Who'sWho in Science and Engineering®

Biographical Titles Currently Published by Marquis Who's Who

Who's Who in America
Who's Who in America Junior & Senior High School Version
Who Was Who in America
  Historical Volume (1607–1896)
  Volume I (1897–1942)
  Volume II (1943–1950)
  Volume III (1951–1960)
  Volume IV (1961–1968)
  Volume V (1969–1973)
  Volume VI (1974–1976)
  Volume VII (1977–1981)
  Volume VIII (1982–1985)
  Volume IX (1985–1989)
  Volume X (1989–1993)
  Index Volume (1607–1993)
Who's Who in the World
Who's Who in the East
Who's Who in the Midwest
Who's Who in the South and Southwest
Who's Who in the West
Who's Who in American Education
Who's Who in American Law
Who's Who in American Nursing
Who's Who of American Women
Who's Who of Emerging Leaders in America
Who's Who in Entertainment
Who's Who in Finance and Industry
Who's Who in Religion
Who's Who in Science and Engineering
Index to Who's Who Books
The *Official* ABMS Directory of Board Certified Medical Specialists

Who'sWho in Science and Engineering®

1994~1995

2nd Edition

MARQUIS
Who'sWho
A REED REFERENCE PUBLISHING COMPANY
121 Chanlon Road
New Providence, NJ 07974 U.S.A.

Marquis Who'sWho
Who'sWho in Science and Engineering®

Senior Vice President, Database Publishing	Peter E. Simon
Vice President & Publisher	Sandra S. Barnes
Managing Director	Leigh C. Yuster-Freeman
Editorial Director	Paul Canning
Senior Managing Editor	Frederick M. Marks
Managing Editor	Samuel J. Dempsey
Senior Editor	Tara Lindquist
Senior Researcher	William J. Cherf

Editorial

Assistant Editors	Hazel C. Conner Daniel F. DeRienzo Kristin Anna Eckes Carl R. Edolo Paul Johnson Eileen McGuinness Stephanie A. Palenque Josh Samber Rebecca Sultzbaugh

Editorial Services

Manager	Nadine Hovan
Supervisors	Debra Krom Mary Lyn K. Sodano
Coordinators	Ellen Bonner Anne Marie C. Calcagno

Editorial Support

Manager	Kevin E. Frank
Coordinator	Sharon L. Gonzalez
Staff	J. Hector Gonzalez

Mail Processing

Manager	Kara A. Seitz
Staff	Shawn Johnston Yaa Okubi-Abebrese Cheryl A. Rodriguez Jill S. Terbell

Quality Assurance & Database Operations

Supervisor	Bob Leonard
Coordinator	Ren Reiner

Research

Data Entry Operators	Lisa A. Heft Rosemarie Iannuzzi

Support Services

Assistants	Jeanne Danzig Denise C. Kreatsas

Published by Marquis Who's Who, A Reed Reference Publishing Company.

Library of Congress Catalog Card Number 92-80582
International Standard Book Number 0-8379-5752-4 (Classic Edition)
0-8379-5753-2 (Deluxe Edition)

Manufactured in the United States of America

Table of Contents

Preface

Scientists, engineers, academics, and researchers in every field are advancing our knowledge each day. As the pace and scope of these advancements continue to grow, communication between scientists of all specializations becomes increasingly necessary. Marquis *Who's Who in Science and Engineering* is a reference tool designed specifically to facilitate such communication. Its paramount purpose is to provide scientists and engineers, as well as other interested readers, with a vehicle for identifying researchers and achievers throughout the scientific community; for gaining an insight into the purpose and progress of their peers' work and accomplishments; and finally, for networking with those of similar or related interests.

This volume contains the biographies of approximately 23,600 scientists, doctors, and engineers; while the majority of them come from the United States, many come from over 115 other nations. Their careers span more than 110 distinct specialties. This edition documents achievements and breakthroughs from the fields of aerospace, civil, environmental, and optical engineering; robotics; genetics; information science; and many others. The comprehensive coverage of medical specialties in this volume includes anesthesiologists, pathologists, surgeons, dentists, and psychiatrists, to name only a few. The physical sciences and the life sciences are represented by over thirty specialties, from astronomers to zoologists.

While the bulk of this volume concerns the technical and physical sciences, appropriate biographies from the "soft sciences" have been included; sketches of select sociologists, economists, psychologists, among other students of the humanities, appear in this work in the belief that their methods, conclusions, and achievements are scientific and noteworthy. Equal representation has also been given those professions which provide support to pure scientific study, such as communications, education, governmental and industrial administration, corporate management, and legal practice.

The reference value of Marquis *Who's Who in Science and Engineering* is derived in part from the inclusion of our Biographees' scientific achievements. Where applicable, sketches contain titles or brief descriptions of patents, patents pending, inventions, invention disclosures, discoveries, designs, pioneering developments, and current research interests. This data is of significant importance, not only in acknowledging the contributions of our Biographees, but in conveying the specific nature of their work to interested readers.

The Geographic, Professional, and Awards indexes are special features within this publication. The Geographic and Professional indexes group Biographees by country, state, and city, as well as by field and specialty. Containing 16 distinct fields and over 100 specializations, these indexes provide interested readers with quick and easy reference to any biography in the book. The new and expanded Awards Index lists some 295 award-granting agencies and over 410 awards bestowed upon approximately 1200 scientists and engineers. This unique index enhances the reference value of Marquis *Who's Who in Science and Engineering*, and attests to Marquis' commitment to the principle of reference value within all our publications. It is organized alphabetically, first by general field of endeavor, and thereafter by award-granting agency, individual award, and award recipient, in descending chronological order. Each recipient has a biographical sketch in Marquis *Who's Who in Science and Engineering*.

Inclusion in Marquis *Who's Who in Science and Engineering* is determined solely by reference value. Positions of responsibility, contributions to the field, and noteworthy individual accomplishment are vital factors in the final selection of listings for the book. As in all Marquis Who's Who publications, the Biographee's own recommendations and ultimate approval ensure the accuracy of the sketches. In those instances where individuals of distinct reference value failed to supply biographical information, Marquis staff members carefully researched the necessary data. Asterisks denote such sketches.

Marquis Who's Who editors and researchers diligently prepare each biographical listing. However, errors may occur—especially in the presentation of advanced scientific terminology. We regret all such errors and invite Biographees and readers to report them to the publisher so that corrections can be made in subsequent editions.

Board of Advisors

Marquis Who's Who gratefully acknowledges the following distinguished individuals who have made themselves available for review, evaluation, and general comment with regard to the publication of the second edition of Marquis *Who's Who in Science and Engineering.* The advisors have enhanced the reference value of this edition by the nomination of outstanding individuals for inclusion. However, the Board of Advisors, either collectively or individually, is in no way responsible for the final selection of names appearing in this volume, nor does the Board of Advisors bear responsibility for the accuracy or comprehensiveness of the biographical information or other material contained herein.

Hojjat Adeli
Professor of Civil Engineering
Ohio State University
Columbus, Ohio

William C. Anderson
Executive Director
American Academy of Environmental Engineers
Annapolis, Maryland

Robert F. Barnes
Executive Vice President
American Society of Agronomy
Madison, Wisconsin

Bruno Boley
Professor of Civil Engineering and Engineering Mechanics
Columbia University
New York, New York

William L. Duax
Research Director
Medical Foundation of Buffalo, Inc. Research Institute
Buffalo, New York

Gordon H. Dunn
Chairman
Joint Institute for Laboratory Astrophysics
University of Colorado
Boulder, Colorado

Herman Feshbach
Professor Emeritus
Massachusetts Institute of Technology
Cambridge, Massachusetts

Walter F. Freiberger
Chairman
University Committee on Statistical Science
Brown University
Providence, Rhode Island

Yuon-Cheng B. Fung
Professor of Bioengineering and Applied Mathematics
University of California
San Diego, California

Pierre M. Galletti
Professor of Medical Science
Brown University
Providence, Rhode Island

Roland D. Glenn
Former Director
Association of Consulting Chemists and Chemical Engineers
New York, New York

C. Bryan Gregor
Former Secretary
The Geochemical Society
Wright State University
Dayton, Ohio

Michael Guillen
Science Editor
ABC-TV
New York, New York
Instructor
Core Curriculum Program
Harvard University
Cambridge, Massachusetts

Leonard C. Klein
Secretary
American Microchemical Society
Princeton, New Jersey

Robert G. McKinnell
Secretary-Treasurer
International Society of Differentiation, Inc.
University of Minnesota
St. Paul, Minnesota

Frederick W. Oehme
Fellow
American Academy of Clinical Toxicology
Kansas State University
Manhattan, Kansas

Jeremiah P. Ostriker
Chairman
Department of Astronomy
Princeton University Observatory
Princeton, New Jersey

Michael S. Teitelbaum
Former Program Officer
Alfred P. Sloan Foundation
New York, New York

Charles A. Wert
Professor Emeritus, Former Head
Department of Metallurgy and Mining Engineering
University of Illinois
Urbana, Illinois

Standards of Admission

Selection of Biographees for Marquis *Who's Who in Science and Engineering* is determined by reference interest, based on involvement in research or administration in pure or applied scientific and technological fields of endeavor.

Among the categories of individuals selected for admission are the following:

Recipients of national and international awards granted by recognized professional scientific and engineering agencies;

Directors of selected university, industrial, governmental, and medical research programs;

Individuals who evince special leadership qualities within their respective scientific or engineering communities;

Selected members of honorary organizations, such as the National Academy of Sciences, the National Academy of Engineering, and the Institute of Medicine;

Heads of research institutes;

Executives of scientific or engineering associations;

Individuals who have made remarkable discoveries or research breakthroughs critical to the development of their chosen field(s).

Key to Information

[1] **STEELE, FLETCHER DAVID,** [2] mechanical engineer; [3] b. Normal, Ill., Jan. 20, 1939; [4] s. Thomas William and Susan (Shobe) S.; [5] m. Julie Ann Walsh, Sept. 8, 1964; [6] children: Honor Elizabeth Carter, Michael Thomas. [7] BSME, Purdue U., 1961; MS, U. Ill., 1965. [8] Registered profl. engr., Ill., Iowa. [9] Asst. engr. Kelly, Kitching, Berendes & Brault, Engrs., Chgo., 1966-67, engr., 1967-71; sr. engr. Kelly, Kitching, Berendes & Brault, Internat., Des Moines, 1971-78, mgr. fluids div., 1979-84, v.p. R & D, 1985—; [10] lectr. Drake U., 1983-88. [11] Contbr. articles to Jour. Biomech. Engring., Jour. Fluids Engring. [12] Asst. troop leader Des Moines coun. Boy Scouts Am., 1992—. [13] Lt. U.S. Army, 1961-63. [14] Fulbright scholar, 1965. [15] Mem. ASME, NSPE, Iowa Mech. Engrs. Assn., Big Sand Lake Club. [16] Republican. [17] Roman Catholic. [18] Achievements include patent for internal piston lock for hydraulic cylinders; design of L16500 Workhorse rotar; research in linear regression analysis for large-lot engine data comparisons. [19] Home: 733 N Ottawa Rd Ankeny IA 50021 [20] Office: 1245 34th St Des Moines IA 50311

KEY

[1] Name
[2] Occupation
[3] Vital statistics
[4] Parents
[5] Marriage
[6] Children
[7] Education
[8] Professional certifications
[9] Career
[10] Career related
[11] Writings and creative works
[12] Civic and political activities
[13] Military
[14] Awards and fellowships
[15] Professional and association memberships, clubs and lodges
[16] Political affiliation
[17] Religion
[18] Achievements information
[19] Home address
[20] Office address

Table of Abbreviations

The following abbreviations and symbols are frequently used in this book.

*An asterisk following a sketch indicates that it was researched by the Marquis Who's Who editorial staff and has not been verified by the Biographee.

A Associate (used with academic degrees only)
AA, A.A. Associate in Arts, Associate of Arts
AAAL American Academy of Arts and Letters
AAAS American Association for the Advancement of Science
AACD American Association for Counseling and Development
AACN American Association of Critical Care Nurses
AAHA American Academy of Health Administrators
AAHP American Association of Hospital Planners
AAHPERD American Alliance for Health, Physical Education, Recreation, and Dance
AAS Associate of Applied Science
AASL American Association of School Librarians
AASPA American Association of School Personnel Administrators
AAU Amateur Athletic Union
AAUP American Association of University Professors
AAUW American Association of University Women
AB, A.B. Arts, Bachelor of
AB Alberta
ABA American Bar Association
ABC American Broadcasting Company
AC Air Corps
acad. academy, academic
acct. accountant
acctg. accounting
ACDA Arms Control and Disarmament Agency
ACHA American College of Hospital Administrators
ACLS Advanced Cardiac Life Support
ACLU American Civil Liberties Union
ACOG American College of Ob-Gyn
ACP American College of Physicians
ACS American College of Surgeons
ADA American Dental Association
a.d.c. aide-de-camp
adj. adjunct, adjutant
adj. gen. adjutant general
adm. admiral
adminstr. administrator
adminstrn. administration
adminstrv. administrative
ADN Associate's Degree in Nursing
ADP Automatic Data Processing
adv. advocate, advisory
advt. advertising
AE, A.E. Agricultural Engineer
A.E. and P. Ambassador Extraordinary and Plenipotentiary
AEC Atomic Energy Commission
aero. aeronautical, aeronautic
aerodyn. aerodynamic
AFB Air Force Base
AFL-CIO American Federation of Labor and Congress of Industrial Organizations
AFTRA American Federation of TV and Radio Artists
AFSCME American Federation of State, County and Municipal Employees
agr. agriculture
agrl. agricultural
agt. agent
AGVA American Guild of Variety Artists
agy. agency
A&I Agricultural and Industrial
AIA American Institute of Architects
AIAA American Institute of Aeronautics and Astronautics
AICE American Institute of Chemical Engineers
AICPA American Institute of Certified Public Accountants
AID Agency for International Development
AIDS Acquired Immune Deficiency Syndrome
AIEE American Institute of Electrical Engineers
AIM American Institute of Management
AIME American Institute of Mining, Metallurgy, and Petroleum Engineers
AK Alaska
AL Alabama
ALA American Library Association
Ala. Alabama
alt. alternate
Alta. Alberta
A&M Agricultural and Mechanical
AM, A.M. Arts, Master of
Am. American, America
AMA American Medical Association
amb. ambassador
A.M.E. African Methodist Episcopal
Amtrak National Railroad Passenger Corporation
AMVETS American Veterans of World War II, Korea, Vietnam
ANA American Nurses Association
anat. anatomical
ANCC American Nurses Credentialing Center
ann. annual
ANTA American National Theatre and Academy
anthrop. anthropological
AP Associated Press
APA American Psychological Association
APGA American Personnel Guidance Association
APHA American Public Health Association
APO Army Post Office
apptd. appointed
Apr. April
apt. apartment
AR Arkansas
ARC American Red Cross
archeol. archeological
archtl. architectural
Ariz. Arizona
Ark. Arkansas
ArtsD, ArtsD. Arts, Doctor of
arty. artillery
AS American Samoa
AS Associate in Science
ASCAP American Society of Composers, Authors and Publishers
ASCD Association for Supervision and Curriculum Development
ASCE American Society of Civil Engineers
ASHRAE American Society of Heating, Refrigeration, and Air Conditioning Engineers
ASME American Society of Mechanical Engineers
ASNSA American Society for Nursing Service Administrators
ASPA American Society for Public Administration
ASPCA American Society for the Prevention of Cruelty to Animals
assn. association
assoc. associate
asst. assistant
ASTD American Society for Training and Development
ASTM American Society for Testing and Materials
astron. astronomical
astrophys. astrophysical
ATSC Air Technical Service Command
AT&T American Telephone & Telegraph Company
atty. attorney
Aug. August
AUS Army of the United States
aux. auxiliary
Ave. Avenue
AVMA American Veterinary Medical Association
AZ Arizona
AWHONN Association of Women's Health Obstetric and Neonatal Nurses

B. Bachelor
b. born
BA, B.A. Bachelor of Arts
BAgr, B.Agr. Bachelor of Agriculture
Balt. Baltimore
Bapt. Baptist
BArch, B.Arch. Bachelor of Architecture
BAS, B.A.S. Bachelor of Agricultural Science
BBA, B.B.A. Bachelor of Business Administration
BBC British Broadcasting Corporation
BC, B.C. British Columbia
BCE, B.C.E. Bachelor of Civil Engineering

BChir, B.Chir. Bachelor of Surgery
BCL, B.C.L. Bachelor of Civil Law
BCLS Basic Cardiac Life Support
BCS, B.C.S. Bachelor of Commercial Science
BD, B.D. Bachelor of Divinity
bd. board
BE, B.E. Bachelor of Education
BEE, B.E.E. Bachelor of Electrical Engineering
BFA, B.F.A. Bachelor of Fine Arts
bibl. biblical
bibliog. bibliographical
biog. biographical
biol. biological
BJ, B.J. Bachelor of Journalism
Bklyn. Brooklyn
BL, B.L. Bachelor of Letters
bldg. building
BLS, B.L.S. Bachelor of Library Science
BLS Basic Life Support
Blvd. Boulevard
BMI Broadcast Music, Inc.
BMW Bavarian Motor Works (Bayerische Motoren Werke)
bn. battalion
B.&O.R.R. Baltimore & Ohio Railroad
bot. botanical
BPE, B.P.E. Bachelor of Physical Education
BPhil, B.Phil. Bachelor of Philosophy
br. branch
BRE, B.R.E. Bachelor of Religious Education
brig. gen. brigadier general
Brit. British, Brittanica
Bros. Brothers
BS, B.S. Bachelor of Science
BSA, B.S.A. Bachelor of Agricultural Science
BSBA Bachelor of Science in Business Administration
BSChemE Bachelor of Science in Chemical Engineering
BSD, B.S.D. Bachelor of Didactic Science
BSEE Bachelor of Science in Electrical Engineering
BSN Bachelor of Science in Nursing
BST, B.S.T. Bachelor of Sacred Theology
BTh, B.Th. Bachelor of Theology
bull. bulletin
bur. bureau
bus. business
B.W.I. British West Indies

CA California
CAA Civil Aeronautics Administration
CAB Civil Aeronautics Board
CAD-CAM Computer Aided Design-Computer Aided Model
Calif. California
C.Am. Central America
Can. Canada, Canadian
CAP Civil Air Patrol
capt. captain
CARE Cooperative American Relief Everywhere
Cath. Catholic
cav. cavalry
CBC Canadian Broadcasting Company
CBI China, Burma, India Theatre of Operations
CBS Columbia Broadcasting Company
C.C. Community College
CCC Commodity Credit Corporation
CCNY City College of New York
CCRN Critical Care Registered Nurse
CCU Cardiac Care Unit
CD Civil Defense
CE, C.E. Corps of Engineers, Civil Engineer
CEN Certified Emergency Nurse
CENTO Central Treaty Organization
CEO chief executive officer
CERN European Organization of Nuclear Research
cert. certificate, certification, certified
CETA Comprehensive Employment Training Act
CFA Chartered Financial Analyst
CFL Canadian Football League
CFO chief financial officer
CFP Certified Financial Planner
ch. church
ChD, Ch.D. Doctor of Chemistry
chem. chemical
ChemE, Chem.E. Chemical Engineer
ChFC Chartered Financial Consultant
Chgo. Chicago
chirurg. chirurgical
chmn. chairman
chpt. chapter
CIA Central Intelligence Agency
Cin. Cincinnati
cir. circuit
Cleve. Cleveland
climatol. climatological
clin. clinical
clk. clerk
C.L.U. Chartered Life Underwriter
CM, C.M. Master in Surgery
CM Northern Mariana Islands
CMA Certified Medical Assistant
CNA Certified Nurse's Aide
CNOR Certified Nurse (Operating Room)
C.&N.W.Ry. Chicago & North Western Railway
CO Colorado
Co. Company
COF Catholic Order of Foresters
C. of C. Chamber of Commerce
col. colonel
coll. college
Colo. Colorado
com. committee
comd. commanded
comdg. commanding
comdr. commander
comdt. commandant
comm. communications
commd. commissioned
comml. commercial
commn. commission
commr. commissioner
compt. comptroller
condr. conductor
Conf. Conference
Congl. Congregational, Congressional
Conglist. Congregationalist
Conn. Connecticut
cons. consultant, consulting
consol. consolidated
constl. constitutional
constn. constitution
constrn. construction
contbd. contributed
contbg. contributing
contbn. contribution
contbr. contributor
contr. controller
Conv. Convention
COO chief operating officer
coop. cooperative
coord. coordinator
CORDS Civil Operations and Revolutionary Development Support
CORE Congress of Racial Equality
corp. corporation, corporate
corr. correspondent, corresponding, correspondence
C.&O.Ry. Chesapeake & Ohio Railway
coun. council
CPA Certified Public Accountant
CPCU Chartered Property and Casualty Underwriter
CPH, C.P.H. Certificate of Public Health
cpl. corporal
CPR Cardio-Pulmonary Resuscitation
C.P.Ry. Canadian Pacific Railway
CRT Cathode Ray Terminal
C.S. Christian Science
CSB, C.S.B. Bachelor of Christian Science
C.S.C. Civil Service Commission
CT Connecticut
ct. court
ctr. center
ctrl. central
CWS Chemical Warfare Service
C.Z. Canal Zone

D. Doctor
d. daughter
DAgr, D.Agr. Doctor of Agriculture
DAR Daughters of the American Revolution
dau. daughter
DAV Disabled American Veterans
DC, D.C. District of Columbia
DCL, D.C.L. Doctor of Civil Law
DCS, D.C.S. Doctor of Commercial Science
DD, D.D. Doctor of Divinity
DDS, D.D.S. Doctor of Dental Surgery
DE Delaware
Dec. December
dec. deceased
def. defense
Del. Delaware

del. delegate, delegation
Dem. Democrat, Democratic
DEng, D.Eng. Doctor of Engineering
denom. denomination, denominational
dep. deputy
dept. department
dermatol. dermatological
desc. descendant
devel. development, developmental
DFA, D.F.A. Doctor of Fine Arts
D.F.C. Distinguished Flying Cross
DHL, D.H.L. Doctor of Hebrew Literature
dir. director
dist. district
distbg. distributing
distbn. distribution
distbr. distributor
disting. distinguished
div. division, divinity, divorce
DLitt, D.Litt. Doctor of Literature
DMD, D.M.D. Doctor of Dental Medicine
DMS, D.M.S. Doctor of Medical Science
DO, D.O. Doctor of Osteopathy
DON Director of Nursing
DPH, D.P.H. Diploma in Public Health
DPhil, D.Phil. Doctor of Philosophy
D.R. Daughters of the Revolution
Dr. Drive, Doctor
DRE, D.R.E. Doctor of Religious Education
DrPH, Dr.P.H. Doctor of Public Health, Doctor of Public Hygiene
D.S.C. Distinguished Service Cross
DSc, D.Sc. Doctor of Science
DSChemE Doctor of Science in Chemical Engineering
D.S.M. Distinguished Service Medal
DST, D.S.T. Doctor of Sacred Theology
DTM, D.T.M. Doctor of Tropical Medicine
DVM, D.V.M. Doctor of Veterinary Medicine
DVS, D.V.S. Doctor of Veterinary Surgery

E, E. East
ea. eastern
E. and P. Extraordinary and Plenipotentiary
Eccles. Ecclesiastical
ecol. ecological
econ. economic
ECOSOC Economic and Social Council (of the UN)
ED, E.D. Doctor of Engineering
ed. educated
EdB, Ed.B. Bachelor of Education
EdD, Ed.D. Doctor of Education
edit. edition
EdM, Ed.M. Master of Education
edn. education
ednl. educational
EDP Electronic Data Processing
EdS, Ed.S. Specialist in Education
EE, E.E. Electrical Engineer
E.E. and M.P. Envoy Extraordinary and Minister Plenipotentiary
EEC European Economic Community
EEG Electroencephalogram
EEO Equal Employment Opportunity
EEOC Equal Employment Opportunity Commission
E.Ger. German Democratic Republic
EKG Electrocardiogram
elec. electrical
electrochem. electrochemical
electrophys. electrophysical
elem. elementary
EM, E.M. Engineer of Mines
EMT Emergency Medical Technician
ency. encyclopedia
Eng. England
engr. engineer
engring. engineering
entomol. entomological
environ. environmental
EPA Environmental Protection Agency
epidemiol. epidemiological
Episc. Episcopalian
ERA Equal Rights Amendment
ERDA Energy Research and Development Administration
ESEA Elementary and Secondary Education Act
ESL English as Second Language
ESPN Entertainment and Sports Programming Network
ESSA Environmental Science Services Administration
ethnol. ethnological
ETO European Theatre of Operations
Evang. Evangelical
exam. examination, examining
Exch. Exchange
exec. executive
exhbn. exhibition
expdn. expedition
expn. exposition
expt. experiment
exptl. experimental
Expy. Expressway
Ext. Extension

F.A. Field Artillery
FAA Federal Aviation Administration
FAO Food and Agriculture Organization (of the UN)
FBI Federal Bureau of Investigation
FCA Farm Credit Administration
FCC Federal Communications Commission
FCDA Federal Civil Defense Administration
FDA Food and Drug Administration
FDIA Federal Deposit Insurance Administration
FDIC Federal Deposit Insurance Corporation
FE, F.E. Forest Engineer
FEA Federal Energy Administration
Feb. February
fed. federal
fedn. federation
FERC Federal Energy Regulatory Commission
fgn. foreign
FHA Federal Housing Administration
fin. financial, finance
FL Florida
Fl. Floor
Fla. Florida
FMC Federal Maritime Commission
FNP Family Nurse Practitioner
FOA Foreign Operations Administration
found. foundation
FPC Federal Power Commission
FPO Fleet Post Office
frat. fraternity
FRS Federal Reserve System
FSA Federal Security Agency
Ft. Fort
FTC Federal Trade Commission
Fwy. Freeway

G-1 (or other number) Division of General Staff
GA, Ga. Georgia
GAO General Accounting Office
gastroent. gastroenterological
GATE Gifted and Talented Educators
GATT General Agreement on Tariffs and Trade
GE General Electric Company
gen. general
geneal. genealogical
geod. geodetic
geog. geographic, geographical
geol. geological
geophys. geophysical
gerontol. gerontological
G.H.Q. General Headquarters
GM General Motors Corporation
GMAC General Motors Acceptance Corporation
G.N.Ry. Great Northern Railway
gov. governor
govt. government
govtl. governmental
GPO Government Printing Office
grad. graduate, graduated
GSA General Services Administration
Gt. Great
GTE General Telephone and Electric Company
GU Guam
gynecol. gynecological

HBO Home Box Office
hdqs. headquarters
HEW Department of Health, Education and Welfare
HHD, H.H.D. Doctor of Humanities
HHFA Housing and Home Finance Agency
HHS Department of Health and Human Services
HI Hawaii

hist. historical, historic
HM, H.M. Master of Humanities
HMO Health Maintenance Organization
homeo. homeopathic
hon. honorary, honorable
Ho. of Dels. House of Delegates
Ho. of Reps. House of Representatives
hort. horticultural
hosp. hospital
HUD Department of Housing and Urban Development
Hwy. Highway
hydrog. hydrographic

IA Iowa
IAEA International Atomic Energy Agency
IATSE International Alliance of Theatrical and Stage Employees and Moving Picture Operators of the United States and Canada
IBM International Business Machines Corporation
IBRD International Bank for Reconstruction and Development
ICA International Cooperation Administration
ICC Interstate Commerce Commission
ICCE International Council for Computers in Education
ICU Intensive Care Unit
ID Idaho
IEEE Institute of Electrical and Electronics Engineers
IFC International Finance Corporation
IGY International Geophysical Year
IL Illinois
Ill. Illinois
illus. illustrated
ILO International Labor Organization
IMF International Monetary Fund
IN Indiana
Inc. Incorporated
Ind. Indiana
ind. independent
Indpls. Indianapolis
Indsl. industrial
inf. infantry
info. information
ins. insurance
insp. inspector
insp. gen. inspector general
inst. institute
instl. institutional
instn. institution
instr. instructor
instrn. instruction
instrnl. instructional
internat. international
intro. introduction
IRE Institute of Radio Engineers
IRS Internal Revenue Service
ITT International Telephone & Telegraph Corporation

JAG Judge Advocate General
JAGC Judge Advocate General Corps
Jan. January
Jaycees Junior Chamber of Commerce
JB, J.B. Jurum Baccalaureus
JCB, J.C.B. Juris Canoni Baccalaureus
JCD, J.C.D. Juris Canonici Doctor, Juris Civilis Doctor
JCL, J.C.L. Juris Canonici Licentiatus
JD, J.D. Juris Doctor
jg. junior grade
jour. journal
jr. junior
JSD, J.S.D. Juris Scientiae Doctor
JUD, J.U.D. Juris Utriusque Doctor
jud. judicial

Kans. Kansas
K.C. Knights of Columbus
K.P. Knights of Pythias
KS Kansas
K.T. Knight Templar
KY, Ky. Kentucky

LA, La. Louisiana
L.A. Los Angeles
lab. laboratory
L.Am. Latin America
lang. language
laryngol. laryngological
LB Labrador
LDS Latter Day Saints
LDS Church Church of Jesus Christ of Latter Day Saints
lectr. lecturer
legis. legislation, legislative
LHD, L.H.D. Doctor of Humane Letters
L.I. Long Island
libr. librarian, library
lic. licensed, license
L.I.R.R. Long Island Railroad
lit. literature
LittB, Litt.B. Bachelor of Letters
LittD, Litt.D. Doctor of Letters
LLB, LL.B. Bachelor of Laws
LLD, L.L.D. Doctor of Laws
LLM, L.L.M. Master of Laws
Ln. Lane
L.&N.R.R. Louisville & Nashville Railroad
LPGA Ladies Professional Golf Association
LPN Licensed Practical Nurse
LS, L.S. Library Science (in degree)
lt. lieutenant
Ltd. Limited
Luth. Lutheran
LWV League of Women Voters

M. Master
m. married
MA, M.A. Master of Arts
MA Massachusetts
MADD Mothers Against Drunk Driving
mag. magazine
MAgr, M.Agr. Master of Agriculture
maj. major
Man. Manitoba
Mar. March
MArch, M.Arch. Master in Architecture
Mass. Massachusetts
math. mathematics, mathematical
MATS Military Air Transport Service
MB, M.B. Bachelor of Medicine
MB Manitoba
MBA, M.B.A. Master of Business Administration
MBS Mutual Broadcasting System
M.C. Medical Corps
MCE, M.C.E. Master of Civil Engineering
mcht. merchant
mcpl. municipal
MCS, M.C.S. Master of Commercial Science
MD, M.D. Doctor of Medicine
MD, Md. Maryland
MDiv Master of Divinity
MDip, M.Dip. Master in Diplomacy
mdse. merchandise
MDV, M.D.V. Doctor of Veterinary Medicine
ME, M.E. Mechanical Engineer
ME Maine
M.E.Ch. Methodist Episcopal Church
mech. mechanical
MEd., M.Ed. Master of Education
med. medical
MEE, M.E.E. Master of Electrical Engineering
mem. member
meml. memorial
merc. mercantile
met. metropolitan
metall. metallurgical
MetE, Met.E. Metallurgical Engineer
meteorol. meteorological
Meth. Methodist
Mex. Mexico
MF, M.F. Master of Forestry
MFA, M.F.A. Master of Fine Arts
mfg. manufacturing
mfr. manufacturer
mgmt. management
mgr. manager
MHA, M.H.A. Master of Hospital Administration
M.I. Military Intelligence
MI Michigan
Mich. Michigan
micros. microscopic, microscopical
mid. middle
mil. military
Milw. Milwaukee
Min. Minister
mineral. mineralogical
Minn. Minnesota
MIS Management Information Systems
Miss. Mississippi
MIT Massachusetts Institute of Technology
mktg. marketing
ML, M.L. Master of Laws
MLA Modern Language Association

M.L.D. Magister Legnum Diplomatic
MLitt, M.Litt. Master of Literature, Master of Letters
MLS, M.L.S. Master of Library Science
MME, M.M.E. Master of Mechanical Engineering
MN Minnesota
mng. managing
MO, Mo. Missouri
moblzn. mobilization
Mont. Montana
MP Northern Mariana Islands
M.P. Member of Parliament
MPA Master of Public Administration
MPE, M.P.E. Master of Physical Education
MPH, M.P.H. Master of Public Health
MPhil, M.Phil. Master of Philosophy
MPL, M.P.L. Master of Patent Law
Mpls. Minneapolis
MRE, M.R.E. Master of Religious Education
MS, M.S. Master of Science
MS, Ms. Mississippi
MSc, M.Sc. Master of Science
MSChemE Master of Science in Chemical Engineering
MSEE Master of Science in Electrical Engineering
MSF, M.S.F. Master of Science of Forestry
MSN Master of Science in Nursing
MST, M.S.T. Master of Sacred Theology
MSW, M.S.W. Master of Social Work
MT Montana
Mt. Mount
MTO Mediterranean Theatre of Operation
MTV Music Television
mus. museum, musical
MusB, Mus.B. Bachelor of Music
MusD, Mus.D. Doctor of Music
MusM, Mus.M. Master of Music
mut. mutual
mycol. mycological

N. North
NAACOG Nurses Association of the American College of Obstetricians and Gynecologists
NAACP National Association for the Advancement of Colored People
NACA National Advisory Committee for Aeronautics
NACU National Association of Colleges and Universities
NAD National Academy of Design
NAE National Academy of Engineering, National Association of Educators
NAESP National Association of Elementary School Principals
NAFE National Association of Female Executives
N.Am. North America
NAM National Association of Manufacturers
NAMH National Association for Mental Health
NAPA National Association of Performing Artists
NARAS National Academy of Recording Arts and Sciences
NAREB National Association of Real Estate Boards
NARS National Archives and Record Service
NAS National Academy of Sciences
NASA National Aeronautics and Space Administration
NASP National Association of School Psychologists
NASW National Association of Social Workers
nat. national
NATAS National Academy of Television Arts and Sciences
NATO North Atlantic Treaty Organization
NATOUSA North African Theatre of Operations, United States Army
nav. navigation
NB, N.B. New Brunswick
NBA National Basketball Association
NBC National Broadcasting Company
NC, N.C. North Carolina
NCAA National College Athletic Association
NCCJ National Conference of Christians and Jews
ND, N.D. North Dakota
NDEA National Defense Education Act
NE Nebraska
NE, N.E. Northeast
NEA National Education Association
Nebr. Nebraska
NEH National Endowment for Humanities
neurol. neurological
Nev. Nevada
NF Newfoundland
NFL National Football League
Nfld. Newfoundland
NG National Guard
NH, N.H. New Hampshire
NHL National Hockey League
NIH National Institutes of Health
NIMH National Institute of Mental Health
NJ, N.J. New Jersey
NLRB National Labor Relations Board
NM New Mexico
N.Mex. New Mexico
No. Northern
NOAA National Oceanographic and Atmospheric Administration
NORAD North America Air Defense
Nov. November
NOW National Organization for Women
N.P.Ry. Northern Pacific Railway
nr. near
NRA National Rifle Association
NRC National Research Council
NS, N.S. Nova Scotia
NSC National Security Council
NSF National Science Foundation
NSTA National Science Teachers Association
NSW New South Wales
N.T. New Testament
NT Northwest Territories
numis. numismatic
NV Nevada
NW, N.W. Northwest
N.W.T. Northwest Territories
NY, N.Y. New York
N.Y.C. New York City
NYU New York University
N.Z. New Zealand

OAS Organization of American States
ob-gyn obstetrics-gynecology
obs. observatory
obstet. obstetrical
Oct. October
OD, O.D. Doctor of Optometry
OECD Organization of European Cooperation and Development
OEEC Organization of European Economic Cooperation
OEO Office of Economic Opportunity
ofcl. official
OH Ohio
OK Oklahoma
Okla. Oklahoma
ON Ontario
Ont. Ontario
oper. operating
ophthal. ophthalmological
ops. operations
OR Oregon
orch. orchestra
Oreg. Oregon
orgn. organization
orgnl. organizational
ornithol. ornithological
OSHA Occupational Safety and Health Administration
OSRD Office of Scientific Research and Development
OSS Office of Strategic Services
osteo. osteopathic
otol. otological
otolaryn. otolaryngological

PA, Pa. Pennsylvania
P.A. Professional Association
paleontol. paleontological
path. pathological
PBS Public Broadcasting System
P.C. Professional Corporation
PE Prince Edward Island
P.E.I. Prince Edward Island
PEN Poets, Playwrights, Editors, Essayists and Novelists (international association)
penol. penological
P.E.O. women's organization (full name not disclosed)
pers. personnel
pfc. private first class
PGA Professional Golfers' Association of America
PHA Public Housing Administration

pharm. pharmaceutical
PharmD, Pharm.D. Doctor of Pharmacy
PharmM, Pharm.M. Master of Pharmacy
PhB, Ph.B. Bachelor of Philosophy
PhD, Ph.D. Doctor of Philosophy
PhDChemE Doctor of Science in Chemical Engineering
PhM, Ph.M. Master of Philosophy
Phila. Philadelphia
philharm. philharmonic
philol. philological
philos. philosophical
photog. photographic
phys. physical
physiol. physiological
Pitts. Pittsburgh
Pk. Park
Pky. Parkway
Pl. Place
P.&L.E.R.R. Pittsburgh & Lake Erie Railroad
Plz. Plaza
PNP Pediatric Nurse Practitioner
P.O. Post Office
PO Box Post Office Box
polit. political
poly. polytechnic, polytechnical
PQ Province of Quebec
PR, P.R. Puerto Rico
prep. preparatory
pres. president
Presbyn. Presbyterian
presdl. presidential
prin. principal
procs. proceedings
prod. produced (play production)
prodn. production
prodr. producer
prof. professor
profl. professional
prog. progressive
propr. proprietor
pros. atty. prosecuting attorney
pro tem. pro tempore
PSRO Professional Services Review Organization
psychiat. psychiatric
psychol. psychological
PTA Parent-Teachers Association
ptnr. partner
PTO Pacific Theatre of Operations, Parent Teacher Organization
pub. publisher, publishing, published
pub. public
publ. publication
pvt. private

quar. quarterly
qm. quartermaster
Q.M.C. Quartermaster Corps
Que. Quebec

radiol. radiological
RAF Royal Air Force
RCA Radio Corporation of America
RCAF Royal Canadian Air Force
RD Rural Delivery
Rd. Road
R&D Research & Development
REA Rural Electrification Administration
rec. recording
ref. reformed
regt. regiment
regtl. regimental
rehab. rehabilitation
rels. relations
Rep. Republican
rep. representative
Res. Reserve
ret. retired
Rev. Reverend
rev. review, revised
RFC Reconstruction Finance Corporation
RFD Rural Free Delivery
rhinol. rhinological
RI, R.I. Rhode Island
RISD Rhode Island School of Design
Rlwy. Railway
Rm. Room
RN, R.N. Registered Nurse
roentgenol. roentgenological
ROTC Reserve Officers Training Corps
RR Rural Route
R.R. Railroad
rsch. research
rschr. researcher
Rt. Route

S. South
s. son
SAC Strategic Air Command
SAG Screen Actors Guild
SALT Strategic Arms Limitation Talks
S.Am. South America
san. sanitary
SAR Sons of the American Revolution
Sask. Saskatchewan
savs. savings
SB, S.B. Bachelor of Science
SBA Small Business Administration
SC, S.C. South Carolina
SCAP Supreme Command Allies Pacific
ScB, Sc.B. Bachelor of Science
SCD, S.C.D. Doctor of Commercial Science
ScD, Sc.D. Doctor of Science
sch. school
sci. science, scientific
SCLC Southern Christian Leadership Conference
SCV Sons of Confederate Veterans
SD, S.D. South Dakota
SE, S.E. Southeast
SEATO Southeast Asia Treaty Organization
SEC Securities and Exchange Commission
sec. secretary
sect. section
seismol. seismological
sem. seminary
Sept. September
s.g. senior grade
sgt. sergeant
SHAEF Supreme Headquarters Allied Expeditionary Forces
SHAPE Supreme Headquarters Allied Powers in Europe
S.I. Staten Island
S.J. Society of Jesus (Jesuit)
SJD Scientiae Juridicae Doctor
SK Saskatchewan
SM, S.M. Master of Science
SNP Society of Nursing Professionals
So. Southern
soc. society
sociol. sociological
S.P.Co. Southern Pacific Company
spl. special
splty. specialty
Sq. Square
S.R. Sons of the Revolution
sr. senior
SS Steamship
SSS Selective Service System
St. Saint, Street
sta. station
stats. statistics
statis. statistical
STB, S.T.B. Bachelor of Sacred Theology
stblzn. stabilization
STD, S.T.D. Doctor of Sacred Theology
std. standard
Ste. Suite
subs. subsidiary
SUNY State University of New York
supr. supervisor
supt. superintendent
surg. surgical
svc. service
SW, S.W. Southwest

TAPPI Technical Association of the Pulp and Paper Industry
tb. tuberculosis
tchr. teacher
tech. technical, technology
technol. technological
tel. telephone
Tel. & Tel. Telephone & Telegraph
telecom. telecommunications
temp. temporary
Tenn. Tennessee
Ter. Territory
Ter. Terrace
Tex. Texas
ThD, Th.D. Doctor of Theology
theol. theological
ThM, Th.M. Master of Theology
TN Tennessee
tng. training
topog. topographical
trans. transaction, transferred
transl. translation, translated
transp. transportation

treas. treasurer
TT Trust Territory
TV television
TVA Tennessee Valley Authority
TWA Trans World Airlines
twp. township
TX Texas
typog. typographical

U. University
UAW United Auto Workers
UCLA University of California at Los Angeles
UDC United Daughters of the Confederacy
U.K. United Kingdom
UN United Nations
UNESCO United Nations Educational, Scientific and Cultural Organization
UNICEF United Nations International Children's Emergency Fund
univ. university
UNRRA United Nations Relief and Rehabilitation Administration
UPI United Press International
U.P.R.R. United Pacific Railroad
urol. urological
U.S. United States
U.S.A. United States of America
USAAF United States Army Air Force
USAF United States Air Force
USAFR United States Air Force Reserve
USAR United States Army Reserve
USCG United States Coast Guard
USCGR United States Coast Guard Reserve
USES United States Employment Service
USIA United States Information Agency
USMC United States Marine Corps
USMCR United States Marine Corps Reserve
USN United States Navy
USNG United States National Guard
USNR United States Naval Reserve
USO United Service Organizations
USPHS United States Public Health Service
USS United States Ship
USSR Union of the Soviet Socialist Republics
USTA United States Tennis Association
USV United States Volunteers
UT Utah

VA Veterans Administration
VA, Va. Virginia
vet. veteran, veterinary
VFW Veterans of Foreign Wars
VI, V.I. Virgin Islands
vice pres. vice president
vis. visiting
VISTA Volunteers in Service to America
VITA Volunteers in Technical Assistance
vocat. vocational
vol. volunteer, volume
v.p. vice president
vs. versus
VT, Vt. Vermont

W, W. West
WA Washington (state)
WAC Women's Army Corps
Wash. Washington (state)
WATS Wide Area Telecommunications Service
WAVES Women's Reserve, US Naval Reserve
WCTU Women's Christian Temperance Union
we. western
W. Ger. Germany, Federal Republic of
WHO World Health Organization
WI Wisconsin
W.I. West Indies
Wis. Wisconsin
WSB Wage Stabilization Board
WV West Virginia
W.Va. West Virginia
WY Wyoming
Wyo. Wyoming

YK Yukon Territory
YMCA Young Men's Christian Association
YMHA Young Men's Hebrew Association
YM & YWHA Young Men's and Young Women's Hebrew Association
yr. year
YT, Y.T. Yukon Territory
YWCA Young Women's Christian Association

zool. zoological

Alphabetical Practices

Names are arranged alphabetically according to the surnames, and under identical surnames according to the first given name. If both surname and first given name are identical, names are arranged alphabetically according to the second given name.

Surnames beginning with De, Des, Du, however capitalized or spaced, are recorded with the prefix preceding the surname and arranged alphabetically under the letter D.

Surnames beginning with Mac and Mc are arranged alphabetically under M.

Surnames beginning with Saint or St. appear after names that begin Sains, and are arranged according to the second part of the name, e.g. St. Clair before Saint Dennis.

Surnames beginning with Van, Von, or von are arranged alphabetically under letter V.

Compound surnames are arranged according to the first member of the compound.

Many hyphenated Arabic names begin Al-, El-, or al-. These names are alphabetized according to each Biographee's designation of last name. Thus Al-Bahar, Mohammed may be listed either under Al- or under Bahar, depending on the preference of the listee.

Parentheses used in connection with a name indicate which part of the full name is usually deleted in common usage. Hence Abbott, W(illiam) Lewis indicates that the usual form of the given name is W. Lewis. In such a case, the parentheses are ignored in alphabetizing and the name would be arranged as Abbott, William Lewis. However, if the name is recorded Abbott, (William) Lewis, signifying that the entire name William is not commonly used, the alphabetizing would be arranged as though the name were Abbott, Lewis. If an entire middle or last name is enclosed in parentheses, that portion of the name is used in the alphabetical arrangement. Hence Abbott, William (Lewis) would be arranged as Abbott, William Lewis.

Who'sWho in Science and Engineering®

Biographies

AARON, JEAN-JACQUES, chemist, educator; b. Fontainebleau, France, Oct. 25, 1939; s. Michel and Marie (Deloustal) A.; m. Paulette Chabriere, Dec. 7, 1961; children: Hélène, Emmanuel. Licence es Sciences, Faculty Sci., Paris, 1961; Doctorate, Faculty Scis., Paris, 1965, Doctorat d'Etat, 1968. Asst., then asst. prof. Faculty Sci., Paris, 1962-70; postdoctoral fellow U. Fla., Gainesville, 1971-72; sci. attaché French Gen. Consulate, Houston, 1972-75; prof. U. Dakar, Senegal, 1976-85; prof. chemistry U. Paris 7 and Marne la Vallée, 1985—. Contbr. chpts. to books, articles to sci. jours. Grantee NATO, Brussels, 1971, 87-93. Mem. Am. Chem. Soc., French Soc. Chemistry. Achievements include patent on protection of metals against corrosion, development of analytical methods for determination of traces of pollutants, new technique for monitoring electropolymerization on solid surfaces. Home: 44 Elysée 2, 78170 La Celle St Cloud France Office: Itodys Univ Paris 7, 1 Rue Guy de la Brosse, 75005 Paris France

ABAJIAN, HENRY KRIKOR, mechanical engineer; b. Aintab, Turkey, Dec. 8, 1909; came to U.S., 1923, naturalized, 1933; s. Hagop H. and Vartouhi (Haleblian) A.; m. Gladys Lucy Mahseregian, Jan. 12, 1942; 1 dau., Carol. ME, Rensselaer Poly. Inst., 1933. Registered profl. engr., Calif. Design engr. Wallace & Tiernan Co., Am. La France Foamite Corp., Heald Machine Co., 1933-40; design specialist Consol. Vultee Aircraft Corp., 1940-46; project engr. Willys Overland Motors, Gen. Tire & Rubber Co., Marquardt Aircraft, Hycon Mfg. Co., 1946-51; v.p., dir., dir. engring. Resdel Industries, 1951-54, pres., chmn. bd., chief exec. officer, 1954—, chmn. bd., 1981; chmn. bd., chief exec. officer Resdel Engring. Corp., 1965—, Mfrs. Assistance Corp., 1970-81, Fanon/Courier Corp., 1971-81, Resdel Internat., 1972-81. Mem. U.S. Power Squadron, Masons, Sigma Xi. Congregationalist (chmn.). Home: 2052 Midlothian Dr Altadena CA 91001-3412 Office: 3333 Michelson Dr Irvine CA 92715-1681

ABALAKIN, VIKTOR KUZ'MICH, astronomer; b. Odessa, USSR, Aug. 27, 1930; s. Kuz'ma Theodore and Mary Vincenc (Rokos-Sokolova) A.; m. Vera Gavrilovna Kuntsevsky, July 9, 1953; children: Alexis, Katherine. Jr. Scientist, Odessa State U., 1953; Candidate Sci., Leningrad (USSR) State U., 1961; PhD, Leningrad (USSR) State U., Leningrad, 1978. Cert. astronomy. Jr. scientist Inst. Geophysics Acad. Sci., Moscow, 1953-56; jr. scientist Inst. Theoretical Astronomy, Leningrad, 1956-57, sr. scientist, 1964—; sr. scientist Astronomical Obs. Odessa State U., 1960-64; dir. Pulkovo Obs. Russian Acad. Sci., St. Petersburg (formerly Leningrad), 1983—. Author: Theory of Artificial Satellite Motion, 1974, Osnovy Efemeridnoi Astronomii, 1978, Rotation of Rigid Earth, 1982. Served to lt. Russian mil. Recipient USSR State prize Com. on State and Lenin Prizes of USSR, 1984. Mem. USSR Acad. Sci., USSR Acad. Astronautics, Astronomische Gesellschaft, Rotary Internat. Russian Orthodox. Home: 162 Nevsky Ave, 193024 Saint Petersburg Russia Office: Pulkovo Obs, Pulkovo M-140, 196140 Saint Petersburg Russia

ABALOS, TED QUINTO, electronics engineer; b. Binmaley, Pangasinan, Philippines, Mar. 23, 1930; came to U.S., 1969; s. Emilio Quinto and Francisca (Quinto) A.; m. Cristina Lamug, Apr. 18, 1955; children: Abraham, Ruben, Marite, Constantine. BSEM, Mapua Inst. Tech., Manila, 1953; intern grad., Naval Material Command Sch., Washington, 1983; postgrad., Nat. U., 1987—. Mine foreman and supr. Lepanto Cons. Mining Co., Philippines, 1957-64, mine safety engr., 1964-68; asst. mine supt. Black Mountain, Inc., Baguio, Philippines, 1968-69; mine supt. FRONTINO, Inc., Makati, Philippines, 1970-72; mine engr. U.S. Borax & Chem. Corp., Boron, Calif., 1975-78; designer piping engring. C.F. Braun & Engrs. Co., Alhambra, Calif., 1979-80; gen. engr. Fleet Analysis Ctr., Corona, Calif., 1980-82, Naval Ocean System Ctr., San Diego, 1982-84; electronics engr. Naval Electronics Engring. Ctr., San Diego, 1984-92, NISE West, San Diego, 1992—. Contbr. articles to profl. jours. Pres. Philippine-Am. Assn. Antelope Valley, Lancaster, Calif., 1977-79, Pangasinan Assn. San Diego City, 1985-87; v.p. Philippine-Am. Community San Diego City, 1983-84; co-chmn. Filipino Mens Forum San Diego, 1988. Named Outstanding Safety Engr., Safety Engrs. Assn. of The Philippines, 1967. Mem. NSPE, KC, Am. Inst. Mining Engrs., Am. Soc. Test Engrs., Calif. Soc. Profl. Engrs., Mapua Alumni Assn. (chmn. 1992—). Republican. Roman Catholic. Avocations: golf, bowling, traveling, playing violin, community activities. Home: 9480 Hiker Hill Rd San Diego CA 92129-2850 Office: NISE West San Diego PO Box 85137 San Diego CA 92186-5137

ABANERO, JOSE NELITO TALAVERA, aerospace engineer; b. Manila, Philippines, Aug. 10, 1960; came to U.S., 1969; s. Nelson Palma and Librada Bate (Talavera) A.; m. Jerilyn Henrietta Montoya, Feb. 21, 1987; children: Elaina Requel Miller, Joshua Ariel. BS in Aero. and Astronautical Engring., U. Ill., 1983. Cert. acquisition profl. USAF. Commd. 2d lt. USAF, 1984, advanced through grades to capt., 1988; flight test engr. USAF, Edwards AFB, Calif., 1984-88; tng. systems engr. USAF, Wright-Patterson AFB, Ohio, 1989-92; completed svc., 1992. Decorated Achievement medal USAF, 1987, Commendation medal USAF, 1988, 1993. Mem. AIAA, Soc. Flight Test Engrs., The Planetary Soc.

ABARBANEL, HENRY DON ISAAC, physicist, academic director; b. Washington, May 31, 1943; s. Abraham Robert and Selma Helen (Kintberger) A.; m. Gwen Lorita Lewis, May 27, 1965 (div. July 1974); m. Beth Leah Levine, Sept. 12, 1982; children: Brett Lillian, Sara Alexis. BS in Physics, Calif. Inst. Tech., 1963; PhD in Physics, Princeton U., 1966. Rsch. assoc. Stanford (Calif.) Linear Accelerator Ctr., 1967-68; asst. prof. physics Princeton (N.J.) U., 1968-72; physicist Theoretical Physics Dept. Fermi Nat. Accelerator Lab., Batavia, Ill., 1972-81; staff scientist Lawrence Berkeley Lab., 1979-82; rsch. physicist Marine Phys. Lab. Scripps Instn. Oceanography, La Jolla, Calif., 1983—; dir. Inst. for Nonlinear Sci. U. Calif. San Diego, La Jolla, 1986—, acting dir. Inst. for Nonlinear Sci., 1986, dir. Inst. for Nonlinear Sci., 1986—, prof. physics in residence, 1986-88, prof. physics, 1988—; cons. Lockheed Palo Alto (Calif.) Rsch. Lab., 1990—; vis. scientist Ctr. Etudes Nucleaires Saclay, France, 1970, Stanford Linear Accelerator Ctr., 1971, 1975-76, Inst. Haute Études Sci., Bures-sur-Yvette, France, 1989, Inst. Applied Physics Acad. Scis. USSR, Gorky, 1990; adj. prof. physics U. Calif. San Diego, 1982; vis. prof. U. Calif., Santa Cruz, 1975-76, San Diego, 1982, 86, Japan Soc. for the Promotion Sci., 1976; vis. assoc. prof. Stanford U., 1975-76; lectr. dept. physics U. Calif., Berkeley, 1980; chmn. steering com. joint program in nonlinear sci. U. Calif.-NASA, 1987-88; lectr. U. Chgo., 1973, U. Calif., Berkeley, 1980; exchange scientist Landau Inst. for Theoretical Physics NAS, Moscow, Lenningrad, 1977; mem. U.S. Physics Del. to People's Republic China, 1973; mem. JASON, 1974—, mem. steering com., 1985-88, participant, project leader; cons. T-div. Los Alamos (N.Mex.) Nat. Sci. Lab., 1980-84; mem. bd. visitors in physics Office Naval Rsch., 1987, mem. bd. visitors in phys. oceanography, 1989-91; mem. bd. tech. advisors Program in Global Climate Change Dept. Energy, 1990—; invited speaker numerous profl. meetings. Co-editor: Encyclopedia of Chaos, 1991; editor-in-chief Springer-Verlag Series in Nonlinear Sci., 1989—; sci. referee various physics jours. Mem. Planning Commn., Del Mar, Calif., 1989-92, mem. city coun., 1992—. NSF predoctoral fellow, 1963-66, NSF postdoctoral fellow, 1966-67, Woodrow Wilson fellow (hon.); grantee Defense Advanced Rsch. and Projects Agy., 1979-83, 83-84, 86-89, 86-91, U.S. Dept. Energy, 1981-83, Office Naval Rsch., 1983-88, NASA, 1986, 88-89, 90-91, Lockheed Palo Alto Rsch. Lab., 1991—, U.S. Army Rsch. Office, 1991-94. Mem. Am. Phys. Soc., Am. Geophys. Union, Soc. for Indsl. and Applied Maths. Office: U Calif San Diego UCSD MC 0402 La Jolla CA 92093-0402

ABATE, JOHN E., electrical/electronic engineer, communications consultant; b. Paterson, N.J., July 25, 1931; s. Joseph and Lucy Abate; m. Mary Ann Parrillo, July 9, 1955; children: John F., Robert J., Mark J., Holly A. MSEE, Stevens Inst. Tech., 1960; ScD in Elec. Engring., N.J. Inst. Tech., 1967. Registered profl. engr., N.J. Astronautic engr. Singer Inc., Little Falls, N.J., 1956-63; mem. tech. staff Bell Labs., Holmdel, N.J., 1963-70; mgr. AT&T Bell Labs., Holmdel, N.J., 1970—; chmn. synchronization standards group ANSI T1X1.3, 1983-86; mem. U.S. NBS Panel for Basic Standards, 1986-89; expert in field of communications network synchronization. Contbr. numerous articles to profl. jours., conf. procs. and mags. Cubmaster Cub Scouts, Holmdel, 1968-70; chmn. ch. coms., Holmdel, 1968-70. 1st lt. USAF, 1954-56. Bell Labs. fellow, 1991; named to Alumni Honor Roll, N.J. Inst. Tech. Alumni Assn., Newark, 1992, Disting. Alumni, NCE, Newark, 1964; recipient commendation Nat. Security Agy., Washington, 1956. Mem. IEEE (sr. mem.). Roman Catholic. Achievements include invention of adaptive delta modulator used in NASA space shuttle communications system. Home: 7 Ardmore Pl Holmdel NJ 07733

ABATE, RALPH FRANCIS, structural engineer; b. Buffalo, May 29, 1945; s. Morris Anthony A. and Lorraine (Wiese) Hunt; m. Barbara Ann Schwelle, June 24, 1967; children: Annette, Michelle. BCE, SUNY, Buffalo, 1968; MCE, SUNY, 1969. Structural engr. Grumman Aerospace Corp., Bethpaige, N.Y., 1969-74; chief structural engr. Krehbiel Assoc., Inc., Tonawanda, N.Y., 1974-79; v.p. DeSerio Engrs., Buffalo, 1980-89; pres. DeSerio-Abate Engrs., Buffalo, 1990-93, Abate Engring. Assocs., Buffalo, 1993—; vice chmn. adv. bd. civil tech. curriculum Erie Community Coll., Buffalo, 1987—. Reviewer: ASCE Jour. Computing in Civil Engring., 1986—. Named Boss of Yr. Nat. Assn. Women in Constrn., Buffalo chpt., 1991, Engring. Mgr. of Yr. N.Y. State Profl. Engring. Soc., Erie-Niagara chpt., 1991. Mem. Am. Concrete Inst. (chpt. pres. 1980-81, McKaig Meml. award 1985), N.Y. State Cons. Engrs. Coun. (chpt. pres. 1983-84), Assn. for Bridge Constrn. and Design (chpt. pres. 1983-84), ASCE (chpt. dir. 1988—). Achievements include first consultant in N.Y. to use high density concrete for bridge overlay. Office: Abate Engring Assocs PC 1888 Statler Towers Buffalo NY 14202

ABBASCHIAN, REZA, materials science and engineering educator; b. Zanjan, Iran, Jan. 23, 1944; came to U.S., 1966; s. Ebrahim and Motahreh Abbaschian; m. Janette S. Johnson, Sept. 6, 1973; children: Lara S., Cyrus E. BS, U. Tehran, Iran, 1965; MS, Mich. Tech. U., 1968; PhD, U. Calif., Berkeley, 1971. Rsch. analyst U.S. Steel Corp., Gary, Ind., 1967; rsch. asst. U. Calif., 1968-71; asst. prof. Shiraz (Iran) U., 1972-74, assoc. prof., 1974-80; with U. Fla., Gainesville, 1980—, acting chmn., 1986-87, chmn., prof., 1987—; vis. assoc. prof. dept. metallurgy and mining engring. U. Ill., Urbana, 1976-78; vis. scientist MIT, Cambridge, 1980, NASA Space Processing Lab., Marshall Space Flight Ctr., Huntsville, Ala., 1981; chmn. Shiraz U., 1974-76. Co-author: Physical Metallurgy Principles, 1992; editor: Grain Refinement in Castings and Welds, 1983, Solidification Processing of Eutectic Alloys, 1988, Modeling and Control of Casting, 1988. NSF grantee, 1984-86. Mem. Univ. Materials Coun. (chmn. policy com., exec. coun. 1990—). Office: U Fla Dept Materials Sci and Engring North-South Dr Gainesville FL 32611

ABBASI, TARIQ AFZAL, psychiatrist, educator; b. Hyderabad, India, Aug. 13, 1946; came to U.S., 1976, naturalized, 1983; s. Shujaat Ali and Salma Khatoon (Siddiqui) A.; m. Kashifa Khatoon, Nov. 10, 1972; children—Sameena, Omar, Osman. B.S., Madrasa-I-Aliya, Hyderabad, 1964; M.B.B.S., Osmania Med. Coll., Hyderabad, 1970; Diploma in Psychol. Medicine, St. John's Hosp., U. Sheffield (Eng.), 1976. Diplomate Am. Bd. Psychiatry and Neurology; diplomate in psychiatry Royal Coll. Physicians of Eng. Sr. house officer St. John's Hosp., Lincoln, Eng., 1972-73, registrar, 1973-76; resident in psychiatry Rutgers Med. Sch., Piscataway, N.J., 1976-79, chief resident, 1979, dir. adult in-patient services Community Mental Health Ctr., Rutgers Med. Sch., also asst. prof. psychiatry, 1979-82; staff psychiatrist Northville Regional Psychiat. Hosp. (Mich.), 1982-83, div. dir., 1983—; cons. psychiatrist Rahway State Prison (N.J.), 1979-82; clin. instr. psychiatry Wayne State U. Med. Sch., Detroit. Mem. Am. Psychiat. Assn., Mich. Psychiat. Soc. Office: Northville Regl Psychiat Hosp 41001 7 Mile Rd Northville MI 48167-2698 also: Personal Dynamics Ctr 23810 Michigan Ave Dearborn MI 48124 also: 999 Haynes Ste 245 Birmingham MI 48008

ABBEY, LELAND RUSSELL, internist; b. N.Y.C., Dec. 18, 1945; s. Solomon and Bettina (Zipper) A. BA magna cum laude, Drew U., 1967; MD, Hahnemann U., 1971. Diplomate Am. Bd. Internal Medicine, Am. Bd. Rheumatology. Intern in internal medicine Mt. Sinai Hosp., N.Y.C., 1971-72, resident in internal medicine, 1972-74, fellow in rheumatology, 1974-76; staff physician, chief sect. rheumatology Bronx (N.Y.) VA Med. Ctr., 1976—, asst. chief med. svc., 1978—; asst. prof. Medicine Mt. Sinai Hosp., N.Y.C., 1991—. Contbr. chpts. to books, articles to profl. jours. Fellow Am. Coll. Rheumatology; mem. AMA, AAAS, Am. Med. Informatics Assn., N.Y. Acad. Scis., Alpha Omega Alpha. Office: Bronx VA Med Ctr 130 W Kingsbridge Rd Bronx NY 10468

ABBONDANZO, SUSAN JANE, research geneticist; b. Plainfield, N.J., Oct. 25, 1962; d. Peter Arthur and Rita Elsie (Towler) Bassett; m. James Vincent Abbondanzo, June 10, 1989; 1 child, Kimberly Marie. BS, Rutgers U., 1986; MS in Biology, Fairleigh Dickinson U., 1991. Lab. asst. environ. sci. dept. Rutgers U., New Brunswick, N.J., 1985-86; rsch. asst. Albert Einstein Coll. Medicine, Bronx, N.Y., 1986-88; assoc. scientist Roche Inst. Molecular Biology, Hoffmann-LaRoche, Nutley, N.J., 1988—. Contbr. articles to Cell, Nature; contbr. chpt. to Methods In Enzymology, 1993. Mem. Sigma Xi. Home: 24 Diana Rd Morris Plains NJ 07950 Office: Roche Inst Molecular Biology 340 Kingsland St Nutley NJ 07110

ABBOTT, ISABELLA AIONA, biology educator; b. Hana, Maui, Hawaii, June 20, 1919; d. Loo Yuen and Annie Patseu (Chung) Aiona; m. Donald P. Abbott, Mar. 3, 1943 (dec.); 1 dau., Ann Kaiue Abbott. A.B., U. Hawaii, 1941; M.S., U. Mich., 1942; Ph.D., U. Calif., Berkeley, 1950. Prof. biology Stanford U., 1972-82; G.P. Wilder prof. botany U. Hawaii, 1978—; vis. research biologist and tchr., Japan and Chile. Author: (with G.J. Hollenberg) Marine Algae of California, 1976, La'au Hawaii, traditional Hawaiian uses of plants, 1992; contbr. articles to profl. jours. Co-recipient N.Y. Bot. Garden award for best book in botany, 1978. Fellow AAAS; mem. Internat. Phycological Soc. (treas. 1964-68), Western Soc. Naturalists (sec. 1962-64, pres. 1977), Phycological Soc. Am., Brit. Phycological Soc., Hawaiian Bot. Soc. Office: U Hawaii Botany Dept 3190 Maile Way Honolulu HI 96822-2279

ABBOTT, JOHN RODGER, electrical engineer; b. L.A., Aug. 2, 1933; s. Carl Raymond and Helen Catherine (Roche) A.; m. Theresa Andrea McQuaide, Apr. 20, 1968. BS with honors, UCLA, 1955; MSEE, U. So. Calif., 1957. Registered profl. engr., Calif. Advanced study engr. Lockheed Missile Systems, L.A., 1955-56; radar systems engr. Hughes Aircraft Co., L.A., 1956-59; devel. engr. Garrett Airesearch Co., L.A., 1959-63, instr. in-plant tng. program, 1962-63; asst. project engr. Litton Industries, L.A., 1963; space power systems engr. TRW Systems, L.A., 1963-65; engr. specialist L.A. Dept. Water and Power, 1965-92; engr. specialist Abtronix, 1992—; frequency coordination chmn. Region X, Utilities Telecommunications Coun., 1977-79, sec.-treas. Utilities Telecommunication Coun., 1979-80; instr. amateur radio course L.A. City Schs., Birmingham High Sch., Van Nuys, Calif., 1965-66, Los Feliz Elem. Sch., Hollywood, Calif., 1990—. Contbr. articles to profl. jours. Mem. IEEE, Am. Radio Relay League (Pub. Svc. award 1971), Tau Beta Pi. Office: Abtronix PO Box 220066 Santa Clarita CA 91322-0066

ABBOTT, REGINA A., neurodiagnostic technologist, consultant, business owner; b. Haverhill, Mass., Mar. 5, 1950; d. Frank A. and Ann (Drelick) A. Student, Pierce Bus. Sch., Boston, 1967-70, Children's Hosp. Med. Ctr. Sch. EEG Tech., Boston, 1970-71. Registered technologist. Tech. dir. electrodiagnostic labs. Salem Hosp., 1972-76; lab. dir. neurophysiology Tufts U. New Eng. Med. Ctr., Boston, 1976-78; clin. instr. EEG program Laboure Coll., Boston, 1977-81; adminstrv. asst. dept. Neurology Mt. Auburn Hosp., Cambridge, Mass., 1978-81; tech. dir. clin. neurophysiology Drs. Diagnostic Service, Virginia Beach, Va.; tech. dir. neurodiagnostic ctr. Portsmouth Psychiatric Ctr., 1981-87; pres., owner Commonwealth Neurodiagnostic Services, Inc., 1986—; co-dir. continuing edn. program EEG tech., Boston, 1977-78; mem. adv. com. sch. neurodiagnostic tech. Laboure Coll., 1977-81, sch. EEG Tech. Children's Hosp. Med. Ctr., Boston, 1980-81; assoc. examiner Am. Bd. Registration of Electroencephalographic Technologists, 1977-83; mem. guest faculty Oxford Medilog Co., 1986; cons. Nihon Kohden, Irvine, Calif., 1981-83; cons., educator Teca Corp., Pleasantville, N.Y., 1981-87; allied health profl. staff mem. Virginia Beach Gen. Hosp., Humana Hosp. Bayside, Virginia Beach. Contbr. articles to profl. jours. EIL scholar, Poland, USSR, 1970. Mem. NAFE, Am. Soc. Electroneurodiagnostic Technologists, New England Soc. EEG Technologists (bd. dirs., sec., tng. and edn. com., faculty tng. and edn.), Epilepsy Soc. of Mass. Avocations: running, gardening, photography, reading, investing.

ABBOUD, FRANCOIS MITRY, physician, educator; b. Cairo, Egypt, Jan. 5, 1931; came to U.S., 1955, naturalized, 1963; s. Mitry Y. and Asma (Habac) A.; m. Doris Evelyn Khal, June 5, 1955; children—Mary Agnese, Susan Marie, Nancy Louise, Anthony Lawrence. Student, U. Cairo, 1948-52; M.B., B.Ch., Ein Chams U., 1955; D. (hon.), U. Lyon, France, 1991. Diplomate Am. Bd. Internal Medicine (bd. govs. 1987—). Intern Demerdash Govt. Hosp., Cairo, 1955; resident Milw. County Hosp., 1955-58; Am. Heart Assn. research fellow cardiovascular labs. Marquette U., 1958-60; Am. Heart Assn. advanced research fellow U. Iowa, 1960-62, asst. prof., 1961-65, assoc. prof. medicine, 1965-68, prof. medicine, 1968—, prof. physiology and biophysics, 1975—, Edith King Pearson prof. cardiovascular rsch., 1988—; dir. cardiovascular div., 1970-76, chmn. dept. internal medicine, 1976—, dir. cardiovascular center, 1974—; attending physician VA Hosp., Iowa City, 1963—, U. Iowa Hosps., 1961—; chmn. rsch. rev. com. Nat. Heart, Lung and Blood Inst., 1978-80. Editor: Circulation Research, 1981-86. Recipient European Traveling fellow French Govt.,1948; NIH career devel. awardee, 1962-71; Dr. Honoris Carusy U. of Lyon France, 1991. Master ACP; mem. Inst. Medicine NAS, AMA, Am. Soc. Clin. Investigation, Central Soc. for Clin. Rsch. (pres. elect 1984-85, pres. 1985-86), Soc. Exptl. Biology and Medicine, Am. Heart Assn. (bd. dirs. 1977-80, past chmn. rsch. coms., award of merit 1982, Disting. Achievement award 1988, CIBA award for hypertension rsch. 1990, pres. elect 1989-90, pres. 1990-91), Am. Fedn. Clin. Rsch. (pres. 1971-72), Assn. Univ. Cardiologists, Assn. Profs. Medicine (Robert H. Williams Disting. Chmn. of Medicine award 1993, bd. dirs. 1993—), Assn. Am. Physicians (treas. 1979-84, councillor 1984-89—, pres. elect 1989-90, pres. 1990-91), Am. Physiol. Soc. (chmn. circulation group 1979-80, chmn. clin. physiology sect. 1979-83, publ. com. 1987-90, Wiggers award 1988), Am. Clin. and Climatological Assn. (councillor 1992), Am. Soc. Pharmacology and Exptl. Therapeutics (award exptl. therapeutics 1972), Sigma Xi, Alpha Omega Alpha (bd. dirs. 1989—). Achievements include research and publications in cardiovascular physiology on neurohumoral control of circulation in humans and animals. Home: 24 Kennedy Pkwy Iowa City IA 52240 Office: U Iowa Coll Medicine Dept Internal Medicine U Iowa Hosps and Clinics Iowa City IA 52242

ABDELHAK, SHERIF SAMY, health science executive; b. Cairo, Egypt, Mar. 14, 1946; came to U.S., 1971; s. Samy Hassan and Aziza (Sobhy) A.; m. Marlynn Singleton; children: S. Jonathan, Matthew H., Derek A., Aaron W. BA in Econs., Polit. Sci., Am. U., Cairo, 1968; MBA, U. Pitts., 1976. cons. Allegheny Health, Edn. and Rsch. Found., Pitts., 1985-86, pres., chief exec. officer, 1986—, bd. dirs.; cons. Price Waterhouse Health Care Div., Pitts., 1985-87; adj. asst. prof. U. Pitts., 1977-89; bd. dirs. Vol. Hosps. of Am., Dallas, 1986—, Juvenile Diabetes Found., 1987—, Hosp. Coun. Western Pa., 1987—; pres., chief exec. officer Greater Canonsburg (Pa.) Health System, 1982-86. Chmn. Hosp. Trustee Forum Prudent Buyer Com., 1989—. Fellow Am. Coll. Healthcare Execs.; mem. Hosp. Assn. Pa. (chmn. uncompensated care com. 1989), Greater Pitts. C. of C. (bd. dirs.). Republican. Presbyterian. Avocation: breeding straight Egyptian horses. Office: Allegheny Health Edn Rsch 320 E North Ave Pittsburgh PA 15212*

ABDELHAMIED, KADRY A., biomedical engineer; b. Cairo, Egypt, July 9, 1954; came to U.S., 1980; s. Abdelmohsen and Kokb (Bayomi) A. BS, Cairo U., 1977; PhD, Ohio State U., 1986. Grad. teaching asst. Cairo U., Egypt, 1977-80, asst. prof., 1986-89; asst. prof. La. Tech U., Ruston, 1989—; cons. IBM Cairo Sci. Ctr., 1987-88, Ohio State U., Columbus, 1988-89. Contbr. articles to profl. jours. Mem. IEEE, Rehab. Engring. Soc. of N.Am., Biomed. Engring. Soc., Internat. Neural Networks Soc., Sigma Xi. Home: 1041 Sybil Dr Ruston LA 71270 Office: Louisiana Tech U Biomed Engring Dept Ruston LA 71272

ABDELLAH, FAYE GLENN, retired public health service executive; b. N.Y.C., Mar. 13, 1919; d. H.B. and Margaret (Glenn) A. BS in Teaching, Columbia U., 1945, MA in Teaching, 1947, EdD, 1955; LLD (hon.), Case Western Res. U., 1967, Rutgers U., 1973; DSc (hon.), U. Akron, 1978, Cath. U. Am., 1981, Monmouth Coll., 1982, Ea. Mich U., 1987, U. Bridgeport, 1987, Georgetown U., 1989; D Pub. Svc. (hon.), Am. U., 1987; LHD (hon.), Georgetown U., 1989, U. S.C., 1991; D Pub. Svc., U. S.C., 1991. RN. Commd. officer USPHS, Rockville, Md., 1949, advanced through grades to rear adm., 1970, asst. surgeon gen., chief nurse officer, 1970-87, dep. surgeon gen., 1981-89, chief nursing edn. br., div. nursing, 1949-59; surgeon gen. USPHS, 1989; chief rsch. grants br. Bur. Health Manpower Edn., NIH, HEW, Rockville, 1959-69; dir. Office Rsch. Tng. Nat. Ctr. for Health Svcs. R & D, Health Svcs. Mental Health Adminstrn., Rockville, 1969; acting dep. dir. Nat. Ctr. for Health Svcs. R & D, Rockville, 1971, Bur. Health Svcs. Rsch. and Evaluation, Health Resources Adminstrn., Rockville, 1973; dir. Office Long-Term Care, Office Asst. Sec. for Health, HEW, Rockville, 1973-80; exec. dir. grad. sch. nursing uniformed svcs. U. Health Scis., Bethesda, Md., 1993; exec. dir., acting Grad. Sch. Nursing Uniformed Svcs. U. Health Scis., Bethesda, Md., 1993; prof. nursing, Emily Smith chair U. S.C., Columbia, 1990-91; dean Grad. Sch. Nursing, Uniformed Srvs. U. Health Scis., 1993—. Author: Effect of Nurse Staffing on Satisfactions with Nursing Care, 1959, Patient Centered Approaches to Nursing, 1960, Better Patient Care Through Nursing Research, 1965, 2d edit., 1979, 3d edit., 1986, Intensive Care, Concepts and Practices for Clinical Nurse Specialists, 1969, New Directions in Patient Centered Nursing, 1972; Contbr. articles to profl. jours. Recipient Mary Adelaide Nutting award, 1983, hon. recognition ANA, 1986, Oustanding Leadership award U. Pa., 1987, 99, Disting. Svc. award, 173-89, Surgeon Gen.'s medal and medallion, 1989, Allied-Signal Achievement award in aging, 1989, Gustav O. Lienhard award Inst. Medicine, NAS, 1992. Fellow Am. Acad. Nursing (charter, past v.p., pres.); mem. Am. Psychol. Assn., AAAS, Assn. Mil. Surgeons U.S., Sigma Theta Tau (Disting. Rsch. Fellow award 1989), Phi Lambda Theta. Home: 3713 Chanel Rd Annandale VA 22003-2024

ABDELNOOR, ALEXANDER MICHAEL, immunologist, educator; b. N.Y.C., Jan. 18, 1941; s. Michael Dib and Evelyn Isber (Massabni) A.; m. May Khalil Achkar, Aug. 25, 1970; children: Natalie, Michael, Cheryl. BS in Pharmacy, Am. U., Beirut, Lebanon, 1964; PhD, U. Mich., 1969. Postdoctoral fellow U. Mich., Ann Arbor, 1969-70; NIH postdoctoral fellow Temple U., Phila., 1970-72; lab. dir. Fanar (Lebanon) Rsch. Inst., 1973-76; from asst. prof. to assoc. prof. Am. U., Beirut, 1977-84, prof., 1984—, chmn. dept. microbiology and immunology faculty medicine, 1992—. Author: (with others) Cellular Antigens, 1972, Benefical Effects of Endotoxins, 1983; contbr. articles to Infection and Immunity, Annals of Saudi Medicine, Jour. Immunology and Immunopharmacology. F.G. Novy fellow U. Mich., Ann Arbor, 1969. Mem. Am. Assn. Immunologists, Am. Soc. Microbiology, N.Y. Acad. Scis., Internat. Endotoxin Soc., Am. Assn. Blood Banks. Achievements include research on structure-function relationship of endotoxins. Office: Am U Beirut 850 3rd Ave New York NY 10022-6222

ABDEL-RAHMAN, MOHAMED, plant pathologist, physiologist; b. Cairo, Egypt, July 3, 1941; came to U.S., 1965; m. Janine Neerdaels, Dec. 21, 1967 (div. 1989); children: Magda, Susan. BS, Ain-Shams U., 1962, MS in Plant Pathology, 1964; PhD in Hort., U. Fla., 1970. Instr. Ain-Shams Coll. Agr., 1962-65; rsch. assoc. U. Fla. Coll. Agrl., Gainesville, 1968-70; product devel. mgr., planning & devel. mgr., bus. mgr. Agway Inc., Syracuse, N.Y., 1972-82; v.p. rsch. devel. Food Source Inc., Larkspur, Calif., 1982-83; dir. Agri-Intelligence Doane Inc., Princeton, N.J., 1983-85; pres. I.A-M Corp., Trenton, N.J., 1985-89; mktg. product mgr. fungicides and herbicides BASF, Inc., Rsch. Triangle Park, N.C., 1989—; sec. gen. PGRWG, 1977-79; pres. I.A.M. Corp., Trenton, N.J., 1985-89. Author: Agway Chemical and Crop-Protection Guides, 1972-82; contbr. over 240 articles to profl. jours. Vol. Flood Relief Orgn., N.Y., 1972. Recipient Nat. award Excellence, Egyptian Govt., 1962. Mem. Am. Soc. Hort. Sci. (program chmn. 1970-76), Am. Phytopathological Soc. (numerous exec. positions 1971—), Plant Growth Regulator Soc. Am. (numerous exec. positions 1973—). Muslim. Achievements include patent for the fungicide glyodin; devel. 2 new plant growth regulators. Home: 23 Beechtree Ct Durham NC 27713-1942 Office: BASF Corp PO Box 13528 Durham NC 27709-3528

ABDUL, ABDUL SHAHEED, hydrogeologist; b. Parika, Essequibo, Guyana, Apr. 20, 1952; came to U.S. 1985; s. Sattaur and Saffiran (Ali) A.; m. Saharoon Nesa, Sept. 24, 1972; children: Elisha S., Shelly N. BS in Physics, U. Guyana, Georgetown, 1976; MS in Engring., U. Guelph, Ont., Can., 1981; PhD in Hydrogeology, U. Waterloo, Ont., Can., 1985. Sr. rsch. scientist GM Rsch. Lab., Warren, Mich., 1985-88, staff rsch. scientist, 1988—, sect. mgr., 1989—. Editorial bd. Ground Water, Dubun, Ohio, 1989—; contbr. articles to profl. jours. Mem. Am. Inst. Hydrology, Am. Geophys. Union, Assn. Ground Water Scientists and Engrs. Achievements include 4 patents on remediation technologies. Home: 4407 Reilly Dr Troy MI 48098 Office: GM Rsch Environ Sci Dept Warren MI 48090-9055

ABDUL, CORINNA GAY, software engineer, consultant; b. Honolulu, Aug. 10, 1961; d. Daniel Lawrence and Katherine Yoshie (Kanada) A. BS in Computer Sci., U. Hawaii, 1984. With computer support for adminstrv. and fiscal svcs. U. Hawaii, Honolulu, 1982-84; mem. tech. staff II test systems and software engr. dept. of Space & Tech. Group TRW Inc., Redondo Beach, Calif., 1985-89; systems software engr. II, Sierra On-Line, Inc., Oakhurst, Calif., 1989-90; sr. programmer, analyst Decision Rsch. Corp., Honolulu, 1990-92; ind. computer cons. Honolulu, Hawaii, 1992—. Home: 1825 Anapuni St Apt 201 Honolulu HI 96822

ABDULLA, MOHAMED, physician, educator; b. Alwaye, Kerala, India, Dec. 8, 1937; s. Mohamed Moulavi and Nacheema Mohamed; m. Nasemma Abdulla, Jan. 16, 1976; children; Nadia, Sabina. BSc, U. Punjab, Labore, Pakistan, 1958; MB, U. Lund, Sweden, 1974, MD, 1978, PhD, 1985. Cert. specialist in clin. chemistry. Sales officer Packages Ltd., Lahore, 1960-62; lab. technician Tottenham County Sch., London, 1962-64; engr. PLM, Malmö, Sweden, 1964-67; researcher McMaster U., Hamilton, Ont., Can., 1967-69; rsch. assoc. Swedish Med. Bd., Dalby, 1970-76; physician, researcher U. Hosp., Lund, 1976-88; prof., chmn. Baqai Med. Coll., Karachi, Pakistan, 1988-90; prof. Hamdard U., New Delhi, 1990—; Organizer internat. sci. mtgs., Sweden, Denmark, Norway, U.S.A., Japan, Portugal, Turkey, Indian and Pakistan; sci. advisor to several pharm. cos. worldwide. Editor: Nutrition and Old Age, 1979, other books, 1975-90; contbr. over 200 articles to profl. jours. Founder, v.p. UNESCO Inst., Lyon, France. Fellow Swedish Med. Soc.; mem. Internat. Soc. Trace Element Rsch. in Humans (sec. 1985-88), Internat. Coll. Nutrition, Internat. Union Elementologists (v.p. 1984—). Avocations: cricket, golf, fishing, music, writing. Home: Harjagersvägen 9, S 232 54 Åkarp Sweden Office: Hamdard U, PO Box 5133, S 220 05 Lund Sweden

ABE, YOSHIHIRO, ceramic engineering educator; b. Nagoya, Aichi-ken, Japan, Jan. 13, 1935; s. Kiichi and Shige Abe; m. Keiko Itoh, Mar. 27, 1966; children: Masanari, Kenshi, Hiroyuki. BS, Nagoya Inst. Tech., Japan, 1958; PhD in Engring., Nagoya U., Japan, 1977. Researcher Nippon Mineral Fibers Co., Kanagawa, Japan, 1955-58; instr. ceramic engring. Nagoya (Japan) Inst. Tech., 1961-66, lectr., 1966-76, assoc. prof., 1976-79, prof., 1979—. Co-author: Topics in Phosphorus Chemistry, 1983, Phosphate Materials, 1989; patent unidirectional crystalli; contbr. over 150 articles to profl. jours. Recipient Bioceramics award Ichimura Found., 1978, Acad. award Ceramic Soc. Japan, 1980, Acad. award Japan Assn. Inorganic Phosphorous Chemistry, 1993. Avocations: fishing, tennis. Home: 6-1705 Higashiyama, Nisshin-cho, Aichi-gun Aichi-Ken 470-01, Japan Office: Nagoya Inst Tech, Gokiso-cho, Showa-ku, Nagoya Aichi 466, Japan

ABEL, ULRICH RAINER, biometrician, researcher, financial consultant; b. Duisburg, Fed. Republic of Germany, Jan. 23, 1952; s. Heinrich and Else (Griebenow) A.; m. Michelle Jacqueline, Mar. 30, 1978; children: Clémentine, Frédéric. Diploma in Math., U. Hannover, Fed. Republic of Germany, 1975, PhD in Human Biology, 1986, Phd in Math., 1977; Habilitation in Biostatistics and Epidemiology, U. Heidelberg, Fed. Republic of Germany, 1989. Asst. prof. faculty Applied Math. U. Bielefeld, Fed. Republic of Germany, 1977-80; biometrician Tumor Ctr. Nat. Cancer Rsch. Ctr., Heidelberg, Fed. Republic of Germany, 1980—; bd. dirs. Boing Vermoegensverwaltung AG, Dusseldorf, N.Y. Broker Deutschland AG, Duesseldorf, INCAM AG, Dusseldorf, Akademie Fuer onkologisch-biologische Forschung, Stuttgart/Heidelberg; v.p. Germania Nichtraucherschutz eV., Hamburg. Contbr. over 120 articles to profl. jours.; author of several books. Recipient prize Johann George Zimmermann Fonds, 1985, Instant Foerderpreis, Inst. Standardisierung und Dokumentation, 1991, Ernst Krokowski prize Gesellschaft Biologische Krebsabwehr, 1991. Mem. Gesellschaft Medizinische Dokumentation and Statistik, Internat. Biometrische Gesellschaft. Avocations: violin, soccer, chess. Home: Rainweg 2, 69118 Heidelberg Germany Office: Deutsches Krebsforschungszentrum, Neuenheimer Feld 280, 69120 Heidelberg Germany

ABELA, GEORGE SAMIH, medical educator, internist, cardiologist; b. Tripoli, Lebanon, Jan. 1, 1950; came to U.S., 1976; s. Anthony George and Maro (Kozma) A.; m. Sonia Zablit, May 14, 1977; children: Oliver George, Andrew John, Scott Anthony. BSc in Chemistry and Biology, Am. U. of Beirut, 1971, MSc in Pharmacology, 1974, MD, 1976. Diplomate Am. Bd. Internal Medicine, Am. Bd. Cardiology. Intern in pathology Emory U., Atlanta, 1976-77, resident in medicine, 1977-80; fellow in cardiology U. Fla., Gainesville, 1980-83, asst. prof. medicine and cardiology, 1983-87, assoc. prof., 1987-93; assoc. prof. Harvard Med. Sch., 1993—; bd. dirs. interventional rsch. Deaconess Hosp., Boston. Assoc. editor Jour. Internat. Cardiology; patentee laser apparatus. Rsch. fellow Am. Heart Assn., 1982-83; Merck fellow Am. Coll. Cardiology, 1982-83; recipient New Investigator Rsch. award NIH, 1983-86, Rsch. Career Devel. award NIH, 1986-91. Fellow Am. Soc. for Lasers in Medicine, Am. Coll. Cardiology, Am. Soc. Laser Medicine and Surgery (chmn. Standards of Tng. and Practice com. 1987-88, bd. dirs., Govs. award 1989); mem. Am. Heart Assn.

ABELLA, ISAAC DAVID, physicist, educator; b. Toronto, Ont., Can., June 20, 1934; s. Samuel A. and Sarah Freida Abella; children: Benjamin, Sarah. BA in Physics, U. Toronto, 1957; MA in Physics, Columbia U., 1959, PhD, 1963. Rsch. assoc. Columbia Radiation Lab., N.Y.C., 1963-65; mem. faculty dept. physics U. Chgo., 1965—, prof., 1986—; cons. Mithras, Inc., Cambridge, Mass., 1966-76; vis. scientist chemistry div. Argonne Nat. Lab., 1978-82, physics div. Atomic Physics Group, 1982—, optical scis. Naval Rsch. Lab., Washinton, 1981-82; mem. tech. rev. com. Office Naval Rsch., Arlington, Va., 1988-90; com. mem. Nat. Acad. Scis., 1992-95. Contbr. to profl. publs. Fellow Am. Phys. Soc., Optical Soc. Am.; mem. AAAS, Sigma Xi.

ABELSON, PHILIP HAUGE, physicist; b. Tacoma, Wash., Apr. 27, 1913; s. Ole Andrew and Ellen (Hauge) A.; m. Neva Martin, Dec. 30, 1936; 1 child, Ellen Hauge Abelson Cherniavsky. BS, Wash. State Coll., 1933, MS, 1935; PhD, U. Calif., 1939; DS, Yale U., 1964, So. Meth. U., 1969, Tufts U., 1976; Duke U., 1981; DHL, U. Puget Sound, 1968. Asst. physicist Carnegie Instn. of Washington, 1939-41, chmn. biophysics sect. dept. terrestrial magnetism, 1946-53, dir. Geophysics Lab., 1953-71, pres. instn., 1971-78; trustee from, 1978—; assoc. physicist Naval Research Lab, Washington, 1941-42, physicist, 1942-44, sr. physicist, 1944-45, prin. physicist, 1945; civilian in charge Naval Research Lab. br. Navy Yard, Phila., 1944-45; resident fellow Resources for the Future Inc., Washington, 1985-88; Chmn. com. on radiation cataracts NRC, 1949-57, sub-com. on shock, 1950-53, mem. Plowshare adv. com., 1959-63; gen. adv. com. AEC, 1960-63; mem. biophysics and biophys. chemistry study sect. Nat. Inst. Arthritis and Metabolic Diseases, NIH, 1956-59, mem. phys. biology tng. grants com. 1958-60, bd. sci. counselors, 1960-63; cons. NASA, 1960-63; sci. advisor AAAS, from 1985, acting exec. dir., 1989. Author: Energy for Tomorrow, 1975, Enough of Pessimism, 1985; mem. adv. bd. Jour Nat. Cancer Inst., 1947-52; editor: Researches in Geochemistry, 1959, Vol. 2, 1967; Energy: Use, Conservation and Supply, 1974, Food: Politics, Economics, Nutrition, and Research, 1975; Materials; Renewable and Nonrenewable, 1976, Electronics: The Continuing Revolution, 1977; co-editor Jour. Geophys. Research, 1959-65; editor Sci. mag., 1962-85, dep. editor for engring. and applied scis., from 1985. Recipient Disting. Civilian Service medal, 1945, ann. award phys. sci. Washington Acad. Sci., 1950, Disting. Alumnus award Wash. State U., 1962, Hillebrand award Chem. Soc. Washington, 1962, Modern Medicine award, 1967, Mellon award Carnegie-Mellon U., 1970, Joseph Priestley award Dickinson Coll., 1973, Sci. Achievement award AMA, 1974, Hon. Scroll award D.C. Inst. Chemists, 1976, Kalinga prize UNESCO, 1972, Disting. Pub. Service award NSF, 1984, Nat. Medal Sci., 1989, PublicWelfare Medal, NAS, 1992. Fellow Am. Phys. Soc., Geol. Soc. Am., Mineral. Soc. Am., Geol. Soc. Washington, Am. Acad. Arts and Scis.; mem. Am. Nuclear Soc., Seismol. Soc. Am., Internat. Union Geol. Scis. (pres. 1972-76), Brit. Biochem. Soc., Brit. Mineral. Soc., Am. Chem. Soc., Am. Philos. Soc., Soc. Am. Bacteriologists, Am. Geophys. Union (pres. 1972-74), Am. Assn. Petroleum Geologists, Geochem. Soc., Washington Acad. Scis., Biophys. Soc., Philos. Soc. Washington, Phi Beta Kappa (senator-at-large 1972—), Sigma Xi, Nat. Acad. Scis., Nat. Inst. Medicine (sr.). Club: Cosmos (pres. 1972). Office: AAAS 1333 H St NW Washington DC 20005-4707

ABENDROTH, REINHARD PAUL, materials engineer; b. St. Louis, Mar. 19, 1931; s. August Paul and Gertrude Marie (Ellersieck) A.; m. Betty Jane Schlake, Aug. 27, 1955; children: Paul M., Nancy J. BS in Metall. Engring., U. Mo., Rolla, 1953, MS in Metall. Engring., 1954, PhD in Metall. Engring., 1957. Registered profl. engr., Ohio. Rsch. metallurgist Union Carbide Corp., Niagara Falls, N.Y., 1957-60; rsch. metallurgist Owens-Ill. Inc., Toledo, 1960-73, glass technologist, 1973-77, sect. chief phys. testing, 1977-82; mgr. lab. svcs. Kimble Glass Inc. subs. of Owens-Ill. Inc., Vineland, N.J., 1982—; Mem. Internat. Commn. on Glass, Istanbul, 1983—. Contbr. articles to profl. jours., chpts. to books. Mem. ASTM (mem. com. 1983—), Internat. Orgn. for Standardization (mem. com. 1983—), Am. Ceramic Soc., Minerals, Metals and Materials Soc., Soc. Glass Tech. Home: 2892 Garwood Ln Vineland NJ 08360 Office: Kimble Glass Inc 537 Crystal Ave Vineland NJ 08360

ABERNATHY, JACK HARVEY, petroleum, utility company and banking executive; b. Shawnee, Okla., June 10, 1911; s. George Carl and Carrie (Howell) A.; m. Mary Ann Staig, June 13, 1932 (dec.); children: Jack Harvey, Carrilee Abernathy Bell; m. Virginia Watson, Dec. 21, 1974 (dec. Aug. 1986). B.S. in Petroleum Engring, U. Okla., 1933. Petroleum engr. Sinclair Oil & Gas Co., 1933-34; chief engr., gen. prodn. supt. Sunray DX OIL Co., Tulsa, Okla., 1935-45; pres. Seneca Oil Co., Oklahoma City, 1959-65, Post Oak Oil Co., Oklahoma City, 1966-72; chmn. exec. com., dir. Entex, Inc., Houston, 1972-85; pres., chmn. dir. Big Chief Drilling Co., Oklahoma City, 1946-82; dir. Entex Petroleum, Inc., Houston, Entex Coal Co., Houston, Hinderliter Industries, 1981-85; dir., mem. exec. com. Liberty Nat. Bank & Trust Co., Oklahoma City, 1967-85; dir., chmn. Southwestern Bank & Trust Co., Oklahoma City, 1964—; mem., past chmn. Nat. Petroleum Council, 1957-83; dir., vice chmn. bd. dirs. Gen. Producing and Drilling Co., Houston, 1981-83. Contbr. numerous tech. articles to profl. jours. Bd. dirs. The Benham Group, Oklahoma City Zool. Soc., 1983—, Oklahoma City Art Mus.; trustee U. Okla. Found., Oklahoma City Community Found, Okla. Ednl. TV Authority, Okla. Found. for Excellence. Recipient numerous awards inlcuding Disting. Engring. Grad award Okla. U., 1990.; elected to Okla. State Hall of Fame, 1971, Nat. Petroleum Hall of Fame, 1990. Mem. Mid-Continent Oil and Gas Assn. (dir., past chmn.), Am. Petroleum Inst. (dir., exec. com. 1964-83), Oklahoma City C. of C. (dir.), All-Am. Wildcatters, Nat. Soc. Profl. Engrs., Soc. Petroleum Engrs., Internat. Assn. Oilwell Drilling Constractors (pres., bd. dirs.), Sigma Tau, Tau Beta Pi, Pi Epsilon Tau, Beta Gamma Sigma. Presbyterian. Clubs: Men's Dinner, Petroleum (Oklahoma City); Oklahoma City Golf and Country. Home: 6208 Waterford # 85 Oklahoma City OK 73118 Office: Southwestern Bank & Trust PO Box 19100 Oklahoma City OK 73144-0100

ABERNETHY, VIRGINIA DEANE, population and environment educator; b. Havana, Cuba, Oct. 4, 1934; d. Bernard Charles and Helen Adele (Arnold) Deane; m. John Benjamin Kendrick II, 1955 (div.); children: Hugh C., Jack B. III, Helen D. Kendrick Campbell, Diana C. Kendrick Untermeyer; m. C. Gregory Smith, Jr., Dec. 24, 1980. BA, Wellesley (Mass.) Coll., 1955; MA, PhD, Harvard U., 1970; MBA, Vanderbilt U., 1981. Rsch. assoc. Harvard Med. Sch., Boston, 1971-72, assoc. in psychiatry (anthropology), 1972-75; asst. prof. Vanderbilt U. Sch. Medicine, Nashville, 1975-76, assoc. prof., 1976-80, prof. psychiatry (anthropology), 1980—; dir. studies in population and family Harvard Med. Sch., 1972-75, dir. med. ethics symposium Vanderbilt Med. Sch., Nashville, 1979, dir. health care resources symposium, 1984; fellow Vanderbilt Inst. for Pub. Policy Studies, Nashville, 1985—. Author: Population Pressur and Cultural Adjustment, 1979, Population Politics: Choices that Shape our Future, 1993; editor, author Frontiers in Medical ethics, 1980; editor-in-chief Population and Environment, 1988—; also articles, chpts. to books. Bd. dirs. Carrying Capacity Network, Washington, 1989—, Fossil Fuels Policy Action Inst., Arcata, Calif., 1989—, Population-Environ. Balance, Washington, 1988—; adv. bd. chair Murphy Sch. for Pregnant Teenagers, Nashville, 1976-79. NIMH grantee; OEO grantee; Rand Corp. grantee. Fellow Am. Anthropology Assn.; mem. AAAS, Sigma Xi (exec. com. Vanderbilt chpt. 1986). Republican. Avocations: tennis, walking, travelling. Home: 6501 Grayson Ct Nashville TN 37205-3033 Office: Vanderbilt Med Sch Dept Psychiatry Nashville TN 37232

ABERTH, WILLIAM HENRY, physicist; b. L.A., Jan. 4, 1933; s. George and Louise (Fuchs) A.; married, May 15, 1954; children: Susan, George, Diane. BS in Physics, CUNY, 1954; MA in Physics, Columbia U., 1957; PhD in Physics, NYU, 1963. Lectr. in physics CUNY, N.Y.C., 1957-60, NYU, N.Y.C., 1960-63; sr. physicist SRI Internat., Menlo Park, Calif., 1963-68, asst. mgr., 1970-75; assoc. prof. physics Sonoma State U., Rhonert Park, Calif., 1969-70; assoc. rsch. prof. Linus Pauling Inst., Palo Alto, Calif., 1976-81; rsch. physicist U. Calif., San Francisco, 1982—; trustee Brownlee Labs., Inc., San Jose, Calif., 1980-84; cons. Genomyx, Inc., South San Francisco, Calif., 1988-90. Contbr. articles to profl. publs., procs. in field. 1st lt. U.S. Army Corps Engrs., 1954-56. Mem. Am. Phys. Soc., Am. Soc. Mass Spectrometry, Bay Area Mass Spectrometry Soc. Achievements include 4 patents in field. Office: Antek 3146 Manchester Ct Palo Alto CA 94303

ABEYESUNDERE, NIHAL ANTON AELIAN, health organization representative; b. Colombo, S.W., Sri Lanka, Feb. 18, 1932; arrived in Bangladesh, 1989; s. Aelian Joseph and Aileen Millicent (Perera) A.; m. Nalini Antoinette Perera, Feb. 15, 1960; children: Nirendra Gerard, Nilani Antoinette. MBBS, Colombo U., Sri Lanka, 1959; Diploma in Pub. Health, U. Sydney, Australia, 1968; MD, Inst. Post Grad. Medicine, Colombo, 1981. Dist. med. asst. Health Svcs. Sri Lanka, Haputale, 1959; dist. med. officer Health Svcs. Sri Lanka, Welimada, Sri Lanka, 1960-61; med. officer Govt. Hosp., Panadura, Sri Lanka, 1961-64, Health Svcs., Horana, Sri Lanka, 1964-68; regional officer Ante Malaria Campaign, Kurnegala, Sri Lanka, 1969; epidemiologist Ante Malaria Campaign, Colombo, 1970-73; dep. supt. Ante Malaria Campaign, 1973-74, supt., head, 1974-81; malariologist WHO, Nepal, Bangladesh, 1982-88; rep., chief of mission WHO, Dhaka, Bangladesh, 1989—; mem. expert adv. com. on malaria, WHO, 1980-82; cons. in field. Contbr. 15 articles to profl. jours. 1st violinist Colombo Symphony Orchn., 1946-50. Mem. sri Lanka Med. Assn. (mem. coun. 1976-80), Lions (pres. 1967, dist. club cabinet 1968-70). Home: # 3 Rd 95, Gulshan DIT 2, Dhaka Bangladesh Office: World Health Orgn, GPO Box 250, Dhaka 1000, Bangladesh

ABEYSUNDARA, URUGAMUWE GAMACHARIGE YASANTHA, civil engineer; b. Marara, South, Sri Lanka, Mar. 15, 1964; s. Dharmadasa and Jayaseeli (Danthanarayana) Urugamuwe Gamacharige. BSc in Engring., U. Moratuwa, Sri Lanka, 1990. Engr. trainee Irrigation Dept., Amparai, Sri Lanka, 1984-85, Road Devel. Authority, Matara, Sri Lanka, 1985-86, Cen. Engring. Consultancy Bur., Kotmale, Sri Lanka, 1986-87; civil engr. Civmecon Engring. (Pvt.) Ltd., Nugegoda, Sri Lanka, 1988-90, Brown & Co. Ltd., Colombo, 1990-91, State Engring. Corp. of Sri Lanka, Colombo, 1991—; cons. engr. Civmecon Engring. (Pvt.) Ltd., Nugegoda, 1990—; divsnl. bldgs. engr. Hambantota Dist., Sri Lanka, 1992—. Inventor D.O.E. method of design for local concrete mixers, 1990. Mem. Sri Lanka Eye Donation Orgn. Mem. ASCE (assoc.), Instn. Engrs. (Sri Lank chpt.). Avocation: photography. Office: Divsnl Bldgs Engr's Office, Hambantota Sri Lanka

ABIKO, TAKASHI, peptide chemist; b. Sendai, Miyagi, Japan, Nov. 8, 1941; s. Buichi and Takako A.; m. Yasuko Abiko, Oct. 20, 1968; children: Keiichi, Harumi. PharmM, Tohoku Coll. Pharmacy, Sendai, 1967; PharmD, Tohoku Coll. Pharmacy, 1975. Lectr. Tohoku Coll. Pharmacy, Sendai, 1967-76; vice dir. Kidney Ctr. in Sendai Ins. Hosp., 1977-85; rsch. dir. Kidney Rsch. Lab. in Kojinkai, Sendai, 1986—. Contbr. articles to profl. jours. Avocations: swimming, movies, travel, driving. Home: 42-12 Asahigaoka, 1-chome Aoba-ku, Sendai 981, Japan Office: Kidney Rsch Lab Kojinai, 1-6 Tsutsujigaoka 2-chome, Miyagino-ku, Sendai 980, Japan

ABILDSKOV, J. A., cardiologist, educator; b. Salem, Utah, Sept. 22, 1923; s. John and Annie Marie (Peterson) A.; m. Mary Helen McKell, Dec. 4, 1944; children—Becky, Alan, Mary, Marilyn. B.A., U. Utah, 1944, M.D., 1946. Diplomate Am. Bd. Internal Medicine. Intern Latter-day Saints Hosp., Salt Lake City, 1947-48; resident Charity Hosp. La., New Orleans, 1948-51; instr. Tulane U., New Orleans, 1948-54; asst. prof. to prof. SUNY-Syracuse, 1955-68; prof. medicine U. Utah, Salt Lake City, 1968—; dir. Nora Eccles Harrison Cardiovascular Research and Tng. Inst., Salt Lake City, 1970—. Contbr. articles to profl. jours. Served to capt. USAR, 1954-56. Recipient Disting. Research award U. Utah, 1976. Fellow Am. Coll. Cardiology; mem. Assn. Am. Physicians, Am. Soc. Clin. Investigation (emeritus), Assn. Univ. Cardiologists (founding), Western Assn. Physicians, Venezuelan Cardiology Soc. (hon.), Cardiology Soc. Peru (corresponding). Republican. Mormon. Home: 1506 Canterbury Dr Salt Lake City UT 84108-2833 Office: U Utah Bdlg 500 Salt Lake City UT 84112

ABOITES, VICENTE, physicist; b. Salamanca, Mexico, Dec. 31, 1958; s. Gilberto and Chopy (Manrique) A.; m. Margarita Ortega, June 20, 1988; 1 child, Ivo. BS in Physics, U. Metropolitana, Mexico, 1981; PhD in Physics, U. Essex, Eng., 1985; Attestation d'Etudes Approfondies d'Electrolique, U. Pierre et Marie Curie, France, 1986; diploma in philosophy, U. Iberoamericana, Mex., 1992. Doctoral student Rutherfor Lab., Oxford, Eng., 1982-85; stage employee Ecole Sup. De Physique Et Chimie De Paris, Paris, 1985-86; postdoctoral fellow Technische Univeristat Berlin, 1989-90; researcher Centro De Investigaciones En Optica, Leon, Mexico, 1986—; acad. vis. Imperial Coll., London, 1992-93; prof. philosophy and electronic engring. Univ. Iberoamericana, Leon, 1988—. Author: El Laser, Una Introduccion, 1991, El Laser, 1991, Opus III, 1991; contbr. articles to profl. jours. Recipient acad. grant Conacyt, Eng., 1981, sci. grant Rotary Found., Eng., 1982, scientific exangegrant DAAD, Germany, 1989, Marie Curie sci. grant, Commn. European Community, 1992. Mem. Sistema Nacional de Investigadores, European Soc. Physics, Soc. Francaise de Physique. Achievements include discovery of way to produce q-switched laser pulses with repetitive and constant temporal and spatial profile using phase conjugation by stimulated brillouin scattering. Office: Centro de Investigaciones, En Optica APDO Post 948, 37000 Leon Mexico

ABOU-SAMRA, ABDUL BADI, biomedical researcher, physician; b. Aleppo, Syria, Oct. 10, 1954; came to U.S., 1984; s. Ahmad Abou-Samra and Fatoum El-Zein; m. Raja Bankasli Abou-Samra, Jan. 6, 1980; children: Ahmad, Rouba, Tareq, Reem. MD, Aleppo U., 1979; PhD, Lyon I (France) U., 1983. Intern in medicine Aleppo U. Hosp., 1978-79; resident in medicine Antiquaille Hosp., Lyon, 1979-80; fellow in medicine Lyon-I U., 1980-82, asst. prof., 1982-84; vis. scientist NIH, Bethesda, Md., 1984-86; asst. prof. Harvard Med. Sch., Boston, 1986—; asst. in medicine Mass. Gen. Hosp., Boston, 1988—. Contbr. articles to profl. jours. and conf. procs. NIH rsch. grantee, 1992. Mem. Endocrine Soc., Am. Fedn. Clin. Rsch., Am. Soc. Bone and Mineral Rsch. Achievements include patent for molecular cloning of parathyroid hormone receptor sequence information, clinical application. Office: Massachusetts Gen Hosp 1 Fruit St Boston MA 02114

ABRAHAM, GEORGE, research physicist, engineer; b. N.Y.C., July 15, 1918; s. Herbert and Dorothy (Jacoby) A.; m. Hilda Mary Wenz, Aug. 26, 1944; children: Edward H., Dorothy J., Anne H., Alice J. Sc.B., Brown U., 1940; S.M., Harvard U., 1942; Ph.D., U. Md., 1972; postgrad., MIT, George Washington U. Registered profl. engr., D.C. Chmn. bd., pres. bd. Intercollegiate Broadcasting System, N.Y., 1940—; radio engr. RCA, Camden, N.J., 1941; with Naval Research Lab., Washington, 1942—; head sci. edn., head exptl. devices and microelectronics sects. Naval Research Lab., 1945-69, head systems applications Office of Dir. Research, 1969-75, research physicist Office Research and Tech. Applications, cons., 1975—; lectr. U. Md., 1945-52, George Washington U., 1952-67, Am. U., 1979; indsl. cons.; mem. D.C. Bd. Registration Profl. Engrs. Contbr. chpts. to books, articles to profl. jours. Chmn. bd. Canterbury Sch., Accokeek, Md.; mem. schs. and scholarships com. Harvard U.; active PTA, Boy Scouts Am. Served to capt. USNR, World War II. Recipient Group Achievement award Fleet Ballistic Missile Program U.S. Navy, 1963, Edison award Naval Research Lab., 1971, Navy Research Publ. award, 1974, 84, Patent awards, 1959-75, D.C. Sci. citation, 1982, Govt. Microcircuit Applications Conf. Founders award, 1986, Outstanding Service award Soc. Profl. Engrs., 1986, D.C. Council Engring. and Archtl. Soc. Nat. Capital Engr. of Yr. award, 1987, Desert Storm Science and Engring. award USN, 1991. Fellow IEEE (Harry Diamond award 1981, Centennial award 1984, Disting. Svc. award Capital area coun. 1991), AAAS, Washington Acad. Scis. (pres. 1974-75), N.Y. Acad. Scis.; mem. AAUP, Am. Phys. Soc., Am. Assn. Physics Tchrs., Am. Soc. Naval Engrs., Washington Soc. Engrs. (pres. 1974, award 1981), Philos. Soc. Washington, Sierra Club, Sigma Xi, Sigma Pi Sigma, Tau Beta Pi, Sigma Tau, Eta Kappa Nu, Iota Beta Sigma. Clubs: Cosmos (Washington), Harvard (Washington); Appalachian Mountain (Boston); Sierra (San Francisco). Achievements include patents in field. Home: 3107 Westover Dr SE Washington DC 20020-3719 Office: Naval Rsch Lab Washington DC 20375-5000

ÀBRAHÀM, GYÖRGY, mechanical engineering educator; b. Budapest, Hungary, May 20, 1948; s. Lajos and Gizella Kaczor (Lajosnè) A.; m. Àgnes Szabò, Jan. 27, 1976; children: Kata, Peter. MSc in Mech. Engring., Tech. U. Budapest, 1972, D degree, 1978; PhD, Hungarian Sci. Acad., Budapest, 1984. Diplomate mech. engring. Asst. prof. Tech. U. Budapest, 1972-75, adj. prof., 1975-84, assoc. prof., 1984-90; optical cons. Hungarian Indsl. Ministry, Budapest, 1984—. Editor Image and Sound Tech. Jour., 1989—. Recipient Silver Work award Hungarian govt., 1986. Mem. Optical Soc. Hungary (adv. bd. 1985—), Internat. Soc. for Optical Engring. Achievements include patent for correction of human color blindness, method for measuring the space telescope for Halley comet program. Office: Tech U Budapest, Egry U 20 22, H 1521 Budapest Hungary

ABRAHAM, JACOB A., computer engineering educator, consultant; b. Kerala, India, Dec. 8, 1948; came to U.S., 1970; s. Jacob and Annamma (Chacko) A.; m. Ruth Anne Dick, July 19, 1975; children—Nathan Thomas, Sarah Anne. B.S., U. Kerala, 1970; M.S., Stanford U., 1971, Ph.D., 1974. Acting asst. prof. Stanford U., Calif., 1974-75; asst. prof. computer engring. U. Ill., Urbana, 1975-80, assoc. prof., 1980-83, prof., 1983—; prof. and Cockrell Family Regents Chair in Engring. #8 U. Tex., Austin, 1988—, dir. Computer Engring. Rsch. Ctr., 1989—; cons. Aerospace Corp., Digital Equipment Corp., GE, GTE, Hewlett-Packard Co., IBM Corp., Intel, Sperry, 1979—; dir. rsch. program in reliable very large scale integration architectures U. Ill., 1984-88; dir. Computer Engring. Rsch. Ctr., U. Tex., Austin,1989—. Assoc. editor JETTA, 1992—; advisory editor Asken Assocs. Pub., 1987—; contbr. over 200 articles to profl. confs., jours. and books. Fellow IEEE (assoc. editor transactions on computer-aided design of integrated circuits and systems 1984-86, assoc. editor transactions on very large scale integrates systems 1992, chair Computer Soc. Tech. Com. on Fault-Tolerant Computing 1992); mem. Assn. Computing Machinery, Sigma Xi. Mem. Ch. of S. India. Achievements include 1 patent. Avocations: flying (instrument-rated pilot), tennis, backpacking. Office: U Tex Computer Engring Rsch Ctr 2201 Donley Dr Ste 395 Austin TX 78758-4538

ABRAHAM, TONSON, chemist, researcher; b. Bombay, Dec. 21, 1948; came to U.S., 1970; s. Thykadavil Jorge and Annie (Joseph) A.; m. Iona Marianne Joseph, June 17, 1978; children: Akash, Kavi. B in Tech., Indian Inst. Tech., Kanpur, India, 1970; PhD in Organic Chemistry, Cath. U. Am., 1976. Fellow NRC, Washington, 1976-78; postdoctoral fellow No. Ill. U., DeKalb, 1978-79; vis. asst. prof. Ill. State U., Normal, 1979-80; polymer scientist Wright-Patterson Air Force Base, Fairborn, Ohio, 1980-86; sr. scientist Owens-Corning Fiberglas, Granville, Ohio, 1986; R & D assoc. B.F. Goodrich Co., Brecksville, Ohio, 1987—. Contbr. articles to profl. jours. Fellow Am. Inst. Chemists; mem. Am. Chem. Soc. Roman Catholic. Achievements include patents in synthetic polymer and organic chemistry; research in the synthesis and applications of water soluble polymers, synthesis of high temperature, oil resistent elastomers, fluoroelastomers, hydrogenation of polymers, homogeneous hydrogenation catalysts, thermosetting resin precursors for aerospace composites and in the chemistry of indoles. Home: 16936 Deer Path Dr Strongsville OH 44136-6260 Office: BF Goodrich Co 9921 Brecksville Rd Cleveland OH 44141-3289

ABRAHAMSON, GEORGE R., civilian military scientist; b. Painesville, Ohio, Aug. 31, 1927; s. George and Gudrun Abrahamson; m. Mildred B. Bratton, Aug. 20, 1948; children: Christine, Karl, Thomas, Norman. BS in Engring. Sci., Stanford U., 1955, PhD in Engring. Mechanics, 1958. Student engr. Poulter Lab., Menlo Park, Calif., 1953-58, dir., dept. engring. mechanics, 1959-68; staff devel. leave Norwegian Defense Rsch. Establishment, 1968-69; dir. Poulter Lab., Menlo Park, Calif., 1969-88; v.p., phys. sci. divsn. SRI Internat., Menlo Park, Calif., 1980-91, sr. v.p., scis. group, 1988-91; chief scientist Dept. U.S. Air Force, Washington, 1991—; chmn. NSF Mech. Engring. and Applied Mechanics Adv. Com., Washington, 1985-86; adv. bd. mem. U.S. Air Force Sci. Bd., Washington, 1988-89. Recipient Applied Mechanics award Am. Soc. Mech. Engrs., 1991. Fellow Am. Soc. Mechanical Engrs.; mem. Am. Nuclear Soc., Am. Assn. Advancement Sci. Office: HQ US Air Force/ST 1060 Air Force Pentagon Washington DC 20330-1060

ABRAHAMSON, JAMES ALAN, retired air force officer; b. Williston, N.D., May 19, 1933; s. Norval S. and Thelma B. (Helle) A.; m. Barbara Jean Northcott, Nov. 7, 1959; children: Kelly Anne, James A. B.S. in Aero. Engring., MIT, 1955; M.S. in Aerospace Engring., U. Okla., 1961. Commd. 2d lt. U.S. Air Force, 1955; advanced through grades to major general US Air Force, 1978; advanced through grades to maj. gen. U.S. Air Force, 1978; flight instr. U.S. Air Force, Bryan AFB, Tex., 1957-59; spacecraft project officer Vela nuclear detection saltellite program U.S. Air Force, Los Angeles AF Sta., 1961-64; figher pilot Tactical Air Command, 1964; astronaut USAF Manned Orbiting Lab., 1967-69; mem. staff NACS, Exec. Office of Pres., 1969-71; program dir. Maverick Program, 1971-73; comdr. 495th text wing U.S. Air Force, Wright Patterson AFB, Ohio, 1973-74; insp. gen. Air Force Systems Command, 1974-76; system program dir., multinat. F-16 program U.S. Air Force, 1976—, promoted to general, 1986. Decorated D.S.M.; decorated Legion of Merit with one cluster, Air medal with cluster, Meritorious Service medal; recipient award Daedalian Weapon System Mgmt. Assn., Aerospace Power award Air Force Assn., Dayton, Von Karmen Lectr. Astronautics AIAA, 1993. also: Hughes Aircraft Company PO Box 45066 Los Angeles CA 90045

ABRAHAMSON, SCOTT DAVID, mechanical engineer, educator; b. Tacoma, Wash., Jan. 18, 1959; s. Richard Clark A. and Joanne Alberta (Schweifier) Trueblood; m. Candace Anne White, Sept. 18, 1988; 1 child, Mark. MS in Mech. Engring., Stanford U., 1983, PhD, 1987. Design engr. Shugart Assocs., Sunnyvale, Calif., 1983-84; rsh. asst. Stanford (Calif.) U., 1984-87, postdoctoral researcher, 1987-89; asst. prof. U. Minn., Mpls., 1989—. Contbr. articles to profl. jours. Mem. AIAA, ASME, Am. Phys. Soc., Sigma Xi. Office: U Minn 110 Union St SE Minneapolis MN 55455

ABRAMCZUK, TOMASZ, computer image processing specialist; b. Warsaw, Poland, Aug. 4, 1954; arrived in Sweden, 1972; s. Kazimierz and Krystyna (Bozek) A.; m. Barbara Dunia, May 12, 1984; 1 child, Monika. MSc, Royal Inst. Tech., Stockholm, 1979; lic. in technology, Royal Inst. Tech., 1989. Rsch. engr. Royal Inst. Tech., Stockholm, 1982-83; researcher Royal Inst. Tech., 1983-85, sr. researcher, 1985-87; chief exec. of R & D Sydat Automation, AB, Stockholm, 1988 ; also bd. dirs. Sydat Automation, AB; chief exec. officer Poltra AB, Stockholm, 1991—, also bd. dirs.; cons. in field. Author: Image Processing, 1984; contbr. articles to profl. publs. Mem. IEEE, Swedish Elec. Engring. Orgn., Internat. Assn. Polish Experts and Cons. (co-founder 1991, bd. dirs. 1991—), Royal Inst. Tech. Radio Club. Avocations: sailing, motor sports. Office: SYDAT Automation AB, Gökottestigenh 17, 163 60 Spänga Sweden

ABRAMOWICZ, JACQUES SYLVAIN, obstetrician, perinatologist; b. Paris, Dec. 5, 1948; s. Theodore Dov and Sara Ethel (Cukiernik) A.; m. Annie Sternelicht, Aug. 1, 1972; children: Shelly, Ory. MD, Sackler Sch. Medicine, Tel-Aviv, 1975. Diplomate Israel Bd Obstetrics and Gynecology. Rotating intern Tel-Aviv Mcpl. Med. Ctr., 1973-74; resident dept. ob-gyn. Sapir Med. Ctr., Kfar-Saba, Israel, 1978-85; rsch. registrar ultrasound dept. ob-gyn. King's Coll. Hosp., London, 1981; resident dept. gen. surgery Sapir Med. Ctr., 1982-83, resident dept. urology, 1983; cons. Timsit Inst. Reproductive Medicine, Tel-Aviv, 1986-87; dir. clin. rsch. Div. Maternal-Fetal Medicine, Ea. Va. Med. Sch., Norfolk, 1987-89; assoc. researcher Jones Inst. Reproductive Medicine, Norfolk, 1989; dir. perinatal ultrasound, asst. prof. dept. ob-gyn. U. Rochester Med. Ctr., 1990-93, assoc. prof., 1993—. Contbr. articles to profl. jours. including Am. Jour. Ob-Gyn., Obstet. Gynecology, Jour. Ultrasound Medicine, Prenatal Diagnosis, Am. Jour. Perinatology, Fetal Therapy, Jour. Perinatal Medicine, also chpts. to books; referee various jours. Maj. Israel Def. Forces, 1974-78. Mem. Am. Inst. Ultrasound in Medicine (sr., internat. rels. com. 1988-91, standards com. 1991—, mfrs. commendation panel 1991-93, chair mfrs. commendation panel 1993—), N.Y. Acad. Sci., Soc. Perinatal Obstetricians, Internat. Perinatal Doppler Soc., Internat. Fetal Med. and Surg. Soc. Jewish. Achievements include research in prenatal diagnosis and therapy, ultrasound, doppler velocimetry. Office: U Rochester Dept Ob/Gyn Box 8668 601 Elmwood Ave Rochester NY 14642-8668

ABRAMS, SCOTT IRWIN, immunologist; b. Bklyn., Mar. 11, 1959; s. Daniel and Sheila (Mintz) A.; m. Debra Segneri Abrams, Nov. 23, 1991. BS in Biology summa cum laude, Delaware Valley Coll., 1981; PhD in Immunology, Ind. U., Indpls., 1987. Postdoctoral fellow Washington U. Sch. Medicine, St. Louis, 1987-91; sr. staff fellow immunology Nat. Cancer Inst., NIH, Bethesda, Md., 1991—. Contbr. articles to profl. jours., chpts. to books. Recipient fellowships, NIH grants, Nat. Rsch. Svc. awards; recipient 1st pl. rsch. competition Autumn Immunology Conf., 1989. Mem. Am. Assn. Immunologists, Sigma Xi. Achievements include research in cytotoxic cells: generation, triggering, effector functions, and methods. Office: Nat Cancer Inst NIH Lab Tumor Immunol & Biol Rm 8B07 9000 Rockville Pike Bldg 10 Bethesda MD 20892

ABRAMSON, HYMAN NORMAN, engineering and science research executive; b. San Antonio, Mar. 4, 1926; s. Nathan and Pearl (Westerman) A.; m. Idelle Rebecca Ringel, Apr. 20, 1947; children—Phillip David, Mark Donald. B.S.M.E., Stanford U., 1950, M.S. in Engring. Mechanics, 1951; Ph.D. in Engring. Mechanics (So. Fellowship Fund fellow), U. Tex., Austin, 1956. Engr. U.S. Naval Air Missile Test Center, Point Mugu, Calif., 1947-48; project engr. Chance Vought Aircraft Co., Dallas, 1951-52; assoc. prof. aero. engring. Tex. A&M U., 1952-55; sect. mgr., dept. dir. S.W. Research Inst., San Antonio, 1956-72, v.p. div. engring. scis., 1972-85, exec. v.p., 1985-91, also bd. dirs.; mem. many research adv. coms. U.S. Govt.; bd. dirs. Broadway Nat. Bank. Author: An Introduction to the Dynamics of Airplanes, 1958, reprinted, 1971; contbr. numerous articles to profl. publs.; editor: (with others) Applied Mechanics Surveys, 1966, The Dynamic Behavior of Liquids in Moving Containers, 1966; assoc. editor: (with others) Applied Mechanics Revs, 1954-85; editorial adv. bd.: (with others) Jour. Computers and Structures, 1970—, Aeros. and Astronautics, 1975-80. Mem. Greater San Antonio C. of C., and City of San Antonio Market Sq. Adv. Com., 1973-77; mem. U.S. Bicentennial Com. of San Antonio, 1975-76; mem. adv. bd. dirs. U.S. Alamo, Inc., 1985—. Served with USN, 1943-45. Fellow AIAA (Disting. Service award 1973, dir., Structures, Structural Dynamics and Materials award 1991), ASME (v.p., gov., hon. mem.); mem. Nat. Acad. Engring., Soc. Naval Architects and Marine Engrs., Nat. Acad. Engring. Mexico, AAAS, Sigma Xi. Republican. Jewish. Home: 1511 Spanish Oaks San Antonio TX 78213-1635 Office: SW Research Inst 6220 Culebra Rd PO Drawer 28510 San Antonio TX 78228-0510

ABRON, LILIA A., engineer; b. Memphis, Mar. 8, 1945; d. Ernest and Bernice (Wise) A.; children: Fredeick, Ernest, David. BS in Chemistry, Lemoyne Coll., 1966; MS in Sanitary Engring., Washington U., 1968; PhD in Chem. Engring., U. Iowa, 1971. Profl. engr. Free lance cons. Washington, 1971-74; asst. prof. Howard U., Washington, 1974-81; chief environ. div. Delon Hampton & Assocs., Washington, 1975-78; pres. Peer Cons., Rockville, Md., 1978--; com. mem. Nat. Coun. Examiners, Clemson, S.C. Pres. Jack & Jill Am., Inc., D.C. Chpt., 1990-92; bd. dirs. Bapt. Home for Children, Washington. Recipient Women Owned Bus. Enterprise award DOT, 1988, Balti. Outstanding Minority Bus. award Fed. Exec. Bd., 1987; named Alumnus of Yr. Lemoyne Owen Coll., 1988. Mem. AAAS, Water Environ. Fedn., Am. Soc. Civil Engrs., Am. Water Works Assn., Sigma Xi. Office: PEER Cons 12300 Twinbrook Pkwy Ste 410 Rockville MD 20852

ABRUMS, JOHN DENISE, internist; b. Trinidad, Colo., Sept. 20, 1923; s. Horatio Ely and Clara (Apfel) A.; m. Annie Louise Manning, June 15, 1947; children: Louanne C. Abrums Sargent, John Ely. BA, U. Colo., 1944; MD, U. Colo., Denver, 1947. Diplomate Am. Bd. Internal Medicine. Intern Wisc. Gen. Hosp., Madison, 1947-48; resident in internal medicine VA Hosp., Albuquerque, 1949-52, attending physician, 1956-80; mem. staff Presbyn. Hosp. Ctr., Albuquerque; cons. staff physician St. Joseph Hosp., Albuquerque, 1957-85; attending physician U. N.Mex. Hosp. (formerly Bernalillo County Med. Ctr.), Albuquerque, 1954-86; cons. physician A.T. & S.F. Meml. Hosp., Albuquerque, 1957-83; clin. assoc. in medicine U. N.Mex.; mem. N.Mex. Bd. of Med. Examiners. Bd. dirs. Blue Cross/Blue Shield, 1962-76. Brig. gen. M.C., U.S. Army, ret., N.Mex. Nat. Guard. Fellow ACP (life), AMA, Am. Soc. Internal Medicine (trustee 1976-82, pres. 1983-84), N.Mex. Soc. Internal Medicine (pres. 1962-64), N.Mex. Med. Soc. (pres. 1980-81), Nat. Acads. Practice (disting. practitioner), Albuquerque and Bernalillo County Med. Assn. (bd. govs. 1959-61, chmn. pub. rels. com. 1959-61), Am. Geriatric Soc., 1992—. Brig. gen. M.C., U.S. Army, ret. Republican. Episcopalian. Office: N Mex Med Group PC 3205 Berkeley Pine Albuquerque NM 87106

ABSE, DAVID WILFRED, psychiatrist, psychologist; b. Cardiff, Wales, Mar. 15, 1915; came to U.S. 1952; s. Rudolphn and Kate (Shepherd) A.; m. Elizabeth Jean Smith, July 26, 1961; children: Edward, Nathan. BSc, U. Wales, 1935; M.B.B.Ch., Med. Coll. Wales, 1938; DPM, U. London, 1940, MD, 1948. Cert. psychoanalyst D.C.; lic. physician U.K., Brit. Commonwealth, N.C., Va. Hon. clin. asst. Westminster Hosp., London, 1942-48; chief asst. psychiatrist Med. Sch., Charing Cross Hosp., London, 1949-50; lectr. psychology City Coll., London, 1950-51; clin. asst. prof. psychiatry U. N.C., Chapel Hill, 1952-53, assoc. prof. psychiatry, dir. inpatient svc., 1957-59, dir. postgrad. edn. psychiatry, 1957-59, prof. psychiatry, 1958-62; prof. psychiatry U. Va., Charlottesville, 1962-80, clin. prof. psychiatry, 1981—, prof. emeritus, 1990—; attending psychiatrist St. Albans Psychiat. Hosp., Radford, Va., 1985-90; dep. med. supt. Monmouthshire Mental Hosp., Abergavenny, U.K., 1946-48; faculty Washington Psychoanalytic Inst., 1962—, Washington Sch. Psychiatry, 1971-75; vis. prof. psychiatry St. Bartholomew's Hosp., London, 1979; dir. psychiat. edn. St. Albans Psychiat. Hosp., 1980-83; sr. psychiat. cons. David C. Wilson Hosp., Charlottesville, Va., 1983-85. Contbr. numerous articles to profl. jours. Maj. RAMC, 1942-47. Fellow Royal Coll. Psychiatrists U.K. (founding), Brit. Psychol. Assn., Royal Soc. Medicine, Am. Psychiat. Assn., Am. Acad. Psychoanalysis; mem. AMA, Brit. Med. Assn., Am. Psychoanalytic Assn., Group-Analytic Soc. London, AAAS, Va. Psychiat. Soc. Home: 1852 Winston Rd Charlottesville VA 22903

ABTS, DANIEL CARL, computer scientist; b. La Crosse, Wis., Feb. 2, 1956. BS in Computer Sci., U. Wis., La Cross, 1978. Mgr. systems and network U. Wis., 1978—. Achievements include planning and performing technical work in making University of Wisconsin, La Cross, part of Bitnet, WiscNet and Internet. Office: U Wis 1725 State St La Crosse WI 54601-3767

ABUZAKOUK, MARAI MOHAMMAD, civil aviation engineer; b. Benghazi, Libya, June 5, 1936; s. Mohammad Abdallah Abuzakouk and Esaida Ibrahim Aljihani; m. Lotfia Mahmoud, Aug. 8, 1961; 1 child, Mohammad. Dipl. in Air Traffic Control, U.K. Coll. Air Traffic Control, Bournemouth, 1968; BSc, Columbia Pacific U., 1985, MSc, 1986, PhD, 1987. Morse operator Ministry of Communication and Transport, Benina Airport, Libya, 1951-67; air traffic contr. Ministry of Comm. and Transp., Benina Airport, Libya, 1968-70, mgr. tech. svc., 1970-74; dir. gen. civil aviation and meteorology engr. Ministry of Comm. and Transp., Tripoli, Libya, 1974-87, chmn. civil aviation and met., 1987—; lectr. U. Tripoli, 1988-92; adviser t Ministry of Comm. and Transp., 1988-92. Contbr. articles to profl. publs. Recipient State prize, Govt. of Libya, 1989. Fellow Royal Meteor. Soc.; mem. Royal Aero. Soc., AIAA (sr.), Canadian Air Space Inst., Canadian Aero. and Astron. Soc. Achievements include development of plans for building airports, telecommunication facilities. Home: PO Box 14149, Tripoli Libya

ACAR, YALCIN BEKIR, civil engineer, soil remediation technology executive, educator; b. Ankara, Turkey, July 7, 1951; arrived in U.S., 1980, naturalized.; s. Mahmut Bedri and Serife Bedriye (Benek) A. BSCE, Robert Coll., Istanbul, Turkey, 1973, MSCE, 1975; PhD in Civil Engring., Bogazici U., Istanbul, 1980. Registered profl. engr., La., Nat. Coun. Engring. Examiners. Asst. site engr. Companie Industriel et Travaux Emile Blaton, Brussels, 1970, 72; rsch. asst. dept. civil engring. Bogazici U., 1973-75, rsch. assoc. Earthquake Engring. Rsch. Inst., 1975-77, rsch. engr. dept. civil engring-Earthquake Engring. Inst., 1977-80; rsch. engr., Fugro postdoctoral fellow dept. civil engring. La. State U., Baton Rouge, 1980-81, asst. prof., 1981-86, assoc. prof., 1986-92, prof., 1992—; presenter in field, condr. seminars; mem. com. on physiochem. phenomena in soils Transp. Rsch. Bd., NRC, 1989—. Patentee in field; contbr. articles to profl. jours. Mem. ASCE (publs. com. 1988, environ. geotechnics com. 1989), ASTM (com. on soil and rock), U.S. Nat. Soc. of Internat. Soc. for Soil Mechanics and Found. Engring. (com. on environ. geotechnics), Assn. Drilled Shaft Contractors, Internat. Geotextile Soc., Masons, Sigma Xi. Republican. Avocations: snow skiing, bicycling, swimming. Home: 7956 Jefferson Place Blvd # A Baton Rouge LA 70809-7622 Office: La State U Civil Engring Dept CEBA Bldg Baton Rouge LA 70803

ACEVEDO, EDMUND OSVALDO, physical education educator; b. Worcester, Mass., July 25, 1960; s. Iris Perry; m. Tracy Arrowood, Aug. 12, 1989. MS, U. Md., 1985; PhD, U. N.C., 1989. Asst. prof. William Paterson Coll., Wayne, N.J., 1989-90; asst. prof. kinesiology Kans. State U., Manhattan, 1990—; cons. Sport/Fitness Risk Mgmt., Denver, 1992—. Named to Outstanding Young Men of Am.; summer rsch. fellow Kans. State U., Manhattan, 1991-92. Mem. Am. Coll. Sports Medicine, Assn. for Advancement of Applied Sport Psychology, N. Am. Soc. for Psychology of Sport and Phys. Activity, Sigma Xi. Achievements include published rsch. in field. Home: 1716 Poyntz Ave Manhattan KS 66502 Office: Kansas State Univ Manhattan KS 66506

ACHENBACH, JAN DREWES, engineer; b. Leeuwarden, Netherlands, Aug. 20, 1935; came to U.S., 1959, naturalized, 1978; s. Johannes and Elizabeth (Schipper) A.; m. Marcia Graham Fee, July 15, 1961. Candidate engr., Delft U. Tech., 1959; PhD, Stanford U., 1962. Preceptor Columbia U., 1962-63; asst. prof. Northwestern U., Evanston, Ill., 1963, assoc. prof., 1966-69, prof. dept. civil engring., 1969—, Walter P. Murphy prof. civil engring., mech. engring. and applied math., 1981—, dir. Ctr. for Quality Engring. and Failure Prevention, 1986—; vis. assoc. prof. U. Calif., San Diego, 1969; vis. prof. Tech. U. Delft, 1970-71; prof. Huazhong Inst. Sci. and Tech., 1981; mem. at large U.S. Nat. Com. Theoretical and Applied Mechanics, 1972-78, 86—. Author: Wave Propagation in Elastic Solids, 1973, A Theory of Elasticity with Microstructure for Directionally Reinforced Composites, 1975, (with A.K. Gautesen and H. McMaken) Ray Methods for Waves in Elastic Solids, 1982, (with Y. Rajapakse) Solid Mechanics Research for Quantitative Non-Destructive Evaluation, 1987; editor: (with J. Miklowitz) Modern Problems in Elastic Wave Propagation, 1978 (with S.K. Datta and Y.S. Rajapakse) Elastic Waves and Ultrasonic Nondestructive Testing, 1990; editor-in-chief: Wave Motion, 1979—. Recipient award C. Gelderman Found., 1970, C.W. McGraw Rsch. award Am. Soc. Engring. Edn., 1975, Tempo All-Professor Team, Sciences, Chicago Tribune, 1993. Fellow ASME (Timoshenko medal 1992), Am. Acad. Mechanics (pres. 1978-79); mem. AAAS, U.S. Nat. Acad. Scis., Acoustical Soc. Am., Am. Geophys. Union, U.S. Nat. Acad. Engring., Am. Soc. Nondestructive Testing. Home: 574 Ingleside Park Evanston IL 60201-1738 Office: Northwestern U Room 324 2137 N Sheridan Catalysis Bldg Evanston IL 60208

ACHESON, SCOTT ALLEN, research biochemist; b. Independence, Mo., July 2, 1958; s. Allen Morrow and Mary Jeanne (Baird) A.; m. Dimitra Dedopoulou, Aug. 20, 1988. BA in Chemistry, Grinnell Coll., 1981; PhD in Chemistry, U. Iowa, 1986. Anesthesiology asst. U. Kans. Med. Ctr., Kansas City, 1981-82; NIH postdoctoral fellow dept. biochemistry, biophysics U. N.C., Chapel Hill, 1986-89; rsch. fellow Glaxco Rsch. Inst., Research Triangle Park, N.C., 1989-91; rsch. biochemist Pacific Hemostasis Fisons, Huntersville, N.C., 1992—. Contbr. articles to sci. jours. Mem. Sierra Club, Chapel Hill, 1987-91. Mem. Am. Chem. Soc., Young Dems. Presbyterian. Home: 525 Olmsted Park Pl Apt C Charlotte NC 28203-5686 Office: Pacific Hemostasis 11515 Vanstory Dr Huntersville NC 28078

ACHTERBERG, CHERYL LYNN, nutrition educator; b. Lynwood, Calif., Sept. 19, 1953; d. David and Margaret Rita Baker; m. Larry Harold Achterberg, May 17, 1975; 1 child, Jerusha. MS, U. Maine, 1981; PhD, Cornell U., 1986. Asst. food technologist food sci. dept. U. Maine, Orono, 1977-79, instr. human devel., 1981-82; grad. student div. nutritional sci. Cornell U., Ithaca, N.Y., 1982-85; asst. prof. nutrition dept. Penn State U., University Park, Pa., 1985-91; vis. prof. U. Guelph, Guelph, Ontario, Can., 1990; assoc. prof. nutrition dept. Penn State U., University Park, 1991—; dir. Penn State Nutrition Ctr., University Park, 1992—; cons. advisor Kraft-Gen. Foods, Chgo., 1992—; task force mem. Am. Heart Assn., Dallas, 1986—; advisor cons. Nat. Dairy Coun., Chgo., 1987—; mem. editorial bd. Jour. of Nutrition Edn. Author: (monograph) Effective Nutrition Communication for Beh Change, 1992; contbr. articles to profl. jours. and chpts. to books. Grantee USDA, 1987, 88, Nat. Dairy Coun., 1988, 91, Howard Heinz Endowment, 1989, 92; named Oustanding Faculty advisor Interhellenic Coun., 1989; recipient Coll. Teaching award Pa. State U., 1992, Mead Johnson Rsch. award, 1993. Mem. Soc. for Nutrition Edn. (dir. N.E. 1992-95), Am. Dietetic Assn., Am. Inst. Nutrition, Sigma Xi. Achievements include svc. as nat. advisor to rsch. on Food Guide Pyramid; leadership in theory-based rsch. in nutrition edn. and domestic nutrition edn. policy on nat. level; svc. as rapporteur for UNESCO/WHO conf. in Paris. Home: 2027 Pinecliff Rd State College PA 16801 Office: Penn State Nutrition Ctr 417 E Calder Way University Park PA 16801-5663

ACHTERBERG, ERNEST REGINALD, mining engineer; b. Little Rock, Sept. 1, 1925; s. Reginald Ernest and Anna Rosa (Schneyer) A.; m. Mollie Elizabeth Mohnkern, Dec. 30, 1949; children: Milton, Kay. BS, MO. Sch. Mines, 1953; MS, U. Ark., Fayetteville, 1971. Registered profl. engr., Mo. Svc. engr. Dowell Inc., Abilene, Tex., 1953, 54-55; engr. Grinnell Corp., Waverly, Ohio, 1953-54; engr. NL Industries, Potosi, Mo., 1955-68, Malvern, Ark., 1968-70; mining engr. Mine Safety and Health Adminstrn., Rolla, Mo., 1971-78, Bur. of Land Mgmt., Tulsa, 1978-91; ret., 1991. Mem. Okla. Mining Commn., Oklahoma City, 1991—. With USAAF, 1943-46. Mem. NSPE (life, chpt. pres. 1966-67), Soc. Mining Engrs. (chpt. bd. dirs. 1965-66), Am. Bowling Congress (city sec. 1963-68). Home: 10641 E 33d St Tulsa OK 74146-1810

ACKERMAN, ROY ALAN, research and development executive; b. Bklyn., Sept. 9, 1951; s. Jack A. and Estelle (Kuchlik) A.; m. Janet Sharon Ostrow, July 4, 1974 (div. 1984); children: Shanna Avrah, Batya; m. Kathleen T. Smith, 1989; 1 child, Daniel Jacob. BSChemE, Poly. Inst. of N.Y., 1972; MSChemE, MIT, 1974; PhD, U. Va./U. B.H., 1986. Chem. engr. Tri-Flo Rsch. Labs., Bellmore, N.Y., 1972-74; sr. project engr. Thetford Corp., Ann Arbor, Mich., 1975; dir. R & D Applied Sci. Through Rsch. and Engring. (now ASTRE), Ann Arbor, 1976-77, ASTRE Cons. Corp., Charlottesville, Va., 1978-81; tech. dir. ASTRE Corp. Group, Charlottesville, 1981-89, Alexandria, Va., 1989—; bd. dirs. Indsl. Microgenics Ltd., Charlottesville; chmn. Bicarbolyte Corp., Alexandria, Va. Author: Water Reuse and Recycle, 1981; patentee in field. Lay leader Congregation Beth Israel, Charlottesville, 1979-89. Scholar Samuel Ruben Found., 1968-72. Fellow Am. Inst. Chemists; mem. Water Pollution Control Fedn., Am. Soc. Artificial Internal Organs, Am. Assn. Rsch. Cos., Sigma Xi, Tau Beta Pi. Avocations: reading, swimming, tennis, politics, dancing. Office: ASTRE Corp Group 809 Princess St Alexandria VA 22314-2223

ACOSTA, NELSON JOHN, civil engineer; b. Newark, N.J., July 8, 1947; s. Pedro Nelson and Bertha Maud (Williams) A.; m. Twyla Liasine Flaherty, June 19, 1970; children: Jeffrey Thomas, Stephen Patrick, Bryan Edward. BCE, Ga. Inst. Tech., 1969, MSCE, 1970. Registered profl. engr., Ill.; API cert. aboveground storage tank inspector. Design engr. Chgo. Bridge and Iron Co., Birmingham, Ala., 1970-73; sales estimator Chgo. Bridge and Iron Co., Oak Brook, Ill., 1973-74; contracting engr. Chgo. Bridge and Iron Co., Atlanta, 1975-79, CBI Constructors, Ltd., London, 1979-80, Arabian CBI Ltd., Al Khobar, Saudi Arabia, 1980-84, CBI Na-Con, Inc., Fontana, Calif., 1984-88; mgr. spl. projects and estimating HMT, Inc., Cerritos, Calif., 1989—. Recipient traineeship NSF, Atlanta, 1969. Mem. ASCE. Republican. Roman Catholic. Office: HMT Inc 13921 Artesia Blvd Cerritos CA 90701

ACRIVOS, ANDREAS, chemical engineering educator; b. Athens, Greece, June 13, 1928; s. Athanasios and Anna (Besi) A.; m. Juana Vivo, Sept. 1, 1956. BSChemE, Syracuse U., 1950; MS, U. Minn., 1951, PhD, 1954. Instr. U. Calif., Berkeley, 1954-55, asst. prof., 1955-59, assoc. prof., 1959-62; prof. Stanford (Calif.) U., 1962-88; Einstein prof. CCNY, 1988—. Contbr. articles to profl. jours. Guggenheim Found. fellow, 1959, 76. Fellow Am. Physical Soc. (fluid dynamics prize 1991), Am. Inst. Chem. engrs. (awards 1963, 68, 84); mem. NAS, NAE, Am. Chem. Soc. Office: CCNY Levich Inst 138th St at Convent Ave New York NY 10031

ACS, JOSEPH STEVEN, transportation engineering consultant; b. Budapest, Hungary, Apr. 26, 1936; came to U.S., 1957; s. Gyula Istvan and Gizella (Sztanek) A.; m. Eva Hegedus, Apr. 18, 1960; 1 child, Joseph S. Jr. BS in Civil Engring., Kvassay Jeno Kozlekedes Ipari Technikum, Budapest, 1956. With Balt. & Ohio R.R. Co., Cin., 1957-68; project engr. Balt. & Ohio Chgo. Terminal R.R. Co., 1968-75, terminal engr., 1975-88; pub. projects engr. CSX Transp., Chgo., 1988-92; prin. Allied Consulting Svcs., Ltd., Chgo., 1992—. Contbr. articles to profl. jours. Mem. NSPE, ASCE, Am. Ry. Engring. Assn. (chmn. com. 14/yard and terminals 1989-92, Profl. Svcs. award 1992), Am. Ry. Bridge and Bldg. Assn., Am. Right of Way Assn., Roadmasters and Maintenance of Way Assn., Chicago Heights Country Club, Eagles, Lions. Republican. Roman Catholic. Home: 18851

Hood Ave Homewood IL 60430-4030 Office: Allied Consulting Svcs Ltd PO Box 277945 Chicago IL 60627-7945

ADAIKAN, GANESAN PERIANNAN, pharmacologist, medical scientist; b. Kuala Lumpur, Selangor, Malaysia, Feb. 20, 1944; arrived in Singapore, 1973; s. Periannan and Alagammai Adaikan; m. Selvamani Arumugam, May 17, 1970; children: Kala, Kalpana, Sangeetha, Siva Sanjeevkumar. M in Biology, Inst. Biology, London, 1976; MSc, U. Singapore, 1979; PhD, Nat. Univ. Singapore, 1985; DSc, Inst. Applied Rsch., London, 1992. Research asst. Makerere U., Kampala, Uganda, 1971-73; research asst. U. Singapore, 1973-79, research fellow, 1980-82; clin. biochemist Nat. Univ. Singapore, 1982-85, clin. scientist, 1986-88, sr. rsch. fellow, 1989—; presented papers at numerous profl. convs. and symposia; chmn.-elect VI World Meeting on Impotence, Singapore, 1994. Contbr. 119 articles to sci. jours., 1972—; cons. editor: Drugs of Today, 1984, Drugs of the Future, 1984—, Asia Pacific Jour. of Pharmacology, 1986—, Internat. Jour. of Impotence Rsch. Fellow Inst. Biology, London, 1987—; recipient Jean-Francois Ginestie award The Internat. Soc., Prague, Czechoslovakia, 1986; co-recipient Surg Soc. Research award, U.K., 1973. Mem. Singapore Planned Parenthood Fedn. (coun. mem.), Obstet. and Gynecol. Soc. Singapore (Benjamin H. Sheare Meml. Lecture award 1987), Internat. Soc. of Impotence Rsch., Asia-Pacific Soc. Impotence Rsch. Hindu. Avocations: art, cartooning. Home: 109 Jalan Hitam Manis, Singapore 1027, Singapore Office: Nat U Singapore, Nat Univ Hosp Dept Ob-Gyn, Lower Kent Ridge Rd, Singapore 0511, Singapore

ADAIR, LILA MCGAHEE, physics educator; b. Griffin, Ga., Jan. 8, 1947; d. Henry Grady and Lila (Smith) McGahee; m. Terry Wayne Adair, July 21, 1973; 1 child, James. BS in Biology, Oglethorpe U., 1967; MAT in Physics, Ga. State U., 1978; Diploma for Advanced Study of Teaching in Ednl. Leadership, Emory U., 1990. Cert. tchr. physics, leadership/supervision, Ga. Tchr., sci. dept. chmn. Lindley Jr. High, Mableton, Ga., 1967-72; tchr. Hohenwald (Tenn.) Elem. Sch., 1972-74; tchr. physics, chemistry South Gwinnett High, Snellville, Ga., 1974-79; tchr., physics, sci. dept. chmn. Cen. Gwinnett High, Lawrenceville, Ga., 1979—; instr. NSF Inst., Emory U., Atlanta, 1985; sci. demonstrator in field. Inventor/dir. Gwinnett County Sci. Bowl, 1980-92. Adv. bd. SciTrek, Atlanta, 1988-91; tchr. Boy Scouts Am., Snellville, Ga., 1990—; instr. Summerscape, Ga. Tech, Atlanta, 1990—; mag. evaluator Coun. for Agrl. Sci. and Tech., Athens, Ga., 1984. Recipient Presdl. award for Excellence in Sci. Teaching, Pres. Reagan, Washington, 1985, Tchr. Hall of Fame awards Gwinnett County Schs., Lawrenceville, 1984, STAR Tchr. award Gwinnett County Schs., 1980, 82, 85, 87, 93; named Tchr. of Yr. Gwinnett County Schs., 1985, 88. Mem. Am. Assn. Physics Tchrs. (sect. v.p. 1991, pres. 1992, exec. bd. 1988-90, physics teaching resource agt. 1986—, planning com., internat. com. 1989-91), Ga. Jr. Acad. Sci. (state dir. 1985), Ga. Sci. Tchrs. Assn. (dist. bd. dirs. 1981-82, sec. 1979-80, Ga. Sci. Tchr. of Yr. 1985), Am. Phys. Soc. (co-chmn. local physics alliance conf. 1991), Ga. Acad. Sci. (symposium speaker 1990), Sigma Pi Sigma, Alpha Delta Kappa. Democrat. Baptist. Avocations: reading, travel, sewing, family. Home: 1994 Skyland Glen Dr Snellville GA 30278-3862 Office: Ctrl Gwinnett High School 564 W Crogan St Lawrenceville GA 30245-4796

ADAIR, ROBERT KEMP, physicist, educator; b. Fort Wayne, Ind., Aug. 14, 1924; s. Robert Cleland and Margaret (Wiegman) A.; m. Eleanor Reed, June 21, 1952; children—Douglas McVeigh, Margaret Guthrie, James Cleland. Ph.B., U. Wis., 1947, Ph.D., 1951. Instr. physics U. Wis., Madison, 1950-53; physicist Brookhaven (N.Y.) Nat. Lab., 1953-58; mem. faculty Yale U., New Haven, 1958—, prof. physics, 1961-72, Eugene Higgins prof. physics, 1972-88, Sterling prof. physics, 1988—, chmn. dept. physics, 1967-70, dir. div. phys. scis., 1977-80; assoc. dir. for high energy and nuclear physics Brookhaven Nat. Lab., Upton, N.Y., 1987-88; physicist Nat. Baseball League, 1987-89. Author: (with Earle C. Fowler) Strange Particles, 1963; Concepts in Physics, 1969, The Great Design, 1987, The Physics of Baseball, 1990; assoc. editor Phys. Rev., 1963-66; assoc. editor Phys. Rev. Letters, 1974-76, editor, 1978-84. Served with inf. AUS, 1943-46. Guggenheim fellow, 1954; Ford Found. fellow, 1962-63; Sloane Found. fellow, 1962-63. Fellow Am. Phys. Soc. (chmn. div. particles and fields 1972-73); mem. NAS (chmn. physics sect. 1986-89, sec. class phys. scis. 1989—). Home: 50 Deepwood Dr Hamden CT 06517-3415 Office: Yale U J W Gibbs Lab New Haven CT 06520

ADAM, STEVEN JEFFREY, chemical engineer; b. Long Beach, Calif., July 2, 1957; s. Martin George and Mary Jean (Campbell) A.; m. Karen Pavilonus, June 26, 1987. AA, AS in Chemistry, Rancho Santiago, 1980; BS in Chemistry, Calif. State U., Fullerton, 1986. Engr. Beckman Instruments, Fullerton, 1979-84, Rockwell Internat., Anaheim, Calif., 1984-89; lead engr., scientist McDonnell Douglas Space Systems Co., Huntington Beach, Calif., 1989—; prof. Univ. Calif., Irvine, 1991—. Contbr. articles to profl. jours. Recipient Recognition award NASA, 1992. Mem. Soc. for Advancement of Materials and Process Engring., Am. Soc. Metals, Am. Chem. Soc. Achievements include patents for gas stream purifier and RGA-coupon permeation measurement apparatus. Home: 4504 E Silverleaf Ave Orange CA 92669 Office: McDonnell Douglas Space 5301 Bolsa Ave Huntington Beach CA 92647

ADAMOVICS, ANDRIS, toxicologist, consultant; b. Riga, Latvia, Nov. 21, 1932; came to U.S., 1951, naturalized, 1963; s. Fricis Janis and Liza Adamovics; m. Vija Mara Zarins, Aug. 26, 1972; children: Andrejs Janis, Maris Valdis. B.S. Portland State Coll., 1958; B.S., U. Wash.-Seattle, 1962, M.S., 1965; Ph.D., Ohio State U., 1975. Pharmacologist Warren-Teed Pharms., Columbus, Ohio, 1966-75; toxicologist SRI Internat. Menlo Park, Calif., 1975-80; freelance cons., San Jose, Calif., 1980—; toxicologist Northview Pacific Labs., Berkeley, Calif., 1985-90; rsch. scientist Shaman Pharms., San Carlos, Calif., 1990-91; rsch. toxicologist Pilkington Barnes Hind, Sunnyvale, Calif., 1991—. Contbr. articles to profl. jours. Mem. N.Y. Acad. Scis., Genetic and Environ. Toxicology Assn. No. Calif., Soc. Toxicology, Contact Lense Inst. (tech. subcom. in vitro alternatives to animal use), Latvian-Am. Assn. Univ. Profs. and Scientists. Lutheran. Home: 957 Sweet Ave San Jose CA 95129-2343

ADAMS, ALFRED BERNARD, JR., environmental engineer; b. Asbury Pk., N.J., Oct. 15, 1920; s. Alfred Bishop and Julia Ruth (Wiseman) A.; m. Claudia Neff, Dec. 28, 1942; children: Alfred B. III, Tamara Adams Dohn, Carla Adams York. BSChemE, Ga. Inst. Technol., 1943; postgrad., Wayne State U., 1946-48, U. Ala., 1986-88, Jefferson State Community Coll, 1989-91. Registered profl. engr., Ala., Mich., Fla., Ga., N.C.; Diplomate in Am. Acad. Environ. Engrs. Project engr. Pennwalt, Wyandotte, Mich., 1946-50; sales mgr., design engr. Goslin-Birmingham Div., Birmingham, Ala., 1950-61; field engr. & Sales Elmco Corp., Birmingham, Ala., 1961-62; prin. engr. Morton-Thiokol Corp., Brunswick, Ga., 1962-64; tech. mgr. Rust Internat., Birmingham, 1964-86; pres., owner Adams Cons. & Engring. Svcs., Birmingham, 1986—; cons. in field. Contbr. tech. papers to profl. publs. Pres. Woodhaven Lakes Property Owners Assn., Pinson, Ala., 1980-82; mem. Pub. Health Com., Birmingham, 1975-78. 2d lt. U.S. Army Chem. Corps, 1943-53. Mem. Air & Waste Mgmt. Assn., Am. Chem. Soc., Tech. Assn. Pulp & Paper Industries. Presbyterian. Avocations: travel, photography, golf. Home: 1824 Lake Park Ln Birmingham AL 35215-5748 Office: Adams Cons & Engring Svcs 1824 Lake Park Ln Birmingham AL 35215-5748

ADAMS, CHESTER Z., sales engineer; b. Pitts., Sept. 4, 1908; s. Zigmond and Rose (Ruptic) A.; m. Mollie Zimbo, Oct. 5, 1935; children: Eugene, Edward Alan; m. Jean Roberts, Dec. 28, 1964. BS in Constrn., Carnegie Tech, 1930. Sales engr. ILG Electric Ventilation Co., N.Y.C., 1930-32; br. mgr. ILG Electric Ventilation Co., Greensboro, N.C., 1934-46; pres. Chet Adams Co., Greensboro, N.C., 1946-84, chmn. bd., 1984-86; pres. Am. Soc. Heating and Ventilation, N.C., 1946, Piedmont Sales Execs., Greensboro, 1964. Mem. Profl. Engrs. N.C., Greensboro Engrs. Club, KC, Greensboro Country Club. Republican. Roman Catholic. Home: 1700 Beechtree Rd Greensboro NC 27408 Office: 1052 Grecade Greensboro NC 27408

ADAMS, DAVID PARRISH, historian, educator; b. Jacksonville, Fla., Aug. 2, 1958; s. David Parrish and Gloria Ann (Nesmith) A.; m. Teri Ann Becker, Aug. 31, 1985; 1 child, Morgan Becker. BA, Emory U., 1980; AM, Washington U., St. Louis, 1982; PhD, U. Fla., 1987. Resource faculty dept. family medicine U. Fla., Gainesville, 1983-87; assoc. prof. dept. humanities Columbus (Ohio) State C.C., 1987—; adj. faculty dept. history Ohio Dominican Coll., Columbus, 1989; adj. asst. prof. dept. family medicine Ohio State U., Columbus, 1990—, dept. family practice, Ponce Sch. Medicine, Ponce, P.R., 1993; cons. in field 1992. Author: The Greatest Good to the Greatest Number: Penicillin Rationing on the American Home Front, 1940-45, 1991; contbr. articles to profl. jours. Med. Humanities fellow U. Ill., Chgo., 1992; grant-in-aid U. Wis., 1985; grantee NIH, 1989, Ohio Acad. Family Physicians, 1991. Mem. APHA, Am. Assn. History of Medicine, Am. Hist. Assn., Orgn. Am. Historians, Soc. Tchrs. Family Medicine. Democrat. Mem. United Ch. of Christ. Achievements include research in general practice in rural Florida; evolution of family practice; penicillin, dentistry and SBE; wartime penicillin policy; community oriented primary care. Home: 2044 Lublin Dr Apt E Reynoldsburg OH 43068 Office: Columbus State C C Humanities Dept 550 E Spring St Columbus OH 43216-1609

ADAMS, DEE BRIANE, hydrologist, civil engineer; b. Provo, Utah, Feb. 6, 1942; s. Dee B. and Helen Beth (Henrichsen) A.; m. Julie Dian Herbert, June 15, 1962; children: Andrew Briane, Sarah, Aaron Thomas. Student, Snow Coll., Ephraim, Utah, 1960-61; BS cum laude, U. Utah, 1971; MS, MIT, 1974. Engr. technician Thiokol Chem. Corp., Brigham City, Utah, 1962-64; hydrologic technician water resources div. U.S. Geol. Survey, Salt Lake City, 1964-71, civil engr., 1971-73; hydrologist U.S. Geol. Survey, Denver, 1974-77; sr. hydrologist U.S. Geol. Survey, Grand Junction, Colo., 1977-82; chief hydrologic studies U.S. Geol. Survey, Tampa, Fla., 1982-87; dist. chief for Ala. U.S. Geol. Survey, Tuscaloosa, 1987-90; for S.E. region U.S. Geol. Survey, Atlanta, 1990—; also nat. tech. cons. U.S. Geol. Survey; com. mem. Water, Energy and Biogeochemical Budgets Global Change Rsch. program. Contbr. tech. reports and articles to profl. jours. Leader Boy Scouts Am., Colo., Fla., Ala., Ga., 1977—; bd. dirs. Colo. Fed. Credit Union, Grand Junction, 1981-82, Panorama Improvement Dist., Grand Junction, 1981-82; tech. com. mem. Ala. Gov.'s Water Resources Study Commn., 1989-91; chmn., mem. exec. com. So. Appalachian Man and the Biosphere Coop.; gen. chmn. Applications and Mgmt. of Geographic Info. Systems in Hydrology and Water Resources internat. symposium, Mobile, Ala., 1993. With U.S. Army, 1966-68. U.S. Geol. Survey fellow, 1973-74. Mem. ASCE (mem. com. "Model Water Code", 1991—), ASTM (vice chmn., mem. com. D18.01.08 on global change 1991—), Am. Water Resources Assn. (coms. 1987—, organizer, pres. Ala. sect. 1988-89, chmn. promotion com.), Am. Geophys. Union, Am. Inst. Hydrology (registered profl., pres. Fla. sect. 1986-87, conf. orgn. com. 1986), Univ. Club, Tau Beta Pi, Chi Epsilon. Mem. LDS Ch. Avocations: outdoor sports, car building, woodworking. Home: 2272 Westridge Dr Snellville GA 30278 Office: Office of Regional Hydrologist Spalding Woods Office Park Ste 160 3850 Holcomb Bridge Rd Norcross GA 30092

ADAMS, DOLPH O., pathologist, educator; b. Montezuma, Ga., Apr. 12, 1939; s. J.F. and Frances O. Adams. AB in Chemistry, Duke U., 1960; MS in Anatomy, Med. Coll. Ga., 1963, MD, 1965; PhD in Exptl. Pathology, U. N.C., 1969. Lic. Ga. State Bd. Med. Examiners, N.C. State Bd. Med. Examiners; diplomate Am. Bd. Pathology. Dir. autopsy svc. Med. Ctr, Duke U., 1972-82, chief, 1983-92; program leader Comprehensive Cancer Ctr., Duke U., Durham, N.C., 1982-92, program dir., 1992—; dir. grad. studies Med. Ctr., Duke U., Durham, N.C., 1986-89, assoc. prof. immunology, 1986-92, dir. Lab. Cell & Molecular Biology of Leukocytes, 1986-92, vice chmn. pathology, 1991-93, prof. pathology, 1981—; vis. prof. The Rockefeller U., N.Y., 1982-83; cons. Ctr. Demographic Studies Duke U., 1973-80, VA Hosp., 1977-82; mem. exec. com. Integrated Program in Toxicology Duke U., Durham, N.C., 1981-90. Editor: (book) Methods for Studying Mononuclear Phagocytes, 1981, Contemporary Topics in Immunobiology, 1984. Major U.S. Army Med. Corps, 1970-72. Recipient Bausch and Lomb Metal award Am. Soc. Clin. Pathology, 1965, Rsch. Recognition award Noble Found., 1985, 2 Pub. Health Svc. awards NIH, Bethesda, Md., 1990-95. Mem. Am. Soc. Pathologists, Am. Soc. Cell Biology, Am. Assn. Immunologists, Am. Assn. Cancer Rsch., Soc. Leukocyte Biology (coun. pres. 1990-91). Achievements include research in the role of macrophages in cellular immunology and tumor biology. Home: 4209 New Bern Pl Durham NC 27707 Office: Duke U Med Ctr Dept Pathology Box 3712 Erwin Rd Durham NC 27710

ADAMS, DONALD E., mechanical engineer; b. Burlington, Vt., Nov. 11, 1938; s. Thurston M. and Dorothy (Johnson) A.; m. Rosemarie Vara, June 5, 1965; children: David E., Thomas J., Emily M. BS, U. Vt., 1960; MS, Cornell U., 1963. Rsch. asst. Cornell U./NSF, Ithaca, N.Y., 1960-62; assoc. engr. Cornell Aero. Lab., Inc., Buffalo, N.Y., 1962-66; rsch. engr. Cornell Aero. Lab., Inc., Buffalo, 1966-71; sr. engr. Calspan Corp., Buffalo, 1971-76, prin. engr., 1976—; jr. engr. Naval Ordnance Lab., Silver Spring, Md., 1961. Contbr. articles to profl. jours. Mem. Burlington (Vt.) Jr. Downtown Athletic Club, 1960-92, Town Recycling Com., Clarence, N.Y., 1991-92. Mem. ASME (chair Buffalo sect. 1972), Sigma Xi. Presbyterian. Achievements include patent on radiation heat transfer measuring instrument. Office: Calspan Corp 4455 Genesee St Buffalo NY 14225

ADAMS, DONALD EDWARD, biology researcher; b. Pitts., June 4, 1964; s. George Oliver and Norma Shirley (Whetton) A.; m. Daria Ann Novekosky, Oct. 21, 1989. BS in Chemistry and Biology, Geneva Coll., 1986; PhD in Microbiology, U. Ala., Birmingham, 1991. Lic. radioactive materials mgr., Ala. Teaching asst. Geneva Coll., Beaver Falls, Pa., 1985-86; rsch. scientist Sch. Medicine, U. Ala., Birmingham, 1986-91; patent examiner U.S. Dept. Commerce, Patent & Trademark Office, Arlington, Va., 1991—; adj. prof. sci. Marymount U., 1991—; judge Patent and Trademark Office Regional Sci. Fair. Contbr. articles to scholarly and profl. jours. NIH grantee, 1986; U. Ala. fellow, 1987. Mem. AAAS, Am. Inst. Biol. Scis., N.Y. Acad. Sci., Patent and Trademark Office Soc., Am. Soc. Microbiology, Phi Sigma (grad. student advisor 1990-91). Achievements include discovery of a novel cell cycle regulated protein kinase and its potential application for the diagnosis of acute lymphoblastic leukemia; discovery of a novel pathway for control of cell division utilizing a glycosyltransferase associated protein kinase. Office: US Dept Commerce Patent and Trademark Office Crystal Mall 1 Rm 9A01 Arlington VA 22202

ADAMS, DONALD SCOTT, research scientist, pharmacist; b. Covington, Ky., Apr. 29, 1960; s. Donald Ray and Joan Marie (Hunter) A.; m. Carol Jean Hausfeld, Mar. 30, 1985; children: Samuel Scott, Michelle Elizabeth. A Applied Sci. in Pharm., U. Cinn., 1980, A Sci. magna cum laude, 1991. Pharmacy tech. Jewish Hosp., Cin., 1980-81; rsch. tech. Procter & Gamble Co., Cin., 1981-90, rsch. scientist, 1991—. Mem. Progressive Citizens Orgn., St. Bernard, Ohio, 1985—. Mem. Assn. Pharmacy Tech. (advisor 1984—, treas. 1980-82), Ohio Valley Football Ofcls. Assn., Eagles, Alpha Sigma Lambda, Delta Tau Kappa, Delta Mu Delta, Golden Key Nat. Hon. Soc. Achievements include 3 patents applied for. Home: 4705 Chalet Dr Cincinnati OH 45217 Office: Procter & Gamble 6060 Center Hill Rd Cincinnati OH 45224

ADAMS, DUNCAN DARTREY, medical researcher, physician; b. Hamilton, Waikato, New Zealand, Apr. 18, 1925; s. H. Dartrey C. and Eliza Veitch (Duncan) A.; m. Yvonne Joan Macfarlane, Apr. 2, 1958 (dec. 1988); children: Christopher, Julia. MB, BCHir, Otago U., Dunedin, New Zealand, 1949, MD, 1962, DSc, 1962. Lectr. J.B. Collip's Lab., London, Ont., 1954; rsch. fellow Med. Rsch. Coun., New Zealand, 1951-53, rsch. officer, 1955-70, dir. autoimmunity rsch. unit, 1971-85; rsch. officer dean's dept. Otago Med. Sch., Dunedin, 1986—; vis. prof. Univ. Sheffield, Eng., 1964; vis. scientist Clin. Rsch. Centre, Med. Rsch. Coun., London, 1974-75. Contbr. 122 med. rsch. papers to profl. jours. With Merchant Navy, 1944. Recipient Van Meter prize Am. Goiter Assn., 1958, Organon Rsch. award Endocrine Soc., 1976, Mallinkrodt prize 8th Internat. Thyroid Congress, 1980, Medal for Contbns. to Biomed. Sci., Univ. Pisa, 1986; Am. Acad. Microbiology fellow, 1985. Fellow Royal Australasian Coll. Physicians (travelling scientist 1989). Avocations: hunting, show jumping, skiing, music, bridge. Home: 123 Cliffs Rd, St Clair, Dunedin New Zealand Office: Med Sch Deans Dept, Box 913, Dunedin New Zealand

ADAMS, EARLE MYLES, technical representative; b. Little Rock, May 9, 1942; s. Louis William and Maida Ora (Moran) A.; m. Jill Lenora Stuckey, Dec. 30, 1966 (div. May 7, 1979); children: Stephen Myles, Steffany Jill; m. Donna Kay Pettey, May 9, 1980. Student, U. Ark., 1960-64. Estimator, asst. mgr. Dyke Bldg. Specialties, North Little Rock, 1964-65; estimator, detailer Lashlee Steel Co., Benton, Ark., 1962-68; v.p. Knox Gill Co., Little Rock, 1968-82; roof cons. EMA, Inc., Little Rock, 1982—; estimator, project mgr. Kullander Constrn., Little Rock, 1984-85; roof cons. Cromwell Architects Engrs., Little Rock, 1985-90; tech. rep. BITEC, Inc., Morrilton, Ark., 1990—. Home: 30 Kensington Dr Conway AR 72032 Office: BITEC Inc # 2 Industrial Pk Dr Morrilton AR 72032-7224

ADAMS, JAMES ALFRED, natural science educator; b. Columbia, Miss., Dec. 17, 1949; s. Joseph Quincy and Bernice (Jackson) A.; m. Beverly Ann Colwell, June 28, 1978 (div. Feb. 1982); 1 child, Jasmine Denise. BS, Alcorn State U., 1970; MS, U. So. Miss., 1971; PhD, U. Pitts., 1975. Asst. prof. biology Tenn. State U., Nashville, 1975-80, assoc. prof. biology, 1981-86, prof. biology, 1986-88; chmn. dept. natural sci. U. Md.-Eastern Shore, Princess Anne, 1988—; mem. exec. com. grad. toxicology program U. Md., Balt., 1988—; mem. exec. com. Minority Inst. Marine Sci. Program, Jackson, Miss., 1989—; cons. State U. System of Fla., Tallahassee, 1992—. Contbr. articles to profl. jours. Mem. Men Inspiring Students to Enjoy Reading, Salisbury, Md., 1991—. Macy fellow Corp. Marine Biol. Lab., 1976, 77, 79, Westinghouse fellow, 1990; grantee NIH-Nat. Inst. Environ. Health Sci., 1991, NSF, 1992. Fellow Tenn. Acad. Scis.; mem. AAAS, N.Y. Acad. Scis., 100 Black Men of Md./Am. Democrat. Bapist. Achievements include discovery that Reagent grade Acetone can induce metamorphosis in Anuran Amphibian, that glutathione can induce budding in Hydra. Office: U Md Eastern Shore Dept Natural Scis Princess Anne MD 21853

ADAMS, JAMES DEREK, construction engineer; b. Newark, Mar. 3, 1962; s. Jimmie Lee and Hazel (Barden) A.; m. Geniece Gary, Sept. 3, 1989. BSCE, Villanova U., 1984. Registered profl. engr., N.J. Prin. Adams Minority Enterprises, Atlantic City, 1984; constrn. engr. RBA Group, Morristown, N.J., 1984-86, N.J. Hwy. Authority, Woodbridge, N.J., 1986—. Pres. Legacy, Woodbridge, 1992; chmn. budget com. Roger Williams Bapt. Ch., Passaic, N.J., 1992. Mem. Am. Concrete Inst. (N.J. chpt.), N.J. Soc. Asphalt Technologists (cert.). Home: 1244B N Broad St Hillside NJ 07205 Office: NJ Hwy Authority PO Box 5050 Woodbridge NJ 07095-5050

ADAMS, JAMES HENRY, chemical engineer; b. Montgomery, Ala., Aug. 16, 1957; s. James Henry and Edith Elizabeth (Toney) A. BS in Chemistry, Tuskegee Inst., 1980, BSChemE, 1981; postgrad., U. Ala., Tuscaloosa, 1984-85. Cert. engr.-in-tng. Process engr. Phillips Petroleum Co., Sweeny, Tex., 1981-82; project engr. CH2M Hill, Montgomery, 1985—. Mem. Am. Inst. Chem. Engrs. (program chmn. 1989-90, chmn. cen. Ala. sect. 1990-92), Am. Chem. Soc. Methodist. Home: 4102 Ardmore Dr Montgomery AL 36105 Office: CH2M Hill 2567 Fairlane Dr Montgomery AL 36116

ADAMS, JAMES LEE, poultry scientist; b. Reading, Pa., May 9, 1953; s. Ralph Donald and Eleanor Isabella (Bobbenmoyer) A.; m. Karen Lee Conner, June 24, 1972; children: Timothy James, Molly Beth, Kelly Lee. BS in Poultry Tech. and Mgmt., Pa. State U., 1980. Dir. R&D Wengers Feed Mill, Inc., Rheems, Pa., 1980—. With U.S. Army, 1975-78. Mem. Poultry Sci. Assn., N.Y. Acad. Sci., Animal Sci. Assn., Coun. for Agrl. Sci. and Tech. Office: Wengers Feed Mill Inc PO Box 26 Rheems PA 17570

ADAMS, JAMES MILLS, chemicals executive; b. Sioux Falls, S.D., Aug. 4, 1936; m. Sherrell D.; 2 children. BSChemE, S.D. Sch. Mines and Tech., 1958; MS in Engring., U. Wash., 1961, PhD in Chem. Nuclear Engring., 1962. Sr. engring. specialist aerophysics rsch. Aerojet-Gen. Corp., Sacramento, 1962-68; sr. spectroscopist Hoffmann La Roche, Inc., Nutley, N.J., 1968-70, sr. scientist applied scis. dept., 1970-73, mgr. CVA engring., 1974-76; plant mgr. aroma chem. plant Haarmann and Reimer Corp., Springfield, N.J., 1976-79, v.p., gen. mgr. aroma chem. div., 1978-85, exec. v.p., 1979-80, pres., 1980—, CEO, 1985—; mem. adv. bd. Cook Coll., Rutgers U., 1985-92; corp. v.p. Miles, Inc., 1989-91; chmn. bd. Creations Aromatiques, Inc., 1990—. Assoc. editor Pyrodynamics, 1966-69; contbr. over 40 articles to profl. jours.; patentee in fields of med. instrumentation, emission spectrometry, pyrometry, remote sensing, others. Mem. Charleston County Aviation Commn., 1978-79; internat. adv. coun. Monell Ctr., 1990—. H.L. Doherty Ednl. Found. scholar 1954-58; W. Alton Jones fellow, 1959-60; recipient Centennial 100 Alumni award, S.D. Sch. Mines and Tech., 1985. Mem. AAAS, Am. Phys. Soc., Am. Mgmt. Assn., Flavor and Extract Mfrs. Assn. (bd. govs. 1985—, v.p., sec. 1989—, pres. elect 1990, pres. 1992-93), Fragrance Materials Assn. (bd. dirs. 1985—, pres. 1988-92), Rsch. Inst. for Fragrance Materials (chmn. bd. dirs. 1984-89, vice-chmn. 1989-90), N.J. C. of C., Pres. Assn., Optimists Club (pres. Watchung chpt. 1984-85), Sigma Xi (pres. Roche Rsch. Club 1973-74). Home: 171 Hillcrest Rd Watchung NJ 07060 Office: Haarmann & Reimer Corp PO Box 175 Springfield NJ 07081-0175

ADAMS, JAMES WILLIAM, retired chemist; b. Conover, Wis., Oct. 29, 1921; s. Aldred Henry and Pauline (Evertson) A.; m. Joyce Marie Braatz, Oct. 27, 1944 (div. Oct. 1986); children: Judy R. Adams Swank, Neal J.; m. Barbara A. Backlund, Apr. 4, 1987. BS in Chemistry, U. Wis., 1943. Analytical chemist U.S. Rubber Co., Institute, W.Va., 1943-47; rsch. chemist U.S. Rubber Co., Naugatuck, Conn., 1947-52; sr. scientist Marathon Corp., Rothschild, Wis., 1952-53, rsch. group leader, 1953-57; sr. rsch. assoc. Am. Can Co., Rothschild, 1957-80; rsch. fellow Reed Lignin Inc., Rothschild, 1980-89; retired, 1989; chemistry specialist J&B Cons., Schofield, Wis., 1989—. Contbr. articles to Chem. and Engring. Progress, Indsl. and Engring. Chemistry, Applied Polymer Symposium, Radiochem. Radioanalytical Letters. Alderman Schofield City Coun., 1958-74; vol. Wausau Nordic Ski Club. Recipient Meritorious Svc. award City of Schofield, 1976, Vol. of Yr. award Badger State Games, 1991. Mem. Am. Chem. Soc. (tour speaker 1986—), Am. Inst. Chemists, N.Y. Acad. Scis. Achievements include 20 patents for using Wood Pulping Liquor in Animal Feed, Process for Making Superabsorbent Fibers, Modified Wood Pulp Fibers for Plant Growth Medium, and others on industrial products and processes; development of test for measuring road wear qualities of tire rubber. Home: 2008 Clarberth St Schofield WI 54476-1211

ADAMS, JOHN EWART, chemistry educator; b. Jefferson City, Mo., Oct. 13, 1952; m. Carol Ann Deakyne, May 25, 1991. BS in Chemistry, U. Mo., Rolla, 1974; PhD, U. Calif., Berkeley, 1979. Postdoctoral fellow Los Alamos (N.Mex.) Nat. Lab., 1979-81; asst. prof. chemistry U. Mo., Columbia, 1981-87, assoc. prof. chemistry, 1987—. Recipient Undergrad. Teaching award Amoco Found., 1987; NSF grad. fellow, 1974-77. Mem. Am. Chem. Soc. (chair, sec.-treas. U. Mo. sect.), Am. Phys. Soc., Alpha Chi Sigma (faculty advisor). Office: U Mo Dept Chemistry Columbia MO 65211

ADAMS, JOSEPH BRIAN, engineer, mathematics educator; b. Lancaster, Pa., Apr. 23, 1961; s. Laurence John and Ann (Onufrak) A. BS in Nuclear Engring., Pa. State U., 1983, M Engring. Sci., 1987; postgrad, U. Del. Registered profl. engr., Pa. Radiol. engr. Phila. Electric Co., Phila., 1983-88; instr. math. U. Del., Newark, 1988—; adj. math. and engring. faculty Widener U., Chester, Pa., 1987—; Pa. State U., 1992—. Mem. Math. Assn. Am. Roman Catholic. Home: 5 Haskell Dr Lancaster PA 17601-4343 Office: U Delaware Dept Ops Rsch Newark DE 19716

ADAMS, LARRY DELL, mining engineering executive; b. Madisonville, Ky., Sept. 9, 1963; s. Dell Hayden and Judith Ann (Mullennix) A.; m. Debra Kay Lohman, June 1, 1985;1 child, Rachel Anne. BS, Pa. State U., 1985. Registered profl. engr., Ky. Student engr. Consolidated Coal Co., Library, Pa., 1981, Moundsville, W.Va., 1982; project engr. Adams Engrs., Inc., Lexington, Ky., 1985-88; project engr. Mining Cons. Svcs., Inc., Lexington, 1989-92, v.p., 1992—. Mem. Ky. Cols., 1984. Mem. NSPE, Soc. for Mining, Metallurgy and Exploration, Ky. Soc. Profl. Engrs., Profl. Engrs. in Mining Ky. (bd. dirs. 1991-92), Triangle frat. Home: 3892 Gladman Way Lexington KY 40514 Office: Mining Cons Svcs Inc 340 S Broadway Ste 200 Lexington KY 40508

ADAMS, NANCY R., nurse, military officer. BSN, Cornell U., N.Y. Hosp. Sch. Nursing; MSN, Cath. U. Am.; grad., Command and Gen. Staff Coll., U.S. Army War Coll. Commd. Nurse Corps, U.S. Army, 1968, advanced through grades to brig. gen.; chief nurse Army Regional Med. Ctr., Frankfurt, Germany; nursing adminstr. various locations; nurse consultant to U.S. Surgeon Gen.; now chief Nurse Corps, U.S. Army. Author textbooks; contbr. articles to profl. jours. Mem. ANA, Assn. Mil. Surgeons of U.S.,

Am. Orgn. Nurse Execs., Sigma Theta Tau. Office: Army Nurse Corp Office of the Surgeon Gen 5109 Leesburg Pike Falls Church VA 22041-3258

ADAMS, ONIE H. POWERS (ONIE H. POWERS), retired biochemist; b. Augusta, Ga., July 28, 1907; d. William Perley and Lucy Jane (Ingram) Hixson; m. George Arthur Powers, May 5, 1948 (dec. 1955); m. Elliott T. Adams, Apr. 20, 1963 (dec. 1989). BSHE, U. Ga., 1928, MSPH, 1931; postgrad., various colls., Boston. Sci. tchr. various schs. and hosps., 1928-40; rsch. fellow U. Ga. Med. Sch., Augusta, 1929-31; tchr. Children's Hosp., Augusta, 1929-31; sr. med. technician, adminstr. Southside Community Hosp., Farmville, Va., 1932-33; med. technician Lynn (Mass.) Hosp., 1934-40; seriologist, instr. USPHS Hosp., Boston, 1940-53; biochemist USPHS CLin. Ctr., Bethesda, Md., 1953-54, NIH Cancer Inst., Bethesda, 1954-67, Med. Libr. Detailed Info. Scis., Bethesda, 1967; retired, 1967. Contbr. articles to Jour. Nat. Cancer Inst. Mem. AAAS, Am. Chem. Soc. (emeritus), Clin. Chemistry Assn. (emeritus), N.Y. Acad. Scis. (life), NIH Alumni Assn. (life), Sigma Delta Epsilon Grad. Women in Sci., Inc.(life). Achievements include research in metabolism of tumors, changes with chemical agents on various enzymes in tumors. Home: 56 Morse Rd Newton MA 02160-2455

ADAMS, PAUL ALLISON, biologist, educator; b. Davenport, Iowa, May 30, 1940; s. George Miller and Irene Henrietta (Johnson) A.; m. Kathleen Belle Hagan, Jan. 2, 1965; children: Nathan, Natalie. AB, Calvin Coll., 1962; PhD, U. Mich., 1969. Rsch. assoc. Mich. State U./AEC Plant Rsch. Lab., East Lansing, 1968-70; asst. prof. U. Mich., Flint, 1970-74, assoc. prof., 1974—, chmn. dept. biology, 1985-88. Mem. AAAS, Am. Scientific Affiliation, Am. Soc. Plant Physiologists, Sigma Xi. Home: 2418 Gold Ave Flint MI 48503 Office: U Mich 303 E Kearsley St Flint MI 48502

ADAMS, ROBERT EDWARD, journalist; b. Geneseo, Ill., Apr. 27, 1941; s. Horace Mann and Florence (Beidelman) A. B.S., U. Ill., 1963. Reporter Champaign-Urbana Courier, Ill., 1962-64; reporter, city staff St. Louis Post-Dispatch, 1966-72, Washington corr., 1972—; bur. chief St. Louis Post-Dispatch, Washington, 1983—; Washington commentator Sta. KMOX, St. Louis, 1984—; founding mem. St. Louis Journalism Rev., 1970. Recipient reporting award Nat. Civil Service League, 1975, polit. reporting award Lincoln U., Jefferson City, Mo., 1984, Raymond Clapper Meml. award for Washington Corr., 1987, citation for excellence Overseas Press Club, for series on Soviet Union, 1988; co-recipient Fgn. Corr. award Overseas Press Club Am., 1984, Nat. Headliner award, 1986. Mem. White House Corr.'s Assn., Com. to Protect Journalists, Washington Ind. Writers, The Gridiron Club, Sigma Delta Chi (Outstanding Young Reporter award St. Louis chpt. 1969). Roman Catholic. Home: 2500 Wisconsin Ave NW Washington DC 20007-4505 Office: St Louis Post-Dispatch Washington Bur 1701 Pennsylvania Ave NW Washington DC 20006-5805

ADAMS, WILLIAM B., consultant; b. Richmond, Va., Feb. 4, 1941; s. William B. and Jeanette (Griffiths) A.; m. Gail Taylor, June 4, 1966; children: Robert Taylor, Suzanne Valerie. BS, U. Md., 1962, postgrad., 1962-67. Cons. specializing in communications, computers, networks, security, software engring., artificial intelligence, computer sci., stats., math. Springfield, Va., 1962—. League dir. Braddock Rd. Youth Club, Springfield, 1982—, mem. exec. com., 1982—. U. Md. fellow, 1963-64. Mem. Math. Assn. Am., Soc. Indsl. and Applied Math., Am. Math. Soc., European Soc. for Theoretical Computer Sci., Am. Assn. for Artificial Intelligence, Pi Mu Epsilon. Libertarian. Office: PO Box 1467 Springfield VA 22151

ADAMS, WILLIAM JOHN, JR., mechanical engineer; b. Riverdale, Calif., Feb. 9, 1917; s. William John and Florence (Dodini) A.; m. Marijane E. Leishman, Dec. 26, 1939; children—W. Michael, John P. B.S., Santa Clara U., 1937. Registered profl. mech. engr., agrl. engr., Calif. Design and project engr. Gen. Electric Co., Schenectady, 1937-45; project engr. FMC Corp., San Jose, Calif., 1946-47; chief engr. Bolens div., Port Washington, Wis., 1947-53; asst. gen. mgr. Central Engring. Labs., Santa Clara, 1953-71; dir. planning and ventures, advanced products div. Central Engring. Labs., 1971-76, mgr. new bus. ventures, 1976-80, cons. to mgmt., 1980—; regent Santa Clara U., 1986—. Chmn. Santa Clara United Fund drive, 1967, Santa Clara Indsl. Citizens Bd., 1965-66; bd. dirs. Santa Clara County council Boy Scouts Am., 1962—, Santa Clara Valley Sci. Fair, Eagle Scout Assn. Recipient Ignatian award disting. svc. to humanity Santa Clara U., 1990, Inaugural Disting. Engring. Alumnus award Santa Clara U., 1991. Fellow Am. Soc. Agrl. Engrs. (dir. Pacific Coast region); mem. ASME (life, del. Silicon Valley Engring. Coun.), Soc. Automotive Engrs., Joint Coun. Sci. and Math. Edn., Tau Beta Pi, Pi Tau Sigma. Achievements include patents in field U.S., Can., numerous fgn. countries. Home and Office: 7940 Caledonia Dr San Jose CA 95135-2113

ADAMS-HILLIARD, BEVERLY LYNN, chemist; b. San Francisco, Oct. 28, 1965; d. Lee F. and Velnet (Johnson) A. BS, Alcorn State U., 1987, MS, 1991. Radiochemist Entergy Ops., Port Gibson, Miss., 1989—. Home: 4417 Sun Drop Ct Jackson MS 39212

ADAMSON, DAN KLINGLESMITH, association executive; b. Vernon, Tex., Oct. 12, 1939; s. Earl Larkin and Edith (Klinglesmith) A.; m. Eva Diane Pope, Aug. 18, 1962; children—Larkin, Rebecca, Amy, Sarah. Student, U. Mo., 1958-59; B.A. in History, Southwestern U., Georgetown, Tex., 1962. Tchr. pub. schs., Jefferson County, Colo., 1962-64; asst. dir. Soc. Petroleum Engrs., Dallas, 1964-67, editor jour., 1967-71, gen. mgr., 1972-79, exec. dir., 1979—. Mem. Am. Soc. Assn. Execs., Council Engring. Sci. Soc. Execs. Republican. Methodist. Office: Soc Petroleum Engrs PO Box 833836 222 Palisades Creek Dr Richardson TX 75080

ADAMSON, JOHN WILLIAM, hematologist; b. Oakland, Calif., Dec. 28, 1936; s. John William and Florence Jean Adamson; m. Susan Elizabeth Wood, June 16, 1960; children: Cairn Elizabeth, Loch Rachael; m. Christine Fenyvest, Sept. 1, 1989. BA, U. Calif., Berkeley, 1958; MD, UCLA, 1962. Intern then resident in medicine U. Wash. Med. Ctr., Seattle, 1962-64, clin. and rsch. fellow hematology, 1964-67, mem. faculty, 1969-90, prof. hematology, 1978-90, head div. hematology, 1981-89; pres. N.Y. Blood Ctr., N.Y.C., 1989—; Josiah Macy Jr. Found. scholar, vis. scientist Nuffield dept. clin. medicine, U. Oxford, Eng., faculty medicine, 1976-77. Author papers in field, chpts. in books. With USPHS, 1967-69. Recipient Rsch. Career Devel. award NIH, 1972-77, rsch. grantee, 1976—. Fellow AAAS; mem. Am. Soc. Hematology, Assn. Am. Physicians, Am. Soc. Clin. Investigation, Western Assn. Physicians. Office: NY Blood Ctr 310 E 67th St New York NY 10021-6295

ADAMSON, RICHARD HENRY, biochemist; b. Council Bluffs, Iowa, Aug. 9, 1937; s. Holger Nels and Mary Carolyn (Dengle) A.; m. Charlene Denham, Oct. 25, 1963; children: Kristin, Kara. B.A., Drake U., 1957; M.S., U. Iowa, 1959, Ph.D., 1961; M.A., George Washington U., 1968. Fellow U. Iowa Coll. Medicine, Iowa City, 1958-61; commd. officer USPHS, NIH, Bethesda, Md., 1961-63; sr. investigator lab. chem. pharmacology Nat. Cancer Inst., Bethesda, Md., 1963-69, head pharmacology and exptl. therapeutics sec., 1969-73, acting chief lab. chem. pharmacology, 1973-76, chief lab. chem. pharmacology, 1976-81, dir. divsn. cancer etiology, 1981—; lectr. physiology George Washington U., Washington, 1963-70; Fulbright vis. scientist St. Mary's Hosp. Med. Sch., London, 1965-66; sr. policy analyst Office Sci. and Tech. Policy Exec. Office of Pres., 1979-80. Author: numerous publs. in field; mem. editorial bd.: Cancer Treatment Reports, 1972-75, Xenobiotica, 1971-84, Cancer Research, 1980-87, Jour. Biolchem. Toxicology, 1984—, Regulatory Toxicology and Pharmacology, 1984—, Health and Environment Digest, 1986—, Japanese Jour. Cancer Research (Gann), 1986—, In Vivo, 1990—, Teratogenesis, Carcinogenesis and Mutagenesis, 1991—. Recipient USPHS Superior Svc. award, 1976, 82, Spl. Achievement award EEO, 1982, Presdl. Meritorious Exec. Rank award, 1989, Arnold J. Lehman award Soc. Toxicology, 1989, PHS Spl. Recognition award, 1992, Leadership for Combined Fed. Campaign award NIH, 1993. Mem. AAAS, Am. Assn. Cancer Research, Biochem. Soc., Am. Soc. Pharmacology and Exptl. Therapeutics, Toxicology Soc., Am. Soc. Internat. Law, Comparative Research Leukemia, Toxicology Forum (Anderson award 1990). Office: NIH-National Cancer Institute Cancer Etiology Rm 11A03 9000 Rockville Pike Bldg 31 Bethesda MD 20892

ADAMSON, SANDRA LEE, anthropologist; b. Vancouver, Wash., Mar. 2, 1948; d. Roland Lee and Mary Lee (Drum) A. BA, U. Wash., 1969; MA, U. Va., 1980. Social worker County of San Mateo, Redwood City, Calif., 1970-73; Wash. rep. L5 Soc., Tucson, 1985-86; sr. analyst Ctr. for Space and Advanced Tech., Fairfax, VA., 1986—; dir. Congl. rsch. Spacepac, Santa Monica, Calif., 1986. S.W. Ariz. coord. for John Glenn Campaign, Tucson, 1984; vol. Washington Park Zoo, Portland, Oreg., 1981-82. Mem. Nat. Space Soc. (sec. 1986-87, exec. v.p. 1990, v.p. 1988-89, 91—), Space Cause (bd. dirs., v.p. 1985—), Am. Soc. Gravitational and Space Biology, Am. Anthrop. Assn., Am. Astro. Soc., Nat. Assn. Practicing Anthropology, Am. Ethnology Soc., Women in Aerospace. Democrat. Home: 4141 N Henderson Rd Apt 214 Arlington VA 22203 Office: Ctr for Space/Advanced Tech 9302 Lee Hwy Ste 400 Fairfax VA 22031

ADANK, JAMES P., physicist, administrator; b. Fountain City, Wis., May 24, 1963; s. Roger W. and Patricia (Schafner) A. BS in Physics and Math., U. Wis., 1983; MS in Physics, U. Minn., 1985. Staff scientist Philips Med. Systems, Shelton, Conn., 1989-90, prin. magnet resonance physicist, 1990-93, clin. sci. mgr., 1993—. Contbr. Magnet Resonance Physics, 1989—. Mem. Am. Assn. Physics in Medicine, Soc. Magnetic Resonance in Medicine. Office: Philips Med Systems 710 Bridgeport Ave Shelton CT 06484

ADDICOTT, FREDRICK TAYLOR, retired botany educator; b. Oakland, Calif., Nov. 16, 1912; s. James Edwin and Ottilia Katherine Elizabeth (Klein) A.; m. Alice Holmes Baldwin, Aug. 11, 1935; children: Donald James, Jean Alice, John Fredrick, David Baldwin. AB in Biology, Stanford U., 1934; PhD in Plant Physiology, Calif. Inst. Tech., 1939. Instr. to asst. prof. Santa Barbara (Calif.) State Coll., 1939-46; assoc. physiologist emergency rubber project USDA, Salinas, Calif., 1942-44; asst. prof. to prof. UCLA, 1946-60; prof. agronomy U. Calif., Davis, 1961-72, prof. botany, 1972-77, prof. emeritus, 1977—; vis. prof. U. Adelaide, Australia, 1966, U. Natal, Pietermaritzburg, Republic South Africa, 1970. Author: Abscission, 1982; editor, author: Abscisic Acid, 1983; contbr. articles to profl. jours. Fulbright rsch. scholar Victoria U., N.Z., 1957, Royal Bot. Garden, U.K., 1976; vis. fellow Australian Nat. U., 1983. Fellow AAAS; mem. Am. Soc. Plant Physiologists (Charles Reid Barnes Life Membership award 1990), Australian Soc. Plant Physiologists, Bot. Soc. Am., Internat. Plant Growth Substance Assn., Internat. Soc. Plant Morphologists, South African Assn. Botanists. Avocations: backpacking, cabinet making. Home: 1003 Pine Ln Davis CA 95616-1764 Office: U Calif Botany Dept Davis CA 95616-8537

ADDINGTON, WILLIAM HAMPTON, civil engineer; b. Berkeley, Calif., Feb. 10, 1934; s. William Hampton and Kathrine Irene (Dwyer) A.; m. Carmel Mary Gurbada, Nov. 20, 1954; children: William Matthew, Christina Anne Miller, Diana Mary Addington. Student, various colls. and univs. Registered profl. engr., Calif.; lic. land surveyor, Calif. Project engr. Neste Bruding & Stone, San Bernardino, Calif., 1965-69; project mgr. J.F. Davidson Assocs., Riverside, Calif., 1969-77; v.p., chief engr. J.F. Davidson Assocs., Colton, Calif., 1977-90; v.p. regional mgr. Engring. Svc. Corp., Culver City, Calif., 1991-92; prin. chief engr. Addington Cons. Engring., Grand Terrace, Calif., 1992—. With USN, 1951-55. Mem. Cons. Engrs. and Land Surveyors of Calif. (sec.-treas. 1992-93), Calif. Coun. of Civil Engrs. and Land Surveyors (v.p. 1992), Civil Engrs. and Land Surveyors of Riverside and San Bernardino Counties (pres. 1982), Rotary Club (pres. 1985-86), Am. Soc. of Civil Engrs., Am. Cons. Engrs. Coun., Calif. Land Surveyors Assn. Home: 12055 Westwood Ln Grand Terrace CA 92324 Office: Addington Cons Engring Ste 2 22737 Barton Rd Grand Terrace CA 92324

ADDISON, WALLACE LLOYD, aerospace engineer; b. Greenville, S.C., June 26, 1965; s. Lloyd Brandford and Patricia Ann (Priester) A. BS in Aerospace Engring., U. Ala., 1988. Mgr. reliability cost program ALD/OAF, Wright-Patterson AFB, Ohio, 1989-90; mgr. F-16 reliability/cost program ASD/YPLI, Wright-Patterson AFB, Ohio, 1990-91, mgr. F-16 reliability and maintainability program, 1991-92; mgr. Korean fighter program ASC/YPX-KFP, Wright-Patterson AFB, Ohio, 1993—. Active Boy Scouts Am., 1975—. 1st lt. USAF, 1988—. Recipient Air Force Orgnl. award, 1990. Mem. AIAA, U. Ala. Nat. Alumni Assn., Capstone Engring. Soc., Nat. Eagle Scout Assn., K.C. Roman Catholic. Home: 5010 Worchester Dr Dayton OH 45431-1108 Office: ASC/YPX-KFP Wright-Patterson AFB Dayton OH 45433-6503

ADEDIRAN, SUARA ADEDEJI, chemistry educator; b. Ibadan, Oyo, Nigeria, Jan. 27, 1948; s. Braimoh Adepoju and Wulemot Olalompe (Yusuf) A.; m. Olusola Abeni Ogunyemi, Dec. 6, 1975; children: Adekemi Olaide, Ayodeji Olaniyi, Adeyemi Olayiwola, Adegoke Oladele. BSc in Chemistry, U. Ibadan, 1973, PhD in Chemistry, 1979. Chartered chemist Royal Soc. Chemistry London. Demonstrator in chemistry U. Ibadan, 1974-79; lectr. chemistry Sch. Med. Lab. Tech., Ibadan, 1976-79, Ogun State Poly., Abeokuta, Nigeria, 1979; postdoctoral rsch. fellow U. Alta., Edmonton, Can., 1981-82; vis. rsch. assoc. U. Kans. Coll. Health Scis., Kansas City, 1985-86; lectr. II and I chemistry U. Ilorin (Nigeria), 1979-86, sr. lectr. chemistry, 1986—; peer reviewer NSF, Washington, 1986, 90. Referee Nigerian Jour. Sci., Ibadan, 1986—; contbr. articles to profl. jours. including European Jour. Biochemistry, Arch. Biochem. Biophys., Jour. Biol. Chemistry, Biochimie. Pres. Fate Lions Club, Ilorin, 1990. Travel fellow Internat. Union Biochemistry, Prague, Czechoslovakia, 1988; rsch. grantee U. Ilorin, 1989, 91, Royal Soc. Chemistry, London, 1992. Mem. Royal Soc. Chemistry, Am. Chem. Soc., Sci. Assn. Nigeria (coun. 1988-92), Chem. Soc. Nigeria. Achievements include research on a mechanism of formation of horseradish peroxidase compounds, on the use of indicator dyes as mechanistic probes for enzymatic reactions. Office: U Ilorin, Chemistry Dept, Ilorin Kwara, Nigeria

ADEGBOLA, SIKIRU KOLAWOLE, aerospace engineer, educator; b. Ibadan, Nigeria, Jan. 21, 1949; came to U.S., 1971; s. Lasisi and Moriamo Abeke (Akinyemi) A. BSME, Calif. State U., Fullerton, 1974; MBA, Calif. State U., Dominguez Hills, 1988; MSME, U. Ariz., 1975; MS in Applied Mechanics, U. So. Calif., 1977; PhD in Engring., Calif. Coast U., 1983. Registered profl. mech. engr., Calif., Ariz. Rsch. engr. Jet Propulsion Lab., Pasadena, Calif., 1976-78; stress analyst Bechtel Power Corp., Norwalk, Calif., 1978-87; engring. mem. tech. staff structural analysis dept. Space Systems div. Rockwell Internat., Downey, Calif., 1987—; prof. engring. Calif. State U., Fullerton, 1984—. Leopold Schepp Found. fellowship, 1972-74. Mem. ASME, NSPE, Calif. Soc. Profl. Engrs., Nat. Mgmt. Assn. Home: PO Box 345 Downey CA 90241-0345 Office: Rockwell Internat Corp Space Systems Divsn 12214 Lakewood Blvd Downey CA 90242-2693

ADELBERGER, REXFORD EARLE, physics educator; b. Cleve., Mar. 30, 1940; s. Erwin G. and Emma M. (Zeschky) A.; m. Patricia Carvalho, Nov. 26, 1966; children: Marc E., Kirstin J. BS, Coll. of William and Mary, 1961; PhD, U. Rochester, 1968. Asst. prof. SUNY, Geneseo, N.Y., 1968-73; assoc. prof. Guilford Coll., Greensboro, N.Y., 1973-80, prof. physics, 1980—. Editor: Jour. of Undergrad. Rsch. in Physics, 1981—. Mem. Kiwanis, Greensboro, 1975—. Mem. Am. Assn. Physics Tchrs., Am. Phys. Soc., Coun. Undergrad. Rsch. Religious Soc. of Friends. Office: Guilford Coll Physics Dept Greensboro NC 27410

ADELI, HOJJAT, civil engineer, computer scientist, educator; b. Langrood, Iran, June 3, 1950; came to U.S., 1974; s. Jafar and Mokarram (Soofi) A.; m. Nahid Dadmehr, Mar. 1979; children: Amir, Anahita, Mona, Cyrus Dean. MSCE, U. Teheran, Iran, 1973; Ph.D. in Civil Engring., Stanford U., 1976. Asst. prof. Northwestern U., Evanston, Ill., 1977; asst. prof. U. Teheran, 1978-81, assoc. prof., 1981-82; assoc. prof. U. Utah, Salt Lake City, 1982-83; assoc. prof. Ohio State U., Columbus, 1983-88, prof., 1988—, chmn. structures faculty, 1988-91; cons. Atomic Orgn. Iran, Teheran, 1978-79, Iran Ministry Housing, Teheran, 1970-82, U.S. Army Constrn. Engring. Rsch. Lab., 1988; keynote lectures in Italy, 1989, Mex., 1989, Japan, 1991, China, 1992, Can., 1992, Portugal, 1992, U.S.A., 1993, Germany, 1993; organizer or lectr. at more than 90 nat. and internat. confs. Author: Interactive Microcomputer-Aided Structural Steel Design, 1988; co-author: Expert Systems for Structural Design: A New Gerneration, 1988, Parallel Processing in Structural Engineering, 1993; editor: Expert Systems in Construction and Structural Engineering, 1988, Microcomputer Knowledge-Based Expert Systems in Civil Engineering, 1988, Parallel and Distributed Processing in Structural Engineering, 1988, Knowledge Engineering, vols. 1 & 2, 1990, Supercomputing in Engineering Analysis, 1992, Parallel Processing in Computational Mechanics, 1992, Advances in Design Optimization, 1993; co-editor: Mechanics Computing in the 1990's and Beyond, vols. 1 & 2, 1991; editor-in-chief, founder Internat. Jour. Microcomputers in Civil Engring., 1986—, Integrated Computer-Aided Engring., 1993—; editor-in-chief Heuristics: The Jour. of Knowledge Engring., 1991-93; assoc. editor Control Engring. Practice; mem. editorial bd., editorial adv. bd. 20 sci. engring. jours.; contbr. over 250 publs. Recipient First Degree medal of Knowledge, Iran Ministry of Higher Edn., 1973, Disting. Leadership award Am. Biog. Inst., 1986, Honor medal, 1987, Rsch. award NSF, USAF Flight Dynamics Lab., Cray Rsch., Ind., Bethlehem Steel Corp., Ohio Thomas Edison Prog., Am. Inst. Steel Construction, Am. Iron and Steel Inst.; named Man of Yr. Am. Biog. Inst., 1990, Internat. Man of Yr. IBC, 1991-92. Mem. ASCE, IEEE, Am. Iron and Steel Inst., Am. Inst. Steel Constrn., Am. Assn. for Artificial Intelligence, Assn. for Computing Machinery, Earthquake Engring. Rsch. Inst. Home: 1540 Picardae Ct Powell OH 43065-9791

ADELMAN, RICHARD CHARLES, gerontology educator, researcher; b. Newark, Mar. 10, 1940; s. Morris and Elanor (Wachman) A.; m. Lynn Betty Richman, Aug. 18, 1963; children—Mindy Robin, Nicole Ann. A.B., Kenyon Coll., 1962; M.A., Temple U., 1965, Ph.D., 1967. Postdoctoral fellow Albert Einstein Coll. Medicine, Bronx, N.Y., 1967-69; from asst. prof. to prof. Temple U., Phila., 1969-82, dir. inst. aging, 1978-82; prof. biol. chemistry U. Mich., Ann Arbor, 1982—, dir. inst. gerontology, 1982—; mem. study sect. NIH, 1975-78; adv. coun. VA, 1981-85; chmn. Gordon Rsch. Conf. Biol. Aging, 1976; adv. com. VA, 1981-91; chmn. VA Geriatrics and Gerontology Adv. Com., 1987-91. Mem. various editorial bds. biomed. research jours., 1972—. Bd. dirs. Botsford Continuing Care Ctrs., Inc., Farmington Hills, Mich., 1984-88. Recipient Medalist award Intrasci. Research Found., 1977; grantee NIH, 1970—; established investigator Am. Heart Assn., 1975-78. Fellow Gerontol. Soc. Am. (v.p. 1976-77, pres. elect 1986-87, Kent award 1990); mem. Am. Soc. Biol. Chemists, Gerontol. Soc. Am. (pres. 1986-87), Am. Chem. Soc., AAAS, Phila. Biochemists (pres.), Practictioners in Aging. Jewish.

ADELSTEIN, S(TANLEY) JAMES, physician, educator; b. N.Y.C., Jan. 24, 1928; s. George and Belle (Schild) A.; m. Mary Charlesworth Taylor, Sept. 20, 1957; children—Joseph Burrows, Elizabeth Dunster. B.S., M.I.T., 1949, M.S., 1949, Ph.D. in Biophysics (Nat. Found. fellow), 1957; M.D., Harvard U., 1953. Med. house officer Peter Bent Brigham Hosp., Boston, 1953-54; sr. asst. resident physician Peter Bent Brigham Hosp., 1957-58, chief resident, 1959-60; fellow Howard Hughes Med. Inst., 1957-58, Henry A. and Camilus Christian fellow, 1959-60; Moseley Traveling fellow Harvard U., 1958-59; instr. anatomy Harvard Med. Sch., 1961-65, asst. prof., 1965-68, asso. prof. radiology, 1968-72, prof., 1972-89, Paul C. Cabot prof. med. biophysics, 1989—, dean for acad. programs, 1978—; mem. Nat. Council for Radiation Protection Measurements, 1978—, dir., 1980—, v.p., 1982—; cons. Med. Found. fellow, 1960-63. Mem. editorial bd. Investigative Radiology, 1972-80l, Postgrad. Radiology, Radiation Rshc., 1990—; assoc. editor Jour. Nuclear Medicine, 1975-81; contbr. articles to profl. jours. Trustee Am. Bd. Nuclear Medicine, 1972-78, sec., 1975-78. NIH Career Devel. awardee, 1965-68; Fogarty Sr. Internat. fellow, 1976. Fellow Am. Coll. Nuclear Physicians, AAAS; mem. Am. Chem. Soc., Biophys. Soc., Assn. for Radiation Research, Radiation Research Soc. (councillor 1975-78), Soc. Nuclear Medicine (trustee 1970-74, Aebersold award 1986), Boylston Med. Soc., Inst. Medicine, Sigma Xi, Tau Beta Pi, Alpha Omega Alpha. Office: Harvard Med Sch 25 Shattuck St Boston MA 02115-6092

ADENIJI-FASHOLA, ADEYEMO AYODELE, mechanical engineer, consultant; b. Lagos, Nigeria, May 22, 1952; arrived in Germany, 1991; s. Christopher Adewusi and Ibidunni Olufunke (Odujomein) A.-F.; m. Jummai Elizabeth Umaru, July 23, 1977; children: Adebunmi, Adetomiwa, Adekanmbi. BME, Ahmadu Bello U., Zaria, Nigeria, 1975; M of Applied Sci., U. Toronto, 1978, PhD, 1981; MBA, Internat. Inst. Mgmt. Devel., Lausanne, Switzerland, 1990. Lectr. U. Ilorin, Nigeria, 1981-85, sr. lectr., 1985-86; rsch. assoc. NASA-Marshall Space Flight Ctr., Huntsville, Ala., 1986-89; mgmt. and tech. cons. Ploenzke Cons., Wiesbaden, Germany, 1991—; rsch. assoc. Turbulent Two-Phase Flow Rsch., Huntsville, 1986-89. Contbr. articles to Acta Mechanica, Internat. Jour. Heat Mass Transfer, Internat. Com. Heat Mass Transfer. Nat. Acad. Sci.-NRC rsch. assoc., 1986; Can. Commonwealth scholar, 1976. Mem. ASME, AIAA. Evangelical. Achievements include modeling of turbulent particulate flows to account for polydisperse particulate phase and include two-way coupling between the dispersed phase and the continuous phase turbulence; use of stochastic tools to model physical as well as business and economic phenomena; modeling/prediction of domain delineation of continuous and discontinuous change. Office: Ploenzke Cons, Kreuzberger Ring 62, Wiesbaden Germany D-6200

ADEYEYE, SAMUEL OYEWOLE, nutritional biochemist; b. Ilesha, Nigeria, Africa, May 21, 1953; came to U.S., 1989; s. Isaac Adeyeye and Alice Taiwo (Ojengbede) A.; m. Olayinka Fajembola, Apr. 28, 1979; children: Olufemi, Olufunmilayo, Temitope. MSc, U. Ibadan, Nigeria, 1984; PhD, U. Ibadan, 1988. Rsch. assoc. Internat. Inst. Tropical Agrl., Ibadan, 1987-89; postdoctoral rsch. assoc. Tukegee (Ala.) U., 1989-91, postdoctoral rsch. assoc./lectr., 1991-92, rsch. assoc. prof., 1992—. Contbr. articles to profl. jours. including Nutrition Reports Internat. and Agronomy Jour. Mem. AAAS, Japan Soc. for Bioscience, Biotechnology and Agrochemistry, N.Y. Acad. Scis., Sigma Xi. Home: 307 B Katherine Dr Tuskegee AL 36088 Office: Tuskegee U Thomas Campbell Bldg Tuskegee AL 36088

ADHOLEYA, ALOK, biotechnologist, researcher; b. Gwalior, India, Aug. 30, 1962; s. Hari Kant and Uma Saxena; m. Radhika Jan. 26, 1991; 1 child, Abhimanyu. BSc, Jiwaji Univ., 1981, MSc, 1986, PhD, 1988. Rsch. assoc. Tata Energy Rsch. Inst., New Delhi, 1986-92; fellow Tata Energy Rsch. Inst., New Delhi, 1992—; mem. tech. adv. com. Mycorrhiza Network Asia, New Delhi, 1989—; speaker and presenter at various convs. Contbr. articles to profl. jours. Recipient numerous grants. Mem. Mycological Soc. Am., British Mycological Soc., Assn. Mycorrhixologist. Achievements in the area of axenic culturing of vesicular arbusculat mycorrhixal fungi using transformed hairy root cultures. Home: C-130 Sarvodya Enclave, New Delhi India 110017 Office: Tata Energy Rsch Inst, 158 Jorbagh, New Delhi India 110003

ADINOFF, BRYON HARLEN, psychiatrist; b. Waynesville, N.C., Mar. 9, 1953; s. Bernard and Madeline (Shiller) A.; m. Patricia Jean Holland, Apr. 29, 1984; children: Zack Nathan, Holland Annette. B in Gen. Studies, U. Mich., 1974; MD, Mich. State U., 1979. Intern, resident Tulane U. Affiliate Hosp., New Orleans, 1979-83; from med. staff fellow to dir. outpatient unit Lab. Clin. Studies, Nat. Inst. Alcohol Abuse & Alcoholism, Bethesda, Md., 1983-88; asst. prof. dept. psychiatry and behavioral scis. Med. U. S.C., Charleston, 1988-92, assoc. prof., 1992—; dir. Substance Abuse Treatment Ctr. VA Med. Ctr., Charleston, 1988—. Contbr. articles to Biol. Psychiatry, Arch Gen. Psychiatry, Am. Jour. Psychiatry, Am. Jour. Addiction. Nat. Inst. Alcohol Abuse and Alcoholism fellow, 1983-86. Mem. Am. Assn. Psychiatrists on Alcoholism and Addictions, Rsch. Soc. on Alcoholism, Am. Psychiat. Assn. Home: 330 Bampfield Dr Mount Pleasant SC 29464 Office: VA Medical Center 109 Bee St Charleston SC 29401

ADJAMAH, KOKOUVI MICHEL, physician; b. Lome, Togo, West Africa, Jan. 21, 1942; s. Carl Yao and Antoinette Afi (Tourné) A.; m. Francoise Legendre, July 30, 1975; children: Isabelle, Celine. MS, U. Paris, 1970, MD, 1973, MPH, 1974, M of Human Ecology, 1975, M of Clin. Toxicology-Pharmacology, 1976. Pub. health physician Dreux, France, 1973-74; gen. practice medicine, homeopathy, acupuncture, holistic medicine Vernouillet, France, 1976; expert in human ecology Paris, 1976, cons. clin. pharmacology, 1977—. Author: Onchocerchosis in West Africa, 1973; contbr. articles to profl. jours. Fellow Rabelais Med. Brotherhood; mem. French Com. for Med. Lexicon, Soc. Functional Medicine, French Soc. Pharmacology-Toxicology (assoc.), Social Medicine Club, Human Rights Assn., XVIII Century Studies Assn., History of Justice Assn. Quaker. Avocations: musicology, piano. Home: 25 rue de Chailloy, Garnay Vernouillet, 28500 Eure et Loir France Office: 5 Ter rue Louis Jouvet, Vernouillet, 28500 Eure et Loir France

ADJEI, ALEX ASIEDU, internist, pharmacologist; b. Kumasi, Ghana, Apr. 21, 1955; came to U.S., 1989; s. James Kwaku and Juliana (Owusua) A. MBChB, U. Ghana, 1982; PhD, U. Alta., Can., 1989. Diplomate Am. Bd. Internal Medicine. Med. officer Ghana Med. Sch., Accra, 1983-84; clin. rsch. fellow U. Alberta, Edmonton, Can., 1987-89; med. resident Howard U. Hosp., Washington, 1989-91, chief med. resident, 1991-92; sr. clin. fellow Oncology Clin. Ctr. Johns Hopkins Hosp., Balt., 1992—; mem. Grad. Student's Coun., Edmonton, 1984-86. Contbg. author book in field, 1988. Named Best Student, Ghana Med. Sch., 1979, Resident of Yr., D.C. Gen. Hosp., 1991; recipient Travel award Nat. Cancer Inst., Can., 1988, rsch. scholarship Alberta Heritage Found., Edmonton, 1985-89. Mem. AAAS, ACP, Am. Assn. Cancer Rsch., Am. Soc. Internal Medicine, N.Y. Acad. Scis. Home: 6608 Waning Moon Way Columbia MD 21045 Office: Johns Hopkins Hosp Oncology Ctr 600 N wolfe St Baltimore MD 21205

ADLER, EUGENE VICTOR, forensic toxicologist, consultant; b. Detroit, June 28, 1947; s. Leonard Eugene and Beatrice Doris (Goldfarb) A.; m. Patricia Ann Vangeloff, Sept. 7, 1969; children: Kimberly, Emily. BS in Chemistry with honors, Wayne State U., 1969. Diplomate Am. Bd. Forensic Toxicology. Analyst crime lab. Mich. Health Dept., Lansing, 1970-72; criminalist crime lab. Ariz. Dept. Pub. Safety, Phoenix, 1972—; cons., lectr. U.S. Dept. of Transp., Washington, 1990—; advisor, lectr. Ariz. Govs. Office of Hwy. Safety, Phoenix, 1989—; lectr. Northwestern U., Traffic Safety Inst., 1991, and various traffic safety confs., 1988—. Editor The DRE newsletter/jour., 1989—. Recipient Commendation award Phoenix City Prosecutors Office, 1989, 92. Mem. Internat. Assn. Forensic Toxicologists, Calif. Assn. Toxicologists, Southwestern Assn. Forensic Scientists, Camelback Toastmasters, U.S. Corr. Chess Team (player). Achievements include establishment, validation and promotion of a scientific method for detecting drug impaired drivers; development of Ariz.'s Drug Recognition Expert (DRE) program. Office: Ariz Dept Pub Safety Crime Lab PO Box 6638 2310 W Encanto Phoenix AZ 85005-6638

ADLER, FREDERICK RUSSELL, mathematical ecologist; b. Boston, Mar. 15, 1963; s. Sheldon and Barbara (Gitter) A. BA, Harvard U., 1984; PhD, Cornell U., 1991. Postdoctoral rsearcher Ctr. for Population Biology U Calif, Davis, 1991-92; asst. prof. depts. of Math. and Biology U. Utah, Salt Lake City, 1993--. Office: U Utah Dept math Salt Lake City UT 84112

ADLER, STEPHEN LOUIS, physicist; b. N.Y.C., Nov. 30, 1939; s. Irving and Ruth (Relis) A.; children: Jessica Wendy, Victoria Stephanie, Anthony Curtis. A.B. summa cum laude, Harvard U., 1961; Ph.D., Princeton U., 1964. Jr. fellow Soc. of Fellows Harvard U., 1964-66; research assoc. Calif. Inst. Tech., 1966; mem. Inst. for Advanced Study, Princeton, N.J., 1966-69; prof. Sch. Natural Scis., Inst. for Advanced Study, 1969-79, N.J. Albert Einstein prof., 1979—; vis. lectr. dept. physics Princeton U., 1969—; cons. in field. Author: (with R.F. Dashen) Current Algebras, 1968; contbr. articles to profl. jours. Recipient J.J. Sakuai prize Am. Phys. Soc., 1988. Fellow Am. Acad. Arts and Scis., AAAS, Am. Phys. Soc.; mem. Nat. Acad. Scis., Phi Beta Kappa, Sigma Xi. Home: 287A Nassau St Princeton NJ 08540-4618 Office: Inst for Advanced Study Sch Natural Scis Olden Ln Princeton NJ 08540

ADNEY, JAMES RICHARD, physicist; b. Indpls., May 15, 1946; s. Frank Brown and Mary Alice (Claycombe) A.; m. Melissa Gay Kepner, June 8, 1974; 1 child, Laura Elizabeth Kepner-Adney. BA, DePauw U., 1968; MA, U. Wis., 1970. Bd. dirs. Yellow Jersey Coop., Madison, Wis., 1970-72; framebuilder Masi Bicycle Co., Carlsbad, Calif., 1973-74; mechanic Yellow Jersey Coop., Madison, 1974-77; physicist Nat. Electrostatics Corp., Middleton, Wis., 1977--. Contbr. articles to profl. jours. Achievements include design and debugging of heavy ion beam probe accelerator with 500 kilovolt solid state power supply, development of RF tuning techniques for subnanosecond pulsing and bunching systems for heavy ion accelerators. Home: 4018 Yuma Dr Madison WI 53711 Office: Nat Electrostatics Corp PO Box 620310 Middleton WI 53562

ADOM, EDWIN NII AMALAI, psychiatrist; b. Accra, Ghana, West Africa, Jan. 12, 1941; came to U.S., 1960; s. Isaac Quaye and Julianna Adorkor (Brown) Adom; m. Margaret Odarkor Lamptey; children: Edwin Nii Nortey Jr., Isaac Michael Nii Nortei. BA, U. Pa., 1963; MD, Meharry Med. Coll., 1968; FRSH (Eng.), Royal Soc. Health, 1974. Diplomate Am. Bd. Psychiatry and Neurology. Intern Pa. Hosp., Phila., 1968-69; resident in psychiatry Thomas Jefferson U. Med. Ctr., Phila., 1969-72; cons. psychiatrist St. Joseph Hosp., Phila., 1976-80, Stephen Smith Home for the Aged, Phila., 1976-85, Mercy Douglas Human Svcs. Ctr., Phila., 1977-79, St. Ignatius Home for the Aged, Phila., 1978-85; attending psychiatrist Phila. Psychiat. Ctr., 1987—; cons. psychiatrist psycho-social dept. Horizon House Rehab. Ctr., Phila., 1987-89; cons. psychiatrist U.S. Dept. Labor, Workmen's Compensation Div., Phila., 1987—, House Staff of the U. Pa. Hosp., Phila., 1989—; cons. The Grad. Hosp., Phila., 1976—; faculty clin. assoc. psychiatry U. Pa. Sch. Medicine, Phila., 1972—; with State of Pa. Bur. Disability Determination, 1975—; attending psychiatrist West Phila. Mental Health Consortium, 1972—, med. dirs., 1991—. Ghana Govt. scholar U. Pa., 1960-68; recipient Citizen Citation Chapel of the Four Chaplains, 1974. Mem. Fellow Royal Soc. Health (Eng.); mem. Am. Psychiatric Assn., Am. Acad. Psychiatry and the Law, Nat. Med. Assn., Black Psychiatrists Am. (exec. mem. 1975-77), Pa. Psychiatric Soc., Med. Soc. Ea. Pa., Phila. Psychiatric Soc., N.Y. Acad. Sci., World Fedn. Mental Health, Psychiatric Alliance, Phila. Acad. Family Psychiatrists, Nat. Geographic Soc. Presbyterian. Avocations: traveling, world affairs, music, cultural affairs. Office: Med Towers 255 S 17th St # 2704 Philadelphia PA 19103

ADVANI, SUNDER, engineering educator, university dean. BSME with first class honors, Bombay U., 1961; MSME, Stanford U., 1962, PhD, 1965. Design engr. Carco Electronics, Menlo Park, Calif., 1962; teaching and rsch. asst. mech. engring. Stanford U., 1962-64; sr. rsch. engr., mem. tech. staff Biodynamics Lab. Northrop Corp., Calif., 1965-67; assoc. prof., prof. mech. engring. and mechanics W.Va. U., 1967-78, assoc. chmn. grad. studies dept. mech. engring. and mechanics, 1975-78; prof., chmn. dept. engring. mechanics Ohio State U., 1978-91, assoc. dean acad. affairs Coll. Engring., 1984-91; dean Coll. Engring. and Applied Sci. Lehigh U., Bethlehem, Pa., 1991—; mem. U.S. nat. rock mechanics com. NRC, 1983-89; assoc. editor ASME JOMAE, 1987-91, JERT, 1986-91; bd. dirs. Ohio State U. Rsch. Found., 1986-90; pres. Assn. Chmn. Dept. Mechanics, 1990-91. Author: (with others) Crashworthiness in Transportation Systems, 1978, Applied Physiological Mechanics, 1979, Electrical Properties of Bone and Cartilage, 1979, Human Body Dynamics, 1982, Finite Elements in Fluids, 1984; editor: (with C.H. Popelar and A.W. Leissa) Developments in Mechanics, 1985; contbr. numerous articles to profl. jours. Recipient Ralph P. Boyer Meritorious award, 1980, George Westinghouse award ASEE, 1985. Fellow ASME, Am. Acad. Mechanics. Achievements include research in geomechanics, computational methods, material property characterization, and biomechanics. Office: Lehigh U Coll of Engring & Applied Sci 308 Packard Laboratory Bethlehem PA 18015

AF EKENSTAM, ADOLF W., mathematics educator; b. Västerås, Sweden, Jan. 19, 1932; s. Wilhelm H. and Elsa (Palmi) af E.; m. Barbara von Tresckow, Oct. 4, 1935; children: Ebba, Karin, Nils. PhD, U. Uppsala, Sweden, 1960. Tchr. Upper Secondary Sch., Linköping, Sweden, 1961-66; lectr. Tchr. Tng. Coll., Linköping, Sweden, 1966-77; sr. lectr. U. Linköping, 1977—. Author textbooks in math. edn.; co-editor Jour. Math. Behavior; contbr. articles Swedish, English, German profl. jours. Home: Ettöresgatan 9, 58266 Linkoping Sweden Office: U Linkoping, 58183 Linköping Sweden

AFFATICATI, GIUSEPPE EUGENIO, telecommunications, instrumentation engineer; b. Cortemaggiore, Piacenza, Italy, July 25, 1932; s. Paolo Giovanni and Maria Anna (Sozzi) A.; Perito Industriale, Istituto Tecnico G Marconi, Piacenza, 1955. With AGIP, 1958—, instrumentation project engr. engring. dept., Gela, Italy, 1961-79, telecommunications, instrumentation project engr. Milan, Italy, 1979—; cons. engring. Served to lt., arty. Italian Army, 1956-57. Roman Catholic. Home: 9 Via Manfredi, 29100 Piacenza Italy Office: AGIP Oil Co, Milanofiori C4, 20090 Assago Italy

AFFLECK, IAN KEITH, physics educator; b. Vancouver, B.C., Can., July 2, 1952; s. William Burchill and Evelyn Mary (Colense) A.; m. Glenda Ruth Harman, July 2, 1977; children: Geoffrey Roger, Ingrid Katherine. BS, Trent U., 1975; AM, Harvard U., 1976, PhD, 1979. Asst. prof. physics Princeton (N.J.) U., 1981-87; rsch. scientist Centre d'Etudes Nucléaire, Saclay, France, 1984-85; prof. U. B.C., Vancouver, 1987—. Contbr. articles on physics to profl. jours. Recipient Steacie prize Nat. Rsch. Coun. Can., 1988, Hertzberg medal Can. Assn. Physicists, Rumford medal Royal Soc., 1991. Fellow Royal Soc. Can. (Rutherford medal 1991), Can. Inst. for Advanced Rsch., Harvard Soc. Fellows (jr.); mem. Aspen Inst. Physics, Can. Inst. Neutron Scattering (trustee 1990). Office: U BC, Physics Dept, Vancouver, BC Canada V6T 121

AFSARMANESH, HAMIDEH, computer science educator, research scientist; b. Tehran, Iran, Aug. 23, 1953; came to U.S., 1977; d. Esmail Afsarmanesh and Mehrangiz Hoveida; m. Farhad Arbab, Mar. 8, 1977; children: Taraneh, Mandana. BBA, Tehran Bus. Sch., 1975; MS, Aryamehr Tech. U., Tehran, 1977; MSc, UCLA, 1980; PhD, U. So. Calif., 1985. Rsch. scientist SDC (now subs. Unisys), L.A., 1985-86; assoc. prof. dept. computer sci. Calif. State U., Dominguez Hills, 1986—; vis. rsch. prof. U. Amsterdam, 1990—; cons. dept. elec. engring. U. So. Calif., 1987—. Contbr. chpts. to books. Rsch., Scholarship and Creative Activities Program fellow Calif. State U., 1989. Mem. IEEE (chairperson various confs. 1988—), Assn. Computing Machinery, Sigma Xi.

AFZAL, MOHAMMAD, biologist; b. Talhar, Pakistan, Jan. 15, 1949; s. Abdul Aziz and Sardar Begum Aziz; m. Nadra Jabeen, Oct. 23, 1978; children: Humaira, Hammad, Imran, Aamir. MS in Marine Zoology, U. Karachi, 1972, MS in Entomology, 1974, PhD in Zoology, 1983; MPhil in Animal Physiology, Quaid-e-Azam U., 1984. Biologist Invertebrate Reference Mus. U. Karachi, Pakistan, 1972-73; rsch. officer dept. zoology U. Karachi, Pakistan, 1974-79; rsch. assoc. Quaid-E-Azam U., Islamabad, Pakistan, 1980-82; scientific officer Nat. Agriculture Rsch. Ctr., Islamabad, 1983; assoc. curator Pakistan Mus. Natural History, Islamabad, 1983, curator, dir. in charge, 1983-90; dir. Pakistan Sci. and Tech. Info. Ctr., Islamabad, 1990—; sec. Standing Com. on Informatics, Pakistan, Com. on 2d T.V. Channel for Edn., Pakistan, mem. steering com. for computerization. Contbr. articles to Chemosphere, Tropical Ecology, Folia Biology, Sabrao Jour., Jour. Stored Product Rsch. Fellow Natural History Mus. France, 1989-90. Fellow Entomol. Soc. Pakistan, Zool. Soc. Pakistan, Fisheries Soc. Pakistan, Scientific Soc. Pakistan. Achievements include research in systematics and biology of Pentatomomorphous super families Coreoidea and Pentatomodea of Pakistan, rice insects of Pakistan, food legume improvement, pulses entomology, zoogeographical studies on diptera in S.W. Asia. Office: Pastic Nat Ctr, Quaid-E-Azam U Campus, Islamabad Pakistan

AGAPITO, J. F. T., mining engineer, mineralogist. ACSM in Mining Engring., Camborne Sch. Mines, Eng.; MS in Mine Ventilation/Mining Engring., U. Mo., Rolla; PhD in Rock Mechanics/Mining Engring., Colo. Sch. Mines. Profl. engr., Colo., Wash. Miner, technician Beralt Wolfram & Tin, Ltd., Portugal, 1958-60, South Crofty, Ltd., Eng., 1958-60; ventilation engr. White Pine (Mich.) Copper Co., 1964-66, mine rsch. engr., 1966-68; instr. Colo. Sch. Mines, Golden, 1968-72; sr. rock mechanics engr. Atlantic Richfield Co., Grand Valley, Colo., 1972-74; assoc. Golder Assocs., Inc., Grand Junction, Colo., 1974-76; ind. cons. mining engr., 1976-78; pres. J. F. T. Agapito & Assocs., Inc., Grand Junction, 1978—. Contbr. articles to profl. jours. Recipient Rock Mechanics award Soc. Mining, Metallurgy and Exploration, 1992. Fellow Inst. Mining and Metallurgy (Eng.); mem. AIME, Internat. Soc. Rock Mechanics. Office: J. F. T. Agapito & Assoc Inc 715 Horizon Dr Ste 340 Grand Junction CO 81506*

AGAR, JOHN RUSSELL, JR., school district supervisor; b. Camden, N.J., July 25, 1949; s. John R. and Evva L. (Wilhelm) A.; m. Beatrice A. B.; children: Rebekah A., Sarah L. BA with high honors, Rutgers U., 1971; MS, U. Pa., 1973, MS in Edn., 1975; EdD with distinction, Temple U., 1983; postgrad., U. Pa., 1989. Cert. secondary educator, supr., prin., dist. supt., Pa., N.J. Lectr. in chemistry U. Pa., Phila., 1974-75; sci. dept. head West Cath. Girls' High Sch., Phila., 1974-79; chemistry tchr. Moorestown (N.J.) Friends' Sch., 1979-82, West Deptford High Sch., Westville, N.J., 1982-84; visiting asst. prof. Temple U., Phila., 1983-88; lectr. in edn. U. Pa., Phila., 1988-90; sci. supr. Marple Newtown Sch. Dist., Newtown Sq., Pa., 1984—; mem. CEPUP staff Lawrence Hall of Sci. U. Calif., Berkeley, 1991—; mem. tchrs. industry environment com. Pa. Chem. Industry Coun., 1992—. NIH fellow U. Pa., 1973-74, CEPUP/CHEM/NSF fellow U. Calif., Berkeley, 1990-91; recipient Nat. Tchr. award CEPUP, U. Calif., Berkeley, 1992. Mem. ASCD, Nat. Sci. Suprs. Assn., Nat. Sci. Tchrs. Assn., Phi Lambda Upsilon, Phi Delta Kappa, Mensa. Avocations: travel, backpacking, scuba diving, cross-country skiing, weightlifting.

AGARWAL, DULI CHAND, mechanical engineer; b. Tankri, Haryana, India, Sept. 22, 1950; came to U.S., 1980; s. Jai Narain and Narbada Devi (Nargada Kumari) A.; m. Chanda Devi, Feb. 9, 1976; children: Ruby, Reena, Pankaj. BSME, Regional Inst. Tech Jamshedpur, Bihar, India, 1976; MS in Engring. Mgmt., U. Md., 1990. Registered profl. engr., Md., D.C. Engr. Bechtel Power Corp., Gaithersburg, Md., 1980-89; sr. mech. engr. U. Md., College Park, 1989-90; nuclear engr. U.S. Dept. Energy, Germantown, Md., 1990—. Merit scholar Regional Inst. Tech., Jamshedpur, 1970-75. Mem. ASME, Am. Soc. for Quality Control. Home: 12331 Quince Valley Dr North Potomac MD 20878

AGARWAL, NEERAJ, anatomy and biology educator; b. Dhampur, India, Nov. 13, 1955; came to U.S., 1982; s. Jagdish Saran Gupta and Kanti (Devi) Agarwal; m. Rajnee Agarwal, Feb. 20, 1981; children: Priyanka P., Meena M. PhD in Biochemistry, Postgrad. Inst. Med. Edn. Rsch, Chandigarh, India, 1982. Postdoctoral fellow U. So. Calif. Sch. Medicine, L.A., 1982-84, Yale U. Med. Sch., New Haven, 1984-87; asst. prof. U. Tex. Health Sci. Ctr., San Antonio, 1987-92; asst. prof. dept. anatomy and cell biology U. North Tex. Health Sci. Ctr., Ft. Worth, 1992—. Rsch. fellow Indian Dept. Sci. and Tech., 1977, Indian Coun. Med. Rsch., 1980, James Hudson Brown fellow, 1985; Fight for Sight grantee Nat. Soc. to Prevent Blindness, 1990. Mem. Am Soc. Cell Biologists, Assn. Rsch. for Vision and Ophthalmologists. Office: Tex Coll Osteo Medicine Anatomy & Cell Biology 3500 Camp Bowie Blvd Fort Worth TX 76107

AGARWAL, SANJAY KRISHNA, chemical engineer; b. Bombay, India, Aug. 18, 1965; came to U.S., 1986; s. Krishna Murari and Saroj (Garg) Goyal. BSChemE, U. Bombay, 1986; MSChemE, U. Pitts., 1988, PhD in Chem. Engr., 1989; MBA, U. N.C., 1993. Rsch. engr. Rsch. Triangle Inst., Durham, N.C., 1990—. News Brief Editorial bd. Applied Catalysis Jour., 1992—. Mem. Am. Inst. Chem. Engrs. (sec. Ea. N.C. sect. 1993—), Am. Chem. Soc., Am. Soc. Quality Control, Sigma Xi, Beta Gamma Sigma. Achievements include research in natural gas upgradings, organic compounds destruction, NOX control, and alternate fuels for automobile applications. Office: Rsch Triangle Inst PO Box 12194 Research Triangle Park NC 27709-2194

AGBA, EMMANUEL IKECHUKWU, mechanical engineering educator; b. Zaria, Kaduna, Nigeria, June 30, 1960; s. Lawrence Okoli and Caroline Uchechukwu (Ezagu) A. M.Eng., U. Benin, Benin City, Nigeria, 1988; PhD, Fla. Atlantic U., 1991. Maintenance engr. Ringin Motors Ltd., Kano, Nigeria, 1984-85; prodn. mgr., maintenance engr. Starline Nigeria Ltd., Aba, 1987-88; rsch. asst. Fla. Atlantic U., Boca Raton, Fla., 1988-91; asst. prof. mech. engring. Miss. State U., Mississippi State, 1991—; cons. PCi, Iuka, Miss., 1992. Contbr. articles to profl. publs. Newell doctoral fellow, 1989, 90, Harbor Br. Oceanographic Inst. fellow, 1990, minority student fellow, 1988. Mem. IEEE, ASME (assoc.), Nigerian Soc. Engrs. Office: PO Drawer ME Mississippi State MS 39762

AGGARWAL, JAGDISHKUMAR KESHORAM, electrical and computer engineering educator, research director; b. Amritsar, India, Nov. 19, 1936; came to U.S., 1960; s. Keshoram J. and Harkaur A.; m. Shanti Seth, July 1965; children: Mala, Raj. BS, U. Bombay, 1957; B in Engring., U. Liverpool, Eng., 1960; MS, U. Ill., 1961, PhD, 1964. Registered profl. engr., Tex. Rsch. asst. Marconi's Rsch. Lab., Chelmsford, Eng., 1959; fellow U. Ill., Urbana, 1960-61, rsch. asst. coordinated sci. lab., 1961-64; asst. prof. elec. engring. U. Tex., Austin, 1964-68, assoc. prof. elec. engring., 1968-72, prof. elec. and computer engring., 1972—, John J. McKetta Energy prof., 1981-90, Cullen Trust for Higher Edn. Endowed prof. No. 2, 1990—; dir. Computer and Vision Rsch. Ctr., Austin, 1985—; cons. IBM Corp., Shell Devel. Corp. Co-author: Non-Linear Systems: Stability Analysis, 1977, Digital Signal Processing, 1979, Deconvolution of Seismic Data, 1982, Motion Understanding, 1988. Recipient Outstanding Contbn. award Pattern Recognition Soc., 1985, 86, Disting. Alumnus award U. Ill., 1986, Alumni Honor award U. Ill. Coll. Engring., 1987, Am. Soc. Engring. Edn. Sr. Rsch. award, 1992. Fellow IEEE (editor Expert jour. 1986-89, editor IEEE Trans. Parallel and Dist. Systems, 1992—); mem. IEEE Computer Soc. (chmn. tech. com. on pattern recognition and machine intelligence 1987-89), Internat. Assn. for Pattern Recognition (rep. 1985—, treas. 1989-92, pres. 1992—, computer vision program chmn. 1990 Conf.), Austin Yacht Club. Avocation: sailing. Office: Univ of Tex ECE Dept Computer & Vision Rsch Ctr Coll Engring Austin TX 78712-1084

AGHAJANIAN, JOHN GREGORY, electron microscopist, cell biologist; b. Boston, Nov. 2, 1947; s. William R. and Anna Delores (Infante) A.; 1 child, Gregory D. BA, U. N.H., 1970; MS, L.I. U., 1972; PhD, U. N.C., 1977. Rsch. assoc. U. N.C., Chapel Hill, 1977-79, Harvard Med. Sch., Boston, 1980-82, Dana Farber Cancer Inst., Boston, 1979-82; instr. biology Quinsigamond C.C., Worcester, Mass., 1984-87; sr. electron microscopist Worcester Found. for Exptl. Biology, Shrewsbury, Mass., 1982—; dir., cons. E.M. Visions, Putnam, Conn., 1986—. Contbr. articles to Jour. of Phycology, Protoplasma, others. Cubmaster Boy Scouts Am., Quincy, Mass., 1986-88, scoutmaster, 1988-90. NSF fellow, 1969, 71. Fellow AAAS; mem. Electron Microscopy Soc. Am., Am. Soc. for Cell Biology, Phycological Soc. Am., New Eng. Soc. for Electron Microscopy (pres. 1988, dir. biol. scis. 1984-86), Masons, Sigma Xi. Home: 29 Pleasant St Putnam CT 06260 Office: Worcester Found Exptl Biology 222 Maple Ave Shrewsbury MA 01545-2795

AGID, STEVEN JAY, aerospace engineer; b. N.Y.C., May 9, 1952; s. Jacob and Marion (Popick) A. BS in Space Sci., Fla. Inst. Technol., 1974. Spacecraft controller RCA Am. Communications, Vernon, N.J., 1976-81; spacecraft analyst RCA/GE Am. Communications, Vernon, N.J., 1981-88; sr. engr. McDonnell Douglas Space Systems Co., Kennedy Space Ctr., Fla., 1988—; cons. in field. Author: (poem) Challenger Memorial Poem, 1986; contbr. articles to mags. Recipient Right Stuff award Ala. Space and Rocket Ctr., 1985, Men of Achievement award Internat. Biographical Ctr., 1979; named Men & Women of Distinction, 1979. Fellow Brit. Interplanetary Soc.; mem. AIAA (sr.), Nat. Space Soc. (charter life), Smithsonian/Air and Space Mus. Avocations: photography, calligraphy, model making, art, poetry. Home: 1310 Plum Ave Merritt Island FL 32953 Office: McDonnel Douglas Space Systems Co Kennedy Space Center Orlando FL 32899

AGNEW, HAROLD MELVIN, physicist; b. Denver, Mar. 28, 1921; s. Sam E. and Augusta (Jacobs) A.; m. Beverly Jackson, May 2, 1942; children: Nancy E. Agnew Owens, John S. AB, U. Denver, 1942; MS, U. Chgo., 1948, PhD, 1949; PhD (hon.), Coll. Santa Fe, 1980, U. Denver, 1992. With Los Alamos Sci. Lab., 1943-46, alt. div. leader, 1949-61, leader weapons div., 1964-70, dir., 1970-79; pres. Gen. Atomics, San Diego, 1979-85, also bd. dirs., 1985—; sci. adviser Supreme Allied Comdr. in Europe, Paris, 1961-64; chmn. Army Sci. Adv. Panel, 1965-70, San Diego County adv. bd.; mem. aircraft panel President's Sci. Adv. Com., 1965-73; mem. USAF Sci. Adv. Bd., 1957-69, Def. Sci. Bd., 1965-70, Gov. of N.Mex.'s Radiation Adv. Coun., 1959-61; sec. N.Mex. Health and Social Svcs., 1971-73; chmn. gen. adv. com. ACDA, 1974-77, mem., 1977-81; mem. aerospace safety adv. panel NASA, 1964-70; mem. U.S. Army Sci. Bd., 1978-80, White House Sci. Coun., 1982-89; adj. prof. U. Calif., San Diego, 1988—. Mem. council engring. NRC, 1978-82; mem. Los Alamos Bd. Ednl. Trustees, 1950-55, pres., 1955; trustee San Diego Mus. Art, 1983-87; mem. Woodrow Wilson Nat. Fellowship Found., 1973-80; N.Mex. State senator, 1955-61; sec. N.Mex. Legis. Council, 1957-61; chmn. N.Mex. Senate Corp. Commn., 1957-61; mem. Fed. Emergency Agy., 1982-88; bd. dirs. Fedn. Rocky Mountain States, Inc., 1975-77, Charles Lee Powell Found., 1993—. Recipient Ernest Orlando Lawrence award AEC, 1966; Enrico Fermi award Dept. Energy, 1978. Fellow Am. Phys. Soc., AAAS; mem. Nat. Acad. Scis., Nat. Acad. Engring., Council on Fgn. Relations, Phi Beta Kappa, Sigma Xi, Omicron Delta Kappa. Home: 322 Punta Baja Dr Solana Beach CA 92075-1720

AGORASTOS, THEODOROS, obstetrics and gynecology educator; b. Drama, Macedonia, Greece, Dec. 31, 1951; s. Agorastos and Paraskevi (Kassimidou) A.; m. Ioanna Ikonomou, July 27, 1977; 1 child, Agorastos. Diploma in medicine, Aristotelian U. Thessaloniki, Greece, 1975; MD U. Aachen, Germany, 1979. Jr. registrar dept. ob-gyn U. Aachen, 1975-81; lectr. Aristotelian U. Thessaloniki, 1982-87, asst. prof., 1987—; sec. gen. 4th Greek Congress Ob-Gyn, 1988. Author, editor: Fetale Epidermis und Vernix c., 1989, Handbook of Cardiotocography, 1991. With Greek Navy, 1981-82. Scholar Aristotelian U. Thessaloniki, 1970-71, German Acad. Exch. Svc., 1984, 86; fellow Alexander von Humboldt Found., Graz, Austria, Aachen, Stockholm, 1989-90; grantee Volkswagen Found., 1991. Mem. European Soc. Human Reprodn. and Embryology, European Assn. Gynecologists and Obstetricians, German Soc. Ob-Gyn, Greek Soc. Ob-Gyn (assoc. editor Hellenic Ob-Gyn 1985—), Greek Soc. Gynecologic Oncology. Avocations: reading, painting, music. Home: 87 Mitropoleos St, 546 22 Thessaloniki Greece Office: Hipprokrateion Hosp O-G Cl, 50 Papanastassiou St, 546 39 Thessaloniki Greece

AGRANOFF, BERNARD WILLIAM, biochemist, educator; b. Detroit, June 26, 1926; s. William and Phyllis (Pelavin) A.; m. Raquel Betty Schwartz, Sept. 1, 1957; children: William, Adam. MD, Wayne State U., 1950; BS, U. Mich., 1954. Intern Robert Packer Hosp., Sayre, Pa., 1950-51; commd. surgeon USPHS, 1954-60; biochemist Nat. Inst. Neurol. Diseases and Blindness, NIH, Bethesda, Md., 1954-60; mem. faculty U. Mich., Ann Arbor, 1960—, prof. biochemistry, 1965—; R.W. Gerard prof. of neurosci. in psychiatry, 1992; rsch. biochemist Mental Health Rsch. Inst., 1960—, assoc. dir., 1977-83, dir., 1983—; vis. scientist Max Planck Inst. Zellchemie, Munich, 1957-58, Nat. Inst. Med. Rsch. Mill Hill Engr., 1974-75; Henry Russel lectr. U. Mich., 1987; cons. pharm. industry, govt. Contbr. articles to profl. jours. Fogarty scholar-in-residence NIH, Bethesda, Md., 1989—; named Mich. Scientist of Yr., Mus. of Sci., Lansing, 1992. Fellow Am. Coll. Neuropsychopharmacology; mem. Am. Soc. Biochemistry and Molecular Biology, Am. Soc. Biol. Chemists, Am. Chem. Soc., Inst. Medicine, Internat. Soc. Neurochemistry (treas. 1983-89, chmn. 1989-91), Am. Soc. Neurochemistry (pres. 1973-75). Achievements include research in brain lipids, biochem. basis of learning, memory and regeneration in the nervous system, human brain imaging. Home: 1942 Boulder Dr Ann Arbor MI 48104-4164 Office: U Mich Neurosci Lab 1103 E Huron St Ann Arbor MI 48104-1687

AGRAWAL, CHANDRA MAULI, biomaterials engineer, educator; b. Allahabad, U.P., India, May 13, 1959; came to U.S., 1983; s. Jagdish Kumar and Raj Bala (Mital) A.; m. Susan Robinson, June 11, 1988. BTech. in Mech. Engring., Indian Inst. Tech., Kanpur, India, 1982; MS in Mech. Engring., Clemson U., 1985; PhD in Materials Sci., Duke U., 1989. Registered profl. engr., Tex. Engr. automotive div. Ashok-Leyland, Ltd., Madras, India, 1982-83; teaching asst. Clemson (S.C.) U., 1983-86; teaching, rsch. asst. Duke U., Durham, N.C., 1986-89; image processing system mgr. dept. psychology Duke U., Durham, 1987-89, adj. asst. prof. dept. mech. engring. material sci., 1989-91, rsch. asst. prof. dept. biomed. engring., 1990-91; asst. prof. dept. orthopedics U. Tex. Health Sci. Ctr., San Antonio, 1991—; dir. Orthopedic Biomaterials Lab. U. Tex. Health Sci. Ctr., San Antonio, 1991—; adj. asst. prof. dept. biomed. engring. U. Tex., Austin, 1992—; asst. dir. for rsch. NSF Ctr. for the Enhancement of the Biology-Biomaterials Interface, 1992—; mem. S.W. Rsch. Consortium Biomaterials-Bioassessment Initiative Adv. Com., 1991—. Contbr. articles to Biomaterials, Jour. Material Sci., Physiology and Behavior, Jour. Polymer Sci., Jour. Polymer Engring. and Sci., Inv. Radiology Design Graphics Jour.; also abstracts of articles pub. in profl. jours. Mentor John Glen Sch. Mentorship Program, San Antonio, 1992. Mem. ASTM, Soc. Biomaterials, Orthopaedic Rsch. Soc., ASM Internat. Achievements include patents pending on biodegradable bone plate for fracture Fixation, biodegraeable multiple drug release system, enhanced total hip prothesis; development of biodegradable intavascular stent. Office: U Tex Health Sci Ctr 7703 Floyd Curl Dr San Antonio TX 78284

AGRAWAL, VIMAL KUMAR, structural engineer; b. Kanpur, U.P., India, Dec. 21, 1962; came to U.S., 1987; s. Jagdish Chandra and Sudha Agrawal; m. Alpana Gupta, Mar. 24, 1989. B Tech in Civil Engring., Indian Inst.

Tech., Kanpur, 1986; MS in Civil and Environ. Engring., U. R.I., 1989. Structural engr. GEI Cons., Winchester, Mass., 1989—. Mem. ASCE (assoc.), Soc. Civil Engrs. Indian Inst. Tech. (pres. 1985-86), Boston Soc. Civil Engrs. Home: 58 Mill St # 8 Woburn MA 01801 Office: GEI Cons Inc 1021 Main St Winchester MA 01890

AGRIOS, GEORGE NICHOLAS, educator; b. Galarinos, Halkidiki, Greece, Jan. 16, 1936; s. Nicholas and Olga (Kotsioudi) A.; m. Annette Braynard, Nov. 11, 1962; children—Nicholas, Anthony, Alexander. B.S. in Horticulture, U. Thessaloniki, Greece, 1957; Ph.D. in Plant Pathology, Iowa State U., 1960. Prof. plant pathology U. Mass., Amherst, 1963-88; prof., chmn. plant pathology dept. U. Fla., 1988—. Author: Plant Pathology, 1969, 2d edit., 1978, 3d edit., 1988; contbr. articles to profl. jours.; editor-in-chief APS Press Books, 1984-87. Served to 2d lt. Engring. Corps, Greek Army, 1961-62. Fellow Am. Phytopathol. Soc.; mem. Can. Phytopathol. Soc., Am. Phytopathological Soc. (v.p., pres.-elect, pres., 1988-91), AAAS. Greek Orthodox. Avocations: reading; outdoor activities. Office: U Fla Dept Plant Pathology Fifield Hall Gainesville FL 32611

AGRUSS, NEIL STUART, cardiologist; b. Chgo., June 2, 1939; s. Meyer and Frances (Spector) A.; B.S., U. Ill., 1960, M.D., 1963; m. Teresa Marie Stafford; children—David, Lauren, Michael, Joshua, Susan. Intern, U. Ill. Hosp., Chgo., 1963-64, resident in internal medicine, 1964-65, 67-68; fellow in cardiology, Cin. Gen. Hosp., 1968-70, dir. coronary care unit, 1971-74, dir. echocardiography lab., 1972-74; dir. cardiac diagnostic labs., Central DuPage Hosp., Winfield, Ill., 1974—; asst. prof. medicine, U. Cin., 1970-74, Rush Med. Coll., 1976—. Chmn. coronary care com. Heart Assn. DuPage County, 1974-76; active Congregation Beth Shalom, Naperville, Ill. Served to capt. M.C. U.S. Army, 1965-67. Diplomate Am. Bd. Internal Medicine. Fellow ACP, Am. Coll. Cardiology, Am. Coll. Chest Physicians, Council Clin. Cardiology, Am. Heart Assn.; mem. AMA, DuPage County, Ill. State Med. Socs., Am. Fedn. Clin. Research, Chgo. Heart Assn. Author and co-author publs. in field. Office: 454 Pennsylvania Ave Glen Ellyn IL 60137-4496

AGUIAR, ADAM MARTIN, chemist, educator; b. Newark, Aug. 11, 1929; s. Joaquim Ramahlo and Emilea Andrada (Nunes) A.; m. Laura E. Brand, Sept. 2, 1980; children: Justine Diane, David Laurence, Adam Albert, Erick Arthur, Aaron Benjamin, Evan Joaquim. B.S., Fairleigh Dickinson U., 1955; M.A., Columbia U., 1957, Ph.D., 1960. Chemist Otto B. May, Newark, 1948-55; asst. prof. Fairleigh Dickinson U., Rutherford, N.J., 1959-63; asst. prof. chemistry Tulane U., New Orleans, 1963-65, assoc. prof., 1965-67, prof., 1967-72, head dept. chemistry Newcomb Coll. div., 1970; dean grad. and research programs William Paterson Coll., Wayne, N.J., 1972-73; research prof. Rutgers U., Newark, 1973-75; prof. chemistry Fairleigh Dickinson U., Madison, N.J., 1975-93, chmn. dept. chemistry/geol. scis., 1984-89; pres. Seltox Corp., N.J., 1980—; adj. prof. chem. Monmouth Coll., West Long Branch, N.J., 1993—; cons. chem. firms in, La. and N.J. Contbr. articles to profl. jours. Union Carbide fellow, 1957; NIH fellow, 1959; recipient other grants. Mem. AAUP, Am. Chem. Soc., AAAS, N.Y. Acad. Sci., Ctr. for Profl. Advancement, Sigma Xi, Phi Lambda Epsilon, Phi Omega Epsilon. Home: 37 Wyncrest Ln Neptune NJ 07753-7421 Office: Monmouth Coll West Long Branch NJ

AGUILERA, JORGE, electrical and solar energy engineer; b. Granada, Andalucia, Spain, June 3, 1962; s. Ramiro and Vicenta (Tejero) A.; m. Ana Huertas, Aug. 7, 1988. MSEE, Universidad Politécnica de Madrid, 1987. Supr. Instituto Nacional de Técnica Aerospacial- European Space Agy., Madrid, 1987-88; rsch. assoc. Instituto de Energia Solar-Universidad Politécnica de Madrid, 1988-89, 90—; mgr. ICI, La Paz, Bolivia, 1989-90; mgr. Sociadad de Especialistas del Desarrollo, Madrid, 1990-92. Achievements include research in solar energy in developing countries, rural electrification, solar energy in Bolivia high-plateau, transfer tech. Office: ETSI Telecomm, Ciudad Universitaria, 28040 Madrid Spain

AHEARNE, JOHN FRANCIS, scientific research society director, researcher; b. New Britain, Conn., June 14, 1934; s. Daniel Paul and Balbena Marian (Baloski) A.; m. Barbara Helen Drezek, June 19, 1956; children: Thomas, Paul, Mary Ann, Robert, Patricia. B of Engring. Physics, Cornell U., 1957, MS in Physics, 1958; MA, Princeton U., 1963, PhD, 1966. Nuclear weapons analyst USAF, 1959-61; assoc. prof. physics USAF Acad., 1964-69; adj. prof. U. Colo., 1966-69; lectr. Colo. Coll., 1966-69; analyst Office Asst. Sec. Def. for Systems Analysis, 1969-70, dir. tactical air, 1970-72; dep. asst. sec. def. for gen. purpose programs, 1972-74, prin. dep. sec. def. manpower and res. affairs, 1974-76; staff White House Energy Office, 1977; dep. asst. sec. Dept. Energy, 1978; commr. U.S. Nuclear Regulatory Commn., 1978-83, chmn., 1980-81; mgmt. cons. to Comptr. Gen of U.S., 1983-84; v.p., sr. fellow Resources for the Future, 1984-89; exec. dir. Sigma Xi, The Sci. Rsch. Soc., Research Triangle Park, N.C., 1989—; adj. fellow Resources for the Future, 1992—; chmn. risk perception and comm. com. NAS, 1987-89, chmn. future nuclear power com., 1990-92, mem. bd. on radioactive waste mgmt., 1993—; chmn. adv. com. on nuclear facility safety U.S. Dept. Energy, 1988-91; mem. pres.'s coun. for nat. labs. U. Calif., 1992—. Vice-chair U.S. Com. for ILASA, 1992—; mem. reactor panel NAS Comm. on Uses of Weapons Plutonium, 1992—; bd. dirs. Woodstock Theol. Ctr., chmn., 1980-85. Served with USAF, 1959-70. Gen. Electric Coffin fellow, 1957-58; recipient Dept. Def. Disting. Civilian Service medal and bronze palm, Sec. Def. Meritorious Svc. medal; named Boss of Year D.C. chpt. Nat. Secs. Assn., 1976. Fellow AAAS; mem. Am. Phys. Soc., Am. Nuclear Soc., Soc. Risk Analysis, Sigma Xi. Democrat. Roman Catholic.

AHERN, JOHN EDWARD, mechanical engineer, consultant; b. Portland, Maine, Apr. 7, 1921; s. Henry Robert and Eva Irene (Legere) A.; m. Cora Marie Wilhelm, Sept. 2, 1950; children: Thomas, Maureen, Corinne, Kathleen, Timothy, Jeannine. BSME, Northeastern U., 1943; MSME, Columbia U., 1952. Registered profl. mech. engr., Calif. Contract engr. Babcock & Wilcox Co., N.Y.C., 1947-52, asst. project engr. Pratt and Whitney Aircraft, East Hartford, Conn., 1952-59; prin. rsch. scientist The Marquardt Corp., Van Nuys, Calif., 1959-67; sr. tech. specialist Aerojet Electro Systems Co., Azusa, Calif., 1967-88; cons. Extech, Glendora, Calif., 1988—; cons. Douglas Aircraft Co., Long Beach, Calif., 1988-89, Aerojet Electro Systems, Co., Azusa, 1988-91. Author: The Exergy Method of Energy Systems Analysis, 1980; contbr. articles to AIAA Jour. 1st lt. USAF, 1943-46. Fellow ASME; mem. AIAA. Roman Catholic. Achievements include four patents on cryogenics and heat exchangers. Home and office: 738 Parkbrook Ln Glendora CA 91740

AHLUWALIA, DHARAM VIR, physicist; b. Fatehpur, India, Oct. 20, 1952; came to U.S., 1976; s. Bikram Singh and Bimla A.; m. Regina Rita Fuchs, Mar. 6, 1982; children: Jugnu, Vikram, Shanti, Wellner. PhD in Physics, Tex. A&M U., 1991. Postdoctoral fellow medium energy physics div. Los Alamos (N.Mex.) Nat. Lab., 1992—; presenter workshops, symposia, lectrs. at profl. confs. Contbr. articles to Phys. Letters, Phys. Rev., other profl. publs. Achievements include research in spacetime symmetries and their violation. Office: Los Alamos Nat Lab Medium Energy Physics Div Los Alamos NM 87545

AHLVIN, RICHARD GLEN, civil engineer; b. Joliet, Ill., Dec. 23, 1919; s. Martin Victor and Mabel Edwardine (Alder) A.; m. Maridel Brown, Aug. 22, 1942; children: Richard Brown, Alix Elisabeth Ahlvin Moro, Martin Charles, Alder Frank. BCE, Purdue U., 1941, MCE, 1949. Registered profl. engr., Miss., Ind. Instr. in engring. Purdue U., West Lafayette, Ind., 1946-49; rsch. engr. U.S. Army Engrs. Waterways Exptl. Sta., Vicksburg, Miss., 1949-80; cons. civil engr. Vicksburg, 1980—. Co-author: Construction Guide for Soils and Foundations, 1988; contbr. over 100 articles to profl. jours. Capt. U.S. Army, 1943-46. Recipient Arch T. Colwell Merit award Soc. Automotive Engrs., 1968, Commendation FAA, 1977; named Disting. Engr. Alumnus Purdue U., 1971. Fellow ASCE (br. pres., sect. dir. 1960, Robert Horonjeff award 1988), Nat. Soc. Profl. Engrs. (hon.), Soc. Am. Mil. Engrs. (hon.), ASTM. Republican. Home and Office: 4 Shadow Wood Dr Vicksburg MS 39180-9741

AHMAD, JAMEEL, civil engineer, researcher, educator; b. Lahore, Punjab, Pakistan, May 22, 1941; came to U.S. 1962; s. Naseer and Iftikhar (Dean) Bakhsh; m. Rosalba Quiroz, March 31, 1983; 1 child, Monica. BSc, Punjab U., Lahore, 1962; MS, U. Hawaii, 1964; PhD, U. Pa., 1967. East-west ctr. fellow U. Hawaii, Honolulu, 1962-65; rsch. fellow U. Pa., Phila., 1965-67; asst. prof. Widener U., Chester, Pa., 1967-68; asst. prof. Cooper Union, N.Y.C., 1968-71, assoc. prof., 1971-80, chmn. civil engring., 1980—, prof. civil engring., 1979—; dir. rsch. Cooper Union Rsch. Found., N.Y.C., 1983—; dir. High Techs., Inc., N.Y.C., 1986—. V.p. Vilmanor Community Assn., N.Y.C., 1992, West Side Community Assn., N.Y.C., 1976. Mem. Am. Soc. Civil Engrs. (Outstanding Svc. award 1985). Achievements include patents for fleximech reinforcement system, asphalt reinforcement system. Office: Cooper Union Coll 51 Astor Pl New York NY 10003

AHMAD, KHALIL, medical practitioner; b. Sheikhupura, Punjab, Pakistan, Feb. 28, 1950; s. Mohammad and Rashida Ibrahim; m. Tehmina Khanum, Feb. 13, 1992; 1 child Sameera. MB, B of Surgery, King Edward Med. Coll., Lahore, Pakistan, 1973; Diploma Tuberculous and Chest Diseases, Postgrad. Med. Inst., Lahore, Pakistan, 1975. Med. officer Kafue Gorge (Zambia) Hosp. Ministry of Health, 1978-84; postgrad. student Whittington Hosp., London, 1985; clin. attachment Royal Shrewsbury (Eng.) Hosp. North, 1985; med. officer Health Dept. Govt. of the Punjab, Lahore, 1989-92; chest specialist Dist. Hdqs. and Tb. Hosps., Sheikhupura, Pakistan, 1992—; physician and chest specialist 6, Huma Block, Allama Iqbal Town, Lahore, 1986-92. Recipient Silver medal Bd. Intermediate & Secondary Edn., 1965, Cert. of Honor, King Edward Med. Coll., 1970, Gold medal U. of the Punjab, 1975. Mem. The N.Y. Acad. Scis., The New Eng. Jour. Medicine. Muslim. Avocations: reading sci. and edn., listening English broadcasts. Home: 69-A Wali St, Hassan Town Multan Rd, Lahore Pakistan Office: Chest Specialist, Dist Hdqs Hosp, Sheikhupura Pakistan

AHMAD, SALAHUDDIN, nuclear scientist; b. Sylhet, Bangladesh, Nov. 25, 1954; arrived in Can., 1978; came to U.S., 1990; s. Jalal and Mamtaz (Begum) A.; m. Munawar Sultana, June 1, 1978; 1 child, Nahid Rubaba. MSc, Dacca U., Bangladesh, 1975; PhD, U. Victoria, B.C., Can., 1981. Lectr. Dacca U., 1978; postdoctoral rsch. assoc. U. Victoria, 1981; rsch. scientist U. Paris South, Orsay, France, 1982-83; profl. rsch. assoc. U. Sask., Saskatoon, Can., 1983-84; rsch. assoc. Triumf Nat. Lab., Vancouver, B.C., 1984-86, U. B.C., Vancouver, 1987-89; faculty fellow Rice U., Houston, 1990—; sci. assoc. Brookhaven Nat. Lab., 1990—, CERN, Geneva, 1991—; mem. Solenoidal Detector Collaboration at the Superconducting Super Collider, Solenoidal Tracker at RHIC, CEBAF Large Acceptance Spectrometer Collaboration. Contbr. over 100 articles to sci. jours. and conf. procs., including Physics Letters, Phys. Rev., Phys. Rev. Letters. Pres. Bangladesh-Can. Cultural Assn. B.C., Vancouver, 1988-89, Bangladesh-Am. Lit., Art and Cultural Assn., Houston, 1992—. Raja Kalinarayan scholar U. Dacca, 1974-75; fellow Can. Commonwealth Fellowship Com., 1978-81. Mem. Am. Phys. Soc. Office: Rice U Bonner Lab Houston TX 77251-1892

AHMADINEJAD, BEHROUZ, chemist, educator; b. Tehran, Iran, Apr. 4, 1967; s. Mostafa and Tayebeh (Yarjou) A. BS in Chemistry, Azad U., Tehran, 1991. Trainee Chemidaru Co., Tehran, 1988-89, Pars Oil Co., Tehran, 1989, Varko Enamel Co., Tehran, 1990-91; dir. Kimia Publ. Co., Tehran, 1991—; dir. Ctr. Rsch. and Tech. in Mil. Instrns., Tehran, 1991—. Mem. Am. Chem. Soc., Iranian Chem. Soc., Iranian Petroleum Inst. Home: PO Box 11365-1975, Tehran Iran

AHMADJIAN, MARK, chemist; b. Worcester, Mass., Mar. 22, 1951; s. Azad and Diana A. (Boyajian) A. MS, U. R.I., 1987, PhD, 1992. Rsch. assoc. U. R.I., Kingston, 1973-80; chemist USAF Phillips Lab., Hanscom AFB, Mass., 1985—. Contbr. articles to profl. jours. Capt. USAF, 1980-84, USAFR, 1984—. Mem. AIAA, Am. Geophysical Assn., Reserve Officers Assn. U.S. Achievements include patent on oil spill identification and measurement of space shuttle glow. Home: 130 Boon St Narragansett RI 02882 Office: USAF Phillips Lab PL/GPOB Hanscom AFB MA 01731-5000

AHMED, ABU, economics educator; b. Feni, Bangladesh, Feb. 1, 1949; s. Nur and Azifa (Khatoon) A.; m. Shahida Sarwath, Sept. 20, 1975 (div.); m. Dilruba Ahmed, June 6, 1977; children: Farzana, Tahmina, Lumaya. BA with hons., Dhaka U., Bangladesh, 1970; MS, Islamabad U., Pakistan, 1972; MA, York U., Toronto, 1974. Teaching asst. Northeastern U., Boston, 1982-84; lectr. U. Mass., Boston, 1984, U. Sierra Leone, 1975-76, Dhaka U., Bangladesh, 1977-78; asst. prof. Dhaka U., 1978-85, assoc. prof., 1985—; bd. dirs. Econ. Rsch. Unit, 1989-91, Kohinoor Chem. Co. Ltd., 1989—; pres. Bangladesh Share Investors Forum, 1990—. Author: The Spirit of Democratic Capitalism by Michael Novak, 1986, A Guide to Investment in Financial Assets, 1990; guest columnist for newspapers; referee Jour. Social Sci., Dkaha U.; contbr. articles to profl. jours. Office: Dhaka U, Econ Dept, Dhaka 1000, Bangladesh

AHMED, HASSAN JUMA, nuclear engineer; b. Suwaira, Kut, Iraq, July 1, 1942; s. Juma Ahmed Al-Khafaji and Naima Ahmed; m. Janet Susan Van-Camerik, Sept. 15, 1965; children: Ali, Nedda. BSc, NYU, 1965, MSc, 1973. Engr. Nuclear Rsch. Inst., Baghdad, Iraq, 1967-70; supr. Amstar Corp., N.Y.C., 1970-72; sr. engr. Westinghouse Electric, Columbia, S.C., 1973-86, prin. engr., 1986-91, fellow engr., 1991—; grad. asst. Ga. Inst. Tech., Atlanta, 1973; adj. prof. U. S.C., Columbia, 1983. 2d lt. Iraqi Air Force, 1965-68. Mem. Am. Nuclear Soc. Achievements include 14 patents for nuclear measurements, image processing, automation and inspection. Home: 125 Winding Rd Irmo SC 29063 Office: Westinghouse Electric 5801 Bluff Rd Columbia SC 29205

AHMED, IMTHYAS ABDUL, computer scientist; b. India, June 5, 1958; came to U.S., 1982; children: Soofia, Saadia. BS in Physics, Madras U., India, 1978, MS in Physics, 1980; MS in Computer Sci., Fla. Inst. Tech., 1985. Software engr. W.R. B. Assocs., Troy, Mich., 1985-88; systems engr. Electronic Data Systems, Troy, Mich., 1988—. Home: 3817 Mark Troy MI 48083

AHMED, IQBAL, psychiatrist, consultant; b. Tumkur, Karnataka, India, Aug. 23, 1951; came to U.S., 1976; s. Rahimuddin Ahmed and Arifa (Banu) Rahimuddin; m. Lisa Suzanne Rose, Oct. 9, 1983; children: Yasmin, Jihan. BS, MB, St. John's Med. Coll., Bangalore, India, 1975. Diplomate in gen. psychiatry and geriatric psychiatry Am. Bd. Psychiatry and Neurology. Intern St. Martha's Hosp., Bangalore, India, 1974-75; resident in psychiatry U. Nebr. Med. Ctr., Omaha, 1976-79; fellowship in consultation Boston U. Sch. Medicine, 1979-81; staff psychiatrist in consultation liaison psychiatry Boston City Hosp., 1981-87, staff psychiatrist, geriatric psychiatry, 1983-85, dir. geriatric neuropsychiatry unit, 1985-87, dir. geriatric psychiatry, 1988-92; assoc. dir. consultation liaison psychiatry New England Med. Ctr., Boston, 1989-92; chief spl. svc. Hawaii State Hosp., 1992—; asst. prof. psychiatry Sch. Medicine, Boston U., 1981-87, Sch. Medicine, Tufts U., Boston, 1987-92; dir. med. student edn. in psychiatry Boston City Hosp., 1981-87; chief spl. svcs. Hawaii State Hosp., 1992—; assoc. prof. John A. Burn's Sch. Medicine, U. Hawaii, 1992—. Contbr. articles to profl. jours. Mem. Mass. State Dem. Party Minority Caucus, Boston, 1983. Mem. AMA, Am. Psychiat. Assn., Am. Neuropsychiatry Assn., Royal Coll. Psychiatrists, Acad. Psychosomatic Medicine, Am. Acad. Geriatric Psychiatry. Democrat. Avocation: computer hacking. Office: Hawaii State Hosp 45-710 Keaahala Rd Kaneohe HI 96744

AHMED, KAZEM UDDIN, technical staff member, consultant; b. Rangpur, Bangladesh, Nov. 1, 1948; came to U.S., 1984; S. Basir Uddin and Kohinoor Begum Ahmed; m. Almoon Nahar, Dec. 5, 1975; children: Naheed, Nafis. MSc, Rajshahi Univ., 1972; MS, Ariz. State Univ., 1988. Lectr. govt. colls. Univ. Dhaka, Bangladesh, 1974-78; lectr. Univ. Ibadan, Nigeria, 1979-83; teaching asst. dept. math. Ariz. State U., Tempe, 1985-88, teaching asst. dept. computer sci., 1988-90; mem. tech. staff Wavephore, Inc., Tempe, 1990—. Contbr. articles to profl. jours. Mem. IEEE, ACM, SPIE. Office: Wavephore Inc 2601 W Broadway Tempe AZ 85282

AHMED, OSMAN, mechanical engineer; b. Berhampore, India, Feb. 20, 1956; came to U.S., 1984; s. Sayed and Anjuman (Begum) A.; m. Aftab Zaman, July 16, 1983; children: Aadil S., Aaqib S. BSME, Bangladesh U. Engring. & Tech., 1979; MSME, U. Windsor, 1983; postgrad., U. Wis. Design engr. Aziz & Co., Dhaka, Bangladesh, 1978-79; cons. Bangladesh Mgmt. Devel. Ctr., 1979-80; rsch. & tech. asst. U. Windsor (Ont., Can.), 1980-83; lectr., teaching asst. U. Toledo, 1984-85; design engr. Beling Cons., Peoria, Ill., 1985-88; prin. applications engr. Landis & Gyr Powers, Buffalo Grove, Ill., 1988—. Tech. Merit scholar, Bangladesh U. Engring. & Tech., 1975-79, U. Windsor, 1982. Mem. Am. Soc. Heating, Refrigeration & Air Conditioning Engrs. (assoc.). Achievements include patents for laboratory fume hood control aparatus, entitled apparatus for controlling the ventilation of laboratory fume hoods, a system for controlling the differential pressure of rooms with fume hoods, a method for determining the open area of multiple moveable doors fume hood. Office: Landis & Gyr Powers 1000 Deerfield Pky Buffalo Grove IL 60089

AHMED, S. BASHEER, research company executive, educator; b. Kurnool, Andhra, India, Jan. 1, 1934; s. S. M. and K.A. (Bee) H.; m. Alice Cordelia Pearce; 1 child, Ivy Amina. BA, Osmania Coll., Kurnool, 1955; MA, Osmania U., Hyderabad, India, 1957; MS, Tex. A&M U., 1963, PhD, 1966. Asst. prof. Tenn. Tech. U., Cookeville, 1966-68, Ohio U., Athens, 1968-70; vis. fellow Princeton U., N.J., 1977-78; prof. Western Ky. U., Bowling Green, 1970-80; prof. Mgmt. Scis. Lubin Grad. Sch. Bus., dir. doctoral program Pace U., N.Y.C., 1982-92; pres. Princeton Econ. Rsch., Inc., 1980—; cons. Oak Ridge (Tenn.) Nat. Lab., 1969-77, Inst. for Energy Analysis, Oak Ridge, 1975, Honeywell Corp., Mpls., 1985—; bd. dir. doctoral programs Lubin Grad. Sch. Bus. Pace U., N.Y.C. Author: Quantitative Methods for Business, 1974, Nuclear Fuel and Energy Policy, 1979; author, editor: Technology, International Stability, and Growth, 1984. Recipient Achievement award Oak Ridge Nat. Lab., 1977, IEEE Centennial Medal, 1983. Fellow AAAS, Systems, Man, and Cybernetics Soc. (pres. 1980-82). Republican. Moslem. Home: 401 Knoll Way Rocky Hill NJ 08553-1016

AHSANULLAH, OMAR FARUK, computer engineer, consultant; b. Suri, India, Feb. 20, 1964; came to U.S., 1981; s. Mohammad and Masuda (Gowas) A.; m. Angela Rahim, August 4, 1991. BS, Rensselaer Poly. Inst., 1986, MS in Computer Systems Engring., 1987. Coop. student IBM, Atlanta, 1984-85; coop. student Factron, Latham, N.Y., 1985-87, design engr., 1987-88; lead design engr. Schlumberger, Simi Valley, Calif., 1988-90, new product introduction project mgr., 1990-92, product mgr. diagnostic systems, 1992-93, product mgr. diagnostic systems and well svcs., 1993—; con. Omega Rsch. & Applications, Valencia, Calif., 1989—; instr. Calif. Luth. U., 1991—. Recipient scholarship Rensselaer Poly. Inst., 1981-85, Nat. Honor Soc., 1981. Mem. Am. Prodn. and Inventory Control Soc., World Affairs Coun., Eta Kappa Nu. Office: Schlumberger Techs 85 Moreland Rd Simi Valley CA 93065-1662

AHTCHI-ALI, BADREDDINE, chemical engineer; b. Annaba, Algeria, Sept. 5, 1956; s. Abdeldjelil and Faffani Ahtchi-Ali; m. Pamela Ahtchi-Ali, Aug. 18, 1986; children: Badreddine II, Sofian. MS, Rutgers U., 1983, PhD, 1985. Rsch. scientist Haut Commissariat a la Recherche, Algiers, 1985-86, dept. head, 1986-87; tech. support mgr. Prosys Tech., Florham Park, N.J., 1987-90; engring. scientist Unilever Rsch. U.S., Edgewater, N.J., 1990-92, sr. engring. scientist, 1993—; cons. DuPont, Wilmington, 1988-90; mem. scholarship com. Hart Commissariat a la Recherche, Algiers, 1987-87. Contbr. articles to profl. jours.; author: Biochemical Engineering for 2001, 1992. Recipient Grad. Studies scholarship Algerian Govt., Rutgers, N.J., 1980, Bklyn. Union Gas Project, Rutgers, 1987. Mem. AICE (Best Rsch. Work 1984). Democrat. Muslim. Achievements include devel. of new process for new chems. in the area of consumer products; devel. new numerical method to solve multiple steady state problems in trickle-bed reactors; devel. new dynamic parameter estimation method for reaction kinetics in batch and semi-batch reactions. Office: Unilever Rsch US 45 River Rd Edgewater NJ 07020

AHUJA, ANIL, energy engineer; b. Gujranwala, Punjab, Pakistan, Apr. 6, 1943; came to U.S., 1976; s. Behari Lal and Sheila (Malik) A.; m. Meena Choudhery, Sept. 13, 1969 (div.); children: Siddarth, Avantika; m. Meenakshi Mona Kalia, Dec. 4, 1986; 1 child, Nitasha. BS in Mech. Engring., Punjab U., 1963; MBA, U. Tenn., 1978. Mktg. mgr. Indian Oil Corp., New Delhi, 1970-76; energy cons. Brinks Inc., Knoxville, 1978-80; energy mgr. L.A. City Schs., 1981-91; energy cons. APJ Energy Environ. Group, L.A., 1991—. Maj. Indian Army, 1964-69. Recipient nat. and state awards for energy innovation. Mem. Assn. Energy Engrs., Assn. Profl. Energy Mgrs. Democrat. Hindu. Achievements include cutting 130 million dollars in utility costs for L.A. city schools. Home: 15768-6 Midwood Dr Granada Hills CA 91344 Office: APJ Energy & Environ Group 16005 Sherman Way # 106 Van Nuys CA 91406

AIDAR, NELSON, civil engineer, consultant; b. São Paulo, Brazil, Oct. 7, 1953; came to U.S., 1980; s. Orlando J. and Valentine (Shick) A. BS in Civil Engring. with honors, FEFAAP, São Paulo, Brazil, 1979; MS in Civil Engring. summa cum laude, Catholic U. Am., 1985. Registered profl. engr., Md., D.C., Va., Brazil; cert. on sediment and erosion control, Md. Civil, structural engr. EMJ Electrack, Inc., Hyattsville, Md., 1980-86; civil, utility engr. Sheladia Assocs., Inc., Riverdale, Md., 1986; project mgr. STV/Lyon Assocs., Inc., Balt., 1987; sr. utilities engr. Deleuw Cather Co, Washington, 1988; sr. civil engr. Daniel, Mann, Johnson & Mendenhall, Balt., 1988-90; project engr. Fleming Corp., Washington, 1990—. Mem. ASCE, Md. Soc. Profl. Engrs. (Potomac chpt. dir. 1993-95), Nat. Soc. Profl. Engrs., Am. Concrete Inst. Home: 1008 Hobbs Dr Colesville MD 20904-6211

AIDUN, CYRUS KHODARAHM, mechanical engineering educator; s. Khodarahm and Paridokht Aidun; m. Sarvar Khosravi, Dec. 29, 1981; children: Kevin Cyrus, Armin Khosrow. BS in Mech. Engring., Rensselaer Poly. Inst., 1978, MS, 1981; PhD, Clarkson U., 1985; postdoctoral, Cornell U., 1985-87. Prin. scientist Battelle Rsch. Inst., Columbus, Ohio, 1987-88; asst. prof. Inst. of Paper Sci., Appleton, Wis., 1988-92; assoc. prof. Inst. of Paper Sci., Ga. Inst. Tech., Atlanta, 1992—. Math. Scis. Rsch. fellow Cornell U., 1987; recipient George Olmsted award API, 1990, Young Investigator award NSF, 1992. Mem. TAPPI (Found. award 1991). Office: Inst Paper Sci & Tech 500 10th St NW Atlanta GA 30318

AIELLO, PIETRO, cardiologist; b. Rome, Dec. 13, 1939; s. Rocco and Angela (Fusaro) A.; m. Maria Cristina Perugini, Oct. 25, 1971; children: Alessandro, Andrea. MD, U. La Sapienza, Rome, 1965; M in Cardiology, U. Turin, Italy, 1967; M in Rheumatology, U. Rome, 1968; M in Internal Medicine, U. Parma, Italy, 1970. Diplomate in Cardiology. Vol. asst. San Giovanni Hosp., Rome, 1965-68; asst. in cardiology San Camillo Hosp., Rome, 1975-86, asst. to head cardiologist, 1986—; vis. lectr. Accademia Lancisiana, Rome, 1975—; staff lectr. Scuola Medica Ospedaliera, Rome, 1983—. Contbr. over 20 articles to scholarly jours. Med. officer Italian Air Force, 1966-67. Recipient Cavaliere al Merito della Republica, Pres. of Italian Republic, 1980. Mem. Nat. Assn. Hosp. Cardiologists. Roman Catholic. Avocations: tennis, motor sailing, underwater fishing. Home: Via Umberto Saba 96, Rome 00144, Italy Office: Viale Europa 300, Rome 00144, Italy

AIHARA, KAZUYUKI, mathematical engineering educator; b. Kitakyusyu, Fukuoda, Japan, June 23, 1954; s. Kazuo and Tokiwa (Mine) A.; m. Chieko Takagi, Mar. 29, 1981; children: Ikkyu, Kazusa, Chiari. B in Engring., Tokyo U., 1977, M in Engring., 1979, PhD, 1982. Postdoctoral fellow Tokyo U., 1982-83; asst. Tokyo Denki U., 1983-86, asst. prof., 1986-88, assoc. prof., 1988-93; assoc. prof. Tokyo U., 1993—. Author: Neural Computers, 1988; editor: Chaos, 1990, Nuero, Fuzzy and Chaos Computing, 1993—, Chaos in Neural Systems, 1993. Mem. Inst. Elec. Engrs. of Japan, Inst. Electronics, Info. and Communication Engrs., Internat. Neural Network Soc., Inst. Elec. Installation Engrs. of Japan, Japan Soc. Indsl. and Applied Maths. Office: Fac Engring Tokyo U, 7-3-1 Hongo, Bunkyo-ku, Tokyo 113, Japan

AIKA, KEN-ICHI, chemistry educator; b. Chiba, Japan, Mar. 17, 1942; s. Kunihiko and Yuki (Tsuyusaki) A.; m. Chigusa Fujita, June 25, 1972; children: Tomohiko, Hidehiko. B of Engring., Tokyo Inst. Tech., Japan, 1964, M of Engring., 1966, DEng, 1969. Rsch. assoc. Tokyo Inst. Tech., Japan, 1969-74, Princeton (N.J.) U., 1974-75, Tex. A&M U., College Sta., 1975-77; rsch. assoc. Tokyo Inst. Tech., Yokohama, Japan, 1977-81, assoc. prof., 1981-92; prof. Tokyo Inst. Tech., Yokohama, 1992—; lectr. Tokai U., Hiratsuka, Japan, 1988—. Patentee in field; author: Catalytic Activation of N2, 1981, (handbook) Ammonia Syntesis, 1991; editor Environmental Protection, 1991. Mem. Am. Chem. Soc., Chem. Soc. Japan, Catalysis Soc. Japan, Petroleum Soc. Japan. Avocations: tennis, jogging, touring, karaoke. Office: Tokyo Inst Tech, 4259 Nagatsuta Midoriku, Yokohama 227, Japan

AIKAWA, JERRY KAZUO, physician, educator; b. Stockton, Calif., Aug. 24, 1921; s. Genmatsu and Shizuko (Yamamoto) A.; m. Chitose Aihara, Sept. 20, 1944; 1 son, Ronald K. A.B., U. Calif., 1942; M.D., Wake Forest Coll., 1945. Intern, asst. resident N.C. Baptist Hosp., 1945-47; NRC fellow in med. scis. U. Calif. Med. Sch., 1947-48; NRC, AEC postdoctoral fellow in med. scis. Bowman Gray Sch. Medicine, 1948-50, instr. internal medicine, 1950-53, asst. prof., 1953; established investigator Am. Heart Assn., 1952-58; exec. officer lab. service Univ. Hosps., 1958-61, dir. lab. services, 1961-83, dir. allied health program, 1969—, assoc. dean allied health program, 1983—, pres. med. bd.; assoc. dean clin. affairs asst. prof. U. Colo. Sch. Medicine, 1953- 60, asso. prof. medicine, 1960-67, prof., 1967—, prof. biometrics, 1974—, assoc. dean clin. affairs, 1974—; Pres. Med. bd. Univ. Hosps. Fellow ACP, Am. Coll. Nutrition; mem. Western Soc. Clin. Research, So. Soc. Clin. Research, Soc. Exptl. Biology and Medicine, Am. Fedn. Clin. Research, AAAS, Central Soc. Clin. Research, AMA, Assn. Am. Med. Colls., Phi Beta Kappa, Sigma Xi, Alpha Omega Alpha. Home: # 115 7222 E Gainey Ranch Rd Scottsdale AZ 85258-1530 Office: U Colo Sch Medicine 4200 E 9th Ave Denver CO 80262-0001

AILLONI-CHARAS, MIRIAM CLARA, interior designer, consultant; b. Veere, The Netherlands, July 31, 1935; came to U.S., 1958; d. Maurits and Elzina (De Groot) Taytelbaum; m. Dan Ailloni-Charas, Oct. 8, 1957; children: Ethan Benjamin, Orrin, Adam. Degree in Interiors, Pratt Inst., 1962; BSc, SUNY, Albany, 1978. Interior designer S.J. Miller Assocs., N.Y.C., 1960-63; interior design cons. Rye Brook, N.Y., 1963-88, 90—; exec. v.p. Contract 2000 Inc., Port Chester, N.Y., 1988-90. Treas. Temple Guild, Congregation Emanu-El, Rye, N.Y., 1979-88, co-chair, 1988—, trustee, 1986-92. Recipient Cert. of Merit, U.S. Jaycees, 1962, March of Dimes, 1989, 91. Mem. Am. Soc. Interior Designers, Allied Bd. Trade, Westchester Assn. Women Bus. Owners (bd. dirs. 1988-93), Nat. Trust for Historic Preservation, Westchester C. of C. (area devel. coun. 1988-90). Home and Office: 23 Woodland Dr Port Chester NY 10573-1797

AIRD, ROBERT BURNS, neurologist, educator; b. Provo, Utah, Nov. 5, 1903; s. John William and Emily Dawn (McAuslan) A.; m. Ellinor Hill Collins, Oct. 5, 1935 (dec. 1988); children: Katharine (dec. 1992), Mary, John, Robert. BA, Cornell U., 1926; MD, Harvard Med. Sch., 1930. Diplomate Nat. Bd. Med. Examiners, Am. Bd. Psychiatry and Neurology. Intern Strong Meml. Hosp-U. Rochester (N.Y.) Sch. Medicine, 1930-31, resident, 1931-32; rsch. assoc. U. Calif., San Francisco, 1932-35, instr., 1935-39, from asst. to assoc. prof., 1939-49, prof. neurology, 1949-71, founder dept. neurology, 1947, emeritus prof., 1971—, established Electroencephalographic Lab., 1940, dir. Electroencephalographic Lab., 1940-71; trustee Deep Springs Coll., Inyo County, Calif., 1959-71, dir. coll., 1960-66, hon. trustee 1971—; founder No. Calif. chpt. Multiple Sclerosis Soc. Author: Foundations of Modern Neurology: A Century of Progress, 1993; sr. author: Management of Epilepsy, 1974, The Epilepsies--A Critical Review, 1984; author: (with others) Clinical Neurology, 1955, Textbooks of Medicine, 1959, 62; contbr. more than 275 articles to profl. jours. Mem. coms. NIH, Washington, 1953-63. Comdr. Order of Hipolito Unanaae, Govt. of Peru, 1963; Fulbright scholar, 1957-58. Mem. AAAS, Am. Epilepsy Soc. (hon., pres. 1959-60, Lennox award 1970), Am. Electroencephalographic Soc. (charter mem., pres. 1953-54), Am. Neurol. Assn. (v.p. 1955, 69, 72, sr. mem.), Calif. Acad. Sci. (life), Calif. Acad. Medicine, San Francisco Neurol. Soc. (co-founder), Gold Headed Cane Soc. (hon.), Commonwealth Club Calif. Achievements include establishment of importance of blood-brain barrier in cerebral concussions, EST and epileptogenesis of CNS lesions; seizure-inducing mechanisms for better control of epilepsy; research on CNS effects of subarachnoid-injected agents, on the effect of thyrotropic hormone in production of malignant exophthalmos. Office: U Calif Dept Neurology M-794 PO Box 0114 San Francisco CA 94143-0001

AIRST, MALCOLM JEFFREY, electronics engineer; b. Toronto, Ont., Can., Jan. 12, 1957; s. Herman and Faye (Stulberg) A.; m. Victoria Kessler, Feb. 27, 1981. Student, DeVry Inst., Toronto, 1977-80; BSEE, George Mason U., 1987. Assoc. engr. Satellite Bus. Systems, McLean, Va., 1983-84; lead engr. COMSAT, Fairfax, Va., 1984-87, MITRE Corp., McLean, 1987—; with tech. lead embedded hardware standards specialty group VITA-VME Industry Trade Assn.; mem. subcom. VITA Live insertion, Space and Naval Warfare VME com. USN, fixed and mobile standards com. Ballistic Defense Missile Orgn. Pres. McLean Greens Homeowners Assn., 1990—. Mem. IEEE, Armed Forces Communications and Electronics Assn. Achievements include instrumental role in the adoption of electronic hardware standards by the U.S. Dept. Defense. Office: MITRE Corp 7525 Colshire Dr Mc Lean VA 22102-7500

AISNER, JOSEPH, oncologist, physician; b. Munich, Jan. 5, 1944; came to U.S., 1948; s. Philip and Faye Aisner; m. Seena Feldman, Aug. 31, 1969; children: Dara Lianna, Leon Andrew. BS in Chemistry, Wayne State U., 1965, MD, 1970. Intern Sinai Hosp. Detroit, 1970-71; resident Georgetown U. Hosp., Washington, 1971-72; commd. med. officer USPHS, 1972, advanced through grades to rank 05; clin. assoc. Nat. Cancer Inst., Balt., 1972-75, sr. investigator, 1975-78, chief med. oncology, 1978-81; resigned USPHS, 1981; chief med. oncology U. Md. Cancer Ctr., Balt., 1981-92, dep. dir. clin. affairs, 1982-88, ctr. dir., 1988-93; prof. medicine U. Md., 1982—, prof. oncology, 1982, prof. pharmacology, 1985, prof. clin. pharmacology, 1987, prof. epidemiology preventive medicine, 1993—. Editor 6 books; contbr. numerous chpts. to books and articles and abstracts to profl. jours. Bd. dirs. Md. chpt. Am. Cancer Soc., 1988—, Am. Assn. Cancer Edn., 1990; exec. com. Md. Cancer Consortium, chmn. breast cancer sect., 1992-93, chmn. 1993—; mem. Gov.'s Coun. Cancer Prevention, 1991, exec. com., 1991—; bd. dirs. Md. Children's Cancer Found., 1991—. Nat. Cancer Inst. grantee, 1984-93. Fellow ACP; mem. Am. Fedn. Clin. Rsch., Am. Soc. Clin. Oncology (dir. edn. program 1985-86, bd. dirs. 1991—), Am. Assn. Cancer Rsch., Cancer Leukemia Group B (bd. dirs. 1982—, vice chair breast sect. 1980-86), Am. Radium Soc. (sci. program. com. 1993—). Home: 1404 Berwick Rd Baltimore MD 21204-6508 Office: U Md Cancer Ctr 22 S Greene St Baltimore MD 21201-1544

AISNER, MARK, internist; b. Boston, Feb. 27, 1910; s. Jacob and Sadie (Solomon) A.; m. Helen Cashman, June 27, 1948; children: Jonathan Alan, Susan Jane. AB, Harvard Coll., 1931; MD, Tufts U., 1935. Diplomate Am. Bd. Internal Medicine. Rsch. fellow physiology Tufts U. Sch. Medicine, Boston, 1935; residency Boston City Hosp., 1936-38; asst. prof. medicine Tufts U. Sch. Medicine, Boston, 1942-55, assoc. clin. prof. medicine, 1955-77, head course in phys. diagnosis, 1946-75; head course in physiology Tufts Dental Sch., Boston, 1938-43; chief cardiology Hahnemann Hosp., Boston, 1975-90; cons. staff Newton-Wellesley Hosp., Newton, Mass., 1991—, Faulkner Hosp., Boston, 1991—, Beth Israel Hosp., Boston, 1991—; vis. physician Holy Ghost Hosp. (now Yonville Hosp.), Cambridge, Mass., 1943-53; attending physician Boston VA Hosp., 1947-75; chief cardiology Winthrop Community Hosp., 1950-75; active staff to cons. staff Faulkner Hosp., 1963—, Newton-Wellesley Hosp., 1938—; assoc. staff to cons. staff Beth Israel Hosp., 1938—. Mem. editorial bd. New Eng. Jour. of Medicine, 1954-73; editor Disease-A-Month, 1954-59, editor emeritus, 1959—; contbr. articles to profl. jours. Trustee Countway Med. Libr., Boston, 1991—; mem. publs. com. Boston Med. Libr., 1992—; mem. Bd. of Registration in Medicine, Boston, 1981-84. Maj. M.C., U.S. Army, 1943-46. Recipient Bronze star U.S. Army, 1945, Army Commendation medal U.S. Army, 1946. Fellow Mass. Med. Soc., Am. Coll. Physicians; mem. Am. Coll. Physicians (Mass. Internist of Yr. 1991), Coun. Clin. Cardiology, Am. Heart Assn., AOA Honor Med. Soc. Home: 35 Evelyn Rd Waban MA 02168

AIZAWA, KEIO, biology educator; b. Okaya, Japan, Feb. 13, 1927; s. Yukio and Chikae (Hanaoka) A.; m. Atsumi Aruga, Jan. 8, 1955; 1 child, Jun Aizawa. B, U. Tokyo, 1950, PhD, 1959. Insect pathologist, chief virus lab. Sericultural Expt. Sta., Ministry Agr. and Forestry, Tokyo, 1950-64; prof. biology Kyushu U., Fukuoka, Japan, 1964-90, prof. emeritus, 1990—; prof. biology Teikyo U., Utsunomiya, Japan, 1990—; subcom. mem. Internat. Com. on Taxonomy of Viruses, 1957-78, Internat. Com. on Systematic Bacteriology, 1973-78; mem. WHO Steering Com. on Biol. Control of Vectors, 1984-87. Author, contbr. 22 books on silkworm diseases, insect pathology, microbial control and microbial insecticides. Recipient prize Japanese Sci., Soc. Agrl. Scis., Tokyo, 1968, Louis Pasteur prize Commn. Séricicole Internat., La Mulatière, France, 1990, Sericultural Sci. prize Japanese Assn. Sericulture, 1991. Mem. Soc. Invertebrate Pathology (trustee 1968-72, hon.). Home: Aoba 3-23-5 Higashi-ku, Fukuoka 813, Japan

AIZAWA, MASUO, bioengineering educator; b. Yokohama, Aug. 31, 1942; s. Kaiji and Shin Aizawa; m. Hideoko Aizawa, Oct. 1, 1971; 1 child, Gaku. BS in Engring., Yokohama Nat. U., 1966; D of Engring., Tokyo Inst. Tech., 1971. Rsch. assoc. Tokyo Inst. Tech., 1971-80; rsch. fellow Lehigh U., Bethlehem, Pa., 1974-75; assoc. prof. U. Tsukuba, Japan, 1980-86; assoc. prof. Tokyo Inst. Tech., 1985-86, prof., 1986—; coord. chmn. MITI Future Generation Project Bioelectronic Devices, Tokyo, 1991—. Author: Electrochemical Measurements, 1985, Electroenzymology, 1988; author, editor: Biocomputer, 1991. Mem. Soc. Chem. Sensors (pres. 1991—), Internat. Soc. Mol. Elect. Biocomputer (pres.-elect 1994—). Home: 2-19-14 Amanuma, Suginami-ku, Tokyo 167, Japan Office: Tokyo Inst Tech, Dept Bioengring Nagatsuta, Midori-ku Yokohama 227, Japan

AIZENMAN, MICHAEL, mathematics and physics educator; b. Aug. 28, 1945; m. Marta Beatriz Gershanik; children: Nurith Celina, Ya'ir Gideon. BS, Hebrew U., Jerusalem, Israel, 1969; PhD, Yeshiva U., 1975. Postdoctoral vis. mem. Courand Inst. Math. Scis. NYU, 1974-75, prof. Courant Inst. Math. Scis., physics dept., 1987-90; postdoctoral position Princeton (N.J.) U., 1975-77, asst. prof. physics dept., 1977-82, prof. math. and physics, 1990—; from assoc. prof. to prof. math. and physics Rutgers U., New Brunswick, N.J., 1982-87; vis. prof. Institut des Hautes Etudes Scientifiques, Bures-sur-Yvette, U. Paris, 1984-85, Inst. Advanced Study, 1991. Mem. editorial bd.: Comm. in Math. Physics, Jour. Statis. Physics, Jour. Math. Physics, Comm. Pure and Applied Math., Annals of Probability, Reports in Math. Physics. Sloan fellow, 1981-84, Guggenheim fellow, 1984-85; Fairchild scholar, 1992; recipient Giudo Stampacchia prize Scuola Normale Superior di Pisa, 1982, Excellence in Rsch. award Rutger U. Bd. Trustees, 1987, Norbert Wiener award Am. Math. Soc. and Soc. Indsl. and Applied Math., 1990. Achievements include rsch. in physics and math. with focus on math. analysis of issues arrising in statis. mechanics, quantum field theory, theory of Schrödinger operators and disorder effects. Office: Princeton U Jadwin Hall PO Box 708 Princeton NJ 08544-0708 Office: Princeton U. Princeton NJ 08544*

AJMERA, KISHORE TARACHAND, civil engineer; b. Karanchi, Pakistan, Mar. 1, 1941; came to U.S. 1970; s. Tarachand J. and Labhuben G. (Ghelani) A.; m. Gita Kishore Doshi, Dec. 7, 1972; children: Ashish, Tejal. BE in Civil Engring., U. Poona, India, 1966; MSCE, U. Tex., Arlington, 1977. Registered profl. engr., Tex., Calif. Engr. City of Bombay, India, 1966-70; design engr. I.W. Santry Inc., Dallas, 1970-73, Rady & Assocs., Ft. Worth, Tex., 1973-75; engr. Tex. Dept. Water Resources, Austin, 1975-78; sr. project engr. Tex. Dept. Health, Austin, 1978-81; sr. engr. Radian Corp., Austin, 1981-86; sr. staff engr. Radian Corp., El Segundo, Calif., 1986—. Mem. NSPE, ASCE. Office: Radian Corp 300 N Sepulveda Blvd # 1000 El Segundo CA 90245

AJMERA, PRATUL KUMAR, electrical engineer; b. Bhavnagar, Gujarat, India, Oct. 4, 1945; came to U.S., 1968; s. Kantilal Jagjivandas and Hansa (Parekh) A.; m. Meena Trivedi, July 2, 1971; children: Sameer, Sujata. B-Tech with honors, Indian Inst. Tech., Kharagpur, West Bengal, India, 1968; PhD, N.C. State U., 1975. Asst. prof. U. Notre Dame, Notre Dame, Ind., 1976-82; assoc. prof. La. State U., Baton Rouge, 1982-92, prof., 1992—; cons. Teltech Resource Network, Mpls., 1989-92, Miles Labs., Ames, Div., Elkhart, Ind., 1980, Wheelabrator Frye, Inc., South Bend, Ind., 1978. Contbr. articles to profl. jours. Co-founder, v.p. Hindu Samaj of Baton Rouge, Inc., 1992-94. Grantee Air Force Office of Sci. Rsch., 1977-79, Solar Energy Rsch. Inst., 1983-88, Army Res. Office, 1984-88, Dept. Def., 1985, La. Edn. Quality Support Fund, 1987-88. Mem. IEEE, Materials Rsch. Soc., Sigma Xi, Phi Kappa Phi. Hinduism. Achievements include tech. presentations at organized conferences and supervision of five doctoral and 19 masters' student rsch. Office: La State U Elec Engring Baton Rouge LA 70803

AKASOFU, SYUN-ICHI, geophysicist; b. Nagano-Ken, Japan, Dec. 4, 1930; came to U.S., 1958, naturalized, 1986; s. Shigenori and Kumiko (Koike) A.; m. Emiko Endo, Sept. 25, 1961; children: Ken-Ichi, Keiko. B.S., Tohoku U., 1953, M.S., 1957; Ph.D., U. Alaska, 1961. Sr. research asst. Nagasaki U., 1953-55; research asst. Geophys. Inst., U. Alaska, Fairbanks, 1958-61, mem. faculty, 1961—, prof. geophysics, 1964—, dir. Geophys. Inst., 1986—. Author: Polar and Magnetospheric Substorms (Russian edit. 1971), 1968, The Aurora: A Discharge Phenomenon Surrounding the Earth (in Japanese), 1975, Physics of Magnetospheric Substorms, 1977, Aurora Borealis: The Amazing Northern Lights (Japanese edit. 1981), 1979; co-author: Sydney Chapman, Eighty, 1968, Solar-Terrestrial Physics (Russian edit. 1974); editor: Dynamics of the Magnetosphere, 1979; co-editor: Physics of Auroral Arc Formation, 1980—, The Solar Wind and the Earth, 1987; editorial bd.: Planet and Earth Sci; co-editor: Space Sci. Revs. Recipient Chapman medal Royal Astron. Soc., 1976, award Japan Acad., 1977; named Disting. Alumnus U. Alaska, 1980, Centennial Alumnus Nat. Assn. State Univs. and Land Grant Colls., 1987. Fellow Am. Geophys. Union (John Adam Fleming medal 1977); mem. AAAS, Sigma Xi.

AKAY, ADNAN, mechanical engineer educator. BS, N.C. State U., 1971, MME, 1972, PhD, 1976. Rsch. fellow Nat. Inst. Environ. Scis., Research Triangle Park, N.C., 1976-78; from asst. prof. to DeVlieg prof. Wayne State U., Detroit, 1978-92; prof., head Mech. Engr. Dept. Carnegie-Mellon U., Pitts., 1992—. Contbr. articles to profl. jours. Mem. ASME, Acoustical Soc. Am. Office: Carnegie-Mellon U Dept Mech Engring Pittsburgh PA 15213

AKBAR, SHEIKH ALI, materials science and engineering educator; b. Faridpur, Bangladesh, Jan. 1, 1955; came to U.S., 1980; s. Abdus Sattar and Achia (Khatun) Sk; m. Parveen Tahmina Akbar, Sept. 6, 1981; children: Nafisa T., Nusrat T., Naeema T. BS, MS, U. Sofia, Bulgaria, 1980; PhD, Purdue U., 1985. Asst. prof. materials sci. and enging. Ohio State U., Columbus, 1988—. Contbr. articles to profl. jours. Rsch. grantee U. Chgo., 1991, Edison Materials Tech. Ctr., Kettering, Ohio, 1991, 92, NSF, Washington, 1992. Mem. AIME, Am. Ceramic Soc., Electrochem. Soc. Achievements include development of sensors for CO gas. Office: Ohio State U 494 Watts Hall Columbus OH 43210

AKEL, OLLIE JAMES, oil company executive; b. Harlan, Ky., Aug. 14, 1933; s. William M. and Jameleh (Raffih) A.; m. Mona, June 11, 1966; children: Omar James, Amanda Dalal, Roanna Lyn. BSME, U. Ky., 1954; M in Aero. Engring., Renssalear Polytech. Inst., 1955; MS in Mgmt., Mass. Inst. Tech., 1967. Thermodynamic engr. North Am. Aviation, Columbus, Ohio, 1958-59; engr. Middle East Airlines, Beirut, Lebanon, 1959-65, Exxon Corp., N.Y., London, Arabia, 1967-80; pres. Exxon Chem. Mideast and Africa, Brussels, 1981-86, Exxon Chem. Belgium, Brussels, 1986-88; dir. corp. comm. Exxon Chem. Internat., Brussels, 1988-89; pres. Exxon Saudi Arabia, Riyadh, 1989-92, Exxon Mexicana, Mex., 1993—. Dir. United Way, Brussels, 1988-89. 2d Lt. U.S. Army, 1956-58. Mem. Am. C.C. Mex. (bd. dirs.), Am. Businessmen's Group of Riyadh (steering com., 1979-81, 90-92), Tau Beta Pi, Pi Tau Sigma. Protestant. Office: Exxon Mexicana, Aristoteles 77, Mexico

AKESSON, NORMAN BERNDT, agricultural engineer, emeritus educator; b. Grandin, N.D., June 12, 1914; s. Joseph Berndt and Jennie (Nonthen) A.; m. Margaret Blasing, Dec. 14, 1946; children—Thomas Ryan, Judith Elizabeth. B.S. in Agrl. Engring., N.D. State U., 1940; M.S. in Agrl. Engring, U. Idaho, 1942. Registered profl. engr., Calif. Research fellow U. Idaho, 1940-42; physicist U.S. Navy, Bremerton, Wash., 1942-47; asst. prof. agrl. engring. U. Calif., Davis, 1947-56; assoc. prof. U. Calif., 1956-62, prof., 1962-84, prof. emeritus, 1984—; cons. United Fruit Honduras, 1959, Israel, 1968, WHO Mosquito Control, 1969-84, FAO Aircraft in Agr., 1971-84, Japan, 1972, Egypt, 1980, China, 1985, Can. Forest Svc. Herbicide Application, 1987, U. Fla. Aircraft Application Herbicides, 1987; chmn. expert com. on vector control equipment WHO, 1976; chmn. com. on aircraft for agr. Coun. for Agrl. Sci. and Tech., 1982; pres. Calif. Weed Control Conf., 1966. Author: The Use of Aircraft in Agriculture, 1974, Pesticide Application Equipment and Techniques, 1979, Aircraft Use for Mosquito Control, 1981; contbr. over 330 articles to profl. jours. Recipient research and devel. award FAO, 1973-74, research and devel. award WHO, 1978; Fulbright fellow, Eng. and East Africa, 1957-58. Fellow Am. Soc. Agrl. Engrs. (chmn. Pacific region 1965, dir. 1972-74, assoc. editor tech. publs. 1983-93); mem. ASTM (chair E-35 1982-84), Nat. Agrl. Aviation Assn., Calif. Agrl. Aviation Assn., Nat. Mosquito Control Assn., Calif. Mosquito Assn., Entomol. Soc. Am., Weed Sci. Soc. Am. (editorial bd. 1968-70), Western Weed Soc. (hon.) Brit. Soc. Agrl. Engring., Farmers Club (London). Republican. Home: 748 Elmwood Dr Davis CA 95616-3517 Office: U Calif Agrl Engring Dept Davis CA 95616-5294

AKIL, HUSEIN AVICENNA, acoustical engineer; b. Cibatu, Garut, Indonesia, Apr. 11, 1956; arrived in Eng., 1988; s. Achjar Mansur and Otih (Djuariah) A.; m. Ida Triastuti Suhidi, Apr. 15, 1984; children: Seinda Nurinawati, Wenda Averroes. Insinyur, Bandung Inst. Tech., 1983; MSc, U. Salford, Eng., 1991; student, U. Liverpool, Eng., 1991—. Rsch. officer Ctr. for Calibration Instrumentation and Metrology, Indonesian Inst. of Scis., Puspiptek, Indonesia, 1983-88. Contbr. articles to profl. jours. Mem. Acoustical Soc. Am., Inst. of Acoustics (Great Britian), Assn. of Indonesian Muslim Intellectual (coord. organizational and instnl. sect. 1992). Muslim. Home: Komplek Puspiptek Blok IA9, Serpong 15310, Indonesia Office: Indonesian Inst Scis, Komplek Puspiptek, Serpong 15310, Indonesia

AKINMUSURU, JOSEPH OLUGBENGA, civil engineer; b. Ileoluji, Nigeria, Oct. 28, 1948; s. Isaiah and Julianah (Akinsiku) A.; m. Bosede Olayinka Adeeko, Dec. 16, 1978; children: Toyin, Tosin, Tobi, Tomi. BS, U. Lagos, Nigeria, 1970; MS, Ahmadu Bello U., Zaria, Nigeria, 1973; PhD, U. Sheffield, Eng., 1978. Registered profl. engr. Civil engr. Pub. Svc. Western State, Ibadan, Nigeria, 1970-71; prof. civil engr. U. Ife, Ile-Ife, Nigeria, 1978-87; vis. prof. U. Mich., Ann Arbor, 1987-88; civil engring. prof. Cleve. State U., 1988-90, Lawrence Tech. U., Southfield, Mich., 1990--; cons. rural rds. Army Engr. Corps Nigerian Army, Lagos, 1980-87; cons. rural infrastructures Fed. Gov. Nigeria, 1980-87. Vol. facilitator, mentor LINK Program Cleve. State U., 1988-90; role model, motivation speaker Boys Scouts Am., Detroit, 1990--. Mem. Am. Soc. Civil Engrs., Am. Soc. Testing and Materials, Internat. Soc. Soil Mechanics and Found. Engring., Assn. Geoscientists for Internat. Devel., Nigerian Soc. Engrs. (mem. governing coun. 1983-85). Achievements include successful application of the reinforced soil technology to problems of rural roads, slope stability and erosion control, as well as to the construction of rural houses and farmsteads. Home: 2121 Alice St Ann Arbor MI 48103 Office: Lawrence Tech U Dept Civil Engring Southfield MI 48075

AKIYAMA, MASAYASU, chemistry educator; b. Okayama, Japan, June 28, 1937; s. Shizuo and Teruko (Tokuda) A.; m. Hiroko Matsuda, June 12, 1969; children: Takuo, Yuko. BS, Okayama, 1960; MS, Tokyo Inst. Tech., 1962, DSc, 1965. Rsch. assoc. Tokyo Inst. Tech., 1965-70; postdoctoral fellow Northwestern U., Evanston, Ill., 1968-70; assoc. prof. Tokyo U. of Agriculture & Tech., Koganei, Tokyo, 1970-82, prof., 1982—, chief libr. faculty br., 1983-85; vis. scholar Harvard U., Cambridge, Mass., 1986; chmn. dept. biol. & chem. sci., dept. grad. course Tokyo U of Agriculture & Tech., 1991-92; mem. editorial bd. Japan Chem. Soc., Tokyo, 1988-90. Co-author: Bioorganic Chemistry of Enzymatic Catalysis, 1992; contbr. articles to profl. jours. Mem. AAAS, Japan Chem. Soc., Am. Chem. Soc., Royal Soc. Chemistry. Achievements include polymerization of p-cyanobenzonitrile N-oxide; preparation of N-Hydroxymaleimide catalytic activity of 2-substituted imidazoles for hydrolysis of acyl derivatives; design and synthesis of artificial siderophores, synthesis and properties of N-hydroxy Peptides. Home: 3-20-11 Nanyodai, Hachioji 192-03, Japan Office: Tokyo U Agriculture & Tech, 2-24-16 Naka-cho, Koganei 184, Japan

AKKARA, JOSEPH AUGUSTINE, biochemist; b. Kerala, India, Feb. 22, 1938; came to U.S., 1964; naturalized, 1980; s. Augustine Aippu Akkara and Theresa Anthony Kolapran; PhD, Sch. Medicine, U. Mo., 1969; m. Mary Ann Malaickel, Aug. 18, 1969; children: Augustine Viju, Jeena Theresa. Lab. asst. Med. Coll., Trivandrum, Kerala, India, 1959-61; tech. asst. Cen. Food Technol. Rsch. Inst., Mysore, India, 1961-64; grad. asst., rsch. assoc. Sch. Medicine U. Mo., Columbia, 1964-69; rsch. assoc. Rockefeller U., N.Y.C., 1969-71; rsch. assoc. Brookdale Hosp. Med. Ctr., Bklyn., 1971-73, chief radioassay, 1973-80; sr. scientist Med. Rsch. Inst., Worcester, Mass., 1980-81; biochemist stat. Toxicology Svc. Boston, 1981-84; rsch. chemist U.S. Army Natick Rsch. and Engring. Ctr., 1984—; adj. asst. prof. Richmond Coll., CUNY. Active Boy Scouts Am. Lic. dir. clin. labs. N.Y.C. Dept. Health. Mem. Materials Rsch. Soc., Am. Chem. Soc., N.Y. Acad. Sci., Kerala Assn. New Eng. (pres. 1986-87), Indian Assn. Greater Boston (sec. 1986-88, 1st v.p. 1988-89), Sigma Xi, Lions Club. Roman Catholic. Achievements include patents, research and publications in synthesis and characterization of bioengineered materials for electro-optic and high performance applications, enzymology, nutrition, endocrinology, analytical biochemistry. Home: 18 Temi Rd Holliston MA 01746-1220 Office: US Army Natick Rsch and Engring Ctr Natick MA 01760-5020

AKKERMANN, SCHAIA, mechanical, electrical and civil engineer; b. Rio de Janeiro, Brazil, Mar. 28, 1928; s. Schaia and Mina A.; m. Bella Ruth Schenker, Oct. 4, 1953; children: Sergio, Davi, Ethel. BSc in Engring., U. São Paulo, Brazil, 1952. Civil engr. Eucatex Indsl. Commn., São Paulo, 1955-62; civil and elec. engr. Departamento de Águas e Energia Elétrica, São Paulo, 1962-82; civil, mech., elec. engr. Acustica Engenharia S/C Ltd., São Paulo, 1962—. Mem. Acoustical Soc. Am., Inst. Noise Control Engring. (assoc.). Office: Acoustica Engenharia S/C, Rua Raggio Nóbrega 63, 01441-010 São Paulo Brazil

AKPAN, EDWARD, metallurgical engineer; b. Kalamazoo, Jan. 19, 1952; s. Edward and Akon Akpan; m. Mildred Jimmeh; children: Ime, Kaema. BS in Metall. Engring., Western Mich. U., 1976, MS in Metall. Engring., 1977. Materials engr. Ford Motor Co., Indpls., 1978; process engr. Hughes Aircraft Co., Tucson, 1979-80; sr. engr. GE, Cleve., 1980-83; lab. mgr. GE, Florence, S.C., 1983-85; sr. engr. Reliance Electric Co., Greenville, S.C., 1985—. Author: Secondary Operations in Powder Metallurgy, 1993; contbr. papers in field. Active Big Bros./Bis Sisters, Cleve., 1982; founding mem. Edn./Bus. Partnership, Greenville, 1987. Mem. ASTM, Am. Powder Metallurgy Inst. (treas. 1987, vice chmn. 1988, chmn. 1988-91, bd. dirs. 1991—, mem. com. Am. Powder Metallurgy Inst. Metal Powder Iindustries Fedn. World Congress, 1991-92). Republican. Mem. Charismatic Ch. Achievements include patent for bonding of refractory metal. Home: 205 Gilderbrook Rd Greenville SC 29615 Office: Reliance Electric 6040 Ponders Ct Greenville SC 29615

AKSELSSON, KJELL ROLAND, environment technology educator, researcher; b. Torhamn, Sweden, Jan. 31, 1940; s. Axel Emanuel Kjell and Evy Elise (Alfredsson) A.; m. Rose-Marie Andersson, Jan. 11, 1964; children: Lisa Maria Magdalena, Anna Cecilia Terese. MSc, Lund (Sweden) U., 1963, PhD, 1973; MB, Lund U., 1979. Rsch. and teaching asst. Lund Inst. Tech. Lund U., 1964-71, prof. working environment tech. Lund Inst. Tech., 1983—, chmn. bd. study program in indsl. hygiene, 1984-87, chmn. dept. indsl. mgmt., 1990—, vice dean Sch. Mech. engring., 1990—, co-founder ctr. environ. measuring tech. Lund Inst. Tech., 1990, rsch. assoc. dept. environ. health and dept. nuclear physics, 1972-73; rsch. assoc. Fla. State U., Tallahassee, 1974-75; rsch. assoc. Lund U., 1975-77, bd. dirs. Ctr. Work Scis., 1991—; rsch. scientist Swedish Bd. Natural Scis., Lund, 1977-83; bd. dirs. internat. com. Swedish Work Environ. Fund, Stockholm, 1986-91. Editorial bd. Jour. Aerosol Sci., 1987-91; contbr. articles to profl. jours. and procs.; co-patentee in personal sample for airborne particles; co-inventor particle induced x-ray emission spectroscopy; prin. investigator multi-disciplinary projects. Grantee Swedish Bd. for Tech. Devel., Swedish Work Environ. Fund, Commn. of European Communities. Mem. Ergonomics Soc. Sweden (bd. dirs. 1987-91), Human Factor Soc. (Europe chpt. assoc.), Nordic Ergonomics Soc., Gesellschaft Aerosolforschung, Nordic Soc. Aerosol Rsch. Internat. Commn. Occupational Health, Am. Indsl. Hygiene Assn. Avocations: orienteering, music. Home: Steglitsv 7A, S-227 32 Lund Sweden Office: Lund Inst Tech, Dept Working Environ Tech, Box 118, S 22100 Lund Sweden

AL-AFALEQ, ELJAZI, organic chemistry educator, administrator; b. Feb. 13, 1956; d. Ibrahim A. and Dalal S. Al-A.;. BS in Chemistry Edn., Grils Edn. Coll., Riyadh, Saudi Arabia, 1978, MSc in Organic Chemistry, 1984; PhD in Organic Chemistry, Girls Sci. Coll., Dammam, 1989. Prof. Girls Sci. Coll., Dammam, Saudi Arabia, 1978-91; rector Girls Sci. Coll., Dammam,

1991—. Recipient Ea. Province Govs. award, Dammam, 1990. Achievements include research to compare extraction of protein and essential amino acids from cardamon; research to extract protein and essential animo acids from defatted phaseolus mango seeds by sudium salts; studies in heterocycles, synthesis and special studies of new thiopyridazine derivatives with possible antimicrobial activity, synthesis and antimicrobial activity of new thiopyridazine derivatives, novel synthesis of biological active pyridazine derivatives, synthesis of triazolo (493-b) pyridazines. Home and Office: Ibrahim A Afaleq, Aramco PO Box 1670, E Prov Dhahran 31311, Saudi Arabia

ALAGIĆ, SUAD, computer science educator, researcher, consultant; b. Derventa, Bosnia, Feb. 15, 1946; m. Mara Karanović, June 10, 1972; children: Irena, Gorjan. BSEE, U. Sarajevo, Bosnia, 1970; MSc in Computer Sci., U. Mass., 1972, PhD in Computer Sci., 1974. Grad. asst. dept. computer and info. sci. U. Mass., Amherst, 1971-74; postdoctoral fellow dept. computer sci. U. Edinborough, Scotland, 1977; asst. prof. dept. computer sci. and informatics U. Sarajevo, 1975-80, assoc. prof., 1980-86, prof., 1986-93, chmn. computer sci. Faculty Elec. Engring., 1989-91, vice rector, 1989-91; prof., chair dept. computer sci. Wichita (Kans.) State U., 1993—; cons. UN Environ. Program, London, 1983, Athens, 1984, Munich, 1990; vis. prof. U. Vt., Burlington, 1991-93. Author: The Design of Well-Structured and Correct Programs, 1978 (transl. into Japanese 1980, Russian 1982, Polish 1985), Relational Database Technology, 1986, Object-Oriented Database Programming, 1989. Recipient gold medal U. Sarajevo, 1971, Univ. medal, 1989; Sci. Achievement award State of Bosnia and Herzegovina, 1985. Mem. Assn. for Computing Machinery. Office: Wichita State U Dept Computer Sci 1845 Fairmount CB 83 Wichita KS 67260-0083

ALANKO, MATTI LAURI JUHANI, animal reproduction educator; b. Jääski, Finland, July 7, 1935; s. Lauri and Hellä (Mure) A.; m. Solveig Nybonde, May 6, 1961; children: Jouko, Marko, Jarko. DVM, Vet. Coll., Oslo, 1961; PhD, Coll. Vet. Medicine, Helsinki, 1974. Pvt. practice Mäntyharju, Finland, 1968-74; rsch. asst. dept. ob-gyn Coll. Vet. Medicine, Hautjärvi, Finland, 1962-68, assoc. prof., 1975-81, prof., 1981—. Contbr. numerous articles to sci. and ednl. jours. Lt. inf. Finnish Army, 1961-62. Mem. Nordic Vet. Assn. for Animal Reprodn. (pres. Finnish sect. 1978—), Nordic Assn. for Andrology (v.p. 1988—), Finnish Vet. Assn., Assn. U. Profs. Lutheran. Avocations: ornithology, outdoors, photography. Home: Metsapolku 6, SF-04600 Mäntsälä Finland Office: Coll Vet Medicine, SF-04840 Hautjärvi Finland

ALARCON, MINELLA CLUTARIO, physics educator, researcher; b. Manila, Aug. 10, 1948; d. Mario and Asuncion (Mazo) Clutario; m. Francisco Martin Alarcon, June 23, 1973; children: Sandra Francesca, Maria Jennifer. BS in Physics, U. Philippines, Quezon City, 1969, PhD in Physics, 1987; MS in Physics, Ateneo de Manila U., 1981. Instr. physics Centro Escolar U., Manila, 1969-70, U. Phillipines, 1970-73, De La Salle U., Manila, 1973-74, Coll. of Holy Spirit, Manila, 1978-80; instr. physics Ateneo de Manila U., 1974-76, 80-87, asst. prof., 1987-90, assoc. prof., 1990—, supr. Laser Lab., 1987—, dept. chair, 1992—; vis. researcher Rsch. Inst. Elec. Communication, Tohoku U., Sendai, Japan, 1985, 88, Fukui (Japan) U., 1990-91. Scholar Philippine Dept. Sci. and Tech., 1982-86, rsch. grantee, 1989—; Bursar grantee Brit. Coun., 1983-84; fellow Japanese Soc. for Promotion Sci., 1985, 88, Asahi Shimbun, 1990-91. Mem. Philippine Phys. Soc. (councilor 1981-82, sec. 1982-83, v.p. 1988—), Am. Phys. Soc., Optical Soc. Am. Roman Catholic. Avocations: reading, singing, listening to music, films, badminton. Office: Ateneo de Manila U Physics, Dept, Loyola Heights, Quezon City 1108, The Philippines

ALARCON, ROGELIO ALFONSO, physician, researcher; b. Yungay, Nuble, Chile, Feb. 14, 1926; came to U.S., 1954; s. Alfredo and Carmen Rosa (Carrasco) A. BS, U. Chile, Concepcion, 1943; MD, U. Chile, Santiago, 1950. Staff physician internal medicine U. Chile Hosp. Salvador, Santiago, 1951-52, Hosp. Gonzalez Cortez, Santiago, 1952-54; resident medicine Meml. Ctr. for Cancer and Allied Diseases, N.Y.C., 1955-56; fellow internal medicine George Washington U. Hops., George Washington Sch. Medicine, Washington, 1956-57; resident internal medicine Lemuel Shattuck Hosp., Boston, 1957-58; rsch. fellow pathology Children's Cancer Rsch. Found., Children's Hosp. Med. Ctr., Boston, 1958-60; rsch. assoc. Children's Cancer Rsch. Found., Boston, 1960-74, Harvard Med. Sch., Boston, 1962-76, Cancer Rsch. Inst., New Eng. Deaconess Hosp., Boston, 1974-76; staff physician Boston Children's Hosp. Med. Ctr., Wrentham, Mass., 1977-79; staff physician VA Med. Ctr., Phila., 1979-80, Bedford, Mass., 1980—. Contbr. articles to profl. jours. Mem. Am. Assn. for Cancer Rsch., N.Y. Acad. Scis., Nat. Assn. VA Physicians. Roman Catholic. Achievements include discovery of the enzymatic generation of acrolein, a highly cytotoxic aldehyde, from biogenic polyamines; development of a fluorometric method to measure minimal amounts of acrolein; research in the growth inhibitory effects of oxidized spermine on mammalian cells, research involving acrolein in cell growth regulation, and identification of acrolein as a metabolite of cyclophosphamide and related chemotherapeutic agents. Home: 33 Pond Ave Apt B-915 Brookline MA 02146-7128 Office: Bedford VA Med Ctr 200 Springs Rd Bedford MA 01730-1114

ALAUPOVIC, PETAR, biochemist, educator; b. Prague, Czechoslovakia, Aug. 3, 1923; married, 1947; 1 child. ChemE, U. Zagreb, 1948, PhD in Chemistry, 1956; DHC (hon.), U. Lille, France, 1987. Rschr. pharms. rsch. lab. Chem Corp, Prague, 1948-49; rschr. organic lab. Inst. Indsl. Rsch., Yugoslavia, 1949-50; asst. agrl. facility U. Zagreb, 1951-54, asst. chem. inst. med. facility, 1954-56; rsch. biochemist U. Ill., 1957-60; with cardiovascular sect. Okla. Med. Rsch. Found., Oklahoma City, 1960—, head lipoprotein lab., 1972—, also head lipoprotein and atherosclerosis rsch. program; prof. rsch. biochemistry, sch. med. U. Okla., 1960—. Assoc. editor Lipids, 1974-78. Grantee NIH, 1961-91. Mem. AAAS, Am. Soc. Biol. Chemists, Am. Chem. Soc., Am. Heart Assn., Am. Oil Chemistry Soc. Achievements include research on chemistry of naturally occuring macromolecular lipid compounds such as serum and tissue lipoproteins and bacterial endotoxins, on biochemistry of red cell membranes; isolation and characterization of tissue lipases. Office: OK Med Rsch Found Lipoprotein & Artherosclerosis 825 NE 13th St Oklahoma City OK 73104*

ALBACH, CARL RUDOLPH, consulting electrical engineer; b. Bayonne, N.J., Feb. 21, 1907; s. George J. and Mary (Bollier) A.; m. Anne Avery, Sept. 15, 1934 (dec.); children—Lyndon Carl, Karen Joy Albach Antikajian. EE, Rensselaer Poly. Inst., 1928; MA in Psychology, U. Buffalo, 1939. Registered profl. engr. N.Mex. (ret.). Outside plant engr. AT&T, N.Y.C., 1928-30, Buffalo, 1930-38; field asst. social security bd. N.Y.C. and Buffalo, 1939-42; engr. U.S. Engrs. Office, Buffalo, 1942-44; dial office engr. Western Electric Co., Chgo., 1944-47; elec. designer W.C. Kruger Co., Santa Fe, 1947-50; cons. elec. engr. various govt. and pvt. sch., ch., hosp. bldgs., 1950-73; elec. engring. cons. U. N.Mex., 1955-70; elec. and lighting cons. sch. plant planning com. N.Mex. Dept. Edn., 1963-70; elec. engr. Lewis Poe & Assocs., 1977-87; lectr. illumination various orgns., 1957-64, 72; vice chmn. N.Mex. Bd. Profl. Engrs., 1953-55; mem. Sweet's Nat. Adv. Bd., 1965-70, Statutes of Limitations Commn., 1966-67; mem. N.Mex. Elect. Code Com., 1967-76. Author: Engineered Projects and Other Topics 1950-1986, 1991; contbr. 45 articles to profl. jours. Mem. deve. council Rensselaer Poly. Inst. Fellow Am. Cons. Engrs. Coun. (life), Illuminating Engring. Soc. (past chmn. N.Mex. chpt., vice chmn. 4th and 18th intermountain regional confs., nat. dir. 1962-64, mem. nat. allied arts com. 1957-64, Meritorious Svc. award); mem. IEEE (sr., life), Cons. Engrs. Coun. N.Mex. (life, pres. 1956-58, bd. dirs. 1958-59, 61-64, 66-67, exec. dir. 1967-85, 30 Yr. Svc. award 1987). Achievements include design of electrical system for Albuquerque Civic Auditorium, of lighting systems for the Univ. N.Mex's. Johnson Gymnasium, Student Union, Fine Arts Center, and Walking Utility Tunnel System, of Sante Fe Outdoor Opera in 1957 and redesign in 1967 after fire, of electrical system for Portales, N.Mex's. Automated Water Supply System. Address: 3471 Cerrillos Rd # 55 Santa Fe NM 87501

ALBACH, HORST, economist; b. Essen, Germany, July 6, 1931; s. Karl Albach; m. Renate Gutenberg; children: Rolf, Karin, Dirk. Student bus. adminstrn., econs. and law, U. Cologne, 1952-56; D of Econs., Bowdoin Coll., 1958. Privatdozent (lectr.) Cologne, 1960, prof., 1961; mem. scientific adv. com. Fed. Econs. Ministry, 1967; head econs. dept. Cologne U., 1968-73; dep. chmn. spl. com. Kosten U. Finanzierung d. berufl. Bild., 1971-74; sr. lectr. Wissenschaftlicher Rat, 1974-77; spl. advisor Begutachtung der gesamtwirtschaftlichen Entwicklung, 1978; pres. Berlin Acad. Scis., 1987—; dir. Inst. Ops. Rsch. and Mgmt., 1987—. Author: Wirtschaftlichkeitsrechn. bei unsich. Erwartungen, 1959, Investitionen u. Liquidität, 1962 (also in Japanese), Degressive Abschreibung, 1967, Beitr. z. Unternehm. plan., 1969, 2d edit., 1978, Steuersystem u. unternehm. Investitionspolitik, 1970, Als-Ob-Konzept und zeitl. Vergleichsmarkt, 1976; co-author numerous publications; contbr. numerous articles to profl. jours. Hon. doctorate Helsinki, Stockholm. Mem. Rhineland-Westphalian Acad. Scis. Office: Academy of Sciences & Technology, Griegstr 5-7, 1000 Berlin 33, Germany*

ALBAGLI, LOUISE MARTHA, psychologist; b. Queens, N.Y., Jan. 15, 1954; d. Meyer Nathan and Leah (Bleier) Greenberg; m. Eli S. Albagli, July 31, 1977. BA in Psychology summa cum laude, CUNY, 1976; D of Clin. Psychology, Rutgers U., 1983. Clin. psychology intern Postgrad. Ctr. Mental Health, N.Y.C., 1980-81; staff psychologist Queens County Neuropsychiat. Inst., Jackson Heights, N.Y., 1981-83, Bklyn. Community Counseling Ctr., 1981-84; sr. clin. psychologist Richard Hall Community Mental Health Ctr., Bridgewater, N.J., 1984-86; pvt. practice specializing in women's reproductive health issues cen. N.J., 1985—; adj. faculty mem. Rutgers U., 1990—. Editor Women's Reproductive Health Newsletter, A Place to Be Heard. Mem. Nat. Register Health Care Providers, Am. Psychol. Assn., N.J. Psychol. Assn. (com. inter-profl. rels.), Internat. Childbirth Edn. Assn., RESOLVE, Raritans, Phi Beta Kappa.

ALBALA, DAVID MOIS, urologist, educator; b. Chgo., Dec. 29, 1955; s. Maurice Mois and Betty (Pancratz) A. BA, Lafayette Coll., Easton, Pa., 1978; postgrad., Brown U., 1979-81; MD, Mich. State U., 1983. Gen. surg. intern Dartmouth-Hitchcock Med. Ctr., Hanover, N.H., 1983-84, resident in gen. surgery, 1984-85, resident and rsch. resident in urology, 1986-89, 90; sr. resident in pediatric urology Mass. Gen. Hosp., Boston, 1989; fellow in endourology and stone disease Barnes Hosp.-Washington U. Sch. Medicine, St. Louis, 1990-91; clin. instr. urology Dartmouth Med. Sch., 1989-90; asst. prof. dept. urology Loyola U. Med. Sch., Maywood, Ill., 1991—; staff urologist VA Hosp., Hines, Ill., 1992—. Contbr. chpts. to books, numerous articles to profl. jours. Recipient Jour. Endourolgy award, 2nd pl. Essay Contest, 1991; recipient numerous grants in field. Mem. ACS, AMA, AAAS, Am. Urol. Assn., Am. Pub. Health Assn., Chgo. Urol. Soc., Endourol. Soc., Soc. for Minimally Invasive Therapy. Office: Loyola U Med Ctr Dept Urology 2160 S 1st Ave Maywood IL 60153

ALBANIS, TRIADAFILLOS ATHANASIOS, chemistry educator, researcher; b. Thessaloniki, Greece, Mar. 3, 1956; s. Athanassios and Dimitra (Papazoglou) A.; m. Maria-Eleni Lekka, Dec. 4, 1984; children: Orestis, Dimitra. 1st deg. in Chemistry, Poly. Sch., Thessaloniki, 1979; Diplome, Mediterranean Agr. Inst., Zaragosa, Spain, 1983; PhD in Chemistry, U. Ioannina, 1987. Asst. researcher Poly. Sch., Thessaloniki, 1980-81; asst. researcher U. Ioannina, Greece, 1981-86, lectr., 1987-89, asst. prof., 1992—. Editor: Modelling in Environment Chemistry, 1991; contbr. articles to profl. jours. Postdoctoral fellow U. Tex., San Antonio, 1989-90. Mem. Greek Chem. Soc., Am. Chem. Soc. (agrochemicals div.). Achievements include research in influence of fly ash on pesticide late in the environment and simulation of pesticide movement in succesive soil layers using an SCTR cascade model. Office: Univ Ioannina, Dept Chemistry, 45332 Ioannina Greece

ALBAUGH, KEVIN BRUCE, chemical engineer; b. Syracuse, N.Y., Sept. 28, 1958; s. Judson Knight and Carolyn Jeanette (Darling) A.; m. Cheryl Diane Cade, June 7, 1980; children: Kurt William, James Patrick. B-SChemE, Clarkson Coll., 1980; PhD in Chem. Engring., Clarkson U., 1987. Profl. engr., Vt. Assoc. engr. IBM, Essex Junction, Vt., 1980-83, sr. assoc. engr., 1983-87, staff engr., 1987-89, devel. mgr., 1989—; mentor Sematech Ctr. of Excellence, Tucson, 1989—; mem. focus tech. adv. bd. Sematech, Austin, Tex., 1992—; bd. dirs. Ctr. for Advanced Materials Processing, Potsdam, N.Y. Contbr. articles to profl. jours. Vol. examiner Am. Radio Relay League, Newington, Conn., 1986—; sec. Burlington (Vt.) Amateur Radio Club, 1986-90; dist. emergency coord. Amateur Radio Emergency Svc., Essex Junction, 1986-90; mem. choir St. James Ch., Essex Junction, 1990—. Recipient Cert. of Merit Am. Radio Relay League, 1988. Mem. Am. Ceramic Soc., Electrochem. Soc., Inst. Environ. Sci. (co-chair tech. program 1992—), Sigma Xi, Omega Chi Epsilon. Episcopalian. Achievements include development of reactor analysis techs. for microcontamination reduction in semiconductor mfr.; demonstrated electrochem. mechanisms for anodic bonding of glass to metals. Office: IBM Corp 1000 River St Essex Junction VT 05452

ALBEE, ARDEN LEROY, geologist, educator; b. Port Huron, Mich., May 28, 1928; s. Emery A. and Mildred (Tool) A.; m. Charleen H. Ettenheim, 1978; children: Janet, Margaret, Carol, Kathy, James, Ginger, Mary, George. B.A., Harvard, 1950, M.A., 1951, Ph.D., 1957. Geologist U.S. Geol. Survey, 1950-59; prof. geology Calif. Inst. Tech., 1959—; chief scientist Jet Propulsion Lab., 1978-84, dean grad. studies, 1984—, project scientist Mars Observer Mission, 1978—; cons. in field, 1950; chmn. lunar sci. rev. panel NASA, 1972-77, mem. space sci. adv. com., 1976-84. Assoc. editor Jour. Geophys. Rsch., 1976-82, Ann. Rev. Earth Space Scis., 1978—; contbr. numerous articles to profl. jours. Mem. bd. regents L.A. Chiropractic Coll., 1990—. Recipient Exceptional Sci. Achievement medal NASA, 1976. Fellow Mineral Soc. Am. (assoc. editor Am. Mineralogist 1972-76), Geol. Soc. Am. (assoc. editor bull. 1972-89, councilor 1989-92), Am. Geophys. Union. Office: Calif Inst Tech Mail Code 02 31 Pasadena CA 91125

ALBERS, JAMES ARTHUR, engineering executive; b. Alton, Ill., Oct. 5, 1941; s. Arthur John and Elisabeth Ann (Wickenhauser) A.; children: James, Joseph; m. Thalia Sorgi, Oct. 28, 1989. B.S. in Aero. Engring., Parks Coll., St. Louis, 1962; M.S. in Engring. Sci., U. Toledo; Ph.D. in Mech. Engring., Mich. State U., East Lansing, 1971; M.B.A. in Mgmt., Golden Gate U., San Francisco, 1979. Chief Propulsion Systems NASA Dryden, Edwards AFB, Calif., 1975-78; candidate exec. devel. program NASA Hdqrs., Washington, 1979-80; chief project engring. office NASA Dryden, 1980-81, asst. chief Dryden Aero. Projects, 1981; asst. dir. aerospace systems NASA Ames Ctr, Moffett Field, Calif., 1982-85; dep. dir. aerospace systems Calif. State U., 1985-91. Author reports. Contbr. articles to profl. jours. Recipient spl. achievement award NASA Lewis, Cleve., 1974, Equal Opportunity award NASA Dryden, 1976, Group Achievement award, YF-12 symposium team, 1978, Exceptional Service medal NASA Ames, 1985. Mem. AIAA, Am. Helicopter Soc. Roman Catholic. Home: 3488 Shafer Dr Santa Clara CA 95051-4617 Office: NASA Ames MS-200-3 Moffett Field CA 94035

ALBERT, GERALD, clinical psychologist; b. N.Y.C., Nov. 13, 1917; s. Andrew I. and Eleanor (Walder) A.; divorced; m. Norma Holm Haskell, 1983; children: Jay Harvey, Laurie Ellen Albert Moxham. BA, CCNY, 1938; MA, New Sch. for Social Research, 1958; EdD, Columbia U., 1964; Cert. psychoanalytic tng. program, L.I. Inst. Mental Health, Queens, N.Y., 1964. Editor Vulcan and Creston Pubs., N.Y.C., 1939-45; nat. dir. advt., pub. relations Universal Pictures, div. ednl. films, N.Y.C., 1945-50; exec. dir. Advt. Enterprises and Continental Research Inst., Queens, N.Y., 1951-64; asst. to full prof. LIU, 1964-85, prof. Emeritus, 1985—; dir. L.I.U. C.W. Post Counseling Ctr., 1964-70; supervising psychologist, clin. dir. L.I. Consultation Ctr., until 1986—, clin. cons., 1986—; pvt. practice marriage and individual therapy, 1958—. Author: (cassette) How to Choose and Keep a Marriage Partner, 1980, The Wonderful Magic of No-Fault Living, 1990; editor-in-chief Jour. Contemporary Psychotherapy, 1985-87; contbr. articles to profl. jours., author booklets. Recipient 1st prize Most Effective Communications/Newsletters Community Agys. Pub. Rels. Assn., 1983. Fellow Am. Assn. for Marriage and Family Therapy (L.I. Family Therapist of Yr. 1993); mem. APA, N.Y. Soc. Clin. Psychologists, Soc. Clin. and Exptl. Hypnosis. Office: 271 Merrick Ave East Meadow NY 11554-1549

ALBERT, MILTON JOHN, retired chemist, microbiologist; b. St. Louis, Oct. 27, 1917; s. Fredrich and Mary (Kuchenmeister) A.; m. Arline Louise Bolle, Mar. 30, 1946; children: Mary Louise, Kathleen Marie, Barbara Jean, Patricia Lee. BS, U. Ill., 1948. Plant mgr. Lucky Club Flavors, St. Louis, 1950-53; quality control Vestal Labs., St. Louis, 1953-57; microbiologist Rexall Drug Co., St. Louis, 1957-64; quality control Falstaff Brewing Co., St. Louis, 1964-67; dept. head Morton Frozen Food, Russellville, Ark., 1967-81; quality control Ga.-Pacific, Crossett, Ark., 1981-82; cons. Russellville, 1982—; cons. Russellville, 1982—; vol. Vols. in Tech. Assistance, Arlington, Va., 1981—; tech. advisor Ark. Valley Vocat. Tech. Sch., Ozark, 1985-87. With U.S. Army, 1942-46, World War II. Mem. Am. Soc. Microbiology, Inst. Food Tech., Am. Chem. Soc. Home: RR 7 Box 259 Russellville AR 72801-9113

ALBERT, ROY ERNEST, environmental health educator, researcher; b. N.Y.C., Jan. 11, 1924; m. Abigail L. Lewin, 1945; 4 children. AB, Columbia U., 1943; MD, NYU, 1946. Med. officer health and safety lab. AEC, N.Y.C., 1952-54, from asst. chief to chief med. br., div. biology and medicine, 1954-56; asst. dir. radioisotope lab., asst. clin. prof. medicine George Washington U. Sch. Medicine, 1956-59; assoc. prof. Inst. Environ. Medicine, NYU Med. Ctr., 1959-66, prof., vice chmn., dept. dir., 1966-85; dep. asst. adminstr. for health and ecology EPA, Washington, 1975-76; Jacob G. Schmidlapp prof. environ. health, chmn. dept. U. Cin. Med. Ctr., 1985—; chmn. Carcinogen Assessment Group EPA, 1976-85. Recipient medal EPA, 1983, Jerry Stara Meml. award, 1993. Mem. AAAS, Am. Assn. for Cancer Rsch., Radiation Rsch. Soc., Soc. for Occupational and Environ. Health, Soc. for Risk Analysis (Disting. Contbn. award 1984), Alpha Omega Alpha. Office: U Cin Dept Environ Health 3223 Eden Ave Cincinnati OH 45267-0001

ALBERTHAL, LESTER M., JR., information processing services executive; b. Corpus Christi, Tex., Feb. 27, 1944; married. BBA, U. Tex., 1967. With EDS subs. Gen. Motors Corp., Dallas, 1968—; v.p. ins. group Electronic Data Systems Corp. subs. Gen. Motors Corp., Dallas, 1979-84, v.p. bus. ops., from 1984, pres., 1986—, chief exec. officer, 1987—, also dir.; chmn. EDS subs. Gen. Motors Corp., Dallas, 1989—. Office: Electronic Data Systems Corp 7171 Forest Ln Dallas TX 75230-2399

ALBERTINI, RICHARD JOSEPH, molecular geneticist, educator; b. Racine, Wis., Mar. 15, 1935; married; 4 children. BS, U. Wis., 1960, MD, 1963, PhD in Med. Genetics, 1972. Instr. medicine U. Wis., Madison, 1970-72; from asst. prof. to assoc. prof. Coll. Medicine U. Vt., 1972-79, prof. medicine, 1979—, adj. prof. microbiology and molecular genetics, 1980—; sci. dir. Vt.-N.H. Red Cross Blood Ctr., 1983—; dir. Vt. Regional Cancer Ctr., Genetics Lab. U Vt., 1984—, Genetics Unit Dept. Medicine, 1988—; pres. Vt. divsn. Am. Cancer Soc., 1989-90; mem. environ. health sci. review com. NIH, 1988; vis. prof. oncology and pediatrics U. Wis., Madison, 1990-91. Editor-in-chief Environ. & Molecular Mutagenesis, 1988—. Recipient St. George medal Am. Cancer Soc., 1990. Mem. Am. Assn. Cancer Rsch., Am. Assn. Clin. Oncology, Am. Soc. Human Genetics, Environ. Mutagen Soc. (pres. 1984-85, Alexander Hollaender award 1990). Achievements include research in nature and consequences of somatic cell gene mutation that arise spontaneously or as a result of mutagen/carcinogen exposure in humans, mutations that involve immunocompetent lymphocytes and their effects in vivo. Office: University of Vermont Genetics Lab 32 N Prospect St Burlington VT 05401*

ALBERTS, ALLISON CHRISTINE, biologist; b. San Francisco, Oct. 13, 1960; d. Walter Watson and Marilyn Janice (West) A.; m. Michael Benton Worley, Apr. 26, 1992. AB, U. Calif., Berkeley, 1982; PhD, U. Calif., San Diego, 1989. Teaching asst. U. Calif., San Diego, 1983-89; postdoctoral fellow Tex. A&M U., College Station, 1989-91; rsch. fellow Zool. Soc. San Diego, San Diego, 1991—, assoc. physiologist, 1993—; environ. cons. Odgen Environ./Michael Brandman Assn./Dudeck Assn., 1991—. contbr. articles to profl. jours. NSF grantee, 1992—. Mem. Animal Behavior Soc. (mem. edn. com. 1991—), Am. Soc. Zoologists, Internat. Soc. Chem. Ecology, Soc. for Conservation Biology, Am. Soc. of Ichthyologists & Herpetologists, Am. Assn. of Zool. Parks and Aquariums (mem. lizard adv. group 1991—). Democrat.

ALBERTS, BRUCE MICHAEL, foundation administrator; b. Chgo., Apr. 14, 1938; s. Harry C. and Lillian (Surasky) A.; m. Betty Neary, June 14, 1960; children: Beth L., Jonathan B., Michael B. AB in Biochemical Scis. summa cum laude, Harvard Coll., 1960; PhD in Biophysics, Harvard U., 1965. Postdoctoral fellow NSF Institut de Biologie Moleculaire, Geneva, 1965-66; asst. prof. dept. chemistry Princeton (N.J.) U., 1966-73, assoc. prof. dept. biochemical scis., 1971-73, Damon Pfeiffer prof. life scis., 1973-76; prof., vice chmn. dept. biochemistry and biophysics U. Calif., San Francisco, 1976-81, Am. Cancer Soc. Rsch. prof., 1981-85, prof., chmn., 1985-90, Am. Cancer Soc. Rsch. prof. of biochemistry, 1990-93; pres. NAS, Washington, 1993—; trustee Cold Spring Harbor Lab., 1972-75; adv. panel human cell biology NSF, 1974-76; adv. coun. dept. biochemical scis. and molecular biology Princeton U., 1979-85; chmn. vis. com. dept. biochemistry and molecular biology Harvard Coll., 1983-86; chmn. mapping and sequencing the human genome Nat. Rsch. Coun. Com., 1986-88; bd. sci. couns. divsn. arthritis and metabolic diseases NIH, 1974-78, molecular cytology study sect. 1982-86, chmn. 1984-86; program adv. com. NIH Human Genome Project, 1988-91; sci. adv. bd. Jane Coffin Childs Meml. Fund for Med. Rsch., 1978-85, Markey Found., 1984—, Fred Hutchinson Cancer Rsch. Ctr., Seattle, 1988—; com. mem. corp. vis. dept. biology MIT, 1978—, dept. embryology Carnegie Inst., Washington, 1983—; faculity rsch. lectr. U. Calif., San Francisco, 1985; sci. adv. com. Marine Biological Lab., Woods Hole, Mass., 1988—; bd. dirs. Genentech Rsch. Found., Fed. Am. Socs. for Experimental Biology; adv. bd. Bethesda Rsch. Labs. Life Tech. Inc., Nat. Sci. Resources Ctr., Smithsonian Inst., 1990—; com. mem. adolescence and young adulthood/sci. standards, Nat. Bd. Profl. Teaching Standards, 1991—. Co-author The Molecular Biology of the Cell, 1989; editor: Mechanistic Studies of DNA Replication and Genetic Recombination, 1980; editorial bd. Jour. Biological Chemistry, 1976-82, Jour. Cell Biology, 1984-87; assoc. editor Annual Reviews Cell Biology, 1984—; essay editor Molecular Biology of th Cell, 1991—; contbr. numerou articles to profl. jours. including Saunders Sci. Publ., Current Sci., Ltd. Fellow NSF, 1960-65; recipient Eli Lilly award in biological chemistry Am. Chemical Soc., 1972, Baxter award for Disting. Rsch. in Biomedical Scis. Assn. Am. Med. Colls., 1992; named Lifetime Rsch. Prof. Am. Cancer Soc., 1990, Outstanding Vol. Coord. Calif. Sch. Vol. Partnership, 1993. Fellow AAAS; mem. NAS (commn. life scis. Nat. Rsch. Coun. 1988—, chmn. 1988-93, adv. bd. Nat. Sci. Resources Ctr. 1990—, Nat. Com. Sci. Edn. Standards and Assessment 1992—, com. mem. Nat. Edn. Support System for Tchrs. and Schs. 1992—, U.S. Steel Found. award 1975), Am. Chemical Soc., Am. Soc. for Cell Biology, Am. Soc. for Microbiology, Genetics Soc. Am., Am. Soc. Biochemistry and Molecular Biology (councilor 1984—), European Molecular Biology Orgn. (assoc.), Phi Beta Kappa. Office: NAS Office of Pres 2101 Constitution Ave NW Washington DC 20418

ALBERTS, JAMES JOSEPH, scientist, researcher; b. Chgo., May 23, 1943; s. Joseph James and Lilyan (Matas) A. AB in Biology and Chemistry, Cornell Coll., 1965; MS in Organic Chemistry, Dartmouth Coll., 1967; PhD in Chem. Oceanography, Fla. State U., 1970. Rsch. assoc. Kans. State Geologic Survey, U. Kans., 1970; rsch. assoc. dept. biology U. Ga., 1970-72, asst. prof. dept. zoology, 1972-74; asst. chemist radiological and environ. rsch. div. Argonne Nat. Lab., 1974-77; assoc. rsch. ecologist Savannah River Ecology Lab., 1977-84; dir. and sr. rsch. scientist U. Ga. Marine Inst., 1984—; adj. prof. dept. biol. scis. Clemson U., 1985—, mem. coll. scis. adv. com., 1984-87; adj. prof. dept. zoology U. Ga., 1986-93, adj. prof. dept. marine sci., 1993—. Contbr. articles to publs. and chpts. to books. Mem. Marine div. Estuarine sub-com. Nat. Assn. State Univs. and Land Grant Colls., 1985-89, co-chmn., 1987-89; mem.-at-large exec. com. Southeastern Estuarine Rsch. Soc., 1986-88; pres. Southeastern Assn. Marine Labs., 1988; mem. Ga. dept. natural resources task force on rsch priorities Sapelo Island Nat. Estuarine Rsch. Reserve, 1988—; mem. exec. com. SAML, 1989; mem. NSF adv. panel Equipment and Facilities for Biol. Field Stas. and Marine Labs., 1989-90, BBS Rsch. Tng. Groups, 1991; mem. steering com. Coun. on Ocean Affairs, 1989-91, Nat. Assn. Marine Labs. Recipient Alexander von Humboldt Found. Sr. U.S. Scientist award, 1989; Petroleum Rsch. Fund fellow, 1968-69. Mem. AAAS, Am. Soc. Limnology and Oceanography, Internat. Assn. on Water Pollution Rsch. and Control, Internat. Humic Substance Soc., Estuarine Rsch. Fedn., Socs. Internationalis Limnologiae Theoreticae et Applicatae, Soc. for Environ. Geochemistry and Health, Geochem. Soc., Oceanography Soc., Phi Beta Delta, Phi Kappa Phi. Office: U Ga Marine Institute Sapelo Island GA 31327

ALBERTY, ROBERT ARNOLD, chemistry educator; b. Winfield, Kans., June 21, 1921; s. Luman Harvey and Mattie (Arnold) A.; m. Lillian Jane

Wind, May 22, 1944; children—Nancy Lou, Steven Charles, Catherine Ann. B.S., U. Nebr., 1943, M.S., 1944, D.Sc., 1967; Ph.D., U. Wis., 1947; D.Sc., Lawrence U., 1967. Engaged in research blood plasma fractionation for U.S. Govt., 1944-46; mem. faculty U. Wis., 1947-67, prof. chemistry, 1955-67, assoc. dean letters and sci., 1961-63; dean U. Wis. (Grad. Sch.), 1963-67; prof. chemistry MIT, 1967-91, prof. emeritus, 1991—, dean Sch. Sci., 1967-82; cons. NSF, 1958-83, NIH, 1962-72; chmn. commn. on human resources NRC, 1974-77; dir. Colt Industries, 1978-88, Inst. for Def. Analysis, 1980-86; pres. phys. chemistry div. Internat. Union Pure and Applied Chemistry, 1991-93. Co-author: Physical Chemistry, 1992, Experimental Physical Chemistry, 3d edit., 1970. Guggenheim fellow Calif. Inst. Tech., 1950-51; recipient Eli Lilly award biol. chemistry, 1955. Fellow AAAS; mem. NAS, Inst. Medicine, Am. Chem. Soc. (chmn. com. on chemistry and pub. affairs 1978-80), Am. Acad. Arts and Scis. (coun. 1991—), Phi Beta Kappa, Sigma Xi. Home: 7 Old Dee Rd Cambridge MA 02138-4633

ALBIN, LESLIE OWENS, biology educator; b. Spur, Tex., Jan. 8, 1940; s. John Leslie and Ottie Maude (Lassetter) A.; m. Monta Kay Gragg, Sept. 3, 1961 (div. 1982); children: Leslie Susan Albin Gann, Kimberly Ann Albin. BA, McMurry Coll., Abilene, 1962; MA, N. Tex. State U., 1969. Instr. biology E. Cen. State U., Ada, Okla., 1969-71; rsch. assoc. M.D. Andrson Hosp. & Tumor Inst., Houston, 1971; asst. prof. biology Western Tex. Coll., Snyder, 1971-74, assoc. prof. biology, 1974-77; instr. and dept. head biology Austin (Tex.) Community Coll., 1977-78, chmn. div. natural scis., 1978—. NDEA fellow, 1968. Mem. Faculty Assn. Western Tex. Coll. (pres. 1973-74), Faculty Assn. Austin Community Coll. (pres. 1987-88), Tex. Jr. Coll. Tchrs. Assn., AAAS, Tex. Acad. Sci., Nat. Sci. Tchrs. Assn., Alpha Chi. Office: Austin Community Coll 1212 Rio grande Box 140587 Austin TX 78714-0587

ALBONE, ERIC STEPHEN, chemist, educator; b. Hanworth, London, Feb. 21, 1940; s. Frank Charles and Dorothy (Kentish) A.; m. Kum-Yul Song, July 19, 1969; children: Michael, Paul. BA, Oxford U., U.K., 1962, MA, D of Philosophy, 1966; postgrad. cert. in edn., U. Bristol. Postdoctoral rsch. fellow Dept. Chemistry, SUNY, Stony Brook, 1966-68; rsch. assoc. Ctr. for Biology Natural Systems, Wash. U., St. Louis, 1968-69; rsch. assoc., fellow Sch. of Chemistry and Sch. of Vet. Sci., U. Bristol, 1970-79; chemistry educator, dir. rsch. Clifton Coll., Bristol, 1980—; hon. rsch. assoc. U. Bristol, 1981—; founding trustee Clifton Scientific Trust, Bristol, 1991—. Author: Mamalian Semiochemistry: The Investigation of Chemical Signals Between Mammals, 1984; contbr. articles to profl. jours. Recipient numerous grants. Fellow Royal Soc. of Arts; mem. Royal Soc. of Chemistry (coun. edn. div., other coms.). Achievements include first in the emerging interdisciplinary field of mammalian semiochemistry, sci. edn., interfacing edn. with profl. sci. Office: Clifton Scientific, Clifton Coll, Bristol BS8 3JH, England

ALBRECHT, ALBERT PEARSON, electronics engineer, consultant; b. Bakersfield, Calif., Aug. 23, 1920; s. Albert Waldo and Elva (Shuck) A.; m. Muriel Elizabeth Grenell, June 15, 1942 (dec. Apr. 1943); m. Edith J. Dorner, July 18, 1944. BSEE, Calif. Inst. Tech., 1942; MSEE, U. So. Calif., L.A., 1947. Registered profl. engr., Calif. Rsch. assoc. radiation lab. MIT, Cambridge, Mass., 1942-43; chief engr. Gilfillan Bros., L.A., 1943-58; v.p. Space Gen. Corp., El Monte, Calif., 1958-68; exec. v.p. Telluran Cons., Santa Monica, Calif., 1968-72; dir. systems evaluation Office of Asst. Sec. of Def. for Intelligence, Washington, 1972-76; assoc. adminstr. FAA, Washington, 1976-86; cons., prin. AP Albrecht-Cons., Alexandria, Va., 1986—; mem. exec. bd. RADIO Tech. Commn. for Aeronautics, Washington, 1980-86; mem. aeronautics adv. com. NASA, Washington, 1980-90. Co-author: Electronic Designers Handbook-Design Compendium, 1957, 2d edit., 1974. Fellow AIAA (adv. com. Aerospace Am. 1984—), IEEE (Engr. Mgr. of the Yr. 1989). Achievements include technical leadership of the replacement and automation of the nation's air traffic control system. Home and Office: 3224 Eagleridge Dr Bellingham WA 98226

ALBRECHT, ALLAN JAMES, computer scientist; b. Pittston, Pa., Feb. 6, 1927; s. Arthur Carl and Edna Louise (Boldt) A.; m. Jean Carol Herman, Jan. 30, 1949; children: Drew Edward, Dale Allan. Student, Bucknell U. Jr. Coll., Wilkes-Barre, Pa., 1944-45, Wilkes Coll., 1947-48; BSEE, Bucknell U., 1949. Field engr. IBM, York, Pa., 1949-52; design engr. IBM Engring. Lab., Poughkeepsie, N.Y., 1952-55; devel. engr. IBM Fed. Systems Div., Kingston, N.Y., 1955-60; sr. engr. IBM Fed. Systems Div., L.A., 1960-69; project mgr. IBM Data Processing Svcs., L.A., 1969-72, Portland, Oreg., 1972-74; program mgr. IBM Data Processing Svcs., White Plains, N.Y., 1974-79; sr. tech. staff mem. IBM Corp., Armonk, N.Y., 1980-89; cons. Orleans, Mass., 1989—; study group mem. Pa. State U., State College, 1956, USAF/Mitre Corp., McLean, Va., 1968. Author: Function Points Definition, 1984; inventor: Function Points Analysis, 1979. Pres. Mayflower Point Assn., Orleans, 1990—. With USN, 1945-46.

ALBRECHT, HELMUT HEINRICH, medical director; b. Trier, Fed. Republic Germany, Mar. 2, 1955; came to U.S., 1988; s. Heinrich Friedrich Paul and Ingeborg (Bräcker) A.; m. Gay Anne Wood, Oct. 12, 1984; children: Ingo, Ashley. MD, U. Hamburg and Heidelberg, Fed. Republic Germany, 1983; MS in Mgmt. and Policy, SUNY, Stony Brook, 1991, advanced cert. in health care mgmt., 1992. Intern in medicine and surgery Gen. Hosp. Hamburg-Altona, Dermatology U. Hosp., Eppendorf, 1982-83; rsch. asst. in pathology U. Heidelberg, 1983-84; asst. in gynecology and obstetrics City Hosp. Wiesbaden, Fed. Republic Germany, 1984-85; regional med. dir. BYK Gulden Pharms., Konstanz, Fed. Republic Germany, 1985-88; med. dir. Byk Gulden Pharma Group, Altana Inc., Melville, N.Y., 1988-92; med. dir. G.I. category Procter & Gamble Co., Cin., 1992—; Co-author, editor: Shopping Guide for Heidelberg, 1980. With Med. Svc., Fed. Republic Germany Air Force, 1974-76. Mem. AMA, Am. Acad. Dermatology, Internat. Soc. Chronobiology, N.Y. Acad. Scis., German Assn. Physicians in Pharm. Industry. Home: 9459 Hunters Creek Dr Cincinnati OH 45242 Office: Procter & Gamble Co 11370 Reed Hartman Hwy Cincinnati OH 45241

ALBRETHSEN, ADRIAN EDYSEL, metallurgist, consultant; b. Carey, Idaho, June 20, 1929; s. Norman Carl and Dollie Gustina (Brown) A.; m. Joan Alice Phelan, July 8, 1961; children: Thomas, Eric, Carl. BS in Mining Engring., U. Idaho, 1952, MSMetE, 1958; PhD in Mineral Engring., MIT, 1963. Analytical chemist Bunker Hill Co., Kellogg, Idaho, 1954-55; mining engr. Anaconda Co., Butte, Mont., 1955-57; rsch. asst. MIT, Cambridge, 1958-63; sr. engr. GE, Richland, Wash., 1963-65; sr. rsch. engr. Battelle Meml. Inst., Richland, Wash., 1965-66, ASARCO, Inc., South Plainfield, N.J., 1966-86; plant metallurgist Nord Ilmenite Corp., Jackson, N.J., 1989-92; cons. pvt. practice, Bridgewater, N.J., 1986—; cons. Bridgewater, N.J., 1986—. 1st lt. USAF, 1952-54, Korea. Mem. ASM Internat., Soc. Mining Engrs., Electrochem. Soc. (chmn. met. N.Y. sect. 1987-88), Assn. Consulting Chemists and Chemical Engrs., Inc., Sigma Xi. Avocation: gardening. Home: 485 Vicki Dr Bridgewater NJ 08807-1941

ALBRIGHT, DEBORAH ELAINE, emergency physician; b. Springfield, Ill., May 21, 1958; d. Jacob Eugene and Josephine (Ciotti) A.; m. William Ray Dudleston, Dec. 29, 1978; 1 child, Victoria Josephine. AA, Springfield Coll., 1978; BS, U. Ill., 1980; MD, So. Ill. U., 1984. Diplomate Am. Bd. Pediatrics, Am. Bd. Internal Medicine. Lab. asst. dept. microbiology Sch. Medicine So. Ill. U., Springfield, 1977-78; resident in internal medicine, pediatrics So. Ill. U. Affiliated Hosps., Springfield, 1984-88; emergency rm. physician St. Vincent's Hosp., Taylorville, Ill., 1987-88, Abraham Lincoln Meml. Hosp., Lincoln, Ill., 1988; pediatrician locum tenens Dr. Anthony Agatucci, Springfield, 1988, Dr. Ann Pearson, Springfield, 1988; staff physician dept. emergency medicine St. John's Hosp., Springfield, 1988—; clin. asst. prof. dept. pediatrics Sch. Medicine So. Ill. U., 1988—; part-time staff physician McFarland Mental Health Ctr., Springfield, 1990, State Ill. Dept. Disability Svcs., Springfield, 1991—. Mem. St. Aloysius Parish, Springfield, 1958—; vol. dept. radiology Mercy Hosp., Champaign, Ill., 1980; vol. Roman Cultural Women's Aux., Springfield, 1985—, Camp CoCo, Springfield, 1988-90; sponsor Sta. WILL, Springfield, 1987-91, Children's Miracle Network Telethon, Springfield, 1988—, Springfield Symphony Orch., 1992—. Fellow Am. Acad. Pediatrics; mem. Am. Coll. Physicians, Am. Acad. Pediatrics, Am. Coll. Emergency Physicians, Phi Theta Kappa, Alpha Epsilon Delta. Republican. Avocations: tennis, crocheting, following professional baseball box scores. Home: 2125 Shabbona Dr Springfield IL 62702-1338

ALBRITTON, DANIEL L., aeronomist; b. Camden, Ala., June 8, 1936. BS in Elec. Engring., Ga. Inst. Tech., 1959, MS in Physics, 1963, PhD in Physics, 1967. Dir. Aeronomy Lab. Environ. Rsch. Labs. NOAA, Boulder, Colo.; leader atmospheric chemistry project Climate and Global Change Program NOAA, v.p. commn. atmospheric chemistry and global pollution; co-chmn. sci. assessments of stratospheric ozone U.N. Environment Program; mem. sci. working group Intergovtl. Panel on Climate Change; former mem. editorial adv. bd. Jour. Molecular Spectroscopy, Jour. Atmospheric Chemistry., co-editor; lectr. in field. Contbr. 150 articles to profl. jours. Recipient Sci. Freedom and Responsibility award AAAS, 1993, Sci. Assessments award Am. Meteorol. Assn. Fellow Am. Phys. Soc., Am. Geophys. Union. Achievements include research in laboratory investigation of atmospheric ion-molecule reactions and theoretical studies of diatomic molecular structure, investigation of atmospheric trace-gas photochemistry. Office: Dept Commerce Aeronomy Lab Bldg 24 325 Broadway St Boulder CO 80303-3328*

ALBRITTON, GAYLE EDWARD, structural engineer; b. Foley, Ala., Jan. 28, 1939; s. Kernice Edward and Quita Fae (Gay) A.; m. Rose Annette Starling, Aug. 30, 1959; children: Jeffrey, Erik, Misty. BS, La. State U., 1962. Rsch. structural engr. U.S. Army Engr. Waterways Ext. Sta., Vicksburg, Miss., 1962-73, 78-86; supervisory rsch. structural engr. USAE Waterways Expt. Sta., Vicksburg, Miss., 1986—; project mgr. Miss. State Hwy. Dept., Jackson, 1973-76, asst. rsch. and devel. engr., 1976-78. Contbr. articles and tech. reports to profl. jours. Commr. Little League Baseball, Clinton, Miss., 1976-80; mem. City Recreation Com., Clinton, 1980. Mem. ASCE (bd. mem. 1990-91), NSPE, Soc. Am. Mil. Engrs., Sertoma Club of Clinton (pres. 1976-77). Achievements include research in structural engineering. Office: US Army Engr Waterways Expt Sta 3909 Halls Ferry Rd Vicksburg MS 39180-6199

ALBRIZIO, FRANCESCO, chemical consultant, chemistry educator; b. Vittorio Veneto, Treviso, Italy, Feb. 6, 1947; s. Tomaso and Olga (Peconi) A.; m. Silvanda Covre, Aug. 10, 1975; children: Laura, Silvia. Degree in chemistry, Padua U., Italy, 1973. Chemistry tchr. Tech. High Sch., Conegliano Veneto, Italy, 1973-77; properties and quality of comml. products tchr. Comml. High Sch., Conegliano Veneto, Italy, 1978-91; cons. Ct. of Justice, Treviso-Venezia-Vicenza, Italy, 1980—; cons. on pollution control local municipalities, Italy, 1980—. Contbr. (mag.) The Lancet, 1989. Councillor local com. Vittorio Veneto, Italy, 1985-90. Mem. Associazione Italiana Igienisti Industriali. Roman Catholic. Office: Via A Volta 6, Treviso, 31029 Vittorio Veneto Italy

ALCAZAR, ANTONIO, electrical engineer; b. Tampico, Mex., Aug. 22, 1965; s. Antonio and Victoria (Toribio) A. BS in Physics Engring., ITESM, Monterrey, Mex., 1987; MS in Electrical Engr., Ariz. State U., 1989; M in Internat. Mgmt., Am. Grad. Sch. of Internat., Mgmt., 1993. Product devel. engr. Rogers Corp., Tempe, Ariz., 1987-89, process engr., 1989-90, new product devel. engr., 1990-92; sr. product devel. engr. Circuit Components, Inc., Tempe, 1992—, program mgr., 1992-93. Ariz. U. scholar, 1992. Roman Catholic. Achievements include one patent.

ALDAY, JOHN HANE, mechanical engineer; b. Oklahoma City, Oct. 19, 1964; s. James Malcolm and Caroline Sloan (Easley) A. BSME magna cum laude, Va. Poly. Inst. and State U., 1988, MSME, 1991. Field engr. Schlumberger Well Svcs., Webster, Tex., 1988-89; devel. engr. Ingersoll-Rand Co., Davidson, N.C., 1991—. Co-author tech. paper. Mem. ASME, AIAA. Episcopalian. Achievements include developement of technique for quantifying randomness in a flow-field (axial flow fan), of performance modeling program for rotary screw compressors. Office: Ingersoll Rand Co 800A Beaty St Davidson NC 28036

ALDAY, PAUL STACKHOUSE, JR., mechanical engineer; b. Camden, N.J., May 31, 1930; s. Paul Stackhouse and Amanda (Knocke) A.; m. Ethel Humes O'Connor, Nov. 29, 1952; children: Amy Jane, Paul Stackhouse III, Sarah Jean. BS in ME, Drexel U., 1953, MS in ME, 1957. Registered profl. engr., N.J. Engr. Naval Shipyard/Burroughs Corp., Phila., 1953-56; rsch. engr. Franklin Rsch. Labs., Phila., 1956-57, Univac, Phila., 1957-59; sr. engr. RCA Corp., Camden, 1959-68; sr. design engr. Univac/Burroughs/Control Data, southeastern Pa., 1968-74; project engr. Campbell Soup Co., Camden, 1974-90; cons. Budd Co., Phila., 1990-91; mail processing equipment U.S. Post Office, Phila., 1991—. Drexel U. scholar, 1948. Mem. Sigma Xi. Achievements include stress analysis, supports and preliminary bearing test on design of Enrico Fermi Nuclear Reactor; concept and design of digital data recorder for Gemini Spacecraft; concepts and designs of video recorder mechanisms used in surveillance satelites, digital computer input/output and memory devices. Home: 5759 Rogers Ave Pennsauken NJ 08109

ALDEMIR, TUNC, nuclear engineering educator; b. Istanbul, Turkey, Mar. 17, 1947; s. Ahmet Cevat and Hatice (Oner) A.; m. Kristine Weichieh King, July 9, 1977. Dipl. in Physics, U. Istanbul, 1970; MS in Nuclear Engring., U. Ill., 1975, PhD, 1978. Researcher Cekmece Nuclear Rsch. & Tng. Ctr., Istanbul, 1971-73, sr. researcher, 1978-83; asst. prof. nuclear engring. Ohio State U., Columbus, 1983-89, assoc. prof., 1989—; cons. Internat. Atomic Energy Agy.; cons. atomic energy rsch. reactor design, optimization and safety to govts. Columbia, 1988, 90, Mexico, 1990; dir. NATO advanced rsch. workshop on the reliability and safety analysis of dynamic process systems, Kusadasi, Turkey, 1992; session organizer and chmn. various sci. confs. in field. Reviewer Reliability Engring. & System Safety, 1988—, IEEE Transactions on Reliability, 1990—, Nuclear Sci. and Engring, 1990—; contbr. articles to profl. jours. Rsch. grantee, U.S. Dept. Energy, 1985-88, 88-91, NSF, 1987-90. Mem. Am. Nuclear Soc. Achievements include research in dynamic system reliability; small, compact reactor core design and analysis; probabilistic and deterministic nuclear reactor safety. Home: 5219 Brynwood Dr Columbus OH 43220-2273 Office: Ohio State U 206 W 18th Ave Columbus OH 43210-1154

ALDEN, INGEMAR BENGT, pharmaceutical executive; b. Stockholm, Feb. 23, 1943; s. Bengt Erik and Agnes (Eriksson) A.; m. Estelle Cuni Skrabanek, June 18, 1977; children: Lars, Sonja, Ingela. M.Social and Bus. Sci., Stockholm U., 1969. Field supr. Astra Lakemedel Sweden, Sodertalje, 1970-71, nat. sales mgr., 1971-72, mgr. mktg. and sales, 1973-74; internat. mktg. mgr. Astra Pharms., Sodertalje, 1975-76; dir. pharm. div. Astra Ltd., Watford, Eng., 1977-78; mng. dir. Merck Sharp & Dohme, Sweden, 1979-89; chief exec. officer Aldenco AB, 1989-91; dir. Pharma/Agro/Vet div. Svenska Hoechst AB, 1991—; bd. mem. Assn. Fgn. Pharm. Cos., Kronans Droghandel AB; chmn. Aldenco AB; ptnr. Midway Internat. Co. Club: Rotary, RVC.

ALDER, BERNI JULIAN, physicist; b. Duisburg, Germany, Sept. 9, 1925; came to U.S., 1941, naturalized, 1944; s. Ludwig and Ottilie (Gottschalk) A.; m. Esther Berger, Dec. 28, 1956; children—Kenneth, Daniel, Janet. B.S., U. Calif., Berkeley, 1947, M.S., 1948; Ph.D., Calif. Inst. Tech., 1951. Instr. chemistry U. Calif., Berkeley, 1951-54; theoretical physicist Lawrence Livermore Lab., Livermore, Calif., 1955-93; prof. dept. applied sci. U. Calif., Davis, 1987-93, prof. emeritus, 1993; van der Waals prof. U. Amsterdam, Netherlands, 1971; prof. associé U. Paris, 1972; G.N. Lewis lectr. U. Calif., Berkeley, 1984, Hinshelwood prof., Oxford, 1986, Lorentz prof., Leiden, 1990, Kistiakowsky lectr. Harvard U., 1990, Royal Soc. lectr., 1991. Author: Methods of Computational Physics, 1963; editor: Jour. Computational Physics, 1966-91. Served with USN, 1944-46. Guggenheim fellow, 1954-55; NSF sr. postdoctoral fellow, 1963-64, Japanese Promotion of Sci. fellow, 1989. Fellow Am. Phys. Soc.; mem. Nat. Acad. Scis., Am. Chem. Soc. (Hildebrand award 1985). Republican. Jewish. Office: PO Box 808 Lawrence Livermore Lab Livermore CA 94550

ALDERETE, JOSEPH FRANK, psychiatrist, medical service adminstrator; b. Las Vegas, N.Mex., Sept. 10, 1920; s. Jose P. and Adela R. (Armijo) A. ; m. Christine Krajewski, June 24, 1964; children—Joseph Frank, Sarah A. BS in Chemistry and Biology, Tex. Western Coll., 1950; MD, Nat. U. Mex., 1959. Intern USPHS Hosp., Balt., 1959-60; resident in psychiatry USPHS Hosp., Lexington, Ky., 1960-62; sr. resident in psychiatry U. Hosp. U. Okla. Sch. of Medicine, Oklahoma City, Ky., 1962-63; practice medicine specializing in psychiatry Balt., 1963-65, Springfield, Mo., 1965-67, Atlanta, 1968—; staff psychiatrist USPHS Hosp., Balt., 1963-65; chief of psychiat. service U.S. Med. Center for Fed. Prisoners, Springfield, Mo., 1965-67; clin. and research fellow in electroencephalography Mass. Gen. Hosp.-Harvard, Boston, 1967-68; clin. instr. psychiatry Emory U. Sch. of Medicine, Atlanta, 1968-84, asst. prof. medicine, 1986—; chief med. officer, hosp. dir. U.S. Penitentiary Hosp., Atlanta, 1968-78; asst. regional flight surgeon So. Dist., ARTCC, Atlanta, 1978—; evaluation sect. VA Hosp., Decatur, Ga., 1983—; psychiat. cons. to Student Health Office, Okla. State U., Stillwater, 1962-63, U.S. Fed. Reformatory, El Reno, Okla., 1962-63. Contbr. articles to profl. jours. Served with USAF, 1944-48, to capt. AUS, 1948-50. Fellow Am. Acad. Psychosomatic Medicine; mem. AMA, Am. Psychiat. Assn., Am. Soc. Clin. Hypnosis, So. EEG Soc., Am., Internat. acads. of law and psychiatry, Acad. of Psychosomatic Medicine, Atlanta Med. Soc., Clin. Soc. of USPHS, Phi Rho Sigma. Home: 4130 E Brockett Creek Ct Tucker GA 30084-6407 Office: VA Hosp 1670 Clairmont Rd Decatur GA 30033-4098

ALDERMAN, MINNIS AMELIA, psychologist, educator, small business owner; b. Douglas, Ga., Oct. 14, 1928; d. Louis Cleveland Sr. and Minnis Amelia (Wooten) A. AB in Music, Speech and Drama, Ga. State Coll., Milledgeville, 1949; MA in Supervision and Counseling Psychology, Murray State U., 1960; postgrad. Columbia Pacific U., 1987—. Tchr. music Lake County Sch. Dist., Umatilla, Fla., 1949-50; instr. vocal and instrumental music, dir. band, orch. and choral Fulton County Sch. Dist., Atlanta, 1950-54; instr. English, speech, debate, vocal and instrumental music, dir. drama, band, choral and orch. Elko County Sch. Dist., Wells, Nev., 1954-59; tchr. English and social studies Christian County Sch. Dist., Hopkinsville, Ky., 1960; instr. psychology, counselor critic prof. Murray (Ky.) State U., 1961-63, U. Nev., Reno, 1963-67; owner Minisizer Exercising Salon, Ely, Nev., 1969-71, Knit Knook, Ely, 1969—, Minimimeo, Ely, 1969—, Gift Gamut, Ely, 1977—; prof. dept. fine arts Wassuk Coll., Ely, 1986-91, assoc. dean, 1986-87, dean, 1987-90; counselor White Pine County Sch. Dist., Ely, 1960-68; dir. Child and Family Ctr., Ely Indian Tribe, 1988—, Family and Community Ctr., Ely Shoshone Indian Tribe, 1989—; adv. Ely Shoshone Tribal Youth Coun., 1990—, Budge Stanton Meml. Scholarship, 1991—, Budge Stanton Meml. Living Mus. and Cultural Ctr., 1991—; fin. aid contracting officer Ely Shoshone Tribe, 1990—; supr. testing Ednl. Testing Svc., Princeton, N.J., 1960-68, Am. Coll. Testing Program, Iowa, 1960-68, U. Nev., Reno, 1960-68; chmn. bd. White Pine Sch. Dist. Employees Fed. Credit Union, Ely, 1961-69; psychologist mental hygiene div. Nev. Pers., Ely, 1969-75, dept. employment security, 1975-80; sec.-treas. bd. dirs. Gt. Basin Enterprises, Ely, 1969-71; speaker at confs. Pvt. instr. piano, violin, voice and organ, Ely, 1981—; bd. dirs. band Sacred Heart Sch., Ely, 1982—. Author various news articles, feature stories, pamphlets, handbooks and grants in field. Pres. White Pine County Mental Health Assn., 1960-63, 78—; mem. Gov.'s Mental Health State Commn., 1963-65, Ely Shoshone Tribal Youth Camp, 1991-92, Elys Shoshone Tribal Unity Conf., 1991-92, Tribal Parenting Skills Coord., 1991; bd. dirs. White Pine County Sch. Employees Fed. Credit Union, 1961-68, pres., 1963-68; 2d v.p. White Pine Community Concert Assn., 1965-67, pres., 1967, 85—, treas., 1975-79, dr. chmn., 1981-85; chmn. of bd., 1984; bd. dirs. White Pine chpt. ARC, 1978-82; mem. Nev. Hwy. Safety Leaders Bd., 1979-82; mem. Gov.'s Commn. on Status Women, 1968-74, Gov.'s Nevada State Juvenile Justice Adv. Commn., 1992—, White Pine Overall Econ. Devel. Plan Coun., 1992—; sec.-treas. White Pine Rehab. Tng. Ctr. for Retarded Persons, 1973-75; mem. Gov.'s Commn. on Hwy. Safety, 1979-81, Gov's. Juvenile Justice Program, White Pine County Juvenile Problems Cabinet; dir. Ret. Sr. Vol. Program, 1973-74; vice chmn. Gt. Basin Health Coun., 1973-75, Home Extension Adv. Bd., 1977-80; sec.-treas. Great Basin chpt. Nev. Employees Assn.; bd. dirs. United Way, 1970-76; vice chmn. White Pine Coun. on Alcoholism and Drug Abuse, 1975-76, chmn., 1976-77; grants author 3 yrs. Indian Child Welfare Act, originator Community Tng. Ctr. for Retarded People, 1972, Ret. Sr. Vol. Program, 1974, Nutrition Program for Sr. Citizens, 1974, Sr. Citizens Ctr., 1974, Home Repairs for Sr. Citizens, 1974, Sr. Citizens Home Assistance Program, 1977, Creative Crafters Assns., 1976, Inst. Current World Affairs, 1989, Victims of Crime, 1990-92; bd. dirs. Family coalition, 1990-92, Sacred Heart Parochial Sch., 1982—, dir. band, 1982—; candidate for diaconal ministry, 1982—; dir. White Pine Community Choir, 1962—, Ely Meth. Ch. Choir, 1960-84; choir dir., organist Sacred Heart Ch., 1984—. Precinct reporter ABC News 1966; speaker U.S. Atty. Gen. Conf. Bringing Nev. Together. Fellow Am. Coll. Musicians, Nat. Guild Piano Tchrs.; mem. NEA (life), Nat. Fedn. Ind. Bus. (dist. chair 1971-85, nat. guardian coun. 1985—, state guardian coun.ells br. 1957-58, pres. White Pine br. 1965-66, 86-87, 89-91, bd. dirs. 1965-87, rep. edn. 1965-67, implementation chair 1967-69, area advisor 1969-73, 89-91), Nat. Fedn. Bus. and Profl. Women (1st v.p. Ely chpt. 1965-66, pres. Ely chpt. 1966-68, 74-76, 85—, bd. dirs. Nev. chpt. 1966—, 1st v.p. Nev. Fedn. 1970-71, pres. Nev. chpt. 1972-73, nat. bd. dirs. 1972-73), Mensa (supr. testing 1965—), Delta Kappa Gamma (br. pres. 1968-72, state bd. 1967—, chpt. parliamentarian 1974-78, state 1st v.p. 1967-69, state pres. 1969-71, nat. bd. 1969-71, state parliamentarian 1971-73), White Pine Knife and Fork Club (1st v.p. 1969-70, pres. 1970-71, bd. dirs. 1979—). Home: 945 Avenue H PO Box 150457 East Ely NV 89315-0457 Office: Ely Shoshone Tribe 16 Shoshone Cir Ely NV 89301-2055 also: 1280 Avenue F East Ely NV 89315

ALDOORI, WALID HAMID, researcher; b. Baghdad, Iraq, June 21, 1953; came to U.S., 1989; s. Hamid Salman Aldoori and Sahira Amin Abda Lghafour; m. Randa Tukan, Jan. 27, 1989. MD, Cairo U., 1977; diploma, London U., 1981; MPA, Harvard U., 1990. House officer Cairo U. Med. Sch., 1978-79; researcher faculty medicine Jordan U., Amman, 1982, lectr., 1982-89; researcher Harvard Sch. Pub. Health, Boston, 1992—; cons. UNESCO, Amman, 1988, Ministry of Health, Ministry of Social Devel., Jordan U. Hosp., Amman, 1982-89. Author: Family Health and Nutrition, 1988. Recipient Middle East Ednl. Fellowship Harvard U. J.F. Kennedy Sch. Govt., Boston, 1989, Internat. Scholarship Harvard Sch. Pub. Health, Boston, 1990, 91. Mem. APHA, Am. Inst. Nutrition, Am. Soc. Clin. Nutrition, Royal Soc. Pub. Health U.K., Physicians for Human Rights. Office: Harvard Sch Pub Health Nutrition Dept 665 Huntington Ave Boston MA 02115

ALDREDGE, RALPH CURTIS, III, mechanical engineer, educator; b. Houston, Calif., Apr. 23, 1964; s. Ralph Curtis and Joyce (Miller) A.; m. Teresa Walker, June 22, 1991; children: Danielle Lashaun, Simone Fay. BSME, Carnegie-Mellon U., 1985; MA, Princeton U., 1987, PhD, 1990. Summer preprofl. IBM, East Fishkill, N.Y., 1983, engr., 1984; mem. tech. staff AT&T Bell Labs., Murray Hill, N.J., 1985, 88; postdoctoral fellow U. Calif., San Diego, 1990-91; rsch. fellow Calif. Inst. Tech., Pasadena, 1991-92; asst. prof. U. Calif., Davis, 1992—. Jour. referee Combustion Sci. & Tech., Reading, U.K., 1990—, AIAA, N.Y.C., 1990—; contbr. articles to profl. jours. AT&T fellow, 1985, U. Calif., LaJolla fellow, 1990, James Irvine Rsch. fellow, 1991. Mem. ASME (assoc.), Combustion Inst., Soc. for Indsl. and Applied Math., Sigma Xi, Tau Beta Pi. Lutheran. Achievements include identification of flow field modification upstream and downstream from laminar premixed flames wrinkled by turbulence. Office: U Calif Engr MAME Davis CA 95616-5294

ALDRICH, NANCY ARMSTRONG, psychotherapist, clinical social worker; b. Taylorville, Ill., Oct. 4, 1925; d. Guy L. and Alice Irene (Hicks) Armstrong; m. Paul Harwood Aldrich, Sept. 30, 1949; children: Gregory Paul, Mark Douglas, Alice Ann Aldrich White, Ruth Lynne Aldrich Sammis. AB with highest honors, U. Ill., 1947, BS in Chemistry, 1948, MS in Chemistry, 1949; MSS, Bryn Mawr Coll., 1986. Lic. clin. social worker, Del., Pa. Parole bd. mem. State of Del., Dover, 1970-74; instr. continuing edn. U. Del., Newark, 1976-78, program specialist, 1978-83; founder Acad. Lifelong Learning; v.p. Aldrich Assocs. Inc., Landenberg, Pa., 1983—, psychotherapist, 1987—; psychotherapist Family Community Service Del. County, Media, Pa., 1986, Tressler Ctr. for Human Growth, Wilmington, Del., 1987-93; clin. affiliate Personal Performance Cons., 1990—, Acorn, 1990—, CMG Health, 1991—, Achievement and Guidance, 1991—, DuPont EAP Program, 1992—; coord. human resources devel. program Tressler Ctr. for Human Growth, 1983-84. Pres. YWCA New Castle County, Wilmington, 1974-76; mem. Statewide Health Coordinating Council, Del., 1978-79; bd. dirs., com. mem. United Way Del., Wilmington, 1975-84; trustee Unitarian Universalist House, Phila., 1992—. Mem. NASW, AAUW (pres. Wilmington br. 1968-70, nat. resoultions com. 1971-72, Fellowship gift named in her honor 1970), Del. Soc. Lic. Clin. Social Workers (pres. 1990-91), Assn. for Humanistic Psychology, Del. Gerontol. Soc., Mental Health Assn. Del., Internat. Soc. Bioenergetic Analysis, Phi Beta Kappa, Phi Kappa Phi, Iota Sigma Pi. Office: 625 Chambers Rock Rd Landenberg PA 19350

ALDRIDGE, DANIEL, aerospace engineer; b. Pitts., Oct. 25, 1960; s. Raymond and Margret Marie (Oblinger) A. BS in Aerospace Engring., Pa. State U., 1983. Engr. USAF SA-ALC, San Antonio, 1984—. Mem. AIAA. Office: USAF SA-ALC/LDIE Kelly AFB San Antonio TX 78241

ALDRIDGE, EDWARD C., JR., aerospace transportation executive; b. Houston, Tex., Aug. 18, 1938. BS, Tex. A&M U., 1960; MS, Ga. Inst. Tech., 1962. Mgr. missile and space div. Douglas Aircraft Co., Santa Monica, Calif., 1962-67, Washington, 1962-67; dir. strategic def. div. Dept. Def., 1967-72, dep. asst. sec. for strategic programs, 1974-76; dir. planning and evaluation Office of Sec. Def., 1976-77; sr. mgr. LTV Aerospace Corp., Dallas, 1972-73; sr. mgmt. assoc. Office Mgmt. and Budget, Washington, 1973-74; v.p. Strategic Systems Group System Planning Corp., Arlington, Va., 1977-81; undersec. Dept. Air Force, 1981-86; sec. of Air Force, 1986-88; pres. McDonnell Douglas Electronic Systems Co., McLean, Va., 1988-92; pres., c.e.o. Aerospace Corp., El Segundo, Calif., 1992—; formerly advisor Strategic Arms Limitation Talks, Helsinki and Vienna. Recipient George M. Low Space Transp. award AIAA, 1990. Office: Aerospace Corp 2350 E El Segundo El Segundo CA 90245-4691

ALDRIN, BUZZ, former astronaut, science consultant; b. Montclair, N.J., Jan. 20, 1930; children from previous marriage: James Michael, Janice Ross, Andrew John; m. Lois Driggs, 1988. BS, U.S. Mil. Acad., 1951; ScD in Astronautics, MIT, 1963; ScD (hon.), Gustavus Adolphus Coll., 1967, Clark U., 1969, U. Portland, 1970, St. Peter's Coll., 1970; LittD (hon.), Montclair State Coll., 1969; HHD (hon.), Seton Hall U., 1970. Commd. officer USAF, 1951, advanced through grades to col.; served as fighter pilot in Korea, 1953; pilot Gemini XII orbital rendezvous space flight, Nov. 11-15, 1966; lunar module pilot on first manned lunar landing Apollo XI; comdr. Aerospace Rsch. Pilots Sch., Edwards AFB, Calif., 1971-72; ret. USAF, 1972; with Ctr. for Aerospace Scis. U. N.D., Grand Forks, 1989; sci. cons. Beverly Hills Oil Co., Inforex Computer Co., Laser Video Corp., Mut. of Omaha Ins.; founder Starcraft Enterprises, 1988. Author: Return to Earth, 1973, Men From Earth, 1989. Decorated D.S.M., Legion of Merit, D.F.C. with oak leaf cluster, Air medal with 2 oak leaf clusters; recipient numerous awards including Presdl. medal of Freedom, 1969. Fellow AIAA; mem. Nat. Space Soc. (chmn.), Soc. Exptl. Test Pilots, Royal Aero. Soc. (hon.), Sea Space Symposium; charter Internat. Acad. Astronautics (corr.), Sigma Xi, Tau Beta Pi. Club: Masons (33 degree). Established record over 7 hours and 52 minutes outside spacecraft in extravehicular activity.

ALDROUBI, AKRAM, mathematician, researcher; b. Homs, Syria, May 20, 1958; came to U.S., 1982; s. Samir and Amira (Azhari) A. Diploma in elec. engring., Swiss Inst. Tech., Lausanne, 1982; PhD in Math., Carnegie-Mellon U., 1987. Assoc. researcher Va. Poly. Inst. and State U., Blacksburg, 1987-88; staff fellow NIH, Bethesda, Md., 1988—. Contbr. chptrs. to books and articles to profl. jours. Mem. Am. Math. Soc., Soc. Indsl. and Applied Math. Achievements include advances in mathematics and applied mathematics. Home: 4901 Battery Ln Apt 2 Bethesda MD 20814 Office: NIH BEIP/NCRR Bldg 13 Rm 3W13 Bethesda MD 20892

ALEKMAN, STANLEY LAWRENCE, chemical company executive; b. N.Y.C., Mar. 21, 1938; s. Harry and Evelyn (Bernstein) A.; m. Alice Finkelstein; children: Rachel, Eric, Elizabeth. BA, CCNY, 1962; PhD, U. Del., 1968. Rsch. chemist E.I. DuPont de Nemours & Co., Wilmington, Del., 1968-79; tech. dir. Air Products & Chemicals Inc., Allentown, Pa., 1979-83; v.p. rsch. and devel. Hüls Am. Inc., Piscataway, N.J., 1983—. With U.S. Army, 1962-64. Mem. Am. Chem. Soc., N.Y. Acad. Scis., Phila. Acad. Sci. Jewish. Avocations: chess, opera, tennis. Home: One Macintosh Ct East Brunswick NJ 08816 Office: Huls America Inc PO Box 456 80 Centennial Ave Piscataway NJ 08855-0365

ALEKSANDROV, LEONID NAUMOVITSH, physicist, educator, researcher; b. Dnepropetrovsk, U.S.S.R., Sept. 27, 1923; s. Naum Lvovitsh and Vera (Markovna) A.; m. Julia Makarovna Melnik, Aug. 19, 1953; children: Svetlana, Andrej. Degree, U. Dnepropetrovsk, 1950; candidate sci., Sci. Coun., Dnepropetrovsk, 1954; DSc, Sci. Coun., Moscow, 1964. Head of chair solid state physics U. Saransk, U.S.S.R., 1953-65; head rsch. lab. Inst. Light Sources, Saransk, 1958-65; head thin film rsch. lab. Inst. Semiconductor Physics, Acad. Sci., Novosibirsk, U.S.S.R., 1965-77; chief sci. staff Inst. Semiconductor Phsyics, Acad. Sci., Novosibirsk, Russia, 1977—; prof. microelectronics Electrotechnical Inst. Technical Univ., Novosibirsk, 1968—; meridet sci. and engring. worker of Russia Technical Univ., 1983—; prof., cons. U. Saransk, 1965-90, UNESCO, U. Habana, Cuba, 1975-76; guest prof. physics U. Vienna, Austria, 1991. Author: Transition region in Epitaxial Semiconductor Films, 1978 (Acad. Sci. award 1979), Growth of Crystalline Semiconductor Materials on Crystal Surfaces, 1984, Crystallization and Regrowth of Semiconductor Films, 1986 (Branch award 1986); mem. editorial bd. internat. jours. Thin Solid Films and Physica Status Solidi. Mem. Dep. Soviet Coun. of Region, Novosibirsk, 1969—; mem. Coun. of Vets., Novosibirsk, 1987—; mem. Znanja, Novosibirsk, 1950—. Lt. Soviet Army, 1943-45. Recipient prize for thin film rsch. Siberian Acad., 1982, Gold Medal of State award, Moscow Govt., 1988. Mem. Munich Bundes Republik Deutschland Exhbn. (mem. scis. com. electronics, medal 1989). Avocations: photography, philately, touring, radio, light athletics. Home: Vojevodskogo 5 w 2, 630090 Novosibirsk Russia Office: Acad Sci USSR, Inst Semiconductor Physics, 630090 Novosibirsk Russia

ALEXANDER, BEVERLY MOORE, mechanical engineer; b. Portsmouth, Va., Apr. 11, 1947; d. Julian Morgan and Ezefferlee (Griffin) Moore; m. Ronald Lee Rutherford, Dec. 21, 1969 (div. Dec. 1977), m. Larry Ray Alexander, Mar. 4, 1978. BS, Aero. Engring., Va. Poly. Inst. and State U., 1969; postgrad., U. New Orleans. Registered profl. engr., La. Assoc. engr. McDonnell Douglas Corp., St. Louis, 1969-74; design engr. Bell Aerospace Textron, New Orleans, 1974-81; supr. systems integration, New Orleans, 1981-83, chief interface activities, 1983-84; chief engr. Bell Aerospace Textron, New Orleans, 1984-85, dir. engring. planning and control, 1985-86, chief engr. engring. svcs., 1986-88, asst. chief engr., supr. of shipbuilding USN, New Orleans, 1988—. Mem. La. Engring. Soc., NAFE, ASNE (sect. chmn. 1992, vice chmn. 1993), SNAME. Republican. Episcopalian. Office: SUPSHIP C201 Naval Support Activity Bldg 16 New Orleans LA 70142

ALEXANDER, BYRON ALLEN, insect systematist; b. Austin, Tex., Apr. 14, 1952; s. Harold Edwin and Elizabeth Ann (Rowe) A. MS, Colo. State U., 1983; PhD, Cornell U., 1989. Rsch. asst. Stanford Outdoor Primate Facility, Calif., 1974-75; field asst. African Primate Project U. Stirling, Parc National du Nickolo-Koba, Senegal, 1976-77; park naturalist Capitol Reef Nat. Pk., Great Sand Dunes Nat. Monument, 1977-81; postdoctoral fellow Smithsonian Instn., Natural History Mus., Washington, 1990; asst. prof., curator Snow Entomol. Mus., U. Kans., Lawrence, 1989—; Contbr. articles to profl. jours.; illustrator: Natural History and Behavior of North American Beewolves, 1988. NSF grantee, 1987-89; Smithsonian Instn. fellow, 1990, Cornell U. fellow, 1983-89; recipient Outstanding Teaching Asst. award Cornell U., Dept. Entomology, 1987. Mem. AAAS, Entomol. Soc. Am., Animal Behavior Soc., Sigma Xi, Alpha Omega. Achievements include phylogenetic analysis of honey bee species is consistent with von Frisch-Lindauer scenario for the evolution of the dance language but does not decisively rule out alternative scenarios; suggested that Philanthinae may be the major lineage of sphecid wasps that is more closely related to bees than to other sphecid wasps. Office: U Kans Snow Entomol Mus Snow Hall Lawrence KS 66045

ALEXANDER, CARL ALBERT, ceramic engineer; b. Chillicothe, Ohio, Nov. 22, 1928; s. Carl B. and Helen E. A.; m. Dolores J Herstenstein, Sept. 4, 1954; children—Carla C., David A. B.S., Ohio U., 1953, M.S., 1956; Ph.D., Ohio State U., 1961. Mem. staff Battelle Columbus Labs., 1956—, research leader, 1974—, mgr. physico-chem. systems, 1976—; mem. faculty Ohio State U., 1963—, prof. ceramic and nuclear engring., 1977—; sr. research leader, chmn. tech. council of Biol. and Chem. Scis. Directorate, 1987—, chief scientist, 1987; prof. materials sci. and engring, 1988—;. Author; patentee in field. Served to lt. (j.g.) USNR, 1951-54. Recipient Merit award NASA, 1971, IR-100 award, 1987, R&D-100 award, 1988; citations Dept. Energy, citations AEC, citations ERDA. Mem. Am. Soc. Mass Spectrometry, Keramos, Sigma Xi. Home: 4249 Haughn Rd Grove City OH 43123-3216 Office: 505 King Ave Columbus OH 43201-2681

ALEXANDER, CHRISTINA LILLIAN, pharmacist; b. N.Y.C., Dec. 25, 1942; d. Stanley Urich and Roselyn Helen (Joseph) A. BS in Pharmacy, Fordham U., 1965; MS in Pharmacology, St. John's U., Jamaica, N.Y., 1977. Chief pharmacist Holy Family Hosp., Bklyn., 1966; asst. chief pharmacist N.Y. Polyclinic Hosp., N.Y.C., 1967-69; pharmacist Mt. Sinai Hosp. Med. Ctr., N.Y.C., 1969; night pharamcist Maimonides Med. Ctr., N.Y.C., 1969-71; pharmacist N.Y. U. Hosp. and Med. Ctr., 1971-72; night pharamacist, drug info. specialist Brookdale (N.Y.) Med. Ctr. and Hosp., 1972-80; asst. clin. prof. St. John's U. Coll. Pharmacy and Allied Health Professions, 1981-82; dir. drug info. ctr. St. John's U. and L.I. Jewish-Hillside Med. Ctr., N.Y.C., 1981-82; pres. Internat. Pharm. Cons., 1982—; asst. dir. pharmacy Burke Rehab. Ctr., White Plains, N.Y., 1984-89, dir., 1989-90; pres., dir. pharm. svcs. Alexander Assocs. Health Cons., 1990—; editor, pub. Internat. Pharm. Cons. Newsletter, 1984—; clin. instr. in pharmacy Rutgers U., New Brunswick, N.J., 1975, clin. cons. Coll. Pharmacy, 1977-79; sec. internat. adv. coun. Am. Bd. Pharmacy, Internat. Contbr. to N.Y. Carib. News. Recipient The Assembly of the State of N.Y. Citation, 1986, Scholarship Incentive award N.Y. State, 1960-65. Mem. AAAS, Am. Mus. Natural History (life), Am. Bd. Diplomates in Pharmacy, Internat., Federation Internationale Pharmaceutique, Fedn. Am. Scientists, Am. Pharm. Assn., Am. Soc. Hosp. Pharmacists, Fordham U. Coll. Pharmacy Alumni Fedn. (dir., pres. 1987—), Caribbean Cultural Ctr. N.Y.C., Lambda Kappa Sigma, Rho Chi. Democrat. Roman Catholic. Home: 3333 Henry Hudson Pky # 22F Bronx NY 10463-3224

ALEXANDER, DALE EDWARD, materials scientist; b. Feb. 22, 1962; s. Edward Francis Alexander and Helen Louise (Jessing) Flynn; m. Denise Anne Lobsinger, Aug. 16, 1986. BS in Nuclear Engring., U. Mich., 1984, MS, 1986, PhD, 1990. Cons. KMS Fusion Inc., Ann Arbor, Mich., 1984-86; asst. materials scientist Argonne (Ill.) Nat. Lab., 1990—. Contbr. articles to Phys. Rev., Jour. Applied Physics, Jour. Nuclear Materials. Active Hands-on Children's Mus., Ann Arbor, 1986. DOE Nuclear Engring. fellow Oak Ridge Associated U., 1984-88. Mem. Am. Phys. Soc., Am. Nuclear Soc., Materials Rsch. Soc., Electron Microscopy Soc. Am., The Metallurgical Soc., Sigma Xi. Roman Catholic. Achievements include research in ion irradiation enhanced diffusion-induced grain boundary migration. Office: Argonne National Lab 9700 S Cass Ave Bldg 212 Argonne IL 60439

ALEXANDER, DUANE FREDERICK, pediatrician, research administrator; b. Balt., Aug. 11, 1940; s. Fred Lucas and Christiana H. (Showacre) A.; m. Marianne Ellis, June 23, 1963; children: Keith Duane, Kristin Marianne. B.S., Pa. State U., 1962; M.D., Johns Hopkins U., 1966. Diplomate: Am. Bd. Pediatrics. Intern Johns Hopkins Hosp., Balt., 1966-67, resident, 1967-68, fellow, 1970-71; commd. officer USPHS, 1968—, now rear adm.; clin. assoc. Nat. Inst. Child Health and Human Devel., NIH, Bethesda, Md., 1968-70, asst. to sci. dir., 1971-74, asst. to dir., 1978-82, dep. dir., 1982-86; dir. Nat. Inst. Child Health and Human Devel., NIH, 1986—; staff pediatrician Nat. Commn. for Protection of Human Subjects of Research, 1974-78. Contbr. articles to profl. jours. Recipient Commendation medal USPHS, 1970, Meritorious Service medal USPHS, 1985, Spl. Recognition medal USPHS, 1985, Surgeon Gen's. Exemplary Svc. medal, 1990, Irving B. Harris Lectureship award, 1991. Fellow Am. Acad. Pediatrics, Soc. Devel. Pediatrics, Am. Pediatric Soc., Assn. for Retarded Citizens. Methodist. Office: Nat Inst Child Health Bldg 31 9000 Rockville Pike Rm 2A03 Bethesda MD 20892-0001

ALEXANDER, HAROLD, bioengineer, educator; b. N.Y.C., Nov. 12, 1940; s. Jack and Freda (Koltun) A.; m. Sheila M. Eisner, Dec. 20, 1964; children: Robin, Andrea. B.S., NYU, 1962, M.S., 1963, Ph.D., 1967. Asst. research scientist NYU, Bronx, 1966-67, assoc. research scientist, 1967-68; asst. prof. Stevens Inst. Tech., Hoboken, N.J., 1968-71, assoc. prof., 1971-77, co-dir. med. engring. lab., 1973-77, head lab. balloon tech., 1968-74; assoc. prof. dept. surgery, dir. G.L. Schultz Labs. for Orthopedic Research N.J. Med. Sch., Newark, 1977-81, prof., 1981-86; dir. dept. bioengring. Hosp. for Joint Diseases Orthopaedic Inst., N.Y.C., 1986—; lectr. in pediatrics Mt. Sinai Sch. of Medicine, N.Y.C., 1975-77; cons. Johnson & Johnson Research Labs., New Brunswick, N.J., 1975-76; vis. prof. Coll. Engring., Rutgers U., 1975-76; adv. on fabrication of balloons USAF, 1968-74; v.p. C.A.S., Inc., 1974-78, pres., 1978-83; dir. Orthomatrix, Inc., 1984-86; sr. advisor CAS Med. Systems, 1984; prof. orthopaedic surgery NYU Sch. Medicine, 1987—, asst. prof. occupational health and safety, 1987—. Associate Biomaterials sect. editor Jour. Biomed. Materials Research, 1987-89; editor: Jour. Applied Biomaterials, 1990—; mem. editorial bd. Jour. Investigative Surgery, 1987—; contbr. articles to profl. jours., chpts. to books; researcher cardiovascular and orthopedic bioengring. Mem. ASME, Soc. Biomaterials (Founders award 1987), N.J. Orthopedic Soc., Orthopedic Research Soc., ASTM. Achievements include being instrumental in development of new system for measurement of infant blood pressure and absorbable composites for orthopedic implant use. Home: 47 Elmwood Pl Short Hills NJ 07078-3320 Office: Hosp for Joint Diseases/Orthopedic Inst Dir Dept Bioengring 301 E 17th St New York NY 10003-3804

ALEXANDER, JOHN CHARLES, editor, writer; b. Lincoln, Nebr., Jan. 25, 1915; s. John Merriam Alexander and Helen (Abbott) Boggs; m. Ruth Edna McLane, Aug. 20, 1955. Student, U. Nebr., 1933-37, Chouinard Art Inst./Ben Bard Playhouse Sch., L.A., 1937-38, Pasadena Playhouse, 1939-42, UCLA, 1945-47. Aircraft assembler N. Am. aviation, Inglewood, Calif., 1941-42; engring. writer Lockheed-Vega Aircraft, Burbank, Calif., 1942-45; prodn. mgr/actor Gryphon Playhouse, Laguna Beach, Calif., 1947-49; asst. producer/writer Young & Rubicam/ABC, Hollywood, Calif., 1949-51; editor-in-chief Grand Cen. Aircraft, Tucson, 1952-53; sr. writer/editor various cos., Calif., 1953-60; sr. editor/writer, sec. Sci. Guidance Rsch. Coun. Stanford Rsch. Inst., U.S. Army Combat Devel. Command, Menlo Park, Calif., 1962-66; editor-in-chief Litton Sci. Support Lab. USACDC, Fort Ord, Calif., 1966-70; editorial dir./sec. The Nelson Co., Film and Video Prodn., Tarzana, Calif., 1971—; editorial cons., dir. Human Resources Rsch. Office George Washington U., The Presidio, Monterey, Calif., 1960-62; book editor The Dryden Press, Hinsdale, Ill., 1971-72; book editor/adaptor Gen. Learning Press, Silver Burdette Co., Morristown, N.J., 1972-74; contbg. editor West Coast Writers Conspiracy mag., Hollywood, Calif., 1975-77; contbg. editor/book reviewer Santa Ynez Valley Times, Solvang, Calif., 1976-77; editorial cons. to author of Strangers in Their Land: CBR Bombardier, 1939-45; participant Santa Barbara Writers Conf., Montecito, Calif., 1974, 75. Author: (TV plays) Michael Has Company for Coffee, 1948, House on the Hill, 1958, (radio drama) The Couple Next Door, 1951; co-author nine films for U.S. Dept. Justice: Under the Law, Parts I and II, 1973; co-author 10 films for Walt Disney Ednl. Media Co.: Lessons in Learning, Parts I and II, 1978-81; author: (with others) The American West Anthology, 1971; editorial cons. Strangers in Their Land: CBI Bombardier, 1939-45, 1990-92. Recipient award for short story, Writer's Digest, 1960, 61, Gold award, The Festival of the Americas, Houston Internat. Film Festival, 1977. Mem. Nat. Cowboy Hall of Fam, Nat. Geog. Soc., Nat. Soc. Lit. and Arts, Am. Film Inst., Western Hist. Soc., Calif. Acad. Sci., Nat. Air and Space Mus., Smithsonian Instn., Woodrow Wilson Internat. Ctr. for Scholars, Aircraft Owners and Pilots Assn., Air Force Assn., Sigma Nu, Alpha Phi Omega. Avocations: scale model building, environmental/wildlife conservation, aviation, science, fgn. affairs, intelligence. Home: 23123 Village 23 Camarillo CA 93012-7602

ALEXANDER, JOHN ROBERT, hospital administrator, internist; b. Tulsa, July 28, 1936; s. Hiram Marshall and Roberta Alice (Greene) A.; m. Marjorie Louise Okeson, Aug., 1958; children: Stephanie Maine, Paul Fulton, James Marshall, Cynthia Ann, Karen Louise, Robert Thomas. BS, U. Okla., 1958; MD, U. Okla., Oklahoma City, 1961. Intern St. John's Hosp., Tulsa, 1961-62, 1961-62; resident Meth. Hosp. of Dallas, Tex., 1962-65; resident in internal medicine Methodist Hosp. of Dallas, 1962-65; pres. Tulsa Internists, Ltd., 1974-89, Wheeling Med. Group, Inc., Tulsa, 1989-90; mem. staff St. John Med. Ctr., Tulsa, chief of staff, 1977-78, v.p. med. affairs, 1991—; vice chmn., bd. dirs. Physicians Liability Ins. Co., Oklahoma City; bd. dirs. adv. coun. Okla. Bd. Nurse Registration and Nursing Edn., Oklahoma City; clin. prof. U. Okla. Coll. Medicine, 1974. Editor (report) Med. Edn. in Tulsa County, 1981. Pres. Tulsa County Heart Assn., 1970-71; elder Kirk of the Hills Presbyn. Ch., Tulsa, 1970—; med. dir. Wright City (Okla.) Free Health Clinic, 1975-80; chmn. bd. dirs. Tulsa County Health Dept., 1982. Recipient Disting. Svc. award Am. Heart Assn., 1971; named Friend of Nursing, Okla. Nurses Assn., 1990. Fellow ACP; mem. AMA (del. 1985—), Okla. State Med. Assn. (pres. 1989-90), Tulsa County Med. Soc. (pres. 1983), Tulsa Internists Soc.(pres. 1972), U. Okla. Alumni Assn. (pres. 1982-83), Rotary Club Will Rogers (pres. 1973-74), Alpha Omega Alpha. Republican. Avocations: racquetball, photography. Home: 6733 S Gary Ave Tulsa OK 74136-4515 Office: St John Med Ctr 1923 S Utica Ave Tulsa OK 74104-6502

ALEXANDER, JONATHAN, cardiologist, consultant; b. N.Y.C., Nov. 29, 1947; s. Josef and Hannah (Margolis) A.; m. Karen Deborah Einhorn, Aug. 8, 1971; children: Jessica Beth, Daniel Lewis, Benjamin Joel. BA, Harvard U., 1968; MD, Albert Einstein Coll. Medicine, 1973. MD. Intern, resident Yale-New Haven Hosp., 1973-76; fellow dept. cardiology Sch. Medicine Yale U., New Haven, 1976-78, asst. clin. prof. medicine, 1978-83, assoc. clin. prof. medicine, 1983—; attending physician Danbury (Conn.) Hosp., 1978—, West Haven (Conn.) Vets. Hosp., 1978—, New Milford (Conn.) Hosp., 1980; dir. cardiac rehab. unit and nuclear cardiology Danbury (Conn.) Hosp., 1978—. Recipient Samuel Kushlan award Yale-New Haven Hosp., 1974, Revlon award 11th Internat. Congress Chemotherapy, 1983. Fellow ACP, Am. Coll. Cardiology (gov.-elect for Conn. 1992—), Am. Coll. Chest Physicians, Am. Heart Assn. (coun. clin. cardiology); mem. Soc. Nuclear Medicine, Alpha Omega Alpha. Jewish.

ALEXANDER, JOSEPH KUNKLE, JR., physicist; b. Staunton, Va., Jan. 9, 1940; s. Joseph Kunkle and Charlotte (Harper) A.; m. Diana Lenore Titolo, Sept. 22, 1962; children: Kathryn, Stephen, David. BS in Physics, Coll. William and Mary, 1960, MA in Physics, 1962. Physicist Nat. Bur. Standards, 1960; research asst. Coll. William and Mary, Williamsburg, Va., 1960-62; physicist Goddard Space Flight Ctr., NASA, Greenbelt, Md., 1962-85, head planetary magnetospheres br., 1976-84; dep. chief scientist NASA, Washington, 1985-87, asst. assoc. adminstrs. space sci. and applications, 1987-93; assoc. dir. space scis. Goddard Space Flight Ctr., NASA, Greenbelt, Md., 1993—; vis. scientist U. Colo., 1973-74; sr. policy analyst White House Office Sci. and Tech. Policy, Washington, 1984-85; assoc. chief Lab. Extraterrestrial Physics, 1985, acting dir. life scis., 1992-93. Contbr. articles to sci. and tech. jours. Mem. Am. Geophys. Union, Am. Astronomical Soc., Internat. Astron. Union. Office: NASA Goddard Space Flight Ctr Code 600 Greenbelt MD 20771

ALEXANDER, MARTIN, educator, researcher; b. Newark, Feb. 4, 1930; s. Meyer and Sarah (Rubinstein) A.; m. Renee Rafaela Wulf, Aug. 26, 1951; children: Miriam H., Stanley W. B.S., Rutgers U., 1951; M.S., U. Wis., 1953, Ph.D., 1955. Asst. prof. Cornell U., Ithaca, N.Y., from 1955, now L.H. Bailey prof; advisor agys. fed. govt., Washington, 1965—, UN agys., Kenya, France, Italy, 1963—; mem. coms. Nat. Acad. Sci., Washington, 1971—; cons. Author: Microbial Ecology, 1971, Introduction to Soil Microbiology, 1977; editor: Advances in Microbial Ecology, 5 vols., 1977-81. Recipient Indsl. Research 100 award, 1968, Fisher award Am. Soc. Microbiology, 1980, Superior Svc. award USDA, 1989. Fellow Am. Acad. Microbiology, AAAS, Internat. Inst. Biotechnology, Am. Soc. Agronomy (Soil Sci. award 1964). Home: 301 Winthrop Dr Ithaca NY 14850-1736 Office: Cornell U Bradfield Hall Ithaca NY 14853

ALEXANDER, MELVIN TAYLOR, quality assurance engineer, statistician; b. Greensboro, N.C., June 2, 1949; s. Melvin Taylor and Sabina Mae (Anglin) A.; m. Karen Gwendolyn Davenport, Aug. 22, 1973 (div. 1982); children: Asia Trinicia, Sabina, Melvin Taylor III; m. Lucia Antoinette Ward, Apr. 23, 1983. Student, Guilford Coll., 1967-70; BS in Math., N.C. A&T State U., 1972; MSPH in Biostats., U. N.C., 1979. Registered quality engr. Instr. math. N.C. A&T State U., Greensboro, 1975-77; grad. asst. biostatis. dept. U. N.C., Chapel Hill, 1977-79; rsch. assoc. Sch. Pub. Health, Chapel Hill, 1980-81, jr. statis. analyst, 1981-82; engring. staff asst. Westinghouse Electronic Systems Group, Balt., 1982-83, sr. engr., 1983—; cons. N.C. Dept. Adminstrn., Raleigh, 1979-80, S.C. Conf. Black Mayors, Gifford, 1980. Co-author: Managing Industrial Processes, 1984. USPHS grantee, 1977; U. N.C. Minority Student fellow, 1978. Mem. Am. Statis. Assn., Am. Soc. Quality Control (sec. Balt. sect. 1990-91, vice chmn. 1991-92, chmn. 1992-93), Internat. Soc. for Hybrid Microelectronics. Democrat. Presbyterian. Avocations: music; photography; computers. Office: Westinghouse Electronic Systems Group 7323 Aviation Blvd Baltimore MD 21203-0746

ALEXANDER, NANCY JEANNE, health science facility administrator, researcher; b. Cleve., Dec. 1, 1939; d. Ralph Stanley and Dorotha Mae (Hunt) Hilditch; m. David Frederick Staat, Nov. 28, 1970; children: Benjamin Alexander, Alex Roy. BS, Miami U. of Ohio, 1960, MA, 1961; PhD, U. Wis., 1965. Prof. anatomy, microbiology, immunology, ob-gyn., urology Oreg. Health Scis. U., Portland, 1980-86; scientist Oreg. Primate Rsch. Ctr., Beaverton, 1978-86; prof. dept. ob-gyn. Ea. Va. Med. Sch., Norfolk, Va., 1986-93; expert contraceptive devel. br. nat. inst. child health and human devel. NIH, Bethesda, Md., 1990-93, chief contraceptive devel. br. nat. inst. child health and human devel., 1993—. Mem. Am. Soc. Andrology (pres. 1979-80), Soc. for the Study of Reproduction (bd. dirs. 1980-83), Am. Fertility Soc. (bd. dirs. 1986-89). Office: NICHD Contraceptive Devel 6100 Executive Blvd 8B13 Bethesda MD 20892

ALEXANDER, THERON, psychologist, writer; s. Theron and Mary Helen (Jones) A.; m. Marie Bailey; children: Thomas, Mary. B.A., Maryville Coll., 1935; M.A., U. Tenn., 1939; postgrad. Naval tng., Princeton U., 1943, Harvard U., 1944; Ph.D., U. Chgo., 1949. Asst. prof. psychology Fla. State U., 1949-54; dir. Mental Health Clinic, 1954-57; assoc. prof. psychology in pediatrics U. Iowa, to 1965; prof. psychology in pediatrics U. Miami (Fla.), 1965-66; research prof. Community Studies Ctr. of Temple U., Phila., 1966-68; dir. Child Devel. Research Ctr., 1966-69; prof. human devel., 1966-80; pres. Alexander Assocs., 1980-86; vis. scholar Hoover Instn., Stanford U., 1987—; dir. study tour human devel. programs, facilties and rsch. in govt., industry and univs. Temple U., Holland, France, Switzerland, Italy, Yugoslavia, Germany, England, Sweden, Denmark, 1969, univ. study leave for travel and writing, England, 1972, study tour in Soviet Union, 1974; invited lectr. Internat. Symposium, Brazil, 1977. Author: Psychotherapy in Our Society, 1963, Children and Adolescents, 1969, Human Development in an Urban Age, 1973, El Desarrollo Humano en la Epoca del Urbanismo, 1978, (with others) Developmental Psychology, 1980; sr. author: Psicologia evolutiva, 1984; contbr. articles to Ency. of Psychology; rsch. paper Stanford Sch. Medicine, 1989, 93. Staff of commdr. USN, World War II, PTO. Recipient cert. Gov. State of Sao Paulo (Brazil), 1977; Legion of Honor, 1979. Mem. APA (cert. disting. contbn. clin. div. 186), Am. Psychol. Soc. (charter fellow). Address: 350 Sharon Park Dr C3 Menlo Park CA 94025

ALEXANDER, WAYNE ANDREW, product engineer; b. Phila., Feb. 26, 1951; s. George H. and Lillie I. (Radcliffe) A.; m. Cheryl P. Walker, June 26, 1976; children: Michelle Christina, Michael Andrew. BS in Mfg. Engring. Tech., Spring Garden Coll., 1987. Moldmaker Linder-Koeger, Feasterville, Pa., 1978-82; chief inspector Oppenheimer Precision, Horsham, Pa., 1982-84; systems engr. Rapistan divsn. LSI, Willow Grove, Pa., 1984-89; devel. specialist Stabilus, Colmar, Pa., 1989—. Elder Faith Community Ch., Roslyn, Pa., 1992—; bd. dirs. Mission Projects Fellowship, Willow Grove, 1993—. With USN, 1970-72. Mem. Soc. Plastics Engrs., Tau Alpha Pi, Alpha Chi. Democrat. Achievements include invention, design and prodn. of prototype of volume compensation separator for hydraulic damper. Home: 902 Truman Ct Warrington PA 18976 Office: Stabilus 92 County Line Rd Colmar PA 18915

ALEXANDER, WILLIAM ROBERT, chemist; b. Huntington, W.Va., Jan. 27, 1967; s. Norman Robert and Letitia Anne (Chamberlain) A. BS in Physics, Bethany Coll., 1989; MS in Phys. Sci., Marshall U., 1992. Chemist Ashland Petroleum Co., Ashland, Ky., 1991--. Mem. Am. Chem. Soc., Soc. for Applied Spectroscopy, Soc. Amateur Radio Astronomers. Office: Ashland Petroleum Co PO Box 391 Ashland KY 41114

ALEXANDER-BRIDGES, MARIA CARMALITA, medical researcher; b. Detroit, June 12, 1952; d. Archie and Gladys Maria (Boykin) A.; m. Kenneth Roland Bridges, Aug. 9, 1975; 1 child, Camille Maria. BA magna cum laude, Mt. Holyoke Coll., 1973; MD, Harvard Med. Sch., 1980, PhD, 1982. Intern Johns Hopkins Hosp., Balt., 1980-81, resident in internal medicine, 1981-82; rsch. fellow Mass. Gen. Hosp., Boston, 1982-84; instr. in medicine Harvard Med. Sch., Boston, 1984-87, asst. prof. medicine, 1987-92, assoc. prof., 1992—; cons., awardee Lilly Rsch. Labs., Indpls., 1989—. Contbr. articles to profl. jours. Co-chair Hinton-Wright Soc., Boston, 1990—.

Recipient Kaiser/NMF Merit fellowship Kaiser Found., 1980, Eli Lilly award, 1989; named Rsch. Assoc., Howard Hughes Med. Inst., 1983-85, Asst. investigator, 1985—. Mem. AAAS, Am. Diabetes Assn., Internat. Diabetes Fedn. Congress, Juvenile Diabetes Found. Internat., Endocrine Soc. Democrat. Methodist. Achievements include cloning of insulin-sensitive transcription factor. Home: 36 Greylock Rd Newtonville MA 02160 Office: Howard Hughes Med Inst Mass Gen Hosp Fruit St Boston MA 02114

ALEXANDRE, GILBERT FERNAND A.E., surgeon; b. Couillet, Belgium, Sept. 19, 1944; s. Fernand Clement and Yvonne (Everard) A.; m. Monique Debienne; 1 child, Luc. MD, Liege U., Belgium, 1969, Surgeon, 1975, Lic. Med. de Exper., 1987. Resident supr., cons. surgeon Liege U. Hosp., 1970-82; gen. orthopedic surgeon Clinique St. Rosalie, Liege, 1978—, Clinique St. Joseph, Liege, 1975—, Clinique de Hermalle, Liege, 1987—. Fellow Internat. Coll. Angiology; mem. Group for Advancement Microsurgery, Soc. Belge Medicine Trafic, Belgium Orthopedic Soc., Belgian Hand Group, Collegium Internat. Chirurgiae Digestivae, Am. Back Soc., Internat. Arthroscopy Assn., Am. Back Surgery Soc. Avocations: music, computing. Home: Rue de Lyser 33, 4430 Ans Belgium Office: Clinique St Rosalie, Rue Des Wallons 72, 4000 Liège Belgium

ALEXANDROPOULOS, NIKOLAOS, physics educator; b. Ithaki, Greece, Aug. 9, 1934; s. Gerassimos N. and Ekaterini G. (Zaverdinou) A.; m. Tina Chatzigeorgiou, Dec. 28, 1964; children: Gerassimos, Karima. Diploma in physics, U. Athens, Greece, 1960, DSc, 1964. Rsch. assoc. Syracuse (N.Y.) U.; asst. prof. physics Poly. Inst. Bklyn.; mem. tech. staff Aerospace Corp., L.A.; prof. physics U. Ioannina, Greece, 1977—; vis. assoc. prof. physics U. Houston; vis. assoc. porf. material sci. Rice U., Houston; prof. adj. physics U. Estadua de Campinas, Sao Paulo, Brazil; adj. rsch. prof. physics Poly. U., Bklyn., 1988; cons. Bell Labs., Murray Hill, N.J., 1971-72, 88; guest scientist Brookhaven Nat. Lab. (NSLS), Upton, N.Y., 1988; SERC visitor U. Warwick (Eng.), 1984; dir. solid state div. U. Ioannina, 1984—. Mem. Athens Acad. (corr.), European Syncrotron Rad. Soc. (exec. com. 1991), Sigma Xi. Orthodox. Home: Chaonon 10, 45221 Ioannina Greece Office: U Ioannina Dept Physics, PO Box 1186, 45110 Ioannina Greece

ALFANO, MICHAEL CHARLES, pharmaceutical company executive; b. Newark, Aug. 8, 1947; s. Michael Ferdinand and Anne Marie (Barrington) A.; m. JoAnn Mary Coletta, Mar. 30, 1969; children: Michael Anthony, Kristin Lynn. Student, Rutgers U., 1965-67; DMD, U. Medicine and Dentistry of N.J., 1971; postgrad. in periodontics, Harvard U., 1971-74; PhD, MIT, 1975. Asst. prof. dentistry Fairleigh Dickinson U., Hackensack, N.J., 1974-77, assoc. prof., 1977-80, prof. with tenure, 1980-82, dir. Oral Health Research Ctr., 1977-82, asst. dean grad. affairs and research, 1981-82; v.p. dental research Block Drug Co., Inc., Jersey City, 1982-84, sr. v.p. R&D, 1987—, bd. dirs., 1988—, pres. dental products div., 1985-88, cons. office of chief exec., 1990—; cons. Nat. Inst. Dental Research, Bethesda, Md., 1976-82; apptd. vis. prof. Nat. Dairy Council, Chgo., 1981; vis. sr. scientist Fairleigh Dickinson U., 1982-88; adj. prof. U. Medicine and Dentistry of N.J., Newark, 1985—; mem. sci. adv. council Office of Gov., State of N.J., 1981-84. Editor: Symposium on Nutrition, 1976; contbr. articles to profl. jours. and chpts. to books; patentee in field. Trustee Found. of U. Medicine and Dentistry of N.J., 1988—; adv. bd. Columbia U. Sch. Dental and Oral Surgery, 1990—; mem. program com. Am. Fund for Dental Health, 1991-93; bd. overseers Forsyth Dental Ctr., Boston, 1992—, U. Pa. Coll. Dental Medicine, 1992—. Recipient Leadership citation Newark YMCA, 1966, Disting. Alumnus award U. Medicine and Dentistry of N.J., 1986; NIH research grantee, 1974-82; NIH postdoctoral fellow, 1971-74. Fellow Am. Coll. Dentists, Am. Coll. of Prosthodontists (hon. fellow); mem. ADA (cons., Nat. Achievement award 1978), Internat. Assn. for Dental Rsch., Am. Assn. for Dental Rsch. (pres. N.J. chpt. 1985), Am. Inst. Nutrition. Democrat. Roman Catholic. Achievements include 5 patents, 2 patents pending; discovery of role of Vitamin C in mucous membrane barrier function. Home: 954 Arapaho Trl Franklin Lakes NJ 07417-2258 Office: Block Drug Co Inc 257 Cornelison Ave Jersey City NJ 07302-3113

ALFONSECA, MANUEL, computer scientist, educator; b. Madrid, Apr. 24, 1946; s. Manuel and Carmen (Moreno) A.; m. Maria Angel Cubero, Nov. 24, 1971; children: Maria de los Angeles, Enrique. Dr. in Electronics Engring., Poly. U. Madrid, 1971, MS in Computer Sci., 1976. Researcher Ministry of Edn., Madrid, 1970-71; computer scientist U. Computing Ctr., Madrid, 1971-72; asst. prof. Poly. U., Madrid, 1971-86; researcher IBM, Madrid, 1972-86; prof. Poly. U., Madrid, 1986—; sr. tech. staff mem. IBM, Madrid, 1986—; sci. collaborator La Vanguardia newspaper, Barcelona, Spain, 1983—, Santillana Pub., Madrid, 1985-87. Author 12 books children's lit., 1986— (Lazarillo award 1988), 7 books computer sci., 1971—, 13 software products, 1979—; contbr. articles to profl. jours. Recipient Nat. Graduation award Spanish Govt., 1970, Cross of Alfonso X El Sabio award Spanish Govt., 1970. Mem. IEEE, N.Y. Acad. Scis., Spanish Assn. Telecommunication, Spanish Assn. Sci. Journalists, Assn. Computer Machinery, British APL Assn., IBM Internat. Tech. Liaison. Roman Catholic. Avocations: piano playing, classical music, writing for children. Home: Belianes 2, Madrid 28043, Spain Office: IBM Spain, Santa Hortensia 26-28, Madrid 28002, Spain

ALFONSI, WILLIAM E., interior designer, funeral industry consultant; b. Niles, Ohio, Jan. 27, 1923; s. Pacifico Tobia and Carmela (D'Angelo) A.; m. Adrell A. Alfonsi; children: Gary William, Pamela Jane, Scott Allan, Kevin Lawrence. Student, Miami U., Oxford, Ohio, 1942. Interior design asst. Carey W. Sims Inc., Cleve., 1946-47; interior designer Masticks, Inc., Cleve., 1947-49; dir. interior planning and decoating div. Superior Funeral Supply, Cleve., 1949-69; pres. Country Furniture Store, Inc., Kinsman, Ohio, 1969—, Custom Planned Funeral Interiors, Kinsman, Ohio, 1969—, William Alfonsi, Inc., Kinsman, Ohio, 1969—, Heritage Funeral Equipment, Kinsman, Ohio, 1969—. Sgt. U.S. Army, 1942-45. Mem. Am. Legion, VFW. Democrat. Methodist. Avocation: working with young people with drug, alcohol or related problems. Home: 9234 N Kingsville Rd Farmdale OH 44417-9750 Office: William Alfonsi Inc PO Box 77 Kinsman OH 44428-0077

ALFORD, JOSEPH SAVAGE, JR., chemical engineer, research scientist; b. Lynn, Mass., Oct. 18, 1943; s. Joseph Savage and Mary Elizabeth (Pearson) A.; m. Martha Louise Boyd, June 15, 1968; children: Michael, David, Randy. BS in Chem. Engring., Purdue U., 1966; MS, U. Cin., 1972, PhD, 1972. Registered profl. engr., Ind. Chem. engr. Eli Lilly & Co., Indpls., 1968, sr. scientist, 1972-79, rsch. scientist, 1979-86, sr. rsch. scientist, 1986—. Contbr. chpts. to books, articles to Can. Jour. Microbiology, Biotech. and Bioengring., Pharm. Tech., Jour. Indsl. Microbiology, others. Deacon, Sunday sch. tchr. Southport Presbyn. Ch., Indpls., 1979—; cubmaster, troop leader Boy Scouts Am., Indpls., 1980-90; soccer, baseball and basketball coach, Indpls., 1978-88. Lt. comdr. USNR, 1966-76, Viet Nam. Dupont fellow, NASA fellow, 1968-72. Mem. Instrument Soc. Am., Sigma Xi. Achievements include development of a hierarchical process control/host computer system which has become the company's standard for fermentation processes; invention of a new parameter, called Ro, used in fermentation industry and development of applications using real time artificial intelligence. Office: Eli Lilly & Co Lilly Corp Ctr Indianapolis IN 46285

ALFREDSSON, MATS LENNART, physics researcher, administrator; b. Gothenburg, Sweden, June 1, 1946; s. Sune Lennart and Marta Henrietta (Sjögren) A.; m. Sara Ingalill Johansson; children: Anna, Frida, Julia, Maria. Diploma mkt. economics, U. Gothenburg, 1973. Lab. engr. dept. electronic physics U. Tech., Gothenburg, 1967-70; rsch. engr. Onsala Space Observatory Chalmers U. Tech., Rao, Ousala, Sweden, 1971-75; mng. dir. Analytical Standards AB, Kungsbacka, Sweden, 1976-87, Referensmaterial AB, Ulricehamn, Sweden, 1988—. Mem. ASTM, Swedish Soc. for Materials Tech., Swedish Chem. Soc. Home: Lobeliav 6, S-523 33 Ulricehamn Sweden Office: Referensmaterial AB, Lobeliav 6, S-523 33 Ulricehamn Sweden

ALFVÉN, HANNES OLOF GOSTA, physicist; b. May 30, 1908. Ph.D., U. Uppsala, 1934. Prof. theory of electricity Royal Inst. Tech., Stockholm, 1940-45, prof. electronics, 1945-63, prof. plasma physics, 1963-73; prof. dept. applied physics, electrical engring. and info. sci. U. Calif., San Diego, 1967—; mem. Swedish Sci. Adv. Council, 1963-67; past mem. Swedish AEC; past gov. Swedish Def. Research Inst., Swedish Atomic Energy Co.; past sci. adv. Swedish Govt.; pres. Pugwash Confs. on Sci. and World Affairs, 1970-75; mem. panel on comets and asteroids NASA. Author: Cosmical Electrodynamics, 1950; On the Origin of the Solar System, 1954; Cosmical Electrodynamics: Fundamental Principles, 1963; Worlds-Antiworlds, 1966; The Tale of the Big Computer, 1968; Atom, Man and the Universe, 1969; Living on the Third Planet, 1972; Evolution of the Solar System, 1976; Cosmic Plasma, 1981. Recipient Nobel prize for physics, 1970; Lomonsov gold medal USSR Acad. Scis., 1971; Franklin medal, 1971, Bowie Gold medal Am. Geophysical Union, 1987. Fellow Royal Soc. (Eng.); mem. Swedish Acad. Scis., Akademia NAUK (USSR), NAS (fgn. assoc.), others. Office: Royal Inst Tech, Alfvén Lab Dept Plasma Physics, S-100-44 Stockholm 70, Sweden

ALGORA DEL VALLE, CARLOS, physics educator, researcher; b. Madrid, Jan. 11, 1962; s. Dionisio and Aurora (del Valle) Algora. Degree in Physics, Facultad de Fisicas, Madrid, 1985, PhD, 1990. Rschr. Inst. Energia Solar, Madrid, 1985-91, project leader, 1992—. Contbr. articles to profl. jours. Cpl. Spanish Infantry, 1981-83. Recipient grant Spanish Govt., 1986-88. Achievements include research in high efficiency GaAs solar cells (among highest efficiencies recorded in Europe) and developments in GaAs IRED's for non-guided optical communications. Office: Inst Energia Solar, Ciudad Universitaria s/n, Madrid 28040, Spain

ALHADEFF, JACK ABRAHAM, biochemist, educator; b. Vallejo, Calif., May 9, 1943. BA, U. Chgo., 1965; PhD in Biochemistry, U. Oreg., 1972. Fellow U. Calif., San Diego, 1972-74, asst. rsch. neuroscientist, 1974-75, from asst. prof. to assoc. prof., 1975-82; prof. biochemistry Lehigh U., 1982—. Recipient Rsch. Career Devel. award NIH, 1978. Mem. AAAS, Am. Chem. Soc., Biochem. Soc., Sigma Xi. Achievements include research in biochemical studies on glycoconjugate metabolism in normal and pathological (cancer, diabetes, cystic fibrosis) human tissues. Office: Lehigh U Inst Health Sciences Bethlehem PA 18015*

AL-HAKIM, ALI HUSSEIN, chemist; b. Najaf, Iraq, Oct. 21, 1951; came to U.S., 1988; s. Hussein and Najat (Moussa) Al-H.; m. Juliet Hilda Pacey, July 21, 1984; 1 child, Adam. Diploma, U. East Anglia, Norwich, Eng., 1979, PhD, 1983. Scientist John Innes Rsch. Inst., Norwich, 1985-88; rsch. assoc. U. Iowa, Iowa City, 1988-91; sr. scientist Wyeth-Ayerst Rsch., Rouses Point, N.Y., 1991—. Co-author: Biocatalysis for Industry, 1991, Biotechnology and Polymers, 1991; contbr. articles to sci. jours. Mem. Am. Chem. Soc., Am. Electrophoresis Soc., Royal Soc. Chemistry. Achievements include patents for non-radioactive nucleic acid hybridization probes, new method for sequencing carbohydrate on gel using charged fluorescent conjugate. Home: 1-D Sandra Ave Plattsburgh NY 12901-2409 Office: Wyeth-Ayerst Rsch 64 Maple St Rouses Point NY 12979

ALHANATI, SHELLEY, clinical psychologist; b. N.Y.C., July 21, 1959. MS in Biochemistry, UCLA, 1983; PhD in Psychology, Calif. Sch. Profl. Psychology, 1989. Rsch. biochemist Molecular Biology Inst., UCLA, 1981-83; genetic engr. Salk Inst. Biotech., La Jolla, 1983-85; clin. psychologist Reiss Davis Child Study Ctr., L.A., 1988—; pvt. practice in psychotherapy Beverly Hills, 1987—; guest speaker So. Calif. Conf. on Genetic Engring., 1982. Fellowship UCLA, 1981-83. Mem. APA, L.A. County Psychol. Assn. Achievements include development of bioreactor tech. for synthesis of neurotransmitters and their analogs, new strain of Pseudomonas for use in oil spill cleanups, novel shuttle vector for use in cloning expts. on biosynthetic regulatory mechanisms, molecular genetics of psychiatric disorders. Office: 435 N Bedford Dr # 404 Beverly Hills CA 90210

AL-HASHIMI, IBTISAM, oral scientist, educator; b. Karbala, Iraq, July 1, 1951; came to U.S., 1981; d. Hadi A. and Rabab H. Al-H.; m. Alkhafaji, Aug. 26, 1988. B Dental Sci., Sch. Dentistry, Baghdad, 1973; MS, SUNY, Buffalo, 1985, PhD, 1989. Diplomate in Oral Surgery. Registrar Sch. Dentistry, Baghdad, 1975-81; postdoctoral assoc. SUNY, Buffalo, 1984-88, asst. prof., 1988-89; asst. prof. U. Pacific, San Francisco, 1989-90; dir. stomatology lab. Baylor Coll. Dentistry, Dallas, 1991—, dir. salivery dysfunction clinic, 1992—; adv. com. SS Found. (western N.Y. chpt.) Buffalo, 1985-89, Dallas-Ft. Worth chpt., 1992—. Author: Proceeding of the Second Dows Symposium, 1987; contbr. articles to profl. jours. Mem. Am. Assn. Dental Rsch., Internat. Assn. Dental Rsch., N.Y. Acad. Sci., Salivery Rsch. Group, Sigma Xi. Muslim. Achievements include research on molecular mechanisms of salivary gland diseases; characterization of a major salivary enzyme inhibitor in the mouth; identification of the principal protein components that participate in the formation of the protective coat of the teeth of healthy subjects. Office: Baylor Coll Dentistry 3302 Gaston Ave Dallas TX 75246

ALI, ASHRAF, civil engineer; b. Solemanpur, Bangladesh, June 4, 1955; s. Romjan and Hawa (Khatun) A.; m. Shayla Sheikh, Apr. 22, 1979; 1 child, Aleef Nayla. MS, Rensselaer Poly. Inst., 1983; DSc, Washington U., St. Louis, 1986. Registered profl. engr., Pa. Lectr. Engring. Univ., Dhaka, Bangladesh, 1979-81; rsch. asst. Rensselaer Poly. Inst., Troy, N.Y., 1981-83, Washington U., St. Louis, 1983-86; rsch. engr. Ill. Dept. Transp., Springfield, 1987-88; sr. rsch. engr. Swanson Analysis Systems, Inc., Houston, Pa., 1988—; presenter confs. Structural Stability Rsch. Coun., 1985, Internat. Conf. on Steel and Aluminum Structures, 1987, Internat. Symposium on BEM, 1989, Ga. Inst. Tech., 1990, 2d Internat. Congress on Recent Devel. in Air and Structure-borne Sound and Vibration, 1992. Contbr. articles to Internat. Jour. Solids and Structures, Jour. Engring. Mechs., Jour. Structural Engring., Earthquake Engring. and Structural Dynamics, Internat. Jour. for Numerical Methods in Engring., Comms. in Applied Numerical Methods, Jour. Acoustical Soc. Am. founder Bangladesh Devel. Initiative, Pitts., 1988—. Mem. ASCE (assoc.), Internat. Soc. for Boundary Elements, Internat. Assn. for Boundary Element Methods. Muslim. Home: 6042 Pennwood Ct Bethel Park PA 15102 Office: Swanson Analysis Systems PO Box 65 Johnson Rd Houston PA 15342

ALI, KHWAJA MOHAMMED, aeronautical engineer; b. Lahore, Punjab, Pakistan, June 13, 1948; came to U.S., 1989; s. Khwaja Mohammed and Rashida (Omar) Savi; m. Yasmin Khan, Aug. 7, 1976; children: Mehvash, Babar. B Aero. Engring., U. Karachi, Pakistan, 1970; MBA in Aviation Mgmt., Dowling Coll. Registered profl. engr., Pakistan. Supr. Pakistan Internat. Airlines, Karachi, 1970-74, aircraft engr., 1974-75, devel. engr., 1975-78; sta. engr. Pakistan Internat. Airlines, Islamabad, Pakistan, 1979-82; programs engr. Pakistan Internat. Airlines, Karachi, 1982-89; maintenance mgr. Pakistan Internat. Airlines, N.Y.C., 1989—; project advisor for engring. univ. students on Hovercraft and Cockpit Armour Protection. Contbr. articles to profl. publs. Bd. dirs. Adventure Found. Pakistan, Karachi, 1986-88; pres. Marine Club, Karachi, 1988-89; leader Indus River Expedition, Pakistan, 1978, Swat White River Expedition, Pakistan, 1981. Fellow Instn. Engrs. (vice chmn. 1984-89); mem. AIAA (sr. mem.), Royal Aero. Soc. (aero. program sec. 1985-89), Delta Mu Delta. Islam. Achievements include design and fabrication of single seater aircraft. Home: 57-47 226 St Bayside NY 11364 Office: Pakistan Internat Airlines JFK Airport Bldg 52 Jamaica NY 11430

ALI, MOHAMMAD FARHAT, chemistry educator; b. Hyderabad, India, Mar. 14, 1938; s. Mohammad Ishaq and Amirunnissa Ali; m. Syeda, Aug. 24, 1969; children: Akbar, Qudsia, Azam. BS, Osmania U., Hyderabad, 1958; MS, Karachi (Pakistan) U., 1960; PhD, St. Andrews (Scotland) U., 1968. Chemist Hunting Tech. Svcs., Herts, Eng., 1961-64, Imperial Chem. Industries, Runcorn, Eng., 1964-65; lab. mgr. Nat. Refinery Ltd., Karachi, 1968-77; asst. prof. U. Petroleum & Minerals, Dhahran, Saudi Arabia, 1977-80, assoc. prof., 1980-86, prof., 1987—; dir. oil testing ctr. King Fahd U. Petroleum and Minerals, Dhahran, 1977—. Contbr. articles to profl. jours. Fellow Inst. Petroleum London; mem. Am. Chem. soc., ASTM. Avocations: reading, photography, gardening. Home: KFUPM Campus, Dhahran Saudi Arabia Office: King Fahd U Petroleum, KFUPM Box 76, Dhahran 31261, Saudi Arabia

ALI, NAUSHAD, physicist; b. Dubra, India, Mar. 10, 1953; came to U.S., 1986; s. Abdul and Khairun A.; m. Donna Mae, Aug. 20, 1983; children: Nadia, Diana. MS, Meml. U. Can., 1978; PhD, U. Alberta, 1984. Asst. prof. Southern Ill. U., Carbondale, 1986-90, assoc. prof., 1990—. Contbr. articles to profl. jours. Postdoctoral fellow McMaster U., Hamilton, Can., 1985-86. Mem. Am. Phys. Soc. Office: Southern Ill U Dept Physics Carbondale IL 62901

ALI, SYED IBRAHIM, civil engineer; b. Rampur, U.P., India, Apr. 29, 1941; s. Syed Yad and Rashida (Begum) A.; m. Begum Parveen Ibrahim, Feb. 14, 1971; children: Syed Ashar, Syeda Ashi, Syeda Mona Jabeen, Syed Adnan. BS, Aligarh (India) Muslim U., 1961, BS in Civil Engring., 1966. Asst. civil engr. Airports Devel. Agy., Karachi, Pakistan, 1967-73, exec. engr., 1973-75, planning engr., 1976; airport engr. Ministry Communication, Binghazi, Libya, 1975-76; div. mgr. quality control Abbasi-Ada-NC Joint Venture, Dhahran, Saudi Arabia, 1976-78, mgr. quality control, 1978-79; resident mgr. ind. engring. testing lab. Alhuseini Ada (SA) Ltd., Alkhobar, Saudi Arabia, 1979-80; chief engr. quality control housing project Alhuseini Ada (SA) Ltd., Hafar Al Batin, Saudi Arabia, 1980-83; project dir. rd. constrn. Alhuseini Ada (SA) Ltd., Shedgum, Saudi Arabia, 1983-84; resident dir. ind. engring. labs. and materials cons. svcs. Alhuseini Ada (SA) Ltd., Alkhobar, Saudi Arabia, 1984-92. Mem. ASTM, Am. Concrete Inst., Soc. Am. Mil. Engrs., S.E. Asian Soc. Soil Engrs., Pakistan Inst. Engrs., Pakistan Soc. Profl. Engrs., Pakistan Engring. Coun. (life). Avocations: music, cricket, reading, photography, driving. Home: 4 A 8/8 Nazimabad, Karachi Sind, Pakistan Office: Samina Decor Works, PO Box 21146, Sharjah EP 31932, Saudi Arabia

ALI, SYED WAJAHAT, physical chemist, researcher; b. Allahabad, India, July 7, 1937; s. Syed Asghar and Matinun Nisa; m. Akhter Jehan Zubairi, July 29, 1977. BSc with honors, U. Karachi, Pakistan, 1958, PhD, 1981. Tech. asst. Pakistan Coun. Sci. and Indsl. Rsch., Karachi, 1960-63, rsch. physicist, 1963-68, rsch. officer, 1968-76, sr. rsch. officer, 1976-85, prin. rsch. officer, 1985-86; dir. Solar Energy Rsch. Centre, Hyderabad, Pakistan, 1986—. Contbr. articles to Pakistan Jour. Sci. and Indsl. Rsch., others. Mem. Royal Soc. Chemistry, Internat. Solar Energy Soc., Pakistan Inst. Chem. Engring., Pakistan Assn. Sci. and Sci. Profession. Muslim. Achievements include establishment of solar energy research center and solar water desalination plants. Home: 117-B Pechs Block # 2, Karachi 75400, Pakistan Office: Solar Energy Rsch Ctr, PO Box 1021 Latifabad, Hyderabad 71800, Pakistan

AL-IBRAHIM, HUSSAM SADIQ, electrical engineer; b. Baghdad, Iraq, Dec. 6, 1962; came to U.S., 1981; s. Sadiq and Saadiah (Al-safar) A.; m. Lisa Ann Brizendine, Dec. 17, 1989. BS in Sci., U. Louisville, 1988, M. Engring., 1989. Instr. U. Louisville, 1986-91; engr. Safetran Systems Corp., Louisville, 1989—; computer cons. P.C. Solutions, Louisville, 1990—. Mem. IEEE, Ky. Profl. Engrs. Soc. Office: Safetran Systems Corp 1044 E Chestnut St Louisville KY 40209

ALIC, JOHN ANTHONY, engineer, policy analyst; b. Oak Park, Ill., Nov. 24, 1941; s. Anthony and Louise Alic. B of Mech. Engring., Cornell U., 1964; MS, Stanford U., 1965; PhD, U. Md., 1972. Case writer Stanford U., Calif., 1965-66; instr. U. Md., College Park, 1966-72; from asst. to assoc. prof. Witchita (Kans.) State U., 1972-78; project dir. and sr. assoc. Office of Tech. Assessment, Washington, 1979—. Co-author: Beyond Spinoff: Mitary and Commercial Technologies in a Changing Wood, 1992. Mem. Soc. Auto. Engrs. (Wright Bros. medal 1975). Office: US Congress Technology Assessment Washington DC 20510-8125

ALICATA, JOSEPH EVERETT, microbiology researcher, parasitologist; b. Carlentini (Siracusa), Italy, Nov. 5, 1904; came to U.S., 1919, naturalized, 1926; s. Antonio and Concetta (Vaccaro) A.; m. Hannah L. Davis, Jan. 23, 1929 (div. 1954); children: Betty Mae, William D.; m. Earleen E. Moyer, June 30, 1958. AB, Grand Island Coll., 1927; MA, Northwestern U., 1929; PhD, George Washington U., 1934. Jr. zoologist Bur. Animal Industry USDA, Washington, 1928-35; parasitologist Hawaii Agrl. Exptl. Sta. U. Hawaii, Honolulu, 1935-37, head dept. animal scis. Parasitology Lab., 1937-70, prof. emeritus, 1970—; parasitologist Hawaii Dept. Health, Honolulu, 1936-37; parasitologist pub. health commn. Honolulu C. of C., 1940-42, Hawaii Sugar Planters' Assn., Honolulu, 1943; sr. scientist div. internat. health USPHS, Amman, Jordan, 1953-54. Author: (with others) Advances in Parasitology, 1965, Parasites of Man and Animals in Hawaii, 1969, Trichinosis in Man and Animals, 1970; editor: Angiostrongylosis in the Pacific and Southeast Asia, 1970; contbr. numerous articles to profl. jours. With USN, 1953-54. Fulbright scholar, 1950-51; NIH rsch. fellow, 1949-50, grantee. Fellow AAAS; mem. Am. Soc. Parasitologists, Am. Soc. Tropical Medicine and Hygiene, Hawaiian Acad. Scis., Washington Acad. Scis. (hon.), Sigma Xi (hon.), Phi Kappa Phi (hon.). Republican. Achievements include first demonstration of presence of leptospirosis among sugar cane plantation workers in Hawaii; of the efficacy of Daraprim as a suppressive medication for field control of malaria in Jordan; first determination that trichinosis in man in Hawaii derived from consumption of dried sausage prepared from wild pigs; first determination that encysted trichina larvae in pork muscle became sterile when exposed to a dose of 10,000 R or more of gamma irradiation; first diagnosis of human heterophyiasis in Hawaii; first confirmation that rat lungworm was etiological agent of human eosinophilic meningitis; research in epidemiology of endemic typhus in man and rats. Home: 1434 Punahou St Apt 736 Honolulu HI 96822-4729

ALI MOHAMED, AHMED YUSUF, chemistry educator, researcher; b. Manama, Bahrain, Mar. 23, 1951; s. Yusuf Ali Mohamed and Zamzam Ahmed Shiban; m. Fatima A. Rasool, Nov. 2, 1988; 1 child, Reham. BSc, U. Kuwait, 1970-74; PhD, U. Wales, 1979-82; invited scholar, Nagoya (Japan) City U., 1992. Sta. chemist Sitra Power and Desalination Plant, Bahrain, 1974-77; asst. chmn. dept. chem. U. Coll. Arts, Sci. and Edn., Bahrain, 1986-88; asst. prof. U. Bahrain (formerly U. Coll. Arts, Sci. and Edn.), Bahrain, 1983-90; chmn. dept. chem. U. Bahrain, 1988-92, assoc. prof., 1990—. Contbr. articles to profl. jours. Mem. Am. Chem. Soc., Royal Soc. Chemistry, Internat. Assn. Water Quality. Achievements include research on antibacterial activity of a number of cobalt (III) complexes and their inhibitory effects in bacteria, activities of nicotine complexes of rhodium (III) estimation of inorganic particulate matter in the atmosphere. Office: U Bahrain, P O Box 32038, Isa Town Bahrain

ALIPOUR-HAGHIGHI, FARIBORZ, mechanical engineer; b. Tehran, Iran, Oct. 13, 1947; came to U.S., 1977; s. Mansour and Zievar (Khazaie) A.-H.; m. Sorour Hamzepour, May 20, 1976; children: Mohammed, Paimon. MS in Mech. Engring., U. Tehran, 1971; PhD in Energy Engring., U. Iowa, 1981. Instr. U. Tehran, 1972-77; researcher U. Iowa, Iowa City, 1982—, adj. faculty mem., 1985—. Contbr. articles, papers to profl. publs.; author computer program. Grantee NIH, 1992, Nat. Ctr. Supercomputing Applications, 1992. Mem. ASME, Acoustical Soc. Am. Office: U Iowa 334 E SHC Iowa City IA 52242

ALLAIRE, PAUL ARTHUR, office equipment company executive; b. Worcester, Mass., July 21, 1938; s. Arthur E. Allaire and Elodie (LePrade) Murphy; m. Kathleen Buckley, Jan. 26, 1963; children: Brian, Christiana. BSEE, Worcester Poly. Inst.,, 60; MSIA, Carnegie-Mellon U., 1966. Fin. analyst Xerox Corp., Rochester, N.Y., 1966-70; dir. fin. analysis Rank Xerox Ltd., London, N.Y., 1970-73; dir. internat. ops. fin. Xerox Corp., Stamford, Conn., 1973-75; chief staff officer Rank Xerox Ltd., London, 1975-79, mng. dir., 1979-83; sr. v.p., chief staff officer Xerox Corp., Stamford, Conn., 1983-86, pres., 1986-91, chmn., 1991, chmn. bd., 1991—; also bd. dirs. Rank Xerox Ltd., London, Conn.; mem. investment policy adv. com. U.S. Trade Rep.; bd. dirs. Rank Xerox Ltd., Fuji Xerox Co. Ltd., Sara Lee Corp., N.Y. Stock Exch., Crum & Forster, Morristown, N.Y. City Ballet, Catalyst, N.J. Nat. Planning Assn., Washington. 1986. Bd. dirs., mem. bus. adv. coun., trustee Grad. Sch. Indsl. Adminstrn. Carnegie Mellon U.; trustee Worchester Poly. Inst. Mem. Tau Beta Pi, Eta Kappa Nu. Democrat. Office: Xerox Corp PO Box 1600 800 Long Ridge Rd Stamford CT 06904-1600

ALLARD, JUDITH LOUISE, secondary education educator; b. Rutland, Vt., Feb. 21, 1945; d. William Edward and Orilla Marion (Trombley) A. BA, U. Vt., 1967, MS, 1969. Tchr. math., sci. Edmunds Jr. High Sch., Burlington, Vt., 1969-73; biology tchr. Edmunds Jr. High Sch., Burlington, 1973-78, sci. dept. chair, 1975-78; biology tchr. Burlington (Vt.) High Sch., 1978—; instr. environ. studies U. Vt., Burlington, 1988-89; adviser Nat. Honor Soc., 1986—. Contbg. author Favorite Labs of Outstanding Tchrs., 1991. Active Amnesty Internat., 1985—; mem. Discovery Mus., Essex Junction, Vt., 1986—, Lake Champlain Com., Burlington, 1987—. Recipient Presdl. Sci. Teaching award NSF, 1983; named Outstanding Vt. Educator, U. Vt., 1983, Outstanding Vt. Sci. Tchr., Sigma Xi Soc., 1984; Tandy Tech.

scholar, 1990. Mem. ASCD, NEA (bd. dirs. Vt. chpt.), Vt. Sci. Tchrs. Assn. (bd. dirs. 1980-92, treas. 1985-92), Burlington Edn. Assn. (exec. bd. 1989-92), Burlington Profl. Standards Bd. (chair 1991—), Parents and Friends of Edn. (trustee) Nat. Assn. Biology Tchrs. (dir. Vt. Outstanding Biology Tchr. award program 1977—, Outstanding Biology Tchr. award 1975), Assn. Presdl. Awardees in Sci. Roman Catholic. Avocations: needlework, fishing, music. Home: 221 Woodlawn Rd Burlington VT 05401-2453 Office: Burlington High Sch 52 Institute Rd Burlington VT 05401-2789

ALLARD, ROBERT WAYNE, geneticist, educator; b. L.A., Sept. 3, 1919; s. Glenn A. and Alma A. (Roose) A.; m. Ann Catherine Wilson, June 16, 1944; children: Susan, Thomas, Jane, Gillian, Stacie. B.S., U. Calif., Davis, 1941; Ph.D., U. Wis., 1946. From asst. to assoc. prof. U. Calif., Davis, 1946—, prof. genetics, 1955—. Author books; contbr. articles to profl. jours. Served to lt. USNR. Recipient Crop Sci. award Am. Soc. Agronomy, 1964, DeKalb Disting. Career award Crop Sci. Soc. Am., 1983; Guggenheim fellow, 1954, 60; Fulbright fellow, 1955. Mem. Nat. Acad. Scis., Am. Acad. Arts and Scis., Am. Soc. Naturalists (pres. 1974-75), Genetics Soc. Am. (pres. 1983-84), Am. Genetics Assn. (pres. 1989), Phi Beta Kappa, Sigma Xi, Alpha Gamma Rho, Alpha Zeta. Democrat. Unitarian. Home: PO Box 185 Bodega Bay CA 94923 Office: U Calif Davis 133 Hunt Davis CA 95616-2322

ALLARDICE, DAVID JOHN, fuel technologist; b. Melbourne, Victoria, Australia, June 26, 1941; s. Kenneth John and Melva Jean (Brewster) A.; m. Alison Nethercote, Aug. 21, 1965; children: Geoffrey John, Lachlan David. BSc with honors, U. Melbourne, 1962, MSc with honors, 1964, PhD, 1967. Chartered chemist, chartered engr. Rsch. assoc. Pa. State U., State College, 1967-70; rsch. scientist State Electricity Commn. of Victoria, Melbourne, 1970-82; mgr. devel. Victorian Brown Coal Coun., Melbourne, 1982-85; mgr. bus. devel. and rsch. Coal Corp., Victoria, Melbourne, 1985—; mem. Nat. Energy Coun., Canberra, ACT, 1989-91; dir. Australian Coal Industry Rsch. Labs., Sydney, 1987-91. Fellow Australian Inst. Energy (pres. 1988-89), Royal Australian Chem. Inst., Inst. Energy. Achievements include research in brown coal science and utilization technology; internationally recognized expert on low rank coal. Office: Coal Corp Victoria, 128 Exhibition St, Melbourne Victoria, Australia 3000

ALLARDICE, JOHN MCCARRELL, coatings manufacturing company executive; b. Balt., May 30, 1940; s. James Barclay and Rebecca Jane (McCarrell) A.; m. J. Ann Benjamin, May 30, 1962 (div. 1979); children: John McCarrell Jr., Scott, Julie; m. Susan Bryson Miller, Aug. 15, 1981; stepchildren: Ben, Ted. Student, Washington and Jefferson Coll., 1958-61; BS in Chemistry, U. Pitts., 1963. Salesman chem. div. PPG, Pitts., 1964; silicone div. GE, Waterford, N.Y., 1965; salesman Stauffer Chem. Co., Adrian, Mich., 1965-69; salesman, sales mgr. Fre Kote, Inc., Boca Raton, Fla., 1969-78; pres. Releasomers, Inc., Bradford Woods, Pa., 1978—. Patentee bladder lubricants. Republican. Mem. Unity Ch. Avocations: golf, tennis, acting, singing, old cars. Office: Releasomers Inc PO Box 82 PO Box 82 Bradford Woods PA 15015-0082

ALLARDT, ERIK ANDERS, academic administrator, educator; b. Helsingfors, Finland, Aug. 9, 1925; s. Arvid and Marita I. (Heikel) A.; m. Sagi E. Nylander; children: Jörn, Monica, Barbro. MA, U. Helsinki, Finland, 1947, PhD, 1952; PhD (hon.), U. Stockholm, 1978, Abo Acad., Finland, 1978, U. Uppsala, Sweden, 1984. Instr. sociology U. Helsinki, 1948-53; rsch. asst. bur. of applied social rsch. Columbia U., 1954; rsch. dir. Helsinki Sch. of Social Scis., 1955-57; prof. U. Helsinki, 1958-70, 80-85; rsch. prof. Acad. of Finland, 1970-80, pres., 1986-91; chancellor Abo Acad. Univ., 1992—; vis. prof. U. Calif., Berkely, 1962-63, U. Ill., Urbana, 1966-67, U. Wis., Madison, 1970, U. Mannheim, Fed. Republic of Germany, 1985; vis. scholar Wilson Ctr., Washington, 1978-79. Author: Drinking Norms and Drinking Habits, 1957, About Dimensions of Welfare, 1973, Implications of the Ethnic Revival in Industrialized Society, 1979, (with Lysgaard and Sorensen) Sociologin i Sverige, vetenskap, miljo och organisation, 1988; author, editor: (with Y. Littunen) Cleavages, Ideologies and Party Systems, 1964, (with S. Rokkan) Mass Politics, 1970, Acta Sociologica, 1968-71, Scandinavian Polit. Studies, 1975-76. Chmn. Finnish sect. Amnesty Internat., Finland, 1977-78, Tampere Inst. for Peace and Conflict Research, Finland, 1977-78, Coms. for Devel. of Future Studies, Finland, 1988-89. Mem. The Norwegian Acad. Sci. and Petters (fgn. mem. 1986—), European Sci. Found. (exec. coun. 1987-92, v.p. 1990-92), Academia Europaea (founder mem. 1988—), Finnish Soc. Scis. and Letters (hon. mem. 1988—, chmn. 1985-86), Scandinavia-Japan Sasakawa Fond. (bd. dirs. 1987—), Royal Swedish Acad. Letters, History and Antiquities (fgn. 1992—). Home: Unionsgatan 45 B 40, SF-00170 Helsinki Finland Office: U Helsinki Rsch Group, Comparative Sociology, Hämeentie 68, SF-00550 Helsinki Finland

ALLEN, BURKLEY, mechanical engineer; b. Marianna, Ark., Mar. 24, 1958; d. Alonzo Greenlaw and June (Beasley) Mann; m. Newton Perkins, Jr., June 19, 1982; children: Sarah, Newton Perkins III, Mary Lobdell. BS in Physics, Davidson Coll., 1980; MSME, U. Va., 1982. Registered profl. engr., Tenn. Design engr. I. C. Thomasson Assoc., Nashville, 1982-84, solid waste dept., 1984-89, process engr., 1990-91; cons. self employed Nashville, 1991—. Sec. Hillsboro West End Neighborhood, Nashville, 1984-91; elder Second Presbyn. Ch., Nashville, 1989-93. Mem. ASHRAE (sec. 1990-91, v.p. 1991-92, pres.-elect. 1992-93, pres. 1993-94).

ALLEN, CHARLES EUGENE, college administrator, agriculturist; b. Burley, Idaho, Jan. 25, 1939; s. Charles W. and Elsie P. (Fowler) A.; m. Connie J. Block, June 19, 1960; children: Kerry J., Tamara S. BS, U. Idaho, 1961; MS, U. Wis.-Madison, 1963, PhD, 1966. NSF postdoctoral fellow Sydney, Australia, 1966-67, asst. prof. of agr. U. Minn., St. Paul, 1967-69, assoc. prof., 1969-72, prof., 1972—, dean Coll. Agr., assoc. dir. agrl. expt. sta., 1984-88, acting v.p., 1988-90, v.p. agriculture, forestry and home econs., dir. Minn. Agr. Expt. Sta., 1990—; vis. prof. Pa. State U., 1978; cons. to industry; C. Glen King lectr. Wash. State U., 1981; Univ. lectr. U. Wyo., Laramie, 1984; adj. prof. Hassan II U., Rabat, Morocco, 1984. Contbr. numerous chpts. to books, articles to sci. jours. on growth and metabolism of muscle and adipose tissue, meat quality. Recipient Horace T. Morse-Amoco Found. award in undergrad. edn. U. Minn., 1984; Disting. Tchr. award U. Minn. Coll. Agr., 1984. Fellow Inst. Food Tech.; mem. Am. Meat Sci. Assn. (dir. 1970-72; Research award 1980, Signal Service award, 1985), Am. Soc. Animal Sci. (Exceptional Research Achievement award 1972, Research award 1977), Am. Inst. Nutrition, Sigma Xi. Avocations: bowling, photography, reading, outdoor sports, golf.

ALLEN, CLAYTON HAMILTON, physicist, acoustician; b. Whitinsville, Mass., June 2, 1918; s. Charles Aaron and Edith Gertrude (Peck) A.; m. Doris Elizabeth LeClaire, Dec. 7, 1981. BS in Physics, Worcester Poly. Inst., 1940; MS in Physics and Math., Pa. State U., 1942, PhD in Physics and Math., Phys.-Chemistry, 1950. Grad. asst. Pa. State U., 1940-42; grad. assoc. Pa. State U., State College, 1945-50; acoustical communications researcher Aircraft Radio Lab., Wright-Patterson AFB, Ohio, 1942-45; ultrasonic researcher Corning (N.Y.) Glass Works, 1950-54; cons. on noise control Bolt Beranek and Newman Inc., Cambridge, Mass., 1954-74; v.p. noise control Sci. Applications Inc., La Jolla, Calif., 1974-75; pres. The Clayton H. Allen Corp., Chebeague Island, Maine, 1975—; cons. in acoustics, vibration and noise control design and devel.; lectr. profl. confs. and meetings. Contbr. articles to Jour. Acoustical Soc. Am., Jr. Cellular and Comparative Physiology, Noise Control, Mech. Contractor, Heating, Piping and Air Conditioning, Am. Indsl. Hygiene Assn., Sound, chpts. to ref. books. Fellow Acoustical Soc. Am.; mem. AAAS, Am. Phys. Soc., Acad. Applied Sci., Innovation Group, Am. Foundrymen's Soc., Nat. Coun. Acoustical Cons., Mass. Coun. Acoustical Cons., Mass. Engrs. Coun., Inst. Noise Control Engring. (past mem. bd. dirs. and bd. examiners), Sigma Xi, Sigma Pi Sigma. Achievements include patents (with others) for High Frequency Siren, Method of and Apparatus for Acoustic Silencing, Apparatus for Silencing Vibrational Energy, Method of Phonograph Record Fabrication, Phonograph System, Nonlinear Acoustic Ear Muff and the like. Office: 80 South Rd Chebeague Island ME 04017

ALLEN, DAVID WOODROFFE, computer scientist; b. Hampton, Iowa, Sept. 20, 1944; s. Edward DeWalt and Julia Woodroffe (Lamb) A.; m. Barbara Ann Schneider, Sept. 15, 1973. BA, Grinnell Coll., 1967; MS, U. Pitts., 1974. Assoc. engr. Westinghouse Electric Corp., Sharon, Pa., 1967-70, engr., 1970-79, sr. engr., 1979-84; sr. computer scientist Westinghouse Electric Corp., Pitts., 1984-90, prin. engr., 1990—. Contbr. articles to profl. jours.; patentee in field. Recipient George Westinghouse Signature award of excellence, 1989, 92, George Westinghouse Innovation award, 1993. Mem. IEEE (sect. sec.-treas. 1981-82, referee tech. papers for Computer jour.), Assn. for Computing Machinery, Silicon Graphics User's Group We. Pa. (treas. 1991—), Digital Equipment Computer User's Soc. Democrat. Achievements include research in current transformer transient performance, magnetic and electric field computation and analysis, computer-aided geometric design. Avocations: woodworking, gardening, reading. Home: 2637 Rossmoor Dr Pittsburgh PA 15241-2572 Office: Westinghouse Electric Corp PO Box 18249 Pittsburgh PA 15236-0249

ALLEN, DUFF SHEDERIC, JR., retired chemist; b. St. Louis, Dec. 8, 1928; s. Duff Shederic and Mildred Lucille (Burns) A.; m. Iris Meitus, Feb. 8, 1960 (dec. 1971); children: Cori, Tana; m. Mary Menzies, Jan. 23, 1976. BA, Princeton U., 1949; postgrad., Washington U., 1949-51; PhD, U. Wis., 1960. Rsch. chemist Am. Cyanamid Co., Bound Brook, N.J., 1960-62; group leader Am. Cyanamid Co., Pearl River, N.Y., 1963-65, dept. head organic chemistry, 1965-75, mgr., 1975-85, sr. rsch. scientist, 1985-91; ret. Mem. Greenwich (Conn.) Rep. Town Meeting, 1968-76. Mem. AAAS, N.Y. Acad. Sci., Sigma Xi. Home: 25 Upland Ave Falmouth MA 02540

ALLEN, FREDERICK GRAHAM, consulting engineer; b. Boston, Feb. 2, 1923; s. Frederick Warren and Agnes (Horner) A.; m. Susan Kate Morse, Sept. 21, 1949; children: Warren Morse, Katherine Holden, Peter Graham. B in Mechanical Engring., Cornell U., 1944; MS, Harvard U., 1955, PhD in Engring. Sci., 1957. Cert. profl. engr., Calif. Mem. tech. staff Bell Tel. Labs., Murray Hill, N.J., 1957-66; dept. head Bellcomm., Inc., Washington, 1966-69; dept. head, prof. U. Calif., L.A., 1969-91; expert witness legal Cases as Cons., Calif., 1972--. Contbr. articles to profl. jours. Fellow Am. Phys. Soc.; mem. IEEE (sr. mem.). Democrat. Home: 823 Westholme Ave Los Angeles CA 90024

ALLEN, HENRY WESLEY, biomedical researcher; b. Louisville, Oct. 16, 1927; s. John Turk and Irene Victoria (Slater) A.; m. Evelyn Chen, Dec. 29, 1968 (div. Dec. 1988); children: Lillian Chen, Rosaniline Chen, Dianne Chen. Student, U. Louisville, 1945-46, U. Chgo., 1946-47, U. So. Calif., 1960-61. Rschr. Loma Linda (Calif.) U., 1962-77, Am. Biologics, Chula Vista, Calif., 1977—. Author: International Protocols in Cancer Management, 1983, The Study of Reactive Oxygen Toxic Species and Their Metabolism, 1985, The Biochemistry of Live Cell Therapy, 1986; contbr. articles to Jour. of Theoretical Biology, Analytical Biochemistry, Nature, others. Achievements include patents in field. Office: Am Biologics 1180 Walnut Ave Chula Vista CA 91911

ALLEN, JAMES HARMON, JR., civil engineer; b. Pratt, Kans., Apr. 8, 1948; s. James Harmon and Glenda Rosena (Hackenberg) A.; m. Betty June Schlegel, July 11, 1970; children: James III, Christine, Benjamin. BCE, Kans. State U., 1971. Registered profl. engr., Okla. Asst. supt. Richards spur Dolese Bros. Co., Elgin, Okla., 1980-82, supt., 1982-84; asst. gen. supt. Dolese Bros. Co., Oklahoma City, 1984, chief engr., 1984-87, asst. aggregate ops. mgr., 1987-88, gen. mgr. aggregate and prestress ops., 1988-91, gen. mgr. ops., 1991—; chmn. surface mining adv. coun. Dept. of Mines, Oklahoma City, 1988-92; mem. ashphalt task force Okla. Dept. Transp., Oklahoma City, 1988. 1st lt. U.S. Army, 1971-74. Mem. Nat. Stone Assn. (chmn. com. 1990-92), NSPE, Okla. Soc. Profl. Engrs. Republican. Presbyterian. Home: 12717 Saint Andrews Dr Oklahoma City OK 73120 Office: Dolese Bros Co PO Box 677 20 NW 13th Oklahoma City OK 73101

ALLEN, JAMES L., electrical engineer; b. Graceville, Fla., Sept. 25, 1936. BEE, Ga. Inst. Tech., 1959, MSEE, 1961, PhD in Elec. Engring., 1966. Sr. engr. Sperry Microwave Electronics, 1961-63; instr. elec. engring. Ga. Inst. Tech., 1963-66; rsch. staff cons. Sperry Microwave Electronics, 1966-68; assoc. prof. elec. engring. U. So. Fla., 1968-70, Colo. State U., 1970-72; assoc. prof. U. So. Fla., 1972-74, prof., 1974-81; pres. Lamar Allen Enterprises, Inc., 1981-83; chief engr., v.p. Kaman Sci. Corp., 1983—; chief editor Transactions Microwave Theory and Techniques, IEEE, 1977-79. Fellow IEEE. Achievements include research in electromagnetic interference, microprocessor applications, microwave theory and techniques, high power microwave weapon systems. Office: Kaman Scis Corp PO Box 7463 Colorado Springs CO 80933*

ALLEN, JAMES MADISON, family practice physician, consultant; b. Columbus, Ohio, Nov. 14, 1944; s. D.C. Allen and Edith (Melvin) Eckfeld; m. Elizabeth Wolfe, Dec. 30, 1972; children: Elaine, Michelle, Katherine, James Jr. BA, U. Ga., 1969, MA, 1970; BS in Medicine, U. N.D., Grand Forks, 1977; MD, U. Ala., Birmingham, 1980; postgrad., Birmingham Sch. Law, Birmingham, 1990—. Med. officer U.S. Indian Health Svc., Phila., Miss., 1981-82; clinic dir. Rush Hosp. Clinics, Phila., 1983-85; v.p., bd. dirs. Am. Family Care, Birmingham, 1986-88; pvt. practice AMI Brookwood Family Med. Ctr., Birmingham, 1988—; cons. Riverchase Clin. Rsch., Birmingham, 1990—; pres. U.S. Physicians Inc., Birmingham, 1982—; cons. DiGiorno Foods div. Kraft Foods, Birmingham, 1988—, So. Nuclear Oper. Co., 1993—. Commr. Boy Scouts Am., Birmingham, 1988—. With USMC, 1962-66, Viet Nam. Recipient Award of Merit, B'Nai-Brith, 1970. Mem. Am. Acad. Family Physicians, Med. Assn. Ala., Assn. Am. Indian Physicians, Shriners (Noble 1988), Masons (Master 1966, 32 degree 1978), Phi Gamma Delta (Ga. chpt. pres. 1967-68), Sigma Delta Kappa. Avocations: flying, model railroading, musician, collector of scout patches. Home: 857 Tulip Poplar Dr Birmingham AL 35244 Office: Riverchase Convenient Care 4515 Southlake Pkwy Ste 104 Birmingham AL 35244

ALLEN, JERRY PAT, aviation science educator; b. Enid, Okla., Apr. 20, 1932; s. Albert Easley and LaRena Jean (McGinnis) A.; m. Lou Ann Schmidt, Mar. 18, 1955; children: Melissa Ann, Jerry Pat Jr. AS in Engring, Okla. State U., 1959; BA, Sch. of Ozarks, 1975; MS, Cen. Mo. State U., 1986. Engr.'s asst. Boeing Aircraft Co., Wichita, 1959-64; sales and svc. rep. Cessna Aircraft Co., Wichita, 1964-71; prof. aviation sci. Coll. of Ozarks, Pt. Lookout, Mo., 1971—. Mem. Profl. Aviation Mechanics Assn., Future Aviation Profls., Mo. Pilots Assn. (pres. Ozark chpt. 1975), Rotary (bull. editor 1974-82), Lions, Alpha Eta Rho. Baptist. Home: PO Box 282 Point Lookout MO 65726 Office: Coll of the Ozarks Dept Aviation Sci Point Lookout MO 65726

ALLEN, JOHN RYBOLT L., chemist, biochemist; b. Indpls., Sept. 14, 1926. BA, Ball State Tchrs. Coll., 1949; PhD in Biochemistry, U. Ill., 1954. Rsch. assoc. biochemistry Northwestern U., 1953-56; asst. prof. Coll. Med. Baylor U., 1956-59; sr. scientist Warner-Lambert Pharm. Co., N.J., 1959-60; rsch. assoc. Dental Sch. Wash. U., 1960-62; prof. chemistry, head. dept. Union Coll., Ky., 1962-64; clin. assoc. clin. chemistry U. Hosp. Case Western Reserve U., 1964-65; asst. prof. pathology and radiology coll. medicine Ohio State U., 1965-68; clin. chemist St. John's Mercy Hosp., St. Louis, 1968-69, Decatur Meml. Hosp., Ill., 1969-70, San Diego Inst. Pathology, 1970, San Bernardino County Hosp., 1970-75; instr. chemistry Phoenix Coll., 1975-80. Recipient G.K. Warren prize Nat. Acad. Scis., 1990. Fellow AAAS, Am. Assn. Clin. Chemistry, Am. Chemistry Soc., Acad. Clin. Lab. Physicians & Scientists, Am. Inst. Chemistry. Achievements include research in quality control and methods, creating phosphokinase, vitamin E deficiency, lipip metabolism and structure. Home: 4645 N 22nd St Apt 220W Phoenix AZ 85016*

ALLEN, JONATHAN DEAN, chemist, researcher; b. Ft. Worth, Oct. 10, 1956; s. Donald Horace Allen and Helen Marie (Wheeless) Grantham. BS in Chemistry, U. Sci. and Arts Okla., 1985. Chemist dept. geology and geophysics U. Okla., Norman, 1988—. Contbr. articles to profl. jours. Home: 511 Chickasha St Norman OK 73071 Office: U Okla Dept Geology and Geophysics 100 E Boyd St Rm S-104 Norman OK 73019

ALLEN, LEATRICE DELORICE, psychologist; b. Chgo., July 15, 1948; d. Burt and Mildred Floy (Taylor) Hawkins; m. Allen Moore, Jr., July 30, 1965 (div. Oct. 1975); children—Chandra, Valarie, Allen; m. Armstead Allen, May 11, 1978 (div. May 1987). A.A. in Bus. Edn., Olive Harvey Coll., Chgo., 1975; B.A. in Psychology cum laude, Chgo. State U., 1977; M.Clin. Psychology, Roosevelt U., 1980; MA in Health Care Adminstrn., Coll. St. Francis, Joliet, Ill., 1993. Clk., U.S. Post Office, Chgo., 1967-72; clin. therapist Bobby Wright Mental Health Ctr., Chgo., 1979-80; clin. therapist Community Mental Health Council, Chgo., 1980-83, assoc. dir., 1983—; cons. Edgewater Mental Health, Chgo., 1984—, Project Pride, Chgo., 1985—; victim services coordinator Community Mental Health Council, Chgo., 1986-87; mgr. youth family services Mile Square Health Ctr., Chgo., 1987-88; coord. Evang. Health Systems, Oakbrook, Ill., 1988—. Scholar Chgo. State U., 1976, Roosevelt U., 1978; fellow Menninger Found., 1985. Mem. Nat. Orgn. for Victim Assistance, Ill. Coalition Against Sexual Assault (del. 1985—), Soc. Traumatic Stress Studies (treatment innovations task force), Chgo. Sexual Assault Svcs. Network (vice-chair, bd. dirs.). Avocations: aerobics; reading; theatre; dining.

ALLEN, LESLIE, physics educator; b. London, Oct. 22, 1935; s. James Herbert and Beatrice May (Blunden) A.; m. Barbara Russell, July 22, 1957 (div. Nov. 1989); children: Michael John, Carol Ann, Jennifer Lesley, David Philip; m. Polly Barnes, Dec. 23, 1991. BSc, U. London, 1957; DIC, diploma Imperial Coll., PhD, 1960; DSc, U. London, 1973; assoc., Royal Coll. Sci., 1957. Chartered physicist. Asst. lectr. Royal Holloway Coll., U. London, 1960-62; lectr. U. Sussex, Brighton, 1962-69, reader, 1969-85; pro rector Polytechnic East London, London, 1986-91; vis. prof. U. Leiden, The Netherlands, 1991-92; vis. lectr. U. Ife, Nigeria, 1966; vis. sr. rsch. assoc. U. Rochester, N.Y., 1968-69, 72, 74, 77, 79; vis. prof. Indian Inst. Tech., Delhi, 1975, U. Sussex, 1989-92, U. Essex, 1992—; rsch. fellow Leverhulme Trust, 1992-93; emeritus prof. U. East London, 1992—; physics advisor Assn. Commonwealth Univs. Scholrship Commn., 1987-91; mem. sci. bd., nuclear physics bd. Sci. and Engring. Rsch. Coun., 1989-92; mem. com. for rsch. Coun. Nat. Acad. Awards, 1987-91; treas. Brit. Pugwash Group, London, 1989—; coord. SERC non-linear optics initiative, 1992—. Author: Essentials of Lasers, 1969; co-author: Principles of Gas Lasers, 1967, Optical Resonance and Two-Level Atoms, 1975, 87, Concepts of Quantum Optics, 1983; editorial advisor Progress in Optics, 1975-88; contbr. over 80 articles to profl. jours. Mem. exec. com. Coun. for Edn. in Commonwealth, 1987-88. Sci. and Engring. Rsch. Coun. rsch. grantee, 1971, 77, 78, 82, 85; Royal Soc. travel grantee. Fellow Royal Soc. Arts, Inst. Physics; mem. Prehistoric Soc., Surrey County Cricket Club, Millwall Football Club. Avocations: theatre, art, opera, archaeology, jazz. Home: 11 W Arbour St, London E1 OPQ, England Office: U Essex Phys Dept, Wivenhoe Park, Colchester England

ALLEN, MARSHALL BONNER, JR., neurosurgeon, consultant; b. Long Beach, Miss., Oct. 19, 1927; s. Marshall Bonner and Annie Marie (Martin) A.; m. Mildred Ann Reynolds, June 1949 (div. Feb. 1957); m. Dorothy Odell Herron, Oct. 11, 1957; children: Lori Lynn Allen Athey, Marshall Bonner III, George Michael. BA, U. Miss., 1949; MD, Harvard U., 1953. Diplomate Am. Bd. Neurosurgery. Intern Jefferson Med. Coll. Hosp., 1953-54; resident in gen. surgery U. Hosp., Jackson, Miss., 1956-57, resident in neurosurgery, 1957, 61; chief neurosurgery VA Hosp., Jackson, Miss., 1962-64; prof. surgery, chief neurosurgery Med. Coll. Ga., Augusta, 1965—; cons. VA Hosp., Augusta, 1965—, Univ. Hosp., Augusta, 1965—, Dwight David Eisenhower Army Med. Ctr., Ft. Gordon, Ga., 1968—, Cen. State Hosp., Milledreville, Ga., 1967—. Author: A Manual of Neurosurgery, 1975, Polytimography of the Brain, 1975; editor: The Pituitary, A Current Reivew, 1968, Essentials of Neurosurgery, 1993; contbr. articles to profl. jours. Mem. Evening Optimist Club of Ga., Augusta, 1978—, pres. 1985. With U.S. Army, 1946-48, Korea; lt. commdr. USNR, 1964-66. Fellow Am. Coll. Surgeons; mem. Soc. Univ. Neurosurgeons (sec., pres. 1978-82), Ga. Neurosurgical Svc. (pres. 1969). Methodist. Office: Med Coll Ga Sect of Neurosurgery Augusta GA 30912

ALLEN, MERRILL JAMES, marine biologist; b. Brady, Tex., July 16, 1945; s. Clarence Francis and Sara Barbara (Finlay) A. BA, U. Calif., Santa Barbara, 1967; MA, UCLA, 1970; PhD, U. Calif., San Diego, 1982. Cert. jr. coll. tchr., Calif. Asst. environ. specialist So. Calif. Coastal Water Rsch. Project, El Segundo, 1971-77; postdoctoral assoc. Nat. Rsch. Coun., Seattle, 1982-84; oceanographer Nat. Marine Fisheries Svc., Seattle, 1984-86; sr. scientist MBC Applied Environ. Scis., Costa Mesa, Calif., 1986-93; prin. scientist So. Calif. Coastal Water Rsch. Project, Long Beach and Westminster, Calif., 1993—; tech. adv. com. Santa Monica Bay Restoration Project, Monterey Park, Calif., 1989—; affiliate asst. prof. sch. fisheries U. Wash., Seattle, 1985-89. Mem. AAAS, Am. Inst. Fisheries Rsch. Biologists (dir. So. Calif. dist. 1991-93), Am. Fisheries Soc., Am. Soc. Ichthyologists and Herpetologists. Achievements include development of most comprehensive atlas of marine fishes from Bering Sea to Mexico; description of state of contamination of Santa Monica Bay. Office: So Calif Coastal Water Rsch Project 7171 Fenwick Ln Westminster CA 92683

ALLEN, NINA STRÖMGREN, biology educator; b. Copenhagen, Denmark, Sept. 17, 1935; came to U.S. 1951; d. Bengt G.D. and Sigrid S.C. (Hartz) Strömgren; m. Robert Jackson Williams, July 18, 1958 (div. 1970); children: Erik Robert, Harriet Hopf; m. Robert Day Allen, Sept. 11, 1970 (dec.); 1 child, Barbara Sigrid. BS, U. Wis., 1957; MS, U. Md., 1970, PhD, 1973. Teaching asst. botany U. Wis., 1957-58; teaching asst. botany U. Md., 1964-67, NIH predoctoral fellow, 1967-70; rsch. asst. SUNY, Albany, 1970-93, postdoctoral fellow, 1973-75; summer rschr. Marine Biol. Lab., Woods Hole, Mass., 1970-93; vis. asst. prof., rsch. assoc. Dartmouth Coll., Hanover, N.H., 1975-76, asst. prof., 1976-83; assoc. prof. biology Wake Forest U., Winston-Salem, 1984-92, prof. biology, 1993—; NSF vis. prof. for women Stanford U., 1990-91; lectr. in field; ad hoc reviewer Biomed. Rsch. Tech. Rev. Com., 1988, study sect. mem., 1989—; trustee, mem. exec. com. Marine Biol. Labs. Editorial bd. Cell Motility, 1980-85; series editor, founder Plant Biology, Alan R. Liss, Inc., 1985—; editorial bd. BioTechniques, 1989—; contbr. articles to profl. jours.; producer sci. films and videotapes. Fellow AAAS; mem. Am. Soc. Cell Biology, Am. Soc. Plant Physiologists, N.E. Algal Soc. (founding mem.), Phycol. Soc. Am., N.C. Acad. Sci., Am. Soc. for Gravitational and Space Biology, Electron Microscopical Soc. Am., Royal Microscopical Soc. (fellow 1985), Phi Beta Kappa, Sigma Xi. Democrat. Unitarian Universalist. Achievements include patent on development of AVEC microscopic methods; co-inventor video microscopy. Home: 1970 Faculty Dr Winston-Salem NC 27106 Office: Wake Forest U Dept Biology Winston Salem NC 27106

ALLEN, PAUL C., physicist; b. Mpls., July 14, 1952; s. Edgar F. and Alice (Nachbar) A.; m. Meredith L. Pederson, July 3, 1981; children: Summer E., Marc D., Matthew V., Sarah C. BA in Physics, Carleton Coll., 1972; MS in Physics, U. Ill., 1973, PhD in Physics, 1977. Postdoctoral rsch. asst. U. Calif. San Diego, La Jolla, 1977-80; process engring. mgr. Advanced Micro Devices, Sunnyvale, Calif., 1980-84; dir. optical devel. ATEQ Corp., Beaverton, Oreg., 1984-88, dir. concept and feasibility, 1988-91; dir. lithography tech. ETEC Systems, Inc., Beaverton, 1991—. Contbr. articles to profl. jours. Mem. IEEE, Soc. Photo-Instrumentation Engrs., Sigma Xi. Achievements include 2 patents on thermodynamic apparatus, 4 patents on microlithography. Office: Etec Systems Inc 9100 SW Gemini Dr Beaverton OR 97005

ALLEN, RHOUIS ERIC, electrical engineer; b. Jonesboro, Ark., July 11, 1964; s. Rhouis Arnold and Loretta Faye (Cooper) A.; m. Jamie Ray Hardin, June 27, 1987. BSEE, Memphis State U., 1987. Design engr. Riddick Engring., Little Rock, 1987-88; elec. instrumentation and control design engr. Ark. Power & Light, Little Rock, 1988-90, Entergy Ops., Little Rock, Russellville, 1990—. Mem. employee polit. action com. Ark. Nuclear One, Russellville. Mem. IEEE, Tau Beta Pi, Gamma Beta Phi, Phi Eta Sigma. Baptist. Home: 102 Stream Rd Russellville AR 72801 Office: Entergy Ops Akr Nuclear One Rt 3 Box 137 G Russellville AR 72801

ALLEN, RYNE CUNLIFFE, electrical engineer; b. Hartford, Conn., Jan. 21, 1962; s. James Cunliffe and Susan Webster (Conlin) A. BSEE, Northeastern U., 1985, MSEE, 1989. Jr. engr. Prime Air, Inc., West Hartford, Conn., 1981-82, Mass. Mfg. Corp., Cambridge, Mass., 1982-84; jr. engr. electronics rsch. lab. Northeastern U., Boston, 1984-85, rsch. asst. plasma sci. lab., 1985-88, rsch. engr., lab. mgr. plasma sci. lab., 1988-89, sr. rsch. engr. plasma sci. lab., 1989—; v.p. AIS, Inc., Newton, Mas., 1991; initated Northeastern U. Electron Beam Lithography Ctr. for the Plasma Sci. and Microelectronics Rsch. Lab., 1986—; past mem. Chem. Hygiene Com., Boston, 1990; cons. electromagnetic interference/electromagnetic compatibility, 1986—. Contbr. articles to jours. on plasma dynamics, interatcions between magnetic fields, electron temperature, low current anocic arc phenomena, coatings, others. Eagle Scout Troop 6, Brookline, Mass., 1976;

mem. Internat. Taekwon-Do Fedn., U.S. Taekwon-Do Union, 1988—; Black Belt, asst. instr. Northeastern Tae Kwon-Do Club, 1988—; cadet comdr. CAP, Boston, 1979-82. Mem. IEEE, ASM Internat., Am. Geophys. Union, Am. Vacuum Soc., Am. Poetry Assn. (pub. poet 1987—), Soaring Soc. Am. Achievements include patent for environmental treatment and anodic/cathedic arc coatings. Office: Northeastern U Plasma Lab 360 Huntington Ave Boston MA 02115-5096

ALLEN, STEPHEN D(EAN), pathologist, microbiologist; b. Linton, Ind., Sept. 8, 1943; s. Wilburn and Betty (Moffett) A.; m. Vally C. Autrey, June 17, 1964; children: Christopher D., Amy C. BA, Ind. U., 1965, MA, 1967; MD, Ind. U., Indpls., 1970. Diplomate Am. Bd. Pathology; cert. in anatomic and clin. pathology and med. microbiology. Intern in pathology Vanderbilt U. Hosp., Nashville, 1970-71; resident in pathology Vanderbilt U. Hosp., 1971-74; clin. asst. prof. pathology Emory U., Atlanta, 1974-77; asst. prof. clin. pathology Ind. U., Indpls., 1977-79; asst. prof. pathology Ind. U., 1979-81, assoc. prof. pathology, 1981-86, prof. pathology, 1986—, assoc. dir. div. clin. microbiology, dept. pathology, 1977—, dir. grad. progam pathology, 1986—, sr. assoc. chmn. dept. pathology, 1990-91, dir. div. clin. microbiology dept. pathology, 1992—. Co-author: Color Atlas of Diagnostic Microbiology, 4th edit. 1992; contbr. articles to profl. jours. With USPHS, 1974-77. Fellow Coll. Am. Pathologists, Am. Acad. Microbiology, Am. Soc. Clin. Pathologists (coun. mem. microbiology 1983-89), Infectious Diseases Soc. of Am., Soc. Sigma Xi; mem. Am. Bd. Pathology (microbiology test com.), Masons (32nd deg.), Shriner. Avocations: music, electric bass and trumpet, fishing. Office: Ind U Hosp Rm 4587 550 N University Blvd Indianapolis IN 46202-5283

ALLEN, STEPHEN LOUIS, electrical engineer, educator; b. St. Louis, Sept. 5, 1956; s. John Joseph and Mary Ann (Schick) A.; m. Judy Marie Luechtefeld, July 3, 1982; children: Lauren, Richard. BEE, U. Mo., Rolla, 1978, PhD in Engring. Mgmt., 1978. Engring. mgr. Schlumberger Ltd., 1978-86; prof. mgmt. N.E. Mo. State U., Kirksville, 1989—. Contbr. articles to Regional Bus. Rev., Engring. Mgmt. Jour. V.p. Parish Sch. Bd., Kirksville, Mo., 1991-92. Mem. Prodn. and Ops. Mgmt. Soc., Am. Soc. for Engring. Mgmt., K.C. (advocate 1992, 93), Delta Sigma Pi. Roman Catholic. Office: NE Mo State U 119 Violette Hall Kirksville MO 63501

ALLEN, STEVEN PAUL, microbiologist; b. Oak Park, Ill., Nov. 6, 1958; s. Paul Samuel and Rosemary (Bieber) A.; m. Wendy Dianne Gunia, Sept. 11, 1982; children: Samantha Rose, Matthew David. BS, U. Ill., 1980, PhD, 1990; MS, Iowa State U., 1982. Rsch. asst. Iowa State U., Ames, 1980-82; rsch. specialist U. Ill., Chgo., 1982-85; rsch. asst. U. Ill., Urbana, 1985-90; postdoctoral assoc. Monsanto Corp. Rsch., St. Louis, 1990—. Contbr. numerous articles to profl. jours. Mem. Am. Soc. for Microbiology, Inst. of Food Technologists (fellowship 1989-90, merit fellowship 1987), Phi Kappa Phi, Gamma Sigma Delta, Sigma Xi. Achievements include development of first intact-cell DNA transformation protocol for the bacterium Clostridium perfringens. Office: Abbott Labs Dept 9FC Bldg AP31 1 Abbott Park Rd Abbott Park IL 60064

ALLEN, THOMAS E., obstetrician/gynecologist; b. Bairdford, Pa., July 2, 1919; s. Emerson Ray and Lillie Mabel (McIntyre) A.; children: Catherine, Christine, Cynthia, Carolyn, Thomas J. Candace. BS, U. Pitts., 1940, MD, 1943. Diplomate Am. Bd. Ob-Gyn. Rotating intern U. Pitts., 1944, assoc. clin. prof. ob-gyn. Sch. Medicine; resident in gynegology Magee Hosp., Pitts., 1944-45, resident in ob-gyn., 1948-51; gen. practice medicine Oakmont, Pa., 1947-48; practice medicine specializing in ob-gyn. Pitts., 1951—; med. dir., co-founder Women's Health Service, Inc., Pitts., 1973—; cons. ob-gyn Russelton Med. Group, New Kensington, Pa., 1953-73. Pres. Oakmont Sch. Bd., 1962-71; pres. bd. dirs. Am. Wateways Wind Orch., Pitts., 1972—; bd. dirs. ACLU, Pitts., 1972-90. Served to capt. U.S. Army, 1945-47. Am. Legion and Buhl scholar, 1937. Fellow ACS, Am. Coll. Obstetricians and Gynecologists, Pan Pacific Surg. Assn., Pitts. Ob-gyn. Soc.; mem. AMA, county and state med. assns. Democrat. Club: Oakmont Country. Avocations: cooking, music, reading, golf. Home: 301 Halket St Pittsburgh PA 15213-3104 Office: 3433 Bates St Pittsburgh PA 15213-3900

ALLEN, WILLIAM CECIL, physician, educator; b. LaBelle, Mo., Sept. 8, 1919; s. William H. and Viola O. (Holt) A.; m. Madge Marie Gehardt, Dec. 25, 1943; children: William Walter, Linda Diane Allen Deardeuff, Robert Lee, Leah Denise Rogers. A.B., U. Nebr., 1947, M.D., 1951; M.P.H., Johns Hopkins U., 1960. Diplomate Am. Bd. Preventive Medicine, Am. Bd. Family Practice. Intern Bishop Clarkson Meml. Hosp., Omaha, 1952; practice medicine specializing in family practice Glasgow, Mo., 1952-59; specializing in preventive medicine Columbia, Mo., 1960—; dir. sect. chronic diseases Mo. Div. Health, Jefferson City, 1960-65; asst. med. dir. U. Mo. Med. Ctr., 1965-75; assoc. coordinator Mo. Regional Med. Program, 1968-73, coordinator health programs, 1969—, clin. asst. prof. community health and med. practice, 1962-65, asst. prof. community health and med. practice, 1965-69, assoc. prof., 1969-75, prof., 1975-76, prof. dept. family and community medicine, 1976-87, prof. emeritus, 1987—; cons. Mo. Regional Med. Program, 1966-67, Norfolk Area Med. Sch. Authority, Va., 1965-66; governing body Area II Health Systems Agy., 1977-79, mem. coordinating com., 1977-79; founding dir. Mid-Mo. PSRO Corp., 1974-79, dir., 1976-84. Contbr. articles to profl. jours. Mem. Gov.'s Adv Council for Comprehensive Health Planning, 1970-73; trustee U. Mo. Med. Sch. Found., 1976—. Served with USMC, 1943-46. Fellow Am. Coll. Preventive Medicine, Am. Acad. Family Physicians (sci. program com. 1972-75, commn. on edn. 1975-80), Royal Soc. Health; mem. Mo. Acad. Family Physicians (dir. 1956-59, 76-82, alt. del. 1982-87, pres. 1985-86, chmn. bd. 1986-87), Mo. Med. Assn., Howard County Med. Soc. (pres. 1958-59), Boone County Med. Soc. (pres. 1974-75), Am. Diabetes Assn. (pres. 1978, dir. 1974-77), Mo. Diabetes Assn. (pres. 1972-73), Soc. Tchrs. Family Medicine, AMA, Mo. Public Health Assn., Am. Heart Assn. (program com. 1979-82), Am. Heart Assn. of Mo. (sec. 1980-81), Mo. Heart Assn. (sec. 1979-82, pres.-elect 1982-84, pres. 1984-86). Methodist. Club: Optimists. Office: U Mo M234 Medical Ctr Columbia MO 65203

ALLENDE, JORGE EDUARDO, biochemist, molecular biologist; b. Catargo, Costa Rica, Nov. 11, 1934; s. Octavio Allende and Amparo Rivera; m. Catherine C. Connelly, 1961; 4 children. Student, La. State U., Yale U. Rsch. assoc. Rockefeller U., N.Y.C., 1961-62; asst. prof. biochemistry U. Chile, Santiago, 1963-68, assoc. prof., 1968-71, prof. biochemistry and molecular biology, 1972—; regional coord. Latin Am. Network Biol. Scis., 1975—; pres. Pan Am. Assn. Biochem. Socs., 1976; mem. exec. com. Internat. Cell Rsch. Orgn., 1976—, Internat. Union Biochemistry, 1982-91; mem. exec. bd. Internat. Coun. Sci. Unions, 1986-90; fgn. assoc. inst. of medicine NAS. Contbr. articles to profl. jours. Fogarty scholar-in-residence NIH. Fellow Third World Acad. Scis.; mem. Chilean Acad. Scis. (pres. 1991—), Chilean Acad. Medicine (hon.). Office: Academia Chilena de Ciencias, Clasificador 1349, Correo Central Santiago Chile*

ALLERTON, SAMUEL ELLSWORTH, biochemist; b. Three Rivers, Mich., Aug. 21, 1933; s. Sanford Ellsworth and Virginia Mary (Dickenson) A.; m. Theresa Mary Pawlak, Aug. 20, 1966; children: Adam Sanford, Eve Samantha. BA summa cum laude, Kalamazoo (Mich.) Coll., 1955; PhD, Harvard U., 1962. Teaching fellow Harvard U. Med. Sch., Boston, 1957-61; rsch. assoc. Rockefeller U., N.Y.C., 1961-65; asst. prof. U. So. Calif., L.A., 1965-69, assoc. prof., 1969—; cons. Woodroof Labs., Santa Ana, Calif., 1978-89. Contbr. articles to profl. jours. Bd. dirs. Huntington Beach (Calif.) Community Clinic, 1990-92. Named Outstanding Young Man of Am., Jaycee's, 1966. Mem. N.Y. Acad. Scis., Am. Coll. of Nutrition, Elks, Sigma Xi, Omicron Kappa Upsilon. Anglican. Achievements include rsch. on phys.-chem. characterization of proteins, tumor products, studies on absorption of copper. Office: Dentistry U So Calif University Park MC-0641 Los Angeles CA 90089-0641

ALLES, RODNEY NEAL, SR., information management executive; b. Orleans, Nebr., Aug. 24, 1950; s. Neal Stanley and Evelyn Dorothy (Zelske) A.; m. Diana Kay Koenig, Nov. 25, 1978; children: Rodney Neal Jr., Jennifer E., Victoria E. BS in Indsl. Engring., U. Okla., 1973, MBA, 1977. Asst. to the pres. Skytop Brewster Co., Inc., Houston, 1978-79, mgr. planning and mfg., 1979-83; v.p. adminstrn. Internat. Meter Co. Inc., Arkansas City, Kans., 1983-84, v.p., 1984-85; dir. info. mgmt. McAlester (Okla.) Army Ammunition Plant, 1987—. Lt. USN, 1973-76. Mem. Am. Inst. Indsl. Engrs., Fed. Info. Processing Coun., PE-ET, Loyal Knights of Old Trusty, Omicron Delta Kappa, Tau Beta PI, Sigma Tau, Alpha Pi Mu. Democrat. Lutheran. Avocations: golf. Office: Army Info Systems Commd ATTN: SMCMC-IM Mcalester OK 74501-5000

ALLEVA, JOHN JAMES, research biologist; b. Norristown, Pa., Apr. 17, 1928; s. John James and Margaret Helen (DelCollo) A.; m. Margaret Christine Martin, June 11, 1960; children: Brian, David, Lynn, Diane. AB, U. Pa., 1950; MS, U. Mo., 1952; PhD, Harvard U., 1959. Sr. scientist Smith Kline & French Labs., Phila., 1959-62; instr. Albany (N.Y.) Med. Coll., 1962-63; rsch. biologist FDA, Washington, 1963—. Contbr. sci. articles to profl. jours. 1st lt. USAF, 1952-56. Mem. Soc. for Rsch. on Biol. Rhythms, Internat. Soc. for Chronobiology. Home: 9511 Spode Ct Fairfax VA 22032 Office: FDA (HFD-472) 200 C St SW Washington DC 20204

ALLEY, E. ROBERTS, environmental engineer; b. Greenville, Miss., Sept. 17, 1938; s. Ernest Hayes and Thelma Mignon (Roberts) A.; m. Marion Catherine Spalta, Jan. 29, 1966; children: E. Roberts Jr., Laura Elizabeth, Aime Catherine, Emma Lynne. B in Engring., Vanderbilt U., 1960, MS, 1973. Registered profl. engr., Tenn., Ala., Ga., Miss., Fla., S.C., N.C., Va., Ky., Ark., Mo., La., Tex., Ohio, Mich., Ill., Ind.; diplomate Am. Acad. Environ. Engrs. Pres. Alley & Brown, Inc., Nashville, Tenn., 1968-72; v.p. Hart Freeland Roberts, Inc., Nashville, 1972-74; pres. E. Roberts Alley & Assocs., Inc., Brentwood, Tenn., 1974-80, 91, Alley, Young & Baumgartner, Inc., Brentwood, 1980-91, Pickering Alley, Inc., Brentwood, 1991-92, Alley & Assocs., Inc., Brentwood, 1992—; adj. prof. Vanderbilt U., Nashville, 1981-86; instr. George Washington U., Washington, D.C., London, San Diego, 1987—, Inst. for Mgmt. Tech., Singapore, Jakarta, Indonesia, 1990—. Contbr. articles to profl. jours. 2d lt. U.S. Army, 1960-61. Mem. ASCE, Am. Water Works Assn., NSPE, Water Pollution Control Fedn. Presbyterian. Home: 6030 Sherwood Ct Nashville TN 37215 Office: Alley & Assocs Inc 230 Wilson Pike Circle Brentwood TN 37027

ALLEY, MARCUS M., agronomy educator. BS, Berea Coll., 1969; MS in Agronomy, Va. Poly. Inst. and State U., 1971, PhD in Agronomy, 1975. Agronomist Park Forest Farms, Inc., Baskerville, Va., 1975; rsch. rep. Ohio and Ky. divsns. Ciba-Geigy Corp., 1975-77; asst. prof. soil and crop mgmt. Va. Poly. Inst. and State U., 1977-83, assoc. prof., 1983-89, prof., 1989—; advisor to exec. bd. Va. Agrl. Chem. and Soil Fertility Assn., 1989—; mem. project com. Best Mgmt. Practices Manual for Wheat Growers, 1993—; invited participant ARS-CSRS Soil Nitrate Workshop, Washington, 1991; mem. nitrogen, phosphorus, waste and metals work com. So. Region review USDA Funded Projects, 1992. Contbr. over 30 papers and abstracts to profl. and refereed jours., 3 chpts. to books. Recipient Werner L. Nelson award Fluid Fertilizer Found., 1986, Robert E. Wagner award Potash and Phosphate Inst./Am. Soc. Agronomy, 1990, Rsch. award Va. Small Grains Assn. Mem. Am. Soc. Agonomy (aronomic achievement awards com. 1989-90, Agronomic Achievement award 1988), Soil Sci. Soc. Am. (assoc. editor jour. 1990-93, chair-elect divsn. S-4, 1993). Achievements include development of restoration strategies for prime farmlands disturbed by mineral sands mining in the Coastal Plain of the Eastern U.S.; research in tissue and soil testing for increasing the efficiency of nitrogen utilization by wheat, efficient placement of N and P fertilizer for corn production, increasing nitrogen use efficiency in barley production, program for increasing barley yields and quality; evaluation of SPAD-502 chlorophyll meter for predicting optimum growth stage 30 nitrogen fertilizer rates for winter wheat. Office: Virginia Polytechnic Inst Blacksburg VA 24061*

ALLGAIER, GLEN ROBERT, electronics engineer, researcher; b. Washington, Sept. 23, 1940; s. Earl LeRoy and Esther Jane (Carlson) A.; m. Cynthia Margaret Allgaier, Apr. 10, 1963; children: Alene, Allison, April, Amber, Ayrih, Amy, Adam, Austin, Ammon. BS, Princeton, 1965; PhD, U. Ariz., 1971. Project systems engr. Naval Ocean Systems Ctr., San Diego, 1971-76, command and control br. head, 1976-81, command and control divsn. head, 1981-84, artificial intelligence br. head, 1984-87; tech. dir. Eaton Corp., Milw., 1987-90, exec. mgr. tech. devel., 1990—. Scoutmaster Boy Scouts Am., Tucson, Ariz., 1971, 72, Brookfield, Wis., 1987—; Mormon bishop, La Mesa, Calif., 1973-80. Office: Eaton Corp 4201 N 27th St Milwaukee WI 53216

ALLI, RICHARD JAMES, SR., electronics executive, service executive; b. McKeesport, Pa., June 16, 1932; s. James and Elizabeth (Hallas) A.; m. Margaret Ann Coursin, Mar. 17. 1950 (div. Dec. 1973); children: Richard James Jr., Deborah Elaine, Stephen John; m. D. Joan Love, Sept. 17, 1976. Cer. in electronics, Devry Inst., 1962. Co-owner R&R Auto Trim, MCkeesport, 1962-68, Poolside Motel, Geneva-on-the-Lake, Ohio, 1966—; structural steel insp. State of Ohio, 1968-79; pres. Pyramid 7 Corp., Warren, Ohio, 1979—; pres., chmn. bd., owner Corflex Internat., Inc., Warren, 1984—. V.p. Teenage Enterprises, Warren, 1979. Mem. KT, Shriners, Internat. Order of Merit. Republican. Avocations: swimming, surfing, boating, art purchasing, water and snow skiing. Home: 734 Kinsman St NW Warren OH 44483-3114 Office: Corflex Internat Inc PO Box 4324 Warren OH 44482-4324

ALLINGER, NORMAN LOUIS, chemistry educator; b. Alameda, Calif., Apr. 6, 1928; s. Norman Clarke and Florence Helen (Young) A.; m. Irene Saez; children: Alan Louis, Ilene Suzanne, James Augustus, Maritza Ivonne Quinn, Vilma Ivelise Veveles, Aida Irene Quinones. BS, U. Calif., Berkeley, 1951; Ph.D., UCLA, 1954. Research fellow U. Calif. at Los Angeles, 1954-55, Harvard, 1955-56; asst. prof. chemistry Wayne State U., 1956-59, assoc. prof., 1959-60, prof., 1960-69; prof. chemistry U. Ga., 1969—. Editor: Jour. Comp. Chemistry; contbr. articles to profl. jours. Served with AUS, 1946-48. Alfred P. Sloan fellow, 1959-63; Arthur C. Cope scholar, 1988. Mem. NAS, Am. Chem. Soc. (Herty medal Atlanta sect. 1988, James Flack Norris award 1989), Chem. Soc. London. Office: U Ga Comp Ctr Molecular Structure and Design Dept Chemistry Athens GA 30602

ALLISON, DAVID BRADLEY, psychologist; b. N.Y.C., Feb. 2, 1963; s. Ronald L. and Bernice C. (Goldschlager) A. BA in Psychology, Vassar Coll., 1985; MA in Clin. and Sch. Psychology, Hofstra U., 1987, PhD in Clin. and Sch. Psychology, 1990. Lic. clin. psychologist, N.Y.; cert. sch. psychologist, N.Y.; cert. instr. phys. intervention and restraint N.Y. State Office Mental Retardation and Devel. Disabilities. Behavior therapist Physician's Weight Loss Ctr., Hicksville, N.Y., 1985-86; child care worker Blueberry Treatment Ctr., Bklyn., 1985-86; psychiat. asst. St. Francis Hosp., Roslyn, N.Y., 1986-87; mental health asst. Mercy Hosp. Community Residence, Wantagh, N.Y., 1987-88; intern psychologist Farmingdale (N.Y.) Sch. Dist., 1987-88, Astor Child guidance Ctr., Bronx, N.Y., 1988-89; applied behavior specialist Plus Group Homes, Westbury, N.Y., 1988-90; postdoctoral fellow Johns Hopkins U. Sch. Medicine, Balt., 1990-91, Obesity Rsch. Ctr. Columbia U. Coll. Physicians and Surgeons, N.Y.C., 1991—; mem. adj. faculty St. John's U., Queens, N.Y., 1990, 92—, Nassau C.C., Garden City, N.Y., 1990, Hofstra U., Hempstead, N.Y., 1990; mem. Obesity & Health Task Force on Weight Loss Abuse; mem. adv. bd. Betty Jane Rehab. Ctr. Ohio; cons. to NIH grant; statis. cons. for doctoral students Hofstra U., Johns Hopkins U., Albert Einstein Coll. Medicine; mem. selection com. for TV entries for AAAS-Westinghouse Sci. Journalism 1990 awards. Ad hoc reviewer Jour. Applied Behavior Analysis, 1991, Children and Youth Svcs. Rev., 1991, Measurement and Evaluation in Counseling and Devel., 1992, Appetite, 1992, Jour. Cons. and Clin. Psychology, 1992, Psychol. Assessment, 1992, Internat. Jour. Obesity, 1992; contbr. articles to profl. jours. Recipient Clin. Nutrition Fellows award Am. Soc. Clin. Nutrition, 1992; granteeNIH, NATO, Am. Psychol. Assn., Orgnl. Behavior Mgmt. Network. Mem. AAAS, APA, Am. Psychol. Soc., Am. Assn. Applied and Preventive Psychology, Soc. Behavioral Medicine (Merit Citation 1992), N.Am. Assn. for Study Obesity, European Assn. for Study Obesity, Soc. for Study Ingestive Behavior, Assn. for Advancement Behavior Therapy, Am. Statis. Assn., The Biometric Soc., Inst. Math. Stats., Classification Soc. N.Am., Assn. for Measurement and Evaluation in Couseling and Devel., N.Y. Acad. Sci. Achievements include research on methodology for genetic studies of twins and families, relative effectiveness and cost-effectiveness of coop., competitive and ind. monentary incentive systems for improving staff performance. Office: Columbia U Coll Phys and Surgs Obesity Rsch Ctr 411 W 114th St # 3D New York NY 10025

ALLISON, JOHN MCCOMB, aeronautical engineer, retired; b. Guthrie, Okla., Nov. 27, 1901; s. John McComb and Mary Ann (Miller) A.; m. Dorothy Louise Olson, Nov. 15, 1931; children: John, Mary Ann, David. BSME, U. Ka., 1928. Aero. engr. Corman Aircraft Corp., Dayton, Ohio, 1928-29; airship engr. U.S. Naval Air Sta., Lakehurst, N.J., Akron, 1929-33; rsch. engr. Nat. Adv. Com. for Aeronautics, Langley Field, Va., 1933-38; engr. Bur. of Ordnance, Washington, 1945-46; head missile test dept U.S. Naval Air Missile Test Ctr., Pt. Mugu, Calif., 1946-50; head specifications Goodyear Aircraft Corp., Akron, 1957-59; proposal engr. Rockwell Internat., Columbus, Ohio, 1959-69, ret., 1969; cons. Allison Airplane Co., North Miami Fla., 1977-85; founder Langley Fed. Credit Union, 1936. Contbr. articles to profl. jours. Del. Nat. Rep. Planning Com., Washngton, 1992. Capt. USNR, 1939-46. Fellow AIAA (assoc.). Immanuel Lutheran. Achievements include patent for a wrap ground rocket to supply boost for another rocket engine or ram jet sustainer, thus shortening missile; a hybrid turbo rocket engine that can house solid booster propellant in its normal combustion chamber; patent disclosure for a retractable hydrofoil at the normal step position of a flying boat to improve take-off; design of Stinson Tri-Motor.

ALLISON, MERLE LEE, geologist; b. Phila., Jan. 15, 1948; s. Merle Raymond and Lois Loretta (Lynch) A. BS, U. Calif., Riverside, 1970; MS, San Diego State U., 1974; PhD, U. Mass., 1986. Registered geologist, Calif. Oil & gas engr. Calif. Divsn. Oil & Gas, Inglewood, 1971-72; geologist Chevron Corp., San Francisco, 1974-79; consulting geologist Amherst, Mass., 1980-84; scientist II Jet Propulsion Lab.-NASA, Pasadena, Calif., 1981; geologist Sohio Petroleum-BP, Dallas & Houston, 1984-87; sr. geologist U. Utah Rsch. Inst., Salt Lake City, 1987-89; dir., state geologist Utah Geol. Survey, Salt Lake City, 1989—; pres. Western Earth Sci. Techs. Inc., Casper, Wyo., 1988-93; chair Utah Geog. Info. Coun., Salt Lake City, 1992-93. Editor: Energy & Mineral Resources of Utah, 1991; contbr. articles to profl. jours. Press sec., mgr. Caprio for Congress, San Diego, 1972, 74; city precinct mgr. Unruh for Gov., Riverside, 1978; energy advisor Clifford for Congress, Houston, 1990. Mem. Am. Assn. Petroleum Geologists, Geol. Soc. Am., Soc. Profl. Well Log Analysts, Am. Geophys. Union, Rocky Mountain Assn. Geologists, Utah Geol. Assn., Assn. for Women Geoscientists, Wyo. Geol. Assn., Nev. Petroleum Soc. Achievements include research on geology of Ganymede; diplog interpretation, various of in-situ crustal structures. Home: 3037 S Plateau Dr Salt Lake City UT 84109 Office: Utah Geological Survey 2363 S Foothill Dr Salt Lake City UT 84109

ALLMAN, MARK C., physicist; b. Rochester, Pa., Aug. 4, 1958; s. Crawford Marcus and Darl Terresa (Hazenstab) A.; m. Mary Beth Decker, Apr. 30, 1983 (div. 1987); m. Janice Kay Hempleman, Dec. 8, 1989. BSBA, Robert Morris Coll., 1980; MS in Phys. Sci., U. Houston Clear Lake, 1991. Programmer/analyst Transcomm Data Systems, Inc., Pitts., 1980-81; ind. cons. Pitts., 1981-82; programmer/analyst, cons. ComTech Systems, Inc., Columbus, Ohio, 1982-83; systems programmer, mgr., project leader DataCom, Inc., Columbus, 1983-86; systems engr. R&D Discovery Systems, Dublin, Ohio, 1986-88; systems analyst On-Line Computer Libr. Ctr., Dublin, 1988-89; engr., physicist Rockwell Space Ops. Co., Houston, 1989—; mem. data collection team Allegheny Obs., U. Pitts., 1981-82; presenter at profl. confs. Co-author: Introduction to Astrodynamics, 1992. Mem. campaign coun. Rep. Nat. Com., Washington, 1991—; pres. Haw Rang Do Kung-Fu Martial Arts Team, 1991-93. Mem. AIAA, Am. Astron. Soc. (assoc.), Coun. Fgn. Rels., Wu Shu Kung Fu Fedn. (2nd degree black belt). Republican. Roman Catholic. Home: Apt 202 16460 Hwy 3 Webster TX 77598 Office: Rockwell Space Ops Co 600 Gemini St Houston TX 77058

ALLSBROOK, OGDEN OLMSTEAD, JR., economics educator; b. Wilmington, N.C., July 1, 1940; s. Ogden Olmstead Sr. and Elizabeth Barringer (Warren) A. BA, Wake Forest U., 1962; PhD, U. Va., 1966. Ops. rsch. analyst Dep. Def., Washington, 1966-68; asst. prof. econs. U. Ga., Athens, 1968-73, dir. grad. studies econs., 1971-81, assoc. prof., 1974—. Author: Utilization of Military Resources, 1969; contbr. articles to profl. jours. Capt. U.S. Army, 1966-68. Mem. AAUP, Nat. Soc. SAR (pres. Athens chpt. 1992—), Cape Fear Club, Atlanta Econ. Club, Pond Soc., So. Econ. Assn. Lutheran. Avocation: motor sports. Home: 115 Tillman Ln Athens GA 30606-4115

ALLSTOT, DAVID JAMES, electrical and computer engineering educator; b. Brookings, S.D., Jan. 22, 1947; s. Elmer C. and Ollie C. (Hillerud) A.; m. Vickie G. Randall, Dec. 21, 1968; children: Kevin J., Emily G. BS, U. Portland, 1969; MS, Oreg. State U., 1974; PhD, U. Calif., Berkeley, 1979. Rsch. asst. U. Calif., Berkeley, 1975-79; mem. tech. staff Tex. Instruments Inc., Dallas, 1979-80; design engr. Mostek, Inc., Carrollton, Tex., 1980-81; pvt. cons. Allstot and Assocs., Carrollton, Tex., 1981-85; vis. prof. U. Calif., Berkeley, 1985-86; prof. Oreg. State U., Corvallis, 1986-90; prof. elec. and computer engring. Carnegie Mellon U., Pitts., 1990—. Fellow IEEE (bd. govs. 1990-93, W.R.G. Baker award 1980); mem. Sigma Xi, Eta Kappa Nu. Democrat. Lutheran. Achievements include patents for current-steering CMOS logic family, MOS folded source-coupled logic. Office: Carnegie Mellon U Dept Elec Enging Pittsburgh PA 15213

ALM, ROGER RUSSELL, chemist; b. Kansas City, Mo., Nov. 19, 1945; s. Russell Arthur and Jean Grace (Frantz) A.; m. Laura Linnea Kiscaden, Oct. 14, 1983. BChemE, U. Minn., 1967; MS in Chemistry, Northwestern U., 1968. Adv. chemist Speciality Chems. divsn. 3M Co., St. Paul, 1968-72, sr. chemist, 1972-82, devel. specialist, 1982-93, patent liaison, 1993—. Contbr. articles to profl. jours. Singer 3M Male Chorus, St. Paul, 1970-85; bd. dirs., arranger Sundown, St. Paul, 1970-85; tutor sci. 3M Vol. Program, 1988—. Mem. Am. Chem. Soc., Alpha Chi Sigma (treas. 1982-84). Democrat. Unitarian Universalist. Achievements include patents and patents pending in fire-fighting foams for class B (hydrocarbon) fires, stabilized foam covers for sanitary land fills, waste sites, polymerizable fluorochemical surfactants. Home: 5187 Jamaca Ave N Lake Elmo MN 55042-9581 Office: 3M Co 3M Ctr Bldg 236-2B Saint Paul MN 55144-1048

ALMAN, TED IRWIN, engineer, consultant; b. Bklyn., July 26, 1946; s. Harry and Gertrude (Karp) A.; m. Renee Marcia Siegel, Aug. 30, 1969; children: Lee Jason, David Lawrence. AAS in Constrn. Tech., CUNY, 1967; BA in History, L.I. U., 1969; MS in Planning, Pratt Inst., 1972. Lic. comml. pilot, FAA; CFP. Draftsman Syska & Hennessey Engrs., N.Y.C., 1966-67; airport planner TransPlan, Inc., N.Y.C., 1972-73, Va. Dept. Aviation, Richmond, 1973-86; mgr. aviation planning, aviation divsn. N.C. Dept. Transp., Raleigh, 1986—; mgr. AeroPave divsn. The AeroGroup, Ltd., Raleigh, 1989—. Scout master Temple Beth El troop Boy Scouts Am., Richmond, 1982-86; pres. Three Chopt Recreation Assn., Richmond, 1985-86. Mem. ASTM, ASCE, Am. Inst. Cert. Planners (cert.), Transp. Rsch. Bd. Democrat. Jewish. Office: The AeroGroup Ltd PO Box 98274 Raleigh NC 27624-8274

ALMAZAN, AUREA MALABAG, biochemist; b. Manila, Philippines, Mar. 10, 1944; d. Cipriano Ayeras and Filomena (Malabag) A. BS in Chemistry, U. Philippines, 1965; MS in Biochemistry, U. Nebr., 1971, PhD in Biochemistry, 1974. Postdoctoral fellow dept. biochemistry Baylor Coll. Medicine, Houston, 1974; postdoctoral fellow dept. chemistry Internat. Rice Rsch. Inst., Los Banos, Philippines, 1975; asst. prof. dept. chemistry U. Philippines, Quezon City, 1975-76; researcher U. Philippines, Los Banos, 1976; sr. rsch. specialist Del Monte, Philippines, Davao City, 1977-82; biochemist, food technologist Internat. Inst. Tropical Agriculture, Ibadan, Nigeria, 1983-89; rsch. asst. prof. Tuskegee (Ala.) U., 1990—. Contbr. articles to profl. jours. UNESCO fellow Tokyo Inst. Tech., 1976-77. Mem. Inst. Food Tech. (profl.), Am. Chem. Soc., Sigma Xi. Office: Tuskegee U 100 Campbell Hall Tuskegee AL 36088

AL-MOHAWES, NASSER ABDULLAH, electrical engineering educator; b. Riyadh, Saudi Arabia, 1951; s. Abdullah Nasser and Lolwah Hamad Al-Suwailem; m. Fatma A., 1976; chldren: Hisham, Hend, Yaser, Riyadh. BSEE, Riyadh U., Saudi Arabia, 1976; MS in Nuclear Engring., MIT, 1979; PhD in Nuclear Engring., Purdue U., 1982. Teaching asst. elec. engring. dept. King Saud U., Riyadh, Saudi Arabia, 1976-82, asst. prof., 1982-87, assoc. prof., 1988—; dir. rsch. ctr. coll. engring. King Saud U., Riyadh, 1985-88, dean coll. engring., 1989-93; gen. mgr. Aseer Shareholder Co., 1993—; chmn. Third Saudi Engring. Conf., 1991. Contbr. numerous articles to profl. jours. Recipient Scholarship for Grad. Study, Grant for

Rsch. Mem. IEEE, Am. Nuclear Soc. Home: PO Box 85715, Riyadh 11612, Saudi Arabia Office: PO Box 800, Riyadh 11421, Saudi Arabia

ALMOND, MATTHEW JOHN, chemistry educator, researcher, writer; b. Blackburn, Lancashire, Eng., May 20, 1960; s. Raymond and Barbara (Crossley) A. BS in Chemistry with honors, U. Reading, Eng., 1981; DPhil in Inorganic Chemistry, U. Oxford, Eng., 1984. Rsch. fellow U. Munster, Fed. Republic Germany, 1984-85, U. Oxford, 1985-86; lectr. U. Reading, 1986—; chmn. bd. studies for geochemistry U. Reading, 1992—. Author: Spectroscopy of Matrix-Isolated Species, 1989, Short-Lived Molecules, 1990; contbr. articles to profl. jours. Recipient Eli Heyworth prize Borough of Blackburn, 1978; Sci. and Engring. Rsch. Coun. grantee, 1988—. Mem. Royal Soc. Chemistry (chmn. Thames Valley sect. 1988—, mem. coun. 1992—). Anglican. Avocations: cricket, walking, history, archaeology, running. Home: 35 Hilmanton, Lower Earley Reading RG6 4HN, England Office: U Reading, Dept Chemistry, Reading RG6 2AD, England

ALMQUIST, DONALD JOHN, retired electronics company executive; b. Elwood, Ind., Aug. 30, 1933; s. Elliott John and Gladys Ione (Jones) A.; m. Charline Gail Mull, Dec. 17, 1955; children: Gregory John, Tracy Gail. B in Indsl. Engring., Gen. Motors Inst., 1955. Supr. Delco Remy div. Gen. Motors Co., Anderson, Ind., 1956-64, plant supt., 1964-69, planr mgr., 1969-72, asst. dir. personnel, 1972-74, mgr. mfg., 1974-78, gen. mgr. mfg., 1978-82, gen. mgr., 1982-84; gen. mgr. Delco Electronics div. Gen. Motors Corp., Kokomo, Ind., 1984-86; v.p., gen. mgr. Delco Electronics Corp., Kokomo, 1986-89, chmn., pres., chief exec. officer, 1989-93; exec. v.p. GM Hughes Electronics, 1989-93; mem. Ind. Telecommunications Network Commn., Indpls., 1985-89; bd. dirs. Ind. Corp. for Sci. Tech., Indpls. Bd. mgrs. Rose Hulman Inst. Tech., Terre Haute, Ind., 1984—; bd. dirs. St. Johns Med. Ctr., Anderson, 1982-84, Ind. Vocat. Tech. Coll., Kokomo, 1984-86, St. Joe Hosp., Kokomo, 1984-86. Mem. Ind. C. of C. (bd. dirs. 1984—). Methodist. Lodge: Elks. Home: 5605 Four Mile Hill Dr Kokomo IN 46901-3829 Office: Gen Motors Corp Delco Electronics Corp 1 Corporate Ctr Kokomo IN 46902-4000

AL-OHALI, KHALID SULIMAN, aeronautical engineer, researcher, industrial development specialist; b. Riyadh, Saudi Arabia, Nov. 12, 1964; s. Suliman Abdullah Al-Ohali and H.M. Al-Saud. BS in Aero. Engring. with honors, U. Petroleum and Minerals, Dahran, Saudi Arabia, 1988. Gen. engr. Riyadh Petromin, 1987; aircraft engr. Saudi Arabian Airlines, Jeddah, Makkah, Saudia Arabia, 1988-91; liaison engr. Alsalam Aircraft Co., Riyadh, 1991-92; coord. mil./comml. programs devel. Alsalam Aircraft Co., 1992-93; project engr. Tornado Aircraft Maj. Svc. & Modification, 1993—. Contbr. articles to profl. jours. Organizer Handicap Inst., Riyadh, 1989; vol. Security Patrol during Gulf War, Riyadh, 1990-91. Mem. ASME, ASTM, Am. Soc. Metal Internat. (seminar com. Materials Engring. Inst. 1990), Am. Soc. Automotive Engrs. Moslem. Office: Alsalam Aircraft Co Ltd, PO Box 8012, Riyadh Saudi Arabia 11482

ALOIMONOS, YIANNIS JOHN, computer sciences educator; b. Sparta, Laconia, Greece, Aug. 17, 1957; came to U.S., 1982; PhD, Rochester U., 1986. Asst. prof. dept. computer sci. U. Md., 1986—; head Computer Vision Lab. Ctr. for Automation Rsch./U. Md., 1990—. Author: Integration of Visual Modules, 1989. Recipient Presdl. Young Investigator award White House/NSF, Washington, 1990, Maer prize hon. mention IEEE, London, 1987; grantee DARPA, U.S. Army Engring. Office: U Md Computer Vision Lab Ctr for Automation Research College Park MD 20742-1743

ALONSO, JOSE RAMON, physicist; b. San Francisco, July 26, 1941; s. Jose Maria and Irene (Windecker) A.; m. Carol Travis, July 26, 1969; children: Laura, Christopher. SB, MIT, 1962, PhD, 1967. Instr. MIT, Cambridge, 1967-68; rsch. scientist Yale U., New Haven, 1968-72; physicist Lawrence Berkeley (Calif.) Lab., 1972—, mgr. Bevalac ops., 1981-89, asst. dir. accelerator and fusion rsch. div., 1989-92; mem. rev. coms. various labs. Dept. Energy; mem. rev. adv. coms. Loma Linda (Calif.) U. Med. Ctr., 1987-91; mem. bd. Proton Therapy Coop. Group, 1986—. Editor conf. procs. Workshop on Highly Charged Ions, 1989. Recipient R&D 100 award, 1987. Mem. Am. Phys. Soc. Achievements include co-discovery of element 106; development of techniques for heavy ion acceleration, particle beams for cancer radiation therapy. Home: 91 Loma Vista Orinda CA 94563 Office: Lawrence Berkeley Lab MS 64-121 1 Cyclotron Rd Berkeley CA 94720

ALONSO-FERNANDEZ, JOSÉ RAMÓN, chemist; b. Orense, Spain, Mar. 16, 1946; s. Bernardino Alonso Fernandez and Felisa Fernandez Morais. Degree in chemistry, U. Santiago de Compostela, Spain, 1971. Asst. lectr. in analytical chemistry U. Santiago de Compostela, 1972-76, asst. lectr. in phys. chemistry, 1977-78, asst. lectr. in pediatrics, faculty of medicine, 1978-86, assoc. prof. pediatrics, 1986—; head metabolopathy sect. Galicia Gen. Hosp., Santiago de Compostela, 1979; coord. metabolopathy screening lab. Galicia Directorate, Gen. Pub. Health, Santiago de Compostela, 1989—; cons. in biotechnology. Contbr. papers to profl. publs.; patentee in field. Recipient prize II Premios Galacia De Saúde Investigation. Fellow Galician Coll. Chemistry; mem. Spanish Soc. Clin. Chemistry, Am. Chem. Soc., Chromatographic Soc., Internat. Soc. for Neonatal Screening, others. Home: Rua do Franco 52-2, 15702 Santiago Compostela Spain Office: Hosp Xeral de Galicia, 15705 Santiago Spain

ALOUANI, MEBAREK, physicist, research specialist; b. Sétif, Algeria, Mar. 1, 1958; came to U.S., 1989.; s. Lakhdar and Lalahom (Hamid) A.; m. Anne Marie Catherine Zint, Feb. 28, 1991; children: Emily, Adam. B in Math, U. Constantine, Algeria, 1978; M, U. Louis Pasteur, Strasbourg, France, 1983; PhD in Physics, U. Strasbourg, France, 1986. Postdoctoral position Max-Planck Inst., Stuttgart, Germany, 1986-89, Los Alamos (N.Mex.) Nat. Lab., 1989-91; rsch. specialist Ohio State U., Columbus, 1991—; cons. U. Nancy, France, 1991—, Los Alamos Nat. Lab., 1992—. Contbr. articles to Solid State Comms., Phys. Review B, Phys. Review Letters. Mem. Am. Phys. Soc., French Phys. Soc. Achievements include research in electronic structure of the MX halogen-bridged transition-metal linear chain compound and optical properties of semiconductors and semiconductor superlattices. Office: Ohio State U 174 W 18th Ave Columbus OH 43210

ALPEN, EDWARD LEWIS, biophysicist, educator; b. San Francisco, May 14, 1922; s. Edward Lawrence and Margaret Lilly (Shipley) A.; m. Wynella June Dosh, Jan. 6, 1945; children: Angela Marie, Jeannette Elise. B.S., U. Calif., Berkeley, 1946, Ph.D., 1950. Br. chief, then dir. biol. and med. scis. Naval Radiol. Def. Lab., San Francisco, 1952-68; mgr. environ. and life scis. Battelle Meml. Inst., Richland, Wash., 1968-69, assoc. dir., then dir. Pacific N.W. div., 1969-75; dir. Donner Lab., U. Calif., Berkeley; also assoc. dir. Lawrence Berkeley Lab., 1975-87; prof. biophysics U. Calif., Berkeley, 1975—; prof. radiology U. Calif., San Francisco, 1976—; dir. U. Calif. Study Ctr., London, 1988-90; councillor, dir. Nat. Council Radiol. Protection, 1969—; mem. Gov. Wash. Council Econ. Devel., 1973-75; bd. dirs. Wash. Bd. Trade, 1973-76. Author papers, abstracts in field. Served to capt. USNR, 1942-46, 50-51. Recipient Navy Sci. medal, 1962, Disting. Service medal Dept. Def., 1963, Sustaining Members medal Assn. Mil. Surgeons, 1971; fellow Guggenheim Found., 1960-61; sr. fellow NSF, 1958-59. Fellow Calif. Acad. Scis.; mem. Bioelectromagnetics Soc. (pres. 1979-80), Am. Physiol. Soc., Radiation Research Soc., Soc. Exptl. Biology and Medicine, Biophys. Soc., Brit. Inst. Radiology, Am. Philatelic Soc., Sigma Xi. Episcopalian. Office: Neutron Tech 1182 Miller Ave Berkeley CA 94708

ALPER, ANNE ELIZABETH, professional association executive; b. Montreal, Que., Can., Nov. 7, 1942; d. John S. and Emma (Flynn) Fairhurst; m. Howard Alper, June 4, 1966; children: Ruth and Lara (twins). BS in Chemistry with honors, Marianopolis Coll., Montreal, 1962; PhD in Organic Chemistry, McGill U., Montreal, 1966. Rsch. assoc. Textile Rsch. Inst., Princeton, N.J., 1968-69; mgr. publs. Chem. Inst. Can., Ottawa, Ont., 1981-86, exec. dir., 1986—. Office: Chem Inst Can, 130 Slater St Ste 550, Ottawa, ON Canada K1P 6E2

ALPERT, WARREN, oil company executive, philanthropist; b. Chelsea, Mass., Dec. 2, 1920; s. Goodman and Tena (Horowitz) A. B.S. Boston U, 1942; M.B.A., Harvard U., 1947. Mgmt. trainee Standard Oil Co. of Calif., 1947-48; financial specialist The Calif. Oil Co., 1948-52; pres. Warren Petroleum Co., 1952-54; now chmn. bd.; founder, pres., chmn. bd. Warren Equities, Inc., from 1954; chmn. Ritz Tower Hotel; chmn. bd. Kenyon Oil Co., Inc., Mid-Valley Petroleum Corp., Puritan Oil Co., Inc., Drake Petroleum Co., Inc.; mem. of U.S. Com. for UN, 1958; exec. com. Small Bus. Adminstrn., 1958; adminstr. for adminstrn. U.S. AID, 1962; former trustee, mem. exec. com. Boston U.; trustee Emerson Coll.; former v.p. Petroleum Marketing Edn. Found.; bd. dirs. Assocs. of Harvard Bus. Sch., Mass. Life; mem. com. for resource and devel. Harvard Med. Sch., bd. fellows. Bd. dirs. World Coun. Synagogues; bd. overseers Albert Einstein Med. Sch.; founder Warren Alpert Found.; bd. fellows Harvard Med. Sch.; former trustee Boston U., Emerson Coll. Andrew Wellington Cordier fellow Sch. Internat. Affairs, Columbia U.; Harvard Med. Sch. Rsch. Ctr. Bldg. named in his honor, 1993. Mem. Am. Petroleum Inst. (dir. mktg. divsn.), Harvard Bus. Sch. Club (exec. com., dir., bd. govs., pres. 1960-61), Am. Petroleum Industry 25 Year Club, Young Presidents Orgn. (past dir.), Harvard Club (N.Y.C., mem. house com.), Marco Polo Club, Met. Club, University Club. Home: 465 Park Ave New York NY 10022-1902 Office: Warren Equities Inc 375 Park Ave New York NY 10152

ALPHER, RALPH ASHER, physicist; b. Washington, Feb. 3, 1921; s. Samuel and Rose (Maleson) A.; m. Louise Ellen Simons, Jan. 28, 1942; children: Harriet Alpher Lebetkin, Victor. B.S. George Washington U, 1943, M.S., 1945, Ph.D., 1948; ScD honoris causa, Union Coll., 1992; ScD Honoris Causa, Renssalaer Poly. Inst., 1993. Physicist Bur. Ordnance and Naval Ordnance Lab., U.S. Navy, Washington, 1940-44, Applied Physics Lab., Johns Hopkins U., Silver Spring, Md., 1944-55, Gen. Electric Research and Devel. Ctr., Schenectady, 1955-86; disting. research prof. of physics Union Coll., Schenectady, 1986—; adj. prof. aero. engring. Rensselaer Poly. Inst., 1968-63, adj. prof. physics, 1986-92. Contbr. articles to books and profl. jours. in fields astrophysics, cosmology, physics of fluids. Bd. dirs. Mohawk-Hudson Council for Ednl. TV, 1974-80, 82-87, chmn., 1978-80, 86-87; bd. dirs. Dudley Obs., Union U., Albany, N.Y., 1968-72, 80-86, v.p. 1983-86, adminstr., disting. sr. scientist, 1987—. Recipient Magellanic Premium Am. Philos. Soc., 1975, Georges Vanderlinden prize Belgian Royal Acad. Scis., Letters and Fine Arts, 1975, John Price Wetherill medal Franklin Inst., 1980, Phys. and Math. Scis. prize N.Y. Acad. Scis., 1981, Disting. Alumnus award George Washington U., 1987, Henry Draper medal NAS, 1993. Fellow Am. Phys. Soc. (councillor-at-large 1979-82, exec. com. 1980-81), AAAS (sect. B physics steering com. 1982-86), Am. Acad. Arts & Scis.; mem. Fedn. Am. Scientists, Am. Astron. Soc., Sigma Xi. Club: Internat. Torch. Home: 2159 Orchard Park Dr Niskayuna NY 12309-2218 Office: Union Coll Dept Physics Schenectady NY 12308

ALPHER, VICTOR SETH, clinical psychologist, consultant; b. Washington, Oct. 20, 1954; s. Ralph Asher and Louise Ellen (Simons) A. BA, U. Pa., 1976; PhD, Vanderbilt U., 1985. Lic. psychologist, Tex., Tenn. Grad. fellow Vanderbilt U., Nashville, 1981-85; asst. prof. U. Tex. Health Sci. Ctr., Houston, 1986-88, clin. asst. prof., 1989—; cons. Rsch. Inst. on Addictions, Buffalo, 1990—, Meml. Geriatric Evaluation and Resource Ctr., Houston, 1991—. Contbr. articles to profl. jours., including Jour. Personality Assessment, Jour. Psychopathology and Behavioral Assessment, Psychotherapy. Mem. APA, Gerontol. Soc. Am., Soc. Psychotherapy Rsch., Soc. Personality Assessment, Sigma Xi.

AL-QADI, IMAD LUTFI, civil engineering educator, researcher; b. Nablus, Palestine, Feb. 5, 1962; came to U.S., 1985; s. Lutfi A. and Fatemah (Abdulmajeed) M.; m. Muna M. Atout, Dec. 24, 1987; children: Dana I., Nora I. BSCE, Yarmouk U., Irbid, Jordan, 1984; M in Engring., Pa. State U., 1986, PhD, 1990. Registered profl. engr., Pa. Project engr. Nablus, 1984-85; instr., rsch. engr. Pa. State U., University Park, 1988-90; asst. prof. Va. Tech. U., Blacksburg, 1990—; cons. AMACO, Atlanta, 1992, Charles Barger & Son, Inc., Lexington, Va., 1992—; mem. Va. Bituminous Adv. Rsch. Com., Charlottesville, Va., Va. Transp. Tech. Transfer Ctr.; founding mem. Va. Tech. Ctr. Infrastructure Assessment & Mgmt. Contbr. articles to profl. jours. including Transp. Engring./ASCE, IEEE Transaction of Geosci. and Rem. Sens., Jour. of Sci. and Engring. Corrosion/NACE, SHRP/Nat. Rsch. Coun. Grantee NSF, 1992, Ctr. Innovative Tech., 1992—, Ctr. Adhesive & Sealant Sci., 1991—. Mem. ASCE, ASTM, Transp. Rsch. Bd., Am. Soc. Engrs. Edn., Assn. of Asphalt Paving Technologists. Achievements include development of a dynamic triaxial apparatus to simulate stopping mechanism of arrester beds, of a new dynamic normal/shear testing technique to evaluate sealants, of a methodology to evaluate membranes used on bridge decks, of different techniques to abate steel corrosion in reinforced concrete; pioneering use of electromagnetic waves in hot-mix asphalt and concrete.

AL-SALQAN, YAHYA YOUSEF, computer science researcher; b. Nablus, West Bank, Mar. 7, 1962; came to U.S., 1989; s. Yousef Idrees and Maryam Yahya (Al-Jayeh) Al-S. BSEE, Birzeit U., Palestine, 1986; MS, Am. U., 1990; postgrad., U. Ill.-Chgo., 1991—. Registered profl. engr., Ill, Washington. Computer engr. Soudah Computer Systems, Ramallah, West Bank, 1986-89; network supr. Am. U., Washington, 1990-91; teaching asst., rsch. asst. U. Ill., Chgo., 1991—. Author papers in field. Fellow Am. U., Washington, 1989-90. Mem. IEEE, Sigma Xi. Avocations: soccer, public speaking, horseback riding, reading. Home: 3738 N Saint Louis Chicago IL 60618 Office: U Ill Chgo EECS Dept M/C 154 851 S Morgan St Rm 1120 SEO Chicago IL 60607

AL-SARI, AHMAD MOHAMMAD, data processing executive; b. Al-Mukalla, South Yemen, Feb. 22, 1947; s. Mohammad Salem and Fatima Daoud (Al-Jilani) Al-S.; student U. Petroleum and Minerals, Dhahran, Saudi Arabia, 1965-67; B.Sc. in Chem. Engring., U. Tex., Austin, 1970. Lab. asst. Center Hwy. Research, Austin, 1969-70; programmer/analyst Data Processing Center, U. Petroleum and Minerals, 1970-76, dir. data processing center, 1976-80; pres. Al-Khaleej Computers & Electronic Systems, Riyadh, Saudi Arabia, 1980—; dir. Saudi-Am. Systems Engring. Co., Steria Al-Khaleej Co., United Computer Svcs. Co., Dar Al-Salam Chocolate Co.; co-founder Al-Falak Electronic Equipment and Supplies Co., United Computer Svcs. Co., United Systems Engring. Co., Steria Al-Khaleej Co.; cons. various Saudi Arabian ministries; chmn. 1st Nat. Computer Conf., 1974. Mem. Assn. Computing Machinery U.S. (assoc.), Saudi Computer Soc. (founding mem.). Editor Procs. Symposium on Arabic Code Standards, Saudi Arabian Standards Orgn., 1981, The DPC Bull., 1972-76. Office: PO Box 16091, Riyadh 11464, Saudi Arabia

ALSHARIF, NASER ZAKI, toxicologist, educator, researcher; b. Nablus, Palestine, Aug. 19, 1960; came to U.S., 1980; s. Rouda A. Younes. PharmD, U. Nebr., Omaha, 1987; PhD, Creighton U., 1987. Staff pharmacist Bergan Mercy Med. Ctr., Omaha, 1988—; lectr. Creighton U., Omaha, 1988—, asst. prof. toxicology, 1992—; vis. scientist Epply Cancer Inst., Omaha, 1992—. Contbr. articles to Xenobiotica, Toxicology Applied Pharmacology, Biol. Reactions Intermed. Mem. Am.-Arab Anti-Discrimination Com., Washington, 1989—. Scholar Nat. Assn. Bds. Pharmacy, 1987. Mem. Soc. Toxicology, Nebr. Pharmacy Assn., Am.-Arab Heritage Soc. (chmn. social and sport com. 1992), Sigma Xi (grantee 1989, travel grants 1988-91). Achievements include emphasis on role of oxidative stress in the overall toxicity of TCDD; establishment that decrease in membrane fluidity increased in lipid peroxidation and enhanced production of reactive oxygen species by peritoneal exudate cells; tumor necrosis factor (TNF-2) was implicated in the production of an oxidative stress state following acute exposure to TCDD;; dexamethasone, antioxidants, scavengers of reactive oxygen species and a monoclonal antibody against TNF-2 protected against the acute toxicity of TCDD. Home: 4519 S 145th St Omaha NE 68137 Office: Epply Cancer Inst 600 S 42d St Omaha NE 68144

ALSTON, STEVEN GAIL, physicist, educator; b. Searcy, Ark., July 20, 1953; s. Robert and Charlotte Gertrude (Dye) A.; m. Carol Anne Hilt, May 20, 1977 (div. 1980); m. Kay Lynne Bredthauer, Jan. 3, 1983; children: Erin, Hope. BS in Physics, Math., Ft. Hays State U., 1976; PhD in Physics, U. Nebr., Lincoln, 1982. Rsch. and teaching assoc. U. Freiburg, Germany, 1983-85; rsch. assoc. Joint Inst. Lab. Astrophysics, Boulder, Colo., 1985-87; asst. prof. physics Pa. State U., Wilkes-Barre, Lehman, 1987-93; assoc. prof. physics Pa. State U., Wilkes-Barre, 1993—. Contbr. articles to Phys. Rev., Jour. Phys. Grantee Dept. Energy, 1989—. Mem. Am. Phys. Soc. Achievements include development of strong-potential born perturbative technique for electron capture, Faddeev theory representation of double-scattering mechanism of capture at high velocities. Office: Pa State U Wilkes Barre Campus Lehman PA 18627

ALTAMIRANO, ANIBAL ALBERTO, physiologist; b. la Rioja, Argentina, Nov. 14, 1954; came to U.S. 1985; s. Lasaro Anibal and Maria Teresa (Semeraro) A. Medico cirujano, U. Nacional, Cordoba, Argentina, 1979, D in Medicine, 1984. Rsch. assoc. U. Tex. Med. Br., Galveston, 1985-92, Med. Coll. Pa., Phila., 1992--. Contbr. articles to profl. jours. Office: Med Coll Pa 2900 Quenn Ln Philadelphia PA 19129

ALTAMURA, MICHAEL VICTOR, physician; b. Bklyn., Sept. 28, 1923; s. Frank and Theresa (Inganamorte) A.; m. Emily Catherine Wandell, Sept. 21, 1948; children: Michael Victor, Robert Frank. BS, LIU, 1949; MA, Columbia U., 1951; DO, Kirksville Coll., 1961; MD, Calif. Coll. Medicine, 1962. Diplomate Am. Bd. Family Practice. Intern Los Angeles County Gen. Hosp., 1961-62; practice medicine specializing in family practice Sunnyvale, Calif., 1962—; staff El Camino Hosp., chief family practice dept., 1972-73; preceptor family practice Stanford Sch. Medicine, 1972-73, clin. asst. prof., 1974-81, clin. assoc. prof., 1982—; assoc. prof. family medicine Calif. Coll. Osteopathic Medicine, 1985—; preceptor family practice Davis (Calif.) Sch. Medicine, 1974-75. Author: (with Mary Falconer and Helen Behnke) Aging Patients: A Guide for Their Care. Served to 1st lt. AUS, 1942-45, 51-53; ETO. Recipient Order of Golden Sword, Am. Cancer Soc., 1973. Fellow Am. Acad. Family Physicians (pres. Santa Clara County chpt. 1972-73, Calif. del. Santa Clara chpt. 1991), Calif. Acad. of Family Physicians (bd. dirs. 1987-90), Royal Soc. Health, Am. Geriatric Soc.; mem. AMA, Calif., Santa Clara County socs., Internat. Platform Assn. Republican. Lutheran. Office: 500 E Remington Dr Sunnyvale CA 94087-2657

ALTAN, TAYLAN, engineering educator, mechanical engineer, consultant; b. Trabzon, Turkey, Feb. 12, 1938; came to U.S., 1962; s. Seref and Sadife (Baysal) Kadioglu; m. Susan Borah, July 18, 1964; children—Peri Michele, Aylin Elisabeth. Diploma in engring., Tech. U., Hannover, Fed. Republic Germany, 1962; M.S. in Mech. Engring., U. Calif.-Berkeley, 1964, Ph.D. in Mech. Engring., 1966. Research engr. DuPont Co., Wilmington, Del., 1966-68; research scientist Battelle Columbus Labs, Ohio, 1968-72, research fellow, 1972-75, sr. research leader, 1975-86; prof. mech. engring., dir. engring. rsch. ctr. Ohio State U., Columbus, 1985—; chmn. sci. com. N.Am. Mfg. Rsch. Inst. Soc. Mfg. Engrs., Detroit, 1982-86, pres., 1987; dir. Ctr. for Net Shape Mfg. Co-author: Forging Equipment, 1973, Metal Forming, 1983, Metal Forming and the Finite Element Method, 1989; assoc. editor Jour. Materials Processing Tech., Eng., 1978—; contbr. over 150 tech. articles to profl. jours. Fellow Am. Soc. Metals (chmn. forging com. 1978-87), Soc. Mfg. Engrs. (Gold medal 1985), ASME. Avocations: languages, travel.

ALTER, BLANCHE PEARL, physician, educator; b. Toronto, Ont., Can., May 1, 1941; came to U.S., 1959; d. Sol and Florence (Andrews) A. AB, Radcliffe Coll., 1963; MD, Johns Hopkins U., 1967. Intern, resident Johns Hopkins Sch. Medicine, Balt., 1967-70; fellow Harvard Med. Sch., Boston, 1970-73, asst. prof., 1974-78, assoc. prof., 1978-79; fellow MIT, Cambridge, 1972-74; assoc. prof. Mt. Sinai Sch. Medicine, N.Y.C., 1979-83; prof., 1983-92; prof., chief pediatric hematology/oncology U. Tex. Med. Br., Galveston, 1992—. Editor: Methods in Hematology-Perinatal Hematology, 1989, Aplastic Anemia-Acquired and Inherited, 1993; contbr. 140 articles to profl. jours., chpts. to books. Recipient Basil O'Connor Starter Rsch. award March of Dimes, 1974, Rsch. Career Devel. award NIH, 1975, Irma T. Hirschl Career Scientist award Hirschl Trust, N.Y.C., 1980.

ALTING, LEO LARSEN, manufacturing engineer, educator; b. Aalborg, Denmark, Apr. 15, 1939; s. Edvard and Caroline Marie (Larsen) A. M.Sc. in Mech. Engring., Tech. U. Denmark, 1965, Ph.D., 1969. Registered profl. engr.; m. Greta Jakobsen, Nov. 6, 1965; children: Rasmus, Caroline. Postdoctoral rsch. fellow U. Denver, 1970-71; head lab. process and prodn. engring. Tech. U. Denmark, 1971—, prof. mfg. engring., 1976—; div. dir. Inst. Product Devel.; bd. dir. AGA Inc., Industriharderiet A/S, CIM Cons. ApS, Surfcoat Aps.; co-dir. mfg. consortium Brigham Young U. With Danish Army, 1965-67. Recipient Prof. Wilkens award, 1967, SKF Indsl. award, 1974. Mem. Danish Acad. Sci., Royal Swedish Acad. of Engring. Scis., Soc. Mfg. Engrs. (Internat. Edn. award 1988), Danish Metall. Soc., Danish Engring. Soc., Danish Welding Soc., Sci. Rsch. Coun., Internat. Product Engring. Rsch. Inst., Inst. Mfg. Engring. Tech. U. Denmark (chmn.). Author: Manufacturing Engineering Processes, 1974,, 2nd edit., 1990, English edit. (Marcel Dekker), 1981, 2nd edit., 1993, Tool and Die Design, 1975, (with H. C. Ehang) Computerized Process Planning, 1993; contbr. articles to profl. jours. Home: 60 Harreshojvej, DK 3080 Tikob Denmark Office: Tech U, Inst Mfg Engring, Bldg 425, DK 2800 Lyngby Denmark

ALTMAN, DAVID WAYNE, geneticist; b. Portsmouth, Va., June 6, 1951; s. Ivan Stuart and Elaine Louise (Koll) A.; m. Kathleen Mary Healy, Dec. 16, 1972; children: Erin Healy, Regan Lewisia, Devin Armand. BA, Vanderbilt U., 1972; BS, Oreg. State U., 1978, MS, 1981; PhD, U. Minn., 1983. Cert. secondary sch. tchr. Vol. prof. Peace Corps, Chott Mariem, Tunisia, 1974-75; curriculum developer dept. crop sci. Oreg. State U., Corvallis, 1977-78, rsch. asst. dept. crop sci., 1978-80; rsch. asst. dept. agronomy and plant genetics U. Minn., St. Paul, 1980-83, postdoctoral specialist dept. agronomy and plant genetics, 1983; rsch. geneticist USDA Agrl. Rsch. Svc., College Station, Tex., 1983-92; prof. dept. plant breeding & biometry Cornell U., Ithaca, N.Y., 1992—; vice chair faculty of genetics Tex. A&M U., College Station, 1989-91, adj. assoc. prof., 1989; faculty Met. State U., Mpls., 1981-83; pres. Internat. Svc. for Acquisition of Agri-Biotech Applications, 1992—. Assoc. editor: In Vitro Cellular and Developmental Biology, 1990—, edit. adv. bd. CAD International, AgBiotech News and Information; contbr. articles to Jour. Heredity, Genome, Plant Cell Reports, Crop Science, others; contbr. chpts. to books. Pres. Unitarian Fellowship, College Station, 1988; v.p. Aggie Swim Club, College Station, 1987; v.p. Falcon Heights Elem. Sch. PTA, St. Paul, 1979. Oreg. Seed Trade Commn. scholar, 1978, DuPont Corp. fellow, 1979, 80; State of Tex. grantee, 1987, 89, USDA grantee, 1990, 91. Mem. Tissue Culture Assn. (exec. coun. 1990—), Crop Sci. Soc. Am., Am. Genetic Assn., Sigma Xi (coord. task force 1989-91). Achievements include co-achievement commercialization for genetically-engineered cotton; co-discoverer of the range of terpenoid production in cotton plants, including the aspects of their inheritance, basic tissue culture methods for cotton which are the foundation of modern biotechnology for this crop. Office: Cornell U Dept Plant Breeding and Biometry 262 Emerson Hall Ithaca NY 14853-1902

ALTMAN, SAMUEL PINOVER, mechanical engineer, research consultant; b. Atlantic City, Apr. 15, 1921; s. Ben and Beatrix (Pinover) A.; m. Francine Danish, Oct. 5, 1943; children: Ellen Beatrix, Sharon Anita. BSME, CCNY, 1942. Registered profl. engr., Ohio. System engr. USAF, Lear, Lockheed, Bendix, various, 1943-58; prin. engr. Martin Co., Waterton, Colo., 1958-61; supr. space scis. United Aircraft Systems Ctr., Farmington, Conn., 1961-63; cons. engr. GE, Missile & Space Div., King of Prussia, Pa., 1963-69; sr. staff engr. System Devel. Corp., Santa Monica, Calif., 1969-72; dir. space mechanics Can. Govt., Dept. Communications, Ottawa, Ont., Can., 1972-85; rsch. cons. Ottawa, 1985—. Author: Orbital Hodograph Analysis, 1965. Capt. USAF, 1943-53. Fellow Can. Aero. and Space Inst. (assoc.); mem. AIAA, IEEE, Am. Astronautical Soc. Achievements include 3 patents on aircraft inertial instruments; development of math. methods for selection/optimization of orbital parameters for 2-impulse orbital transfer, of algebraic transformations of orbital parameters between position, velocity and acceleration states, of alternative hodographic formulations for orbit analysis, based upon orbital hodograph parameters. Home and Office: 465 Richmond Rd # 2107, Ottawa, ON Canada K2A 1Z1

ALTMAN, SIDNEY, biology educator; b. Montreal, Que., Can., May 7, 1939. BS, MIT, 1960; PhD in Biophys., U. Colo., 1967; DSc (hon.), McGill U., Montreal, 1991. Teaching asst. Columbia U., 1960-62; Damon Runyon Meml. Fund cancer rsch. fellow in molecular biology Harvard U., 1967-69; Anna Fuller Fund fellow, then Med. Rsch. Coun. fellow Med. Rsch. Coun. Lab. Molecular Biology, 1969-71; from asst. to assoc. prof. Yale U., New Haven, 1971-80, prof. biology, 1980—, chmn. dept., 1983-85; dean Yale Coll., from 1985; tutor Radcliffe Coll., 1968-69; researcher effects of acidrines on T4 DNA replications, mutants, precursors of RNA processing by catalytic RNA and ribonuclease function. Recipient Nobel prize in chemistry, 1989. Fellow AAAS; mem. Am. Soc. Biol. Chemists, Genetics Soc. Am. Office: Yale U Dept Biology PO Box 6666 New Haven CT 06520

ALTMANN, JEANNE, zoologist, educator. BA in Math., U. Alta., Can., 1962; MAT, Emory U., 1970; PhD, U. Chgo., 1979. Rsch. assoc., co-investigator U. Alta., Can., 1963-65, Yerkes Regional Primate Rsch. Ctr., Atlanta, 1965-67, 69-70; rsch. assoc. dept. biology U. Chgo., 1970-85, assoc. prof. dept. ecology and evolution, 1985-89, prof., 1989—; rsch. curator, assoc. curator primates Chgo. Zool. Soc., 1985—; hon. lectr. dept. zoology U. Nairobi, Africa, 1989-90; chair com. evolutionary biology U. Chgo., 1991—; bd. sci. dirs. Karisoke Rsch. Ctr., Rwanda, 1980-82, 86-89, acting chairperson, 1980; mem. biosocial perspectives on parent behavior and off spring devel. com. Social Sci. Rsch. Coun., 1984-91; mem. adv. coun. dept. ecology and evolutionary biology Princeton (N.J.) U., 1991—; mem. rev. com. dept. zool. rsch. Nat. Zool. Park, Smithsonian Inst., Washington, 1992; mem. com. Internat. Ethol. Congress, 1992—; mem. vis. com. dept. anatomy and biol. anthropology Duke U., Durham, N.C., 1993; reviewer manuscripts various jours. Author: (with S. Altmann) Baboon Ecology: African Field Research, 1970, Baboon Mothers and Infants, 1980; editor: Animal Behavior, 1978-82; consulting editor: Am. Jour. Primatology, 1981—; mem. editorial panel: Monographs in Primatology, 1982-90; mem. editorial bd. Bioscience, 1983-88, ISI Reviews in Animal Science, 1988, Human Nature, 1989-92, Internat. Jour. Primatology, 1990—, Am. Naturalist, 1991—; contbr. articles to profl. jours. Fellow Ctr. Advanced Study in Behavioral Scis., 1990-91. Fellow Animal Behavior Soc. (mem. exec. com 1978-82, 84-87, mem, nominating com. 1987-89, pres. 1985-86), Animal Behavior Soc.; mem. NSF (mem. sci. adv. panel psychobiology program 1983-86, mem. adv. panel for vis. professorships for women 1987, 88, mem. adv. panel conservation and restoration biology 1990, mem. task force behavioral, biol. and social scis. Looking Toward the 21st Century 1990-91, mem. adv. coun. directorate for social, behavioral and econ. scis. 1992—), Internat. Primatol. Soc. (v.p. conservation, mem. exec. com.). Home: 1507 E 56th St Chicago IL 60637 Office: U Chgo Ecol and Evolution Dept 940 E 57th St Chicago IL 60637

ALTOMARI, MARK G., clinical psychologist; b. Ft. Monroe, Va., July 13, 1947; s. Guido and Mary Ann (Zalos) A.; m. Susan Alice Gross, Mar. 22, 1969; children: Alicia, Devin, Paul. BA, Villanova U., 1974; MA, West Chester State U., 1978; MS, Va. Poly. Inst. and State U., 1982, PhD, 1984. Lic. psychologist. Traffic hearing examiner Bur. Traffic Safety, Commonwealth of Pa., Harrisburg, 1973-78; behavioral specialist Community Svcs. Inc., Lancaster, Pa., 1977-78; counselor New River Valley Coun. on Alcoholism, Christiansburg, Va., 1981-82; psychology technician VA Med. Ctr., Salem, Va., 1981-82; case mgr. New River Valley Alcohol Safety Action Project, Christiansburg, 1982; clin. psychologist Va. Poly. Inst., Blacksburg, 1982-83; clin. psychology supr. Fulton (Mo.) State Hosp., 1984-86; pvt. practice clin. psychologist Columbia, Mo., 1986—; cons. Disability Determinations, Jefferson City, Mo., 1986—, Archdiocese of Jefferson City, 1987—, Fulton Police Dept., 1988—; cons. cert., forensic examiner Fulton State Hosp., 1986—; dir. of counseling William Woods Coll., Fulton, 1985—. Contbr. articles to profl. jours. Coach, mgr. Columbia Soccer Club, 1985; cons. Mo. Water Patrol, 1990—; fund raiser Clearview Neighborhood Assn., Columbia, 1990; assoc. chief justice Grad. Honor Ct.; pres. Strategy and Tactics Soc., Pottstown, Pa., 1975-80; dep. Pa. Athletic Commn., 1972. Senatorial scholar Commonwealth of Pa., 1971; rsch. grantee Gen. Motors, 1983. Mem. Am. Psychol. Assn. (divsn. clin. psychology, divsn. clin. hypnosis, psychology-law soc., neuropsychology divsn., divsn. psychologists in pub. svc., divsn. psychologists in ind. practice, divsn. women's issues, divsn. consulting), Mo. Psychol. Assn. Avocations: military history, sports, chess, wargaming, physical fitness. Office: 916 N College Ave Columbia MO 65201-4784

ALTON, CECIL CLAUDE, computer scientist; b. Annapolis, Md., June 15, 1943; s. Howard Walter and Helen Mullaney (Simmons) A.; m. Ieda Mendonca Constantino, July 6, 1970; children: Valerie, Robert, Richard. BBA, U. Tex., 1965; M Computer Sci., Navy Postgrad. Sch., Monterey, Calif., 1977. From commd. ensign aviator to lt. comdr. USN, 1965-85, ret., 1985; mgr. Electronic Data Systems, Herndon, Va., 1985—. Mem. Data Processing Mgrs. Assn. Republican. Lutheran. Home: 15510 Laurel Ridge Rd Dumfries VA 22026 Office: Electronic Data Systems 13600 EDS Pl Herndon VA 22171

ALTSHILLER, ARTHUR LEONARD, physics educator; b. N.Y.C., Aug. 12, 1942; s. Samuel Martin and Betty Rose (Lepson) A.; m. Gloria Silvern, Nov. 23, 1970 (div. 1975); m. Carol Heiser, Aug. 16, 1980. BS in Physics, U. Okla., 1963; MS in Physics, Calif. State U., Northridge, 1971. Elec. engr. Garrett Corp., Torrance, Calif., 1963-64, Volt Tech. Corp., Phoenix, 1965; engr., physicist Aerojet Gen. Corp., Azusa, Calif., 1966-68; elec. engr. Magnavox Rsch. Labs., Torrance, 1968-69; sr. engr. Litton Guidance & Control, Canoga Park, Calif., 1969; physics tchr. L.A. Unified Sch. Dist./Van Nuys Math/Sci. Magnet High Sch., 1971—; math. instr. Valley Coll., Van Nuys, Calif., 1986—; part-time physics/chemistry tchr. West Coast Talmudical Sem., L.A., 1978-88. Mesa Club sponsor Math.-Engring. Sci. Achievement L.A. High Sch. and U. So. Calif., 1984-87. Recipient Cert. of Honor Westinghouse Sci. Talent Search, 1990. Mem. AAAS, Am. Assn. Physics Tchrs., Nat. Coun. Tchrs. of Math., So. Calif. Striders, Santa Monica Astron. Soc., United Tchrs. L.A. Democrat. Jewish. Avocations: tennis, table tennis, shot put, swimming, tournament chess. Home: 6776 Vickiview Dr Canoga Park CA 91307-2751 Office: Van Nuys High Sch 6535 Cedros Ave Van Nuys CA 91411-1599

ALTSHULER, DAVID THOMAS, computer scientist; b. Boston, Dec. 14, 1961; s. John H. and Barbara A. Altshuler; m. Sharman Buechner, July 14, 1990. BA in Polit. Economy, Williams Coll., 1984; MBA, Pa. State U., 1993. Rsch. assoc. Williams Coll., Williamstown, Mass., 1984-86; researcher Internat. Med. Industries, Denver, 1986-87; dir. MIS Internat. Investors, N.Y.C., 1987-90; pres. Mut. Analytics Corp., Dorset, Vt., 1989—; adviser Apollo Adv. Group, N.Y.C., 1989—. Designer software Mutual Marketscope, 1991. Treas. Dorset (Vt.) Fire Dept., 1989-91. Recipient Gold Medal Rsch. award Colo.-Wyo. Acad. Sci., 1980. Mem. Investment Co. Inst. (operation com. 1987-90). Achievements include co-invention of PapPrint device. Office: Mutual Analytics/Empath Corp 3721 Chestnut St Philadelphia PA 19104

ALTURA, BELLA T., physiologist, educator; b. Solingen, Federal Republic Germany; came to U.S., 1948; d. Sol and Rosa (Brandstetter) Tabak; m. Burton M. Altura, Dec. 27, 1961; 1 child, Rachel Allison. BA, CUNY, 1953, MA, 1962, PhD, 1968. Instr. exptl. anesthesiology Albert Einstein Coll. Medicine, Bronx, 1970-74; asst. prof. physiology SUNY Health Sci Ctr., Bklyn., 1974-82, assoc. prof. physiology, 1982—; vis. prof. Beijing Coll. of Traditional Chinese Medicine, 1988, Jiangsu (China) Med. Coll., 1988, Tokyo U. Med. Sch., 1993; mem. Nat. Coun. on Magnesium and Cardiovascular Disease, 1991—; cons., Niché pharm. cons. Protina GmbH, Munich, 1992—. Contbr. over 250 articles to profl. jours. Fellowship NASA, 1966-67, CUNY, 1968; recipient Gold-Silver medal French Nat. Acad. Medicine, 1982, Silver medal Mayor of Paris, 1982. Mem. Am. Physiol. Soc., Am. Soc. Pharmacol. and Exptl. Therapy, Am. Soc. for Magnesium Rsch. (founder, treas. 1984—), Phi Beta Kappa. Achievements include first measurement ionized magnesium with ion selective electrode in blood, serum and plasma; demonstration that substances of abuse can cause cerebrovasospasm and stroke. Office: SUNY Health Sci Ctr Box 31 450 Clarkson Ave Brooklyn NY 11203

ALTURA, BURTON MYRON, physiologist, educator; b. N.Y.C., Apr. 9, 1936; s. Barney and Frances (Dorfman) A.; m. Bella Tabak, Dec. 27, 1961; 1 child, Rachel Allison. BA, Hofstra U., 1957; MS, NYU, 1961, PhD (USPHS fellow), 1964. Teaching fellow in biology NYU, N.Y.C., 1960-61, instr. exptl. anesthesiology Sch. Medicine, 1964-65, asst. prof. Sch. Medicine, 1965-66; asst. prof. physiology and anesthesiology Albert Einstein Coll. Medicine, N.Y.C., 1967-70, assoc. prof., 1970-74, vis. prof., 1974-78; prof. physiology SUNY Health Sci Ctr., Bklyn., 1974—, prof. medicine, 1992—; rsch. fellow Bronx Mcpl. Hosp. Center, 1967-76; mem. spl. study sect. on toxicology Nat. Inst. Environ. Health Scis., 1977-78; mem. Alcohol Biomed. Rsch. Rev. Com., Nat. Inst. Alcohol Abuse and Alcoholism, 1978-83; adj. prof. biology Queens Coll., CUNY, 1983-84; cons. NSF, VA Grants Rev. Commn, Nat. Heart, Lung and Blood Inst., CUNY, Miles Inst., Nat. Inst. Drug Abuse, Merck, Sharpe and Dohme, Millipore Corp., Internat. Ctr. Disabled, Upjohn Co., Bayer AG, Ciba-Geigy, Zyma SA., Genentech, Nova Biomed., Parke, Davis & Co., Chem. Def. Unit, Brit. Govt., Schering-Key Corp., Sterling Drug Co., Searle Pharm. Co., Niché Pharm. Inc., Chem. Def. Establishment, U.K., Am. Speech and Hearing Assn., Protina GmbH; hon. pres. Internat. Symposium on Interactions of Magnesium and Potassium on Cardiac and Vascular Muscle, Montbazon, France, 1984; organizer, condr. symposia; organizer workshop Nat Inst. Alcohol Abuse and Alcoholism, 1992; condr., chmn. Gordon Rsch. Conf. on Magnesium in Biochem. Processes and Medicine, 1984, symposium Am. Soc. Nephrology, 1993; v.p. Fourth Internat. Symposium on Magnesium, Blacksburg, 1985; judge Am. Inst. Sci. and Tech., 1984, 85, 86, 88, 89, 90, 91, 93, Jr. Acad. N.Y. Acad. of Scis., 1987, 89, 90; mem. adv. council Nat. Found. for Addictive Drugs, 1986—; vis. prof. Yamaguchi U., Japan, 1988, 93, Beijing Coll. Traditional Chinese Medicine, People's Republic of China, 1988, Jiangsu Med. Coll., 1988, Beijing Med. U., 1988, Mass. Gen. Hosp., Harvard U. Med. Sch., 1989, U. Tokyo, 1993, Kokura Meml. Hosp, Kiyushi U. Japan, 1993, Yamaguchi U. Hosp., Japan, 1993, Tokyo U., 1993, Kyoto Sch. Medicine, 1993, Kumamoto U. Sch. Medicine, 1993; hon. prof. Yamaguchi U. Hosp., Japan, 1988; mentor Aaron Diamond fellowships, 1990—; vis. prof., hon. lectr. Inst. for Water, Soil and Air Hygiene, Fed. Health Inst., Berlin, 1991; vis. prof., hon. lectr. Max Planck Inst., Dortmund, Ger., 1992; mem. working group convened by Congressman Durbin III, 1991; mem. Nat. Coun. on Magnesium and Cardiovascular Disease, 1991—; hon. lectr. Yamonouchi Co. Ltd., Japan, 1993, Searle Co., Japan, 1993. Author: Microcirculation, 3 vols., 1977-80, Vascular Endothelium and Basement Membranes, 1980, Pathophysiology of the Reticuloendothelial System, 1981, Ionic Regulation of the Microcirculation, 1982; Handbook of Shock and Trauma, Vol. 1: Basic Science, 1983, Magnesium and the Cardiovascular System, 1985, Cardiovascular Actions of Anesthetic Agents and Drugs Used in Anesthesia, vol.,1986, vol. II, 1987, Magnesium, Stress and the Cardiovascular System, 1986, Magnesium in Biochemical Processes and Medicine, 1987, Magnesium in Clinical Medicine and Therapeutics, 1992, Magnesium in Clinical Medicine and Therapeutics, 1992; editor in chief: Physiology and Patho-physiology Series, 1976-81, Microcirculation, 1980-84, Magnesium: Exptl. and Clin. Research, 1981-89, Microcirculation, Endothelium and Lymphatics, 1984—, Magnesium and Trace Elements, 1990—; mem. editorial bd.: Jour. Circulatory Shock, 1973-85, Advances in Microcirculation, 1976—, Jour. Cardiovascular Pharmacology, 1977-84, Prostaglandins, Leukotrienes and Fatty Acids, 1978—, Substance and Alcohol Actions/Misuse, 1979-84, Alcoholism: Clin. and Exptl. Research, 1982-87; assoc. editor: Jour. of Artery, 1974—; assoc. editor: Microvascular Research, 1978-'85, Agents and Actions, 1981-88, Biogenic Amines, 1985-88, Jour. Am. Coll. Nutrition, 1982—; over 650 articles to profl. jours. Recipient Rsch. Career Devel. award USPHS, 1968-72; Silver medal for furthering French-U.S. sci. rels., mayor of Paris, 1984; Medaille Vermeille, French Acad. Medicine, 1984; recipient travel awards NIH, 1968, Am. Soc. Pharm. and Exptl. Therapeutics, 1969; grantee NIH, 1968—, NIMH, 1974-78, Nat. Heart Lung Blood Inst., 1974-86, Nat. Inst. Drug Abuse, 1979-83, Nat. Inst. Alcohol Abuse and Alcoholism, 1990—. Fellow Internat. Coll. Angiology, Am. Coll. Angiology, Am. Inst. Chemists, Am. Heart Assn. (mem. coun. on stroke 1973—, coun. basic sci. 1969—, coun. on thrombosis 1971—, coun. on circulation 1978—, coun. on high blood pressure 1978—, coun. on cardiopulmonary circulation, 1987—, cardiovascular A study sect. 1978-81), Am. Coll. Nutrition, Am. Physiol. Soc. (mem. circulation group 1971—, pub. info. com. 1980-84, symposium organizer); mem. AAUP, AAAS, Microcirculatory Soc. (past mem. exec. council, mem. nominating com. 1973-74), Soc. Exptl. Biology and Medicine (editorial bd. 1976-83), Am. Assn. for Clin. Chemistry (hon. lectr. 1989), Am. Diabetes Assn., Am. Pub. Health Assn., Am. Chem. Soc. (div. medicinal chemistry, div. analytical chemistry), Am. Soc. Pharm. and Exptl. Therapeutics (symposium organizer), Endocrine Soc., Harvey Soc., Am. Coll. Toxicology, Rsch. Soc. on Alcoholism (organizer several symposia), Soc. for Critical Care Medicine, Am. Thoracic Soc., Soc. for Neurosci., Shock Soc. (founder, hon. lectr.), Am. Fedn. Clin. Rsch., Microscopy Soc. Am., European Conf. Microcirculation (symposium organizer, hon. lectr.), Internat. Anesthesia Rsch. Soc., Fedn. Am. Soc. Exptl. Biology (pub. info. com. 1981-86), Am. Inst. Nutrition (organizer minisymposium), Am. Assn. Pathologists, Am. Soc. Microbiology, Internat. Soc. Thrombosis and Haemostasis, Internat. Soc. Biomed. Rsch. on Alcoholism (founding mem.), Internat. Soc. Biorheology, Soc. Leukocyte Biology, Soc. Environ. Geochemistry and Health, Soc. Neurosci., Soc. Cardiovascular Pathology, Reticuloendothelial Soc., Internat. Soc. Exposure Analysis, Soc. of Parenteral and Enteral Nutrition, Soc. Nutrition Edn., Soc. Scholarly Pub., Gerontol. Soc., Internat. Platform Assn., Am. Assn. Lab. Animal Sci., Am. Inst. Biol. Sci., Assn. Gnotobiotics, Am. Microscopical Soc., Am. Soc. Zoologists, The Oxygen Soc., Am. Soc. Cell Biology, Am. Soc. Bone and Mineral Rsch., Am. Soc. Magnesium Rsch. (founder, pres., exec. dir. 1984—, symposium, workshop organizer), N.Y. Acad. Scis., Am. Pub. Health Assn., N.Y. Heart Assn., N.Y. Soc. Electron Microscopy, Coun. Biology Editors, Soc. for Scholarly Pub., Internat. Anesthesia Soc., Internat. Soc. for Hypertension, Am. Soc. Hypertension (founding mem.), Am. Soc. Microbiology, Am. Assn. Pharm. Scis., Nat. Coun. for Magnesium and Cardiovascular Disease, Am. Assn. Clin. Chemistry, Soc. for Critical Care Medicine, Am. Med. Writers Assn., Am. Speech and Hearing Assn., Sigma Xi. Office: 450 Clarkson Ave Brooklyn NY 11203-2098

ALVAREZ, GUILLERMO A., research scientist; b. Pacora, Caldas, Colombia, Oct. 30, 1950; came to U.S., 1982; s. Guillermo and Nelly (Lopez) A. BS in Chemistry, Essex U., 1975, PhD in Polymer Sci., 1978. Rsch. asst. U. Freiburg, Germany, 1979-82; rsch. assoc. U. Wis., Madison, 1982-84; rsch. chemist Exxon Chem. Co., Linden, N.J., 1984-85; rsch. scientist CR Inds., Elgin, Ill., 1986—. Mem. Soc. Rheology. Home: 234 S Emmett St Apt 2E Genoa IL 60135 Office: CR Inds 900 N State St Elgin IL 60123

ALVAREZ, PABLO, neuroscientist; b. Madrid, Spain, Feb. 5, 1964; came to U.S., 1986; s. Jose Luis and Mercedes (Royo-Villanova) A. BS in Biochemistry, U. Autonoma de Madrid, Madrid, Spain, 1986; MS in Neurosci., U. Calif. San Diego, 1989, PhD, 1993. Teaching asst. U. Calif. San Diego, La Jolla, 1987, 89, rsch. asst., 1986-93, postdoctoral fellow Salk Inst., 1993—. Contbr. articles to profl. jours. Recipient 1st Pl. award Spanish Math. Olympics, 1981, fellowship U.S.-Spain Joint Cultural and Ednl. Com., 1987-89, McDonnell-Pew Ctr. for Cognitive Neurosci., 1992-94. Mem. AAAS, Soc. for Neurosci., Internat. Neural Network Soc., Sociedad Española de Neurociencia. Office: U Calif San Diego Dept Neuroscience 0608 La Jolla CA 92093

ALVING, AMY ELSA, aerospace engineering educator; b. Miami, Fla., Oct. 29, 1962; d. Ralph Eric and Therese (Fongeallaz) A. BSE, Stanford U., 1983; MA, Princeton U., 1985, PhD, 1988. Postdoctoral rsch. fellow Technische U. Berlin, 1988-90; asst. prof. aerospace engring. and mechanics U. Minn., Mpls., 1990—. Fellow NSF, 1983-86, Hertz Found., 1986-88, Airlift Found., 1988-89. Mem. AIAA, Am. Phys. Soc., Phi Beta Kappa, Tau Beta Pi. Office: U Minn Aerospace Engring/Mechanics 110 Union St SE Minneapolis MN 55455

ALWAN, ABEER ABDUL-HUSSAIN, electrical engineering educator; b. Baghdad, Iraq, Feb. 26, 1959; came to U.S., 1981; d. Abdul-Hussain Alwan Shlash and Amina Wahab Mashta. BSEE, Northeastern U., 1983; SMEE, MIT, 1986, EE, 1987, PhD in Elec. Engring., 1992. Intern Concord Data Systems, Waltham, Mass., 1982-83; rsch. asst., teaching asst. dept. elec. engring. MIT, Cambridge, Mass., 1982-92; asst. prof. dept. elec. engring. UCLA, 1992—; cons. Digital Equipment Corp., Waltham, 1990. Contbr. articles to profl. publs. Named one of Outstanding Young Women Am. Mem. IEEE, Acoustical Soc. Am., Sigma Xi, Tau Beta Pi. Office: U Calif 405 Hilgard Ave Los Angeles CA 90024-1594

ALZOFON, JULIA, laboratory administrator; b. Santa Barbara, Calif., Aug. 27, 1954; d. Frederick Efriam and Norma Dorothy (Shaw) A. BS in Med. Tech., U. Utah, 1976; MS in Med. Tech., U. Bridgeport, 1979; PhD in Biol. Scis., Fordham U., 1984. Clin. lab. dir. N.Y., Conn.; med. technologist. Med. technologist Norwalk (Conn.) Hosp., 1977-79; teaching fellow Fordham U., Bronx, 1980-83; postdoctoral rsch. fellow Am. Health Found., Valhalla, N.Y., 1984-86; hematology/blood bank supr. FDR VA Med. Ctr., Peekskill, N.Y., 1986-90; hematology lab. supr. Yonkers Gen. Hosp., 1990-91; clin. lab. co-dir. Med. Lab. of Stamford (Conn.), 1992—; lab. inspection team mem. Coll. Am. Pathologists, United Hosp. Lab., Port Chester, N.Y., 1989—. Contbr. articles to profl. jours. Schering-Plough fellow, 1983. Mem. Entomol. Soc. Am., Am. Soc. Clin. Pathologists.

AL-ZUBAIDI, ALI ABDUL JABBAR, engineer; b. Baghdad, Mar. 17, 1952; s. Abdul Jabbar and Hasiba (Ibrahim) Al-Z.; m. Latif Nidhal, June 26, 1982; children: Jumana, Hussain. BSc in Engring., London U., 1976; PhD, Brunel U., Eng., 1982. Cons. engr. Ministry Electricity and Water, Kuwait, Kuwait, 1983-90; regional mgr. SGS U.K., Camberley, Eng., 1990—. Contbr. articles on desalination, water treatment and corrosion to profl. publs. Mem. ASTM, Am. Water Works Assn., Inst. Chem. Engrs. Moslem. Office: SGS UK, 227-221 London Rd, Camberley Surrey GU153EY, England

AL-ZUBAIDI, AMER AZIZ, physicist, educator; b. Najaf, Iraq, June 10, 1945; came to U.S., 1974; s. Aziz Allawi and Shahai Ali (Al Fortousi) A.; m. Haifa M. Al-Zubaidi, Aug. 24, 1972; children: Samer, Akrum. BS in Physics, U. Baghdad, Iraq, 1966; MS in Physics, Pa. State U., 1976, post-grad., 1977, 81; postgrad., Va. Poly. Inst. and State U., 1977-82. High sch. tchr. Inst. for Tchrs., Riyadh, Suadi Arabia, 1966-68; high sch. tchr. physics, math., and related scis. Saudi Ministry of Edn., Riyadh, 1966-68; high sch. tchr. physics, math., mem. phys. lab. supplies and equipments com. Agrl. Vocat. Sch., Iraqi Ministry Edn., Baghdad, 1968-74; grad. teaching asst. Va. Poly. Inst. and State U., Blacksburg, 1976-82, rsch. sci. nuclear physics, 1982—. Mem. Union of Concerned Scientists, Sigma Xi, Sigma Pi Sigma. Home: 2319 10th St Ext NW Roanoke VA 24012

AMABILE, MICHAEL JOHN, mechanical engineer, consultant; b. Belmar, N.J., Dec. 6, 1968; s. A. Thomas and Regina (Chaput) A. BSME, Princeton U., 1991. Software engr. David Sarnoff Rsch. Ctr., Princeton, N.J., 1989-91; info. cons. Andersen Cons., Florham Park, N.J., 1991—. Mem. Sigma Xi, Phi Beta Kappa, Tau Beta Pi. Roman Catholic. Home: 106 Fairacres Dr Toms River NJ 08753 Office: Andersen Cons 100 Campus Dr Florham Park NJ 07932

AMADO, RALPH DAVID, physics educator; b. Los Angeles, Nov. 23, 1932; s. Richard Joseph and Suzanne (Nahoum) A.; m. Carol Stein, May 28, 1961; children—Richard Lewis, David Philip. B.A., Stanford, 1954; Ph.D. (Rhodes scholar), Oxford U., 1957. Research asso. U. Pa., 1957-59, asst. prof., 1959-62, asso. prof., 1962-65, prof. physics, 1965—; Cons. Arms Control and Disarmament Agy., 1962-65, Los Alamos Sci. Lab., 1965—. Fellow Am. Phys. Soc., AAAS. Home: 509 Latmer Rd Merion Station PA 19066-1811

AMADOR, ARMANDO GERARDO, medical educator; b. Mexico City, Feb. 2, 1953; came to U.S., 1979; s. Mario Armando and Maria (Meza-Calix) A.; m. Gabriela L. Velazquez, July 8, 1977; children: Santiago, Diego V. MD, U. Nacional Autonoma de Mexico, 1976. Rsch. intern Inst. Nacional de la Nutricion, Mexico City, 1975; Inst. Nacional de la Nutricion; lectr. U. Nacional Autonoma de Mexico, Mexico City, 1976-79, 84, chair biol. sci., 1978-79; post-doctoral fellow U. Tex. Health Sci. Ctr., San Antonio, 1979-83; assoc. scientist So. Ill. U. Sch. Medicine, Carbondale, 1985-89; assoc. prof. So. Ill. U. Sch. Medicine, Springfield, 1989—; sr. cons., sci. dir. ReproGen, Mexico City, 1991—. Contbr. over 120 articles to profl. jours. Mem. Am. Soc. Andrology, Genetics Soc. Am., Am. Soc. Human Genetics, Internat. Mammalian Genome Soc. Achievements include first description of the importance of genetic regulation of testicular luteinizing hormone receptors and its correlation in endocrine disease. Office: So Ill U Sch Medicine 801 N Rutledge Springfield IL 62794-9230

AMADOR, JOSE MANUEL, plant pathologist, research center administrator; b. Calimete, Matanzas, Cuba, Mar. 3, 1938; came to U.S., 1957; s. Luis Felipe and Blanca Rosa (Muñiz) A.; m. Silvia G. Garcia, Nov. 25, 1965; children: Silvia G. Amador Bibb, Marian L., Daniel J. BS in Agronomy and Soil Chemistry, U. Havana, Cuba, 1960; MS in Botany, Plant Pathology and Breeding, La. State U., 1962, PhD in Plant Pathology and Biochemistry. Rsch. asst. in plant pathology La. State U., Baton Rouge, 1960-65; estension plant pathologist Tex. A&M U. Extension, Weslaco, 1965-91, dir. agrl. rsch. and extenion ctr., 1991—; mem. extension futures task force Tex. A&M U. Extension, 1988, internat. task force agrl. complex, 1989; cons. Rio Grande Sugar Growers, Inc., Santa Rosa, Tex., Big-B Ranch, Belle Glade, Fla., US/AID/U. Fla.-El Salvador, Internat. Planning Svcs., Inc., US/AID Mission, Panama, Citrus Devel. Corp., Chiqueta Brands Internat., XAFRA, Inc., Veracruz, Mex. Contbr. articles to profl. and sci. publs. Recipient Svc. to Agriculture award Hidalgo Farm Bur., 1993. Mem. Am. Phytopathological Soc. (long standing, adv. bd. office internat. programs 1989, rep. to Internat. Soc. Plant Pathology 1989, immediate past chmn. tropical plant pathology com. 1989, past chmn. internat. cooperation com. 1980-90, mem. coun., past mem. extension com., counselor Carribean divsn. 1980-89, Excellence in Extension award 1990), Tex. Vegetable Assn. (bd. dirs.), Tex. Assn. Plant Pathologists and Nematologists, Lower Rio Grande Valley Hort. Soc. (past treas.), Gamma Sigma Delta, Epsilon Sigma Phi. Roman Catholic. Home: 1400 Yucca Mcallen TX 78504 Office: Texas A & M Univ Agricultural Rsch & Extension Ctr 2415 E Hwy 83 Weslaco TX 78596

AMARAL, JOSEPH FERREIRA, surgeon; b. Pawtucket, R.I., Aug. 9, 1955; s. Joseph and Rosa (Ferreira) A.; m. Linda Watson, June 6, 1981; children: Courtney, Ashley, Gregory. BS in Biology summa cum laude, Providence Coll., 1977; MD, Brown U., 1981. Diplomate Am. Bd. Surgery, Am. Bd. Med. Examiners. Intern R.I. Hosp., Providence, 1981-82, resident, 1982-83; surg. rsch. fellow Brown U./R.I. Hosp., Providence, 1983-86; sr. surg. resident R.I. Hosp., Providence, 1986-88, adminstrv. chief surg. resident, 1988-89, ccord. surg. residency, asst. surgeon, asst. prof. Brown U., 1989-91, coord. surg. residencey, dir. laparoscopic surgery, 1991-92, dir. laparoscopic surgery, asst. surgeon, asst. prof., 1991-93, assoc. prof., surgeon, 1993—; treas. R.I. Hosp. Staff Assn., 1991—; sec. R.I. Hosp. Surg. Found., 1992—; bd. dirs. R.I. Hosp. PHO; vis. surgeon hosps. in Australia, Argentina, Portugal, Austria, Rome, Singapore and Brazil. Contbr. articles to numerous profl. jours.; numerous internat., nat. and regional presentations; various scientific exhibits. Recipient Merck Clin. Achievement award, 1981, Haffenraffer Surg. Rsch. fellowship, 1983-85, 16th ACS scholarship, 1984-86, Young Investigators award Shock Soc., 1986, Residents Rsch. award Surg. Infection Soc., 1986. Fellow ACS, Internat. Coll. Surgeons; mem. AMA, AAAS, R.I. Med. Soc., Providence Surg. Soc., New Eng. Surg. Soc., Soc. Laproendoscopic Surgeons, Assn. Surg. Edn., Ctrl. N.Y. Surg. Soc. (hon.), Soc. Minimally Invasive Therapy, Am. Soc. Gastrointestinal Endoscopy, Am. Biatric Soc., Surg. Infection Soc., N.Y. Acad. Scis., Wound Healing Soc., Am. Soc. Eternal and Parenteral Nutrition, Shock Soc., Assn. Acad. Surgeons, Brown Med. Alumni Assn., Sigma Xi, Phi Sigma Tau, Sigma Pi Sigma. Office: R I Hosp Dept Surgery 593 Eddy St Providence RI 02903-4923

AMATEAU, MAURICE FRANCIS, materials scientist, educator; b. N.Y.C., Dec. 3, 1935; s. Edward I. and Lucy (Russo) A.; m. Dorothy Elaine Letow, Nov. 1, 1959; children: Geoffrey Allan, Yvonne Gabrielle. BMetE, Ohio State U., 1957, MSc in Metall., 1963; PhD in Metall., Case We. Res. U., 1968. Prin. metall. Battelle Meml. Inst., Columbus, Ohio, 1957-63; rsch. engr. TRW Inc., Euclid, Ohio, 1963-68; mgr. metal composites Aerospace Corp., El Segundo, Calif., 1968-79; dir. materials Internat. Harvester, Hinsdale, Ill., 1979-83; prof. engring. sci. Pa. State U., University Park, 1983—. Author: (book chpt.) Encyclopedia on Composites, 1990; editor: Mechanical Properties of Metal Composites, 1983. Mem. AIME (Metall. sect. So. Calif. sect. chmn. 1976-77), ASME Internat., Am. Ceramic Soc., Soc. Advanced Materials and Process. Achievements include patent for Thermomechanical processing of machine elements, developed laminated ceraamic composite materials for cutting tools, researched tribological behavior of metal composites. Home: 629 Sunset Rd State College PA 16803 Office: Pa State U 227 Hammond Bldg University Park PA 16802

AMATO, VINCENT EDWARD, geotechnical engineer, consultant; b. Celina, Ohio, Sept. 18, 1962; s. Augustine Joseph and Rosemarie Phyllis (Art) A.; m. Jeanne Ann DeMatteis, June 11, 1988. BS, Ohio State U., 1984, MS, 1986. Registered profl. engr., Ohio, W.Va., Ky. Geotech. engr. Burgess & Niple, Ltd., Columbus, Ohio, 1986-90, dir. soils and founds., 1990—. Mem. ASCE (editor newsletter 1989), Internat. Soc. Soil Mechanics and Found. Engrs., Delta Tau Delta. Office: Burgess & Niple Ltd 5085 Reed Rd Columbus OH 43220

AMBROSE, ROBERT MICHEAL, application engineer, consultant; b. Cleve., Dec. 22, 1960; s. John Thomas and Anne Marie (Neroni) A.; m. Terri Ann Fazio, June 14, 1986; 1 child, Joseph Robert. BSME, U. Cin., 1985. Sales engr. York Internat., Cin., 1985-90; svc. mgr. York Internat., Richmond, Va., 1990-92; constrn. and engring. mgr. Trane Internat., Cleve., 1992—. Mem. ASHRAE (assoc.), Assn. Energy Engrs. (assoc.), Cleve. Coun. Sml. Bus. Enterprises, ASME (assoc.). Roman Catholic. Home: 1187 Homestead St Cleveland OH 44121 Office: Trane PO Box 39280 Solon OH 44139

AMBRUSKO, JOHN STEPHEN, retired surgeon, county official; b. North Tonawanda, N.Y., May 3, 1913; s. Joseph and Magdalena (Isky) A.; m. Phyllis Eusterman, Sept. 21, 1946; children: Therese Ambrusko Carlson, Gretchen Ambrusko Schaeffer, Mary Ambrusko Bennett, Sara Ambrusko Tokars, Joni Ambrusko Crane, Karen Ambrusko Wilcox, Krissy. MD, SUNY, Buffalo, 1937. Diplomate Am. Bd. Surgery. Intern Buffalo Gen. Hosp., 1937-38; asst. resident surgeon Buffalo Gen. Hosp., Buffalo Children's Hosp., 1938-39, resident pathology, 1939-40; fellow in gen. surgery Mayo Found., Rochester, Minn., 1940-42, 46-47; asst. surgeon Mayo Clinic, Rochester, 1947-48; instr. surgery SUNY, Buffalo, 1948-56; chief surgery, chmn. dept. surgery Kenmore (N.Y.) Mercy Hosp., 1950-70, chief cons. surgeon, 1970-90; med. exec. dir. Manatee County Pub. Health Dept., Bradenton, Fla., 1977-90; ret., 1990; clin. asst. prof. U. South Fla. Med. Sch., Tampa, 1979-89; surgical cons. quality assurance program Manatee Meml. Hosp., Brandenton, 1991. Mem. Erie County Alcoholic Beverage Control Commn., 1965-75; trustee Rosary Hill Coll., Buffalo, 1965-75; bd. visitors Roswell Park Cancer Rsch. Hosp., Buffalo, 1965-75. Lt. comdr. USN, 1942-46, PTO. Doctor John Ambrusko Pub. Health Ctr. named in his honor Mannatee County Bd. of County Commrs., 1988; recipient Outstanding Achievement award Western N.Y. Jr. C. of C., 1949, Sci. Achievement award N.Y. State Med. Soc., 1952, Cancer Rsch. award Roswell Park Cancer Rsch. Hosp., 1974, Outstanding Achievement award Nat. Assn. Counties, 1989. Fellow ACS, Plaza Club (bd. dirs. 1985-89), Buffalo Club. Republican. Roman Catholic. Avocations: swimming, reading, boating. Home: 3804 Bayside Dr Bradenton FL 34210-4109 Office: Dr John Ambrusko Pub Health Ctr 410 6th Ave E Bradenton FL 34208-1986

AMBUR, DAMODAR REDDY, aerospace engineer; b. Puthalpet, India, Sept. 30, 1947; came to U.S., 1977; s. Subramanyam Reddy and Sakuntalamma Ambur; m. Manjula Yaramakala, June 8, 1973; children: Sumanth, Vishnu Vardhan. B Engring., S.V. Un., Tirupathi, India, 1968; M Tech., Indian Inst. Tech., Madras, 1972; PhD, Ga. Inst. Tech., 1980. Cert. profl. mgr. Sr. engr. Indian Space Rsch. Orgn., Trivandrum, 1972-77; postdoctoral fellow Ga. Inst. Tech., Atlanta, 1980-82, sr. rsch. engr., 1982-85; engr. specialist Lockheed Aero. Systems Co., Marietta, Ga., 1985-87; design specialist Lockheed Aero. Systems Co., Burbank, Calif., 1989; dep. gen. mgr. Hindustan Aeronautics Ltd., Bangalore, India, 1987-88; sr. aerospace engr. NASA Langley Rsch. Ctr., Hampton, Va., 1989—; cons. in field, 1980—. Contbr. articles to numerous profl. publs. Treas. Telugu Assn. Metro Atlanta, 1982, pres., 1986. Assoc. fellow AIAA; mem. Indian Soc. for Advancement of Materials and Process Engring., Inst. Cert. Profl. Mgrs. Avocations: social service, tennis, ping-pong, swimming. Home: 108 Flag Crk Yorktown VA 23693-2331

AMEMIYA, CHRIS TSUYOSHI, geneticist, biomedical scientist; b. Wahiawa, Hawaii, Dec. 21, 1959; s. Keiji and Setsuko (Matsumiya) A. BS, Purdue U., West Lafayette, Ind., 1981; PhD, Tex. A&M U., College Station, 1987. Rsch. and teaching asst. Tex. A&M U., College Station, 1981-87; postdoctoral fellow Tampa Bay Rsch. Inst., St. Petersburg, Fla., 1987-90; rsch. fellow Lawrence Livermore (Calif.) Nat. Lab., 1990-93, Boston U. Sch. Medicine, 1993—; reviewer Analytical Biochemistry, Molecular Immunology, Devel. Immunology, Copeia Jour., 1987—, NSF/NIH Lab. Agys., 1987—, J. Immunology Jour., 1990. Contbr. 40 jour. articles, 13 book chpts., 42 pub. abstracts. Recipient Stoye award Am. Soc. Ichthyologists, Knoxville, Tex., 1985; Tom Slick Grad. Fellow Tex. A&M U., 1986-87; Nat. Inst. Health rsch. support grantee Showa Inst., St. Petersburg, Fla., 1988, fellow 1988; rsch. grantee, fellow Alfred Sloan Found., 1988-90. Mem. Assn. Advancement Sci., Soc. Study of Evolution, Genetics Soc. of Am., Am. Soc. of Zoologists, Am. Soc. Ichthyologists and Herpetologists. Office: Lawrence Livermore Nat Lab 7000 East Ave Livermore CA 94550-9244

AMEMIYA, KENJIE, toxicologist; b. Camden, N.J., Nov. 8, 1958; Hiroshi and Michiyo (Kaneko) A.; m. Jacqueline Inez Lewis, July 18, 1987; 1 child, Jamie Lewis. BS, U. Calif., Davis, 1982, PhD, 1987. Reproductive and gen. toxicologist CIBA-Geigy Pharms., Summit, N.J., 1987—. Contbr. articles to Teratology. Mem. Mid Atlantic Reproductive Toxicology Assn. (mem. steering com. 1991—), Am. Inst. Nutrition, Behavioral Teratology Soc., Teratology Soc. Republican. Achievements include copyright for training videotapes on fetal rat/rabbit anomaly examination. Office: CIBA-Geigy Pharms 556 Morris Ave Summit NJ 07901

AMER, MAGID HASHIM, physician; b. Cario, June 5, 1941; came to U.S. 1971; s. Hashim and Zeinab (Iskander) A.; m. Sabah El Sayed Shehata, Mar. 12, 1973; children: Sophi, Mona, Hoda. M.B.B.Ch., Cairo U. Med. Sch., 1963. Rotating intern Cairo U. Hosp., 1964-65; resident gen. surgery Northampton (UK) Gen. Hosp., 1968-69; resident internal medicine Worcester City Hosp., Mass., 1971-73, Lemuel Shattuck Hosp., Boston, 1973-74; fellow med. oncology Wayne State U., Detroit, 1974-76, asst. prof., 1977-78; cons. oncologist King Faisal Specialist Hosp., Riyadh, Saudi Arabia, 1978-84, head div. med. oncology, 1984-91, chmn. rsch. ethics com., 1986-91; pvt. practice oncology Maadi-Cairo, 1992—. Contbr. articles to profl. jours. Fellow Royal Coll. Physicians Can., ACP, Royal Coll. Surgeons Edinburgh; mem. Am. Soc. Clin. Oncology, Am. Assn. Cancer Research, AAAS. Muslim. Home: 24 Rd 12, Maadi-Cairo 11431, Egypt

AMERINE, MAYNARD ANDREW, enologist, educator; b. San Jose, Calif., Oct. 30, 1911; s. Roy Reagan and Tennie (Davis) A. B.S., U. Calif.-Berkeley, 1932, Ph.D. in Plant Physiology, 1936. Mem. faculty U. Calif., Davis, 1935—, prof. enology, enologist Exptl. Sta., 1952-74; emeritus U. Calif. and Expt. Sta., Davis, 1974—; chmn. dept. viticulture and enology U. Calif., Davis, 1957-62; cons. Wine Inst., 1974-85. Author: (with M. A. Joslyn) Table Wines: The Technology of Their Production in California, 1951, 2d edit., 1970, (with Louise Wheeler) A Check-List of Books and Pamphlets on Grapes and Wines and Related Subjects, 1951, A Short Check-List of Books and Pamphlets in English on Grapes, Wine and Related Subjects, 1949-1959, 1959, (with others) The Technology of Wine Making, 4th edit., 1980, (with G. L. Marsh) Wine Making at Home, 1962, (with M.A. Joslyn) Dessert, Appetizer and Related Flavored Wines: The Technology of Their Production, 1964, (with V.L. Singleton) Wine: An Introduction for Americans, 1965, 2d edit., 1977, (with Rose M. Pangborn and E. B. Roessler) Principles of Sensory Evaluation of Food, 1965, A Check List of Books on Grapes and Wines, 1960-68, (with supplement for), 1949-59, 1969, (with G.F. Stewart) Introduction to Food Science and Technology, 1973, 2d edit., 1982, (with C.S. Ough) Wine and Must Analyses, 1974, 80, 2d edit., 1988, (with E.B. Roessler) Wines: Their Sensory Evaluation, 1976, 2d edit., 1983; (with H. Phaff) Bibliography of Univeristy of California Publications, 1876-1980, on Grapes, Wines, and Related Subjects, 1986; editor and contbr.: Wine Production Technology in the U.S, 1981; co-editor and contbr. (with D. Muscatine and B. Thompson) The University of California/Sotheby Book of California Wine, 1984. Served to maj. AUS, 1942-46. Decorated Chevalier de Merite Agricole (France), 1947, officier Ordre National du Merite (France), 1976; recipient diplôme d'honneur Office Internat. du Vin, 1952, 65, 84, 89, 2d prize Oberly award A.L.A., 1953, Merit award Am. Soc. Enologists, 1967, Am. Wine Soc., 1976, Man of Year award Les Amis du Vin, 1976, The Wine Spectator, 1985, Adams award Wine industry Tech. Seminar, 1989, award of Distinction, U. Calif. at Davis, 1990; Guggenheim fellow, 1954-55. Mem. Am. Soc. Enologists (pres. 1958-59), AAAS, Am. Chem. Soc., Inst. Food Technologists. Republican. Baptist. Club: Bohemian (San Francisco). Home and Office: PO Box 208 Saint Helena CA 94574

AMERO, SALLY ANN, molecular biologist, researcher; b. Pitts., Feb. 24, 1952; d. Robert Clayton and Grace I. (Kunselman) A. BS, Indiana U. Pa., 1974; PhD, W.Va. U., 1979. Postdoctoral rsch. in chemistry U. Va., Charlottesville, 1979-80, postdoctoral rsch. in biology, 1980-84, rsch. asst. prof., 1987-91; postdoctoral rschr. Washington U., St. Louis, 1984-87; asst. prof. Loyola Med. Sch., Maywood, Ill., 1991—. Contbr. articles to profl. jours. Mem. AAAS, Am. Soc. Cell Biology, Genetics Soc. Am., Am. Soc. Microbiology. Office: Loyola U Med Ctr 2160 S 1st St Maywood IL 60153

AMES, BRUCE N(ATHAN), biochemistry and molecular biology educator; b. N.Y.C., Dec. 16, 1928; s. Maurice U. and Dorothy (Andres) A.; m. Giovanna Ferro-Luzzi, Aug. 26, 1960; children: Sofia, Matteo. BA, Cornell U., 1950; PhD, Calif. Inst. Tech., 1953. Chief sect. microbial genetics NIH, Bethesda, Md., 1953-68; prof. biochemistry and molecular biology U. Calif., Berkeley, 1968—, chmn. biochemistry dept., 1983-89; mem. Nat. Cancer Adv. Bd., 1976-82. Research, publs. on bacterial molecular biology, histidine biosynthesis and its control, aging, mutagenesis, detection of environ. mutagens and carcinogens, genetic toxicology, oxygen radicals and disease. Recipient Flemming award, 1966, Rosensteil award, 1976, Fedn. Am. Soc. Exptl. Biology award, 1976, Felix Wankel award, 1978, John Scott medal, 1979, Corson medal, 1980, N.B. lectureship Am. Soc. Microbiology, 1980, Mott prize GM Cancer Rsch. Found., 1983, Gairdner award, 1983, Tyler prize Environ. Achievement, 1985, Gold medal Am. Inst. Chemists, 1991, Glenn Found. award, 1992. Fellow Acad. Toxicol. Scis., Am. Acad. Microbiology, Gerontol. Soc. Am.; mem. NAS, Am. Soc. Biol. Chemists, Am. Soc. Microbiology, Environ. Mutagen Soc. (award 1977), Genetics Soc., Am. Assn. Cancer Rsch., Soc. Toxicology, Am. Chem. Soc. (Eli Lilly award 1964, Gustavus John Esselen award 1992), Royal Swedish Acad. Scis., Am. Acad. Arts and Scis. Home: 1324 Spruce St Berkeley CA 94709-1435 Office: U Calif 401 Barker Hall Berkeley CA 94720

AMEY, EARLE BARTLEY, materials engineer; b. Easton, Pa., Jan. 3, 1942; s. Earl Bartley and Mary Ria (Altenbach) A.; m. Frances Ellen Caldwell, Aug. 28, 1971; children: Sharon, Angela, Jessica. BS, U. Md., 1966, MS, 1968, PhD, 1974. Chem. rsch. engr. Bur. of Mines, College Park, Md., 1966-75; prin. investigator Bur. of Mines, Boulder City, Nev., 1975-78; staff engr. Bur. of Mines, Washington, 1978-80, tech. liaison officer, 1980-82, staff engr., 1982-90, asst. br. chief, 1990—; adj. prof. U. Nev., Las Vegas, 1978-79; alternate EE officer Bur. of Mines, Washington, 1989—; adv. bd. mem. SETAC, Pensacola, Fla., 1992—. Editor: New Materials Soc., Vol. 3, 1991, (proceedings) Canadia Waste Management Conf., 1991; author, editor: (brochure) Information and Analysis Materials Program, 1991. Vol. fireman Boulder City Fire Dept., 1975-78. Recipient Spl. Achievement award Bur. of Mines, 1990, 91. Mem. AICE, Nat. Fire Protection Assn., Profl. Engr. Republican. Roman Catholic. Achievements include patents in field. Home: 1508 Crofton Pky Crofton MD 21114 Office: Bur of Mines 810 Seventh St NW Washington DC 20240

AMICK, CHARLES L., electrical engineer, consultant; b. Clarinda, Iowa, Nov. 21, 1916; s. Charles Henry and Zoa Marie (Nixon) A.; m. Janet Robertson Campbell, Nov. 25,1939; children: Joan A. Toomey, Carol C. Amick, Charles J. Amick (dec.). BSEE, Iowa State U., 1937, profl. elec. engr., 1946. Registered profl. engr. Ohio, Mo. Test engr. GE, Schenectady, N.Y., 1937-38; illuminating engr. GE, Cleve., 1938-55; dir. engring. svcs. Day-Brite Lighting, Inc., St. Louis, 1955-80; lighting cons. Emerson Elec., Day-Brite Lighting Divsn., Tupelo, Miss., 1980-89, Thomas Industries, Inc., Tupelo, 1989—; pres. Illuminating Engring. Soc. of N.Am., N.Y.C., 1964-65. Author: (book) Fluorescent Lighting Manual, 1942 (reprinted 1947, 1960); contbr.numerous articles to profl. jours. Water commr. Village Dist. of Eastman, Grantham, N.H., 1988—; auditor Grantham Village Sch. Dist., 1987-91. Fellow and gold medal recipient Illuminating Engring. Soc. of N.Am., 1955, 75; fellow Chartered Inst. of Bldg. Svcs. Engrs., 1974; recipient Profl. Achievement Citation Iowa State U., 1976. Life mem. Internat. Commn. on Illumination (pres. 1983-87); mem. Tau Beta Pi, Sigma Delta Chi, Eta Kappa Nu, Pi Mu Epsilon. Avocations: photography, woodworking, reading. Home and Office: C L Amick Lighting Cons 12 Wedgewood Dr Box 233 Grantham NH 03753-0233

AMICK, S. EUGENE, military engineer; b. Scottsburg, Ind., Oct. 12, 1936; s. Willoughby Fern and Vera Maxine (Peacock) Schrink; m. Karen Yvonne Miller, Oct. 2, 1960; children: Scott Eugene, Steven Eric. BSEE, Rose-Hulman, 1959. Registered profl. engr., Ark.; cert. energy mgr. Missile elec. engr. to elec. engr. 314 Civil Engring. Squadron, Little Rock AFB, 1962-68, chief constrn. mgmt., 1971-75, 78-79, 89, chief design engring., 1979-80, chief engring. and constrn., 1980-81, missile civil engr. to chief missile engring., 1981-87, chief EMCS, 1989-92, elec./electronic engr., 1992-93; elec. engr. Directorate of Engring Hdqs. 5th AF, Fuchu AS, Japan, 1968-71, Engring. and Svcs. Directorate Hdqrs. USAF Europe, RAF West Ruislip, England, 1975; sr. elec. engr. Engring. and Svcs. Directorate Hdqrs. USAF Europe, Ramstein AB, Germany, 1975-78; dep. base civil engr. 43 Civil Engring. Squadron, Andersen AFB, Guam, 1987-89. Mem. Assn. Energy Engrs. (sr.). Home: 4909 Lakeview Rd North Little Rock AR 72116

AMIEL, DAVID, orthopaedic surgery educator; b. Alexandria, Egypt, Oct. 25, 1938; came to U.S., 1962; s. Eli and Inez (Bokey) A.; m. Nancy Joy Lyons, Nov. 27, 1966; 1 child, Michael Eli. B Math., Lycee Francais, Alexandria, 1955; PhD in Chem. Engring., U. Brussels, Belgium, 1962. Chem. engr. Polymers Lab. Boeing Aeorspace, Renton, Wash., 1962-63; assoc. in orthopaedics U. Wash., Seattle, 1964-66; chief chemist Laucks Testing Lab., Seattle, 1966-68; assoc. orthopaedic specialist U. Calif., La Jolla, 1968-75, orthopaedic specialist, 1975-83, from asst. prof. to assoc. prof. surgery, 1983-91, prof. orthopaedics, 1992—, mem. exec. com. sch. medicine, 1990-91; dept. head biochemistry M&D Coutts Inst., San Diego, 1984—; bd. dirs. Am. Coll. Sports Medicine, Indpls., 1989-92. Reviewer Am. J. Physiol., 1989-92, JBJS, JOR; contbr. chpts. to books and articles to profl. jours. Grant reviewer Arthritis Soc. Can. Recipient Award Excellence Basic Sci. Rsch., Am. Orthopaedic Soc. Sports Medicine, 1983, 86, Merit award NIH, 1989-99. Grant revewier Arthritis Soc. Can. Achievements include patents for Continuous Passive Motion Machine used as a rehabilitation device for post-ligamentous injuries and post total joint replacements (with others); among first to describe the response to stresses of periarticular connective tissues such as ligaments and tendons.

AMIES, ALEX PHILLIP, civil engineer; b. Canberra, Australia, Jan. 6, 1967; s. William and Barbara (Westren) A.; m. Cui Lan Fang, Sept. 15, 1990. B of Civil Engring., U. NSW, 1991; MS of Civil Engring., Stanford U., 1992. Geotech. engr. Soil Mechanics Ltd., Eng., 1991; rsch. asst. Stanford (Calif.) U., 1992; engr. Woodward-Clyde Cons., Santa Ana, Calif., 1993—. Contbr. articles to Jour. Geotech. Engring. Assn. Mem. ASCE, Inst. Engrs. (Australia). Office: Woodward-Clyde Cons Santa Ana CA 92705

AMINO, NOBUYUKI, endocrinologist, educator; b. Kobe, Hyogo-ken, Japan, Dec. 4, 1940; s. Ichie and Ei (Yoshikawa) A.; m. Masae Tanaka, Oct. 17, 1969; children: Shingo, Ikuko. MD, Osaka U. Med. Sch., Japan, 1965. Clin. fellow Osaka U. Hosp., Japan, 1966-71; clin. asst. Osaka U. Hosp., 1974-78; rsch. fellow U. Chgo., 1971-73; asst. prof. Med. Sch. Osaka U., 1978-89, assoc. prof., 1989-92; prof. Med. Sch. Osaka U., 1992—. Recipient 1st Prize of Kozakai Nozomu, Clin. Pathology Rsch. Found., 1991, Annual award Japan Med. Assn., 1989, Prize, Asia and Oceania Thyroid Assn., 1989, 1st Annual award Japan Endocrine Soc., 1981, 6th Annual award Japan Thyroid Assn., 1977. Avocations: swimming, bird watching. Home: 5-60-38 Nampeidai, Takatsuki Osaka 569, Japan Office: Osaka U Med Sch, 2-2 Yamadaoka, Suita Osaka 565, Japan

AMIR-MOEZ, ALI REZA, mathematician, educator; b. Teheran, Iran, Apr. 7, 1919; s. Mohammad and Fatema (Gorgestani) A.-M.; BA, U. Teheran, 1942; MA, UCLA, 1951, PhD, 1955. Came to U.S., 1947, naturalized, 1961. Instr. math. Teheran Tech. Coll., 1942-46; asst. prof. math. U. Idaho, 1955-56, Queens Coll., N.Y.C., 1956-60, Purdue U., 1960-61; assoc. prof. U. Fla., Gainesville, 1961-63; prof. math. Clarkson Coll., Potsdam, N.Y., 1963-65, Tex. Tech U., Lubbock, 1965-88, prof. emeritus, 1988—. Author: Elements of Linear Space, 1961; (play) Kaleeheh & Demneh, 1962; Three Persian Tales, 1961, Matrix Techniques Trigonometry and Analytic Geometry, 1964; Mathematics and String Figures, 1966; Classes Residues et Figures avec Ficelle, 1968; Extreme Properties of Linear Transformations and Geometry in Unitary Spaces, 1971; Elements of Multilinear Algebra, 1971; Linear Algebra of the Plane, 1973; contbr. articles to math. jours. on proper and singular values of linear operators and matrices. 2d lt. Persian Army, 1936-38. Decorated Honor emblem Persian Royal Ct., medal Pro Mundi Beneficio Academia Brasileira de Ciencias Humanas. Mem. Am. Math. Soc., Math. Assn. Am., Sigma Xi, Pi Mu Epsilon. Office: Tex Tech U Math Dept Lubbock TX 79409

AMLADI, PRASAD GANESH, management consulting executive, health care consultant, researcher; b. Mudhol, India, Sept. 12, 1941; came to U.S., 1967, naturalized, 1968; s. Ganesh L. and Sundari G. Amladi; m. Chitra G. Panje, Dec. 20, 1970; children: Amita, Amol. B in Engring. with honors, Indian Inst. Tech., Bombay, 1963; MS in Indsl. Engring., Ops. Rsch., Stanford U., 1968; MBA with high distinction U. Mich., 1975. Sr. rsch. engr. Ford Motor Co., Dearborn, Mich., 1968-75; mgr. strategic planning Mich. Consol. Gas Co., Detroit, 1975-78; mgr. planning services The Resources Group, Bloomfield Hills, Mich., 1978-80; project mgr., sr. cons. Mediflex Systems Corp., Bloomfield Hills, 1980-85; mgr. strategic planning services Mersco Corp., Bloomfield Hills, 1985-86, mgr. corp. planning and rsch. Diversified Techs., Inc., New Hudson, Mich., 1986-87; mgr. planning and rsch. Blue Cross & Blue Shield of Mich., Detroit, 1987—. Contbr. papers to profl. publs. Recipient Kodama Meml. Gold medal, 1957; India Merit scholar Govt. of India, 1959-63, K.C. Mahindra scholar, 1967, R.D. Sethna Grad. scholar, 1968. Mem. Inst. Indsl. Engrs. (sr.), N.Am. Soc. Corp. Planning, Econ. Club Detroit, Beta Gamma Sigma. Office: Blue Cross Blue Shield of Mich 441 E Jefferson Ave # J738 Detroit MI 48226-4322

AMMAR, REDA ANWAR, computer science engineering educator; b. Giza, Egypt, Aug. 22, 1950; s. Anwar Mohamed and Saadia Khalil (Ahmed) A.; m. Tahany Abd El-Monsef, July 15, 1976; children: Rabab, Doaa, Haytham. BSEE, Cairo U., 1973; BS in Math., Ain Shames U., Cairo, 1975; MS in Computer Sci., U. Conn., 1981, PhD in Computer Sci., 1983. Demonstrator Cairo U., Giza, 1973-78; teaching asst./rsch. asst. U. Conn., Storrs, 1978-83; ass.t prof. Cairo U., Giza, 1984-85; vis. asst. prof. The Am. U., Cairo, 1984-85; sr. analyst Cons. Engring. Grp., Cairo, 1984-85; vis. asst. prof. U. Conn., Storrs, 1985-87; asst. prof. computer sci. and egring. U. Conn., 1987-92; assoc. prof. computer sci. and engring., 1992—. Contbr. articles to profl. jours. U. Conn. fellow, 1980-83; Naval Underwater Sys. Ctr. grantee, 1987-89, U. Conn. grantee, 1987, 90, NSF grantee, 1989. Mem. IEEE, ACM, Internat. Soc. Mini-Micro Computers. Office: U Conn Dept Computer Sci and Engring 260 Glenbrook Rd # 155U Storrs Mansfield CT 06268

AMON, CRISTINA HORTENSIA, mechanical engineer, educator; b. Montevideo, Uruguay, Oct. 12, 1956; came to the U.S., 1983; d. Mirko and Marisa (Kovacic) A.; m. Carmelo Parisi, Dec. 6, 1980; children: Andreina, Gabriel. Degree in mech. engring., U. Simon Bolivar, 1981; MS, DSc in Engring., MIT, 1988. Rsch. and teaching asst. MIT, Cambridge, 1984-88; assoc. prof. Carnegie Mellon U., Pitts., 1988—; cons. Aavid, Laconia, N.H., 1986, Vanzetti Systems, Stoughton, Mass., 1986, Tex. Instruments, Calif., 1987, Bally Design, Inc., Pitts., 1988. Contbr. numerous articles to Cooling Tech. for Electronic Equipment, Physics of Fluids A., Jour. Thermodynamics and Heat Transfer, ASME Jour. Electronic Packaging, Numerical Heat Transfer, Quar. Yugoslav Acad. Scis., Internat. Jour. Heat and Mass Transfer, Jour. Heat Transfer, AIAA Jour., several conf. proceedings. Recipient Rsch. Initiation award NSF, 1989, G.T. Ladd award CIT, 1991. Mem. ASME, AIAA, Am. Soc. Engring. Educators, Soc. Women Engrs., SAE, Sigma Xi. Achievements include development of spectral element-Fourier method for Navier-Stokes and energy equations. Home: 1457 Beulah Rd Pittsburgh PA 15235 Office: Carnegie Mellon U Dept Mech Engring Pittsburgh PA 14213

AMORNMARN, LINA, chemist; b. Bangkok, Thailand; came to U.S., 1981; BS in Chemistry, Chulalongkorn U., Bangkok, 1978; MS in Chemistry, Fairleigh Dickinson U., 1985; AAS in Computer with high honors, County Coll. Morris, 1990. Chemist YKK Zipper Co., Ltd., Bangkok, 1978-81, Atlantic Industries Inc., Nutley, N.J., 1985-88, York Labs., Whippany, N.J., 1988-89; coop. computer scientist Sandoz Pharm., East Hanover, N.J., 1987-88; asst. scientist Hoffmann La-Roche, Nutley, 1989-91, assoc. scientist, acting supr. pharm. quality and control, 1991—. Mem. Am. Chem. Soc., Alpha Beta Gamma. Home: 360 Park Rd Parsippany NJ 07054 Office: Hoffmann La-Roche 340 Kingland St Nutley NJ 07110

AMOROSO, MARIE DOROTHY, retired medical technologist; b. Phila., Jan. 16, 1924; d. Salvatore and Clorinda (Gaudio) A. Med. Lab. Tech., Hahnemann Hosp., Phila., 1943; postgrad., Temple U., Phila., 1945-48, U. Pa., Phila., 1947-48, 1950. Registered EEG Technologist; cert. registered EEG Technologist. EEG technician Hahnemann Med. Coll., Phila., 1943-53, Phila. Gen. Hosp., 1953-62; histology technician Temple Med. Coll. Temple U., Phila., 1962-63; allergy technician Harry Rogers, M.D., Phila., 1963; EEG technologist Haverford (Pa.) State Hosp., 1963-85, Irvin M. Gerson, MD, Haverford, 1985-88; EEG technologist to pvt. physician Haverford State Hosp., 1985-88; ret., 1988; instr. EEG Osteopathic Med. Ctr. Sch. Allied Health, Phila., 1978-85. Editor: The Eastern Breeze, 1977-79; contbr. articles to profl. jours.; patentee in field. Mem. Am. Soc. Electroneurodiagnostic Technologists Inc., Western Soc. Electrodiagnostic Technologists, So. Soc. EEG Technicians Inc., Ea. Soc. EEG and Neurodiagnostic Technicians (sec. 1977-79), Phila. Regional EEG Technician's Assn. (exec. bd. 1967, sec.. 1969), Electro-Physiological Technologists Assn. Gt. Britain (subscriber mem.), Ea. Assn. Electroencephalographers (subscriber mem.). Avocations: writing musical compositions, poetry. Home: 477 Brookfield Rd Drexel Hill PA 19026-1198

AMOS, DENNIS B., immunologist; b. Bromley, Eng., Apr. 16, 1923; s. Benjamin and Vera (Oliver) A.; m. Solange M. Labesse, Aug. 25, 1949 (dec. 1980); children: Susan V., Martin D., Christopher I., Nigel P., Irene C.; m. Kay B. Veale, Mar. 9, 1984. MBBS, Guy's Hosp., London, 1951, MD, 1963. House officer Guy's Hosp., London, 1951-52, rsch. pathologist, 1952-55; prin. cancer rsch. scientist Roswell Park Inst., Buffalo, 1955-62; prof. immunology, exptl. surgery Duke U., Durham, N.C., 1962—; cons. NIH, Bethesda, Md., 1957—. Mem. Nat. Acad. Scis., Inst. of Medicine. Office: Duke Med Ctr PO Box 3010 Durham NC 27710-0001

AMOUZADEH, HAMID R., toxicologist; b. Firouzkouh, Iran. MS, Okla. State U., 1986, PhD, 1991. Teaching assoc. Okla. State U., Stillwater, 1985-91; post-doct. fellow NIH, Bethesda, Md., 1991—. Solvay Resident grantee Solvay Am., 1986; recipient Grad. Rsch. Excellence award Okla. State U., 1991. Office: Nat Heart Lung and Blood Inst Lab Chem Pharmacology Bldg 10 Rm 8-N-104 Bethesda MD 20892

AMR, ASAD TAMER, environmental engineer, consultant; b. Beirut, Sept. 21, 1941; came to U.S., 1969; s. Tamer A. and Nimreh K. (Harakeh) A.; m. Thea Brodski, July 4, 1965; children: Hady A., Tammir, J. Dean. BSCE, U. Tehran, 1965; MSCE, Teheran (Iran) U., 1966; MS in Environ. Engring., Drexel U., 1973, PhD in Environ. Engring., 1975. Registered profl. engr., N.J., Va. Instr. Maquasid Coll., Beirut, 1966-67; site engr. Ministry Pub. Works, Beirut, 1967-69; environ. engr. Gen. Pub. Utility, Parsippany, N.J., 1969-75; project engr. Exxon Chem. Co., Florham Park, N.J., 1975-76; dept. staff Mitre Corp., McLean, Va., 1976-80; systems engr. Mitre Corp., Mc Lean, 1987—; prog. mgr. Amartech Ltd., Jubail, Saudi Arabia, 1980-87. Author: Energy Systems, 1981; contbr. articles to profl. jours. Soccer coach Potomac Kiwanis, Arlington, Va., 1976-79. Scholar govts. of Iran and Lebanon, 1960. mem. ASCE. Avocations: soccer, tennis, swimming, travel. Home: 3612 N Woodstock St Arlington VA 22207-4323 Office: Mitre Corp 7525 Colshire Dr Mc Lean VA 22102-7500

AMSTERDAM, JAY D., psychiatry educator, researcher; b. Phila., Feb. 10, 1949. BA, Syracuse U., 1970; MD, Jefferson Med. Coll., 1974. Diplomate Am. Bd. Med. Examiners, Am. Bd. Psychiatry and Neurology; lic. physician, Pa., N.J. Intern in ob-gyn. Upstate Med. Ctr., Syracuse, N.Y., 1974; resident in psychiatry Thomas Jefferson U. Hosp., Phila., 1974-77, chief resident in psychiatry, 1976-77; sr. resident psychiat. rsch. svc. depression rsch. unit VA Med. Ctr., Phila., 1977-78; NIMH postdoctoral fellow in neuropsychopharmacology VA Med. Ctr./U. Pa. Hosp., Phila., 1978-79; asst. prof. psychiatry Thomas Jefferson U., Phila., 1978-83, adj. assoc. prof. psychiatry, 1983-88; dir. depression rsch. unit U. Pa. Hosp., Phila., 1979—; asst. prof. psychiatry U. Pa., Phila., 1979-86, assoc. prof. psychiatry, 1986-92, prof., 1992—; mem. instl. review bd. com. on studies involving human beings U. Pa., 1983—, faculty senate exec. com., 1987-88, univ. coun., 1987-88, undergraduate admissions and fin. aid, 1987-90, univ. facilities com., 1988-89, studebt affairs com., 1989-91, univ. disability bd., 1990—; mem. student-faculty interaction com. U. Pa. Hosp. and Med. Sch., 1987-89, continuing med. edn. lecture series, 1986—, ad hoc com. on irgan transplantation, 1988, pharmacy and therapeutics com., 1989-93, investigational drug task force, 1991-92, mem. clin. svcs. steering com., 1985-86, residency tng. com., 1985-86, rsch. com., 1985-90, Sachar Rsch. Award com., 1985—; adj. assoc. prof. Wistar Inst. Anatomy, Phila., 1988—; mem. program planning com. IV World Congress Biol. Psychiatry, Phila., 1984-85; sci. adv. com. 2nd Internat. Conf. on Viruses, Immunity and Mental Health, Montreal, Que., Can., 1988; program dir. organizing com. 1st Internat. Conf. on Refractory Depression, Phila., 1988; mem. organizing com. 2nd Internat. Conf. on Refractory Depression, Amsterdam, 1990-92; program co-dir. 3rd Internat. Conf. on Refractory Depression, 1992—. Editor: Pharmacology of Depression: Applications for the Outpatient Practitioner, 1990, Refractory Depression, 1991, (with J. Mendels) Psychobiology of Affective Disorders, 1980; assoc. editor Jour. Affective Disorders, 1986—, referee; asst. editor Psychosomatics, 1987—; referee Archives Gen. Psychiatry, Am. Jour. Psychiatry, Biol. Psychiatry, Jour. Clin. Psychopharmacology, Jour. Neuropsychiatry and Clin. Neuroscience, Psychoneuroendocrinology, Psychiatry Rsch., Psychosomatics, Psychosomatic Medicine, Psychobiology; contbr. 113 articles to med. and sci. jours., 196 presentations, abstracts, and invited papers in field. Grantee NIMH, 1980-81, 1980-85, 89—, Jack Warsaw Fund, 1984—; Biomedical Rsch. Support grantee NIH, 1989-90. Fellow Am. Psychiat. Assn. (advisor mood disorders work group for DSM-IV 1989—, Marie H. Eldredge award 1986), Am. Bd. Med. Psychotherapists; mem. Am. Fedn. for Clin. Rsch., Am. Assn. History Medicine, Internat. Soc. Psychoneuroendocrinology, Internat. Soc. for Investigation of Stress, Soc. Biol. Psychiatry (membership com. 1988-91), Pa. Psychiat. Soc., Phila. Psychiat. Soc. (exec. coun. 1985—, chair clin. rsch. resource com. 1989—). Home: PO Box 1931 Cherry Hill NJ 08034 Office: U Pa Univ Sci Ctr Depression Rsch Unit 3600 Market St 8th Fl Philadelphia PA 19104-2649

AMTOFT-NIELSEN, JOAN THERESA, physician, educator, researcher; b. Reading, Pa., Jan. 31, 1940; children: Andre Christian, Nikolaj Johan, Anja. BS, Kutztown (Pa.) State U., 1960; MD, Ansalt U. Munchen, Fed. Republic Germany, 1965; DC, Nat. Coll., 1968; MD, U. Copenhagen, 1978; postgrad., Harvard U., 1989-90, 91. Regional dir. Pa. Acad. Sci., Reading, 1961; intern Cook County Hosp., Chgo., 1966-68; clin. instr. U. Copenhagen, 1975-80; proctor N.C. Coalition Health, Durham, 1985-87; founder, cons. Triangle PMS Ctr., Cary, N.C., 1987—, also bd. dirs. Contbr. articles to profl. jours. Bd. dirs. shelter St. Francis Ho., Chapel Hill, N.C., 1989—; bd. dirs., grant coordinator N.C. Coalition Chs., Raleigh; v.p. Danish Red Cross, 1975-80; cons. physician Handicapped Encounter in Christ, Raleigh, 1984-87. NSF grantee, 1961; recipient award Sardoni Found., 1964, Walter Morris Found., 1957, Community Svc. award K.C., 1989. Mem. Am. Acad. Holistic Physicians, European Acad. Preventative Medicine, AAUW (v.p. Raleigh chpt. 1987—), NAFE, Scandinavian Club. Republican. Roman Catholic. Avocations: swimming, bridge. Home: 218 Rosebrooks Dr Cary NC 27513-3609

AMUNDSON, CLYDE HOWARD, engineering educator, researcher; b. Wood County, Wis., Aug. 15, 1927; s. Howard John and Fern Lydia (Ross) A.; m. Marilyn Ann Amundson, Aug. 11, 1951; children—Nora Lynn, Clyde Eric. B.S. in Food Sci., U. Wis., 1955, M.S. in Food Engring., 1956, Ph.D. in Food Sci., Biochemistry, 1960. Instr. food sci. U. Wis., Madison, 1956-60, asst. prof., 1960-65, assoc. prof., 1965-70, prof., 1970—, dir. Aquaculture Rsch Ctr, 1980—, chmn. dept., 1980-87, dir. food engring., 1974—, prof. oceanography and limnology, Coll. Engring., 1980—, prof. agrl. engring., 1982—. Contbr. numerous articles to profl. jours. Patentee in field. Recipient DRInc. award, 1970, Pfizer Inc. Research award, 1987; Inst. Food Tech. fellow 1987. Fellow N.Y. Acad. Scis., Wis. Acad. Sci., Arts and Letters, AAAS; mem. Am. Dairy Sci. Assn., Inst. Food Tech., Am. Oil Chemistry Soc., ASHRAE, World Mariculture Soc., Sigma Xi, Phi Tau Sigma. Office: U Wis Dept Food Sci 1605 Linden Dr Madison WI 53706-1565

AMUNDSON, MERLE EDWARD, pharmaceuticals executive; b. Sioux Falls, S.D., Aug. 21, 1936; s. Edward Percival and Marjorie Chloe (Lifto) A.; m. Avis Eyvonne Schneekloth, June 5, 1958; children: Eric, Paul, Peter. BS in Pharmacy, S.D. State U., 1958; PhD in Pharmacy, Mass. Coll. Pharmacy, 1961. Sr. chemist Eli Lilly & Co., Indpls., 1961-67, rsch. scientist, 1967-68, mgr., 1968-75, dir., 1975-84, exec. dir., 1984—; mem. adv. coun. Coll. of Pharmacy, S.D. State U., 1992—. Contbr. articles to profl. jours. Bd. dirs. Boys and Girls Club, Greenfield. Am. Found. Pharmacy Edn. fellow, 1958-61; recipient Medallion, Boys Club Greenfield, 1984, Disting. Alumnus award S. D. State U., 1989. Mem. Am. Chem. Soc., N.Y. Acad. Scis., Greenfield Sertoma. Lutheran. Home: 1414 Sherwood Dr Greenfield IN 46140 Office: Eli Lilly and Co Lilly Corp Ctr Indianapolis IN 46285

AMUSIA, MIRON YA, physics educator; b. Lenningrad, Russia, Nov. 18, 1934; s. Yankel Amusia and Golda Gurevich; m. Aneta Kominarova, Apr. 15, 1961; 1 child, Vladislav. MSc, Leningrade State U., 1958, PhDin Theoretical Nuclear Physics, 1963, D in Physical and Math. Scis., 1973. Prin. sci. A.F. Ioffe Physical-Tech. Inst., St. Petersburg, Russia; prof. theoretical physics St. Petersburg Maritime Tech. Inst.; vis. prof. Boris Kidric Inst. Nuclear Physics, Belgrade, Yugoslavia, 1972, 73, Daresbury Lab., 1988, Kaiserlautern U, Germany, 1989, Tohoky U., Tokyo, 1990, Flinders U., Adelaide, Australia, 1990, Inst. for Theoretical Atomic and Molecular Physics, Harvard U., 1990, 91, 93, Inst. for Theoretical Physics, Frankfurt U., Germany, 1991, 92, Manne Siegbahn Inst. of Physics, Stockholm, 1991, Helsinki U. of Tech., Finland, 1991, Argonne Nat. Labs., 1993; vice chmn., mem. orgn., program coms. Internat., Nat. (All Union) Confs. Atomic Physics, 1977, U. Paris-Sud, 1992, Argonne Nat. Lab., 1993; mem. adv. bd. Internat. Confs. Vacuum Ultraviolet radiation physics, 1989—; mem. Internat. Sci. Com. X-Ray and Inner Shell Processes Confs., 1993—; chmn. workshop Today and Tomorrow in Photoionization, Leningrad, 1990. Author: Many-body Effects in Electron Atomic Shells, 1968, (with L.V. Chernysheva) Automatic System for Numerical Investigation of the Atomic structure-ATOM, 1983, Atomic Photoeffect, 1987, rev. edit. 1990, Bremsstrahlung, 1990; (with V.M. Buimistrov, B.A. Zon, V.N. Tsitovitch) Polarizational Bremsstrahlung of Atoms and Particles, 1987, others; edit. bd. Jour. of Physics B, 1991-92., Comments on Atomic and Molecular Physics, 1976—; adv. panel atomic and molecular physics "Adam Hilger" Pubs., 1990—; pub. more than 400 sci. papers. Recipient Humbold award, Germany, 1990; Argonne Nat. Lab. fellow, U.S., 1992. Mem. St. Petersburg Assn. Scis. (mem. coord. com., chmn. internat. commn.), Acad. Scis. (sci. coun. electronic, atomic collisions, commn. on synchroton radiation of Presidium, commn. on spectroscopy). Achievements include development of many-body theory of nucleus, atoms, multiatomic formations and electron gas; discovery of the collective nature of the atomic photoionization and prediction of collectivization of the few electron shells under the action of many electron neighboring shells; suggestion of a new mechanism of bremsstrahlung ("atomic" Bremsstrahlung) and its extensive investigation; prediction of a new mechanism leading to the ordered motion of electrons and atoms due to light absorption: evaluation of quadrupole moments of atomic shells in their J=1/2 states; derivation of dispersion relation for the elastic electron-hydrogren atom forward scattering amplitude; quantum mechanical formulation of the Post-Collision phenomena; description of correlational and cooperative effects in vacancies decay. Home: Orbeli Str 20 Apt 64, 194223 Saint Petersburg Russia Office: A F Ioffe Phys Tech Inst, Acad Sci, 194021 Saint Petersburg K 21, Russia

ANAEBONAM, ALOYSIUS ONYEABO, pharmacist; b. Udi, Nigeria, June 25, 1955; came to the U.S., 1980; s. George Nwoye and Maria Nneka (Ofoedu) A.; m. Nneka Chinyere Esimai, Feb. 21, 1992. BS in Pharmacy, U. Nigeria, 1978; PhD, Mass. Coll. Pharmacy, 1986. Registered pharmacist, Nigeria. Rsch. scientist Pfeiffer Pharm. Sci. Labs., Boston, 1981-86; product devel. scientist Fisons Corp., Bedford, Mass., 1986-89; mgr. analytical devel. Fisons Corp., Rochester, N.Y., 1989-90; dir. product devel. and quality control Ascent Pharms., Inc., Billerica, Mass., 1991—; cons. Al-Consult, Rochester, 1990—. Author: Chewable Tablets, 1990, 89; contbr. articles to profl. jours. Mem. Am. Assn. Pharm. Sci., Rho Chi, Beta Simga (gov. 1976-77). Roman Catholic. Achievements include patent for stabilization of pentamidine isethionate solutions. Home: 301 Arboretum Way Burlington MA 01803-3829

ANAGNOSTOPOULOS, STAVROS ARISTIDOU, earthquake engineer, educator; b. Megalo Horio, Evrytania, Greece, Jan. 8, 1946; came to U.S., 1968; s. Aristides A. and Agapi (Katsiyanni) A.; m. Panagiota Papacosta, Aug. 31, 1975; children—Aristides, Haralampos, Demetrios. Diploma Nat. Tech. U. Athens, 1968; M.S., MIT, 1970, Sc.D., 1972. Research assoc. MIT, Cambridge, 1975-76; research engr. Shell Devel., Houston, 1976-81, sr. research engr., 1981; dir. Inst. Engring. Seismology and Earthquake Engring., Thessaloniki, Greece, 1981-86; prof. dept civi. engring. U. Patras, Greece, 1986—; chmn. bd. dirs. Greek Earthquake Planning and Protection Orgn.; pres. Hellenic Assn. for Earthquake Engring., 1991—; mem. sci. com. European Ctr. for Earthquake Prediction and Prevention; mem. European Com. Drafting Eurocode #8 for Earthquake Resistant Design. Contbr. articles to profl. jours. Greek state fellow, Athens, 1964-68; doctoral fellow Greek Ministry Coordination, Athens, 1969-72. Mem. ASCE, Seismol. Soc. Am., Earthquake Engring. Rsch. Inst. Office: U Patras, Dept Civil Engring, 26110 Patras Greece

ANAND, SURESH CHANDRA, physician; b. Mathura, India, Sept. 13, 1931; came to U.S., 1957, naturalized, 1971; s. Satchit and Sumaran (Bai) A. m. Wiltrud, Jan. 29, 1966; children: Miriam, Michael. MB, BS, King George's Coll., U. Lucknow (India), 1954; MS in Medicine, U. Colo., 1962. Diplomate Am. Bd. Allergy and Immunology. Fellow pulmonary diseases Nat. Jewish Hosp., Denver, 1957-58, resident in chest medicine, 1958-59, chief resident allergy-asthma, 1960-62; intern Mt. Sinai Hosp., Toronto, Ont., Can., 1962-63, resident in medicine, 1963-64, chief resident, 1964-65, demonstrator clin. technique, 1963-64, U. Toronto fellow in medicine, 1964-65; rsch. assoc. asthma-allergy Nat. Jewish Hosp., Denver, 1967-69; clin. instr. medicine U. Colo., Denver, 1967-69; internist Ft. Logan Mental Health Ctr., Denver, 1968-69; pres. Allergy Assocs. & Lab., Ltd., Phoenix, 1974—; mem. staff Phoenix Bapt. Hosp., chmn. med. records com., 1987; mem. staff St. Joseph's Hosp., St. Luke's Hosp., Humana Hosp., Chandler Regional Hosp., Valley Luth. Hosp., John C. Lincoln Hosp., Good Samaritan Hosp., Phoenix Children's Hosp., Tempe St. Luke Hosp., Desert Samaritan Hosp., Mesa Luth. Hosp., Scottsdale Meml. Hosp., Phoenix Meml. Hosp., Chandler (Ariz.) Regional Hosp., Valley Luth. Hosp., Mesa, Ariz.; pres. NJH Fed. Credit Union, 1967-68. Contbr. articles to profl. jours. Mem. Camelback Hosp. Mental Health Ctr. Citizens Adv. Bd., Scottsdale, Ariz., 1974-80; mem. Phoenix Symphony Coun., 1973-90; mem. Ariz. Opera Co., Boyce Thmpson Southwestern Arboretum; mem. Ariz. Hist. Soc., Phoenix Arts. Mus., Smithsonian Inst. Fellow ACP, Am. Coll. Chest Physicians (critical care com.), Am. Acad. Allergy, Am. Assn. Cert. Allergists, Am. Coll. Allergy and Immunology (aerobiology com., internat. com., pub. edn. com.); mem. AAAS, AMA, Internat. Assn. Allergy and Clin. Immunology, Ariz. Med. Assn., Ariz. Allergy Soc. (v.p. 1988-90, pres. 1990-91), Maricopa County Med. Soc., West Coast Soc. Allergy and Immunology, Greater Phoenix Allergy Soc (v.p. 1984-86, pres. 1986-88, med. adv. team sports medicine Ariz. State U.), Phoenix Zoo, N.Y. Acad. Scis., World Med. Assn., Internat. Assn. Asthmology, Assn. Care of Asthma, Ariz. Thoracic Soc., Nat. Geographic Soc., Village Tennis Club, Ariz. Club, Ariz. Wild Life Assn. Office: 1006 E Guadalupe Rd Tempe AZ 85283-3044 also: 6641 Baywood Ave Mesa AZ 85206 also: Ste 350 7331 E Osborn Dr Scottsdale AZ 85253

ANANDAN, MUNISAMY, physicist; b. Perampattu, T.Nadu, India, July 1, 1939; came to U.S. 1987; s. Ramasamy Munisamy and Sali; m. Visalakshi Anandan, May 10, 1968; children: Sharadamani, Yesheswini. M.Tech., Indian Inst. Tech., Bombay, 1967; PhD, Indian Inst. Sci., Bangalore, 1979. Devel. engr. Bharat Electronics Ltd., Bangalore, 1967-69, sr. devel. engr., 1969-74, dep. mgr. devel., 1974-79, mgr. display devices, 1979-87; vis. scientist Bell Communications Rsch., Red Bank, N.J., 1987-88; rsch. scientist Thomas Electronics, Inc., Wayne, N.J., 1988—. Contbr. articles to profl. jours. Recipient R & D Excellence 1st prize Electronic Components Industries Assn., India, 1985, R & D 100 award R & D Mag., 1992. Mem. IEEE (sr.), Am. Phys. Soc., Soc. for Info. Display (chmn. 1993—), Illuminating Engring. Soc. Achievements include patents for Thick Film Integrated Flat Fluorescent Lamp, Coplanar Electrode Electric Gas-Discharge Device (India), Method of Manufacturing True Analogue Multi-column Bar-Graph Liquid Crystal Display (India); research for self-aligned process for the preparation of suspended electrodes for a single substrate AC color plasma display. Office: Thomas Electronics Inc 100 Riverview Dr Wayne NJ 07470-3104

ANANIA, WILLIAM CHRISTIAN, podiatrist; b. Long Branch, N.J., May 11, 1958; s. Joseph John and Marie (Forgione) A.; m. Pamela Capone, Dec. 18, 1982; 1 child, William Christian Jr. BS in Biology, Villanova U., 1980; D of Podiatric Medicine, Ohio Coll. Podiatric Medicine, 1984. Diplomate Nat. Bd. Podiatry Examiners; diplomate in podiatric surgery and medicine Am. Podiatric Med. Specialties Bd. Resident James C. Giuffré Med. Ctr., Phila., 1984-86; pvt. practice Middletown, N.J., 1986—; cons. Fern Med., Boston, 1991. Chmn. editorial review bd. The Contemporary Podiatric Physician; assoc. editor: Jour. Current Podiatric Medicine, 1986-89; contbr. numerous articles to profl. jours. Named Dr. of the Month Jour. Current Podiatric Medicine, 1986. Fellow Am. Soc. Podiatric Medicine, Am. Soc. Podiatric Dermatology, Nat. Soc. Conscious Sedation (med. advisor 1989—); mem. Am. Coll. Foot Surgeons, Am. Podiatric Med. Assn., Middletown Area C. of C. (bd. dirs. 1988-90, 2d v.p. 1990-91, pres. 1993). Avocations: Civil War reenactor, medical and military antique collectibles. Home: 112 Tindall Rd Middletown NJ 07748-2327 Office: 112 Tindall Rd # 673 Middletown NJ 07748-2327

ANAS, JULIANNE KAY, administrative laboratory director; b. Detroit, Oct. 31, 1941; d. Theodore John and Lorraine (Comment) Knechtges; m. Donald Cartwright, Jan. 25, 1965 (div. June 1968); m. Daniel James Anas, Jan. 6, 1979. BS, Ea. Mich. U., 1969; MA, Cen. Mich. U., 1978. Cert. specialist in chemistry and med. tech. Am Soc. Clin. Pathologists. Med. technologist W.A. Foote Hosp., Jackson, Mich., 1962-63; med. technologist PCHA Annapolis Hosp., Wayne, Mich., 1964-65, supr. spl. chemistry and nuclear medicine, 1969-71; med. technologist Herrick Hosp., Tecumseh, Mich., 1965, Emma L. Bixby Hosp., Adrian, Mich., 1965-68; asst. clin. chemist Peoples Community Hosp. Authority, Wayne, 1971-81; adminstrv. lab. dir. Health Alliance Plan, Henry Ford Health System, Detroit, 1981—; adv. panel Medicalab Observor mag., 1988—. Contbr. articles to profl. publs. Mem. Am. Soc. Med. Tech. (bd. dirs. Mich. sect. 1972-73, pres. Detroit sect. 1972), Hosp. Lab. Mgrs. Assn. (membership chmn. 1984, 85, 90), Detroit Soc. Med. Technologists (Med. Technologist of Yr. 1975), Am. Assn. Clin. Chemists (nominations chair Mich. sect. 1992), Founders Art Inst. Republican. Avocations: boating, gardening, art, reading. Home: 30774 Bobrich St Livonia MI 48152-3410

ANASTASIO, THOMAS JOSEPH, neuroscientist, educator, researcher; b. Washington, Dec. 7, 1958; s. Albert Thomas and Giovanna Grace (Russo) A.; m. Anne E. McKusick, Sept. 2, 1990; 1 child: Albert Thomas. BS, McGill U., Montreal, Que., Can., 1980; PhD, U. Tex. Med. Br., Galveston, 1986. NASA fellow Vestibular Rsch. Facility, Moffett Field, Calif., 1982; predoctoral rsch. fellow U. Tex. Med. Br., Galveston, 1980-86; postdoctoral fellow Johns Hopkins U. Sch. Medicine, Balt., 1986-88; rsch. asst. prof. dept. otolaryngology U. So. Calif., L.A., 1988-91; asst. prof. dept. physiology and biophysics U. Ill., Urbana, 1991—; presenter seminars; reviewer Biological Cybernetics, others. Contbr. chpts. to books, 10 articles to peer-reviewed jours.; author abstracts. Mem. Internat. Brain Rsch. Orgn., Internat. Neural Network Soc., Soc. Neurosci. Achievements include research on linear, nonlinear and distributed aspects of signal processing in neural control of eye movements. Office: U Ill Beckman Inst 405 N Mathews Ave Urbana IL 61801

ANBAR, MICHAEL, biophysics educator; b. Danzig, June 29, 1927; came to U.S., 1967, naturalized, 1973; s. Joshua and Chava A.; m. Ada Komet, Aug. 11, 1953; children: Ran D., Ariel D. MSc, Hebrew U., Jerusalem, 1950, PhD, 1953. Instr. chemistry U. Chgo., 1953-55; sr. scientist Weizmann Inst. Sci., 1955-67; prof. Frienberg Grad. Sch., Rehovoth, Israel, 1960-67; sr. rsch. assoc. NASA Ames Rsch. Ctr., 1967-68; dir. phys. sci. SRI Internat., Menlo Park, Calif., 1968-72; dir. mass spectrometry research ctr. SRI Internat., 1972-77; prof. biophysical sci., chmn. dept. Sch. Medicine, SUNY, Buffalo, 1977-90, Faculty prof., dir. Interdeptl. Clin. Biophysics Group, 1990—, exec. dir. Health Instrument and Device Inst., 1983-85; assoc. dean applied research Sch. Medicine, SUNY, 1983-85; v.p. R & D AMARA Inc, Amherst, N.Y., 1992—. Author: The Hydrated Electron, 1970, The Machine of the Bedside—Strategies for Using Technology in Patient Care, 1984, Clinical Biophysics, 1985, Computers in Medicine, 1986; editor-in-chief: Thermology, 1993; contbr. articles to profl. jours. With Israeli Air Force, 1947-49. Grantee in field. Mem. IEEE, AAAS, IEEE Computer Soc., Assn. Am. Med. Colls., N.Y. Acad. Scis., Am. Chem. Soc., Biophys. Soc., Am.Inst. Ultrasound in Medicine, Am. Assn. Clin. Chemistry, Internat. Assn. Dental Rsch., Radiation Rsch. Soc., Am. Assn. Dental Rsch., Am. Assn. Mass Spectrometry, Assn. Advancement Med. Instrumentation, Am. Acad. Thermology, Am. Assn. Med. Systems Informatics, Engring. in Medicine and Biology Soc., Internat. Med. Informatics Assn., Internat. Soc. Optical Engring. Office: SUNY 118 Cary Hall Buffalo NY 14214

ANCES, I. G(EORGE), obstetrician/gynecologist, educator; b. Balt., July 3, 1935; s. Harry and Fanny A.; m. Marlene Roth, Oct. 23, 1966; 1 son, Beau Mark. B.S., U. Md., 1956, M.D., 1959. Diplomate: Am. Bd. Obstetrics and Gynecology. Intern Ohio State U. Hosp., 1959-60; resident in obstetrics and gynecology Univ. Hosp., Balt., 1960-61, 63-65; mem. faculty U. Md. Med. Sch., Balt., 1966—; prof. obstetrics and gynecology U. Md. Med. Sch., 1975-83, dir. labs. obstetrics and gynecol. research and clin. labs., 1967-83, dir. div. adolescent obstetrics and gynecology and family planning, 1981-83;; prof. ob-gyn. Rutgers U. Sch. Medicine,, Camden,, N.J.,, 1983—; chmn. dept., 1983—. Contbr. chpts. to books, articles to profl. jours. Capt. sustaining fund drive Balt. Symphony Orch., Opera Co. Phila.; med. adv. com. Fire Dept. Balt. City. Served with USAF, 1961-63. Fellow Am. Coll. Obstetrics and Gynecology; mem. Endocrine Soc., Soc. Gynecol. Investigation, Soc. Study Reprodn. (charter), Internat. Soc. Rsch. in Biology Reprodn. (charter), Md. Obstetrics and Gynecol. Soc. (sec. 1978-81, dir. 1979—), Med. and Chirurgical Soc. Md., Soc. Adolescent Medicine, Douglas Obstet. and Gynecol. Soc. (pres. 1984—), N.J. State Med. Soc. (chmn. neo-natal coop. So. Jersey 1986—), English Speaking Union, Cooper Found., N.J. Conservation Coun., Harbour League Club, Sigma Xi. Clubs: Maryland, Towson Golf and Country. Home: 1 Lane Of Acres Haddonfield NJ 08033-3504 Office: Rutgers U Sch Medicine Dept Ob-Gyn 3 Cooper Plz Camden NJ 08103-1438

ANCHETA, CAESAR PAUL, software engineer; b. Manila, June 1, 1947; s. Carlos Fortunato and Rosalinda (Huliganga) A.; m. Ruth Segalman, June 1, 1969; children: Rebecca E., Amy L. BS in Physics, U. Tex., 1969; MS in Physics, UCLA, 1971. Mem. tech. staff Hughes Aircraft Co., Culver City, Calif., 1969-78; sr. staff engr. Fairchild Camera and Instrument, Simi Valley, Calif., 1978-82; software engr. Internat. Remote Imaging Systems, Chatsworth, Calif., 1982-84, Teradyne, Inc., Woodland Hills, Calif., 1984-86; sr. scientist Internat. Remote Imaging Systems, Chatsworth, 1986-88; rsch. scientist Teledyne Industries, Northridge, Calif., 1988-89; sr. systems engr. Hughes Aircraft Co., Long Beach, Calif., 1989-90; sr. software engr. GE, Milw., 1990—. Author: (publs.) Proceedings of the Society of Photo Optical Instrumentation Engineers, 1978, Proceedings of the International Test Conference, 1981. Mem. The Elfun Soc., Milw., 1992. Named New Elfun of Yr., Milw. chpt. The Elfun Soc., 1993; Hughes fellow Hughes Aircraft Co., 1969-71; Stevens scholar U. Tex., El Paso, 1965-68. Mem. AAAS. Achievements include patent pending (with Arthur F. Griffin) on Rectilinear Object Matcher; development of white blood cell recognition machine vision project, pap smear scanning machine vision project.

AN DER HEIDEN, WULF-UWE, mathematics educator; b. Marburg, Hessia, Germany, Dec. 24, 1942; s. Werner and Brunhilde (Redler) Hoch; m. Doris Hauser; children: Matthias, Iris, Judith. Student, U. Cologne, Fed. Republic Germany, 1964-70; PhD, U. Göttingen, Fed. Republic Germany, 1972. Sci. asst. U. Kiel, Fed. Republic Germany, 1971-79, U. Tübingen, Fed. Republic Germany, 1971-79; pvt. docent U. Bremen, Fed. Republic Germany, 1980-85, prof. theoretical biology, 1986; prof. math. and theory of complex systems U. Witten/Herdecke, Fed. Republic Germany, 1987—; mem. editorials Acta Biotheoretica, Systeme, Jour. Math. Biology, Neural Networks. Author: Analysis of Neural Networks, 1980; editor: Temporal Disorder of Human Oscillatory Systems, 1987; contbr. articles to sci. publs. Mem. Soc. Math. Biology, Biometric Soc., Deutsche Math.-Vereinigung, Gesellschaft Angewandte Math. and Mech., Soc. Indsl. and Applied Math. Home: Kirchender Dorfweg 163, D-58313 Herdecke Germany Office: U Witten/Herdecke, Stockumer Str 10, D-58448 Witten Germany

ANDERMAN, IRVING INGERSOLL, dentist; b. N.Y.C., June 10, 1918; s. Louis and Regina (Mandel) A.; m. Georgia Gordon, Feb. 14, 1943; children: Susan, Robert. BS, CCNY, 1938; DDS, NYU, 1942. Diplomate Am. Bd. Oral Electrosurgery. Pvt. practice Middletown, N.Y., 1946—; clin. instr. NYU, N.Y.C., 1947-48; vis. lectr. dental sch. NYU, 1965-80, 9th Dist. Teaching Ctr., Hawthorne, N.Y., 1965—; adj. prof., chmn. dental hygiene adv. bd. Orange County Community Coll., Middletown, 1965—; mem. Orange County Bd. Health, 1985—. Co-author: Electrosurgery in Dental Practice, 1976, Dental Clinics of North America, 1982; contbr. articles to profl. jours. Chmn. Middletown Housing Authority, 1955—; disaster chmn. ARC, 1960-62; mem. exec. bd., dist. commr. Boy Scouts Am., Orange County, 1947—. Recipient gold medal 9th Dist. Dental Soc., 1984. Fellow Am. Coll. Dentists, Am. Acad. Oral Medicine (acad.), Internat. Coll. Dentists, Soc. Physiology and Occlusion (pres.), Acad. Gen. Dentistry; mem. AAAS, Internat. Assn. Dental Rsch., N.Y. Acad. Scis., Kiwanis Internat. (pres., disting. svc. award 1979). Republican. Jewish. Achievements include invention of vital pulpotomy technique by means of electrosurgery. Office: 3 Linden Pl Middletown NY 10940-4887

ANDERS, WILLIAM ALISON, aerospace and diversified manufacturing company executive, former astronaut, former ambassador; b. Hong Kong, Oct. 17, 1933; s. Arthur Ferdinand and Muriel Florence (Adams) A.; m. Valerie Elizabeth Hoard, June 26, 1955; children: Alan Frank, Glen Thomas, Gayle Alison, Gregory Michael, Eric William, Diana Elizabeth. BS, U.S. Naval Acad., Annapolis, 1955; MS in Naval Engring., U.S. Inst. Tech., Wright-Patterson AFB, 1962. Commnd. 2d lt. U.S. Air Force, 1955, pilot, engr., 1955-69; astronaut NASA-Johnson Space Ctr., Houston, 1963-69, Apollo 8, 1st lunar flight, 1968; exec. sec. Nat. Aero. and Space Council, Washington, 1969-72; commr. AEC, Washington, 1973-74; chmn. Nuclear Regulatory Commn., Washington, 1975-76; U.S. Ambassador to Norway, 1976-77; v.p., gen. mgr. nuclear energy products div. Gen. Electric Co., 1977-80; v.p., gen. mgr. aircraft equipment div. Gen. Electric Co., DeWitt, N.Y., 1980-84; sr. exec., v.p. ops. Textron Inc., Providence, R.I., 1984-89; vice chmn. Gen. Dynamics, St. Louis, 1990-91; chmn., CEO Gen. Dynamics, 1991-93, chmn. bd. dirs., 1993—; bd. dirs. Enron Corp. Trustee Battell Meml. Inst., Washington U., St. Louis. Maj. gen. USAFR, 1983-88. Decorated various mil. awards; recipient Wright, Collier, Goddard and Arnold flight awards; co-holder several world flight records. Mem. Soc. Exptl. Test Pilots , Nat. Acad. Engring., Tau Beta Pi. Office: PO Box 1618 Eastsound WA 98245-1618

ANDERSEN, HANS CHRISTIAN, chemistry educator; b. Brooklyn, N.Y., Sept. 25, 1941; m. June Jenny, June 17, 1967; children—Hans Christian, Albert William. S.B., MIT, 1962, Ph.D., 1966. Jr. fellow Soc. Fellows Harvard U., Cambridge, 1965-68; asst. prof. chemistry Stanford U., Calif., 1968-74; assoc. prof. Stanford U., 1974-80, prof., 1980—; vis. prof. chemistry Columbia U., N.Y.C., 1981-82; co-dir. Stanford Ctr. for Materials Rsch., 1988-89, dep. dir., 1989—; mem. allocation com. San Diego Supercomputer Ctr., 1986-89, chmn., 1988-89; vice chmn. Gordon Rsch. Conf. on Physics and Chemistry of Liquids, 1989, chmn. 1991. Mem. editorial com. Ann. Rev. of Phys. Chemistry, 1983-87; mem. editorial bd. Jour. of Chem. Physics, 1984-86, Chem. Physics, 1986—; mem. adv. bd. Jour. of Phys. Chemistry, 1987-92. Recipient Gores Award for Excellence in Teaching Stanford U., 1973; Sloan fellow, 1972-74; Guggenheim fellow, 1976-77. Fellow AAAS, Am. Acad. Arts and Scis., Am. Phys. Soc.; mem. NAS, Am. Chem. Soc. (chmn. phys. chemistry div. 1986, Joel Henry Hildebrand award 1988). Office: Stanford U Dept Chemistry Stanford CA 94305

ANDERSEN, LEIF PERCIVAL, physician; b. Copenhagen, Apr. 27, 1951; s. John Eivin Percival and Inger Viola Andrea (Wiberg) A.; m. Annemarie Skovshoved Petersen, July 23, 1977; children: Tine Skovshoved, Henrik Percival, Karsten Wiberg. MD, U. Copenhagen, 1982. Physician dept. organ surgery Cen. Hosp., Hillerød, Denmark, 1983; physician internal medicine Sundby Hosp., Copenhagen, 1984-86; med. doctor infectious diseases dept. medicine Rigshospitalet, Copenhagen, 1985-86; med. registrar clin. microbiolgoy Cen. Hosp., Hillerød, 1985; med. doctor dept. virology

Statens Seruminstitut, Copenhagen, 1986-88; med. sr.registrar dept. clin. microbiology Rigshopitalet, Copenhagen, 1988-91; med. doctor dept. clin. microbiology Statens Seruminstitut, Copenhagen, 1992—; mem., coord. European study group on cellular immunology in helicobacter pylori infections. Grantee Brochades pharma Denmark, 1988-92, Danish Med. Rsch. Coun., 1989-93, Found. for Promotion of Med. Rsch., 1990-92; recipient Hoechst Denmark's Antibiotic award, 1992. Mem. Danish Med. Soc., Danish Helicobacter Pylori Study Group (head), Danish Soc. for Clin. Microbiology, Scandinavian Soc. for Antimicrobial Chemotherapy, European Soc. for Clin. Microbiology and Infectious Diseases, European Helicobacter Pylori Study Group (organizing com.), Am. Soc. Microbiology, Danish Soc. Gastroenterology. Office: Rigshosp Dept Clin Microbio, Juliane Mariesvej 28 2, DK 2100 Copenhagen 0, Denmark also: Statens Seruminstitut, Artillerivej 5, 2300S Copenhagen Denmark

ANDERSEN, THEODORE SELMER, engineering manager; b. N.Y.C., Dec. 4, 1944; s. Selmer and Irene Frances (McManus) A.; m. Elva Glenna Layden, June 19, 1965; children: Elva Irene, Theodore Christian, Caroline Elizabeth. BChemE, Cooper Union, 1965; MSChemE, U. Pitts., 1968, PhDChemE, 1971, MBA, 1977. Registered profl. engr., Pa. From engr. to mgr. compensation, evaluation and tng. Bettis Atomic Power Lab. Westinghouse, West Mifflin, Pa., 1965-77; mgr. emerging systems programs Advanced Energy Systems Div Westinghouse, Waltz Mill, Pa., 1978-84, mgr. energy program, 1985-86, mgr. stragetic program mktg., 1987-88; dep. dir. AP600 Program Nuclear and Advanced Tech. Div. Westinghouse Elec., Pitts., 1989—. Mem. Am. Inst. Chem. Engrs., Am. Wind Energy Assn. (pres. 1981-84), Am. Nuclear Soc. Republican. Methodist. Home: 5170 Caste Dr Pittsburgh PA 15236-1646 Office: Westinghouse Nuclear Advance PO Box 355 Pittsburgh PA 15230-0355

ANDERSEN, TORBEN BRENDER, optical researcher, astronomer; b. Naestved, Denmark, May 17, 1954; came to U.S., 1983; s. Bjarne and Anna Margrethe (Brender) A.; m. Alice Louise Palmer, Nov. 3, 1990. PhD, Copenhagen U., Denmark, 1979. Rsch. fellow Copenhagen U., 1980-82, sr. rsch. fellow, 1982-85; optical cons. Nordic Optical Telescope Assn., Roskilde, Denmark, 1985; optical systems analyst Telos Corp., Santa Clara, Calif., 1985-88; rsch. scientist Lockheed Missiles and Space Co., Palo Alto, Calif., 1988—, staff scientist, 1993—; vis. scholar Optical Scis. Ctr., U. Ariz., Tucson, 1983-85. Editor: Astronomical Papers Dedicated to Bengt Strömgren, 1978; contbr. articles to Jour. Quantitative Spectroscopy Radiation Transfer, Applied Optics, Astronomische Nachrichten. Mem. Optical Soc. Am., Internat. Astron. Union, Soc. Photo-Optical Instrumentation Engrs. Achievements include development of method for computing optical aberration coefficients to arbitrarily high orders; discovery of set of differential equations for the Voigt function. Office: Lockheed Rsch Labs O/97-20 3251 Hanover St # 254G Palo Alto CA 94304-1191

ANDERSLAND, ORLANDO BALDWIN, civil engineering educator; b. Albert Lea, Minn., Aug. 15, 1929; s. Ole Larsen and Brita Kristine (Okland) A.; m. Phyllis Elaine Burgess, Aug. 15, 1958; children—Mark, John, Ruth. B.C.E., U. Minn., 1952; M.S.C.E., Purdue U., 1956, Ph.D., 1960. Registered profl. engr., Minn., Mich. Staff engr. Nat. Acad. Sci., Am. Assn. State Hwy. Ofcls. Road Test, Ottawa, Ill., 1956-57; research engr. Purdue U., West Lafayette, Ind., 1957-59; faculty Mich. State U., East Lansing, 1960—, prof. civil engring., 1968—. Co-author: Geotechnical Software for the IBM, PC, 1987, Geotechnical Engineering and Soil Testing, 1992; co-editor: Geotechnical Engineering for Cold Regions, 1978; contbr. chpt. Ground Engineer's Handbook, 1987; contbr. articles to profl. jours. Served to 1st lt. C.E., U.S. Army, 1952-55. Decorated Nat. Def. Svc. medal; UN Svc. medal; Korean Svc. medal; recipient Best Paper award Assn. Asphalt Paving Technologists, 1956, Disting. Faculty award Mich. State U., 1979; postdoctoral fellow Norwegian Geotech. Inst., 1966; grantee NSF, EPA. Fellow ASCE; mem. ASTM, Internat. Soc. Soil Mechanics and Found. Engring., Am. Soc. Engring. Edn., Sigma Xi, Chi Epsilon, Tau Beta Pi. Lutheran. Home: 901 Woodingham Dr East Lansing MI 48823-1855 Office: Mich State U Dept Civil and Environ Engring East Lansing MI 48824

ANDERSON, ALAN JULIAN, database consultant; b. Atlanta, Aug. 2, 1953; s. James Louis and Mary Grace (Burson) A.; m. Elfreida Marie Grimes, June 27, 1985 (div. 1986); m. Kristine Lavian Renee Jones, May 17, 1991. BS in Computer Sci., Morehouse Coll., Atlanta, 1975. Engr. IBM, Atlanta, 1967—. Author two relational database products, 1981-86. Bd. treas. Nat. Urban League, N.Y.C., 1990; adv. bd. Langston U., 1990. Recipient Lockheed Corp. award, 1975, others. Avocations: sports officiating, community service, sports, reading. Home: 2312 Renaissance Way NE Atlanta GA 30308-2462 Office: Fulton County Bd Edn 786 Cleveland Ave SW Atlanta GA 30315

ANDERSON, ALFRED OLIVER, mathematician, consultant; b. Marmon, N.D., May 18, 1928; s. Frederick Gustav and Minnie Petrine (Jensen) A. BS, Ore. State U., 1953. Systems programmer U.S. Army Ballistics Research Lab., Aberdeen (Md.) Proving Ground, 1953-83; cons. Aberdeen, 1983—; investment specialist, Aberdeen, 1983—. Mem. Pi Mu Epsilon. Democrat. Lutheran. Avocations: wood working, investment analysis. Home and Office: 717 Plater St Aberdeen MD 21001-3023

ANDERSON, BRENDA JEAN, biological psychologist; b. Moundridge, Kans., Jan. 11, 1961; d. John Dale and Jean Carolyn (Heim) A.; m. William John Arloff, May 23, 1992. BS, Emporia State U., 1983, MS, 1985; PhD, U. Ill., 1993. Rsch. asst. Emporia (Kans.) State U., 1982-85, U. Ill., Urbana, 1985-92; lab. technician Parsons (Kans.) State Hosp., 1984; postdoctoral researcher Ind. U., Bloomington, 1992—. Author:Instructor's Manual and Test Item File, 1984; contbr. to profl. publs. Grantee NSF, 1987, 88, 89, 90. Mem. Soc. Neurosci. Achievements include research on neural basis of learning and memory, effects of exercise and learning on neuronal morphological plasticity. Office: Ind U Dept Psychology Bloomington IN 47405

ANDERSON, BROR ERNEST, chemist; b. Mazeppa, Minn., Nov. 24, 1914; s. Ernest and Huldah (Celine) A.; m. Ruth Mae Vruwink, June 14, 1941; children: Erik, Bonnie, Kent. BA, Gustavus Adolphus Coll., 1935; PhD, U. Minn., 1940. Chemist Northwest Paper Co., Cloquet, Minn., 1936-37; tech. mgr. filter products Johnson & Johnson, Chgo., 1940-46; mgr. paper and fiber rsch. A.B. Dick Co., Chgo., 1946-55, mgr. stencil rsch., 1955-58; tech. dir. Wyomissing Corp., West Reading, Pa., 1958-65; v.p., tech. dir. Weber Marking Systems, Arlington Heights, Ill., 1965-79; owner RAMCOR, Yonkers, N.Y., 1979—; cons. Diagraph-Bradley Co., Herrin, Ill., 1979-86. Contbg. author books on filtration and on paper coating. Chair Berks County Sci. Fair, Reading, Pa., 1963. Achievements include about 50 patents worldwide in fields of duplicating, non-woven fabrics, and thermographic coatings; involvement in pollution reduction by finding uses for reclaimed solvents that generator cannot reuse. Home: 582 Scarsdale Rd Yonkers NY 10707 Office: RAMCOR 582 Scarsdale Rd Yonkers NY 10707

ANDERSON, BRUCE MORGAN, computer scientist; b. Battle Creek, Mich., Oct. 8, 1941; s. James Albert and Beverly Jane (Morgan) A.; m. Jeannie Marie Hignight, May 24, 1975; children: Ronald, Michael, Valerie, John, Carolyn. BEE, Northwestern U., 1964; MEE, Purdue U., 1966; PhD in Elec. Engring., Northwestern U., 1973. Rsch. engr. Zenith Radio Corp., Chgo., 1965-66; assoc. engr. Ill. Inst. Tech. Rsch. Inst., Chgo., 1966-68; sr. electronics engr. Rockwell Internat., Downers Grove, Ill., 1973-75; computer scientist Argonne (Ill.) Nat. Lab., 1975-77; mem. group tech. staff Tex. Instruments, Dallas, 1977-88; sr. scientist BBN Systems and Techs., Cambridge, Mass., 1988-90; sr. systems engr. Martin Marietta, Denver, 1990—; lectr. computer sci. U. Tex.-Arlington and Dallas; adj. prof. computer sci. N. Tex. State U.; vis. indsl. prof. So. Meth. U.; computer systems cons. Info. Internat., Culver City, Calif., HCM Graphic Systems, Gt. Neck, N.Y.; computer cons. depts. geography, transp., econs., sociology and computer sci. Northwestern U., also instr. computer sci.; expert witness for firm Burleson, Pate and Gibson. Contbr. articles to tech. jours. NASA fellow Northwestern U., 1973. Mem. IEEE Computer Soc. (chmn. Dallas 1984-85), Am. Assn. Artificial Intelligence, Assn. Computing Machinery (publs. chmn. 1986 fall joint computer conf. IEEE and Assn. Computing Machinery), Toastmasters Internat., Sigma Xi, Eta Kappa Nu, Theta Delta Chi. Home: 3473 E Euclid Ave Littleton CO 80121-3663 Office: Martin Marietta Mail Stop XL4370 700 W Mineral Ave Littleton CO 80120-4511

ANDERSON, CHARLES ROSS, civil engineer; b. N.Y.C., Oct. 4, 1937; s. Biard Eclare and Melva (Smith) A.; m. Susan Breinholt, Aug. 29, 1961; children: Loralee, Brian, Craig, Thomas, David. BSCE, U. Utah, 1961; MBA, Harvard U., 1963. Registered profl. engr.; cert. land surveyor. Owner, operator AAA Engring. and Drafting, Inc., Salt Lake City, 1960—; mem. acad. adv. com. U. Utah, 1990-91. Mayoral appointee Housing Devel. Com., Salt Lake City, 1981-86; bd. dirs., cons. Met. Water Dist., Salt Lake City, 1985—; bd. dirs., pres., v.p., sec. bd. Utah Mus. Natural History, Salt Lake City, 1980-92; asst. dist. commr. Sunrise Dist. Boy Scouts Am., Salt Lake City, 1985-86; fundraising coord. architects and engrs. United Fund; mem. Sunstone Nat. Adv. Bd., 1980-88; bd. dirs. Provo River Water Users Assn., 1986—. Fellow Am. Gen. Contractors, Salt Lake City, 1960; recipient Hamilton Watch award, 1961. Mem. ASCE, Am. Congress on Surveying and Mapping, U. Utah Alumni Assn. (bd. dirs. 1989-92), Harvard U. Bus. Sch. Club (pres. 1970-72), The Country Club, Bonneville Knife and Fork Club, Rotary (v.p. 1990-91, chmn. election com. 1980-81, vice chmn. and chmn. membership com. 1988-90), Pi Kappa Alpha (internat. pres. 1972-74, trustee endowment fund 1974-80, Outstanding Alumnus 1967, 72), Phi Eta Sigma, Chi Epsilon, Tau Beta Pi. Avocations: fly fishing, golfing, foreign travel. Home: 2689 Comanche Dr Salt Lake City UT 84108-2846 Office: AAA Engring & Drafting Inc 1865 S Main St Salt Lake City UT 84115-2045

ANDERSON, CHRISTINE LEE, analytical chemist; b. Akron, Ohio, Apr. 17, 1963; d. Bruce Walter and Janet Harriet (Keckler) A. BS in Chemistry, Urbana U., 1987. Mgr. chemistry lab. Midwest Testing Labs., Piqua, Ohio, 1983-85; chief chemist Tri-mark Inc., Piqua, 1985—. Mem. Am. Chemical Soc., Am. Standard for Testing and Materials, Order of Ea. Star. Republican. Methodist. Home: 909 Marlboro Ave Piqua OH 45356 Office: Tri-Mark Inc 8585 Industry Park Dr Piqua OH 45356

ANDERSON, CHRISTINE MARLENE, software engineer; b. Washington, D.C., Nov. 19, 1947. BS in Math., U. Md., 1969. Mathematician Naval Oceanographic Office, Suitland, Md., 1969-71; computer researcher USAF Avionics Lab., Dayton, Ohio, 1971-74; sr. analyst System Devel. Corp., Colorado Springs, Colo., 1974-76; chief computer tech. USAF Wright Lab./Armament Directorate, Ft. Walton Beach, Fla., 1981-82; dir. space software rsch. ctr. USAF Phillips Lab., Albuquerque, 1992—; co-chmn. on Ada computer programming lang. Am. Nat. Standards Inst., 1989—; editor Ada standard Internat. Standards Orgn., 1991—. Co-author: Aerospace Software Engineering, 1991; contbr. articles to profl. jours. Fellow AIAA (chair software systems tech. com. 1987-89, bd. dirs. 1989—, Aerospace Software Engring. award 1991). Office: Air Force Phillips Lab PL/UTES Kirtland AFB NM 87117

ANDERSON, DAVID PREWITT, university dean; b. Twin Falls, Idaho, Sept. 14, 1934; m. Janice Gale Schmied, Dec. 21, 1962; children: Kathryn Lynn, Christopher Kyle. Student, U. Idaho, 1952-54; B.S., Wash. State U., 1959, D.V.M., 1961; M.S., U. Wis., 1964, Ph.D., 1965. NIH trainee U. Wis., 1961-62, asst. prof. vet. sci., asst. dir. biotron, 1965-69; prof. med. microbiology, dir. Poultry Disease Research Center, U. Ga., 1969-71, asso. dean research and grad. affairs Coll. Vet. Medicine, 1971-73, dean, 1975—; mem. com. animal health Nat. Acad. Sci., 1977-80; mem. animal health sci. rsch. adv. bd. USDA, 1978-85, mem. nat. adv. com. on meat and poultry inspection, 1990-92; mem. adv. com. Ctr. for Vet. Medicine, FDA, 1984-88. Editor: Avian Diseases. Mem. AVMA, Am. Coll. Vet. Microbiologists (diplomate), Am. Coll. Poultry Vets. (diplomate), Am. Assn. Avian Pathologists (pres. 1988-89), Nat. Acads. Practice. Home: 190 Harris St Winterville GA 30683-9710

ANDERSON, DON LYNN, geophysicist, educator; b. Frederick, Md., Mar. 5, 1933; s. Richard Andrew and Minola (Phares) A.; m. Nancy Lois Ruth, Sept. 15, 1956; children: Lynn Ellen, Lee Weston. B.S., Rensselaer Poly. Inst., 1955; M.S., Calif. Inst. Tech., 1959, Ph.D., 1962. With Chevron Oil Co., Mont., Wyo., Calif., 1955-56; with Air Force Cambridge Research Center, Boston, 1956-58, Arctic Inst. N.Am., Boston, 1958; mem. faculty Calif. Inst. Tech., Pasadena, 1962—, assoc. prof. geophysics, 1964-68, prof., 1968—, dir. seismol. lab., 1967-89, Eleanor and John R. McMillan prof. in geophysics, 1990—; prin. investigator Viking Mars Seismic Expt.; mem. various coms. NASA; chmn. geophysics research forum NAS, chmn. Arthur L. Day award com., also mem. various coms.; mem. adv. com. Purdue U., U. Chgo., U. Tex., Stanford U., U. Calif. Berkeley, Carnegie Instn. of Washington, U. Paris. Asso. editor: Jour. Geophys. Research, 1965-67, Tectonophysics, 1974-77; editor: Physics of the Earth and Planetary Interiors. Recipient Exceptional Sci. Achievement award NASA, 1977; Sloan Found. fellow, 1965-67; Emil Wiechert Medal German Geophys. Soc., 1986. Fellow AAAS, Am. Geophys. Union (James B. Macelwane award, 1966, pres. tectonophysics sect. 1971-72, chmn. Macelwane award com. 1975, mem. Bowie medal cCom. 1985, pres. elect 1986-88, pres. 1988-90, Bowie medal 1990), Geol. Soc. Am. (assoc. editor bull. 1971—, Arthur L. Day medal 1987, mem. Penrose medal com. 1989, Arthur L. Day medal com. 1989-90, long range planning com. 1990—); mem. NAS (chmn. seismology com. 1975, chmn. Geophysics Research Forum 1984-86), Am. Philos. Soc., Royal Astron. Soc. (Gold medal 1988), Seismol. Soc. Am., Sigma Xi. Home: 669 Alameda St Altadena CA 91001-3001 Office: Calif Inst Tech Divsn Geology & Planetary Scis Pasadena CA 91109

ANDERSON, DONALD MARK, biological oceanographer; b. Milw., Sept. 7, 1948; s. Josephine (Stanley) A.; m. Kathleen A. Hudock, May 27, 1978; children: Brian, Eric, Lauren. BS, MIT, 1970, MS, 1975, PhD, 1977. Postdoctoral investigator biology dept. Woods Hole (Mass.) Oceanographic Inst., 1978-79, asst. scientist, 1979-83, assoc. scientist, 1983-91, sr. scientist, 1991—; chmn. SCOR working group on harmful algae, 1992—; sci. advisor U.S. Delegation to Joint IOC/FAO Intergovtl. Panel on Harmful Algae, 1992; chmn. IOC ad hoc group of experts on harmful algae blooms, 1989-92. Contbr. over 70 sci. papers to profl. publs.; edited conf. proceedings. Recipient Rsch. fellowship NSF, 1989-90. Mem. AAAS, Am. Soc. Limnology and Oceanography, Am. Phycological Soc. Achievements include patent pending in field. Office: Woods Hole Oceanographic Inst Red Field 3-32 Biology Dept Woods Hole MA 02543

ANDERSON, DONALD MORGAN, entomologist; b. Washington, Dec. 27, 1930; s. John Kenneth and Alice Cornelia (Morgan) A. B.A., Miami U., Oxford, Ohio, 1953; Ph.D., Cornell U.l, 1958. Grad. teaching asst. Cornell U., 1954-57; asst. prof. sci. SUNY-Buffalo, 1959-60, rsch. fellow, 1960; rsch. entomologist Dept. Agrl., Washington, 1960-90, rsch. collaborator, 1990—; rsch. assoc. Buffalo Mus. Sci., 1972—, Smithsonian Instn., 1978—. Contbr. articles to profl. jours., chpts to books. Sigma Xi grantee, 1959. Mem. Entomol. Soc. Washington (corr. sec. 1963-65, pres. 1985), Entomol. Soc. Am., Coleoperists Soc., Am. Inst. Biol. Sci., St. Andrews Soc. Washington, Clan Anderson Soc. (editor 1979-84, treas. 1985-90, pres. 1990-92), Sigma Xi, Phi Kappa Phi. Home: 1900 Lyttonsville Rd Apt 804 Silver Spring MD 20910-2238 Office: Mus Natural History Dept Agr Nat Systematic Entomology Lab Washington DC 20560

ANDERSON, DONALD NORTON, JR., retired electrical engineer; b. Chgo., Aug. 15, 1928; s. Donald Norton and Helen Dorothy (Lehmann) A.; B.S., Purdue U., 1950, M.S., 1952. With Hughes Aircraft Co., Culver City and El Segundo, Calif., 1952-84, sect. head, sr. project engr., 1960-65, tech. mgr. Apollo program, 1965-66, mgr. visible systems dept., 1966-69, 70-73, project mgr., 1969-70, mgr. space sensors lab., 1973-79, mgr. space electro-optical systems labs., 1979-80, mgr. space electro-optical systems labs., 1980-84, ret., 1984. Recipient Apollo Achievement award, 1970; Robert J. Collier Landsat award, 1974. Mem. Research Soc. Am., Nat. Speleological Soc., Am. Theatre Organ Soc., Sigma XI (sec. Hughes Labs. br. 1974-75), Eta Kappa Nu, Sierra Club. Home: 2625 Topanga Skyline Dr Topanga CA 90290-9543

ANDERSON, DONALD THOMAS, zoologist, educator; b. Eng., Dec. 29, 1931; s. Thomas and Flora A.; m. Joanne Trevathan Claridge; 1 son. BSc, U. London, 1953, PhD, 1956, DSc, 1966; DSc, U. Sydney, NSW, Australia, 1983. Lectr. zoology U. Sydney, 1958-61, sr. lectr., 1962-66, reader, 1967-71, prof., 1972-83, Challis prof. biology, 1984—. Author: Embryology and Phylogeny in Annelids and Arthropods, 1973; contbr. articles on zoology, evolution to profl. jours. Trustee Australian Mus., 1982-87. Decorated officer Order of Australia; recipient Clarke medal Royal Soc. NSW, 1979, Excellence in Rsch. award Crustacean Soc., 1990. Fellow Royal Soc., Linnean Soc. London, Australian Inst. Biology; mem. Linnean Soc. NSW (pres. 1966), Australian Marine Scis. Assn. (pres. 1980-82), Australian and New Zealand Assn. for the Advancement Sci. (pres. sect. 12 zoology 1981). Home: 52 Spruson St, Neutral Bay 2089, Australia Office: U Sydney, Dept Zoology A08, Sydney 2006, Australia

ANDERSON, DONALD THOMAS, JR., environmental consultant; b. Springfield, Ill., Jan. 19, 1937; s. Donald Thomas and Lucille Ann (Leonard) A.; m. Judith Lynn Skiles, Oct. 28, 1967. BS, Purdue U., 1959; MA, Sangamon State U., 1978; PhD, So. Ill. U., 1984. Tchr. sci. Springfield Sch. Dist. 186, 1959-69; sales engr. Simplex, Inc., Springfield, 1969-71; mktg. rep. Motorola, Inc., Springfield, 1971-73; energy specialist Ill. State Bd. Edn., Springfield, 1973-84; state asbestos dir. Ill. Dept. Pub. Health, Springfield, 1984-87; environ. cons. Asbestos Profl. Svcs., Inc., Breese, Ill., 1987—; cons. U.S. EPA, Washington, 1986-87; seminar dir. The Environ. Inst., Marietta, Ga., 1986—; condr. human resource mgmt. seminars various profl. orgns. Mem. Am. Mgmt. Assn., Nat. Asbestos Coun., Ill. Assn. Sch. Bus. Ofcls. Avocation: public speaking. Home: 92 Golf Rd Springfield IL 62704-3142 Office: Assessment Profl Svcs 501 N 2nd St Breese IL 62230-9998

ANDERSON, DOUGLAS WARREN, optics scientist; b. Wooster, Ohio, June 2, 1950; s. Earl Godfrey and Elsie Aileen (Heath) A.; m. Virginia Southwell Hunt, Aug. 3, 1974; children: Charles Alford, Julia Heath. BA in Physics, Ohio Wesleyan U., 1972; MS, U. Ariz., 1974, PhD in Optical Scis., 1978. Physicist Air Force Human Resources Lab., Wright-Patterson AFB, Ohio, 1976-79; lectr. U. Tex., Dallas, 1989-90; optical designer Tex. Instruments, Dallas, 1979—. Treas. Richardson (Tex.) Crop Walk, 1987—; den leader Cub Scout Pack 720, Richardson, 1987-92. Mem. Optical Soc. Am., Soc. Photo-Optical Instrumentation Engrs. Achievements include one patent in field; design of optical systems for visible and infrared military products, commercial display and laser-based systems. Office: Tex Instruments PO Box 660246 M/S 3122 Dallas TX 75266

ANDERSON, DUWAYNE MARLO, earth and polar scientist, university administrator; b. Lehi, Utah, Sept. 9, 1927; s. Duwayne LeRoy and Fern Francell (Fagan) A.; m. June B. Hodgin, Apr. 2, 1980; children by previous marriage: Lynna Nadine, Christopher Kent, Lesleigh Leigh. B.S., Brigham Young U., 1954; Ph.D. (Purdue Research Found. fellow), Purdue U., 1958. Prof. soil physics U. Ariz., Tucson, 1958-63; research scientist, chief earth scis. br. (Cold Regions Research and Engring. Lab.), Hanover, N.H., 1963-76; chief scientist, div. polar programs NSF, Washington, 1976-78; mem. Viking sci. team NASA, 1969-76; dean faculty natural scis. and math. SUNY Buffalo, 1978-84; assoc. provost for research and grad. studies Tex. A&M U., College Station, 1984-92, prof. Coll. Geoscis., 1992—; also councilor Tex. A&M Research Found., 1984—; Pegrum lectr. SUNY, 1980; v.p. Assn. Tex. Grad. Schs., 1990-91, pres., 1991-92; bd. dirs. Buffalo Mus. Sci.; cons. NASA, 1964, NSF, 1979-81, U.S. Army Cold Regions Research and Engring. Lab., Hanover, N. H.; sr. U.S. rep., Antarctica, 1976, 77; bd. dirs., mem. exec. com. Houston Area Research Ctr., 1984-89; vis. prof., lectr., cons. numerous univs. Editor: (with O.B. Andersland) Geotechnical Engineering for Cold Regions, 1978; Cons. editor: Soil Sci, 1965-81, (with O.B. Andersland) Cold Regions Sci. and Tech, 1978-82; Contbr. numerous sci. and tech. articles to profl. jours. Bd. dirs. Ford K. Sayre Meml. Ski Council, Hanover, 1969-71; bd. dirs. Grafton County Fish and Game Assn., 1965-76, pres., 1968-70; bd. dirs. Hanover Conservation Council, 1970-76, v.p., 1970-73; bd. dirs. Buffalo Mus. Sci., 1980-84, v.p., 1982-84. Served in USAF, 1946-49. Recipient Sci. Achievement award Cold Regions Research and Engring. Lab., 1968; co-recipient Newcomb Cleveland award AAAS, 1976; Sec. of Army Research fellow, 1966. Fellow Geol. Soc. Am., Am. Soc. Agronomy; mem. Internat. Glaciological Soc., Am. Polar Soc., Am. Geophys. Union (spl. task force on cold regions hydrology 1974-84), AASS, Soil Sci. Soc. Am., Niagara Frontier Assn. Research and Devel. Dirs. (pres. 1983-84), Mem. Licensing Execs. Soc., Buffalo Mus. Sci. (bd. dirs.), NASA Teams (Viking, Skylab & Planetary Geology & Geophysics Working Group), Comet Rendevous/Asteroid Flyby Mission Team, Arctic Rsch. Consortium U.S. (exec. com.), Sigma Xi, Sigma Gamma Epsilon, Phi Kappa Phi. Republican. Home: 8720 Bent Tree Dr College Station TX 77845-5558 Office: Tex A&M U Dept Geology 267 Halbouty Blvd College Station TX 77843-3115

ANDERSON, FRANCES SWEM, nuclear medical technologist; b. Grand Rapids, Mich., Nov. 27, 1913; d. Frank Oscar and Carrie (Strang) Swem; m. Clarence A.F. Anderson, Apr. 9, 1934; children: Robert Curtis, Clarelyn Christine (Mrs. Roger L. Schmelling), Stanley Herbert. Student, Muskegon Sch. Bus., 1959-60; cert., Muskegon Community Coll., 1964. Registered nuclear med. technologist Am. Registry Radiol. Technologists. X-ray file clk., film librarian Hackley Hosp., Muskegon, Mich., 1957-59, radioisotope technologist and sec., 1959-65; nuclear med. technologist Butler Meml. Hosp., Muskegon Heights, Mich., 1966-70; nuclear med. technologist Mercy Hosp., Muskegon, 1970-79, ret., 1979. Mem. Muskegon Civic A Capella choir, 1932-39; mem. Mother-Tchr. Singers, PTA, Muskegon, 1941-48, treas. 1944-48; with Muskegon Civic Opera Assn., 1950-51, office vol. Alive '88 Crusade, mem. com. for 60th High Sch. class reunion. Soc. Nuclear Medicine Cert. nuclear medicine technologist Soc. Nuclear Medicine; active Forest Park Covenant Ch., mem. choir 1953-79, 83—, choir pres 1992—, choir sec. 1963-69, Sunday sch. tchr. 1954-75, supt. Sunday sch. 1975-78, treas. Sunday sch. 1981-86, sec., 1993, chmn. master planning coun., coord. centennial com. to 1981, ch. sec. 1982-84, 87, 91, registrar vacation Bible sch., 1988-89, 90, 91, sec. Sunday Sch. 1991, 92, mem. Sunday Sch. support team, sec., 1993; co-chmn. Jackson Hill Old Timers Reunion, 1982, 83, 85. Mem. Am. Registry Radiologic Technologists. Home: 5757 Sternberg Rd Fruitport MI 49415-9740

ANDERSON, FREDERIC SIMON B., physicist; b. Adelaide, Australia, Apr. 14, 1953; came to U.S., 1981; s. Frederic Brian and Lorna Doreen (Schultz) A.; m. Marie Helen Willoughby, Feb. 28, 1954 (div. 1987); children: Tom Adam F., Jesse David S., Arwen Eleanor M. BS with honors, Flinders U., South Australia, 1977, PhD, 1981. Postdoctoral researcher U. Wis., Madison, 1981-83, rsch. assoc., 1984—; rsch. cons. Oak Ridge (Tenn.) Nat. Lab., 1988-92; co-prin. investigator DOE, 1992—. Author: Nuclear Fusion, 1989, Reviews of Scientific Instruments, 1990. Scout leader Boy Scouts of Am., Madison, 1986-88. Mem. Am. Phys. Soc. Achievements include patent (with other) for three axis machining tool for toroidal machining. Home: 116 Proudfit St Madison WI 53715 Office: U Wis 1415 Johnson Dr Madison WI 53706

ANDERSON, FREEDOLPH DERYL, gynecologist; b. Bismarck, N.D., Aug. 14, 1933; s. Freedolph E. and Olga (Mayer) A.; m. Harlean Ruth Peterson; children: Robert Allan, Daniel David, Rebecca Lynn. AB in Biology, Stanford Univ., 1955; MD/BS in Med., Sch. Med. U. Minn., 1959. Diplomate Am. Bd. Ob-Gyn., Am. Bd. Med. Examiners. Intern San Diego County Gen. Hosp., 1959-60; resident in obstetrics and gynecology Salt Lake County Gen. Hosp., Salt Lake City, Utah, 1960-61, Boston City Hosp., 1961-62; Chief Resident Salt Lake County Gen. Hosp., 1962-63; pvt. practice Missoula, Mont., 1963-75; interim health officer Missoula City-County Health Dept., Missoula, Mont., 1975; ob-gyn. FHP, Long Beach, Calif., 1975; assoc. med. dir. ob-gyn. FHP, Guam, 1975-78, Gynecology and Infertility, Missoula, 1978-82; asst. dir. clinical rsch. Ortho Pharm. Corp., Raritan, N.J., 1982-84, assoc. dir., clinical rsch., 1984-85, dir., clinical rsch., 1985-86, Group dir, clinical rsch and devel., 1986-88, exec. dir., clinical rsch. and devel. R.W. Johnson Pharm. Rsch. Inst., 1988-90; dir. product devel. ctr., assoc. prof. clinical rsch. The Jones Inst. Reproductive Med., Ea. Va. Med. Sch., Norfolk, Va., 1990—. Contrb. articles to profl. jours. Fellow Am. Coll. Ob-Gyn; mem. AMA, APHA, Am. Assn. Advancement of Sci., Am. Assn. Pharm. Scientists, Am. Fertility Soc., Assn. Reproductive Health Profls., Drug Information Assn., Med. Soc Va., Nat. Assn. Family Planning Doctors, Norfolk Acad. Med. Office: Ea Va Med Sch Jones Inst Reproductive Med 601 Colley Ave Rm 427 Norfolk VA 23507

ANDERSON, GARY ALAN, waste water plant executive; b. Winchester, Ind., Oct. 29, 1955; s. Donald Rex and Judith (Gaddis) A.; m. Melanie Kolp, Sept. 24, 1976; children: Joshua Alan, Todd Michael. Student, Ball State U., 1974-76. Farmer, 1976-92; plant mgr. Winchester (Ind.) Waste Water, 1985—. Mem. edn. com. IWPCA, Ind., 1991-92; musician Winchester Community Orch.; coach Little League, Winchester; v.p. Winchester Wrestling Club. Mem. Ind. Water Pollution Control Assn. (Lab.

Excellence award 1991). Presbyterian. Home: RR 4Box 250 Winchester IN 47394 Office: Winchester Waste Water 901 N West St Winchester IN 47394

ANDERSON, GARY ARLEN, agricultural engineering educator, consultant; b. Spencer, Iowa, Sept. 15, 1953; s. Donal Arlen and Shirley Annette (Westegard) A.; m. Joanne Carol Puetz, Jan. 7, 1984; children: Donnell, Janelle, Nichelle. BS in Agrl. Engring., S.D. State U., 1975; MS in Agrl. and Structural Engring., Iowa State U., 1985, PhD in Agrl. Engring., 1987. Asst. prof. freshman engring. Iowa State U., Ames, 1982-85; asst. prof. agrl. engring. S.D. State U., Brookings, 1987-92, assoc. prof., 1992—; cons. Pella (Iowa) Co., 1991—, Fiber-Tech Industries, Inc., Spokane, Wash., 1992-93. Contbr. articles to profl. jours., chpts. to monograph. 1st lt. U.S. Army, 1976-80, Capt. USAR, 1980-89. Grantee FABRAL divsn., Alcon Aluminum Corp., 1985, Granite City Steel, 1986, Wick Bldg., 1987, Mendards, 1989, Metal Sales Mfg. Corp., 1991, Fiber-Tech Industries, Inc., 1991; named to Outstanding Young Men of Am., 1989. Assoc. mem. ASHRAE, ASCE, Am. Soc. Agrl. Engrs. (chmn. diaphrams action com. 1991—, Outstanding Reviewer award 1991), Sigma Xi (chair banquet com., 1990) Alpha Epsilon, Gamma Sigma Delta. Achievements include pioneering research in light-gage metal on lumber frame diaphragms; developer test procedure for diaphragms, diaphragm design procedures for buildings, model for predicting contaminant generation from stored waste, model for heat generation and loss to environment of an animal. Home: 614 11th Ave Brookings SD 57006 Office: SD State U PO Box 2120 Brookings SD 57007

ANDERSON, GERALDINE LOUISE, laboratory scientist; b. Mpls., July 7, 1941; d. George M. and Viola Julia-Mary (Abel) Havrilla; m. Henry Clifford Anderson, May 21, 1966; children: Bruce Henry, Julie Lynne. BS, U. Minn., 1963. Med. technologist Swedish Hosp., Mpls., 1963-68; hematology supr. Glenwood Hills Hosp. lab., Golden Valley, Minn., 1968-70; assoc. scientist dept. pediatrics U. Minn. Hosps., Mpls., 1970-74; instr. health occupations and med. lab. asst. Suburban Hennepin County Area Vocat. Tech. Ctr., Brooklyn Park, Minn., 1974-81, 92, St. Paul Tech. Vocat. Inst., 1978-81; rsch. med. technologist Miller Hosp., St. Paul, 1975-78; rsch. assoc. Children's and United Hosps., St. Paul, 1979-88; sr. lab. analyst Cascade Med. Inc., Eden Prairie, Minn., 1989-90; lab. mgr. VA Med. Ctr., Mpls., 1990; technical support scientist INCSTAR Corp., 1990—; mem. health occupations adv. com. Hennepin Tech. Ctrs., 1975-90, chairperson, 1978-79; mem. hematology slide edn. rev. bd. Am. Soc. Hematology, 1976—; mem. flow cytometry and clin. chemistry quality control subcoms. Nat. Com. for Clin. Lab. Standards, 1988-92; cons. FCM Specialists, 1989—. Mem. rev. bd. Clin. Lab. Sci., 1990-91, The Learning Laboratorian Series 1991; contbr. and presenter In Svc. Rev. in Clin. Lab. Sci., audio taped study program for ASMT, 1992; contbr. articles to profl. jours. Mem. Med. Lab. Tech. Polit. Action Com., 1978—; charter orgns. rep. troop #534 Boy Scouts Am., Viking Coun., 1988-90; resource person lab. careers Robbinsdale Sch. Dist., Minn., 1970-79; del. Crest View Home Assn., 1981—; mem. sci. and math. subcom. Minn. High Tech. Council, 1983-88, mem. Women Scientists Speakers Bur., 1989-92.Recipient svc. awards and honors Omicron Sigma. Mem. AAAS, AAUW, NAFE, Nat. Assn. Women Cons., Inc., Minn. Emerging Med. Orgns., Minn. Soc. Med. Tech. (sec. 1969-71), Am. Soc. Profl. and Exec. Women, Am. Soc. Med. Tech. (del. to ann. meetings 1972—, chmn. hematology sci. assembly 1977-79, nomination com. 1979-81, bd. dirs. 1985-88), Twin City Hosp. Assn. (speakers bur. 1968-70), Assn. Women in Sci., World Future Soc., Minn. Med. Tech. Alumni, Am. Soc. Hematology, Soc. Analytical Cytology, Great Lakes Internat. Flow Cytometry Assn. (charter mem. 1992), Sigma Delta Epsilon (corr. sec. Xi chpt. 1980-82, pres. 1982-84, membership com. 1990-92, nat. nominations chair 1991-92, nat. v.p. 1992—), Alpha Mu Tau. Lutheran. Office: INCSTAR Corp PO Box 285 1990 Industrial Blvd Stillwater MN 55082

ANDERSON, GLORIA LONG, chemistry educator; b. Altheimer, Ark., Nov. 5, 1938; d. Charley and Elsie Lee (Foggie) L.; divorced, 1977; 1 child, Gerald Leavell. BS, Ark. Agr. Mech. & Normal Coll., 1958; MS, Atlanta U., 1961; PhD, U. Chgo., 1968. Instr. S.C. State Coll., Orangeburg, 1961-62, Morehouse Coll., Atlanta, 1962-64; teaching and rsch. asst. U. Chgo., 1964-68; assoc. prof., chmn. Morris Brown Coll., Atlanta, 1968-73, Callaway prof., chmn., 1973-84, acad. dean, 1984-89, United Negro Coll. Fund disting. scholar, 1989-90, Callaway prof. chemistry, 1990—; interim pres. Morris Brown Coll., 1992—. Contbr. articles to profl. jours. Bd. dirs. Corp. for Pub. Broadcasting, Washington, 1972-79, vice chmn. 1977-79; Pub. Broadcasting Atlanta, 1980—; mem. Pub. Telecommunications Task Force, Atlanta, 1980. Postdoctoral rsch. fellow NSF, 1969, faculty industry fellow, 1981, faculty rsch. fellow Southeastern Sci. Ctr. for Elec. Engring and Edn., 1984. Mem. Am. Chem. Soc., Am. Inst. Chemists, Nat. Sci. Tchrs. Assn., Sigma Xi. Baptist. Home: 560 Lynn Valley Rd SW Atlanta GA 30311-2331 Office: Morris Brown Coll 643 M L King Jr Dr NW Atlanta GA 30314

ANDERSON, GORDON MACKENZIE, petroleum service contractors executive; b. Los Angeles, Mar. 25, 1932; s. Kenneth C.M. and Edith (King) A.; m. Elizabeth Ann Pugh, Mar. 21, 1959; children: Michael James, Greg Mark, Jeffrey Stevens. AA, Glendale Coll., 1951; BSME, U. So. Calif., 1954; grad., Officers Candidate Sch., Newport, R.I., 1955; student, various Navy Schs. including CIC Sch. Mgr. Santa Fe Drilling Co., Chile, 1960-63, Libya, 1963-67; mgr. contracts adminstrn. Santa Fe Drilling Co., Calif., 1967-70; pres. Santa Fe Drilling Co., Alhambra, Calif., 1970-80, from 87; exec. v.p. Santa Fe Internat. Corp., Alhambra, Calif., 1974-80, pres., chief operating officer, 1980-87, exec. v.p., from 1987, also bd. dirs., pres.; bd. dirs. Baker Hughes, Houston. Mem. adv. bd. U. So. Calif. Sch. Engring.; bd. dirs. St. Jude Hosp., Fullerton, Calif. Served to lt. (j.g.) USN, 1955-58. Mem. Young Pres.'s Orgn. (chmn. 1978-79), Internat. Assn. Oilwell Drilling Contractors. Office: Santa Fe Drilling Co Box 4000 1000 S Fremont Ave Alhambra CA 91802-4000

ANDERSON, GREGORY MARTIN, medical company representative; b. Spokane, Wash., Mar. 3, 1959; s. Norman Clarence and Jean Marie (Huggar) A. BS, Ea. Wash. U., 1986, MS, 1988. Med. rep. pharms. div. CIBA Geigy Corp., Yakima, Wash., 1989—; mem. adv. bd. Walk in the Wild Zoo, Spokane, 1986-88. Chpt. advisor Order of De Molay, Spokane, 1986; deacon 7th Day Adventist Ch. Ea. Wash. U. fellow, 1987, 88. Mem. Am. Inst. Biol. Sci., Eastern Wash. Lodge of Rsch., Manito Masons (jr. warden Spokane chpt. 1986), Sigma Xi. Home: 1800 River Rd Apt 40 Yakima WA 98902-6212

ANDERSON, HOLLY GEIS, medical clinic executive, radio personality; b. Waukesha, Wis., Oct. 23, 1946; d. Henry H. and Hulda (Sebroff) Geis; m. Richard Kent Anderson, June 6, 1969. BA, Azusa Pacific U., 1970. CEO Oak Tree Antiques, San Gabriel, Calif., 1975-82; pres., founder, CEO Premenstrual Syndrome Treatment Clinic, Arcadia, Calif., 1982—, Hormonal Treatment Ctrs., Inc., Arcadia, 1992—; Lectr. radio and TV shows, L.A.; on-air radio personality Women's Clinic with Holly Anderson, 1990—. Author: What Every Woman Needs to Know About PMS (audio cassette), 1987, The PMS Treatment Program (video cassette), 1989, PMS Talk (audio cassette), 1989. Mem. NAFE, The Dalton Soc. Republican. Avocations: writing, poetry, travel, hiking, antique restoration. Office: PMS Treatment Clinic 150 N Santa Anita Ave Ste 755 Arcadia CA 91006-3113

ANDERSON, JACK ROY, health care company executive; b. Mansfield, Ohio, Feb. 14, 1925; s. Roy L. and Katherine (Munson) A.; m. Rose-Marie J. Garcia, June 24, 1950; children—Gail Ellen, Neil Robert, Barbara Ann. B.S., Miami U., Oxford, Ohio, 1947; M.S., Columbia U. Grad. Sch. Bus., 1949. Acctg. mgr. Time, Inc., N.Y.C., 1950-59; asst. to controller W.R. Grace & Co., N.Y.C., 1959-62; v.p., treas. Hartford Publs., Inc., N.Y.C., 1962-65; controller McCall Corp., N.Y.C., 1965-68; v.p. Reliance Group, Inc., N.Y.C., 1968-70; pres., dir. Hosp. Affiliates Internat., Inc., Nashville, 1970-76, chmn. bd., dir., 1977-81; chmn. INA Health Care Group, Dallas, 1978-81; pres. Manor Care, Inc., Silver Spring, Md., 1981-82, Calver Corp., Dallas, 1982—; adj. faculty Owen Grad. Sch. Mgmt., 1978-79; bd. dirs. Manor Care, Inc., Med. Care Am., Inc., TakeCare, Inc., Navistar Internat. Corp., United Dental Care, Inc. Author: The Road to Recovery, 1976. Vis. com. Vanderbilt Owen Grad. Sch. Mgmt., 1973-77; trustee Nat. Com. for Quality Health Care, 1979-87, vice chmn., 1979-82; mem. bus. adv. coun. Miami U., 1975-78, chmn., 1978. Lt. (j.g.) USNR, 1943-46. Mem. Columbia Bus. Assocs., Fin. Execs. Inst. (chmn. employee benefits com. 1980-82), Preston Trail Golf Club (Dallas), John's Island Club (Vero Beach, Fla.), Desert Forest Golf Club (Carefree, Ariz.), Desert Mountain Country Club (Scottsdale, Ariz.), Greenwich Country Club, Stanwich Club (Conn.), Sky Club (N.Y.C.), Reform Club (London), Sigma Chi, Beta Alpha Psi. Office: Calver Corp 14755 Preston Rd 515 Dallas TX 75240

ANDERSON, JAMES GILBERT, chemistry educator. BS, physics, U Washington, Seattle; Ph.D., physics, astrogeophysics, U Colorado, Boulder, 1970. Prof. Harvard U., Cambridge, Mass., 1978—; now Philip S. Weld prof. atmospheric chemistry Harvard U. Recipient Gustavus John Esselen awd., Amer. Chem. Soc., 1993. Mem. NAS. Achievements include research in stratospheric physics and chemistry central to the understanding of atmospheric ozone and the ozone hole above the Antarctic. Office: Harvard U Ctr Earth & Planetary Physics Cambridge MA 02138*

ANDERSON, JAMES HENRY, university dean; b. Odum, Ga., Jan. 11, 1926; s. James Tillman and Mamie (Aspinwall) A.; m. Dorothy Allen, Dec. 29, 1951; children: Alicia Carol, Laurie Beth, James Hampton, Sue Ellen, John Allen. B.S., U. Ga., 1949; M.S., N.C. State U., 1955; Ph.D., Iowa State U., 1957. Bar: Registered profl. engr., Miss. Prof., head agrl. engring. dept U. Tenn., Knoxville, 1960-61; prof., head agrl. engring. dept. Miss. State U., Starkville, 1961-68; dean resident instrn. Coll. Agr., 1967-68; dir. Miss. Agrl. and Forestry Exptl. Sta., Starkville, 1969-77; dean Coll. Agrl. and Natural Resources, Mich. State U., Lansing, 1977—; now also vice provost Coll. Agrl. and Natural Resources, Mich. State U. Contbr. articles to various publs. Mem. exec. com S. Bapt. Conv., 1973-77. Served with U.S. Army, 1944-46. Fellow Am. Soc. Agrl. Engrs. (pres. 1984-85). Baptist. Club: Rotary. Office: Michigan State U 104 Agriculture Hall East Lansing MI 48824-1039

ANDERSON, JAY LAMAR, horticulture educator, researcher, consultant; b. Madison, Wis., Apr. 22, 1931; s. Melvin Eliason and Ruth (Crittenden) A.; m. Geraldine Olsen, Oct. 17, 1955; children: Marc Olsen, Denise, Kraig LaMar, Kurt David. BS, Utah State U., 1955; PhD, U. Wis., 1961. Asst. prof. Utah State U., Logan, 1961-67, assoc. prof., 1967-75, prof., 1975—; vis. prof. U. Calif., Riverside, 1971-72; hort. cons. Bur. Indian Affairs U.S. Dept. Justice, 1986—. Author, editor Acta Horticulturae symposium reports; contbr. articles to profl. jours. including Jour. Am. Soc. Hort. Sci., Hort. Sci., Weed Sci. 1st lt. U.S. Army, 1955-57, Korea. Grantee numerous chem. cos., 1966, USDA, 1976. Fellow Am. Soc. Hort. Sci. (pres. western region 1978-81), Western Soc. Weed Sci. (bus. mgr. 1965-89, merit award 1974, Disting. Svc. award 1989); mem. Am. Pomological Soc. (Paul Howe Shepard award 1964), Weed Sci. Soc. Am., Internat. Soc. Hort. Sci., Utah State U. Faculty Assn. (pres. 1974-75). Republican. Mem. LDS Ch. Achievements include research in anatomical and ultrastructural effects of herbicides, on modeling of deciduous fruit tree phenology, on bloom delay of deciduous fruit trees, and on effects of plant growth regulators on fruit set and development. Home: 856 Juniper Dr Logan UT 84321-3622 Office: Utah State U Dept Plants Soils Biometeorology Logan UT 84322-4820

ANDERSON, JEFFREY LANCE, cardiologist, educator; b. Salt Lake City, Oct. 27, 1944; s. Aldon Jr. and Virginia (Weilenmann) A.; m. Kathleen Tadje, Aug. 18, 1967; children: Russell, Nathan, Derek, Megan. BA magna cum laude, U. Utah, 1968; MD cum laude, Harvard U., 1972. Diplomate Am. Bd. Internal Medicine, Am. Bd. Cardiovascular Diseases. Resident in internal medicine Mass. Gen. Hosp., Boston, 1972-74; staff assoc. NIH, Bethesda, Md., 1974-76; fellow in cardiology Stanford (Calif.) U., 1976-78; asst. prof. medicine U. Mich., Ann Arbor, 1978-80; asst. prof. medicine U. Utah, Salt Lake City, 1980-83, assoc. prof. medicine, 1983-89, prof. medicine, 1989—; presenter in field. Author over 200 med./sci. papers in field; author, editor 48 book chpts. and books in field; contbr. articles to profl. jours. Recipient numerous fed., local and indsl. grants, 1980—. Mem. ACP (gov. 1993—), Am. Coll. Cardiology (gov. 1986-88), Am. Heart Assn. (pres. Utah chpt. 1985). Mem. LDS Ch. Achievements include first randomized trial to show benefit of thrombolytic therapy in heart attacks, to show possible benefit of beta blockers in heart failure; investigation of land mark cardiac arrhythmia suppression trial; development of multiple pharmacologic agents to treat arrhythmias, heart attacks, heart failure. Home: 795 18th Ave Salt Lake City UT 84103 Office: LDS Hosp 8th Ave C St Salt Lake City UT 84143

ANDERSON, JERRY WILLIAM, JR., technical and business consulting executive, educator; b. Stow, Mass., Jan. 14, 1926; s. Jerry William and Heda Charlotte (Petersen) A. ; m. Joan Hukill Balyeat, Sept. 13, 1947; children: Katheleen, Diane. BS in Physics, U. Cin., 1949, PhD in Econs., 1976; MBA, Xavier U., 1959. Rsch. and test project engr. Wright-Patterson AFB, Ohio, 1949-53; project engr., electronics div. AVCO Corp., Cin., 1953-70, program mgr., 1970-73; program dir. Cin. Electronics Corp., 1973-78; pres. Anderson Industries Unltd., 1978—; chmn. dept. mgmt. and mgmt. info. svcs. Xavier U., 1980-89, prof. mgmt., 1989—; lectr. No. Ky. U., 1977-78; tech. adviser Cin. Tech. Coll., 1971-80. Contbr. articles on radar, lasers, infrared detection equipment, air pollution to govt. publs. and profl. jours.; author 3 books in field. Mem. Madeira (Ohio) City Planning Commn., 1962-80; founder, pres. Grassroots, Inc., 1964; active United Appeal, Heart Fund, Multiple Sclerosis Fund. With USNR, 1943-46. Named Man of Year, City of Madeira, 1964. Mem. Am. Mgmt. Assn., Assn. Energy Engrs. (charter), Internat. Acad. Mgmt. and Mktg., Assn. Cogeneration Engrs. (charter), Assn. Environ. Engrs. (charter), Am. Legion (past comdr.), Acad. Mgmt., Madeira Civic Assn. (past v.p.), Omicron Delta Epsilon. Republican. Home and Office: 7208 Sycamorehill Ln Cincinnati OH 45243-2101

ANDERSON, JOHN BAILEY, electrical engineering educator; b. Canandaigua, N.Y., May 30, 1945; s. Jack Richard and Martha (Wommer) A.; m. Janet Smith, June 8, 1968; children: Katherine E., Alix C. BSEE, Cornell U., 1967, MS, 1969, PhD, 1972. Asst./assoc. prof. dept. elec. engring. McMaster U., Hamilton, Ont., Can., 1972-80; assoc., later full prof. elec., computer and systems engring. Rensselaer Poly. Inst., Troy, N.Y., 1981—; vis. assoc. prof. elec., computer and systems engring. U. Calif., Berkeley, 1978-79; guest researcher info. theory dept. Chalmers Tech. U., Gothenburg, Sweden, 1987; vis. prof. elec. engring. Queens U., Kingston, Ont., 1987; vis. scientist Inst. Communications Engring., German Aerospace Rsch., Munich, 1991-92. Author: Digital Phase Modulation, 1986, Modern Electrical Communications, 1988, Source and Channel Coding, 1991. Pres. McMaster Symphony Orch., Hamilton, 1973-78. Recipient Humboldt Sr. Scientist award, 1991. Fellow IEEE (publs. bd. 1989-90); mem. Info. Theory Soc. of IEEE (pres. 1985, v.p. 1983-84), Am. Fedn. Musicians, Sigma Xi. Avocation: semi-profl. musician. Office: Rensselaer Poly Inst Dept Elec Computer & Systems Engring Troy NY 12180

ANDERSON, JOHN ROY, grouting engineer; b. Culberson, N.C., June 22, 1919; s. Oscar Garfield and Lula Adeline (Russell) A.; m. Rheba Ulma Nichols, Dec. 31, 1951 (dec. Oct. 1989); children: Richard Allen, John Steven, Mark Garfield. Student, Berea Coll., 1950. From clk. to field inspector, then constrn. engr. Govt. Agys., 1941-51; project engr., mgr. Intrusion Prepakt, Cleve., 1951-58, 64-65; regional mgr./project mgr. Lee Turzillo Contracting Co., Breaksville, Ohio, 1957-60; engr. foundations, ops. mgr. Harza Engring. Co., Chgo., 1960-61, 65-68, 1972-86; foundations engr. Tippetts Abbett McCarthy Stratton, N.Y.C., 1961-64, 68-72. Mem. ASCE. Baptist. Achievements include foundation engineering on dams, bridges, tunnels and high rise buildings throughout the world. Home and Office: 7770 Skipper Ln Tallahassee FL 32311-9534

ANDERSON, KARL ELMO, educator; b. Buffalo, May 16, 1940; s. Richmond Karl and Cleo (Holland) A.; m. Carol Jean DeMambro, Mar. 31, 1983; children: Matthew, Gillian, Jeffrey. BA, Johns Hopkins U., 1962, MD, 1963. Bd. cert. internal medicine & gastroenterology. Intern, resident Vanderbilt U. Hosp., Nashville, 1965-67; resident internal medicine N.Y. Hosp.-Cornell Med. Ctr., N.Y.C., 1967-68, fellow gastroenterology, 1968-70; asst. prof., assoc. prof. Rockefeller U., N.Y.C., 1973-85; prof. N.Y. Med. Coll., Valhalla, 1985-87, U. Tex. Med. Br., Galveston, 1987—; study sect. mem. NIH, Bethesda, Md., 1980-82; mem. scientific adv. bd. Am. Prophyria Found., Montgomery, Ala., 1982—; gastroenterology adv. panel U.S. Pharm. Conv., Rockville, Md., 1991—. Contbr. over 100 articles to profl. jours. Lt. comdr. USN, 1971-73. Fellow Am. Coll. Physicians. Achievements include development of new therapies for human porphyrias; demonstration that diet influences drug metabolism in humans. Office: U Tex Med Br 700 The Strand Galveston TX 77555-1009

ANDERSON, KENNETH EDWIN, technical writer; b. N.Y.C., Sept. 7, 1931; s. Murray Simon and Louise (Merchant) Anderson; m. Therese Catherine Tully, Apr. 18, 1955; children: Karen A. Dow, Kim A. Payne, Deborah A. Doggett, Barbara A. Stuehm. BS with honors, U. Fla., 1969, MA, 1970; PhD, Okla. State U., 1975. Instr. Eastern Ill. U., Charleston, 1970-71; editor engring. publs. Okla. State U., Stillwater, 1971-74, dir. Native Am. Tech. Program, 1974-78; pres. Anderson Petroleum Svcs., Stillwater, 1978-87; tech. program designer TAD Tech. Svcs., Norman, Okla., 1987-89; instrnl. systems designer U. Okla., Norman, Okla., 1989-92; ptnr. Berger & Anderson Consultants, Yale, Okla., 1977—, Anderson-Baldwin Consultants, Tulsa, 1986—. Co-author: Modern Petroleum, 1st edit., 1978, 3rd edit., 1992, Basic Processing Knowledge, 1979, Refinery Operations, 1979, Gas Handling and Field Processing, 1980; editor: Manual of Practical Pipeline Construction, 1982; contbr. more than 100 articles to profl. jours. Mem. Rock Knoll Homeowners Assn., Oklahoma City, 1991—. Sgt. Corps of Engrs., 1957. Named Most Disting. Grad., Coll. of Journalism, 1969; recipient Grad. fellowship Poynter Found., 1970, Okla. Heritage Assn., 1974. Mem. Soc. Profl. Journalists, Phi Kappa Phi, Omicron Delta Kappa. Democrat. Roman Catholic. Achievements include conducting 10 environ. impact assessments for Dept. of Interior covering more than a quarter-million acres of fed. land, from 1981-87. Home and Office: 6505 Eastbourne Ln Oklahoma City OK 73132-2006

ANDERSON, LLOYD LEE, animal science educator; b. Nevada, Iowa, Nov. 18, 1933; s. Clarence and Carrie G. (Sampson) A.; m. Janice G. Peterson, Sept. 7, 1958 (dec. Dec. 1966); m. JaNelle R. Hall, June 15, 1970; children: Marc C., James R. Student, Simpson Coll., 1951-52, Iowa State U., 1952-53; BS in Animal Husbandry, Iowa State U., 1957, PhD in Animal Reproduction, 1961. NIH postdoctoral fellow Iowa State U., Ames, 1961-62, asst. prof., 1961-65, assoc. prof., 1965-71, prof. animal sci., 1971—, Charles F. Curtiss Disting. prof. agr., 1992—; Lalor Found. fellow Station de Recherches de Physiologie Animale, Institut National de Recherche Agronomique, Jouy-en-Josas, France, 1963-64; rschr. physiology of reproduction; mem. reproductive biology study sect. NIH, 1984-88, Nat. Insts. Health Reviewers Res. (NRR), 1988-92; mem. peer rev. panel animal health spl. rsch. grants on beef and dairy cattle reproductive diseases USDA, 1986-88; Honor lectr. representing Iowa State U., Mid-Am. State Univs. Assn., 1989-90. Mem. editorial bd. Biology Reproduction, 1968-70, 86-90, Jour. Animal Sci., 1982-87, Animal Reproduction Sci., 1978—, Inst. for Sci. Info. Atlas of Sci., 1987-90, Domestic Animal Endocrinology, 1992-95, Endocrinology, 1993-96; contbr. articles to profl. jours. With U.S. Army, 1953-55. USDA grantee, 1978—. Fellow Am. Soc. Animal Sci. (hon., Animal Physiology and Endocrinology award 1988, Nat. Pork Prodrs. Coun. Innovation award in basic rsch. 1993); mem. AAAS, VFW, Endocrine Soc., Am. Physiol. Soc., Am. Assn. Anatomists, Soc. for Study of Reproduction, Soc. for Exptl. Biology and Medicine (mem. coun. 1980-83), Brit. Soc. for Study of Fertility, Am. Legion, Sigma Xi, Gamma Sigma Delta. Methodist. Home: 2812 Valley View Rd Ames IA 50014 Office: Iowa State U Dept Animal Sci 11 Kildee Hall Ames IA 50011-3150

ANDERSON, MARK EDWARD, mining engineer, consultant; b. Greenville, Pa., Mar. 21, 1957; s. John Waldemer and Evelyn Frances (Speir) A.; m. Toni Renee Putt, Aug. 2, 1986; 1 child, Stephenie Grace. BS in Mineral Econs., Pa. State U., 1979. Registered profl. engr., Pa.; lic. geologist, N.C., S.C. Staff geologist L. Robert Kimball and Assocs., Ebensburg, Pa., 1979-82, staff engr., 1982-84, project engr., 1984-86, project mgr., 1986-89, mgr. solid waste projects, 1989—. Contbr. articles to profl. publs. Past pres., treas. Highland Fed. Credit Union, Ebensburg, 1982-86; exec. adviser Jr. Achievement, Johnstown, Pa., 1982-86. Mem. AIME, Pa. Mining Profls. (auditor 1982). Office: L Robert Kimball & Assocs PO Box 1000 615 W Highland Ave Ebensburg PA 15931

ANDERSON, MARTIN CARL, economist; b. Lowell, Mass., Aug. 5, 1936; s. Ralph and Evelyn (Anderson) A.; m. Annelise Graebner, Sept. 25, 1965. AB summa cum laude, Dartmouth Coll., 1957, MS in Engring., MSBA; PhD in Indsl. Mgmt., MIT, 1962. Asst. to dean, instr. engring. Thayer Sch. Engring. Dartmouth Coll., Hanover, N.H., 1959; research fellow Joint Ctr. for Urban Studies MIT and Harvard U., Cambridge, 1961-62; asst. prof. fin. Grad. Sch. Bus. Columbia U., N.Y.C., 1962-65, assoc. prof. bus., 1965-68; sr. fellow Hoover Inst. on War, Revolution and Peace Stanford (Calif.) U., 1971—; spl. asst. to Pres. of U.S. The White House, 1969-70, spl. cons. for systems analysis, 1970-71, asst. for policy devel., 1981-82; syndicated columnist Scripps Howard News Svc., 1993—; mem. Pres.' Fgn. Intelligence Adv. Bd., 1982-85, Pres.' Econ. Policy Adv. Bd., 1982-88, Pres.' Gen. Adv. Com. on Arms Control and Disarmament, 1987—; pub. interest dir. Fed. Home Loan Bank San Francisco, 1972-79; mem Commn. on Crucial Choices for Ams., 1973-75, Def. Manpower Commn., 1975-76, Com. on the Present Danger, 1977—. Author: The Federal Bulldozer: A Critical Analysis of Urban Renewal, 1949-62, 1964, Conscription: A Select and Annotated Bibliography, 1976, Welfare: The Political Economy of Welfare Reform in the U.S., 1978, Registration and the Draft, 1982, The Military Draft, 1982, Revolution, 1988, Impostors in the Temple, 1992; columnist Scripps-Howard News Svc. Dir. research Nixon presdl. campaign, 1968; issues adviser Reagan presdl. campaign, 1976, 80; trustee Ronald Reagan Presdl. Found., 1985-88. 2d lt. AUS, 1958-59. Mem. Am. Econ. Assn., Mont Pelerin Soc., Phi Beta Kappa. Club: Bohemian. Office: Stanford U Hoover Instn Stanford CA 94305-6010

ANDERSON, PAMELA BOYETTE, quality assurance professional; b. Wilson, N.C., June 9, 1957; d. Robert D. and Janice (Pearson) Boyette; 1 child, Courtney Lauren Anderson. BS, Atlantic Christian Coll., 1979. Quality assurance chemist Miles Lab., Clayton, N.C., 1979-88; quality assurance chemistry specialist Kabi Pharmacia Inc, Clayton, N.C., 1988-90, quality assurance chemistry supr., 1990-92, quality assurance compliance mgr., 1992—. Sec. PTO, Kenly, N.C., 1987; mem. adv. coun. Johnston County Sch. Bd. of Edn., Kenly, 1989-92. Mem. Am. Soc. Quality Control (cert. quality auditor), N.C. Pharm. Discussion Group. Office: Kabi Pharmacia Inc 1899 Hwy 70 E Clayton NC 27520

ANDERSON, PAUL, product management executive; b. N.Y.C., June 10, 1963; s. Percy and Myrna (Young) A. BA in Econs., UCLA, 1986; postgrad., Mich. Bus. Sch., 1992—. Cert. Intel tech. sales specialist. Tech. sales rep. Arrow Electronics, Chatsworth, Calif., 1986-88, Permacel, Buena Park, Calif., 1988-89, GE Plastics (Silicones), Brea, Calif., 1989-92; product mgr. Kellogg's, Battle Creek, Mich., 1993—. Home: 95 Old Broadway New York NY 10027

ANDERSON, PETER GLENNIE, research pathologist, educator; b. Oxford, N.Y., Dec. 28, 1954; s. Robert Jr. and Ruth (Fyfe) A.; m. Joan Cherre, June 25, 1988; 1 child, Robert Cherre Anderson. BA in Zoology, U. Wash., 1977; DVM, Wash. State U., 1981; PhD in Pathology, U. Ala., Birmingham, 1986. Asst. prof. depts. pathology and comparative medicine U. Ala., Birmingham, 1986—, mem. grad. faculty, 1986—; vis. rsch. scientist U. Ariz., Tucson, 1988-89; invited chmn. ann. meetings Internat. Acad. Pathology, Atlanta, 1992, Fedn. Am. Socs. for Exptl. Biology, Anaheim, Calif., 1992. Contbr. articles to profl. jours. Mem. spl. peer rev. com. Fla. Am. Heart Assn., Tampa, 1991. Recipient NIH 1st award Heart Lung Blood Inst., 1989—; grantee Ala. Heart Assn., 1992-93. Mem. Soc. for Cardiovascular Pathology (councilor 1990-93), Internat. Soc. Heart Rsch., Am. Assn. Pathologists, Group for Rsch. in Pathology Edn. Achievements include patent pending for Monoclonol Antibodies Conjugates for Coronary Artery Angioplasty Restenosis. Office: U Ala Birmingham Volker Hall G038 1670 University Blvd Birmingham AL 35294-0019

ANDERSON, PHILIP WARREN, physicist; b. Indpls., Dec. 13, 1923; s. Harry W. and Elsie (Osborne) A.; m. Joyce Gothwaite, July 31, 1947; 1 dau., Susan Osborne. B.S., Harvard U., 1943, M.A., 1947, Ph.D., 1949; D.Sc. (hon.), U. Ill., 1979; DSc (hon.), Rutgers U., 1991. Mem. staff Naval Research Lab., 1943-45; mem. tech. staff Bell Telephone Labs., Murray Hill, N.J., 1949-84; chmn. theoretical physics dept. Bell Telephone Labs., 1959-60, asst. dir. phys. research lab., 1974-76, cons. dir., 1976-84; researcher in quantum theory, especially theoretical physics of solids, spectral line broadening, magnetism, superconductivity; Fulbright lectr. U. Tokyo, 1953-54; Loeb lectr. Harvard U., 1964; prof. theoretical physics Cambridge (Eng.) U., 1967-75; prof. physics Princeton U., 1975—; Overseas fellow Churchill Coll., Cambridge U., 1961-62; fellow Jesus Coll., 1969-75, hon. fellow,

1978—; Bethe lectr., 1984; external prof., Santa Fe Inst., 1990—. Author: Concepts in Solids, 1963, Basic Notions of Condensed Matter Physics, 1984. Recipient Oliver E. Buckley prize Am. Physical Soc., 1964; Dannie Heinemann prize Göttingen (Ger.) Acad. Scis., 1975; Nobel prize in physics, 1977; Guthrie medal Inst. of Physics, 1978; Nat. Medal Sci., 1982. Fellow AAAS, Am. Phys. Soc., Am. Acad. Arts and Scis., Japan Acad. Scis. (fgn.), Indian Acad. Scis. (fgn.); mem. Nat. Acad. Scis., Royal Soc. (fgn.), Accademia Lincei, Am. Philos. Soc., N.Y. Acad. Scis. (hon. life). Office: Princeton U Dept Physics Princeton NJ 08544

ANDERSON, RAYMOND HARTWELL, JR., metallurgical engineer; b. Staunton, Va., Feb. 25, 1932; s. Raymond Hartwell and Virginia Boatwright (Moseley) A.; m. Dana Bratton Wilson, Sept. 5, 1959; children: Kathryn, Margaret, Susan. BS in Ceramic and Metall. Engring., Va. Poly. Inst. and State U., 1957, MSMetE, 1959. Registered profl. engr. Asst. prof. metall. engring. Va. Poly. Inst. and State U., Blacksburg, 1957-59; metall. engr. Gen. Dynamics Corp., Ft. Worth, 1959-61; sr. engr. Babcock & Wilcox Co., Lynchburg, Va., 1961-65; tech. specialist McDonnell Douglas Astronautics Co., Huntington Beach, Calif., 1965-88, sr. engring. specialist space sta. div., 1988—; tchr. materials sci. and chemistry U. Calif., Irvine, Calif., 1990—; cons. in field Los Angeles Area, 1967-71. Author, patentee Roll Diffusion Bonding of Beryllium, 1970-71, Increased Ductility of Beryllium, 1971-72. Served to 1st lt. U.S. Army, 1954-56. Mem. Am. Soc. Metals (lectr. 1968-70), Nat. Soc. Corrosion Engrs., Bolting Tech. Coun., Am. Ceramic Soc., Am. Welding Soc. (space welding adv. bd. 1991—). Republican. Avocations: gardening, antique cars, music, bicycling. Home: 1672 Kenneth Dr Santa Ana CA 92705-3429 Office: McDonnell Douglas Space Sys Co Space Sta Divsn 5301 Bolsa Ave Huntington Beach CA 92647-2099

ANDERSON, REBECCA COGWELL, psychologist; b. San Antonio, Tex., June 28, 1948; d. Alvy Lynn and Georgia (Earles) Cogwell; m. Kim Edward Anderson, May 29, 1976. BS, U. Ctrl. Ark., 1972, MSE, 1976; PhD, Marquette U., 1988. Chemistry tchr. Pulaski County Schs., Little Rock, 1971-72; grad. rsch. asst. U. Ctrl. Ark., Conway, 1972-73; indsl. hygienist U.S., Ark. Depts. Labor, Little Rock, 1974-77; instr. health edn. U. Ctrl. Ark., Conway, 1977-85; asst. dir. Marquette U. Parenting Ctr., Milw., 1987-92; asst. clin. prof. Med. Coll. of Wis., Milw., 1991—; psychol. examiner Psychol. Corp., Milw., 1987-88; adj. faculty Marquette U., 1988—; wellness cons., 1987-91. Author: Worksite Wellness: a Guide to Program Planning, 1986; author: 2 book chpts., 1988, 89; contbr. articles to profl. jours. Chairperson Gov.'s Safety Conf. health mgmt. sect., Hot Springs, Ark., 1984, 85; judge Ark. State Sci. Fair, Conway, 1985; mem. Elmbrook Schs. human growth and devel. com., Brookfield, Wis., 1992. Recipient Schmitt fellowship, 1987. Mem. APA, Assn. for Advancement of Behavior Therapy, Wis. Psychol. Assn., Internat. Soc. for Traumatic Stress, Alphi Chi. Home: 15145 Bending Brae Ct Brookfield WI 53005 Office: Med Coll of Wis 9200 W Wisconsin Ave Milwaukee WI 53226

ANDERSON, RICHARD MCLEMORE, internist; b. Gainesville, Fla., Mar. 3, 1930; s. Montgomery Drummond and Myrtle (McLemore) A.; m. Leewood Shaw, Mar. 21, 1959; children: Richard McLemore Jr., Bruce Dexter. BS, U. Fla., 1951; MD, Emory U., 1958. Diplomate Am. Bd. Internal Medicine. Chief of staff Alachua Gen. Hosp., Gainesville, Fla., 1973-75; internist Gainesville, Fla., 1962—; Chmn. of bd. Santa Fe Health Care, Gainesville, 1984-91; vice chmn. AV Med., 1992—. Pres. Rotary Club of Gainesville, 1980-81. Capt. USAF, 1951-54. Mem. AMA, Alachua County Med. Soc. (v.p. 1972), Fla. Med. Assn., Am. Soc. Internal Medicine. Presbyterian. Avocation: tennis. Office: 106 SW 10th St Gainesville FL 32601-6201

ANDERSON, RICHARD VERNON, ecology educator, researcher; b. Julesburg, Colo., Sept. 9, 1946; s. Vernon Franklin and Charolett Iona (Jeppesen) A.; m. Arline June Rosentreter, Jan. 23, 1971; children: Rustle R., Michael C., Theodore F. Student, Chadron State Coll., 1964-66, Western State Coll., 1970; BS, No. Ill. U., 1974, MS, 1975; PhD, Colo. State U., 1978. Grad. teaching asst. No. Ill. U., DeKalb, 1974-75; grad. rsch. asst. Colo. State U., 1975-78, postdoctoral fellow Nat. Resource Ecology Lab., 1978-79; asst. prof. Western Ill. U., Macomb, 1979-82, assoc. prof., 1982-87, prof., dir. Kibbe Life Scis. Field Sta., 1987—; vis. asst. prof. inst. for. environ. studies Water Resources Ctr., U. Ill., 1980; mem. assoc. faculty Argonne Nat. Lab., 1985—; assoc. supportive scientist Ill. Natural History Survey, 1985—; proposal reviewer ecology, ecosystem studies, regulatory biology, divsn. internat. programs NSF, 1981—, mem. proposal panel for equipment and facilities grants, 1987; proposal reviewer U.S./Israel Binational Sci. Found., 1981-82, Natural Environ. Rsch. Coun., Eng., 1983-84; environ. cons. aquatic sect. Environ. Cons. and Planners, DeKalb, 1974. Reviewer Natural Resource Ecology Lab., 1977-81, Jour. Nematology, 1977-81, Archives Environ. Contamination and Toxicology, 1978-81, Ecology, 1978-85, Argonne Nat. Lab., 1980—, Pedobiologia, 1982-87, Jour. Freshwater Ecology, 1982—, Freshwater Invertebrate Ecology, 1982—; contbr. over 225 sci. articles, reports, papers and abstracts; presenter papers in field. Grantee NSF, 1972, 73, 82, 83, 84, 85, (two grants), 86, (two grants), 87, 88, Western Ill. U., 1980 (two grants), 81, Upper Miss. River Basin Comm./U. Ill., 1980, Abbott Labs., 1981, Ill. Dept. Transp., 1981, 85, Ctrl. Ill. Light Co., 1981, 82, 83, 84, Nat. Fish and Wildlife Svc., 1983, Ill. Dept. Conservation, 1985, 87, 88, 89, 91, U.S. Fish and Wildlife Svc./Ill. Dept. Conservation, 1988, 89, 90, 91, Environ. Cons. and Planners, Inc., 1988, 89, 91, Booker Assocs., Inc., 1989 (two grants), Ill. Natural History Survey, 1989, Wetlands Rsch., Inc., 1989, 90, 91, 92, USDA/U. Ill., 1991, 92. Mem. Entomol. Soc. Am., N.Am. Benthological Soc. (program. com. 1982-83, reviewer jour. 1990—), Ecol. Soc., Soc. Nematologists (ecology com. 1981-82, systematic resources com. 1981-82, Internat. Congress Ecologists, Ill. State Acad. Sci., Miss. River Rsch. Consortium (mem. exec. bd. 1981-82, v.p. bd. dirs. 1991-92, pres. bd. dirs. 1992-93), Xerces Soc., Sigma Xi.(Rschr. of Yr. award 1984), Phi Kappa Phi. Achievements include research in invertebrate ecology, aquatic biology with an emphasis on large river ecosystems, aquatic invertebrates and freeliving nematodes, the effects of invertebrates on nutrient cycling. Home: 704 S Randolph Macomb IL 61455 Office: Western Ill U Dept Biol Scis Macomb IL 61455

ANDERSON, ROBERT, environmental specialist, physician; b. Ft. Sill, Okla., May 22, 1944. MD, George Washington U., 1970; MPH, U. Tex., 1974. Chief med. svcs. TWA, Kansas City, Kans., 1976-81; med. dir. Air Can., Montreal, Que., 1981-82; corp. med. dir. Manville Corp., Denver, 1982-87, v.p. health, safety and environ., 1987-89, v.p. sci. and tech., 1989-90, sr. v.p. sci. and tech., 1990—. Bd. dirs. Rocky Mountain Multiple Sclerosis Ctr., Denver, 1991. Maj. USAF, 1971-76. Mem. AMA, Am. Occupational Health Assn., Am. Coll. Preventive Medicine, Aerospace Med. Assn. Office: Manville Corp PO Box 5108 Denver CO 80217-5108

ANDERSON, ROGER CLARK, biology educator; b. Wausau, Wis., Oct. 30, 1941; s. Jerome Alfred and Virginia Stella (Hoffman) A.; m. Mary Rebecca Blocher, Aug. 5, 1967; children: John Allen, Nancy Lynn. BS magna cum laude, La Crosse State Coll., 1963; MS, U. Wis., 1965, PhD, 1968. Asst. prof. So. Ill. U., Carbondale, 1968-70; arboretum dir. U. Wis., Madison, 1970-73, assoc. prof., 1970-73; assoc. prof. Cen. State U., Edmond, Okla., 1973-76; prof. Ill. State U., Normal, 1976—; mem. Ill. Nature Preserves Commn., 1985-90; mem., chmn. PARKNET adv. com. Fermilab, Batavia, Ill., 1986—. Editorial bd. Jour. Restoration Ecology, 1992—; author: Enviromental Biology, 1970; author: (with others) Fire in North American Tallgrass Prairie, 1990, Grasses and Grasslands Systematics and Ecology, 1982, Phenology and Seasonality Modeling, 1974; contbr. 75 articles to profl. jours. Pres. Parkland Found. Bd., McLean County, Ill., 1987—. Named McMullen lectr. Monmouth Coll., 1983. Fellow Ill. Acad. Sci.; mem. Ecol. Soc. Am., Soc. for Ecol. Restoration, Am. Bot. Soc., Kappa Delta Pi. Achievements include research on the role of fire in native grassland and savannas, on the relationships between native prairie plants and mycorrhizae fungi. Home: 14 McCormick Blvd Normal IL 61761 Office: Ill State U Biology Dept Normal IL 61761

ANDERSON, ROY ALAN, chemical engineer; b. Freeport, Tex., Oct. 23, 1949; s. Roy Orville and Rose Louise (King) A.; m. Donna Carol Massey, May 27, 1978; children: Roy Brian, Laura Michelle. BS in Chem. Engring., U. Tex., 1973. Plant engr. Pennwalt Corp., Houston, 1975-78; environ. engr. Temple-Eastex, Dibol, Tex., 1978-80; project devel. engr. ARAMCO, Dhahram, Saudi Arabia, 1980-85; project engr. C&I Engring., Cin., 1985-87; pres. Tek-Arts, Loveland, Ohio, 1987-90; project mgr. Jacobs Engring., Cin., 1990-92; pres. TRACO INC, Loveland, 1992—; lectr. in field. Author tech. manuals: Refinery Turnaround Manual, 1990, Engineering Administrative Procedures Manual, 1989, Fire and SAfety Procedures Manual, 1988. Computer cons. Children's Meeting House-Montessori Sch., Loveland, 1989; soccer coach Soccer Assn. for Youth, Loveland, 1992; discus/shotput coach ARAMCO Schs., Dhahran, 1985. Mem. Am. Inst. Chem. Engrs., Project Mgmt. Inst., Am. Assn. Individual Investors. Achievements include numerous design/construct projects in fields of chemical specialties, plastics, food ingredients; research on interdisciplinary design coordination resulting in improved project management techniques. Office: TRACO INC 1049 Red Bird Rd Loveland OH 45140

ANDERSON, ROY EVERETT, electrical engineering consultant; b. Batavia, Ill., Oct. 30, 1918; s. Elof and Nellie Amanda Anderson; m. Gladys Marie Nelson, Aug. 22, 1943; children: Paul V., David L., Barbara J., Dorothy M. BA in Physics, Augustana Coll., Rock Island, Ill., 1943; MSEE, Union Coll., Schenectady, 1952. Instr. physics Augustana Coll., 1943-44, 46-47; cons. engr. GE, Schenectady, 1947-83; co-founder, v.p. Mobile Satellite Corp., Malvern, Pa., 1983-88; owner, mgr., cons. Anderson Assocs., Glenville, N.Y., 1988-93; pres. Rega Assocs., Inc., Glenville, 1993—; cons. Am. Mobile Satellite Corp., Washington, 1988-91; participant nat. and internat. regulatory and tech. orgns. leading to establishment generic mobile satellite svc. Contbr. over 125 articles to profl. jours.; patentee indsl. electronic measurement and quality control instruments, tone code ranging technique for position surveillance using satellites; developer Doppler radio direction finder. Trustee Dudley Obs., Schenectady, 1975-83, 90—, chmn. bd. trustees, 1980-83, 90. With USN, 1944-46. GE Coolidge fellow, 1970. Fellow IEEE, AAAS, Radio Club Am.; mem. AIAA, Inst. Navigation, Soc. Satellite Profls. Internat. Home and Office: PO Box 2531 Glenville NY 12325-0531

ANDERSON, RUSSELL KARL, JR., physicist, horse breeder; b. Passaic, N.J., Jan. 5, 1943; s. Russell Karl and Hilda (Bartles) A.; m. Jane Louise Blair, Apr. 20, 1973; children: Christina Lynn, Melissa Jane. BS in Physics, Fairleigh Dickinson U., 1965; MS in Physics, Rensselaer Poly. Inst., 1967; cert., Penn Tech. Inst., Pitts., 1969; MBA, Duquesne U., 1973. Scientist Westinghouse Bettis Atomic Power Lab., West Mifflin, Pa., 1967-73; engr. Westinghouse Nuclear & Advanced Tech. Div., Ctr., Westinghouse Energy Systems Div., Energy Ctr., Monroeville, Pa., 1973—. NSF grantee, 1969. Mem. Am. Quarter Horse Assn., Nat. Snaffle Bit Assn., Western Pa. Quarter Horse Assn., Phi Omega Epsilon. Home: Fern Valley Farm PO Box 12 Fenelton PA 16034 Office: Westinghouse Electric Corp PO Box 355 Pittsburgh PA 15230-0355

ANDERSON, SAMUEL WENTWORTH, research scientist; b. Norwood, Mass., Sept. 11, 1929; s. Samuel Miner and Hope (Dennis) A.; m. Sylvia Phelops, June 2, 1958; 1 child, Roger Dennis Anderson. BA, George Washington U., 1958; PhD, Harvard U., 1965. Statistician Bur. of Census-Dept. Commerce, Washington, 1958-60, rsch. psychologist, 1960-64; asst. prof. psychology Wesleyan U., Middletown, Conn., 1965-71; rsch. scientist N.Y. State Psychiat. Inst., 1969—; rsch. assoc. Dept. Psychiatry, Columbia U., N.Y.C., 1976—; dir. Comm. Scis. Lab., Columbia U., N.Y.C., 1982—; survey advisor Louis Harris Assoc., Internat. Ctr. for Disabled, N.Y.C., 1988-89. Author: (with others) Language Related Asymmetries in Evoked Potentials, 1977, Integer Concept Among School Children, 1970; co-editor: Cochlear Prosthese International Symposium, 1983; editorial bd. Jour. of Psycholinguistic Rsch., 1970—; contbr. articles to profl. jours. Advisor Nat. Coun. on the Handicapped, Washington, 1984-87, Manhattan Boro Pres. Disability Adv. G.P., N.Y., 1991-92.; mem. N.Y. State Task Force on Tech. and Disability, Albany, 1987. Fellowship N.Y. Acad. Scis., 1985. Mem. Phi Beta Kappa, Sigma Xi. Achievements include devel. of computer program to extract and time vowels in spontaneous speech. Office: NY Psychiat Inst Unit # 108 722 W 168th St New York NY 10032

ANDERSON, STEPHEN THOMAS, computer design engineer; b. Pasadena, Calif., Apr. 6, 1949; s. Thomas Anderson and Barbara (Ives) Stencel; m. Dale Kimberly, Apr. 6, 1993. Student motion picture dept., Brooks Inst. Photography, Santa Barbara, Calif., 1993—. Field engr. Applicon Inc., Burlington, Mass., 1976-79; asst. systems engr. Pacific Missile Test Ctr. Grumman Data Systems, Port Mugu, Calif., 1979-80; R & D engr. Omnidata Corp., Westlake Village, Calif., 1980-82; engr. STS program Martin Marietta, Vanderberg AFB, Calif., 1982-85; systems design engr. Quintron Systems, Vanderberg AFB, 1985-87; R & D engr. Indsl. Computer Design, Westlake Village, 1987-88; owner, engr. Have Computer Will Travel, Westlake Village, 1988-92; prin. Anderson Digital Imaging, Santa Barbara, 1993—; cons. in field. Author: (novel) Medic! Medic!, 1991. With U.S. Army, 1966-69, Vietnam. Decorated Purple Heart. Achievements include integration of the first complete CAD-CAM link in U.S.; design and development of Net Selectable Jack Sta. communication device, Integration Safety Info. System database mgmt. system; established software verification program for enabling testing of all software used in STS (Shuttle Transp. System) program, of the Facility Checkout and Maintenance Control System. Home and Office: Anderson Digital Imaging PO Box 40376 Santa Barbara CA 93140

ANDERSON, TERRY MARLENE, civil engineer; b. Honolulu, Sept. 26, 1954; d. Stanley Dale and Anna Clara (Heigert) A.; m. Jack Willard Steinberg, Feb. 29, 1980 (div. May 1983). Student, U. San Diego, 1971-72, U. Calif., San Diego, 1972-74; BS in Biol. Scis., U. Calif., Davis, 1974, BS in Aquacultural Engrs., 1979. Registered civil engr., Calif., Colo. Project mgr. John Carollo Engrs., Walnut Creek, Calif., 1979-85; assoc. civil engr. Grice Engring. Inc., Salinas, Calif., 1985; self-employed Durango, Colo., 1985-86; BS in Aquacultural Engring. Charpier, Martin & Assocs., Sacramento, Calif., 1986-87; project engr. CWC-HDR, Inc., Cameron Park, Calif., 1987; sr. civil engr. El Dorado County Dept. Transp., Placerville, Calif., 1987-90; dep. dir. pub. works Medocino County Dept. Pub. Works, Ukiah, Calif., 1990—; enging. div. mgr. pub. works Sonoma County. Vol. Office Emergency Svcs., Santa Rosa, 1987—. Recipient Resolution of Appreciaiton, City Coun., City of Gonzales, 1983. Mem. Woman's Transp. Seminar, ASCE. Republican. Office: Sonoma County dept Pub Works 575 Administration Dr Santa Rosa CA 95403

ANDERSON, THEODORE ROBERT, physicist; b. Lodi, Ohio, Jan. 30, 1949; s. Robert Anderson and LaVaughn (Mitchell) Gillotti. BS in Physics, Fla. State U., 1971; postgrad. in math. physics, U. Geneva, Switzerland, 1973, 75; MS in Physics, NYU, 1979, MS in Applied Sci., 1983, PhD in Physics, 1986. Nuclear engr. Gibbs & Hill Inc., N.Y.C., 1980-83; rsch. physicist elec. boat div. Gen. Dynamics, Groton, Conn., 1983-88; rsch. physicist Naval Underwater Systems Ctr., New London, Conn., 1988—; adj. prof. mech. engring., astronomy U. Conn., Storrs, Groton, 1983—; adj. prof. math. Mitchell Coll., New London Conn., 1985, U. Hartford, 1990—; adj. prof. mech. and aeronautical engring. U. Bridgeport, 1989—, Hunter Coll.; adj. prof. physics and astronomy CUNY, 1979-83; adj. prof. physics L.I. Univ., 1980-83; adj. prof. elec. and mech. engring. Rensellaer Poly. Inst., Hartford, 1986—; adj. prof. Sch. Bus. U. New Haven, 1989—, mech. engring., 1983—, elec. engring., 1983—; instr. Cooper Union Sch. Engring., N.Y.C., 1980. Rsch. in fluid dynamics, acoustics and atomic physics; contbr. articles to profl. jours. Active Met. Opera Guild, N.Y.C., 1986—, Mus. Modern Art, N.Y.C., 1985—, Met. Mus. Art, N.Y.C., 1984—, Am. Mus. Natural History, N.Y.C., 1987—, N.Y. Shakespeare Festival, 1987—, N.Y. Zool. Soc., 1988—, Ea. Nat. Park and Monument Assn., 1990—. Recipient Spl. Achievement award USN, 1989, 90. Mem. Nat. Geographic Soc., Nat. Parks and Conservation Assn., Am. Phys. Soc., AIAA, Soc. Rheology. Achievements include research in fluid dynamics, acoustics and atomic physics. Home: 2 Lake Dr Old Lyme CT 06371-1211

ANDERSON, TIMOTHY LEE, electrical engineer; b. Seoul, Republic of Korea, May 20, 1969; came to U.S., 1974; s. Burl Eldon and Rita Kay (Maxwell) A. BA in Physics, Berea Coll., 1992. Co-op elec. engr. IBM Corp., Lexington, Ky., 1992—. Mem. IEEE, Sigma Pi Sigma, Pi Mu Epsilon. Republican. Nazarene. Home: 370 Rose St Apt 5 Lexington KY 40508

ANDERSON, VICTOR CHARLES, applied physics educator; b. Shanghai, China, Mar. 31, 1922; s. Elam Johnathan and Colina (Michael) A.; m. Anne Dowden, May 9, 1943; children: Victor C., Judith. B.A., U. Redlands, 1943; Ph.D., UCLA, 1953. Research technician U. Calif. Radiation Lab., Berkeley, 1943-46; research asst. Scripps Instn. Oceanography, La Jolla, 1947-53; postdoctoral fellow Harvard U. Acoustics Research Lab., 1954-55; research physicist Scripps Instn. Oceanography, U. Calif., 1955—, prof. applied physics, 1969—; chmn. dept. elec. engring. and computer sci. Scripps Instn. Oceanograpy, U. Calif., 1981—; dep. dir. marine Phys. Lab. Scripps Instn. Oceangraphy, U. Calif., 1969—. Patentee digital multibeam steering, delay line time compressor. Recipient Disting. Civilian Service award Office of Naval Research, 1976. Fellow Acoustical Soc. Am. (silver medal 1993); mem. IEEE, Sigma Pi Sigma. Presbyterian. Home: 2325 Poinsettia Dr San Diego CA 92106-1224 Office: U Calif-San Diego EECS Dept La Jolla CA 92093

ANDERSON, WAYNE ARTHUR, electrical engineering educator; b. Jamestown, N.Y., May 20, 1938; s. Arthur Charles and Flora Mary (Funicello) A.; m. Marilyn Mae Anderson, July 28, 1964; children—Wayne P., Leslie M. B.A., SUNY-Buffalo, 1961, M.S., 1965, Ph.D., 1970. Research engr. Great Lakes Carbon, Niagara Falls, N.Y., 1961-65; instr. SUNY-Buffalo, 1965-70, prof. elec. engring., 1978—; prof. Rutgers U., New Brunswick, N.J., 1970-78; cons. Exxon, Linden, N.J., 1972-76, Amerace Corp., N.J., 1974-76; reviewer NSF, 1979—; dir., Ctr. for Electronic and Electro-Optic Materials, 1986—, ECE dept. chmn. 1989—. Contbr. more than 100 articles to profl. jours. Chmn. Internat. Students Com., 1983—; deacon 1st Baptist Ch., 1984-87, Westerly Rd. Ch., 1973-76. Grantee Solar Energy Research Inst., 1975-83, NSF, 1972-74, 80-83, 92—, Intelstat, 1980-84, ONR 1986-89. NYSERDA, 1991-93, NREL, 1993—. Mem. IEEE (sr., bd. dirs. 1981-84), Materials Rsch. Soc., Am. Inst. Physics. Republican. Avocations: tennis; swimming; biking. Home: 39 Sleepy Hollow Ln Orchard Park NY 14127-4617 Office: SUNY Elec and Computer Engring 217 C Bonner Hall Buffalo NY 14260

ANDERSON, WILLIAM CARL, association executive, environmental engineer; b. Vinton, Iowa, Sept. 24, 1943; s. Ivan D. and Lois B. (Schlotterback) A.; m. Elizabeth A. Dingman, Nov. 12, 1966; children: William Carl III, Erica Dawn. BSCE, Iowa State U., 1967. Registered profl. engr., N.Y., N.J., Pa., Iowa; diplomate Am. Acad. Environ. Engrs. Dir. environ. health Cayuga County Health Dept., Auburn, N.Y., 1969-73; ptnr. Pickard & Anderson, Auburn, 1973—; trustee Am. Acad. Environ. Engrs., Annapolis, Md., 1982-85, exec. dir., 1985—. Editor: The Diplomate, 1985—. Gen. chmn. Cayuga County United Way, 1982, exec. com., 1982-84, bd. dirs., 1981-84; health and safety com. Cayuga County council Boy Scouts Am., 1969-83; parish council Sacred Heart Parish, 1981-82; bd. dirs. YMCA-WEIU Cayuga County, 1982-85. Served with USNR, 1967-69. Recipient Recognition award United Way, 1982; named Honorable Conceptor, Mich. Cons. Engrs. Council, 1983. Fellow ASCE (exec. com., mgmt. group "D" 1983-87, tech. activities com. 1985-86, nat. water policy council 1986-87, Outstanding Service award 1981, 86); mem. Am. Water Works Assn., Air Waste Mgmt. Assn., Assn. Environ. Engring. Profs., NSPE, N.Y. Soc. Profl. Engrs., N.Y. Water Pollution Control Assn. (Lewis Van Carpenter award 1974), Water Pollution Control Fedn. (Philip F. Morgan medal 1973), Chi Epsilon. Republican. Roman Catholic. Office: Am Acad Environ Engrs 130 Holiday Ct Ste 100 Annapolis MD 21401-7032

ANDERSON, WILLIAM R., biologist, educator, curator, director. BS Botany, Duke U., 1964; MS Systematic Botany, U. Mich., 1965, PhD, 1971. Assoc. curator N.Y. Botanical garden, 1971-74; from asst. prof. to prof. Hebarium, dept. Biology U. Mich., 1974—, also assoc. curator, 1974-86, curator, 1986—, dir. Herbarium, 1986—; field work in Jamaica, 1963, 66, Hawaii, 1964, Mexico, 1965, 66, 68, 70, 81, 83, 88, Costa Rica, 1969, 90, Brazil, 1972-76, 78, 82, 90, Argentina, 1982, 90, Venezuela, 1984. Gen. editor numerous vols., chpts. in field; contbr. articles in field to profl. jours. Office: U of Mich Herbarium N University Building Ann Arbor MI 48109-1057

ANDERSON, WILLIAM ROBERT, physicist; b. Moline, Ill., Nov. 7, 1950; s. Clair Howard and Mary Louise (Tingle) A.; m. Judy Ann Reber, June 16, 1983; 1 child, Jessica K. BA, Augustana Coll., 1972; PhD, Tex. A&M U., 1977. Postdoctoral assoc. Ballistic Rsch. Lab., Aberdeen Proving Ground, Md., 1977-78; rsch. physicist Chem. Systems Lab., Aberdeen Proving Ground, 1978-79, Army Rsch. Lab., Aberdeen Proving Ground, 1979—. Contbr. articles to Jour. Chem. Physics, Combustion and Flame, other sci. publs. Mem. Am. Phys. Soc., Combustion Inst., Phi Lambda Upsilon (chpt. v.p. 1974-75, pres. 1975-76). Achievements include development of detection methods for various flame molecules, study of detailed combustion chemistry of nitrogen oxides. Office: AMSRLC-WT-PC Aberdeen Proving Ground MD 21005

ANDERSSON, STIG INGVAR, educator; b. Göteborg, Sweden, Oct. 25, 1945; s. Jack Ivan Ebbe and Ingegerd (Andreasson) A. Fil.kand., U. Göteborg, 1969, fil.dr., 1975. Vis. researcher Acad. Sci., Warsaw, Poland, 1968; Royal Soc. fellow Royal Soc., London, 1969-70; rsch. fellow Royal Soc. Stockholm, 1974-76; vis. prof. Inst. Hautes Etudes, Sci., Bures, Yvette, France, 1976-79; vis. prof. Calif. Inst. Tech., Pasadena, 1979-80; prof. math. physics U. Clausthal, Fed. Republic Germany, 1982-86; dir. rsch. group of math. immunology Chalmers Sci. Park, Göteborg, 1986—; vis. prof. U. Toulouse, France, 1987-88, vis. prof. U. Vienna, 1989-90; cons. in field. Author: Non-abelian Cohomology Theory and Applications, 1985; editor conf. procs.; contbr. articles to profl. jours. Alexander von Humboldt rsch. fellow, 1980. Fellow Royal Soc. London; mem. Am. Math. Soc., Swedish Soc. for Physicists, Deutsche Mathematiker-Vereinigung. Office: Chalmers Sci Park, S-412 88 Göteborg Sweden

ANDERSSON, TOMMY EVERT, researcher; b. Karlskrona, Sweden, June 10, 1956; s. Evert Jonny and Inga Maria (Hakansson) A.; m. Anita Marianne Sjölander, Aug. 10, 1985; children: Jonatan, Kristina. PhD, U. Uppsala, Sweden, 1983. Rsch. assoc. Dept. Med. Microbiology, U. Linköping, 1984-88, assoc. prof. Dept. Cell Biology, U. Linköping, 1988—. Contbr. articles to profl. jours. including Jour. Biol. Chem. and Proc. Nat. Acad. Fogarthy fellowship NIH, 1990; research Fernström Found., 1988, Odd Fellow Found., 1987. Mem. Am. Soc. for Cell Biology, Am. Soc. for Biochemistry and Molecular Biology, N.Y. Acad. Scis., Swedish Soc. Medicine. Home: Majorsgatan 3, S582 63 Linköping Sweden Office: U Linköping US, Dept Cell Biology, S 581 85 Linköping Sweden

ANDERSSON, ULF GÒRAN CHRISTER, biochemist; b. Malmö, Sweden, Oct. 25, 1960; s. Stig Borje and Birgitta (Persson) A. MSc in Chem. Engring., U. Lund, 1986. Mng. dir. Ingenjörsfirma Perulf, Malmo, 1983-86; mktg. mgr. AC Biotechnics AB, Lund, 1986-87; mktg. dir. SILAB-MILAB, Malmo, 1987-90; v.p. Medscand Diagnostics AB, Malmo, 1990-92, Euro-Diagnostica (formerly Ferring Diagnostica), Malmo, 1992—; exec. officer Medscand Ingeny, Rijswijk, Holland, 1990-91. Capt. Swedish Army, 1982—. Mem. Lions, Bivrast Men's Club Malmo (sec. 1987—). Office: Euro-Diagnostica AB c/o Ferring AB, PO Box 30047, 20061 Malmö Sweden

ANDES, CHARLES LOVETT, museum executive, technology association executive; b. Phila., Sept. 23, 1930; s. Charles Lovett and Gladys (Stead) A.; m. Dorothea Roberta Abbott, Aug. 25, 1961; children: Elizabeth, Susan, Karen, Page. Student, Swarthmore Coll., 1948-50; B.A., Syracuse U., 1952. Pres. Adtech Industries, Phila., 1954-68; exec. v.p., dir. The Franklin Mint Corp., Franklin Ctr., Pa., 1969-73, pres., 1972-73, chmn. bd., CEO, 1973-85; chmn., CEO The Franklin Inst., Phila., 1985-91, chair, 1991—; pres., CEO Tech. Coun. Greater Phila., 1991—; bd. dirs. O'Brien Environ. Energy Systems, ptnr. Tech. Leaders Venture Capital Fund; mem. Pa. Tech. Coun. Pres. Inst. Biotech. and Advanced Molecular Medicine; bd. dirs. Pa. Acad. Fine Arts, Clin. Nutritional Found., Fidelity Bank,; vice chmn. Pa. Intergovtl. Coop. Authority. Presbyterian. Clubs: Phila. Country (Phila.), Union League (Phila.), Merion Cricket (Phila.); Johns Island (Fla.). Office: The Franklin Inst 20th & The Benjamin Philadelphia PA 19103 also: Tech Coun Greater Phila 435 Devon Park Dr Wayne PA 19087-1900

ANDIS, MICHAEL D., urban entomologist; b. Amarillo, Tex., June 6, 1954; s. Billie Ralph and B.J. (Pyron) A. BS, Tex. A&M Univ., 1976; MS, La. State Univ., 1980. Diplomate Am. Bd. Entomologist. Rsch. tech. USDA Cotton Insects Rsch. Lab., College Station, Tex., 1975-76; entomologist Tex. Agrl. Experiment Station, College Station, Tex., 1976-78;

rsch. asst. dept. entomology La. State Univ., Baton Rouge, 1978-80, rsch. asssoc., 1980-85; medical entomologist New Orleans Mosquito Control Bd., 1985-87; biology mgr. Roussel UCLAF Corp., Englewood Cliffs, N.J., 1987-92; product devel. mgr. EcoScience Corp., Worcester, Mass., 1992—; adv. La. Dept. Agrl., Baton Rouge, 1982-85; cons. La. State Police, Baton Rouge, 1983-85; team mem. mgmt. Roussel UCLAF Corp., 1985-92. Author: (training manual) Emergency Vector-Borne Disease Control, 1985, (tech. manual) Biorational Insect Management Strategies, 1992; contbr. articles to profl. jours. Recipient Outstanding Rsch. award La. State Univ. Entomology Dept., 1982, Excellence award Ctr. for Disease Control, 1986. Mem. Am. Assn. Advancment Sci., Entomological Soc. Am., N.Y. Acad. Sci., Gamma Sigma Delta. Achievements include 3 insecticidal process patents; developed and implemented vector borne disease control program for City of New Orleans; received EPA/USDA grant to investigate nonchemical means of controlling disease vectors; conducted forensic entomological investigations to assist law enforcement agencies in ascertaining post-mortem intervals; developed numerous commercial products for environmentally compatible and commercially feasible insect control. Office: EcoScience Corp One Innovation Dr Worcester MA 01605

ANDO, SHIGERU, biology educator; b. Nagoya, Japan, Oct. 26, 1929; s. Kenji and Kikue (Ikai) A.; m. Youko Ishii, Apr. 8, 1958; children: Satoshi, Atsushi. BS, Nagoya U., 1954, MS, 1956, DSc, 1961. Instr. Nagoya U., 1959-68; assoc. prof. biology (biotelemetry) Aichi Kenritsu U., Nagoya, 1968-78, prof., 1978—; cons. on harmful birds Chuubu Electric Power Co. Ltd., Nagoya, 1988-90; dir. investigation com. on natural nomument of cormorant, Nagoya, 1983. Co-author: Telemeter and Stimulator for Living, 1980, Sociology before Man, 1990; editor: Wild Herd, 1979. Mem. Com. Natural Properties, Aichi Prefecture, Nagoya, 1984—. Recipient award Found. for Sci. in Tokai, 1965, Ishida Found. for Sci., 1983; grantee Sakkoukai, 1964-66. Mem. Ecol. Soc. Japan. Avocations: computer programming, hobby electronics, model railways. Home: 2-229 Shirotuchi Haruki, Tougoucho Aichi-gun 470-01, Japan Office: Aichi Kenritsu U, 3-28 Takada-cho, Nagoya 467, Japan

ANDO, TAKASHI, chemistry educator; b. Okazaki, Aichi, Japan, Apr. 1, 1937; s. Masaharu and Toshiko (Yamada) A.; m. Kazuko Shimada, Apr. 16, 1967; children: Yuka, Rika, Mika. BS, Osaka U., 1959, PhD, 1965. Rsch. assoc. Inst. Sci. Ind. Rsch., Osaka U., Sakai, 1965-73, assoc. prof., 1973-83; rsch. fellow Dept. Chemistry, Harvard U., Cambridge, 1969-71; prof. Dept. Chemistry, Shiga U. Med. Sci., Otsu, Japan, 1983—. Contbr. articles to profl. jours. including Jour. Am. Chem. Soc., Jour. Organic Chemistry, Jour. Phys. Chemistry, others. Recipient Divisional award Chem. Soc. Japan, 1991. Achievements include rsch. on kinetic isotope effects, supported reagents, sonochemistry. Office: Shiga U Med Sci Dept Chemistry, Seta Tsukinowa, Otsu 520-21, Japan

ANDOSCA, ROBERT GEORGE, engineering educator; b. Boston, Apr. 9, 1967; s. Francis George and Mary Louise (Brennan) A.; m. Katherine Marie Flores, June 30, 1990; 1 child, Chantelle Ann-Marie. BS, U. N.H., Keene State coll., 1989; MS, U. Vt., 1991. Tutor KSC: Spl. Acad. Svcs., Keene, N.H., 1986-88; teaching asst. KSC: Physics Dept., Keene, N.H., 1987-89; engring. asst. Kimball Physics, Inc., Wilton, N.H., 1989; rsch. asst. elec. engring. dept. U. Vt., Burlington, 1989-91, rsch. asst., teaching fellow physics dept., 1991—. Contbr. articles to Jour. of Applied Physics, 1992, Chemical Perspectives of Microelectronic Materials II, 1990, Am. Phys. Soc. Bulletin, 1989, 88. Recipient N.H. Gov.'s Success grant, 1988-89, Teaching fellowship U. Vt., 1991—. Mem. Acoustical Soc. Am., Materials Rsch. Soc. Achievements include study of bulk properties of silicon dioxide thin films deposited by electron cyclotron resonance plasma enhanced chem. vapor deposition. Office: U Vt Physics Dept Cook Phys Sci Bldg Burlington VT 05405

ANDRADE, FRANCISCO ALVARO CONCEICAO, chemistry educator; b. Simao Dias, Sergipe, Brazil, Oct. 4, 1943; s. Alvaro Lucindo and Josefa Conceição (Fortunato) A.; m. Maria Auxiliadora Carvalho dos Anjos, Jan. 11, 1951; children: Paulo, Marilia, Mariana. B. in Chemistry, Fed. U. of Bahia, Salvador, Brazil, 1968; PhD, U. São Paulo (Brazil), 1979. Chemist JOANES Indsl., Salvador, 1972-73; asst. prof. Fed. U. Bahia, Salvador, 1973-81, assoc. prof., 1981—; head organic chemistry dept. U. Bahia, Salvador, 1984-86; rsch. assoc. U. Ga., Athens, 1989-91. Contbr. articles to profl. jours. including Phosphours and Sulfur, An. Acad. Brasil. Ciênc., Internat. Jour. Sulfur Chemistry, Biooorganic Medicinal Chemistry Letters. Mem. Am. Chem. Soc., Brazilian Chem. Soc. Achievements include research on relative basicities measurements, on sterochemistry of some enzymatic reactions, on synthetic organic reactions and intermediates. Office: Inst de Quimica da UFBA, Campus Universitario, 40170-280 Salvador Brazil

ANDRADE-GORDON, PATRICIA, biological scientist; b. Cali, Colombia, Jan. 19, 1956; came to U.S., 1975; d. Victor Hugo and Sylvia (Franco) A.; m. David A. Gordon, Dec. 11, 1982; 1 child, Andrea E. BA in Chemistry, Spanish cum laude, Barnard Coll., 1979; PhD in Pharmacology, SUNY, 1984. Postdoct. rsch. SUNY, Stony Brook, 1985-89; assoc. rsch. scientist Yale U., New Haven, Conn., 1989-90, asst. prof., 1990-91; sr. scientist RW Johnson Pharm. Rsch. Inst., Spring House, Pa., 1991-92, prin. scientist, 1992—. Contbr. articles to profl. jours. Recipient Predoct. Trainee award NIH, Stony Brook, 1979-80. Mem. Internat. Soc. thrombosis and Hemostasis, N.Y. Acad. Scis., Thrombosis Coun. Am. Heart Assn., Phi Lambda Upsilon. Achievements include 2 patents in field. Home: 301 Windy Run Doylestown PA 18901 Office: RW Johnson Pharm Rsch Inst Welsh & McKean Rds Spring House PA 19477

ANDREA, MARIO IACOBUCCI, engineer, scientist, gemologist, appraiser; b. Haverhill, Mass., May 21, 1917; s. Andrea and Lucia (Antolini) Iacobucci; m. Muriel Grace Litchfield, June 30, 1940 (div. Dec. 1947); children: Gail, Patricia; m. Elizabeth Dwight (Bowes) Bray, Dec. 31, 1949 (div. Jan. 1986); children: Marjorie, Lucia, Janet; m. Elma Williams, Nov. 29, 1986. BSc, Webb Inst., Glen Cove, N.Y., 1939; grad., Oak Ridge Sch. Tech., 1958; MSE, Cath. U. Am., 1967; PhD, Pacific Western U., 1984. Grad. gemologist; registered profl. engr., Md. Application engr. GE Co., Schenectady, 1948-52; marine engr. Mil. Sea Transp. Svc., Washington, 1952-54, supervisory naval architect, Yokosuka, Japan, 1954-56; nuclear and gen. engr. R & D, Maritime Adminstrn., Washington, 1956-74; grad. gemologist, appraiser The Gem Tree, Bethesda, Md., 1974—. Patentee helical ship hull form. Pres., treas. Maritime Recreation Assn., Washington, 1970. Lt. comdr. USNR, 1941-61. Decorated naval medals. Mem. Nat. Assn. Jewelry Appraisers (sr.), Gemol. Inst. Am. Alumni Assn. (life). Episcopalian. Avocations: chess, bridge, golf, gardening.

ANDREASEN, NANCY COOVER, psychiatrist, educator; d. John A. Sr. and Pauline G. Coover; children: Robin, Susan. BA summa cum laude, U. Nebr., 1958, PhD, 1963; MA, Radcliffe Coll., 1959; MD, U. Iowa, 1970. Instr. English Nebr. Wesleyan Coll., 1960-61, U. Nebr., Lincoln, 1962-63; asst. prof. English U. Iowa, Iowa City, 1963-66; resident U. Iowa, 1970-73; asst. prof. psychiatry U. Iowa, Iowa City, 1973-77, assoc. prof., 1977-81, prof. psychiatry, 1981—, dir. Mental Health Clin. Rsch. Ctr., 1987—; sr. cons. Northwick Pk. Hosp., London, 1983; acad. visitor Maudsley Hosp., London, 1986. Author: The Broken Brain, 1984, Introductory Psychiatry Textbook, 1991; editor: Can Schizophrenia be Localized to the Brain?, 1986, Brain Imaging: Applications in Psychiatry, 1988; book forum editor: Am. Jour. Psychiatry, 1988-93, dep. editor, 1989-93, editor, 1993—. Woodrow Wilson fellow, 1958-59, Fulbright fellow Oxford U., London, 1959-60. Fellow Royal Coll. Physicians Surgeons Can. (hon.), Am. Psychiat. Assn., Am. Coll. Neuropharmacologists; mem. Am. Psychopathol. Assn. (pres. 1989-90), Inst. of Medicine of NAS. Office: U Iowa Hosps & Clinics 200 Hawkins Dr Iowa City IA 52242-1009

ANDREEV, VACHESLAV MIKCHAYLOVITCH, physicist; b. Astrakhan, Russia, Sept. 18, 1941; s. Mikchail and Anna Andreeva; m. Natalia Georgievna Gagena, July 30, 1980; children: Vladimir, Aleksandra, Olga. Dr., Ioffe Phys. Tech. Inst., 1970. Prof. Ioffe Phys. Tech. Inst., St. Petersburg, Russia, 1980—; head lab. Ioffe Phys. Tech. Inst., St. Petersburg, 1981—. Author: (in Russian) Liquid Phase Epitaxy, 1975, Photovoltaic Conversion of Concentrated Sunlight, 1989; contbr. over 160 scientific articles to profl. jours. Mem. Russian Acad. Scis. (head photovoltaic sect. 1985). Achievements include 56 inventions. Office: Ioffe Physico-Tech Inst, Polytechnicheskay 26, Russia Saint Petersburg Russia

ANDREOTTI, RAYMOND EDWARD, chemical engineering technologist; b. Milford, Mass., Sept. 6, 1940; s. Raymond and Congetta Josephine (Fino) A.; m. Joanne Sue Lemon, Sept. 12, 1964; children: Raymond Michael, Gina Marie. Student, Tufts U., 1959, Carnegie Inst. of Med. Tech., 1962-63. Phys. sci. technologist U.S. Army Natick (Mass.) Rsch. and Devel. Command, 1959-75; biochem. technologist U.S. Army Natick Rsch. Devel., Engring. Ctr., 1975-85, chem. engring. technologist, 1985—; membership com. Sigma Xi Natick Ctr., 1975-88; mem. Lab. Robotics Group, Hopkinton, Mass., 1990—. Contbr. articles to profl. jours. Mem., past chair Hopedale (Mass.) Town Fin. Com., 1972—; lectr Sacred Heart Ch., Hopedale, 1966—; mem. Mass. Arts Lottery Adv. Bd., Boston,1 982. Recipient U.S. Army Achievement award, 1968, Army Civilian Performance awards, 1987-92. Mem. Am. Watercolor Soc., Acad. Artists Assn., Sigma Xi. Achievements include patent in chromatography column injector; devel. of safe test methods using lab. robotics for evaluating chem. warfare protective clothing. Home: 116 Freedom St Hopedale MA 01747 Office: US Army Natic R & D Ctr Kansas St Natick MA 01760

ANDRESEN, MARK NILS, electrical engineer; b. Patuxent River, Md., Mar. 21, 1957; s. Ronald N. and Nancy R. (Foster) A.; m. Silia M. Andrews, Nov. 6, 1982; children: Lauren, Jonathan, Jaqueline, Kimberly. BSEE, Va. Tech., 1980. Design engr. U.S. Army Corps Engrs., Norfolk, Va., 1977-79; R & D engr. Hewlett Packard, Cupertino, Calif., 1980-85; applications engr. Oneac Corp., Sunnyvale, Calif., 1985-86; pvt. practice cons. San Jose, Calif., 1989; tech. instr. Basic Measuring Instruments, Foster City, Calif., 1989-91, product mgr., 1991-92, sr. applications engr., power quality specialist, 1992-93; dir. of edn., 1993—; presenter in field, 1989—. Mem. Calvary Community Ch., San Jose, Calif., pastor staff, 1986-89; mem. Cupertino Foursquare Ch., 1991—. Mem. IEEE (editor publ., chpt. chair), Assn. Energy Engrs., Nat. Fire Protection Agy. Republican. Home: 10344 N Portal Ave Cupertino CA 95014 Office: Basic Measuring Instruments 335 Lakeside Dr Foster City CA 94404

ANDREW, JOHN WALLACE, medical physicist; b. Charlottetown, P.E.I., Can., June 10, 1946; s. Wallace Jenkins and Georgia Florence Matheson) A.; m. Christine Mary McCleave, June 20, 1970; children: Melissa, Alan. BSc, Dalhousie U., Halifax, N.S., Can., 1968, MSc, 1969; PhD, U. Alta., Can., 1975. Diplomate Am. Bd. Med. Physics. Med. rsch. fellow Princess Margaret Hosp., Toronto, Ont., Can., 1975-78; asst. prof. U. Toronto, 1978-79; med. physicist N.S. (Can.) Cancer Ctr., Halifax, 1979-83; sr. med. physicist N.S. (Can.) Cancer Ctr., 1983-89, dir. med. physics, 1989—; asst. prof. Dalhousie U., Halifax, 1979-90, assoc. prof., 1990—; exec. mem. Med. and Biol. Physicists, Can., 1982-86, chmn., 1984-85. Author booklet: Medical Physics in Canada, 1992; contbr. articles to profl. publs. Leader Boy Scouts Can., Halifax, 1990—, cub scout leader, 1985-88; coord. United Way, Halifax, 1988-89. 2d lt. Can. Army, 1964-67. Fellow Can. MRC, 1975-78; NRC scholar, 1971-74. Fellow Can. Coll. Physicists in Medicine (bd. dirs. 1988—, registrar 1990—); mem. Can. Orgn. Med. Physicists, Am. Assn. Physicists in Medicine, Can. Assn. Radiation Oncologists. Achievements include development of compensator systems for radiotherapy; contributions to beam ortimization in radiotherapy. Office: NS Cancer Ctr, 5820 University Ave, Halifax, NS Canada B3H 1V7

ANDREWS, ANGUS PERCY, mathematician, writer; b. North Chatham, N.H., Dec. 30, 1937; s. Angus Irvine and Dora Angette (Jones) A.; m. Geraldine Arden Betschick, June 12, 1959; children: Eric Brace, Margaret Collins, Erin Louise. SB, MIT, 1960; MA, UCLA, 1965, PhD, 1968. Tech. trainee Std. Oil of Calif., Houston, 1960; rsch. engr. Gen. Dynamics, San Diego, 1961; tech. staff Rockwell Internat. (formerly N.Am. Aviation), Downey, Calif., 1962-66, Anaheim, Calif., 1967-78, Thousand Oaks, Calif., 1978—; mem. ednl. coun. MIT, Cambridge, 1979—. Co-author: (textbook) Kalman Filtering Theory and Practice, 1993; contbr. over 30 articles to profl. jours. Mem. exec. com. Sierra Club, Conejo Group, Thousand Oaks, 1990—; mem. LWV, Thousand Oaks, 1992, Friends of Libr., Thousand Oaks, 1986—. Mem. IEEE, AIAA, Soc. Indsl. and Applied Math., Math. Assn. Am. Achievements include patent in field; 2 patents pending; discovery of unknown landmark tracking; of methods for square root Kalman filtering. Office: Rockwell Sci Ctr 1049 Camino Dos Rios Thousand Oaks CA 91360

ANDREWS, BETHLEHEM KOTTES, research chemist; b. New Orleans, Sept. 18, 1936; d. George Leonidas and Anna Mercedes (Russell) Kottes; B.A. with honors in Chemistry, Newcomb Coll., Tulane U., 1957; m. William Edward Andrews, May 9, 1959; children—Sharon Leslie, Keith Edward. Chemist wash wear investigation, So. Regional Research Center, Sci. and Edn. Adminstrn., Dept. Agr., New Orleans, 1958-63, research chemist wash wear investigation, cotton textile chemistry lab., 1968-70, research chemist spl. products research, cotton textile chemistry lab., 1976-83, sr. research chemist cotton chem. reactions research, 1983-85, lead scientist textile finishing chemistry research, 1985—; scientist-supr. Grace King High Sch. Lab. Tech. Tng. Program; U.S. del. ISO Meeting on Textiles, 1984. Recipient outstanding professionalism citation New Orleans Fedn. Businessman's Assn., 1977, Disting. Service award in med./sci. category, 1983, named Women of Yr. award in profl. category, 1978; La. Heart Assn. grantee, 1957. Mem. Am. Chem. Soc., Am. Assn. Textile Chemists and Colorists (exec. com. on research, Olney medal 1992), Fiber Soc., Phi Beta Kappa, Sigma Xi, Phi Mu. Democrat. Roman Catholic. Clubs: P.E.O., Southern Yacht. Contbr. chpts. to books, articles to sci. jours.; patentee. Office: So Regional Rsch Ctr Agrl Rsch Svc Dept Agriculture 1100 Robert E Lee Blvd New Orleans LA 70124

ANDREWS, DAVID CHARLES, energy and power consultant; b. London, Mar. 16, 1950; s. Alan Charles and Maisie Joy (Brien) A.; m. Janet Alison Hooker; children: Joseph James, Annabel Jane. BS in Engring. with honors, U. Cardiff, 1973. Chartered engr. Engr. Mouchel & Ptnrs., Bath, Eng., 1974-78; researcher Open Univ., Milton Keynes, Eng., 1978-82; area mgr. Ellis Tylin, London, 1982-84, Applied Energy Systems, Watford, Eng., 1984-86; mgr. sales and mktg. CHP Conversions Ltd., Llantrisant, Eng., 1986—; prin. David Andrews Assocs., Bath, 1988—; mng. dir. Power Gasifiers Internat., Ltd., Silent Clean Power Ltd.; sr. cons. power project Leverton Caterpillar, Windsor, U.K., 1988—. Author: The IRG Solution, 1984, The Hidden Manager, 1985; inventor info. routing group concept. Fellow Inst. Plant Engrs.; mem. Assn. Ind. Electricity Producers (founder mem.), Instn. Diesel and Gas Turbine Engrs. (coun. 1990—), Inst. Energy. Achievements include research in domestic sterling engined co-generation for gas and electricity industry.

ANDREWS, GLEN K., biochemist; b. Emporia, Kans., Oct. 1, 1948; s. Theodore Francis and Mae B. (Bryant) A.; m. D. Janell Stevenson, May 25, 1968; 1 child, Hillary Kristen. BS, Emporia State U., 1971, MS, 1973; PhD, Baylor Coll. of Medicine, 1978. Postdoctoral fellow U. Calgary, Alberta, Can., 1978-82; sr. staff fellow Nat. Inst. Environ. Health Scis., Research Triangle Park, N.C., 1982-84; asst. prof. Kans. U. Med. Sch., Kansas City, Kans., 1984-88, assoc. prof., 1988-93; prof., 1993—; study sect. mem. NIH, Washington, 1991—; mem. editorial bd. Endocrinology, Washington, 1993—. Contbr. articles to profl. jours. and chpts. to books. Grantee USDA, 1986-92, NIH, 1988-92, 91-96, 91-93. Mem. AAAS, Endocrine Soc., Am. Soc. Biochemists and Molecular Biologists, Am. Soc. Cell Biologists. Office: Kans U Med Ctr Dept Biochemistry 39th and Rainbow Kansas City KS 66160-7421

ANDREWS, HARVEY WELLINGTON, medical company executive; b. Stowe Twp., Pa., Sept. 9, 1928; s. Robert W. and Theresa R. (Reis) A.; B.B.A. cum laude, U. Pitts., 1952; M.B.A., Harvard, 1957; m. Jane Garland, Aug. 9, 1969; children: Marcia Lynne, Glynis Suzanne, Elizabeth Jane. With Gen. Electric Co., Syracuse, N.Y., 1952-55, Scovill Mfg. Co., Waterbury, Conn., 1957; comptr. Alcon Labs., Inc., Ft. Worth, 1958-61, comptr., treas., 1961-65, v.p. finance, 1964-68; founder, pres. Medimation, Inc., Ft. Worth, 1968—; bd. dirs. Med. Scis. Computor Corp., 1st Clin. Labs. ; pres., chmn. bd. Dalworth Med. Labs. Bd. dirs., mem. exec. com. Ft. Worth Opera Assn. Served with AUS, 1946-48. Named to Alcon Labs. Hall of Fame, 1988. Mem. AAAS, Am. Acad. Polit. & Social Scis., Ft. Worth C. of C., Soc. Advancement Mgmt. Order Artus, Scabbard & Blade, Golden Eagle Assn., Sigma Alpha Epsilon. Lutheran. Masons (32 deg.). Clubs: Rotary, Met. Knife and Fork, Tex. Christian U. Pres.'s Round Table Assn., Colonial Country, Century II. Home: 3124 Chaparral Ln Fort Worth TX 76109 Office: PO Box 1786 Fort Worth TX 76101-1786

ANDREWS, JUDY COKER, electronics company executive; b. Hot Springs, Ark., Dec. 19, 1940; d. Leon G. and Bobbie (Randles) Coker; m. William Campbell Andrews, June 27, 1961; children: Alan Campbell, Theresa Lee Andrews Mills. BSE, Henderson State U., 1961; MEd, U. N.C., 1973. Instr. math. High Point (N.C.) Coll., 1973, Greensboro (N.C.) Pub. Schs., 1974, Richland Coll., Dallas, 1974-78; systems analyst J.C. Penney Co., Dallas, 1978-80; systems analyst Texas Instruments, Dallas, 1980-83, info. ctr. mgr., 1983-85, customer svc. mgr., 1986-90, bus. devel. mgr., 1990-92; acct. mgr. Tex. Instruments, Dallas, 1993—. Inventor: system and method for securing cellular telephone access through a cellular telephone network using voice verification (pat. pending). Adult advisor Explorer Post 444, Boy Scouts Am., Richardson, Tex., 1980-82. U. N.C. fellow, 1971-73. Mem. Assn. for Systems Mgmt. (chpt. pres. 1984-85, chair internat. ann. conf. 1989, chair internat. corp. ptnr. com. 1992, Disting. Svc. award 1992). Avocation: golf. Office: Texas Instruments 6550 Chase Oaks Blvd Plano TX 75086

ANDREWS, KEVIN PAUL, physical chemist; b. Lakewood, Ohio, Sept. 7, 1965; s. Chester Arthur and B. Jean (Etling) A.; m. Anna Catherine Ploplis, Aug. 15, 1987. BA in Physics, Coll. Wooster, 1987; MS in Chem. Physics, U. Md., 1990. Rsch. asst. U. Md., College Park, 1990—. Trustee Emmanuel United Meth. Ch., Beltsville, Md., 1990-93. Wooster Coll. scholar, 1983-87, Edward B. Westlake scholar, 1983-87, Ohio Acad. scholar, 1983-87. Mem. AAAS, Am. Chem. Soc., Am. Phys. Soc., Neutron Scattering Soc. Am., Assn. for Women in Sci., Sigma Xi, Phi Beta Kappa. Office: U Md Dept Chemistry College Park MD 20742

ANDREWS, MARK ANTHONY WILLIAM, physiologist, educator; b. Brownsville, Pa., Apr. 25, 1959; s. Lester Vincent and Virginia Eileen (Krajeski) A. BS in Biology, Chemistry, St. Vincent Coll., 1981; MS in Exercise Physiology, U. Pitts., 1985; PhD in Physiology, Biophysics, Med. Coll. Ga., 1989. Postdoctoral fellow Med. Coll. Ga., Augusta, 1989-90; rsch. asst. prof. U. Heidelberg, Fed. Republic Germany, 1990-91; rsch. assoc. U. Vt. Sch. Medicine, Burlington, 1991-92; asst. prof. physiology N.Y. Coll. Osteo. Medicine, Old Westbury, N.Y., 1992—; cons. Pro Fitness, Inc., Burlington, 1991-92, head cons., Albertson, N.Y., 1992—. Contbr. articles to Jour. Gen. Physiology, Biophys. Jour., others. Mem. Greenpeace, 1980—. Grantee Internat. Union Pure and Applied Biophysics, 1990. Mem. Am. Physiol. Soc., Am. Coll. Sports Medicine, N.Y. Acad. Scis. Biophys. Soc., Mensa, Sigma Xi. Roman Catholic. Achievements include study of activation of contractile mechanism of muscle cells, particularly in protein thermodynamics, physical chemistry and fatigue mechanisms; development of theory of muscle fatigue involving muscle protein destabilization. Office: NY Coll Osteo Medicine Dept Physiology Old Westbury NY 11568

ANDREWS, SALLY MAY, healthcare administrator; b. Westfield, Mass., Feb. 29, 1956; d. Roger N. and Dorothy M. (Goodhind) A. Student, U. Conn., 1974-76; BA, Simmons Coll., Boston, 1978; MBA, Boston U., 1986. Payroll clk. Children's Hosp., Boston, 1978-79, asst. payroll supr., 1979-81, staff analyst dept. medicine, 1981-83, asst. adminstr. dept. medicine, 1983-86, adminstr. dept. medicine, 1986—. Bd. overseers Lasell Coll. for Women, Newton, Mass. Mem. Am. Mgmt. Assn., Adminstrs. of Internal Medicine, Assn. Adminstrs. in Acad. Pediatrics. Congregationalist. Office: Children's Hosp Dept Medicine 300 Longwood Ave Boston MA 02115-5737

ANDREWS, WALLACE HENRY, microbiologist; b. Biloxi, Miss., Oct. 6, 1943; s. Wallace Henry and Rita Louise (Wentzell) A. BA in Biology with distinction, U. Miss., 1965, MS in Microbiology, 1967, PhD in Microbiology, 1969. Microbiologist FDA, Dauphin Island, Ala., 1969-71; rsch. microbiologist FDA, Washington, 1971—; cons. Food and Agr. Orgn. of UN, Rome, 1983—, Pan Am. Health Orgn., WHO, Washington, 1989—. Assoc. editor: Official Methods of Analysis, 1984, 2d edit., 1990; mem. editorial bd. Jour. Food Protection, 1984—; contbr. articles to profl. jours. NASA fellow, 1966-69, Assn. Ofcl. Analytical Chemists fellow, 1993—. Mem. Am. Soc. for Microbiology, Assn. Ofcl. Analytical Chemists (Gen. Referee of Yr. 1986, 91), Internat. Assn. Milk, Food, and Environ. Sanitarians, Inc., Phi Kappa Phi. Roman Catholic. Achievements include development of methods used by the FDA for detection of Salmonella in foods. Office: FDA 200 C St SW Washington DC 20204-0002

ANDROUTSELLIS-THEOTOKIS, PAUL, civil engineer; b. Corfu, Ionian Islands, Greece, Feb. 27, 1939; s. Arsene P. and Alice (De Vida) A-T.; m. Carla Cerruti; children: Stephanos, Andreas. D of Civil Engring., Turin (Italy) Poly. U., 1967, postgrad., 1967; postgrad., Technoecon. formation, Techn. Chamber Gr., Athens, Greece, 1973. Bridge calculator Euroconsult, Turin, Italy, 1967; engr. A. Despotopoulos Contrs., Corfu, Greece, 1967; static calculator Diamantaras Despotopoulos Engring., Athens, 1967-68; with Ministry of Coordination, Athens, 1968-69; chief engr. Edok-Eter Contrs., Athens, 1969-78; vice chmn. bd. dirs. Geoerevna Soil Mechanics, Athens, 1971-78; tech. cons. Generali Ins. Co., Athens, 1978—; pres., gen. mgr. Exte Tourist Devel., Athens, Corfu, 1979-82; coord. Androutsellis Engring. Cons., Athens, 1983—; tech. cons. San Paolo Bank, 1990—, Cariplo Bank, 1991, Fidel Group, 1991—. Contbr. articles to profl. jours. on rsch. and application of membranal theory. Mem. Empros Study of Nat. Problems, Athens, 1982-89. Mem. Tech. Chamber of Greece, Corfu Reading Soc., Athens Tennis Club, Athens Ski Club, Yachting Club of Greece. Avocations: classical music, tennis, sailing. Home: 14 Ardittou Str, 11636 Athens Greece Office: Androutsellis Engineers, 4 Charitos St, 10675 Athens Greece

ANDRUS, W(INFIELD) SCOTT, scientist, consultant; b. N.Y.C., Aug. 10, 1938; s. Winfield and Julia M. (Arduino) A.; m. Marjorie Stevenson, July 13, 1974. AB, Wagner Coll., 1960; MA, SUNY, Stony Brook, 1965, PhD, 1967. Rsch. assoc. AMMRC, Watertown, Mass., 1968-69; sr. scientist Am. Sci. and Engring., Cambridge, Mass., 1969-71; project scientist Mass. Gen. Hosp., Boston, 1972-77; staff scientist Photo Metrics, Lexington, 1977-81; mgr. of rsch. Rorer Group, Andover, 1981-86; dir. laser devel. USCI Div. C.R. Bard, Billerica, 1986-91; pvt. practice cons. Lexington, 1991-92; prin. rsch. scientist Am. Med. Systems, Minnetonka, Minn., 1992—. Contbr. articles to profl. jours. Office: 11001 Bren Rd E Minnetonka MN 55343

ANDRY, STEVEN CRAIG, electrical engineer, educator; b. Bklyn., Dec. 11, 1959; s. Daniel and Carolyn Ann (Bindhamer) A. BS, Polytech. Inst. N.Y., 1982, MS, 1987. Elec. engr. Grumman Aerospace Corp., Bethpage, N.Y., 1982-89; mem. tech. staff AT&T Bell Labs., Whippany, N.J., 1989—. Contbr. articles to profl. jours. Mentor sumer sci. student program, Whippany, 1991. Mem. IEEE. Achievements include devel. of filters for real time processing systems, undersea mil. transmission systems, digital cellular communications systems, feed forward amplifier development. Home: 2069 57th St Brooklyn NY 11204 Office: AT&T Bell Labs 67 Whippany Rd Whippany NJ 07981

ANDRZEJEWSKI, CHESTER, JR., immunologist, research scientist; b. Bridgeport, Conn., Mar. 20, 1953; s. Chester Sr. and Helen (Sholomicky) A.; m. Kathleen Marie O'Connor, Aug. 7, 1976; children: Nicholas Chester, Michael Yuri, Danielle Natalya. ScB, Brown U., 1975; PhD, Tufts U., 1981, MD, 1984. Diplomate Am. Bd. Pathology, Nat. Bd. Med. Examiners. Resident, fellow Hosp. U. Pa., Phila., 1984-88; staff pathologist Wilford Hall USAF Med. Ctr., Lackland AFB, Tex., 1988-92; med. dir., transfusion medicine and clin. immunology Wilford Hall USAF Med Ctr., Lackland AFB, 1989-92; med. dir. transfusion medicine svcs. Baystate Med. Ctr., Springfield, Mass., 1992—. Contbr. articles to profl. jours. Sub-deacon Orthodx Ch. Am., 1977. Maj. USAF, 1980—. Recipient Rsch. Fellowship award Mass. Lupus Found., 1979, Internat. Disting. Dissertation award Coun. Grad. Students in U.S. and Univ. Microfilms, 1981, Excellence Rsch. award Roche Labs., 1987. Mem. Am. Assn. Clin. Pathologists, Am. Assn. Blood Banks (rsch. grantee 1987), N.Y. Acad. Scis., Clin. Immunology Soc., Coll. Am. Pathologists. Mem. Orthodox Ch. Am. Achievements include discovery of the first hybridoma monoclonal anti-DNA autoantibodies from a murine SLE model, the first anti-idiotypes to such antibodies;demonstration of the immunochemical and genetic relatedness of these autoantibodies.

Home: 19 Eagle Brook Dr Somers CT 06071-9999 Office: Baystate Med Ctr Dept Pathology Springfield MA 01199

ANEL, ALBERTO, biochemist; b. San Sebastian, Guipuzcoa, Spain, Aug. 25, 1963; came to U.S., 1991; s. Valentin Anel and Pilar Bernal; m. Elvira Cebollero, Nov. 16, 1990. BS in Organic Chemistry, U. Zaragoza, Spain, 1986, PhD in Biochemistry, 1990. Postdoctoral fellow Institute de Recherches Scientifiques sur le Cancer, Paris, 1990, Med. Biology Inst., La Jolla, Calif., 1991—. Contbr. articles to AIDS Rsch. and Human Retroviruses, Leukemia, Biochemistry, other sci. jours. Mem. Amnesty Internat., 1989-91, Greenpeace, 1989, Assoc. for Protection of Aragon Pyrenees, 1990-91. Grantee Diputacion Gen. de Aragon, 1986, fellow, 1987; fellow Ministry Edn. y Ciencia, 1991. Mem. N.Y. Acad. Scis. Achievements include improvements in the understanding of lipid and fatty acid metabolism in T-lymphocytes, either normal, tumoral or HIV-infected. Office: Med Biology Inst 11077 N Torrey Pines Rd La Jolla CA 92037

ANFIMOV, NIKOLAI, aerospace researcher; b. Moscow, Mar. 29, 1935; s. Apollon Anfimov and Maria Butovsky; m. Anna Golovachev, Sept. 22, 1956; 1 child, Valery. Engr. physicist in thermodynamics, Phys.-Tech. Inst., Moscow, 1958; candidate scis., Rsch. Inst. Thermal Processes, Moscow, 1962; DSc, Rsch. Inst. Thermal Processes, 1973. Engr., sr. researcher, chief heat transfer dept. Rsch. Inst. Thermal Processes, Moscow, 1958-73; chief thermal div., dep. dir., first dep. dir. Ctrl. Rsch. Inst. Machine-Bldg., Kaliningrad, Russia, 1974—; prof. Phys.-Tech. Inst., Moscow, 1974—; head expert coun. Russian Attestation Commn., Moscow, 1988—; editorial bd. Gas and Fluid Mechanics Jour., 1986—. Sci. editor Russian Translations; contbr. articles to profl. jours. Recipient Zuckovsky award for radiation gas dynamic rsch., 1970, USSR State award, 1980, Yu Gagarin medal USSR Cosmonautics Fedn., 1981, M. Yangel medal, 1984, S. Korolev medal, 1985, K. Tsiolkovsky medal, 1989. Mem. AIAA, Russia's Acad. Scis. (corr. mem.), Russia's Acad. Cosmonautics (honour mem.). Achievements include research in heat and mass transfer and thermal protection of spacecraft and rockets, aerodynamic and thermal design, analysis and ground testing of spacecraft and rockets. Home: 9-4 Rizsky Proezd, Moscow 129278, Russia Office: Tsniimash, 4 Pionerskaya St, Kaliningrad 141070, Russia

ANFINSEN, CHRISTIAN BOEHMER, biochemist; b. Monessen, Pa., Mar. 26, 1916; s. Christian Boehmer and Sophie (Rasmussen) A.; m. Florence Bernice Kenenger, Nov. 29, 1941 (div. 1978); children: Carol Bernice, Margot Sophie, Christian Boehmer; m. Libby Shulman Ely, 1979. B.A., Swarthmore Coll., 1937, D.Sc., 1965; M.S., U. Pa., 1939; Ph.D., Harvard, 1943; D.Sc. (hon.), Georgetown U., 1967, N.Y. Med. Coll., 1969, Gustavus Adolphus Coll., 1975, Brandeis U., 1977, Providence Coll., 1978; M.D. (hon.), U. Naples Med. Sch., 1980, Adelphi U., 1987. Asst. prof. biol. chemistry Harvard Med. Sch., 1948-50, prof. biochemistry, 1962-63; chief lab. cellular physiology and metabolism Nat. Heart Inst., Bethesda, Md., 1950-62; chief lab. chem. biology Nat. Inst. Arthritis and Metabolic Diseases, Bethesda, 1963-82; prof. biology Johns Hopkins U., Balt., 1982—; vis. prof. Weizmann Inst. Sci., Rehovot, Israel, 1981-82, bd. govs. 1962. Author: The Molecular Basis of Evolution, 1959; contbr. articles to profl. jours. Am. Scandinavian fellow Carlsberg Lab., Copenhagen 1939; sr. cancer research fellow Nobel Inst., Stockholm, 1947; Markle scholar 1948; Guggenheim fellow Weizmann Inst., 1958; recipient Rockefeller Pub. Service award, 1954-55, Nobel prize in chemistry, 1972. Mem. Am. Soc. Biol. Chemists (pres. 1971-72), Am. Acad. Arts and Scis., Nat. Acad. Scis., Washington Acad. Scis., Am. Philos. Soc., Fedn. Am. Scientists (treas. 1958-59, vice chmn. 1959-60, 73-76), Pontifical Acad. Sci. Home: 4 Tanner Ct Baltimore MD 21208-1332 Office: Johns Hopkins U Dept Biology 34th and Charles Sts Baltimore MD 21218-2685

ANFOSSO, CHRISTIAN LORENZ, analytical chemist; b. Oklahoma City, Feb. 25, 1963; s. Frank John Jr. and Laura Jane (Faatz) A. BS in Chemistry cum laude, Stephen F. Austin State U., 1985. Analyst Shell Westhollow R&D, Houston, 1986; chemist resin div. Ga. Pacific Chems., Lufkin, Tex., 1986-89; analytical chemist polyethylene lab. Union Carbide Chems. and Plastics, Port Lavaca, Tex., 1989—. Mem. Am. Chem. Soc., Am. Taekwondo Assn. (trainee instr., black belt), Alpha Chi, Phi Eta Sigma, Xi Epsilon. Achievements include analytical work with pyrophoric catalyst, particularly XRF. Home: Spt 131 209 Forest Hills St Rockport TX 78382 Office: Union Carbide Chems/Plastic Bldg 542 PO Box 186 Port Lavaca TX 77979

ANGEL, AUBIE, physician, academic administrator; b. Winnipeg, Man., Can., Aug. 28, 1935; s. Benjamin and Minnie (Kaplan) A.; m. Esther-Rose Newhouse; children: Jennifer, Jonathan, Suzanne, Steven, Michael. BSc in Medicine, U. Man., 1959, MD, 1959; MSc, McGill U., 1963. Jr to sr. intern in medicine Winnipeg Gen. Hosp., 1959-61; speciality resident in diabetes and endocrinology Montreal Gen. Hosp., 1961-62; postgrad. dept. exptl. medicine McGill U., 1962-63; asst. resident in medicine Royal Victoria Hosp., Montreal, 1963-64; asst. prof. pathology McGill U., Montreal, Que., Can., 1965-68; staff physician Royal Victoria Hosp., Montreal, 1965-68; sr. endocrinologist Toronto Gen. Hosp., 1968-90; asst. prof. medicine U. Toronto, Ont., Can., 1968-72, assoc. prof., 1972-81, prof. medicine, 1981-90, dir. Inst. Med. Sci., 1983-90; prof., head dept. medicine U. Man., 1991—; physician in chief Health Sci. Ctr., Winnipeg, Man., 1991—; vis. scientist U. Calif., San Diego, 1977-78, Hammersmith Hosp., London, 1978. Editor: (with C.H. Hollenberg and D.A.K. Roncari) The Adipocyte and Obesity: Cellular and Molecular Mechanisms, (with J. Frohlich) Lipoprotein Deficiency Syndromes: Advances in Experimental Medicine and Biology, 1986. Project dir. Can. Internat. Devel. Agy., Toronto and Costa Rica, 1987—. Recipient Outstanding Svc. award Heart and Stroke Found. Ont., 1985; U. Toronto Med. Rsch. Coun. scholar, 1965-71; Trinity Coll., Toronto, fellow, 1989—. Fellow Royal Coll. Physicians and Surgeons, Royal Soc. Medicine, London; mem. Can. Soc. Endocrinology and Metabolism (pres. 1980-82), Coll. Physicians and Surgeons Costa Rica (hon.), N.Am. Assn. Study Obesity (pres. 1986-87), Can. Soc. Clin. Investigation (councillor 1977-80), Am. Soc. Clin. Investigation, Can. Inst. Acad. Medicine (founding pres. 1990-92), Internat. Assn. Study Obesity (mem. bd. govs. 1986—), Juvenile Diabetes Found. Internat. (mem. hon. bd. 1987—). Office: Univ Manitoba Dept Internal Med, HE Sellers Prof & Chmn, Rm GC 430 Health Sci Ctr, 700 William Ave, Winnipeg, MB Canada R3E 0Z3

ANGEL, JAMES ROGER PRIOR, astronomer; b. St. Helens, Eng., Feb. 7, 1941; came to U.S., 1967; s. James Lee and Joan (Prior) A.; m. Ellinor M. Goonan, Aug. 21, 1965; children—Jennifer, James. B.A., Oxford (Eng.) U., 1963, D.Phil., 1967; M.S., Calif. Inst. Tech., 1966. From rsch. assoc. to assoc. prof. physics Columbia U., 1967-74; vis. assoc. prof. astronomy U. Tex., Austin, 1974; mem. faculty U. Ariz., Tucson, 1974—; prof. astronomy U. Ariz., 1975—, prof. optical sci., 1984—, Regents prof., 1990—. Sloan fellow, 1970-74. Fellow Royal Soc., Royal Astron. Soc.; mem. Am. Astron. Soc. (v.p. 1987-90, Pierce prize 1976). Achievements include research on white dwarf stars, quasars, astronomical instruments, telescopes, and adaptive optics. Office: Univ Ariz Steward Obs Tucson AZ 85721

ANGELIDES, DEMOSTHENES CONSTANTINOS, civil engineer; b. Thessaloniki, Greece, June 18, 1947; came to U.S., 1973; s. Constantinos D. and Chrysavgi (Papatsa) A.; m. Chryssanthi Koutsandrea, Dec. 25, 1991; 1 child, Constantine. Diplom. ingenieur, Aristoteles U., Greece, 1970; MSCE, MIT, 1975, PhD, 1978. Registered profl. engr., Tex. Rsch. asst. MIT, Cambridge, 1973-78; supervising structural engr. Brian Watt Assocs., Inc., Houston, 1978-80; sr. cons. structural engr. McDermott, Inc., New Orleans, 1980-83, Hudson Engring./McDermott, Inc., Houston, 1983-85; sr. cons. engr. McDermott Internat., Inc., New Orleans, 1985-90, mgr. total quality, 1990—; mem. tech. com. Am. Petroleum Inst., New Orleans, 1983-85; McDermott rep. Indsl. Liaison program MIT, New Orleans, 1981-89. Co-author: Offshore Structures, 1991; contbr. articles to Earthquake Engring. and Structural Dynamics, Jour. Engring. Mechanics, Jour. Structural Div., Jour. Offshore Mechanics and Arctic Engring. Mem. La. Fin. com. Dukakis for Pres., 1988; pres. Hellenic Arts Soc., New Orleans, 1986-88; bd. dirs. Trustees of Holy Trinity Greek Orthodox Ch., New Orleans, 1989-90; 2d lt. corps. engrs. Hellenic Army, Greece, 1971-73. Postdoctoral fellowship Coun. for Sci. and Indsl. Rsch., 1978; Govtl. scholar Aristoteles U., Thessaloniki, 1965-70. Mem. ASCE, ASME (chmn. computer tech. comm. of offshore mechanics and arctic engring. div. 1985-88), Am. Assn. Artificial Intelligence, Am. Geophys. Union, Sigma Xi. Achievements include development and design of pioneering concepts of offshore oil platforms, applications of artificial intelligence in optimization of manufacturing and engineering design, planning and implementation of statistical methods for improving quality of products and processes in manufacturing and software organizations. Home: 3915 St Charles Ave Apt 606 New Orleans LA 70115-4661 Office: McDermott Internat Inc 1010 Common St New Orleans LA 70112-2401

ANGELL, JAMES BROWNE, electrical engineering educator; b. S.I., N.Y., Dec. 25, 1924; s. Robert Corson and Jessie (Browne) A.; m. Elizabeth Isabelle Rice, July 22, 1950; children: Charles Lawrence, Carolyn Corson. S.B., S.M., MIT, 1946, Sc.D. in Elec. Engring, 1952. Research asst. MIT, 1946-51; mgr. solid-state circuit research, research div. Philco Corp., Phila., 1951-60; mem. faculty Stanford U., 1960—, prof. elec. engring., 1962-89, prof. emeritus, 1990—, dir. Solid-State Electronics Lab., 1964-71, assoc. dept. chmn., 1970-89; cons. to industry and govt., 1960—; Mem. electronics adv. group for comdg. gen. U.S. Army Electronics Command, 1964-74; mem. U.S. Army Sci. Adv. Panel, 1968-74; Carillonneur Stanford, 1960-91. Author sect. book. Area chmn. town incorporation com. Portola Valley, Calif., 1963-64; Bd. dirs. Portola Valley Assn., 1964-67. Fellow IEEE (life, chmn. internat. solid state circuits conf. 1964); mem. Am. Guild Organists, Guild Carillonneurs in N. Am. (dir. 1969-75, rec. sec. 1970-75). Home: 30 Shoshone Pl Portola Valley CA 94028-7632 Office: Stanford U Dept Elec Engring Stanford CA 94305

ANGELO, GAYLE-JEAN, mathematics and physical sciences educator; b. Winchester, Mass., Nov. 27, 1951; d. John William and Josephine Marie (Tavano) A. BA in Physics with honor, Northeastern U., 1975, MEd in Curriculum and Instrn. of Sci. and Math., 1978; MS in Applied Stats., Columbia U., 1984, postgrad., 1984—. Cert. secondary tchr., Mass.; cert. community coll. tchr., Calif. Clin. chemist Boston Med. Lab., Inc., 1971-73; exptl. physicist Northeastern U., Boston, 1975-76; tchr. natural scis., head sci. dept. Girls Cath. High Sch., Malden, Mass., 1977-78; rsch. and teaching asst. Columbia U., N.Y.C., 1979-80, rsch. assoc., 1982-83; rsch. scientist Air Force Rocket Propulsion Lab., Edwards AFB, Calif., 1980-82; cons. stats., 1982—; R&D analyst, engr. Varian-Extrion Div., Gloucester, Mass., 1984-86; asst. prof. math. Imperial (Calif.) Valley Coll., 1986—; engring. technician Imperial Irrigation Dist., 1991-92; instr. math. Golden Gate U., Cerro Coso Community Coll., 1981-82, Columbia U., 1982-83; instr. chemistry North Shore Community Coll., 1984-86; curriculum cons. Calif. State Dept. Edn., 1988-90; lectr. edn. San Diego State U., 1989-90. With USAF, 1980-82; with Air N.G., 1982-85; with USAFR, 1985-87. Mem. Am. Phys. Soc., Am. Statis. Assn., Nat. Coun. Tchrs. Math., Nat. Sci. Tchrs. Assn., Soc. Coll. Sci. Tchrs., Mensa, Sigma Xi, Phi Delta Kappa, Sigma Pi Sigma, Sigma Delta Epsilon, Kappa Delta Pi. Home: PO Box 565 Imperial CA 92251-0565 Office: Imperial Valley Coll Sci Math Engring Dept 380 E Aten Rd Imperial CA 92251-0158

ANGELO, MICHAEL ARNOLD, information scientist; b. Oregon City, Oreg., Nov. 25, 1944; s. Floyd Frederick and Alice Alison (Hammans) A.; m. Victoria Sophie Mann, Apr. 16, 1978; 1 child, Sierra Victoria. BS in Chemistry, Oreg. State U., 1966; BS, U. Alaska, 1969. Lab. asst. U.S. Forest Rsch. Lab., Corvallis, Oreg., 1964-66; grad. asst. U. Alaska, Fairbanks, 1966-68; chemist U.S. Environ. Prodn. Agy., College, Alaska, 1968-73; computer programmer R&M Engring. Cons., Fairbanks, 1973-75; laborer Trans-Alaskan Pipeline, Fairbanks, 1975-77; analyst programmer dept. Transp. State of Alaska, Juneau, 1977-92, database specialist, 1992—. Recipient Appreciation of Svc. award Gov. Alaska, Juneau, 1987, Exceptional Svc. award Dept. Environ. Cons., 1984. Mem. AAAS, Ancient Free and Accepted Masons. Office: Alaska Dept Transp 3132 Channel Dr Juneau AK 99801

ANGELOTTI, RICHARD H., banker, science administrator; b. Erie, Pa., Dec. 12, 1944; s. Henry and Ann (DiPlacido) A.; m. Carol A. Flaherty, Feb. 14, 1944; children: Kimberly, Nicole. BA, U. Notre Dame, 1966; JD, Loyola U., Chgo., 1969. Assoc. Brown Fox and Blumberg, Chgo., 1972-75, Angelotti and Cesario, Hinsdale, Ill., 1975-82; dir. mktg., v.p. Northern Trust Bank of Fla., Sarasota, 1983—, also bd. dirs.; chair. Mote Marine Laboratory Inc., Sarasota, Fla. Author: Estate Planning Techniques, 1975. Trustee, treas. Mote Marine Lab., Sarasota, 1987—; bd. dirs. United Way of Sarasota, 1985—, Suncoast Heart Assn., 1986—, Sarasota Family YMCA, 1988; pres., bd. dirs. Asolo Theater, Sarasota, 1986—. Mem. Am. Inst. Bank Marketers. Republican. Roman Catholic. Office: Mote Marine Laboratory Inc 1600 Thompson Pky Sarasota FL 34236

ANGHAIE, SAMIM, nuclear engineer, educator; b. Mallayer, Iran, Sept. 18, 1949; came to U.S., 1978; s. Tavakol and Raffieh (Khademi) Onghai; m. Sousan Novnejad, June 20, 1976; children: Amir-Ali, Amir-Hamid. BS, Pahlavi U., Shiraz, Iran, 1972, MS, 1974; PhD, Pa. State U., 1982. Instr. Pahlavi U., Iran, 1974-75; instr., lectr. Coll. Engring. U. Baluchistan, Zalredan, Iran, 1975-78; research asst. U. Fla., Gainesville, 1980-81, assoc. researcher, 1981-82, asst. prof., 1982-84, assoc. prof., 1986-92, prof., 1992—, assoc. dir. Innovative Nuclear Space Power and Propulsion Inst., 1986—; asst. prof. Oreg. State U., Corvallis, 1984-86; cons. Fla. Nuclear Assn., Gainesville, 1982—, United Techs., Jacksonville, Fla., 1983-84, Exxon Nuclear Co., Richland, Wash., 1985-86, Battelle Pacific Northwest Lab., Richland, Wash., 1986, Rocketdyne div. Rockwell Internat., Canoga Park, Calif., 1991—, United Techs., Pratt and Whitney, West Palm Beach, Fla., 1991—. Contbr. articles to nuclear, mech. and aerospace engring. jours.; patentee in field. Mem. adv. bd. Vols. of Am., Gainesville, 1982—; bd. dirs. Childcare Resources, Gainesville, 1989—. Grantee NSF, 1987, Elec. Power Rsch. Inst., 1989, NASA, 1990. Mem. ASM Internat., ASME, AIAA, Am. Nuclear Soc., Sigma Xi. Mem. Bahai Ch. Achievements include patent for Differential Gama Scattering Spectroscopy Method; development of self collider fusion reaction concept; nuclear droplet core concept; development of compatible materials for high temperature gaseous and liquid fuel nuclear reactors. Home: 2435 NW 36th Ter Gainesville FL 32605-2633 Office: U Fla 202 NSC Gainesville FL 32611

ANGHILERI, LEOPOLDO JOSÉ, researcher; b. Buenos Aires, Aug. 22, 1928; s. José and María (Orlando) A.; m. Akemi Itoh, Jan. 8, 1983; children: María Luján, Noelle, Juan Manuel. D in Chemistry, U. Buenos Aires, 1957. Fellow Johns Hopkins Med. Instns., Balt., 1967-69; asst. prof. U. Denver, 1969-70; rschr. German Cancer Ctr., Heidelberg, Germany, 1966-67, Tumor Rsch. Ctr., Essen, Germany, 1970-75; fellow Inst. du Radium, Paris, 1968-69, Inst. Nat. Rsch. Med., Paris, 1975-80; rsch. dir. lab. biophysics U. Nancy, France, 1980-91, ret., 1991. Editor, author: General Processes Radiotracer Localization Vol. 2, 1982, Hyperthermia in Cancer Treatment, Vol. 3, 1986, Role of Calcium in Biological Systems Vol. 5, 1982-90; contbr. 250 articles to profl. jours. Mem. Am. Chem. Soc. Home: 95 rue de Mareville, 54520 Laxou France

ANGUS, SARA GOODWIN, environmental engineer; b. Columbia, S.C., Dec. 28, 1943; d. John Edward and Sara Elizabeth (Sharp) Goodwin; m. Henry Brantlin Angus, Nov. 23, 1990; children: Kenneth Dale Fitzner, John Fredrick Biskup. BS in Chemistry, N.Mex. State U., 1969; MS in Engring., U. Tex., El Paso, 1981. Registered profl. engr., Tex., registered chemist Am. Soc. Clin. Pathologists. Chemist Providence Meml. Hosp., El Paso, Tex., 1970-75; chief chemist El Paso Water Utility, 1975-82; environ. engr. U.S. Army Corps Engrs., Ft. Worth, 1982-92, Washington, 1992—. Contbr. articles to SW and Tex. Waterworks Jour., Water Engring. and Mgmt. Mem. ASCE, Soc. Mil. Engrs., Am. Soc. Profl. Engrs. Home: 5713 Eliot Ct # 272 Alexandria VA 22311 Office: US Army Corps Engrs 20 Massachusetts Ave NW Washington DC 20314-1000

ANNI, ELENI (HELEN ANNI), biochemistry and biophysics educator; b. Athens, Greece, June 16, 1955; d. Athanassios and Julia (Stamatelatou) A. BSc in Biology, U. Athens, 1978; PhD in Biology, U. Patras, Greece, 1984. Predoctoral fellow Nat. Rsch. Ctr. Demokritos, Athens, 1977-81; postdoctoral fellow U. Pa., Phila., 1984-86, rsch. assoc., 1986-88; vis. scholar Okazaki Nat. Rsch. Insts., Okazaki, Japan, 1988-89, Kyoto (Japan) U., 1989-90; lectr. Agrl. U. Athens, 1990-91; rsch. asst. prof. U. Pa., Phila., 1991—. Author: Peroxidases in Chemistry and Biology, 1990, Metal Ions in Biological Systems, 1991. Greek Atomic Energy Commn. fellow, Athens, 1977, EC Sci. and Tech. fellow, Japan, 1988. Mem. AAAS, Assn. Women in Sci., Biophys. Soc., N.Y. Acad. Scis. Greek Christian Orthodox. Achievements include research in structural characterization of cytochrome c peroxidase, a mitochondrial hemeprotein that catalyzes the decomposition of hydrogen peroxide, as a five-coordinated high-spin species by electronic absorption, electron paramagnetic, nuclear magnetic and resonance Raman scattering spectroscopy, and functional determination of its active and inactive forms by stopped-flow kinetics, in relation to freezing and isolation artifacts. Office: U Pa Med Sch Dept Biochemistry & Biophysics 37th and Hamilton Walk Philadelphia PA 19104-6089

ANNIS, PATRICIA ANNE, textile scientist, educator, researcher; b. Council Bluffs, Iowa; d. Maurice Hamilton and Margaret Jean (Parquet) Wilson; m. Michael Clifford Annis, Sept. 4, 1971; 1 child, Paul Michael. MS, U. Ark., 1980; PhD, Kans. State U., 1988. Instr. Keokuk (Iowa) Community Schs., 1974-77; rsch. and teaching asst. U. Ark., Fayetteville, 1978-80, instr., 1980-81; rsch. asst. Kans. State U., Manhattan, 1982-87; asst. prof. dept. textile scis. U. Ga., Athens, 1988-93, assoc. prof., 1993—; expert witness on textile trace evidence; rsch. cons. to corp. and fed. orgns. related to the textile industry, 1990—. Contbr. articles to Textile Rsch. Jour., Jour. Forensic Scis., Textile Chemist and Colorist, Am. Dyestuff Reporter, Colourage, Canadian Textile Jour. Fellow Kappa Omicron Nu, 1986, Phi Upsilon Omicron, 1986, 87; grantee Hoechst-Celanese Corp., 1992, 93, Milliken Rsch. Corp., 1993, Advanced Technology Devel. Ctr., 1993. Mem. ASTM, Assn. of the Nonwovens Fabric Industry, Am. Assn. Textile Chemists and Colorists, Sigma Xi, Gamma Sigma Delta. Achievements include patent for material wear testing devices and techniques. Office: U Ga 309 Dawson Hall Athens GA 30602

ANODINA, TATYANA GRIGORYEVNA, aviation expert; b. Leningrad, 1939. Grad., L'vov Polytech. Inst. Dir. Aviation Rsch. Inst.; chmn. Interstate Aviation Com. (MAK), 1991—. Office: Rsch & Exper Ctr of Aviation Tech, Volokolamskoye Shosse 26, 123182 Moscow Russia*

ANSELL, GEORGE STEPHEN, metallurgical engineering educator, academic administrator; b. Akron, Ohio, Apr. 1, 1934; s. Frederick Jesse and Fanny (Soletsky) A.; m. Marjorie Boris, Dec. 18, 1960; children: Frederick Stuart, Laura Ruth, Benjamin Jesse. B. in Metall. Engring., Rensselaer Poly. Inst., 1954, M. in Metall. Engring., 1955, PhD, 1960. Physical metallurgist USN Research Lab., Washington, 1957-58; mem. faculty Rensselaer Poly. Inst., Troy, N.Y., 1960-84, Robert W. Hunt prof., 1965-84, chmn. materials div., 1969-74, dean engring., 1974-84; pres. Colo. Sch. Mines, Golden, 1984—, now pres., chancellor; bd. dirs. Norwest Bank, Cyprus Minerals Co. Editor books; patentee in field; contbr. over 100 articles to profl. jours. Served with USN, 1955-58. Recipient Hardy Gold Medal AIME, 1961, Curtis W. McGraw award Am. Soc. Engring. Edn., 1971, Souzandrade Gold Medal of Univ. Merit Fed. U. Maranhao, 1986. Fellow Metall. Soc. (pres. 1986-87), Am. Soc. Metals (Alfred H. Geisler award 1964, Bradley Stoughton award 1968); mem. NSPE, Am. Soc. Engring. Edn. (Curtis W. McGraw award 1971), Sigma Xi, Tau Beta Pi, Phi Lambda Upsilon. Club: Denver. Office: Colo Sch of Mines 1500 Illinois St Golden CO 80401-1887*

ANSON, FRED COLVIG, chemistry educator; b. Los Angeles, Feb. 17, 1933; m. Roxana Anson; children: Alison, Eric. BS, Calif. Inst. Tech., 1954; MS, Harvard U., 1955, PhD, 1957. Instr. chemistry Calif. Inst. Tech., Pasadena, 1957-58, asst. prof., 1958-62, assoc. prof., 1962-68, prof. chemistry, 1968—, chmn. div. chemistry and chem. engr., 1984—. Contbr. numerous articles to profl. jours. Fellow J.S. Guggenheim Found. U. Brussels, 1964, Alfred P. Sloan Found., 1965-69; scholar Fulbright-Hays Found. U. Florence, Italy, 1972, A. von Humboldt Found. Fritz Haber Inst., Berlin, 1984-86. Mem. AAAS, Nat. Acad. Sci., Am. Chem. Soc., Am. Electrochem. Soc., Internat. Soc. Electrochemistry, Soc. Electroanalytical Chemistry, Tau Beta Pi. Office: Calif Inst Tech Divsn Chemistry and Chem Engring MS 127-72 Pasadena CA 91125

ANTAYA, TIMOTHY ALLEN, physicist; b. Flint, Mich., May 5, 1956; s. John Clifford and Patricia Lee (Kennedy) A.; m. Ann Louise Martyn, June 4, 1976; children: Claire Louise, Andrew Martyn. BA in Physics, U. Mich., Flint, 1977; PhD in Physics, Mich. State U., 1985. Sr. physicist Nat. Superconducting Cyclotron Lab., East Lansing, Mich., 1984—; presenter in field. Contbr. articles to Nuclear Instr. Meth., Rev. Sci. Instr. Recipient 4th Pl. biology div. Internat. Sci. and Engring. Fair, San Diego, 1973. Mem. Am. Phys. Soc. Achievements include research in the area of highly charged ion beams. Office: Mich State U Cyclotron Lab East Lansing MI 48824

ANTHES, CLIFFORD CHARLES, retired mechanical engineer, consultant; b. Buffalo, Aug. 2, 1907; s. Edward Charles and Ethel Elmira (Cliff) A.; m. Theresa Roselyn Bischof, Sept. 2, 1931 (dec. Dec. 1982); children: Carol Louise, Clifford Charles Jr.; m. Ursula Elizabeth O'Leary, Apr. 7, 1984. Student, U. Buffalo, 1928-32, Newark Coll. Engring., 1934-39. Technician Linde div. apparatus devel. lab. Union Carbide Corp., Tonawanda, N.Y., 1928-33; project engr. Union Carbide Corp., Newark, 1933-64; sr. project engr. Union Carbide Corp., Florence, S.C., 1964-70; cons. Union Carbide Corp., Piscataway Twp., N.J., 1970-76; cons. So. Meth. U., Dallas, 1973, Gray Corp., Dallas, 1973, Brown & Brown Architects, Tucson, 1977, CF&I Steel Corp., Pueblo, Colo., 1982. Patentee in field; contbr. articles to profl. jours. and tech. papers to profl. soc. Vol. Civil Def., Union, N.J., 1939-42; com. mem., explorer, advisor Boy Scouts Am., Union, 1940s; trustee, mem. exec. bd. Florence Mus., 1960s. Mem. Eastern Fedn. Gem and Mineral Soc. (pres. 1967, citation for advancement of earth scis. 1972), Newark Gem and Mineral Soc. (pres. 1960s), Tucson Mineral Soc., Masons (Master 1949). Republican. Lutheran. Avocations: collecting minerals, cutting and faceting gems, mountain climbing. Home: 5711 W Rafter Cir Tucson AZ 85713-4444

ANTHONISEN, NICHOLAS R., respiratory physiologist; b. Boston, Oct. 12, 1933. AB, Dartmouth Coll., 1955; MD, Harvard U., 1958; PhD in Exptl. Medicine, McGill U., 1969. Intern medicine N.C. Meml. Hosp., 1958-59, jr. asst. resident, 1959-60; sr. asst. resident respiratory dept. Royal Victoria Hosp., 1963-64; demonstrator medicine McGill U., 1964-66, from asst. to assoc. prof. exptl. medicine, 1969-73, prof., 1973-75; prof. medicine U. Man., Winnipeg, 1975—. Scholar Med. Rsch. Coun. Can., 1969-71. Mem. Can. Soc. Clin. Investigation, Can. Thoracic Soc., Am. Physiol. Soc., Am. Soc. Clin. Investigation, Am. Thoracic Soc. Achievements include research in chest disease, pulmonary physiology, physiologic aspects of respiratory disease. Office: U Man, 753 McDermot Ave, Winnipeg, MB Canada R3E 0W3*

ANTHONY, DAMON SHERMAN, biologist, researcher; b. Hazleton, Pa., Sept. 28, 1960; s. George Sherman and Sophie Effie (Popovich) A. BS in Biology, King's U., 1986; MS in Biology, East Stroudsburg Coll., 1992. Order entry clk., then asst. technician Connaught Labs., Inc., Swiftwater, Pa., 1986-88, technician, then rsch. asst., 1988-92, clin. rsch. assoc., 1992—. Mem. AAAS, Am. Soc. Microbiology, Parasitology Soc. Republican. Lutheran. Achievements include development of influenza-bearing liposome vaccine currently in human trial. Office: Connaught Labs Inc Rte 611 Box 187 Swiftwater PA 18370-0187

ANTHONY, ETHAN, architect; b. Iowa City, Oct. 14, 1950; s. Frank and Carol (Kessler) A.; m. Luz Eugenia, Feb. 18, 1984; children: Winston Eugene, Alexandra Luce, Edward Rey. Student, Boston Arch. Ctr., 1971-77; BArch, U. Oregon, 1980. Project architect Payette Assocs., Boston, 1980-83; prin. Anthony Assocs., Boston, 1983-90; ptnr. Hoyle, Doran and Berry, Inc., Boston, 1991—; cons. architect Springfield (Vt.) Hosp., 1984—; instr. design Roger Williams Coll., Bristol, R.I., 1984-89; thesis advisor Boston Archtl. Ctr., 1985-87; speaker in field. Mem. Nat. Trust for Hist. Preservation. Mem. AIA, S.Am. Exploreres Club, U. Oreg. Alumni Assn. Avocations: painting, history. Home: 427 Elm St Concord MA 01742 Office: Hoyle Doran and Berry Inc 585 Boylston St Boston MA 02116-3609

ANTHONY, FRANK ANDREW, biochemist; b. Arlington, Va., Nov. 30, 1957; s. David Banks and Frances Stowe (Hart) A. BS, Tex. Christian U., 1979, PhD, 1984. Teaching asst. Tex. Christian U., Ft. Worth, 1980-82; rsch. fellow St. Jude Children's Rsch. Hosp., Memphis, 1984-87, rsch. asst., 1987-89; biologist II Schering-Plough Health Care Products, Memphis, 1989-91, sr. biologist, 1991—; adj. faculty Memphis State U., 1991—; radiation safety com. Schering-Plough Health Care Products, Memphis, 1992—. Contbr. articles to profl. jours. Mem., treas., past pres. Germantown (Tenn.) Symphony Orch., 1985—; mem. Rivercity Community Band, Memphis, 1987—. Recipient Nat. Rsch. Svc. award NIH, 1984-87, Chemistry of Behavior Rsch. fellowship Tex. Christian U., 1980-83, Shering-Plough Sci. Achievement award 1992; named to Outstanding Young Men of Am., 1985, 87. Mem. Tex. Soc. for Electron Microscopy, Am. Soc. for Photobiology, N.Y. Acad. Scis., Sigma Xi. Home: 3789 Kenwood Memphis TN 38122 Office: Schering Plough Healthcare 3030 Jackson Ave Memphis TN 38151-0001

ANTHONY, GREGORY MILTON, infosystems executive; b. Rochester, N.H., June 22, 1958; s. Milton and Effie K. (Pallas) A. BS, U. N.H., 1980. Sales rep. G.T. Johnson Co., Burlington, Mass., 1981-84; sales rep. Honeywell, Bedford, N.H., 1984-88, sr. sales rep., 1988-90, account mgr., 1990-92. Scoutmaster Boy Scouts of Am., Nashua, N.H., 1986-92. Recipient Silver Beaver award Boy Scouts of Am., 1990, Dist. Award of Merit, 1988. Mem. Assn. of Energy Engrs. Home: 35 Donovan Ct Merrimack NH 03054

ANTHONY, THOMAS RICHARD, research physicist; b. Pitts., June 27, 1941; s. Harry Louis III and Evelyn Gertrude (Fischer) A.; m. Angela Ute Klugert, Jan. 26, 1966; children: Wendy Christine, Jason Wayne. BS, U. Fla., 1962; MS, Harvard U., 1964, PhD, 1967. Rsch. physicist GE Corp. R&D Ctr., Schenectady, N.Y., 1967—. Assoc. editor Diamonds & Related Materials, 1991—; contbr. articles to profl. jours.; patentee in field. With USAF, 1959-63. Recipient IR. 100 award IRD mag., 1977, U.S. Patent and Commerce Assn. medal, 1990, John A. Thornton Meml. award Am. Vacuum Soc., 1992, Best Product award Popular Sci., 1992, Best Product award Bus. Week, 1990; Coolidge fellow GE, 1978; NAE fellow, 1990. Mem. Materials Rsch. Soc. (David Turnbull Lectureship award 1992), Phi Beta Kappa, Phi Kappa Phi. Republican. Office: GE Corp R&D Ctr Bldg K-1 Rm 586 PO Box 8 1 River Rd Schenectady NY 12301

ANTOCH, ZDENEK VINCENT, electrical engineering educator; b. Prague, Czechoslovakia, Oct. 16, 1943; came to U.S., 1950; s. Zdenek Antoch and Marta (Smidova) Frank; m. Maureen O. Shaw, June 24, 1968 (div.); 1 child, Anna Marie. BS, Portland State U., 1971, postgrad. in Engring., 1971-73, postgrad. in Physics, 1973-75, MS, 1989, postgrad., 1989—. Research asst. Portland (Oreg.) State U., 1972-75; electronics instr. Portland (Oreg.) Community Coll., 1975-80, 81—; part time instr. Portland (Oreg.) State U., 1989—. Mem. IEEE, Am. Soc. Engring. Edn. Democrat. Avocation: sailing. Office: Portland Community Coll 12000 SW 49th Ave Portland OR 97219-7197

ANTOLOVICH, STEPHEN DALE, engineering educator. Dir., mechanical properties lab. Georgia Inst. Tech., Atlanta. Recipient M. Eugene Merchant Mfg. medal ASME/SME, 1990. Office: Ga Inst Tech Mechanical Properties Rsch Lab 778 Atlantic Dr Atlanta GA 30332-0245

ANTON, WALTER FOSTER, civil engineer; b. Alameda, Calif., Jan. 24, 1936; s. Leroy Fredrick and Dorothy Belle (Siebe) A.; m. Kristin Ingrid Palmquist, June 15, 1957 (div. 1984); children: Mark, Julie, Elisabeth; m. Diane Elisabeth Stern, Sept. 29, 1984; stepchildren: Marc, Eric. BCE, U. Calif., Berkeley, 1957, MCE, 1962. Registered profl. engr., Calif., Wash., Oreg., Ill. From asst. to asst. gen. mgr. Kaiser Engrs., Oakland, Calif., 1959-61; from project engr. to asst. to v.p. Harza Engring. Co., Chgo., 1962-69; project mgr. Tudor Engring. Co., San Francisco, 1969-71, v.p., mgr. water resources, 1981-85; dir. engring., asst. gen. mgr. and chief engr. East Bay Mcpl. Utility Dist., Oakland, 1971-81; div. dir. and chief engr. Seattle Water Dept., 1986—. Pres. bd. dirs. Vols. Am., Seattle, 1987-88; v.p. bd. dirs. Salem Luth. Home, Oakland, 1978-84. Lt. USN, 1957-59. Fellow ASCE (pres. San Francisco sect., chair nat. water policy com., tech. coun. lifeline earthquake engring., dir. Seattle sect., H.J. Brunnier award 1984); mem. Am. Water Works Assn., Am. Acad. Environ. Engrs., U.S. Com. on Large Dams. Presbyterian.

ANTONACCI, ANTHONY EUGENE, food corporation engineer; b. Sept. 21, 1949; s. Salvatore Natali and Odile Estella (Stanton) A.; m. Sherry Lee Kessler, Mar. 6, 1971; children: Don Warren, Lance Anthony. Student U.S. Air Force Acad., 1968-69; Assocs. in Sci., Forest Park Coll., St. Louis, 1971. Lic. power engr. Asst. supr. data processing ops. 1st Nat. Bank, St. Louis, 1969-71; engr. Installation and Service Engring. (Mech. and Nuclear) div. Gen. Electric Corp., St. Louis, 1971-76; engr. Anheuser-Busch Corp., St. Louis, 1976—; software author. Trustee, treas. Antonette Hills Trusteeship, Affton, Mo., 1976-80. Recipient Spl. Performance awards Gen. Electric Co., 1972, 74. Mem. Brewers and Maltsters Local 6 (del. 1982, 83), Nat. Aerospace Edn. Council, Apple Programmers and Developers Assn., Am. Legion. Republican. Roman Catholic. Avocations: classic auto restoration, music (trumpet). Home: 8971 Antonette Hills Dr Saint Louis MO 63123-6503

ANTONIOU, PANAYOTIS A., surgeon; b. Patras, Achaia, Greece, Feb. 17, 1962; s. Antonios Antoniou and Maria Flogera; m. Argiro Paloumpi, May 17, 1986. MD, Nat. U. Athens (Greece), 1986. Resident surgery Army Share Fund Hosp., Athens, 1986-89; resident surgery Air Force Gen. Hosp., 1990-91, resident urology, 1991; pvt. practice Athens, 1991-92; rural community work Korinthos Gen. Hosp., 1992-93; resident in surgery Air Force Gen. Hosp., 1993—. Contbr. articles to med. jours., 30 presentations to med. congresses. Mem. European Assn. Endoscopic Surgery, Soc. for Minimally Invasive Therapy, Internat. Gastro-Surg. Club, Greek Assn. Endoscopic Surgery and other Interventional Techniques, Assn. Med. Grads. Athens U., Hellenic Action Against Cancer. Orthodox. Home: 25th Martiou 28, Pefki, GR 15121 Athens Greece

ANTONS, PAULINE MARIE, mathematics educator; b. Monticello, Iowa, Jan. 15, 1926; d. Henry and Eliza (Zimmerman) Tobiason; m. Richard William Antons, Aug. 13, 1950; children: Sharon Kay, Karen Lyn. BS, U. Dubuque, 1948. Cert. secondary tchr., Iowa. Tchr. math. Elkader (Iowa) Community Sch., 1948-50, Onslow (Iowa) Ind. Schs., 1950-60, Midland Community, Wyoming, 1960-90, Kirkwood Coll., Cedar Rapids, Iowa, 1982—; mem. scholarship adv. bd. Jones County Health Assn., Anamosa, Iowa, 1983—; asst. soil commr. Iowa, County Soil and Water Dist. Asst. commr. County of Jones; treas. Evang. Luth. Ch. Women. asst. soil commr. Iowa County Soil and Water Conservation Dist.; mem. exec. com, treas. Wayne Zion Evang. Luth. Ch. Women. Recipient Pres. award for excellence, 1988; Pres.'s scholar U. Dubuque, 1945-48, NSF scholar Drake U., 1967, Clarke Coll., 1968, U. Iowa, 1969. Mem. Altrurian Study Club (officer, Outstanding Club Woman award 1961), Jones County Conservation Edn. Coun. (Conservation Tchr. award, asst. soil commr.), Delta Kappa Gamma (past officer). Lutheran. Avocations; gardening, reading, travel. Home and Office: RR 1 Box 3 Center Junction IA 52212-9702

ANTTILA, SAMUEL DAVID, environmental scientist; b. New Castle, Pa., Nov. 5, 1942; s. William Samuel and Mildred Francis (Dengler) A.; m. Mary Ellen Freeman, Oct. 22, 1966; 1 child, Yetive Ann Anttila Himes. BS in Forestry, Youngstown State U., 1975; AS in Mech. and Elec. Tech., C.C. USAF, 1992. Gen. foreman environ. control Sharon Steel Corp., Farrell, Pa., 1975—; bd. dirs. Sharon Steel Mgmt. Club, 1989—. Ch. coun. Christ Luth. Ch., New Castle, 1978-91; bd. dirs. Sharon Steel Credit Union, 1989—. Sgt. USAFR, 1960-66, Vietnam. Decorated D.S.M., Presdl. citation. Mem. VFW, Eagles, Moose. Democrat. Home: RD 1 Box 263 A Edinburg PA 16116 Office: Sharon Steel Corp PO Box 291 Sharon PA 16146-0291

ANTUNES, CARLOS LEMOS, electrical engineering educator; b. Lisbon, Portugal, Dec. 29, 1951; s. Arlindo Lemos and Carolina (Afonso) A.; m. Maria Teresa Guimaraes; children: Daniel, André, Ricardo. Diploma in engring., U. Porto, Portugal, 1974; PhD, U. London, 1981; diploma (hon.), U. Coimbra, 1981; PhD, U. Coimbra, Portugal, 1982. Jr. lectr. U. Coimbra, 1974-78, auxiliar prof., 1982-85, assoc. prof., head elec. engr. dept., 1985-87, prof. elec. engring., 1991—, dir. dept. elec. engring. lab., 1985—; gen. mgr. Enaco, Ltd., Coimbra, 1988—; sci. dir. Frapil Ltd., Aveiro, Portugal, 1982-85; chmn., sci. reviewer various confs. in field. Contbr. articles to profl. publs. Lt. Portuguese Army. Grantee Found. Gulbenkian, 1978-81, INIC, 1978-81. Mem. Ordem dos Engrs. Roman Catholic. Avocations: football, music, tennis. Home: Picoto dos Barbados, 3000 Coimbra Portugal Office: U Coimbra Dept Elec Engring, Largo Marques de Pombal, 3000 Coimbra Portugal

ANTZELEVITCH, CHARLES, research center executive; b. Israel, Mar. 25, 1951; came to U.S., 1959; s. Chaim and Frida (Hassman) A.; m. Brenda Reisner, June 24, 1973; children: Daniel Avi, Lisa Rachel. BA, Queens Coll., 1973; PhD, SUNY, Syracuse, 1977. Postdoctoral fellow Masonic Med. Rsch. Lab., Utica, N.Y., 1977-80, rsch. scientist, 1980-83, sr. rsch. scientist, 1984, dir., 1984—; asst. prof. SUNY Health Scis. Ctr. Pharmacology, Syracuse, N.Y., 1980-83, assoc. prof., 1983-86; prof. of Pharmacology SUNY Health Scis. Ctr., Syracuse, N.Y., 1987—. Mem. editorial bd. Jour. Cardiovascular Electrophysiology, 1990, NASPETAPES; contbr. articles to profl. jours. Bd. dirs. Clin. Med. Network, Utica, 1987, Jewish Community Ctr., Utica, 1987-92, Royal Arch Masons Med. Rsch. Found., 1989, Cen. N.Y. Heart Assn., 1989, Temple Beth El, Utica, 1991—; com. mem. N.Y. State Heart Assn., Syracuse, 1982-87l; mem. instnl. rev. bd. Faxton Hosp., Utica, 1990—. Recipient Van Horne award Cen. N.Y. Heart Assn., 1981-84, numerous grants 1986—; Gordon K. Moe scholar, 1987—. Mem. AAAS, Am. Heart Assn., N.Y. Acad. Scis., Internat. Soc. for Heart Rsch., Cardiac Electrophysiol. Soc., N.Am. Soc. Pacing and Electrophysiology. Lodges: Masons. Avocation: swimming. Office: Masonic Med Rsch Lab 2150 Bleecker St Utica NY 13504*

ANYANWU, CHUKWUKRE, alcohol and drug abuse facility administrator; b. Ogbor-Ugiri, Nigeria, Apr. 14, 1943; came to U.S., 1963; s. Peter Ebo and Eunice Ikwuaha (Madu) A.; m. Ngozi G. Nwaike, Jan. 10, 1980; children: Okechukwu-Pat, Adaku Cathy, Ikechukwu-Uzo, Uremegbulem, Kingsley-Ugo, Ucheckukwu. BS in Biology and Chemistry, St. Joseph's Coll., 1971; MS in Biochemistry, Fairleigh Dickenson U., 1972; postgrad., Temple U., 1979; MD, Cetec U., Dominican Republic, 1981. Postdoctorate Temple Hosp.; diplomatic envoy Nigeria, 1973-75; extern various hosps., Phila. area, 1977-79; obstetrician-gynecologist, cons. Lagos U. Teaching Hosp., Nigeria, 1983-84; cons. psychiatry St. Mary's Hosp., Phila., 1981-82; rsch. nuclear medicine Temple U. Hosp., Phila., 1980-81; chmn. A-B Assocs. Inc., Phila., 1970—; chief exec. officer, owner, founder A-B Assocs., Inc., Phila., 1989—; virolog rsch. A-B Assocs. Inc., Phila., 1979-82, owner, chief exec. officer, dir.; mem. staff dept. of psychiatry JFK Mental Health/Retardation, Phila., 1985-88; mem. staff dept. of drug and alcohol addiction Giuffré Med. Ctr., 1988-89; counselor in psychiatry Misericordia Hosp., Phila., 1987-88; mem. staff addiction svcs. Guiffre Med. Ctr., Phila., 1988—; founder, chief exec. officer AB Assocs. Am. Beats Addiction, Inc., Phila., 1989—; peer rev. cons. NIH, Alcohol, Drug Abuse and Mental Health Adminstrn. Author numerous poems; contbr. articles profl. jours. Senate candidate Imo State Govt., Nigeria, 1983; mem. free standing steering com., pub. policy com. & providers com. Healthy Start Initiatives-Phila. Dept. Pub. Health; Olympian athlete competing in pole vault, 1500 meters and 400 meter hurdles, Mex., 1968. Mem. Am. Coll. Healthcare Execs., Pa. Cert. Addiction Counselors, Orgn. Nigerian Profs. USA (chmn. jud. com.), Fedn. Police Law Enforcement, Phila. Fraternal Order Police, Interagency Coun. Homeless. Democrat. Roman Catholic. Office: America Beats Addiction Inc PO Box 38127 Philadelphia PA 19140

AOKI, ICHIRO, theoretical biophysics educator, researcher; b. Takefu, Fukui-ken, Japan, Sept. 15, 1935; s. Shoichi and Fumiko (Yamazaki) A.; m. Atsuko Ookawara, Dec. 5, 1970; children: Kyoko, Keiko. BS, Kyoto (Japan) U., 1958, MS, 1960, DSc, 1988. Assoc. prof. theoretical biophysics Osaka Med. Coll., Takatsuki, Japan, 1965—. Contbr. articles to profl. jours. Grantee Yukawa Meml. Found., 1963-65, Japan Soc. for Promotion Sci., 1986-87, 88. Mem. Internat. Soc. Ecol. Modelling (editorial bd. 1989—). Home: 10-604 Ginkakujimae-cho, Sakyo, Kyoto 606, Japan Office: Osaka Med Coll, 2-41 Sawaragi-cho, Takatsuki Osaka 569, Japan

AOKI, JUNJIRO, chemical engineer; b. Japan, Feb. 17, 1910; s. Ihei and Sada (Nohara) A.; B.Eng., Waseda U., 1934; m. Sumie Takanoha, Dec. 3, 1939; children: Hajime, Minoru, Arata. Head research tech. research and devel. dept. Fujikura Rubber Works Co., Tokyo, 1934-64; head staff KPE Co., Tokyo, 1965-75; dir. Chubu Kogyo Co., Nagoya, Japan, 1965-77; chem. engr. Aoki Chem. Lab., Tokyo, 1965—; adviser Nitto Shoji, Fujikagaku Shi, Takada Co., Mikasa Communication Parts Co., Japan Unisys Supply Co., Courier Internat. Corp.; Nippon Sheet Glass Co., Nippon Steel Co. Served with Japanese Army, 1944-45. Mem. Adhesion Soc. Japan, Japanese Assn. Leather Tech., Japan Soc. Colour Material, Soc. Rubber Industry Japan, Soc. Powder Tech. Japan, Soc. Surface Sci. Japan, Soc. Polymer Sci. Japan, Sampe Japan, Soc. Complex Sci. Japan. Club: Tokyo Chofu Lions. Research on polyamid and polyurethane artificial leather, coupling agts. application in inorganic and organic composites, magnetic tapes and desk, tonner and carrier, color hard copy and print. Home and Office: 4 20 8 Higashi Nogawa, Komae City, Tokyo 201, Japan

AOKI, KEIZO, chemistry educator; b. Toyohashi, Aichi, Japan, Nov. 11, 1941; s. Isoshi Ito and Teru Aoki; m. Shizuko Kato, Sept. 1, 1976; children: Hirokazu, Michiko. B Engring., Nagoya U., 1964, M Engring., 1966, D Engring., 1969. Rsch. assoc. Nagoya (Japan) U., 1969—. Mem. Chem. Soc. Japan, Am. Chem. Soc., Royal Soc. Chemistry. Avocations: hiking, music. Office: Nagoya U, Furo-cho, Chikusa-ku, Nagoya 464-01, Japan

AOKI, MASAMITSU, chemical company executive; b. Ohshima, Yamaguchi, Japan, Dec. 19, 1945; s. Sadao and Toshiko (Toyotake) A.; children: Mariko, Masahiro, Kazuhiro, Akemi. B in Chemistry, Kyushu Inst. of Technoloy, Tobata, Japan, 1968. Chem. engr. Toshiba Corp., Kawasaki, Japan, 1968-74; group leader Toshiba Chem. Corp., Kawasaki, 1974-76, asst. tech. mgr., 1977-82, tech. mgr., 1982-85; tech. mgr. Toshiba Chem. Corp., Kawaguchi, Japan, 1985-87; sr. specialist Toshiba Chem. Corp., Kawasaki, 1987-91; engring. adminstrv. mgr. Toshiba Chem. Corp., Tokyo, 1991-92, sr. mgr., 1992—; cons. Japan Printed Circuit Assn., 1987-88; mem. Underwriters Lab. Industry Adb. Group, 1992—. Author: Electronic Packaging Technology, 1985, Advanced Technology for PCB, 1987, PCB Technical Handbook, 1987, Electronic Engineering, 1989, SMT Handbook, 1990, Polymide Resin, 1991, High Density PWB Technology, 1991, Printed Circuit Technical Handbook, 1993. Mem. Inst. for Interconnecting and Packaging Electronic Circuits, Japan Thermosetting Plastics Industry Assn., Japan Inst. Printed Circuit (Pres. award 1992). Home: Maeda Heights 1141, 511-2 Maeda-cho, Totsuka-ku Yokohama 244, Japan Office: Toshiba Chem Corp, 3-3-9 Shimbashi, Minatoku Tokyo 105, Japan

AONUMA, TATSUO, management sciences educator; b. Nagano, Japan, Aug. 17, 1933; s. Kiyoshi and Kazuyo (Tokunaga) A.; m. Mami Matsuki, Apr. 8, 1961; children: Hidehiro, Keiko. BS, Tokyo U. Edn., 1956; MS, Tokyo Inst. Tech., 1958, PhD in Ops. Rsch. and Mgmt. Sci., 1989. System analyst Mitsubishi Oil Co., Tokyo, 1958-63; assoc. prof. Kobe (Japan) U. Commerce, 1963-70, prof., 1970—, dir. info. system ctr., 1988-92. Author: Theory of Mathematical Programming, 1973. Fellow Ops. Rsch. Soc. Japan; mem. Inst. Mgmt. Sci., Spl. Interest Group in APL, Assn. for Computing Machinery. Avocations: tennis, computer language. Home: 2-1-23-503 Takakura-cho, Suma, Kobe Hyogo 654, Japan Office: Kobe U Commerce Mgmt Scis, Nishi, Kobe Hyogo 651-21, Japan

AOYAMA, HIROYUKI, structural engineering educator; b. Shinjuku, Tokyo, Japan, July 14, 1932; s. Hidesaburo and Sadako (Nishimura) A.; m. Kikuko Sugiura, Apr. 16, 1960; children: Masako, Nobuyuki. B in Engring., U. Tokyo, 1955, M in Engring., 1957, DEng., 1960. Registered first class architect. Lectr. U. Tokyo, 1960-64, assoc. prof., 1964-78, prof., 1978-93; prof. Nihon U., 1993—; vis. rschr. U. Ill., Urbana, 1961-63, vis. prof., 1971-72; vis. prof. U. Canterbury, Christchurch, N.Z., 1980-81. Fellow Am. Concrete Inst., New Zealand Nat. Soc. Earthquake Engring.; mem. ASCE, Archtl. Inst. Japan (award 1976), Japan Concrete Inst. (award 1975), Japan Soc. Civil Engrs. Home: 4-2-13 Takadanobaba, Shinjuku-ku, Tokyo 169, Japan Office: Aoyama Lab, Bunkyo-ku, 1-13-14 Sekiguchi, Tokyo 112, Japan

APEL, JOHN PAUL, civil engineer; b. Columbus, Ohio, May 19, 1932; s. John Henry and Helen Mackenzie (Aitken) A.; m. Nancy Elizabeth Stephenson; children: John Andrew, Janice Elizabeth, Carol Ann, Daniel Eric. BS in Civil Engring., Ohio State U., 1956. Registered profl. engr., Ohio, land surveyor, Ohio. Engr. Thomas Engring. and Surveying Co., Columbus, 1956-66; civil engr. Columbus & So. Ohio Electric Co., 1966-69, supr. civil engring., 1969-71, staff asst. environ. programs, 1971-74, mgr. environ. div., 1974-76, v.p. environ., 1976-80; v.p. regulatory affairs dept. fuel supply Am. Electric Power Svc. Corp., Lancaster, Ohio, 1980-83, v.p. govtl. affairs dept. fuel supply, 1983—. Mem. steering com. City of Columbus Comprehensive Plan. Capt. U.S. Army Corps Engrs., 1956-65. Recipient Outstanding Civil Engring. Alumni award Ohio State U. Civil Engring. Alumni Assn., 1992. Fellow ASCE (officer 1966-71); mem. Am. Ground Water Trust (bd. dirs. 1986-92), Nat. Coal Assn. (bd. dirs. 1990-92). Lutheran. Achievements include patent for tower design. Home: 2801 Zollinger Rd Columbus OH 43221

APELIAN, DIRAN, materials scientist, provost; b. Cairo, Egypt, Oct. 28, 1945; s. Jack Hagop and Santina (Zarmanian) A.; m. Seta Masseredjian, Aug. 21, 1975; children: Teni, Lara. BS, Drexel U., 1968; ScD, MIT, 1972. Teaching asst. dept. metallurgy and materials sci. MIT, Cambridge, mass., 1969-72; product rsch. Bethlehem (Pa.) Steel Corp., 1972-75; vis. prof., asst. prof. dept. materials engring. Drexel U., Phila., 1975-76, 76-79, assoc. prof., 1979-83, prof. dept. materials engring., dept. head, 1983-87, Howmet prof. materials engring., 1987-90, assoc. dean coll. engring., 1987-89, assoc. v.p. acad. affairs/grad. studies, 1989-90; provost, Howmet prof. materials engring. Worcester (Mass.) Poly. Inst., 1990—; chair Blue Ribbon Task Force, Worcester Poly. Inst., 1992—, acad. planning and student affairs, 1990—, inst. budget com., 1990—, com. on governance, 1990—, fin. and adminstrv. policy com., 1990—. Contbr. numerous articles to profl. jours. Recipient Champion H. Mathewson Gold medal, 1992, Kabakjian award, 1990, Scientific Merit award, 1990, Howe medal for Best Papers in Metall. Transactions, 1989, Howmet Chair Professorship, Drexel U., 1987, Howard Taylor Gold medal, 1987, Univ. Rsch. Scholar award, 1987, Lindback Teaching award, 1985, Am. Soc. of Metals' Bradley-Stoughton award, 1980, DOW Outstanding Young Faculty award, 1979. Mem. AAAS, AICE, Am. Powder Metallurgy Inst., Am. Foundrymen's Soc., Am. Inst. Metall. Engrs., The Metals Soc., The Iron and Steel Soc., Am. Soc. Metals, Am. Soc. Engring. Educators, Hist. Metallurgy Soc., Materials Rsch. Soc., Nat. Assn. Scholars, Societe Francaise de Metallurgie, Sigma Mu Epsilon, Sigma Xi, Alpha Sigma Mu, Tau Beta Pi. Home: 15 Regent St Worcester MA 01609 Office: Worcester Poly Inst 100 Institute Rd Worcester MA 01609

APELIAN, VIRGINIA MATOSIAN, psychologist, assertiveness training instructor, lecturer, consultant; b. Yoghun-Oluk, Turkey, Dec. 3, 1934; came to U.S., 1950; d. Hagop M. and Christina (Atamian) Matosian; m. Henry M. Apelian, Apr. 4, 1959; children: Arminée, Gregory, Christopher and David (twins). AA in Liberal Arts, Union County Coll., 1973; BA in Psychology, Douglass Coll., Rutgers U., 1975. From clk. to v.p. N.J. Bank & Trust Co., Paterson, N.J., 1955-59; freelance artist, 1966—; adminstrv. aide dist. 22 N.J. State Assembly, Clark, Trenton, N.J., 1976-78; elected councilwoman Township Coun., Clark, 1978; office supr. Electronic Corp. Am., Springfield, N.J., 1979-81; pres. Township Council, Clark, 1979-82; dir. sr. citizens group Union County Dept. Parks, Clark, 1983-84; coord. youth programming Cen. Presbyn. Ch., Summit, N.J., 1984-85; tchr., cons. Union County Adult Edn. System, Clark, 1975-87; psychologist, assertiveness trainer, lectr. Union County Coll., Cranford, N.J., 1987—; elected to bd. govs. Union County Coll., 1992; mem. Juvenile Conf. Com., Clark, N.J., 1983-86, 90—, chair, 1990-93, N.J. Coun. for the Social Studies, 1990-93; lectr. in field. Co-editor The Presbyn., 1987-90; contbr. articles to profl. jours. Apptd. mem. Ethnic Adv. Coun., State of N.J., 1992—; tchr. Christian edn. various Presbyn. chs., Clark and Paterson, 1955-91; mem., tchr., budget and evangelism com. mem., pulpit nominating com., mediator counselor, chmn. mission and Christian edn. coms. Osceloa Presbyn. Ch., Clark; Christian edn. commr. Fanwood (N.J.) Presbyn. Ch.; liaison Elizabeth Presbytery and Armenian Missionary Assn. Am. for Armenian Earthquake Relief Fund of Dec. 1988 Catastrophe; co-chair Heart Fund Drive, Clark, 1977; chair and coord. cancer crusade Am. Cancer Soc., Clark, 1978; legis. com., chair local pub. schs. and high schs., Clark, 1975-80; lectr. on drug abuse at local schs., Clark, 1975-80; lectr. on human rights issues; mem. pub. rels. coord., lectr. fund raiser Mayor's Com. on Drug Abuse, Clark, 1973-80; mem. Clark Little League Aux., 1970-80; mem., research chair environ. health bd. Union County, 1983-84, consumer affairs bd., 1980-84; mem., sec. Union County Juvenile Adv. Bd., Elizabeth Superior Ct., 1983-86, 90—, elected chair, 1990; trustee Clark Pub. Libr. Bd., 1988-89; commr. Christian edn. dept. Fanwood (N.J.) Presbyn. Ch.; bd. govs. Union County Coll. Recipient spl. plaque Mayor and Township Council, Clark, 1982, Disting. Alumna award Union County Coll., 1983; named in N.J. State Assembly resolution in honor of continued disting. svc. to the community, 1984. Mem. Profl. Women's Assn. (spl. cert. recognition 1988), N.J. Assn. for Elected Women Ofcls. (charter, treas., 1st v.p.), Armenian Relief Soc. (adv. bd. mem. 1955—, sec. 1955-60, pres. 1960-70, Gold Pin 1985), N.J. Coun. for the Social Studies, Gov.'s Ethnic Adv. Coun. N.J., Clark Hist. Soc. (cultural programs 1980—, pres. 1993—), Union County Coll. Alumni Assn. (pres. 1993—), Clark Rep. Club (v.p. 1970-75, pres. 1976-82). Republican. Presbyterian. Avocations: painting, writing, gardening, interior design, fashion design, travel, meeting people. Home: 85 Rutgers Rd Clark NJ 07066-2729

APELOIG, YITZHAK, chemistry educator, researcher; b. Buchara, USSR, Sept. 1, 1944; arrived in Israel, 1947; s. Israel and Pola (Medalion) A.; m. Zipora Zaltzberg; children: Shai, Noa. BSc in Chemistry, Hebrew U., Israel, 1967, PhD in Chemistry, 1973. Rsch. assoc. Hebrew U., Jerusalem, 1967-74; rsch. fellow Princeton (N.J.) U., 1974-76; asst. prof. Technion, Haifa, Israel, 1976-82, assoc. prof., 1982-88, prof., 1988—, chair, 1993—; chmn. 10th Internat. Union Pure and Applied Chemistry Conf., Israel, 1990. Author: (with others) Theoretical Calculations of Organoslicon Compounds, 1990; editorial bd. Jour. Chem. Soc., 1991—; contbr. numerous articles to profl. jours. Sgt.-Maj. Paratrooper, 1962-64, Israel. Recipient Deutscher Akedemischen Austauschdienst award, Germany, 1979, 85, 91, Japan Soc. for Promotion of Sci. award, 1991. Mem. Am. Chem. Soc., Israel Chem. Soc., Royal Soc. Chemistry. Home: 17 Shazar St, Haifa 34861, Israel Office: Technion, Technion City, Haifa 32000, Israel

APLAN, FRANK FULTON, metallurgical engineering educator; b. Boulder, Colo., Aug. 11, 1923; s. Frank Fulton Sr. and Helen Elizabeth (Fischer) A.; m. Clare Marie Donaghue, July 30, 1955; children: Susan M., Peter D., Lucy A. BS, S.D. Sch. Mines and Tech., 1948; MS, Mont. Sch. Mines, 1950; ScD, MIT, 1957; hon. degree in mineral engring., Mont. Coll. Mineral Sci. and Tech., 1968. Mill engr. Climax Molybdenum Co., Climax, Colo., 1950-51, 53; asst. prof. U. Wash., Seattle, 1951-53; sr. scientist Kennecott Copper Corp., Salt Lake City, 1957; group mgr. mineral engring. R & D Mining and Metals div., Union Carbide Corp., Niagara Falls, Tuxedo, N.Y., 1957-67; prof. metallurgy and mineral processing Pa. State U., University Park, 1968—, Disting. prof., 1990, head dept. mineral preparation, 1968-71; chmn. mineral processing sect. Pa. State U., University Pk., 1971-77; chmn. metallurgy sect. Pa. State U., University Park, 1973-75; bd. dirs. Engring. Found., N.Y.C., 1977-90, chmn. 1985-87. Contbr. articles to profl. jours.; patentee in field. T/Sgt. U.S. Army, 1942-46, ETO. Decorated Bronze Star; recipient Antoine M. Gaudin award Soc. Mining, Metallurgy and Exploration, 1991. Mem. NAE, AIME (hon. mem., Robert H. Richards award 1978, Mineral Industry Edn. award 1992), AICE, Am. Soc. for Metals, Archeol. Inst. Am., Am. Filtration Soc., Am. Chem. Soc., Soc. Mining Engrs. (bd. dirs. 1973-76, chmn. mineral processing divsn. 1972-73, Arthur F. Taggart award 1985, Disting. Mem. award 1978), Minerals, Metals and Materials Soc., Sigma Xi. Home: 432 W Fairmount Ave State College PA 16801-4612 Office: Pa State U Dept Mineral Engring 155 Mineral Scis Bldg University Park PA 16802

APOSTLE, CHRISTOS NICHOLAS, social psychologist; b. N.Y.C., Nov. 14, 1935; s. Nicholas Christos and Maria (Katsaros) A. BS, U. Colo., 1958; postgrad., New Sch. Social Rsch., CCNY, 1959, 69; MS in Social Psychology, U. Md., 1963. Interviewer Columbia U. Sch. Pub. Health, N.Y.C., 1962; rsch. supr. Nat. Opinion Rsch. Ctr., U. Chgo., N.Y.C., 1963; instr. Wagner Coll., S.I., N.Y., 1964, Hunter Coll., Bronx, N.Y., 1964, Hofstra U., Hempstead, N.Y., 1965; asst. prof. SUNY, Albany, 1965-68; founder, dir. rsch. Inst. Temporal and Durational Studies, Albany, 1970—; prin. scientist Booz-Allen Applied Rsch., Ft. Monmouth, N.J., 1968-70; instr. Rutgers U., Newark, 1968; bd. dirs. Effective Advt., Albany; arbitrator Better Bus. Bur., 1981-92. Author: Getting Through College Using Sociological Principles, 1966; editor Indian Sociol. Bull., 1966-69; contbr. articles to profl. jours. Bd. dirs. Albany Colonie C. of C., 1975-85, YMCA, Albany, 1979-82; committeeman Town of Colonie (N.Y.) Dem. Com., 1987-91. NYU fellow, 1963-65. Fellow Am. Sociol. Assn.; mem. N.Y. Acad. Scis., World Assn. Pub. Opinion Rsch. Democrat. Greek Orthodox. Achievements include development of concept of temporal sociology, establishment of influence of Karl Jaspers on Talcott Parsons suggesting possible plagarism; development of belted landing field for airplanes. Office: Inst Temporal/Durational Studies 6 Pine St Albany NY 12207

APOSTOLIDES, ANTHONY DEMETRIOS, economist, educator; b. Salonika, Greece; came to U.S., 1956; s. Demos Demetrios and Kalliopi (Papadourakis) A. BA with honors, U. Cin., 1965; MA, U. Pitts., 1966; PhD, U. Oxford, England, 1970. Economist The Conf. Bd., N.Y.C., 1972-74; assoc. econ. affairs officer UN, Geneva, Switzerland, 1975-78; economist Inst. Internat. Law Econ. Devel., Washington, 1978-79, Jack Faucett Assocs., Chevy Chase, Md., 1979-80; asst. prof., econs. Mary Washington Coll., Fredericksburg, Va., 1981-85; economist III Md. Dept. Health, Balt., 1985-88; asst. prof. Ind. U., South Bend, 1988—. Author: Overseas R & D by U.S. Multinationals, 1976, (with others) Energy Consumption in Manufacturing, 1974, (pamphlet) Public Health Services in Elkhart County, 1989. Capt. U.S. Army, 1970-72. Nu Tone Inc. scholar, Cin., 1962-65; Faculty Summer fellow, Ind. U., 1989. Mem. APHA, Am. Econ. Assn. Midwest Econ. Assn., Am. Assn. for Budget and Program Analysis, Hertford Soc., Omicron Delta Epsilon. Avocation: travel. Office: Ind U 1700 Mishawaka Ave South Bend IN 46634

APPEL, STANLEY HERSH, neurologist; b. Boston, May 8, 1933; married; 2 children. AB, Harvard U., 1954; MD, Columbia U., 1960. Diplomate Am. Bd. Psyciatry & Neurology. Intern medicine Mass. Gen. Hosp., 1960-61; resident neurology Mt. Sinai Hosp., 1961-62; rsch. assoc. Lab. Moleculat Biology NIH, 1962-64; chief rsch. assoc. Sch. Medicine U. Pa., 1965-66, asst. prof., 1966-67; assoc. of neurology Med. Ctr. Duke U., 1964-65, from assoc. prof. to prof. neurology, 1967-77, assoc. prof. biochemistry, 1968-77, chief divsn. neurology, 1969-77; prof. neurology, dept. chmn. Baylor Coll. Medicine, 1977—, chmn. program neurosci., 1977—, dir. Jerry Lewis Neuromuscular Disorder Rsch. Ctr., 1977—. Recipient Rsch. Career Devel. award USPHS, 1965-70. Mem. Am. Acad. Neurology, Am. Soc. Biol. Chemistry, Am. Soc. Clin. Investigation, Am. Neurol. Assn. Achievements include research in molecular neurobiology, neurochemistry, synapse function, muscle membranes and disease. Office: Baylor Coll Medicine J Lewis Neuromuscular Desease Rsch Ctr 6051 Fannin Houston TX 77030*

APPELBAUM, BRUCE DAVID, physician; b. Lincroft, N.J., Apr. 24, 1957; s. John S. and Shirley B. (Wolfson) A. BS in pharmacy, Rutgers Coll., 1980; MS in pharmacology, Emory U., 1983, PhD in pharmacology, 1985; MD, Medical Coll. Ga., 1989. Diplomate Nat. Bd. Med. Examiners. Rsch. assoc. Emory U. Dept. Pharmacology, Atlanta, 1985; resident physician U. Calif. Dept. Psychiatry, Irvine, Calif., 1989—; cons. Avalon Med. Group, Garden Grove, Calif, 1990—; com. mem. quality assurance, Fountainblue Nursing Facility, Anaheim, Calif., 1990—. Contbr. articles to profl. jours. Recipient Nat. Rsch. Svc. award Nat. Inst. Health, 1982-83, Eastern Student Rsch. Forum U. Miami Medical Sch., 1984, Nat. Student Rsch. Forum, 1987. Mem. Am. Medical Assn., Am. Psychiatric Assn., Orange County Psychiatric Soc., Sigma Xi. Democrat. Jewish. Avocations: traveling, photography, bicycling, reading. Home: 18602 Creek Ln Huntington Beach CA 92648-1629 Office: U Calif Irvine Medical Ctr 101 City Dr South Orange CA

APPELL, GEORGE NATHAN, social anthropologist; b. York, Pa., Aug. 8, 1926; s. Louis Jacob and Helen (Pfaltzgraff) A.; m. Laura Wolcott Reynolds, May 25, 1957; children: Laura Parker Appell Warren, Amity Cheney Pfaltzgraff Appell Doolittle, Charity Reynolds Appell Wheelock. AB, Harvard Coll., 1949, MBA, 1952; AM, Harvard U., 1957; PhD, Australian Nat. U., Canberra, 1966. Rsch. assoc. Brandeis U., Waltham, Mass., 1968-78; collaborator Smithsonian Instn., Washington, 1986-87; pres. Borneo Rsch. Coun., Phillips, Maine, 1986—; project dir. Sabah (Malaysia) Oral Lit. Project, 1986—; sr. rsch. assoc. Brandeis U., Waltham, 1979—; fieldwork among the Rungus of Sabah, 1959-60, 61-63, 86, 90, 92, Bulusu Dayak & Punan of Kalimantan Timur, Indonesia, 1980-81; officer Susquehanna-Pfaltzgraff Co., York. Mem. internat. editorial bd. Studies in Third World Socs., 1973—; coord. editor Lang. Atlas, Pacific Area, Canberra, Australia, 1983; editor: The Societies of Borneo, 1976, Modernization and the Emergence of a Landless Peasantry, 1985; author: Dilemmas and Ethical Conflicts in Anthropological Inquiry, 1978; co-editor: Choice and Morality in Anthropological Perspective, 1988. Grantee NSF, Sabah, 1965-66, Am. Coun. Learned Socs.-Social Sci. Rsch. Coun., Sabah, 1968-69, NSF, Kalimantan, Indonesia, 1980-81, Ford Found., Kalimantan, Wenner-Gren Found., Sabah, 1989. Fellow Borneo Rsch. Coun.; mem. Maine Sociol. Assn. (pres. 1973-74), N.E. Anthropol. Assn. (pres. 1976-77), Fund for Astrophys. Rsch. (pres. 1966-70). Taoist. Achievements include research on female and male in Borneo and discovery of the absence of rape and homosexuality in a Bornean society. Office: Borneo Rsch Coun Phillips ME 04966

APPLE, DAVID JOSEPH, ophthalmology educator; b. Alton, Ill., Sept. 14, 1941; s. Joseph Bernard and Margaret Josephine (Bearden) A. BS, Northwestern U., 1962; MD, U. Ill., 1966. Intern and resident in pathology Charity Hosp. La., La. State U., New Orleans, 1966-71; ophthalmology resident U. Iowa, Iowa City, 1977-80; prof. ophthalmology and pathology O'Brien Lab. Ocular Pathology, Tulane U. Sch. Medicine, New Orleans, 1980-81, U. Utah Health Scis. Ctr., Salt Lake City, 1981-88; dir. Ctr. for Intraocular Lens Rsch. U. Utah Sch. Medicine, 1984-88; prof. ophthalmology dept. Storm Eye Inst., Med. U. S.C., Charleston, 1988—, chmn. dept., 1988—, dir. Ctr. Intraocular Lens Rsch., 1988—; assoc. prof. ophthalmology and pathology U. Ill. Eye and Ear Infirmary, Chgo., 1974-76; vis. rsch. prof. eye clinics U. Tuebingen, 1975-77, U. Bonn, U. Munich, 1981, Fed. Republic Germany; lectr. U.K. Intraocular Soc., Turnberry, Scotland, 1989; cons. McGraw Hill Pub. Co., N.Y.C., WHO Prevention Blindness Program. Author: Pathology of the Eye, 1980 (German and Japanese edits.), Ocular Pathology, 4th edit., 1991, Intraocular Lenses, 1989, 7 other books, numerous book chpts. and articles. Recipient Témoignage d'Honneur Can. Implant Soc., Montreal, 1986, Ridley medal, lectr. 6th European Intraocular Coun., Copenhagen, 1988, Choyce medal. Mem. Am. Acad. Ophthalmology (Binkhorst lectr. and medal 1988, Walter Wright lectr., Toronto, 1993), Internat. Acad. Pathology, Am. Assn. Ophthalmic Pathologists, Internat. Soc. Eye Rsch., Theobald Soc., Verhoeff Soc., Lions, Nu Sigma Nu, Phi Kappa Epsilon. Home: 93 E Bay St Charleston SC 29401-2544 Office: Med U SC Storm Eye Inst Dept Ophthalmology 171 Ashley Ave Charleston SC 29425-2236

APPLEGARTH, RONALD WILBERT, engineer; b. Keefeton, Okla., Feb. 17, 1934; s. George Wilbert and Elsie Viola (Farmer) A.; m. Glennie Mabel Graham, June 23, 1957; children: Verna Mae and Leona Faye (twins), Ronald Dwain. AS, Connors State Coll., 1954; student, U. Okla., 1954-56. Clk., sr. clk. Okla. Gas and Electric Co., Oklahoma City, 1958-63, engr. technician, 1963-70, sr. engr. technician, 1970-82, engr., 1982-84, project engr., 1984-87; project engr. Okla. Gas and Electric Co., Norman, 1987—. Mem. Norman Utilities Commn., 1991-93. Mem. NSPE, Okla. Soc. Profl. Engrs. (sec.-treas. Canadian Valley chpt. 1992-93, v.p. membership, pres.-elect Canadian Valley chpt. 1993-94), Engring. Club Oklahoma City. Mem. Ch. of Christ. Office: Okla Gas and Electric Co PO Box H1 Norman OK 73070-7107

APRIGLIANO, LOUIS FRANCIS, metallurgist, researcher; b. Bklyn., Jan. 17, 1950; s. Pasquale Anthony and Margarita (Ingrisano) A.; m. Patricia Anne Lakeman, June 20, 1971; children: Christina, Amy. BS in Metall. Engring., Poly. Inst. N.Y., 1971; ScD, George Washington U., 1987. Metallurgist Carderock div. Naval Surface Warfare Ctr., Bethesda, Md., 1971—. Author govt. reports on high temperature gas turbine blade coatings and superconductors; contbr. articles to profl. jours. Pres. Annapolis Landing Homeowners Assn., Riva, Md., 1989; v.p. parish coun. Our Lady of Perpetual Help Ch., Edgewater, md., 1992. Mem. ASTM, Sigma Xi. Achievements include 6 patents for gas turbine blade coatings and superconductor fabrication, development of corrosion resistant coating for marine gas turbine blades, test method to simulate corrosion found on marine gas turbine blades. Home: 354 Westbury Dr Riva MD 21140 Office: Naval Surface Warfare Ctr Code 612 Annapolis MD 21402

APTER, NATHANIEL STANLEY, psychiatrist, educator, retired, researcher; b. N.Y.C., May 10, 1913; s. Louis Leo and Sadie (Friedman) A.; m. Julia Esther Tutelman, Dec. 13, 1941 (dec. Apr. 9, 1979); children: Marion Hope Apter Quinn, Terry Eve Apter Newbery; m. Valerie Vivienne Van, July 18, 1980; 1 child, Jessie Irene Cooley. AB, Cornell U., 1933; MD, SUNY, Buffalo, 1938. Diplomate Am. Bd. Psychiatry and Neurology. Asst. in psychiatry Johns Hopkins U., Balt., 1942-44; asst. prof. psychiatry U. Chgo., 1946-50, head div. psychiatry, 1950-54; sr. attending and rsch. psychiatrist Michael Reese Hosp., Chgo., 1954-83; C.I. schizophrenic rsch. Manteno (Ill.) State Hosp., 1950-64; prof., lectr. psychiatry U. Chgo., 1954-83; resident adj. prof. marine biology Oceanographic Ctr. Nova U., Ft. Lauderdale, Fla., 1983—; mem., cons. ethics com. APA, Washington, 1976-80; disting. cons. Ill. State Psychiat. Inst., Chgo., 1958-78. Capt. Aus, 1942-44. Fellow AAAS; mem. Ill. Psychiat. Soc. (pres. 1958-59). Office: Nova U Oceanographic Ctr 8000 N Ocean Dr Dania FL 33004

AQUINO, JOSEPH MARIO, clinical psychologist; b. N.Y.C., Nov. 21, 1947; s. Joseph and Rose (Nasi) A.; m. Kathleen Ann Ryan, Oct. 6, 1990. BA in English, So. Ill. U., 1969, MS in Secondary Edn., 1976; PhD in Clin. Psychology, St. John's U., Jamaica, N.Y., 1987. Lic. psychologist, N.Y. Tchr. English Wappingers Cen. Schs., Wappingers Falls, N.Y., 1969-79; intern psychology Maimonides Med. Ctr., Bklyn., 1983-84; specialist in applied behavior sci. Builders for Family and Youth, Bklyn., 1984-85; trainee psychology and psychologist St. Vincent's Svcs., Bklyn., 1984-89; psychologist St. Christopher-Ottilie Svcs., Sea Cliff, N.Y., 1989—; pvt. practice psychology N.Y.C., 1988—; guest lectr. St. John's U., 1990. Co-author: Situational Leadership for Principals, 1983; mem. editorial bd. Jour. Urban Psychiatry, 1982-84; contbr. articles to profl. jours. Recipient citation VFW, Wappingers Falls, N.Y., 1977; Bethany House Achievement award Bethany House II, 1991; psychology teaching fellow St. John's U., 1981; cited in article Emergency mag., 1991. Mem. Am. Psychol. Assn. (div. 12). Office: 10 Rye Ridge Plz Ste 213 Rye Brook NY 10573-2828

ARABYAN, ARA, mechanical engineer; b. Istanbul, Turkey, Mar. 11, 1953; came to U.S., 1979; s. Varujan and Lusanus (Bogosyan) A.; m. Colette Djeredjian, May 1, 1985. BSc, Tex. A&M U., 1980; MSc, U. So. Calif., 1982, PhD, 1986. Asst. prof. U. Ariz., Tucson, 1986-92, assoc. prof., 1992—; panel mem. NSF, Washington, 1989. Reviewer NSF, and jours., 1987—; contbr. articles to profl. jours. Recipient Presdl. Young Investigator award NSF, 1990. Mem. ASME. Office: U Ariz AME Dept Tucson AZ 85721

ARAKAWA, KASUMI, physician, educator; b. Toyohashi, Japan, Feb. 19, 1926; came to U.S., 1954, naturalized, 1963; s. Masumi and Fayuko (Hattori) A.; m. Juen Hope Takahara, Aug. 27, 1956; children: Jane Riet, Kenneth Luke, Amy Kathryn. M.D., Tokyo Med. Coll., 1953; Ph.D., Showa U. Sch. Med., Tokyo, 1984. Diplomate: Am. Bd. Anesthesiology. Intern Iowa Meth. Hosp., Des Moines, 1954-56; resident U. Kans. Med. Ctr., Kansas City, 1956-58, instr. anesthesiology, 1961-64, asst. prof., 1964-71, assoc. prof., 1971-77, prof., 1977—, Arakawa Disting. prof. anesthesiology, 1990; clin. assoc. prof. U. Mo.-Kans. City Sch. Dentistry, 1973—; dir. Kansas City Health Care, Inc. Fulbright scholar, 1954; civilian cons. USAF. Recipient Outstanding Faculty award Student AMA, 1970. Fellow Am. Coll. Anesthesiology; mem. Assn. Univ. Anesthetists, Acad. Anesthesiology (pres. 1986-87), Japan-Am. Soc. Midwest (v.p. 1965, 71). Home: 7917 El Monte St Shawnee Mission KS 66208-5048 Office: Univ Med Ctr 3901 Rainbow Blvd Kansas City KS 66160-7415

ARAKI, HIROSHI, chemist; b. Aichi, Japan, Aug. 19, 1949; s. Sadasuke and Hatsuko A.; m. Reiko Ohta, Apr. 23, 1979; children: Takahiro, Tomohiro. BS, Nagoya U., 1972, MS, 1974, PhD, 1977. Rsch. assoc. N.C. State U., Raleigh, 1978-80; vis. researcher Swedish Pulp and Paper Rsch. Inst., Stockholm, 1980-81; assoc. rsch. scientist Japan Pulp and Paper Rsch. Inst., Tsukuba, 1981-87, prin. rsch. scientist, 1987—. Co-author: Dictionary of Analytical Chemistry, 1992; contbr. articles to profl. jours. Mem. Japan Tech. Assn. Pulp and Paper Industry (publ. com. 1991-92, Japan Tappi award 1991). Achievements include research in paper physics and wood chemistry. Office: Japan Pulp and Paper Rsch, 5-13-11 Tokodai, Tsukuba 300-26, Japan

ARAKI, TAKAHARU, editor; b. Kyoto, Japan, Dec. 22, 1929; came to U.S., 1965; s. Shiro and Kiyo (Ohmori) A.; m. Motoko Yoshizawa, Nov. 23, 1958. MS, Kyoto U., Japan, 1957, DSc, 1961. Rsch. assoc. Kyoto U., Japan, 1960-62; sr. chemist Tekkosha Corp., Mitaka, Tokyo, Japan, 1962-65, 68-70; rsch. fellow U. Minn., Mpls., 1965-67, 70-71; sr. rsch. scientist U. Chgo., 1971-82; sr. rsch. assoc. McGill U., Montreal, Can., 1983-85; sr. assoc. editor Chem. Abstracts Svc., Columbus, Ohio, 1985—. Contbr. articles to profl. jours. Fellow Mineralogical Soc Am.; mem. Am. Ceramic Soc., Am. Chem. Soc., Am. Crystallographic Assn. Home: 97 Fitz Henry Blvd Columbus OH 43214-1611 Office: Chem Abstracts Svc 2540 Olentangy River Rd Columbus OH 43210

ARAKI, TAKEO, chemistry educator; b. Kyoto, Japan, Jan. 10, 1934; s. Choji Tanaka and Fumi Araki; m. Akiko Nakatani, Apr. 9, 1962; children: Makoto, Atsushi. BS, Kyoto U., 1957, MS, 1959, PhD in Sci., 1963. Asst. prof. Osaka (Japan) U., Toyonaka, 1962-66, lectr., 1966-71, assoc. prof., 1971-80; postdoctoral fellow Manchester (Eng.) U., 1967-68; prof. Shimane U., Matsue, Japan, 1980-89; prof. of chemistry Kyoto Inst. Tech., 1989—, dean faculty textile engring., 1992—; vis. prof. Kyoto U., 1990-92; councilor Shimane U., 1984-88, Kyoto Inst. Tech., 1991—. Author, editor: Liquid Membranes, 1990; author 12 books in field; contbr. articles to profl. jours.; patentee in field. Mem. Chem. Soc. Japan (editorial Chemistry and Chem. Industry, 1983-86, Chemistry Letters, 1989-92, conf. chmn. 1987), Polymer Sci. Japan, Am. Chem. Soc. (polymer chemistry div.), Kinki Chem. Soc. Japan, Synthetic Organic Chemistry Japan, Internat. Union Pure and Applied Chemistry (award for poster session 1986). Buddhist. Avocations: photography, breeding butterflies, mountains, music, poetry. Home: 3-5-3 Sakura-ga-oka, Minoo, Osaka 562, Japan Office: Kyoto Inst Tech, Goshokaido, Matsugasaki, Kyoto 606, Japan

ARAL, MUSTAFA MEHMET, civil engineer; b. Ankara, Turkey, Feb. 26, 1945; came to U.S., 1978; s. Faruk and Bedia A.; m. Sevg H. (div. 1991); 1 child, Sinan K. MSCE, Ga. Inst. Tech., PhD. Asst. prof. Mid. East Tech. U., Ankara, Turkey, 1971-76, assoc. prof., 1976-78; assoc. prof. Ga. Inst. Tech., Atlanta, 1978—; bd. scientific advisors ASTDR/DHHS, Atlanta, 1990-92; mem. edit. bd. Jou. Engring. Sci. and Health, Atlanta, 1990-92. Author: Ground Water Modeling in Multilayer, 1990; contbr. articles to profl. jours. Advisor, coach YMCA, Atlanta, 1980-90. NATO fellow, 1974, 78. Mem. ASCE, Am. Geophys. Union, Am. Water Resources Assn., Am. Inst. Hydrology. Office: Ga Tech Inst Sch Civil Engring Atlanta GA 30332

ARAMBEWELA, LAKSHMI SRIYANI RAJAPAKSE, chemist; b. Colombo, Sri Lanka, May 29, 1946; d. Don Elbin and Chandrawathi (Rodrigo) Rajapakse; m. Ranjit Parakrama de Silva Arambewela, 1976; children: Maulee, Chamat. BSc in Chemistry, U. Ceylon, Colombo, 1970; PhD, U. Sri Lanka, Colombo, 1976. Rsch. officer Ceylon Inst. Sci. and Indsl. Rsch., 1972—; coun. mem. Inst. Chemistry Ceylon, 1990—. Contbr. articles to Jour. Chromatography, Phytochemistry, Recent Advance in Natural Products Chemistry, Jour. of Natural Products, Fitoterapia, Jour. Nat. Sci. Coun. of Sri Lanka. Grantee Internat. Found. for Sci., 1982, 88. Fellow Inst. Chemistry Ceylon (treas. 1991); mem. Sri Lanka Assn. Advancement Sci., Am. Soc. Pharmacognosy. Home: Castle St, 14 Sarasavi Ln, Castle St, Colombo 8, Sri Lanka Office: Ceylon Inst Sci Indsl Rsch, PO Box 787, 363 Bauddhaloka Mawatha, Colombo 7, Sri Lanka

ARANGO, RICHARD STEVEN, architect, graphic and industrial designer; b. Bogota, Colombia, June 30, 1953; s. Jorge Arango Sanin and Judith (Wolpert) Arango; m. Maria Francesca Violich, Aug. 1977; children: Ruy Rafael, Antonia. BA in Architecture with honors, U. Calif., Berkeley, 1976, MArch, 1980. Registered architect. Prin. Richard Arango Design, Berkeley, 1976-85, Richard Arango Architects, Coral Gables, Fla., 1986-88, Seckinger Arango Architects, Coral Gables, 1988—; cons. designer Herman Miller Corp., Berkeley, 1984; John K. Branner traveling fellow U. Calif.-Berkeley grad. div., 1980. Designer greeting cards Mus. Modern Art, 1991; contbr. articles to profl. jours. Mem. Bd. Architects, City of Coral Gables, 1988-90. Recipient Commendation, Progressive Architecture Mag., 1982. Mem. AIA (chair architecture in the schs. 1988-90, editor, chair Miami chpt. Newsletter 1990—). Democrat. Unitarian. Office: Seckinger Arango Architects 121 Majorca Ave Miami FL 33134-4508

ARANTES, JOSÉ CARLOS, industrial engineer, educator; b. Itamogi, Brazil, May 10, 1955; came to U.S., 1986; s. Antonio A. and Parizina (Marinzeck) A.; m. Nadia Maria Monti, July 26, 1986. MSc in Indsl. Mgmt., Katholieke U. Leuven, Belgium, 1982; PhD in Indsl. Engring., U. Mich., 1991. Product engr. Kodak Co., Brazil, 1979-81; instr., cons. U. Campinas, Brazil, 1983-86; rsch. asst., teaching asst. U. Mich., Ann Arbor, 1987-90; asst. prof. U. Cin., 1991—; cons. Criminal Justice Task Force, Cin., 1992. Author: Degeneracy in Gereralized Networks, 1990; contbr. to profl. jours. 2d lt. Brazilian armed forces, 1973-75. Grantee Westinghouse Environ., Cin., 1992, County of Hamilton, Cin., 1992, Fernald Environ. Mng. Co., Cinn., 1993. Mem. Sci. Rsch. Soc., Inst. Indsl. Engrs., Inst. Mgmt. Sci., Ops. Rsch. Soc. Am. Home: 6939 Lynnfield Ct. #143 Cincinnati OH 45243-1732 Office: U Cin Dept Engring Mail Location 116 Cincinnati OH 45221-0116

ARAUJO, MARCIO SANTOS SILVA, chemical engineer; b. Rio de Janeiro, Jan. 26, 1946; s. Manoel Silva Araujo and Augusta do Carmo S. Silva Araujo; m. Vera Lucias, Nov. 6, 1972 (div. Aug. 1990); children: Marcia Lucia, Carlos Gustavo; m. Vilma Maria Luna S. Silva Araujo, Oct. 6, 1990. Degree in chem. engring., Chemistry Sch., Rio de Janeiro, 1968. Researcher Inst. de Quimico URFJ, Rio de Janeiro, 1968-72; tchr. U. Fed. Fluminense, Rio de Janeiro, 1971-80; lab. chief Demillus, Rio de Janeiro, 1974-76; lab. mgr. CIBA-GEIGY Quimica S.A., Rio de Janeiro, 1976-80; tech. mgr. Temana Produtos Consumo, São Paulo, 1980-83; plant mgr. Leiner Brasil Gelatines, São Paulo, 1983-84; ops. mgr. Stanley Home Products, São Paulo, 1984-86; mfg. mgr. Alba Quimica Industries Com. LTDA, São Paulo, 1986—. With Brazilian Army, 1964-65. Fellow Paineiras Club. Avocations: swimming, jogging, snooker, playing cards, tennis. Home: Oscar Freire St 1961 Apt 81, 05409 São Paulo Brazil Office: Alba Quimica Ind Com LTDA, Rod Raposo Tavares KM 28,5, 06700 Cotia Brazil

ARAVAS, NIKOLAOS, mechanical engineering educator; b. Thessaloniki, Greece, Dec. 6, 1957; m. A. Zerva, Aug. 19, 1990. Diploma mech. engring., Aristoteleion U., Thessaloniki, 1980; MS in Theoretical Applied Mechanics, U. Ill., 1982, PhD in Theoretical Applied Mechanics, 1984. Asst. engr. Deutsche Bundesbahn, Nurnberg, Germany, 1977, Esso-Papas Oil Refinery, Thessaloniki, 1978; teaching asst. U. Ill., Urbana-Champaign, 1980-82, rsch. asst., 1982-84; sr. engr. Hibbitt, Karlsson & Sorenson, Inc., Providence, 1985; asst. prof. U. Pa., Phila., 1986-92, assoc. prof., 1992—; engr. cons. IBM-T.J. Watson Rsch. Ctr, Yorktown Heights, N.Y., 1987—, Braddock, Dunn & McDonald, Ford Aerospace Co., Washington, 1990—; presdl. young investigator NSF, 1987. Contbr. articles to profl. jours. Mem. ASME, Am. Acad. Mechanics, Phi Kappa Phi, Pi Tau Sigma. Achievements include research in fracture mechanics, plasticity, metal forming, computational mechanics. Office: U Pa 220 S 33rd St Philadelphia PA 19104

ARBEIT, ROBERT DAVID, physician; b. Jersey City, Aug. 16, 1947; s. Sidney Robert and Marie (Gluck) A.; m. Susan Abelson, Dec. 20, 1970; children: Jeffrey, Miriam. BA, Williams Coll., 1968; MD, Yale U., 1972. Diplomate Am. Bd. Internat. Medicine, Am. Bd. Infectious Disease. Intern then resident Yale-New Haven Hosp., New Haven, 1972-74; clin. assoc. Nat. Cancer Inst., Bethesda, Md., 1974-76; fellow Sidney Farber Cancer Inst., Boston, 1976-79; staff physician VA Med. Ctr., Boston, 1979—, asst. chief med. svcs., 1989—, dir. infectious diseases rsch., 1991—, assoc. chief of staff, rsch., 1991—; asst. prof. Sch. Med. Boston U., 1979-87, assoc. prof. Sch. Med., 1987—. Contbr. articles to profl. jours. Fellow Infectious Diseases Soc. of Am., Am. Coll. Physicians; mem. Am. Soc. for Microbiology, Phi Beta Kappa, Alpha Omega Alpha. Avocation: personal computers. Office: VA Med Ctr 150 S Huntington Ave Jamaica Plain MA 02130-4820

ARBER, WERNER, microbiologist; b. Gränichen, Switzerland, June 3, 1929; married; 2 children. Ed., Aargau (Switzerland) Gymnasium, Eidgenössische Technische Hochschule, Zurich. Asst. Lab. Biophysics, U. Geneva, 1953-58, docent, then extraordinary prof. molecular genetics, 1962-70; research assoc. dept. microbiology U. So. Calif., 1958-59; vis. investigator dept. molecular biology U. Calif., Berkeley, 1970-71; prof. microbiology U. Basel (Switzerland), 1971, rector, 1986-88. Co-recipient Nobel prize for physiology or medicine, 1978. Mem. Nat. Acad. Scis. (fgn. assoc.). Office: Biozentrum der Universität, 70 Klingelbergstrasse, CH-4056 Basel Switzerland

ARBIN, ASTRID VALBORG, chemist; b. Ström, Sweden, Jan. 31, 1942; d. Anders Mikael and Eva Vilhelmina Mikaelsson; m. Stig Mattias Arbin, Mar. 6, 1971; children: Eva Katarina, Karl-Anders. Pharm. Chemist, Pharm. Inst., Stockholm, 1967; D Pharm., U. Uppsala, 1980. Lab. pharmacist Ctrl. Lab. Sweden Pharm., Solna, 1967-72; mgr. bioanalysis ACO Läkemedel AB, Solna, 1972-80, mgr. analy. chemistry, 1980-88; mgr. analy. chemistry Kabi Vitrum AB, Solna, 1988-91, Kabi Pharmacia AB, Solna, 1991—. Contbr. articles to profl. jours. Mem. Swedish Acad. Pharm. Sci. (sec. bd.), European Pharm. Com Group 16 (Coun. of Euorpe). Achievements include research in bioanalysis, alkylation, analysis, migration, plastics, polymers. Office: Kabi Pharmacia AB, S-112 87 Stockholm Sweden

ARBITELL, MICHELLE RENEÉ, clinical psychologist; b. Trenton, N.J., Oct. 24, 1962; d. John A. and Adele M. (Klama) A.; m. Michael F. Belinc, Aug. 7, 1993. BA, Lehigh U., 1984; MA, Indiana U. of Pa., 1986, D Psychology, 1988. Lic. psychologist, Pa., cert. counselor Office Vocat. Rehab. Clin. pscyhology intern Geisinger Med. Ctr., Danville, Pa., 1987-88; dir. behavioral medicine and neuropsychology Rehab. Hosp. Altoona (Pa.), 1988—; pvt. practice psychology, cons., Altoona, State College, Pa., 1991—; invited lectr. Pa. State U., University Park, 1991—; adv. bd. mem. Blair County Pain Support Groups, Altoona, 1988—; presenter in field. Author: (with others) On Spouse Abuse..., 1985; (with others) Bulimics' Perceptions..., 1991. Mem. APA, Nat. Acad. Neuropsychology, Pa. Psychol. Assn., Soc. Behavioral Medicine, Phi Beta Kappa, Psi Chi (v.p./sec. 1980-84). Avocations: fitness, music, cooking. Office: Rehab Hosp Altoona 2005 Valley View Blvd Altoona PA 16602

ARCA, GIUSEPPE, mathematician, educator; b. Cagliari, Italy, Oct. 23, 1949; s. Sebastiano and Annamaria (Arca) A.; m. Maria Rosaria Contu, Feb. 7, 1988; 1 child, Elena. Laurea in Math., U. Cagliari (Italy), 1971. Asst. prof. U. Cagliari, 1971-73, lectr., 1975-79, assoc. prof., 1982—; researcher U. Warwick, U.K., 1974; maître de conférences Faculté des Sci. U. Sfax, Tunisia, 1980-81; reviewer Math. Reviews, 1982, Zentralblatt für Mathematik, 1981. Author: Lezioni di Algebra Lineare, 1987, Geometria Euclidea Elementare, 1990; contbr. articles to profl. jours. Nat. Rsch. Coun. grantee, 1974, NATO rsch. grantee, 1980. Mem. Am. Math. Soc., Tensor Soc., Nat. Rsch. Coun., Italian Math. Union (Cagliari br. mem. pres. office 1975—). Roman Catholic. Home: Via Cagna 60, 09126 Cagliari Italy Office: U Cagliari, Dept Math Fac Eng, Viale Merello 93, 09123 Cagliari Italy

ARCE, PEDRO EDGARDO, chemical engineering educator; b. Nogoya, Entre Rios, Argentina, Feb. 27, 1952; came to U.S., 1983; s. Pedro Ismael and Julia Celina (Traverso) A.; m. Maria Beatriz Trigatti, Feb. 9, 1978; children: Maria Paula, Andrea Lucia. Diploma in Chem. Engring., U. Nacional del Litoral, Santa Fe, Argentina, 1977; MSChemE, Purdue U., 1987, PhDChemE, 1990. Grad. fellow Coun. for Sci. Rsch., Santa Fe, Argentina, 1978-84; lectr. Universidad Nacional del Litoral, Santa Fe, Argentina, 1980-87; grad. student Purdue U., West Lafayette, Ind., 1984-90; asst. prof. Fla. A&M U./Fla. State U. Coll. Engring., Tallahassee, 1990—; rsch. staff mem. CONICET, Argentina, 1984-90. Contbr. articles to profl. jours. Recipient Fellowship, CONICET, 1978, 80, 84, Univ. of Queensland, Australia, 1982, Purdue U., 1988. Mem. AAAS, AIChE (student chpt. Prof. of Yr. 1990), Am. Chem. Soc., Am. Phys. Soc., Phi Lambda Upsilon, Sigma Xi. Achievements include rsch. in corona discharge in liquid phase for waste treatment; discovery of pattern formation in catalytic reactors with implications for selectivity and yield improvement; new approaches in computer mathematics; development of The Colloquial Approach teaching technique. Office: FAMU/FSU Coll of Engring Chem Engring Dept PO Box 2175 Tallahassee FL 32316-2175

ARCE-CACHO, ERIC AMAURY, solar energy engineer, consultant; b. Morovis, P.R., Sept. 24, 1940; s. Eduardo and Celia Arce-Cacho; BSEE; m. Carmen Ruth Gonaalez, Nov. 19, 1960; children: Eric Edmaury, Ruth Dagmar. Student, Coll. Engring., Myz, P.R., 1960, U.S. Air Force Adminstrn. Sch., 1963; BBA, InterAm. U., Bay, P.R., 1970, MA in Econs., 1972; M.A., Ch. of God Sch. of Theology, 1985, postgrad. Intercept control tech Dept. of Def., San Juan, P.R., 1963-77, instr. air control scis., 1967-79; safety engring. cons., Bay, 1978-80; cons., researcher Energy Saving Equipment Inc., Bay, 1980—; pres. ESE Reliable Svcs., Inc.; dir., founder World Christian Embassies, 1986. Author: (poems) Soledad, 1979. Editor articles. Mem. Am. Soc. Safety Engrs., Internat. Platform Assn. Clubs: Community (Fort Buchanan, P.R.), Gulf Course, Disabled Am. Vets. Assn., Am. Legion. Avocations: UFO-OVNI enigma sci. rsch. Home: C-17 Forest Hills Bay PR 00959-2000

ARCEO, THELMA LLAVE, energy engineer; b. Manila, Philippines, Apr. 19, 1956; d. Paulino Marasigan and Nellie (Llave) A. BS in Biology, U. Philippines, 1978; MS in Energy Mgmt., N.Y. Inst. Tech., 1987. Cert. asbestos investigator. Rsch. analyst Philippine Nat. Oil Co., Quezon City, Philippines, 1983-84; asst. dir. East Manhattan Sch., N.Y.C., 1985-87; energy specialist/dir. N.Y. Urban Coalition Housing Group, N.Y.C., 1987—; cons. CONSERVE, Inc., N.Y.C., 1991-92, ConEdison Consumer Edn., N.Y.C. Contbr. articles to profl. jours. Mem. AAAS, Assn. of Energy Engrs., Illuminating Engring. Soc. Achievements include rsch. on comml. biogas digester designs, use of Sargassum substrate for biogas prodn. Home: 98-50 67 Ave Forest Hills NY 11374 Office: NY Urban Coalition Housing 99 Hudson St New York NY 10013

ARCHAMBEAU, CHARLES BRUCE, physics educator, geophysics research scientist; b. Chisholm, Minn., Sept. 1, 1933; s. Charles and Amelia (Jackse) A.; m. Eleanor Williams, June 22, 1956 (div. 1973); children: Richard, Annette, Jane. BS, U. Minn., 1955; PhD, Calif. Inst. Tech., 1965. Seismologist United Electrodynamics, Pasadena, Calif., 1962-64, Teledyne Industries, Inc., Alexandria, Va., 1964-66; assoc. prof. geophysics Calif. Inst. Tech., Pasadena, 1966-69, prof., 1969-74; prof. geol. scis. U. Colo., Boulder, 1975-92, prof. physics, 1992—; sr. vis. fellow Coop. Inst. for Rsch. in Environ. Scis. U. Colo./NOAH, 1974-75, mem., 1976-82, rsch. assoc., 1982-92; cons. Systems, Sci. & Software-Maxwell Ind., La Jolla, Calif. and Reston, Va., 1970-92, Amoco Prodn. Rsch. Lab., Tulsa, 1986-89, Sci. Applications Internat., Dept. Energy, Las Vegas, Nev., 1990-91; bd. dirs. Tech. and Resource Assessment Corp., Boulder. Contbr. articles to profl. jours., chpts. to books. Advisor on arms control affairs Parliamentarians for Global Action, N.Y.C., 1984-88, Natural Resources Def. Coun., Washington, 1985-87. MacArthur Found. fellow, 1988. Mem. AAAS, Am. Geophys. Union, Seismol. Soc. Am., Fedn. Am. Scientists (Pub. Svc. award 1986), N.Y. Acad. Scis. Democrat. Roman Catholic. Avocations: art, literature, outdoor sports. Office: U Colo Campus Box 449 Boulder CO 80309

ARCHER, GREGORY ALAN, clinical psychologist, writer; b. Euclid, Ohio, Oct. 27, 1957; Robert Dale and Martha Jane (Heller) A. BS, Ohio State U., 1980; D in Psychology, Wright State U., 1984. Psychologist Cigna Health Plan, Tucson, 1984-90; psychologist, pvt. practice Archer, Searfoss and Assocs. Inc., Phoenix, 1990—; cons. Paradise Valley Schs. Steroid Abuse, Phoenix, 1991, Boys and Girls Clubs Met. Phoenix, 1992. Author: Big Kids: Parents Guide to Weight Control for Children, 1989, Rites of PAssage, 1992; (screenplays) Total Teen Fitness with Dr. A., 1991, NEtworking, 1992. Cons. Big Brothers, Big Sisters, Phoenix, 1992. Recipient Rsch. award Midwest Psychol. Assn., 1980. Mem. SAG. Am. Psychol. Assn., Ariz. Psychol. Assn. Achievements include the development of innovative fitness programs for teens. Home and Office: Archer Searfoss and Assocs 9433 N 19th St Phoenix AZ 85020

ARCHER, STEVEN RONALD, ecology educator; b. Sioux Falls, S.D., Sept. 23, 1953; s. Ronald Franklin and Carol Margurite (Wright) A.; m. Gewndolyn Maureen Allen, Aug. 18, 1975; children: Jordan, Cadie. BA, Augustana Coll., Sioux Falls, S.D., 1975; PhD, Colo. State U., 1983. Rsch. assoc. Augustana Coll., Sioux Falls, 1976-78; grad. rsch. asst. Colo. State U., Ft. Collins, 1978-81, instr., 1981-83; asst. prof. Tex. A&M Univ., Coll. Station, 1983-88, assoc. prof., 1988—; vis. scientist Nat. Ctr. for Atmospheric Rsch., Boulder, Colo., 1991-92. Contbr. articles to Ecology, Am. Naturalist, Oikos, Jour. of Vegetation Sci. Grantee: USDA, Dept. Energy, NSF, NASA. Mem. Internat. Assn. for Vegetation Sci., Soc. for Range Mgmt., Am. Geophysical Union, Ecol. Soc. Am., Wilderness Soc. Office: Tex A&M U Dept Rangeland Ecology & Mgmt College Station TX 77843-2126

ARCHER, THOMAS JOHN, aeronautical engineer; b. Bangor, Northern Ireland, Nov. 29, 1927; came to U.S., 1956; s. Beresford Robert and Sarah (Andrews) A.; m. Mary Pollock, Sept. 3, 1949; children: Christine Anne, David Robert. Grad. in Mech. Engring., Coll. of Tech., Belfast, 1950. Design engr. Short Bros. & Harland Ltd., Belfast, Northern Ireland, 1943-56, Lockheed Aircraft, Marietta, Ga., 1956-58; supr., mgr. Lockheed Missiles & Space, Sunnyvale, Calif., 1958-87; v.p. Target Constructors Inc., Fresno, Calif., 1987—. Contbr. articles to profl. jours. Recipient Fleet Ballistic Missile award USN, 1978. Office: Target Constructors Inc 4055 W Shaw Ave Fresno CA 93722

ARCHIBALD, DAVID WILLIAM, virologist, dentist; b. Melrose, Mass., Mar. 2, 1953; s. Ralph Strong and Shirely (Clark) A.; m. Marilyn Davis, June 8, 1985; children: Cameron, Hunter. BS, Tufts U., 1975; DMD, Harvard U., 1979, DSc, 1986. Oral oncology fellow Harvard Sch. of Dental Medicine, Boston, 1980-81; rsch. fellow cancer biology Harvard Sch. Pub. Health, Boston, 1982-86; asst. prof. oral pathology dental sch. U. Md. Microbiology and Immunology Med. Sch., Balt., 1986-91; assoc. prof. oral pathology U. Md., Balt., 1991-93; ad hoc reviewer NIH, Bethesda, Md., 1990—. Contbr. articles to profl. jours. Fellow Acad. Gen. Dentistry; mem. Am. Assn. Dental Rsch. (councillor 1991-93), Am. Acad. Oral Pathology, Infectious Diseases Soc. Am., Omicron Kappa Upsilon.

ARCHIBALD, FREDERICK RATCLIFFE, chemical engineer; b. Huron County, Ont., Can., Jan. 30, 1905; came to U.S. 1942; s. Andrew and Margaret S. (Wallace) A.; m. Doris Pearl Kindree, July 24, 1928; children: Doreen A. Messenger, Margaret Ruth Hafely. MA in Chemistry, Queen's U., 1934. Registered profl. engr., Ont. From rsch. chemist to chief metallurgist Ventures Ltd. & Assocs. Cos. and Panaminas Inc., Toronto, N.Y.C., 1934-47; chem. engr. Oliver Iron Mining Co., Duluth, Minn., 1948-50; from chief metallurgist to v.p. Falconbridge Ltd. & Assoc. Cos., Toronto, Ont., 1951-70; chem. engring. cons. Naples, Fla., 1970-92; ret. Contbr. articles to profl. jours. Com. mem. Solid Waste Advisory, Collier County, Fla., 1980-85. Recipient Disting. Svc. medal Can. Inst. Mining and Metallurgy, Montreal, 1971. Mem. Fla. Engring. soc., Assn. of Profl. Engrs. Ont. Presbyterian. Achievements include patents on various metallurgical processes. Home: 412 Bentley Dr Naples FL 33963

ARCHIBALD, JAMES DAVID, biology educator, paleontologist; b. Lawrence, Kans., Mar. 23, 1950; s. James R. and Donna L. (Accord) A. B.S. in Geology, Kent State U., 1972; Ph.D. in Paleontology, U. Calif., Berkeley, 1977. Gibb's instr. dept. geology Yale U., New Haven, 1977-79, asst., then assoc. prof. dept. biology, 1979-83; curator of mammals Peabody Mus. Natural Hist., New Haven, 1979-83; assoc. prof., then prof. dept. biology San Diego State U., 1983—; extensive field expeditions in Mont., Colo., N.Mex., Pakistan, 1973—. Author: A Study of Mammalia and Geology Across the Cretaceous-Tertiary Boundary, 1982; contbr. articles to profl. jours. Trustee San Diego Natural History Mus. Scholar Yale U., San Diego State U.; fellow Alcoa Found., U. Calif.-Berkeley; grantee Sigma Xi, Nat. Geog. Soc., NSF, Petroleum Research Found., San Diego State U. Mem. Soc. Vertebrate Paleontology, Geol. Soc. Am., Paleontol. Soc., Soc. Systematic Zoologists, Am. Soc. Mammalogists, Willi Hennig Soc., Sigma Xi. Office: San Diego State U Dept Biology San Diego CA 92182

ARCINO, MANUEL DAGAN, microbiologist, consultant; b. Manila, May 9, 1941; came to U.S., 1959; s. Francisco Villaneuva and Felicidad (Dagan) A.; m. Ofelia Caponpon Chavez, July 11, 1970; children: Jennifer Eillen, Michelle Monel, Catherine Anne, Mary Beth, Melissa Christy. BS in Bacteriology, Kans. State U., 1964; cert. in med. tech., St. Francis Hosp., 1966; MS in Microbiology, U. Bridgeport, Conn., 1974. Bench microbiologist Wesley Med. Ctr., Wichita, Kans., 1966-68; head microbiologist Mercy Hosp., Muskegon, Mich., 1968-70; specialist in microbiology Bridgeport Hosp., 1970-76; supervising microbiologist Lenoir Meml. Hosp., Kinston, N.C., 1976-81; cons. microbiologist Kinston Clinics, 1981—; teaching fellow, rsch. asst. Wichita State U., 1966-67; lectr. in microbiology Housatonic Coll., Bridgeport, 1973-74; med. lab. cons. Kinston Clinics/Tri-County Med. Health Ctr., 1981-91; presenter in field; coord. workshop/seminar. Member choir, cantor, eucharistic minister Holy Trinity Ch., Kinston, 1976-91, vice chmn. pastoral coun., 1989—; founder Evangelization program, Kinston, 1985; chmn. Evangelization Commn. & Spiritual Life Commn., renew coord. welcome com., bldg. com., liturgy commn., tchr. childrens' liturgy; facilitator Faith Sharing Group, Kinston, 1985-91. Wichita State U. fellow, 1967; recipient Greater Bridgeport Heart Assn., 1973, Teaching Recognition award Christ the King Sch., 1980. Mem. Am. Soc. Microbiology, Am. Soc. Clin. Pathologists, Assn. Practitioners in Infection Control, Southeastern Assn. Clin. Microbiologists, N.C. chpt. Am. Soc. Microbiology. Republican. Roman Catholic. Achievements include research on intestinal parasites among Puerto Ricans, differential diagnosis of Vibrio parahaemolyticus and its pathogenicity, yeast like fungi, family enterobacteriaceae, rapid diagnostic procedures in clinical microbiology, antimicrobial agents. Home: 2703 Fairfax Rd Kinston NC 28501-1154

ARDASH, GARIN, mechanical engineer; b. Detroit, July 14, 1963; s. Berge and Lucy Alice (Souldourian) A. BSME, U. Mich., 1986, MME, 1988. Grad. rsch. asst. U. Mich. Coll. Engring., Ann Arbor, 1986-87, Los Alamos (N.Mex.) Nat. Lab., 1987; rsch./analysis engr. materials tech. dept. Bettis Atomic Power Lab. Westinghouse Electric Co., West Mifflin, Pa., 1989—. U. Mich. Coll. Engring. fellow, 1986-87; State Mich. Coop. Scholar, 1982-83. Mem. ASTM, ASME (assoc.), Soc. Mfg. Engrs., Internat. Legion Intelligence, Mensa, Pocatello Ski Assn., Idaho Falls Men's Soccer League, Pitts. South Soccer Assn. Avocations: photography, skiing, chess. Home: 1220 Royal Dr Apt 266 Library PA 15129-8613 Office: Westinghouse Electric Co Bettis Atomic Power Lab M/S 14B PO Box 79 West Mifflin PA 15122-0079

ARDEN, BRUCE WESLEY, computer science and electrical engineering educator; b. Mpls., May 29, 1927; s. Wesley and Clare Montgomery (Newton) A.; m. Patricia Ann Joy, Aug. 25, 1951; children: Wayne Wesley, Michelle Joy. Student, U. Del., 1944; B.S. in Elec. Engring., Purdue U., 1949; postgrad., U. Chgo., 1949; M.A., U. Mich., 1955, Ph.D., 1965. Detail engr. Allison div. Gen. Motors Corp., Indpls., 1950-51; asst. prof. dept. computing and communication scis. U. Mich., Ann Arbor, 1965-67, assoc. prof., 1967-70, prof., 1970-73, chmn. dept., 1971-73, from research asst. to assoc. dir. Computing Facilities, 1951-73; prof., chmn. dept. elec. engring. and computer sci. Princeton U., 1973-85, Arthur Le Grand Doty prof. engring., 1981-86; prof. elec. engring., computer sci., dean engring. and applied sci. U. Rochester, 1986—, vice provost computing, 1992—, William F. May Prof. Engring., 1993—; vis. prof. U. Grenoble, France, 1971-72; guest prof. Siemens Research, Munich, Germany, 1983, also cons.; cons. to Gen. Motors Corp., Ford Corp., Westinghouse Co., RCA, Xerox Data Systems, IBM.; mem. sci. council USRA Inst. for Computer Applications in Sci. and Engring., 1973-79, 82-88; mem. sci coun. USRA Inst. Advanced Computer Sci., 1982-88; chmn. com. on anti-ballistic missile data processing Nat. Acad. Sci., 1966-71; mem. panel Inst. Computer Sci. and Tech., 1980-86; mem. acad. adv. council Wang Inst., 1978-87; mem. study sect. NIH, 1985-88; reviewer Guggenheim Found., 1985-91. Author: An Introduction to Digital Computing, 1963; (with K. Astil) Numerical Algorithms: Their Origins and Applications, 1970; editor: What Can Be Automated?, 1980. Served with USNR, 1944-46, 49-50. Fellow AAAS; mem. IEEE (sr.), Assn. for Computing Machinery, Univs. Space Research Assn. (bd. dirs. 1982-88), Sigma Xi, Tau Beta Pi, Eta Kappa Nu. Office: U Rochester Coll Engring & Applied Sci Rochester NY 14627

AREF, HASSAN, fluid mechanics educator; b. Alexandria, Egypt, Sept. 28, 1950; s. Moustapha and Jytte (Adolphsen) A.; m. Susanne Eriksen, Aug. 3, 1974; children: Michael, Thomas. Cand.Sci., U. Copenhagen, Denmark, 1975; PhD, Cornell U., 1980. Asst. prof. Brown U., Providence, 1980-85, assoc. prof., 1985; assoc. prof. fluid mechanics U. Calif., San Diego, 1985-88, prof. fluid mechanics, 1988-92; chief scientist San Diego Supercomputer Ctr., 1989-92; prof., head dept. theoretical and applied mechanics U. Ill., Urbana-Champaign, 1992—; Corrsin lectr. Johns Hopkins U., Baltimore, 1988; Westinghouse disting. lectr. U. Mich., Ann Arbor, 1991; lectr. Midwest Mechanics, 1991. Contbr. articles to profl. jours.; assoc. editor Jour. Fluid Mechanics, 1984—; editor Cambridge Texts in Applied Math., 1987—. Recipient Presdl. Young Investigator award, NSF 1985. Fellow Am. Physical Soc.; mem. Soc. Indsl. and Applied Math. Office: U Ill Dept Theoretical and Applied Mechanics 104 S Wright St Urbana IL 61801-2935

ARENA, BLAISE JOSEPH, research chemist; b. Chgo., Nov. 17, 1948; s. Joseph R. and Margaret H. (Shogrin) A.; m. Kathleen M. Ganch, July 21, 1973; children: Evan, Eve, Carmen. BS in Biology, Ill. State U., 1971; MSc in Chemistry, Northea. Ill. U., 1977. Rsch. specialist UOP Rsch. Ctr. UOP Inc., Des Plaines, Ill., 1978—. Contbr. articles Jour. Chem. Edn., Trans. Ill. State Acad. Sci. Mem. AAAS. Achievements include 30 patents in Catalysis and Carbohydrate Reactions. Office: UOP Inc 25 E Algonquin Rd Des Plaines IL 60016-6100

ARESKOG, DONALD CLINTON, chiropractor; b. Bklyn., Aug. 6, 1926; s. Andrew Albert and Jennie Margaret (Dickson) A.; m. Julia Catherine Koskela, May 15, 1954. D Chiropractic, Logan Coll., St. Louis, 1950; Philosopher of Chiropractic, Atlantic States Chiropractic Coll. Pvt. practice Bklyn., 1952-56, Wappingers Falls, N.Y., 1956-61, Poughkeepsie, N.Y., 1961-89; retired, 1989; bd. govs. Atlantic States Chiropractic Coll., Bklyn., 1954; research in field. Mem. Am. Chiropractic Assn. (speakers bur. 1964), Ednl. Rsch. Soc., Internat. Basic Rsch. Inst., Internat. Platform Assn., Wappingers Falls C. of C. (treas. 1959), Toastmasters. Achievements include research on the removal of emotional pressures that interfere with the complete mental, physical and social well-being of the individual. Home: 330 SE 20th Ave Apt 514 Deerfield Beach FL 33441-5181

ARGENTO, VITTORIO KARL, environmental engineer; b. Naples, Italy, Mar. 7, 1937; came to U.S., 1937; s. Valentino and Sophie Caroline (Ness) A.; m. Emma Frances Raley, Aug. 16, 1959. BS, San Diego State U., 1964; MS, U. Tex. at Dallas, 1976, PhD, 1989. Registered profl. engr., Tex. Aerospace engr. Calif., 1964-71; chief City of Dallas Air Pollution Control Sect., Dallas, 1971-78; sr. lectr. U. Tex., Arlington, 1978-89, assoc. prof., 1989—, dir. ctr. for environ. rsch. and tng., 1986—; vice-chair North Ctrl. Tex. Coun. of Govts., Arlington, 1991—; chair Dallas Environ. Health Com., Dallas, 1988-91; engring. mem. Tex. Air Control Bd., Austin, Tex., 1979-86. Contbr. articles to profl. jours. Chair-elect Envrion. Health and Safety Tng. Div. Nat. Univ./Continuing Edn. Assn., Washington, 1993; assoc. dir. Environ. Inst. for Tech. Transfer, Arlington, 1990—; mem. Air and Waste Mmgt. Assn., Pitts., 1972—. With USAF, 1955-59. Mem. Am. Acad. Environ. Engrs., Sigma Xi. Office: U Tex at Arlington CERT Box 19021 Arlington TX 76019

ARGIBAY, JORGE LUIS, information systems firm executive and founder; b. Montevideo, Uruguay, May 17, 1953; s. Candido Argibay and Blanca Martinez; m. Stella Gonzalez, Feb. 20, 1974 (div. Aug. 1981); children: Laura, Andres; m. M. Ines Sencion, Mar. 22, 1982; 1 child, Nicolas. Ingeniero de Sistemas, U. de la Republica, 1978. Cert. engring. Researcher laser optics Univ. Inst. of Physics, Montevideo, 1971-74; researcher operating systems Univ. Computing Div., Montevideo, 1975-77; prof. automata theory Faculty of Engring., Montevideo, 1977-81, prof. low level langs., 1977-81; pres. Swann S.A., Montevideo, 1980-91, Fla. Swan, Inc., Miami, 1991—; cons. Supreme Ct. of Justice, Montevideo, 1987—. Contbr. articles to profl. jours.; co-inventor DACOL programming lang. Avocation: boat sailing. Home: 4726 NW 97th Ct Miami FL 33178-1977 Office: 8390 NW 53d St Ste 318 Miami FL 33166

ARHAR, JOSEPH RONALD, chemist; b. Lakewood, Ohio, Aug. 9, 1964; s. Ronald Joseph and Joan Carol (Kruteck) A.; m. Kathryn Sara Neth, Mar. 7, 1987; children: Kayla Elizabeth, Carolen Sarah, Rebecca Eleanor, Annmarie Josephine. BS, Ohio No. U., 1986; MA, SUNY, 1989. Assoc. tech. engr. Clevite Elastomers, Milan, Ohio, 1989-90, tech. svc. engr., 1990-92, materials engr., 1992—. Contbr. articles to profl. jours. Village councilman Wakeman (Ohio) Village Coun., 1992; mem. St. Mary's Parish Coun.; head maintenance St. Mary's Ch., Wakeman, Ohio. Mem. Am. Chem. Soc. Republican. Roman Catholic. Home: 30 River St Wakeman OH 44889

ARIAS, PEDRO LUIS, chemical engineering educator, researcher; b. Baracaldo, Vizcaya, Spain, Apr. 7, 1959; s. Jaime Americo Arias and Maria Soledad Ergueta. Degree in indsl. engring., U. Bilbao, Spain, 1981, DEng, 1984. Rsch. scientist Labein Rsch. Labs., Bilbao, 1981-84; asst. prof. Sch. Engring., Bilbao, 1984-90, prof., 1990—; postdoctoral assoc. Imperial Coll., London, 1986, MIT, Cambridge, Mass., 1988; cons. Cementos Portland Lemona (Spain) S.A., 1985-87, Formica Española S.A., Galdacano, 1990-91; dir. coun. Labeim Rsch. Labs., Bilbao, 1986—, Cadem, Bilbao, 1991—. Contbr. articles to profl. jours. Bd. dirs. U. of the Basque Country, Leioa, Spain, 1991. Mem. Computing Div. IEEE, Am. Inst. Chem. Engrs., Am. Computing Machinery Assn., Engring. Indsl. Coll. (chartered). Roman Catholic. Avocations: reading, climbing, swimming. Home: Aizpuru 2 1, 48910 Sestao Vizcaya, Spain Office: Sch Engring, Alameda Urquijo SN, 48013 Bilbao Vizcaya, Spain

ARIEFF, ALLEN IVES, physician; b. Chgo., Sept. 30, 1938. BS in Math. and Chemistry, U. Ill., 1960; MS in Physiology, Northwestern U., 1964, MD, 1964. Intern Phila. Gen. Hosp., 1964-65; resident SUNY, Bklyn., 1967-68; renal fellow U. Colo., Denver, 1968-69; rsch. and edn. assoc., clin. investigator Wadsworth VA Med. Ctr., L.A., 1970-74; asst. prof. medicine, rsch. scientist UCLA Med. Ctr., 1971-74; asst. prof. medicine, dir. hemodialysis U. Calif. VA Med. Ctr., San Francisco, 1975-76, assoc. prof. medicine, dir. nephrology sect., 1976-83, prof. medicine, chief clin. nephrology, 1983-86, prof. medicine, dir. rsch. & edn. geriatrics, 1986—; cons. and speaker in field. Contbr. numerous articles to profl. jours. Fellow ACP; mem. Am. Soc. Nephrology, Am. Fedn. Clin. Rsch., Am. Diabetes Assn., Am. Psychol. Soc., Am. Soc. Neurochemistry, Am. Soc. Clin. Investigation, Am. Soc. Bone and Mineral Rsch., Assn. Am. Physicians, Western Assn. Physicians, Western Soc. Clin. Rsch., Internat. Soc. Nephrology. Office: VAMC/UCSF 4150 Clement St (111)G San Francisco CA 94121

ARIMA, AKITO, academic administrator; b. Osaka, Japan, Sept. 13, 1930; s. Jyoji and Kazuko Arima; m. Hiroko Aota, Mar. 16, 1957; children: Yoshihito, Akiko. BS, U. Tokyo, 1953, DSc, 1958; DSc (hon.), Glasgow (Scotland) U., 1984; DrS (hon.), Drexel U., 1992. Dir. computer ctr. U. Tokyo, 1981-85, dean faculty of sci., 1985-87, v.p., 1987-89, pres., 1989—. Author: The Interacting Boson Model, 1987; contbr. articles to profl. publs. Recipient prize Nishina Found., 1978, award Humboldt Found., 1987, Grosse Verdienstkreuz award German Govt., 1991, award of Comdr. in Order of Orange Nassau, Her Majesty Queen Beatrix of the Netherlands, 1991, Tom W. Bonner prize Nuclear Physics Am. Physical Soc., 1993. Mem. Phys. Soc. Japan (pres. 1981-82), Haijin Assn. (prize 1988), Japan Writers' Assn., Japan Pen Club. Avocations: reading, painting. Home: Sangenjaya 2-23-6, Setagaya-ku 154, Japan Office: U Tokyo, Hongo 7-3-1 Bundyo-ku, Tokyo 13, Japan

ARIMOTO, RICHARD, atmospheric chemist; b. Wilmington, Del., May 1, 1952; s. Fred S. and Amy E. (Inouye) A.; m. Leslie Anne Diggs, Oct. 2, 1982. MS, U. Del., 1977; PhD, U. Conn., 1981. Marine rsch. assoc. U. R.I., Narragansett, 1982-84; asst. marine scientist U. R.I., narragansett, 1984-87, assoc. marine scientist, 1987-92, assoc. rsch. prof., 1992--. Author: Jour. Atmos. Chemistry, 1992, Chem. Oceanog., 1989, Jour. Geophys. Rsch., 1990, Jour. Great Lakes Rsch., 1989, Global Biogeochem. Cycles, 1991; assoc. editor: Jour. Geophys. Rsch. Atmospheres, 1988—, Environ. Sci. and Engring. Fellow AAAS, Am. Geophys. Union. Office: U RI South Ferry Rd Narragansett RI 02882-1197

ARIMURA, AKIRA, biomedical research laboratory administrator, educator; b. Kagoshima, Japan, Dec. 26, 1923; came to U.S., 1965; s. Jyojiro and Kiyoko (Kajiwara) A.; m. Katsuko Yamashita, July 31, 1957; children: Jerome J., Mark M., Margaret M. BS, 7th Nat. Coll., Kagoshima, 1943; MD, Nagoya (Japan) U., 1951, PhD, 1957. James Hudson Brown postdoctoral fellow Yale U., New Haven, Conn., 1956-58; instr., rsch. assoc. Hokkaido U., Sapporo, Japan, 1961-65; instr. Tulane U., New Orleans, 1958-61, asst. prof., 1965-68, assoc. prof., 1968-73; prof. medicine Tulane U., New Orleans, 1973—; dir. U.S.-Japan Biomedical Rsch. Lab. Tulane U., Belle Chasse, La., 1985—; rsch. physician VA Hosp., New Orleans, 1965-80; mem. Endocrine Study Sect., NIH, 1978-82; adj. prof. anatomy Tulane U., New Orleans, 1979—, Physiology, 1989—, founder, dir. clin. RIA Lab., Tulane U. Med. Ctr., 1980-87, molecular neuroendo and diabetes lab. Belle Chasse, 1980-85, dir., 1985; vis. prof. Keio U., Tokyo, 1990—; founder U.S.-Japan Biomed. Rsch. Labs., Belle Chasse, 1985—; reviewer Jour. Clin. Endocrinology and Metabolism, Am. Jour. Physiology, Jour. Clin. Investigation, Sci., Life Sci., Procs. Soc. Exptl. Biology and Medicine, others. Mem. editorial bd. Peptides, Turkish Jour. Med. and Biol. Rsch.; contbr. articles to scholarly and profl. jours. Planner, initiator student exchange program Tulane U. and Keio U., New Orleans and Tokyo, 1986. Fulbright scholar, 1956. Mem. AAAS, Internat. Soc. Neuroendocrine, Endocrine Soc. U.S.A., Japan Endocrine Soc. (hon.), Am. Physiology Soc., Am. Soc. Neuroscience, Soc. Exptl. Biology and Medicine, N.Y. Acad. Sci. Achievements include co-development of LHRH, somatostatin, Interleukin-1, pituitary adenylate cyslau activating polypeptide; discovery of PACAP. Office: Tulane U Herbert Rsch Ctr US-Japan Biomed Rsch Labs 3705 Main St Belle Chasse LA 70037*

ARISMAN, RUTH KATHLEEN, environmental manager; b. Camp Atterbury, Ind., Dec. 14, 1942; d. Paul Mason and Ruth Kathleen (Pruitt) Ambrose; children: Caroline Frances Arisman Melton, Phillip Ambrose. BS in Chemistry magna cum laude, U. S.C., 1971, PhD in Chemistry, 1975. Rsch. asst. U. S.C., Columbia, 1975-76; phys. chemist GE Capacitor Products, Hudson Falls, N.Y., 1976-79, mgr. environment and safety, 1979-83; mgr. environ. protection GE Capacitor Products, Ft. Edward, N.Y., 1983-84, mgr. environment and analytical, 1984-85; mgr. environ. programs GE Plastics, Burkville, Ala., 1985-91; mgr. environ. programs GE Plastics Group, Parkersburg, W.Va., 1991-92, mgr. CMA programs, 1992—; mem. air toxics com. ADEM, Montgomery, Ala., 1985-90. Youth leader YPF Ch. of the Ascension, Montgomery, 1985-89. NDEA grantee U. S.C., 1971-74. Mem. Soc. Environ. Toxicology and Chemistry (membership chair 1979-80, bd. dirs. 1979-84, com. chair 1980-84, historian 1987-90), N.Y. Acad. Scis., Am. Chem. Soc., Alpha Kappa Mu. Episcopalian.

ARISTODEMOU, LOUCAS ELIAS, industrial engineer, manufacturing company executive; b. Paphos, Cyprus, Oct. 17, 1946; s. Elias and Eleni (Alexandrou) A.; m. Mary Heraclides, Dec. 28, 1975; childre: Eliagne L., Constantinos L. MSc in Engring., Poly. Inst. Bucharest, Romania, 1972, PhD in Indsl. Mgmt., 1977. Asst. prof. Poly. Inst., Bucharest, 1976-77; gen. mgr. Couvas Bros. Ltd., Limassol, Cyprus, 1977-78, Nemitsas Industries Ltd., Limassol, 1978-80, Metalco (Heaters) Ltd., Nicosia, Cyprus, 1980—; cons. Commonwealth Secretariat, London, 1983-87; adv. com. industry and tech. Govt. of Nicosia, 1990—; mem. Cyprus Standards Orgn., Nicosia, 1991—; mem. syllabus com. Higher Tech. Inst., Nicosia, 1986—; assoc. judge Cyprus Labor Ct., Nicosia, 1988—; bd. dirs. Cyprus Employers and Industrialists Fedn., Nicosia, 1992. Editor, co-author Ency. of Math., 1975; co-author: Informatic and Activity Analysis of Enterprises, 1976; chief editor Tech. Word, 1990—; contbr. to tech. publs. Mem. Cyprus Profl. Engrs. Assn. (pres. 1992), Cyprus Tech. and Sci. Chamber (gen. sec. 1992), Am. Mgmt. Assn., Internat. Solar Energy Soc. Home: 3 Doridos St, 143 Nicosia Cyprus Office: Metalco (Heaters) Ltd, PO Box 1307, 143 Nicosia Cyprus

ARIYOSHI, TOSHIHIKO, toxicologist, educator; b. Fukuoka, Japan, July 1, 1930; m. Nobu Yasutake, Oct. 30, 1958; children: Sakiko, Noritaka. B in Pharmacy, Kyushu U., Fukuoka, 1954, PharmM, 1956, PharmD, 1964. Pharmacist Tachiarai Hosp., Fukuoka, 1958-60; from rsch. assoc. to assoc. prof. Nagasaki (Japan) U., 1960-74, prof., 1975—, dean of students, 1986-88. Author: Drug Metabolism, 1986; contbr. 110 articles to profl. jours. Humboldt fellow Tubingen U. 1966-68; recipient award Japanese Min. Environment Agy. 1991. Mem. Coun. Environ. Pollution Control of Nagasaki Prefecture (chmn. 1990—). Office: Nagasaki U, 1-14 Bunkyo-Machi, Nagasaki 852, Japan

ARIZPE, LOURDES, anthropologist, researcher; b. Mexico City, Apr. 10, 1946; d. Jesus and Luisa (Schlosser) Arizpe. MA in Ethnology, Escuela Nat.

Anthropology, Mexico City, 1970; PhD in Anthropology, London Sch. Econs., 1975. With Escuela Nat. de Anthropologia, 1966-69; ethnologist Nat. Anthropology Mus., Mexico City, 1967-69; with London Sch. of Econs., 1970-75; researcher El Colegio de Mex., Mexico City, 1973-85; prof. Nat. U., 1985—; lectr. Rutgers U., N.J., 1978; cons. UNESCO/ILO/UN, Geneva, 1978-87, Indian Orgn., Mexico, 1975-85; dir. Nat. Mus. Popular Culture Mexico, Mexico City, 1985-88, Inst. Anthrop. Rsch. Nat. U., 1991—; mem. steering com. Devel. Alternatives for Women, Brazil, 1985; com. mem. Human Dimensions of Global Change, Paris, 1989; com. mem. Soc. Scis. Rsch. Coun., N.Y., 1986-89; sec. Mexican Nat. Sci. Acad., 1992—. Author: Economy & Kinship, 1972, Indian Migrants, 1975, Migration and Ethnicity, 1978, Peasant Migration, 1984, Women in Development, 1987, Culture and Development, 1989, Culture and Global Change, 1993. Recipient medal Nat. Indianist Inst., Mexico, 1978, Sahagun prize Nat. Anthrop. and Hist. Inst., 1992, Mahatma Gandhi prize Coll. of William and Mary, William sburg, Va., 1993; Fulbright-Hays grantee, 1978, Guggenheim grantee, 1982. Fellow Inst. Devel. Studies; mem. Internat. Union Anthrop. Scis. (pres. 1988-93), Internat. Social Sci. Coun. (exec. com. 1990-92, v.p. 1992—), Royal Anthrop. Soc., Latin Am. Studies Assn., Internat. Sociol. Assn. (officer 1986-88), Mexican Anthrop. Assn. (pres. 1984-86), Soc. Internat. Devel. (v.p. 1991—), Mexican Soc. for Anthropology (sec. 1992—). Home: Campestre 54, 01060 Mexico City Mexico

ARJAS, ELJA, statistician; b. Tampere, Finland, Feb. 9, 1943; s. Olavi and Kaja (Kivekäs) A.; m. Pirkko Molander, Nov. 23, 1968 (div. 1992); children: Inari, Aino. MSc, U. Helsinki, Finland, 1965, Lic. Philosphy, 1970, PhD, 1972. Lab. engr. Helsinki U. Tech., Espoo, Finland, 1965-70; rsch. asst. The Acad. Finland, Helsinki, 1970-72; rsch. assoc. Ctr. for Ops. Rsch. and Econometrics, Louvain, Belgium, 1972-73; acting assoc. prof. U. Helsinki, 1973-74; prof., dept. chmn. U. Oulu, Finland, 1975—; rsch. prof. The Acad. Finland, 1992—; vis. prof., Can., U.S., Australia. Assoc. editor Stochastic Process Applications, 1980-93, Math. Ops. Rsch., 1987-92; assoc. editor Scandinavian Jour. Stats., 1986-90, editor, 1991—; contbr. articles to profl. jours. Fellow Inst. Math. Stats.; mem. ISI, Bernoulli Soc. (program chmn. European region 1988, program coord. 1989-91). Office: Univ Oulu, Linnanmaa, Oulu SF-90570, Finland

ARKADAN, ABDUL-RAHMAN AHMAD, electrical engineer, educator; b. Saidon, Lebanon, June 1, 1956; came to U.S., 1977; s. Ahmad Abdul-Rahman and Amina A. (Yaman) A.; m. Maha M. Habli, Aug. 31, 1986; children: Farah, Nour. BSEE magna cum laude, U. Miss., 1980; MSEE, Va. Poly. Inst. and State U., 1981; PhD in Elec. Engring., Clarkson U., 1988. Registered profl. engr., Wis. Rsch. asst. Va. Poly. Inst. and State U., Blacksburg, 1980-81; quality control engr. Saudi Oger Ltd., Saudi Arabia, 1981-83; project engr. AIM Ltd., Saudi Arabia, 1983-84; devel. engr. Sundstrand Aerospace Corp., Rockford, Ill., summers of 1985, 86; teaching & rsch. asst. Clarkson U., Potsdam, N.Y., 1984-88; asst. prof. dept. elec. and computer engring. Marquette U., Milw., 1988-93, assoc. prof., 1993—; cons. MacNeal Schwendler Corp., Fleck Controls, Inc., Sundstrand Aerospace Corp., Outboard Marine Corp.; presenter at profl. confs. Contbr. articles to tech. publs. Grantee Nat. Sci. Found, Elec. Power Rsch. Inst., 1992—, various corps. Mem. IEEE (sr.), IEEE Power Engring. Soc., IEEE Magnetics Soc., IEEE Indsl. Electronics Soc., Am. Soc. Engring. Edn., Sigma Xi, Phi Kappa Phi, Tau Beta Pi, Eta Kappa Nu, Pi Mu Epsilon. Office: Marquette U Dept Elec and Computer Engring 1515 W Wisconsin Ave Milwaukee WI 53233

ARKILIC, GALIP MEHMET, mechanical engineer, educator; b. Sivas, Turkey, Mar. 10, 1920; came to U.S., 1943, naturalized, 1960; s. Sabir Mehmet and Zahra Fatima (Hocazade) A.; m. Ann A. Bryan, Mar. 31, 1956; children: Victor, Dennis, Layla, Errol. BME, Cornell U., 1946; MS, Ill. Inst. Tech., 1948; PhD, Northwestern U., 1954. Registered profl. engr., Va. Mech. engr. Miehle Printing Press and Mfg. Co., Chgo., 1948-49, analyst, 1954-56; research and devel. engr. Mech. and Chem. Industries, Turkey, 1949-52; asst. prof. Pa. State U., University Park, 1956-58; assoc. prof. dept. civil engring. George Washington U., Washington, 1958-63, prof. engring. and applied sci., 1963—, prof. emeritus, 1990—, chmn. dept. engring. mechanics, 1966-69, asst. dean, 1969-74. Contbr. articles to sci. jours. Vice pres. Courtland Civic Assn., Arlington, Va., 1965-66; pres. Am. Turkish Assn., Washington, 1967-71. Served to 2d lt. Turkish Army, 1939-41. Recipient Disting. Leadership award Am. Turkish Assn., 1972; Recognition of Service award Sch. Engring. and Applied Sci., George Washington U., 1976, Spl. Appreciation award Engring. Alumni Assn., George Washington U., 1990; Air Force Office of Sci. Research grantee, 1963-69. Mem. ASME, AAUP, Am. Acad. Mechanics, Math. Assn. of Am., Am. Math. Soc., Wash. Soc. Engrs., Sigma Xi. Club: George Washington U. (Washington). Home: 8403 Camden St Alexandria VA 22308-2111 Office: George Washington U Washington DC 20052

ARKIN, GERALD FRANKLIN, agricultural research administrator, educator; b. Washington, Sept. 16, 1942; s. Hyman Paul and Lottie (Werbitzky) A.; m. Karen Judith Horwitz, Aug. 28, 1966; children: Melissa, Marleigh. BS, Cornell U., 1966; MS, U. Ga., 1968; PhD, U. Ill., 1971. Lic. profl. engr., Tex. Asst. prof. Tex. A&M U., College Station, 1972-76, assoc. prof., 1976-80, prof., 1980-87; resident dir. Blackland Rsch. Ctr. Tex. A&M U., Temple, 1983-87; prof. U. Ga., Athens, 1987—, assoc. dir. Ga. Agrl. Expt. Stas., 1987—. Co-editor: Modifying the Root Environment to Reduce Crop Stress, 1981. Mem. AAAS, Am. Soc. Agrl. Engrs., Am. Soc. Agronomy, Soil Sci. Soc. Am., Crop Sci. Soc. Am., Coun. for Agr. and Tech. Office: Ga Experiment Sta 1109 Experiment St Griffin GA 30223-1797

ARKIN, JOSEPH, mathematician, lecturer; b. Bklyn., May 25, 1923; s. Ben and Helen (Heller) A.; m. Judith H. Lobel, Aug. 28, 1954; children: Helen, Aviva, Jessica, Sarah. Ph.D. (hon.), Brantridge Sch., Eng., 1967. Vis. lectr. Orange Community Coll., Middletown, N.Y., 1962-67; lectr. Nanuet Pub. Schs., Rockland County, N.Y., 1962-67; sr. lectr. U.S. Mil. Acad.; researcher various profs., 1965—; math. reviewer Am. Math. Soc., R.I., 1976—. Contbr. articles to profl. jours. Active Mus. Village, Monroe, N.Y., 1978. Served with U.S. Army, 1942-43. Mem. N.Y. Acad. Scis., Internat. Congress Math (Can.), AAAS, Fibonacci Assn. (charter mem.), Math. Assn. Am., Am. Math. Soc., Can. Math. Soc., Calcutta Math Soc., Soc. Indsl. and Applied math., Am. Legion, DAV, NCO Club (West Point). Home: 197 Old Nyack Tpke Spring Valley NY 10977-5304 Office: US Mil Acad Dept Math West Point NY 10996

ARLING, BRYAN JEREMY, internist; b. Mpls., Dec. 10, 1944; s. Leonard Swenson and Marion (Schroeder) A.; m. Donna Dickson; children: Elissa, Jeremy, Timothy. BA summa cum laude, U. Minn., 1965; MD, Harvard U., 1969. Diplomate, Am. Bd. Internal Medicine. Intern Stanford (Calif.) Affiliated Hosps., 1969-70, resident in medicine, 1970-71; spl. asst. to adminstr. Health Sci. Mental Health Adminstrn. USPHS, Rockville, Md., 1971-73; instr., chief resident medicine George Washington U. Hosp., Washington, 1973-74; asst. prof. medicine George Washington U. Hosp., 1974-77; pvt. practice gen. internal medicine Washington, 1977—; clin. prof. medicine George Washington U., 1988—. Mem. adminstrv. bd., Chevy Chase United Meth. Ch.; mem. devel. com., Maret Sch., 1985—, co-chair ann. giving campaign, 1986-87, 87-88, bd. trustees 1991—; question relevance reviewer ABIM, 1991-92; com. mem. ABIM Certifying and Recertifying Exam, 1992—. Mem. ACP, Am. Soc. Internal Medicine, AMA, D.C. Med. Soc., Smithsonian Assocs., Friends of Kennedy Ctr., Harvard Club Washington, Nat. Trust for Historic Preservation, Friends of nat. Zoo, Common Cause, ACLU, Physicians for Social Responsibility, Columbia Country Club, Bahamas Air-Sea Rescue Assn. Home: 3803 Taylor St Bethesda MD 20815-4117 Office: 2440 M St NW Ste 817 Washington DC 20037-1404

ARLINGHAUS, WILLIAM CHARLES, mathematics educator; b. Detroit, July 17, 1944; s. Francis Anthony and Blanche Therese (Stolinski) A.; m. Sandra Judith Lach, Sept. 3, 1966; 1 child, William Edward. BS summa cum laude, U. Detroit, 1964; PhD, Wayne State U., 1979. Mathematician GM Tech. Ctr., Warren, Mich., 1967-74; tchr. math. U. Detroit High Sch., 1974-75; lectr. Eastern Mich. U., 1975-77, Wayne State U., 1975-77; lectr. math. Ohio State U., Columbus, 1977-79; asst. prof. math. Loyola U., Chgo., 1979-83, U. Detroit, 1983-85; asst. prof. math., computer sci. Lawrence Technol. U., Southfield, Mich., 1985-90, assoc. prof., 1990-92, prof., 1992—, chair dept. math., computer sci., 1990—; vis. asst. prof. math. U. Mich., Dearborn, 1982-83; bus. mgr. Inst. Math. Geography, Ann Arbor, Mich., 1985—, adv. bd. for monograph series, 1985—. Author: The Classification of Minimal Graphs with Given Abelian Automorphism Group, 1985; mem. editorial bd. Solstice, 1990—; contbr. articles to internat. profl. publs. Mem. pastoral coun. St. Thomas the Apostle Ch., Ann Arbor, 1990—, v.p., 1991, pres., 1992; mem. local arrangements com. World Jr. Bridge Championship, 1991. Mem. Am. Math. Soc., Math. Assn. Am., N.Y. Acad. Scis., Assn. Am. Geographers, Am. Contract Bridge League (silver life master), Mich. Bridge Assn. (bd. dirs., tournament chmn. 1974, 87, chmn. 1976, 89, pres. 1975, 88), Alpha Sigma Nu. Roman Catholic. Avocations: duplicate bridge, golf. Home: 2790 Briarcliff St Ann Arbor MI 48105-1429 Office: Lawrence Technol Univ 21000 W 10 Mile Rd Southfield MI 48075-1058

ARLOOK, THEODORE DAVID, dermatologist; b. Boston, Mar. 12, 1910; s. Louis and Rebecca (Sakansky) A.; BS, U. Ind. Sch. Medicine, 1932, M.D., 1934; postgrad. dermatology U. So. Calif., 1946-47. Diplomate Am. Bd. Dermatology. Intern, Luth. Meml. Hosp., Chgo., 1934-35; resident in dermatology Indpls. Gen. Hosp., 1947-49; practice medicine specializing in dermatology, Elkhart, Ind., 1950—; mem. staff Elkhart Gen. Hosp.; assoc. mem. dermatology dept. Wishard Meml. Hosp., Indpls, 1950-86, Regenstrief Hosp., Indpls., 1987—. Pres., Temple Israel, Elkhart, 1963-64; pres. B'nai B'rith, 1955. Served to capt. M.C. AUS, 1941-46; PTO. Mem. AMA, Ind. State Med. Assn., Am. Acad. Dermatology, Elkhart County Med. Soc. (pres. 1967), Noah Worcester Dermatol. Soc. Contbr. articles to med. jours. Office: 1825 Rainbow Bend Blvd Elkhart IN 46514-1402

ARMANIOS, ERIAN ABDELMESSIH, aerospace engineer, educator; b. Cairo, July 6, 1950; came to the U.S., 1980; s. Abdelmessih Armanios; m. Mahera S. Philobos, May 2, 1980; children: Daniel, Laura. BS in Aero. Engring., Cairo U., 1974, MS in Aero. Engring., 1979; PhD in Aerospace Engring., Ga. Inst. Tech., 1985. Teaching asst. U. Cairo, 1974-79, asst. lectr., 1979-80; grad. rsch. asst. Ga. Inst. Tech., Atlanta, 1980-84, rsch. engr. I, 1985-86, asst. prof., 1986-91, assoc. prof., 1991—; cons. Bell Helicopter Textron Inc., Ft. Worth, 1984-85, Rolls-Royce Inc., Atlanta, 1989—; judge Ga. Sci. and Engring. Fair, Atlanta, 1987; judge space sci. student program NASA, Atlanta, 1988—; dir. Ga. Space Grant Consortium, 1991—. Editor: Interlaminar Fracture of Composites, 1989; contbr. articles to profl. jours.; mem. editorial bd. Jour. Composites Tech. and Rsch., 1992. Recipient Teaching Excellence award Ctr. for Enhancement of Teaching and Learning Amoco Found., 1990, Outstanding Paper award Jour. Aerospace Engring., 1990. Fellow AIAA (assoc.); mem. ASTM (com. on high modulus fibers and composites 1988); Am. Soc. for Composites, Am. Helicopter Soc. (com. on structures and materials). Office: Ga Inst Tech Sch Aerospace Engring Atlanta GA 30332-0150

ARMBRUSTER, BARBARA LOUISE, botanist; b. Oceanside, N.Y., May 27, 1952; d. John William and Dorthy Matilda (Carter) A.; m. Robert David Vincent, May 22, 1985. BA in Biology, Hamilton-Kirkland Coll., 1974; PhD in Botany, Duke U., 1980. Postdoctoral fellow Biozentrum, U. Basel, Switzerland, 1980-83, Med. Sch., Washington U., St. Louis, 1983-85; sr. rsch. biologist Monsanto Co., St. Louis, 1985-89, rsch. specialist, 1989—; indsl. fellow U. Minn. Ctr. for Interfacial Engring., Mpls., 1992—; adj. prof. U. Mo., 1991—. Contbr. articles to profl. jours. Rsch. grantee Nat. Inst. Allergies and Infectious Diseases, Washington U., 1984-85. Mem. AAAS, Cen. States Electron Microscopy Soc. (pres. 1993—), Am. Soc. Cell Biology, Microscopy Soc. Am. Achievements include research on fungal, microbial and plant ultrastructure, fungal sporogenesis, STM/AFM studies of metal cluster formation, low temperature embedding and immunolocalization techniques. Office: Monsanto Co 800 N Lindbergh Blvd Saint Louis MO 63167

ARMBRUSTER, WALTER JOSEPH, foundation administrator; b. Cin., July 18, 1940; s. Joseph Anthony and Nellie (Barry) A.; m. Helen Gartner, Feb. 27, 1987; 1 child, Sean W. BS, Purdue U., 1962, MS, 1964; PhD, Oreg. State U., 1970. Agrl. economist USDA Econ. Rsch. Svc., Washington, 1968-76; staff economist USDA Agrl. Mktg. Svc., Washington, 1976-78; assoc. mng. dir. Farm Found., Oak Brook, Ill., 1978-91, mng. dir., 1991—; mem. adv. bd. Childhood Agrl. Injury Prevention Symp. Nat. Coun., 1991—, Annenburg/CPB Social Sci. Telecourse, 1990—, Ill. Inst. for Rural Affairs, 1987—. Editor, author: Federal Marketing Programs in Agriculture, 1983, Economic Efficiency in Agricultural and Food Marketing, 1987, Research on Effectiveness of Agricultural Commodities Promotion, 1985, (pamphlets) Generic Agricultural Commodity Advertising and Promotion, 1988, Federal Agricultural Marketing Programs, 1988. V.p., bd. dirs. Nat. Farm-City Coun., 1992-93; bd. dirs. North Ctrl. Regional Ctr. for Rural Devel.; mem. So. Rural Devel. Ctr.; asst. supt. Future Farmers of Am. Farm Bus. Mgmt. Contest Com., Kansas City, Mo., 1983—. Recipient Cert. of Appreciation, HHS, 1991. Mem. Am. Agr. Econs. Assn. (bd. dirs. 1988-91), Am. Agr. Law Assn. (bd. dirs. 1988-91), Internat. Assn. Agr. Econs. (sec.-treas. 1991—), The Chgo. Farmers (pres. 1992—), Alpha Zeta. Office: Farm Found 1211 W 22d St Oak Brook IL 60521

ARMENANTE, PIERO M., chemical engineering educator; b. Avezzano, Aquila, Italy, June 2, 1953; came to U.S., 1979; s. Euclide and Maria (Antonini) A.; m. Annemarie Aigner, Oct. 21, 1983. Laurea in chem. engring., U. Rome, 1977; PhD in Chem. Engring., U. Va., 1983. Rsch. asst. Internat. Inst. for Applied Systems Analysis, Laxenburg, Austria, 1978, U. Lund (Sweden), 1978-79; engring. specialist UN Indsl. Devel. Orgn., Vienna, Austria, 1979-87; process engr. Farmitalia Carlo Erba, Milan, Italy, 1985-87; asst. prof. N.J. Inst. Tech., Newark, 1984-91, assoc. prof., 1991-93; prof., 1993—; cons. UN Indsl. Devel. Orgn., Vienna, 1978-86; presenter in field. Author: Contingency Planning for Industrial Emergencies, 1991; author: (with others) Risk Assessment and Risk Management for the Chemical Process Industry, 1991; editor: Biotechnology Applications in Hazardous Treatment, 1989; contbr. articles to profl. jours.; author reports; peer reviewer Chem. Engring. Sci., Can. Jour. Chem. Engring., Biotech. Progress, Chem. Engring. Communications. Sec. North Am. Mixing Forum, 1990—. Grantee NSF, 1991, EPA, 1989-90, 91-93, Exxon Edn. Found., 1991, Schering-Plough, Inc., 1992-93, Hazardous Substance Mgmt. Rsch. Ctr., 1988-91, 91-92, Ctr. for Mfg. Engring. Systems, Schering-Plough, Inc., 1990, 90-91, Industry/Univ. Coop. Ctr. for Hazardous Substance Mgmt., 1986-89, P.M. Armenant Inst. Tech., 1989, N.J. Inst. Tech., 1984-85. Mem. AICE (chmn. north Jersey sect. 1992-93), Am. Chem. Soc., Am. Soc. Engring. Edn., Order of the Engr., Sigma Xi, Tau Beta Pi. Office: NJ Inst Tech Dept Chem Engring Chemistry and Environ Sci Newark NJ 07102

ARMENIADES, CONSTANTINE D., chemical engineer, educator; b. Thessaloniki, Greece, May 29, 1936; came to U.S., 1954; s. Demosthenes C. and Eleni (Vantis) A.; m. Mary Sophia Orsini, Feb. 14, 1959; children: Christopher, Eleni. BS in Chem. Engring., Northeastern U., Boston, 1961; MS in Engring., Case Inst. of Tech., 1967; PhD, Case Western Reserve U., 1969. Profl. staff mem. Arthur D. Little, Inc., Cambridge, Mass., 1961-65; prof. Chem. Engring. Rice U., Houston, 1969—; master of Will Rice Coll., Rice U., Houston, 1976-82; editorial bd. mem. Jour. Applied Polymer Sci., 1987-91; cons. and expert witness several maj. law firms, Houston, N.Y.C., 1975—. Contbr. numerous articles to profl. jours. Mem. Soc. Plastics Engrs., N.Y. Acad. of Scis., Sigma Xi. Achievements include invention of patented Kenics Mixer, a static mixer used in process industries; co-invention of patented ophthalmic surgical instruments. Office: Rice U 6100 S Main St Houston TX 77005

ARMITAGE, KENNETH BARCLAY, biology educator, ecologist; b. Steubenville, Ohio, Apr. 18, 1925; s. Albert Kenneth and Virginia Ethel (Barclay) A.; m. Katie Lou Hart, June 5, 1953; children—Carol, Keith, Kevin. B.S. summa cum laude, Bethany Coll., W.Va., 1949; M.S., U. Wis.-Madison, 1951, Ph.D., 1954. Instr. U. Wis.-Green Bay, 1954-55; instr. U. Wis.-Wausau, 1955-56; asst. prof. biology U. Kans., Lawrence, 1956-62, assoc. prof., 1962-66, prof., 1966—, William J. Baumgartner disting. prof., 1987—, chmn. dept. systematics & ecology, 1982-88, dir. environ. studies program, 1976-82, dir. exptl. and applied ecology program, 1974—; vis. prof. U. Modena, Italy, 1989; mem. com. examiners Grad. Record Exam. Biology Test, 1986-92, chmn., 1988-92; sr. investigator Rocky Mountain Biol. Lab., Gothic, Colo., 1962—, trustee, 1969-86, pres. bd. trustees, 1985-86. Author: (lab. manual) Investigations in General Biology, (with others) Principles of Modern Biology; contbr. articles to profl. jours.; mem. editorial bd. Ethology, Ecology and Evolution, 1989—. Pres. Douglas County chpt. Zero Population Growth, 1969-71; bd. dirs. Children's Hour, Inc., Lawrence, 1969-70. Served with U.S. Army, 1943-46, ETO. Recipient Antarctic medal NSF, 1968, Edn. Service award U. Kans., 1979, Alumni Achievement award Bethany Coll., 1989. Fellow AAAS; mem. Am. Soc. Naturalists (treas. 1984-86), Am. Inst. Biol. Scis. (mem. task force for 90s), Ecol. Soc. Am., Am. Soc. Zoologists, Orgn. Biol. Field Stations (v.p. 1986-87, pres. 1988-89), Sigma Xi, Phi Beta Kappa. Avocations: stamp collecting; gardening; hiking; jogging; natural history. Home: 1616 Indiana St Lawrence KS 66044-4046 Office: U Kans Kansas Ecological Reserves Lawrence KS 66045-2106

ARMOR, JOHN N., chemical company research manager; b. Phila., Sept. 14, 1944; s. Lloyd N. and Cornelia Armor; m. Connie B. Korzuch, Dec. 17, 1966; children: Kimberly, Gregory, Jennifer. BS in Chemistry, Pa. State U., 1966; PhD, Stanford U., 1970. Asst. prof. chemistry Boston U., 1970-74; group leader Allied Signal Corp., Morristown, N.J., 1974-85; prin. rsch. assoc. Air Products and Chems. Inc., Allentown, Pa., 1985—; chmn. Inorganic Gordon Rsch. Conf., New London, N.H., 1988. Editor Applied Catalysis, 1987—; mem. editorial bd. Catalysis Today, Chemistry of Materials, Microporous Solids; contbr. over 70 articles to profl. jours. Mem. Am. Chem. Soc. (organizer symposium on environ. catalysis 1993), Am. Ceramic Soc., Materials Rsch. Soc., N.Y. Acad. Scis. (chmn. catalysis sect. 1983-85), Catalysis Soc. (bd. dirs., treas. 1993—), Catalysis Club Phila., Catalysis Club N.Y. (bd. dirs.). Achievements include over 30 patents. Home: 1608 Barkwood Dr Orefield PA 18069-8923 Office: Air Products & Chem Inc 7201 Hamilton Blvd Allentown PA 18195-9642

ARMSTRONG, ANDREW THURMAN, chemist; b. Haslet, Tex., May 26, 1935; s. Andrew Thurman and Ila (Kitchen) A.; m. Kay Francis Masters, May 30, 1958; children: Michael Andrew, Marion Kay, Benjamin Niel. BS, North Tex. State U., 1958, MS, 1959; PhD, La. State U., 1967. Diplomate Am. Bd. Criminolistics. Instr. West Tex. State U., 1959-61, La. State U., 1963-66; postgrad. researcher UCLA, 1966-67; asst. prof. U. Tex., Arlington, 1968-72, assoc. prof., 1972-84; prin. Armstrong Forensic Lab., Arlington, 1984—; vis. asst. prof. La. State U., 1967-68; bd. dirs. Tarrant County Adv. Coun. on Arson, 1981-87, treas., 1985-87; lectr. in field; presenter at profl. confs.; presenter short courses. Contbr. numerous articles to sci. publs. Fellow Am. Inst. Chemists (cert.), Am. Acad. Forensic Scis.; mem. ASTM (task force on gas chromatography/mass spectrometry in fire residue analysis 1990—, mem. forensic sci. com.), Am. Chem. Soc. (sect. treas. 1972-73, chmn. edn. com. 1976-84), Am. Indsl. Hygiene Assn., Internat. Assn. Arson Investigators (forensic sci. com. 1990—), Coblentz Soc., North Tex. Assn. Arson Investigators, Phi Lambda Upsilon, Alpha Chi Sigma. Achievements include research in physical-analytical chemistry, instrumentation, industrial hygiene.

ARMSTRONG, DAVID WILLIAM, biotechnologist, microbiologist; b. Ottawa, Ont., Can., Jan. 29, 1954; s. Robert Crosby Armstrong and Margaret Theresa (Larose) Shepherd; m. J. Elaine Robertson, June 10, 1978; 1 child, Laura Lynne. BSc with honors, U. Ottawa, Ontario, Can., 1978, MSc, 1980; PhD, Carleton U., Ottawa, 1984. Rsch. scientist Nat. Rsch. Coun., Ottawa, 1979—; head Cell Culture Facility Inst. Biol. Scis. NRC, Ottawa, 1985—; advisor Biotech. and Biomedicine, Canadian Space Agy., Ottawa, 1985-88. Named Ontario scholar, 1973; recipient Ontario Grad. scholarship, 1978-80. Mem. Can. Coll. Microbiologist (registered specialist), Tissue Culture Assn. Mem. United Ch. Achievements include patent for unique cell culture system for cultivation of human/animal cells; development of processes for specialty chemicals production by manipulation of cell metabolic regulation. Office: Nat Rsch Coun Biol Scis, 100 Sussex Dr, Ottawa, ON Canada K1A OR6

ARMSTRONG, DOUGLAS, organic chemist, educator; b. New Albany, Ind., Jan. 4, 1941; s. Robert Edwin and Mabel Gladys (McIntosh) A.; m. Moonyean Devine, June 8, 1970; children: Paul, Mary. BS in Chemistry with honors, Ind. U., 1963; PhD in Organic (Medicinal) Chemistry, U. Iowa, 1968. Rsch. assoc. MIT, Cambridge, Mass., 1968-69; asst. prof. Mass. Coll. Pharmacy, Boston, 1969-74; various edn. and indsl. positions, 1974-85; assoc. prof. Olivet Nazarene U., Kankakee, Ill., 1985-88, prof., 1988—; participant acad.-industry polymer edn. program Tenn. Eastman Co., 1985; Smith Kline Beecham vis. rsch. fellow U. Iowa, 1987; cons. Armour Pharm. Co., Kankakee, 1990—, Rsch. Biochemicals Internat., 1992—. Contbr. articles to Jour. Medicinal Chemistry, Jour. Pharm. Exp. Ther., Nature. Mem. Internat. Soc. Heterocyclic Chemistry, Am. Chem. Soc. (faculty travel grantee organic chemistry divsn. 1989). Nazarene. Office: Olivet Nazarene U Dept Chemistry Kankakee IL 60901-0592

ARMSTRONG, EDWARD BRADFORD, JR., oral and maxillofacial surgeon, educator, naval officer; b. Teaneck, N.J., Sept. 24, 1928; s. Edward Bradford and Ruth Elizabeth (Fippinger) A.; A.B., U. Pa., 1950; D.D.S., N.Y.U., 1954; m. Dusanka Vladimirovna Jakovljevic, Nov. 5, 1960; children: Edward Bradford, III, James B., Hugh B. Commd. lt. j.g. U.S. Navy, 1954, advanced through grades to capt. 1971; intern oral surgery Roosevelt Hosp., N.Y.C., 1958, assoc. attending oral surgery, 1959—, attending oral surgeon out-patient dept., 1959—, chmn., moderator Oral Surgery Staff Confs., 1963-70; resident Carle Hosp., Urbana, Ill., 1959; assoc. attending oral surgeon Flower and Fifth Ave. hosps., N.Y.C., 1960-78; asst. attending oral surgeon Hackensack (N.J.) Hosp., 1963-65; adminstrv. officer Naval Res. Dental Co. 3-2, 1965-68, exec. officer, 1968-71, comdg. officer, 1971-73; comdt.'s rep. 3d Naval Dist., Naval Acad., 1972-78, 3d Naval Dist for Dentistry, 1973-75, group staff officer for dentistry and medicine, 1973-75, Ready Res. Unit 502, 1975-77, VTU 0207, 1977-79, ret., 1979; assoc. clin. prof. oral surgery N.Y. Med. Coll., 1963—; adj. assoc. clin. prof. oral surgery Columbia U. Sch. Dentistry, 1973-89; chmn. bd. E. & R. Armstrong, Inc., Albany, N.Y., 1966-77; pres. Edward B. Armstrong, P.C., N.Y.C., 1979-90; dir. Songtime, Inc., Boston; dir., mem. exec. com. PGP Internat. Corps, Inc. Bd. dirs., trustee Christian Mission Farms of Paraguay, Inc., 1974-84; pres., trustee Central Bible Chapel, Palisades Park, N.J.; area rep., ann. giving U. Pa., 1960-68; Blue and Gold officer Naval Acad. Admissions Com., sec. bd. dirs., trustee Boys' Club of N.Y. Health Services, Inc. Diplomate Am. Bd. Oral Surgery. Fellow N.Y. Acad. Dentistry (sec., dir., pres. 1979-80), Am., Internat. colls. dentists, Am. Coll. Oral and Maxillofacial Surgeons (founding), Am. Dental Soc. Anesthesiology (hon. life); mem. Am. Soc. Oral Surgeons (N.J. rep. Ho. of Dels. 1963-65), N.Y. Soc. Oral Surgeons (chmn. audit and budget com. 1972-79), ADA, First Dist., N.Y., Bklyn., Yokosuka (hon.) dental socs., Assn. Mil. Surgeons U.S., Mil. Order World Wars, Naval Res. Assn. (life), Union League (chmn. art com. 1973-76, bd. govs. 1974-77, 82-84, v.p. 1977-80, 85-88), Met. Club (bd. gov. 1992), N.Y.C., U. Pa. Club, U. Pa. Club of Met. N.J. (dir. 1982—), Acacia, Xi Psi Phi, Psi Omega (hon.), Delta Sigma Delta. Mem. Plymouth Brethren Ch. Home: 110 Broad Ave Leonia NJ 07605-2093

ARMSTRONG, ELMER FRANKLIN, information systems and data processing educator; b. Pray, Mont., Dec. 12, 1931; s. Elmer F. and Mary Naomi (Conlin) A.; m. Vera E. Browning, May 12, 1952; children: James R., Alma N. BS, Mont. State Coll., 1953; MS, U. Mont., 1972; PhD, Colo. State U., 1979. Cert. Data Processor. Various edn. positions pub. schs., Mont., 1953-66; pvt. practice Helena, Mont., 1965-68; instr. computer programming and acctg. Helena (Mont.) Vol. Tech. Ctr., 1969-74; prof. data processing Mayville (N.D.) State Coll., 1975-84; prof. math. Valley City (N.D.) State U., 1984-91; computer cons., 1991—. Author automated acctg. system, 1969, nutritional rsch., 1990, computer programs, 1970-78. Mem. Data Processing Mgmt. Assn., Pi Omega Pi, Phi Kappa Phi.

ARMSTRONG, GENE LEE, retired aerospace company executive; b. Clinton, Ill., Mar. 9, 1922; s. George Dewey and Ruby Imald (Dickerson) A.; m. Lael Jeanne Baker, Apr. 3, 1946; children—Susan Lael, Roberta Lynn, Gene Lee. BS with high honors, U. Ill., 1948, MS, 1951. registered profl. engr., Calif. With Boeing Aircraft, 1948-50, 51-52; chief engr. astronautics div., corp. dir. Gen. Dynamics, 1954-65; chief engr. Def. Systems Group TRW, Redondo Beach, Calif., 1956-86; pvt. cons. systems engring. Def. Systems Group TRW, 1986—; Mem. NASA Research Adv. Com. on Control, Guidance & Navigation, 1959-62. Contbr. chpts. to books, articles to profl. publs. Served to 1st lt. USAAF, 1942-45. Decorated Air medal; recipient alumni awards U. Ill., 1965, 77;. Mem. Am. Math. Soc., AIAA, Nat. Mgmt. Assn., Am. Def. Preparedness Assn., Masons. Home: 5242 Bryant Cir Westminster CA 92683-1713 Office: Armstrong Systems Engring Co PO Box 86 Westminster CA 92684-0086

ARMSTRONG, GERALD CARVER, mining engineer; b. Kansas City, Mo., May 15, 1934; s. Thomas Jack and Cuba Athelia (Hall) A.; m. Susan Elaine Erickson, May 3, 1956 (div. 1972); children: Gerald C. II, Willard J., James R., Robert D.; m. Katherine Ann Kentner, Oct. 12, 1974. BS in Mining Engring., Mo. Sch. Mines, 1958. Registered profl. engr., Mo., Ill. Quarry engr. U.S. Gypsum Co., Chgo., 1958-59; tech. instr. South Plains Coll., Levelland, Tex., 1959-60; sales engr. The Bristol Co., Waterbury, Conn., 1960-63; sr. devel./evaluation engr. Monsanto Co., St. Louis, 1963-66; corp. instrument engr. Chemplex Co., Rolling Meadows, Ill., 1966-72; project mgr. Mgmt. & Tng. Systems, St. Louis, 1972-81; measurement/loss control supt. Phillips Pipeline Co., Bartlesville, Okla., 1981—; internat. tng. com. cons. Instrument Soc. Am., Chapel Hill, N.C., 1981-83. Coantbr. articles to profl. jours. Instr. Jr. Achievement, Odessa, Tex., 1988. With USMC, 1953-55. Fellow Internat. Soc. Philos. Inquiry; mem. Instrument Soc. Am. (Permian Basin pres. 1988-89), Permian Basin Measurement Soc. (founding mem., charter v.p. 1986-87), Am. Mensa. Republican. Presbyterian. Achievements include design/development of self supervising switching system, sequenced time simulator; assisted development for AC2 control systems; developed high pressure reactor dump system based on temperature/pressure rate of rise. Home: 13908 W County Rd 123 Odessa TX 79765 Office: Phillips Pipeline Co 4001 E 42nd St Odessa TX 79762

ARMSTRONG, JOANNE MARIE, clinical psychologist, family mediator; b. Cooperstown, N.Y., Nov. 26, 1956; d. William John and Joan Alice (Larsen) A.; m. Brian Joseph Yore, July 31, 1983; children: Mackensie A., Campbell A. BA, Trinity U., San Antonio, 1978; MA, U. Louisville, 1982, PhD, 1987. Lic. psychologist, Wis. Mgmt. trainee, adminstrv. asst. Gentec Hosp. Supply Co., San Antonio, 1978-79; rsch. asst. U. Louisville, 1980-81; therapist Seven Counties Svcs., Louisville, 1981-82; mental health profl. Head Start, Louisville, 1982-83; dir. Kaufman County Outreach Clinic, Tex. Dept. Mental Health-Mental Retardation, Terrell, 1984-85; clin. psychologist Nicolet Clinic/La Salle Clinic, S.C., Menasha, Wis., 1987-89; pvt. practice, Neenah, 1989-93; pvt. practice Spartanburg, S.C., 1993—; cons. Wellness Counseling Ctr., Appleton, Wis., 1989-93, Fox Valley Hosp., Green Bay, Wis., 1990-91. Mem. bd. Birthing Network, Neenah, 1988-93; mem. Citizens for Better Environ., Neenah, 1989-93. Rsch. fellow U. Louisville, 1985-86. Mem. Am. Psychol. Assn., Nat. Register Health Svc. Providers in Psychology, Wis. Psychol. Assn., Fox Valley Psychol. Assn. (founding), Acad. Family Mediators (cert., assoc.). Episcopalian. Avocations: renovating old houses, bicycling, tennis, skiing, piano. Office: 390 E Henry St Ste 206 Spartanburg SC 29302

ARMSTRONG, KEITH BERNARD, engineering consultant; b. Ashford, Eng., Sept. 8, 1931; s. George Roland and Doris Maud (Weatherley) A.; m. Dulcie Marina Sampson, Mar. 31, 1956; children: Bernard Paul, Gillian Ruth, Derek George Andrew. H.N.C., Kingston (Eng.) Poly., 1952; MSc, City U., London, 1978, PhD, 1990. Aviation apprentice Vickers-Armstrongs Ltd., Weybridge, Eng., 1948-52, design draughtsman, 1952-55; sr. draughtsman Vickers-Armstrongs (Aircraft) Ltd., Weybridge, 1958-60; exptl. officer Nat. Phys. Lab., Teddington, Eng., 1960-67; devel. engr. Brit. Airways, Heathrow, London, 1967-77, sr. devel. engr., 1977-88, prin. devel. engr., 1988-91; cons., 1991—; chmn. Internat. Air Transport Assn. Composite Repair Task Force, Montreal, Can., 1988-91. Youth leader Meth. Ch., Ashford, Middlesex, Eng., 1952-55, Ashford Common, Middlesex, Eng., 1961-64. Flying officer RAF, 1955-58. Fellow Plastics & Rubber Inst., Instn. Mech. Engrs., Royal Aero Soc. Avocations: flying, family history. Home: 20 Homewaters Ave, Sunbury on Thames TW16 6NS, England

ARMSTRONG, NEIL A., former astronaut; b. Wapakoneta, Ohio, Aug. 5, 1930; s. Stephen Armstrong; m. Janet Shearon; children: Eric, Mark. B.S. In Aero. Engring., Purdue U., 1955; M.S. in Aero. Engring., U. So. Calif. With Lewis Flight Propulsion Lab., NACA, 1955; then aero. research pilot for NACA (later NASA, High Speed Flight Sta.), Edwards, Calif.; astronaut Manned Spacecraft Center, NASA, Houston, 1962-70; command pilot Gemini 8; comdr. Apollo 11; dep. assoc. adminstr. for aeros. Office Advanced Research and Tech., Hdqrs. NASA, Washington, 1970-71; prof. aerospace engring. U. Cin., 1971-79; chmn. bd. Cardwell Internat., Ltd., 1980-82; chmn. CTA, Inc., 1982-92. Mem. Pres.'s Commn. on Space Shuttle, 1986, Nat. Commn. on Space, 1985-86. Served as naval aviator USN, 1949-52, Korea. Recipient numerous awards, including Octave Chanute award Inst. Aero. Scis., 1962, Presdl. Medal for Freedom, 1969, Exceptional Service medal NASA, Hubbard Gold medal Nat. Geog. Soc., 1970, Kitty Hawk Meml. award, 1969, Pere Marquette medal, 1969, Arthur S. Fleming award, 1970, Congl. Space Medal of Honor, Explorers Club medal. Fellow AIAA (hon., Astronautics award 1966), Internat. Astronautical Fedn. (hon.), Soc. Exptl. Test Pilots; mem. Nat. Acad. Engring.

ARMSTRONG, PETER BROWNELL, biologist; b. Syracuse, N.Y., Apr. 27, 1939; s. Philip Brownell and Marian Louise (Schmuck) A.; m. Margaret Tryon, Sept. 22, 1962; children: Katharine, Elisabeth, Philip. BS, U. Rochester, 1961; PhD, Johns Hopkins U., 1966. Asst. prof. Biology U. Calif., Davis, 1966-72, assoc. prof. Biology, 1972-80, prof. Biology, 1980—; sci. trustee Marine Biol. Lab., Woods Hole, Maine, 1986-90. Contbr. articles to profl. jours. Office: U Calif Dept Molecular and Cell Biology Davis CA 95616-8755

ARMSTRONG, RICHARD SCOTT, mechanical engineer; b. Teaneck, N.J., Mar. 31, 1942; s. Richard Thomas and Edna Mae (Scott) A.; m. Paula Kelley (div.); children: R. Scott, Kelley, Amy; m. Marsha Marie Major, June 15, 1981. BSME, Fairleigh Dickinson U., 1969, MBA in Mgmt., 1974. Registered profl. engr. Alaska, HAwaii, Calif., Wash. Quality control analyst Ford Motor Co., Mahwah, N.J., 1961-67; quality assurance engr. Singer-Kearfott, Little Falls, N.J., 1967-74; constrn. supt. Alyeska Pipeline Svc. Col, Anchorage, 1974-75, support svcs. mgr., 1975-79; dir. design & constrn. div. State of Alaska Dept. Transp., Anchorage, 1979-84; pres. RSA Engring., Inc., Anchorage, 1984—; mem. Architects, Engrs. Land Surveyors Bd. of Registration, Juneau, Alaska, 1992—. Mem. Am. Soc. Heating, Refrigerating and Air Conditioning Engrs., Internat. Conf. Bldg. Ofcls., Assn. Energy Engrs., Am. Arbitration Assn., Alaska Soc. Profl. Engrs. Office: RSA Engring 2522 Arctic Blvd Anchorage AK 99503

ARMSTRONG, ROBERT DON, toxicologist; b. Sligo, Pa., July 21, 1928; s. Vivian Dake and Kathryn Jane (Mahle) A.; m. Joan Marshall Van Scoyk, Sept. 1, 1951; children: Wanda, Kimble, Claudia, Robin, Matthew, Russel. BA, Coll. of Wooster, 1952; MS, U. Rochester, 1958. Tech. rsch. assoc. Atomic Energy Project, U. Rochester, N.Y., 1952-63; behavioral toxicologist Chem. Rsch. Devel. & Engring. Ctr. U.S. Army APG-EA, Aberdeen Proving Ground, Md., 1963—. Mem. AAAS, N.Y. Acad. Sci., Sigma Chi. Methodist. Achievements include design and operation of the first behavioral toxicology lab at U. of Rochester; created pigeon model of chronic mercury poisoning. Home: 324 Clyde Ct Abingdon MD 21009-1560 Office: US Army CRDEC Toxicology Div SMCCR-RST-C Aberdeen Proving Ground MD 21010-5423

ARMSTRONG, ROBIN LOUIS, university official, physicist; b. Galt, Ont., Can., May 14, 1935; s. Robert Dockstader and Beatrice Jenny (Grill) S.; m. Karen Elisabeth Feilberg Hansen, July 8, 1960; children: Keir Grill, Christopher Drew. B.A., U. Toronto, 1958, M.A., 1959; Ph.D., 1961; Ph.D. Rutherford Meml. fellow, Oxford (Eng.) U., 1961-62; FRSC, 1979. Mem. faculty U. Toronto, 1962-90, prof. physics, 1971-90, chmn. dept., 1974-82, dean Faculty of Arts and Sci., 1982-90; pres. U. N.B., Fredericton and Saint John, 1990—; pres. Can. Inst. for Neutron Scattering, 1986-89; mem. rsch. coun. Can. Inst. for Advanced Rsch., 1982—; mem. coun. NSERC, 1991—. Co-author: Mechanics, Waves and Thermal Physics, 1970, Electromagnetic Interaction, 1973; contbr. articles to profl. jours. Recipient Commemorative medal for 125th Anniversary of Can. Confedn., 1992. Mem. Can. Assn. Physicists (Herzberg medal 1973, v.p. 1989-90, pres. 1990-91, medal for achievement 1990), Internat. Soc. Magnetic Resonance, Internat. Soc. Magnetic Resonance in Medicine, Rsch. Coun. Can. Inst. Advanced Rsch. Home: 58 Waterloo Row, Fredericton, NB Canada E3B 1Y9 Office: U NB, PO Box 4400, Fredericton, NB Canada E3B 5A3

ARN, KENNETH DALE, physician, city official; b. Dayton, Ohio, July 19, 1921; s. Elmer R. and Minna Marie (Wannagat) A.; m. Vivien Rose Fontini, Sept. 24, 1966; children—Christine H. Hulme, Laura P. Hafstad, Kevin D., Kimmel R. B.A., Miami U., Oxford, Ohio, 1943; M.D., U. Mich., 1946. Intern Miami Valley Hosp., Dayton, Ohio, 1947-48; resident in pathology U. Mich., 1948-49, fellow in renal research, 1949-50; fellow in internal medicine Cleve. Clinic, 1950-52; pvt. practice specializing in internal medicine, pub. health and vocat. rehab. Dayton, 1952—; commr. of health City of Oakwood, Ohio, 1953—; assoc. clin. prof. medicine Wright State U., 1975—; mem. staffs Kettering Med. Ctr., Dayton, Miami Valley Hosp.; adj. assoc. prof. edn. Wright State U.; field med. cons. Bur. Vocat. Rehab., 1958—, Bur. Svcs. to Blind, 1975—; med. dir. Ohio Rehab. Svcs. Commn., 1979-87; mem. Pres.'s Com. on Employment of Handicapped, 1971—; chmn. med. adv. com. Goodwill Industries, 1960-75, chmn. bd. trustees 1985-87, chmn. rehab. com. 1987—; mem., chmn. lay adv. com. vocat. edn. Dayton Pub. Schs., 1973-82; exec. com. Gov.'s Com. on Employment Handicapped; bd. dirs. Vis. Nurses Assn. Greater Dayton; chmn. profl. adv. com. Combined Gen. Health Dist. Montgomery County. Trustee Luth. Social Svc. of Miami Valley, 1982—. Named City of Dayton's Outstanding Young Man, Jr. C. of C., 1957; 1 of 5 Outstanding Young Men of State, Ohio Jr. C. of C., 1958; Physician of Yr., Pres.'s Com. on Employment of Handicapped, 1971; Bishop's medal for meritorious service Miami U., 1972. Mem. AMA, Ohio Med. Assn., Montgomery County Med. Soc. (chmn. com. on diabetic detection 1955-65, chmn. polio com. 1954-58), Nat. Rehab. Assn., Am. Diabetes Assn., Am. Profl. Practice Assn., Am. Heart Assn., Am. Pub. Health Assn., Ohio Pub. Health Assn., Aerospace Med. Assn., Fraternal Order Police, Dayton Country Club, Kiwanis, Royal Order Jesters, Masons (past potentate), Shriners, K.T., Scottish Rite (33 deg.), Nu Sigma Nu, Sigma Chi. Lutheran. Home: 167 Lookout Dr Dayton OH 45409-2238 Office: 30 Park Ave Dayton OH 45419-3426

ARNAL, MICHEL PHILIPPE, mechanical engineer, consultant; b. Inglewood, Calif., Sept. 8, 1955; s. Robert Emile and Frances Annabelle (Leuthold) A.; m. Anne-Kathrin Meyer, Mar. 31, 1989; 1 child, Hannah Madeleine. MS, U. Calif., Berkeley, 1981-83, PhD in Engring. Sci., 1985-88. Rsch. asst. dept. mech. engring. U. Calif., Berkeley, 1985-88; rsch. scientist Inst. Fluid Mechanics Tech. U., Munich, Germany, 1989-92; design engr. ABB Power Generation Ltd., Baden, Switzerland, 1992—; cons. Advanced Scientific Computing, Holzkirchen, Germany, 1991-92. Contbr. articles to profl. jours. Doct. Study fellow IBM, Berkeley, 1986-87, German Acad. Exch. Svc. Exch. fellow German Govt., Erlangen, 1983-85. Mem. ASME, Am. Phys. Soc. Office: ABB Power Generation Ltd, KWDT31 Haselstrasse 16, CH-5401 Baden Switzerland

ARNASON, BARRY GILBERT WYATT, neurologist, educator; b. Winnipeg, Man., Can., Aug. 30, 1933; s. Ingolfur Gilbert and Elsie (Wyatt) A.; m. Joan Frances Morton, Dec. 27, 1961; children—Stephen, Jon, Eva. M.D., U. Man., 1957. Intern Winnipeg Gen. Hosp., 1956-57, resident in medicine, 1957-58; resident in neurology Mass. Gen. Hosp., Boston, 1958-62; asst. prof. neurology Harvard Med. Sch., Boston, 1965-71, assoc. prof., 1971-76; prof., chmn. dept. neurology U. Chgo. Pritzker Sch. Medicine, 1976—, dir. Brain Research Inst., 1985—; mem. med. adv. bd. Nat. Multiple Sclerosis Soc., 1977—, Amyotrophic Lateral Sclerosis Soc. Am., 1976-79, Hereditary Disease Found., 1977-85. Contbr. articles to med. jours. Nat. Multiple Sclerosis Soc. rsch. fellow, 1959-61, 62-64; grantee NIH, other founds. Mem. Am. Soc. Clin. Investigation, Am. Assn. Immunology, Am. Neurol. Assn., Am. Acad. Neurology, Am. Assn. Neuropathology, Soc. Neurosci. Home: 4832 S Ellis Ave Chicago IL 60615-1810 Office: U Chgo Brain Rsch Inst Dept Neurobiology Box 425 5841 S Maryland Ave Chicago IL 60637-1470*

ARNDT, BRUCE ALLEN, mechanical engineer; b. Sault Ste. Marie, Mich., July 3, 1956; s. Charles John and Mary Lee (Marby) A.; m. JoAnn Wilkins, May 2, 1982. AS, Marion Tech. Coll., 1980. Engr. Marion (Ohio)-Dresser, 1979-84; project engr. Rotec Indsl., Elmhurst, Ill., 1984-88; chief application engr. Flender Corp., Elgin, Ill., 1988—. Mem. ASME, Fluid Power Soc. Office: Flender Corp 950 Tollgate Rd Elgin IL 60123

ARNDT, JULIE ANNE PREUSS, aeronautical engineer; b. Midland, Mich., Nov. 5, 1968; d. Roland Julius and Betty (Linse) P. BS in Aero./Astron. Engring., Purdue U., 1991. Co-op engr. Gen. Dynamics, Ft. Worth, 1987-90; field engr. GE Indsl. and Power, Columbia, Md., 1991—. Mem. AIAA, U.S. Rowing Assn. Lutheran. Home: 5721 Ridge View Dr Alexandria VA 22310 Office: 6905D Oakland Mills Rd Columbia MD 21045

ARNDT, NORBERT KARL ERHARD, mechanical engineer; b. Heide, Holstein, Germany, Jan. 3, 1960; s. Erhard Karl Wilhelm and Johanna Eva Edith (Kruschka) A. MSc, Calif. Inst. Tech., 1984, PhD, 1988. Devel. engr. Motoren-Turbinen-Union, Munich, 1988-91; mgr. low pressure system BMW Rolls-Royce, Lohhof, Germany, 1991—. Contbr. articles to profl. jours. Rotary Found. scholar, 1983-84; Calif. Inst. Tech. Spl. Tuition award, 1984-88. Mem. ASME, AIAA, Verein Deutscher Ingenieure. Home: 55 Richildenstr, 8000 Munich Federal Republic of Germany Office: BMW Rolls-Royce GmbH, Carl-von-Linde Str 25, 8044 Lohhof Germany

ARNDT, ROGER EDWARD ANTHONY, hydraulic engineer, educator; b. N.Y.C., May 25, 1935; s. Ernest Otto Paul and Olive (Walters) A.; m. Jane Elizabeth Pfund, Dec. 1, 1990; children from previous marriage: Larysa Tamara, Tanya Sofia. B.C.E., CCNY, 1960; S.M., M.I.T., 1962, Ph.D., 1967. Chemist Consol. Testing Labs., New Hyde Park, N.Y., 1956-57; jr. civil engr. N.Y.C. Dept. Public Works, 1960; research engr. Allegheny Ballistics Lab., Cumberland, Md., 1962-63; sr. research engr. Lockheed Calif. Corp., Burbank, 1963-64; assoc. prof. aerospace engring. Pa. State U., 1967-77; prof. hydromechanics, dir. St. Anthony Falls Hydraulic Labs., U. Minn., Mpls., 1977-93; mem. Gov.'s Commn. on Cold Weather Research; 1st Theodor Ranov disting. lectr. SUNY, Buffalo, 1979; cons. in field. Editor several books on fluid mechanics and hydropower; contbr. articles in field, chpts. in books; also spl. publs. mech. engring. Recipient George Taylor Teaching award U. Minn., 1978, Lorenz G. Straub award, 1968, Fluids Engring. award ASME Fluids, 1993; NASA fellow, 1965-67. Fellow AIAA (assoc., Outstanding Faculty Adv. award 1971, 72, 73, 74), ASME (Fluids Engring. award 1993); mem. Internat. Assn. Hydraulic Rsch., Acoustical Soc. Am., ASCE, ASTM, Am. Water Resources Assn., N.Y. Acad. Scis., Twin City Cloud 7 Club, Sigma Xi. Home: 1820 N Ham Lake Dr Anoka MN 55304-5651 Office: U Minn St Anthony Falls Hydraulic Lab Miss River at 3rd Ave SE Minneapolis MN 55414

ARNESON, HAROLD ELIAS GRANT, manufacturing engineer, consultant; b. Mpls., Dec. 18, 1925; s. Theodore John and Ella Marie (Eliason) A.; m. Lorna May Mullen, Nov. 25, 1950 (dec. May 1991); 1 child, Grant. Student, U. Minn. Inst. Tech., 1943-46. Ptnr. Profl. Instruments Co., Mpls., 1947-77; cons. Precision Engring., Ft. Myers Beach, Fla., 1977—. With USAF, 1943-45. Mem. AAAS, Am. Soc. Precision Engrs., Soc. Mfg. Engrs. Achievements include research in precision mechanics and 10 patents in this field ,the most significant of which relates to the hydrodynamics of thin films as applied to externally pressurized gas bearings. Home and Office: 131 Andre Mar Dr Fort Myers Beach FL 33931

ARNETT, EDWARD MCCOLLIN, chemistry educator, researcher; b. Phila., Sept. 25, 1922; s. John Hancock and Katherine Williams (McCollin) A.; m. Sylvia Gettmann, Dec. 10, 1970; children—Eric, Brian; stepchildren—Elden, Byron, Colin Gatwood. B.S., U. Pa., 1943, M.S., 1946, Ph.D., 1949. Research dir. Max Levy and Co., Phila., 1949-53; asst. prof. Western Md. Coll., Westminister, 1953-54, assoc. prof., 1954-55; research fellow Harvard U., Cambridge, Mass., 1955-57; asst. prof. chemistry U. Pitts., 1957-61, assoc. prof., 1961-64, prof., 1964-80; R. J. Reynolds Corp. prof. Duke U., Durham, N.C., 1980-92, prof. emeritus, 1992—; vis. lectr. U. Ill., 1963; vis. prof. U. Kent, Canterbury, Eng., 1970; dir. Pitts Chem. Info. Ctr., 1967-70; mem. adv. bd. Petroleum Research Fund, 1968-71; mem. com. on chem. info. NRC, 1969-71. DuPont fellow, 1948-49; Guggenheim fellow, 1968-69; Mellon Inst. adj. sr. fellow, 1964-80; Inst. Hydrocarbon Chemistry sr. fellow, 1980. Fellow AAAS; mem. Am. Chem. Soc. (James Flack Norris award 1977, Pitts. award Pitts. chpt. 1976, Petroleum Chemistry award 1985), Nat. Acad. Scis., The Chem. Soc., Sigma Xi, Phi Lambda Upsilon. Author 220 papers in field.

ARNIZAUT DE MATTOS, ANA BEATRIZ, veterinarian; b. Rio de Janeiro, Oct. 6, 1959; came to U.S., 1984; d. Fabio F. and Emilia (Vasconcellos) A.; m. Francisco J. Vilella, Sept. 24, 1988; 1 child, Isabela Beatriz. DVM, U. Fed. Rural Rio de Janeiro, Brazil, 1982; MSc, La. State U., 1988. Natural resources specialist Dept. of Natural Resources, San Juan, P.R., 1990-92; aviary coord. Puerto Rican Parrot project Dept. of Interior-U.S. Fish and Wildlife Svc., Palmer, P.R., 1992—. Contbr. articles to profl. jours. Fellowship Brazillan Govt., 1984. Mem. Assn. of Avian Vets., Wildlife Disease Assn., Soc. for Tropical Vet. Medicine (reg. council vet. medicine of Rio de Janeiro 1982—), Am. Soc. Tropical Vet. Medicine. Democrat. Roman Catholic. Office: USFWS Puerto Rican Parrot Field Office 1-E Fernandez Garcia St Luquillo PR 00773

ARNOLD, ARTHUR PALMER, neurobiologist; b. Phila., Mar. 16, 1946; s. Wiley Ellsworth and Leona Arnold; m. Caroline Scheaffer, June 24, 1967; children: Jennifer, Matthew. AB, Grinnell Coll., 1967; PhD, Rockefeller U., 1974. Instr. Cen. State U., Wilberforce, Ohio, 1968-69; prof. UCLA, 1976—. Contbr. articles to profl. jours. Mem. AAAS, Soc. Neurosci., Soc. for Study Reproduction, Phi Beta Kappa. Office: UCLA Dept Psychology 405 Hilgard Ave Los Angeles CA 90024-1563

ARNOLD, DAVID DEAN, electrical engineer; b. Wichita, Kans., May 20, 1958; s. Ronald and Sharon A.; m. Janet Snyder, May 30, 1981; children: Lindsey, Nathan. BSEE, Kans. State U., 1981. Registered profl. engr., Kans. Student engr. Kans. Power and Light, Manhattan, 1980-81, divsn. engr., 1981-84; divsn. supt. Kans. Power and Light, Salina, 1984-85, KPL Gas Svc., Salina, 1985-90; mgr. distbn. ops. KPL Gas Svc., Topeka, 1990-92; mgr. meters and svc. KG&E (A Western Resources Co.), Wichita, 1992—. Bd. moderator Belmont Christian Ch., Salina, 1989. Mem. NSPE, Kans. Engring. Soc. (multiple offices 1988-90, Outstanding Young Engr. of Smoky Valley chpt. 1989). Republican. Office: KG&E 1900 C Central PO Box 208 Wichita KS 67201

ARNOLD, DONALD SMITH, chemical engineer, consultant; b. Cuyahoga Falls, Ohio, Sept. 14, 1920; s. Elton Dewey and Esther Anna (Schmid) A.; m. Eleanor Ann Webster, July 9, 1944; children: Ann A., Jane D., Elaine S., Dale F., David W., Douglas E. BSChemE, Ohio State U., 1942, MSChemE, 1947, PhD in Chem. Engring., 1949. Profl. engr., Ohio, Calif., Okla. Asst. prof., instr. dept. chem. engring. N.C. State U., Raleigh, 1947-53; head chem. dept. Nat. Lead Co. of Ohio, Fernald, 1953-59; head high energy fuels sect. Am. Potash & Chem. Corp., Henderson, Nev., 1959; mgr. Trona rsch. Am. Potash & Chem. Corp., Trona, Calif., 1959-67; dir. cen. engring. Am. Potash & Chem. Corp., L.A., 1967-69; from engring. specialist to sr. tech. advisor Kerr-McGee Corp., Oklahoma City, 1969-91; pvt. cons. Bethany, Okla., 1991—; mem. tech. com. Fractionation Rsch. Inc. Contbr. articles to profl. jours. Mem. County Svc. Area Com,. Trona, 1960-62; assoc. advisor Explorer Post, Boy Scouts Am., Trona, 1960-64; mem. gen. coun. Washita Presbytery, Oklahoma City, 1979-80. 1st lt. U.S. Army C.E., 1942-46, PTO, ETO. Named Disting. Alumnus Ohio State U. Coll. Engring., 1970. Fellow AICE (section chmn., com. mem. Design Inst. for Phys. Property Data), Am. Inst. Chemists, Sigma Xi, Tau Beta Pi; mem. AAAS, NSPE, Am. Chem. Soc. (sect. chmn.), Am. Inst. Mining & Metall. Engrs., Am. Soc. for Engring. Edn., Armed Forces Def. Preparedness Assn. Achievements include patent in field. Home and Office: 2005 N Briarcliff Ave Bethany OK 73008-5656

ARNOLD, JACK WALDO, physicist; b. Rossville, Ga., Apr. 14, 1935; s. Irvin Otis and Eula Louise (Kenimer) A.; m. Martha Ann Ford, May 15, 1955; children: Anna Marie, Jack Waldo II, William Thomas, Bradley Scott. BS in Physics, U. Tenn., Chattanooga, 1976. Lic. chief engr., commd. engr. examiner, commd. tech. instr. Nat. Inst. for Uniform Licensing of Power Engrs. Radiographer boilers Combustion Engring., Inc., Chattanooga, 1957; trainee to shift engr. TVA, 1957-76; sect. mgr./sr. engr. Gibbs & Hill, N.Y.C., 1976-77; div. mgr. Gen. Physics Corp., Chattanooga, 1977-81, Gilbert-Commonwealth, Inc., Jackson, Mich., 1981-83; group mgr. United Energy Svcs. Corp., Atlanta, 1983-88; pres. Arnold & Assocs., Inc., Roswell, Ga., 1988—. With USN, 1952-56. Mem. Ga. State Assn. Power Engrs. (state sec. 1992, 93), Masons, Sigma Pi Sigma. Republican. Methodist. Office: Arnold & Assocs Inc 11100 Bowen Rd Roswell GA 30075-2241

ARNOLD, JAY, engineering executive; b. Balt., Jan. 1, 1936; s. Otto Joseph and Margaret (Flannery) A.; m. Harriet Mary Metzbower, July 4, 1959; children: Kelly Marie Arnold Wood, Philip Driscoll Arnold, Michael Flannery Arnold. BS, Loyola Coll., Balt., 1965; MBA, Loyola Coll., Potomac, Md., 1977; postgrad., George Washington U., 1980-81, Berlitz Inst., Washington, 1987-90. Software and systems engring. positions including mgt. NASA's Manned Space Program IBM, 1962-78; vis. IBM prof. Morgan State U., Balt., 1978-79; planner of automation strategy Fed. Systems div. IBM, Gaithersburg, Md., 1979-81; sr. mgr. systems design depts. USAF Data Systems Modernization Fed. Systems div. IBM, Gaithersburg, 1981-83, sr. mgr. systems design depts. FAA Advanced Automation System, 1983-87; dir. network mgmt. and control Comsat Systems div. Communications Satellite Corp., Clarksburg, Md., 1987-88, sr. dir. Deutsche Fermelde Satellite program, 1988-90, sr. dir. MOSCOM program, 1990, sr. dir. engring. advanced systems, 1991—; speaker, instr. and lectr. in computer and communication field, 1973—. Caregiver Frederick County Hospice, 1984-87; club leader Frederick County 4-H, 1975-80; pres./v.p. Frederick County Sheep Breeders Assn., 1983-84; chmn. bd. govs. Am. Bouviers Des Flandres Club, 1981-82; mem. St. Peter's Ch. Parish Coun., 1991—, St. Peter's Parish Men's Club, 1990-91. With USAF, 1958-62. Recipient parenting awards Future Farmers of Am., 1978-80, Award for Advancement of Human Rights UN Assn., 1984; named Alumni of Yr. Mt. St. Joseph Coll. High Sch., 1989. Mem. Am. Assn. for Retired Persons, Armed Forces Communications and Electronics Assn., Johnsville Ruritan Club. Democrat. Roman Catholic. Avocations: farming, breeding quarter horses, raising sheep, personal computing, boating. Home: Heaven Sent Farm 11131 Repp Rd Union Bridge MD 21791 Office: COMSAT 22300 Comsat Dr Clarksburg MD 20871-9470

ARNOLD, KENT LOWRY, electrical engineer; b. Washington, Sept. 14, 1960; s. Anthony and Ruth (Lowry) A.; m. Doreen McMahon, Nov. 10, 1990; children: Kathleen, John. BA in Physics, Dartmouth Coll., 1982, MBA in Ops., 1989. Mem. tech. staff Hughes Aircraft Co., El Segundo, Calif., 1982-85, Avco Everett (Mass.) Rsch. Lab., 1985-87, Mitre Corp., McLean, Va., 1990—; v.p. Picture Conversion, Inc., Falls Church, Va., 1989-90. Author papers in field. Mem. IEEE, AIAA (sr.), Inst. Navigation, U.S. Naval Inst., Armed Forces Comm. and Electronics Assn. Office: Mitre Corp 7525 Colshire Dr Mc Lean VA 22102

ARNOLD, TONI LAVALLE, designer; b. N.Y.C., Nov. 29, 1947; d. Aldo Peter and Margaret E. (Tessitore) Lavalle; m. Asbury Rembert Arnold, July 26, 1975. Student, Marymount Coll., 1965-67. Electro-mech. drafter PRD Electronics, Syosset, N.Y., 1973-74; electro-mech. designer/drafter Cadre Corp., Atlanta, 1974-79; EDA libr. resource mgr. Harris Corp., Palm Bay, Fla., 1982—. Stained glass artist represented in galleries in Fla., N.Y. Vol., leader Camp Fire Sunshine Coun., Lakeland, Fla., 1984-87, bd. dirs., 1993—; mem. Brevard County Dem. Exec. Com., 1993—; vol. South Brevard Habitat for Humanity. Recipient scholarship N.Y. State Bd. Regents, 1964, Blue Ribbon award Camp Fire Nat., Kansas City, 1985. Democrat. Avocations: stained glass art, Appalachian basketweaving, wilderness camping, jewelry design, beachcombing. Home: 203E 6th Ave Melbourne Beach FL 32951-2337 Office: Harris Corp Govt Com Divsn PO Box 91000 Melbourne FL 32902-3001

ARNOLD HUBERT, NANCY KAY, writer; b. Kalamazoo, Mich., May 9, 1951; d. Byron Lyle and Ada (Doorlag) Arnold; m. Louis Scott Hubert, May 5, 1989. BFA in Painting, Western Mich. U., 1983, postgrad., 1985-86. Writer Advanced Systems & Designs, Inc., Farmington Hills, Mich., 1987-89; pres., owner TechWrite, Kalamazoo, 1989—. Author: (poetry) Tetragonal Pyramids, 1982; exhibited in group shows, Kalamazoo, 1983, Western Mich. U., 1982, 85. Mem. ACLU, NAFE, Humane Farming Assn. Am., People for Ethical Treatment of Animals, Greenpeace. Libertarian. Avocations: bike riding, reading, piano, singing, cross-country skiing.

ARNONE, MARY GRACE, radiology technologist; b. Bronx, N.Y., Dec. 28, 1961; d. Anthony Rocco and Mary Helen (Doring) A. AA, Acad. Health Sci., U.S. Army, 1982. Lic. radiologist, lic. mammographer, N.Y. Radiology technologist New Rochelle (N.Y.) Hosp., 1986-90, Our Lady of Mercy Hosp., Yonkers, N.Y., 1988—. With U.S. Army, 1982-86.

Democrat. Lutheran. Office: Our Lady of Mercy Hosp PO Box 566 Yonkers NY 10704-0566

ARNOTT, ERIC JOHN, ophthalmologist; b. Sunningdale, Berkshire, Eng., June 12, 1929; s. Robert and Cynthia Emita Amelia (James) A.; m. Vernica Mary Langue, Nov. 19, 1960; children: Stephen John, Tatiana Amelia, Robert Laureston John. BA, Trinity Coll., Dublin, 1953, M.B., B.Ch., BAO, 1954; DO, U. London, 1956, FRCS, 1963, FC Ophth., 1989; FC Oph. (hon.), 1989. Houseman Royal Victoria Eye & Ear Hosp. and Adelaide Hosp., Dublin, 1953-54; resident surg. officer Moorfields Eye Hosp., London, 1959-60; sr. registrar U. Coll. Hosp., London, 1961-63; cons. ophthalmologist surgeon Royal Eye Hosp., London, 1965-74, Charing Cross Hosp., London, 1971—, Royal Masonic Hosp., London, 1971—; faculty U. London, 1966—, Royal Coll. Opthalmologists, London, 1966—. Co-author: Emergency Surgery, 1983, Intraocular Lens Implantation, 1987; contbr. articles to profl. jours.; developer, inventor totally encircling loop intraocular lens for implantation into the capsular bag of the eye after surgery; patentee in field. Founder, chmn. The Arnott Trust, The Great London Treasure Hunt. Fellow Royal Coll. Surgeons, Royal Coll. Ophthalmologists, Am. Acad. Ophthalmologists; mem. Order of St. John, Outpatient Ophthalmic Surgery Soc., London Med. Soc., Royal Soc. Medicine, European Phaco & Laser Soc. (pres. 1986—), Kildare St. Club (Dublin), RAC, The Garrick Club (London). Achievements include research on excimer laser for correction of short sight. Home: Trottsford Farm, Headley, Nr Bordon, Hampshire England GU358TF Office: Arnott Ophthalmic Clinic, 11 Milfors House 7 QUeen Anne St, London W1M 9FD, England Office: The Cromwell Hosp, The Arnott Ophthalmic Unit, Cromwell Rd, London England SW5 O2U

ARNOWITZ, LEONARD, engineering executive; b. Paterson, N.J.; s. Nathan and Esther (Liboff) A.; m. Shirley Sarah Goteiner; children: Edward Bruce, Gary Mark, Mitchell Charles, Joanbeth. BSEE, N.J. Inst. Tech.; MS in Engring., Johns Hopkins U. Div. chief Goddard Space Flight Ctr. NASA, Greenbelt, Md., 1967-88; cons. Israel Aircraft Industries, Yehud, Israel, 1988-92; com. mem. Israel Aircraft Industries, Tel Aviv, 1988-92; CEO, pres. GAR Aerospace, Daytona Beach, Fla., 1991—; exec. v.p. IDEA, Beltsville, Md., 1992—. Program chmn. Vanguard 25th Anniversary Commemoration. With U.S. Signal Corps, 1944-46. Assoc. fellow AIAA (bd. dirs. nat. capital sect. 1985-88). Home: 100 E Wind Ln Fern Park FL 32730

ARNSDORF, MORTON FRANK, cardiologist, educator; b. Chgo., Aug. 7, 1940; s. Selmar N. and Irmgard C. (Steinmann) A.; m. Mary Hunter Tower, Dec. 26, 1963 (div. 1982); m. Rosemary Crowley, Dec. 27, 1986. BA magna cum laude, Harvard U., 1962; MD, Columbia U., 1966. Diplomate Am. Bd. Internal Medicine. House staff officer U. Chgo., 1966-69; fellow cardiology Columbia-Presbyn. Med. Ctr., N.Y.C., 1969-71; asst. prof. medicine U. Chgo., 1973-79, assoc. prof., 1979-83, prof., 1983—, chief sect. cardiology, 1981-90; mem. pharmacology study sect. NIH, 1981-84. Contbr. articles to profl. jours. Served to maj. USAF, 1971-73. Recipient Rsch. Career Devel. award NIH, 1976-81; rsch. grantee Chgo. Heart Assn., 1976-78, NIH, 1977—, NIH Merit award, 1989—. Fellow ACP, Am. Coll. Cardiology (mem. editorial bd. JACC 1983-87, 90—, gov.-elect Ill. 1990-91, gov. Ill. 1991—, pres. Ill. chpt. 1991—); mem. AMA, Am. Heart Assn. (dir. 1981-83, chmn. exec. com. basic sci. council 1981-83, steering com. 1983-86, mem. rsch. program and evaluation com. 1989-91, assoc. editor circulation Rsch. 1986-91), Am. Heart Assn. Met. Chgo. (v.p. 1986, pres.-elect 1987-88, pres. 1988-89, bd. govs., chmn. rsch. council 1981, chmn. program coun. 1986-88), Am. Fedn. Clin. Research, Assn. Univ. Cardiologists, Cen. Soc. Clin. Research (chmn. cardiovascular council 1986-87, sec.-treas. 1991—), Assn. Profs. Cardiology (founding mem., bylaws com. 1989-90), Chgo. Cardiology Group (pres. 1990-92), Chgo. Med. Soc., Ill. Med. Soc., Cardiac Electrophysiology Soc. (sec.-treas. 1984-86, pres. 1986-88). Club: Quadrangle. Office: U Chgo Hosps and Clinics Sect Cardiology MC 6080 5841 S Maryland Ave Chicago IL 60637-1470

ARO, GLENN SCOTT, environmental and safety executive; b. Balt., Jan. 18, 1948; s. Raymond Charles Sr. and Elizabeth Virginia (Coppage) A.; m. Marlene Rose Lefler, Jan. 8, 1972 (div. June 1987); children: Vincent Wade, Marlena Irene. BS in Mech. Engring., Gen. Motors Inst., Flint, Mich., 1972; MBA in Fin., Wayne State U., 1980. Registered environmental assessor, Calif. From engr. to supr. GM, Detroit, Balt., L.A., 1966-84; environ. specialist New United Motor, Fremont, Calif., 1984-86; environ. engring. mgr. Def. Systems FMC Corp., San Jose, Calif., 1986-89; cons./exec. sales rep. Gaia Systems, Menlo Park, Calif., 1990; corp. environ. & safety mgr. Ampex Corp., Redwood City, Calif., 1990-92; audit programs mgr. Hughes Environ. Systems, Manhattan Beach, Calif., 1992—; lectr. colls. and seminars Environ. Regulatory Issues, 1988—. Author: Developing a National Environmental Policy in a Global Market, 1989; contbd. articles to profl. jours. Panel mem. Toxics Awareness Project, San Francisco, 1989—; com. mem. Environ. Working Group, Sacramento, 1986-88. Mem. Peninsula Indsl. & Bus. Assn. (bd. dirs., v.p. 1988-91). Republican. Roman Catholic. Avocations: running, reading, travel, baseball, basketball. Home: 2836 Palos Verdes Dr W Palos Verdes Estates CA 90274

ARONHIME, PETER BYRON, electrical engineering educator, researcher; b. Louisville, Apr. 21, 1940; s. Samuel Howard and Mary (Patterson) A.; m. Rose Spalding, June 6, 1964; children: Reagan, Anna, Samuel, Emily, Sarah. BEE, U. Louisville, 1962; MSEE, Colo. State U., 1964, PhD in Elec. Engring., 1971. Electronics engr. Bell Telephone Labs., Winston-Salem, N.C., 1962; mem. tech. staff Hughes Aircraft, Fullerton, Calif., 1964-65; from asst. prof. to assoc. prof. Tri-State U., Angola, Ind., 1965-74; asst. prof. Ill. Inst. Tech., Chgo., 1974-76, from assoc. prof. to prof. U. Louisville, 1976—; NSF sci. faculty fellow, 1969-70, ASEE summer faculty fellow NASA, Houston, 1974, 76; vis. prof. Colo. State U., 1987; cons. Inst. for Gas Tech., Chgo., 1975; mem. organizing com. Southeastcon '84, Louisville; co-chmn. 28th Midwest Symposium on CAS, Louisville, 1985. Contbr. articles to profl. jours. NASA grantee, 1977-79; NSF grantee, 1988-91. Mem. IEEE (sr.), Circuits and Systems Soc. of IEEE, Instrumentation and Measurement Soc., Edn. Soc. Achievements include research in network theory, electronics, synthesis and computer-aided testing of electrical systems. Office: U Louisville Dept Elec Engring Eastern Pkwy Louisville KY 40292

ARONOFF, GEORGE RODGER, medicine and pharmacology educator; b. Peoria, Ill., Mar. 6, 1950. BA in Chemistry with distinction, Ind. U., 1972; MD with honors, Ind. U., Indpls., 1975, MS in Pharmacology, 1984. Diplomate Am. Bd. Internal Medicine; diplomate Am. Bd. Internal Medicine Nephrology. Intern in internal medicine Ind. U., Indpls., 1975-76, resident, 1976-77, clin. fellow div. nephrology, 1977-78, chief resident in internal medicine Wishard Meml. Hosp., 1978-79, rsch. fellow div. nephrology, 1979-80, instr. phys. diagnosis, 1977-78, instr. medicine, 1978-79, from asst. prof. to assoc. prof. medicine, 1980-87, assoc. prof. pharmacology, 1985-87; prof. medicine, prof. pharmacology U. Louisville, 1987—; mem. staff Ind. U. Hosp., Indpls., 1980-87, Wishard Meml. Hosp., Indpls., 1980-87; mem. staff VA Med. Ctr., Indpls., 1980-87, Louisville, 1987—; mem. staff Jewish Hosp., Louisville, 1987—, Norton's Hosp., Louisville, 1987—, Meth. Evang. Hosp., Louisville, 1987—, Humana Hosp. U. Louisville, 1987—, Clark County Hosp., Jeffersonville, Ind., 1987—; fellow in clin. pharmacology Eli Lilly & Co., Indpls., 1979-80; acting dir. med. ICU Wishard Meml. Hosp., 1978-79, mem. exec. coun., 1977-78; mem. exec. coun. Med. Ctr. Ind. U., 1977-78, mem. adv. com. clin. rsch. ctr., 1981-87, mem. pharmacy & therapeutics com., 1983-87, mem. biomed. rsch. com., 1983-87, others; mem. R&D com. VA Med. Ctr., Indpls., 1981-84, Louisville, 1987—; mem. faculty coun. Ind. U.-Purdue U., Indpls., 1982-84, mem. athletic affairs com., 1982-84; mem. core group study com. dept. medicine Sch. Medicine U. Louisville, 1987—, mem. residency evaluation com., 1988—; mem. pharmacy and therapeutics com. Humana Hosp. U. Louisville, 1989—, chmn., 1989; William N. Creasy vis. prof. clin. pharmacology U. Oreg., Portland, 1986; mem. U.S. Pharmacopeial Conv., 1985—; presenter numerous profl. meetings. Mem. editorial bd. Am. Jour. Kidney Disease, 1981—, Antimicrobial Agents and Chemotherapy, 1981—, Seminars in Dialysis, 1990—; reviewer Kidney Internat., 1983—; cons. The Med. Letter, Inc., 1981—; contbr. U.S. Pharmacopeia Dispensing Info., 1981—, numerous articles and abstracts to profl. jours. Mem. Nat. Kidney Found. Ind., 1979-87, bd. dirs., 1985-87, mem. exec. com., 1985-87, bd. dirs. cen. Ind. chpt., 1985-87, chmn. fundraising, 1985-87; bd. dirs. Nat. Kidney Found. Ky., 1987—, exec. com., 1988—, pres. med. and sci. adv. bd., 1989—. Fellow ACP; mem. AAAS, Am. Fedn. Clin. Rsch., Am. Soc. Clin. Pharmacology and Therapeutics (med. edn. rev. com. 1984-86), Am. Soc. Nephrology, Cen. Soc. Clin. Rsch., Ky. State Med. Assn., Inter-Am. Soc. Chemotherapy (chmn. div. medicinal chemistry and pharmacology 1982-87), Jefferson County Med. Soc. (editorial bd. Louisville Medicine 1989—, editor 1990), Phi Eta Sigma, Phi Lambda Upsilon, Phi Beta Kappa, Alpha Omega Alpha, Sigma Xi. Achievements include research on pharmacokinetics of common antibiotics used in CAPD, on acid base disorders, on clinical pharmacology and antidepressant action of specific serotonin uptake blockers, on clinical pharmacology and physiologic dispositon of fluoxetine, on guidelines for drug therapy in dialysis, on disorders of potassium balance, on antimicrobial therapy in impaired renal function, on effect of captopril on gentamicin induced acute renal failure, and on effect of non-steroidal anti-inflammatory agents on renal function. Office: U Louisville Kidney Disease Program 615 S Preston St Louisville KY 40292-0001

ARONOWITZ, FREDERICK, physicist; b. N.Y.C., July 3, 1935; s. Nathan and Beatrice (Harris) A.; m. Marguerite Aronowitz; children: Malica, Michelle, Jacqueline. BS in Physics, Polytechnic Inst. Bklyn., 1956; PhD in Physics, NYU, 1969. Rsch. scientist Honeywell Inc., Mpls., 1962-69, staff scientist, 1969-78, sect. chief, 1978-82, program mgr., 1982-83; mgr. Raytheon Co., Sudbury, Mass., 1983-84; chief scientist Rockwell Internat., Anaheim, Calif., 1984—; cons. Ea. Rsch. Group, N.Y.C., 1960-62. Author: (with others) The Laser Gyro, 1971, Theory of a Traveling Wave Optical Maser, 1965. Recipient Prize for Indsl. Applications of Physics Am. Inst. Physics, 1983, Elmer A. Sperry award ASME, IEEE, Soc. Auto. Engrs., 1984, Kirshner award IEEE, 1988, Tech. Achievement award Internat. Soc. Optical Engring., 1990, Albert A. Michelson award Navy League, 1990. Achievements include first in the development of the ring laser gyroscope; theoretical/experimental studies provided the foundation for the development of laser gyro technology worldwide; numerous patents relating to laser gyro technology. Home: 32542 Adriatic Dr Monarch Beach CA 92629 Office: Rockwell Internat 3370 Miraloma Ave Anaheim CA 92803

ARONSON, CASPER JACOB, physicist; b. Canisteo, N.Y., Sept. 1, 1916; s. Aaron Julius and Rhyda (Fybush) A.; m. Eleanor Jaffray Gould, Nov. 11, 1943; 1 child, Rhyda Anne Aronson Conant. BS, U. Rochester, 1938, MS, 1940. Physicist dept. terr. magnetism Carnegie Inst., Washington, 1940-41, Naval Ordnance Lab., Silver Spring, Md., 1941-42, 44-74, USN Bur. Ordnance, Washington, 1942-44; cons. VSE Corp., Alexandria, Va., 1980-86. Recipient Flemming award U.S. Jr. C. of C., Washington, 1955. Mem. Philos. Soc. Wash. (treas. 1970-72), Phi Beta Kappa, Sigma Xi. Home: 3401 Oberon St Kensington MD 20895

ARONSON, DAVID, chemical and mechanical engineer; b. Bklyn., Sept. 24, 1912; s. Oscar and Amy (Maas) A.; m. Hannah Unger, Feb. 11, 1945; children: Deborah, Judith. B.S., Cooper Union Sch. Engring., 1936; B.S. Ch.E., Poly. Inst. Bklyn., 1944. Chem. engr. Sanderson & Porter, Pine Bluff, Ark., 1942-43, Kellex Corp., N.Y.C., 1943-45, Elliott Co., Jeannette, Pa., 1945-51; staff engr. Worthington Corp., Harrison, N.J., 1951-54; partner Deutsch & Loonam, N.Y.C., 1954-55; cons., dir. rsch. and devel. Worthington Corp., Harrison, N.J., 1955-70; cons. engr. David Aronson Assocs., Upper Montclair, N.J., 1970—. Contbr. articles to profl. jours. Chmn. community relations com. Congregation Shomrei Emunah, 1971—. Fellow ASME (rep.-coord. N.J. State Legis. 1989—); mem. Am. Inst. Chem. Engrs., N.Y. Acad. Scis., Sigma Xi, Tau Beta Pi. Achievements include numerous patents on cryogenic systems power generating systems using low grade heat, and on large capacity water chillers both mechanical compression and chemical absorption; development of simplified techniques used to control production barrier for separation of uranium hexafluoride isotopes. Address: 9 Riverview Dr W Upper Montclair NJ 07043

ARONSON, MIRIAM KLAUSNER, gerontologist, consultant, researcher, educator; b. N.Y.C., July 12, 1940; d. Joseph and Martha (Sklower) Klausner; children: Eric, Andrew, Elliott. AB, Barnard Coll., 1961; EdM, Columbia U., 1970, EdD, 1980. Cons., researcher geriatric facilities N.Y, N.J., 1969-75; dir. geriatric program Soundview-Throgs Neck Community Mental Health Ctr., Bronx, N.Y., 1975-78; chief services to elderly Bronx-Lebanon Hosp. Community Mental Health Ctr., Bronx, 1978-79; dir. longterm care Gerontol Ctr. Albert Einstein Coll. Medicine, Bronx, 1979-91, from asst. prof. to prof. neurology and psychiatry, 1980-90, clin. assoc. prof. epidemiology and social medicine and psychiatry, 1991-92; dir. Inst. on Aging Bergen Pines County Hosp., Paramus, N.J., 1993—; prin. investigator senile dementias, risks and course, 1984—; dir. long-term care svcs. Jewish Guild for the Blind, Yonkers, N.Y., 1991-92; cons. gerontologist, dir. Inst. on Aging, Bergen Pines County Hosp., Paramus, N.J. Author, dir.: film series Teaching Series on Alzheimers Disease, 1980; author (with R. Bennett and B. Gurland) The Acting Out Elderly, 1983; editor, contbr. Understanding Alzheimers Disease, 1988. Mem. Hillsdale (N.J.) Bd. Health, 1975-82, pres., 1977-81; mem. planning and policy com. Outreach Health Svc. Program, Bergen County, N.J., 1976; mem. long range planning com. Bergen Pines County Hosp., 1989-92; co-chair task force on AIDS Bronx Mcpl. Hosp., 1989-91. N.Y. State Regents scholar, 1957-61; adminstrn. Aging grantee, 1968-70, 74-75; Nat. Council Community Mental Health Ctrs. Best Outreach Program award, 1976; Alzheimers Disease Soc. Greater N.Y. award, 1982. Fellow Gerontol. Soc. Am. (dir. task force on long term care 1981-85), Am. Orthopsychiat. Soc. (program com. 1985-88, co-chair study group on aging 1988—, N.Y.C. mayor's alzheimers disease adv. com.); mem. Am. Geriatrics Soc., Nat. Alzheimers Disease and Related Disorders Assn. (dir. edn. and pub. awareness com. 1979-84, cons. 1985-90, Founders award 1987), Western Gerontol. Soc. (bd. dirs. 1983-85). Office: 230 E Ridgewood Ave Paramus NJ 07632

ARORA, SAM SUNDER, mechanical engineer; b. Lyallpur, India, Jan. 4, 1942; came to U.S., 1981; s. Ram Das and Vidya Wati (Taneja) A. B of Engring. U. Delhi, 1963; M of Engring., U. Toronto, 1969. Registered profl. engr. N.H., Ont. Mech. engr. Salzgitter (Germany) Stahlbau, 1964-67; design engr. Ont. (Can.) Hydro, 1969-81; consulting engr. United Engrs. & Constrn., Phila., 1981-86; sr. mech. engr. Gibbs & Hill, N.Y.C., 1986-87, Bechtel Corp., San Francisco, 1987-88; sr. staff mem. Advanced Tech. Inc., Portland, Oreg., 1988-91; procurement engr. Nuclear Energy Svcs., San Clemente, Calif., 1992—. Ontario Grad. fellow, 1968; U. Delhi scholar, 1958-63. Mem. NSPE, ASME, Am. Soc. Quality Control, Am. Nuclear Soc. Home: PO Box 5767 San Clemente CA 92674-5767

ARORA, SARDARI LAL, chemistry educator; b. Lahore, Pakistan, June 4, 1929; came to U.S., 1964; s. Uttam Chand and Kushal Devi Arora; m. Sunita Chawla, May 9, 1960; children: Nita, Nalini. MSc, Lucknow (India) U., 1953, PhD, 1959. Chief chemist, dir. R & D, Internat. Liquid Crystal Co., Cleve., 1971-74; rsch. assoc. Liquid Crystal Inst., Kent (Ohio) State U., 1966-71, researcher, 1983—, casual asst. prof. chemistry, 1975-77, vis. asst. prof., 1977-80, asst. prof., 1980-86, assoc. prof., 1986—; cons. Crystaloid Electronic Corp., 1976-78, Timex Corp., 1976-78, Liquid Crystal Applica-tion, Inc., 1983; presenter in field. Contbr. articles to sci. jours.; inventor, patentee in field. Fellow Coun. Sci. Indsl. Rsch, Govt. of India, 1957; fellow Aerospace Med. Rsch Lab., 1966, NASA, 1968; grantee NSF, 1983, indsl. grantee, 1986. Fellow Am. Inst. Chemists; mem. AAAS, Am. Chem. Soc., Internat. Union Pure and Applied Chemistry, Sigma Xi. Achievements include patents for Field Effect Light Shutter Employing Low Temperature Nematic Liquid Crystals; for Liquid Crystal Materials; research in development of polymer and other new liquid crystal materials, their characterization and technical applications. Home: 162 Steeplechase Ln Munroe Falls OH 44262 Office: Kent State U 6000 Frank Ave NW Canton OH 44720-7599

ARP, DANIEL J., biochemistry educator; b. Henderson, Nebr., Mar. 14, 1954; s. Jack Jr. and Delores Lucille (Brown) A.; m. Wanda Hofman, Aug. 10, 1974; children: Sarah, James. BS in Chemistry with distinction, U. Nebr., 1976; PhD in Biochemistry and Bacteriology, U. Wis., Madison, 1980. Lab. asst. dept. agrl. biochemistry U. Nebr., Lincoln, 1973-76; grad. rsch. asst. biochemistry dept. U. Wis., Madison, 1976-80; NATO postdoctoral fellow U. Erlangen, West Germany, 1980-82; from asst. prof. to assoc. prof. dept. biochemistry U. Calif., Riverside, 1982-89; assoc. prof., dir. Lab. for Nitrogen Fixation Rsch., Dept. Botany and Plant Pathology Oreg. State U., Corvallis, 1990-92, prof., dir. Lab. for Nitrogen Fixation Rsch., Dept. Botany and Plant Pathology, 1992—, dir. molecular biology program, 1993—; adminstr. dept. botany and plant pathology Oreg. State U., chair grad. studies com., 1990-93, adv. com. 1990-92, affiliate mem. dept. biochemistry and biophysics, mem. molecular and cellular biology grad. program, recruitment com., 1990-92, admissions com., 1992-93, mem. plant physiology grad. program, mem. Ctr. for Gene Rsch. and Biotech., participant hazardous waste management program; grant panel mem. Dept. Energy, Basic Energy Scis., 1988, Dept. Agriculture, Competitive Rsch. Grants Office, Nitrogen Fixation Program, 1989, 91; mem. rev. team Dept. Navy, Naval Med. Rsch. and Devel. Command, 1991; lectr. Am. Soc. Microbiology, 1991, Ctr. for Gene Rsch. and Biotech., 1991, Western Oreg. State Coll., 1991, Reed Coll., 1991, Advanced Sci. & Tech. Inst., 1992, Western Region Hazardous Substances Rsch. Ctr., 1992, U. Nebr., 1992, U. Georgia, 1992, other presentations and seminars. Contbr. numerous chpts. to books, articles to profl. jours. including Can. Jour. Microbiology, Jour. Bacteriology, Biochemistry, Jour. Biol. Chemistry. Vol. demonstrator of sci. concepts to elem. schs., youth groups; co-founder Partnership for Sci. Edn. Com. Grantee NSF, 1983-85, 85-88, 86-87, DOE, 1984-86, 86-88, 88-91, 91-94, USDA, 1984-86, 88-90, 90-94, 91-94, WRRI, 1990-91, EPA, 1990-93; Regent's scholar U. Nebr., 1972; Evelyn Steenbock fellow, 1977, Wharton fellow, 1979. Mem. AAAS, Am. Soc. for Microbiology, Am. Chem. Soc., Am. Soc. for Biochemistry and Molecular Biology, Phi Beta Kappa. Achievements include research in biochemistry and physiology of the microbial N cycle, biological N2 fixation, H2-utilizing microorganisms, biochemistry, physiology and molecular biology of nitrification, enzymology of gas-utilizing metalloenzymes; enzyme inhibitors as probes of enzyme mechanism and physiological function; bioremediation. Home: 1999 NW Lantana Corvallis OR 97330 Office: Oreg State U Dept Botony and Plant Path Cordley 2082 Corvallis OR 97331-2902

ARPS, DAVID FOSTER, electronics engineer; b. Napoleon, Ohio, July 28, 1948; s. Fred B. and Melba Lavern (Harrison) A.; m. Jacqueline A. Vollmar, Feb. 16, 1973 (div. June 1978); m. Vickie Lee Westrick, Mar. 19, 1982; children: Derek, Elizabeth. BS in Astronomy, Case Inst. Tech., 1970; MAT in Physics, Bowling Green State U., 1975; MS in Atmospheric Physics, U. Nev., Reno, 1977. Cert. secondary edn. and community coll. tchr. Astronomy instr. U. Toledo, Ohio, 1970; physics tchr. Napoleon (Ohio) High Sch., 1970-74; teaching asst. Bowling Green (Ohio) State U., 1974-75; rsch. fellow Desert Rsch. Inst., Reno, 1975-78; mech. engr., physicist Naval Air Warfare Ctr., Aircraft Div., Indpls., 1978-84, electronics engr., failure analyst, 1984—; astronomical rschr. Ritter Obs., U. Toledo, Ohio, 1970; solar radiation rschr. Desert Rsch. Inst., Reno, 1975-78. Mem. PTA, Mt. Comfort (Ind.) Elem., 1989-93; asst. coach Mt. Comfort (Ind.) Elem. Sports, 1992-93. Recipient Rsch. fellowships Desert Rsch. Inst., Reno, 1975-78. Mem. Nat. Geog. Soc., Ind. Astron. Soc., Wilderness Soc., Sigma Pi Sigma. Methodist. Achievements include numerous technical reports on component failure analysis, surface analysis and x-ray microanalysis. Home: 7041 Glendale Ln Greenfield IN 46140 Office: Naval Air Warfare Ctr Aircraft Div 6000 E 21st St Indianapolis IN 46219-2189

ARRATIA-PEREZ, RAMIRO, chemistry educator; b. Santiago, Chile, Aug. 19, 1949; s. Cristina Perez; m. Lucia Hernandez Arratia. Degree chemistry, U. Chile, Santiago, 1975, licenciate, 1976; PhD in Chemistry, U. Calif., Davis, 1983. Vis. scientist U. Santa Clara, Calif., 1983-84; postdoctoral fellow Simon Fraser U., Burnaby, Can., 1984-86, vis. prof., 1987-89; vis. prof. U. Tex., Arlington, 1986-87; asst. prof. U. Catolica, Santiago, Chile, 1989-90, assoc. prof., 1990—. Contbr. articles to Jour. Chem. Physics, Jour. Magnetic Resonance, Phys. Rev. B., Jour. Phys. Chemistry. DAAD fellowship, 1976, Fulbright fellowship, 1979. Mem. Am. Chem. Soc., Chilean Chem. Soc. (v.p.), U. Catolica (assoc.). Roman Catholic. Office: U Catolica De Chile, Vicuna Mackenna 4860, 306 Santiago Chile

ARRIGO, SALVATORE JOSEPH, biologist; b. Washington, Nov. 26, 1960; s. Salvatore Joseph and Elizabeth Jane (Phelps) A.; m. Kelly Colleen Greer, Oct. 1, 1988; 1 child, Caitlin. BA in Biology, U. Va., 1982; PhD in Biology, Johns Hopkins U., 1988. Postdoctoral fellow UCLA Sch. Medicine, 1988-91; asst. prof. microbiology and immunology Med. U. S.C., Charleston, 1991—. Rsch. grantee Am. Found. for AIDS Rsch., 1991-93, NIH, 1992—; Leukemai Soc. Am. fellow, 1989-91. Mem. AAAS. Achievements include research on involvement of the HIV-1 Rev protein in the expression of viral proteins. Office: Med U SC 171 Ashley Ave 203 BSB Charleston SC 29425

ARROW, KENNETH JOSEPH, economist, educator; b. N.Y.C., N.Y., Aug. 23, 1921; s. Harry I. and Lillian (Greenberg) A.; m. Selma Schweitzer, Aug. 31, 1947; children: David Michael, Andrew. B.S. in Social Sci., CCNY, 1940; M.A., Columbia U., 1941, Ph.D., 1951, DSc (hon.), 1973; LL.D. (hon.), U. Chgo., 1967, City U. N.Y., 1972, Hebrew U. Jerusalem, 1975, U. Pa., 1976, Washington U., St. Louis, 1989; D.Social and Econ. Scis. (hon.), U. Vienna, Austria, 1971; LL.D. (hon.), Ben-Gurion U. of the Negev, 1992; D.Social Scis. (hon.), Yale, 1974; Doctor (hon.), Université René Descartes, Paris, 1974, U. Aix-Marseille III, 1985; Dr.Pol., U. Helsinki, 1976; M.A. (hon.), Harvard U., 1968; D.Litt., Cambridge U., 1985. Research assoc. Cowles Commn. for Research in Econs., 1947-49; asst. prof. econs. U. Chgo., 1948-49; acting asst. prof. econs. and stats. Stanford, 1949-50, assoc. prof., 1950-53, prof. econs., statistics and ops. research, 1953-68; prof. econs. Harvard, 1968-74, James Bryant Conant univ. prof., 1974-79; exec. head dept. econs. Stanford U., 1954-56, acting exec. head dept., 1962-63, Joan Kenney prof. econs. and prof. ops. research, 1979-91, prof. emeritus, 1991—; economist Council Econ. Advisers, U.S. Govt., 1962; cons. RAND Corp. Author: Social Choice and Individual Values, 1951, Essays in the Theory of Risk Bearing, 1971, The Limits of Organization, 1974, Collected Papers, Vols. I-VI, 1983-85; co-author: Mathematical Studies in Inventory and Production, 1958, Studies in Linear and Nonlinear Programming, 1958, Time Series Analysis of Inter-industry Demands, 1959, Public Investment, The Rate of Return and Optimal Fiscal Policy, 1971, General Competitive Analysis, 1971, Studies in Resource Allocation Processes, 1977, Social Choice and Multicriterion Decision Making, 1985. Served as capt. AUS, 1942-46. Social Sci. Research fellow, 1952; fellow Center for Advanced Study in the Behavioral Scis., 1956-57; fellow Churchill Coll., Cambridge, Eng., 1963-64, 70, 73, 86; Guggenheim fellow, 1972-73; Recipient John Bates Clark medal Am. Econ. Assn., 1957; Alfred Nobel Meml. prize in econ. scis., 1972, von Neumann prize, 1986. Fellow AAAS (chmn. sect. K. 1983), Am. Acad. Arts and Scis. (v.p. 1979-81, 91-93), Econometric Soc. (v.p. 1955, pres. 1956), Am. Statis. Assn., Inst. Math. Stats., Am. Econ. Assn. (exec. com. 1967-69, pres. 1973), Internat. Soc. Inventory Rsch. (pres. 1983-90); mem. NAS (mem. coun. 1990-93), Internat. Econs. Assn. (pres. 1983-86), Am. Philos. Soc., Inst. Mgmt. Scis. (pres. 1963, chmn. coun. 1964, Finnish Acad. Scis. (fgn. hon.), Brit. Acad. (corr.), Western Econ. Assn. (pres. 1980-81), Soc. Social Choice and Welfare (pres. 1991-93). Office: Stanford U Dept Econs Stanford CA 94305-6072

ARROWSMITH-LOWE, THOMAS, federal agency administrator, medical educator; b. San Angelo, Tex., Sept. 2, 1948; s. James Thomas Lowe and Martha Beatrice Archer; m. Janet Arrowsmith, Sept. 15, 1990; children: Dylan Thomas, Victoria Archer. BA, U. North Tex., 1971; DDS, U. Tex., Houston, 1975; MPH, U. Minn., 1980. Cert. dentist, Tex., 1975, N.Mex., 1976. Commd. lt. USPHS, 1976, advanced through grades to capt., 1989; dental dir. Centro Campesino de Salud, Española, N.Mex., 1976-77; dir. Mille Lacs Indian Health Ctr., Onamia, Minn., 1977-78; dep. chief dental br. Navajo Area Indian Health Svc., Window Rock, Ariz., 1978-80, chief dental br., 1980-82, chief med. officer, 1982-84; dep. dir. office health affairs FDA, 1985—; AIDS coord. FDA/Ctr. Devices and Radiol. Health, 1985—; mem. standing com. on contraception and sexually transmitted diseases NIH, surgeon gen.'s adv. group AIDS and condoms; bd. dirs. Washington Free Clin., Episcopal Caring Response to AIDS, chair edn. com.; mem clin. operating com., HIV med. staff Whitman-Walker Clin.; assoc. prof. Health Scis. Ctr., U. Tex., San Antonio; health system cons. The Gambia, West Africa, 1983. Contbr. articles to profl. jours., chpts. to books. Mem. vestry St. Thomas' Episcopal Ch., Washington, 1992—. Decorated Commendation medal, two Outstanding Svc. medals; recipient Exemplary Svc. medal U.S. Surgeon Gen., 1992. Mem. ADA, N.Y. Acad. Sci., Internat. AIDS Soc., USPHS Commd. Officers Assn. Address: 1905 New Hampshire Ave NW Washington DC 20009

ARROYO, CARMEN MILAGROS, research chemist; b. Rio Piedras, P.R., May 7, 1958; d. Tomas Arroyo and Ana (Felicita) Rivera; m. Alasdair John

Carmichael, Feb. 1, 1989; children: Anita Yvonne, Ian William. BS in Chemistry, U. P.R., 1978, BS in Math. cum laude, 1978; PhD in Physical Chemistry, U. Ala., 1983. Lab. asst. U. P.R., Rio Piedras, 1977-78; rsch. asst. dept. chemistry U. Ala., Tuscaloosa, 1980-83; radiation protection officer Letterman Army Med. Ctr., San Francisco, 1983-84; prin. investigator radiation sci. dept. Armed Forces Radiobiology Rsch. Ints., Bethesda, Md., 1984-86; rsch. assoc. dept. medicine, div. experimental medicine Goerge Washington U., Washington, 1986-87, rsch. asst. prof., sr. rsch. scientist, 1987-88; rsch. assoc. VA/U. Md., Balt., 1989-91; rsch. chemist U.S. Army Med. Rsch. Inst. Chem. Defense, Physiology Br., Aberdeen Proving Ground, Md., 1991—; vis. scientist Tokai U. Sch. Medicine Dept. Physiology, Boheseidai, Japan, 1989, 90, 91. Contbr. articles to profl. jours. Capt. U.S. Army, 1983-86. Rsch. fellow U.S. Dept. Energy, 1979-82. Mem. Internat. EPR Soc., Southeastern Magnetic Resonance Conf., Oxygen Club Greater Washington, Sigma Xi. Roman Catholic. Achievements include research on magnetic resonance: electron paramagnetic resonance, electron nuclear double resonance and nuclear magnetic resonance; flourescence spectroscopy; techniques in free radical rsch. Office: USA Med Rsch Inst Chem Def Aberdeen Proving Ground MD 21010-5425

ARROYO-VAZQUEZ, BRYAN, wildlife biologist; b. N.Y.C., Apr. 28, 1964; s. Teodoro and Sonia (Vazquez) A.; m. Enidza Segarra, Jan. 20, 1990; 1 child, Enidza Nicole. BS, Cath. U. P.R., 1987; MS, U. Ark., 1991. Night filler K-Mart Corp., Ponce, P.R., 1985-88; teaching asst. U. Ark., Fayetteville, 1988-90; student trainee U.S. Fish and Wildlife Svc., Arlington, Tex., 1990; wildlife biologist U.S. Fish and Wildlife Svc., Arlington, 1991, Austin, Tex., 1991—; teaching asst. U. Ark., Fayetteville, 1991. Contbr. articles to Nova Hedwigia, Sci.-Ciencia, Wilson Bulletin, Oritholigia Neotropical 2; compiler (booklet) Threatened and Endangered Species of Tex., 1992. Registration official, Junta Electoral de P.R., Ponce, 1982; election official Partido Popular Democratico, Ponce, 1984; pres. Tri-Beta Biol. Soc., 1985-86. Recipient Minority fellowship Univ Ark., Fayetteville, Ark., 1989, Nat. Hispanic fellowship, Nat. Hispanic Found., Boulder, Colo. 1990, Dr. David Causey award, David Causey Meml. Fund, Fayetteville, Ark., 1990; grantee: Audubon Soc. Ark, Little Rock, 1989, Zoology Dept., U. Ark., Fayetteville, 1990. Mem. Ornithological Socs. Am., Soc. for Conserv. Biology, Sigma Xi (assoc.). Roman Catholic. Achievements include first recording of avian nesting among suspended leaf litter; first description of Elfin Woods Warbler (Dendroica angelae) nest and breeding behavior. Home: 306 W Rundberg Apt 251 Austin TX 78753 Office: Fish and Wildlife Svc 611 E 6th St Rm 407 Austin TX 78701

ARRUDA-NETO, JOÃO DIAS DE TOLEDO, nuclear physicist; b. São Paulo, Brazil, Aug. 30, 1943; s. Ruy de Toledo and Ida F. de Toledo (Forte) A.; m. Cristina Magalhães, Dec. 15, 1964 (div. Dec. 1981); 1 child, Flóra-Cristina; m. Sonia Maria de Almeida Perri, Sept. 26, 1982; children: Tayguara, Guacyara, Aymberê. BS, U. São Paulo, 1968, MS, 1973, PhD, 1976. Asst. master U. São Paulo, 1969-76, asst. prof., 1976-79; rsch. affiliate Stanford (Calif.) U., 1979-81; assoc. prof. U. São Paulo, 1982-85; prof. Tohoku U., Sendai, Japan, 1984, 87, 91, U. Ill., Champaign, 1985-86, U. Gent (Belgium), 1988, 90, U. São Paulo, 1986—; vis. scientist Inst. Atomic Energy, Beijing, China, 1987; rsch. leader Nuclear Fission Rsch. group, Sendai, 1984—; coord. Photonuclear Reactions Group, São Paulo, 1982—; dir. Lab. Nuclear Microanalysis, São Paulo, 1989—; collaborator Inst. Physics & Tech., Kharkov, Ukraine, 1990—. Contbr. articles to profl. jours. Japan Soc. Promotion of Sci. grantee, 1984, USSR Acad. Scis. grantee, 1990. Mem. Am. Physical Soc., Japan Physical Soc., Brazilian Physical Soc., N.Y. Acad. Scis. Social. Democrat. Avocations: pipe collecting, jogging, movies. Office: U São Paulo, Caixa Postal 20516, 01498 São Paulo Brazil

ARSENAULT, RICHARD JOSEPH, materials science and engineering educator; b. Champion, Mich., June 16, 1935; s. Eugene and Eleanor (Chipman) A.; m. Elizabeth Kisiel, Aug. 1959 (dec. Nov. 1976); children: Joseph, Carol, Robert, David; m. Louise Marie Brusco, May 23, 1980. BS, Mich. Technol. U., 1957; PhD, Northwestern U., Evanston, Ill., 1962. Assoc. engr. Westinghouse, Pitts., 1957-59; rsch. metallurgist Oak Ridge (Tenn.) Nat. Lab., 1962-66; prof. U. Md., College Park, 1966—; vis. prof. U. Liverpool, Eng., 1973-74; assoc. program dir. NSF, Washington, 1978-79; prin. scientist Erich Schmid, Leoben, Austria, 1982; mem. sci. adv. bd. USAF, 1990—. Author: Metal Matrix Composites, 1989; contbr. more than 200 articles to profl. jours. Fellow ASM Internat.; mem. Am. Inst. Metall. Engrs., Nat. Assn. of Corrosion Engring. Office: U Md Dept Mat and Nuclear Engrng College Park MD 20742-2115

ARSHAM, HOSSEIN, information scientist educator; b. Mashhad, Iran, Mar. 28, 1947; came to U.S., 1978; s. Gholam Reza and Habebeh (Babai) A.; m. Elaheh-Naaze Khoshghadam, Dec. 20, 1986; 1 child, Aryana. MSc, Cranfield Inst., Eng., 1977; DSc, George Washington U., 1982. Cert. info. scientist. Rschr. Internat. Water Resources Inst., Washington, 1982-83; asst., assoc. prof. U. Balt., 1983—; assoc. faculty advanced studies Calif. Nat. U., 1991—; tech. lectr. Bethlehem Steel Co., Balt., 1983-84. Assoc. editor: Computational Stats. and Data Analysis; editorial bd. mem. Jour. of End User Computing; contbr. articles to profl. jours. Recipient Black & Decker Corp. Rsch. award, 1987, 88. Fellow Royal Statis. Soc., Operational Rsch. Soc., Inst. Combinatorics and Applications; mem. AAAS, IEEE, Am. Math. Soc., Internat. Assn. Math. and Computer Modeling, Internat. Forecasting Soc., Am. Statis. Assn., Assn. for Computing Machinery, Digital Equipment Computer Users Soc., Info. Resources Mgmt. Assn., Math. Assn. Am., Ops. Rsch. Soc. Am., Soc. Indsl. and Applied Math., Soc. for Info. Mgmt., N.Y. Acad. Scis., Beta Gamma Sigma, Omega Rho. Office: U Balt 1420 N Charles St Baltimore MD 21201-5779

ARTERBURN, DAVID ROE, mathematics educator; b. Norfolk, Nebr., Dec. 29, 1939; s. David Allen and Vivianne Joan (Cranwell) A.; m. Kathleen Diane Weaver, Aug. 31, 1963; children: Autumn, David Alan. BS, So. Meth. U., 1961; MS, N.Mex. State U., 1963; PhD, S, 1964. Asst. prof. math. U. Mont., Missoula, 1964-67; asst. prof. math. N.Mex. Inst. of Mining and Tech., Socorro, 1967-68, assoc. prof. math., 1968—, math. chair, 1987-90; Author: E.I.T. Exam Study Guide; contbr. articles to profl. jours. Mem. AAUP, AAAS, Math. Assn. Am. (gov. 1984-87, 91—). Avocations: welding, cabinet making. Home: Rte 1 Box 39A Lemitar NM 87823 Office: New Mexico Inst Mining Math Dept Socorro NM 87801

ARTHUR, JAMES HOWARD, mechanical engineering educator; b. Lynchburg, Va., Feb. 2, 1958; s. Howard Board and Minnabelle (Williams) A.; m. Cindy Patriece May, Dec. 10, 1983; children: James Michael, David Benjamin, Peter Timothy. BS, U. Va., 1980, MS, 1982, PhD, 1988. Registered profl. engr., Va. Mech. engr. Babcock & Wilcox Co., Lynchburg, Va., 1980-87; asst. prof. Va. Mil. Inst., Lexington, 1988—; assoc. rsch. engr. U. Va., Charlottesville, 1988, 89; rsch. cons. U. Va., Charlottesville, 1990. Mem. ASME (assoc., treas. Shenandoah sect. 1991-93), ASHRAE, Sigma Xi, Tau Beta Pi. Achievements include research in numerical heat transfer, analysis of natural convection in a vertical, asymmetrically heated, perforated walled channel including radiative exchange at the boundaries. Office: Va Mil Inst Mech Engring Dept Lexington VA 24450

ARTS, HENRY ALEXANDER, otolaryngologist; b. Buffalo, Feb. 19, 1956; s. Henry Francis and Betty Jo (Campbell) A.; m. Susan G. Wichhart, Apr. 14, 1990; children: Henry William, Alexander Nicholas. BSME, Rice U., Houston, 1978; MD, Baylor U., 1983; MSE, U. Wash., 1990. Diplomate Am. Bd. Otolaryngology. Resident gen. surgery U. Wash., Seattle, 1983-85, resident otolaryngology, 1985-90; fellow neuro-otology U. Va., Charlottesville, 1990-91, asst. prof. otolaryngology, 1992-93; otolaryngologist Indian Health Svc., Gallup, N.Mex., 1991; asst. prof. Otolaryngology U. Mich., Ann Arbor, 1993—; mem. adv. bd. Lions of Va. Hearing Found., Charlottesville, 1991—. Fellow Am. Acad. Otolaryngology (head and neck surgery); mem. AMA, Am. Neurotology Soc., Acoustical Soc. Am. Home: 2670 Appleway Ann Arbor MI 48104 Office: U Mich Dept Otolaryngology Box 0312 1500 E Med Ctr Dr Ann Arbor MI 48109-0312

ARTURI, ANTHONY JOSEPH, engineering executive, consultant; b. Paterson, N.J., Sept. 6, 1937; s. Emanuel and Mary (Territo) A.; m. Betty Jane Hanner, July 14, 1962; children: Anthony David, Dawn Elizabeth. Degree in mech. engring., Stevens Inst. Tech., 1959, MS in Math., 1966. Registered profl. engr., N.J. Project engr. Gen. Precision-Kearfott, Wayne, N.J., 1962-64, Bendix Corp., Teterboro, N.J., 1964-66; engr. supr. Singer Kearfott Divsn., Wayne, 1966-74; project mgr. Lummus Co., Bloomfield, N.J., 1974-77, GE Info. Svcs., N.Y.C., 1977-82; pres. ARTECH Assocs., Wayne, 1982—; trustee Stevens Inst. Tech., Hoboken, N.J., 1986-88; adj. faculty Fairleigh Dickinson U., Teaneck, N.J., 1968-75. Presenter in field. Organizer, panelist Stevens Enterprise Forum, 1985-90. Recipient Alumni Achievement award Stevens Alumni Assn., 1979. Mem. IEEE, Drug Info. Assn., Stevens Entrepreneurs Club (organizer small bus. network 1986), Lions. Home: 13 Miller Rd Wayne NJ 07470-3620 Office: ARTECH Assocs 1341 Hamburg Tpke Wayne NJ 07470-4042

ARTUSHENIA, MARILYN JOANNE, internist, educator; b. Glen Ridge, N.J., Feb. 16, 1950; d. Gregory and Julia (Markewicz) A.; A.B., Boston U., 1970; M.D., Hahnemann Med. Coll., 1974. Intern, Mount Sinai Hosp., N.Y.C., 1974-75, resident medicine, 1975-77; fellow nephrology Bronx (N.Y.) VA Hosp., also Mt. Sinai Hosp., 1977-79; research fellow endocrinology Bronx VA Hosp., 1979-80, research cons. endocrinology, 1980—; asst. attending physician in medicine and psychiatry Elmhurst Gen. Hosp., N.Y., 1980-85; attending staff St. Joseph's Hosp., 1982—; instr. medicine Mt. Sinai Hosp., 1980—; instr. advanced cardiac life support and cardiopulmonary resuscitation Am. Heart Assn., 1980—; practice medicine specializing in internal medicine and nephrology, Torrington, Conn. Mem. A.C.P., AAAS, AMA, N.Y. Acad. Scis., Queens County Med. Soc., Internat. Platform Assn., Phi Beta Kappa, Alpha Omega Alpha. Home and Office: 732 Weigold Rd Torrington CT 06790-2037

ARTZT, ALICE FELDMAN, mathematics educator; b. Flushing, N.Y., June 8, 1946; d. Robert and Martha (Weintraub) Feldman; m. Russell Michael Artzt, July 6, 1968; children: Michele, Julie, Gregory. BA in Math., CUNY, Flushing, 1968, MS in Math. Edn., 1973; PhD in Math. Edn., NYU, Manhattan, 1983. Math. tchr. Herricks (N.Y.) High Sch., 1968, North Shore High Sch., Glen Head, N.Y., 1968-70; computer sci. tchr. Bais Yaakov High Sch. of Queens (N.Y.), 1973-79; prof. math. and math. edn. Queens Coll., CUNY, Flushing, 1983—, dir. math. edn. program, 1985—. Author: (with C. Newman) How to Use Cooperative Learning in the Mathematics Teacher, 1990; book reviewer Math. Tchr., 1990—, column co-editor 1990-91; contbr. articles and columns to profl. jours. Recipient award NSF, 1986-88, Rsch. award CUNY, N.Y.C., 1988-89. Mem. ASCD, Nat. Coun. Tchrs. Math. (reviewer, mem. ednl. materials com. 1993—), Math. Assn. Am., Am. Ednl. Rsch. Assn., Assn. Computers in Math. and Sci. Teaching, Internat. Assn. for Study Coop. in Edn., Assn. Math. Tchrs. N.Y. State (reviewer). Jewish. Avocations: sewing, singing, tennis. Home: 45 Rolling Hill Ln Old Westbury NY 11568-1028 Office: Queens Coll CUNY 65-30 Kissena Blvd Flushing NY 11367

ARUMUGANATHAN, KATHIRAVEPILLAI, plant flow cytometrist, cell and molecular biologist; b. Point Pedro, Sri Lanka, Oct. 12, 1950; came to U.S., 1983; s. A. and Sivapaikium Kathiravepillai; m. Vanmathy, Jan. 30, 1978. MSc, U. Wales, Aberystwyth, 1983; PhD, Ohio U., 1988. Asst. lectr. U. Colombo, Sri Lanka, 1975, U. Jaffna, Thirunelvely, Sri Lanka, 1975-81; Brit. coun. scholar U. Wales, Aberystwyth, 1981-83; teaching assoc. Ohio U., Athens, 1983-87; postdoctoral assoc. Cornell U., Ithaca, N.Y., 1988-91, rsch. assoc. II, 1991-92; rsch. asst. prof., flow cytometry lab. mgr. U. Nebr., Lincoln, 1992—. Achievements include research in isolation, flow cytometric analysis and sorting of plant chromosomes; construction of chromosome-specific DNA Libraries of important crop plants. Office: U Nebr Ctr Biotech Lincoln NE 68588-0159

ARVANITOYANNIS, IOANNIS, polymer chemistry researcher; b. Thessaloniki, Greece, Dec. 6, 1962; s. Sotirios and Victoria (Vergidou) A.; m. Eleni Psomiadou, Dec. 26, 1991. BSc in Chemistry, U. Thessaloniki, 1984, PhD in Polymer Chemistry, 1990. Rsch. asst. Canning Food Co. Zanae, Thessaloniki, 1984; rsch. asst., demonstrator Dept. Chemistry, Aristotle U., Thessaloniki, 1984-87; rsch. asst. Plastics divsn. Ciba-Geigy, Fribourg, Switzerland, 1986-89; tchr. chemistry and physics High Sch., Thessaloniki, 1988; quality control chemist Four-F Hellas, Chalastra, Greece, 1989; postdoctoral rsch. fellow U. Nottingham & Loughboro, Eng., 1990—. Contbr. numerous articles to profl. jours. With Greek Army, 1987-88. Scholarship Nat. Inst., 1977, Rotary, 1977, Nat. Inst. of Scholarships, 1981-87, Onassis Inst., 1990-92; fellowship John Coppeck Meml. Inst., 1991. Mem. Union of Greek Chemists, Union of Chemists of North Greece, Am. Chem. Soc. Achievements include research on polyamides, polyesters, polystyrene, polymethylmethacrylate and gutta-percha (trans-polyisoprene). Home: 14 Clifford Rd, Loughborough LE11 0NG, England Office: Loughborough U of Tech, Dept Chemistry, Loughborough LE 113TU, England

ARVIZU, DAN ELIAB, mechanical engineer; b. Douglas, Ariz., Aug. 23, 1950; s. Walter and Ella (Rodriguez) A.; m. Patricia Ann Brady, Feb. 23, 1980; children: Joshua, Angela, Elizabeth, Kayley, Tecia. BSME, New Mexico State U., 1973; MSME, Stanford U., 1974, PhD in Mech. Engring., 1981. Mfg. engring. asst. Texas Instruments, Dallas, 1969-72; mem. tech. staff Bell Telephone Labs., Denver, 1973-77; mem. solar thermal tech. staff Sandia Nat. Labs., Albuquerque, 1977-81, mem. solar photovoltaic tech. staff, 1981-86, supr. photovoltaic cell rsch., 1986-88, mgr. tech. transfer, 1988-91, dir. tech. transfer, 1991-93; dir. adv. energy tech. Sandia Nat. Labs., 1993—; mem. tech. transfer steering com. Nat. Ctr. for Mfg. Scis., Ann Arbor, Mich., 1992; mem. tech. transfer mgrs. adv. bd. Nat. Tech. Transfer Ctr., Wheeling, W.Va., 1992. Contbr. articles to profl. jours. Named Disting. Engring. Alumnus N.Mex. State U., 1988, Ingeniero Eminente, 1990; named Rising Star in Sci. Albuquerque Tribune newspaper, 1989. Mem. ASME (nat. lab. tech. transfer com. 1990—), IEEE, IEEE Electronic Device Soc. (adminstrv. com. 1986-91), Tech. Transfer Soc. Achievements include leadership of national laboratory negotiating teams that resulted in Department of Energy policy changes to improve U.S. Goverment/ Industry partnership agreements, and management of research effort that developed 30 percent solar to electric conversion efficiency solar cell. Office: Sandia Nat Labs Adv Energy Tech Ctr 6200 PO Box 5800 Albuquerque NM 87185

ARWASHAN, NAJI, structural engineer; b. Damascus, Syria, Feb. 23, 1965; came to U.S., 1989; s. Michel and Georgette (Diab) A. MS, Ecole Nat. des Ponts et Chaussées, Paris, 1988; PhD, U. Mich., 1992. Structural engr. Europe Etudes, Paris, 1988-89; summer asst. The World Bank for Reconstruction and Devel., Washington, 1990; rsch. asst. U. Mich., Ann Arbor, 1990-91; structural engr. Carl Walker Engrs., Inc., Kalamazoo, Mich., 1992—; adj. asst. prof. engring. tech. Western Mich. U., Kalamazoo, 1992—. Recipient Fellowship Arab Student Aid Internat., U.S.A., 1988-90, Scholarship Akram Ojjeh Found., Paris, 1989-92. Mem. ASCE (assoc.). Greek Catholic. Achievements include research in structural analysis and structural safety. Office: Carl Walker Engrs Inc 445 W Michigan Ave Ste 101 Kalamazoo MI 49007

ARY, T. S., federal official; b. Eldorado, Ill., Mar. 30, 1925; s. McKinley and Emma (Busby) A.; m. Martha K. Metz, Dec. 23, 1945; 1 child, David Metz. Student, Evansville (Ind.), 1942-43; BS in Mineral Sci., Stanford U., 1947. Registered geologist, Calif. Football, basketball, baseball coach Jacksonville (Fla.) Naval Air Sta., 1946-47; asst. football coach jr. varsity Stanford (Calif.) U., 1947-49; shift boss, asst. supt. Anaconda Copper Co., Butte, Mont., 1951-53; mining engr. Union Carbide Corp. (U.S. Vanadium Co.), Rifle, Colo., 1953-55; asst. mgr. exploration Union Carbide Corp. (U.S. Vanadium Co.), Grand Juction, Colo., 1955-57, mgr. domestic exploration, 1957-62; land mgr. Union Carbide Nuclear Co., N.Y.C., 1962-67, v.p. mineral exploration mining and metals div., 1967-74; mgr. devel., v.p. mineral exploration Utah Internat. Inc., San Francisco, 1974-80; pres. mineral exploration, pres. resource div. Kerr-McGee Corp., Oklahoma City, 1980-87; dir. Bur. Mines U.S. Dept. Interior, Washington, 1988-93; chmn. Nat. Critical Materials Coun., 1989-93; bd. dirs. Mineral Info. Inst., Denver; mem. Dept. State Adv. Com. to Task Force Com. of UN Law of the Sea, Washington, 1966-77, Internat. Atomic Energy Program Adv. Com., Vienna, Austria, 1970-75, Nat. Strategic Materials and Minerals Program Adv. Com., Washington, 1984-88; com. chmn. Am. Mining Congress, Washington, 1960-88; mem. minority staff Senate Energy and Natural Resource Com., 1993—. Assoc. editor, mem. editorial bd. Jour. Resource Mgmt. and Tech., 1982-88; author over 100 published articles on mineral resources, pub. lands mgmt., pub. land law, pub. policy, internat. bus. mgmt., fly fishing techniques, sports, religion. Sec.-treas. Lakehurst Homeowners Assn., Oklahoma City, 1983-88; dir. Last Frontier Coun. Boy Scouts Am., Oklahoma City, 1980-86; trustee Westminister Presbyterian Ch., Oklahoma City, 1984-87; telethon chmn. Oklahoma Soc. for Crippled Children, 1985-86; pres. Lido Isle Homeowners Assn., Foster City, Calif., 1978-80, sec. 1976-78; dir. Internat. Student Svcs. Internat. YMCA, N.Y.C., 1973-75, Sch. Bd. Dist. 51, Grand Junction, Colo., 1959-65; committeeman We. Colo. Boy Scout Coun., Grand Junction, Colo., 1960-64; dir. Grand Mesa Ski Corp., Grand Junction, 1960-65, pres. 1963-64; regional dir. Nat. Ski Patrol Assn., Denver, 1952-65; dir. Colo. Expenditure Coun., Denver, 1961-65, South Rocky Mountain Ski Assn., Denver, 1955-65, Butte (Mont.) Ski Club, 1952-53; active Boy Scouts Am., 1935-65. Lt. (j.g.) USN, 1943-47, ETO. Recipient Disting. Svc. award Rocky Mountain Coal Mining Inst., 1988, Disting. Svc. award Nat. Ind. Coal Operators, 1988, numerous athletic awards USN, AP, Nat. Coaches Assn., Nat. Athletic Scholastic Soc., 1938-48; named Outstanding Miner of Yr., Idaho Mining Assn., 1989; inducted into Am. Mining Hall of Fame, 1992; Paul Harris fellow Oklahoma City Rotary Internat. Club. Mem. Am. Inst. Profl. Geologists, Circum-Pacific Coun. Energy and Mineral Resources (program chmn. 1978-80), Am. Assn. Profl. Landmen, Wyo. Mining Assn. (Outstanding Man of Yr. 1991), Colo. Mining Assn. (Edn. Found. award 1990), N.W. Mining Assn. (Mining Man of Yr. 1988), Calif. Mining Assn., N.Mex. Mining Assn. (Mining Man of Yr. 1988), Ariz. Mining Assn., AIME (sec., treas., vice-chmn., chm. 1955-65, Disting. Svc. award Wyo. sect. 1990, Pres. citation 1992, Robert Earll McConnell award 1993), Rocky Mountain Mineral Law Found., Mining Club of N.Y., Commonwealth Club of San Francisco, Forum on Fgn. Affairs (San Francisco), Nat. Assn. Mfrs. (chmn. natural resource com. 1985-87), Sigma Nu. Republican. Avocations: Bible studies, fly fishing, snow skiing, reading, rose competition. Home: 3301 N Nottingham St Arlington VA 22207-1345

ARYA, SATYA PAL SINGH, meteorology educator; b. Mavi Kalan, Dist Meerut, India, Aug. 24, 1939; came to U.S., 1965; BE (Civil), U. Roorkee (India), 1961, ME (Civil), 1964; PhD, Colo. State U., 1968. Asst. engr. Irrigation Dept., Lucknow, India, 1961-62; lectr. U. Roorkee, 1963-65; rsch. asst. Colo. State U., Ft. Collins, 1965-68, rsch. assoc., 1968-69; rsch. asst., assoc. prof. U. Wash., Seattle, 1969-76; assoc. prof. N.C. State U., Raleigh, 1976-81, prof. meteorology, 1981—, acting head MEAS dept., 1982-83; vis. prof. Indian Inst. Tech., Delhi, 1983-84. Author: Introduction to Micrometeorology, 1988; contbr. sci. articles to jours. of atmospheric scis., applied meteorology, fluid mechanics, others. Fellow AAAS, Am. Meteorol. Soc.; mem. Am. Geophys. Union, Sigma Xi. Achievements include research in atmospheric sciences, applied meteorology, fluid mechanics, atmospheric environment, geophysical research, environmental pollution. Office: NC State U Dept Marine Earth and Atmospheric Sci Raleigh NC 27695-8208

ARZBAECHER, ROBERT C(HARLES), research institute executive, electrical engineer, researcher; b. Chgo., Oct. 28, 1931; s. Hugo L. and Caroline G. A.; m. Joan Collins, June 16, 1956; children: Carolyn, Robert, Mary Beth, Jean, Thomas. B.S., Fournier Inst., 1953; M.S., U. Ill., 1958; Ph.D., 1960. Asst. prof. elec. engring. Christian Bros. Coll., Memphis, 1960-63, assoc. prof., 1963-67; assoc. prof. elec. engring. U. Ill.-Chgo., 1967-70, prof., 1970-76; chmn. dept. elec. engring. U. Iowa, Iowa City, 1976-81; dir. Pritzker Inst. Med. Engring., Ill. Inst. Tech., Chgo., 1981—; v.p. U. Iowa Research Found., 1978-81; pres. Arzco. Inc., Chgo., 1980-87. Contbr. articles to profl. jours.; inventor Arzco pill electrode. Trustee Ill. Cancer Council, Chgo., 1981—. Fellow IEEE, Am. Coll. Cardiology, Am. Inst. Med. Biol. Engring. Home: 5757 N Sheridan Rd Chicago IL 60660-4746 Office: Ill Inst Tech Pritzker Inst Med Engr 10 W 32nd St Chicago IL 60616

ASADA, TOSHI, seismologist, educator; b. Tokyo, Dec. 15, 1919; s. Shunsuke and Sumi (Asakura) A.; m. Teruko Uchida, Nov. 29, 1955; children: Takashi, Satoshi, Yuko. BS, U. Tokyo, 1944, DSc, 1958. Rsch. asst. U. Tokyo, 1944-55, mem. faculty, 1955—, prof. seismology, 1966-80, prof. emeritus, 1980—; prof. geophysics Inst. Rsch. and Devel., Tokai U., 1980—; chmn. coordinating com. for earthquake prediction Geog. Survey Inst., 1980-90; chmn. earthquake assessment com. Japan Meteorol. Agy., 1980-90; chmn. Geodesy coun. Ministry of Edn., Sci. & culture. Sr. fellow Carnegie Instn., Washington, 1960-61. Mem. Seismol. Soc. Japan, Am. Geophys. Union. Author papers on microearthquakes, explosion seismology, ocean bottom seismometers. Home: 3-13-17 101 Shimo-ochiai, Shinjuku-ku, Tokyo 161, Japan Office: Inst Rsch and Devel Tokai U, 1117 Kitakaname, Hiratsuka 259-12 Kanagawa-ken, Japan

ASAJIMA, SHOICHI, economics educator; b. Tokyo, Mar. 11, 1931; m. Kinuko Yagi, Nov. 14, 1957; children: Sumiko, Mieko. BA in Econs., Tokyo U., 1953, D in Econs., 1972. With Sumitomo Trust & Banking Co., Osaka, Japan, 1953-73, dir. Tokyo rsch. dept., 1973-77; prof. bus. history Senshu U., Tokyo, 1977—. Author: History of Japanese Trust Business, 1969, Studies of Japanese Trust Business Laws, 1980, Business History of Sumitomo Zaibatsu, 1921-1945, 1983, Financial Structure of Mitsubishi Zaibatsu, 1986, Financial History of Japanese Life Insurance Companies, 1991. Mem. Sci. Coun. Japan, Bus. History Soc. Japan (bd. dirs. 1987-90). Home: Higashi 2-14-36, Kunitachi Tokyo 186, Japan Office: Senshu U, Higashimita 2-1-1, Tamaku Kawasaki 214, Japan

ASAKURA, HITOSHI, internal medicine educator; b. Yokohama, Kanagawa, Japan, Feb. 21, 1937; s. Takeshi and Kito A.; m. Hiroko Suzuki, May 18, 1975; children: Tomoko, Makoto. Student, Keio Gijuku U., Tokyo, 1963; MD, Keio Gijuku U., 1968, postgrad., 1964-68. Medical diplomate, Japan. Intern Keio Gijuku U. Hosp., 1963-64; med. dr. Keio Gijuku U. Hosp., Tokyo, 1968-73; asst. prof. internal medicine Keio Gijuku U. Sch. Medicine, Tokyo, 1973-88; prof. internal medicine Niigata (Japan) U. Sch. Medicine, 1988—. Fellow Internat. Coll. Angiology; mem. N.Y. Acad. Sci., Am. Gastroent. Assn., Japanese Soc. Gastroenterology, Japanese Soc. Internal Medicine. Home: 7-26-14 Koonandai, Koonan-ku, Yokohama Kanagawa 233, Japan Office: Niigata Univ Sch Medicine, 757 Ichibancho/ Asahimachidori, Niigata Niigata 951, Japan

ASANUMA, HIROSHI, physician, educator; b. Kobe, Japan, Aug. 17, 1926; s. Kisaburo and Yukiko (Takahashi) A.; m. Reiko Shimazu, Dec. 15, 1953; children—Chisato, Mari. M.D., Keio U., Tokyo, 1952; D.M.S., Kobe Med. Coll., 1959. Instr., Kobe Med. Coll., 1953-59; asst. prof. Osaka City U., Japan, 1959-65; guest investigator Rockefeller Inst., N.Y.C., 1961-63; assoc. prof. N.Y. Med. Coll., N.Y.C., 1965-71, prof., 1971-72; prof. Rockefeller U., N.Y.C., 1972—. Contbr. articles to profl. jours. and books. Mem. Am. Physiol. Soc., Soc. for Neurosci., Harvey Soc., Japanese Physiol. Soc. Home: 505 E 79th St New York NY 10021-0709 Office: Rockefeller U Dept of Motorphysiology 1230 York Ave New York NY 10021-6341

ASAWA, TATSURŌ, chemistry researcher; b. Anzan, Manchuko, Dec. 2, 1932; arrived in Japan, 1946; s. Saburō and Tae (Asawa) A.; m. Seiko Akino, Oct. 12, 1960; children: Makiko, Akio, Yukiko. M. in Chemistry, Tokyo U., 1958. Lab. dir. Asahi Glass Co., Ltd., Yokohama, Japan, 1978-84; dir. devel. Asahi Glass Co., Ltd., Tokyo, 1984-90; dir. tech. Rsch. Inst. Tech. for the Earth, Tokyo, 1990—. Co-author: Ullman's Encyclopedia, 1985. Mem. tech. com. Japan Soda Assn., Tokyo, 1988-90; mem. frontier com. Assn. for New Chemistry, Tokyo, 1988-90. Mem. Japan Polymer Soc. (Tech. award 1981), Japan Electrochem. Soc. (Tech. award 1980). Avocations: classic music, philately, gardening. Home: 5-15-2 Hiyoshihoncho, Kohokuku, Yokohama 223, Japan

ASBURY, ARTHUR KNIGHT, neurologist, educator; b. Cin., Nov. 22, 1928; s. Eslie and Mary (Knight) A.; m. Carolyn Holstein, May 17, 1980; children by previous marriage: Dana, Patricia Knight, William Francis. Grad., Phillips Acad., Andover, Mass., 1946; student, Stanford, 1947-48; B.S., U. Ky., 1951; M.D., U. Cin., 1958; M.A. (hon.), U. Pa., 1974. Intern in medicine Mass. Gen. Hosp., Boston, 1958-59; resident Mass. Gen. Hosp., 1959-63, fellow, 1963-65, staff neurologist, 1965-69; chief neurology San Francisco VA Hosp., 1969-74; prof. dept. neurology U. Pa., Phila., 1974—, chmn. dept. neurology, 1974-82; Van Meter prof. neurology U. Pa., 1983—; acting dean, exec. v.p. U. Pa. Sch. Medicine, 1988-89, vice dean for rsch., 1990—; prof. The Wistar Inst., 1990—; teaching fellow Harvard Med. Sch., 1958-65, instr., 1965-68, assoc., 1968-69; assoc. prof. neurology U. Calif. at San Francisco, 1969-73, vice-chmn., 1969-74, prof., 1973-74. Sr. editor Internat. Med. Rev. Series-Neurology, Butterworth & Co., London, 1980—; assoc. editor: Archives of Neurology, 1975-76, Annals of Neurology, 1976-81, chief editor, 1985-93; mem. editorial bd. Muscle and Nerve, 1977-89,

Neurology, 1981-85, Jour. Neuropathology and Exptl. Neurology, 1981-83, Jour. Neurol. Scis., 1989—; contbr. chpts. to med. textbooks, articles to med. jours. V.p., bd. dirs. Forest Retreat Farms Inc., Carlisle, Ky., 1970-92. With AUS, 1951-53. Recipient Daniel Drake medal U. Cin., 1988; inducted Inst. of Medicine Nat. Acad. Scis., 1993; grantee USPHS, 1967—, Muscular Dystrophy Assn., 1974—. Fellow Am. Acad. Neurology (v.p. 1977-79); mem. Am. Neurol. Assn. (councillor 1976-81, pres. 1982-83), Am. Assn. Neuropathologists (v.p. 1983-84), Soc. Neurosci., Assn. Univ. Prof. Neurology (pres. 1980-82), World Fedn. Neurology (v.p. 1989-93), European Neurol. Soc. (hon.). Episcopalian (vestryman). Home: 408 S Van Pelt St Philadelphia PA 19146-1233 Office: U Pa Hosp Dept Neurology 36th & Hamilton Walk Philadelphia PA 19104

ASBURY, CHARLES THEODORE, JR., civil engineering consultant; b. Abbington, Pa., July 21, 1938; s. Charles T. Asbury and Janet (Scott) Asbury Murray; divorced; children: Charles T. III, Heather Elizabeth. Student, Rutgers U., 1964-67; BSCE, N.Mex. State U., 1970; postgrad., U. Houston, 1970-72. Registered profl. engr., N.Mex., Tex., Ariz., Colo. Staff engr. Continental Oil Co., Houston, 1970-72; ops. mgr. William Matotan & Assoc., Inc., Albuquerque, 1972-75; pres. Dean-Hunt-Asbury, Inc., Albuquerque, 1975-77; sec.-treas. Andrews, Asbury & Robert, Inc., Albuquerque, 1977—; chmn. N.Mex. Bd. Registration For Profl. Engrs. and Surveyors, Santa Fe, 1985-91, N.Mex. Joint Practice Com.-Architects/Engrs./Surveyors, Albuquerque, 1988-91; mem pres.'s coun. various engring. socs.; N.Mex. del. Nat. Conf. Examiners Engrs. and Surveyors. Investigator rsch. publs., patent applications; designer system design publ. Legislator N.Mex. Ho. of Reps., Santa Fe, 1981-84; candidate U.S. Ho. of Reps., Washington, 1984, N.Mex. State Senate, Santa Fe, 1986. Cpl. USM, 1956-63. Recipient Colonel Aid-de-Camp award Gov. of N.Mex., 1983. Mem. ASCE, NSPE (bd. dirs. 1991-93, chmn. 1991-93), N.Mex. Cons. Engrs. Coun. (sec.-treas. 1993-94), Greater Albuquerque C. of C. (bd. dirs.), Kiwanis Internat. (v.p.), Blue Key, Sigma Tau, Chi Epsilon. Democrat. Roman Catholic. Achievements include research in moving bed filter for B.O.D. and particulate removal, acidic waste stabilization using oyster shells, chromic acid precipitation using SO2 wastes, high technology for small communities. Office: Andrews Asbury & Robert Inc 149 Jackson St NE Albuquerque NM 87108

ASCH, DAVID KENT, medical educator; b. Neligh, Nebr., Nov. 6, 1958; s. Karl Edward and Marlynn Gail (Brodie) A.; m. Joanna Lynn, June 13, 1981; children: Stephanie, Jeffery. MS, Creighton U., Omaha, 1984; PhD, U. Kans., Kansas City, 1991. Lab. technician St. Joseph's Hosp., Omaha, 1982-83; teaching asst. Kans. U. Med. Ctr., Kansas City, 1983-88, rsch. asst., 1988-89; postdoctoral assoc. U. Ga., Athens, 1989-92; asst. prof. Youngstown (Ohio) State U., 1992—. Contbr. articles to profl. jours. Mem. AAAS. Republican.

ASCHENBRENNER, FRANK ALOYSIOUS, former diversified manufacturing company executive; b. Ellis, Kans., June 26, 1924; s. Philip A. and Rose E. Aschenbrenner; m. Gertrude Wilhelmina DeBie, Nov. 15, 1946; children: Richard David, Robert Wayne, Mary Lynne. BS with high honors, Kans. State U., 1950; PhD in Physics, M.I.T., 1954. Mgr. physics and math. Gen. Electric, Cin., 1958-61; asst. dir. space div. Rockwell Internat., Downey, Calif., 1961-69; corp. dir. tech. Rockwell Internat., Pitts., 1969-71; v.p., gen. mgr. div. yarn machinery Rockwell Internat., Charlotte, N.C., 1971-75; pres. COR, Inc., Charlotte, 1975-77; v.p. research and devel. and engring. Ball Corp., Muncie, Ind., 1977-86; pvt. bus. cons. Poway, Calif., 1986—; chmn. bd. RAMZ Corp., Dunkirk, Ind., 1985—; nat. bd. advisors Rose-Hulman Inst., Terre Haute, Ind., 1984—, U. Tenn. Space Inst., Tullahoma, 1982—. Served with USN, 1943-47. Mem. AIAA, Am. Phys. Soc., Naval Res. Assn., San Diego Venture Group. Achievements include design of measurement technique which was needed to determine threshold energy for nuclear fission in order to improve efficiency of nuclear bombs; pioneering of computerized nuclear radiation shielding design and analysis for nuclear reactors. Home and Office: 14258 Palisades Dr Poway CA 92064-6443

ASFOURY, ZAKARIA MOHAMMED, physician; b. Port Said, Egypt, Aug. 5, 1921; arrived in Eng., 1946; s. Mohammed El Asfouri; m. Fadia Katamesh. BSc with hons., Cairo U., 1944, MB BCh, 1946; Msc, London U., 1952; PhD, Heliopolice & Georgia U., 1965. Hon. demonstrator in medicine Liverpool (Eng.) U., 1946-49; hon. demonstrator, rsch. fellow London U. Medicine, 1949-52; curator mus., asst. lectr. Cairo U., 1952-55; pvt. practice Cairo, 1955-69, London, 1969—; owner Brislington Pvt. Nursing Home and Hosp., Bristol, Eng.; rep. London Hosp., London U. 5th Internat. Anatomical Congress-Oxford, 1950; cons. various hosps., U.K.; lectr. Oxford, Cambridge, others. Author: Sympathectomy and the Innervation of the Kidney, 1971; contbr. numerous articles and rsch. papers to various profl. publs. Mem. Royal Coll. Physicians London (lic.), Royal Coll. Surgeons Eng., Brit. Med. Assn., Royal Soc. Medicine, Renal Assn. Britain, Anatomical Soc. U.K., Geriatrics Soc. Britain, Hunterian Soc. Britain, Internat. Cultural Exch., Egyptian Med. Union in World (chmn.), Ea. Carpet Soc., Egyptian Scholars Abroad (chmn.). Achievements include first demonstration of detailed innervation of whole kidneys, of microscopic details of sympathetic and Vagus nerve distributions above and below the diaphragm in human embryos; pioneer in breaking of kidney stones by very high frequency currents, treatment of bilharsia, amoeba, malaria, etc. by daithermy, treatment of heart ischemia by very high frequency currents, treatment of psoriasis; discovery of collagenase enzyme and anticollagenase in man and animals. Office: 27 Devonshire Pl, London W1N 1PD, England

ASH, MAJOR MCKINLEY, JR., dentist, educator; b. Bellaire, Mich., Apr. 7, 1921; s. Major McKinley Sr. and Helen Marguerite (Early) A.; m. Fayola Foltz, Sept. 2, 1947; children: George McKinley, Carolyn Marguerite, Jeffrey LeRoy, Thomas Edward. BS, Mich. State U., 1947; DDS, Emory U., 1951; MS, U. Mich., 1954; Doctoris Medicine Honoris Causa, U. Bern, 1973. Instr. sch. dentistry Emory U., Atlanta, 1952-53; instr. U. Mich., Ann Arbor, 1953-56, asst. prof., 1956-59, assoc. prof., 1959-62, prof., 1962—, chmn. dept. occlusion, sch. dentistry, 1962-87, dir. stomatognathic physiology lab., sch. dentistry, 1969-87, dir. TMJ/oral facial pain clinic, sch. dentistry, 1983-87, Marcus L. Ward prof. dentistry, 1984-89, prof. emeritus, rsch. scientist emeritus, 1989—; cons. N.E. Regional Dental Bd., 1989—; vis. prof. U. Bern, 1989, U. Tex., San Antonio, 1990—; pres. Basic Sci. Bd., State of Mich., 1962-74; cons. over the counter drugs FDA, Washington, 1985-89. Author, co-author 30 textbooks, 1958—; editor 4 books; contbr. 160 articles to profl. jours. Served to tech. sgt. Signal Corps, U.S. Army, 1942-45, ETO. Nat. Inst. Dental Research grantee, 1962-85. Fellow Am. Coll. Dentists, Internat. Coll. Dentists, European Soc. Craniomandibular Disorders; mem. N.Y. Acad. Scis., AAAS, Am. Dental Assn. (cons. council on dental therapeutics 1982—), Washtenaw Dist. Dental Soc. (pres. 1963-64), Phi Kappa Phi. Presbyterian. Avocations: photography, bird watching. Home: 1206 Snyder Ave Ann Arbor MI 48103-5328 Office: U of Mich Sch of Dentistry Ann Arbor MI 48109

ASH, PHILIP, psychologist; b. N.Y.C., Feb. 2, 1917; s. Samuel Kieval and Estella (Feldstein) A.; m. Ruth Clyde, Sept. 16, 1945 (div. Dec. 1972); children—Peter, Sharon; m. Judith Nelson Cates, June 6, 1973; 1 son, Nelson E. B.S. in Psychology, City U. N.Y., 1938; M.A. in Personnel Adminstrn, Am. U., 1949; Ph.D. in Psychology, Pa. State U., 1949. Diplomate: Indsl. Psychlogy Am. Bd. Profl. Psychology. Analyst to unit chief occupational research Dept. Labor, 1940-47; research fellow Pa. State U., 1947-49, asso. prof., 1949-52; asst. to v.p. indsl. relatons Inland Steel, 1952-68; prof. psychology U. Ill., Chgo., 1968-80; prof. emeritus U. Ill., 1980—; dir. rsch. John E. Reid Assocs., Chgo., 1969-87; v.p. rsch. Reid Psychol. Systems, 1985-87; cons. London House, Inc., Park Ridge, Ill., 1987-91; dir. Ash., Blackstone & Cates, Blacksburg, Va. Author: Guide for Selection and Placement of Employees, 2d edit., 1977, Volunteers for Mental Health, 1973, The Legality of Preemployment Inquiries, 1989, The Construct of Employee Theft Proneness, 1991, Preparing for Retirement: Guidelines and Information Sources, 1993, also other books, monographs and articles; editor: Forensic Psychology and Disability Evaluation, 1972. Mem. public adv. com. Chgo. Commn. Human Relations, 1957-80; retirement com. Chgo. Commn. Sr. Citizens, 1960-80; chmn. Ill. Psychologist Examining Com., 1963-72. Fellow AAAS, Am. Psychol. Assn. (pres. div. indsl. psychology 1968-69); mem. Ill. Psychol. Assn. (pres. 1963-64), Chgo. Psychol. Assn., Va. Psychol. Assn., Va. Applied Psychology Acad. (pres. 1992-93), Midwest Psychol. Assn., Am. Pers. and Guidance Assn., Acad. for Criminal Justice Scis., Am. Criminology Assn., Internat. Assn. Applied Psychology, Sigma Xi, Phi Beta Kappa, Psi Chi. Home: 817 Hutcheson Dr Blacksburg VA 24060-3211

ASHBACHER, CHARLES DAVID, computer programmer, educator; b. Fort Riley, Kans., Sept. 24, 1954; s. Rudolph Carl and Paula Louis (Enos) A.; m. Valencia Sue Ashbacher, Oct. 27, 1973 (div. May 1984); m. Mary L. Rhiner, Dec. 14, 1991; 1 child, Katrina; stepchildren: Nicholas Rhiner, Jay Rhiner. AS, Kirkwood Community Coll., Cedar Rapids, Iowa, 1978; BS, Mount Mercy Coll., 1980. Tchr. math, computer sci. Mount Mercy Coll., Cedar Rapids, 1983-89; tchr. computers Kirkwood Community Coll., Cedar Rapids, 1990—; rsch. programmer U. Iowa, Iowa City, 1990-92; rsch. scientist Geog. Decison Systems, Cedar Rapids, Iowa, 1993—; rev. panelist Math. and Computer Edn., Hicksville, N.Y., 1990—; mem. editorial bd., revs. editor Jour. Recreational Math. Editorial bd. Jour. Recreational Math., 1991, Recreational and Ednl. Computing, 1989—; PC software revs. editor Math. and Computer Edn.; contbr. articles to profl. jours. Basketball coach YMCA, 1990—; active local PTA. Mem. AAAS, IEEE, Am. Math .Soc. Am. Assn. Artificial Intelligence, Am. Math. Assn. Two Yr. Colls. (rev. panelist 1990—), Assn. Automated Reasoning, Assn. Computing Machinery, Math. Assn. Am. Avocations: reading, languages, solving problems, travel. Home: 1615 K Ave NE Cedar Rapids IA 52402

ASHBAUGH, SCOTT GREGORY, mechanical engineer, consultant; b. Grand Forks AFB, N.D., Sept. 14, 1968; s. Dennis Michael and Pamela Sue (Pierce) A.; m. Lesha Ann Goodchild, June 29, 1991; 1 child, Brandon Pierce. BS in Aero. Engring., Calif. Poly. State U., 1991. Cert. engr.-in-tng., Calif. Staff engr. Sci. Applications Internat. Corp., Albuquerque, 1991—. Mem. AIAA, System Safety Soc. Office: Sci Applications Internat 2109 Airpark Rd SE Albuquerque NM 87106

ASHBY, EUGENE CHRISTOPHER, chemistry educator, consultant; b. New Orleans, Oct. 30, 1930; s. Anthony and Ida (Bruno) A.; m. Carolyn Turner, Sept. 13, 1952; children: Chris, Stephen, Terry, Marie, Angela, Julie, Rachel. BS in Chemistry, Loyola U., New Orleans, 1951; MS in Chemistry, Auburn U., 1953; PhD in Chemistry, U. Notre Dame, 1956. Rsch. chemist Ethyl Corp., Baton Rouge, 1956-59, rsch. assoc., 1959-63; asst. prof. Ga. Inst. Tech., Atlanta, 1963-65, assoc. prof., 1965-69, prof., 1969-73, Regents prof., 1973—; cons. Ethyl Corp., 1980—, Conoco, Ponca City, Okla., 1972-76. Contbr. over 250 articles to profl. publs. Recipient Lavoisier medal French Chem. Soc., 1971, Sigma Xi rsch. award, 1968-75, Herty medal Am. Chem. Soc., 1983, Disting. Prof. award Ga. Inst. Tech., 1988. Avocations: tennis, cattle farming. Home: 2516 Flair Knolls Dr NE Atlanta GA 30345-1316 Office: Ga Inst Tech Sch Chemistry Atlanta GA 30332

ASHBY, MICHAEL FARRIES, engineering educator; b. Bristol, Eng., Nov. 20, 1935; s. Eric and Helen Ashby; m. Maureen Stewart; children: Benjamin, Zachary, Victoria. Student, Campbell Coll., Belfast, No. Ireland, 1950-53, Queens Coll., No. Ireland, 1953-54; BA, U. Cambridge, 1961, MA, 1962, PhD, 1964; MA, Harvard U., 1970. Royal Soc. Rsch. prof. U. Cambridge, 1975. Author: Deformation Maps, 1982, Engineering Materials I, 1980, Engineering Materials II, 1986, Cellular Solids, 1989, Materials Selection in Engineering Design, 1992. Recipient Von Hippel award Materials Rsch. Soc., 1992. Fellow Royal Acad. Engring., London, The Royal Soc. London; mem. U.S. Acad. Engring., Swedish Acad. Engring. (fgn.), Acad. of Sci. (Gottingen, Ger.). Home: 51 Maids Causeway, Cambridge England CB5 8DE Office: Dept of Engring U Cambridge, Trumpington St, Cambridge CB2 1PZ, England

ASHEN, PHILIP, chemist; b. Bklyn., Nov. 5, 1915; s. Joel and Fannie (Hirt) A. BA in Chemistry, Bklyn. Coll., 1936; MBA, NYU, 1957, PHD, 1968. Cert. profl. chemist. Chief chemist Alco Mfg. Corp., Bklyn., 1936-48; mgr. chem. div. M.W. Hardy & Co., Inc., N.Y.C., 1948-63, v.p., 1963-77, pres., chief exec. officer, 1977—; lectr. grad. sch. NYU, N.Y.C., 1954-56; lectr. chem. warfare U.S. Citizens Def. Corps., 1940-45, gas reconnaissance officer, 1940-45; cons. in field, 1957—. Author: Foreign Chemical Companies Engaged in International Trade, 1957, The American Selling Price Method of Valuation in U.S. Chemical Imports, 1969. Bd. dirs. Bklyn. Coll. Alumni Assn., Bklyn., 1969-86. Recipient citation, U.S. Treasury Dept., 1944, Nat. War Fund Commn., 1945, Civilian Def. Vol. Office, 1945, War Finance Com. U.S. Treasury Dept., 1945, award ARC, 1945, Founders Day award NYU, 1969, Roosevelt medal Theodore Roosevelt Assn.. Fellow Am. Inst. Chemists (chmn. profl. accreditation com. N.Y. chpt. 1964-68, treas. N.Y. chpt. 1970-76, councillor N.Y. chpt. 1964-70), AAAS; mem. SACI, Am. Chem. Soc., N.Y. Acad. Sci., Chemistry Alumni Soc.-Bklyn. Coll. (pres. 1966—), Chemists Club (resident). Home: 2315 Avenue I Brooklyn NY 11210-2825 Office: M W Hardy & Co Inc 111 Broadway New York NY 10006-1901

ASHFORD, NICHOLAS ASKOUNES, technology and policy educator; b. Chgo., Jan. 9, 1938; s. Theodore Askounes and Venette (Tomaras) A.; 1 child, Androniki. BA in Chem., Washington Univ., 1959; PhD in physical chem., Univ. Chgo., 1965; JD, The Law Sch., 1972. Rsch. scientist Ill. Inst. Tech., Chgo., 1966-72, instr., 1966-71; adj. assoc. prof. Boston Univ. Sch., 1978—; sr. rsch. assoc. Ctr. for Policy Alt. MIT, Cambridge, Mass., 1972-78, assoc. prof. tech. and policy, 1978-92, prof. tech. and policy, 1992—; adj. faculty Harvard, Boston, 1984—; cons. U.S. Environmental Protection Agency, Washington, 1978-84, Consumer Product Safety Commn., Washington, 1979-81. Author: Crisis in the Workplace: Occupational Disease and Injury, 1976; (with others) Monitoring the Community for Exposure and Disease, 1993, Technology Law and the Working Environment, 1991, Chemical Exposures: Low Level and High Stakes, 1991, Monitoring the Worker for Exposure and Disease, 1990. Mem. adv. bd. State of Mass. Energy, 1983-85; chmn. com. on tech. EPA Nat. Adv. Coun., 1988-92; chmn., mem. Nat. Adv. Com. Occupational Safety, Washington, 1976-82; sci. adv. bd. mem. EPA, Washington, 1978-82. Recipient Macedo award Am. Assn. World Health, Washington, 1990, Lawrence R. Klein award Monthly Labor Review, Washington, 1975; fellow Woodrow Wilson Found., 1959-60. Fellow AAAS; mem. APHA (gov.), Am. Chem. Soc., Soc. Occupational and Environ. Health (v.p. 1985-86), Phi Beta Kappa, Sigma Xi. Democrat. Greek Orthodox. Office: Mass Inst Tech E40-239 77 Massachusetts Ave Cambridge MA 02139

ASHFORD, ROBERT LOUIS, computer professional; b. Meridian, Miss., Sept. 8, 1938; s. Walter and Bertha (Edmonds) A.; m. Ruth L. Sypert, May 16, 1992. Student, Tougaloo Coll., 1956-58. Programmer, analyst State of Calif., San Francisco, 1964-68; programming mgr. Control Data Corp., Palo Alto, Calif., 1968-73; office tech. cons. Hewlett-Packard Co., Palo Alto, 1973—; advisor, cons. NAACP, Palo Alto, 1988—. Mem. Legal Def. Fund: Com. of 100, N.Y.C. With U.S. Army, 1961-64. Mem. Space Studies Inst., Search for Extraterristrial Intelligence, Union Concerned Scientists, Astron. Soc. Pacific (advisor, cons. 1989—), Group 70 (bd. dirs. pub. rels. large amateur telescope project). Democrat. Home: 3005 Breen Ct San Jose CA 95121-2412 Office: Hewlett-Packard Co 100 Mayfield Ave Mountain View CA 94043-4158

ASHKENAZI, JOSEF, physicist; b. Rehovot, Israel, July 29, 1944; came to U.S., 1988; s. Itzhak and Eva (Romer) A.; m. Dalia Sadeh, Aug. 23, 1966; children: Avivit, Oded. BSc, Hebrew U., 1968, MSc, 1970, PhD, 1975. Charge' des recherches U. Geneva, Switzerland, 1976-81; asst. prof. Technion, Haifa, Israel, 1981-88; assoc. prof. U. Miami, Coral Gables, Fla., 1988—; advisor Weizmann Inst. of Sci., Rehovot, 1982-88, Armament Devel. Authority, Israel, 1983-88; guest scientist, cons. Naval Rsch. Lab., Washington, 1985—. Contbg. author: High Temperature Superconductivity, 1992; contbr. articles to profl. jours. Grantee Israel Acad. of Scis., 1982-84, 87-88, U.S.-Israel Binational Scis. Found., 1982-85, 87-88, Office of Naval Rsch., 1991. Mem. Israeli Phys. Soc., European Phys. Soc., Am. Phys. Soc., Nat. Geog. Soc. Jewish. Achievements include contributions to realistic theory of solid-state physics, including the subjects of metal-insulator transitions in transition metal oxides; electron-phonon coupling and superconductivity, high temperature superconductivity and others. Office: U of Miami Dept of Physics PO Box 248046 Miami FL 33124

ASHLER, PHILIP FREDERIC, international trade and development advisor; b. N.Y.C., Oct. 15, 1914; s. Philip and Charlotte (Barth) A.; m. Jane Porter, Mar. 4, 1942 (dec. 1968); children: Philip Frederic, Robert Porter, Richard Harrison; m. Elise Barrett Duvall, June 21, 1969; stepchildren: Richard Edward Duvall, Jeffries Harding Duvall. B.B.A. cum laude, St. Johns Coll., 1935; M.B.A., Harvard U., 1937; grad., Indsl. Coll. Armed Forces, 1956; Sc.D., Fla. Inst. Tech., 1969; LL.D. (hon.), U. West Fla., 1969; postgrad., U. Oxford, Eng., 1988, 89, 91. Enlisted USMCR, 1932; commd. ensign USN, 1938, advanced through grades to rear adm., 1959; served in Normandy, So. France, Iwo Jima, Korea; dir. Office Small Bus., Dept. Def., Washington, 1948-49; mem. joint staff Joint Chiefs Staff, 1957-59; ret., 1959; dir. devel. Pensacola Jr. Coll., 1960-68; vice chancellor adminstrn. State Univ. System Fla., 1968-70, exec. vice chancellor, 1970-75; treas., ins. commr., fire marshal State of Fla., 1975-76, sec. of commerce, 1977-79; pres. Philip F. Ashler & Assos., Tallahassee, 1979—; chmn. bd. Cambridge Community Care, Inc., Tallahassee, 1981-86, Circle Seven Internat., Tampa, 1988—; past dir. Fidelity Guaranty Life Ins. Co., Balt., U.S. Fidelity & Guaranty Co., 1st Fla. Bank N.A., Tallahassee; sec., dir. Fringe Benefits Mgmt. Co., Tallahassee, 1987—; mem. Fla. Edn. Council, 1967-68; commr. from Fla. Edn. Commn. States, 1967-68; mem. U.S. Dept. Commerce Dist. Export Council, 1978-92; pres., dir. Fla. Internat. Vol. Corps., 1988—; mem. legis. adv. council So. Regional Edn. Bd., 1966-68; mem. Fla. Bd. Ind. Colls. and Univs., 1971-75, mem. adv. council for mil. edn., 1980-85; bd. advisors Ctr. Profl. Devel., Fla. State U., 1988—; chmn. Fla. Civil Def. Adv. Council, 1966-69; mem. Fla. Council Internat. Devel., 1973-92, vice chmn., 1979-80, chmn., 1980-82, chmn. emeritus, 1990—; mem. Select Council on Post High Sch. Edn., 1967-68; chmn. Fla. Med. Liability Ins. Commn., 1975-76, Fla. Task Force on Auto and Workers Compensation, 1975-76; mem. Yugoslavia Adv. Council, 1976-87, InterAm. Congress on Psychology, Bogota, Colombia, 1974, NATO Advanced Sci. Inst., W.Ger., 1973; guest lectr. U. Belgrade, Yugoslavia, 1973; adviser econ. devel. to gov. Fla., 1977-78; mission leader Japan/S.E. U.S. Assn., Tokyo, 1977; trustee Fla. Council on Econ. Edn., 1979-81; mem. services policy adv. com. Office of U.S. Trade Rep., Exec. Office of Pres., Washington, 1980-85; mem. Republic of China/U.S.A. Econ. Council, 1979-92. Mem. Fla. Ho. of Reps., 1963-68; chmn. bd. dirs. Fla. Heart Assn., 1969-71; bd. dirs., treas. Internat. Cardiology Found.; bd. dirs. Tallahassee Meml. Hosp., Easter Seal Soc., 1963-68; bd. dirs., mem. exec. com. Am. Heart Assn., 1971-77, Internat. Cardiology Fedn., Geneva, 1975-77; founding chmn. Tallahassee Symphony Orch., 1981-82; trustee So. Ctr. Internat. Studies, Atlanta, 1988-91; mem. adv. bd. Fla./China Inst., Miami, Fla./Japan Inst., Tampa, Fla./Brazil Inst. Decorated Bronze Star with Combat V, Korean Presdl. citation; recipient Internat. Distinguished Service award Kiwanis Internat., 1965; Distinguished Service award Am. Heart Assn., 1965, 71; Distinguished Achievement award, 1975; Legislative award St. Petersburg Times, 1967. Mem. Fla. Med. Malpractice Joint Underwriting Assn. (chmn. bd. govs. 1975-76), Nat. Assn. Ins. Commrs. (vice chmn. exec. com. 1976), Internat. C. of C. (U.S. coun. 1979-87), U.S. S.E./Japan Assn. (chmn. 1981-83), S.E. U.S./Korea Econ. Coop. Coun. (bd. dirs.), Capital Tiger Bay Club (chmn. bd. dirs.), Govs. Club (bd. govs. 1989-93, v.p. for fin. 1992-93), Econ. Club Fla. (chmn. 1987-90, chmn. emeritus 1991—), Masons (32 degree), Shriners, Rotary, Kappa Delta. Episcopalian (lic. lay eucharistic minister). Home: 2115 E Randolph Cir Tallahassee FL 32312-3325 also: 11 Riad Sultan Kasbah, Tangier Morocco Office: Fringe Benefits Mgmt Co PO Box 1878 Tallahassee FL 32302

ASHLEY, ELLA JANE (ELLA JANE RADER), medical technologist; b. Dewitt, Ark., Mar. 6, 1941; d. Clayton Ervin and Emma Mae (Coleman) Funderburk; m. Albert Ashley, Sept. 27, 1957 (div. Nov. 1962); 1 child, Cynthia Gayle. Student, Westark Community Coll. Cert. clin. lab. technologist, clin. lab. scientist. Lab. asst. U. Ark. Med. Ctr., Little Rock, 1966-67; lab. technician II Ark. State Hosp., Little Rock, 1967-68; staff technologist Cooper Clinic, Ft. Smith, Ark., 1969-71; asst. chief technologist Nat. Health Labs (formerly Bioassay/Am. Biomed.) div. Revlon, Ft. Smith, 1972—; mem. profl. adv. panel Med. Lab. Observer, 1976—. research in lithium carbonate. Mem. Am. Soc. Med. Technology. Lutheran. Avocations: travel, theater, concerts, painting. Home: 1310 S Houston St Fort Smith AR 72901-7271 Office: Nat Health Labs 500 Lexington Ave Fort Smith AR 72901-4641

ASHLEY, HOLT, aerospace scientist, educator; b. San Francisco, Jan. 10, 1923; s. Harold Harrison and Anne (Oates) A.; m. Frances M. Day, Feb. 1, 1947. Student, Calif. Inst. Tech., 1940-43; BS, U. Chgo., 1944; MS, MIT, 1948, ScD, 1951. Mem. faculty MIT, 1946-67, prof. aero., 1960-67; prof. aeros. and astronautics Stanford U., Palo Alto, Calif., 1967-89, prof. emeritus, 1989—; spl. rsch. aeroelasticity, aerodynamics; cons. govt. agys., rsch. orgns., indsl. corps.; dir. office of exploratory rsch. and problem assessment and div. advanced tech. applications NSF, 1972-74; mem. sci. adv. bd. USAF, 1958-80, rsch. adv. com. structural dynamics NASA, 1952-60, rsch. adv. com. on aircraft structures, 1962-70, chmn. rsch. adv. com. on materials and structures, 1974-77; mem. Kanpur Indo-American program Indian Inst. Tech., 1964-65, governing bd. Nat. Rsch. Coun., 1988-91; AIAA Wright Bros. lectr., 1981; bd. dirs. Hexcel Corp. Co-author: Aeroelasticity, 1955, Principles of Aeroelasticity, 1962, Aerodynamics of Wings and Bodies, 1969, Engineering Analysis of Flight Vehicles, 1974. Recipient Goodwin medal M.I.T., 1952; Exceptional Civilian Service award U.S. Air Force, 1972, 80; Public Service award NASA, 1981; named one of 10 outstanding young men of year Boston Jr. C. of C., 1956; recipient Ludwig-Prandtl Ring, West German DGLR, 1987, Spirit of St. Louis Medal, ASME, 1992. Fellow AIAA (hon., assoc. editor jour., v.p. tech. 1971, pres. 1973, Structures, Structural Dynamics and Materials award 1969), Am. Acad. Arts and Scis., Royal Aero. Soc. (hon.); mem. AAAS, NAE (aeros. and space engring. bd. 1977-79, mem. coun. 1985-91), Am. Meterol. Soc. (profl., 50th Ann. medal 1971), Phi Beta Kappa, Sigma Xi, Tau Beta Pi. Home: 475 Woodside Dr Woodside CA 94062-2363

ASHLEY, KEVIN EDWARD, research chemist; b. Hammond, Ind., Aug. 31, 1958; s. Kenneth Edward Ashley and Judith Diane (Theisen) O'Neal; m. Diane Barbara Parry, Oct. 8, 1988; children: Amanda Diane, Elizabeth Lynn. BS in Biology, U. Ariz., 1980; MS in Chemistry, No. Ariz. U., 1984; PhD in Chemistry, U. Utah, 1987. Vis. faculty scientist IBM Almaden Rsch. Ctr., San Jose, Calif., 1988; vis. asst. prof. U. Utah, Salt Lake City, 1989; asst. prof. San Jose (Calif.) State U., 1988-91, assoc. prof., 1991-93; rsch. chemist Nat. Inst. Occupational Safety & Health, Cin., 1991—; Sci. fair judge local jr. high and high schs., Cin. Contbr. over 25 articles to sci. jours., 1985-93. Rsch. grantee Petroleum Rsch. Fund, San Jose State U., 1988, 90, Rsch. Corp., San Jose State U., 1989, NIH-Minority Biomed. Rsch. Support, Keck Found., San Jose State U., 1991. Mem. ASTM (subcom. vice-chair 1991—), Am. Chem. Soc. (analytical div.), Soc. Electroanalytical Chemistry, Sierra Club. Democrat. Achievements include rsch. in infrared spectroelectrochem. studies of ionic absorption on electrode surfaces, electrochem. sensors, mechanistic studies of reactions at interfaces and surface electrochemistry. Office: Nat Inst Occupational Safety & Health 4676 Columbia Pkwy MSR-7 Cincinnati OH 45226

ASHLEY, SHARON ANITA, pediatric anesthesiologist; b. Goulds, Fla., Dec. 28, 1948; d. John H. Ashley and Johnnie Mae (Everett) Ashley-Mitchell; m. Clifford K. Sessions, Sept. 1977 (div. 1985); children: Cecili, Nicole, Erika. BA, Lincoln U., 1970; postgrad., Pomona Coll., 1971; MD, Hahnemann Med. Sch., Phila., 1976. Diplomate Am. Bd. Pain Mgmt. Intern pediatrics Martin Luther King Hosp., L.A., 1976-77, resident pediatrics, 1977-78, resident anesthesiology, 1978-80, mem. staff, 1981—. Named Outstanding Tchr. of Yr., King Drew Med. Ctr., Dept. Anesthesia, 1989, Outstanding Faculty of Yr., 1991. Mem. Am. Soc. Anesthesiologists, Calif. Med. Assn., L.A. County Med. Soc., Soc. Regional Anesthesia, Soc. Pediatric Anesthesia. Democrat. Baptist. Avocations: reading, crocheting, sailing. Office: Martin Luther King Hosp 12021 Wilmington Ave Los Angeles CA 90059-3099

ASHLEY, WILLIAM HILTON, JR., software engineer; b. Laurel, Miss., Jan. 27, 1954; s. William Hilton and Kay Kroh (Forrester) A.; m. Mary Frances Presnell, May 23, 1981; children: William Hilton III, Victoria Kay. BA in Math., Southwestern U., 1976; MS in Math., Numerical Analysis, Tex. A&M U., 1978. Grad. teaching asst. dept. math. Tex. A&M U., College Station, 1976-78; programmer IBM Shuttle SImulator Math. Models, Houston, 1979-80; programmer, designer, linkage editor IBM Shuttle Relink, Houston, 1980-82; software designer IBM Shuttle Configuration Mgmt. Databases, Houston, 1982-83; lead software architect, designer IBM Shuttle Reconfiguration Databases, Houston, 1983-86; founder, chair tech. insertion IBM On-board Shuttle, Houston, 1986-92; systems engr. IBM Space Sta. Program, Houston, 1992—; presenter at profl. confs., presenter workshops.

Contbr. to profl. publs. Founder, chair Heritage Park Neighborhood Crime Watch, Friendswood, Tex., 1991-92; sect. chmn. Brookwood Neighborhood Crime Watch, Houston, 1992—; bd. dirs. Blue Marlins Swim Team, Houston, 1992—. Mem. AIAA, Phi Mu Epsilon. Achievements include invention of processes for organizational change including technology ownership life cycle, production-oriented information processes, production-oriented education processes, technology insertion process. Office: IBM Fed Sector Co 3700 Bay Area Blvd Houston TX 77058

ASHMEAD, ALLEZ MORRILL, speech-hearing-language pathologist, orofacial myologist, consultant; b. Provo, Utah, Dec. 18, 1916; d. Laban Rupert and Zella May (Miller) M.; m. Harvey H. Ashmead, 1940; children: Harve DeWayne, Sheryl Mae Harames, Zeltha Janeel Henderson, Emma Allez Broadfoot. BS, Utah State U., 1938; MS summa cum laude, U. Utah, 1952, PhD summa cum laude, 1970; postgrad., Idaho State U., Oreg. State Coll., U. Denver, U. Utah, Brigham Young U., Utah State U., U. Washington, U. No. Colo. Cert. secondary edn., remedial reading, spl. edn., learning disabilities; cert. ASHA clin. competence speech pathology and audiology; profl. cert. in orofacial myology. Tchr. pub. schs. Utah, Idaho, 1938-43; speech and hearing pathologist Bushnell Hosp., Brigham City, Utah, 1943-45; sr. speech correctionist Utah State Dept. Health, Salt Lake City, 1945-52; dir. speech and hearing dept. Davis County Sch. Dist., Farmington, Utah, 1952-65; clin., field supr. U. Utah, Salt Lake City, 1965-70, 75-78; speech pathologist Box Elder Sch. Dist., Brigham City, 1970-75, 78-84; teaching specialist Brigham Young U., Provo, 1970-73; speech pathologist Primary Children's Med. Ctr., Salt Lake City, 1975-77; pvt. practice speech pathology and orofacial myology, 1970-88; del. USSR Profl. Speech Pathology seminar, 1984, 86; participant numerous internat. seminars. Author: Physical Facilities for Handicapped Children, 1957, A Guide for Training Public School Speech and Hearing Clinicians, 1965, A Guide for Public School Speech Hearing Programs, 1959, Impact of Orofacial Myofunctional Treatment on Orthodontic Correction, 1982, Meeting Needs of Handicapped Children, 1975, Relationship of Trace Minerals to Disease, 1972, Macro and Trace Minerals in Human Metabolism, 1971, Electromotive Potential Differences Between Stutterers and Non-stutterers, 1970, Learning Disability, An Educational Adventure, 1969, New Horizons in Special Education, 1969, Developing Speech and Language in the Exceptional Child, 1961, Parent Teacher Guidance in Primary Stuttering, 1951, numerous others; contbr. research articles to profl. jours. Student Placement chair Am. Field Service, Kaysville, Utah, 1962-66; ednl. del. Women's State Legis. Council, Salt Lake City, 1958-70; chairwoman fund raising Utah Symphony Orch., Salt Lake City, 1970-71; sec., treas. Utah chpt. U.S. Council for Exceptional Children, 1958-62, membership com. chair, 1962-66, program com. chair, 1966-68. Recipient Scholarship award for Higher Edn. U. Utah, Salt Lake City, 1969; Delta Kappa Gamma scholar, 1968; rsch. grantee Utah Dept. Edn., 1962. Mem. NEA, Utah Ednl. Assn., Am. Speech, Lang. Hearing Assn. (life, continuing edn. com. 1985, Ace award for Continuing Edn. 1984), Western Speech Assn., Internat. Assn. Orofacial Myology (life, bd. examiners, Sci. Contribution award 1982), Utah Speech, Hearing and Lang. Assn. (life, sec., treas. 1956-60), AAUW (Utah state bd. chair status of women 1959-62, Kaysville br. 1957-60, bd. dirs. Kaysville-Davis br. 1987-92, chair internat. rels. 1987-91, chair cultural interests Kaysville-Davis br. 1991-92), Delta Kappa Gamma (state scholarship award 1968, del. Woman's State Legis. Coun. 1958-70, profl. affairs chair 1963-67, tchr. of yr. award 1978), AAUW (bd. dirs. internat. rels. Kaysvile-Davis br., 1988-91), Sigma Alpha Eta, Theta Alpha Phi, Psi Chi, Zeta Phi Eta, Phi Kappa Phi. Republican. Mormon. Lodges: Daus. Utah Pioneers (parlimentarian Kaysville chpt. 1980-92, historian 1975-80, lesson leader 1992—), Soroptimist Internat. (charter mem. 1954, bd. dirs. 1954-56, pres. Davis County chpt. 1965-69, treas. 1956-54, Rocky Mountain regional bd. dirs. 1965-70, community service award 1968, pub. service award 1970). Avocations: international travel, reading, boating, sports, fine and performing arts. Home: 719 E Center St Kaysville UT 84037-2138

ASHMEAD, HARVE DEWAYNE, nutritionist, executive, educator; b. Brigham City, Utah, June 6, 1944; s. Harvey Harold and Allez (Morrill) A.; m. Eugele Baird, June 24, 1966; children: Stephen, Jilane, Brett, Angelique, Heidi. BS, Weber State Coll., 1969; PhD, Pacific Inst., 1970; PhD magna cum laude, Donsbach U., 1981. Cert. nutritional cons. With Ch. Jesus Christ of Latter Day Saints, Paris, 1963-66; v.p. Albion Labs., Ogden, Utah, 1966-71, exec. v.p., Clearfield, Utah, 1971-82, pres., 1982—, also bd. dirs.; pres. Albion Internat.; adj. prof. Weber State Coll., also adv. council; former advisor Weber County Sch. Dist.; bd. dirs. Albion Internat., Zions Bank, Albion Labs., Inc., Unilabco, Inc., Albion Middle East, Albion Europe, Rhondell Labs.; guest lectr. Adv. Fruit Heights City (Utah); pres. PTA. Fellow Am. Coll. Nutrition; mem. Am. Soc. Animal Sci., Am. Assn. Nutrition and Dietary Cons., Internat. Acad. Nutritional Cons., Am. Assn. Nutritional Cons., Am. Acad. Applied Health Sci., AAAS, Am. Biographical Inst. (bd. govs.), Clearfield C. of C. (bd. dirs.), Delta Sigma Pi. Mormon. Author: Chelated Mineral Nutrition, 1981, Mineral Absorption Mechanisms, 1981, Chelated Mineral Nutrition in Plants, Animals and Man, 1982, A New Era in Plant Nutrition, 1982, Intestinal Absorption of Metal Ions and Chelates, 1985, Foliar Feeding of Plants with Amino Acid Chelates, 1986, In Search of a Rainbow, 1988, Amino Acids in Animal Nutrition, 1991; Conversations on Chelation and Mineral Nutrition, 1989, The Roles of Amino Acid Creates in Animal Nutrition, 1993; contbr. numerous articles to profl. jours. Office: Albion Labs 101 N Main St Clearfield UT 84015-2243

ASHMORE, ROBERT WINSTON, computational pharmacologist; b. Boston, Apr. 7, 1955; s. James Elwin and Lottie John (Holt) A.; m. Jacqueline Stephens, May 26, 1979; children: John James, Lillian Rosa. BA, Ind. U., 1977; MS, Johns Hopkins U., 1988. Asst. supr. Ind. St. Dept. Toxicology, Indpls., 1977-84; sr. lab. technician Cal Health Rsch. Inst., Indpls., 1984-85; mass spectro rsearcher Walter Reed Army Inst. Rsch., Washington, 1985-88; scientific applications analyst Program Resources, Inc., Frederick, Md., 1988-90; scientific systems analyst Eli Lilly And Co., Indpls., 1990—; v.p. Gain, Inc., Indlps., 1992. Contbr. articles to profl. jours. Mem. Adminstr. Bd. Epworth UM, Indpls., 1992. Mem. USENIX, Am. Soc. for Mass Spectrometry, Am. Electrophoresis Soc., Assn. for Computing Machinery. Methodist. Achievements include development of thermospray LC/MS analysis of metabolites of antileishmanial 8-aminoquinoline generated in isolated perfused rat liver and development of various biomedically oriented data acquistion mgmt. systems. Office: Eli Lilly and Co Lilly Corporate Ctr Indianapolis IN 46285

ASKER, JAMES ROBERT, journalist; b. Louisville, June 4, 1952; s. James Edwin and Virginia Ann (Thornburrow) A.; m. Jane Roberta Gurin, July 28, 1985; children: Nathaniel Leif, Eric Louis. BA in Policy Scis., Rice U., 1974; postgrad., MIT, 1987-88. Reporter sci. tech. and space Houston Post, 1974-88; Knight fellow MIT, Cambridge, 1987-88; reporter, editor Boston, 1989; mng. editor Electronic Bus., Boston, 1989; sr. space tech. editor Aviation Week & Space Tech., Washington, 1989—. Contbr. articles to Colliers Ency. and World Book Sci. Yr. Mem. AAAS, Soc. Profl. Journalists (dir. freedom of info. Tex. Gulf Coast chpt. 1978-86, ethics com. 1986), Nat. Assn. Sci. Writers, Investigative Reporters and Editors, D.C. Sci. Writers Assn., Aviation/Space Writers Assn. Home: 2436 Drexel St Vienna VA 22180 Office: Aviation Week & Space Tech 1200 G St NW Ste 922 Washington DC 20005

ASKEW, THOMAS RENDALL, physics educator, researcher, consultant; b. Geneva, Ill., June 11, 1955; s. Thomas Addelbert and Jean Mary (Somerville) A.; m. Mary Louise Kazmaier, July 15, 1978; 1 child, Steven Thomas. MS, U. Ill., 1982, PhD, 1984. Mem. tech. staff DuPont Rsch., Wilmington, Del., 1984-91; asst. prof. physics Kalamazoo (Mich.) Coll., 1991—; vis. scientist Argonne (Ill.) Nat. Lab., 1992; cons. to numerous U.S. corps. Contbr. articles to profl. jours. Recipient Ball Meml. scholarship Gordon Coll., 1977, Mac Arthur scholarship John D. and Catherine T. Mac Arthur Found., 1991-93. Mem. Am. Phys. Soc., Material Rsch. Soc. Achievements include patent in superconductivity. Office: Kalamazoo Coll Physics Dept Kalamazoo MI 49006

ASKEW, WILLIAM EARL, chemist, educator; b. Maysville, N.C., Aug. 31, 1943; s. Carl Lee and Sally Chinese (Pope) A. BA in Chemistry, U. N.C., 1965; MA in Biology, East Carolina U., 1968; PhD in Biophys. Sci., U. Houston, 1973. Rsch. assoc. Baylor Coll. Medicine, Houston, 1973-77, Vets.' Hosp., Houston, 1973-77; instr. chemistry Houston C.C. System, 1977—. With U.S. Army, 1968-70. Mem. Am. Chem. Soc., Am. Acad. Sci., Tex. Jr. Coll. Tchrs. Assn., 2-Yr. Chemistry Soc. Office: Northwest Coll Houston CC System 901 Yorkchester St Houston TX 77079

ASKEY, RICHARD ALLEN, mathematician; b. St. Louis, June 4, 1933; s. Philip Edwin and Bessie May (Yates) A.; m. Elizabeth Ann Hill, June 14, 1958; children: James, Suzanne. B.A., Washington U., St. Louis, 1955; M.A., Harvard U., 1956; Ph.D., Princeton U., 1961. Instr. in math. Washington U., 1958-61, U. Chgo., 1961-63; asst. prof. math. U. Wis., Madison, 1963-65; asso. prof. U. Wis., 1965-68, prof., 1968-86, Gabor Szego prof., 1986—. Author: Orthogonal Polynomials and Special Functions, 1975; editor: Theory and Application of Special Functions, 1975, Collected Papers of Gabor Szego, 1982. Guggenheim fellow, 1969-70. Fellow Indian Acad. Sci. (hon.), Am. Acad. Arts and Scis.; mem. Am. Math. Soc., Math. Assn. Am., Soc. Indsl. and Applied Math. Home: 2105 Regent St Madison WI 53705-3941 Office: Van Vleck Hall U Wis Madison WI 53706

ASLAM, MUHAMMED JAVED, physician; b. Shillong, India, June 27, 1938; came to U.S., 1963; m. Tasnim Qadir, Feb. 5, 1967; children: Anissa, Shaazia, Sohail. MBBS, King Edward Med. Coll., Lahore, Pakistan, 1962. Diplomate Am. Bd. Internal Medicine, Am. Bd. Hematology; Fellow Royal Coll. Physicians/Can. Hematologist Winnipeg (Can.) Clinic, 1971-77; pvt. practice, hematologist Houston, 1977—; pres. Tess Data Systems, Houston, 1985—. Author: (computer software) Dietician, 1980, Tess System One, 1985, Tess System Two, 1987, Tess System Three, 1990; co-author: (computer software) Dietician Dietwae, 1984. Mem. Tex. Med. Assn., Harris County Med. Soc., N.Y. Acad. Scis. Office: Tess Data Systems Inc 14340 Torrey Chase Blvd Ste 340 Houston TX 77014

ASMAR, CHARLES EDMOND, structural engineer, consultant; b. Beirut, Lebanon, Oct. 6, 1958; came to U.S., 1984; s. Edmond Eid and Constantina (Canarelli) A.; m. Suad Omer Mohamud; 1 child, Karina. BSCE, U. Toledo, 1986. Registered profl. engr., N.Y. Inspector Paul Zein Cons., Jounieh, 1976-82; structural engr. Parsons Brinckerhoff, Hartford, Conn., 1987-89, N. Massand P.C., Bayside, N.Y., 1989-91, Goodkind & O'Dea, Inc., Rutherford, N.J., 1991-93; cons. pvt. practice, Stamford, Conn., 1993—. Mem. ASCE. Home and office: 95 Liberty St A-8 Stamford CT 06902

ASMUSSEN, JES, JR., electrical engineer; b. Milw., June 12, 1938; s. Jes and Anita (Weltzien) A.; m. Judith Adele Knopp, June 18, 1960 (div. Mar. 1980); m. Colleen Cooper, Jan. 4, 1987; children: Kirsten, Jes III, Stig; stepchildren: Scott Cooper, Jill Cooper. BSEE, U. Wis., 1960, MSEE, 1964, PhD in Elec. Engring., 1967. Design and devel. engr. Louis Allis Co., Milw., 1960-62; asst. prof. elec. engring. Mich. State U., East Lansing, 1967-71, assoc. prof. elec. engring., 1971-75, prof. elec. engring., 1975—, acting assoc. dir. div. engring. rsch., 1983-84, chairperson dept. elec. engring., 1990—; cons. to industry. Contbr. over 200 articles to profl. jours. Recipient Disting. Faculty award Mich. State U., 1988, Disting. Svc. citation Coll. Engring. U. Wis.-Madison, 1993; grantee NSF, Dept. Energy, NASA, Def. Advanced Rsch. Projects Agy., 1971—. Fellow IEEE; mem. AIAA, AAAS, Am. Vacuum Soc., Material Rsch. Soc., Sigma Xi, Eta Kappa Nu. Lutheran. Achievements include 11 patents in field for microwave, multipolar, electron cyclotron resonance plasma and ion source technology used in thin film deposition and submicron etching applications, microwave plasma assisted diamond thin film technology; research in microwave discharges (plasma) as applied to microchip technology and electric propulsion engines for spacecraft propulsion, microwave technology for materials processing, and wind power engineering. Home: 3811 Viceroy Dr Okemos MI 48864-3844 Office: Mich State U Dept Elec Engring East Lansing MI 48824

ASOMOZA, RENE, physicist; b. Puebla, Mexico, June 29, 1948; s. Vicente and Concepcion (Palacio) A.; m. Martha Torres-Hernandez Asomoza, June 26, 1972. BS in Physics, Faculty Physics Math.-Nat. Polytech. Inst., Mexico City, 1972; degree, Paris U., Orsay, France, 1975, PhD in Physics, 1980. Assoc. prof. U. Paris-SUD, 1976-80; assoc. prof. Ctr. for Rsch. and Advanced Studies, Nat. Poly. Inst., Mexico City, 1980-81, prof., 1981—, head solid state electronics sect., 1987—. Contbr. 34 articles to profl. jours. Mem. Am. Vaccum Soc., Materials Rsch. Soc., Mexican Surface Sci. Soc. (rsch. award 1991), Mexican Acad. of Materials. Achievements include rsch. on basic properties of amorphous semiconductors, in particular amorphous hydrogenated carbon films. Home: Salamina 318 Col Lindavista, 07300 Mexico City Mexico Office: Cinvestav, AV IPN 2508, 07300 Mexico City Mexico

ASSANIS, DENNIS N. (DIONISSIOS ASSANIS), mechanical engineering educator; b. Athens, Greece, Feb. 9, 1959; came to U.S., 1980; s. Nicholas and Kyriaki (Gyftakis) A.; m. Helen Stavrianos, Aug. 25, 1984; children: Nicholas, Dimitris. BSc in Marine Engring. with distinction, Newcastle U., U.K., 1980; SMME, SM in Naval Arch. Marine Engring., MIT, 1982, PhD in Power and Propulsion, 1985, SM in Mgmt., 1986. Asst. prof. mech. engring. U. Ill., Urbana-Champaign, 1985-90, assoc. prof. mech. engring., 1990—; assoc. prof. Nat. Ctr. for Supercomputing Applications, 1990—, head thermal scis./systems divsn., 1992—; part-time rsch. staff energy and environ. systems divsn. Argonne Nat. Lab., 1987-92; cons. in field. Contbr. over 30 articles to Internat. Jour. Vehicle Design, Jour. Heat Recovery Systems and Combined Heat and Power, ASME Transactions: Jour. Fluids Engring, ASME Transactions: Jour. Engring. for Gas Turbines and Power, SAE Transactions: Jour. Engines, Internat. Jour. Materials and Product Tech., Jour. Computational Physics, Internat. Jour. Numerical Methods in Fluids, Numerical Heat Transfer: Fundamentals, SAE Transactions: Jour. Passenger Cars, SAE Transactions: Jour. Materials, Internat. Jour. Heat and Mass Transfer, also to conf. proceedings. Univ. scholar, 1991-94, Athens Coll. Acad. scholar, 1967-77; recipient IBM Rsch. award, 1991, NSF Presdl. Young Investigator award, 1988-93, NSF Engring. Initiation award, 1987, NASA Cert. of Recognition for Creative Devel. of a Tech. Innovation, 1987; Lilly Endowment Teaching fellow, 1988. Mem. ASME (faculty advisor U. Ill. student sect. 1989-91, ASME/Pi Tau Sigma Gold Medal award 1990, Internal Combustion Engine Divsn. Speaker award 1993), Soc. Automotive Engrs. (Ralph Teetor award 1987, Russell Springer Best Paper award 1991), Am. Soc. for Engring. Edn., Combustion Inst., Sigma Xi. Achievements include development of comprehensive models of internal combustion engine processes. Office: U Ill 1206 W Green St Urbana IL 61801

AST, DIETER GERHARD, materials science educator; b. Stuttgart, Germany, Mar. 14, 1939; came to U.S., 1966; s. Reinhold and Hildegard (Rau) A.; m. Carol Patricia Abitabilo, Aug. 21, 1971; children; Ingrid, Karen. Dipl. Phys., U. Stuttgart, Germany, 1966; PhD, Cornell U., 1970. Rsch. scientist IBM T.Y. Watson Lab., Yorktown Heights, Calif., 1976-77; vis. scientist Hewlett Packard Corp. Labs., Palo Alto, Calif., 1983-84, 91-92; cons. 3M, Mpls., 1987-89, Electronic Display Systems, Hatfield, Pa., 1984-89. Contbr. articles to profl. jours. Achievements include patents on non-lattice matched growth and rapidly solidified composites; research in active matrix flat panel displays, defects in semiconductors and non-lattice matched growth of semiconductors. Office: Cornell U Materials Sci Bard Hall Ithaca NY 14853-1501

ASTILL, BERNARD DOUGLAS, environmental health and safety consultant; b. Nottingham, Eng., Feb. 11, 1925; came to U.S., 1952; s. Bernard and Constance (Harriet) A.; m. Norma Sarah Di Lauro, Apr. 9, 1955; children: Paul, Alexandra. BS, U. Nottingham, Eng., 1950, PhD in Organic Chemistry, 1953. Postdoctoral Fulbright fellow U. Rochester, N.Y., 1952-55; biochemist Eastman Kodak Co., Rochester, N.Y., 1955-64; clin. instr. U. Rochester Med. Sch., Rochester, N.Y., 1960-74; rsch. assoc. Eastman Kodak Co., Rochester, 1964-72, supr. biochemistry health & environ. lab., 1972-85, dir. regulatory affairs, life scis., 1985-87; cons. Bernard D. Astill Assocs., Rochester, 1987—; cons. in field; com. mem. NAS/NRC Div. Toxicology, Washington, 1980-85, 88-91. Author: (book chpt.) Patty's Toxicology & Indsl. Hygiene, 1981; contbr. articles to profl. jours. Chmn. Rochester Cosmopolitan Club, 1957-59; co-chmn. new mem. campaign Rochester Philharmonic Orch., 1976-78. Petty officer Royal Navy, 1943-46, South Pacific. Mem. Soc. Toxicology, Am. Chem. Soc. (chmn. Rochester sect. 1983), Am. Guild Organists. Avocations: piano, organ, choral dir., gardening, theatre. Home and Office: 195 Lyell St Spencerport NY 14559-9536

ASTOR, PETER H., environmental consultant, mathematician; b. N.Y.C., Jan. 16, 1944; s. Jack I. and June (Solot) A.; m. Harriet Ann Swartz, Mar. 20, 1965; children: Gregory E., Aaron M., James E. BS, Stevens Inst. Tech., 1964, MS, 1966, PhD, 1970. Cert. environ. auditor. Instr. N.C. State U., Raleigh, 1968-70; asst. prof. Drexel U., Phila., 1970-73; supervising engr. Ebasco Svcs., Inc., Lyndhurst, N.J., 1973-85; dir. N.J. Inst., Newark, 1985-86; project mgr. Woodward-Clyde Cons., Wayne, N.J., 1986-89; sr. project mgr., dir. Louis Berger & Assocs., Inc., East Orange, N.J., 1989—. Author: Systems Analysis in Ecology, 1975; contbr. articles to profl. jours. Recipient Harold Fee Alumni award Stevens Inst. Tech., 1989. Mem. Stevens Alumni Environ. Profls. (founder, pres. 1990-93), Hazardous Material Control Rsch. Inst., Stevens Alumni Coun. (v.p. 1990—). Achievements include development of techniques to include environmental/human health variables in siting studies and government program decisions, techniques used to site numerous industrial facilities including U.S. Dept. of Energy's management operations nuclear waste nationwide. Home: 52 Clinton Ave Maplewood NJ 07040 Office: Louis Berger and Assocs Inc 100 Halsted St East Orange NJ 07019

ASTWOOD, WILLIAM PETER, psychotherapist; b. N.Y.C., May 18, 1940; s. Henry Kenneth and Rose Margit (Eastby) A.; m. Sharon Lisa Sprung, June 10, 1979; 1 child, Jesse Jack. BA, CUNY, 1962; MA, NYU, 1967, PhD, 1975. Case worker, supr. dept. social services City N.Y., 1964-67; community orgn. trainer Block Communities, Inc., N.Y.C., 1967-68; field rep. Office Econ. Opportunity, N.Y.C., 1968-70, U.S. Dept. Health, Edn., Welfare, N.Y.C., 1970-71; pvt. practice Bklyn., 1971—; dir. family therapy div. DiMele Ctr. for Psychotherapy, N.Y.C., 1990—; bd. dirs. South Beach Psychiat. Ctr., Bklyn., 1976-78, N.Y. Group for Comprehensive Family Therapy, Mineola, 1988—; exec. bd. Met. Ctr. for Psychotherapy, N.Y.C., 1969-72. Co-author: Practicing Psychotherapy, 1980. Exec. bd. Social Service Employees Union, N.Y.C., 1965-67. Staff sgt. USANG, 1963-69. Mem. N.Y. Acad. Scis., Assn. for Humanistic Psychology. Home: 394 Atlantic Ave Brooklyn NY 11217-1703 Office: 163 Clinton St Brooklyn NY 11201-4601

ATAIIFAR, ALI AKBAR, economist, educator; b. Khoy, Azerbijan, Iran, July 2, 1952; came to U.S., 1970; s. Ayoob and Khadijeh Ataiifar. BS, Internat. Iranzamin Coll., Tehran, 1977; MBA, Northrop U., 1979; MA, New Sch. for Social Rsch., N.Y.C., 1989; ABD, New Sch. for Social Rsch., 1990. Lectr. Kean Coll. N.J., Union, 1983-87, NYU, 1988-89; asst. prof. Delaware County Community Coll., Media, Pa., 1990—. Avocations: speaking, reading and writing languages. Office: Delaware County CC Media PA 19063

ATAL, BISHNU SAROOP, speech research executive; b. Kanpur, Uttar Pradesh, India, May 10, 1933; came to U.S., 1961; s. Jagannath Prasad and Lakshmi Devi (Lakshmi) A.; m. Kamla Atal, July 3, 1959; children: Alka, Namita. BS with honors, U. Lucknow, India, 1952; elec. engring. degree, Indian Inst. Sci., Bangalore, 1955; PhD in Elec. Engring., Poly. Inst. Bklyn., 1968. Sr. rsch. asst. Indian Inst. Sci., Bangalore, 1955-56, lectr., 1957-60; sr. rsch. fellow Cen. Elec. Engring. Rsch. Inst., Pilani, Rajasthan, India, 1960-61; mem. tech. staff AT&T Bell Labs., Murray Hill, N.J., 1961-85, head acoustics rsch., 1985-90, head speech rsch., 1990—. Contbr. articles to various publs. Fellow Acoustical Soc. Am., IEEE (Acoustics, Speech and Signal Processing Sr. Tech. Achievement award 1975, ASSP Sr. award 1980, Centennial medal 1984, Morris N. Liebman Meml. Field award 1986); mem. NAE, NAS. Office: AT&T Bell Labs 600 Mountain Ave Rm 2D-535 Murray Hill NJ 07974

ATANASOFF, JOHN VINCENT, physicist; b. Hamilton, N.Y., Oct. 4, 1903; s. John and Iva Lucina (Purdy) A.; m. Lura Meeks, June, 1926 (div. 1949); children: Elsie Whistler, Joanne Gathers, John Vincent; m. Alice Crosby, June 17, 1949. BSEE, U. Fla., 1925, DSc (hon.), 1974; MS, Iowa State U., 1926; PhD, U. Wis., 1930, DSc (hon.), 1987; DSc (hon.), Moravian Coll., Bethlehem, Pa., 1981; LittD, Western Md. Coll., 1984; LHD (hon.), Mount St. Mary's Coll., 1990. Asst. prof. to assoc. prof. physics Iowa State U., Ames, 1930-42; chief acoustics div. Naval Ordnance Lab., White Oak, Md., 1942-49; dir. fuses Naval Ordnance Lab., White Oak, 1951-52; sci. advisor Chief Army Field Forces, Fort Monroe, Va., 1949-51; pres. Ordnance Engring. Corp., Frederick, Md., 1952-56; v.p. Aerojet Gen. Corp., Azusa, Calif., 1956-61; pres. Cybernetics Inc., Frederick, 1961-82; cons. scientist Stewart-Warner Corp., Chgo., 1961-63, Honeywell Inc., Mpls., 1968-71, Control Data Corp., Washington, 1968-71. Contbr. articles to acoustical, phys. and seismol. jours.; inventor 1st electronic digital computer, 1935, binary alphabet, 1943; patentee in field. Decorated Order of Cyril and Methodius 1st class (Bulgaria), 1970; recipient Disting. Civilian Service award U.S. Navy, 1945; citation Seismol. Soc., 1947, citation Bur. Ordnance, 1947; Disting. Achievement award Iowa State U. Alumni, 1983; Computing Appreciation award EDUCOM, 1985, IEEE elec. engring. milestone, 1990; named to Iowa Inventors Hall of Fame, 1978, Computer Pioneer medal IEEE, 1984, Iowa Gov.'s Sci. medal, 1985, medal of Bulgaria, 1985, Holley medal ASME, 1985, Nat. medal Technology U.S. Dept. Commerce Tech. Adminstrn., 1990. Mem. Bulgarian Acad. Sci. (fgn. mem.), Phi Beta Kappa, Pi Mu Epsilon, Tau Beta Pi. Democrat. Club: Cosmos (Washington). Home: 11928 E Baldwin Rd Monrovia MD 21770-9714

ATCHISON, ARTHUR MARK, industrial, research and development engineer; b. Cleve., Aug. 22, 1944; s. James Edward and Zella Katherine (Beecher) A.; m. Lora Suzanne Ferlet (div.); 1 child, Tiara Lynne; m. Patricia Fay Jones, July 9, 1983; 1 child, James Edward II. AA, Cuyahoga C.C., Parma, Ohio, 1971; BS in Indsl. Mgmt., U. Akron, 1973; MS in Indsl. Engring. and Ops. Rsch., Va. Poly. Inst. and State U., 1980; PhD in Indsl. Engring., Kennedy-Western U., 1992. Supr. indsl. engring. Firestone Tire and Rubber Co., Akron, Ohio, 1973-76; divisional indsl. engr. Firestone Internat. Co., Akron, 1976-77; regional ops. engr. Reynolds Aluminum Co., Richmond, Va., 1981-82; plant engr. AMP, Inc., Harrisonburg, Va., 1982-83; supr. engring. lab. E.R. Carpenter Co., Richmond, 1977-81; mgr. corp. devel. engring. E.R. Carpenter Co., Elkhart, Ind., 1983—; co. advisor Jr. Achievement, Akron, 1976. Vol. for emergency communications, Kodak Liberty Bike-a-thon, Richmond, 1986. Vol. lic. examiner FCC. With USN, 1964-70, Vietnam. Mem. Am. Inst. Indsl. Engrs. (sr.), Am. Soc. Quality Control, Amateur Radio Relay League, Richmond Amateur Telecom. Soc., Internat. Amateur Radio Soc., Greater Richmond Sailing Assn., Delta Sigma Pi (life). Republican. Episcopalian. Avocations: amateur radio, photography, coin collecting, sailing, art collecting. Office: ER Carpenter Co 195 County Rd 15 Elkhart IN 46516-9630

ATCHISON, WILLIAM DAVID, pharmacology educator; b. Madison, Wis., Feb. 25, 1952; s. William D. and Barbara A. (Breitenbach) A.; m. Mary Jo Schultz, Sept. 29, 1979; children: Douglas Kyle, Bradley Joseph. BS with honors, U. Wis., 1974, MS, 1978, PhD, 1980. Rsch. asst. U. Wis., Madison, 1979-80; NIEHS postdoctoral fellow Northwestern U. Med. Sch., Chgo., 1981-82; asst. prof. Pharm. Dept., Mich. State U., East Lansing, 1982-87, assoc. prof., 1987-91, prof., 1991—, dir. neurosci., 1992—; instr. Lansing (Mich.) C.C., 1983-91; mem. study sect. toxicology 2, NIH, Bethesda, 1990—. Assoc. editor Toxicology and Applied Pharmacol., Neurotoxicology. Recipient Travel award ASPET, 1979, 84. Mem. Myasthenia Gravis Assn. (med. adv. bd.). Office: Mich State U Dept Pharmacology/Toxicology B331 Life Scis East Lansing MI 48824-1713

ATCHLEY, ANTHONY ARMSTRONG, physicist, educator; b. Lebanon, Jan. 23, 1957. BS, U. South, 1979; MS, NMex. Tech., 1982; PhD in Physics, U. Miss., 1984. Assoc. prof. physics Naval Post Grad. Sch., 1985—. Recipient R. Bruce Lindsay award Acoustical Soc., 1992. Mem. Am. Physics Soc., Acoust. Soc. Am., Am. Assn. Physics Tchrs. Achievements include research in Thermoacoustics heat transport. Office: Naval Post-Graduate School Dept of Physics Monterey CA 93943*

ATHAPPILLY, KURIAKOSE KUNJUVARKEY, computers and quantitative methods educator; b. Mala, Kerala, India, Oct. 22, 1937; came to U.S., 1974; s. Kunjuvarkey I. and Monica A.; m. Sossa K. Kallingal, Nov. 5, 1972; children: George, Geena, Geetha. BSc, MSc, U. Kerala, 1964, 67; MBA, U. Guam, Agana, 1976; EdD, Western Mich. U., 1978, postgrad., 1979-81. Lectr. math St. Michael's Sch., Patna, India, 1970-72, St. Cambridge Sch., Bhopal, India, 1972-73; asst. prof. math. Western Mich. U., Kalamazoo, 1979-84, assoc. prof., 1984-88, prof. quantitative methods and mgmt. info. systems, 1988—, area coord. computer info. systems, 1982-89, mentor Coll. Bus., 1987-89; vis. prof. Mich. State U., Lansing, 1990-91; cons. Ronningen Rsch., Vicksburg, Mich., 1989—; vis. prof. Indian univs.; liaison Indian bus. ventures. Author: Programming, 1985, Vax Basic, 1989; mem. editoral rev. bd. Data Management, 1984-88, Computerage, 1985-87, Jour. of Computer Info. Systems, 1989—, Info. Exec., 1989—; contbr. articles to profl. publs. Amb. Kerala Cath. Ch., Kalamazoo, 1989—. Mem. AAUP, Soc. Computer Educators, Info. System Edn. Assn., Data Processing Mgmt. Assn. Avocations: music, painting, tennis, basketball. Office: Western Mich U BIS Dept Kalamazoo MI 49008

ATIF, MORAD RACHID, architect; b. Algiers, Algeria, Mar. 3, 1962; came to U.S. 1984; s. Arezki and Fatma-Zohra (Aktouche) A.; m. Zakia Amrouche, Dec. 25, 1987; 1 child, Sara. BArch, Poly. Sch. Architects, Algeria, 1984; MArch, UCLA, 1987; PhD in Arch., Tex. A&M U., 1992. Architect/intern Urban Innovation Group/Charles Moore, L.A., 1985-86; space surveyor UCLA, 1986-87; rsch. asst. Tex. A&M U., College Station, 1987-88, Caudill rsch. fellow, 1988-89, 90-91, lectr. archtl. design, 1991—. Contbr. articles to profl. jours. Mem. ASHRAE, Illuminating Engring. Soc., Phi Beta Delta, Tau Sigma Delta. Achievements include development of method by which architects/engineers can size thermal mass and glass of atrium buildings to optimize daylighting without excess of thermal loads. Home: PO Box 626 College Station TX 77841 Office: Nat Rsch Coun Can, Bldg Performance Bldg M-24, Ottawa, ON Canada K1A 0R6

ATIYEH, ELIA MTANOS, civil engineer; b. Kattineh, Homs, Syria, Mar. 20, 1960; came to U.S., 1986; s. Mtanos Elia and Kathrina Abdul Karim Atiyeh. B Civil Engring., U. Okla., 1989. Registered profl. engr., Okla. Project engr. The Breisch Engring., Sand Springs, Okla., 1990—. Mem. ASCE, Okla. Soc. Profl. Engrs. Roman Orthodox. Office: The Breisch Engring 16 S Main Sand Springs OK 74063

ATKINS, PETER WILLIAM, chemistry educator; b. Amersham, U.K., Aug. 10, 1940; s. William Henry and Ellen (Edwards) A.; m. Judith Ann Kearton, Aug. 22, 1964 (div. 1983); 1 child, Juliet Louise Tiffany; m. Susan Adele Greenfield, Mar. 31, 1991. BSc, U. Leicester, 1961, PhD, 1964; MA, U. Oxford, 1965; DSc, U. Utrecht, Netherlands, 1992. Harkness fellow UCLA, 1964-65; lectr. U. Oxford, U.K., 1965—; fellow, tutor Lincoln Coll., Oxford, 1965—. Author: Molecular Quantum Mechanics, 2d edit., 1983, The Second Law, 1984, Molecules, 1987, Physical Chemistry, 1990, Inorganic Chemistry, 1990, Quanta, 2d edit., 1991, Atoms, Electrons and Change, 1991, General Chemistry, 2d edit., 1992, The Elements of Physical Chemistry, 1992, Creation Revisited, 1992. Recipient Meldola medal Chem. Soc., 1969. Office: Lincoln Coll, Oxford England OX1 3DR

ATKINSON, DAVID JOHN, computer scientist; b. Ann Arbor, Mich., Feb. 2, 1959. BA, U. Mich., 1980; MS, Yale U., 1982, MPhil, 1984; PhD, Chalmers U. Tech., Goteborg, Sweden, 1992. Mem. tech. staff Caltech-Jet Propulsion Lab., Pasadena, 1984-89, program mgr., tech. sect. mgr., 1989—; vis. scientist Chalmers U. of Tech., 1992-93. Contbr. articles to profl. jours. Recipient Exceptional Svc. medal NASA, 1990. Fellow IEEE Computer Soc.; mem. ACM, Am. Assn. Artificial Intelligence. Achievements include patent pending for Star Tool, a system for expert systems implmentation; developed intelligent monitoring and diagnosis software used in Voyager Spacecraft mission during flyby of Neptune; research on mobile robots for Mars; founding of knowledge systems program at NASA JPL. Office: Jet Propulsion Lab 4800 Oak Grove Dr Pasadena CA 91109

ATKINSON, HAROLD WITHERSPOON, utilities consultant, real estate broker; b. Lake City, S.C., June 12, 1914; s. Leland G. and Kathleen (Dunlap) A.; BS in Elec. Engring., Duke, 1934; MS in Engring., Harvard U., 1935; m. Pickett Rancke, Oct. 6, 1946; children: Henry Leland, Harold Witherspoon. Various positions in sales, engring. Cambridge Electric Light Co. (Mass.), 1935-39, 46-73, asst. mgr. power sales dept., 1946-49, gen. mgr., 1957-73, dir., 1959-84, exec. v.p., 1972-73; mgr. Pee Dee Electric Membership Corp., Wadesboro, N.C., 1939-46; gen. mgr. Cambridge Steam Corp., 1951-73, v.p., 1959-73, dir. 1955-84. Chmn., Cambridge Traffic Bd., 1962-73; pres. Cambridge Ctr. Adult Edn., 1962-64; v.p. Cambridge Mental Health Assn.; chmn. allocations com. Greater Boston United Community Svcs., 1971-72; chmn. Cambridge Commn. Svcs., 1955-56; adv. bd. Cambridge Coun. Boy Scouts Am.; mem. corp., chmn. camping com. Cambridge YMCA, 1964-71; chmn. Cambridge chpt. ARC, 1969-71; trustee of trust funds Town of Harrisville, N.H., 1976-83; treas. North Myrtle Beach Citizens Assn., 1982-84. Served from pvt. to capt. AUS, 1942-45. Registered profl. engr., Mass. Mem. IEEE (sr.), Mass. Soc. Profl. Engrs., Elec. Inst. (pres. 1971), Harvard Engring. Soc., Cambridge C. of C. (pres. 1957-58). Newcomen Soc. N.Am., Phi Beta Kappa, Tau Beta Pi, Pi Mu Epsilon. Clubs: Cambridge Boat (treas. 1962-65), Cambridge (pres. 1972-73), Sandpiper Bay Golf, Bay Tree Golf; Plantation; Civitan (pres. Wadesboro 1940-41); Rotary (pres. Cambridge club 1959-60, v.p. North Myrtle Beach, S.C. club) Home: 705 Holloway Circle N North Myrtle Beach SC 29582-2613 Office: 710 17th Ave S North Myrtle Beach SC 29582

ATKINSON, HOLLY GAIL, physician, journalist, author, lecturer; b. Detroit, Oct. 20, 1952. BA in Biology magna cum laude, Colgate U., 1974; MD, U. Rochester, 1978; MS in Journalism, Columbia U., 1981. Diplomate Nat. Med. Bds. Intern in internal medicine Strong Meml. Hosp., Rochester, N.Y., 1978-79; rschr. Walter Cronkite's Universe show CBS News, N.Y.C., 1981-82; med. reporter CBS Morning News, N.Y.C., 1982-83; on-air co-host Bodywatch health show PBS, 1983-88; contbg. editor and health columnist New Woman mag., 1983—; on-air corr., med. editor Lifetime Med. TV, 1985-89, sr. v.p. programming and med. affairs, 1989—; assoc. editor Journal Watch, 1986-90; med. corr. Today show NBC News, N.Y.C., 1991—; mem. trustee's com. U. Rochester, 1983-90. Author: Women and Fatigue, 1986. Vol. nat. and local level Am. Heart Assn., 1984-91; bd. dirs., chairperson Nat. Comms. Com., 1987-91. Commd. officer USPHS, 1979-80. Recipient Young Achievers award Nat. Coun. of Women, 1986. Phi Beta Kappa. Office: Lifetime Med TV 36-12 35th Ave Astoria NY 11106

ATKINSON, PAUL HENRY, cell biologist; b. Auckland, New Zealand, Aug. 31, 1943; s. Henry and Edna May (Witten) A.; m. Anne Catherine, Nov. 21, 1966; children: Harry Maxwell, Catherine Marianne. BSc with first class honors, U. Canterbury, New Zealand, 1965; PhD, U. Auckland, New Zealand, 1968. Asst. prof. dept. pathology/devel. biology and cancer Albert Einstein Coll. Medicine N.Y., 1971-77, assoc. prof., 1977-81, assoc. prof. dept. devel. biology and cancer, 1977-83, prof., 1983-92; sci. mgr. Wallaceville Animal Rsch. Ctr., Upper Hutt, New Zealand, 1991-92, gen. mgr. sci. and tech., 1992—; postdoctoral fellow dept. microbiology and immunology Albert Einstein Coll. Medicine N.Y., 1969-71, spl. asst. to dean, 1989-91, assoc. dir. Cancer Rsch. Ctr., 1990-92; established investigator Am. Heart Assn., 1975-80. Author 58 original papers and 8 book chpts. Treas. 320 Riverside Apts. Corp., N.Y.C., 1983-86. Postgrad. fellow New Zealand Univ. Grants Com., WHO, 1966-68; postdoctoral fellow Internat. Agy. for Rsch. on Cancer, 1969-70, Damon Runyon fellow, 1970-71. Mem. Am. Soc. Cell Biology, Am. Heart Assn. (coun. basic sci.), Am. Chem. Soc., Soc. Complex Carbohydrates. Achievements include patents for rotavirus antigens: a method to solubilize simian rotavirus rotavirus neutralising antigens of use in combatting viral diarrhoea. Home: 6 Puketea St Eastbourne, Wellington 6304, New Zealand Office: AgRsch Wallaceville, Ward St, Upper Hutt 6400, New Zealand

ATLAN, PAUL, gynecologist; b. Algiers, Algeria, France, June 13, 1942; s. Gabriel Nathan and Rosine (Chemaoun) A.; m. Liliane Haddad, Dec. 26, 1968; children: Sophie, Catherine. MD, U. Paris, 1969, degree gynecology, 1973. Resident Psychiatric Hosp., Epinay, 1971-74; planning ctr. dir. Fontenay Sous Bois (France), 1973—; gynecologist St. Anne Psychiatric Hosp., 1980—, Hosp. Antoine Beclere; cons. RTL radio station, Paris, 1979-86; attaché First Paris Hosp., 1980—; dir. Internat. Conf. on Jewish Ethics and New Med. Procreation, conf. on Laws and Ethics for New Med. Procreations in French Nat. Assembly, 1988. Mem. French Psychosomatic Gynecological Soc. Avocations: reading, movies. Office: 105 Rue de la Convention, 75015 Paris France

ATLAS, RONALD M., microbiologist educator, ecologist; b. N.Y., Oct. 19, 1946. BS, SUNY, 1968; MS, Rutgers U., 1970, PhD in Microbiology, 1972. Rsch. assoc. jet propulsion lab Calif. Inst. Tech., 1972-73, from asst. prof. to assoc. prof., 1973-81; prof. biology U, Louisville, 1981—; nat. lectr. Sigma Xi, 1981—; rsch. grantee Off Naval Res., 1973-79, Nat. Oceanic & Atmospheric Adminstrn., 1975—, Dept. Energy, 1976-81, Environ. Microbiol. Com., Am. Soc. Microbiol., 1981—. Recipient ASM award in Applied and Environmental Microbiology Am. Soc. Microbiology, 1991. Fellow Am. Acad. Microbiol., Soc. Indsl. Microbiol.; mem. AAAS, Am. Soc. Microbiol. Achievements include research in microbial degradation of hydrocarbons and other organic pollutants, numerical taxonomy of marine bacteria, effects of pesticides and other organic compounds on microorganisms, ecology of soil and marine microorganisms. Office: Dept of Biology Univ of Louisville Louisville KY 40292*

ATLEE, JOHN LIGHT, III, physician; b. Lancaster, Pa., Feb. 22, 1941; s. John Light Jr. and Ann (Stevens) A.; m. Barbara Sheaffer, June 20, 1964 (dec. Apr. 14, 1967); m. Barbara Sanford, Feb. 3, 1968; children: Sarah Sanford, John Light. BA, Franklin & Marshall Coll., 1963; MD, Temple U., 1967, MS in Pharmacology, 1971. Diplomate Am. Bd. Anesthesiology. Intern Germantown Hosp., Phila., 1967-68; resident in anesthesiology Temple U. Hosp., Phila., 1968-70; postdoctoral rsch. fellow pharmacology Temple U. Grad. Sch. Medicine, 1970-71; staff anesthesiologist U.S. Naval Hosp, Bethesda, Md., 1971-73; asst. prof. anesthesiology U. Wis., Madison, 1973-78, assoc. prof. anesthesiology, 1978-85, prof. anesthesiology, 1985-88; prof. anesthesiology Med. Coll. Wis., Milw., 1988—; referee/cons. peer rev. jours. Anesthesia & Analgesia, Am. Jour. of Physiology, Anesthesiology, Jour. of Cardiothoracic and Vascular Anesthesia, 1980—; cons. Marquette Electronics, Milw., 1989—, ARZCO Med. Electronics, Vernon Hills, Ill., 1989-92, Anaquest, Murray Hill, N.J., 1987—. Author: Perioperative Cardiac Arrhythmias, 1985, 2d edit., 1990; editor: Perioperative Management of Pacemaker Patients, 1991, Jour. Cardiothor. and Vascular Anesthesia, 1987-92; contbr. articles to profl. jours. Lt. comdr. USN, 1971-73. NIH grantee, 1978—. Fellow Am. Coll. Cardiology and Anesthesiology; mem. Am. Soc. Anesthesiologists, Assn. Univ. Anesthesiologists, N.Am. Soc. Pacing and Electrophysiology, Soc. Register Assn., Sigma Xi. Republican. Episcopalian. Achievements include development of new transeosphageal stimulation and recording technology for use in anesthesia and emergency medicine and intensive care. Home: N71 W29436 Tamron Ln Hartland WI 53209 Office: Med Coll Wis 8700 W Wisconsin Ave Milwaukee WI 53226-3595

ATLURI, SATYA N(ADHAM), aerospace engineering educator; b. Gudivada, Andhra, India, Oct. 7, 1945; came to U.S., 1966, naturalized, 1976; s. Tirupati Rao and Tulasi (Devi) A.; m. Revati Adusumilli, May 17,, 1972; children: Neelima, Niroupa. B.E., Andhra U., Vizag, 1964; M.E., Indian Inst. Sci., Bangalore, India, 1966; DSc, MIT, 1969; DSc honoris causa, Nat. U. Ireland, Dublin, 1989. Researcher MIT, Cambridge, 1966-71; Jerome Clarke Hunsaker vis. prof. aeronautics and astronautics MIT, 1990-91; asst. prof. U. Wash., Seattle, 1971-74; assoc. prof. engring. sci. and mech. Ga. Inst. Tech., Atlanta, 1974-77, prof., 1977-79, Regents' prof. mechanics, 1979—, dir. Ctr. for Advancement Computational Mechanics, 1980—, inst. prof., 1991—, Regents' prof. aerospace engring., 1991—; White House nominee Com. for Evaluation Nat. Medal of Tech. Dept. Commerce, 1992—; co-chmn. Internat. Conf. on Computational Engring. Sci., Tokyo, 1986, Atlanta, 1988, Melbourne, Australia, 1991, Patras, Greece, 1991, Hong Kong, 1992; gen. lectr., invited keynote speaker over 150 internat. tech. confs.; adv. prof. Southwestern Jiaotong U., Emei, Sichuan, China, 1988; bd. dirs. FAA Ctr. for Excellence in Computational Modeling of Aerospace Structures, Ga. Inst. Tech., 1992; White House nominee Evaluation Com. for Nat. Medal of Tech., U.S. Dept. Commerce, 1992. Contbr. over 400 articles to profl. jours.; gen. editor: Springer Verlag Series on Computational Mechanics, 1988-91, 91, Structural Intgegrity of Aging Airplanes, 1991, Frontiers in Computational Mechanics, 1989—; author 25 books, including Computational Methods in the Mechanics of Solids and Structures, 1984; editor: books including Hybrid and Mixed Finite Element Methods, 1983, Computational Methods in the Mechanics of Fracture, 1985 (Russian transl. 1989), Handbook of Finite Elements, 1986, Dynamic Fracture Mechanics, 1986, Computational Mechanics 86, 1986, Large-Space-Structurs: Dynamics and Control, 1987, Computational Mechanics '88, 1988, Computational Mechanics '91, 1991, Frontiers in Computational Mechanics, 1989, Computational Mechanics '91, 1991, Structural Integrity of Aging Airplanes, 1991, Durability of Metal Aircraft Structures., 1992, Nonlinear Computational Mechanics in Aerospace Engineering, 1992; editor-in-chief Internat. Jour. Computational Mechanics; mem. editorial bd. Computers and Structures, Engring. Fracture Mechanics, Internat. Jour. Plasticity, Internat. Jour. for Numerical Methods in Engring., Acta Mecanica Solida Sinica, also others. Grantee NSF, 1975—, USAF Office Sci. Research, 1973—, Office Naval Research, 1978—, Air Force Rocket Propulsion Lab., 1976-79, NASA, 1980—, NRC, 1978-80, Dept. Transp., 1987—, ARO, 1988—, FAA, 1991—; recipient V.K. Murti Gold medal Andhra U., India, 1964, Roll of Honors award Indian Inst. Sci., 1966, Disting. Alumnus award, 1991, Class of 1934 Disting. Prof. award for 1986 Ga. Inst. Tech., Outstanding Faculty Research award Ga. Inst. Tech., 1986, 91, Survey Paper Citation AIAA Jour., 1988, Monie Ferst Meml. award for Sustained Research, Sigma Xi, 1988, Outstanding Rsch. award, 1991; fellow Japan Soc. for Promotion Sci., 1987—, Computational Mechanics Div. medal Japan Soc. Mech. Engrs., 1991, ICES Gold medal, 1992; named Southwest Mechanics lectr., 1987, Midwestern Mechanics lectr., 1988. Fellow Internat. Congress on Fracture (hon.), ASME (chmn. com. computing in applied mechanics 1983-85, assoc. editor Applied Mech. Revs.), AIAA (assoc. editor AIAA Jour. 1983—, Structures Dynamics and Materials award 1988), Am. Acad. Mechanics, Aero. Soc. India; mem. ASCE (assoc. editor Jour. Engring. Mechanics 1982-84, Aerospace Structures and Materials award 1986), Internat. Assn. for Computational Mechanics (founding mem.), U.S. Assn. for Computational Mechanics (founding mem.). Home: 871 Springdale Rd NE Atlanta GA 30306-4617 Office: Ga Inst Tech Ctr Computational Modeling and Infrastructure Rehab Atlanta GA 30332-0356

ATLURU, DURGAPRASADARAO, veterinarian, educator; b. Tenneru, India, Oct. 15, 1949; came to U.S., 1976; s. Veeraraghavaiah and Kanakabasavamma (Boyapati) A.; m. Subbayamma Gorijavolu, Mar. 18, 1984; 1 child, Sreevalli. DVM, Coll. Vet. Sci., Hyderabad, India, 1971, MSc, 1974; MS, U. Minn., St. Paul, 1980. Vet. officer Dairy Devel. Corp., Hyderabad, 1974-76; rsch. asst. dept. large animal clin. scis. U. Minn., St. Paul, 1977-81; rsch. assoc. dept. microbiology U. Auburn, Ala., 1982; NIH postdoctoral fellow dept. medicine U. N.Mex., Albuquerque, 1983-85; instr. dept. medicine Med. Coll. Wis., Milw., 1985-87; asst. prof. dept. anatomy and physiology Kans. State U., Manhattan, 1987-90; dir. immunology Regional Kidney Disease Program, Mpls., 1990—. Contbr. rsch. articles to sci. publs. Active Kans. affiliate Am. Heart Assn. Grantee NIH, 1987-93, USDA, 1988-92, Am. Heart Assn., 1988-90. Mem. AAAS, Am. Assn. Immunologists (travel award 1989), Am. Assn. Vet. Immunologists. Achievements include research in organ transplants. Home: 653 White Birch Dr Shoreview MN 55126 Office: Regional Kidney Disease Program 701 Park Ave Minneapolis MN 55415-1829

ATREYA, ARVIND, mechanical engineering educator; b. Jan. 15, 1954. BTech, Indian Inst. Tech., New Delhi, 1975; MScE, U. N.B., Can., 1977; SM, Harvard U., 1978, PhD, 1983. Mech. engr. Nat. Engring. Lab. for Fire Rsch., Nat. Bur. Standards, Gaithersburg, Md., 1983-84; asst. prof. dept. mech. engring. Mich. State U., East Lansing, 1983-87; scientist on sabbatical leave Nat. Inst. of Stds. and Tech., Gaithersburg, 1991; assoc. prof. mech. engring. Mich. State U., East Lansing, 1987-92, U. Mich., Ann Arbor, 1993—; cons. Hughes Assocs., Inc., Wheaton, Md., 1990-91, Hughes, Hubbard & Reed, N.Y.C., 1988-92; mem. program. subcom. Internat. Symposium for Combustion, 1986—. Contbr. articles to Combustion and Flame, ASME Jour. of Heat Transfer, others. Recipient Presdl. Young Investigator award NSF, 1986-91. Mem. ASME, The Combustion Inst., Internat. Assn. for Fire Safety Sci., Sigma Xi. Achievements include significant and internationally respected contributions in fire safety science and industrial burners. Office: U Mich Dept Mech Engring/Applied Mechanics 2158 GG Brown Bldg Ann Arbor MI 48109-2125

ATTALLA, ALBERT, chemist; b. Cuyahoga Falls, Ohio, Sept. 29, 1931; sm. Mary Jane Carroll, Dec. 29, 1956; children: Ruth Anne, Suzanne Marie, Jessanne Carroll. BA in Chemistry, Kent State U., 1955, MA in Chemisty and Biochemistry, 1959; PhD in Phys. Chemistry, U. Cin., 1962. Chemist Corning (N.Y.) Glass Works, 1962-63, Monsanto Rsch. Group, Miamisburg, Ohio, 1963-88, EG&G Mound Applied Technologies, Inc., Miamisburg, 1988--. Office: EG&G Mound Applied Tech Miamisburg OH 45342

ATTAWAY, DAVID HENRY, federal research administrator, oceanographer; b. Sterling, Okla., June 9, 1938; s. Fred John and Minnie Ora (Yandell) A. BS in Chemistry, U. Okla., 1960, PhD in Chemistry, 1968. Physical oceanographer U.S. Naval Oceanographic Office, Suitland, Md., 1962-65; postdoctoral scholar, Marine Sci. Inst. U. Tex., Port Aransas, 1968-69; chief, geochemistry U. Kans., State Geol. Survey, Lawrence, 1969-71; chem. oceanographer USCG, Washington, 1971-72; rsch. and grants adminstr. Nat. Sea Grant Coll. Program, Washington, 1972—. Contbr. articles to profl. jours. Mem. Am Chem. Soc., AAAS, Inst. Food Tech., Marine Tech. Soc., Am. Geophysical Union, Sigma Xi, Phi Lambda Upsilon, Phi Sigma. Home: 609 7th St NW Washington DC 20001-3713 Office: Nat Sea Grant Coll Program 1335 East-West Hwy Silver Spring MD 20910

ATTLES, LEROY, aerospace engineer; b. Plainfield, N.J., June 12, 1966; s. LeRoy Sr. and Henrietta S. (Evans) A. BS in Aerospace Engring., Tuskegee U., 1990. Coop. engr. NASA Ames Dryden, Edwards, Calif., 1987; engr. Rockwell Space Operation Co., Houston, 1990—. Mem. AIAA, Tuskegee Alumni Club. Home: 15123 Hillside Park Way Cypress TX 77429-5612

ATTWOOD, DAVID THOMAS, physicist, educator; b. N.Y.C., Aug. 15, 1941; s. David Thomas and Josephine (Banks) A.; divorced; children: Timothy David, Courtney Catherine, Kevin Richard; m. Linda Jean Geniesse, Aug. 3, 1991. BS, Hofstra U., 1963; MS, Northwestern U., 1964; D in Engring. Sci., NYU, 1972. Physicist Lawrence Livermore Nat. Lab., Livermore, Calif., 1972-83, Lawrence Berkeley Lab., Berkeley, Calif., 1983—; prof. in residence U. Calif., Berkeley, 1989—, chair applied sci. and tech., 1991—; founder, dir. Ctr. for X-ray Optics Lawrence Berkeley Lab. Editor: (with B.L. Henke) X-Ray Diagnostics, (with J. Bokor) Short Wavelength Coherent Radiation; reviewer numerous sci. jours.; contbr. numerous articles to profl. publs. Mem. AAAS, Am. Phys. Soc., Optical Soc. Am. Achievements include research on laser-plasma interactions, x-ray optics, synchrotrons, and partially coherent x-rays. Office: Lawrence Berkeley Lab Ctr X-ray Optics Berkeley CA 94720

ATUTIS, BERNARD P., manufacturing company executive; b. Phoenix, Jan. 24, 1933; m. Ursula Igna Weiss, May 22, 1954; children: Diana, Bernard E., Robert A., Bernardine E., Mark A. Student, Cornell U., 1966-77, SUNY, Binghamton, 1972. Sr. programmer IBM, White Plains, N.Y., 1955-72; with Control Data, Mpls., 1977-79; assoc. programmer Karsten Mfg. Corp., Phoenix, 1979-81, CAD/CAM sales staff, 1982-85, project coord., 1985-92, prin. engr. GPS vehicle tracking, 1992—; grad. asst. Dale Carnegie Tng./Sales, Binghamton, 1961-69; cropland rschr. Agr. Ext. Svc., Binghamton, 1960-80; soil water chmn. U.S. Dept. Agr./ASC, Little Falls, N.Y., 1954. Contbr. articles to profl. jours. Instl. rep. Boy Scouts Am., Binghamton, 1955-60; active 4-H Club, Ithaca, 1961-72. With USAF, 1950-54. Mem. Am. Legion, Moon Valley Golf Course. Republican. Roman Catholic. Achievements include development of computerized integrated technology system for golf courses and agriculture crops. Home: 4109 W Banff Phoenix AZ 85023 Office: Karsten Mfg Corp 2201 W Desert Cove Phoenix AZ 85029

ATWATER, TANYA MARIA, marine geophysicist, educator; b. Los Angeles, Aug. 27, 1942; d. Eugene and Elizabeth Ruth (Ransom) A.; 1 child, Alyosha Molnar. Student, MIT, 1960-63; BA, U. Calif., Berkeley, 1965; PhD, Scripps Inst. Oceanography, 1972. Vis. earthquake researcher U. Chile, 1966; research assoc. Stanford U., 1970-71; asst. prof. Scripps Inst. Oceanography, 1972-73; U.S.-USSR Acad. Scis. exchange scientist, 1973; asst. prof. MIT, 1974-79, assoc. prof., 1979-80, research assoc., 1980-81; prof. dept. geoscis. U. Calif., Santa Barbara, 1980—; chairperson ocean margin drilling Ocean Crust Planning Adv. Com.; mem. pub. adv. com. on law of sea U.S. Dept. State, 1979-83; mem. tectonics panel Ocean Drilling Project, 1990—; Sigma Xi lectr., 1975-76. Sci. cons.: Planet Earth: Continents in Collision (R. Miller), 1983; contbr. articles to profl. jours. Sloan fellow, 1975-77; recipient Newcomb Cleveland prize AAAS, 1980; named Scientist of Yr. World Book Ency., 1980. Fellow Am. Geophys. Union (fellows com. 1980-81, Ewing award subcom. 1980), Geol. Soc. Am. (Penrose Conf. com. 1978-80); mem. AAAS, Assn. Women in Sci, Am. Geol. Inst., Phi Beta Kappa, Eta Kappa Nu. Office: U Calif Dept Geoscis Santa Barbara CA 93106

ATWONG, MATTHEW KOK LUN, chemical engineer; b. Hong Kong, Sept. 28, 1950; came to U.S., 1973; s. Tsi Cheong and Wai Fong (Leong) Wong; m. Catherine C. Tang, Dec. 19, 1987; 1 child, Andrew Philip. BS in Chem. Engring., U. Mich., 1975, MS in Chem. Engring., 1977; MBA, Grand Valley State U., 1981. Registered profl. engr., Mich. Project and process engr. BASF, Inc., Holland, Mich., 1977-85; engring. mgr., prin. engr. Johnson Matthey, Inc., Wayne, Pa., 1985-90; project mgr., sr. project engr. Miles, Inc., Elkhart, Ind., 1990-91; mgr. sterilization ops. and engring. McGaw, Inc., Irvine, Calif., 1991—. Mem. AIChE, ACS, NSPE, Internat. Soc. Pharm. Engring., Mich. Soc. Profl. Engrs., Lions (charter pres. Chesterbrook club 1986, pres. Holland club 1983, hearing conversation chmn. Pa. dist. 1987, zone chmn. Mich. dist. 1984), Phi Lambda Upsilon, Delta Theta Phi. Roman Catholic. Achievements include publication in technical journals and patent for apparatus for purifying copper phthalocyanne pigment, inventions infield of catalytic emission control system, electronic emission control system, electroless copper plating and chemical process design and devel. Office: McGaw Inc 2525 McGaw Ave Irvine CA 92714

ATWOOD, THEODORE, marine engineer, researcher; b. Cambridge, Mass., Nov. 7, 1925; s. Raymond Loring and Pauline B. (Stoughton) A.; m. Jeanne Carol Sterling, Sept. 10, 1955; children: Sterling, Susan, Alan. BS in Marine Transp., Mass. Inst. Tech., 1949. Marine engr. United Fruit Co., N.Y.C., 1949-53; engr. Merchants Refrigerating Co., N.Y., 1953-60; engr. in tech. svc. Allied-Signal, Morristown, N.J., 1960—; engr. in R & D Allied-Signal, Buffalo, 1981—. Contbr. articles to profl. jours. Fellow ASHRAE (Best Tech. Paper Yr. award 1971), Refrigeration Svc. Engrs. Soc. (mem. svc. adv. coun.), Internat. Inst. Refrigeration. Achievements include development of variable capacity refrigeration system. Home: 1520 N Forest Rd Williamsville NY 14221 Office: Allied-Signal R & D Lab 20 Peabody St Buffalo NY 14210

AUBERT, ALLAN CHARLES, aerospace engineer, consultant; b. Arlington, Mass., Apr. 24, 1957; s. Eugene James and Dorothy Marion (Stephens) A.; m. Sandra Jean Remington, May 30, 1981; children: Jeremy, Nathan, Ryan. BS in Aerospace Engring., U. Mich., 1980; SM in Aeronautics and Astronautics, MIT, 1983, Engr. of Aeronautics and Astronautics, 1983. Engr. trainee U.S. Army Corps of Engrs., Detroit, 1976-77, NASA, Houston, 1977-79; assoc. engr. Northrop Svcs. Corp., Houston, 1979; engr. Airflow Scis. Corp., Livonia, Mich., 1980-81; assoc. scientist Kaman Avi Dyne, Burlington, Mass., 1983-84; scientist Cambridge (Mass.) Collaborative, Inc., 1984-88, sr. scientist, 1988-89; sr. engr. Atlantic Applied Rsch. Corp., Burlington, Mass., 1989-92; ind. software developer Vienna, Va., 1992—. Contbr. articles to profl. jours. Mem. AIAA, Inst. Noise Control Engrs. Achievements include development of Force-State Mapping technique for measuring space structure joint properties, of various computer codes for modeling of vibration and acoustics; expert in Statistical Energy Analysis method of vibro-acoustic modeling and research and development. Home and Office: 300 Bowfin Ct Oakland CA 94619

AUCHINCLOSS, PETER ERIC, water quality improvement executive, consultant; b. Manchester, Conn., Nov. 8, 1960; s. Eric E. and Barbara J. (Perkin) A.; m. Gerri Coady Stephenson, May 10, 1981; 1 child, Sara P. Sales rep. Culligan, Wakefield, R.I., 1982-84; gen. mgr. Schumacher & Seiler, Balt., 1984-88; v.p. Hydrosource, Inc., Balt., 1986-90, pres., chief exec. officer, 1990-92; pres. Watermark Corp., Balt., 1992—. Contbr. articles to profl. jours. Pres. Greater Jacksonville (Md.) Assn., 1989-90; bd. dirs. Mayor's Adv. Coun., Balt., 1992—. Mem. Better Bus. Bur. Md. (bd. dirs. 1992—). Achievements include 2 trademarks, 4 patents, numerous copyrights. Home: 3806 Juniper Rd Baltimore MD 21218 Office: Watermark Corp 14 E Eager St Baltimore MD 21202

AUCIELLO, ORLANDO HECTOR, physicist; b. Cordoba, Argentina, Dec. 16, 1945; s. Domingo and Maria Magdalena (Cairo) A.; m. Hildegard Dolores Klenk, Oct. 5, 1973; children: Stephen, Elizabeth. Diploma/Electronic Engr., U. Cordoba, 1970; MS, U. Cuyo, Bariloche, Argentina, 1973, PhD, 1976. Postdoctoral rsch. assoc. McMaster U., Hamilton, Can., 1977-

79; rsch. assoc. Inst. Aerospace Studies/Univ. Toronto, Can., 1979-81, sr. rsch. assoc., 1982-84; assoc. prof. dept. nuclear engring. N.C. State U., Raleigh, 1985-88; assoc. faculty researcher Argonne (Ill.) Nat. Lab., 1988—; materials scientist MCNC Ctr. for Microelectronics, Research Triangle Park, 1989—; adj. prof. N.C. State U., Raleigh, 1989—; demonstrator Thin Films Formation, McMaster U., 1977, assistantship electronic properties of materials, 1977, assistantship introduction to materials sci., 1976; mem. grad. student adv. coms.; guest scientist various orgns., invited lectr. various orgns. Editor: (book series) Plasma-Materials Interaction, 1988—, Plasma Diagnostic, Vols. 1 and 2, 1989-90, Plasma-Surface Interaction, 1990, Ion Bombardment Modification of Surfaces, 1984. Lt. Argentinan Army, 1959-63. Recipient Outstanding Inventor recognition, U.S. Dept. Energy, 1991. Mem. AAAS, Am. Vacuum Soc., Materials Rsch. Soc., Planetary Soc., Sigma Xi. Achievements include patent for automation of ion beam and laser beam sputter-ablation deposition system for thin film synthesis; group producer of YBaCUO superconducting thin films by laser ablation deposition; research in atomic collision with gases. Office: MCNC Ctr Microelectronics 3021 W Cornwallis Rd Durham NC 27709-2889

AUDIN, LINDSAY PETER, energy conservation professional; b. N.Y.C., Sept. 3, 1946; s. Gabriel Jean and Mildred (Kettner) A.; m. Sondra Schechter; 1 child, Samantha. Student, Rensselaer Poly. Inst., 1964-69. Cert. energy mgr.; cert. lighting efficiency profl. Lab. technician Hudson Wire Co., Ossining, N.Y., 1970-72; owner People's Printing Co., Ossining, 1972-75; editor Avco Pub. Co., Scarborough, N.Y., 1975-76; ptnr. Homeland Industries, Bklyn., 1976-78, Omedia Comms., Ossining, 1978-79; dept. head Goldman Copeland Assoc. Engrs., N.Y.C., 1979-88; energy conservation mgr. Columbia U., N.Y.C., 1988—. Co-author: The Next Nuclear Gamble, 1983. Named Lighting Designer of Yr., Metal Optics Corp., 1991, Energy Mgr. of Yr., Energy User News, 1991, N.Y. Assn. Energy Engrs., 1992; recipient Renew America Environmental Achievement award, 1993. Mem. ASHRAE (mem. lighting standards panel 1991—), Assn. Energy Engrs. (mem. lighting certification bd. 1990—, Internat. Energy Mgr. of Yr. award 1993), Illuminating Engring. Soc. (Cert. of Merit 1992). Office: Columbia U B-230 Central Mail Rm New York NY 10027

AUERBACH, STANLEY IRVING, ecologist, environmental scientist, educator; b. Chgo., May 21, 1921; s. Abraham and Carrie (Friedman) A.; m. Dawn Patricia Davey, June 12, 1954; children: Andrew J., Anne E., Jonathan B., Alison M. BS, U. Ill., 1946, MS, 1947; PhD, Northwestern U., 1949. Instr., then asst. prof. Roosevelt U., Chgo., 1950-54; assoc. scientist, then scientist health physics divsn. Oak Ridge (Tenn.) Nat. Lab., 1954-59, sr. scientist, sect. leader, 1959-70, dir. ecol. sci. divsn., 1970-72, dir. environ. scis. divsn., 1972-86, sr. rsch. advisor, 1986-90; adj. prof. ecology U. Tenn., 1965-90; adj. rsch. prof. radiation ecology U. Ga., 1964-90; mem. U.S. exec. com. Internat. Biol. Program, co-chmn. program coordinating com., dir. deciduous forest biome project, 1969-74; mem. exec. com. Sci. Adv. Bd., U.S. EPA, 1986-92; adv. com. Sci. and Tech. NSF, 1989-91; environ. adv. bd. U.S.C.E., 1989-93; mem. NAS Adv. Com. on Research to Sec. Agr., 1969-70, NAE Power Plant Siting Program Commn., 1970-71, mem. bd. energy studies, 1974-77, chmn. com. on energy and environ., 1974-77, chmn. environ. studies bd., 1983-86, mem. com. on phys. scis., math. and resources, 1982-83, mem. com. on natural resources, 1979-82; chmn. archtl. rev. com. Oak Ridge Nat. Lab., 1971-84; mem. ecol. adv. bd. Bur. Reclamation; mem. NAS-NAE Bd. on Energy Studies, 1974-77; mem. C.E. bd. environ. cons. Tenn.-Tombigbee Waterway, 1975-86; mem. ad hoc com. on transuranic burial ERDA (Dept. Energy), 1976-78; mem. Pres.'s Spl. Com. on Health and Environ. Effects of Increasing Coal Utilization, 1977-78, Resources for the Future Research Adv. Com., 1978-81; mem. commn. natural resources NRC, 1979-81; mem. adv. council Water Resources Research Center, U. Tenn., 1980—; Mem. Tenn. Citizens Wilderness Planning, bd. dirs., 1968; trustee Inst. Ecology, 1971-74. Ecology editor Environ. Internat., 1979—; mem. adv. bd. Environ. and Exptl. Botany, 1967-92; mem. bd. editors Radiation Rsch., 1975-77. 2nd lt. AUS, 1942-44. Recipient Dist. Assoc. award U.S. Dept. Energy, 1987, Comdr.'s award U.S. Dept. Army, 1990. Fellow AAAS; mem. Am. Inst. Biol. Scis. (bd. govs.), Soc. Zoology (chmn. ecology div.), Am. Soc. Agronomy, Brit. Ecol. Soc., Health Physics Soc., Entomol. Soc. Am., Soc. Systematic Biology, Ecol. Soc. Am. (chmn. com. radioecology 1963-65, sec. 1964-69, chmn. fin. com. 1969, pres. 1971-72, Disting. Service award 1985), Internat. Union Radioecology (pres. 1984-87, mem. of honour 1991), Sigma Xi (pres. Oak Ridge br. 1972-73, chmn. admissions 1980-82), Alpha Epsilon Pi. Achievements include research on ecology centipedes, radioecology and radioactive waste disposal; on environmental behavior of radionuclides and ecosystem analysis. Home: 24 Wildwood Dr Oak Ridge TN 37830-8622

AUERBACHER, PETER, cancer research organization administrator; b. St. Louis, Mar. 14, 1950; s. Ernest and Patricia (Mayer) A.; m. Josephina Francisca Somers, June 18, 1988; children: David, Daniel. BA, U. Colo., 1975. V.p. ITG Internat., Brussels, 1982-86; v.p., dir. ITG Internat., Brussels, Wilmington, Del., 1986-88; exec. dir. European Orgn. for Rsch. and Treatment of Cancer, Brussels, 1989-91; ind. mgmt. and fund raising cons. Mem. Am. C. of C. in Belgium, Am. Club Brussels (bd. govs.). Home: 127 Avenue Montjoie, 1180 Brussels Belgium

AUGERSON, WILLIAM SINCLAIR, internist; b. Denton, Tex., Mar. 13, 1927; s. Harold Wilbur and Bernice (Sinclair) A.; m. Virginia Benham, Aug. 29, 1953; children: Christopher C., Elizabeth C. BS cum laude, Bowdoin Coll., 1949; MD, Cornell U., 1955. Diplomate Am. Bd. Internal Medicine; med. lic. Mass., S.C. Commd. 2d lt. U.S. Army, 1955, advanced through grades to maj. gen., 1977; flight surgeon Space Task Group NASA, Langley AFB, Va., 1958-62; intern Brooke Gen. Hosp., Ft. Sam Houston, Tex., 1955-60; resident, fellow dept. medicine Walter Reed Army Med. Ctr., Washington, 1964-67; mil. asst. life sci. Dir. Def. Rsch. and Engring., Washington, 1971-74; chief profl. svcs. 2d Gen. Hosp., Landstuhl, Germany, 1974-75; commanding gen. U.S. Army Med. R&D, Frederick, Md., 1976-79; dep. asst. sec. def. Office of the Sec. Def. (Health), Washington, 1979-82; ret. U.S. Army, 1982; v.p. Arthur D. Little Inc., Cambridge, Mass., 1982—; cons. Def. Sci. Bd., Washington, 1979-82, Legis. Task Force Hazardous Materials, Boston, 1985-86; dir. Matritech, Cambridge, Mass., 1988—. Author: Treatment Protocols Chemical/Biological Casualties, 1987; author: (monographs) Nerve Agents, Mustard Agents, Lewisite, Phosgene Oxine, Tricothecene Mycotoxins, Heat; contbr. 16 articles to profl. jours. Decorated Silver Star, Def. Superior Svc. medal, Disting. Svc. medal, Outstanding Svc. medal. Fellow Aerospace Med. Assn., Mass. Med. Soc.; mem. AMA (Honor Citation-Space Medicine 1962), Am. Coll. Occupational Medicine. Episcopalian. Achievements include multiple contributions to the development and safety of manned space flight (Mercury "Apollo"); demonstration of positive pressure breathing for hi G protection; use of epidemiologic techniques to lower disease injuries in US forces in combat; co-founder of Tri-Service Medical System (TRIMIS) and program coordinator; Restructuring of U.S. Medical R and D Command along with application areas, starting health hazards of systems program; Improvement of U.S. Forces, chemical defense, better treatments for toxic injuries. Home: 37 Pilgrim Dr Winchester MA 01890 Office: Arthur D Little Inc Acorn Park Cambridge MA 02140

AUGHENBAUGH, NOLAN BLAINE, engineering educator, consultant; b. Akron, Ohio, July 29, 1928; s. Russell Lowell and Virginia Elena (Squires) A.; m. Barbara Alinora Hill, Feb. 1959; children: Debra Jean, Lowell Doric, Amy Elena. BSCE, Purdue U., 1955, PhD, 1963; MS in Geology, U. Mich., 1959. Registered profl. engr. Ind., Ill., Miss. Civil engr. Cold Regions Rsch. and Engring. Lab., U.S. Corps Engrs., Greenland, 1955; engr., geologist, glaciologist U.S. Internat. Geophys. Yr., Antarctica, 1956-59; instr., asst. prof. civil engring. Purdue U., West Lafayette, Ind., 1959-66; prof., dept. chair mining, petroleum and geologic engring. U. Mo., Rolla, 1966-83; prof., dean sch. mineral engring. U. Alaska, Fairbanks, 1983-85; prof. sch. engring. U. Miss., University, 1985—; cons. in field, 1959—. Contbr. articles to profl. jours. Mountain in Antarctica named in his honor U.S. Bd. Geog. Names; rock type named in his honor USGS. Mem. ASCE, Soc. Mining Engrs. (various offices), Assn. Engring. Geologists (mem. various coms.), Soc. Profl. Engrs. (various local chpt. offices). Achievements include development of laboratory tests for moisture activity index and flexibility index; design of apparatus to evaluate the flexibility of soils; exploration and mapping of approximately 300,000 square miles of previously unknown land in Antarctica. Home: 311 Dogwood Dr Oxford MS 38655 Office: U Miss Dept Geology and Geol Engring University MS 38677

AUGUST, RUDOLF See SCHLOEMANN, ERNST FRITZ

AUGUSTINE, BRIAN HOWARD, electronics researcher; b. Kingston, N.Y., Aug. 29, 1968; s. Roger Allen and Judith (Furst) A.; m. Kristin Michelle Van Housen, June 6, 1992. BA Chemistry, SUNY, Geneseo, 1990; postgrad. in chemistry, U. N.C., 1990—. Analytical chemist IBM, East Fishkill, N.Y., 1989, process engr., 1990; grad. fellow U. N.C., Chapel Hill, 1990—. Mem. Ch. of the Good Shepherd, Durham, N.C., 1991—. Fellow Dept. Edn., U.S. Dept. Edn., 1991-93; microelectronic fellow Microelectronic Ctr. of N.C., Research Triangle Park, 1990-91, Charles Reilley fellow U. N.C. Dept. Chemistry, 1990, SUNY Alumni fellow, Geneseo, 1988-89, SUNY Presdl. scholar, 1989-90. Mem. Am. Phys. Soc., Optical Soc. Am., Phi Lambda Upsilon. Office: U NC Chapel Hill Dept Chemistry CB # 3290 Venable Hall Chapel Hill NC 27599-3290

AUGUSTINE, MARGRET L., architect, designer; b. Buffalo, Mar. 18, 1953; d. Allen and Patricia A.; 1 child, Starr L. BA in Arch., Inst. of Ecotechnics, 1976. Project mgr. Inst. Ecotechnics, Oakland, Calif., 1974-75; treas., project design mgr. Synopco Corp., Santa Fe, N.Mex., 1975-77; with design dept. Project and Design Cons. Pty., Ltd., Singapore, 1978; mng. dir. Sarbid Ltd., London, 1978-91; v.p. Sarbid Corp., Ariz., 1991-92; pres. Biosphere Design, Inc., 1992—; pres., CEO Space Biospheres Ventures, Oracle, Ariz., 1984—; dir. Decisions Team Ltd., Hong Kong, Planetary Design Corp., Tucson, Internat. Space U. Contbr. articles to profl. jours. Office: Space Biospheres Ventures PO Box 689 Oracle AZ 85623

AUGUSTINE, NORMAN RALPH, industrial executive; b. Denver, July 27, 1935; s. Ralph Harvey and Freda Irene (Immenga) A.; m. Margareta Engman, Jan. 20, 1962; children: Gregory Eugen, René Irene. B.S.E. magna cum laude, Princeton U., 1957, M.S.E., 1959; D of Engring. (hon.), Rensselaer Poly. Inst., 1988; DSc (hon.), U. Colo., 1989; ED(hon.), Western Md. Coll., 1990; DEng (hon.), U. Md., 1992; D Aerospace Mgmt. (hon.), Embry Riddle U., 1992; D Engring. (hon.), Stevens Inst., 1993. Research asst. Princeton U., 1957-58; program mgr., chief engr. Douglas Aircraft Co., Inc., Santa Monica, Calif., 1958-65; asst. dir. def. research and engring. U.S. Govt., Office of Sec. Def., Washington, 1965-70; v.p. advanced systems Missiles and Space Co., LTV Aerospace Corp., Dallas, 1970-73; asst. sec. army The Pentagon, Washington, 1973-75; undersec. army The Pentagon, 1975-77; v.p. ops. Martin Marietta Aerospace Corp., Bethesda, Md., 1977-82; pres. Martin Marietta Denver Aerospace Co., 1982-85, sr. v.p., info. systems, 1985, pres., chief operating officer, 1986-87, vice chmn., chief exec. officer, 1987-88, chmn., chief exec. officer, 1988—, also bd. dirs.; bd. dirs. Phillips Petroleum Co., Procter & Gamble Co., Riggs Nat. Bank Corp. , New Am. Schs. Devel. Corp.; cons. office Sec. of Def., 1971—, Exec. Office Pres., 1971-73, Dept. Army, Dept. Air Force, Dept. Navy, FAA, Dept. Energy, Dept. Transp.; mem. USAF Sci. Adv. Bd.; chmn. Def. Sci. Bd.; mem. NATO Group Experts on Air Def., 1966-70, NASA Rsch. and Tech. Adv. Coun., 1973-75; chmn. NASA Space Sytems and Tech. Adv. Bd., 1985—; mem. Chief of Naval Ops. Exec. Bd., 1989—; chmn. def. policy adv. com. on trade, 1988—. Author: Augustine's Laws; co-author: The Defense Revolution, 1990; mem. adv. bd. Jour. Def. Rsch., 1970—; assoc. editor Def. Systems Mgmt. Rev., 1977-82; mem. editorial bd. Astronautics and Aeros.; contbr. articles to profl. jours. Trustee Johns Hopkins U., Princeton U.; chmn. nat. program evaluation com., coun. v.p., exec. v.p. Boy Scouts Am, 1990—; chmn. ARC; mem. Immanuel Presbyn. Ch., McLean, Va., Policy Coun. Bus. Roundtable, 1988—, Bus. Coun., 1989—. Recipient Meritorious Svc. medal Dept. Def., 1970, 4 Disting. Civilian Svc. medals Dept. Def., James Forrestal medal Nat. Security Indsl. Assn., 1988, Nat. Engring. award Am. Assn. Engring. Socs., 1991. Fellow IEEE, AIAA (hon., bd. dirs. 1978-85, pres. 1983-84, Goddard medal 1988), Am. Astron. Soc., Am. Helicopter Soc. (dir. 1974-75); mem. NAE (Arthur M. Bueche award 1991), Am. Acad. Arts and Scis., Internat. Acad. Astronautics, Assn. U.S. Army (pres. 1980-84, chmn. 1990—), Nat. Security Indsl. Assn. (Forrestal medal 1988), Indsl. Coll. Armed Forces (Eisenhower award 1990), Armed Forces Communications and Electronics Assn. (Sarnoff medal 1990), Nat. Space Club (Goddard Trophy 1991), Rotary (Nat. Space Trophy 1992), Phi Beta Kappa, Sigma Xi, Tau Beta Pi. Office: Martin Marietta Corp 6801 Rockledge Dr Bethesda MD 20817-1836

AULD, BERNIE DYSON, civil engineer, consultant; b. Austin, Tex., Oct. 5, 1958; s. George Ernest and Lois Ann (Justillian) A.; m. Cynthia Lynn Smith, June 15, 1985; children: Jamie Meredith, Kirsten Marlee, Lucas Austin. BSCE, Lamar U., 1982; MSCE, U. Tenn., 1992. Registered civil engr., Ky., Miss., Tenn. Civil engr. City of San Angelo, Tex., 1982-84, Arlington (Tex.) Engring. Co., 1984-85; civil project mgr. Barge, Waggoner, Sumner & Cannon, Nashville, 1985-89; civil dept. mgr. Lockwood Greene Engrs., Nashville, 1989-92; pres. BA Engring., Mt. Juliet, Tenn., 1992—. Athletic scholar Lamar U., 1977-82. Mem. ASCE, Engrs. Assn. Nashville (pres. 1993). Home and Office: 2026 Sanford Dr Mount Juliet TN 37122

AULD, ROBERT HENRY, JR., biomedical engineer, educator, consultant, author; b. Akron, Ohio, Sept. 19, 1942; s. Robert Henry Sr. and Elsie Mae (Rollans) A.; m. Karen Kay Atkinson, June 28, 1968 (div. Dec. 1981); children: Sheila Kay, Jason Craig. BSBA, Biomed. Engr., U. San Francisco, 1978. Registered profl. egnr., calif.; cert. clin. engr. Reg. svc. mgr. scientific products div. AHSC, Sunnyvale, Calif., 1963-68; owner, gen. mgr. Lab. Instrument Svc., Campbell, Calif., 1968-77; nat. mgr. biomed Honeywell, Inc., Denver, 1977-79; profl. engr. Robert Auld Enterprises, San Jose, Calif., 1979-86; dir. clin. engring. St. Louis Reg. Med. Ctr., 1987-89; engring. mgr. Robert Auld Enterprises, Imperial, Mo., 1989—; engring. advisor St. Louis Reg. Career Access Ctr., 1987-89. Contbr. articles to profl. jours. Del. at large Rep. Legion of Merit, Imperial, 1990-93. With USN, 1959-61. Recipient Disting. Leadership award Am. Biographical Inst., Raleigh, N.C., 1988. Mem. N.Y. Acad. Scis., IEEE, Am. Soc. Hosp. Engrs., NSPE, Mo. Soc. Profl. Engrs. (chmn. 1988-89, chmn. minority Math Counts pilot project 1987-89). Republican. Achievements include development of device for equilibrating gases in a liquid or blood for measurement of gases in blood; patent pending for dual halogen colormetric light source. Home and Office: 869 Country Glen Dr Imperial MO 63052

AULITZKY, HERBERT, retired erosion and avalanche control educator; b. Innsbruck, Tyrol, Austria, Feb. 25, 1922; s. Karl and Ida (Demetz) A.; m. Franziska Wolf, Aug. 9, 1947; children: Helmut, Wolfgang, Herbert, Walter, Elisabeth. Grad. engr., U. Bodenkultur, Vienna, Austria, 1948, D, 1950; Senator Forestry, U. Bodenkultur, 1974. Chief asst. Austrian Erosian and Avalanche Control Svc., Innsbruck, Tyrol, 1949-53, chief researcher, 1953-63, dist. chief, 1963-71, state chief, 1971-72; lectr. in bioclimatology U. Bodenkultur, Vienna, Austria, 1967; full prof. erosion and avalanche control U. Bodenkultur, Vienna, 1972-90, ret., 1990; cons. European Coun., Strasbourg, France, 1969-75, Austrian Parliament, Vienna, 1985-90. Author: Austrian Erosion Control-method, 1983; editor: Erosion and Avalanche Control in Tyrol, 1975. 2d lt. Austrian Inf., 1940-45. Recipient Verdienstkreuz des Landes Tyrol award Tiroler Landesregierung, Innsbruck, 1971, Grosz Silberne Ehrenzeichen für Verdienste um die Republik Oesterreich award Bundesministerium für Wissenschaft und Forschung, Vienna, 1983, Ehrenkreuz für Wissenschaft und Kunst 1.Klasse, 1992. Mem. Engrs. for Torrent Control of Austria, Assn. Italiana di Idronomia (hon.), Forschungsgesellschaft fur Vorbeugende Hochwasserbekämpfung (hon.), Chinese Soc. for Erosion Control Beijing (hon.). Roman Catholic. Avocations: writing, torrent and avalanche control, silviculture, bioclimatology, general planning, human ecology ethics. Home: Karlweisgasse 18, A-1180 Vienna Austria Office: U Bodenkultur, Peter-Jordanstr 82, A-1190 Vienna Austria

AUNIS, DOMINIQUE, neurobiologist; b. Caen, France, May 3, 1948; s. Jean and Anne (Jeanne) A.; m. Michelle Ganter, Oct. 28, 1972; children: Coline, Camille. Ingenieur, INSA, Lyon, France, 1969; DSc, Univ., Strasbourg, 1976. Asst. Medicine Faculty, Strasbourg, France, 1969-72; rsch. asst. CNRS, Strasbourg, France, 1972-76; rsch. asst. Inserm U.44, Strasbourg, France, 1976-83, dir. rsch., 1983-89; head Inserm unit Inserm U. 338, Strasbourg, France, 1989—; cons. in field. Contbr. articles to profl. publs. Recipient French Acad. Scis. award, 1989. Mem. Societe des Neurosciences (treas. 1989-92), Internat. Soc. for Neurochemistry. Achievements include rsch. on molecular mechanisms of neurotransmitter release and on cell communication. Office: Inserm U 338, 5 rue Blaise Pascal, 67084 Strasbourg France

AUPPERLE, ERIC MAX, data network center administrator, research scientist, engineering educator; b. Batavia, N.Y., Apr. 14, 1935; s. Max Karl and Hedwig Elise (Haas) A.; m. Nancy Ann Jach, June 21, 1958; children: Bryan, Lisa. BSEE, U. Mich., 1957, BSE in Math, 1957, MSE in Nuclear Engring., 1958, Instm.E., 1964. Registered profl. engr., Mich. Lectr. dept. electrical engring. U. Mich., 1972, lectr. dept. indsl. & ops. engring., 1973-74, 1963—, from asst. rsch. engr. to rsch. engr. Cooley Electronics Lab., 1957-69, rsch. engr. Inst. Sci. & Tech., 1969-74, rsch. scientist Inst. Sci. & Tech., 1974—; project leader Merit Computer Network of Inst. Sci. & Tech., U. Mich., 1969-73, assoc. dir., 1973-74, dir., 1974-78; assoc. dir. comm. Computing Ctr. U. Mich., 1981-89, interim dir. Info. Tech. Divsn. Network Systems, 1990-92, pres. Merit Network, Inc. Info. Tech. Divsn. Network Systems, 1988—; alt. mem. senate assembly U. Mich., 1975, coll. engring. rep. senate assembly, 1976-79, mem. univ. hierarchical computing study com., 1978-79, mem. univ. com. computer policy and utilization, 1979-80, chmn., 1980-84; guest lectr. computer sci. sect. dept. math. Wayne State U., 1975; cons. Cholette, Perkins & Buchanan, Computer Tech. Mgmt. Svcs., Reliance Electric Co., Votrax, Prentice Hall, IMB, Owens-Ill., Donnelly Mirror, Inc., others; panelist instructional sci. equipment program NSF, 1978; mem. program com. Nat. Electronic Conf., 1962-65, 70-73, chmn. student activities com., 1966, chmn. intensive refresher seminar com. 1967, faculty mem. profl. growth seminar series, 1969-74. Editorial bd. Spectrum, 1975-80. Mem. IEEE (sr., dir. Southeastern Mich. sect. 1963-64, 65-67, mem. aid to Apelscor com. 1967-76, mem. Jackson ednl. com. 1967—, Southeastern Mich. sect. treas., 1972, Southeastern Mich. sect. sec. 1973, Southeastern Mich. sect. vice chmn. 1974, Southeastern Mich. sect. chmn. 1975, jr. past chmn. 1976, sr. past chmn. 1977), U. Mich. Sci. Rsch. Club, Phi Eta Sigma, Eta Kappa Nu, Tau Beta Pi, Sigma Xi, Pi Kappa Phi. Home: 3606 Chatham Way Ann Arbor MI 48105 Office: Merit Network Inc U Mich 5115 IST Bldg Ann Arbor MI 48109

AURILIA, ANTONIO, physicist; b. Napoli, Italy, May 14, 1942; came to U.S., 1986; s. Clemente and Assunta (Ligesto) A.; m. Elizabeth Christine Adams, Dec. 1, 1972; children: Darius Matthew, Alexandra Rebecca. Laurea in Physics, U. Naples, Italy, 1966; PhD in Physics, U. Wis., Milw., 1970. Postdoctoral fellow dept. physics U. Alta., Edmonton, 1970-72; rsch. assoc. dept. physics Syracuse (N.Y.) U., 1972-74; rsch. scientist Internat. Ctr. Theoretical Physics, Trieste, Italy, 1974-75, Nat. Inst. Nuclear Physics, Trieste, 1975-86; prof. dept. physics Calif. State Poly. U., Pomona, 1986—; vis. scientist Imperial Coll., London, 1977-78, U. Alta., Edmonton, 1980-82, U. Toronto, Ont., 1985-86. Referee Founds. Physics, Can. Jour. Physics; contbr. articles to profl. jours. Recipient Faculty Scholarship Assoc. Western U., Stanford U., 1987-89, Calif. State U. System, Pomona, 1988-90; rsch. grantee NRC, Italy, 1978, Nat. Scis. and Engring. Rsch. Coun., Can., 1983-86. Mem. Am. Phys. Soc., Am. Assn. Physics Tchrs., N.Y. Acad. Sci., Sigma Xi. Democrat. Roman Catholic. Achievements include rsch. in theory of high spin fields, spontaneous symmetry breaking, theory of relativistic extended objects and generalized Maxwell fields. Office: Calif State U Dept Physics 3801 W Temple Ave Pomona CA 91768

AURNER, ROBERT RAY, II, oil company, auto diagnostic and restaurant franchise and company development executive; b. Madison, Wis., Mar. 24, 1927; s. Robert Ray and Kathryn (Dayton) A.; m. Phyllis Barrett, 1951 (div. 1966); children: Sheryl, Roxanne, Kathryn, Suzanne, Robert III; m. Deborah Marion Lucas, Jan. 31, 1976; children: William Lucas, Christopher Ray. AA, Monterey (Calif.) Peninsula Coll., 1948; BA, Calif. State U., Fresno, 1950; postgrad., U. Calif., U. Iowa, Duquesne U., Pitts. Lic. real estate broker, Calif., Pa., N.Y.; registered investment advisor. Sr. sales rep. retail svc. stas. Shell Oil Co., San Francisco, 1952-60; mgr. west coast regional real estate Gulf Oil Corp., San Francisco and Oakland, Calif., 1960-67; real estate mgr., dir. Midwest ops. Sunray DX Oil Co. (merger Sunoco), Tulsa, 1967-71, Milex Auto Diagnostic Franchise, Inc., Plymouth Meeting, Pa., 1972-74; dir. real estate Pitts. divsn. Atlantic & Pacific Tea Co. Supermarkets, 1974-76; real estate adminstr. N.E. region Steak and Ale Restaurants Divsn. Pillsbury Cos., Dallas, 1978-80; real estate mgr. N.Y. and Phila. regions Burger King Corp. Div. Pillsbury Cos., 1980-86; real estate mgr. Ky. Fried Chicken div. (formerly owned by Heublein) and Pizza Hut div. Pepsi-Cola Inc., N.Y.C. Metro and SMSA; corp. dir. real estate and franchising Nathan's Famous Restaurants, Inc., N.Y.C., 1989-90; bd. dirs. Bristlecone Trading and Devel. Inc., Carmel, Calif., Surf City, N.J.; pres., CEO Aurner and Assocs., Carmel, 1987-90; chmn. of bd., 1990—. With USNR, PTO. Named to Hon. Order Ky. Col., Gov. of Ky., Commodore in Okla. Navy Gov. of Okla. Mem. USS Yellowstone Assn. (USNR), U. Iowa and Calif. State U. Fraternity, Sigma Alpha Epsilon, Buccaneer Club (pres. N.Y. and Conn.), Elks, Rotary. Republican. Episcopalian. Avocations: golf, precious metals conniseuer, Civil War buff. Office: 908 Long Beach Blvd Surf City NJ 08008 also: 3855 Via Nona Marie Ste 203A Carmel CA 93923

AUSES, JOHN PAUL (JAY), technical specialist; b. Johnstown, Pa., Jan. 28, 1949; s. Frank Frederick and Stephanie (Pilot) A.; m. Christine Wank, Aug. 11, 1973; children: John A., Julia C., Kevin F. BS in Chemistry, St. Francis Coll., 1970; MS in Analytical Chemistry, W.Va. U., 1974. Tech. specialist Aluminum Co. Am., Alcoa Center, Pa., 1974—; assoc. devel. dir. St. Joseph High Sch., Natrona Heights, Pa., 1991—. Inst. CPR, ARC, New Kensington, Pa., 1977—; com. mem. Pitts. Conf. Analytical Chemistry and Applied Spectroscopy, 1976—, treas., 1986, show mgr., 1988, pres., 1992. Mem. Soc. Analytical Chemists Pitts. (treas. 1983-84, chmn. 1987-88), Tri-City Jaycees (treas. 1974-77). Roman Catholic. Home: 1817 Victoria Ave Arnold PA 15068 Office: Alcoa Tech Ctr 100 Technical Dr Alcoa Center PA 15069-0001

AUSTIN, CHARLES LOUIS, software engineer; b. Seattle, Jan. 26, 1948; s. Louis Claude and Jean (Harbaugh) A. BA, U. Puget Sound, 1970; MBA, Golden Gate U., 1975; postgrad., MIT, 1988. Dep. dir. Mobilization Systems/Office of Sec. of Defense, Pentagon, Washington, 1983-85, dir., 1986-87; dir. Tech. Mgmt./Office of Sec. of Defense, Pentagon, Washington, 1985-86, Program Analysis.Office of Sec. of Defense, Pentagon, Washington, 1987-89, Command Systems Integration Agy., Pentagon, Washington, 1989-92; program exec. officer Standard Army Mgmt. Info. Systems, Ft. Belvoir, Va., 1992—. Capt. USAF, 1970-77. Mem. Armed Forces Comms. and Electronics Assn., Assn. of the U.S. Army. Achievements include development of numerous state of the art command and control technologies. Office: PEO Stamis AAE Program Exec Bldg 1465 SFAE-PS Stop C-3 Fort Belvoir VA 22060-5895

AUSTIN, FRANK HUTCHES, JR., aerospace physician, educator; b. 1924. Intern Long Beach Naval Hosp.; resident in aerospace medicine U. Calif, Berkeley, 1960-63; assoc. clin. prof. sch medicine Wright State U. Recipient Louis H. Bauer Founders award Aerospace Medical Assn., 1985, Jeffries Med. Rsch. award AIAA., 1991. Home: 1409 Trap Rd Vienna VA 22182-1642*

AUSTIN, GEORGE LYNN, surgeon; b. Eagle Pass, Tex., Aug. 19, 1944; s. George Nicole and Mary Annyce (Brisco) A.; m. Linda Gitlitz, Mar. 6, 1971; children: Juliane Beth, Kristine Nicole. BS, Johns Hopkins U., 1966; MD, U. Md., 1970. Diplomate Am. Bd. Surgery. Intern Duke U. Hosp., Durham, N.C., 1970-71, resident, 1971-72, 74-76, fellow in surgery, 1976-77; resident Med. Coll. Va., Richmond, 1977-80, instr. in surgery, 1980; pvt. practice surgery Greensburg, Pa., 1980—. Contbr. sci. papers to profl. publs. Maj. U.S. Army, 1971-74. Fellow ACS; mem. Pitts. Surg. Soc. Soc. Critical Care. Office: 562 Shearer St Ste 302 Greensburg PA 15601

AUSTIN, INEZ J., nuclear scientist. Student, Idaho State U., Columbia Basin Coll. Cert. vocat. tchr., Wash., plumbers and steamfitters journeyman, Wash., cross connection control specialist 1, Wash. Staff engr. West Tank Farm Ops. Westinghouse Hanford Co./Rockwell Hanford Ops., Richland, Wash., sr. engr. single emphasis task team, sr. engr., single shell tank process engr., phase II congnizant engr. B/BX/BY farms, sr. engr. data processing application, acting mgr. MEA&A, inspector quality control, monitor radiation. Recipient Hugh M. Hefner First Amendment award for Individual Conscience, 1991, Scientific Freedom and Responsibility award AAAS, 1992. Office: Hanford Nuclear Reservation Battelle Blvd PO Box 999 Richland WA 99352

AUSTIN, JAMES ALBERT, healthcare executive, obstetrician-gynecologist; b. Phoenix, Sept. 23, 1931; s. Albert Morris and Martha Lupkin (Mercer) A.; m. Margaret Jeanne Arnold, July 26, 1952 (div. 1978); children: Cynthia Milee Ludgin, Lauri Jeanne Fuller, Wendy Patrice Rhea; m. Sandra Lee Marsh, Jan. 3, 1979 (div. 1992). BA, U. So. Calif., 1952; MD, George Wash. U., 1956; MBA, Pepperdine U., 1991. Diplomate Am. Bd. Ob-Gyn., Am. Bd. Med. Mgmt. Intern U.S. Naval Hosp., Bethesda, Md., 1956-57, resident in ob-gyn, 1957-60; ob-gyn. Washington Gynecologists, Washington, 1966-69; pres. Ariz. Obstetrics and Gynecology Ltd., Phoenix, 1969-79; chmn. dept. ob-gyn. USN, Agana Hgts., Guam, 1979-81; ob-gyn. Sanger Med. Group, Coronado, Calif., 1981-83; chmn. ob-gyn. FHP Corp., Salt Lake City, 1983-84, assoc. med. dir., 1984-85; hosp. med. dir. FHP Corp., Fountain Valley, Calif., 1985-86; assoc. v.p. med. affairs FHP Corp., Fountain Valley, 1987-90; chief exec. officer Ultra Link Nationwide HMO Network, Costa Mesa, Calif., 1990-93; chief med. officer Downey (Calif.) Community Hosp., 1993—; clin. prof. ob-gyn. George Wash. U., Georgetown, Washington, 1966-69; asst. clin. prof. U. Calif. San Diego, 1981-83, U. Utah, Salt Lake City, 1983-85. Rear adm. USNR, 1956-88. Fellow Am. Coll. Ob-Gyn.; mem. AMA, Am. Acad. Med. Dir., Ariz. Med. Assn. (bd. dirs. 1978), Am. coll. Physician Execs. Republican. Presbyterian. Home: PO Box 1450 16811 S Pacific Ave Sunset Beach CA 90742-1450 Office: Downey Community Hosp PO Box 7010 11500 Brookshire Ave Downey CA 90241-7010

AUSTIN, LISA SUSAN COLEMAN, mechanical engineer; b. Queens, N.Y., Aug. 5, 1963; d. George Joseph and Lucille Duncan (Saxton) C. BSME, U. Rochester, 1985. Mech. engr. ISC Cardion Electronics, Woodbury, N.Y., 1985-87, Grumman Aerospace, Bethpage, N.Y., 1987-90, EG&G Fla., Kennedy Space Ctr., 1990—. Tchr. art Roslyn (N.Y.) Sch. Painting, 1987-90; math. tutor Bus./Edn. Ptnrship, Brevard County, Fla., 1991—. Mem. ASME, Soc. Allied Weights Engring. Office: EG&G Fla BOC 125 PO Box 21267 Kennedy Space Center FL 32815

AUSTIN, RALPH LEROY, chemicals executive; b. Potwin, Kans., Dec. 3, 1929; s. Dorse Thomas and Lydia Julia (Weber) A.; m. Wanda Lea McWilliams, Apr. 21, 1950 (div. Apr. 1969); children: Stephen M., Joy S. Austin Cooper, Jane L. Austin Duke, Mary K. Austin Coward; m. Cheryl Dianne Ryan, June 10, 1988. AS in Engring., El Dorado Jr. Coll., 1949; student, U. Okla., 1949-50. Sr. engr. Vickers Petroleum Co., Inc., Potwin, 1950-62; project mgr. Refining Chem. Corp., Odessa, Tex., 1962-64; v.p. Refining Chem. Corp., El Paso, Tex., 1969-71; mgr. engring. Ehrhart & Assocs., San Francisco, 1964-66; Gulf Coast mgr. Ehrhart & Assocs., Houston, 1966-69; exec. v.p. Jacob's Group Co., Houston, 1971-78; pres., chief exec. officer Blackburn, Inc., Houston, 1978-87, now bd. dirs.; pres., chmn. Blackburn Group Inc., Houston, 1987—; bd. dirs. Am. Gen. Constructors, Houston, Sixth Properties, Houston, Uniq - Con, Houston. Founder, pres. Butler County Substance Abuse Found., El Dorado, Kans., 1980-85; adv. bd. dirs. South Cen. Mental Health Clinic, El Dorado, 1987—. Mem. El Dorado C. of C. (pres., bd. dirs. 1982-84), Butler County Hist. Soc., El Dorado Country Club. Republican. Avocations: wood carving, collecting antiques.

AUSTIN, ROBERT BRENDON, civil engineer; b. West Point, N.Y., Aug. 10, 1956; s. Thomas and Margaret Ann (Hart) A. BS, U. Conn., 1979; M Civil Engring., U. Tex., Arlington, 1992. Structural engr. Stone & Webster Engring. Corp., Dallas, 1979-89; rsch. assoc. U. Tex., Arlington, 1990-92; resident engr. Stone & Webster Engring. Corp., Denver, 1992—; cons. Dallas, 1990-92; instr., vis. prof. U. Tex., Arlington, 1991-92. Contbr. articles to profl. jours. Recipient Constrn. Innovation award Stone & Webster Engring., Boston, 1987. Mem. Am. Soc. Civil Engrs., Am. Concrete Inst. Office: Stone & Webster Engr Corp 4411 Worcola St Dallas TX 75206

AUSTIN, SAM M., physics educator; b. Columbus, Wis., June 6, 1933; s. A. Wright and Mildred G. (Rheinhard) A.; m. Mary E. Herb, Aug. 15, 1959; children: Laura Gail, Sara Kay. BS in Physics, U. Wis., 1955, MS, 1957, PhD, 1960. Rsch. assoc. U. Wis., Madison, 1960; NSF postdoctoral fellow Oxford U., Eng., 1960-61; asst. prof. Stanford U., Calif., 1961-65; assoc. prof. physics Mich. State U., East Lansing, 1965-69, prof., 1969-90, univ. disting. prof., 1990—, chmn. dept., 1980-83; assoc. dir. Cyclotron Lab. Mich. State U., 1976-79, rsch. dir., 1983-85, co-dir., 1985-89, dir., 1989-92; guest Niels Bohr Inst., 1970; guest prof. U. Munich, 1972-73; scientific collaborator Saclay and Lab. Rene Bernas, 1979-80. Author, editor: The Two Body Force in Nuclei, 1972, the (p,n) Reaction and Nucleon-Nucleon Force, 1980; editor Phys. Rev. C, 1988—; assoc. editor Atomic Data and Nuclear Data Tables, 1990—; mem. editorial com. Ann. Rev. Nuclear and Particle Sci. Fellow NSF, 1960-61, Alfred P. Sloan Found., 1963-66; recipient Mich. Assn. of Governing Bds. Disting. Prof., 1992. Fellow AAAS, Am. Phys. Soc. (vice chmn. nuclear physics div. 1981-82, chmn. 1982-83, exec. com., 1983-84, 86-89, coun. 1986-89, coun. exec. com. 1987-88); mem. Am. Assn. Physics Tchrs., Sigma Xi (Sr. Rsch. award 1977). Achievements include research in nuclear physics, nuclear astrophysics and nitrogen fixation. Home: 1201 Woodwind Trl Haslett MI 48840-8956 Office: Mich State U Nat Supercondr Cyclotron Lab East Lansing MI 48824

AUSTRIAN, ROBERT, physician, educator; b. Balt., Apr. 12, 1916; s. Charles Robert and Florence (Hochschild) A.; m. Babette Friedmann Bernstein, Dec. 29, 1963; stepchildren: Jill Bernstein, Toni Bernstein. AB, Johns Hopkins U., 1937, MD, 1941; DSc honoris causa, Hahnemann Med. Coll., 1980, Phila. Coll. Pharmacy and Sci., 1981, U. Pa., 1987. Diplomate: Am. Bd. Internal Medicine. House officer Johns Hopkins Hosp., 1941-50, asst. dir. med. out-patient dept., 1951-52; assoc. prof. medicine, then prof. medicine SUNY Coll. Medicine, 1952-62; John Herr Musser prof., chmn. rsch. medicine U. Pa. Sch. Medicine, 1962-86, prof. emeritus, chmn. emeritus, 1986—; attending physician Hosp. U. Pa.; Tyndale vis. lectr. and prof. Coll. Medicine U. Utah, 1964; spl. research on infectious diseases, bacterial genetics; mem. Meningococcal Infections Com., 1964-72, Commn. on Acute Respiratory Disease, 1963-72, Commn. Streptococcal and Staphylococcal Diseases, 1970-72, Armed Forces Epidemiol. Bd.; cons. surg. gen. U.S. Army Research and Devel. Command, 1966-69; mem. subcom. streptococcus and pneumococcus Internat. Com. Bacteriol. Nomenclature; mem. allergy and immunology study sect. Nat. Inst. Allergy and Infectious Diseases, 1965-69, mem. bd. sci. counselors, 1967-70, chmn., 1969-70; mem. Who Expert adv. panel Acute Bacterial diseases, 1979—. Mem. editorial bd.: Jour. Bacteriology 1964-69, Am. Rev. Respiratory Diseases, 1963-66, Bacteriol. Rev., 1967-71, Jour. Infectious Diseases, 1969-74, Antimicrobial Agents and Chemotherapy, 1972-86, Infection and Immunity, 1973-81, Revs. of Infectious Diseases, 1979-89, Vaccine, 1983—. Trustee Johns Hopkins U., 1963-69. Served to capt. M.C. AUS, 1943-45. Recipient U.S. Typhus Commn. medal, 1947; Albert Lasker Clin. Med. Research award, 1978; Phila. award, 1979; Willard O. Thompson award Am. Geriatric Soc., 1981, others. Fellow ACP (master, James D. Bruce Meml. award 1979), N.Y. Acad. Scis., Am. Acad. Microbiology, AAAS (chmn. sect. on med. scis. 1975); mem. Assn. Am. Physicians, Am. Soc. Clin. Investigation, Am. Clin. and Climatol. Assn. (pres. 1984), Am. Soc. Microbiology (v.p. N.Y. br. 1961-62), Am. Philos. Soc., Nat. Acad. Scis., Soc. Exptl. Biology and Medicine, Harvey Soc., Am. Fedn. Clin. Research, Inst. Medicine (sr.), Balt. Med. Soc., Am. Assn. Immunologists, N.Y. Acad. Medicine (sec. sect. microbiology 1961-62), Phila. County Med. Soc. (Strittmatter award 1979), Coll. Physicians Phila. (Meritorious Service award 1980, pres.-elect 1986, pres. 1988-90), Interurban Clin. Club (pres. 1970), Infectious Disease Soc. Am. (pres. 1971, Maxwell Finland lecture award 1974, Bristol award 1986), Johns Hopkins Soc. Scholars, Phi Beta Kappa, Sigma Xi, Alpha Omega Alpha, Omicron Delta Kappa. Club: 14 W. Hamilton Street (Balt.). Achievements include demonstration of the continuing importance of lobar pneumonia as a cause of death despite treatment with antibiotics and of the efficacy of polyvalent pneumococcal vaccine in preventing such illness leading to its relincensure. Office: U Pa Sch Medicine Dept Rsch Medicine 36th & Hamilton Walk Philadelphia PA 19104

AUWERX, JOHAN HENRI, molecular biologist, medical educator; b. Diepenbeek, Limburg, Belgium, Oct. 4, 1958; s. Rene Auwerx and Maria Gilissen. B of Medicine, Cath. U., Leuven, Belgium, 1978, MD, 1982, PhD, 1990, Specialist in Endocrinology, 1987. Specialist internal medicine. Intern Royal Victoria Hosp., Montreal, Can., 1981; resident in medicine Cath. U., Leuven, 1982-84, fellow in endocrinology, 1984-86; rsch. dir. CNRS, Nice, 1991—; assoc. prof. Cath. U., Leuven, 1989—; sr. rsch. fellow U. Wash., Seattle, 1986-88; cons. Janssen Pharm. Industry, Beerse, Belgium. Co-author/editor seven scientific books; contbr. articles to profl. jours. Recipient Horlait-Dapsens prize Belgium, 1985, Fulbright award, 1986, J. Fogarty Rsch. award NIH, 1987, ILSI Rsch. award Internat. Life Sci. Inst., USA/Europe, 1989, Morgagni Young Investigator award European Metabolic soc., Italy, 1991, F. De Waele award, Belgium, 1989, Boehring award, Belgium, 1991, Danone award, France, 1993. Mem. AAAS, Am. Fedn. Clin. Rsch., Am. Heart Assn. (corrs. fellow coun. on atherosclerosis 1989), Am. Soc. Microbiology, European Atherosclerosis Soc., Belgian Lipid Club. Avocations: sailing, swimming, travel. Office: Centre de Biochimie, Parc Valrose, 06108 Nice France

AVASTHI, RAM BANDHU, physician, general practitioner; b. Siswan, Punjab, India, July 30, 1941; arrived in England, 1967; s. Lekh Raj and Paramajot (Sharma) A.; m. Promila Devi Vasudeva, Apr. 10, 1971; 1 child, Durga Arti. MBBS, Med. Coll., Armitsar, Punjab, India, 1964; MD, Ednl. Coun. Fgn. Med. Grads., Ill., U.S., 1972. Resident in gen. surgery and orthopaedics Hartlepool, Birmingham, Eng., 1967-70; registrar in gen. surgery Nuneaton Hosps., England, 1970-73; registrar in cardiovascular and thoracic surgery Gen. Infirmary, Leeds, Eng., 1973-74, Castle Hill Hosp., Hull, Eng., 1974-76, Royal Infirmary and City Hosps., Edinburgh, Scotland, 1976-78; pvt. practice gen. medicine Blackpool, Eng., 1978—; hon. med. officer to British Limbless Ex-Sevicemen Assn. Home, Blackpool, 1979-87. Justice of Peace, Magistrates' Bench, Blackpool, 1984—; fund raiser Park Sch., Blackpool, 1984. Fellow Royal Coll. Surgeons; mem. Royal Coll. Physicians (licentiate), Lancashire Local Med. Com. Mem. Conservative Party. Hindu. Avocations: motor cycling, studying religions. Home: 55 Saint David's Rd N, FY8 2BT Saint Annes-on-sea England Office: The Surgery, 194 Lytham Rd, FY1 6EU Blackpool England

AVENDANO, TANIA, software engineer; b. La Habana, Cuba, Sept. 13, 1963; arrived in PR, 1970.; d. Tito Joel and Nilda (Ulloa) A.; m. Orlando Rodriguez, Oct. 10, 1987. Bs in Computer Engring., Recinto Universitario de Mayaguez, PR, 1987. Registered profl. computer engr. Product support engr. Digital Equipment Corp., Aguadilla, PR, 1987-88, software engr., 1988-91, sr. software engr., 1991-93; systems software engr. Computer Engring. Assocs., San Juan, PR, 1993—. Mem. Computer Soc. of IEEE. Achievements include design of Decserver 200 value engineering, Decserver 250 (parallel printer server) manufacturing diagnostics developer, DecAgent 90 (simple network management server) firmware project leader.

AVERILL, BRUCE ALAN, chemistry educator; b. Bucyrus, Ohio, May 19, 1948; s. Kenneth L. Averill and Mildred (Reid) Krug; m. Patricia Ann Eldredge, Aug. 23, 1986; children: Lindsay Patricia, Alan Eldredge. BS, Mich. State U., 1969; PhD, MIT, 1973. Asst. prof. Mich. State U., East Lansing, 1976-81, assoc. prof., 1981-82; assoc. prof. U. Va., Charlottesville, 1982-88, prof. chemistry, 1988—; mem. biophysics adv. panel NSF, Washington, 1985-88; mem. faculty forum for sci. rsch. U. Va., Charlottesville, 1984-88. Contbr. over 100 articles to sci. jours. A.P. Sloan fellow, 1981-83; recipient creativity award NSF, 1991. Mem. AAAS, Am. Soc. Biochemistry and Molecular Biology, Am. Chem. Soc., Royal Soc. Chemistry, Sigma Xi. Office: U Va Dept Chemistry Mccormick Rd Charlottesville VA 22904-1000

AVERY, CHARLES CARRINGTON, forestry educator, researcher; b. Syracuse, N.Y., July 22, 1933; s. Edward Carrington and Elizabeth Amelia (Boorum) A.; m. Valeen Tippetts, Sept. 13, 1961 (div. 1986); children: Christopher E., Maureen E., Nathan C., Theodore C. BS, Utah State U., 1961; cert., French Nat. Sch. Forests and Waters, Nancy, France, 1962; MF, Duke U., 1963; PhD, U. Wash., 1974. Forestry technician Fremont Nat. Forest/Pacific N.W. Experiment Sta., Paisley and Portland, Oreg., 1955-61; forester Medicine Bow Nat. Forest/Wenatchee Nat. Forest, Laramie, Wy. and Ellensburg, Wis., 1961-66; rschr. Rocky Mountain Experiment Sta., Flagstaff, Ariz., 1966-74; assoc. prof. No. Ariz. U., Flagstaff, 1974-80, prof., 1980—; cons. in hydrology, 1980—; vis. prof. Free. U. Brussels, 1985, Utah State U., 1986. Co-author: Watershed Management for Hydrologists, 1990; contbr. articles to sci. jours. Advisor Resource Ctr. for Environ. Edn., Flagstaff, 1986-90; moderator Gov.'s Commn. on Ariz. Environment, Prescott, 1988; dir. Commn. Mental Health, The Guidance Ctr., Flagstaff, 1989—. Cpl. U.S. Army, 1953-55, France. Fullbright scholar 1961. Fellow Ariz./Nev. Acad. Scis. (pres. 1978-79); mem. Soc. Am. Foresters (chair San Francisco Peaks chpt. 1984-85), Am. Geophysical Union, Am. Inst. Hydrology (registered profl. hydrologist), Ariz. Hydrological Soc. (pres. Flagstaff chpt. 1988-89, bd. dirs. 1989-90, 92-93), N.Y. Acad. Sci., Sigma Xi (pres. No. Ariz. chpt. 1990-91), Phi Kappa Phi. Democrat. Home: PO Box 22514 Flagstaff AZ 86002-2514 Office: No Ariz U Sch Forestry Box 15018 Flagstaff AZ 86011

AVERY, MARY ELLEN, pediatrician, educator; b. Camden, N.J., May 6, 1927; d. William Clarence and Mary (Miller) A. AB, Wheaton Coll., Norton, Mass., 1948, DSc, 1974; MD, Johns Hopkins U., 1952; DSc (hon.), Trinity Coll., 1976, U. Mich., 1975, Med. Coll. Pa., 1976, Albany Med. Coll., 1977, Med. Coll. Wis., 1978, Radcliffe Coll., 1978; MA (hon.), Harvard U., 1974; LHD (hon.), Emmanuel Coll., 1979, Northeastern U., 1981, Russell Sage Coll., 1983, Meml. U., Newfoundland, 1993. Intern Johns Hopkins Hosp., 1953-54, resident, 1954-57; research fellow in pediatrics Boston, 1957-59, Balt., 1959-69; assoc. prof. pediatrics Johns Hopkins U., 1964-69; prof., chmn. dept. pediatrics McGill U. Med. Sch., 1969-74; prof. pediatrics Harvard U., 1974—; physician-in-chief Montreal Children's Hosp., 1969-74, Children's Hosp. Med. Center, Boston, 1974-85; mem. Med. Rsch. Coun. Can.; mem. study sect. NIH, 1968-71, 84-88. Author: The Lung and Its Disorders in the Newborn Infant, 4th edit., 1981, (with A. Schaffer) Diseases of the Newborn, 1971, 6th edit. (with H.W. Taeusch and R. Ballard), 1991; (with G. Litwack) Born Early, 1984; author, editor: (with L. First) Pediatric Medicine, 1988, 2nd edit., 1993; also articles; mem. editorial bd. Pediatrics, 1965-71, Am. Rev. Respiratory Diseases, 1969-73, Am. Jour. Physiology, 1967-73, Jour. Pediatrics, 1974-84, Medicine, 1985, Johns Hopkins Med. Jour., 1970-82, Clin. and Investigative Critical Care Medicine, 1990, New Eng. Jour. Medicine, 1990. Trustee Wheaton Coll. (1965-85), Radcliffe Coll., Johns Hopkins U., 1982-88. Recipient Mead Johnson award in pediatric rsch., 1968, Trudeau medal Am. Thoracic Soc., 1984, Nat. Medal of Sci. NSF, 1991; Markle scholar in med. scis., 1961-66. Fellow AAAS (dir. 1989), Internat. Pediatric Assn. (standing com. 1986-89), Am. Acad. Pediatrics, Am. Acad. Arts and Scis., Royal Coll. Physicians and Surgeons Can.; mem. Can. Pediatric Soc., Am. Physiol. Soc., Soc. Pediatric Rsch. (pres. 1972-73), Brit. Pediatric Assn. (hon.), Inst. Medicine (coun. 1987), Am. Pediatric Soc. (pres. 1990), Phi Beta Kappa, Alpha Omega Alpha. Office: 221 Longwood Ave Boston MA 02115-5817

AVERY, SUSAN KATHRYN, electrical engineering educator, researcher; b. Detroit, Jan. 5, 1950; d. Theodore Peter and Alice Jane (Greene) Rykala; m. James Paul Avery, Aug. 12, 1972; 1 child, Christopher Scott. BS in Physics, Mich. State U., 1972; MS in Physics, U. Ill., 1974, PhD in Atmospheric Sci., 1978. Asst. prof. elec. engring. U. Ill., Urbana, 1978-83; fellow CIRES U. Colo., Boulder, 1982—, assoc. prof. elec. engring., 1985-92, assoc. dean rsch. and grad. edn. Coll. Engring., 1989-92, prof. elec. engring., 1992—; adv. com. chair Elec. and Communications Div. NSF, Washington, 1991-93, adv. panel atmospheric scis. program, 1985-88, steering com. CEDAR program, 1986-87, adv. com. engring. directorate, 1991-93, vis. professorship, 1982-83; working group ionosphere, thermosphere, mesosphere NASA, Washington, 1991—; mem.-at-large USNC/URSI NRC, Washington, 1991-93, com. on solar-terrestrial rsch., 1987-90; trustee Univ. Corp. for Atmospheric Rsch., 1991—, vice chair bd. trustees, 1993, sci. programs evaluation com., 1989-91; working group on tides in mesosphere and lower thermosphere Internat. Commn. Meteorology of Upper Atmosphere, 1981-86; mesosphere-lower thermosphere network steering com. Internat. STEP Program, 1989—, equatorial mid. atmosphere dynamics steering com., 1990—. Contbr. articles to Radio Sci., Adv. Space Rsch., Jour. Atmosphere Terrestrial Physics, Jour. Geophys. Rsch., others. Recipient Faculty Award for Women, NSF, 1991, Outstanding Publication award NCAR, 1990; vis. fellow Coop. Inst. for Rsch. in Environ. Scis., 1982-83. Mem. IEEE, Am. Meteorol. Soc. (com. on mid. atmosphere 1990—), Am. Geophys. Union (com. edn and human resources 1988—), Am. Soc. Engring. Edn., Sigma Xi. Achievements include research on the dynamics of the mesosphere, stratosphere and troposphere with emphasis on unifying observational analyses and theoretical studies, on wave dynamics including the coupling of the atmosphere/ocean and interactions between large-scale and small-scale motions, on the use of ground-based doppler radar techniques for observing the clear-air atmosphere and use of new signal processing algorithms for radar data analysis. Office: U Colo Engring Rsch Ctr CB 423 Boulder CO 80309-0425

AVGOUSTI, MARIOS, chemical engineer; b. Nicosia, Cyprus, Aug. 9, 1966; came to U.S., 1988; s. Charalambos and Maria (Georgiadou) A. M in Engring., Imperial Coll., 1988; PhD, U. Del., 1992. Rsch. student Unilever Rsch., Port Sunlight, U.K., 1986; rsch. asst. U. Del., Newark, 1988—, teaching asst., 1991; asst. prof. Stevens Inst. Tech., Hoboken, N.J., 1992. Contbr. articles to Procs. Royal Soc. London, Internat. Jour. Numerical Methods in Fluid Mechanics, Jour. Non-Newtonian Fluid Mechanics. Mem. AICE, Soc. Indsl. and Applied Math., Inst. Chem. Engrs. (Eng. and Wales), Soc. Rheology, Sigma Xi. Christian Orthodox. Home: 730 Hudson St Apt 20 A Hoboken NJ 07030 Office: Chem Engring Dept Stevens Inst Tech Hoboken NJ 07030

AVISSAR, YAEL JULIA, molecular biology educator; b. Budapest, Hungary, July 29, 1946; came to U.S., 1986; d. Istvan and Iren (Steiner) Bauer; m. Jacob Avissar, Sept. 26, 1967; children: Oded, Michele. MSc, Ben Gurion U., Beer Sheva, Israel, 1973; PhD, Mich. State U., 1978. Rsch. assoc. Mich. State U., East Lansing, 1978-79; lectr. Ben Gurion U., 1979-85; vis. prof. Pasteur Inst., Paris, 1985-86; rsch. assoc. Brown U., Providence, 1986-90; asst. prof. molecular biology R.I. Coll., Providence, 1990—. Contbr. articles to profl. jours. Mem. Am. Soc. Plant Physiologists, Sigma Xi. Jewish. Achievements include explanation of the biochemistry and molecular biology of the early steps of tetrapyrrole biosynthesis in bacteria and protists. Home: 76 Edgehill Rd Providence RI 02906 Office: RI Coll 600 Mt Pleasant Ave Providence RI 02908

AVIV, DAVID GORDON, electronics engineering executive; b. Tel-Aviv, Israel, Dec. 18, 1928; came to U.S., 1939; s. Mendel and Clare (Gordon) A.; m. Rena Rod, Aug. 23, 1959; children: Jonathan Enoch, Oren Rod, Robert Eli. BEE, Pratt Inst., 1948; MA in Math., Columbia U., 1951, MSc in Elec. Engring., 1953. Sr. mem. rsch. staff Rand Corp., Santa Monica, Calif., 1965-67; study mgr. Aerospace Corp., El Segundo, Calif., 1968-79; engring. specialist Rockwell Internat., Downey, Calif., 1980-85; program mgr. Lockheed, Plainfield, N.J., 1985-90; lead engr. Mitre Corp., Eatontown, N.J., 1990-92; pres. ARC Inc., N.Y.C., 1992—. Contbr. articles to profl. jours. Mem. IEEE (sr., chmn. edn. N.Y. sect. 1992), N.Y. Acad. Sci., Internat. Laser Comm. Soc. (founding mem.), Soc. Photo Instrumentation Engrs. and Scientists, Columbia U. Engring. Sch. Alumni Orgn. (pres. 1972-85). Achievements include patent for AC and DC Operational Amplifier with Infinite Memory; development of security systems, intelligent vehicles and highways, semi-active identification friend or foe (IFF) system, synthetic aperture radar identifying vibrating systems, digital encrypted speech interpolation system, early warning fire alarm system for civilian applications, laser communications systems, electroencephalometer and other medical electronic systems. Home: 150 W 56th St New York NY 10019

AVIV, JONATHAN ENOCH, otolaryngologist, educator; b. N.Y.C., Aug. 24, 1960; s. David Gordon and Rena (Rod) A.; m. Caryn Lee Schacht, Sept. 11, 1986. BA, Columbia U., 1981, MD, 1985. Diplomate Am. Bd. Otolaryngology, Nat. Bd. Med. Examiners. Resident dept. surgery Mount Sinai Med. Ctr., N.Y.C., 1985-87, resident dept. otolaryngology, 1987-90, fellow microvascular surgery, 1990-91; asst. prof. dept. otolaryngology Coll. Physicians and Surgeons, Columbia U., N.Y.C., 1991—; Contbr. articles to profl. jours. Mem. AMA, Am. Acad. Otolarynology, Am. Acad. Facial Plastic and Reconstructive Surgery, N.Y. Acad. Scis., Assn. for Rsch. in Otolarynology. Achievements include development of implantable laryngeal electrode platform to stimulate electrically the muscles of the voice box; use of microvascular free tissue tranfer for head and neck reconstruction. Office: Columbia-Presbyn Med Ctr Dept Otolaryngology 630 W 168th St New York NY 10032-3702

AVOLIO, JOHN, chemist, inventor; b. Bklyn., Feb. 22, 1958; s. John Joseph Jr. and Angelina (Felitti) A.; m. Catherine Rosario, Dec. 31, 1991. BS in Chemistry, CUNY, 1984. Pres., CEO, co-founder Chiral Innovations, Inc., Somerset, N.J., 1990—. Somerset county coord. Perot Campaign, 1992, United We Stand Am., 1992—. Mem. Am. Chem. Soc. (cert. lab. safety), Sports Car Club Am. Roman Catholic. Achievements include patents pending for Chiral scientific instruments. Office: Chiral Innovations Inc 23 Kingsberry Dr Somerset NJ 08873

AVOURIS, PHAEDON, chemical physicist; b. Athens, Greece, June 16, 1945; came to U.S., 1970.; s. Dionisios and Ourania (Nomikos) A.; m. Alice Laura Dearden, Oct. 7, 1976; 1 child, Ann. BS, Aristotle U., Thessaloniki, greece, 1968; PhD, Mich. State U., 1974. Postdoctoral fellow U. Calif., L.A., 1975-77; rsch. assoc. AT&T Bell Labs., Murray Hill, N.J., 1978; rsch. staff IBM Watson Rsch. Ctr., Yorktown Heights, N.Y., 1978-84; mgr. chem. physics IBM Watson Rsch. Ctr., Yorktown Heights, 1984—; panel for chem. sci. Nat. Rsch. Coun., Washington, 1990—. Editor: (book) Atomic and Nanoscale Modifications of Materials; contbr. articles to profl. jours. including Phys. Rev., Sci., Jour. Chem. Physics. Fellow Am. Phys. Soc.; mem. Am. Chem. Soc. (adv. editorial bd. 1990—). Achievements include pioneering the study of surface chemistry on atomic scale with scanning tunneling microscopy, the manipulation of individual atoms; contbutions to understanding of electronically excited states at surfaces. Office: IBM Watson Rsch Ctr Rt 134 Yorktown Heights NY 10598

AWAN, AHMAD NOOR, civil engineer; b. Chakwal, Punjab, Pakistan, June 2, 1942; came to U.S., 1969; s. Ghulam Hussain and Sayada Awan; m. Nargis Parveen Janjua, Dec. 24, 1972; children: Monazza, Shujah, Noureen, Farah. BSc in Civil Engring., U. Engring., Lahore, Pakistan, 1965; MS in Civil Engring., U. Pa., Phila., 1971; cert. project mgmt., Poly. Inst. N.Y., 1976. Civil engr. Water & Power Devel. Authority of Govt. Pakistan, Lahore, 1965-66; project resident engr., cons. Govt. Libya, El Beida, 1966-68; sr. structural engr. Stone & Webster Engring. Corp., N.Y.C., 1971-79; sr. engr., project mgmt. cons. U.S. Army C.E. Middle East, Saudi Arabia, 1979-83; sr. staff engr. project Port Authority of N.Y. and N.J., N.Y.C., 1985—; Mem. internat. roster of experts in fields of engring., constrn. bldg., fin. and contracts and tenders Habitat, UN Centre for Human Settlements, 1980. Mem. ASCE, Am. Concrete Inst. Achievements include development of computerized project management system for U.S. Army Corps of Engineers; managed major restoration team after New York World Trade Center bombing, 1993. Home: 6 Silver Hollow North Brunswick NJ 08902

AWAN, GHULAM MUSTAFA, economist, political scientist, educator; b. Shikarpur, Sindh, Pakistan, June 1, 1940; s. Saeed Rasul and Saeedah Begum Awan; m. Saeedah Awan, Feb. 25, 1960; children: Fatimah, Tariq Mustafa, Zahida, Riffat, Khalid, Noor, Rabiah. BA with honors, Sindh U., Hyderabad, Pakistan, 1960, MA in Econs., 1965, LLB, 1966, BEd, 1968, MA in Polit. Sci., 1968; MS in Econ. Devel., Inst. Social Studies, The Hague, The Netherlands, 1978; MS in Def. and Strategic Studies, Quaid-i-Azam U., Islamabad, Pakistan, 1991; PhD in Econs., Pacific Western U., L.A., 1992. Prof. econs. Govt. Postgrad. Studies Ctr., Sukkar, Sindh, Pakistan, 1965-71; asst. econ. adviser Office Ministry of Fin. Govt. Pakistan, Islamabad, 1972-76; dep. chief economic group Ministry of Planning and Devel., Islamabad, 1976-83; chief nat. planning commn. Govt. Pakistan, Islamabad, 1984-89, 89-90; mem. directing staff Ministry of Def. Nat. Def. Coll. Armed Forces, Govt. Pakistan, Rawalpindi, 1990-91; chief programming Ministry Planning and Devel., Islamabad, Pakistan, 1992—; joint chief economist Plan Commn., 1993—. Author: Role of Statistics in Planning, 1972, Regional Cooperation for Development, 1978, Cooperation in Pakistan and Foreign Lands, 1980, Role of Banking in Pakistan, 1983, Federal Provincial Relationship, 1990, Interest Free Banking in Pakistan, 1993; contbr. 85 articles to profl. jours. Municipality counselor, Shikarpur, 1960, mayor, 1962; patron-in-chief Awan Assn., Shikarpur, 1964. Recipient Sermon award Muslin Fed. (The Netherlands), 1979; King Abdulaziz fellow U. Jeddah, South Arabia, 1984, U. Pitts. fellow, 1991, USA. Fellow Adminstrv. Staff Coll. India, Higher Productivity Orgn. Japan; mem. Am. Econ. Assn., Soc. Pakistan Economists (pres.), Islamabad Club, Army Golf Club. Avocations: hockey, golf, philately, travel, lecturing on economics. Home: House # 35 Category 1, St 11 1-8/1, Islamabad Pakistan Office: Nat Planning Commn, "P" Block Pakistan Secretariat, Islamabad Pakistan

AWERBUCH, SHIMON, research consultant; b. Tel Aviv, Israel, May 19, 1946; s. Erich Eli and Lilly Leah (Drabin) A.; came to U.S., 1956, naturalized, 1963; BS, Rensselaer Poly. Inst., 1968, MS in Urban and Environ. Studies, 1969, PhD in Urban and Environ. Studies, 1975. Ops. rsch. analyst asst. chief of staff, USAF, Washington, 1969-71; com. counsel Standing Com. on Environ. Conservation, N.Y. State Assembly, Albany, 1971-73; sr. cons. Mgmt. Cons. Svc., Ernst & Whinney, Washington, 1974-75; dir. policy analysis and planning project N.Y. State Edn. Dept., Albany, 1975-76; policy analyst Gov.'s Econ. Devel. Bd., Albany, 1976-77; mng. ptnr. and co-founder Tibbits Assocs., Troy, 1977-80; chief econ. and policy studies Exec. Dept., Utility Intervention Office, State of N.Y., 1980-86; prof. fin. Coll. Mgmt. Sci., U. Lowell, Mass., 1986—; cons. in utility rate-case procs.; rsch. cons. Mass. Energy Dept., 1986—, Sandia Nat. Labs. U.S. Dept. Energy, Nat. Renewable Energy Lab, various regional telephone cos., electric utilities, 1988—. With USAF, 1969-71. Mem. Am. Fin. Assn., Fin. Mgmt. Assn., Inst. Mgmt. Sci., Sigma Xi. Author: (with William A. Wallace) Policy Evaluation for Community Development: Decision Tools for Local Government, 1976, (with D. Freireich) Nuclear Cancellations, Award Papers in Public Utility Economics and Regulations, 1983, Efficient Income Measures and the Partially Regulated Firm: Deregulating Utilities, 1989; contbr. articles on fin., regulatory econs., public policy to profl. jours. Office: U Mass Coll Mgmt Sci 1 University Ave Lowell MA 01854-2881

AWSCHALOM, DAVID DANIEL, physicist; b. Baton Rouge, Oct. 11, 1956; s. Miguel and Evelyn A.; m. Nancy L. Kawalek, Aug. 6, 1988. BS in Physics, U. Ill., 1978; PhD in Physics, Cornell U., 1982. Exxon rsch. fellow Cornell U., Ithaca, N.Y., 1981-82; postdoctoral fellow IBM Watson Rsch. Ctr., Yorktown Heights, N.Y., 1982-83, rsch. staff mem., 1984-89, mgr. nonequilibrium physics dept., 1989-92; prof. physics U. Calif., Santa Barbara, 1992—; mem. Nat. Rsch. Coun. Panel on Magnetic Semiconductors, 1990, Nat. Rsch. Coun. Panel on Naval Rsch., 1991, NSF Ctr. for Quantized Electronics Structures, 1992; seminar speaker in field. Contbr. over 70 articles to profl. jours. Named James scholar U. Ill., 1976-78, Exxon Predoctoral fellow, 1981; recipient Lyman Physics prize U. Ill., 1978. Fellow Am. Phys. Soc.; mem. Materials Rsch. Soc. (Outstanding Investigator prize 1992). Achievements include development and application of ultrafast optical technique for exploring electronic and magnetic interations in quantum systems; invented new time-resolved magnetic spectroscopies using superconduction devices. Office: U of California Dept of Physics Santa Barbara CA 93106

AXEL, LEON, radiologist, educator; b. Lakewood, N.J., Nov. 1, 1947; s. Milton and Alice (Terry) A.; m. Katherine Lorber, 1979 (div.); m. Linda Susan Koenigsberg, Oct. 15, 1983; 1 child, Nathaniel Irving. PhD, Princeton U., 1971; MD, U. Calif., San Francisco, 1976. Diplomate Am. Bd. Radiology. Intern U. Calif., San Francisco, 1976-77, resident, 1977-80, fellow, 1980-81; mem. staff Hosp. of U. Pa., Phila.; asst. prof. radiology U. Pa., Phila., 1981-86, assoc. prof., 1986-90, prof. radiology, 1990—. Mem. Soc. Magnetic Resonance Medicine (mem. bd. trustees), Soc. Magnetic Resonance Imaging (bd. dirs.), Am. Coll. Radiologists (mem. commn. on magnetic resonance), Am. Assn. Physicists in Medicine. Achievements include 3 patents for methods of measurement of regional heart motion with magnetic resonance imaging; development of methods for study of blood flow with magnetic resonance imaging. Office: Hosp of Univ of Pa 36th & Hamilton Walk 308 Stemmler Philadelphia PA 19104-6086

AXELROD, JULIUS, biochemist, pharmacologist; b. N.Y.C., May 30, 1912; s. Isadore and Molly (Leichtling) A.; m. Sally Taub, Aug. 30, 1938; children: Paul Mark, Alfred Nathan. BS, CCNY, 1933; MA, NYU, 1941, DSc (hon.), 1971; PhD, George Washington U., 1955, LLD (hon.), 1971; DSc (hon.), U. Chgo., 1965, Med. Coll. Wis., 1971, Med. Coll. Pa., 1974, U. Pa., 1986, Hahnemann U., 1987; LLD (hon.), CCNY, 1972; D honoris causa, U. Panama, 1972, U. Paris (Sud), 1982, Ripon Coll, 1984, Tel Aviv U., 1984; DSC (hon., McGill U., Montreal, 1989. Chemist Lab. Indsl. Hygiene, 1935-46; research assoc. 3d N.Y. U. research divsn. Goldwater Meml. Hosp., 1946-49; assoc. chemist sect. chem. pharmacology Nat. Heart Inst., NIH, 1949-50, chemist, 1950-53, sr. chemist, 1953-55; acting chief sect. pharmacology Lab. Clin. Sci. NIMH, 1955, chief sect. pharmacology, 1955-84; guest worker Lab. Cell Biology NIMH, 1984—; Otto Loewi meml. lectr. N.Y. U., 1963; Karl E. Paschkis meml. lectr. Phila. Endocine Soc., 1966; NIH lectr., 1967; Nathanson meml. lectr. U. So. Calif., 1968; James Parkinson lectr. Columbia U., 1971; Wartenberg lectr. Am. Acad. Neurology, 1971; Arnold D. Welch lectr. Yale U., 1971; Harold Carpenter Hodge distinguished lectr. toxicology U. Rochester, 1971; Bennett lectr. Am. Neurol. Assn., 1971; Harvey lectr., 1971; Mayer lectr. Mass. Inst. Tech., 1971; distinguished prof. sci. George Washington U., 1972; Salmon lectr. N.Y. Acad. Medicine, 1972; Eli Lilly lectr., 1972; Mike Hogg lectr. U. Tex., 1972; Fred Schueler lectr. Tulane U., 1972; numerous other hon. lectures; vis. scholar Herbert Lehman Coll. City U. N.Y., 1973; professorial lectr. George Washington U., 1959—; panelist U.S. Bd. Civil Service Examiners, 1958-67; mem. research adv. com. United Cerebral Palsy Assn., 1966-69; mem. psychopharmacology study sect. NIMH, 1970-74; mem. Internat. Brain Research Orgn.; mem. research adv. com. Nat. Found.; vis. com. Brookhaven Nat. Lab., 1972-76; bd. overseers Jackson Lab., 1974-88. Mem. editorial bd. Jour. Pharmacology and Exptl. Therapeutics, 1956-72, Jour. Medicinal Chemistry, 1962-67, Circulation Research, 1963-71, Currents in Modern Biology, 1966-72; mem. editorial adv. bd. Communication in Behavioral Biology, 1967-73, Jour. Neurobiology, 1968-77, Jour. Neurochemistry, 1969-77, Jour. Neurovisceral Relation, 1969, Rassegna di Neurologia Vegetativa, 1969—, Internat. Jour. Psychobiology, 1970-75; hon. cons. editor Life Scis, 1961-69; co-author: The Pineal, 1968; contbr. papers in biochem. actions and metabolism of drugs, hormones, action of pineal gland, enzymes, neurochem. transmission to profl. jours. Recipient Meritorious Rsch. award Assn. Rsch. Nervous and Mental Diseases, 1965; Gairdner award disting. rsch., 1967; Nobel prize in med. physiology, 1970; Alumni Disting. Achievement award George Washington U., 1968; Superior Service award HEW, 1968; Disting. Svc. award, 1970; Claude Bernard professorship and medal U. Montreal, 1969; Disting. Svc. award Modern Medicine mag., 1970; Albert Einstein award Yeshiva U., 1971; medal Rudolf Virchow Med. Soc., 1971; Myrtle Wreath award Hadassah, 1972; Leibniz medal Acad. Sci. East Germany, 1984; Salmon medal N.Y. Acad. Medicine, Bristol-Myers award for disting. rsch. in neurosci., 1989, Thudicum medal Brit. Biochem. Soc. (lectr.), 1989, Gerard medal Soc. Microsci., 1991. Fellow AAAS, Am. Acad. Arts and Scis., Am. Soc. Neuropsychopharmacology; mem. German Pharmacol. Soc. (corr.), Am. Chem. Soc., Am. Soc. Pharmacology and Exptl. Therapeutics (Torald Sollmann award 1973), Nat. Acad. Scis., Am. Neurol. Assn. (hon.), Royal Soc. London (fgn.), Inst. Medicine (sr.), Deutshe Academie Naturfoucher (East Germany) Sigma Xi, Am. Psychopathol. Assn. (hon.). Home: 10401 Grosvenor Pl Rockville MD 20852-4646 Office: NIH Dept Health Edn and Welfare Bldg 36 Rm 3A-15 Bethesda MD 20892

AXFORD, ROY ARTHUR, nuclear engineering educator; b. Detroit, Aug. 26, 1928; s. Morgan and Charlotte (Donaldson) A.; m. Anne-Sofie Langfeldt Rasmussen, Apr. 1, 1954; children: Roy Arthur, Elizabeth Carole, Trevor Craig Charles. B.A., Williams Coll., 1952; B.S., Mass. Inst. Tech., 1952, M.S., 1955, Sc.D., 1958. Supr. theoretical physics group Atomics Internat., Canoga Park, Calif., 1958-60; assoc. prof. nuclear engring. Tex. A&M, 1960-62, prof., 1962-63; assoc. prof. nuclear engring. Northwestern U., 1963-66; assoc. prof. U. Ill., Urbana, 1966-68, prof., 1968—; cons. Los Alamos Nat. Lab., 1963—. Vice-chmn. Mass. Inst. Tech. Alumni Fund Drive, 1970-72, chmn., 1973-75; sustaining fellow MIT, 1984. Recipient cert. of recognition for excellence in undergrad. teaching U. Ill., 1979, 81; Everitt award for teaching excellence, 1985. Mem. AIAA, ASME, Am. Nuclear Soc. (Excellence in Undergrad. Teaching award 1990, Disting. faculty Alpha Nu Sigma 1991), SAR (sec.-treas. Piankeshaw chpt. 1975-81, v.p. chpt. 1982-83, pres. chpt. 1984-86), Kiwanis (charter life patron fellow 1992), Sigma Xi, Phi Kappa Phi, Tau Beta Pi. Home: 2017 S Cottage Grove Ave Urbana IL 61801-6353

AXON, KENNETH STUART, chemist; b. Manchester, Eng., Jan. 1, 1940; s. Frederick Alan and Ada Marjorie (Baldwin) A.; m. Margaret Jones, July 27, 1963; children: Katherine Jane, Joanna Elizabeth, Timothy Alan. MA honors, Cambridge U., Eng., 1962. Prodn. mgr., devel. engr. Bakelite Xylonite Ltd., Manningtree, Essex, Eng., 1962-70; sr. cons. Inbucon-AIC Mgmt. Cons., London, 1970-75; gen. mgr. Harshaw Chems. Ltd., Glasgow, Scotland, 1975-80; group mng. dir. Daniel C. Griffith & Co. Ltd., Witham, Essex, Eng., 1980-85; mng. dir. Associated Lab. Svcs. Ltd., Bocking, Essex, Eng., 1985—; bd. dirs. Cotecna Inspection SA, Geneva, 1985—, Kitto Labs. Ltd., Bocking, 1987—. Avocations: mountaineering, sailing, running, music. Home: Wych Elm Mayes Ln, Danbury CM3 4NJ Essex, England Office: Associated Lab Svcs Ltd, Christy House Church Ln, Bocking CM7 5RX Essex, England

AYAD, JOSEPH MAGDY, psychologist; b. Cairo, Egypt, May 21, 1926; s. Fahim Gayed and Victoria Gabour (El-Masri) A.; came to U.S., 1949, naturalized, 1961; B.A. in Social Scis., Am. U., Cairo, 1946; M.A. in Clin. Psychology (Univ. scholar), Stanford U., 1952; Ph.D. in Clin. Psychology (Univ. scholar), U. Denver, 1956; m. Widad Fareed Bishai, May 29, 1954; children—Fareed Merritt, Victor Maher, Michael Joseph, Mona Elaine. Lectr., Fitzsimmons Army Hosp., Denver, 1953-54; staff psychologist Cons. Psychol. Services, Denver, 1954-55; psychologist, Denver, 1956-57, High Plains Neurol. Center, Amarillo, Tex., 1957—. Pres. JMA Cattle Co., Amarillo, 1973—; v.p., treas. Filigon Inc., Amarillo, 1962-75, pres., 1976—; cons. psychologist Tex. Dept. Pub. Welfare. Mem. profl. adv. bd. Amarillo Mental Health Assn., 1968-69. Mem. Amarillo Child Welfare Bd., 1961-63; area chmn. U. Denver Fund Raising Campaign, 1963; mem. profl. adv. bd. St. Paul's Meth. Ch. Sch. for Children with Learning Disabilities, Amarillo, 1969-70. Recipient Grad. Sr. award in Philosophy Am. U. at Cairo, 1946. Mem. Am. Psychol. Soc., Am. Psychol. Assn., Internat. Assn. Applied Psychology, Am. Assn. Marriage and Family Therapists, Potter-Randall County (Tex.) Psychol. Soc. (pres. 1974), Tex. Psychol. Assn., Calif. Psychol. Assn. Presbyn. Club: Amarillo Country. Contbr. articles to profl. jours. Home: 4239 Erik Ave Amarillo TX 79106-6008 Office: 2301 W 7th Ave Amarillo TX 79106-6601

AYALA, JUAN ALFONSO, molecular biology research, biochemistry educator; b. Cartagena, Spain, Nov. 3, 1951; s. Emilio Ayala and Ines Serrano; m. Dolores Vicente, July 14, 1979; children: Emilio, Elena, Laura. Lic., U. Complutense, 1974, PhD, 1978. Cert. sci. researcher. Dept. head Labs. Normon S.A., Madrid, 1978-79; postgrad. U. Paris-SUD, 1980-81; postdoct. fellow Centro Biologia Molecular, Madrid, 1982-85, researcher, 1986—; cons. Prensa Cientifica, Madrid, 1987—. Author: The Target of Penicillin, 1983, The Antibiotics, 1985; contbr. articles to profl. jours. Recipient Best Fellow award Ministerio Edn., Madrid, 1974. Mem. Am. Chem. Soc., Am. Soc. Microbiology, Spanish Soc. Biochemistry, N.Y. Acad. Sci. Office: Centro Biologia Molecular, U A M Canto-Blanco, 28049 Madrid Spain

AYBAR, CHARLES ANTON, aviation executive; b. N.Y.C., Sept. 27, 1956; s. Louis Adolf and Elisabeth A. (Schwarz) A.; m. Deborah Ann Benson, May 1, 1988; 1 child, Heidi Brita. AS in aeronautics, Embry-Riddle Aero. U., 1987; BS in Aviation Mgmt., Pacific-Western U., 1988, MBA in Mktg., 1988, PhD in Mktg., 1993. Lic. airline transport pilot; cert. FAA flight instr. and aircraft dispatcher. Mdse. mgr. Korvettes, Inc., N.Y.C., 1976-79; gen. mgr. Family Games Ctr., Inc., Bklyn., 1979-81; pres. N.Am. Sch., Inc., Bklyn., 1979-81; exec. dir. of acad. Laces, Inc., New Hyde Park, N.Y., 1981-85; pilot Air Sedona (Ariz.) Airlines, Inc., 1986-88; v.p. Ruidoso (N.Mex.) Airlines, Inc., 1988; pres. S&S Aircraft, Inc., Plant City, Fla., 1989-91; dir. flight ops. Plant City Airport, Inc., 1991-92; chief flight instr. airline prep. program Scottsdale (Ariz.) Aeromech, Inc., 1992—; written test examiner FAA, Orlando, Fla., 1989—, accident prevention counselor, 1991—; cons. Laces, Inc., 1981-84, Hillsborough Aviation Authority, Tampa, Fla., 1989—, CBS TV Network, 1979-80. Inventor children's toothpaste; contbr. articles to periodicals. Founder Children's Air Mus., Ariz., 1993. Recruiter USAF-CAP, Prescott, Ariz., 1987. Named Flight Instr. of Yr. FAA, Orlando, Fla., 1990. Mem. Tampa Bay Super Bowl Task Force, Assn. Ind. Airmen, Roller Skating Rink Operators Assn. Avocations: roller skating, ice skating, swimming, electronics. Home: 5640 E Bell Rd Townhouse 1073 Scottsdale AZ 85254 Office: Scottsdale Aeromech Inc 14605 N Airport Dr Scottsdale AZ 85260

AYBAR, ROMEO, architect; b. Buenos Aires, Argentina, Feb. 8, 1930; came to U.S., 1960, naturalized, 1965; s. Aristobulo Romeo and Maria Sara (Figoli) A.; m. Rose Delia Caceres, Oct. 18, 1954; children: Patricia Monica Aybar Smith, Viviana Sylvia Aybar Pugaczewski, Cynthia Jenny Aybar Giordano. B.Arch., U. Buenos Aires, 1954. Lic. architect, N.J., Pa., Del., Md., Vt., Va.; registered planner, N.J.; cert. Dept. Def. fall-out shelter analyst. Pvt. practice architecture Buenos Aires, 1955-60; sr. draftsman Widersum Assocs., N.Y.C., 1960-61; job capt. Mahony Troast, Clifton, N.J., 1961-63; project mgr. R. Cadien Architect, Cliffside Park, N.J., 1963-67; ptnr. Cadien & Aybar, Cliffside Park, N.J., 1968-69; sr. ptnr. The Aybar Partnership-Architects and Planners, Ridgefield, N.J., 1969—; organizer, dir. First Fed. Bank, Clifton, N.J.; mem. adv. bd. archtl. drafting course The Plaza Sch., Paramus, N.J., 1971—; lectr. Ft. Lee High Sch., Ridgefield High Sch., N.J. Sch. Architecture, N.J. Inst. Tech., others; mem. adj. faculty Montclair State Coll., 1971-74. Mem. Indsl. Safety Council N.J., 1973-78; mem. Ridgefield Zoning Bd. Adjustments, 1969-71, chmn., 1972-73; mem. Hudson Riverfront Planning Commn. State of N.J., 1979-81; acting bldg. insp. City of Ridgefield, 1968; mem. planning commn. Ellis Island & Statue of Liberty Restoration Master Plan, 1979-80. 1st lt., pilot N.J. wing CAP, 1978. Recipient Dir.'s award Architects League N.J., 1971, 84, 86, Vegliante Meml. award Architects League N.J., 1973, Outstanding Excellence in Design award N.J. Soc. Architects, 1971, 73, citation for Outstanding Svcs. Am. Concrete Assn., 1979; named Jerseyan of Week Star Ledger Publs., 1979. Fellow AIA (N.J. regional dir. 1981-83, 125th Anniversary Presdl. citation 1982, Presdl. citation 1982, citation Dedicated Svcs. 1983); mem. Archtl. League No. N.J. (pres. 1975), N.J. Soc. Architects (dir. 1971, established Romeo Aybar Scholarship 1973, treas. 1974-75, pres. 1979, award of Honor 20 Yrs. Meritorious Accomplishments 1988, regional dir. 1983), Aircraft Owners and Pilots Assn., Nat. Pilots Assn. Republican. Club: Ridgefield Exchange (pres.) (1972-73); Ridgefield Exchange (dir. N.J. dist.) (1973-74). Home: 2150 Center Ave Fort Lee NJ 07024-5806 Office: Aybar Partnership 605 Broad Ave Ridgefield NJ 07657-1604

AYERS, JACK DUANE, metallurgist; b. Nampa, Idaho, Feb. 25, 1941; s. Robert Henry and Ruth (Nowland) A.; m. Sharon Ann Denny, Feb. 23, 1963; children: Gayle Marie, Bryan Douglas, Stuart Duncan. BS in Metall. Engring., U. Wash., 1966; PhD, Carnegie-Mellon U., 1970. Metallurgist Naval Rsch. Lab., Washington, 1971-78, supervisory metallurgist, 1978—. Contbr. articles to profl. jours. With USNR, 1960-62. NATO postdoctoral fellow, Oxford U., 1971. Mem. The Materials Soc., Am. Soc. for Materials Internat. Achievements include 7 patents. Office: Naval Rsch Lab 4555 Overlook Ave Washington VA 20375

AYERS, JOSEPH WILLIAMS, chemical company executive; b. Easton, Pa., Jan. 6, 1904; s. Charles Pierson and Emma Cottman (Williams) A.; B.Chemistry, Cornell U., 1927, postgrad., 1927-28; m. Caroline Brooke Stone, Oct. 6, 1934; children: Katherine, Phyllis Ayers Harmon. Research dir. C.K. Williams & Co., Easton, 1930-48, v.p., 1945-62; gen. mgr. minerals pigments and metals div. Pfizer, Inc., N.Y.C., 1962-68, pres., 1968-69; pres. Calcium Chem. Corp., Adams, Mass., 1937-48; pres. Agrashell, Inc., Los Angeles, 1939-76, chmn. bd., 1976—; pres. J.W. Ayers, & Co., Easton, 1955-61, The Ayers Co., Easton, 1971—; pres., treas. Joseph Ayers, Inc., Bethlehem, Pa., 1973-86; dir. New Eng. Lime Co., Foote Mineral Co. Chmn. Pa. Hosp. and Health Council, 1971-74; v.p. Northampton County Citizens for Regional Progress, 1973-78; trustee Emma Willard Sch., 1957-65, Greater Valley Council Girl Scouts U.S.; treas. Bach Choir Bethlehem, Pa, 1976-84, now trustee. Fellow Am. Inst. Chemists, N.Y. Acad. Scis., Am. Inst. Chem. Engrs.; mem. AAAS, ASTM (D-1 com. 1932-48), Am. Chem. Soc., Soc. Chem. Industries (Eng.), N.Y. Chemist Club, Zeta Psi. Republican. Presbyterian. Clubs: Cornell, Union League (N.Y.C.); Northhampton County Country; Pomfret (Easton); Oyster Harbors, Wianno, Wianno Yacht (Osterville, Mass.); Beach (Craigsville, Mass.), Los Angeles Athletic. Contbr. articles to profl. jours. Patentee in field. Home: 22 N 14th St Easton PA 18042-3216 Office: 5934 Keystone Dr Bath PA 18014-8848

AYLESWORTH, JOHN RICHARD, software professional; b. Manhattan, N.Y., July 9, 1962; s. John Banzley and Nancy Lee (Eberle) A.; m. Natalie Jane Herrebrugh, June 20, 1987. AA in Bus., Santa Monica Coll., 1987. Asst. mgr. tech. support Arrays, Inc./Continental Software, L.A., 1984-86; prod. support engr. C.ITOH Electronics, Irvine, Calif., 1986-88; supr., software support Ashton-Tate Corp., Torrance, Calif., 1988-91; product group mgr. Borland Internat., Scotts Valley, Calif., 1991—. Officer Civil Air Patrol, USAF Aux., 1983—. Democrat. Lutheran. Home: 1324 S Winchester Bl # 72 San Jose CA 95128

AYLWARD, GLEN PHILIP, psychologist; b. Hoboken, N.J., July 30, 1950; s. Philip E. and Marion E. (Attubato) A.; m. Deborah E. Bellini, Sept. 9, 1973; children: Shawn C., Megan B., Brandon S. BA, Rutgers U., 1972; MA, Fairleigh Dickinson U., 1974; PhD, Ga. State U., 1979. Diplomate Nat. Register of Health Svc. Providers in Psychology. Asst. prof. psychology So. Ill. U., Springfield, 1979-86, assoc. prof. psychology, 1986—, dir. Div. Developmental and Behavioral Pediatrics, 1984—; cons./data policy adv. bd. NIH-NHLBI, Bethesda, Md., 1977-84. Author: Developmental and Psychological Testing: A Reference Handbook for Practitioners, 1993; author (test) Bayley Infant Neurodevelopmental Screen, 1993; contbr. articles to profl. jours. Fellow APA (Div. 37 treas.), Am. Acad. Cerebral Palsy and Developmental Medicine; mem. Soc. for Behavioral Pediatrics (exec. coun.). Office: So Ill Univ Sch Medicine/Pediatrics PO Box 19230 Springfield IL 62794-9230

AYOUB, AYOUB BARSOUM, mathematician, educator; b. Cairo, Egypt, May 22, 1931; came to U.S., 1975; s. Barsoum Ayoub and Linda (Naguib) Rizk; m. Germaine Hozayen Saad, Feb. 5, 1972; children: Sameh, Mariane. BSc in Math., Ain-Shams U., Cairo, 1951; MA in Math., Temple U., Phila., 1977, PhD in Math., 1980. Tchr. Tawfikia High Sch., Cairo, 1951-55; instr. Ain-Shams U., Cairo, 1955-75; teaching asst. Temple U., Phila., 1975-77, rsch. asst., 1977-80, vis. asst. prof., 1982-83; asst. prof. Ain-Shams U., Cairo, 1980-82; asst. prof. math. Pa. State U.-Ogontz Campus, Abington, 1983-90; assoc. prof. Pa. State U., Abington, 1990—, coord. math. dept., 1992—. Referee Math. Mag., Coll. Math. Jour., referee, reviewer Math. Tchr.; contbr. articles to profl. jours. Chmn. United Way Campaign, Pa. State, Ogontz Campus, 1990. Mem. Math. Assn. Am., Nat. Coun. Tchrs. Math., Pa. State Math. Assn. Two Yr. Colls., Pa. Coun. Tchrs. Math., Assn. Math. Tchrs. N.Y. State. Avocation: travel. Office: Pa State U Ogontz Campus Abington PA 19001

AYRES, RAYMOND MAURICE, mechanical engineer; b. Leicester, Eng., Aug. 25, 1945; s. Maurice Harold and Kathleen Mary (Young) A.; m. Claire Pidgeon, Dec. 16, 1979. BSc, Nottingham (Eng.) U., 1967; MSc, London U., 1969; PhD, Leicester (Eng.) U., 1979. cert. chartered engineer. Student apprentice A.A. Jones-Shipman Ltd., Leicester, 1964-69; devel. engr. Rolls-Royce Ltd., Derby, Eng., 1969-71; rsch. fellow U. Leicester, 1971-75; devel. engr. London Borough of Hammersmith, 1975-79; cons. Systems Designers Ltd., Camberley, Eng., 1979-90; tech. dir. Silicon Valley Systems Ltd., Camberley, Eng., 1990—; dir. Eayre & Smith Ltd., Derby. Freeman City of Leicester, 1966—. Mem. AIAA, Inst. Mech. Engrs., Brit. Inst. Mgmt., Coun. Engring. Instns., Chartered Engrs. Achievements include development of parachute aerodynamic theory; world authority on the tuning theory and dynamic structural analysis of bells rung for change-ringing; consultation in application of computers to command and control for air defence and air traffic control applications; integrated logistical support of large high technology systems. Home: 239 Upper Chobham Rd, GU15 1HB Camberley Surrey, England Office: Silicon Valley Systems Ltd, High St Sandhurst, GU17 8DY Camberley Surrey, England

AYYADURAI, V.A. SHIVA, software engineer; b. Bombay, India, Dec. 2, 1963; came to the U.S., 1970; s. Vellayappa and Meenakshi (Sivasundarum) A. MS, MIT, 1990, postgrad., 1990—. Sr. software engr. Lotus Devel. Corp., Cambridge, Mass., 1986-90; sr. scientist Dataware Techs., Cambridge, 1990—; participant various confs., Seattle. Mem. IEEE (chmn. sci. visualization group 1990-91), Sigma Xi (Outstanding Achievement award), Tau Beta Pi, Phi Delta Theta. Achievements include development of first user-friendly networkwide electronic mail system. Home: 440 Massachusetts Ave Ste 6 Cambridge MA 02139 Office: MIT 77 Massachusetts Ave Rm 3-3637 Cambridge MA 02138

AYYUB, BILAL MOHAMMED, civil engineering educator; b. Shweikeh, Tulkaram, Palestine, Jan. 5, 1958; came to U.S., 1980; s. Mohammed S. and Thuraya Ayyub; m. Deena L. Ziadeh, June 27, 1987; children: Omar, Rami, Samar. BSCE, U. Kuwait, 1980; MSCE, Ga. Inst. Tech., 1981, PhD, 1983. Registered profl. engr., Md. Asst. prof. dept. civil engring. U. Md., College Park, 1983-88, assoc. prof., 1988-93; prof. U. Md., 1993—; cons. USCG, Groton, Conn., 1988—, USN, Crystal City, Va., 1990—, ASME, Washington, 1990—; mem. adv. bd. to internat. jours. and Naval Engrs. Jour., 1989—; gen. chmn. Internat. Symposium on Uncertainty Modeling and Analysis, 1990, 93. Editor: Analysis and Management of Uncertainty, 1992. Grantee NSF, Washington, 1985—, Md. State Hwy. Adminstrn., Balt., 1986-90, USN, Washington, 1990. Mem. ASCE (Outstanding Rsch. Oriented Paper award 1988, Edmund Friedman award 1989), ASME (polit. action com. 1990—), Am. Soc. Naval Engrs. (life, Jimmie Hamilton award 1986, 1993), Soc. Naval Architects and Marine Engrs., Am. Concrete Inst. Achievements include risk and uncertainty analysis in engineering, design guidelines for posttensioned composite bridges, gen. guidelines for risk-based inspection, structural reliability assessment using variance reduction techniques, uncertainty modeling and analysis in engring., fuzzy logic in civil engring. Office: U Md Dept Civil Engring College Park MD 20742

AZAM, FAROOQ, chemist, researcher; b. Punjab, Pakistan, June 17, 1956; came to U.S., 1981; s. Abdul and Khurshid (Begum) Hameed. MS, Quaid.E.Azam U., Pakistan, 1981, U. Scranton, 1983; PhD, SUNY, Albany, 1993. Rsch. assoc. U. Scranton, Pa., 1981-83; analytical chemist Roco Medilabs, N.Y.C., 1983-88; rsch. assoc. N.Y. State Dept., Albany, 1988—; acad. com. Sch. Pub. Health, SUNY, Albany, 1988-90. Rep. Pakistan Assn., Albany, 1990—; pres. Pakistan Student Assn., SUNY, Albany, 1992—. Muslim. Office: NY State Dept Labs and Rsch Div Empire State Plz Albany NY 12201

AZARKHIN, ALEXANDER, mechanical engineer, mathematician; b. Kharkov, Ukraine, Sept. 22, 1939; came to U.S., 1981; s. Moisei and Bertha (Dvorchik) A.; m. Alla Georgiev, June 17, 1963 (div. Apr. 1985); 1 child, Olga Hedler. DSc in Civil Engring., Transp. Inst., Novosihirsk, USSR, 1967; PhD in Applied Mechanics, U. Mich., 1985. Rsch. assoc. Metallurgy Rsch. Inst., Kharkov, USSR, 1967-74; prof. Polytech. Inst., Krasnoyarsk, USSR, 1974-81; engring. assoc. Aluminum Co. Am., Alcoa Center, Pa., 1985—; editorial cons. Am. Math. Soc., Ann Arbor, Mich., 1982—. Rackham fellow U. Mich., Ann Arbor, 1984. Mem. ASME, Sigma Xi. Achievements include research on convergence of iterative methods in elasticity theory, plastic deformations for combined loading with free surfaces, various wear and friction mechanisms, contact problems and cutting tool modeling. Home: 118 McMasters Dr Monroeville PA 15146 Office: Aluminum Co Am Rd 780 Alcoa Center PA 15069

AZARYAN, ANAHIT VAZGENOVNA, biochemist, researcher; b. Yerevan, Armenia, Jan. 9, 1950; came to U.S., 1991; d. Vazgen Kh. and Dazy T. (Mirzoyan) A.; m. David B. Akopian, Jan. 30, 1981; 1 Child, Tigran. MD, Med. Sch., Yerevan, 1972; PhD, Inst. Molecular Biology, Moscow, 1979; Dr. Sc., Inst. Biochemistry, Yerevan, 1988. Prin. investigator Inst. Biochemistry, Yerevan, 1980-84, head proteolysis group, 1984-90; rsch. program vis. scientist Nat. Inst. Child Health and Human Devel./NIH, Bethesda, Md., 1991; rsch. assoc. bochemistry dept. Uniformed Svcs. Univ. of Health Scis., Bethesda, 1992—. Author: Brain Peptide Hydrolases and Their Biological Functions, 1989; contbr. articles to Jour. Biol. Chem., Neurochem. Rsch., Jour. Neurosci. Rsch., others. Recipient fellowship Martin Luther U., Halle, Germany, 1983, fellowship N. Kline Inst. Psychiat. Rsch., N.Y.C., 1988, Travel award Internat. Soc. Neurochem.-FIDIA, Washington, 1991; Fogarty Internat. Ctr. Rsch. fellow, 1991. Mem. Internat. Soc. for Neurochemistry, European Soc. for Neurochemistry, Am. Soc. for Neuroscience, N.Y. Acad. Scis. Achievements include characterization of ATP-Ubiquitin-dependent protease in brain proves the existence of cytosolic ATP-Ub-dependent proteolysis pathway in that tissue; characterization of YAP3, a novel yeast aspartic potease involved in the activation of prohormones and proneuropeptides.

AZHAR, BARKAT ALI, economic adviser, researcher; b. Faisalabad, Punjab, Pakistan, June 30, 1927; s. M. Mahabat Khan; m. Mahmuda Azhar, May 23, 1956; 1 child, Rafay. BSc, Agrl. Coll., Lyallpur, Punjab, 1948; MS, U. Ill., 1951, MA, 1952, PhD, 1954. Prof. econs. Govt. Coll., Khairpur, 1955-57; resident economist Fin. Svcs. Acad., Lahore, 1957-64, dept. dir., 1964-65; dean agrl. econs. U. Agr., Faisalabad, 1965-66; dir. rsch. Cen. Bd. Revenue, Islamabad, 1966-71; joint econ. advisor Govt. of Pakistan, Islamabad, 1971-74; sr. fiscal adviser UNDP/IBRD, Khartoum, Sudan, 1974-

84; dir. rsch. Cen. Bd. Revenue, Islamabad, 1984-87; mng. dir. Ctr. for Devel. Studies, Lahore, 1987—; mem., sec. Taxation Commn., Islamabad, 1971-74; expert, agr. taxation Fact Finding Com., Lahore, 1962-63; mem. Study Group on Mobilization of Rural Incomes, Islamabad, 1968-69, Agrl. Policy, Islamabad, 1969-70. Editor Pakistan Agrl. Dev. Rev., 1988—; contbr. articles to profl. jours. Sec. Econ. Welfare Com., Islamabad, 1966, Budget Adv. Coun., Islamabad, 1972; rsch. adviser Pakistan Inst. Devel. Econs., Islamabad, 1971-74; mem. acad. coun. U. Agr., Faisalabad, 1965-73. Booth fellow U. Ill., 1951-54, Cento fellow Brit. Coun., 1961. Mem. Pakistan Soc. Agrl. Econs., Pakistan Fiscal Assn., Pakistan Assn. Agrl. Social Scientists, Pakistan Econ. Assn. (v.p. 1973-77), Punjab Econ. Rsch. Inst. (bd. govs. 1990—), Gamma Sigma Delta. Avocations: reading, writing, research, group discussions, meditation. Home and Office: 37-A/2 Gulberg-III, Lahore Pakistan

AZIZ, KHALID, petroleum engineering educator; b. Bahawalpur, Pakistan, Sept. 29, 1936; came to U.S., 1952; s. Aziz Ul and Rshida (Atamohammed) Hassan; m. Mussarrat Rizwani, Nov. 12, 1962; children: Natasha, Imraan. BS in Mech. Engring., U. Mich., 1955; BSc in Petroleum Engring., U. Alta., 1958, MSc in Petroleum Engring., 1961; PhD in Chem. Engring., Rice U., 1966. Jr. design engr. Massey-Ferguson, 1955-56; various position to asst. prof. petroleum engring. U. Alta., 1960-62; various positions, chmn. bd. Neotech. Cons. Ltd., 1972-85; mgr., dir. Computer Modelling Group, Calgary, Alta., 1977-82; various positions to chief engr. Karachi (Pakistan) Gas Co., 1958-59, 62-63; various positions to prof. chem. and petroleum engring. U. Calgary, 1965-82; prof. petroleum engring. dept. Stanford (Calif.) U., 1982—, assoc. dean rsch. Sch. Earth Scis., 1983-86, chmn. petroleum engring. dept., 1986-91, Otto N. Miller prof. in earth scis., 1989. Author: Flow of Complex Mixtures in Pipes, 1972, Petroleum Reservoir Simulation, 1979; contbr. articles to profl. jours. Recipient Diploma of Honor, Pi Epsilon Tau, 1991; Chem. Inst. Can. fellow, 1974, Killam Resident fellow U. Calgary, 1977. Mem. AICE, Soc. Petroleum Engrs. (disting., Ferguson award 1979, Reservoir Engring. award 1987, Lester C. Uren award 1988, Disting. Achievement award for Petroleum Engring. faculty 1990), Soc. Indsl. and Applied Math., Assn. Profl. Engrs., Geologists and Goephysicists Alta, Sigma xi. Muslim. Achievements include rsch. in multiphase flow of oil/gas mixtures & steam in pipes & wells, multiphase flow in porous media, reservoir simulation (black-oil, compositional, thermal, geothermal), natural gas engring., hydrocarbon fluid phase behavior. Home: 112 Peter Coutts Circle Stanford CA 94305 Office: Stanford U Dept Petroleum Engring Stanford CA 94305

AZIZ, SALMAN, chemical engineer, company executive; b. Lahore, Pakistan, Apr. 2, 1955; came to U.S., 1977; s. Abdus and Suriya Salam; m. Aisha (Teresa) Hennes, Mar. 26, 1982; children: Saad, Amna, Hassan. B-SChemE, U. Wash., 1979, MS in Pulp and Paper Engring., 1981. Rsch. scientist U. Wash., Seattle, 1981-82; process engr. Biol. Energy Corp., Valley Forge, Pa., 1982-85; mgr. pulping and bleaching Inst. Paper Chemistry, Appleton, Wis., 1985-89; v.p. tech. Integrated Paper Svcs., Inc., Appleton, 1989—. Contbr. articles to profl. jours. Mem. adv. com. Montessori Adventure Schs., Appleton, 1986—. Grantee Wis. Dept. Natural Resources, 1991-92. Mem. TAPPI (bd. dirs. Lake State chpt. 1987-89, mem. alkaline com. 1990—, v.p. solvent pulping com. 1991—, bd. dirs. TMCC 1991—). Islam. Achievements include evolution of commercial feasibility of solvent pulping technology; environmentally friendly bleaching technologies, simplification of extended kraft pulping, deinking of waste paper with solvents, use of expert system in paper processing. Home: 14 Sunray Ct Appleton WI 54915 Office: Integrated Paper Svcs 101 W Edison Ave Ste 250 Appleton WI 54912

AZIZKHAN, RICHARD GEORGE, pediatric surgeon; b. London, Aug. 10, 1953; came to U.S., 1964; s. Reza George and Helga Marianne (Behnke) A.; m. Jane Elizabeth Clifford, May 8, 1976; children: Richard Anthony, Kathryn Marie, Christine Elizabeth Ann. BS with honors, Dickinson Coll., Carlisle, Pa., 1972; MD, Pa. State U., 1975. Diplomate Am. Bd. Surgery in Pediatric Surgery and Surgical Critical Care. Resident in surgery U. Va., Charlottesville, 1976-78, 80-83; rsch. fellow in pediatric surgery Harvard Med. Sch. Boston Children's Hosp., 1978-80; fellow in pediatric surgery Johns Hopkins Univ., Balt., 1983-85; chief pediatric surgery U. N.C., Chapel Hill, 1985-93; surgeon-in-chief Child Hosp., Buffalo, 1993—; mem. surgical adv. bd. Smith, Kline, Beecham, Phila., 1990—. Author (med. text) Congential Malformations: Prenatal Diagnosis and Management, 1990; contbr. over 70 articles to profl. jours. Recipient Upjohn Achievment award, U. Va. Sch. Medicine, 1981; Schering scholarship, Am. Coll. Surgeons, Chgo., 1982, Hugh J. Warren Teaching award U. Va., Sch. Medicine, Charlottesville, 1983, Smith, Kline & French fellowship, Am. Coll. Surgeons, 1986, Battle Disting. Excellence in Teaching award, U. N.C. Sch. Medicine, Chapel Hill, 1988. Fellow Am. Coll. Surgeons, Am. Acad. Pediatrics, Am. Pediatric Surgery Assn. (program com. 1990-93); mem. Assn. of Acad. Surgery (exec. coun. 1986-89), Alpha Omega Alpha. Roman Catholic. Achievements include research that helped illucidate the importance of heparin in the growth of new blood vessels (angiogenesis); development of novel technique utilizing fiberoptic laser to treat bronchial stenosis in infants. Office: Childrens Hosp Buffalo Burnett-Womack 229 H 219 Bryant St Buffalo NY 14222

AZOLA, MARTIN P., civil engineer, construction manager; b. Elmhurst, Ill., Jan. 12, 1947; s. Joseph Ramon and Lillian Alice (Zeeman) A.; m. Lone Tidemand, June 29, 1968; children: Anthony, Matthew, Kirsten. BSCE, Va. Polytechnic Inst., 1968, MSCE, 1969. Registered profl. engr., Md. V.p. J.R. Azola and Assocs., Balt., 1973-80, pres. M.P. Azola, Inc., Balt., 1980-90, Azola and Assocs., Inc., Balt., 1990—; commr. Balt. County Landmarks Com., Towson, Md., 1983-93; chmn. Nat. Remodelers Coun., Washington, 1988-89; instr. Goucher Coll., Towson, 1992-93. Contbr. articles to profl. jours. Pres. Charles St. Assoc., Balt., 1978; dir. Soc. For Preservation of Md. Antiques, Balt., 1988; trustee Children's Hosp., Balt., 1992. Recipient Preservation Svc. award State of Md., 1988, Profl. Achievement award Profl. Builder, Washington, 1987, Renaissance Grand award Nat. Assn. Home Builders, Washington, 1991; named Builder of Yr. Home Builders Assn. Md., 1988, Remodeler of Yr., 1991. Mem. ASCE, Nat. Soc. Profl. Engrs., Soc. Am. Mil. Engrs., Engring. Soc. Balt. (dir. 1980), Omicron Delta Kappa (pres. 1967), Chi Epsilon. Office: Azola and Assocs PO Box 140 Brooklandville MD 21022

AZOPARDI, KORITA MARIE, mathematics educator; b. Galveston, Tex., Jan. 30, 1941; d. John and Cecil Marie (Kierbow) Hamilton; m. Benny Lee Azopardi, May 27, 1960; children: Connie, B. Lee Jr. BS in Edn. and Math., S.W. Tex. State U., 1961; MA in Math., St. Louis U., 1970. Tchr. Navarro Pub. Schs., Geronimo, Tex., 1961-62, Victoria (Tex.) Ind. Sch. Dist., 1962-65, St. Louis Ind. Sch. Dist., 1965-67, Normandy Ind. Sch. Dist., St. Louis, 1967-69, Mary Inst., St. Louis, 1969-71, Corpus Christi (Tex.) Ind. Sch. Dist., 1979-84, Calallen Ind. Sch. Dist., Corpus Christi, 1984—. Mem. NEA, Nat. Coun. Tchrs. of Math., Math. Assn. Am., Tex. Coun. Tchrs. of Math., Tex. Tchrs. Assn., Tex. Computer Edn. Assn. Republican. Presbyterian. Avocation: ranching. Home: 4705 Gayle Dr Corpus Christi TX 78413-3322 Office: Calallen Ind Sch Dist 4001 Wildcat Dr Corpus Christi TX 78410-5197

AZUMA, TAKAMITSU, architect, educator; b. Osaka, Japan, Sept. 20, 1933; s. Yoshimatsu and Yoshiko (Ikeda) A.; m. Setsuko Nakaoka, Mar. 17, 1957; 1 child, Rie. BArch, Osaka U., 1957, DArch, D. of Engring, 1985. Designer, Ministry of Postal Svcs., Osaka, 1957-60; chief designer Junzo Sakakura Architect & Assocs., Osaka, 1960-63; chief designer Junzo Sakakura Architect and Assocs., Tokyo, 1963-67; prin. Takamitsu Azuma Architect & Assocs., Tokyo, 1967-85; instr. Tokyo U. Art and Design, 1976-78, Tokyo Denki U., 1980-82, Tokyo U., 1983-85; instr. Osaka U., 1981-85, prof., 1985—; instr. Osaka Art U., 1985-87; architecht Azuma Architects & Assocs., 1985—; vis. prof. Sch. Architecture Washington U., St. Louis, 1989. Recipient 1st prize Kinki Br., Japan Inst. Architects Competition, 1957. Mem. Archtl. Inst. Japan, Japan Architects Assn. Author: Reevaluation of the Residence, 1971, On the Japanese Architectural Space, 1981, Takamitsu Azuma-Contemporary Japanese Architects Series, 1982, Philosophy of Living in the City, 1983, Device from Architecture, 1986, Space Analysis of Urban Residence, 1986, 100 Chpt. for Children's Place, 1987, White Book about Tower House, 1987. Home: 3-39-4 Jingumae, Shibuya-ku, Tokyo 150 Japan Office: Azuma Architects & Assocs, 3-6-1 Minami-Aoyama Minato-ku, Tokyo 107, Japan

AZZI, DANIEL W., mathematician; b. Tripoly, Feb. 25, 1966; s. Amal A. and Nelly (Issa) A. BS, Am. U. Beirut, 1986; MS, U. Ill., 1989. Software engr. U.S. Army C.E., Champaign, Ill., 1987-90, researcher, 1990—; teaching asst. Am. U. Beirut, Lebanon, 1986-87. Contbr. articles to profl. jours. Vol. tchr. of gifted students Benjamin Franklin Sch., Champaign, 1990—. Recipient rsch. grant U.S. Army Corp Engrs., 1990-92. Mem. Am. Math. Soc. Home: 1304 Christopher Cir # 4 Urbana IL 61801-1282 Office: Math Dept Univ Ill 273 Altgeld Hall Green St Urbana IL 61801

AZZOPARDI, MARC ANTOINE, astrophysicist, scientist; b. Philippeville, Algeria, Oct. 28, 1940; s. Antoine Philippe and Marie Madeleine (Grech) A.; m. Aug. 27, 1966; children: Pauline, Mathilde, Marceau. Lic. es sci., Alger and Marseilles U., Algeria, 1963; DSc, Toulouse U., France, 1981. Astronomy aide U. Toulouse, 1964-81, adj. astronomer, 1981-83; sci. assoc. European South Obs., Garching, Fed. Republic of Germany, 1983-87; sci. adv. coun. Can.-France-Hawaii Telephone Co., Garching; chmn. user's com. European South Obs., Garching; astronomer 2nd class U. Marseilles, France, 1987-92, astronomer 1st class, 1992—; guest prof. European So. Obs., Garching, 1992-93; vis. scholar U. Tex., Austin, 1982-83. Contbr. articles to profl. jours. Recipient NSF-Ctr. Nat. Sci. Rsch. award, 1982-83. Mem. Soc. French Specialists in Astronomy, Am. Astron. Soc., European Astron. Soc. Roman Catholic. Office: Obs Marseilles, 2 Place Le Verrier, Marseilles 13248, France

BAALMAN, ROBERT JOSEPH, biology educator; b. Grinnell, Kans., Oct. 28, 1939; s. Francis B. and Sarah M. (Murphy) B.; m. Judith Ann Oates, Aug. 22, 1964; children: Janet E., Joseph A., Gary E. BS in Biology, Ft. Hays (Kans.) State U., 1960, MS in Botany, 1961; PhD in Botany, U. Okla., 1965. Sci. tchr. Damar (Kans.) High Sch., 1961-62; teaching asst. U. Okla., Norman, 1962-63, NSF fellow, 1963-65; from asst. prof. to assoc. prof. biology Calif. State U., Hayward, 1965-74; prof., 1974—; cons. in field. Contbr. articles to profl. jours. Mem. AAAS, Am. Inst. Biol. Scis., Ecol. Soc. Am., Southwestern Assn. Naturalists, Calif. Native Plant Soc. Office: Calif State Univ Dept Biology Hayward CA 94542

BABA, YOSHINOBU, biotechnologist; b. Hitoyoshi, Kumamoto, Japan, Sept. 21, 1958; s. Kazumi and Kayoko (Iwasaki) B.; m. Mikiko Sato, May 4, 1986; children: Tomohiro, Takahiro. BS in Chemistry, Kyushu U., Fukuoka, Japan, 1981, DSc in Chemistry, 1986. Postdoctoral fellow Kobe Women's Coll. of Pharmacy, Japan, 1986; rsch. assoc. Oita U., Japan, 1986-88, lectr. to asst. prof., 1988-90; lectr. to asst. prof. Kobe Pharm. U., Japan, 1990—. Author: Computer-Assisted HPLC, 1990, Flow Injection Analysis, 1989, Trends in Organic Chemistry, 1990, Capillary Electrophoresis, 1993. Mem. AAAS, Am. Chem. Soc., Chem. Soc. Japan, Pharm. Soc. Japan (Young Pharm. Scientist award), Japanese Soc. Analytical Chemistry, Soc. for Chromatographic Sci., N.Y. Acad. Scis. Avocation: tennis. Office: Kobe Pharm U, Higashinada, Kobe 658, Japan

BABCOCK, ELKANAH ANDREW, geologist; b. Elizabeth, N.J., Nov. 23, 1941. BS, Union Coll., N.Y., 1963; MS, Syracuse U., 1965; PhD in Geology, U. Calif., Riverside, 1969. Asst. prof. geology U. Alta., 1969-75; dir. natural resources divsn. Alta. Rsch. Coun., Ottawa, 1975-80, v.p., 1980—. Mem. AAAS, Geol. Assn. Can., Can. Soc. Petroleum Geologists. Achievements include research in structural geology of the Salton Trough, Southern California, terrain analysis and geologic mapping applications of remote sensing, fracture phenomena in rock. Office: Geol Survey of Canada, 601 Booth St, Ottawa, ON Canada K1A 0E4*

BABCOCK, HORACE W., astronomer; b. Pasadena, Calif., Sept. 13, 1912; s. Harold Delos and Mary Geddie (Henderson) B.; children: Ann Lucille, Bruce Harold, Kenneth L. B.S., Calif. Inst. Tech., 1934; Ph.D., U. Calif., 1938; D.Sc. (hon.), U. Newcastle-upon-Tyne (Eng.), 1965. Asst. Lick Obs., Mt. Hamilton, Calif., 1938-39; instr. Yerkes and McDonald Obs., Williams Bay, Wis., Ft. Davis, Tex., 1939-41; with Radiation Lab., MIT, 1941-42, Rocket Project, Calif. Inst. Tech., 1942-45; staff mem. Mt. Wilson and Palomar Obs., Carnegie Instn. of Washington, Calif. Inst. Tech., Pasadena, 1946-80; dir. Mt. Wilson and Palomar Obs., 1964-78. Author sci. and tech. papers in profl. jours. Recipient USN Bur. Ordnance Devel. award, 1946, Draper medal NAS, 1957, Eddington medal Royal Astron. Soc., 1958, Gold medal, 1970; Bruce medal Astron. Soc. Pacific, 1969, Rank prize, 1993. Mem. NAS (councilor 1973-76), Royal Astron. Soc. (assoc.), Societé Royale des Sciences de Liege (corr. mem.), Am. Philos. Soc., Am. Acad. Arts and Scis., Am. Astron. Soc. (councilor 1956-58, George Ellery Hale award 1990), Astron. Soc. Pacific, Internat. Astron. Union. Home: 2189 N Altadena Dr Altadena CA 91001-3533 Office: Obs of Carnegie Instn Washington 813 Santa Barbara St Pasadena CA 91101-1232

BABCOCK, JANICE BEATRICE, health care coordinator; b. Milw., June 2, 1942; d. Delbert Martin and Constance Josephine (Dworschack) B. BS in Med. Tech., Marquette U., 1964; MA in Healthcare Mgmt. and Supervision, Cen. Mich. U., 1975. Registered med. technologist and microbiologist., clin. lab. scientist, epidemiologist; cert. bioanalytical lab. mgr. Intern St. Luke's Hosp., Milw., 1963-64; microbiologist St. Michael's Hosp., Milw., 1964-65; supr. clin. lab. svc. VA Regional Office, Milw., 1965-66; hosp. epidemiologist VA Ctr., Milw., 1966-74, supr. anaerobic microbiology and rsch. lab., 1974-78, adminstrv. officer, chief med. tech., 1978-83, quality assurance coord., 1983-86, asst. to chief of staff profl. svcs., 1986-92; coord. vet. affairs outpatient clinic, acting coord. VHA Nat. Cost Containment Ctr., Milw., 1992—; lectr. Marquette U., 1966-86, U. Wis., 1966-86, Med. Coll. Wis., 1966-86. Contbr. numerous articles to profl. jours. Rec. sec. Wis. Svc. League, 1989-92, corr. sec., 1991. Recipient Wood VA Fed. Woman's award, 1975, Profl. Achievement award Lab. World jour., 1981, Disting. Alumni award Cen. Mich. U., 1986. Fellow Royal Soc. Health, Am. Acad. Med. Adminstrs. (Wis. state Dir. of the Yr. award 1989, Diplomate 1989, mem. editorial bd. Exec. jour. 1987—, regional dir. 1992—); mem. Internat. Acad. Healthcare Mgmt., Internat. Soc. of Tech. Assessment in Health Care, Am. Soc. Microbiology, Am. Coll. Healthcare Execs., Am. Soc. Med. Tech. (Nat. Sci. Creativity award 1974, Nat. Microbiology Sci. Achievement award 1978, Mem. of the Yr. award 1979, Profl. Achievement Lectureship award 1981, French Lectureship award 1983), Assn. for Health Svcs. Rsch., Assn. Marquette U. Women (bd. dirs. 1987-93, v.p., sec.), Assn. Mil. Surgeons U.S. (lifetime), Nat. Assn. Med. Staff Svcs. (mem. editorial bd. Overview Jour. 1990-93), Wis. Assn. Med. Staff Svcs., Wis. Hosp. Assn., Wis. Svc. League (corr. sec., recording sec.), Fed. Execs. Assn. (Milw. 1983—), Alpha Mu Tau (pres. 1984-85), Alpha Delta Theta, Sigma Iota Epsilon, Alpha Delta Pi (Alumni Honor award 1979). Home: 6839 Blanchard St Milwaukee WI 53213-2853 Office: VA Med Ctr NCCC 5000 W National Ave Milwaukee WI 53295-0002

BABCOCK, MICHAEL WARD, economics educator; b. Bloomington, Ill., Dec. 10, 1944; s. Bruce W. and Virginia (Neeson) B.; B.S.B.A., Drake U., 1967; M.A. in Econs., U. Ill., 1971, Ph.D. in Econs., 1973; m. Virginia Lee Brooks, Aug. 4, 1973; children: John, Karen. Teaching asst. U. Ill., Urbana, 1968, 71, research asst., 1972; prof. econs. Kans. State U., Manhattan, 1972—; con. Santa Fe, Burlington Northern, and Union Pacific R.R.; United Transp. Union, Kans. Dept. Transp., Kans. Dept. Agr., U.S. Dept. Agr., Kans. Dept. Commerce. Contbr. articles to profl. jours., newspapers, mags. With U.S. Army, 1969-71. Fed. R.R. Adminstrn. grantee, 1976-78; U.S. Army C.E. grantee, 1978-79; USDA grantee, 1978-79, 80-82, 84-85; Kans. Dept. Agrl. grantee, 1987; Kans. Wheat Commn. grantee, 1989, 92, 93; Midwest Transp. Ctr. grantee, 1989, 92, 93; Kans. Dept. Transp. grantee, 1991-93; recipient A.T. Kearney award Transp. Research Forum, 1987, 89, UPS Found. award, 1990, Edgar S. Bagley award Kans. State U., 1989, 93. Mem. Am. Assn. Agrl. Economists, Missouri Valley Econ. Assn., Mid-Continent Regional Sci. Assn., So. Regional Sci. Assn., Nat. Assn. Bus. Economists, Transp. Research Forum, Transp. Rsch. Bd., Coun. Logistics Mgmt., So. Econs. Assn., Western Econs. Assn., Beta Gamma Sigma, Omicron Delta Epsilon. Club: Optimist. Home: 720 Harris Ave Manhattan KS 66502-3614 Office: Kans State Univ Econs Dept Manhattan KS 66506

BABIĆ, DAVORIN, electrical engineer, researcher; b. Karlovac, Croatia, Apr. 17, 1957; s. Marko and Vilma B.; m. Adrienne R. Hekster, May 25, 1990. MSEE, U. Pa., 1984, PhD, 1988. Rsch. engr. Rade Končar, Zagreb, Croatia, 1980-81; rsch. fellow U. Pa., Philadelphia, 1981-88; rsch. engr. Bosch Power Tool Corp., New Bern, N.C., 1988-90; postdoctoral rsch. assoc. U. N.C., Charlotte, 1991—. Contbr. articles to profl. jours. Grantee N.C. Supercomputing Ctr., 1991. Mem. IEEE, Am. Phys. Soc. Office: U NC Charlotte Charlotte NC 28223

BABIN, STEVEN MICHAEL, atmospheric scientist, researcher; b. Lawton, Okla., Sept. 6, 1954; s. Cleveland Victor Jr. and Delys Lilian (Lowry) B.; m. Pamela Gail Nee, June 23, 1990. BS in Engring. Physics with special distinction, U. Okla., 1976; MD, U. Okla., Okla. City, 1980; MSEE, U. Pa., 1983. Diplomate Am. Bd. Med. Examiners. Assoc. instr. pathology and lab. medicine U. Pa. Hosp., Phila., 1980-82; sr. engr. Applied Physics Lab. Johns Hopkins U., Laurel, Md., 1983—; Presenter in field. Contbr. articles to profl. jours. Engring. scholar Frontiers Sci. Found., 1972, Spl. scholar Nat. Merit Found., 1972. Mem. IEEE, Am. Meteorol. Soc., Am. Geophysics Union (life), Am. Mensa (life), Tau Beta Pi, Alpha Epsilon Delta, Phi Eta Sigma. Achievements include investigation of meteorological effects on microwave propagation in the marine boundary layer; design and development of data acquisition and analysis systems in use on helicopters, rocketsondes, buoys, etc.; development of optical waveguide pH sensor; design and creation of working proportional counter for exo-electron research. Office: Applied Physics Lab Johns Hopkins U Johns Hopkins Rd Laurel MD 20723-6099

BABIUK, LORNE ALAN, virologist, immunologist, research administrator; b. Canora, Sask., Can., Jan. 25, 1946; s. Paul and Mary (Mayden) B.; m. Betty Lou Carol Wayar, Sept. 29, 1973; children: Shawn, Kimberley. BSA, U. Sask., Saskatoon, 1967, MSc, 1969, DSc, 1987; PhD, U. B.C., Vancouver, 1972. Postdoctoral fellow U. Toronto, Ont., Can., 1972-73; asst. prof. We. Coll. Vet. Medicine, Saskatoon, Sask., 1973-75, assoc. prof., 1975-79, prof., 1979—; assoc. dir. rsch. Vet. Infectious Disease Orgn., Saskatoon, 1984-93, dir., 1993—; dir. sci. affairs Biostar, Saskatoon, 1993—; cons. Molecular Genetics, Mpls., 1980-84, Genetech., San Francisco, 1981-84, Ciba Geigy, basel, Switzerland, 1984-91. Contbr. 300 artciles to refereed publs., 40 chpts. to books. Recipient award Can. Soc. Microbiology, 1990, Am. Vet. Immunology, 1992, Xerox-Can. Forum., 1993. Achievements include six patents in field. Home: 245 East Pl, Saskatoon, SK Canada S7J 2Y1 Office: Vet Infectious Disease Orgn, 124 Veterinary Rd, Saskatoon, SK Canada S7N 0W0

BACALOGLU, RADU, chemical engineer; b. Bucharest, Romania, July 21, 1937; came to the U.S., 1985; s. Dan and Elena (Lazarescu) B.; m. Ilze Irina Schiff, Aug. 29, 1963; 1 child, Radu Dan. MSChE, Poly. Inst., Romania, 1959, PhD in Organic Chemistry, 1968. Asst. prof. Poly. Inst., Timisoara, Romania, 1959-79, prof. chemistry, 1979-85; rsch. assoc. U. Calif., Santa Barbara, 1985-89; asst. prof. Rutgers U., New Brunswick, N.J., 1989-91; rsch. scientist Witco Corp., Oakland, N.J., 1991—; dir. rsch. group of homogeneous and enzymatic catalysis Inst. Chem. Energetics, Bucharest, 1981-85. Author 2 books, over 150 papers; contbr. articles to Rev. Roumaine Chemie, Tetrahedron, Spectrochimica Acta, Jour. Am. Chem. Soc., Jour. Phys. Chem., others. Recipient 1st prize for scientific rsch. Ministry of Edn., Romania, 1965, G. Spacu award for rsch. in chemistry Romanian Acad., 1981. Mem. Am. Chem. Soc., N.Y. Acad. Scis. Achievements include 12 patents in organic chemistry; research in physical and synthetic organic chemistry, kinetic study of degradation and stabilization of polymers, especially polyvinyl chloride. Home: 27 Burlington Ct Hamburg NJ 07419 Office: Witco 100 Bauer Dr Oakland NJ 07436

BACANI, NICANOR-GUGLIELMO VILA, civil and structural engineer, consultant, real estate investor; b. Dagupan City, Pangasinan, Philippines, Jan. 10, 1947; s. Jose Montero and Felisa Lomibao (Vila) B.; m. Julieta, June 24, 1972; children: Julinor, Jazmin, Joymita, Normina, Nicanor Jr., Noel-John. BCE, U. Philippines, 1968, MSE, 1973. Registered profl. engr., Philippines. Structural engr. FR Estuar, PhD. Assocs., Quezon City, Philippines, 1970-72; civil structural engr. BestPhil Cons., Dagupan City, 1972-73; engring. mgr. Supreme Structural Products, Inc., Manila, 1974; chief engr. Tecphil Cons., Quezon City, 1974-76; v.p. Erectors, Inc., Makati, Philippines, 1977-81; pres. NGV Bacani & Assocs., various locations, 1981—; advisor, cons. Met. Manila Office of Commr. Planning, 1980—; profl. lectr. U. Manila Grad. Sch., 1982—; resource person Nat. Engring. Ctr. U.P., Quezon City, 1983—; cons. Geo J. Fosdyke Assocs., L.A., 1985—, Victor Constrn. & Devel., 1986—, H.A. Simons Internat., 1988-90, Azlon Devel. Corp., 1990—; pres. Mgmt. Design & Investment Co., 1987—, Stanley Assocs. Internat., 1988; sr. structural cons. Seismic Engring. Ltd., 1990—; sr. cons. InterCoast Cons. Ltd., 1991—, Davey Gibson Cons., 1991—; pres. Bestphil Can., 1992—, Seismic Cons. Author: A Reference for Engineers and Builders, 1983. Mem. Am. Mgmt. Assn., Internat. Assn. Bridge and Structural Engrs., Prestressed Concrete Inst., U. Philippines Alumni Engrs. Assn. (life), Assn. Structural Engrs. Philippines (bd. dirs. 4 terms). Avocations: guitar playing, swimming, jogging.

BACH, GÜNTHER, organic chemist, researcher; b. Scheibenberg, Erzgebirge, Germany, Feb. 27, 1928; s. Otto and Frieda (Thierfelder) B.; m. Irmintraut Hess, June 30, 1957; children: Thomas, Renate. Diploma, U. Leipzig, Germany, 1951, Dr. rer. nat., 1953. Rsch. chemist Filmfabrik Wolfen, Germany, 1951-55, dept. head, 1955-91. Contbr. articles to profl. jours.; patentee polymethine dyes (50). Recipient Wöhlerpreis, Chem. Gesellschaft der Deutschen Demokratischen Republik, 1961, Deutscher Nationalpreis der Deutschen Demokratischen Republik, 1965. Home: Luxemburgstr 5, O6846 Dessaŭ Germany

BACH, LARS, wood products engineer, researcher; b. Aalborg, Denmark, May 19, 1934; arrived in Canada, 1965; s. Berg and Tove (Kjaer) B.; m. Mary Doreen Rebagliati, mar. 9, 1968; children: John Berg, Jennifer Mari. MSc, R.V.A. U., Denmark, 1960; PhD, SUNY, Syracuse, 1966. Registered profl. engr., Denmark, B.C., Alta. Rsch. asst., fellow dept. wood product engring. SUNY, Syracuse, 1963-65; rsch. scientist Western Forest Products Lab., Vancouver, B.C., Can., 1965-67; project engr. Bach & Egmose Ltd., Aalborg, 1967-68; assoc. rsch. prof. Tech. U. Denmark, Copenhagen, 1968-74; sect. head MacMillan Bloedel Rsch. Ltd., Vancouver, 1974-80; wood products engr. Alta. Rsch. Coun., Edmonton, Can., 1980—; adj. prof. U. Alta., Edmonton, 1982—; prin. Bach & Assocs., Gambier Island, B.C., 1968—; bd. dirs. Aktie-Plantageselkabet, Aalborg. Contbr. 69 articles to tech. publs. Sgt. Danish army, 1960-62. NSERC grantee govt. of Can., 1986—. Mem. Forest Products Rsch. Soc. (Markwardt Wood Engring. award 1991), Can. Soc. Civil Engring., Sigma Xi. Achievements include invention of corrugated waferboard, development of equipment for grading wood panels, molded wood composites, 19 patents. Home: 7912 147th St, Edmonton, AB Canada T5R 0X7 Office: Alta Rsch Coun, 250 Karl Clark Rd, Edmonton, AB Canada T6H 5X2

BACH, STEPHAN BRUNO HEINRICH, chemistry educator; b. Knoxville, Tenn., Nov. 4, 1959; s. Bernd Bruno and Marianne Christina (Woit) B.; m. Patricia Cosper, May 28, 1988. BA, U. Cin., 1977, BS, 1983, cert. bus. adminstrn., 1983; PhD, U. Fla., 1987. Asst. prof. chemistry U. Tex., San Antonio, 1987—; rsch. assoc. NRC/Naval Rsch. Lab., Washington, 1989-91. U. Fla. postdoctoral rsch. fellowship, 1987. Mem. AAAS, Am. Chem. Soc., Am. Soc. Mass Spectrometry. Achievements include research in Ag7: Pentagonal Bipyramid; ionization potentials of gas-phase species via charge transfer reaction (specifically clusters); direct laser vaporization of aluminum nitride for the prodn. of aluminum clusters. Office: U Tex Div Earth & Phys Sci San Antonio TX 78249-0663

BACHAS, LEONIDAS GREGORY, chemistry educator; b. Chios, Greece, Oct. 9, 1958; s. Gregory and Maria (Exadaktylou) B.; m. Sylvia Daunert, 1986; children: Stephanie, Philip. MS in Oceanic Scis., U. Mich., 1985, PhD in Chemistry, 1986. Rsch. asst. U. Mich., Ann Arbor, 1982-85; asst. prof. U. Ky., Lexington, 1986-91, assoc. prof., 1991—. Contbr. articles to profl. jours. Recipient Am. Cyanamid award, 1992, grant Soc. Analytical Chemistry Pitts., 1988, First award NIH, 1988, Rsch. grant NSF, 1992. Mem. Am. Chem. Soc. (councilor 1990-93, Petroleum Rsch. Fund grant 1987), Am. Assn. Clin. Chemistry (archives com. 1990), Soc. Electroanalytical Chemistry (Young Investigator award 1993), Ky. Acad. Sci. Office: Univ Ky Dept Chemistry Lexington KY 40506-0055

BACHICHA, JOSEPH ALFRED, obstetrician/gynecologist, educator; b. Rock Springs, Wyo.; s. Alfred and Helen B. BA, Stanford U., 1977; MD, Boston U., 1982. Diplomate Am. Board of Obstetrics and Gynecology. Intern St. Luke's-Roosevelt Hosp., N.Y.C., 1982-83; resident in ob-gyn Stanford U. Hosp., Palo Alto, Calif., 1983-86; pvt. practice medicine Chgo., 1986—; cons. WHO, UN Family Planning Assn.; asst. prof. Northwestern U., Chgo., 1986—; dir. low-risk obstetrics, coord. undergrad. med. edn. Prentice Women's Hosp., Chgo. Contbr. articles to med. jours. Mem. Chgo. Coun. Fgn. Rels. Rotary Found. grad. fellow, 1980. Fellow Am. Coll. Ob-Gyn, Assn. Profs. Gynecology & Obstetrics, Internat. Coll. Surgeons; mem. AMA, Am. Assn. Maternal & Neonatal Health, Am. Fertility Soc., APHA, Chgo. Gynecol. Soc., Stanford U. Alumni Assn., Boston U. Sch. Medicine Alumni Assn., Phil Delta Epsilon. Roman Catholic. Avocations: mystery books, cross-country skiing, weight tng., long distance running, aerobics. Home: 215 E Chicago Ave Bldg 2608 Chicago IL 60611-2610 Office: Printers Row Med Office 715 S Dearborn Chicago IL 60605

BACHKOSKY, JOHN M. (JACK BACHKOSKY), federal agency administrator; b. Taylor, Pa., Nov. 1, 1939; m. Helen Bachkosky; children: Arlene, Janice, Karen, John. B in Engring. Physics, Pa. State U. Aircraft engr. USN, 1963-82; tech dir. direct energy weapon devel., dir. advanced tech. on the army staff Office Dir. Def. for Rsch. and Engring., Washington, 1982-86, dep. dir., 1991—; from various tech. and mgmt. pos. to chief of staff Def. Nuclear Agency, Washington, 1986-91. Office: Dept of Defense Defense Research & Engineering The Pentagon Rm 3E1045 Washington DC 20301-3030*

BACHMAN, CLIFFORD ALBERT, engineering specialist, technical consultant; b. St. Louis, Jan. 8, 1958; s. Verman Clifford and Shirley Marie (Lishen) B.; m. Elizabeth Appelbaum, Nov. 24, 1984; children: Blake Elizabeth, Paul Henry, Ellie Marie. Cert. automotive tech., Ranken Tech. Inst., 1978, cert. auto body, 1983; AAS magna cum laude, Jefferson Coll., 1987; AT, cert. supervisory mgmt. tech., Ranken Tech. Coll., 1991. Mechanic Soutar Motors Inc., St. Louis, 1978-82, Gamer Auto Body Inc., St. Louis, 1983; machinist LP Tool & Die, Inc., St. Louis, 1984-85; tool room supr. Jefferson Coll., Hillsboro, Mo., 1985-87; machinist Spl. Parts, Inc., St. Louis, 1987-88; lab. technician Sunnen Products Co., 1988-92, engr., 1993—. Mem. Soc. Mfg. Engrs. (cert. mfg. technologist). Republican. Mem. Evangel. Free Ch. Am. Avocations: automobile & motorcycle restoration, macrophotography, metrology, modelmaking, woodworking. Office: Sunnen Products Co 7910 Manchester Rd Saint Louis MO 63143-2793

BACHMANN, FEDOR WOLFGANG, hematology educator, laboratory director; b. Zurich, Switzerland, May 23, 1927; s. Theodor E. and Maria (Isler) B.; m. Edith I. Derendinger, Oct. 17, 1957; 1 child, Christian M. MD, U. Zurich, 1954. Diplomate Swiss Bd. Internal Medicine and Hematology. Intern, resident Med. Sch. U. Zurich, 1955-61; trainee USPHS Med. Sch. Washington U., St. Louis, 1961-64, asst. prof. Med. Sch., 1964-68; assoc. prof. Med. Sch. Rush-Presbyn.-St. Luke's Hosp., Chgo., 1968-73; dir. med. rsch. Schering Corp., USA, Lucerne, Switzerland, 1973-76; prof. medicine Med. Sch. U. Lausanne, Switzerland, 1976-92, prof. emeritus, 1992—; dir. hematology labs. U. Lausanne Med. Ctr., 1976-92, acting chmn. Dept. Medicine, 1980-81, provost, 1986-91. Editor: Progress in Fibrinolysis, vol. 6, 1983; assoc. editor Fibrinolysis, Clin. Lab. Haematology, Internat. Jour. Haematology; contbr. over 200 articles to profl. jours. Recipient Biannual prize Internat. Com. on Fibrinolysis, 1988; rsch. grantee NIH, 1964-73, Swiss Nat. Found., 1977-93. Fellow Am. Coll. Physicians; mem. Internat. Com. Thrombosis and Haemostasis (chmn. 1984-86), Internat. Soc. Thrombosis and Haematology (exec. coun. 1988—), Swiss Soc. Haematology (pres. 1986-88). Avocations: reading, skiing. Home: Chemin Praz-Mandry 20, 1052 Le Mont Switzerland

BACHMEYER, THOMAS JOHN, fundraising executive; b. Kalamazoo, Mich., July 7, 1942; s. Thomas Loften Bachmeyer and Bess Bruce Kies; divorced; 1 child, Thomas John Jr.; m. Ellen Christine Rau, Sept. 2, 1979; children: Justin Daniel, Jenna Kristen. BA, Yale U., 1964; BD, U. Chgo., 1967, MA, 1968, PhD, 1971. From asst. prof. to assoc. prof. St. Francis Sch. Pastoral Ministry, Milw., 1970-75; prin., cons. Prospect Counseling & Consultation Svcs., Milw., 1975-83; dir. planned giving Sharp Hosps. Found., San Diego, 1983-85; dir. devel. New Eng. Bapt. Hosp., Boston, 1985-88; v.p. Boca Raton (Fla.) Community Hosp. Found., 1988-92; dir. devel. The Miami Project to Cure Paralysis U. Miami Sch. Medicine, 1992—. Co-author: Disciplines in Transformation: A Guide to Theology and the Behavioral Sciences, 1979; contbr. articles to profl. jours. Mem. Assn. for Healthcare Philanthropy, Rotary Internat. Episcopalian. Avocations: golf, tennis, swimming, reading. Office: U Miami Sch Medicine The Miami Project to Cure Paralysis 1600 NW 10th Ave R 48 Miami FL 33136

BACHUS, BENSON FLOYD, mechanical engineer, consultant; b. LeRoy, Kans., Aug. 10, 1917; s. Perry Claude and Eva Marie (Benson) B.; m. Ruth Elizabeth Beck, May 31, 1942; children: Carol Jean Schueler, Bruce Floyd, Linda Ruth Gadway. Degree, Hemphill Diesel Sch., Chgo., 1937; student, Sterling Coll., 1937-39; BSME, Kans. State U., 1942; postgrad., Ohio State U., 1961, Stevens Inst., 1964; MBA, Creighton U., 1967. Registered profl. engr., Ariz., Ill., Nebr. Researcher, mech. engr. Naval Ordnance Rsch. Lab., Washington, 1942-43; jr. product engr. Western Electric Co., Inc., Chgo. and Eau Claire, Wis., 1944-46; sr. devel. engr. Western Electric Co., Inc., Chgo., 1946-56; devel. engr. Western Electric Co., Inc., Omaha, 1960-66; product engr. mgr. Century Electronics and Instruments, Inc., Tulsa, Okla., 1956-60; sr. staff engr. Western Electric Co. div. AT&T Techs., Phoenix, 1966-85; cons. in field, 1985-93; cons. in field, Phoenix, 1985—; chmn. energy conservation AT&T Techs., Inc., 1973-85; advisor to student engrs. Ariz. State U., 1967-87. Patentee in field. Trustee, Village of Westchester (Ill.), 1949-53; sec.-treas. Westchester Broadview Water Commn., 1949-53; Sunday Sch. supr. Westchester Community Ch., 1949-56; vol. campaign worker, precint committeeman Phoenix Rep. Party, 1986—. Named Westchester Family of Yr., Westchester Community Ch., 1952; recipient Centennial medal Am. Soc. Engrs., 1979. Fellow ASME (state legis. coord. 1985-86, 88-93, treas. Ariz. sect. 1971-72, sec. 1972-73, vice chmn. 1973-74, chmn. 1974-75, 50-Yr. Membership award, President's Dedicated Svc., Devotion, Leadership, Performance award 1992, Dedicated Svc. award 1993); mem. TAPPI, NSPE (Engr. of Yr. award 1979), Soc. Profl. Engrs. (editor mag. 1972-86), Ariz. Coun. Engring. and Sci. Assn., Am. Security Coun., Soc. Plastics Engrs., Weoma Sci. Club (pres. 1963-66), Tel. Pioneers Am., Order of Engrs., Elks. Avocations: woodworking, hiking, fishing, tennis, writing. Home and Office: 5229 N 43d St Phoenix AZ 85018

BACHUS, BLAINE LOUIS, pilot, aircraft mechanical engineer; b. Chgo., July 30, 1954; s. James Franklin and Jean (Clifton) B.; m. Kathleen Ann Hager, May 1, 1979; children: Briana Kathleen, Brenna Christine. BS in Aero., St. Louis U., 1976. Commd. 2d lt. USAF, 1977, advanced through grades to maj., various positions, 1977-85; aircraft comdr. KC-135 909th Air Refueling Squadron, Kadena AB, Japan, 1985-86, asst. fligh comdr., 1986-87; chief aircrew tng. devices 376th Strategic Wing, Kadena AB, 1987-88; U-2/TR-1 aircraft comdr. 95th Reconnaissance Squadron, RAF Alconbury, U.K., 1988-89, fligh comdr., 1989-90; chief standardization and evalution 17 Reconnaissance Wing, RAF Alconbury, 1990-91; flight test ops. officer USAF Plant 42 (U-2 Flight test), Palmdale, Calif., 1991—. Decorated 3 Air medals. Mem. Blackbird Assn. Achievements include flying first high altitude reconaissance mission over Iraq during the Gulf war. Office: USAF OL Det 8FT Edwards AFB CA 93523

BACKENROTH-OHSAKO, GUNNEL ANNE MAJ, psychologist, educator, researcher; b. Grums, Värmland, Sweden, June 12, 1951; parents: K. Hugo and Maj E.I.S. (Joelsson) B. BA, Göteborgs U., Gothenberg, Sweden, 1974; MA, Umeå U., Sweden, 1976; PhD, Stockholm U., 1983. Lic. psychologist, Sweden. Child psychologist Barn-och ungdomspsykiatriska Kliniken, Lidköping, Sweden, 1973-76; researcher, program leader Stockholm U., 1976-79, 1981—; psychologist, program leader Parental Assn. for Cooperation, Stockholm, 1979-81; univ. lectr. Stockholm U., 1988, assoc. prof., 1989; lectr. various univs.; cons. Stockholms läns Lansting, 1985-91, Sveriges Dövas Riksförbund, 1991-92, Nämnden för vårdartjänst, 1992—. Contbr. numerous articles to profl., acad. jours. Lecture award Singapore Assn. for the Deaf, 1984. Mem. Swedish Assn. for Psychologists (sci. adv. council), Internat. Assn. Human Relations Lab. Tng. (hon., life), Univ. Assn. Stockholm, Assn. for Female Researchers, Scandinavian Assn. for Research of the Deafs Mental Health, Internat. Round. Table for the Advancement of Counseling, Swedish Rorschach Assn., Internat. Assn. of Applied Psychology. Avocations: foreign cultures, photography, music, art, literature. Home: Skidbacken 4, S-17245 Sundbyberg Sweden Office: Stockholm U, Dept Psychology, S-10691 Stockholm Sweden

BACKENSTOSS, HENRY BRIGHTBILL, electrical engineer, consultant; b. Washington, Sept. 28, 1912; s. Ross Elwood and Susan Catherine (Brightbill) B.; m. Violet Pentleton, Jan. 23, 1942 (div. 1952); m. Bernadette Humbert, Sept. 24, 1954; 1 child, Martine Susan. BSEE, MSEE, MIT, 1935. Registered profl. engr., Pa., Mass., Conn. Project mgr. Jackson & Moreland, Engrs., Boston, 1945-59; prof. power tech. Am. U. Beirut (Lebanon), 1959-61; spl. cons. Gen. Pub. Utilities Corp., N.Y.C., 1961-62; v.p. Jackson & Moreland Internat., Beirut, 1962-68; sr. cons. Gen. Pub. Utilities Svc. Corp., Reading, Pa., 1970-77; cons. Devel. Analysis Assocs., Cambridge, Mass., 1977-82; Govt. Saudi Arabia, 1962-69; panelist fuel crisis and power industry IEEE Tech. Conf., 1973. Contbr. articles to profl. publs. Bd. dirs. Reading Symphony Orch. (Pa.), 1975-86, Berks County Conservancy (Pa.), 1984-92, Reading Mus. Found., 1986—, pres., 1988-91, chmn., 1991—, Am. Soc. Utility Investors, New Cumberland, Pa., 1989-93. Mem. IEEE (life sr., power system engring. com. 1952-87, system econs. subcom. 1952-76), Am. Soc. Utility Investors (New Cumberland chpt. 1982—), Nat. Soc. Profl. Engrs., Pa. Soc. Profl. Engrs., Sigma Xi (assoc.), Tau Beta Pi. Congregationalist. Achievements include research in preparing economic studies leading to bulk power generation at mine-mouth in western Pennsylvania with transmission at 500 kv. to eastern markets. Home: 408 S Tulpehocken Rd Reading PA 19601-1030

BACKMAN, ARI ISMO, electronics company executive; b. Helsinki, Finland, Dec. 19, 1961; came to U.S., 1990; s. Jan Egon Sigfrid and Edit Tellervo (Liukkonen) B. BS in Econs. and MS in Econs., Sch. Econs., 1988. Systems analyst Orion Pharms. Corp., Helsinki, Finland, 1982-85; pres. NBV Advisors Corp., Helsinki, Finland, 1986-92, IB Innovation Corp., Helsinki, Finland, 1985-89; acct. mgr. Digital Equipment Corp., Helsinki, Finland, 1989-90, Chgo., 1990—. With Nylands Brigade, 1981-82. Hewlett PAckard grantee, 1987. Mem. Finnish-Am. C. of C. (bd. dirs.). Lutheran. Home: 1824 N Lincoln Park W #201 Chicago IL 60614 Office: Digital Equipment Corp 225 W Washington St Chicago IL 60614

BACKUS, CHARLES EDWARD, engineering educator, researcher; b. Wadestown, W.Va., Sept. 17, 1937; s. Clyde Harvey and Opal Daisy (Strader) B.; m. Judith Ann Clouston, Sept. 1, 1957; children: David, Elizabeth, Amy. B.S. in Mech. Engring., Ohio U., 1959; M.S., U. Ariz.-Tucson, 1961, Ph.D., 1965. Supr., system engr. Westinghouse Astronuclear, Pitts., 1965-68; asst. prof. engring. Ariz. State U., Tempe, 1968-71, assoc. prof., 1971-76, prof., 1976—, asst. dean research, 1979-90, assoc. dean research, 1990-91, dir. Ctr. for Research, 1980-91, interim dean Coll. Engring. and Applied Sci., 1991-92, assoc. dean Indsl. and Profl. Devel., 1992—. Contbr. chpts. to books, articles to profl. jours. Mem. Ariz. Solar Energy Commn., Phoenix, 1975-87. Fellow IEEE; mem. AAAS, Am. Nuclear Soc., ASME, Am. Soc. Engring. Edn., Mesa C. of C., Sigma Xi, Phi Mu Epsilon, Tau Beta Pi. Methodist. Lodge: Rotary. Office: Ariz State U Coll Engring & Applied Sci Ctr Rsch Tempe AZ 85287-5506

BACKUS, ELAINE ATHENE, entomologist, educator; b. L.A., Sept. 22, 1956; d. Henry Floyd and Penelope (Mihalakis) B.; m. Ned M. Gruenhagen, July 30, 1988. BS, Brigham Young U., 1978; PhD, U. Calif., Davis, 1983. Postdoctoral assoc. U. Calif., Davis, 1983-84; asst. prof. U. Mo., Columbia, 1984-90, assoc. prof. entomology, 1990—. Contbr. articles to profl. publs., chpt. to book. Mem. AAAS, Entomol. Soc. Am. (Outstanding Grad. Student of Pacific Br., Comstock award 1983, Pres.'s Prize for Student Paper, 1982, Provost's award for Outstanding Jr. Faculty Teaching, 1990), Sigma Xi. Achievements include discovery and study of the sensory organs that mediate feeding in leafhoppers, understanding how the feeding of Empoasca leafhoppers cause damage to crop plants to develop resistant crops via genetic engineering. Office: U Mo Dept Entomology 1-87 Agriculture Bldg Columbia MO 65211

BACKUS, JOHN, computer scientist; b. Phila., Dec. 3, 1924; m. Una Stannard, 1968; children: Karen, Paula. BS, Columbia U., 1949, AM, 1950; D.Univ. (hon.), U. York, Eng., 1985; DSc (hon.), U. Ariz., 1988; Docteur honoris causa, Université de Nancy 1, France, 1989; DSc (hon.), U. Ind., 1992. Programmer IBM, N.Y.C., 1950-53, mgr. programming rsch., 1954-59; staff mem. IBM T.J. Watson Rsch. Ctr., Yorktown Heights, N.Y., 1959-63; IBM fellow IBM Rsch., Yorktown Heights and San Jose, Calif., 1963-91; mgr. functional programming IBM Almaden Rsch. Ctr., San Jose, 1980-91; cons., 1991—. With AUS, 1943-46. Recipient W. Wallace McDowell award IEEE, 1967; Nat. medal of Sci., 1975; Harold Pender award Moore Sch. Elec. Engring., U. Pa., 1983; Achievement award Indsl. Research Inst., Inc., 1983. Fellow Am. Acad. Arts and Scis.; mem. NAS, NAE (recipient Charles Stark Draper prize, 1993), Assn. Computing Machinery (Turing award 1977). Achievements include system design of IBM 704, Fortran programming lang., Backus-Naur Form Lang., funcion-level programming; mem. design group ALGOL 60 lang. Home: 91 St Germain Ave San Francisco CA 94114-2129

BACKUS, ROBERT COBURN, biophysical chemist; b. Carroll, Iowa, Aug. 25, 1913; s. Roy Eugene and Ethel (Coburn) B.; m. Beverly Helen Torwelle, July 5, 1940; children: Byron Torwelle, Robley Dean. BS, Dakota Wesleyan U., 1937; MS in Biochemistry, U. Mich., 1944, PhD in Bacteriology/Immunology, 1951. Instr. rsch. Sch Pub. Health, 1942-45; rschr. physics dept. U. Mich., Ann Arbor, 1946-50; rschr. Virus Lab. U. Calif., Berkeley, 1950-56; cons. rsch. grants, contracts Am. Cancer Soc., N.Y.C., 1957; rsch. grants adminstr. Nat. Cancer Inst.-NIH, Bethesda, Md., 1958-61, Nat. Inst. Allergy, Infectious Disease, NIH, Bethesda, Md., 1962-66; office of dir. NIH, Bethesda, Md., 1965-87; ret.; biophysicist U.S. Dept. Agrl. European Commn. Foot and Mouth Disease, Pirbright, Eng., 1951; staff Surgeon Gen.'s Ad Hoc Adv. Com. on Quarantine, Washington, 1967; staff HEW Tuskegee Syphilis Study Ad Hoc Adv. Panel, Washington, 1972-73; acting exec. sec. HEW Ethics Adv. Bd., 1978; bd. trustees Am. Type Culture Collection, 1962-68; mem. Civil Svc. Exam. Bd. for Health Scientist Adminstrn. Applicants, 1962-68. Contbr. articles to profl. jours. Mem. Electron Microscope Soc. Am. (bd. dirs. 1954-57), Sigma Xi (life). Achievements include electron microscopic counts of virus particles and determination of macromolecular weights; correlation of bacteriophage counts with infectivity; invention of field aligning capsule centrifugation; others. Home: 5305 Roosevelt St Bethesda MD 20814

BÄCKVALL, JAN-ERLING, chemist; b. Malung, Dalarna, Sweden, Dec. 7, 1947; s. Sigvard Aron and Karin Lily (Matsson) B.; m. Berith Linnea Carlsson, Sept. 4, 1971; children: Helena, Fredrik. MSc, Royal Inst. of Tech., Stockholm, 1971, PhD, 1975. Rsch. assoc. Royal Inst. of Tech., 1976-77, docent, 1977, assoc. prof., 1977-86; prof. chemistry U. Uppsala, Stockholm, 1986; mem. adv. bd. Internat. Symposia on Homogenous Catalysis, 1988—, Synlett, 1991—, Jour. Chem. Soc. Perkin Transaction, 1992—. Mem. Swedish Chem. Soc. (Arrhenius medal 1986), Am. Chem. Soc., Royal Chem. Soc. Office: U Uppsala, Dept Chemistry, Box 531, S-75121 Uppsala Sweden

BACON, JOHN STUART, biochemical engineer; b. Washington, June 8, 1959; s. Edward Jennings and Martha Ora (Landefeld) B. BS in Chemistry, U. Mich., 1982; MS in Chemical Engring., New Mex. State U., 1989. Student scientist NIH Lab. of Molecular Biology, Bethesda, Md., 1980; rsch. asst. U. Mich. Medical Sch., ANn Arbor, 1982-84; grad. asst. dept. of chemical engring. New Mex. State Univ., Las Cruces, 1984-87; process devel. engr. ChemGen Corp., Gaithersburg, Md., 1987-90; rsch. engr. Red Star Yeast Univ. Foods Corp., Milw., 1990-92, sr. rsch. engr., 1992—; adv. com. New Mex. grad. student, 1986-87; mem. Engrs. Coun., Las Cruces, 1987. Contbr. articles to profl. jours. Mem. Grand Jury Dona Ana County, Las Cruces, 1985. Mem. AICE, Am. Chem. Soc., Phi Kappa Phi, Soc. Omega Chi Epsilon. Home: N7673 Hwy 175 Theresa WI 53091-9762 Office: Universal Foods Corp 6143 N 60th St Milwaukee WI 53218

BACON, VICKY LEE, lighting services executive; b. Oregon City, Oreg., Mar. 25, 1950; d. Herbert Kenneth and Lorean Betty (Boltz) Rushford; m. Dennis M. Bacon, Aug. 7, 1971; 1 child, Randene Tess. Student, Portland Community Coll., 1974-75, Mt. Hood Community Coll., 1976, Portland State Coll., 1979. With All Electric Constrn., Milwaukie, Oreg., 1968-70, Lighting Maintenance Co., Portland, Oreg., 1970-78; svc. mgr. GTE Sylvania Lighting Svcs., Portland, 1978-80, br. mgr., 1980-83; div. mgr. Christenson Electric Co. Inc., Portland, 1983-90, v.p. mktg. and lighting svcs., 1990-91, v.p. svc. ops. and mktg., 1991—. Mem. Illuminating Engring. Soc., Nat. Assn. Lighting Maintenance Contractors. Office: Christenson Electric Co Inc 111 SW Columbia St Ste 480 Portland OR 97201-5886

BADEER, HENRY SARKIS, physiology educator; b. Mersine, Turkey, Jan. 31, 1915; came to U.S., 1965, naturalized, 1971; s. Sarkis and Persape Hagop (Koundakjian) B.; m. Mariam Mihran Kassarjian, July 12, 1948; children: Gilbert H., Daniel H. M.D., Am. U., Beirut, Lebanon, 1938. Gen. practice medicine Beirut, 1940-51; asst. instr. Am. U. Sch. Medicine, Beirut, 1938-45; adj. prof. Am. U. Sch. Medicine, 1945-51, asso. prof., 1951-62, prof. physiology, 1962-65, acting chmn. dept., 1951-56, chmn., 1956-65; research fellow Harvard U. Med. Sch., Boston, 1948-49; prof. physiology Creighton U. Med. Sch., Omaha, 1967-91, emeritus prof., 1991—; acting chmn. dept. Creighton U. Med. Sch., 1971-72; vis. prof. U. Iowa, Iowa City, 1957-58, Downstate Med. Center, Bklyn., 1965-67; mem. med. com. Azounieh Sanatorium, Beirut, 1961-65; mem. research com. Nebr. Heart Assn., 1967-70, 85-88. Author textbook Spanish translation; contbr. chpts. to books, articles to profl. jours. Recipient Golden Apple award Students of AMA, 1975, Disting. Prof. award, 1992; Rockefeller fellow., 1948-49; grantee med. research com. Am. U. Beirut, 1956-65. Mem. Internat. Soc. Heart Rsch., Am. Physiol. Soc., Internat. Soc. for Adaptive Medicine (founding mem.). Home: 2808 S 99th Ave Omaha NE 68124-2603 Office: Creighton U Med Sch 2500 California St Omaha NE 68178-0224

BADEN, MICHAEL M., pathologist, educator; b. N.Y.C., July 27, 1934; s. Harry and Fannie (Linn) B.; m. Judianne Densen-Gerber, June 14, 1958; children—Trissa, Judson, Lindsey, Sarah. BS, CCNY, 1955; MD, NYU, 1959. Diplomate Am. Bd. Pathology. Intern, first med. div. Bellevue Hosp., N.Y.C., 1959-60, resident, 1960-61, resident in pathology, 1961-63, chief resident in pathology, 1963-64, fellow in pathology, 1964-65; pvt. practice in pathology N.Y.C.; asst. med. examiner City of N.Y., 1961-65, jr. med. examiner, 1965-66, assoc. med. examiner, 1966-70, dep. chief med. examiner, 1970-78, 79-81, 83-86, chief med. examiner, 1978-79; dep. chief med. examiner, dir. labs. Suffolk County, N.Y., 1981-83; dep. chief med. examiner Suffolk County, 1983-86; dir. forensic scis. unit N.Y. State Police, 1986—; instr. in pathology NYU, N.Y.C., 1964-65, asst. prof. pathology, 1966-70, assoc. prof. forensic medicine, 1970-89; adj. prof. law N.Y. Law Sch., N.Y.C., 1975-88, John Jay Coll. Criminal Justice, N.Y.C., 1989-90, 93; vis. prof. pathology Albert Einstein Sch. Medicine, N.Y.C., 1975—; lectr. pathology Coll. Physicians and Surgeons, Columbia U., N.Y.C., 1975—, adj. prof. pathology and lab. medicine, 1993—; asst. vis. pathologist Bellevue Hosp., N.Y.C., 1965—; lectr. Drug Enforcement Adminstrn., Dept. Justice, 1973—; vis. lectr. Fairleigh Dickinson Dentistry, Hackensack, N.J., 1968-70; spl. forensic pathology cons. N.Y. State Organized Crime Task Force, 1971-75; chmn. forensic pathology panel U.S. Ho. of Reps. select coms. on assassinations of Pres. John F. Kennedy and Dr. Martin Luther King, Jr., 1977-79; mem. med. adv. bd. Andrew Menchell Infant Survival Found., 1969-74; mem. cert. bd. Addiction Svcs. Agy., N.Y.C., 1966-69; preceptor health research tng. program N.Y.C. Dept. Health, 1968-79; v.p. Coun. for Interdisciplinary Communication in Medicine, 1967-69; forensic pathology cons. N.Y. State Police, 1985—. Author: Alcohol, Other Drugs and Violent Death, 1978, Unnatural Death, 1989; contbr. articles on forensic medicine to profl. jours.; mem. editorial bd. Am. Jour. Drug and Alcohol Abuse, 1973—, Internat. Microfilm Jour. Legal Medicine, 1969-73, Contemporary Drug Problems, 1971. Active N.Y. adv. bd. Odyssey House, Inc., 1966-76; bd. dirs. N.Y. Coun. on Alcoholism, sec., 1969-79; bd. dirs. Belco Scholarship Found., Inc., 1971-87. Recipient Great Tchr. award NYU, 1980. Fellow Coll. Am. Pathologists (chmn. toxicology subcom. 1972-74), Am. Soc. Clin. Pathologists (mem. drug abuse task force 1973—), Am. Acad. Forensic Scis. (program chmn. 1971-72, sec. sect. pathology and biology 1970-71, exec. com. 1971-74, v.p. 1982-83); mem. Med. Soc. County N.Y. (mem. pub. health com. 1966-76), Soc. Med. Jurisprudence (corr. sec. 1971-78, v.p. 1979-81, pres. 1981-85, chmn. bd. 1985—), Nat. Assn. Med. Examiners, N.Y. Path. Soc., N.Y. State Med. Soc., AMA, Internat. Royal Coll. Health. Office: 142 E End Ave New York NY 10028-7503

BADER, KEITH BRYAN, chemical engineer; b. Columbus, Ohio, Oct. 7, 1956; s. Meinhardt and Delores (Wolf) B.; m. Karen Helen Koppang, May 27, 1978; children: Kenneth, Lauren. BS in Chem. Engring., U. N.D., 1978. Fire protection engr. Indsl. Risk Insurers, Houston, 1978-79; prodn. engr. Pennwalt-Lucidol, Houston, 1979-81; process/product engr. GE Plastics, Mt. Vernon, Ind., 1981-87, project mgr., 1987—. Webelos leader Boy Scouts Am., Mt. Vernin, 1981—. Home: 624 Raintree Cir Mount Vernon IN 47620

BADER, WILLIAM ALAN, computer engineer; b. Bethlehem, Pa., Sept. 10, 1964; s. Morris and Sophie Karen (Roberts) B. BS in Computer Engring., Lehigh U., 1985, MSEE, 1986. Sr. analyst Software Cons. Svcs., Nazareth, Pa., 1982—. Office: Software Cons Svcs 3162 Bath Pike Nazareth PA 18064

BADETTI, ROLANDO EMILIO, health science facility administrator; b. Istanbul, Turkey, Mar. 25, 1947; s. Umberto and Iole (Bianchi) B.; m. Emanuela Ponte, Oct. 29, 1973; children: Barbara, Fabiana. PhD in Pharmacy, U. Padua, Italy, 1971. Mfg. supr. Gruppo Lepetit, El Jadida, Morocco, 1971-76, plant mgr., 1977-79; quality assurance mgr. Gruppo Lepetit, Milan, 1980-81, material mgr., 1982-83; tech. dir. Lirca Synthelabo, Milan, 1983-91; gen. mgr. asst. Arval, Milan, 1991-92; tech. dir. Laboratori UCB, Turin, 1992—. Office: Laboratori UCB SpA, Via Praglia 15, 10044 Pianezza Italy

BADGLEY, JOHN ROY, architect; b. Huntington, W. Va., July 10, 1922; s. Roy Joseph and Fannie Myrtle (Limbaugh) B.; m. Janice Atwell, July 10, 1975; 1 son, Adam; children by previous marriage: Dan, Lisa, Holly, Marcus, Michael. AB, Occidental Coll., 1943; MArch, Harvard, 1949; postgrad., Centro Internazionale, Vincenza, Italy, 1959. Pvt. practice, San Luis Obispo, Calif., 1952-65; chief architect, planner Crocker Land Co., San Francisco, 1965-80; v.p. Cushman & Wakefield Inc., San Francisco, 1980-84; pvt. practice, San Rafael, Calif., 1984—; tchr. Calif. State U. at San Luis Obispo, 1952-65; bd. dirs. Ft. Mason Ctr., Angel Island Assn. Served with USCGR, 1942-46. Mem. AIA, Am. Arbitration Assn., Golden Gate Wine Soc. Home and Office: 1356 Idylberry Rd San Rafael CA 94903-1074

BADINGER, MICHAEL ALBERT, computer applications designer, inventor, consultant; b. New Orleans, Aug. 5, 1954; s. Albert Francis and Myrl Ann Badinger. Cert. in aviation electronics, Coll. Air Force, 1978; cert. in elec. sci., Butler Coll., 1978. Cert. comml. pilot, aircraft repairman. Instr. Delgado Coll., New Orleans, 1978-79; supr. aircraft elec. dept. Pan Air Corp., New Orleans, 1979-80; pvt. practice comml. taxi owner, operator New Orleans, 1979-81; sr. engr. Martin Marietta, New Orleans, 1981-92; pvt. practice cons., inventor Slidell, La., 1992—; committeeman La. Aircraft Repair Apprenticeship Program, New Orleans, 1978-79; mem. material review bd. Martin Marietta/NASA, New Orleans, 1981-84. Contbr. articles to profl. jours. Pres. Ozone Woods Homeowners Assn., Slidell, 1987. Sgt. USAF, 1974-78. Recipient Cert. of Recognition NASA, 1990, 91. Mem. IEEE, Am. Soc. for Quality Control, Air Force Sgts. Assn. Achievements include patents on high voltage ion pump using electro hydro dynamic principle, on a computer based inspection.

BADZIAN, ANDRZEJ RYSZARD, physicist; b. Zamosc, Poland, May 6, 1938; came to U.S., 1985; s. Janusz and Nadzieja (Chodak) B.; m. Teresa Kieniewicz, Oct. 5, 1963. MS in Chemistry, Warsaw U., 1961; PhD, Polish Acad. of Sci., 1969, DSc in Physics, 1985. Rsch. asst. Polish Acad. of Scis., Warsaw, Poland, 1962-69; head dept. Rsch. Ctr. for Crystals, Warsaw, Poland, 1969-76, Inst. for Electronic Materials, Warsaw, Poland, 1976-85; assoc. prof. Penn State U., University Park, Pa., 1985—. Contbr. articles to profl. jours. Grantee Office of Naval Rsch., 1992, NSF, 1991. Mem. Am. Phys. Soc., Materials Rsch. Soc. Roman Catholic. Achievements include discovery of diamond-cubic boron nitride solid solutions; discovery of method of cubic boron nitride synthesis; discovery of selection rules for thermal scattering of x-rays in crystals; new crystal form of tetrahedral carbon; Gem quality crystals of diamond, grown by chemical vapor deposition; demonstration of feasibility of high temperature diamond electronics. Office: The Penn State U 271 Materials Rsch Lab University Park PA 16802

BAE, BEN HEE CHAN, microbiologist; b. Pusan, Korea, Dec. 18, 1939; came to the U.S., 1966; s. Kuyeol and Jaesoon (Suh) B.; m. Inha Ahn, Oct. 30, 1971; children: Susan, Shane. BS in Biology, Yonsei U., 1965; PhD in Microbiology, Pa. State U., 1971. Rsch. assoc. Pa. State U., University Park, 1971-72; postdoctoral trainee Mount Sinai Hosp., N.Y.C., 1972-73; chief microbiology lab. L.I. Coll. Hosp., Bklyn., 1973—. Contbr. articles to Applied Microbiology, Jour. Bacteriology, Canadian Jour. Microbiology, Jour. Clin. Microbiology, Am. Jour. Clin. Pathology. Mem. Am. Soc. for Microbiology, AAAS, N.Y. Acad. Scis. Office: LI Coll Hosp 340 Henry St Brooklyn NY 11201

BAEK, SE-MIN, plastic surgeon; b. Busan, Republic of Korea, Apr. 13, 1943; s. Young-Woo and In-Soon (Yoo) B.; m. Boyung Hah, Apr. 30, 1968; children: Siunna, William, Shivonne. MD, Seoul Nat. U., 1967. Diplomate Am. Bd. Plastic Surgery, Am. Bd. Surgery. Intern Sisters of Charity Hosp., Buffalo, 1968-69; resident in plastic surgery Mt. Sinai Hosp., N.Y.C., 1970-75, St. Louis U., 1975-77; prof., chmn. dept. plastic surgery Kyung Hee U., Seoul, 1982-83; chief plastic surgery Bronx (N.Y.) VA Hosp., 1977-82; prof., dir. Inst. Plastic Surgery Korea U., Seoul, 1983-88; chmn. dept. plastic surgery Inje U., Seoul, 1988—, dir. Plastic Facial Deformity Ctr., 1988—; acting chief plastic surgery Mt. Sinai Hosp., 1980-82; dir. plastic surgery Mt. Sinai Sch. Medicine, 1980-82. Contbr. articles to profl. jours. Fellow ACS; mem. Am. Soc. Plastic and Reconstructive Surgery (1st prize Ednl. Found. 1982), Soc. Head and Neck Surgeons, Am. Soc. Maxillofacial Surgeons, Am. Soc. Head and Neck Surgery, Am. Soc. Aesthetic Surgery, Am. Assn. Hand Surgery, N.Y. Acad. Scis. Office: Chung Ryang, PO Box 256 Dongdaemun-ku, Seoul Republic of Korea

BAER, LEDOLPH, oceanographer, meteorologist; b. Monroe, La., Nov. 21, 1929; s. Leo M. and Leonora (Lieber); m. Inge Rosenbaum, Dec. 22, 1957; children: Teresa, Margaret, Leonard. Student, Tulane U., 1946-48; BS, La. Poly., 1950; MS, Tex. A&M U., 1955; PhD, NYU, 1962. Meteorologist Gulf Cons., Houston, 1955-57; oceanographer Lockheed Corp., Sunnyvale, Burbank, San Diego, Calif., 1959—, Nat. Oceanic and Atmospheric Adminstrn., Rockville, Md., 1974—. Contbr. articles to profl. jours. Served to sgt. USAF, 1950-54. Fellow AAAS; mem. Am. Meteorol. Soc. (profl.), Am. Geophys. Union. Home: 9920 Bedfordshire Ct Rockville MD 20854-2015 Office: NOAA/NOS Office Ocean Earth Scis 1305 East-West Hwy Silver Spring MD 20910

BAER, RUDOLF LEWIS, dermatologist, educator; b. Strasbourg, France, July 22, 1910; came to U.S., 1934, naturalized, 1940; s. Ludwig and Clara (Mainzer) B.; m. Louise Jeanne Grumbach, Nov. 6, 1941; children: John Reckford, Andrew Rudolph. MD, U. Basel, Switzerland, 1934; postgrad. dermatology, N.Y. Postgrad. Med. Sch., 1937-39; MD (hon.), U. Munich, 1981. Diplomate: Am. Bd. Dermatology (mem. 1964-72, pres. 1967-70). Intern Beth Israel Hosp., N.Y.C., 1934-35; resident dermatology Montefiore Hosp., N.Y.C., 1936-37; faculty Columbia U. Sch. Medicine, 1939-48; dir. dept. dermatology Univ. Hosp., 1961-81; faculty NYU Sch. Medicine, 1948—, prof. dermatology, 1961—, chmn. dept. dermatology, 1961-81, George Miller MacKee prof., 1961-81; dir. dept. dermatology Bellevue Hosp. Center, 1961-81; cons. Elizabeth A. Horton Meml. Hosp., Middletown, N.J., Hackensack (N.J.) Hosp.; surgeon gen. U.S. Army, FDA; mem. Internat. Com. Dermatology, 1967-82, pres., 1972-77; mem. com. on revision U.S. Pharmacopeia, 1970-75; mem. commn. cutaneous diseases Armed Forces Epidemiologic Bd., 1967-72; hon. pres. 18th World Congress Dermatology, N.Y.C., 1992; chmn. Dermatology Found., 1974-78. Editor: Office Immunology, 1947, Atopic Dermatitis, 1955, Year Book Dermatology, 1955-65; also past mem. numerous editorial bds.; Author over 300 articles. Chmn. bd. Dermatology Found., 1974-77; bd. dirs. Rudolf L. Baer Found. for Skin Diseases, 1975—. Decorated Order of the Rising Sun (Japan), 1991; Dohi lectr. and recipient Dohi medal Japanese Dermatol. Soc., 1965; Von Zumbusch lectr. Munich, 1967, Hellerstrom lectr. Stockholm, 1970, O'Leary lectr. Mayo Clinic, Rochester, Minn., 1971, Robinson lectr. U. Md., 1972, Barrett Kennedy lectr., 1973, Louis A. Duhring lectr., 1974, Samuel M. Bluefarb lectr., 1975, Frederick J. Novy Jr. vis. scholar, 1978, Morris Samitz lectr., 1979, Ruben Nomland-Robert Carney lectr., 1979, Barrett Kennedy meml. lectr., 1980, A. Harvey Neidorff lectr., 1985, Ferdinand von Hebra meml. lectr., 1988, Alexander A. Fisher lectr., 1989, Stuart B. Fisher lectr., 1991, Tamotsu Imaeda lectr., 1991, Hermann Pinkus lectr., 1992; recipient von Hebra medal U. Vienna, 1988, Discovery award Dermatology Found., 1993. Fellow N.Y. Acad. Medicine (chmn. sect. dermatology 1963-64), Am. Acad. Dermatology (pres. 1974-75, Dome lectr. 1976, Gold medal 1978, hon. mem. 1980), Am. Acad. Allergy, Am. Coll. Allergists; mem. AMA (chmn. sect. dermatology 1965-66), Am. Dermatol. Assn. (pres. 1977, hon. mem. 1992), Soc. Investigative Dermatology (pres. 1963-64, Stephen Rothman medal 1973, hon. mem. 1980), Am. Contact Dermatitis Soc. (hon.), Bronx Dermatol. Soc. (pres. 1952), N.Y. Dermatol. Soc. (pres. 1982-83, hon. mem. 1991), N.Y. Allergy Soc., N.Y. Acad. Scis., N.Y. County and State Med. Soc., World Congress Dermatology (hon. pres. 1992), Internat. League Dermatol. Socs. (pres. 1972-77, Alfred Marchionini Gold medal 1977); hon. mem. Argentinian, Austrian, Brit., Brazilian, Danish, Finnish, German, Iranian, Israeli, Italian, Japanese, Mexican, Polish, Swedish, Yugoslav, Venezuelan dermatol. socs., Brazilian Nat. Acad. Medicine; corr. mem. Pacific, Cuban, French, dermatol. socs., French Allergy Soc. Home: 1185 Park Ave New York NY 10128-1308 Office: 566 1st Ave New York NY 10016-6402

BAER, STEPHEN COOPER, manufacturing executive; b. L.A., Oct. 13, 1938; s. Herman Lewis and Beatrice Adah (Cooper) B.; m. Holiday Pedotti, May 21, 1960; children: Audrey, José. Student, Amherst Coll., 1956-59, Swiss Fed. Inst. Tech., 1963-65. Pres. Zomeworks Corp., Albuquerque, 1969—. Author: Dome Cookbook, 1968, Zome Primer, 1970, Sunspots, 1975. With U.S. Army, 1960-63. Recipient Passive Solar Pioneer award Solary Energy Soc., U.S. sect., 1990, Solar Hall of Fame, Harry Thomason, 1991. Achievements include inventions and patents in fields of structures, passive solar controls and heat engines. Home: Box 422 Corrales NM 87048 Office: Zomeworks 1011 Sawmill Rd Albuquerque NM 87104

BAERG, RICHARD HENRY, podiatrist, educator, consultant; b. L.A., Jan. 19, 1937; s. Henry Francis and Ruth Elizabeth (Loven) B.; children from previous marriage: Carol Elizabeth, William Richard, Michael David; m. Yvonne Marie Estrada, 1987; children from previous marriage: Yvette Marie. AA, Reedley Coll., 1956; BS, Calif. Coll. Podiatric Medicine, 1965, DPM, 1968, MSc in Foot Surgery, 1970; MPH in Med. Adminstrn., U. Calif., Berkeley, 1971; ScD (hon.), N.Y. Coll. Podiatric Medicine, 1980; LittD (hon.), Ohio Coll. Podiatric Medicine, 1984. Diplomate Am. Bd. Podiatric Surgery, Am. Bd. Podiatric Orthopedics and Primary Podiatric Medicine, Am. Bd. Podiatric Pub. Health. Intern Highland Gen. Hosp., Oakland, Calif., 1969; resident in surgery Calif. Podiatry Hosp. (Pacific Coast Hosp.), San Francisco, 1970; acad. dean N.Y. Coll. Podiatric Medicine, N.Y.C., 1971-74; v.p., dean Calif. Coll. Podiatric Medicine, San Francisco, 1974-76; chief podiatric medicine Los Angeles County-U. So. Calif. Med. Ctr., 1976-78; dir. So. Calif. Podiatric Med. Ctr., 1976-78; pvt. practice Beverly Hills, Calif., 1976-78; pres. Ill. Coll. Podiatric Medicine, Chgo., 1978-79; mem. spl. med. adv. group to sec. Dept. Vets. Affairs, Washington, 1976-79, dir. podiatric service, dept. medicine and surgery, 1979-84, acting dir., 1984-86; chief podiatry VA Med. Ctr., Loma Linda, Calif., 1984-89; exec. v.p., med. dir. Dr. Footcare Corp., Montclair, Calif., 1988-90; faculty podiatry U. N.C. Hosps., Chapel Hill, 1992—; clin. prof. Sch. of Podiatric Medicine Barry U., Miami, Fla., 1993—; clin. prof. Med. Sch., U. N.C., 1992; assoc. clin. prof. Stanford U. Med. Sch., 1974-76; clin. prof. Inst. Ednl. Mgmt., Harvard U., 1975, Pa. Coll. Podiatric Medicine, 1979-86, U. Osteo. Medicine and Health Sci., 1984-92; pres. Baerg & Assocs.; mem. podiatry adv. panel NAS-Inst. Medicine, 1974; mem. bd. podiatric medicine Calif. dept. Consumer Affairs, 1989-90, chmn. residency, edn. and hosp. inspection com. Contbg. author: (text) Podiatric Medicine and Public Health, 1987; editorial bd.: Jour. Podiatric Edn., Yearbook of Podiatric Medicine and Surgery; contbr. articles to profl. jours., chpts. to textbooks. Served with M.C. U.S. Army and U.S. Navy, 1958-61. Mead-Johnson fellow, 1968-69. Fellow Am. Podiatric Med. Assn. (com. on pub. health 1971-84, coun. podiatric edn., chmn. profl. edn. com. 1977-78, com. on hosp. 1980-85, Kenison award 1984, cert. appreciation 1990, com. on pub. health and preventive medicine), Am. Coll. Foot and Ankle Surgeons, Am. Coll. Foot Orthopedics and Podiatric Medicine (exec. dir. 1980-90), Am. Acad. Podiatric Adminstrns. (exec. dir. 1990-91), Acad. Ambulatory Foot Surgery; mem. Am. Public Health Assn. (governing coun. 1977-80, chmn. podiatric health sect. 1991—), Nat. Bd. Podiatric Med. Examiners (bd. dirs.), Assn. Podiatrists in Fed. Svc., Am. Assn. Colls. Podiatric Medicine (exec. com. 1973, pres. 1980-81), Assn. Mil. Surgeons U.S., Nat. Acads. of Practice (podiatric medicine 1985), Sigma Pi Epsilon, Pi Delta. Republican. Clubs: Masons, Scottish Rite (32 degree), Commonwealth of Calif. Home: 211-21 Melville Loop Chapel Hill NC 27514 Office: Chapel Hill Podiatry Clinic 1777 Fordham Blvd # 203 Chapel Hill NC 27514

BAGCHI, AMALENDU, environmental engineer; b. Katwa, India, Oct. 4, 1946; came to U.S., 1978; s. Kamakhya Kumar and Shanti Rani (Sanyal) B.; m. Sujata Chakravorti, Feb. 7, 1973; 1 child, Sudeshna. BTech with honors, Indian Inst. Tech., 1968, MTech, 1970; MS, U. Wis., 1982. Registered profl. engr., Wis. Cons. engr. Civil Svc., Calcutta, India, 1970-74; site engr., design engr., indsl. coord. Pile Found. Constrn. Co., Calcutta, 1974; lectr. Bengal Engring. Coll., Howrah, India, 1975-78; environ. engr. Wis. Dept. Natural Resources, Madison, 1979—; panelist Internat. Symposium Environ. Geotech., Allentown, Pa., 1987; editorial bd. ASCE Nat. Conv., N.Y.C., 1992. Author: Design, Construction and Monitoring of Sanitary Landfill, 1990; contbr. articles to profl. publs. Mem. ASCE, Assn. Indians in Am. Achievements include development of design method for natural attenuation type LF landfills, blow-out of clay liners in landfill, field method for soil blending for liner construction, computer model for leachate apportionment in lined landfill. Home: 222 Saint Croix Ln Madison WI 53705 Office: Wis Dept Natural Resources PO Box 7921 Madison WI 53707

BAGCHI, KALLOL KUMAR, computer science and engineering educator, researcher; b. Calcutta, India, Nov. 3, 1951; arrived in Denmark, 1987; s. Saroj and Champa Bagchi. MSc in Math., Calcutta U., 1976; postgrad. diploma in computer sci., Jadavpur U., Calcutta, 1978, PhD in Computer Sci., 1988. Rsch. scholar Jadavpur U., 1979-83; sr. systems engr. Webel Computers, Calcutta, 1983-85; edn. and tng. officer CMC India Ltd., Calcutta, 1985-86; project mgr. CCC Software Profls., Oulu, Finland, 1986-87; asst./assoc. prof. computer sci. and engring. Aalborg (Denmark) U., 1987—; organizer, session chmn. internat. confs. in field. Guest editor jours. in field; assoc. editor Internat. Jour. in Computer Simulation; contbr. articles to profl. jours. Mem. IEE Computer Soc., Assn. for Computing Machinery, N.Y. Acad. Scis., Soc. for Computer Simulation (bd. dirs.). Hindu. Avocations: writing, music, literature, travel, meditation. Office: Aalborg U, Fredrick Bajers Vej 7, DK-9220 Aalborg Denmark

BAGGERLY, JOHN LYNWOOD, nuclear engineer; b. Richmond, Va., Sept. 18, 1956; s. John Thomas and Shirley Evelyn (Bryant) B.; m. Susan Kaye Brown, March 23, 1985; children: Anne Bryant, Laura Marie. BS in Biology, Va. Poly. Inst. & State U., 1978, BSME, 1982; MBA, Old Dominion U., 1993. Rsch. asst. biology dept. Va. Poly. Inst., Blacksburg, 1978-79; nuclear test engr. nuclear engring. dept. Norfolk Naval Shipyard, Portsmouth, Va., 1982-88; chief test engr. nuclear engring. dept. Norfolk Naval Shipyard, Portsmouth, 1988—. Recipient Antarctic Svc. medal of USA, NSF, Washington, 1979. Mem. Naval Civilian Mgrs. Assn. Achievements include being member of first underwater research dive team to perform studies of freshwater lakes on the Antarctic continent.

BAGHAI, NINA LUCILLE, geology researcher; b. Pitts., Nov. 19, 1954; d. Ali Asghar and Betty Jane (Lehman) B. MS in Geology/Paleobotany, U. Idaho, 1983; MS in Geol. Sci., U. Rochester, 1989. Lectr. in earth sci. Wassuk Coll., Hawthorne, Nev., 1986; geology rschr. Calif. Acad. Scis., San Francisco, 1983-84; seasonal ranger, interpreter Nev. State Parks, Berlin, 1985-88 summers; seasonal ranger II Nev. State Parks, Panaca, Nev., 1989-90 summers; NSF geology rschr. U. Rochester, 1987-88; geology rsch. asst. Bur. of Econ. Geology, Austin, 1990—. Contbr. articles to profl. jours. including Biochemical Systematics and Biology, Am. Jour. of Botany, Jour. Sedimentary Petrology, Jour. Paleontology. Pipe organist Austin Ward Ch. LDS, Austin, 1990—, music chmn., 1989—. Grantee Gulf Coast Geology, 1992, Geol. Soc. Am., 1990, 92, Paleontol. Soc., 1991; Featherstone scholarship U. Idaho, 1982. Mem. Am. Assn. for Women Geoscientists (co-chmn. 1991—), Am. guild of Organists (Austin student chpt. sec. 1992-93), Soc. Vertebrate Paleontology, Sigma Xi (grant), Phi Kappa Phi. Republican. Achievements include research on Miocene Liriodendron, Upper Jurassic ichthyosaurs from North America, Upper Cretaceous plants from Big Bend National Park, Pleistocene mammoths and Bison from San Francisco, Miocene plant megafossils and microfossils. Home: Apt B 1918 Robbins Pl Austin TX 78705 Office: Dept Botany U Tex Austin TX 78712

BAGLEY, BRIAN G., materials science educator, researcher; b. Racine, Wis., Nov. 20, 1934; s. Wesley John and Ethel (Rasmussen) B.; m. Dorothy Elizabeth Olson, Nov. 20, 1959; children: Brian John, James David, Kristin Marie. BS, U. Wis., 1958, MS, 1959; AM, Harvard U. , 1964, PhD, 1968. Mem. tech. staff Bell Telephone Labs., Inc., Murray Hill, N.J., 1967-83, Bell Communications Rsch. Inc., Red Bank, N.J., 1984-91; NEG endowed chair, dir. Eitel Inst., prof. physics U. Toledo, 1991—. Served to 1st lt. AUS, 1960-61. Xerox predoctoral fellow Harvard U., 1964-66, Robert J. Painter predoctoral fellow Harvard U., 1966-67. Mem. Am. Phys. Soc., Am. Vacuum Soc. (chpt. chmn. 1991), Materials Rsch. Soc., Sigma Xi, Sigma Pi Sigma. Home: 16474 West River Rd Bowling Green OH 43402-9469 Office: U Toledo Dept Physics and Astronomy 2801 W Bancroft St Toledo OH 43606-3390

BAGLIO, VINCENT PAUL, aeronautical engineer; b. Patchogue, N.Y., Feb. 18, 1960; s. Lorenzo and Nancy (Morello) B.; m. Donna Marie Cappo, Sept. 8, 1985. BS, Princeton U., 1982; MS, Poly. U., Bklyn., 1986. Sr. engr. aircraft systems div. Grumman Aerospace Corp., Bethpage, N.Y., 1982—. Contbr. articles to profl. jours. Mem. alumni schs. com. Princeton (N.J.) U. Mem. AIAA (com. chmn. 1988-91), Soc. Automotive Engrs. (indsl. lectr. 1990-91), Internat. Coun. Aero. Scis. (program com. 1989-93), Friends Princeton Football. Avocations: golf, running. Office: Grumman Aerospace Corp MS B21-35 Bethpage NY 11714

BAGRODIA, SHRIRAM, chemical engineer; b. Gwalior, India, Sept. 24, 1952; came to the U.S., 1980; s. Vishwanath and Shanti (Bhageria) B.; m. Rajni Khemka, June 18, 1978; children: Mridula, Aditya. B in Tech., IIT Kanpur, 1973; MS in Engring., Princeton U., 1975; PhD, Va. Poly. Inst. and State U., 1984. Sr. engr. Orient Paper Mills, Amlai, India, 1975-80; sr. rsch. engr. Monsanto, Pensacola, Fla., 1984-86; prin. rsch. chem. engr. Eastman Chem. Co., Kingsport, Tenn., 1986—. Contbr. articles to profl. jours. Active Meals on Wheels, Kingsport. Mem. Am. Chem. Soc., Soc. Plastic Engrs., Fiber Soc., Phi Kappa Phi. Achievements include 10 U.S. patents for novel fibers and processes and polymer blends; co-discovery of spontaneously wettable fibers; research in polymer science and engineering and the structure-property relationship in polymers. Home: 2649 Suffolk St Kingsport TN 37660 Office: Eastman Chem Co Rsch Labs Kingsport TN 37660

BAGWELL, KATHLEEN KAY, infosystems specialist; b. Pratt, Kans., Oct. 8, 1951; d. Robert Wilfred and Mildred Henrietta (Koch) Ahrens; m. Scott Bagwell, Aug. 17, 1991. BS, Emporia State U., 1973; AA, Washburn U., 1984. Tchr., coach Peabody (Kans.) High Sch., 1973-75, Shawnee Heights High Sch., Tecumseh, Kans., 1975-84; programmer Excel Corp., Wichita, Kans., 1984-85, analyst, mem. standards com., 1985-88, mem. fitness and activity com., 1985—, chmn. standards com., 1987-88, coord. quality assurance, 1988-89, product mgr. acctg. systems, 1988-90, project leader yields, quality and tng. systems, 1990-92, project mgr. distbn. systems, 1991-92; project mgr. order processing, sales and mktg., 1993—. Coach Wichita West YMCA Volleyball Clinic, 1985-90. Named Area Coach of Yr. Topeka Volleyball Coaches Assn., 1978. Mem. Wichita IBM System 3X/400 Users Group, U.S. Tennis Assn., Sedgwick County Zool. Soc. Avocations: tennis, bicycling, running. Home: 1042 Coolidge St Wichita KS 67203-3019 Office: Excel Corp 151 N Main St Wichita KS 67202-1413

BAHADUR, BIRENDRA, display specialist, liquid crystal researcher; b. Gorakhpur, India, July 7, 1949; came to Can., 1981; s. Bijai Bahadur and Shakuntala Srivastva; m. Urmila Bahadur, May 29, 1970; children: Shivendra, Shachindra. BSc in Physics, Chemistry and Math., Gorakhpur U., 1967, MSc in Physics, 1969, PhD, 1976. Rsch. scholar physics dept. Gorakhpur U., 1969-76, asst. prof. physics dept., 1976-77; sr. sci. officer Nat. Phys. Lab. India, New Delhi, 1977-81; v.p. R&D Data Images, Ottawa, Ont., Can., 1981-85; mgr. R&D Litton Data Images, Ottawa, 1985-91; mgr. liquid crystal display material and process Litton Systems, Can., Etobicoke, Ont., 1988—; mem. sci. com. 13th Internat. Conf. on Liquid Crystals, Vancouver, B.C., Can., 1990; participant numerous profl. meetings. Author: Liquid Crystal Displays, 1984; editor: Liquid Crystals--Applications and Uses, vol. I, 1990, vol. II, 1991, vol. III, 1992; mem. editorial bd. Displays, 1993—; mem. abstracting panel Liquid Crystal Abstracts, 1978-80; contbr. numerous articles to profl. publs. V.p. nat. capitol region India Can. Assn., 1989-90, pres., 1990-91. Grantee Industrial Rsch. Assistance Program, NRC Can., 1982-85, 84-87, 88-91, Wright Patterson AFB, 1991. Mem. Optical Soc. Am., Soc. Info. Displays (Spl. Recognition award 1993), Inst. Physics, Soc. de Chimie Physique. Achievements include patent for Process for Production of Printed Electrode Pattern for Use in Electro-Optical Display Devices (India); patent application for Novel Vacuum Bag Sealing Technique for Producing Extremely Uniform Displays (U.S., Can.), and Day Night Displays; co-development of technology of various liquid crystal displays. Home: 21 Arborwood Dr, Etobicoke, ON Canada M9W 6W5 Office: Litton Systems Display Sys Eng, 25 City View Dr, Etobicoke, ON Canada M9W 5A7

BAHADUR, KHAWAJA ALI, mechanical engineer, consultant; b. Rawalpindi, Punjab, Pakistan, Oct. 14, 1930; s. Elahi and Jan (Azeema) Bukhsh; m. Hussan Ara Fazal, Oct. 31, 1965; children: Asad Ali, Babar Ali, Sadia Ali. Indsl. cert., Chamber of Industry & Commerce, Fed. Republic of Germany, 1955; grad., Inst. Mech. Engrs., London, 1957. Cert. engr. Engr. Pakistan Ordnance Factories, 1958-64; prodn. mgr. Singer Industries Ltd., Pakistan, 1965-72; chief mech. engr. Bendix-Siyanco, Saudi Arabia, 1973-78; project mgr. Nat. Constrn. Co., Saudi Arabia, 1982-83, Almihdar Binladen, Saudi Arabia, 1982-83; plant engr. Nat. Indsl. Corp., Saudi Arabia, 1991-92; mng. dir. Super Foods, Islamabad, Pakistan, 1993—; cons. various engring. firms, Pakistan, 1984-90. Avocations: chess, bridge, reading. Home: St No 24 H No 187, Sector I/9-1, Islamabad Pakistan Office: Super Foods, I & T Centre, Plot No 34 Sector G/9/4, Islamabad Pakistan

BAHCALL, NETA ASSAF, astrophysicist; b. Israel, Dec. 16, 1942; d. Yehezkel Oscar and Gita (Zilberstein) Assaf; m. John Norris Bahcall, Mar. 21, 1966; children: Ron Assaf, Dan Ophir, Orli Gilat. BS, Hebrew U., Jerusalem, 1963; MS, Weizmann Inst. Sci., Israel, 1965; PhD, Tel Aviv U., 1970. Rsch. asst. astrophysics Calif. Inst. Tech., 1965-67; rsch. fellow Calif. Inst. Tech., 1970-71; rsch. assoc. at observatory Princeton U., 1971-74, rsch. staff mem. 1974-75, rsch. astronomer, 1975-79, sr. rsch. astronomer, 1979-83, chief gen. observer br., from 1983; with Space Telescope Sci. Inst., Balt.; prof. dept. astronomy Princeton (N.J.) U., 1990—. Contbr. articles to profl. jours. Mem. Am. Astron. Soc. Office: Princeton U Dept Astro-Physics Payton Hall Princeton NJ 08544

BAHN, GILBERT SCHUYLER, retired mechanical engineer, researcher; b. Syracuse, N.Y., Apr. 25, 1922; s. Chester Bert and Irene Eliza (Schuyler) B.; BS, Columbia U., 1943; MS in Mech. Engring., Rensselaer Poly. Inst., 1965; PhD in Engring., Columbia Pacific U., 1979; m. Iris Cummings Birch, Sept. 14, 1957 (dec.); 1 child, Gilbert Kennedy. Chem. engr. Gen. Electric Co., Pittsfield, Mass., 1946-48, devel. engr., Schenectady, 1948-53; sr. thermodynamics engr. Marquardt Co., Van Nuys, Calif., 1953-54, rsch. scientist, 1954-64, rsch. cons., 1964-70; engring. specialist LTV Aerospace Corp., Hampton, Va., 1970-88; ret.; freelance rsch. FDR at Nadir, 1988—. Mem. JANNAF Performance Standardization Working Group, 1966-83, thermochemistry working group, 1967-72; propr. Schuyler Tech. Libr., 1952—. Active Boy Scouts Am., 1958-78. Served to capt. USAAF, 1943-46. Recipient Silver Beaver award Boy Scouts Am., 1970. Registered profl. engr., N.Y., Calif. Mem. ASME, Combustion Inst. (sec. western states sect. 1957-71), Soc. for Preservation Book of Common Prayer. Episcopalian (vestryman 1968-70). Author: Reaction Rate Compilation for the H-O-N System, 1968; Blue and White and Evergreen: William Byron Mowery and His Novels, 1981; Oliver Norton Worden's Family, 1982; Studies in American Historical Demography to 1850, Vol. 1, 1987; Overall Population Trends, Age Profiles, and Settlement, Vol. 2, 1987; The Wordens, Representative of the Native Northern Population, 1987; The Ancient Worden Family in America: A Story of Growth and Migration, 1988. Founding editor Pyrodynamics, 1963-69; proceedings editor Kinetics, Equilibria and Performance of High Temperature Systems, 1960, 63, 67; contbr. articles to profl. jours.; discoverer free radical chem. species diboron monoxide, 1966. Home: 4519 N Ashtree St Moorpark CA 93021-2156 Office: 238 Encino Vista Dr Thousand Oaks CA 91362-2537

BAHR, JANICE MARY, reproductive physiologist; b. LaCrosse, Wis., Feb. 14, 1935; d. Frank and Elizabeth (Schmitz) B. BA, Viterbo Coll., 1964; MS, U. Ill., 1968, PhD, 1974. Instr. biology Viterbo Coll., LaCrosse, 1968-70; predoctoral fellow U. Ill., Urbana-Champaign, 1970-74, asst. prof., 1974-79, assoc. prof., 1979-83, prof., 1983—; mem. panel NSF, Washington, 1981-86, USDA, Washington, 1987, NIH, Washington, 1988-92. Author: catecholamines as Hormone Regulators, 1985, Reproduction in Domestic Aminals, 1991; contbr. articles to sci. jours. Bd. dirs. McKinley Family YWCA, Champaign, 1989—, Covenant Med. Found., Urbana, 1991—. Recipient Outstanding Alumina award Viterbo Coll., 1988. Fellow AAAS; mem. Soc. for Study of Reproduction (treas. 1988-91, pres. 1993—), Endocrine Soc., Am. Soc. for Animal Sci., Poultry Sci. Assn., Champaign Rotary (pres. 1992-93), Sigma Xi, Phi Kappa Phi. Achievements include first report that male germ cells produce estrogen; discoveries in ovarian function of the domestic hen; research in comparative endocrinology. Office: U Ill 1207 W Gregory Dr Urbana IL 61801

BAI, SUNGCHUL CHARLES, nutritionist; b. Keo Jae, Korea, Feb. 23, 1954; s. Ki Kwon and Seong Rae (Park) B.; m. Heesock Sue Na, Apr. 29, 1981; children: Seokhyoun Ezekiel, Daniel Seokil. MA in Agri., Calif. State U., Fresno, 1984; PhD, U. Calif., Davis, 1990. Grad. rsch. asst. Wash. State U., Pullman, 1984-85; grad. rsch. asst. Univ. Calif., Davis, 1986-89, post-grad. researcher, 1990; rsch. assoc. Tex. A&M U., College Station, Tex., 1990-92; adj. asst. prof. Ohio State U., Columbus, 1993—. Contbr. articles to profl. jours. Recipient scholarship Kon-Kuk U., Seoul, Korea, 1976, Spl. Grad. award Mil. Med. Sch., Taegu, Korea, 1978, scholarship Calif. State U., Fresno, 1983; grantee Baker Commodities, Inc., L.A., 1983. Mem. Am. Inst. Nutrition, Am. Fisheries Soc., World Aquaculture Soc., Sigma Xi. Office: Ohio State U Sch Natural Resources 2021 Coffey Rd Columbus OH 43210

BAI, TAEIL ALBERT, research physicist; b. Muan, Jeonnam, Korea, July 16, 1945; came to U.S., 1972; s. Jong Hun and Soo-Bong (Suh) B.; m. Sue, 1973; children: Samuel, Jean, Helen. BS, Kyung Hee U., 1967; MS, U. Md., 1974, PhD, 1977. Asst. rsch. scientist U. Calif. San Diego, 1978-82; sr. rsch. scientist Stanford (Calif.) U., 1982—. Contbr. articles to The Astrophys. Jour., The Ann. Rev. Astron. Astrophysics, Solar Physics. Recipient Donald E. Billings award in Astro-Geophysics, 1978. Mem. Am. Phys. Soc., Am. Astron. Soc., Am. Geophys. Union. Achievements include original contbns. on classification of solar flares and the 154 day periodicity of solar activity; discovery that 25.5 days is the fundamental period of the sun. Office: Stanford U Stanford CA 94305

BAIER, AUGUSTO CARLOS, plant researcher; b. Bagé, Rio Grande do Sul, Brazil, May 10, 1941; s. Karl Fabian and Johanna (Bossard) B.; m. Selma Mielke; children: Luciane Mielke, Auro Augusto. Bel. Agronomy, E.A Eliseu Maciel, Pelotas, Brazil, 1968; D in Agr., Tech. U., Munich, 1972; postgrad., U. Mo., 1993—. Wheat breeder EMBRAPA/CNPT, Passo Fundo, Brazil, 1972-77; triticale breeder EMBRAPA/CNPT, Passo Fundo, 1977—, coordinator nat. triticale research program, 1981—, lupin breeder, 1982-91; coordinator, organizer Nat. Wheat Rsch. Ctr., Brazil, 1974; organizer 2d Internat. Triticale Symposium, 1990. Author: As Lavouras de Inverno I and II, 1988; contbr. articles to profl. jours. Recipient Medalha Marechal Hermes award Ministro do Exército, Bagé, 1960. Mem. Am. Soc. Agronomy, Soc. Brasileira P/Prog.da Ciência, Soc. Brasileira de Genética, Internat. Tritacale Assn. (pres. 1986-90), Rotary. Lutheran. Club: Passo Fundo (pres. 1982-83), Loj. Luz do Planalto (ven. master 1986-87). Avocation: gardening. Home: R Independencia 2245, 99025 Passo Fundo RS, Brazil Office: Nat Wheat Research Ctr, Cx Postal 569, 99001 Passo Fundo RS, Brazil

BAILAR, JOHN CHRISTIAN, III, public health educator, physician, statistician; b. Urbana, Ill., Oct. 9, 1932; married; 4 children. BA, U. Colo., 1953; MD, Yale U., 1955; PhD in Stats., Am. U., 1973. Intern U. Colo. Med. Ctr., Denver, 1955-56; field investigator, biometry br. Nat Cancer Inst., NIH, Bethesda, Md., 1956-62, head demography sect., 1962-70, dir. 3d nat. cancer survey, 1967-70, dep. assoc. dir. for cancer control, 1972-74; editor-in-chief JNCI, 1974-80; dir. research service VA, Washington, 1970-72; lectr. in biostats. Harvard U., Cambridge, Mass., 1980-87; prof. McGill U., Montreal, Que., Can., 1987—, chair dept. epidemiology and biostats., 1993—; sr. scientist Office Disease Prevention and Health Promotion, Dept. HHS, Washington, 1983-92; sr. scientist health and environ. rev. div. EPA, 1980-83; lectr. epidemiology and pub. health Yale U., New Haven, 1958-83; mem. faculty math. and stats. USDA Grad. Sch., Washington, 1966-76; vis. prof. stats. SUNY, Buffalo, 1974-80; professorial lectr. George Washington U., Washington, 1975-80; cons. in biostats. and epidemiology Dana-Farber Cancer Inst., Boston, 1977-83; vis. prof. Harvard U., 1977-79; spl. appointment grad. faculty U. Colo. Med. Ctr., Denver, 1979-81; scholar in residence NAS, 1992—. Mem. editorial adv. bd. Cancer Rsch., 1968-72; statis. cons. New Eng. Jour. Medicine, 1980-91; mem. bd. editors New England Jour. Medicine, 1992—; contbr. numerous articles to profl. jours. John D. and Catherine T. MacArthur Found. grantee, 1990—. Fellow AAAS, Am. Coll. Epidemiology, Am. Statis. Assn. (chair-elect and chair biometric sect. 1979-81, chair sect. stats. and environment 1990); mem. Inst. of Medicine, Internat. Statis. Inst., Coun. Biology Editors (chair publishing policy com. 1983-89, pres.-elect, pres., past pres. 1986-89), Soc. Risk Analysis (founding chair Boston chpt. 1985-86). Office: McGill U Sch Med, Dept Epidem and Biostats, Montreal, PQ Canada H3A 1A2

BAILE, CLIFTON A., biologist, researcher; b. Warrensburg, Mo., Feb. 8, 1940; s. Harold F. and Salome (Mohler) B.; m. Beth Lucile Hoover, Aug. 21, 1960; children: Christopher A., Marisa B. BS in Agr., Bus., Cen. Mo. State U., 1962; PhD in Nutrition, U. Mo., 1965; MA (hon.), U. Pa., 1979. NIH rsch. fellow Sch. Pub. Health Harvard U., Boston, 1964-66, from. instr. to asst. prof. Sch. Pub. Health, 1966-71; mgr. neurobiol. rsch. SmithKline Animal Health, Phila., 1971-75; from assoc. prof. to prof. Sch. Vet. Medicine U. Pa., Phila., 1975-82; disting. fellow, dir. R & D Monsanto Agrl. Co., St. Louis, 1982—; adj. prof. nutrition Sch. Medicine Washington U., St. Louis, 1982—; adj. prof. dept. animal sci. U. Mo., 1982—; presentor numerous seminars and symposiums. Contbr. over 225 articles and 200 abstracts to sci. jours. Ralston Purina rsch. fellow, 1962-64, NIH spl. postdoctoral fellow, 1969. Mem. Am. Soc. Animal Sci. (bd. dirs. 1990-93, animal growth and devel. award 1989), Am. Physiol. Soc., Am. Inst. Nutrition, Am. Dairy Sci. Assn. (Am. Feed Mgmt. award 1979), Soc. Neurosci. Achievements include 12 patents in field; research in control and feed intake and regulation of energy balance. Office: Monsanto Co BB3F 700 Chesterfield Village Pky Saint Louis MO 63198-0002

BAILEY, ARTHUR EMERY, natural products chemist; b. Niagara Falls, N.Y., Sept. 24, 1953; s. Randal E. and Ethel J. (Dorchak) B.; m. Karen C. Worman, July 16, 1988; 1 child, Evan Michael. BA, Duke U., 1975; PhD, U. Wyo., 1985. Post-doctoral Monell Inst., Phila., 1985-86; rsch. chemist Organon Tecknika, West Chester, Pa., 1986-87; program mgr. Pepsi Cola, Valhalla, N.Y., 1987-90; sr. rsch. chemist Lederle Labs., Pearl River, N.Y., 1990—. Lt. j.g. USN, 1975-79. Mem. Am. Chem. Soc.

BAILEY, DON MATTHEW, aerospace company executive; b. Pitts., Jan. 2, 1946; s. William and Vera (Mitchell) B.; m. Linda Reed, Sept. 15, 1967; children: Don Matthew Jr., Kirsten Paige, Terrance Reed. BSME, Drexel U., 1968; MS in Ops. Rsch., U. So. Calif., 1971; MBA, Pepperdine U., 1986. Programmer Naval Air Systems Command, Washington, 1963-69; engr. Rockwell Internat., Anaheim, Calif., 1969-71; program mgr. Logicon, Inc., San Pedro, Calif., 1971-80; sec., v.p. corp. devel. Comarco, Inc., Anaheim, 1980—; bd. dirs. Devel. Disabilities Ctr. Orange County, Anaheim, Calif., Perspective Instructional Communications, San Diego. Mem. Assn. for Corp. Growth. Office: Comarco Inc 160 S Old Springs Rd Anaheim CA 92808-1246

BAILEY, LEONARD LEE, surgeon; b. Takoma Park, Md., Aug. 28, 1942; s. Nelson Hulburt and Catherine Effie (Long) B.; m. Nancy Ann Schroeder, Aug. 21, 1966; children: Jonathan Brooks, Charles Connor. BS, Columbia Union Coll., 1960-64; postgrad., NIH, 1965; MD, Loma Linda U., 1969. Diplomate Am. Bd. Surgery, Am. Bd. Thoracic Surgery. Intern Loma Linda U. Med. Ctr., 1969-70, resident in surgery, 1970-73, resident in thoracic and cardiovascular surgery, 1973-74; resident in pediatric cardiovascular surgery Hosp. for Sick Children, Toronto, Ont., Can., 1974-75; resident in thoracic and cardiovascular surgery Loma Linda U. Med. Sch., 1975-76, asst. prof. surgery, 1976-86, prof. surgery, 1986—, dir. pediatric cardiac surgery, 1976—, chief div. cardiothoracic surgery, 1988, chair dept. surgery, 1992. Mem. AMA, ACS, AAAS, Calif. Med. Assn., San Bernadino County Med. Soc., Am. Coll. Cardiology, Tri-County Surg. Soc., Western Thoracic Surg. Assn., Soc. Thoracic Surgery, Western Soc. Pediatric Rsch., L.A. Transplant Soc., Internat. Soc. for Heart Transplantation, Lyman A. Brewer III Internat. Surg. Soc., Am. Heart Assn., Internat. Assn. for Cardiac Biol. Implants, Am. Soc. for Artificial Internal Organs, So. Calif. Transplant Soc., Pacific Coast Surg. Assn., Western Assn. Transplant Surgeons, Internat. Soc. for Cardiovascular Surgery, United Network for Organ Sharing, The Transplant Soc. Democrat. Seventh-day Adventist. Office: Loma Linda U Med Ctr Rm 2560 Loma Linda CA 92354

BAILEY, MARK WILLIAM, electrical engineer; b. Milw., July 3, 1945; s. Hamilton Tower and Marion Lois (Burtch) B.; m. Gail E. Mathews; children, Alexander, Hugh. BEE, U. Wis., 1968. Registered profl. engr., Ill. Engr. Barber Colman, Rockford, Ill., 1968-70; field engr. Commonwealth Edison, Rockford, 1970-73; engr. Commonwealth Edison, Chgo., 1974-81; prin. engr. Commonwealth Edison, Zion, Ill., 1981-88; sr. design engr. Fluor-Daniel, Chgo., 1988—. Active planning comm. City of Round Lake Beach, Ill., 1984—, Storm Water Mgmt. Tech. Adv. Com., Lake County, Ill., 1989—. Mem. IEEE, Masons. Home: 1413 N Pleasant Dr Round Lake Beach IL 60073 Office: Fluor Daniel 200 W Monroe Chicago IL 60606

BAILEY, NORMAN ALISHAN, economist; b. Chgo., May 22, 1931; s. Percival and Yevnige (Bashian) B.; m. Lorraine Baillargeon, Sept. 1, 1962 (div. Feb. 1966); m. Suzin Robbins, July 8, 1966; children: Stacy, Anthony, Samara, Gabrielle. AB, Oberlin (Ohio) Coll., 1953; MA, Columbia U., 1955, PhD, 1962; LLD, Hanyang U., Seoul, Korea, 1983. Economist Mobil Oil Co., N.Y.C., 1960-62; prof. CUNY, Queens, 1962-83; pres. Bailey, Tondu, Warwick & Co., N.Y.C., 1962-75; spl. asst. to Pres. Reagan The White House, Washington, 1981-83; pvt. practice Washington, 1984—. Author 8 books; contbr. numerous articles to profl. jours. Cons. Rep. Campaign, N.Y., Washington, 1980, 84, 88. With U.S. Army, 1956-58. Named Knight of the Order of Our Lady of the Conception of Vila Vicosa, 1988. Mem. The Univ. Club (Washington), Columbia Club of N.Y., Clube Micaelense (Ponta Delgada, Azores), Phi Beta Kappa. Avocations: swimming, boating, writing. Office: Norman A Bailey Inc 1912 Sunderland Pl NW Washington DC 20036-1608

BAILEY, THOMAS EVERETT, engineering company executive; b. Atlantic, Iowa, Mar. 30, 1936; s. Merritt E. and Clara May (Richardson) B.; m. Elizabeth Jane Taylor, Sept. 9, 1956; children: Thomas E., Douglas L., Steven W. BS, U. Iowa, 1959. Reg. profl. engr., environ. assessor, expert witness, arbitrator. Engr. Calif. Dept. Water Resources, Sacramento, 1960-67; sr. engr. Calif. Water Quality Control Bd., San Luis Obispo, 1967-72; asst. div. chief, dir. water quality planning State Water Resources Control Bd., Sacramento, 1972-75, chief div. planning rsch., 1975-77, chief tech. support br., 1977-79; sr. tech. advisor Yemen Arab Republic, Sana'a, 1979-81; chief Calif. superfund program Calif. Dept. Health Svcs., Sacramento, 1982-86; prin., v.p. Kleinfelder Inc., Walnut Creek, Calif., 1986-92; also bd. dirs. Kleinfelder Inc., Walnut Creek; pres. Bailey Environ., Sacramento, 1992—; cons. engr., arbitrator Calif. Hazardous Substance Cleanup Arbitration Panel. Contbr. articles to profl. jours. Mem. San Luis County Obispo Rep. Ctrl. Com., 1969-72, vice-chmn., 1970-71, chmn., 1971-72; vice-chmn. bd. trustees Meth. Ch., San Luis Obispo, 1970-72; mem. Contra Costa County Hazardous Materials Com., 1988-89, Calif. Remedial Action Group, co-chmn. 1991-92; With U.S. Army, 1959-60. Mem. ASCE, Cons. Engrs. and Land Surveyors of Calif. (chmn. hazardous waste mgmt. acad. 1992—), Am. Consulting Engrs. Coun., Hazardous Waste Action Coalition (chmn. bus. practices com. 1991-93, bd. dirs. 1992-93). Office: Bailey Environ Engring 7064 Riverside Blvd Sacramento CA 95831

BAILEY, WILLIAM NATHAN, systems analyst; b. Thomasville, N.C., Nov. 30, 1955; s. Charlie Franklin and Bonnie Mae (West) B.; m. Joy Linda Wagner, June 10, 1978 (dec. Feb. 1987); m. Belinda Carol Church, Aug. 8, 1988; 1 child, Benjamin Franklin. BFA, U. N.C., Greensboro, 1978. Sr. programmer CIBA-GEIGY Corp., Greensboro, N.C., 1986-88, sr. system analyst, 1990—; system mgr. N.C. A&T State U., Greensboro, 1988-90; chmn. DECUS PT-LUG, Greensboro, 1987-88, 90—, asst. chmn., 1988-89. Contbr. articles to profl. jours. Tchr. Faith & Victory Ch., Greensboro, 1988-92. Mem. IEEE, Assn. Computing Machinery, Digital Equipment Computer Users Soc. Republican. Office: CIBA-GEIGY Corp 410 Swing Rd Greensboro NC 27419

BAILLE, DAVID L., biologist, educator. Prof. dept. biology Simon Fraser U., Burnaby, B.C., Can. Recipient Luigi Provasoli award Phycological Soc. Am., 1990. Office: Dept of Biology, Simon Fraser University, Burnaby, BC Canada V5A 1S6*

BAINS, ELIZABETH MILLER, aerospace engineer; b. Phila., Feb. 13, 1943; d. Robert Keck and Grace Elizabeth (Clemons) Miller; m. James Albert Bains, Jr., Sept. 16, 1972. BS, Lebanon Valley Coll., 1964; MA in Coll. Teaching, U. Tenn., 1968, PhD, 1972. Asst. prof. Alcorn (Miss.) State U., 1976-79; engr. Lockheed Engring. Svcs. Corp., Houston, 1979-88, Johnson Space Ctr.-NASA, Houston, 1988—; dep. br. chief System Simulation br. Johnson Space Ctr.-NASA, Houston, 1990—. NSF grantee, 1977, summer student sci. tng. grantee, 1977. Assoc. fellow AIAA; mem. Nat. Coun. of Systems Engrs., Soc. for Computer Simulation. Achievements include patent on hopping mechanism for pipe and coupling inspection probe. Office: NASA/JSC MS ET5 Houston TX 77058

BAINUM, PETER MONTGOMERY, aerospace engineer, consultant; b. St. Petersburg, Fla., Feb. 4, 1938; s. Charles J. Bainum and Mildred (Trincher) Salyer; m. Carmen Cecilia Perez, Sept. 7, 1968; 1 child, David P. BS, Tex. A&M U., 1959; SM, MIT, 1960; PhD, Cath. U., 1967. Asst. engr. MIT Naval Supersonic Lab., Cambridge, Mass., 1959-60; sr. engr. Martin Co., Orlando, Fla., 1960-62; staff engr. IBM Fed. Systems Div., Bethesda, Md., 1962-65; sr. staff, aerospace engr., cons. Johns Hopkins U. Applied Physics Lab., Laurel, Md., 1965-69, 69-72; assoc. prof. Howard U., Washington, 1969-73, prof., 1973-90, disting. prof., 1990—; v.p. rsch., cons. WHF & Assocs., Bethesda, 1977-86; mem. NASA/PSN Tether Applications Simulation Working Group, 1987—; lectr. various internat. univs., rsch. ctrs. and confs.; hon. vis. prof. Universidad Francisco Marroquin, Guatemala, 1991. Editor, co-editor 12 books, 1981-92; author tech. reports and conf. proceedings; contbr. numerous articles to profl. jours. Judge, D.C. Sci. Fair, Washington, 1973. Recipient Ralph R. Teetor award Soc. Automotive Engrs., 1971. Fellow AIAA (capital sect. community action com. 1975-76), Am. Astronautical Soc. (v.p. internat. 1986—, Brouwer award 1990), AAAS, British Interplanetary Soc.,; mem. Internat. Acad. Astronautics, Sigma Xi. Office: Howard Univ Dept of Mechanical Engr Washington DC 20059

BAIR, WILLIAM ALOIS, engineer; b. Bklyn., Aug. 13, 1931; s. Henry Auchu and Anna Margaret (Zidar) B.; m. Patricia Anne Doyle, July 23, 1955; children: William A. Jr., Joseph M. Student, Pa. State U., 1949-51; BS in Engring., U.S. Naval Acad., 1955; BS in Civil Engring., Rensselaer Poly. Inst., 1958; MS in Nuclear Engring., U. Calif., 1966; grad. advanced mgmt. program, Wharton Sch., 1987. Registered profl. engr., N.Y., N.J., Pa., Conn., Md., Del., Va., S.C., Ga., D.C. Commd. ensign USN, 1955, advanced through grades to comdr., 1969; with USN Civil Engr. Corps, 1955-77; ret. USN, 1977; project mgr. Ebasco Svcs. Inc., Princeton, N.Y., 1977-85; dir. program planning and devel. Ebasco Svcs. Inc., N.Y.C., 1985—; appointed mem. spl. 3 man NATO tech. com. to evaluate effectiveness of European Airfield Phys. Protection program to counter damage from attack by Warsaw Pact Nations. Author: Helium 3 Neutron Spectrometer, 1966; contbr. articles to profl. jours. Scoutmaster Boy Scouts Am., Rockville, Md., 1969-70; coun. mem. European br., CAsteau, Belgium, 1971-75. Decorated Legion of Merit, Bronze Star with V; Cross of Gallantry, Medal of Honor 1st class (Vietnam). Mem. ASCE, NSPE, Am. Nuclear Soc., Soc. Am. Mil. Engrs., Nat. Contract. Mgmt. Assn., Am. Legion. Republican. Roman Catholic. Achievements include research on decontamination and demolition of radioactive structures. Home: 21 Lorrie Ln #15 Lawrenceville NJ 08648-5112 Office: Ebasco Svcs Inc Princeton Plasma Physic Lab PO Box 451 Forrestal Campus Princeton NJ 08543-0451

BAIR, WILLIAM J., radiation biologist; b. Jackson, Mich., July 14, 1924; s. William J. and Mona J. (Gamble) B.; m. Barbara Joan Sites, Feb. 16, 1952; children: William J., Michael Braden, Andrew Emil. B.A. in chemistry, Ohio Wesleyan U., 1949; Ph.D. in Radiation Biology, U. Rochester, 1954. NRC-AEC fellow U. Rochester, 1949-50, research asso. radiation biology, 1950-54; biol. scientist Hanford Labs. of Gen. Electric Co., Richland, Wash., 1954-56, mgr. inhalation toxicology sect., biology dept., 1956-65; mgr. inhalation toxicology sect., biology dept. Battelle Meml. Inst., 1965-68; mgr. biology dept. Pacific Northwest Labs., Richland, Wash., 1968-74, dir. life scis. program, 1973-75, mgr. biomed. and environ. research program, 1975-76, mgr. environ. health and safety research program, 1976-86, mgr. life scis. ctr., 1986-93; demonstrated toxicology of plutonium and carcinogenesis of radioactive particles in lung; lectr. radiation biology Joint Ctr. Grad. Study, Richland, 1955-75; cons. to adv. com. on reactor safeguards Nuclear Regulatory Commn., 1971-87; mem. several coms. on plutonium toxicology; mem. subcom. inhalation hazards, com. pathologic effects atomic radiation NAS, 1957-64, mem. ad hoc com. on hot particles of subcom. biol. effects ionizing radiation NAS-NRC, 1974-76, vice chmn. com. on Biol. effects of ionizing radiation, IV Alpha radiation, 1985-88, chmn. task force on biol. effects of inhaled particles Internat. Commn. on Radiol. Protection, 1970-79, mem. com. 2 on permissible dose for internal radiation, 1973-93, chmn. task group on respiratory tract models, 1984-93; mem. Nat. Coun. on Radiation Protection and Measurements, 1974-92, hon. mem., 1992, bd. dirs., mem. com. of radionuclides on maximum permissible concentrations for occupational and nonoccupational exposure, 1970-74, mem. com. basic radiation protection criteria, 1975-92, chmn. ad hoc com. on hot particles,, 1974, chmn. ad hoc com. internal emitter activities, 1976-77, mem. com. on internal emitter standards, 1977-92. Author 200 books, articles, reports, chpts. in books. Recipient E.O. Lawrence Meml. award AEC, 1970; cert. of appreciation AEC, 1975; Alumni Disting. Achievement citation Ohio Wesleyan U., 1986. Fellow AAAS, Health Physics Soc. (bd. dirs. 1970-73, 83-86, pres. elect 1983-84, pres. 1984-85, Disting. Sci. Achievement award 1991); mem. Radiation Rsch. Soc., N.Y. Acad. Sci., Soc. Exptl. Biology and Medicine (vice chmn. N.W. chpt. 1967-70, 74-75), Reticuloendothelial Soc., Soc. Occupational and Environ. Health, Sci. Soc. Pres. (coun. 1984-85), Sigma Xi. Club: Kiwanis (dir.). Home: 102 Somerset St Richland WA 99352-1966 Office: Battelle Pacific NW Labs PO Box 999 Richland WA 99352-0999

BAIRAM, ERKIN IBRAHIM, economics educator; b. Nicosia, Cyprus, Apr. 7, 1958; arrived in New Zealand, 1987; s. Ibrahim and Ayshe Kahya. BA in Econs. with honors, Essex U., Colchester, Eng., 1980; MA in Econometrics, Hull (Eng.) U., 1982, PhD in Econs., 1986. Asst. lectr. Hull U., 1985-86; lectr. Otago U., Dunedin, New Zealand, 1987-88, sr. lectr., 1989, assoc. prof., 1990, chairperson dept. econs., 1990, prof., 1991—. Author: Applied Econometrics, 1988, 91, 93; contbr. over 60 articles to refereed internat. jours. Recipient Keio Econ. Soc. award Keio U., Japan, 1988, 89, 90. Mem. Econometric Soc., Royal Econ. Soc., Am. Econ. Assn. Avocations: chess, movies. Office: Otago U, Box 56, Dunedin New Zealand

BAIRD, HAYNES WALLACE, pathologist; b. St. Louis, Jan. 28, 1943; s. Harry Haynes and Mary Cornelia (Wallace) B.; m. Phyllis Jean Tipton, June 26, 1965; children—Teresa Lee, Christopher Wallace, Kelly Wallace. BA, U. N.C., 1965, MD, 1969. Diplomate Am. Bd. Pathology. Radio announcer, disc jockey, 1961-63; intern N.C. Meml. Hosp., Chapel Hill, 1969-70, resident in pathology, 1970-72, chief resident in pathology, 1972-73; assoc. pathologist Moses H. Cone Meml. Hosp., Greensboro, N.C., 1973—; practice medicine, specializing in pathology Greensboro, 1973—; clin. asst. prof. U. N.C., Chapel Hill, 1978—; clin. lectr. chemistry U. N.C., Greensboro, 1973—. Mem. adminstrv. bd. West Market St. United Meth. Ch., 1985-88, usher, 1988—; bd. dirs. Greensboro unit Am. Cancer Soc., 1980-81. Fellow Coll. Am. Pathologists (ho. of dels. 1983-85); mem. AMA, So. Med. Assn., Am. Assn. for Clin. Chemistry, Am. Soc. Cytology, Am. Soc. Clin. Pathologists, Internat. Acad. Pathology, N.C. Med. Soc., Guilford County Med. Soc., N.C. Soc. Pathologists (sec.-treas. 1977-79), Greensboro Acad. Medicine. Methodist. Home: 2805 New Hanover Dr Greensboro NC 27408-6705 Office: 1200 N Elm St Greensboro NC 27401-1020

BAIRD, ROSEMARIE ANNETTE, pharmacist; b. Kingston, Jamaica, Oct. 30, 1956; came to U.S., 1981; d. Canute Egbert and Kathleen Gloria (Miles) B. BS cum laude, L.I. U., 1983, postgrad. in pharmacy and bus. adminstrn., 1984—. Registered pharmacist, N.Y., N.J. Grad. asst. Dean Undergrad. Bus. Sch. L.I. (N.Y.) U., 1983-85; pharmacist Nat. Pharmacy, Elmwood, N.J., 1985-86; pharmacist II Manhattan (N.Y.) Devel. Ctr., 1986-87, Kingsboro Psychiat. Ctr., Bklyn., 1987—; sr. assoc. pharmacist Kings County Hosp., Bklyn., 1986—. Continuing Edn. scholar; Columbia Pharmacy Sch. scholar. Mem. Am. Soc. Hosp. Pharmacists, N.Y. State Coun. Hosp. Pharmacists, Optima-Biology Honor Soc., Caribbean-Am. C. of C., Empire Pharm. Soc., Rho Chi, Pharmacy Honor Soc. Anglican. Avocations: reading, arts, crafts. Home: 1200 E 53rd St Apt 6D Brooklyn NY 11234-2340

BAIRD, WILLIAM MCKENZIE, chemical carcinogenesis researcher, biochemistry educator; b. Phila., Mar. 23, 1944; s. William Henry Jr. and Edna (McKenzie) B.; m. Elizabeth A. Myers, June 21, 1969; children: Heather Jean, Elizabeth Joanne, Scott William. BS in Chemistry, Lehigh U., 1966; PhD in Oncology, U. Wis., 1971. Postdoctoral fellow Inst. Cancer Research, London, 1971-73; from asst. to assoc. prof. biochemistry Wistar Inst., Phila., 1973-80; assoc. prof. medicinal chemistry Purdue U., West Lafayette, Ind., 1980-82, prof., 1982—, Glenn L. Jenkins prof. medicinal chemistry, 1989—, dir. Cancer Ctr., 1986—; faculty participant cancer ctr., biochemistry program Purdue U., 1980—; adv. com. on biochemistry and chem. carcinogenesis Am. Cancer Soc., 1983-86; mem. chem. pathology study sect. NIH, 1986-90. Contbr. articles to profl. jours.; assoc. editor Cancer Rsch. Grantee NCI. Mem. ISSX, AAAS, Am. Assn. Cancer Rsch., Am. Soc. Biol. Chemists, Am. Chem. Soc., Am. Soc. Biochemistry and Molecular Biology, Environ. Mutagen Soc. Office: Purdue U Cancer Ctr Life Scis Rsch Bldg West Lafayette IN 47907

BAISCH, STEVEN DALE, pediatrician; b. Glendive, Mont., May 30, 1955; s. Maynard Jack and Edith Maxine (Milne) B.; children: Christopher, Rebecca. BA with honors, Jamestown Coll., 1977; Calculus fellow, Harvard U., 1975; MD, U. N.D., 1981. Pediatric resident U. N.D. Med. Ctr., 1981-84, asst. chief resident, 1983-84; ptnr., pres. Panhandle Pediatric Clinic, P.C., Scottsbluff, Nebr., 1984-88; pres. Region West Pediatric Svcs., P.C., Scottsbluff, 1988—; fellow pediatric critical care medicine U. Minn., Mpls., 1991—; mem. staff Children's Hosp. of St. Paul, 1992—; instr. pediatrics U. Nebr. Med. Ctr., Omaha, 1984—; med. dir. Camp Cosmos-Diabetic Camp, Scottsbluff, 1984—; dir. Asthma Care Tng. Program, Scottsbluff, 1986—; med. staff Regional West Med. Ctr. Active 1st United Meth. Ch., Scottsbluff, Scottsbluff Little League Baseball/Football. U. Minn. Dept. Pediatrics Mark Snelling Meml. Outstanding fellow teaching award, 1993; U. Minn. fellow, 1991—. Fellow Am. Acad. Pediatrics (PREP award 1990). Office: U Minn Dept Ped Critical Care Med Box 737 UMHC Minneapolis MN 55455

BAJPAI, PRAMOD KUMAR, chemical and biochemical engineer; b. Kanpur, Utter Prad, India, Jan. 1, 1951; s. Kamal Narain and Shiv Kumari (Tripathi) B.; m. Pratima Pande, June 30, 1975. BSc in Chem. Engring., Harcourt Butler Tech. Inst., Kanpur, India, 1971, MSc in Chem. Engring. Practice, 1973; PhD in Chem. Engring., Indian Inst. Tech., Kanpur, India, 1977. Rsch. assoc. U. Western Ontario, London, Ontario, Can., 1981-82; lectr. U. Rorkee, Rorkee, U.P; rsch. engr. and head chem. engring div. Thapar Cor. R&D Ctr., Patiala, Punjab, India, 1984—; adj. prof. Environ. Engring. Thapar Inst. Engring. and Tech., Patiala, 1989-90, prof chem. engring, 1991—; vis. prof. microbial biotech., U. Waterloo, Ont., 1990-91. Contbr. articles to Canadian Jour. Chem. Engring., Biotech. and Bioengring., Biotech. Technique, Enzyme and Microbial Tech., Process Biochemistry. Recipient postdoctoral fellowship U. Saskatchewan, Saskatoon, 1978. Mem. AICE, Indian Inst. Chem. Engrs., Indian Soc. for Tech. Edn. (life), Bioenergy Soc. India. Hindu. Achievements include patents on a novel method of immobilization of yeast or bacteria; a process for enhanced production of thermostable alpha-anaylase enzyme; research in design of bioreactors, immobolized cell fermentation, membrane technology, environmental engineering. Office: Thapar Corp R&D, Ctr, Punjab Patiala 147001, India

BAK, MARTIN JOSEPH, biomedical engineer; b. Washington, Apr. 15, 1947; s. Anthony Frank and Irene Louise (Hutton) B.; m. Tina Mauree Hight, Jan. 22, 1972; children: Mauree, Brian, Cheryl, David. BSEE, U. Md., 1972. Biomed. engr. NIH, Bethesda, Md., 1970—; prin. investigator NIH, Bethesda, Md., 1979—; cons. Micro Probe, Inc., Clarksburg, Md., 1984—, Multichannel Concepts, Inc., Gaithersburg, Md., 1991—. Trustee Living Word Fellowship, Gaithersburg, 1987-92. Cpl. U.S. Army Res., 1969-75. Mem. AAAS, IEEE, Soc. Neurosci., N.Y. Acad. Scis. Achievements include patent for thumbtack microelectrode and method of making same; development of floating microelectrode which is presently being implanted chronically into the visual cortex of blind patient vols. to test the feasibility of a visual prosthesis for the blind. Office: NIH LNLC Bldg 49 Rm 3A50 Bethesda MD 20892

BAKALIAN, ALEXANDER EDWARD, public health engineer; b. Beirut, July 25, 1957; s. Edward Dikran and Verkin Bedros (Arpadjian) B. BCE, Am. U., Beirut, 1979; PhD in Environ. Engring., Johns Hopkins U., 1987; MBA, George Washington U., 1992. Project engr. Saulex Engrs. and Contractors, Alkhobar, Saudi Arabia, 1979-80; environ. specialist World Bank, Washington, 1986—. Contbr. articles to profl. jours. Bd. dirs. Miriam's Kitchen Feeding Program for Homeless, Washington, 1990—; vol. ARC, Balt. 1984-86. Mem. Armenian Network Am. (v.p. 1990-91, pres. 1992—), World Affairs Coun., Water Pollution Control Fedn., Am. WaterWorks Assn. Armenian Orthodox. Home: 10303 Montrose Ave Bethesda MD 20814-4151 Office: World Bank 1818 H St NW Washington DC 20433-0002

BAKAY, ROY ARPAD EARLE, neurosurgeon, educator; b. Chgo., Mar. 5, 1949; s. Archie Joseph and Marjorie (Jordal) B.; m. Joann P. Feiertag; children: Mark, Scott, Candace, Jacqueline. BS, Beloit Coll., 1971; MD, Northwestern U., 1975. Diplomate Am. Bd. Med. Examiners, Am. Bd. Neurol. Surgeons. Intern U. Mich., Ann Arbor, 1975-76; resident in neurosurgery U. Wash., Seattle, 1976-82; acting instr., asst. in neurosurgery U. Wash. Med. Sch., Seattle, 1980-82, NIH fellow, 1981-82; asst. prof. sect. neurol. surgery Emory U. Med. Sch., Atlanta, 1982-88, dir. neurol. surgery resident rsch., 1984—, assoc. prof., 1988-93, prof., 1993—; mem. R & D Com. VA Med. Ctr., Decatur, Ga., 1982-86, sect. chief neurol. surgery, 1982—; affiliate scientist neurobiology Yerkes Regional Primate Rsch. Ctr., Atlanta, 1982—. Author: (with others) Yearbook of Science and Technology, 1989; abstractor Jour. Surg. Gynecology and Obstetrics, 1978—; mem. editorial Bd. Jour. Contemporay Neurosurgery, 1987—; contbr. articles to profl. jours., chpts. to books. Chmn. profl. adv. bd. Ga. chpt. Epilepsy Found. Am., 1987-88; mem. adv. panel U.S. Congl. Office Tech. Assessment, Washington, 1988-90; profl. rep. Am. Cancer Soc., Atlanta, 1987-90. Recipient Resident Rsch. award Western Neurosurgery Soc., 1979, No. Pacific Soc. Neurology and Psychiatry, 1979, Soc. Neurology Anesthesists and Neurology Supportive Care, 1981; named one of Outstanding Athletes of Am., 1971. Mem. AAAS, Soc. for Neurosci., Am. Stereotactic and Functional Neurosurgeons (v.p. 1988-91, pres. 1991—), Am. Assn. Neurol. Surgeons (chmn. GRAFT Registry Com. 1987—), Congress Neurol. Surgeons (v.p. joint com. 1988-91, pres. 1991—). Presbyterian. Avocations: hiking, camping, skiing, fishing, team sports. Office: The Emory Clinic 1327 Clifton Rd NE Atlanta GA 30307-1013

BAKER, BRUCE EDWARD, orthopedic surgeon, consultant; b. Oswego, N.Y., Mar. 22, 1937; s. Elbert J. and Reatha (Hartranft) B.; m. Patricia Therese Gormel, Aug. 19, 1961; children: Brett, Clayton, Sean, Reatha. BSME, Syracuse U., 1959; MD, SUNY-Syracuse, 1965. Intern State U. Iowa, Iowa City, 1965-66, asst. resident, 1966-67; resident orthopaedics SUNY-Upstate Med. Ctr., Syracuse, 1969-72, NIH orthopaedic rsch. fellow, 1972-73, asst. prof. orthopaedic surgery, 1973-79, assoc. prof., 1979-86, prof., 1986-89; dir. univ. sports medicine service div. dept.

orthopaedic surgery 1980-89; team physician, dir. sports medicine athletic dept., Syracuse U., 1973—, orthopaedic cons. Student Health Ctr., 1973—; staff SUNY Hosp., Syracuse, 1973-89, Syracuse VA Hosp., 1973-89, A.C. Silverman Pub. Health Hosp., 1973-77, Crouse-Irving Meml. Hosp., 1973—; cons. in field. Contbr. numerous articles to profl. jours. Capt. USAF, 1967-69. Recipient AMA Physicians Recognition award, 1978, Bronze medal award Am. Roentgen Ray Soc., 1980, Gold medal award Sound Slide Prodn. Conditioning, 1977; Syracuse U. scholar, 1955; N.Y. State Regents scholar, 1955-59; USPHS grantee, 1973-74; Hendricks Research fund grantee, 1973-75; NIH grantee, 1974-76, 76-77. Fellow ACS, Am. Acad. Orthopaedic Surgeons; mem. AME, Med. Soc. State N.Y., Onondaga County Med. Soc., Orthopaedic Rsch. Soc., Am. Coll. Sports Medicine, Am. Orthopaedic Soc. for Sports Medicine, N.Y. Soc. Orthopaedic Surgeons, Royal Soc. Medicine, Internat. Arthroscopy Assn., Arthroscopy Assn. N.Am., Biolelec Repair and Growth Soc. Office: 475 Irving Ave Ste 108A Syracuse NY 13210-1756

BAKER, BRUCE S., molecular biologist. Professor Stanford University Dept. of Biological Sciences, Stanford, CA. Recipient NAS Molecular Biology award, Nat. Acad. Sci., 1992. Office: Stanford Univ Dept of Biol Scis Stanford CA 94305

BAKER, CARL GWIN, science educator; b. Louisville, Ky., Nov. 27, 1920; s. Edward Forrest and Naomi (Taylor) B.; m. Lois Eleane Oxsen, Mar. 24, 1949 (div. May 1975); children: Cathryn, Jeanette; m. Catherine Valerie Smith, May 23, 1975. AB in Zoology, U. Louisville, 1942, MD, 1944; MA in Biochemistry, U. Calif., Berkeley, 1949; DSc, U. Louisville, 1980. Lic. med. practice, Ky., Calif. Rsch. investigator NIH, Bethesda, Md., 1949-55; asst. to the assoc. dir. NIH, 1958-61; assoc. dir. program Nat. Cancer Inst., NIH, Bethesda, Md., 1961-67, sci. dir. etiology, 1967-69, dir., 1969-72; dir. program policy staff Health & Human Svcs. Adminstr., Rockville, Md., 1975-76; med. dir. Ludwig Inst. Cancer Rsch., Zurich, Switzerland, 1977-85, ret., 1985; adj. instr. U. Md., College Park, 1989—; governing coun. Internat. Agy. for Cancer Rsch., Lyon, France, 1969-72. Assoc. editor Jour. of the Nat. Cancer Inst., 1954-55; mem. editorial adv. bd. Cancer Jour., 1965-73; contbr. articles to jours. Biochemsitry, Immunology, Mgmt. Sci. Del. State Bd. Edn., Annapolis, Md., 1957; mem. exec. com. adv. panel on health Am. Revolution Bicentenial Commn., Washington, 1970-72; v.p. 10th Internat. Cancer Congress, Houston, 1970. Asst. surgeon gen. USPHS, 1970—. Decorated PHS Meritorious Svc. medal; Jane Coffin Childs Fund fellow, 1946-48, Spl. fellow NCI, NIH, 1949. Mem. Am. Assn. Cancer Rsch. (bd. dirs. 1972-76), Am Chem. Soc. (div. biol. chemistry, sec. 1955-57, councillor 1958-61), Am. Soc. Biochemistry and Molecular Biology, Cosmos Club. Achievements include research in application of systems analysis and planning to strategic planning in medical research and laying the foundations for development of national cancer plan. Home: 19408 Charline Manor Rd Olney MD 20832

BAKER, CAROLYN ANN, research biochemist; b. Fort Wayne, Ind., Mar. 2, 1936; d. Howard L. and Edith L. Wisner; m. Nicholas Baker, June 5, 1965 (dec. July 25, 1991); children: Barbara L., Wilte H. BS, Ind. U., 1958; MS in Microbiology, U. Mich., 1960. Registered med. tech. Bacteriology rsch. asst. Ind. U., Bloomington, 1955-57; med. tech. Thorton-Haymond Med. Lab., Indpls., 1957-58; teaching asst. dept. microbiology U. Mich., Ann Arbor, Mich., 1959-60; virologist The Upjohn Co., Kalamazoo, Mich., 1960-66; med. tech. Bronson Hosp., Kalamazoo, Mich., 1967-75; rsch. biochemist The Upjohn Co., Kalamazoo, Mich., 1975—. Contbr. articles to Jour. Investigative Dermatology, Life Sci. Mem. Soc. for Investigative Dermatology. Home: 5817 N Westnedge Kalamazoo MI 49004 Office: The Upjohn Co 7240-209-6 301 Henrietta St Kalamazoo MI 49001

BAKER, CHARLES H., engineering company executive; b. 1951. With Rohm & Haff Seed Co., 1972-81, dir., 1981—; with North Am. Plant Breeders, 1972-81, dir., 1981—; pres. Biotechnica Agriculture, Inc., 1987—, CEO. Office: Biotechnica Internat Inc 7300 W 110th St Overland Park KS 66210*

BAKER, D. JAMES, federal agency administrator; m. Emily Lind. BS in Physics, Stanford Univ.; PhD in Physics, Cornell Univ. Post doctoral fellow Univ. Calif., Berkeley, Univ. R.I.; co-founder, dean Coll. Ocean and Fishery Scis., Univ. Wash., Seattle; pres. Joint Oceanographic Institutions Inc. (JOI), Wash., 1983-93; under sec. of commerce Dept. Commerce, Oceans and Atmosphere, Wash., 1993—; adminstr. Dept. Commerce, NOAA, Wash., 1993—; leader deep sea physics group NOAA Pacific Marine Environ. Lab.; faculty mem. Harvard Univ.; advisor White House, Nat. Acad. Scis. Author: Planet Earth--the View from Space; contbr. over 80 articles to profl. jours. Am. Assn. Advancement Sci. fellow, 1988. Mem. The Oceanography Soc. (co-founder, past pres.), Coun. Ocean Affairs (develoeped and formed orgn.). Achievements include research in ocean circulation and air sea interaction, initiating and coordinating major programs in marine science from ocean drilling to satellites; participation in research expeditions to most of the major oceans and holds a joint patent for a deep sea pressure gauge. Office: Dept of Commerce (NOAA) 14th & Constitution Ave NW Washington DC 20230*

BAKER, DALE EUGENE, soil chemist; b. Marble Hill, Mo., May 5, 1930; s. William Emroe and Daisy Almiranda (Wilkinson) B.; m. Doris LaDonna Scheper, Sept. 10, 1949; children: Carol Sue, Kenneth Lee, Donna Kay. BS in Agr., U. Mo., 1957, MS in Soils, 1958, PhD in Soil Chemistry, 1960; postgrad., Purdue U., 1968. Asst. prof. and asst. supr. U. Minn., N.E. Expt. Sta., Duluth, 1960-61; asst. prof. Pa. State U., University Park, 1961-65, assoc. prof., 1965-70, prof. soil chemistry, 1970—; pres. Land Mgmt. Decisions, Inc., State College, Pa., 1988—; vis. scientist Am. Soc. Agronomy, 1965-67. Contbr. over 200 articles to Soil Sci. Plant Analysis, Trace Substances in Environ. Health, others. Chmn. Dem. Party, State College, 1976-78. Rsch. awardee, Gamma Sigma Delta, 1970, N.E. Br. Am. Soc. Agronomy, 1972, U. Mo. Alumni Assn., 1985, Water Pollution Control Assn. Pa., 1986. Fellow AAAS (exec. com. sect. O 1985-90), Am. Soc. Agronomy (chmn. budget and fin. 1978, pres. N.E. br. 1983), Soil Sci. Soc. Am. (chmn. soil chemistry div. 1977), Am. Coll Ash Assn. (life), Lions (pres. 1990-91), Sigma Xi, Gamma Sigma Delta, Phi Epsilon Phi, Phi Lambda Upsilon. Democrat. Baptist. Achievements include research in soil physical chemistry applications in land management leading to development of the trademarked Baker Soil Test, in physical chemistry applications relating plant physiological responses to properties of soils, spoils, and wastes for integrated waste management for land reclamation and environmental quality, in soil-plant-animal health interrelationships. Home: 1429 Harris St State College PA 16803-3024 Office: Land Mgmt Decision Inc 3048 Research Dr State College PA 16801-2782

BAKER, DAVID B., environmental scientist; b. Akron, Ohio, May 29, 1936; s. Arlus Roosevelt Baker and Mary Eleanor Brown; m. Margaret Jane Swinehart, June 18, 1960; children: Sarah Jane, Mark David, Susan Carol. BS, Heidelberg Coll., 1958; MS in Botany, U. Mich., 1960, PhD in Botany, 1963. Asst. prof. Rutgers U., New Brunswick, N.J., 1964-66; prof. Heidelberg Coll., Tiffin, Ohio, 1966—; bd. dirs. Water Quality Lab., Tiffin, 1970—. Contbr. articles to profl. jours. Fellow Soil and Water Conservation Soc. (honor award 1984), Ohio Acad. Sci.; mem. Am. Chem. Soc., Agronomy Soc., Ohio Acad. Sci. Presbyterian. Achievements include development and operation of major water quality monitoring program for Ohio tributaries to Lake Erie, development of major voluntary private well monitoring program; research includes pesticide risk assessment in drinking water supplies, operation of large scale, long term agricultural ecosystem studies. Office: Heidelberg College 310 East Market St Tiffin OH 44883

BAKER, DONALD JAMES, federal official, oceanographer, administrator; b. Long Beach, Calif., Mar. 23, 1937; s. Donald James and Lillian Mae (Pund) B.; m. Emily Lind Delman, Sept. 7, 1968. B.S., Stanford U., 1958; Ph.D., Cornell U., 1962; LHD (hon.), Nova U., 1993. Postdoctoral fellow Grad. Sch. Oceanography, U. R.I., Kingston, 1962-63; NIH fellow in chem. biodynamics Lawrence Radiation Lab., U. Calif., Berkeley, 1963-64; research fellow, asst. prof., assoc. prof. phys. oceanography Harvard U., 1964-73; research prof. dept. oceanography, sr. oceanographer, applied physics lab. U. Wash., Seattle, 1973-77; sr. fellow Joint Inst. for Study Atmosphere and Ocean, 1977-86; group leader deep sea physics group Pacific Marine Environ. Lab., NOAA, Seattle, 1977-79; prof. dept. oceanography Joint Inst. for Study Atmosphere and Ocean, 1979-86, chmn. dept. oceanography, 1979-81; dean Coll. Ocean and Fishery Scis., 1981-83; under sec. NOAA, Washington, DC, 1993—; bd. govs. Joint Oceanographic Instns., Inc., 1979-93, pres., 1983-93; disting. vis. scientist Jet Propulsion Lab., Calif. Inst. Tech., 1982-93; undersec. for oceans and atmosphere and adminstr. Nat. Oceanic and Atmospheric Adminstrn. U.S. Dept. Commerce, 1993—; co-chmn. exec. com. Internat. So. Ocean Studies (NSF project), 1974-84; vice-chmn. joint panel, global weather experiment NAS, mem. ocean scis. bd., ocean scis. policy bd., ocean studies bd., 1987-91; mem. com. on atmospheric scis., 1978-81, mem. climate rsch. com., 1979-90; mem. space and earth sci. adv. com. NASA, 1982-86, mem. space sci. bd., 1984-87, chmn. com. on earth scis., 1984-87; mem. environ. panel Navy Rsch. Com., 1983-86, mem. NASA earth system scis. com., 1983-86; officer Joint Sci. Com. for World Climate Rsch. Program, 1987-93; mem. U.S. nat. com. for internat. coun. sci. unions NAS, 1987-89, com. on global change rsch., 1987-93, com. on environ. rsch., 1991-93; mem. U.S. Sci. Steering Com. for World Ocean Circulation Experiment, 1985-90; chmn. NASA Ctr. Sci. Assessment Team, 1987-88; chmn. Internat. World Ocean Circulation Experiment Sci. Steering Group, 1988-92; mem. earth sci. and applications adv. com. NASA, 1989-90; chmn. tech. com. on ocean processes and climate Intergovtl. Oceanographic Commn., 1989-93; bd. dirs. Coun. of Ocean Law, 1989-91; mem. panel on megaprojects Pres.'s Coun. of Advisors on Sci. and Tech., 1991-93; mem. NASA Earth Observing System Engring. Rev. Com., 1990-92; mem. com. on the future of the U.S. Space Program NASA, 1990-92; mem. adv. panel on climate and global change Nat. Oceanic and Atmospheric Adminstrn., 1989-93; mem. V.P.'s Space Policy Adv. Bd., 1991-93; mem. adv. panel Earth Observing System Data and Info. System., 1992-93, adv. com. divsn. of polar programs, NSF, 1992—; liaison mem. Pres.'s Commn. on Sustainable Devel., 1993—. Author: Planet Earth: The View from Space, 1990; mem. editorial bd. Oceanus Mag., 1992-93; co-editor in chief: Dynamics of Atmospheres and Oceans Jour., 1975-79; contbr. sci. articles to profl. jours. Fellow AAAS, Explorers Club; mem. Am. Geophys. Union, Am. Meteorol. Soc. (coun. 1982-88), The Oceanography Soc. (pres. 1988-92), Marine Tech. Soc., Nat. Assn. State Univs. and Land Grant Colls. (bd. dirs. marine div. 1989-92), Sigma Xi. Achievements include patent for deep-sea pressure gauge. Office: NOAA US Dept Commerce Washington DC 20230

BAKER, EDWARD GEORGE, retired mechanical engineer; b. Freeport, N.Y., Oct. 20, 1908; s. Edward George and Mary (Dunham) B.; m. Mary Louise Freer, Feb. 7, 1931; children—Edward Clark, Marna Larson, Ellen Freer (Mrs. George W. Lewis), John Durrin, Bruce Robert. B.A., Columbia Coll., 1930, M.A., 1931, Ed.D., 1938. Assoc. prof. math. Newark Coll. Engring., N.J., 1930-42; mem. tech. staff Am. Bur. Shipping, N.Y.C., 1942-73. Author: First Course in Mathematics, 1942. Contbr. articles on marine engring. to profl. jours. Pres. Nutley (N.J.) Symphony Soc., 1939-41; chmn. zoning bd. of adjustment, Pine Knoll Shores, N.C., 1979-84. Recipient Order of Long Leaf Pine award State of N.C., 1982. Mem. Am. Math. Soc., ASME, Soc. Naval Architects and Marine Engrs., N.Y. Acad. Sci., Phi Beta Kappa. Republican. Episcopalian. Home: 106 Carob Ct Pine Knoll Shores NC 28512-6205

BAKER, GEORGE HAROLD, III, physicist; b. Cheverly, Md., Mar. 23, 1949; s. George Harold Jr. and Betty (Fost) B.; m. Donna Prillaman, Jun 21, 1975; children: Matthew C., Jeffrey P., Virginia E. MS, U. Va., 1974; PhD, USAF Inst. Tech., Dayton, Ohio, 1987. Teaching asst. U. Va., Charlottesville, 1971-73; physicist Harry Diamond Labs., Adelphi, Md., 1973-77; physicist Def. Nuclear Agy., Alexandria, 1977-87, group leader, 1987-89, asst. for program devel., 1989—. Contbr. articles to profl. jours. Tchr. Agape Christian Fellowship, Chantilly, Va., 1974—; music and youth leader New Life Fellowship, Annandale, Va., 1979-83; canvasser Citizens for Sensible County Planning, Fairfax County, Va., 1989—. Fellow Nuclear Electromagnetic Soc. (chmn. program com. 1984); mem. Am. Phys. Soc., IEEE (session chmn. 1987, 92), Forum for Mil. Application of Directed Energy, Phi Delta Theta. Achievements include patent for optically coupled differential voltage probe, 1976; co-developer sea-going nuclear emp simulator concept, 1979; initiated Def. Nuclear Agy. EMP underground test program, 1983, High Power Microwave program, 1984. Office: Def Nuclear Agy 6801 Telegraph Rd Alexandria VA 22310

BAKER, GREGORY RICHARD, mathematician; b. Johannesburg, Transvaal, Republic South Africa, Nov. 9, 1947; came to U.S., 1972; s. Mervyn Colin and Valerie Rita (Deary) B.; M. Joanne Broker, Nov. 5, 1971 (div. Apr. 1978); 1 child, Kim; m. Maryellen Asgeirsson, Oct. 7, 1979; 1 child, Kathryn Ann. BS, U. Natal, Durban, Republic South Africa, 1970, MS, 1973; PhD, Calif. Inst. Tech., 1977. Rsch. fellow Calif. Inst. Tech., Pasadena, 1976-77; instr. MIT, Cambridge, 1977-79, asst. prof., 1979-81; assoc. prof. U. Ariz., Tucson, 1981-86; rsch. math. Exxon Rsch. and Engring. Co., Annondale, N.J., 1986-88; eminent scholar Ohio State U., Columbus, 1988—; cons. Cambridge Hydrodynamics Inc., Princeton, N.J., 1978-86, ICASE, NASA-Langley, 1980-81; mem. applied math. rev. panel for Dept. of Engery, 1993. Contbr. articles to profl. jours. including Physics of Fluids, Jour. Fluid Mechanics. Recipient Presdl. Young Investigator award, NSF, 1984. Mem. Soc. Indsl. and Applied Math., Am. Physics Soc. Achievements include development of reliable numerical methods for studies of evolution of free-surfaces in incompressible fluid flow. Office: Ohio State Univ Math Dept Columbus OH 43210

BAKER, HELEN MARIE, health services executive; b. Tulsa, Oct. 12, 1946; d. Joseph Donald and Caroline Emma (Nelson) Waldhelm; m. Lewis Edward Browder, 1964 (div. 1966); m. Lawrence Selden Baker, Nov. 23, 1978; children: Lawrence Nelson, Marjorie Lyn. Student, U. Tex., 1965-66. Staff asst. to pres. White House, Washington, 1970-73; v.p. Mgmt., Systems, Sales, Inc., Washington, 1973-74, Inter-Am. Svcs., Inc., Washington and Tex., 1974-83; v.p. Med. Diversified Svcs. Inc., San Antonio, 1983-90, exec. v.p., 1990-92, also bd. dirs., pres., CEO, 1992—. Editor newsletter Physician and Family, 1983-86. Elder St. Andrew Presbyn Ch., San Antonio, 1986-89. Mem. San Antonio Mus. Assn., Club of Sonterra. Republican. Avocations: reading, singing, walking. Office: Med Diversified Svcs 15600 San Pedro Ave Ste 107 San Antonio TX 78232-3700

BAKER, HERBERT GEORGE, botany educator; b. Brighton, Eng., Feb. 23, 1920; came to U.S., 1957; s. Herbert Reginald and Alice (Bambridge) B.; m. Irene Williams, Apr. 4, 1945; 1 dau., Ruth Elaine. B.S., U. London, 1941, Ph.D., 1945. Research chemist, asst. plant physiologist Hosa Research Labs., Sunbury-on-Thames, Eng., 1940-45; lectr. botany U. Leeds, Eng., 1945-54; research fellow Carnegie Instn., Washington, 1948-49; prof. botany U. Coll. Ghana, 1954-57; faculty U. Calif., Berkeley, 1957—, assoc. prof. botany, 1957-60, prof., 1960-90, prof. integrative biology emeritus, 1990—, dir. bot. garden, 1957-69. Author: Plants and Civilization, 1965, 70, 78 (translated into Spanish and Japanese); editor: (with G. L. Stebbins) Genetics of Colonizing Species, 1965; series editor: Bot. Monographs, 1971-84; contbr. articles to sci. jours. Fellow Assn. Tropical Biology, AAAS (past pres. Pacific div.); mem. Am. Acad. Arts and Sci., Am. Philos. Soc., Am. Inst. Biol. Sci., Brit. Ecol. Soc. (hon. mem.), Soc. Econ. Botany (Disting. mem. award), Internat. Assn. Botanic Gardens (past v.p.), Ecol. Soc. Am., Soc. for Study Evolution (past pres.), Bot. Soc. Am. (past pres.), Sigma Xi. Home: 635 Creston Rd Berkeley CA 94708-1239

BAKER, JAMES BURNELL, systems engineer; b. Reidsville, N.C., Mar. 6, 1930; s. Numa Reid and Carrie Lee (Mitchell) B.; m. Marie Whisant Richardson, Dec. 26, 1952; children: Sara Anne, James Richardson, William Reid. BSEE, N.C. State U., 1952; MS, MIT, 1954. Elec. engr. AT&T, Burlington, N.C., 1952-53; program mgr. Honeywell, Inc., Clearwater, Fla., 1957—. Author: Control of Hydrofoil, 1954. Bd. dirs. Pinellas County chpt. Nat. Safety Coun., 1993—. Capt. USAF, 1952-57. Recipient Tony Janus award AIAA, 1967. Mem. Rotary Club of Clearwater East (treas. 1971-78, bd. dirs. 1978, Paul Harris fellow 1985), Sigma Xi. Democrat. Methodist. Home: 3170 San Pedro Clearwater FL 34619 Office: Honeywell 13350 US Hwy 19 North Clearwater FL 34624

BAKER, JAMES GILBERT, optics scientist; b. Louisville, Nov. 11, 1914; s. Jesse Blanton and Hattie May (Stallard) B.; m. Elizabeth Katherine Breitenstein, Jan. 1, 1938; children: Kirby Alan, Dennis Graham, Neal Kenton, Brenda Sue. A.B., U. Louisville, 1935, Sc.D. (hon.), 1948; A.M. (Townsend Scholar), Harvard, 1936, Ph.D. (mem. Soc. Fellows 1937-42), 1942. Lowell lectr., 1940; dir. Optical Research Lab., Harvard, 1941-46, assoc. prof., 1946-48; research fellow Harvard Obs., 1942-46, research assoc., 1949—; research assoc. (Lick Obs.), 1949-60; pres. Spica, Inc., 1955-60; cons. optical physics Air Force, 1946-57; cons. Polaroid Corp., 1966—, Perkin-Elmer Corp., 1943-55; chmn. U.S. nat. com. Internat. Commn. Optics, 1956-59, internat. v.p., 1959-62; sci. adv. bd. USAF, 1952-57. Author: (with George Z. Dimitroff) Telescopes and Accessories, 1945. Trustee The Perkin Fund, 1970—. Recipient Woodcock Soc. medal U. Louisville, 1935; Presdl. Medal of Merit for War Work, 1947, Magellanic medal for contbns. to astron. optics Am. Philos. Soc., 1952, Exceptional Civilian Svc. award USAF, 1957, Elliott Cresson medal Franklin Inst., 1962; Alan Gordon award SPIE, 1976, Gold medal, 1978; Internat. Optical Design Conf. award, 1990, Joseph Fraunhofer award Optical. Soc. Am., 1991. Mem. NAS, NAE, Am. Philos. Soc., Am. Astron. Soc. (councillor 1956-59), Am. Optical Soc. (Adolph Lomb medal for contbns. to optics 1942, pres. 1960, Frederick Ives medal 1965, Fraunhofer award 1991), Am. Acad. Arts and Scis. (councillor 1957-59), Explorers Club, Gamma Alpha (pres. Harvard chpt. 1939), Sigma Xi (sec.-treas. Harvard chpt. 1946-48). Home: 14 French Dr Bedford NH 03110-5717 Office: Harvard Coll Obs 60 Garden St Cambridge MA 02138-1596*

BAKER, JAMES REGINALD, aerospace company executive, engineer; b. Newberry, S.C., Oct. 28, 1945; s. C.K. and Mary Evoline (Nobles) B.; m. Joyce Canady, Aug. 2, 1986; 1 child, James Christopher. BSEE, Clemson U., 1968; MS in Systems Mgmt., Fla. Inst. Tech., 1971. Prin. engr. Harris Corp. GESD, Melbourne, Fla., 1968-81; dir. E Systems, Falls Church, Va., 1981-82; product line dir. Harris Corp. GISD, Melbourne, 1982-87; bus. area mgr. GTE Govt. Systems, Mt. View, Calif., 1987-90; v.p. engring. Comsat Systms Divsn., Clarksburg, Md., 1990-91; dir. European networks Brit. Aerospace Commn., Stevenage, Eng., 1991; v.p. Loral WDL Bus. Devel., San Jose, Calif., 1991—. Mem. IEEE, Armed Forces Communications Electronics Assn. Home: 12323 Benson Branch Rd Ellicott City MD 21042 Office: Loral WDL Strategic Systems 3200 Zanker Rd San Jose CA 95134

BAKER, JOHN E., cardiac biochemist, educator; b. London, Dec. 12, 1954; came to U.S., 1984; s. Edward D. and Florence I. (Dobson) B.; m. Mary E. Zurawski, Oct. 29, 1988; children: David J.. Elizabeth A. BSc, Poly. Wolverhampton, Eng., 1977; PhD, St. Thomas' Med. Sch., London, 1984. Sr. biochemist Cen. Pathology Labs., London, 1977-78; rsch. asst. St. Thomas' Hosp. Med. Sch., London, 1978-84; rsch. fellow Med. Coll. Wis., Milw., 1984-86, vis. prof., 1986-87, asst. prof. cardiothoracic surgery, 1987-92; assoc. prof., 1992—. Contbr. rsch. med. articles to profl. jours. Grantee NIH, 1989, 90, Culpeper Found., 1987, Ronald McDonald Children's Charities, 1989, 91. Mem. Am. Heart Assn. (mem. coun. on basic sci., mem. peer rev. rsch. com. Wis. affiliate 1989-93). Methodist. Avocation: walking, music. Office: Med Coll Wis 8701 W Watertown Plank Rd Milwaukee WI 53226-4801

BAKER, JOHN STEVENSON (MICHAEL DYREGROV), writer; b. Mpls., June 18, 1931; s. Everette Barrette and Ione May (Kadletz) B. BA cum laude, Pomona Coll., Claremont Colls., 1953; MD, U. Calif. at Berkeley and San Francisco, 1957. Writer, 1958—; book cataloger Walker Art Center, Mpls., 1958-59; editor, writer neurol. rsch. articles Louis E. Phillips Psychobiol. Rsch. Fund, Mpls., 1960-61. Contbr. articles and poetry to various publs. in Eng. and U.S.; author 65 pub. poems, 21 short essays and 10 sets of aphorisms. Donor numerous species of native plants and seeds to Minn. Landscape Arboretum, U.S. Nat. Arboretum and Arnold Arboretum, Harvard U., papers of LeRoi Jones and Hart Crane to Yale U., Brahms recs. to Bennington Coll., several others. Recipient Disting. Service award Minn. State Hort. Soc., 1976; Cert. of Appreciation U.S. Nat. Arboretum, 1978; property registered as a Minn. Natural Area Minn. chpt. Nature Conservancy, 1990. Mem. Nu Sigma Nu. Office: PO Box 16007 Minneapolis MN 55416-0007

BAKER, KERRY ALLEN, household products company executive; b. Selmer, Tenn., Sept. 21, 1949; s. Austin Clark and Betty Ann (Brooks) B.; m. Ellen Fleming. BIE, Ga. Inst. Tech., 1971; MBA, Ga. State U., 1973; JD, Memphis State U., 1987. With dept. law State of Ga., 1971-73; div. engr. N.W. Ga. div. Gold Kist Inc., Ellijay, 1977-80; sr. mfg. engr. Plough, Inc., Memphis, 1980-82; mgr. indsl. engring. Plough, Inc., 1983-86, supr. mfg. engr., 1986-90; plant bus. mgr. Clorox Co., Dyersburg, Tenn., 1990—. Bd. dirs. Dyersburg Caummunity Concerts Assn., Adopt-a-Sch. Capt. U.S. Army, 1973-77. Decorated Order of St. Barbara. Mem. Inst. Indsl. Engrs., Soc. for Advancement Mgmt., Am. Inst. Plant Engrs., Soc. Am. Mil. Engrs., Am. Prodn. and Inventory Control Soc., Nat. Fire Protection Assn., Scabbard and Blade, Rotary, Masons, Sigma Phi Epsilon, Phi Delta Phi, Alpha Phi Omega. Methodist. Home: 727 N Sampson Ave Dyersburg TN 38024-3961

BAKER, LAURENCE HOWARD, oncology educator; b. Bklyn., Jan. 14, 1943; s. Jacob and Sylvia (Tannenbaum) B.; m. Maxine V. Friedman, July 25, 1964; children: Mindy, Jennifer. BA, Bklyn. Coll. of CUNY, 1962; DO, U. Osteo. Medicine and Surgery, Des Moines, 1966. Diplomate Am. Bd. Internal Medicine. Rotating intern Flint Osteo. Hosp., Flint, Mich., 1966-67; med. resident Detroit Osteo. Hosp., 1967-69; fellow in oncology Wayne State U., Detroit, 1970-72; asst. prof. medicine, dept. oncology Wayne State U. Sch. Medicine, Detroit, 1972-76, assoc. prof. medicine, dept. oncology, 1978-79, prof. medicine, dept. oncology, 1979-82, assoc. chmn., dept. oncology, 1980-82, prof. medicine, dir. div. med. oncology, dept. internal medicine, 1982-86, prof. medicine, dir. div. hematology and oncology, dept. internal medicine, 1986-93, asst. dean for cancer programs, 1988—; Bd. dirs. Meyer L. Prentis Comprehensive Cancer Ctr. of Met. Detroit, Children's Leukemia Soc. Mich. (med. rev. bd.), Mich. Cancer Consortium, Dept. Pub. Health, Mich. Cancer Found., US Bioscis. Sci. Bd.; assoc. chmn. Southwest Oncology Group; mem. search coms., Wayne State U. Med. Sch.; cons. Upjohn Co., Kalamazoo, Mich., 1986, USPC, Rockville, Md., 1986, U. Chgo. Clin. Ctr., 1986, Nat. Cancer Inst., Bethesda, Md., 1987, others; site visits Comprehensive Cancer Ctr., Harper Hosp., Detroit, 1986, 87, Southwest Oncology Group, Seattle, 1987, U. Mich., 1987. Author or co-author 120 pub. articles, 28 books, 15 case reports, 64 abstracts, 35 presentations; mem. editorial adv. bd. Primary Care and Cancer; assoc. editor New Agents and Pharmacology; reviewer Cancer Rsch., Cancer Treatment Reports, Cancer, Am. Jour. Clin. Oncology, JAMA, Investigational New Drugs. Major U.S. Army, 1968-70, Vietnam; USAR, 1970-74. Recipient Faculty Ednl. Devel. award Bur. Health Manpower NIH, 1973, grants Southwest Oncology Group, 1989, Intergroup Sarcoma Contract, 1989, Burroughs Wellcome Phase I Evaluation of BW A 77OU Mesylate on a Single Dose Infusion Schedule, 1986-89, Cancer Ctr., 1989-90, Clin. Therapeutics, Kasle Trust, 1988-89, Marilyn J. Smith Breast Cancer Rsch. Fund, 1986—, Program Project, New Drug Devel., 1989-90. Mem. Am. Soc. Cancer Rsch., Am. Soc. Clin. Oncology, Am. Soc. for Clin. Pharmacology and Therapeutics, Am. Assn. Clin. Rsch., Am. Assn. Cancer Edn., Am. Coll. Osteo. Internists, Cen. Soc. Clin. Rsch. Democrat. Jewish. Office: Harper Hosp 3990 John R St Detroit MI 48201-2018

BAKER, LEE EDWARD, biomedical engineering educator; b. Springfield, Mo., Aug. 31, 1924; s. Edward Fielding and Oneita Geneva (Patton) B.; m. Jeanne Carolyn Ferbrache, June 20, 1948; children: Carson Phillips, Carolyn Patton. BEE, U. Kans., 1945; MEE, Rice U., 1960; PhD in Physiology, Baylor U., 1965. Registered profl. engr., Tex. Asst. prof. electrical engring. Rice U., Houston, 1960-64; asst. prof. physiology Baylor U. Coll. Medicine, Houston, 1965-69, assoc. prof., 1969-75; prof. biomed. engring. U. Tex., 1975-82; Robert L. Parker Sr. Centennial Prof. Engring. U. Tex., Austin, 1982—. Co-author: Principles of Applied Biomedical Engineering, 1968, 3d edit., 1989; author, co-author scientific papers. Served to lt. USN, 1943-46, PTO, 1951-53. Spl. research fellow NIH, 1964-65. Fellow Am. Inst. of Med. and Biol. Engring., Royal Soc. of Medicine; mem. IEEE (sr.), Biomed. Engring. Soc. (sr.), Assn. for Advancement Med. Instrumentation, Am. Physiol. Soc., N.Y. Acad. Scis. Episcopalian. Avocation: gardening. Office: U Tex Biomed Engring Program Engring Sci Bldg 610 Austin TX 78712

BAKER, LENOX DIAL, orthopaedist, genealogist; b. DeKalb, Tex., Nov. 10, 1902; s. James D. and Dorothy Hamilton (Lenox) B.; m. Virginia Flowers, Aug. 22, 1933 (dec.); children: Robert Flowers, Lenox Dial; m. Margaret Copeland, Apr. 22, 1967. Student, St. Edwards Coll., Austin, Tex., 1912-13, Pierce Sch. Bus. Adminstrn., Phila., 1920-21, Carver Chiropractic Coll., 1922-24, U. Tenn., 1925-29, Sch. Medicine, U. N.C., 1929-30; M.D., Duke U., 1934. Diplomate: Am. Bd. Orthopaedic Surgery. Athletic trainer U. Tenn., 1925-29, asst. in zoology, 1927-29; athletic trainer

Duke U., 1929-33; ofcl. So. Football Conf., 1933-40; orthopaedic intern Johns Hopkins Hosp., 1933-34, surg. intern, 1934-35, asst. resident orthopaedics, 1935-36, resident orthopaedics, 1936-37; asst., instr. orthopaedic surgery, sch. med. Johns Hopkins U., 1935-37; asst. orthopaedics Duke U., Durham, N.C., 1937-38, assoc., 1938-39, asst. prof., 1940-42, assoc. prof., 1942-46, prof., 1947-72, emeritus, 1972—, Pres.'s assoc., 1974—; orthopaedist Duke Hosp., 1937-72, founder and dir. div. phys. therapy, 1943-62; co-op. orthopaedic surgeon crippled children's div. N.C. Bd. Health; also vocational rehab. div. N.C. Dept. Pub. Instrn., 1937-74; orthopaedist Lincoln Hosp., 1937-74, trustee, 1939-74, exec. com., 1941-74, chmn. exec. com., 1951-74; vis. orthopaedist Watts Hosp., 1937—; faculty div. pub. health and soc. work U. NC., 1938-41; founder, med. dir. Lenox Baker Children's Hosp. N.C., Durham, 1949-72; established Virginia Flowers Baker chair of Orthopaedic Surgery Duke U., 1968; mem. gov.'s cabinet, sec. human resources, State of N.C., 1972; orthopaedic cons. to several hosps., founds., sanitaria, govtl. agys.; active in cerebral palsy work, pres. League for Crippled Children, 1941-43; mem. N.C. Bd. Health, 1956-72, pres., 1963-68, v.p., 1968-72. Author: Treatment of Minor Injuries of Baseball, Bone Tumors, 1952, (with others) History of Medicine in North Carolina, 1972; Mem. editorial com.: (with others) Jour. Bone and Joints Surgery, 1960-61; trustee (with others), 1967—; Contbr. (with others) articles to profl. publs. Recipient U.S. President's Physician's award, 1958, Citizenship award Triangle chpt. Nat. Football Hall of Fame, 1969, Am. Legion 50th Anniversary Physician of Half-Century award, 1969, Service to Athletics award Atlantic Coast Sportswriters, 1970, N.C. Gov.'s Baseball award, 1979, N.C. Order of the Long Leaf Pine, 1979, 88, Derby Day Dedication award Sigma Chi, 1983, Cannon Cup award N.C. History and Preservation Soc., Disting. Alumnus award Duke U. Sch. Medicine, 1989, Mr. Sports Medicine award Am. Orthopaedic Soc. Sports Medicine, 1989, Disting. Svc. award So. Med. Assn., 1990, Svc. award Am. Assn. for State and Local History, 1991, Hope award Nat. Multiple Sclerosis Soc., 1991, Disting. Alumnus award Duke U., 1992, Disting. Alumnus award Texarkana Tex. Ind. Sch. Dist., 1993; named Durham Father of Yr., Exchange Club Book of Golden Deeds; named N.C. Tar Heel of Wk.; named to Duke U. Sports Hall of Fame, 1979, N.C. Sports Hall of Fame, 1983, East Tenn. chpt. Nat. Football Found. Hall of Fame, 1992, Hon. Order Ky. Cols.; Ortho Clinic Duke Med. Ctr. named in his honor, 1966, Lenox Baker Lectureship established at Duke Med. Ctr., 1966. Mem. AMA (chmn. orthopaedic sect. 1958-59), and other nat., regional, state, local profl. and sci. orgns., including Am. Acad. Cerebral Palsy (pres. 1954-55), Am. Orthopaedic Assn. (pres. 1963-64), So. Med. Assn. (editorial com. 1960—, past. chmn. orthopaedic sect., Disting. Svc. award 1990), So. Surg. Assn., Med. Soc. N.C. (pres. 1959), N.C. Orthopaedic Assn. (pres. 1947, Outstanding Svc. award 1990), Tex. Orthopaedic Assn. (hon.), Internat. Cerebral Palsy Soc. (spl. mem.), Guatemala Orthopaedic Assn. (hon.), N.C. Geneal. Soc. (dir. 1974-76, v.p. 1978, pres. 1982-83), Friends of Archives of N.C. (pres. 1980—), Ky. Col., Soc. of the Cincinnati, Wake Forest Monogram Club (hon.), Kappa Sigma, Nu Sigma Nu, Alpha Omega Alpha. Presbyerian. Clubs: Hope Valley Country (Durham, N.C.); Sertoma (hon.); Wake Forest U. Monogram (hon.). Home: 1 Hastings The Valley Durham NC 27707-3643 Office: Duke Hosp Box 3706 Durham NC 27710

BAKER, LOUIS COOMBS WELLER, chemistry educator, researcher; b. N.Y.C., Nov. 24, 1921; s. F(rancis) Godfrey and Marion Georgina (Weller) B.; m. Violet Eva Simmons, June 28, 1964; children—William W.S., Godfrey A.S. A.B., Columbia Univ., 1943; M.S., U. Pa., 1947, Ph.D., 1950; LHD honoris causa Georgetown U., 1988. Asst. instr. chemistry Towne Sci. Sch. U. Pa., Phila., 1943-50, instr. chemistry The College, U. Pa., 1945-50, assoc. in chemistry The Johnson Found., U. Pa., Phila., 1950-51; instr. Martin Coll. and Rittenhouse Coll., Phila., 1945-48; asst. prof. to assoc. prof., head inorganic div. Boston U., 1951-62; prof. chemistry Georgetown U., Washington, 1962—, chmn. dept., 1962-84; co-project dir. OPRD high thermal efficiency engine project, N.Y.C. and Boundbrook, N.J., 1943-45; chmn. Internat. Symposium on Heteropoly Electrolytes, Am. Chem. Soc., 1956; chmn. Nat. Acad. Sci. Com. on Recommendations to U.S. Army for Basic Sci. Research, 1974-78; plenary lectr. Internat. Conf. Coordination Chemistry, Moscow, 1973; sci. mem. com. on accreditation Coll. and Univs. Middle States Assn. Coll. and Secondary Schs.; sci. mem. vis. com. Ferdowsi Univ., Mashhad, Iran, 1974-75; speaker Gordon Research Confs., 1956, 1967; cons. in field. Contbr. articles on inorganic chemistry to profl. jours. Co-inventor High Thermal Efficiency Internal Combustion Engine, 1943-45. Guggenheim fellow, 1961—; recipient Tchugaev medal USSR Acad. Scis., 1973; Vicennial Gold medal Georgetown U., 1983, Pres.'s medal Disting. Service, 1984; grantee NSF, NIH, numerous others. Fellow Washington Acad. Scis.; mem. Am. Chem. Soc. (chmn. com. rev. inorganic papers 1957-59, councilor, bd. mgrs.), Sigma Xi (past pres. Georgetown chpt.). Quaker. Club: Cosmos (Washington). Office: Georgetown U Dept Chemistry Washington DC 20057

BAKER, M(ERVIN) DUANE, healthcare, biotechnology marketing and business consultant; b. Greensburg, Ind., Mar. 2, 1949; s. Mervin Loren and Barbara Ellen (Marlow) B.; m. Jill Ruth Gardner, May 30, 1976; children: Carrie, Cheri. BS in Microbiology and Organic Chemistry, Purdue U., 1971; MBA in Mktg., Northwestern U., 1976. Internat. mktg. dir. Hollister, Inc., Libertyville, Ill., 1977-82; dir. mktg. Gambro, Inc., Lincolnshire, Ill., 1982-85; v.p. mktg. and sales Genesis Funding Group, Inc., Northbrook, Ill., 1985-90; sr. market cons. Govt. of Ont. (Can.), Chgo., 1990—; mem. health-care sector com. Govt. of Ont. (Can.), Toronto, Ont., 1990—; mem. Health Econ. Devel. Group of Ont., Ministry of Health, 1992—. Contbg. editor newspaper Ont. Report, 1990—. Chmn. election com. Northbrook Caucus Party, 1987-88; trustee Northbrook Pub. Libr., 1988—. Paul Harris fellow, 1988. Mem. Rotary Internat., Mensa Internat. Republican. Home: 1957 Redwood Ln Northbrook IL 60062-3626 Office: Govt of Ont (Can) 221 N La Salle St Ste 2700 Chicago IL 60601-1504

BAKER, SCOTT MICHAEL, industrial engineer; b. Oklahoma City, Aug. 28, 1960; s. Earl Andrew and Earline Goldie (Wilmoth) B.; m. Patricia Caroline Spenner, July 19, 1985. BS in Indsl. Engring., U. Nebr., 1983; MS in Indsl. Engring., Purdue U., 1988. Registered profl. engr., Tex. Indsl. engr. AT&T-Western Electric, Oklahoma City, 1983-85; devel. engr. AT&T Network Systems, Oklahoma City, 1985-88; engring. mgr. AT&T Network Systems, Morristown, N.J., 1988-90; sales mgr. AT&T Network Systems, Irving, Tex., 1990—. Contbr. articles to profl. publs. Exec. advisor Jr. Achievement, Oklahoma City, 1986. Named Competent Toastmaster, 1987. Mem. Inst. Indsl. Engring. (pres. Oklahoma City chpt. 1987-88). Republican. Methodist. Home: 4122 Harvestwood Blvd Grapevine TX 76051

BAKER, SCOTT PRESTON, computer engineer; b. Dallas, Feb. 4, 1955; s. Calvin E.D. and Billie L. B.; m. Victoria Gayle Shreeve, June 23, 1979. A.Avionics Tech., Mountain View Jr. Coll., Dallas, 1978; BS in Computer Sci., U. North Tex., 1986. Systems cons. Logic Works, Flower Mound, Tex., 1986—. Chmn. Planning and Zoning Commn., Town of Flower Mound, 1988—; pres. Summit Club-Non-Profit Community svc., 1990—. Office: Logic Works 4904 Wolf Creek Trail Flower Mound TX 75028-1954

BAKER, SCOTT RALPH, toxicologist; b. Winnipeg, Canada, Jan. 13, 1950; came to the U.S., 1975; s. Harvey and Sally (Handleman) B. BS, U. Manitoba, 1971, MS, 1975; PhD, Iowa State U., 1978. Resident rsch. assoc. Nat. Rsch. Coun./Walter Reed Army Med. Ctr., Washington, 1978-79; sr. staff officer Nat. Acad. Scis., Washington, 1979-83; sci. advisor, asst. adminstr. for R&D U.S. EPA, Washington, 1983-88; dep. dir. risk focus div. Versar, Inc., Springfield, Va., 1988-92; dir. health scis. group EA Engring. Sci & Tech. Inc., Silver Spring, Md., 1992—; chmn. Interagy. Task Force on Environ. Cancer, Heart and Lung Disease, Washington, 1983-88. Editor: Environmental Toxicity and the Aging Process, 1987, Effects of Pesticides on Human Health, 1990; author: Methods for Assessing and Reducing Injury from Chemical Accidents, 1989, Methods for Assessing Adverse Effects of Pesticides on Non-Target Organisms, 1992. Mem. Soc. Toxicology, Am. Coll. Toxicology, Phi Kappa Phi, Sigma Xi. Home: 2423 McCormick Rd Rockville MD 20850 Office: EA Engring Sci Tech Inc 8401 Colesville Rd Ste 500 Silver Spring MD 20910

BAKER, STEPHEN HOLBROOKE, quality engineering executive; b. Concord, Mass., May 31, 1952; s. Morton H. and Mary E. (Gillette) B.; m. Heide Y. Wegener, Mar. 31, 1973; children: Kimberly M., Michelle E. BS mech. engr., Univ. Lowell, 1981. Cert. quality engr. Nuclear machinist instr. U. S. Navy, Saratoga Springs, N.Y., 1974-77; reactor operator Univ. Lowell (Mass.) Nuclear Engr., 1977-79; cons. asst. Arthur D. Little, Inc., Cambridge, Mass., 1980-81; quality engr. Avco Lycoming Greer, Greer, S.C., 1982-86; metrology mgr. Texton Lycoming Greer, Greer, S.C., 1986-88, supplier quality mgr., 1989-91, mgr. quality engr., 1991—. Commr. Brookfield Spl. Tax Dist., Greenville, S.C., 1989—; bd. mem. Greenville Eagles Soccer Club, 1984-89. With U. S. Navy, 1972-77. Mem. Am. Soc. Quality Control. Achievements include patent on geometric simulator for coordinate measuring machines. Home: 5 Doverdale Ct Greenville SC 29615 Office: Textron Lycoming Greer 400 S Buncombe Rd Greer SC 29652

BAKER, SUZON LYNNE, mathematics educator; b. Sacramento, Calif., Mar. 29, 1943; d. Thomas Kestell and Dorothy (Espinosa) Lockart; m. Carl Leroy Baker, June 13, 1966; children: Michele, Eric. BA, Sacramento State U., 1965; MS, St. Francis Coll., Ft. Wayne, Ind., 1974. Cert. tchr., Ind. Math. tchr. Richland-Bean Blossom Sch., Ellettsville, Ind., 1966-68, Ft. Wayne Community Schs., 1968-70, Garrett-Keyser-Butler (Ind.) Schs., 1977—. Mem. Nat. Coun. Tchrs. Math., Ind. Tchrs. Math. Methodist. Avocations: sewing, yard work. Home: 4 Pinetree Rd Garrett IN 46738-9772 Office: Garrett High Sch 801 E Houston St Garrett IN 46738-1699

BAKER, THADDEOUS JOSEPH, electronics engineer; b. Washington, Nov. 3, 1921; s. John Thaddeus and Angelina Elvira (Rappa) B.; m. Mary Frederick, Dec. 14, 1946; children: Fred, Jim. BA, U. Okla., 1950. Engr. power plant Boeing Airplane Co., Seattle, 1942-44; engr. Airforce OCMA, Midwest City, Okla., 1950-55; engr. Collins Radio Co., Dallas, 1955-57, Richardson, Tex., 1960-71; engr. Temco Aircraft Co., Garland, Tex., 1957-60, Telephony Internat., Richardson, 1979—. Achievements inlcude development of open clock phase encoded non-contact magnetic recording, development of missile range tracking device; patent for balanced constnat current telephone circuit. Office: Telephony Internat 710 Presidential Dr Richardson TX 75081-2964

BAKER, TIMOTHY ALAN, healthcare administrator, educator, consultant; b. Myrtle Point, Oreg., July 30, 1954; s. Farris D. and Billie G. (Bradford) B.; 1 child, Amanda Susann. BS in Mgmt. with honors, Linfield Coll., McMinnville, Oreg., 1988; MPA in Health Adminstrn. with distinction, Portland State U., 1989, PhD in Pub. Adminstrn. and Policy, 1992. Registered emergency med. technician. Gen. mgr. Pennington's, Inc., Coos Bay, Oreg., 1974-83; dep. dir. Internat. Airport Projects Med. Svcs., Riyadh, Saudi Arabia, 1983-87; adminstrv. intern Kaiser Sunnyside Hosp., Portland, Oreg., 1988-89; grant mgr. Oreg. Health Sci. U., Portland, 1989-90; dir. health sci. program Linfield Coll., Portland, Oreg., 1992—; rsch. assoc. Portland State U., 1990—; instr. S.W. Oreg. C.C., Coos Bay, 1980-83; pres. Intermed. Inc., Portland, 1987—; sr. rschr. small area analysis Oreg. Health Sci. U., 1990, The Oreg. Helath Plann Project, 1990-91; developer, planner, prin. author trauma system devel. S.W. Wash. EMS and Trauma System, 1991-93; cons. ednl. defense Min. Civil Defense, Riyadh, Saudi Arabia, 1992. Pub. Jour. Family Practice. Planner mass disaster plan King Khaled Internat. Airport, 1983; EMS planner Emergency Med. Plan, Province of Cholburi, Thailand, 1985; bd. dirs. Coos County Kidney Assn., 1982, Coos Bay Kiwanis Club, 1979; regional adv. com. EMS and Trauma, State Wash. Dept. Health, 1990—. Recipient Pub. Svc. award Am. Radio and Relay League, 1969, Med. Excellence award KKIA Hosp., 1985; named Fireman of Yr. Eastside Fire Dept., 1982, Adminstr. of Yr., Wash. Dept. Health, 1993. Mem. Am. Soc. Pub. Asminstrn. (doctoral rep. to faculty senate 1990, Portland State U.), Am. Pub. Health Assn., Am. Coll. Healthcare Execs. Avocations: flying, scuba diving, photography, racquetball, amateur radio. Home: 12008 N Jantzen Beach Ave Portland OR 97217-8151 Office: Linfield Coll Portland Campus 2255 NW Northrup Portland OR 97210

BAKER, WILLIAM OLIVER, research chemist, educator; b. Chestertown, Md., July 15, 1915; s. Harold May and Helen (Stokes) B.; m. Frances Burrill, Nov. 15, 1941; 1 son, Joseph Burrill. BS, Washington Coll., 1935, ScD, 1957; PhD, Princeton U., 1938; ScD, Georgetown U., 1962, U. Pitts., 1963, Seton Hall U., 1965, U. Akron, 1968, U. Mich., 1970, St. Peter's Coll., 1972, Poly. Inst. N.Y., 1973, Trinity Coll., Dublin, Ireland, 1975, Northwestern U., 1976, U. Notre Dame, 1978, Tufts U., 1981, N.J. U. Medicine and Dentistry, 1981, Clark U., 1983, Fairleigh Dickinson U., 1983, Rockefeller U., 1990; DEng., Stevens Inst. Tech., 1962, N.J. Inst. Tech., 1978; LLD (hon.), U. Glasgow, 1965, U. Pa., 1974, Kean Coll., N.J., 1976, Lehigh U., 1980, Drew U., 1981, Monmouth Coll., 1973, Clarkson Coll. Tech., 1974, Princeton U., 1993. With AT&T Bell Labs., 1939-80, in charge polymer research and devel., 1948-51, asst. dir. chem. and metall. research, 1951-54, dir. research, phys. scis., 1954-55, v.p. research, 1955-73, pres., 1973-79, chmn. bd., 1979-80; bd. dirs. Summit Trust Co., Gen. Am. Investors, Inc.; dir. Health Effects Inst., 1980—; vis. lectr. Northwestern U., Princeton U., Duke; Schmitt lectr. U. Notre Dame, 1968; Harrelson lectr. N.C. State U., 1971; Herbert Spencer lectr. U. Pa., 1974; Charles M. Schwab Meml. lectr. Am. Iron and Steel Inst., 1976; NIH lectr., 1958, Metall. Soc. Am. Inst. Mining Engrs./Am. Soc. Metals disting. lectr., 1976; Miles Conrad Meml. lectr. Nat. Fedn. Abstracting and Indexing Services, 1977; Wulff lectr. MIT, 1979; Mayo Found. lectr., 1980; Logue lectr., 1981; Whitehead lectr. U. Ga., 1985; Lazerow lectr. U. Pitts., 1984; Taylor lectr. Pa. State U., 1984; other lectureships; cons. Office Sci. and Tech., 1977-81; mem. Princeton Grad. Council, 1956-64; bd. visitors Tulane U., 1963-82; mem. commn. sociotech. systems NRC, 1974-78, also chmn. adv. bd. on mil. personnel supplies, 1964-78; mem. com. on phys. chemistry of div. chemistry and chem. tech., 1963-70; also steering com. Pres.'s Food and Nutrition Study Commn. Internat. Relations Nat. Acad. Scis.-NRC, 1975; mem. panel on phys. chemistry Office Naval Research, 1948-51; past mem. Pres.'s Sci. Adv. Com., 1957-60; nat. sci. bd. NSF, 1960-66; past chmn. Nat. Sci. Info. Council, 1959-61; mem. sci. adv. bd. Nat. Security Agy., 1959-76, cons., 1976—; cons. Dept. Def., 1958-71; cons. to spl. asst. pres. for sci. and tech., 1963-73; cons. Panel of Ops. Evaluation Group, USN, 1960-62; mem. N.J. Bd. Higher Edn., 1967—, exec. com., 1970—, vice chmn., 1970-72, 82-84; mem. liaison com. for sci. and tech. Library of Congress, 1963-73; mem. Pres.'s Fgn. Intelligence Adv. Bd., 1959-77, 81-90; chmn. diplomatic telecommunication systems policy bd. Dept. State, 1984—; chmn. Pres.'s Adv. Group Anticipated Advances in Sci. and Tech., 1975-76; vice chmn. Pres.'s Com. Sci. and Tech., 1976-77; bd. regents Nat. Library Medicine, 1969-73; bd. visitors Air Force Systems Command, 1962-73; mem. mgmt. adv. council Oak Ridge Nat. Lab., 1970-78; mem. Nat. Commn. on Libraries and Info. Scis., 1971-75, Commn. on Critical Choices for Ams., 1973-75, Nat. Cancer Adv. Bd., 1974-80, Nat. Commn. on Excellence in Edn., 1981-83, Nat. Commn. on Jobs and Small Bus., 1985-87, Nat. Commn. on Role and Future State Colls. and Univs., 1985-87; mem., vice chmn. Commn. on Sci. and Tech. N.J. 1985—; mem. Carnegie Forum on Edn. Sci. Tech. and The Economy, 1985—; co-chmn. nat. coun. on sci. and tech. edn. AAAS, 1985—; mem. panel adv. Inst. Materials Rsch., Nat. Bur. Standards, 1966-69; mem. Council Trends and Perspectives, U.S. C. of C., 1966-74; chmn. tech. panels adv. to Nat. Bur. Standards, Nat. Acad. Scis.-NRC, 1969-78; mem. Nat. Council Ednl. Research, 1973-75; mem. energy research and devel. adv. council Energy Policy Office, 1973-75; mem. Project Independence adv. com. Fed. Energy Adminstrn., 1974-75, Gov.'s Com. to Evaluate Capital Needs N.J., 1974-75; mem. governing bd. Nat. Enquiry into Scholarly Communication, 1975-79;uncil N.J. Regional Med. Library, 1975—, Spl. Libr. Assn., 1985—, Fed. Emergency Mgmt. Adv. Bd., 1980-93, Gas Rsch. Inst. Adv. Bd., 1978-85; mem. adv. bd. N.J. Sci./Tech. Center, 1980-86; mem. sci. adv. bd. Robert A. Welch Found., 1968—; vis. com. for chemistry Harvard, 1959-72; mem. council Marconi Fellowships, 1978—; vis. com., div. chemistry and chem. engring. Calif. Inst. Tech., 1969-72; vis. com. on scis. and math. Drew U., 1969—; assoc. in univ. seminar on tech. and social change Columbia, 1969-80; vis. com., dept. materials sci. and engring. MIT, 1973-76; bd. overseers Coll. Engring. and Applied Sci. U. Pa., 1975—; bd. dirs. Council on Library Resources, 1970—, Health Effects Inst., 1980—, Clin. Scholar Program Robert Wood Johnson Found., 1973-76, Third Century Corp., 1973-76, EDUCOM, 1985-92; organizer labs. for numerous companies; originator nat. tech. means of satellite survey Nat. Security Orgn. Fed. Telecommunications System; co-sponsor Nat. Cancer Plan, Nat. Materials Program; co-founder Aerospace Corp., 1961, Health Effects Inst., 1980—, N.J. Commn. on Sci. and Tech., 1985—. Contbr.: High Polymers, 1945, Symposium on Basic Research, AAAS, 1959, Rheology, Vol. III, 1960, Technology and Social Change, 1964, Science: The Achievement and the Promise, 1968, Ann. Rev. Materials Sci, 1976, Advancing Materials Research, 1987, various other books.; mem. editorial adv. bd. Jour. Info. Sci; past mem. adv. editorial bd. Chem. and Engring. News; hon. editorial adv. bd. Carbon; contbr. numerous articles to tech. jours. Trustee Urban Studies, Inc., 1960-78, Aerospace Corp., 1961-76, Carnegie-Mellon U., 1967-87, now emeritus, Princeton, 1964-86, now emeritus, Fund N.J., 1974—, Harry Frank Guggenheim Found., 1976—, Gen. Motors Cancer Rsch. Found., 1978—, Charles Babbage Inst., 1978—, Newark Mus., 1979-89; trustee Rockefeller U., 1960—, chmn., 1980-90, chmn. emeritus, 1990—; trustee Andrew W. Mellon Found., 1965—, chmn., 1975-90, chmn. emeritus, 1990—. Named 1 of 10 top scientists in U.S. industry, 1954; recipient Perkin medal, 1963; Honor scroll Am. Inst. Chemist, 1962; award to execs. ASTM, 1967; Edgar Marburg award, 1967; Indsl. Research Inst. medal, 1970; Frederik Philips award IEEE, 1972; Indsl. Research Man of Year award, 1973; Procter prize Sigma Xi, 1973; James Madison medal Princeton U., 1975; Mellon Inst. award, 1975; Soc. Research Adminstrs. award for disting. contbns., 1976; von Hippel award Materials Research Soc., 1978; Fahrney medal Franklin Inst., 1977; N.J. Sci/Tech. medal, 1980; Harvard U. fellow, 1937-38; Procter fellow, 1938-39; recipient Jefferson medal N.J. Patent Law Assn., 1981; David Sarnoff prize AFCEA, 1981; Vannevar Bush prize Nat. Sci. Bd., 1981; Pres.'s Nat. Security medal, 1983; Baker medal Security Affairs Support Assn., 1984; Disting. Svc. award Nat. Assn. Gov. Bd., 1993; co-recipient Nat. Medal Tech., 1985, Thomas Alva Edison Sci. medal State of N.J., 1987, Nat. Materials Advancement award Fedn. Materials Socs., 1987, Nat. Medal Sci., 1988. Fellow Am. Phys. Soc., Am. Inst. Chemists (Gold medal 1975), Franklin Inst., Am. Acad. Arts and Scis.; mem. Dirs. of Indsl. Research, Am. Chem. Soc. (past mem. com. nat. def., cons., past mem. com. chemistry and pub. affairs, Priestley medal 1966, Parsons award 1976, Willard Gibbs award 1978, Madison Marshall award 1980), Am. Philos. Soc., Nat. Acad. Scis. (council 1969-72, com. sci. and pub. policy 1966-69), Nat. Acad. Engring. (Bueche prize 1986), Inst. Medicine (council 1973-75), Indsl. Research Inst. (dir. 1960-63, medal 1970), Sigma Xi, Phi Lambda Upsilon, Omicron Delta Kappa. Clubs: Chemists of N.Y. (hon.), Cosmos, Princeton of Northwestern N.J. Achievements include 13 patents; research on semiconducting polymers, on solid state structure of linear polymers-polamides and polyesters, on influence of microstructure on mechanical and engineering properties of rubbers and plastics, on development of polyethylene for cable sheathing and microwave dielectric, on synthesis and properties of polymer carbon-patents as resistor, on microphonic and composite (fibrous) material, on dynamic mechanics of polymers, and on relaxation times of dilute macromolecules. Office: AT&T Bell Labs 600 Mountain Ave Murray Hill NJ 07974-2010

BAKINTAS, KONSTANTINE, civil, environmental design engineer; b. Heraklion, Crete, Greece, Sept. 24, 1961; came to U.S., 1970; s. Savas and Dimitra (Kouvoukas) B. BSCE, U. Utah, 1985; MSCE, U. Tex., Arlington, 1990. Engring. asst. Horrocks/Corollo Engrs., Salt Lake City, 1984-85; civil engr. Yandell & Hiller Inc., Ft. Worth, 1985—, prin., 1992%. Mem. ASCE, Nat. Soc. Profl. Engrs., Chi Epsilon, Tau Beta Pi. Office: Yandell & Hiller Inc 512 Main St Ste 1500 Fort Worth TX 76102-3999

BAKSHI, PRADIP M., physicist; b. Baroda, Gujarat, India, Aug. 21, 1936; s. Manharlal M. and Rama M. (Shroff) Baxi; m. Hansika P. Parekh, Sept. 14, 1967; children: Vaishali, Ashesh. BSc, Bombay U., 1955; AM, Harvard U., 1957, PhD, 1962. Postdoctoral fellow Harvard U., Cambridge, Mass., 1963; sr. rsch. physicist Air Force Cambridge Rsch. Labs., Hanscom AFB, Mass., 1963-66, Brandeis U., Waltham, Mass., 1966-68; vis. assoc. prof. Brandeis U., Boston, 1968-70; rsch. assoc. prof. Boston Coll., 1970-75, rsch. prof., 1975—. Contbr. articles to sci. and profl. jours. Grantee NSF, NATO, DOE, Air Force Office of Sci. Rsch., Office of Naval Rsch., Army Rsch. Office, Air Force Cambridge Rsch. Lab./Air Force Geophysics Lab. Office: Boston Coll Dept Physics Chestnut Hill MA 02167

BALABAN, EDWARD PAUL, physician; b. Pitts., Apr. 20, 1951; s. Edward Paul and Rose (Sich) B.; m. Deborah K. Balaban; children: Matthew, Eric. BS, U. Pitts., 1973; DO, Phila. Coll. Osteo. Medicine, 1977. Diplomate in hematology and med. oncology. Intern Detroit Osteo. Hosp., 1977-78; resident Allegheny Gen. Hosp., 1978-81; fellow U. Tex. Southwestern, 1981-84; asst. prof. U. Tex. Southwestern Med. Ctr., Dallas, 1988—; cons. staff Am. Running and Fitness Assn., 1991—. Contbr. articles to profl. jours. Mem. planning coun. Hearts and Hammers, Dallas, 1992. Fellow ACP; mem. Am. Soc. Hematology, Am. Soc. Clin. Oncology, Am. Osteo. Assn. Achievements include dissection of clinical role of protein ferriten within the red cell; clinical studies in acute leukemia, lung cancer, and colorectal neoplasia.

BALABANIAN, NORMAN, electrical engineering educator; b. New London, Conn., Aug. 13, 1922; s. Adam B. and Elizabeth (Seklemian) B.; m. Jean Tajerian, Aug. 16, 1947 (div. 1977); children: Karen J., Doris R., Gary N., Linda C.; m. 2d, Rosemary Lynch, Jan. 19, 1979. BSEE, Syracuse U., 1949, MSEE, 1951, PhD, 1954. From instr. to prof. Syracuse U., 1949-91, prof. emeritus, 1991—; mem. tech. staff Bell Labs., Murray Hill, N.J., 1956, IBM Devel. Lab, Poughkeepsie, N.Y., 1962; vis. prof. U. Calif., Berkeley, 1965-66; mem. UNESCO field staff Inst. Politecnico Nacional, Mexico City, 1969-70; Fulbright fellow U. Zagreb, Zagreb, Jugoslavia, 1974-75; acad. advisor Inst. Nat. d'Elec. et d'Elec., Boumerdes, Algeria, 1977-78; chmn. Dept. of Elec. & Computer Engring. Syracuse U., 1983-90; vis. scholar MIT, 1990—, Tufts U., 1990—. Author: Network Synthesis, 1958, Fundamentals of Circuit Theory, 1961, Fourier Series, 1976, Ensenanza Programada en la Education Activa (in Spanish), 1974, Activne RC Mreze (in Serbo-Croatian), 1977, Electric Circuits, 1993; co-author: Linear Network Analysis, 1959, Electrical Network Theory, 1969, Electrical Science: Resistive Networks, 1970, Electrical Science: Dynamic Networks, 1973, Linear Network Theory, 1981; editor: Undergraduate Physics and Mathematics in Electrical Engineering, 1960, Electrical Engineering Education, 1961; editor (jour.) IEEE Transactions on Circuit Theory, 1963-65, (mag.) IEEE Technology and Society, 1979-86, 93—. Dist. commr. Dem. Party, Syracuse, N.Y., 1959-61; pres. Cen. N.Y. Civil Liberties Union, Syracuse, 1963-64, 79-80 (Civil Liberties award 1966); congl. candidate Liberal Party, People's Peace Party, Syracuse, N.Y., 1966. S/Sgt. Army AC, 1943-46. Recipient peace award Syracuse Peace Coun., 1966. Fellow IEEE (life mem., Centennial award 1984), IEEE Soc. Implications Tech. (v.p., pres. 1988-91), AAAS; mem. Am. Soc. for Engring. Edn. (life mem., pres. EE div. 1966-67), AAUP (pres. Syracuse U. chpt. 1964-65). Office: Tufts U EE Dept Medford MA 02156-1240

BALACHANDRAN, SWAMINATHAN, industrial engineering educator; b. Coimbatore, India, Nov. 6, 1946; s. Ardhanari Swaminathan and Karunambal (Chettiar) B.; m. Lalitha Kathiresan, Dec. 1, 1976; children: Jay Shankar, Dave Kumar. BME, U. Madras, India, 1968; M in Aerospace Engring., Ind. Inst. Sci., 1970; PhD in Indsl. Engring., Va. Poly. Inst. and State U., 1984. Grad. rsch. asst. Va. Poly. Inst. and State U., Blacksburg, 1974-76, instr. indsl. engring., 1976-80, asst. prof., 1980-85; from assoc. prof. to prof. Va. Poly. Inst. and State U., Plateville, 1985—; chmn. dept. indsl. engring.; program evaluator ABET; cons. in field. Reviewer profl. jours. NSF grantee, 1988. Mem. Inst. Indsl. Egnring. (sr.), Soc. Mfg. Engrs., Am. Soc. Engr. Educators, Am. Soc. Quality Control (sr.), Am. Prodn. and Inventory Control Soc., Phi Kappa Phi, Alpha Pi Mu. Hindu. Avocations: collecting stamps, playing tennis, swimming, playing table tennis. Home: 270 Flower Ct Platteville WI 53818-1914 Office: U Wis Indsl Engring Dept 1 University Plz Platteville WI 53818-3024

BALADI, VIVIANE, mathematician; b. Tour de Peilz, Vaud, Switzerland, May 23, 1963; arrived in Switzerland, 1991.; d. André and Adrienne Sylvia (Barben) B. MSc in Math. and Computer Sci., U. Geneva, 1986, PhD in Math., 1989. Asst. U. of Geneva, 1985-90; postdoctoral fellow IBM Rsch., Yorktown Heights, N.Y., 1990-91; chargée de recherche CNRS, Lyons, France, 1990—; assistenzprofessorin für mathematik Technische Hochschule Zurich, Switzerland. Contbr. articles to Jour. of Statis. Physics, Communications in Math, Physics, Nonlinearity, La Gazette des Maths, Ergodic Theory and Dynamical Systems. Mem. Am. Math. Soc., Soc. Math. Suisse, Soc. Math. de France, Femmes et Maths. Office: Technische Hochschule Zurich, ETH Zentrum, CH8094 Zurich Switzerland

BALAGUER, JOHN P., aircraft manufacturing executive; b. Paris, July 1, 1935; came to U.S., 1940; m. Veronica Glasser; children: John, Michelle. B in Aero. Engring., U. Detroit, 1958; MME, U. Bridgeport. Flight test engr. Sikorsky Aircraft divsn. United Techs., 1960-80, v.p. S-76 program Sikorsky Aircraft divsn., 1980-82; exec. v.p. mfg. divsn. Pratt & Whitney, East Hartford, Conn., 1982-85; pres. Govt. Engines & Space Propulsion divsn.

United Techs. Pratt & Whitney, West Palm Beach, Fla., 1985—. With U.S. Army, 1960. Office: Pratt & Whitney Government Govt Engines & Space Propulsion PO Box 109600 West Palm Beach FL 33410*

BALAKRISHNA, SUBASH, chemical engineer; b. Madras, India, Sept. 25, 1967; came to U.S., 1988; s. Balakrishnan and Jayalakshmi (Raman) B. B Tech. with honors, Indian Inst. Tech., Kharagpur, 1988; PhD in Chem. Engring., Carnegie Mellon U., 1992. Grad. fellow Carnegie Mellon U., Pitts., 1988-92; advanced engr. Mobil R&D Corp., Princeton, N.J., 1992—. Contbr. articles to profl. publs. Mem. Sigma Xi. Achievements include development of technique for reactor network synthesis and optimal process flow sheet integration. Home: 189 Harrison St Princeton NJ 08540 Office: Mobil R&D Co PO Box 1026 Princeton NJ 08540

BALAKRISHNAN, KRISHNA (BALKI BALAKRISHNAN), biotechnologist, corporate executive; b. Chelakkara, Kerala, India, June 30, 1955; came to U.S., 1977.; s. S. Krishna Iyer and C.K. Parvathy (Ammal); m. Sheela Kalyanakrishnan, Dec. 12, 1984; children: Karthik, Purnima. MS in Chem., Indian Inst. Tech., 1977; PhD in Biophysical Chem., Stanford U., 1982. Tchr., rsch. asst. dept. biophysical chem. Stanford U., Calif., 1977-82; staff sci. DNAX, Ltd., 1982-84; sr. sci. Biogenex, Inc., 1984-85; dir. hybridoma scis. Berkeley Antibody Co., Calif., 1985—, v.p. rsch. & devel., 1988—; guest lectr., sci. adv. biotech. program Contra Costa Coll., 1992—; indsl. ptrn. Stanford-NIH grad. tng. program biotech., 1992—. Contbr. articles to profl. jours.; speaker in field; patent applications. Mem. AAAS, Am. Chem. Soc. Office: Berkeley Antibody Co 4131 Lakeside Dr Ste B Richmond CA 94806-1965

BALALE, EMANUEL MICHAEL, chemical engineer; b. Cleve., Dec. 25, 1956; s. Michael and Stella (Bymakos) B. BS in Chem. Engring., Case Western Res. U., 1978. Lic. profl. engr., Pa., N.J. Process engr. Gulf Oil Co., Phila., 1978-86; project engr. Allstate Engring. Co., Newark, 1986-88; lead process engr. Atlantic Refining Co., Phila., 1988-89; sr. process engr. Sun Oil Co., Phila., 1989—. Achievements include numerous design projects on waste energy recovery, high octane gasoline production, jet fuel production, petroleum coke handling, heavy oil storage. Office: Sun Oil Co 1501 Blueball Ave Linwood PA 19061

BALARAS, CONSTANTINOS AGELOU, mechanical engineer; b. Athens, Greece, Sept. 20, 1962; s. Agelos and Christina Balaras. BS in Mech. Engring. with honors, Mich. Tech. U., 1984; MSME, Ga. Inst. of Tech., 1985, PhD, 1988. Cert. engr., Europe. Cons. engr. Am. Combustion Inc., Norcross, Ga., 1987-88; sr. mech. engr. Am. Standards Testing Bur., Inc., Phila., 1988-89; rsch. scientist Nat. Def. Rsch. Ctr., Athens, 1989-90, Nat. U. Athens, 1990-91; seminar lectr. Inst. of Tech. Applications, Athens, 1990—; sr. mech. engr. Protechna Ltd., Athens, 1991—; asst. prof. mech. engring. dept. T.E.I., Pireaus, 1991—; rsch. assoc. U. Athens, 1992—. Contbr. over 30 articles to profl. jours. and confs. Fgn. student scholarship Mich. Tech. U., 1980-84. Mem. ASME, ASHRAE, The Scientific Rsch. Soc., Internat. Soc. of Solar Energy, Tech. Chamber of Greece, Nat. Soc. of Lic. Mech. Elec. Engrs., Sigma Xi. Achievements include research in renewable energy sources, thermal sciences, energy conservation, passive heating and cooling of buildings, indoor air quaity, numerical applications. Home: Sotiriou Charalabi 5, GR114 72 Athens Greece Office: Protechna Ltd, Themistokleous 87, GR10683 Athens Greece

BALASUBRAMANIAN, AIYLAM SUBRAMANIAIER, biochemistry educator; b. Perumbavoor, Kerala, India, July 15, 1937; s. Aiylam Venkateswar and Seetha Subramaniaier; m. Jyothi Balasubramanian, Mar. 30, 1970; children: Ravi, Raju. MSc, Kerala U., 1959, PhD, 1964. Rsch. assoc. U. So. Calif. Med. Sch., L.A., 1966-68; reader Christian Med. Coll., Vellore, India, 1969-74, assoc. prof., 1974-77, prof. biochemistry, chief neurochemistry lab., 1977—; mem. med. sci. rsch. com. Coun. of Scientific and Indsl. Rsch., New Delhi, 1990—. Contbr. over 106 articles to profl. jours. Recipient Prof. Shadaksharaswami Endowment award Soc. of Biol. Chemists, 1988. Mem. Am. Chem. Soc., Biochem. Soc. London, Internat. Soc. for Neurochemistry, Sigma Xi. Achievements include first to report amine sensitive aryl acylamidase activity in acetylcholinesterase and butyrylcholinesterase and the deficiency of arylsulfatase A in metachromatic leukodystrophy. Home and Office: Neurochemistry Lab, CMC Hosp, Vellore 632004, India

BALASUBRAMANIAN, KRISHNA, chemistry educator, consultant; b. Madras, India, Nov. 25, 1945; s. Narasimhan and Anandammal Krishnan B.; m. Thirupurasundari Balasubramanian, Feb. 5, 1975; children: Anand, Lalitha. MSc, Madras U., India, 1968, PhD, 1974. Faculty mem. Madras U., 1968-73, Wayne State U., Detroit, 1974-76, Temple U., Phila., 1976-77, Drexel U., Phila., 1977-78; sr. project mgr. CLRI, Madras, 1979—; tech. mgr. AL Hotystanger Ltd., Khobar, Saudi Arabia, 1991—; bd. dirs. Indian Jour. Chemistry, New Delhi, 1979—; cons. various cos., India, 1970—. Contbr. articles to Jour. Am. Chem. Soc., Indian Jour. Chemistry, Synthesis (Fed. Republic Germany). Mem. Am. Chem. Soc., Internat. Soc. Magnetic Resonance, Indian Soc. Mass Spectrometry, Sob. Biol. Chemists, Madras Sci. Assn. Achievements include pending Indian patent for synthesis of novel polyurethane; discovery of a new dimerisation reaction based on oxaziridines; synthesized the natural products deoxyschizandrin, squalene, brittonin A, 1, 4 Bis-(3, 4-dimethoxyphenyl)-2,3-Dimethylbutane; carried out the controlled oxidation of collagen with pyridinium chlorochromate. Home: Plot 60 Ashtalakshmi Nagar, 600116 Porur Madras, India Office: Al Hoty Stanger Ltd, PO Box 1122, Khobar Alkhobar 31952, Saudi Arabia

BALATSKY, ALEXANDER VASILIEVITCH, physicist; b. Pushkin, Russia, Oct. 19, 1961; came to U.S., 1989; s. Vasiliy Efimovitch and ELizaveta (Pliskina) B.; m. Galina Ivanovna Martemyanova; 1 child, Kirill. MS, Moscow Phys.-Tech. Inst., 1984; PhD, Landau Inst., 1987. Researcher Landau Inst., Moscow, 1987-89; J.R. Oppenheimer fellow Los Alamos (N.Mex.) Nat. Lab., 1991—; vis. asst. prof. U. Ill., Urbana, 1990-91. Contbr. articles to profl. jours. Postdoctoral fellow U. Ill., 1989-90. Mem. Am. Phys. Soc., Rotary. Achievements include theoretically predicted principally new kind of superconducting order parameter; research interests include superconductivity, superfluidity, quantum hall effect. Office: Los Alamos Nat Lab K765 CMS Los Alamos NM 87545

BALCÁZAR, JOSÉ LUIS, computer scientist, educator; b. Valladolid, Spain, Aug. 26, 1959; s. José Luis and Delfina (Navarro) B.; 1 child, Adrian. Lic. en math., U. Valladolid, U. Complutense, Madrid, Spain, 1981; D in Info., U. Politecnica de Catalunya, Barcelona, 1984. Encargado de curso d U. Politecnica de Catalunya, 1981-82, colaborador, 1982-84, titular contratado, 1984-86, prof. titular de universidad, 1986-88, catedratico de universidad, 1988—; head theoretical computer sci. sect. Leenguatges Sistemes Informatics, U. Politecnica de Catalunya, 1988—. Co-author: Structural Complexity I, 1988, Structural Complexity II, 1990; contbr. articles to profl. jours. Grantee Comite Conjunto Hispano-Norteamericano, 1982. Mem. European Assn. for Theoretical Computer Sci., Assn. for Computing Machinery, Spl. Interest Group for Automata and Computability Theory. Avocations: music theory and composing, piano playing, fencing, skiing. Office: U Politecnica, Dept LSI, 08071 Barcelona Spain

BALCERZAK, MARION JOHN, mechanical engineer; b. Balt., Oct. 28, 1933; s. Marion Frank and Cecilia V. (Mazur) B.; m. Mary Joan Kenny, June 21, 1958; children—Stephanie, Susan, Jennifer, Jeffrey. B.Mech. Engring. magna cum laude, U. Detroit, 1956; M.S. (transp. fellow 1957-60), Northwestern U., 1958, Ph.D., 1961. Research engr. Borg-Warner Corp., 1960-62; with GATX Corp., Chgo., 1962-84; dir. research and devel. GATX Corp., 1980-84; tech. dir., then v.p., gen. mgr. subs. GARD, Inc., 1969-80, pres., 1980-84; with Chamberlain Mfg. Corp. (a unit of Duchossois Industries, Inc.), 1984—; v.p., gen. mgr. GARD div. Duchossois Industries, 1984-85, v.p. tech. group, 1985—. Served with AUS, 1961. Mem. ASME (Charles T. Main award 1956), Am. Mgmt. Assn., AIAA, ASTM, Am. Def. Preparedness Assn., Research Dirs. Assn. of Chgo. Home: 2750 Crabtree Ln Northbrook IL 60062-3460 Office: Duchossois Indust Inc 845 N Larch Ave Elmhurst IL 60126-1114

BALCERZAK, STANLEY PAUL, physician, educator; b. Pitts., Apr. 27, 1930. B.S., U. Pitts., 1953; M.D., U. Md., 1955. Diplomate Am. Bd. Internal Medicine, Am. Bd. Hematology, Am. Bd. Oncology. Instr. medicine U Chgo., 1959-60; instr. medicine U. Pitts., 1962-64, asst. prof., 1964-67; assoc. prof. medicine Ohio State U., Columbus, 1967-71, prof., 1971—, dir. div. hematology and oncology, 1969—, dep. dir. Ohio State U. Comprehensive Cancer Ctr., 1984—, assoc. chmn. dept. medicine, 1984—, dir. Hemophilia Ctr., 1975-79, 1981—; mem. clin. rev. com. Am. Cancer Soc., N.Y.C., 1976-82. Contbr. chpts. to books, numerous articles to profl. jours. Served to capt. U.S. Army, 1960-62. Recipient numerous grants. Fellow ACP; mem. Central Soc. for Clin. Research (chmn. subsplty. council in hematology 1980-81, councillor 1980-83), Am. Soc. for Clin. Oncology, Am. Assn. for Cancer Research, Am. Soc. Hematology, Phi Beta Kappa, Alpha Omega Alpha. Home: 3113 3B's and K Rd Sunbury OH 43074 Office: Ohio State U Divsn Hematology Oncology 10 N Doan Hall Columbus OH 43210-1228

BALDESCHWIELER, JOHN DICKSON, chemist, educator; b. Elizabeth, N.J., Nov. 14, 1933; s. Emile L. and Isobel (Dickson) B.; m. Marlene R. Konnar, Apr. 15, 1991; children from previous marriage: John Eric, Karen Anne, David Russell. B. Chem. Engring., Cornell U., 1956; Ph.D., U. Calif. at Berkeley, 1959. From instr. to asso. prof. chemistry Harvard U., 1960-65; faculty Stanford (Calif.) U., 1965-71, prof. chemistry, 1967-71; chmn. adv. bd. Synchrotron Radiation Project, 1972-75; vis. scientist Synchrotron Radiation Lab., 1977; dep. dir. Office Sci. and Tech., Exec. Office Pres., Washington, 1971-73; prof. chemistry Calif. Inst. Tech., Pasadena, 1973—; chmn. div. chemistry and chem. engring. Calif. Inst. Tech., 1973-78; OAS vis. lectr. U. Chile, 1969; spl. lectr. in chemistry U. London, Queen Mary Coll., 1970; vis. scientist Bell Labs., 1978; mem. Pres.'s Sci. Adv. Com., 1969—, vice chmn., 1970-71; mem. Def. Sci. Bd., 1973-80, vice chmn., 1974-76; mem. carcinogenesis adv. panel Nat. Cancer Inst., 1973—; mem. com. planning and instl. affairs NSF, 1973-77; adv. com. Arms Control and Disarmament Agy., 1974-76; mem. NAS Bd. Sci. and Tech. for Internat. Devel., 1974-76, ad hoc com. on fed. sci. policy, 1979, task force on synfuels, 1979, Com. Internat. Security and Arms Control, 1992—; mem. Pres.'s Com. on Nat. Medal of Sci., 1974-76, pres., 1986-88, Pres.'s Adv. Group on Sci. and Tech., 1975-76; mem. governing bd. Reza Shah Kabir U., 1975-79; mem. Sloan Commn. on Govt. and Higher Edn., 1977-79, U.S.-USSR Joint Commn. on Sci. and Tech. Coop., 1977-79; vice chmn. del. on pure and applied chemistry to China, 1978; mem. com. on scholarly communication with China, 1978-84; chmn. com. on comml. aviation security NAS, 1988—, mem. def. sci. bd. task force on 'operation desert shield', 1990-91, mem. com. on internat. security and arms control, 1991—; mem. rsch. adv. coun. Ford Motor Co., 1979—, mem. chem. and engring. adv. bd., 1981-83; vis. lectr. Rand Afrikaans U., Johannesburg, South Africa, 1987, Found. Rsch. and Devel., Pretoria, South Africa, 1989. Mem. editorial adv. bd. Chem. Physics Letters, 1979-83, Jour. Liposome Rsch., 1986—. Served to 1st lt. AUS, 1959-60. Sloan Found. fellow, 1962-64, 64-65; recipient Fresenius award Phi Lambda Upsilon, 1968, Tolman award ACS, 1989. Mem. NAS, Am. Chem. Soc. (award in pure chemistry 1967, William H. Nichols medal 1990), Council on Sci. and Tech. for Devel., Am. Acad. Arts and Scis., Am. Philos. Soc. Home: PO Box 50065 Pasadena CA 91115-0065 Office: Calif Inst Tech Divsn Chemistry & Chem Engring # 127-72 Pasadena CA 91125

BALDO, GEORGE JESSE, biophysicist, physiology educator; b. Herkimer, N.Y., Aug. 14, 1952; s. Sullivan George and Dellalouise (Crane) B.; m. Linda Kathryn Brown, Aug. 22, 1981; 1 child, Jesse Sullivan. BS in Biology, Union Coll., 1974; PhD in Physiology, SUNY, Stony Brook, 1982. Tchr. asst. in spl. edn. BOCES II, Patchogue, N.Y., 1974-76; rsch. instr. dept. physiology SUNY, Stony Brook, 1982-85, rsch. asst. prof. dept. physiology, 1985—. Contbg. author: Modern Cell Biology, vol. 7, 1988; contbr. articles to Jour. of Physiology, Biophys. Jour., Investigative Ophthalmology and Visual Sci., Am. Jour. Physiology, others; editor Setauket United Meth. Ch. Wesleyan. Fight for Sight/Nat. Soc. to Prevent Blindness grantee, 1989-90. Mem. Assn. for Rsch. in Vision and Ophthalmology, Biophysics Soc., Am. Physiol. Soc., Sigma Xi. Republican. Roman Catholic. Achievements include research in synaptic transmission; research in intercellular communication in syncytial tissues via gap junctional coupling, notably the ocular lens and ventricular muscle. Home: 18 Fox Rd Setauket NY 11733 Office: SUNY Stony Brook Dept Physiology Health Scis Ctr Stony Brook NY 11794-8661

BALDWIN, BETTY JO, computer specialist; b. Fresno, Calif., May 28, 1925; d. Charles Monroe and Irma Blanche (Law) Inks; m. Barrett Stone Baldwin Jr.; two daughters. AB, U. Calif., Berkeley, 1945. With NASA Ames Rsch. Ctr., Moffett Field, Calif., 1951-53, math tech. 14' Wind Tunnel, 1954-55, math analyst 14' Wind Tunnel, 1956-63, supr. math analyst Structural Dynamics, 1963-68, supervisory computer programmer Structural Dynamics, 1968-71, computer programmer Theoretical Studies, 1971-82, adminstrv. specialist Astrophys. Experiments, 1982-85, computer specialist, resource mgr. Astrophysics br., 1985—; v.p. B&B Baldwin Farms, Bakersfield, Calif., 1978—. Mem. IEEE, Assn. for Computing Machinery, Am. Geophys. Union, Am. Bus. Womens Assn. (pres., v.p. 1967, one of Top 10 Women of Yr. 1971). Presbyterian. Avocations: reading, bridge, hiking. Office: NASA Ames Rsch Ctr Mail Stop 245-6 Moffett Field CA 94035-1000

BALDWIN, DANIEL FLANAGAN, mechanical engineer; b. Fort Collins, Colo., Jan. 4, 1965; s. Lionel Vernon and Kathleen Mae (Flanagan) B.; m. Kristen Jean Schamberger, Aug. 1989; 1 child, Kelsey Rae. BS in Engring. summa cum laude, Ariz. State U., 1988; MS, MIT, 1990, postgrad. Engr. in tng., Ariz. Software analyst Colo. State U., Fort Collins, 1984, 85, rsch. asst., 1986; engring. intern Mitsubishi Electric Corp., Kamakura, Japan, 1987; Draper fellow C.S. Draper Lab., Cambridge, Mass., 1988-90; rsch. mgr. MIT, Cambridge, 1990—. Referee Robotics and Computer-Integrated Mfg., 1992-93; contbr.: Computer-Aided Mechanical Assembly Planning, 1991; contbr. articles to IEEE Transactions in Robotics and Automation. Mem. ASME, IEEE, Am. Soc. Engring. Edn., Soc. Mfg. Engrs., Soc. Plastics Engrs., Sigma Xi, Pi Tau Sigma, Tau Beta Pi, Phi Kappa Phi. Achievements include a patent pending for supermicrocellular foamed materials, gas-assisted injection molding of microcellular and supermicrocellular plastics; research in materials processing, manufacturing systems, design, and process control. Office: MIT Rm 35-009 77 Massachusetts Ave Cambridge MA 02139

BALDWIN, DAVID E., physicist; b. Bradenton, Fla., June 12, 1936; m. Eleanore McCord Haggard, June 24, 1961; children: Kenneth, Johanna. BS in Physics, MIT, 1958, PhD in Physics, 1962. Rsch. assoc. Stanford U., 1962-64, Culham Lab., Eng., 1964-66; asst. prof. engring. and applied sci. Yale U., 1966-68, assoc. prof., 1968-70; fusion theory program physicist Lawrence Livermore Nat. Lab., Magnetic Fusion Divsn., 1970-78; assoc. theory program leader Lawrence Livermore Nat. Lab, Magnetic Fusion Divsn., 1978-83, deputy assoc. dir., MFE, 1983-88, acting assoc. dir., MFE, 1988, assoc. dir., MFE, 1991-92, assoc. dir. energy, 1992—; prof. physics U. Tex., Austin, 1988-91, dir. inst. for fusion studies, 1988-91; various programs, coms. office fusion energy, Dept. Energy. 1983—; vis. com. nuclear engring. dept., MIT, 1987-93; magnetic fusion adv. bd. Los Alamos Nat. Lab., 1988-91; vis. com. applied sci. dept., Yale U., 1990-93; Tokamak Physics Experiment coun. Princeton Plasma Physics Lab, 1992—. Assoc. editor: Physical Fluids, 1975-78, Physical Review Letters, 1976-77, Review of Modern Physics, 1980-88. Fellow Am. Physical Soc. Achievements include rsch. in theoretical aspects of plasma physics and magnetic fusion energy, patent in Generating End Plug Potentials in Tandem Mirrors by Heating Thermal Particles Out of Low Density End Plugs. Office: Lawrence Livermore Nat Lab L-640 PO Box 808 Livermore CA 94551

BALDWIN, GEORGE MICHAEL, industrial marketing professional; b. Koza City, Okinawa, Oct. 8, 1960; s. George Ramon and Haruko (Toyama) B.; 1 child, Corey. BSBA, Appalachian State U., 1983; MBA, Campbell U., 1988. Cert. cogeneration profl. Comml. mktg. coord. N.C. Natural Gas Corp., Fayetteville, 1983-88, mgr. merchandising, 1988-91, mgr. indsl. sales, 1991—. Mem. ASHRAE, N.C. Agri-Bus. (bd. dirs. 1992—), Eastern Chamber of N.C. (bd. dirs. 1992—), Am. Heart Assn. of Cumberland County (bd. dirs. 1992—), N.C. Econ. Developers Assn. (legis. com. 1992—), Assn. of Energy Engrs. (v.p. Sandhill chpt. 1989-90), West Fayetteville Rotary Club (sec. 1992),. Republican. Presbyterian. Home: 3244-D Turtlepoint Dr Fayetteville NC 28304 Office: NC Natural Gas Corp 150 Rowan St Fayetteville NC 28302

BALDWIN, JOHN CHARLES, surgeon, researcher; b. Ft. Worth, Sept. 23, 1948; s. Charles Leon and Anabel (West) B.; m. Christine Janet Stewart, Mar. 31, 1973; children: Alistair Edward Stewart, John Benjamin West, Andrew Christian William. BA summa cum laude, Harvard U., 1971; MD, Stanford U., 1975; MA Privatim (hon.), Yale U., 1989. Diplomate Am. Bd. Internal Medicine, Am. Bd. Surgery, Am. Bd. Thoracic Surgery. Fellow in medicine Harvard Med. Sch., Boston, 1975-77, fellow in surgery, 1977-81; resident in cardiothoracic surgery Stanford (Calif.) U., 1981-82, chief resident cardiothoracic surgery, 1983, asst. prof., 1984-87, head heart-lung transplantation, co-dir. exptl. lab. Dept. Cardiovascular Surgery,, 1986-87; prof. surgery and chief cardiothoracic surgery Yale Univ., New Haven, 1988—; cardiothoracic-surgeon-in chief Yale-New Haven Hosp.; dir. thoracic surgery residency program Yale-New Haven Hosp.; vis. lectr. Yale U.; cons. gen. thoracic surgery Waterbury (Conn.) Hosp. Health Ctr.; bd. dirs. United Network for Organ Sharing, 1984—; mem. clin. rsch. com. ad hoc rsch. grant rev. Cystic Fibrosis Found.; trustee New Eng. Organ Bank, 1988; mem. solid organ transplant com. Blue Cross & Blue Shield of Conn., 1990—. Co-editor: Thoracic Surgery, Oxford Textbook of Surgery, 1989—; assoc. editor Jour. Applied Cardiology, 1985-92; editorial bd. Jour. Thoracic and Cardiovascular Surgery, 1990—, Transplantation, 1990—, Transplantation Sci., 1992—; contbr. numerous articles and book chpts. in field. Mem. Harvard Club Schs. Com., Harvard Coll. Fund, Harvard U. Undergrad. Admissions Interview Com.; fellow Timothy Dwight Coll. Yale U., Yale U. Art Gallery Assocs.; mem. appointments and promotions com. Sch. Medicine, Yale U., 1991—, clin. scis. bldg. planning com., 1990—; bd. dirs. Neighborhood Music Sch. New Haven. John Harvard scholar, 1969, 70, Wendell scholar Harvard U., 1969, Rhodes scholar Oxford U., Alumni scholar Stanford Sch. Medicine, 1974 ; medalist Gothenburg (Sweden) Thoracic Soc., 1985; recipient Medaille de la Ville de Bordeaux French Thoracic Soc., 1987, travelling lectureship, 1988, Master Tchr. award Cardiovascular Revs. & Reports, 1990; travelling fellow Australia and New Zealand chpt., ACS, 1989; traveling lectureship, 1989. Fellow ACP, Royal Coll. Surgeons (Eng., traveling lectr. 1989), Am. Coll. Angiology, Am. Coll. Cardiology (mem. transplantation com. 1991—, chmn. task force cardiac donor procurement Bethesda Conf. 1992), Am. Coll. Surgeons (bd. govs.), Am. Coll. Chest Physicians, Mass. Med. Soc.; mem. AMA, Am. Assn. Thoracic Surgery (mem. com. grad. edn. thoracic surgery 1992—, Evarts A. Graham Meml. Traveling Fellowship com. 1993—), Am. Soc. Transplant Surgeons (com. on heart transplantation 1986—, adv. com. in issues 1989—, chmn. subcom. on heart transplantation, physician payment reform commn. 1989—), Assn. Acad. Surgery, Am. Physiol. Soc., Am. Heart Assn. (mem. rsch. grant peer rev. subcom 1984—), Am. Surg. Assn., Am. Thoracic Soc., Am. Soc. Artificial Internal Organs, Am. Soc. Extracorporeal Tech., Internat. Soc. Heart Transplantation (chmn. program com. 1988), New Eng. Surg. Soc., Pan Am. Med. Assn. (council on organ transplantation), Societe Internat. de Chirurgie, Royal Soc. Medicine, Soc. Univ. Surgeons, Thoracic Surgery Dirs. Assn., Transplantation Soc., Assn. Alumni of Magdalen Coll. Oxford U., Assn. Rhodes Scholars, Calif. Med. Assn., Calif. Thoracic Soc., Conn. Med. Soc., Conn. Soc. Am. Bd. Surgeons, Mass. Med. Soc., New Haven County Med. Soc., Harvard Med. Alumni Assn. (assoc.), San Francisco Surg. Soc., Santa Clara Med. Soc., Stanford Med. Alumni Assn., Stanford Club Conn., Harvard Clubs San Francisco, Peninsula, N.Y.C., So. Conn., Mory's Assn., New Haven Lawn Club, Inner Squad Stanford U., The Hasty Pudding Club - Inst. 1770, Quinnipiack Club, Yale Club New Haven. Office: Yale U Sch Medicine Dept Surgery 333 Cedar St New Haven CT 06510-3289

BALDWIN, RANSOM LELAND, animal science educator; b. Meriden, Conn., Sept. 21, 1935; s. Ransom Leland and Edna (Thurrot) B.; m. Mary Ellen Burns, June 1, 1957; children: Ransom Leland VI, Cheryl Lee, Robert Ryan. BS, U. Conn., 1957; MS, Mich State U., 1958, PhD, 1963. Research asst. Mich. State U., East Lansing, 1957-61; from asst. to assoc. prof. U. Calif., Davis, 1963-70, prof., 1970—. Assoc. editor Jour. Nutrition, 1971-73, 83—; contbr. research articles to profl. jours. Guggenheim fellow; Fullbright fellow; NSF fellow; recipient Borden award, 1980, Am. Feed Mfrs. Assn. award, 1970. Fellow AAAS; mem. Am. Inst. Nutrition, Am. Diary Sci. Assn., Soc. Animal Sci., Sigma Xi. Home: 2101 Amador Ave Davis CA 95616-3014 Office: U Calif Dept Animal Sci Davis CA 95616

BALDWIN, WENDY HARMER, social demographer; b. Phila., Aug. 29, 1945; B.A. magna cum laude, Stetson U., DeLand, Fla., 1967; M.A., U. Ky., Lexington, 1970, Ph.D. (NDEA fellow, spl. grantee Population Council), 1973. Research asst. Colombian Assn. Med. Faculties, Bogatá, 1971; research asst. sociology U. Ky., 1971-72; health scientist adminstr. behavioral scis. br. Center Population Research, Nat. Inst. Child Health and Human Devel., NIH, 1972-79, chief demographic and behavioral scis. br., 1979-91, dep. dir., 1991—, acting dep. dir. Extramural Rsch., 1993. Recipient Merit award NIH, 1978; USPHS Superior Service award, 1985; Carl S. Schultz award population amd family planning sec., Am. Pub. Health and Planning Assn. Mem. Population Assn. Am. (dir. 1978-80, 2d v.p. 1984), Am. Sociol. Assn. (sec. population sect. 1977-80, chmn. 1985), So. Sociol. Assn., Phi Beta Kappa. Author articles in field. Office: NIH/ NICHD/OD Bldg 31 Rm # 2A03 9000 Rockville Pk Bethesda MD 20892

BALEJA, JAMES DONALD, biochemist, educator; b. Carman, Man., Can., Dec. 23, 1961; s. Roy and Vera B. BSc in Biochemistry with honors, U. Man., Winnipeg, 1983, MSc in Chemistry, 1984; PhD in Biochemistry, U. Alta., Edmonton, Can., 1990. Postdoctoral fellow Harvard U., Cambridge, Mass., 1990-92; asst. prof. Tufts U., Medford, Mass., 1992—. Contbr. articles to Nature, Biochemistry, other jours. Recipient scholarship Nat. Sci. and Engring. Rsch. Coun. Can., 1983-85, Med. Rsch. Coun. Can., 1985-89; fellow Med. Rsch. Coun. Can., 1990-92; recipient Gov. Gen. Gold medal, Edmonton, 1990. Achievements include research in macromolecular structure determination using nuclear magnetic resonance spectroscopy; basis of protein-DNA recognition. Office: Tufts U Dept Biochemistry 136 Harrison Ave Boston MA 02111

BALESTRA, CHESTER LEE, electrical engineer, educator; b. Decatur, Ala., June 21, 1943; s. Lee and Mildred (Droter) B.; m. Patricia Souza, Aug. 27, 1966; children: Scott Anthony, Mark Andrew, Joel Alan. BS, MIT, 1966, ScD, 1971. Rsch. chemist Eastman Kodak Corp., Rochester, N.Y., 1971-74; surface physicist Thermo Electron Corp., Waltham, Mass., 1974-78; mem. tech. staff Tex. Instruments, Inc., Dallas, 1978-81; engring. mgr. Solid State Sci. Corp., Willow Grove, Pa., 1981-82, EG&G, Salem, Mass., 1982-83; chief electronics engr. McDonnell Douglas Corp., St. Louis, 1983—. Mem. IEEE, Electrochem. Soc., Am. Vacuum Soc., Sigma Xi. Achievements include invention of monolithic semiconductor active waveguide optical crossbar switch. Home: 2030 Medicine Bow Dr Ellisville MO 63011 Office: McDonnell Douglas Corp Box 516 Saint Louis MO 63166

BALETTIE, ROGER EUGENE, aerospace engineer; b. St. Joseph, Mo., Jan. 27, 1964; s. Eugene Valerio and Suzanne (Dixon) B.; m. Gail Marie Sottosanti, Feb. 8, 1986; 1 child, Andrew Joseph. BS in Aero. Engring., U. Tex., 1985; MS in Phys. Sci., U. Houston, 1992. Applications software team leader Rockwell/NASA Johnson Space Ctr., Houston, 1986-90, flight dynamics officer, 1990—. Mem. Young Conservatives of Tex. Republican. Episcopalian.

BALFE, ALAN, biochemist; b. Dublin, Ireland, Dec. 13, 1956; s. Andrew and Eithne (McEveney) B. BS, U. Coll. Dublin, Dublin, 1978, PhD, 1985. Rsch. biochemist Mater Misericordiae Hosp., Dublin, 1978-82; biochemist St. James's Hosp., Dublin, 1982-87, 90—; rsch. fellow depts. neurology and biol. chemistry Kennedy Inst., Johns Hopkins Hosp., Balt., 1987-89. Contbr. articles to profl. jours. Vol. leader Lady Md. Found., Balt., 1988-89. Rsch. fellow St. Luke's Cancer Fund, 1979, Med. Rsch. Coun. Ireland, 1980-82, Fogarty Ctr. NIH, 1987-89. Mem. Assn. Clin. Biochemists Ireland, Assn. Clin. Biochemists London, Soc. for Study of Inborn Errors of Metabolism, Royal Coll. Pathologists (assoc.). Avocations: sailing, photography, rugby, literature, music. Home: 105 Whitecliff Rathfarnham, Dublin 16, Ireland Office: Saint James Hosp, James St, Dublin 8, Ireland

BALIGA, BANTVAL JAYANT, electrical engineering educator, consultant; b. Madras, India, Apr. 28, 1948; came to U.S., 1969; s. Bantval Vittal and Sanjivi (Rao) B.; m. Pratima Nayak, Dec. 25, 1975; children: Avinash, Vinay. B in Tech., Indian Inst. Tech., Madras, 1969; MS, Rensselaer Poly. Inst., 1971, PhD, 1974. Mem. staff GE Rsch. Ctr., Schenectady, N.Y., 1974-

78, program mgr., 1979-88, Coolidge fellow, 1983-88; prof. N.C. State U., Raleigh, 1988—, dir. power semiconductor rsch. ctr., 1991—; cons. power semiconductor industry, 1988—, EPRI, Palo Alto, Calif., 1989-92. Author: Modern Power Devices, 1987; author, editor: Epitaxial Silicon Tech, 1986; editor: Power Transistors, 1984, High Voltage ICs, 1988; contbr. articles to profl. jours.; patentee in field. Recipient Dushman award GE, 1983; named among 100 Brightest Scientists in Am., Science Digest, 1984, Disting. Scientist, Asian Indians in N.Am., 1984. Fellow IEEE (assoc. editor Transactions on Electron Devices 1984—, William Newell award 1991, Morris Liebman award 1993); mem. Nat. Acad. Engring. Avocations: tennis, stamp collecting, travel. Office: NC State U Power Semiconductor Rsch Ctr Box 7924 Ctrl Campus Raleigh NC 27695

BALL, EDNA MARION, cardiac rehabilitation nurse educator; b. Sault Ste Marie, Mich., Oct. 22, 1944; d. Stanley Thomas and Clara P. (Lewis) Piteau; m. Richard J. Ball, July 25, 1963; children: Richard Jr., Robin. AS in Nursing, Southwestern Jr. Coll., 1974; BS in Nursing with distinction, U. N. Fla., 1977; postgrad., U. West Fla. RN, Fla.; cert. advanced cardiac life support provider, basic CPR instr. trainer, caridac rehab. nurse therapist, advanced coronary care, exercise specialist. Staff nurse, med./surg. fl., intensive critical care unit Beaches Hosp., Jacksonville Beach, Fla.; weight control for life counselor Bapt. Hosp., Pensacola, Fla., nurse educator, diabetes edn. program, cardiac rehab. and pacer clinic coord.; staff nurse, critical care unit West Fla. Hosp. Coll., Pensacola; clin. instr., adj. faculty mem. Pensacola Jr. Coll. Mem. Am. Assn. Critical Care Nurses (nat. and local chpts.), Pensacola Area Continuing Edn. Resource System, Fla. Assn. Cardiovascular and Pulmonary Rehab., Am. Assn. Cardiovascular and Pulmonary Rehab., Am. Heart Assn. (bd. dirs. local chpt., sec. to bd. dirs. local chpt. 1988-89, Cert. Appreciation 1983, 84), Mended Hearts (exec. bd. local chpt., adv. com., Merit award), Calif. Jr. Coll. Honor Scholarship Soc., Alpha Gamma Sigma (Gamma Pi chpt.).

BALL, WILLIAM JAMES, physician; b. Charleston, S.C., Apr. 16, 1910; s. Elias and Mary (Cain) B.; B.S., U. of South, 1930; M.D., Med. Coll. S.C., 1934; m. Doris Hallowell Mason, July 9, 1938. Intern, Roper Hosp., Charleston, 1934-35; resident dept. pediatrics U. Chgo. Clinics, 1935-37; instr. pediatrics Med. Coll. S.C., 1938-42; practice medicine specializing in pediatrics, Charleston, 1938-42, Northwest Clinic, Minot N.D., 1946-51, Aurora, Ill., 1951-70; physician student Health Service No. Ill. U., 1970-72; mem. staff Copley Meml., Mercy Ctr. Health Care Services; assoc. prof. Sch. Nursing, No. Ill. U., 1971-72. Mem. Bd. Health, Aurora, Ill., 1958-62; pediatrician, div. services for crippled children U. Ill., 1952-86; pediatric cons. sch. dists. 129 and 131, Aurora, 1972-85, DeKalb County Spl. Edn. Assn., 1972-81, Sch. Assn. Spl. Edn. Dupage County, 1980-83, Mooseheart, Ill., 1970-83, Northwestern Ill. Assn. Handicapped Children; chmn. adv. com. Kane County Health Dept., 1986—; pres. Kane County sub-area council Health Systems Agy., Kane, Lake, McHenry Counties, 1977-78, sec., 1978-79. Served as capt. M.C., AUS, 1942-46; maj., 1946 to col., 1963, ret. 1970. Diplomate Am. Bd. Pediatrics. Fellow Royal Soc. Health, Am. Acad. Pediatrics; mem. AMA, Kane County Med. Soc. (pres. 1962), Am. Heart Assn., Am. Sch. Health Assn., Am. Cancer Soc., Juvenile Protective Assn. of Aurora , The Ret. Officers Assn. (west suburban Chgo. chpt.), Phi Beta Kappa, Phi Chi, Pi Kappa Phi. Republican. Rotarian. Address: 433 S Commonwealth Ave Aurora IL 60506

BALL, WILLIAM JAMES, JR., biochemistry educator; b. Hornell, N.Y., Sept. 5, 1942; s. William James and Elizabeth (Brady) B.; m. Elvira Ponce, June 8, 1987; children by previous marriage: Valerie, Darek; stepchildren: Margarita Barrios, Celina Barrios. MS in Chemistry, San Diego State U., 1968; PhD in Biochemistry, UCLA, 1973. Postdoctoral fellow Sloan-Kettering Inst. Cancer Rsch., N.Y.C., 1972-75; instr. dept. pharmacology Baylor Coll. Medicine, Houston, 1976; asst. prof. dept. pharmacology U. Cin. Coll. Medicine, 1978-84, assoc. prof., 1984—; grant rev. com. Am. Heart Assn., Ohio affiliate, 1987—. Nat. Cancer Inst. fellow UPSHS, 1973-75; recipient Nat. award USPHS, 1978, Established Investigator award Am. Heart Assn., 1984-89. Mem. Am. Heart Assn., Am. Soc. Biochemistry and Molecular Biology, Biophys. Soc., Sierra Club. Achievements include research in mechanisms of ion transport and digitalis-receptor interactions. Office: U Cin Coll Medicine Dept Pharmacology 231 Bethesda Ave Cincinnati OH 45267-0575

BALL, WILLIAM PAUL, physicist, engineer; b. San Diego, Nov. 16, 1913; s. John and Mary (Kajla) B.; m. Edith Lucile March, June 28, 1941 (dec. 1976); children: Lura Irene Ball Raplee, Roy Ernest. AB, UCLA, 1940; PhD, U. Calif., Berkeley, 1952. Registered profl. engr. Calif. Projectionist, sound technician studios and theatres in Los Angeles, 1932-41; tchr. high sch. Montebello, Calif., 1941-42; instr. math. and physics Santa Ana (Calif.) Army Air Base, 1942-43; physicist U. Calif. Radiation Lab., Berkeley and Livermore, 1943-58; mem. tech. staff Ramo-Wooldridge Corp., Los Angeles, 1958-59; sr. scientist Hughes Aircraft Co., Culver City, Calif., 1959-64; sr. staff engr. TRW-Def. Systems Group, Redondo Beach, Calif., 1964-83, Hughes Aircraft Co., 1983-86; cons. Redondo Beach, 1986—. Contbr. articles to profl. jours.; patentee in field. Bd. dirs. So. Dist. Los Angeles chpt. ARC, 1979-86. Recipient Manhattan Project award for contbn. to 1st atomic bomb, 1945. Mem. AAAS, Am. Phys. Soc., Am. Nuclear Soc., N.Y. Acad. Scis., Torrance (Calif.) Area C. of C. (bd. dirs. 1978-84), Sigma Xi. Home and Office: 209 Via El Toro Redondo Beach CA 90277-6561

BALLAL, DILIP RAMCHANDRA, mechanical engineering educator; b. Nagpur, India, Jan. 16, 1946; came to U.S., 1979; s. Ramchandra Govind Ballal and Padma (Balwant) Zadkar; m. Shubhangi Sadashiv Ayachit, Dec. 17, 1975; children: Rahul, Deepti. BSME, Coll. Engring., Bhopal, India, 1967; PhD, Cranfield (U.K.) Inst. Tech., 1972, DSc in Engring. (hon.), 1983. Registered profl. engr., Ohio. Lectr. mech. engring. Cranfield Inst. Tech., 1972-79; sr. staff engr. GM Rsch. Labs., Warren, Mich., 1979-83; prof. mech. engring. U. Dayton (Ohio), 1983—; cons. GMR Labs. and GE Aircraft, Warren, Cin., 1987—. Author: (with others) Combustion Measurements and Modern Development in Combustion, 1990, 91; contbr. about 100 articles on combustion, turbulence, heat transfer and pollution to profl. jours. Project leader Engrs. Club Dayton, 1986, 88, 90; judge, organizer "Odyssey of Mind" Sch. Contest, Dayton, 1985, 87, 88; vice chmn. edn. com. Miami Valley Sch., Dayton, 1988, 90. Named Outstanding Engr., Engrs. Club, Dayton, 1988. Fellow ASME (vice chmn. combustion and fuels com. 1993—, Best Rsch. award 1986, 92), AIAA (Energy Systems award 1993). Achievements include patents on Ignitor Plug for Jet Engine Combustor. Home: 5519 Knollcrest Ct Dayton OH 45429-5913 Office: U Dayton KL 465 300 College Park Dayton OH 45469-0140

BALLANTYNE, JOSEPH MERRILL, science educator, program administrator, researcher; b. Tucson; s. Alando Bannerman and Annie Hyde (Merrill) B.; m. Martha B. Cox; children: Joseph, Elizabeth, Catherine, Mary Joy, Annie, Richard, Merrill, Leonora. BS, BSEE, U. Utah, 1959; SM, MIT, 1960, PhD, 1964. Assoc. prof. Cornell U., Ithaca, N.Y., 1968-75, prof., 1975—, dir. elec. engring., 1980-84, v.p. rsch. adv. studies, 1984-89; dir. SRC ctr. of excellence in microscience and tech. Cornell U., 1992—; cons. various cos. and univs., 1961—; bd. dirs. N.Y. Photonics Devel. Corp., Rome., vis. assoc. prof. Stanford Univ., Calif., 1970-71, vis. scientist IBM Watson Rsch. Ctr., Yorktown Heights, N.Y., 1978-79, vis. prof. U. Calif., Santa Barbara, 1990, vis. prof. Tech. Univ. Aachen, Germany, 1990; acting dir. Nat. Nanofabrication Facility, Ithaca, 1977-78. Contbr. articles to profl. jours; patentee semiconductor devices. Bishop LDS Ch., Ithaca, 1972-77; v.p. bd. dirs. Tompkins County Area Devel. Corp., Ithaca, 1984-89, trustee Associated Univs. Inc., Washington, 1984-89, pres. Cornell Rsch. Found., Ithaca, 1984-89; mem. High Tech. Adv. Com., N.Y. State Urban Devel. Corp., N.Y.C., 1984-87; bd dirs. Coun. on Rsch. & Tech., Washington, 1987-89, Univ. Industry Partnership for Econ. Growth Waverly, N.Y., 1987-89. Recipient George Emery Fellows medal Phi Kappa Phi, 1959, Whitney fellow MIT, 1959-63, Schlumberger fellow MIT, 1959-63, NSF sr. fellow, 1970. Fellow IEEE. Avocation: music performance. Office: Cornell U 217 Phillips Hall Ithaca NY 14853

BALLARD, LOWELL DOUGLAS, mechanical engineer; b. Seiling, Okla., June 27, 1933; s. Auty Wayne and Mabel (Henderson) Haynes; B.S., U. Md., 1962. Mech. engr. Rabinow Inc., Rockville, Md., 1962; mech. engr. Nat. Bur. Standards, 1962-82; export licensing officer Dept. Commerce, Washington, 1981-88; cons., 1989—; panel mem. Nat. Elec. Code, 1975 edit.; chmn. Joint Bd. on Sci. and Engring. Edn., 1981-82. V.p. South Townhouse Assn., 1974-77, pres., 1978. With USAF, 1954-58. Fellow Washington Acad. Sci.; mem. IEEE (sr.), Am. Def. Preparedness Assn., Philos. Soc. Washington, Optical Soc. Am. Presbyterian. Club: Toastmasters (area gov. 1981-82). Home: 7823 Mineral Springs Dr Gaithersburg MD 20877-3822

BALLARD, RICHARD OWEN, rocket propulsion engineer; b. Lincoln, Nebr., Jan. 29, 1962; s. J.H. Curtis and Barbara Helen (Lloyd) B.; m. Amy Lynn Pedigo, Oct. 19, 1991. BS in Aerospace Engring., Tex. A&M U., 1986. Propulsion engr. Martin Marietta Manned Space Systems, Huntsville, Ala., 1987-89, Sverdrup Tech./MSFC Group, Huntsville, 1989—; lectr. U.S. Space Camp, Huntsville, 1990—. Recipient Group Achievement award NASA, Marshall Space Flight Ctr., 1988. Home: 5308 Panorama Dr Huntsville AL 35801 Office: Sverdrup Tech/MSFC Group 620 Discovery Dr Huntsville AL 35806

BALLARD, ROBERT DUANE, marine scientist; b. Wichita, Kans., June 30, 1942; s. Chester Patrick and Harriet Nell (May) B.; m. Marjorie C. Jacobsen (div.), July 1, 1966; children: Todd (dec.), Doug; m. Barbara Earle, Jan. 1991. BS, U. Calif., Santa Barbara, 1965; postgrad., U. Hawaii, 1965-66, U. So. Calif., 1966-67; PhD in Marine Geology and Geophysics, U. R.I., 1974. Asst. scientist Woods Hole (Mass.) Oceanographic Instn., 1974-76, assoc. scientist, 1976-83, sr. scientist, 1983—; pvt. cons. Benthos, Inc., North Falmouth, Mass., 1982-83; cons. dep. chief naval ops. for submarine warfare, 1983—; vis. scholar Stanford U., 1979-80, cons. prof., 1980-81, dir. Deep Submergence Lab., 1983—; bd. dirs., founder Marquest Group, The Jason Found. for Edn. Author: Exploring Our Living Planet, 1983, Discovery of the Titanic, 1989, Discovery of the Bismarck, The Wreck of the ISIS, 1990. With U.S. Army, 1965-67; with USN, 1967-70. Recipient Sci. award Underwater Soc. Am., 1976, Newcomb Cleveland prize AAAS, 1981, Cutty Sark Sci. award, 1982, Centennial Awd. Nat. Geog. Soc., 1988, Westinghouse award AAAS, 1990, Golden Plate award Am. Acad. Achievement, 1990, U.S. Navy Robert Dexter Conrad award for Sci. Achievement, 1992. Mem. Geol. Soc. Am., Marine Tech. Soc. (Compass Disting. Achievement award 1977), Am. Geophys. Union, Explorers Club. Achievements include being leader first and second expeditions to reach sunken ship Titanic, 1985, 86. Office: Woods Hole Oceanographic Water St Woods Hole MA 02543

BALLENGER, HURLEY RENÉ, electrical engineer; b. Jacksonville, Ill., Nov. 26, 1946; s. Leonard Hurley and Katherine Natalie (Daniel) B.; m. Sandra Ann Rubley, Dec. 9, 1986. Student, Ill. Coll., 1964-65, 75. Technician electronics div. Hughs Aircraft, Inc., Tucson, 1973; maintenance supr. Fiatallis N.Am., Springfield, Ill., 1973-75, project engr., 1975-83, plant engr., 1983-86; tech. advisor CNC/CAM Fiatallis Europe, Lecce, Italy, 1986-87; plant engr. Illini Tech., Inc., Springfield, Ill., 1988, plant and mfg. engr., 1988—. Mem. career adv. bd. Lincoln Land Community Coll., Springfield, 1983-85. Served to staff sgt. USAF, 1965-72, Vietnam. Lutheran. Avocations: photography, home computing. Office: Illini Tech Inc 3430 Constitution Dr Springfield IL 62707-9402

BALLINGALL, JAMES MCLEAN, III, physicist; b. Washington, Oct. 3, 1955; s. James McLean and Phylis (Dodson) B.; m. Brigitte Charreyron Wintermute, July 24, 1982; children: James McLean IV, Cameron Christopher, Connor Steven. BS in Engring. Physics, U. Calif., Berkeley, 1978; MS, Cornell U., 1980, PhD in Applied Physics, 1982. Scientist McDonnell Douglas Rsch. Labs., St. Louis, 1982-86; mgr. epitaxial tech. GE Electronics Lab., Syracuse, N.Y., 1986-91, mgr. III-V materials lab., 1991—; program com. mem. numerous internat. confs. on electronic materials, 1989—; lectr. in field. Contbr. over 100 articles to profl. jours. Co-recipient Nelson P. Jackson Aerospace award Nat. Space Club, 1990. Mem. IEEE (sr.), Materials Rsch. Soc., Tau Beta Pi. Achievements include co-development of high electron mobility transistor technology for microwave and millimeter wave radar and satellite communication; development of molecular beam epitaxial materials for infrared sensors and microwave/millimeter wave transistors. Home: 8009 Summerview Dr Fayetteville NY 13066 Office: Martin Marietta Electronics Lab PO Box 4840 EP3 Syracuse NY 13221-4840

BALLOD, MARTIN CHARLES, civil engineer, mechanical engineer; b. Abington, Pa., Nov. 21, 1950; s. Martin Albert and Ethel Mary (Paul) B.; m. JoAnn Shirley Wyant, Sept. 16, 1972; 1 child, Christopher Eric. BSCE, Pa. State U., 1973; MCE, Villanova U., 1981. Registered profl. engr., Pa., La., Tex., Ohio, Md., Va., Ala., N.J., Utah, Calif., Minn., Mass. Engr.-in-tng. FMC Corp., Colmar, Pa., 1973-74, design engr., sr. design engr., 1974-79, project engr., 1980-88; quality assurance mgr., 1985-90; sr. project engr. FMC Corp., Chalfont, Pa., 1988—. Asst. coach Northampton (Pa.) Little League, 1984; lector St. Vincent Cath. Ch., Richboro, Pa., 1986-88. Mem. Am. Welding Soc., Am. Concrete Inst., Tau Beta Pi. Home: 47 Bobbie Dr Warminster PA 18974-1638 Office: FMC Corp 400 Highpoint Dr Chalfont PA 18914-3924

BALLOU, CHRISTOPHER AARON, civil engineer; b. Hartsville, Tenn., Apr. 20, 1963; s. Bobby Aaron and Doris Anne (Day) B.; m. Jovita Denise Minchey Ballou, Mar. 15, 1985; 1 child, Mary Anne Ballou. BS in Civil Engring., Tenn. Tech. U., Cookeville, 1985-88. Engr.-in-Tng., Tenn. Draftsman Tenn. Dept. Transp., Nashville, 1982-83; surveyor Tenn. Dept. Transp., Gallatin, 1985; project engr. Gresham, Smith and Ptnrs., Nashville, 1988—; mem. Chi Epsilon Engr. Hon. Soc., 1987—; assoc. mem. Am. Soc. Civil Engrs., 1988—. Recipient Merit award for Outstanding Svc. Gresham, Smith & Ptnrs., 1991. Baptist. Home: 4906 Packard Dr Nashville TN 37211 Office: Gresham Smith & Ptnrs 3310 W End Ave PO Box 1625 Nashville TN 37202

BALLWEBER, HETTIE LOU, archaeologist; b. Pitts., Dec. 27, 1944; d. Nicholas George and Harriett Elizabeth (Tucker) Beresh; m. Walter David Boyce, Aug. 24, 1963 (div. 1984); children: Michael David, Steven Todd; m. William Arterberry Ballweber, Nov. 8, 1986. BA summa cum laude, Calif. U., Pa., 1985; M. Applied Anthropology, U. Md., 1987. Cons. archaeologist Monongahela, Pa., 1980-85; archaeologist archeology div. Md. Geol. Survey, Balt., 1985-86; dir. Md. New Directions, Balt., 1987; cons. Columbia, Md., 1987—; prin. ACS Cons., Columbia, Md., 1987—; bd. dirs. Alternative Directions, Inc., Balt. Author: First People of Maryland, 1985; contbr. articles to profl. jours. State publicity chmn. Pa. Congress Parents and Tchrs., Harrisburg, Pa., 1981-84, regional v.p., 1984. With USN, 1979-87. Fellow Soc. Applied Anthropology; mem. Mon-Yough Archaeol. Soc. (pres. 1983-84), Westmoreland Archaeol. Soc. (v.p. 1982-83), Coun. Md. Archeology (pres. 1990-91), Washington Assn. Profl. Anthropologists, Soc. Hist. Archaeology, Shriners, Order Eastern Star. Home and Office: 3212 Peddicoat Ct Woodstock MD 21163

BALLY, ALBERT W., geology educator. PhD, U. Zurich, Switzerland, 1953. Harry Corothers Weiss prof. geology Rice U., Houston. Office: Rice U Dept Geology Houston TX 77251

BALLY, LAURENT MARIE JOSEPH, software engineering company executive; b. Fort-de-France, Martinique, France, Dec. 23, 1943; s. Raoul and Therese (de Gentile) B.; m. Iris Kirstaetter, Oct. 8, 1971; 1 child, Laetitia. Diploma in engring., Inst. Superieur d'Electronique, Paris, 1966. Systems analyst NCR Corp., London, 1966-67; systems engr. Siemens A.G., Karlsruhe, Fed. Republic Germany, 1968-71; tech. mgr. Sesa-Deutschland GmbH, Frankfurt, Fed. Republic Germany, 1971-78; head of dept. Siemens A.G., Munich, 1979-84; head of div. Sesa S.A., Paris, 1984-88; dep. gen. mgr. Cap Sesa Tertiaire, Paris, 1989-92; gen. mgr. Cap Sesa Telecom, Paris, 1992—. Contbr. articles to profl. jours. Office: Cap Sesa Telecom, 30 Quai de Dion Bouton, 92806 Puteaux France

BALSARA, PORAS TEHMURASP, electrical engineering educator; b. Bombay, India, Jan. 9, 1961; s. Tehmurasp Erachshah and Roshan Tehmurasp (Balsara) B.; m. Pearl Poras, May 25, 1989. Diploma in Electronics, Victoria Jubilee Tech. Inst., Bombay, 1980; BEEE, U. Bombay, 1983; MS in Computer Sci., Pa. State U., 1985, PhD in Computer Sci., 1989. Inplant trainee, rsch. and devel. div. Tata Electr. Cos., Bombay, 1979; student trainee Nat. Electronics and Radio Corp., Bombay, 1981, Tata Inst. Fundamental Rsch., Bombay, 1982; grad. asst. Pa. State U., University Park, 1983-89; asst. prof. of elec. engring. U. Tex. at Dallas, Richardson, 1989—. Contbr. articles to profl. jours. Recipient Rsch. Initiation award NSF, 1990, J.N. Tata scholarship J.N. Tata Endowment for Higher Edn., Bombay, 1983. Mem. IEEE, IEEE Computer Soc., Assn. Computing Machinery, Sigma Xi. Office: Univ of Tex at Dallas Dept Elec Engring PO Box 830688 EC 33 Richardson TX 75083

BALSCHI, JAMES ALVIN, medical educator; b. Bloomsburg, Pa., Aug. 18, 1950; s. Alvin James and Elizabeth (Petusky) B.; m. So Ling Chan, Aug. 26, 1983; children: Ian Waicue, Sean Wingcue. BA in Anthropology, Temple U., 1973; PhD in Chemistry, SUNY, Stony Brook, 1984. Tech. dir. nuclear magnetic resonance lab. Harvard Med. Sch., Boston, 1984-88, rsch. assoc., 1984-86, instr. in medicine, 1986-88; asst. prof. dept. medicine U. Ala., Birmingham, 1988—. Contbr. articles to profl. jours. NIH grantee, 1991—. Mem. AAAS, Am. Heart Assn. (basic sci. coun.), Soc. Magnetic Resonance in Medicine. Achievements include patent in field. Office: U Ala Ctr Nuclear Imaging Rsch CMRL Bldg Birmingham AL 35294-4470

BALSLEY, HOWARD LLOYD, economist; b. Chgo., Dec. 3, 1913; s. Elmer Lloyd and Katherine (McGlashing) B.; m. Irol Verneth Whitmore, Aug. 24, 1947. A.B., Ind. U., 1946, M.A., 1947, Ph.D., 1950; postgrad., Johns Hopkins U., 1947-48, U. Chgo., summer 1948. Asst. prof. econs. U. Utah, Salt Lake City, 1949-50; assoc. prof. econs., dir. Sch. Bus., Russell Sage Coll., Troy, N.Y., 1950-52; asso. prof. econs. Washington and Lee U., Lexington, Va., 1952-54; prof. bus. stats., head dept. bus. and econ. research La. Tech. U., Ruston, 1954-65; prof. bus. adminstrn. and stats. Tex. Tech U., Lubbock, 1965-75; head dept. econs. and fin., prof. econs. and stats. U. Ark., Little Rock, 1975-80; adj. prof. econs. and stats. Hardin-Simmons U., Abilene, Tex., 1980-81; adj. prof. math. U. South Fla., Tampa, 1992. Author: (with James Gemmell) Principles of Economics, 1953, Readings in Economic Doctrines, vols. 1 and 2, 1961, Introduction to Statistical Method, 1964, Quantitative Research Methods for Business and Economics, 1970, (with Vernon Clover) Business Research, 1974, 2d edit., 1979, 3d edit. 1984, 4th edit. 1988, alt. 4th edit. 1992, Basic Statistics for Business and Economics, 1978, Data for Decision: Statistics in a Dynamic Economy, 1989, 2nd edit., 1992, (with James Conway) Acquiring a Fortune: Financial Novice to Millionaire, 1991. Served with USAAF, 1943-46. Mem. So. Econ. Assn., S.W. Fedn. Adminstrv. Disciplines, , Phi Beta Kappa, Beta Gamma Sigma. Home: 6501 15th Ave W Bradenton FL 34209-4528

BALTACIOGLU, MEHMET NECIP, civil engineer; b. Ankara, Turkey, Sept. 4, 1954; s. Salih Zeki and Semiha Necla Baltacioglu; m. Yaprak Erem, Nov. 24, 1977; 1 child, Sinan Alp. BS, Mid. East Tech., Ankara, 1976; MS, Bhosporous U., Istanbul, Turkey, 1977; PhD, Carleton U., Ottawa, Ont., Can., 1986. Registered profl. engr., Ont. Structural engr. CAG Ltd., Istanbul, 1977-80; pres. Cube Systems Inc., Ottawa, 1986-90, Winsoft Software Inc., Ottawa, 1990—. Author: (software) FINESSE, 1989, PCA-Frame, 1991, CONCISE, 1992, PCA-CAD, 1993. IRAP grantee Nat. Rsch. Coun., Can., 1988. Mem. Can. Soc. for Civil Engring. (Whitman Wright excellence in CAD design award 1992). Achievements include patent pending for polarized three dimensional vision drivers, patent applied for stereoscopic vision drivers using bi-coloured lenses. Office: Winsoft Software Inc, PO Box 187 Stn B, Ottawa, ON Canada K1P 6C4

BALTAZAR, ROMULO FLORES, cardiologist; b. Naga, Camarines Sur, Philippines, Oct. 15, 1941; came to U.S., 1970; s.Melecio Perez and Socorro (Flores) B.; m. Ophelia Zarzuela, June 6, 1970; children: Maria Cristina, Romulo Jr. BA, U. Philippines, 1961, MD, 1966. Intern U. Philippines, Philippine Gen. Hosp., 1965-66; resident medicine Philippine Gen. Hosp., 1966-69, chief resident medicine, 1969-70; instr. medicine U. Philippines, 1969-70; resident medicine Sinai Hosp., Balt., 1970-71, resident cardiology, 1971-73, assoc. cardiology, 1975-87, dir. non-invasive cardiology, 1987—; resident pediatric cardiology Johns Hopkins Hosp., Balt., 1971-72; fellow cardiology Maimonides Med. Ctr., N.Y.C., 1973-75; instr. medicine Johns Hopkins U. Sch. Medicine, Balt., 1977-87, asst. prof. medicine, 1987—; referee Archives of Internal Medicine, 1991—, JAMA, 1988. Contbr. articles to med. jours. Fellow ACP, Am. Coll. Cardiology, Am. Coll. Chest Physicians. Roman Catholic. Avocations: listening to popular music, playing chess, reading, travel, photography. Office: Sinai Hosp Balt Div Cardiology Baltimore MD 21215

BALTIMORE, DAVID, microbiologist, educator; b. N.Y.C., N.Y., Mar. 7, 1938; s. Richard I. and Gertrude (Lipschitz) B.; m. Alice S. Huang, Oct. 5, 1968; 1 dau., Teak. BA with high honors in Chemistry, Swarthmore Coll., 1960; postgrad., MIT, 1960-61; PhD, Rockefeller U., 1964. Research assoc. Salk Inst. Biol. Studies, La Jolla, Calif., 1965-68; assoc. prof. microbiology MIT, Cambridge, 1968-72, prof. biology, from 1972, Am. Cancer Soc. prof. microbiology, 1973-83, dir. Whitehead Inst. Biomed. Rsch., 1982-90; pres. Rockefeller U., N.Y.C., 1990-91, now prof. Mem. editorial bd. Jour. Molecular Biology, 1971-73, Jour. Virology, 1969-90, Sci., 1986—, New Eng. Jour. Medicine, 1990—. Bd. govs. Weizmann Inst. Sci., Israel; bd. dirs. Life Sci. Rsch. Found.; co-chmn. Commn. on a Nat. Strategy of AIDS; ad hoc program adv. com. on complex genome, NIH. Recipient Gustav Stern award in virology, 1970; Warren Triennial prize Mass. Gen. Hosp., 1971; Eli Lilly and Co. award in microbiology and immunology, 1971; U.S. Steel Found. award in molecular biology, 1974; Gairdner Found. ann. award, 1974; Nobel prize in physiology or medicine, 1975. Fellow AAAS, Am. Med. Writers Assn. (hon.); mem. NAS, Am. Acad. Arts and Scis., Inst. Medicine, Pontifical Acad. Scis., Royal Soc. (Eng.) (fgn.). Office: MIT Ctr Cancer Res Dept Biology 77 Massachusetts Ave Cambridge MA 02139

BALTZ, RICHARD ARTHUR, chemical engineer; b. Red Bud, Ill., Aug. 1, 1959; s. Arthur A. and Arlou M. (McDonald) B. BS in Chem. Engring., U. Mo., Rolla, 1981. Process design engr. Corp. Engring. dept. Monsanto, St. Louis, 1981-83; process engr. Nitro Plant Monsanto, Nitro, W.Va., 1983-89; process engring. specialist W.G. Krummrich Plant Monsanto, Sauget, Ill., 1989--. Mem. Am. Inst. Chem. Engrs. Roman Catholic. Home: 3749-J Huntington Valley Dr Saint Louis MO 63129 Office: Monsanto Co 500 Monsanto Ave Sauget IL 62206-1198

BALVE, BEBA CARMEN, research center administrator; b. Rosario, Santa Fe, Argentina, Aug. 3, 1931; d. Dalmiro Carmelo and Asunta (Apolito) B.; m. Rodriguez Fidel, 1989. Lic. in philosophy, Nat. Litoral U., Rosario, 1962. Prin. investigator Latin Am. Inst. Econ. and Social Planning, Buenos Aires, 1964-68; investigator Proyecto Marginalidad, Ford Found. Inst. Torcuato Di Tella, 1969-71; prin. investigator Lab. Indsl. Sociology, U. Paris-Francia, Buenos Aires, 1967-69; prof. Nat. U. de La Plata Faculty Agronomy, Buenos Aires, 1971-74; founder, dir. Social Sci. Rsch. Ctr., Buenos Aires, 1966—; dir. Seminario de Investigación, Nat. U. Buenos Aires, 1987-93; mem. nat. com. World Univ. Svc., 1980-82. Author: Lucha Calles. Lucha de Clases, 1971, El !69. Huelga política de Masas, 1989, Algunas Consideraciones Acerca de la Temática de los Movmientos Sociales, 1989, Ciencias Sociales y Sujeto Social, 1992, La Relación que Guarda el Proceso Social y la Actividad de los Partidos Políticos Acerca de la Teoría de la Organización Social, 1992. Nat. dir. Socialist Party Argentina, 1954-70. Rsch. grantee Social Sci. Rsch. Coun., 1976-77.. Mem. Consejo Latin Am. Investigatin Para La Paz Am. Latina (directive com. 1979-82), Latin Am. Sociology Assn. (directive com. 1988-93), Consejo Latin Am. Ciencias Sociales (directive com. 1987-91). Avocations: flying, golf. Office: CICSO, Defensa 665-5-C, 1065 Buenos Aires Argentina

BALZHISER, RICHARD EARL, research and development company executive; b. Wheaton, Ill., May 27, 1932; s. Frank E. and Esther K. (Merrill Werner) B.; m. Christine Karnuth, 1951; children: Gary, Robert, Patricia, Michelle. B.S. in Chem. Engring., U. Mich., 1955, M.S. in Nuclear Engring., 1956, Ph.D. in Chem. Engring., 1961. Mem. faculty U. Mich., Ann Arbor, 1961-67; White House fellow, spl. asst. to sec. Dept. Def., Washington, 1967-68; chmn. dept. chem. engring. U. Mich., 1970-71; assoc. dir. energy, environ. and natural resources White House Office of Sci. and Tech., Washington, 1971-73; dir. fossil fuel and advanced systems Electric Power Rsch. Inst., Palo Alto, Calif., 1973-79, sr. v.p. R&D, 1979-87, exec. v.p. R&D, 1987-88, pres., chief exec. officer, 1988—; chmn. U.S. Energy R&D Exch. with USSR, 1973-74; mem. EPCOT Ctr., Orlando, Fla., 1977-80, NAS acad. industry program, 1988—, U. Mich. Coll. Engring. nat. adv. com., 1989-92; energy rsch. adv. bd. U.S. Dept. Energy, 1988-89, mem. innovative control tech. adv. panel, 1989; adv. coun. U. Tex. Natural Sci. Found., Austin, 1990—; mem. NRC Energy Engring. Bd., 1991-92; bd. dirs. Atlantic Coun. U.S., 1991—, Pres.'s Coun. Competitiveness, 1992, U.S. Energy Assn.,

1993—; adv. bd. Woods Hole (Mass.) Oceanographic Instn., 1992—. Editorial bd.: Sci. mag., 1977—; co-author: Chemical Engineering Thermodynamics, 1972, Engineering Thermodynamics, 1977. Mem. Ann Arbor City Coun., 1965-67, mayor pro tem, 1967. Charles M. Schwab Meml. lectr. Am. Iron And Steel Inst., 1983. Mem. AAAS, ASME, Am. Inst. Chem. Engrs., Cosmos Club (Washington), Sigma Chi. Republican. Lutheran. Office: Electric Power Rsch Inst PO Box 10412 3412 Hillview Ave Palo Alto CA 94304-1395

BAMBERGER, JOSEPH ALEXANDER, mechanical engineer, educator; b. Hamburg, Germany, Nov. 21, 1927; came to U.S., 1940; s. Seligman and Else (Buxbaum) B.; m. Dorothy Frank, Dec. 24, 1950; children: David, Michael. BME, CUNY, 1949; MME, NYU, 1954. R & D engr. Kramer Trenton Co., Trenton, N.J., 1949-59; mech. engr., scientific staff Brookhaven Nat. Lab., Upton, N.Y., 1959-82; prof. mech. tech. Suffolk Community Coll., Selden, N.Y., 1982—; cons. Typhoon Air Conditioning, Div. Hupp Corp., Bklyn., 1952-59. Contbr. articles to ASHRAE Jour., Advances in Cryogenic Engring., Cryogenics, ASME Transactions, Jour. Vacuum Sci. and Tech., Nuclear Instruments and Methods. Dir. Temple Beth El, Patchogue, N.Y., 1962—; chmn. Beth El Sch., Patchogue, 1968-70; chmn. Cryogenic Safety Com., Brookhaven Lab., 1980-82. Mem. N.Y. Acad. Sci., AAAS, ASHRAE. Achievements include patent for Electrically Insulating Feed-through for Cryogenic Applications; research in low temperature cooling systems for superconducting magnets, cryogenic pumping systems, liquid hydrogen bubble chamber design and operation.

BAMBURG, JAMES ROBERT, biochemistry educator; b. Chgo., Aug. 20, 1943; s. Leslie H. and Rose A. (Abrahams) B.; m. Alma Y. Vigo, June 7, 1970 (div. Dec. 1984); children: Eric Gregory, Leslie Ann; m. Laurie S. Minamide, June 22, 1985. BS in Chemistry, U. Ill., 1965; PhD, U. Wis., 1969. Project assoc. U. Wis., Madison, 1968-69; postdoctoral fellow Stanford U., Palo Alto, Calif., 1969-71; from asst. to full prof. Colo. State U., Ft. Collins, 1971—, acad. coordinator cell and molecular biol. program, 1975-78, interim chmn. dept. biochemistry, 1982-85, 88-89, assoc. dir. neuronal growth and devel., 1986-90, dir. neuronal growth and devel., 1990—; vis. prof. MRC Molecular Biol. Lab., Cambridge, Eng., 1978-79, MRC Cell Biophysics Unit, London, 1985-86, Children's Med. Rsch. Inst., Sydney, Australia, 1992-93; mem., chmn. NIH Biomed. Scis. Study Sect., Bethesda, Md., 1980-85. Contbr. articles to sci. jours. Fellow NSF, 1964-65, Nat. Multiple Sclerosis Soc., 1969-71, J.S. Guggenheim Found., 1978-79, Fogarty Ctr., 1985-86, 92-93. Mem. Am. Chem. Soc., Am. Soc. Cell Biology, Am. Soc. Biol. Chemists, Internat. Neurochem. Soc., Sigma Xi (pres. CSU chpt. 1989). Home: 2125 Sandstone St Fort Collins CO 80524-1825 Office: Colo State U Dept Biochemistry MRB Rm 235 Fort Collins CO 80523

BAMBURY, RONALD EDWARD, polymer chemist; b. Aberdeen, S.D., Dec. 26, 1932; s. Daniel L. and Ruby L. (Kline) B.; m. Laura M. Ukasick, Sept. 11, 1954; children: Kevin, Catherine, Jennifer. B Chemistry, U. Minn., 1955; PhD, U. Nebr., 1960. Chemist Am. Cyanamid, Princeton, N.J., 1960-64; mgr. Richardson Merrell, Inc., Cin., 1964-76; group leader Diamond Shamrock Corp., Painesville, Ohio, 1976-83; mgr. polymer chemistry Bausch and Lomb, Inc., Rochester, N.Y., 1984—. Contbr. chpts. to profl. books, articles to jours. Mem., chmn. Montgomery Sch. Adv. Bd., Cin., 1973-76. Mem. Am. Chem. Soc. (chmn. local sect. 1976). Achievements include 25 U.S. patents and 27 publications in the fields of anti-infective agents and contact lens polymers. Office: Bausch and Lomb Inc 1400 Goodman St N Rochester NY 14692-0450

BAMFORD, DONALD LAWRENCE (LARRY BAMFORD), computer scientist; b. Dayton, Ohio, Aug. 12, 1957; s. Harold E. and Barbara Jean (Ihrig) B.; m. Ann Marie Thompson, May 13, 1984; children: Thomas, Michael. BA in Psychology, U. Va., 1980, M Computer Sci., 1982. Grad. rsch. asst. Grad. Sch. Bus. U. Va., Charlottesville, 1981-82; programmer/analyst Vector Rsch. Co., Inc., Bethesda, Md., 1983-84; rsch. computer scientist Fed. Jud. Ctr., Washington, 1984-87; computer specialist Adminstrv. Office U.S. Cts., Washington, 1987—, Internet postmaster, tech. contact, 1990—. Contbr. articles to UNIX/World. Mem. vestry Ch. of Apostles Episc. Ch., Fairfax, Va., 1991—. Republican. Achievements include development of software in federal judiciary. Home: 3978 Wilcoxson Dr Fairfax VA 22031 Office: Adminstrv Office US Cts OAT-STD SEB 1 Columbus Cir NE Washington DC 20544

BAN, SADAYUKI, radiation geneticist; b. Fukui, Japan, May 26, 1949; s. Yasobei and Sadako B.; m. Kumiko Kato, May 26, 1979; children: Yuriko, Tomoaki. BS, Kanazawa U., Ishikawa, Japan, 1973, MS, 1975; DMSc, Kyoto U., Japan, 1980. Rsch. fellow Japan Sci. Promotion Soc., Kyoto, Japan, 1979; rsch. sci. Radiation Effects Rsch. Found., Hiroshima, Japan, 1980-83; vis. scientist Brookhaven Nat. Lab., L.I., N.Y., 1983-85; rsch. sci. Radiation Effects Rsch. Found., Hiroshima, 1985—. Author: Modification of Radiosensitivity in Cancer Treatment, 1984; contbr. over 30 sci. papers to internat. jours. Mem. AAAS, Japan Radiation Rsch. Soc., Japan Tissue Culture Soc. (councilor 1978-79, editor of correspondence 1979-81), Japanese Soc. Biomed. Gerontology (councilor 1986-87), N.Y. Acad. Scis., Planetary Soc., Am. Assn. Cancer Rsch. (corr.). Avocations: theatergoing, stamp collecting. Office: Radiation Effects Rsch Found, Minami-ku Hijiyama Park 5-2, Hiroshima 732, Japan

BANAS, EMIL MIKE, physicist, educator; b. East Chicago, Ind., Dec. 5, 1921; s. John J. and Rose M. (Valcicak) B.; ed. Ill. Benedictine Coll., 1940-43; B.A. (U.S. Rubber fellow), U. Notre Dame, 1954, Ph.D., 1955; m. Margaret Fagyas, Oct. 9, 1948; children—Mary K., Barbara A. Instr. math. and physics Ill. Benedictine Coll., Lisle, 1946-48, adj. faculty mem., 1971-82, trustee, 1959-61; with Civil Service, State of Ind., Hammond, 1948-50; lectr. physics Purdue U., Hammond, 1955-60; staff research physicist Amoco Corp., Naperville, Ill., 1955-82; cons., 1983—. Served with USNR, 1943-46. Mem. Am. Philatelic Soc., Ill. Benedictine Coll. Alumni Assn. (dir. hon., named alumnus of yr., 1965, pres. 1959-60), U. Notre Dame Alumni Assn. (sec. grad. physics alumni), Am. Legion, Sigma Pi Sigma. Roman Catholic. Clubs: Soc. of Procopians. Contbr. articles to sci. jours. Home: SW 325 Clarkson Ct Apt 4 Pullman WA 99163

BANCROFT, GEORGE MICHAEL, chemical physicist, educator; b. Saskatoon, Sask., Can., Apr. 3, 1942; s. Fred and Florence Jean B.; m. Joan Marion MacFarlane, Sept. 16, 1967; children: David Kenneth, Catherine Jean. B.Sc., U. Man., 1963; M.Sc., 1964; Ph.D., Cambridge (Eng.) U., 1967, M.A., 1970, Sc.D. (E.W. Staecie fellow), 1979. Univ. demonstrator Cambridge U.; then teaching fellow Christ Coll.; mem. faculty U. Western Ont., London, now prof., chmn. chemistry dept. Author: Mössbauer Spectroscopy, 1973; also articles in photoelectron spectroscopy, synchrotron radiation studies; revs. Mössbauer Spectroscopy. Recipient Harrison Meml. prize, 1972, Meldola medal, 1972, Rutherford Meml. medal, 1980, Alcan award, 1990, Herzberg award, 1991; Guggenheim fellow, 1982-83. Fellow Royal Soc. Can.; mem. Royal Soc. Chemistry, Can. Chem. Soc., Can. Geol. Soc., Can. Physics Soc. Mem. United Ch. Can. Clubs: Curling, Tennis (London). Office: U Western Ont, Chemistry Dept, London, ON Canada N6A 5B7

BANDOPADHYAYA, AMITAVA (AMIT BANDO), economist, consultant, educator; b. Mombasa, Kenya, Oct. 2, 1957; (parents Indian citizens); came to U.S., 1980; s. Parry Mohan and Neelima (Chatterjee) B.; m. Carolyn A. Berry, Aug. 15, 1988; 1 child, Nikhil Alexander. BA with honors, St. Stephen's Coll., 1977; MA, Delhi Sch. Econs., 1979; PhD, U. Minn., 1988. Researcher Indo-Canadian Inst., Delhi, 1977-80; teaching asst. U. Minn., Mpls., 1980-85; asst. prof. New Mex. State U., Las Cruces, N.M., 1985-88; economist sect. econs. and law U. Chgo./Argonne (Ill.) Nat. Lab., 1988-92; adj. prof. Lewis U., Romeoville, Ill., 1990-92; v.p. RCF, Inc., Chgo., 1990-92; mgr. internat. environ. div. RCG/Hagler, Bailly, Inc., Arlington, Va., 1992—; cons. nat. and internat. energy/environ. issues U.S. Govt., fgn. govts. and multilateral banks, including World Bank, ABD; cons. energy/environ./market analysis and forecasting U.S., Australasia, Europe, Africa, Latin Am.; bd. dirs., cons. environ. issues in Ea. Europe, Asian Pacific and Mid. East, Abacus Internat. Group, Brookfield, Ill., 1989—; bd. dirs. Ctr. Econometric Modeling and Forecasting. Contbr. articles to numerous jours.; editor: Business Forecaster, 1987-90, Nat. Social Sci. Jour., 1988—. Sloan Found. fellow, 1984-85, Frost Found. fellow, 1986-88. Mem. AAAS, Am. Econ. Assn., Western Econ. Assn., Internat. Assn. for Energy Econs. Office: RCG/Hagler Bailly 1530 Wilson Blvd Ste 900 Arlington VA 22209-2406

BANDURSKI, BRUCE LORD, ecological and environmental scientist; b. Waterbury, Conn., June 28, 1940; s. Stanley Alexander Bandurski and Virginia Ann (VanRennselaer) Bandurski Hinckley; m. Ruth Anne Fuhrman Sudar, June 28, 1990; stepchildren: Adam Sudar, Karl Sudar. BS with honors, Mich. State U., 1962; postgrad., George Washington U., 1964-65, U.S. Dept. Agr. Grad. Sch., 1965-66. Park ranger Nat. Park Service, 1962-63; sci. reference analyst USPHS, Washington, 1963-65; intelligence ops. specialist U.S. Army, Washington, 1965-66; analyst planner U.S. Dept. Interior, Washington, 1966-74, coord., br. chief, Nat. Environ. Policy Act officer, 1974-83; on detail as ecologist, ecomgmt. advisor Internat. Joint Commn. U.S. and Can., Washington, 1983-85, sr. ecomgmt. advisor, ecologist, 1985—; mem. faculty U.S. Dept. Agr. Grad. Sch., 1968—; guest lectr. No. Va. Community Coll., U. Wis., Bucknell U., Am. U., U. Pitts.; mem. subcom. Fed. Interagy. Com. on Edn., 1967—; watch dir., dep. and acting mission dir. U.S. Man-in-Sea program, St. John, V.I., 1970; chmn. Conservation Roundtable of Washington, 1970-71; chmn. com. on definitions, spl. com. on environ. protection U.S. nat. com. World Energy Conf., Washington, 1981-85; initiator, dir. Binat. Workshop on Transboundary Monitoring, 1984; mem. exec. com. Great Lakes Sci. Adv. Bd., 1986-92; liaison Coun. Great Lakes Rsch. Mgrs.; mem. steering com. Great Lakes-St. Lawrence Ecosystem Model Framework; participant in ECE Seminar on Ecosystems Approaches to Water Management UN; mem. Lake Superior Biodiversity Project Adv. Com. Nat. Wildlife Fedn.; initiator multi year project Ecological Com. Great Lakes Adv. Bd., 1990—. Writer planning and recreation impact mgmt. series, 1967-73; author U.S. Bur. Land Mgmt. Environ. Mgmt. Procedures, 1976-84 (Achievement award 1978, 79, 84), Steering Group on Marine Environ. Monitoring, Commn. on Engring. and Tech. Studies, NRC, 1986-87, Complimentaries between holism and reductionism as they pertain to governance of human/environ. rels.; co-author The Ecosystem Approach: Theory and Ecosystem Integrity, 1990—. MEM. AAAS, Ecol. Soc. Am. (charter Met. Washington chpt.), Internat. Assn. for Ecology, Am. Soc. Naturalists, The Wildlife Soc., Am. Soc. Mammalogists, Fed. Profl. Assn., Wash. Soc. Engrs., Outdoor Ethics Guild, Nature Conservancy, Maine Coast Heritage Trust, Island Inst., Earthwatch, Assn. Ecosystem Rsch. Ctrs., Alpha Zeta, Beta Beta Beta. Home: Bandura/Point of Maine/Starboard Bucks Harbor ME 04618

BANEGAS, ESTEVAN BROWN, agricultural biotechnology executive; b. Hatch, N.Mex., May 10, 1941; s. Estevan Vera Banegas and Josephine (Brown) Crew; m. Amanda Martin, Sept. 5, 1970. BS, N.Mex. U., 1964; MBA, Wake Forest U., 1978. Sales mgr. agr. div. Ciba-Geigy Corp., San Juan, P.R., 1968-73; mktg. mgr. agr. div. Ciba-Geigy Corp., Greensboro, N.C., 1974-80; dir. corp. planning Ciba-Geigy Corp., Ardsley, N.Y., 1980-81; dir. product mgmt. agr. div. Ciba-Geigy Corp., Greensboro, 1981-83, dir. strategic planning agr. div., 1983-85; pres. joint venture Union Carbide Corp. & DNA Plant Tech. Agri-Diagnostics Assocs., Cinnaminson, N.J., 1985-92, bd. dirs., 1985-92; pres. Techshare, Inc, Greensboro, 1992—; pres., CEO Dominion BioScis., Inc., Blacksburg, Va., 1993—; speaker mktg. biotech. products Agbio Conf., 1989; vis. faculty joint ventures and strategic partnering Internation Rsch. Inst., 1992. Bd. advisors U. Minn., St. Paul, 1985-88, N. Mex. State U. Leadership Project, 1993—; bd. dirs. agrl. devel. bd. Ohio State U., Columbus, 1987-93. Capt. USMC, 1964-67, Vietnam. Decorated Cross of Gallantry with silver star Govt. of South Vietnam, 1966. Mem. Am. Chem. Soc. (speaker mktg. strategies 1988), Am. Phytopathology Soc., Golf Course Supts. Am., Sedgefield Country Club, Rotary. Republican. Roman Catholic. Avocations: golf, gardening, hiking, church activities. Office: Techshare Inc 3558 Old Onslow Rd Greensboro NC 27407

BANERJEE, AJOY KUMAR, engineer, constructor, consultant; b. Dacca, Bangladesh, Apr. 23, 1945; came to U.S., 1966; s. Kalidas and Anjali (Mukherjee) B.; m. Teri Sandra Siegel, Aug. 23, 1970; children: Shonali Misha, Monisha Jenni. B Tech. in Civil Engring. with honors, Indian Inst. Tech., Kharagpur, 1966; M of Engring., U. Detroit, 1967; PhD in Structural Engring., Cornell U., 1973. Registered profl. engr., N.Y., Mass., Va., Mo. Mgr. of projects Stone & Webster Engring. Corp., Boston, 1973—. Contbr. articles to Nuclear Safety Jour., Nuclear Engring. and Design Jour., Jour. Structural Div. ASCE, Am. Nuclear Soc. Proceedings, ASME Proceedings. Fellow ASCE; mem. ASME, IEEE (com. on probabilistic risk assessment 1981-82), Am. Nuclear Soc. (power div. program com., vice chmn. tech. and pub. issues com.), Internat. Assn. Structural Mechanics in Reactor Tech., Nuclear Mgmt. and Resources Coun. (utility tech. group on life extension, alt. com. on design basis documentation). Achievements include development of programs and methodologies for nuclear power plant life extension and increasing power rating. Office: Stone & Webster Engring Co 245 Summer St Boston MA 02210-1116

BANERJEE, AMIYA KUMAR, biochemist; b. Rangoon, Burma, May 3, 1936; came to U.S., 1965, naturalized, 1976; d. Phanindra Nath and Bibhati (Ghosal) B.; m. Sipra Datta, Jan. 23, 1965; children: Antara, Arjun. MSc, Calcutta (India) U., 1958, PhD, 1965, DSc, 1970. Staff assoc. Roche Inst. Molecular Biology, Nutley, N.J., 1969-71, asst. mem., 1971-74, assoc. mem., 1974-80, mem., 1980-87; chmn. dept. molecular biology Cleve. Clinic Found., 1987—, vice chmn. Rsch. Inst., 1990—; adj. prof. NYU Med. Ctr., N.Y.C., 1981—, SUNY Health Sci., Bklyn., 1986—, Case Western Res. U., Cleve., 1989—. Assoc. editor Virology, 1983—; mem. editorial bd. Jour. Virology, 1988—. Pres. Tagore Sco., N.Y.C., 1978-82. Recipient Phoebe Weinstein award NIH, Washington, 1977, Prof. S.C. Roy Commemoration medal Calcutta (India) U., 1983. Achievements include major contributions to mechanism of gene expression of negative strand animal RNA viruses. Home: 60 Lochspur Ln Chagrin Falls OH 44022-2310 Office: The Cleve Clinic Found Dept Molecular Biology Rsch Inst 9500 Euclid Ave # 2124C Cleveland OH 44195-0002

BANERJEE, BEJOY KUMAR, mechanical engineer; b. Rangoon, Burma, Jan. 21, 1931, s. Lalit Chandra and Hem (Mukherjee) B.; m. Runu, Mar. 12, 1962; children: Roopa, Polly, Sanjay, Moni. AS, U. Rangoon, Burma, 1950; BSCE, Govt. Tech. Inst., Rangoon, Burma, 1954. Asst. engr. Elec. Power Corp., Rangoon, Burma, 1954-56, engr., 1956-60, asst. exec. engr., 1960-68, exec. engr., 1968-78; sr. engr. Ebasco Svcs. Inc., New Orleans, 1978-88; sr. lead engr. Ebasco Svcs. Inc., Commanche Peak, Tex., 1988; lead engr. Ebasco Svcs. Inc., Fort St. Vrain, Colo., 1988-90; sr. prin. engr. EG & G Rocky Flats, Inc., Golden, Colo., 1990—. Mem. AMSE. Home: 5133 W 69th Pl Westminster CO 80030

BANERJEE, PRASHANT, industrial engineering educator, researcher; b. Calcutta, West Bengal, India, Apr. 15, 1962; came to U.S., 1986; s. Prabhat K. and Bani Banerjee; m. Madhumita Banerjee, Dec. 11, 1987; 1 child, Jay. BSME, Indian Inst. Tech., Kanpur, India, 1984; MS in Indsl. Engring., Purdue U., 1987, PhD, 1990. Indsl. engr. Tata Steel Co., Jamshedpur, India, 1984-85; rsch. scientist Engring. Rsch. Ctr. Intelligent Mfg. Systems, Purdue U., West Lafayette, Ind., 1986-90; asst. prof. U. Ill., Chgo., 1990—; cons. Kraft Cheese Co., Nobesville, Ind., 1988, Caterpillar Inc., Peoria, Ill., 1992. Author: Automation and Control of Manufacturing Systems, 1991, Object-oriented Technology in Manufacturing, 1992; contbr. articles to profl. jours. Equipment grantee Digital Equipment Corp., 1990, NSF rsch. grantee, 1992. Mem. ASME, Inst. Indsl. Engrs., Am. Assn. Artificial Intelligence, Inst. Mgmt. Scis., Soc. Mfg. Engrs. Avocations: sports, current events, religious discussions. Home: 197 Brookwood Ln W Bolingbrook IL 60440 Office: Univ Ill Chicago IL 60680-4348

BANERJEE, SAMARENDRANATH, orthopaedic surgeon; b. Calcutta, India, July 12, 1932; s. Haridhone and Nihar Bala (Mukherjee) B.; m. Hima Ganguly, Mar. 1977; 1 child, Rabindranath. M.B. B.S., R.G. Kar Med. Coll., Calcutta, 1957; postgrad., U. Edinburgh, 1965-66. Intern R.G. Kar Med. Coll., Calcutta, 1956-58; resident in surgery Bklyn. Jewish Hosp. Med. Ctr., 1958-60, Brookdale Med. Ctr., Bklyn., 1960-61, Jersey City Med. Ctr., 1961-63; orthopaedic registrar Royal Postgrad. Med. Sch., Hammersmith Hosp., London, 1966-67, Heatherwood Orthopaedic Hosp., Ascot, Eng., 1967-68; research fellow Hosp. for Sick Children, U. Toronto, Ont., Can., 1968-69; practice medicine specializing in orthopedics Sault Ste. Marie, Ont.; past pres. med. staff, chmn. exec. com. Gen. Hosp., Sault Ste. Marie, Ont., chief dept. surgery, mem. adv. com., 1980-88, chief div. orthopaedic surgery, 1980—; cons. orthopaedic surgeon Gen. Hosp. Plummer Meml. Hosp., Crippled Children Ctr., Ministry Nat. Health and Welfare, Dept. Vets. Adminstrn; civilian orthopaedic surgeon to 44th Div. Armed Forces Base Hosp., Kaduna, Nigeria, 1969. Trustee, Gen. Hosp., Sault Ste. Marie, 1975-76. Miss Betsy Burton Meml. fellow N.Y. U. Med. Ctr., 1963-64. Fellow ACS, Royal Coll. Surgeons Can., Royal Coll. Surgeons Edinburgh; mem. Am. Fracture Assn. (regional v.p. Can. chpt., bd. govs. 1991—), Can. Orthopaedic Assn., Can. Med. Assn., Ont. Orthopaedic Assn., N.Y. Acad. Sci. Home: 50 Alworth Pl, Sault Sainte Marie, ON Canada P6B 5W5 Office: 125-955 Queen St E, Sault Sainte Marie, ON Canada P6A 2C3

BANEY, RICHARD NEIL, physician, internist; b. Phila., Apr. 13, 1937; s. Robert Emmet and Mary Elizabeth (Hedges) B.; m. Carolyn Vern Kurey, Feb. 17, 1962; children: Richard N. Jr., Michael D., Marisa V., Brian E. BS, Georgetown U., 1958; MD, U. Pitts., 1963. Diplomate Am. Bd. Internal Medicine, Am. Bd. Rheumatology. Intern V.A. & Parkland Hosp., Dallas, 1963-64; resident U. Pitts., 1967-70; internist Jess Parrish Hosp., Titusville, Fla., 1971-76, Melbourne (Fla.) Internal Med. Assocs., Holmes Regional Med Ctr., 1976—; trustee Holmes Regional Med. Ctr., Melbourne, 1984—; founding dir., chmn. bd. Reliance Bank of Fla., Melbourne, 1985—; chief med. staff Jess Parrish Hosp., Titusville, 1974-76; dir. DBA Systems Inc. Trustee Fla. Inst. Tech., Melbourne, 1986—, mem. exec. com., 1987—, vice chmn. bd. trustees, 1991—; pres. Canaveral chpt. Am. Heart Assn., Rockledge, Fla., 1973-74; chmn. bd. trustees Sea Pines Rehab. Hosp., Melbourne, 1992—. Fellow ACP; mem. Am. Coll. Rheumatology, Am. Soc. Nuclear Medicine, Underwater and Hyperbaric Med. Soc., Brevard County Med. Soc. (pres. 1977-78), Navy League U.S., Eau Gallie Yacht Club (commodore 1985-86), Coast Club (bd. dirs. 1985-91, chmn. bd. 1989-91). Republican. Roman Catholic. Avocations: tennis, jogging, bicycling, travel, collecting antique maps, golf. Office: Melbourne Internal Med Assn 200 E Sheridan Rd Melbourne FL 32901-3182

BANFIELD, WILLIAM GETHIN, physician; b. Hartford, Conn., Mar. 2, 1920; s. William Gethin and Mabel (Dean) B.; m. Joan Sanders; children: William Gethin, Peter Dean, Sarah Morgan. BS, R.I. State Coll., 1941, MS, 1943; MD, Yale U., 1946, DNB, 1947; JD, Am. U., 1974. Diplomate Nat. Bd. Med. Examiners. Intern Mt. Auburn Hosp., Cambridge, Mass., 1946-47; Am. Cancer Soc. fellow dept. pathology Yale U. Sch. Medicine, New Haven, 1949-52, asst. prof. dept. pathology, 1954; assoc. pathologist Grace New Haven Community Hosp., 1952-54; med. officer lab. of pathology NIH, Bethesda, Md., 1954-70, med. officer comparative oncology sect., 1970-80; ret., 1980; cons. electroprobe microanalysis, Rockville, Md., 1988—. Contbr. articles to profl. publs. Capt. Med. Corps, AUS, 1947-49. Mem. Microbeam Analysis Soc., Microscopy Soc. Am. Achievements include research in cellular transmission on lymphoma through mosquito vector; first pictures of polyome virus in thin section; early application of electron probe in biology. Home and Office: 15715 Avery Rd Rockville MD 20855-1718

BANGS, ALLAN PHILIP, computer scientist, software engineer, consultant; b. Seattle, Jan. 10, 1930; s. Scholer and Helen (Hutton) B.; m. Adelaide Harris, Aug. 31, 1957; children: Ruth E., Leonard A. BA in Physics, San Francisco State U., 1960. Sr. system analyst Aerojet Gen. Corp., System Devel. Corp., Rockwell Internat. Space Divsn., Lockheed Aircraft Corp., Calif., N.Y., 1960-78; mem. tech. staff Hughes Aircraft Co., L.A., 1978-89, Jet Propulsion Lab., Pasadena, Calif., 1990-91; mgr., CEO, ASSET Cons., L.A., 1983—; referee IEEE Computer and Software jours., 1990-92. With U.S. Army, 1950-53. Recipient contbn. of merit award NASA Space Exploration Initiative, 1990. Fellow AIAA (assoc., vice chmn. pub. affairs L.A. 1980-82, nat. pub. policy com. 1983); mem. Am. Phys. Soc. (sr.). Achievements include being member of development teams in space programs (Apollo, shuttle, Magelan).

BANIK, NIRANJAN CHANDRA-DUTTA, physicist, researcher; b. Barisal, Bangladesh, Jan. 17, 1946; came to U.S., 1973; s. Benode Behari and Mahamaya Banik; m. Alo Rani, July 24, 1970; children: Sanjoy, Anuva. MSc., Dacca U., Dhaka, Bangladesh, 1968; PhD, Purdue U., 1978. Prof. and head, dept. physics Comilla Victoria Coll., Bangladesh, 1969-71; lectr. physics Dhaka U., 1971-73; postdoctoral rsch. asst. Case Western Res. U., Cleve., 1978-79; mem. rsch. staff Xerox Corp., Webster, N.Y., 1979-81; rsch. geophysicist Gulf Oil Corp., Pitts., 1981-84; sr. rsch. physicist ARCO Oil and Gas, Plano, Tex., 1984-86; rsch. assoc. T.D. Williamson, Inc., Tulsa, 1986-91, sr. scientist, 1991—. Reviewer Soc. Exploration Geophysics, 1981—; contbr. over 40 articles to profl. jours. Pres. Bengali Assn. Tulsa, 1989; sec. India Assn. Tulsa, 1990, pres. 1991. Merit Scholarship grantee E. Pakistan Edn., 1964. Mem. Am. Soc. Nondestructive Testing, Am. Phys. Soc., Soc. Exploration Geophysics, Sigma Chai. Achievements include resolution of depth anomalies in the North Sea through anisotropy; descriptions of phase transitions in piezo-electric PVF2 polymer; made EMAT transducers work for pipeline corrosion inspection. Home: 6211 E 77th St Tulsa OK 74136-8539

BANKS, BRUCE A., engineer, physicist, researcher; b. Cleve., Mar. 12, 1942; s. Clarence F. and Marie E. (Schaffer) B.; m. Judith E. Sturgeon, Aug. 28, 1965; children: Michael A., Thomas J., Eric B., Cynthia E. BS in Physics, Case Inst. Tech., 1964; MS in Physics, U. Mo., Rolla, 1966. Devel. physicist GE, Nela Park, Ohio, 1964; mem. faculty mech. engring. dept. U. Mo., Rolla, 1964-65, Atomic Energy Commn. fellow, 1966; chief electro-physics br. NASA Lewis Rsch. Ctr., Cleve., 1966—; mem. adj. faculty Ohio Aerospace Inst., Cleve., 1990—. Contbr. 83 articles to profl. jours., 2 chpts. to books; editor 1 book; author 12 NASA tech. briefs. V.p. City Sch. Dist. Bd. Edn., North Olmsted, Ohio, 1980-88. Mem. Materials Rsch. Soc., Kiwanis, Tau Beta Pi. Achievements include 25 patents, including Piezoelectric Deicing Device, Ion Beam Sputteretching, Texturing Polymer Surfaces by Transfer Casting, Deposition of Diamondlike Carbon Films, Oxidation Protection Coatings for Polymers, others. Office: NASA Lewis Rsch Ctr MS 302-1 21000 Brookpark Rd Cleveland OH 44135

BANKS, EPHRAIM, chemistry educator, consultant; b. Norfolk, Va., Apr. 21, 1918, s. Israel and Ada (Gezunsky) B.; m. Libby Kohl, Mar. 17, 1943 (dec. May 14, 1985); children: Thomas Israel, Jay Lewis. BS, CCNY, 1937; PhD, Poly. Inst. of Bklyn., 1949. Jr. metallurgist N.Y. Naval Shipyard, Bklyn., 1941-46; rsch. fellow Poly. Inst. of Bklyn., N.Y., 1946-49, rsch. assoc., 1949-50; instr. to prof. Poly. Inst. of Bklyn. (now Poly. U.), 1951-87, prof. emeritus, 1987—; cons. Westinghouse, GM, Mallinckrodt Chem., 1966—. Contbr. over 120 articles to profl. publs. including Jour. Am. Chem. Soc., Jour. Electrochem. Soc., Sci., Jour. Solid State Chemistry, Jour. Phys. Soc., Jour. Chem. Physics, and others. With USNR, 1944-46. Weizmann fellow, 1963-64; NSF Faculty fellow, 1971-72. Fellow AAAS, N.Y. Acad. Sci., Mineral. Soc. Am.; mem. Am. Chem. Soc. (mem. exec. com. inorganic div. 1959-62), Am. Phys. Soc., Electrochem. Soc. (assoc. editor jour. 1956-85), Am. Crystallography Assn. Achievements include research in luminescent materials, magnetic oxides, oxide bronzes, high temperature superconductors, semiconducting materials. Office: Poly U 333 Jay St Brooklyn NY 11201-2990

BANKS, JAMES DANIEL, chemical engineer; b. Port Arthur, Tex., Apr. 26, 1947; s. James Daniel and Frances Audrey (King) B.; m. Nancy Vickrey, May 16, 1970; children: Emily Vickrey, Laura Elizabeth. BA, Rice U., 1970, MChemE, 1990. Registered profl. engr., Okla. Process engr. Fluor Engrs., Houston, 1970-71, Mobil Oil Co., Beaumont, Tex., 1971-76; applications engr. John Zink Co., Tulsa, 1976-80; bus. mgr. IT-McGill PCS, Tulsa, 1980-92; dir. engring. John Zink Co., Tulsa, 1992—. Mem. AICHE (sec. 1989-90, chmn. 1990-91). Home: 3715 E 55 St Tulsa OK 74135 Office: John Zink Co 11920 E Apache Tulsa OK 74116

BANKS, PETER MORGAN, electrical engineering educator; b. San Diego, May 21, 1937; s. George Willard and Mary Margaret (Morgan) B.; m. Paulett M. Behanna, May 21, 1983; children by previous marriage: Kevin, Michael, Steven, David. M.S. in E.E, Stanford U., 1960; Ph.D. in Physics, Pa. State U., 1965. Postdoctoral fellow Institut d'Aeronomie Spatiale de Belgique, Brussels, Belgium, 1965-66; prof. applied physics U. Calif., San Diego, 1966-76; prof. physics Utah State U., 1976-81, head dept. physics, 1976-81; vis. assoc. prof. Stanford U., 1972-73, prof. elec. engring., 1981-90, dir. space, telecommunications and radiosci. lab., 1982-90, dir. ctr. for aeronautics and space info. systems, 1983-90; prof. atmospheres, oceans, and space sci. U. Mich., 1990—; dean Coll. Engring., U. Mich., 1990—; pres. Earth Data Corp., 1985-86; vis. scientist Max Planck Inst. for Aeronomie,

Ger., 1975; pres. La Jolla Scis. Inc., 1973-77, Upper Atmosphere Research Corp., 1978-82; chmn. NASA adv. com on sci. uses of space sta., 1985-87, prin. investigator space shuttle experiments, 1982, 85, 91; mem. Jason Group, 1983—; bd. dirs. Ctr. for Space and Advanced Tech., Vienna, Va., Indsl. Tech. Inst., Ann Arbor, Mich., Mich. Instrnl. TV Network, Lansing, Tecumseh Products Corp., Consortium for Internat. Earth Sci. Data Networking, Saginaw, Mich., Great Lakes Environ. Svcs., Inc., Rsch. Environ. Industries, Inc., 1993—; chmn. bd. trustees, 1991—, Consortium Internat. Earth Sci. Info. Networks. Author: (with G. Kockarts) Aeronomy, 1973, (with J.R. Doupnik) Introduction to Computer Science, 1976; assoc. editor: Jour. Geophys. Research, 1974-77; assoc. editor: Planetary and Space Sci, 1977-83, regional editor, 1983-86; contbr. numerous articles in field to profl. jours. Mem. space sci. adv. council NASA, 1976-80. Served with U.S. Navy, 1960-63. Recipient Appleton prize Royal Soc. London, 1978, Space Sci. award AIAA, 1981, NASA Disting. Service medal, 1986; Alumni fellow Pa. State U., 1982. Fellow Am. Geophys. Union; mem. Internat. Union Radio Sci., Nat. Acad. Engring. Episcopalian. Club: Cosmos. Home: 3485 Narrow Gauge Way Ann Arbor MI 48105-2576 Office: Univ Mich 2401 EECS Bldg Ann Arbor MI 48109

BANKS, ROBERT EARL, family practice physician; b. Kansas City, Mo., Mar. 21, 1929; s. Earl Wesley and Mary Margaret (Driskill) B.; m. Shirley Nadine Smith, Aug. 6, 1950; children: Robert Emory, Larry Earl, Donald Edward, Gregory Eldon. BA, Kansas U., 1951, MD, 1955. Diplomate Am. Bd. Family Practice. Intern San Diego Naval Hosp., 1955-56; pvt. practice Paola, Kans., 1958—; chief of staff Miami County Hosp., Paola, 1961—; mem. adv. bd. Baehr Found, Paola, 1970—, postgrad. medicine Kans. U. Med. Ctr., adv. bd. Osawatonie State Hosp. Lt. USN, 1955-58. Republican. Methodist. Home: PO Box 298 Paola KS 66071-0298

BANKS, RONALD ERIC, chemistry educator. Prof. chemistry dept. U. Manchester, Eng. Recipient ACS award for Creative work in Flourine Chemistry, Am. Chemical Soc., 1993. Office: U Manchester Inst Techn Sci, Dept Chemistry, Manchester M15 6BH, England*

BANKS, RUSSELL, chemical company executive; b. N.Y.C., Aug. 2, 1919; s. Thomas and Fay (Cowen) B.; m. Janice Reed, June 19, 1949; 1 son, Gordon L. B.B.A., CCNY, 1936-40; J.D., N.Y. Law Sch., 1960. Bar: N.Y. 1961. Sr. acct. Selverne, Davis Co., N.Y.C., 1940-45; pvt. practice N.Y.C., 1945-61; exec. v.p. Met. Telecommunications Corp., Plainview, N.Y., 1961-62; pres., chief exec. officer Grow Group, Inc. (formerly Grow Chem. Corp.), N.Y.C., 1962—; also dir. Grow Group, Inc. (formerly Grow Chem. Corp.); chmn. bd. dirs. GVC Venture Corp. Editor: Managing the Small Company. Recipient award of achievement Sch. of Bus. Alumni Soc. of CCNY, 1977; Winthrop-Sears medal Chem. Industry Assn., 1980. Mem. Nat. Paint and Coatings Assn. (past pres.), Am. Mgmt. Assn. (gen. mgmt. planning coun. 1966—, former trustee, exec. com.), Met. Club, Annabel's Club. Home: 1000 Park Ave New York NY 10028-0934 Office: Grow Group Inc 200 Park Ave New York NY 10166-0005

BANNING, RONALD RAY, systems engineer; b. Aberdeen, S.D., May 6, 1960; s. LaVern Calvin and Arlene Ann (Knutsen) B.; m. Jill Marie Garris, Apr. 30, 1983; children: Matthew, Samuel. BSEE, S.D. Sch. of Mines and Tech., 1981; MSEE, U. N.Mex., 1985. Systems engr. Honeywell Def. Avionics Systems Divsn., Albuquerque, 1992—. Mem. AIAA, Tau Beta Pi, Etta Kappa Nu. Achievements include design of flight control systems for C-135/B-52/C-130 autopilot replacement programs. Home: 4024 71st St NW Albuquerque NM 87120-1652 Office: Honeywell DASD 9201 San Mateo NE Albuquerque NM 87113

BANNISTER, LANCE TERRY, applications engineer; b. Indpls., Feb. 2, 1954; s. Roy Edwin and Mary Louise (Grayson) B.; m. Rona Jaye Cleveland, Aug. 1, 1973; children: Andrew, Michael. Student, Ind. U. Purdue Indpls., 1981-85. Technician USAF, 1973-80; technician Wavetek, Indpls., 1981-82, supr., 1982-89, applications engr., 1989—. Mem. Soc. Cable TV Engrs. Office: Wavetek 5808 Churchman By-Pass Indianapolis IN 46203

BANNON, GARY ANTHONY, molecular geneticist; b. Dubuque, Iowa, June 11, 1954; s. Raymond Anthony and Therese Marie (Campbell) B.; m. Linda Marie Zagrocki, July 2, 1977; children: Sean, Kevin. BA, Wabash Coll., 1976; PhD, Iowa State U., 1981. Postdoctoral fellow U. Rochester, N.Y., 1981-84; asst. prof. U. Ark. Med. Sch., Little Rock, 1984-89, assoc. prof., 1989—; dir. DNA lab. Ark. Genetics Program, Little Rock, 1990—; cons. NSF, Washington, 1986—, Dep. Prosecutor's Office, Little Rock, 1990. Contbr. articles to Molecular and Cellular Biology, 1988, 90, Am. Jour. of Obstetrics and Gynecology; book chpts. to Clinical Practice of Gynecology, 1989, Histone Genes and Histone Gene Expression, 1984. Grantee NSF, 1985, 88, 91. Mem. AAAS, Am. Soc. Microbiology, Sigma Xi (pres. 1992-93). Roman Catholic. Avocations include first to determine that temperature specific expression of ciliate surface proteins was controlled by message stability; first to use DNA polymorphisms for determination of paternity. Office: Univ of Ark for Med Sci 4301 W Markham Little Rock AR 72205

BANNON, GEORGE, economics educator, department chairman; b. Phila., May 25, 1925; s. Joseph Aloysius and Violet May (McCartney) B.; m. Rosemary Ann Chirico, Aug. 19, 1950; children: Patricia Ann, Christina Ann, Terence George. Student, U. Ga., 1944, N.C. State U., 1944; AB, Muhlenberg Coll., 1947; MBA, Lehigh U., 1967. Contr. Overseas Underwriters Ltd., Nassau, Bahamas, 1957-61; internal auditor Bethlehem (Pa.) Steel Corp., 1961-68, sr. systems and procedure analyst, 1968-72, adminstrv. asst., 1972-81; vis. assoc. prof. Moravian Coll., Bethlehem, 1981-85; asst. prof. Muhlenberg Coll., Allentown, Pa., 1985-88; chmn. dept. econs. Muhlenberg Coll., Allentown, 1988—; official Ea. Collegiate Football Officials Assn., Princeton, N.J., 1955-65; Pa. Interscholastic Football Officials Assn., Harrisburg, Pa., 1952-71. Organizer Allentown Area Luth. Parish, Luth. Ch. in Am., 1964-67; bd. dirs. Allentown Area Luth. Parish, 1984-87. Recipient Outstanding Official award Pa. Interscholastic Football Officials Assn., 1971. Mem. Nat. Assn. Accts. (rsch. com. 1979-84, mktg. com. 1985-86, bd. dirs. 1987—), Am. Mgmt. Assn., Fin. Exec. Inst., Allentown C. of C., Am. Assn. Collegiate Schs. Bus., West Allentown Kiwanis Club (pres. 1986-87). Avocations: photography, coin collecting, gardening. Home: 4254 Winchester Rd Allentown PA 18104-1952 Office: Muhlenberg Coll 24th & Chew St Allentown PA 18104

BANSAL, NAROTTAM PRASAD, ceramic research engineer; b. Narnaul, Haryana, India, Sept. 26, 1946; s. Ram Narayan and Santara Bansal; m. Shashi Bansal, May 9, 1974; children: Gaurav, Saurabh. MS, BITS, India, 1968; PhD, Delhi U., India, 1972. Rsch. fellow Delhi U., 1968-71; postdoctoral fellow U. Alberta, Edmonton, Can., 1975-77; asst. prof. Delhi U., 1971-75, 78-82; rsch. assoc. material engring. dept. Rensselaer Poly. Inst., Troy, N.Y., 1982-85; sr. rsch. assoc. Lewis Rsch. Ctr. NASA, Cleve., 1985-87; prin. rsch. scientist Case Western Res. U., Cleve., 1987-90; sr. rsch. engr. Lewis Rsch. Ctr. NASA, Cleve., 1990—; mem. tech. com. Internat. Commn. Glass, 1986—; mem. com. IUPAC, 1985—; symp. chair Ceramic-Matrix Composites Am. Ceramic Soc., 1992-93. Co-author: Handbook of Glass Properties, 1986, Physical Properties Data Energy Storage, 1979; co-author, co-editor: Gases in Molten Salts, 1991; editor: Advances in Ceramic-Matrix Composites, 1993; contbr. articles to profl. jours. and chpts. to books. Recipient Tech. Innovations awards NASA. Mem. Am. Ceramic Soc. (sec.1992—, symp. chair ceramic-matrix composites 1992-93). Achievements include 2 patents and 4 patents pending. Office: NASA Lewis Rsch Ctr 21000 Brookpark Rd Cleveland OH 44135

BANSAL, SATISH KUMAR, civil engineer, educator; b. Sunam, India, Dec. 10, 1960; s. Shambhu Nath and Darshana (Devi) B.; m. Deepika Bansal, Feb. 20, 1988; 1 child, Digvijay. BCE, T.I.E.T., 1983; MCE, PEC, 1986. Design engr. Unites Ltd., Delhi, 1983-84; jr. engr. I.A.A.I., Delhi, 1984-85; lectr. R.E.C., Kurukshetra, India, 1986-88; with P.W.D., Patiala, India, 1988-91; asst. prof. civil engring. Engring. Coll., Bathinda, India, 1991—; cons. engr. H.A.U., Hisar, India, 1987. Merit scholar Govt. of Punjab, 1988, 90-92. Mem. I.S.T.E. Delhi, Inst. Engring. Calcutta (assoc.). Home: AP-9 Staff Colony Engring, Bathinda 151001, India Office: Engring Coll, Dept Civil Engring, Bathinda 151001, India

BANSCHBACH, VALERIE SUZANNE, biologist; b. Cin., June 22, 1964; d. Jerome Lee and Shirley Ann (Bishop) B.; m. James Duncan Rowe, Apr. 24, 1992. BA, Pomona Coll., 1986; PhD, U. Miami, 1992. Rsch. assoc. U. Miami, Coral Gables, 1988-89, teaching asst., 1987-92; postdoctoral rsch. assoc. U. Vt., Burlington, 1992--; manuscript reviewer Animal Behaviour, 1992—, Environmental Entomology, 1992—. Mentor Assn. for Women in Sci., Burlington, 1992; scientist Sci. by Mail, Orlando, Fla., 1992. Named Outstanding teaching asst. U. Miami, 1992; Fla. Entomological Soc. grantee, 1989, Sigma Xi grantee, 1990. Mem. Animal Behavior Soc., Internat. Soc. for Behavior Ecology, Internat. Union for the Study of Social Insects, Entomological Soc. Am. Democrat. Achievements include first demonstration of lack of risk sensitivity in a foraging animal the honey bee. Office: U Vt Dept Zoology Burlington VT 05405

BANUK, RONALD EDWARD, mechanical engineer; b. Brockton, Mass., Oct. 22, 1944; s. Joseph John and Leocadia Marilyn (Gusciora) B.; m. Patricia Audrey Ryan, July 4, 1969; children: Kim, Lance. BSME, Northeastern U., 1967; MSME, San Diego State U., 1971. Design and stress engr. in advanced systems Ryan Aero. Co., San Diego, 1967-76; sr. tech. specialist Northrop Corp., Pico Rivera, Calif., 1976-93, program mgr., 1987-89; structures tech. area mgr. Northrop Corp., Pico Rivera, 1991, prin. investigator in advanced structure and foam devel., 1986-93. Author: Design Considerations for Foam and Honeycomb Structure, 1993, On Saleem, 1993, Mary: Past, Present and Future. 1993. Mem. Soc. Adv. Material and Process Engring. Republican. Avocation: writing on religion. Home: 6441 Ringo Cir Huntington Beach CA 92647-3323 Office: Adv Structural Design & Devel T235/GK 8900 E Washington Blvd Pico Rivera CA 90660-3783

BANVILLE, DEBRA LEE, research chemist; b. Astoria, N.Y., Mar. 25, 1959; d. George Joseph and Marilyn Donna (Johansson) B.; m. Kevin Paul Scott, Sept. 2, 1989. BS, Brandeis U., 1981; PhD, Emory U., 1986; postgrad., U. Calif., San Francisco, 1986-90. Asst. nuclear magnetic resonance facility U. Calif., 1987-90; rsch. chemist ICI Pharms., Wilmington, Del., 1990—. Contbr. articles to Biochemistry, Jour. Am. Chem. Soc., Jour. Phys. Chemistry. Mem. Am. Chem. Soc. (sec. Del. sect. 1991-92, environ. chmn. 1993—), Sigma Pi, Pi Alpha. Unitarian. Achievements include discovery of unusally high sequence specificity of actinomycin and a series of metalloporphyrin intercalators for synthetic DNA polymers. Office: ICI Pharms Concord Pike Wilmington DE 19897-2500

BANWELL, MARTIN GERHARDT, chemistry researcher; b. Lower Hutt, New Zealand, Nov. 24, 1954; s. Congreve John and Margot (Hormes) B.; m. Catherine Louise Beckwith, Dec. 19, 1981; 1 child, James. BS with honors, Victoria U. Wellington, 1977, PhD, 1979. Postdoctoral fellow dept. chemistry Ohio State U., Columbus, 1979-80; sr. teaching fellow U. Adelaide, South Australia, 1980-81; lectr. in chemistry U. Auckland, New Zealand, 1982-86; lectr. in organic chemistry U. Melbourne, Australia, 1986-88, sr. lectr., 1989-92, assoc. prof., 1993—; chmn. organic chemistry group RACI, Melbourne, 1989-91. Contbr. chpt.: Advances in Strain in Organic Chemistry, 1991; contbr. articles to Australian Jour. Chemistry. Mem. Am. Chem. Soc., Royal Australian Chem. Inst. (Rennie Meml. medal 1986), Royal Soc. Chemistry. Achievements include development of new methods for synthesis of biologically active compounds, most notably the alkaloid colchicine and its congeners. Office: U Melbourne Sch Chemistry, Parkville, Melbourne 3052, Australia

BAO, JOSEPH YUE-SE, orthopaedist, microsurgeon, educator; b. Shanghai, China, Feb. 20, 1937; s. George Zheng-En and Margaret Zhi-De (Wang) B.; m. Delia Way, Mar. 30, 1963; children: Alice, Angela. MD, Shanghai First Med. Coll., 1958. Intern affiliated hosps. Shanghai First. Med. Coll.; resident Shanghai Sixth People's Hosp., orthopaedist, 1958-78, orthopaedist-in-charge, 1978-79, vice chief orthopaedist, 1979-84; rsch. assoc. orthopaedic hosp. U. So. Calif., L.A., 1985-90, vis. clin. assoc. prof. dept. orthopaedics, 1986-89; coord. microvascular svcs. Orthopaedic Hosp., L.A., 1989-91; clin. assoc. prof. dept. orthopaedics U. So. Calif., L.A., 1989—; attending physician Los Angeles County and U. So. Calif. Med. Ctr., L.A., 1986, 90—; cons. Rancho Los Amigos Med. Ctr., Downey, Calif., 1986. Contbr. articles to profl. jours., chpts. to books. Mem. Internat. Microsurgical Soc., Am. Soc. for Reconstructive Microsurgery, Orthopaedic Rsch. Soc. Home: 17436 Terry Lyn Ln Cerritos CA 90701-4522 Office: LA County USC Med Ctr Dept Orthopaedics 2025 Zonal Ave COH 3900 Los Angeles CA 90033-4526

BAO, ZHENLEI, physicist; b. Tianjin, China, Nov. 5, 1963; s. Yazheng Wang and Shuhui Bao; m. Yili Zhang, Dec. 24, 1987. BS, Peking U., Beijing, 1985; MA, SUNY, Stony Brook, 1992. Rsch. asst. Peking U., Beijing, 1986-89, SUNY, Stony Brook, 1990—. Contbr. articles to profl. jours. Mem. Am. Physics Soc. Achievements include research on high and low temperature superconductors. Office: Dept Physics SUNY Stony Brook NY 11794

BAPTIST, JAMES NOEL, biochemist; b. Shelbyville, Ill., June 6, 1930; s. Noel A. and Vivian L. (Lockhart) B. BS, Case Inst. Tech., 1952; PhD, U. Ill., 1957; postgrad., U. Mich., 1957-59. Chemist Phillips Petroleum Co., Bartlesville, Okla., 1952-54; biochemist W.R. Grace & Co., Clarksville, Md., 1959-63; microbiologist Internat. Minerals and Chemicals Co., Skokie, Ill., 1963-65; self-employed biochemist Bradenton, Fla., 1965-68; biologist M.D. Anderson Hosp., Houston, 1968-79; biology rschr. Kerrville, Tex., 1980-87, El Paso, Tex., 1987—. Contbr. articles to profl. jours. Republican. Lutheran. Achievements include research in hydrocarbon oxidation by cell free bacterial enzymes, microbial taxonomy by zone electrophoresis of enzymes, bacterial mutations visualized by enzyme zone electrophoresis, protein purification. Home and Office: 2620 Lake Victoria El Paso TX 79936

BAR, ROBERT S., endocrinologist; b. Gainesville, Tex., Dec. 2, 1943; s. Samuel and Mildred Emma (Kaplan) B.; m. Laurel Ellen Burns, June 23, 1970; children: Katharine June, Matthew Thomas. BS, Tufts Univ., 1964; MS in Biochemistry, Ohio State U., 1970, MD, 1970. Medicine intern Pa. Hosp., Phila., 1970-71; medicine resident Ohio State Univ., Columbus, 1971-72; asst. prof., dept. medicine Univ. Iowa, Iowa City, 1977-82, assoc. prof., dept. medicine, 1982-86, prof., dept. medicine, 1986—; acting dir. divsn. of endocrinology and metabolism, U. Iowa, 1985-90; dir. diabetes-endocrinology rsch ctr., U. Iowa, 1986—, nat. rsch. svc. award in endocrinology, 1984—, endocrinology fellowship program, 1979—, divsn. of endocrinology and metabolism, 1990—; mem. ad hoc study sect. NIH, 1985; mem. editorial bd. Jour. of Clin. Endocrinology and Metabolism, 1984-87; mem study sect. Nat. Veterans Adminstrn., 1984-87; v.p. rsch. Nat. Am. Diabetes Assn., 1987-88; mem. orgn. com. Endothelium and Diabetes Symposium, Melbourne, 1988; mem. study sect. numerous assns. and coms.; guest reviewer numerous jours. Editor Endocrinology, 1987-89, Advances in Endocrinology and Metabolism, 1989—. Mem. Am. Diabetes Assn., Am. Soc. for Clin. Investigation, Assn. Am. Physicians, Endocrine Soc., Ctrl. Soc. for Clin. Rsch., Sigma Xi. Office: U Iowa Hwy 6 West 3E19 VA Iowa City IA 52246

BARAGIOLA, RAUL ANTONIO, physicist; b. Santa Fe, Argentina, Mar. 31, 1945; came to U.S., 1988; s. Jorge Alberto and Apolonia Manuela (Torres) B.; m. Beatriz Carolina Pfister, July 3, 1970; children: Verena Gisela, Valeria Monica, Pablo Antonio. Lic. Physics, Inst. Balseiro, Bariloche, Argentina, 1969, D of Physics, 1971. Rschr. Argentine Atomic Energy Commn., Bariloche, 1971-88; founder, CEO Altec S.E., Bariloche, 1985-87; v.p. rsch. Itron S.A., Bariloche, 1986-87; prof. and dir. Lab. for Atomic and Surface Physics U. Va., Charlottesville, 1990—; vis. scientist Rutgers U., Piscataway, N.J., 1988-90; advisor Rio Negro Sec. Sci. and Tech., Bariloche, 1984-88; referee Physics Rev., NYC, 1980—. Editor: Ionization of Solids by Heavy Particles, 1992; contbr. articles to profl. jours. Recipient Morales Interamerican prize Orgn. Am. States, 1983, Argentine Physics prize Ex-Internos de la Fraternidad, 1979, Coca-Cola Arts Scis. prize Coca-Cola Co., 1982. Mem. Am. Phys. Soc., Am. Vacuum Soc., Bömisches Phys. Soc. Achievements include patent for Time of Flight Fast-particle Velocity Detector ; research on scaling law for electron emission by ions, origin of auger electrons from ion-solid interactions, dielectric breakdown of solid rare gases under irradiation. Home: 1808 Wakefield Rd Charlottesville VA 22901 Office: U Va Thornton Hall Charlottesville VA 22901

BARANOWSKI, PAUL JOSEPH, nuclear instrumentation technician; b. Norwich, Conn., July 29, 1950; s. Joseph Baranowski Jr. and Margaret Olive (Croteau) Momut; m. S. Rose Bottom, Sept. 3, 1977; 1 child, Bettyann Cole. AS in Indsl. Electronics, Thames Valley State Coll., 1982; cert., Inst. Nuclear Power Ops., 1992. Cert. control technician Nat. Acad. Nuclear Tng. Welder Gen. Dynamics Elec. Boat, Groton, Conn., 1973-74; automotive technician Goodyear, Norwich, 1974-75, Mallon Chevrolet, Norwich, 1975-77; maintainence mechanic Wyre Wynd, Jewett City, Conn., 1977-79; engring. technician Victor Elec. Wire & Cable, Westerly, R.I., 1982-84; instrumentation and controls technician, mech. tech. various nuclear facilities, 1984-90; instrumentation and controls technician Calvert Cliffs Nuclear Power Plant Balt. Gas & Elec. Co., Lusby, Md., 1990—. With USN, 1968-73. Mem. Am. Legion. Democrat. Avocations: fishing, collecting old bottles, antiques, gardening. Home: PO Box 1667 Solomons MD 20688-1667

BARASSI, DARIO, management consultant; b. Milan, Italy, June 1, 1940; s. Giovanni and Antonietta (Croci) B.; m. Gerlinde Brenner, Jan. 10, 1969; children: Sebastiano, Lorenzo. Diploma in Econs. and Commerce, U. Cattolica, Milan, 1962; MBA, Econs. Sch., Rotterdam, The Netherlands, 1971. Mgr. Unilever Italy, Milan, 1962-75; sr. cons. Hay Italiana, Milan, 1976-77; sr. ptnr. Unit Mgmt. Cons., Milan, 1977—; pres. Svc. Idea, Milan, 1989—; Bd. dirs. Associazione Disegno Industriale. Author: La Service Idea, 1988, Divenire dellž Impresa, 1993, Forme Libere, 1994; contbr. articles to mags.; lectr. for confs. Avocation: contemporary music editor. Home: Viale Monte Nero 70, 20135 Milan Italy Office: Unit Mgmt Cons SrL, Via Saffi 25, 20123 Milan Italy

BARBA, EVANS MICHAEL, civil engineer; b. N.Y.C., July 27, 1950. BECE, Cooper Union, 1974; MSCE, Poly. Inst. N.Y., 1979. Registered profl. engr., N.Y., N.J., Conn., Mich., Pa. Resident/soils lab. engr. Meuser, Rutledge, Wentworth, Johnston, N.Y.C., 1972-74; civil/sanitary engr. Malcolm Pirnie, White Plains, N.Y., 1974-78; v.p. Hill Internat., Willingboro, N.J., 1978-82; pres., chief exec. officer Arkhon Corp., Cherry Hill, N.J., 1983-90; chmn., chief exec. officer Barba Internat. Inc., Cherry Hill, 1991-92; chmn., CEO Barba-Arkhon Internat. Inc., Mt. Laurel, N.J., 1993—. Contbr. articles to Constrn. Litigation Reporter, Hotel and Motel Mgmt., Constrn. Briefings, Constrn. Law Reports; contbg. author: Handbook of Construction Law and Claims, 1982. Mem. ASCE, Constrn. Mgmt. Assn. Am., Am. Arbitration Assn. (nat. panel), NSPE. Office: Barba-Arkhon Internat Inc Ste 300 East 10000 Midlantic Dr Mount Laurel NJ 08054

BARBALAS, LORINA CHENG, chemist; b. Forest Hills, N.Y., July 9, 1958; d. Benjamin K. and Laura (Kwan) Cheng; m. Michael Peter Barbalas, Oct. 10, 1981. BA cum laude, Cornell U., 1979; MA, Columbia U., 1981. Mem. tech. staff AT&T Bell Labs., Murray Hill, N.J., 1982-85; lectr. Lanzhou (Peoples Republic of China) U., 1985; cons. Mgmt. Techs., Internat., Tianjin, Peoples Republic of China, 1986-91; adminstr. Friends of China Found., Hong Kong, 1991—. Summer rsch. grantee Exxon, 1978. Mem. Am. Chem. Soc. Home: 44 Country Squire Rd Old Tappan NJ 07675 Office: Friends of China Foundation, PO Box 887 Shatin Post Office, Tai Wai Hong Kong

BARBARÁN, FRANCISCO RAMÓN, educator, researcher; b. Salta, Argentina, Sept. 2, 1960; s. Medardo Ramón and María Graciela (Zigarán) B. Licenciate, U. Nacional de Salta, Argentina, 1986. Licenciate in natural resources. Adscript asst. student Facultad de Ciencias Naturales, U. Nacional de Salta, Argentina, 1983-85, asst. student 2d category, 1985-86, asst. tchr. 1st category, 1986—; researcher Centro de Investigaciones Ecológicas del Chaco, Salta, 1988—; cons. Constituent Mcpl. Conv., 1988; dir. Commercialization of Furs and Skins of Wildlife in Salta, 1989. Author: How to Prepare a Professional Thesis, 1992, Politica, Desarrollo y Medio Ambiente: la Nueva Síntesis, Vida Silvestre 1(2), 1991, Commercialization Statistics of the Blue-Fronted Amazon (Amazona aestiva) in Salt Province, Argentina, 1977-92; contbr. articles to profl. jours. Superior coun. mem. U. Nacional de Salta, 1984-85. Grantee Inst. Nacional de Tecnologia Agropecuaria, 1986, Iguana Colorada Project, 1989, Sistema Para el Apoyo a la Investigacion y Desarrollo de la Ecologia en la República Argentina, 1990, Banff (Can.) Ctr. for Mgmt., 1991; recipient Can. Embassy award, 1990. Avocations: reading, travel. Home: Pachi Gorriti 1780, 4400 Salta Argentina

BARBAT, ALEX HORIA, civil engineer; b. Brasov, Romania, Mar. 5, 1947; s. Alexandru and Ana (Droc) B.; m. Eugenia Ileana Vlad, Oct. 5, 1973. Civil engr., Tech. U. Iasi, Romania, 1970; D of Civil Engring., Tech. U. Iasi (Romania), 1978. Asst. prof. Tech. U. Iasi, 1970-73; assoc. prof. Tech. U Iasi, 1974-79; assoc. prof. Tech. U. Catalonia, Barcelona, Spain, 1980-85, rsch. assoc. prof., 1986-90, rsch. prof., 1990—; cons. structural engring. Indus, S.A., Barcelona, 1979-80, DPI, S.A., Barcelona, 1980-81; tech. dir. ESPES, S.A., Barcelona, 1981-85. Author: Seismic Analysis of Structures, 1982, Structures Subjected to Seismic Actions, 1988, 2d edit., 1993, Structural Response Computations in Earthquake Engineering, 1989; contbr. to book, monographs and papers to profl. jours. Mem. Internat. Assn. Computational Mechanics, Earthquake Engring. Rsch. Inst., N.Y. Acad. Scis., Seismol. Soc. of Am. Home: Estapé 46 Atico 4a, 08190 Sant Cugat del Vallés, Barcelona Spain Office: Tech U Barcelona, Gran Capitán s/n, 08034 Barcelona Spain

BARBATIS, CALYPSO, histopathologist; b. Kefallonia, Greece, July 24, 1946; d. Dimitrios and Krystalia (Nentopoulou) B.; m. Mr. Petropoulos, Dec. 11, 1982; 1 child, Michaela. Grad. high sch., Kefallinia; grad., Athens U., Greece, 1970. Registrar Radcliffe Infirmary Hosp., Oxford, Eng., 1975, temp. clin. lectr., 1976; sr. registrar Southampton (Eng.) Gen. Hosp., 1976-78; clin. lectr. Southampton U., 1978-79, John Radcliffe Hosp./Oxford U., 1979-82; cons. Lewisham Hosp., London, 1982-89; tchr. Guy's and St. Thomas United Med. Schs., London, 1982-89; cons. Red Cross Hosp., Athens, Greece, 1989—; dir. pathology Red Cross Hosp., 1990—. Contbr. articles to profl. jours. Athens U. scholar, 1965-70; Guy's Hosp. grantee, 1984-88. Mem. Assn. Clin. Pathologists, Soc. Hellenic Pathol. Anatomy (exec. com.), Pathol. Soc. Great Britain and Ireland, European Soc. Pathology, Internat. Gastro-Surg. Club, Internat. Acad. Pathology, Hellenic Gastro-Enterology. Avocations: reading, swimming. Home: Parnithos 37 P.Psychikon, Athens Greece Office: Red Cross Hosp, Red Cross St, Athens Greece

BARBATO, JOSEPH ALLEN, writer; b. N.Y.C., Feb. 23, 1944; s. Joseph Michael and Florence (Kelly) B.; m. Augusta Ann DeLait, Oct. 23, 1965; children: Louise, Joseph. BA, NYU, 1964, MA, 1969. Newswriter NYU, N.Y.C., 1964-68, dir. alumni communications, 1969-74, sr. devel. writer, 1974-78; staff writer Shell Oil Co., N.Y.C., 1968-69; ind. writer N.Y.C., 1978-90; editorial dir. The Nature Conservancy, Arlington, Va., 1990—; mem. edit. bd. Small Press mag., N.Y.C., 1984-86; communications cons. univs., hosps., etc., 1978-90. Co-author: You Are What You Drink, 1989; editor: What We Really Know About Mind-Body Health, 1991; contbg. author: The Book of the Month, 1986; columnist edn., health, lit. numerous mags. and newspapers including Smithsonian, N.Y. Times, Village Voice, Christian Sci. Monitor, others. Mem. Authors Guild, Nat. Book Critics Cir., Soc. Profl. Journalists. Home: 1418 Juliana Pl Alexandria VA 22304-5935 Office: The Nature Conservancy 1815 N Lynn St Arlington VA 22209-2003

BARBEE, JOE ED, lawyer; b. Pharr, Tex., Feb. 27, 1934; s. Archie Allen and Concha (Leal) B.; m. Yolanda Margaret Atonna, Feb. 17, 1962; children—Cynthia M., Adam A., Walter J. BSEE, U. Ariz., 1961; JD, Western New Eng. Coll., 1973. Bar: Mass. 1973, U.S. Patent Office 1973, U.S. Ct. Appeals (fed. cir.) 1982. Engr. Gen. Electric Co., Pittsfield, Mass., 1961-73; patent atty. Fort Wayne, Ind., 1973-75, Magnavox, Fort Wayne, 1975-76, Motorola, Inc., Phoenix, 1976—. Sgt. U.S. Army, 1953-56. Recipient Outstanding Performance award U.S. Civil Svc., 1960. Mem. ABA, Am. Patent Law Assn., Am. Intellectual Property Law Assn. Republican. Methodist. Avocations: tennis, hunting, fishing. Home: 7611 N Mockingbird Ln Paradise Valley AZ 85253-3126 Office: Motorola Inc 8220 E Roosevelt B3 Scottsdale AZ 85257

BARBEE, ROBERT WAYNE, cardiovascular physiologist; b. Greensboro, N.C., Feb. 6, 1956; s. Wendell Wayne and Rogers Carlene (McNeal) B.; m. Mary Suzanne Poulton, Aug. 2, 1981. MS, U. Fla., 1981; PhD, La. State U.,

1986. Rsch. and teaching asst. Dept. Physiology, U. Fla., Gainesville, 1979-81; pre-doctoral fellow Dept. Physiology, La. State U., New Orleans, 1982-86, asst. prof. (part-time), 1988—; postdoctoral fellow Hypertension Rsch., Ochsner Med. Found., New Orleans, 1986-88, staff rsch. scientist, 1988-90; staff rsch. scientist Cardiology sect., Ochsner Med. Found., New Orleans, 1990—; adj. asst. prof. Dept. Physiology, Tulane Med. Sch., New Orleans, 1988—; chmn. Ochsner Animal Care and Use COm., New Orleans, 1989—, mem. Oschner Basic Rsch. Adv. Com., 1990—, Mem. Oschner Safety Com., 1988—. Author 2 book chpts.; contbr. over 18 articles to profl. jours. Subscription mgr. 20/20 Vision, New Orleans, 1991—; donating mem. Union of Concerned Scientists, Washington, 1988—. Recipient Am. Chem. Soc. award Am. Chem. Soc., 1973. Mem. Am. Fedn. for Clin. Rsch., Am. Physiol. Soc., Internat. Soc. for Heart Rsch. Achievements include rsch. on role of atrial natiuretic peptide in fluid and pressure homeostasis. Office: Div Rsch Alton Ochsner Med Found 1520 Jefferson Hwy New Orleans LA 70121

BARBER, ANN MCDONALD, internist; b. Washington, Jan. 14, 1951; d. Charles Finch and Lois Helen (LaCroix) B. BS in Math., Stanford U., 1974, MS in Math., 1974; MD, Northwestern U., Chgo., 1981. Mathematician NIH, Bethesda, Md., 1974-76; program analyst engr. II Mass. Gen. Hosp., Boston, 1976-77; resident physician Northwestern U. Med. Ctr., Chgo., 1981-84; med. staff fellow NIH, Bethesda, 1984-87, sr. staff fellow, 1987-91; computer scientist DOE, 1991-92; attending physician Providence Hosp., Washington, 1992—; peer reviewer Annuals of Internal Medicine, ACP, Phila., 1986—; cons. Inst. for New Generation Computer Tech., Tokyo, 1991—; guest scientist NIH, 1991—. Contbr. articles to profl. jours. Vol. Zacchaeus Med. Clinic, Washington, 1990—. Recipient Physicians Recognition award, AMA. Fellow Royal Soc. Medicine London; mem. Am. Med. Info. Assn. Office: NCI NIH Lab Math Biology 9000 Rockville Pike Bethesda MD 20892-0001

BARBER, EDMUND AMARAL, JR., retired mechanical engineer; b. East Providence, R.I., Oct. 20, 1916; s. Edmund Amaral and Clara Veronica (Amaral) B.; Sc.B., Brown U., 1938; postgrad. M.I.T., 1951, Syracuse U., 1955-57; m. Marion McKelvy Yost, Mar. 15, 1941; children—Marion Elizabeth Barber Goodrich, Jean Claire Barber Keehan. Designer, New Eng. Butt Co., Providence, 1936-38; with IBM, 1938-71, mgr. research, Endicott, N.Y., 1952-55, engring. mgr., Owego, N.Y., 1955-59, lab. adminstrn. mgr., Owego, 1959-64, spl. asst., Owego, 1964-66, avionics systems adminstrn. mgr., 1966-67, staff adminstrn. mgr., 1967-68, facility plans mgr., 1968-69, tech. staff mem., 1969-70, data mgr., 1970-71; ret., 1971; mem. Ithaca (N.Y.) chpt. Service Corps Ret. Execs., 1976—. Mem. Town of Owego Planning Bd., 1956-72. Pres. Tioga County Indsl. Devel. Corp., 1970-72, sec.-exec. dir., 1972, dir., 1967-72; adminstrv. dir. Tioga County Indsl. Devel. Agy., 1972; v.p. N.Y.-Penn Health Planning Council, 1969-72, dir., 1968-72; mem. exec. com., dir. N.Y.-Penn Health Mgmt. Corp., 1972; mem. exec. com. Tioga Gen. Hosp., Waverly, N.Y., 1962-72, dir., 1961-72, chmn. planning com., 1964-72; v.p. Owego Boys' Club, 1963-66, dir., 1962-66; pres. Christmas League, Owego, 1963-64, dir., 1961-70; bd. mgrs. Tompkins County Hosp., Ithaca, 1976, sec., 1976, chmn. planning com., 1976. Mem. ASME, Nat., N.Y. socs. profl. engrs., Sigma Xi, Tau Beta Pi, Elks. Republican. Roman Catholic. Patentee in field of data processing machine, med. research equipment, music instrn. devices. Home: 42 Fairview Sq Ithaca NY 14850-4911 also: 2650 Pearce Dr Apt 311 Clearwater FL 34624-1133

BARBER, SUSAN CARROL, biology educator; b. Anson, Tex., Oct. 2, 1952; d. Raymond Reginald and Loreta (Judkins) Barber; m. Robert Joseph Mulholland Jr., Oct. 17, 1981 (dec. 1989); m. David P. Nagle Jr., Oct. 11, 1991. BS, Howard Payne U., 1974; MS, Okla. State U., 1975; PhD, U. Okla., 1980. Asst. prof. biology Sam Houston State U., Huntsville, Tex., 1979-82; asst. prof. biology Okla. City U., 1983-88, assoc. prof. and chair dept. biology, 1988-92, prof. and chair dept. biology, 1992—; adj. prof. U. Okla., Norman, 1982—; cons. Okla. Biol. Survey, Norman, various other orgns.; presenter wildflower workshops, Green Earth Series, Myriad Gardens, Oklahoma City, 1990. Recipient Sears Teaching 2d pl. award Oklahoma City U., 1991, United Meth.-related Instn. Exemplary Teaching award, 1992. Mem. Bot. Soc. Am., Am. Soc. Plant Taxonomists, Beta Beta Beta (advisor 1985—). Home: 724 Oakbrook Dr Norman OK 73072 Office: Oklahoma City Univ Dept Biology 2501 N Blackwelder Oklahoma City OK 73106

BARBERO, GIULIO JOHN, physician, educator; b. Mt. Vernon, N.Y., Oct. 13, 1923; s. Armando and Mary (Celoria) B.; m. Margaret Goff, May 30, 1947; children: Paul, Christopher, Mary, Peter, Claudia, David. BA, U. Maine, 1943; MD, U. Pa., 1947. Intern Hosp. U. Pa., 1947-49; resident Children's Hosp. Phila., 1949-50, chief resident, 1951-52; mem. faculty U. Pa. Med. Sch., 1953-67, asso. prof. pediatrics, 1963-67; prof., chmn. dept. pediatrics Hahnemann Med. Coll., Phila., 1967-72, U. Mo. Med. Sch., Columbia, 1972-90; prof. U. Mo. Med. Sch., 1991—; Chmn. gen. med. and sci. council Nat. Cystic Fibrosis Research Found., 1971-74; mem. Nat. Heart, Lung, Blood Council, 1977-78. Served with AUS, 1950-51, Korea. Decorated Bronze Star; sr. fellow Nat. Polio Found., 1952-54; recipient Bernard Wenrich award for research in cystic fibrosis, 1962. Mem. Phi Beta Kappa, Sigma Xi, Pi Kappa Phi. Home: 408 S Glenwood Ave Columbia MO 65203-2716

BARBERO, JOSÉ ALFREDO, physics researcher; b. Mendoza, Argentina, June 1, 1950; s. Benito Antonio and Rosa Matilde (Becker) B.; m. Mónica Dolores Sosa, Sept. 22, 1976; children: Sebastián Alfredo, Fernando Álvaro, José Alfredo Jr., María Andrea. MSc, Nat. U. Cuyo, Bariloche, Argentina, 1974, PhD in Physics, 1984. Dean rsch. Nat. U. Lomas de Zamora, Buenos Aires, 1984-86, jr. researcher Nat. Atomic Energy Commn., Bariloche, 1975-77, Buenos Aires, 1977-85; sr. researcher, electrochemistry group leader Nat. Atomic Energy Commn., Bariloche, 1986—; jr. researcher Alta. (Can.) Rsch. Coun., Edmonton, 1981-82; cons. in instrumentation Gekmece Nuclear Rsch. Ctr., Istanbul, Turkey, 1989—; mem. Sci. Investigation Commn., Province Buenos Aires, 1984-86, Nat. Rsch. Coun., Buenos Aires, 1984-86. Contbr. articles to sci. jours. Scholar Nat. U. del Sur, 1969-71, Nat. Atomic Energy Commn., 1971-74, 75-77. Mem. Argentine Assn. Nuclear Tech., Argentine Soc. Physics and Chemistry. Avocation: long-distance running. Office: Nat Atomic Energy Commn, Avda E Bustillo Km 9.500, 8400 Bariloche Rio Negro, Argentina

BARBI, JOSEF WALTER, engineering, manufacturing and export companies executive; b. Melk, Noe, Austria, Sept. 26, 1949; s. Walter and Hermine (Mayr) B.; m. Yolanda Rojas, Aug. 29, 1981; 1 child, Anna Katherina. Student, U. Saskatoon, Sask., Can., 1974, Kans. State U., 1982. Mech. engr. Zizala Metalwarenfabriken, Melk, 1963-65, Austrian Farmers Coop., Pochlarn, Austria, 1970-72; area mgr. Internat. Systems & Controls Corp., Regina, Sask., Can., 1974-75; mng. dir. Bakem Agro-Indsl. C.A., Caracas, Venezuela; gen. mgr. Intercon. Agro Indsl. Devel. Inc., Hialeah, Fla., 1977-81; internat. mktg. mgr. MEC Co., Kans., 1982-84; adviser internat. ops. Calif. Pellet Mill Co., I.R., San Francisco, 1984-91; CEO ASIMA Corp., Independence, Kans., 1984—; pres. Internat. Nutrition Techs., Independence, 1987—, Engineered Systems & Equipment, Inc., Caney, Kans., 1988—; cons. govts. of Venezuela, 1976-77, fish farm coops., Europe, 1987; speaker at internat. profl. confs. Contbr. articles to profl. jours. Bd. dirs. Internat. Independence Community Coll., 1987-89, Jr. Achievement, Independence, 1987-90, AFIA Com., 1993—. Mem. World Aquaculture Soc., C. of C., Rotary Internat. Home: RR 4 Box 194E Independence KS 67301-9169

BARBOUR, ERIC S., electrical engineer; b. Pinehurst, N.C., Apr. 24, 1966; s. Earl G. and Susan C. Barbour. BSEE, N.C. State U., 1989, MS in Mgmt., 1991. Registered engr.-in-tng., N.C. Utilities engr. BASF Corp., Asheville, N.C., 1991—. Mem. IEEE, Nat. Soc. Profl. Engrs., Jaycees. Home: 1205 Turtle Creek Dr Asheville NC 28803 Office: BASF Corp Utilities Dept Sandhill Rd Enka NC 28728

BARBOUR, ROBERT CHARLES, technology executive; b. Seymour, Ind., Sept. 1, 1935; s. Robert Henry and Audrey Silence (Trueblood) B.; m. Rose G. Dienes, Mar. 18, 1960; children: Eric M., Timothy A. BS, Purdue U., 1957. Supr. new products B.F. Goodrich Co., Akron, Ohio, 1965-70; v.p., gen. mgr. structures div. Irvin Industries, Inc., Greenwich, Conn., 1970-79; v.p. mktg. shielding group Bairnco/Keene, Norwalk, Conn., 1979-84; gen. mgr. SRS. div. Bairnco/Keene, Monrovia, Calif., 1985; ind. cons. Darien, Conn., 1986; v.p. mktg. Haskon Corp., Taunton, Mass., 1987-92; chmn. bd. dirs., CEO Haskon Internat., Inc., Taunton, 1992—. Pres. Area Neighborhood Assn., Darien, 1984. Mem. AIAA. Office: Haskon Internat Inc 336 Weir St Taunton MA 02780

BARBUY, BEATRIZ, astronomy educator; b. Sao Paulo, Brazil, Feb. 16, 1950; d. Heraldo and Belkiss (Silveira) B. Student, U. Sao Paulo, 1972, M in Astronomy, 1976; PhD in Astrophysics, U. Paris, 1982. Prof. U. Sao Paulo, 1982-87, assoc. prof., 1987—. Contbr. articles to profl. jours. Mem. Am. Astron. Soc., Internat. Astron. Union, Brazilian Astron. Soc. (pres. 1992-94), Royal Astron. Soc., Soc. Francaise Specialistes D'Astronomie, Academia Ciências Estado de São Paulo. Office: U São Paulo Inst Astronomy, Av Miquel Stefano 4200, 04301-904 São Paulo Brazil

BARCHI, ROBERT LAWRENCE, neuroscience educator, clinical neurologist, neuroscientist; b. Phila., Nov. 23, 1946; s. Henry John and Elizabeth (Pesci) B.; m. Joan E. Mollman, Sept. 20, 1976; children: Jonathan Robert, Jennifer Elizabeth. BS, Georgetown U., 1968, MS, 1969; PhD, U. Pa., 1972, MD, 1973. Diplomate Am. Bd. Neurology and Psychiatry, Am. Bd. Med. Examiners. Resident in neurology U. Pa. Hosp., 1973-75; asst. prof. biochemistry U. Pa. Med. Sch., Phila., 1974-75, asst. prof. neurology and biochemistry, 1975-78, assoc. prof., 1978-81, prof., 1981—, David Mahoney prof. neurosci., 1985—, chmn. neurosci. grad. program, 1983-89, dir. Mahoney Inst. Neurol. Scis., 1983—; vice-dean rsch. sch. medicine U. Pa. Med. Sch., 1989-91, chmn. dept. neurosci., 1992—; mem. med. adv. bd. Muscular Dystrophy Assn., 1982—; adv. bd. Caphalon Inc., 1992, Phila. Ventures Inc., 1992. Author: (with R. Lisak) Myasthenia Gravis; contbr. chpts. to textbooks, numerous articles to profl. jours.; mem. editorial bd. Muscle & Nerve jour. 1981—, Jour. Neurochemistry, 1981-90, Jour. Neurosci., 1985—, Ion channels, 1988—, Current Opinion Neurology and Neurosurgery, 1992—. Recipient Lindback award U. Pa., 1979, Javits award NIH, 1985. Fellow Am. Acad. Neurology; mem. NAS, Inst. Medicine, Am. Neurol. Assn. (bd. councillors 1992—), Biophys. Soc., Soc. for Neurosci. (pub. lectr. 1985), Am. Soc. Clin. Investigation, Assn. Am. Physicians, Phi Beta Kappa, Alpha Omega Alpha. Avocation: antiquarian horology. Office: U Pa David Mahoney Inst Neurol Scis 215 Stemmler Hall 452 Med Edn Bldg Philadelphia PA 19104

BARCLAY, ROBERT, JR., chemist; b. Mt. Vernon, N.Y., Apr. 1, 1928; s. Robert and Emma Josephina (Neher) B. AB, Cornell U., 1948; PhD, U. Md., 1957. Chemist Barrett Div. Allied Chem. Corp., Edgewater, N.J., 1948-51, Am. Cyanamid Co, Linden, N.J., 1951-52; project scientist Union Carbide Corp., Bound Brook, N.J., 1956-69; sr. rsch. scientist Chem. Div. Morton Thiokol, Trenton, N.J., 1969-79; sect. head Hydrocarbon Rsch. Inc., Lawrenceville, N.J., 1979-86; cons. Amoco Performance Products Inc., Bound Brook, 1986-90. Contbr. chpt. to book Condensation Monomers, 1972. Fellow Am. Inst. Chemists; mem. Am. Chem. Soc. (sec. Trenton sect. 1979-81, alt. councillor 1983-84). Roman Catholic. Achievements include 13 patents for synthesis of high performance condensation polymers and ultraviolet cured urethane acrylate polymers, and others. Home: 6 Berrywood Dr Trenton NJ 08619-1906

BARCLAY, STANTON DEWITT, engineering executive, consultant; b. Pa., Apr. 27, 1899; s. James Arthur Barclay and Elsie Arvilla Gore; married; children: Stanton D. Jr., Gail Dee (dec.). BS, Pa. State U., 1922, M in Engring., 1926; LHD (hon.), Lycoming Coll., 1986. Lic. profl. engr. Instr. engring. Rensselaer Poly. Inst., 1922-25; asst. head dept. mech. engring. Pratt Inst., Bklyn., 1922-31; founder, pres. Barclay Chem. Co., Watertown, Mass., 1931—. Pres. local PTA. With SATC, 1918. Mem. ASME, Charlton R.R. Assn. (gen. mgr. 1991), Pioneer Valley R.R. Assn., Rotary, Masons. Republican. Achievements include patent for use of amines to adjust pH level of reconditioned fuel oil; 2 patents pending for automatic fuel saver. Home: 21 Pleasant St Newton MA 02165-1230

BARD, ALLEN JOSEPH, chemist, educator; b. Dec. 18, 1933; m. Fran; children: Eddie, Sara. BSc in Chemistry summa cum laude, CCNY, 1955; MA in Chemistry, Harvard U., 1956, PhD in Chemistry, 1958. Instr. chemistry The U. Tex., Austin, 1958-60, asst. prof., 1960-62, assoc. prof., 1962-67, prof., 1967—, Jack S. Josey Professorship Energy Studies, 1980-82, Norman Hackerman Prof. Chemistry, 1982-85, Hackerman-Welch Regents Chair Chemistry, 1985—; cons. SACHEM, ClearFlow, E.I. DuPont, Electric Power Rsch. Inst., IGEN; lectr. numerous univs., 1969-93; mem. U.S. nat. com. IUPAC Nat. Rsch. Coun., 1983—, chair 1988-89, energy engring. bd. 1983-86, bd. chem. scis. and tech. 1982-87, co-chair 1985-87, nat. materials adv. bd. com. on electrochem. aspectr of energy conservatin and prod., 1985, com. on chem. scis. and ad hoc panel on DOE rsch. 1980-84, NAS, NRC liaison com. on high temp. sci. and tech., 1984; pres. Internat. Union Pure and Applied Chemistry, 1991-93; mem. adv. bd. Dept. Energy and Energy Rsch., panel on Cold Fusion, 1989, chem. adv. com. NSF, 1981-84, external adv. bd. Beckman Inst., 1989—. Author: Chemical Equilibrium; co-author: Electrochemical Methods; editor: Electroanalytical Chemistry, 18 vols., Eneyclopedia of the Electrochemistry of the Elements, 16 vols., (with others) Standard Potentials in Aqueous Solution; editorial and adv. bd. mem.: Journal of American Chemical Soc. (editor-in-chief 1982—), Electrochimica Acta (div. editor 1978-80), Dictionary of Modern Science and Technology, Encyclopedia of Scientific Instrumentation, Encyclopedia of Physical Science and Technology, Encyclopedia of Science and Technology, Yearbook of Science and Technology, Analytical Letters, Catalysis Letters, Chemical Instrumentation, Chemical Physics Letters, Critical Reviews in Analytical Chemistry, Journal of Photoacoustics, New Journal of Chemistry, Jour. Supercritical Fluids, Organic Thin Films and Surfaces. Recipient Outstanding Achievement in Fields of Analytical Chemistry award Eastern Analytical Symposium, 1990, Townsend Harris medal City Coll. N.Y., 1989, Edward Mack award Ohio State U., 1989, Math. and Phys. Scis. award N.Y. Acad. Scis., 1986, Docteur Honoris Causa award U. de Paris-VII, 1986, Bruno Breyer Meml. award Royal Australian Chem. Inst., 1981, Scientific Achievement award City Coll. N.Y., 1983, Sherman Mills Fairchild scholar Calif. Inst. Tech., 1977, Ward Medal in Chemistry, 1955, Luigi Galvani medal Societa Chimica Italiana, 1992. Fellow Electrochem. Soc. (Olin-Palladium medal 1987, Henry Linford award 1986, Carl Wagner Meml. award 1981); mem. AAAS (coun. del. 1992-95), Am. Chem. Soc. (G.M. Kosolapoff award 1992, Oseper award Cin. sect. 1987, Analytical Chemistry award 1988, Willard Gibbs award Chgo. sect. 1987, Fisher award in Alaytical Chemistry 1984, Harrison Howe award Rochester sect. 1980), Nat. Acad. Scis., Am. Acad. Arts and Scis. (award 1990), Internat. Soc. Electrochemists, Assn. Harvard Chemists, Sigma Xi. Achievements include research involving application of electrochemical methods to study of chemical problems and include investigations in electroanalytical chemistry, electron spin resonance, electro-organic chemistry, high resolution electrochemistry, electrogenerated chemiluminescence and photoelectrochemistry. Office: Dept Chemistry Lab Electrochem U Tex Austin TX 78712

BARD, JONATHAN ADAM, molecular biologist; b. Bayside, N.Y., June 8, 1958. BS, SUNY, Albany, 1980; PhD, U. Chgo., 1985. Staff scientist Synaptic Pharm. Corp., Paramus, N.J., 1990-92; group leader Synaptic Pharm. Corp., Paramus, 1992—. Named postdoctoral fellow Irvington House for Med. Rsch., N.Y., 1986-89, postdoctoral fellow NIH, 1986, 90. Mem. AAAS. Achievements include patents in field. Office: Synaptic Pharm Corp 215 College Rd Paramus NJ 07652

BARD, JONATHAN F., mechanical engineering and operations research educator; b. N.Y.C., Dec. 11, 1946; s. Soli and Jewel (Fenichel) B. BS, Rensselaer Poly. Inst., 1968; MS, Stanford U., 1969; DSc, George Washington U., 1979. Systems analyst The Mitre Corp., Bedford, Mass., 1969-72; project mgr. Booz, Allen & Hamilton, Bethesda, Md., 1972-75; program mgr. The Aerospace Corp., Washington, 1975-79; asst. prof. U. Mass., Boston, 1979-81, Northeastern U., Boston, 1981-83; assoc. prof. U. Calif., Berkeley, 1983-84; prof. dept. mech. engring. U. Tex., Austin, 1984—; cons. Office of Tech. Assessment, Washington, 1981-82, Jet Propulsion Lab., Pasadena, Calif., 1984-86, U.S. Postal Svc., Washington, 1989—, Tex. Instruments, Austin, 1986—, Am. Airlines, Dallas, 1985—. Author: Project Management: Engineering, Technology and Implementation, 1993; editor several jours.; contbr. more than 100 articles to profl. jours. Lady Davis fellow, 1989-90; NASA-Am. Soc. Engring. Faculty fellow, 1983, 85; Johnson Space Ctr. grantee, 1988-89; NSF grantee, 1980-83, others. Fellow Indsl. Properties Corp.; mem. IEEE (sr., pres. Austin chpt. 1985-86, 91—), Inst. Indsl. Engrs. (sr., Applications award 1993), Ops. Rsch. Soc. Am. (geog. sects. com. 1991-92), Inst. Mgmt. Scis. (treas. Boston chpt. 1981-82). Jewish. Achievements include work in hierarchical optimization and the decomposition of multilevel systems; research in the analysis and control of manufacturing systems. Office: U Tex ETC 5-160 Dept Mech Engring Austin TX 78712-1063

BARDA, JEAN FRANCIS, electronics engineer, corporate executive; b. Gargilesse, France, June 26, 1940; s. Ernest and Clothilde (Darmon) B.; m. Monique Marie Vianey, July 28, 1980; children: Nathalie, Xavier, Louis. Diploma, Lycee St. Louis in Paris, 1959; diploma in engring., Ecole Nationale de l'Aviation Civile, Paris-Orly, 1963. Registered electronic engr. Sect. mgr. Service Technique de la Navigation Aërienne Civil Aviation, Paris, 1965-70; tech. mgr. Utilisations Nouvelles de L'informatique et de la Television Sarl, Paris, 1970-83, Ateliers Techniques de Gargilesse Sarl, 1983-85; gen. mgr. Logiciel, Materiel et Applications de la Videographie Sarl (AVELEM) SA, Gargilesse, 1985—; tchr. math., Paris, 1967-70, electronics, Chambre de Commerce, Chateauroux, France, 1982-85. Patentee in field. Recipient 1st prize Conseil Gen. Indre, 1983, Innovation prize Region Centre, 1990. Mem. Soc. Motion Picture and TV Engrs. Home and Office: La Billardiere, 36190 Gargilesse France

BARDE, DIGAMBAR KRUSHNAJI, manufacturing executive; b. Kopra, India, Jan. 15, 1940; came to U.S., 1967; s. Krushnaji V. and Gaya (Keote) B.; m. Kumudini Kale, Dec. 5, 1970; children: Anupam, Soniya. BS in Chem. Engring., Indian Inst. Tech., Madras, 1967; MS in Chem. Engring., Mich. Tech. U., 1969. Registered professional engineer, Ill., Ala., Mass., Maine. Project engr. Copeland Systems, Oakbrook, Ill., 1977-78; project engr. HPD, Inc., Naperville, Ill., 1978-81, project mgr., then sr. project mgr., 1981-89; mgr. corp. engring. Lincoln (Maine) Pulp and Paper (subs. Eastern Pulp and Paper), 1989-91, v.p. engring., 1991—; v.p. engring. Ea. Fine Paper (subs. Eastern Pulp and Paper), Brewer, Maine, 1991—. Mem. TAPPI. Office: Eastern Pulp and Paper 100 University Dr Amherst MA 01002

BARDELL, PAUL HAROLD, JR., electrical engineer; b. Casper, Wyo., Feb. 2, 1935; s. Paul Harold and Grace Adalee (Hooser) B.; m. Dorothy Estelle Chandler, June 9, 1956 (div. 1974); children: Paul Harold III, Renée Grace; m. Adrienne Marie Pati, Jan. 10, 1976. BSEE, U. Colo., 1956; MSEE, Stanford U., 1962, PhD, 1965. With IBM, 1956-93; sr. tech. staff mem. IBM, Poughkeepsie, N.Y., 1982-91, fellow, 1991-93; chief scientist Virginin Laser Tech., Inc., Carmel, N.Y., 1993—; mem. indsl. adv. bd. elec. and computer engring. dept. U. Mass., Amherst, 1987-90; mem. curriculum coun. on elec. engring. Nat. Tech. U., 1990-92; ind. cons., 1993—. Co-author: Built-In Test for VSLI, 1987; contbr. articles to IEEE Transactions on Computers; contbr. articles, mem. editorial bd. Jour. Electronic Testing, Theory and Applications. Lt. USNR, 1956-58. Fellow IEEE; mem. AAAS, Am. Phys. Soc., N.Y. Acad. Sci. Achievements include patents for Industrial Circuits; for Built-In Self Test. Home and Office: 46 Wellington Dr Carmel NY 10512

BARDEN, ROBERT CHRISTOPHER, psychologist, educator, lawyer; b. Richmond, Va., June 7, 1954; s. Elliott Hatcher and Jane Elizabeth Cole (Ferris) B.; m. Robin Jones, Nov. 14, 1987. BA summa cum laude, U. Minn., 1976, PhD in Clin. Psychology, 1982; postgrad., U. Calif., Berkeley, 1977; JD cum laude, Harvard U., 1992. Lic. consulting psychologist, Minn., Tex. Project asst. NSF, 1978-79; intern in psychology VA Med. Ctr., Stanford Med. Ctr., Palo Alto, Calif., 1979-80; dir. psychology Internat. Craniofacial Surg. Inst., Dallas, 1980-87; corp., litigation, family and health law atty. Lindquist and Vennum, Mpls., 1992—; asst. prof. psychology So. Meth. U., Dallas, 1980-84; asst. prof., dir. child clin. psychology U. Utah, Salt Lake City, 1984-87, rsch. faculty dept. surgery, 1987-93; vis. faculty, asst. prof. psychology Gustavus Adolphus Coll., St. Peter, Minn., 1988; pres. Optimal Performance Systems, Inc., Cambridge, 1989—; cons. in field. Consulting editor Devel. Psychology, 1989; contbr. to profl. publs. Project dir. ch. community svc. projects, Mpls. and Cambridge, 1988—. Fellow NSF 1978, NIMH 1976, 77; Recipient Young Scholar award Found. for Child Devel., faculty scholar award W.T. Grant Found. 1987-89. Mem. ABA, APA, Soc. for Rsch. in Child Devel., Internat. Soc. Clin. Hypnosis, Harvard Law Sch. Soc. Law and Medicine, Lowell House Commons Rm. Harvard U., Sigma Xi, Phi Beta Kappa. Avocations: tennis, martial arts, mountain climbing, music. Home: 4025 Quaker Ln N Plymouth MN 55441-9999 Office: Lindquist and Vennum 4200 IDS Ctr Minneapolis MN 55402-2205

BARDIN, CLYDE WAYNE, biomedical researcher and developer of contraceptives; b. McCamey, Tex., Sept. 18, 1934; s. James A. and Nora Irene (Barnett) B.; m. Dorothy Kreiger, Aug. 11, 1978 (dec. Apr. 1985); m. Beatrice MacDonald, June 12, 1987; children: Charlotte E., Stephanie F. BA in Biology, Rice U., 1957; MS with honor, Baylor U., 1962, MD with honor, 1962; Docteur honoris cause, Université de Caen, France, 1990. Cert., licensed MD, Tex., Pa., N.Y. Resident in medicine N.Y. Hosp., N.Y.C., 1962-64; clin. assoc. NIH, Bethesda, Md, 1964-67, sr. investigator, 1967-70; assoc. prof. Milton S. Hershey Med. Ctr., Pa. State U., Hershey, 1970-72, prof. medicine, 1972-78; v.p. The Population Coun., N.Y.C., 1978—; adj. prof. Rockefeller U., N.Y.C., 1978—, Cornell Med. Ctr., N.Y.C., 1985—; cons. WHO, 1972-73; chmn. bd. sci. counselors NICHD, Bethesda, 1982-83; chmn. endocrine study sect. NIH, Bethesda, 1977-79; mem. nat. prostate cancer task force Nat. Cancer Inst., 1973-78. Editor 8 books on medicine and endocrinology; author over 400 sci. papers; mem. editorial bd. 14 sci. jours. Advisor internat. div. Ford Found., N.Y.C., 1975-79. Decorated Order of Comdr. of Lion (Finland); recipient Transatlantic medal Brit. Endocrine Socs., 1988; fellow Josiah Macy Jr. Found., 1976-77. Mem. Am. Assn. Physicians, Am. Soc. Clin. Investigation, Am. Soc. Andrology (coun., v.p., pres. 1984-89, Serono award 1984, Disting. Andrologist award 1992), Endocrine Soc. (coun. 1976-79, pres. 1993-94), Internat. Soc. Andrology (exec. coun. 1981-85), Internat. Assn. Axel Munthe Awards (bd. dirs. 1902-92), Internat. Com. Contraception Rsch. (chmn. 1978—), Inst. Medicine. Democrat. Avocations: music, sports, ballet, theater. Office: The Population Coun 1230 York Ave New York NY 10021-6341

BARER, SOL JOSEPH, biotechnology company executive; b. Windsheim, Germany, Apr. 20, 1947; came to U.S., 1949; s. Isaac and Hela Barer; m. Meri I. Barer, Aug. 17, 1969; children: Jennifer, Lori, Ilyssa, Joshua. BS, CUNY, Bklyn., 1968; MS, PhD, Rutgers U., 1974. Sr. rsch. chemist Celanese Co., Summit, N.J., 1974-78, supr. R&D, 1978-82, mgr. chem. R&D, 1982-84; dir. Chem. Systems, Tarrytown, N.Y., 1984-87; v.p. tech. Celgene (biotechnology co.), Warren, N.J., 1987-90, exec. v.p., gen. mgr., 1990—. Contbr. articles to profl. jours. NSF undergrad. fellow NSF, Bklyn., 1968, 70, NDEA grad. fellow Dept. Def., Rutgers U., 1970-72. Mem. Am. Chem. Soc., Am. Soc. Microbiology. Achievements include a wide variety of developments in chemical and biological industries. Home: 625 Westfield Ave Westfield NJ 07090 Office: Celgene 7 Powder Horn Dr Warren NJ 07059

BARES, WILLIAM G., chemical company executive; b. 1941; married. B.S. in Chem. Engring., Purdue U., 1963; M.B.A., Case Western Res. U., 1969. Process devel. engr. Lubrizol Corp., Wickliffe, Ohio, 1963-67; group leader, pilot plant Lubrizol Corop., Wickliffe, Ohio, 1967-71; asst. dept. head Lubrizol Corp., Wickliffe, Ohio, 1971-72, dept. head., 1972-78, asst. to pres., 1978, v.p., 1978-80, exec. v.p., 1980-82, pres., dir., 1982—, chief operating officer, 1987—. Office: Lubrizol Corp 29400 Lakeland Blvd Wickliffe OH 44092-2298*

BARFIELD, WALTER DAVID, physicist; b. Gainesville, Fla., Nov. 25, 1928; s. Walter Hugh and Myrtle (May) B. BS, U. Fla., 1950, MS, 1951; PhD, Rice U., 1961. Staff mem. Los Alamos (N.Mex.) Sci. Lab., 1951-63, Inst. for Def. Analyses, Washington, 1963-67; staff mem. Los Alamos Nat. Lab., 1968-90, guest scientist, 1990—. Author papers on photoionization cross sects., gamma rays from nuclear reactions, mesh generators for hydrodynamic calculations, others. Mem. Am. Phys. Soc. Home: 4647 Ridgeway Dr Los Alamos NM 87544-1963

BARGER, JAMES EDWIN, physicist; b. Manhattan, Kans., Dec. 28, 1934; s. Edgar Lee and Carolyn Marie (Grantham) B.; m. Mary Elizabeth Rupp, Aug. 24, 1957; children—Elaine Marie, Carolyn Ruth, James Rupp, Corinne

Elizabeth. B.S., U. Mich., 1957; M.S., U. Conn., 1960; Ph.D., Harvard U., 1964. Teaching asst. Harvard U., Cambridge, 1961-64; v.p. Bolt Beranek & Newman, Inc., Cambridge, Mass., 1965-75; chief scientist Bolt Beranek & Newman, Inc., 1975—; trustee Winchester Savs. Bank. Mem. Methods and Procedures Com., Town of Winchester, 1967-71; trustee Winchester Hosp., 1972—; corp. mem. Mt. Vernon House, 1979—. Served with USNR, 1957-63. NSF fellowship, 1960-64. Fellow AAAS, Acoustical Soc. Am.; mem. Marine Tech. Soc., Winchester Country Club, Cosmos Club, Tau Beta Pi, Pi Tau Sigma. Congregationalist (deacon). Home: 3 Lakeview Rd Winchester MA 01890-3801 Office: Bolt Beranek & Newman Inc 70 Fawcett St Cambridge MA 02138-1110

BARHAM, WARREN S., horticulturist; b. Prescott, Ark., Feb. 15, 1919; s. Clint A. and Hannah Jane (Sandusky) B.; m. Margaret Alice Kyle, Dec. 27, 1940; children: Barbara E., Juanita S., Margaret Ann, Robert W. BS in Agr., U. Ark., 1941; PhD, Cornell U., 1950. Grad. asst. in plant breeding Cornell U., Ithaca, N.Y., 1942-45; assoc. prof. horticulture N.C. State U., Raleigh, 1949-58; dir. raw material R & D Basic Vegetable Products, Inc., Vacaville, Calif., 1958-76; prof., head dept. hort. sci. Tex. A&M U., College Station, 1976-82; v.p. Castle & Cook Techniculture, Watsonville, Calif., 1982-84; dir. watermelon R & D Tom Castle Seed Co., Morgan Hill, Calif., 1984-86; pres. Barham Seeds Inc., Gilroy, Calif., 1987—; cons. Basic Vegetable Products, Inc., Vacaville, 1976-78, U.S. AID, Central Am., 1977, Egypt and U.S., 1980-82, Gentry Foods & Gilroy Foods, 1978—, Fed. Republic Germany Govt., Ethiopia, 1984; industry rep. adv. com. Onion Rsch. Program USDA, 1960-70. Contbr. articles to profl. jours. Bd. dirs., pres. Vacaville Sch. Bd., 1964-74. Sgt. USAF, 1942-45, ETO. Fellow Am. Soc. Hort. Sci. (pres. 1982, bd. dirs. 1979-83); mem. Sons in Retirement (elected), Rotary Internat. (bd. dirs. 1964). Achievements include development of 34 varieties and hybrids of onions, of 8 hybrids of watermelon, of 2 cucumber varieties. Home: 7401 Crawford Dr Gilroy CA 95020-5421

BARKER, CLYDE FREDERICK, surgeon, educator; b. Salt Lake City, Aug. 16, 1932; s. Frederick George and Jennetta Elizabeth (Stephens) B.; m. Dorothy Joan Bieler, Aug. 11, 1956; children—Frederick George II, John Randolph, William Stephens, Elizabeth Dell. BA, Cornell U., 1954, MD, 1958. Diplomate Am. Bd. Surgery. Intern Hosp. U. Pa., Phila., 1958-59, resident in surgery, 1959-64, fellow in vascular surgery, 1964-65; fellow in med. genetics U. Pa. Sch. Medicine, Phila., 1965-66, assoc. in surgery, 1964-68, assoc. in med. genetics, 1966-72; attending surgeon Hosp. U. Pa., Phila., 1966—; chief div. transplantation U. Pa. Sch. Medicine, Phila., 1966—, asst. prof. surgery, 1968-69, assoc. prof. surgery, 1969-73, prof. surgery, 1973—, J. William White prof. surg. research, 1978-82, chief div. vascular surgery, 1982—, Guthrie prof. surgery, 1982—, John Rhea Barton prof. surgery, 1983—, chmn. dept. surgery, 1983—; chief surgery Hosp. U. Pa., Phila., 1983—; dir. Harrison Dept. Surgery research U. Pa., Phila., 1983—; mem. immunobiology study sect. NIH; chmn. clin. practices U. Pa., 1987-89. Mem. editorial bd. Jour. Transplantation, 1977—, Clin. Transplantation, 1988—, Jour. Surg. Rsch., 1979-85, Jour. Diabetes, 1981-86, Archives of Surgery, 1987—, Surgery, 1991—, Cell Transplantation, 1991—, Postgrad. Gen. Surgery, 1991—; contbr. articles to profl. jours. and textbooks. Markle Found. Scholar, 1968-74; NIH grantee, 1974—; recipient Merit award NIH, 1987—. Fellow NAS (Inst. Medicine), ACS (com. Forum on Fundamental Surg. Problems 1983-88, vice chmn. 1987-88, pres. elect Phila. chpt. 1990-91, pres. 1991-92), Coll. Physicians Phila.; mem. AMA, Soc. Univ. Surgeons, Am. Surg. Assn. (recorder 1991—), Soc. Clin. Surgery (chmn. membership 1984-85), Halsted Soc. (chmn. membership 1984-85, v.p. 1985-86, pres. 1986-87), Surg. Biology Club II, Soc. Vascular Surgery, Internat. Cardiovascular Soc., Internat. Surg. Group (treas. 1988—), Transplantation Soc. (councilman 1978-84), Am. Soc. Transplant Surgeons (chmn. membership 1980-81, treas. 1988-91, pres. 1992-93), Assn. Acad. Surgery, Am. Diabetes Assn., Am. Soc. Artificial Internal Organs, Am. Fedn. Clin. Rsch., ACS (pres. elect Phila. chpt. 1990-91, pres. 1991-92), Phila. Acad. Surgery (program chmn. 1984-86, v.p. 1986-88, pres. 1988-89), Greater Del. Valley Soc. Transplant Surgeons (pres. 1978-80), Juvenile Diabetes Found. Clubs: Merion Cricket, Phila. Home: 3 Coopertown Rd Haverford PA 19041-1012 Office: U Pa Dept Surgical Rsch 313 Stemmler Bldg Philadelphia PA 19104

BARKER, DEE H., chemical engineering educator; b. Salt Lake City, Mar. 28, 1921; s. John Henry and Christina Selina (Heaton) Barker; m. Catherine Thompson, Apr. 24, 1945; children: DeeAnn, Lynn, Craig, Gary, Pamela. BS, U. Utah, 1948, PhD, 1951. Research engr. E.I. DuPont de Nemours & Co., Inc., Wilmington, Del., 1951-54; reactor engr. E.I. DuPont de Nemours & Co., Inc., Baton, S.C., 1954-59; prof. chem. engring. Brigham Young U., Provo, Utah, 1959—; cons. Brila Inst. Tech. & Sci., Rajasthan, India, 1966-78; Chonnam Nat. U. fellow, 1980-81, 87; prof. emeritus Brigham Young U., Provo, 1986—. Active Boy Scouts Am., Salt Lake City. With USN, 1944-46. Fellow Am. Inst. Chem. Engrs.; mem. Am. Soc. Engring. Educators, Nat. Council Engring. Examinations, Kiwanis. Avocations: photography, reading, scouting. Home: 1398 Cherry Ln Provo UT 84604-2851

BARKER, MICHAEL DEAN, nuclear engineer; b. Lampasas, Tex., Jan. 21, 1960; s. Hughby Frank Barker and Georgia Ann (Bales) Alsbrooks; m. Joy Ann Lively, May 19, 1984; children: Sarah Elizabeth, Michael Austin, Matthew Hamilton. B in Nuclear Engring., Ga. Inst. Technology, 1983. Coop. engring. student Ga. Power Co., Waynesboro, 1978-83; sr. nuclear engr. Ga. Power Co., Atlanta, 1983-88; sr. engr. SONOPCO, Birmingham, 1988-90; sr. project mgr. Inst. Nuclear Power Ops., Atlanta, 1991—. Candidate U.S. Congress, Rep. Party, Ala., 1990; elected mem. Rep. Exec. Com., Shelby County, Ala., 1989-91. Mem. Am. Nuclear Soc. Republican. Presbyterian. Office: Inst of Nuclear Power Ops 700 Galleria Pkwy Atlanta GA 30339

BARKER, ROBERT, biochemistry educator; b. Northumberland, Eng., Sept. 21, 1928; came to U.S., 1955, naturalized, 1966; s. Albert E. S. and Hannah M. (Ferry) B.; m. Kazuko Yamanaka, June 18, 1955; children: Hana, Robin. BA, U. B.C., Vancouver, Can., 1952, MA (B. C. Sugar Refineries scholar), 1953; PhD, U. Calif., Berkeley, 1958. Technician Fisheries Rsch. Bd. Can., 1953-55; Atlas Powder Co. postdoctoral fellow in chemistry Washington U., St. Louis, 1958-59; vis. scientist NIH, 1959-60; asst. prof. biochemistry U. Tenn., 1960-63, assoc. prof., 1963; assoc. prof. U. Iowa, 1963-67, prof., 1967-74; prof., chmn. dept. biochemistry Mich. State U., 1974-79; prof., dir. div. biol. scis. Cornell U., 1979-83, v.p. rsch. and advanced studies, 1983-84, provost, 1984-89, sr. provost, 1989-91; dir. Ctr. for the Environment Cornell U., Ithaca, N.Y., 1991—; vis. prof. U. Minn., 1968, Duke U., 1970-71; mem. Nat. Bd. Med. Examiners, 1967-79; cons. to govt. agys. Author: Organic Chemistry of Biological Compounds, 1971; contbr. numerous articles to profl. jours. Recipient Career Devel. award NIH, 1965-70. Mem. Am. Soc. Biol. Chemists, Am. Chem. Soc. (chmn. div. biol. chemistry 1978-79). Office: Cornell U 425 Hollister Hall Ithaca NY 14853

BARKLEY, LINDA KAY, chemical analyst, spectroscopist; b. Hicksville, Ohio, Feb. 8, 1953; d. Charles Jacob and Audrey Elizabeth (Williams) B. AAS in Chem. Engring. Tech., Pellissippi State Coll., 1988. Quality control lab. technician Gold Bond Bldg. Products, Gibsonburg, Ohio, 1975-78; lab. technician Martin Marietta Refractories, Woodville, Ohio, 1978-79; lab. analyst Galbraith Labs., Knoxville, Tenn., 1988-89; sr. lab. analyst Martin Marietta Energy Systems, Oak Ridge, Tenn., 1989—. Mem. Am. Soc. Cert. Engring. Technicians (treas. 1987-88, Presdl. award of Excellence 1988), Knoxville Gem and Mineral Soc. Republican. Nazarene. Office: Martin Marietta Energy Systems K25 Site Mail 7446 Oak Ridge TN 37830

BARKLEY, THEODORE MITCHELL, biology educator; b. Modesto, Calif., May 14, 1934; s. Theodore W. and Faye (Mitchell) B.; m. Carolyn Adair, Aug. 1955 (div. 1976); children: Theodore A., Stephen M., Elaine A.; m. Eileen K. Schofield, Feb. 7, 1981. BS, Kans. State U., 1955; MS, Oreg. State U., 1957; PhD, Columbia U., 1960. Instr. Occidental Coll., L.A., 1960-61; asst. prof. Kans. State U., Manhattan, 1961-67, assoc. prof., 1967-73, prof., 1973—, coord. Konza Prairie Rsch. Natural Area, 1988-90, 91-92. Author: Field Guide to the Weeds of Kansas, 1983; editor, co-author: Atlas Flora Great Plains, 1977, Flora of the Great Plains, 1986 (Gleason award 1987). Pres., sec. Manhattan Arts Coun., 1986-89. Mem. Am. Soc. Plant Taxonomists, Internat. Assn. Plant Taxonomists, Bot. Soc. Am., AAAS, Torrey Bot. Club, New Eng. Bot. Club. Home: 1414 McCain Ln Manhattan KS 66502-4621 Office: Kans State Univ Ackert Hall Div of Biology Manhattan KS 66506

BARLOW, HORACE BASIL, physiologist; b. Chesham Bois, Eng., Dec. 8, 1921; s. James Alan and Emma Nora Barlow; m. Miranda Weston-Smith, June 28, 1980; children: Oscar Hugh, Ida Lucy, Pepika Elizabeth; children by previous marriage: Rebecca Nora, Natasha Helen, Naomi Jane, Emily Anne. BA, Cambridge (Eng.) U., 1943; MD, Harvard U., 1946; MBB, Univ. Coll. Hosp. of London, 1947. Fellow Trinity Coll., 1950-54; demonstrator, asst. dir. research physiology lab. King's Coll., 1954-63; Royal Soc. research prof. Cambridge U., 1973-87; prof. physiol. optics U. Calif., Berkeley, 1963-73. Editor Jour. of Physiology, 1972-77; contbr. articles to profl. jours. Recipient Edward D. Tillyer award Optical Soc. Am., 1992. Mem. Physiol. Soc., Exptl. Psychology Soc., Brain Research Assn. Avocations: walking, music. Home: 9 Selwyn Gardens, Cambridge CB3 9AX, England Office: U Cambridge Physiol Lab, Downing St, Cambridge CB2 3EG, England

BARLOW, JOEL WILLIAM, chemical engineering educator; b. Burbank, Calif., May 2, 1942; married; 2 children. BS, U. Wis., 1964, MS, 1965, PhD in Chem. Engring., 1970. Fellow materials and thermodynamics Washington U., 1968-70; research engr. in plastics Union Carbide Corp., Bound Brook, N.J., 1970-73; from asst. prof. to assoc. prof. U. Tex., Austin, 1973-83, prof. chem. engring., 1983-90, Z.D. Bonner prof. chem. engring., Cullen Trust for Higher Edn. Endowed Professorship Number 5, 1984—. Mem. Am. Chem. Soc., Am. Inst. Chem. Engrs., Soc. Plastics Engrs. Rsch. interests include polymer processing and physics, thermodynamics of polymer blends, polymer flammability characterization, reaction injection molding, selective laser sintering. Office: Univ Tex Dept Chem Engring CPE 3.466 Austin TX 78712

BARLOW, KENT MICHAEL, mechanical engineer; b. Grosse Pointe Park, Mich., Mar. 16, 1935; s. Jewett Dunton and Bernadine Margaret (Culkins) B.; m. Patricia Ann Limmex, Sept. 1, 1962; children: Daniel J., Carole J., Timothy P., Renée B., Andrew M., Jonathan R. BS, Mich. Coll. Mining and Tech., 1957. Registered profl. engr., Wis. Jr. engring. asst. Madison (Wis.) Gas and Electric Co., 1961-62, results engr., 1962-64, asst. supt., 1964-66, supt., 1966-74, mgr., 1974-81, v.p., 1981-87, sr. v.p., 1988—; treas. Wis. Ctr. for Demand Side Rsch., Madison, 1988—. Author papers. Mem. Dane County Recycling Task Force, Madison, 1986-87. 1st lt. U.S. Army, 1959-61. Mem. ASME, Wis. Soc. Profl. Engrs., Madison Downtown Kiwanis. Achievements include development of first successful combustion of municipal waste (refuse derived fuel) in an investor owned utility boiler. Home: 5909 Driftwood Ave Madison WI 53705 Office: Madison Gas and Elec Co PO Box 1231 133 S Blair St Madison WI 53701-1231

BARLOW, NADINE GAIL, planetary geoscientist; b. La Jolla, Calif., Nov. 9, 1958; d. Nathan Dale and Marcella Isabel (Menken) B.; m. Michael Ewing Zolensky, Apr. 23, 1989. BS, U. Ariz., 1980, PhD, 1987. Instr., planetarium lectr. Palomar Coll., San Marcos, Calif., 1982; grad. rsch. asst. U. Ariz., Tucson, 1982-87; postdoctoral fellow Lunar and Planetary Inst., Houston, 1987-89; NRC assoc. NASA/Johnson Space Ctr., Houston, 1989-91, vis. scientist, 1991-92, support scientist exploration programs office, 1992; vis. scientist Lunar and Planetary Inst., Houston, 1992—; assoc. prof. U. Houston, Clear Lake, 1991—; co-dir. intern program Lunar and Planetary Inst., 1988-89. Editor (slide set) A Guide to Martian Impact Craters, 1988; assoc. editor Encyclopedia of Earth Sciences; contbr. articles to profl. jours. Named among Outstanding Women and Ethnic Minorities Engaged in Sci. and Engring., Lawrence Livermore Nat. Lab., 1991. Mem. AIAA, AAUW (pres. Clear Lake chpt. 1991-93, program v.p. 1993—, v.p. interbr. coun. 1990-91, chmn. Tex. task force on women and girls in sci. and math, 1991-92, dir. state pub. policy, 1992—, Tex. Woman of Yr. 1992, chmn. steering com. Tex. ednl. equity roundtable, 1991), Am. Astron. Soc. (press officer div. planetary scis, 1993—, status of women in astronomy com. 1987-90), Am. Geophys. Union, Geol. Soc. Am., Assn. Women in Sci., Assn. Women Geoscientists. Achievements include research and compilation of primary data source on 42,283 impact craters on Mars. Office: Lunar and Planetary Inst 3600 Bay Area Blvd Houston TX 77058

BARLOW, ROBERT DUDLEY, medical biochemist; b. New Romney, Kent, Eng., July 21, 1954; s. Dudley Brian and Patricia Alice (Butters) B.; m. Iris Heather Wheeler, June 29, 1974; children: Michelle Rose, Charlotte Patricia. BSc with honors, U. Birmingham (Eng.), 1976, MSc, 1978; PhD, U. London, 1988. Cert. clin. biochemist; chartered chemist. Biochemist dept. ob.-gyn. U. Oxford (Eng.), 1978-84; biochemist dept. environ. and preventive medicine St. Bartholomew's Hosp. Med. Coll., London, 1984-88; prodn. mgr. European Diagnostic Products Corp., Witney, Eng., 1988-90, divsn. mgr., 1990—; cons. Diagnostic Products Corp., L.A., 1985-88, European Media Resource Svc., Ciba Found., London, 1991—. Contbr. articles to major sci. jours. Regional coord. Dr. Billy Graham's Mission Eng., 1989. Mem. Royal Soc. Chemistry (C. Chem. MRSC award 1990), Assn. Clin. Biochemists. Anglican. Avocations: squash, swimming, music, reading. Home: 82 Dinerth Rd, Colwyn Bay LL28 4YH, England Office: Euro/DPC Ltd, Glyn Rhonwy, North Wales LLSS 4EL, England

BARMACK, NEAL HERBERT, neuroscientist; b. N.Y.C., Aug. 23, 1942; married, 1964; 2 children. BS, U. Mich., 1963; PhD, U. Rochester, 1970. Asst. lectr. psychology U. Rochester, 1968-69, rsch. assoc. to sr. rsch. assoc. neurophysiology dept ophthalmology, 1969-75, assoc. scientist, 1975-80; sr. scientist Neurological Sci. Inst., Good Samaritan Hosp. & Med. Ctr., Portland, 1980-81; assoc. prof. biological sci. U. Conn., 1981-82; with R S Dow Neurological Sci. Inst., Good Samaritan Hosp. & Med. Ctr., 1982—. Mem. Soc. Neuroscience, Am. Physiological Soc., Assn. Rsch. Vision, Internat. Brain Rsch. Organization, Nat. Eye Inst. Achievements include research in Neural control of eye movements; plasticity of reflexive eye movements; the cellular and biochemical basis of cerebellar modulation of reflex function. Office: R S Dow Neurological Sci Instit Good Sammaritan Hospital & Med 1120 NW 20th AVe Portland OR 97209*

BARNARD, MAXWELL KAY, inventor, computer systems executive; b. Hackensack, N.J., Mar. 8, 1946; s. Jack D. and Carolyn (Mastenbrook) B.; m. Kathy A. Luch, Oct. 11, 1982. BS in Elec. Engring., U. Wash., 1968. Systems engr. Boeing Co., Seattle, 1966-71; rsch. engr. VA Hosp., Seattle, 1972-75; scientific programmer Stanford U., Palo Alto, Calif., 1976-78; chmn. Barnard Systems Co., Port Townsend, Wash., 1978—; tchr. Portland, Oreg., 1980-81, Seattle C.C., 1971-72. Author: Without Me You're Nothing, 1980. Mem. Am. GO Assn. Achievements include patents for Inertial Energy Interchange System, 1985, Computer Programming System, 1987, Self-Fairing Wind Vane, 1987. Home and Office: 1933 San Juan Ave Port Townsend WA 98368

BARNARD, PETER DEANE, dentist; b. Canberra, ACT, Australia, Apr. 25, 1932; s. Colin and Elsie Joyce (Deane) B.; children: Pauline Elizabeth, Peter Hahn. BDS, U. Sydney (N.S.W., Australia), 1954, MDS, 1967, DDSc, 1991; MPH, U. Mich., 1956. Dental officer Inst. Dental Rsch., Sydney, 1954-55; rsch. assoc. U. Rochester (N.Y.), 1957-58; pvt. practice dentistry London, 1958-61; sr. lectr. U. Sydney, 1961-69, assoc. prof., 1970—, head dept. preventive dentistry, 1989—; dir. preventive dentistry Westmead (Australia) Hosp. Dental Clin. Sch., 1984—; cons. WHO, India, Indonesia, Malaysia, Geneva, 1972, 73, 74, 75, 78, 81, 83, 84, South Pacific Commn., Samoa, Noumea, 1976, 78, 80, 90. Contbr. articles to profl. jours. Lt. col. Dental Corps, Australian mil., 1982-87. Fellow APHA, Internat. Coll. Dentists, Royal Australasian Coll. Dental Surgeons; mem. Australian Dental Assn. (cons. 1978—, Meritorious Svc. award 1985), Fedn. Dentaire Internat. (life, vice chair pub. dental health svc. 1970-76), Asian Pacific Dental Fedn. (life, chair pub. dental health 1977-79). Office: U Sydney C 24, Sydney NSW 2006, Australia

BARNARD, WILLIAM HOWARD, JR., biologist, educator; b. Montclair, N.J., Apr. 26, 1945; s. William Howard and Jean (Ackerly) B.; married, June 22, 1968; children: Craig, Ethan. BA, Franklin Coll., 1968; PhD, Ind. U., 1979. Asst. prof. dept. biology Norwich U., Northfield, Vt., 1974-81, assoc. prof., 1981-92, prof., 1992—, Dana prof., 1988; dir. New Eng. Hawk Watch; mem. Vt. Endangered Species Com., Vt. Fragile Areas Com., Sci. Adv. Group on Reptiles and Amphibians, Sci. Adv. Group on Birds. Editor: Vermont Hawk Watch Results; contbr. articles to profl. publs. Grantee Vt. Fish and Wildlife Dept., 1988, 90. Mem. Ecol. Soc. Am., Am. Ornithologists' Union, Assn. Field Ornithologists, Wilson Ornithol. Soc., Cooper Ornithol. Soc., Hawk Migration Assn. N.Am., Northeastern Bird-Banding Assn., Ea. Bird-Banding Assn. Achievements include research in blood parasites of birds in central Vermont, hawk migrating in New England, Atlas of Vermont Reptiles and Amphibians, Gray Jay population dynamics, species distribution of turtles in Vermont. Home: Rte 2 Box 110 Northfield VT 05663 Office: Dept Biology Norwich Univ Northfield VT 05663

BARNARD, WILLIAM MARION, psychiatrist; b. Mt. Pleasant, Tex., Dec. 17, 1949; s. Marion Jaggers and Med (Cody) B. BA, Yale U., 1972; MD, Baylor U., 1976. Diplomate Am. Bd. Psychiatry and Neurology. Resident NYU/Bellevue Med. Ctr., 1976-79; fellow L.I. Jewish/Hillside Med. Ctr., 1979-80; chief, liaison, consultation psychiatrist Queens (N.Y.) Med. Ctr., 1980-83; liaison, consultation psychiatrist Mt. Sinai Med. Ctr., N.Y.C., 1983-84; clin. asst. prof., adminstrn. and emergency psychiatrist VA. Med. Ctr., NYU Med. Sch., N.Y.C., 1984-87; pvt. practice Pasadena, Calif., 1987—; chief psychiat. svc. Las Encinas Hosp., Pasadena, 1989, chief staff, 1990, med. dir. gen. adult. psychiat. svc., 1990-92, asst. med. dir., 1992; med. dir. CPC Alhambra Hosp., Rosemead, Calif., 1992—. Chmn. mental health com. All Saints AIDS Svc. Ctr., Pasadena, 1990—, bd. dirs., 1991—; bd. dirs. Pasadena Symphony, 1989—, Whiffenpoof Alumni, New Haven, 1991—. Wilson scholar Yale U., 1973. Mem. Am. Psychiat. Assn., NYU-Bellevue Psychiat. Assn., Am. Soc. Addiction Medicine, Amateur Comedy Club, Met. Opera Club, Yale Club N.Y.C. Republican. Episcopalian. Avocations: singing, fitness, bicycling. Office: 2810 E Del Mar Blvd Pasadena CA 91107

BARNES, BETTY RAE, counselor; b. Wichita, Kans., June 24, 1932; d. Henry Charles and Vivian Augusta (Lamberth) Archer; m. Orland Eugene Barnes, Mar. 18, 1953; children: Terry Lee, Steven Gregory. BA, Our Lady of the Lake, San Antonio, 1986, MS in Counseling Psychology, 1989. Cert. profl. sec., lic. profl. counselor; lic. marriage and family therapist. Adminstrv. asst. S.W. Rsch. Inst., San Antonio, 1975—; counselor Community Clinic, Inc., San Antonio, 1989—, counseling coord., 1991—; counselor Community Counseling Ctr., Our Lady of the Lake U., San Antonio, 1989-91. Recipient Outstanding Achievement award Sch. Bus. and Pub. Adminstrn., Our Lady of the Lake U., 1984. Mem. AACD, San Antonio Mus. Assn., Delta Mu Delta. Avocations: piano, reading, collecting pewter. Office: Community Clinic Inc 210 W Olmos Dr San Antonio TX 78212-1956

BARNES, CHARLES D., neuroscientist, educator; b. Carroll, Iowa, Aug. 17, 1935; s. Jack Y. and Gladys R. (Beckwith) B.; m. Leona Gladys Wohler, Sept. 8, 1957; children: Tara Lee, Teagen Yale, Kalee Meion, Kyler Alan. BS, Mont. State U., 1958; MS, U. Wash., 1961; PhD, U. Iowa, 1962. Asst. prof. dept. anatomy and physiology Ind. U., Bloomington, 1964-68, assoc. prof., 1968-71; prof. Ind. U., Terre Haute, 1971-75; prof., chmn. dept. physiology Tex. Tech. U. Health Scis. Ctr., Lubbock, 1975-83; chmn. dept. vet. and comparative anatomy, pharmacology and physiology Wash. State U., Pullman, 1983—; vis. scientist Instituto di Fisiologia Umana della U. di Pisa, Italy, 1968-69. Author: (with L. Eltherington) Drug Dosage in Laboratory Animals, 1964, 2d edit., 1973, (with C. Kircher) Readings in Neurophysiology, 1968, (with Davies) Regulation of Ventilation and Gas Exchange, 1978, (with Hughes) Neural Control of Circulation, 1980, (with Orem) Physiology in Sleep, 1980, (with Crass) Vascular Smooth Muscle, 1982, (with McGrath) Air Pollution-Physiologic Effects, 1982, Brainstem Control of Spinal Cord Function, 1984, (with Janssen) Cardiovascular Shock, 1985, (with Ritter and Ritter) Feeding Behavior: Neural and Humoral Controls, 1986, (with Harding, Wright and Speth), Angiotensin and Blood Pressure Regulation, 1986, (with Kalivas) Sensitization in the Nervous System, 1988, (with Borchard and Elthrington) Drug Dosage in Laboratory Animals, 1990, (with Pompeiano) Neurobiology of the Locus Coeruleus, 1991, (with Johnston) Brain-Gut Peptides and Reproductive Function, 1991, (with Meyers) Platelet Amine Storage Granule, 1992. Recipient NIH Career Devel. award, 1967-72. Mem. AAAS, Am. Inst. Biol. Scis., Am. Physiol. Soc., Am. Soc. Pharmacology and Exptl. Therapeutics, Assn. Chmn. Depts. Physiology, Assn. Anat. Chmns., Internat. Brain Research Orgn., Nat. Caucus Basic Biomed. Sci. Chairs, Soc. Exptl. Biology and Medicine, Soc. Neuroscis., Soc. Gen. Physiologists, Am. Assn. Anatomists, Am. Assn. Vet. Anatomists, Am. Soc. Vet. Physiology and Pharm., Sigma Xi. Office: Wash State U Dept VCAPP Pullman WA 99164-6520

BARNES, DAVID BENTON, school psychologist; m. Cheryle Kirkland; children: David, Matthew, Bryan. BSc with honors, Springfield Coll., 1958; MEd, U. Maine, 1962; EdD, Rutgers U., 1970. Cert. tchr., Maine. Asst. football coach Boston U., 1964-66; dir. Counselling Ctr., asst. prof. edn. Acadia U., Wolfville, N.S., Can., 1966-71, acting dean Sch. Edn., 1969-70, assoc. prof., 1970; chief psychologist Fundy Med. Health Ctr., Wolfville, 1971-73; psychologist Atlantic Child Guidance Ctr., 1974-77; supr. spl. svcs. Cape Breton County Sch. Bd., 1977-82, Lunenburg County Dist. Bd., 1982-87; sch. psychologist, spl. edn. adminstr. Bennington-Rutland Supervisory Union, Vt., 1987-90; psychologist N.W. Psycho-Ednl. Program, Rome, Ga., 1990—; mem. adminstrv. task force for spl. edn. N.S. Dept. Edn.; mem. N.S. Adv. Coun. on Tchr. Edn.; mem. Met. Mental Health Planning Bd.; mem. Aqua Percept Nat. Adv. Bd.; founder Camp Recskill; founder, bd. dirs. Cape Breton Child Guidance Ctr.; adj. faculty mem. U. Coll. of Cape Breton, Acadia U., Walden U. Co-author: Special Educator's Survival Guide: Practical Techniques and Materials for Supervision and Instruction. Grantee Can. Govt., 1978-87, Province of N.S., 1978-79, N.S. Tchrs. Union, 1980-81, Internat. Youth Yr., 1985, Donner Found., Laidlaw Found., Windsor Found. Mem. Can. Univ. Counselors Assn., Atlantic Inst. Edn. (steering com. for counselor edn.), Assn. Profl. Staffs of Community Mental Ctrs. (sec.-treas.), Can. Assn. for Children With Learning Disabilities (v.p.), Provincial Assn. for Children with Learning Disabilities (bd. dirs.), N.S. Mental Health Assn. (bd. dirs.), Dartmouth Mental Health Assn. (v.p.), Cape Breton Mental Health Assn. (bd. dirs.), Coun. Exceptional Children, Assn. Psychologists N.C., Nat. Assn. Sch. Psychologists. Home: 309 Rigas Rd Americus GA 31709

BARNES, FRANK STEPHENSON, electrical engineer, educator; b. Pasadena, Cal., July 31, 1932; s. Donald Porter and Thedia (Schellenberg) B.; m. Gay Dirstine, Dec. 17, 1955; children: Stephen, Amy. BS, Princeton U., 1954; MS, Stanford U., 1955, PhD, 1958. Fulbright prof. Coll. Engring., Baghdad, Iraq, 1957-58; rsch. assoc. Colo. Rsch. Corp., Broomfield, 1958-59; assoc. prof. U. Colo., Boulder, 1959-65, prof. dept. elec. engring., 1965—; chmn. dept. U. Colo., 1964-81, faculty rsch. lectr., 1965, acting dean Coll. Engring. and Applied Sci., 1980-81; mem. bd. advisors Columbine Ventures. Regional editor Electronics Letters of Brit. Inst. Elec. Engrs, 1970-75. Bd. dirs. Accreditation Bd. Engring. and Tech., 1980-82. Recipient Cert. of Merit, Internat. Communications Assn., 1989. Fellow AAAS, IEEE (editor Student Jour. 1967-70, mem. G-Ed Adcom 1970-77, v.p. publ. activities 1974-75, pres. device soc. 1974-75, mem. ednl. activities bd. 1976-82, editor IEEE Transactions on Edn. 1988—, mem. press bd. 1989-90, ednl. activities bd., cert. of merit, Centennial medal), Soc. Lasers in Medicine, Engrs. Coun. Profl. Devel. (dir. 1976-82, chmn. com. on advanced level accreditation 1976-78), Bioelectromagnetics Soc. (bd. dirs. 1982-84), Engring. Info. (bd. dirs. 1984-90). Home: 225 Continental View Dr Boulder CO 80303-4516

BARNES, HUBERT LLOYD, geochemistry educator; b. Chelsea, Mass., July 20, 1928; s. George Lloyd and Mary Eileen (MacPherson) B.; m. Mary Talbot Westergaard; children—Roy Malcolm, Catherine Patricia. B.S., MIT, 1950; Ph.D., Columbia U., 1958. Resident geologist Peru Mining Co., Hanover, N.Mex., 1950-52; lectr. geology Columbia U., N.Y.C., 1952-54; postdoctoral fellow Geophys. Lab. Carnegie Inst., Washington, 1956-60; prof. Pa. State U., University Park, 1960—, dir. ore deposits rsch. sect., 1969—; vis. prof. Mineralogy-Petrology Inst., Heidelberg, 1974, Academia Sinica, 1983; Crosby lectr. MIT, Cambridge, 1983; mem. geophysics rsch. bd. NRC, 1976-80; mem. U.S. Nat. Com. on Geology, 1983-87; cons. numerous corps.; dir. NATO Advanced Study Inst., Salamanca, 1987; gen. chmn. Goldschmidt Conf., Balt., 1988; chmn., sec. Internat. Symposium on Hydrothermal Reactions, Pa. State U., 1985. Author: Uranium Prospecting, 1956. Editor: Geochemistry of Hydrothermal Ore Deposits, 1967, 79; co-editor: Hydrothermal Experimental Techniques, 1987. N.L. Britton scholar, 1955-56; Guggenheim fellow, 1966-67; recipient Sr. Humboldt prize Humboldt Found. West Germany, 1988; named Disting. Prof. Geochemistry

Pa. State U., 1990; Can. Inst. Mining and Metallurgy lectr., 1969, C.F. Davidson lectr., St. Andrews, Scotland, 1971. Fellow Mineral Soc. Am., Geol. Soc. Am.; mem. Geochem. Soc. (councillor 1970-73, v.p. 1983, pres. 1984-85), Soc. Econ. Geologists (councilor 1981-84, Thayer Lindsley lectr. 1980-81), Am. Geologic Inst. (governing bd. 1981-83), U.S. Nat. Geochemistry Com. (chmn. 1983-85). Home: 213 E Mitchell Ave State College PA 16803-3655 Office: Pa State U 235 Deike Bldg University Park PA 16802

BARNES, JOHN MAURICE, plant pathologist; b. Washington, Apr. 22, 1931; s. Lester George and Lucille Jane (Scott) B.; m. Pauline Marie Boari, Sept. 7, 1957; children: John A., Charlotte L., Stephanie L., Brian G. BS, U. Md., 1954; MS, Cornell U., 1957, PhD, 1960. Biologist Resources Rsch./Hazelton Lab., Leesburg, Va., 1961-67; prin. plant pathologist U.S. Dept. Agr., Washington, 1967-91; sr. scientist Coun. on Environ. Quality, Washington, 1991—; profl. soc. rep. Com. on Life Scis. NAS, Washington, 1988—. Assoc. editor Acad. Press, Ltd., 1987—; co-editor: Bridging Interdisciplinary Strengths, 1986. Canvasser Am. Diabetes Assn., Montgomery County, Md., 1992. Recipient Certs. of Merit U.S. Dept. Agr., Washington, 1972, 84, 85, 88. Mem. AAAS, Am. Phytopath. Soc., Soc. Nematologists. Democrat. Roman Catholic. Achievements include initiation of interdisciplinary rsch. and planning focus in U.S. Dept. Agr., thrusts to establish nat. ultraviolet monitoring network. Home: 4406 Morgal St Rockville MD 20853

BARNES, MELVER RAYMOND, chemist; b. nr. Salisbury, N.C., Nov. 15, 1917; s. Oscar Lester and Sarah Albertine (Rowe) B. AB in Chemistry, U. N.C., 1947; D of Physics (hon.), World U., 1983; DSc in Chemistry (hon.), Assoc. Univs., 1987, PhD in Chemistry (hon.), 1990; PhD in Chemistry (hon.), Albert Einstein Internat. Acad. Found. and Associated Univs., 1990. Chemist Pitts. Testing Labs., Greensboro, N.C., 1948-49, N.C. State Hwy. and Pub. Works Commn., Raleigh, 1949-51, Edgewood (Md.) Arsenal, 1951-61, Dugway (Utah) Proving Ground, 1961-70. Recipient Albert Einstein Bronze medal, 1988, Albert Einstein Acad. Found. Cross of Merit, 1992. Mem. AAAS, Am. Statistical Assn., Am. Chem. Soc., Am. Phys. Soc., Am. Math. Soc., Soc. Indsl. and Applied Math, Math. Assn. Am. Home: RR 1 Box 646 Linwood NC 27299-9753

BARNES, PAUL RANDALL, electrical engineer; b. Lexington, Ky., Jan. 14, 1943; s. Joel R. and Pauline (Whitlock) B.; m. Joan P. Shames; children: Christopher R., Matthew J. BSEE, U. Ky., 1967; MSEE, U. N.Mex., 1971. Registered profl. engr., Tenn. Mem. rsch. staff Oak Ridge (Tenn.) Nat. Lab., 1972—. Contbr. articles on nuclear elecgtromagnetic pulse and solar energy to profl. publs. Capt. USAF, 1968-72. Mem. IEEE (sr.), Internat. Electrotech. Commn. (mem. subcom.). Achievements include patent on reflective insulating blinds. Office: Oak Ridge Nat Lab PO Box 2008 Oak Ridge TN 37831-6070

BARNES, PETER JOHN, food scientist; b. Blackburn, Lancashire, U.K., Feb. 3, 1947. BSc in Applied Biology with honors, Liverpool (Eng.) Poly., 1972, PhD, 1975. Trainee Boots Chemists, Ltd., Darwen, U.K., 1964-66; technician Lion Brewery, Blackburn, U.K., 1966-68; postdoctoral fellow Brewing Rsch. Found., Redhill, U.K., 1975-76, U. Bristol, U.K., 1976-78; head of food chemistry RHM Rsch. & Engring., Ltd., High Wycombe, U.K., 1978-88; prin. cons. P.J. Barnes & Assocs., High Wycombe, U.K., 1988-90; sr. editor Elsevier Advanced Tech., Oxford, Eng., 1990—; mem. Soc. Chem. Industry, Oils and Fats Group, U.K., 1973-90, Inst. of Brewing, U.K., 1972-88. Author and editor: (book) Lipids in Cereal Technology, 1983; editor Food Safety & Security, 1990—, Food, Cosmetics and Drug Packaging, 1990—, Lipid Technology, 1990—. Mem. Inst. Food Technologists, Am. Oil Chemists Soc. Achievements include 40 scientific research papers and reviews. Office: Elsevier Advanced Tech, Mayfield House 256 Banbury Rd, Oxford OX2 7DH, England

BARNES, ROBERT F., agronomist; b. Estherville, Iowa, Feb. 6, 1933; s. Chester Arthur and Pearl Adella (Stoelting) B.; m. Bettye Jeanne Burrell, June 25, 1955; children: Bradley R., Rebecca L. Reinalda, Roberta K. Nixon, Brian L. AA, Estherville Jr. Coll., 1953; BS, Iowa State U., 1957; MS, Rutgers U., 1959; PhD, Purdue U., 1963. Rsch. agronomist USDA-Agrl. Rsch. Svc., West Lafayette, Ind., 1959-70; lab. dir. USDA-Agrl. Rsch. Svc., University Park, Pa., 1970-75; staff scientist nat. program staff USDA-Agrl. Rsch. Svc., Beltsville, Md., 1975-79; assoc. dep. adminstr. So. region USDA-Agrl. Rsch. Svc., New Orleans, 1979-84, dep. adminstr. So. region, 1984-86; exec. v.p. Am. Soc. of Agronomy, Madison, Wis., 1986—, also fellow; asst. prof. Purdue U., West Lafayette, 1963-66; assoc. prof. Pa. State U., University Park, 1966-70, adj. prof., 1970-75; pres. Internat. Grassland Congress, Lexington, Ky., 1981. Editor: Forages, 1985; contbr. articles to profl. jours. With U.S. Army, 1953-55, Germany. Recipient H.S. Stubbs Meml. Lecture award Tropical Grassland Soc., Brisbane, Australia, 1984, Henry A. Wallace award Iowa State U., 1991. Fellow AAAS, Crop Sci. Soc. Am. (pres. 1984-85); mem. Am. Forage and Grassland Coun. (medallion 1981), Grazing Lands Forum (pres. 1986-87), Forage and Grassland Found. (pres. 1993—). Avocations: walking, reading. Office: Am Soc of Agronomy 677 S Segoe Rd Madison WI 53711-1048

BARNES, RONALD FRANCIS, mathematics educator; b. Fillmore, N.Y., June 4, 1942; s. Ramon Charles and Frances (Coughlin) B. BS, St. Bonaventure U., 1964; MS, Syracuse U., 1966, PhD, 1972. Asst. prof. math. SUNY, Brockport, 1971-77; assoc. prof. math. U. Houston-Downtown Campus, 1977-90, prof. math., 1990—, lectr. in stats. Hilton Coll. Hotel Mgmt., 1990—; cons. to oil and gas industries, Dallas, 1992—; lectr. Chinese Assn. for Sci. and Tech., Shanghai, Nanjing, Hangzhou, People's Republic China, 1990; speaker in field; math. cons. Guangxi Normal U., Guilin, People's Republic China, 1992. Co-editor conf. proc. in field; reviewer Coll. Math. Jour., 1984-86; mem. editorial bd. UMAP Jour. Applications in Math., 1990—; author: Statistics Study Guide, 1985; contbr. articles, reports to profl. publs. Judge Houston Area Sci. Fairs, 1989—. Grantee SUNY, 1972, 74, Exxon Ednl. Found., 1990—; NASA summer faculty fellow, 1989, 90. Mem. Am. Statis. Assn. (Houston chpt., v.p. 1984-85, pres. 1985-86, nat. coun. chpts. rep. 1986-88, sec.-treas. statis. cons. sect. 1988—, instnl. rep. comap consortium coun. 1988—), Math. Assn. Am. (vis. lectr., bd. lectrs. 1985-90, vis. cons. 1990—, sect. coord. student chpts. 1988—, sect. coord. Tex. sect. 1988—). Avocations: jogging, marathons, tennis. Office: U Houston Downtown 1 Main St Houston TX 77002

BARNESS, LEWIS ABRAHAM, physician; b. Atlantic City, N.J., July 31, 1921; s. Joseph and Mary (Silverstein) B.; m. Elaine Berger, June 14, 1953 (dec. Jan. 1985); children: Carol, Laura, Joseph; m. Enid May Fischer Gilbert, July 5, 1987; stepchildren: Mary, Elizabeth, Jennifer, Rebecca. A.B., Harvard U., 1941, M.D., 1944; M.A. (hon.), U. Pa., 1971. Intern Phila. Gen. Hosp., 1944-45; resident Children's Med. Center, Boston, 1947-50; asst. chief, then chief dept. pediatrics Phila. Gen. Hosp., 1951-72; vis. physician U. Pa. Hosp., 1952-57, acting chief, then chief, 1957-72; mem. faculty U. Pa. Sch. Medicine, 1951-72, prof. pediatrics, 1964-72; chmn. dept. U. So. Fla. Med. Sch., Tampa, 1972-88, prof. pediatrics, 1988—; vis. prof. Univ. Wis., 1987-92, prof. emeritus, 1993—. Author: Pediatric Physical Diagnosis Yearbook, edits. 1-6, 1957—; editor: Advances in Pediatrics, 1976—, Pediatric Nutrition Handbook, 3d edit., 1991; asst. editor Pediatric Gastroenterology and Nutrition, 1981-91; editorial bd. Cons., 1960-84, Pediatrics, 1978-83, Core Jour. Pediatrics, 1980—, Contemporary Pediatrics, 1984—, Jour. Clin. Medicine and Nutrition, 1985—, Nutrition Rev., 1985-87. Served to capt. AUS, 1945-46. Recipient Lindback Teaching award U. Pa., 1963; Borden award nutrition, 1972; Noer Disting. Prof. award, 1980, Joseph B. Goldberger award in clin. nutrition, 1984, Joseph St. Geme Leadership award 7 pediatric socs., 1991. Mem. AAAS, Am. Pediatric Soc. (recorder-editor 1964-75, pres. 1985-86, John Howland award 1993), Soc. Pediatric Research, Am. Acad. Pediatrics (chmn. com. on nutrition 1974-81, Abraham Jacobi award 1991), Am. Coll. Nutrition, Am. Inst. Nutrition, Dietary Guidelines Adv. Commn., USDA, Sigma Xi, Alpha Omega Alpha. Home: 1115 W Virginia Ave Tampa FL 33603-4538 Office: Univ of S Florida Dept of Pediatrics 12901 Bruce B. Downs Blvd Tampa FL 33612

BAR-NESS, YEHESKEL, electrical engineer, educator; b. Baghdad, Iraq, Apr. 28, 1932; arrived in Israel, 1950; came to U.S., 1978; m. Varda Bar-Ness, Aug. 21, 1952; children: Yael, Yaron, Yegal. BEE, Technion U., Haifa, Israel, 1958, MEE, 1963; PhD, Brown U., 1969. Chief engr. Elscint Inc., Haifa, 1971-75; assoc. prof. Tel-Aviv U., 1973-78; vis. prof. Brown U., 1978-79, U. Pa., Phila., 1979-81; prof. elec. engring. Drexel U., Phila., 1981-83; tech. staff mem. AT&T Bell Lab., Holmdel, N.J., 1983-85; disting. prof. elec. and computer engring. N.J. Inst. Tech., Newark, 1985—, dir. ctr. communication and signal processing rsch., 1985—; vis. prof. Brown U., Providence, R.I., 1978-79. Recipient Kaplan Prece award Gov. of Israel, 1974. Fellow IEEE; mem. Communication Soc. of IEEE (sec. communications systems engring. com. 1985-87, vice chmn., 1987-89, chmn. 1990-91, editor IEEE transaction on communication). Home: 2 Etna Ct Marlboro NJ 07746-1307 Office: NJ Inst of Tech 323 Kings Blvd Newark NJ 07102

BARNETT, CRAWFORD FANNIN, JR., internist, educator, cardiologist; b. Atlanta, May 11, 1938; s. Crawford Fannin and Penelope Hollinshead (Brown) B.; m. Elizabeth McCarthy Hale, June 6, 1964; children: Crawford Fannin III, Robert Hale. Student Taft Sch., 1953-56, U. Minn., 1957; AB magna cum laude, Yale U., 1960; postgrad. (Davison scholar) Oxford (Eng.) U., 1963; MD (Trent scholar), Duke, 1964. Intern internal medicine Duke U. Med. Center, Durham, N.C., 1964-65, resident, 1965; resident internal medicine Wilmington (Del.) Med. Ctr., 1965-66; dir. Tenn. Heart Disease Control Program, Nashville, 1966-68; pvt. practice medicine specializing in internal medicine, Atlanta, 1968—; dir. Travel Immunization Ctr., Atlanta; mem. staff Crawford Long, Northside, Grady Meml., West Paces Ferry, North Fulton, hosps. (all Atlanta); mem. teaching staff Vanderbilt Med. Ctr., Nashville, 1966-68, Crawford Long Meml. Hosp., 1969—; clin. instr. internal medicine, dept. medicine Emory U. Med. Sch., Atlanta, 1969—. Bd. govs. Doctors Meml. Hosp., 1971-80; bd. dirs. Atlanta Speech Sch., 1976-80, 92—, Historic Oakland Cemetery, 1976-86, So. Turf Nurseries, 1977-92, Tech Industries, 1978-92; bd. dirs. Am. Chestnut Found., 1990. Served as surgeon USPHS, 1966-68. Fellow Am. Geog. Soc., Royal Soc. of Tropical Medicine and Hygiene, Royal Geog. Soc., Royal Soc. Medicine, Explorers Club (life, N.Y.C.); mem. Am. Soc. Tropical Medicine and Hygiene, Am. Fedn. Clin. Rsch., Coun. Clin. Cardiology, AMA, Ga. Med. Assn., Atlanta Med. Assn., Am. Heart Assn., Ga. Heart Assn., Am. Soc. Internal Medicine, Ga. Soc. Internal Medicine, Am. Assn. History Medicine, Ga. Hist. Soc. Atlanta Hist. Soc. (dir. 1976-84), Ga. Trust for Hist. Preservation, Nat. Trust Hist. Preservation, Internat. Hippocratic Found. Soc. (Greece), Faculty of History of Medicine and Pharmacy Worshipful Soc. Apothecaries of London, Atlanta Com. on Fgn. Relations (chmn. exec. com. 1972-88), So. Council Internat. and Public Affairs, Newcomen Soc., Atlanta Clin. Soc., Wilderness Med. Soc., Internat. Soc. Travel Medicine (founding), Travelers Century Club, Circumnavigators Club, South Am. Explorers Club, Victorian Soc. Am. (bd. advisers Atlanta chpt. 1971-86), Mensa, Gridiron, Piedmont Driving Club, Yale Club (dir. 1970-74), Nine O'Clocks Club, Pan Am. Doctors Club, Phi Beta Kappa. Episcopalian. Contbr. articles to profl. publs. Home: 2739 Ramsgate Ct NW Atlanta GA 30305-2830 Office: 3250 Howell Mill Rd NW 205 Atlanta GA 30327-4187

BARNETT, HENRY JOSEPH MACAULAY, neurologist; b. New Castle upon Tyne, Eng., Oct. 2, 1922; s. Thomas William Barnett and Sadie Banks Macaulay; m. Kathleen Barnett, Feb. 23, 1946; children: Ann, William, Jane, Ian. MD, U. Toronto, Ont., Can., 1944; LLD (hon.), Dalhousie U., Halifax, N.S., Can., 1984; DSc (hon.), N.Y. Inst. Tech., 1985. Jr. rotating intern Toronto Gen. Hosp., 1944, sr. med. resident, 1947-48, resident in neurology, 1949-50, physician, 1952-67; fellow in pathology Banting Inst. U. Toronto, 1946-47, clin. tchr., 1952-54, assoc. in medicine, 1954-63, asst. prof., 1963-66, assoc. prof., 1966-69; sr. med. resident in neurology Sunnybrook Hosp., Toronto, 1948-49, chief divsn. neurology, 1966-69; mem. house staff Nat. Hosp., Queen Sq., London, 1950-51; rsch. asst. Dept. Neurology, Oxford, Eng., 1951; chief divsn. neurology Victoria Hosp., London, Ont., Can., 1969-72; prof. neurology U. Western Ont., London, Ont., Can., 1969, chmn. dept. clin. neurol. scis., 1974-84, emeritus prof., 1991; chief divsn. neurology Univ. Hosp., London, Ont., Can., 1972-84, chief dept. clin. neurolog. scis., 1974-84; pres., sci. dir. John P. Robarts Rsch. Inst., London, Ont., Can., 1984-92, scientist, 1992; clin. prof. dept. clin. epidemiology and biostatistics McMaster U., Hamilton, Ont., Can., 1989—; cons. neurology Sunnybrook Hosp., Toronto, 1953-66, Clarke Inst. Psych., Toronto, 1953-68, Toronto Hosp., Weston, Ont., 1954-66; past chmn. Royal Coll. Com. in Neurology, 1966-74; juror St. Vincent Internat. Med. Prize Jury, Italy, 1979, 83; hon. pres. Bishin-kai, Rsch. Inst. for Brain and Blood Vessels, Akita, Japan, 1983; past pres. Med. Staff Assn., Univ. Hosp.; mem. adv. com. on clin. applications of nuclear magnetic resonance Ont. Ministry of Health, 1986-87, monitoring com. European Carotid Surgery Trial, protocol & evaluation com. The PACK Group, Ctr. for Thrombosis & Vascular Rsch., data monitoring com. Multicenter Acute Stroke Trial; mem. Stroke Data Bank Ad Hoc Adv. Com., NIH; sci. advisor Nat. Inst. Neurol. Disorders and Stroke Project; assoc. Can. Brain Bank, Med. Rsch. Coun., 1983; mem. WHO Task Force on Stroke and Other Vascular Disorders; chmn. safety com. Sandoz Pharm.; adv. com. Antiphospholipid Antibody in Stroke Study Group, Propranolol for Patients with AAA; bd. dirs. Heart and Stroke Found. Can.; project dir. orgn. and incorporation Can. Stroke Soc. (J.P. Bickle Found.), Nat. Insts. Neurol. Diseases and Stroke (NIH); vis. prof. U. Newcastle, England, 1976, Barrow Neurol. Inst., Phoenix, Ariz., 1981, U. Toronto, 1981, Washington U., St. Louis, 1983, Mass. Gen. Hosp., Boston, 1985, Vanderbilt U. Med. Ctr., Nashville, 1986, Koret, San Francisco, 1991; lectr. McGill U., Montreal, 1978, U. Belfast, Ireland, 1979, Ohio State U., Columbus, 1982, Sunnybrook Hosp., Toronto, 1983, U. Minn., 1983, Toronto Gen. Hosp., 1983, Jewish Gen. Hosp., Toronto, 1984, Australian Assn. Neurologists, 1985, U. Texas, Houston, 1987, Dalhousie U., Halifax, N.S., 1988, Albert Einstein Coll. Medicine, N.Y.C., 1989, U. Rochester, N.Y., 1991, U. Ottawa, Ont., 1991, U. Wash., Seattle, 1991, U. Tex. Med. Sch., Houston, 1991, Fairfax Hosp., Va., 1992, others; prin. investigator various rsch. grants and studies. Mem. editorial bd. Neurology, 1973-76, Jour. of Neurol. Scis., 1970-84, Can. Jour. of Neurol. Scis., 1980-92, Neuroepidemiology, 1981-84, Current Opinion in Neurology and Neurosurgery, 1988—, Advances in Neurology, 1992—, Stroke, 1974—, editor-in-chief, 1981-86, cons. editor, 1987—. Bd. dirs. London (Ont.) Convention Ctr., 1993. Decorated Officer of Order of Can., 1984; recipient Royal Bank award (with C.G. Drake), 1983, Disting. Svc. award Heart and Stroke Found. Ont., 1985, Sir Izaak Walton Killam award in Medicine, Can. Coun., 1988, FNG Starr award Can. Med. Assn., 1991, Commemorative medal 125th Anniversary of Confedn., 1993; Annual Henry Barnett Stroke Lectureship of Heart and Stroke Found. of Ont. created in his honor, 1986. Fellow ACP, Royal Coll. Physicians (Londoed. Soc. London; mem. Royal Coll. Physicians Can. (mem. coun. 1981-84, chmn. rsch. com. 1983-84, royal coll. lectr.), Am. Heart Assn. (exec. com., stroke coun., publs. com. 1981-86, Disting. Svc. award 1989, Disting. Achievement award 1990, Willis lectr. 1988), Am. Assn. Neurol. Surgeons (Donaghy lectr. joint sect. on cerebrovascular surgery 1993), Am. Neurol. Assn. (hon., v.p. 1980-81, 1st v.p. 1982), Royal Soc. Medicine (hon. mem. sect. neurology 1991), Hungarian Neurosurg. Soc. (hon.), Neurol. Assn. Costa Rica (hon.), Australian Assn. Neurologists (hon.), L'Inst. des Neuroscis. Cliniques de Bordeaux (hon.), Hong Kong Neurol. Soc. (hon.), Assn. Brit. Neurologists (hon. foreign), New Orleans Acad. Internal Medicine (hon.), European Neurol. Soc. (hon.), Russian Acad. Med. Scis. (hon. foreign), Can. Neurol. Soc. (sec. 1964-68, v.p. 1974-75, pres. 1975-76), Toronto Neurol. Soc. (founding pres. 1968), World Fedn. Neurology (rsch. group cerebrovascular disease), N.Y. Acad. Scis., Internat. Med. Soc. Paraplegia, London Neuroscis. Assn. (past pres.), Can. Stroke Soc. (founding pres. 1976-80, pres. 1990—), Can. Heart Found. (mem., dir. med. adv. com. 1977-81, dir. 1991—), Ont. Heart Found. (dir. 1972-86, chmn. stroke com. 1972-84, mgmt. com. 1982, Murray Robertson Meml. lectr. 1984), Ont. Deafness Rsch. Found. (past chmn. sci. adv. com.), Internat. Stroke Soc. (pres. 1989-92), World Congress Neurology (pres. 1990—), World Wildlife Fund (dir. Can. 1992—), AOA. Office: John P Robarts Rsch Inst, 100 Perth Dr, London, ON Canada N6A 5K8

BARNETT, PETER RALPH, health science facility administrator, dentist; b. Bklyn., Oct. 21, 1951; s. Seymour and Betty Natalie (Cobbs) B.; m. Susan Clay, Jan. 27, 1990; children: Regina, Alexis, Alana. AB, Colgate U., 1973; DMD, U. Pa., 1977, MBA, 1979. Lic. dentist, Pa., N.J., N.Y. Dir. mgmt. systems U. Pa. Sch. Dental Med., Phila., 1979-81, asst. dir. clinic mgmt., 1981-84; dir. profl. affairs Pearle Dental, Inc., Dallas, 1984-86, v.p. and dir. dental ops., 1986-87; dir. vision benefits Pearle Health Svcs., Inc., Dallas, 1987-88, v.p. managed vision care, 1988-91, sr. v.p. franchising and sales, 1991-92, sr. v.p. quality and franchising, 1992-93. Author and co-author several profl. articles on health care mgmt., fin., and mktg. Bd. dirs. PTA Brinker Elementary Sch. Plano, Tex. 1988-89. Mem. ADA, Beta Beta Beta. Democrat. Jewish. Avocations: running, reading, tennis, karate, gardening. Home: 4619 Creekmeadow Dr Dallas TX 75287-6814

BARNETT, ROBERT NEAL, physicist, researcher; b. Littlefield, Tex., Nov. 28, 1947; s. Robert Lee and Norma Elizabeth (Bates) B.; m. Stacy Irene Brown, Feb. 20, 1987; children: Daniel Lee, Matthew Neal. BS in Engring. Physics, Tex. Tech. U., 1975; PhD in Physics, U. Kans., 1980. Postdoctoral Ga. Inst. Tech., Sch. Physics, Atlanta, 1980-82, rsch. scientist II, 1982-93, sr. rsch. scientist, 1993—. Contbr. articles to Phys. Rev. Letters, Phys. Rev. B., Jour. Chem. Physics, others. With U.S. Army, 1967-70. Mem. Am. Phys. Soc. Office: Ga Inst Tech Sch Physics Atlanta GA 30332

BARNETT, TIM P., meteorologist. BA in Physics, Pomona Coll., 1960; MS in Physical Oceanography, Scripps Inst. of Oceanography, 1962, PhD in Physical Oceanography, 1966. Rsch. asst. Scripps Inst. Oceanography, La Jolla, Calif., 1960-64; oceanographer, cons. Marine Advisors, La Jolla, Calif., 1961-64; oceanographer U.S. Naval Oceanographic Office, Washington, 1964-66; sr. scientist, mgr. ocean physics Westinghouse Ocean Rsch. Lab., San Diego, 1966-71; rsch. marine physicist, acad. adminstr. Scripps Inst. Oceanography, 1971—; co-originator and U.S. lead participant in Joint North Sea Wave Project, 1968-72, Sea Wave Modeling Project, 1979-81; principal sci. and co-originator first U.S. Exptl. Climate Forcaster ctr., 1981—; convenor nat. mtg. El Niño CRC/NAS, 1982-83; mem. review panel NCAR Oceanography program, NOAA Pacific Marine Environmental Lab. Author or co-author 113 refereed publs. plus 57 additional book chpts. Recipient Special Creativity award NSF, 1991, 92, Harald Ulrick Sverdrup Gold medal Am. Meterological Soc., 1993. Fellow Am. Meteorological Soc.; mem. Internat. Assn. for the Phys. Scis. of the Ocean (com. on Tides and Mean Sea Level, 1983-86). Office: Scripps Inst of Oceanography Climate Research Division UCSD Mail Code 0224 La Jolla CA 92093*

BARNHARDT, ROBERT ALEXANDER, college dean; b. Jenkins Township, Pa., Sept. 21, 1937; s. Daniel T. and Janet A. (MacCartney) B.; married. BS in Textile Engring., Phila. Coll. Textiles and Sci., 1959; MS, Inst. Textile Tech., 1961; MEd, U. Va., 1970, EdD, 1974. Assoc. prof. fabric tech. Phila. Coll. Textiles and Sci., 1961-64, chmn. dept. textiles, 1964-66; dir. edn. Inst. Textile Tech., Charlottesville, Va., 1966-69, dean and dir. edn., 1972-76, dir. rsch. and edn., 1977-78, v.p. rsch. and edn., 1978-84, exec. v.p., chief oper. officer, 1984-87; dean Coll. Textiles N.C. State U., Raleigh, 1987—; bd. dirs. Textile/Clothing Tech. Corp., Raleigh, Harriet & Henderson Yarns, Inc., N.C., So. Textile Assn. Fellow Textile Inst. Gt. Britain (medal 1988); mem. Am. Soc. Engring. Edn., Nat. Coun. for Textile Edn. (pres. 1990—), Internat. Conf. Textile Edn., Phi Kappa Phi. Episcopalian. Avocations: tennis, snow skiing, singing, golf, gardening. Office: NC State U Coll Textiles Box 8301 4700 Hillsborough St Raleigh NC 27695-8301

BARNUM, MARY ANN MOOK, information management manager; b. Arlington, Va., Apr. 3, 1946; d. Conrad Payne and Barbara Heer (Held) Mook; m. William Douglas Barnum, Aug. 10, 1968. BS in Mathematics, Radford U., 1967. Cert. tchr., Va., N.J., N.Mex. Math. tchr. Prince William County Schs., Woodbridge, Va., 1967-68; mathematician RCA Svc. Co., Andros Island, Bahamas, 1968-70; math. tchr. Cinnaminson (N.J.) Schs., 1970-73, Alamagordo (N.Mex.) Sch. System, 1973-74; data svcs. supr. A.M. Best Co., Oldwick, N.J., 1975-78; assoc. mgr. AT&T Communications, Piscataway, N.J., 1978-86; mgr. AT&T Info. Mgmt. Svcs., Piscataway, N.J., 1986-90, AT&T Bus. Comm. Svcs., Somerset, N.J., 1990-91; mem. tech. staff AT&T Network Systems, Berkeley Heights, N.J., 1991—. Sec. Cherry Hill (N.J.) Jaycettes, 1972-73; sec. bd. trustees Friends of Clarence Dillon Libr., Bedminster, N.J., sec., 1989-90, pres., 1990-92, vol., 1986—; mem. Far Hills Environ. Commn., 1990-92, chmn., 1992—; founder Somerset Hills Environ. Commn. Mem. IEEE Computer Soc., Am. Soc. Quality Control, DAR, Descendants of Washington's Army at Valley Forge (capt. of the guard 1988-90, dep. adjutant gen. 1990-92, adjutant gen. 1992—), Soc. of the Descendants of the Mayflower, Kappa Delta Pi. Presbyterian. Home: PO Box 893 Lake Rd Far Hills NJ 07931

BARON, DAVID HUME, science journalist; b. Phila., Mar. 31, 1964; s. Charles Hillel and Irma (Frankel) B. BS in Physics and Geology, Yale U., 1986. Freelance sci. reporter nat. pub. radio, BBC, CBC, CNN, Boston Globe, 1985—; sci. reporter WBUR-FM, Boston, 1987—; Knight sci. journalism fellow MIT, Boston, 1989-90; sci. writing fellow Marine Biol. Lab., Woods Hole, Mass., 1991. Recipient Nat. Med. Reporting award AMA, 1992, Nat. Sci. Reporting award AAAS-Westinghouse, 1991; Corp. Pub. Broadcasting grantee, 1990. Mem. AAAS, Assn. Inds. in Radio, Nat. Assn. Sci. Writers. Office: WBUR-FM 630 Commonwealth Ave Boston MA 02215

BARON, GINO VICTOR, chemical engineering educator; b. Roeselare, Belgium, Oct. 26, 1948; s. Marcel J. Baron and Francine Calmeyn; m. Anne C. Lauwers, July 26, 1971; children: Ilan, Eline. Degree in chem. engring., Vrije U. Brussels, 1971; DSc in Engring., Technion, Haifa, Israel, 1976. Cert. chem. engr. Instr. chem. engring. Technion, 1972-74; Asst. Vrije U. Brussels, 1971-72, asst., 1974-81, lectr., 1981-89, prof., 1989—. Contbr. articles to profl. jours.; inventor, patentee in field. Recipient I. Akerman award NSF, 1983. Fellow Royal Flemish Inst. Engrs. Avocations: running, tennis, gardening. Home: Bleuckeveldlaan 34, B 3080 Tervuren Belgium Office: Vrije Universiteit Brussels, Dept CHIS, Pleinlaan 2, B1050 Brussels Belgium

BARON, HAZEN JAY, dental scientist; b. Detroit, Mar. 12, 1934; s. Hazen Cornelius and Lauretta (Schoenfeld) B.; m. Julie Christine Enders, Nov. 26, 1953; children: Hazen Lawrence, Melvin Philip. BS, Wayne State U., 1954; DDS, Northwestern U., 1958, PhD, 1968. Pvt. practice gen. dentistry Waukegan, Ill., 1959-62; lctr. Northwestern Dental Sch., Chgo., 1962-64; asst. dir. Warner Lambert Co., Morris Plains, N.J., 1964-69, dir., 1969-80; dir. Johnson and Johnson, New Brunswick, N.J., 1980-87, v.p., 1987-91; freelance cons. Morristown, 1991--; mgt. bd. Johnson and Johnson, New Brunswick, 1987-89; trustee advisor Am. Fund Dental Health, Chgo., 1988-90; vis. com. Northwestern Dental Sch., Chgo., 1987-91. Editor: Tri-County Dental Soc., 1965-69; co-author: Oral Pathology, 1965; contrb. articles to profl. jours. Rsch. fellow NIH, 1959-62. Fellow Am. Acad. Oral Pathology; mem. Am. Acad. Periodontology, Internat. Assn. for Dental Rsch. Home and Office: 41 Springbrook Rd Morristown NJ 07960

BARONDES, SAMUEL HERBERT, psychiatrist, educator; b. Bklyn., Dec. 21, 1933; s. Solomon and Yetta (Kaplow) B.; m. Ellen Slater, Sept. 1, 1963 (dec. Nov. 22, 1971); children: Elizabeth Francesca, Jessica Gabrielle. AB, Columbia U., 1954, MD, 1958. Intern, then asst. resident in medicine Peter Bent Brigham Hosp., Boston, 1958-60; sr. asst. surgeon USPHS, NIH, Bethesda, Md., 1960-63; resident in psychiatry McLean and Mass. Gen. hosps., Boston, 1963-66; asst. prof., then assoc. prof. psychiatry and molecular biology Albert Einstein Coll. Medicine., Bronx, N.Y., 1966-69; prof. psychiatry U. Calif., San Diego, 1969-86; prof., chmn. dept. psychiatry, dir. Langley Porter Psychiat. Inst. U. Calif., San Francisco, 1986—; pres. McKnight Endowment Fund for Neurosci., 1989; mem. Inst. Med. NAS, 1990, sci. adv. com. Charles E. Culpeper Found.; mem. extramural sci. adv. bd. NIMH; mem. bd. sci. advisors Buck Ctr. for Rsch. in Aging; sci. adv. com. Rsch.!Am.; mem. fellowship rev. panel Howard Hughes Med. Inst. Mem. editorial bds. profl. jours.; contbr. numerous articles to profl. publs. Recipient Rsch. Career Devel. award USPHS, 1967; Fogarty Internat. scholar NIH, 1979. Fellow AAAS, Am. Coll. Neuropsychopharmacology; mem. Am. Soc. Biol. Chemistry, Am. Psychiat. Assn. Am. Soc. Cell Biology, Soc. Neurosci., Am. Soc. Neurochemistry, Psychiat. Rsch. Soc., Soc. for Complex Carbohydrates, Phi Beta Kappa, Alpha Omega Alpha. Office: U Calif-San Francisco Langly Porter Psychiat Inst 401 Parnassus Ave San Francisco CA 94143-0001*

BARPAL, ISAAC RUBEN, technology and operations executive; b. Argentina, Feb. 21, 1940; came to U.S., 1964; s. David and Gala (Trajtengertz) B.; children: David, Daniel, Donna. BSEE, BS in Math., Calif. State Poly. U., 1967; MSEE, PhDEE, U. Calif., Santa Barbara, 1970. Registered profl. engr., Pa., Calif., Fla., Brazil. Asst. engring. mgr. Westinghouse Electric Co., Pitts., 1971-74, v.p. sci. and tech., 1987-93; exec. mgr. Westinghouse Electric Co., Sao Paulo, Brazil, 1974-80; pres. Westinghouse de Venezuela, Caracas, 1980-83, Westinghouse S. Latin Am., Sao Paulo, 1983-87; sr. v.p., chief tech. officer Allied Signal Inc., Morristown, NJ, 1993—. Bd. dirs. Ben Franklin in Advanced Tech. Ctr., Pitts., 1987-93, Carnegie Sci. Ctr., Pitts., 1987-93, Pitts. Symphony Soc., 1989-93; mem. adv. coun. Sch. Engring. Calif. Poly.

State U., San Luis Obispo, 1988—. Recipient Honored Alumnus award Calif. Poly. State U., 1988. Fellow IEEE; mem. IEEE Engring. Mgmt. Soc. (Engring. Mgr. of Yr. award 1991), NSPE. Office: Allied Signal Inc 101 Columbia Rd Morristown NJ 07962-1021

BARRANCO, SAM CHRISTOPHER, biologist, researcher; b. Beaumont, Tex., Nov. 17, 1938; s. Sam and Leola (LaRocca) B.; m. Barbara Reese, Feb. 25, 1967; children: Rhys S., Melissa E. BS, Tex. A&M U., 1960, MS, 1962; PhD, Johns Hopkins U., 1969. Postdoctoral fellow/asst. prof. M.D. Anderson Tumor Inst., Houston, 1969-72; prof. U. Tex. Med. Br., Galveston, 1972-87, Ea. Va. Med. Sch., Norfolk, 1987-90; pres., CEO Prism Diagnostics & Devel. Corp., Virginia Beach, Va., 1990—, Quest Biomed. Rsch. Ctr., Virginia Beach, Va., 1990—. Contbr. over 100 articles to profl. jours., chpts. to books. Capt. U.S. Army, 1963-64. Nat. Cancer Inst. predoctoral fellow, 1964-69, postdoctoral fellow, 1969-70, rsch. grantee, 1972—. Mem. Am. Assn. Cancer Rsch., Soc. Analytical Cytology, Cell Kinetics Soc. (treas. 1991-93, membership com. 1990-91). Achievements include 1 patent pending. Home: 1005 Dartford Mews Virginia Beach VA 23452 Office: Prism Diagnostics Ste 201 841 Seahawk Cir Virginia Beach VA 23452

BARRANTES, DENNY MANNY, biochemist; b. Karata, Zelaya, Nicaragua, July 30, 1948; s. Luis and Margie (Angus) P.; m. Camilla Joan Binette, May 11, 1974; children: Laura Michele, Sonia Maria, Elena Celeste. BS, U. N.H., 1972; MS, George Washington U., 1974. Prof. biochemistry Nat. U. Nicaragua, Leon, 1974-76; rsch. assoc. Va. Mason Rsch. Ctr., Seattle, 1976-83; chemistry instr. Seattle Cen. Community Coll., 1981-82; rsch. scientist Atlantic Antibodies, Scarborough, Maine, 1983-86, chief scientist, 1987-89; sr. scientist Binax, Inc., South Portland, Maine, 1989-90; mgr., rsch. scientist Midland Bioproducts Corp., Scarborough, 1990—; cons. Midland Bioproducts Corp., Boone, Iowa, 1989-90. Co-author: Agents and Actions, 1973, Biochemical Pharmacology, 1982, International Jour. of Tissue Reactions, 1983, Clinical Chemistry, 1987. Sr. deacon State St. Congl. Ch., Portland, Maine, 1990-91; soccer coach Community Svcs., Scarborough, 1986-92; mem. Community Involvement Group of Beacon Schs. grant for improvement of math. and sci. skills in grades K-5, Scarborough, 1993—. Recipient Laspau scholarship, Cambridge, Mass., 1970-74, Juarez scholarship, Mexican govt., 1970. Mem. AAAS, Am. Chem. Soc., Am. Assn. Clin. Chemistry, N.Y. Acad. Scis. Democrat. Mem. United Ch. of Christ. Achievements include devel. of liquid-stabilized lipoprotein calibrators and controls, stable, low/aggregate-free enzyme-antibody conjugate chemistries. Home: 3 Green Needle Dr Scarborough ME 04074-9505

BARRANTES, FRANCISCO JOSE, biochemist, educator; b. Buenos Aires, Mar. 13, 1944; s. Guillermo Horacio and Dolores Barrantes; m. Phyllis Johnson, Sept. 3, 1973; children: Alexandra Nadine, Caroline Roxana, Diego Christian. MD, U. Buenos Aires, 1968, PhD, 1971. Rsch. fellow Nat. Rsch. Coun., Buenos Aires, 1969-71; rsch. assoc. dept. biochemistry U. Ill., Urbana, 1972-73; staff mem. dept. molecular biology Max Planck Inst., Goettingen, Germany, 1974-79; sect. co-head biophys. chemistry dept. Max Planck Inst., Goettingen, 1979-83; dir. inst. biochemistry U. South, Bahia Blanca, Argentina, 1983—; prof. dept. biochemistry U. South, Bahia Blanca, 1983—; vis. prof. dept. neurobiology SUNY, Stony Brook, 1986-87, Neuroscis. Inst., N.Y.C., 1986, Weizmann Inst., Rehovot, Israel, 1987. Author: Goytia Award, 1971; external reviewer Rsch. Coun. Argentina, 1985-87, Rsch. Coun. Chile, 1989—. Fellow John S. Guggenheim Meml. Found. 1990, Royal Soc. London 1991; recipient award Argentina Acad. Advanced Scis. 1971, Bernardo Houssay award Argentina Sci. Rsch. Coun. 1987, Biology award Third World Acad. Scis. 1989. Fellow Third World Acad. Sci. (corr.); mem. Internat. Union Pure and Applied Biophysics (mem. spl. commn. on cell and membrane biophysics 1987—), Biochem. Soc. London, Am. Biophys. Soc., German Soc. Biol. Chemistry, Deutsche ForschungsGemeinschaft. Avocations: woodwork, music, reading, sports. Office: Inst of Biochemistry, Camino de la Carrindanga km 7, 8000 Bahía Blanca Argentina

BARREIRO, ELIEZER JESUS, medicinal chemistry educator; b. Rio de Jan, Brazil, May 23, 1947; s. Marcelino Jesus and Carmen Sylvia B.; m. Lais Angela Tortorella, June 15, 1971 (div. 1983); m. Iolanda Margheritta, June 23, 1985; children: Matheus F., Joana, Gabriela. Grad., Anglo-Copaczbanz Coll., Rio de Janeior, 1966. Cert. pharmacist. Asst. prof. U. Fed. Rio de Janeiro, 1973-74, U. Fed. de São Carlos (Brazil), 1979-83; assoc. prof. U. Fed. Rio De Janeiro, 1983-86, prof., 1986—; cons. Brazilian Rsch. Coun., Brasilia, 1978—. Contbr. articles to profl. jours. including Tetrahedron Letters, Jour. Organic Chemistry, Jour. Heterocycl. Chem., Braz. Jour. Biol. Med. Rsch. Recipient 1st award pharm. rsch. Laboratorio Farmaceutico de Pernambuco, Recife, Brazil, 1991. Fellow Am. Chem. Soc.; mem. Brazilian Assn. Pharmacists (sci. dir. 1992). Roman Catholic. Achievements include patent for the process to new prostaglandin derivatives synthesized from natural hidnocarpic acid. Office: U Fed Rio de Janeiro, Ilha do Fundao Cidade Univ, 21941370 Rio de Janeiro Brazil

BARRETT, BERNARD MORRIS, JR., plastic and reconstructive surgeon; b. Pensacola, Fla., May 3, 1944; s. Bernard Morris and Blanche (Lischkoff) B.; BS, Tulane U., 1965; MD, U. Miami, 1969; m. Julia Mae Prokop, Nov. 26, 1972; children: Beverly Frances, Julie Blaine, Audrey Blake, Bernard Joseph. Surg. intern Meth. Hosp. and Ben Taub Hosp., Houston, 1969-70; resident in gen. surgery Baylor Coll. Medicine, Houston, 1970-71, UCLA, 1971-73; resident in plastic surgery U. Miami Affiliated Hosps., Fla., 1973-75, chief resident in plastic surgery, 1975; fellow in plastic surgery Clinica Ivo Pitanguy, Rio de Janeiro, Brazil, 1973; instr. surgery Baylor Coll. Medicine, 1970-71, clin. instr. plastic surgery, 1977-80, clin. asst. prof., 1980-90, clin. assoc. prof., 1991—; instr. surg. emergencies L.A. County Paramedics, 1972-73; plastic surgery coordinator for jr. med. students Sch. Medicine U. Miami, 1975; practice medicine specalizing in plastic and reconstructive surgery, Houston, 1976—; pres., chmn. bd. dirs. Plastic and Reconstructive Surgeons, P.A., Houston, 1978—; chmn. Tex. Inst. Plastic Surgery, Houston; assoc. chief plastic surgery St. Luke's Episcopal Hosp., Houston, 1991—; attending physician Jr. League Clinic, Tex. Children's Hosp., Houston, 1977—; active staff St. Luke's Hosp., Houston, Meth. Hosp., Houston; clin. assoc. in plastic surgery U. Tex. Med. Sch., Houston, 1976—; instr. surg. emergencies Harris County Community Coll.; dir. Am. Physicians Ins. Exchange, Austin, API Life Ins. Co., Austin, past chief of staff, chief plastic surgery Travis Centre Hosp., Houston, 1985—; cons. physician Houston Oilers, 1978—; attending physician Ontario Motor Speedway, Calif., 1972-73. Bd. dirs. Plastic Surgery Ednl. Found., Chgo; mem. Fed. Coun. on Aging, Washington, 1991. Served to lt. comdr., M.C., USNR, 1969-74. Surg. exchange scholar to Royal Coll. Surgeons, London, 1968; hon. dep. sheriff Harris County, Tex. (Houston); diplomate Am. Bd. Plastic Surgery. Fellow ACS; mem. Am. Soc. Plastic and Reconstructive Surgeons, Royal Soc. Medicine, Michael E. DeBakey Internat. Cardiovascular Surg. Soc., Am. Soc. for Aesthetic Plastic Surgery, Denton A. Cooley Cardiovascular Surg. Soc., Tex. Med. Assn., Tex. Soc. Plastic Surgery, Harris County Med. Assn., Lipoplasty Soc. N.Am., Houston Soc. Plastic Surgery, D. Ralph Millard Plastic Surg. Soc. (pres.-elect 1992-93, v.p. 1977-79, sec., treas. 1975-77, historian 1980—), U. Miami Sch. Medicine Nat. Alumni Assn. (bd. dirs. 1975-77), Alpha Kappa Kappa (pres., 1968-69). Clubs: Houston City, Houstonian; Royal Biscayne Racquet; Commodore (Key Biscayne, Fla.), Coral Beach and Tennis Club (Bermuda). Author: Patient Care in Plastic Surgery, 1982; Manuel de Ciudados en Cirugia Plastica, 1985. Contbr. articles to med. publs., presentations to profl. confs.; inventor Barrett sterling surgigrip. Office: 6624 Fannin St Ste 2200 Houston TX 77030-2334

BARRETT, EUGENE J., researcher, medical educator, physician; b. Jersey City, N.J., May 22, 1946; s. Joseph Francis and Margaret (Harney) B.; m. Paul Marie Quiricani, Jan. 31, 1976; children: Nora, Matthew. BS, St. Peters Coll., 1968; MD, U. Rochester, 1975, PhD, 1975. Asst. prof. Yale U., New Haven, 1980-85, assoc. prof., 1985-91; prof. U. Va., Charlottesville, 1991—; dir. diabetes unit Yale U. Sch. Medicine, 1987-91; dir. diabetes ctr. U. Va., 1991—. Contbr. over 70 articles to profl. jours. Recipient Rsch. Career award NIH, 1981-85. Mem. Am. Diabetes Assn., Am. Heart Assn. (Established Investigator 1986-90), Am. Fedn. Clin. Rsch. Roman Catholic. Avocations: sailing, tennis. Office: U Va Sch Medicine Diabetes Rsch Ctr MR4 Box 5116 Charlottesville VA 22908

BARRETT, GREGORY LAWRENCE, environmental scientist; b. Chgo., Jan. 4, 1952; s. Joseph E. and Helen (Urek) B.; d. Sarah J. Foster, Sept. 30, 1978; 1 child, Danielle. BS, Ill. State Univ., 1973; MS, Univ. Ill., 1976. Lab. supr. Univ. Ill., Urbana, 1976-79; analyst chemist Nalco Chemical Co., Naperville, Ill., 1979-82, Stuart-Ironsides, Willowbrook, Ill., 1982-84; quality anaylst supr. Allied-Signal Inc., Danville, Ill., 1984-88, environ. supr., 1988-89; compliance mgr. Ecolab Inc., Joliet, Ill., 1989-92; environ. scientist Argonne (Ill.) Nat. Lab., 1992—. Mem. Am. Chem. Soc., AF & AM Lodge, Phi Eta Sigma, Phi Sigma. Achievements include three year research effort on environmental effects of sewage sludge disposal. Assisted in developing mechanical integrity testing procedures for underground injection well with Illinois EPA and USEPA Region V. Office: Argonne Nat Lab 9700 S Cass Ave Argonne IL 60439

BARRETT, JAMES EDWARD, biology educator, research; b. Camden, N.J., Aug. 9, 1942; s. Thomas T. and Ruth E. (Taylor) B.; m. Maura Dean Bird, June 10, 1962; children; Jennifer, Andrea, Stephanie. BS, U. Md., 1966; PhD, Pa. State U., 1971. Prof. U. Md., College Park, 1972-79, Uniformed Svcs. Univ., Bethesda, Md., 1979-91; scientist Eli Lilly & Co., Indpls., 1991-92; dept. head Lederle Labs., Pearl River, N.Y., 1992—; adv. bd. N.E. Regional Primate Ctr., Harvard Med. Sch., Boston, 1990—, Bowman Gray Sch. of Medicine, Winston-Salem, N.C., 1989—; mem. NIDA IRG, Rockville, Md., 1989-92; adj. prof. of behavioral biology in psychiatry Columbia U., 1993—. Editor Jour. of the Exptl. Analysis of Behavior, 1985-92, Advances in Behavioral Pharmacology, 1987—; mem. editorial bd. Jour. of Exptl. Analysis of Behavior, 1993. With U.S. Army, 1960-63. Recipient Solvay-Duphar award, 1991, Nat. Inst. Drug Abuse, 1982—. Mem. Am. Coll. Neuropsychopharmacology, Soc. for Neurosci., Behavioral Pharmacology Soc. (pres. 1987-89), Am. Soc. Pharm. Exptl. Therapy. Office: Lederle Labs 200/3608 401 N Middletown Rd Pearl River NY 10965

BARRETT, MICHAEL JOHN, anesthesiologist; b. Milw., Feb. 27, 1954; s. Walter Joseph and Valerie Clara (Wisniewski) Baclawski; m. Joan Marie Rowley, May 28, 1983; children: Michael J. Jr., Jessica Marie, Monica Jane. BS in math. with honors, U. Wis., 1974; MD, Med. Coll. Wis., 1981. Diplomate Am. Bd. Anesthesiology, Nat. Bd. Medicine and Surgery, Nat. Bd. Med. Examiners, Am. Acad. Pain Mgmt. Intern Med. Coll. Wis. Affiliated Hosps., Milw., 1981, resident in anesthesiology, 1982-84; dir. anesthesiology Putnam Community Hosp., Palatka, Fla., 1984-92, dir. Putnam Pain Ctr., 1985-92; pres. Putnam Anesthesia Assocs., Palatka, 1985-92; staff anesthesiologist St. Vincent Med. Ctr., Toledo, 1992—. Bd. dirs. Round Lake Park Homeowners Assn., Palatka, 1986-88. Walter Zeit fellow. Mem. AMA, Internat. Anesthesia Rsch. Soc., Am. Soc. Anesthesiologists, Am. Soc. Regional Anesthesiologists, Ohio Med. Assn., Acad. Medicine of Toledo and Lucas County, Ohio Soc. Anesthesiologists, Fla. Soc. Anesthesiologists, Fla. Med. Assn., So. Med. Soc., Putnam County Med. Soc. (pres. 1989-91), Putnam Soc., Putnam C. of C., Phi Beta Kappa, Phi Kappa Phi. Republican. Roman Catholic. Avocations: boating, private pilot, swimming. Home: 7017 Westwind Dr Sylvania OH 43560 Office: Assoc Anesthesiologists 2409 Cherry St #4 Toledo OH 43608

BARRETT, MICHAEL WAYNE, product development company executive; b. Tacoma, Wash., June 13, 1955; s. Dean Garnet and Phyllis Emily (Martin) B. Assembler Rocket Rsch., Redmond, Wash., 1982-83; technician Spectra Tech., Bellevue, Wash., 1983-88; mfg. mgr. Synrad, Inc., Bothell, Wash., 1988-92, plant mgr., 1992—; pres. BMW Enterprises, Seattle, 1993—; pres. Vacuum Systems, Seattle, 1988—. Sgt. USAF, 1978-81. Achievements include method for manufacture of reliable ultra high vacuum, high temperature seals for caf2, mgf2, Lif, vuv, windows for lasers; patent for sealed RF excited gas laser and method for manufacture; invention of cork screw attachment. Home: 3705 S Ainsworth St Tacoma WA 98408 Office: Synrad 4056 NE 174th St Seattle WA 98155-5514

BARRETT, RONALD MARTIN, aerospace engineer; b. Warner-Robins AFB, Ga., Nov. 15, 1965; s. Ronald Paul and Joy Anne (Gonzalez) B.; m. Karin Harmien Van Dijk, Feb. 2, 1990. BS with distinction, U. Kans., 1988; MS, U. Md., 1990; PhD with honors, U. Kans., 1993. Flight test engr. Skytrader Corp., Kansas City, Mo., 1988; instr. aerospace U. Kans., Lawrence, 1990-91; cons. Barrett Aerospace, Lawrence, 1990—; asst. prof. Auburn (Ala.) U., 1993—; smart structures cons. Gen. Dynamics, 1990-91, McDonnell Douglas, St. Louis, 1991. Contbr. tech. papers. Officer Internat. Student Assn., Lawrence, 1990-92, Internat. Club, Lawrence, 1987-89. Grad. fellow U. Md., 1988-90, NASA Space Grant fellow U. Kans., 1991—. Mem. AIAA (Paper awards 1987, 88), Soc. Hispanic Profl. Engrs., Soc. Advancement Materials and Processing Engrs., Internat. Soc. for Optical Engring., Tau Beta Pi. Achievements include patent on method of achieving different stiffnesses in desired directions through directional attachment; discovery of branch of structural mechanics called directional attachment. Office: Auburn U Aerospace Engring 211 Aerospace Engring Bldg Auburn AL 36849

BARRETTE, JEAN, physicist, researcher; b. Montreal, Quebec, Canada, May 1, 1946; s. Bertrand and Marguerite Ducharme B. ScB, U. Montréal, 1967, MSc, 1968, PhD, 1974. Postdoctoral fellow Max-Planck Inst., Heidelberg, Germany, 1974-76; physicist Brookhaven Nat. Lab., Upton, N.Y., 1976-82; engring. physicist Commissariat a l'energie Atomique, Saclay, France, 1982-87; prof. McGill U., Montréal, 1987—; dir Foster Radiation Lab., Montréal, 1988—. Mem. Am. Physical Soc., Can. Assn. of Physicists (divsn. chair). Achievements include research in nucleus-nucleus reactions and heavy-ion physics with particular interest in the study of reaction mechanism at intermediate and relativistic bombarding energies. Office: McGill Univ-Foster Radiation Lab, 3610 University St, Montreal, PQ Canada H3A 2B2

BARRON, CHARLES THOMAS, psychiatrist; b. Hattiesburg, Miss., May 2, 1950; s. Palmer H. and Eleanor Clarice (Sherman) B. BS, U. So. Miss., 1972; MD, U. Miss., 1976. Diplomate Am. Bd. Psychiatry and Neurology. Resident psychiatry St. Vincent's Hosp. and Med. Ctr. N.Y., N.Y.C., 1976-79; fellow inpatient psychiatry St. Vincent's Hosp. and Med. Ctr. N.Y., 1979-80; physician-in-charge psychiatry/substance abuse Beth Israel Med. Ctr., N.Y.C., 1980-84; physician-in-charge inpatient svcs. (psychiatry) Beth Israel Med. Ctr., 1984-88; instr. Mt. Sinai Sch. Medicine, N.Y.C., 1980-87; asst. clin. prof. psychiatry Mt. Sinai Sch. Medicine, 1987-88, 91—; assoc. dir. psychiatry Gouverneur Hosp., N.Y.C., 1988-90; clin. asst. prof. psychiatry NYU, 1989-90; dir. inpatient svcs. (psychiatry), assoc. dir. psychiatry Mt. Sinai Svcs., Elmhurst, N.Y., 1990—; author/presenter presentation World Psychiat. Assn., 1981; presenter Ottawa Child & Adolescent Conf., 1988. Mem. Am. Psychiat. Assn. (presenter 1990), Internat. AIDS Soc., Am. Orthopsychiat. Assn. Office: CIO-21 79-91 Broadway Elmhurst NY 11373

BARRON, RONALD MICHAEL, applied mathematician, educator, researcher; b. Windsor, Ont., Can., July 12, 1948; s. Bernard Matthew and Mary Alice (Marontate) B.; m. Coral Lynn Mailloux, Aug. 9, 1969; 1 child, Ronald Melvin Jr. BA with honors, U. Windsor, 1970, MS, 1971; PhD, Carleton U., Ottawa, Ont., 1974; MS in Engring., Stanford U., 1982. Asst. prof. U. Windsor, 1975-79, assoc. prof., 1979-84, head dept. math. and stats., 1986-89, dir. Fluid Dynamics Rsch. Inst., 1987-90; vis. prof. Bangalore (India) U., 1989-90; prof. U. Windsor, 1984—, dir. Fluid Dynamics Rsch. Inst., 1993—. Contbr. articles to profl. jours. Mem. AIAA, Can. Applied Math. Soc., Soc. Indsl. and Applied Math., Can. Aeronautics and Space Inst. Roman Catholic. Achievements include development of efficient grid generators and flow solvers for computational fluid dynamics applications. Office: Univ Windsor, 401 Sunset Ave, Windsor, ON Canada N9B 3P4

BARRON, SUSAN, clinical psychologist; b. Chgo., May 13, 1940; d. Earl and Trixie (Chernoff) B.; m. Eugene Pratt, Jan. 18, 1975 (div. 1983). BBA, CCNY, 1960, MA, 1963; PhD, CUNY, 1973. Lic. psychologist. Intern psychologist Bellevue Psychiat. Hosp., N.Y.C., 1964-65, psychologist, 1966-67; teaching fellow CUNY, 1965-66; staff psychologist Lighthouse, N.Y. Assn. for the Blind, N.Y.C., 1968-71, sr. clin. psychologist, 1971-74; dir. psychol. counseling svcs. Peninsula Ctr. for the Blind, Palo Alto, Calif., 1974-75; cons. psychologist N.Y. State Commn. for Blind and Visually Handicapped, N.Y.C., 1975-78, 86—; dir. psychol. svcs. Thoms Rehab. Hosp., Asheville, N.C., 1978-79; state coord. psychol. svcs. N.Y. State Office Vocat. Rehab., Albany, 1979-85; founder, dir. Family Support Program ICU N.Y. Infirmary-Beekman Downtown Hosp., N.Y.C., 1982-84; cons. clin. psychologist N.Y. Hosp. Cornell U. Med. Ctr., 1987—; pvt. practice, 1987—; behavioral scientist diabetes control and complications NIH Cornell U. Med. Ctr., N.Y.C., 1987—; Mem. Nat. Human Svcs. Adv. Bd.-Retinitis Pigmentosa Found., Balt., 1975-82; cons. Del. State Commn. for Blind, 1975-78, Am. Found. Blind, 1974-82, Calif. Dept. Rehab., 1974-82, Hawaii State Svcs. Blind, 1974-82, Ariz. State Svcs. Blind, 1974-82, Nev. State Svcs. Blind, 1974-82; speaker Nat. Multiple Disabilities Conf., 1982, NAS, 1981; mem. adv. bd. doctoral psychology internship program Rusk Inst. of Rehab. Medicine, NYU Med. Ctr., 1979-84; behavioral scientist Diabetes Control and Complications Trial NIH-Cornell U. Med. Ctr., 1987—. Contbr. articles to profl. jours. Recipient Leadership award Alumni Assn. CCNY, 1960, 62, Rsch. award Retinal Dystrophy Soc., Australia, 1975. Fellow Am. Orthopsychiat. Assn.; mem. APA, AAAS, Calif. State Psychol. Assn., N.Y. Acad. Sci. Office: NY Hosp Cornell U Med Ctr 515 E 71st St # S102 New York NY 10021-4895

BARROW, ARTHUR RAY, chemical engineer; b. Mt. Clemens, Mich., June 9, 1959; s. Daniel Glidwell and Anna Viola (Kemp) Galloway; m. Kelly Lynn McCarty, June 19, 1982; children: Brittney Nichole, Danielle Jean. BSCE, U. Tex., Austin, 1982, MS in Engring., 1993. registered profl. engr. Tex. Coatings/traffic mgr. engr. Tex. Dept. Transp., Austin, 1983—; advisor AICE, 1985-86. Vol. Capitol 10,000 Workforce, Austin, 1982-87. Named Outstanding Chemistry Student Tex. Chem. Soc., 1977. Mem. Am. Soc. Testing & Materials, Fedn. Socs. for Coating Techs., Steel Structures Painting Coun., Austin Golf Assn., Alpha Chi Sigma. Achievements include research in traffic materials, coatings and application technologies; trouble shooting coating related problems/failures. Home: 2205 Shiloh Dr Austin TX 78745 Office: Tex Dept Transp 125 E 11th St Austin TX 78701

BARROW, THOMAS DAVIES, oil and mining company executive; b. San Antonio, Dec. 27, 1924; s. Leonidas Theodore and Laura Editha (Thomson) B.; m. Janice Meredith Hood, Sept. 16, 1950; children—Theodore Hood, Kenneth Thomson, Barbara Loyd, Elizabeth Ann. B.S., U. Tex., 1945, M.A., 1948; Ph.D., Stanford U., 1953; grad. advanced mgmt. program, Harvard U., 1963. With Humble Oil & Refining Co., 1951-72; regional exploration mgr. Humble Oil & Refining Co., New Orleans, 1962-64, sr. v.p., 1967-70, pres., 1970-72, also dir.; exec. v.p. Esso Exploration, Inc., 1964-65; sr. v.p. Exxon Corp., N.Y.C., 1972-78, also dir.; chmn., chief exec. officer Kennecott Corp., Stamford, Conn., 1978-81; vice chmn. Standard Oil Co., 1981-85; investment cons. Houston, 1985—; chmn. GX Tech., Houston, 1990—; also bd. dirs. and chmn. GX Technology, Houston; sr. chmn. GeoQuest Internat. Holdings, Inc., Houston, 1990—; with GPS Tech. Corp., Houston, McDermott Internat. Inc., New Orleans, Am. Gen. Corp., Houston; mem. commn. on natural resources NRC, 1973-78, commn. on phys. sci., math. and natural resources, 1984-87, bd. on earth scis., 1982-84; trustee Woods Hole Oceanographic Instn., 20th Century Fund-Task Force on U.S. Energy Policy. Pres. Houston Grand Opera, 1985-87, chmn., 1987-91; trustee Am. Mus. Natural History, Stanford U., 1980-90, Tex. Med. Ctr., 1983—, Geol. Soc. Am. Found., 1982-87; trustee Baylor Coll. Medicine, 1984—, vice chmn bd. trustees, 1991—. Served to ensign USNR, 1943-46. Recipient Disting. Achievement award Offshore Tech. Conf., 1973, Disting. Engring. Grad. award U. Tex., 1970, Disting. Alumnus, 1982, Disting. Geology Grad., 1985, Disting. Natural Sci. Grad., 1990; named Chief Exec. of Yr. in Mining Industry, Fin. World, 1979. Fellow N.Y. Acad. Scis.; mem. Nat. Acad. Engring., Am. Mining Congress (bd. dirs. 1979-85, vice chmn. 1983-85), Am. Assn. Petroleum Geologists, Geol. Soc. Am., Internat. Copper Research Assn. (bd. dirs. 1979-85), Nat. Ocean Industry Assn. (bd. dirs. 1982-85), AAAS, Am. Soc. Oceanography (pres. 1970-71), Am. Geophys. Union, Am. Petroleum Inst., Am. Geog. Soc., Sigma Xi, Tau Beta Pi, Sigma Gamma Epsilon, Phi Eta Sigma, Alpha Tau Omega. Episcopalian. Clubs: Houston Country, The Hills, Petroleum, River Oaks Country, Ramada. Office: 4605 Post Oak Place Dr Ste 207 Houston TX 77027-9728

BARRY, BRENDA ELIZABETH, respiratory biologist; b. Pawtucket, R.I., Dec. 18, 1950; d. James Joseph and Catherine Theresa (Callahan) B. BS, U. R.I., 1973, MS, 1975; PhD, Duke U., 1983. Sr. electron microscopy technician Duke U., Durham, N.C., 1975-78, lab. rsch. analyst, 1978-79; staff scientist Health Effects Inst., Cambridge, Mass., 1989—. Fellow Parker B. Francis Found., 1986-89. Mem. AAAS, Am. Thoracic Soc., Air and Waste Mgmt. Assn., Appalachian Mountain Club (Boston chpt. bicycle chairperson 1991—). Office: Health Effects Inst 141 Portland St # 7300 Cambridge MA 02139

BARRY, DAVID WALTER, infectious diseases physician, researcher; b. Nashua, N.H., July 19, 1943; s. Walter and Claire B.; m. Gracia Chen; children: Christopher, Jennifer. BA in French literature with highest honors magna cum laude, Yale Coll., 1965; student, Sorbonne, Paris, 1963-64; MD, Yale U., 1969. Cert. State of Conn. Med. Examining Bd., State of Md. Med. Examining Bd., State of N.C. Bd. Med. Examiners; diplomate Nat. Bd. Med. Examiners, Am. Bd. Internal Medicine, Am. Bd. Infectious Diseases. Intern, then resident Yale-New Haven Hosp., 1969-72; staff assoc., dir., acting dep. dir. divsn. virology, bur. biologics FDA, 1972-77; head anti-infectives sect., dept. clin. investigation, med. divsn. Burroughs Wellcome Co., Research Triangle Park, N.C., 1977-78, head dept. clin. investigation, med. divsn., 1978-85, head dept. virology, Wellcome rsch. labs., 1983-89, dir. divsn. clin. investigation, 1985-86, v.p. rsch. Wellcome rsch. labs., 1986-89, v.p. rsch., devel. and med. affairs, Wellcome rsch. labs., 1989—, also bd. dirs.; dir. influenza vaccine task force, bur. biologics FDA, 1976-77, mem. rsch. human subjects com.; adj. prof. sch. medicine Duke U., 1977—; mem. com. pub.-pvt. sector rels. vaccine innovation, inst. medicine NAS, 1983, roundtable drugs and vaccines against AIDS, 1989, industry liaison panel, 1992; mem. AIDS task force NIH, 1986, AIDS program adv. com., 1988; deans coun. Yale Sch. Medicine, 1989—; cons., lectr. in field; bd. dirs. Burroughs Wellcome Fund, Wellcome Found., Family Health Internat. Mem. editorial bd. AIDS rsch. and Human Retroviruses, AIDS Patient Care; contbr. articles to profl. jours. Active N.C. Indsl. (Vaccine) Commn. Sr. surgeon USPHS, 1972-77. Vis. fellow U. Md., 1975-76. Fellow ACP, Infectious Diseases Soc. Am., Royal Soc. Medicine; mem. AAAS, AMA, Am. Soc. Virology, Am. Soc. Microbiology, Am. Fedn. Clin. Rsch., Pharm. Mfrs. Assn. (med. and sci. sect., com. AIDS 1989, AIDS task force 1989, R & D steering com. 1990, commn. treatment drug dependence and abuse 1990—), N.C. Med. Soc., Durham-Orange Counties Med. Soc., Venezuelan Soc. Internal Medicine (hon.), Alpha Omega Alpha. Achievements include 12 patents for treatment of viral infections, patent for the Use of Azidothymidine to Treat AIDS and ARC, for the Treatment of Idiopathic Thrombocytopaenic Purpura. Office: Burroughs Wellcome Co 3030 Cornwallis Rd Research Triangle Park NC 27709

BARRY, RICHARD WILLIAM, chemist, consultant; b. Meriden, Conn., Dec. 16, 1934; s. Joseph F. and Irene Antoinette (Baraldi) B.; m. Mary Jane Tylec, July 29, 1967; children: Eliza, Todd, Sean. BS, Cen. Conn. State Coll., 1966; MA, U. R.I., 1973. Tchr. sci. New Haven Pub. Schs., 1965-67, Clinton (Conn.) Pub. Schs., 1967-79; mgr. quality control, cons. Bridgeport (Conn.) Testing Lab., 1979-85; sr. chemist Turbine Components Corp., Branford, Conn., 1985-93; process engr. Light Metals Coloring Co., Southington, Conn., 1993—; cons. Casco Corp., Bridgeport, 1985, C. Cowles & Co., New Haven, 1985-86, Bridgeport Testing Lab., 1985-86. Bd. dirs. Clinton Improvement Assn., 1972-79; deacon Branford 1st Congl. Ch., 1983-89. Mem. Am. Soc. Metals. Achievements include development of several processes for removal of worn out protective coatings from turbine components, including a chemical process for the removal of chrome carbide. Home: 23 Meadow Wood Rd Branford CT 06405-6412 Office: Light Metals Coloring Co 270 Spring St Southington CT 06489

BARTEK, GORDON LUKE, radiologist; b. Valpraiso, Nebr., Dec. 27, 1925; s. Luke Victor and Sylvia (Buner) B.; m. Ruth Evelyn Rowley, Sept. 10, 1949; children: John, David, James. BSc, U. Nebr., 1948, MD, 1949. Diplomate Am. Bd. Radiology. Intern Bishop Clarksen Hosp., Omaha, 1949-50; resident in medicine Henry Ford Hosp., Detroit, 1952-53, resident in radiology, 1953-56; staff radiologist Ferguson Hosp., Grand Rapids, Mich., 1956-76, Holland City Hosp., Mich., 1956-76, Logan Hosp., Utah, 1976-78, St. Lawrence Hosp., Lansing, Mich., 1978—; asst. clin. prof. dept. radiology Mich. State Univ. Coll. Medicine, 1977—, asst. prof. radiology, 1993—; dir. Accord Ins. Co., Cayman Islands, 1983-90. Served to lt. USN, 1949-52. Fellow Am. Coll. Radiology; mem. Mich. Radiology Practice Assn. (bd. dirs. 1984—, chmn. western Mich. sect. 1970-71), Am. Coll. Radiology (councilor 1972-76). Republican. Roman Catholic. Club: Manhattan Tennis (pres.). Avocations: flying, photography, skiing, snorkeling. Home: 1350 Briarcliff Dr SE Grand Rapids MI 49546-9737

BARTER, JAMES T., psychiatrist, educator; b. South Portland, Maine, May 31, 1930; married; 3 children. BA, Antioch Coll., 1952; MA, U. Ariz., 1955; MD, U. Rochester, 1961. Asst. prof. psychiatry Sch. Medicine U. Colo., 1965-69; assoc. chief Colo. Psychiat. Hosp., 1968-69; dep. dir. Sacramento Mental Health Svcs., 1969-73, dir., 1973-77; assoc. clin. prof. psychiatry U. Calif., Davis, 1978-85; dir. Ill. State Psychiat. Inst., Chgo., 1985—; cons. USPHS Indian Health Svc., 1968-71. Fellow Am. Psychiat. Assn., Am. Coll. Psychiatry; mem. AMA. Achievements include research in cross cultural psychiatry, drugs and drug abuse, adolescents, suicide, administration. Office: Ill State Psychiatric Inst 1601 W Taylor St Chicago IL 60612*

BARTH, CAROLYN LOU, hospital administrator, microbiologist; b. Elmhurst, Ill., Oct. 30, 1936; d. Victor Christian and Luella W. (Senf) B. BS in Biology and Chemistry, Elmhurst (Ill.) Coll., 1958; MS in Med. Tech., Wayne State U., 1960, PhD in Microbiology, 1974. Grad. teaching asst. microbiology dept. Wayne State U., Detroit, 1967, 69-73; chief med. technologist hematology div. Henry Ford Hosp., Detroit, 1960-67, rsch. assoc. Ophthalmology Rsch. Lab., 1970-73, technologist, then sr. med. technologist pathology dept., 1974-75, supr. immunopathology div., 1975-87, ops. mgr. pathology dept., 1987—; mem. eyes on classics com. Detroit Inst. Ophthalmology; presenter at nat. meetings, 1963—. Contbg. author: Cell Wall-Deficient Bacteria, 1982; contbr. articles to Clin. Rsch., Annals N.Y. Acad. Scis., Jour. Reticuloendothelial Soc., Henry Ford Hosp. Med. Jour., Jour. Immunological Methods. Past bd. dirs. Children's Mus. Friends; mem. staff parish rels. com. United Meth. Ch. Mem. Am. Soc. Clin. Pathologists (cert. med. technologist), Am. Soc. Microbiology, AAUW (chmn. Ednl. Found., past pres.), Wayne State U. Alumni Assn., Women of Wayne State U. Alumni Assn. (pres.), Sigma Xi, Beta Sigma Phi (past pres.). Home: 1011 S Renaud Grosse Pointe Woods MI 48236

BARTH, DAVID VICTOR, computer systems designer; b. Tulsa, Sept. 23, 1942; s. Vincent David and Norma (Bell) B. BS summa cum laude, Met. State Coll., Denver, 1977; MS, U. No. Colo., 1982. Programming mgr. Am. Nat. Bank, Denver, 1967-72; cons. Colo. Farm Bur. Ins. Corp., Denver, 1972; systems analyst Mid-Continent Computer Services, Denver, 1972-73; programming mgr. Bayly Corp., Denver, 1973-75; project leader Cobe Labs. Inc., Denver, 1976-84; part-time tchr. Met. State Coll., 1982-83; systems analyst Affiliated Banks Service Co., Denver, 1985-87; real estate broker Van Schaack & Co., Denver, 1985; tech. supr. Affiliated Banks Svc. Co., Denver, 1987-89; software engr. Computer Data Systems, Inc., Aurora, Colo., 1990-91; sr. computer systems designer Martin Marietta Corp., Golden, Colo., 1991-92; computer systems designer and salesman Computer Shop, Lakewood, Colo., 1992—; freelance flight instr., 1977—. Vol. Am. Red Cross, 1987—; Served with USN, 1961-66. Mem. Soc. for Info. Mgmt. (editor newsletter 1983), Exptl. Aircraft Assn. (editor newsletter chpt. 660, 1989-91), Aircraft Owners and Pilots Assn., Flying Circus Skating Club. Republican. Avocations: ice skating, amateur radio, flying, creative writing. Home: 509 S Cody St Lakewood CO 80226-3047 Office: Computer Shop 6591 W Colfax Ave Denver CO 80214-1803

BARTH, DELBERT SYLVESTER, environmental studies educator; b. Lawrenceburg, Ind., June 6, 1925. BS in Mil. Engring., U.S. Mil. Acad., 1946; MS in Nuclear Physics, Ohio State U., 1952; MS in Solid State Physics, Stevens Inst. Tech., 1960; PhD in Biophysics, Ohio State U., 1962. Health physics trainee Oak Ridge Nat. Lab., Tenn., 1947-49; asst. prof. dept. physics and chemistry U.S. Mil. Acad., West Point, N.Y., 1956-60; staff officer evaluation and planning sect. rsch. br. div. radiological health U.S. Pub. Health Svc., Dept. Health, Edn. and Welfare, 1960-61; investigator, staff officer experimental radiobiology program rsch. br. DRH, Rockville, Md., 1962-63; dir. bioenviron. rsch. program Southwestern Radiological Health Lab., U.S. Pub. Health Svc., Las Vegas, 1963-69; dir. bur. criteria and standards Nat. Air Pollution Control Adminstrn., DHEW, Durham, N.C., 1969-71; dir. Bur. Air Pollution Scis., Research Triangle Park, N.C., 1971; dir. Nat. Environ. Rsch. Ctr., Research Triangle Park, 1971-72, Las Vegas, 1972-76; dep. asst. adminstr. health and ecological effects ORD U.S. Environ. Protection Agy., 1976-78; vis. prof. biophysics U. Nev., Las Vegas, 1978-82; sr. scientist environ. rsch. ctr. U. Nev., 1982-88, dir. environ. rsch. ctr., 1989-92, prof. environ. studies program, 1992—. Mem. sub-com. on environ. effects, mem. adv. com. to fed. radiation, mem. com. on hearing, bioaccoustics, biomechs.; chmn. adv. com. Nat. Air Quality Criteria; environ. monitoring advisor, mem. environ. com. ecological scis. divsn. Inst. Environ. Scis.; mem. Army Sci. Bd., 1984-90; mem. awards bd. EPA. Recipient PHS Disting. Svc. medal. Mem. AAAS, Sigma Xi. Office: U Nev Las Vegas Environ Studies Program 4505 S Maryland Pky Las Vegas NV 89154-4030

BARTHOLOMEW, DONALD DEKLE, engineering executive, inventor; b. Atlanta, Aug. 2, 1929; s. Rudolph A. and Rubye C. (Delke) B.; m. Paula Hagood; children: John Marshall, Barbara Ann, Deborah Paige, Sandra Dianne. Student in Physics, Ga. Inst. Tech., 1946-48, 55-58. Owner Happy Cottons and Jalopy Jungleland, Atlanta, 1946-48, Beach Hotel Supply, Miami Beach, Fla., 1949-50; engr. Sperry Microwave Electronics, Clearwater, Fla., 1958-61; v.p., owner Draft Pak, Inc., Tampa, Fla., 1961-65, Merit Plastics, Inc., East Canton, Ohio, 1966-79; pres., owner Modern Tech., Inc., Marine City, Mich., 1979—; owner, officer and dir. various internat. mfg. companies, 1981—. Patentee in field. Served as sgt. USAF, 1951-54. Mem. Soc. Automotive Engrs., Soc. Plastics Engrs. (dir. 1982), Soc. Mfg. Engrs., Holiday Isles Jr. C. of C. (founding dir.). Republican.

BARTHOLOMEW, RICHARD WILLIAM, mechanical engineer; b. Montclair, N.J., June 27, 1954; s. Richard William and Gladys Jane (Van Wort) B.; m. Barbara Ann Sari, Aug. 26, 1983; children: Kellie Anne, Richard William III. BS, Rensselaer Poly. Inst., 1976, MS, 1977; PhD, U. Mich., 1982. Asst. prof. Mich. State U., East Lansing, 1982-87; sr. thermodynamicist Nat. Inst. Stds. & Tech., Gaithersburg, Md., 1987—. Contbr. articles to profl. jours. Mem. ASME, SAE, Sigma Xi, Tau beta Pi. Office: Nat Inst Stds & Tech Bldg 411 Rm A115 Gaithersburg MD 20899

BARTKY, IAN ROBERTSON, physical chemist; b. Chgo., Mar. 15, 1934; s. Walter and Elizabeth Inrig (Robertson) B.; m. Elizabeth Louise Hodgins, July 30, 1960; children: David John, Anne Robertson. BS, Ill. Inst. Tech., 1955; PhD, U. Calif., Berkeley, 1962. Phys. chemist Nat. Bur. Stds., Washington, 1961-68; sci. asst. Nat. Bur. Stds., Gaithersburg, Md., 1968-83; phys. sci. adminstr. Hdqs. U.S. Army Materiel Command, Alexandria, Va., 1983-86, Hdqs. U.S. Army Lab. Command, Adelphi, Md., 1986-92; cons. Bethesda, Md., 1992—; staff fellow subcom., interstate and fgn. commerce com. U.S. Ho. Reps., Washington, 1973-74; mng. editor Nat. Climate Program Office, Rockville, Md., 1979-80. Contbr. articles to profl. jours. Mem. local PTA, Bethesda, 1970's. Nat. Inst. Pub. Affairs fellow, 1967-68, Dept. Commerce Sci. and Tech. fellow, 1973-74. Mem. AAAS, Am. Chem. Soc., Soc. for the History Tech., Sigma Xi. Achievements include development of studies and recommendations to Congress on Daylight Saving Time's effects and proposed legislation. Home and Office: 7804 Custer Rd Bethesda MD 20814

BARTLETT, ELIZABETH EASTON, interior designer; b. Cleve., Apr. 1, 1937; d. Walter James Easton and Elizabeth (Scott) Easton Sullivan; m. Peter B. Bartlett, Nov. 24, 1956 (div. Sept. 1987); children: Elizabeth Kimberley Bartlett Kernan, Christopher, Katherine. Student, Skidmore Coll., 1959. Model Cluett, Peabody & Co., N.Y.C., 1958-65; pvt. practice N.Y.C., 1978—; buyer, bd. dirs. Boutique de Noël, N.Y.C., 1976-87. Trustee, vice chmn. St. Barnabas Hosp., Bronx, N.Y., 1978—; v.p. N.Y. Soc. for Prevention of Cruelty to Children, N.Y.C., 1979; trustee, bd. dirs. Youth Counseling League, N.Y.C., 1974. Mem. Rolling Rock Club. Episcopalian. Home and Office: 30 E 72nd St New York NY 10021-4248

BARTLETT, ELSA JAFFE, neuropsychologist, educator; b. Pitts., July 20, 1935; d. Aaron and Hortense (Greenberg) Jaffe; m. Jonathan Bartlett (div.). BFA, Columbia U., 1957; MS, Bank St. Coll., 1966; EdD, Harvard U., 1974. Editor various pub. cos., N.Y.C., 1957-62; tchr. N.Y.C. (N.Y.) Pub. Schs., 1962-64; publs. assoc. Bank St. Coll., N.Y.C., 1964-66; rsch. asst. Harvard U., Cambridge, Mass., 1966-72; rsch. assoc. The Rockefeller U., N.Y.C., 1972-80; asst. prof. NYU Med. Ctr., N.Y.C., 1980—; bd. advisors Children's Television Workshop, N.Y.C., 1972-75; cons. Tchrs. and Writers Collaborative, NYU, 1974-78, Nat. Assessment of Ednl. Progress, 1981-84; outside reviewer NSF, Washington, 1980-85. Contbr. articles to Jour. Cerebral Blood Flow, Psychiatry Rsch., Brain and Lang., Jour. Verbal Learning, Jour. Child Lang. Named post doctoral fellow NIH/Nat. Inst. Neurol. Disorders and Stroke, Washington, 1980-83. Achievements include research in language acquisition, word learning in young children, brain imaging, effects of haloperidol on cerebral metabolism. Office: NYU Med Ctr Psychiatry Dept 550 1st Ave New York NY 10016

BARTLETT, FRED MICHAEL PEARCE, structural engineer; b. Ottowa, Ont., Can., Oct. 29, 1957; s. David Wittington and Elizabeth Agnes (Pearce) B.; m. Doreen Joan, Sept. 18, 1982. BS in Civil Engring., Queen's U. Kingston, Ont., 1979; M of Applied Sci., U. Waterloo, Ont., 1982. registered profl. engr., Alta., B.C., Yukon territory. From engr.-in-tng. to project engr. Buckland and Taylor, Ltd., North Vancouver, B.C., 1982-89; Mem. Can. Highway Bridge Design Code Subcom. on Evaluation and Rehab. Contbr. numerous articles to profl. jours. NSERC postgraduate scholar 1989-90, 90-91; Isaac Walton Killam Meml. scholar The Killam Trust, 1993; recipient Ralph Steinhauser award of Distinction, Alta. Heritage Scholarship Fund, 1991-92, 92-93, Andrew Stewart Meml. Graduate prize, 1992. Mem. ASCE, Can. Soc. Civil Engring (Donald Jamieson fellow 1990, 91, vice chmn. Edmonton sect. 1993—, Le Prix P. L. Pratley award 1992), Am. Concrete Inst. (contbr. materials jour. 1993), Assn. Preservation Tech. Home: 8534 106A St, Edmonton, AB Canada T6E 4J9

BARTLETT, JEFFREY STANTON, molecular biologist; b. Boston, Oct. 3, 1962; s. Frederick Stewart and Elizabeth (Stanton) B.; m. Kirsten Marie Kenyherz, Aug. 4, 1991. AB with highest honors, Middlebury Coll., 1984; PhD, U. Pitts. Sch. Medicine, 1991. Fellow U. Pitts. Sch. Medicine, 1984-91; rsch. assoc. U. N.C., Chapel Hill, 1991—. Contbr. articles to profl. jours. Recipient Tng. award Nat. Cancer Inst., 1985; dean's fellow U. Pitts. Sch. Medicine, 1984. Mem. AAAS, Am. Soc. for Microbiology. Office: Dept Biochemistry and Biophysics CB 7260 Chapel Hill NC 27599

BARTLETT, PAUL DOUGHTY, chemist, educator; b. Ann Arbor, Mich., Aug. 14, 1907; s. George Miller and Mary Louise (Doughty) B.; m. Mary Lula Court, June 24, 1931; children: Joanna Court (Mrs. Stephen D. Kennedy), Geoffrey McSwain, Sarah Webster (Mrs. James Hester). A.B., Amherst Coll., 1928, Sc.D. (hon.), 1953; M.A., Harvard U., 1929, Ph.D., 1931; Sc.D. (hon.), U. Chgo., 1954; Sc.D. Dr. honoris causa, U. Montpellier, 1967, U. Paris, 1968, U. Munich, 1977. NRC fellow Rockefeller Inst., 1931-32; instr. chemistry U. Minn., 1932-34; mem. faculty Harvard U., 1934-75, prof. chemistry, 1946-75, Erving prof. chemistry, 1948-75, Erving prof. emeritus, 1975—, chmn. dept., 1950-53; Robert A. Welch prof. chemistry Tex. Christian U., 1974-85, prof. emeritus, 1985—; George Fisher Baker lectr. Cornell U., spring 1949; vis. prof. UCLA, 1950; Walker-Ames lectr. U. Wash., 1952; guest lectr. U. Munich, Germany, 1957; speaker 15th Internat. Congress Pure and Applied Chemistry, Paris, France, 1957; Karl Folkers lectr. U. Ill., 1960; Spl. Univ. lectr. U. London, Eng., 1961; lectr. Japan Soc. for Promotion of Sci., 1978; mem. div. com. math., phys. and engring. scis. NSF, 1957-61. Author: Nonclassical Ions, 1965; also chpts. in textbooks, numerous research papers.; Mem. editorial bd.: Jour. Am. Chem. Soc., 1945-55, Jour. Organic Chemistry, 1954-57, Tetrahedron. Recipient award in pure chemistry Am. Chem. Soc., 1938; August Wilhelm von Hofmann gold medal German Chem. Soc., 1962; Roger Adams award organic chemistry, 1963; Willard Gibbs medal, 1963; Theodore William Richards medal, 1966; Nat. Medal of Sci., 1968; James Flack Norris award in phys. organic chemistry Am. Chem. Soc., 1969; John Price Wetherill medal, 1970; Linus Pauling award, 1976; Nichols medal, 1976; James Flack Norris award in teaching chemistry, 1978; Alexander von Humboldt sr. scientist award U. Freiburg, Germany, 1976; Alexander von Humboldt sr. scientist award U. Munich, 1977; Wilfred T. Doherty award, 1980; Max Tishler award Harvard U., 1981; Robert A. Welch award, 1981; Guggenheim and Fulbright fellow, spring 1957. Hon. fellow Royal Soc. Chemistry (London; Centenary lectr. 1969, Ingold lectr. 1975); mem. Swiss Chem Soc. (hon.), Chem. Soc. Japan (hon.), Nat., N.Y. acads. scis., Am. Acad. Arts and Scis., Am. Philos. Soc., Franklin Inst. (hon.), Am. Chem. Soc. (chmn. Northeastern sect. 1953-54, Southwest regional award 1984), Internat. Union Pure and Applied Chemistry (pres. organic div. 1967-69, program chmn. 23d internat. congress 1971), Deutsche Akademie der Naturforscher Leopoldina, Phi Beta Kappa, Sigma Xi, Phi Lambda Upsilon. Achievements include research in kinetics and mechanism organic reactions. Address: Brookhaven A-311 1010 Waltham St Lexington MA 02173

BARTLETT, PETER GREENOUGH, engineering company executive; b. Manchester, N.H., Apr. 22, 1930; s. Richard Cilley and Dorothy (Pillsbury) B.; Ph.B., Northwestern U., 1955; m. Jeanne Eddes, July 8, 1954 (dec. 1980); children: Peter G., Marta, Lauren, Karla, Richard E.; m. Kathleen Organ, July 21, 1984. Engr., Westinghouse Electric Co., Balt., 1955-58; mgr. mil. communications Motorola, Inc., Chgo., 1958-60; pres. Bartlett Labs., Inc., Indpls., 1960-63; assoc. prof. elec. engring. U. S.C., Columbia, 1963-64; dir. research Eagle Signal Co., Davenport, Iowa, 1964-67; div. mgr. Struthers-Dunn, Inc., Bettendorf, Iowa, 1967-74; pres. Automation Systems, Inc., Eldridge, Iowa, 1974-89; pres., chmn. Cybertronics, Inc., Davenport, 1989—. Mem. IEEE. Republican. Presbyterian. Patentee in field. Home and Office: 2336 E 11th St Davenport IA 52803-3701 Office: Cybertronics Inc Davenport IA 52802

BARTLETT, RANDOLPH W., engineer; b. Balt., Oct. 6, 1954; s. Robert B. and Mary E. (Hail) B.; m. Lee Harrel, Apr. 19, 1980; children: Patrick D., Charles T. BS, Va. Tech U., 1976. Registered profl. engr., Va. Design engr. Norfolk (Va.) DPW, 1976-78, head planning and engring., 1978-81; design egnr. William and Tazell, Norfolk, 1980-81; dir. pub. work City of Bedford, Va., 1981-83, Town of Blacksburg, Va., 1983-89; ops. div. chief Dept. of Pub. Works, Arlington, Va., 1989-91; div. chief water, sewer and sts. Arlington County, 1991—. Office: Water Sewer and Sts 4200 S 28th St Arlington VA 22206

BARTLETT, ROBERT WATKINS, academic dean, metallurgist; b. Salt Lake City, Jan. 8, 1933; s. Charles E. and Phyllis (Watkins) B.; m. Betty Cameron, Dec. 3, 1954; children: John C., Robin Parmley, Bruce R., Susanne. BS, U. Utah, 1953, PhD, 1961. Registered profl. engr., Calif. Scientist Ford Aerospace, 1961-64; group leader ceramics SRI Internat., Menlo Park, Calif., 1964-67; assoc. prof. metallurgy Stanford U., Palo Alto, Calif., 1967-74; mgr. hydrometallurgy Kennecott Minerals Co., Salt Lake City, 1974-77; dir. materials lab. SRI Internat., Menlo Park, Calif., 1977-80; v.p. research Anaconda Minerals Co., Tucson, 1980-85; mgr. materials tech. Idaho Sci. and Tech. Dept., Idaho Falls, 1985-87; dean Coll. Mines and Earth Resources, U. Idaho, Moscow, 1987—; dir. Idaho Geol. Survey, Moscow. Contbr. approx. 70 research publs. in metallurgy; 9 patents in field. Served to lt. (j.g.) USN, 1953-56. Recipient Turner award Electrochem. Soc., 1965, McConnell award AIME, 1985. Mem. Metall. Soc. (pres. 1989), Soc. Mining Engrs. (disting. mem.), Sigma Xi, Tau Beta Pi. Office: U Idaho Dept Metall & Mining Moscow ID 83844-3025

BARTLEY, THOMAS LEE, electronics engineer; b. San Fernando, Calif., Nov. 6, 1942; s. George R. Bartley and Collette (Drury) Wescott; children: Cara, Christina. BSEE, Stanford U., 1965, MSEE, 1966. Assoc. elec. engr. IBM, San Jose, Calif., 1966-67; engr. to prin. engr. Gen. Dynamics, San Diego, 1967-80; mktg. mgr. Cubic, San Diego, 1980-82, AMEX, San Diego, 1982-83, SAIC, San Diego, 1983-85; product mktg. mgr. Loral, San Diego, 1985-89; owner Tom Bartley Ideas, San Diego, 1989—; sr. scientist Netrologic, 1993—. Mem. IEEE, Stanford Alumni Assn., Assn. Old Crows (pres., bd. dirs. San Diego chpt. 1985-87). Achievements include rsch. in parallel computer broadcast bus architecture.

BARTOLO, DONNA M., hospital administrator, nurse; b. Springfield, Ill., Mar. 21, 1941; d. Elmer Ralph Bartolomucci and Zoe (Rose) Cavatorta. Diploma in nursing, St. John's Sch. Nursing, Springfield, Ill., 1962; BS, Milliken U., 1976; MS, Sangamon State U., 1978. Pediatric nurse Springfield Clin., 1962-64, physician's asst., 1972-74; gynecol. nurse Watson Clin., Lakeland, Fla., 1964-66; cons. state sch. nurses Office of Edn. State of Ill., Springfield, 1974-78; assoc. dir. operating rm. svcs. Cedars-Sinai Med. Ctr., L.A., 1978-82, co-dir. div. nursing, 1981-82; surg. nurse Emory U. Hosp., Atlanta, 1966-70, asst. dir. of nursing, dir. surg. svcs., 1982—; asst. prof. Nell Hodgson Woodruff Sch. Nursing Emory U. Mem. editorial bd. Perioperative Nursing Quarterly; contbr. articles to nursing jours. Mem. Org. Nurse Execs., Ga. Assn. Nurse Exec. (pres. elect, pres. 1992), Assn. Operating Rm. Nurses, Sigma Theta Tau (sec. 1990—). Home: 1328 Mill Glen Dr Dunwoody GA 30338

BARTON, COLE, psychologist, educator; b. Rule, Tex., Oct. 17, 1946; s. Roger Darl and Ruth Lou (Cole) B.; m. Sydney Theurer, July 20, 1968; children: Bret Cole, Luke Colin, Amanda. BS, U. Utah, 1974, MS, 1981, PhD, 1982. Lic. clin. psychologist, N.C. Asst. prof. psychology U. Utah, Salt Lake City, 1982-83; asst. prof. psychology Davidson (N.C.) Coll., 1983-87, assoc. prof. psychology, 1988—; mem. adv. bd. Family Support Ctr., Charlotte, N.C., 1984-87. Contbr. articles to profl. jours., chpts. to books. Rsch. fellow MacArthur Found., 1986, U. Utah, 1980-81. Mem. APA, Soc. of Behavioral Medicine, Internat. Soc. for Study of Traumatic Stress. Democrat. Presbyterian. Office: Davidson Coll Dept Psychology PO Box 1719 Davidson NC 28036

BARTON, DEREK HAROLD RICHARD, chemist; b. Eng., Sept. 8, 1918; s. William Thomas and Maude Henrietta (Lukes) B.; m. Jeanne Wilkins, Dec. 20, 1944; 1 son, William Godfrey Lukes; m. Christiane Cosnet, Nov. 5, 1969. B.Sc. with 1st class honours, Imperial Coll., London, 1940, Ph.D. in Organic Chemistry (Hofmann prize), 1942; D.Sc., U. London, 1949; D.Sc. (hon.), U. Montpelier, France, 1962, U. Dublin, Ireland, 1964, U. St. Andrews, Scotland, 1970, Columbia U., Scotland, 1970, U. Coimbra, Portugal, 1971, Oxford U., 1972, Manchester U., 1972, U. South Africa, 1973, U. La Laguna, Tenerife, Spain, 1975, U. Western Va., 1975, U. Sydney, Australia, 1976; D (h.c.), U. Valencia, 1979, U. Sheffield, 1979, U. Western Ont., 1979, U. Metz, 1979, Weizmann Inst. Sci., 1979. Govt. research chemist, 1942-44; research chemist Messrs. Albright & Wilson, Birmingham, 1944-45; asst. lectr., then ICI research fellow Imperial Coll., 1945-49; vis. lectr. Harvard U., 1949-50; reader organic chemistry Birkbeck Coll., London, 1950-53, prof. organic chemistry, 1953-55, fellow, 1970-85, emeritus prof., 1978—; Regius prof. chemistry Glasgow U., 1955; Max Tishler lectr. Harvard U., 1956; Aub lectr. Med. Sch., 1962; prof. organic chemistry Imperial Coll., 1957-78; dir. Inst. of Chemistry and Natural Substances, Gif-sur-Yvette, France, 1978-85; prof. chemistry Tex. A&M U., College Station, 1985—; Arthur D. Little prof. MIT, 1958; Karl Folkers vis. prof. univs. Ill. and Wis., 1959; Falk-Plaut lectr. Columbia U., 1961; Renaud lectr. Mich. State U., 1962; inaugural 3M's lectr. U. Western Ont. (Can.), 1962; 3m's lectr. U. Minn., 1963; Sandin lectr. U. Alta., 1969; Graham Young lectr., Glasgow, 1970; Rose Endowment lectr. Bose Inst., Calcutta, 1972; Stieglitz lectr. U. Chgo., 1974; Bachman lectr. U. Mich., 1975; Woodward lectr. Yale U., 1972; 1st Smissman lectr. U. Kans., 1976; Priestley lectr. Pa. State U., 1977; Cecil H. and Ida Green vis. prof. U. B.C., 1977; Benjamin Rush lectr. U. Pa., 1977; Firth vis. prof. U. Sheffield, 1978-79; Romanes lectr., Edinburgh, 1979; mem. Council Sci. Policy U.K., 1965—. Created knight bachelor, 1972; decorated Order Rising Sun 2d class (Japan), 1972; chevalier Legion of Honor (France), officier, 1986, 1974; knight of Mark Twain, 1975; recipient Nobel prize in chemistry, 1969, Royal medal, 1972, B.C. Law Gold medal Indian Assn. Cultivation Scis., 1972, medal Soc. Cosmetic Chemistry Gt. Britain, 1972, medal Union Sci. Workers Bulgaria, 1978, medal U. Sofia, 1978, medal Acad. Scis. Bulgaria, Chemical Pioneer award Am. Inst. Chemists, 1993. Fellow Chem. Soc. (1st Corday-Morgan medal 1951, Tilden lectr. 1952, 1st Simonsen meml. lectr. 1958, Hugo Muller lectr. 1963, Pedler lectr. 1967, Robert Robinson lectr. 1970, 1st award natural product chemistry, 1961, Longstaff medal 1972; pres. Perkin div. 1971; nat. pres. 1973), Royal Soc. (Davy medal 1961), Royal Soc. Edinburgh; fgn. fellow Am. Chem. Soc. (Fritzsche medal 1956, 1st Roger Adams medal 1959, 2d Centennial Priestly Chemistry award 1974), Indian Nat. Sci. Acad.; fgn. assoc. Nat. Acad. Scis.; fgn. hon. mem. Am. Acad. Arts and Scis., Pharm. Soc. Japan, Royal Acad. Scis. Spain; hon. fellow Deutsche Akademie der Naturforscher Leopoldina, Indian Chem. Soc.; hon. mem. Soc. Quimica de Mex., Belgium Chem. Soc., Chilean Chem. Soc., Acad. Pharm. Scis. U.S., Danish Acad. Scis. and Letters, Nacional Acad. Exact, Phys. and Natural Scis. Argentina, Hungarian Acad. Scis., Soc. Italiana per il Progresso delle Scienze, Acad. Scis. France; corr. mem. Argentinian Chem. Soc.; fgn. mem. Acad. des Ciencias da Lisboa, Acad. Nazionale dei Lincei; mem. Brit. Assn. Advancement Sci. (pres. Sect. B, 1969), Internat. Union Pure and Applied Chemistry (pres. 1969). Office: Tex A&M U Chemistry Dept College Station TX 77843-3255 also: Imperial Coll Sci & Tech, Prince Consort Rd, S Kensington, London SW7, England*

BARTON, HENRY DAVID, operations research analyst; b. Wichita Falls, Tex., Jan. 13, 1944; s. Henry Will and Margaret Helen (Cline) B.; m. Salley Elizabeth Reynolds, Aug. 31, 1968 (div. Dec. 1977); 1 child, Sarah Jessica; m. Ingrid Maria Crown, Aug. 10, 1985; 1 child, Katherine Michelle Barton. MA, George Mason U., 1982; PhD, U. Tex. at Austin, 1972. Project asst. sociolinguist Ford Found: Survey of Lang., East Africa, Dares Salaam, East Africa, Dar es Salaam, 1969-70; rsch. psychologist, sociolinguist U.S. Army Rsch. Inst., Alexandria, Va., 1972-79; ops. rsch. analyst, rsch. psychologist U.S. Office of Pers. Mgmt., Washington, 1979-87; ops. rsch. analyst Dept. Def. Inspector Gen., Arlington, Va., 1987—; mem., sec. GS-13 and GM-14 Coun., DoDIG, Arlington, 1989-92. Author: (with others) Language in Tanzania, 1979. Capt. U.S. Army, 1971-74. Doctoral Rsch. grantee NSF, 1971-72, NDEA Title VI fellowship U. Tex., 1966-69. Mem. Ops. Rsch. Soc. Am., Phi Beta Kappa. Achievements include analysis of Fed. Alternative Work Schedules Experiment program evaluation and report to the Pres. and the Congress; analysis and authorship of U.S. Office of Personnel Management internal evaluation of the first two full years of Merit Pay System implementation. Home: PO Box 1106 Fairfax VA 22030 Office: Dept Def Inspector Gen 400 Army Navy Dr Rm 801 Arlington VA 22202

BARTON, JAMES CLYDE, JR., hematologist, medical oncologist; b. Knoxville, Tenn., Oct. 28, 1945; s. James Clyde and Katherine Louise (Jackson) B.; m. Nancy Elaine Huston, June 16, 1977; children: Ellen, Clay. BS summa cum laude, U. Tenn., 1968, MD, 1971. Diplomate Am. Bd. Internal Medicine, Am. Bd. Med. Oncology, Am. Bd. Hematology. Intern and resident dept. medicine U. Tenn./City of Memphis Hosps., 1972-74; chief med. resident U. Tenn.-City of Memphis Hosp., 1974-75; instr. medicine U. Ala., Birmingham, 1976-77, prof. hematology, 1977-90, mem. grad. faculty, 1986—; chief hematology rsch. VA Med. Ctr., Birmingham, 1983-90; dir. bone marrow transplant Brookwood Med. Ctr., Birmingham, 1990—; expert scientist OECD, Paris, 1985-86, Internat. Commn. Radiol. Protection, Chilton, Eng., 1986—; bd. dirs. Iron Overload Diseases Assn., Inc., West Palm Beach, Fla., 1987—, co-founder, med. dir. S.E. unit, 1987. Author books on metal metabolism, clin. hematology; contbr. articles, revs., abstracts to profl. publs. Recipient fellowship Nat. Rsch. Svc., NIH, 1978. Fellow ACP; mem. So. Soc. Clin. Investigation, Am. Fedn. Clin. Rsch., Am. Soc. Hematology, Am. Soc. Clin. Oncology, Histochem. Soc., Phi Beta Kappa, Phi Kappa Phi. Achievements include development of animal model for quantification of absorption and organ distribution of metals, methods for heavy metal and metal-binding protein analysis in cells and tissues, discovery of beneficial effects of hepatitis in acute leukemia, predictability of type of blast transformation in chronic myelogenous leukemia. Office: Brookwood Med Ctr G 105 ACC 2022 Medical Center Dr Birmingham AL 35209

BARTON, LYNDON O'DOWD, mechanical engineer; b. Buxton, Guyana, S.Am.; came to U.S., 1961; s. Lyndon O. and Clarice A. (Stewart) B.; m. Olive Barton, June 27, 1964; children: Rhonda, Loren, Carol, Leon. BSME cum laude, Howard U., 1966; M in Mech. and Aerospace Engring., U. Del., 1972. Process engr. E.I. DuPont de Nemours, Wilmington, Del., 1966-70, 74-79, mech. engr., 1970-74; project engr. E.I. DuPont de Nemours, Wilmington, 1979-86, sr. engr., 1986—; adj. instr. Delta Tech. & C.C., Stanton, 1974—; adj. lectr. in engring. mechanics Del. Tech. and C.C., 1974—, Widner U., Chester, Pa., 1993. Author: Mechanism Analysis, 1984, 2d edit., 1993; contbr. articles to profl. jours. Mem. curriculum com. Christina Sch. Dist., Newark; bd. mem. troop 257 Boy Scouts Am., Newark. Recipient award for contbn. to arts, Com. to Improve Buxton, Washington, 1985. Mem. Am. Soc. Inventors, Howard U. Alumni Assn. (chpt. pres. 1977-79, v.p.), Tau Beta Pi (life, chpt. pres. 1965-66), Beta Kappa Chi, Pi Mu Epsilon. Achievements include patents in mechanical timers for competitors in concentration games such as chess, checkers and scrabble and a game board to teach beginners the rules of bidding as they play bridge. Home: 26 Shull Dr Newark DE 19711 Office: E I duPont de Nemours & Co Louviers Bldg Newark DE 19714

BARTON, MARK QUAYLE, physicist; b. Kansas City, Mo., June 5, 1928; s. Paul and Ruth Florence (Hoffman) B.; m. Lois Joan Schneeberger, Dec. 18, 1954; children: Linda Sue, Mark Alan. AB, Cen. Meth. U., 1950; PhD, U. Ill., 1956. Physicist Brookhaven Nat. Lab., Upton, N.Y., 1955-87; rsch. staff mem. IBM, Yorktown Heights, N.Y., 1987-91; ret., 1991—; mem. high energy physics adv. panel DOE, Washington, 1974-78. Active various musical groups, L.I., N.Y. Recipient Lindgren award Soc. Econ. Geologists, 1992. Fellow IEEE (chair tech. com. particle accelerators Nuclear and Plasma Scis. Soc. 1987-90), Am. Phys. Soc. Achievements include research on intensity limits of particle accelerators, design of controls and diagnostics for accelerators and mgmt. of accelerator facilities. Home: 6 Academy Ln Bellport NY 11713-2702

BARTON, META PACKARD, business executive, medical science research executive; b. Balt., Dec. 2, 1928; d. Charles Lee and Dorothy (Levering) Packard; m. David W. Barton Jr., July 4, 1951 (div. 1989); children: Blair Lee, Meta Walker, Priscilla Taylor, Emilie Packard. AB, Vassar Coll., 1950; MA, Loyola Coll., Balt., 1977. Bus. rsch. Balt. Assn. Commerce, 1950-51; faculty Bryn Mawr (Md.) Sch., 1952-62; staff psychologist Epoch House, Essex, Md., 1973-81; employee benefits coordinator, profit sharing plan adminstr., treas. Barton-Gillet Co., Balt., 1980-90; pres. Friends Med. Sci. Rsch. Ctr., Balt., 1981—, Namaste Enterprises Inc., Balt., 1987—. Mem. APA, Am. Evaluation Assn., Md. Vassar Club (treas.). Office: Friends Med Sci Rsch Ctr 2330 W Joppa Rd Ste 103 Lutherville MD 21093

BARTON, NICK, rock mechanics engineer; b. Leamington, Eng., Aug. 10, 1944; s. John Ryland and Phyl (Fox) B.; children: David, Andrew, Laurence, Karsten. BSc with honours, King's Coll., London, 1966; PhD in Rock Mechanics, Imperial Coll., London, 1971. Sr. engr. Norwegian Geotech. Inst., Oslo, 1971-80; sr staff cons. Terratek Inc., Salt Lake City, 1980-83, mgr. geomechanics, 1983-84; div. dir. Norwegian Geotech. Ins., Oslo, 1984-89, tech. advisor, 1989—; adj. prof. U. Utah, Salt Lake City, 1983-84, U. Luleå, Sweden, 1986-90; prin. investigator for rock mechanics NEA/OECD Internat. Stripa Project, Sweden, 1986-92; geotech. cons. UK Nirex Ltd., Sellafield Radioactive Waste Repository, 1990—; cavern engring. cons. Hong Kong Geotech. Control Office, 1989-91; cons. Xiaolangdi Hydro Project, World Bank, People's Republic China, 1990—, Mingtan Hydro Project, Sinotech, Republic of China, 1987-90, UN Devel. Project, India, 1988, Gjovik Olympic Cavern, Norway, 1991. Contbr. over 120 articles to tech. jours., conf. procs. Recipient award U.S. Nat. Com. Rock Mechanics, 1975, E.B. Burwell Jr. Meml. award Geol. Soc. Am., 1978, Lauritz Bjerrum Meml. lecture, Oslo, 1985, Manuel Rocha Meml. lecture, Lisbon, 1987. Mem. Internat. Soc. for Rock Mechanics (com. for rock joints, com. for scale effects in rock masses, com. for failure mechanisms in underground openings 1988-92, coord. working party recommendations for quantitative description of discontinuities in rock masses 1977-78). Achievements include development of principal international method of characterizing rock masses for designing tunnel and rock cavern support (Q-system, Barton, Lien and Lunde 1974); co-development of Barton-Bandis joint constitutive model for computer modelling of rock mass response to excavation, 1981. Office: Norwegian Geotech Inst, 72 Sognsveien, 0806 Oslo Norway

BARTON, RUSSELL WILLIAM, psychiatrist, author; b. London, Apr. 21, 1923; s. Charles William and Muriel Marguerite (Hart) B.; m. Katherine Grizel Maitland-Makgill-Crichton, July 24, 1954; children: Karen Elizabeth, Sarah Muriel. M.B., BS, U. London, 1949; MD, SUNY, NY, 1990. Diplomate Am. Bd. Neurology and Psychiatry; lic. MD, N.Y. House physician, registrar, psychiat. registrar Westminster Hosp., London, 1948-53; registrar Maudsley Hosp., London, 1953-56; physician supt. Severalls Hosp., Colchester, Eng., 1960-71; dir. Rochester (N.Y.) Psychiat. Center, 1970-77; cons. psychiatrist WHO, 1964, Minn. Dept. Pub. Welfare, 1965; dir. edn. Pilgrim State Hosp., Brentwood, N.Y., 1969, 70; clin. prof. psychiatry N.Y. Sch. Psychiatry, N.Y.C., 1968—; assoc. clin. prof. U. Rochester, 1971—; mem. select coms. mental hosps., med. edn., psychogeriatrics Ministry Health, London, 1963-66. Author: Institutional Neurosis, 1959, 3d edit., 1976, transl. Greek, French, German, Spanish, Dutch, Italian, Japanese, Science and Psychiatry, 1963, A Short Practice of Clinical Psychiatry, 1975, Diabetes Insipidus and Obsessional Neurosis, 1986. Served with Brit. Red Cross, 1945, Belsen Concentration Camp; with Royal Navy, 1948-50. Fellow Royal Coll. Physicians (Can.), Royal Coll. Psychiatrists, Royal Soc. Medicine, Am. Psychiat. Assn., A.C.P.; mem. Royal Coll. Physicians (London). Home and Office: 2322 Clover St Rochester NY 14618-4125

BARTON, WILLIAM ELLIOTT, biologist, administrator; b. Carlisle, Pa., Feb. 20, 1956; s. Frederick Lowell and Dorothy Anne (Haigh) B.; m. Susan Diane Bell, Aug. 5, 1978; 1 child, Lorrin Diane. BS in Biology, Psychology, Grove City Coll., 1978; MEd, U. Md., 1980, PhD in Human Devel., 1990. Co-founder, exec. v.p. Stellar Bio Systems, Inc., Columbia, Md., 1982—. Bd. dirs. Greencastle Lakes Community Assn., Burtonsville, Md., 1986-91, pres., 1991. Mem. Am. Psychol. Soc. Office: Stellar Bio Systems Inc 9075 Guilford Rd Columbia MD 21046

BARTOO, RICHARD KIETH, chemical engineer, consultant; b. Potter County, Pa., June 3, 1938; s. Raymond Eldon and Norma Grace (Butler) B.; m. Nancy Jo Hoebler, Oct. 15, 1988 (div.); children: Scott Lee, Roy Keith. BS in chem. engr., Carnegie Inst. Tech., 1962; MS in chem. engr., Carnegie Mellon Univ., 1972. Process engr. Atlas Chemical Industries, Wilmington, Del., 1962-64, Central Romana By-Products Co., La Romana, Dominican Republic, 1964-66; process engr. rsch. div. Consol Coal Co., Library, Pa., 1966-68; pilot plant engr. The Benfield Corp., Mt. Lebanon, Pa., 1968-72, sr. process engr., 1972-81; sr. process engr. Union Carbide Corp. (now UOP), Tarrytown, N.Y., 1989—. Author: (chapter in book) Acid and Sour Gas Treating Processes, 1984; contbr. articles to profl. jours. With U.S. Army Res., 1956-60. Mem. Am. Inst. Chemical Engrs., Am. Youth Hostel (trip leader). Methodist. Achievements include design, construction, operations, and troubleshooting acid gas scrubbing process (a chemical process called "Hot Potassium Carbonate Process", or Benfield Process used for removal of Co2 and H2S from industrial gas streams). Office: UOP/Union Carbide Corp 777 Old Sawmill River Rd Tarrytown NY 10591

BARTSCH, DAVID LEO, environmental engineer; b. Wheeling, W.Va., Sept. 27, 1953; s. Leo Donald and Mary Anne (Shepard) B.; m. Helen Louise Mauck, Aug. 23, 1975; children: Timothy Aaron, Therese Marie, Jonathan Daniel. BS in Engring. of Mines, W.Va. U., 1976; MS in Mining Engring., Pa. State U., 1980. Registered profl. engr., Pa., Ohio, Conn.; cert. mine foreman, Ohio, W.Va., Pa. Mining engr. So. Appalachian Coal Co., Julian, W.Va., 1979-81; mine-indsl. engr. GM & W Coal Co., Inc., Jennerstown, Pa., 1981-82; ventilation engr./dir. Quarto Mining Co., Powhatan Point, Ohio, 1983-87; prin. engr. TRC Environ. Cons., East Hartford, Conn., 1988; project engr., permitting dir. Ohio Valley Coal Co., Alledonia, 1988—; presenter 2nd Ann. Ventilation Symposium, 1985. Lay leader prayer group St. Mary Ch., St. Clairsville, Ohio, 1984—. Mem. Soc. Mining Engrs. (chmn. health safety com. 1989-90). Republican. Roman Catholic. Avocations: running, hunting, fishing. Office: Ohio Valley Coal Co 56854 Pleasant Ridge Rd Alledonia OH 43902

BARTSCH, DIRK-UWE GUENTHER, research bioengineer, consultant; b. Detmold, Lippe, Germany, June 6, 1965; came to U.S., 1987; s. Guenther and Inge Bartsch. BS, Technische Hochschule, Darmstadt, Germany, 1986; MS, U. Calif.-San Diego, La Jolla, 1989, PhD, 1992. Rsch. asst. U. Karlsruhe, Germany, 1987; rsch. asst. U. Calif.-San Diego, 1988-92, rsch. bioengr., 1992—; cons. to ophthalmic imaging co., 1992—; speaker U. Tex. Health Sci. Ctr., San Antonio, 1992. Contbg. author: Scanning Laser Ophthalmoscopy and Tomography, 1989, Scanning Laser Ophthalmology, Tomography and Microscopy, 1992; contbr. article to Am. Jour. Ophthalmology. Scholar Konrad Adenauer Stiftung, 1985-90, Rotary Internat. Found., 1987-88. Fellow Inst. for Biomed. Engring.; mem. Assn. for Rsch. in Vision and Ophthalmology. Office: U Calif Shiley Eye Ctr 9500 Gilman Dr MS 0946 La Jolla CA 92093-0946

BARZ, RICHARD L., microbiologist; b. Rockford, Ill., Mar. 22, 1955; s. William Edward Barz and Rosemary Alice (Easton) Scott; m. Nancy Ellen Kozakewich, May 14, 1976 (div. Nov. 1985); m. Susan Jane Hennefent, May 12, 1989; children: Megan, Richard L. Jr. BS in Microbiology, Colo. State U., 1975. Microbiologist Leprino Foods, Denver, 1975-76, rsch. technician, 1976-78, mgr. quality assurance, 1978-82, dir. quality assurance, researcher, 1982-86, v.p., 1986—. Mem. Am. Dairy Assn. Achievements include patents in field in IQF Freezing Mozzarella Cheese; Coating Application of Mozzarella Cheese; Same Day Manufacture of Mozzarella Cheese. Home: 5368 E Mineral Circle Littleton CO 80122 Office: Leprino Foods 1830 W 38th Ave Denver CO 80211

BASAPPA, SHIVANAND, computer engineer; b. Kolar, India, Dec. 2, 1954; came to U.S., 1984; s. Basappa Naganna Chandrashekara and Sharadamma (Sharadamma) B.; m. Manasa Shivanand, June 22, 1987. MS in Aerospace Engring., Indian Inst. Sci., Bangalore, 1978; MS in Computer Sci., Johns Hopkins U., 1992. Sr. scientist and engr. Indian Space Rsch. Orgn., Bangalore, 1978-82, project engr., 1982-84; cons. Burroughs Corp., Camarillo, Calif., 1984-85; sr. analyst STX-Hughes Corp., Lanham, Md., 1985-91; sr. engr. Comsat Corp., Washington, 1991-92; computer scientist Computer Scis. Corp., Laurel, Md., 1992—. Recipient Achievement award NASA. Mem. IEEE, AIAA. Home: 10346 Hickory Ridge Rd 426 Columbia MD 21044 Office: Computer Scis Corp 1100 West St Laurel MD 20707

BASAR, RONALD JOHN, research engineer, engineering executive; b. Kingston, Pa., Mar. 28, 1950; s. John and Sophie (Turowski) B.; divorced; children: Amber Lynn, Kimberly Ann. BS in Chem. Engring., Wayne State U., 1972; MS in Electrical Engring., Rochester Inst. Tech., 1976, MBA in Fin., 1979. Registered profl. engr., N.Y. Rsch. engr., mgr. Eastman Kodak Co., Rochester, N.Y., 1972-87; dir. strat. market devel. BAT Industries, Glassboro, N.J., 1987-89; dir. engring. and ops. AM Multigraphics Divsn. AM Internat., Chgo., 1989—, sr. dir. engring. and facilities; cons. RJB Fin. Mktg. Coach Buffalo Grove (Ill.) Swim Team. Mem. Rochester Inst. Tech. Alumni Club. Republican. Achievements include 4 patents for market introduction of line of Green Chemicals, 2 patents for laser imaging master. Home: 912 Shambliss Ln Buffalo Grove IL 60089 Office: AM Internat Inc AM Multigraphics Divsn 1800 W Central Rd Mount Prospect IL 60056

BASAR, TANGUL ÜNERDEM, electrical engineering educator, researcher; b. Dikili, Turkey, May 30, 1951; came to U.S., 1980; d. Hüsnü and Kevser (Yilmaz) Ünerdem; m. Tamer Basar, Dec. 27, 1975; children: Gözen, Elif. PhD, Bogazici U., Istanbul, Turkey, 1978. Rsch. engr. Marmmara Rsch. Inst., Istanbul, 1975-76; rsch. asst. Bogazici U., 1976-78; rsch. fellow Twente U. Tech., The Netherlands, 1978-79; asst. prof. Tech. U. Istanbul, 1979-80; assoc. prof. Ill. Inst. Tech., Chgo., 1986-89; lectr. elec. engring. U. Ill., Urbana, 1981-86, 89—, rsch. asst., 1989—. Mem. IEEE, Sigma Xi. Achievements include research on optimum linear causal coding schemes in the presence of correlated jammings. Office: U Ill ECE Dept 155 Everitt Lab 1406 W Green St Urbana IL 61801-2991

BASARAN, OSMAN A., chemical engineer; b. Istanbul, Turkey, July 31, 1956; came to U.S., 1974; s. Turgut H. and Leyla E. (Balan) B.; m. Sonya M. Swintek, Aug. 12, 1983; children: Benjamin, Eric. BS, MIT, 1978; PhD, U. Minn., 1984. Rsch. engr. Air Products and Chems., Inc., Allentown, Pa., 1984-88; group leader Oak Ridge (Tenn.) Nat. Lab., 1988--; mem. NASA Drops and Colloids Panel, Washington, 1992. Mem. Am. Inst. Chem. Engrs. (Best Paper award 1991, mem. nat. fluid mechanics com. 1992—), Am. Phys. Soc., Soc. Indsl. and Applied Math. Achievements include research on drop and fluid dynamics. Office: Oak Ridge Nat Lab PO Box 2008 Oak Ridge TN 37831-6224

BASAVANHALLY, NAGESH RAMAMOORTHY, opto-electronics packaging engineer; b. Mysore, Karnataka, India, Dec. 9, 1952; came to U.S., 1974; s. Ramamoorthy and Sithalakshamma Basavanhally; m. Jayanthi Basavanhally, June 14, 1982; children: Ajay, Naveen. BE, U. Mysore, 1974; PhD, U. Pitts., 1980. Registered profl. engr., Pa. Head of mechanics Schneider Cons. Engrs., Bridgeville, Pa., 1979-84; mem. tech. staff AT&T Bell Labs., Princeton, N.J., 1984—. Contbr. articles to Internat. Jour. of Solids and Structures and Procs. of 41st, 42d, 43d, Elec. Components and Tech. Conf. Mem. ASME. Achievements include research on optical fiber alignment method, optical fiber alignment apparatus, optical laser connector, electronic device manipulating apparatus and methods, methods and apparatus for inserting pins in a substrate. Home: 3 Moro Dr Mercerville NJ 08619 Office: AT&T Bell Labs Carter Rd Hopewell NJ 08525

BASCH, MICHAEL FRANZ, psychiatrist, psychoanalyst, educator; b. Berlin, Germany, May 5, 1929; came to U.S., 1939; s. Martin and Lilli (Hess) B.; m. Carol Gay Shoff, Dec. 14, 1957; children: Gail Marie, Thomas Michael, John Steven. BS, Loyola U., Chgo., 1949; MD, Loyola U., 1953; cert., Inst. Psychoanalysis, Chgo., 1966. Diplomate Am. Bd. Psychiatry. Intern Michael Reese Hosp., Chgo., 1953-54, resident, 1956-59; Pvt. practice Chgo., 1959—; clin. prof. psychiatry Rush Med. Coll., Chgo., 1984-92, Cynthia Oudejans Harris M.D. prof. psychiatry, 1992—; prof. psychiatry U. Chgo., 1979-83. Author: Doing Psychotherapy, 1980, Understanding Psychotherapy, 1988, Practicing Psychotherapy, 1992; contbr. 60 articles to profl. jours. Lt. USNR (Med. Corps), 1954-56. Fellow Am. Psychiat. Assn. Office: M F Basch MD 55 E Washington St Chicago IL 60602

BASCOM, WILLARD NEWELL, research engineer, scientist; b. N.Y.C., Nov. 7, 1916; s. Willard Newell and Pearle (Boyd) B.; m. Rhoda Nergaard, Apr. 15, 1946; children: Willard, Anitra. Student, Colo. Sch. Mines; D in Natural Scis. (hon.), U. Genoa, Italy, 1992. Registered profl. engr., Fla., D.C. Research engr. U. Calif., Berkeley, 1945-50, Scripps Inst. Oceanography, La Jolla, Calif., 1950-54; exec. sec., dir. Mohole Project Nat. Acad. Scis., Washington, 1954-62; pres. Ocean Sci. and Engring., Inc., Washington, 1962-72; dir. Coastal Water Research Project, Long Beach, Calif., 1973-85; cons. to govt. and industry; leader 1st Russian-Am. archaeol. expdn. to Black Sea, 1991, 1st systematic search of deep waters of Sea of Marmara for ancient shipwrecks, Turkey. 1992. Author: Waves and Beaches, 1964, A Hole in the Bottom of the Sea, 1961, Deep Water, Ancient Ships, 1976, The Crest of the Wave, 1988; over 100 articles; patentee deep ocean search/recovery system. Recipient Disting. Achievement medal Colo. Sch. Mines, 1979, Compass Disting. Achievement award Marine Tech. Soc., 1970, Rolex award, 1993. Clubs: Explorers (Explorers medal 1980), Adventurers.

BAS CSATARY, LASZLO KALMAN, anesthesiologist, cancer researcher; b. Kolbasa, Hungary, Dec. 26, 1923; came to U.S., 1957; s. Gyula and Jolan (Szabo) C.; m. Magda M. Horvath, Sept. 1956 (div. 1961); 1 child, Christine; m. Eva Maria Oszter, June 11, 1963. BA, Hunfalvy Gymnasium, Kassa, Hungary, 1942; MD, Pazmany Peter U., 1949. Dep. head dept. Tetenyi Hosp. Ear-Nose-Throat Dept., Budapest, Hungary, 1950-57; resident Washington Hosp. Ctr., 1957-60; anesthesiologist Nat. Orthopedic Hosp., Alexandria, Va., 1960-63, Casualty Hosp., Washington, 1963-65; pvt. practice Washington, 1965-69, Arlington, Va., 1969-82; chmn. bd. United Cancer Rsch. Inst., Alexandria, 1983—. Mem. AMA, Med. Assn. Va., N.Y. Acad. Scis. Achievements include patents for method for treating viral diseases, intra-uterine contraceptive device, sickbed with hammock; development of method to prevent active hepatitis B to proceed into chronic phase. Home: 2100 S Ocean Ln # 2503 Fort Lauderdale FL 33316

BASELT, RANDALL CLINT, toxicologist; b. Chgo., Feb. 12, 1944; s. Benjamin Oliver and Vivian Marie (Rende) B.; m. Lana Mak, June 11, 1966; 1 child, David. BS in Chemistry, U. Ill., 1965; PhD in Pharmacology, U. Hawaii, 1972. Cert. Am. Bd. Forensic Toxicology, Am. Bd. Clin. Chemistry, Am. Bd. Toxicology, forensic alcohol supr., clin. toxicologist technologist, clin. chemist, clin. lab. toxicologist. Forensic toxicologist Office of Coroner, County of Orange, Calif., 1965-69; rsch. fellow dept. pharmacology U. Hawaii Sch. Medicine, Honolulu, 1969-72; NIH postdoctoral rsch. fellow Medizinisch-Chemisches Inst., U. Bern (Switzerland) Sch. Medicine, 1972-73; rsch. toxicologist Office of Coroner, San Francisco, 1973-75; chief toxicologist Office of Med. Examiner, Farmington, Conn., 1975-78; dir. toxicology and drug analysis lab. U. Calif. Med. Ctr., Sacramento, 1978-84; dir. Chem. Toxicology Inst., Calif., 1984—; asst. prof. lab. medicine U. Conn. Health Ctr., Farmington, 1975-78; assoc. prof. pathology U. Calif. Sch. Medicine, Davis, 1978-84; cons. drug abuse USN, 1983—, USAF, 1984—; accredited lab. inspector Nat. Lab. Cert. Program, 1988—. Author: Disposition of Toxic Drugs and Chemicals in Man, 3d. edit., 1989, Biological Monitoring Methods for Industrial Chemicals, 2d edit., 1988, Analytical Procedures for Therapeutic Drug Monitoring and Emergency Toxicology , 2d edit., 1987, (with M. Houts and R.H. Cravey) Courtroom Toxicology, 1980; editor 7 other books; founder, editor Jour. Analytical Toxicology, 1977—; mem. editorial bd. Jour. Forensic Scis., 1983—; contbr. articles to profl. jours. Mem. Am. Soc. Clin. Toxicology, Am. Assn. for Clin. Chemistry, Am. Indsl. Hygiene Assn., Calif. Assn. Toxicologists (past pres.), Internat. Assn. Forensic Toxicologists, Jour. Am. Med. Assn. (peer rev. com. 1985—), Soc. Forensic Toxicologists (bd. dirs. 1978-80, lab. survey com. 1982-83), Soc. Toxicology, Southwestern Assn. Toxicologists. Office: Chem Toxicology Inst 1167 Chess Dr # E Foster City CA 94404-1112

BASH, FRANK NESS, astronomer, educator; b. Medford, Oreg., May 3, 1937; s. Frank Cozad and Kathleen Jane (Ness) B.; m. Susan Martin Fay, Sept. 10, 1960; children—Kathryn Fay, Francis Lee. B.A., Willamette U., 1959; M.A. in Astronomy, Harvard U., 1962; Ph.D., U. Va., 1967. Staff scientist Lincoln Lab. MIT, 1962; assoc. astronomer Nat. Radio Astronomy Obs., Green Bank, W.Va., 1962-64; rsch. asst. U. Va., 1965-67; postdoctoral faculty assoc. U. Tex., Austin, 1967-69, asst. prof. astronomy, 1969-73, assoc. prof., 1973-81, prof., 1981—, Frank N. Edmonds Regents prof., 1985—; chmn. dept. astronomy U. Tex., 1983-86, dir. W.J. McDonald Obs., 1989—; mem. astronomy adv. panel NSF, 1988-91; chmn. vis. com. Nat. Radio Astronomy Obs., 1990, mem., 1990—; mem. vis. com. Aricebo Obs., 1990—; mem. planning com. NASA Astrophys. Data System, 1991—. Author: (with Daniel Schiller and Dilip Balamore) Astronomy, 1977; contbr. articles to profl. jours. Grantee NSF, 1967—, The Netherlands NSF, 1979, W.M. Keck Found., 1988. Mem. Am. Astron. Soc., Internat. Astron. Union, Internat. Sci. Radio Union, Tex. Assn. Coll. Tchrs. (pres. U. Tex. chpt. 1980-82), Tex. Philos. Soc. Club: Town and Gown (Austin). Office: Univ of Texas RLM 15 208 Austin TX 78712

BASHAM-TOOKER, JANET BROOKS, geropsychologist, educator; b. Hampton, Va., Sept. 27, 1919; d. Thomas Westmore and Cora Evelyn Brooks; m. Linwood Cecil Basham (div. 1968); m. Frederick Fitch Tooker. BA cum laude, U. N.C., Greensboro, 1948; ABD in Psychology, Calif. State U., L.A., 1981; MA in Human Devel., Pacific Oaks Coll., 1984. Tchr., Calif. Grad. asst. psychology Duke U., Durham, N.C., 1948-49; tchr. Albuquerque City Schs., 1950-51; tchr. L.A. City Schs., 1953-54, counselor, 1981; lectr. L.A., 1988—; docent Las Angelitas del Pueblo, L.A., 1971-74; active project with autistic children, through Pepperdine Univ., UCLA Neuropsychiatric Inst., L.A., 1974. Author numerous poems. Mem. planning com., women's conf. Commn. on Status of Women, Pasadena, Calif., 1982-85, sr. com. Task Force on Aging, San Marino, Calif., 1986-89, United Way, Arcadia, Calif., 1984-88, Symphony Guild, Fayetteville, 1990; adv. mem. San Gabriel Presbytery Commn. on Aging, 1984-88; mem. grad. studies subcom. Calif. State U., L.A., 1975-78; v.p. San Marino Aux. Meth. Hosp., Arcadia, Calif., 1985-86; docent Duarte Hist. Soc., Calif., 1986-89; moderator sr. adults 1st United Presbyn. Ch., Fayetteville, 1990; facilitator fin. info. program for women AARP, Fayetteville, 1990; vol. in gerontology Fayetteville (Ark.) City Hosp., 1991, Health Care Unit, Butterfield Trail Village, Fayetteville, 1993; adv. com. Single Parent Scholarship Fund, Fayetteville, 1992; mem. League of Women Voters, 1991-93. Recipient Margaret Noffsinger award Va. Intermont Coll., 1937. Mem. AAUW, Am. Soc. Aging, Mental Health Assn., Older Women's League, Flaming Hills Garden Club, League of Woman Voters, Phi Beta Kappa, Phi Theta Kappa. Republican. Presbyterian. Avocations: art, art history, writing poetry.

BASHAW, MATTHEW CHARLES, physicist; b. LaPorte, Ind., Oct. 22, 1963. BS, U. Notre Dame, 1985; MS, Yale U., 1987, MPhil, 1988, PhD, 1991. Postdoctoral fellow Stanford (Calif.) U., 1991—. Contbr. articles to profl jours. including Applied Physics Letters, Phys. Rev. B., Jour. Optical Soc. Am. B., Applied Optics and Optic Letters. Mem. Optical Soc. Am., Am. Phys. Soc., Soc. Francaise D'Optique, Sigma Xi. Achievements include research on new concepts in volume holography, nonlinear optics and optical information systems. Office: Stanford U Dept Electrical Engring Stanford CA 94305-4035

BASHFORD, JAMES ADNEY, JR., experimental psychology researcher, educator; b. Ontario, Oreg., Dec. 17, 1948; s. James Adney and Juanita Marie (Veristain) B.; m. Michael M. MacClarence, Sept. 8, 1973; children: Sean T., Gillian W. BA, Portland State U., 1971; MS, Western Wash. State U., 1973; PhD, U. Wis., 1984. Rschr. in exptl. psychology U. Wis., Milw., 1985—; instr. Alverno Coll., Milw., 1986—. Contbr. articles to Perception and Psychophysics, Jour. of the Acoustical Soc. Am. Mem. Psychonomic Soc., Acoustical Soc. Am., Am. Psychol. Soc. Home: 1153 E Sylvan Whitefish Bay WI 53217 Office: U Wis Milw 2441 E Hartford Milwaukee WI 53211

BASHIAS, NORMAN JACK, software engineer, computer consultant; b. Long Island City, N.Y., Nov. 24, 1962; s. Zaharias Naoum and Helen Maria (Samaras) B. BA, NYU, 1984, MS, 1988, postgrad. Computer specialist U.S. Dept. Treasury, N.Y.C., 1984-87; software engr. Neometrics, Inc., Northport, N.Y., 1987-88, Arthur Young and Co., N.Y.C., 1988-89; computer cons. Strategic Application Systems, Inc., Southport, Conn., 1990-92; adj. lectr. Coll. S.I. CUNY, 1993—. Mem. IEEE Computer Soc., Assn. Computing Machinery, Am. Assn. Artificial Intelligence, N.Y. Acad. Sci., Pan-Macedonian Youth Assn. (sec. 1992—). Home: Apt 2 32-12 54th St Woodside NY 11377

BASHIR, NABIL AHMAD, biochemist, educator; b. Amman, Jordan, Mar. 30, 1954; s. Ahmad Mohammad and Aysha Hassan (Shehada) B.; m. Iman Kamel, June 24, 1983; children: Aseil, Hadeel, Mohammad. PhD, SUNY, Buffalo, 1987. Chmn. med. diag. lab. U. Health Ctr., Irbid, Jordan, 1987-91; assoc. prof. biochemistry, chmn. dept. Med. Sch. Jordan U. Sci. and Tech., Irbid, 1987—; chief biochemistry lab. Princess Basma Teaching Hosp., Irbid, 1987—; mem. biotech. com. Higher Coun. Sci. and Tech., Amman, 1988—. Contbr. articles to profl. jours. Mem. Am. Assn. Clin. Chemistry, Jordan Assn. Clin. Biochemistry. Office: U Sci and Tech Med Sch, Dept Biochemistry, Irbid Jordan

BASILIER, ERIK NILS, software scientist and engineer; b. Töre, Lapland, Sweden, Apr. 7, 1947; came to U.S., 1981; s. Stig Anders Basilier and Anna Britta (Arvidson) Basilier Blau. M in Engring., U. Uppsala, Sweden, 1972, D in Exptl. Physics, 1980. Software engr. Geosource, Houston, 1984-86, Syntech, Dallas, 1986-87; pvt. cons. Houston, 1988; software engr. Computer Assistance, Inc., Dallas, 1988; software engr. Motorola Semiconductor, Phoenix, 1988-90, engr., scientist sr. staff, 1990-92, engr., scientist prin. staff, 1992—; bd. dirs. Cactus Lug: Digital Equip. Corp. Local Users' Group, Phoenix, 1991—. Contbr. articles to Proceedings of Decus. Mem. IEEE, Digital Equip. Corp. Users Soc. Achievements include research in security software. Home: 774 W Chandler Blvd Chandler AZ 85224

BASINGER, WILLIAM DANIEL, computer programmer; b. Washington, Feb. 14, 1952; s. James Samuel and Eleanor (Freeburger) B.; m. Martha Kecskes, July 1, 1978 (div. 1983); m. Mary Teresa Richardson, June 11, 1988. BA in Linguistics, U. Md., 1974; MS in Linguistics, Georgetown U., 1977; MS in Computer Sci., Johns Hopkins U., 1989. Programmer Evaluation Techs., Arlington, Va., 1977-78; programmer, analyst, cons. Vitro Corp., Silver Spring, Md., 1978-84, 87-88; programmer, analyst Tracor Applied Scis., Rockville, Md., 1984-88, Planning Rsch. Corp., McLean, Va., 1988-89; sr. programmer, analyst Systems & Computer Tech. group George Washington U., Washington, 1989—; cons. in applications software dept. Geology George Washington U., Washington, 1990-91. Contbr. articles to profl. jours. Contbr., sponsor Statue of Liberty/Ellis Island Found., N.Y.C., 1985—. Md. State Sen. scholar U. Md., 1970-74. Mem. Assn. Computing Machinery, N.Y. Acad. Sci. Republican. Roman Catholic. Home: 11342 Cherry Hill Rd T-203 Beltsville MD 20705 Office: Systems & Computer Tech George Washington U B-148 801 22d St NW Acad Ctr Washington DC 20052

BASKERVILLE, CHARLES ALEXANDER, geologist, educator; b. Jamaica, N.Y., Aug. 19, 1928; s. Charles H. and Annie M. (Allen) B.; m. Susan Platt, July 5, 1979; children: Mark Dana, Shawn Allison, Charles Morris, Thomas Marshall. BS, CCNY, 1953; MS, NYU, 1958, PhD, 1965. Cert. geologist, Maine, Ind.; cert. profl. geologist. Asst. civil engr. N.Y. State Dept. Transp., Babylon, 1953-66; prof. engring. geology CUNY, N,Y.C., 1966-79, dean sch. of gen. studies, 1970-79, prof. emeritus, 1979—; project rsch. geologist U.S. Geol. Survey, 1979-90; prof. geology Cen. Conn.

State U., New Britain, 1990—, dept. chmn., 1992—; Commonwealth vis. prof. George Mason U., Fairfax, Va., 1987-89; mem. U.S. Nat. Com. on Tunnelling Tech. NAS, chmn. subcom. on edn. and tng.; mem. Am. del. Internat. Tunnelling Assn. to Internat. Colloquium of Tunnelling and Underground Works, Beijing, People's Republic of China, 1984; mem. adv. com. earth scis. divsn. NSF, 1989-92; cons. in field, guest lectr. various colls., 1964—. Author numerous sci. papers. Mem. com. for minority participation in the geosciences U.S. Dept. of Interior, 1972-75; panelist Grad. Fellowship Program, NSF; chmn. Minority Grad. Fellowship Program, 1979-80. Recipient Founders Day award N.Y. U., 1966, 125th Anniversary medal The City Coll., 1973, award for excellence in engring. geology Nat. Consortium Black Profl. Devel., 1978. Fellow Geol. Soc. Am. (mem. com. on minorities in geoscis., chmn. com. on coms. 1989, Nat. Rsch. Coun. (liaison to U.S. nat. com. on tunnelling tech. engring. and tech. systems 1992—); mem. N.Y. Acad. Scis., Geol. Soc. Washington, Am. Inst. Profl. Geologists, Assn. Engring. Geologists (rep. to nat. bd. dirs. 1973-74, chmn. N.Y.-Phila. sect. 1973-74), Internat. Assn. Engring. Geology, Yellowstone-Bighorn Rsch. Assn., Sigma Xi. Office: Cen Conn State U Dept Physics and Earth Scis 1615 Stanley St New Britain CT 06050-4010

BASKHARONE, ERIAN AZIZ, mechanical and aerospace engineering educator; b. Cairo, Egypt, Sept. 28, 1947; came to U.S. 1974; s. Aziz and Mofida (Khlil) B.; m. Samia Lamie Massoud, Sept. 24, 1978; children: Richard, Robert. MS in Aerospace Engring., U. Cin., 1975, PhD in Aerospace Engring., 1979. Registered profl. engr., Tex. Asst. lectr. Coll. Engring., U. Cairo, 1970-74; rsch. asst. U. Cin., 1974-79, postdoctoral fellow, 1979-80; sr. engr. Garrett Turbine Engine Co., Phoenix, 1980-85; assoc. prof. mech. and aerospace engring. Tex. A&M U., College Station, 1985—; cons. Aux. Power Units div. Allied Signal Aerospace Co., Phoenix, 1989—. Contbr. articles to profl. jours. Recipient Disting. Teaching award Amoco Found., 1992, award of Excellence in Engring. Teaching, Gen. Dynamics Corp., 1991, cert. of Recognition by the Inventors and Contbns. Bd., NASA Lewis Rsch. Ctr., 1983. Mem. ASME (turbomachinery com. 1987—), AIAA, Am. Soc. Engring. Edn., Sigma Gamma Tau. Democrat. Orthodox Christian. Achievements include patent for radial inboard pre-swirl system for turboprop engine cooling; research on turbomachinery flow fields and non-traditional perturbation models in the mathematical and application aspects of the finite-element method. Home: 2706 Echo Glen Cir Bryan TX 77803 Office: Tex A&M Univ Dept Mech Engring College Station TX 77843-3123

BASMAJIAN, JOHN VAROUJAN, medical scientist, educator, physician; b. Constantinople, Turkey, June 21, 1921; came to Can., 1923, naturalized, 1927; s. Mihran and Mary (Evelian) B.; m. Dora Belle Lucas, Oct. 4, 1947; children—Haig, Nancy, Sally. M.D. with honors, U. Toronto, Ont., Can., 1945. Intern Toronto Gen. Hosp., 1945; surg. resident Sunnybrook Hosp. and Hosp. for Sick Children, Toronto, 1946-48; from lectr. to prof. U. Toronto, Ont., Can., 1949-57; prof. anatomy, chmn. dept. anatomy Queen's U., Kingston, Ont., Can., 1957-69; prof., dir. regional rehab. research and tng. ctr. Emory U., Atlanta, 1969-77; prof. medicine McMaster U., Hamilton, Ont., Can., 1977-86; prof. emeritus McMaster U., Hamilton, Ont., 1986—; dir. rehab. ctr. Chedoke-McMaster Hosps., Hamilton, 1977-86; exec. sec. Banting Rsch. Found., Toronto, 1954-57; chmn. rsch. com. Fitness Coun. Can., Ottawa, Ont., 1965-69; spl. cons. med. rsch. Ga. Inst. Tech., Atlanta, 1984—; dir. rsch. and tng. grants Ea. Seal Rsch. Inst., Toronto; bd. dirs. Can. Physiotherapy Found., Toronto, 1984—; lectureships in Europe, Asia, South Am., Australia, Japan, others. Author 11 med. sci. and clin. books in multiple edits. and transls., 1953—; editor 6 med. clin. books in multiple edits., and transls., 1977—; series editor: Rehabilitation Medicine Library, 22 vols., 1977—; editorial bd. Am. Jour. Phys. Medicine, 1968—, Am. Jour. Anatomy, 1971-74, Electromyography and Clin. Neurophysiology, 1966—, Electro-diagnostic-therapy, Physiotherapy Can., 1979—, Jour. Motor Behavior, 1980—, Med. Post; assoc. editor Anat. Record, 1970-73, BMA Audiotape Series, 1970-77; contbr. articles to profl. jours., 1950—; producer several motion pictures; inventor sci. and med. devices and techniques. Mem. and chmn. Bd. Edn., Kingston, Ont., 1960-68; founding chmn. bd. govs. St. Lawrence Coll. Applied Arts and Tech., Ont., 1964-69. Served to capt. M.C., Can. Army, 1943-46. Decorated officer Order of Ont., 1991; recipient awards including Starr Gold medal U. Toronto, 1957, Kabakjian award Armenian Youth Fedn., 1967; N.R.C. (Can.) vis. scientist Soviet Acad. Scis., 1963, Henry Gray Laureate, 1991, Order of Ontario, 1991. Fellow Am. Acad. Angiology, Royal Coll. Physicians (Can.), Royal Coll. Physicians and Surgeons (Glasgow, hon.), Coll. Rehabilitative Medicine (Australia, hon.); mem. Am. Assn. Anatomists (pres. 1985-86, Henry Gray Laureate award 1991), Can. Assn. Anatomists (founding, sec. 1965-69, J.C.B. Grant award 1985), Am. Congress Rehab. Medicine (Gold Key award 1977, Coulter lectr. 1988), Biofeedback Soc. Am. (founding, pres. 1978-79), Internat. Soc. Electromyographic Kinesiology (founding, pres. 1955-60); hon. life mem. Order St. John of Jerusalem, Am. Orthopedic Foot Soc., Australian Biofeedback Soc., Venezuelan Biofeedback Soc., Mexican Soc. Anatomy, Colombian Assn. Phys. Medicine. Avocations: travel; music; gardening; writing. Office: McMaster U Med Sch, Box 2000 Sta A, Hamilton, ON Canada L8N 3Z5

BASNIGHT, ARVIN ODELL, public administrator, aviation consultant; b. Manteo, N.C., Sept. 14, 1915; s. Thomas Allen and Mary Meekins Basnight; m. Marjorie Jane Gauthier, Dec. 6, 1942; children: Mary Ann Basnight Wolf, William Gaylord, Michael André. Student in Mech. Engring., N.C. State U., 1932-35; student in Pub. Adminstrn., Am. Univ., 1936-42. Park ranger U.S. Nat. Parks, Kitty Hawk, N.C. and Mesa Verde, Colo., 1938-40; pers. adminstr. CAA, Washington, 1940-42; fin., budget accounts FAA, Washington, 1945-62; adminstr. Washington, Atlanta, L.A., 1962-74; bd. dir., chmn. bd. Palos Verdes (Calif.) Nat. Bank, 1982-92; aviation cons., L.A., 1974-92. Co-chmn. Salute to Doolittle Raiders, Santa Monica, Calif., 1991; pres. Palos Verdes Breakfast Club, 1990. Maj. USAF, 1942-45. Decorated D.F.C.; recipient Disting. Svc. medal U.S. Fed. Aviation, 1966. Mem. Nat. Aeros. Assn. (Honor award 1991), Aero Club So. Calif. (pres., bd. dirs. 1974-92). Achievements include development of air safety programs and certification for Lockheed C-141 and 1011, MacDonald Douglas MD-10, Boeing 737 and 747. Home: 1536 Paseo Del Mar Palos Verdes Estates CA 90274

BASOLO, FRED, chemistry educator; b. Coello, Ill., Feb. 11, 1920; s. John and Catherine (Marino) B.; m. Mary P. Nutley, June 14, 1947; children: Mary Catherine, Freddie, Margaret-Ann, Elizabeth Rose. BE, So. Ill. U., 1940, DSc (hon.), 1984; MS, U. Ill., 1942, PhD in Inorganic Chemistry, 1943; LLD (hon.), U. Turin, 1988. Rsch. chemist Rohm & Haas Chem. Co., Phila., 1943-46; mem. faculty Northwestern U., Evanston, Ill., 1946—, prof. chemistry, 1958—, Morrison prof. chemistry, 1980-90, chmn. dept. chemistry, 1969-72; Charles E. and Emma H. Morrison prof. emeritus Nortwestern U., Evanston, Ill., 1990—; guest lectr. NSF summer insts.; chmn. bd. trustees Gordon Rsch. Conf., 1976; pres. Inorganic Syntheses, Inc., 1979-81; mem. bd. chem. scis. and tech. NRC-Nat. Acad. Scis.; adv. bd. Who's Who in Am., 1983; cons. in field. Author: (with R.G. Pearson) Mechanisms of Inorganic Reactions, 1958, (with R.C. Johnson) Coordination Chemistry, 1964; assoc. editor Chem. Revs., 1960-65, Inorganica Chemica Acta, 1967—, Inorganica Chemica Acta Letters, 1977—; editorial bd. Jour. Inorganic and Nuclear Chemistry, 1959—, Jour. Molecular Catalysis, Chem. Revs.; co-editor Catalysis, Transition Metal Chemistry; editor Inorganic Syntheses XVI; contbr. articles to profl. jours. Recipient Ballar medal, 1972, So. Ill. U. Alumni Achievement award, 1974, Dwyer medal, 1976, James Flack Norris award for Outstanding Achievement in Teaching of Chemistry, 1981, Oesper Meml. award, 1983, IX Century medal Bologna U., 1988, Mosher award, 1990, Padova U. medal, 1991, Chinese Chem. Soc. medal, 1991, G.C. Pimental award, 1992, Chemical Pioneer award, 1992, Gold medal Am. Inst. Chemists, 1993; Guggenheim fellow, 1954-55; NSF fellow, 1961-62; NATO sr. scientist fellow Italy, 1981; Sr. Humboldt fellow, 1992. Fellow NAS, AAAS (chmn. chemistry sect. 1979), Am. Acad. Arts and Scis.; mem. Am. Chem. Soc. (asst. editor jour. 1961-64, chmn. div. inorganic chemistry 1970, pres. 1983, bd. dirs. 1982-84, award for research in inorganic chemistry 1964, Disting. Service award in inorganic chemistry 1975, N.E. regional award 1971, award in chem. edn. 1992, Chem. Pioneer award 1992), Chem. Soc. London, Italian Chem. Soc. (hon.), Acad. Nat. dei Lincei (Italy), Sigma Xi (Monie A. Ferst medal 1992), Phi Lambda Upsilon, Alpha Chi Sigma, Phi Kappa Phi, Kappa Delta Phi, Phi Lambda Theta (hon.). Office: Dept Chemistry Northwestern U 2145 Sheridan Rd Evanston IL 60201

BASOV, NIKOLAI GENNADIEVICH, physicist; b. Usman, nr. Voronezh, USSR, Dec. 14, 1922; s. Gennadiy Fedorovich and Zinaida Andreevna (Molchanova) B.; m. Ksenia Tikhonovna, July 18, 1950; children: Gennadiy, Dmitriy. Grad., Moscow Mech. Inst., 1950, Can. Phys. Math. Sci., 1953, D. Phys. Math. Sci., 1956; LL.D. (hon.), Polish-Mil.-Tech. Acad., 1972, Jena U., 1974, Prague Poly. Inst., 1975, U. Pavia, Italy, 1977, Madrid Poly. U., 1985; Karl Marx Stadt Tech. U., 1988. With P. N. Lebedev Phys. Inst., USSR Acad. Sci., 1948—, vice dir. for sci. work, 1958-73, head lab. quantum radio physics, 1963—; prof. solid state physics Moscow Inst. Phys. Engrs., 1963—; dir. P. N. Lebedev Phys. Inst., 1973-89, dir. quantum radiophysics div., 1989—; mem. expert coun. of prime-minister of Russian Govt., 1991—. Author over 500 works. Research on principle of molecular generator, 1952, realized molecular generator on molecular beam of ammonia, 1955, 3-level system for receiving states with inversal population suggested, 1955, proposed use of semicondrs. for creation lasers, 1958, realized various types of semicondr. lasers with excitation through p-n junctions, electronic and optical pumping, 1960-65, research on obtaining short powerful pulses of coherent light; proposed thermal and chem. methods for laser pumping, 1962, gas dynamic lasers, 1966; research optical data processing, 1965—; proposed, 1961, realized thermonuclear reactions by using powerful lasers, 1968; developed main trends in optical frequency standards, 1967-68; inventor electron-beam pumped semicondr. laser projection TV, 1968; proposed, 1966, realized eximer lasers, 1970; realized stimulation of chem. reactions by infrared laser radiation, 1970; proposed and realized electro-ionization laser, 1971; proposed concept of low-entropy compression of high-aspect ratio multilayer thermonuclear targets, 1974, showed possibility of their stable compression, 1983; realized lasers with long-time stability of 2.10-14, 1982; chief editor Priroda, 1967-90, Kvantovaya Elektronika, 1971—. Chmn. bd. All-Union Soc., Znanie, 1978-90, hon. chmn., 1990—; dep. USSR Supreme Soviet, 1974-89; mem. presidium Supreme Soviet, 1982-89; v.p. exec. coun. World Fedn. Sci. Workers, 1976-83; v.p. WFSW, 1983-90, hon. mem., 1990—. Decorated Order Lenin (5), Order of Patriotic War hero twice Socialist Labour; recipient Lenin prize, 1959, Nobel prize for fundamental rsch. in quantum electronics resulting in creation of masers and lasers, 1964, Gold medal Czechoslovakian Acad. Scis., 1975, A. Volta's Gold medal, 1977, Order of Kirill and Mephodii (Bulgaria), 1981, E. Henkel Gold medal German Dem. Rep., 1986, Commodor's cross Order of Merit, Poland, 1986, Kalinga prize UNESCO, 1986, Gold medal Slovakian Acad. Scis., 1988, M.V. Lomonosov Gold medal USSR Acad. Scis., 1989, State prize of USSR, 1989, Edward Teller medal, 1991. Fellow Optical Soc. Am., Indian Nat. Sci. Acad.; mem. European Acad. Scis. and Arts (Salzburg chpt.), Internat. Acad. Scis. (hon.), USSR Acad. Scis. (presidium 1967-90, advisor 1990—), Acad. Natural Scis. of Russian Fedn. (hon.) Acad. Scis. German Dem. Rep., Polish and Czechoslovakian Acad. Scis., German Acad. Natural Scis. Leopoldina, Bulgarian Acad. Scis., Royal Swedish Acad. Engring., European Acad. Arts, Scis. and Humanities (Paris chpt.). Office: PN Lebedev Phys Inst, 53 Leninsky Prospekt, Moscow Russia

BASS, GEORGE FLETCHER, archaeology educator; b. Columbia, S.C., Dec. 9, 1932; s. Robert Duncan and Virginia (Wauchope) B.; m. Ann Singletary, Mar. 19, 1960; children: Gordon Wauchope, Alan Joseph. MA, Johns Hopkins U., 1955; PhD, U. Pa., 1964; PhD (hon.), Bogazici U., Istanbul, Turkey, 1987. Asst. prof. U. Pa., Phila., 1964-68, assoc. prof., 1968-73; prof. archaeology Tex. A&M U., College Sta., 1976-80, Disting. prof., 1980—, George T. and Gladys H. Abell prof. nautical archaeology, 1986—; dir. excavations of ancient shipwrecks off Turkish coast, 1960-87. Author: Archaeology Under Water, 1966, Cape Gelidonya, 1967, History of Seafaring, 1972, Archaeology Beneath the Sea, 1975, Yassi Ada I, 1982, Ships and Shipwrecks of the Americas, 1988; adv. editor Am. Jour. Archaeology, 1987—, Archaeology, 1987—, Internat. Jour. Nautical Archaeology, 1987—, Nat. Geog. Research, 1987—. Served to lt. U.S. Army, 1957-59, Korea. Recipient Centennial award Nat. Geog. Soc., 1988, La. Gorce Gold medal, 1979, Lowell Thomas award Explorers Club, 1986; named one of Outstanding Young Men of Yr., Jaycees, 1967. Mem. Inst. Nautical Archaeology (pres. 1973-82), Archaeol. Inst. Am. (Gold medal for disting. archaeol. achievement 1986), Soc. for Hist. Archaeology, Nat. Maritime Hist. Soc., Mothers Against Drunk Driving. Presbyterian. Avocation: classical music. Home: 1600 Dominik Dr College Station TX 77840-3623 Office: Tex A&M U Nautical Archaeology PO Drawer HG College Station TX 77841

BASS, MIKHAIL, electrical engineer; b. Bucharest, Rumania, Apr. 3, 1948; came to U.S., 1975; s. Zelman M. and Khasya (Kipnis) B.; m. Irene, Apr. 21, 1973; children: Julia, Gabriale. BSEE, Poly. Coll., 1973. Sr. elec. engr. Concast Inc., Montvale, N.J., 1975-80, Hyward Robinson Inc., N.Y.C., 1980-84; dept. mgr. Frederic R. Harris, N.Y.C., 1984—. Mem. IEEE. Achievements include patents for Method and Apparatus for Regulating the Bath Level in Continuous Casting Mold. Office: Frederic R Harris 300 E 42nd St New York NY 10017

BASS, THOMAS DAVID, biology educator; b. Port Arthur, Tex., June 18, 1956; s. Thomas A. and Edith L. (Linscomb) B.; m. Donna Lynn Pace, May 26, 1979; 1 child, Courtney Megen. BS in Sci. Edn., Lamar U., 1978, MS in Biology, 1979; PhD in Zoology, Tex. A&M U., 1985. Assoc. prof. U. Ctrl. Okla., Edmond, 1985—; resource person Nature Conservancy, Okla. Chpt., 1989—. Contbr. articles to profl. jours. Recipient Faculty Rsch. grants U. Ctrl. Okla., 1986—, Nature Conservancy, Oklahoma Tall Grass Prairie, 1991-92; named Researcher of Yr., U. Ctrl. Okla. Sigma Xi Club, Edmond, 1992. Mem. N.Am. Benthological Soc., Okla. Acad. Sci. (biology chair 1991—), Sigma Xi. Democrat. Office: Univ Ctrl Okla Biology Dept 100 N University Dr Edmond OK 73034-0177

BASSETT, C(HARLES) ANDREW L(OOCKERMAN), orthopaedic surgeon, educator; b. Crisfield, Md., Aug. 4, 1924; s. Harold Reuben and Vesta (Loockerman) B.; m. Nancy Taylor Clark, June 15, 1946; children: Susan, David Clark, Lee Sterling. Student, Princeton U., 1941-43; MD, Columbia U., 1948, ScD, 1955; LHD (hon.), SUNY, 1988. Diplomate Am. Bd. Orthopedic Surgery. Intern, resident St. Lukes Hosp., N.Y.C., 1948-50; asst. resident orthopedic surgery N.Y. Orthopaedic Hosp., 1950, Annie C. Kane fellow, 1953-55; asst. attending orthopedic surgeon Presbyn. Hosp., 1955-60, assoc. attending orthopedic surgeon, 1960-63, attending orthopedic surgeon, 1963-83, cons., 1983-88; instr. orthopedic surgery Columbia U., 1955-59, dir. orthopedic rsch. labs., 1957-86, asst. prof., 1959-61, assoc. prof., 1961-67, prof., 1967-83, prof. emeritus, spl. lectr., 1983-86; dir. Bioelectric Rsch. Ctr., 1986—; cons. Naval Med. Rsch. Inst., Bethesda, Md., 1952-54; spl. cons. NIH, Nat. Inst. Neurol. Diseases and Blindness, Bethesda, 1959-62; career scientist N.Y.C. Health Research Council, 1961-71; vis. scientist Strangeway Research Lab., Cambridge, Eng., 1965-66; cons. div. med. scis., cons., exec. sec. com. on skeletal system NRC-Nat. Acad. Scis., 1963-71; sci. advisors Schwerizerischen Abeitsgemeinschaft fur Osteosynthesefragen, 1959-75; bd. sci. advisors Inst. Calot, Berck-Plage, France, 1969-89; cons. N.Y. State Rehab. Hosp., West Haverstraw, 1968-86; cons. on med. devices FDA, HEW, 1970-77; founder, cons., bd. dirs. Electro-Biology, Inc., 1975-87; chmn. bd. dirs., founder Osteodyne, Inc., 1988—. Contbr. chpts. to books, articles to profl. jours.; patentee in field. Served to lt. (j.g.) USNR, 1950-54. Recipient Nat. award Paralyzed Vets. Am., 1959, Max Weinstein award United Cerebral Palsy, 1960, James Mather Smith prize Columbia Coll. Phys. and Surg., 1971, Galvani award U. Bologna, 1989. Fellow ACS, N.Y. Acad. Scis.; mem. AMA, Am. Acad. Orthopaedic Surgeons, Am. Orthopaedic Assn., Am. Soc. Cell Biology, Bioelectromagnetics Soc. (bd. dirs. 1991—, d'Arsonval medal 1991), Bioelectric Repair and Growth Soc. (Kappa Delta award 1988, pres.-elect 1993), Internat. Soc. Orthopedic Surgery and Traumatology (sci. adv. bd. internat. orthopedics), N.Y. State, N.Y. County med. socs., Orthopaedic Research Soc. (pres.), Can. Orthopaedic Assn. (hon.), Royal Coll. Medicine (Eng.), Royal Micros. Soc., Tissue Culture Assn., Can. Orthopaedic Research Soc. (hon.), Soc. Exptl. Biology and Medicine, Harvey Soc., S.C. Orthopaedic Assn. (hon.), Société Belge de Chirurgie Orthopédique et de Traumatologie (hon.), Sigma Xi (Ann. award), Alpha Omega Alpha. Achievements include 4 patents; identification of physical control mechanisms which modify cell function; of mechanical, stress generated, electric potentials in bone; of the importance of mechanical loading rate in triggering bone formation; of critical nature of frequency content, amplitude, timing and vectoral factors for electromagnetic and electric fields in producing selective bioeffects; development of tubulation methods for neural repair; of specific, pulsed electromagnetic fields as therapeutic agents in un-united fractures, osteonecrosis, osteoporosis, nerve regeneration, wound healing, development of new blood vessels, and degenerative conditions of tendons in humans and animals, without the need for surgery. Home: 108 Midland Ave Bronxville NY 10708-3206 Office: 2600 Netherland Ave Ste 103 Bronx NY 10463-4889

BASSETT, LELAND KINSEY, communication company executive, educator, author; b. May 27, 1945; s. Wilfred George and Vera Agnes (Scheffel) B.; m. Tina Bassett; children: Joshua Alan, Robert Ian. BA, Mich. State U., 1968, postgrad., 1969, 70, 76, 84; postgrad. Wayne State U., 1972, 74, 75, 81, U. Mich., 1970, 76, 77, 78, Harvard U., 1991. Legis. page Mich. Ho. of Reps., Lansing, 1957-58; owner, prin. Communication Assocs., East Lansing, 1969-72; pres. East Lansing-Meridian Area C. of C., 1968-70; project assoc. Mgmt. Assistance Program Mich. State C. of C., Lansing, 1970-71, dir. communication, 1971-72; legis. analyst Greater Detroit C. of C., 1972, v.p. communication div., 1973-76; prin. Leland K. Bassett & Assocs., Detroit, 1976-86; strategic programs developer pub. affairs Detroit Edison Co., 1977-81, dir. communication analysis and plng., 1981-86; chmn. Bassett and Bassett, Inc., 1986—; vis. lectr. Mich. State U., 1985-88. Editor The Detroiter Bus. News, 1974-76; assoc. pub. The Detroiter Mag., 1974-75. Producer, dir., writer (film) Detroit: Our Decisive Moment in History, 1973; creator, producer Alive and Well, 1973, collaborator A Play Half Written: The Energy Adventure (winner numerous awards), 1979, collaborator Radiation ... Naturally (winner numerous awards), 1981. Auctioner WTVS Annual Fund Raising Auction, 1975-76; mem. exec. com. Detroit Bicentennial Commn., 1975-76; area chmn. YMCA Capitol Funds Drive, Lansing, 1969; bd. dirs. Greater Lansing Assn. Retarded Children, 1972; trustee Music Hall Ctr. for Performing Arts, Detroit, 1984—, chmn. planning com., 1984-86, mem. exec. com., 1985—, chmn. nominating com., 1988-92, founding co-chair Producers, 1988—, corp. sec., 1990-92; mem. mgmt. team Mich. Citizens for Jobs and Energy, 1982, Voters for Responsible Govt., 1982; mem. exec. com. to retain Supreme Ct. Justice Archer, 1986. Recipient Mayor's Citation East Lansing, 1972, Outstanding Alumni award Mich. State U. Coll. Comm., Arts & Scis., 1992; named Outstanding Young Man East Lansing-Meridian Area C. of C., 1970. Mem. Greater Detroit C. of C., Mich. State C. of C., Internat. Assn. Bus. Communicators, Internat. Communication Assn., Mich. Speech Assn., Pub. Rels. Soc. Am. (student soc., counselors acad., chair awards adn recognition com. Detroit chpt.), Founders Soc. Detroit Inst. Arts, Rivertown Bus. Assn., Nat. Alliance for the Mentally Ill, Detroit Hist. Soc., Mich. Assn. Retarded Citizens, Nat. Assn. Retarded Citizens, Detroit Assn. Retarded Citizens, Mich. State Trust for Ry. Preservation, Mich. State U. Alumni Assn. (coll. commn. arts and scis., charter mem., past pres., past dir.), Press Club, Renaissance Club (membership com. 1990-92), Econ. Club, Detroit Soc. Clubs, Detroit Athletic Club. Office: 672 Woodbridge St Detroit MI 48226-4302

BASSI, JOSEPH ARTHUR, physician; b. Memphis, Feb. 6, 1951; s. Joseph Arthur Sr. and Sarah (Smith) B.; m. Leslie Lynn Taylor, Dec. 14, 1985; children: Carly, Lauren, Michael, Chad. BA in Biology, St. Louis U., 1973, MD, 1977. Diplomate Am. Bd. Internal Medicine. Rschr. NSF, St. Louis, 1971-72; intern Bapt. Meml. Hosp., Memphis, 1977, resident, 1978-80; physician Jackson Purchase Med. Assocs., Puducah, Ky., 1980—; pres. Lourdes Hosp. Med. Staff, Paducah, 1988; mem., bd. dirs. Lourdes Hosp., Paducah, 1987-92; physician advisor West Ky. Lupus Found., Paducah, 1981-87. Grantee NSF, 1971-72. Mem. McCracken County Med. Soc., Ky. Med. Soc., ARC Assn. Republican. Roman Catholic. Achievements include research in molecular virology-host DNA transmission of viral genome. Home: 7 West Vale Paducah KY 42001 Office: Jackson Purchase Med Assocs 225 Medical Center Dr Paducah KY 42002

BAST, ALBERT JOHN, III, civil engineer; b. Canton, Ohio, Nov. 14, 1948; s. Albert John Jr. and Nicole Agnes (Baboud) B.; m. Susan Elsa Burner, Dec. 12, 1970; 1 child, Nicole Susanne. BSCE, Va. Mil. Inst., 1970; M in Engring., U. Calif., 1977; grad., U.S. Army Command and Gen. Staff Coll., 1987, U.S. Army War Coll., 1991. Commd. 2d lt. U.S. Army, 1970, advanced through grades to maj., 1981; civil engr. C.E. U.S. Army, Fayetteville, N.C., 1970-74; co./installation comdr. U.S. Army, Republic of Korea, 1975-77; aero. facilities project mgr. NASA U.S. Army, Washington, 1977-81, resigned, 1981; lt. col. C.E. USAR, 1981—; resident engr. Parsons Brinckerhoff Constrn. Svcs. Inc., Norfolk, Va., 1981-84; project engr. Euroroute Tech. Audit, Paris, 1985; v.p. bus. devel. Parsons Brinckerhoff Constrn. Svcs. Inc., Herndon, Va., 1985-87; constrn. planner Storebaelt Tunnel Parsons Brinckerhoff Constrn. Svcs. Inc., Copenhagen, 1988; v.p. Parsons Brinckerhoff Internat. Inc., Hong Kong, 1989—; dir. Parsons Brinckerhoff Asia Ltd., Hong Kong, 1992—, sr. v.p., 1993—; guest lectr. Hong Kong Open Forum Series, 1990; constrn. planner Panchiao rwy. undergrounding project, Taipei, 1990; prin. in charge 3d Stage Expwy., Bangkok, 1990-91, Bangkok Elevated Transport System Design Rev., 1992—; dep. regional mgr., Asia, 1989—; country mgr., Japan, 1989—. Mem. ASCE, Constrn. Mgmt. Assn. Am., Va. Mil. Inst. Alumni Assn., U. Calif. Alumni Assn., Berkeley Engring. Alumni Soc., U.S. Army War Coll. Alumni Assn., Am. C. of C. Hong Kong, Am. Club Hong Kong, Kappa Alpha, Chi Epsilon. Avocations: sailing, traveling, hiking. Office: Parsons Brinckerhoff Asia Ltd, 979 King's Rd, Quarry Bay Hong Kong

BAST, ROBERT CLINTON, JR., research scientist, medical educator; b. Washington, Dec. 8, 1943; s. Robert Clinton and Ann Christine (Borland) B.; m. Blanche Amy Simpson, Oct. 21, 1972; 1 child, Elizabeth Simpson Bast. BA cum laude, Wesleyan U., Middletown, Conn., 1965; MD magna cum laude, Harvard U., 1971. Diplomate Am. Bd. Internal Medicine, Am. Bd. Med. Oncology, Am. Bd. Hematology. Predoctoral fellow dept. pathology Mass. Gen. Hosp., Boston, 1967-69; intern Johns Hopkins Hosp., Balt., 1971-72; research assoc. biology br. Nat. Cancer Inst., NIH, Bethesda, Md., 1972-74, research scientist biology br., 1974-75; asst. resident Peter Bent Brigham Hosp., Boston, 1975-76, fellow med. oncology Sidney Farber Cancer Inst., Boston, 1976-77; asst. prof. medicine Harvard U. Med. Sch., Boston, 1977-83, assoc. prof., 1983-84; prof. Duke U. Med. Ctr., Durham, N.C., 1984-92, Wellcome clin. prof. medicine in honor of R. Wayne Rundles, 1992—, co-dir. div. hematology-oncology, 1984—; dir. clin. research programs Duke U. Comprehensive Cancer Ctr., Durham, 1984-87, dir., 1987—; hosp. appointments include asst. in medicine Peter Bent Brigham Hosp., 1976-77; jr. assoc. in medicine Brigham and Women's Hosp., 1977-82; cons. oncologist Boston Hosp. Women, 1978-80; mem. biol. response modifiers decision network com. Nat. Cancer Inst., 1984-87, exptl. immunology study sect., 1983-84, 90-92; mem. grant rev. com. Leukemia Soc. Am., 1985-87, adv. com. oncologic drugs FDA, 1985-89, chmn. 1988-89; bd. dirs. Cancer and Leukemia Group B., 1986-88, Am. Council Transplantation, 1985-87; mem. grant rev. com. Am. Cancer Soc., 1987; numerous other coms.; Edward G. Waters Meml. lectr., 1987; John Ohtani Meml. lectr., 1991; D. Nelson Henderson lectr., 1991; Stolte Meml. lectr., 1992; Arnold O. Beckman Disting. Lectureship, 1993. Contbr. numerous articles on tumor immunology, immunodiagnosis and immunotherapy of cancer and cellular immunology to profl. jours. Served as surgeon USPHS, 1972-75. Recipient Dominus award, 1984, Robert C. Knapp award, 1990; grantee Nat. Cancer Inst., NIH, HHS, 1978—; scholar Leukemia Soc. Am., 1978-83. Fellow ACP; mem. The Reticuloendothelial Soc., Am. Soc. Microbiology, Am. Assn. Cancer Research, Am. Assn. Immunologists, Am. Soc. Clin. Oncology, Am. Fedn. Clin. Research., Am. Soc. Clin. Investigation, Internat. Soc. Immunopharmacology, Soc. Biol. Therapy (bd. dirs. 1984-86), Am. Soc. Hematology, Soc. Gynol. Oncology (assoc.). Achievements include development of monoclonal antibodies to react with human ovarian cancer, leading to CA125 blood test; techniques for selective elimination of tumor cells from human bone marrow; identification of changes associated with malignant transformation of ovarian epithelium. Office: Duke U Med Ctr Comprehensive Cancer Ctr PO Box 3708 Durham NC 27710

BASTIAANS, GLENN JOHN, analytical chemist, researcher; b. Oak Park, Ill., Oct. 25, 1947; s. John Peter and Bernice Ann (Lastovka) B.; m. Mary Jane Elizabeth Singler, Aug. 24, 1974; 1 child, Elizabeth Jane. BS in Chemistry, U. Ill., 1969; PhD in Analytical Chemistry, Ind. U., 1973. Postdoctoral fellow Colo. State U., Ft. Collins, 1973-74; asst. prof. Georgetown U., Washington, 1974-79, Tex. A&M U., College Station, 1979-85; chief tech. office Integrated Chem. Sensors Inc., Newton, Mass., 1985-90; assoc. dir. Analytical Instrumentation Ctr. Iowa State U., Ames, 1990—; cons. GJB Techs., Ames, 1990—. Author: (chpt.) Piezoelectric Biosensors in Chemical Sensor Technology IV, 1992, Piezoelectric Transducers in Chemical Sensors, 1988; mem. editorial bd. PI Quality, 1992—. Mem. AAAS, Am. Chem. Soc., Soc. for Applied Spectroscopy. Achievements include patent in sensor having piezoelectric crystal for microgravimetric immunoassays. Office: Iowa State Univ Analytical Instrumentation 1915 Scholl Rd ASC 2 Ames IA 50011

BASTIDA, DANIEL, data processing facility administrator, educator; b. Sept. 9, 1954; s. Daniel and M. Angels (Obiols) B.; m. Carolina Serra, Oct. 7, 1977; children: Carolina, Daniel. Indsl. engr. degree, U. Politecnica, Barcelona, Spain, 1978; M. in Bus., Inst. Estudios Superiores Empresa, Barcelona, 1988. Engr. Hewlett-Packard, Barcelona, 1978-85; dir. Nat. Ctr. Info. Processing, Andorra, Andorra, 1985—; adminstr. Official Jour. Andorra, 1989—; dir., educator Data Processing Sch., Andorra, 1988—; venue tech. mgr. Olympic Games, Barcelona, 1991-92. Editor Revista de Informatica, 1988; contbr. articles to profl. jours. Mem. Andorra-CEE Com., 1987, Cercle of Economy of Andorra; active Olympic Com., Andorra, 1988—; coord. econ. strategy plan for Andorra, 1992-93. Fellow Latin Data Processing Congress (Barcelona), Sci. Soc. (Andorra); mem. IEEE, ACM, Internat. Fedn. Info. Processing (mem. gen. assembly), Engrs. Assn. (v.p. Andorra, Barcelona), Anyos Club, Reial Automòbil CLub, Automòbil Club D'Andorra, Moto Club Andorra, Esqui Club Ordino-Arcalis, Centre Recerca i Investigació Submarina, French Soc. for Info. and Systems Sci. and Tech., Lions. Roman Catholic. Avocations: squash, windsurf, skiing, motorcycle, photography. Home: La Pleta M, Ordino Andorra Office: CNIA, Av Sta Coloma 91, Andorra Andorra

BASU, JANET ELSE, journalist; b. Schenectady, N.Y., May 17, 1947; d. Robert Alan and Alta (Howard) Else; m. Basab K. Basu, Jan. 27, 1967. AB, U. Calif., Berkeley, 1970, M in Journalism, 1984. Editor Oman Pub., Inc., Mill Valley, Calif., 1970-80; freelance journalist San Francisco, 1980—. Contbg. editor: Hippocrates Mag., 1992—; contbr. articles to Am. Health, Scientist, San Francisco Focus, Science '81 and others. Mem. AAAS, Nat. Assn. Sci. Writers, No. Calif. Sci. Writers Assn. (pres. 1987-90).

BASU, PRITHWISH, mechanical engineering researcher; b. Calcutta, India, Nov. 21, 1956; came to U.S., 1981; s. Chitta Ranjan and Anjali (Ghose) B.;m. Suchismita Bose, Feb. 23, 1981; 1 child, Anasuya. B Tech., Indian Inst. Tech., Kharagpur, 1978; M Engring., Carnegie-Mellon U., 1984, PhD, 1988. Design engr. Devel. Cons. Pvt. Ltd., Calcutta, 1978-81; sr. analyst R & D EG&G Fluid Components Tech. Group, Cranston, R.I., 1987—. Contbr. articles to profl. publs. Mem. ASME, AIAA, Soc. Tribologists and Lubrication Engrs., Sigma Xi. Achievements include developments in field of low hysteresis and extended life brush seals, and non-contacting gas seals. Home: 60 Ridge St Pawtucket RI 02860 Office: EG&G FCTG R&D 50 Sharpe Dr Cranston RI 02920

BASU, PRODYOT KUMAR, civil engineer, educator; b. Lucknow, India, Nov. 15, 1939; came to U.S., 1974; s. Krishna Kamal and Usha Rani (Ghosh) B.; m. Liliya Bhattacharya, Jan. 26, 1966; 1 child, Devraj. MS, Calcutta (India) U., 1963; ScD, Washington U., St. Louis, 1977. Registered profl. engr., Tenn. Lectr., then asst. prof. civil engring. Calcutta U., 1964-74; instr., sr. rsch. engr. Washington U., 1975-79, asst. prof., then assoc. prof. civil engring., 1980-84; assoc. prof., now prof. civil engring. Vanderbilt U., Nashville, 1984—; chair computer applications Structural Stability Rsch. Coun., Lehigh, Pa., 1988—; cons. Ala. A&M U., Huntsville, 1990—. Contbr. articles to tech. publs. Fellow ASCE; mem. Internat. Assn. Computational Mechanics, Assn. Computing Machinery. Achievements include development of p- and hp- version of finite and boundary element methods. Office: Vanderbilt U Box 15 Sta B Nashville TN 37221

BATBAYAR, BAT-ERDENIIN, microbiologist; b. Arkhangai, Mongolian People's Republic, 1955. Student, Mongolian State U., U. London. Tchr. secondary sch. Province of Hentii, Mongolian People's Republic, 1982-84; scientist Inst. Microbiology, Ulan Bator, Mongolian People's Republic, 1984—. Founding mem. Dem. Socialist Movement, Mongolian Social Dems., chair, 1990—. Office: Academy of Sciences, Suhbaatar Sq 3, Ulan Bator 11, Mongolian People's Republic*

BATCHA, GEORGE, mechanical and nuclear engineer; b. Marblehead, Ohio, Oct. 24, 1928; s. John and Anna (Groholy) B.; m. Erika Voelker, Jan. 1, 1982; 1 child, Susan Kolodziejczyk. BA, Bowling Green State U., 1951; MS in Engring. Sci., U. Toledo, 1968; R&D test and evaluation program cert., U.S. Army Logistics Mgmt. Coll. and Assn. for Systems Mgmt.; certs. numerous U.S. Army tng. schs. Registered profl. engr., Ohio, Mich. With Standard Products Co., Port Clinton, Ohio, 1951, A.O. Smith Co. Landing Gear div., Toledo, Ohio, 1951, army rep. at Glenn L. Martin Co., Balt., 1952-54, Cleve. Pneumatic Tool Co., 1954-55, Hardware Stamping div. Ford Motor Co., Sandusky, Ohio, 1955-59; mech. design and test engr. Missile and Def. Engring. divs. Chrysler Corp., Detroit, 1959-62; mech. and nuclear engr. NASA, Lewis Research Ctr., Plum Brook Sta., Sandusky, 1962-74; mech. and system mgmt. engr. Armament Research and Devel. Command, U.S. Army, Dover, N.J., and Rock Island, Ill., 1974-81, mech. engr. Tank Automotive Command, Warren, Mich., 1981—. Author numerous tech. reports. Served with U.S. Army, 1952-54. Scholar, Bowling Green State U., 1948; recipient Apollo Achievement award NASA, 1969, accomplishment awards, 1981, 82, Cost Reduction awards, 1971, 74, Dept. Army Achievement award Tank Automotive Command, 1985, Superior Performance award 1990, 91, 92. Mem. Nat. Soc. Profl. Engrs., Order of Engr., Nat. Council Engring. Examiners (cert.), Am. Acad. Environ. Engrs. (diplomate, radiation protection), Soc. Logistics Engrs. (cert. profl. logistician), U.S. Army Logistics Mgmt. Coll. and Assn. for Systems Mgmt. (cert. in R & D), Assn. U.S. Army, Port Clinton Power Squadron of Ohio, Am. Legion. Byzantine Catholic. Current work: Technical assessment and guidance of developmental programs of all elements of integrated logistics support in tank-automotive weapon system and equipment. Subspecialties: Mechanical engineering; Nuclear engineering. Home: 12851 E Outer Dr Detroit MI 48224-2730

BATCHELOR, ANTHONY STEPHEN, geotechnical engineer, consultant; b. London, June 11, 1948; s. Alfred James and Freda Edith May (Smith) B.; m. Linda Patricia Weller, Aug. 22, 1970; children: James William, Charlotte Louise. BSc, Nottingham (Eng.) U., 1969, PhD, 1972. Cert. engr., Eng. Lectr. Camborne (Eng.) Sch. of Mines, 1972-75, sr. lectr., 1975-78, prin. engr., 1978-80, project dir., 1980-85; mng. dir. GeoSci. Ltd., Falmouth, Eng., 1986—; vis. staff mem. Los Alamos (N.Mex.) Nat. Lab., 1978-87; vis. lectr. MIT, Cambridge, 1984—; supr. Cambridge (Eng.) U., 1990—; dir. Total Flow Cycles Ltd., London, 1990—. Co-author: Applied Geothermics, 1987; contbr. numerous articles to profl. jours. Mem. Soc. Petroleum Engrs., Internat. Geothermal Assn., Geothermal Resources Coun., Inst. Mining and Metallurgy. Avocations: skiing, photography, computer graphics, biking. Home: 1 Clifton Gardens, Truro TR1 3HL, England Office: GeoSci Ltd, Falmouth Bus Park, Falmouth Cornwall TR11 4SZ, England

BATCHELOR, BARRY LEE, software engineer; b. Alexandria, Va., June 10, 1952; s. LeRoy Elsworth and Angeline (Hood) B.; m. Catherine Gray Holmes, Sept. 18, 1982; children: Erin Gray, Kellea Margaret. Student, Va. Poly. Inst. and State U., 1971-76. Technician J.J. Henry Co., Inc., Arlington, Va., 1976-79; mgr. Advanced Marine Enterprises, Arlington, 1979—. Office: Advanced Marine Enterprises # 1300 1725 Jefferson Davis Hwy Arlington VA 22202

BATCHELOR, BILL, civil engineering educator; b. Houston, Feb. 9, 1949; s. William Sr. and Katherine (McClarnen) B.; m. Colleen Jennings, July 7, 1973; children: Brian Jennings Batchelor, Megan Jennings Batchelor. BA, Rice U., 1971, MS, 1974; PhD, Cornell U., 1976. Profl. engr., Tex. Asst. prof. civil engring. Texas A&M U., College Station, 1976-81, assoc. prof., 1981-86, prof., 1986—; bd. dirs. Inst. for Environ. Engring. Tex. A&M Univ. System, College Station; cons. Argonne (Ill.) Nat. Lab., 1992—, PRC Environ. Mgmt., McLean, Va., 1992—, U.S. EPA, Washington, 1985—. Contbr. over 30 articles to profl. jours.; presenter at numerous seminars, symposia & confs. Advancements chair Boy Scouts Am., Bryan, Tex., 1990—; precinct chair Dem. Party, Bryan, 1988, 90. Named Halliburton Prof. Civil Engring. Tex. A&M U., 1986-87. Fellow AAAS; mem. Assn. Environ. Engring. Profs. (sec.-treas. 1984-86), Water Environment Fedn. (subcom. chair 1989-92, Harrison Prescott Eddy medal 1983), Am. Chem. Soc., Am. Soc. Civil Engrs., Am. Water Works Assn. Democrat. Roman Catholic. Achievements include development of model and design procedure for single-sludge denitrification; demonstrated feasibility of autotrophic denitrification using sulfur; demonstrated feasibility of ultra-high lime treatment for recycled cooling water. Office: Texas A & M Univ Environmental & Water Resources Civil Engineering Dept College Station TX 77843

BATCHELOR, JOSEPH BROOKLYN, JR., electronics engineer, consultant; b. Jersey, Ga., Apr. 11, 1922; s. Joseph Brooklyn and Mary Arlie (Reece) B.; m. Clara Owens, July 14, 1940; children: Joseph Brooklyn III, James Alfred, William Owens. Diploma, North Ga. Coll., 1940. Registered profl. engr., Ill., Ga. Owner, pres. JRS Electronics-Svc., Rsch. & Cons., Monroe, Ga., 1946-57; dir. rsch. and engring. Cen. Electronics, Inc., Chgo., 1957-61; rsch. engr. Hallicrafters Corp., Chgo., 1961-63; aircraft devel. engring. specialist Lockheed A/C Corp., Marietta, Ga., 1965-70; pres., chmn. BRECO Corp., Walnut Grove, Ga., 1972-75; cons. engr. Batchelor Labs., Libertyville, Ill., 1963-65; owner, mgr. Batchelor Labs., Walnut Grove, 1970-72, Jersey, 1975—; pres. PATRONIX-Patent Holding Corp., Chgo., 1958-61. Patentee in field; inventor radio-location device for lost student pilots WWII. Chmn. bd. Monroe Christian Ch., 1950-57. Sgt. USAF, 1945-46, World War II. Mem. AAAS, Citizens Adv. Coun. on Energy, Mensa. Avocations: amateur radio, classical music, reading. Home and Office: 101 Main St Jersey GA 30235

BATDORF, SAMUEL B(URBRIDGE), physicist; b. Jung Hsien, China, Mar. 31, 1914; s. Charles William and Nellie (Burbridge) B.; m. Carol Catherine Schweiss, July 19, 1940; children: Samuel Charles, Laura Ann. A.B., U. Calif.-Berkeley, 1934, A.M., 1936, Ph.D., 1938. Assoc. prof. physics U. Nev., 1938-43; aero. rsch. scientist Langley Lab., NACA, 1943-51, chmn. advanced study com., 1946-51, mem. NACA subcom. on aircraft structural metals, 1946-51; dir. devel. Westinghouse Elec. Corp., Pitts., 1951-56; tech. dir. weapons systems Lockheed Missile & Space Co., Palo Alto, Calif., 1956-58; mgr. communication satellites Inst. Def. Analysis, Washington, 1958-59; dir. rsch. in physics, electronics and bionics Aeronutronic, Newport Beach, Calif., 1959-62; prin. staff scientist Aerospace Corp., El Segundo, Calif., 1962-77; Sigma Xi lectr. communication satellites; Disting. prof. Tsing Hua U., Republic of China, 1969; vis. scholar Va. Poly. Inst. and State U., 1984; adj. prof. engring. and applied sci. UCLA, 1973-86; mem. aeromechanics adv. com. Air Force Office Sci. Rsch., 1965-71, tech. assessment panel Engrs. Joint Coun., 1968-71. Contbr. articles to profl. jours. Fellow AIAA (edn., structures and materials coms.), ASME (hon., edn. materials and space structures coms., chmn. applied mechanics div. exec. com. materials and space structures com.), Am. Phys. Soc., Am. Acad. Mechanics (pres. 1982-83); mem. Aerospace Club, Rod and Gun Club, Academia Club, Engrs. Club of Va. Peninsula (pres. 1948), Phi Beta Kappa, Phi Kappa Phi. Republican. Presbyterian. Home: 5536 B Via La Mesa Laguna Hills CA 92653

BATE, BRIAN R., psychologist; b. Cleve., July 4, 1940; s. Paul A. and Claire N. B.; children: Jennifer A., Julia L. BA in English, Case Western Reserve U., 1963, MS in Psychology, 1965, PhD in Psychology, 1972. Lic. psychologist, Ohio. Instr. Cuyahoga Community Coll., Parma, Ohio, 1969, from asst. prof. to prof. of Psychology, 1970—; pvt. practice Cleve., 1972—. Contbr. articles to profl. jours. Nat. Merit Scholar Princeton U., 1958-61, Western Res. U., 1962-63; USPHS fellow, 1963-67. Mem. APA, Am. Fedn. Musicians, Gestalt Inst. of Cleve., Ohio Psychologists for Soc. Responsibility, Edelweiss Ski Club. Buddhist. Achievements include development and teaching of the first underclass-level behavior modification course in USA, 1970. Home and Office: 6511 Mill Rd Cleveland OH 44141-1560

BATEMAN, ROBERT EDWIN, aeronautical engineer; b. Butte, Mont., Apr. 11, 1923; s. Edwin Joseph and Katherine (Bronner) B.; m. Sarah Elizabeth Hayes, Mar. 2, 1947; children: Robert Eugene, Lucy Annette, Paul William. BS in Aero. Engring., Purdue U., 1946; ED (hon.), Purdue, 1992. Aero. staff Boeing Co., Seattle, 1946-59; devel. program mgr. Boeing Aerospace, Seattle, 1959-65, gen. mgr. turbine div., 1965-67, 747 program exec., 1967-71, v.p., mgr. Washington D.C. ops., 1971-75, v.p. gen. mgr. marine systems, 1975-85; v.p. govt. and internat. affairs Boeing Co., Seattle, 1985-88; chmn., bd. dirs. Mus. Flight, Seattle, 1989—. Bd. dirs. Naval War Coll., Newport, R.I., 1976—, Naval Meml. Found., 1983-88; pres. World Affairs Coun., Seattle, 1988—. Lt. comdr. USNR, 1946-66. Sec. Navy Meritorius Pub. Svc. award, 1968, Disting. Pub. Svc. award, 1972; named Disting. Engring. Alumnus Purdue U., 1974, Old Master award, 1988. Fellow (assoc.) AIAA; mem. Navy League U.S. (exec. com., nat. dir. 1965—, Disting. Svc. award 1986, Hall of Fame 1991). Home: 1645 E Boston Ter Seattle WA 98112-2831

BATES, DAVID MARTIN, botanist, educator; b. Everett, Mass., May 31, 1934; s. Leslie Marriner and Ann Louise (Gustafson) B.; m. Jane Sandra Schwarz, Sept. 4, 1956; children: Jonathan David, Leslie Marriner. BS, Cornell U., 1959; PhD, UCLA, 1963. Postdoctoral fellow Brit. Mus., 1962-63; asst. prof. Cornell U., Ithaca, N.Y., 1963-69, assoc. prof., 1969-75, prof., 1975—, dir. L.M. Bailey Hortorium, 1969-83. Author: Hortus Third, Biology and Utilization of the Cucurbitaceae. With U.S. Army, 1954-56. NSF fellow; recipient Gold medal Garden Club Am. Fellow AAAS, Societe de Biogeographie; mem.. Bot. Soc. Am., Am. Soc. Plant Taxonomists (pres. 1983), Internat. Assn. Plant Taxonomy, Soc. for Econ. Botany (pres. 1989), Internat. Palm Soc., Assn. Systematic Collections (pres. 1976-78). Home: PO Box 15 King Ferry NY 13081 Office: Cornell U L H Bailey Hortorium 468 Mann Libr Ithaca NY 14853-4301

BATES, SHARON ANN, plant and soil scientist, educator; b. Murphysboro, Ill., June 18, 1958; d. Bill E. and Lennis M. (Ditzler) B. BS, So. Ill. U., 1980, MS, 1990, postgrad., 1990—. Lab. technician Arch Mineral Corp., Percy, Ill., 1980-82, lab. supr., 1982-85; br. mgr. Comml. Testing & Engring., Percy, 1985; rsch. and teaching asst. So. Ill. U., Carbondale, 1988—. Contbr. chpt. to book. Recipient Philip White Meml. award Tissue Culture Assn., 1992. Mem. Am. Soc. for Hort. Sci. (travel grantee 1992), Sigma Xi, Pi Alpha Xi. Office: So Ill U Plant and Soil Sci Dept Carbondale IL 62901

BATES, STEPHEN CUYLER, research engineering executive; b. Jenkintown, Pa., Aug. 27, 1948; s. Frederick Heston and Barbara Jane (Snedeker) B.; m. Joyce Ethel Carpenter, Apr. 20, 1986; children: Cuyler Nathaniel, Faye Gwendolyn. BS, MIT, 1970, MS, 1971, PhD, 1977. Scientist AVCO Everett (Mass.) Rsch. Lab., 1977, Oak Ridge (Tenn.) Nat. Lab., 1977-82, GA Technologies, San Diego, 1983; engr. Gen. Motors Rsch. Lab., Warren, Mich., 1984-90; sr. scientist Advanced Fuel Rsch., East Hartford, Conn., 1990-93; pres. Thoughtventions Unlimited, Glastonbury, Conn., 1993—. Contbr. articles to profl. jours. Fellow Hertz Found., 1970; grantee DOD, NASA, 1992. Mem. AAAS, Soc. Photo-optical Instrument Engrs., Combustion Inst. Achievements include research in design, construction, operation, and diagnosis of single-cylinder internal combustion visualization engine. Sapphire stand-along cylinder wall and full combustion for multiple cycles for many cases. Developed techniques for using sapphire. Home: 59 Lexington Rd Glastonbury CT 06033 Office: Thoughtventions Unlimited PO Box 1310 Glastonbury CT 06033

BATES, WILLIAM LAWRENCE, civil engineer; b. Leominister, Mass., Dec. 6, 1957; s. William Earl and Imogene (Pope) B.; m. Susan Joy Cook, Aug. 16, 1980; children: Jonathan Edwards, David William, Rebecca Ann. BSCE, Clemson U., 1979. Registered profl. engr., N.C., S.C. Design engr. I Duke Power Co., Charlotte, N.C., 1980-88; structural engr. Republic Contracting Corp., Columbia, S.C., 1989—. Dir. Awana Pioneer Club, Chapel in the Wildwood Bapt. Ch., Springdale, S.C., 1989—. Mem. ASCE, Am. Inst. Steel Constrn. (profl. mem.). Republican. Office: Republic Contracting Corp 829 Pepper St # 9167 Columbia SC 29209-2138

BATESON, MARY CATHERINE, anthropology educator; b. N.Y.C., Dec. 8, 1939; d. Gregory and Margaret (Mead) B.; m. J. Barkev Kassarjian, June 4, 1960; 1 child, Sevanne Margaret. BA, Radcliffe Coll., 1960; PhD, Harvard U., 1963. Instr. Arabic Harvard U., 1963-66; asso. prof. anthropology Ateneo de Manila U., 1966-68; sr. research fellow psychology and philosophy Brandeis U., 1968-69; assoc. prof. anthropology Northeastern U., Boston, 1969-71; researcher U. Tehran, 1972-74; vis. prof. Northeastern U., 1974-75; prof. anthropology, dean grad. studies Damavand Coll., 1975-77; prof. anthropology, dean social sci. and humanities U. No. Iran, 1977-79; vis. scholar Harvard U., 1979-80; dean faculty, prof. anthropology Amherst Coll., 1980-87; Clarence Robinson prof. anthropology and English George Mason U., 1987—; pres. Inst. Intercultural Studies, from 1979. Author: Structural Continuity in Poetry: A Linguistic Study of Five Early Arabic Odes, 1970, Our Own Metaphor: A Personal Account of a Conference on Consciousness and Human Adaption, 1972, 2d edit., 1991, With a Daughter's Eye: A Memoir of Margaret Mead and Gregory Bateson, 1984, Composing a Life, 1989; co-author: Angels Fear: Towards an Epistemology of the Sacred, 1987, Thinking AIDS, 1988; co-editor: Approaches to Semiotics: Anthropology, Education, Linguistics, Psychiatry and Psychology, 1964. Fellow Ford Found., 1961-63, NSF, 1968-69, Wenner-Gren Found., 1972, Bunting Inst., 1983-84, Guggenheim Found., 1987-88. Mem. Am. Anthrop. Assn., Lindisfarne Assn., Phi Beta Kappa.

BATHEY, BALAKRISHNAN R., materials scientist; b. Madurai, Madras, India, May 20, 1940; came to U.S., 1971; s. Ramachary B. and Visalakshi R. B.; m. Meera B. Bathey, Mar. 29, 1973; children: Shrilekha, Shrimathi. BSc, Madras U., 1963; MTech, Indian Inst. Tech., Bombay, 1966. Metallurgist Allied Chems., New Delhi, India, 1966-70; metallurgist, cons. Nat. Defense Lab., New Delhi, 1966-70; quality control engr. Am. Electroplating Co., Cambridge, Mass., 1971-74; rsch. technologist Mobil Solar Energy Corp., Billerica, Mass., 1974—. Contbr. articles to profl. publs. Achievements include patent for method of fabricating solar cells. Home: 4 Rogers Brook E Andover MA 01810 Office: Mobil Solar Energy Corp 4 Suburban Park Dr Billerica MA 01821

BATHIAS, CLAUDE, materials science educator, consultant; b. Bénévent, France, Aug. 5, 1938; s. André and Gabrielle (Thienot) B.; m. Marie-Claude Leveque, Dec. 26, 1964; children: Anne, Claire. French PhD, U. Poitiers (France), 1964; postgrad., MIT, 1972. Engr., cons. Aerospatiale Co., Paris, 1973-85; material engring. dept. head French Ministry for Rsch. and Tech., Paris, 1978-82; prof. U. Technology, Compiègne, France, 1974-88, Conservatoire Nat. des Arts et Métiers, Paris, 1989-93; vis. prof. Ga. Inst. Tech., Atlanta, 1988-92; dir. CNRS Lab. 914, U. Compiègne, 1980-85, Inst. for Tech. and Materials Sci., Paris, 1985-93, dir.; chmn. numerous sci. confs.; del. Japan Prize, Tokyo, 1988, VAMAS Steering Com. Editor: European Newsletter Advanced Materials, books from conf. proc.; contbr. numerous articles to sci. and profl. jours. Advisor Ministry for Rsch., Paris, 1978-82, Ministry for Def., Paris, 1983-86, Ministry for Foreign Affairs, Paris, 1988-91; del. VAMAS-France, Paris, 1983-91. Recipient Chevalier award Ordre Nat. du Merite, Paris, 1983, Oppenheim award Ingénieurs et Scientifiques de France, Paris, 1978, award Nat. Rsch. Inst. for Metals, Tokyo, 1988. Fellow Am. Soc. for Materials; mem. Am. Soc. for Testing and Materials (voting mem.), MIT Club of Paris, CODATA France (staff mem.). Avocations: tennis, music, international cooperation. Home: 80 Boulevard Bourdon, 92200 Neuilly Sur Seine France Office: Conservatoire Nat des Arts, et Métiers, 292 rue Saint Martin, 75141 Paris France

BATHINA, HARINATH BABU, chemist, researcher; b. Eluru, India, Mar. 9, 1940; came to U.S., 1967; s. Ramachandra R. and Sesirekha (Tata) B.; m. Sudha H. (Matta), Apr. 29, 1962; children: Jyothi, Murali, Srinath. MSc, Osmania U., Hyderabad, India, 1960, PhD, 1965; M in Mgmt., Northwestern U., 1976. Rsch. assoc. Northwestern U., Chgo., 1967-73; devel. mgr. Armak, Chgo., 1974-79; R & D dir. IDL Chems., Hyderabad, 1980-82; tech. dir. Southland Corp., Summit, Ill., 1982-88; product devel. assoc. Stepan Co., Northfield, Ill., 1989—; cons. Bathina Assocs., Burr Ridge, Ill., 1988—; rsch. guide Osmania U. Grad. Sch., Hyderabad, 1980—. Contbr. articles to sci. and profl. jours. Trustee Hindu Temple, Lemont, Ill., 1986-90. Fellow Am. Inst. Chemists; mem. Am. Chem. Soc., Soc. Tribologists and Lube Engrs. Home: 159 Circle Ridge Dr Burr Ridge IL 60521-8381

BATLIVALA, ROBERT BOMI D., oil company executive, economist educator; b. Bombay, India, Feb. 17, 1940; came to U.S., 1962, naturalized, 1968; s. Dean Shaw and Rose (Engineer) B.; m. Carole Gretchen Feustel, May 9, 1964; children: Amy, Dina. BS in Geology, Chemistry, St. Xavier Coll., Bombay, Ind., 1960; MBA in Bus., Econs., Loyola U., Chgo., 1970; PHD in Bus., Econs., Ill. Inst. Tech., 1971; post-doctoral studies, U. Chgo., 1972-73. Rsch. chemist Reynolds Metals Co., McCook, Ill., 1962-64; from sales engr. to staff dir. econs. Amoco Corp., Chgo., 1964-1988, dir. antitrust econs., 1988-93; dir. regulatory econs., 1993—; adj. prof. bus. and econs. Rosary Coll., River Forest, Ill., 1976—, Graduate Sch. Bus., 1986—; bd. dirs. Pvt. Bancorp, Inc., Chgo. Contbr. articles to profl. jours., 1971-78. Bd. dirs. Ctr. for Conflict Resolution, 1990—. Recipient Stuart Tuition scholarship, Ill. Inst. Tech., 1970-71, Recognition award Rosary Coll. Grad. Sch. Bus. Alumni Assn., River Forest, 1986. Mem. ABA (assoc.), Nat. Assn. Mfrs. (corp. fin., mgmt. & competition com., regulation, transp. com. 1980—), Am. Econ. Assn., Assn. of Energy Economists, Loyola U. Grad. Bus. Alumni Assn. (pres., sr. v.p. 1971-73, Disting. Alumni award 1975), Oak Park Country Club. Avocations: ancient history, reading, writing, travel, languages. Home: 1106 Keystone Ave River Forest IL 60305-1326 Office: Amoco Corp 200 E Randolph Dr Chicago IL 60601-6401

BATLLE, DANIEL CAMPI, nephrologist; b. Barcelona, Spain, Feb. 11, 1950; came to U.S., 1975; s. Narciso and Francisca (Campi) B.; m. Joan Batlle, Mar. 11, 1983; children: Jordi, Nicholas, Natalie. MD, U. Barcelona, 1973. Diplomate Am. Bd. Internal Medicine and Nephrology. Resident Wayne State U., Detroit, 1975-77; nephrology fellow U. Ill., Chgo., 1977-79, asst. prof. medicine, 1980-85; assoc. prof. medicine Northwestern U., Chgo., 1985-89, prof. medicine, 1989—; chief divsn. nephrology, hypertension Northwestern U., 2d Northwest Meml. Hosp., Chgo., 1992—. Editorial bd. Seminars in Nephrology, 1987—, Am. Jour. of Kidney Diseases, 1988—, Hypertension, 1992—, The Kidney, 1992—. Fellow ACP, Am. Heart Assn. (high blood pressure coun.); mem. Nat. Kidney Found. (scientific adv. bd. 1987—, chmn. nominating com. scientific adv bd. 1989-91, chmn. program com. clin. meeting, 1993—), Am. Soc. for Clin. Investigation, Cen. Soc. for Clin. Rsch., Soc. Gen. Physiologists, Am. Physiol. Soc., Am. Soc. Nephrology, Internat. Soc. Nephrology, Am. Heart Assn., N.Y. Acad. Scis., Inter-Am. Soc. Hypertension, Am. Soc. Renal Biochemistry and Metabolism. Achievements include research on intracellular pH, Ca2 and Na regulation in hypertension, acid-base and potassium physiology and pathophysiology. Office: Northwestern U Med Sch 303 E Chicago Ave Chicago IL 60611

BATLOGG, BERTRAM, physicist; b. Bludenz, Austria, 1950; came to the U.S., 1979; Dipl. Phys., ETH, Zurich, 1975, Dr. rer. nat., 1979. Mem. tech. staff Bell Telephone Labs., Murray Hill, N.J., 1979-86; dept. head AT&T Bell Labs., Murray Hill, 1986—; dir. Consortium for Superconducting Electronics, 1990—. Fellow Am. Phys. Soc. (exec. com. div. condensed material physics 1991—); mem. Materials Rsch. Soc. Office: AT&T Bell Labs Murray Hill NJ 07974

BATSON, SUSAN CATHERINE, chemistry educator; b. McKees Rocks, Pa., Mar. 13, 1957; d. Cyril F. and Zora (Sokolovich) Kurtz; m. W. Brayton Batson, Nov. 22, 1980; children: Francis O., Edward S., Brayton T. BS in Chemistry, Clarion State Coll., 1979; postgrad., Clarion U., 1984, U. Pitts. Cert. tchr. chemistry, earth and space sci., Pa. Tchr. North Hills Sch. Dist., Pitts., 1980—; mem. Integration Team, North Hills High Sch., Pitts., 1991—; dir. North Hills Planetarium, Pitts., 1992—. Leader Cub Scout Pack 258, Bellevue, Pa., 1991—; advisor North Hills Environment Club, Pitts., 1991—. Grantee Spectroscopy Soc. of Pitts., 1992. Mem. NSTA, Pa. Sci. Tchrs. of Am., Am. Chem. Soc. Republican. Roman Catholic. Achievements include development (with others) of sci. and technology course for high school students which is innovative alternative to gen. sci. emphasizing critcal reading, analysis, problem solving-decision making rather than memorization of facts.

BATTAGLIA, FRANCO, chemistry educator; b. Catania, Italy, Dec. 15, 1953; s. Salvatore and Francesca (Nicolosi) B.; m. Manuela Libertini, Jan. 5, 1985. Italian Laurea, U. Catania (Italy), 1979; PhD, U. Rochester (N.Y.), 1985. Rsch. assoc. Max Planck Inst., Gottingen, Germany, 1980-81, U. Rochester, 1985, U. Rome, 1984-87, SUNY, Buffalo, 1987; prof. theoretical chemistry U. Della Basilicata, Potenza, Italy, 1987—; vis. prof. Columbia U., N.Y.C., 1992. Author: Notes in Classical and Quantum Physics, 1990. Recipient awards Sherman Clarke, 1982-85, Elon Huntington Hooker, 1983-84, Consiglio Nazional delle Ricerche, 1983, 86, 92. Mem. Am. Phys. Soc. (life), Italian Assn. Phys. Chemistry. Home: Via Dei Campani 55, Rome 00185, Italy Office: Chemistry Dept, U Della Basilicata, Potenza 85100, Italy

BATTERMAN, BORIS WILLIAM, physicist, educator, academic director; b. N.Y.C., Aug. 25, 1930; children: Robert W., William E., Thomas A. Student, Cooper Union Coll., 1949-50, Technische Hochschule, Stuttgart, Germany; student (Fulbright scholar), 1953-54; S.B., Mass. Inst. Tech.,

1952, Ph.D., 1956. Mem. tech. staff Bell Telephone Labs., Murray Hill, N.J., 1956-65; assoc. prof. Cornell U., 1965-67, prof. applied and engring. physics, 1967—, dir. Sch. Applied and Engring. Physics,, 1974-78, 1986—, dir. Synchrotron Radiation Lab. (CHESS),, 1978—, Walter S. Carpenter Jr. prof. engring., 1985—; cons. x-ray diffraction; mem. U.S.A. Nat. Com. Crystallography, Nat. Acad. Sci., 1969-72. Assoc. editor: Jour. Crystal Growth, 1964-74. Guggenheim fellow, 1971; Fulbright Hayes fellow, 1971; Alexander von Humboldt fellow, 1983. Fellow Am. Phys. Soc. Office: Wilson Lab Cornell U Ithaca NY 14853

BATTERMAN, STEVEN CHARLES, mechanical engineering, bioengineering educator; b. Bklyn., Aug. 15, 1937; s. Jacob and Anna (Abramowitz) B.; m. Judith Wilpon, Mar. 29, 1959; children: Scott David, Risa Karen, Daniel Adam. B.C.E., Cooper Union, 1959; Sc.M. (NSF fellow), Brown U., 1961, Ph.D., 1964; M.A. (hon.), U. Pa., 1971. Mem. faculty U. Pa., 1964—, prof. mech. engring. and applied mechanics, 1974-79; asso. prof. orthopaedic surgery research U. Pa. (Sch. Medicine), 1972-75, prof. orthopaedic surgery research, 1975—; prof. biomechanics in vet. medicine U. Pa Sch. Vet Medicine, 1975-84, prof. bioengring., 1974—; cons. to govt., industry, ins. cos., attys. Contbr. numerous articles to profl. jours.; patentee apparatus for acoustically determining periodontal health. Recipient S.R. Warren Disting. Teaching award U. Pa., 1982. Mem. ASCE, ASME, Am. Acad. Mechanics, Am. Soc. Engring. Edn., Biomed. Engring. Soc., Soc. Exptl. Stress Analysis, Soc. Automotive Engrs., Am. Soc. Safety Engrs., Am. Acad. Forensic Scis. (Founder's award 1992, pres-elect 1993—), Assn. for Advancement Automotive Medicine, Sigma Xi, Tau Beta Pi, Chi Epsilon. Jewish. Home: 109 Charlann Cir Cherry Hill NJ 08003-2906 Office: U Pa 120 Hayden Hall Philadelphia PA 19104

BATTISTA, JERRY JOSEPH, medical physicist; b. Montreal, Que., Can., Jan. 11, 1950; s. Evan and Angelina (Cuccioletta) B.; m. Leigh Barton, May 17, 1975; children: Michael, Susan, Andrea. MSc, U. Western Ont., London, 1973; PhD, U. Toronto, 1977. Jr. physicist Ont. Cancer Inst., Toronto, 1977-79; sr. physicist Cross Cancer Inst., Edmonton, Alb., Can., 1979-88; chief physicist London (Ont.) Regional Cancer Ctr., 1988—; nat. adv. bd. Nat. Rsch. Coun. of Can., Ottawa, 1991-92; mem. profl. adv. com. Ont. Cancer Found., Toronto, 1991-92, rsch. planning com. 1991-92; Can. grants reviewer Med. Rsch. Coun., 1990—. Contbr. articles to profl. jours. Med. Rsch. Coun. scholarship, 1973-76, Nat. Rsch. Coun. scholarship, 1971-73, Loyola Coll. Entrance scholarship, 1967-71; recipient numerous grants. Mem. Can. Organization of Med. Physicists, Can. Assn. of Physicist (chmn. div. of med. and biol. physics 1986-87), Am. Assn. of Physicists in Medicine (conf. program com. 1991—, Farrington-Daniels award 1986, Sylvia Fedoruk awards 1989, 92),. Achievements include devel. of convolution method for x-ray doses in cancer therapy; rsch. includes introduction of Ytterbium-169 for cancer brachytherapy, excellence in teachings of radiation physics and biology. Office: London Regional Cancer Ctr, 790 Commissioners Rd, London, ON Canada N6A 4L6

BATTISTELLI, JOSEPH JOHN, electronics executive; b. Bridgeport, Conn., Oct. 22, 1930; s. Joseph John and Maria (Brunetti) B.; m. Helen Josephine Thompson, Apr. 5, 1961; children: Jay Dominick, Randall Victor. BSEE, U. Conn., 1958; MSEE, U. Ariz., 1960. Registered electrical engr., Ariz., Ohio. V.p. Electro Tech. analysis Corp., Tucson, 1960-68; rsch. engr. Ohio U., Athens, 1968-72; sr. engr. Hughes Aircraft Co., Culver City, Calif., 1972-74; dir. engring. Lockheed Aircraft Co., Ont., Calif., 1974-80; dir. Riyadh area Litton Industries, Beverly Hills, Calif., 1980-91; v.p. Orion Ltd., Reston, Va., 1991—; cons. FAA, Washington, 1962-72, U.S. Army Electronics Command, Ft. Monmouth, N.J., 1962-74, Lochkeed Aircraft Co., Ont., 1980—. Contbr. articles to profl. jours. With U.S. Army, 1952-54. Mem. IEEE, Sigma Xi, Tau Beta Pi, Eta Kappa Nu, Phi Kappa Phi. Office: Litton Ltd, PO Box 7529, Riyadh 11472, Saudi Arabia

BATTISTI, ORESTE GUERINO, pediatrician; b. Fontaine, Belgium, June 10, 1951; s. Luigi and Sirch (Giuseppina) B.; m. Marie Van Pariss, Sept. 15, 1977 (div. 1985); children: Catherine, Anne, Marie, Frank; m. Tessariol Loretta, Mar. 22, 1986; children: Elene, Peolo, Ciril, Ariel. B, Saint Michel, Belgium, 1969; MD, Cath. U. Louvain, Belgium, 1976; grad. in pub. health, U. Nancy, France, 1988. Rsch. fellow Hammersmith Hosp., London, 1979-80; cons. MD Clin. Notre Dame, Belgium, 1980-83, Clin. St. Vincent, Belgium, 1984—; cons.-in-charge social pediatrics P. St. Raphaël, Liege, Belgium; tchr. Numing Sch., Liege, 1988, Brussels, 1988. Author: La Croissance du Prématuré an Alimentation Enterale, 1990. With Belgium Army. Mem. Soc. for Rsch. in Child Devel., Neonatal Soc. (U.K.), Belgian Soc. Neuropediatrics. Avocations: computing, music, soccer. Home: Rue Des Pinsons 50, 4451 Juprelle Belgium Office: Clin Saint Vincent, 207 Rue F Lefebvre, 4000 Liège Belgium

BATTY, J. MICHAEL, geographer, educator; b. Liverpool, Eng., Jan. 11, 1945; s. Jack and Nellie (Marsden) B.; m. Susan Elizabeth Howell, Jan. 4, 1969; 1 child, Daniel Jack. BA, U. Manchester, Eng., 1966; PhD, U. Wales, 1984. Rsch. asst. U. Manchester, 1966-69; lectr. U. Reading, Eng., 1969-74; visiting asst. prof. U. Waterloo, Ontario, Can., 1974-75; reader U. Reading, 1975-79; prof. U. Wales, Cardiff, 1979-84, dean environ. design, 1984-87, chmn. planning, 1987-90; dir. Nat. Ctr. for Geog. Info. and Analysis SUNY, Buffalo, N.Y., 1991—. Author: Urban Modelling, 1976, Microcomputer Graphics, 1987; editor: Systems Analysis in Urban Policy-Making, 1983, Advances in Urban Systems Models, 1986, Cities of the 21st Century, 1991. Recipient rsch. grants Econ. & Social Rsch. Coun., 1974-90, NSF, 1991—. Fellow Royal Town Planning Inst., Royal Soc. Arts, Chartered Inst. Transport. Avocations: travel, reading, drawing, China. Office: Nat Ctr Geog Info Analysis SUNY Buffalo NY 14261

BATZEL, ROGER ELWOOD, chemist; b. Weiser, Idaho, Dec. 1, 1921; s. Walter George and Inez Ruth (Klinefelter) B.; m. Edwina Lorraine Grindstaff, Aug. 18, 1946; children: Stella Lynne, Roger Edward, Stacy Lorraine. B.S., U. Idaho, 1947; Ph.D., U. Calif. at Berkeley, 1951. Mem. staff Lawrence Livermore (Calif.) Lab., 1953—, head chemistry dept., 1959-67, asso. dir. for chemistry, 1961-71, asso. dir. for testing, 1961-64, asso. dir. for space reactors, 1966-68, asso. dir. chem. and bio-med. research, 1969-71, dir. lab., 1971-88, assoc. dir. at large, 1988-89, dir. emeritus, 1989—. Served with USAAF, 1943-45. Named to Alumni Hall of Fame U. Idaho, 1972; recipient disting. assoc. award U.S. Dept. Energy, 1982. Fellow AAAS, Am. Phys. Soc.; mem. Sigma Xi. Office: Lawrence Livermore Lab PO Box 808 Livermore CA 94551-0808

BATZER, MARK ANDREW, molecular geneticist; b. Detroit, Oct. 26, 1961; s. Douglas Herman and Marie (Bean) B.; m. Pamela Elizabeth Richard, Feb. 3, 1990. BS in Zoology, Microbiology, Mich. State U., 1983, MS in Zoology, 1985; PhD in Genetics, Zoology, La. State U., 1988. Rsch. asst., grad. teaching asst. dept. zoology Mich. State U., East Lansing, 1983-85; grad. rsch. asst. dept. zoology La. State U., Baton Rouge, 1985-88; postdoctoral rsch. assoc. dept. biochemistry La. State U. Med. Ctr., New Orleans, 1988-92; biomed. scientist Human Genome Ctr., Lawrence Livermore (Calif.) Nat. Lab., 1992—; edit. bd. Analytical Biochemistry, 1993—, ad hoc reviewer, 1989—, Nucleic Acids Rsch., 1991—, Proc. of the Nat. Acad. of Scis., 1993—. Contbr. articles to profl. jours, peer reviewed jours. Recipient Basic rsch. award Cancer Crusaders, 1990-91, 91-92, Basic Rsch. award Ctr. for Molecular Cytometry, 1993-94. Mem. AAAS, Am. Soc. Microbiology, Am. Soc. Biochemistry and Molecular Biology, Am. Soc. for Cell Biology, Am. Soc. Human Genetics, Human Genome Organisation, Genetics Soc. Am. (travel award 1987). Home: 7320 Park Wood Cir Apt L Dublin CA 94568 Office: Lawrence Livermore Nat Lab Human Genome Ctr L-452 PO Box 808 Livermore CA 94551

BAUCHSPIESS, KARL RUDOLF, physicist; b. Cologne, Germany, Sept. 24, 1955; s. Rudolf and Gertrud (Dostert) B. Diplom, U. Koln, 1982; PhD, Simon Fraser U., Burnaby, B.C., Can., 1990. Vis. scientist IBM Almaden Rsch. Ctr., San Jose, 1989-90, Photon Factory, Nat. Lab. for High Energy Physics, Tsukuba, Ibaraki, Japan, 1990-92; vis. scientist Sch. Math. and Phys. Scis. Murdoch (Western Australia) U., 1992—. Contbr. articles to profl. jours. Mem. Am. Phys. Soc., European Phys. Soc, German Phys. Soc. Roman Catholic. Office: Murdoch U, Sch Math and Phys Scis, Murdoch 6150, Australia

BAUER, A(UGUST) ROBERT, JR., surgeon, educator; b. Phila., Dec. 23, 1928; s. A(ugust) Robert and Jessie Martha-Maynard (Monie) B.; BS, U. Mich., 1949, MS, 1950, MD, 1954; M Med. Sci.-Surgery, Ohio State U., 1960; m. Charmaine Louise Studer, June 28, 1957; children: Robert, John, William, Anne, Charles, James. Intern Walter Reed Army Med. Ctr., 1954-55; resident in surgery Univ. Hosp., Ohio State U., Columbus, also instr., 1957-61; pvt. practice medicine, specializing in surgery, Mt. Pleasant, Mich., 1962-74; chief surgery Ctrl. Mich. Community Hosp., Mt. Pleasant, 1964-65, vice chief of staff, 1967, chief of staff, 1968; clin. faculty Mich. State Med. Sch., East Lansing, 1974; mem. staff St. Mark's Hosp., Salt Lake City, 1974-91; pvt. practice surgery, Salt Lake City, 1974-91; clin. instr. surgery U. Utah, 1975-91. Trustee Rowland Hall, St. Mark's Sch., Salt Lake City, 1978-84; mem. Utah Health Planning Coun., 1979-81. Served with M.C., U.S. Army, 1954-57. Diplomate Am. Bd. Surgery. Fellow ACS, Southwestern Surg. Congress; mem. AMA, Salt Lake County Med. Soc., Utah Med. Assn. (various coms.), Utah Soc. Certified Surgeons, Salt Lake Surg. Soc., Pan Am. Med. Assn. (affiliate), AAAS (affiliate), Sigma Phi Epsilon, Phi Rho Sigma. Episcopalian. Club: Zollinger. Contbr. articles to profl. publs., researcher surg. immunology. Office: PO Box 17533 Salt Lake City UT 84117-0533

BAUER, ERNST GEORG, physicist, educator; b. Schonberg, Germany, Feb. 27, 1928. MS, U. Munich, 1953, PhD in Physics, 1955. Rsch. asst. U. Munich, 1955-58; head crystal physics br. Michelson Lab., China Lake, Calif., 1958-69; prof. Tech. U. Clausthal, Germany, 1969—. Author: Elektronenbeugung, 1958. Recipient Gaede Preis German Vacuum Soc., 1988. Fellow Am. Phys. Soc.; mem. Am. Vacuum Soc. (Welch award 1992), Material Rsch. Soc., German Electron Microscope Soc., Gottingen Acad. Sci. Office: Tech Univ Clausthal, 3392 Clausthal Germany

BAUER, EUGENE ANDREW, dermatologist, educator; b. Mattoon, Ill., June 17, 1942; s. Eugene C. and Madge L. (Armer) B.; m. Gloria Anne Hehman, Feb. 19, 1966; children: Marc A., Christine A., J. Michael, Amanda F. BS, Northwestern U., 1963, MD, 1967. Diplomate Am. Bd. Dermatology, Nat. Bd. Med. Examiners. Intern Barnes Hosp., St. Louis, 1967-68; resident, fellow div. dermatology Washington U. Med. Ctr., 1968-70; instr. Washington U., St. Louis, 1971-72, asst. prof. dermatology, 1974-78, assoc. prof., 1978-82, prof., 1982-88; prof. chmn. Stanford U. Sch. Medicine, 1988—; program dir. Gen. Clin. Rsch. Ctr. Contbr. numerous articles to profl. jours. Served to lt. comdr. USNR, 1972-74. Fellow Am. Acad. Dermatology; mem. Am. Fedn. Clin. Research, Am. Soc. Clin. Investigation, Am. Dermatol. Assn., Soc. Investigative Dermatology (bd. dirs. 1981-86, assoc. editor Jour. Investigative Dermatology 1982-87), Cen. Soc. Clin. Research, Assn. Am. Physicians. Office: Stanford U Med Ctr Dept Dermatology R-144 Stanford CA 94305*

BAUER, MICHAEL ANTHONY, computer scientist, educator; b. Dayton, Ohio, Feb. 18, 1948; married; 2 children. BSc, U. Dayton, 1970; MSc, U. Toronto, 1971, PhD in Computer Sci., 1978. Rschr. artificial intelligence Edinburgh U., 1974-75; assoc. prof. computer sci. U. Western Ont., 1975—, chmn. dept., 1991—; cons. Geac Computers Internat., 1984-88; mem. bd. Can. Info. Processing Soc., 1984-88, Assn. Computer Machinery, 1989—; advisor Internat. Bus. Machine Ctr. Advanement Studies, 1990-91. Achievements include research in distributed computing, especially distributed algorithms, correctness, languages for distributed computing, verification; software engineering, including methodologies, formal specifications, development environments. Office: University of Western Ontario, Middlesex College Rm 355, London, ON Canada N6A 5B7*

BAUER, STEVEN MICHAEL, cost containment engineer; b. Hemet, Calif., Nov. 8, 1949; s. Donald Richard and Jeanne Patricia (Lamont) B.; m. Myung-Hee Min, Sept. 10, 1983; children: Claudia Margaret, Monica Anne. BA in Physics, Calif. State U., San Bernardino, 1971, BS in Physics, 1984, cert. in acctg., 1980, cert. in computer programming, 1986; postgrad., U. Calif., 1974; post grad., Calif. State U., 1982, 87—; cert. in counseling skills, U. Calif. extension, 1991., cert. in alcohol and other drug studies, 1992. Registered engr. in tng., Calif., 1976. Asst. nuclear engr. So. Calif. Edison Co., Rosemead, 1973-76, assoc. nuclear engr., 1976-88, cost containment engr., 1988—; cons. rsch. dept. Jerry L. Pettis Meml. Vets. Hosp., 1978-79, Calif. State U., San Bernardino, 1983—; cons. planning San Bernardino County, 1975-76; cons. alumni rels. Calif. State U., San. Bernardino, 1989-90. Supporter St. Labre Indian Sch., 1984, Asian Relief Fund, 1985—, So. Poverty Law Ctr., Amnesty Internat., Freedom Writer, 1988; mem. Greenpeace, Wilderness Soc., Internat. Platform Assn.; supporter United Negro Coll. Fund., 1985, vol., 1988; vol. counselor San Bernandino Girl's Juvenile Hall, ARC, 1990—; fellow Casa Colina Hosp.; mem. L.A. County Mus. Art; campaign vol. Congressment George E. Brown, 1966; block capt. Neighborhood Watch, sec., bd. dirs., 1992—; chpt. sec. Sierra Club, 1992. Mem. Am. Nuclear Soc. (assoc.), Calif. State U. San Bernardino Alumni Assn. (sec. bd. 1979-80, rep. food com. 1980-82), Nat. Assn. Accts., Astron. Soc. Pacific, Assn. Computing Machinery (assoc.), Ams. for Energy Independence (bd. dirs. 1990—), K.C. (sec., recorder 1989, community dir., Outstanding Svcs. award 1989), Toastmasters, Numismatic Assn. So. Calif., UCLA Alumni (life), Calif. State U. Fullerton Computer Club, Sierra Club (sec. San Garganio chpt. 1992). Avocations: personal computers, reading, swimming, aerobic exercise, hiking. Home and Office: 131 Monroe Ct San Bernardino CA 92408-4137

BAUER, THEODORE HENRY, nuclear engineer; b. N.Y.C., Sept. 16, 1943; s. George Adolf and Dorothy Johanna (Eberling) B.; m. Sandra Kay Crittenden, June 26, 1971; children: Samuel George Henry, Gerard Bradford. BS in Applied Physics magna cum laude, Hofstra U., 1965; PhD in Theoretical Physics, Cornell U., 1970. Sr. rsch. assoc. Daresbury Nuclear Physics Lab., Cheshire, Eng., 1970-73; lectr., postdoctoral rsch. assoc. Cornell U., Ithaca, N.Y., 1973-77; nuclear engr., physicist Argonne (Ill.) Nat. Lab., 1977—. Contbr. articles to profl. publs. NSF trainee, 1965-69; recipient Literary award Material Sci. and Tech. div. Am. Nuclear Soc., 1990. Mem. Sigma Xi (chpt. pres. 1990). Achievements include planning and analysis of large scale in-reactor experiments in support of advanced reactor safety. Office: Argonne Nat Lab 9700 S Cass Ave Argonne IL 60439

BAUER, VICTOR JOHN, pharmaceutical company executive; b. N.Y.C., May 14, 1935; s. Victor and Ottilie (Wild) B.; m. Sonia Witkowski, Sept. 14, 1957; children: Katherine E., Steven E. B.S., MIT, 1956; Ph.D., U. Wis., 1960. Research fellow Harvard U., 1961; research chemist Lederle Labs., Pearl River, N.Y., 1961-71; with Hoechst-Roussel Pharms. Inc., Somerville, N.J., 1971-92; dir. chem. research Hoechst-Roussel Pharms. Inc., 1971-74, v.p. ops., 1974-80, exec. v.p., chief operating officer, 1980-87, exec. v.p., pres. research and mfg. div., 1987—, pres., 1989-92. Trustee N.J. Symphony Orch., 1992—. Mem. Am. Chem. Soc.

BAUGH, CHARLES MILTON, biochemistry educator, college dean; b. Fayetteville, N.C., June 20, 1931; s. John Yewell and Dorothy Ann (Shaw) B.; m. Ebby O. Jonsdottir, Oct. 24, 1953; children: Dorothy Baugh Ledbetter, Barbara Baugh Baumer, Charis Baugh Spyridon, Lisa Baugh Eckert. BS in Biochemistry, U. Chgo., 1958; PhD in Biochemistry, Tulane U., 1962. Instr. Tulane U., New Orleans, 1963-64, asst. prof. biochemistry, 1964-65; asst. prof. medicine and pharmacology Washington U., St. Louis, 1965-66; assoc. prof. medicine and biochemistry U. Ala., Birmingham, 1966-70, prof. pediatrics, medicine and biochemistry, 1970-73; prof. biochemistry U. South Ala., Mobile, 1973—, chmn. dept., 1973-81, assoc. dean basic sci., 1976-87, dean Coll. Medicine, 1987-92; extensive rsch. cons. Australian Nat. Health and Med. Rsch. Coun., 1975—, Med. Rsch. Coun. Can., 1976—; pres. South Ala. Med. Sci. Found., Mobile, 1982-92. Contbr. numerous articles, book chpts. to profl. publs. With USN, 1951-55. Predoctoral fellow NIH, Walter Libby Rsch. fellow Am. Heart Assn., La. chpt.; scholastic scholar U. Chgo.; recipient numerous grants NIH, Am. Cancer Soc., others. Fellow Royal Soc. Medicine (Eng.); mem. Soc. Exptl. Biology and Medicine, Am. Inst. Nutrition, Am. Soc. Biochemistry and Molecular Biology, Ala. Acad. Sci. (pres. 1982), So. Med. Assn. (hon.), Alpha Omega Alpha (hon.). Home and Office: PO Box 586 Salemburg NC 28385

BAUGHMAN, GEORGE WASHINGTON, aeronautical operations research scientist; b. Strasbourg, Can., Feb. 22, 1911; came to U.S., 1923; s. John Allen and Daisy (Lafferty) B.; m. Helen Frances Hille, July 2, 1938; children: Barbara Ann (dec.), Susan Jane (dec.), Charles Allen. BS in Aero. Engring. magna cum laude, Wichita State U., 1931; postgrad., Ga. Tech., 1954-55, Wichita State U., 1933-37. Petroleum geologist Phillips Petroleum Co., Wichita, Kans., 1932-40; project engr. Cessna Aircraft Co., Wichita, 1940-51; staff specialist Lockheed Aircraft Corp., Marietta, Ga., 1951-54, ops. rsch. scientist, 1954-56, asst. project engr., Monticello project, 1956-58; prin. ops. analyst Cornell Aero. Lab., Arlington, Va., 1958-67; head systems environ. group Cornell Aero. Lab., Bangkok, Thailand, 1967-70; cons. future aviation Pres. Aviation Adv. Com., Washington, 1971-72; tech. coord. High Frontier Inc., Arlington, 1990—; mem. task force sr. scientists engrs. AAAS, Washington, 1991—; cons. SST program, Fed. Aviation Agy., Washington, 1965-67. Contbr. over 70 corp. tech. pubs. including aircraft pilot handbooks, maintenance manuals, catalogues, structural repair, aircraft design & performance, noise & sonic boom analysis and others. Bd. dirs. Vis. Nurses Assn. Va., 1976-82, Eldercrafters Inc., Alexandria, 1980-86; v.p. Alexandria Taxpayers Alliance, 1980—; past chmn., pres. several local or nat. aircraft orgns. Recipient Missildine Chemistry prize, Wichita State U., 1928. Mem. Am. Inst. Aeronautics Astronautics (chmn. Atlanta chpt. 1953-54, Nat. counselman 1954), Am. Assn. Petroleum Geologists, Kans. Geol. Soc., Tau Beta Pi. Republican. Achievements include design of airframe and control for 2 cycle engine tractor propeller, radio controlled winged bomb, of new series of composite metal and wood furniture cases, new series of low cost hydraulic actuators and controls for implements, of aerodynamic modifications of a tactical cargo aircraft for improved short field performance; development of aircraft industry standard for the selection and quality control of aircraft spruce during increasing scarcity of supply, standard for primary and secondary aircraft control cable assemblies, standards for a series of commercially available utility fasteners and fittings for secondary non-structural applications, of heat-treating procedures, for maximizing energy absorbing properties of chrome-vanadium steel, of method for compensating automated aircraft navigation data to improve target accuracy of long range aircraft reconnaissance missions of various other improvements in evaluating and projecting uses and efficiencies of aircraft. Home: 316 Crown View Dr Alexandria VA 22314-4802 Office: High Frontier Inc 2800 Shirlington Rd Arlington VA 22206-3601

BAUGHMAN, JOHN THOMAS, data base designer, educator; b. Mishawaka, Ind., Dec. 22, 1943; s. George Thomas and Ida Marie (Forman) B.; m. RoxAnn Henderson, Aug. 22, 1964; children: Michael, Angela. BS, Purdue U., 1966, MAT, 1967. Tchr. Mishawaka (Ind.) High Sch., 1968-69; programmer Assocs. Corp. N. Am., South Bend, Ind., 1969-72; systems analyst Notre Dame (Ind.) U., 1972-74, Am. Airlines, Tulsa, 1974-75; data base adminstr. Notre Dame (Ind.) U., 1975-81, Valero Energy Corp., San Antonio, 1981-88, USAA, San Antonio, 1988—. Mem. Data Processing Mgmt. Assn., Toastmasters. Office: USAA USAA Bldg DIW-A1 San Antonio TX 78288

BAULIEU, ETIENNE-EMILE, endocrinologist; b. Strasbourg, France, Dec. 12, 1926; s. Léon Blum and Thérèse (Lion) B.; Dr. Medicine and Phys. Scis., Faculty Medicine and Scis., U. Paris, 1955; Dr. Honoris Causa Université de Gand, 1991, Tufts U., 1991; m. Yolande Compagnon, Oct. 4, 1947; children: Catherine, Laurent, Frédérique. Intern, Hosp. of Paris, 1951; from lectr. to prof. biol. chemistry U. Reims Sch. Medicine, 1958, U. Rouen Sch. Medicine, 1960, Faculty Medicine, U. Paris, 1961; Claude Bernard lectr.; vis. lectr. Columbia U., N.Y.C., 1961-62; sci. dir. Nat. Inst. Health and Med. Research, 1963—; co-founder Internat. Soc. for Research in Biology of Reprodn., 1967, Karolinska Symposia in Reproductive Endocrinology, 1970; chmn. sci. council Fondation pour la Recherche médicale française, 1973-75; chmn. sci. council Inst. Nat. de la Santé et de la Recherche Médicale, 1975-79; chmn. sci. council Centre Internat. de Recherches Médicales de Franceville (Gabon), 1978-85. Decorated chevalier Order of Merit, commandeur Legion of Honor; recipient Reichstein award Internat. Soc. Endocrinology, 1972, Grand Prix Scientifique, City of Paris, 1973, (with E. Jensen) Roussel prize, 1976, Pincus Meml. award, 1978, 1st European medal English Soc. Endocrinology, 1985, Lasker award, 1989. Mem. French Acad. Scis., French Soc. Biochemistry, French Soc. Endocrinology, Am. Acad. Achievement, Royal Soc. Medicine, Endocrine Soc. (U.S.), Nat. Acad. Sci. (fgn. 1990), N.Y. Acad. Scis. Research on secretion, metabolism, physio-pathology and mechanism of action of steroid hormones, and biology of reprodn. Address: Unité 33 Inserm Hosp de Bicêtre, 80 Rue du General Leclerc, 94276 Bicêtre France

BAUM, BERNARD RENE, biosystematist; b. Paris, Feb. 14, 1937; s. Kurt and Martha (Berl) B.; m. Danielle Habib, May 24, 1961; 1 child, Anat. B.S., Hebrew U., Jerusalem, 1963, M.S., 1963, Ph.D., 1966. Research scientist Agr. Can., Ottawa, Ont., 1966-74, sr. research scientist, 1974-80, prin. research scientist, 1980—; chief vascular plants sect. Biosystematics Research Inst. Agr. Can., 1981—. Author: Oats: Wild and Cultivated, 1977, Monograph of Tamarix, 1978, World Registry of Avena Cultivars, 1972, World Registry of Barley Cultivars, 1985; assoc. editor Can. Jour. Botany, 1986—, Euphytica, 1987—. Fellow Acad. Sci.-Royal Soc. Can.; mem. Can. Bot. Assn. (Lawson medal 1979), Bot. Soc. Am., Am. Soc. Plant Taxonomists, Internat. Assn. Plant Taxonomists, Classification Soc., Linnean Soc. London, Orgn. Plant Taxonomy of the Mediterranean Area. Home: 15 Murray St, Ste 408, Ottawa, ON Canada K1M 9M5 Office: Biosystematics Rsch Ctr, Research Br, Cen Exptl Farm, Ottawa, ON Canada K1A 0C6

BAUM, CARL EDWARD, electromagnetic theorist; b. Binghamton, N.Y., Feb. 6, 1940; s. George Theodore and Evelyn Monica (Bliven) B. BS with honors, Calif. Inst. Tech., 1962, MS, 1963, PhD, 1969. Commd. 2d lt. USAF, 1962; advanced through grades to capt., 1967, resigned, 1971; project officer Phillips Lab. (formerly Air Force Weapons Lab.), Kirtland AFB, N.Mex., 1963-71, sr. scientist for electromagnetics, 1971—; pres. SUMMA Found.; U.S. del. to gen. assembly Internat. Union Radio Sci., Lima, Peru, 1975, Helsinki, Finland, 1978, Washington,1 981, Florence, Italy, 1984, Tel Aviv, 1987, Prague, Czechoslovakia, 1990; mem. Commn. B U.S. Nat. Com., 1975—, Commn. E, 1982—, Commn. A, 1990—. Author: (with others) Transient Electromagnetic Fields, 1976, Electromagnetic Scattering, 1978, Acoustic, Electromagnetic and Elastic Wave Scattering, 1980, Fast Electrical and Optical Measurements, 1986, EMP Interaction: Principles, Techniques and Reference Data, 1986, Lightning Electromagnetics, 1990, Modern Radio Science, 1990, Recent Advances in Electromagnetic Theory, 1990, Direct and Inverse Methods in Radar Polarimetry, 1992; co-author: (with A.P. Stone) Transient Lens Synthesis: Differential Geometry in Electromagnetic Theory, 1991; contbr. articles to profl. jours. Recipient award Honeywell Corp., 1962, R & D award USAF, 1970, Harold Brown award Air Force Systems Command, 1990; Electromagnetic pulse fellow. Fellow IEEE (Harry Diamond Meml. award, 1987, Richard R. Stoddart award, 1984); mem. Electromagnetics Soc. (pres. 1983-85), Electromagnetics Acad., Sigma Xi, Tau Beta Pi. Roman Catholic. Home: 5116 Eastern Ave SE Unit D Albuquerque NM 87108 Office: Phillips Lab/WSR Kirtland AFB NM 87117

BAUM, HOWARD BARRY, physician; b. Passaic, N.J., Feb. 14, 1952; s. Samuel and Ethel (Stuhlbach) B.; m. Carolyn Frey, Sept. 7, 1986; children: Eric, Evan. AB summa cum laude, Dartmouth Coll., 1973; MD, Cornell U. Med. Coll., 1977. Diplomate Am. Bd. Internal Medicine and Gastroenterology. Resident internal medicine Dartmouth-Hitchcock Med. Ctr., Hanover, N.H., 1977-80; fellow in gastroenterology The N.Y. Hosp., Cornell Med. Ctr., N.Y.C., 1980-82; ptnr. Passaic (N.J.) Med. Assocs., PA, 1982—; trustee Passaic (N.J.) Valley Prof. Standards Review Orgn., 1983-84, Passaic (N.J.) Beth Israel Hosp., 1987—; dept. chief gastroenterology, Passaic Beth Israel Hosp., Gen. Hosp. Ctr. at Passaic, 1990-91; governing body Region One Health Planning Consortium, N.J., 1991—. Co-founder Doctors Against Misusing Passaic's Environ. Resources, 1985; trustee Jewish Fedn. Greater Clifton, Passaic, 1987—; steering com. PASS Plan, Passaic County, 1988—; v.p. Assn. Jewish Fedns. N.J., 1990—. Recipient Arthur Palmer prize Cornell Med. Coll., 1977; named Disting. Health Profl. United Passaic Orgn., 1987. Mem. AMA, ACP, Passaic County Med. Soc. (pres. 1993—, v.p. 1991-92), N.J. Med. Soc., N.J. Gastroenterology Soc., Phi Beta Kappa. Office: Passaic Med Assocs 540 Broadway Passaic NJ 07055

BAUM, JANET SUZANNE, architect; b. Chgo., Jan. 31, 1944; d. Lee S. and Gladys M. (Dial) Fantz; 1 child, Rachel L. BS in Archtl. Sci., Washington U., St. Louis, 1966; MArch, Harvard U., 1970. Registered architect, Mass. Prin. Janet Baum Architecture and Planning, Cambridge, Mass., 1973—; staff architect Harvard U. Med. Sch., Boston, 1973-85; prin. Payette Assocs., Inc., Boston, 1985-92, Near Earth Archtl. Rsch., Boston, 1988—; group v.p. Hellmuth Obata Kassabaum, St. Louis, 1993—. Co-author:

Health and Safety Guidelines for Laboratory Design, 1980, Improving Safety in the Chemical Lab, 1987, 2d edit., 1990, Guidelines for Laboratory Design, 1987, 2d edit., 1993, Safe Lab Design, 1990, The Biotechnology Industry, 1991. Mem. AIA (chair com. on biomed. rsch. bldgs. 1986—), Internat. Soc. Pharm. Engrs., Am. Chem. Soc., N.Y. Acad. Sci. Office: Hellmuth Obata Kassabaum 1831 Chestnut St Saint Louis MO 63103

BAUM, LAWRENCE STEPHEN, biologist, educator; b. Scranton, Pa., Mar. 3, 1938; married; 3 children. BS, U. Ala., 1960, MS, 1962, PhD in Cell Physiology, 1965. Asst. prof. biology Extension Ctr. U. Ala., 1965-66; asst. prof. N.E. La. U., 1966-69, assoc. prof. biology, 1969—, dir. Cancer Rsch. Ctr., 1981—. Mem. Am. Physiol. Soc., Sigma Xi. Achievements include research in angiogenesis during tumor development, scanning electron microscope study of surface morphology of neoplastic urothelium, mode of action of various chemical carcinogens, incidence of cancer in Louisiana. Office: Northeasr LA University Cancer Research Ctr 700 Univ Ave Monroe LA 71209*

BAUM, PAUL FRANK, mathematics educator; b. N.Y.C., July 20, 1936; s. Mark and Celia (Frank) B.; m. Barbara Alice Bigelow, June 21, 1961; children: Sarah Alice, Michael Eli, Jessica Louise. AB, Harvard U., 1958; PhD, Princeton U., 1963. Vis. mem. Inst. for Advanced Study, Princeton, N.J., 1964-65, 76-77; asst. prof. math. Princeton U., 1965-67, vis. assoc. prof. math., 1967-68; assoc. prof. math. Brown U., Providence, 1967-71, prof. math., 1971-87; prof. math. Pa. State U., University Park, 1987-91, disting. prof. math., 1991—; speaker at conf., seminar in field. Contbr. articles to profl. publs. Japan Soc. for Promotion of Scis. fellow, 1986; NSF rsch. grantee, 1965—. Mem. Am. Math. Soc. (profl. editorial com. 1983-86). Jewish. Office: Pa State U 206 McAllister Bldg University Park PA 16802

BAUM, PETER SAMUEL, research director; b. Baltimore, June 12, 1944; s. Bernard Baum and Mary (Stalker) Whitenack; m. Naomi Lois Arenberg, Sept. 9, 1990. BS in Math., Shimer Coll., 1966; MS in Math., MS in Computer Sci., Southern Ill. Univ., 1975. Systems programmer Southern Ill. Univ. Med. Sch., Springfield, Ill., 1975-77; sr. software engr. Digital Equipment Corp., Maynard, Mass., 1977-86; rsch. dir. Aesir Rsch., Cambridge, Mass., 1987—. With U.S. Army, 1968-70, Korea. Achievements include patent for Tessellator, Projecting Tessellator. Office: Aesir Research 9 Eustis St Mail Stop 2R Cambridge MA 02140-2226

BAUM, RICHARD THEODORE, engineering executive; b. N.Y.C., Oct. 3, 1919. BA, Columbia U., 1940, BS, 1941, MS, 1948. Registered profl. engr., N.Y., D.C., and 20 other states, Nat. Bur. Engrng. Registration. Engr. Electric Boat Co., Groton, Conn., 1941-43; with Jaros, Baum & Bolles, N.Y.C., 1946—, ptnr., 1958-86, ptnr. emeritus, cons. to firm, 1986—; mem. adv. coun., faculty of engring. and applied sci. Columbia U., N.Y.C., 1972—. 1st lt. USAAF, 1943-46. Egleston medalist Columbia U., 1985. Fellow ASME, ASHRAE, Am. Cons. Engrs. Coun.; mem. NAE (mech. engring. peer com. 1991-93), NSPE, N.Y. Acad. Scis., Nat. Soc. Energy Engrs., NRC (chmn. bldg. rsch. bd. 1987-91), Am. Arbitration Assn. (panel arbitrators 1973—), Coun. on Tall Bldgs. and Urban Habitat (vice chmn. N.Am. chpt.), Univ. Club N.Y.C. Office: Jaros Baum & Bolles 345 Park Ave New York NY 10154-0002

BAUM, STEFI ALISON, astronomer; b. Chgo., Dec. 11, 1958; d. Leonard E. and Julia (Lieberman) B.; m. Christopher Peter O'Dea, Aug. 17, 1985; children: Connor Baum O'Dea, Kieran Peter Baum O'Dea. BA in Physics, Harvard U., Radcliffe, 1980; PhD in Astronomy, U. Md., 1987. Postdoctoral researcher Netherlands Foun. for Rsch. in Astronomy, Dwingeloo, 1987—. Mem. Am. Astron. Soc. (Annie Jump Cannon award astronomy 1993). Avocations: badminton, knitting, swimming, bike riding, visiting zoos. Home: Achter de Hoven 3, 7991 AD Dwingeloo The Netherlands Office: Space Telescope & Science Inst 3700 San Martin Dr Baltimore MD 21218

BAUMAN, DALE ELTON, nutritional biochemistry educator; b. Detroit, Dec. 26, 1942; s. Elton Blaine and Waneta Mary (Taylor) B.; m. L. Marie Vinande, Aug. 28, 1965; children: Rebecca, Todd, Jeffrey. B.S., Mich. State U., 1964, M.S., 1968; Ph.D., U. Ill., 1969. Asst. prof., assoc. prof. U. Ill.-Urbana, 1969-78; vis. prof. Mich. State U., East Lansing, 1978; assoc. prof., then prof. Cornell U., Ithaca, N.Y., 1979—, Liberty Hyde Bailey prof., 1987; mem. U.S. Bd. Agr., U.S. Com. Biotech. Contbr. articles to profl. jours. Leader and scoutmaster Boy Scouts Am., Mich., N.Y., 1978-83. Recipient N.Y. Farmers award, 1982, Alexander von Humboldt award, 1985, USDA Superior Service award, 1986. Mem. Am. Dairy Sci. Assn. (Nat. Student award 1967, Nutrition Research award 1982, Biotech. award 1987), Am. Soc. Animal Sci. (Young Scientist award 1977), Am. Inst. Nutrition, Nat. Acad. Scis. Methodist. Home: 2 Eagleshead Rd Ithaca NY 14850-9659 Office: Cornell U 262 Morrison Ithaca NY 14853

BAUMAN, JAN GEORGIUS JOSEF, cell biologist, histochemist; b. Vlaardingen, The Netherlands, Mar. 1, 1950; s. Petrus W.M. and Maria (van Holstijn) B.; m. Marionne J.A.Th. van de Kruijs, June 14, 1976; children: Sanne Maartje, Eveline Sabine, Nadine Marieke. D of Biology, State U. Leiden, The Netherlands, 1975, PhD, 1980. Doctoral researcher lab. histocytochemistry U. Leiden, 1975-78; rsch. scientist dept. radiobiology Erasmus U., Rotterdam, The Netherlands, 1980-85, Netherlands Cancer Found., Ryswyk, 1985-87; fellow Royal Netherlands Acad. Scis., Amsterdam; appt. at dept. radiobiology Erasmus U., Rotterdam, 1987-92; sect. head cytometry Inst. for Applied Radiobiology and Immunology, Ryswyk, 1990—. Contbr. articles to profl. jours. Recipient Robert Feulgen prize Gesellschaft fur Histochemie, 1983, Eleanor Roosevelt Inst. Cancer Rsch. fellowship Internat. Union Against Cancer, 1990. Mem. Internat. Assn. Histochemists (bd. dirs. 1991—), Internat. Soc. Analytical Cylology, Am. Soc. Cell Biology. Avocations: computer programming, squash, bicycling.

BAUMAN, NATAN, audiologist, acoustical engineer; b. Magnitogorsk, USSR, Sept. 7, 1945; came to U.S., 1970; s. Alter Grzybmacher and Cela (Majerczyk) Bauman; m. Eliza Tabak, July 19, 1971; children: Joel, Noam, Dara-Ann. MS, Poly. Inst., Wroclan, Poland, 1969; EdD, Columbia U., 1984. Sch. audiologist St. Francis de Sales Sch. for the Deaf, Bklyn., 1973-75; clin. audiologist Misericorida Hosp., Bronx, N.Y., 1975-77; rsch. asst. Columbia U. Tchrs. Coll., N.Y.C., 1976-77; head dept. audiology Lincoln Hosp., Bronx, 1977-78; clin. audiologist St. Michael's Hosp., Toronto, Ont., Can., 1978-80; dir. hearing and speech clinic Yale New Haven Hosp., 1980-88; dir. Hearing Balance and Speech Ctr., New Haven, 1988—; cons. Precision Acoustic Lab., N.Y.C., 1970-72, St. Joseph Sch. for the Deaf, Bronx, 1974-75; acoustical engr. Paso Electronics, N.Y.C., 1972-73. Columbia U. fellow, 1973; recipient Spencer Found. award, 1977. Mem. Am. Acad. Audiology, Am. Speech-Lang.-Hearing Assn. (cert. clin. competence in audiology), Acoustical Soc. Am., N.Y. Sci. Acad. Achievements include patent pending for a hand-held hearing device. Office: Hearing Balance and Speech Ctr 2 Church St S New Haven CT 06519-1717

BAUMANN, THEODORE ROBERT, aerospace engineer, consultant, army officer; b. Bklyn., May 13, 1932; s. Emil Joseph and Sophie (Reiblein) B.; m. Patricia Louise Drake, Dec. 16, 1967; children: Veronica Ann, Robert Theodore, Joseph Edmund. B in Aerospace Engring., Poly. U., Bklyn., 1954; MS in Aerospace Engring., U. So. Calif., L.A., 1962; grad., US Army C&GS Coll., 1970, Indsl. Coll. of Armed Forces, 1970, US Army War Coll., 1979, Air War Coll., 1982. Structures engr. Glenn L. Martin Co., Balt., 1954-55; structural loads engr. N.Am. Rockwell, L.A., 1958-67; dynamics engr. TRW Systems Group, Redondo Beach, Calif., 1967-71, systems engr., 1971-75, project engr., 1975-84, sr. project engr., 1984-92; cons. SAAB-Scania Aerospace Div., Linkoping, Sweden, 1981-82; asst. dir. Dir. Weapons Systems, U.S. Army, Washington, 1981-85, staff officer Missile & Air Def. System div., 1975-81. Contbr. articles to Machine Design, tech. publs., tech. symposia. Asst. scoutmaster Boy Scouts Am., Downey, Calif., 1985—; instr. Venice Judo Boys Club, 1966-86. 1st. lt. U.S. Army, 1955-58, col. USAR. Decorated Legion of Merit. Mem. AIAA; mem. Soc. Am. Mil. Engrs (life), Am. Legion, Res. Officers Assn. (life), U.S. Judo Fedn., Nat. Rifle Assn. Republican. Roman Catholic. Achievements include developing a new method for the analysis and classification of random data; contbr. to air force ballistic missile program; devel. procedure for design of prestressed joints and fittings. Office: Theodore R Baumann & Assoc 7732 Brunache St Downey CA 90242

BAUMGARDT, BILLY RAY, university official, agriculturist; b. Lafayette, Ind., Jan. 17, 1933; s. Raymond P. and Mildred L. (Cordray) B.; m. D. Elaine Blain, June 8, 1952; children: Pamela K. Baumgardt Farley, Teresa Jo Baumgardt Adolfsen, Donald Ray. B.S. in Agr., Purdue U., 1955, M.S., 1956; Ph.D., Rutgers U., 1959. From asst. to assoc. prof. U. Wis., Madison, 1959-67; prof. animal nutrition Pa. State U., University Park, 1967-70, head dept. dairy and animal sci., 1970-79, assoc. dir. agrl. expt. sta., 1979-80; dir. agrl. research, assoc. dean Purdue U., West Lafayette, Ind., 1980—. Contbr. chpts. to books, articles to sci. jours. Recipient Wilkinson award Pa. State U., 1979. Fellow AAAS; mem. Am. Dairy Sci. Assn. (Nutrition Rsch. award 1966, pres. 1984-85, award of Honor 1993), Am. Inst. Nutrition, Am. Soc. Animal Sci., Rotary, Sigma Xi. Home: 812 Lazy Ln Lafayette IN 47904-2722 Office: Purdue U West Lafayette IN 47907

BAUMGARTNER, DONALD LAWRENCE, entomologist, educator; b. Chgo., June 4, 1954; s. Lawrence and Gloria Ann (Winkler) B.; m. Claudia Jeanne Zaloudek, Aug. 21, 1982; 1 child, Kenneth. BS, U. Ill., 1979, MS, 1984. Apiary rsch. asst. U. Ill., Chgo., 1977, 78, 81, teaching asst., 1979-84; mgr. Ill. Lyme Disease Project Coll. of Medicine, U. Ill., Rockford, 1990-91; mus. specialist Field Mus. Natural History, Chgo., 1978-80, adult edn. instr., 1981-85; field supr., rsch. coord. N.W. Mosquito Abatement Dist., Wheeling, Ill., 1985-90; life scientist U.S. EPA, Chgo., 1991—; adj. faculty William Rainey Harper Coll., Palatine, Ill., 1987—; entomol. cons. Kane County Health Dept., Aurora, Ill., 1990. Contbr. articles to Jour. Med. Entomology, Jour. Am. Mosquito Control Assn., Gt. Lakes Entomologist, Proceed, Ill. Mosquito and Vector Control Assn. Newsletter; state editor: Vector Control Bull. North Cen. States. Grantee Biosystematic Rsch. Inst., 1981, U. Ill. Rsch. Bd., 1990, Ill. Vector Control, 1991-92. Mem. Entomol. Soc. Am., Am. Mosquito Control Assn., Soc. for Vector Ecologists, Mich. Entomol. Soc., Ill. Mosquito and Vector Control Assn. Democrat. Home: 120 S Walnut St Palatine IL 60067 Office: US EPA SP-14J 77 W Jackson Blvd Chicago IL 60604

BAUMGARTNER, JAMES EARL, mathematics educator; b. Wichita, Mar. 23, 1943; s. Earl Benjamin and Gertrude J. (Socolofsky) B.; m. Yolanda Yen-Hsu Loo, Jan. 29, 1966; children—Eric James, Jonathan David. A.B., U. Calif., Berkeley, 1964, Ph.D., 1970; A.M. (hon.), Dartmouth Coll., 1981. J. W. Young research instr. Dartmouth Coll., Hanover, N.H., 1969-71, asst. prof. math., 1971-76, assoc. prof., 1976-80, prof., 1980-83, J.G. Kemeny prof. math., 1983—; vis. asst. prof. Calif. Inst. Tech. Pasadena 1971-72; cons. Coll. Bd., 1990—. Cons. editor Jour. Symbolic Logic, 1983-90; editor: Axiomatic Set Theory, 1984; contbr. articles to profl. jours., chpts. to books. Mem. Am. Math. Soc. (editor Transactions and Memoirs 1988-92, mng. editor 1992—), Math. Assn. Am., Assn. Symbolic Logic. Home: Lindy Ln Hanover NH 03755-1333 Office: Dartmouth Coll Dept Math Hanover NH 03755

BAUMHEFNER, ROBERT WALTER, neurologist; b. San Francisco, May 28, 1948; s. Clarence Herman and Virginia (Marie) B.; m. Rita Christine Hund, Apr. 26, 1980; children: Adam, Genevieve. BA in Biochemistry, U. Calif., 1970; MD, Northwestern U., Chgo., 1974. Am. Bd. Psychiatry and Neurology. Intern Pacific Med. Ctr., San Francisco, 1974-75; resident Harbor Gen. Hosp., Torrance, Calif., 1975-78; rsch. fellow VA West L.A. Med. Ctr., 1978-80; asst. prof. dept. neurology UCLA Sch. Medicine, L.A., 1979-85; staff physician VA West L.A. Med. Ctr., 1980—; adj. assoc. prof. dept. neurology UCLA Sch. Medicine, L.A., 1985-93; prof. dept. neurology UCLA Sch. Medicine, 1993—; chmn. Med. Records Com., L.A., 1989—; mem. clin. exec. bd., VA West L.A. Med. Ctr., 1989—; qualified med. examiner State of Calif. Dept. Indsl. Rels., 1991—. Author: Multiple Sclerosis Pathology Diagnosis and Management, 1983, Cellular Immunology in Multiple Sclerosis, 1985; contbr. articles to profl. jours. Mem. Am. Profl. Practice Assn., Am. Acad. Neurology, N.Y. Acad. Scis. Republican. Lutheran. Office: Neorology Svc Med Ctr Wilshire and Sawtell Blvds Los Angeles CA 90073

BAUR, WERNER HEINZ, mineralogist, educator; b. Warsaw, Poland, Aug. 2, 1931; came to U.S., 1962; s. Heinrich Ernst and Melanie (Borkowska) B.; m. Renate Grossmann, June 22, 1962; children: Wolfgang, Brigitte. Dr. rer.nat, U. Gottingen, Germany, 1956, privat-dozent, 1961. Sci. officer U. Göttingen, 1956-63; asst. to assoc. prof. U. Pitts., 1963-65; asso. prof. to prof. U. Ill.-Chgo., 1965-86, head dept. geol. scis., 1967-80, asso. dean Coll. Liberal Arts and Scis., 1978-80; prof. crystallography Johann Wolfgang Goethe U., Frankfurt am Main, Fed. Republic Germany, 1986—; postdoctoral fellow U. Berne, Switzerland, 1957; vis. asso. chemist Brookhaven Nat. Lab., 1962-63; vis. prof. U. Karlsruhe, Germany, 1971-72. Editor: Zentralblatt fur Mineralogie, Teil I, 1989—; assoc. editor Crystallography Revs.; contbr. articles to sci. jours. Fellow Mineral. Soc. Am.; mem. Am. Crystallogical Assn., Am. Geophys. Union. Research on crystal chemistry of minerals and inorganic compounds, crystal structure determination, zeolites, computer simulation of crystal structures, empirical theories of chem. bonding, predictive crystal chemistry, ionic conductivity. Office: Institut Kristallographie, Senckenberganlage 30, D-60054 Frankfurt am Main Germany

BAUSELL, R. BARKER, JR., research methodology educator; s. Rufus B. and Nellie (Bowman) B.; m. Carole R. Vinograd, Jan. 6, 1978; children: Jesse T., Rebecca B. BS in Edn., U. Del., 1968, PhD in Ednl. Rsch. and Evaluation, 1975. Rsch. methodologist Med. Coll. Pa., 1975-76; prof., coord. faculty rsch. U. Md., Balt., 1976-91, dir. office rsch. methodology, 1991—; cons., part-time dir. prevention rsch. ctr. Rodale Press, Inc.; presenter numberous seminars and confs. Author: (with C.R. Bausell and N.B. Bausell) The Bausell Home Learning Guide: Teach Your Child to Read, 1980, (with C.R. Bausell and N.B. Bausell) The Bausell Home Learning Guide: Teach Your Child to Write, 1980, (with C.F. Waltz) Nursing Research: Design, Statistics and Computer Analysis, 1981, (with C.R. Bausell and N.B. Bausell) The Bausell Home Learning Guide: Teach Your Child Math, A Practical Guide to Conducting Empirical Research, 1986, An Instructor's Manual for a Practical Guide to Conducting Empirical Research, 1986, (with C. Inlander and M. Rooney) How to Evaluate and Select a Nursing Home, 1988, Advanced Research Methodology: An Annotatoed Guide to Sources, 1991; editor: Evaluation and the Health Professions; author numerous monographs; contbr. over 100 articles to profl. jours. Recipient Outstanding Rsch. award Nat. Wellness Conf., 1986, 87, Gov.'s award Meritorious Svc., 1992, award for Disting. Assessment Project Md. Assessment Resource Ctr., 1993. Achievements include research on documented effects of class size on student learning, effects of teacher experience on student learning, and determinants of health seeking (preventative) behavior. Home: 1311 Doves Cove Rd Baltimore MD 21286 Office: U Md Office for Rsch Methodology 655 W Lombard St Baltimore MD 21201

BAUTISTA, ABRAHAM PARANA, immunologist; b. Davao, Philippines, Mar. 15, 1952; s. Eufronio Bernardo and Loreto (Parana) B. BS in Biology, Far Eastern U., Manila, Philippines, 1972; Diploma in Microbiology, U. Tokyo, 1978; MS, Aberdeen U., Aberdeen, Scotland, 1981, PhD in Immunology, 1984. Sr. researcher, lectr. U. of Santo Tomas, Manila, Philippines, 1976-81; rsch. scholar U. Aberdeen, Scotland, 1979-84; rsch. assoc. East Carolina U., Greenville, N.C., 1984-89; asst. prof. La. State U. Med. Ctr., New Orleans, 1989—; guest editor, reviewer Circulatory Shock, 1991—, Jour. of Leukocyte Biology, 1988—, Am. Jour. of Physiology, 1991—. Contbr. more than 50 sci. articles to profl. jours. Named Internat. fellow UNESCO, 1978, Internat. scholar Brit. Coun., 1979; recipient Rsch. award in Medicine, U. Aberdeen, 1981-84, F.I.R.S.T. award/Rsch. grant NIH, 1991-96, Travel fellowship Am. Assn. for Study of Liver Disease, 1990. Mem. Am. Assn. Immunology, Inst. of Biology, Soc. for Leukocyte Biology, Rsch. Soc. on Alcholism, Sigma Chi. Achievements include first demonstration that endogenous or exogenous interleukin-1 regulates insulin biosynthesis in vivo. Home: 103 Hollow Rock Ct Slidell LA 70461 Office: La State U Med Ctr 1901 Perdido St New Orleans LA 70112-1328

BAUTISTA, RENATO GO, chemical engineer, educator; b. Manila, The Philippines, Mar. 27, 1934; s. Teodulo Herera and Felicidad (Tiongko-Go) B.; m. Elaine Tsang, July 1, 1978; 1 child, Derek Kevin. B.S. in Chem. Engring., U. Santo Tomas, 1955; S.M. in Metallurgy, MIT, 1957; Ph.D. in Metall. Engring., U. Wis., 1961. Registered profl. engr., Iowa, Nev., The Philippines. Research metallurgist Allis Chalmers Corp., Milw., 1957-58; research assoc. chemistry dept. U. Wis., Madison, 1961-63; staff metallurgist A. Soriano Corp., Manila, 1963-67; research assoc. dept. chemistry Rice U., Houston, 1967-69; asst. prof. chem. engring Iowa State U., Ames, 1969-73, assoc. prof., 1973-78, prof., 1978-84; prof. dept. chem. and metall. engring Mackay Sch. Mines U. Nev., Reno, 1984—; vis. prof. Imperial Coll., London, 1976; vis. scientist Warren Spring Lab., Stevenage, Britain, 1975-76; cons. NASA, Asahi Chem. Co., Olin Corp., W.R. Grace, Gen. Motors, Behre-Dolbear-Riverside, Inc., Jamaica Bauxite Inst. Editor: Hydrometallurgical Process Fundamentals, 1984, (with Rolf Wesely) Energy Reduction Techniques in Metal Electrochemical Processes, 1985, (with Rolf Wesely, Gary W. Warren) Hydrometallurgical Reactor Design and Kinetics, 1986, (with J.E. Hoffman, V.A. Ettel, V. Kudryk, R.J. Wesely) The Electrorefining and Winning of Copper, 1986, (with M.M. Wong) Rare Earths, Extraction, Preparations and Applications, 1988, (with Knona C. Liddell, Donald R. Sadoway) Refractory Metals, Extraction, Processing and Applications, 1990, (with Norton Jackson) Rare Earths: Resources, Science, Technology and Applications, 1991, (with V.I. Lakshmanan and S. Somasundaran) Emerging Separation Technologies for Metals and Fuels, 1993; contbr. articles to profl. jours. Mem. Am. Inst. Chem. Engrs., Soc. Mining Engrs., AIME, Metall. Soc. AIME, Am. Chem. Soc., Sigma Xi, Phi Lambda Upsilon. Roman Catholic. Achievements include patents in field. Home: 3622 Big Bend Ln Reno NV 89509-7427 Office: U Nev Mackay Sch Mines Reno NV 89557

BAUTZ, GORDON THOMAS, information scientist, researcher; b. Bridgeport, Conn., June 6, 1942; s. Milton and Janet Bedford (Ballou) B.; m. Delmae Hannemann, June 12, 1965; children: Tracy, Jennifer. BS, Salem (W.Va.) Coll., 1964. Assoc. scientist Mt. Sinai Hosp., N.Y.C., 1965-68; from asst. scientist to scientist Hoffmann-La Roche, Nutley, N.J., 1968-84, rsch. scientist, 1984—. Author: Advances in Neurology, 1974, Central Nervous Effects of Hypothalamic Hormones and Other Peptides, 1979, Protides of the Biological Fluids Colloquium, 1982, Spinal Cord Injury, 1982; contbr. to profl. jours. Mem. Soc. Neurosci., Am. Soc. Pharmacology and Exptl. Therapeutics, N.Y. Acad. Scis., Sigma Xi. Office: Hoffman La Roche Kingsland St Nutley NJ 07110

BAVICCHI, ROBERT FERRIS, construction materials technician, concrete batchplant operator; b. Portsmouth, N.H., Oct. 28, 1957; s. Ferris George and Marga Theresa (Shannon) B.; m. Sandra Rena Tarbox, May 3, 1992. Student, U. N.H. Cert. grade I concrete field testing technician. Dir. materials tech. John Iafolla Co., Inc., Portsmouth, N.H., 1983—; mem., juror N.H. Concrete Certification Com., Concord, 1991-93. Recipient Cert. of Achievement, Portland Cement Assn., 1987. Office: John Iafolla Co Inc Peverly Hill Rd Portsmouth NH 03801

BAXTER, JOHN DARLING, physician, educator, health facility administrator; b. Lexington, Ky., June 11, 1940; s. William Elbert and Genevive Lockhart (Wilson) B.; m. Ethelee Davidson Baxter, Aug. 10, 1963; children: Leslie Lockhart, Gillian Booth. BA in Chemistry, U. Ky., 1962; MD, Yale U., 1966. Intern, then resident in internal medicine Yale-New Haven Hosp., 1966-68; USPHS research assoc. Nat. Inst. Arthritis and Metabolic Diseases, NIH, 1968-70; Dernham sr. fellow oncology U. Calif. Med. Sch., San Francisco, 1970-72; mem. faculty U. Calif. Med. Sch., 1972—, prof. medicine and biochemistry and biophysics, 1979—; dir. endocrine research Howard Hughes Med. Inst., 1976-81, investigator, 1975-81; chief div. endocrinology Moffitt Hosp., 1980—, dir. Metabolic Research Unit, 1981—; attending physician U. Calif. Med. Center, 1972—. Editor textbook of endocrinology and metabolism; Author research papers in field; mem. editorial bd. profl. jours. Recipient George W. Thorn award Howard Hughes Med. Inst., 1978, Disting. Alumni award U. Ky., 1980, Dautrebande prize for research in cellular and molecular biology, Belgium, 1985, Albion Bernstein award N.Y. Med. Soc., 1987; grantee NIH, Am. Cancer Soc., others. Mem. Am. Chem. Soc., Am. Soc. Hypertension, Am. Soc. Clin. Investigation, Am. Thyroid Assn., Assn. Am. Physicians, Am. Fedn. Clin. Research, Endocrine Soc., Western Assn. Physicians, Western Soc. Clin. Research. Office: U Calif Med Sch 671 HSE San Francisco CA 94143

BAXTER, JOSEPH DIEDRICH, dentist; b. New Albany, Ind., Sept. 11, 1937; s. James William, Jr. and Beatrice (Diedrich) B.; A.B., Ind. U., 1959, D.M.D., U. Louisville, 1969; m. Carroll Jane Bell, Dec. 23, 1972. Practice dentistry, New Albany, 1969—. Bd. dirs. Floyd County (Ind.) Econ. Opportunity Corp., 1971-76. Served with AUS, 1960-61. Mem. Floyd County Dental Soc. (pres. 1972-74), Am. Dental Assn., Phi Gamma Delta. Republican. Methodist. Home: 2702 Paoli Pike Apt 150 New Albany IN 47150-5145 Office: Profl Arts Bldg New Albany IN 47150

BAXTER, LORAN RICHARD, civil engineer; b. Sherwood, Oreg., Feb. 3, 1953; s. Vernon W. and Marie (Wotypka) B.; m. Duby Yvonne Allenson, Apr. 7, 1979. BSCE cum laude, U. Wash., 1977; MSCE, Stanford U., 1983. Registered profl. engr., Alaska. Civil engr. in water resources planning U.S. Army Corps Engrs. Alaska Dist., Anchorage, 1978-84; supervisory civil engr. project ops., nav. and flood control U.S. Army C.E. Alaska Dist., Anchorage, 1984-85, supervisory civil engr. Richardson resident office, 1985-87; civil engr., constrn. mgmt. office U.S. Army C.E. Ft. Drum (N.Y.), 1987-89; supervisory civil engr., quality assurance br. U.S. Army C.E. Alaska Dist., Anchorage, 1989-91, civil engr., army project mgr., 1991—; civil engr., civilian exec. officer Hdqrs. U.S. Army C.E., Washington, 1992-93. Bd. dirs. Grace Brethren Missions, Anchorage, 1992—. Mem. ASCE, NRA, Soc. Am. Mil. Engrs., Alumni Assn. Stanford, Safari Club Internat., Found. N.Am. Wild Sheep, Tau Beta Pi.

BAXTER, MERIWETHER LEWIS, JR., gear engineer, consultant; b. Nashville, Dec. 7, 1914; s. Meriwether Lewis and Elizabeth (Young) B.; m. Louise Sweetnam, Aug. 31, 1940 (dec. 1985); children: Paul Douglas, Hugh Harris; m. Phyllis Probst Johnson, May 5, 1990. BS, Yale U., 1935. Engr., then rsch. engr., chief rsch. engr., dir. engring. Gleason Works, Rochester, N.Y., 1935-76; gear cons. Rochester, 1976—; invited lectr. Chinese govt., 1985. Contbg. author: Gear Handbook, 1962; contbr. articles to profl. jours. Mem. Eagle bd. Boy Scouts Am., Rochester, through 1985. Fellow ASME (life, past chmn. Rochester sect.); mem. Indsl. Math. Soc., Rochester Engring. Soc., Sigma Xi. Achievements include development of "Tooth Contact Analysis," making it possible to determine exactly the contact conditions between gear tooth surfaces; first to apply vector-matrix mathematics to three-dimensional gear problems; numerous patents. Home: 86 New Wickham Dr Penfield NY 14526

BAY, NIELS, manufacturing engineering educator; b. Copenhagen, Mar. 13, 1947; s. Børge Henrik and Margaret (Bay) Madsen; m. Ingrid Bjerrum, Jan. 17, 1970; children: Astrid, Rolf. MSc, Tech. U. Denmark, Lyngby, 1972, PhD, 1977, DSc, 1987. Asst. prof. Tech. U. Denmark, Lyngby, 1974-79; head lab. for advanced welding tech. Danish Welding Inst., Glostrup, Denmark, 1979-80; assoc. prof. Tech. U. Denmark, Lyngby, 1980-88, docent, 1988-93, prof., 1993—; cons. Nat. Testing Bd., Denmark, 1982—; chmn. Danish Cold Forging Group, Denmark, 1985—, Internat. Cold Forging Groups Tribology Subgroup, 1985—, Internat. Cold Forging Group, 1991—; Danish del. EEC's CAN for Indsl. & Materials Tech., 1987-92. Contbr. over 60 articles to internat. jour. and conf. proc. Mem. Am. Soc. Mfg. Engrs., N.Am. Mfg. Rsch. Inst., Am. Soc. Exptl. Mechanics, Danish Acad. for Tech. Scis. Avocations: tennis, table tennis, l'hombre, bridge. Home: Hegnstoften 14, DK 2630 Taastrup Denmark Office: Tech U Denmark, Inst Mfg Engring Bldg 425, DK 2800 Lyngby Denmark

BAYER, ARTHUR CRAIG, organic chemist; b. Bklyn., Apr. 5, 1946; s. Joseph and Henrietta (Cetto) B.; m. Barbara Ann Dockendorf, Oct. 23, 1976; children: Craig Andrew, Jessica Ann. BS in Chemistry, Manhattan Coll., 1967; PhD in Organic Chemistry, SUNY, Buffalo, 1972. Rsch. chemist Hooker Chem. Corp., Grand Island, N.Y., 1974-78; sr. rsch. chemist Stauffer Chem. Co., Dobbs Ferry, N.Y., 1978-84; dir. R&D First Chem. Corp., Pascagoula, Miss., 1984-88; sr. group leader CIBA-GEIGY Corp., St. Gabriel, La., 1988—. Referee Jour. Organic Chemistry, Columbus, Ohio, 1986—; contbr. articles to profl. jours. Adult leader Boy Scouts of Am., Baton Rouge, 1989—. Mem. AAAS, Am. Chem. Soc. (officer local sects.), Am. Mgmt. Assn., Sigma Xi. Achievements include U.S. and fgn. patents in agricultural and polymer intermediates and product and process R&D. Home: 12335 Cardeza Ave Baton Rouge LA 70816 Office: CIBA GEIGY Corp PO Box 11 Saint Gabriel LA 70776

BAYER, JANICE ILENE, mechanical engineer; b. N.Y.C., June 22, 1966; d. Joel Stephen and Karen Marjorie (Kurtz) B. BSME, Cornell U., 1988; MS in Engring. Mechanics, Pa. State U., 1991. Rsch. asst. Ctr. for Engring. of Electronic and Acoustic Materials, State College, Pa., 1988-90; rsch. asst spacecraft controls br. NASA Langley, Hampton, Va., 1990-91, rsch. engr. structural acoustics br., 1991; mission planning specialist SpaceTec Ventures, Inc., Hampton, 1991—. Mem. AIAA, Nat. Space Soc. (v.p. Langley chpt. 1992-93), Adventure Club East, Toastmasters, NASA Langley Volleyball Club (sec. 1992-93). Home: # 303 1024 Porte Harbour Arch Hampton VA 23664 Office: SpaceTec Ventures Inc 2713 Magruder Blvd Ste H Hampton VA 23666

BAYES, KYLE DAVID, chemistry educator; b. Colfax, Wash., Mar. 3, 1935; s. George Percy and Goldie Marie (Kopp) B.; m. Jane Higginbotham, June 17, 1961; children: Joseph C., Stephen K. BS, Calif. Inst. Tech., 1956; PhD, Harvard U., 1959. NSF postdoctoral fellow U. Bonn, Fed. Republic of Germany, 1959-60; asst. prof. U. Calif., L.A., 1960-65, assoc. prof., 1965-71, prof., 1971—, chmn. dept. chemistry and biochemistry, 1987-90; Sloan vis. lectr., Harvard U., Cambridge, Mass., 1967; Gastprofessor Bergische Universitat, Wuppertal, Fed. Republic of Germany, 1985; Erskine fellow U. Canterbury, Christchurch, New Zealand, 1988. Mem. Am. Chem. Soc., Am. Phys. Soc. Office: Univ Calif Dept Chemistry and Biochemistry Los Angeles CA 90024

BAYKUT, MEHMET GÖKHAN, chemist; b. Istanbul, Turkey, Feb. 22, 1954; s. Muhittin Fikret and Sacide B. BSCE, U. Istanbul, 1976; PhD in phys. chemistry, U. Frankfurt, Germany, 1980. Asst. prof. U. Istanbul, 1982; rsch. assoc. chemistry dept. U. Fla., Gainesville, 1983-87; sr. scientist Ionspec Corp., Irvine, Calif., 1987-89; assoc. prof. degree U. Istanbul, 1988; R&D scientist Bruker-Franzen Analytik, Bremen, Germany, 1989; product mgr. for environ. sampling systems Bruker-Franzen Analytik, Bremen, 1990—. Inventor in field of mass spectrometric sampling; patents in field; mem. editorial bd. Chimica Acta Turcica, Istanbul, 1983—; contbr. articles on ion physics/chemistry and mass spectrometry to profl. jours. Mem. Am. Chem. Soc., Am. Soc. Mass Spectrometry, Deutsche Gesellschaft fur Luft- und Raumfahrt. Avocations: aerospace technology, flying, black pencil drawing. Home: Senator Balcke Str 56, D-28279 Bremen Germany Office: Bruker-Franzen Analytik, Fahrenheit Str 4, D-28359 Bremen Germany

BAYLIS, WILLIAM THOMAS, systems logistics engineer; b. Bay Shore, N.Y., Oct. 21, 1952; s. William Wood and Viola Elaine (Burtis) B.; m. Milagros Marfisi, July 3, 1988;children: Christopher Thomas, Justin William Andrew. BSBA, U. Tenn., 1981; MBA in Mgmt., Dowling Coll., Oakdale, N.Y., 1984; PhD in Mgmt., Columbia Pacific U., San Raefel, Calif., 1986; postgrad., N.Y. Inst. Tech., 1987—. Asst. mgr. AIL div. Eaton Corp., Deer Park, N.Y., 1984-86, group leader, 1986-88; program mgr. Gen. Instrument Corp., Hicksville, N.Y., 1988-89, mgr. logistics engring., 1988-89; sr. logistics engr., logistics support analysis engring. lead McDonnell Douglas Space Systems Co., Kennedy Space Center, Fla., 1989—; Level II Integrated Database adminstr., security coord. for Space Sta. Freedom Program; pres. WTB Enterprises, Melbourne, Fla., 1990—; part-time instr. space logistics tech. Brevard Community Coll (adv. comm. Logistics Systems Tech. program), tchr. various logistics engring. courses; panel speaker on space station ground operations/logistics integration, Cocoa Beach, Fla., 1991, logistics support analysis speaker 27th Annual Internat. Conf. and Tech. Exposition, Indpls., 1992; mem. tech. adv. bd. SOLE Logistics Engring., 1992-93; speaker in field. Author, editor: Starting a Retail Business, 1988, Trainer's Guide to Task Analysis, 1989, Logistics Engineer's Desk Reference, 1991, 92; author: Training Requirements for Defense Contracts: A Practitioner's Desk Reference, 1989, Developing, Designing and Delivering Productive and Efficient Training, 1992; developer software course Lotus 123, 1989. With USN, 1974-78. Recipient Chmn. award SOLE chpt., 1992, Guest Speaker award SOLE, 1992, Space Congress Achievement award 30th Space Congress, 1992. Mem. AIAA, Am. Mgmt. Assn., Soc. Logistics Engrs. (Splty. award in logistics support analysis 1991, 92), Internat. Soc. Philos. Enquiry, Assn. MBA Execs., Mensa, Intertel, Am. Legion, Phi Eta Sigma. Avocations: developing software, writing, photography, woodworking, reading. Home: 2421 Eden Park Dr Melbourne FL 32935 Office: McDonnell Douglas Space Systems Co Kennedy Space Center FL 32815

BAYLOR, DENIS ARISTIDE, neurobiology educator; b. Oskaloosa, Iowa, Jan. 30, 1940; s. Hugh Murray and Elisabeth Anne (Barbou) B.; m. Eileen Margaret Steele, Aug. 12, 1983; children: Denis Murray, Michael Randel; stepchild Michele Van Tassel. BA Chemistry magna cum laude, Knox Coll., 1961, DS (hon.), 1989; MD cum laude, Yale U., 1965. Post-doctoral fellow Yale Med. Sch., New Haven, Conn., 1965-68; staff assoc. Nat. Inst. Neurological Diseases and Stroke, Bethesda, 1968-70; spl. fellow USPSH Physiological Lab. Cambridge U., Eng., 1970-72; assoc. prof. physiology U. Colo. Med. Sch., Denver, 1972-74; assoc. prof. physiology Stanford U., Calif., 1974-75, assoc. prof. neurobiology, 1975-78, prof. neurobiology, 1978—, chmn. dept. neurobiology, 1992—; First Annual W.S. Stiles lecturer U. Coll., London, England, 1989; Jonathan Magnes lecturer Hebrew U., Jerusalem, Israel, 1990; Woolsey lecturer U. Wis., 1992; mem. NIH Visual Scis. Study Sect., 1984-88, chmn. 1986-88; vis. com. med. scis. Harvard U., 1987—; chmn. Summer Conf. on Vision FASEB, 1989. Author: (with others) Physiological and Biochemical Aspects of Nervous Integration, 1968, Photoreception, 1977, Function and Formation of Neural Systems, 1977, Theoretical Approaches in Neurobiology, 1981, Molecular Mechanisms of Photoreceptor Transduction, 1981, Biomembranes Part H Visual Pigments and Purple Membranes, 1982, Color Vision: Physiology and Psychophysics, 1983, Central and Peripheral Mechanisms of Color Vision, 1985, Sensory Transduction, 1992, Colour, 1993; edit. bd. Jour. Physiology, 1977-84, Neuron, 1988—, Jour. Neurophysiology, 1989—, Visual Neuroscience, 1990—, Jour. Neuroscience, 1991—; contbr. articles to profl. jours. Recipient Sinsheimer Found. award, 1975, Mathilde Solowey award, 1978, Kayser Internat. award Retina Rsch. Found., 1988, Golden Brain award Minerva Found., 1988, Merit award Nat. Eye Inst., 1990, Alcon Rsch. Inst. award, 1991; Rank Optoelectronics prize Rank Orgn., Eng., 1980; Proctor medal Assn. Rsch. Visin & Ophthalmology, 1986. Fellow Am. Acad. Arts and Scis.; mem. NAS, Phi Beta Kappa, Alpha Omega Alpha. Avocations: jogging, woodworking. Office: Stanford U Sch Med Dept Neurobiology Fairchild Bldg Stanford CA 94305

BAYLOR, JILL S(TEIN), electrical engineer; b. Newport News, Va., Dec. 20, 1954; d. Manuel and Bernice (Malkin) Stein; m. Mark D. Baylor, May 30, 1976. BS in Applied Math., U. Va., 1976; MBA, U. N.C., Charlotte, 1979. Registered profl. engr., Colo. Planning engr. Duke Power Co., Charlotte, 1976-81; planning analyst mining and coal div. Mobil Oil Corp., Denver, 1981-84; asst. v.p. Stone & Webster Mgmt. Cons., Denver, 1984-92; prin. RCG/Hagler, Bailly, Boulder, Colo., 1992—; adv. bd. Coal Market Strategies Conf., Denver, 1990—; external adv. bd. U. Colo. at Boulder Women in Engring. Program, 1993—. Contbr. articles to profl. publs. Violinist, bookkeeper Littleton (Colo.) Symphony, 1985—; mem. leadership com. Colo. Women's Leadership Coalition, Denver, 1991-92. Recipient Cert. of Honor, Colo. Engring. Coun., 1990. Mem. IEEE (sr.), Am. Assn. Engring. Socs. (bd. govs., awards com. 1992-94), Soc. Women Engrs. (sr., life mem., pres. 1991-92), Women's Forum of Colo., Tau Beta Pi, Omicron Delta Kappa. Home: 7377 S Hudson Way Littleton CO 80122 Office: RCG/Hagler Bailly PO Drawer O Boulder CO 80306

BAYLOR, JOHN PATRICK, nurse; b. Quincy, Mass., Aug. 30, 1952; s. Garrett Michael and Anna Rita (Timmons) B.; m. Deborah Jean Tolar, May 6, 1978; children: Jennifer, Julia. ASN, Fayetteville (N.C.) Tech. C.C., 1978. RN, N.C. Staff nurse Southeastern Regional Rehab. Ctr., Fayetteville, 1978-79, asst. head nurse, 1979-80, staff nurse, 1982-83, asst. patient care mgr., 1983—; staff nurse Cardinal Cushing Gen. Hosp., Brockton, Mass., 1980-82. Pub. radio vol. announcer Sta. WFSS. With U.S. Army, 1972-75. Mem. Assn. Rehab. Nurses, K.of C. Home: 3733 Masters Dr Hope Mills NC 28348-2224

BAYLOR, SANDRA JOHNSON, electrical engineer; b. Fukuoka, Japan, Sept. 19, 1960; came to the U.S., 1961; d. George Garland and Gloria Dean (Hagger) Johnson; m. Alfred M. Baylor Jr., Feb. 11, 1989. BSEE, So. U., Baton Rouge, La., 1982; MSEE, Stanford U., 1984; PhD, Rice U., 1988. Rsch. staff mem. IBM T.J. Watson Rsch. Ctr., Yorktown Heights, N.Y., 1988—. Mem. IEEE, IEEE Computer Soc., Assn. for Computing Machinery, Computing Rsch. Assn. (com. on status of women in computer sci. 1990—), Delta Sigma Theta. Home: 63 Steven Dr Ossining NY 10562

BAYM, GORDON ALAN, physicist, educator; b. N.Y.C., July 1, 1935; s. Louis and Lillian B.; children—Nancy, Geoffrey, Michael, Carol. A.B., Cornell U., 1956; A.M., Harvard U., 1957, Ph.D., 1960. Fellow Universitetets Institut for Teoretisk Fysik, Copenhagen, Denmark, 1960-62; lectr. U. Calif., Berkeley, 1962-63; prof. physics U. Ill., Urbana, 1963—; vis. prof. U. Tokyo and U. Kyoto, 1968, Nordita, Copenhagen, 1970, 76, Niels Bohr Inst., Copenhagen, 1976, U. Nagoya, 1979; vis. scientist Academia Sinica, China, 1979; mem. adv. bd. Inst. Theoretical Physics, Santa Barbara, Calif., 1978-83; mem. subcom. theoretical physics, physics adv. com. NSF, 1980-81, mem. phys. adv. com., 1982-85; mem. nuclear sci. adv. com. Dept. of Energy/NSF, 1982-86, subcom. on theoretical physics; mem. adv. com. physics Los Alamos Nat. Lab., 1988; mem. nat. adv. com. Inst. Nuclear Theory. Author: Lectures on Quantum Mechanics, 1969, Neutron Stars, 1970, Neutron Stars and the Properties of Matter at High Density, 1977, (with L.P. Kadanoff) Quantum Statistical Mechanics, 1962, (with C.J. Pethick) Landau Fermi Liquid Theory: Concepts and Applications, 1991; assoc. editor Nuclear Physics; mem. editorial bd. Procs. Nat. Acad. Scis., 1986-92. Trustee Assoc. U. Inc., 1986-90. Recipient Alexander von Humboldt Found. sr. U.S. Scientist award, 1983; fellow Am. Acad. Arts and Scis.; Alfred P. Sloan Found. research fellow, 1965-68; NSF postdoctoral fellow, 1960-62. Fellow AAAS, Am. Phys. Soc. (exec. com. div. history of physics 1986-88); mem. NAS, Am. Astron. Soc., Internat. Astron. Union.

BAYNE, JAMES WILMER, mechanical engineering educator; b. Balt., Apr. 2, 1925; s. John Ernest and Marie Jeanette (Sullivan) B.; m. Loretta Catherine Schumacher, Aug. 16, 1948; children: Marilyn Bayne Schroeder, Charles, Cathy, John, Kenneth, Lisa, Robert. BSME, U. Ill., 1946, MSME, 1950. From instr. to assoc. prof. mech. engring. U. Ill., Urbana, 1946-57, assoc. prof., 1957-68, prof., 1969—, prof. emeritus, 1985—, acting head dept. mech. and indsl. engring., 1974-75, assoc. head dept. mech. and indsl. engring., 1970-85. Co-author: Opportunities in Mechanical Engineering. Trustee Holy Cross Ch., 1970-92. Served with USN, 1943-46. Recipient Everitt Undergrad. Teaching Excellence award, 1968, Western Electric-ASEE Outstanding Teaching award, 1972, Pierce award, 1981, Haliburton award, 1985, Disting. Alumni award U. Ill. Dept. Indsl./Mech. Engring., 1985, Coll. Engring. Alumni Honor award U. Ill., 1986. Mem. ASME (exec. com. mech. engring. div., past pres., sec.-treas. U. Ill. br.), Am. Soc. for Engring. Edn. (exec. com.), Assn. Coll. Honor Socs. (pres. 1975-76), U. Ill. Alumni Assn. (Loyalty award 1985), Sigma Tau, Pi Tau Sigma (pres. 1972-75, nat. sec.-treas. 1959-71, editor Condenser 1959-71), Tau Beta Pi, Omicron Delta Kappa, Sigma Chi (Order of Constantine 1986). Roman Catholic. Lodge: KC (grand knight 1958). Avocation: golf. Home: 1209 W Clark St Champaign IL 61821-3236 Office: Univ Ill 140 Mech Engring Bldg 1206 W Green St Urbana IL 61801-2906

BAYSINGER, STEPHEN MICHAEL, air force officer; b. St. Louis, May 9, 1954; s. David Richard and Betty I. (Elledge) B.; m. Robin Ann Weber, Oct. 8, 1977; children: Devin, Derrick, Corey, Jocelyn. BA in Polit. Sci., U. Denver, 1977; MS in Human Resource Mgmt., Troy State U., 1989. Cert. quality auditor. Commd. 2d lt. USAF, 1970, advanced through grades to maj., 1990; maintenance supr. 7th Bombardment Wing, Carswell AFB, Tex., 1979-82, 343d Fighter Wing, Eielson AFB, Alaska, 1982-85; logistics project mgr. Air Force Logistics Mgmt. Ctr., Gunter AFB, Ala., 1985-90; chief maintenance mgmt. 58 Fighter Wing, Luke AFB, Ariz., 1990-91, maintenance supr., 1991, chief quality assurance, 1991-92; mgr. quality assurance, project officer ISO 9001 EIM Co., Houston, 1993—. Contbr. articles to profl. jours. Leader Boys Scouts Am., Luke AFB, 1991-92; bd. dirs. Ala. Am. Diabetes Assn., Montgomery, 1985-90. Mem. Am. Soc. for Quality Control (cert. quality auditor), Soc. Logistics Engrs. Achievements include development of the first-ever automated personnel and equipment performance trend identification and analysis program for USAF, of the automated information center, of the quality division. Home: 108 Cherry Point Dayton TX 77535

BAYYA, SHYAM SUNDAR, ceramic engineer; b. Visakhapatnam, India, Feb. 13, 1964; came to U.S., 1987; s. Suryanarayana V. and Ratnavali (Maddala) B.; m. Madhavi Tadepalli, Dec. 15, 1991. MS, Alfred (N.Y.) U., 1989, PhD, 1992. Rsch. asst. Alfred U., 1987-88, teaching asst., 1989-92; rsch. assoc. Alfred (N.Y.) U., 1993—; vis. scientist G.M. Rsch. Lab., Warren, summer, 1991. Author: (with others) Corrosion of Glass, Ceramic and Ceramic Superconductors, 1991, Superconductivity and its Applications, 1991. Mem. Am. Ceramic Soc., Materials Rsch. Soc., Am. Soc. for Matels, Am. Crystallographic Assn. Achievements include patents pending for molten salt synthesis of superconductors in T1-Ba-Ca-Cu-O systems. Home: 456 Bennett Pkwy 3D Hornell NY 14843 Office: Alfred U BMH Coll of Ceramics Alfred NY 14802

BAZZAZ, FAKHRI A., plant biology educator, administrator; b. Baghdad, Iraq, June 16, 1933; came to U.S., 1958; s. Abdul-Lalif and Munifa B.; m. Maarib Bazzaz, Aug. 25, 1958; children: Sahar, Ammar. B.S., U. Baghdad, 1953; M.S., U. Ill., 1960, Ph.D., 1963; A.M. (hon.), Harvard U., 1984. Prof. U. Ill., Urbana, 1977-84, head dept. plant biology, acting dir. Sch. Life Scis., 1983-84; prof. Harvard U., Cambridge, Mass., 1984—; fellow Clare Hall, Cambridge U., Eng., 1981. Editor: Oecologia, 1983. Guggenheim fellow. Fellow AAAS, Am. Acad. Arts and Scis.; mem. Ecol. Soc. Am., Brit. Ecol. Soc. Office: Harvard U Dept Organismic and Evolutionary Biology 16 Divinity Ave Cambridge MA 02138-2097

BAZZAZIEH, NADER, environmental engineer; b. Tabriz, Iran, Dec. 12, 1947, s. Ali Akbar and Bozorg (Sadaghiani) B.; m. Sholeh Sadaghiani Azarbaijani, June 22, 1973; children: Nava, Saba. BS, U. Tabriz, 1971; MS, Washington State U., 1979. Registered profl. engr., Md., Va.; diplomate Am. Acad. Environ. Engrs. Lectr., researcher U. Tabriz, Iran, 1971-74; project engr. Behrouzan and Ptnrs., Inc., Tehran, 1974-76; lectr. Coll. Technicom, Tehran, 1980-81; chief engr. Min. Energy Tech. Bur. Water, Tehran, 1982-86; project mgr. Associated Engring. Scis., Inc., Hagerstown, Md., 1986-89; design engr. Metcalf and Eddy, Inc., Laurel, Md., 1989-91; pres. Environ. Consulting Engr., Jessup, Md., 1991-93; engring. mgr., sr. engr. Handex, Inc., Odenton, Md., 1993—. Co-author: Design of Hydraulic Structures, 1986. Mem. ASCE, Am. Acad. Environ. Engrs., Water Environ. Fedn. Home: 8707 Clemente Ct Jessup MD 20794

BEACH, ARTHUR THOMAS, mechanical engineer; b. N.Y.C., Dec. 2, 1920; s. Charles Arthur and Elsa Margaret (Nichols) B.; m. KiKi Duggan, Oct. 2, 1948; children: Richard, Elizabeth, Charles, Cynthia, Edward, Peter. BSME, Lafayette Coll., 1942; MS, Cornell U., 1947. Chmn. bd. Abbe Engring. Co., Bklyn., 1950—, Beach-Russ Co., Bklyn., 1946—; dir. First Huntington Securities, Huntington, N.Y., 1982; pres. Huntington Rsch. Corp., 1985. Capt. U.S. Army, 1942-46. Mem. Am. Vacuum Soc. (bd. dirs. 1958-60), Inst. Environ. Engrs., Newcombe Soc., BPOE, Huntington Country Club. Republican. Presbyterian. Achievements include patents in Vacuum Pumping Equipment; Water Vapor Pumping; Condensing Equipment. Office: Beach-Russ Co 544 Union Ave Brooklyn NY 11211

BEACH, DAVID DUNCAN, naval architect; b. New Haven, Conn., May 1, 1918; s. David Duncan and Helen (Gibson) B.; m. Helen Elizabeth Fisher, July 22, 1943; 1 child, David Duncan. Student, U. Mich., 1935-40; Diplomate, Westlawn Inst., Stamford, Conn., 1980. Registered profl. engr., D.C. Naval architect Higgins Boats Corp., New Orleans, 1955-58, Outboard Marine Corp., Waukegan, Ill., 1958-60, McCulloch Corp., L.A., 1960-62, Owens Yachts (Brunswick Corp.), Balt., 1962-66, Nat. Marine Mfrs. Assn., Chgo., 1966-92; cons. F.A.O. of UN, Rome, 1971-72. Fellow Royal Instn. Naval Architects; mem. Soc. Naval Architects and Marine Engrs. (life), Soc. Small Craft Designers (bd. dirs. 1982-92). Office: Nat Marine Mfg Assocs 401 N Michigan Ave Chicago IL 60611-4274

BEACH, ROBERT MARK, biologist; b. Athens, Ga., Oct. 7, 1957; s. Robert Ervin and Frances (Myers) B.; m. Catherine Cesaro, Oct. 3, 1987; 1 child, Katelyn Marie. BS, Clemson U., 1979; PhD, U. Ga., 1985. Rsch. asst. Clemson (S.C.) U., 1979-81, U. Ga., Athens, 1981-85, 1985-87; scientist Crop Genetics Internat., Hanover, Md., 1987-91, project mgr., 1991—. Contbr. articles to profl. jours. Mem. Entomol. Soc. Am., Internat. Orgn. Biocontrol, Soc. Invertebrate Pathologists. Achievements include research on control in the field of an economically significant insect pest using genetically manipulated living microbe. Home: 8382 Montgomery Run Rd Unit B Ellicott City MD 21043-7246 Office: Crop Genetics Internat 10150 Old Columbia Rd Columbia MD 21046

BEACHAM, DOROTHY ANN, medical research scientist; b. Bklyn., Dec. 28, 1957; d. Frederic Charles and Doris Elvira Beacham. BA in Biology/German, SUNY, Binghamton, 1979; PhD in Biochemistry/Molecular Biology, Health Sci. Ctr. Syracuse, 1986. Postdoctoral fellow biochemistry and cell biology U. Mass., Amherst, Mass., 1987-89; rsch. fellow in medicine hematology/oncology div. Brigham and Women's Hosp., Harvard Med. Sch., Boston, 1989-92; rsch. asst. prof. medicine Cardeza Found. for Hematologic Rsch., Jefferson Med. Coll., Phila., 1992—. Contbr. articles to profl. jours., including Biol. Chem., Exptl. Cell Rsch. Affiliate fellow Am. Heart Assn., 1990-91; grant-in-aid Am. Heart Assn., 1993—. Mem. AAAS, Am. Soc. for Cell Biology, N.Y. Acad. Scis. Achievements include research on characterization of endothelial cell adhesion to von Willebrand Factor mediated by endothelial GPIb. Office: Cardeza Found for Hematologic Rsch 1015 Walnut St Rm 711 Philadelphia PA 19107-5099

BEAHRS, OLIVER HOWARD, surgeon; b. Eufaula, Ala., Sept. 19, 1914; s. Elmer Charles and Elsa Katherine (Smith) B.; m. Helen Edith Taylor, July 27, 1947; children: John Randolf, David Howard, Nancy Ann Beahrs Oster. B.A., U. Calif., Berkeley, 1937; M.D., Northwestern U., 1942; M.S. in Surgery, Mayo Grad. Sch. Medicine, 1949. Diplomate Am. Bd. Surgery. Fellow surgery Mayo Grad. Sch. Medicine, Rochester, Minn., 1942, 46-49; prof. surgery, 1966-79, prof. emeritus, 1979—; asst. surgeon Mayo Clinic, 1949-50, head sect. gen. surgery, 1950-79, vice chmn. bd. govs., 1964-75; Bd. dirs. Rochester Meth. Hosp.; trustee Mayo Found.; mem. cancer control and rehab. adv. com. Nat. Cancer Inst., 1975-84; mem. Am. Joint Com. on Cancer, 1975-78, exec. dir., 1980-92. Editor: Surgical Consultations; editorial bd.: Surgery, Surg. Techniques Illustrated; contbr. over 400 articles to profl. jours. Hon. life, bd. dirs. Am. Cancer Soc., 1975—; trustee Rochester Meth. Hosp; adv. bd. Uniform Svcs. Univ. Med. Sch; med. cons. Pres. and Mrs. Reagan. Served to Capt. USNR, 1942-64. Fellow Royal Coll. Surgery in Ireland (hon.), Royal Australasian Coll. Surgery (hon.); mem. AMA, ACS (mem. exec. com., bd. govs., chmn. cen. jud. com., long-range planning com., chmn. bd. govs., chmn. bd. regents, pres. 1988-89), Am. Group Practice Assn. (sec.-treas. 1974-75), Minn. Surg. Soc. (pres. 1960-61), Am. Thyroid Assn., James IV Assn. Surgeons, Am. Surg. Assn. (pres. 1979-80, chmn. com. on issues 1980-83), So. Surg. Assn., Cen. Surg. Assn., Western Surg. Assn., Soc. Head and Neck Surgeons (pres. 1966-67), Am. Assn. Endocrine Surgeons (pres. 1986-87), Am. Assn. Clin. Anatomists (pres. 1986-87), Soc. Surgery Alimentary Tract, Soc. Pelvic Surgeons (pres. 1983-84), Soc. Surg. Oncology, Am. Assn. Clin. Anatomists (pres.), Philippine Coll. Surgeons (hon.), Hellenic Coll. Surgery (hon.), L'Association Française de Chirurgie Française, Northwestern U. Alumni Assn. (merit award), Sigma Xi, Phi Kappa Epsilon, Phi Beta Pi, Theta Delta Chi. Republican. Methodist. Home: 1010 60th Ave SW Rochester MN 55902-8700 Office: 200 1st St SW Rochester MN 55905-0001

BEAK, PETER ANDREW, chemistry educator; b. Syracuse, N.Y., Jan. 12, 1936; s. Ralph E. and Belva (Edinger) B.; m. Sandra J. Burns, July 25, 1959; children: Bryan A., Stacia W. B.A., Harvard U., 1957; Ph.D., Iowa State U., 1961. From instr. to prof. chemistry U. Ill., Urbana, 1961—; cons. Abbott Labs., North Chgo., Ill., 1964—, Monsanto Co., St. Louis, 1969—; G.D. Searle Co., Ill., 1987—. Contbr. articles to profl. jours. A.P. Sloan Found. fellow, 1967-69; Guggenheim fellow, 1968-69. Fellow AAAS; mem. Am. Chem. Soc. (A.C. Cope scholar award 1993, editorial and adv. bds., sec. and divsn. officer 1980). Home: 304 E Sherwin Ave Urbana IL 61801-7130

BEALE, GUY OTIS, engineering educator, consultant; b. Cleve., June 16, 1944; s. Guy Otis and Hilda (Booth) B.; m. Susan Ann Weaver, Dec. 16, 1967; 1 child, Michael Scott. BSEE, Va. Poly. Inst., 1967; MS in Physics, Lynchburg Coll., 1974; PhDEE, U. Va., 1977. Engr. Babcock & Wilcox, Lynchburg, Va., 1971-81; asst. prof. Vanderbilt U., Nashville, 1981-86; assoc. prof. George Mason U., Fairfax, Va., 1986—; cons. David Taylor Rsch. Ctr., Carderock, Md., 1987—, Advanced Control Tech., Nashville, 1987-88, Vanderbilt U. Med. Ctr., Nashville, 1984-85, Syerdrup Tech., Tullahoma, Tenn., 1982-83. Contbr. over 40 articles to profl. jours. 1st lt. U.S. Army, 1968-70. Recipient Teaching Excellence award Tau Beta Pi, 1983. Mem. IEEE (sr. mem.), Sigma Xi, Eta Kappa Nu. Lutheran. Avocation: photography. Office: George Mason U Elec/Computer Engring Fairfax VA 22030

BEALE, WILLIAM TAYLOR, engineering company executive; b. Chattanooga, Tenn., Apr. 17, 1928; s. David and Katherine (Burris) B.; m. Carol Phyllis Brand, June 30, 1959; children: Faith, Daniel, John. BSME, Wash. State Coll., 1950; MSME, Calif. Inst. Tech., 1953. Profl. engr., Ohio. Rsch. engr. Wash. State Coll., Pullman, 1950-52; engr. NACA, Cleve., 1953-56; rsch. engr. MIT, Cambridge, Mass., 1957-58; asst. prof. Boston U., 1958-59; assoc. prof. Ohio U., Athens, 1959-74; pres. Sundowner Inc., Athens, 1974—. Contbr. articles to profl. jours. Democrat. Unitarian/Quaker. Achievements include invention of free piston stirling engine, cooled variable expansion ratio diesel free piston. Office: Sunpower Inc 5 Byard St Athens OH 45701

BEALL, GEORGE HALSEY, ceramic engineer; b. Montreal, Can., Oct. 14, 1935. BS in Physics, Geology with honors, McGill U., 1956, MS in Geology, 1958; PhD in Geology, MIT, 1962. Rsch. geologist/mineralogist Corning Inc., Corning, N.Y., 1962-66, mgr. glass/ceramic rsch. dept., 1966-77, rsch. fellow, 1977—; courtesy prof. Materials Sci. Cornell U., 1980; vis. corp. com. Materials Sci. and Engring, MIT, 1981. Contrb. to over 65 profl. jours. Recipient George D. Morey award Am. Ceramic Soc., 1988, Samuel Geijsbeek award Am. Ceramic Soc., 1993, John Jeppson award Am. Ceramic Soc., 1993. Fellow Am. Ceramic Soc.; mem. Acad. Ceramics. Achievements include research in glass crystallization andglass ceramics which are nonporous, microcrystalline ceramics formed by controlled internal nucleation and crystallization of glass. Current research interest involve melt mixed blends of advanced polymers from glass and the related development of strength and toughness in glass ceramics luminescence in transparent glass ceramics for use in solar collectors, lasers and controlled nucleation of gas phases in glass. Holder of 69 U.S. Patents. (Macor, Cercor, Corelle, Suprema, Cortem, The Counter That Cooks, Visions, Dicor). Home: 106 Woodland Rd Big Flats NY 14814 Office: Corning Inc Glass Ceramic Res Division Sullivan Pk FR 51 Corning NY 14831*

BEALL, JAMES HOWARD, physicist, educator, public policy analyst; b. Grantsville, W.Va., May 12, 1945; s. Judson Harmon and Mary Lenore (Burns) B.; m. Mary Ruth Clance; B.A. in Physics cum laude, U. Colo., 1972; M.S., U. Md., 1975, Ph.D., 1979; 1 dau., Tara Siobhan. Astrophysicist, Goddard Space Flight Ctr., NASA, Greenbelt, Md., 1975-78; Congressional Fellow U.S. Congress Office Tech. Assessment, Washington, 1978-79; project scientist sci. and analysis div. BKD, Rockville, Md., 1979-81; NAS/NRC resident research assoc. Naval Research Lab., Washington, 1981-83; mem. faculty St. John's Coll., 1983—; sr. cons. E. O. Hulburt Ctr. Space Rsch., Washington, 1983—. mem. sci. and engring. adv. bd. High Frontier, Arlington, Va., 1991—; prof. space scis., computational scis., and informatics George Mason U., Fairfax, Va., 1992—; project adminstr. black oral history project Folger Shakespeare Library, Washington, 1981; moderator Library of Congress Symposium, 1981. Dir. edn. Environ. Action Com., Denver, 1971-72; bd. dirs. Partridgeberry Sch., Greenbelt, 1977-78; Poets-in-the Schs. participant Va. Public Schs., 1975—. Served in USAF, 1963-67. Nat. Endowment for Humanities grantee, 1976, 78; recipient Teaching Excellence award U. Md., 1974-75. Mem. AAAS, Am. Phys. Soc., Am. Astron. Soc., Md. Writers Council, Phi Beta Kappa, Sigma Xi, Sigma Pi Sigma. Republican. Author: Hickey, the Days, 1980. Research in theoretical and observational astrophysics, renewable energy resources and public policy; discovered 1st concurrent radio and x-ray variability of active galaxy; made first prediction of inverse compton x-ray emission from supernovae, first prediction of detectable infrared and optical emission from accretion disks around black holes, first detection of a ring of x-ray light around the earth's equator. Home: 15606 Powell Ln Mitchellville MD 20716-1439 Office: Naval Rsch Lab Code 4120 Washington DC 20375

BEALL, ROBERT JOSEPH, foundation executive; b. Washington, May 19, 1943; s. William Joseph and Louise Rachel (Tayman) B.; m. Mary Ellen

O'Connor, June 24, 1967; children: Thomas Joseph, Robert Andrew. B.S., Albright Coll., 1965; M.A., SUNY, Buffalo, 1970, Ph.D., 1970. Asst. prof. dept. physiology Case-Western Reserve U., Cleve., 1971-74; asst. prof. Case-Western Reserve U. (Sch. Dentistry), 1972-74; grants asso. div. research grants NIH, 1974-75; program dir. metabolic diseases program Nat. Inst. Arthritis, Metabolism & Digestive Diseases, 1975-79; med. dir. Cystic Fibrosis Found., Rockville, Md., 1980—; nat. dir. Cystic Fibrosis Found., 1981-84, exec. v.p., 1984—. Recipient Merit award NIH, 1980. Mem. AAAS, N.Y. Acad. Scis., Am. Soc. Human Genetics, Sigma Xi. Presbyterian. Office: Cystic Fibrosis Found 6931 Arlington Rd Bethesda MD 20814-5231

BEAM, JERRY EDWARD, electrical engineer; b. Dayton, Ohio, Apr. 20, 1947; s. William Edward and Alberta Rovene (Morris) B.; m. Mary Hayes Hornbeck, Sept. 9, 1972; children: Amy, Sarah. BSEE, U. Cin., 1975; MS in Materials Engring., U. Dayton, 1979, PhD in Mech. Engring., 1985. Registered engr., Ohio. Mem. staff NCR, Dayton, Ohio, 1970-75; engr. propulsion and power lab. USAF, Dayton, 1975-89, sect. chief thermal control group, 1989—; prof. mech. engring. and heat transfer Wright State U., Dayton, 1985—; mem. com. NRC, 1989—; adj. mem. staff Air Force Inst. Tech., 1992—. Contbr. articles to profl. jours. With USAF, 1967-70. Assoc. fellow AIAA (steering com. 1986-89, thermophysics com. 1986-90, Aerospace Power com. 1992—, Nat. Rsch. Coun. 1989—); mem. SAE (exec. com. Aerospace Atlantic 1990—). Achievements include patents in laser electrostatic bonding, heat pipe wick, LiH thermal storage, heat pipe artery fabric. Home: 1694 Edith Marie Dr Dayton OH 45431-3382

BEAM, WILLIAM WASHINGTON, III, data coordinator; b. L.A., Jan. 21, 1960; s. William Washington and Ada Frances (Towler) B. BS, UCLA, 1982; MA, U. Wash., 1985. Paralegal Arco, L.A., 1985-88, programmer, 1988-90, data coord., 1990—. Mem. Am. Econ. Assn. Office: Arco 515 S Flower St AP-4661 Los Angeles CA 90071

BEAMER, LESA JEAN, molecular biologist; b. Cleve., Dec. 12, 1963; d. James Alfred and Nancy Elizabeth (Bumbarger) Peterson; m. Brock Allen Beamer, June 20, 1987. BS, Kent State U., 1986; PhD, Johns Hopkins U., 1991. Postdoctoral fellow Molecular Biology Inst. UCLA, 1991—. Contbr. articles to Sci., Phys. Chem. Liquid, Jour. Molecular Biology. Mem. AAAS, Am. Crystallographic Assn., Nature Conservancy, World Wildlife Found., Nat. Audubon Soc., Phi Beta Kappa. Democrat. Lutheran. Office: Molecular Biology Inst UCLA 405 Hilgard Ave Los Angeles CA 90024

BEARB, MICHAEL EDWIN, anesthesiologist; b. Beaumont, Tex., June 30, 1956; s. Edwin and Ella Lou (Broussard) B.; m. Joanne Ruth Patterson, Nov. 18, 1989; 1 child, Emily. BS in Psychology with highest honors, Lamar U., 1978; MD, U. Tex., Dallas, 1984. Diplomate Am. Bd. Anesthesiology. Intern St. Paul Hosp., Dallas, 1984-85; resident in anesthesiology Parkland Meml. Hosp., Dallas, 1985-87; fellow in cardio-thoracic anesthesiology The Cleve. Clin. Found., Cleve., 1987-88; instr. in anesthesiology Georgetown U. Hosp., Washington, 1988-90, asst. prof. in anesthesiology, 1990—; chmn. resident selection com. Georgetown U. Hosp., 1990—, coord. cardiovascular lectr. series, 1990-91, attending intensivist cardiovascular ICU, 1989—. Author (with others): Hemoglobinopathies and Anesthetic Care of the Trauma Patient, 1992. Palladian mem. Friends of Hist. Mt. Vernon, Alexandria. Fellow Am. Coll. Angiology; mem. AMA, Internat. Anesthesia Rsch. Soc., Soc. Critical Care Medicine, Soc. Cardiovascular Anesthesiologists, Anesthesia History Assn., Am. Soc. Anesthesiologists, Civil War Soc., Smithsonian Inst., Phi Kappa Phi. Avocations: mil. history, presdl. history, photography, astronomy, short wave radio. Home: 69 Elmhurst Dr Jackson TN 38305 Office: Jackson-Madison County Gen Hosp 708 W Forest Ave Jackson TN 38301

BEARD, ROBERT DOUGLAS, geologist; b. Plattsburgh, N.Y., Aug. 6, 1961; s. John Roland and Nancy Kay (Ware) B.; m. Rosalina Tellez, June 16, 1990. BA in Geology, Calif. State U., Chico, 1983; MS in Geology, U. N.Mex., 1987. Storage tank program mgr. Rust Environ. & Infrastructure, Mechanicsburg, Pa., 1988—; cons. in field. Author: (trade jour.) Pit and Quarry, 1991. Mem. Assn. Groundwater Scientists and Engrs. Office: Rust Environ & Infrastructure 2 Market Plz Way Mechanicsburg PA 17055

BEARER, CYNTHIA FRANCES, neonatologist; b. Morristown, N.J., June 8, 1950; d. Paul Joseph and Dorothy Louise (Hughes) B. BA cum laude, Smith Coll., 1972; PhD, Case Western Res., 1977; MD, Johns Hopkins U., 1982. Diplomate Am. Bd. Perinatal-Neonatal, Am. Bd. Pediatrics. Intern pediatrics Johns Hopkins Hosp., Balt., 1982-83, clin. fellow in pediatrics, 1982-84, resident pediatrics, 1983-84; instr. in pediatrics Wash. U. Sch. Medicine, St. Louis, 1987-89; asst. clin. prof. Dept. Medicine, U. Calif., San Francisco, 1990—; asst. in pediatrics Children's Hosp., Barnes Hosp., and Jewish Hosp., St. Louis, 1987-89; dir. div. of pediatric environ. health Children's Hosp. Oakland (Calif.) Rsch. Inst., 1990-92; assoc. neonatologist Children's Hosp., Oakland, 1990-92, assoc. rsch. biochemist, 1990-92; asst. adj. prof. Sch. Dentistry, U. Calif., San Francisco, 1990-92; dir. neonatology and pediatric environ. health Tod Children's Hosp., 1992—; dir. Ctr. Environ. Health Scis., Youngstown State U., 1992—, affiliated scholar dept. chemistry, 1992—; lectr. in field. Author: (with others) Advances in Cyclic Nucleotide Research, 1980, Kids and the Environment: Toxic Hazards, 1990; contbr. numerous articles to profl. jours. Pub. policy com. March of Dimes Greater Bay Area, 1991-92. Recipient numerous grants in field, 1990-92; postdoctoral rsch. fellow NIH, 1978; Farley fellow Children's Hosp., Harvard Med. Sch., W. Barry Wood Student Rsch. fellow Johns Hopkins U., 1979-80. Fellow Am. Acad. Pediatrics (subcom. environ. hazards no. Calif. chpt. 1990—); mem. AAAS, APHA, Am. Fedn. Clin. Rsch., Nat. Coun. Juvenile and Family Ct. Judges, Teratology Soc. Achievements include research on non-oxidative metabolism of ethanol occurring in fetus; transfusion as source of lead exposure to infants in intensive care nurseries. Office: TOD Children's Hosp 500 Gypsy Ln Youngstown OH 44501

BEARNSON, WILLIAM R., systems engineer; b. Salt Lake City, Feb. 9, 1961; s. LeRoy Wood and Barbara Ruth (Barker) B. BSEE, Brigham Young U., 1989. Spl. projections technician Brigham Young U., Provo, Utah, 1982-83, audio visual supr., 1983-87; broadcast engr. Sta. KBYU-TV/FM, Provo, Utah, 1987-89; system engr. Sattel Techs. Inc., Chatsworth, Calif., 1989-91, project engr., 1991—; spl. effects supr. Jimmy Osmond Prodns., Provo, 1984. Full time mission Ch. of Jesus Christ of LDS, Sydney Australia, 1980-82, ward clk., Provo, 1986-88, ward mission leader, Chatsworth, Calif., 1991-92. Home: 1334 E 2300 N Provo UT 84604 Office: Sattel Techs 9145 Deering Ave Chatsworth CA 91311

BEASLEY, CHARLES ALFRED, mining engineer, educator; b. Summerlee, W.Va., Nov. 3, 1934; s. Emerson McKinley and Maude (Gwinn) B.; m. Frances Sayre, July 23, 1977; children: Carol Ann, David Emerson, Kathryn Ann. BS, MS, Ohio State U., 1957; PhD, U. Minn., 1969; MA, Colgate Rochester-Bexley Hall-Crozer Div. Sch. Seminary, Rochester, N.Y., 1987. Regional dir., then asst. dir. Office of Surface Mining, U.S. Dept. Interior, Washington, 1978-81; spl. asst. to dir. conservation div. U. S. Geol. Survey, Reston, Va., 1981; dir. mining ops. Law Engring. and Testing Co., Atlanta, 1981-83; prof., chair mining engring. U. Mo., Rolla, 1983-85; dir. Ctr. Environ. Rsch. SUNY, Buffalo, 1989-92, chair, prof. tech., 1985—; prin. Geonomic Assocs., Fairbanks, Alaska,1 963-65; prin., v.p. 3R Corp., Denver, 1976-78. Contbr. to profl. publs. Chair bd. govs. Gt. Lakes Rsch. Consortium, Syracuse, N.Y., 1992; mem. conservation task group SUNY, Albany, 1992. Recipient Gold Watch award Coal Mining industry, 1957, NSF award, 1965, Iron Ring award Can. Soc. Engrs., 1974. Mem. AAAS, Am. Assn. Mining, Metall. and Petroleum Engrs. (environ. com. 1983, student affairs com. 1985), Tech. Edn. Assn., Assn. Rsch. Dirs. Achievements include patent for high pressure jet cutting, ground control system, implementation of federal program for abandoned land reclamation. Office: SUNY Buffalo Dept Tech 1300 Elmwood Ave Buffalo NY 14222-1095

BEASLEY, ERNEST WILLIAM, JR., endocrinologist; b. Atlanta, May 7, 1924; s. Ernest William and Arrinda Elizabeth (Eidson) B.; M.D., Georgetown U., 1949; m. Ann Lee Jeffreys, July 1, 1950; children—Janet Ann, Ernest William III, Mary Elizabeth, Barbara Elaine. Intern, Walter Reed Hosp., Washington, 1949-50; resident in internal medicine VA Hosp.-Grady Meml. Hosp.-Emory U. Hosp., Atlanta; practice medicine specializing in family practice, Atlanta, 1955-65, in internal medicine, Atlanta, 1966-75, in endocrinology, Atlanta, 1975—; chief endocrinology and metabolism Ga. Bapt. Med. Center; assoc. dept. internal medicine Emory U.; cons. Ga. Assn. Retarded Children, 1955-65; dir. Diabetes Assn. Atlanta, 1976. Served with AUS, 1943-45, M.C., U.S. Army, 50-52. Diplomate Am. Bd. Internal Medicine, Sub-Bd. Endocrinology, Am. Bd. Family Practice Geriatrics. Mem. A.C.P. Med. Assn. Atlanta, Med. Assn. Ga., AMA, Am. Soc. Internal Medicine, Am. Diabetes Assn. Methodist. Club: Cherokee Country. Address: 315 Boulevard NE Ste 242 Atlanta GA 30312

BEASLEY, MALCOLM ROY, physics educator; b. San Francisco, Jan. 4, 1940; s. Robert Williams and Cora (Miller) B.; m. Jo Anne Horsfall, Sept. 29, 1962; children: Michael, Matthew, Claire. B in Engring. Physics, Cornell U., 1962, PhD in Physics, 1967. Research assoc. Harvard U., Cambridge, Mass., 1967-69, from asst. prof. to assoc. prof. applied physics, 1969-74; from assoc. prof. to prof. Stanford (Calif.) U., 1974—, chmn. dept. applied physics, 1984—. Fellow Am. Phys. Soc.; mem. AAAS. Office: Stanford U Dept Applied Physics Stanford CA 94305

BEATON, JAMES DUNCAN, soil scientist; b. Vancouver, B.C., Can., Aug. 28, 1930; s. James Andrew Beaton and Gertrude Marion Lorimer; m. Doris Irene Ford, Aug. 30, 1952; children: Barbara Ruth, Andrea Irene, Alice Shirley. BSA, U. B.C., 1951, MSA, 1953; PhD, Utah State U., 1957. Soil specialist Agriculture Can., Kamloops, B.C., 1953-57; phys. chemist Agriculture Can., Swift Current, Sask., 1959-61; instr. U. B.C., Vancouver, 1957-59; dir. agrl. rsch. The Sulphur Inst., Washington, 1968-73; head of soil sci. Cominco Ltd., Trail, B.C., 1961-65, sr. agronomist, 1965-67; regional dir. Potash & Phosphate Inst., Calgary, Alta., 1978-86, v.p., pres., 1986—. Author: Soil Fertility and Fertilizers, 5th edit., 1993. Head coach Lake Bonavista Bantam and Midget Box LaCrosse Teams, Calgary, 1973-77. Fellow Am. Soc. Agronomy (Agronomic Svc. award 1983), Soil Sci. Soc. Am., Can. Soc. Soil Sci., Agrl. Inst. Can. (AIC Fellowship award 1990); mem. Nat. Fertilization Solutions Assn. (hon.), Western Can. Fertilizer and Chem. Dealers Assn. (hon.), Western Coop. Fertilizers Ltd. (Agronomy Merit award 1981), Western Can. Fertilizer Assn. (pres. 1977-79, Award of Merit 1993). Achievements include development of forestry grade urea, high analysis degradeable elementals fertilizers, urea-ammonium sulphate 40-0-0-6 (S) granular fertilizer, impregnation of granular fertilizers with herbicides. Home: Potash & Phosphate Inst, 771 McCartney Rd, Kelowna, BC Canada V1Z 1R6 Office: Potash & Phosphate Inst, Ste 704 CN Tower, Midtown Pla, Saskatoon, SK Canada S7K 1J5

BEATTIE, DONALD A., energy scientist, consultant; b. N.Y.C., Oct. 30, 1929; s. James Francis and Evelyn Margaret (Hickey) B.; m. Ann Mary Kean, Mar. 27, 1973; children: Thomas James, Bruce Andrew. A.B., Columbia U., 1951; M.S., Colo. Sch. Mines, 1958. Regional geologist Mobil Oil Co., 1958-63; Apollo lunar expts. program mgr. NASA, 1963-72; dir. NASA energy systems div. NASA, Washington, 1978-82; v.p. Houston ops. BDM Corp., 1983-84;; cons. on energy and space tech., 1984—; pres. Endosat Inc., 1991—; dir. advanced energy research and tech. NSF, 1973-75; dep. asst. adminstr. ERDA, 1975-77; acting asst. sec. Dept. Energy, Washington, 1977-78; solar energy coordinator U.S./USSR Coop. in Sci. and Tech.; U.S. rep. Vienne Inst. for Comparative Econ. Studies Workshop on Energy. Contbr. numerous articles on lunar sci., energy to profl. jours. Active Boy Scouts Am., 1958-71. Served with AC USN, 1951-56. Recipient Exceptional Service medal NASA, 1971, Sr. Exec. Service and Outstanding Performance award, 1980; Superior Achievement award Dept. Energy, 1978. Fellow AAAS; mem. AIAA, Geol. Soc. Am., Am. Astron. Soc., Nat. Space Club. Office: 13831 Dowlais Dr Rockville MD 20853-2630

BEAUCHAMP, GARY KEITH, physiologist; b. Belvidere, Ill., Apr. 5, 1943; married; 2 children. BA, Carleton Coll., 1965; PhD in Biopsychology, U. Chgo., 1971. Fellow chemosensation U. Pa., 1971-73, asst. prof. psychology dept. otorhinolaryngology and human comm. Med. Sch., 1974—; assoc. mem. Monell Chem. Senses Ctr., 1973—, dir., 1990—. Mem. AAAS, Assn. Chemoreception Soc. Achievements include research in investigation of the role of the chemical senses in regulating behavior and physiology in a variety of animal species, including humans. Office: Monell Chemical Senses Ct 3500 Market St Philadelphia PA 19104*

BEAUCHAMP, JEFFERY OLIVER, mechanical engineer; b. Alice, Tex., Jan. 19, 1943; s. Charles Kirkland and Lila Arminda (Calk) B.; m. Toni Ramona Nobler, Sept. 7, 1963. BSME, U. Houston, 1969, MSME, 1973. Registered profl. engr., Tex. Mech. designer Great Lakes Petroleum Service, Houston, 1963-64; mech. engr. Elliott Co. div. Carrier Corp., Houston, 1964-68; research asst. U. Houston, 1968-70; chief engr. Mallay Corp., Houston, 1970-74; project mgr. Fluor Engrs. & Constructors, Houston, 1974-78; pres. INTERMAT Internat. Materials Mgmt. Engrs., Houston, 1978—; cons. in field; speaker, lectr., founding bd. dirs. Westheimer Nat. Bank, 1983-84. Bd. dirs. Houston Dist. Export Council, Dept. Commerce, 1985-87; pres. Leadership Houston Assn., 1985-86, Sci. Engring. Fair of Houston, Inc., 1983-84, Engrs. Council of Houston, 1982-83. Mem. Nat. Soc. Profl. Engrs., Tex. Soc. Profl. Engrs. (Outstanding Young Engr. 1974), Am. Cons. Engrs. Council, Cons. Engrs. Council Tex., ASME, Greater Houston Partnership (internat. and domestic bus. council), Inst. Internat. Edn., Common Cause, Houston Mus. Fine Arts, Contemporary Arts Mus. Houston, Los Angeles County Mus., CEO Roundtable, Smithsonian Assocs., Sigma Xi, Phi Kappa Phi, Pi Tau Sigma. Contbr. articles to profl. jours. Home: 9 Pinehill Ln Houston TX 77019-1111 Office: 9 Greenway Pla Ste # 2400 Houston TX 77046

BEAUCHAMP, JESSE LEE (JACK BEAUCHAMP), chemistry educator; b. Glendale, Calif., Nov. 1, 1942; m. Patricia George; children: Melissa Ann, Thomas Alton, Amanda Jane, Ryan Howell. BS with honors in Chemistry, Calif. Inst. Tech., 1964; PhD in Chemistry, Harvard U., 1967. Arthyr Amos Noyes instr. in Chemistry Calif. Inst Tech., Pasadena, 1967-69, asst. prof. chemistry, 1969-71, assoc. prof. chemistry, 1971-74, prof. chemistry, 1974—; panelist chem. rsch. evaluation Directorate of Chem. Scis. Air Force Office of Sci. Rsch., 1978-81, adv. panelist high energy density materials, 1988-92; exec. com. advanced light source users, LBL, 1984-87; experimental evaluation com. TRIUMPH, U. B.C., 1985-88; grad. fellow selection panel, NSF, 1986-89; postdoctoral selection panel NATO, 1987-89; mem. com. critical techs.: role of Chemistry and Chem. Engring. Nat. Rsch. Coun., 1991-92. Mem. editorial adv. bd. Chemical Physics Letters, 1981-87, Jour. Am. Chem. Soc., 1984-87, Jour Physical Chemistry, 1984-87, Organometallics, 1989-92, Interat. Jour. Chemical Kinetics, 1990—. Woodrow Wilson fellow Harvard U., 1964-65, NAS grad. fellow, 1965-67; fellow Alfred P. Sloan Found., 1967-70; tchr.-scholar Camille and Henry Dreyfus, 1971-76; meml. fellow John Simon Guggenheim, 1976-77. Fellow AAAS; mem. NAS (com. chem. scis., chem. kinetics subgroup 1980-83), Am. Chem. Soc. (award in pure chemistry 1978, exec. com. divsn. physical chem., 1980—), Am. Assn. Mas. Spectometry, Aircraft Owners and Pilots Assn., Soc. Fellows Harvard U. Office: Calif Inst Tech Dept of Chemistry Noyes Lab 127-72 Pasadena CA 91125

BEAUGRAND, MICHEL, physician, educator; b. Paris, Oct. 30, 1945; s. Henri and Christiane (Bertrand) B. BS, 1963; MD, Faculte de Medicine, Paris, 1974. Intern Assistance Publique de Paris, 1969-74; prof. hepatogastroenterology Faculte de Medicine, Paris, 1981—; mem. staff Hosp. Jean Verdier, Bondy, France. Office: Hosp Jean Verdier, Ave 14 Juillet, 93143 Bondy France

BEAULIEU, JACQUES ALEXANDRE, physicist; b. St. Jean, Que., Can., Apr. 15, 1932; s. Armand J. and Alice (Boulais) B.; m. Fleur-Ange Tardif, Feb. 5, 1955; children—Jacqueline, Pierre. B.Sc., McGill U., Montreal, Que., 1953, M.Sc., 1954; Ph.D., U. London, 1969; Ph.D. (hon.), U. Moncton, 1972. Research scientist Dept. Nat. Defense, Valcartier, Que., 1954-71, sr. system officer, 1974-91, ret., 1991; dir. R&D Gentec Co., Quebec, Que., 1971-74; pres. Beaulieu Cons. Inc., Quebec, Que., 1991—. Contbr. articles to profl. jours.; patentee in field. Recipient Dennis Gabor award Soc. Photo-Instrumentation Engrs., 1986, commendation Min. Nat. Def., Can., 1989. Fellow Royal Soc. Can. (T. Eady medal 1978); mem. Can. Assn. Physicists (Outstanding Achievement in Indsl. and Applied Physics award 1993), Assn. Canadienne Francaise pour l'Avancement des Sciences (Armchambeault medal 1971, J.A. Bombardier medal 1980, CAP medal for achievements in indsl. and applied physics 1993), Optical Assn. Am. Home: 2526 Chasse St, Sainte-Foy, PQ Canada G1W 1L9

BEAUPAIN, ELAINE SHAPIRO, psychiatric social worker; b. Boston, Nov. 1, 1949; d. Abraham and Anna Marilyn (Gass) S.; m. Dean A. Beaupain, Feb. 14, 1987; 1 child, Andrew. BA, McGill U., Montreal, Que., 1971, MSW, 1974. Ind. clin. social worker, Mass.; cert. social worker, Maine; cert. social worker with ind. practice lic., Maine; lic. ind. clin. social worker, Mass. Psychiat. social worker Bangor (Maine) Mental Health Inst., 1974-75; outpatient therapist The Counseling Ctr., Bangor, 1975-76, The Counseling Ctr., Millinocket, Maine, 1979-86; asst. core group leader adolescent unit Jackson Brook Inst., Portland, Maine, 1986-87; area dir. Community Health and Counseling Svcs., 1981-86; pvt. practice social work, 1987—, psychotherapy with individuals, couples and families Millinocket and Bangor, 1987—. Mem. AAUW, Nat. Assn. Social Workers, Acad. Cert. Social Workers (diplomate 1992). Republican. Office: 122 Pine St Bangor ME 04401-5216

BEAVEN, THORNTON RAY, physical scientist; b. Evansville, Ind., Jan. 25, 1937; s. Clark Thornton and Rosalma (Kirkpatrick) B.; m. Barbara Sue Porter, Dec. 1961 (div. 1971); 1 child, Eric Craig; m. Sonja Judith Radzynski, July 23, 1971; children: Dominique Maria, Max Clark. AA, Miami Dade Jr. Coll., 1971; BS, U. Cen. Fla., 1974. Biologist water resources div. U.S. Geol. Survey, Miami, Fla., 1974-79; environ. scientist conservation div. U.S. Geol. Survey, Casper, Wyo., 1979-83; environ. scientist Bur. Land Mgmt., Casper, 1983-84; environ. scientist Bur. Land Mgmt., Cheyenne, Wyo., 1984-90, supervisory phys. scientist, 1990—. With U.S. Army, 1957-67. Mem. Am. Chem. Soc. Republican. Baptist. Home: 3512 Central Ave Cheyenne WY 82001 Office: Bur Land Mgmt PO Box 1828 Cheyenne WY 82003

BEAVERS, ROY L., utility executive; b. Joplin, Mo., Apr. 24, 1930; s. Roy L. Sr. and Margarette Nellie (Loughlin) B.; m. Valerie Evelyn Gurney; children: Leslie Anne, Brendan G. BS in Bus., U. Mo., 1952; MA in Polit. Sci., U. Md., 1970. Commd. ens. USN, 1952, advanced through grades to comdr., 1966, retired, 1972; agt., broker ins. agy., Lebanon, Mo., 1972-77; field rep. Nat. Rural Electric Coop. Assn., Washington, 1977-84; mgr. pub. info. and legis. liaison wholesale power coop. KAMO Power, Vinita, Okla., 1984—. Contbr. polit. and mil. essays to newspapers and other publs. State hdqrs. dir. Va. Com. to Re-elect Nixon, Richmond, Va., 1972; mem. Bd. Mo. Community Betterment Edn. Fund, Bd. Okla. Acad. for State Goals. Decorated Bronze, Silver, and Gold medals U.S. Naval Inst., Pres. Merit Svc. medal, Navy Commendation medal. Mem. Mo. Indsl. Devel. Coun., So. Indsl. Devel. Coun., U.S. Naval Inst. Office: KAMO Power Electric Coop Inc PO Box 577 Vinita OK 74301-0577

BECHERER, RICHARD JOHN, architectural educator; b. East St. Louis, Ill., Nov. 8, 1951; s. Adam Jacob and Agnes Evelyn (Baker) B.; m. Charlene Castellano, Aug. 13, 1982. Student Courtauld Inst., U. London, 1973; BA, BArch, Rice U., 1974; MA, Cornell U., 1977, PhD, 1981. Archtl. asst. Colin St. John Wilson and Ptnr., London, 1972-73; designer The Brooks Assn., Houston, 1973-74; grad. asst. Cornell U., Ithaca, N.Y., 1974-80, asst. prof. architecture, 1981; asst. prof. Auburn (Ala.) U., 1980-82, U. Va., Charlottesville, 1982-86; head grad. architecture program Carnegie Mellon U., Pitts., 1986-90, assoc. prof. architecture, 1987—; presenter seminars NEH, 1982, 88, 89, Am. Collegiate Schs. Architecture, 1988, 93; lectr. Centre Canadien d'Architecture, Montreal and various colls., univs. and nat. confs. Author: Science Plus Sentiment; César Daly's Formula for Modern Architecture, 1984, (mus. catalogue and display) Urban Theory and Transformation, 1976, (tourist guidebook) Canandaigua: A Walking Tour, 1977; contbr. articles to profl. jours.; prin. works include interiors Michael P. Kelley House, Belleville, Ill., 1978, Robert Becherer House, Stonybrook, 1990; selected exhibitor Venice Biennale, Prato della Valle, Padua, 1985; exhibitor Heart of the Park, Houston. Grad. fellow Cornell U., 1975-79, Eidlitz fellow, 1978, Soc. for Humanities and Mellon Found. fellow, 1984-85, NEH fellow, 1986; Travel to Collections grantee NEH, 1985; recipient Design Arts award Nat. Endowment for Arts, 1989-90, Graham Found. award, 1993. Mem. AAUP, Soc. Archtl. Historians (session chmn. ann. meeting 1989), Coll. Art Assn., Rice U. Alumni Assn. Democrat. Roman Catholic. Avocations: free-hand drawing, bicycling, ballroom dancing, French cinema. Home: 119 Race St Edgewood PA 15218 Office: Carnegie Mellon Univ Dept Architecture CFA 5000 Forbes Ave Pittsburgh PA 15213

BECHLER, RONALD JERRY, structural engineer; b. Sioux City, Iowa, Feb. 14, 1943; s. Gerald Francis and Lauretta Mae (Stephens) B.; m. Nancy Ann Reed, June 10, 1967; children: Bethany, Lorraine, Cynthia. BSCE, Iowa State U., 1967; MSCE, U. Iowa, 1975. Lic. profl. engr. Engr.-in-tng. Stanley Consultant, Muscatine, Iowa, 1967-71, engr., 1971-80, dept. head, 1980-85; v.p. engring., chief structural engr. Lear-Siegler Cuckler Bldg. Div., Monticello, Iowa, 1985-88; ptnr. Jack C. Miller & Assocs., Structural Engrs., Cedar Rapids, Iowa, 1988—. Mem. ASCE, NSPE, Iowa Engring. Soc., Structural Engrs. Assn. of Eastern Iowa (pres. 1991-92), Iowa Engring. Soc. Home: RR 1 Box 151B Monticello IA 52310 Office: Jack C Miller & Assocs 422 Second Ave SE Cedar Rapids IA 52401

BECHTEL, DONALD BRUCE, biologist, educator, research chemist; b. Paterson, N.J., Aug. 1, 1949; s. Joseph Frederick and Helene Julia (Fiedeldey) B.; m. Kathleen Ann Barnwell, Aug. 7, 1971; children: Michael S., Ryan J. BS, Iowa State U., 1971, MS, 1974; PhD, Kans. State U., 1982. Grad. teaching asst. dept. botany Iowa State U., Ames, 1973-74; chemist U.S. Grain Mktg. Rsch. Lab., Manhattan, Kans., 1974-77; rsch. chemist U.S. Grain Mktg. Rsch. Lab., Manhattan, 1977—; adj. prof. div. biology Kans. State U., Manhattan, 1983—. Editor: New Frontiers in Food Microstructure, 1983; assoc. editor: Cereal Chemistry, 1985-89, Food Structure, 1985-87, 92—; contbr. numerous sci. papers. Com. chmn. Boy Scout Troop 73, Manhattan, 1986-92; bd. dirs. Manhattan (Kans.) Optimist Club, 1990-92. Recipient Rsch. grants Los Alamos (N.Mex.) Stable Isotope Lab., 1980, USDA Agrl. Rsch. Svc. Competitive Grant, Washington, 1982. Mem. Am. Assn. Cereal Chemists (chmn. books com. 1986-91), Botanical Soc. Am., Gamma Sigma Delta, Sigma Xi. Republican. Lutheran. Achievements include developed numerous techniques and methodologies for the study of cereal grain in electron microscopy. Home: 1406 Sharingbrook Dr Manhattan KS 66502 Office: US Grain Mktg Rsch Lab 1515 College Ave Manhattan KS 66502

BECHTEL, STEPHEN DAVISON, JR., engineering company executive; b. Oakland, Calif., May 10, 1925; s. Stephen Davison and Laura (Peart) B.; m. Elizabeth Mead Hogan, June 5, 1946; 5 children. Student, U. Colo., 1943-44; BS, Purdue U., 1946, D. in Engring. (hon.), 1972; MBA, Stanford U., 1948; DSc (hon.), U. Colo., 1981. Registered profl. engr., N.Y., Mich., Alaska, Calif., Md., Hawaii, Ohio, D.C., Va., Ill. Engring. and mgmt. positions Bechtel Corp., San Francisco, 1941-60, pres., 1960-73, chmn. of cos. in Bechtel group, 1973-80; chmn. Bechtel Group, Inc., 1980-90, chmn. emeritus, 1990—; bd. dirs. IBM; former chmn., mem. bus. coun., life councillor, past chmn. conf. bd. Trustee, mem., past chmn. bldg. and grounds com. Calif. Inst. Tech.; mem. pres.'s coun. Purdue U.; adv. coun. Inst. Internat. Studies, bd. visitors, former charter mem., adv. coun. Stanford U. Grad. Sch. Bus. With USMC, 1943-46. Decorated officer French Legion of Honor; recipient Disting. Alumnus award Purdue U., 1964, U. Colo., 1978, Ernest C. Arbuckle Disting. Alumnus award Stanford Grad. Sch. Bus., 1974, Disting. Engring. Alumnus award 1979; named Man of Yr. Engring. News-Record, 1974, Outstanding Achievement in Constrn. award Moles, 1977, Chmn.'s award Am. Assn. Engring. Soc., 1982, Washington award Western Soc. Engrs., 1985, Nat. Medal Tech. from Pres. Bush, 1991, Golden Beaver award 1992, Herbert Hoover medal 1980. Fellow ASCE (Engring. Mgmt. award 1979, Pres. award 1985), AAAS, Instn. Chem. Engrs. (U.K., hon.); mem. AIME, NSPE (hon. chmn. Nat. Engrs. Week 1990), Nat. Acad. Engring. (past chmn.), Calif. Acad. Scis. (hon. trustee),Am. Soc. French Legion Honor (bd. dirs.), Am. Acad. Arts and Scis., Royal Acad. Engring. (U.K., fgn. mem.), Pacific Union Club, Bohemian Club, San Francisco Golf Club, Claremont Country Club, Cypress Point Club, Met. Club (Washington), Chi Epsilon, Tau Beta Pi. Office: Bechtel Group Inc PO Box 193965 San Francisco CA 94119-3965

BECHTEL, STEPHEN E., mechanical engineer, educator. BS in Engring. summa cum laude, U. Mich., 1979; PhD in Engring., U. Calif., Berkeley, 1983. Rsch. asst. U. Calif., Berkeley, 1979-83; assoc. prof. dept. engring. mechanics Ohio State U., Columbus, 1983—; mem. grad. studies com. dept. engring. mechanics Ohio State U., 1986—, dept. engring. mechanics rep. coll.

engring. undergrad. honors com., 1987—, mem. undergrad. com. dept. engring. mechanics, 1988—; reviewer design, mfg. and computer-integrated engring. divsn. NSF, reviewer fluid dynamics and hydraulics directorate; cons. Hoechst Celanese Corp., plastics divsn. Rockwell Internat., Los Alamos Nat. Lab. Referee Jour. Rheology, Jour. Non-Newtonian Fluid Mechanics, others. James B. Angell scholar U. Mich., 1976-79. Mem. ASME (mem. fluid mechanics com. applied mechanics divsn. 1989—, rec. sec. gen. com. 1991-92, rec. sec. exec. com. 1992-93, Henry Hess award 1990), Am. Acad. Mechanics, Soc. Rheology, Tau Beta Pi. Achievements include research in modeling of industrial polymer processing and fiber spinning, viscoelastic fluid flows, free surface flows and instability mechanisms, material characterization, transducer characterization in non-destructive evaluation. Office: Ohio St Univ Dept of Eng Mech 155 W Woodruff Ave Columbus OH 43210*

BECK, AARON TEMKIN, psychiatrist; b. Providence, July 18, 1921; s. Harry S. and Elizabeth (Temkin) B.; m. Phyllis Whitman, June 4, 1950; children: Judith, Daniel, Alice, Roy. B.A., Brown U., 1942, Dr.Med.Sci. (hon.), 1982; M.D., Yale U., 1946. Mem. faculty U. Pa. Med. Sch., 1954—, prof. psychiatry, 1971—, Univ. prof., 1983—; dir. Center Cognitive Therapy, 1965—; mem. rev. panel NIMH, 1965-80, chmn. task force suicide prevention , 1969-80; bd. dirs. West Philadelphia Community Mental Health Consortium, 1975-77. Author: Depression: Causes and Treatment, 1967, Diagnosis and Management of Depression, 1973, Prediction of Suicide, 1973, Cognitive Therapy and the Emotional Disorders, 1976, Cognitive Therapy of Depression, 1979, Anxiety Disorders and Phobias: A Cognitive Perspective, 1985, Love is Never Enough, 1988, Cognitive Therapy of Personality Disorders, 1990; co-author: Cognitive Therapy in Clinical Practice, 1989, Cognitive Therapy with Inpatients, 1992, Cognitive Therapy of Substance Abuse, 1993. Served as officer M.C. U.S. Army, 1952-54. Recipient rsch. award R.I. Med. Soc., 1948, ann. award Phila. Soc. Clin. Psychologists, 1978, Am. Psychopathol. Assn., 1983, Disting. Sci. award APA, 1989, rsch. award Am. Suicide Found., 1991, Albert Einstein Sch. Medicine award, 1992. Fellow Royal Coll. Psychiatry, N.Y. Acad. Medicine (Thomas Salmon award 1992), Am. Psychol. Soc. (rsch. award 1993); mem. Soc. Psychotherapy Rsch. (pres. 1975-76), Am. Psychiat. Assn. (prize rsch. psychiatry 1979), Am. Assn. Suicidology (rsch. prize 1985), Assn. Advancement Behavior Therapy. Office: 3600 Market St 7th fl Philadelphia PA 19104*

BECK, DAVID PAUL, biochemist; b. Wilmington, Del., Aug. 3, 1944; s. David Franklin and Mary Jane (Lazar) B.; m. Jeanne Elaine Crawford, Nov. 19, 1966; children: Jennifer Jeanne, David Andrew. AB, Princeton U., 1966; PhD, Johns Hopkins U., 1971. Postdoctoral fellow Harvard U., Cambridge, Mass., 1971-74; staff scientist Md. Psychiat. Rsch. Ctr., Balt., 1974-77; health scientist, adminstr. NIH, Bethesda, Md., 1977-84; assoc. dir., sec. bd. dirs. Pub. Health Rsch. Inst., N.Y.C., 1984-91; pres. Coriell Inst. Med. Rsch., Camden, N.J., 1991—. Contbr. rsch. articles to sci. jours. Mem., officer various polit. and civic orgns.; mem. Baltimore County Bd. Recreation and Pks., 1977-84; bd. dirs. Hoff-Barthelson Music Sch., Scarsdale, N.Y., 1989-91; bd. dirs. West Jersey Chamber Music Soc., Moorestown, N.J., 1992—. Mem. Assn. Ind. Rsch. Insts. (v.p. 1989-92). Office: Coriell Inst Med Rsch 401 Haddon Ave Camden NJ 08103-1505

BECK, JOHN ROLAND, environmental consultant; b. Las Vegas, N.Mex., Feb. 26, 1929; s. Roland L. and Betty L. (Shrock) B.; m. Doris A. Olson, Feb. 9, 1951; children: Elizabeth J., Thomas R., Patricia L., John William. BS, Okla. A. & M. U., 1950; MS, Okla. State U., 1957; postgrad., U. Tex., 1954, George Washington U., 1965. Registered sanitarian, Ohio, Ariz.; cert. wildlife biologist. Wildlife researcher King Ranch, Kingsville, Tex., 1950-51; faculty Inst. Human Physiology U. Tenn., Martin, 1954-55; rsch. biologist FWS, USDI, Grangeville, Idaho, 1955-57; ctr. dir. Job Corps, OEO, Indiahoma, Okla., 1965-67; supr. animal control biology FWS, USDI, 1953-69; operating v.p. Bio-Svc. Corp., Troy, Mich., 1969-78; pres. BECS Ltd., Prescott, Ariz., 1981-85; spl. asst. USDA - APHIS, Washington, 1986-87; prin. cons. Biol. Environ. Cons. Svc. Inc., Phoenix, 1978—; faculty assoc. Ariz. State U., Tempe, 1980-89. Sr. author: Managing Service for Success, 1987, 2d edit., 1991; columnist mo. column on pest control in 2 mags., 1980-88; contbr. articles to profl. jours. Capt. USAR, 1950-62. Fellow Royal Soc. Health, N.Y. Explorers Soc.; mem. ASTM, Wildlife Soc., Wildlife Disease Assn., nat. Environ. Health Assn., Sigma Xi. Republican. Baptist. Avocations: coaching amateur sports, botany studies, ornithology, mammalogy. Office: PO Box 26482 Prescott Valley AZ 86312

BECK, LOIS GRANT, anthropologist, educator; b. Bogota, Colombia, Nov. 5, 1944; d. Martin Lawrence and Dorothy (Sweet) Grant; m. Henry Huang; 1 dau., Julia. BA, Portland State U., 1967; MA, U. Chgo., 1969, PhD, 1977. Asst. prof. Amherst (Mass.) Coll., 1973-76, Univ. Utah, Salt Lake City, 1976-80; from asst. to assoc. prof. Washington U., St. Louis, 1980-92, prof., 1992—. Author: Qashqa'i of Iran, 1986, Nomad, 1991; co-editor Women in the Muslim World, 1978. Grantee Social Scis. Rsch. Coun., 1990, NEH, 1990-92. Mem. Middle East Studies Assn. (bd. dirs. 1981-84), Soc. Iranian Studies (exec. sec. 1979-82, edit. bd. 1982-91). Office: Washington Univ Dept Anthropology 1 Brookings Dr Saint Louis MO 63130

BECK, MORRIS, allergist; b. Miami, Fla., Oct. 12, 1927; s. Max and Anna (Luks) B.; m. Hollis Schwartz, Aug. 6, 1960; children: Gayle Beck Finan, Anne. BA, UCLA, 1949; MD, U. Zurich, Switzerland, 1957. Diplomate Am. Bd. Allergy and Immunology, Am. Bd. Pediatrics. Intern Queens Hosp. Ctr., 1958, resident in pediatrics, 1959-60; preceptor in allergy U. Miami (Fla.) Med. Sch., 1961 77; pvt. practice pediatrician Miami, 1961 77, pvt. practice allergist, 1978—; chief dept. allergy Miami Childrens Hosp, 1986—. With U.S. Army, 1950-52. Fellow Am. Coll. Allergy & Immunology, Am. Acad. Pediatrics, Am. Assn. Cert. Allergists; mem. Am. Acad. Allergy & Immunology, Am. Coll. Chest Physicians. Republican. Jewish. Avocations: photography, fishing, travel. Home: 12015 SW 68th Ave Miami FL 33156-5406 Office: 7800 SW 87th Ave Miami FL 33173-3570

BECK, ROBERT N., nuclear medicine educator; b. San Angelo, Tex., Mar. 26, 1928; married, 1958. AB, U. Chgo., 1954, BS, 1955. Chief scientist Argonne Cancer Rsch. Hosp., 1957-67, assoc. prof., 1967-76; prof. radiology sci. U. Chgo., 1976—; dir. Franklin McLean Inst., 1977—; cons. Internat. Atomic Energy Agency, 1966-68; mem. Internat. Com. on Radiation Units, 1968—, Nat. Coun. on Radiation, Protection & Measurements, 1970—. Mem. Soc. Nuclear Med., Am. Assn. Physicists in Medicine. Achievements include research in development of a theory of the process by which images can be formed of the distribution of radioactive material in a patient in order to diagnose his disease. Office: U Chgo-Franklin McClean Meml Rsch Inst 5841 S Maryland Ave Chicago IL 60637-1470 also: Abbot Labbs Abbott Park IL 60064*

BECK, VERNON DAVID, physicist, consultant; b. Evergreen Park, Ill., May 19, 1949; s. David John and Violet (Franzen) B.; m. Nancy Jo Katagiri, Sept. 11, 1971; 1 child, Laura Ingrid. BA in Physics, U. Chgo., 1971, MS in Physics, 1971, PhD in Physics, 1977. Postdoctoral researcher IBM Rsch., Yorktown Heights, N.Y., 1971-73, mem. rsch. staff, 1973-84, mgr. display devices, 1984-87; vis. scholar Enrico Fermi Inst., Chgo., 1987-88; sr. engr. IBM East Fishkill, N.Y., 1988-91; pres. Vernon Beck Cons., Inc., Ridgefield, Conn., 1991—; interviewer for fellow selection Fannie and John Hertz Found., Livermore, Calif., 1991—. Contbr. articles to tech. publs. Fellow Fannie and John Hertz Found., 1971-76. Mem. Soc. Info. Display, Electron Microscopy Soc. Am., Sigma Xi (pres. 1982-83). Lutheran. Achievements include 5 patents; research interests include electron optics, lithography, precision engineering. Office: Vernon Beck Cons Inc 1 Hobby Dr Ridgefield CT 06877-1922

BECKER, BROOKS, management consultant, chemist; b. Emporia, Kans., Aug. 12, 1931. PhD, U. Kans., 1959. Sr. rsch. chemist PPG Industries, Akron, Ohio, 1960-61; rsch. scientist AMF, Alexandria, Va., 1961-64; sr. rsch. chemist Gulf R&D Corp., Pitts., 1964-70; dir. bur. air and waste mgmt. State of Wis., Madison, 1970-78; pres., CEO RMT, Inc., Madison, 1978-90; mgmt. cons. Madison, 1990—. Pres. Village of Shorewood Hills, Wis., 1987-89. With U.S. Army, 1953-59. Mem. Am. Inst. Chemists, Am. Chem. Soc., Am. Mgmt. Assn.

BECKER, BRUCE CARL, II, physician, educator; b. Chgo., Sept. 8, 1948; s. Carl Max and Lillian (Podzamsky) B. BS in Aero. and Astron. Engring., U. Ill., 1970; MSME, Colo. State U., 1972; postgrad. Wright State U., 1973-74; MD, Chgo. Med. Sch., 1978; MS in Health Svcs. Adminstrn., Coll. St. Francis, Joliet, Ill., 1984; Diploma in Spanish, U. Chgo., 1988; Diploma in Polish, Coll. of Du Page, 1989. Diplomate Am. Bd. Family Practice. Resident in surgery U. N.C.-Chapel Hill, 1978-79, in family practice St. Mary of Nazareth Hosp. Ctr., Chgo., 1979-81, chmn., program dir. dept. family practice, 1985-90; clin. instr. Chgo. Med. Sch., 1982, affiliate instr., 1982-83, asst. prof., 1983, vice chmn. dept. family medicine, 1983-91; asst. dir. med. edn. St. Mary of Nazareth Hosp. Ctr., Chgo., 1981-82, dir. family practice residency, 1983-90, chief Family Practice Ctr., 1983-85, chmn. dept. family practice, 1985-90, med. dir. Home Health Svc., 1985—, med. dir. HMO-Ill., 1985—, mem. fin. com. governing bd., 1987-91, planning and devel. com. governing bd., 1990—, v.p. med. affairs, 1989—. Contbr. articles to profl. jours. Mem. editorial rev. bd. Postgrad. Medicine, 1987-89. Mem. Pub. Health Svc. Adv. Network Dept. Health & Human Svcs., 1990-91; bd. dirs. Inn Care of Am. Midwest Region, 1991; mem. dinner com. Ill. chpt. Lupus Found. Am., 1991. Capt., USAF, 1970-75. Recipient Literary Key award St. Mary of Nazareth Hosp. Ctr., 1981, 85. Fellow Am. Acad. Family Physicians (rep. to accreditation rev. com. for physician assts. 1989—, chmn. 1991—); mem. AMA, Ill. Acad. Family Physicians (commn. on internal affairs 1986, commn. pub. and govt. policy 1987-89, chmn. 1989-90, bd. dirs. 1988-92, chmn. pub. rels. and info. com. 1988-92, state rep. family practice rev. act com. 1990—, vice speaker, 1991—), Soc. Tchrs. of Family Medicine, Assn. Am. Med. Colls., Alliance Continuing Med. Edn., Am. Coll. Health Care Execs., Am. Coll. Occupational Medicine, Am. Coll. Physician Execs., Am. Acad. Med. Adminstrs., Chgo. Med. Soc. (councilor for Chgo. Med. Sch. 1986—, physicians stress ad hoc com. 1989-90, vice chmn. 1990-91, adv. com. on pub. health policy 1990—, presdl. adv. com. 1991—), Ill. State Med. Soc. (coun. on edn. and manpower 1986-91, chmn. com. on CME 1991, chmn. subcom. physican placement and practice issues 1986-90, third party payment and processes com. IAFP rep. 1990—, chmn. com. edn. 1991—), Phi Delta Epsilon. Lutheran.

BECKER, FREDERICK FENIMORE, cancer center administrator, pathologist; b. N.Y.C., July 23, 1931; s. Louis I. and Ruth (Shurr) B.; m. Mary Ellen Terry, Nov. 23, 1971; 1 dau., Bronwyn Elizabeth. BA, Columbia U., 1952; MD, NYU, 1956. Intern Harvard svc. Harvard service Boston City Hosp., 1956-57; resident Bellevue Hosp., N.Y.C.; pathology trainee NYU Sch. Medicine, N.Y.C., 1957-60, prof., dir. pathology, 1962-75; chmn. dept. pathology U. Tex. Cancer Center, M.D. Anderson Hosp. and Tumor Inst., Houston, 1976-79; v.p. research U. Tex. M.D. Anderson Cancer Ctr., 1979—. Contbr. numerous articles to various publs. Served with USN, 1960-62. Mem. Am. Assn. Pathologists (pres. 1980-81), Am. Assn. Cancer Research, Am. Soc. Cell Biology, Tex. Med. Assn. Club: Athenaeum (London). Office: U Tex MD Anderson Cancer Ctr 1515 Holcombe Blvd Houston TX 77030-4095

BECKER, GARY STANLEY, economist, educator; b. Pottsville, Pa., Dec. 2, 1930; s. Louis William and Anna (Siskind) B.; m. Doria Slote, Sept. 19, 1954 (dec.); children: Judith Sarah, Catherine Jean; m. Guity Nashat, Oct. 31, 1979; children: Michael Claffey, Cyrus Claffey. AB summa cum laude, Princeton U., N.J., 1951; AM, U. Chgo., 1953, PhD, 1955; PhD (hon.), Hebrew U., Jerusalem, 1985, Knox Coll., 1985, U. Ill., Chgo., 1988, SUNY, 1990, Princeton U., 1991, U. Palermo, Buenos Aires, 1993, U Palermo, Argentina, 1993. Asst. prof. U. Chgo., 1954-57; from asst. prof. to assoc. prof. Columbia U., N.Y.C., 1957-60, prof. econs., 1960-68, Arthur Lehman prof. econs., 1968-70; Univ. prof. U. Chgo., 1970-83, Univ. prof. econs. and sociology, 1983—, chmn. dept. econs., 1984-85; Ford Found. vis. prof. econs. U. Chgo., 1969-70; assoc. Econs. Rsch. Ctr. Nat. Opinion Rsch. Ctr., Chgo., 1980—; mem. domestic adv. bd. Hoover Instn., Stanford, Calif., 1973-90, sr. fellow, 1990—; mem. acad. adv. bd. Am. Enterprise Inst., 1987-90; rsch. policy advisor Ctr. for Econ. Analysis Human Behavior Nat. Bur. Econ. Rsch., 1972-78, mem. and sr. research assoc., 1957-79; assoc. mem. Inst. Fiscal and Monetary Policy, Ministry of Japan, 1988—. Author: The Economics of Discrimination, 1957, 3d edit., 1993, Human Capital, 1964, 2d edit., 1975, 3rd edit., 1993 (W.S. Woytinsky award U. Mich. 1967), Human Capital and the Personal Distribution of Income: An Analytical Approach, 1967, Economic Theory, 1971, (with Gilbert Ghez) The Allocation of Time and Goods Over the Life Cycle, 1975, The Economic Approach to Human Behavior, 1976, A Treatise on the Family, 1981, expanded edit., 1991; editor: Essays in Labor Economics in Honor of H. Gregg Lewis, 1976; co-editor: (with William M. Landes) Essays in the Economics of Crime and Punishment, 1974; columnist, Bus. Week, 1985—; contbr. articles to profl. jours. Recipient Profl. Achievement award U. Chgo. Alumni Assn., 1968, Frank E. Seidman Disting. award in Polit. Economy, 1985, MERIT award NIH, 1986, John R. Commons award Omicron Delta Epsilon, 1987, recipient: Nobel prize in Economic Scis., 1992. Fellow Am. Statis. Assn., Econometric Soc., Am. Acad. Arts and Scis., Am. Econ. Assn. (Disting., v.p. 1974, pres. 1987, John Bates Clark medal 1967); mem. NAS, NAE (founding mem., v.p. 1965-67), Am. Philos. Soc., Internat. Union for Scientific Study Population, Mont Pelerin Soc. (exec. bd. dirs. 1985—, v.p. 1989-90, pres. 1990-92), Phi Beta Kappa. Office: U Chgo Dept Econs 1126 E 59th St Chicago IL 60637-1539

BECKER, HERBERT P., mechanical engineer; b. N.Y.C., Oct. 8, 1920; s. David and Jeanette (Solomon) B.; m. Shirley Schneider, Apr. 13, 1947; children: Eileen Lois, Robert Bruce. BME, CCNY, 1942; MME, NYU, 1948. Registered profl. engr., N.Y., N.H., Mass. Sales engr., dir. A.I. McFarlan, N.Y.C., 1950-58; project engr. Syska & Hennessey, N.Y.C., 1958-64; asst. v.p. Michael Baker Jr. N.Y., N.Y.C., 1964-76; energy systems analyst Pope, Evans & Robbins, N.Y.C., 1976-80, Flack & Kurtz, N.Y.C., 1980-81; prin. H.P Becker, P.E., N.Y.C., 1981—; energy crisis statement Congrl. Record, Washington, 1973; tchr. HVAC courses Voorhees C.C., NYU, U. Wis., Ctr. Profl. Advancement; lectr. in field. Contbr. numerous articles to profl. jours. With USN, 1943-45. Recipient Energy Engr. of Yr. award Bklyn Engrs. Club, 1990. Fellow ASHRAE (contbg. author handbook 1973-92, Best Jour. Paper award 1974, Crosby Field award 1974), Assn. Energy Engrs. (Energy Engr. of Yr. 1992). Achievements include research in concept of variable speed pumping systems to the commercial heating ventilating and air conditioning industry. Home: 3901 Independence Ave New York NY 10463 Office: Chervenak Keane & Co 307 E 44 St New York NY 10017

BECKER, JULIETTE, psychologist, marriage and family therapist; b. L.A., Sept. 22, 1938; d. Louis Joseph and Elissa Cecelia (Bevacqua) Cevola; m. Richard Charles Sprenger, Aug. 13, 1960 (div. Dec. 1984); children: Lisa Anne, Stephen Louis, Gina Marie, Paul Joseph, Gretchen Lynette; m. Vance Benjiman Becker, Nov. 7, 1986. BA in Psychology, Calif. State U., Fullerton, 1983; M in Marriage and Family Therapy, U.S. Internat. U., 1985; PhD in Clin. Psychology, William Lyon U., 1988. Therapist Villa Park (Calif.) Psychol. Svcs., 1985-88, psychologist, 1988—. Mem. APA, Am. Assn. Marriage, Family and Child Therapists, Calif. Assn. Marriage, Family and Child Therapists. Avocations: painting, opera, classical piano, interior design. Office: Villa Park Psychol Svcs 17871 Santiago Blvd Ste 206 Orange CA 92667-4131

BECKER, LEWIS CHARLES, cardiology educator. MD, Johns Hopkins U., 1966. Prof. medicine, divsn. cardiology Johns Hopkins U., 1972—. Achievements include research in nuclear cardiology. Office: Johns Hopkins Hosp Ischemic Heart Dis Ctr Rsch 600 N Wolfe St Baltimore MD 21205*

BECKER, SEYMOUR, hazardous materials and wastes specialist; b. Bronx, N.Y., Feb. 14, 1924; m. Ruth Schmitt, Aug. 30, 1958. MS, U. Wis., 1949; PhD, Pacific Western U., 1981. Nationally cert. hazardous materials mgr. and hazardous control mgr. Radiation control insp. Suffolk County Dept. Health Svcs., Hauppauge, N.Y., 1960-81; tech. cons., 1981-83; hazardous materials and wastes cons. Environ. Svcs., Portland, Maine, 1983-85, Mercy Hosp., Portland, 1985—; del. to China, People to People, Spokane, Wash., 1987, del. to Russia and Ukraine, 1992; advisor and cons. State of Maine Hosp. Assn, Augusta, 1988-90, Low Level Radioactive Wastes Authority, Augusta, 1989—, Dept. Environ. Protection, Augusta, 1989—. Contbr. articles to profl. jours. Cons. Emergency Mgmt. Agy., Windham, Maine, 1983—, Local Emergency Planning Com., Windham, 1989—. Fellow APHA; mem. Acad. Hazardous Materials Mgmt., Health Physics Soc., N.Y. Acad. Scis., Maine Pub. Health Assn. Achievements include development of N.Y. State radiation code; initiation of radiation control program in Suffolk County, N.Y. Home: 169 High St Apt 312 Portland ME 04101-2852 Office: Mercy Hosp 144 State St Portland ME 04101-3795

BECKER, UWE EUGEN, physicist; b. Chemnitz, Germany, May 17, 1947; s. Chin-Jung and Hildegard B.; m. Sigrid Hanna Flath, Feb. 2, 1979; children: Markus, Anne-Grit. Diplom, Tech. U., Berlin, Germany, 1971, PhD, 1977, Habilitation, 1985. Rsch. asst. Tech. U. Berlin, Germany, 1971-74; teaching asst. Tech. U. Berlin, 1974-79; rsch. assoc. Free U. of Berlin, 1979-80; vis. scientist Lawrence Berkeley Lab., Calif., 1980-81; asst. prof. Tech. U. of Berlin, Germany, 1981-86; prof. physics U. Würzburg, Germany, 1987; rsch. fellow German Rsch. Coun., Bonn, Germany, 1988-89; rsch. advisor Tech. U. of Berlin, Germany, 1989-90; leader atomic and molecular physics group Fritz-Haber Inst. of Max Planck Soc., Berlin, Germany, 1990—; faculty adv. coun. Tech. U. of Berlin, 1982-86, rep. for cooperation, 1985-86, adj. prof. physics, 1991; volume editor Plenum Press, N.Y.C., London, 1991; mem. Nat. Com. Rsch. Synchrotron Radiation, 1993. Contbr. articles to profl. jours. Recipient Award medal for Outstanding Achievement, Tech. U. Berlin, 1971, Rsch. fellowship German Rsch. Coun., 1986. Mem. German Phys. Soc., European Phys. Soc., Am. Phys. Soc., European Synchrotron Radiation Soc. Coun. Home: Niedstr 6, 1000 Berlin 41 Germany Office: Fritz-Haber Inst, Max Planck Soc, Faradayweg 4-6, 1000 Berlin Germany

BECKER, VIRGINIA GRAFTON, psychologist; b. Goshen, Ind., Oct. 22, 1951; d. Peter E. and Mary Louise (Jessup) Grafton; m. Frederick Hugo Becker, June 18, 1988; 1 child, Guinevere Virginia. MEd, U. Wash., 1976; PsyD, U. Denver, 1987. Inter/instr./therapist Boise (Idaho) State U., 1976-78; therapist Warm Springs Ctr., Boise, 1978-84; intern Am. Lake VA Med. Ctr., Tacoma, Wash., 1986-87; fellow Reproductive Sexual Med. Clinic, U. Wash., Harbor View Hosp., Seattle, 1987-88; psychologist in pvt. practice Seattle, 1988—; chair treatment com. Sexual Abuse Task Force, Boise, 1982-84; provider Snohomish County Juvenile Sex Offender Project, Everett, Wash., 1989-91; mental health person, critical incident stress debriefing, Everett, 1989-92. Named to Outstanding Young Women of Am., 1985. Mem. APA, Wash. State Psychol. Assn. (mem. at large 1993), Am. Assn. Sex Educators, Counselors and Therapists (cert. sex therapist, educator). Achievements include research on infertility and stress. Office: 727 N 182nd St Ste 202 Seattle WA 98133

BECKERMAN, ROBERT CY, pediatrician, educator; b. Bklyn., Oct. 29, 1946; s. Edward and Beatrice (Sherman) B.; children: Sarah, Ashley, Mollye. BS, Dickinson Coll., 1968; MD, Jefferson Med. Coll., 1972. Cert. Nat. Bd. Med. Examiners, Bd. Pulmonary Disease subspecialty Am. Bd. Pediatrics. Intern Riverside Hosp., Newport News, Va., 1972-73; resident in pediatrics Children's Hosp./U. of Cin., 1973-75, devel. disability fellow, 1975-76; pulmonary fellow U. Ariz., Tucson, 1976-78; asst. prof. pediatrics U. Miami (Fla.) Sch. Medicine, 1978-79; dir. pediatric pulmonary and sleep lab. Tulane U. Sch. Medicine, New Orleans, 1980-90, prof. pediatrics, physiology, chief pediatric pulmonology, 1978—; dir. Constance Kaufman Ctr., New Orleans, 1980—; dir. fellowship program pediatric pulmonary sect. Tulane U. Sch. Medicine, New Orleans, 1983—; dir. CF Ctr., chief pediat. pulmonology CF Ctr. La., New Orleans, 1987—; dir. Pediatric Pulmonary Ctr. So. La., New Orleans, 1988—, Pediatric Pulmonary Outreach Program, Lafayette, La., 1988—; chief med. cons. Audubon Zoo, New Orleans; med. cons. Friends of SIDS, La. Chpt. Sudden Infant Death Syndrome Found., New Orleans, Spinal Muscular Atrophy Assn. Exec. Com., New Orleans. Author: (with others) Practical Manual of Pediatrics, 2d edit., 1981, Current Pediatric Therapy, 13th edit., 1990, Respiratory Control Disorders in Infants and Children, 1992; editorial bd.: So. Med. Jour., 1993-96. Fellow Am. Coll. Chest Physicians (pediat. assembly), Am. Acad. Pediatrics; mem. Am. Thoracic Soc., La. Lung Assn., So. Soc. for Pediat. Rsch. Office: Tulane U Sch Medicine 1430 Tulane Ave New Orleans LA 70112-2699

BECKHUSEN, ERIC HERMAN, chemical engineer, consultant; b. Rahway, N.J., June 23, 1922; s. Ernest Henry and Elizabeth (Pilz) B.; m. Jean Eleanor Deitrich, June 14, 1949; children: Elizabeth, Eric, David, John. BS in Chem. Engring., Newark Coll. Engring., 1943; M Chem. Engring., NYU, 1949; MBA, Rutgers U., 1956. Registered profl. engr., N.J. Mgr. plant design GAF Corp., Linden, N.J., 1960-70; sr. project mgr. Jacobs Engring., Mountainside, N.J., 1975-80; dir. process plant design Carbogel, Inc., Livingston, N.J., 1980-85; prin. Beckhusen Assocs., Rahway, 1985—; lectr. seminars, Chgo., Amsterdam, Copenhagen. Contbr. articles to profl. publs. Elder 2d Presbyn. Ch., Rahway, 1961—; v.p. Rahway Bd. Edn., 1974, mem., 1972-75. With USNR, 1944-46, PTO. Mem. AICE. Office: Beckhusen Assocs 602 Grove St Rahway NJ 07065

BECKJORD, ERIC STEPHEN, energy researcher, nuclear engineering educator; b. Evanston, Ill., Feb. 17, 1929; s. Walter Clarence and Mary Amelia (Hitchcox) B.; m. Caroline Wendell Gardner, Feb. 28, 1953; children—Eric H., Amy W., Charles A., Sarah H. A.B. cum laude, Harvard U., 1951; M.S. in Elec. Engring., MIT, 1956; M.B.A., U. Chgo., 1984. Devel. engr. GE, San Jose, Calif., 1956-60; project engr. GE, Pleasanton, Calif., 1960-63; engring. mgr. Westinghouse Electric Corp., Pitts., 1963-70; project dir., mgr. strategic planning-nuclear Westinghouse Electric Corp., 1973-75; v.p. Westinghouse Nuclear Europe, Brussels, 1970-73; dep. dir. FEA, Washington, 1975; dir. div. reactor devel. and demonstration ERDA, Washington, 1976-77; dir. nuclear power devel. Dept. of Energy, Washington, 1977-78; coordinator internat. nuclear study Dept. of Energy, 1978-80; dep. dir. Argonne Nat. Lab., Ill., 1980-84; vis. prof. nuclear engring. MIT, Cambridge, 1984-86; dir. rsch. U.S. Nuclear Regulatory Commn., Washington, 1986—. Author: Boiling Water Reactor Design, 1962; contbr. articles to profl. jours. Committeeman 14th Ward Rep. Com., Pitts., 1968-70, vestryman Calvary Episcopal Ch., Pitts., 1970, 75; vestry St. Alban's Ch., Washington, 1989-93. Lt. (j.g.) USNR, 1951-54. Fellow Am. Nuclear Soc. (v.p. Belgian sect. 1972, chmn. honors and awards 1985); mem. IEEE (sr.), Sigma Xi. Avocations: photography; astronomy. Office: NRC Nuclear Regulatory Rsch Mail Stop NLS007 Rockville MD 20852

BECKLER, DAVID ZANDER, government official, science administrator; b. Detroit, June 29, 1918; s. William J. and Thekla (Levy) B.; m. Harriet Levy, Aug. 1, 1943; children—Stephen, Paul, Rochelle. BSChemE, U. Rochester (N.Y.), 1939; JD, George Washington U., 1943. Bar: D.C. 1942. Patent atty. Pennie, Davis, Marvin & Edmonds, Washington, 1939-42; tech. aide fgn. liaison office Office Sci. R & D, Exec. Office of Pres., Washington, 1942-45; patent atty. Eastman Kodak Co., Rochester, 1946; dep. tech. historian Ops. Crossroads Joint Chiefs of Staff, Washington, 1946; chief tech. intelligence br. R & D Bd., Office of Sec. of Def., Washington, 1947-49; mem. internat. sci. policy survey group Dept. of State, Washington, 1949-50; exec. dir. com. atomic energy R & D Bd., Washington, 1950-52; asst. dir. office indsl. devel. AEC, Washington, 1952-53; exec. officer Pres.'s Sci. Adv. Com., Washington, 1953-73; spl. asst. to dir. Office of Def. Mobilzation, Washington, 1954-57; asst. to spl. asst. to pres. for sci. and tech. The White House, Washington, 1957-62; asst. to dir. Office Sci. and Tech., Exec. Office of Pres., Washington, 1962-73; asst. to pres. NAS, Washington, 1973-76; dir. sci., tech. and industry OECD, Paris, 1976-83; assoc. dir. Carnegie Commn. on Sci., Tech. and Govt., 1988—; cons. sci. and tech. policies, 1983-88. Recipient cert. of appreciation War and Navy Depts., Washington, 1945. Fellow AAAS; mem. Coun. Fgn. Rels., Cosmos Club (Washington). Home: 8709 Duvall St Fairfax VA 22031 Office: Carnegie Commn Sci Tech & Govt 1616 P St NW # 400 Washington DC 20036-1405

BECKMAN, ARNOLD ORVILLE, analytical instrument manufacturing company executive; b. Cullom, Ill., Apr. 10, 1900; s. George W. and Elizabeth E. (Jewkes) B.; m. Mabel S. Meinzer, June 10, 1925; children: Gloria Patricia, Arnold Stone. BS, U. Ill., 1922, MS, 1923; PhD, Calif. Inst. Tech., 1928; DSc (hon.), Chapman Coll., 1965, Whittier Coll., 1971, Clarkson U., 1989, Rockefeller U., 1992; LLD (hon.), U. Calif., Riverside, 1966, Loyola U., L.A., 1969, U. Ill., 1982, Pepperdine U., 1977, Ill. Wesleyan U., 1991; DHL (hon.), Calif. State U., Fullerton, 1984, Ill. State U., 1990. Rsch. assoc. Bell Tel. Labs., N.Y.C., 1924-26; chem. faculty Calif. Inst. Tech., 1926-39; v.p. Nat. Tech. Lab., Pasadena, Calif., 1935-39; pres. Nat. Tech. Lab., 1939-40, Helipot Corp., 1944-58, Arnold O. Beckman, Inc., South Pasadena, Calif., 1946-58; founder, chmn. Beckman Instruments, Inc., Fullerton, Calif., 1940-65, chmn. emeritus, 1988—; vice chmn. SmithKline Beckman Corp., 1984-86; bd. dirs. Security Pacific Nat. Bank, 1956-72, adv. dir., 1972-75, bd. dirs. Continental Airlines, 1956-71, adv. dir., 1971-73.

Author articles in field; inventor; patentee in field. Mem. Pres.'s Air Quality Bd., 1970-74; chmn. System Devel. Found., 1970-88; chmn. bd. trustees emeritus Calif. Inst. Tech.; hon. trustee Calif. Mus. Found.; bd. overseers House Ear Inst., 1981—; trustee Scripps Clinic and Rsch. Found., 1971—; bd. dirs. Hoag Meml. Hosp.; co-founder, bd. dirs. Beckman Laser Inst. and Med. Clinic, 1982—; mem. bd. overseers U. Calif., Irvine, 1982—. With USMC, 1918-19. Benjamin Franklin fellow Royal Soc. Arts; named to Nat. Inventors Hall of Fame, 1987; recipient Nat. Medal Tech., 1988, Presdl. Citizens medal, 1989, Nat. Medal of Sci., 1989, Order of Lincoln award State of Ill., 1991, Bower award for Bus. Leadership, 1992. Fellow Assn. Clin. Scientists; mem. NAM, AAAS, Am. Acad. Arts and Scis., L.A. C. of C. (bd. dir. 1954-58, pres. 1956), Calif. C. of C. (dir., pres. 1967-68), Nat. Acad. Engring. (Disting. Honoree, 1986, Founders Award, 1987), Am. Inst. Chemists (Gold medal 1987), Instrument Soc. Am. (pres. 1952), Am. Chem. Soc., Social Sci. Rsch. Coun., Am. Assn. Clin. Chemistry (hon.), Newcomen Soc., Auto Club So. Calif. (bd. dirs. 1965-73, hon. dir. 1973—), Sigma Xi, Delta Upsilon, Alpha Chi Sigma, Phi Lambda Upsilon. Clubs: Newport Harbor Yacht, Pacific. Office: 100 Academy Dr Irvine CA 92715-3002

BECKMAN, DONALD A., engineer; b. Charleroi, Pa., Feb. 9, 1947; s. Verner Reese and Rosalie Marie (Trozzo) B.; m. Vicki Lynn Cutter, July 17, 1971; 1 child, Erika Leigh. BS, U.S. Merchant Marine Acad., 1969. Engring. officer Nuclear Ship Savannah, N.Y.C., 1969-71; submarine reactor test supr. Newport News (Va.) Shipbuilding, 1971-76; power plant test and ops. supr. Burns & Roe, Inc., Oradell, N.J., 1976-77; engring. mgr., sr. resident inspector U.S. NRC, King of Prussia, Pa., 1977-82; v.p. engring. Energy Cons., Inc., Pitts., 1982-84; v.p., gen. mgr. Shaker Mech. Corp., Cleve., 1984-87; pres., prin. cons. Beckman & Assoc., Inc., Belle Vernon, Pa., 1986—; mem. quality assurance adv. bd. C.C. Alleghany County, Pa., 1980-82. Author, co-author over 200 reports in field. Mem. CAP, Belle Vernon, 1982-92; mem. Mon Valley Progress Coun., Monessen, Pa., 1989-91. Lt. USNR, 1969-76. Mem. Am. Nuclear Soc., Civil Air Patrol, Aircraft Owners and Pilots Assn., Pitts. Aero Club. Achievements include identification of problems with commercial nuclear plant piping and component dynamic analyses which led to major industry safety initiatives.

BECKMAN, JAMES WALLACE BIM, economist, marketing executive; b. Mpls., May 2, 1936; s. Wallace Gerald and Mary Louise (Frissell) B. BA, Princeton U., 1958; PhD, U. Calif., 1973. Pvt. practice econ. cons., Berkeley, Calif. 1962-67; cons. Calif. State Assembly, Sacramento, 1967-68; pvt. practice market rsch. and econs. cons., Laguna Beach, Calif., 1969-77; cons. Calif. State Gov.'s Office, Sacramento 1977-80; pvt. practice real estate cons., L.A. 1980-83; v.p. mktg. Gold-Well Investments, Inc., L.A. 1982-83; pres. Beckman Analytics Internat., econ. cons. to bus. and govt., L.A. and Lake Arrowhead, Calif., 1983—, East European/Middle East Bus. and Govt., 1992—; adj. prof. Calif. State U. Sch. Bus., San Bernardino, 1989—, U. Redlands, 1992—; cons. E European. Contbr. articles on regional & internat. econ. devel. & social change to profl. jours. Maj. USMC 1958-67. NIMH fellow 1971-72. Fellow Soc. Applied Anthropology; mem. Am. Econs. Assn., Am. Statis. Assn., Am. Mktg. Assn. (officer), Nat. Assn. Bus. Economists (officer). Democrat. Presbyterian. Home: PO Box 1753 Lake Arrowhead CA 92352-1753

BECKMAN, JOSEPH ALFRED, research and development administrator; b. Macomb, Ill., Oct. 30, 1937; s. Alfred Jacob and Mary Jeanette (Botts) B.; m. Peggy Ann Miller, Feb. 1, 1938; children: Bruce, Jill. AB in Chemistry, Western Ill. U., 1960; PhD in Organic Chemistry, Iowa State U., 1965. Various research positions The Firestone Tire & Rubber Co., Akron, Ohio, 1964-74; mgr. elastomer research The Firestone Tire & Rubber Co., Akron, 1974-80, asst. dir. research, 1980-87; v.p. research & devel. DSM Copolymer, Inc., Baton Rouge, 1987—; mem. Indsl. Rsch. Adv. Com. La. Bd. Regents, Baton Rouge, 1989-91, Indsl. Adv. Com. Adv. Com. Dept. Polymer Sci., U. So. Miss., Hattiesburg, 1989-93. Patentee in field. Mem. Am. Chem. Soc., Akron Rubber Group. Republican. Avocations: gardening, golf, woodworking, reading, tropical fish. Home: 2925 Dakin Ave Baton Rouge LA 70820-4434 Office: DSM Copolymer Inc PO Box 2591 Baton Rouge LA 70821-2591

BECKMEYER, HENRY ERNEST, anesthesiologist, medical educator; b. Cape Girardeau, Mo., Apr. 13, 1939; s. Henry Ernest Jr. and Margaret Gertrude (Link) B.; m. Suzanne Samberg, June 24, 1961 (div. Nov. 1982); children: Henry, James, Martha; m. Barbara Truitt, Feb. 12, 1983; children: Leigh, Hillary. BA, Mich. State U., 1961; DO, U. Health Scis., 1965. Diplomate Am. Bd. Med. Examiners, Am. Osteo. Bd. Anesthesiology, Am. Acad. Pain Mgmt. Chief physician migrant worker program and op. head start Sheridan (Mich.) Community Hosp., 1967-69; resident in anesthesia Bi-County Community Hosp./DOH Corp., Detroit, 1969-71, chief resident, 1968-69; staff anesthesiologist Detroit Osteo. Hosp./BCCH, 1971-75; founding chmn. dept. anesthesia Humana Hosp. of the Palm Beaches, West Palm Beach, Fla., 1975-79; assoc. prof. Mich. State U., East Lansing, 1979-88, prof. anesthesia, 1988—, chmn. dept. osteo. medicine, 1985—; chief staff Mich. State U. Health Facilities, 1988-90, chmn. med. staff exec. and steering coms., 1988-90; chmn. of anesthesia St. Lawrence Hosp., Lansing, Mich., 1984-90; chief of staff Sheridan Community Hosp., 1968-69; mem. adminstrv. coun. Mich. State U., 1988—, mem. acad. coun., 1992—, mem. faculty coun., 1992—, mem. clin. practice bd., bd. dirs. sports medicine; mem. internal mgmt. com. Mich. Ctr. for Rural Health; cons. Ministry Health, Belize C.A., 1993—; amb. Midwestern Univ. Consortium Internat. Activities, 1993; program chmn. Am. Russian Med. Exch., 1993—; bd. dirs. Belize Med. Partnership. Speaker Sta. WKAR, Mich. State U.; bd. dirs. Boy Scouts Am., W. Bloomfield, Mich., 1973-74, Palm Beach Mental Health, 1977-79, Care Choices HMO, Lansing, 1987-88. Fellow Am. Coll. Osteo. Anesthesiologists; mem. Am. Osteo. Coll. Anesthesiology (chmn. commn. on colls. 1988-89, cert. anesthesiology 1976), Soc. Critical Care Medicine, Internat. Anesthesiology Rsch. Soc., Am. Coll. Physician Execs., Am. Osteo. Assn. (speaker), Am. Acad. Pain Mgmt. (cert. 1991—), Am. Arbitration Assn., Mich. Pain Soc., Mich. Peer Review Orgn., Am. Soc. Regional Anesthesia, Soc. Security Disability Evaluation, Univ. Club. Republican. Avocations: reading, family, exercise, music. Office: Mich State U West Fee Hall East Lansing MI 48824

BECKWITH, CATHERINE S., veterinarian; b. St. Louis, Apr. 8, 1958; d. John P. Sr. and Dolores A. Beckwith. BA in Germanic Langs. & Lit., U. Ill., 1981, BS in Vet. Medicine with honors, 1984, DVM with honors, 1986. Assoc. vet. Cen. Hosp. for Animals, Carterville, Ill., 1986-87, Coble Animal Hosp., Springfield, Ill., 1987-91, Northgate Pet Clinic, Decatur, Ill., 1991-92; postdoctoral fellow in lab. animal medicine U. Mo., Columbia, 1992—. Contbr. articles to profl. jours. Fulbright grantee Tech. U., Munich, 1981. Mem. Am. Assn. for Lab. Animal Sci., Am. Soc. for Lab. Animal Practitioners, Am. Vet. Med. Assn. Avocations: skiing, swimming, running. Home: 406A Southampton Dr Columbia MO 65203 Office: Office Lab Animal Medicine M144 Med Sci Bldg Columbia MO 65212

BECUWE, IVAN GERARD, aeronautical engineer; b. Veurne, Flanders, Belgium, Feb. 28, 1967; s. Roland and Christiane (Decdodt) B. Indsl. engr. electromechanics, KIHWV, Belgium, 1990; MSc in Aeronautics, Ghent U., Belgium, 1991. Engr. trainee Trans European Airways, Brussels, 1989-91; rsch. engr. Kingston U., Kingston upon Thames, U.K., 1992; maintenance instr. European Aviation Tng. Ctr., Brussels, 1992—. Mem. AIAA, Instn. of Flemisch Engrs., Belgian Inst. of Automatic Control (working group in aviation automation). Home: St Elisabethlane 6, B-8660 De Panne Belgium Office: European Aviation Tng Ctr, Bldg 119 Bloc 3, B-1820 Melsbroek Belgium

BEDAPUDI, PRAKASH, aerospace engineer; b. Chittoor, Andhra, India, Aug. 25, 1966; came to U.S. 1989; s. Bhaskar Naidu and Ambujakshi (Bhaskar) B. BS in Auto. Engring., Karnatak U., India, 1987; MS in Aerospace Engring., U. Cin., 1991. Mktg. engr. Avanti KoppElec. Ltd., Hyderabad, India, 1986-87; project engr. Escorts India Ltd., New Delhi, 1987-89; project engr. Cummins Engine Co., Inc., Columbus, Ind., 1990, sr. performance devel. engr., 1991—; rsch. asst. U. Cin., 1989-91; cons. Escorts Rsch. Ctr., New Delhi, 1989—. Author tech. reports on high pressure injection systems. U. Cin. grad. scholar, 1989-91. Mem. ASME, AIAA, India Student Assn., Assn. Auto. Engrs. India (pres. 1987). Achievements include design and development of lean burn combustion system for 2-stroke gasoline engines; designed fuel manifold which solved the start of injection variation problem of a 6-cylinder DI diesel engine. Home: 490 Oakbrook Dr Columbus IN 47201 Office: Cummins Engine Co Inc 1900 McKinley Dr Columbus IN 47203

BEDEAUX, DICK, chemist; b. The Hague, The Netherlands, Sept. 19, 1941; s. Theodoor Marie and Geertje (Lambo) B.; m. Tonia Maria Michon, July 9, 1965; children: Anouk Janneke, Yana Marike, Dietske Geertien, Zenda Manoesja. M in Physics, U. Utrecht, The Netherlands, 1964, PhD, 1969. Postdoctoral fellow U. Calif., Chemistry Dept., San Diego, 1969-71; vis. prof. Wolfgang Goethe U., Inst. Theoretical Phys. Chemistry, Frankfurt, Germany, 1971-72; rsch. assoc. U. Leiden, Inst. Lorentz for Theoretical Physics, The Netherlands, 1972-84, MIT, Dept. Chemistry, Cambridge, Mass., 1976-77; prof. physics N.T.H., Inst. Theoretical Physics, Trondheim, Norway, 1981-83; prof. phys. chemistry dept. macromolecular & phys. chemistry Gorlaeus Labs., U. Leiden, The Netherlands, 1984—. Contbr. articles to profl. jours. Grantee NATO, 1976-79, 89—; recipient Twinning grant EC, 1991-94. Mem. Am. Phys. Soc., European Phys. Soc., Dutch Phys. Soc., Royal Dutch CHem. Soc., IUPAC. Avocations: skating, cross country skiing. Home: Burg Colijnstraat 175, 2771 GL Boskoop The Netherlands Office: Dept Phys & Macromol Chem, PO Box 9502, 2300 RA Leiden The Netherlands

BEDFORD, ANTHONY JOHN, defense science executive; b. Victor Harbor, Australia, July 15, 1943; s. Harold Herbert and Winifred Mary (Limb) B.; m. Margaret Elizabeth Mumme, Jan. 6, 1966; children: Julie Michelle, Susan Elizabeth, David Ronald. B Applied Sci., U. Adelaide, Australia, 1966, BApp.Sci with honors, 1967, PhD, 1971. Cert. profl. engr. Tech. asst. Weapons Rsch. Establishment, Salisbury, South Australia, 1961-63; cadet defence sci. U. Adelaide, South Australia, 1963-67, post grad. student, 1967-70; rsch scientist, sr. rsch. scientist Materials Rsch. Lab., Melbourne, Victoria, Australia, 1971-76; prin. rsch. scientist Defence Sci. & Tech. Orgn., Canberra, Australia, 1976-80; group head Materials Rsch. Lab., Melbourne, 1980-86; counsellor defence sci. Embassy of Australia, Washington, 1986-89; chief explosives div. Materials Rsch. Lab., 1989-91; chief optoelectronics div. DSTO, Salisbury, 1991—. Contbr. articles and reports to sci. jours. Fellow Instn. Engrs. (Australia). Avocations: sailing, golf, tennis, swimming. Address: COED/SRL/DSTO, PO Box 1500, Salisbury 5108, Australia

BEDFORD, KEITH WILSON, civil engineering and atmospheric science educator; b. Schenectady, N.Y., May 5, 1945; s. Alexander Wilson and Elsie Maude (Flickinger) B.; m. Marilyn Kay Bettoney, Aug. 18, 1972; children: Nathaniel Keith, Hilary Alexis. BSME with honors, Union Coll., 1969, MSME, 1971; PhD, Cornell U., 1974. Rsch. asst. Union Coll., 1968-70; rsch. fellow EPA Cornell U., 1970-73, teaching asst. hydaulics/hydrology lab., 1973; asst. prof. civil engring. Ohio State U., 1973-78, assoc. prof., 1978-83, prof. civil engring. and atmospheric sciences, 1983—; fellow Coop. Inst. Limnology and Ecosystem Rsch., NOAA Great Lakes Environ. Rsch. Lab/U. Mich., 1990—; rsch. cons. U.S. Fish and Wildlife Svc., 1977-78, Geotechnics, Inc., 1980-82, U.S. Army Corps. Engrs., 1982-83, 89-90, HRC Corp., 1984-86, UN, 1985-87, Camp, Dresser and McKee, 1991—, others. Contbr. articles to profl. jours. Fed. Water Quality Adminstrn. Rsch. fellow Cornell U., 1970-71, EPA Rsch. fellow, 1971-73. Mem. ASCE (Outstanding Svc. award 1976, 77, Huber prize 1986, Karl Emil Hilgard prize 1989, Review Article of Yr. award hydraulics divsn. 1991), Am. Geophysical Union, Internat. Assn. Great Lakes Rsch. (bd. dirs.), Internat. Assn. Hydraulic Rsch., Sigma Xi, Tau Beta Pi, Chi Epsilon. Unitarian. Achievements include development of Great Lakes Forecasting System, Acoustic Resuspension Measurement System, HANDS high frequency nondestructive particle sizer, Dynamic Water Quality Planning Model; research includes measurements of erosion rates and scour, closure free turbulence models. Office: Ohio State U Dept Civil Engring 2070 Neil Ave Columbus OH 43210

BEDNARZ, JAMES C., wildlife ecologist educator. BS in Wildlife and Fishery Biology, N.Mex. State U., 1976; MS in Animal Ecology, Iowa State U., 1979; PhD in Biology, U. N.Mex., 1986. Wildlife biologist U.S. Fish and Wildlife Svc., 1977-78; wildlife cons., 1979-80, 91—; rsch. assoc. U. N.Mex., 1985-91; dir. higher edn. and rsch. Hawk Mountain Sanctuary Assn., 1987-90; prin. investigator Greenfalk Cons., 1990-91; asst. prof. wildlife ecology Ark. State U., State Univ., 1993—; rsch. asst. part-time Iowa State U., 1977-78; invited mem. Eastside Forests Scientific Panel; referee Jour. Mammalogy, Wilson Bull., Raptor Rsch., NSF, and others. Contbr. articles to profl. jours. Recipient Rsch. award Hawk Mountain, Marcia Brady Tucker Travel award; Ding Darling scholar. Am. Ornithologists' Union, Animal Behavior Soc., Wildlife Soc., Cooper Ornithological Soc., Wilson Ornithological Soc., Raptor Rsch. Found., Acad. Nat. Scis. Phila. (assoc. in ornithology), Hawk Watch Internat. (rsch. adv. com.), Sigma Xi. Achievements include research in conservation biology, the influence on human disturbance on wildlife populations, the effects on habitat alterations on wildlife populations, social behavior of birds and mammals, evolution of mating systems, the influence of ecology on animal social systems, migratory strategies of birds, and the application of basic ecological and evolutionary principles to conservation and wildlife management. Office: Ark State U Dept Biological Scis State University AR 72467

BEDNORZ, J. GEORG, crystallographer; b. May 16, 1950. Grad., U. Munster, Fed. Republic of Germany, 1976; PhD, Swiss Federal Inst. Tech., ETH Zurich, 1982. Researcher IBM Zurich Rsch. Lab., Ruschlikon, Switzerland, 1982—; lectr. Swiss Fed. Inst. Tech. and U. Zurich, 1987—. Co-recipient Thirteenth Fritz London Meml. award, 1987, Nobel Prize in physics Royal Swedish Acad. Soc., 1987; recipient Dannie Heineman prize Minna James Heineman Stiftung, Acad. Scis. Gottingen, Fed. Republic of Germany, 1987, Robert Wichard Pohl prize German Phys. Soc., 1987, Hewlett-Packard Europhysics prize, 1988, Marcel-Benoist prize Marcel-Benoist Found., 1986, APS Internat. prize for new materials research, 1988, Viktor Moritz Goldschmidt prize German Mineralogical Soc., 1987, Otto-Klung prize Otto-Klung Found., 1987. Office: IBM Zurich Rsch Lab, Saumerstrasse 4, CH-8803 Ruschlikon Zurich Switzerland

BEDOYA, MICHAEL JULIAN, veterinarian; b. Mexico City, June 11, 1945; s. Carlos and Irene (Stabenow) B.; m. Gladys Cervera, Sept. 25, 1970; children: Sandra, Julian. DVM, Universidad Nacional Autonoma de Mex., Mexico City, 1969; PhD, U. Ill., 1983. Diplomate Vet. State Medicine/Applied Vet. Pathology. Dir. Regional Diagnostic Lab./Sec. Agr., Queretaro, Mex., 1969-70; tech. dir. Vet. Diagnostic Labs. Network/Sec. Agr., Mexico City, 1972-77; dir. Nat. Animal Health Ctr./Sec. Agr., Mexico City, 1978-79; tech. co-dir. Mex. Am. Foot and Mouth Disease Commn., Mexico City, 1983-85; animal health cons. Interam. Inst. Cooperation Agr., Brasilia, Brazil, 1985-92, Andean Region, Ecuador, 1992—; temp. cons. Pan-Am. Health Orgn./WHO, Cen. Am., 1984, Interam. Inst. Cooperation in Agr., Cen. Am., 1976-77; part-time prof. Universidad Nacional Autonoma de Mex., 1972-78. Contbr. articles to profl. jours. Fellow Brit. Coun., U. Edinburgh, 1970, FAO/Swedish Internat. Devel. Agy., Royal Vet. Coll. Stockholm, 1974. Office: Q 1 05 Cj 9 Bl D, Mariana de Jesus 197 4, 71600 Lapradera Quito Ecuador

BEDRICK, BERNICE, retired science educator, consultant; b. Jersey City, Sept. 29, 1916; d. Abraham Lewis and Esther (Cowan) Grodjesk; m. Emanuel Arthur Bedrick, Dec. 25, 1938 (dec. 1967); children: Allen Paul, Jane Bedrick Abels; m. Samuel Milberger, Sept. 23, 1984 (dec. 1984); stepchildren: Susan Milberger Rafael, Stanford. BS, U. Md., 1938; MA, NYU, 1952. Cert. tchr., N.J. Tchr. Linden (N.J.) Pub. Sch. System, 1950-69, supr. sci. curriculum, 1969-79, sch. prin., 1979-87; ret., 1987. Co-author: A Universe to Explore, 1969; developer program of safety and survival N.J. Dept. Edn., 1975. Founder, mem. Temple Mekor Chayim, Linden; pres. bd. trustees Linden Pub. Libr., 1989-90, v.p., 1991; pres. Friends of Linden Libr., 1987-92. Mem. NEA (life), N.J. Edn. Assn. (life), Am. Fedn. Sch. Adminstrs. (chpt. pres. 1984-86), Linden Edn. Found. (bd. dirs.), N.Y. Acad. Scis., N.J. Prins. and Suprs. Assn., N.J. Sci. tchrs. Assn., Nat. Sci. Tchrs. Assn., Alumni Assn. U. Md. (life), N.J. PTA (life), Hadassah (life), Linden Ceramics Club (sec. 1991-92), Nat. Coun. Jewish Women (life), Alpha Lambda Delta, Phi kappa Phi. Home: 2016 Orchard Terr Linden NJ 07036-3719

BEDRIJ, OREST J., industrialist, scientist; b. Ukraine, May 24, 1933; arrived in U.S., 1949, naturalized, 1955; s. Eustachy and Olga (Banach) B.; m. Oksana Cymbalista, Nov. 10, 1956; children: Orest W., Roksana, Chrystyna Bedrij Stecyk. BSEE, Rochester Inst. Tech., 1956, MS in Humanities; PhD in Theoretical Physics, Columbia Pacific U., 1986. Various mgmt. positions IBM Corp., Poughkeepsie, N.Y. and Los Angeles, 1956-68; IBM tech. dir. Space Flight Facility Jet Propulsion Lab., Calif. Inst. Tech., 1962-63; pres., dir. Securities Council, Inc., 1965-83; founder, pres., dir. Profit Technology, Inc., 1983-89, Griffin Capital Mgmt. Corp., N.Y.C., 1989—; founder, dir. Advance Memory Systems Inc., now with G.E. as Intersil, Inc., Sunnyvale, Calif., 1968, Xytex Corp. (merged with Calcomp Corp.), Boulder, Colo., 1970, Internat. Jour. Nonlinear Math. Physics, Kyyiv, 1992; mem. exec. com., treas., dir. Ukrainian Studies Fund, Harvard U., 1959-72, adviser Ctr. for the Study of World Religions, Harvard U., 1991—; trustee, treas. John E. Fetzer Found., 1987-89. Author: Yes It's Love: Your Life Can Be a Miracle, 1974, One, 1977, 2d rev. edit., 1978, You, 1989; contbr. articles to profl. jours.; patentee in field. With U.S. Army (Res.), 1954-60. Recipient Outstanding Contribution award IBM, 1967. Mem. AAAS, Royal Soc. Arts Mfg. and Commerce, N.Y. Acad. Arts and Scis., Inst. Noetic Scis., Found. Econ. Edn. Achievements include research in physics of the Absolute Principal, determination of fundamental constants in quantum electrodynamics and nucleon-meson dynamics, and on the mathematics of parallel addition in computers. Office: Griffin Captial Mgmt 200 Rector Pl 41st Fl New York NY 10280-1176

BEECHER, STEPHEN CLINTON, physicist; b. Dansville, N.Y., Apr. 15, 1954; s. Hugh Charles and Rita Doris (Lemen) B. BS, SUNY, Geneseo, 1983; PhD, U. Del., 1990. Teaching asst. U. Del. Dept. Physics, Newark, 1983-90; post doctoral fellow U. Del., 1990-92; rsch. scientist Oak Ridge (Tenn.) Nat. Lab., 1992—. Contbr. articles to profl. jours. With USN, 1973-79. Postdoctoral fellow U. Del., 1990-92. Mem. Am. Phys. Soc., Am. Ceramic Soc., Materials Rsch. Soc. Democrat. Roman Catholic. Office: Oak Ridge Nat Lab Bldg 4515 MS 6064 Oak Ridge TN 37831

BEEHLER, BRUCE MCPHERSON, research zoologist, ornithologist; b. Baltimore, Md., Oct. 11, 1951; s. William Henry Jr. and Cary (Baxter) B.; m. Carol Hare, June 7, 1982; children: Grace Bryant, Andrew McPherson. BA, Williams Coll., 1974; MA, Princeton U., 1978, PhD, 1983. Sci. asst. to sec. Smithsonian Instn., Washington, 1981-84, sci. asst. to sec. emeritus, 1984-88, zoologist, 1988-91; assoc. rsch. zoologist N.Y. Zool. Soc., Washington, 1991—; mem. expdns. to Papua New Guinea, 1975-76, 78-80, 81, 82, 83, 84, 86, 87, to India, 1983, 85, 86, 88; rsch. assoc. dept. vertebrate zoology Nat. Mus. Natural History, 1985—. Author: Birdlife of the Adirondack Park, 1978, Upland Birds of Northeastern New Guinea, 1978, A Naturalist in New Guinea, 1991; sr. co-author: Birds of New Guinea; contbr. articles to sci. jours. Thomas J. Watson Found. fellow, 1974; Nat. Geog. Soc. rsch. grantee, 1980, 86, 89, N.Y. Zool. Soc. rsch. grantee, 1986. Mem. Am. Ornithologists Union (elective). Democrat. Avocations: hiking, tennis.Co-discoverer with John P. Dumbacher of the Pitohui, a poisonous genus of bird that uses as a chemical defense a powerful homobatrachotoxin. Home: 6421 Broad St Bethesda MD 20816-2607 Office: Smithsonian Instn Div of Birds Washington DC 20560

BEEM, JOHN KELLY, mathematician, educator; b. Detroit, Jan. 24, 1942; s. William Richard and June Ellen (Kelly) B.; m. Eloise Masako Yamamoto, Mar. 24, 1964; 1 child, Thomas Kelly. A.B. in Math., U. So. Calif., 1963, M.A. in Math., 1965, Ph.D. in Math., 1968. Asst. prof. math. U. Mo., Columbia, 1968-71, assoc. prof., 1971-79, prof., 1979—. Co-author: (with P. Y. Woo) Doubly Timelike Surfaces, 1969, (with P. E. Ehrlich) Global Lorentzian Geometry, 1981; condr. research in differential geometry and gen. relativity. NSF fellow, 1965, 68. Mem. Math. Assn. Am., Am. Math. Soc., Phi Beta Kappa. Home: 5204 E Tayside Circle Columbia MO 65203-5191

BEEMAN, CURT PLETCHER, research chemist; b. Mt. Vernon, Ohio, May 30, 1944; s. Milton Henry and Virginia (Hartsel) B.; m. Elisabeth Anne Bartlett, Apr. 16, 1966; children: Anne Elisabeth, Christopher Pletcher. BS, U. Fla., 1967; MS, Auburn U., 1969, PhD, 1971. Postdoctoral fellow U. Alta., Edmonton, Can., 1971-73; rsch. assoc. Mt. Holyoke Coll., South Hadley, Mass., 1973-77, DWQRC Fla. Internat. U., Miami, 1977-79; sr. rsch. scientist Internat. Paper Co., Mobile, Ala., 1979—. Contbr. articles to profl. jours. Mem. Am. Chem. Soc., Sigma Xi, Phi Lambda Upsilon. Office: Internat Paper Co PO Box 2787 Mobile AL 36652

BEENE, KIRK D., systems engineer; b. Detroit, Sept. 19, 1967; s. Kenneth Dwight and Sharon Elaine (Vaughn) B.; m. Kristine Dawn Sheehy, Mar. 28, 1992. BS in Aerospace Engring., Va. Tech., 1991. Systems engr. DDL Omni Engring. Corp., McLean, Va., 1986—. Contbr. articles to profl. jours. Republican. Baptist. Office: DDL Omni Engring Corp Ste 600 8260 Greensboro Dr Mc Lean VA 22102

BEETON, ALFRED MERLE, laboratory director, limnologist, educator; b. Denver, Aug. 15, 1927; s. Charles Frederick and Edna F. (Smith) B.; m. Mary Eileen Wilcox, July 20, 1945; children—Maureen Ann, Heather Ann, Celeste Nadine; m. Ruth Elizabeth Holland, June 4, 1966; children—Jonathan Eugene, Daniel Paul. B.S., U. Mich., 1952, M.S., 1954, Ph.D., 1958. Fishery biologist U.S. Bur. Comml. Fisheries, Ann Arbor, Mich., 1957-65; chief environ. research U.S. Bur. Comml. Fisheries, 1960-65; prof. zoology U. Wis.-Milw., 1965-76; asst. dir. U. Wis.-Milw. (Center for Gt. Lakes Studies), 1965-69, assoc. dir., 1969-73; assoc. dean U. Wis.-Milw. (Grad. Sch.), 1973-76; dir. Gt. Lakes and Marine Waters Center; prof. U. Mich., Ann Arbor, 1976-86; dir. Gt. Lakes Environ. Research Lab., Nat. Oceanic and Atmospheric Adminstrn. Dept. Commerce, Ann Arbor, 1986—; lectr. biology Wayne State U., 1957-61; lectr. civil engring. U. Mich., 1961-65; mem. Mich. Toxic Substance Control Commn., 1987-89; U.S. chmn. Sci. Adv Bd. Internat. Joint Commn., 1986-91; mem. research adv. council Wis. Dept. Natural Resources; mem. water quality criteria com. Nat. Acad. Scis.; cons. U.S. Army C.E., 1967-73, Met. San. Dist. Chgo., 1968-76, EPA, 1973-83; adviser to Smithsonian Instn. on projects in, Ghana, Laos, Yugoslavia, 1972-82; to WHO/Pan Am. Health Orgn. in, Venezuela, 1978; mem. environ. studies bd. NRC, 1976-82, internat. environ. program com., 1977-82. Contbr. chpts. to books; articles Ency. Brit. Mem. Internat. Assn. Theoretical and Applied Limnology, Am. Soc. Limnology and Oceanography (treas. 1962-81), Internat. Assn. Gt. Lakes Research, Mich. Acad. Sci., Arts and Letters. Home: 2761 Oakcleft St Ann Arbor MI 48103-2247 Office: 2205 Commonwealth Blvd Ann Arbor MI 48105-1593

BEEVER, JAMES WILLIAM, III, biologist; b. Balt., Aug. 17, 1955; s. James William Jr. and Virginia Irene (Ruhlmann) B.; m. Lisa Britt Dodd, May 26, 1990. BS, Fla. State U., 1977, MS, 1979; postgrad., U. Calif., Davis, 1991. Environ. specialist Fla. Dept. of Environ. Regulation, Ft. Myers, 1984-88; resource mgmt. and rsch. coord. South West Fla. Aquatic Preserves, Bokeelia, Fla., 1988-90; biol. scientist III Fla. Game and Fresh Water Fish Commn., Punta Gorda, 1990—; bd. dirs. Calusa Land Trust and Nature Preserve, Bokeelia, 1989—, Agy. on Bay Mgmt., Tampa, Fla., 1990—; tech. adv. bd. Sarasota Bay and Tampa Bay Nat. Estuary Program, Sarasota, 1989—; expert witness in field, 1986—. Author: Lemon Bay Aquatic Preserve Management Plan, 1988, (computer database) Resource Inventory of Species in S.W. Fla., The Cedar Point Syudy, 1992; contbr. articles to profl. jours. Com. mem. G.A.C. Internal Improvement Trust Fund Com., Ft. Myers, 1988—; chair Grad. Student Assn., Davis, 1981-83. Regents fellowship U. Calif, 1983-84; recipient Grad. Rsch. award, 1982-83, Outstanding Profl. Achievements award Fla. DNR, 1989. Mem. Fla. Acad. Sci., Estuarine Rsch. Fedn., Soc. Wetland Scientists, Soc. for Conservation Biology, Ecol. Soc. Am., Phi Beta Kappa, Sigma Xi. Achievements include research on mangrove tree crab and arboreal folivore, mangrove cutting, endangered species protection, red cockaded wood peckers; regional wildlife habitat and wildlife corridor planning; involved in preserve designation and acquisition of Florida ecosystems. Office: Fla Game & Freshwater Fish 29200 Tuckers Grade Punta Gorda FL 33955-2207

BEEVER, LISA BRITT-DODD, environmental planner; b. Alton, Ill., Apr. 16, 1960; d. Ralph Everett and Martha Guinilda (Ebbersten) D.; m. James William Beever III, May 26, 1990. BS in Landscape Architecture, Tex. A&M, 1982, PhD, 1987; MLA, N.C. State U., 1983. Registered landscape architect; cert. planner. Landscape designer Dave Bost Group, Round Rock, Tex., 1983-84; teaching asst. Tex. A&M U., College Station, 1984-85; landscape planner Richardson-Verdoorn, Austin, Tex., 1985; planner Austin Parks and Recreation Dept., 1985-88; prin. planner divsn. planning Lee County, Ft. Myers, Fla., 1988-89; dir. environ. scis. Lee County, Ft. Myers, 1989-92; dir. Charlotte County-Punta Gorda (Fla.) Met. Planning Orgn., 1993—. Author several environ. regulations, Lee County Wildlife Corridor Plan; contbr. articles to profl. jours. Mem., vol. landscape architect Calusa Land Trust, Pine Island, Fla., 1990—; vol. Children's Sci. Ctr., N. Ft. Myers, Fla., 1992, Profl. Placement Network, Ft. Myers, 1993. Recipient County Achievement award Nat. Assn. Counties, 1990, 91, 92, award of excellence Fla. Planning Assn. Mem. Am. Planning Assn., Fla. Native Plant Soc., Am. Inst. Cert. Planners, Fla. Acad. Scis. (chair urban and regional planning sect. 1991—). Democrat. Unitarian. Avocations: gardening, art, travel, dachshunds, hiking. Home: 306 Little Grove Ln Fort Myers FL 33917 Office: Charlotte County-Punta Gorda Met Planning Orgn 28000 Airport Rd A-6 Punta Gorda FL 33982-2411

BEEVER, WILLIAM HERBERT, chemist; b. St. Joseph, Mo., May 14, 1952; s. Herbert Henry and Mildred Peterson (Tootle) B.; m. Deborah Jo McDevitt, Sept. 1, 1972; children: Matthew W., Lara D., Taira B. PhD, Colo. State U., 1979. From rsch. chemist to sect. supr. Phillips Petroleum Co., Bartlesville, Okla., 1980-92; technical mgr. Phillips Petroleum Co., Bartlesville, 1992—. Contbr. articles to profl. jours. Elder Grace Bapt. Ch., Bartlesville, 1988—. Mem. Am. Chem. Soc., Soc. Plastics Engrs. Republican. Achievements include patents in pultrusion of thermoplastics and polymer blends and alloys. Office: Bldg 93-G PRC Phillips Petroleum Co Bartlesville OK 74004

BEEZHOLD, DONALD HARRY, immunologist; b. Chgo., Oct. 17, 1955; s. William Phillip and Harriet Ellen (Kramer) B.; m. Elizabeth Wilma Verwoerd, July 2, 1977; children: Brian James, Eric William, Kevin John. MS, U. Ill. Med. Ctr., 1979, PhD, 1981. Postdoctoral fellow Med. Coll. of Ga., Augusta, 1981-82, instr. anatomy, 1982-84, asst. prof., 1984-88; asst. scientist Guthrie Rsch. Inst., Sayre, Pa., 1988-92, assoc. scientist, 1992—. Contbr. articles to profl. jours. Mem. ASTM, Leukocyte Biology, Am. Assn. Immunologists, N.Y. Acad. Scis., Internat. Endotoxin Soc. Office: Guthrie Rsch Inst 1 Guthrie Sq Sayre PA 18840

BEG, MIRZA UMAIR, toxicologist; b. Lucknow, U.P., India, July 8, 1946; s. Abdul Aziz Beg and Mumtaz Jehan Begum; m. Naheed, Aug. 5, 1970; children: Uzma, Hina, Mansoor. BSc, Lucknow U., 1963, MSc in Biochemistry, 1965, PhD in Biochemistry, 1970. Sr. rsch. fellow Coun. Sci. and Indsl. Rsch., Delhi, 1970-73; scientist B Indsl. Toxicology Rsch. Ctr., Lucknow, 1973-78, scientist C, 1978-84, asst. dir., 1984-90; founder dir. Dubai (United Arab Emirates) Inst. Environ. Rsch., 1990—. Author: Toxicology Map of India, 1990; contbr. articles to profl. jours. WHO fellow, Geneva, 1979-80; rsch. grantee Indian Coun. Med. Rsch., 1986-90, Coun. Sci. and Tech., 1987-90. Mem. AAAS, Indian Soc. Agr. Biochemists (v.p. 1989-91), Soc. Toxicology India (life), Soc. Biol. Chemists (India). Avocations: short story writing, popularisation of science through mass media. Home: 50 Mohammed Ali Ln, Lucknow 226018, India Office: Dubai Inst Environ Rsch, PO Box 19099, Dubai United Arab Emirates

BEGELMAN, MITCHELL C., astrophysicist, educator. AB, AM, Harvard U., 1974; PhD, U. Cambridge (Eng.), 1978. Asst. prof. dept. astrophys., planetary and atmospheric scis. U. Colo., Boulder, 1982-87, assoc. prof., 1987-91, prof., 1991—, fellow Joint Inst. for Lab. Astrophysics, 1984—. Recipient Presdl. Young Investigator award, 1984, Helen B. Warner prize Am. Astron. Soc., 1988; Alfred P. Sloan Found. rsch. fellow, 1987-91. Office: U Colo Joint Inst Lab Astrophysics Campus Box 440 Boulder CO 80309

BEGENISICH, TED BERT, physiology educator; b. Sacramento; s. Robert and Phyllis (Richmond) B. BS in Physics, U. Calif., Davis, 1968; MA in Physics, U. Calif., Irvine, 1971; PhD in Biophysics, U. Md., Balt., 1974. Engr. North Am. Rockwell, Anaheim, Calif., 1968-70; postdoctoral fellow U. Wash., Seattle, 1974-75; asst. prof. physiology U. Rochester (N.Y.), 1975-81, assoc. prof., 1981—, prof., 1993—; cons. Merck Sharp & Dohme, West Point, Pa., 1988-89; editorial bd. Biophysical Jour., 1981-87, Jour. Gen. Physiology, 1990—. Recipient Career Devel. award Pub. Health Svc., 1978-83, rsch. grants NIH, 1978—. Mem. Biophysical Soc., Soc. for Gen. Physiology, Soc. for Neuroscience. Office: U Rochester Physiology Box 642 Rochester NY 14642-8642

BEGGS, JAMES HARRY, electrical engineer; b. Ft. Wayne, Ind., Nov. 12, 1940; s. Harry George and Louise Ann (Simminger) B.; m. Ann Pinckney Buek, Sept. 12, 1970; children: Robert Thomas, Alice Louise. B. Engring., AS, Yale U., 1966. Registered profl. engr., Nev. Mem. tech. staff Sanders Assocs., Nashua, N.H., 1966-70; v.p. Integrated Systems Co. Inc., Las Vegas, Nev., 1970-74; chief engr. Integrated Systems div. Robertshaw Controls, Las Vegas, 1974-89; cons. SESCOM, Inc., Henderson, Nev., 1989-90; pres. Beggs Assocs., Inc., Las Vegas, 1990—. Chair diaconate 1st Congregational Ch., Las Vegas, 1975, chair trustees, 1985, moderator, 1990. With USMC, 1961-64. Mem. IEEE (chmn. 1983), ASHRAE, Nat. Fire Protection Assn. Republican. Achievements include patents in Two-way AC Powerline Communication System, (co-author), claims a system using 1974 techniques below (United Kingdom); Noise Trap Arrangement for AC Comm System, claims a noise-reduction technique for the above patent; Powerline Carrier System (co-author) claims digital communication on neutral-ground path. Home: 2012 N Parkway Las Vegas NV 89106-4819 Office: Beggs Assocs Inc 2012 N Parkway Las Vegas NV 89106-4819

BEGLE, DOUGLAS PIERCE, ichthyologist; b. New Haven, July 22, 1958; s. Edward Griffith and Elsie Alkin (Pierce) B.; m. Christine Marie Garzon Martinez, Oct. 14, 1989. BS, Stanford U., 1980; PhD, U. Mich., 1991. Lectr. U. Mich., Ann Arbor, 1989-90; research ichthyologist Smithsonian Instn., Washington, 1991. Contbr. articles to profl. jours. Walker fellow, 1988, Hinsdale fellow, 1989, fellow Smithsonian Instn., 1991, U. Mich., 1990. Mem. Soc. Systematic Biology, Am. Soc. Ichthyologists and Herpetologists (Stoye award 1988), Willi Henning Soc. Achievements include providing the first empirical analysis of reductive evolution in fishes, theoretical framework for other evolutionary studies of paedomorphosis, first evolutionary analysis of eye evolution in deep-sea fishes. Home: 3621 Newark St NW # 306 Washington DC 20016 Office: Nat Mus Natural History NHB WG-12 Washington DC 20560

BEGLEITER, HENRI, psychiatry educator; b. Nimes, France, Sept. 11, 1935; married, 1963; 2 children. PhD in Phychophysiology, New Sch. Social Rsch., 1967. Rsch. assoc. neurophysiology SUNY, 1964-66, asst. prof. psychiatry and psychophysiology, 1967-72, assoc. prof. psychiatry, 1972-75; dir. Neurodynamics Lab, SUNY, Downstate Med. Ctr., 1966—; prof. psychiatry SUNY, 1975—. Recipient Thorp award, Isaacson prize. Fellow AAAS, Am. Electroencephalogram Soc., Am. Coll. Neurosychopharmacol; mem. Am. Psychology Assn., Soc. Psychophysiology Rsch., N.Y. Acad. Sci. Achievements include research in neurophysiological as it correlates to brain dysfunction. Office: State Univ of New York Dept of Psychiatry 450 Clarkson Ave Box 1203 Brooklyn NY 11203*

BEGLEY, ANTHONY MARTIN, research physicist; b. Stockport, Great Britain, Mar. 20, 1966; came to U.S., 1989; s. Thomas Michael and Patricia (Reid) B.; m. Rachel Janet Ward, Aug. 19, 1989. BS, U. Birmingham, Eng., 1987, PhD, 1990. Postdoctoral rsch. assoc. Fla. Atlantic U., Boca Raton, 1990-91, SUNY, Stony Brook, 1991-93; editorial asst. Phys. Rev. B, Ridge, N.Y., 1993—. Contbr. articles to profl. jours. Mem. Am. Phys. Soc. Achievements include the first detail resolved photomission from rare earth single crystal metals; structural analysis of ultra-thin metal films on transition metal substrates. Office: Am Phys Soc Editorial Office 1 Research Rd PO Box 1000 Ridge NY 11961

BEHBEHANIAN, MAHIN FAZELI, surgeon; b. Kermanshah region, Iran; d. M Jaafar and Ozra (A.) B.; m. Abolfath H. Fazeli, Sept, 4, 1969; children: Pouneh, Pontea. BS, Wilmington (Ohio) Coll., 1961; MD, Med. Coll. of Pa., Phila., 1965; general surgeon, Lankenan Hosp., Phila., 1970. Diplomate Am. Bd. Surgery, 1981. Chief surgery, pres. med. staff Imperial Ct. Hosp., Teheran, Iran, 1971-79; gen. surgery Riddle Meml. Hosp., Media, Pa., 1980—; pvt. practice Chester, Media, Phila., Pa., 1984 ; mem. operating room com. Riddle Meml. Hosp., Media, 1988—, also Emergency room com., utilization com. Editor-in-chief Behkoosh Jour. of Medicine, Teheran, 1976-79. Recipient Gilson Colby Engel award, 1966. Fellow Am. Coll. Surgeons; mem. AMA, Am. Women Surgical Soc., Pa. Med. Soc., Del. County Med. Soc. Office: Riddle Meml Health Care Ctr 1088 W Baltimore Pike Media PA 19063

BEHL, CHARANJIT R., pharmaceutical scientist; b. Nabha, Punjab, India, Dec. 11, 1950; came to U.S., 1971; s. Dev Prakash and Lajwanti Behl (Kohli) B. BS in Pharmacy, Birla Inst. Tech. and Sci., Pilani, India, 1971; MS, Duquesne U., 1975; PhD, U. Mich., 1979. Asst. rsch. scientist U. Mich., Ann Arbor, 1978-81; asst. rsch. group chief Hoffmann-LaRoche, Inc., Nutley, N.J., 1981-83, rsch. fellow, 1983-85, rsch. investigator, 1985-90, rsch. leader, 1990—. Co-contbr. over 6 chpts. to books; contbr. over 40 articles to profl. jours. and over 100 abstracts. Fellow Am. Assn. Pharm. Scientist; mem. Am. Pharm. Assn. (Ebert prize 1990), Am. Chem. Soc., Controlled Release Soc. Achievements include patents in field. Home: 27 Cottage Pl Nutley NJ 07110 Office: Hoffman LaRoche Inc 340 Kingsland St Nutley NJ 07110

BEHM, DENNIS ARTHUR, engineering specialist; b. Grand Rapids, Minn., Jan. 21, 1947; s. Arthur Walter and Sigrid Ingeborg (Lee) B.; m. Annette Lee Norvitch, Oct. 4, 1980. BEE, U. Minn., 1969, PhD, 1982. Rsch. assoc. Space Sci. Ctr., Mpls., 1976-80; sr. physicist 3M Co., St. Paul, 1980-85; applications specialist 3M Co., Austin, Tex., 1985-88; sr. engring. specialist Gen. Dynamics, Ft. Worth, 1988-93, Lockheed, Ft. Worth, 1993—. Contbr. articles to profl. jours. Spl. awards judge Internat. Sci. and Engring. Fair, Pitts., 1989, Orlando, Fla., 1990, Nashville, 1992. Lt. USN, 1969-74. Mem. IEEE, Am. Physical Soc.

BEHNKE, ERICA JEAN, physiologist; b. Rochester, N.Y., Jan. 16, 1957; d. Chester E. and Melitta C. (Schilling) B. BS, Mich. State U., 1978; MS, U. Minn., 1980, PhD, 1987. Dir. in vitro fertilization/reproductive studies Bethesda Hosps., Inc., Cin., 1987-92; dir. Christ Hosp. Ctr. Reproductive Studies, Cin., 1992—; locum tenens King Faisal Hosp. and Rsch. Ctr., Riyadh, Saudi Arabia, 1991; rsch. mem. Ctr. Reproduction Endangered Wildlife, Cin., 1990—; mem. internal rev. bd. Christ Hosp., Cin., 1993—. Named one of Outstanding Young Women of Am., 1988. Mem. Am. Fertility Soc. (abstract com. 1993), Am. Soc. Andrology, Soc. Study Reproduction, N.Y. Acad. Scis., Cin. Acad. Medicine, Sigma Xi, Sigma Delta Epsilon. Avocation: breeding horses. Office: The Christ Hosp Ctr Reproductive Studies 2139 Auburn Ave Cincinnati OH 45219-2706

BEHNKE, JAMES RALPH, food company executive; b. Milw., May 2, 1943; s. Ralph A. and Esther M. (Gruenwald) B.; BS, U. Wis. Madison, 1966, MS, 1968, PhD (NIH, Inst. Food Technologists fellow), 1970; children: Erica, Heather. Group leader, sect. mgr. research and devel. Quaker Oats Co., Barrington, Ill., 1970-74; with Pillsbury Co., Mpls., 1974—, v.p. research and devel., 1979-81, sr. v.p. ops. and tech., 1981-86, corp. sr. v.p. tech., 1986—; dir. Snow Brand-Pillsbury, Inc., Japan. Mem. Inst. Food Technologists, Am. Assn. Cereal Chemists. Club: Minneapolis. Patentee in field. Home: 2300 Parklands Rd Minneapolis MN 55416-3861 Office: Pillsbury Co 200 S 6th St Minneapolis MN 55402-1404*

BEHNKE, ROY HERBERT, physician, educator; b. Chgo., Feb. 24, 1921; s. Harry and Florence Alice (MacArthur) B.; m. Ruth Gretchen Zinszer, June 3, 1944; children: Roy, Michael, Donald, Elise. A.B., Hanover Coll., 1943; Ph.D. (hon.), 1972; M.D., Ind. U., 1946. Diplomate: Am. Bd. Internal Medicine. Intern Ind. U. Med. Center, 1946-47, resident, 1949-51, chief resident medicine, 1951-52; instr. medicine Ind. U. Sch. Medicine, Indpls., 1952-55, asst. prof. medicine, 1955-58, assoc. prof., 1958-61, prof., 1961-72; chief medicine VA Hosp., Indpls., 1957-72; prof. medicine, chmn. dept. U. South Fla. Coll. Medicine, Tampa, 1972—; AMA rep. to residency rev. com. in internal medicine, 1970-75; mem. exec. and adv. com. Inter-Soc. Comm. Heart Disease Resources, 1968-72, chmn. pulmonary study sect., 1969-72; chmn. career devel. com. VA, 1980-83. Mem. Met. Sch. Bd. Washington Twp., 1968-72, pres., 1971; bd. dirs. Southside Community Health Center, 1968; trustee Tampa Gen. Hosp. Found., 1979-85; mem. research coordinating com. Am. Lung Assn., 1983-85, chmn., 1985-87, bd. dirs., 1983-87. Served with AUS, 1943-45, 47-49. Recipient Clin. Tchr. of Year award Ind. U. Sch. Medicine, 1968, 69, 70, Clin. Tchr. of Year award U. South Fla. Coll. Medicine, 1977, Disting. Prof., 1983, also recipient Founders award, 1984; recipient Standard Oil Found. award Ind. U., 1971; Alumni Achievement award Hanover Coll., 1971; John and Mary Markle scholar, 1952, 57. Fellow ACP (gov. Fla. chpt. 1980-84, Laureate award 1991), Am. Coll. Chest Physicians; mem. AMA, Am. Fedn. Clin. Rsch., Cen. Soc. Clin. Rsch., So. Soc. Clin. Rsch., Alpha Omega Alpha. Home: 5111 Rolling Hill Ct Tampa FL 33617-1024 Office: Dept Internal Medicine 12901 N 30th St Box 19 Tampa FL 33612

BEHREND, WILLIAM LOUIS, electrical engineer; b. Wisconsin Rapids, Wis., Jan. 11, 1923; s. Albert and Eva Mae (Barney) B.; m. Manet Louise Whitrock, July 7, 1945; children: Jane Louise, Ann Elizabeth. B.S. in Elec. Engring., U. Wis., 1946, M.S., 1947. Research engr. David Sarnoff Research Ctr., RCA, 1947-64; advanced devel. engr. comml. systems div. RCA, Meadows Lands, Pa., 1964-66, preliminary design and systems analyst, 1966-84; ret., 1984, cons. engr., 1984-90. Contbr. articles on elec. engring. to profl. jours.; patentee in field. Served with USNR, 1944-46. Recipient RCA David Sarnoff Rsch Ctr. Outstanding Rsch. award, 1956, 59, 63, RCA Comml. Systems Div. Outstanding Contbns. to Product Tech. award, 1974. Fellow IEEE (Scott Helt award 1971); mem. AAAS, N.Y. Acad. Scis., Sigma Xi. Achievements include development of television transmitter systems and measurement methods. Address: 479 Carnegie Dr Pittsburgh PA 15243

BEHRENS, RUDOLPH, mechanical engineer; b. Paterson, N.J., Sept. 24, 1933; s. Herman and Yolanda (Matarazzo) B.; m. Barbara Marek, Apr. 19, 1980; children: Courtney, Todd, Derek. BS in Mech. Engring., Rutgers U., 1976, BA in Econs., 1976. Engr. Worthington Pump Corp., Harrison, N.J., 1976-79; chief engr. pumps Sun Oil Corp., Phila., 1979-81; engr. sales Goulds Pump Co., Seneca Falls, N.Y., 1981-84; pres., founder Pumpcom Inc., Collegeville, Pa., 1981—; cons. US Nuclear Regulatory Commn., King of Prussia, Pa., 1985—. Inventor barometric dryer, synergetic geometry as applied to fluid dynamics. Ben Franklin grantee, Phila., 1986. Mem. ASME, Chi Psi. Democrat. Achievements include invention of autonomous fluid-electric generating module. Home and Office: 118 Sycamore Ct Collegeville PA 19426-2910

BEHRMAN, EDWARD JOSEPH, biochemistry educator; b. N.Y.C., Dec. 13, 1930; s. Morris Harry and Janet Cahn (Solomons) B.; m. Cynthia Fansler, Aug. 29, 1953; children—David Murray, Elizabeth Colden, Victoria Anne. B.S., Yale, 1952; Ph.D., U. Calif. at Berkeley, 1957. Research asso. biochemistry Cancer Research Inst., Boston, 1960-64; bd. tutors biochem. scis. Harvard, 1961-64; asst. prof. chemistry Brown U., Providence, 1964-65; mem. faculty Ohio State U., Columbus, 1965—; asso. prof. biochemistry Ohio State U., 1967-69, prof., 1969—. Contbr. articles profl. jours. USPHS fellow, 1955-56, 57-60; NSF grantee, 1966-73; NIH grantee, 1973-81. Mem. Am. Chem. Soc., Royal Soc. Chemistry, Am. Soc. Biol. Chemists, Phi Beta Kappa, Sigma Xi. Home: 6533 Hayden Run Rd Hilliard OH 43026-9642 Office: Ohio State U Dept Biochemistry Columbus OH 43210

BEHRMAN, JACK NEWTON, economist; b. Waco, Tex., Mar. 5, 1922; s. Mayes and Marguerite (Newton) B.; m. Edwina Louise Sims, Sept. 6, 1945; children: Douglas, Gayle, Paul (dec.), Andrea. B.S. with honors in Econs. cum laude, Davidson Coll., 1943, LL.D. (hon.), 1979; M.A., U. N.C., 1945, Princeton U., 1950; Ph.D., Princeton U., 1952. Research asst. ILO, 1945-46; asst. prof. econs. Davidson Coll., 1946-48; research asst. internat. fin. sect. Princeton U., 1950-52; asso. prof. econs. and polit. sci. Washington and Lee U., 1952-57; prof. econs. and bus. adminstrn. U. Del., 1957-61; asst. sec. for internat. affairs U.S. Dept. Commerce, 1961-62, asst. sec. for domestic and internat. bus., 1962-64; co-founder Internat. Exec. Svc. Corps, 1964; prof. internat. bus. Sch. Bus., U. N.C., 1964-77, Luther Hodges Disting. prof., 1977-91, prof. emeritus, 1991—; Drexel research prof. Sch. Bus., U. N.C., 1970-71; Thomas Carroll vis. prof. Harvard Bus. Sch., 1967, chmn. MBA Program, 1971-77, assoc. dean acad. programs, 1983-87, assoc. dean faculty, 1987-88; dir. Ctr. for Internat. Trade and Investment Promotion, Kenan Inst. Private Enterprise, 1990-92; initiator, dir. MBA Enterprise Corps, 1990—; bd. dirs. 1st Union Nat. Bank at Chapel Hill (chmn. 1990-92), Troxler Electronics, Inc., 1980-92; mem. bd. sci. and tech. for internat. devel. NRC, 1973-75; mem. panel sci. and tech. in internat. econs. and trade Nat. Acad. Sci., 1977; dir. research project Dept. State, 1973-74; sr. research advisor Fund for Multinat. Mgmt. Edn., 1974-82, v.p. research and program devel., 1982-85; cons. U.S. Dept. State, Pan Am. Union, Com. Econ. Devel., Econ. Council Can., OAS, Nat. Fgn. Trade Council, So. Regional Council, Nat. Planning Assn., Hudson Inst., Nat. Acad. Sci., U.S. C. of C., UN Ctr. Sci. and Tech. for Devel., UN Ctr. on Transnat. Corps, GAO, N.C. Dept. Agr., also pvt. bus.; prin. investigator Patent, Trademark and Copyright Research Inst., George Washington U., 1955-60; cons. Presdl. Commn. on Internat. Trade and Investments, EPA, Office of Sci. and Tech. Policy, Exec. Office of Pres., 1980-81; mem. adv. com. on fgn. investment State Dept., 1976-77; mem. adv. com. on internat. trade, investment and devel. Undersec. State, 1978-86; research sec. for study on multinat. enterprises Council on Fgn. Relations, 1976-77; mem. panel tech. and internat. competitiveness Nat. Acad. Engring., 1978-79; lectr. Am. Mgmt. Assn., Columbia U., U. Tenn., U. Miami, Salzburg Seminar, USIA, Motorola Exec. Inst., U. N.C. Exec. Program, Young Execs. Inst., Govt. Exec. Program in 55 fgn. countries; mem. Dept. State del. Latin Am. Pre-Ministerial Conf., 1974-75; mem. del. UN Conf. Sci. and Tech. in Devel., Vienna, 1979; bd. dirs. Am. Viewpoint, Inc., Ethics Resource Center, 1977-85, N.C. World Trade Ctr. Co-author: (with Gardner Patterson) Survey of United States International Finance, 3 vols., 1950, 51, 52, (with Wilson E. Schmidt) International Economics, 1957, (with Raymond F. Mikesell) Financing Free World Trade with the Sino-Soviet Bloc, 1958, U.S. Private and Government Investment Abroad, 1962, (with Roy Blough) Regional Integration and the Trade of Latin America, 1968, (with A. Kapoor and J. Boddewyn) International Business-Government Communications, 1975, (with H. Wallender) Transfer of Manufacturing Technology within Multinational Enterprises, 1976, (with W. Fischer) Science and Technology for Development, 1980, Overseas Research and Development Activities of Transnational Companies, 1980, (with Robert E. Grosse) International Business and Governments: Issues and Institutions, 1990; (with James E. Shapiro, William A. Fischer, Simon E. Powell) Investing and Joint Ventures in China, 1991; author: Rise of the Multinational Enterprise, 1969, National Interests and the Multinational Enterprise, 1970, U.S. International Business and Governments, 1971, Multinational Production Consortia, 1971, Role of International Companies in Latin American Integration, 1972, Decision Criteria for Foreign Directnvestment in Latin America, 1974, Conflicting Constraints on the Multinational Enterprise, 1974, Toward a New International Economic Order, 1974, Tropical Diseases: Responses of Pharmaceutical Companies, 1980, Industry Ties with Science and Technology Policies in Developing Countries, 1980, Discourses on Ethics and Business, 1981, Industrial Policies, International Restructuring and Transnationals, 1984, The Rise of the Phoenix: The U.S. in a Restructured World Economy, 1987, Essays on Ethics in Business and the Professions, 1988; editor: (with Robert E. Driscoll) National Industrial Policies, 1983; contbr. 110 articles to profl. jours.; mem. editorial bd.: Jour. Internat. Bus. Studies, 1976-82. Fellow Acad. Internat. Bus.; mem. Council Fgn. Relations, Regional Export Expansion Council (vice chmn. 1971-73), Assn. Edn. in Internat. Bus. (sec. 1959-60, pres. 1966-68), N.C. World Trade Assn. (dir.), Sigma Phi Epsilon, Pi Gamma Mu, Alpha Phi Omega, Beta Gamma Sigma. Democrat. Presbyterian (elder). Home: 1702 Audubon Rd Chapel Hill NC 27514-7605

BEIDEL, DEBORAH CASAMASSA, clinical psychologist, researcher; b. Reading, Pa., Mar. 1, 1955; d. Anthony Joseph and Jean Marie (Lyons) Casamassa; m. Edward M. Beidel, Aug. 20, 1977. BA, Pa. State U., 1976; MS, U. Pitts., 1984, PhD, 1986. Asst. prof. U. Pitts., 1988-92; assoc. prof. Med. U. S.C., Charleston, 1992—. Author: Treating Obsessive-Compulsive Disorder, 1988; mem. editorial bd. Jour. Anxiety Disorders, 1991—, Jour. Psychopathology and Behavioral Assessment, 1988—, Profl. Psychology: Research and Practice, 1989—; contbr. articles to profl. jours. Recipient Pres.'s New Rschr., Assn. for Advancement of Behavior Therapy, 1990. Office: Med Univ South Carolina Dept Psychiatry 171 Ashley Ave Charleston SC 29425

BEIDLEMAN, RICHARD GOOCH, biologist, educator; b. Grand Forks, N.D., June 3, 1923; s. Fred Allen and Olive M. (Gooch) B.; m. Reba E. Rutz, Sept. 5, 1946 (dec. 1990); children: Sterling Kirk, Janet Gail (Mrs. Robson), Carol Aileen (Mrs. Tiemeyer); m. Linda G. Havighurst, June 3, 1991. BA, U. Colo., 1947, MA, 1948, PhD, 1954; DS (hon.), Colo. Coll., 1989. Asst. prof. zoology Colo. State U., Ft. Collins, 1948-56; asst. prof. biology U. Colo., Boulder, 1956-57; from asst. prof. to prof. biology Colo. Coll., 1957-92, chmn. biology, 1968-71, prof. emeritus biology, 1988; Hulbert Endowed chair Southwestern Studies Colo. Coll., Colorado Springs, 1992; mem. Steering com. Biol. Sci. Curriculum Study, Commn. Undergrad. Edn. Biol. Sci., South Front Range Planning Team, Colo. State Planning Dept., 1969; cons., author, instr., and ranger naturalist U.S. Nat. Parks Svc., 1948-86. Co-author BSCS Green Version Biology Textbook, 1968, Interaction of Man and the Biosphere, 1980, Wildlife and Plants of the Southern Rocky Mountains, 1966; co-editor IBP The Grassland Biome, 1969; author booklets in field. Mem. and chmn. Colo. State Parks Bd., Denver, 1976-84; mem. Colo. State Natural Areas Coun., Denver, 1980-84; adv. com. Pikes Peak Area Coun. of Govts. Lt. (j.g.) USN, 1943-46, PTO, lt. USNR, 1946-63. Recipient Environ. Edn. award R.M. Ctr. Environment, 1970, Burlington No. Faculty Achievement award Colo. Coll., 1988, award Beidleman Environ. Ctr. Colo. Springs, 1987, Garden Club of Am. Environ. citation, 1993; Ford Found. grantee, 1954-55; Faculty Improvement rsch. grantee Colo. Coll., Australia, 1971-72. Fellow AAAS; mem. Am. Behavior Soc., Soc. Range Mgmt., Am. Soc. Mammals, Ecol. Soc. Am., Cooper Orn. Soc., Am. Orn. Soc., Soc. Study Evolution, Colo./Wyo. Acad. Sci. (past exec. sec., pres.), History of Sci. Soc., Sigma Xi. Democrat. Home: 766 Bayview Ave Pacific Grove CA 93950

BEIERWALTES, WILLIAM HENRY, physician, educator; b. Saginaw, Mich., Nov. 23, 1916; s. John Andrew and Fanny (Aris) B.; m. Mary Martha Nichols, Jan. 1, 1942; children: Andrew George, William Howard, Martha Louise. A.B., U. Mich., 1938, M.D., 1941. Diplomate: Am. Bd. Internal Medicine and Nuclear Medicine. Intern, then asst. resident medicine Cleve. City Hosp., 1941-43; mem. faculty U. Mich. Med. Center, 1944—, prof. medicine, 1959—; dir. nuclear medicine, also dir. Thyroid Research Lab., 1952-86, cons., 1987—; cons. nuclear medicine depts. St. John Hosp., Detroit, Wm. Beaumont Hosp., Royal Oak and Troy, Mich., The UpJohn Co. Rsch. div., 1952-65, The Abbott Labs. Rsch. div., 1960-67; sr. med. cons. MD (Med. Fedn.), Bagdad, Iraq, 1963; mem. exec. com. Inst. Sci. and Tech., 1963; lectr. Nat. Naval Med. Ctr., 1964-88, Ctr. for Environ. Health Mich. State Dept. Health, 1988-89; Peter Heimann lectr. 34th meeting Internat. Congress Surgery, Stockholm, Sweden, 1991; adv. panel on radionuclide labeled compounds for tumor diagnosis Internat. AEC, 1974-75; mem. Mich. State Radiation Bd., 1980-84; co-chmn. Nat. Coop., Thyroid Cancer Therapy Group, 1978-81. Author: Clinical Use of Radioisotopes, 1957, Manual of Nuclear Medicine Procedures, 1971, also numerous articles.; assoc. editor Jour. Lab. and Clin. Medicine, 1954-60; mem. editorial bd. Jour. Nuclear Medicine, 1959-64; assoc. editor, 1975-81; mem. editorial bd. Jour. Clin. Endocrinology and Metabolism, 1963; mem. adv. bd. Annals of Saudi Medicine, 1986-90; patentee for monoclonal antibodies to HCG, and radionuclide in vivo biochem. imaging of endocrine glands, 1951; first to treat a patient for cancer with radio labeled antibodies, 1951; co-inventor radiopharms, 1971; originator of radioimmunodetection of human cancer, first description of cytogenetic evolution of thyroid cancer, first description of fall of serum antithyroid antibodies during pregnancy with rise after delivery, other med. techniques. Guggenheim fellow, 1966-67; Commonwealth Fund fellow, 1967; Hevesy Nuclear Medicine Pioneer award, 1982; Disting. Faculty award U. Mich., 1982; Johann-Georg-Zimmerman Trust for Cancer Research Sci. prize for greatest contbn. to treatment of thyroid cancer, 1983. Mem. AMA, ACP, Am. Fedn. Clin. Rsch. (pres. 1954-55), Soc. Nuclear Medicine (pres. 1965-66, Disting. Educator's award 1989, The Best Doctors in Am. award 1993), Ctrl. Clin. Rsch. Club (pres. 1958-59), Am. Thyroid Assn. (v.p. 1964-65, 66-67, Disting. Svc. award 1972), Ctrl. Soc. Clin. Rsch. (councillor 1964-67, 67-71), Galens Med. Soc., Assn. Am. Physicians, Mich. Med. Soc., Am. Endocrine Soc., Am. Soc. Clin. Oncology. Home: 917 Whittier Rd Grosse Pointe MI 48230-1850 Office: St John Hosp 22101 Moross Rd Grosse Pointe MI 48236-2172

BEIK, MOSTAFA ALI-AKBAR, laser, electro-optical and fiber optics research engineer; b. Dec. 24, 1958. BSEE and Computer Engrng., Ariz. State U., 1984; MSEE, U. Wyo., 1986, PhD in Elec. Engrng., 1990. Cert. engr.-in-tng. Instr. math. U. Wyo., Laramie, 1985-87, instr. elec. engring., 1987-88, vis. asst. prof. dept. physics and astronomy, 1990-91; rsch. assoc. Atmospheric Scis. Rsch. Ctr. SUNY, Albany, 1991—; elec. engring. con. dept. physics U. Wyo., Laramie, 1987. Contbr. articles to profl. jours. Recipient Global Change Studies grant NASA, 1992, Rsch. grant NOAA, 1990. Mem. IEEE (Lasers and Electro-Optics Soc., chmn. student br. Ariz. State U., chief editor teaching effectiveness com. Ariz. State U.), Optical Soc. Am., Am. Soc. Engring. Edn., Sigma Xi (rsch. grant 1988, 89). Home: PO Box 3938 Albany NY 12203 Office: SUNY Albany Atmospheric Scis Rsch Ctr 100 Fuller Rd Albany NY 12205

BEIMAN, ELLIOTT, research pharmaceutical chemist; b. Manhattan, N.Y., Oct. 25, 1935; s. Jack David and Sally (Kaplan) B.; m. Phyllis Barbara Wissner, June 23, 1957; children: Jacqueline Diane, Lawrence David. BS in Chemistry, CCNY, 1960. Chief chemist Foods Plus Inc., Moonachie, N.J., 1960-64; dir. mfg. Zenith Labs., Northvale, N.J., 1964-79; pres. Deena Corp., Norwood, N.J., 1979-83; dir. mfg. Pvt. Formulations, Edison, N.J., 1983-85; dir. product devel. Barr Labs., Pomona, N.Y., 1985-87, Clonmel (Ireland) Chem. Co., 1987—; mng. dir. Dorotock, Inc., River Vale, N.J., 1987—. Named Honor Citizen Ireland, Irish Govt., Clonmel, 1989. Mem. N.Y. Acad. Scis., Am. Assn. Sci., Assn. Food Technicians. Achievements include patent in Centrifugal Apparatus and Cell Auto Analyzer. Home: 60 Birchwood Rd Bedminster NJ 07921

BEINEKE, LOWELL WAYNE, mathematics educator; b. Decatur, Ind., Nov. 20, 1939; s. Elmer Henry and Lillie Agnes (Snell) B.; m. Judith Rowena Wooldridge, Dec. 23, 1967; children: Jennifer Elaine, Philip Lennox. BS, Purdue U., 1961; MA, U. Mich., 1962, PhD, 1965. Asst. prof. Purdue U., Ft. Wayne, Ind., 1965-68, assoc. prof., 1968-71, prof., 1971—, Jack W. Schrey prof., 1986—; tutor Oxford (Eng.) U., 1974, The Open U., Milton Keynes, Eng., 1974, 75; vis. lectr. Poly. N. London, Eng., 1980-81. Co-author, co-editor: Selected Topics in Graph Theory, 3 vols., 1978, 83, 88, Applications of Graph Theory, 1979; mem. editorial bd., assoc. editor Jour. Graph Theory, 1977-80, 89—; mem. editorial bd. Internat. Jour. Graph Theory, 1991—; co-editor: Congressus Numerantium, Vols. 63-64, 1988; contbr. numerous articles to profl. jours. Corp. mem. Bd. for Homeland Ministries, United Ch. of Christ, N.Y., 1988-91, del. Gen. Synod, 1989, 91. Recipient Outstanding Tchr. award AMOCO Found., 1978, Friends of the Univ., 1992; Fulbright Found. grantee London, 1980-81, rsch. grantee Office Naval Rsch., Washington, 1986-89; fellow Inst. Combinatorics and its Applications, 1990—. Mem. AAUP, Math. Assn. Am. (chairperson Ind. sect. 1987-88, mem. bd. govs. 1990-93), Am. Math. Soc., London Math. Soc., Common Cause, Amnesty Internat., Summit Book Club, Internat. Affairs Forum, Sigma Xi (club pres. 1984-86), Phi Kappa Phi(chpt. pres. 1993—). Achievements include characterization of line graphs and thickness of complete graphs; enumeration of multidimensional trees. Avocations: British culture, reading, gardening, stamp collecting, jogging. Home: 4529 Bradwood Ter Fort Wayne IN 46815-6028 Office: Ind U-Purdue U Dept of Math Scis 2101 E Coliseum Blvd Fort Wayne IN 46805-1445

BEINS, BERNARD CHARLES, psychology educator; b. Toledo, Feb. 4, 1950; s. William Edward and Gladys A. (Hilditch) B.; m. Linda Grossman, Aug. 8, 1949; children: Agatha Meryl, Simon Frederick. AB, Miami U., 1972; PhD, CUNY, 1979. Psychology prof. Thomas More Coll., Crestview Hills, Ky., 1979-85, Ithaca (N.Y.) Coll., 1986—. Editor: Teaching of Psychology: Computers in Teaching, 1987—; author lectr.'s manual: Psychology: Science, Behavior and Life, 1990; contbr. articles to profl. jours. Mem. APA (sec. divsn. 2), Ea. Psychol. Assn., Am. Statis. Assn., Psi Chi, Sigma Xi. Democrat. Home: 47 Cayuga St Trumansburg NY 14886 Office: Ithaca Coll 953 Danby Rd Ithaca NY 14850

BEITLER, STEPHEN SETH, cosmetics company executive; b. N.Y.C., Oct. 1, 1956; s. Stanley Samuel and Arline (Mandell) B.; m. Deborah Joy Gottlieb, Jan. 16, 1982. BA, cert. of Asian Study, Am. U., Washington, 1977; postgrad., U. Chgo., 1977-78; MS, Def. Intelligence Coll., 1986. Legis. aide U.S. Ho. of Reps., Washington, 1975-77; commd. 2d lt. U.S. Army, 1977, advanced through grades to maj., 1989; intelligence briefing officer Office of Chmn. Joint Chiefs of Staff/ Office Sec. Def./Def. Intelligence Agy., Washington, 1984-86; asst. to asst. sec. of def. Office Sec. Def., Washington, 1987-88; asst. to undersec. of def. Office of Sec. of Def., Washington, 1988-89; resigned U.S. Army, 1989; mgr. ops. devel. Helene Curtis, Inc., Chgo., 1989-90, corp. mgr. strategy and devel., 1990—; comdr. 305th psychol. ops. bn. USAR, Arlington Heights, Ill., 1992—; cons. MGA, Inc., Chgo., 1985—. Contbg. author: The Military Intelligence Community, 1986; contbr. articles to profl. publs. Vol. Bus. Vols. for the Arts, Chgo., 1991—. Fellow Inter-Univ. Seminar on Armed Forces and Soc.; mem. Soc. Competitive Intelligence Profls. (bd. dirs. 1991—), Spl. Forces Club (London), Army and Navy Club. Home: 378 Delta Ln Highland Park IL 60035 Office: Helene Curtis Inc 4401 W North Ave Chicago IL 60639

BEJAN, ADRIAN, mechanical engineering educator; b. Sept. 24, 1948; married; 3 children. SB in Mech. Engring., MIT, 1972, SM in Mech. Engring., 1972, PhD Mech. Engring., 1975; PhD (hon.), U. Bucharest, 1992. Staff. engr. Sci. Energy Systems, Inc., Watertown, Mass., 1972; rsch. asst. dept. mech. engring. MIT, 1971-74, lectr., rsch. assoc., dept. mech. engring., 1975-76; fellow Miller Inst. Basic Rsch. Sci., U. Calif., Berkeley, 1976-78; asst. prof., dept. mech. engring. U. Colo., Boulder, 1978-81; Croft prof. Coll. Engring., U. Colo., 1981-82; assoc. prof., dept. mech. engring. U. Colo., Boulder, 1981-84; prof., dept. mech. engring. and materials sci. Duke U., Durham, N.C., 1984-89, J.A. Jones prof., dept. mech. engring., 1989—; mem. Nusselt-Reynolds Prize Bd., 1990-92; cons. Dept. Energy, NSF, Solar Energy Rsch. Inst., Procter & Gamble, The Lord Corp., John Wiley & Sons, McGraw Hill, Harper & Row, Hemisphere Pub. Corp., Rolscreen Co., EVI Inc. Author: Entropy Generation Through Heat and Fluid Flow, 1982, Convection Heat Transfer, 1984, Advanced Engineering Thermodynamics, Heat Transfer, 1992; co-author: Convection in Porous Media, 1992; hon. editorial bd.: International Journal of Heat and Mass Transfer, 1992, International Communications in Heat and Mass Transfer, 1992; bd. editors: International Journal for Engineering Analysis and Design; adv. editor: Heat Transfer Japanese Research, 1990, International Journal of Heat and Fluid Flow, 1988; reviewer manuscrips for numerous jours.; contbr. articles to profl. jours. Recipient James Harry Potter Gold medal Am. Soc. Mech. Engrs., 1990; Gustus L. Larson award Am. Soc. Mech. Engrs., 1988, Ralph R. Teetor award Soc. Automotive Engrs., 1980, De Florez award MIT, 1969; Faculty fellow U. Colo., 1984-85; F. Mosey Vis. scholar U. Western Australia. Fellow Am. Soc. Mech. Engrs.; mem. Am. Acad. Mecanics, Tau Beta Pi, Pi Tau Sigma. Office: Duke Univ Dept Mech Engring Materials Sci Durham NC 27706*

BEKÂROĞLU, ÖZER, chemist, educator; b. Trabzon, Turkey, May 3, 1933; s. Ibrahim Shevki and Halide (Sak) B.; m. Afife Cingilli, July 1, 1969; 1 child, Burak. BSChE, MSChE, U. Istanbul, 1960; PhD, U. Basel, 1963. Researcher Chemische Werke Hüls, Germany, 1961-62; postdoctoral fellow U. Calif., Davis, 1966-68; investment mgr. Eezacibasi Holding, Istanbul, 1968-70; assoc. prof. U. Istanbul, 1970-71; assoc. prof. Tech. U. Istanbul, 1971-75, prof., 1975—; head dept. chemistry TUBITAK Rsch. Ctr., Istanbul, 1984—; dean faculty chemistry Tech. U. Istanbul, 1979-82, faculty chem. and metall. engring., 1985-88, faculty sci. and letters, 1988-91. Author: Coordination Chemistry, 1972, General Chemistry, 1985, Inorganic Chemistry I, 1991. 1st lt. Turkish Army, 1964-66. Mem. Chem. Soc. Turkey, Am. Chem. Soc., New Swiss Chem. Soc., Turkish Chem. Found. Home: Kazim Ozalp Sok, Cingilli Apt 10/10, Kantarci-Istanbul Turkey Office: Tech U Istanbul, Maslak-Istanbul Turkey

BEKESI, JULIS GEORGE, medical researcher; b. Budapest, Hungary, July 2, 1931; came to U.S., 1958; s. Ernest and Katalin (Sztraka) B.; m. Eva, June 18, 1955; children: Michelle G., Richard G. B of Chem. Engring., U. Budapest, 1956; MS, U. Alberta, 1965; PhD, U. Buffalo, 1968. Rsch. scientist U. Alberta, Edmonton, Can., 1958-65; assoc. rsch. prof. Roswell Park Cancer Inst., Buffalo, 1968-73; prof. Mt. Sinai Sch. Medicine, N.Y.C., 1973—. Author: Immunological Approaches to Cancer Therapeutics, 1982; editor: Immunobiology of Cancer and AIDS, 1987, Immune Dysfunction in Cancer and AIDS, 1988, Immunoactive Products in Oncology and Persistent Viral Infections, 1988. Mem. AAAS, Am. Cancer Soc., Prevention of Cancer (dir. 1978-89). Achievements include chemo-immunotherapy of adult Leukemia; early diagnosis of breast cancer and colorectal cancer by LAI-CMI method (patent), recognation of PBB and PCB induced immune dysfunction in human, immunology & treatment of HIV-1. Office: Mt Sinai Sch Medicine 10 E 102d St New York NY 10029

BELAGAJE, RAMA M., biotechnology scientist; b. Belagaje, Karnataka, India, May 9, 1942; s. Subraya and Shankary Belagaje; m. Nalini Belagaje, Jan. 25, 1973; children: Samir, Sudhir. BSc, Mysore (India) U., 1962; MSc, Banaras Hindu U., Varanasi, India, 1965; MS, NYU, 1970, PhD, 1971. Instr. in chemistry St. Philomena's Coll., Puttur, India, 1962-63; rsch. fellow Indian Inst. Sci., Bangalore, India, 1965-67; teaching fellow NYU, N.Y.C., 1971-72; sr. rsch. assoc. MIT, Cambridge, Mass., 1972-79; sr. scientist Eli Lilly & Co., Indpls., 1979—. Contbr. articles to profl. publs. Chmn. devel. com. India Community Ctr., Indpls., 1988-91; bd. dirs. Gita Mandal of Indpls., 1989—; scoutmaster Boy Scouts Am., Indpls., 1988—. Recipient Padma Shri award Govt. of India, 1977. Mem. Am. Chem. Soc., N.Y. Acad. Scis., Acad. Gen. Edn., India Assn. Indpls. (bd. dirs. 1990—). Achievements include patents for cloning vectors for expression of exogenous protein, transcription terminators, inverted (A-C-B) Proinsulin, Bacteriophage Lambda PL Promoters, others, also numerous patents pending. Home: 7821 Mohawk Ln Indianapolis IN 46260 Office: Eli Lilly and Co Dept MC625 Bldg 98C/2318 Indianapolis IN 46285

BELANGER, PIERRE ROLLAND, university dean, electrical engineering educator; b. Montreal, Que., Can., Aug. 18, 1937; s. Pierre Henri and Lucille R. B.; m. Margaret Mary Clark, Aug. 24, 1963; children: Mark, Suzanne, David. B.Eng., McGill U., 1959; S.M., M.I.T., 1961, Ph.D., 1964. Asst. prof. elec. engring. M.I.T., 1964-65; systems analyst The Foxboro (Mass.) Co., 1965-67; assoc. prof. McGill U., 1967-75, prof., chmn. dept. elec. engring., 1975-84, dean Sch. Engring., 1984—. Mem. Nat. Adv. Bd. for Sci. and Tech., 1987-90; vice chmn. Def. Sci. Adv. Bd. Mem. IEEE (v.p. 1981-82), Ordre Ingenieurs du Que., Can. Acad. Engring. (founder). Home: 59 Somerville St, Westmount, PQ Canada H3Z 1J4 Office: McGill Univ, 817 Sherbrooke W, Montreal, PQ Canada H3A 2K6

BELANGER, RONALD LOUIS, chemist; b. Bklyn., Nov. 4, 1955. BA, U. Vt., 1977; MS, NYU, 1983. Assoc. chemist Revlon Health Care, Tuckahoe, N.Y., 1977-80; tech. specialist Gen. Foods U.S.A., Tarrytown, N.Y., 1980—; rep. tech. com. FEMA, Washington, 1990—. Mem. Am. Chem. Soc., Am. Soc. for Mass Spectrometry. Achievements include Gas Chromatography/ Mass Spectrometry, Nuclear Magnetic Resource, and Fourier Transform Infrared Spectroscopy analysis of foods, flavors, and ingredients. Office: Gen Foods USA Tech Ctr T23-1 Tarrytown NY 10625

BELARBI, ABDELDJELIL, civil engineering educator, researcher; b. Tlemcen, Algeria, Apr. 21, 1959; came to U.S., 1983; s. Sid Ahmed and Rabia (Benchouk) B.; m. Samira Bereksi, Aug. 14, 1986; children: Sihem L., Hishem I. BSc, U. Oran, Algeria, 1983; MSc, U. Houston, 1986, PhD, 1991. Rsch. asst. U. Houston, 1984-90, teaching fellow, 1990-91; asst. prof. civil engring. U. Mo., Rolla, 1991—. Contbr. articles to profl. jours. Recipient Outstanding Tchr. award U. Houston, 1991; Algerian Govt. scholar, 1984-90; rsch. grantee NSF, 1992. Mem. ASCE (assoc.), Am. Concrete Inst., Am. Soc. for Engring. Edn., Sigma Xi (scholar 1986), Tau Beta Pi, Chi Epsilon. Islam. Achievements include research on shear and in-plane forces on reinforced concrete, performance and durability of architectural glazing systems under wind and earthquake effects. Home: 1116 Joyce Ave Rolla MO 65401 Office: Dept Civil Engring U Mo Rolla MO 65401

BELCHER, RONALD ANTHONY, nuclear energy educator; b. Mt. Clemens, Mich., Mar. 12, 1950; s. Earnest W. and Helen S. (Schinkai) B.; m. Janice Rose Bathurst Belcher, Aug. 4, 1973; children: Andrew R., Stefanie Lynn. BS, Cen. Mich. U., 1973; MA, Saginaw Valley State U., 1977. Tchr. Swan Valley High Sch., Saginaw, Mich., 1973-81; gen. nuclear instr. Midland (Mich.) Nuclear Plant, 1981-84; simulator instr. tng. dept. Plant E. I. Hatch, Baxley, Ga., 1984-86, sr. simulator instr., 1986-89, plant nuclear instr., 1989—. Named Nat. Outstanding Tchr. Ednl. Publs. Am., 1975. Mem. Am. Nuclear Soc. Home: 219 Amberwood Dr E Vidalia GA 30474-3079 Office: Plant El Hatch Tng Dept PO Box 439 Baxley GA 31513-0439

BELDEN, DAVID LEIGH, professional association executive, engineering educator; b. Mpls., Jan. 9, 1935; m. Lois Marion Lind, June 14, 1956; children: Richard Alan, Grant David. B.Gen. Edn., U. Omaha, 1961; M.S. in Indsl. Engring., Stanford U., 1963, Ph.D., 1969; grad., Indsl. Coll. Armed Forces, 1973; DSc (hon.), Manhattan Coll., 1992. Registered profl. engr., Calif. rated navigator, aviator. Enlisted U.S. Air Force, 1954, commd. 2d lt., 1956, advanced through grades to col., 1973; served Thailand; asst. for procurement mgmt. to Sec. Air Force, Washington; ret., 1976; exec. dir. Inst. Indsl. Engrs., Norcross, Ga., 1976-87, ASME, N.Y.C., 1987—; adj. prof. Far East div. U. Md., 1970; asso. prof. George Washington U., 1974. Author articles in field. Decorated Legion of Merit, Meritorious Service medal, Commendation medal (3). Fellow Am. Inst. Indsl. Engrs., Instn. Prodn. Engrs. (Eng., life); mem. ASME, Am. Assn. Engring. Socs. (bd. govs.), Coun. Engring. and Sci. Soc. Execs. (pres. 1984-85), Am. Soc. Engring. Edn., N.Y. Soc. Assn. Execs., Am. Soc. Assn. Execs. (found. bd.), Australian Inst. Indsl. Engrs. (hon.), Japan Mgmt. Assn. (assoc.), Alpha Pi Mu, Tau Beta Pi. Republican. Home: 6 Bates Farm Ln Darien CT 06820-3500 Office: ASME 345 E 47th St New York NY 10017-2330

BELDING, WILLIAM ANSON, ceramic engineer; b. Boston, Nov. 23, 1944; s. Harwood Seymore and Lola (Selley) B.; m. Mary Jane Inglis (div.); children: Andrew W., Kerry L. BS, Rutgers U., 1966, PhD, 1969. Sr. ceramic engr. Kaiser Aluminum & Chem. Co., Baton Rouge, 1969-72, group leader, 1972-75; sect. head Kaiser Aluminum & Chem. Co., Pleasanton, Calif., 1975-83, mgr. applied chemistry, 1983-85, dir. rsch., 1985-88; dir. rsch. LaRoche Chems., Pleasanton, 1988-89; pres., cons. Innovative Rsch. Enterprises, Danville, Calif., 1989—. Contbr. articles to profl. publs. Mem. Am. Ceramic Soc., Am. Assn. Heating, Refrigeration and Air-Conditioning Engrs., Nat. Inst. Ceramic Engrs. Republican. Achievements include patents in spherical alumina catalyst substrate development, flame retardant composition from bauxite; patents pending on low pressure drop adsorbent filter for removal of air pollutants, desiccant paper for cooling aplications. Home and Office: 84 La Pera Ct Danville CA 94526

BELEFANT, ARTHUR, engineer; b. N.Y.C., June 17, 1927; s. Benjamin and Lena Helen (Roth) B.; m. Rita Myra Sinclair, July 29, 1958; children: Helen Miller, Brian, Sterling. BEE, CCNY, 1949; BA in Mgmt., U. Md., 1964; MS in Mgmt., Fla. Inst. Tech., 1969. Profl. engr. D.C. Fla., Ga., Md., N.Y., Tex. Pres. Belefant Assocs., Inc., Cocoa Beach, Fla., 1966-79; chief electrical engr. Louis Berger Interant., Tel Aviv, Israel, 1979-81; dir. ops. Frank E. Basil, Inc., Washington, 1981-84; asst. project mgr. TRW, Omaha, Nebr., 1984-87; sr. staff engr. TRW, Redondo Beach, Calif., 1987-90; prin. project engr. Walt Disney Imagineering, Glendale, Calif., 1990-91; free lance cons. Melbourne, Fla., 1991--; mem. Canaveral Coun. Tech. Soc., Cape Canaveral, Fla., 1992, Lt. Gov. Com. on the Space Shuttle, Kennedy Space Ctr., 1970. Contbr. articles to profl. jours. Mem. Bd. Adjustment, Melbourne Beach, 1992-93, GSA Citizens Adv. Bd., Atla., 1978. With U.S. Army, 1947-48. Recipient Outstanding Civic Contribution award Fla. Engring. Soc., 1977, Spl. Commendation award Balistic Missile Office, 1989. Mem. IEEE (sr.), Soc. Am. Mil. Engrs. (pres. 1972, Engr. of Yr. 1971), Illuminating Engring. Soc. (libr. light 1970), Am. Soc. Mechanical Engring. Office: 305 Oak St Melbourne Beach FL 32951

BELFIELD, KEVIN DONALD, chemistry educator; b. Utica, N.Y., Oct. 8, 1960; s. Donald E. and Marilyn J. Belfield; m. Jing Sun; 1 child, Dylan Sun. BS in Chemistry, Rochester Inst. Tech., 1982; PhD in Chemistry, Syracuse U., 1988. R&D chemist Bristol-Myers Co., Syracuse, N.Y., 1982-83; sr. chemist Ciba-Geigy Corp., Ardsley, N.Y., 1988-89; postdoctoral assoc. SUNY Coll. Environ. Sci. and Forestry, Syracuse, 1989-90; postdoctoral fellow Harvard U. Dept. Chemistry, Cambridge, Mass., 1990-92; asst. prof. U. Detroit Mercy, 1992—; adj. rsch. fellow Harvard U., JFK Sch. Govt., Ctr. Sci. and Internat. Affairs, Cambridge, Mass., 1990-93; cons. Accustar Corp./Chrysler Corp., Syracuse, 1989-90. Contbr. articles to profl. jours. Jour. Polymer Sci., Chemistry Edn., Jour. of Am. Chem. Soc., Jour. Phys. Organic Chemistry, Jour. Organic Chemistry, Jour. Chromatography. Coadvisor U. Detroit Mercy Chemistry Club, chair Detroit sect. younger chemists com. Lilly Found. grantee, 1992; Summer Rsch. fellow Syracuse U., 1984-86, N.Y. State Regents scholar, 1978, Freshman scholar Rochester Inst. Tech., 1978. Mem. Am. Chem. Soc. Office: U Detroit Mercy PO Box 19900 Detroit MI 48219-0900

BELGAU, ROBERT JOSEPH, aeronautical engineer; b. Miami, Fla., Nov. 18, 1960; s. V. Eugene Belgau and Marie (Siena) Skinner. BS in Aero. Engring., Embry Riddle Aero. U., 1990. Aero. engr. R.T. Aerospace, Miami, 1990—. Mem. AIAA, Soc. Automotive Engrs.

BELGRADER, PHILLIP, molecular biologist; b. N.Y.C., June 16, 1961; s. Mike and Serena DeRose; m. Rita Pirk, June 16, 1990; 1 child, Maria. BS, Syracuse U., 1983; PhD, SUNY, Buffalo, 1990. Postdoctoral fellow Roswell Park Cancer Inst., Buffalo, 1990—. Contbr. articles to profl. jours. Recipient Neuromuscular Disease Rsch. award Muscular Dystrophy Assn., 1992. Mem. AAAS, Am. Heart Assn. Office: Roswell Park Cancer Inst Elm and Carlton Sts Buffalo NY 14263

BELIVEAU, JEAN-GUY LIONEL, civil engineering educator; b. Arthabaska, Que., Can., Aug. 1, 1946; came to U.S. 1955; s. Laurier Joseph and Yvette (Plante) B.; m. Constance Jane Moore, July 6, 1968; children: Michelle, Denise, Chantal, Lia. BSCE, U. Vt., 1968; PhD in Civil Engring., Princeton U., 1974. Registered profl. engr., Vt., Que. Postdoctoral fellow Columbia U., N.Y.C., 1973-74; prof. civil engring. U. Sherbrooke, Que., 1974-84; prof. civil engring. U. Vt., Burlington, 1985—, interim chmn., 1990-93; vis. rsch. engr. UCLA, 1976; vis. rsch. prof. U. Joseph Fourier, Grenoble, France, 1982-83; adv. bd. Region I Univ. Transp. Ctr., MIT, Boston, 1988-90. Editorial bd. Can. Jour. Civil Engring., 1984-91; contbr. articles to profl. jours. 1st lt. U.S. Army, 1972. Recipient Phelps prize U. Vt., 1968; USN/ ASEE summer faculty fellow, 1986-88; recipient Robert Angus medal Can. Soc. Mech. Engrs., 1986. Mem. ASCE (Vt. sect. pres. 1990-92). Roman Catholic. Achievements include research in structural system identification with application to testing of satellite components, stability of buildings and temporary structures, vibrations of bridges, soil-structure interaction of nuclear power plants and dynamic behavior of soils. Home: 1474 RR 1 Charlotte VT 05445 Office: Univ of Vt 213 Votey Bldg Burlington VT 05405

BELK, KEITH E., international marketing specialist, researcher; b. Denver, Aug. 8, 1961; s. James R. and B. Kay (Robinson) B.; m. Jo Ann Mills, Jan. 6, 1990. BS, Colo. State U., 1983, MS, 1986; PhD in Animal and Food Sci., Tex. A&M U., 1992. Quality assurance supr. Safeway Stores, Inc., L.A., 1986; buyer Safeway Stores, Inc., Denver, 1986-88; grad. asst. Tex. A&M U., College Station, 1988-92, rsch. assoc., 1992; internat. mktg. specialist agrl. mktg. svc. USDA, Washington, 1992—; cons. Hoerst-Reussel Agri-Vet, Greeley, Colo., 1988. Contbr. articles to profl. jours. Com. mem. Fall Classic Regional 4-H Contest, Denver, 1983, Nat. Western Intercollegiate Judging Contest, Denver, 1988. Scholar Jr. Colo. Cattlemens Assn., 1979, Douglas County 4-H, 1980. Mem. AAAS, Am. Soc. Animal Sci., Inst. Food Technologists, Am. Meat Sci. Assn., Nat. Cattlemen's Assn. (cons. 1988-92). Achievements include development of foodservice cooking guidelines, research on adipose tissue growth in feeder steers, U.S. beef consumption vs. rain forest depletion, tissue-specific activity of pentose-phosphate enzymes, multivariate analysis. Office: USDA-AMS Livestock-Seed Div 2603 South Bldg PO Box 96456 Washington DC 20090-6456

BELKA, KEVIN LEE, electrical engineer; b. Colby, Kans., Feb. 28, 1949; s. Bernard Albert and Louise Harriet (Havranek) B.; m. Rochelle Ruth Franklin, Apr. 10, 1971; children: Kevin Lee Jr., Laura Franklin. BSEE, U. Nebr., 1972, MS, 1990. Registered profl. engr., Nebr., Calif., S.C., N.C., Ga., Fla. Dist. engr. Cutler-Hammer Inc., Southfield, Mich., 1972-76; owner, contr. K.B. Elec., Lincoln, Nebr., 1976-83; forensic engr. R.L. Large & assocs., Inc., Lincoln, Nebr., 1983-86; sr. engr. Maniktala Assoc., Inc., Lincoln, Nebr., 1986-88; prin., elec. engring. dept. head Archtl. Engring. Assocs., Columbia, S.C., 1988—. IEEE, NSPE, Nat. Fire Protection Assn., Illuminating Engring. Soc. Office: AEA PO Box 11437 Columbia SC 29211

BELKIN, MICHAEL, ophthalmologist, educator, researcher; b. Tel Aviv, Nov. 14, 1941; s. Yerachmeal and Gita (Seligson) B.; m. Ruth Miriam Loeb, Nov. 8, 1967; children: Tamar, Dan, Daphna. MA, Cambridge U., 1965; MD, Hebrew U., 1969. Chief physician, eye dept. Haddasah Univ. Hosp., Jerusalem, 1975-76; chief rsch. and devel. Mil. Med. Corps., Israel, 1976-79; head, Laser Lab. Tel Aviv Univ. Eye Rsch. Inst., 1981—; dir. Eye Rsch. Inst. Tel Aviv Univ. Med. Sch., 1981-91, prof. ophthalmol., 1987—, chmn., dept. ophthalmology, 1988-92; chmn. Israel Eye Rsch. Assn., 1983-86, Adv. com. in Ophthalmology, Israel, 1985—; mem. World Coun. Ergoophthalmology, Stockholm, 1988—; vis. prof. Letterman Army Inst. Rsch., San Francisco, 1979-81, 89, 90. Contbg. author books in field, publs.; contbr. articles to profl. jours. Chmn. Ophthalmic Planning Com. for Nat. Emergencies, Israel, 1982—; mem. Human Use in Sci. Experiment Com., Tel Hashomer, Israel, 1984—; bd. dirs. Ramot Ltd., Tel Aviv, 1988—. Lt. col. Israeli Med. Corps, 1976-84. Recipient Landau Rsch. prize Israel Med. Assn., 1976, Stein Rsch. prize Tel Aviv U., 1985, 88, Merit award and medal Internat. Ophthalmol. Congress, 1990, Beatrice lectr. U.S. Army, 1990. Mem. Internat. Soc. for Eye Rsch., Assn. for Rsch. in Vision and Ophthalmology, Am. Acad. Ophthalmology. Achievements include 10 patents; development of instrument for detection and analysis of intraocular foreign bodies; design of means of eye protection for military and civilian operations. Office: Sheba Med Ctr, Eye Inst, 52621 Tel Hashomer Israel

BELKIND-GERSON, JAIME, gastroenterologist, nutritionist, researcher; b. Mexico City, Sept. 28, 1964; came to U.S., 1988; s. Roberto Belkind and Raquel (Gerson) Shveid. MD, U. Nat. Antonoma Mex., Mexico City, 1988. Resident in pediatrics Mass. Gen. Hosp. Harvard U., Boston, 1991—, clin. fellow in pediatrics, 1991—. Mem. AMA (resident), Am. Acad. Pediatrics (resident), Mass. Med. Soc. (resident), N.Y. Acad. Scis., Harvard U. Mex. Assn. Jewish. Achievements include research in various aspects of pediatric nutrition as well as intestinal and hepatic disease. Home: 55 Wedgewood Rd Newton MA 02165 Office: Mass Gen Hosp Fruit St Boston MA 02114

BELL, CHARLES EUGENE, JR., industrial engineer; b. N.Y.C., Dec. 13, 1932; s. Charles Edward and Constance Elizabeth (Verbelia) B.; B. Engring., Johns Hopkins U., 1954, M.S. in Engring., 1959; m. Doris R. Clifton, Jan. 14, 1967; 1 son, Scott Charles Bell. Indsl. engr. Signode Corp., Balt., 1957-61, asst. to plant mgr., 1961-63, plant engr., 1963-64, div. indsl. engr., Glenview, Ill., 1964-69, asst. to div. mgr., 1969-76, engring. mgr., 1976—; host committeeman Internat. Indsl. Engring. Conf., Chgo., 1984, 92. Served with U.S. Army, 1955-57. Registered profl. engr., Calif. Mem. Am. Inst. Indsl. Engrs. (pres. 1981), Indsl. Mgmt. Club Central Md. (pres. 1964), Nat. Soc. Profl. Engrs., Ill. Soc. Profl. Engrs., Soc. Plastics Engrs. Republican. Roman Catholic. Home: 1021 W Old Mill Rd Lake Forest IL 60045-3749 Office: Signode Corp 3610 W Lake Ave Glenview IL 60025-5800

BELL, CHESTER GORDON, computer engineering company executive; b. Kirksville, Mo., Aug. 19, 1934; s. Roy Chester and Lola Dolph (Gordon) B.; m. Gwendolyn Kay Druyor, Jan. 3, 1959; children: Brigham Roy, Laura Louise. BSEE, MIT, 1956, MSEE, 1957; DEng (hon.), WPI, 1993. Engr. Speech Communication Lab., MIT, Cambridge, 1959-60; mgr. computer design Digital Equipment Corp., Maynard, Mass., 1960-66, v.p. engring., 1972-83; prof. computer sci. Carnegie-Mellon U., 1966-72; vice chmn. Encore Computer Corp., Marlboro, Mass., 1983-86; asst. dir. NSF, Washington, 1986-87; v.p. R & D Stardent Computer, Sunnyvale, Calif., 1987-89; bd. dirs. Inst. Rsch. and Coordination Acoustic Music, Cirrus Logic, Chronologic Simulation, The Bell-Mason Group, Univ. Video Comm., Visix Software, Velox; bd. dirs., trustee Computer Mus., 1982—. Author: (with Newell) Computer Structures, 1971, (with Grason, Newell) Designing Computers and Digital Systems, 1972, (with Mudge, McNamara) Computer Engineering, 1978, (with Siewiorek, Newell) Computer Structures, 1982, (with McNamara) High Tech Ventures, 1991. Recipient 6th Mellon Inst. award, 1972, Nat. medal Tech. U.S. Dept. COmmerce Rechnology Adminstrn., 1991. Fellow IEEE (McDowell award 1975, Eckert-Mauchly award 1982), AAAS (von Neumann medal 1992, AEA award for greatest econ. contbn. to region 1993); mem. Nat. Acad. Engring., Assn. for Computing Machinery (editor Computer Structures sect. 1972-78), Eta Kappa Nu. Home: 450 Old Oak Ct Los Altos CA 94022-2634

BELL, CORINNE REED, psychologist; b. Holly Springs, Miss., July 6, 1943; d. Robert Norris and Laura Kathleen (Robinson) Reed; children: Jeffrey Kenneth, Jennifer Michelle. BA with highest honors, U. Tenn., 1976, MA, 1978, PhD, 1985. Lic. psychologist, Tenn. Rsch. asst. Lakeshore Mental Health Inst., Knoxville, 1976; instr. psychology dept. Roane State C.C., Harriman, Tenn., 1979-80; cons., sch. psychologist dept. spl. edn. U. Tenn., Knoxville, 1979-80; pvt. practice psychology Knoxville, 1979-85; psychologist, ptnr. Clin. & Sch. Assocs., Knoxville, 1985—; profl. supr., adminstr., bus. mgr. Clin. & Sch. Assocs., Knoxville, 1985—; cons. social svcs. Dept. Human Svcs., Tenn., 1985—. Contbr. articles to profl. jours. Mem. adv. bd. John Tarleton Children's Home, Knoxville, 1987—; bd. dirs. Sexual Assault Crisis Ctr., Knoxville, 1987—; mem., cons. Knox County Child Abuse Rev. Team, Knoxville, 1982—. Acad. scholar U. Tenn., 1974; rsch. grantee Knox County Children's Found., 1979. Mem. APA, Tenn. Psychol. Assn. (pub. rels. chair 1992—), Knoxville Area Psychol. Assn. (treas. 1986-88, pub. rels. chair 1989-90, pres. 1990-91), Unified Psychology Coalition (co-founder, legis. chair/spokesperson 1989-92), Tenn. Assn. Sch. Psychologists, Mortar Bd., Phi Kappa Phi, Phi Beta Kappa. Avocations: performing arts, piano, creative writing, physical fitness. Office: Clin and Sch Assocs 5912 Toole Dr Ste B Knoxville TN 37919-4173

BELL, FRANCES LOUISE, medical technologist; b. Milton, Pa., Apr. 28, 1926; d. George Earl and Kathryn Robbins (Fairchild) Reichard; m. Edwin Lewis Bell II, Dec. 27, 1950; children: Ernest Michael, Stephen Thomas, Eric Leslie. BS in Biology cum laude, Bucknell U., 1948; MT, Geisinger Meml. Hosp., 1949. Registered med. technologist. Med. technologist Burlington County Hosp., Mt. Holly, N.J., 1949-50, Robert Packer Hosp., Sayre, Pa., 1950, Carle Hosp./Clinic, Urbana, Ill., 1951-52, St. Joseph Hosp., Reading, Pa., 1972-83. Vol. Crime Watch, City Hall, Reading, 1985-90, Am. Heart Assn., Reading, 1956—, March of Dimes, Reading, 1956-72, Am. Cancer Soc., Reading, 1956-71, Multiple Sclerosis, Reading, 1956-72, Reading Musical Found., 1985-90, Hist. Soc. Berks County; corr. sec. women's aux., 1986-90; fin. sec. women's aux. Albright Coll., 1988—; hospitality co-chmn. women's com. Reading Symphony Orch., 1985-90, co-editor yearbook women's com., 1990-92, editor yearbook women's com., 1992—; chmn. hospitality Reading-Berks Pub. Librs., 1988-91; mem. Friends Reading Mus., Berks County Conservancy. Mem. Woman's Club of Reading (treas. 1986-88, fin. sec. 1991—), AAUW (assoc. editor bull. 1961-63, cultural interests rep. 1967-68), United Meth. Women, Phi Beta Kappa. Republican. Methodist. Avocations: music appreciation, photography, postcard art prints. Home: 1454 Oak Ln Reading PA 19604-1865

BELL, HELEN CHERRY, chemistry educator; b. Halls, Tenn., June 4, 1937; d. William McKinley and Bessie (Riddick) Cherry; m. Tom Bell, Aug. 2, 1959; children: Sandra Claire Bell Baker, Tom Bell II. BS, BA, U. Tenn., Martin, 1959; MS, Memphis State U., 1965; PhD, U. Tenn., 1975; postdoctoral work in gas chromatography, atomic absorption spectrophotometry, UV and IR spectrophotometry, high performance liquid chromatogrphy. Cert. edn. in chemistry, physics, math. Tchr. algebra I and II Halls High Sch., 1959-60; tchr. chemistry, physics and biochemistry Dyersburg (Tenn.) City Schs., 1960-69; supr. instrn. Dyer County Sch. System, Dyersburg; prof. chemistry Dyersburg State Community Coll., 1983—; chmn. faculty senate Dyersburg State C.C., 1989-93; faculty bd. mem. Tenn. Bd. Regents, 1992—. Mem. election commn. Dyer County, Dyersburg, 1985-88, campaign mgr., 1978, mem. exec. com., 1988; mem. Dyer County Consol. Sch. Planning Commn. Recipient Master Tchr. award for excellence in teaching, 1993, Nat. Catalyst award Chem. Mfrs. Assn., 1993. Fellow Am. Inst. Chemists ; mem. Am. Chem. Soc. (bd. dirs. SE region 1985-91, presenter, sponsor student affiliates 1983—), Tenn. Acad. Sci. (bd. dirs., all other offices), AAAS, Nat. Assn. Sci. Tchrs. (presenter). Mem. Ch. of Christ. Achievements include development of a Scientific Instrumentation Program for Tennessee Board of Regents Schools, Colleges and Universities. Home: 225 Tom Bell Rd Friendship TN 38034 Office: Dyersburg State Community Coll Lake Rd Dyersburg TN 38024-3815

BELL, JAMES MILTON, psychiatrist; b. Portsmouth, Va., Nov. 5, 1921; s. Charles Edward and Lucy (Barnes) B. Student, Va. State Coll., 1939-40; BS, N.C. Cen. U. (formerly N.C. Coll.), 1943; MD, Meharry Med. Coll., 1947. Diplomate in psychiatry and child psychiatry Am. Bd. Psychiatry and Neurology (examiner 1980—), Pan. Am. Med. Assn. (coun. psychiatry sect.); cert. N.Y. State Dept. Mental Hygiene. With Harlem Hosp., N.Y.C., 1947-48; asst. physician to clin. dir. Lakin (W.Va.) State Hosp., N.Y.C., 1948-51; fellow gen. psychiatry Menninger Sch. Psychiatry-Menninger Found., Topeka, 1953-56, tng. child psychiatry, 1957-58; resident Winter VA Hosp., Topeka, 1953-56; asst. sect. chief children's unit Topeka State Hosp., 1956-58; clin. teaching staff Menninger Sch. Psychiatry, 1956-58; clin. dir., psychiatrist Berkshire Farm Ctr. and Svcs. for Youth, Canaan, N.Y., 1959-86, sr. child and adolescent psychiatrist, 1986—; clin. asst. to clin. prof. psychiatry Albany Med. Coll., Union U., 1959—, mem. admission com., 1972-79; psychiatrist-in-charge Albany Home for Children, N.Y., 1959-77; staff psychiatrist Parsons Child and Family Ctr., 1977—; asst. dispensary to dispensary psychiatrist Albany Med. Ctr. Clinic, 1960; trainee cons. Albany Child Guidance Ctr. Psychiat. Svc., Inc., 1961. cons. Keller U.S. Army Hosp., U.S. Mil. Acad., West Point, N.Y. Contbr. numerous articles to profl. jours. Cons. Astor Home for Children, Rhinebeck, N.Y., 1965; instrnl. staff Frederick Amman Meml. Inst. Delinquency and Crime, St. Lawrence U., 1965-70; cons. adolescence N.Y. State Div. Youth, 1966-76, mem. med. rev. bd., 1974-76; mem. Child Abuse Adv. Coun., Albany; bd. dirs., mem. com. on proposed policy N.Y. Spaulding for Children, v.p., 1988-89; bd. dirs., exec. com. Guold Farm, Barrington, Mass. Capt. M.C., AUS, 1951-53; col. USAR, 1955-85, ret., 1985. Decorated Army Commendation medal, Meritorious Svc. medal, others; named Disting. Alumnus, Meharry Med. Coll., 1990. Fellow AAAS (life), Am. Psychiat. Assn. (life, chmn. coun. nat. affairs 1973-75, past vice-chmn.), Am. Acad. of Child and Adolescent Psychiatry (life, chmn. com. psychiat. facilities for children and adolescence 1973-75), Am. Orthopsychiat. Assn. (life, past dir.), Am. Soc. Adolescent Psychiatry (life), N.Y. Acad. Scis., Am. Coll. Psychiatrists (past mem. Stanley Dean award com.), Am. Psychopathol. Assn., mem. AMA, NAACP (life), Am. Acad. Polit. and Social Sci., Am. Soc. Addiction Medicine, Black Psychiatrists Am., Group for Advancement of Psychiatry (com. on child psychiatry), Inst. Religion and Health (charter), Council for Exceptional Children, Nat. Assn. Tng. Schs. and Juvenile Agys., Assn. N.Y. Educators of Emotionally Disturbed, Nat., N.Y. State, Columbia Country med. assns., Child Care Workers (bd. dirs. N.Y.), Assn. Psychiat. Treatment of Offenders, N.Y. State Soc. Med. Rsch., N.Y. Capitol Dist. Coun. Child Psychiatry (pres. 1974), Am. Legion (life), Rotary, Alpha Omega Alpha. Home: Hudsonview 175 Old Post Rd N Croton-on-Hudson NY 10520 Office: Berkshire Farm Ctr & Svcs for Youth Canaan NY 12029

BELL, JEROME ALBERT, aerospace engineer; b. Houston, Mar. 24, 1940; m. Judith Louise Traiber; children: Elyse R., Robert F. Student, Northrop Aero. Inst., 1957-58; BS, U. Tex., 1963. Mgr. spl. projects NASA, Houston, 1976-83, tech. asst. for space sta., 1983-85, mgr. customer integration space sta. program office, 1985-87, system engr. for resource allocation, 1987-88, system engr. new initiatives div., 1988-90, mgr. ops. explorations programs office, 1990-92, sr. lead engr. for future programs, 1992—. Sr. mem. AIAA (mem. space and ops. tech. support com.); mem. Lion's Club. Achievements include development of PROX operations concepts and heading up BASA explorations operations group. Home: 6219 S Braeswood Blvd Houston TX 77096-3715

BELL, LAURA JEANE, retired nurse; b. Chgo., Mar. 11, 1922; d. Harold Elwood and Mary Etta (Sprague) Downey; m. David Hoge Bell, Feb. 21, 1943; children: David, Roy, Thomas, John, Ruth, Keith, Mary, Richard, Howard. AA, Blackburn Coll., 1941; diploma in nursing, St. Elizabeths Hosp., Washington, 1946; BS in Nursing, Washington U., St. Louis, 1962; MS in Edn., St. Louis U., 1969. RN, Mo. Relief supr. St. Vincent's Hosp., St. Louis, 1949-58; staff asst. head nurse, head nurse Barnes Hosp., St. Louis, 1958-61, instr., coord. Sch. Nursing, 1962-67; instr. Jefferson Barracks VA Hosp., St. Louis, 1967-70, asst. chief nursing svc. for edn., 1970-71; instr. Jefferson Barracks div. St. Louis VA Med. Ctr., 1971-73, med. supr. John Cochran div., 1973-81, supr. ambulatory care Jefferson Barracks div., 1981-84; ret., 1984. Dist. rep. intercultural programs Am. Field Svc., St. Louis, 1983-91; rec. sec. Overland (Mo.) Hist. Soc., 1988-89, pres., 1990-92; pres. United Meth. Women, Stephan Meml. Ch., Charlack, Mo., 1988-92; mem. Women's Polit. Caucus, St. Louis, 1987—; v.p. St. Louis North Dist. United Meth. Women, 1992—; sec. Mo. Ch. Women United, 1992—; chair Commn. on Status and Role of Women, Mo. East Conf., U. Mo., 1992—; mem. Wesley Found. Bd., St. Louis, 1992—; pres. Met. St. Louis Ch. Women United, 1993—. Named Fed. Employee of Yr., Fed. Exec. Bd., 1970. Mem. ANA (accreditation com. cen. region 1984-88), Mo. League for Nursing (pres. 1978-80, chmn. bylaws 1985-92), Mo. Nurses Assn. (continuing edn. and bylaws com., chmn. nominating com., 3d dist. 1988, regional bd. dirs. 1988-90, Pres.'s award 1989), Mo. Student Nurses Assn. (hon., scholarship named in her honor 1982). Home: 2418 Oakland Ave Saint Louis MO 63114-5016

BELL, PETER MAYO, geophysicist; b. N.Y.C., Jan. 3, 1934; s. Frank Kirkhaugh and Mary Elizabeth (Mayo) B.; m. Norma Joan Erkert, June 20, 1959; children: Peter Mayo, James, Elizabeth, Bradford. B.S., St. Lawrence U., 1956; M.S., U. Cin., 1959; A.M., Harvard U., 1961, Ph.D., 1963. Solid state physicist Office of Aerospace Research, Bedford, Mass., 1963; postdoctoral fellow Carnegie Instn., Washington, 1963-64, staff geophysicist, 1964—; adv. space mission planning NASA, Houston and Washington, 1973-74, lunar sample advisor, 1974-80; mem. adv. bd. Petroleum Research Fund, Washington, 1982—; sci. and tech. advisor Norton-Christensen Inc., Salt Lake City, 1983—; mem. corp. tech. com. Norton Co., Worcester, Mass., 1983—, v.p., corp. tech., 1986-90; v.p. corp. rsch., chief scientist St.-Gobain Corp., Worcester, Mass., 1990—. Author Infrared of Minerals, 1975; editor: EOS, 1980—; mem. editorial bd.: Sci. Mag., 1982-83; mem.: Advances in Geochemistry, 1980—; contbr. articles to profl. jours.; patentee system to make solid H2, 1982. Nat. Capital Area Council Cub Scout leader Boy Scouts of Am., 1964-72; mem. Pres.'s bd. advisors MIT, Pres.'s Com. Worcester Poly. Inst. Served to 1st lt. U.S. Army, 1957. Recipient Medal for Highest Sci. Achievement NASA, 1976; named Guiness Book of Records, Guiness Found., 1981; Guggenheim Found. fellow, 1982; Fairchild Disting. Scholar Calif. Inst. Tech., 1983. Fellow Am. Geophys. Union, Am. Mineral Soc.; mem. Geochem. Soc., Geol. Soc. Washington, Potomac Geophys. Soc. (v.p. 1983-84), Geol. Soc. Am., Am. Chem. Soc., Am. Phys. Soc., Sigma Xi. Republican. Episcopalian. Clubs: Cosmos (Washington); West River Sailing (Galesville, Md.); Royal Bermuda Yacht (Hamilton); Annapolis (Md.) Yacht; Plaza Club (Worcester). Office: Norton Co PO Box 15008 Worcester MA 01615-0008 also: 1 New Bond St Worcester MA 01606-2698

BELL, THOMAS EUGENE, psychologist, educational administrator; b. Okmulgee, Okla., Feb. 20, 1945; s. Wilmer Ordell and Betty Jean (Good) Bell; m. Ramona Kay Ashlock, Aug. 26, 1965; 1 child, Stacie Lane. BA, Cen. State U., Edmond, Okla., 1972, MEd, 1975; postgrad., Okla. State U., 1986-88. Lic. profl. counselor, Okla. Psychometrist Guthrie (Okla.) Pub. Schs., 1975-79, sch. psychologist, 1979-89, dir. counseling, 1989—. Developer Teen Buddies, 1990. With USAF, 1965-67. Recipient Parent Edn. award Okla. Juvenile Justice, Oklahoma City, 1991. Mem. Mensa Internat. (proctor 1979-86), Okla. Psychol. Assn. (rep. 1986-87), Okla. Sch. Psychology Assn. (area rep. 1987-88), Nat. Assn. Sch. Psychologists, Youth Suicide Prevention Assn. (v.p. 1991—). Democrat. Mem. Ch. of Christ. Avocations: scuba diving, swimming, golf, travel. Home: 1101 Apollo Cir Edmond OK 73034-2410 Office: Guthrie Pub Schs 802 E Vilas Ave Guthrie OK 73044-5228

BELLAICHE, CHARLES ROGER, computer company executive; b. Tunis, Tunisia, June 21, 1955; s. Robert and Colette (Sarfati) B.; m. Andrea Mustaff, July 25, 1986; children: Lisa, Mickaël. Degree in civil engring., Ecole des Mines de Paris, 1977. Sales engr. Hewlett-Packard, Boblingen, Germany, 1979-82, product mgr., 1981-85, mktg. mgr. Europe, 1985—. Lt. French Air Force, 1977-79. Office: Hewlett Packard, Rte du Nant d'Avril 150, CH-1217 Meyrin 2 Geneva Switzerland

BELLANGER, BARBARA DORIS HOYSAK, biomedical research technologist; b. Syracuse, N.Y., Oct. 24, 1936; d. Edward George and Bernardine Elizabeth (Blaney) Hoysak; m. Ronald Patrick Bellanger, July 1, 1961; children: Laura Jeanne, Andrea Lynne, Janis Anne. BS, Syracuse U., 1958. Cert. lab. animal technician. Tech. asst. Bur. of Labs., Syracuse, 1958; rsch. scientist Bristol Labs., Syracuse, 1958-63; rsch. assoc. Syracuse Cancer Rsch. Inst., 1973—. Pres. CNS Northstars Band Parents, Inc., Cicero-North Syracuse, N.Y., 1986-87. Mem. Am. Assn. Lab. Animal Sci. (cert. lab. animal technician, sec. Upstate N.Y. br. 1990—, Technician of Yr. award 1992), N.Y. Acad. Scis., Alpha Gamma Delta (pres. Alpha alumnae chpt. 1959-60, treas. 1989—). Home: 410 David Dr North Syracuse NY 13212-1929 Office: Syracuse Cancer Rsch Inst Presidential Pla 600 E Genesee St Syracuse NY 13202-3108

BELLANTI, JOSEPH A., microbiologist, educator; b. Buffalo, Nov. 21, 1934. MD, U. Buffalo, 1958. Diplomate Am. Bd. Pediatrics, Am. Bd. Allergy and Immunology. Intern Millard Fillmore Hosp., Buffalo, 1958-59; resident in pediatrics Children's Hosp., Buffalo, 1959-61; NIH spl. trainee immunology J. Hillis Miller health ctr. U. Fla., 1961-62; rsch. virologist dept. virus disease Walter Reed Army Inst. Rsch., Washington, 1962-64; from asst. to assoc. prof. sch. medicine Georgetown U., Washington, 1963-70, prof. pediatrics and microbiology, 1970—, dir. ctr. interdisciplinary studies immunology, 1975—; mem. growth and devel. com. NIH, 1970-75; mem. med. adv. com. Nat. Kidney Found., 1971—; chmn. Infectious Disease Com., 1972; dir. Am. Bd. Allergy and Immunology, 1975—. Recipient William Peck Sci. Rsch. award, 1966, Sci. Exhibit award Am. Acad. Clin. Pathologists, Coll. Am. Pathology, 1966; Mead Johnson grantee, 1964. Fellow Am. Acad. Pediatrics (E. Mead Johnson award 1970), Am. Acad. Allergy, Am. Acad. Allergists, Am. Assn. Clin. Immunology and Allergy; mem. AMA. Achievements include research in immunologic aspects of facultatively slow virus infections, biochemical changes in human polymorphonuclear leukocytes during maturation. Office: Georgetown U Internat Ctr Studies of Immunology 3800 Reservoir Rd NW Washington DC 20007-2196*

BELLENKES, ANDREW HILARY, aerospace experimental psychologist; b. Bklyn., July 21, 1950; s. William and Thelma (Lackowitz) B.; m. Susanna von Urbanski, June 23, 1990; children: Stephanie, Christoph. BA, C.W. Post Coll., 1973; MA, Fairleigh Dickinson U., 1976; diploma, U. Innsbruck, Austria, 1988. Cert. practicing human factors engr.; registered ergonomics specialist. Aero. exptl. psychologist Naval Aero. Med. Rsch. Lab., Pensacola, Fla., 1982-85; human factors engr. Grumman Aircraft Systems Div., Bethpage, N.Y., 1985-88; dir. human factor engring. Control Data Corp., Govt. Systems, Bloomington, Minn., 1989-90; chief human factors div. U.S. Naval Safety Ctr., Norfolk, Va., 1990—. Contbr. articles to profl. jours. Lt. comdr. USN, 1982—. Decorated D.S.M.; recipient Nat. Def. Medal, 1991, USN Achievement medal, 1985, Second Pl. Grad. Student Paper award, 1978, U. Del. Assistantship Support, 1978-81, U. Del. Grant-in-Aid, 1978, N.Y. State Scholarship Incentive award, 1968. Mem. AIAA (mem. life sci. and systems tech. com., group head young mems. com.), Aero. Med. Assn. (mem. sci. and technology com.), Internat. Astro. Fedn. (observer bioastronautics com., mem. student activities commn., judge student awards com. 1986), Ergonomics Soc. Home: 4712 Hermitage Rd Virginia Beach VA 23455-4032 Office: Naval Safety Ctr NAS Norfolk Norfolk VA 23511-5796

BELLER, MARTIN LEONARD, retired orthopaedic surgeon; b. N.Y.C., Apr. 30, 1924; s. Abraham Jacob and Ida (Fishkin) B.; m. Wilma Gertrude Kjelgaard, June 29, 1947; children: Alan Lewis, Beatrice Ann Beller Foreman, Peter James. A.B. with honors, Columbia U., 1944, M.D., 1946. Diplomate: Am. Bd. Orthopaedic Surgery. Intern Mt. Sinai Hosp., N.Y.C., 1946-47; resident in orthopaedic surgery Hosp. Joint Diseases, N.Y.C., 1949-52; practice medicine specializing in orthopaedic surgery Phila., 1952-87; asst. prof. orthopaedic surgery U. Pa. Sch. Medicine, Phila., 1967-72; assoc. prof. U. Pa. Sch. Medicine, 1972-80, clin. prof., 1980-87; attending orthopaedic surgeon Hosp. U. Pa., 1963-87; assoc. attending orthopaedic surgeon Albert Einstein Med. Center, Phila., 1960-70; chmn. dept. orthopaedic surgery Albert Einstein Med. Center (Daroff div.), 1970-79. Author: (with I. Stein and R. O. Stein) Living Bone in Health and Disease, 1955, (with I. Stein) Clinical Densitometry of Bone, 1970. Vestryman Episcopal Ch., 1966-70, 71-87, 90-93. Capt. M.C., AUS, 1947-49. Am. Orthopaedic Assn. exchange fellow Gt. Britain, 1963. Fellow ACS, Am. Acad. Orthopaedic Surgeons (bd. councilors 1978-81, Pa. rep. commn. on trauma 1984-87), Internat. Soc. Orthopaedic Surgery and Traumatology; mem. Am. Orthopaedic Assn., Pa. Orthopaedic Soc. (pres. 1975-77), Orthopaedic Rsch. Soc., Am. Coll. Rheumatology, N.Y. Acad. Sci., Phi Beta Kappa, Alpha Omega Alpha, Phi Delta Epsilon (nat. pres. 1975-76, chmn. bd. trustees 1984-85, assoc. exec. sec. 1991—). Republican. Home: RR 1 Box 256 B Gaines PA 16921-9768

BELLERO, CHIAFFREDO JOHN, civil engineer; b. Torino, Italy, Aug. 17, 1926; came to the U.S., 1982; s. Francesco and Ebba Olga (Ferrato) B.; m. Anna Cristina Larghi, Jan. 7, 1957; 1 child, Luisa. MCE, Poly. U., Turin, Italy, 1951; postgrad., Poly. U., Guayaquil, Ecuador, 1968. Registered profl. engr., Italy, Ecuador. Mgr. Recchi S.P.A. Costruzioni, Turin, 1958-78; joint gen. mgr. Sideco Am., Buenos Aires, 1978-83; pres., chief exec. officer Sideco North Am., N.Y.C., 1982-84; mng. dir. SEIFRA, Rome, 1985-88; chief exec. officer, sr. v.p. Recchi Am. Inc., Miami, Fla., 1988—. With British Army, 1943-45. Decorated Silver medal Republic of Italy, 1944. Fellow ASCE; mem. Internat. Assn. for Bridges and Structural Engring., Canadian Soc. Civil Engring., Post Tensioning Inst. Roman Catholic. Home: 1280 S Alhambra Cir # 1116 Coral Gables FL 33146 Office: Recchi Am Inc 9200 S Dadeland Blvd # 225 Miami FL 33156

BELLI, FEVZI, computing science educator, consultant; b. Menemen, Turkey, June 14, 1948; arrived in Germany, 1965; s. Murat and Kiymet Belli (Guler) B.; m. Bettina Weyer, June 20, 1973; children: Murat, Aslan. Diploma in Engring., Tech. U. Berlin, 1973, D of Engring., 1978, habil., 1988; cert. in econs., U. Hagen, Germany, 1985. Mem. sci. staff Gesellschaft fuer Mathematik und Datenverarbeitung mbH, Bonn, Germany, 1974-78; systems analyst Elektronik-System-GmbH, Munich, 1978-83; prof. Hochschule Bremerhaven, Germany, 1983-88; univ. prof. U. Paderborn, Germany, 1989—; faculty European divsn. U. Md., 1985-91; cons. software, Munich, Berlin and Istanbul, Turkey, 1978&; bd. sci. adv. dir. Cad Lab. of Siemens Nixdorf U., Paderborn, 1989—; chmn. several internat. sci. confs. Author 8 sci. textbooks on programming and software engring.; editor numerous books, mags.; contbr. over 70 papers to confs., books and profl. jours. Chmn. Internat. Student Orgns., Berlin, 1967-69. With Turkish Army, 1976. Mem. IEEE (founding mem., tech. com. software reliability), German Assn. for Computing (exec. bd.) Assn. for Computing Machinery, Info. Resources Mgmt. Assn. (founding mem., mem. adv. bd.), Computing Soc. Fault Tolerant Computing (bd. dirs. 1982—), Internat. Soc. for Applied Intelligence (founding mem., assoc. v.p.), Rotary (chmn. vocat. svc. 1991—). Avocations: swimming, fishing, jogging, music. Office: U Paderborn Dept Elec Engrg, PO Box 1621, D-33095 Paderborn Germany

BELLINI, FRANCESCO, chemist; b. Ascoli, Piceno, Italy, Nov. 20, 1947; s. Berardino Bellini; m. Marisa; children: Roberto, Carlo. Diplome in Chem. Engring., I.T.I.S., Italy, 1967; BSc in Chemistry, Coll. Loyola, Montreal, Can., 1972; PhD in Organic Chemistry, U. New Brunswick (Can.), 1977. Rsch. asst. Ayerst Labs., 1968-74, postdoctoral fellow, 1977-79, sr. scientist, 1979-81, rsch. assoc., 1981-84; dir. biochems. div. Institut Armand Frappier, Laval, Que., 1984-86; pres., chief exec. officer Biochem Pharma Inc. (formerly IAF Biochem Internat. Inc.), Laval, Que., 1986—; vice chmn. bd. Nava, Inc., Beltsville, 1990—; dir. BioCapital Inc., Montreal, IAF BioVac Inc., Laval. Contbr. numerous articles to profl. jours. Achievements include patents on angiotensin enzyme inhibitors; discovery of 6-(lower alkoxy)-5-(trifluorimenthyl)-1-naphtalene-carboxylic acid, known and Tolrestat, used as an aldose reductase inhibitor; co-author of 20 patents. Office: Biochem Pharma Inc, 2550 b Daniel-Johnson #600, Laval, PQ Canada H7T 2L1

BELLIS, CARROLL JOSEPH, surgeon; b. Shreveport, La.; s. Joseph and Rose (Bloome) B.; m. Mildred Darmody, Dec. 26, 1939; children—Joseph, David. BS, U. Minn., 1930, MS in Physiology, 1932, PhD in Physiology, 1934, MD, 1936, PhD in Surgery, 1941. Diplomate Am. Bd. Surgery. Resident surgery U. Minn. Hosps., 1937-41; pvt. practice surgery Long Beach, Calif., 1945—; mem. staff St. Mary's, Community hosps., Long Beach; cons. surgery Long Beach Gen. Hosp.; prof., chmn. dept. surgery Calif. Coll. Medicine, 1962—; surgical cons. to Surgeon-Gen., U.S. Army. Author: Fundamentals of Human Physiology, 1935, A Critique of Reason, 1938, Lectures in Medical Physiology; contbr. numerous articles in field of surgery, physiology to profl. jours. Served to col. M.C. AUS, 1941-46. Nat. Cancer Inst. fellow, 1934; recipient Charles Lyman Green prize in physiology, 1934; prize Mpls. Surg. Soc., 1938; ann. award Mississippi Valley Med. Soc., 1955. Fellow ACS, Royal Soc. Medicine, Internat. Coll. Surgeons, Am. Coll. Gastroenterology, Am. Med. Writers Assn., Internat. Coll. Angiology (sci. council), Gerontol. Soc., Am. Soc. Abdominal Surgeons, Nat. Cancer Inst., Phlebology Soc. Am., Internat. Acad. Proctology, Peripheral Vascular Soc. Am. (founding); mem. AAAS, Am. Assn. Study Neoplastic Diseases, Mississippi Valley Med. Soc., N.Y. Acad. Scis., Hollywood Acad. Medicine, Am. Geriatrics Soc., Irish Med. Assn., Am. Assn. History Medicine, Pan Pacific Surgical Assn., Indsl. Med. Assn., L.A. Musicians Union (hon.), Pan Am. Med. Assn. (diplomate), Internat. Bd. Surgery (cert.), Internat. Bd. Proctology (cert.), Wisdom Soc. (wisdom award of honor), Sigma Xi, Phi Beta Kappa, Alpha Omega Alpha. Office: 904 Silver Spur Rd Ste 804 Rolling Hills Estates CA 90274

BELLISTON, EDWARD GLEN, medical facility administrator, consultant; b. Upland, Calif., Oct. 20, 1958; s. G. Howard and MaryAnn (Fitzgerald) B.; m. Kristine Marie Holmes, Aug. 12, 1981. BS, Brigham Young U., 1984, MHA, 1987. Cert. EMT. Admitting clk. Utah Valley Regional Med. Ctr., Provo, 1985; adminstrv. resident St. Benedict's Hosp., Holy Cross Health System, Ogden, Utah, 1986; fin. counselor Utah Valley Regional Med. Ctr., Provo, 1986, adminstrv. dir., 1986; regional clin. admminstrv. coord. Intermountain Health Care-IHC Physicians Svcs., Salt Lake City, 1989; sr. phys. cons. Intermountain Health Care-IHC Physicians Svcs., Salt Lake, 1990—. Instr. gospel doctrine LDS Ch., Springville, Utah, 1989-91, bishopric Brigham Young U. 81st ward, 1991—, leader Boy Scouts Am., Springville, 1990. Mem. Med. Group Mgmt. Assn., Am. Coll. Healthcare Execs. Republican. Avocations: fishing, collecting antique medical instruments, gardening, coin collecting. Home: 363 W 1400 N American Fork UT 84003-2794 Office: Intermountain Health Care PO Box 57010 650 E 4500 S Ste # 340 Salt Lake City UT 84157-0010

BELLOBONO, IGNAZIO RENATO, chemist, educator; b. Alexandria, Egypt, Nov. 12, 1932; s. Sebastiano and Paola Lucia (Cifali) B.; m. Maria Letizia Stefanelli, May 7, 1966. Dipl. indsl. chemistry, U. Milan, Italy, 1954; PhD of Chemistry, U. Milan, 1964. Lectr., asst. prof. U. Milan, Italy, 1955-71; prof. chemistry U. Rome, Italy, 1971-72, U. Milan, 1973—. Author: Physical Principles of Chemistry, 1974, General Chemistry, 1964; contbr. articles to profl. jours. Mem. Royal Soc. Chemistry, Am. Chem. Soc., Am. Soc. of Imaging Sci. and Engring. Roman Catholic. Achievements include patents in field; discovered an entirely innovative method for membranes production and their application to environmental technologies. Home: 10 via G Keplero, I-20124 Milan Italy Office: Univ Milan, 19 via C Golgi, I-20133 Milan Italy

BELLSTEDT, OLAF, software engineer; b. Munich, Oct. 23, 1966; came to U.S., 1984; s. Karl-Heinz and Ria (Filsinger) B.; m. Jessica Mae Patterson, Dec. 16, 1989. BS in Space Sci., Fla. Inst. Tech., 1990. Mgr. product devel., support Varimetrix Corp., Palm Bay, Fla., 1990—. Ambulance driver, attendant Harbor City Vol. Ambulance Squad, Melbourne, Fla., 1988-89. Mem. AIAA (vice chmn. 1988-89), Profl. Assn. Diving Instrs. (dive master), Theta Xi. Avocations: tennis, flying. Office: Varimetrix Corp 2350 Commerce Park Dr NE Ste 4 Palm Bay FL 32905

BELLUSSI, GIUSEPPE CARLO, chemical research manager; b. Cremona, Italy, Feb. 25, 1953; s. Paride and Angela Rosa (Serafini) B.; m. Maria Grazia Sabini, July 20, 1980; 1 child, Brando. M in Chemistry, U. Parma, Italy, 1978. Researcher Duco S.p.A., Fombio, Italy, 1980-81, Assoreni, S. Donato, Italy, 1981-85; researcher Eniricerche S.p.A., S. Donato, Italy, 1986-89, sr. scientist, 1989-90, mgr., 1991—; cons. UN Ind. Devel. Orgn., Pune, India, 1989. Contbr. articles to Studies in Surface Sci. and Catalysis, 1986, Jour. of Catalysis, 1992, Applied Catalysis, 1991. Recipient D. Breck award Internat. Zeolite Assn., 1992. Mem. Italian Chem. Soc., Am. Chem. Soc., Internat. Zeolite Assn., Stanford Rsch. Internat. (fellowship 1987-89). Achievements include 47 patents in field and 21 scientific papers. Home: Via Scoto 44, 29100 Piacenza Italy Office: Enirichere SpA, Via F Maritano 26, 20097 San Donato Italy

BELMAN, ANITA LEGGOLD, pediatric neurologist; b. N.Y.C., Feb. 20, 1940; d. Benjamin and Martha (Broyde) Leggold; m. Stefan G. Belman, June

10, 1960; children: Matthew, Benjamin, Melissa Hope. BS, Cornell U., 1960, MS, 1961; MD, NYU, 1979. Diplomate am. Acad. Psychiatry and Neurology. Asst. to assoc. prof. neurology SUNY Sch. Medicine, Stony Brook, 1984-89; assoc. prof. neurology and pediatrics Sch. of Medicine, SUNY, Stony Brook, 1989-93, Albert Einstein Coll. Medicine, Bronx, N.Y., 1992—; cons. NIH, Bethesda, 1987—, NIMH, Rockville, 1988—; advisor WHO, Geneva, Switzerland, 1990. Contbr. articles to profl. jours. Trustee Cold Spring Harbor Libr., 1970-76. Grantee WHO, 1989, 1988. Fellow Am. Acad. Pediatrics, Am. Acad. Neurology. Achievements include research on neurologic involvement in infants and children with HIV-I infection.

BELMANS, RONNIE JOZEF MARIA, foundation adminstrator, researcher; b. Duffel, Belgium, May 19, 1956; m. Mieke Alberte Andre; children Nathalie, Wim. MS in Engring. with great distinction, Katholieke U. Leuven, Belgium, 1979, PhD with great distinction, 1984. Sr. rsch. asst. NSF, 1985-87, rsch. assoc., 1987-91, sr. rsch. assoc., 1991—; vis. assoc. prof. McMaster U., Hamilton, Ont. Can., 1989-90; part time assoc. prof. dept. elec. engring., Katholieke U. Leuven; vis. prof. Imperial Coll., London, 1991; internat. sec. Internat. Conf. Elec. Machines, 1990—; bd. dirs. Brandweer Informatiecentrum Gevaarlijke Stoffen te Geel, 1990—, Belgin Inst. for Normalisation; referee SCIENCE project, 1989-91, NATO Advanced Rsch. Workshop, 1990—; speaker in field; expert witness various legal cases. Co-author: Algemene Elektrotechniek, 1991, Elekrtische aandrijvingen voor gebruikers, 1992; editorial bd. ETEP jour., 1992—; contbr. numerous articles to profl. jours. Fellow Von Humboldt, 1988-89. Mem. IEEE (mem. elec. machines com., elec. drives com.), Royal Flemish Engring. Soc. (mem. electrotechnical com. 1991—), Royal Belgian Soc. Elec. Engrs. (mem scientific com.), Instn. Elec. Engrs., Conférence Internationale des Grands Réseaux (study com.), Société Européenne pour la Formation des Ingénieyrs (bd. dirs. 1992—, treas. 1988-89), Soc. Engrs. Leuven, Vlaamse Leergangen Leuven, Soc. Flemish Profs., European Assn. Elec. Drives (treas 1988-90, pres. 1992—). Office: Katholieke U Leuven Dept EE, Kard Mercierlaan 94, B 3001 Leuven-Heverlee Belgium*

BELMARES, HECTOR, chemist; b. Monclova, Coahuila, Mex., Feb. 21, 1938; s. Armando and Guadalupe (Sarabia) B.; B.Sc., Instituto Tecnológico de Monterrey (Mex.), 1960; Ph.D. (Todd fellow 1961-63), Cornell U., 1963; postdoctoral student Calif. Inst. Tech., 1965; m. Eleanor Johanna Wold, Aug. 28, 1965; children: Michelle Anne, Michael Paul, Elizabeth Myrna, Mary Eleanor. Sr. research chemist Rohm and Haas Co., Phila., 1965-71; gen. mgr. tech. and quality control Fibras Químicas, S.A., Monterrey, Mex., 1972-75; sr. research chemist Centro de Investigación en Química Aplicada, Saltillo Coahuila, Mex., 1976-83, Sola Optical USA Inc., 1984—; mem. adv. panel Modern Plastics Mgmt., 1986-87; cons. on polymers for industry; cons. UN Indsl. Devel. Orgn. Community rep. Against Indsl. Air Pollution, Moorestown, N.J., 1968-70. Mem. Am. Chem. Soc., N.Y. Acad. Scis., AAAS, Sigma Xi. Mem. Christian Evangelical Ch. Patentee Plexiglas 70. Contbr. articles to profl. jours. Office: Sola Optical USA Inc 1500 Cader Ln Petaluma CA 94954-6905

BELSHE, JOHN FRANCIS, zoology and ecology educator; b. Marshall, Mo., Feb. 6, 1935; s. John Sherman and Velma Dee (Robbins) B.; m. Donna Joan Petre, June 2, 1957; children: Jeffrey Dean, Rhonda Lynn. BS in Edn., Ctrl. Mo. State U., 1957; MS in Zoology, U. Miami, 1961, PhD in Zoology, 1967. Cert. secondary tchr., Mo. From instr. to asst. prof. Miami-Dade C.C., Fla., 1961-64; from asst. prof. to assoc. prof. Ctrl. Mo. State U., Warrensburg, 1964-72, prof., 1972—; treas., exec. com. Mo. Acad. Sci., 1989—; presenter papers various profl. meetings. Grantee USDA, 1972. Mem. AAAS, Nat. Assn. Biology Tchrs., Am. Fisheries Soc. (pres. Mo. chpt. 1984), Am. Soc. Zoologists, Soc. Internat. Odonatologica. Democrat. Office: Ctrl Mo State U Biology Dept Warrensburg MO 64093

BELTRACCHI, LEO, engineer; b. Rochester, N.Y., Apr. 13, 1930; s. Cirillo and Cesarina B.; m. Mary Ann, June 27, 1959; children: Michael, Todd, John. B of Mech. Engring., Clarkson U., 1952; M of Mech. Engring., Rensselaer Polytech. Inst., 1957. Control systems engr. Pratt & Whitney Aircraft, East Hartford, Conn., 1954-65; systems engr. Avco Systems Div., Wilmington, Mass., 1965-71; mgr., systems engring. Computer Systems Engring., North Billerica, Mass., 1971-74; systems engr. U.S. Nuclear Regulatory Commn., Bethesda, Md., 1974-87; sr. human factors engr. U.S. Nuclear Regulatory Commn., Rockville, Md., 1987—; mem. NSF panel, Dec. 1990; presenter in field. Author 31 tech. papers; patentee in field. 1st lt. U.S. Army, Korea, 1952-54. Mem. ASME, IEEE, Nat. Computer Graphics Assn., Sigma Xi, Tau Beta Pi, Pi Tau Sigma. Home: 12112 Triple Crown Rd Gaithersburg MD 20878

BEMBEN, MICHAEL GEORGE, exercise physiologist; b. Thunder Bay, Ont., Can., Dec. 18, 1956; s. Michael and Mary (Demeo) B.; m. Debra Anne Arnberg, July 16, 1982. BSc, Lakehead U., Thunder Bay, 1978; MSc, U. Sask., 1981; PhD, U. Ill., 1989. Adj. asst. prof. N.E. Mo. State U., Kirksville, 1988-92; postdoctoral fellow Kirksville Coll. of Osteopathic Medicine, Kirksville, 1989-91, rsch. assoc., 1991-92; asst. prof. dept. physiology U. Okla., Norman, 1992—. Contbr. articles to profl. jours.; manuscript reviewer for Jour. gerontology, Jour. Sports Scis., Jour. Applied Sport Sci. Rsch., Jour. Orthopaedic and Sports Phys. Therapy, Jour. Sports Medicine, Tng. and Rehab., Jour. Osteopathic Sports Medicine. Am. Heart Assn. rsch. fellow, 1989-90, 90-91, Am. Osteopathic Assn./Nat. Osteopathic Found. rsch. grantee, 1988-89, Sask. Sport Rsch. grantee, 1980-81. Mem. AAAS, AAHPERD, Am. Coll. Sports Medicine, Am. Physiol. Soc., Can. Assn. Sports Scis., Exercise Physiology Acad., Nat. Strength and Conditioning Assn., Sigma Xi. Office: Univ of Okla 1401 Asp St Norman OK 73019

BEMENT, ARDEN LEE, JR., engineering educator; b. Pitts., May 22, 1932; s. Arden Lee and Edith Ardella (Bigelow) B.; m. Mary Ann Baroch, Aug. 24, 1952; children: Kristine, Kenneth, Vincent, Cynthia, Mark, David, Paul, Mary. Deg. of Engr. in Metallurgy, Colo. Sch. Mines, 1954; MSMetE, U. Idaho, 1959; PhD, U. Mich., 1963; DEng (hon.), Fenn Sch. Engring., Cleve. State U., 1989; Hon. Doctorate degree, Cleve. State U., 1989. Rsch. metallurgist Hanford Labs., GE, Richland, Wash., 1954-65; sr. rsch. mgr. Pacific N.W. Lab., Battelle Meml. Inst., Richland, 1965-70; prof. nuclear materials MIT, 1970-76; dir. Def. Advanced Rsch. Projects Agy. Office Materials Sci., Dept. Dept., Washington, 1976-79, dep. undersec. rsch. and advanced tech., 1979-80; v.p. tech. resources TRW, Lyndhurst, Ohio, 1980-89, v.p. sci. and tech., 1990-92; Basil Turner disting. prof. engring. Purdue U., West Lafayette, Ind., 1993—; Basil S. Turner disting. prof. engring. Purdue U., 1992—; tech. assistance expert to Mexico UNIAEA, 1974-76; cons. NRC, Taiwan, 1975; mem. Nat. Sci. Bd., 1988—; mem. sci. adv. com. Electric Power Rsch. Inst., 1987—, Advanced Tech. Inc., 1993—; bd. dirs. Keithley Instrument Co., Lord Corp. Author publs. in field; editor: Biomaterials: Structural and Biomedical Bases for Hard Tissue and Soft Tissue Substitutes, 1971; co-editor: Dislocation Dynamics, 1968, Creep of Zirconium Alloys in Nuclear Reactors, 1983; mem. editorial bd. Jour. Nuclear Materials, 1970-77, Materials Tech., 1987—; contbr. articles to profl. jours. Chmn. bd. health Mental Health/Mental Retardation, Benton-Franklin Counties, Wash., 1968-70; pres. Arts Coun., Richland, Pasco and Kennewick, Wash., 1968-70; tbd. dirs. Cleve. Opera Bd., 1991-93, treas., 1982-86, v.p., 1986-81. Lt. col. USAR, 1954-79. Recipient Outstanding Achievement award Colo. Sch. Mines, 1984, Melville F. Coolbaugh award, 1991, Disting. Engr. award UCLA, 1987, Honor Roll award U. Idaho Alumni Assn., 1991, Alumnus of Yr. award U. Mich. Alumni Assn. (Cleve. br., 1992; Ford Found. fellow, 1959-60. Fellow Am. Nuclear Soc., Am. Soc. Metals, Am. Inst. Chemists; mem. NAE, ASTM, AIME, Metals Soc. of AIME (Leadership award 1988), Sigma Xi, Tau Beta Pi, Sigma Gamma Epsilon. Republican. Roman Catholic. Home: 4709 Doe Path Ct Lafayette IN 47905 Office: Purdue Univ 1289 MSEE Bldg Rm 308J West Lafayette IN 47907-1289

BENACERRAF, BARUJ, pathologist, educator; b. Caracas, Venezuela, Oct. 29, 1920; came to U.S., 1939, naturalized, 1943; s. Abraham and Henriette (Lasry) B.; m. Annette Dreyfus, Mar. 24, 1943; 1 child, Beryl. B es L, Lycee Janson, 1940; BS, Columbia U., 1942; MD, Med. Sch. Va., 1945; MA, Harvard U., 1970; MD (hon.), U. Geneva, 1980; DSc (hon.), NYU, 1981, Va. Commonwealth U., 1981, Yeshiva U., 1982, U. Aix-Marseille, 1982, Columbia U., 1985, Adelphi U., 1988, Weizmann Inst., 1989, Harvard U., 1992, U. Bordeaux, 1993. Intern Queens Gen. Hosp., N.Y.C., 1945-46; rsch. fellow dept. microbiology Med. Sch. Columbia U., 1948-50; charge de recherches Centre Nat. de Recherche Scientique Hosp. Broussais, Paris, 1950-56; asst. prof. pathology Sch. Medicine NYU, 1956-58, assoc. prof. Sch. Medicine, 1958-60, prof. Sch. Medicine, 1960-68; chief immunology Nat. Inst. Allergy and Infectious Diseases NIH, Bethesda, Md., 1968-70; Fabyan prof. comparative pathology, chmn. dept. Med. Sch. Harvard U., 1970-91; ret. Med. Sch., Harvard U., Cambridge, Mass., 1991; pres., chief exec. officer Dana-Farber Cancer Inst., 1980-91; pres., chief exec. officer Dana-Farber Inc., 1990; mem. immunology study sect. NIH; pres. Fedn. Am. Socs. Exptl. Biology, 1974-75; chmn. sci. adv. com. Centre d'Immunologie de Marseille. Bd. govs. Weizmann Inst. Medicine; mem. sci. adv. com. Children's Hosp. Boston; mem. award com. GM Cancer Rsch. Found., also chmn. selection com. Sloan prize, 1980. Capt. M.C. AUS, 1946-48. Recipient T. Duckett Jones Meml. award Helen Hay Whitney Found., 1976, Rabbi Shai Shacknai lectr. and prize Hebrew U. Jerusalem, 1974, Waterford award, 1980, Nobel prize, 1980, Corr. Emerite de l'Institut de la Sante et de la Recherche Scientifique, Nat. Medal of Sci. NSF, 1990. Fellow Am. Acad. Arts and Scis.; mem. NAS, Nat. Inst. Medicine, Am. Assn. Immunologists (pres. 1973-74), Brit. Assn. Immunology, French Soc. Biol. Chemistry, Internat. Union Immunology Socs. (pres. 1980-83). Home: 111 Perkins St Jamaica Plain MA 02130-4313 Office: Dana-Farber Cancer Inst 44 Binney St Boston MA 02115-6084

BEN AMOR, ISMÄIL, obstetrician/gynecologist; b. Tunis, Tunisia, Mar. 31, 1937; s. Youssef Ben Mohamed and Khira (Slaïti) B.; m. Janine Bernadette Cheype, Sept. 18, 1964; 1 child, Leyla. Baccalauréat degree, Coll. Sadiki-Bardo, Tunis, 1959; PCB, U. Tunis, 1960; Sterility Cert., U. Paris, 1976; Clin. Thermography Cert., Bobigny (France), 1978; Cert. in Reproduction and Devel. Biology, Paris Sud Kremlin Bicêtre, 1989, Cert. in Andrology, 1989; Cert. in Gynecologic Surgery Endoscopy, Clermont Ferrand, 1990; Cert. in Fetal Medicine, Paris U., 1991—, diploma of Health and Smoking, 1993. Cert. MD, Gynecologist/Obstetrician; cert. Gynecology and Obstetrics Sonography, U. Paris, 1980, Breast Pathology, U. Strasbourg, 1981, Clin. Carcinology, Inst. Gustave Roussy, 1982, Gynaecology Endoscopy, U. Paris, 1985; cert. foetal medicine, Paris U., 1991. Non-resident med. student Paris Hosp., 1965-69, serving resident med. student, 1969-74; Doctor's Thesis Paris U., 1971; attaché Intercommunal Hosp., Créteil, Val de Marne, France, 1974—; pvt. practice Paris, 1977—; attaché Intercommunal Hosp., Créteil; cons. med. Ctr. Etoile, Paris; task force on tobacco dependency Biomedical Saints Pères Rsch. Unit, 1993. Co-author: French Gynecologist Jour. Exptl. Study in EW, 1975; Med. Jour. Colposcopy in Early Diagnosis, 1983, Syn/Obst. Biology Rep. Jour., Uterus Glioma, 1986, Gyneacology, 1987, Detection of Microinvasive Cancer, 1987. Mem. French Soc. Gynecology & Obstetrics, Soc. of Fertility & Sterility, Soc. Sonography (Echography), Assn. Le Val de Seine, Colposcopy Soc., Gynecology Pathology French Soc., French Andrology Assn. Avocation: art. Office: Cabinet Med, 17 Ave d'Italie, 75013 Paris France

BEN-ASHER, JOSEPH ZALMAN, aeronautical engineer; b. Jerusalem, Oct. 9, 1955; s. Asher Dov and Sara Lea (Kalach) Druk; m. Avital Elon, Oct. 23, 1974; children: Noa, Matan, Aya, Sara-Lea. BS, Technion, Haifa, Israel, 1978; MS, Va. Poly. Inst., 1986, PhD, 1988. Project officer Israel Air Force, 1978-84; chief project engr. Israel Mil. Industries, Ramat-Hasharon, Israel, 1984-85, control sect. head, 1988--; adj. prof. electronics dept. Tel-Aviv U., Israel, 1988--. Contbr. articles to profl. jours. Maj. Israeli Def. Forces. Mem. Am. Inst. Aeronautics and Astronautics, Israel Assn. Automatic Control. Office: IMI Sys Div, PO Box 1044-77, Ramat-Hasharon 47100, Israel

BÉNASSY, JEAN-PASCAL, economist, researcher, educator; b. Paris, Dec. 30, 1948; s. Jean and Jeannine Bénassy. Grad., École Normale Supérieure, Paris, 1970; PhD in Econs., U. Calif., Berkeley, 1973. Rsch. assoc. Cepremap, Paris, 1973—; dir. rsch. CNRS, Paris, 1981—; with dept. econs. École Polytechnique, Paris, 1987—. Author: The Economics of Market Disequilibrium, 1982, Macroeconomics: An Introduction to the Non-Walrasian Approach, 1986. Recipient Guido Zerilli Marimo prize Académie des Scis. Morales et Politiques, Paris, 1990. Fellow Econometric Soc. (coun. mem. 1990-92). Office: Cepremap, 142 Rue du Chevaleret, 75013 Paris France

BENAVIDES, JAIME MIGUEL, orthopedist; b. Chuquicamata, Chile, Oct. 20, 1923; came to U.S., 1926; s. Jaime and Elena (Spikula) B.; m. Nela Julieta Montejo, May 14, 1947; children: Suzanne Benavides Egle, Maria, Jaime Manuel. AB, Duke U., 1943; MD, U. Pa., 1947. Diplomate Am. Bd. Orthopedic Surgeons. Intern Luth. Hosp., Cleve., 1947-48, resident in surgery, 1948-49; resident in orthopedic surgery U.S. Naval Hosp., Phila., 1953-55; asst. chief orthopedics U.S. Naval Hosp., Newport, R.I., 1955-56; resident in orthopedic surgery Newington (Conn.) Childrens' Hosp., 1957; asst. chief orthopedics U.S. Naval Hosp., Phila., 1958-61; chief of orthopedics, surg. services U.S. Naval Hosp., Key West, Fla., 1961-66; chief of staff Monroe Gen. Hosp., Key West, 1966-70, Fla. Keys Meml. Hosp., Key West, 1971-78; staff physician DePoo Hosp., Key West, 1971-78; staff physician, vice chmn. med. staff Glades Gen. Hosp., Belle Glade, Fla., 1982-83, sec.-treas. med. staff, 1983-84; med. dir. Woodrow Wilson Rehab. Ctr., Fishersville, Va., 1984-86, staff physician, 1986-89; pvt. practice Key West, 1989—; chmn. bd. Lower Fla. Keys Hosp. Tax Dist., 1970-71; bd. dirs. Monroe County Comprehensive Health Planning Council, 1969-75; profl. advisor Easter Seal Soc., Ctr. of Hope in Key West, 1980; profl. adv. com. Fla. Easter Seal Soc.; mem. adv. com. Health Related Occupation Programs Fla. Keys Community Coll., 1972; team physician Key West High Sch., 1966-80, Mary Immaculate High Sch., 1972-80, Key West Conchs Profl. Baseball Team, 1977-80; mem. med. adv. com. Fla. State Dept. Vocat. Rehab., 1980, chmn. med. subcouncil, 1977-78; chmn. regional adv. council Emory U. Research and Tng. Ctr., 1978-84. Bd. dirs., pres. Beachwood Villas Condominium Assn., Stuart, Fla., 1981-84; mem. Mil. Affairs Com. City of Key West, 1980, spl. population adv. bd. Waynesboro (Va.) Dept. Parks and Recreation; lector St. John's Ch., Waynesboro, 1985-87. Named Advisor of Yr. Fla. Easter Seal Soc., 1972. Fellow ACS, Am. Acad. Orthopedic Surgeons; mem. AMA (numerous Recognition awards), Am. Fracture Assn., Am. Coll. Sports Medicine, Soc. Internat. de Chirurgie Orthopedique et Traumatologie, Soc. Latinoam. de Orthopedia y Traumatologia, Physicians for Automotive Safety, Am. Orthopedic Soc. Sports Medicine, Nat. Rehab. Adminstrn. Assn., Am. Congress Rehab. Medicine (Roy Hoover Physician of the Yr. award 1985), So. Orthopedic Assn., Fla. Orthopedic Soc. (program chmn. 1974-75), So. Med. Soc., Navy League (life, pres. Key West council 1979), U.S. Power Squadron, Internat. Oceanographic Found., U.S. Naval Inst., Am. Mgmt. Assn., Va. Rehab. Assn. (bd. dirs. 1985-88). Republican. Roman Catholic. Lodge: Rotary. Avocation: photography. Office: 1901 Fogarty Ave Key West FL 33040-3607

BENDER, ERWIN RADER, JR., air force officer; b. Waynoka, Okla., Apr. 16, 1956; s. Erwin Rader and Beth Lois (Deweese) B.; m. Melinda Louise Higgins, May 15, 1981; children: Christopher, Elizabeth. BS in Pharmacy, Southwestern Okla. State U., 1979, BA in chemistry, 1979; MA in Computer & Info. Resources Mgmt., Webster U., St. Louis, 1991. Registered pharmacist, Okla. Staff pharmacist Doctor's Hosp., Tulsa, 1980-81, Meml. Hosp., Woodward, Okla., 1981-84; commd. 2nd lt. USAF, 1984, advanced through grades to capt., 1987; officer in charge Refill Pharmacy Wilford Hall USAF Med. Ctr., Lackland AFB, Tex., 1984-86; officer in charge Main Pharmacy Wilford Hall Med. Ctr., Lackland AFB, Tex., 1986-87; chief Bolling AFB Clinic Pharmacy, Washington, 1987-88; chief pharmacy systems br. Composite Health Care System Devel. and Test Evaluation, Pentagon, Va., 1988-93, dir. tech. analysis and testing, 1993—. Pres. bd. dirs. E.P. Clapper Meml. Hosp., Waynoka, Okla., 1982-84. Mem. MUMPS (devel. com.), Soc. Air Force Pharmacists (pres. 1993—), C. of C. Waynoka, Okla. (pres. 1983-84), Am. Mensa Ltd., Masons, Scottish Rite. Office: Devel Test and Evaluation 5401 Westbard Ave Ste 900 Bethesda MD 20816

BENDER, HARVEY A., biology educator; b. Cleve., June 5, 1933; m. Eileen Adelle Teper, June 16, 1956; children: Leslie Carol, Samuel David, Philip Michael. AB in Chemistry, Case Western Res. U., 1954, student, 1954-55; MS, Northwestern U., 1957, PhD, 1959. Diplomate Am. Bd. Medical Genetics (founding). Post-doctoral fellow USPHS U. Calif., Berkeley, 1959-60; asst. prof. biology U. Notre Dame, Ind., 1960-64, assoc. prof., 1964-69, prof., 1969—; adj. prof. law U. Notre Dame, 1974—; dir. No. Ind. Regional Genetics Ctr., Meml. Hosp. South Bend (Ind.), 1979—, Gt. Lakes Regional Genetics Group, 1991—; NSF In-Svc. Inst. prof., fall term 1962-63; vis. prof. human genetics, rsch. assoc. Yale U., 1973-74; vis. prof. zoology So. Ill. U., Carbondale, summer 1978; adj. prof. medical genetics Ind. U., 1979—; vis. prof. natural scis. Washington Coll., Chestertown, Md., 1984; cons. Ednl. Rsch. Coun. Am., 1967-69, Pres.'s Com. on Mental Retardation, 1973, N.J. Inst. Tech., 1975-76, Ind. State Bd. of Health, 1991—, mem. sickle cell commn., 1987—, chronic disease commn., 1989—; genetics cons. Ind. State Bd. Health, 1991—. Editorial reviewer various profl. jours. Bd. dirs. Internat. Rels. Coun., 1961-69, v.p., 1962-64, pres., 1964-65; bd. dirs. Coun. for Retarded of St. Joseph County, 1964-76, 1st v.p., 1967-76; chmn. human rights com. No. Ind. State Hosp., 1980—. Predoctoral fellow USPHS, 1957-59, Cross-disciplinary fellow Yale U., 1973-74; grantee NIH, 1961-67, DOE, 1961—, United Health Svc., 1963-73, NSF, 1978-81, HEW, HHS, others. Fellow AAAS; mem. AAUP, Am. Assn. Mental Deficiency, Am. Inst. Biol. Scientists, Am. Soc. Human Genetics, Genetics Soc. Am., Ind. Acad. Sci., Radiation Rsch. Soc., Soc. Devel. Biology, Soc. for Values in Higher Edn., Sigma Xi (regional lectr. 1977—, mem. nat. com. on sci. and society 1978-89, chmn. 1981-89, mem. nat. com. awards, 1981-86, chmn. 1981-83, dir.-at-large 1980-86, bd. dirs. nat. exec. com. 1983-84, long range planning com. 1986—). Office: U Notre Dame Dept Biol Scis Notre Dame IN 46556-0369

BENDER, MICHAEL DAVID, gastroenterologist; b. Newark, Dec. 7, 1942; s. Abraham and Temi (Bleznak) B.; m. Pearl Wang, May 31, 1981; 1 child, Alex Philip. AB, Brandeis U., 1964; MD, Columbia U., 1968. Diplomate Am. Bd. Internal Medicine, Am. Bd. Gastroenterology. Asst. clin. prof. medicine U. Calif., San Francisco, 1976-83, assoc. clin. prof. medicine, 1983-90, clin. prof. medicine, 1990—; chief of medicine Peninsula Hosp., Burlingame, Calif., 1985-89. Contbr. articles to profl. jours., chpts. to books. Lt. comdr. USN, 1973-75. Fellow ACP; mem. Am. Gastroent. Assn., Am. Soc. Gastrointestinal Endoscopy, No. Calif. Soc. Clin. Gastroenterology (pres. 1983-84), Phi Beta Kappa, Alpha Omega Alpha. Jewish. Achievements include first clinical endoscopic description of glycogenic acanthosis (glycogen plaques) in the human esophagus; summarized and classified diseases of the peritoneum and mesentery. Office: 1828 El Camino Real Burlingame CA 94010

BENDIXEN, HENRIK HOLT, physician, educator, dean; b. Fredriksberg, Denmark, Dec. 2, 1923; came to U.S., 1954, naturalized, 1960; s. Carl Julius and Borghild (Holt) B.; m. Karen Skakke, Dec. 20, 1947 (dec. 1984); children: Nils, Birgitte; m. Lilo M. Laver, May 29, 1985. Cand. Phil., U. Copenhagen, 1943, CM, CChir, 1951, MD (hon.), 1987; MD (hon.), Jagiellonian U., Krakow, Poland, 1985. Diplomate Am. Bd. Anesthesiologists. Intern Copenhagen County Hosp., 1951-52; resident in surgery and anesthesia Denmark and Sweden, 1952-54; resident in anesthesia Mass. Gen. Hosp., Boston, 1954-57, anesthetist, 1957-69; asst. clin. prof. Harvard U., Boston, 1957-69; prof. anesthesia, chief dept. U. Calif., San Diego, 1969-73; med. dir. Univ. Hosp., San Diego, 1971-72; prof. anesthesiology Columbia U., N.Y.C., 1973—, chmn. dept. anesthesiology, 1973-85, acting provost Coll. Physicians and Surgeons, 1980-81, alumni prof., 1984, v.p. health scis., dean faculty medicine, 1984-89, E.M. Papper prof. anesthesiology, 1985-86; sr. assoc. v.p. health scis., sr. assoc. dean medicine, 1989—. Author: Respiratory Care, 1965; contbr. numerous articles to profl. jours. Mem. bd. visitors sch. medicine U. Pitts., 1985; trustee Mary Imogene Bassett Hosp., Cooperstown, N.Y., 1986—. Fellow AAAS, Faculty of Anesthetists, Royal Coll. Surgeons Eng.; mem. Minn. Surg. Soc. (hon.), Belgian Soc. Anesthesiologists (hon.), NAS Inst. Medicine, Scandinavian Soc. Anesthesiologists (hon.). Clubs: Harvard (Boston); Univ., Century (N.Y.C.). Home: Daisy Ln Irvington NY 10533-2015 Office: Columbia U Coll Physicians & Surgeons 630 W 168th St New York NY 10032-3702

BENEDEK, GEORGE BERNARD, physicist, educator; b. N.Y.C., Dec. 1, 1928. B.S., Rensselaer Poly. Inst., 1949; M.A., Harvard U., 1951, Ph.D. in Physics, 1954. Mem. staff joint Harvard-Lincoln Lab. MIT Project, 1953-55; research fellow Harvard U., 1955-57, lectr. in solid state physics, 1957-58, asst. prof. applied physics, 1958-61, assoc. prof., 1961-65; prof. physics MIT, Cambridge, 1965—, now Alfred H. Caspary prof. physics and biol. physics; mem. physics adv. com. NSF, 1983-86. Guggenheim fellow, 1960; profl. fellow Atomic Energy Research Establishment, Harwell, Eng., 1967. Fellow Am. Phys. Soc., Am. Acad. Arts and Scis.; mem. Nat. Acad. Scis. (mem. inst. medicine 1983), Am. Inst. Physics (bd. govs. 1971-74). Office: MIT Dept Physics Rm 13-2005 Cambridge MA 02139

BENEDICT, AUDREY DELELLA, biologist, educator; b. Schenectady, N.Y., Sept. 27, 1951; d. George Peter and Louise Irene (Johnson) DeLella; m. James Bell Benedict, July 14, 1971; stepchildren: William Logan, Robert James. BA, U. Colo., 1983. Founder, dir. Cloud Ridge Naturalists, Ward, Colo., 1979—; cons. biologist. Author: Sierra Club Naturalists' Guide: The Southern Rockies, 1991. Mem. Am. Soc. Mammalogists, Southwestern Assn. Naturalists, Colo. Native Plant Soc., Nature Conservancy, Colo. Authors' League. Democrat. Home: 8297 Overland Rd Ward CO 80481 Office: Cloud Ridge Naturalists 8297 Overland Rd Ward CO 80481

BENEDICT, JEFFREY DEAN, financial and technical consultant, mathematics educator; b. Lynwood, Calif., Mar. 13, 1961; s. Bruce Howard and Beverley Jean (Fiedler) B. BSME, U. Va., 1983; MBA in Fin., U. Pa., 1986. Mech. design engr. Amercom div. Litton Co., College Park, Md., 1983-85; with Questech, Inc., Falls Church, Va., 1985-89; fin. and tech. cons. Mgmt. Cons. & Rsch., Falls Church, 1989—; adj. prof. math. Strayer Coll., Ashburn, Va., 1992—. Mem. Kiwanis (program chmn. McLean, Va. 1988—, v.p 1988), Sigma Xi (assoc.). Republican. Home: 9621 Locust Hill Dr Great Falls VA 22066 Office: Mgmt Cons & Rsch Inc 5113 Leesburg Pike Ste 509 Falls Church VA 22041

BENEFIELD, JENIEFER LEN, software and systems engineer; b. Washington, Apr. 25, 1957; d. Allen W. and Marcus Jean (Avens) B. BSME, George Washington U., 1980, MSME, 1986. Program mgr., engr. Ensco, Inc., Springfield, Va., 1980-88; sr. systems engr. FSC, Inc., Fairfax, Va., 1988-89; v.p. adv. tech. Amron Corp., Arlington, Va., 1989-92; v.p. systems engring. logistics and software devel. divsn. Global Assocs. Ltd., Arlington, 1992—; computer designer BJB, Arlington, 1989-91. Office: Global Assocs Ltd Ste 205 2300 Clarendon Blvd Arlington VA 22201

BENES, SOLOMON, biomedical scientist, physician; b. Iasi, Romania, Mar. 28, 1925; came to U.S., 1978; s. Moritz and Cecilia (Abramovici) B.; m. Liudmila Topor, Mar. 27, 1954. Baccalaureate, Cultura Lyceum, Bucharest, Romania, 1943; MD, Sch. of Medicine, Bucharest, Romania, 1952. Intern microbiology lab. Mil. Hosp., Bucharest, 1949-50, fellow microbiology lab., 1950-51, dir. clin. lab. outpatient dept., 1951-52; dir. rsch. lab. Ctr. for Radiobiology Rsch., Bucharest, 1953-57, 59-66; chief physician microbiology lab. Mil. Hosp., Bucharest, 1967-73; chief physician clin. lab. Ctr. of Haematology, Bucharest, 1973-76; assoc. in medicine Havard Med. Sch., Boston, 1978-81; asst. rsch. scientist, asst. prof. SUNY Health Sci. Ctr., Bklyn., 1982—. Author: (with others) Seminars in Infectious Diseases, 1983; contbr. articles to Sexually Transmitted Diseases, Antimicrobial Agts. and Chemotherapy, Jour. Clin. Microbiology, Proceedings of the 6th Internat. Symposium on Human Chlamydial Infections. Col., Romanian Army Med. Svc., 1946-73. Achievements include discovery that the Trachoma biovar of Chlamydia trachomatis is able to achieve intercellular propagation in cell culture and that, in a proper cell setting, this bacterium spreads from cell to cell in cell culture, contrary to what was generally believed. Home: 2828 Bragg St # 3 Brooklyn NY 11235-1102 Office: SUNY Health Sci Ctr Bklyn 450 Clarkson Ave # 56 Brooklyn NY 11203-2098

BENET, LESLIE ZACHARY, pharmacokineticist; b. Cin., May 17, 1937; s. Jonas John and Esther Racie (Hirschfeld) B.; m. Carol Ann Levin, Sept. 8, 1960; children: Reed Michael, Gillian Vivia. AB in English, U. Mich., 1959, BS in Pharmacy, 1960, MS in Pharm. Chemistry, 1962; PhD in Pharm. Chemistry, U. Calif., San Francisco, 1965; PharmD (hon.), Uppsala U., Sweden, 1987. Asst. prof. pharmacy Wash. State U., Pullman, 1965-69; asst. prof. pharmacy and pharm. chemistry U. Calif., San Francisco, 1969-71, assoc. prof., 1971-76, prof., 1976—, vice chmn. dept. pharmacy, 1973-78, chmn. dept. pharmacy, 1978—, dir. drug studies unit, 1977—, dir. drug kinetics and dynamics ctr., 1979—; mem. pharmacology study sect. NIH, Washington, 1977-81, chmn. 1979-81, mem. pharmacol. scis. rev. com. 1984-88, chmn. 1986-88; mem. generic drugs adv. com. FDA, Washington, 1990—, mem. Sci. Bd., 1992—; mem. sci. adv. bd. SmithKline Beecham

Pharms., 1989-92, Pharmetrix, 1989-92. Editor: Jour. Pharmacokinetics and Biopharmaceutics, 1976—; mem. editorial bd. Pharmacology, 1979—, Pharmacy Internat., 1979-82, Pharmaceutical Research, 1983—, ISI Atlas of Sci.: Pharmacology, 1988-89; editor: The Effect of Disease States on Drug Pharmacokinetics, 1976, Pharmacokinetic Basis for Drug Treatment, 1984, Pharmacokinetics: A Modern View, 1984,, Integration of Pharmacokinetics, Pharmacodynamics and Toxicokinetics in Rational Drug Development, 1992; contbr. articles to profl. jours. Appt. to Forum on Drug Devel. and Regulation, 1988. Fellow Acad. Pharm. Scis. (pres. 1985-86, chmn. basic pharmaceutics sect. 1976-77, mem.-at-large exec. com. 1979-83, Rsch. Achievement award 1982), AAAS (mem.-at-large exec. com. pharm. scis. sect. 1978-81, 91—), Am. Assn. Pharm. Scientists (pres. 1986, treas. 1987, bd. dirs. 1988—, Disting. Pharm. Scientist award 1989); mem. Inst. Medicine NAS (forum on drug devel. and regulation 1988, chmn. com. on antiprogestins, 1993), AAUP, Am. Found. for Pharm. Edn. (bd. dirs. 1987—, Disting. Svc. "Profile" award 1993), Am. Coll. Clin. Pharmacology (Disting. Svc. award 1988), ISSX (councillor 1992—), Am. Pharm. Assn., Am. Soc. Clin. Pharmacology and Therapeutics, Am. Soc. for Pharmacology and Exptl. Therapeutics, Generic Pharm. Industry Assn. (mem. blue ribbon com. on generic medicines 1990), Internat. Pharm. Fedn. (bd. pharm. scis. 1988—), Drug Info. Assn., Am. Assn. Colls. Pharmacy (Volwiler Rsch. Achievement award 1991, pres. 1993—, bd. dirs. 1992—), Sigma Xi, Rho Chi (Ann. Lecture award 1990), Phi Lambda Sigma. Home: 53 Beach Rd Belvedere CA 94920-2364 Office: U Calif San Francisco Dept Pharmacy San Francisco CA 94143-0446

BENETSCHIK, HANNES, mechanical engineer; b. Bonn, Germany, Mar. 7, 1960; s. Bruno and Christa (Niemann) B.; m. Gerwita Hees, 1992. Diploma in engring., Tech. U., Aachen, 1986, D in Engring., 1992. Rsch. engr. Inst. für Strahlantriebe und Turboarbeitsmaschinen, Aachen, 1986-92, chief in computational fluid dynamics, 1992—; rsch. engr. Motoren- und Turbinen Union Muenchen (Germany) GmbH, 1989. Recipient Borchers medal Tech. U., 1992. Roman Catholic. Office: Inst fur Strahlantriebe, Templergraben 55, Aachen D-52062, Germany

BENFENATI, EMILIO, chemist, researcher; b. Milan, Sept. 27, 1954; s. Gualtiero and Aurelia (Alberani) B.; m. Giuseppina Gini, May 13, 1981; children: Chiara, Francesco. Laurea in Chimica, U. Statale, Milan, 1979; Attestato qualificazione, Inst. Mario Negri, Milan, 1985. Chartered chemist. Researcher Inst. Biochimico Italiano, Milan, 1979-81, Inst. Mario Negri, 1981-83; visiting scientist Stanford (Calif.) U., 1983-84; sr. researcher Inst. Mario Negri, 1984-86, unit chief, 1987—; vis. scientist U. Calif., Berkeley, 1984; mem. experts com. Bur. Reference, European Econ. Communities, Brussels, 1986—, Com. for Pollution, Lombardy, Milan, 1986-87; mem. com. U. Verde di Bergam, Italy, 1986—. Contbr. articles to profl. jours. European Econ. Communities fellow, 1983-84; recipient rsch. program directorship Lombardy, Milan, 1986-91. Mem. Am. Soc. Mass Spectrometry, Ordine dei Chimici Italy, Soc. Chimica Italiana. Home: Viale Lombardia 32, Milan 20131, Italy Office: Inst Mario Negri, Via Eritrea 62, Milan 20157, Italy

BENHAM, LINDA SUE, civil engineer; b. Toledo, Oct. 31, 1954; m. William H. Benham; children: William H. IV, Katherine L. BS in Civil Engring., U. Toledo, 1977. Structural engr. Itil and Assocs., Toledo, 1977-78; project and scheduling mgr. Finkbeiner, Pettis and Strout, Ltd., Toledo, 1978—. Trustee Huntington Community Ctr., Sylvania, Ohio, 1990-92. Recipient Spirit of Am. Woman in Bus. award, 1990; Young Engr. of Yr. nomination Toledo sect. Am. Soc. Civil Engrs. Mem. NAFE, Tech. Soc. Toledo, Kiwanis (pres.). Republican. Avocations: flying, sailing, pianist. Office: Finkbeiner Pettis and Strout Ltd 4405 Talmadge Rd Toledo OH 43623-3509

BENI, GERARDO, electrical and computer engineering educator, robotics scientist; b. Florence, Italy, Feb. 21, 1946; came to U.S., 1970; s. Edoardo and Tina (Bazzanti) B.; m. Susan Hackwood, May 24, 1986; children: Catherine Elizabeth, Juliet Beatrice. Laurea in Physics, U. Firenze, Florence, Italy, 1970; PhD in Physics, UCLA, 1974. Research scientist AT&T Bell Labs., Murray Hill, N.J., 1974-77; research scientist AT&T Bell Labs., Holmdel, N.J., 1977-82, disting. mem. tech. staff, 1982-84; prof. elec. and computer engring. U. Calif., Santa Barbara, 1984-91, dir. Ctr. for Robotic Systems in Microelectronics, 1985-91; prof. elec. engring., dir. distbn. robotic system lab. U. Calif., Riverside, 1991—. Founder, editor: Jours. Robotic Systems, 1983 (Jour. of Yr. award 1984); editor: Recent Advances in Robotics, 1985, Vacuum Mechatronics, 1990; contbr. more than 130 articles to tech. jours.; 16 patents in field. Fellow Am. Physics Soc. Office: U Calif-Riverside Coll Engring Riverside CA 92521-4009

BENIDICKSON, AGNES, university chancellor. Chancellor Queen's U. at Kingston, Ont., Can.; bd. dirs. James Richardson & Sons, Ltd., Mut. Life Assurance Can. Office: Queen's U at Kingston, Office of Chancellor, Kingston, ON Canada K7L 3N6

BENITEZ, ISIDRO BASA, obstetrician/gynecologist, oncologist; b. Tacloban, Leyte, The Philippines, June 15, 1927; s. Victoriano Antonio and Juliana (Basa) B.; m. Teresita Limjap Boncan, Mar. 19, 1961; children: Rene, Glenn, Mellissa, Eric. MD, U. of The Philippines, Manila, 1953. Diplomate Philippine Bd. Ob-Gyn. Intern Philippine Gen. Hosp., 1952-53, resident dept. gynecology, 1954-59; cons., oncology svc., dept. ob-gyn. U. Philippines and Philippines Gen. Hosp., 1960—; pvt. practice Makati Med. Ctr., 1969—; prof. dept. ob-gyn. U. Philippines, 1981—; chief of svc. dept. ob-gyn Philippines Gen. Hosp., 1974-85, chmn. dept. ob-gyn, 1985-88; chmn. dept. ob-gyn Coll. Medicine, U. of The Philippines, Manila, 1985-88, mem. pers. bd., 1986-88, vice chair admission com., 1986, mem. rsch. coun., 1987-88, med. educator, rschr.; vice chmn. dept. ob-gyn Makati Med. Ctr., The Philippines, 1989—; mem. editorial Philippine Jour. Ob-Gyn, 1988—; mem. nat. adv. coun. Philippine chpt. World Assn. for Gynecol. Cancer, 1965; chmn. Philippine Bd. Ob-Gyn., 1979. Author: Philippine Physician Board Examination Reviewer, 1962, (manual) Ovarian Cancer, 1975, Carcinoma of the Vulva, 1987; contbr. articles to profl. publs. Fellow ACS, Philippine Obstetrical and Gynecol. Soc., Philippine Coll. Surgeons, Soc. Gynecologic Oncologists of Phillipines (pres. 1990—). Roman Catholic. Avocations: golf, music, plays. Home: 19 Mercedes St Bel-Air, Makati, Manila The Philippines Office: U Philippines Coll Medicine, Pedro Gil, 2801 Manila The Philippines

BENJAMIN, KEITH EDWARD, mechanical engineer; b. Sunderland, Durham, Eng., Oct. 3, 1943; s. Edward Percy and Iris (Marshall) B.; m. Rosella Truglio, June 1965 (div. 1975); m. Violet A. Forker, Aug. 29, 1975; children: Edward, Anthony, Dawn, Matthew, Keith II, Kurt, Eve. BSME, Newark Coll. Engring., 1966; MSIE, NYU, 1967. Designer Corning (N.Y.) Inc., 1967-69; supr. equip. engring. Corning (N.Y.) Inc., Erwin, N.Y., 1969-74; supr. ops. engring. Corning (N.Y.) Inc., Corning, 1974-75, sr. project engr., 1975-87, supr. rsch. and devel. engring. svcs., 1987—. Treas. Presho Meth. Ch., Lindley, N.Y.; umpire Little League, Lindley, 1986-87. Mem. Nat. Rifle Assn. Republican. Methodist. Home: Bell Hill Rd RD 1 Box 65 Painted Post NY 14870 Office: Corning Inc Sullivan Park Rd Painted Post NY 14870-9116

BENJAMIN, STEPHEN ALFRED, veterinary medicine educator, environmental pathologist, researcher; b. N.Y.C., Mar. 27, 1939; s. Frank Benjamin and Dorothy (Zweighaft) Fabricant; m. Barbara Larson, July 25, 1982; children: Jeffrey, Karen, Susan, Douglas. AB, Brandeis U., 1960; DVM, Cornell U., 1964, PhD, 1968. Diplomate Am. Coll. Vet. Pathologists. Fellow pathology Johns Hopkins U., Balt., 1966-67; asst. prof. comparative medicine M.S. Hershey (Pa.) Med. Ctr. of Pa. State U.; exptl. pathologist Inhalation Toxicology Research Inst., Albuquerque, 1970-77; prof. pathology and radiation biology Colo. State U., Ft. Collins, 1977—, dir. collaborative radiol. health lab., 1977—, assoc. dean grad. sch., 1986—, co-dir. ctr. for environ. toxicology, 1991—. Contbr. sci. articles to profl. mags. Mem. Am. Assn. Pathologists, Internat. Acad. Pathology, Radiation Rsch. Soc., Nat. Coun. for Radiation Protection (liver task group). Office: Colo State U Dept Pathology Dept for Environ Toxicology Fort Collins CO 80523

BENJAMINS, JOYCE ANN, neurology educator; b. Bay City, Mich., June 1, 1941; d. John E. and Mary (Buben) Livak; m. David Benjamins, Dec. 27, 1965; children: Mary, Laura. BA in Chemistry, Albion Coll., 1963; PhD in Chemistry, U. Mich., 1967. Rsch. assoc. neurology Johns Hopkins Sch. Medicine, Balt., 1968-69, instr., 1969-70, asst. prof., 1971-73; asst. prof. biochemistry Biol. Sci. Rsch. Ctr., U. N.C., Balt., 1973-75; asst. prof. neurology, assoc. biochemistry Wayne State U. Sch. Medicine, Detroit, 1975-78, assoc. prof. neurology, biochemistry, 1978-85, prof. neurology, assoc. biochemistry, immunology and microbiology, 1985—; mem. sci. adv. com. Amyotrophic Lateral Sclerosis Soc. Am., 1984-87; mem. Nat. Inst. Neurological Diseases and Stroke Neurol. C Study Sect., 1991-95, ad hoc mem., 1990-91. Recipient Agnes Faye Morgan Rsch. award Iota Sigma Pi, 1978, Javits Neurosci. Investigator award, 1987—, Wayne State U. Bd. Govs. Disting. Faculty Fellowship, 1988-90. Mem. Am. Soc. Neurochemistry, Internat. Soc. Neurochemistry, Soc. Neurosci., Sigma Xi. Achievements include research in neurochemistry, neuroscience and cell biology. Office: Wayne State U Sch Medicine 3124 Elliman 421 E Canfield Ave Detroit MI 48201

BENKER, HANS OTTO, mathematics educator; b. Mühltroff, Germany, May 26, 1942; s. Herbert and Helene (Herzmann) B.; m. Doris Kober, Dec. 24, 1965; 1 child, Uta. Mathematician, U. Dresden (Germany), 1967; PhD, U. Merseburg (Germany), 1970, MS, 1974. Asst. prof. U. Merseburg, 1967-74, lectr., 1975-88, prof. math., 1988-93, dir. Inst. Math Analysis, 1990-93; prof. math U. Halle, 1993—; dir. Inst. Math., Tech. U. Merseburg, 1990—. Author 3 books of math., 37 papers math. rsch. Home: Klobikauer Str 139, 06217 Merseburg Germany Office: U Halle, Universitätsplatz 10, Halle Germany

BENKO, JAMES JOHN, chemist; b. Tallahassee, Fla., Feb. 13, 1943; s. John and Helen (Gyure) B.; m. Shirley May Williams, Dec. 1, 1975; children: James David, Daniel Mark. BS, Ill. Inst. Tech., 1968. Lab tech. Sherwin-Williams, Chgo., 1962-66; assoc. chemist McCrone Assocs., Chgo., 1967-68; chemist, lab. mgr. Libby McNeill Libby, Chgo., 1968-72; project leader Nalco Chem. Co., Chgo., 1972-75; sr. chemist Chemetron, Inc., Holland, Mich., 1975—; rsch. assoc. BASF, Holland, 1975—. Contbr. articles to The Microscope, Sci. Probe, Am. Lab., Chem. News, Jour. Water Pollution Control Fedn. Tchr. Grand Valley State U., 1982, event supr. sci. olympiad, 1988-89; Webelos leader Boy Scouts Am., Griffith, Ind., 1973-74. Fellow Am. Inst. Chemists; mem. Am. Chem. Soc. (chmn. Western Mich. sect. 1980-81, councillor 1986-88, treas.-sec. 1977-79), State Micrscopical Soc. Ill. Presbyterian. Achievements include 2 U.S. patents. Home: 916 Knoll Dr Zeeland MI 49464 Office: BASF-Coatings Colorants 491 Columbia Ave Holland MI 49423

BENMARK, LESLIE ANN, chemical company executive; b. Morrison, Ill., Jan. 15, 1944; d. Joseph Allen and Mable (Heiss) Freemon; m. Gary Nyle Benmark, Mar. 1, 1969. BS, U. Tenn., 1967, MS, 1970; PhD, Vanderbilt U., 1976; JD, U. Del., 1984. Registered profl. engr., Tenn., Del.; bar: Pa. 1984, Del. 1986, D.C. 1986, Tenn. 1986. Systems analyst Monsanto Co., St. Louis, 1967-68; systems analyst E.I. Dupont de Nemours and Co., Old Hickory, Tenn., 1968-70, systems analysis supv., 1970-75, design supv., 1975-76, planning/indsl. engring. supv., 1976-79, bus. analysis mgr., 1979-87, bus. strategy mgr., 1987-90, mgmt. systems cons., 1990-93, supply chain systems mgr., 1993—; part-time computer sci. instr. U. Tenn., Nashville, 1973-75; part-time asst. to dean engring., dir. women engring. program Vanderbilt U., Nashville, 1975-79; speaker in field. Mem. editorial bd. Engring. Mgmt. Internat. Jour., 1982-90. Mem. vis com. W.Va. Coll. Engring., 1987—; mem. adv. bd. Gateway Engring. Edn. Coalition, 1993—, Universidad Politecnica de Puerto Rico, San Juan, 1993—; bd. dirs. Swin Early and Live Found., 1989-92; mem. Tenn. consumer panel Tenn. Pub. Svc. Commn. 1975; active Fire Code Bd. Appeals, 1976-79. Recipient Outstanding Tennessean award Gov. Tenn., 1978; named Woman of Yr. by Bus./Profl. Women's Assn., 1977, Tenn. Outstanding Young Woman, 1978, Del. Engr. of Yr., 1992. Fellow Inst. Indsl. Engrs. (rep. to ABET bd. dirs., bd. dirs. State of Del. 1988-90, bd. trustees, mem. exec. com. 1988-90, nat. group v.p. profl. enhancement 1988-90), Inst. Indsl. Engrs. Ireland; mem. NAE, NSPE (mem. industry adv. group 1989-91, mem. engring. 2000 task force 1989-91), Del. Engring. Soc. (New Castle chpt. v.p., pres. 1982-87, chmn. Del. engrs. week banquet 1988, State Del. pres. 1987-90), Am. Soc. Engring. Edn., Del. Assn. Profl. Engrs., Union Panamericana de Asociaciones de Ingenieros (chair total quality engring. com.), Nat. Sci. Found. (mem. adv. com. engring. 1991—, chair presdl. faculty fellow selection panel 1993, mem. engring. edn. coalition rev. panel 1990, 91), Nat. Rsch. Coun. (mem. mfg. studies bd. dirs. 1993—), Moot Ct. Honor Soc., Alpha Pi Mu.

BENNER, RONALD ALLEN, JR., mechanical engineer; b. Pitts., Nov. 30, 1966; s. Ronald Allen Sr. and Margaret Louise (Covert) B.; m. Sherri Marie Cargile, June 20, 1992. BS in Mining Engring., U. Pitts., 1987, BSME, 1989. Project engr. Elec. Boat Divsn. Gen. Dynamics, Groton, Conn., 1989—. Mem. NSPE. Home: 29B Greensboro Blvd Clifton Park NY 12065

BENNETT, ALBERT FARRELL, biology educator; b. Whittier, Calif., July 18, 1944; s. John C. and Edna R. (Lederer) Kidd; m. Rudi C. Berkelhamer, May 14, 1977; children: Hilary J. Arnold, Laura K. Arnold, Mari J., Andrew M. BA in Zoology, U. Calif., Riverside, 1966; PhD in Zoology, U. Mich., 1971. Miller postdoctoral fellow U. Calif., Berkeley, 1971-73, acting asst. prof. zoology, 1973-74; asst. prof. biol. scis. U. Calif., Irvine, 1974-78, assoc. prof., 1978-83, prof., 1983—, chmn. dept. developmental and cell biology, 1984-86, 1988-89, acting dean Sch. Biol. Scis., 1986-88; vis. rsch. assoc., assoc. prof. anatomy U. Chgo., 1981-82; vis. rsch. fellow zoology U. Adelaide, South Australia, 1983-84, vet. physiology U.Nairobi, 1990. Contbr. articles to profl. jours. Recipient hon. fellowship Woodrow Wilson Found., 1966, Career Devel. award NIH, 1978-83; grantee NSF, 1974—. Fellow AAAS; mem. Am. Physiol. Soc. (mem. editorial bd. 1982—, chair com. physiology sect. 1988-91), Am. Soc. Naturalists, Ecol. Soc. Am., Soc. Exptl. Biology, Am. Soc. Zoologists (mem. editorial bd. 1986-91, chair membership com. 1988, pres. 1990), Soc. for Study Evolution, Phi Beta Kappa. Home: 24 Perkins Ct Irvine CA 92715-4043 Office: U Calif-Irvine Dept Ecology Evolutionary Biology Irvine CA 92717

BENNETT, BRIAN O'LEARY, utilities executive; b. Bklyn., Dec. 5, 1955; s. Robert Joseph and Barbara Ashton (Michael) B. BA in Econs., George Washington U., 1982; JD, Southwestern U., 1982. Legis. caseworker U.S. Sen. James L. Buckley, Washington, 1973-77; legis. asst. U.S. Congressman Bob Dornan, L.A., 1977-78; dist. field rep. Congressman Bob Dornan, L.A., 1978-83; dir. comm. Calif. Dept. Housing & Community Devel., Sacramento, 1983-84; chief of staff U.S. Congressman R.K. Dornan, Washington, 1985-89; reg. affairs mgr. So. Calif. Edison Co., Santa Ana, 1989—. Contbr. articles to L.A. Times. Active organizing com. Calif. Bush for U.S. Pres., 1986-88; Calif. del. selection com., 1988, 92; campaign mgr. Dornan for U.S. Congress, 1984, 86, 88; mem. cen. com. Calif. State Rep. Party, mem. platform com., 1988, vice chair proxies and credentials com., 1993-94; del. Rep. Nat. Conv., 1988, 92; mem. Orange County Pro-Life PAC; bd. dirs. World Affairs Coun. of Orange County, Orange County Urban League, Orange County Task Force on Air Quality. Named one of Outstanding Young Men of Am., 1988. Roman Catholic. Avocations: movies, history, skiing, travel, racquetball. Office: So Calif Edison 1325 S Grand Ave Santa Ana CA 92705-4499

BENNETT, BRUCE ANTHONY, civil engineer; b. Providence, Nov. 8, 1950; s. George Sr. and Anne (Rominyk) B.; m. Patricia Anne Matteson, May 29, 1971; children: Paul Jason, Susan Lynn. BSCE, U. R.I., 1973; MSCE, Northeastern U., 1981. Sr. engr. Stone & Webster Engring. Corp., Boston, 1973-81; project mgr. Nat. Hydro Corp., Boston, 1981-86; sr. engr. Heat Exch. Systems, Boston, 1986-89; project mgr. Stowe Engring. Corp., Quincy, Mass.; conf. com. Nat. Sci. Found., Amherst, Mass., 1988. Editor: Pipeline Infrastructure, 1988. Com. mem. Pub. Works Study, Duxbury, Mass., 1985, Solid Waste Study, Duxbury, 1986, Nuclear Matters, Duxbury, 1987. Mem. ASCE (div. chmn. 1981—, conf. chmn. 1988, conf. com. 1993), Tau Beta Pi. Achievements include inventionr of generator cooling system, condenser perf. monitor, most recent discovery proving that substitution of a chilled water process facilitates the achievement of higher electric power generator output. Home: 280 Temple St Duxbury MA 02332 Office: Stowe Engring Corp 1150 Hancock St Quincy MA 02169

BENNETT, CURTIS OWEN, research engineer; b. Williston, N.D., May 21, 1954; s. Owen B. and Adele Louise (Hutchinson) B.; m. Kelly Bennett, July 24, 1992; children: Max, Alexis. BS, U. Tulsa, 1976, MS, 1979, PhD, 1982. Reservoir engr. BP Exploration, Houston, 1990-91; rsch. engr. Amoco Prodn. Co., Tulsa, 1982-90, 91—; mem. industry adv. bd., petroleum engring. dept. U. Tulsa, 1989—. Contbr. articles to profl. jours. Mem. Soc. Petroleum Engrs. (tech. editor 1983-88, rev. chmn. 1988-93), Sigma Xi. Office: Amoco Prodn Co PO Box 3385 Tulsa OK 74102-3385

BENNETT, JAY BRETT, medical equipment company executive; b. Durham, N.C., Dec. 13, 1961; s. James Leonard Jr. and Yoalder Kathleen (Brunson) B.; m. Trisha Helen Folds, Feb. 3, 1990. BA in Econs., Wake Forest U., 1984; M Health Adminstrn., Duke U., 1986. Sr. cons. Ernst and Whinney (now Ernst and Young), Charlotte, N.C., 1986-89; assoc. dir. strategic planning SSI Med. Svcs., Inc., Charleston, S.C., 1989-92, dir. strategic planning, 1992—. Del. N.C. Rep. Conv., Raleigh, 1988, Charleston County Rep. Conv., 1989. Mem. Am. Coll. Healthcare Execs (nominee), Nat. Trust for Hist. Preservation, Am. Hosp. Assn., S.C. Hist. Soc., Nat. Soc. Sons of Am. Revolution, S.C. Coastal Conservation League, Charleston Lib. Soc., Ducks Unltd., Quail Unltd., Trout Unltd. Avocations: outdoors, history of American south, photography. Office: SSI Med Svcs Inc 4349 Corporate Rd Charleston SC 29405-7487

BENNETT, MARTIN ARTHUR, chemist, educator; b. Harrow, Middlesex, United Kingdom, Aug. 11, 1935; s. Arthur Edward Charles and Dorothy Ivy (Bennett) B.; m. Rae Elizabeth Mathews, Dec. 14, 1964; children: Simon, Andrew. BS, Imperial Coll. Sci., 1957, PhD, 1960; DSc, U. London, 1974. Turner and Newall fellow Univ. Coll., London, 1961-63, lectr. in chemistry, 1963-67; fellow Rsch. Sch. Chemistry, Canberra, Australia, 1967-70; sr. fellow Australian Nat. U., 1970-79, professorial fellow, 1979-92, prof., 1992—. Mem. editorial adv. bd.: Organometallics, 1986-88, Inorganica Chimica Acta, 1968—; mem. internat. editorial adv. bd. Dictionary of Organometallic Compounds, 1984—, Dictionary of Inorganic Compounds, 1988—; contbr. 180 articles to Jour. Am. Chem. Soc., Organometallics, Inorganic Chemistry, Angewandte Chemie, others. Fellow Australian Acad. Sci.; mem. Am. Chem. Soc., Royal Soc. Chemistry (Nyholm medal 1991), Royal Australian Chem. Inst. (G.J. Burrows award 1987, H.G. Smith medal 1977). Achievements include discovery or formation of transition metal to carbon bonds in cyclometallated complexes, stabilization of short-lived organic molecules, such as benzyne and cyclohexyne in form of transition metal complexes. Office: Australian Nat U, Rsch Sch Chemistry, Canberra ACT 0200, Australia

BENNETT, OVELL FRANCIS, chemistry educator, researcher, consultant; b. Middleboro, Mass., Nov. 5, 1929; s. Orville Thomas and Elizabeth Marthalene (Dauner) B.; m. Marjorie Claire Devlin, June 25, 1955; children: Thomas Ryan, Carla. BS, Bridgewater (Mass.) Coll., 1953; MS, Boston Coll., Chestnut Hill, Mass., 1955; PhD, Pa. State U., 1958. Rsch. chemist DuPont Co., Gibbstown, N.J., 1958-61; prof. chemistry Boston Coll., 1961—; cons. various orgns., 1963—. Contbr. articles on organic chemistry to sci. jours. Mem. AAAS, AAUP, Am. Chem. Soc., Sigma Xi. Achievements include patents in field of organic chemistry. Home: 10 Erick Rd Apt 30 Mansfield MA 02048-3075 Office: Boston Coll Dept Chemistry Commonwealth Ave Chestnut Hill MA 02167-3848

BENNETT, PETER BRIAN, researcher, anesthesiology educator; b. Portsmouth, Hampshire, Eng., June 12, 1931; s. Charles Risby and Doris Isobel (Peckham) B.; m. Margaret Camellia Rose, July 7, 1956; children: Caroline Susan, Christopher Charles. B.Sc., U. London, 1951; Ph.D., U. Southampton, 1964, D.Sc., 1984. Asst. head surg. sect. Royal Navy Physiol. Lab., Alverstoke, Eng., 1953-56, head inert gas narcosis sect., 1953-66; dep. dir., prin. sci. officer, head pressure physiology sect. Royal Naval Physiol. Lab., Alverstoke, 1968-72; head pressure physiology group Can. Def. and Civil Inst. for Environ. Research, Toronto, Ont., 1966-68; prof. biomed. engring. Duke U., Durham, N.C., 1972-75, assoc. prof. physiology, 1975—, prof. anesthesiology, 1972—, dir. research dept. anesthesiology Med. Ctr., 1973-84, dir. Nat. Divers Alert Network, 1980—; dep. dir. F.G. Hall Lab. Environ. Research, 1973-74, co-dir., 1974-77, dir., 1977-88; sr. dir. Hyperbaric Ctr., 1988—; cons. in field. Author: The Aetiology of Compressed Air Intoxication and Inert Gas Narcosis, 1966; author, editor: The Physiology and Medicine of Diving and Compressed Air Work, 1969, Russian edit., 1987, 4th edit., 1993; contbr. over 200 articles to profl. jours. Served with RAF, 1951-53. Recipient Letter of Commendation Pres. Ronald Reagan, 1981, Sci. award Underwater Soc. Am., 1980, Leonard Greenstone Safety award Nat. Assn. Underwater Instrs., 1985, 1st Prince Tomohito of Mikasa Japan prize, 1990, Craif Hoffman Meml. award, 1992. Mem. Undersea Med. Soc. (exec. com. 1972-75, editor jour. 1976-79, pres. 1975-76, 1st Oceaneering Internat. award 1975, Albert R. Behnke award 1983), Am. Physiol. Soc., European Undersea Biomed. Soc., Aerospace Med. Soc., Marine Tech. Soc. Avocations: scuba diving; swimming; boating. Home: 1921 S Lakeshore Dr Chapel Hill NC 27514-2029 Office: Duke U Med Ctr FG Hall Lab PO Box 3823 Durham NC 27702-3823

BENNETT, STEPHEN CHRISTOPHER, biology educator; b. New Haven, Mar. 20, 1956; s. Emmett Leslie Jr. and Marja Dorothy (Adams) B.; m. Linda Marie Phillips, Aug. 1, 1987; 1 child, Daniel Lawrence. MS in Biology, Yale U., 1985; PHD in Systematics and Ecology, U. Kans., 1991. Adj. prof. dept. biology Washburn U., Topeka, 1992; asst. prof. dept. systematics and ecology U. Kans., Lawrence, 1992—. Contbr. articles to Jour. of Paleontology, Jour. Vertebrate Paleontology, Ichnos, others. Mem. Soc. Vertebrate Paleontology, Paleontological Soc., Will Hennig Soc. Office: U Kans Mus of Natural History Lawrence KS 66045

BENNETT, THOMAS PETER, museum director, educator, biologist; b. Lakeland, Fla., Oct. 8, 1937; s. Thomas Edward and Hazel Dean (Smith) B.; m. Gudrun Dorothea Staub, Sept. 1, 1962; children: Vanessa Hildegard, Alexander Staub. AB, Fla. State U., 1959; PhD, Rockefeller U., 1965. Asst. prof. biology Harvard U., Cambridge, 1967-71; prof. biology U. Ky., Lexington, 1971-72; prof. biology, chmn. dept. Fla. State U., Tallahasee, 1972-76; spl. asst. to pres., acting v.p. Fla. State U., Tallahassee, 1973-76; pres. Acad. Nat. Scis., Phila., 1976-86; dir. Fla. Mus. Nat. History, U. Fla., Gainesville, 1986—; cons. biology edits W.H. Freeman Co., 1971; bd. dirs. Wistar Inst., Phila., 1979—; chmn. Pa. Mus. Coalition, Phila., 1979-81. Author: Graphic Biochemistry, 1968, Elements of Protein Synthesis, 1969; co-author: Modern Topics in Biochemistry, 1966, Biology Today, Physical Basis of Life, Biochemistry, 1979, 81. Bd. dirs. coun. Boy Scouts Am., Phila, 1977-85, Cultural Alliance, Phila., 1976-86; founder, pres. Friends of Logan Sq. Found., Inc., Phila., 1983-86; founder, chmn. Mus. Assocs. Pa., Phila., 1978-84. Mem. Assn. Nat. Sci. Instns. (chmn. 1984-86), Assn. Sci. Mus. Dirs. (pres. 1986-90), Linnean Soc., The Explorers Club (bd. dirs.), Phi Beta Kappa, Sigma Xi, Phi Eta Sigma, Phi Kappa Phi. Avocations: fern gardening, stamp collecting, poetry. Office: U Fla Fla Mus Nat History Gainesville FL 32611

BENNICI, ANDREA, botany educator; b. Livorno, Tuscany, Italy, Dec. 18, 1942; s. Salvatore Bennici and Concetta Di Falco; m. Paola Battaglia, July 25, 1971; 1 child, Marco. Diploma in geometry, Inst. Tech. Amerigo Vespucci, Livorno, 1962; D in Agrl. Scis. cum laude, U. Pisa, Italy, 1966. Asst. prof. botany Faculty Agrl. Scis., U. Pisa, 1966-74, lectr., 1974-80; prof., chairperson botany Faculty Math., Phys. and Natural Scis., U. Genoa, Italy, 1980-84, Faculty Agrl. and Forestal Scis., U. Florence, Italy, 1984—. Author: Development in Higher Plants, 1978; contbr. numerous articles to nat. and internat. sci. jours. Recipient award for studies on control of differentation in plants Acad. Nat. dei Lincei, Rome, 1974. Fellow Italian Soc. Botany (head nat. assn. for differentiation and plant tissue and cell culture 1973-78), Internat. Assn. Plant Tissue and Cell Culture, Internat. Hort. Soc. Avocation: religious studies. Office: U Florence Dept Plant Biol, Piazzale della Cascine 28, 50144 Florence Italy

BENNION, JOHN STRADLING, nuclear engineer, consultant; b. Salt Lake City, Sept. 19, 1954; s. Mervyn S. Jr. and Larrie (Stradling) B. BS in Chemistry, U. Utah, 1987, BSChemE, 1987, MS in Nuclear Engring., 1990, postgrad., 1991—. Registered profl. engr., Utah, radiation protection technologist Nat. Registry of Radiation Protection Technologists; lic. sr. reactor operator U.S. Nuclear Regulatory Commn. Carpenter various cos. Utah, 1974-86; sr. reactor engr. U. Utah Nuclear Engring. Lab., Salt Lake City, 1987—; instr. mech. engring. dept. U. Utah, Salt Lake City, 1992—; mem. reactor safety com. U. Utah, Salt Lake City, 1987—. Author: (with others) Irradiation Damage Assessment of Electronics, 1991; author tech.

reports. Merit badge counselor Boy Scouts Am., Salt Lake City, 1990. Mem. AAAS, ASME, NSPE, IEEE, Nuclear and Plasma Scis. Soc. of the IEEE, Am. Nuclear Soc. (student br. pres. 1988-90), Am. Soc. Quality Control, Health Physics Soc., Internat. Soc. Radiation Physics, Utah Acad. Arts & Scis., Phi Kappa Phi, Alpha Nu Sigma, Pi Tau Sigma, Tau Beta Pi, Sigma Xi. Republican. Mem. LDS Church. Achievements include research on maintenance regulations and risk assessment in the nuclear energy industry, atmospheric dispersion, responce of thermoluminescent dosimetry to mixed neutrongamma radiation fields and radiation hardness testing of electronic components. Office: U Utah Nuclear Engring Lab 1205 Merrill Engring Bldg Salt Lake City UT 84112

BENNISON, ALLAN PARNELL, geological consultant; b. Stockton, Calif., Mar. 8, 1918; s. Ellis Norman Lambly and Cora Mae (Parnell) B.; m. DeLeo Smith, Sept. 4, 1941; children: Victor, Christina, Mary. BA, U. Calif., Berkeley, 1940. Cert. petroleum geologist, cert. profl. geologist. Geology fellow Antioch Coll., Yellow Springs, Ohio, 1940-42; photogrammetrist U.S. Geol. Survey, Arlington, Va., 1942-45; stratigrapher, asst. chief geologist Companias Unidas de Petroleos, Cartagena, Colombia, 1945-49; staff stratigrapher Sinclair Oil & Gas Co., Tulsa, 1949-69; geol. cons. Tulsa, 1969—; cons. in field. Editor: Tulsa's Physical Environment, 1973; compiler maps; contbr. articles to profl. jours. Fellow AAAS, Geol. Soc. Am., Explorers Club; mem. Am. Assn. Petroleum Geologists (trustee assoc., Disting. Svc. award 1986), Soc. Econ. Paleontologists and Mineralogists (Disting. Svc. award 1990), Tulsa Geol. Soc. (pres. 1965), Tulsa Astronomy Club (v.p. 1965), Sigma Xi. Republican. Episcopalian. Avocations: photography, astronomy, reading, travel. Home: 1410 Terrace Dr Tulsa OK 74104-4626 Office: 125 W 15th St # 401 Tulsa OK 74119-3810

BENOIT, JEAN-PIERRE ROBERT, pneumologist, consultant; b. Cotonou, Dahomey, May 19, 1930; s. Samuel Pierre and Renée (Meffre) B.; m. Isabelle Rappard, Apr. 10, 1969; children: Laurence, Arnaud. MD in Pneumo-phtisiology, Paris U., 1964, PhD in Econs., 1968. Intern in medicine Paris Hosp.; resident in pneumology-indsl. medicine Corbeil (France) Hosp.; sr. cons., dept. head Pneumology Hosp. Fontenoy, Chartres, France, 1970—; v.p. Ligue contre le Cancer, Chartres, 1990—. Contbr. articles to profl. jours. Officer French Med. Svc., 1957-59. Mem. Soc. Pneumology de Langue Francaise, N.Y. Acad. Scis., Nat. Geog. Soc., European Respiratory Soc., Imagery Thoracic Soc. Avocations: viola, bridge, music composition, archeology. Office: Hôpital Fontenoy Ctr, Hospitalier, Chartres 28018, France

BENORDEN, ROBERT ROY, utility executive; b. Columbus, Nebr., Apr. 21, 1932; s. Roy Fredrick and Lucille Otillia (Grossnicklaus) B.; m. Edith Lorraine Homan, Sept. 27, 1959 (div. 1969); 1 child, Rodney Roy; m. Shirley Jean Sanburn, Jan. 8, 1970; children: Debra Sue, Kevin Charles, Gina Marie. Grad. high sch., Loup City, Nebr. Farmer, rancher Nebr., 1955-69; parts mgr. Hwy. Implement, Inc., Geneva, Nebr., 1969-71, Green Country, Inc., Geneva, 1971-74, Welch Implement Co., York, Nebr., 1975-83; utility supt. Village of Fairmont, Nebr., 1984-91, Village of Grafton, Nebr., 1991—. Sgt. USAF, 1951-55, Korea. Mem. A.m Legion. Republican. Lutheran. Home: Box 485 745 9th Ave Fairmont NE 68354 Office: Village of Grafton Box 71 205 N Washington St Grafton NE 68365

BENOS, DALE JOHN, physiology educator; b. Cleve., Sept. 30, 1950; s. John George and Corinne Rachele (Casini) B.; m. Kimberlee Webb, May 30, 1992. BA, Case Western Res. U., 1972; PhD, Duke U., 1976. Asst., prof. Harvard Med. Sch., Boston, 1978-83, assoc. prof., 1983-85; assoc. prof. U. Ala., Birmingham, 1983-87, prof. physiology, 1987—; rsch. scientist Cystic Fibrosis Rsch. Ctr., Birmingham, 1985—; sr. scientist Nephrology Rsch. and Tng. Ctr., Birmingham, 1985—; editor cell physiology Am. Jour. Physiology, Birmingham, 1990—; sr. scientist UAB AIDS Ctr., Birmingham. Contbr. articles to profl. jours.; author: Membrane Transport in Biology, 1991. Andrew Mellon Found. grantee, 1978-85, NIH grantee, 1979—, 87—, 89—; Coca-Cola Student Rsch. Program summer scholar, 1991-92. Mem. Am. Physiol. Soc., Am. Soc. Cell Biology, Biophys. Soc., Soc. for Gen. Physiologists. Office: Univ of Ala BHSB 706 Physiology Dept Birmingham AL 35294-0005

BENOWITZ, STEVEN IRA, science-medical writer; b. Phila., Jan. 24, 1960; s. Edward and Pauline (Malin) B. BS in Biology, Pa. State U., 1982, MA in Journalism, 1991. Med.-sci. writer Ohio State U., Columbus, 1985-90, U. Chgo., 1990-91, Hershey Med. Ctr. of Pa. State, 1991—; intern Nat. Cancer Inst., Bethesda, Md., 1984, Sci. News, Washington, 1984. Recipient Gold Medal award Coun. for Advancement and Support Edn., 1987, Bronze Medal award, 1988, 91, Silver Medal award, 1988, Grand Gold Medal award, 1990. Mem. AAAS, Nat. Assn. Sci. Writers, Phila. Area Sci. Writers Assn., Penn State Sci. Writers Assn. Home and office: 15 Chevy Chase Hershey PA 17033

BENSON, ALAN JAMES, avaiation medical doctor; b. Liverpool, Eng., Oct. 25, 1929. BSc with honours, Manchester U., 1951, MSc, 1952, MB, ChB, 1955. House officer Manchester Royal Infirmary, 1955-56; svc. med. officer RAF Inst. Aviation Medicine, Farnborough, Eng., 1956-59, civilian med. officer, 1959-91; cons. RAF Inst. Aviation Medicine, Farnborough, 1991—; doctor RAF Inst. Aviation Medicine, Hartz, U.K.; cons. European Space Agy., Paris, 1981-88, Matra, Toulouse, France, 1991—. Contbr. over 100 articles to profl. jours., chpts. to books. Flt. lt. RAF, 1956-59. Recipient W.S. Smith prize Inst. Mech. Engrs. (U.K.), 1988, Yearsley medal Royal Soc. Medicine, 1988, Eric Liljencrantz award Aerospace Med. Assn., 1992. Fellow Royal Aeronautical Soc., Aerospace Med. Soc. (Longacre award 1975, Liljenkrantz award 1992); mem. Physiol. Soc., Exptl. Psychol. Soc. Achievements include research on spatial disorientation in flight, motion sickness, perception of motion, vestibular function in weightlessness. Office: RAF Inst Aviation Medicine, Farnborough GU14 6SZ, England

BENSON, ALLEN B., chemist, educator, consultant; b. Sioux Rapids, Iowa, Oct. 1, 1936; s. Bennett and Freda (Smith) B.; m. Marian Richter, Aug. 24, 1959; children: Bradley Gerard, Jill Germaine. BS in Secondary Edn. magna cum laude, Western Mont. U., 1960; postgrad., U. Mont., Missoula, 1960-61, Seattle U., 1962-63; M in Natural Sci., Highlands U., 1965; postgrad., Ill. Inst. Tech., 1969; PhD in Chemistry, U. Idaho, 1970. Chemistry instr. U. Wis., Whitewater, 1968-69, Spokane (Wash.) Falls Community Coll., 1969—; mem. steering com. Hanford Edn. Action League, Spokane, 1984-86; energy and nuclear cons., 1970—; mem. Hanford Health Effects Panel, Richland, Wash., 1986; numerous speeches, interviews and pub. articles on energy and nuclear issues, including speaker nat. conv. Physicians for Social Responsibility, Denver, 1990; lead sci. cons. Hanford Radiation Litigation Lawsuit for Hanford Downwinders against GE, Du-Pont and Rockwell, Wash., 1991; sci. conf. leader UNLV on radiation and health effects, 1992. Author: Hanford Radioactive Fallout: Are There Observable Health Effects?, 1989. Active Spokane County Dem. Platform Com., 1980, 84. Served as pvt. U.S. Army, 1955-57. Roman Catholic. Avocations: golf, philosophy. Home: 4528 N Windsor Dr Spokane WA 99205-2052 Office: Spokane Falls Community Coll Spokane WA 99204

BENSON, ALVIN K., geophysicist, consultant, educator; b. Payson, Utah, Jan. 25, 1944; s. Carl William and Josephine Katherine (Wirthlin) B.; m. Connie Lynn Perry, June 17, 1966; children: Alauna Marie, Alisa Michelle, Alaura Dawn. BS, Brigham Young U., 1966, PhD in Physics, 1972. Cert. environmentalist. Nuclear group physicist Phillips Petroleum Co., Arco, Idaho, 1966; assoc. prof. physics Ind. U., New Albany, 1972-78, head physics dept., 1976-78; sr. rsch. geophysicist Conoco, Inc., Ponca City, Okla., 1978-81, supr. geophysical rsch., 1981-85; geophysics rsch. assoc. DuPont, Ponca City, 1985-86; prof. geophysics Brigham Young U., Provo, Utah, 1986—; cons. Dames and Moore Engring., Salt Lake city, 1987-88, DuPont, Ponca City, 1989-91, Kuwait U., 1991-92, Coleman Rsch., Laurel, Md., 1991, Centennial Mine, Boise, 1990-91, Certified Environ., Salt Lake City, 1991-92, EPA, Washington, 1992; developer vis. geoscientist program Bringham Young U. Author: Seismic Migration, 1986, Theory and Practice of Seismic Imaging, 1988; contbr. articles to over 90 publs. including Geophysics, Jour. Computational Physics, Geophys Prospecting, Engring. Geology. Bishop LDS Ch., New Albany, 1976-78, Stake High Coun., Tulsa, 1979-81; active polit. adv. com. Rep. Party, Provo, 1990; polit. cons. Guatemala, 1991-92. Recipient Hon. Sci. award Bausch and Lomb, Rochester, N.Y., 1966; geophysics grantee Rotary, Provo, 1987, Am. Assn. Petroleum Geologists, Tulsa, 1988, Geol. Soc. Am., Boulder, Colo., 1988. Mem. Am. Phys. Soc., Am. Geophys. Union, Soc. Exploration Geophysicists (referee 1980-93), Utah Geol. Assn. Achievements include development of a stable, explicit seismic depth imaging algorithm, a residual depth imaging algorithm for seismic data, a linearized elastic wave inversion process for seismic data, an aperture compensated migration-inversion process for seismic data, a modified self-consistent quantum field theory; research in ground penetrating radar and electrical resistivity methods for delineating hazardous materials in the subsurface. Home: 249 W 1100 S Orem UT 84058-6709 Office: Brigham Young U University Hill Provo UT 84602

BENSON, ANDREW ALM, biochemistry educator; b. Modesto, Calif., Sept. 24, 1917; s. Carl Bennett and Emma Carolina (Alm) B.; m. Ruth Carkeek, May 22, 1942 (div. 1969); children: Claudia Benson Matthews, Linnea; m. Dorothy Dorgan Neri, July 31, 1971. BS, U. Calif., Berkeley, 1939; PhD, Calif. Inst. Tech., 1942; Phil D h.c., U. Oslo., 1965; Docteur h.c., U. Paris, 1986. Instr. chemistry U. Calif., Berkeley, 1942-43, asst. dir. Bio-organic group Radiation Lab., 1946-54; rsch. assoc. dept. chemistry Stanford U., 1944-45; assoc. prof. agrl. biol. chemistry Pa. State U., 1955-60, prof., 1960-61; prof.-in-residence biophys./physiol. chemistry UCLA, 1961-62; prof. Scripps Instn. Oceanography, U. Calif., San Diego, 1962-88, prof. emeritus, 1988—; Fulbright vis. prof. Agrl. Coll. Norway, 1951-52. Contbr. articles on biochem. rsch. on photosynthesis, lipids, coral metabolism, arsenic metabolism, methanol application in agr. to profl. jours. Trustee Found. for Ocean Rsch., San Diego, 1970-88; mem. adv. council The Costeau Soc., 1976—; mem. adv. bd. Marine Biotech. Inst. Co. Ltd., Tokyo, 1990—. Recipient Sugar Rsch. Found. award, 1950, Ernest Orlando Lawrence Meml. award, 1962, Rsch. award Supelco/Am. Oil Chemists Soc., 1987; Sr. Queen's fellow Australia, 1979. Fellow NAS, AAAS, Am. Acad. Arts and Sci.; mem. Royal Norwegian Soc. Sci. and Letters, Am. Oil Chemists Soc., Am. Chem. Soc. (emeritus), Am. Soc. Plant Physiologists (Stephen Hales award 1972), Am. Soc. Biochemistry and Molecular Biology, Inst. Marine Biology, Far East Br., Acad. Sci. Russia (hon.). Home: 6044 Folsom Dr La Jolla CA 92037 Office: Scripps Instn Oceanography La Jolla CA 92093-0202

BENSON, BARTON KENNETH, civil engineer; b. Chickasha, Okla., Mar. 15, 1938; s. Oscar Gains and Eunice Marie (Price) B.; m. Jessie Joan Bandy, July 20, 1956; children: Barton Kenneth Jr., Benny James, Virginia Kay, Dennis John, Vincent Keith. BA in Math., Cameron Coll., 1973; BSCE, U. Ark., 1981, MSCE, 1985. Registered profl. engr., Ark. Field constrn. engr. Ark. Hwy. and Transp. Dept., Van Buren, 1975-85; rsch. engr. Ark. Hwy. and Transp. Dept., Little Rock, 1985, specifications engr., 1986—. Author: Resident Engineers Manual, 1986; author, editor: Standard Specifications for Highway Construction, 1991, 2d rev. edit, 1993, also various rsch. reports. Minister of music Berry St. Bapt. Ch., Springdale, Ark., 1979-81. Capt. U.S. Army, 1967-75. Fed. Hwy. Adminstrn. scholar, 1979; Fed. Hwy. Adminstrn. fellow, 1984. Mem. ASCE, Tau Beta Pi (life). Achievements include design of computer systems for field use in the construction division, electronic data collection system for weight tickets; research on use of fly ash in bridge concrete, bridge deck joints, electronic data processing. Office: Ark Hwy and Transp PO Box 2261 Little Rock AR 72203-2261

BENSON, D(AVID) MICHAEL, plant pathologist; b. Dayton, Ohio, Aug. 28, 1945; s. Phillip Wayne and Edna Mae (Yowler) B.; m. Patricia D. Miller, Jan. 28, 1967; children: Julie Ann, Jeremy M., Jamie M. BS, Earlham Coll., Richmond, Ind., 1967; MS, Colo. State U., 1968, PhD, 1973. Postdoctoral fellow U. Calif., Berkeley, 1973-74; prof. plant pathology N.C. State U., Raleigh, 1974—. Contbr. articles to profl. jours.; editor Phytopathology, 1988-90. Fellow Am. Phytopathol. Soc.; mem. Sigma Xi (v.p. 1987, Young Rschr. award 1980), Gamma Sigma Delta (treas. 1991-93). Office: N C State Univ Dept Plant Pathology Box 7616 Raleigh NC 27695

BENSON, RICHARD CHARLES, chemist; b. Kalamazoo, Mich., Jan. 23, 1944; s. Harry B. and Helen (Clayton) B.; m. Jill Robinson, July 22, 1972; children: Michelle E., Ross C. BS, Mich. State U., 1966; PhD in Phys. Chemistry, U. Ill., 1972. Sr. chemist applied physics lab. Johns Hopkins U., Laurel, Md., 1972-80, prin. profl. staff, 1980—, group supr., 1990—. Contbr. articles to tech. jours. Mem. IEEE, Am. Phys. Soc., Am. Vacuum Soc., Materials Rsch. Soc. Achievements include research in surface science, properties of microelectronic materials, stability of spacecraft materials, molecular spectroscopy, mass spectrometry, gas chromatography and application of optical techniques to surface science. Office: Johns Hopkins Univ Applied Physics Lab Johns Hopkins Rd Laurel MD 20723

BENSON, ROBERT JOHN, computer science educator, administrator; b. Michigan City, Ind., Sept. 2, 1943; s. John Richard and Dorothy Ethel (Lorimer) B.; m. Nancy Louise White, Aug. 6, 1966; children: John Richard, David Paul. BS, Washington U., St. Louis, 1965; JD, Washington U., 1968. Affiliate assoc. prof. computer sci. Washington U., 1968—, assoc. vice chancellor, 1975—, dir. computing facilities, 1973—, founder, dir. Ctr. for Study of Data Processing, 1980—; cons. in field, St. Louis, 1968—; pres., founder Regional Consortium for Edn. and Tech., St. Louis, 1984—. Mem. Data Processing Mgmt. Assn., Assn. for Computing Machinery, Assn. for Systems Mgmt. Home: 10300 Richview Dr Saint Louis MO 63127-1433 Office: Washington U PO Box 1080 Saint Louis MO 63188-1080

BENSON, RONALD EDWARD, JR., environmental engineer; b. Lansing, Mich., Sept. 6, 1957; s. Ronald Edward and Donna Kathryn (Ellis) B.; m. Susann Boulware, Oct. 27, 1979; children: Ronald Edward III, Marie Kathryn. BSCE, Ohio No. U., 1978; MS in Sanitary Engring., Ga. Inst. Tech., 1979; PhD in Civil Engring., Clemson U., 1985. Registered profl. engr., Fla., Ga., Ohio, Ind. Asst. project engr. Roy F. Weston, Inc., Decatur, Ga., 1979-80; project engr. Robert and Co. Assoc., Atlanta, 1980-82; asst. prof. The Citadel, Charleston, S.C., 1982-85, Rose-Holman Inst. Tech., Terre Haute, Ind., 1985-8; assoc. v.p. Flood Engrs., Inc., Jacksonville, Fla., 1988-90; sr. project mgr. Hole, Montes & Assoc., Inc., Naples, Fla., 1990—; rsch. engr. USAE Waterways Experiment Sta., Vicksburg, Miss., 1986-88; adj. prof. Jacksonville U., 1989-90; presenter in field. Contbr. articles to profl. publs. Leader Boy Scouts Am., Jacksonville, 1988-90, Naples, 1991—. Mem. ASCE, NSPE, Water Environ. Fedn., Rotary (pres. 1992-93). Office: Hole Montes & Assoc Inc 715 10th St S Naples FL 33940

BENSON, SIDNEY WILLIAM, chemistry researcher; b. N.Y.C., Sept. 26, 1918; m. Anna Bruni, 1986; 2 children. A.B., Columbia Coll., 1938; A.M., Harvard U., 1941, Ph.D., 1941; Docteur Honoris Causa, U. Nancy, France, 1989. Rsch. asst. Gen. Electric Co., 1940; rsch. fellow Harvard U., 1941-42; instr. chemistry CCNY, 1942-43; group leader Manhattan Project Kellex Corp., 1943; asst. prof. U. So. Calif., 1943-48, assoc. prof., 1948-51, prof. chemistry, 1951-64, Disting. prof. chemistry, 1976-89, Disting. prof. emeritus, 1989—, dir. chem. physics program, 1962-63; dir. dept. kinetics and thermochemistry Stanford Rsch. Inst., 1963-76; sci. dir. Hydrocarbon Rsch. Inst. U. So. Calif., 1977-90, sci. dir. emeritus, 1991—; rsch. assoc. dept. chemistry and chem. engring. Calif. Inst. Tech., 1957-58; vis. prof. UCLA, 1959, U. Ill., 1959; hon. Glidden lectr. Purdue U., 1961; vis. prof. chemistry Stanford U., 1966-70, 71, 73; mem. adv. panel phys. chemistry Nat. Bur. Standards, 1969-72, chmn., 1970-71; hon. vis. prof. U. Utah, 1971; vis. prof. U. Paris VII and XI, 1971-72, U. St. Andrews, Scotland, 1973, U. Lausanne, Switzerland, 1979; Frank Gucker lectr. U. Ind., 1984; Brotherton prof. in phys. chemistry U. Leeds, 1984; cons. G.N. Lewis; lectr. U. Calif., Berkeley, 1989. Author: Foundations of Chemical Kinetics, 1960, Thermochemical Kinetics, 1968, 2d edit., 1976, Critical Survey of the Data of the Kinetics of Gas Phase Unimolecular Reactions, 1970, Atoms, Molecules, and Chemical Reactions, 1970, Chemical Calculations, 3d edit., 1971; founder, editor-in-chief Internat. Jour. Chem. Kinetics, 1967-83; mem. editorial adv. bd. Combustion Sci. and Tech., 1973—; mem. editorial bd. Oxidation Communications, 1978—, Revs. of Chem. Intermediates, 1979-87; mem. Hydrocarbon Letters, 1980-81; mem. editorial bd. Jour. Phys. Chemistry, 1981-85. Recipient Polanyi medal Royal Soc. Eng., 1986; faculty rsch. award U. So. Calif., 1984, Presdl. medal, 1986; Guggenheim fellow, 1950-51, Fulbright fellow, France, 1950-51, fellow NSF, 1957-58, 71-72. Fellow AAAS, Am. Phys. Soc.; mem. NAS, Am. Chem. Soc. (Tolman medal 1977, Hydrocarbon Chem. award 1977, Langmuir award 1986, Orange County award 1986), Faraday Soc., Indian Acad. Sci., Phi Beta Kappa, Sigma Xi, Pi Mu Epsilon, Phi Lambda Upsilon, Phi Kappa Phi. Home: 1110 N Bundy Dr Los Angeles CA 90049-1513 Office: U So Calif University Pk MC-1661 Los Angeles CA 90089

BENSON, WILLIAM HAZLEHURST, environmental toxicologist; b. Phila., Nov. 21, 1954; s. Henry Edward and Marian (Owens) B.; m. Mary Them, Apr. 29, 1977; children: Ian Hazlehurst, William Owens. BS in Biology, Fla. Inst. Tech., 1976; MS in Toxicology, U. Ky., 1980, PhD, 1984. Sr. scientist Cannon Labs., Inc., Reading, Pa., 1976-78; dir. toxicology N.E. La. U., Monroe, 1984-88; coord. environ. toxicology U. Miss., Oxford, 1988—; Dow-Corning Disting. lectr. in environ. toxicology Mich. State U., 1989. Contbr. articles to profl. jours. Fellow SETAC, 1983. Mem. Soc. Toxicology, Soc. Environ. Toxicology and Chemistry (bd. dirs. 1991-94), Am. Chem. Soc. Achievements include research in assessment of acute and chronic health effects of environmental contaminants, reproduction of aquatic organisms, use of biomarkers in environmental monitoring, residue health effects, pesticide toxicology. Office: U Miss Sch Pharmacy University MS 38677

BENTLEY, WILLIAM ARTHUR, engineer, electro-optical consultant; b. Jan. 21, 1931; s. Garth Ashley and Helen (Dieterle) B.; m. Erika Bernadette Seuthe, Nov. 17, 1956; children: David Garth, Barbara Elizabeth. BS in Physics, Northwestern U., 1952; MS in Systems Engring., Calif. State U., Fullerton, 1972. Engr. N. Am. Aircraft, Downey, Calif., 1956-69; chief engr. Fairchild Optical, El Segundo, Calif., 1969-72; sr. staff engr. Advanced Controls, Irvine, Calif., 1975-80; prin. Instrument Design Cons., Stanta Ana, Calif., 1978—; mgr. mfg. research and devel. Xerox Electro-Optical, Pomona, Calif., 1980-83; cons. Kasper Industries, Sunnyvale, Calif., 1977-78, Lincoln Laser Co., Phoenix, 1983—, Coopervision, Irvine, 1984-88, Triad Microsystems, 1985-88, Baxter, 1988-90, Indsl. Dynamics Co., 1990—. Patentee in field including 1st automatic optical printed wiring board inspector. Served with U.S. Army, 1952-54. Mem. Soc. Photo-optical Instrumentation Engrs., Optical Soc. Am., Mensa. Democrat. Unitarian. Home: 170 The Masters Cir Costa Mesa CA 92627-4640 Office: Instrument Design Cons PO Box 2203 Santa Ana CA 92707-0203

BENTLY, DONALD EMERY, electrical engineer; b. Cleve., Oct. 18, 1924; s. Oliver E. Bently and Mary Evelyn (Conway) B.; m. Susan Lorraine Pumphrey, Sept. 1961 (div. Sept. 1982); 1 child, Christopher Paul. BSEE with distinction, State U. of Iowa, 1949; MSEE, U. Iowa, 1950; DS (hon.), U. Nev., 1987. Registered profl. engr., Calif., Nev. Pres. Bently Nev. Corp., Minden, 1961-85, chief exec. officer, 1985—; chief exec. officer Bently Rotor Dynamics and Research Corp., Minden, 1985—; also chmn. bd. dirs. Bently Nev. Corp., Minden; chief exec. officer Gibson Tool Co., Carson City, Nev., 1978—; bd. dirs. Sierra Pacific Resources, 1982-83. Contbr. articles to profl. jours.; developer electronic instruments for the observation of rotating machinery, and the algorithm for rotor fluid-induced instability; inventor in field. Served with USN, 1943-46, PTO. Named Inventor or Yr., State of Nev. Invention and Tech. Coun., 1983; recipient first Decade award Vibration Inst. Mem. ASME, Am. Petroleum Inst., Engrs. Club of San Francisco, St. Petersburg (Russian Fedn.) Acad. Engring., Sigma Xi, Eta Kappa Nu, Tau Beta Pi, Sigma Alpha Epsilon. Episcopalian. Avocations: skiing, hiking, biking. Office: Bently Nev Corp PO Box 157 Minden NV 89423-0157

BENTON, PETER MONTGOMERY, business development and information science consultant; b. St. Paul, Apr. 9, 1948; s. James and Nancy (Bell) B.; m. Margaret Drain, May 22, 1973; children: Jonah, Susannah. BS, City Coll. Engring., 1973; MBA, Bradley U., 1979. Bus. analyst Caterpillar Tractor Co., Peoria, Ill., 1973-79; cons., mgr. Arthur Andersen & Co., N.Y.C., 1979-81; bus. systems planner McGraw-Hill Inc., N.Y.C., 1981-83, tech. exec., 1983-87, bus. devel. v.p., 1987-90, chief scientist, v.p., 1990-91; cons., owner Informed Decisions, Bklyn., 1984—; mem. Bd. of Standards Rev. Am. Nat. Standards Inst., N.Y.C., 1992—. Contbr. articles to sci. and profl. jours. Mem. Assn. Computing Machinery, IEEE, Am. Assn. for Artificial Intelligence, Am. Soc. for Information Sci. Office: Informed Decisions 524 10th St Brooklyn NY 11215

BENTON, STEPHEN RICHARD, civil and mechanical engineer; b. Brawley, Calif., Aug. 26, 1952; s. Homer Grabill and Blanche Carolyn (Saxe) B.; m. Diane Gordon Brooks, June 19, 1976; children: Matthew Richard, Sarah Ruth, Carolyn Brooks. BS, U.S. Mil. Acad., 1974; MSCE, MSME, Stanford U., 1982. Registered prof. engr., Va. Commd. 2d lt. U.S. Army, 1974, advanced through grades to lt. col., 1991; exch. officer Sch. Mil. Engring., Casula, New South Wales, Australia, 1988-90; mil. planning officer Office of Chief Engrs., Washington, 1990-92; dep. dist. engr. U.S. Army C.E., San Juan, P.R., 1992—. Decorated Meritorious Svc. medal with 3 oak leaf clusters, Army Commendation medal with 3 oak leaf clusters, Army Achievement medal with 2 oak leaf clusters, Nat. Def. Svc. medal with bronze svc. star; recipient Bronze de Fleury medal Army Engr. Assn., 1992. Mem. ASCE, ASME, Soc. Am. Mil. Engrs., Phi Kappa Phi. Office: U.S. Army CE 400 Fernandez Juncos Ave San Juan PR 00901-3299

BENTZ, BRYAN LLOYD, chemist; b. York, Pa., June 6, 1953; s. Lloyd Oscar and Margaret (Netling) B.; m. Paula Jane Gale, July 6, 1985; 1 child, Elizabeth Wyatt Gale-Bentz. BS in Chemistry with high distinction, U. Va., 1975, PhD in Chemistry, 1980. Guest scientist Max Planck Institut für Plasmaphysik, Garching, Fed. Republic Germany, 1979-81; mem. tech. staff RCA Labs., Princeton, N.J., 1982-87, David Sarnoff Rsch. Ctr., Princeton, 1987—. Contbr. articles to profl. publs.; mem. editorial bd. Internat. Jour. Mass Spectrom. Ion Processes. Mem. Am. Soc. for Mass Spectrometry, Am. Vacuum Soc., Am. Chem. Soc., Bohmische Phys. Soc. (sci.), Sigma Xi. Achievements include development of first quadrupole GDMS instrument; research in organic mass spectrometry, ion optics, surface analysis using static SIMS, trace element analysis and atomic beams. Office: David Sarnoff Rsch Ctr CN 5300 Princeton NJ 08543-5300

BEN-YAACOV, GIDEON, computer system designer; b. Bney Brack, Israel, July 26, 1941; came to U.S., 1979; s. Abraham and Henda (Natel) B-Y.; m. Miriam R. Schultz, May 11, 1967; children: David, Saul. BSEE, Technion Israel Inst. Tech., Haifa, 1966. R&D engr. Israeli Ministry of Def., Haifa, 1967-69; sci. asst. U. of the Witwatersrand, Johannesburg, Republic of South Africa, 1959-71; head office engr. ESCOM, Johannesburg, 1971-79; staff engr. Gibbs & Hill, Omaha, 1979-82; head process computer engring. HDR, Omaha, 1982-83; cons. engr. Power Utility Process Computer Engring., Omaha, 1983-92; sr. engr. advanced techs. MFS Network Techs., Omaha, 1992—. Contbr. tech. articles to Process Computer Systems Engring.; author tech. papers. Mem. IEEE (sr.), Instrument Soc. Am. (sr., Philip P. Sprague Application award for devel. advanced operator interface, 1980). Achievements include research on human-factors deficiencies in power plants; development of operator interface terminals for power plant computer systems, introduction of application of distributed controls for electrostatic precipitators. Home: 1870 Mayfair Dr Omaha NE 68144

BENZ, DONALD RAY, nuclear safety engineer, researcher; b. Carbondale, Ill., Feb. 16, 1950; s. Raymond and Estella (Ebersohl) B.; m. Betty Jo Ring, Dec. 30, 1972; children: Kathy Lynn, Jeffrey Alan. BSEE, So. Ill. U., 1972; MA in Bus. Adminstrn., Sangamon State U., 1982. Registered profl. engr., Ill. Design engr. Moldovan & Assocs. Cons. Engrs., Salem, Ill., 1973-75; elec. estimator McWilliams Electric Co., Vandalia, Ill., 1975-77, Volle Electric, Springfield, Ill., 1977-78; evaluation engr. Ill. Capitol Devel. Bd., Springfield, 1978-82; sect. head engring. div. Ill. Dept. Nuclear Safety, Springfield, 1982—; program dir. Nat. Clean Air com. Conf. Radiation Control, 1984-85; presenter in field, 1990. Bldg. com. mem. Faith Evang. Luth. Ch., Jacksonville, Ill., 1988-90; precinct com. mem. Rep. party, Prentice-Sinclair, Ill., 1982-85; vol. ARC, 1993 floods. Am. Nuclear Soc. (chair. cen. mid-west sect. 1986-87, cert. of governance 1987), Ill. Soc. Profl. Engrs. (pres. 1986-87, pres's. award Capital chpt. 1987), Health Physics Soc. (charter, Prairieland chpt. by-laws com. 1989-90), Am. Contract Bridge League (sect. master 1989). Achievements include the design, development and installation of the Ill. Dept. of Nuclear Safety's continuous radiation monitoring system consisting of isotopic stackmonitors, environmental detection network and liquid effluent monitors for nuclear power plants, the establishment of programs and goals for the development of a research and testing center for radiation monitoring systems. Home: RR 1 Box 90A Ashland IL 62612-9801

BENZ, ROBERT DANIEL, toxicologist; b. Indpls., Apr. 4, 1947; s. Robert Raymond and Mary Lou (Rasico) B.; m. Bernadette Ann Kuhn, Feb. 22, 1969; children: Zachary Oliver, Nicholas Joseph. BS, Ill. Inst. Tech., 1969;

PhD, U. Calif., Berkeley, 1974. Sta. mgr. Radio Sta. WIIT, Chgo., 1967-69; postdoctoral fellow York Univ., Downsview, Ont., Can., 1974-76; adj. asst. prof. Univ. Calif., Irvine, 1976-80; asst. scientist Brookhaven Nat. Lab., Upton, N.Y., 1980-82; assoc. scientist Brookhaven Nat. Lab., Upton, 1982-86; rev. toxicologist U.S. FDA, Washington, 1987—, group leader, 1991—. Contbr. articles to profl. jours. V.p. PTO, Miller Place, N.Y., 1983-87; first v.p. Civic Assn., Miller Place, N.Y., 1986-87; mem. Nat. Assn. R.R. Passengers, Washington, 1976—. Mem. AAAS, Assn. Govt. Toxicologists (dir. 1991-93, pres. elect 1993—), Genetic Toxicology Assn. (dir. 1989-93, sec. 1991-92, chmn. 1992-93). Achievements include discovery that 60Hz electromagnetic fields do not cause toxicological effects in mice; rsch. in families of repeated DNA sequences not evenly spread over hamster chromosomes, in plastic implants causing DNA damage in mouse bone marrow. Office: USFDA CFSAN DHEE HFS-227 200 "C" St SW Washington DC 20204

BENZ, SAMUEL PAUL, physicist; b. Dubuque, Iowa, Dec. 4, 1962; s. Frank Leonard and Joyce Elaine (Schumann) B.; m. Elizabeth Ann Roberts, Oct. 15, 1988; 1 child, Paul Alexander. BA, Luther Coll., 1985; PhD, Harvard U., 1990. Rsch. asst. dept. physics Harvard U., Cambridge, Mass., 1985-90; postdoctoral fellow Nat. Inst. of Standards and Tech., Boulder, Colo., 1990-92; physicist NIST, Boulder, Colo., 1992—. Author: Dynamical Properties of Two-dimensional Josephson Junction Arrays, '1990; contbr. articles to profl. publs. Officer Luth. Brotherhood Br. 8659, Boulder, 1991-93. Achievements include patent on voltage-tunable high-frequency superconducting oscillator; expertise in superconducting electronics. Office: NIST Div 814-03 325 Broadway St Boulder CO 80303-3328

BENZER, SEYMOUR, neurosciences educator; b. N.Y.C., Oct. 15, 1921; s. Mayer and Eva (Naidorf) B.; m. Dorothy Vlosky, Jan. 10, 1942 (dec. 1978); children: Barbara Ann Benzer Freidin, Martha Jane Benzer Goldberg; m. Carol A. Miller, May 11, 1980; 1 child, Alexander Robin. B.A., Bklyn. Coll., 1942; M.S., Purdue U., 1943, Ph.D., 1947, D.Sc. (hon.), 1968; D.Sc., Columbia U., 1974, Yale U., 1977, Brandeis U., 1978, CUNY, 1978, U. Paris, 1983, Rockefeller U., N.Y.C., 1993. Mem. faculty Purdue U., 1945-67, prof. biophysics, 1958-61, Stuart distinguished prof. biology, 1961-67; prof. biology Calif. Inst. Tech., 1967-75, Boswell prof. neurosci., 1975—; biophysicist Oak Ridge Nat. Lab., 1948-49; vis. asso. Calif. Inst. Tech., Pasadena, 1965-67. Contbr. articles to profl. jours. Rsch. fellow Calif. Inst. Tech., 1949-51; Fulbright rsch. fellow Pasteur Inst., Paris, 1951-52; sr. NSF postdoctoral fellow Cambridge, Eng., 1957-58; recipient Award of Honor Bklyn. Coll., 1956, Sigma Xi rsch. award Purdue U., 1957, Ricketts award U. Chgo., 1961, Gold medal N.Y. City Coll. Chemistry Alumni Assn., 1962, Gairdner award of merit, 1964, McCoy award Purdue U., 1965, Lasker award, 1971, T. Duckett Jones award, 1975, Prix Leopold Mayer French Acad. Scis., 1975, Louisa Gross Horwitz award, 1976, Harvey award Israel, 1977, Warren Triennial prize Mass. Gen. Hosp., 1977, Dickson award, 1978, Rosenstiel award, 1986, T.H. Morgan medal Genetics Soc. Am., 1986, Karl Spencer Lashley award, 1988, Gerard award Soc. Neurosci., 1989, Helmerich award, 1990, Wolf Found. Prize (in medicine), Israel, 1991, Bristol-Myers Squibb Neurosci. award, 1992, Crafoord prize Royal Swedish Acad. Scis., 1993. Fellow Indian Acad. Scis. (hon.); mem. Nat. Acad. Scis., Am. Acad. Arts and Scis., Am. Philos. Soc. (Lashley award 1988), Harvey Soc., N.Y. Acad. Scis., AAAS, Royal Soc. London (fgn. mem.). Home: 2075 Robin Rd San Marino CA 91108-2831

BERANEK, LEO LEROY, scientific foundation executive, engineering consultant; b. Solon, Iowa, Sept. 15, 1914; s. Edward Fred and Beatrice (Stahle) B.; m. Phyllis Knight, Sept. 6, 1941 (dec. Nov. 1982); children: James Knight, Thomas Haynes; m. Gabriella Sohn, Aug. 10, 1985. A.B., Cornell Coll., 1936, D.Sc. (hon.), 1946; M.S., Harvard U., 1937, D.Sc., 1940; D.Eng. (hon.), Worcester Poly. Inst., 1971; D.Comml. Sci. (hon.), Suffolk U., 1979; LL.D. (hon.), Emerson College, 1982; Dr. Pub. Service (hon.), Northeastern U., 1984. Instr. physics Harvard U., 1940-41, asst. prof., 1941-43, dir. rsch. on sound, 1941-45; dir. Electro-Acoustics and Systems Rsch. Labs., 1945-46; assoc. prof. communications engring. MIT, 1947-58, lectr., 1958-81; tech. dir. Acoustics Lab., 1947-53; pres., dir., chief exec. officer Bolt Beranek & Newman, Cambridge, Mass., 1953-69, chief scientist, 1969-71, dir., 1953-84; pres., chief exec. officer, dir. Boston Broadcasters, Inc., 1963-79, chmn. bd., 1980-83; pres. Am. Acad. Arts and Scis., Cambridge, 1989—; part-owner WCVB-TV, Boston, 1972-82; chmn. bd. Mueller-BBM GmbH, Munich, 1962-86; bd. dirs. Tech. Integration Inc., Bedford, Mass., 1987—. Author: (with others) Principles of Sound Control in Airplanes, 1944, Acoustic Measurements, 1949, 2d edit., 1986, Music, Acoustics and Architecture, 1962; editor, contbr. (with others) Noise Reduction, 1960, Noise and Vibration Control, 1971, 2d edit., 1988, Noise and Vibration Control Engineering, 1992; contbr. articles on acoustics, audio and TV communications systems to tech. publs. Mem. Mass. Gov.'s Task Force on Coastal Resources, 1974-77; charter mem. bd. overseers Boston Symphony Orch., 1968-80, chmn., 1977-80, trustee, 1977-87, v.p., 1980-83, chmn. bd. trustees, 1983-86, hon. chmn., 1987; vis. com. biology and related rsch. facilities Ctr. Behavioral Scis. Harvard U., 1971-77, 86-91, vis. com. physics dept., 1983-90, vis. com. Bus. Sch., 1983-90, bd. overseers, 1984-90; mem. council for arts MIT, 1972—; hon. chmn. bd. dirs. Opera Co. of Boston, 1987-89; mem. Mass. Commn. on Jud. Conduct, 1986-88, others in past. Guggenheim fellow, 1946-47; recipient Presdl. certificate of merit, 1948; Cornell Coll. Alumni Citation, 1953; 1st Silver medal le Groupement des Acousticiens de Langue Francaise Paris, 1966; Abe Lincoln TV award So. Bapt. Conv., 1975; Media award NAACP, 1975; named Sta. WCRB Person of the Yr., 1987. Fellow Acoustical Soc. Am. (Biennial award 1944, exec. council 1944-47, v.p. 1949-50, pres. 1954-55, assoc. editor 1946-60, Wallace Clement Sabine Archtl. Acoustics award 1961, Gold medal award 1975), NAE (bd. dir. marine bd., com. pub. engring. policy, aeros. and space engring. bd.),, Am. Phys. Soc., AAAS, Am. Acad. Arts and Scis., Audio Engring. Soc. (pres. 1967-68, Gold medal 1971, gov. 1966-71), IEEE (chmn. profl. group audio 1950-51); mem. Inst. Noise Control Engring. (charter pres. 1971-73, dir. 1973-75), Am. Standards Assn. (chmn. acoustical standards bd. 1956-68, dir. 1963-68), Mass. Broadcasters Assn. (bd. dir. 1973-80, pres. 1978-79, Disting. Service award 1980), Boston Community Media Council (treas. 1973-76, v.p. 1976-77), Cambridge Soc. Early Music (pres. 1963-71, dir. 1961-79), Acad. Disting. Bostonians, Greater Boston C. of C. (dir. 1973-79, v.p. 1976-79, Disting. Community Service award 1980, 83), Phi Beta Kappa, Sigma Xi, Eta Kappa Nu. Episcopalian. Clubs: Mass. Inst. Tech. Faculty, St. Botolph, Harvard, City. Home: 975 Memorial Dr Apt 804 Cambridge MA 02138-5755 Office: 136 Irving St Cambridge MA 02138

BERBARY, MAURICE SHEHADEH, physician, military officer, hospital administrator, educator; b. Beirut, Lebanon, Jan. 14, 1923; came to U.S., 1945, naturalized, 1952; s. Shehadeh M. and Marie K. Berbary; children: Geoffrey Maurice, Laura Marie. BA, Am. U., Beirut, 1943; MD, U. Tex., 1948; MA in Hosp. Adminstrn., Baylor U., 1970; diploma, Army Command and Gen. Staff Coll., Leavenworth, Kan., 1963, Air Force Sch. Aerospace Medicine, San Antonio, 1964, Army War Coll., Carlisle, Pa., 1969. Diplomate Am. Bd. Ob-Gyn. Intern Parkland Meml. Hosp., Dallas, 1948-49, resident in ob-gyn., gen. surgery and urology, 1949-53; resident in ob-gyn. Walter Reed Army Hosp., Washington, 1955-57; fellow in obstetric and gynecologic pathology Armed Forces Inst. Pathology, Washington, 1959-60; practice clin. medicine in ob-gyn., 1953—; capt. MC U.S. Army, 1952, advanced through grades to col., 1968; chief dept. ob-gyn. U.S. Army Hosp., Ft. Polk, La., 1957-59, Womack Army Hosp., Ft. Bragg, N.C., 1960-62; div. surgeon 1st. inf. div., Ft. Riley, Kans., 1963-64, 3d. Armored div., Germany, 1964-65; corps surgeon V. Corps, Germany, 1965-67, 24th Army Corps, S. Vietnam Theater of Operation, 1970; comdr., hosp. adminstr. U.S. Army Hosp., Teheran, Iran, 1954-55; comdr. 43d Hosp. Group Complex, Vietnam, 1969-70; command surgeon U.S. Armed Forces Command and U.S. Army South, U.S. C.Z., Panama, 1970-73; comdr. 5th Gen. Hosp., Stuttgart, West Germany, 1973-77, Munson Army Hosp., Ft. Leavenworth, Kans., 1977-81; sr. staff officer dept. ob-gyn William Beaumont Army Med. Ctr., Ft. Bliss, Tex., 1981-83; ret., 1983, cons. health care adminstrn. and med.-legal affairs, 1984—; vis. lectr. ob-gyn. and pathology Duke U. Med. Ctr., Durham, N.C., 1960-62; clin. instr. dept. ob-gyn. U. Kans. Coll. of Medicine, Kansas City, 1963-80, advanced to clin. asst. prof. dept. ob-gyn., 1980—; instr. 5th Army NCO Acad., Fort Riley, Kans., 1963-64. Decorated Legion of Merit with three oak leaf clusters, Bronze Star medal, Army Commendation medal, Combat Air medal. Fellow ACS, Am. Coll. Obstetricians and Gynecologists, Am. Coll. Health Care Execs.; mem. AMA, Assn. of Mil. Surgeons, Soc. of U.S. Army Flight Surgeons, Internat. Platform Assn., Am. Hosp. Assn., N.Y. Acad. Scis., Dallas County Med. Soc., Tex. State Med. Assn., Mason (32 deg.). Home and Office: 7923 Abramshire Ave Dallas TX 75231-4712

BERBENICH, WILLIAM ALFRED, electronics engineer; b. Atlanta, Sept. 24, 1961; s. Charles Joseph and Dorothy June (Albers) B. Student, Chapman Coll., 1985-86, Ga. State U., 1987—. Exec. v.p. Silicon Solutions USA, Atlanta, 1988—; mem. rsch. staff Ga. Inst. Tech., Atlanta, 1988-93; radio frequency engr. PacTel Cellular, Atlanta, 1993—; propr. S.E. News Photo, Atlanta, 1992—. Vol. Ga. Games, 1990—, Scottish Rite Children's Med. Ctr., 1990—. Decorated D.S.M., Navy Achievement medal, Meritorious Svc. medal. Mem. VFW (life), NRA (life). Libertarian. Avocations: skiing, sailing, amateur radio, golf, outdoors. Office: PacTel Cellular Ste 300 4151 Ashford Dunwoody Rd Atlanta GA 30319

BERCAW, JOHN EDWARD, chemistry educator, consultant; b. Cin., Dec. 3, 1944; s. James Witherow and Mary Josephine (Heywood) B.; m. Teresa Diane Ingram, July 10, 1965; children—David Lawrence, Karin Elizabeth. B.S. in Chemistry, N.C. State U., 1967; Ph.D. in Chemistry, U. Mich., 1971. Postdoctoral U. Chgo., 1971-72; A.A. Noyes fellow Calif. Inst. Tech., Pasadena, 1972-74, asst. prof. chemistry, 1974-77, assoc. prof. chemistry, 1977-79, prof. chemistry, 1979—, Shell Disting. prof., 1985-90; cons. Exxon Corp., Annandale, N.J., 1979—. Fellow Am. Acad. Arts and Scis.; mem. NAS, AAAS, Am. Chem. Soc. (chmn. div. inorganic chemistry 1988—, organometallic subdiv. chair 1980, recipient award in pure chemistry 1980, award in organometallic chemistry 1990, vice chmn. Gordon Rsch. Conf. on Organometallic Chemistry 1990). Home: 985 N Chester Ave Pasadena CA 91104-2942 Office: Calif Inst Tech 1201 E California Blvd Pasadena CA 91125-0001

BERCEL, NICHOLAS ANTHONY, neurologist, neurophysiologist; b. Budapest, Hungary, Aug. 20, 1911; came to U.S., 1940; s. Desiré and Julia (Kapos) B.; children: Diana, Anthony, Christopher, Patrick, Yvette. MD, U. Rome, 1936., 1940; Resident U. Rome, 1936-38, U. Paris, 1938-40; intern Swedish Hosp., Mpls., 1958—; assoc. prof. in physiology U. So. Calif., L.A., 1948-67; mem. staff St. John Hosp., Santa Monica, 1954-84; mem. staff dept. neurodiagnosis Queen of Angels Hosp., L.A., 1960-80; neuro-psychiat. cons. Social Security Adminstrn., West Los Angeles, 1958—. Author: Textbook on Etiology of Schizophrenia, Psychopathology, 1959; contbr. numerous articles to profl. publs., including Diseases of the Nervous System, Jour. of Neuropsychiatry, Jour. AMA, Calif. Medicine, Am. Jour. Med. Scis., others. Republican. Achievements include research on experimental epilepsy for testing the comparative activity of anticonvulsants; schizophrenic serum influence on spider beahvior; schizophrenic model psychoses induced with LSD-25.

BERCLAZ, THEODORE M., chemistry educator; b. Nyon, Switzerland, June 27, 1949; s. Armand and Germaine M. (Andreoli) B.; m. Nadia M. Berclaz, Mar. 12, 1979; 1 child, Matthieu. Diplome, U. Geneva, 1974, PhD, 1979. Asst. U. Geneva, 1974-78; rsch. assoc. Stanford (Calif.) U., 1979-81; chef de travaux U. Geneva, 1981-86, maître enseignement et de recherche, 1986—. Contbr. articles to profl. jours. Swiss NSF grantee 1979-80. Mem. Am. Chem. Soc., Am. Biophysical Soc., Soc. Suisse de Chimie, Soc. Chimique de Geneve (sec. 1986). Roman Catholic. Office: U Geneve, Dep Chimie Physique, 1211 Geneva Switzerland

BERENS, RANDOLPH LEE, microbiologist, educator; b. L.A., June 13, 1943. BS, U. Calif., Irvine, 1967, PhD in Cell Biology, 1975. Rsch. asst. cell biology U. Calif., Irvine, 1971-75; fellow sch. medicine Washington U., 1975-78; asst. prof. physiology sch. med. St. Louis U., 1978—; asst. prof. medicine and microbiology health sci. ctr. U. Colo., Denver; instr. physiology Cerritos Coll., 1974-75. Grantee WHO, 1979-82, NIH, 1981—. Mem. AAAS, Am. Soc. Protozoologists, Am. Soc. Microbiologists, Am. Soc. Parasitologists, N.Y. Acad. Sci. Achievements include research in various intermediate metabolic pathways in pathogenic protozoans with the primary goal of finding metabolic differences between these organisms and their mammalian hosts. Office: University of Colorado 4200 E 9th Ave Box 168 Denver CO 80262*

BERENSON, GERALD SANDERS, physician; b. Bogalusa, La., Sept. 19, 1922; s. Meyer A. and Eva (Singerman) B.; m. Joan Seidenbach, Mar. 7, 1951; children—Leslie, Ann, Robert, Laurie. B.S., Tulane U., 1943, M.D., 1945. Intern U.S. Navy Hosp., Great Lakes, Ill., 1945-46; practice medicine specializing in cardiology New Orleans; mem. staff Charity Hosp., Hotel Dieu; instr. dept. medicine Tulane U., 1949-52, prof. medicine, pediatrics Sch. Pub. Health, 1992—; asst. prof. medicine La. State U. Med. Sch., 1954-58, assoc. prof., 1958-63, prof., 1963-92, Boyd prof., 1988-92, prof. emeritus, 1992—; prof. medicine and pediatrics Tulane U. Sch. Pub. Health, New Orleans, 1992—; dir. Specialized Ctr. Rsch. Arteriosclerosis, New Orleans, 1972-87, Nat. Rsch. and Demonstration Ctr. in Arteriosclerosis, 1984-87, Nat. Ctr. Cardiovascular Health, Sch. Pub. Health and Tropical Medicine Tulane U., 1992—; sr. vis. physician Charity Hosp. La., New Orleans, 1948—; cons. Touro Infirmary, 1967—; cons. medicine Hotel Dieu, 1962—. Contbr. articles to profl. jours. Served with USNR, 1945-48. USPHS fellow U. Chgo., 1952-54. Mem. Am. Coll. Cardiology (gov. La. 1985-88, trustee 1988, chmn. prevention com. 1990-93), So. Soc. Clin. Investigation (pres. 1969), La. Heart Assn. (pres. 1971), New Orleans Acad. Internal Medicine (pres. 1966), Musser-Burch Soc. (pres. 1981), Sigma Xi, Alpha Omega Alpha. Home: 505 Northline St Metairie LA 70005-4435 Office: Tulane Sch Pub Health Nat Ctr Cardiovascular Health 1501 Canal St New Orleans LA 70112

BERES, JOEL EDWARD, electrical engineer, consultant; b. Meadville, Pa., Jan. 6, 1961. BSEE cum laude, W.Va. U., 1982. Registered profl. engr., N.Y.; cert. sr. reactor operator. Systems analyst GE Aerospace Bus. Group, Springfield, Va., 1988-89; nuclear licensing engr. Niagara Mohawk Power Corp., Syracuse, N.Y., 1989-91; cons. engr. Tenera L.P., Monticello, Minn., 1991—. Lt. USN, 1982-88. Mem. Am. Nuclear Soc., Tau Beta Pi. Home: 9739 Gilbert Ave NE Monticello MN 55362 Office: MNGP Hwy 75 Monticello MN 55362

BERES, MILAN, surgeon; b. Trebisov, Slovak Republic, Jan. 13, 1936; came to U.S., 1968, naturalized, 1974; s. Juraj and Barbara (Hrinova) B.; grad. Slovak U., 1955; M.D., P.J. Safarik U. (Slovak Republic), 1959; m. Terezia Marcinova, Nov. 17, 1962; children: Stephen, Milan Jr. Otolaryngologist, Univ. Hosp., Kosice, Slovak Republic, 1962-68; fellow in otolaryngology Cleve. Clin. Ednl. Found., 1970-71; resident physician in otolaryngology and facial plastic and reconstructive surgery U. Conn. Health Center, Farmington, 1971-74, sr. attending physician, 1974-87, chief otolaryngology sect., 1987—, Bridgeport (Ct.) Hosp.; elected pres. Slovak-Am. Cultural Ctr., N.Y.C., 1991. Served with Czechoslovakian Army, 1960-61. Fellow ACS; mem. AMA (physicians recognition awards), Conn. Med. Assn., Fairfield County Med. Assn., Greater Bridgeport Med. Assn., Am. Acad. Otolaryngology, ACS, Internat. Corr. Soc. Ophthalmologists and Otolaryngologists, Am. Acad. Facial Plastic and Reconstructive Surgery, New England Otolaryn. Soc., Pan Am. Assn. Oto-rhino-laryngology Head and Neck Surgery. Roman Catholic. Contbr. articles to med. jours. Home: 31 Isinglass Ter Trumbull CT 06611-4038 Office: 1681 Barnum Ave Stratford CT 06497-5302

BERETTA, GIORDANO BRUNO, computer scientist, researcher; b. Brugg, Aargau, Switzerland, Apr. 14, 1951; came to U.S., 1984; PhD, ETH, Zurich, Switzerland, 1984. Mem. rsch. staff Xerox Palo Alto (Calif.) Rsch. Ctr., 1984-90; charter mem., sr. scientist Canon Info. Systems, Palo Alto, 1990—. Contbr. articles to profl. jours. Mem. IEEE, Assn. Computing Machinery, Soc. Imaging Sci. and Tech., Soc. Info. Display, Inter-Soc. Color Coun. Office: Canon Info Systems 4009 Miranda Ave Palo Alto CA 94304-1218

BERG, BERND ALBERT, physics educator; b. Delmenhorst, Germany, Aug. 23, 1949; came to U.S., 1985; s. Max and Irmgard (Tetzlaff) B.; m. Ursula A. Schroder, Mar. 26, 1975; 1 child, Felix. Dr rer Nat., Free U., Berlin, 1977. Postdoctoral fellow Free U., Berlin, 1977-78; postdoctoral fellow Univ. Hamburg, Germany, 1978-80, asst. prof., 1982-85; fellow CERN, Geneva, 1980-82; assoc. prof. Fla. State U., Tallahassee, 1985-88, prof., 1988—; vis. prof. U. Bielefeld, Germany, 1990-91; fellow Inst. for Advanced Study, Berlin, 1992-93. Contbr. articles to profl. jours. Mem. Am. Phys. Soc. Achievements include work in quantum field theory and computational phsyics, in particular introduction of the multicanonical ensemble. Office: Fla State U Tallahassee FL 32306

BERG, CLYDE CLARENCE, research agronomist; b. Meriden, Kans., Nov. 2, 1936; s. Clarence W. and Edna G. (Jensen) B.; m. Rebecca M. Hoskinson, June 8, 1958; children: Judy E. West, Karen L. Duncan, Larry K. BS, Kans. State U., 1958; MS, Okla. State U., 1960; PhD, Wash. State U., 1965. Rsch. agronomist USDA-ARS, Univ. Park, Pa., 1965—. Contbr. articles to profl. jours. Scoutmaster Boy Scouts Am., State College, Pa., 1982-91. Mem. Am. Soc. Agronomy, Crop Sci. Soc. Am., Am. Genetics Assn., Genetics Soc. Can. Republican. Methodist. Home: 272 Bradley Ave State College PA 16801 Office: USDA ARS US Regional Pasture Rsch Lab University Park PA 16801

BERG, DEAN MICHAEL, applications engineer; b. Rochester, Minn., Mar. 17, 1968; s. Lyle Richard and Rose Marie (Ellis) B. BS, U. Wis., Stout, 1991. Registered profl. engr., Minn. Customer support engr. Camax Systems, Mpls., 1991—. Mem. NSPE (Engr. in Tng. award 1990), AIAA (Minn. br.), Soc. Mfg. Engrs. Home: 910 Wescott Trail # 201 Eagan MN 55123

BERG, GEORGE G., toxicologist; b. Warsaw, Poland, May 27, 1919; came to U.S. 1938; s. Leon and Stefa (Goldman) Gliksberg; m. Olga Aronowitz, (dec. 1988); children: Peggy Ann, Carl Lester, Lora Jane; m. Raiza Tuchman Bolgla, May 5, 1991. PhD in Zoology, Columbia U., 1954. Lectr. biology U. Rochester, N.Y., 1955-63, asst. then assoc. prof. biophysics, 1960-88; vis. prof. Zhejiang U., Hangzhou, China, 1989; emeritus prof. toxicology U. Rochester, N.Y., 1988--; coun. mem. Finger Lakes Health Systems Agy., Rochester, 1976-82. Co-author: The Encyclopedia of Microscopy and Microtechnique, 1973; editor: Measurement of Risks, 1981. Founder, bd. dirs. The Rochester Com. for Scientific Info., 1964-88; fellow Scientists Inst. for Pub. Info., N.Y.C., 1969-80. Sgt. U.S. Army, 1942-46, ETO. Recipient Feinstone Environ. award Coll. Environ. Sci. and Forestry, 1987, Linder Conservation award Conservation Coun., 1985, Cert. Appreciation Genesee Region Health Planning Coun., 1976. Mem. Soc. for Risk Analysis, Soc. Toxicology, Am. Assn. for the Advancement Sci., Fedn. Am. Scientists. Jewish. Achievements include development of inhibitor partition kinetics for mercuric salts and the localization of polyphosphate hydrolases by chelate removal. Home and Office: 1242 Wildcliff Pkwy Atlanta GA 30329

BERG, IVAR ELIS, JR., social science educator; b. Bklyn., Jan. 3, 1929; s. Ivar Elis and Hjordis (Holmgren) B.; m. Calli J. Smallwood, Feb. 16, 1991; 1 child, Geoffrey Sverre. AB, Colgate U., 1954; postgrad., U. Oslo, Norway, 1954-55; PhD, Harvard U., 1959; MA (hon.), U. Pa., 1979. Asst. prof. to prof. sociology Columbia U., N.Y.C., 1959-75; dean faculties Columbia U., 1969-71; prof. sociology Vanderbilt U., Nashville, 1975-79; Justin Potter prof. bus. Vanderbilt U., 1983-84; prof. and chmn. dept. sociology U. Pa., Phila., 1979-83, prof. sociology/dean of coll., 1984-89, dean social sci., 1989-91; cons. Chancellor of Higher Edn., Trenton, N.J., 1982-89, Pres.'s Commn. on crime, Washington, 1966-67; chmn. coll. svcs. Coll. Bd., N.Y.C., 1989-91. Author: Great Training Robbery, 1970, Managers and Work Reform, 1978, Work and Industry, 1987; contbr. articles to profl. jours. Conciliator Ad Hoc Com. on Pub. Edn., Hastings-on-Hudson, N.Y., 1967-69. Maj. USMC, 1946-65; ATO. Guggenheim fellow, 1973-74, Rockefeller fellow, 1975-76, Woodrow Wilson fellow, 1954-55. Fellow AAAS, N.Y. Acad. Sci., Internat. Acad. Mgmt.; mem. Am. Sociol. Assn. (coun. mem. 1989-91), Ea. Sociol. Soc. (v.p. 1991), Harvard Club (N.Y.C.), Pres.'s Club of Colgate U., Phi Beta Kappa. Presbyterian. Avocations: tennis, stamp collecting. Home: 2501 Christian St # 405 Philadelphia PA 19146-2322 Office: U Pa 113 McNeil Bldg Philadelphia PA 19104

BERG, JAN MIKAEL, science educator; b. Stockholm, Sweden, Jan. 5, 1928; arrived in Germany, 1969; s. Curt Per Assar and Eva Elisabet (Ekstroem) B.; m. Catarina Eva Eklund, Apr. 22, 1959; 1 child, Jenny Elisabet. BA, U. Stockholm, 1950, MA, 1956, PhD, 1962. Teaching asst. U. Stockholm, 1957-62, asst. prof., 1964-69; asst. prof. U. Minn., Mpls., 1962-64; prof. Tech. U. Munich, 1969—. Author: Bolzano's Logic, 1962, Ontology Without Ultrafilters and Possible Worlds, 1992, Mathematische Logik. Eine Bibliometrische Untersuchung, 1993; editor: Collected Works of Bernard Bolzano, 23 vols., 1972-93; contbr. articles to profl. jours. Mem. Assn. Symbolic Logic, Philosophy of Sci. Assn. Home: Paul-Hey-Strasse 25, D-82131 Gauting Germany Office: Munich Tech Univ, Lothstrasse 17, D-80335 Munich Germany

BERG, JEREMY M., chemistry educator. Prof., dir. dept. biophysics Johns Hopkins U., Balt. Recipient ACS Pure Chemistry award Am. Chemical Soc., 1993. Office: Johns Hopkins Univ Sch Dept Biophysics 720 Rutland Ave Baltimore MD 21205*

BERG, LILLIAN DOUGLAS, chemistry educator; b. Birmingham, Ala., July 9, 1925; d. Gilbert Franklin and Mary Rachel (Griffin) Douglas; m. Joseph Wilbur Berg, June 26, 1950; children: Anne Berg Jenkins, Joseph Wilbur III, Frederick Douglas. BS in Chemistry, Birmingham So. Coll., 1946; MS in Chemistry, Emory U., 1948. Instr. chemistry Armstrong Jr. Coll., Savannah, Ga., 1948-50; rsch. asst. chemistry Pa. State U., University Park, 1950-54; instr. chemistry U. Utah, Salt Lake City, 1955-56; prof. chemistry No. Va. Community Coll., Annandale, 1974—. Mem. Am. Chem. Soc., Am. Women in Sci., Mortar Bd. Soc., Iota Sigma Pi, Sigma Delta Epsilon, Phi Beta Kappa. Avocation: music. Home: 3319 Dauphine Dr Falls Church VA 22042-3724 Office: No Va Community Coll 8333 Little River Tpke Annandale VA 22003-3796

BERG, MARY JAYLENE, pharmacy educator, researcher; b. Fargo, N.D., Nov. 7, 1950; d. Ordean Kenneth and Anna Margaret (Skramstad) B. BS in Pharmacy, N.D. State U., 1974; PharmD, U. Ky., 1978. Lic. pharmacist, N.D., Ky., Iowa. Fellow in pharmacokinetics Millard Fillmore Hosp./SUNY, Buffalo, 1978-79; asst. prof. U. Iowa, Iowa City, 1980-85, assoc. prof., 1985—; with dept. clin. rsch., clin. pharmacology/pharmacokinetics F. Hoffmann-La Roche, Ltd., Basel, Switzerland, 1992. Reviewer Cin. Pharmacy, 1984—, Epilepsia, 1987—; editor: Internat. Leadership Symposium, The Role of Women in Pharmacy, 1990, Pharmacy World Congress '91: Women-A Force in Pharmacy Symposium, 1992; contbr. articles to Drug Intelligence & Clin. Pharmacy, New Eng. Jour. of Medicine, Jour. Forensic Scis., Therapeutic Drug Monitoring, Epilepsia. Advisor Kappa Epislon, Iowa City, 1980—; pres. Mortar Bd. Alumnae, Iowa City, 1986-88. NIH grantee, 1984, Nat. Insts. on Drug Abuse grantee, 1986; recipient Career Achievement award Kappa Epsilon, 1985. Mem. Am. Assn. Pharm. Scientists, Am. Soc. Hosp. Phrmacists (chair spl. interest group of clin. pharmacokinetics 1987-89), Am. Epilepsy Soc., Am. Pharm. Assn., Internat. Forum for Women in Pharmacy (U.S. contact), Fedn. Internat. Pharmaceutique, Leadership Internat., Women in Pharmacy (bd. dirs. 1991—), Sigma Xi, Rho Chi, Kappa Epsilon, Phi Beta Delta. Lutheran. Avocations include research in multiple doses of oral activated charcoal to clear totally absorbed drug, pharmacokinetics of drug-drug interaction between phenytoin and folic acid, also women's health research. Office: U Iowa Coll of Pharmacy Iowa City IA 52242

BERG, PAUL, biochemist, educator; b. N.Y.C., June 30, 1926; s. Harry and Sarah (Brodsky) B.; m. Mildred Levy, Sept. 13, 1947; 1 son, John. BS, Pa. State U., 1948; PhD (NIH fellow 1950-52), Western Res. U., 1952; DSc (hon.), U. Rochester, 1978, Yale U., 1978, Wash. U., St. Louis, 1986, Oreg. State U., 1989. Postdoctoral fellow Copenhagen (Denmark) U., 1952-53; postdoctoral fellow sch. medicine Washington U., St. Louis, 1953-54; Am. Cancer Soc. scholar cancer research dept. microbiology sch. medicine Washington U., 1954-57, from asst. to assoc. prof. microbiology sch. medicine, 1955-59; prof. biochemistry sch. medicine Stanford U., 1959—, Sam, Lula and Jack Willson prof. biochemistry sch. medicine, 1970, chmn. dept. sch. medicine, 1969-74; dir. Stanford U. Beckman Ctr. for Molecular and Genetic Medicine, 1985—; non-resident fellow Salk Inst., 1973-83; adv. bd. NIH, NSF, MIT; vis. com. dept. biochemistry and molecular biology Harvard U.; bd. sci. advisors Jane Coffin Childs Found. Med. Rsch., 1970-80; chmn. sci. adv. com. Whitehead Inst., 1984-90; internat. adv. bd. Basel Inst. Immunology; chmn. nat. adv. com. Human Genome Project, 1990-92. Contbr. profl. jours.; Editor: Biochem. and Biophys. Research Communications, 1959-68; editorial bd.: Molecular Biology, 1966-69. Trustee Rockefeller U.,

1990-92. Served to lt. (j.g.) USNR, 1943-46. Recipient Eli Lilly prize biochemistry, 1959; V.D. Mattia award Roche Inst. Molecular Biology, 1972; Henry J. Kaiser award for excellence in teaching, 1974; Disting. Alumnus award Pa. State U., 1972; Sarasota Med. awards for achievement and excellence, 1979; Gairdner Found. annual award, 1980; Lasker Found. award, 1980; Nobel award in chemistry, 1980; N.Y. Acad. Sci. award, 1980; Sci. Freedom and Responsibility award AAAS, 1982; Nat. Medal of Sci., 1983; named Calif. Scientist of Yr. Calif. Museum Sci. and Industry, 1963; numerous disting. lectureships including Harvey lectr., 1972, Lynen lectr., 1977, Priestly lectrs. Pa. State U., 1978, Dreyfus Disting. lectrs. Northwestern U., 1979, Lawrence Livermore Dir.'s Disting. lectr., 1983. Fellow AAAS; mem. NAS, Inst. Medicine, Am. Acad. Arts and Scis., Am. Soc. Biol. Chemists (pres. 1974-75), Am. Soc. Microbiology, Am. Philos. Soc., Japan Biochem. Soc. (elected fgn. mem. 1978), French Acad. Sci. (elected fgn. mem. 1981), Royal Soc. (elected fgn. mem. 1992). Office: Stanford Sch Medicine Beckman Ctr B-062 Stanford CA 94305-5425

BERG, ROBERT RAYMOND, geologist, educator; b. St. Paul, May 28, 1924; s. Raymond F. and Jennie (Swanson) B.; m. Josephine Finck, Dec. 22, 1946; children: James R., (dec.), Charles R., William R. B.A., U. Minn., 1948, Ph.D., 1951. Geologist, Calif. Co., Denver, 1951-56; cons. Berg and Wasson, Denver, 1957-66; prof. geology, head dept. Tex. A&M U., 1967—, Michel T. Halbouty prof. geology, 1982—; dir. univ. research Tex. A & M U., 1972—; cons. petroleum geology, 1959—. Contbr. papers in field. Served with AUS, 1943-46. Fellow Geol. Soc. Am.; mem. Am. Assn. Petroleum Geologists (disting. lectr. 1972, hon. mem. 1985, Sidney Powers Meml. award 1993), Am. Inst. Profl. Geologists (pres. 1971, hon. mem. 1988), Nat. Acad. Engring. Home: 414 Brookside Bryan TX 77801-3701 Office: Texas A&M Univ Geology Dept College Station TX 77843

BERG, STANTON ONEAL, firearms and ballistics consultant; b. Barron, Wis., June 14, 1928; s. Thomas C. and Ellen Florence (Nedland) Silbaugh; m. June K. Rolstad, Aug. 16, 1952; children: David M., Daniel L., Susan E., Julie L. Student U. Wis., 1949-50; LLB, LaSalle Extension U., 1951; postgrad. U. Minn., 1960-69. Claim rep. State Farm Ins. Co., Mpls., Hibbing and Duluth, Minn., 1952-57, claim supt., 1957-66, divisional claim supt., 1966-70; firearms cons., Mpls., 1961—; regional mgr. State Farm Fire and Casualty Co., St. Paul, 1970-84; bd. dirs. Am. Bd. Forensic Firearm and Tool Mark Examiners, 1980—; instr. home firearms safety, Mpls., 1975—; cons. to Sporting Arms and Ammunition Mfrs. Inst., 1974—; internat. lectr. on forensic ballistics. Adv. bd. Milton Helpern Internat. Ctr. for Forensic Scis., 1975—; mem. bd. cons. Inst. Applied Sci., Chgo., 1974—; cons. for reexam. of ballistics evidence in Robert Kennedy assasination/Sirhan case Superior Ct. L.A., 1975; ct. expert witness in most state cts., Mil. Gen. Ct. Martial and U.S. Dist. Cts., and the Supreme Ct. of Ontario, Can.; mem. Nat. Forensic Ctr., 1979—, internat. study group in forensic scis., 1985—; chmn. internat. symposiums on forensic ballistics, Edinburgh, Scotland, 1972, Zurich, 1975, Bergen, Norway, 1981, Dusseldorf, Germany, 1993. With CIC, RA, 1948-52. Fellow Am. Acad. Forensic Sci.; mem. ASTM, Assn. of Firearm and Tool Mark Examiners (exec. council 1970-71, charter mem., life mem., Disting. Mem. and Key Man award 1972, exam. and standards com. 1975-76, spl. honors award 1976, nat. peer group on cert. of firearms examiners 1978—), Forensic Sci. Soc., Internat. Assn. Forensic Scis., Internat. Assn. for Identification (mem. firearms subcom. of sci. and practice com. 1961-74, 86-93, chmn. firearm subcom. 1964-66, 69-70, 91-93, lab. rsch. and techniques subcom. 1980-81, life and disting. mem. 1947—, life charter mem. Minn. dvsn. 1963—), Internat. Wound Ballistics Assn., Western Conf. Criminal and Civil Problems (sci. adv. com.), Am. Legion, Army Counter-Intelligence Corp. Vets. Assn., Browning Arms Collectors Assn. (life 1988—), Am. Ordnance Assn. (life), NRA (life mem. 1957—), Minn. Weapons Collectors, Internat. Cartridge Collectors Assns. (life mem.), Internat. Reference Orgn. Forensic Medicine and Scis, Internat. Assn. Bloodstain Pattern Analysts, Assn. Firearms and Toolmark Examiners (editorial com. AFTE jour. 1989-92), Am. Nat. Standards Inst., Am. Soc. Testing and Materials (standards devel. subcom. 1989—). So Contbg. editor Am. Rifleman mag., 1973-84; mem. editorial bd. Internat. Microform Jour. Legal Medicine and Forensic Scis., 1979—, Am. Jour. Forensic Medicine and Pathology, 1979-91; contbr. articles on firearms and forensic ballistics to profl. publs. Address: 6025 Gardena Ln NE Minneapolis MN 55432

BERGAMINI, EDUARDO WHITAKER, electrical engineer; b. Ribeirao Preto, Sao Paulo, Brazil, Apr. 30, 1944; s. Fausto A.B. and Eudoxia (Whitaker) B.; m. Zuleica Maria Nogueira, May 3, 1969; children—Ana Luiza, Ana Cristina. Elec. Engr., Escola Politecnica da Universidade de Sao Paulo, 1967; M.Sc. in Space Sci., Instituto de Pesquisas Espacias, 1969; M.Sc. in Elec. Engring., Stanford U., 1971, Ph.D. in Elec. Engring., 1973, postdoctoral, 1973. Asst. research Instituto de Pesquisas Espaciais Brazil, S. J. Campos, Sao Paulo, 1968-69; research asst. elec. engring. dept. Stanford U., 1971-73, research assoc., 1973; head group Instituto Nacional de Pesquisas Espaciais, São José dos Campos, Sao Paulo, Brazil, 1974-78, head div., 1978-82, head dept., 1982-88, resp. line of research and devel., 1988— ; cons. Financiadora de Estudios e Projetos, Conselho Nacional de Desenvolvimento Cientifico e Tecnologico, Brazil, 1984; head del. Consultative Com. for Space Data Systems, Washington, 1982—; exec. vice-dir. Laboratorio Nacional de Redes de Computadores, Sao Paulo, 1983-90. NASA fellow Stanford U., 1969-71, Conselho Nacional de Desenvolvimento Cientifico e Tecnologico fellow Stanford U., 1969-72, Coordenacao do Aperfeicoamento Pessoal de Nivel Superior fellow Ministry Edn. Brazil, Stanford U., 1971-73. Mem. IEEE Computer Soc. (sr.), IEEE Aerospace and Electronics Soc., AIAA (sr., subcom. space quality IAA), Internat. Astronautical Fedn. (space exploration com.), Sociedade Brasileira de Computacao, Sociedade Brasileira de Telecomunicacoes, Stanford U. Alumni Assn. Roman Catholic. Club: Tenis (Sao Jose Campos, Sao Paulo). Avocations: tennis; swimming. Home: Rua Beatriz Sa de Toledo 38, San Jose dos Campos, São Paulo Brazil 12243-050 Office: Instituto de Pesquisas Espacias, Avenida dos Astronautas 17 58, San Jose dos Campos São Paulo Brazil 12225

BERGEMAN, GEORGE WILLIAM, mathematics educator, software author; b. Ft. Dodge, Iowa, July 16, 1946; s. Harold Levi and Hilda Carolyn (Nuhn) B.; m. Clarissa Elaine Hellman, Oct. 24, 1968; 1 child, Jessica Ann. BA, U. Iowa, 1970, MS, 1972; postgrad., Va. Inst. Tech., 1978-83. Teaching asst. U. Iowa, Iowa City, 1970-72; coll. instr. Peace Corps, Liberia, 1972-75; asst. prof. math. No. Va. C.C., Sterling, 1975—; software author George W. Bergeman Software, Round Hill, Va., 1984—; cons. Excel Corp., Reston, Va., 1983-84; developer software including graphics and expert systems. Author: (software, book) 20/20 Statistics, 1985, 2d edit., 1988, (software) MathCue, 1987, 2d edit., 1991, 93, Maxis, 1989, MathPath, 1989, 2d edit., 1991, Graph 2D/3D, 1990, 93, MathCue Solution Finder, 1991, 2d edit., 1992, F/C Graph, 1993, F/C/P Graph, 1993. Community worker VISTA, cen. Fla., 1968-69. Named Outstanding Educator Phi Theta Kappa, 1987. Mem. Math. Assn. Am., Am. Math. Soc., Am. Math. Assn. Two-Yr. Colls., Phi Kappa Phi. Home: RR 1 Box 499 Round Hill VA 22141-9304 Office: No Va Community Coll 1000 Byrd Hwy Sterling VA 22170

BERGEN, ROBERT LUDLUM, JR., materials scientist; b. Islip, N.Y., Oct. 29, 1929; s. Robert Ludlum and Alice (D'Oench) B.; m. Grace-Elizabeth Field, June 11, 1951; children: Beryl F., Alice D'Oench, Robert Ludlum III, Jennifer U. AB cum laude, Williams Coll., 1951; MS, Cornell U., 1953, PhD, 1955. Various tech. assignments Uniroyal Chem. div. Uniroyal, Inc., Naugatuck, Conn., 1955-68; mgr. plastics and fibers rsch. corp. R&D Uniroyal, Inc., Wayne, N.J., 1969-72; various mgmt. assignments Uniroyal Chem. div. Uniroyal, Inc., Naugatuck, 1972-75; mgr. elastomers R & D Uniroyal, Inc., Wayne, N.J.; group mgr. chems. and polymers R & D Uniroyal Chem. div. Uniroyal, Inc., Naugatuck, 1975-79; dir. corp. R & D Uniroyal, Inc., Middlebury, Conn., 1979-81, dir., rsch., devel. and engring., Engineered Products Group, 1981-84; dir. corp. engring. Uniroyal, Inc., Middlebury, 1985; adj. prof. math. U. New Haven, 1986—; cons. Bethany, Conn., 1986—; mem. adv. bd. Inst. Materials Sci., U. Conn., 1979—; adj. prof. chemistry U. New Haven, 1964-69; chmn. Soc. Plastic Engrs., Engring. Properties, 1970-71. Author: Testing of Polymers-Stress Relaxation Tests, 1966, various publs., 1954-68. Pres. Bethany Conservation Trust, 1979-82; moderator New Haven Assn. of United Ch. of Christ, 1991—. Fellow AAAS; mem. Am. Chem. Soc., Sigma Xi. Achievements include patents on improving stress cracking resistance of plastics; development of specialized impact test for plastics, of correlations between long term mech. properties of plastics and environ. stress cracking. Home and Office: 79 Lebanon Rd Bethany CT 06524-3033

BERGER, BRUCE WARREN, physician, urologist; b. Auburn, N.Y., Sept. 25, 1942; m. Toni M. LeRoy, Aug. 27, 1966; children: Jill, David. BA, Cornell U., 1964; MD, Upstate Med. Ctr., 1968. Diplomate Am. Bd. of Urology, Nat. Bd. of Med. Examiners. Surg. intern Hosp. U. of Pa., Phila., 1968-69; surgery resident NYU Hosp.-Bellevue (N.Y.) Med. Ctr., 1969-70; resident in urology Johns Hopkins Hosp., Balt., 1972-76; urologist Cohen, Berger, Epstein & Jaskulsky, P.A., Balt., 1976—; assoc. prof. clin. surgery urology U. Md., Baot., 1976—; attending, pres. med. staff Sinai Hosp., 1989-90, bd. dirs., 1989-93; assoc. attending Balt. Med. Ctr., Union Meml. Hosp.; bd. dirs. geriatric ctr. Maj. USAR, 1970-72, Vietnam. Mem. AMA, Balt. Med. Soc., Am. Assn. Clin. Urologists, Md. Urologist Assn., Am. Urologist Assn., The Associated Jewish Community Fedn. (chmn. physicians div. 1991—), Alpha Omega Alpha. Avocations: tennis, golf, skiing. Office: Cohen Berger Epstein et al 2411 W Belvedere Ave Ste 305 Baltimore MD 21215-5213

BERGER, FREDERICK JEROME, electrical engineer, educator; b. Szatmar, Hungary, Nov. 26, 1916; came to U.S. 1929; s. Joseph and Goldie (Weiss) B. BS, CCNY, 1959, BEE, 1961; MEE, NYU, 1964; LLD, Frank Ross Stuart U., 1981; DSc, Capitol Coll., Laurel, Md., 1986. Tool and die maker Brewster Aero. Co., 1935-39, chief tool, gauge and plant engr., 1939-45; process engr. Arma Co., 1946-51; entrepreneur Elec. Electronic Communication Systems and Machine Shop Equipments, 1952-61; prof., dep. chmn., chmn. and engring. sci. coord. CUNY, 1962-82; evaluator Accrediting Bd. Engring., 1962-81; cons. NSF, 1969-80. Editor Jour. of Tau Alpha Pi, 1975—. With U.S. Army, WWII. Fellow Am. Soc. Engring. Edn. (Frederick J. Berger ann. scholarship award 1990—, James H. McGraw award in Engring. Tech. Edn. 1992, Centennial cert. and medallion 1993); mem. IEEE (life, Engring. Svc. award 1964-81), Am. Nuclear Engring. Soc., Instrument Soc. Am. (life), Masons, Tau Alpha Pi (founding exec. dir. 1973—), Tau Beta Pi. Office: Tau Alpha Pi PO Box 266 Bronx NY 10471-0266

BERGER, HERBERT, retired internist, educator; b. Bklyn., Dec. 14, 1909; s. Louis and Augusta (Feldman) B.; m. Sylvia Berger, Oct. 1934; children: Leland S. (dec.), Shelby L. (Mrs. William Jakoby). BSc, NYU, 1929; MD, U. Md., 1932. Diplomate: Am. Bd. Internal Medicine. Intern Morrisania City Hosp., Bronx, N.Y., 1932-34; resident U.S. Naval Hosps., 1941-45; pvt. practice medicine, 1934-88; cons. cardiologist Seaview Hosp., S.I., 1934-88; attending physician Flower-Fifth Ave. Hosp., Met. Hosp.; cons. USPHS Hosp.; prof. medicine N.Y. Med. Coll., 1962-90, prof. medicine emeritus, 1990—; pres. med. staff, dir. medicine emeritus Richmond Meml. Hosp., 1975-88; bd. dir. emeritus Group Health Ins., Inc.; med. lectr. in over 96 countries. Author: Did This Really Happen?, 1991; cons. editor: Internat. Jour. Addictions; contbr. chpts. to med. textbooks and over 200 articles to med. jours. and local newspapers. Comdr. USNR, 1942-45. Recipient gold medal U. Md., 1978, established scholarship Tottenville High Sch.; Karlinski scholar, 1929; named Ednowed Lectr. various institutions. Fellow ACP, Am. Coll. Chest Physicians; mem. Internat. Coll. Angiology, N.Y. Acad. Medicine (v.p., mem. coun., chmn. sect. on medicine, vice chmn. med. edn. com.), Brit. Soc. Health Edn., Richmond County Med. Soc. (past pres.), N.Y. C. Med. Soc. (past pres.), Med. Soc. State N.Y. (past v.p.), Blood Banks Assn. (past pres.), Am. Soc. to Study Addictions (past pres.), N.Y. Cardiol. Soc., N.Y. State Soc. Internal Medicine (founder, past pres.), Am. Astronomers Assn., Amateur Astronomers Inc. Republican. Jewish. Clubs: Richmond County Country, Richmond County Yacht, Circumnavigators. Home: 25 Bloomingdale Rd Staten Island NY 10309-2813

BERGFIELD, GENE RAYMOND, engineering educator; b. Granite City, Ill., July 11, 1951; s. Walter Irvin Bergfield and Venie Edith (Sanders) Nolen; m. Juanita Pauline Kapp, Sept. 19, 1970; children: Gene Raymond Jr., Timothy Shawn. BA in Applied Behavioral Scis., Nat. Coll. Edn., Chgo., 1988. Field engr. Westinghouse PGSD, St. Louis, 1979-81; instr. Westinghouse PGSD, Phila., 1982-84; asst. resource mgr. Westinghouse PGSD, Chgo., 1984-89; power plant instr. Westinghouse PGPD, Orlando, Fla., 1989—. With USN, 1971-79.

BERGGREN, THAGE, automotive executive; b. 1932. Chalmers U. Tech., Gothenberg, Sweden, U. Wis., 1964. Mfg. engr. A B Volvo, Sweden, 1960, head truck ops., 1960, prodn. coord. truck divsn., 1966; chief engr. Volvo Trucks Corp., 1978; mng. dir. Volvo Europe, 1979; pres. Volvo White Trucks Corp., Greensboro, N.C., 1979-81; pres., CEO, dir. Volvo GM Heavy Trucks Corp., 1987—. Recipient Eli Whitney Productivity award Soc. Mfg. Engrs., 1990. Office: Volvo GM Heavy Truck Corp PO Box 26115 7825 National Service Rd Greensboro NC 27409-9667

BERGGREN, WILLIAM ALFRED, geologist, research micropaleontologist, educator; b. N.Y.C., Jan. 15, 1931; s. Wilhelm Fritjof and Lilly Maria (Skog) B.; m. Lois Albee, June 19, 1954 (div. July 1981); children—Erik, Anna Lisa, Anders, Sara Maria; m. Marie Pierre Aubry, June 19, 1982. B.S., Dickinson Coll., 1952; M.Sc., U. Houston, 1957; Ph.D., U. Stockholm, 1960, D.Sc., 1962. Research micropaleontologist Oasis Oil Co., Tripoli, Libya, 1962-65; asst. scientist Woods Hole Oceanographic Inst., Mass., 1965-68; assoc. scientist Woods Hole Oceanographic Inst., 1968-71, sr. scientist, 1971—; adj. prof. Brown U., Providence, 1968—. Editor: (with others) Catastrophes and Earth History, 1984; contbr. articles to sci. jours. Fellow Geol. Soc. Am., Geol. Soc. London (hon.); mem. NAS (Mary Clark Thompson medal 1982), Am. Assn. Petroleum Geologists, Soc. Econ. Paleontologists and Mineralogists, Paleontol. Soc. Am. (co-editor jour. 1980-84), Am. Geophys. Union, Geol. Soc. Switzerland. Avocation: skiing. Office: Woods Hole Oceanographic Inst Water St Woods Hole MA 02543

BERGLES, ARTHUR EDWARD, mechanical engineering educator; b. N.Y.C., Aug. 9, 1935; s. Edward H. and Victoria (Winkelmann) B.; m. Priscilla Lou Maule, June 19, 1960; children: Eric, Dwight. S.B., S.M., MIT, 1958, Ph.D., 1962. Registered profl. engr., Mass. Research staff Nat. Magnet Lab., Cambridge, Mass., 1962-69; asst. prof. to assoc. prof. mech. engring. MIT, Cambridge, 1963-69, assoc. dir. heat transfer lab., 1966-69; prof. mech. engring. Ga. Inst. Tech., Atlanta, 1970-72; prof., chmn. dept. mech. engring. Iowa State U., Ames, 1972-83, prof., dir. heat transfer lab., 1983-86; Clark and Crossan prof. engring., dir. heat transfer lab. Rensselaer Poly. Inst., Troy, N.Y., 1986—, dean of engring., 1989-92; U.S. rep. Internat. Heat Transfer Conf., 1978-82; chmn. U.S. group heat transfer U.S./USSR Agreement, Washington, 1979-82; cons. to industry, mem. numerous adv. groups. Co-author: Tho-Phase Flow and Heat Transfer in the Power and Process Industries, 1981; co-editor: Two-Phase Heat Exchangers, 1988, others; editor: Heat Transfer in Electronic and Microelectronic Equipment, 1990; mem. editorial adv. bd. 17 jours.; contbr. numerous articles to tech. jours. Scoutmaster Boy Scout Am., Ames, 1976-84; bd. dirs. Ames Soc. for Arts, 1975-79. Fulbright fellow Technische Hochschule, Munich, Fed. Republic Germany, 1958-59; recipient U.S. Sr. Scientist award Alexander von Humboldt Found., U. Hanover, Fed. Republic Germany, 1979-80, Faculty Achievement award in research Iowa State U., 1986; named Anson Marston Disting. prof. engring., Iowa State U., 1981. Fellow ASHRAE, AAAS, NAE, ASME (v.p. 1981-85, chmn. heat transfer div. 1982-83, bd. govs. 1985-89, pres. 1990-91, Heat Transfer Meml. award 1979, Dedicated Svc. award 1984), Internat. Ctr. Heat and Mass Transfer (exec. com. 1984—), Am. Soc. Engring. Edn. (Lamme award 1987, Centennial Certificate and Medallion 1993); mem. AIAA, Soc. Automotive Engrs. (Ralph R. Teetor award 1987), Am. Inst. Chem. Engrs. (Donald Q. Kern award 1990), Union of Mechanical and Electrical Engrs. and Technicians of Yugoslavia (hon.), Polish Soc. Theoretical and Applied Mech. (fgn.), Theta Chi. Republican. Lutheran. Lodge: Rotary (Troy). Office: Rensselaer Poly Inst Sch Engring Troy NY 12180-3590

BERGMAN, CARLA ELAINE, hydrologist, consultant; b. Ross, Calif., Dec. 24, 1962; d. Kenneth Leroy and Mary Alice (Biddle) B. BS in Civil Engring., Tex. A&M U., 1985, MS in Civil Engring., 1987. Registered profl. engr., Mo. Rsch. asst. Tex. Water Resources Inst., College Station, 1985-87; water resources engr. Burns & McDonnell Engring., Kansas City, Mo., 1988-90; groundwater hydrologist Burns & McDonnell Waste Cons., Overland Park, Kans., 1990-92; hydrologist, project engr. HNTB Corp., Kansas City, 1992—. Co-author several tech. publs. Recipient W.G. Mills fellowship Tex. Water Resources Inst., 1986, 87. Home: 601 E 63rd Terr Kansas City MO 64110 Office: HNTB Corp 1201 Walnut Ste 700 Kansas City MO 64106

BERGMAN, LAWRENCE ALAN, engineering educator; b. N.Y.C., Oct. 14, 1944; s. Henry and Sylvia (Kirsch) B.; m. Jane Howell, May 24, 1969; 1 child, Sarah Michelle. BSME, Stevens Inst. Tech., Hokoben, N.J., 1966; MS in Civil Engring., Case Western Res. U., 1978, PhD, 1980. Mem. tech. staff TRW Inc., Cleve., 1966-68, Lord Corp., Erie, Pa., 1968-75; cons. engr. J.D. Stevenson Cons., Cleve., 1975-77, Woodward-Clyde Cons., Cleve., 1977-79; asst. prof. Engring. U. Ill., Urbana, 1979-85, assoc. prof. Engring., 1985—; cons. with various orgns., 1980—. Assoc. editor (jour.) Transactions of ASME, 1990—; contbr. articles to profl. jours.; patentee in field. Assoc. fellow AIAA; mem. ASCE (State of the Art in Civil Engring. award 1983), ASME, Am. Acad. of Mechanics. Office: U Ill AAE dept 104 S Wright St Urbana IL 61801

BERGMAN, ROBERT GEORGE, chemist, educator; b. Chgo., May 23, 1942; s. Joseph J. and Stella (Horowitz) B.; m. Wendy L. Street, June 17, 1965; children: David R., Michael S. BA cum laude in chemistry, Carleton Coll., 1963; PhD (NIH fellow), U. Wis., 1966. NATO fellow in chemistry Columbia U., N.Y.C., 1966-67; Arthur Amos Noyes instr. chemistry Calif. Inst. Tech., Pasadena, 1967-69; asst. prof. chemistry Calif. Inst. Tech., 1969-71, assoc. prof. chemistry, 1971-73, prof., 1973-77; prof. chemistry U. Calif. at Berkeley, 1977—, asst. dean Coll. Chemistry, 1987-91; Miller Rsch. prof. U. Calif., Berkeley, 1982-83, 93—; Sherman Fairchild Disting. scholar Calif. Inst. Tech., 1984; mem. panel NIH bioinorganic and metallobiochemistry study sect. NIH, 1977-80; chmn. Gordon conf. in organometallic Chemistry, 1991; cons. EI Du Pont de Nemours, 1982-85, Chevron Rsch. Co., 1983-89, Union Carbide Corp., 1977-81, 90—. Mem. editorial bd. Chem. Revs., Jour. Am. Chem. Soc., Organometallics; contbr. articles to profl. jours. Alfred P. Sloan Found. fellow, 1970-72; recipient Camille and Henry Dreyfus Found. Tchr. Scholar award, 1970-75; Excellence in Teaching award Calif. Inst. Tech., 1978, award in organometallic chemistry Am. Chem. Soc., 1986, Edgar Fahs Smith award ACS Pa. sect., 1990, Ira Remsen award ACS Balt. sect., 1990, Merit award NIH, 1991; Arthur C. Cope scholar Am. Chem. Soc., 1987. Mem. Nat. Acad. Scis., AAAS, Phi Beta Kappa, Sigma Xi, Phi Lambda Upsilon. Home: 501 Coventry Rd Kensington CA 94707-1316 Office: U Calif Dept Chemistry Berkeley CA 94720

BERGO, CONRAD HUNTER, chemistry educator; b. Evanston, Ill., Jan. 5, 1943; s. Arthur Conrad and Mary Margret (Hunter) B.; m. Nancy Wallace, Mar. 12, 1977; children: Stacey Lynn, Fred Monteabaro. BA, St. Olaf Coll., 1965; PhD, U. Minn., 1972. Asst. prof. Chieng Mai (Thailand) U., 1972-75; rsch. assoc. dept. pharmacology U. Ky., Lexington, 1975-77; asst. prof. Alliance Coll., Cambridge Springs, Pa., 1977-80; assoc. prof. East Stroudsburg (Pa.) U., 1980—; exec. dir. Pa. State Coll. Chemistry Consortium, 1991—; book reviewer McGraw-Hill, Freeman, John Wiley and West Pub. Pres. bd. dirs. Burnley Workshop, Stroudsburg, 1993—. Recipient Cert. of Citizen Svc., Commonwealth of Pa., 1989, award Beyond War, 1990. Mem. Am. Chem. Soc., Pa. Acad. Sci., Sigma Xi. Office: East Stroudsburg Univ Chemistry Dept East Stroudsburg PA 18301

BERGOLD, ORM, medical educator; b. Nuremberg, Germany, Apr. 30, 1925; s. Friedrich and Wilhelmine (Schering) B.; M.D., Chgo. Med. Coll., 1974; D.Chemistry, Benjamin Franklin Inst., N.Y.C., 1976; M.Acupuncture, Old Chinese Acupuncture Acad., Hong Kong, 1978; D.Sc. (hon.), St. Andrew's Coll., London, 1965; m. Sylvia Patricia Sanchez, 1983; children: Heike, Timm. Pres., Orm Bergold Chemie, Langlau and Bochum, Fed. Republic Germany, 1953-63; pres. Inst. Med. Biophysics and Biochemistry, Campione, Switzerland, 1963-82; pres. Inst. Med. Biophysics and Biochemistry, San Jose, Costa Rica, 1982—, Inst. Biocybernetic and Natural Therapy, San Jose, 1985—; pres. Stress and Aging Control Inc., Panama, Costa Rica, 1986—; pres. AIDS Control, Inc., Panama, 1989—; prof. cybernetic medicine Academia Gentium Pro Pace, Rome, 1977—, senator, 1979—; prof. extraordinary U. Francisco Marroquin, Guatemala City, 1979—, senator, 1980—. Named Hon. Pres. Acad. for Biocybernetic Holistic Medicine, Rheda, Germany, 1990—; decorated grand cross Ordre Equestre de la San Croix de Jérusalem; chevalier du Tastevin. Author: Kybernetische Medizin, 1977, Cancer prophylaxis: a problem of early recognition and treatment, 1980, Cancer Treatment with Human Fibroblast Interferon, 1982, Cancer Treatment by Natural Remedies, 1983, Stress, Cortisol and Stress Diseases, 1989, AIDS Treatment, 1989; also articles. Home: PO Box 359-1250, Escazu Costa Rica Office: PO Box 257, 1005 Barrio Mexico, San Jose Costa Rica

BERGRUN, NORMAN RILEY, aerospace executive; b. Green Camp, Ohio, Aug. 4, 1921; s. Theodore and Naomi Ruth (Stemm) B.; m. Claire Michaelson, May 23, 1943; children: Clark, Jay, Joan. BSME, Cornell U., 1943; LLB, LaSalle U. Ext., 1955; DSc, World U., 1983. Registered profl. mech. engr. Thermodynamicist Douglas Aircraft Co., El Segundo, Calif., 1943-44; rsch. scientist NACA Ames Rsch. Lab., Mt. View, Calif., 1944-56; mgr. analysis Lockheed Missile & Space Co., Sunnyvale, Calif., 1956-67, staff scientist, 1967-69; dir. mgmt. systems Nielsen Engring. and Rsch., Mt. View, 1969-71; CEO, scientist Bergrun Rsch. and Engring., Los Altos, Calif., 1971—; advisor to bd. CSPE Edn. Found., Sacramento, Calif., 1985-92. Author: Ringmakers of Saturn, 1986, Tomorrow's Technology Today, 1972; photographer including The Sir Francis Drake Collection, 1990; contbr. more than 80 articles and reports to profl. jours. Incorporator Aurora Sigers Found., Palo Alto, Calif., 1989. Chief USN, 1944-46. Named Man of Yr., Am. Biog. Assn.; recipient Archimedes award, 1988, Cert. of Appreciation, Eglin AFB, 1961. Assoc. fellow AIAA (sr. judge 7th and 8th Grade Essay Contest, Bay Area, Calif., 1992, 93); mem. CSPE Edn. Found. (Engr. of Yr., CEO 1985-86, Appreciation award 1986), Profl. Engrs. Soc. (pres. 1988-89, Integrity award 1989), L'Academie Europeene (speaker 1987). Achievements include discoveries of existence of large, mobile cylindrical objects, identified at Saturn, Miranda, Iapetus, Mars, Neptune and the Sun; patents for Cyclic Electric Thermal Ice-Prevention System for Airplanes. Office: Bergrun Rsch and Engring 26865 Saint Francis Rd Los Altos CA 94022

BERGSTEINER, HARALD, architect; b. Munich, Germany, May 30, 1944; m. Gayle Christine Avery, Aug. 12, 1947. B.Arch., U. Sydney, Australia, 1967; Dipl.Town & Country Plg., U. Sydney, 1970. Chartered architect. Architect Commonwealth Dept. Wks., Syndey, 1968-71; head spl. projects City of Bankstown, Sydney, 1971-73; dir. Bergsteiner, McInnes & Rigby P/L Sydney, 1974-79; mng. dir. H. Bergsteiner & Assocs., Sydney, Munich, N.Y.C., 1980—. Decorated Knight Gov. of Order of St. Joseph, 1984. Mem. Royal Australian Inst. Architects, Royal Australian Planning Inst., German Castle Soc. Home: Hollandstrasse 2, 80805 Munich Germany

BERGSTRÖM, K. SUNE D., biochemist; b. Stockholm, Jan. 10, 1916; s. Sverker B. and Wera (Wistrand) B.; m. Maj Gernandt, July 30, 1943. Docent physiol. chemistry, MD, Karolinska Inst., Stockholm, 1944, D. Med. Sci. in Biochemistry, 1944; D h.c., U. Basel, Switzerland, 1960, U. Chgo., 1960, Harvard U., 1976, Mt. Sinai Med. Sch., 1976, Med. Acad. Wroclaw, Poland, 1976, McMaster U., Hamilton, Can., 1988. Rsch. fellow U. London, 1938, Columbia U., N.Y.C., 1940-41, Squibb Inst. Med. Rsch., New Brunswick, N.J., 1941-42; asst. biochem. dept. Med. Nobel Inst., Karolinska Inst., Stockholm, 1944-47; rsch. fellow U. Basel, 1946-47; prof. physiol. chemistry U. Lund, Sweden, 1947-58; prof. chemistry Karolinska Inst., 1958-80, dean med. faculty, 1963-66, rector, 1969-77; chmn. bd. dirs. Nobel Found., Stockholm, 1975-87; pres. Royal Swedish Acad. Scis., 1983-85; chmn. WHO Adv. Com. Med. Research, Geneva, 1977-82; La Madonnina lectr., Milan, Italy, 1972; Dunham Lectr. Harvard U., 1972; Dohme lectr. Johns Hopkins U., 1972-73; Merrimon lectr. U. N.C., Chapel Hill, 1973; V.D. Mattia lectr. Roche Inst., 1974; Harvey lectr. Harvey Soc., N.Y.C., 1974; Gen. Amir Chand orator All India Inst., New Delhi, 1978; Cairlton lectr. U. Tex. Health Sci. Ctr., Dallas, 1979; mem. Swedish Med. Rsch. Coun., 1952-58, 64-70, Swedish Natural Sci. Rsch. Coun., 1955-62. Contbr. articles to sci. jours. Decorated Grand Officier de l'Ordre du Mérite, Paris, 1984; recipient Anders Jahre Med. prize, Oslo, 1972, Gairdner award U. Toronto, 1972, Louisa Gross Horwitz prize Columbia U., 1975, Francis Amory prize Am. Acad. Arts and Scis., 1975, Albert Lasker Basic Med. Rsch. award, N.Y.C., 1977, Robert A. Welch award, Houston, 1980, Nobel prize, 1982. Mem. Royal Swedish Acad. Scis., Swedish Acad. Engring. Scis., Am. Acad. Arts and Scis., Am. Philos Soc. (Benjamin Franklin medal 1988), Am. Soc. Biol. Chemists, Acad. Scis. USSR, Academia Leopoldina (German Democratic Republic), Royal Soc. Edinburgh, Med. Acad. USSR, Finska Vetenskaps-Societeten, Swedish Soc. Med. Scis., Inst. of Medicine (sr.), NAS, fgn. assoc. NAS, Pontifical Acad. Scis., Città del Vàticano. Office: Karolinska Institutet, Nobelkansliet Box 270, 171 77 Stockholm Sweden

BERGSTRÖM, STEN RUDOLF, psychology educator; b. Uppsala, Sweden, Dec. 10, 1934; s. Rudolf Vilhelm and Maud Margit (Kihlberg) B.; m. Karin Birgitta Ahlstrand, Dec. 29, 1962 (div. 1966); children: Erik, Eleonora; m. Inga Lena Ahlstrand, June 13, 1970. BA, U. Uppsala, 1961, MA, 1965, PhD, 1969. Amanuensis Inst. Psychology U. Uppsala, 1961-65, asst. prof. Inst. Psychology, 1965-90, dir. studies Inst. Psychology, 1965-69, head of dept. Inst. of Psychology, 1969-86, head of dept Inst. of Applied Psychology, 1972-85, docent Faculty of Social Scis., 1969—; advisor Swedish Nat. Bd. Health and Welfare, Stockholm, 1970-75. Contbr. articles to sci. publs.; patentee in med. equipment field. Churchwarden Ch. of Sweden, Häggeby, 1977—; mem. Coun. of Archbishops Diocese, Uppsala, 1989—, Bd. of Archbishops Diocese, 1989—. Mem. Swedish Soc. Medicine, Swedish Soc. for Traffic Medicine, Rotary. Lutheran. Avocations: lit., history, gardening, pistol and MG shooting. Home: Häggeby prästgård, S-74694 Bålsta Sweden

BERKELMAN, KARL, physics educator; b. Lewiston, Maine, June 7, 1933; s. Robert George and Yvonne (Langlois) B.; m. Mary Bowen Hobbie, Oct. 10, 1959; children: Thomas, James, Peter. BS, U. Rochester, N.Y., 1955; PhD, Cornell U., 1959. From asst. prof. to prof. physics Cornell U., Ithaca, N.Y., 1961—, dir. lab. nuclear studies; sci. assoc. DESY, Hamburg, Fed. Republic of Germany, 1974-75, CERN, Geneva, 1967-68, 81-82, 91-92. Office: Cornell U Newman Lab Nuclear Studies Ithaca NY 14853

BERLE, PETER ADOLF AUGUSTUS, lawyer, association executive; b. N.Y.C., Dec. 8, 1937; s. Adolf Augustus and Beatrice (Bishop) B.; m. Lila Sloane Wilde, May 30, 1960; children: Adolf Agustus, Mary Alice, Beatrice Lila, Robert Thomas. B.A. (Knox fellow), Harvard U., 1958, LL.B., 1964; LL.D., Hobart Smith Coll., 1977; LLB, North Adams Tchrs. Coll., 1988. Bar: N.Y. Assoc. Paul, Weiss, Rifkind, Wharton & Garrison, N.Y.C., 1964-71; ptnr. Berle, Butzel & Kass, N.Y.C., 1971-76; N.Y. state commr. environ. conservation, 1976-79; ptnr. Berle, Butzel, Kass & Case, 1979-85; pres., CEO (pub. Audubon mag.) Nat. Audubon Soc., 1985—; trustee Twentieth Century Fund, Inc., 1971—, chmn., 1982-87; teaching fellow econs. Harvard Coll., Cambridge, Mass., 1963-64; assoc. adj. prof. Sch. Urban Affairs Hunter Coll., 1974, 84; vis. prof. environ. sci. and forrestry SUNY, 1986. Author: Does the Citizen Stand a Chance, 1974. Mem. N.Y. State Assembly, 1968-74; chmn. N.Y. Gov.'s Transition Task Force on Environment, 1974-75; commr. N.Y. State Moreland Act Commn. on Nursing Homes, 1975-77; bd. dirs. 20th Century Fund, 1971—, Clean Sites Inc., 1986—; chmn. Commn. on the Adirondacks in the 21st Century, 1989-90; mem. EPA adv. group on biotech., 1989-92, EPA adv. group air quality; mem. nat. com. environ., 1991-92, nat. commn. superfund, 1992—. Served to 1st lt. USAF, 1959-61. Decorated Commendation medal; named Outstanding Legislator Eagleton Inst. Politics, 1971. Mem. Assn. Bar City N.Y. (mem. environ. law com., profl. responsibility com., energy policy com., internat. human rights com.), Adirondack Mountain Club (bd. govs. 1972-76). Episcopalian. Home: 530 E 86th St New York NY 10028-7535 Office: Nat Audubon Soc 950 Broadway New York NY 10022

BERLIN, CHARLES I., otolaryngologist, educator; b. N.Y.C., Dec. 26, 1933; married, 1958; 4 children. BS, U. Wis., 1953, MA, 1954; PhD in Hearing and Speech, U. Pitts., 1958. Audiologist, speech pathologist U.S. VA Hosp., Calif., 1959-61; asst. prof. otolaryngology Johns Hopkins Hosp., 1963-67, assoc. prof., 1967-70; dir. Kresge Hearing Rsch. Lab. of South, 1968—; prof. otolaryngology La. State U. Med. Ctr., New Orleans, 1970—. Recipient Rsch. Career Devel. award Nat. Inst. Neurol. Disease and Blindness, 1963-67; fellow Johns Hopkins Hosp., 1962-63. Fellow Acoustical Soc. Am., Am. Speech and Hearing Assn.; mem. AAAS, Am. Acad. Audiol., Sigma Xi. Achievements include research in hearing and speech sciences, evoked potentials, speech perception, communication in pathological states in humans and animals, otoacoustic emissions, hearing disorders and hearing aids. Office: La State U Kresge Hearing Rsch Lab 2020 Gravier St Ste A New Orleans LA 70112-2272*

BERLINER, HANS JACK, computer scientist; b. Berlin, Germany, Jan. 27, 1929; came to U.S., 1937, naturalized, 1943; s. Paul and Theodora (Lehfeld) B.; m. Araxie Yacoubian, Aug. 15, 1969. B.A., George Washington U., 1954; Ph.D., Carnegie Mellon U., 1975. Systems analyst U.S. Naval Rsch. Lab., 1954-58; group head systems analysis Martin Co., Denver, 1959-60; adv. systems analyst IBM, Gaithersburg, Md., 1960-69; sr. rsch. scientist Carnegie-Mellon U., Pitts., 1974—. Mem. editorial bd. Artificial Intelligence, 1976—, Pitman: Research Notes in Artificial Intelligence, 1984—. Served with AUS, 1951-53. Awarded title Internat. Grandmaster Corr. Chess, 1968; inducted into U.S. Chess Hall of Fame, 1990. Fellow Am. Assn. for Artificial Intelligence; mem. Internat. Joint Conf. Artificial Intelligence, U.S. Chess Fedn., Internat. Computer Chess Assn. Among leading chess players U.S., 1950—, N.Y. State champion, 1953, Southwest Open champion, 1960, So. Open champion, 1949, U.S. Open Corr. Chess champion, 1955, 56, 59, World Corr. Chess champion, 1968-72. Developed 1st computer program to defeat a world champion at his own game (backgammon), 1979; co-developer Hitech, first chess computer to become a U.S. Chess Fedn. sr. master; among .5% of all registered tournment chess players; discovered B* tree search algorithm, 1975, SNAC method of constructing polynomial evaluation functions, 1979. Home: 657 Ridgefield Ave Pittsburgh PA 15216-1141

BERMAN, MARCELO SAMUEL, mathematics and physics educator, cosmology researcher; b. Buenos Aires, Apr. 10, 1945; s. Bernardo and Rosa (Soifer) B.; m. Geni Lima, June 21, 1986; children: Albert, Paula. EE, ITA, San José Campos, Brazil, 1967, MSc in Physics, 1981; DSc in Physics, Univ. Fed. Rio De Janeiro, 1988. Ptnr. Plasti-Tact Ltd., Curitiba, Brazil, 1974-79; ptnr., fin. dir. Constr. Gustavo Berman, Curitiba, 1979-88; prof. math. dept. exact scis. Fundaçao Educacional da Regiao de Joinville, Brazil, 1986-90; postdoctoral assoc. dept. astronomy U. Fla., Gainesville, 1989-90; asst adj. prof. dept. Physics and Astronomy U. Ala., Tuscaloosa, 1991; mng. ptnr. Editora Albert Einstein Ltda, Curitiba, Brazil. Co-author: Tensor Calculus and General Relativity, 1987, Relativistic Cosmology, 1988; contbr. over 40 articles to profl. jours. and 40 articles in newspapers. Ministry Aeros. scholar, 1963-67, Pro-Nuclear Brazil scholar, 1980-81, CNPQ scholar, 1989-91. Mem. AAAS, IEEE (sr.), Internat. Astronomical Union, Am. Assn. Physics Tchrs., Am. Phys. Soc., Am. Astron. Soc., N.Y. Acad. Scis., Gen. Relativity and Gravitation Soc. Jewish. Achievements include presentation of models with constant deceleration parameter in cosmology, which cover and accept inflationary scenario. Office: Champagnat Exec Ctr, Rua Candido Hartman 528 Y25, 80730-220 Curitiba Brazil

BERMAN, MARLENE OSCAR, neuropsychologist, educator; b. Phila., Nov. 21, 1939; d. Paul Oscar and Evelyn (Hess) Oscar; m. Michael Brack Berman, June 23, 1963 (div. Feb. 1980); 1 son, Jesse Michael. BA, U.Pa., 1961; MA, Bryn Mawr Coll., 1964; PhD, U. Conn., 1968; postgrad., Harvard U., 1968-70. Research assoc. Boston VA Med. Ctr., 1970-72, clin. investigator, 1973-76, research psychologist, 1976—; assoc. prof. neurology Boston U. Sch. Medicine, 1975-82, prof. neurology and psychiatry, 1982—, dir. Neuropsychology Lab., dept. psychiatry, 1981—; mem. Com. for Protection Human Participants in Rsch., 1979-82, chmn., 1983-85; affiliate prof. psychology Clark U., Worcester, Mass., 1975—; mem. biomed. rsch. initial rev. group Nat. Inst. Alcohol Abuse and Alcoholism, 1987-91, chmn. 1990-91. Contbr. articles to profl. jours. Coordinator Newton Community Schs. (Mass.), 1978-80. Recipient rsch. scientist devel. award Nat. Inst. Neurol. and Communicative Disorders and Stroke, 1976-81, Nat. Inst. Alcohol Abuse and Alcoholism, 1981-86, clin. investigator award VA, 1973-76; grantee USPHS and HHS, 1964—; Fulbright sr. scholar, 1991. Fellow Mass. Psychol. Assn., Am. Psychol. Assn. (sec.-treas. 1981-83); mem. Acad. Aphasia, Soc. Neurosci., Internat. Neuropsychol. Soc., Psychonomic Soc., Huntington's Disease Soc. Am., New Eng. Psychol. Assn., Mass. Neuropsychol. Soc., N.Y. Acad. Scis., Eastern Psychol. Assn. Democrat. Jewish. Office: Boston U Lab Neuropsych Dept Psychiatry M-9 80 E Concord St Roxbury MA 02118-2394

BERMAN, MERVYN CLIVE, biochemist; b. Cape Town, South Africa, Jan. 21, 1934; s. Joseph and Hena Nancy (Stolpinsky) B.; m. Jacqueline Wendy Gersh, Dec. 9, 1956; children: Michael, Neil, Lara. BSc, U. Cape Town, 1953, MBChB, 1957, MMed in Pathology, 1961, PhD, 1964. Lic. med. practitioner. Sr. lectr. U. Cape Town, 1967-69, prin. specialist, 1970-73, assoc. prof., 1974-77, prof., head dep. chem. pathology, 1977—, head div. pathology, 1978-80, 83; dir. med. rsch. coun. Biomembran Rshc. Unit, Cape Town, 1979—. Contbr. articles to Jour. Biol. Chemistry, Biophys. Biochim Acta. Mem. panel Wellcome/MRC Award for Med. Rsch., South Africa. Recipient Millipore Gold medal, 1987, John F. W. Herschel medal Royal Soc. South Africa, 1992; elected fellow U. Cape Town, 1976. Mem. N.Y. Acad. Scis., Biophys. Soc. U.S. Jewish. Office: U Cape Town, Dept Chem Pathology, Anzio Rd Obs, Cape Town 7925, South Africa

BERMAN, MICHAEL ALLAN, biomedical clinical engineer; b. Detroit, July 25, 1957; s. George Arthur and Pearl Sarah (Zion) B.; m. Elizabeth Ann Mattingly; children: Jennifer, Jessica, Rachel, Mo. BS in Elec. Tech., Purdue U., 1985; MBA, Ind. Wesleyan U., 1989. Cert. Biomed. Equipment technician. Electroncis technician Purdue U., West Lafayette, Ind., 1977; lead technician CTS Microelectronics, West Lafayette, Ind., 1978; chief biomed. equipment technician Home Hosp., Lafayette, Ind., 1978-81; supr. clin. engring. St. Elizabeth Hosp., Lafayette, Ind., 1981—; assov. faculty Ind. Vocat. Tech. Coll., Lafayette, 1982—; educator radiology Project Hope, Millwood, Va., 1991—; mem. laser safety com. Purdue U. Vet Sch., West Lafayette, 1993. Author: Diagnostic Radiology Module, 1992. Den leader Boy Scouts Am., West Lafayette, 1992-93. Mem. Am. Soc. Hosp. Engring., Assn. Advancement Med. Instrumentation. Office: St Elizabeth Hosp Med Ctr PO Box 7501 1501 Hartford St Lafayette IN 47904

BERMAN, NEIL SHELDON, chemical engineering educator; b. Milw., Sept. 21, 1933; s. Henry and Ella B.; m. Sarah Ayres, June 3, 1962; children—Jenny, Daniel. B.S., U. Wis., 1955; M.S., M.A., U. Tex., Austin, 1961, Ph.D., 1962. Engr. Standard Oil Co. Calif., Los Angeles, 1955-62; research engr. E.I. DuPont Co., Wilmington, Del., 1962-64; from asst. prof. to prof. chem. engring. Ariz. State U., 1964—, Grad. Coll. Disting. Rsch. prof., 1984-85; cons. air pollution, fluid dynamics; mem. Phoenix Air Quality Maintenance Area Task Force, 1976-77. Contbr. articles on fluid dynamics of polymer solutions, air pollution, thermodynamics and chem. engring. edn. to profl. jours. Served to capt. M.S.C. USAR, 1956-58. Recipient numerous grants for research in fluid dynamics and air pollution. Fellow Am. Inst. Chem. Engrs. (chmn. Ariz. sect. 1978-79), AAAS, Ariz.-Nev. Acad. Sci. (corr. sec. 1981-88, pres.-elect 1988-89, pres. 1989-90); mem. ASME, Am. Chem. Soc., Am. Phys. Soc., Ariz. Council Engring. and Sci. Assns. (chmn. 1980-81), Soc. Rheology, Am. Soc. Engring. Edn., Am. Acad. Mechanics, Sigma Xi, Tau Beta Pi, Phi Kappa Phi. Home: 418 E Geneva Dr Tempe AZ 85282-3731 Office: Ariz State U Dept Chem Engring Tempe AZ 85287-6006

BERN, MURRAY MORRIS, hematologist, oncologist; b. Montgomery, Ala., Feb. 26, 1944; s. Hymie and Ruth Edith (Schaeffer) B.; m. Nancy Frazee, Nov. 23, 1967; 1 child, Alan. BA, Vanderbilt U., 1966; MD, Tulane U., 1970. Diplomate Am. Bd. Internal Medicine, Am. Bd. Hematology, Am. Bd. Oncology. Intern, then resident New Eng. Deaconess Hosp., Boston, 1970-71; resident in medicine Boston City Hosp., 1971-73; Am. Cancer Soc. fellow Ctr. for Blood Rsch., Boston, 1973-75; sr. staff, sect. chief hematology New Eng. Deaconess Hosp., Boston, 1975-86; dir. hematology, lab. dir. Cancer Ctr. Boston, Boston, South Dartmouth, Plymouth, Mass., 1986—; prin., founder and dir. stem cell support care Cancer Ctr. of Boston, 1986—; asst. prof. medicine Harvard U., 1980-87, asst. clin. prof. medicine, 1987—. Author, editor: Urinary Track Bleeding, 1985, Hematologic Disorders in Maternal and Fetal Medicine, 1990. Bd. med. advisors Am. Cancer Soc. Mass., 1976-80, fellow, 1973-75. Recipient Tullis award for rsch. Fellow ACP; mem. Am. Soc. Hematology, Am. Soc. Clin. Oncology. Avocations: camping, fishing. Office: Cancer Ctr Boston 125 Parker Hill Rd Boston MA 02120

BERNARD, EDDIE NOLAN, oceanographer; b. Houston, Nov. 23, 1946; s. Edward Nolan and Geraldine Marie (Dempsey) B.; m. Shirley Ann Fielder, May 30, 1970; 1 child, Elizabeth Ann. B.S., Lamar U., 1969; M.S., Tex. A&M U., 1970, Ph.D., 1976. Geophysicist Pan Am. Petroleum Co., 1969; research asst. oceanographic research Tex. A&M U., College Station, Tex., 1969-70; researcher Nat. Oceanic and Atmospheric Adminstrn. (NOAA), 1970-73; dep. dir. NOAA Pacific Marine Environ. Lab., Seattle, 1980-82; researcher Joint Tsunami Research Effort, 1973-77; dir. Nat. Tsunami Warning Ctr., 1977-80; dir., Pacific Marine Environ. Lab., Seattle, 1982—; dir. NOAA hydrothermal vents program, fisheries oceanography program; mem. adminstrv. bd. Joint Inst. for Study of Atmosphere and Ocean, U. Wash., Seattle; mem. exec. com. Coop. Inst. for Marine Resource Studies Oreg. State U.; mem. adminstrv. bd. Joint Inst. Marine and Atmospheric Research, U. Hawaii; chmn. Internat. Union of Geodesy and Geophysics Tsunami Commn., 1987—; mem. Panel on Wind and Seismic Effects U.S.-Japan Coop. Program in Nat. Resources, 1981—; mem. Internat. Recruitment Investigations in the Subarctic Council, 1982—; mem. Washington Sea Grant Steering Com., 1987—; mem. sci. coun. Joint Inst. for Marine Observations, Scripps Instn. of Oceanography, 1992—, bd. dirs. Pacific Northwest Reg. Marine Rsch. Program, 1992—. Editor: Tsunami Hazard: A Practical Guide for Tsunami Hazard Reduction, 1991; mem. editorial adv. bd. Natural Hazards Jour.; contbr. articles to profl. jours. Recipient Best of New Generation 1984 Register award Esquire Mag., 1984. Mem. Internat. Union of Geodesy and Geophysics (chmn. Tsunami commn. 1987—), Am. Geophys. Union, Am. Meteorol. Soc. Office: Pacific Marine Environ Lab 7600 Sand Point Way NE Seattle WA 98115-6349

BERNARD, HERBERT FRITZ, aerospace engineer; b. Bremen, Germany, Aug. 8, 1957; s. Herbert Albert Friedrich and Margarete Henriette (Obhues) B.; m. Annette Margarethe Ursel Mund, July 8, 1983. MS, Tech. U. Berlin, 1983, PhD, 1990. Devel. engr. Kurt Eichweber Co., Hamburg, Germany, 1983-85; lectr. aeronautics Tech. U. Berlin, 1985-90; rsh. scientist NASA Am. Rsch. Ctr., Moffett Field, Calif., 1990-92; tech. mgr. City Line Simulator & Tng., Berlin, 1992—; cons. Munich Airport Co., 1989-90, Vienna (Austria) Airport Co., 1989-90. Contbr. articles to profl. jours. Rsch. Assoc. award Nat. Rsch. Coun. Wash., 1990-92. Mem. AIAA. Achievements include research in human factors in man-machine-interfaces. Home: Dieffenbachstr 77, 10967 Berlin Germany Office: City Line Simulator & Tng, Schuetzenstr 10, 12526 Berlin Germany

BERNARD, JOHN P., chemical engineer; b. Buffalo, Apr. 8, 1958; s. John F. and Loretta L. (Schall) B.; m. Barbara S. Vail, Oct. 17, 1981. BS in Chem. Engring., SUNY, Buffalo, 1980. Engr. plant systems design Linde div. Union Carbide, Tonawanda, N.Y., 1980-81, asst. staff engr. systems design, 1981-85, staff engr. mech. equipment, 1985-88, sr. engr. mech. equipment, 1988, project engr. process engring., 1988-93; project engr. equipment engring. Praxair, Tonawanda, N.Y., 1993—. Mem. Town of Orchard Park (N.Y.) Planning Bd., 1989-92, chair, 1992—; ad hoc chmn. Citizen's Com. for Town of Orchard Park, 1988—. Mem. Compressed Gas Assn. (subcom. for liquid oxygen pumps 1992—, subcom. for storage of large vols. of cryogenic liquids 1993—), Am. Inst. Chem. Engrs. (bd. dirs. 1991-92, sec. 1992-93, vice chair 1993—), Lions Club of Orchard Park (sec. 1992—). Home: 6760 Powers Rd Orchard Park NY 14127-3223 Office: Praxair 175 E Park Dr Tonawanda NY 14151-0044

BERNARD, RONALD ALLAN, computer performance analyst; b. Dover, N.H., Sept. 28, 1953; s. Robert Ronald and Joyce (Bodwell) B.; children: Laura Jean, Jessica Diane. BS, U. Vt., 1975. Characterization engr. IBM, Essex Junction, Vt., 1979-83; diffusion engring. group leader IBM, 1983-85, evaporation engring. group leader, 1985-87, VM performance analyst, 1987-89, performance group leader, 1989-91; AiX support IBM, Essex Junction, Vt., 1991—; task force mem. IBM N.E. Region Info. and Telecommunications Support Svcs. Consolidation, Endicott, N.Y., 1988-89; speaker VM Internal Tech. Exch., 1988, SHARE, 1988-89, Latin Am. Guide Group, 1989; task force mem. World Wide LAN Security Task Force, White Plains, N.Y., and Toronto, Ont., Can., 1992-93. Bd. dirs. Royal Parke Assn., 1986—, v.p. 1987-91. Mem. Racquet Edge Club (Essex, Vt.), Amnesty Internat. (writer 1988—). Roman Catholic. Avocations: racquetball, nautilus, swimming. Home: 1F 86 Pinecrest Dr Essex Junction VT 05452

BERNARDONE, JEFFREY JOHN, podiatrist; b. Southbridge, Mass., Feb. 10, 1958; s. John Paul Jr. and Lina (Bonadies) B.; m. Janet Rae Bolea, June 3, 1990; 1 child, Jeffrey Michael. BA, Assumption Coll., Worcester, Mass., 1980; D Podiatric Medicine, N.Y. Coll. Podiatric Medicine, 1984. Preceptorship N.Y. Coll. Podiatric Medicine, 1986; pvt. practice Quincy, Mass., 1986—; cons. to various hosps.; podiatrist Boston Marathon. Mem. Am. Podiatric Med. Assn., Mass. Podiatric Med. Assn., Am. Diabetes Assn. Office: 1157 Hancock St Quincy MA 02169-4329

BERNASCONI, CHRISTIAN, chemical engineer; b. Paris, Oct. 29, 1946; s. Maurice and Germaine (Leboeuf) B.; m. Ginette Besse, Sept. 13, 1973; children: Christelle, Marilyne. Degree in chem. engring., Ecole Superieure Chimie, Lyon, France, 1971; PhD, U. Lyon, 1978. Assoc. prof. chemistry U. Lyon, 1971-82; chem. engr. Elf Antar France, Solaize, France, 1982—; speaker at internat. confs. and symposiums. Contbr. articles to Chemistry, Photochemistry, Photobiology, Petroleum Products Chemistry, Physico-Chemistry. Active Croix Rouge Francaise, Lyon, 1965; camp trainer, France, 1965-71. With French Army, 1973-74. Mem. Assn. Francaise des Techniciens du Pétrole, Soc. Automotive Engrs., Internat. Assn. Stability and Handling of Liquid Fuels (steering com. 1990—). Roman Catholic. Achievements include several French and foreign patents for energetical products, petroleum products, processes and additives. Home: 168 Chemin du Bois Contal, 69390 Charly France Office: Elf Antar France Rsch Ctr, BP22 Rue du Canal, 69360 Saint Symphorien Ozo France

BERNATOWICZ, FELIX JAN BRZOZOWSKI, mechanical engineer, consultant; b. Warsaw, Poland, Sept. 12, 1920; came to U.S., 1956; s. Jan and Leokadia (Malinowska) Brzozowski-Bernatowicz; m. June 26, 1948 (div. 1973); 1 child, Monica D. Fulton. Dipl. Ing., Swiss Fed. Inst. of Tech., Zurich, 1946; MSME, U. Conn., 1962. Asst. to the dir. Inst. of Exptl. Physics U. Bern, Switzerland, 1946-49; tech. officer New South Wales U. Tech., Sydney, Australia, 1950-56; R & D engr. Armzen Co., Waterbury, Conn., 1956-58, Bristol Co., Naugatuck, Conn., 1958-66; analytical engr. Colt Chandler Evans Control Systems Div., West Hartford, Conn., 1966-68; sr. engr. Gen. Dynamics Electric Boat Div., Groton, Conn., 1969-90, ret., 1990; pvt. practice cons. New London, Conn., 1990—. Vol. Rep. Party, New London, 1972, 80. Cpl. French Army, 1940, ETO. Mem. AAAS, IEEE, ASME, Am. Assn. for Artificial Intelligence, Am. Def. Preparedness Assn., N.Y. Acad. Scis., Polish Inst. Arts and Scis. of Am., Inc., U.S. Naval Inst. Roman Catholic. Avocations: classical music, modern history, sailing, walking. Home: 91 Crown Knoll Ct Ste 124 Groton CT 06430-6247

BERNAYS, PETER MICHAEL, retired chemical editor; b. N.Y.C., July 19, 1918; s. Murray C. and Hella Freud B.; m. Marie Rasmusson, Apr. 5, 1947; children: Lynda, Sally, Michael. SB, MIT, 1939; MS, U. Ill., 1940, PhD, 1942. From instr. to asst. prof. Ill. Inst. Tech., Chgo., 1946-50; assoc. prof. Southwestern La. Inst., Lafayette, 1950-54; from assoc. editor to asst. dept. head Chem. Abstracts Svc., Columbus, Ohio, 1954-88, retired, 1988. Maj. U.S. Army, 1942-46; col. U.S. Army Reserve, 1946-72. Mem. AAAS, Am. Chem. Soc. Home: 2391 Eastcleft Columbus OH 43221

BERNHARD, MICHAEL IAN, pharmaceutical company executive; b. N.Y.C., Apr. 29, 1944; s. Seymour Jean and Ella (Mackler) B.; m. Deborah Gutcheon; children: Jessica Lauren, Adam David. BS, Tufts U., 1966; MS, NYU, 1969; PhD, Cornell U., 1976. Post-doctoral fellow Sloan-Kettering Inst., N.Y.C., 1976-78; asst. prof. U. Va., Charlottesville, 1978-80; acting chief monoclonal program NIH, Frederick, Md., 1980-82; assoc. dir. Coulter Immunology, Hialeah, Fla., 1982-83, Key Pharms., Miami, Fla., 1983-86, Schering Rsch., Inc., Miami and N.J., 1986-90; v.p. pharm. product devel. Block Drug Co., N.J., 1990—; cons. in field. Contbr. numerous articles to profl. jours.; speaker in field. Recipient Outstanding Inst. Research award Sloan-Kettering Inst. for Cancer Research, 1975, Young Investigator award N.Y. State Health Research Council, 1976, 1978, Scientist of Yr. award Richard Molin Meml. Found. for Cancer Research, 1976, 1977, Pratt Found. award U. Va., 1979. Achievements include research in use of monoclonal antibodies to image and treat tumors. Home: 11 Harvey Dr Summit NJ 07901-1204

BERNIERI, FRANK JOHN, social psychology educator; b. Bklyn., May 2, 1961; s. Gene J. and Rose (Autunnale) B.; divorced; 1 child, Jennifer. BA, U. Rochester, 1983; PhD, Harvard U., 1988. Asst. prof. Oreg. State U., Corvallis, 1988-93, assoc. prof., 1993—. Author: (with others) Coordinated Movement in Human Interaction, 1991; mem. editorial bd. Jour. Nonverbal Behavior, 1990—; contbr. articles to profl. jours. Fellow Harvard U., 1987; grantee NIH, 1988, Oreg. State U. Coll. Liberal Arts, 1990; NSF Young Investigator awardee, 1992. Mem. AAAS, APA, Am. Pschol. Soc., Soc. for Personality and Social Psychology. Democrat. Office: Oreg State Univ Dept Psychology Moreland Hall Corvallis OR 97331

BERNIUS, MARK T., research physicist; b. Bklyn., Oct. 24, 1957; s. Edward and Loretta B. BSc, Polytechnic Inst. N.Y., 1979; MS, SUNY, Stony Brook, 1981; PhD, Cornell U., 1987. Postdoctoral physicist Jet Propulsion Lab., Pasadena, Calif., 1987-89; sr. rsch. fellow Calif. Inst. Tech., Pasadena, 1989-91; sr. rsch. physicist Dow Chem. Co., Midland, Mich., 1991—. Assoc. editor Review Scientific Instruments; contbr. articles to profl. jours. Mem. Am. Phys. Soc., Am. Chem. Soc., Pasadena Jaycees (bd. dirs. 1988-89, exec. bd. dirs. 1989-90). Achievements include 2 patents; research in mass analyzed secondary ion microscopy, polymer optics and characterization. Office: Dow Chem Co Ctrl Rsch 1712 Midland MI 48674

BERNOCO, DOMENICO, immunogeneticist, educator; b. Cherasco, Cuneo, Italy, Apr. 6, 1935; s. Giuseppe and Lucia (Merlo) B.; m. Marietta Magdelene von Diepow, July 20, 1972. DVM, U. Torino, Italy, 1959; lic. vet. medicine, Rome, 1961; Libera Docenza, Ministry Pub. Instrn., Rome, 1971. Asst. prof. med. genetics U. Torino, 1961-70; mem. staff Basel (Switzerland) Inst. Immunology, 1970-76; assoc. rsch. immunologist dept. surgery UCLA, 1977-81; assoc. prof. vet. medicine reproduction U. Calif., Davis, 1981—. Contbr. 98 articles to profl. jours. Fellow Italian Nat. Coun. Rsch., 1962-63, Italian Ministry for Pub. Instrn., 1963-64, fellow for fgn. countries NATO, 1967-68. Mem. Am. Assn. Immunologists, Internat. Soc. Animal Genetics, Am. Soc. Histocompatibility and Immunogenetics. Avocations: gardening, bicycling, hiking, wildlife photography, travel. Home: 1002 Deodara Ct Davis CA 95616-5037 Office: U Calif Sch Vet Medicine Dept Reproduction Davis CA 95616-8743

BERNS, DONALD SHELDON, research scientist; b. N.Y.C., June 27, 1934; s. Benjamin and May (Shapiro) B.; m. Sylvia Schleicher, Feb. 5, 1956; children: Brian Keith, Neil Gary, Amy Sue. BS in Chemistry, Wilkes Coll., 1955; PhD in Phys. Chemistry, U. Pa., 1959. Postdoctoral fellow Yale U., New Haven, Conn., 1959-61; resident rsch. assoc. Argonne (Ill.) Nat. Lab., 1961-62; sr. rsch. scientist Wadsworth Ctr. N.Y. State Dept. Health, Albany, 1962-66, dir. biophysics, 1967-82; dir. clin. scis. div. N.Y. State Dept. Health, Albany, 1983-89; dir. med. rsch. orgn. Ministry of Health, Jerusalem, 1989—; adj. prof. chemistry Rensselaer Poly. Inst., Troy, N.Y., 1971-89; prof. Sch. Pub. Health SUNY, Albany, 1984—, Hebrew U. Hadassah Med. Sch., Jerusalem, 1989—. Contbr. numerous articles to profl. jours. Mem. Am. Soc. Biochemistry and Molecular Biology, Am. Chem. Soc., AAAS. Jewish. Achievements include patents in field of spectrophotometric assay determination of lipid content of biological fluids; elucidated aggregation states of biliprotein phycocyanin; supervised lab. work of extensive HIV-Sero prevalence study published in Am. Jour. Pub. Healt. Office: Israel Ministry Health, 2 Ben Tabai, Jerusalem Israel

BERNS, KENNETH IRA, physician; b. Cleve., June 14, 1938; s. Charles and Delnet (Cohn) B.; m. Laura Louise Lawless, June 26, 1964; children: Jonathan Charles, Deborah Louise. Student, Harvard U., 1956-59; A.B., Johns Hopkins U., 1960, Ph.D., 1964, M.D., 1966. Intern Johns Hopkins Hosp., 1966-67; asst. prof. microbiology Johns Hopkins U. Sch. Medicine, 1970-74, asst. prof. pediatrics, 1970-76, asso. prof. microbiology, 1974-76; dir. Johns Hopkins U. Sch. Medicine (Year I program), 1973-76; prof., chmn. dept. immunology and med. microbiology, prof. pediatrics U. Fla. Coll. Medicine, Gainesville, 1976-84; R.A. Rees Pritchett prof., chmn. dept. microbiology Cornell U. Med. Coll., 1984—; Howard Hughes med. investigator, 1970-75; mem. microbiology test com. Nat. Bd. Med. Examiners, 1979-82, chmn., 1983-86, mem. exec. bd., 1986—; mem. Recombinant DNA adv. com. NIH, 1980-83; mem. genetic biology panel NSF, 1981-84; Fogarty sr. internat. fellow virology dept. Weizmann Inst. Sci., Rehovot, Israel, 1982-83; ad hoc mem. Bd. Sci. Counselors, Nat. Inst. Allergy and Infectious Diseases, 1981-82, permanent mem., 1992—; del. U.S.-Japan Coop. Program on Recombinant DNA, 1981; mem. Internat. Com. Taxonomy of Viruses, 1981—; mem. virology study sect. NIH, 1985-89; mem. virology and microbiology adv. com. Am. Cancer Soc., 1985-89, liason com. on med. edn., 1989—. Served with USPHS, 1967-70. Recipient faculty research award Am. Cancer Soc., 1975-76; grantee NIH, 1970-76, 80—; grantee NSF, 1973-75, 77-80; grantee Am. Cancer Soc., 1970-72; Shell Oil fellow, 1963-64. Mem. AAAS, NAS, Am. Acad. Microbiology, Am. Soc. Biol. Chemists, Am. Soc. Microbiology (bd. pub. and sci. affairs), Assn. Med. Sch. Microbiology Chmns. (counselor 1980-83, chmn. com. pub. policy 1979, pres. 1985), Am. Soc. Virology (pres. 1988-89), Soc. Gen. Microbiology, Soc. Pediatric Rsch., Internat. Union Microbiol. Socs. (v.p. 1990—), Inst. Medicine, Phi Beta Kappa, Sigma Xi. Home: 915 Sylvan Ln Mamaroneck NY 10543-3954 Office: Cornell U Med Coll Dept Microbiology 1300 York Ave New York NY 10021

BERNS, MICHAEL W., cell biologist, educator; b. Burlington, Vt., Dec. 1, 1942; married, 1963; 2 children. BS, Cornell U., 1964, MS, 1966, PhD in Biology, 1968. Assoc. dir. laser biology Pasadena Found. Med. Rsch. Calif., 1969-70; asst. prof. zoology U. Mich., 1970-72, assoc. prof. devel. biology and cell biology, 1972-75; prof. devel. biology and cell biology U. Calif. Irvine, 1975—, also chmn. dept., 1975—; co-prin. investigator laser microbeam studies on mitosis NSF, 1973-75; mem. ad hoc com. lasers in biomedical rsch. NAS, 1977; chmn. Gordon Conf. Lasers in Medicine and Biology, 1978. Grantee NIH, GMS, 1970, 72, 80, NIH Heart and Lung Inst., 1971-80, Nat. Cancer Inst., 1975-80, NSF, 1975, USAF, 1977-79; recipient Internat. Union Against Cancer ICRETT award U.S.-USSR Joint Program Establishment, 1977-80. Mem. AAAS, Am. Soc. Cell Biology, Am. Soc. Photobiology, Tissue Culture Assn., Soc. Devel. Biology. Achievements include laser microbeam studies on chromosomes; research on nucleoli and mitochondria of tissue culture cells; studies on mitosis of cells; laser instrumentation for biomedical research; research on cellular and embryonic development. Office: U Calif Laser Microbeam Program Dept Surgery Irvine CA 92717*

BERNSTEIN, BERNARD, engineering executive; b. N.Y.C., Dec. 6, 1915; s. Sam and Mamie (Meltzer) B.; m. Mitzi Steinhaus, 1939; children: Abby Marcia, Debbie Marilyn. BS, George Washington U., 1942, BME, 1947; MS, U. Md., 1954. Registered profl. engr., D.C. Gen. mgr. Gulton Industries, Metuchen, N.J., 1956-59; mgr. undersea warfare dept. IBM Corp., Bethesda, 1959-63; mgr. spl. project dept. United Electrodynamics, Alexandria, Va., 1963-64; mgr. marine system dept. Philco-Ford, Blue Bell, Pa., 1964-67; dep. dir. Dept. of Navy, Washington, 1967-77; pres. BB Mgmt. Svcs., Bethesda, 1978—. Marine Tech. Soc. (chmn. data engring. com.), Am. Geophys. Union (corp. mem.), Am. Ordnance Assn. (underwater ordnance div., detection and control sect.), Acoustical Soc. Am., Washington Acad. Scis., Sigma Tau. Democrat. Jewish. Achievements include 2 patents for split ring electroacoustic transducer; contbr. articles to profl. jours. Home: 7420 Westlake Ter #608 Bethesda MD 20817

BERNSTEIN, ELLIOT ROY, chemistry educator; b. N.Y.C., Apr. 14, 1941; s. Leonard H. Bernstein and Geraldine (Roman) Goldberg; m. Barbara Wyman, Dec. 19, 1965; children—Jephta, Rebecca. A.B., Princeton U., 1963; Ph.D., Calif. Inst. Tech., 1967. Postdoctoral fellow U. Chgo., 1967-69; asst. prof. Princeton U., N.J., 1969-75; assoc. prof. Colo. State U., Ft. Collins, 1975-80, prof. chemistry, 1980—; cons. Los Alamos Nat. Lab., 1975-83, Philip Morris, 1984—, Du Pont Corp., 1985—. Contbr. articles to profl. jours. NSF fellow, 1961-62; Woodrow Wilson fellow, 1963-64. Fellow Am.Phys. Soc.; mem. Am. Chem. Soc., Am. Phys. Soc., Sigma Xi. Office: Colo Cate U Dept Chemistry Condensed Matter Scis Lab Fort Collins CO 80523

BERNSTEIN, HAROLD SETH, pediatric cardiologist, molecular geneticist; b. N.Y.C., Oct. 6, 1959; s. Wallace Carl and Naomi (Oldak) B.; m. Patricia Margaret Foster. AB, Harvard Coll., 1982; MPhil, CUNY, 1985, PhD, 1986; MD, Mt. Sinai Sch. Med., 1990. Diplomate Nat. Bd. Med. Examiners. Postdoctoral fellow div. med. & molecular genetics Mt. Sinai, N.Y.C., 1986-88; intern U. Calif., San Francisco, 1990-91, resident in pediatrics, 1991-93; clin., rsch. fellow div. pediatric cardiology Cardiovascular Rsch. Inst., U. Calif., San Francisco, 1993—. Contbr. articles to profl. jours. Harvard Coll. scholar, 1980; NIH fellow in med. genetics, 1982-86, pediatric cardiology, 1993—; recipient Disting. Performance in Rsch. award Associated Med. Schs. N.Y., 1989, Achievement award for clin. excellence Upjohn, 1990. Fellow Am. Acad. Pediatrics; mem. AAAS, Am. Soc. Human Genetics, Am. Fedn. Clin. Rsch., Alpha Omega Alpha. Achievements include rsch. in cloning and sequencing of the first human CDNA encoding galactosidase A; first to identify molecular defect in the human galactosidase A gene resulting in Fabry Disease. Office: Univ Calif Div Pediatric Cardiology Box 0544 San Francisco CA 94143

BERNSTEIN, JEREMY, physicist, educator; b. Rochester, N.Y., Dec. 31, 1929. BA, Harvard U., 1951, MA, 1953, PhD in Physics, 1955. Rsch. assoc. physics, cyclotron lab. Harvard U., 1955-57, mem. inst. advanced study, 1957-59; assoc. Brookhaven Nat. Lab., 1960-62; assoc. prof. NYU, 1962-67; prof. physics Stevens Inst. Tech., 1967—; fellow NSF, 1959-61; Ferris prof. Princeton U., 1980-82; Rabi vis. prof. Columbia U., 1983-84. Recipient Westinghouse Writing prize AAAS, 1964, US Steel Found. Sci. Writing prize Am. Inst. Physics, 1970, Brandeis award, 1979, Brittanica award, 1987, Germont award Am. Inst. Physics, N.Y., 1990. Fellow Am. Physics Soc., Franklin Royal Soc. Arts. Achievements include research in elementary particles weak interactions, and cosmology. Office: Stevens Inst of Tech Dept of Physics Castle Pt Sta Hoboken NJ 07030*

BERNSTEIN, LARRY HOWARD, clinical pathologist; b. Highland Park, Mich., Dec. 28, 1941; s. David Mordecai and Lillian Cecilia (Schwartz) B.; m. Audrey Jean Mellen, Dec. 20, 1969; children: Rachel Laura, Naomi Beth. BS, Wayne State U., 1963, MS, 1966, MD, 1968. Intern pathology Kans. U. Med. Ctr., Kansas City, 1968-69; resident and fellow in pathology U. Calif.-San Diego, La Jolla, 1970-73; pathologist Armed Forces Inst. Pathology, Washington, 1973-75; asst. prof. pathology U. South Fla., Tampa, 1975-77; assoc. prof. pathology U. South Ala., Mobile, 1977-78; dir. chemistry Iowa Meth. Med. Ctr., Des Moines, 1979-80, United Health Svcs., Binghampton, N.Y., 1981-82; dir. chemistry and blood bank Bridgeport (Conn.) Hosp., 1983—; cons. Beckman, Boehringer Mannheim, Brea, Calif. and Indpls., 1985-92; Nat. Com. Clin. Lab. Scis. rev. com., Chgo., 1988-92. Contbr. articles to Nutrition, Clin. Chemistry, Cancer, Arch. Pathol. Lab. Medicine, Jour. Biol. Chemistry, Brit. Jour. Cancer, Jour. Molecular Cellular Cardiology. Fellow Am. Assn. Clin. Chemistry (lectr., program chmn. nat. meetings 1985—), Coll. Am. Pathologists; mem. ASTM, Clin. Lab. Mgmt. Assn. (lectr., nat. meetings 1985). Democrat. Jewish. Achievements include patents for lactate dehydrogenase method, malate dehydrogenase mthod; rsch. in effect of nutritional states; rsch. in determining decision values for laboratory tests using truth-table comprehension and quality management using data classification and analysis; rsch. in diagnosis of acute myocardial infarction (heart attack), and in cancer markers in serum and body fluids. Office: Bridgeport Hosp Dept Pathology 267 Grant St Bridgeport CT 06610

BERNSTEIN, LORI ROBIN, biochemist, molecular biologist, educator; b. Hartford, Conn., July 22, 1958; d. Ernest and Barbara S. (Goldberg) B. BA in Biochem. Scis. magna cum laude, Harvard U., 1981; postgrad., Johns Hopkins U., 1983, MS in Biology, 1985, PhD in Biology, 1989. Teaching asst. organic chemistry Johns Hopkins U., Balt., 1982-83, teaching asst. molecular biology, 1984-85; biologist cell biology sect. Lab. Viral Carcinogenesis Nat. Cancer Inst., Frederick, Md., 1987-89, Intramural Rsch. Tng. fellow cell biology sect. Lab. Viral Carcinogenesis, 1989-92, Intramural Rsch. Tng. fellow Lab. Pathology, 1993—; mem. faculty biology and biochemistry Essex C. C., 1983; presenter Gordon Conf., 1990, 93; speaker Lab. Exptl. Carcinogenesis NIH, 1991, Coriell Inst. for Med. Rsch., Camden, N.J., 1992, dept. anatomy and cell biology Georgetown U. Med. Ctr., 1992, dept. chem. biology and pharmacognosy Coll. Pharmacy, Piscataway, N.J., Lab. for Cancer Rsch., Rutgers U., 1992, Nat. Inst. Aging Gerontol. Rsch. Ctr., Francis Scott Key Med. Ctr., Balt., 1992. Contbr. articles to Sci., Environ. Health Perspective, Molecular Carcinogenesis, Jour. Cell Sci., Current Opinions in Oncology. Active Md. Libertarian party, 1990—, Nat. Libertarian party, Washington, 1990—. Grantee NIH, 1985; fellow NIH, 1989—. Mem. AAAS, Am. Assn. Cancer Rsch. Jewish. Achievements include identification of novel protein related to transcription factors inducible by tumor promoters in cells resistant to neoplastic transformation; discovery that transcription factor AP-1 is specifically active in cells susceptible to tumor promoter induced transformation but inactive in resistant cells, found modified C-Jun protein specifically in cells susceptable to tumor promoter-induced transformation. Office: Nat Cancer Inst Bldg 10 Rm 2A33 Bethesda MD 20892

BERNSTEIN, SHELDON, biochemist; b. Milw., Mar. 23, 1927; s. Jacob Louis and Tillie (Lewis) B.; m. Estelle Lou Katz, June 27, 1948; children: Bradley Alan, Richard Neal, Jodi Lynn. BS in Chemistry, U. Wis., 1949, PhD in Physiological Chemistry, 1952. NIH rsch. fellow U. Wis., Madison, 1949-52; sr. rsch. assoc. The Upjohn Co., Kalamazoo, 1952-54; v.p., dir. rsch. Amber Labs and Milbrew, Inc., Milw., 1954-67, pres., 1967-83; dir. tech. devel. Universal Foods Corp., Milw., 1983-89; pres. Bernstein Assocs., Inc., North Miami Beach, Fla., 1989--; bd. visitors U. Wis. Coll. Agr. and Life Scis., madison, 1988--. Contbr. articles to profl. jours. Dir., officer Wis. Biotech Assn., Milw, Madison, 1986-89. With USNR 1944-46. Fellow Am. Inst. Chemists; mem. Am. Chem. Soc. (chmn. biotech div. 1984, James VanLanen Disting. Svc. award 1987). Home: 3600 Mystic Pointe Dr #913 North Miami Beach FL 33180 Office: Bernstein Assocs Inc 3600 Mystic Pointe Dr #913 North Miami Beach FL 33180

BERNSTEIN, SOL, cardiologist, medical services administrator; b. West New York, N.J., Feb. 3, 1927; s. Morris Irving and Rose (Leibowitz) B.; m. Suzi Maris Sommer, Sept. 15, 1963; 1 son, Paul. A.B. in Bacteriology, U. Southern Calif., 1952, M.D., 1956. Diplomate Am. Bd. Internal Medicine. Intern Los Angeles County Hosp., 1956-57, resident, 1957-60; practice medicine specializing in cardiology L.A., 1960—; staff physician dept. medicine Los Angeles County Hosp. U. So. Calif. Med. Center, L.A., 1960—, chief cardiology clinics, 1964, asst. dir. dept. medicine, 1965-72; chief profl. services Gen. Hosp., 1972-74; med. dir. Los Angeles County-U So. Calif. Med. Center, L.A., 1974—; med. dir. central region Los Angeles County, 1974-78; dir. Dept. Health Services, Los Angeles County, 1978; assoc. dean Sch. Medicine, U. So. Calif., L.A., 1986—, assoc. prof., 1968—; cons. Crippled Childrens Svc. Calif., 1965—. Contbr. articles on cardiac surgery, cardiology, diabetes and health care planning to med. jours. Served with AUS, 1946-47, 52-53. Fellow A.C.P., Am. Coll. Cardiology; mem. Am. Acad. Phys. Execs., Am. Fedn. Clin. Research, N.Y. Acad. Sci., Los Angeles, Am. heart assns., Los Angeles Soc. Internal Medicine, Los Angeles Acad. Medicine, Sigma Xi, Phi Beta Phi, Phi Eta Sigma, Alpha Omega Alpha. Home: 4966 Ambrose Ave Los Angeles CA 90027-1756 Office: 1200 N State St Los Angeles CA 90033-4525

BERNTHAL, FREDERICK MICHAEL, federal agency administrator; b. Sheridan, Wyo., Jan. 10, 1943; s. Erwin John and Erna Emma (Kregar) B.; m. Heather A. Lancaster; 1 son, Justin. B.S., Valparaiso U., 1964; Ph.D., U. Calif.-Berkeley, 1969. Research staff Yale U., New Haven, 1969-70; prof. Mich. State U., East Lansing, 1970-80; legis. asst. Senator Howard Baker, Washington, 1978-80, chief legis. asst., 1980-83; mem. U.S. Nuclear Regulatory Commn., Washington, 1983-88; asst. sec. oceans, environment, and sci. Dept. of State, Washington, 1988-90; dep. dir. NSF, Washington, 1990-93, acting dir., 1993—. Contbr. 45 articles to sci. jours. NATO Sr. Scientist fellow U. Copenhagen, 1977; Congl. Sci. fellow Am. Phys. Soc., 1978-79. Mem. Am. Phys. Soc., Am. Chem. Soc., Sigma Xi. Republican. Lutheran. Office: Nat Sci Found Office of Dir 1800 G St NW Washington DC 20572

BERNTSON, GARY GLEN, psychiatry, psychology and pediatrics educator; b. Mpls., June 16, 1945; s. Edward Mathias and Meryle (Nelson) B.; m. Sarah Till Boysen, Mar. 5, 1984. BA, U. Minn., 1968, PhD, 1971. Postdoctoral fellow Rockefeller U., N.Y.C., 1971-73; asst. prof. dept. psychology Ohio State U., Columbus, 1973-77, assoc. prof., 1977-81, prof., 1981—, prof. dept. pediatrics, 1983—, prof. of psychiatry, 1988—; affiliate scientist Yerkes Regional Primate Rsch. Ctr., Emory U., Atlanta, 1984—; mem. initial rev. group ADAMHA, Washington, 1989—; mem. fellowship rev. panel NSF, Washington, 1991. Contbr. over 50 articles to profl. jours., 8 chpts. to books. Fellow NSF, 1969, USPHS, 1972. Mem. Soc. for Neurosci., Internat. Soc. for Developmental Psychobiology, Soc. for Psychophysiol. Rsch. Achievements include discovery of methods of studying cognition in chimpanzees, novel concepts of control of the autonomic nervous system. Office: Ohio State U 1885 Neil Ave Columbus OH 43210-1281*

BERQUEZ, GÉRARD PAUL, psychiatrist, psychoanalyst; b. Chateauroux, France, Nov. 23, 1945; s. Berquez Maurice and Josette (Bredy) B.; m. Rosemarie Saez, Aug. 22, 1970; children: Michael, Fabien. MD, U. Montpellier, France, 1978; diploma in psychopathology and psychoanalysis, U. Paris VII, 1978; PhD, U. Paris X, 1992. Spl. studies cert. in psychiatry and childhood psychiatry, 1978. Intern Navarre Psychiat. Hosp., Evreux, France, 1974-77, Ville Evrard Psychiat. Hosp., Neuilly, France, 1978-80; psychiatrist Medico-Educative Inst., Ecouis, France, 1977-83, Children's Social Service Aid, Villepinte, France, 1980; asst. lectr. U. Paris-X-Nanterre, 1983—; mem. splty. and establishment commn. U. Paris VII, 1985-91. Author: Early Infantile Autism, 1983; contbr. articles to med. jours. Mem. French Analytical Psychology Soc., Evolution Psychiatrique (corr.), Univ. Tchrs. Psychology Assn. Home: 4 Rue Crevaux, 75116 Paris France Office: U Paris X Nanterre, 200 Ave de la Republique, 92001 Nanterre France

BERRE, ANDRE DIEUDONNE, federal agency executive, oil company executive; b. Lambarene, Moyen-Ogooue, Gabon, Sept. 3, 1940; s. Andre Marie and Germaine (King) B.; m. Michele Bernadette Delachatre; children: Eugenie, Henri, Emmanuel, Stephan, Germain. BS, Faculte de Scis., Poitiers, France, 1965, MS, 1966, D in Geology, 1970. Lic. petroleum geologist. Mem. sr. staff exploration dept. Shell Gabon, Port-Gentil, 1970-73, dep. exploration mgr., 1973-74, dep. gen. mgr., 1974-77, gen. mgr., 1977—, chmn., 1982—; now min. commerce, industry and sci. rsch. Gabon Fed. Govt., Libreville, 1993—. Spl. advisor to Pres. of Gabon, Libreville, 1979—; pres. Confedn. Patronale Gabonaise, Libreville, 1988—. Decorated Commdr. Equatorial Star Gabon, Officer of Merit Gabon, Medal of Honor Gabonese Gendarmerie, Chevalier de Legion d'Honneur (France). Mem. Rotary. Avocations: fishing, music. Office: Minister Commerce Industry & Sci Rsch, BP 3906 Libreville Gabon*

BERRIDGE, MARC SHELDON, chemist, educator; b. Marblehead, Mass., May 17, 1954; s. Charles A. and Peggy Berridge; m. Catherine Gillespie, Mar. 12, 1977; children: Kevin, Dennis. BS, Carnegie-Mellon U., 1976; PhD, Washington U., 1980. Exch. scientist Commissarsat a l'energie Atomique, France, 1980-82; asst. prof. U. Tex. Med. Sch., Houston, 1982-86; assoc. prof., head of radiochemistry Case-Western Res. U., Cleve., 1986—; cons. in radiochemistry various hosps. and mfrs. Contbr. articles to profl. jours. U.S./France Exch. grantee NSF, 1980-82; Louderman fellow Washington U., 1976-78; recipient First award NIH, 1989. Mem. AAAS, Am. Chem. Soc., Soc. Nuclear Medicine (Young Investigator award 1979), Kappa Sigma (v.p., sec. 1974-76, Rollie Bradford award 1976). Achievements include patent for process for preparation of O-15 butanol for positron tomagraphy. Office: Univ Hosps Cleve Div Radiology 2074 Abington Rd Cleveland OH 44106

BERRIOS, JAVIER, electrical engineer, consultant; b. San Francisco, July 2, 1961; s. Enrique and Francisca (Curto) B.; m. Ani Lesbet Gonzalez, July 11, 1992. MSEE, Union Coll., Schenectady, 1988; PhDEE, Worcester Poly. Inst., 1992. Mktg. rep. Inteldata, S.A., Barcelona, Spain, 1982-83; quality control engr. EFOSA, Barcelona, 1983-86; lab. engr. SAINEL, Barcelona, 1986; grad. fellow Union Coll., 1987-88; rsch. and teaching asst. Worcester (Mass.) Poly. Inst., 1988-92, postdoctoral fellow, 1992—; cons. engr. GE, Schenectady, 1988. Contbr. articles to profl. jours. Mem. IEEE, Sigma Xi, Eta Kappa Nu. Home: 96 Elm St Worcester MA 01609 Office: Worcester Poly Inst 100 Institute Rd Worcester MA 01609

BERRY, CHESTER RIDLON, physicist; b. Boston, Aug. 15, 1919; m. Ruthe Marcia McKusick, Sept. 7, 1940; children: Linda Berry Corcoran, Charles, Lillian Berry Brown. AB, Dartmouth Coll., 1940; PhD, Cornell U., 1946. Tech. supr. Manhattan Project, Oak Ridge, Tenn., 1944-45; rsch. physicist Eastman Kodak, Rochester, N.Y., 1946-75; pvt. practice cons. South Orleans, Mass., 1976—; cons. Fuji Photo Film, Tokyo, 1976-79. Contbr. articles to profl. publs. Mem. Darthmouth Alumni Assn. (pres. Rochester chpt. 1952-53, Cape Cod chpt. 1982-83), Cape Cod Men's Club (sec. 1991—), Sigma Xi (pres. Rochester chpt. 1971-72). Achievements include 3 patents in field.

BERRY, JONI INGRAM, hospice pharmacist, educator; b. Charlotte, N.C., June 6, 1953; d. James Clifford and Patricia Ann (Ebener) Ingram; m. William Rosser Berry, May 29, 1976; children: Erin Blair, Rachel Anne, James Rosser. BS in Pharmacy, U. N.C., 1976, MS in Pharmacy, 1979. Lic. pharmacist, N.C. Resident in pharmacy Sch. Pharmacy, U. N.C., Chapel Hill, 1977-79, adj. asst. prof., 1985—; pharmacist Durham County Gen. Hosp., Durham, N.C., 1977-79; coord. clin. pharm. Wake Med. Ctr., Raleigh, N.C., 1979-80; co-dir. pharmacy edn. Wake Area Health Edn. Ctr., Raleigh, 1980-85; pharmacist cons. Hospice of Wake County, Raleigh, 1980—. Mem. editorial adv. bd. Hospice Jour., 1985-91, Jour. Pharm. Care in Pain and Symptom Mgmt., 1992—; reviever Am. Jour. Hospice Care, 1986—; contbr. articles to profl. jours. Troop leader Girl Scouts U.S., Raleigh, 1987—, trainer, 1989-91, mgr. svc. unit, 1990—; tchr. Sunday sch. St. Phillips Luth. Ch., Raleigh, 1990-92. Recipient Silver Pinecone award Girl Scouts U.S., 1991, Golden Rule award J.C. Penney Co., 1991. Mem. Am. Pharm. Assn. (hospice pharmacist steering com. 1990—), Am. Soc. Hosp. Pharmacists, Nat. Hospice Assn., N.C. Pharm. Assn. (Don Blanton award 1985, mem. continuing edn. com. 1986-87, com. chairperson 1981-84), N.C. Soc. Hosp. Pharmacists (bd. dirs. 1984-86, program com. 1988-91), Wake County Pharm. Assn. (sec. 1982-85), Rho Chi. Democrat. Avocations: gardening, weight lifting, aerobics. Office: Hospice Wake County 4th Fl 4513 Creedmoor Rd Raleigh NC 27612

BERRY, KIM LAMAR, energy engineer; b. Atlanta, Sept. 13, 1956; s. James Daniel and Lorraine Burg (Huistra) B.; m. Diane Lee Drumm, Oct. 1, 1983; children: Jennifer Danielle, Jason Daniel. B of Mech. Engring., Ga. Inst. Tech., 1979; MBA, Ga. State U., 1985. Cert. energy mgr.; cert. lighting efficiency profl. Div. indsl. engr. Ga. Power Co., MAcon, 1979-81; dist. indsl. engr. Ga. Power Co., Atlanta, 1981-82; sr. dist. power engr. Ga. Power Co., Tucker, 1982-87; program coord. Ga. Power Co., Atlanta, 1987-90, program mgr., 1990—; mem. electrification coun. Indsl. Heat Pump Adv. Group., Washington, 1987-90. Pres. Stone Mt. (Ga.) Indsl. Park Assn., 1987-88; acct. exec. United Way, Atlanta, 1982-84, sect. chmn., 1985-86. Am. Environ. Engrs. Soc., Assn. Energy Engrs. (Atlanta chpt.), Demand Side Mgmt. Soc.

BERRY, MICHAEL JAMES, chemist; b. Chgo., July 17, 1947; s. Bernie Milton and Irene Barbara (Lentz) B.; m. Julianne Elward, Apr. 28, 1967; children—Michael James, II, Jennifer Anne; m. Patricia Gale Hackerman, July 7, 1984. B.S. in Chemistry, U. Mich., 1967; Ph.D. (NSF predoctoral fellow), U. Calif., Berkeley, 1970. Asst. prof., then assoc. prof. chemistry U. Wis., Madison, 1970-76; mgr. photon chemistry dept., corp. research center Allied Chem. Corp. Morristown, N.J, 1976-79; Robert A. Welch prof. chemistry Rice U., Houston, 1979-92; pres. Antropix Corp., Houston, 1982—; dir. Laser Applications Research Ctr., Houston Area Research Ctr., 1984-90; chief scientist Emmetropix Corp., Woodlands, Tex., 1991-92; dir. cornea program Sunrise Technologies, Fremont, Calif., 1992—. Author research papers in field; patentee in field. Recipient Fresenius award Phi Lambda Upsilon, 1982; grantee Air Force Office Sci. Rsch., 1972-76, 85-88, Office Naval Rsch., 1972-74, NSF, 1975-76; tchr.-scholar Camille and Henry Dreyfus Found., 1974-76; Alfred P. Sloan rsch. fellow, 1975-76, John Simon Guggenheim Meml. Found. fellow, 1981-82. Fellow AAAS; mem. Am. Chem. Soc. (Pure Chemistry award 1983), Am. Phys. Soc., Optical Soc. Am., Am. Soc. Photobiology, Am. Soc. Laser Medicine and Surgery, Assn. Rsch. Vision and Opthalmology, Internat. Soc. Eye Rsch., Internat. Soc. Refractive Keratoplasty, Sigma Xi. Home: PO Box 1421 Pebble Beach CA 93953 Office: Sunrise Technologies 47257 Fremont Blvd Fremont CA 94538

BERRY, RICHARD STEPHEN, chemist; b. Denver, Apr. 9, 1931; s. Morris and Ethel (Alpert) B.; m. Carla Lamport Friedman, Sept. 4, 1955; children: Andrea, Denise, Eric. AB, Harvard U., 1952, AM, 1954, PhD, 1956. Instr. chemistry Harvard U., 1956-57, U. Mich., 1957-60; asst. prof. Yale U., 1960-64; assoc. prof. U. Chgo., 1964-67, prof., 1967—, James Franck Disting. Svc. prof., 1989—; Arthur D. Little prof. MIT, 1968; Phillips lectr. Haverford Coll., 1968; Lôwdin lectr. Uppsala U, 1989; cons. Avco-Everett Research Labs., 1964-83, Argonne Nat. Lab., 1976—, Oak Ridge Nat. Labs., 1978-81, Los Alamos Sci. Lab., 1975—, mem. adv. com. theory; vis. prof. U. Copenhagen, 1967, 79; mem. adv. panel for chemistry NSF, 1971-73; mem. rev. com. radiol. and environ. research div. Argonne Nat. Lab., 1970-76; mem. evaluation panel measures for air quality Nat. Bur. Standards; mem. numerical data adv. bd. NRC, 1978-86, chmn., 1981-86; mem. steering com. panel on environ. monitoring, mem. com. on atomic and molecular sci., 1984-89, com. on chem. scis. Nat. Acad. Scis.-NRC, 1977-79; mem. adv. panel on health of sci. and tech. enterprise, mem. adv. panel on nat. labs. Office Tech. Assessment; mem. adv. bd. Environ. Health Resource Center, Inst. for Theoretical Physics, Santa Barbara, 1989-91; mem. vis. com. div. applied physics Harvard U., 1977-81; mem. adv. panel dept. chemistry Princeton U., 1978-81; Hinshelwood lectr. Oxford U., 1980; prof. associé U. Paris-Sud, 1979-80; Newton Abraham prof. Oxford U., 1986-87; Phi Beta Kappa lectr. 1989-90; pres. Telluride Summer Rsch. Ctr., 1989-93. Author: Understanding Energy, 1991; co-author: TOSCA, The Total Social Cost of Fossil and Nuclear Power, 1979, Physical Chemistry, 1980; assoc. editor: Jour. Chem. Physics, 1971-74, Accounts Chem. Rsch., 1975-90, Revs. Modern Physics, 1983—, Phys. Rev. A, 1986-92; bd. dirs. Bull. Atomic Scientists, 1974-83; adv. editor: Resources and Energy; contbr. articles to porfl. jours. Alfred P. Sloan fellow, 1962-66; Guggenheim fellow, 1972-73; MacArthur prize fellow, 1983; Humboldt Rsch. awardee, 1993. Fellow AAAS (chmn. chemistry sect. 1993-94), Am. Phys. Soc., Japan Soc. Promotion of Scis., Am. Acad. Arts and Scis. (v.p. 1987-90); mem. NAS, Am. Chem. Soc., Nat. Coun. Lawyers and Scientists, Royal Danish Acad. Arts and Letters (fgn.), Sigma Xi (nat. lectr. 1976, 77).

BERRY, WILLIAM BENJAMIN NEWELL, geologist, educator, former museum administrator; b. Boston, Sept. 1, 1931; s. John King and Margaret Elizabeth (Newell) B.; m. Suzanne Foster Spaulding, June 10, 1961; 1 child, Bradford Brown. A.B., Harvard U., 1953, A.M., 1955; Ph.D., Yale U., 1957. Asst. prof. geology U. Houston, 1957-58; asst. prof. to prof. paleontology U. Calif., Berkeley, 1958—; prof. geology, 1991—; curator Mus. of Paleontology U. Calif., Berkeley, 1960-75; dir. Mus. of Paleontology U. Calif., 1975-87, chmn. dept. paleontology, 1975-87; marine scientist Lawrence Berkeley Lab., 1989—; cons. U.S. Geol. Survey., Environ. Edn. to Ministry for Environ., Catalonia, Spain. Author: Growth of a Prehistoric Time Scale, 1968, revised edit., 1987, Principles of Stratigraphic Analysis, 1991; assoc. editor Paleoceanography; contbr. numerous articles on stratigraphic and paleontol. subjects to profl. jours.; editor publs. in geol. scis. Guggenheim Found. fellow, 1966-67. Fellow Calif. Acad. Scis.; mem. Paleontol. Soc., Geol. Soc. Norway, Internat. Platform Assn., Explorers Club, Commonwealth Club Calif. Home: 1366 Summit Rd Berkeley CA 94708-2139 Office: U Calif Dept Geology and Geophysics Earth Scis Bldg Berkeley CA 94720

BERRY, WILLIAM BERNARD, engineering educator, researcher; b. Shelby, Ohio, July 23, 1931; s. Dorwin James and Norma Clare (Laux) B.; m. Lois Georgia Langford, June 25, 1955; children: Elizabeth Maria, William Joseph, Mary Suzanne, Thomas James. BSEE, U. Notre Dame, 1953, MSEE, 1957; PhD, Purdue U., 1964. Profl. engr., Ind. Instr., asst. prof. Marquette U., Milw., 1957-61; asst. prof. U. Notre Dame, Ind., 1963-67, assoc. prof., 1967-70, prof., 1970—, asst. dean for rsch., 1974-85, program mgr. cold weather transit tech., 1979—, co-dir. Energy Analysis and Diagnostics Ctr., 1990—; vis. rsch. assoc. NASA-Lewis Rsch. Ctr., Cleve., 1972-73; vis. profl. Solar Energy Rsch. Inst., Golden, Colo., 1986-87. Author: Electronic Materials Properties and Solid State Energy Conversion, 1968. With U.S. Army, 1953-55. Mem. IEEE, Am. Soc. for Engring. Edn., Electrochem. Soc. Office: U Notre Dame Coll Engring Notre Dame IN 46556

BERSON, ELIOT LAWRENCE, ophthalmologist, medical educator; b. Boston, 1937. MD, Harvard U., 1962. Intern Calif. Hosp., San Francisco, 1962-63; resident in ophthalmology Barnes and McMillan Hosps., St. Louis, 1963-66; clin. assoc. ophthalmologist Nat. Inst. Neurol. Diseases and Blindness, Bethesda, Md., 1966-68; asst. Mass. Eye and Ear Infirmary, Boston, 1968-73, asst. surgeon, 1974-78, dir. Berman-Gund Lab. for Study of Retinal Degenerations, Harvard Med. Sch., 1974—, assoc. surgeon in ophthalmology, 1979-84, surgeon in ophthalmology, 1984—; instr. Harvard U. Sch. Medicine, Boston, 1968-70, asst. prof., 1971-76, assoc. prof. ophthalmology, 1976-82, Chatlos prof. ophthalmology, 1982—. Surgeon USPHS, 1966-68. Mem. AMA, Assn. for Rsch. in Vision and Ophthalmology, Am. Acad. Ophthalmology, Am. Ophthal. Soc. Office: Berman-Gund Lab Mass Eye and Ear Infirmary 243 Charles St Boston MA 02114-3004

BERSON, JEROME ABRAHAM, chemistry educator; b. Sanford, Fla., May 10, 1924; s. Joseph and Rebecca (Bernicker) B.; m. Bella Zevitovsky, June 30, 1946; children: Ruth, David, Jonathan. B.S. cum laude, CCNY, 1944; M.A., Columbia U., 1947; Ph.D., 1949. NRC postdoctoral fellow Harvard U., 1949-50; asst. chemist Hoffmann-LaRoche, Inc., Nutley, N.J., 1944; asst. prof. U. So. Calif., 1950-53, asso. prof., 1953-58, prof., 1958-63; prof. U. Wis., 1963-69; prof. Yale U., 1969-79, Irénée du Pont prof., 1979-92, Sterling prof., 1992—, chmn. dept. chemistry, 1971-74, dir. div. phys. sci. and engring., 1983-90; vis. prof. U. Calif., U. Cologne, U. Western Ont., U. Lausanne; Fairchild Disting. scholar Calif. Inst. Tech.; cons. Riker Labs., Goodyear Tire & Rubber Co., Am. Cyanamid Co., IBM, Cord Labs., SCM Corp., B.F. Goodrich Corp.; mem. medicinal chemistry study sect. NIH, 1969-73; mem. adv. panel chemistry NSF, 1964-70. Mem. editorial adv. bd.: Jour. Organic Chemistry, 1961-65, Accounts of Chemical Research, 1971-77, Nouveau Journal de Chimie, 1977-85, Chem. Revs., 1980-83, Jour. Am. Chem. Soc., 1988—; contbr. articles to profl. jours. Served with AUS, 1944-46, CBI. Recipient Alexander von Humboldt award, 1980, Townsend Harris medal Alumni Assn. CCNY, 1984, Merit award NIH, 1989; John Simon Guggenheim fellow, 1980. Fellow Am. Acad. Arts and Scis.; mem. NAS, Am. Chem. Soc. (Calif. sect. award 1963, James Flack Norris award 1978, Nichols medal 1985, Roger Adams award 1987, Arthur C. Cope scholar 1992, chmn. div. organic chemistry 1971), Chem. Soc. London, Phi Beta Kappa, Sigma Xi, Phi Lambda Upsilon. Home: 45 Bayberry Rd Hamden CT 06517-3401 Office: Yale U Dept Chemistry PO Box 6666 New Haven CT 06511-8118

BERSTEIN, IRVING AARON, biotechnology and medical technology executive; b. Providence, Oct. 11, 1926; s. Robert Louis and Laura (Sperber) B.; m. Suzanne D'Amico, Apr. 16, 1972; children: Jonathan, Robert Laurance. ScB, Brown U., 1947; PhD, Cornell U., 1951. Assoc. tech. dir., sr. scientist Tracer Lab Inc., 1951-57; pres., tech. dir. Controls for Radiation, Inc., Cambridge, Mass., 1957-69, Controls for Radiation Inc. (aquired by Teledyne Inc.), Cambridge, Mass., 1969; v.p. Isotopes Inc. (subs. Teledyne Inc.), Cambridge, Mass., 1969-70; dir. med. div., v.p. AGA Corp., Secaucus, N.J., 1970-71; asst. dir. rsch. program devel. div. health sci. and tech. Harvard U.-MIT, 1972-86; founder and chmn. bd. Hygeia Scis. Inc., 1980-87; pres. Tambrands, Inc. (merged Hygeia Scis. Inc. into Tambrands, Inc.), 1985-87; sr. sci. advisor Hygeia Scis., Inc., 1988-90; chmn. bd. Endogen, Inc., Boston, 1990—; pres. Berstein Tech. Corp., 1980—; cons. for Med. & Biotechnology Investment, Corp. Devel. Francis Wayland scholar; Cornell U. fellow. Mem. World Pres.'s Orgn., Forty-Niners, Harvard Club Boston, Cornell Club Boston, Sigma Xi. Home and Office: 42 Buckman Dr Lexington MA 02173-6040 Office: Endogen Inc 68 Fargo St Boston MA 02210-1950

BERT, CHARLES WESLEY, mechanical and aerospace engineer, educator; b. Chambersburg, Pa., Nov. 11, 1929; s. Charles Wesley and Gladys Adelle (Raff) B.; m. Charlotte Elizabeth Davis (June 29, 1957); children: Charles Wesley IV, David Raff. B.S. in Mech. Engring, Pa. State U., 1951, M.S., 1956; Ph.D. in Engring. Mechanics, Ohio State U., 1961. Registered profl. engr., Pa., Okla. Jr. design engr. Am. Flexible Coupling Co., State Coll., Pa., 1951-52; aero. design engr. Fairchild Aircraft div. Fairchild Engine and Airplane Corp., Hagerstown, Md., 1954-56; prin. M.E. Battelle Inst., Columbus, Ohio, 1956-61; sr. research engr., 1961-62, program dir., solid and structural mechanics research, 1962-63, cons., 1964-65; assoc. prof. U. Okla., 1963-66, prof., 1966—; dir. Sch. Aerospace and Mech. Engring., 1972-77, 90—, Benjamin H. Perkinson Chair prof. engring., 1978—; instr. engring. mechanics Ohio State U., Columbus, 1959-61; cons. various indsl. firms.; bd. dirs. Midwestern Mechanics Conf., 1971-79, chmn., 1973-75; Honor lectr. Mid-Am. State Univs. Assn., 1983-84; seminar lectr. Midwest Mechanics, 1983-84; Plenary lectr. 4th Internat. Conf. on Composite Structures, Paisley, Scotland, 1987. Mem. editorial bd. Composite Structures Jour., 1982—, Jour. Sound & Vibration, 1988—, Advanced Composite Materials, 1991—, Composites Engring., 1991—, Mechanics of Composite Materials and Structures, 1993—; assoc. editor Exptl. Mechanics, 1982-87, Applied Mechanics Revs., 1984-87; contbr. chpts. in books. and articles to profl. jours. Served from 2d lt. to 1st lt. USAF, 1952-54. Recipient Disting. Alumnus award Ohio State U. Coll. Engring., 1985. Fellow AAAS, AIAA (nat. tech. com. structures 1969-72, vice chmn. Cen. Okla. chpt. 1965-66, chmn. 1966-67), ASME (Cen. Okla. chpt. exec. com. 1973-78, 90—, sec. 1990-91, region X mech. engring. dept. heads com. 1972-77, 90—, chmn. 1975-77), Am. Acad. Mechs. (bd. dirs. 1979-82), Soc. Exptl. Mechanics (monograph com. 1978-82, chmn. 1980-82, sec. Mid-Ohio chpt. 1958-59, chmn. 1959-60, adv. bd. 1960-63); mem. NSPE, Soc. Engring. Sci. (bd. dirs. 1982-88), N.Y. Acad. Scis., Okla. Acad. Sci., Okla. Soc. Profl. Engrs., Scabbard and Blade, Pa. State U. Alumni Assn. (Outstanding Engring. Alumnus award 1992), Sigma Xi, Sigma Tau, Pi Tau Sigma, Sigma Gamma Tau (Disting. Engr. award), Tau Beta Pi (Disting. Engr. award). Achievements include co-development of world's smallest pressure transducer capable of measuring both steady and fluctuating pressures; first general solution of cylindrically orthotropic plates of radially varying thickness under arbitrary body forces; origination of several minimum-weight optimal designs for multicell cylindrical pressure vessels, experimental techniques and associated data reduction equations for determining residual stresses in both flat-sheet and thick-walled cylindrical specimens of composite materials; first successful application of Kennedy-Pancu system identification method to shell structures, noninteger polynomial version of Rayleigh's method to heat conduction; first application of differential quadrature method to static structural problems, structural vibration problems and non-linear structural problems; first application of noninteger polynomial method to finite element methods; first dynamic stability analysis of unicycles and monocycles; origination of concept of stress gages for composite materials; research on sandwich structures with bimodular facings, prediction of ply steer behavior of automobile tires, non-linear flutter of laminated composite panels; many others. Home: 2516 Butler Dr Norman OK 73069-5059 Office: U Okla Sch Aerospace and Mech Engring 865 Asp Ave Norman OK 73019-0601

BERTELE, WILLIAM, environmental engineer; b. Phila., Aug. 7, 1935; s. William J. and Viola H. (DeFant) B.; children: William Bradford, Theodore, Amanda. BS in Commerce and Engring., Drexel U., 1963. Registered profl. engr., N.J., Pa. Field engr. Sam P. Wallace Co., Santurce, P.R., 1960-65; ptnr. Engineered Products Co., Santurce, 1965-68; sr. application engr. Am. Air Filter, Louisville, 1969-78; v.p. engring. Recon Systems Inc., Raritan, N.J., 1978-92; pvt. practice New Hope, Pa., 1992—. With USAR, 1960-65. Mem. CAP (dep. comdr. 1985-90, comdr. 1991, commendation 1984), NSPE, Air and Waste Mgmt. Assn., Cen. Bucks C. of C. Avocations: music, children, hiking, dancing, gardening. Home and Office: 163 N Sugan Rd New Hope PA 18938

BERTHOLF, ROGER LLOYD, chemist, toxicologist; b. Roanoke, Va., Aug. 21, 1955; s. Max Erwin and Nancy Jane (Layman) B.; m. Marsha Gay Frazelle, Sept. 21, 1985; children: Aaron Lloyd, Abby Gay. BS, James Madison U., 1977; MS, U. Va., 1981, PhD, 1985. Diplomate Am. Bd. Clinical Chemistry. Fellow in neuropathology U. Va. Med. Ctr., Charlottesville, 1985-86, rsch. assoc., 1986-88; asst. prof. U. Fla. Coll. Medicine, Gainesville, 1988—. Contbr. over 30 papers and 25 abstracts to profl. jours. and 11 chpts. to books. Active Gainesville Civic Chorus, 1991-92. Rsch. grantee Am. Health Assistance Found., 1989. Fellow Am. Inst. Chemists; mem. Am. Assn. for Clin. Chemistry, AAAS, Omicron Delta Kappa, Tau Kappa Alpha. Methodist. Office: U Fla Coll Medicine Dept Pathology PO Box 100275 Gainesville FL 32610-0275

BERTHOUEX, PAUL MAC, civil and environmental engineer, educator; b. Olwein, Iowa, Aug. 15, 1940; s. George Albert and LaVadia Fay (McBride) B.; m. Susan Jean Powell, Sept. 8, 1962; 1 child, Stephanie Fay. BSCE, U. Iowa, 1963, MSCE, 1964; PhD, U. Wis., 1969. Registered profl. engr. Instr. U. Iowa, Iowa City, 1964-65; asst. prof. civil engring. U. Conn., Storrs, 1965-67; chief rsch. engr. GKW Cons., Mannheim, Fed. Republic of Germany, 1969-71; prof. civil engring. U. Wis., Madison, 1971—. Author: Strategy of Pollution Control, 1978; contbr. over 100 articles to profl. jours. Recipient Radebaugh Prize, CSWPCA, 1989, 91. Mem. ASCE (Rudolf Herring medal 1974, 92), Water Environ. Fedn. (Eddy medal 1971), Internat. Assn. Water Pollution Rsch. and Control, Am. Water Wks. Assn., Assn. Environ. Engring. Profs. Office: U Wis 3204 Engineering Bldg 1415 Johnson Dr Madison WI 53706

BERTI, ARTURO LUIS, sanitary engineer; b. Bocono, Trujillo, Venezuela, Aug. 5, 1912; s. Arturo Berti and Virginia Marquez; m. Olga Cupello de Berti, Aug. 24, 1942; children: Elena, Arturo, Beatriz Margarita. B, Liceo Andres Bello, Caracas, 1936; PhD, Cen. U., Caracas, 1936; MS, Tex. A&M, 1939. Registered profl. engr., Venezuela. Field engr. Malariology Min. Sanidad, Pto. Cabello, Venezuela, 1936-37; chief zona Malariology Min. Sanidad, Maracay, Venezuela, 1936-40, chief antimalaria engr., 1941-50, chief div., 1950-60; dir. Bur. of Malariology, Maracay, Venezuela, 1960-70; pres. Acad. Ciencias, Fisicas, Mat. y Nat., Caracas, 1989-91, As. Mundial Vivienda Rural, Caracas, 1974—. Mem. Am. Soc., Rotary. Roman Catholic. Avocations: reading. Home: Calle 14, Urb La Boyera, Caracas 1081, Venezuela Office: AS Mundial De Vivienda Rural, Av Libertador, Caracas 1050, Venezuela*

BERTINO, JOSEPH ROCCO, physician, educator; b. Port Chester, N.Y., Aug. 16, 1930; s. Joseph and Madaleine (Posillipo) B.; m. Mary Patricia Hagemeyer, Sept. 29, 1956; children—Frederick, Amy Marie, Thomas Allen, Paul Phillip. Student, Cornell U., 1947-50; M.D., Downstate Med. Center N.Y., 1954. USPHS Research fellow U. Wash. Sch. Medicine, Seattle, 1958-61; mem. faculty Yale U. Sch. Medicine, 1961-87, assoc. prof. pharmacology and medicine, 1964-67, prof., 1967-87, Am. Cancer Soc. prof., 1975—; head program molecular pharmacology and therapeutics Sloan Kettering Ctr., 1987—; prof. medicine and pharmacology Cornell U. Sch. Medicine, N.Y.C., 1987—; cons. USPHS, 1966—; N.Y. State scholar for medicine, 1950-54. Contbr. articles to profl. jours. Recipient Honor medal Am. Cancer Soc., 1992. Mem. Am. Soc. for Clin. Investigation, Am. Soc. Hematology, Biol. Chemists, Pharmacology and Therapeutics. Home: 117 Sunset Hill Rd Branford CT 06405-6419 Office: Meml Sloan Kettering Cancer Ctr 1275 York Ave New York NY 10021-6094

BERTNESS, KRISTINE ANN, physicist; b. Appleton, Minn., Sept. 28, 1959; d. Charles H. and Joyce M. (Harms) B.; m. John C. Price, Aug. 24, 1985. BA in Physics, Oberlin Coll., 1981; MS, PhD, Stanford U., 1987. Engr. Varian Rsch. Ctr., Palo Alto, Calif., 1987-89; staff scientist Nat. Renewable Energy Lab., Golden, Colo., 1989—. Contbr. articles to Applied Physics, Physics Rev., articles to conf. proceedings. Victim advocate, crisis line worker Boulder (Colo.) County Safehouse, 1991-92. Mem. Am. Phys. Soc., Am. Assn. Crystal Growth, Materials Rsch. Soc., Phi Beta Kappa. Democrat. Episcopalian. Achievements include design and construction of materials growth system for solar cells. Office: Nat Renewable Energy Lab 1617 Cole Blvd Golden CO 80401-3393

BERTOLATUS, JOHN ANDREW, physician, educator; b. Jersey City, Apr. 6, 1951; s. John Richard and Mabel (Mayer) B.; m. Dianne Lee Atkins, June 22, 1974; children: Matthew, David. BA, Johns Hopkins U., 1972, MD, 1976. Diplomate Am. Bd. of Internal Medicine. Intern U. Ky. Med. Ctr., Lexington, 1976, resident, 1977-79, fellow in nephrology, 1979-80; fellow in nephrology U. Iowa, Iowa City, 1980-82, assoc. prof. nephrology, 1982-84, asst. prof., 1984-92; assoc. prof. U. Iowa, 1992—. Mem. Am. Soc. Transplant Physicians, Am. Soc. Nephrology, Am. Fedn. Clin. Rsch., Ctrl. Soc. Clin. Rsch. Baptist. Office: U Iowa Dept Internal Med E 300 GH Univ Hosps Iowa City IA 52242

BERTOLETT, CRAIG RANDOLPH, mechanical engineer consultant; b. Richmond, Va., Oct. 16, 1936; s. Arthur Disbrow and Marget (Richardson) B.; m. Sarah G. Swaim, June 27, 1959 (div. July, 1978); children: Craig R. Jr., Christopher Robert, Margo Elizabeth; m. Barbara Frances Bertolett, Dec. 9, 1978. BS, U.S. Military Acad., 1959; MS in Mech. Engring., New Mex. State U., 1965; MBA, Loyola Coll., Balt., 1974. Registered profl. engr., Tex.; cert. product safety engr.; cert. hazard control mgr. Internat. Cert. Bd. Commd. 2nd lt. U.S. Army, 1959, advanced through grades to major, 1970, resigned, 1970; dir. test and evaluation Black & Decker Mfg. Co., Towson, Md., 1970-76; v.p. engring. spl. products divsn. Emerson Elect. Co., St. Louis, 1976-80; pres., prin. CR Bertolett Assocs., Inc., San Antonio, 1980—. Contbr. articles to profl. jours. Mem. Mayor's Blue Ribbon Com., Castle Hills, Tex., 1989—, vestry St. Luke's Episc. Ch., 1989-92; mem. Castle Hill Water Policy Com. 1992. Major U.S. Army, 1959-70, Korea; col. USAR, 1970-86; ret. 1986. Fellow Nat. Acad. Forensic Engrs.; mem. ASME, Am. Nat. Standards Inst., Human Factors Soc., Nat. Safety Coun., Nat. Soc. Profl. Engrs., Profl. Engrs. in Pvt. Practice, Survival and Flight Equipment Assn., Soc. Automotive Engrs., Soc. Am. Military Engrs., Tex. Soc. Profl. Engrs., The Ordnance Corps. Assn., Correspondence Group of ANSI Z535 Safety Signs and Colors. Avocations: sailing, hunting, woodworking.

BERTOLINI, FERNANDO, mathematics educator, Italian embassy cultural administrator; b. Florence, Italy, Feb. 25, 1925; s. Ottorino and Francesca (Da Villa) B.; m. Adriana Ribella, Oct. 20, 1956; children—Giovanni, Francesca, Alessandro, Gabriella, Paolo, Enrico, Luciano, Lorenzo. D.Math., U. Rome, 1947. Instr. math. U. Rome, 1947-61; from assoc. prof. to prof. U. Pitts., 1961-68; prof. math. analysis U. Parma, Italy, 1968—; vis. prof. math. Fordham U., N.Y.C., 1969, Somali Nat. U., Mogadishu, 1982, 83; visitor, cons. Escuela Superior Politecnica de Chimborazo, Riobamba Ecuador, 1983; dir. cultural centre Italian embassy, New Delhi, 1984-91. Contbr. articles to profl. jours. Mem. Unione Matematica Italiana, Circolo Matematico di Palermo, Am. Math. Soc., Sigma Xi. Roman Catholic. Home: Galleria Polidoro 7, 43100 Parma Italy Office: U Parma Math Dept, via M D'Azeglio 85, 43100 Parma Italy

BERTRAND, WILLIAM ELLIS, public health educator, international health center administrator; b. Midland, Tex., May 24, 1944; s. Alvin Lee and Mary Nick (Ellis) B.; m. Jane Trowbridge; children: Katherine, Jacob. BA, La. State U., 1966; PhD, Tulane U., 1972. Instr. to assoc. prof. Delgado Community Coll., New Orleans, 1966-70; instr. to prof. dept. biostats. and epidemiology Tulane U., New Orleans, 1974—, chmn. dept., 1981-86, postdoctoral fellow, 1972-75, also adj. asst. prof. to adj. prof. sociology dept., 1974-78, Wisner prof. public health, 1986—; dir. Tulane Internat. Devel. Ctr. and Tulane Ctr. for Internat. Resource Devel., 1988—; chmn. Dept. Internat. Health and Devel., Tulane SPHTM, 1992—; vis. lectr. Sch. Social Work, 1977-80; vis. lectr. Universidad del Valle, Cali, Colombia, 1969-71, 73; dir. and founder Inst. for Study Change in the Americas, New Orleans, 1984-86; co-dir. and prin. investigator Zaire Sch. Pub. Health U. Kinshasa, 1985—; co-prin. investigator Kenya Ministry Health, Nairobi, 1988—, Niger Ministry Health, Niamey, 1988—; prin. investigator Famine Early Warning System, Burkina Faso, Mali, Mauritania, Niger, Sudan, Chad, 1988—. Contbr. numerous articles to profl. publs., book chpts. Mem. State and City AIDS Task Force, New Orleans, 1985. Grantee U.S. Agy. for Internat. Devel., Zaire, 1985, Kenya, 1988, Niger, 1988, Sudan, 1988. Mem. Am. Coll. Epidemiology, Am. Pub. Health Assn., So. Sociol. Soc., Applied Anthrop. Assn., PanAm. Health and Edn. Found. (bd. dirs. 1990—), Delta Omega. Democrat. Baptist. Avocations: tennis, fishing. Office: Tulane U SPH & TM 1501 Canal St Ste 1300 New Orleans LA 70112-2817*

BERVEN, NORMAN LEE, counselor, psychologist, educator; b. Des Moines, May 14, 1945; s. Arthur N. and Ruth N. (Sharp) B.; m. Estella Stone, Oct. 11, 1969; 1 child, Jennifer. BS, U. Iowa, 1967, MA, 1969; PhD, U. Wis., 1973. Lic. psychologist; cert. rehab. counselor. Rehab. counselor San Mateo County Mental Health Svc., San Mateo, Calif., 1969-71; rsch. assoc. Internat. Ctr. for Disabled, N.Y.C., 1973-75; asst. prof. counseling and spl. svcs. Seton Hall U., South Orange, N.J., 1975-76; asst. prof. to prof. rehab. psychology, program chair U. Wis., Madison, 1976—; cons. to univ., govt. and pvt. non-profit programs. Editor: Rehab. Counseling Bull., 1985-92, assoc. editor, 1982-85, editorial bd., 1980-82, 92—; editorial bd. Rehab. Psychology, 1981—, Vocat. Evaluation and Work Adjustment Bull., 1980—, Assessment in Rehab. and Exceptionality, 1992—; contbr. articles to profl. jours., chpts. to books. U.S. Dept. Edn. tng. grantee, 1986-89, 89-92, 90-93, 93—, rsch. grantee Spencer Found., 1981-82, Wis. Alumni Rsch. Found., 1979-80. Fellow APA (rehab., counseling and evaluation, measurement and stats div.); mem. ACA (rsch. award 1986), Am. Rehab. Counseling Assn. (bd. dirs. N.J. chpt. 1975-76, disting. profl. award 1990, rsch. award 1981, 84, 86, 92, 93), Nat. Rehab. Counseling Assn. (bd. dirs. Wis. chpt. 1981-83, Meritorious Svc. award 1992, Calif. chpt. 1971), Nat. Rehab. Assn. (bd. dirs. S.W. Wis. chpt. 1980—, San Mateo chpt. 1969-71, grad. lit. award 1968), Assn. for Counselor Edn. and Supervision, Assn. for Assessment in Counseling, Assn. for Specialists in Group Work, Vocational Evaluation and Work Adjustment Assn. Home: 10 Southwick Cir Madison WI 53717-1415 Office: U Wis Madison Rehab Psychology 432 N Murray St Madison WI 53706-1496

BERWIG, NEWTON URBANO, aerospace executive; b. Taquara, Brazil, June 27, 1934; s. Adolfo and Maria (Santos) B.; m. Ana Maria Netto, Dec. 20, 1958; 1 child, Newton A. Jr. BS in Physics, UCLA, 1955. Commd. 2d lt. USAF, 1955, advanced through grades to capt., 1962, served as pilot various locations, 1955-64, resigned, 1964; mgr. S. Am. ops. Piper Aircraft Corp., Lock Haven, Pa., 19645-79; pres. Embraer Aircraft Corp., Ft. Lauderdale, Fla., 1979—; bd. dirs. Delta Nat. Bank, Miami, Fla., Brazil-U.S. C. of C., Miami. Office: Embraer Aircraft Corp 276 SW 34th St Fort Lauderdale FL 33315

BERZINS, ERNA MARIJA, physician; b. Latvia, Nov. 27, 1914; d. Arturs and Anna (Steckenbergs) Meilands; came to U.S., 1951, naturalized, 1956; M.D., Latvian State U., 1940; m. Verners Berzins, Aug. 24, 1935; children—Valdis, Andis. Mem. pediatric faculty Latvian State U., 1940-44; intern Good Samaritan Hosp., Dayton, Ohio, 1951-52; resident in pediatrics Children's Hosp. of Mich., Detroit, 1953-55; practice medicine specialising in pediatrics, Detroit, 1956-60; with ARC, Cleve., 1961-63; physician pediatric outpatient dept. Cleve. Met. Gen. Hosp., 1963-84; asst. prof. emeritus Case-Western Res. U., Cleve. Mem. AMA, Ohio Med. Assn., Acad. Medicine, No. Ohio Pediatric Soc., Am. Women's Med. Assn., Am. Med. Polit. Action Com. Lutheran. Address: 5460 Friar Cir Cleveland OH 44126

BESCH, EVERETT DICKMAN, veterinarian, university dean emeritus; b. Hammond, Ind., May 4, 1924; s. Ernst Henry and Carolyn (Dieckmann) B.; m. Mellie Darnell Brockman, Apr. 3, 1946; children: Carolyn Darnell, Ceryl Lynn, Cynthia Lee, Charlotte Ann, Everett Dickman. D.V.M., Tex. A&M Coll., 1954; M.P.H., U. Minn., 1956; Ph.D., Okla. State U., 1963. Instr. U. Minn., 1954-56; asst. prof. Okla. State U., 1956-64, prof., head dept. vet. parasitology and pub. health, 1964-68; dean Sch. Vet. Medicine, La. State U., 1968-88, prof., 1988-89; sec.-treas. Assn. Am. Vet. Med. Colls., 1974-78, sec. coun. deans, 1976-80, chmn. coun. deans, 1980-81; mem. Nat. Adv. Coun. Health Professions Edn., 1982-86; treas. Am. Vet. Med. Assn. Found., 1991-93, v.p., 1993—. Author articles, chpt. in book. Served with USN, 1942-48. Mem. AVMA (ho. of dels. 1988-91, mem. exec. bd. 1991—), Assn. Tchrs. Vet. Pub. Health and Preventive Medicine (pres. 1968), Am. Vet. Med. Assn. (exec. bd. 1991—), La. Vet. Med. Assn., Am. Soc. Parasitologists, Conf. Pub. Health Vet. (pres. 1971-72), Am. Assn. Food Hygiene Vet. (pres. 1976-77), Am. Assn. Vet. Parasitologists (pres. 1964-65). Research interests: arthropod vectors of disease, internal parasites of ruminants. Home: 1453 Ashland Dr Baton Rouge LA 70806-7838

BESCH, HENRY ROLAND, JR., pharmacologist, educator; b. San Antonio, Sept. 12, 1942; s. Henry Roland and Monette Helen (Kasten) B.; m. Frankie R. Drejer; 1 child, Kurt Theodore. B.Sc. in Physiology, Ohio State U., 1964, Ph.D. in Pharmacology (USPHS predoctoral trainee 1964-67), 1967; USPHS postdoctoral trainee, Baylor U. Coll. Medicine, Houston, 1968-70. Instr. ob-gyn. Ohio State U. Med. Sch., Columbus, 1967-68; asst. prof. Ind. U. Sch. Medicine, Indpls., 1971-73, assoc. prof., 1973-77, prof., 1977, Showalter prof. and chmn. pharmacology and toxicology, 1977—; dir. Ind. State Dept. Toxicology, Indpls., 1991—; Can. Med. Research Council vis. prof., 1979, investigator fed. grants, mem. nat. panels and coms., cons. in field. Contbr. numerous articles pharm. and med. jours.; mem. editorial bds. profl. jours. Fellow Brit. Med. Research Council, 1970-71; Grantee Showalter Trust, 1975—. Fellow Am. Coll. Cardiology; mem. AAAS, Am. Assn. Clin. Chemistry, Am. Fed. Clin. Rsch., Am. Heart Assn., Am. Physiol. Soc., Am. Soc. Biochem. Molecular Biol., Am. Soc. Pharmacology and Exptl. Therapeutics, Assn. Med. Sch. Pharmacologists, Biochem. Soc., Cardiac Muscle Soc., Internat. Soc. Heart Rsch. (exec. com. Am. sect.), Nat. Acad. Clin. Biochemistry, N.Y. Acad. Scis., Sigma Xi. Office: Ind U Sch Medicine 635 Barnhill Dr Indianapolis IN 46202-5126

BESING, JOAN MARIE, audiology and hearing science educator; b. Albuquerque, Jan. 9, 1955; d. Carroll Earl and Mary Marcella (O'Laughlin) B. BA, U. Iowa, 1977; MS, Ill. State U., 1979; PhD, La. State U., 1988. Cert. clin. competence in audiology. Clin. audiologist Ctr. for Craniofacial Anomalies U. Ill., Chgo., 1979-81; clin. audiologist W.G. Theilemann, M.D., Bloomington, Ill., 1981-84; postdoctoral fellow MIT, Cambridge, 1988-91; asst. prof. La. State U., Baton Rouge, 1991—; instr., clin. supr. Ill. State U., Normal, 1981-84. Contbr. articles to Jour. Acoustical Soc. Am. Mem. pet therapy program Tiger Hats, Baton Rouge, 1992—. Grantee NIMH, 1987, Sertoma Club, 1992, 93, Deafness Rsch. Found., 1992. Mem. Am. Speech, Lang. and Hearing Assn., Am. Audiotory Soc., Acoustical Soc. Am., La. Speech, Lang. and Hearing Assn. Office: La State U Dept Comm Scis & Disorders 163 M&DA Bldg Baton Rouge LA 70803

BEST, MELVYN EDWARD, geophysicist, researcher; b. Victoria, B.C., Can., Mar. 8, 1941; s. Herbert and Irene Jessie (Kelly) B.; m. Virginia Marie Pignato, July 19, 1970; children: Lisette Anne, Aaron Michael. BSc, U. B.C., 1965, MSc, 1966; PhD, MIT, 1970. Rsch. assoc. dept. physics McGill U., Montreal, Que., 1970-72; geophysicist minerals Shell Can. Ltd., Calgary, Alta., 1972-78; head non-seismic rsch. Royal Dutch Shell Exploration & Prodn. Lab., Rijswijh, The Netherlands, 1978-80; divn. geophysicist minerals Shell Can. Ltd., Calgary, 1980-82, mgr. petroleum engring. rsch., 1982-85; geophys. advisor Teknica Resource Devel. Ltd., Calgary, 1985-86; head basin analysis subdivsn. Atlantic Geosci. Ctr., Geol. Survey of Can., Dartmouth, N.S., 1986-90; dir. Pacific Geosci. Ctr., Geol. Survey of Can., Sidney, B.C., 1990—; mem. oil and gas com. Can.-Newfoundland Offshore Petroleum Bd., Calgary, Alta., 1990-93; mem. sci. com. Geol. Survey Can., Ottawa, Ont., 1990—. Author: Resistivity Mapping and Electromagnetic Imaging, 1992; editor: A Geophysical Handbook for Geologists, 1990; editor Can. Jour. Exploration Geophysics, 1993—. Com. mem. Resource Assessment Panel (Hydrocarbon) Jeanne d'Arc Basin, Ottawa, 1980-90; com. mem. Panel of Energy, Rsch. and Devel. (Petroleum Geology), Calgary, 1987-92. Mem. Can. Soc. Exploration Geophysicists (exec. com.), Soc. Exploration Geophysicists (geophys. rsch. com. 1989—), Assn. Profl. Engrs. Geologists & Geophysicists of Alta. Achievements include development of Sweepem Em (Airborne) System. Home: 5288 Cordova Bay Rd, Victoria, BC Canada V8Y 2L4 Office: Pacific Geoscience Ctr, Box 6000, Sidney, BC Canada V8L 4B2

BETANCOURT, HECTOR MAINHARD, psychology scientist, educator; b. Chile, Sept. 1, 1949; came to U.S. 1979; s. Hector and Eleonora (Mainhard) B.; m. Bernardita Sahli; children: Paul, Daniel. BA, Cath. U., Santiago, Chile, 1976; MA, UCLA, 1981, PhD in Psychology, 1983. From asst. prof. to assoc. prof. psychology Cath. U., Santiago, Chile, 1977-79, 83-85; assoc. prof. psychology Loma Linda U., Riverside, Calif., 1985-91; prof. psychology, 1991-93; chmn., dept. psychology Loma Linda (Calif.) U., 1990-93, chmn. dept. psychology grad. sch., 1993—. Editor Interam. Psychologist, 1962-86; mem. editorial bd. Jour. Community Psychology, 1986-89, Spanish Jour. Social Psychology, 1986—; contbr. articles to profl. jours. Recipient Rotary Found. award for Internat. Understanding, Rotary Internat., 1976-77; Fulbright fellow, UCLA, 1979-80. Mem. APA, Internat. Soc. Polit. Psychology, Internat. Soc. Cross-Cultural Psychology (exec. com. 1984-86), Interam. Soc. Psychology (sec. gen. 1983-87), Am. Psychol. Soc., Soc. for Psychol. Study Social Issues, Soc. Personality and Social Psychology. Avocations: internat. politics, literature, philosophy. Office: Loma Linda U Grad Sch Loma Linda CA 92350

BETHE, HANS ALBRECHT, physicist, educator; b. Strassburg, Alsace-Lorraine, Germany, July 2, 1906; came to U.S., 1935; s. Albrecht Theodore and Anna (Kuhn) B.; m. Rose Ewald, 1939; children: Henry, Monica. Ed. Goethe Gymnasium, Frankfurt on Main, U. Frankfurt; Ph.D., U. Munich,

1928; D.Sc., Bklyn. Poly. Inst., 1950, U. Denver, 1952, U. Chgo., 1953, U. Birmingham, 1956, Harvard U., 1958. Instr. in theoretical physics univs. of Frankfort, Stuttgart, Munich and Tubingen, 1928-33; lectr. univs. of Manchester and Bristol, Eng., 1933-35; asst. prof. Cornell U., 1935, prof., 1937-75, prof. emeritus, 1975—; dir. theoretical physics div. Los Alamos Sci. Lab., 1943-46; Mem. Presdl. Study Disarmament, 1958; mem. Pres.'s Sci. Adv. Com., 1956-60. Author: Mesons and Fields, 1953, Elementary Nuclear Theory, 1957, Quantum Mechanics of One-and Two-Electron Atoms, 1957, Intermediate Quantum Mechanics, 1964; Contbr. to: Handbuch der Physik, 1933, Reviews of Modern Physics, 1936-37, Phys. Rev. Recipient A. Cressy Morrison prize N.Y. Acad. Sci., 1938-40; Presdl. Medal of Merit, 1946; Max Planck medal, 1953; Enrico Fermi award AEC, 1961; Nobel Prize in physics, 1967; Nat. Medal of Sci., 1976; Vannevar Bush award NSF, 1985, Einstein Peace prize Albert Einstein Peace Prize Found., 1993. Fgn. mem. Royal Soc. London; mem. Am. Philos. Soc., NAS (Henry Draper medal 1968), Am. Phys. Soc. (pres. 1954), Am. Astron. Soc. Office: Cornell U Nuclear Studies Lab Ithaca NY 14853

BETHJE, ROBERT, general surgeon, retired; b. Braunschweig, Fed. Republic of Germany, Nov. 15, 1922; came to U.S., 1923; s. Robert Paul and Elisabeth Augusta (Lieder) B.; m. Maria Vatral, June 11, 1955; children: Susan Leslie, Robert Eric, Alan Randolph. BS, CUNY, 1945; MD, N.Y. Med. Coll., 1949. Diplomate Nat. Bd., 1950, Am. Bd. Surgery, 1958. Asst. treas. Broome County Med. Soc., Binghamton, N.Y., 1964, v.p., 1965, pres., 1966; pres. med. staff Ideal Hosp., Endicott, N.Y., 1973-76, chief of surgery, 1971-77; chief of surgery Wilson Meml. Hosp., Johnson City, N.Y., 1979-80. Bd. dirs. Broome-Tioga Assn. for Retarded Children, Binghamton, 1983—. Capt. U.S. Army Med. Corps, 1951-53. Fellow Am. Coll. Surgeons; mem. Rotary (Endicott v.p. 1980-81, dir. 1981-84, pres. 1985-86). Avocations: painting, photography, fishing, hunting, gardening. Home: 4 Ivanhoe Rd Binghamton NY 13903-1424

BETINIS, EMANUEL JAMES, physics and mathematics educator; b. Oak Park, Ill., Oct. 31, 1927; s. James Emanuel and Ioanna Helen (Kallas) B.; children: Demetrios, Joanna, Markos. BS in Chemistry and Math., Northwestern U., 1950; MS in Applied Math., U. Ill., 1952; MS in Physics, U. Chgo., 1979. Aerodynamicist Northrop Aviation, Hawthorne, Calif., 1953-54; theoretical reactor physicist Atomics Internat., Canoga Park, Calif., 1954-57; applied sci. rep. IBM, Chgo., 1957-61; math. cons. Math. Cons. Svc., Chgo., 1961-81; adj. prof. math. and physics IIT, Roosevelt U., Chgo., 1981-88; mathematician Batelle Meml. Labs., Willowbrook, Ill., 1988-89; asst. prof. physics Elmhurst (Ill.) Coll., 1990—. Contbr. articles to Jour. Geophysical Rsch., Jour. Brit. Interplanetary Soc., Hadronic Jour., Matrix, Lensor Soc. Great Britain. Mem. PTO. With U.S. Army, 1946-47. Fellow Brit. Interplanetary Soc.; mem. Am. Nuclear Soc., Sigma Pi Sigma, Pi Mu Epsilon. Republican. Orthodox. Achievements include patent in golf ball trajectory with lift and drag; research in analytic solution of boundary-value problems in arbitrary geometry, special relativity, quantum mechanical proof of speed of light limitation, analytic solution of 3 dimensional heat conduction equation in arbitrary geometry, nuclear potential and prediction of 470MeV elementary particles, analytic solution of non-linear hydrodynamics equations. Office: Elmhurst Coll Dept Physics Box 47 190 Prospect Ave Elmhurst IL 60126-3296

BETLACH, MARY CAROLYN, biochemist, molecular biology researcher; b. Madison, Wis., June 12, 1945; d. William Thompson Stafford and Carolyn Jesse Gillette McCormick; m. Charles J. Betlach, Nov. 14, 1970 (div. 1978); children: John F., Melanie Carolyn. Student, U. Wis., 1963-68; PhD, U. Calif., San Francisco, 1992. Staff rsch. assoc. dept. pathology and biochemistry U. Wis., Madison, 1967-69; staff rsch. assoc. dept. biol. scis. U. Calif., Santa Barbara, 1969-70; staff rsch. assoc. dept. pediatrics U. Calif., San Francisco, 1970-72, staff rsch. assoc. dept. microbiology/biochemistry, 1972-83, rsch. specialist dept. biochemistry, 1983—; cons. Codon, Inc., Brisbane, Calif., 1984; mem. various grant rev. panels. Contbr. chpts. to books, articles to Gene, Microbiology, Nucleic Acids Rsch., Biochemistry, Jour. Bacteriology, others. NIH grantee, 1983—; Am. Vaccine Inc. grantee, 1993. Mem. AAAS, Biophys. Soc., Am. Soc. for Microbiology. Achievements include patent pending on high level expression system for heterologous membrane proteins in halobacteria; development of recombinant DNA technology and early cloning vectors. Office: U Calif Dept Biochemistry San Francisco CA 94143

BETTA, MARY BETH, engineer; b. Bloomfield, N.J., Aug. 19, 1961. MS, Rutgers U., 1987. Engr. IMD Picatinny Arsenal, Dover, N.J., 1988—; cons. IMD Picatinny Arsenal, Dover, N.J., 1987-88. Office: IMD Picatinny Arsenal Dover NJ 07806-5000

BETTI, RAIMONDO, civil engineering educator; b. Rome, Oct. 22, 1959; came to U.S. 1986; s. Renato and Anna Maria (Fais) B.; m. Karen Lisa Whale, Oct. 26, 1991. Laurea, U. Rome, 1985; MSCE, U. So. Calif., 1988, PhD, 1991. Rsch. assocs. U. Rome, 1985-86; asst. prof. Columbia U., N.Y.C., 1991--. Mem. Am. Soc. Civil Engrs., Earthquake Engring. Rsch. Inst., Sigma Xi. Office: Columbia U 640 SW Mudd Bldg New York NY 10027

BETTINGER, JEFFRIE ALLEN, chemical engineer; b. Detroit, Apr. 4, 1954; s. Robert Richard and Genevieve Elizabeth (Maletz) B.; m. Nancy Anne Swanberg, Aug. 9, 1980; 1 child, Jonathan Robert. BS in Chem. Engring., Mich. Tech. Univ., 1982. Registered profl. engr., Mass. Process engr. Stone and Webster Engring. Corp., Boston, 1982-86, mgr. bus. devel., 1986—. Contbr. articles to profl. jours.; author tech. reports. Chmn. Water Protection Task Force, Cohasset, Mass., 1984, 85; mem. Conservation Commn., Cohasset, 1985; pres. Friends of Webb State Park, Weymouth, Mass., 1990, 91, 93. With USN, 1972-77. Mem. AICE, Soc. Mil. Engrs. (tech. advancement com. 1991—), Cohasset Sailing Club (trustee, treas. 1984-91, SVC. award 1992). Democrat. Methodist. Achievements include participation in passage of watershed protection and hazardous material bylaws in Town of Cohasset. Office: Stone and Webster Engring 245 Summer St Boston MA 02210

BETTS, ALAN KEITH, atmospheric scientist; b. Southend, Eng., Sept. 10, 1945; came to U.S., 1970; s. Ronald Frank Henry and Jean Mary (Clegg) B.; m. Karen James, Apr., 1993; children by previous marriage: Elizabeth, Heather. BA, MA, Cambridge U., Eng., 1967; PhD, London U., 1970. Assoc. prof. Colo. State U., Ft. Collins, 1970-78; pvt. practice atmospheric rsch. Pittsford, Vt., 1978—. Contbr. articles to profl. jours. Fellow Royal Meteorol. Soc. (L.F. Richardson prize 1973); mem. Am. Meteorol. Soc. (spl. award 1977, Meisinger award 1978). Office: Atmospheric Rsch RR 3 Box 3125 Pittsford VT 05763

BETTS, BURR JOSEPH, biology educator; b. Denver, May 5, 1945; s. Charles Julius and Virginia (Dare) B.; m. Donna Lea Palmer, Sept. 14, 1968; children: Amy Jeanette, Erika Anne. BS, Purdue U., 1967; PhD, U. Mont., 1973. Prof. Ea. Oreg. State Coll., La Grande, 1975—; dir. Wallowa Mountain Field Sta., La Grande, 1977-85, George Ott Wildlife Rsch. Area, La Grande, 1983—. Author: (jour.) Northwestern Naturalist, 1990, '91, The Wilson Bulletin, 1991. Mem. budget com. La Grande Sch. Dist., 1978-81; bd. mem. Union County Ednl. Svc. Dist., 1981—. Recipient Predoctoral fellowship NSF, 1967-70, 1970-72; grantee Oreg. Nongame Wildlife Fund, 1984. Mem. N.W. Sci. Assn. (trustee 1989-94, pres. 1992-93), Soc. for Northwestern Vertebrate Biology, Animal Behavior Soc., Sigma Xi. Home: 2112 Second St La Grande OR 97850 Office: Ea Oreg State Coll 1410 L Ave La Grande OR 97850

BETTS, DOUGLAS NORMAN, civil engineer; b. Watopa Twp., Minn., Oct. 26, 1932; s. Everet Ernest and Hazel Julia Edel (Erickson) B.; m. Lillian Eva Von Osten, June 23, 1957; children: Kay Marie, Karen Ann, Ann Marie. AA, Rochester Jr. Coll., 1954; B in Civil Engring., U. Minn., 1957. Registered profl. engr., Minn., Mont., Iowa, S.D. Project engr. K.M. McGhie, Rochester, Minn., 1951-59, Wenzel & Co., Great Falls, Mont., 1959-65; v.p. McGhie & Betts, Inc., Rochester, 1966-72, pres., 1972-88, cons., v.p., 1988--. Mem. MARLS, NSPE, ASCE, Minn. Soc. Profl. Surveyors (pres. 1980-81), Minn. Soc. Profl. Engr. (chpt. pres.). Lutheran. Home: PO Box 3073 Bozeman MT 59772-3073 Office: McGhie & Betts Inc 1648 Third Ave SE Rochester MN 55904

BETTS, JAMES GORDON, biology educator; b. Atlanta, Dec. 14, 1954; s. W.M. and Annie Lee (Hobbs) B.; m. Janelle Marie Keith, Dec. 15, 1983; 1 child, Hutson Keith. BS, Stephen F. Austin State U., 1981; MS, Tex. A&M U., 1983, PhD, 1987. Postdoctoral assoc. Med. U. S.C., Charleston, 1987-89, U. Fla., Gainesville, 1989-91; asst. prof. biology East Tex. State U., Commerce, 1991—. With USN, 1975-79. Mem. Lions (bd. dirs. 1992). Democrat. Methodist. Office: East Tex State U Biology Dept Commerce TX 75429

BETZ, NORMAN L., science educator, consultant; b. Baton Rouge, Jan. 23, 1938; s. Louis F. and Doretta M. (Vallee) B.; m. Betty Jane Jarreau, Nov. 19, 1960; children: Stephanie, Sherrel, Sondra, Norman A., Sharise. BS, La. State U., 1961, MS, 1963, PhD, 1966. Investigator La. Petroleum Refiners Waste Control, Baton Rouge, 1956-62, USDA, Baton Rouge, 1962-66; scientist Mallinckrodt Chem. Co., St. Louis, 1968-72; sr. scientist Ralston Purina Co., St. Louis, 1972-85; pvt. cons. Technol. Resources, Ltd., St. Louis, 1985-90; prof., Parker Chair of Excellence U. Tenn., Martin, 1990—; dir. BNB Tec S.A. de C.V., Monterrey, Mex., 1985—; ptnr. BNB Techs., Inc., St. Louis, 1985—; pres. Bio-Aide, Inc., St. Louis, 1987—. Author: Nutrient Help for Children With Lymphoblastic Leukemia, 1991; contbr. Presentation of Leukema Work jour. Active St. Jude Cath. Ch., 1990—. Mem. Am. Assn. Cereal Chemists (chmn. 1982-83, Geddes Meml. award 1980). Republican. Achievements include 14 patents in the areas of taste smell, and appetite control. Home: Rt 1 Box 670C Dresden TN 38225

BEUBE, FRANK EDWARD, periodontist, educator; b. Kingston, Ont., Can., July 1, 1904; came to the U.S., 1930; naturalized, 1937; s. Gabriel and Fannie Bessie (Florence) B. L.D.S., D.D.S., U. Toronto, 1930. Diplomate Am. Bd. Periodontology; m. Edith Schweitzer, Oct. 5, 1930; children: Eric, Stephen. Clin. asst. div. periodontology Sch. Dental and Oral Surgery, Columbia, 1930-37, instr., 1937-41, asst. prof., 1941-46, assoc. prof., 1946-53, head div., 1948-68, clin. prof. dentistry, 1953-84, clin. prof. emeritus, 1984—, emeritus prof.-spl. lectr., 1984—; head dept. periodontology Presbyn. Hosp., N.Y.C., 1941-70; lectr. dept. periodontology, Dental Sch. N.Y. U., 1973—; found. mem. Hebrew U. Recipient William J. Gies award, 1979, Disting. Alumnus award Columbia U. Periodontal Alumni Assn., 1984, Isidore Herschfeld award, 1990. Fellow AAAS, Am. Coll. Dentists, Am. Acad. Periodontology (councilman 1962, chmn. edn. com. 1963, chmn. com. on coms. 1964, pres. 1964-65, chmn. exec. council 1965-66, Pres. award 1988); mem. ADA (chmn. periodontia sect. 1964-65), Western Soc. Periodontology (hon.), Acad. Oral Pathology, So. Acad. Periodontology (hon.), Internat. Assn. Dental Rsch., First Dist. Dental Soc. (past pres. pathodontia sect.), Sigma Xi. Author: Periodontology: Diagnosis and Treatment, 1953; Prevention of Periodontal Diseases, 1956; Gingivectomy in the treatment of Periodontal Diseases, 1957; Disadvantages of Surgical Techniques, 1960; contbr. chpts. to books, articles to dental jours. Rsch. in study of healing of cementum and bone, periodontal diseases and their treatment. Home: 701 Pelham Rd New Rochelle NY 10805 Office: 933 5th Ave New York NY 10021-2603

BEUTLER, LARRY EDWARD, psychology educator; b. Logan, Utah, Feb. 14, 1941; s. Edward and Beulah (Andrus) B.; m. M. Elena Oró, Feb. 25, 1977; children: Jana, Kelly, Ian David, Gail. BS, Utah State U., 1965, MS, 1966; PhD, U. Nebr., 1970. Diplomate Am. Bd. Clin. Psychology. Asst. prof. psychology Duke U., Ashville, N.C., 1970-71; asst. prof. Stephen F. Austin State U., Nacogdoches, Tex., 1971-73; assoc. prof. Baylor Coll. Medicine, Houston, 1973-79; prof. U. Ariz., Tucson, 1979-90, U. Calif., Santa Barbara, 1990—. Author: Eclectic Psychotherapy, 1983; editor Jour. Cons. Clin. Psychology, 1990—. Fellow Am. Psychology Assn., Internat. Acad. Eclectic Psychotherapy; mem. Soc. Psychotherapy Research (pres. 1986-88). Home: 7602 Hollister # 301 Goleta CA 93117 Office: U Calif Santa Barbara GSE Santa Barbara CA 93106

BEVINS, ROBERT JACKSON, agricultural economics educator; b. Concord, Tenn., Mar. 13, 1928; s. Samuel Robert and Elizabeth Leucretia (Jackson) B.; m. Priscilla Ruth LeBaron, Nov. 5, 1955; children: Colleen, Brian. BS, U. Tenn., 1949, MS, 1955; PhD, Mich. State U., 1960. Tchr. agriculture Vevay, Ind., 1949-50; ind. farmer Concord, 1950-51; agrl. specialist U. Tenn., Jackson, 1955-56; asst. prof. agrl. econs. Mich. State U., East Lansing, 1960-61, Kans. State U., Manhattan, 1961-67; prof. U. Mo., Columbia, 1967—. Sgt. U.S. Army, 1951-54, Korea. Anglican. Office: U Mo Dept Agrl Econs 316 Mumford Hall UMC Columbia MO 65211

BEYENE, WENDEMAGEGNEHU TSEGAYE, electrical engineer; b. Addis Ababa, Showa, Ethiopia, Aug. 27, 1963; came to U.S. 1984; s. Tsegaye and Elifinesh Gebretsadik Beyene. BS, Columbia U., 1988, MS, 1991. Undergrad. rsch. asst. Microelectronics Lab., Columbia U., N.Y.C., 1986-88; engr. IBM Corp., Hopewell Junction, N.Y., 1988—. Mem. IEEE, Internat. Soc. Hybrid Microelectronics, Tau Beta Pi, Eta Kappa Nu. Orthodox Christian Ch. Home: 135 Violet Ave Poughkeepsie NY 12601 Office: IBM Corp 1580 Rt 52 Hopewell Junction NY 12533

BEYER-MEARS, ANNETTE, physiologist; b. Madison, Wis., May 26, 1941; d. Karl and Annette (Weiss) Beyer. B.A., Vassar Coll., 1963; M.S., Fairleigh Dickinson U., 1973; Ph.D., Coll. Medicine and Dentistry N.J., 1977. NIH fellow Cornell U. Med. Sch., 1963-65; instr. physiology Springside Sch., Phila., 1967-71; teaching asst. dept. physiology Coll. Medicine & Dentistry N.J., N.J. Med. Sch., 1974-77, NIH fellow dept. ophthalmology, 1978-80; asst. prof. dept. ophthalmology U. Medicine and Dentistry N.J., N.J. Med. Sch., Newark, 1979-85, asst. prof. dept. physiology, 1980-85, assoc. prof. dept. physiology, 1986—, assoc. prof. dept. ophthalmology, 1986—; cons. Alcon Labs. Contbr. articles in field of diabetic lens and kidney therapy to profl. jours. Chmn. admissions No. N.J., Vassar Coll., 1974-79; mem. minister search com. St. Bartholomew Episcopal Ch., N.J., 1978, fund-raising chmn., 1978, 79; del. Episc. Diocesian Conv., 1977, 78; long range planning com. Christ Ch., Ridgewood, N.J., 1985-87. Recipient NIH Nat. Rsch. Svc. award, 1978-80, Found. CMDNJ Rsch. award, 1980; grantee Juvenile Diabetes Found., 1985-87, NIH, NEI grantee, 1980—, Pfizer, Inc. grantee, 1985-87. Mem. Am. Physiol. Soc., N.Y. Acad. Scis., Soc. for Neurosci., Am. Soc. Pharmacology and Exptl. Therapeutics, Assn. for Rsch. Vision & Ophthalmology, Internat. Soc. for Eye Research, AAAS, The Royal Soc. Medicine, Internat. Diabetes Found., Am. Diabetes Assn., Aircraft Owners and Pilots Assn., Civil Air Patrol, Sigma Xi. Office: NJ Med Sch Dept Physiology 185 S Orange Ave Newark NJ 07103-2714

BEYLKIN, GREGORY, mathematician; b. St. Petersburg, USSR, Mar. 16, 1953; came to U.S., 1980; naturalized citizen, 1985; s. Jacob and Raya (Pripshtein) B.; m. Helen Simontov, 1974; children: Michael, Daniel. Diploma in Math., U. St. Petersburg, Leningrad, 1975; PhD in Math., NYU, 1982. Assoc. rsch. sci. NYU, 1982-83; mem. profl. staff Schlumberger-Doll Research, Ridgefield, Conn., 1983-91; prof. program in applied math. U. Colo., Boulder, 1991—. Contbr. articles to profl. jours. Mem. Am. Math. Soc., Soc. for Indsl. and Applied Math., Soc. Exptl. Geophysicists. Home: 2738 Winding Trail Pl Boulder CO 80304 Office: Program in Applied Math Univ Colo at Boulder University Of Colorado CO 80309-0526

BEYROUTY, CRAIG A., agronomist, educator. Prof. dept. agronomy U. Ark., Fayetteville. Recipient CIBA-GEIGY Agronomy award Am. Soc. Agronomy, 1991. Office: Univ of Arkansas Dept of Agronomy Maple St Fayetteville AR 72701*

BEYSTER, JOHN ROBERT, engineering company executive; b. Detroit, July 26, 1924; s. John Frederick and Lillian Edith (Jondro) B.; m. Betty Jean Brock, Sept. 8, 1955; children: James Frederick, Mark Daneil, Mary Ann. B.S. in Engring., U. Mich., 1945, M.S., 1948, Ph.D., 1950. Registered profl. engr., Calif. Mem. staff Los Alamos Sci. Lab., 1951-56; chmn. dept. accel. physics Gulf Gen. Atomic Co., San Diego, 1957-69; pres., chmn. bd. Sci. Applications, Inc., La Jolla, Calif., from 1969, now chmn. bd., chief exec. officer; mem. Joint Strategic Target Planning Staff, Sci. Adv. Group, Omaha, 1978—; panel mem. Nat. Measurement Lab. Evaluation panel for Radiation Research, Washington, 1983—; dir. Scripps Bancorp, La Jolla, 1983. Co-author: Slow Neutron Scattering and Thermalization, 1970. Served to lt. comdr. USN, 1943-46. Fellow Am. Nuclear Soc., Am. Phys. Soc.; mem. NAE. Republican. Roman Catholic. Home: 9321 La Jolla Farms Rd La Jolla CA 92037-1126 Office: Science Applications Inter Corp 10260 Campus Point Dr San Diego CA 92121-1522

BEZDEK, HUGO FRANK, scientific laboratory administrator; b. Washington, Feb. 28, 1936; s. Hugo F. and Louise Bezdek. B.S. in Physics, N.Mex. State U., 1965; Ph.D. in Physics, U. Colo., 1970. Research scientist Scripps Inst. Oceanography, La Jolla, Calif., 1970-74; dep. dir. Ocean Sci. Research Office Devel., Arlington, Va., 1974-80; dir. Atlantic Oceanographic and Meteorol. Labs., Miami, Fla., 1980—. Contbr. numerous articles to profl. jours. Mem. Acoustical Soc. Am., Am. Geophys. Union. Office: Atlantic Oceanographic and Meteorol Labs 4301 Rickenbacker Causeway Virginia Key Miami FL 33149

BEZOARI, MASSIMO DANIEL, chemistry educator, writer; b. Glasgow, Scotland, Sept. 11, 1952; came to U.S., 1975; s. Ubaldo Indo and Marta (Agresti) B.; m. Charlotte Anne Pope, May 30, 1980 (div. Dec. 1992); 1 child, Marco Daniel. BSc, U. Glasgow, 1974; PhD, U. Ala., Tuscaloosa, 1981. Postdoctoral fellow U. Wis., Milw., 1980-81; sr. rsch. chemist Dow Chem. USA, Baton Rouge, 1981-85, project leader, 1985-88; asst. prof. chemistry Coker Coll., Hartsville, S.C., 1988-89, assoc. prof. chemistry, 1989-90; assoc. prof. chemistry Livingston (Ala.) U., 1990-93; summer rsch. fellow U. S.C., Columbia, 1991; assoc. prof. chemistry Huntingdon (Ala.) Coll., 1993—. Author: General Chemistry Lab. Practice, 1st and 2d edits. 1991, 3d edit., 1992, (rev. chpts.) Magill's Survey of Science-Physical Science Series, 1991, Magill's Survey of Science-Life Science Series, 1991, Nobel Prize Winners in Chemistry, 1990, Great Lives from History-20th Century Chemists, 1990, "Cyclophanes", Organic Chemistry-A Series of Monographs, 1983; contbr. articles to profl. jours. Sci. fair judge Dow Chem. USA, Baton Rouge, 1985-88, Livingston U., 1991. Grad. Coun. Rsch. fellow U. Ala., 1977-80; Andrew Carnegie Trust Fund scholar, 1971-74, Dean's scholar, 1976-77. Fellow Am. Inst. Chemists; mem. Am. Chem. Soc., Ala. Acad. Sci. Achievements include patents for anionic polymerization, for organophosphazenes, for chemistry of chlorine-containing polymers. Home: 2617 The Meadows Montgomery AL 36116-1129 Office: Huntingdon Coll 1500 E Fairview Ave Montgomery AL 36106-2148

BHAGAT, ASHOK KUMAR, chemical engineer; b. Dehradun, India, Sept. 22, 1942; came to U.S., 1964; s. Shanti Swaroop (Sahgal) and Saroj B.; m. Susheila Raghavan, Apr. 16, 1975; 1 child, Vikram R. MSChemE, U. Calif., Berkeley, 1966; MS in Mgmt., Renssalaer (N.Y.) Poly. Inst., 1971. Process engr. GE Co., Schenectady, N.Y., 1968-71; cons. Condoctoras Monterey, Mex., 1971-72; project mgr. I.C.I. Ams., Wilmington, Del., 1973—. Mem. AICE. Home: 489 Stella Dr Hockessin DE 19707 Office: ICI Americas Concord Plz Wilmington DE 19897

BHAGAT, PHIROZ MANECK, mechanical engineer; b. Poona, India, Oct. 28, 1948; came to U.S., 1970; s. Maneck Phirozshaw and Khorshed Eduljee (Batliwala) B.; m. Patricia Jane Steckler, Oct. 13, 1979; children: Kay, Sarah. B.Tech., Indian Inst. Tech.-Bombay, 1970; M.S. in Engring., U. Mich., 1971, Ph.D., 1975. Rsch. fellow in applied mechanics Harvard U., Cambridge, Mass., 1975-77; asst. prof. engring. Columbia U., N.Y.C., 1977-81, adj. asst. prof., 1981-84; staff engr. Exxon Rsch. & Engring. Co., Florham Park, N.J., 1981-83, sr. staff engr., 1983—, head sci. computing group, 1988-90; mng. dir. Janus Enterprise Internat., 1992—. Contbr. articles to profl. jours. K.C., Mahindra scholar, 1970, J.N. Tata scholar, 1970; Horace Rackham predoctoral fellow, 1973-74, 74-75. Mem. N.Y. Acad. Scis., Am. Inst. Chem. Engrs., ASME, Tau Beta Pi, Sigma Xi. Rsch. devel. of Neural Nets to engring. scis., pattern recognition and forecasting, on application of thermal scis. to model petrochemical processes, sci. computing, heat transfer, fluid mechanics, thermodynamics, computer modeling, business global computer simulation. Home: 519 Alden St Westfield NJ 07090-3040 Office: Exxon Rsch & Engring Co Florham Park NJ 07932

BHAGWAN, SUDHIR, computer industry and research executive, consultant; b. Lahore, West Pakistan, Aug. 9, 1942; came to U.S., 1963; s. Vishan and Lakshmi Devi (Arora) B.; m. Sarita Bahl, Oct. 25, 1969; children: Sonia, Sunil. BSEE, Punjab Engring. Coll., Chandigarh, India, 1963; MSEE, Stanford U., 1964; MBA with honors, Golden Gate U., 1977. Engr. Gaylor Products, North Hollywood, Calif., 1964-68, Burroughs Corp., Pasadena, Calif., 1968-70; engring. mgr. Burroughs Corp., Santa Barbara, Calif., 1970-78; engring. mgr. Intel Corp., Hillsboro, Oreg., 1978-81, chmn. strategic planning, 1981-82, gen. mgr., 1983-88; pres., exec. dir., bd. dirs. Oreg. Advanced Computing Inst., Beaverton, 1988-90; strategic bus. mgr. INTEL Corp., Hillsboro, Oreg., 1990-92, gen. mgr. bus. multimedia products, 1992—; speaker to high tech. industry, Oreg., 1988—; mem. organizing com. Distributed Memory Computing Conf., 1989-90, gen. chmn., 1990-91; chmn. computer tech. adv. bd. Oreg. Mus. Sci. and Industry, 1991—. Cons. Oreg. Econ. Devel. Dept., 1988—; bd. dirs. St. Mary's Acad., Portland, Oreg., 1989-92. Mem. Am. Electronics Assn. (higher edn. com. Oreg. chpt. 1989-90, exec. com. 1990). Avocations: electronics, photography, tennis, art. Home: 13940 NW Harvest Ln Portland OR 97229-3653 Office: INTEL Corp 5200 NE Elam Young Pky Hillsboro OR 97124-6497

BHALLA, DEEPAK KUMAR, cell biologist, toxicologist, educator; b. Kasauli, India, Aug. 31, 1946; s. Khazan Chand and Shyama Bhalla; m. Lilly Bhalla; 1 child, Neel. BS, Punjab U., India, 1968, MS, 1969; PhD, Howard U., Washington, 1976. Postdoctoral fellow Harvard U., Boston, 1976-79; asst. rsch. cell biologist U. Calif., San Francisco, 1979-82; asst. prof. U. Calif., Irvine, 1982-86, assoc. prof., 1986—; speaker in field. Contbr. articles and revs. to profl. jours. NIH grantee, 1985-88, 88—, Calif. Air Resources Bd. grantee, 1990—. Mem. AAAS, Am. Thoracic Soc., Am. Soc. for Cell Biology. Office: U Calif Community & Environ Medicine Irvine CA 92717

BHARGHAVA, VIJAY, engineer. Recipient John B. Sterling medal Engring. Inst. Can., 1990. Office: care Engring Inst of Canada, 202 280 Albert St, Ottawa, ON Canada K1P 5G8*

BHASKAR, RAMAMOORTHI, artifical intelligence scientist, researcher; b. Bangalore, Karnataka, India, Jan. 14, 1951; came to U.S., 1973; s. Appu Sastai and Mangalam Ramamoorthi; m. Mahzarin R. Banaji, Nov. 25, 1983. B in Engring., Nat. Inst. Engring., Mysore, India, 1972; PhD, Carnegie Mellon U., 1978. Tech. asst. Ministry Def., Govt. India, Bangalore, 1972-73; asst. prof. Ohio State U., 1977-82; mem. rsch. staff IBM Watson Rsch. Ctr., Yorktown Heights, N.Y., 1982—. Contbr. articles to sci. jours. Office: IBM Watson Rsch Ctr PO Box 218 Yorktown Heights NY 10598

BHAT, BAL KRISHEN, geneticist, plant breeder; b. Srinagar, India, May 3, 1940; came to U.S. 1989; s. Justice Janki Nath and Dhanwati (Kaul) B.; m. Sarla Kaul, Sept. 23, 1966; children: Arun Bhat, Anupama Bhat. MSc, Indian Agrl. Rsch. Inst., New Delhi, 1963; PhD, I.A.R.I., New Delhi, 1967. Rsch. assoc. Rockefeller Found., New Delhi, 1967; plant breeder in charge of rsch. Birla Inst. of Sci. Rsch., Rupar, Punjab, India, 1967-68; scientist "C" Reg. Rsch. Lab. Coun. of Sci. and Indsl. Rsch., Srinagar, India, 1968-74, head, 1972-79, 87-89, scientist "E I", 1974-79, scientist "E II", 1981-85, scientist "F" (dep. dir.), 1985-89; v.p., dir. rsch. Bot. Resources, Inc., Independence, Oreg., 1989—; rsch. fellow U. Tasmania, Hobart, Australia, 1979-81, sr. rsch. fellow, 1981-86; cons. in field. Contbr. over 100 articles to profl. jours. Named Scientist of the Yr., Reg. Rsch. Lab., Srinagar, 1976. Fellow Indian Soc. Genetics and Plant Breeding; mem. Am. Soc. Agronomy, Crop Sci. Soc. Am., Soc. for Advancement of Breeding Rsch. in Asia and Oceania, Coun. for Agrl. Sci. and Tech. Achievements include introduction and organization of commercial production of crops to new lands such as hops in India, pyrethrum in Australia, others; evolved a number of new cultivars in hops, pyrethrum, and medicinal and aromatic plants. Office: Bot Resources Inc 5465 Halls Ferry Rd Independence OR 97351-9616

BHAT, HARI KRISHEN, biochemical toxicologist; b. Srinagar, Kashmir, India, Jan. 15, 1954; came to U.S., 1986; s. Radha Krishen and Somavati (Bakshi) B.; m. Nimee Kotha, Nov. 4, 1981; children: Abhinav, Simi. MS, U. Kashmir, 1976; PhD, U. Tex., 1990. Sr. scientific asst. Coun. Scientific and Indsl. Rsch., Jammu, India, 1977-81, sr. scientist, 1981-86; rsch. assoc. U. Tex., Galveston, 1990. Contbr. articles to profl. jours. Recipient Reynolds Meml. award U. Tex. Med. Br., Galveston, 1989, 90, Elias Hochman award, 1989, 90; Kempner fellow, Galveston, 1991; Soc. Toxicology grantee, 1990. Mem. Soc. Toxicology, Am. Cancer Soc., Sigma Xi. Achievements include demonstration that environ. pollutants can interact with body lipids and be retained in the body for a long time and may contribute to long-term toxicity; rsch. interests include demonstrating molecular mechanisms in hormonal cancer. Office: U Tex Med Branch Dept Pharmacology 10th and Market Galveston TX 77555

BHATIA, DEEPAK HAZARILAL, engineering executive; b. Bareilly, India, Jan. 11, 1953; came to U.S., 1970; s. Hazarilal and Radharani Bhatia; m. Kokila Kapur, Feb. 9, 1980; children: Chirag, Ankur. BSME, Tenn. Tech. U., 1973, MSME, 1976. Engr. Keene Corp., Cookeville, Tenn., 1973; design engr. Sq. D. Co., Peru, Ind., 1976-79; sr. engr. nuclear qualification Wyle Labs., Huntsville, Ala., 1979-80; sr. engr., nuclear cons. NUS Corp., Clearwater, Fla., 1980-81; dir. EGS Corp., Hutnsville, Ala., 1982-86; supr. Impell Corp., Ft. Worth, 1987-89; pres., owner Rainbow Tech. Svcs., Ft. Worth, 1989—; equipment qualification specialsit, cons. Tuelectric Co.'s, Glen Rose, Tex., 1989-91. Contbr. articles to jours. Am. Nuclear Soc., ASME. Sec. Jaycees, Peru, Ind., 1977-78. Mem. Pi Tau Sigma, Tau Beta Pi. Achievements include research in optimizing equipment qualification programs, life extension and preventive maintenance during the life of plants; replacements of critical equipment. Home: 8004 Morning Ln Fort Worth TX 76123-1923 Office: Rainbow Tech Svcs 307 W 7th St Ste 1215 Fort Worth TX 76102

BHATTACHARYA, PALLAB KUMAR, electrical engineering educator, researcher; b. Calcutta, West Bengal, India, Dec. 6, 1949; came to U.S., 1978; s. Promode Ranjan and Sipra (Chatterjee) B.; m. Meena Mukerji, Aug. 11, 1975, children: Ramona, Monica. BSc with honors, U. Calcutta, 1968, B of Tech., 1970, M of Tech., 1971; M of Engring., U. Sheffield, Eng., 1976, PhD, 1978. Sr. rsch. asst. Radar and Communication Ctr., Kharagpur, India, 1972-73; asst. stores officer Hindustan Steel Ltd., Rourkela, India, 1973-75; asst., then assoc. prof. dept. elec. engring. Oreg. State U., Corvallis, 1978-83; assoc. prof. dept. elec. engring. and computer sci. U. Mich., Ann Arbor, 1984-87, prof., 1987—, dir. Solid State Electronics Lab., 1991—; invited prof. Swiss Fed. Inst. Tech., Lausanne, 1981-82. Contbr. articles to profl. jours. Fellow IEEE; mem. Am. Phys. Soc. Avocations: photography, music. Office: U Mich Dept Elec Engring & Comp Sci 2228 EECS Bldg Ann Arbor MI 48109-2122

BHATTACHERJEE, PARIMAL, pharmacologist; b. Chittagong, Bangladesh, Feb. 14, 1937; s. Charu Ranjan and Kusum Bhattacherjee; m. Pratima Choudhuri, June 18, 1962; children: Charbak, Vasker. BS, Chittagong Govt. Coll., 1955; MS, Dhaka U., Bangladesh, 1959; PhD, London U., 1972. Prodn. mgr., with quality control pharm. industry Bangladesh, 1960-67; research asst. Inst. Ophthalmology, London, 1967-72; lectr. pharmacology Inst. Ophthalmology, 1972-78; vis. asst. prof. ophthalmology Coll. Physicians and Surgeons, Columbia U., N.Y.C., 1978-79; sr. lectr. pharmacology Inst. Ophthalmology, London, 1978-79; sr. scientist Wellcome Research Labs., Beckenham, England, 1979-84; sect. head biologicals dept. pharmacology Wellcome Research Labs., 1984-87, sr. research scientist dept. mediator pharmacology, 1987-90; vis. prof. Ky. Lions Eye Research Inst., U. Louisville Med. Ctr., Louisville, 1990; prof., 1974-79; tchr., bd. studies pharmacology U. London, 1974-79; organizer Symposium on Ocular Inflammation, 5th Internat. Congress for Eye Research, Eindoven, Netherlands, 1982, 6th Congress for Eye Research, Alicante, Spain, 1984; organizer 8th Internat. Congress for Eye Research, San Francisco, 1988; organizer Symposiums on the Inflammatory Process and Mediators, 9th Internat. Congress for Eye Research, Helsinki, 1990, 92. Co-author: The Leukotrienes, 1984, Ocular Effects of Eicosanoids and Related Substances, 1989, Interleukin 1 Inflammation and Disease, 1989, Lipid Mediators in Eye Inflammation, 1989; contbr. articles to profl. jours. UNESCO fellow, Belgium, 1963, Alexander Pigott Wernher Meml. Trust fellow Med. Research Coun., England, 1972; research grantee Med. Research Coun., England. Mem. British Pharmacological Soc., Assn. for Eye Research (Europe), Internat. Soc. Eye Research, Assn. for Research in Vision & Ophthalmology, Experimenta Eye Research (editorial bd.), Soc. for Drug Research, Inst. of Biology. Hindu. Avocations: playing flute, classical music, playing squash. Office: U Louisville Ky Lions Eye Rsch Inst 301E E Muhammad Ali Blvd Louisville KY 40292-0001

BHATTI, IFTIKHAR HAMID, chiropractic educator; b. Lahore, Pahjab, Pakistan, Sept. 15, 1929; came to U.S., 1954; s. Hamid Khan and Akhtar (Hussain) B.; m. Sabahat Rafi, Dec. 17, 1961; children: Tanees, Tahir. BSc, U. Panjab, 1949, MSc, 1951; PhD, U. S.D., 1971. Prof. zoology Forman Christian Coll., Lahore, 1951-54, 57-61; biology instr. Swarthmore (Pa.) Coll., 1954-55, Haverford (Pa.) Coll., 1955-57; teaching fellow Woman's Med. Coll., Phila., 1961-63; biology prof. Buena Vista Coll., Storm Lake, Iowa, 1963-76; anatomy prof. Palmer Coll. Chiropractic, Davenport, Iowa, 1976—, chmn. dept., 1977-83, v.p. acad. affairs, 1988-91; dean of rsch. and grad. studies Palmer Coll. Chripractic, Davenport, Iowa, 1991—. Author: Human Embryology, 1988; contbr. articles to profl. publs. Fellow Duke U., 1969. Mem. Davenport C. of C., Davenport Club, Davenport Country Club. Avocations: reading, gardening, mechanics. Home: 1419 Tanglefoot Ln Bettendorf IA 52722-2413 Office: Palmer Coll Chiropractic 1000 Brady St Davenport IA 52803-5287

BHATTI, NEELOO, environmental scientist; b. New Delhi, Jan. 30, 1955; arrived in Can., 1958, came to U.S., 1982; d. Daljeet Singh and Abnash (Singh) B.; m. James Joseph McAndrew, Sept. 14, 1985. MES, Yale U., 1984, PhD, 1988. Rsch. asst. McGill U., Montreal, Que., Can., 1976-78; teaching asst. Yale U., New Haven, 1983; rsch. intern Cary Arboretum, Millbrook, N.Y., 1983; postdoctoral fellow Argonne (Ill.) Nat. Lab., 1989-90, energy and environ. scientist, 1990—; environ. effects specialist, cons. World Bank, Washington, 1989—. Author: Dispelling the North American Acid Rain Clouds, 1988, Responding to Threat of Global Warming: Options for Asia and Pacific, 1989. F.C.A.C. scholar Govt. Que., 1983-87; Yale U. fellow, New Haven, 1984-88. Mem. Am. Chem. Soc., Am Soc. Foresters, Sigma Xi. Achievements include identification of specific regions within Asia at highest risk from acid deposition; identified basis of pollution problems in Romania, Poland, and in Ankara, Turkey; contributed to NAPAP research; compiled emissions inventory for mercury in Great Lakes region and for SO2 in Asia; assisted in development of climate change strategy for government of People's Republic of China. Home: 15425 Purley Ct Lockport IL 60441 Office: Argonne Nat Lab 9700 Cass Ave Argonne IL 60439

BHAYANI, KIRAN LILACHAND, environmental engineer, programs manager; b. Bhavnagar, Gujarat, India, Dec. 2, 1944; came to U.S., 1968, naturalized; s. Lilachand Premchand and Rasila (Chhotalal Shah) B.; m. Chandra Vasantlal Gandhi, June 24, 1971; children: Nikhil K., Mihir K. B.Engring. with honors, U. Bombay, India, 1965, M.Engring., 1968; MS, U. R.I., 1970. Diplomate Am. Acad. Environ. Engrs.; registered profl. engr., Va., Ga., Utah. San. engr. Greeley & Hansen, N.Y.C., 1971-72, Hayes, Seay, Mattern & Mattern, Roanoke, Va., 1972-77; environ. engr. Hussey, Gay & Bell, Inc., Savannah, Ga., 1977-80; engring. mgr., Utah Div. Water Quality, Dept. Environ. Quality, Salt Lake City, 1980—; tech. transfer and sludge mgmt. coord., 1982—; mem. fair employment com. Dept. Health, Salt Lake City, 1982-90, adv. 1991—, chmn., 1988-89, cons., 1989-91; mem. Utah Engrs. Coun., 1989—, vice-chmn., 1992-93, chmn. 1993—; chmn. Engr's. Week, 1992; v.p. Gujarati Samaj of Utah. Reviewer (practice manual) Financing Sewer Projects, 1984; design of Municipal Wastewater Treatment Plants, 1990-91. Fellow ASCE (profl. coordination com. 1981-88, reviewer Jour. Environ. Engring. Div., Proceedings ASCE 1988—); mem. NSPE, Am. Acad. Environ. Engrs. (state chmn. 1988—), Am. Water Works Assn., Internat. Assn. Water Quality, Water and Environ. Fedn. (internat. com. 1984, mem. tech. rev. com. for manual of practice, 1990—), MATHCOUNTS (chmn. 1985-88, bd. govs. 1988—, regional coord. 1988—). Office: Utah Div Water Quality PO Box 144870 288 N 1460 W Salt Lake City UT 84114

BHIDE, DAN BHAGWATPRASAD DINKAR, chemical engineer; b. Bombay, India, Apr. 4, 1964; came to U.S., 1985; s. Dinkar V. and Vijaya D. B.; m. Monali D., Jan. 23, 1990. B of Chem. Engring., U. Bombay, 1985; MS in Chem. Engring., Syracuse U., 1987, MS in Environ. Engring., 1989, PhD in Chem. Engring., 1991; MBA, Dowling Coll., 1993. Project mgr. Pall Corp., Hauppauge, N.Y., 1991—. Mem. AICE, N.Am. Membrane Soc., Sigma Xi. Office: Pall Corp 225 Marcus Blvd Hauppauge NY 11788

BHUGRA, BINDU, microbiologist, researcher; b. New Delhi, India, Aug. 11, 1963; d. Om Parkash and Chandra Kanta (Verma) B. BS in Chemistry, Delhi U., India, 1981-84; MS in Biorganic Chemistry, IIT, Delhi, India, 1984-86; PhD in Microbiology, UAB, 1992. Postdoctoral U. Ala. at Bham, 1992—. Pres. Assn. Indian Students, Bham, 1990-92; sec. Indian Cultural Ad of B'ham, 1989-91; mem. Cahaba Riut Soc., B'ham, 1991-92; v.p. Phi Beta Delta, B'ham, 1992—. Mem. Am. Soc. Microbiology, Internat. Orgn. Mycoplasmology, Indian Cultural Assn., Phi Beta Delta. Hindu. Avocations: painting, canoeing, reading, sewing. Office: U Ala Dept Microbiology VH 422 Birmingham AL 35294

BHUNIA, ARUN KUMAR, microbiologist, immunologist, researcher; b. Garania, Midnapur, India, Jan. 12, 1960; came to U.S., 1985; s. Nandalal and Bangabala (Bera) B. BS in Vet. Medicine, BCKV, India, 1984; PhD, U. Wyo., 1989. Rsch. asst. U. Wyo., Laramie, 1985-89; postdoctoral rsch. assoc. U. Ark., Fayetteville, 1989—, rsch. assoc., dir. Hybridoma Lab., 1990—, vis. asst. prof. Biol. Scis., 1992—; mem. grad. student com. Biol. Scis., U. Ark., Fayetteville, 1992—. Contbr. articles to profl. jours. Recipient Young Investigator award USDA, 1992. Mem. AAAS, Am. Soc. Microbiology, Inst. Food Technologist (1st place poster award 1986), Gama Sigma Delta, Sigma Xi. Achievements include development of a specific monoclonal antibody probe for listeria monocytogenes detection; research in characterization and mode of action of antimicrobial peptide from pediococcus acidilactici; tissue culture assay to detect pathogenicity of L. monocytogenes. Office: U Ark Food Scis Dept 272 Young Ave Fayetteville AR 72703

BHUSHAN, BHARAT, mechanical engineer; b. Jhinjhana, India, Sept. 30, 1949; came to U.S., 1970, naturalized, 1977; s. Narain Dass and Devi (Vati) B.; m. Sudha Bhushan, June 14, 1975; children: Ankur, Noopur. B.Engring. with honors in Mech. Engring., Birla Inst. Tech. and Sci., 1970; M.S. in M.E. (Ford Found. fellow), MIT, 1971; M.S. in Mechanics, U. Colo, 1973, Ph.D. in Mech. Engring., 1976, MBA Rensselaer Polytechnic Inst., 1980, DSc U. Trondheim, Norway, 1990. Mem. research staff dept. mech. engring. MIT, Cambridge, 1970-72; research asst., instr. dept. mech. engring. U. Colo., Boulder, 1972-76; expert investigator Automotive Specialists, Denver, 1973-76; program mgr. Mech. Tech. Inc., Latham, N.Y., 1976-80; research scientist SKF Industries, Inc., King of Prussia, Pa., 1980-81; adv. engr. IBM, Tucson, 1981-85, devel. engr., mgr., 1985-86, sr. engr., mgr. head-disk interface Almaden Research Ctr. IBM, 1986-91; Ohio Eminent scholar prof., Ohio State U., 1991—. Contbr. numerous articles to profl. jours. Recipient George Norlin award U. Colo., 1983, Disting. Service award, 1985, Alfred Noble prize ASCE, IEEE, ASME, AIME, Western Soc. Engrs., 1981, Tech. Excellence award Am. Soc. Engrs. India, 1989, Cert. Appreciation award NASA, 1987; grantee U.S. Navy, NASA, Dept. Energy, DuPont Co., USAF, Chrysler Corp. Fellow ASME (cert. of recognition Design Engring. Conf., Henry Hess award 1980, Burt L. Newkirk award 1983, Gustus L. Larson Meml. award 1986, tribology div. Best Paper award 1989, Melville medal for best current original paper 1992), N.Y. Acad. Scis.; mem. NSPE, Am. Soc. Lubrication Engrs., Am. Acad. Mechanics, Internat. Humanists Soc., Tri-City India Assn., Sigma Xi, Tau Beta Pi. Hindu. Lodge: Rotary. Home: 10235 Widdington Close Powell OH 43065 Office: Ohio State U 206 W 18th Ave Columbus OH 43210

BIANCHI, THOMAS STEPHEN, biology and oceanography educator; b. Richmond Hill, N.Y., Nov. 24, 1956; s. Thomas and Rita (Radigan) B.; m. JoAnn Marie Guiffre, June 16, 1984; 1 child, Christopher. BA in Biology, Dowling Coll., Oakdale, N.Y., 1978; MA in Biology, SUNY, Stony Brook, 1981; PhD in Marine Sci., U. Md., 1987. Rsch. assoc. Inst. Ecosystem Studies, Millbrook, N.Y., 1988-90; asst. prof. biology Lamar U., Beaumont, Tex., 1990—; reviewer NSF, 1989—, Marine Ecology Progress Series, Ger., 1986—, Jour. Marine Rsch., 1987—. Contbr. articles to profl. jours. Cons. S.E. Tex. Wetlands Com., Beaumont, 1991—; tchr. Govs. Program for Gifted Students, Austin, 1991. Bermuda Biol. Sta. Sterrer fellow, 1989-90; rsch. grantee Hudson River Found., 1988-90, Tex. Higher Edn. Bd., 1992—, Dept. Energy, 1992—, NSF, 1992—; Fulbright scholar, 1993. Mem. Ecol. Soc. Am., Soc. for Limnology and Oceanography, Sigma Xi. Roman Catholic. Achievements include research on application of plant pigments as tracers of organic carbon in coastal and wetland ecosystems; novel biomarkers that can be used to understand carbon cycling in aquatic ecosystems research. Office: Lamar U Martin Luther King Jr Dr Beaumont TX 77710

BIANCHINE, JOSEPH RAYMOND, pharmacologist; b. Albany, N.Y., Sept. 7, 1929; s. Nunzie and Rose (Gallela) B.; m. Josette Woel, Oct. 10, 1956; children: Peter Joseph, Christine Rose. B.S., Siena Coll., 1951; Ph.D., Albany Med. Coll., 1959; M.D., State U. N.Y., Syracuse, 1960. Intern Johns Hopkins Hosp., Balt., 1960-61; resident, fellow Johns Hopkins Hosp., 1960-65; instr. medicine and pharmacology Sch. Medicine, Johns Hopkins U., Balt., 1965-68; asst. prof. Sch. Medicine, Johns Hopkins U., 1968-70, assoc. prof., 1970-72; prof., chmn. dept. pharmacology, prof. medicine Tex. Tech U., Lubbock, 1972-74, Sch. Medicine, Ohio State U., Columbus, 1974-83; v.p. med. research Hoffmann-LaRoche Inc., Nutley, N.J., 1983-85; v.p. med. research and devel. Am. Critical Care Corp., McGaw Park, Ill., 1985-86, DuPont Critical Care Corp., McGaw Park, Ill., 1986-87; sr. v.p., dir. med. rsch. ctr. Adria/Erbamont, Columbus, Ohio, 1987—. Contbr. research articles to various jours. Served with USPHS, 1962-63. Mem. Am. Soc. Pharmacology and Exptl. Therapeutics, Am. Fedn. Clin. Research, Am. Assn. Study Headache, Am. Assn. Univ. Pharmacologists, Am. Soc. Clin. Parmacology and Therapeutics, Am. Pharm. Assn., Franklin County Med. Assn., Ohio Med. Assn. Office: Adria Labs Inc PO Box 16529 Columbus OH 43216-6529

BICEHOUSE, HENRY JAMES, health physicist; b. New Castle, Pa., Feb. 8, 1943; s. Roy Hamilton and Carrie (Krepps) B.; m. Susan Evelyn Tannahill, Jan. 27, 1967; 1 child, Julia Anne. BS in Biochemistry, Pa. State U., 1965; BS in Biology, Incarnate Word Coll., 1969; MS in Radiol. Health, Temple U., 1970. Health and safety supr. ARCO Radiation Process Ctr., Quehanna, Pa., 1970-72; radiol. health physicist Pa. Dept. of Environ. Resources, Reading, Pa., 1972-78; radiol. engr. Westinghouse Hanford Co., Richland, Wash., 1978-82; health physicist U.S. Nuclear Regulatory Commn., King of Prussia, Pa., 1982-88; dir. regulatory compliance Quadrex Recycle Ctr., Oak Ridge, Tenn., 1988—; corp. RSO Quadrex Corp., Oak Ridge, 1990—; chmn. task force on mamography Conf. Radiation Control Dirs., Reading, 1975-77. Vol. Knox County Sr. Citizen Ctr., Knoxville, 1990—; civil def. dir. Centre County, Pa., 1971-72. With USAF, 1965-68. Mem. AAAS (life), Am. Chem. Soc., Health Physics Soc. Avocations: model railroading, toy train collecting. Home: 5805 Penshurst Ct Powell TN 37849-4969 Office: Quadrex Corp Recycle Ctr 109 Flint Rd Oak Ridge TN 37830-7033

BICK, KATHERINE LIVINGSTONE, scientist, international liaison consultant; b. Charlottetown, Can., May 3, 1932; came to U.S., 1954; d. Spurgeon Arthur and Flora Hazel (Murray) Livingstone; m. James Harry Bick, Aug. 20, 1955 (div.); children: James A., Charles L. (dec.); m. Ernst Freese, 1986 (dec. 1990). BS with honors, Acadia U., Can., 1951, MS, 1952; PhD, Brown U., 1957; DSc (hon.), Acadia U., 1990. Research pathologist UCLA Med. Sch., 1959-61; asst. prof. Calif. State U., Northridge, 1961-66; lab. instr. Georgetown U., Washington, 1970-72, asst. prof., 1972-76; dep. dir. neurol. disorder program Nat. Inst. Neurol. and Communicative Disorders and Stroke, NIH, Bethesda, Md., 1976-81, acting dep. dir., 1981-83, dep. dir., 1983-87; dep. dir. extramural research Office of Dir. NIH, 1987-90; sci. liaison Centro Studio Multicentrico Internazionale Sulla Demenza, Washington, 1990—; cons. Nat. Rsch. Coun., Italy, 1991—. Editor: Alzheimer's Disease: Senile Dementia and Related Disorders, 1978, Neurosecretion and Brain Peptides, Implications for Brain Functions and Neurol. Disease, 1981, The Early Story of Alzheimer's Disease, 1987, Alzheimer Disease, 1993; contbr. articles to profl. jours. Pres. Woman's Club, McLean, Va., 1968-69; bd. dirs. Fairfax County (Va.) YWCA, 1969-70; pres. Emerson Unitarian Ch., 1964-66, Bethesda Place Community Coun., 1993. Recipient Can. NRC award Acadia U., 1951-52, NIH Dir.'s award, 1978, Spl. Achievement award NIH, 1981, 83, Superior Svc. award USPHS, 1986, Presdl. Rank award meritorious sr. exec., 1989; Universal Match Found. fellow Brown U., 1956-57, Fed. Exec. Inst. Leadership fellow, 1980. Mem. AAAS, Am. Neurol. Assn., Am. Acad. Neurology, Assn. for Research in Nervous and Mental Disease, Internat. Brain Research Orgn., World Fedn. Neurology Research Group on Dementias (exec. sec. Am. region 1984-86, chmn. 1986—), Soc. for Neuroscience. Office: Centro SMID USA 7300 Greentree Rd Bethesda MD 20817-1552

BICKEL, JOHN FREDERICK, consulting engineer; b. Cleona, Pa., Feb. 25, 1928; s. Frederick Phillip and Margaret Catherine (Miller) B.; m. Dorothy Marie Saunders, Oct. 23, 1954; children: Richard George, Jane Dorothy. BS in Engring., Drexel U., 1962, MS in Engring. Mgmt., 1968. Registered profl. engr., Pa., Del., N.J.; profl. planner, N.J. Engr. Fischer & Porter Co., Warminster, Pa., 1956-66; chief engr. W.F. Keegan & Co., Havertown, Pa., 1966-73; cons. engr. JBPE Cons. Engrs., Hatboro, Pa., 1973-88, J & B Design Engring., Hatboro, 1988—. With U.S. Army, 1950-51. Mem. Inst. Transp. Engrs., Cons. Coun. ITE, Expert Witness Coun. ITE, Nat. Soc. Profl. Engrs. Avocation: home improvements. Home: 3700 Meyer Ln Hatboro PA 19040-3720 Office: J & B Design Engring 126 S York Rd Hatboro PA 19040-3327

BICKEL, PETER JOHN, statistician, educator; b. Bucharest, Roumania, Sept. 21, 1940; came to U.S., 1957, naturalized, 1964; s. Eliezer and P. Madeleine (Moscovici) B.; m. Nancy Kramer, Mar. 2, 1964; children: Amanda, Stephen. AB, U. Calif., Berkeley, 1960, MA, 1961, PhD, 1963; PhD (hon.), Hebrew U. Jerusalem, 1988. Asst. prof. stats. U. Calif., Berkeley, 1964-67, assoc. prof., 1967-70, prof., 1970—, chmn. dept. stats., 1976-79, dean phys. scis., 1980-86; vis. lectr. math. Imperial Coll., London, 1965-66; fellow J. S. Guggenheim Meml. Found., 1970-71; NATO sr. sci. fellow, 1974. Author: (with K. Doksum) Mathematical Statistics, 1976; assoc. editor Annals of Math. Statistics, 1968-76, 1986—; contbr. articles to profl. jours. Fellow John D. and Catherine T. MacArthur Found., 1984. Fellow Inst. Math. Stats. (pres. 1980), Am. Statis. Assn., AAAS; mem. Royal Statis. Soc., Internat. Statis. Inst., Nat. Acad. Sci., Am. Acad. Arts Scis., Bernoulli Soc. (pres. 1990). Office: U Calif Dept Stats Evans Hall Berkeley CA 94720

BICKEL, WARREN KURT, psychiatry and psychology educator; b. East Meadow, N.Y., June 16, 1956; s. Walter Warren and Catherine (Skelenko) B.; m. Connie Jo Clay, Dec. 24, 1987; children: Keefer Rurik Clay, Corena Keely Clay. MA, U. Kans., 1981, PhD, 1983. Postdoctoral fellow Sch. Medicine U. N.C., Chapel Hill, 1983-84; postdoctoral fellow Johns Hopkins U., Balt., 1984-85; asst. prof. Albert Einstein Coll., N.Y.C., 1985-87; rsch. asst. prof. U. Vt., Burlington, 1987-89, asst. prof., 1989-92, assoc. prof. psychiatry and psychology, 1992—; mem. spl. rev. com. Nat. Inst. Drug Abuse, Washington, 1987-91; bd. editors Jour. Exptl. Analysis of Behavior, 1990—, Drug and Alcohol Dependence, 1991—. Contbr. articles to profl. publs. Fellow APA (Young Psychopharmacologist award 1988); mem. Assn. Behavior Analysis, Behavioral Pharmacology Soc. Achievements include study of applied behavioral economics to drug-taking behavior, demonstration of efficacy of buprenorphine as medication to treat opioid addicts. Office: U Vt 38 Fletcher Pl Burlington VT 05401-1419

BIDDLE, DAVID, neurologist; b. Phila., Nov. 17, 1944; s. Benjamin and Anna (Spector) B.; m. Trisha Fae Karagannis, Feb. 14, 1971; children: Erika Lauren, Jeremy Justin. BA, LaSalle Coll., 1966; MD, Jefferson Med. Coll., 1970. Diplomate Am. Bd. Psychiatry and Neurology. Intern medicine Long Island (N.Y.) Jewish Hosp., 1970-71, resident medicine, 1971-72; resident neurology L.I. (N.Y.) Jewish Hosp., 1974-77; pvt. practice Drs. Biddle-Blau-Mallin, Lake Success, N.Y., 1977—; mem. adv. bd. L.I. chpt. Myasthenia Gravis Found., 1989—. Contbr. articles to Neurology, Archives of Neurology. Bd. dirs. Kennilworth Owners' Assn., Kings Point, N.Y., 1988—. Maj. U.S. Army, 1972-74. Fellow Am. Acad. Neurology, Coll. Phila. Physicians; mem. Am Acad. Electodiagnostic Medicine, Royal Soc. Internal Medicine, L.I. Jewish Staff Soc. (pres. 1989-90). Office: 3003 New Hyde Park Rd Ste 202 New Hyde Park NY 11042-1165

BIDDLE, DONALD RAY, aerospace company executive; b. Alton, Mo., June 30, 1936; s. Ernest Everet and Dortha Marie (McGuire) B.; m. Nancy Ann Dunham, Mar. 13, 1955; children: Jeanne Kay Biddle Bednash, Mitchell Lee, Charles Alan. Student El Dorado (Kans.) Jr. Coll., 1953-55, Pratt (Kans.) Jr. Coll., 1955-56; BSME, Washington U., St. Louis, 1961; postgrad. computer sci. Pa. State U. Extension, 1963; cert. bus. mgmt. Alexander Hamilton Inst., 1958. Design group engr. Emerson Elec. Mfg., St. Louis, 1957-61; design specialist Boeing Vertol, Springfield, Pa., 1962; cons. engr. Ewing Tech. Design, Phila., 1962-66; chief engr. rotary wing Gates Learjet, Wichita, Kans., 1967-70; dir. engring./R & D BP Chems., Inc. Advanced Materials Div., Stockton, Calif., 1971-93; prin. Biddle & Assocs., Consulting Engrs., Stockton, 1993—. Guest lectr. on manrated structures various univs. and tech. socs. Cons. engr. Scoutmaster, counselor, instl. rep. Boy Scouts Am., St. Ann, Mo., 1958-61; mem. Springfield Sch. Bd., 1964. Mem. ASME, ASTM, AIAA, Am. Helicopter Soc. (sec.-treas. Wichita chpt. 1969), Am. Mgmt. Assn., Exptl. Pilots Assn., Soc. for Advancement of Metals and Process Engring. Republican. Methodist (trustee, chmn. 1974-76, 84-86, staff parish 1987—). Patentee landing gear designs, inflatable rescue system, glass retention systems, adjustable jack system, cold weather start fluorescent lamp, paper honeycomb core post-process systems. Home: 1140 Stanton Way Stockton CA 95207-2537 Office: Biddle & Assocs 1140 Stanton Way Stockton CA 95207

BIEBER, FREDERICK ROBERT, medical educator; b. Regina, Sask., Can., Feb. 9, 1950; s. Frederick John and Marjorie (Davidson) B.; m. Jane Marie McNamara, June 23, 1973. BA, SUNY, Oswego, 1972; MS, U. Rochester, 1976; PhD, Med. Coll. Va., 1981. Diplomate Am. Bd. Med. Genetics. Asst. prof. Harvard Med. Sch., Boston, 1985-91, assoc. prof., 1992—. Author: The Malformed Fetus and Stillbirth, 1988; editorial bd. Clin. Dysmorphology; contbr. articles to profl. jours. Bd. dirs. Greyhound Friends, Hopkinton, Mass., 1992. Office: Brigham & Women's Hosp 75 Francis St Boston MA 02115

BIEDERMAN, EDWIN WILLIAMS, JR., petroleum geologist; b. Stamford, Conn., June 30, 1930; s. Edwin Williams and Thelma Frances (Morrow) B.; m. Margaret Jane White, Aug. 23, 1958; children: Robert, Mary, Jane, James. BA, Cornell U., 1952; PhD, Pa. State U., 1958. Cert. petroleum geologist, Pa. Project leader Cities Svc. Co., Tulsa, 1958-68, pres. staff, Cranbury, N.J., 1968-72; asst. dir. Pa. Tech. Assistance program, University Park, Pa., 1972-77, tech. specialist, 1980—; field ctr. dir. NSF Chautauqua Courses, University Park, 1977-80. Author: Atlas of Oil and Gas Reservoir Rocks From North America, 1986; contbr. articles to profl. jours.; holder 5 patents for geochem. exploration, in situ acidulation of phosphate rock, grate for vertical oil shale kiln, fire retardant foam, lightweight cement for oil wells. Served to 1st lt. USAF, 1952-54. Pa. State U. scholar, 1956-58; Am. Assn. Petroleum Geologists grantee, 1957; recipient First Place award Project of Yr. Nat. Assn. Mgmt. and Tech. Assistance Ctrs., 1985. Mem. AAAS, Am. Assn. Petroleum Geologists, Soc. Econ. Paleontologists and Mineralogists, Geochem. Soc., Assn. Profl. Geol. Scientists. Office: Pa State U 232 Mineral Scis Bldg University Park PA 16802

BIEDLER, JUNE L., oncologist; b. N.Y., June 24, 1925. BA, Vassar Coll., 1947; MA, Columbia U., 1954; PhD in Biology, Cornell U., 1959. Exchg. investigator Inst. Gustave-Roussy, France, 1959-60, rsch. fellow, 1959-60, rsch. assoc., 1960-62, assoc., 1962-72, sect. head, 1966-72, assoc. mem., 1972-78; mem Sloan-Kettering Inst. Cancer Rsch., 1978—, lab head, 1975—; assoc. prof. biology, Sloan-Kettering Divsn. Grad. Sch. Med. Sci. Cornell U., 1973—; researcher Meml. Sloan-Kettering Cancer Ctr., N.Y. Recipient Rsch. Career Devel. award USPHS, 1963, G.M.A. Clowes Meml. award Am. Assn. for Cancer Rsch., Phila., 1992. Fellow AAAS; mem. Am. Assn. Cancer Rsch., Am. Soc. Cell Biology, Genetics Soc. Am., N.Y. Acad. Sci. Achievements include research in cytogenetics and somatic cell genetics, tumor biology, chromosome structure-function relationships, drug resistance of mammalian cells. Office: Meml Sloan-Kettering Cancer Ctr 1275 York Ave New York NY 10021*

BIEGEN, ELAINE RUTH, psychologist; b. N.Y.C., June 25, 1939; d. Louis J. and Rose (Drazin) Ebert; children: Richard S., Peter T. Student, Syracuse U., 1956-58; BA, N.Y.U., 1960; MS in Edn., CUNY, N.Y.C., 1970; PhD, CUNY Grad. Ctr., N.Y.C., 1985. Cert. sch. psychologist, N.Y., nationally certified sch. psychologist, 1989; lic. psychologist, N.Y. Sch. psychologist Mt. Vernon (N.Y.) Pub. Schs., 1970-73; sch. psychologist Dobbs Ferry

(N.Y.) Pub. Schs., 1973-75, coordinator of pupil personnel svcs., 1974-75; psychologist Behavioral Cons., Jericho, N.Y., 1975-87, Plainview (N.Y.) Mental Health Ctr., 1975-87, Shield Devel. Evaluation Ctr., N.Y.C., 1976-79; cons. psychologist Lexington Sch. for Deaf and Lexington Ctr., Inc., N.Y.C., 1987-89; psychologist Western Queens Cons. Ctr., N.Y.C., 1987—; pvt. practice Beechurst, Flushing, N.Y., 1985—; adj. asst. prof. CUNY, LaGuardia, N.Y.C., 1985—; psychology cons. N.Y.C. Bd. Edn., 1988—. Designer/creator Transitional Program from Elem. to Middle Sch., Dobbs Ferry, N.Y., 1975. Speaker Lexington Ctr., Inc., 1987-88; liberal Dem. party worker for E. McCarthy Campaign, N.Y.C., poll-watcher; mem. Guggenheim Mus., N.Y.C., Mus. Modern Art, N.Y.C. Mem. NOW, NAFE, Am. Psychol. Assn., Nassau County Psychol. Assn., Nat. Assn. Sch. Psychologists, Smithsonian, Greenpeace, Pub. Broadcasting. Avocations: music, dance, art, lit., theatre. Home: 162-41 Powells Cove Rd Beechhurst Flushing NY 11357-1449

BIELBY, GREGORY JOHN, electrical engineer; b. Adelaide, Australia, Jan. 19, 1955; arrived in Canada, 1983; s. Robert John and Pauline Joyce (Dawes) B.; m. Karen Lesley Walker, Aug. 2, 1986; children: Christopher Mark, Steven Scott. BS, U. Adelaide, 1975, BEE with first class honors, 1976, PhD in Engring., 1983. With Telecom Australia, Adelaide, 1976, Philips Allied Industries Australia, Adelaide, 1977; part-time tutor U. Adelaide, 1977-83; mem. scientific staff Bell-No. Rsch., Montreal, Quebec, Canada, 1983-90, mgr. interactive applications, 1990—. Contbr. articles to conf. proceedings. Commonwealth scholar Australian U., 1972. Mem. IEEE, N.Y. Acad. Scis., Acoustical Soc. Am. Office: Bell No Rsch, 16 Place du Commerce, Verdun, PQ Canada H3E 1H6

BIELER, RUDIGER, biologist, curator; b. Hamburg, Germany, Apr. 9, 1955; came to U.S., 1985; s. Wolfgang K.H. and Erna J. (Meier) B.; m. Petra Sierwald, Jan. 29, 1987; 1 child, Anke. MS in Biology and Geography, U. Hamburg, 1982, PhD in Zoology, 1985. Lectr. zoology U. Hamburg, 1982-85; postdoctoral rsch. fellow Smithsonian Instn., Washington, 1985-86; postdoctoral fellow Smithsonian Marine Sta., Ft. Pierce, Fla., 1986-87, NATO postdoctoral rsch. fellow, 1987-88; asst. curator, divsn. head Del. Mus. Natural History, Wilmington, 1988-90, acting exec. dir., 1990; asst. curator Field Mus. Natural History, Chgo., 1990-93, assoc. curator, 1993—; rsch. assoc. Mus. of Comparative Zoology Harvard U., Cambridge, Mass., 1988—, Nat. Mus. Natural History, Washington, 1989-92; lectr. com. on evolutionary biology U. Chgo., 1992—. Editor-in-chief Monographs of Marine Mollusca, 1990—; mem. editorial bd. Nemouria, 1987—, The Nautilus, 1988, The Malacologia, 1991—; contbr. articles to Ann. Rev. Ecology and Systematics, Jour. of Morphology, Jour. Molluscan Studies. Bd. dirs. Del. Mus. Natural History, Wilmington, 1991—. 1st lt. German Army, 1974-76. Mem. Am. Soc. Zoologists, Am. Malacol. Union (councillor 1989-91, v.p. 1993—), Soc. Systematic Biology, Coun. Systematic Malacologists (sec. 1991—), Willi Hennig Soc., Sigma Xi. Achievements include research in evolutionary biology of Mollusca. Office: Field Mus Natural History Dept Zoology Roosevelt Rd at Lake Shore Chicago IL 60605

BIENENSTOCK, ARTHUR IRWIN, physicist, educator; b. N.Y.C., Mar. 20, 1935; s. Leo and Lena (Senator) B.; m. Roslyn Doris Goldberg, Apr. 14, 1957; children—Eric Lawrence, Amy Elizabeth (dec.), Adam Paul. B.S., Poly. Inst. Bklyn., 1955, M.S., 1957; Ph.D., Harvard U., 1962. Asst. prof. Harvard U., Cambridge, Mass., 1963-67; mem. faculty Stanford (Calif.) U., 1967—, prof. applied physics, 1972—, vice provost faculty affairs, 1972-77, dir. synchrotron radiation lab., 1978—; mem. U.S. Nat. Com. for Crystallography, 1983-88, sci. adv. com. European Synchrotron Radiation Facility, 1988-90, 93—. Author papers in field. Bd. dirs. No. Calif. chpt. Cystic Fibrosis Research Found., 1970-73, mem. pres.'s adv. council, 1980-82; trustee Cystic Fibrosis Found., 1982-88. Recipient Sidhu award Pitts. Diffraction Soc., 1968, Disting. Alumnus award Poly. Inst. N.Y., 1977; NSF fellow, 1962-63. Fellow Am. Phys. Soc. (gen. councilor 1993—), AAAS; mem. Am. Crystallographic Assn., N.Y. Acad. Scis., Materials Rsch. Soc. Jewish. Home: 967 Mears Ct Palo Alto CA 94305-1041 Office: Synchrotron Radiation Lab Bin 69 PO Box 4349 Palo Alto CA 94309-4349

BIENENSTOCK, JOHN, physician, educator; b. Budapest, Hungary, Oct. 6, 1936; s. Maurice and Anne (Horn) B.; m. Dody Sanders, Nov. 24, 1961; children: Jimson Andrew, Adam Sebastian, Robin Anne. MB, BS, Westminster Med. Sch., London, 1960; postgrad., Harvard Med. Sch., 1964-66, SUNY, Buffalo, 1966-68. Fellow Harvard U. Med. Sch., Boston, 1964-66; Buswell fellow SUNY, Buffalo, 1966-68, asst. rsch. prof. medicine, 1967-68; asst. prof. medicine McMaster U., Hamilton, Ont., Can., 1968-74, assoc. dean rsch., 1972-78, prof. medicine and pathology, 1974—, chmn. dept. pathology, 1978-89, v.p. health scis., 1989—, dean health scis., 1992—; founder AB Biol. Supply Inc., 1977, Agritech Rsch. Inc., 1980; D.W. Harrington lectr. SUNY, Buffalo, 1986, Rayne vis. prof. U. Western Australia, Perth, 1987; cons. WHO, Geneva, 1970—; also cons. various pharm. cos. Editor: Immunology of Lung, 1984, Mast Cell Differentiation, 1986, Recent Advances in Mucosal Immunology, 1987; contbr. over 350 articles to sci. jours. Chmn. bd. Dundas Valley (Ont.) Sch. Art, 1984-86; chmn., bd. dirs. Can. Red Cross Soc., mem. adv. com. nat. blood svcs. 1985-90. Recipient Purkynje medal Assn. Czechoslovak Socs., Prague, 1989, Ross A. McIntyre gold medal U. Nebr., Omaha, 1989. Fellow Royal Coll. Physicians (Can.), Royal Soc. (Can.), Royal Coll. Physicians (London); mem. Swiss Soc. Allergy and Immunology (hon.), Can. Soc. Immunology (pres. 1985-87), Assn. Am. Physicians, Am. Soc. Clin. Investigation, Am. Thoracic Soc., Internat. Union Immunological Socs. (mem. coun.). Jewish. Avocation: painting. Home: 50 Albert St, Dundas, ON Canada L9H 2X1 Office: McMaster U Faculty Health Scis, 1200 Main St W, Hamilton, ON Canada L8N 3Z5

BIENHOFF, DALLAS GENE, aerospace engineer; b. Holdrege, Nebr., Apr. 1, 1952; s. Vernon Henry and Betty Lee (Pettite) B.; m. Yolanda Kathryn Hymes, Sept. 12, 1976; children: Erik Ryan Hern, Danielle Geannine Hern. BSME, Fla. Inst. Tech., 1974; MS in Engring., Calif. State U., Northridge, 1985. Assoc. engr. Martin Marietta, Denver, 1974-75; mem. tech. staff The Aerospace Corp., El Segundo, Calif., 1979-82; mem. tech. staff Rocketdyne divsn. Rockwell Internat., Canoga Park, Calif., 1975-79; mem. tech. staff Space Systems divsn. Rockwell Internat., Downey, Calif., 1982-87, supr. Space Systems divsn., 1987-89, project mgr. Space Systems divsn., 1989—. Bd. dirs. Trinity Luth. Ch., Reseda, Calif., 1980-88, Holy Cross Luth. Ch., Cypress, Calif., 1989-92. Mem. AIAA (chairperson 1992-93). Office: Rockwell Internat Space Systems divsn 12214 Lakewood Blvd Downey CA 90241

BIENIAWSKI, ZDZISLAW TADEUSZ, mineral engineer, educator, consultant; b. Cracow, Poland, Oct. 1, 1936; came to U.S., 1978, naturalized; married; 3 children. Student, Gdansk (Poland) Tech. U., 1954-58; BS in Mech. Engring., U. Witwatersrand, Johannesburg, Republic South Africa, 1961, MS in Engring. Mechanics, 1963; PhD in Mining Engring., U. Pretoria, South Africa, 1968. Rsch. engr. Atomic Energy Bd., 1963-66; dir. geomechanics divsn. Coun. for Sci. and Indsl. Rsch., Pretoria, 1966-78; prof. mineral engring., sr. mem. grad. sch. faculty Pa. State U., University Park, 1978—; dir. Pa. Mining and Mineral Resources Rsch. Inst. Pa. State U., 1980-90; vis. prof. U. Karlsruhe, Germany, 1972, Stanford U., 1985, Harvard U., 1990; chmn. U.S. Nat. Com. on Tunneling Tech., 1984-85; U.S. rep. to Internat. Tunnel Assn.; exec. com. U.S. Nat. Com. for Rock Mechanics, 1983-84. Author: Rock Mechanics Design in Mining and Tunneling, 1984, Strata Control in Mineral Engineering, 1987, Aiming High-A Collection of Essays, 1988, Engineering Rock Mass Classifications, 1989, Design Methodology in Rock Engineering, 1992; editor: Tunneling in Rock, 1974, Exploration for Rock Engineering, 1976; contbr. over 130 articles to profl. jours.; prod., dir. ednl. film Stressing the Point in Rock Mechanics, 1970. Proclamation mayor City of State Coll. Dick Bieniawski day, 1983. Mem. ASCE (mem. com. on rock mechanics 1985-89), Soc. Mining Engrs. of AIME (Rock Mechanics award 1984), Am. Soc. Engring. Edn., Internat. Soc. Rock Mechanics (v.p. 1974-79), Toastmasters Internat. (Disting. Toastmaster award 1974), Elks, Sigma Xi. Avocations: research specialties rock mechanics, engring. design, tunneling, tennis, photography, genealogy. Home: 1352 Sandpiper Dr Nittany Hills State College PA 16801-7713 Office: Pa State U 122 Mineral Scis Bldg University Park PA 16802

BIERMAN, PHILIP JAY, physician, reseracher, educator; b. St. Louis, Sept. 7, 1955; s. Marvin and Sylvia R. B.; m. Mary Wampler, Aug. 8, 1988. BA, U. Mo., Kansas City, 1977, MD, 1979. Diplomate Am. Bd. Internal Medicine, Am. Bd. Med. Oncology, Am. Bd. Hematology. Resident in medicine U. Nebr. Med. Ctr., Omaha, 1979-83, fellow in oncology, 1983-85, asst. prof., 1987—; fellow in hematology City of Hope Nat. Med. Ctr., Duarte, Calif., 1985-86, staff physician, 1986-87. Home: 6423 S 106th Cir Omaha NE 68127 Office: U Nebr Med Ctr 600 S 42d St Omaha NE 68198

BIESELE, JOHN JULIUS, biologist, educator; b. Waco, Tex., Mar. 24, 1918; s. Rudolph Leopold and Anna Emma (Jahn) B.; m. Marguerite Calfee McAfee, July 29, 1943 (dec. 1991); children: Marguerite Anne, Diana Terry, Elizabeth Jane; m. Esther Aline Eakin, Mar. 9, 1992. B.A. with highest honors, U. Tex., 1939, Ph.D., 1942. Fellow Internat. Cancer Research Found., U. Tex., 1942-43, Barnard Skin and Cancer Hosp., St. Louis, also; U. Pa., 1943-44, instr. zoology, 1943-44; temporary research assoc. dept. genetics Carnegie Instn. of Washington, Cold Spring Harbor, 1944-46; research assoc. biology dept. Mass. Inst. Tech., 1946-47; asst. Sloan-Kettering Inst. Cancer Research, 1946-47, research fellow, 1947, assoc., 1947-55, head cell growth sect., div. exptl. chemotherapy, 1947-58, mem., 1955-58, assoc. scientist div., 1959-78; asst. prof. anatomy Cornell U. Med. Sch., 1950-52; assoc. prof. biology Sloan-Kettering div. Cornell U. Grad. Sch. Med. Scis., 1952-55, prof. biology, 1955-58; prof. zoology, mem. grad. faculty U. Tex., Austin, 1958-78; also mem. faculty U. Tex. (Coll. Pharmacy), 1969-71, prof. edn., 1973-78; prof. emeritus zoology U. Tex., Austin, 1978—; cons. cell biology M.D. Anderson Hosp. and Tumor Inst., U. Tex. at Houston, 1958-72; dir. Genetics Found., 1959-78; mem. cell biology study sect. NIH, 1958-63; Sigma Xi lectr. N.Y. U. Grad. Sch. Arts and Scis., 1957; Mendel lectr. St. Peter's Coll., Jersey City, 1958; Mendel Club lectr. Canisius Coll., Buffalo, 1971; mem. adv. com. rsch. etiology of cancer Am. Cancer Soc., 1961-64, pres. Travis County unit, 1966, mem. adv. com. on personnel for rsch., 1969-73; counsellor Cancer Internat. Rsch. Coop., Inc., 1962-90; mem. cancer rsch. tng. com. Nat. Cancer Inst., 1969-72; Gen. chmn. Conf. Advancement Sci. and Math. Teaching, 1966. Author: Mitotic Poisons and the Cancer Problem, 1958; mem. editorial bd. Year Book Cancer, 1959-72; mem. editorial adv. bd. Cancer Rsch., 1960-64, assoc. editor, 1969-72; cons. editor: Am. Jour. Mental Deficiency, 1963-68; mem. editorial bd. The Jour. of Applied Nutrition, 1987-91; contbr. articles to profl. jours. Research Career award NIH, 1962, 67, 72, 77. Fellow N.Y., Tex. acads. scis., AAAS; mem. Am. Assn. Cancer Research (dir. 1960-63), Am. Soc. Cell Biology, Am. Inst. Biol. Scis., Phi Beta Kappa, Sigma Xi (pres. Tex. chpt. 1963-64), Phi Eta Sigma, Phi Kappa Phi. Achievements include provision of early evidence for abnormal chromosome numbers in cancer cells, for occasional excessively multiple-stranded state of cancer chromosomes; demonstration of a direct relation of chromosomal size in mammalian tissues and organs to the local metabolic activity, as evidenced by the local content of B vitamins, of differential toxicity of certain antimetabolites to cancer cells in culture; research on concept of hyper-replication of certain genes in the resistance to chemotherapy developed by some cancer cells. Home: 2500 Great Oaks Pky Austin TX 78756-2908

BIESIOT, PATRICIA MARIE, biology educator, researcher; b. St. Cloud, Minn., Mar. 23, 1950; d. Anthony Carroll and Rosemary Helen (Lenz) B.; m. Shiao Yu Wang, May 16, 1988; 1 child, Henry Carson Wang. BS, Bowling Green State U., 1972, MS, 1975; PhD in Biol. Oceanography, MIT/Woods Hole Oceanog. Instn., 1986. Biotechnician Gulf Coast Rsch. Lab., Ocean Springs, Miss., 1973-78, lab. supr., 1978-80; grad. fellow MIT/Woods Hole Oceanog. Instn., Mass., 1980-86; postdoctoral fellow Oak Ridge (Tenn.) Nat. Lab., 1986-88; asst. prof. U. So. Miss., Hattiesburg, 1989—; peer reviewer EPA, Gulf Breeze, Fla., 1991—. Contbr. articles to profl. jours. Grantee Miss. Soil Conservation Svc., 1990-93, Sea Grant, 1991-93, NSF, 1991-93. Mem. Am. Soc. Zoologists, Crustacean Soc., World Aquaculture Soc., Western Soc. Naturalists, Miss. Acad. Scis. (divsn. chair 1992-93), Sigma Xi. Democrat. Office: U So Miss SS Box 5018 Hattiesburg MS 39406-5018

BIGELEISEN, JACOB, chemist, educator; b. Paterson, N.J., May 2, 1919; s. Harry and Ida (Slomowitz) B.; m. Grace Alice Simon, Oct. 21, 1945; children: David M., Ira S., Paul E. A.B., NYU, 1939; M.S., Wash. State U., 1941; Ph.D., U. Calif., Berkeley, 1943. Rsch. scientist Manhattan Dist., Columbia, 1943-45; rsch. assoc. Ohio State U., Columbus, 1945-46; fellow Enrico Fermi Inst., U. Chgo., 1946-48; sr. chemist Brookhaven Nat. Lab., Upton, N.Y., 1948-68; prof. chemistry U. Rochester, N.Y., 1968-78; chmn. dept. U. Rochester, 1970-75; Tracy H. Harris prof. U. Rochester (Coll. Arts and Scis.), 1973-78; v.p. research, dean grad. studies SUNY, Stony Brook, 1978-80; Leading prof. chemistry SUNY, 1978-89, Disting. prof., 1989, Disting. prof. emeritus, 1989—; vis. prof. Cornell U., 1953; NSF sr. fellow, vis. prof. Eidgen Techn. Hochschule, Switzerland, 1962-63; chmn. Assembly Math. and Phys. Scis., NRC-Nat. Acad. Scis., 1976-80. Mem. editorial bd. Jour. Phys. Chemistry, Jour. Chem. Physics. Trustee Sayville Jewish Center, 1954-68. Recipient Nuclear award Am. Chem. Soc., 1958, Gilbert N. Lewis lectr., 1963, E.O. Lawrence award, 1964, Disting. Alumnus award Wash. State U., 1983; John Simon Guggenheim fellow, 1974-75. Fellow Am. Phys. Soc., Am. Chem. Soc., AAAS, Am. Acad. Arts and Sci.; mem. Nat. Acad. Scis. (councilor 1982-85), Phi Beta Kappa, Sigma Xi, Phi Lambda Upsilon. Achievements include research in photochemistry in rigid media, semiquinones, cryogenics, chemistry of isotopes, quantum statistics of gases, liquids and solids. Home: PO Box 217 Saint James NY 11780-0217

BIGGERSTAFF, RANDY LEE, sports medicine consultant; b. Buffalo, Feb. 13, 1951; s. Dever Poole and Mary Martha (Smith) B.; m. Sue Ann Knobeloch, Nov. 26, 1977; children: Nicholas Lee, Amy Elizabeth. BS, U. Mo., 1973. Dist. athletic tng. tchr. Granite City (Ill.) Community Sch. Dist., 1973-77; athletic trainer St. Louis Hummers, Profl. Softball Team, Valley Park, Mo., 1978-79; founder-ptnr., clinic dir. St. Louis Sports Medicine Clinic, Chesterfield, Mo., 1977-82; founder, clin. dir. Iowa Orthopedic Sports Medicine Clinic, Urbandale, 1982-84; clinic dir. St. Louis Orthopedic Sports Medicine Clinic, Chesterfield, 1984-86; ptnr., v.p. St. Louis Rehab. Sports Clinic, Crystal City, Mo., 1986-88; adminstr., regional dir. St. Louis Orthopedic Sports Medicine Clinic, Chesterfield, 1989-90; coord., trainer, cons. St. Luke's Hosp., Chesterfield, 1990-92; v.p. D. P. Biggs Cons. Ltd., Inc., 1992-93; pres. Phoenix Sports Med. Systems, St. Louis, 1993—; cons. Brentwood & Creve Coeur Skating, St. Louis, 1986—, Gateway Athletics, St. Louis, 1984—; med. coord. Show-Me-Bowl, St. Louis, 1979-82, Summer Biathalon Series, Essex Junction, Vt., 1989—. Contbr. articles to profl. jours. Sec. bd. overseers Lindenwood Coll., St. Charles, Mo., 1992-93, vice chmn., 1993—. Mem. Nat. Athletic Trainers Assn. (cert., clin. corp. com.), Mo. Athletic Trainer Assn. (registered, chair Hall of Fame com.). Methodist. Avocations: running, cycling, fitness, hiking, spectator sports. Home: 82 Shirecreek Ct Saint Charles MO 63303-5432 Office: Phoenix Sports Med Systems 13357 Olive St Rd Chesterfield MO 63017

BIGIANI, ALBERTINO ROBERTO, electrophysiologist, researcher; b. Modena, Italy, Jan. 24, 1959; came to U.S. 1989; s. Enrico and Maria (Palazzoni) B.; m. Silva Gavioli, Dec. 8, 1991; children: Lorenzo, Stefano. Degree in Biol. Scis., U. Modena, 1984; PhD in Neurosci., U. Pisa, Italy, 1992. Rschr. Colo. State U., Ft. Collins, 1989-92, U. Modena, 1992—. Mem. AAAS, Soc. for Neurosci., Assn. for Chemoreception Scis. Achievements include research in potassium channels as molecular detector of calcium ions in taste receptor cells. Office: Colo State U Dept Anat & Neurobiology Cente Ave Fort Collins CO 80523

BIGLER, WILLIAM NORMAN, chemistry educator, biochemist; b. Oakland, Calif., Aug. 29, 1937; married, 1961; 3 children. AB, U. Calif., Berkeley, 1960; MS, San Jose State Coll., 1963; PhD in Biochemistry, U. Colo., Boulder, 1968. Acting asst. prof. biochemistry U. Calif., L.A., 1970; from asst. to assoc. prof. chem. Rochester Inst. Tech., 1970-81, dir. clinical chem., 1976-81; prof. chemistry and dir., Ctr. Advanced Med. Tech., San Francisco State U., 1981—; vis. assoc. prof. biochemistry and biophysics U. Hawaii, 1977-78. NIH fellow U. Calif., L.A., 1968-70. Mem. AAAS, N.Y. Acad. Sci., Am. Assn. Clinical Chemists, Am. Chem. Soc. Achievements include research in biochemistry of cell division, regulatory enzymes; effects of ionizing radiation and slime molds; high pressure liquid chromatography; von Hipple-Lindau syndrome. Office: San Francisco State U Dept Clin Sci 1600 Holloway Ave San Francisco CA 94132-1722*

BIGLEY, WILLIAM JOSEPH, JR., control engineer; b. Union City, N.J., May 8, 1924; s. William Joseph and Mary May (Quigley) B.; B.M.E., Rensselaer Polytech. Inst., 1950; M.S. in Elec. Engring., N.J. Inst. Tech., 1962, M.S. in Computer Sci., 1973; Ed.D., Fairleigh Dickinson U., 1984; children: Laura C., William Joseph IV, Susan J. Project engr. Tube Reducing Corp., Wallington, N.J., 1953-58, Flight Support, Inc., Metuchen, N.J., 1958-59, Airborne Accessories, Inc., Hillside, N.J., 1959-61; prin. staff engr. in control engring. Lockheed Electronics Co. div. Lockheed Aircraft, Inc., Plainfield, N.J., 1961-90; pres. Systems Engring. Corp. N.J. Inst. Tech., Newark, 1990—; prof. engring. control systems, George Washington U., 1989—; prof. cons. engr. Automatic Control Systems, 1958—. Named Engr.-Scientist of Yr., Lockheed Electronics Co., Inc., 1980, recipient Robert E. Gross award for tech. excellence, 1980; Achievement Honor Roll award N.J. Inst. Tech. Alumni Assn., 1982; registered profl. engr., N.Y., N.J., Calif. Mem. Nat. Soc. Profl. Engrs., IEEE, ASME, NRA, Tau Beta Pi (eminent engr. 1986). Contbr. articles to profl. jours. Office: Systems Engring Corp Enterprise Devel Ctr 240 Martin Luther King Blvd Newark NJ 07102

BIKALES, NORBERT M., chemist, science administrator; b. Berlin, Jan. 7, 1929; came to U.S., 1946; s. Salomon and Bertha (Bander) B.; m. Gerda V. Bierzonski, Apr. 28, 1951; children: Marguerite Sarlin, Edward A. BS in Chemistry, CCNY, 1951; MS in Chemistry, Polytech. U., 1956; PhD in Chemistry, Poly. U., 1961. Rsch. chemist Am. Cyanamid Co., Stamford, Conn., 1951-62; tech. dir. Gaylord Assocs., Newark, 1962-65; pres. N.M. Bikales & Co., Cons., Livingston, N.J., 1965-76; prof. chemistry, dir. continuing edn. in scis. Rutgers U., New Brunswick and Newark, N.J., 1973-79; dir. polymers program NSF, Washington, 1976—; trustee Gordon Rsch. Conf., 1990—. Editor Encyclopedia of Polymer Science and Technology, 1962-77; mem. editorial bd. Encyclopedia of Polymer Science and Engineering, 1982-90; contbr. articles to profl. jours., chpts. to books. Pres., Friends of Livingston (N.J.) Libr., 1968-72, Livingston Symphony Orch., 1970-76; judge Internat. Tech. Film '89 Festival, Pardubice, Czechoslovakia, 1989. Recipient award Twp. of Livingston, 1976, Great Medal City of Paris, 1985, Disting. Alumnus award Poly. U., Bklyn., 1986, Disting. Lectr. award Soc. Polymer Sci., Tokyo, 1986, Chevalier des Palmes Academiques award French Govt., 1993. Fellow AAAS, Am. Phys. Soc., N.Y. Acad. Sci.; mem. Am. Chem. Soc. (councilor 1987-89, chmn. polymer div. 1983), Internat. Union Pure and Applied Chemistry (titular mem., sec. 1979-87, chmn. commn. on recycling of polymers 1993—), Soc. Plastics Engrs. (sr. bd. dirs. 1979-82), Polish Chem. Soc. (hon.). Achievements include 26 patents in materials, chemicals and chemical processes. Office: NSF 1800 G St NW 4201 Wilson Blvd Arlington VA 22230

BILBRO, GRIFF LUHRS, electronics engineering educator; b. New Orleans, Nov. 3, 1948; s. Griff Wofford and Frieda (Luhrs) B.; m. Carla Diane Savage, Mar. 15, 1982; children: Rebecca, Lucas. BS, Case Inst. Tech., 1973; PhD, U. Ill., 1977. Sci. programmer Applied Rsch. Labs., Austin, 1977-78; sr. systems analyst Rsch. Triangle Inst., Rsch. Triangle Park, N.C., 1978-81; systems programmer Found. Computer Systems, Cary, N.C., 1982-84; rsch. engr. Ctr. Communications and Signal Processing N.C. State U., Raleigh, 1985-90, vis. asst. prof. dept. electrical and computer engring., 1990-93. Contbr. articles to profl. jours. Grad. fellow NSF, 1977-81. Mem. IEEE, Sigma Xi. Achievements include invention of mean field annealing as global optimization technique; development of early theory of high temperature superconductivity, learning theory based on principle of maximum entropy. Home: 1205 Bancroft St Raleigh NC 27612-4703 Office: Box 7911 NC State U Raleigh NC 27695-7911

BILBY, CURT, computer systems executive; b. Noblesville, Ind., Mar. 21, 1960; s. Joe L. and G. Gloria (Harris) B.; m. Diane R. Kluck, Feb. 10, 1990. BS, Rose-Hulman Inst. Tech., 1982; MS, Auburn U., 1984; PhD, U. Tex., 1992. Ops. engr. Lewis Rsch. Ctr. NASA, Cleve., 1982; rsch. asst. Auburn (Ala.) U., 1982-84; NASA rsch. fellow U. Tex., Austin, 1984-86; prin. investigator Large Scale Programs Inst., Austin, 1985-88; gen. mgr. KDT Industries, Inc., Austin, 1988-91; v.p., COO Arrowsmith Techs., Inc., Austin, 1991—. Author: Modeling: Simulation of Advanced Space Programs, 1991; contbr. articles to profl. jours. Mem. AIAA, ASME, Inst. Navigation, Soc. Cable TV Engrs. Republican. Presbyterian. Home: 3000 Bryker Austin TX 78703 Office: Arrowsmith Techs Inc 1301 W 25th Ste 300 Austin TX 78705

BILELLO, JOHN CHARLES, materials science and engineering educator; b. Bklyn., Oct. 15, 1938; s. Charles and Catherine (Buonadonna) B.; m. Mary Josephine Gloria, Aug. 1, 1959; children: Andrew Charles, Peter Angelo, Matthew Jonathan. B.E., NYU, 1960, M.S., 1962; Ph.D., U. Ill., 1965. Sr. research engr. Gen. Telephone & Electronics Lab., Bayside, N.Y., 1965-67; mem. faculty SUNY, Stony Brook, 1967-87, asst. prof., 1967-71, assoc. prof., 1971-75, prof. engring., 1975-87, dean, 1977-81; dean Sch. Engring and Computer Sci., prof. mech. engring. Calif. State U., Fullerton, 1986-89; prof. materials sci. and engring., prof. applied physics U. Mich., Ann Arbor, 1989—; guest scientist Brookhaven Nat. Labs., 1975—; vis. prof. Poly. of Milan, 1973-74; vis. scholar King's Coll., London U., 1983; vis. NATO exchange scholar Oxford U., 1986; project dir. synchroton topography project Univ. Consortium, 1981-86. Assoc. editor Jour. Materials Sci. and Engring., 1984—. NATO sr. faculty fellow Enrico Fermi Center, Milan, Italy, 1973. Fellow Am. Soc. for Metals; mem. AIME, Am. Phys. Soc., Materials Rsch. Soc. Office: U Mich Dept Material Sci Engring Ann Arbor MI 48109

BILES, JOHN ALEXANDER, pharmaceutical chemistry educator; b. Del Norte, Colo., May 4, 1923; s. John Alexander and Lillie (Willis) B.; m. Margaret Pauline Off, June 19, 1943; children: Paula M. (Mrs. Patrick Murphy), M. Suzanne. B.S., U. Colo., 1944, Ph.D. (AEC fellow), 1949. Prof. pharm. chemistry Midwestern U., 1949-50; asst. prof. pharmacy Ohio State U., 1950-52; asst. prof. pharm. chemistry U. So. Calif., Los Angeles, 1952-53; assoc. prof. U. So. Calif., 1953-57, prof., 1957—, dean, prof. pharm. scis., 1968—, John Stauffer dean's chair in pharmacy, 1988—; bd. dirs. Marion Labs. 1984-89, Marion Merrell Dow 1989—, Am. Found. for Pharm. Edn. 1991-92, Am. Assn. Colls. Pharmacy 1989-92; cons. Allergan Pharms., 1953-68, bd. dirs., 1968-80; cons. Region IX, Bur. Health Manpower Edn., Health Resources Adminstrn., 1973, Region X, 1974, Region VI, 1975, VA Central Office Pharmacy Services.; Mem. Nat. Adv. Council, Edn. for Health Professions, 1970-71, Nat. Adv. Council, Edn. for Health Professions (Nat. Study Commn. on Pharmacy), 1972-75; mem. adv. panel on pharmacy for study costs of educating profls. Nat. Acad. Scis., Inst. Medicine, 1973; mem. interdisciplinary tng. in health scis. com. Bur. Health Manpower Edn., 1972, post constrn. evaluation com., 1972, health facilities survey com., 1971; mem. adv. council Howard U. Coll. Pharmacy, 1985-90. Reviewer: Jour. of AMA, 1982-90. Recipient Lehn and Fink Scholarship award, 1945, S.C. Assos. award for excellence in teaching, 1962. Fellow Acad. Pharm. Scis., Am. Assn. Pharm. Scientists; mem. Am., Cal. pharm. assns., Am. Cancer Soc. (mem. sci. adv. com. Los Angeles County), Am. Assn. Colls. Pharmacy (study commn. on pharmacy 1973-75, pres. 1990-91), Nat. Adv. Health Svcs. Coun. (bur. health svcs. rsch. 1974), Phi Kappa Phi. Office: U So Calif Sch Pharmacy 1985 Zonal Ave Los Angeles CA 90033-1086

BILETT, JANE LOUISE, clinical psychologist; b. Queens, N.Y., Oct. 20, 1944; d. Mark Matthews and Dorothy Kors; m. Joseph S. Bilett, Mar. 28, 1965 (div. June 1971); m. E. Brian Potter, Aug. 13, 1975; children: Matthew David, Jessica Jamie. BA, U. Denver, 1965, MA, 1972, PhD, 1974. Lic. psychologist, Colo. Psychology trainee Denver VA Hosp., 1968-72; rsch. psychologist U. Colo. Med. Ctr., Denver, 1972-74; psychology cons. Headstart, Denver, 1972-74; staff psychologist Arapahoe Mental Health Ctr., Englewood, Colo., 1974-76, Aurora (Colo.) Psychotherapy Assocs., 1976-80; psychologist pvt. practice, Aurora, 1980-92; staff psychologist Affiliated Therapeutic Svcs., Aurora, 1992; cons. Ctr. for Emotional Growth, Aurora, 1985-86. Mem. Am. Psychol. Assn., Colo. Psychol. Assn., Denver Assn. Gifted and Talented. Home: 3190 S Holly St Denver CO 80222 Office: Affiliated Therapeutic Svcs 6 Abilene Ste 301 Aurora CO 80011

BILIMORIA, KARL DHUNJISHAW, aerospace engineer; b. Bombay, June 27, 1960; s. Dhun K. and Najoo D. (Motafram) B. BTech in Aero. Engring., Indian Inst. Tech., 1982; MS in Aero. Engring., Va. Poly. Inst. and State U., 1984, PhD in Aero. Engring., 1986. Grad. asst. Va. Poly. Inst. and State U., Blacksburg, 1982-86; asst. prof. Ariz. State U., Tempe, 1987-91, rsch. scientist, 1991—; referee Jour. Aircraft, Jour. Guidance, Control and Dynamics. Contbr. articles to profl. jours. Mem. AIAA (sr.), Sigma Gamma Tau. Achievements include research in optimal control of aerospace vehicles, dynamics of hypersonic flight vehicles. Office: Ariz State U Aero Rsch Ctr Tempe AZ 85287-8006

BILINSKY, YAROSLAV, political scientist; b. Lutsk, Ukraine, USSR, Feb. 26, 1932; s. Peter Bilinsky and Natalia (Balabaj) Bilinska; m. Wira Rusaniwskyj, Feb. 18, 1962; children: Peter Yaroslav, Sophia Vera Yaroslava, Nadia Yaroslava, Mark Paul Yaroslav. A.B. magna cum laude, Harvard U., 1954, postgrad. in Soviet affairs, 1956-57; Ph.D., Princeton U., 1958. Asso. Harvard U. Russian Research Center, 1956-58; instr. polit. sci. Douglass Coll., Rutgers U., New Brunswick, N.J., 1958-61; asst. prof. U. Del., Newark, 1961-65; asso. prof. U. Del., 1965-69, prof., 1969—; vis. instr. U. Pa., 1961; vis. prof. Columbia U., 1976. Author: The Second Soviet Republic: The Ukraine after World War II, 1964. Corr. sec. Peter and Paul Ukrainian Orthodox Ch., Wilmington, Del., 1965-66, trustee, 1967-71. Mem. Am. Polit. Sci. Assn., Am. Assn. Advancement Slavic Studies (pres. Mid-Atlantic Slavic Conf. 1992-93), Ukrainian Acad. Arts and Scis. in U.S. (pres. 1987-90). Home: 2 Mimosa Dr Newark DE 19711-7523 Office: U Del Polit Sci Polit Sci and Internat Rels Newark DE 19716-2574

BILISOLY, ROGER SESSA, statistician; b. Seattle, Jan. 16, 1963; s. Walter and Hattie Bilisoly; m. Charlotte Lynn Saavedra, Aug. 8, 1988; children: Chad, Avesta. BA, U. Chgo., 1984; MS, Ohio State U., 1992. Food relocation engr. Kroger, Westerville, Ohio, 1984-90; pres. Sessa Systems, Westerville, 1992—. Tech. advisor short film Twinkie Bomb, 1992. Mem. Phi Beta Kappa, Sigma Xi. Libertarian. Office: Sessa Systems 42 Pleasant Ave Westerville OH 43081

BILJETINA, RICHARD, chemical engineering researcher; b. Austria, Apr. 2, 1947; U.S. citizen, 1959; m. Mary Patricia McCarthy, Oct. 18, 1969; children: Eric, Christine. BSChemE, Ill. Inst. Tech., 1969; MBA, U. Chgo., 1974. From mem. staff to mgr. R & D Inst. Gas Tech., Chgo., 1969-89, asst. v.p., 1989-92, v.p. product tech. R & D, 1992—. Contbr. articles to profl. jours. Mgr. Little League, Skokie, Ill., 1978-85; scout master Boy Scouts Am., Skokie, 1980-83. Mem. Automatic Meter Reading Assn., Internat. Assn. Energy Economists. Achievements include patent for novel anaerobic digester. Office: Inst Gas Tech 3424 S State St Chicago IL 60616-3896

BILLEAUD, MANNING FRANCIS, civil engineer; b. Lafayette, La., July 23, 1927; s. Manning F. and Jeanne (Mouton) B.; m. Mildred Marie Martin, Sept. 1, 1949; children: Martin, Jeanne, Louise, Robert Manning, Suzanne Andre, Valerie, Edward, Marguerite, Therese, Lawrence, Claire, Philip, Germaine, Jacques, Nicole. BCE, La. State U., 1948. Registered profl. engr., La. Field engr. J.B. Mouton & Sons Inc., Lafayette, 1948-60, sec.-treas., 1960-63, v.p. constrn. and estimating, 1963-82, pres., 1982—. Chmn. Metro Code Bd. Bldg. Standards, Lafayette, 1974-92; active judiciary commn., La., 1983-87. With USN, 1945-46, PTO. Mem. Assn. Gen. Contractors (nat. dir. 1980—). Democrat. Roman Catholic. Home: 145 Girard Woods Dr Lafayette LA 70503 Office: JB Mouton & Sons Inc 202 Toledo Dr Lafayette LA 70506

BILLERA, LOUIS J(OSEPH), mathematics educator; b. N.Y.C., Apr. 12, 1943; s. Joseph James and Florence Ann (Lombardi) B.; m. Jeanne Marie Kebba, June 20, 1964; children: John L., Mark A. BS, Rensselaer Poly. Inst., 1964; postgrad., Princeton U., 1964-65; MA, CUNY, 1967, PhD, 1968. Asst. prof. Cornell U., Ithaca, N.Y., 1968-73; postdoctoral fellow Hebrew Univ., Jerusalem, Israel, 1969; assoc. prof. Cornell U., Ithaca, N.Y., 1973-80, prof. math., 1980—; prof. math. Rutgers U., New Brunswick, N.J., 1985-89; assoc. dir. DIMACS Ctr. for Discrete Math. and Theoretical Computer Sci., New Brunswick, N.J., 1988-89; vis. rsch. assoc. Brandeis U., Waltham, Mass., 1974-75; prof. invité CORE, U. Catholique de Louvain, Belgium, 1980. Contbr. over 50 articles to profl. jours. Fellow Ednl. Testing Svc., 1964-65, NDEA, 1965-68, NSF, 1969. Office: Cornell U Dept Math White Hall Ithaca NY 14853-7901

BILLIG, FREDERICK STUCKY, mechanical engineer; b. Pittsburgh, Pa., Feb. 28, 1933; s. Thomas Clifford and Melba Helen (Stucky) B.; m. Margaret Rose Pelicano, Nov. 30, 1933; children: Linda Ann Baumler, Donna Marie Bartley, Frederick Thomas, James Richard. B. Engring., Johns Hopkins U., 1955; MS, U. Md., 1958, PhD, 1964. Assoc. engr. Applied Physics Lab., Johns Hopkins U., Laurel, Md., 1955-58, engr., 1958-63, project engr., 1963-73, group supr., 1974-77, asst. dept. supr., 1977-87, assoc. dept. supr., chief scientist, 1987—; lectr. U. Md., College Park, 1965—, Space Inst., U. Tenn., Tullahoma, 1965-69, 84, UCLA, 1987-89, Purdue U., 1986-89, SUNY, Buffalo, 1987; lectr. hypersonics short course, Munich, London, Paris, Rome, 1988; NATO Adv. Group for Aerospace Rsch. Devel. lectr. Ramjet and Ramrocket propulsion system for missiles short course, Monterey, Calif., London, Munich, 1984; pres., chmn. bd. Pyrodyne, Inc., 1977-82, 1984—; mem. consultants panel Project SQUID, Office of Naval Rsch., 1972-74; cons. propulsion directorate USAF Nat. Aerospace Plane Program. Contbr. articles to profl. jours.; editorial bd. Johns Hopkins APL Tech. Digest, 1981-89; patentee (with others) cooled leading edges, fuel injector pilons, high reactivity fuels for supersonic combustion ramjets (a supersonic combustion missile). Mem. Joint Army-Navy-NASA-Air Force Working Group on Combustion, 1971-74, diffuser and ramjets panel Com. on Aeroballistics, Naval Bur. Weapons, 1958-60; mem. U.S. nat. com. Internat. Airbreathing Engines Com., 1972-74, chmn., 1980—; mem. hypersonic propulsion peer rev. group NASA, 1980, High Enthalpy Aerothermal Testing aerothermal testing feasibility adv. com., 1983—; mem. sci. adv. bd. USAF, 1988—, Sci. Adv. Bd. aerospace vehicles standing panel, 1988—. Recipient Silver medal Combustion Inst., 1970, Nat. Aerospace Plane Program Pioneer award, 1989, M. M. Bondaruck award USSR Acad. Sci. and Aviation Sport Fedn. Fellow AIAA (tech. com. on airbreathing propulsion 1966-68, chmn. judges AIAA Middle Atlantic student conf. 1968, tech. com. on propellants and combustion 1970-72, membership chmn. coun. nat. capital sect. 1971-73, standing com. on membership 1973-84, treas. nat. capital sect. 1973-74, sec. nat. capital sect. 1974-75, standing com. on publs. 1974-82, bd. dirs. dir. region 1 1974-81, v.p. bd. dirs. mem. exec. 1981-82, Hugh L. Dryden lectr. in rsch. 1992), Combustion Inst., Md. Acad. Sci., Pi Tau Sigma, Phi Kappa Phi. Republican. Avocations: golf, fishing, hunting. Home: 14522 Faraday Dr Rockville MD 20853-1940 Office: Johns Hopkins U Applied Physics Lab Johns Hopkins Rd Laurel MD 20723-6099

BILLIG, ROBERT EMMANUEL, psychiatric social worker; b. N.Y.C.; s. Benjamin and Pearl (Kwiat) B. B.A., McKendree Coll., 1966; M.S., Fort Hays (Kans.) State Coll., 1969; M.S.W., Marywood Coll., 1974. Diplomate Clin. Social Work Am. Bd. Examiners in Clin. Social Work; cert. hypnotherapist Nat. Bd. Hypnotherapist Examiners. Psychiat. social worker S.I. Devel. Center, 1974-76, Queens and Bklyn. Devel. Centers, 1976-79, Bellevue Psychiat. Hosp., N.Y.C., 1979-88; pvt. practice, 1988—; adminstr. Bellevue Community Support Systems, 1983-86. Mem. Acad. Cert. Social Workers, Nat. Assn. Social Workers, N.Y. State Soc. Clin. Social Work Psychotherapists, Am. Soc. Group Psychotherapy and Psychodrama, Am. Bd. Hypnotherapy, Am. Assn. Profl. Hypnotherapist, Am. Assn. Behavioral Therapists, Mensa. Jewish. Home: 10 Park Ter E New York NY 10034-1524

BILLINGHAM, RUPERT EVERETT, zoologist, educator; b. Warminster, Eng., Oct. 15, 1921; s. Albert E. and Helen (Green) B.; m. Jean Mary Morpeth, Mar. 29, 1951; children: John David, Peter Jeremy, Elizabeth Anne. BA, Oriel Coll., Oxford, Eng., 1943, MA, 1947, DPhil, 1950, DSc, 1957; DSc (hon.), Trinity Coll., 1965, U. Pa., 1992. Lectr. zoology U. Birmingham, Eng., 1947-51; research fellow Brit. Empire Cancer Campaign; hon. research asso. dept. zoology Univ. Coll., London, 1951-57; mem. Wistar Inst.; Wistar prof. zoology U. Pa., Phila., 1957-65; prof., chmn. dept. med. genetics, dir. Henry Phipps Inst., U. Pa. Med. Sch., 1965-71; prof., chmn. dept. cell biology and anatomy U. Tex. Health Sci. Center at Dallas, 1971-86, prof. emeritus, 1990—; mem. allergy and immunology study sect. NIH, 1959-62; mem. transplantation immunology com. Nat. Inst. Allergy and Infectious Diseases, NIH, 1968-70, 71-73, mem. council, 1980-83; mem. sci. adv. bd. St. Jude Children's Research Hosp., Memphis, 1965-70; mem. sci. adv. com. Mass. Gen. Hosp., 1976-79. Contbr. articles to profl. jours.; Editorial bd.: Transplantation, 1980-82; adv. editorial bd.: Placenta, 1980-85; adv. editor: Jour. Exptl. Medicine, 1963-84; assoc. editor: Am. Jour. Reproductive Immunology, 1981-86, Jour. Immunology, 1964-72, Cellular Immunology, 1970-86, Jour. Exptl. Zoology, 1976-80; hon. editorial bd.: Developmental and Comparative Immunology, 1977-80. Served to lt. Royal Navy, 1942-46. Recipient Alvarenga prize Coll. Physicians, Phila., 1963; hon. award Soc. Plastic Surgeons, 1964; Fred Lyman Adair award Am. Gynecol. Soc., 1971. Fellow Royal Soc. (London), N.Y. Acad. Scis., Am. Acad. Arts and Scis.; mem. Am. Assn. Immunologists, Transplantation Soc. (pres. 1974-76), Am. Assn. Transplant Surgeons (hon.), Internat. Soc. Immunology of Reprodn. (pres. 1983-86), British Transplantation Soc. (hon. 1988). Home: RR 3 Box 86P Vineyard Haven MA 02568-9802

BILLINGS, LINDA, writer, analyst; b. Binghamton, N.Y., May 1, 1951; d. John Arthur and Hermina (Kubik) Meskunas; m. Richard N. Billings, Oct. 23, 1981 (div. Dec. 1990). BA, SUNY, Binghamton, 1974; postgrad., George Mason U., 1991—. Editor Space Bus. News, Washington, 1983-85; rsch. assoc. Nat. Commn. on Space, Washington, 1985; sr. editor Air and Space/Smithsonian Mag., Washington, 1986-90, Lockheed Engring., Washington, 1988-90; analyst BDM Internat., Washington, 1990—. Co-author: First Contact: The Search for Extraterrestrial Intelligence, 1990. Vol. Women's Ctr. No. Va., Vienna, 1990—. Recipient Media award Washington Space Bus. Roundtable, 1985, Outstanding Achievement award Women in Aerospace, 1992. Mem. AIAA, Nat. Space Soc., Planetary Soc., Women in Aerospace (bd. dirs. 1990-93), Am. Astron. Soc. (chmn. program com. 1991-92). Democrat. Unitarian Universalist. Home: 5409 N 21st St Arlington VA 22205 Office: BDM Internat 409 3d St SW # 340 Washington DC 20024

BILLINGS, THOMAS NEAL, computer and publishing executive, management consultant; b. Milw., Mar. 2, 1931; s. Neal and Gladys Victoria (Lockard) B.; m. Barta Hope Chipman, June 12, 1954 (div. 1967); children: Bridget Ann, Bruce Neal; m. Marie Louise Farrell, Mar. 27, 1982. AB with honors, Harvard U., 1952, MBA, 1954. V.p. fin. and adminstrn. and technol. innovation Copley Newspapers Inc., La Jolla, Calif., 1957-70; group v.p., dir. tech. Harte-Hanks Comm. Inc., San Antonio, 1970-73; exec. v.p. United Media, Inc., Phoenix, 1973-75; asst. to pres., dir. corp. mgmt. systems Ramada Inns, Inc., Phoenix, 1975-76; exec. dir. NRA, Washington, 1976-77; pres. Ideation Inc., N.Y.C., 1977-81; chmn. Bergen-Billings Inc., N.Y.C., 1977-80; pres. The Assn. Svc. Corp, San Francisco, 1978-91; pres. Recorder Printing and Pub. Co. Inc., San Francisco, 1980-82; v.p. adminstrn. Victor Techs. Inc., Scotts Valley, Calif., 1982-84; mng. dir. Saga-Wilcox Computers Ltd., Wrexham, Wales, 1984-85; chmn. Thomas Billings & Assocs., Inc., Reno, 1978—, Intercontinental Travel Svc. Inc., Reno, 1983—, Oberon Optical Character Recognition, Ltd., Hemel-Hemstead, Eng., 1985-86; bd. dirs. 5M Corp., San Francisco, Intercontinental Rsch. Coun., London, Corp. Comm. Coun., Alameda; dir., chief exec. officer Insignia Software Solutions group, High Wycombe, England, Cupertino, Calif., 1986-89; chmn. Intercontinental News Svc. Inc., London and Alameda, Calif., 1989—; v.p. Cromer Equipment Co., Oakland, Calif., 1989—; bd. dirs. Digital Broadcasting Corp., Mountain View, Calif., Lenny's Restaurants Inc., Wichita, Kans., Tymyndr Corp., Dover Del., Zzyzzyx Corp., Reno, Harrod's Hotel & Casino Corp., Las Vegas, Pandemonium Pictures, Inc., San Mateo, Calif., Bonanza Corp., Virginia City, Nev., Quillmill Ltd., London; speaker and seminar leader. Bd. dirs. Nat. Allergy Found., 1973—, The Wilderness Fund, 1978—, San Diego Civic Light Opera Assn., 1965-69; ; chief exec. San Diego 200th Anniversary Expn., 1969. Served with U.S. Army, 1955-57. Recipient Walter F. Carley Meml. award, 1966, 69. Fellow U.K. Inst. Dirs.; mem. Am. Newspaper Pubs. Assn., Inst. Execs. Inc. (dir.), Inst. Newspaper Fin. Officers, Sigma Delta Chi. Republican. Clubs: West Side Tennis, LaJolla Country; Washington Athletic; San Francisco Press; Harvard (N.Y.C.); Elks. Author: Creative Controllership, 1978, Our Credibility Crisis, 1983, Non-Euclidean Theology, 1987, Ruminations on Meta Mentality, 1990, Fixing our Broken System, 1992; editor: The Vice Presidents' Letter, 1978—; pub. The Microcomputer Letter, 1982—, Synthetic Hardware Update, 1987—; editor: Intercontinental News Svc., London, England, and Alameda Calif., 1985—. Office: PO Drawer I Alameda CA 94501-0262 Office: 4701 Oakport St Oakland CA 94601

BILLINGS, WILLIAM DWIGHT, ecology educator; b. Washington, Dec. 29, 1910; s. William Pence and Mabel (Burke) B.; m. Shirley Ann Miller, July 29, 1958. BA, Butler U., 1933, DSc, 1955; MA, Duke U., 1935, PhD, 1936. Instr. botany U. Tenn., 1936-37; instr. biology U. Nev., 1938-40, asst. prof., 1940-43, assoc. prof., 1943-49, prof., chmn. biology dept., 1949-52; assoc. prof. botany Duke U., 1952-58, prof., 1958-67, James B. Duke prof., 1967—; prin. rsch. assoc. Inst. Arctic Biology, U. Alaska, 1984—; mem. adv. panels NSF-AEC, Washington, 1954-58; adj. rsch. prof. Desert Rsch. Inst., U. Nev., 1982. Author: Plants and the Ecosystem, 1964, 78, Plants, Man and the Ecosystem, 1970, Vegetation and the Environment, 1974, The Vegetation of North America, 1988; editor Ecology, 1952-57, Ecol. Monographs, 1969; mem. editorial bd. Ecol. Studies, 1975-93, Arctic and Alpine Rsch., 1975-82. Fulbright rsch. scholar, N.Z., 1959. Fellow Arctic Inst. N.Am., Explorers Club, Am. Acad. Arts and Scis.; mem. Ecol. Soc. Am. (v.p. 1960, pres. 1978-79, Mercer award 1962, Disting. Svc. award 1981, Nevada medal 1989, Eminent Ecologist award 1991), Brit. Ecol. Soc. (hon. fgn. mem.), Bot. Soc. Am. (chmn. ecology sect. 1976, Cert. of Merit 1960), Inst. Arctic and Alpine Rsch. (mem. sci. adv. com. 1975-89). Achievements include research on arctic, alpine and desert ecology. Home: 1628 Marion Ave Durham NC 27705 Office: Duke U Biol Sci Bldg Rm 142 Durham NC 27708-0338

BILLINGSLEY, DAVID STUART, researcher; b. Chgo., June 6, 1929; s. Archibald Stuart and Helen Wilson (Murdoch) B. BSChemE, Tex. A&M U., 1950, BSME and BS in Indsl. Engring., 1956, MSChemE, 1958, PhD ChemE, 1961. System engr. IBM, Houston, 1961-63, with scientific staff, 1963-76; with scientific staff IBM, Palo Alto, Calif., 1976-82; internat. petroleum ctr. staff IBM, Houston, 1982-85; ind. rschr. Houston, 1985—; referee AIChE Jour., 1972-85, Canadian Jour. Chem. Engring., 1972. Author: Descendants of Richard Billingsley, 1992; contbr. articles to profl. jours. 1st lt. U.S. Army, 1951-53. Home: Apt 708 7520 Hornwood Houston TX 77036-4326

BILLMEYER, FRED WALLACE, JR., chemist, educator; b. Chattanooga, Aug. 24, 1919; s. Fred W. and Eleanor (Salmon) B.; m. Annette M. Trzcinski, Aug. 4, 1951; children: Eleanor A., Dean W., David M. B.S., Calif. Inst. Tech., 1941; Ph.D., Cornell U., 1945. With plastics dept. E.I. du Pont de Nemours & Co., 1945-64; lectr. high polymers dept. chemistry U. Del., 1951-64; Vis. prof. chem. engring. Mass. Inst. Tech., 1960-61; prof. analytical chemistry Rensselaer Poly. Inst., Troy, N.Y., 1964-84, prof. emeritus, 1984—; cons. various coms. Internat. Commn. Illumination (CIE), 1964—; mem. U.S. Nat. Com. CIE, 1968—, v.p., 1975-79, mem. for life, 1983—. Author: Textbook of Polymer Chemistry, 1957, Textbook of Polymer Science, 1961, 3d edit., 1984, Synthetic Polymers, 1972, (with Max Saltzman) Principles of Color Technology, 1966, 2d edit., 1981, (with E.A. Collins and J. Bares) Experiments in Polymer Science, 1973, (with R. N. Kelley) Entering Industry, 1975; also articles; editorial adviser: Optical Spectra, 1967-80; editor-in-chief: Color Research and Application, 1976-86. Trustee Munsell Color Found., sec., 1975-83. Recipient Bruning award Fedn. Socs. Coatings Tech., 1977. Fellow Am. Phys. Soc., Optical Soc. Am., Soc. Plastics Engrs., ASTM (award of Merit 1990), AAAS; mem. Am. Chem. Soc., N.Y. Soc. Coatings Tech., Inter-Soc. Color Coun. (pres. 1968-70, sec. 1970-82, Macbeth award 1978, Service award 1983, Godlove award 1993)., Coun. Optical Radiation Measurements (sec. 1979-83), Sigma Xi, Phi Kappa Phi. Home: 1294 Garner Ave Schenectady NY 12309-5746

BILODEAU, JOHN LOWELL, engineer; b. Millinocket, Maine, Apr. 30, 1941. BS in Physics, Auburn (Ala.) U.; MSEE, Naval Postgrad. Sch. Engr. Firestone Synthetic Fiber, Hopewell, Va., 1981-82; instr. Va. State U., Petersburg, 1982-83; mgr. tech. svcs. BMY, York, Pa., 1983-90; mgr. engring. Chamberlain MRC, Hunt Valley, Md., 1990—. Maj. USMC, 1961-81. Office: Chamberlain MRC 336 Clubhouse Ln Cockeysville Hunt Valley MD 21031

BILOFSKY, HOWARD STEVEN, scientist; b. N.Y.C., June 21, 1943. AB, CUNY, 1965; PhD, Worcester (Mass.) Poly. Inst., 1972. Sr. mgr. life sci. dept. Bolt, Beranek & Newman, Inc., Cambridge, Mass., 1973-90; exec. planning officer European Bioinformatics Inst. European Molecular Biology Lab., Heidelberg, Germany, 1990-93; v.p., dir. core sci. info. techs. R & D SmithKline Beecham Pharms., King of Prussia, Pa., 1993—. Contbr. articles to profl. publs. Mem. AAAS, Am. Computing Machinery, Am. Chem. Soc. Office: SmithKline Beecham Pharms R & D 709 Swedesford Rd King Of Prussia PA 19406

BILPUCH, EDWARD GEORGE, nuclear physicist, educator; b. Connellsville, Pa., Feb. 10, 1927; s. John and Elizabeth (Kochisco) B.; m. Marilyn Jean Strohkorb, Sept. 6, 1952. B.S., U. N.C., 1950, M.S., 1952, Ph.D. (Morehead scholar), 1956; Dr. phil. nat. honoris causa, Johann Wolfgang Goethe-U., Frankfurt am Main, Germany, 1992. Research assoc. Duke U., 1956-59, asst. prof., 1960-65, assoc. prof., 1966-70, prof. physics, 1970—; dep. dir. Triangle Univs. Nuclear Lab., Duke Sta., Durham, N.C., 1966-78; dir. Triangle Univs. Nuclear Lab., 1978-92; vis. prof. U. Frankfurt (West Germany), 1972, 74, Fudan U., Shanghai, China, spring 1983; chmn. NBS Evaluation Panel for Radiation Research, 1983-85. Contbr. articles to profl. jours. Served with USNR, 1945-46. Sr. U.S. scientist Humboldt award Fed. Republic of Germany, 1983-84; named Disting. Prof. Duke U., 1987, hon. prof. Fudan U., Shanghai, Peoples Republic of China, 1987; recipient Disting. Alumnus award U. N.C., 1988, Jesse W. Beams award Southeastern Sect. Am. Phys. Soc., 1992. Fellow NAS (physics survey com.), Am. Phys. Soc. (Jesse W. Beams award Southeastern sect. 1992), Phi Beta Kappa, Sigma Xi; mem. AAAS. Home: 106 Cherokee Cir Chapel Hill NC 27514-2718 Office: Duke U Triangle U Nuclear Lab PO Box 90308 Durham NC 27708-0308

BINDER, KURT, physicist, educator; b. Korneuburg, Austria, Feb. 10, 1944; s. Eduard B. and Anna; m. Marlies Ecker; children: Martin, Stefan. Degree, Tech. U. Vienna, 1967, PhD, 1969. Prof. Saarland U., Saarbrücken, 1974-77, U. Cologne, 1977—; dir. Inst. of Solid State Rsch. Nuclear Rsch. Sta., Jülich, %; prof. dept. physics U. Mainz, Germany. Contbr. numerous articles to sci. jours. Recipient Max-Planck-Medaille, Deutsche Physikalische Gesellschaft, 1993. *

BINGHAM, J. PETER, electronics executive; married; 2 children. BS in Physics cum laude, Polytechnic Inst., N.Y.; MS in Exptl. Physics, U. Md., PhD in Elec. Engring. With RCA Consumer Electronics, David Sarnoff Rsch. Ctr.; exec. v.p., tech. Thomson Consumer Electronics; v.p., engring. CEC, and Magnavox CATV Systems; with consumer electronics divsn. Philips Lab., 1982-91, pres., 1991—. Recipient David Sarnoff award, RCA Lab. Achievements award; Naned in his honor Bingham Peak in Antarctica, Arctic Inst. of North Am. Office: Phillips Lab 345 Scarborough Rd Briarcliff Manor NY 10510*

BINGHAM, PARIS EDWARD, JR., electrical engineer, computer consultant; b. Aurora, Colo., Sept. 26, 1957; s. Paris Edward and Shirley Ann (Blehm) B.; m. Laurie Sue Piersol, May 9, 1981 (div. Sept. 1987); m. Helen Naef, Aug. 7, 1993. BS in Elec. Engring. and Computer Sci., U. Colo., 1979. Mem. tech. staff Western Electric Co., Aurora, 1979-81, system engr., 1981; mem. electronic tech. staff Hughes Aircraft Co., Aurora, 1981-83, staff engr., 1983-86, sr. staff engr., 1986-93, scientist, engr., 1993—; cons. RJM Assocs., Huntington, N.Y., 1987—; cons. Aurora, 1988—. Mem. AAAS, IEEE, Assn. for Computing Machinery, Math. Assn. Am. Republican. Presbyterian. Achievements include research on migration to and integration of OSI networking protocols, distributed, networked computing, and neural net applications. Office: Hughes Aircraft Co 16800 E Centretech Pky Aurora CO 80011-9046

BINI, DANTE NATALE, architect, industrial designer; b. Castelfranco Emilia, Modena, Italy, Apr. 22, 1932; came to U.S., 1981; s. Giovanni and Maria (Cavallini) B.; m. Adria Vittoria Moretti, June 27, 1963; children: Stefano Alec, Nicolo Guiseppe. Grad., L.S.A. Righi, Bologna, Italy, 1952; PhD in Architecture, U. Florence, 1962. Chmn. Societa' Anonima Immobiliare Castelfranco Emilia, Castelfranco Emilia, 1960-64, Vedova Bini, Castelfranco Emilia, 1960-64; founder, chmn. Unipack, Old Home, Bologna, 1961-65; founder, exec. v.p. Binishell Spa, Bologna, 1966-69; cons. Dept. Pub. Works New South Wales, Sydney, Australia, 1972-74, Jennings Industries Ltd., Melbourne, Australia, 1975-80; founder Bini Cons. Australia; founder, pres., chmn. Binistar, Inc., San Francisco, 1981—; external cons. Bechtel Nat., Inc., San Francisco, 1985-86; founder, pres., chmn. Pak-Home, Inc. (now Napa Valley Corp.), San Francisco, 1986—; cons. Shimizu Corp., Tokyo, 1989-93; spl. cons. to UN, Rome, 1968, Shimizu Tech. Ctr. Am., 1989, Shimizu Co., Tokyo, 1991-92; lectr. Moscow Expocenter, 1986; vis. lectr. NASA Ames Rsch. Ctr., 1989, vis. lectr. univs. Italy, Australia, U.S., Mex., Venezuela, Brazil, Argentina, Peru, Eng., USSR, Fed. Republic Germany. Contbr. articles and papers to profl. jours., also conf. procs.; patentee self-shaping structures for low-cost, sport and indsl./comml. bldgs., designer of a self-shaping mega-structure for a new city's infrastructure and framework, 1989; developer new automated constrn. system for multi-storied bldg.; co-researcher for self-shaping, self-sinking lunar habitat. Decorated Order of Commendatore Pres. Italian Republic, 1989; recipient Eurostar award European Inst. Packaging, 1964, Excellence in Engring. Design award Design News Mag., 1968, Best Idea of Yr. award European Design News Mag., 1968, Excellence in Indsl. Design award I.E.S. Australia, 1976. Mem. Bd. Architects Emilia e Romagna, Assn. Architects Bologna (co-founder 1963), Italian Assn. Indsl. Design, Italian Inst. Packaging Design (Oscar award 1961-63), Royal Australian Inst. Architects New South Wales, Am. Assn. Mil. Engrs. Roman Catholic. Avocations: golf, tennis, skiing (Univ. World Championship contestant 1957).

BINNEY, JAMES JEFFREY, physicist; b. London, Eng., Apr. 12, 1950; s. Harry Augustus Roy and Barbara (Poole) B. BA, Cambridge U., 1971; PhD, Oxford U., 1975. Rsch. fellow Magd Coll., Oxford, Eng., 1975-79; lectr. in theoretical physics Oxford U., 1981-90, reader in theoretical physics, 1990—; vis. asst. prof. Princeton (N.J.) U., 1979-81; mem. theory panel bd. U.K. Serc, London, 1986-88. Co-author: Galactic Astronomy, 1981, Galactic Dynamics, 1988, Pick for Humans, 1990, The Theory of Critical Phenomena, 1992; contbr. articles to profl. jours. Recipient Maxwell prize Inst. Physics, London, 1986; Fairchild Disting. scholar, 1983-84. Fellow Royal Astron. Soc.; mem. Am. Astron. Soc. Office: Oxford U, Dept Theoretical Physics, 1 Keble Rd, Oxford OX1 3NP, England

BINNIG, GERD KARL, physicist; b. July 20, 1947; m. Lore Binnig, 1969; 2 children. Diploma in Physics, Goethe U., Frankfurt, Fed. Republic Germany, PhD, 1978. Rsch. staff mem. IBM Zurich Rsch. Lab., 1978—, group leader, 1984—; with Stanford U., 1985-86; hon. prof. physics U. Munich, 1987—; vis. prof. Stanford U., 1986-88; mem. supervisory bd. Daimler Benz Holding AG, 1990; mem. tech. coun. IBM Acad., adv. bd. Bild der Wissenschaft. Author: Aus dem Nichts, 1989; editor Natural Sci.; mem. editorial bd. Rev. Sci. Instruments. Co-recipient Nobel prize in physics, 1986; recipient physics prize German Phys. Soc., 1982, Otto Klung prize, 1983, Joint King Faisal Internat. prize for sci., Hewlett-Packard Europhysics prize, 1984, Elliot Cresson medal Franklin Inst., 1987, Grosses Verdienstkreuz mit Stern und Schulterband des Verdienstordens, 1987, Minnie Rosen award Ross U., 1988. Fellow Royal Microscopical Soc. (hon. 1988); Acad. Scis. (fgn. assoc. 1987). Avocations: music, tennis, soccer, golf. Office: IBM Rsch Div Physics Group, Schellingstrasse 4, 80799 Munich 40, Germany

BINNS, WALTER ROBERT, physics researcher; b. Ottawa, Kans., Nov. 7, 1940; s. Willard Russel and Dorothy Ethel (Moore) B.; married; children: Cynthia, Martha. BS, Ottawa U., 1962; PhD, Colo. State U., 1969. Scientist McDonnell Douglas Rsch. Lab., St. Louis, 1969-80; rsch. prof. physics Washington U., St. Louis, 1980—. Republican. Office: Washington U Dept Physics 1 Brookings Dr Saint Louis MO 63130

BINNS, WILLIAM ARTHUR, clinical psychologist; b. Kansas City, Kans., June 8, 1925; s. Ralph Clare and Kathryn Miller (Hetherington) B.; m. Erika Anna MAria Thoelldte, Jan. 31, 1948; children: Kathryn May Binns Jackson, Sandra Helen Binns Willey. BA, U. Kans., 1949; postgrad., Menninger Sch. Psychiatry, Topeka, 1950-56. Clinical trainee Winter VA. Hosp. Menninger Found., Topeka, 1949-51; clin. psychologist Shawnee Ctr. Menninger Clinic and CF Menninger Hosp., Topeka, 1951-56, Watkins Meml. Hosp., U. Kans., Lawrence, 1956-87; pres. Accumulative Investor's Inc., Kansas City, Mo., 1987-90; chmn. endowment fund com. Community Mental Health Clinic, Lawrence, 1986—, also bd. dirs., 1987—; bd. dirs. emeritus Bert Nash Mental Health Clinic; v.p. Kansas City chpt. Am. Assn. Individual Investors, 1987-90, pres., 1991-93, dir., 1993—. Contbr. articles to profl. jours. Bd. dirs. Trinity Foster Home, Lawrence; chmn. City

Human Rels. Commn., Lawrence, 1961-67, Ea. Cen. Kans. Econ. Opportunity Coun., Lawrence and Ottawa, 1968-74, Univ. Ecumenical Ctr. Bd., Lawrence, 1975-78; selection officer Peace Corps, Washington, 1960-61. Mem. NASD (bd arbitrators), Am. Assn. Individual Investors. Democrat. Episcopalian. Avocations: study of economic theory and investments. Home: 1014 Wellington Rd Lawrence KS 66049-3029

BIONDO, RAYMOND VITUS, dermatologist; b. N.Y.C., June 13, 1936; s. Joseph Pernice and Bena (Schwartz) B.; m. Mary McKinnon, Dec. 24, 1976. BA in Biology, U. No. Colo., 1960; MS in Biochemistry, U. Ark., 1963, BS in Medicine, 1967, MD, 1967. Diplomate Am. Bd. Dermatology. Asst. mgmt. analyst 389th USAF Hosp., Francis E. Warren AFB, Wyo., 1954-58; rsch. trainee NIH at U. Ark., Little Rock, 1961-63; rsch. biochemist VA Hosp., Little Rock, 1963-65; intern U. Cin. Med. Ctr., 1967-68; resident in dermatology U. Ark., Little Rock, 1968-71, asst. clin. prof. dermatology, 1971-90; pres. North Little Rock (Ark.) Dermatology Clinic, 1971-90. Contbr. rsch. articles to profl. publs. Mem. nat. adv. com. on scouting for the handicapped Boy Scouts Am., 1976-81, nat. chmn. med. exploring com., 1981-84, nat. coun., 1977—, nat. exploring com., 1977-92, nat. urban field svc. com. 1990—, nat. Jewish com. on scouting, 1981-84; mem. Ark. Kidney Disease Commn., 1979-83; bd. dirs. Cen. Ark. Health Systems Agy., 1984, Congregation B'nai Israel, Little Rock, 1982-84, Jewish Fedn. Ark., 1989-92; founder, pres. Am. Red Magen David for Israel, Ark., 1987-92. Recipient Outstanding Alumnus award U. No. Colo., 1977, William H. Spurgeon and Whitney M. Young awards Boy Scouts Am., 1981, Silver Antelope award, 1983, Silver Beaver award Boy Scouts Am., 1977, Shofar award, 1992, Vol. Action award The White House, 1982, Gov.'s Vol. Excellence award Gov. Bill Clinton, Ark., 1983, cert. of appreciation for pub. svc. Gov. Bill Clinton, 1992, Father Joseph H. Biltz award Ark. Coun. of NCCJ, 1990. Fellow Am. Acad. Dermatology (adv. bd. nat. program dermatology of Chgo. 1973-75); mem. AMA (cert. appreciation 1984, Physician's Recognition award 1971-90), Ark. Med. Soc., Ark. Dermatol. Soc. (pres. 1977, ho. of dels. 1971—), Pulaski County Med. Soc. Jewish. Home: PO Box 6361 North Little Rock AR 72116-0761

BIRBARI, ADIL ELIAS, physician, educator; b. Ziguinchor, Senegal, May 26, 1933; s. Elias George and Sophia George (Nasrallah) B.; m. Micheline Michel Ghosn, Feb. 4, 1978; children: Yolande, Sophia. Baccalaureat II, Internat. Coll., Beirut, 1952; BS, Am. U., Beirut, 1955, MD, 1959. Fellow in hypertension Peter Bent Brigham Hosp., Boston, 1963-65, assoc. dir. hypertension lab, 1965-66; asst. prof. medicine and physiology Am. U., Beirut, 1967-72, assoc. prof. medicine and physiology, 1972-77; prof., chmn. dept. medicine Lebanese U., Beirut, 1987-90. Author: Kidney and Genetic Diseases, 1986, Manual of Clinical Hypertension; contbr. 55 articles to profl. jours. Fellow high blood pressure coun. Am. Heart Assn.; pres. Lebanese Hypertension League, Beirut, 1986; v.p. Lebanese Health Soc., Beirut, 1986; active Camille Shamoun Found., Beirut, 1992. Grantee Nat. Coun. Sci. Rsch., 1973-91. Fellow ACP, Internat. Coll. Angiology; mem. Lebanese Assn. Advancement of Sci. (sec. gen. 1974-92), Lebanese Soc. Nephrology and Hypertension (pres. 1975—, editor-in-chief Jour. 1990), Lebanese Hypertension League (pres.), Internat. Soc. Nephrology, Internat./European Soc. Hypertension, Am. Soc. Hypertension, Nat. Coun. for Sci. Rsch. (Lebanon). Avocations: photography, gardening. Office: Am Univ Beirut 850 3rd Ave Fl 18 New York NY 10022

BIRD, HARRIE WALDO, JR., psychiatrist, educator; b. Detroit, Sept. 21, 1917; s. Harrie Waldo and Ann Josephine (Tossy) B.; m. Della Mae Clemmer, Jan. 4, 1943; children: Harrie Waldo, Kathleen Bird Steinhour, Deborah Bird Hall, Mark Henry, Matthew Alexius, Liza George-Aidan Browning. A.B., Yale U., 1939; postgrad., U. Mich. Med. Sch., 1939-41; M.D., Harvard U., 1943. Intern Phila. Gen. Hosp., 1943-44; resident Menninger Sch. Psychiatry, Topeka, 1946-48; chief infirmary sect. Winter VA Hosp., Topeka, 1946; psychiatrist Adult Psychiat. Clinic, Detroit, 1949; acting dir. Adult Psychiat. Clinic, 1950; psychiat. cons. Mich. Epilepsy Center, Detroit, 1950-55; clin. instr. psychiatry Wayne State U., Detroit, 1952-55; asso. prof. psychiatry U. Chgo., 1955-56, U. Mich., Ann Arbor, 1956-63; asst. dean Med Sch., 1959-61; prof. psychiatry, asso. dean St. Louis U. Sch. Medicine, 1965-68, clin. prof., 1970—, dir. The Family Psychiat. Ctr., 1972-93; lectr., cons. in field. Bd. dirs. Mich. Epilepsy Ctr., 1956-63, Wayne County Mental Health Soc., 1956-63, Mich. Epilepsy Assn., 1956-63, El Paso Mental Health Assn., 1969-70, Cranbrook Sch., 1961-63. Served with M.C. AUS, 1944-46. Recipient Mental Health Inst. award St. John's U., 1966. Fellow Am. Psychiat. Assn. (life); mem. Am. Family Therapy Assn. (charter), AMA, Group for Advancement Psychiatry, Mo. Med. Soc. (hon.), St. Louis Met. Med. Soc. (life), Ea. Mo. Psychiat. Soc., Phi Beta Kappa. Address: 62 Conway Ln Saint Louis MO 63124-1203

BIRD, MARY LYNNE MILLER, association executive; b. Buffalo, Feb. 25, 1934; d. Joseph William and Mildred Dorothy (Wallette) Miller; m. Thomas Edward Bird, Aug. 23, 1958; children: Matthew David, Lisa Bronwen. AB magna cum laude, Syracuse U., 1956; postgrad., Columbia U., 1956-58. Mem. rsch. staff Ctr. for Rsch. in Personality, Harvard U., Cambridge, Mass., 1959-62, Ctr. Internat. Studies, Princeton (N.J.) U., 1962-66, Inst. Internat. Social Rsch., Princeton, 1965, Sch. Internat. Affairs, Columbia U., N.Y.C., 1966-67, Coun. Fgn. Rels., N.Y.C., 1967-69, Twentieth Century Fund, N.Y.C., 1969-72; asst. to pres. World Policy Inst., N.Y.C., 1972-74; dir. devel. Fund for Peace, N.Y.C., 1974-78; dir. fellows program Exec. Council Fgn. Diplomats, N.Y.C., 1978-79; dir. devel. Assn. Vol. Surgical Contraception, N.Y.C., 1979-83; exec. dir. Am. Geog. Soc., N.Y.C., 1983—; cons. Fedn. Am. Scientists, Washington, 1974-75. Trustee Bel Canto Opera Co., N.Y.C., 1975—. Maxwell Citizenship scholar Syracuse U., 1952-56. Mem. Assn. Am. Geographers, Soc. Woman Geographers, Inst. for Current World Affairs, Nat. Coun. Geographic Edn., Internat. Geographical Union (liason mem. U.S. nat. com.), Conf. Latin Americanist Geographers, Planning Com. for Nat. Assessment on Ednl. Progress in Geography, N.Y. Acad. Sci., St. David's Soc. (v.p.), English Speaking Union, Phi Beta Kappa, Phi Kappa Phi, Eta Pi Upsilon. Avocations: singing, sailing. Office: Am Geog Soc Ste 600 156 Fifth Ave New York NY 10010-7002

BIREWAR, DEEPAK BABURAO, chemical engineer, researcher; b. Wani, India, Oct. 29, 1962; came to U.S., 1985; s. Baburao V. and Alka B. (Kondawar) B.; m. Shobanaa Raman, Aug. 22, 1991. BTech, Indian Inst. Tech., Bombay, 1985; PhD, Carnegie Mellon U., 1989. Tech. dir. Multi Organics Pvt Ltd., Bombay, 1990-91; rsch. engr. NuKem Devel., Houston, 1991-92; staff engr. DuPont Polymers, Wilmington, Del., 1992—. Contbr. articles to profl. jours. Hindu. Office: DuPont CR&D Exptl Sta Bldg 1 Rm 221 PO Box 80101 Wilmington DE 19880-0101

BIRGE, ROBERT RICHARDS, chemistry educator; b. Washington, Aug. 10, 1946; s. Robert Bowen and Dorothy (Richards) B.; m. Constance A. Reed, Aug. 3, 1993; children: Jonathan Richards, David Porter. B.S. in Chemistry, Yale U., 1968; Ph.D. in Chem. Physics, Wesleyan U., Middletown, Conn., 1972. NIH postdoctoral fellow Harvard U., Cambridge, Mass., 1973-75; asst. prof. dept. chemistry U. Calif.-Riverside, 1975-81, chmn. com. on research, 1981-82, assoc. prof. dept. chemistry, 1981-84; Weingart sabbatical fellow Calif. Inst. Tech., Pasadena, 1982-83; prof., head dept. chemistry Carnegie-Mellon U., Pitts., 1984-87, dir. Ctr. Molecular Electronics,, 1984-87; prof. chemistry, dir. W.M. Keck Ctr. Molecular Electronics Syracuse (N.Y.) U., 1988—, dir. grad. biophysics program, 1989-93; rsch. dir. N.Y. State Ctr. for Advanced Tech. in Computer Applications and Software Engring., 1992—; NATO prof. Advanced Study Inst., Maratea, Italy, 1983; permanent mem. molecular and cellular biophysics study sect. NIH, Bethesda, Md., 1984-89; bd. dirs. West Penn Hosp. Rsch. Found., 1987-88; co-chmn. adv. com. molecular electronics NAS, 1987. Editorial bd. Jour. Nanotechnology, Brit. Inst. Physics, 1990—; contbr. chpts. to books, over 100 articles to profl. jours. Treas. council Carnegie Inst. Natural History, Pitts., 1985-87. Served to 1st lt. USAF, 1972-73. Recipient Nat. Sci. award Am. Cyanamid Corp., 1964; Regents fellow U. Calif., 1976. Mem. Am. Chem. Soc., Am. Phys. Soc., Biophys. Soc. Home: 346 Summerhaven Dr East Syracuse NY 13057 Office: Syracuse U Dept Chemistry W M Keck Ctr Molecular Electronics Syracuse NY 13244

BIRKEDAL-HANSEN, HENNING, dentist, educator. PhD in Biochemistry, Royal Dentistry Coll., Denmark, 1977. Prof. dentistry Sch. Dentistry, U. Ala., Birmingham, 1983—. Office: U Ala Rsch Ctr in Oral Biology UAB Sta Birmingham AL 35294*

BIRKS, JOHN WILLIAM, chemistry educator; b. Vinita, Okla., Dec. 10, 1946; s. Melvin Young Birks and Mary Ann (Garman) Brooks; m. Kathy Lou Rowlen, July 21, 1990; children: Krishna D., Alicia S., Shannon L. BS, U. Ark., 1968; MS, U. Calif., Berkeley, 1970, PhD, 1974. Asst. prof. U. Ill., Urbana-Champaign, 1974-77; assoc. prof. U. Colo., Boulder, 1977-84, prof., 1984—; vis. scientist Max Planck Inst. Chemistry, Mainz, Germany, 1981-82. Editor: Chemiluminescence and Photochemical Reaction Detection in Chromatography, 1989; co-editor: Hidden Dangers: Environmental Consequences of Preparing for War, 1990, CHEMRAWN VII: The Chemistry of the Atmosphere, Its Impact on Global Change, 1992; contbr. over 100 articles to publs. Recipient Leo Szilard award Am. Phys. Soc., 1985, Witherspoon Peace and Justice award Witherspoon Soc., 1986; Alfred P. Sloan fellow, 1979-81, John Simon Guggenheim fellow, 1986. Mem. AAAS, Am. Chem. Soc. (Colo. Sect. award 1990), Sigma Xi, Alpha Chi Sigma, Phi Beta Kappa. Achievements include co-development of nuclear winter theory. Office: U Colo Dept Chemistry Campus Box 216 Boulder CO 80309-0216

BIRK-UPDYKE, DAWN MARIE, psychologist; b. Saddle Brook, N.J., July 9, 1963; d. George Benty and Constance (Pepsis) Birk; m. James A. Updyke, Aug. 19, 1989; children: William O., Elizabeth N. BA, Colo. Coll., 1985; PhD, Utah State U., 1989. Lic. psychologist, Md. Psychology specialist clin. svcs. Devel. Ctr. for Handicapped Persons, Logan, Utah, 1987-89; postdoctoral fellow Kennedy Inst./Johns Hopkins U., Balt., 1989-90; lic. psychologist Franklin Square Hosp., Balt., 1990-93, Southwestern C.M.H.C., Balt., 1993, Ea. Mont. Mental Health Ctr., Miles City, 1988-89; cons. Utah Sch. for Deaf, Ogden, 1988-89, Cache Instrnl. Workshop, Logan, 1987-89, Meridian Heritage Nursing Ctr., Dundalk, Md., 1990-93. Contbr. articles and abstracts to profl. jours. Mem. Am. Psychol. Assn., Assn. for Behavior Analysis. Office: Eastern Montana Mental Health Ctr 2508 Wilson PO Box 1530 Miles City MT 59301

BIRMAN, JOSEPH LEON, physics educator; b. N.Y.C., May 21, 1927; s. Max and Miriam Ida (Meyerson) B.; m. Joan Sylvia Lyttle, Feb. 22, 1950; children: Kenneth, Deborah, David. BS, CCNY, 1947; MA, Columbia U., 1950, PhD, 1952; Docteur ès Sciences honoris causa, U. Rènnes, France, 1974. Sr. physicist, head luminescence sect. GTE Research Labs., N.Y., 1952-62; Mary Amanda Wood vis. prof. U. Pa., 1960; assoc. prof. physics NYU, 1962-64, prof., 1964-74; Henry Semat prof. physics CCNY, 1974-88, Disting. prof. physics, 1987—; cons. research labs.; vis. prof. U. Paris, Ecole Normale Superieure, 1969-70, Japan Soc. for Promotion of Sci., Research Inst. for Fundamental Physics, U. Kyoto, Japan, 1978, 80, Inst. Hautes Etudes Scientifiques, Bures/Yvette, France, 1976, 78, 80, 82, 86, 87-88, 91, U. Regensburg, 1983, 84, 85, 87; vis. prof. dept. theoretical physics Oxford (Eng.) U., 1981, 84, 85, 86; Lady Davis vis. prof. Technion, Israel, 1981; vis. prof. Peking, Fudan, Nanking, Xian Univs., 1980, 82, 85, U. Stuttgart, 1986, U. de Paris VI, 1987-88, 91; Meyerhoff vis. prof. Weitzman Inst., Rehovoth, 1988; founder, chmn. Am. coordinating com. Chinese Scholars Program, joint program of Am. Phys. Soc. and Chinese Acad. Sci./Chinese State Com. Edn., 1983-86. Author: Theoretical Physics, 1952, Handbuch der Physik, Vol. 25/2b, 1974, reprinted 1984 (Russian transl. 1978); editor: Light Scattering in Solids, 1976, 79; co-editor: Laser Optics of Condensed Matter, 1988; mem. editorial bds., cons. Springer Verlag, Plenum Press, Nova Pubs., World Sci. Press, Oxford U. Press; contbr. over 250 articles to profl. jours. Served with USNR, 1945-46. Research grantee NSF, Army Research Office, Aerospace Research Labs., Dept. Def.; J.S. Guggenheim Meml. Found. fellow, 1980-81. Fellow Am. Phys. Soc. (com. on internat. freedom of scientists 1991-93, chmn. 1993), AAAS (com. on sci. freedom and responsibility 1991-93); mem. Com. Concerned Scientists (nat. bd. dirs.), N.Y. Acad. Scis. (human rights com. 1980—, chair 1993—, gov.-at-large 1989-90, v.p. 1991-92). Home: 100 Wellington Ave New Rochelle NY 10804-3708 Office: CCNY Physics Dept 138th St and Convent Ave New York NY 10031

BIRMAN, VICTOR MARK, mechanical and aerospace engineering educator; b. Leningrad, Russia, Jan. 13, 1950; came to U.S., 1984; s. Mark Samuel and Sima (Pesenson) B.; m. Anna Irene Rabkin, Apr. 9, 1977; children: Michael, Shirley. MS, Shipbuilding Inst., Leningrad, 1973; PhD, Technion, Haifa, Israel, 1983. Engr. Steel Structures Design Inst., Leningrad, 1973-78; grad. teaching asst. Technion, Haifa, 1979-82, rsch. fellow, 1983; engr. Isarel Aircraft Industries, Lod, 1984; asst. prof. U. New Orleans, 1984-87, assoc. prof., 1987-89; assoc. prof. U. Mo.-Rolla, St. Louis, 1989—; mem. summer faculty Air Force Office of Sci. Rsch., Wright-Patterson AFB, 1992, NASA Lewis Ctr., 1993; vis. scientist Air Force Inst. Tech., 1993; cons. to numerous cos., orgns. Assoc. editor Composites Engring. jours., 1991—; translator; reviewer profl. jours., 1989—; contbr. rsch. papers to profl. jours., papers to profl confs. Recipient Award for Excellence in Rsch., U. New Orleans Alumni Assn., 1987; summer scholar U. New Orleans, 1986. Fellow AIAA (assoc.); mem. ASME (mem. composite materials com., mem. structures and materials com., mem. design and analysis com.). Achievements include rsch. in dynamic stability of composite structures, imperfection-sensitivity, thermoelasticity and divergence instability of stiffened composite shells and smart composite structures. Office: U Mo-Rolla Engring Edn Ctr 8001 Natural Bridge Rd Saint Louis MO 63121

BIRNBAUM, JEROME, pharmaceutical company executive; married; 2 children. BS in Biology, CUNY, 1961, postgrad., 1961-63; MS in Microbiology, U. Cin., 1964, PhD, 1966. Sr. rsch. microbiologist dept. fermentation rsch. Merck Sharp and Dohme Rsch. Labs., 1966-69, assoc. dir., then dir. dept. fermentation rsch., 1969-74; sr. dir. basic microbiology Merck Inst. Therapeutic Rsch., 1974-75, exec. dir. basic biol. scis., 1975-80; v.p. microbiology and agrl. rsch. Merck Sharp and Dohme Rsch. Labs., Merck & Co., Inc., 1981-87; sr. v.p., therapeutic area dir. Bristol-Myers Co., Princeton, N.J., 1987-89; exec. v.p. rsch. Bristol-Myers Squibb Co., Princeton, 1989-90, sr. v.p. pharm. devel., 1990—; adv. bd. biotech. program Rutgers U., Newark, 1986-87, sci. adv. coun. Waksman Inst. Microbiology, Rutgers U., New Brunswick, 1983-88; mem. AIDS task force Pharm. Mfrs. Assn., 1988—; mem. adv. bd. internat. AIDS rsch. Inst. Medicine, Nat. Acad. Sci., 1990—; adv. bd. Nat. Inst. Community Health Edn., 1990—; invited lectr. numerous confs., ednl. instns. Editorial bd. Jour. Antibiotics, 1982-90, Environ. and Applied Microbiology, 1976-79, Applied Microbiology, 1974-76; contbr. articles, abstracts to profl. publs. Bd. dirs. R.W. Johnson U. Hosp., New Brunswick, N.J., 1992—; trustee New Brunswick Cultural Ctr., 1992—. NSF rsch. fellow U. Cin., 1963-66; recipient Award for Excellence, U. Cin., 1986. Mem. AAAS, Infectious Disease Soc. Am., Am. Soc. Microbiology (vice-chmn. fermentation sect. 1971-72, chmn. 1972-73, exec. bd. 1972-78, chmn. nominations com. 1978), Theobald Smith Soc. N.J. (councilor 1974-77, program chmn. 1976, Waksmon award 1983), Am. Chem. Soc. (microbial tech. div.), N.Y. Acad. Scis., Soc. Indsl. Microbiology, Internat. Soc. Antiviral Rsch. Achievements include 8 patents in microbial process development. Office: Bristol Myers Squibb Co Pharm Rsch Inst PO Box 4000 Princeton NJ 08543

BIRON, CHRISTINE ANNE, medical science educator, researcher; b. Woonsocket, R.I., Aug. 8, 1951; d. R. Bernard and Theresa Priscilla (Sauvageau) B. BS, U. Mass., 1973; PhD, U. N.C., 1980. Rsch. technician U. Mass., Amherst, 1973-75; grad. researcher U. N.C., Chapel Hill, 1975-80; postdoctoral fellow Scripps Clinic and Rsch., La Jolla, Calif., 1980; fellow U. Mass. Med. Sch., Worcester, 1981-82, instr., 1983, asst. prof., 1984-87; vis. scientist Karolinska Inst., Stockholm, 1984; asst. prof. Brown U., Providence, R.I., 1988-90, assoc. prof., 1990—; mem. AIDS and related rsch. study sect. 3 NIH, 1991—. Assoc. editor Jour. Immunology, 1990—; contbr. articles, revs. to sci. jours. Leukemia Soc. Am. fellow, 1981, Spl. fellow, 1983, scholar, 1987; grantee NIH, 1985—; rsch. grantee MacArthur Found., 1991—. Mem. AAAS, Am. Assn. Immunologists (co-chmn. symposium 1990), Am. Soc. Virology, Sigma Xi. Office: Brown U Biomed Ctr Box G-B618 Providence RI 02912

BISCHOFF, JOYCE ARLENE, information systems consultant, lecturer; b. Chgo., Apr. 1, 1938; d. Carl Henry and Gertrude Alma (Lohn) Winterberg; m. Kenneth B. Bischoff, June 6, 1959; children: Kathryn Ann, James Eric. BS in Math., Ill. Inst. Tech., 1959; cert. computer tech., U. Del., 1979. Programmer, analyst Inst. of Gas Tech., Chgo., 1959-60, U. Ghent, Belgium, 1960-61; database adminstr. Med. Ctr. Del., Wilmington, 1979-84; sr. database analyst ICI Ams., Wilmington, 1984-87; sr. cons. CSC Ptnrs., Malvern, Pa., 1987-90; pres. Bischoff Cons., Inc., Hockessin, Del., 1990—; chairperson, founder Del. Valley DB2-SQL/DS Users Group, Phila., 1986—; task force leader DB2 performance task force Guide Internat., Chgo., 1987-90; speaker, mem. conf. planning com. Internat. DB2 Users Group. Contbr. articles to profl. jours. Recipient McGrath award Del. Valley DB2-SQL/DS Users Group, 1990, Quality award Guide Internat., 1989. Mem. Internat. Platform Assn., N.Y. Acad. Scis., Assn. for Computing Machinery (Del. Valley chpt. pres. 1986-87, program chair 1985-86), Data Processing Mgmt. Assn. (Wilmington chpt.), Network of Women in Computer Tech., Sigma Kappa (pres. 1958-59). Avocations: travel, music, miniatures. Home and Office: Bischoff Cons Inc 1007 Benge Rd Hockessin DE 19707-9242

BISCHOFF, KENNETH BRUCE, chemical engineer, educator; b. Chgo., Feb. 29, 1936; s. Arthur William and Evelyn Mary (Hansen) B.; m. Joyce Arlene Winterberg, June 6, 1959; children: Kathryn Ann, James Eric. B.S., Ill. Inst. Tech., 1957, Ph.D., 1961. Asst. to assoc. prof. U. Tex., Austin, 1961-67; assoc. prof., then prof. U. Md., 1967-70; Walter R. Read prof. engring. Cornell U., 1970-76, dir. Sch. Chem. Engring., 1970-75; Unidel prof. biomed. and chem. engring. U. Del., 1976—, chmn. dept. chem. engring., 1978-82; mem. NRC Bd. on Chem. Scis. and Tech., 1984-86, various coms., 1984—; cons. Exxon Rsch. and Engring., NIH, Gen. Foods Corp., W.R. Grace Co., Koppers Co., DuPont Co. Author: (with D.M. Himmelblau) Process Analysis and Simulation, 1968, (with G.F. Froment) Chemical Reactor Analysis and Design, 1979, 2d edit., 1989; editor: (with R.L. Dedrick and E.F. Leonard) The Artificial Kidney, Proc. 1st. Internat. Symposium Chem. Reaction Engring., 1970, (with R.M. Koros and T.R. Keane) Proc. 9th Symposium, 1986, 87; mem. editorial bd. Advances in Chemistry Series, 1973-76, 78-81, Jour. Bioengring., 1976-80, Jour. Pharmacokin, Biopharmaceutics, 1975-92, Biotech. Progress, 1987—, Advances in Chem. Engring., 1981—. Recipient Ebert prize Acad. Pharm. Scis., 1972, Founders award Chem. Indsl. Inst. Toxicology, 1992; Shell Found. fellow, 1959, NSF fellow, 1960, U. Ghent fellow, 1960-61, NAE fellow. Fellow AAAS, Am. Inst. Chem. Engrs. (dir. 1972-74, chmn. food, pharm. and bioengring. divsn. 1985, chmn. nat. program com. 1978, Profl. Progress award 1976, Food Pharm. and Bioengring. divsn. award 1982, 34th Ann. Inst. lectr. 1982, R.H. Wilhem award 1987); mem. Am. Inst. Chemists, Am. Chem. Soc., Am. Soc. Artificial Internal Organs, Engrs. Coun. for Profl. Devel. (bd. dirs. 1972-78), Coun. Chem. Rsch. (governing bd. 1981-84, chmn. 1985), Catalysis Soc., AAUP, N.Y. Acad. Scis., Sigma XI, Tau Beta Pi, Phi Lambda Upsilon, Omega Chi Epsilon, Alpha Chi Sigma. Home: 1007 Benge Rd Hockessin DE 19707-9242

BISH, DAVID LEE, mineralogist; b. Arlington, Va., Mar. 5, 1952; s. Henry L. and Rosemary M. (Osborn) B.; m. Karen Ainsworth, July 15, 1981; 1 child, Rebecca L. BS, Furman U., 1974; PhD, Pa. State U., 1977. Rsch. assoc. Harvard U., Cambridge, Mass., 1977-80; tech. staff mem. Los Alamos (N.Mex.) Nat. Lab., 1980—. Assoc. editor: Am. Mineralogist, 1987-91; author: Crystal Structures and Cation Sites, 1988; author, editor: Modern Powder Diffraction, 1989, Thermal Analysis in Clay Science, 1990; contbr. articles to profl. jours. Panel mem. Los Alamos Community Pride, 1985. Recipient Award for Materials Sci. Rsch. Xerox Corp., 1977. Fellow Mineralogical Soc. Am.; mem. Clay Minerals Soc. (mem. coun. 1987-90), Am. Geophys. Union, Internat. Assn. for Study Clay Minerals, Phi Beta Kappa. Achievements include discovery of phenomenon of anion exchange in mixed hydroxide minerals; development of quantitative analysis by X-ray powder diffraction using the Rietveld method. Home: 394 Richard Ct Los Alamos NM 87544 Office: Los Alamos Nat Lab Mail Stop D469 Los Alamos NM 87545

BISH, ROBERT LEONARD, applied mathematician, metallurgist, researcher; b. Suva, Fiji, Nov. 30, 1941; s. Leonard Oswald and Frances Susan (Poulton) B.; m. Patricia Irene Field, Aug. 16, 1969; children: Joanna, Rebecca, Peter. BSc, Sydney (Australia) U., 1964. Tech. asst. 2 Def. Standards Lab., Sydney, 1965-69, expt. officer 1, 1969-70, exptl. officer 2, 1970-74; exptl. officer 2 Materials Rsch. Lab., Melbourne, Australia, 1974—. Contbr. articles to profl. jours. Mem. Am. Math. Soc. Avocations: mechanics, metallography. Office: Materials Rsch Lab, Cordite Ave Victoria, Ascot Vale 3032, Australia

BISHARA, AMIN TAWADROS, mechanical engineer, technical services executive; b. Cairo, Oct. 22, 1944; came to U.S., 1973; s. Tawadros and Fakha (Boules) B.; m. Suzi Guirguis, Aug. 27, 1977; children: James A., Robert A. BSME, Ain Shams U., Cairo, 1968; MSME, Poly. U. N.Y., 1976. Registered profl. engr., N.Y., Tex., Ark., Fla. Field engr. Gen. Engring. Co., Cairo, 1968-71; mech. engr. Engring. Co. for Indsl. Enterprises, Cairo, 1971-73; project engr. Cosentini Assocs., N.Y.C., 1973-76; sr. engr. Ebasco Svcs., Inc., N.Y.C., 1976-79, lead engr., 1979-84; chmn., chief exec. officer PTS Tech. Svcs., Inc., Hurst, Tex., 1985—; mem. adv. bd. Entrepreneurship Inst., Ft. Worth, 1990—. Mem. Ft. Worth Civic Leaders Assn., 1988, North Tex. Commn., Dallas, 1988, Dallas Coun. of World Affairs, 1988. Named Hon. Citizen, State of Tex., 1987. Mem. Internat. Soc. Nuclear Air Tech. (v.p. 1990), ASME (vice chmn. subgroup 1986—, mem. main com. nuclear gas treatment 1990—), Nat. Soc. Profl. Engrs., Masons, Moslah Temple of Ft. Worth. Roman Catholic. Home: 2625 Brookridge Dr Hurst TX 76054-2761 Office: PTS Tech Svcs Inc 500 Grapevine Hwy Ste 488 Hurst TX 76054-2766

BISHOP, DAVID JOHN, physicist; b. Montgomery, Ala., Oct. 6, 1951; s. Cleo Merton Bishop and Dorothy Johanna Rielly; m. Vanessa Joy Levin, Aug. 22 1982; 1 child, Noah Samuel Bishop. BS in Physics magna cum laude with honors, Syracuse U., 1973; MS in Physics, Cornell U., 1977, PhD, 1978. Postdoctoral mem. tech. staff AT&T Bell Lab., Murry Hill, N.J., 1978-79, mem. tech. staff, 1979-88, head, microstructure physics rsch. dept., 1988—; adjunct prof. Physics SUNY, Buffalo, N.Y. Contbr. to profl. jours. Recipient Bausch and Lomb Hon. Sci. award. Fellow Am. Physical Soc.; mem. Phi Beta Kappa. Office: AT&T Bell Labs Room 1D-231 PO Box 636 New Providence NJ 07974-0636

BISHOP, DAVID NOLAN, electrical engineer; b. Memphis, Jan. 14, 1940; s. Robert Allen Bishop and Sara Frances (Gammon) Marett; m. Lois Margaret Baudouin, Nov. 16, 1963; children: Julie Frances Bishop Malouse, Anne Marie. BSEE, Miss. State U., 1962, MSEE, 1965. Registered profl. engr., La., Miss., Tex., Fla. Constrn. engr. Chevron Oil Co., New Orleans, 1964-70, lead constrn. engr., 1970-72, sr. constrn. engr., 1972-78; staff elec. engr. Chevron U.S.A., Inc., New Orleans, 1978-82, sr. staff elec. engr., 1982-85, elec. engring. cons., 1985-92; facilities engring. tech. specialist Chevron U.S.A. Prodn. Co., Houston, 1992—; offshore safety and anti-pollution equipment rep. Am. Petroleum Inst., 1982-86, mem. various coms.; mem. Nat. Elec. Code Panel 14, 1989—; instr. sch. electrical systems oil and gas prodn. facilities, Petroleum Ext. Svc., U. Tex. Austin, 1990-92; mem. instrumentation craft com. New Orleans Regional Vocat. Tech. Inst., 1985-92; guest lectr. in field. Author: (book) Electrical Systems for Oil and Gas Production Facilities, 1988, 2d edit. 1992; contbr. articles to profl. jours. Asst. scoutmaster Boy Scouts Am., Metairie, La., 1964-69; adminstrv. bd. St. Matthew's United Meth. Ch., Metairie, 1974-76; adv. com. U. Tex., Austin, Petroleum Extension Tex., 1980—; engring. adv. com. Miss. State U., 1982—, chmn. curriculum and rsch. subcom., 1986-89, sec., 1989-91, vice-chmn., 1992—. With U.S. Army, Miss. Nat. Guard. Named Coll. Alumnus of Yr., Coll. Engring. Miss. State U., 1990, Disting. Engring. fellow, 1992, Chpt. Alumnus of Yr. Miss. State U. Alumni Assn., 1990. Mem. IEEE, Instrument Soc. Am. (sr., sect. pres. 1977-78, dist. v.p. 1983-85, chmn. coun. dist. v.p. 1984-85, v.p. standards and practices 1988-90, host com. chmn. ISA/90, 1989-90, pres.-elect, sec. 1990-91, pres. 1991-92, New Orleans Sect. Commendation award 1983-84, Standards & Practices Dept. Recognition Achievement award 1987, 90), Tau Beta Pi, Phi Eta Sigma, Eta Kappa Nu (student sect. pres. 1961-62), Omicron Delta Kappa. Republican. Avocations: hunting, gardening, crossword puzzles. Home: 10 Mistflower PL The Woodlands TX 77381 Office: Chevron USA Prodn Co PO Box 1635 Houston TX 77251

BISHOP, JOHN MICHAEL, biomedical research scientist, educator; b. York, Pa., Feb. 22, 1936; married 1959. AB, Gettysburg Coll., 1957; MD, Harvard U., 1962. Intern in internal medicine Mass. Gen. Hosp., Boston, 1962-63, resident, 1963-64; rsch. assoc. virology NIH, Washington, 1964-66, sr. investigator, 1966-68; from asst. prof. to assoc. prof. U. Calif. Med. Ctr., San Francisco, 1968-72, prof. microbiology and immunology, 1972—, prof. biochemistry and biophysics, 1982—; dir. G.W. Hooper Rsch. Found., 1981—. Recipient Nobel prize in physiology or medicine, 1989, Biomed.

Rsch. award Am. Assn. Med. Colls., 1981, Albert Lasker Basic Med. Rsch. award, 1981, Armand Hammer Cancer award, 1984, GM Found. Cancer Rsch. award, 1984, Gairdner Found. Internat. award, Can., 1984, Medal of Honor, Am. Cancer Soc., 1984; NIH grantee, 1968—. Fellow Salk Inst. (trustee 1991—); mem. NAS, Inst. Medicine. Achievements include research in biochemistry of animal viruses, replication of nucleic acids, viral oncogenesis, molecular genetics. Office: U Calif Medical Ctr Dept Microbiology Box 0552 San Francisco CA 94143-0552

BISHOP, ROBERT HAROLD, aerospace engineering educator; b. Vicenza, Italy, June 8, 1957; s. William Robert and Anna Maria (DiPietro) B.; m. Lynda R. Ferrera, July 27, 1985; children: Robert Emerson, Joseph Taylor. BS, Tex. A&M U., 1980, MS, 1980; PhD, Rice U., 1990. Registered profl. engr., Tex. Tech. staff mem. Charles Stark Draper Lab., Cambridge, Mass., 1980-90; asst. prof. aerospace engring. and engring. mechanics U. Tex., Austin, 1990—. Author: Modern Control Systems: Analysis and Design Using Matlab, 1993. Mem. AIAA, IEEE, Am. Astronautical Soc., Am. Soc. Engring. Educators. Office: U Tex Dept Aerospace Engring Austin TX 78712

BISHOP, THOMAS RAY, retired mechanical engineer; b. Hutchinson, Kans., Oct. 26, 1925; s. Orren E. and Myrtle (Dale) Bish; m. Mary Lou Nesmith, Sept. 1, 1951; children: Thomas Ray II, Frances Joann. Student California (Pa.) State Tchrs. Coll., 1947-48; BS, U. Houston, 1953; postgrad. U. Wash., 1960-61; grad. Alexander Hamilton Bus. Inst., 1972. Rsch. engr. Boeing Co., Seattle, 1953-64, rsch. engr. Apollo program, 1964-69; asst. chief engr. Product div. Bowen Tools, Inc., Houston, 1969-75, chief engr., 1975-77, chief product engr., 1977-86; chief engr. rsch. and devel.; founder, pres. Tom Bishop Enterprises, 1986—, oil field tool design and quality control programs, quality control engr. cons. Precinct committeeman King County (Wash.) Democratic Com., 1960. With USMCR, 1944-46. Decorated Purple Heart with Gold Star; named Engr. of Year, Boeing Aerospace Co., 1966; recipient Excellence in Engring. citation A.I.S.I., 1975. Registered profl. engr., Ala., La., Tex. Mem. Am. Soc. Quality Control, Tex. Soc. Profl. Engrs. Unitarian. Mason. Contbr. articles to profl. jours.; multi-patentee (16) oil field equipment field. Home and Office: 2202 Viking Dr Houston TX 77018-1728

BISHOP, WALTON BURRELL, science writer, consultant; b. Leroy, Ill., Aug. 25, 1917; s. Minor and Tressie Mae (Staley) B.; m. Marjorie Nelle Hanson, June 7, 1940 (div. 1959); 1 child, Gregory; m. Anne-Marie Delachapelle, Dec. 16, 1959; children: Chantal, Véronique. B of Edn. in Physical Sci., Ill. State U., 1939; MA in Math., Boston U., 1950; MSEE, Northeastern U., 1954; MA in Journalism, U. Md., 1983, PhD in Pub. Comm., 1990. Tchr., coach Milford (Ill.) Twp. High Sch., 1939-41, Lincoln (Ill.) Community High Sch., 1941-42; electronics engr. Air Force Cambridge Rsch. Labs., Hanscom AFB, Mass., 1946-66; electronics engr. Naval Rsch. Lab., Washington, 1966-80, cons., sci. writer, 1980—; lectr. math Northeastern U. Grad Sch., Boston, 1960-66. Adv. editor Jour. Microelectronics and Reliability, 1963-80; contbr. articles to profl. jours. Mem. AAAS, IEEE, Soc. of Profl. Journalists, Sigma Xi. Democrat. Achievements include 22 patents for IFF and reliability; proposed passive electronic guidance device for blind. Home: 6 Balmoral Dr East Oxon Hill MD 20745-1011 Office: Naval Rsch Lab 4555 Overlook Ave SW Washington DC 20375-5336

BISMARCK-NASR, MAHER NASR, aeronautical engineering educator; b. Cairo, Oct. 18, 1940; arrived in Brazil, 1970; s. Bismarck Nasr and Alice (Mikhael) Mansour; m. Maria Eunice Marinho, Jan. 2, 1971; children: Paula Christina, Elizabeth Maria. BSc in Aero. Engring., Cairo U., 1964, MSc in Aero. Engring., 1967; DSc, U. Paris, 1970. Engr. Embraer Empresa Brasileira de Aeronautica, Sáo Jose dos Campos, Sao Paulo, Brazil, 1970-78, head aeroelasticity sect., 1978-79, asst. tech. dir., 1979-89; prof. Instituto Technologicode Aeronautica, Sáo Jose dos Campos, 1989—. Contbr. articles to profl. publs. Mem. AIAA. Roman Catholic. Home: Av Cidade Jardim 3141 Qa47, 12228900 Sao Jose dos Campos Sao Paulo, Brazil Office: Instituto Tech de Aero, CTA/ITA/IEA, 12228-900 Sao Jose dos Campos Brazil

BISNAR, MIGUEL CHIONG, water pollution engineer; b. Cebu City, Philippines, June 24, 1953; s. Alberto Caballes and Barda (Chiong) B.; m. Librada Aquino Sudaria; children: Michelle, Michael-Albert, Ramil. Grad. chem. engring., U. Santo Tomas, Manila, Philippines, 1976; MS in Sanitary Engring., Internat. Inst. Hydraulics, Delft, the Netherlands, 1983. Lic. chem. engr. Project leader air pollution monitoring project Nat. Pollution Control Commn., Manila, 1978-79, pollution control engr., 1979-84, tech. asst. dep. commr. on standards setting, 1984-86; prin. engr. environ. mgmt. dept. Nat. Power Corp., Quezon City, Philippines, 1986—; chief pollution monitoring sect. Nat. Power Corp., Quezon City. Mem. Environ. Impact Assessment, Trainer's Network Manchester (U.K.). Office: Nat Power Corp, BIR Rd Cor Quezon Ave, Quezon City 1100, The Philippines

BISSADA, NABIL KADDIS, urologist, educator, researcher, author; b. Cairo, Egypt, Sept. 2, 1938; s. Kaddis B. and Negma Bissada; m. Samia Shafik Henain, July 23, 1967; children: Sally, Nancy, Mary, Amy, Andrew. M.D., Cairo U., 1963. Diplomate Am. Bd. Urology. Intern Cairo Univ. Hosp., 1964-65; resident in surgery Babelsharia Hosp., 1965-69; resident in urology U. N.C. Hosp., 1970-73; asst. prof. urology U. Ark., 1973-77, assoc. prof., 1977-79; cons. urologist King Faisal Specialist Hosp., Riyadh, Saudi Arabia, 1979-87; prof. urology Med. U. S.C., 1987—, chief urologic oncology, 1988—; chief urologic surgery Ralph H. Johnson Med. Ctr., 1987—; co-chmn. Div. U., U.S. Sect. Internat. Coll. Surgeons, 1989-91, chmn., 1991—, frequent speaker to regional, nat. and internat. med. groups. Author: Lower Urinary Tract Function and Dysfunction: Diagnosis and Management, 1978; Pharmacology of the Urinary Tract and the Male Reproductive System, 1982; contbr. to hundreds of articles and books; pioneer, developer of several significant surgical and med. urologic treatment methods. Fellow ACS, Internat. Coll. Surgeons; mem. AMA, Am. Urol. Assn., Soc. Internat. D'Urologie, Urologiconcology Soc., Urodynamic Soc., Egypt Am. Urol. Assn. (pres. 1989-91), Sigma Xi. Address: Med U of SC Med Ctr 171 Ashley Ave Charleston SC 29425-2280

BITSCH-LARSEN, LARS KRISTIAN, anaesthesia and intensive care specialist; b. Herning, Denmark, May 12, 1945; s. Immanuel and Valborg (Larsen) B.-L.; m. Kate Lund Jensen; children: Kristian, Simon. MD, Copenhagen U., Denmark, 1973; Degree, Tech. Inst., 1976; diploma in health econs., 1993. Diplomate European Acad. Anaesthesia. Residency Växsjö, Lasaret, Sweden, 1984; specialist tng. Växsjö Lassaret, Sweden, 1984; specialist tng. Haukeland U., Bergen, Norway, 1985, cons. anaesthesia, 1986-88; dir. anaesthesia Kalundborg Sygehus, Denmark, 1989—; vice-chmn. staff com. Kalundborg Sugehus, Denmark, 1990-93, chief adminstrv. dr., 1993—; sci. advisor for developing monitoring systems Nimorod, 1992. Author: Phsyologi in Anaesthesi, 1986, 2d edit., 1987; editor: pratical Respirator Therapy, 1986, 2d edit., 1987; contbr. articles to profl. jours. Chmn. County Adv. Com., Vestsjaelland, 1989—, Computer Implimentation Com., Kalundborg, 1990—. Grantee Vestsjaellands County, 1990, Comers & Industry Bd., 1991, 92. Mem. Engrs. Assn. (div. for risk evaluation), European Acad. Anaesthesia, European Cardiothoracis Anesthetist Assn., European Soc. Computing and Tech. in Anaesthesia, Norti Neuroanesthist Assn., Mensa. Avocation: sailing ships. Office: Kalundborg Sygehus, Noerrre Alle 27, DK 4400 Kalundborg Denmark

BITTENBENDER, BRAD JAMES, environmental safety and industrial hygiene manager; b. Kalamazoo, Dec. 4, 1948; s. Don J. and Thelma Lu (Bacon) B.; m. Patricia Stahl Hubbell, June, 1992. BS, Western Mich. U., 1972; Cert. Hazardous Material Mgmt., U. Calif., Irvine, 1987; Cert. Environ. Auditing, Calif. State U., Long Beach, 1992. Cert. safety profl. of the Ams.; cert. hazardous materials mgr. Supr. mfg. Am. Cyanamid, Kalamazoo, 1973-77; supr. mfg. Productol Chem. div. Ferro Corp., Santa Fe Springs, Calif., 1977-79, environ. adminstr., 1979-80; sr. environ. engr. Ferro Corp., Los Angeles, 1980-87, mgr. environ. safety and indsl. hygiene dept., 1988-91; mgr. environ. safety and indsl. hygiene dept. Structural Polymer Systems, Inc., Montedison, Calif., 1991—; bd. dirs., mem. adv. bd. safety and health extension program U. Calif. Irvine, 1985—. Bd. dirs. adv. com. hazardous materials Community Right to Know, Culver City, Calif., 1987—; mem. Calif. Mus. Found., L.A., 1985—, Mus. Contemporary Art, L.A., 1985—; founding sponsor Challenger Ctr. Mem. Am. Inst. Chem. Engrs., Nat. Assn. Environ. Mgmt., Acad. Cert. Hazardous Materials Mgrs., Suppliers of Advanced Composites Materials Assn. (mem. environ. health and safety com. 1989-92), Am. Indsl. Hygiene Assn., Am. Soc. Safety Engrs., Nat. Fire Protection Assn., Beta Beta Beta. Republican. Presbyterian. Avocations: camping, mountaineering, skiing, distance running, reading. Office: Structural Polymer Systems Inc 5915 Rodeo Rd Los Angeles CA 90016-4381

BITTNER, RONALD JOSEPH, computer systems analyst; b. Schenectady, N.Y., July 30, 1954; s. Richard Joseph and Catherine (Stepnowski) B.; m. Elayne Louise Simpson, May 14, 1983; 1 child, Krysten Elayne. AS in Chemistry, Orange County C.C., Middletown, N.Y., 1978; BA in Bus. Mgmt., Herbert Lehman Coll., Bronx, N.Y., 1983. Internal cons. Internat. Paper Co., Tuxedo, N.Y., 1978-87; systems analyst McGraw-Hill News, N.Y.C., 1987-89; sr. systems analyst Orange County Info. Svcs., Goshen, N.Y., 1989-91, Columbia Tristar Home Video, N.Y.C., 1991—; MIS mgr. Post Perfect, N.Y.C., 1993—; cons. in magic Shawnee Playhouse, Poconos, Pa., 1985. Contbr. articles to newspapers and mags., 1980-88. Mem. Variety Clubs Internat., Soc. Am. Magicians (pres. 1985-87), Internat. Brotherhood of Magicians, Microcomputer Mgrs. Assn. (program coord.), Acad. of Magical Arts Soc. Am. Magicians (assembly coord., 1993). Avocations: scuba diving, underwater photography, cooking, bowling, chess. Home: RR 2 Box 173B Wallkill NY 12589-9802

BIXBY, MARK ELLIS, city official; b. Beloit, Wis., Mar. 21, 1955; s. Milton W. and Esther M. (Olsen) B.; m. Stacey Marie French, Jan. 19, 1980; children: Sarah, Lindsey, Andrew. BS, So. Ill. U., 1977; postgrad., No. Ill. U., 1979—. Energy auditor State of Ill., Springfield, 1986—; Ihwap coord. Ill. Home Weatherization Program, Ill., 1987—; mem. adv. bd. Ill. Home Weatherization Program, State of Ill., Springfield, 1985-89. Pres. Rock River ValleyPantry, Rockford, 1991, 92, City of Rockford Employment Coun., 1990—. Mem. Assn. Energy Engrs. (charter mem. demand side mgmt.), Lions. Home: 5601 Tasselbury Close Rockford IL 61114 Office: City of Rockford 3010 Kishwaukee St Rockford IL 61109

BIZZIGOTTI, GEORGE ORA, environmental chemist, consultant; b. Manhasset, N.Y., Sept. 29, 1957; s. Raymond Anthony and Edna June (Becker) B.; m. Mary Gretchen Allen, Aug. 30, 1986. AB, Rutgers U., 1979, PhD, 1982. Post doctoral fellow Univ. Louis Pasteur, Strasbourg, France, 1982-84; asst. prof. Auburn U., 1984-87; technical staff mem. The Mitre Corp., McLean, Va., 1987—. Contbr. articles to Jour. Am. Chem. Soc., Jour. of Organic Chemistry, Tetrahedron Letters; chpt. to Biomemetic Chemistry of Functional Vesicles and Micelles, 1983. Mem. AAAS, Am. Chem. Soc., Phi Beta Kappa. Achievements include consultation with U.S. EPA to implement revised hazard ranking system for Superfund hazardous waste sites; teaching of use of ranking to EPA staff and state agys; review of application to 20 hazardous sites. Office: Mitre Corp 7525 Colshire Dr Mc Lean VA 22102

BJERKE, ROBERT KEITH, chemist; b. Eau Claire, Wis., Apr. 21, 1941; s. Robert L. and Keitha E. (Thompson) B.; m. JoAnn L. Balsiger, June 12, 1965; children: Amy, Susan, Robert W. BS in Chemistry, Wis. State U., Eau Claire, 1963. Ammunition chemist Twin City Army Ammo Plant, Mpls., 1965-70; cosmetics chemist LaMaur, Inc., Mpls., 1971-78; explosives chemist Blount, Inc., Lewiston, Idaho, 1978-90, chemistry mgr., 1990—. Author: Proprietary Thermodynamics of Small Arms Primers, 1991. 1st lt. U.S. Army, 1963-65. Achievements include U.S., British and Italian patents and patents pending in area of explosives for improved small arms primers; adaptation of NASA equilibrium thermochemistry computer program for initiating compositions. Home: 709 Cedar Ave Lewiston ID 83501 Office: Blount Inc PO Box 856 Lewiston ID 83501

BJORCK, JEFFREY PAUL, psychology educator, clinical psychologist; b. Hackensack, N.J., Nov. 17, 1960; s. Walter Jr. and Irene Louise (Cubberley) B.; m. Sharon-Rose McConnell, Nov. 10, 1990. BA in Psychology summa cum laude, Colgate U., 1983; MA in Clin. Psychology, U. Del., 1987, PhD in Clin. Psychology, 1991. Lic. clin. psychologist. Adj. instr. psychology dept. U. Del., Newark, 1986-87; psychotherapist Cecil County Mental Health Ctr., Elkton, Md., 1987-88; psychol. asst. Del. Guidance Svcs., Wilmington, Del., 1988-90; psychology assoc. Bayer & Langsdorf, PA, Elkton, 1989-90, A.D. Hart, PhD, Pasadena, Calif., 1990-93; asst. prof. Fuller Grad. Sch. Psychology, Pasadena, Calif., 1990—; lic. clin. psychologist, 1993—. Guest editorial reviewer: (jour.) Research in the Social Scientific Study of religion, 1992—; contbr. articles to profl. jours. Mem. adv. bd. Parent Alert, San Marino, Calif., 1991; mem. Christ Community Ch., Arcadia, 1991—. Mem. APA (divs. 12 and 36), Calif. Psychol. Assn., Soc. for Scientific Study of Religion, Pasadena Area Psychol. Assn. Evangelical Presbyterian. Office: Fuller Grad Sch Psychology 180 N Oakland Ave Pasadena CA 91101

BJORKMAN, DAVID JESS, gastroenterologist, educator; b. Salt Lake City, Oct. 28, 1952; s. Jesse Harold and Violet Maureen (Neese) B.; m. Kaye Hansen, Aug. 20, 1975; children: D. James, Michael. BA, U. Utah, 1976, MD, 1980. Diplomate Am. Bd. Internal Medicine, Am. Bd. Gastroenterology. Intern Brigham and Womens Hosp., Harvard U. Med. Sch., 1980-81, resident in internal medicine, 1981-83; clin. fellow, rsch. fellow Harvard U. Med. Sch., Boston, 1983-85; instr. medicine U. Utah Sch. Medicine, Salt Lake City, 1985-88, asst. prof. medicine, 1988-92, assoc. prof. medicine, 1992—; sci. rev. com. Nat. Cancer Inst., Bethesda, Md., 1991; exec. com. Utah chpt. Am. Liver Found., Salt Lake City, 1988—. Contbr. articles to profl. jours. Fellow ACP, Am. Coll. Gastroenterology; mem. Utah State Med. Assn. (legis com. 1990—), Am. Soc. Laser Medicine and Surgery (com. mem. 1991—), Phi Beta Kappa, Alpha Omega Alpha (bd. dirs. 1979-82). Achievements include laser identification of colonic cancer using photoactive agent, research on changes in intestinal membrane composition and fluidity. Office: Univ Med Ctr 50 N Medical Dr Salt Lake City UT 84132

BJORKMAN, GORDON STUART, JR., structural engineer, consultant; b. N.Y.C., May 22, 1944; s. Gordon Stuart Sr. and Siri (Peterson) B.; m. Loris Cecilia Lovelace, June 18, 1977; children: Gordon, Eliseo, Loris. BSCE, Princeton U., 1966; MS in Structural Engring., Cornell U., 1968; PhD of Applied Mechanics, U. Del., 1975. Registered profl. engr. Engr. Grumman Aircraft, Bethpage, N.Y., 1967-69; asst. prof. Drexel U., Phila., 1975-78; cons. United Engrs., Phila., 1978-81, Cygna Energy Svcs., Boston, 1981-86; tech. mgr. ABB Impell, Boston, 1986-91; sr. cons. EQE Internat., San Francisco, 1991—. Contbr. articles to profl. jours. Grantee NSF 1978, 82. Mem. ASCE (Structural Computations com. 1991—, Optimun Structural Design com. 1991—), Sigma Xi. Achievements include discovery of harmonic holes and inclusions. Home: RR 1 Box 354 Woodstock VT 05091 Office: EQE Internat 139 R Portsmouth Ave Stratham NH 03885

BJØRN-ANDERSEN, NIELS, information systems researcher; b. Frederiksberg, Denmark, Feb. 29, 1944; s. Hans Christian and Karen Bjørn (Mauritzen) A.; m. Alice Bjørn, Sept. 30, 1967; children: Tine, Mette. BS, Copenhagen Bus. Sch., 1966, MS, 1969, PhD, 1973. Trainee Western Pacific R.R., San Francisco, 1966; systems analyst Danish Unilever, Copenhagen, 1967-69; rsch. asst. Copenhagen Bus. Sch., 1969-71, asst. prof., 1971-73, assoc. prof., 1974-87, prof., 1987—; vis. prof. Manchester Bus. Sch., 1973-74, U. Calif., Irvine, 1987; bd. dirs. Schou Mgmt. Cons., Copenhagen. Author: (book) IS for Decision Making, 1974, Impact of Systems Change in Organizations, 1979, Managing Computer Impact, 1986; editor: (book) Human Side of Information Processing, 1980. Recipient Tietgen Gold medal Soc. for Advancement of Bus. Edn., Copenhagen, 1973; Internat. Fedn. for Info. Processing Outstanding Svc. award, 1988. Mem. Internat. Conf. on Info. Systems (chmn., exec. coun. mem. 1989-91), Internat. Fedn. Info. Processing Socs. (Danish nat. rep. TC 8 1979—). Avocations: scuba diving, golf, bridge. Home: Philip Schousvej 19,3, DK2000 Frederiksberg Denmark Office: Copenhagen Bus Sch, 60 Howitzvej, DK2000 Frederiksberg Denmark

BJORNDAHL, DAVID LEE, electrical engineer; b. Rock Island, Ill., June 19, 1927; s. Richard Gideon and Olive Muriel (Winter) B.; m. Clara Mae Buck, Feb. 16, 1952; children: William, Jay, Jan, Jill. PhD in Elec. Engring., Purdue U., 1956. Sr. engr. Litton Guidance & Control Systems, Beverly Hills, Calif., 1956-58; project engr. Litton Guidance & Control Systems, Woodland Hills, Calif., 1958-62, dir. advanced programs, 1962-66; mgr. Martin-Marietta, Denver, 1966-67; dir. advanced programs Litton Aero. Products, Woodland Hills, 1967-74; v.p. engring. Litton Aero. Products, Moorpark, Calif, 1974-86, chief scientist, 1986-93; part-time cons., Chatsworth, Calif., 1993—. Contbr. articles to profl. jours. Mem. AIEE, Inst. Navigation, Simga Chi, Eta Kappa Nu. Republican. Achievements include design of various electronic navigation systems for aircraft.

BJØRNØ, LEIF, industrial acoustics educator; b. Svendborg, Denmark, Mar. 30, 1937; s. Svend Aage Valdemar and Elna Marie Jensen. Student, U. Roskilde (Denmark), 1956; MSc in Mech. Engring., Tech. U. Denmark, Lyngby, 1962, PhD, 1967; diploma, Imperial Coll., London, 1971. Asst. prof. Tech. U. Denmark, Lyngby, 1967-69, assoc. prof., 1970-78, prof. indsl. acoustics, 1978—; vis. prof. Imperial Coll., London, 1969-70; chmn. mech. engring., v.p. Acad. Tech. Scis., 1989—; chmn. bd. dirs. Brunata A/S, Reson System A/S, Reson Inc. USA, Sensor Tech. System A/S, Syd-Tek, Food Tech A/S. Author: Fluid Mechanics, 1972; author or co-author 8 books, 1 translation; editor Ultrasonics, 1973—; contbr. 230 articles to internat. jours., conf. proceedings and books. Mem. Coun. of the Ch., Taastrup, 1977-80. Named Knight of the Order of Dannebrog, Her Majesty the Queen, Denmark, 1991. Fellow Acoustical Soc. Am., Inst. Acoustics U.K., South African Acoustical Inst., IEEE; mem. Soc. Exploration Geophysicists, Acoustical Soc. Japan, Internat. Soc. Optical Engrs., U.S. Naval Inst., Instr. Noise Control Engring. (corr.), Spanish Acoustical Soc. (hon.), Sigma Xi. Office: Tech U Denmark, Dept Indsl Acoustics, 2800 Lyngby Denmark

BJORNSSON, JOHANNES, pathologist; b. Reykjavik, Iceland, Nov. 13, 1947; came to U.S. 1980; s. Bjorn and Una (Johannesdottir) Sigurdsson; m. Margret Ingvarsdottir, Sept. 11, 1971; children: Bjorn, Una Bjorg. BA, Hamilton Coll., Clinton, N.Y., 1969; MD, U. Iceland, 1977. Diplomate Am. Bd. Pathology. Fellow U. Hamburg, Germany, 1978-80, Mayo Clinic, Rochester, Minn., 1980-84; cons. dept. pathology U. Iceland, Reykjavik, 1984-92; sr. assoc. cons. Mayo Clinic, Rochester, 1992—; assoc. prof. pathology U. Iceland, 1987-92, Mayo Med. Sch., Rochester, 1992—. Contbr. articles to profl. jours. Fulbright scholar, 1967-69. Mem. AMA, Coll. Am. Pathologists, German Soc. Pathology, European Soc. Pathology, Internat. Acad. Pathology, N.Y. Acad. Scis., Sigma Xi.

BJORO, EDWIN FRANCIS, JR., nuclear engineer; b. Chgo., Nov. 14, 1928; s. Edwin Francis and Florence Laverne (Francis) B.; m. Barbara Anne Hanrahan, Aug. 2, 1952; 1 child, Michael Edwin. BA, Columbia U., 1950; B in Aero. Engring., NYU, 1957. Asst. dept. mgr. Titan III program United Tech. Corp., Sunnyvale, Calif., 1963-64; mem. tech. staff Rsch. Analysis Corp., McLean, Va., 1964-65; program mgr. Vitro Labs., Silver Spring, Md., 1965-69; mgr. reliability skylab program Martin Marietta, Washington, 1969-74, tech. dir., cons., 1974-76; with U.S. Dept. of Energy, Washington, 1976—, acting dir. nuclear energy quality assurance, 1990—. Contbr. articles to profl. jours. cons., data rev. bd., reliability audit com., artificial heart program Nat. Heart, Lung, and Blood Inst., NIH, Washington, 1987-92. 1st lt. USAF, 1951-55. Mem. Am. Soc. Quality Control, Tau Beta Pi. Avocations: swimming, tennis, reading. Home: 4206 Pebble Branch Rd Ellicott City MD 21042 Office: US Dept of Energy Washington DC 20585

BJORVATTEN, TOR ANDERSON, chemical engineer; b. Vegaarshei, Norway, Mar. 14, 1929; s. Anders and Karen (Lindstol) B.; m. Helene Winther, June 30, 1958; children: Terje, Marie Kristine. Degree, U. Oslo, 1959. Rsch. fellow U. Oslo, 1959-66; rsch. scientist Norwegian Def. Rsch. Establishment, Kjeller, 1967—. Contbr. articles to Acta Chemica Scandinavica. Office: Norwegian Def Rsch Establis, 2007 Kjeller Norway

BLACK, CLINTON JAMES, engineering company executive; b. Beatrice, Ala., Dec. 19, 1938; s. Billie and V. Oweda (Bradley) B.; m. Rhutelia McCants, Aug. 19, 1962; children: Michael, Michelle, Marcia. BS, Tuskegee U., 1963; MBA, NY Inst. Tech., 1976. Commd. 2d lt. U.S. Army, 1964; advanced through grades to col., 1986, ret., 1992; COO Stephens Engring. Co., Inc., Lanham, Md., 1992—. Chmn. Meth. Men St. Paul United Meth. Ch., Oxon Hill, Md., 1992, chmn. stewardship, 1991, 92. Decorated Bronze Star, 1971, Legion of Merit, 1992. Mem. Armed Forces Comms. Elec. Assn., A & AF Mut. Aid Soc., Alumni Assn. U.S. AWC, Omega Psi Phi (pres. 1984). Home: 4709 Cedell Pl Camp Springs MD 20748 Office: Stephens Engring Co Inc Ste 300 4601 Forbes Blvd Lanham MD 20748

BLACK, CRAIG CALL, museum administrator; b. Peking, China, May 28, 1932; s. Arthur and Mary (Nichols) B.; children—Christopher Arthur, Lorna Varn; m. Mary elizabeth King, Jan. 4, 1986. A.B., Amherst Coll., 1954; M.A. (Simpson fellow), Johns Hopkins U., 1957; Ph.D. (NIH fellow), Harvard U., 1962. Geologist Okla. Geol. Survey, summer 1956; asst. curator Carnegie Inst., Pitts., 1960-62; curator Carnegie Inst., 1962-70; prof. biology U. Kans., 1970-72; dir. Mus. of Tex. Tech. U., Lubbock, 1970-75, Carnegie Inst., 1975-82, L.A. Mus. Natural History, 1982—; co-leader John F. Kennedy U. Western Mus. Conf. Seminar, 1983; faculty Am. Law Inst.-ABA Course of Study in Legal Problems of Mus. Adminstrn., 1980, 82; co-dir. Mus. Mgmt. Inst., U. Calif., Berkeley, summers, 1979, 80; adj. prof. U. Mus. and dept. geology U. Colo., 1965, Dept. of Marine Scis. U. So. Calif., 1987—; assoc. com. on evolution UCLA, 1987—. Contbr. articles to profl. jours. Presdl. appointee Nat. Mus. Svc. Bd., 1985-90; rev. panel Inst. of Mus. Svcs., Washington, 1978-79, policy panel, 1978-80. Simpson fellow, 1954-55; Kellog fellow, 1956-59; NIH precototal fellow, 1959-60. Fellow AAAS; mem. Am. Assn. Mus. (pres. 1980-82, mem. commn. on mus. for a new century 1982-84, mem. council 1977—). Club: Cosmos (Washington); Calif. (Los Angeles). Home: 625 S Muirfield Rd Los Angeles CA 90005-3832 Office: Nat Hist Mus Los Angeles County 900 Exposition Blvd Los Angeles CA 90007-4000

BLACK, DAVID CHARLES, astrophysicist; b. Waterloo, Iowa, May 14, 1943. BS, U. Minn., 1965, MS, 1967, PhD in Physics, 1970. Fellow NAS, 1970-72; rsch. scientist theoretical astrophysics Ames rsch. ctr. NASA, Houston, 1972—; chief scientist space sta. NASA, 1985—. Mem. AAAS, N.Y. Acad. Sci. Achievements include research in theoretical studies of the formation and evolution of stars and planetary systems, interpretation of rare gas isotopic data from meteorites and lunar samples. Office: Lunar & Planetary Institute 3303 NASA Rd 1 Houston TX 77058*

BLACK, SIR JAMES (WHYTE), pharmacologist; b. June 14, 1924. MB, ChB, U. St. Andrews; MD (hon.), U. Edinburgh, 1989; DSc (hon.), U. Glasgow, 1989. Asst. lectr. physiology U. St. Andrews, 1946; lectr. physiology U. Malaya, 1947-50; sr. lectr. U. Glasgow Vet. Sch., 1950-58; with ICI Pharms. Ltd., 1958-64; head biol. rsch., dep. rsch. dir. Smith, Kline & French, Welwyn Garden City, 1964-73; prof., chmn. dept. pharmacology Univ. Coll., London, 1973-77; dir. therapeutic rsch. Wellcome Rsch. Labs., 1978-84; prof. analytical pharmacology King's Coll. Hosp. Med. Sch., U. London, 1984—; chancellor elect Dundee (Scotland) U., 1992—. Created knight, 1981; recipient Nobel prize for medicine, 1988. Fellow Royal Coll. Physicians, Royal Soc. (Mullard award 1978); mem. Royal Coll. Vet. Surgeons (hon. assoc.). Office: U Dundee, Dundee DD1 4HN, Scotland

BLACK, JONATHAN, bioengineering educator; b. London, June 12, 1939; came to U.S., 1947; s. Max and Michal Mabel (Landesberg) B.; m. Susan Starr Williams, June, 1960 (annulled 1962); m. Toni Louise Rogers, June 1, 1963; children: David Lionel, Christina Louise, Matthew Leonard. BS, Cornell U., 1961; ME, Pa. State U., 1968; PhD, U. Pa., 1972. Physicist Carborundum Corp., Niagara Falls, N.Y., 1961-62; devel. engr. RCA, Harrison, N.J., 1962-64; rsch. leader Melpar, Inc., Falls Church, Va., 1964-65; engr. GE, Valley Forge, Pa., 1965-68; assoc. prof. Sch. Medicine U. Pa., Phila., 1975-80, prof., 1980-88; Hunter prof. bioengring. Clemson (S.C.) U., 1988—; chmn. Osteonics Sci. Adv. Bd., Allendale, N.J., 1993—. Author: Biological Performance of Materials, 1981, 2d edit., 1992, Electrical Stimulation, 1987, Orthopaedic Biomaterials, 1988; asst. editor Jour. Biomed. Materials Rsch., 1978—; contbr. over 100 articles to sci. jours. Recipient Gold medal Brit. Orthopaedic Assn., 1987. Mem. Am. Acad. Orthopedic Surgeons (assoc.), Soc. for Biomaterials (founding, pres. 1983-84), Bioelectric Growth and Repair Soc. (founding, Kappa Delta award 1987). Home and Office: 409 Dorothy Dr King Of Prussia PA 19406

BLACK, KRISTINE MARY, physicist; b. St. Paul, July 11, 1953; d. Jaurd Oliver and Dorothy Helen (Amos) B. B in Physics, U. Minn., 1975, MS in Cell Biology, 1978, MS in Metallurgy and Materials Sci., 1981. Analytical physicist Cardiac Pacemakers, St. Paul, 1978, qualifications engr., 1978-81;

biomaterials engr. St. Jude Med., Inc., St. Paul, 1981-83; mgr. quality assurance Unisys Semicondr. Ops., St. Paul, 1983-88; systems assurance sect. mgr. Unisys, 1988-90, quality assurance mgr. SMPO div., San Diego, 1990-91; mgr. reliability and quality assurance Carborundum Co., Phoenix, 1991-93; pvt. practice quality cons., Phoenix, 1993—. Contbr. articles to profl. jours. Mem. IEEE, Am. Soc. for Quality Control, Am. Soc. Metals, U. Minn. Inst. Tech. Alumni Soc. (dir. 1980-87, v.p. 1986-87, pres. 1987-88).

BLACK, PATRICIA JEAN, medical technologist; b. Milw., Oct. 22, 1954; d. Dale B. and Geraldine L. (Milligan) Heywood; m. Robert S. Black, Oct. 14, 1978. BS, Millikin U., 1978; degree in med. tech., St. Mary's Hosp., 1978. Med. technologist Mercy Hosp., Urbana, Ill., 1978-85; biol. lab. technician No. Regional Rsch. Ctr., USDA Agrl. Rsch. Svc., Peoria, Ill., 1985-88; lab. mgr. Chapman Cancer Ctr., Joplin, Mo., 1989—. Patentee in field. Mem. AAUW, Am. Soc. Clin. Pathologists (cert., assoc.), Clin. Lab. Mgmt. Assn., Zeta Tau Alpha (scholar chmn. 1976, house mgr. 1977), Sigma Zeta. Avocations: fishing, arts and crafts.

BLACK, PAUL HENRY, medical educator, researcher; b. Boston, Mar. 11, 1930; s. Samuel Louis and May (Goldberg) B.; m. Sandra Merkin, June 2, 1962; children: Scott, Marc, Jeffrey. A.B., Dartmouth Coll., 1952; M.D., Columbia U., 1956. Diplomate Am. Bd. Internal Medicine. Intern Mass. Gen. Hosp., Boston, 1956-57, asst. resident in medicine, 1957-58, clin. and research fellow, 1958-60, resident in medicine, 1960-61; sr. asst. surgeon Lab. Infectious Diseases USPHS Nat. Inst. Allergy and Infectious Diseases, NIH, Bethesda, Md., 1961-63; sr. surgeon Lab. Infectious Diseases USPHS Nat. Inst. Allergy and Infectious Diseases, U. Glasgow Inst. Virology, Scotland, 1963-64, Nat. Inst. Allergy and Infectious Diseases, NIH, Bethesda, Md., 1964-67; asst. prof. medicine Harvard U. Med. Sch., Boston, 1967-70, assoc. prof. medicine, 1970-80; asst. physician Mass. Gen. Hosp., Boston, 1967-70, assoc. physician, 1970-80, hon. physician, 1980—; chmn., prof. microbiology, research prof. surgery, prof. medicine Boston U. Sch. Medicine, 1979—; dir. Hubert H. Humphrey Cancer Research Ctr. Boston U., 1979-83; cons. Roswell Park Meml. Inst., Buffalo, 1976-80 , Monsanto Chem. Corp., St. Louis, 1976-82, Collaborative Research, Inc., Lexington, Mass., 1984—, Nat. Cancer Adv. Bd. Subcom. on the Evaluation of Cancer Ctrs., Bethesda, 1975-80 ; sci. cons. U.S.-Israel Binat. Sci. Found., Jerusalem, Israel, 1974—; mem. NIH Study Sect. Virology, 1968-72, Tumor Virus Detection Segment, Spl. Virus Cancer Program, Bethesda, 1972-76; mem. subcom. on environ. carcinogens, Am. Cancer Soc. Task Force on Cancer Prevention, 1975-82 , sci. adv. bd. Worcester Found. for Exptl. Biology, Mass., 1976-78, sci. adv. bd. Dartmouth-Hitchcock Med. Ctr., Hanover, N.H., 1976-80, Gov.'s Task Force on AIDS, Commonwealth of Mass., Boston, 1983—; chmn. spl. virus cancer program contract rev. com., Nat. Cancer Inst., 1977-79. Author monograph; contbr. articles to profl. jours., chpts. to books. Nat. Cancer Inst. grantee, 1967-87. Mem. Am. Soc. Clin. Investigation, Infectious Diseases Soc., Am. Soc. Microbiology, Am. Soc. Virology, Am. Assn. Med. Sch. Microbiology Chmn., AAAS, Soc. Gen. Microbiology, Sigma Xi. Democrat. Jewish. Home: 21 Dawes Rd Lexington MA 02173-5926 Office: Boston U Sch Medicine 80 E Concord St Boston MA 02118-2394

BLACK, PERRY, neurological surgeon, educator; b. Montreal, P.Q., Can., Oct. 2, 1930; came to U.S., 1959, naturalized 1979; s. Ovido and Rose (Vasilevsky) B.; children—Daniel Ovid, Julie Miriam, Amy Rose. B.Sc., McGill U., Montreal, 1951; M.D., C.M., McGill U., 1956. Intern, asst. resident in medicine and gen. surgery Jewish Gen. Hosp., Montreal, 1956-58; asst. resident in neurology Montreal Neurol. Inst., 1958-59; resident in neurosurgery Johns Hopkins Hosp., Balt., 1959-63, neurosurgeon, 1964-79; NIH fellow in physiology Johns Hopkins U. Sch. Medicine, 1961-62, instr. neurol. surgery, 1964-67, asst. prof., 1967-69, assoc. prof., 1969-79, asst. prof. psychiatry, 1967-70, assoc. prof., 1970-79; prof., chmn. dept. neurosurgery, dir. pain treatment program Hahnemann U. Sch. Medicine, Phila., 1979—, dir. brain tumor program, 1983—; dir. child head injury project, dept. neurol. surgery Johns Hopkins Hosp., 1963-79; dir. lab. neurol. scis., chmn. central research authority Friends Med. Sci. Research Ctr., Balt., 1972-79; mem. neurology study sect. NIH, 1973-77; council neurosurgical rep. Johns. Hopkins U. Med. Sch., 1977-78, council vice-chmn., 1978-79; hon. dir. Friends Med. Sci. Research Ctr. Editor: Drugs and the Brain, 1969, Physiological Correlates of Emotion, 1970, Brain Dysfunction in Children: Etiology, Diagnosis, and Management, 1981; contbr. articles to profl. jours. Bd. dirs. Epilepsy Assn. Central Md., 1966-77, chmn. profl. adv. bd., 1973-75; mem. com. of fifty Epilepsy Found. Am., 1970-76, state coordinator for Md., 1973-76. Recipient Residents Paper award So. Neurol. Soc., 1963, Volvo award World Fedn. of Neurosurgical Socs., 1985. Mem. AAAS, Congress Neurol. Surgeons (chmn. sci. and edn. com. 1969-72, chmn. sci. program com. 1971-72, editor newsletter 1972-75, exec. com. 1972-75, nominating com. 1975-77, chmn. internat. com. 1975-81, jour. assoc. editor 1976-82, jour. internat. neurosurgery editor 1976-87, Disting. Svc. award 1977), Am. Assn. Neurol. Surgeons (Harvey Cushing Soc.) (subcom. on continuing edn. 1974-78), Am. Pain Soc.Soc. Neurol. Surgeons, AAUP, Am. Soc. Stereotactic and Functional Neurosurgery, Soc. for Neurosci., Research Soc. Neurol. Surgeons, Am. Epilepsy Soc., Am. Neurol. Assn., Internat. Assn. for Study of Pain, Phila. County Med. Soc., Phila. Neurol. Soc. (2d. v.p. 1982-83), Pa. Neurosurgical Soc. (mem. coun. 1989—, sec., treas. 1990—, pres., 1992), AMA, Mid-Atlantic Neurosurgical Soc. Office: Hahnemann U Dept Neurosurgery Broad & Vine Sts Philadelphia PA 19102-1192

BLACK, SUZANNE ALEXANDRA, clinical psychologist, clinical neuropsychologist, researcher; b. N.Y.C., May 6, 1958; d. Lawrence E. and Aline R. (Amsellem) B. BA in Psychology, Clark U., 1980; MA in Gen. Psychology, Yeshiva U., 1984, PsyD in Clin. Psychology, 1987; postgrad., Inst. Contemporary Psychoanalysis, 1992—. Lic. psychologist, Calif., 1989. Rsch. assoc. Inst. for Study of Exceptional Children Roosevelt Hosp. Ctr., 1980-82; rsch. cons. Sch. Pub. Health Columbia U., N.Y.C., 1982-83; clin. psychology extern Albert Einstein Coll. of Medicine Bronx Psychiat. Ctr., 1983-84; clin. psychology extern Jewish Bd. Family and Children's Svcs., N.Y.C., 1984-85; clin. psychology, neuropsychol. extern NYU Med. Ctr./Bellevue Hosp., N.Y.C., 1985-86; pre-doctoral clin. psychology/neuropsychology intern Rusk Inst. Rehabilitation NYU Med. Ctr., N.Y.C., 1986-87; post-doctoral clin. psychology fellow in emergency room and adult inpatient psychiatry Harbor/UCLA Med. Ctr., Torrance, 1987-89; inpatient and outpatient pvt. practice clin. psychology and neuropsychology Torrance, 1989—; rsch. assoc. depts. neurology, psychiatry and medicine Harbor/UCLA Med. Ctr., 1991—; dir. clin. svcs. New Life Adult Inpatient Psychiat. unit Suncrest Hosp. of South Bay, Torrance, Calif., 1992; clin. asst. prof. psychology Fuller Grad. Sch. Psychology, Pasadena, Calif., 1987-88; clin. supr. psychiat. residents and psychology externs Harbor/UCLA Med. Ctr., 1987-89; lectr. in field. NIMH grantee, 1986-87. Mem. Am. Psychol. Assn. (div. mem. women com., clin. neuropsychology com.), Calif. Psychol. Assn., Group Psychotherapy Assn. So. Calif., Assn. Humanistic Psychology. Avocations: travel, photography, artist, hiking, languages. Office: 25550 Hawthorne Blvd Ste 212 Torrance CA 90505-6832

BLACK, WILLIAM CORMACK, IV, insect geneticist, statistician; b. Denver, Jan. 6, 1957; s. William Cormack III and Katherine (Marshall) B.; m. Nancy Marie DuTeau, Aug. 15, 1981; children: Christine, Emma. BA, Grinnell Coll., 1979; MS, Duke U., 1981; PhD, Iowa State U., 1985. Psotdoctoral rsch. fellow U. Notre Dame, South Bend, Ind., 1985-88; asst. prof. Kans. State U., Manhattan, 1988-91; asst. prof. dept. microbiology Colo. State U., Ft. Collins, 1991—; Contbr. articles to Theoretical/Applied Genetics, Genetical Rsch., Heredity, Genetics, Bull. Entomol. Rsch. Democrat. Unitarian. Achievements include, in ecological genetics, use of allozymes and molecular genetic markers to monitor patterns of breeding in insect populations; molecular systematics of Deltocephaline leafhoppers, ticks and aedine mosquitoes. Office: Colo State U Dept Microbiology Fort Collins CO 80523

BLACKBURN, CHARLES LEE, oil company executive; b. Cushing, Okla., Jan. 9, 1928; s. Samuel and Lillian (Beall) B.; m. Mary Ann Bullock Colburn, July 14, 1984; children: Kern A., Alan J., Charles L., Derek R. BS in Engring. Physics, U. Okla., 1952. With Shell Oil Co., 1952-86, exec. v.p. exploration and prodn., 1976-86; pres., chief exec. officer Diamond Shamrock Corp. (name changed to Maxus Energy Corp., 1987), Dallas, 1987—; chmn., pres., chief exec. officer Maxus Energy Corp., 1987—. Served with U.S. Army, 1946-48. Mem. Am. Petroleum Inst. (dir.), Soc. Petroleum Engrs., Mid-Continent Oil and Gas Assn., Tau Beta Pi, Sigma Tau. Methodist. Clubs: Houston Country, Coronado; Dallas Country, Dallas Petroleum. Office: Maxus Energy Corp 717 N Harwood St Dallas TX 75201-6505

BLACKBURN, ELIZABETH HELEN, molecular biologist; b. Hobart, Australia, Nov. 26, 1948; 1 child. BS, U. Melbourne, Australia, 1970, MS, 1971; PhD in Molecular Biology, Cambridge (Eng.) U., 1975; DSc (hon.), Yale U., 1991. Fellow in biology Yale U., New Haven, 1975-77; fellow in biochemistry U. Calif., San Francisco, 1977-78; from asst. prof. to prof. molecular biology U. Calif., Berkeley, 1978-90; prof. U. Calif., San Francisco, 1990—. Recipient Eli Lilly award in microbiology, 1988, NAS award in molecular biology, 1990. Mem. AAAS (elected 1991), NAS (fgn. assoc.), Royal Soc. of London (England 1992). Office: U Calif Microbiology/Biochemistry Depts San Francisco CA 94143-0414

BLACKBURN, JAMES KENT, astrophysicist; b. Chattanooga, Tenn., July 15, 1957; s. James McKinley Blackburn and Myra Sue (Douglas) Martin. BS in Physics, U. N.C., 1979; PhD in Physics, U. Fla., 1990. Sr. engr., physicist Rudolph Rsch., Inc., Flanders, N.J., 1990-91; sr. scientist Data Link, Inc., Herndon, Va., 1991; sr. scientist Hughes STX, Inc., Lanham, Md., 1991-92, prin. scientist, 1992—; sci. cons. Rudolph Rsch., Inc., Flanders, 1990; support scientist NASA Goddard Space Flight Ctr., Greenbelt, Md., 1991—. Contbr. articles to profl. jours. Recipient DSR Rsch. fellow U. Fla., 1987, 88. Mem. AAAS, Am. Physics Soc., Am. Astron. Soc. Achievements include development of "FTOOLS", a FITS data analysis package for NASA's High Energy Astrophysics Community; development of "GRPP", a genaral relativity and tensor analysis computer language; development of "Star Traveler", a relativistic star travel simulation; discovery of numerical models for binary black hole orbits and gravitational radiation. Home: 14027 Vista Dr No 191-B Laurel MD 20707 Office: NASA Code 664.0 Goddard Space Flight Greenbelt MD 20771

BLACKBURN, RUTH ELIZABETH, biomedical research scientist; b. Arlecdon, Cumbria, Eng., Feb. 26, 1963; came to U.S., 1989; d. Walter and Harriet (McDonald) B. BSc in Applied Biology with honors, U. Thames, London, 1985; PhD in Neuroendocrinology, U. Cambridge, Eng., 1989. Rsch. asst. Pub. Health Lab. Svc., Salisbury, U.K., 1983-84; postdoctoral fellow Nuffield Found., Cambridge, U.K., 1989; rsch. assoc. dept. medicine and behavioral neurosci. U. Pitts., 1989—; participant Internat. Workshop on Fluid Balance, Kitakyushu, Japan, 1988. Referee Jour. Neuroendocrinology, Oxford, Eng.; contbr. articles to Jour. Endocrinology, Jour. Physiology, Jour. Neuroendocrinology, Am. Jour. Physiology, other refereed jours. Named Young Scientist Brit. Nuclear Fuels Ltd., 1988; recipient travel award Agrl. and Food Rsch. Coun., Eng., 1988. Mem. Brit. Neuroendocrine Group, Soc. Endocrinology (Young Endocrinologist 1988), Physiol. Soc., Endocrine Soc., Soc. Neurosci. Office: U Pitts Sch Medicine E1140 Biomed Sci Tower Pittsburgh PA 15261

BLACKEY, EDWIN ARTHUR, JR., geologist; b. Tamworth, N.H., Oct. 19, 1927; s. Edwin Arthur and Flora (Whipple) B.; m. Patricia Ann Matthews, Jan. 22, 1955; children: Mark Edwin, Janet Angove. B.S., U. N.H., 1951; postgrad. Worcester (Mass.) Poly. Inst., 1955-56. Geologist, N.E. div. Corps Engrs., Waltham, Mass., 1951-72, div. geologist, 1972-82, cons. engring. geologist, 1982—. Chmn., Hist. Dists. Commn. Sudbury (Mass.); trustee Sudbury Hist. Soc., pres., 1988-89; mem. Earth Removal Bd. Sudbury; past pres. Sudbury Jr. Ski Program. Served with AUS, 1946-47. Cert. geologist, Maine. Recipient Meritorious Civilian Svc. award C.E., 1983. Mem. Assn. Engring. Geologists (nat. chmn. bldg. codes com., nat. bd. dirs., 1982, dir. New Eng. sect., chmn. New Eng. sect. 1982-85, nat. exec. dir. 1987—), Am. Geologic Inst. Episcopalian. Club: U.S. Eastern Ski. Home and Office: 62 King Philip Rd Sudbury MA 01776-2363 Office: Assoc of Engin Geologists 323 Boston Post Rd Ste 2D Sudbury MA 01776

BLACKMAN, EDWIN JACKSON, software engineer; b. Pulaski, Tenn., Nov. 19, 1947; s. Alley J. and Martha (Williams) B.; m. Nancy Kamin, Mar. 11, 1982 (div. Mar., 1990); m. Michelle Fautz, May 25, 1990. AA, Martin Coll., Pulaski, Tenn., 1972; BS, Tenn. Tech. U., 1974. State auditor State of Tenn., Nashville, 1974-76; internal auditor Firestone Tire & Rubber Co., Akron, Ohio, 1976-79; mgr. internal audit Leewards Creative Crafts, Elgin, Ill., 1979-81; acct. mgr. Mgmt. Sci. Am., Oakbrook, Ill., 1981-85; sales cons. Mgmt. Sci. Am., Oakbrook, 1985-88, sr. mktg. rep., 1988-90; customer svc. mgr. Dun & Bradstreet Software, Columbus, Ohio, 1990-92, acct. exec., 1993—. With U.S. Navy, 1966-70, Vietnam. Mem. Moose, Young Ams. for Freedom, Elks, Exchange Club. Methodist. Avocations: sailing, golf. Home: 13824 Bainwick Dr NW Pickerington OH 43147-8722

BLACKMON, JAMES BERTRAM, aerospace engineer; b. Charlotte, N.C., Dec. 6, 1938; s. James B. and Willa Myra (Suits) B.; m. Marcia Fritz, June 1963 (div. 1981); children: James Craig, Jeffrey Scott; m. Christina Marie Almgren, Jan. 29, 1984; children: Laura E. Dyer, Sean N. Dyer. BS in Engring., Calif. Tech., 1961; MS in Engring. and Applied Sci., UCLA, 1967, PhD in Engring. and Applied Sci., 1972. Program mgr. MDAC Energy Systems, 1977-84; mgr. advanced space systems McDonnell Douglas Space Systems Co., Huntington Beach, Calif., 1984-89; sr. mgr. advanced space systems McDonnell Douglas Space Systems Co., Huntington Beach, 1989-91; dir. advanced program devel. and prodn. support McDonnell Douglas Space Systems Co., Huntsville, Ala., 1992—; chair solar energy rev. panel Dept. Energy, Washington, 1992. Contbr. articles on solar energy systems to profl. jours. Asst. scoutmaster Boy Scouts Am., Costa Mesa, Calif., 1978-90. Named McDonnell Douglas Corp. fellow, 1990; recipient Orange County (Calif.) Engrin. Merit award Orange County Engring. Coun., 1991. Fellow Inst. for Advancement of Engring.; mem. AIAA (tech. aerospace power com. mem. 1990-92), ASTM (com. chmn. 1981-92). Methodist. Achievements include patents in dielectrophoretic propellant orientation apparatus and method for calculation of shape factors; in LDR centrifugal collector apparatus; in moving belt radiator heat exchanger; in turbulent droplet generator with boom; in advanced survivable radiation. Office: McDonnell Douglas Aerospace 689 Discovery Dr Huntsville AL 35806

BLACKMON, MARGARET LEE, pharmaceutical chemist; b. Clinton, N.C., Dec. 23, 1951; d. Edgar Lee and Eunice Margaret (Wilson) B. BA in Chemistry, U. N.C., 1990; MDiv, Shaw Div. Sch., Raleigh, N.C., 1993. Cert. quality auditor Am. Soc. Quality Control. Control scientist II Burroughs Wellcome Co., Greenville, N.C., 1975-77, devel. scientist I-IV, 1977-91, assoc. sect. head Analytical Devel. Labs., 1991—. Min. African Meth. Episc. Zion Ch., also dir. Christian edn., dist. sect. of young women, woman's home and overseas missionary soc., deaconess, tchr. Sunday sch; bd. dirs. Second Chance Ministries. Mem. Am Chem. Soc., Drug Info. Assn., Am. Assn. Pharm. Scientists, Alpha Chi Sigma. Home: 218 Belvedere Dr Greenville NC 27834-6804 Office: Burroughs Wellcome Co Hwys 11 & 264 PO Box 1887 Greenville NC 27834

BLACKWELL, JOHN, polymer scientist, educator; b. Oughtibridge, Sheffield, Eng., Jan. 15, 1942; came to U.S., 1967; s. Leonard and Vera (Brook) B.; m. Susan Margaret Crawshaw, Aug. 5, 1965; children: Martin Jonathan, Helen Elizabeth. B.Sc. in Chemistry, U. Leeds, Eng., 1963, Ph.D. in Biophysics, 1967. Postdoctoral fellow SUNY-Syracuse Coll. Forestry, 1967-69; vis. asst. prof. Case Western Res. U., Cleve., 1969-70; asst. prof. Case Western Res. U., 1970-74, asso. prof., 1974-77, prof. macromolecular sci., 1977—; chmn. dept., 1985—; F. Alex Nason prof., 1991—; vis. prof. Kennedy Inst. Rheumatology, London, 1975, Centre National de Recherche Scientifique, Grenoble, France, 1977, U. Frieburg, Fed. Republic Germany, 1982; chmn. Gordon Conf. on Liquid Crystalline Polymers, 1992; cons. in field. Author: (with A.G. Walton) Biopolymers, 1973; mem. editorial bd. Macromolecules, 1989-92; adv. bd. Jour. Macromolecular Sci.-Physics, 1986—; internat. adv. bd. Acta Polymerica, 1992—; contbr. articles to profl. jours. Recipient award for disting. achievement Fiber Soc., 1981, Sr. Scientist award Alexander von Humboldt Found., Max Planck Inst. for Polymer Rsch., Mainz, Fed. Republic Germany, 1991, Rsch. Career Devel. award, 1973-77; grantee NSF, 1970—, NIH, 1970—, Dept. Def., 1976-82. Fellow Am. Phys. Soc. (exec. com. div. high polymer physics 1986—, vice-chmn. 1987-88, chmn. 1988-89); mem. Am. Chem. Soc., Am. Crystallography Soc. (chmn. fiber diffraction spl. interest group 1993-94), Biophys. Soc. (chmn. biopolymer subgroup 1975-76), Fiber Soc. Episcopalian. Home: 2951 Atleboro Rd Cleveland OH 44120-1815 Office: Dept Macromolecular Sci Case Western Res U Cleveland OH 44106-7202

BLACKWELL, JOHN ADRIAN, JR., computer company executive; b. Tulsa, Aug. 1, 1940; s. John Adrian and Daisy Edith (Webb) B. MusB, Westminster Choir Coll., 1962, MusM, 1963. Minister of music 1st Presbyn. Ch., Warren, Ohio, 1963-68, Oklahoma City, 1968-79; artistic dir. Okla. Choral Assn., Oklahoma City, 1980-82; pres. Okla. Digital Technologies Inc., Oklahoma City, 1987-92; ptnr. JJ Enterprises (now Megabyn Assocs., Inc.); pres., co-owner JJ Enterprises (now Megabyn Assocs., Inc.), Oklahoma City, 1992—; cons. Union Oil Co. Calif., Oklahoma City, 1989—; conductor Warren (Ohio) Symphony Orch., 1965-68; choral dir. NBC-TV Stars & Stripes Shows, Oklahoma City, 1975-76. Commd. ch. worker Presbyn. Ch. in the U.S.A., 1965. Recipient Paul Harris award Rotary Found., 1993. Mem. Rotary Internat. Home: 2413 NW 112th Ter Oklahoma City OK 73120-7202 Office: Megabyn Assocs 3120 Chaucer Dr Oklahoma City OK 73120-2228

BLACKWELL, NEAL ELWOOD, agricultural and mechanical engineer, researcher; b. Chase City, Va., Oct. 3, 1955; s. Elwood Toy and Nancy (Smith) B. BS in Agrl. Engring., Va. Poly. Inst. & State U., 1979; MS in Agrl. Engring., VPI & SU, 1982, PhD in Environ. Sci. and Engring., 1986. Engr. in tng., Va. Project engr. Alternative Energy Corp., Rsch. Triangle Pk., N.C., 1986-87, Environ. Techs., Inc., Raleigh, N.C., 1987; pvt. cons. Leading Edge, Inc., Miami, Fla., 1987; mech. engr. Belvior Rsch. and Devel. Ctr., Fort Belvoir, Va., 1987—. Contbr. articles to profl. jours. Design engr. Engring. Ministries Internat., Haiti, 1986, 87. Pratt Presdl. fellow VPI & SU, Blackburg, Va., 1981, 84. Mem. Am. Soc. Agrl. Engrs. (referee transactions 1988, student pres. 1978-79, student honor 1980), Sigma Xi, Alpha Epsilon, Gamma Sigma Delta. Achievements include copyright on wall jet computer model for ceiling fan applications; development of mobile test stand for large peanut drying fans. Office: Belvoir Rsch and Devel Ctr SATBE-FED Fort Belvoir VA 22060-5606

BLADYKAS, MICHAEL P., mechanical engineering consultant; b. Glen Cove, N.Y., Mar. 13, 1959; s. Francis and Velma Grace (Vogtlander) B.; m. Susan Ann Kahler, June 24, 1989; children: John P. Riordan, Paul C. Riordan, Lois Ann, Julia Kahler. BSME summa cum laude, N.Y. Inst. Tech., 1984. Profl. engr., N.Y. Mech. engr. Jaros Baum & Bolles, N.Y.C., 1984-89; mech. engr. Lizardos Engring. Assocs., PC, Albertson, N.Y., 1989-90, prin., 1990-93; engring. mgr. The Sear-Brown Group, Lake Success, N.Y., 1993—. Mem. ASHRAE, NSPE, Assn. of Energy Engrs., Am. Soc. of Plumbing Engrs. Office: The Sear-Brown Group 2300 Marcus Ave Lake Success NY 11042

BLAESSER, GERD, theoretical physics researcher; b. Berlin, Mar. 25, 1933; arrived in Italy, 1961; s. Udo Herbert and Herta Lilly (Krueger) B.; m. Ingrid Renate Loepelmann, Dec. 21, 1957; children: Gabriele Monika, Sven Sergio. Diploma in physics, Free U. Berlin, 1956; Dr Rer. Nat., Ruprecht-Karl U., Heidelberg, Germany, 1958. Reactor physics rschr. Nuclear Rsch. Ctr., Karlsruhe, Germany, 1956-61; reactor physics rschr. Joint Rsch. Ctr., Ispra, Italy, 1961-67, solid state physics rschr., 1967-73, atmospheric physics rschr., 1973-77; photovoltaics rschr. Joint Rsch. Ctr., Ispra, 1975—, head European Solar Test Installation, 1987—; cons. mem. Internat. Electrochemical Commn., 1984—. Contbr. over 30 articles to profl. jours., chpts. to books. Achievements include development of reliable method for on-site power measurements on large PV-generators. Home: Via Verdi 5, I-21018 Osmate Italy Office: Joint Rsch Ctr of CEC, I-21020 Ispra Italy

BLAEVOET, JEFFREY PAUL, mechanical engineer; b. London, June 28, 1959; came to U.S., 1986; s. Roger Yves Gaston and Patricia Mary (Young) B.; m. Susanne Marie Los, Oct. 6, 1990. BS in Agrl. Engring., U. Newcastle Upon Tyne, 1980; MS in Energy Conservation & Environ., Cranfield Inst. Tech., 1985. Registered profl. engr. Calif., N.Y. Mech. engr. Ove Arup & Ptnrs., London, 1984-86; assoc. Ove Arup & Ptnrs., San Francisco, 1986-90, Flack & Kurtz Consulting Engrs., San Francisco, 1990—. Lt. British Army, 1980-83. Serc Advanced scholar, 1984. Mem. Am. Soc. Heating, Refrigerating and Air-Conditioning Engrs., Am. Soc. Hosp. Engrs., Assn. Energy Engrs., Inst. Mech. Engrs., Chartered Inst. Bldg. Svcs. Engrs., Inst. Energy. Office: Flack & Kurtz Cons Engrs 343 Sansome St #450 San Francisco CA 94104-1309

BLAIR, DAVID WILLIAM, mechanical engineer; b. Santa Barbara, Calif., Oct. 5, 1929; s. David Sutherland and Norah Mildred (Higgins) B.; m. Rosemary Constance Miles, Jan. 30, 1954; children: Karen E., Barbara A., M. Maria, Amanda M., David B. O., Rachel P. BS, Oreg. State U., 1952; MS, Columbia U., 1954, PhD, 1961. From asst. to instr. mech. engring. Columbia U., N.Y.C., 1952-58; rsch. assoc. Princeton (N.J.) U., 1958-61; rsch. scientist AeroChem Rsch. Labs., Princeton, 1961-62; postdoctoral fellow Royal Norwegian Coun. Indsl. and Engring. Rsch., Kjeller, Norway, 1962-63; assoc. prof. Polytechnic Inst. Bklyn., 1962-69; engring. assoc. Corp. Rsch. Labs. Exxon Rsch. and Engring. Co., Linden, N.J., 1969-83; pres. Princeton Sci. Enterprises, Inc., 1985—. Contbr. articles to Handbook of the Engring. Scis., AIAA Jour., Jour. Quantitative Spectroscopy and Radiative Transfer, Environ. Sci. and Tech.; patentee for multi-stage process for combusting fuels containing fixed-nitrogen chemical species, efficient high temperature radiant furnace. Mem. Princeton Township Com., 1975-82, Princeton Joint Commn. on Civil Rights, 1975-87. Mem. Am. Inst. Chem. Engrs., Am. Phys. Soc., Combustion Inst., Sigma Xi, Tau Beta Pi, Sigma Tau, Phi Kappa Phi, Pi Mu Epsilon, Pi Tau Sigma. Democrat. Achievements include patent for multi-stage process for combusting fuels containing fixed-nitrogen chemical species, efficient high temperature radiant furnace. Home and Office: Princeton Sci Enterprises Inc 1108 Kingston Rd Princeton NJ 08540-4132

BLAIR, KIM BILLY, mechanical engineer, researcher; b. Kearney, Nebr., Dec. 12, 1958; s. Billy Dick and Kay Elaine (Bauman) B.; m. Cheryl Jakoby, Sept. 5, 1992. BA in Psychology, U. Nebr., 1981, BSME, 1983; MSME, Purdue U., 1987, PhD in Aero. & Astro. Engring., 1992. Project engr. Alpo Petfoods Inc., Crete, Nebr., 1983-85; teaching asst. Purdue U., West Lafayette, Ind., 1985-92; aerospace engr. NASA-Johnson Space Ctr., Houston, 1987-91; sr. staff scientist Moldyn, Inc., Cambridge, Mass., 1992—. Author conf. papers. Mem. AIAA, ASME. Home: 325 Franklin St # 307 Cambridge MA 02139 Office: Moldyn Inc 1033 Massachusetts Ave Cambridge MA 02138

BLAIR, PATRICIA WOHLGEMUTH, economics writer; b. N.Y.C., Nov. 30, 1929; m. James P. Blair, Aug. 13, 1964; children: David A., Matthew W. BA with honors, Wellesley Coll., 1950; MA, Haverford Coll., 1952. Officer U.S. Agy. Internat. Devel., New Delhi, 1953-55, 63-64; editor Carnegie Endowment for Internat. Peace, N.Y.C., 1956-63, Devel. Digest, Nat. Planning Assn., Washington, 1965-68; staff assoc. Commn. on Internat. Devel., World Bank, Washington, 1969-70; ind. cons., writer, editor, 1970—. Editor: Health Needs of the World's Poor Women, 1980; contbr. articles to profl. publs. Mem. adv. com. Unitarian-Universalist Holdeen India Fund, Washington, 1984—; bd. dirs. Equity Policy Ctr., Washington, 1980-85. Mem. Soc. Internat. Devel. (internat. governing coun. 1975-79), Assn. Women in Devel., Asia Soc., World Affairs Coun. Home and Office: 1411 30th St NW Washington DC 20007-3141

BLAIR, ROBERT EUGENE, physicist, researcher; b. Upper Darby, Pa., June 30, 1949; s. William James and Lenore (Talbot) B.; m. Teresa Tarallo, May 11, 1976; 1 child, Byron. BS, Carnegie-Mellon U., 1971; PhD, Calif. Inst. Tech., 1982. Asst. prof. Columbia U., N.Y.C., 1981-86; physicist Argonne (Ill.) Nat. Lab., 1986—. Democrat. Achievements include research in the rate of increase of the neutrino cross sections with energy, monitoring and calibration systems for neutrino flux measurements in a high-energy dichromatic beam, simulating SSC event detection, measurement of prompt photon production in pp interactions at square root of s = 1800 Gev, diphoton production in anti-proton proton at square root of s = 1800 Gev. Home: 1 South 235 Lloyd Ave Lombard IL 60148 Office: Argonne Nat Lab 9700 S Cass Ave Bldg 362 Rm E277 Argonne IL 60439

BLAIS, ROGER NATHANIEL, physics educator; b. Duluth, Minn., Oct. 3, 1944; s. Eusebe Joseph and Edith Seldina (Anderson) B.; m. Mary Louise Leclerc, Aug. 2, 1971; children: Christopher Edward, Laura Louise. BA in Physics and French Lit., U. Minn., 1966; PhD in Physics, U. Okla., 1971; cert. in computer programming, Tulsa Jr. Coll., 1981; cert. in bus., UCLA, 1986. Registered profl. engr., Okla. Instr. physics Westark Community

Coll., Ft. Smith, Ark., 1971-72; asst. prof. physics and geophys. scis Old Dominion U., Norfolk, Va., 1972-77; asst. prof. engring. physics U. Tulsa, 1977-81, assoc. prof., 1981—; assoc. dir. Tulsa U. Artificial Lift Projects, 1983—, chmn. physics, 1986-88, vice provost, 1989-92, acting provost, v.p. acad. affairs, 1990-91. Contbr. articles to profl. jours. Mem. AAAS, AAUP, NSPE, Am. Phys. Soc., Am. Geophys. Union, Soc. Petroleum Engrs., Instrument Soc. Am. (dir.-elect test measurement divsn. 1993), N.Y. Acad. Scis., Iron Wedge Soc., Am. Assn. Physics Tchrs., Am. Soc. Engring. Edn., Phi Beta Kappa, Sigma Xi, Sigma Pi Sigma, Tau Beta Pi, Phi Kappa Phi. Home: 5348 E 30th Pl Tulsa OK 74114-6314 Office: U Tulsa Provost 600 S College Ave Tulsa OK 74104-3189

BLAKE, JEANNETTE BELISLE, psychotherapist; b. Manchester, N.H., Aug. 1, 1920; d. Emile Henry and Mathilda Cecelia (Martin) Belisle; m. Roland Oscar Royer, Sept. 6, 1937 (div. 1948); 1 child, Dorothy Marie Royer Lyman; m. Albert Willard Blake Sr., Aug. 11, 1979. Cert. profl. counselor. Cons. Al Blake Advt. Cons., Manchester, N.H., 1959-68; pvt. practice Manchester, 1968—; founder, dir. N.H. Metaphys. Establishment, Manchester, 1976—; presenter workshops in field. Recipient medal N.H. Metaphysicians, 1978; cert. Greater Manchester Mental Health Ctr., 1985, 86. Mem. N.H. Assn. for Counseling and Devel., N.H. Assn. of Family Counselors, Am. Assn. of Mental Health Counselors (N.H. br.), Therapeutic Touch Healing (Manchester chpt.), Soc. for Psychic Rsch. of N.h. and Mass. (adv. bd.). Roman Catholic. Avocations: harness horse racing, oil painting, workshops on Holistic health. Home and Office: 131 Russell St Manchester NH 03104-3769

BLAKE, JOHN CHARLES, JR., environmental engineer; b. Roanoke, Va., Dec. 19, 1956; s. John Charles and Marguerite Ellen (Cahill) B.; m. Brenda Mae Williams, Jan. 7, 1984; children: Jonathan David, Zachary Aaron. BSCE, Va. Mil. Inst., 1980. Registered engr.-in-tng., Va. Engr. trainee Commonwealth Gas Distbn. Co., Petersburg, Va., 1980-82; asst. dist. engr. Office Water Programs, Va. Dept. Health, Abingdon, 1984—. Mem. Water Environ. Fedn., Am. Water Works Assn., Va. Water Environment Assn. (safety com. 1991—), Civitan Club. Methodist. Home: 221 Carter St Bristol VA 24201 Office: Va Dept Health Office Water Programs 454 E Main St Abingdon VA 24210

BLAKE, WILLIAM BRUCE, aerospace engineer; b. Phila., Oct. 8, 1960; s. Bruce Borsum and Margaret Eleanor (Stebbins) B.; m. Valerie Kaye Bailey, May 19, 1990. BS in Aerospace Engring., Rensselaer Poly. Inst., 1982; MS in Aerospace Engring., U. Dayton, 1992. Aerospace engr. USAF Wright Lab., Wright Patterson AFB, 1987—. Contbr. articles to profl. jours. including AIAA Jour. Aircraft, Tech. Rev., Vertiflite, The Skeptical Inquirer. Trustee Archimedes Rotorcraft Mus., Brookville, Ohio, 1991—. Capt. USAF, 1982-86. Decorated USAF Commendation medal, Sci. Achievement award. Mem. AIAA (sr.), Am. Helicopter Soc. (Dayton chpt., v.p. 1988-89, pres. 1989-90, bd. dirs. 1990—). Achievements include research in prediction methods for aerodynamics/stability and control. Office: USAF Wright Lab WL/FIGC Wright Patterson AFB OH 45433-6553

BLAKELEY, ROGER WILLIAM GEORGE, engineering executive; b. Wellington, New Zealand, Nov. 13, 1945; s. Philip William and Ida May (Rogers) B.; m. Lynnette Marie Steedman, Aug. 22, 1970; children: Nicholas Dylan, Marissa Jade. B in Civil Engring. with honors, U. Canterbury, Christchurch, New Zealand, 1968, PhD, 1972; MS in Mgmt., Stanford U., 1981. Asst. engr. Ministry of Works and Devel., Wellington, 1971-74, engr., 1974-76, sr. engr., 1976-81; dist. civil engr. Ministry of Works and Devel., Wanganui, 1981-84; gen. mgr. State Coal Mines, Wellington, New Zealand, 1984-86; sec. for the environ. Ministry for the Environment, Wellington, 1986—; chairperson OECD Environ. Com., Paris, 1989-91; adv. bd. Climate Inst., Washington, 1990—; chairperson Pub. Svc. Chief Execs.' Forum, Wellington, 1992; New Zealand rep. to internat. confs. including Earth Summit, Rio de Janeiro, 1992; bd. mem. Inst. Policy Studies, Victoria U. Contbr. chpts. to books, papers and articles to profl. jours. and conf. procs. Dir. Wellington Samaritans, Inc., 1977-80. Harkness fellow Commonwealth Fund N.Y., 1980-81; Sloan fellow Stanford U. Grad. Sch. Bus., 1980-81; recipient New Zealand 1990 Commemoration medal. Fellow Instn. Profl. Engrs. New Zealand (Fulton Downer gold medal 1977, Freyssinet award 1975, 76, Structural award 1988), New Zealand Nat. Soc. Earthquake Engring. (Otto Glogau award 1983), Engrs. for Social Responsibility. Methodist. Achievements include development of seismic design of prestressed concrete beam-column joints, procedures for seismic design of base isolated bridges. Home: 25 Maire St Woburn, Lower Hutt New Zealand Office: Ministry for the Environ, 84 Boulcott St, Wellington New Zealand

BLAKLEY, BRENT ALAN, polymer chemist, computer programmer; b. Columbus, Ohio, Feb. 28, 1958; s. Charles Jack and Wilma (Smith) B.; m. Kathy Hursey, Aug. 23, 1980 (div. Sept. 1983); m. Carol Lea Taylor, Oct. 26, 1985; children: Alexandra, Bradley. BS, Otterbein Coll., 1985. Chem. technician Ashland Chem. Co., Dublin, Ohio, 1983-85, chemist, 1985-1989, product coding specialist, 1989—. Mem. Am. Chem. Soc. Achievements include patent on vapor permeation curable coatings comprising polymercaptan resins and multi-isocyanate curing agents, in situ quaternary ammonium catalyst formation for curing polymeric isocyanates, stabilized moisture curable polyurethane coating compositions. Home: 1017 Newfields Ln Westerville OH 43081-3570 Office: Ashland Chem Co Inc 5200 Blazer Pky Dublin OH 43017-5309

BLANC, THEODORE VON SICKLE, physicist; b. N.Y.C., Dec. 18, 1942; s. Francis James and Verlie Elizabeth (Heimsath) B. BA in Physics, Hanover (Ind.) Coll., 1966; MS in Physics, Ball State U., 1968. Rsch. physicist Naval Rsch. Lab., Washington, 1971—. Mem. Royal Meteorol. Sco., Am. Meteorol. Soc. Achievements include research in boundary-layer meteorology, phys. oceanography, atmospheric and ocean tech. Office: Naval Rsch Lab 4555 Overlook Ave SW Washington DC 20375

BLANCH, ROY LAVERN, electrical engineer; b. Burley, Idaho, Nov. 5, 1946; s. James Donald and Georgia Elaine (Egan) B.; m. Garland Hill Smith, Jan. 23, 1982; children: Steven James, Edith Ann. BSEE, Brigham Young U., 1977. Elec. engr. asst. Burroughs Corp., San Diego, 1977-79, elec. engr., 1979-81; elec. engr. Burroughs Corp., Flemington, N.J., 1981-83, sr. elec. engr., 1983-84, project elec. engr., 1984-86; project elec. engr. Unisys, Flemington, 1986-89; staff engr. Unisys, San Jose, Calif., 1989—. Patentee in field. Sgt. USAF, 1970-74. Mormon. Avocations: skiing, flying, reading. Office: Unisys Corp 2700 N 1st St M/S 18-004 San Jose CA 95134-2028

BLANCHARD, ROBERT LORNE, engineering physicist; b. Montreal, Que., Can., May 8, 1919; came to U.S., 1946; s. Lorne Albert and Ida Bernadette (Ganter) B. BSc, McGill U., Montreal, 1941; SM, Harvard U., 1948. Sr. engr. Raytheon Mfg. Co., Waltham, Mass., 1946-47; pvt. cons. Newton, Mass., 1948-49; sr. engr. Trans-Sonics Inc., Bedford, Mass., 1949-52; v.p., dir. rsch. and systems Trans-Sonics Inc., Burlington, Mass., 1952-74, also bd. dirs.; v.p., dir. rsch. and mktg. Foxboro/Trans-Sonics, Burlington, Mass., 1974-78; mgr. tech. product planning Foxboro/Analytical, Burlington, Mass., 1978-84; pres. Arlby Co., Lexington, Mass., 1984—; bd. dirs. Inst. Guilfoyle, Belmont, Mass. Contbr. articles to AIAA Jour., Am. Rocket Soc., others, and confs. Maj. Can. Army Signal Corps, 1942-46. Fellow AIAA (assoc.); mem. AAAS. Achievements include patents for Liquid Measuring System, Fluid Level Meter, Densimeter, Heater for Chromatography Air Bath Oven, Transfer Calibration System, Measurement Calibration; development of Navy ground-air backpack communications receiver, C.W. airborne radar system to track and score machine-gun bullets and missiles, propellant gauging systems for Saturn V booster rocket and for Lunar Excursion Module on the Apollo program, custody transfer gauging system for LNG ships, an LNG cargo system simulator for ship crew training. Home and Office: 1310 Massachusetts Ave Lexington MA 02173-3809

BLANCHARD, WILLIAM HENRY, psychologist; b. St. Paul, Mar. 25, 1922; s. Charles Edgar and Ethel Rachael (Gurney) B.; m. Martha Ida Lang, Aug. 11, 1947; children: Gregory Marcus, Mary Lisa. Diploma in Sci. Mason City Jr. Coll., 1942; BS in Chemistry, Iowa State U., 1944; PhD in Psychology, U. So. Calif., 1954. Lic. clin. psychologist, Calif. Shift chemist B.F. Goodrich Chem. Co., Port Neches, Tex., 1946-47; court psychologist L.A. County Gen. Hosp., 1954-55; psychologist, dir. rsch. So. Reception Ctr. and Clinic, Calif. Youth Authority, Norwalk, 1955-58; social scientist Rand Corp., 1958-60, System Devel. Corp., 1960-70; mem. faculty Calif. State U.-Northridge, L.A., 1970; assoc. prof. UCLA, 1971; faculty group leader urban semester U. So. Calif., L.A., 1971-75; sr. rsch. assoc. Office of Chancellor, Calif. State U., L.A., 1975-76; sr. rsch. fellow Planning Analysis and Rsch. Inst., Santa Monica, Calif., 1976—; pvt. practice psychologist, Calif., 1976—; clin. assoc. dept. psychology U. So. Calif., 1956-58. Author: Rousseau and the Spirit of Revolt, 1967; Aggression American Style, 1978; Revolutionary Morality, 1984. Contbr. articles to profl. jours. Mem. com. on mental health West Area Welfare Planning Council, L.A., 1960-61; bd. dirs. L.A. County Psychol. Assn., 1969; commr. Bd. Med. Examiners, Psychology Exam. Com., State of Calif., 1969; v.p. Parents and Friends of Mentally Ill Children, 1968—, pres., 1966-68, trustee, 1968—. Mem. Am. Psychol. Assn., Internat. Soc. Polit. Psychology, AAAS, Brit. Psychol. Assn. Home: 4307 Rosario Rd Woodland Hills CA 91364-5546

BLANCHE, JOE ADVINCULA, aerospace engineer; b. Rizal, Santa, Ilocos Sur, Philippines, Sept. 11, 1954; came to U.S. 1976; s. Emilio Peralta and Concepcion (Advincula) B.; m. Albine Selerio Lansangan, Oct. 9, 1982; children: Emmanuel Joseph, Earl Jordan. Cert. in mil. sci., U. Philippines, 1973; BS in Math., Adamson U., Manila, 1976; postgrad., Calif. State U., Long Beach, 1982-85; AAS in Avionics Systems Tech., Community Coll. Air Force, Maxwell AFB, Ala., 1990; cert. in mgmt., Cen. Tex. Coll., 1990; PhD in Mgmt., Pacific Western U., 1993. Lic. real estate broker, Calif.; registered tax preparer, Calif. Assoc. engr./scientist McDonnell Douglas Corp., Long Beach, Calif., 1981-84; engr./scientist McDonnell Douglas Corp., 1984-86, engr./scientist specialist, 1987-88, sr. engr./scientist, 1988—; lead aerospace engr. Sikorsky Aircraft-UTC, Stratford, Conn., 1986-87; avionics maint. inspector USAF, 1983-86, 87—. With USAF, 1976-80. Bur. Forestry grantee and scholar U. Philippines, 1971-73. Mem. AIAA, Nat. Notary Assn., NRA, So. Calif. Profl. Engrs. Assn., Corona-Norco Bd. Realtors, Internat. Soc. Allied Weight Engrs. (sr. mem.), Santanians USA Inc. (bd. dirs. 1984-87), Marinduque Assn So. Calif., Fil-Am. Assn. Corona (auditor). Republican. Roman Catholic. Home: 2179 Tehachapi Dr Corona CA 91719-1138 Office: McDonnell Douglas Corp 3855 N Lakewood Blvd Long Beach CA 90846-0001

BLANCHET, THIERRY ALAIN, mechanical engineering educator; b. Huntington, N.Y., Dec. 18, 1963; s. Alain Maurice and Mary Ann (Langr) B. BSME magna cum laude, U. Vt., 1986; MS in Engring. Sci., Dartmouth Coll., 1988, PhD in Engring. Sci., 1992. Summer intern Champlain Cable Rsch. and Devel., Colchester, Vt., 1983-86, Dow Chem. Ctrl. Rsch., Midland, Mich., 1988; summer grad. fellow NASA Lewis Rsch. Ctr., Cleve., 1989-91; post-doctoral rsch. engr. Rensselaer Polytech. Inst., Troy, N.Y., 1992-93; asst. prof. Rensselaer Poly. Inst., Troy, N.Y., 1993—. Contbr. articles to profl. jours. Mem. ASME, Soc. Tribologists and Lubrication Engrs. (Al Sonntag award best publ. in the field of solid lubrication), Sigma Xi, Tau Beta Pi. Achievements include work on solid lubrication for high-temperature ceramic engines. Home: 26 Glenwood Dr Ballston Lake NY 12019 Office: Rensselaer Polytech Inst Dept Mech Engring Troy NY 12180

BLANCHET-FINCHER, GRACIELA BEATRIZ, physicist; b. Buenos Aires, May 7, 1951; came to U.S., 1976; d. Rodolfo and Rosa (Vignolo) Blanchet; m. Curtis R. Fincher Jr., Mar. 27, 1982; children: Katrina, Curtis Lee. MSc, U. Buenos Aires, 1976; PhD, Brown U., 1980. Postdoctoral fellow U. Pa., Phila., 1980-82, U. Calif., Santa Barbara, 1982-83; rsch. physicist E.I. DuPont de Nemours Co., Wilmington, Del., 1983-86, sr. rsch. physicist, 1986-92, rsch. assoc. Cen. Rsch. div., 1992—. Contbr. articles to sci. publs. Achievements include patents in photopolymer for electrostatic proofing applications; deposition of polymer films by laser aflation high temperature superconductors and polymeric membranes. Office: DuPont Cen Rsch Exptl Sta Wilmington DE 19898

BLANCHET-SADRI, FRANCINE, mathematician; b. Trois-Rivieres, Quebec, Can., July 25, 1953; came to U.S., 1990; d. Jean and Rolande (Delage) B.; m. Fereidoon Sadri, July 28, 1979; children: Ahmad, Hamid, Mariamme. BSc in Math., U. du Quebec a Trois-Rivieres, Quebec, Can., 1976; MS, Princeton U., 1979; PhD, McGill U., 1989. Rsch., teaching asst. U. du Quebec, Trois-Rivieres, Quebec, Can., 1974-76; lectr. U. du Quebec, 1976; rsch. asst. Princeton (N.J.) U., 1978; lectr. U. Tech. Isfahan, Iran, 1982-84, McGill U., Montreal, Quebec, 1988-89; asst. prof. U. N.C., Greensboro, 1990—. Contbr. articles to profl. jours. Recipient Rsch. Excellence award, 1991; Natural Scis. and Engring. Coun. Can. postgrad. fellow, 1976-80, Fonds pour la Formation de Chercheurs et L'aide a la Recherche doctoral fellow, 1985-87, Natural Scis. and Engring. Rsch. Coun. Can. postdoctoral fellow, 1990; new faculty grantee U. N.C., Greensboro, 1990-91, NSF grantee, 1991-94. Mem. Am. Math. Assn. of Am., Am. Math. Soc., Assn. for Computing Machinery. Achievements include discovery that the dot-depth of a generating class of aperiodic monoids is computable. Office: U NC Dept Math Greensboro NC 27412

BLANCO, FRANCES REBECCA BRIONES, chemist, medical technologist; b. Manila, Philippines, Nov. 2, 1958; d. Jose Capulong and Fe Reynoso (Briones) B. BS, U. Philippines, 1980; MS, Tufts U., 1989. Rsch. assoc. Nat. Rsch. Coun. of Philippines, Manila, 1980-86; sr. sci. rsch. specialist Philippine Coun. for Health, Rsch. and Devel., Manila, 1984-86; med. technologist Mass. Gen. Hosp., Boston, 1988. Author and co-author numerous articles to profl. jours. Active Assocs. Boston Classical Orch., 1992—. Mem. NAFE, Philippine Biochem. Soc., Philippine Environ. Mutagen Soc. Home: 5570 Ilaya St, Makati, Metro Manila 1200, Philippines Office: Tufts U Chemistry Dept 62 Talbot Ave Medford MA 02155

BLANCO, LUCIANO-NILO, physicist; b. Havana, Cuba, May 28, 1932; s. Luciano and Maria Teresa (Zayas) B.; m. Noemi de los A. Vitier, Dec. 16, 1956; 1 child, Marina Margarita. Student, U. Havana, 1949-54; fellow, Pa. State U., MIT, 1954-55, PhD in Physics, U Havana, Acad. Scis., 1962, 67. Inspector Chas. Martin Co. of Cuba, Havana, 1953-54; researcher Co. Rayonera Cubana, Matanzas, 1955-59; rsch. scientist Comision de Fomento Nacional, Havana, 1959-63; instr., physics prof. U. Havana, 1959-65; dir., rsch. physicist Acad. Scis., Havana, 1963-70; dir. phys. lab. and operation rsch. Avon, SA, Madrid, 1970-76; rsch scientist, mem. faculty physics U. Miami, Coral Gables, Fla., 1976—; prof. physics, dir. Inst. Theoretical Rsch., Coral Gables, 1990—; sci. advisor Internat. Yrs. of the Quiet Sun, Havana, 1964-65; cons. Clean Energy Rsch. Inst. U. Miami, Coral Gables, 1980—; cons. physicist, U.S.A., Spain, 1970—. Editor: Energias No-Convencionales, 1982; editor Boletin de Geofisica, 1965-67, EnergyNotes and EnergyLetters, 1990—; contbr. articles to profl. jours. Fellow Fgn. Ops. Adminstrn. and U.S. Weather Bur., Washington, 1954, Clean Energy Rsch. Inst., Coral Gables, 1982. Mem. AAAS, Am. Phys. Soc., Internat. Energy Soc. (pres. 1982—), N.Y. Acad. Scis., Sigma Xi. Achievements include research in solar-terrestrial relationships, neutrino physics and astrophysics, fundamental principles in energy, theoretical physics, biophysics; patent in field. Office: Inst Theoretical Rsch PO Box 248514 Miami FL 33124-8514

BLANDINO, RAMON ARTURO, psychologist, consultant, researcher; b. Santo Domingo, Dominican Republic, Oct. 22, 1956; came to U.S., 1990; s. Ramon Arturo Blandino and Yolanda Gomez; m. Esther Wong, 1980 (div. 1984); 1 child, Solange; m. Aida Ripley, Sept. 21, 1986; children: Ramon, Enrique, Aidat, George. BA in Clin. Psychology, U. Autonoma Santo Domingo, 1987, MA in Community Psychology, 1990. Lic. psychologist, Dominican Republic. Assoc. producer Radio HIN and TV (Rahintel), Dominican Republic, 1974-76; editor Pubs. Decada/Genesis, Dominican Republic, 1976-78; v.p. Promuca, Dominican Republic, 1978-85, Audiolab, Dominican Republic, 1983-88, Musicor/CBS, Dominican Republic, 1985-89; cons. Programa Control ETS and SIDA (Procets), Dominican Republic, 1988-90; vis. researcher, intern in psychology Bellevue Hosp., NYU, 1990—; cons. A.V. Blandino Ltd., Dominican Republic, 1989—; chief psychol. intern NYU-Robert Clemente Family Guidance Ctr.; vis. researcher N.Y.C. Health & Hosp. Corp./Inst. for Family & Community Care; program dir. Health Industry Resources Enterprises, Inc., 1992, dep. dir., 1993. Mem. task force on learning Assn. Scouts Dominicanos, Dominican Republic, 1983. Mem. APA, AAAS, N.Y. Acad. Sci., Internat. Coun. Psychologists, Assn. Dominicana Psicologos (Dominican Republic) (bd. dirs.), N.Y. State Psychol. Assn., Assn. of Hispanic Mental Health Profl. Avocations: reading, anthropology research.

BLANK, MERLE LEONARD, biochemist, researcher; b. Kanawha, Iowa, Nov. 8, 1933; s. Cecil L. and Marie L. (Krambeer) B.; m. Dorothy M. Siegle, Sept. 25, 1954; children: Nancy, Janice, David, Linda, James. BS, Bemidji State Univ., 1958. Asst. scientist Hormel Inst., Austin, Minn., 1958-65; sr. scientist Kraft Foods, Glenview, Ill., 1965-68, Oak Ridge (Tenn.) Assn. Univ., 1968—. Contbr. articles to profl. jours. With U.S. Army, 1954-56. Recipient Citation Classic award Inst. Sci. Info., 1980, Basic Rsch. award Nat. Glycerin Producer Assn., 1961. Mem. Am. Soc. Biochemistry Mol. Biology, Am. Legion. Achievements include discovery of one of the most active biological (natural) substances found to date Aklylacetylglycero Phosphocholine (1979). Co-holder of two U.S. patents on this and a related compound. Home: 118 Andover Cir Oak Ridge TN 37830 Office: ORAU/ORISE Medical Sci Div P O Box 117 Oak Ridge TN 37831-0117

BLANKENSHIP, ROBERT EUGENE, chemistry educator; b. Auburn, Nebr., Aug. 25, 1948; s. George Robert and Jane (Kehoe) Leech; m. Elizabeth Marie Dorland, June 26, 1971; children: Larissa Dorland, Samuel Robert. BS, Wesleyan U., Nebr., 1970; PhD, U. Calif., Berkeley, 1975. Postdoctoral fellow Lawrence Berkeley Lab., Berkeley, 1975-76, U. Washington, Seattle, 1976-79; asst. prof. Amherst (Mass.) Coll., 1979-85; assoc. prof. Ariz. State U., Tempe, 1985-88, prof., 1988—, dir. Ctr. Study of Early Events in Photosynthesis, 1988-91. Editor-in-chief Photosynthesis Rsch., 1988—; cons. editor Advances in Photosynthesis, 1991—; contbr. numerous articles to sci. jours. Recipient Alumni award Nebr. Wesleyan U., 1991, Disting. Rsch. award Ariz. State U., 1992. Mem. AAAS, Am. Chem. Soc., Union of Concerned Scientists. Democrat. Avocations: bicycling, hiking, camping, cooking, travel. Home: 1806 S Paseo Loma Cir Mesa AZ 85202-5528 Office: Ariz State U Dept Chemistry and Biochemistry Tempe AZ 85287-1604

BLANTON, JOHN ARTHUR, architect; b. Houston, Jan. 1, 1928; s. Arthur Alva and Caroline (Jeter) B.; m. Marietta Louise Newton, Apr. 10, 1953 (dec. 1976); children: Jill Blanton Lewis, Lynette Blanton Rowe, Elena Diane. BA, Rice U., 1948, BS in Architecture, 1949. With Richard J. Neutra, Los Angeles, 1950-64; pvt. practice architecture, Manhattan Beach, Calif., 1964—; lectr. UCLA Extension, 1967-76, 85, Harbor Coll., Los Angeles, 1970-72. Mem. Capital Improvements Com., Manhattan Beach, 1966, city commr. Bd. of Bldg. Code Appeals; mem. Bd. Zoning Adjustment Planning Commn., chmn., 1990. Served with Signal Corps, U.S. Army, 1951-53. Recipient Best House of Year award C. of C., 1969, 70, 71, 83, Preservation of Natural Site award, 1974, design award, 1975, 84. Mem. AIA (contbr. book revs. to jour. 1972-76, recipient Red Cedar Shingle/AIA nat. merit award 1979). Six bldgs. included in A Guide to the Architecture of Los Angeles and Southern California; works featured in L'architettura mag., 1988; design philosophy included in American Architects (Les Krantz), 1989. Office: 1456 12th St # 4 Manhattan Beach CA 90266-6113

BLANTON, ROGER EDMUND, mechanical engineer; b. Sherman, Tex., Apr. 20, 1955; s. Ray Edwin Blanton and Shirley Warlene (Ball) Gallion; m. Brenda Kay, Oct. 22, 1976; children: Ryan E., Rachel E., Rebekah E. BSME with honors, U. Tulsa, 1980, postgrad., 1981—. Registered profl. engr., Okla. Sr. project engr. John Zink Co., Tulsa, 1980-84; sr. applications engr., sr. oil console engr. Dresser-Rand, Broken Arrow, Okla., 1984-90; dir. incineration divsn. Radco, Tulsa, 1990—. Deacon 1st Bapt. Ch., Tulsa, 1989. Mem. ASME, NSPE, Tau Beta Pi. Republican. Baptist. Achievements include development of solutions to unique hazardous waste incineration problems; development and design of specialty burners and incinerators; testing of destruction efficiency for hazardous waste incineration. Home: 7186 S 75th E Ave Tulsa OK 74133

BLANTON, ROY EDGAR, sanitation agency executive; b. El Centro, Calif., Apr. 30, 1933. BSBA, La Verne U., 1975. Field supr. water and sewerage treatment City of Thousand Oaks, Calif., 1966-72; gen. mgr. Goleta (Calif.) Sanitary Dist., 1972-80, Selma-Kingsburg-Fowler (Calif.) County Sanitation Dist., 1980—; mem. operator tng. and cert. com. State Water Resources Control Bd., Sacramento, Calif., 1979-80, 89-92. Sgt. U.S. Army, 1952-55; staff sgt. USAF, 1957-63. Mem. Am. Water Works Assn., Calif. Water Pollution Control Assn., Rotary (pres. Kingsburg club 1984-85), Elks. Republican. Roman Catholic. Office: Selma-Kingsburg-Fowler County Sanitation Dist PO Box 95 Kingsburg CA 93631

BLASCHKE, LAWRENCE RAYMOND, utility company professional; b. Elgin, Ill., Feb. 24, 1950; s. Raymond Otto and Margaret Irma (Palm) B.; m. Diane Charlotte Hartwell, Apr. 12, 1974 (dec. 1986); children: Matthew Robert, Bryan Raymond; m. Karen Juliann Larson, Feb. 14, 1987. AS, William Rainey Harper Coll., 1973; student, Valparaiso U., 1974—. Jr. engr., then assoc. engr. No. Ind. Pub. Svc. Co., Hobart, 1974-79, dist. engr., 1979-84, project engr., 1984-87; project engr. No. Ind. Pub. Svc. Co., Gary, 1987-92; engr. level III Merrillville, Ind., 1992—. Chmn. bd. social ministry Immanuel Luth. Ch., Valparaiso, sec. bd. evangelism; cubmaster Valparaiso area Boy Scouts Am., 1984-88, merit badge counselor, com. mem. and asst. scoutmaster Troop 963, 1992—; supervisory com. No. Ind. Fed. Credit Union, chmn., 1987-88, 90-91, 91—; treas. Montessori Sch. Porter County, 1981-83; sec., treas. Quality Devel., Inc., 1990—; mem. Nat. Arbor Day Found. Recipient Edward A. Filene award Ind. Credit Union League and Credit Union Nat. Assn., Inc., 1991; listed as owner, info., plans, interiors and landscaping of geodesic dome home complex in directory Dome Mag. Mem. IEEE, Instrument Soc. Am., Nat. Parks and Conservation Assn., Alpha Phi Omega. Republican. Avocations: table tennis, stereo audio equipment and recording, carpentry, electronics. Home: 396 W Southfield Ln Valparaiso IN 46383-9633 Office: No Ind Pub Svc Co 801 E 86th Ave Merrillville IN 46410

BLASKÓ, ANDREI, chemist, researcher; b. Baia Mare, Maramures, Romania, Aug. 4, 1952; s. Stefan and Elizabeth (Mezinger) B.; m. Magdalena Kardos, Aug. 30, 1986. PhD, Poly. Inst. Iasi, Romania, 1987. Registered profl. engr., Calif. Chem. engr. Petrochem. Works Solventul, Timisoara, Romania, 1976-81; rsch. scientist Inst. Chem. Biochem. Energetics, Timisoara, 1981-88; rsch. assoc. U. Calif., Santa Barbara, 1988—; cosn. Norsc Assocs., Newbury Park, Calif., 1990-91. Contbr. articles to Jour. Phys. Organic Chemistry, Jour. Organic Chemisty, Jour. Am. Chem. Soc., Jour. Phys. Chem., Langmuir, Polyhedron, Carbohydrate Rsch., Inorganic Chemistry, Bioorganic Chemistry, Tetrahedron, Procs. Nat. Acad. Sci. USA. Mem. Am. Chem. Soc., Royal Soc. Chem. (U.K). Greek Orthodox. Office: U Calif Chemistry Dept Santa Barbara CA 93106

BLASS, GERHARD ALOIS, physics educator; b. Chemnitz, Germany, Mar. 12, 1916; came to U.S., 1949, naturalized, 1955; s. Gustav Alois and Anna (Mehnert) B.; m. Barbara Siegert, July 16, 1945; children—Andrew, Marcus, Evamaria, Annamaria, Peter. Abitur, Oberrealschule Chemnitz, 1935; Dr. rer. nat., Universität Leipzig, 1943. Asst. Institut für Theoretische Physik, Leipzig, 1939-43; research cons. Siemens & Halske, Berlin, 1943-46; dozent math. and physics Oberrealschule, Nuremberg, 1946-47, Ohm Polytechnikum, Nuremberg, 1947-49; prof. physics Coll. St. Thomas, St. Paul, 1949-51; prof. physics U. Detroit, 1951-81, chmn. dept., 1962-71; guest prof. U. Baroda, India, spring, 1967. Author: Theoretical Physics, 1962, "Weil Hierseinvielist" Poems in German, 1987. Fellow AAAS; mem. Soc. Asian and Comparative Philosophy, Esperanto League N.Am., Sigma Pi Sigma. Roman Catholic. Home: 4441 Stewart Rd Metamora MI 48455-9777

BLASS, JOHN PAUL, medical educator, physician; b. Vienna, Austria, Feb. 21, 1937; s. Gustaf and Jolan (Wirth) B.; m. Birgit Annelise Knudsen, Dec. 20, 1960; children—Charles, Lisa. A.B. summa cum laude, Harvard U., 1958; Ph.D., U. London, 1960; M.D., Columbia U., 1965. Postdoctoral fellow Am. Cancer Soc., Columbia U., 1962-63; intern Mass. Gen. Hosp., Boston, 1965-66; resident in medicine Mass. Gen. Hosp., 1966-67; research assoc. Nat. Heart and Lung Inst., Bethesda, Md., 1967-70; asst. prof. psychiatry and biol. chemistry UCLA, 1970-76, assoc. prof., 1976-78; mem. staff UCLA Hosps. Clinics, 1970-78; Winifred Masterson Burke prof. neurology, prof. medicine Cornell U. Med. Center, 1978—; attending neurologist N.Y. Hosp.; mem. NBS-1 rev. com. NIH, 1981-84; councilor Nat. Inst. Aging, 1986-89; chmn. Nat. Adv. Panel on Alzheimers's Disease U.S. Congress, 1987-91, mem., 1993—. Mem. editorial bd. Jour. Neurochemistry, 1981-86, Neurochem. Rsch., Neurochem. Pathology, Neurobiol. Aging, Jour. Neurol. Scis.; assoc. editor Jour. Am. Geriatric Soc.,

1982-87; co-editor: Principles of Geriatric Medicine and Gerontology, Caring for Alzheimer's Patients, Familial Alzheimer's Disease, Treatment of Alzheimer's Disease; contbr. articles to profl. jours. Mem. sci. adv. bd. Will Rogers Inst., Allied Signal Aging Award Com. Served as asst. surgeon USPHS, 1967-70. Marshall scholar, 1958-60. Mem. Soc. Neurosci. (chmn. social issues com.), Biochem. Soc., Am. Soc. Biol. Chemists, Am. Soc. Neurochemistry (council, chmn. public policy com.), Internat. Soc. Neurochemistry (council, chmn. clin. com.), Am. Soc. Clin. Investigation, Am. Geriatrics Soc., Am. Fedn. Aging Research (v.p., chmn. research com. 1982-87), Assn. Alzheimers and Related Disease (sci. adv. bd. 1982-86), Am. Chem. Soc., Phi Beta Kappa, Sigma Xi, Alpha Omega Alpha. Jewish. Home: 1 Orchard Pl Bronxville NY 10708-2509 Office: Cornell U 785 Mamaroneck Ave White Plains NY 10605-2523

BLATCHLEY, BRETT LANCE, computer engineer; b. Syracuse, N.Y., Oct. 31, 1961; s. Barry Rene and Olive Janet (Nelson) B.; m. Judith Ann Wall, Nov. 11, 1989; 1 child, Joshua. Student, Sandhill Coll., 1979-82. Computer operator, programer Moore Meml. Hosp., Pinehurst, N.C., 1979-83; svc. engr. Carolina Computer Ctr., Columbia, S.C., 1986-87; lead computer engr. Quick Assocs., Columbia, 1988-91, MDO/Soft Talk, Columbia, 1983-93, CMS, Inc., Winston-Salem, N.C., 1993—; co-founder, engr. The Telesis Group, Columbia, 1988-89, The HOPE Lab., Columbia, 1992—. Author, publisher Life Learner's Jour., 1992; contbr. articles to profl. jours. Bd. mem. Servants' Missionary Svc., Columbia, 1991—; computer coord., disaster svc. vol. ARC, Columbia, 1992—. Recipient Eagle Scout with 3 Palms, Boy Scouts Am., Aberdeen, N.C., 1977. Mem. IEEE, Assn. Computing Machinery. Republican. Achievements include creation of "Perfect View" a nationally marketed business management system, "SoftTools" a series of Unix and filePro tools which are nationally marketed, "lower-CASE" tool called "FPLine" which is nationally marketed. Home: 4755 Country Club Rd #117-F Winston-Salem NC 27104 Office: CMS Inc 9400 B Two Notch Rd 150 Stratford Rd Ste 500 Winston Salem NC 27104

BLATT, STEPHEN ROBERT, systems engineer; b. N.Y.C., Apr. 13, 1956; m. Faith E. Minard, June 29, 1986. ScB, MIT, 1977; PhD, Yale U., 1981. Sr. engr. SSE, Sanders Assocs., Nashua, N.H., 1983-85; prin. engr. SSE and ASWD, Lockheed Sanders, Nashua, N.H., 1985-91; sr. prin. engr. ASWD, Lockheed Sanders, Nashua, N.H., 1992—. Mem. AAAS, Am. Phys. Soc. Office: MAN 6-2100 Lockheed Sanders PO Box 868 Nashua NH 03061

BLATTNER, MEERA MCCUAIG, computer science educator; b. Chgo., Aug. 14, 1930; d. William D. McCuaig and Nina (Spertus) Klevs; m. Minao Kamegai, June 22 1985; children: Douglas, Robert, William. B.A., U. Chgo., 1952; M.S., U. So. Calif., 1966; Ph.D., UCLA, 1973 . Research fellow in computer sci. Harvard U., 1973-74; asst. prof. Rice U., 1974-80; assoc. prof. applied sci. U. Calif. at Davis, Livermore, 1980-91, prof. applied sci., 1991—; adj. prof. U. Tex., Houston, 1977—; vis. prof. U. Paris, 1980; program dir. theoretical computer sci. NSF, Washington, 1979-80. Co-editor: (with R. Dannenberg) Multimedia Interface Design, 1992. NSF grantee, 1977-81, 93—. Mem. Soc. Women Engrs., Assn. Computing Machinery, IEEE Computer Soc. Contbr. articles to profl. jours. Office: U Calif Davis/Livermore Dept Applied Sci Livermore CA 94550

BLATTNER, WOLFRAM GEORG MICHAEL, meteorologist; b. Nuremberg, Germany, Sept. 28, 1940; came to U.S., 1969; s. Richard and Margarete (Zirngibl) Blattner; m. Brunhilde Klara Wey, Oct. 31, 1969; children: Michelle (dec.), Paul. Pre-diploma Meteorology, U. Mainz, Germany, 1964, diploma in meteorology, 1968. Rschr. NATO, Hanscom Field, Mass., 1968-69; project scientist Radiation Rsch. Assocs., Ft. Worth, 1969-83, dir. atmospheric optics, 1983-85; engring. specialist, Electro-Optics LTV, Dallas, 1985-90; sr. analyst Mission Analysis Loral Vought Systems, Dallas, 1990—; cons. SciTec, Inc., Princeton, N.J., 1983-91. Contbg. author: Radiation in the Atmosphere, 1977; contbr. articles to profl. jours. Mem. referee com. U.S. Soccer Fedn., Chgo., 1990—. Mem. Am. Meteorol. Soc. Roman Catholic. Achievements include research in twilight radiation and radiative transfer. Home: 611 Joyce Weatherford TX 76086 Office: Loral Vought Systems WT-52 PO Box 650003 Dallas TX 75265-0003

BLAUERT, JENS PETER, acoustician, educator; b. Hamburg, Germany, June 20, 1938; s. Werner and Hedwig (Mueller) B.; m. Brigitte Raape, 1965; children: Heike, Michael. Diploma engring., Tech. U., Aachen, Germany, 1964, DEng, 1969; habilitation, Inst. Tech., Berlin, 1973. Sci. asst.Comm. Engring. Inst. Tech., Aachen, Germany, 1964-74; dean Ruhr U., Bochum, 1976-77, senator 1985-87; prof. gen. elec. engring. and acoustics 1974—; cons. pvt. practice Aachen, Bochum, 1968—, Bell Labs., Murray Hill, N.J., 1976; Guest scientist, cons. U. Calif., Berkeley, 1983, CNRS, Marseille, France, 1989-90, NTT Rsch. Labs., Tokyo, 1990; mem. environ. protection coun. State of North Rhine Westphalia, 1985—; chmn. com. German Standard Assn., 1986—. Author: Spatial Hearing, 1983; editorial bd. Acta Acustica, Presence, Applied Acoustics; advisory bd. to editors Acustica; contbr. articles to profl. jours. Grantee German Rsch. Found., 1975-92, Min. Sci. and Rsch. North-Rhine Westphalia, 1976-89; fellow NATO-Scientific Commn., 1984; recipient Silver Needle Verein Deutscher Elektrotechniker, 1986, Silver medal Soc. Francaise d'Acoustique, 1990. Fellow Acoustical Soc. Am.; mem. IEEE (sr.), German Acoustical Soc. (bd. dirs.), European Speech Communication Soc. (bd. dirs.), Info. Tech. Gesellschaft, Deutsches Inst. Normung (adv. bd., com. chmn.), Audio-Enging. Soc., Rotary. Achievements include 5 patents; research in auditory virtual environment and telepresence, binaural technology, models of binaural hearing, architectural acoustics, noise and product-sound engineering, speech technology. Office: Ruhr U, D-44780 Bochum Germany

BLAYDES, JAMES ELLIOTT, ophthalmologist; b. Bluefield, W.Va., Feb. 26, 1927; s. James Elliott Sr. and Mabel Lucetta (Hill) B.; children: James Elliott IV, William Mitchell, Stephen Hill, Elizabeth Boyd Blaydes Lewis; m. Anita G. Shrader, Sept. 25, 1976; 1 child, Jaime Brittany. AB, Princeton U., 1950; MD, U. Pa., 1954; postgrad., NYU, 1956. Intern Pa. Hosp., Phila., 1954-55; resident N.Y. Eye & Ear Infimary, N.Y.C., 1955-58, chief resident, 1957-58; asst. Blaydes Clinic, Bluefield, W.Va., 1950-58, ophthalmologist, 1958—, dir., owner, 1972—; bd. dirs. Blaydes Found., Bluefield, Humana Ophthalmology Ctr. of Excellence, Bluefield; clin. prof. ophthalmology W.Va. U., Morgantown, 1987—; assoc. clin. prof. Marshall U., Huntington, W.Va., 1977—; cons. Norfolk So. Rlwy. Cons. editor Ophthalmology Mgmt., 1983—; mem. editorial adv. bd. Ocular Surgery News-Internat., 1982—, Phaco & Foldables; author: (with others) The Second Report on Cataract Surgery, 1971, Current Concepts in Cataract Surgery, 1974, rev. edits., 1976, 80, 82; contbr. articles to profl. jours. Named N.Y. Eye and Ear Infirmary Alumnus of Yr., 1990. Fellow Am. Coll. Eye Surgeons, Am. Acad. Ophthalmology (Hon. award 1979, Sr. Hon. award 1992); mem. Mercer County Med. Soc. (pres. 1967), W.Va. State Med. Assn., AMA, W.Va. Acad. Ophthalmology and Otolaryngology, Outpatient Ophthalmic Surg. Soc., Contact Lens Assn., Ophthalmologists, Am. Soc. Cataract and Refractive Surgery, China Vision Project Med. Adv. Bd., So. Med. Assn., PanAm. Med. Assn., Am. Assn. Ophthalmology, Am. Assn. Physicians and Surgeons, Soc. Eye Surgeons, N.Y. Eye and Ear Imfirmary Alumni Assn. Home: 908 Edgewood Rd Bluefield WV 24701-4209 Office: The Blaydes Clinic PO Box 1380 Bluefield WV 24701-1380

BLAYLOCK, NEIL WINGFIELD, JR., applied statistician, educator; b. Ft. Smith, Ark., Aug. 18, 1946; s. Neil Wingfield Sr. and Phyllis Catherine (Brown) B.; m. Naomi Josephine Smith, Aug. 25, 1968; children: Neil Wingfield III, Scott Allen, Adrian Philip, Paul Alexander. BA in Math., St. Mary's U., 1968; MS in Stats., U. Tex., San Antonio, 1987. Cert. tchr., Tex. High sch. math. tchr. San Antonio Ind. Sch. Dist., 1968-70; sr. engr. Martin Marietta Aerospace Corp., Orlando, Fla., 1972-79; prin. analyst S.W. Rsch. Inst., San Antonio, 1979—; evening faculty mem. U. Tex., San Antonio, 1987—. Contbr. articles to profl. jours. Com. chmn. Boy Scouts Am., San Antonio, 1982—. With U.S. Army, 1970-72. Fellow AIAA (assoc., chmn. S.W. Tex. sect. 1983, dep. dir. region IV 1991—, Spl. Svc. award 1984, Disting. Achievement award 1989), Hypervelocity Impact Soc. (co-founder, nat. membership chmn. 1988-92), Am. Statis. Assn. Roman Catholic. Achievements include development of specialty military devices, quantitative risk mgmt. techniques, regression techniques for nondimensional scale modeling experiments, advanced devel. of nonnuclear Pershing missile. Home: 7111 Moss Creek Dr San Antonio TX 78238 Office: SW Rsch Inst 6220 Culebra Rd San Antonio TX 78238

BLAZEK, JIRI, aerospace engineer, researcher; b. Prague, Czechoslovakia, July 31, 1961; arrived in Germany, 1979; s. Vladislav and Marie (Jenacek) B.; m. Aug. 13, 1960. MSc, Inst. Tech., Aachen, Germany, 1989; postgrad., U. Braunschweig, Germany, 1989—. Rschr. German Aerospace Rsch. Establishment, Braunschweig, 1990—; participant profl. confs. Mem. AIAA. Achievements include development of upwind implicit residual smoothing method, simplified upwind prolongation; research in computational fluid dynamics, particularly in hypersonic flows. Office: German Aerospace Rsch Est, Flughafen, D-38108 Braunschweig Germany

BLAZIC, MARTIN LOUIS, manufacturing engineer; b. Euclid, Ohio, Aug. 19, 1959; s. Louis Bernard and Isabella (Szabo) B.; m. Anne Louise Ward, Apr. 13, 1985; children: David Ward, Andrew Martin. B Indsl. Engring., Cleve. State U., 1982; Assoc. Elec. Engring. Tech., Franklin U., 1988. Engr.-in-tng., Ohio. Engring. technician lighting bus. group GE, Twinsburg, Ohio, 1980-82; process engr. appliance control dept. GE, Morrison, Ill., 1982-83, materials specialist appliance control dept., 1983; mfg. engr. med. systems group GE, Milw., 1983, prodn. foreman med. systems group, 1983-84; quality engr. lighting bus. group GE, Circleville, Ohio, 1984-87, mfg. engr. lighting bus. group, 1987-89, advanced process engr. lighting bus. group, 1989; sr. mgr. engr. Tosoh SMD, Inc., Grove City, Ohio, 1989-91, mfg. engring. mgr., 1991—. Mem. Soc. Mfg. Engrs. Achievements include patents for sputtering target wrench and sputtering target design, automatic brushplating machine, sputtering target mounting design. Home: 13844 Stonehenge Cir Pickerington OH 43147 Office: Tosoh SMD Inc 3600 Gantz Rd Grove City OH 43123

BLECK, PHYLLIS CLAIRE, surgeon, musician; b. Oak Park, Ill., Mar. 10, 1936; d. William Fred and Mildred A. (Jones) B. BS, U. Ill., 1958; MM, Northwestern U., 1968; DMA, U. So. Calif., 1970; postgrad., Autonoma U., Guadalajara, Mex., 1973-76; MD, Rush Med. Coll., 1979; MS in Surgery, U. Ill., 1983. Diplomate Am. Bd. Surgery, Am. Bd. Thoracic Surgery. Prin. trumpet Fla. Symphony Orch., 1960-66, Orch. Sinfonica Nat. de Peru, 1965; instr. Thornton Jr. Coll., 1966-68; lectr. U. So. Calif., 1969-73; asst. prof. Whittier Coll., 1973; intern Rush Presbyn. St. Luke's Med. Ctr., Chgo., 1979-80, resident, asst. in gen. surgery, 1980-82, instr. gen. surgery, 1982-84; resident in cardiothoracic surgery U. Medicine and Dentistry N.J., 1984-87; pvt. practice medicine specializing in cardiothoracic surgery, Aurora, Ill., 1987—. Editor: Mozart Divertimento for Winds; research on vascular ischemia. Fellow ACS, Am. Coll. Chest Physicians, Ill. Thoracic Surg. Soc., Ill. Surg. Soc.; mem. AAAS, Soc. Thoracic Surgeons, Kappa Delta Pi, Pi Kappa Lambda, Sigma Alpha Iota. Office: 1315 N Highland Ave Aurora IL 60506-1400

BLECK, THOMAS PRITCHETT, neurologist, neuroscientist, educator; b. Michigan City, Ind., Dec. 18, 1951; s. Donald Charles Bleck and Ruth Elizabeth (Pritchett) Trudeau; m. Jane Washburn, Aug. 3, 1985; children: W. Jordan, Holly Jane. BA, Northwestern U., Evanston, Ill., 1972; MD, Rush Med. Coll., Chgo., 1977. Diplomate Am. Bd. Internal Medicine, Am. Bd. Psychiatry and Neurology, Am. Bd. Clin. Neurophysiology. Intern then resident in internal medicine Rush-Presbyn.-St. Luke';s Med. Ctr., Chgo., 1980-83; asst. prof., then assoc. prof. Rush Med. Coll., Chgo., 1983-90, asst. dean, 1989-90; assoc. prof. neurology and neurol. surgery U. Va., Charlottesville, 1990—, Nerancy prof. neurology, 1993—. Assoc. editor: Critical Care Medicine jour., 1992—, Clin. Neuropharmacology, 1992—. Fellowship Royal Soc. Medicine, London, 1991. Fellow ACP, Am. Acad. Neurology (sci. issues com. 1989—), Am. Coll. Chest Physicians, Am. Coll. Critical Care Medicine, Am. Electroencephalographic Soc.; mem. Infectious Diseases Soc. Am. Office: U Va Dept Neurology PO Box 394 Charlottesville VA 22908

BLEDSOE, WOODROW WILSON, mathematics and computer sciences educator; b. Maysville, Okla., Nov. 12, 1921; s. Thomas Franklin and Eva (Matthews) B.; m. Virginia Norgaard, Jan. 29, 1944; children: Gregory Kent, Pamela Nelson, Lance Woodrow. B.S. in Math., U. Utah, 1948; Ph.D. in Math., U. Calif., 1953. Lectr. in math. U. Calif.-Berkeley, 1951-53; mathematician, staff mem. Sandia Corp., Albuquerque, 1953-60, head math. dept., 1957-60; mathematician, researcher Panoramic Research Inc., Palo Alto, Calif., 1960-65, pres., 1963-65; prof. math. computer sci. U. Tex., Austin, 1966—, acting chmn. dept. math., 1967-69, chmn. dept. math., 1973-75, Ashbel Smith prof. math. and computer sci., 1981-84; on leave U. Tex., 1984-87; Peter O'Donnell Jr. Centennial chair in computing systems U. Tex., Austin, 1987—; v.p., dir. artificial intelligence Microelectronics and Computer Tech. Corp., Austin, 1984-87; gen. chmn. Internat. Joint Conf. Artificial Intelligence, MIT, Cambridge, Mass., 1975-77; trustee Internat. Joint Conf. on Artificial Intelligence, 1978-83; mem. subcom. for computer sci. Adv. Com. Math. and Computer Sci., NSF, 1979-82, 88-90; vis. prof. MIT, 1970-71, Carnegie-Mellon U., Pitts., 1978. Editor: (with Donald Loveland) Automated Theorem Proving, 1984; bd. editors Internat. Jour. of Artificial Intelligence, 1972—, also rev. editor, 1973-77; author numerous tech. papers in refereed jours., confs. Vice-pres. Capitol Area council Boy Scouts Am., Austin, 1979-83. Served to capt. U.S. Army, 1940-45, ETO. NSF research grantee, 1972—; NIH research grantee, 1967-72. Mem. Am. Math. Soc., Assn. Computing Machinery, Am. Assn. Artificial Intelligence (pres. 1984-85). Mem. LDS Ch. Office: U Tex Dept Computer Sci Taylor Hall 4.140B Austin TX 78712

BLEIBERG, LEON WILLIAM, surgical podiatrist; b. Bklyn., June 9, 1932; s. Paul Pincus and Helen (Epstein) B.; m. Beth Daigle, June 7, 1970; children: Kristina Noel, Kelley Lynn, Kimberly Ann, Paul Joseph. Student, L.A. City Coll., 1950-51, U. So. Calif., 1951, Case Western Res. U., 1951-53; DSc with honors, Temple U., 1955; PhD, U. Beverly Hills, 1970. Served rotating internship various hosps., Phila., 1954-55; resident various hosps., Montebello, L.A., 1956-58; surg. podiatrist So. Calif. Podiatry Group, Westchester (Calif.), L.A., 1956-75; health care economist, researcher Drs. Home Health Care Svcs., 1976—; podiatric cons. U. So. Calif. Athletic Dept., Morningside and Inglewood (Calif.) High Schs., Internet Corp., Royal Naval Assn., Long Beach, Calif. Naval Sta.; lectr. in field; healthcare affiliate Internat. div. CARE/ASIA, 1987; pres. Medica, Totalcare, Cine-Medics Corp., and World-Wide Health Care Svcs.; exec. dir. Internat. Health Trust, developer Health Banking Program; adminstr. Orthotic Concepts, 1993. Producer (films) The Gun Hawk, 1963, Terrified, Day of the Nightmare; contbr. articles to profl. jours. Hon. Sheriff Westchester 1962-64; commd. mem. Rep. Senatorial Inner Circle, 1984-86; co-chmn. health reform com. United We Stand Am., Thousand Oaks, Calif.; mem. exec. coun. State of Calif., United We Stand Am.; active 1st Security and Safety, Westlake Village, Calif., 1993—; lt. comdr. med. svcs. corps Brit.-Am. Sea Cadet Corps, 1984—; track coach Westlake High Sch., Westlake Village. With USN, 1955-56. Recipient Medal of Merit, U.S. Presdl. Task Force. Mem. Philippine Hosp. Assn. (Cert. of Appreciation 1964, trophy for Outstanding Svc. 1979), Calif. Podiatry Assn. (hon.), Am. Podiatric Med. Assn. (hon.), Acad. TV Arts and Scis., Royal Soc. Health (Eng.), Western Foot Surgery Assn., Am. Coll. Foot Surgeons, Am. Coll. Podiatric Sports Medicine, Internat. Coll. Preventive Medicine, Hollywood Comedy Club, Sts. and Sinners Club, Westchester C. of C., Hals Und Beinbruch Ski Club, Beach Cities Ski Club, Orange County Stamp Club, Las Virgenes Track Club, Masons, Shriners. Home: 30556 Agoura Rd # J-16 Agoura Hills CA 91301

BLEICHER, SHELDON JOSEPH, endocrinologist, medical educator; b. N.Y.C., Apr. 9, 1931; s. Max and Fannie (Klieger) B.; m. Diane D. Cole, Aug. 1990; children from previous marriages: Erick Max, Phillip Thaddeus Samuel, Deborah Ann Cole, Sandra Lynn Cole, Jodie Lisa Cole. A.B., NYU, 1951; M.S., Western Ill. U., 1952; M.D., SUNY Downstate Med. Center, Bklyn., 1956. Intern L.I. Jewish Hosp. Ctr., New Hyde Park, N.Y., 1956-57; resident Boston City Hosp., 1959-60; research fellow in medicine Harvard-Thorndike Meml. Lab., Boston, 1960-63; chief metabolic research unit Jewish Hosp. Med. Center, Bklyn., 1963-67, chief div. endocrinology and metabolism, 1967-77; pvt. practice specializing in endocrinology and diabetes Woodbury, N.Y., 1990—; prof. medicine SUNY Downstate Med. Center, 1975—; chmn. dept. internal medicine Bklyn.-Cumberland Med. Center, 1978-83, Bklyn.-Caledonian Med. Ctr., 1983-90; cons. IAEA, Vienna, 1966—; mem. attending staff Syosset Community Hosp., Cen. Gen. Hosp. Mem. editorial bd. Diabetes in News, Practical Diabetes; contbr. articles to profl. jours. Vice pres. Locust Valley Central Sch. Bd., 1981-82, pres., 1982-85. Served to capt. M.C., USNR, 1957-92, ret. NIH fellow, 1960-63; NIH research career devel. award, 1970-75; recipient Torch of Liberty award Anti-Defamation League of B'nai Brith, 1982 . Fellow ACP; mem. AMA, AAAS, Am. Soc. Internal Medicine, Am. Diabetes Assn. (bd. dirs. 1979-85, achievement award 1986, 90), N.Y. Diabetes Assn. (bd. dirs. 1965—, pres. 1976-78), L.I. Diabetes Assn. (pres. 1978-81), N.Y. State Soc. Internal Medicine (state bd. dirs., treas. Bklyn. chpt., chmn. continuing edn. com.), Bklyn. Soc. Internal Medicine (treas. 1983-85, sec. 1985-87, pres. 1987-89), Endocrine Soc., Am. Assn. Clin. Endocrinologists, Internat. Diabetes Fedn., European Assn. for Study of Diabetes, Juvenile Diabetes Found. Internat., Harvey Soc., Sagamore Yacht Club (L.I., fleet surgeon 1983-86). Jewish. Office: 165 Froehlich Farm Blvd Woodbury NY 11797

BLELLOCH, PAUL ANDREW, mechanical engineer; b. Milan, Italy, Dec. 9, 1959; came to U.S., 1970; s. Robert Andrew and Paola (Panizza) B.; m. Mia Caprice Guess, July 20, 1985; children: Amy Nicole, Christopher Paul. BS, MIT, 1981; MSc, UCLA, 1984, PhD, 1986. Lead engr. advanced dynamics SDRC Engring. Svcs., San Diego, 1986—. Contbr. articles to profl. jours. Mem. AIAA. Home: 4350 Rous St San Diego CA 92122 Office: SDRC Engring Svcs 11995 El Camino Real # 200 San Diego CA 92130

BLESSING, EDWARD WARFIELD, petroleum company executive; b. Glenridge, N.J., Oct. 6, 1936; s. Jess Edward Himes and Laura Louise (Warfield) Blessing; adopted s. Donald L. Blessing; m. Cynthia Harris, July 1, 1961 (div. 1969); m. Jeanne Kyle, Jan. 19, 1970 (div. 1980); 1 child, Megan Louise; m. Debra Jean Wayne, July 12, 1986. BA, San Diego State U., 1960; MBA, Harvard U., 1965. Rep. Shearson, Hammill & Co, La Jolla, Calif., 1961-63; cons. McKinsey & Co. Inc., San Francisco and L.A., 1965-68; assoc. mng. dir. Canadawide Investments, Vancouver, B.C. and Calgary, Alta., 1968-69; misc. investor energy and fin. related activities, 1969-75; mng. ptnr. Dexer Assocs., L.A. and Sharjah, United Arab Emirates, 1975-78; exec. v.p. Okla. Oil & Gas Co., Oklahoma City, 1978-80; pres. Blessing Petroleum Co., Blessing Oil Co., Oklahoma City, 1980-87; dir., pres., chief exec. officer Strategic Petroleum, Inc., Dallas, 1987-89; pres. Blessing Corp., Dallas, 1989—; Vis. instr., adj. prof. U. Okla. Grad. Sch. Bus. Adminstrn., Oklahoma City, 1983-84. Res. dep. sheriff Okla. County Sheriff's Dept., Oklahoma City, 1986-87, Dallas County Sheriff's Dept., 1987-89; mem. Mayor's Adv. Com. on Crime, Dallas, 1988-91; Rep. candidate Calif. 79th Assembly Dist., 1960; hon. dir., chmn. bd. dirs. Calif. Pediatric Ctr., L.A., 1973—; mem. energy subcom. Okla. Dept. Commerce, Oklahoma City, 1987; mem. planning com. Okla. Gov.'s 1987 Energy Conf., Oklahoma City. With USMC, 1960-61. Mem. Am. Assn. Petroleum Landmen, Ind. Petroleum Assn. Am. (exec. com., fin. com., v.p., dir., Roustabout, mem. econs. policy com., crude oil policy com., econ. task force 1980—), Okla. Ind. Petroleum Assn., Tex. Ind. Producers and Royalty Owners Assn., Dallas C. of C. (energy subcom. 1987-90), Harvard Bus. Sch. Alumni Assn. (sponsor), Oklahoma City C. of C. (chmn. energy coun. 1982-87), Tex. Mid-Continent Oil and Gas Assn., North Tex. Oil and Gas Assn., Hard Hatters, Dallas Wildcat Com., Dallas Petroleum Club (bd. dirs.). Episcopalian. Home: 6582 Briarmeade Dr Dallas TX 75240-7912 Office: Blessing Oil Co 1600 Three Lincoln Ctr 5430 LBJ Fwy Dallas TX 75240-2601

BLEWETT, JOHN PAUL, physicist; b. Toronto, Ont., Can., Apr. 12, 1910; s. George John and Clara Marcia (Woodsworth) B.; m. Hildred Hunt, 1936 (div. 1966); m. Joan N. Warnow, 1983. B.A., U. Toronto, 1932, M.A., 1933; Ph.D., Princeton U., 1936. Mem. staff research lab Gen. Electric Co., Schenectady, 1937-46; with Brookhaven Nat. Lab., Upton, N.Y., 1947-78; cons. on nuclear physics and elec. engring. Author: Particle Accelerators. Contbr. articles to profl. jours. Fellow Am. Phys. Soc. (R.R. Wilson award 1993), IEEE, N.Y. Acad. Scis., AAAS. Home: 310 W 106th St New York NY 10025-3429

BLITZER, ANDREW, otolaryngologist, educator; b. Pitts., Apr. 25, 1946; s. Martin Hollander and Lyrene Iris (Lave) B.; m. Patricia Volk, Dec. 21, 1969; children: Peter Morgen, Polly Volk. BA, Adelphi U., 1967; DDS, Columbia U., 1970; MD, Mt. Sinai Sch. Medicine, 1973. Diplomate Am. Bd. Otolaryngology. Resident in gen. surgery Beth Israel Hosp., N.Y.C., 1973-74; resident in otolaryngology Mt. Sinai Hosp., N.Y.C., 1974-77; asst. prof. otolaryngology Coll. Phys. & Surg., Columbia U., N.Y.C., 1977-82, assoc. prof. otolaryngology and oral surgery, 1982-84, prof. clin. otolaryngology and oral surgery, 1984—, prof. clin. otolaryngology in neurology, 1993—, vice chmn. dept. otolaryngology, 1983—, acting chmn., dir., 1991—; dir. div. head and neck surgery Columbia-Presbyn. Med. Ctr., N.Y.C., 1980—, acting chmn. dept. otolaryngology, dir. otolaryngology svc., Columbia-Presbyn. Med. Ctr., 1991—, dir. residency edn., 1978—; lectr. dept. otolaryngology Mt. Sinai Sch. Medicine, N.Y.C., 1977—. Co-Author several books; assoc. editor: Oncology Times, Controversies in Otolaryngology; mem. editorial rev. bd. The Laryngoscope, Otolaryngology-Head and Neck Surgery, Jour. Otolaryngology; contbr. chpts. to books, articles to profl. jours. Recipient award for excellence Am. Assn. Orthodontists, 1970, Tchr.-Investigator award Nat. Inst. Neurol. Communicative Disorders and Strokes, 1978-83, Maxwell Abramson Meml. award Excellence in Resident Teaching, 1993. Fellow ACS, N.Y. Acad. Medicine, Am. Soc. Head and Neck Surgery, Am. Acad. Facial Plastic and Reconstructive Surgery, Am. Laryngol. Assn., Am. Laryngol., Rhinol. and Otol. Soc., Am. Acad. Otolaryngology-Head and Neck Surgery (honor award), Am. Broncho-esophagological Assn. Home: 1136 5th Ave New York NY 10128-0122 Office: Columbia U Coll Physicians & Surgeons Dept Otolaryngology 630 W 168th St New York NY 10032-3702

BLIX, HANS MARTIN, international atomic energy official; b. Uppsala, Sweden, June 28, 1928; s. Gunnar and Hertha (Wiberg) B.; m. Eva Kettis, Mar. 17, 1962; children—Marten, Goran. LL.B., U. Uppsala, 1951; Ph.D., Cambridge U., 1959; LL.D., Stockholm U., 1960. Assoc. prof. U. Stockholm, 1960; legal adviser Ministry Fgn. Affairs, Stockholm, 1963-76, under sec. of state in charge of internat. devel. corp., 1976-78, 79-81; minister fgn. affairs Sweden, 1978-79; dir. gen. Internat. Atomic Energy Agy., Vienna, Austria, 1981—; mem. Swedish Del. UN Gen. Assembly, N.Y., 1961-81, Swedish Del. Conf. Disarmament, Geneva, 1962-78. Author: Treaty Making Power, 1959; Statsmyndigheternas Internationella Forbindelser, 1964; Sovereignty, Aggression and Neutrality, 1970; The Treaty Maker's Handbook, 1974. Mem. Inst. de Droit Internat. Office: IAEA, Wagramerstr 5 POB 100, A-1400 Vienna Austria*

BLOCH, ANTOINE, cardiologist; b. Lausanne, Switzerland, Aug. 9, 1938; s. Paul and Herta (Sonnenfeld) B.; M.D., U. Lausanne, 1963; m. Josee Sánchez, Aug. 25, 1973. Intern, U. Lausanne Hosp., 1964-66; med. resident St. Antonius Hosp., Utrecht, Netherlands, 1966-67, univ. hosps., Lausanne and Geneva, 1967-70; chief resident Univ. Cardiac Center of Geneva, 1970-73, physician, 1975-80; cardiac fellow Mass. Gen. Hosp., Boston, 1973-75; privat-docent Geneva Med. Sch., 1975-80, charge de cours, 1980—; chief cardiac unit Hopital de la Tour, Geneva, 1981—. Swiss Nat. Fund grantee, 1977-79. Fellow Am. Coll. Cardiology, European Soc. Cardiology; mem. Am. Heart Assn., Am. Soc. Echocardiography, Swiss Med. Assn., Swiss Soc. Cardiology, French Soc. Cardiology, Swiss Soc. Intensive Care, Swiss Soc. Ultrasound. Author books, including: L'echocardiographie, 1978; L'infarctus du myocarde, 1979; contbr. numerous articles to profl. publs. Home: 33 Crêt-de-Choully, CH-1242 Choully Switzerland Office: Hosp de la Tour, Cardiac Unit, Geneva CH-1217, Switzerland

BLOCH, EDWARD HENRY, scientist, retired anatomy educator; b. Berlin, Fed. Republic of Germany, Feb. 1, 1914; s. Ernst Bloch and Louise Ehmer. MD, U. Tenn., 1945; PhD, U. Chgo., 1949. Intern Michael Reese Hosp., Chgo., 1946; established investigator Am. Heart Assn., Cleve., 1950-55; acting chmn. dept. anatomy Case Western Res. U., Cleve., 1980-82, emeritus prof., 1982—. Contbr. over 145 articles to profl. jours. Recipient numerous grants NIH, Rockefeller Found. and others. Mem. Am. Assn. Anatomists (emeritus), Am. Assn. Tropical Medicine (emeritus), Am. Assn. Immunologists (emeritus), Microcirculatory Soc. (emeritus, past pres., co-founder), Handerson Med. History Soc. (past pres.), Allen Meml. Libr. (trustee, past pres.), Rowfant Club (past pres.) Sigma Xi. Republican. Lutheran. Office: Case Western Res U 2109 Adelbert Rd Cleveland OH 44106

BLOCH, ERICH, electrical engineer, former science foundation administrator; b. Sulzburg, Germany, Jan. 9, 1925; came to U.S., 1948, naturalized, 1952; s. Joseph and Tony B.; m. Renee Stern, Mar. 4, 1948; 1 dau., Rebecca Bloch Rosen. Student, Fed. Poly. Inst., Zurich, Switzerland, 1945-48; BSEE,

U. Buffalo, 1952; hon. degrees, U. Mass., George Washington U., Colo. Sch. Mines, SUNY Buffalo, U. Rochester, Oberlin Coll., U. Notre Dame, Ohio State U., Rensselaer Poly. Inst., 1989, Washington Coll., 1989, CUNY, N.Y.C., 1991; hon. degree, Poly. U., Bklyn., N.Y., 1993. With IBM, 1952-84; v.p. gen. mgr. IBM, East Fishkill, N.Y., 1975-80; v.p. tech. personnel devel. IBM, Armonk, N.Y., 1980-84; mem. com. computers in automated mfg. NRC, 1980-84; dir. NSF, Washington, 1984-90; fellow Coun. on Competitiveness, 1990—; bd. dirs. Motorola Inc., Convex Computers, Quality Edn. for Minorities Network. Patentee in field. Recipient U.S. medal of tech., 1985, Computer World/Smithsonian award for innovation, 1991. Fellow IEEE (Founders award 1990), AAAS; mem. NAE, Am. Soc. Mfg. Engrs. (hon.).

BLOCH, JAMES PHILLIPS, clinical psychologist; b. Louisville, Oct. 8, 1946; s. William Austin and Mary Frances (Jordan) B.; m. Kathleen Agnes Bowen (div.); m. Jane Ann Blemker, Oct. 8, 1980; 1 child, Alexander Armstrong. BA, Shimer Coll., 1967; MA, U. Louisville, 1971, PhD, 1974. Lic. psychologist, Ky.; cert. psychologist, Ind. Staff psychologist Quinco Cons. Ctr., Columbus, Ind., 1972-80; clin. supr. Seven Counties Svcs., Louisville, 1980-89; pvt. practice Louisville, 1980—; asst. prof. Spalding U., Louisville, 1986-88. Author: Assessment and Treatment of Multiple Personality and Dissociative Disorder, 1991; contbr. articles to profl. jours. Mem. APA, Ky. Psychol. Assn., Am. Soc. of Clin. Hypnosis, Internat. Soc. for the Study of Multiple Personality and Dissociation. Office: Bloch/Harpenau Assocs Ste 307 2100 Gardiner Ln Louisville KY 40205

BLOCH, KONRAD EMIL, biochemist; b. Neisse, Germany, Jan. 12, 1912; came to U.S., 1936, naturalized, 1944; s. Frederick D. and Hedwig (Steimer) B.; m. Lore Teutsch, Feb. 15, 1941; children—Peter, Susan. Chem. Engr., Technische Hochschule, Munich, 1934; Ph.D., Columbia U., 1938. Asst. prof. biochemistry U. Chgo., 1946-50, prof., 1950-54; Higgins prof. chemistry Harvard U., Cambridge, Mass., 1954—. Recipient Nobel prize in physiology and medicine, 1964, Ernest Guenther award in chemistry of essential oils and related products, 1965, Nat. Medal of Sci., 1988. Fellow AAAS; mem. Nat. Acad. Scis., Am. Philos. Soc. Office: Harvard U Dept Chemistry 12 Oxford St Cambridge MA 02138-2900

BLOCK, BARBARA ANN, biology educator; b. Springfield, Mass., Apr. 25, 1958; d. Merrill and Myra (Winograd) B. BA, U. Vt., 1980; PhD, Duke U., 1986. Postdoctoral fellow U. Pa., Phila., 1986-88; asst. prof. organismal biology U. Chgo., 1988-93; asst. prof. biol. sci. Stanford U., 1993—; chmn. sci. adv. bd. Pacific Ocean Rsch. Found., Kona, Hawaii, 1992—. Contbr. articles to profl. jours. Recipient Presdl. Young Investigator award NSF, 1989. Mem. AAAS, Am. Soc. Zoologists, Biophys. Soc. Democrat.

BLOCK, JACK, psychology educator; b. N.Y.C., Apr. 28, 1924; s. Isadore and Tessie (Goldberg) B.; m. Jeanne L. Humphrey, Oct. 7, 1950 (dec. Dec. 1981); children: Susan, Judith, David, Carol. MA, U. Wis., 1947; PhD, Stanford U., 1950. Prof. psychology U. Calif., Berkeley, 1957—; sci. rev. com. NIMH, Rockville, Md., 1966. Author: The Q-Sort Method in Personality Assessment, 1961, The Challenge of Response Sets, 1965, Lives Through Time, 1971. NIMH rsch. grantee, Rockville, 1955—. Fellow APA (G. Stanley Hall award 1991). Office: U Calif Psychology Dept Berkeley CA 94720

BLOCK, RICHARD ATTEN, psychology educator; b. Evanston, Ill.. BA, U. Mich., 1968; PhD, U. Oreg., 1973. Vis. asst. prof. SUNY, Plattsburgh, 1973-74; asst. prof. Mont. State U., Bozeman, 1974-79, assoc. prof., 1979-85, prof. psychology, 1985—, head dept. psychology, 1986—. Editor: Cognitive Models of Psychological Time, 1990; contbr. chpts. to books. Mem. Am. Psychol. Soc., Psychonomic Soc., Internat. Soc. for Study of Time (treas. 1989—). Office: Mont State U Dept Psychology Bozeman MT 59717

BLOCK, ROBERT I., psychologist, researcher, educator; b. Irvington, N.J., Jan. 30, 1951; s. Milton and Harriet (Safier) B. BA with honors, Shimer Coll., 1969; MS, Harvard U., 1972, Rutgers U., 1977; PhD, Rutgers U., 1981. Teaching asst. psychology dept. Rutgers U., New Brunswick, N.J., 1975-76; psychologist Lafayette Clinic, Detroit, 1982-84; rsch. assoc. psychiatry dept. Wayne State U., Detroit, 1982, instr., 1982-84; assoc. rsch. scientist dept. anesthesia U. Iowa, Iowa City, 1984-88, asst. prof. dept. anethesia, 1988—; cons. State of Mich., Lafayette Clinic, Detroit, Hoffmann La-Roche, Inc.; reviewer Psychopharmacology and Anesthesiology; mem. faculty senate Sch. of Medicine, Wayne State U., Detroit, 1982-84. Contbr. articles to Anesthesiology, Brit. Jour. Anaesthesia, Psychopharmacology, Pharmacol. Biochem. Behavior. Recipient fellowship Rutgers U., New Brunswick; grantee Nat. Inst. on Drug Abuse, 1987-91. Mem. AAAS, Collegium Internat. Neuro-Psychopharmacologicum, Am. Psychol. Assn. Achievements include research on effects of nitrous oxide, benzodiazepines, marijuana, and other drugs on human associative processes, memory, and cognition. Home: 2029 Waterford Dr Coralville IA 52241-2734 Office: U Iowa Dept Anesthesia Westlawn Bldg Iowa City IA 52242

BLODGETT, FORREST CLINTON, economics educator; b. Oregon City, Oreg., Oct. 6, 1927; s. Clinton Alexander and Mabel (Wells) B.; m. Beverley Janice Buchholz, Dec. 21, 1946; children: Cherine (Mrs. Jon R. Klein), Candis Melis, Clinton George. BS, U. Omaha, 1961; MA, U. Mo., 1969; PhD, Portland State U., 1979. Joined C.E. U.S. Army, 1946, commd. 2d lt., 1946, advanced through grades to lt. col., 1965, ret., 1968; engring. assignments U.S. Army, Japan, 1947-49, U.K., 1950-53, Korea, 1955-56, Alaska, 1958-60, Vietnam, 1963; staff engr. 2d Army Air Def. Region U.S. Army, Richards-Gebaur AFB, Mo., 1964-66; base engr. Def. Atomic Support Agy., Sandia Base, N.Mex., 1966-68; bus. mgr., trustee, asst. prof. econs. Linfield Coll., McMinnville, Oreg., 1968-73, assoc. prof., 1973-83, prof., 1983-90, emeritus prof. econs., 1990—; pres. Blodgett Enterprises, Inc., 1983-85; founder, dir. Valley Community Bank, 1980-86, vice chmn. bd. dirs. 1985-86. Commr., Housing Authority of Yamhill County (Oreg.), chmn., 1980-83; mem. Yamhill County Econ. Devel. Com., 1978-83; bd. dirs. Yamhill County Found., 1983-91. Decorated Army Commendation medal with oak leaf cluster; recipient Joint Service Commendation medal Dept. of Def. Mem. Soc. Am. Mil. Engrs. (pres. Albuquerque post 1968), Am. Econ. Assn., Western Econ. Assn. Internat., Nat. Retired Officers Assn., Res. Officers Assn. (pres. Marion chpt. 1976), SAR (pres. Oreg. soc. 1985-86, v.p. gen. Nat. Soc., 1991—), Urban Affairs Assn., Pi Sigma Epsilon, Pi Gamma Mu, Omicron Delta Epsilon (Pacific NW regional dir. 1978-88), Rotary (pres. McMinnville club 1983-84). Republican. Episcopalian. Office: Linfield Coll Mcminnville OR 97128

BLOEMBERGEN, NICOLAAS, physicist, educator; b. Dordrecht, The Netherlands, Mar. 11, 1920; came to U.S., 1952, naturalized, 1958; s. Auke and Sophia M. (Quint) B.; m. Huberta D. Brink, June 26, 1950; children: Antonia, Brink, Juliana. BA, Utrecht U., 1941, MA, 1943; PhD, Leiden U., 1948; MA (hon.), Harvard U., 1951; D of Sci. (hon.), Laval U., 1987, U. Conn., 1988, U. Hartford, 1991. Teaching asst. Utrecht U., 1942-45; research fellow Leiden U., 1948; mem. Soc. Fellows Harvard U., 1949-51, assoc. prof., 1951-57, Gordon McKay prof. applied physics, 1957—, Rumford prof. physics, 1974, Gerhard Gade univ. prof., 1980, prof. emeritus, 1990; vis. prof. U. Paris, 1957, U. Calif., 1965, Collège de France, Paris, 1980; Lorentz guest prof. U. Leiden, 1973; Raman vis. prof. Bangalore, India, 1979; Fairchild Disting. scholar Calif. Inst. Tech., 1984; von Humboldt Sr. Scientist, Munic, Fed. Republic Germany.; hon. prof. Fudan U., Shanghai, People's Republic of China. Author: Nuclear Magnetic Relaxation, 1948, Nonlinear Optics, 1965; also articles in profl. jours. Recipient Buckley prize for solid state physics Am. Phys. Soc., 1958, Dirac medal U. New South Wales (Australia), 1983, Stuart Ballantine medal Franklin Inst., 1961, Half Moon trophy Netherlands Club N.Y., 1972, Nat. medal of Sci., 1975, Lorentz medal Royal Dutch Acad., 1978, Frederic Ives medal Optical Soc. Am., 1979; von Humboldt sr. scientist award Munich, 1980, von Humboldt medal, 1989, Nobel prize in Physics, 1981; Guggenheim fellow, 1957. Fellow Am. Phys. Soc., Am. Acad. Arts and Scis., IEEE (Morris Liebmann award 1959, Medal of Honor 1983, Indian Acad. Scis.(hon.); mem. Optical Soc. Am. (hon.), Nat., Royal Dutch acads. scis., Nat. Acad. Engring., Am. Philos. Soc., Deutsche Akademie der Naturforscher Leopoldina, Koninklyke Nederlandse Akademie von Wetenschappen (corr.), Paris Acad. Scis. (fgn. assoc.). Office: Harvard U Div Applied Scis Pierce Hall Cambridge MA 02138

BLOMHOFF, RUNE, biochemist educator, researcher; b. Drammen, Norway, Jan. 22, 1955; s. Thor and Randi (Olsen) B.; m. Heidi Kiil, Aug. 16, 1978; children: Maia Kiil, Henrik Kiil. BS in Biochemistry, U. Oslo, 1981, PhD in Biochemistry, 1985. Rsch. fellow Norwegian Cancer Soc., Oslo, 1981-85, sr. rsch. scientist, 1986-91; prof. Inst. Nutrition Rsch. U. Oslo, 1992—; mem. Med. Rsch. Coun. unit Norwegian Rsch. Coun. for Sci. and Humanities, 1990—. Contbr. numerous articles to profl. jours. Recipient Anders Jahres prize for med. scis., 1991. Mem. Am. Inst. Nutrition (Mead Johnson award 1988), Norwegian Biochem. Soc., Biochem. Soc. U.K. Home: Frognerseterveien 5B, 0387 Oslo 3, Norway Office: U Oslo Inst Nutrition Rsch, PO Box 1046, 0316 Oslo 3, Norway

BLOMMEL, SCOT ANTHONY, environmental engineer; b. Dayton, Ohio, July 23, 1966; s. John William and Joan Francis (Roberts) B.; m. Stephanie Anne Lush, Aug. 24, 1991. BS in Environ. Sci., Bowling Green State U., 1988. Summer intern GM, Dayton, Ohio, 1986, assoc. engr., 1988-90, project engr., 1990-91, environ. engr., 1991—. Mem. Soc. Automotive Engrs. Achievements include 2 patents for Wiper Blades. Home: 833 Xenia Ave #3 Yellow Springs OH 45387 Office: Delco Products Mail Code 02 2 PO Box 1224 Dayton OH 45401-1224

BLOMQUIST, MICHAEL ALLEN, civil engineer; b. Milw., Mar. 28, 1953; s. Glen Allen and Dorothy Ann (Baumgartner) B.; m. Laurel Ann Schacherl, Sept. 16, 1972 (div. Dec. 18, 1981); m. Donna Lynn, Nov. 13, 1982; children: Ryan, Megan. BSCE, U. Wis.-Milw., 1976. Registered profl. engr.; lic. real estate broker. Engr. Eskenazi & Farrell, Chgo., 1983-85; design engr. Beer, Gorski & Graff, Chgo., 1985-88; structural engr. Zimmerman Design Group, Milw., 1988-89; project engr. Graef, Anhalt & Schloemer, Milw., 1985-89; project mgr. Klug & Smith Co., Milw., 1989—. Chmn., trustee Emanujel Ch., Hales Corners, Wis., 1991-92, exec. coun., 1991-92. Named Engr. of the Yr., Wis. Soc. Profl. Engrs., 1991, Engr. of the Yr. in Constrn., Wis. Soc. Profl. Engrs., 1991. Mem. Wis. Soc. Profl. Engrs. (pres. 1990-91). Republican. Lutheran. Office: Klug & Smith Co 4425 W Mitchell Milwaukee WI 53214

BLOOM, ALAN ARTHUR, gastroenterologist, educator; b. Elizabeth, N.J., Dec. 14, 1930; s. Harry and Frances (Aronowitz) B.; m. Roslyn Sadihoff, June 29, 1952; children: Jordan, Sherry, Lee. BA, Syracuse U., 1952; MD, U. Chgo., 1956. Diplomate Am. Bd. Internal Medicine, Am. Bd. Gastroenterology. Intern Montefiore Hosp., Bronx, N.Y., 1956-57, resident, 1957-59; fellowship Manhattan Veterans Hosp., N.Y.C., 1961-63; dir. gastroenterology Bronx (N.Y.)-Lebanon Hosp., 1963—; assoc. prof. medicine Albert Einstein Coll. Medicine, Bronx, 1974—. Fellow Am. Coll. Physicians; mem. Am. Gastroent. Assn. Democrat. Jewish. Office: Bronx Lebanon Hosp 1650 Grand Concourse Bronx NY 10457

BLOOM, WALLACE, psychologist; b. N.Y.C., Apr. 9, 1916; s. Irving and Agnes (Weinstein) B.; m. Riselle Levis, Dec. 24, 1950; 1 child, Michael. BBA, CUNY, 1936; MS, Trinity U., San Antonio, 1951; PhD, U. Tex., 1964. Commd. 2d lt. USAF, 1937, advance through grades to lt. col., 1946, ret., 1960; human factors United Tech. Corp., Sunnyville, Calif., 1963-64; clin. psychologist Porterville (Calif.) State Hosp., 1964-66; rsch. clin. psychologist Wilford Hall USAF Med. Ctr., San Antonio, 1966-86; pvt. practice San Antonio, 1986—. Author: Manual Bloom Sentence Completion Survey, 1975; author: Shift Work and Human Efficiency, 1962, Changes Made Lesson Learned After Mental Health Screening, 1983; assoc. editor: Personality Assessment Jour., 1982-84; contbr. articles to profl. jours.; author more than 70 profl. papers. Fellow Cerebral Palsy Assn., N.Y.C., 1960-62, NDEA, Washington, 1962-63. Fellow Soc. Personality Assessment; mem. APA. Home: 133 Twinleaf Ln San Antonio TX 78213

BLOOM, WILLIAM MILLARD, industrial design engineer; b. New Kensington, Pa., Aug. 10, 1925; s. William Lewis and Natalie Tillbrook (McMillin) B.; m. Judith Ann Callen, May 23, 1953; children: Kimberly Ann, Stacey Ellen. BA, Geneva Coll., 1951; BSME, Carnegie Inst. Tech., 1951. Registered profl. engr., Pa. Fuel engr. maintenance dept. Brackenridge (Pa.) Plant, Allegheny Ludlum Steel, 1951-56; fuel engr. gen. engring. divsn. Allegheny Ludlum Steel Corp., Brackenridge, 1956-59, sr. engr. furnaces and fuels, gen. engring. divsn., 1959-61; chief engr. furnaces and fuels gen. engring. divsn. Allegheny Industries, Pitts., 1961-71; asst. to v.p. engring. spl. assignments Allegheny Ludlum Steel Corp., Brackenridge, 1971-81; mgr. furnace design engring., mfg. engring. Allegheny Ludlum Steel Corp., Pitts., 1981-92; pvt. practice cons. indsl. furnaces Pitts., 1992—; cons. Alloy Rods Corp., Hanover, Pa., 1989, Timet Corp., Henderson, Nev., Toronto, Ohio, IPM Corp., Ridgeway, Pa., Columbus, Ohio, Tube Turn Corp., Louisville, True Temper, Geneva, Ohio, Arnold Engring., Chgo., Altech, Dunkirk, N.Y., Posco, Korea, Kuhlman Electric, Lexington, Ky., 1961-92. With U.S. Army, 1944-46, ETO. Mem. NSPE, Assn. Iron and Steel Engrs. (life, bd. dirs., chmn. combustion com., AISE-KELLY award 1st pl. 1979), 70th Divsn. Assn. (life), Theta Xi (life). Republican. Methodist. Achievements include patents for Bar Furnace Seals, Annealing Apparatus, Coil Quench, Conveyor Roll, Tunnel Furnace, Annealing Furnace, Steel Scrap Preheater, Apparatus Scrap Preheater, Roll Turner/Remover, Jet Heat Reucperator, Replaceable Ladle Heater Seals, High Temp Fan Plug, Hot Strip Mill Cover Heat Retention; developed high temperature hydrogen anneal tunnel furnace for grain oriented silicon steels that significantly lowered watt losses/pound to develop class of steel, jet heat recuperators that reduce continous anneal furnaces fuel input by 50% and increases production 50%. Home: 1522 King John Dr Pittsburgh PA 15237

BLOSSER, HENRY GABRIEL, physicist; b. Harrisonburg, Va., Mar. 16, 1928; s. Emanuel and Leona (Branum) B.; m. Priscilla May Beard, June 30, 1951 (div. Oct. 1972); children: William Henry, Stephan Emanuel, Gabe Fawley, Mary Margaret; m. 2d, Mary Margaret Gray, Mar. 16, 1973. BS, U. Va., 1951, MS, 1952, PhD, 1954. Physicist Oak Ridge (Tenn.) Nat. Lab., 1954-56, group leader, 1956-68; assoc. prof. physics Mich. State U., East Lansing, 1950-61, prof., 1961-90, Univ. Disting. prof., 1990—, dir. Cyclotron Lab., 1961-89; cons. U. Mich., Ann Arbor, 1960-61, Washington U., St. Louis, 1961-62, Lawrence Radiation Lab., 1962, U. Md., 1962-65, Princeton U., 1965—, others. Bd. dirs. Midwest Univs. Rsch. Assocs., 1960-63. With USNR, 1946-48. NSF postdoctoral fellow, 1966-67; Guggenheim fellow, 1973-74. Mem. Am. Phys. Soc. (Bonner prize 1992), Sigma Xi, Phi Beta Kappa, Kappa Alpha. Home: 609 Beech St East Lansing MI 48823-3405 Office: Mich State U Cycltron Lab Dept Physics & Astronomy East Lansing MI 48824

BLOSSOM, NEAL WILLIAM, chemical engineer; b. Great Falls, Mont., Jan. 3, 1961; s. Richard Conrad and Isabel Verna (Clelland) B.; m. Kathie Maria Roos, Jan. 30, 1988. BSChemE, Mont. State U., 1984. Registered profl. engr., Mont. Process engr. Stauffer Chem. Co., Butte, Mont., 1984-88; rsch. engr. Am. Chemet Corp., East Helena, Mont., 1988—. Author seminar publ. Advances in Powder Metallurgy, 1990. Mem. Toastmasters (pres. 1991 Lewis & Clark chpt., area gov. 1992). Achievements include development of dispersion strenghtened copper powders and high green strength copper powders. Office: Am Chemet Corp 1 Smelter Rd East Helena MT 59635

BLOUNT, ALICE MCDANIEL, museum curator; b. Carbondale, Ill., Aug. 18, 1942; d. Wilbur Charles and Claribel Wilhelmina (McClanahan) McDaniel; m. John F. Blount, Aug. 2, 1968. BS in Geology, U. Mo., 1964; MS, U. Wis., 1966, PhD in Geology, 1970. Curator earth sci. The Newark Mus., 1968—; cons. in indsl. minerals Newark, 1973—; mem. grad. faculty Rutgers U., Newark, 1973—. Contbr. articles to profl. jours. Fellow Soc. Econ. Geologists (mem. Thayer Lindsley Vis. Lectrs. Com.); mem. Clay Minerals Soc., Mineralog. Soc. Am., Geol. Soc. Am. Achievements include research in geology, mineralogy and methods of exploration of talc and other nonmetallic mineral deposits. Office: The Newark Mus PO Box 540 Newark NJ 07101-0540

BLOUT, ELKAN ROGERS, biological chemistry educator, university dean; b. N.Y.C., July 2, 1919; s. Eugene and Lillian B.; m. Joan E. Dreyfus, Aug. 27, 1939; children: James E., Susan L., William L.; m. Gail A. Ferris, Mar. 29, 1985. A.B., Princeton U., 1939; Ph.D., Columbia U., 1942; A.M. (hon.), Harvard U., 1962; D.Sc. (hon.), Loyola U., 1976. With Polaroid Corp., Cambridge, Mass., 1943-62, successively rsch. chemist, assoc. dir. rsch., 1948-58, v.p., gen. mgr. rsch., 1958-62; rsch. assoc. Harvard U., 1950-52, 56-60, lectr. on biophysics, 1960-62, prof. biol. chemistry, 1962-90, Edward S. Harkness prof. biol. chemistry, 1964-90, Edward S. Harkness prof. emeritus, 1990—, head dept. biol. chemistry, 1965-69; dean for acad. affairs Harvard Sch. Pub. Health, 1978-89, chmn. dep. environ. sci. and physiology, 1986-88, dir. div. biol. scis., prof., 1987-91; prof. emeritus Harvard Sch. of Pub. Health, Boston, 1991—; cons. Polaroid Corp., 1962—; rsch. assoc. Children's Hosp. Med. Center, Boston, 1950-52, cons. chemistry, 1952—; mem. conseil de surveillance Compagnie Financiére du Scribe, 1975-81; trustee Bay Biochem. Rsch., Inc., 1973-83; mem. exec. com. div. chemistry and chem. tech. NRC, 1972-74, mem. assembly of math. and phys. scis., 1979-82; mem. sci. adv. com. Ctr. for Blood Rsch., Inc., 1972-92, emeritus trustee, 1992—; also mem. bd. dirs.; mem. rsch.adv. com. Children's Hosp. Med. Ctr., 1976-80, 84-90, chmn., 1987-90; mem. sci. adv. com. Mass. Gen. Hosp., 1968-71, Rsch. Inst., Hosp. for Sick Children, Toronto, Ont., Can., 1976-79; mem. adv. coun. dept. biochem. scis. Princeton U., 1974-83, chmn. adv. coun. program in biology, 1983—; mem. vis. com. dept. chemistry Carnegie-Mellon U., 1968-72; bd. visitors Faculty Health Scis., SUNY, Buffalo, 1968-70; overseer Boston Mus. Sci.; trustee, v.p. Boston Biomed. Rsch. Inst., 1990—; bd. govs. Weizmann Inst. Sci., Rehovot, Israel, 1978—; bd. dirs. Nat. Health Rsch. Found., ESA, Inc.; bd. dirs., sec.-treas. Nat. Acads. Corp.; gen. ptnr. Gosnold Investment Fund Ltd. Partnership, 1985—; bd. dirs., investment mgr. Auburn Investment Mgmt. Corp., 1985—; sci. advisor Affymax Rsch. Inst., 1988—; sr. adviser sci. FDA, 1991—; mem. sr. adv. bd. The Encyclopedia of Molecular Biological, 1991—; mem. council visitors Marine Biol. Lab., 1992—. Mem. adv. bd. Jour. Polymer Sci, 1956-62; mem. editorial bd. Biopolymers, 1963-85, hon. founding editor, 1985—; mem. editorial bd. Am. Chem. Soc. Monograph Series, 1965-72, Internat. Jour. Peptide and Protein Rsch., 1978-89; mem. editorial adv. bd. Macromolecules, 1967-70, Jour. Am. Chem. Soc., 1978-82; contbr. articles to profl. jours. Recipient Princeton Class of 1939 Achievement award, 1970, Nat. Med. Sci. award NSF, 1990; NRC fellow Harvard U., 1942-43. Fellow AAAS, Am. Acad. Arts and Scis. (fin. com. 1977-84, com. on investments 1984—, chmn. budget com. 1988-92, treas. 1992—), N.Y. Acad. Arts and Scis., Optical Soc. Am. (past pres. New Eng. sect.); mem. NAS (treas. 1980-92, treas. emeritus 1992—, fin. com. 1976-92, adv. com. USSR and Eastern Europe 1979-84, mem. Com. Sci. Engring. and Pub. Policy, 1992—), Inst. Medicine, USSR Acad. Scis. (fgn. mem.), Am. Chem. Soc. (nat. councillor 1958-61, Ralph F. Hirschmann award 1991), The Chem. Soc., Am. Soc. Biol. Chemists (fin. com. 1973-82), Biophys. Soc., Commn. on Phys. Scis., Math., and Resources of NRC, Internat. Orgn. Chem. Scis. in Devel. (council 1981—, chmn. fin. com. 1982—, v.p., treas. 1985—, bd. dirs. 1985—), Fedn. Am. Socs. Exptl. Biology (investments adv. com. 1981-85). Achievements include patents in field. Home: 1010 Memorial Dr Apt 12A Cambridge MA 02138-4856 Office: Harvard U Med Sch Dept Biol Chemistry Molecular Pharmacology Boston MA 02115 also: Harvard Sch Pub Health Div Biol Scis 677 Huntington Ave Boston MA 02115

BLUE, JEFFREY KENNETH, neuromuscular and skeletal researcher; b. Albany, Ga., Feb. 24, 1956; s. Daniel Monroe and Margaret (Von Hindber) B. BS in Sociology, Coll. of the Ozarks, 1983; MS in Biology, Cen. Mo. State U., 1990. Project/curriculum asst. NASA Space Life Scis. Tng. Program, Kennedy Space Ctr., Fla., 1991; rsch. asst., therapist Midwest Back and Neck Care Ctr., Topeka, 1992—. Sec. Humane Treatment of Rsch. Animals, Warrensburg, Mo., 1988-89; search rescue pilot Civil Air Patrol, Saline County, Mo., 1986-90; election bd. County Election Bd., Burlington County, N.J., 1984. With USMCR, 1983-89. Rsch. grant Cen. Mo. State Grad. Sch., 1989-90. Mem. AIAA, Mo. Acad. Sci., Topeka Rowing Assn., College of the Ozarks Alumni Assn., Sigma Xi (rsch. grant 1989-92). United Methodist. Achievements include rsch. on integrating human excercise physiology in the controlled ecol. life support system biomass prodn. chamber on the Kennedy Space Ctr. Home: Apt 3 317 SW 15th St Topeka KS 66612

BLUE, JOSEPH EDWARD, physicist; b. Quitman, Miss., Sept. 29, 1936; s. Edward Lee and Allie Belle (Corley) B.; m. Neva Rosetta Deal, Apr. 14, 1962; children: Tracy Marie, Gina Lynn. BS in Physics, Miss. State U., 1961; MS in Engring. Sci., Fla. State U., 1966; PhD in Mech. Engring., U. Tex., 1971. Physicist Navy Mine Def. Lab., Panama City, Fla., 1961-68; rsch. sci. engr. U. Tex., Austin, 1968-71; rsch. physicist Naval Rsch. Lab., Orlando, Fla., 1971-73, Meas br. head, 1973-81, supt., 1981—. Author: (with others) Benchmark Papers in U/W Acoust, 1975; contbr. articles to Jour. Acoustical Soc. Am. Fellow Acoustical Soc. Am. Democrat. Methodist. Achievements include patents for Low Frequency Acoustic Source, Color Sonar Display, Time Internal to Pulse Height Converter; research in resonant scattering, parametric depth sounder using water's nonlinearity and substantial sound pressure from tow-powered sources. Office: Naval Rsch Lab PO Box 8337 Orlando FL 32856

BLUE, REGINALD C., psychologist, educator; b. Cleve., Oct. 6, 1942; s. Walter L. and Emma L. Blue; divorced; children: Melonn, Monique, Veronica, Toussaint. MA, John Carroll U., 1969; PhD, Ohio State U., 1974. Tchr. Cleve. Pub. Schs., 1965-69, counselor, 1969-71; intern sch. psychologist Lorain City Schs, 1971-72; supr. learning disabilities/behavioral disorders, sch. psychologist Springfield (Ohio) City Schs., 1973-75; dir. pupil pers. Shaker Heights (Ohio) City Schs., 1975—; psychologist Blue & Assocs., Shaker Heights, 1977—; bd. dirs. Cuyahoga County Youth Svcs., Cleve., 1981-87, Divorce Equity, Inc., Cleve., 1985-89, Home Instrn. Program for Presch., Warrensville Heights, Ohio, 1992—. Mem. Rotary. Baptist. Home: 14411 Onaway Rd Shaker Heights OH 44120-2840 Office: Blue & Assocs 20310 Chagrin Blvd Ste 6 Shaker Heights OH 44122-4973

BLUHER, GRIGORY, computer scientist, mathematician; b. Odessa, Ukraine, May 9, 1960; came to U.S., 1979; s. Froim and Alla (Shvetz) Blyukher; m. Antonia Rose Wilson, May 25, 1986; children: Andrew Emmanuel, Julia Elizabeth, Sarah Elena. MA in Math. with deptl. and gen. honors., Johns Hopkins U., 1983; PhD in Math., Princeton U., 1988; MS in Computer Sci., UCLA, 1992. Asst. prof. Trenton (N.J.) State Coll., 1987-88, Whittier (Calif.) Coll., 1988-89; programmer The Software Toolworks, L.A., 1989-90; researcher computer sci. dept. UCLA, 1990-92; staff programmer IBM, San Jose, 1992-93; project leader ORACLE, Redwood City, Calif., 1993—. Translator: Introduction to the Classical Theory of Abelian Functions, 1990. Interviewer alumni coun. Johns Hopkins U., Balt., 1985-89. IBM scholar, 1983. Mem. IEE-CS, Math. Assn. Am., Assn. for Computing Machinery, Phi Beta Kappa. Home: 6463 Broadway Ave Newark CA 94560-4011 Office: Box 659411 500 Oracle Pkwy Redwood City CA 94065

BLUHM, TERRY LEE, materials scientist, chemist; b. Niagara Falls, N.Y., June 22, 1947; s. Milford Charles and Vera Jane (Perry) B.; m. Ellen C. Manwaring, Apr. 3, 1971; 1 child, Sylvia. PhD, SUNY Coll. Environ. Sci., Syracuse, 1976. Postdoctoral fellow inst. for molecular chemistry U. Freiburg, Germany, 1976-78; rschr. Xerox Rsch. Centre Can., Mississauga, Ont., 1979-92; project mgr. Xerox Corp., Webster, N.Y., 1992—. Contbr. articles on polymer sci. and materials characterization to profl. jours. Recipient Polymer Sci. award Soc. Plastics Engring., 1970. Mem. AAAS, Am. Chem. Soc., Materials Rsch. Soc. Achievements include patents in xerographic materials. Office: Xerox Corp 800 Phillips Rd 114-42D Webster NY 14580

BLUITT, KAREN, computer program manager; b. N.Y.C., Oct. 25, 1957; d. James Bertrand and Beatrice (Kaufman) B.; m. Kenneth Mark Curry, Nov. 24, 1979. BS, Fordham U., 1979; MBA, Calif. State Poly. U., 1982. Software engr. Hughes Aircraft Co., Fullerton, Calif., 1979-81; microprocessor engr. Beckman Instruments Co., Fullerton, 1981-82, Singer Co., Glendale, Calif., 1982-83; sr. software engr. Sanders Assoc., Nashua, N.H., 1983-85; software project mgr. GTE Corp., Billerica, Mass., 1985-86; sr. software engr. Wang Labs, Lowell, Mass., 1986-87; project task leader Vanguard Rsch., Lexington, Mass., 1987-88; program mgr. Applied Rsch. & Engring., Bedford, Mass., 1989-91; program mgr. Sparta, McLean, Va., 1992—. 1st lt. USAR, 1979-88. Scholar Gov. N.Y. Scholarship Com., 1975-79; Beta Gamma Sigma scholar, 1978—. Mem. IEEE, NAFE, NOW, LWV, AAUW, Am. Brokers Network, Armed Forces Comm. and Electronics Assn., Am. Def. Preparedness Assn., Ops. Rsch. Soc., Soc. Women Engrs. Office: Sparta Inc 7926 Jones Branch Dr Mc Lean VA 22102

BLUM, GREGORY LEE, civil engineer; b. Dubuque, Iowa, Nov. 4, 1960. BSCE, U. Iowa, 1983. Registered profl. engr. Engr. Cook Incorp., Bloomington, Ind., 1984—.

BLUM, NORMAN ALLEN, physicist; b. Boston, Dec. 29, 1932; s. John William and Natalie (Levine) B.; m. Elaine Ruth Grossman, Sept. 8, 1957; children: Scott Michael, Wendy Sue, Andrew Paul. AB, Harvard U., 1954; PhD, Brandeis U., 1964. Mem. rsch. staff Bitter Magnet Lab. MIT, Cambridge, 1960-70; sr. physicist Electronics Rsch. Ctr., NASA, Cambridge, 1966-70; prin. physicist Applied Physics Lab., Johns Hopkins U., Laurel, Md., 1970—, mgr. Microelectronics Lab., 1992—; mem. Presdl. Awards Panel; mem. peer rev. panels NSF. Contbr. over 75 articles to sci. jours., chpts. to books.. Lt. USN, 1954-57. Mem. IEEE, Am. Phys. Soc. (vis. com. for physics dept. evaluation), Am. Vacuum Soc. Office: Johns Hopkins U Applied Physics Lab Laurel MD 20723

BLUMBERG, BARUCH SAMUEL, academic administrator; b. N.Y.C., July 28, 1925; s. Meyer and Ida (Simonoff) B.; m. Jean Liebesman, Apr. 4, 1954; children: Anne, George, Jane, Noah. BS, Union Coll., Schenectady, 1946; MD, Columbia U., 1951; PhD, Oxford (Eng.) U., 1957; 20 hon. doctoral degrees. Intern, then resident Columbia div. Bellevue Hosp., N.Y.C., 1951-53; fellow in medicine Columbia-Presbyn. Med. Ctr., N.Y.C., 1953-55; chief geog. medicine and genetics sect. NIH, Bethesda, Md., 1957-64; assoc. dir. clin. rsch. Fox Chase Cancer Ctr., Phila., 1964-86, v.p. population oncology, 1984-89, disting. scientist, 1989—; master Balliol Coll., Oxford U., 1989—; prof. medicine and anthropology U. Pa.; George Eastman vis. prof. Balliol Coll., Oxford U., 1983-84; Raman vis. prof. Indian Inst. Scis., Bangalore, 1986; Ashland vis. prof. U. Ky., Lexington, 1986, 87; disting. vis. Nat. U. Singapore, 1992; sr. advisor to pres. Fox Chase Cancer Ctr., 1989—. Contbr. articles to profl. jours. Lt. USNR, 1943-46. Recipient Albion O. Berstein, M.D. award Med. Soc. State of N.Y., 1969, Grand Sci. award Phi Lambda Kappa, 1972, Ann. award Eastern Pa. br. Am. Soc. Microbiology, 1972, Passano award Williams & Wilkens Co., 1974, Modern Medicine Disting. Achievement award, 1975, Internat. award Gairdner Found., 1975, Karl Landsteiner Meml. award Am. Assn. Blood Banks, 1975, Nobel prize in physiology or medicine, 1976, Scopus award Am. Friends of Hebrew U., 1977, Strittmatter award Philadelphia County Med. Soc., 1980, Disting. Service award Pa. Med. Soc., 1982, Zubrow award Pa. Hosp., 1986, Achievement award Sammy Davis Jr. Nat. Liver Inst., 1987, John P. McGovern award Am. Med. Writers Assn., 1988, Gov.'s Award in the Scis. Commonwealth of Pa., 1989, John Blundell award Brit. Blood Transfusion Soc., 1989, Gold Medal award Can. Liver Found. and Can. Assn. Study of Liver, 1990, elected to Nat. Inventor Hall of Fame, 1993. Fellow ACP, Royal Coll. Physicians; mem. NAS, Assn. Am. Physicians, Am. Soc. Clin. Investigation, Am. Soc. Human Genetics, Am. Assn. Phys. Anthropologists, John Morgan Soc., Chesapeake and Ohio Canal Soc., United Oxford and Cambridge (London), Explorers Club N.Y., Athenaeum (London). Office: Fox Chase Cancer Ctr 7701 Burholme Ave Philadelphia PA 19111-2497

BLUMBERG, BENJAMIN MAUTNER, virologist; b. St. Louis, Apr. 30, 1942; s. Morris Burgheim B. and Dorothy (Mautner) Cordes; m. Bonnie Birtwistle, JUne 15, 1974 (div. 1977). BA, Harvard Coll., 1963; PhD, U. Chgo., 1974. Rsch. asst. U. Geneva, Switzerland, 1980-84; rsch. chemist VA, East Orange, N.J., 1984-89; assoc. prof. U. Rochester (N.Y.) Med. Ctr., 1989—; cons. VA, 1988-89. Contbr. articles to PNAS, Virology, Cell. Donor Rochester Philharmonic Orch., 1989—, WXXI Rochester, 1989—. Mem. AAAS, N.Y. Acad. Sci., Am. Soc. for Virology, Harvard Club Rochester. Achievements include patent pending for xenograft model for HIV-1 infection of human brain; research in replication mechanism of vesicular stomatitis virus, polymerase structure of paramyxoviruses, HIV-1 infection of astrocytes in vivo. Office: U Rochester Med Ctr Dept Neurology 601 Elmwood Ave Rochester NY 14642

BLUMBERG, HERBERT HASKELL, psychology educator; b. Phila., Dec. 8, 1941; s. Daniel and Sara Frieda (Peiper) B.; m. Alison Jean Britton, Oct. 9, 1980; 1 child, Joanna Britton. BA, Haverford Coll., 1963; PhD, Johns Hopkins U., 1967. Lectr., asst. prof. Wilson Coll., Chambersburg, Pa., 1967-69; rsch. psychologist, addiction rsch. unit Inst. of Psychiatry, U. London, 1969-77; lectr., sr. lectr. Goldsmiths Coll., U. London, 1977—; vis. scholar Harvard U., 1988, 90; rsch. assoc. Ctr. for Nonviolent Conflict Resolution, Haverford (Pa.) Coll., 1979-71, vis. prof., 1992-93; mem. exec. com. Initiative for Peace Studies in the U. London, 1989-92; mem. editorial cons. bd. Cahiers Internationaux de Psychologie Sociale, Liege, Belgium, 1989—. Co-author: Small Group Research: A Handbook, 1993; co-editor: Nonviolent Direct Action, 1969, Liberation Without Violence, 1977, Small Groups and Social Interaction, 2 vols., 1979, Peace: Abstracts of the. . .Behavioral Literature, 1992. Niwano Peace Found. grantee, 1993. Fellow Brit. Psychol. Soc. (assoc.); mem. APA, Am. Psychol. Soc., Brit. Sociol. Assn., Scientists for Global Responsibility, Soc. for the Psychol. Study of Social Issues, Psychologists for Social Responsibility. Achievements include preparation of a viable taxonomy of peace-psychology research. Home: 71 Harvist Rd, London England NW6 6EX Office: Haverford Coll Haverford PA 19041

BLUME, HORST KARL, manufacturing engineer; b. Berlin, Dec. 9, 1927; s. Karl and Luise (Leidel) B.; m. Gertrud Schweigert, May 19, 1924. Chem./technician, Coll. Analytic Chemistry, Berlin, 1948; EE, Barth Coll. Engring., Berlin, 1951; Assoc., Free Univ. Berlin, 1952. Engr. Honeywell, Inc., Phila., 1956-58; pres. Phoenix Precision Instrument Co., Phila., 1958-70; owner TCS Med. Products Co., Huntingdon Valley, Pa., 1970—; cons. Tech. Consulting Svcs., Southampton, Pa., 1972—. Patentee in field. Mem. Republican Nat. Com., Washington, 1980. Mem. Schlaraffia Club (v.p. 1978—). Home: 228 Brookdale Dr Huntingdon Valley PA 19006-2429 Office: TCS Med Products Co 2793 Philmont Ave Huntingdon Valley PA 19006-5303

BLUME, JOHN AUGUST, consulting civil engineer; b. Gonzales, Calif., Apr. 8, 1909; s. Charles August and Vashti (Rankin) B.; m. Ruth Clarissa Reed, Sept. 14, 1942 (dec. 1984); m. Jene Frances Osborn, Aug. 28, 1985. A.B., Stanford, 1932, C.E., 1934, Ph.D., 1966. Constrn. engr. San Francisco-Oakland Bay Bridge, 1935-36; individual practice civil and structural engring. San Francisco, 1945-57; pres. John A. Blume & Assos. (Engrs.), San Francisco, 1957-81; chmn., sr. cons. John A. Blume & Assos. (Engrs.), 1980-85; now pvt. cons. Hillsborough, Calif.; past mem., chmn. adv. council Sch. Engring., Stanford U.; past chmn. adv. com. Earthquake Engring. Research Center, U. Calif. at Berkeley.; cons. prof. civil engring. Stanford U. Author: A Machine for Setting Structures and Ground into Forced Vibration, 1935, Structural Dynamics in Earthquake Resistant Design, 1958, A Reserve Energy Technique for the Design and Rating of Structures in the Inelastic Range, 1960, Dynamic Characteristics of Multistory Buildings, 1969; co-author: Design of Multistory Reinforced Concrete Buildings for Earthquake Motions, 1961, An Engineering Intensity Scale for Earthquakes and Other Ground Motion, 1970, The SAM Procedure for Site-Acceleration-Magnitude Relationships, 1977; Contbr. articles to profl. jours. John A. Blume Earthquake Engring. Center at Stanford U. named in his honor. Mem. Nat. Acad. Engring., Structural Engrs. Assn. Calif. (pres. 1949), Cons. Engrs. Assn. Calif. (pres. 1959), ASCE (hon.; pres. San Francisco sect. 1960, Moisseiff award 1953, 61, 69, Ernest E. Howard award 1962), Seismol. Soc. Am. (medal 1986), N.Y. Acad. Scis. (hon. life), Soc. Am. Mil. Engrs., Internat. Assn. Earthquake Engring. (hon.), Earthquake Engring. Research Inst. (hon.; pres. 1977-81, medal 1991), Sigma Xi, Tau Beta Pi. Home and Office: 85 El Cerrito Ave Hillsborough CA 94010-6805

BLUMENREICH, MARTIN SIGVART, medical educator; b. Oslo, Norway, Dec. 1, 1949; came to U.S., 1975; s. Sane and Bluma (Nomberg) B.; m. Patricia Estela Dulman, Dec. 23, 1978; children: Hannah Vardit, Arnina Mirit. MD, U. Uruguay, Montevideo, 1975. Diplomate Am. Bd. Internal Medicine, Am. Bd. Med. Oncology. Intern Jewish Hosp. and Med. Ctr. Bklyn., 1975-76, resident, 1976-78; fellow med. oncology Meml. Sloan-Kettering Cancer Ctr., N.Y.C., 1978-81; asst. prof. medicine U. Louisville, 1981-85, assoc. prof., 1988—; asst. prof. medicine N.J. Med. Sch., Newark, 1985-88. Fellow ACP; mem. Am. Assn. Cancer Rsch., Soc. Clin. Oncology. Office: U Louisville J G Brown Cancer Ctr 529 S Jackson St Louisville KY 40292-0001

BLUMRICH, JOSEF FRANZ, aerospace engineer; b. Steyr, Austria, Mar. 17, 1913; s. Franz and Maria Theresia (Mayr) B.; m. Hildegard Anna Schmidt-Elgers, Nov. 7, 1935; children: Michael Sebastian, Christoph, Stefan. BS in Aero. and Mech. Engring., Ingenieurschule Weimar (Germany), 1934. Engr., Gothaer Waggonfabrik A.G., Gotha, Germany, 1934-44; ct. interpreter U.S. Mil. Ct., Linz, Austria, 1946-51; dep. chief hydraulics dept. United Austrian Iron and Steel Works, Linz, 1951-59; structural design engr. Army Ballistic Missile Agy., Huntsville, Ala., 1959-61; chief structural engring. br. G.C. Marshall Space Flight Ctr., NASA, Huntsville, 1961-69, chief systems layout br., 1969-74; cons. in field, 1974—. Served with German Army, 1944-45. Recipient Apollo Achievement award NASA, 1969, Exceptional Service medal, 1972. Author: The Spaceships of Ezekiel, 1974; Kasskara, 1979; editorial cons. on space sci. and rocketry Scribner-Bantam English Dictionary, 1977; contbr. articles to profl. jours. Patentee in field. Home: PO Box 433 Estes Park CO 80517-0433

BLUNDELL, THOMAS LEON, scientist, science administrator; b. Brighton, Eng., July 7, 1942; s. Horace Leon and Marjorie (Davis) B.; m. Lesley Ratcliff, July 8, 1964 (div. 1972); 1 child, Richard; m. Bancinyane Lynn Sibanda, May 22, 1987; children: Sichelesile, Samkeliso. BA, Oxford (Eng.) U., 1964, D. Philosophy, 1967. Rsch. fellow Oxford U., 1967-72; lectr. biology Sussex U., 1973-76; prof. Birkbeck Coll., London, 1976-90; dir. gen. Agrl. & Food Rsch. Coun., Swindon, Eng., 1991—; hon. dir. Imperial Cancer Rsch. Fund Unit, London, 1989—; coun. mem. Sci. Engring. Rsch. Coun., 1989-90, Agrl. & Food Rsch. Coun., 1985-90, Adv. Coun. Sci. Tech., London, 1988-90, Adv. Bd. Rsch. Coun., 1991—. Author: Protein Crystallography, 1976; editor Progress in Biophysics. Councilor Oxford City Coun., 1970-73, chmn. planning, 1972-73. Recipient Gold medal Inst. Biotechs., 1988—, Ciba medal Biochem. Soc. UK, 1987; hon. fellow Brasenose Coll., Oxford, 1989, Linacre Coll., Oxford, 1991. Fellow Royal Soc. Office: AFRC, Polaris House, Swindon SN2 1UH, England

BLUSH, STEVEN MICHAEL, nuclear scientist, safety consultant; b. Parkville, Mo., June 1, 1948; s. William Edwin Blush and Jeanne Arlene Harrington; m. Lynn Francine Rusten; children: Sarah Courtney Blush, Hilary Dale Florance. BA in Anthropology, U. Calif., 1970; postgrad., Boston Coll., San Francisco State U. Investigator Korean-Am. relations, subcommittee on internat. orgns. U.S. House Of Representatives, Washington, 1977-79; task group leader, investigator, Three Mile Island subcommittee on Nuclear Regulation U.S. Senate, Washington, 1979-80; chief investigator Pres. Nuclear Safety Oversight Com., Washington, 1980-81; pvt. cons., 1981-83; sr. study dir. of nuclear safety-related policy issues NRC, NAS, NAE, Inst. Med., Washington, 1984-90; special cons. to the sec. of energy U.S. Dept. Energy, Washington, 1989-90; dir. Office of Nuclear Saftey, U.S. Dept. Energy, Washington, 1990-93; pres. Steve Blush Cons., Inc., Bethesda, Md., 1993—; mem. Ill. Power Co. Nuclear Review and Audit Group, 1993—; tech. advisor NOVA; researcher Nuclear Control Inst. Adolph Kersten scholar Univ. Calif., 1966. Mem. Am. Soc. Quality Control, Senior Exec. Svcs. Home: 5914 Harwick Rd Bethesda MD 20816

BLY, CHARLES ALBERT, nuclear engineer, research scientist; b. Winchester, Va., Jan. 11, 1952; s. Theodore and Nancy Irma (Fisher) B.; m. April Marie Monnen, July 24, 1976. BS in Nuclear Engring., U. Va., 1978, MS in Nuclear Engring., 1983; student, Nat. Acad. Nuclear Tng., 1992-93. Nuclear reactor operator Nuclear Reactor Facility of the U. Va., Charlottesville, 1977-80, sr. reactor operator, 1980-83, rsch. engr., 1981-83; vis. engr. British Nuclear Fuels Ltd., Springfields, Lancashire, Eng., 1983; nuclear engr. Comml. Nuclear Fuel div. Westinghouse Electric, Pitts., 1983-92, Beaver Valley Power Sta. Duquesne Light Co., Shippingport, Pa., 1992—. Inventor fusion and hybrid fission/fusion fuel rod, combined cycle steam turbine, gas turbine nuclear power plants; discoverer neutrino-driven nucleon fission chain reactions; contbr. numerous articles to profl. jours. Candidate Shenandoah County (Va.) Bd. of Supervisor, 1975; mem. Ad Hoc Com. to Prevent Extension of I-66 Hwy. Through George Washington Nat. Forest, Strasburg, Va., 1979, Ad Hoc Com. to Preserve the Pitts. Aviary, 1991. Mem. ASME, IEEE, ASTM, AAAS, Am. Nuclear Soc., Am. Phys. Soc., ASM Internat., Assn. Energy Engrs, The Engring. Soc., Profl. Engr's. Soc., Fedn. Am. Scientists, Engr's. Soc. Western Pa., N.Y. Acad. Scis., Internat. Platform Assn. Democrat. Lutheran. Avocations: sci. rsch., hang gliding, hiking, camping, travel. Home: 908 William Penn Ct Pittsburgh PA 15221 Office: Duquesne Light Co Beaver Valley Power Sta PO Box 4 Mail Drop SBX-W Shippingport PA 15077

BOADO, RUBEN JOSE, biochemist; b. Buenos Aires, Argentina, Feb. 8, 1955; came to U.S., 1985; s. Osvaldo Ruben and Lucia B.; m. Adriana Graciela Swiecicki, Jan. 11, 1980; children: Augusto Ruben, Lucrecia Adriana. MS, U. Buenos Aires, 1979, Diploma in Biochemistry, 1980, PhD, 1982. Rsch. fellow endocrinology Nat. Coun. Scientific Rsch., Buenos Aires, 1979-81, postdoctoral rsch. fellow in endocrinology, 1981-83, established investigator, 1983-89; internat. fellow UCLA Sch. Medicine, 1985-88, asst. rsch. endocrinologist, 1988-91, asst. prof. medicine, 1991—. Author numerous scientific publs. Recipient Best Scientific Paper award Internat. Assn. Radiopharmacology, Chgo., 1981, Cross-Town Endocrine Soc., L.A., 1988. Mem. AAAS, Argentine Soc. Clin. Rsch., Am. Thyroid Assn. (travel award 1987), Endocrine Soc. (travel award 1984), Brain Rsch. Inst., Soc. Neurosci. Office: UCLA Dept Medicine/Endocrin Rsch Labs C-Lot Rm 104 Los Angeles CA 90024-1682

BOARDMAN, GREGORY DALE, environmental engineer, educator; b. Montpelier, Vt., Dec. 12, 1950; s. Theodore Robert and June Irene (Rogers) B.; m. Gail Cynthia Bedell, June 6, 1970 (div. Dec. 1986); children: Heather Eve, Kristina Marie, Jessica Anne; m. Shelley Ann Mitchell, Aug. 28, 1987; 1 child, Courtney Dale. MS, U. N.H., 1973; PhD, U. Maine, 1976. Registered profl. engr., Va. Asst. prof. civil engring. Va. Poly. Inst. and State U., Blacksburg, 1976-83, assoc. prof., 1983—; mem. bd. Dept. Commerce, Richmond, Va., 1987—; cons. to numerous cos., 1976—. Author 3 book chpts., 2 manuals; contbr. numerous papers to profl. jours. Chmn. Montgomery County Community Shelter, Va., 1986-91; mem. planning commn. Town of Blackburg, 1989-93. Rsch. grantee EPA, NIH, NOAA, Water Rsch., numerous others, 1976—. Mem. ASCE (mem. coms.), Soc. Environ. Toxicology and Chemistry, Water Environment Fedn., Internat. Assn. on Water Quality, Assn. Environ. Engring. Profs., Sigma Xi, Tau Beta Pi, Phi Kappa Phi. Achievements include research on industrial waste treatment and development of short-term toxicity tests. Office: Va Poly Inst and State U Dept Civil Engring 322 Norris Hall Blacksburg VA 24061

BOARDMAN, ROSANNE VIRGINIA, logistics consultant; b. Twin Falls, Idaho, Oct. 4, 1946; d. Gordon Ross and Garnet Othalia (Peterson) Tobin; m. Lowell Jay Boardman, May 12, 1973; 1 child, Christina Garnet. BA cum laude, Occidental Coll., 1968; MA with honors, Columbia U., 1969; postgrad., U. Calif., Irvine, 1971-72, U. Calif., L.A. and Santa Barbara, 1969, 73-74. Cert. jr. coll. tchr., Calif., cert. secondary tchg., Calif. Instr. U. Calif., Irvine, 1971-72, Ventura (Calif.) Community Coll., 1973-77; tech. writer Raytheon Svc. Co., Ventura, 1977-78; engring. analyst John J. McMullen Co., Ventura, 1978-80; sr. logistics specialist Raytheon Co., Ventura, 1977-78, 80-83; civilian tech. writer, editor USN, Port Hueneme, Calif., 1983-84, civilian logistics mgr., 1984-88; cons. Support Mgmt. Systems, Oxnard, Calif., 1988—. Author numerous manuals and logistics guides. Internat. fellow Occidental Coll., 1967; recipient Outstanding Performance award Naval Ship Weapon Systems Engring. Sta., 1985, 86. Mem. Soc. Logistics Engrs., Phi Beta Kappa.

BOBLETT, MARK ANTHONY, civil engineering technician; b. Beckley, W.Va., Jan. 21, 1959; s. Murriel Garner and Meredith Genevieve (Sheppard) B.; m. Susan Renee Walker, June 26, 1982; children: Miranda Lauren, Adrienne Lisbeth. AS in Civil Engring. Tech., W.Va. Inst. Tech., 1983, AS in Bldg. Constrn. Tech., 1983. Quality control technician Pittsburgh Testing Lab., Houston, 1984, Elmo Greer & Sons, Beckley, 1984-86; quality control coord. Green Constrn. Co., Beckley, 1986-88; assoc. agt. Nationwide Ins., Beckley, 1987-88; lab. mgr. Law Engring. Inc., Raleigh, N.C., 1988—. Baptist. Home: 111 Triple Crown Run Louisburg NC 27549 Office: Law Engring Inc 3301 Atlantic Ave Raleigh NC 27640-1695

BOCHKAREV, NIKOLAI GENNADIEVICH, astrophysics researcher; b. Moscow, May 19, 1947; s. Genneady Afanasievich and Elena Konstantinovna (Scherbakova) B.; m. Eugenia Alekseevna Karitskaya, Mar. 14, 1969; 1 child, Yury Nikolaevich. PhD in Math. Sci., Moscow State U., 1974, Dr. Math. Sci., 1988. Cert. astronomer, astrophysicist. Jr. researcher Sternberg Astronomical Inst., Moscow, 1974-86, sr. researcher, 1987-88, leading researcher, 1989—; prof., lectr. astrophysics dept. Moscow State U., 1975—; chmn. Soviet Working Group on Interstellar Matter, Moscow, 1979—; vis. French Govt. Meudon Obs, 1976, U. Calif., Berkeley, 1984, U. Wisc., Madison, 1991. Author: Magnetic Fields in Space, 1985, Local Interstellar Matter, 1990, Basic Physics of the Interstellar Matter, 1992; co-author: Engineering Calculations of Telescope Designing, 1983, 2nd. edit., 1985; editor: Physics of the Interstellar Matter, 1979; editor-in-chief jour. Astronomical and Astrophysical Transactions, 1991—, cons. editor Soviet Encyclopedia Pub., Moscow, 1983—. Mem. All-Union Astronom. Geodesical Soc., Internat. Astronom. Union (invited lectr. Argentina 1991), Soviet Astronom. Soc. (co-chmn. 1990—), European Astronom. Soc., Am. Astronom. Soc., Russian Union Sci. Socs. (v.p.). Achievements include discovery of x-ray emission from nebulae blown by stellar winds. Office: Sternberg Astronomical Inst, Universitetskij prosp 13, 119899 Moscow Russia

BOCHNER, BRUCE SCOTT, immunologist, educator. BA in Natural Scis. with honors, Johns Hopkins U., 1978; MD with honors, U. Ill., Chgo., 1982. Diplomate Am. Bd. Internal Medicine, Am. Bd. Allergy and Immunology. Intern, then resident in internal medicine U. Ill. Hosps., Chgo., 1982-85; fellow in clin. immunology Johns Hopkins U., Balt., 1985-88, instr. in medicine, 1988-89, asst. prof., 1989—; mem. immunology coun. Johns Hopkins U., vice chmn. seminars, 1993-94; lectr. in field. Author: (with others) Progress in Allergy and Clinical Immunology, 1989, 92, Late Phase Allergic Reactions, 1990, Focus on Pulmonary Pharmacology and Toxicology, 1990, Current Therapy in Allergy, Immunology, and Rheumatology, 1992; assoc. editor, mem. editorial bd. Jour. Allergy & Clin. Immunology, 1993—; contbr. articles to profl. jours.; reviewer various jours. New Investigator Rsch. grantee Am. Lung Assn., 1988-90; grantee NIH, 1989-94, 91—, Pfizer, Inc., 1991-93; recipient Developing Investigator award Asthma & Allergy Found. of Am., 1990, Developing Investigator award Burroughs Wellcome Fund, 1992. Mem. Am. Assn. Immunology, Am. Acad. Allergy and Immunology (Pres.'s Grant-in-Aid award 1987, Charles Reed lectureship 1993), Md. Asthma and Allergy Soc. (v.p. 1992-93), Alpha Omega Alpha Med. Honors Soc. Office: Johns Hopkins U Sch Med 720 Rutland Ave Baltimore MD 21205*

BOCKHOP, CLARENCE WILLIAM, retired agricultural engineer; b. Paullina, Iowa, Mar. 28, 1921; s. Fred Henry and Sophie Dorothea (Laue) B.; m. Virginia Buhman, July 9, 1949; children—Barbara Lucille, Nancy Jeanne, Bryan William, Karl David. B.S. in Agrl. Engring, Iowa State U., 1943, M.S. in Agrl. Engring, 1955, Ph.D. in Agr. Engring. and Theoretical and Applied Mechanics, 1957. Mgr. service and edn. Stewart Co., Dallas, 1948-53; mem. faculty Iowa State U., Ames, 1953-57, 60-80, prof. agrl. engring., 1960-80, head dept. agrl. engring., 1962-80; prof., head dept. agrl. engring. U. Tenn., 1957-60; head dept. agrl. engring. Internat. Rice Research Inst., Los Banos, The Philippines, 1980-86; vis. prof. U. Ghana, 1969-70. Gen. reporter, VIth Internat. Congress Agrl. Engring., Lausanne, Switzerland, 1964; Author articles in field. Served to capt. AUS, 1943-48. Fellow Am. Soc. Agrl. Engrs. (chmn. Tenn. sect. 1958-59, chmn. mid-central sect. 1960-61, chmn. Iowa sect. 1963-64, chmn. edn. and research div. 1966-67, dir. 1973-75); mem. Am. Soc. Engring. Edn. (chmn. agrl. engring. div. 1966-67), Sigma Xi, Gamma Sigma Delta, Phi Kappa Phi, Phi Mu Alpha, Tau Beta Pi. Lutheran. Address: 424 Hide-A-Way Ln E Lindale TX 75771

BOCKIAN, JAMES BERNARD, computer systems executive; b. Jersey City, Sept. 16, 1941; s. Abraham and Evelyn (Skner) B.; m. Donna M. Hastings; children: Vivian Shifra, Adrian Adena, Lillian Tova. BA, Columbia U., 1953; MPA, U. Mich., 1955; MA, Yale U., 1957. Vice-consul, 3d sec. Embassy Dept. State, Washington, 1957-61; sr. systems analyst J.C. Penney Co., N.Y.C., 1961-67; mgr. systems svcs., head dept. systems projects McDonnell Douglas Automation Co., East Orange, N.J., 1967-76; prin. JBBA, Inc. (formerly James B. Bockian & Assocs., Inc.), Morristown, N.J., 1976—; v.p. MIS Thomas Cook, Inc., 1980-83, exec. cons. to Thomas Cook Group; lectr. in field. Author: Management Manual for Systems Development Projects, 1979, Project Management for Systems Development, 1981, AT&T User Guide to Information Systems Development, 1980; contbr. treatises and articles to profl. publs. Mem. Internat. Assn. Cybernetics, Assn. Computing Machinery, Data Processing Mgmt. Assn., Am. Mgmt. Assn., Systems and Procedures Assn., Yale Club (N.Y.C.). Home: 26 Farmhouse Ln Morristown NJ 07960-3022 Office: JBBA Inc Olde Forge E Ste 26-5B Morristown NJ 07960-3022

BOCKIUS, THOMAS JOHN, mechanical engineer; b. Wilmington, Del., June 16, 1968; s. John Cochran and Isabella (Shields) B. BS in Mech. Engring., U. Del., 1990. Project engr. Ametek Inc. Haveg Divsn., Wilmington, 1991-92; process engr. Ametek Inc. Specialty Metals, Eighty Four, Pa., 1992-93; project engr. Plymouth products divsn. Ametek Inc., Sheboygan, Wis., 1993—. Home: 2227 Terrace View Dr A2-D Sheboygan WI 53081 Office: Ametek Inc Plymouth Products Divsn 502 Indiana Ave Sheboygan WI 53081

BOCKSERMAN, ROBERT JULIAN, chemist; b. St. Louis, Dec. 20, 1929; s. Max Louis and Bertha Anna (Kremen) B.; m. Clarice K. Kreisman, June 9, 1957; children: Michael Jay, Joyce Ellen, Carol Beth. BSc, U. Mo., 1952, MSc, 1955. Chemist Sealtest Corp., Peoria, Ill., 1955-56; prodn. mgr. Allan Drug Co., St. Louis, 1957-59; rsch. chemist Monsanto Co., St. Louis, 1960-65; purchasing agt. Monsanto Co., Sauget, Ill., 1966-67; founder, pres. Pharma-Tech Industries, Inc., Union, Mo., 1967-84; tech. dir. Overlock-Howe Consulting Group, St. Louis, 1984-85; founder, pres. Conatech Consulting Group, Chesterfield, Mo., 1985—; sec., mem. industry packaging adv. com. Sch. of Engring., U. Mo., Rolla, 1979—; adj. prof. dept. food sci./nutrition U. Mo., Columbia; adj. prof. dept. engring. mgmt. U. Mo., Rolla; vis. lectr. U. Mo., Clayton, Northwestern U., Evanston, Ill., and various programs. Tech. reviewer Jour. Inst. of Packaging Profls., Jour. Packaging Tech. Mem. Mo. Waste Control Coalition; mem. stormwater engring. com. City of Creve Coeur, Mo. With U.S. Army, 1952-54, Korea. SBIR grantee. Mem. ASTM, Cons. Packaging Engring. Coun., Inst. Packaging Profls., Am. Technion Soc., Inst. Food Technologists Arrangements (St. Louis), Nat. Forensic Ctr., Teltech Resource Network, Sigma Xi. Achievements include research on toxicological effects of additives from packaging materials upon foodstuffs, on biological and photo degradation of polymers, on technology of form/fill/seal packaging engineering. Home: 54 Morwood Ln Saint Louis MO 63141-7621 Office: Conatech Cons Group 744-7 Spirit of Saint Louis Blvd Chesterfield MO 63005-1024

BOCKWOLDT, TODD SHANE, nuclear engineer; b. Spirit Lake, Iowa, July 31, 1967; s. Larry Ray and Gale Glee (Bobzien) B.; m. Margery Pitzer, June 9, 1990. BS in Nuclear Engring., Ga. Tech, 1989, MS in Nuclear Engring., 1990. Grad. rsch. asst. Ga. Inst. Technology, Atlanta, 1989-90; S5W (submarines) and A1G (carriers) fleet reactor engr. DOE/USN Naval Reactors Hdqrs., Arlington, Va., 1990—; tech. program chmn. Am. Nuclear Soc. Student Conf., Atlanta, 1988; nuclear engring. rep. Mech. Engring. Student Adv. Com., Ga. Tech, 1988-89. Lt. USN, 1990—. Scholar NROTC, 1985-89, MCDAC, 1985-89, Am. Soc. Naval Engrs. scholar, 1987-89; recipient Gold medal Soc. Am. Mil. Engrs., 1988, Ga. Tech Honor award Soc. Am. Mil. Engrs., 1989, Outstanding Coll. Students of Am. award, 1989. Mem. Am. Nuclear Soc. (grad. scholar 1989-90), Tau Beta Pi, Mensa, Alpha Nu Sigma. Lutheran. Home: 7113 Vantage Dr Alexandria VA 22306-1251

BOCVAROV, SPIRO, aerospace and electrical engineer; b. Stip, Macedonia, Aug. 15, 1962; came to U.S. 1988; s. Risto and Elena (Andonovic) B. BS, U. Beograd, Yugoslavia, 1985, MS, 1988; PhD in Aerospace Engring., Va. Poly. Inst. and State U., 1991. Rsch. engr. Inst. Mihailo Pupin, Beograd, 1985-88; grad. asst. aerospace/ocean engring. dept. Va. Poly. Inst. and State U., Blacksburg, 1988-91; rsch. assoc. aerospace/ocean engring. dept. Ctr. for Applied Math., Va. Poly. Inst. and State U., Blacksburg, 1991—. Contbr. articles to profl. jours. Yugoslav Acad. Sci. fellow, 1986-87. Mem. AIAA (Young Profl. Paper Competition award), IEEE, Tau Beta Pi. Achievements include research in methods for analysis of optimal tactical maneuvering problems for supermaneuverable combat aircraft; real-time control system design. Office: Va Poly Inst and State U Interdisciplinary Ctr Applied Math Blacksburg VA 24061-0531

BOCZKAJ, BOHDAN KAROL, structural engineer; b. Kowel, Poland, Nov. 14, 1930; came to U.S., 1973; s. Walenty and Anna (Sarnecka) B.; m. Teresa Marcela Bioniosek, Aug. 23, 1955; 1 child, Boleslaw. MS in Civil Engring., Tech. U. of Silesia, Gliwice, Poland, 1962; PhD, Tech. U. Lodz (Poland), 1969. Registered profl. engr., Pa. Asst. prof. Tech. U. of Silesia, 1969-73, Tech. U., Rzeszow, Poland, 1970-71; sr. engr. Dravo Engrs., Pitts., 1973-83; vis. prof. Birzeit U., West Bank, Israel, 1984-86; prin. engr. Schneider Engrs., Bridgeville, Pa., 1986-88; design engr. Rust Internat., Pitts., 1988-90; prin. engr. S.E.I. Engrs. and Cons., Pitts., 1990-91; cons., 1992—. Contbr. articles on prestressed concrete and theory of plates, concret fatigue, structure on subsiding area to profl. jours. Teaching grantee Fulbright Found. Coun. for Internat. Exch. of Scholars, Birzeit U., 1985-86. Mem. ASCE, Am. Concret Inst. Roman Catholic. Achievements include patent on Coke Oven Machinery. Home: 728 Riehl Dr Pittsburgh PA 15234-2511

BODDU, VEERA MALLU, chemical engineering researcher; b. Vetapalem, India, July 12, 1958; came to U.S., 1982; s. Chenchu Nadam and Venkata Subbamma (Doguparthi) B.; m. Prabha Vathi Doguparthi, Aug. 14, 1981; children: Sivali Chenchu, Nathan Kailas. MTech, Indian Inst. Tech., Kanpur, 1982; PhD, U. Mo., 1988. Registered engr.-in-tng., Mo. Teaching asst. dept. chem. engring. U. Mo., Columbia, 1982-87, rsch. asst. environ. labs., 1987-88, rsch. assoc. in chem. engring., 1988-91; rsch. engr. U.S. Army C.E., Champaign, Ill., 1991—. Contbr. articles to Jour. Am. Ceramic Soc., Jour. Chem. Thermodynamics, Jour. Chem. and Engring. Data, Chem. Physics, Jour. Water Pollution Control Fedn., Jour. Envring. Engring. Govt. scholar Andhra Pradesh State, India, 1975-80; fellow Indian Inst. Tech., 1980-82. Mem. AICE, Am. Chem. Soc., Am. Ceramic Soc., Sigma Xi. Hindu. Achievements include development of nomograph for estimating thermal conductivity of organic liquids; method for extracting pesticides from biological tissues and soils using supercritical fluid extraction; obtained virial coefficients and pressure-volume-temperature properties for hydrogen, carbon dioxide, carbon monoxide at temperatures 50-150 degrees C and pressures up to 65 bars; binder removal in electronic ceramics through steam oxidation. Office: US Army CE 2902 Newmark Dr Champaign IL 61826-9005

BODE, CHRISTOPH ALBERT-MARIA, cardiology educator, researcher; b. Cologne, Germany, Aug. 15, 1955; s. Johannes and Bétrice (Engel) B.; m. Birgit Susanne Halle, May 24, 1983; children: Michael, Martin, Christine. MD, Cologne Med. Sch., 1981. Rsch. fellow Biomed. Inst. Cologne Med. Sch., 1982-83; rsch. fellow Mass. Gen. Hosp./Harvard U. Med. Sch., Boston, 1984-86; clin. fellow cardiac unit U. Heidelberg, 1986, assoc. prof., 1992—; internist Heidelberg (Fed. Republic of Germany) Med. Sch., 1991—; Author books and articles; inventee in field. Mem. Lebensrecht Alle, Munich, 1986—. With Germany Army, 1981-82. Internat. Soc. for Thrombosis and Hemostasis grantee, 1989. Mem. Am. Heart Assn. (thrombosis coun.), German Soc. Heart and Circulation, German Soc. Internal Medicine, German Soc. Thrombosis and Hemostasis. Roman Catholic. Avocations: travel, tennis. Office: U Heidelberg Med Clinic III, Bergheimerstrasse 58, Heidelberg 69115, Germany

BODINE, PETER VAN NEST, biochemist; b. Syracuse, N.Y., Mar. 14, 1958; s. George Edward and Mary Rachel (Goeble) B.; m. Judith Marie La Londe, Dec. 27, 1986. BS in Biology, Syracuse U., 1980; PhD in Biochemistry, Temple U., 1988. Rsch. asst. Syracuse U., Syracuse, 1980-82; postdoctoral fellow Fels Rsch. Inst., Phila., 1988-91; asst. prof. pharmacology Thomas Jefferson U., Phila., 1991-92; fellow Mayo Clinic & Found., Rochester, Minn., 1992—; mem. adj. faculty Phila. Coll. Osteopathic Medicine, Phila., 1991. Contbr. articles to profl. jours. Coach Syracuse Chargers Track Club, 1977-80; deacon Park Central Presbyn. Ch., Syracuse, 1980-82. Swern fellow Temple U., 1985; recipient Freedman award Temple U., 1984, grad. rsch. award N.Y. Acad. Sci., N.Y.C., 1988; Kendall-Mayo fellow, 1992—. Mem. Am. Assn. for Cancer Rsch., The Endocrine Soc., Am. Soc. for Biochemistry & Molecular Biology, Sigma Xi. Democrat. Achievements include identification of new biological regulator of steroid hormone receptors and protein kinase C. Office: Mayo Clinic & Found 200 First St SW Rochester MN 55905

BODINGTON, CHARLES E., chemical engineer; b. Alameda, Calif., Aug. 23, 1930; s. Harold Pierce and Mercedes Veronica (Jackson) B.; m. Helen Champlin Lohman, June 28, 1952; children: Jeffrey, Celia. BS, Stanford U., 1952; MS, MIT, 1954. With Chevron Rsch. Co., Richmond, Calif., 1954-86, sr. engring. assoc., 1968-86; ind. cons. San Anselmo, Calif., 1986—. Contbr. articles on gasoline mfg. and petroleum industry to profl. jours. Mem. Am. Inst. Chem. Engrs. (session chair 1989, 90, 91, 93). Achievements include patents on chem. recovery and dewaxing process. Office: PO Box 275 San Anselmo CA 94979

BODKIN, LAWRENCE EDWARD, inventor, research development company executive, gemologist; b. Sapulpa, Okla., May 17, 1927; s. Clarence Elsworth and Lillie (Moore) B.; m. Ruby Emma Pate, Jan. 15, 1949; children: Karen Bodkin Snead, Cinda, Lawrence Jr. Student, Fla. State U., 1947-50; grad., Gemological Inst., 1969. Chief announcer, program dir., mgr. various radio stations, Winter Haven, Fla., Tallahassee and Jacksonville, Fla., 1947-60; ind. jewelry salesman and appraiser Underwood Jewelers, 1961-87; pres. Bodkin Jewelers and Appraisers, Jacksonville, 1984—, Telanon, Jacksonville, 1981—, Bodkin Co., Jacksonville, 1974—; chmn., chief exec. officer Bodkin Corp., Jacksonville, 1975—; dir. safety, R&D Innovative Designer Products div. Brooke Shields Beauty Care, Kendall Park, N.J., 1989-92; cons. gem and mineral groups, Jacksonville, 1960—, numerous corps. and industries (on inventions); lectr. in field. Author: Dual Imagery of Ultra Speed Bodies, 1971, Miniatures, 1976; contbr. articles to sci. publs.; inventor Universal-Fault Circuit-Interrupter, 1973, TIP (tested immersion protection), 1992; holder more than 15 patents. Mem. Jacksonville Mus. Sci. and Hist., 1981—, Jacksonville Symphony Assn., 1985—, Cummer Gallery Art, Jacksonville, 1985—. Served with U.S. Army, 1945-47, ETO. Mem. Am. Gem Soc. (cert.), Fla. State U. Alumni Assn, Mensa Internat. Clubs: San Jose Country (Jacksonville); Ponte Vedra Country (Fla.). Avocations: fossil collecting, beach combing, philosophy, writing, theoretical physics. Home: 1149 Molokai Rd Jacksonville FL 32216-3273

BODLAJ, VIKTOR, electrical engineer; b. Kranj, Slovenia, Apr. 19, 1928; arrived in Germany, 1959; s. Josef and Katherina (Gros) B.; m. Maria Ravnikar, Apr. 7, 1956 (div. Aug. 1964); 1 child, Darij; m. Theodora Maria Rogac, Jan. 21, 1965; 1 child, Robert Roland. Diploma in engring., U. Ljubljana, Slovenia, 1956; DSc, U. Munich, 1970. Rsch. engr. Inst. Electronics, Ljubljana, 1956-59; devel. engr. Rohde & Schwarz, Munich, 1959-60; devel. engr. Cen. Devel. Lab., Siemens AG, Munich, 1960-65, rsch. engr. corp. rsch., 1965-71, sci. mgr., 1971-75, sr. sci. mgr., 1975—; inventor, 1986—. Co-author: Industrial Applications of Laser; contbr. articles on laser tech. and electronics to sci. jours.; patentee on laser tech. and electronics in Germany and fgn. countries. Roman Catholic. Avocations: inventing, popular science, jogging. Home: Hans-Schweikart Strasse 14, D-81739 Munich Germany Office: Siemens AG Corp Rsch, Otto-Hann-Ring 6, D-81739 Munich Germany

BODMER, WALTER FRED, cancer research administrator; b. Frankfurt-am-Main, Fed. Republic of Germany, Jan. 10, 1936; s. Ernest Julius and Sylvia Emily B.; m. Julia Gwynaeth Pilkington, Aug. 11, 1956; children: Mark William, Helen Clare, Charles Walter. BA, U. Cambridge, Eng., 1956, PHD, 1959; laurea honoris causa, U. Bologna, Italy, 1987; DSc (hon.), U. Oxford, 1988, U. Bath, Eng., 1988, U. Edinburgh, 1990, U. Surrey, 1990, U. Bristol, 1991; Dr. honoris causa, U. Leuven, 1992; LLD (hon.), U. Dundee, 1993; DSc (hon.), U. Loughborough, 1993. Rsch. fellow Clare Coll., U. Cambridge, Eng., 1958-60; fellow Clare Coll., U. Cambridge, 1961; demonstrator Dept. of Genetics, U. Cambridge, 1960-61; vis. asst. prof. to prof. Dept. of Genetics Stanford U. Sch. of Medicine, Palo Alto, Calif., 1961-70; prof. Dept. Genetics U. Oxford, Eng., 1970-79; dir. rsch. Imperial Cancer Rsch. Fund, London, 1979-91, dir. gen., 1991—; hon. fellow Keble Coll., Oxford, 1981, Clare Coll., Cambridge, 1989. Co-author (with others) The Genetics of Human Populations, 1971, Our Future Inheritance - Choice or Chance?, 1974, Genetics Evolution and Man, 1976; contbr. numerous articles to profl. jours. Recipient the William Allen Memorial award, Am. Soc. Human Genetics, 1980, The Conway Evans Prize, the Royal Coll. of Physicians, 1982, the Rabbi Shai Shacknai Meml. lectureship in immunology and cancer rsch., 1983, the John Alexander Meml. prize and lectureship , U. Pa. Med. Sch., 1984, The Rose Payne Disting. Scientist lectureship, Am. Soc. for Histocompatability and Immunogenetics, 1985, Ellison Cliffe lecture and medal, Royal Soc. Medicine, 1987; named Knight Batchelor, 1986, many others. Fellow Royal Soc., Royal Coll. of Pathologists, Royal Coll. of Surgeons (hon.), Royal Coll. Physicians (hon.), Internat. Inst. Biotechnology; mem. Acad. Europea, The Assn. for Sci. Edn. (pres. 1989-90), Brit. Assn. for the Advancement of Sci. (pres. 1987-88, v.p. 1989—), Brit. Soc. for Histocompatibility and Innunogenetics (pres. 1990-91), Am. Acad. Arts and Scis. (foreign hon. mem.), U.S. Nat. Acad. Sci. (assoc.), Am. Assn. Immunologists, Czechoslovak Acad. Scis., Human Genome Org. (pres. 1990-92). Home and Office: Imperial Cancer Rsch Fund, PO Box 123 Lincoln's Inn Fields, London WC2A 3PX, England

BODOR, NICHOLAS STEPHEN, medicinal chemistry researcher, educator, consultant; b. Satu Mare, Transylvania, Romania, Feb. 1, 1939; came to U.S., 1968, naturalized, 1976; s. Miklos Sandor and Berta (Horvath) B.; m. Gyongyver Gorog, Sept. 21, 1961 (div. 1971); 1 son, Miklos; m. Sheryl Lee Reimann, Nov. 26, 1971; children: Nicole, Erik. BS, MS in Organic Chemistry, Bolyai U., Cluj, Romania, 1959; D. Chemistry, Babes-Bolyai U., Supreme Council of Acad. Sci., Bucharest, Romania, 1965; DSc (hon.), Tech. U., Budapest, 1989, Med. U., Debreseu, 1990. Prin. investigator Chem. and Pharm. Research Inst., Cluj, 1961-68, 69-70; R.A. Welch postdoctoral fellow U. Tex., Austin, 1968-69, 70-72; sr. research scientist Alza Co., Lawrence, Kans., 1972-73; dir. medicinal chemistry INTERx Research Corp., Lawrence, Kans., 1973-79; adj. prof. U. Kans., 1974-79; prof. medicinal chemistry, chmn. dept. medicinal chemistry U. Fla., Gainesville, 1979-83, grad. research prof. medicinal chemistry, 1983—, chmn. dept., 1989—; cons. to pharm. cos.; v.p., dir. research Pharmatec, Inc., 1983—; dir. Ctr. for Drug Design and Delivery, 1986—. Author numerous publs. in field; N.Am. editor Jour. Biopharm. Sci.; editorial bd. several jours. Named Fla. Scientist of Yr., 1984; grantee NIH, 1976—. Fellow Acad. Pharm. Sci., Am. Assn. Pharm. Sci. (Rsch. Achievement award 1989), AAAS, Hung Chem. Soc. (hon.), Panhellenic Soc. Pharmacy (hon.); mem. Am. Chem. Soc., Am. Pharm. Assn. (Rsch. Achievement award 1989). Home: 6219 SW 93rd Ave Gainesville FL 32608-6305 Office: U Fla Dept Medicinal Chemistry PO Box 485J Gainesville FL 32602-0485

BOEHM, BARRY WILLIAM, computer science educator; b. Santa Monica, Calif., May 16, 1935; s. Edward G. and Kathryn G. (Kane) B.; m. Sharla Perrine, July 1, 1961; children: Romney Ann, Tenley Lynn. BA, Harvard U., 1957; PhD, UCLA, 1964. Programmer, analyst Gen. Dynamics, San Diego, 1955-59; head infosci. dept. Rand Corp., Santa Monica, 1959-73; chief scientist TRW Def. Systems Group, Redondo Beach, Calif., 1973-89; dir. infosci. and tech. office Def. Advanced Rsch. Agy. Dept. Def., Arlington, Va., 1989-92, dir. software and computer tech. office, dir. def. rsch. and engring., 1992; prof. software engring. U. So. Calif., L.A., 1992—; co-chmn. Fed. Coordinating Coun. Sci., Engring. and Tech. High Performance Computing WG, Washington, 1989-91; chmn. DOD Software Tech. Plan WG, Arlington, 1990-92, NASA G & C/Infosystems Adv. Com., Washington, 1973-76; guest lectr. USSR Acad. Sci., 1970. Author: ROCKET, 1964, Software Engineering Economics, 1981; co-author: Characteristics of Software Quality, 1978, Software Risk Management, 1989; co-editor: Planning Community Information Utilities, 1972. Recipient Wariner prize Soc. Software Analysts, 1984, Freiman award Internat. Soc. Parametric Analysts, 1988, Award for Excellence Office of Sec. of Def., 1992. Fellow AIAA (chair TC computers 1968-70, Information Systems award 1979), IEEE (gov. bd. computer sci. 1981-82, 86-87). Office: U So Calif Computer Sci Dept Los Angeles CA 90089-0781

BOEHM, ERIC WALTER ALBERT, molecular mycology researcher; b. Carthage, Tunisia, May 6, 1960; s. Walter William and Clotilde (Amzalag) B. BS, U. Calif., Santa Cruz, 1986; MS, U. Minn., 1988, PhD, 1992. Postdoctoral assoc. Inst. Fla. Agricultural Sciences U. Fla., Gainesville, 1992—. Contbr. articles to profl. publs. Mem. Am. Phytopathological Soc., Mycological Soc. Am., Sigma Xi. Democrat. Buddhist. Achievements include combined usage of epifluorescence microscopy, transmission electron microscopy and pulsed field gel electrophoresis to characterize and quantify phytopathogenic fungal chromosome numbers. Home: 1324-D SW 13th St Gainesville FL 32608 Office: U Fla IFAS Plant Pathology 1453 Fifield Hall UF Gainesville FL 32611

BOEHM, GÜNTHER, pediatrician; b. Gerstungen, Germany, Oct. 24, 1946; s. Heinz Werner and Ingeborg (Fräbel) B.; m. Margaret Kessner, Apr. 7, 1968 (div. 1980); children: Andreas, Steffen; m. Heidi Dippe, July 2, 1981; 1 child, Alexander. MD, U. Leipzig, Fed. Republic of Germany, 1972, PhD, 1986. Pediatrician U. Leipzig, Fed. Republic of Germany, 1972-86, sr. physician dept. neonatology, 1986-92; docent of pediatrics U. Leipzig, Germany, 1989-92; guest prof. Gondar (Ethopia) Coll. Med. Scis., 1981-82; vis. prof. Nat. Rsch. Coun. Italy Inst. for Infant Nutrition, Milan, 1992—. Patentee in field; contbr. numerous articles to profl. jours. Recipient prize European Assn. Perinatal Medicine, 1986, Virchow prize, 1989; neonatology fellow U. Milan, 1987, 90, 91, 92; pediatrics fellow U. Lund, Sweden, 1984, 86-92; WHO fellow. Mem. Pediatric Soc. Germany, Soc. Perinatal Medicine of Germany, European Soc. Pediatric Gastroenterological Nutrition, German Soc. Pediatric Gastroenterological Nutrition. Office: Inst Infant Nutrition, Via Macedonio Melloni 52, 1-20129 Milan Italy

BOEHM, LAWRENCE EDWARD, psychologist, educator; b. Columbus, Ohio, Sept. 19, 1961; s. Lawrence Emil and JoAnn (Frey) B.; m. Annette Cooper, June 22, 1985; children: Lauren Elizabeth, Alexander Matthew. BS, Ohio State U., 1984; PhD, Ohio U., 1988. Rsch. psychologist U.S. Bur. Labor Statistics, Washington, 1988-89; asst. prof. psychology Thomas More Coll., Crestview Hills, Ky., 1989—. Contbr. articles to Jour. Exptl. Social Psychology, Jour. Behavioral Decision Making, Proceedings of Am. Statis. Assn. Mem. Am. Psychol. Soc., Midwestern Psychol. Assn., Coun. Tchrs. Undergrad. Psychology. Achievements include research in repetition and perceived validity, implicit memory, frequency judgments, decision-making. Office: Thomas More Coll Dept Psych 333 Thomas More Pkwy Crestview Hills KY 41017

BOEKELHEIDE, VIRGIL CARL, chemistry educator; b. Cheslea, S.D., July 28, 1919; s. Charles F. and Eleonor (Toennies) B.; m. Caroline Barrett, Apr. 7, 1924; children: Karl, Anne, Erich. AB magna cum laude, U. Minn., Mpls., 1939, PhD, 1943. Instr. U. Ill., Urbana, 1943-46; asst. prof. to prof. U. Rochester, 1946-60; prof. dept. chemistry U. Oreg., Eugene, 1960—. Contbr. articles to profl. jours. Recipient Disting. Achievement award U. Minn., 1967; recipient Alexander von Humboldt award W.Ger. Govt., 1974, 82, Centenary Lectureship Royal Soc. G.B., 1983, Coover award Iowa State U., 1981; Disting. scholar designate U.S.-China Acad. Sci., 1981; Fulbright Disting. prof. Yugoslavia, 1972. Mem. NAS, Pharm. Soc. Japan (hon.). Home: 2017 Elk Ave Eugene OR 97403-1788 Office: U Oreg Dept Chemistry Eugene OR 97403

BOER, F. PETER, chemical company executive; b. 1941. AB, Princeton U., 1961; PhD, Harvard U., 1965. With Tex Div. Lab. Dow Chem. Co., 1965-78, dir; v.p., mgr. R & D Am. Can Co., 1978-83; v.p., pres. rsch. div., corp. tech. group W.R. Grace & Co., from 1983; sr. v.p., until 1989, now exec. v.p. corp. tech. group. Office: W R Grace & Co 1 Town Ctr Rd Boca Raton FL 33486-1010

BÖER, KARL WOLFGANG, physicist, educator; b. Berlin, Mar. 23, 1936; came to the U.S., 1961; s. Karl and Charlotte (Gruhlke) B.; m. Renate Schröder, May 18, 1967; children: Ralf Reinhard, Katarina Karlotta. Dipl. physics, Humboldt U., 1949, dr. rer. nat., 1952, dr. rer. nat. habil., 1955. Docent Humboldt U., Berlin, 1955-58, prof., chair, 1958-61; prof. physics U. Del., Newark, 1962-71, prof. physics and engring., 1971-92, disting. prof. physics and solar energy, 1993—, dir. inst. energy, 1972-75; chmn. bd. SES, Inc., Newark, 1972-81. Author: Survey of Physics, Vols. I and II, 1990, 92; editor: Advances of Solar Energy, Vols. I-VII, 1980-92; contbr. articles to profl. jours. Fellow Am. Phys. Soc.; mem. IEEE (sr.), Am. Solar Energy Soc. (pres. 1976-77, Abbott award 1981). Achievements include patents for solid state devices; first to measure Franz-Keldysh effect; first proposed Bose-Einstein condensation of excitons. Home: 239 Bucktoe Hills Rd Kennett Square PA 19348 Office: Univ Del Material Sci Newark DE 19716

BOERI, RENATO RAIMONDO, neurologist; b. Milan, Italy, May 15, 1922; s. Giovanni Battista and Pierina (Martinelli) B.; m. Cini Mariani, Sept. 14, 1950 (div. 1969); children: Sandro, Stefano, Tito; m. Maria Grazia Casiraghi, Sept. 15, 1978. Liceo, Beccaria, Milan, 1940; M.D., U. Milan, 1947. Diplomate Neurology and Psychiatry, Univ. Milan, 1951. Intern, resident Istituto Neurologico Carlo Besta, Milan; clin. and sci. dir. Instituto Neurologico Carlo Besta, Milan, 1977-87; pres. Consulta di Bioetica, 1990—. Editor Italian Jour. Neurol. Scis., 1980-93. Mem. Italian Soc. Neurology (v.p. 1982), Am. Acad. Neurology, Societe Francaise de Neurologie, N.Y. Acad. Scis. Home: Corso Porta Nuova 20, 20121 Milan Italy Office: Via Sant'Andrea N5, 20121 Milan Italy

BOERMA, HENRY ROGER, agronomist, educator. Prof. agronomy U. Ga., Athens. Recipient Agronomic Achievement award Am. Soc. Agronomy, 1992. Office: Univ of Georgia Athens GA 30602*

BOESE, MARK ALAN, forensic scientist, chemist, educator; b. Chgo., Sept. 26, 1960; s. Robert Alan and June Carol (Franke) B.; m. Deborah I. Sullivan, June 25, 1983; 1 child, Anthony Robert. BS, U. Ill., Chgo., 1984. Lab. technician B&W Cons. Forensic Chemists, Inc., Downers Grove, Ill., 1979-84, chemist, analyst, 1984-87, v.p., 1984—, tech. coord., 1987—; mem. Ill. Adv. Com. on Arson Prevention, 1984—; field instr. Fire Svcs. Inst., U. Ill., Champaign, 1988—; guest lectr. profl. orgns. Contbr. chpt. to book; author, presenter scientific papers on toolmarks, forensic chemistry, thermometry and incendiary devices. Pres. high adventure explorer post Chgo. Area coun. Boy Scouts Am., 1978-79, post advisor, 1979-82, 1st mate Sea Explorer Ship, 1981-82; supr. Ray Graham Assn. for Handicapped, Autism Program, 1987-88; aide Ray Graham Family Summer Camp, George Williams Coll., Lake Geneva, Wis., 1987, supr., 1988; coach No. Ill. Regional Spl. Olympics, 1987. Mem. ASTM, Am. Chem. Soc., Am. Acad. Forensic Scis., Am. Soc. for Metals, Am. Inst. Chemists, Internat. Assn. Identification (assoc.), Internat. Assn. Arson Investigators (forensic sci. com., 1986-90, chmn. forensic sci. and engring. com. Ill. chpt. 1986—), Evidence Photographers Internat. Coun., Internat. Fire Photographers Assn., Internat. Union Pure and Applied Chemistry (affiliate), Chgo. Chromatography Discussion Group, Assn. Firearm and Tool Mark Examiners. Achievements include design of a custom adaptation of a gas chromatographic system for the analysis of LP and natural fuel gases. Office: B&W Cons Forensic Chemists 2901 Finley Rd Ste 101 Downers Grove IL 60515-1041

BOESE, ROBERT ALAN, forensic chemist; b. Chgo., Mar. 30, 1934; s. Fred W. and Adeline B. (Kondrad) B.; m. June C. Franke, Dec. 10, 1955; children—Mark A., Brian A. A.A. in Chemistry, Wright Jr. Coll., 1960; B.S. in Chemistry, Ill. Inst. Tech., 1969, M.P.A., 1974. Patrolman Chgo. Police Dept., 1956-58, investigator crime lab., 1958-60, firearms examiner, 1960-63, sr. firearms examiner, 1963-65, chemist, crime lab., 1965-71, chief chemist, 1971-74, tech. coordinator, 1974-83, asst. dir. crime lab. div., 1983-86, ret., 1986; now pres. B and W Cons. Forensic Chemists, Inc.; prof. U. Ill.-Chgo.; mem. Ill. Adv. Com. on Arson Prevention; mem. part-time faculty St. Xavier Coll., 1978-79; lectr. Northwestern U., Ill. Inst. Tech., U. Notre Dame, Loyola U., Chgo., Chgo. Police Acad.; state instr. in use of breathalyzer; mem. Task Force for Evaluation of Ill. Crime Lab. System, 1974-75; mem. Ill. Arson Adv. Com. Expert witness in firearms identification and forensic chemistry mcpl., county, fed. cts. Contbr. articles to profl. jours. Fellow Am. Acad. Forensic Scis. (criminalistics sect.); mem. Am. Chem. Soc., Internat. Assn. Arson Investigators (chmn. forensic lab. com. Ill. chpt.), Midwestern Assn. Forensic Scientists (founder, pres. 1976-77), Am. Firearms and Tool Marks Examiners (charter), Chgo. Gas Chromatography Discussion Group. Lodges: Masons, Shriners. Home: 5657 S Mason Ave Chicago IL 60638-3604 Office: B&W Cons Forensic Chemists Inc 2901 Finley Rd Ste 101 Downers Grove IL 60515-1041

BOFILL, RANO SOLIDUM, physician; b. Panay, Capiz, The Philippines, Mar. 14, 1942; came to U.S., 1969; s. Saturnino Bernas and Consoladora (Solidum) B.; m. Judy Libo-On, May 27, 1972; children: Lora, Rano Libo-on II, Mariju. MD, U. Sto. Tomas, Manila, 1966. Pvt. practice Romney, W.Va., 1973-79; resident radiology Episcopal Hosp., Phila., 1980-82; chief resident radiology Germantown Hosp., Phila., 1982-83; assoc. radiologist Pleasant Valley Hosp., Point Pleasants, W.Va., 1984-89; chmn. credentials com., pres. exec. com., chmn. A.R. Hosp., Man, W.Va., 1984—, pres. med. staff, 1984—. Editor: newsletter, 1984—. Coord. health topic com. A.R. Hosp., Man, 1988. Lt. col. USAR, 1989. Fellow Am. Acad. Family Physicians, Am. Coll. Internat. Physicians (trustee 1985-87, treas. 1987—), Internat. Coll. of Surgeons; mem. Assn. Philippine Physicians Am. (del. 1989—), W.Va. Am. Coll. Internat. Physicians (founding pres. Point Pleasant 1984-85), Philippine Med. Assn. W.Va. (pres. Man 1989—). Roman Catholic. Avocations: musical, entertaining nursing homes, traveling. Home: 309 W Avis Ave Man WV 25635-1132

BOGDAN, JAMES THOMAS, secondary education educator, electronics researcher and developer; b. Kingston, Pa., Aug. 14, 1938; s. Fabian and Edna A. (Spray) B.; m. Carolyn Louetta Carpenter, May 5, 1961; 1 child, Thomas James. BS in Edn., Wilkes U., Wilkes-Barre, Pa., 1960. Cert. chemistry and physics tchr., Calif. Tchr. Forty Fort (Pa.) Sch. Dist., 1960-63; tchr., chmn. sci. dept. L.A. Unified Sch. Dist., 1963—; owner, mgr. Bogdan Electronic Rsch. & Devel., Lakewood, Calif., 1978—; cons. Lunar Electronics, San Diego, 1978-83, T.E. Systems, L.A., 1988-89. Author, pub. The VHF Reporter newsletter, 1967-76. Tng. officer L.A. County Disaster Communications, 1968-91, UHF & microwave systems staff officer, 1991—; pin chmn. Tournament of Roses Communications Group, Pasadena, Calif., 1985—. Republican. Achievements include development of specialized electronic test equipment for automotive and marine magneto ignition systems, of electronic bomb disposal equipment, of portable military satellite communication antenna. Office: PO Box 62 Lakewood CA 90714-0062

BOGDANOV, NIKITA ALEXEEVICH, geology educator; b. Astrakhan, Russia, July 23, 1931; s. Alexey Alexeevich and Irina Vladimirovna (Butkevich) B.; m. Olga Andreevna Dmitrieva, Oct. 6, 1971; children: Vladimir, Alexandr. Student, Moscow Geol. Prospecting Inst., 1954; cand. sci., Geol. Inst. of USSR Acad. Sci., 1962, DSc, 1972. Jr. scientist to sr. scientist dept. tectonics and geology Ocean Geol. Inst. USSR Acad. Sci., 1954-78; head dept. oceanic lithosphere, dep. dir., dir. Inst. Lithosphere, Russian Acad. Sci., 1978—; prof. Moscow U., 1990—; mem. coun. of dirs. Circum-Pacific Coun. for Energy and Mineral Resources, Houston, 1980—; mem. steering com. Internat. Geol. Congress, 1984—. Author: Stratigraphy and Geological Correlation, 1992—, Izvestiya AN SSSR, 1984-91, Island Arcs; mem. editorial bd. Interperiodica Jour.; contbr. more than 200 articles to profl. jours., chpts. to books. Recipient Internat. Spendiarov prize in geology, 1984. Mem. Russian Acad. Scis., Australian Geol. Soc. (hon.). Home: Inst of Lithosphere, Garibaldi Str 15 Bldg 1, 109180 Moscow 117335, Russia Office: Inst of Lithosphere, Staromonetny per 22, Moscow 109180, Russia

BOGDONOFF, MAURICE LAMBERT, physician; b. Chgo., May 11, 1926; s. Harry A. and Mary Ivy (Grogan) B.; m. Diana Edith Rauschkolb, June 29, 1956; children: Vivian, Gregory, Audrey. BS, Tufts U., 1948; MD, Yale U., 1952. Intern U. Ill. Rsch. and Edn. Hosp., Chgo., 1952-53; resident in internal medicine Boston City Hosp., 1953-54; resident in radiology Columbia-Presbyn. Med. Ctr., N.Y., 1955-57; asst. prof. to assoc. prof. radiology to prof. U. Ill., Chgo., 1958-69; attending radiologist Rush-Presbyn.-St. Luke's Med. Ctr., Chgo., pres. med. staff, 1975-77; prof. radiology, medicine Rush Med. Coll., Chgo., 1970—; cons. Argonne (Ill.) Nat. Lab., 1963-88; cons., health dir. Canal Zone Panama, 1973-80; vis. lectr. nuclear power engring. Maine Maritime Acad., 1989. Contbr. articles to profl. jours. Pres. Wheaton (Ill.) Sch. Bd., 1964-67; bd. visitors Coll. of DuPage Radio and TV System, Glen Ellyn, Ill., 1987—. With USN, 1944-46. Fellow Am. Coll. Radiology, Inst. Med. Sch., also others. Republican. Avocations: boating, astronomy, classics. Home: 203 W Willow Ave Wheaton IL 60187-5238

BOGDONOFF, SEYMOUR MOSES, aeronautical engineer; b. N.Y.C., N.Y., Jan. 10, 1921; s. Glenn and Kate (Cohen) B.; m. Harriet Eisenberg, Oct. 1, 1944; children: Sondra Sue, Zelda Lynn, Alan Charles. B.S., Rensselaer Poly. Inst., 1942; M.S., Princeton U., 1948. Asst. sect. head fluid and gas dynamics sect. Langley Meml. Aero. Lab., NACA, 1942-46; research assoc. aero. engring. dept. Princeton U., 1946-53, asso. prof., 1953-57, prof.,

1957-63, Henry Porter Patterson prof. aero. engring., 1963-89, prof. emeritus and sr. rsch. scholar, 1989—, chmn. dept. mech. and aerospace engring., 1974-83; head gas dynamics lab. James Forrestal Research Campus, Princeton U., sr. rsch. scholar, 1989—; cons. aero. engr.; mem. adv. council NASA; mem. sci. adv. bd. Dept. Air Force, 1958-76, 80-84; Von Karman Meml. lectr., Israel, 1992. Recipient Exceptional Civilian Svc. award Dept. Air Force, 1968, 86; Rensselaer Alumni fellow, 1993. Fellow AIAA (dir., Fluid and Plasma Dynamics award 1983, Hugh L. Dryden lectr. in rsch. 1990); mem. NAE, ASME, Am. Phys. Soc., Internat. Acad. Astronautics of Internat. Astronautical Fedn. (corr.), French Nat. Acad. of Air and Space (hon.), Sigma Xi, Tau Beta Pi. Home: 39 Random Rd Princeton NJ 08540-4065

BOGER, DALE L., chemistry educator; b. Hutchinson, Kans., Aug. 22, 1953; s. Lester W. and Elizabeth (Korkish) B. BS in Chemistry, U. Kans., 1975; PhD in Chemistry, Harvard U., 1980. Asst. prof. U. Kans., Lawrence, 1979-83, assoc. prof., 1983-85; assoc. prof. Purdue U., West Lafayette, Ind., 1985-87, prof., 1987-91; Richard and Alice Cramer chair chemistry Scripps Rsch. Inst., La Jolla, Calif., 1991—. Recipient Career Devel. award NIH, 1983-88; NSF fellow, 1975-78, Alfred P. Sloan fellow, 1985-89; Searle scholar, 1981-84. Mem. Am. Chem. Soc. (A.C. Cope scholar 1989). Home: 2819 Via Posada La Jolla CA 92037-2205 Office: Scripps Rsch Inst 10666 N Torrey Pines Rd La Jolla CA 92037-1027

BOGER, DAN CALVIN, economics educator, statistical and economic consultant; b. Salisbury, N.C., July 9, 1946; s. Brady Cashwell and Gertrude Virginia (Hamilton) B.; m. Gail Lorraine Zivna, June 23, 1973; children: Gretchen Zivna, Gregory Zivna. B.S. in Mgmt. Sci., U. Rochester, 1968; M.S. in Mgmt. Sci., Naval Postgrad. Sch., Monterey, Calif., 1969; M.A. in Stats., U. Calif.-Berkeley, 1977, Ph.D. in Econs., 1979. Cert. cost analyst, profl. estimator. Research asst. U. Calif.-Berkeley, 1975-79; asst. prof. econs. Naval Postgrad. Sch., Monterey, Calif., 1979-85, assoc. prof., 1985-92, prof., 1992—; bd. dirs. Evan-Moor Corp., 1992—; cons. econs. and statis. legal matters CSX Corp, others, 1977—. Assoc. editor The Logistics and Transportation Rev., 1981-85, Jour. of Cost Analysis, 1989-92; mem. editorial rev. bd. Jour. Transp. Research Forum, 1987-91; contbr. articles to profl. jours. Served to lt. USN, 1968-75. Flood fellow Dept. Econs., U. Calif.-Berkeley, 1975-76; dissertation research grantee A.P. Sloan Found., 1978-79. Mem. Am. Econ. Assn., Am. Statis. Assn., Econometric Soc., Math. Assn. Am., Inst. Mgmt. Sci., Ops. Research Soc. Am. (sec., treas. mil. applications sect. 1987-91), Sigma Xi. Home: 61 Ave Maria Rd Monterey CA 93940-4407 Office: Naval Postgrad Sch Code AS/Bo Monterey CA 93943

BOGGS, JOSEPH DODRIDGE, pediatric pathologist, educator; b. Bellefontaine, Ohio, Dec. 31, 1921; s. Walter C. and Birdella Z. (Coons) B.; m. Donna Lee Shoemaker, June 12, 1964; 1 son, Joseph Dodridge. A.B., Ohio U., 1941, Litt.D., 1966; M.D., Jefferson Med. Coll., 1945. Intern Jefferson Med. Coll. Hosp., Phila., 1945-46; resident Peter Bent Brigham Hosp., Boston, 1946-48; asso. pathologist Peter Bent Brigham Hosp., 1947-51; instr. pathology Harvard Med. Sch., Boston, 1948-51; with Children's Meml. Hosp., Chgo., 1951—; dir. labs. Children's Meml. Hosp., 1951—; prof. pathology Northwestern U., Chgo., 1952—; dir. BSP Ins. Co., Phoenix. Contbr. articles to profl. jours. Mem. med. adv. bd. Ill. Dept. Corrections, Springfield, 1971-77; bd. dirs. Blood Svcs., Phoenix, 1972, Community Hosp., Evanston, Ill., 1958-61, Lorretto Hosp., Chgo., 1971-72; chmn. Chgo. Regional Blood Program, 1978-80; bd. dirs. Ben Venue Labs., 1985—. Capt. M.C., U.S. Army, 1948-51. Mem. Am. Soc. Study of Liver Disease, N.Y. Acad. Scis., Midwest Soc. Pediatric Research, Inst. Medicine, Ill. Soc. Pathologists (pres. 1965), Ill. Assn. Blood Banks (pres. 1969-70). Office: 1448 N Lake Shore Dr Chicago IL 60610-1625

BOGHOSIAN, PAULA DER, educator, consultant; b. Watervileit, N.Y., Nov. 11, 1933; d. Harry and Osgi (Piligian) der B. BS magna cum laude, Syracuse U., 1964, MS, 1967; postgrad., SUNY, Oswego, 1972, SUNY, Albany, 1974. Cert. profl. sec., 1974. Asst. prof. Cazenovia (N.Y.) Coll., 1964-73; instr. Bd. of Coop., Syracuse, N.Y., 1973-76, dir. bus. careers, 1976-92; cons. computer bus., prin. Syracuse, 1984—. Zonta scholar, 1964; Jessie Smith Noyes grantee Syracuse U., 1965. Mem. Assn. Info. Systems Profl. (com. chmn.), Bus. Tchrs. Assn. of N.Y. State, Adminstrv. Mgmt. Soc., Eastern Bus. Tchrs. Assn., Assn. for Supervision and Curriculum Devel., Assn. of Am. Jr. Colls., Assn. of Am. U. Profs., Nat. Assn. for Armenian Studies and Rsch. Harvard U., Internat. Tng. Communications (v.p. 1985-86), Delta Pi Epsilon, Beta Gamma Sigma, Phi Kappa Phi, Pi Lambda Theta, Sigma Lamda Delta. Republican. Mem. Armenian Apostolic. Avocations: music, golf, water colors, designer, travel. Home: 3181 Bellevue Ave Apt B6 Syracuse NY 13219-3165

BOGORAD, BARBARA ELLEN, psychologist; b. N.Y.C.; d. Albert Lyon and Miriam Ida (Serlin) B. BA, CUNY, 1969; MS, Rutgers U., 1972, Yeshiva U., 1981; PsyD, Yeshiva U., 1983. Lic. psychologist, N.Y.; diplomate Am. Bd. Profl. Psychol., 1992. Psychotherapist South Shore Ctr. Psychotherapy, Merrick, N.Y., 1978-82; psychology intern Birch Ctr. Exceptional Children, Queens, N.Y., 1980-81; clinical intern Long Island Jewish Hosp., Glen Oaks, N.Y., 1981-82; clin. intern South Oaks Hosp., Amityville, N.Y., 1982-83; staff psychologist St. John's Episc. Hosp., Far Rockaway, N.Y., 1984-86, St. Charles Hosp., Port Jeff, N.Y., 1987-88; pvt. practice Amityville, 1985—; staff psychologist South Oaks Hosp., Amityville, 1988—, dir. sexual abuse recovery program, 1991—; speaker in field; radio and TV appearances 1990—. Vol. crisis relief worker Nassau and Suffolk counties, N.Y., 1990—. Mem. APA, Ea. Psychol. Assn., N.Y. State Psychol. Assn., Nassau County Psychol. Assn., Suffolk County Psychol. Assn., Am. Assn. Psychiat. Svcs. Children., Psychologists in Hosp. Practice, Am. Profl. Soc. Abuse of Children, Nat. Assn. Childcare Resource and Referral Agys. (aux.). Avocations: photography, gardening, travel, choral singing. Office: South Oaks Hosp 400 Sunrise Hwy Amityville NY 11701-2508

BOGORAD, LAWRENCE, biologist; b. Tashkent, U.S.S.R., Aug. 29, 1921; came to U.S., 1922; s. Boris and Florence (Bernard) B.; m. Rosalyn G. Sagen, June 29, 1943; children—Leonard Paul, Kiki M. Lee. B.S., U. Chgo., 1942, Ph.D., 1949. Instr. botany U. Chgo., 1948-51, asst. prof. dept. botany, 1953-57, assoc. prof., 1957-61, prof., 1961-67; prof. biology Harvard U., Cambridge, Mass., 1967—; chmn. dept. biology Harvard U., 1974-76; dir. Maria Moors Cabot Found. Harvard U., Cambridge, Mass., 1976-87; Maria Moors Cabot prof. biology Harvard U., 1980—; vis. investigator Rockefeller Inst., N.Y.C., 1951-53; mem. com. on sci. and public policy Nat. Acad. Scis., 1977-81; mem. Assembly of Life Scis., NRC; mem. joint council on food and agrl. scis. Dept. Agr., 1978-82. Assoc. editor Bot. Gazette, 1958; mem. editorial com. Ann. Rev. Plant Physiology, 1963-67, Ann. Rev. Cell Biology, 1984-88; mem. editorial bd. Plant Physiology, 1965-66, Biochimica Biophysica Acta, 1967-69, Jour. Cell Biology, 1967-70, Jour. Applied and Molecular Genetics, 1981-85, Plant Molecular Biology, 1981-85, Plant Cell Reports, 1981-85; editor, chmn. editorial bd. Proc. Nat. Acad. Scis., 1991—. Served with AUS, 1943-46. Merck fellow, 1951-53; Fulbright fellow, 1960; recipient Career Research award NIH, 1963. Fellow Am. Acad. Arts and Scis.; mem. NAS (chmn. botany sect. 1974-77, mem. coun. 1989-92, editor proc., chmn. editorial bd. Proceedings 1991—), AAAS (bd. dirs. 1982-86, pres. 1986-87, chmn. bd. 1987), Am. Philos. Soc., Am. Soc. Biol. Chemistry, Am. Soc. Cell Biology, Am. Soc. Plant Physiologists (pres. 1968-69, Stephen Hales award 1982), Royal Danish Acad. Scis. and Letters (fgn.), Soc. Developmental Biology (pres. 1984). Office: Harvard U Dept of Biology 16 Divinity Ave Cambridge MA 02138-2097

BOGREN, HUGO GUNNAR, radiology educator; b. Jönköping, Sweden, Jan. 9, 1933; came to U.S., 1970; s. Gunnar Hugo and Signe Victoria (Holmström) B.; m. Elisabeth Faxén, Nov. 1, 1956 (div. 1976); children: Cecilia, Niclas, Joakim; m. Gunilla Lady Whitmore, July 2, 1988. MD, U. Göteborg, Sweden, 1958, PhD, 1964. Diplomate Swedish Bd. Radiology. Resident, fellow U. Göteborg, 1958-64, asst. to assoc. prof. radiology, 1964-69; from assoc. prof. to prof. radiology and internal medicine U. Calif. Davis, Sacramento, 1972—; vis. assoc. prof. U. San Francisco, 1970-71; vis. prof. U. Kiel, Fed. Republic Germany, 1980, magnetic resonance unit U. London, 1986-87, 93-94; participant in med. aid fact finding mission, Bangladesh, 1992. Contbr. numerous articles to profl. jours., chpts. to books. Sr. Internat. Fogarty fellow NIH, London, 1986-87. Fellow Am. Heart Assn., Am. Coll. Cardiology, Radiol. Soc., N.Am. Soc. Cardiac Radiology, Cardiac Angiographic Soc., Assn. Univ. Radiologists, Soc. Thoracic Radiology, Soc. Magnetic Resonance Imaging, Soc. Magnetic Resonance in Medicine, Swedish Assn. Med. Radiology; mem. Royal Gothenburg Sailing Club (Sweden), Swedish Cruising Club, Rotary (del.). Lutheran. Avocations: ocean sailing, skiing, classical music. Office: U Calif Davis Med Ctr Div Diagnostic Radiology 2516 Stockton Blvd TICON II Bldg Sacramento CA 95817

BOGUE, SEAN KENNETH, civil engineer; b. Saskatoon, Sask., Can., Aug. 21, 1969; came to U.S., 1981; s. Kenneth George and Carol-Ann (Johnson) B.; m. Kristin Elisabet Johnson, June 12, 1993. BSCE, U. Va., 1991; postgrad., Cornell U., 1993—. Assoc. engr. II Parsons Deleuw, Inc., Washington, 1991-93. Mem. ASCE (assoc., treas. U. Va. chpt. 1990-91). Home: 1808 N Quinn St # 522 Arlington VA 22209

BOHI, DOUGLAS RAY, economist; b. Pocatello, Idaho, Sept. 9, 1939; s. Clarence R. and Florence E. (Karstad) B.; m. Marjorie A. Brenner; children: Heidi, James. BS in Econs., Idaho State U., 1962; PhD in Econs., Wash. State U., 1967. Economist Caterpiller Co., Peoria, Ill., 1969-70; prof. econs. So. Ill. U., Carbondale, 1970-78; sr. fellow Resources for the Future, Washington, 1978-87, div. dir., 1988—; chief economist FERC, Washington, 1987-88; vis. prof. Monash U., Melbourne, Australia, 1982. Author: Analyzing Demand Behavior, 1981; co-author: Limiting Oil Imports, 1978, Oil Prices, Energy Security, and Import Policy, 1982, Analyzing Nonrenewable Resource Supply, 1984. Capt. U.S. Army, 1967-69. Fulbright scholar, Erasmus U., Rotterdam, Netherlands, 1977. Mem. Am. Econs. Assn. Office: Resources for Future 1616 P St NW Washington DC 20036-1434

BOHM, ARNO RUDOLF, physicist; b. Stettin, Poland, Apr. 26, 1936; came to U.S., 1966; m. Darlene Wiley; children: Anita, Rudolf, Joseph. PhD, Univ., 1962, Dr.rer.nat., 1966. Instr. Univ., Karlsruhe, 1963-64; prof. U. Tex., Austin, 1966—. Author 4 books and contbr. 100 articles to profl. jours. Jukrauhova Ctr. Rsch. fellow, 1964-66, Mortimer and Raymond Sachler fellow, 1980; recipient Humboldt prize, 1981. Fellow Am. Phys. Soc. Office: U Tex Physics Dept Austin TX 78712

BOHN, DENNIS ALLEN, electrical engineer, consultant, writer; b. San Fernando, Calif., Oct. 5, 1942; s. Raymond Virgil and Iris Elouise (Johnson) B.; 1 dau., Kira Michelle; m. Patricia Tolle, Aug. 12, 1986. BSEE with honors, U. Calif., Berkeley, 1972, MSEE with honors, 1974. Engring. technician Gen. Electric Co., San Leandro, Calif., 1964-72; research and devel. engr. Hewlett-Packard Co., Santa Clara, Calif., 1973; application engr. Nat. Semicondr. Corp., Santa Clara, 1974-76; engring. mgr. Phase Linear Corp., Lynnwood, Wash., 1976-82; v.p. research and devel., ptnr. Rane Corp., Mukilteo, Wash., 1982—; founder Toleco Systems, Kingston, Wash., 1980. Suicide and crisis ctr. vol., Berkeley, 1972-74, Santa Clara, 1974-76. Served with USAF, 1960-64. Recipient Am. Spirit Honor medal USAF, 1961. Math. Achievement award Chem. Rubber Co., 1962-63. Editor: We Are Not Just Daffodils, 1975; contbr. poetry to Reason mag.; tech. editor Audio Handbook, 1976; contbr. articles to tech. jours.; columnist Polyphony mag., 1981-83; 2 patents in field. Mem. IEEE, Audio Engring. Soc., Tau Beta Pi. Office: Rane Corp 10802 47th Ave W Mukilteo WA 98275-5098

BOHN, HORST-ULRICH, physicist; b. Schw. Hall, Fed. Republic Germany, May 22, 1946; s. Erich Karl and Lydia Maria (Ottenbacher) B.; m. Monika Franziska Gwosdek, Sept. 24, 1977; children: Andreas-Constantin, Marc-Dominic. MS in Physics, U. Wurzburg, Fed. Republic Germany, 1974, PhD in Physics, 1981. Rsch. asst. Inst. Astronomy and Astrophysics, Wurzburg, 1974-80, scientist, 1980-81, asst. prof. astrophysics, 1981-86; staff physicist Fraunhofer Ges. für ang. Forschung, Munich, 1986-90; mng. dir. Nat. Instruments Germany GmbH, Munich, 1991—; cons. engr. Data Aquisition and Image Processing, Wurzburg, 1974-80. Contbr. articles to Astronomy & Astrophysics; author: (with others) Activity in Red Dwarf Stars, 1983, Lecture Notes in Physics, 1986. Mem. Internat. Astron. Union, Am. Astron. Soc., Deutsche Physikalische Gesellschaft, Astronomische Gesellschaft. Office: Nat Instruments Germany, Konrad-Celtis-Str 79, D-81369 Munich Germany

BOHNE, JEANETTE KATHRYN, mathematics and science educator; b. Quincy, Ill., June 7, 1936; d. Anton Henry and Hilda Wilhelminia (Ohnemus) B. BA, Ursuline Coll., Louisville, 1961; MA, St. Louis U., 1962. Cert. math. and chemistry tchr., N.D., Ill., Mo. Math. tchr. Ryan High Sch., Minot, N.D., 1962-66, Althoff Cath. High Sch., Belleville, Ill., 1966-72, St. Francis Borgia High Sch., Washington, Mo., 1974-77; math. tchr. St. Louis Pub. Schs., 1977—, head dept. math., 1977-85; speaker in field. Treas. Welcome Wagon Club, Washington, 1974-76; pres. Bus. & Profl. Women's Club, Washington, 1978-79; active Animal Protective Assn., Zoo Friends of St. Louis Zoo. Fellow AAUW, Mo. State Tchrs. Assn., Mo. NEA, St. Louis Tchrs. Union, Nat. Coun. Tchrs. Math., Am. Assn. Univ. Women, Math. Educators Group St. Louis, Mo. Coun. Tchrs. Math., St. Louis Sci. Ctr., St. Louis Zoo Friends. Avocations: collecting thimbles, raising plants. Home: PO Box 2252 5625 Reber Pl Saint Louis MO 63109-0252 Office: St Louis Pub Schs 911 Locust St Saint Louis MO 63101-1471

BOHR, AAGE NIELS, physicist; b. June 19, 1922; s. Niels and Margrethe (Norlund) B.; m. Marietta Bettina Soffer (dec. 1978); 3 children: m. Bente Meyer Scharff, 1981. Ph.D., U. Copenhagen, Denmark, 1954; D honoris causa, Manchester U., 1961; hon. degrees, Oslo U., 1969, Heidelberg U., 1971, Trondheim U., 1972, Uppsala U., 1975. Jr. sci. officer Dept. Sci. and Indsl. Research, London, 1943-45; research assoc. Inst. Theoretical Physics U. Copenhagen, 1946—, prof. physics, 1956—; dir. Niels Bohr Inst., 1962-70; mem. bd. Nordita, 1958-74, dir., 1975-81. Author: Rotational States of Atomic Nuclei, 1954; (with Ben R. Mottelson) Nuclear Structure, Vol. 1, 1969, Vol. 2, 1975. Recipient Dannie Heineman prize, 1960; Pope Pius XI medal 1963; Atoms for Peace award, 1969; H.C. Ørsted medal, 1970; Rutherford medal, 1972; John Price Wetherill medal, 1974; Nobel prize in physics, 1975; Ole Rõmer medal, 1976. Mem. Danish, Norwegian, Yugoslavian, Polish, Swedish acads. scis., Royal Physiograph. Soc. Lund, Sweden, Am. Acad. Arts and Scis., Nat. Acad. Scis. (U.S.), Deutsche Akademie der Naturforscher Leopoldina, Am. Philos. Soc., Finska Vetenskaps-Societeten, Pontifical Acad. Research quantum physics; specialist nuclear physics. Office: Niels Bohr Inst, Blegdamsvej 15-17, DK-2100 Copenhagen Denmark

BOIKE, SHAWN PAUL, aerospace project/design engineer; b. Detroit, June 20, 1964; s. Charles Paul and Catherine Diane (Supanich) B.; m. Christine Helen Chrzandiuski, July 28, 1990. BSME, Mich. State U., 1982; San Diego U., 1989. Registered mech. engr., aerospace design engr. Automation/robotics designer Boico Engring. Inc., Sterling Hgts., Mich., 1981-85; design engr. Northrop Aircraft Corp., Hawthorne, Calif., 1985-89; project leader McDonnell Douglas Aircraft Co., Long Beach, Calif., 1987-89; project engr. Gen. Dynamics-Space Systems, San Diego, 1989-90; sr. producibility engr. Teledyne Ryan Aero., San Diego, 1990-91; design engr. Boeing Comml. Aircraft, Seattle, 1992—. Author: (proposal) Laws-Laser Assembly Welding System, 1989, SAND-FLEA=An Advance VTOL Aircraft, 1991, ATAC-The Automatic Tank Assembly Cell-Space Craft Manufacturing System, 1990. Participating mem. The Way Back Inn Found., Renton, Wash., 1992. Recipient Innovative Ideas award McDonnell Douglas Aircraft, 1987. Fellow Soc. for Advancement of Materials and Process Engring., Indsl. Modernization and Incentive Program; mem. AIAA. Republican. Achievements include patents pending for hybred electric car using solar, wind and rep. braking with battery mgmt.; for proposed flying disc using counter-rotating cowled fan blades; for automatic anti-terrorist device; for laser welding method. Office: Solution Vehicles Co PO Box 70322 Bellevue WA 98007

BOILLAT, GUY MAURICE GEORGES, mathematical physicist; b. Pontarlier, France, May 18, 1937; s. Georges Paul Charles and Lucie Marguerite Charlotte (Jubin) B. Licence scis., U. Besançon, France, 1959; postgrad., Inst. Henri-Poincaré, Paris, 1959-60, Inst. Theoretical Physics, Copenhagen, 1960-62, Norwegian Tech. U., Trondheim; DSc, Sorbonne U., Paris, 1964. Assoc. prof. dept. math. U. Clermont, Aubière, France, 1966-69, prof., 1969—; lectr., Italy, 1970—; researcher U. Messina, U. Catania, U. Bologna, Italy, 1970—. Contbr. over 80 rsch. articles to profl. jours. Dep. mem. Internat. Parliament for Safety and Peace. Mem. Maison Internat. Intellectuels, Acad. M.I.D.I. Roman Catholic. Avocations: arts, bibliophilism, horology. Home: 16 rue Ronchaux, 25000 Besançon France

BOIS, PIERRE, former medical research organization executive; b. Oka, Que., Can., Mar. 22, 1924; s. Henri and Ethier (Germaine) B.; m. Joyce Casey, Sept. 8, 1953; children: Monique, Marie, Louise. M.D., U. Montreal, Que., 1953, Ph.D., 1957; hon. doctorate, U. Ottawa, Ont., 1982, U. Man., 1985, U. Sherbrooke, 1986. Research fellow pathology U. Montreal, 1957-58, asst. assoc. prof. dept. pharmacology, 1960-64, prof., head dept. anatomy, 1964-70, dean faculty medicine, 1970-81; pres. Med. Research Council of Can., 1981-91; asst. prof. histology, Ottawa, Ont., Can., 1958-60. Contbr. over 130 publs. to profl. jours. Decorated chevalier de l'Ordre do Méite scientifique de la République de France, 1988; recipient Disting. Svc. award Can. Soc. Clin. Investigation, 1986, Francisco Hernandez award, 1986. Fellow Royal Soc. Can., Royal Coll. Physicians and Surgeons Can.; mem. Am., French Canadian assns. anatomists, N.Y. Acad. Scis., AAAS, Can. Fedn. Biol. Socs., Can. Soc. Clin. Investigation, Can. Soc. Acad. Medicine. Research and numerous publs. on morphological effects of hormones, histamine and mast cells in magnesium deficiency, muscular dystrophy, exptl. thymic tumors.

BOISJOLY, ROGER MARK, structural engineer; b. Lowell, Mass., Apr. 25, 1938; s. Joseph Antonio and Isabelle Evelyn (St Cyr) B.; m. Roberta Gibb Malcolm, Apr. 21, 1962; children: Norma, Darlene. BS, Lowell Tech. Inst., 1960. Design engr. Autometics, Anaheim, Calif., 1962-63; mech. design engr. aerospace divsn. Hamilton Standard, Broad Brook, Conn., 1965-66; sr. design engr. Elektron Standard, South Windsor, Conn., 1966-67; asst. prin. engr. Celesco Industries Inc. formerly Atlantic Rsch. Corp., Costa Mesa, Calif., 1963-65, 67-74; lead engr. MTS-4 space divsn. Rockwell Internat., Downey, Calif., 1974-76; cons. engr. Hughes Helicopter and Rockwell Internat., 1976-77; applied mechanics sr. engr. Garrett Airesearch Corp., Torrance, Calif., 1977-79; prin. engr. Brunswick Def. Divsn., Costa Mesa, 1979-80; project engring. mgr. Morton Thiokol, Brigham City, Utah, 1980-87; cons. forensic engr., expert witness Mesa, Ariz., 1987—; lectr. ethics, Mesa, 1987—. Contbr. articles to The Scientist, jours. of ASME and ASCE, Careers and the Engr., Electronic Engring. Times. Mayor City of Willard, Utah, 1982-83. Recipient Cert. of Appreciation, NASA, 1982, 86, Presdl. award Nat. Space Soc. for Profl. Integrity and Courage, 1987, Sci. Freedom and Responsibility award AAAS, 1988, Cavallo prize Cavallo Found., 1990. Mem. U. Lowell Alumni Assn. (Disting. Alumni 1987), IEEE (sr.), Nat. Acad. Forensic Engrs., NSPE. Ch. LDS. Home and Office: 3047 E Menlo St Mesa AZ 85213

BOISVERT, WILLIAM ANDREW, nutritional biochemist, researcher; b. Seoul, May 7, 1962; came to U.S., 1980; s. Robert Andrew and Jung Sup (Song) B. BS in Biochemistry, U. Oreg., 1985; PhD in Nutritional Biochemistry, Tufts U., 1992. Rsch. asst. Tufts U., Boston, 1986-87, rsch. assoc., 1991-93; postdoctoral fellow Cardiovascular Rsch. Inst. U. Calif., San Francisco, 1993—; with Human Nutrition Rsch. Ctr., USDA, Boston. Contbr. articles to sci. jours. Recipient community svc. award Elderly Soc. Jocotenango, Guatemala, 1990; scholar Nutrasweet, Inc., 1986; rsch. grantee USDA, 1988. Mem. Sigma Xi. Achievements include synthesis of several novel heterocyclic organic compounds that may have anti-malarial activity; determination of nutritional requirement of vitamin B2 (riboflavin) in the elderly population. Office: U Calif San Francisco Cardiovascular Rsch Inst 505 Parnassus Ave San Francisco CA 94143

BOK, SONG HAE, biotechnologist, researcher; b. Chongyang, Chungnum, Korea, Apr. 25, 1943; s. Chichan and Soon (Yoo) B.; m. Yon Hee Kim, Sept. 2, 1972; children: Jayne, Eugene H., Jonathan. BS in Biology, Seoul (Korea) Nat. Univ., 1966; MS in Biochem. Engring., MIT, 1972; PhD in Microbiology, Pa. State U., 1976. Sr. microbiologist A.E. Staley Mfg. Co., Decatur, Ill., 1976-79; rsch. fellow Hoffman La Roche, Inc., Nutley, N.J., 1980-85; sr. rsch. specialist Monsanto Co., St. Louis, 1986-87; dir. biotech. Korea Rsch. Inst. Chemical Technology, Taejeon, 1987—. Recipient New Product Discovery award, Korean Ministry Sci. and Engring., Seoul, 1987. Mem. Am. Chem. Soc., Am. Soc. for Microbiology, Soc. for Indsl. Microbiology. Achievements include development of microbial pesticides coated in biodegradable polymers. Office: Korea Rsch Inst Chem Tech, Biotech Dept GERI, PO Box 9 Daedong Dangi, Taejeon Republic of Korea

BOKSAY, ISTVAN JANOS ENDRE, geriatric psychiatrist; b. Unguar, Hungary, June 22, 1940; came to U.S., 1974; s. Endre Bela and Sylvia (Budai) B.; m. Patricia Kamins, Dec. 31, 1979; children: Eszter, Ana, Endre, Alexandra. MD, Med. Sch., Budapest, Hungary, 1964; PhD, Med. Sch., Frankfurt, Germany, 1971. Pharmacologist von Heyden, Regensburg, Germany, 1965-68; sr. pharmacologist Hoechst, Wiesbaden, Germany, 1968-74; assoc. dir. Hoechst-Roussel, Somerville, N.J., 1974-78; clin. dir. Ayerst Lab., N.Y.C., 1978-80; resident psychiatrist Montefiore, Bronx, N.Y., 1981-84; attending psychiatrist, rsch. psychiatrist NYU Med. Ctr., N.Y.C., 1984—. Contbr. articles to profl. jours. Home: 7 N Crescent Maplewood NJ 07040 Office: Aging and Dementia Rsch 550 First Ave NH314 New York NY 10016

BOLAND, LOIS WALKER, retired mathematician and computer systems analyst; b. Newton Center, Mass., Sept. 14, 1919; d. Charles Nelson and Nell Flora (Kruse) Walker; m. Ralph Montrose Boland, June 2, 1943; children: Charles Montrose, William Ralph (dec.), Ann Helen Boland Garner, Mark Alan Boland (dec.). BS, Stetson U., 1940; grad. fellow U. Ala., 1940-41; postgrad. U. Fla., 1948, 51-52, U. Mich., 1958, 65, U. Colo., 1966, 68. Physics tchr., Lakeland Fla. High Sch., 1941-42; chemist, IM&CC, Mulberry, Fla., 1942-43; elec. engring. draftsman Tampa Ship Corp., Fla., 1943-46; math. tchr. Plant High Sch., Tampa, 1947-50; mathematician/computers, data reduction div. Patrick AFB, Fla., 1951-54; mathematician, computer supr. data reduction div. White Sands Missile Range, N.Mex., 1954-63; ops. research analyst Peterson AFB, Colo. Springs, Colo., 1963-78, ret., 1978. Author: AUPRE Computer Program Manual, 1963, 77; Askania Photheodolite Computer Manual, 1956; editor Q-Point mag., 1966. Elected mem. Democratic exec. com. Volusia County, Fla., 1984-89. Pioneer mathematician Pioneer Group White Sands Missile Range, 1985; math. fellow U. Ala., 1940-41. Mem. Math. Assn. Am. (emeritus), AAUW (pres. DeLand chpt. 1984-86), Mensa, Am. Contract Bridge League. Clubs: Halifax, Lake Beresford Yacht. Avocations: writing; pianist; travel; duplicate bridge; swimming. Home: PO Box 215 Cassadaga FL 32706-0215

BOLDT, HEINZ, aerospace engineer; b. Schönfeld Krs. Friedeberg, Germany, July 12, 1923; s. August and Marie (Hamann) B.; m. Christa Friebel, Mar. 25, 1965; children: Pierre, Manon. Diploma in engring., Technische Universität, Berlin, 1951; student Wirtschaftsakademie, Berlin, 1953-57. Tech. dir. Borsig AG, Berlin, 1951-66; gen. mgr., dir. Messerschmitt-Werke Flugzeug-Union Sud, München-Augsburg, Fed. Republic Germany, 1967-70; gen. proxi Klöckner-Humboldt-Deutz, Köln, Fed. Republic Germany, 1970-72; mem. exec. bd. for devel., constrn. and prodn. FAHR AG, Gottmadingen, Fed. Republic Germany, 1970-72; pres. VDI-Bodenseebezirksverein, Friedrichshafen, Germany, 1971-76 (Verein Deutscher Ingenieure) ; mem. exec. bd. Dornier GmbH, Munich, 1972-77; pres. Deutsche Industrieanlagen Gesellschaft mbH, Berlin, 1978-82; rep. Machinoexport. Holder over 100 patents in field. Served with German Army Air Force, 1942-45. Recipient Ring for Honour VDI-Ehrenring, 1962. Mem. Am. C. of C. Club: Club der Luftfahrt. Home: Paartalweg 4 Merching, Bayern 86504, Germany

BOLDT, PETER, chemistry educator, researcher; b. Berlin, Dec. 9, 1927; s. Lothar and Marianne (Fischer) B.; m. Inge Grubert, Aug. 2, 1963 (div. Oct. 1985); children: Michael, Thomas. Diploma in Chemistry, U. Göttingen, Fed. Republic Germany, 1955, PhD, 1958, Habilitation, 1966. Lectr., asst. prof. U. Göttingen, 1969-72; prof. Tech. U. Braunschweig, Fed. Republic Germany, 1972—; dean of faculty, Tech. U. Braunschweig, 1983-85. Author: Extended Quinones, 1988; inventor in field of non linear optics and optical data storage. With German mil., 1944-45. Mem. Gesellschaft Deutscher Chemiker, Am. Chem. Soc., Workshop Inst. for Living Learning Internat. Avocations: trainer for group dynamics. Home: Zeppelin Str 3, D-38106 Braunschweig Germany Office: Tech U Inst Organic Chemistry, Hagenring 30, D-3300 Braunschweig Germany

BOLEBRUCH, JEFFREY JOHN, sales executive; b. Gloversville, N.Y., Jan. 13, 1963; s. John George and Peggy Anne (Spawn) B.; m. Abagail Trainor, July 16, 1988. BS in Geography (Environ. Sci.), U.S. Mil. Acad., 1985. Commd. 2d lt. U.S. Army, 1985, advanced through grades to capt.,

1989; served with 1st Cavalry Div., 4th Bn., 5th Air Defense Arty. U.S. Army, Ft. Hood, Tex., 1985-90; tech. sales rep. Blasch Precision Ceramics, Schenectady, N.Y., 1990—. Mem. Am. Inst. Chem. Engrs., Am. Ceramic Soc. Republican. Home: 217 Hillview Rd Gloversville NY 12078 Office: Blasch Precision Ceramics 99 Cordell Rd Schenectady NY 12304

BOLEN, TERRY LEE, optometrist; b. Newark, Ohio, Sept. 16, 1945; s. Robert Howard and Mildred Irene (Hoover) B.; BS, Ohio U., 1968; postgrad. Youngstown State U., 1973; O.D., Ohio State U., 1978; m. Debbie Elaine Thompson, Mar. 23, 1985. Quality control inspector ITT Grinnell Corp., Warren, Ohio, 1973, jr. quality control engr., 1974; pvt. practice optometry, El Paso, Tex., 1978-80, Dallas, 1980-81, Waco, Tex., 1981-85, Hewitt, Tex., 1983-89; comdr. U.S. Pub. Health Svc., 1989—; bd. dirs. Am. Optometric Found., 1975-77; nat. pres. Am. Optometric Student Assn., 1977-78; pres. El Paso Optometric Soc., 1980. Vol. visual examiner, Juarez, Mex., 1979—; chmn. Westside Recreation Ctr. Adv. Com., El Paso, 1979, Lions Internat. Sight Conservation and Work With the Blind Chmn. award, 1989 . Served to lt. USN, 1969-72; capt. USAIRNG, 1987-89. Recipient pub. svc. award, City of El Paso, 1980. Mem. Am. Optometric Assn., Tex. Optometric Assn., Assn. Mil. Surgeons of U.S. (life, USPHS HSO liaison 1990—, edn. coord. optometry section, 1992), Reserve Officers Assn., Optometric Assn. (clin. assoc. Optometric Extension Program Found. 1978-89), Heart of Tex. Optometric Soc. (sec.-treas. 1984-85, pres.-elect 1986, pres. 1987), North Tex. Optometric Soc., USPHS (pres. No. Nev. chpt. Commd. Officers Assn. 1989-91), Epsilon Psi Epsilon (pres., 1977-78), Lions (3rd v.p. Coronado El Paso, svc. award, 1978, 79, Hewitt pres. 1985, v.p. W.Tex. Lions Eye Bank, 1980, 2d v.p. Cen. Tex. Lions Eye Bank, 1988, Hewitt Lion of Yr. 1987). Republican. Mem. Christian Ch. (Disciples of Christ). Home: 750 E Stillwater Ave # 173 Fallon NV 89406-4058

BOLES, JEFFREY OAKLEY, biochemist, researcher; b. Cookeville, Tenn., Dec. 31, 1961; s. Larris Oakley and Noreen Gail (Moore) B.; m. Tammy Hatfield, June 18, 1988; children: Kathleen Gail, Jarrod Oakley. BS, Tenn. tech. U., 1986, MS, 1988; PhD, U. S.C., 1992. Post doctoral fellow inst. biosciences and tech. Tex. A&M U., Houston, 1992-93. Contbr. articles to Sci. Mag., Synthetic Comms., Biochemistry, Jour. Biol. Chem. Mem. Sigma Xi. Baptist. Home: 6802 Los Tios Houston TX 77083 Office: Tex A&M U Inst Biosciences and Tech 2121 Holcombe Houston TX 77030

BOLEY, BRUNO ADRIAN, engineering educator; b. Gorizia, Italy, May 13, 1924; came to U.S., 1939, naturalized, 1945; s. Orville F. and Rita (Luzzatto) B.; m. Sara R. Kaufman, May 12, 1949 (dec. Sept. 1983); children: Jacqueline Boley Acquaviva, Daniel L. B.C.E., CCNY, 1943, D.Sc. hon., 1982; M. in Aero. Engring., Poly. Inst. Bklyn., 1945, D.Sc. in Aero. Engring., 1946. Asst. dir. structural research, aero. engring. dept. Poly. Inst. Bklyn., 1943-48; engring. specialist Goodyear Aircraft Corp., 1948-50; assoc. prof. aero. engring. Ohio State U., 1950-52; assoc. prof. civil engring. Columbia U., 1952-58, prof., 1958-68, dir. postdoctoral preceptor program, 1962-68; Joseph P. Ripley prof. engring., chmn. theoretical and applied mechanics Cornell U., Ithaca, N.Y., 1968-72; dean Technol. Inst., Walter P. Murphy prof. Northwestern U., Evanston, Ill., 1973-86, dean emeritus, Walter P. Murphy prof. emeritus, 1986—; prof. civil engring. and engring. mechanics Columbia U., N.Y.C., 1987—; mem. adv. com. George Washington U., Princeton U., Yale U., Cornell U., FAMU.FSU Inst. Engring., Duke U., Lehigh U., Nat. Cheng Kung U., Republic of China, Istanbul Tech. U., Rowan Coll. N.J.; mem. sci. adv. coun. Internat. Ctr. forMech. Scis., Udine, Italy, 1980—, Istanbul Tech. U.; chmn. Midwest Program for Minoroties in Engring., 1975-82; bd. govs. Argonne Nat. Lab., 1983-86; bd. advisors Who's Who in Sci. and Engring. Author: Theory of Thermal Stresses, 1960, High Temperature Structures and Materials, 1964, Thermoinelasticity, 1970, Crossfire in Professional Education, 1976; also articles, numerous tech. papers; editor-in-chief: Mechanics Research Communications; bd. editors Jour. Thermal Stresses, Bull. Mech. Engring. Edn., Internat. Jour. Computers and Structures, Internat. Jour. Engring. Sci., Internat. Jour. Fracture Mechanics, Internat. Jour. Mech. Engring. Scis., Internat. Jour. Solids and Structures, Jour. Applied Mechanics, Jour. Structural Mechanics Software, Letters in Applied and Engring. Sci., Nuclear Engring. and Design. NATO fellow, 1964-65, NSF fellow, 1965, Japan Soc. Promotion of Sci. Rsch. fellow, 1987; recipient Disting. Alumnus award Poly. Inst. N.Y., 1974, Townsend Harris medal, 1981, Commendation Ill. Ho. of Reps., 1986, Theodore von Karman medal ASCE, 1991. Fellow AIAA, AAAS, Am. Acad. Mechanics (pres. 1974, Disting. Svc. medal 1987); hon. mem. ASME (exec. com., pres. applied mechanics div. 1975, bd. govs. 1984-86, Worcester Reed Warner medal 1991), NAE (life, chmn. task force engring. edn. 1979-80, edn. adv. bd. 1982-86, editorial bd. The Bridge 1986-90, mem. com. 1984-88), Soc. Engring. Scis. (pres. 1975, Disting. Svc. medal 1987, life), Assn. Chairmen Depts. Mechanics (founder, pres. 1970-72), Internat. Assn. Structural Mechanics in Reactor Tech. (chmn. 1977, adv.-gen. 1979—), Internat. Union Theoretical and Applied Mechanics (sec. Congress com. 1976-84, mem. bur. 1988—, treas. 1992—, personal mem. Gen. Assembly 1980—, treas. 1992—), Am. Soc. Engring. Edn (project bd. 1987, Centennial award 1993), N.Y. Acad. Scis. (named Outstanding Educator of Am. 1971), U.S. Nat. Com. Theoretical and Applied Mech. (chmn. 1975-79, personal mem. gen. assembley 1980—), Ill. Council Energy Research and Devel. (chmn. 1979-84), Eng. Found. (conf. com. 1986-88). Office: Columbia U Dept Civil Engring/Engring Mechanics 610 SW Mudd Bldg New York NY 10027

BOLEY, MARK S., physicist, mathematician; b. Carthage, Ill., Feb. 23, 1967; s. Delbert Lawrence and Ruth Ann (McHargue) B.; m. Leah Grace Starbuck, May 12, 1991. BS in Physics summa cum laude, Western Ill. U., 1987, MS in Physics summa cum laude, 1989; postgrad., U. Mo., Columbia. Undergrad. asst. Western Ill. U., Macomb, 1985-87, grad. teaching asst., 1987-89, instr., researcher, 1989-90; rsch. asst. U. Mo., Columbia, 1990—, teaching asst., 1991—; cons. Materials Rsch. Soc., Boston, 1989—; mem. rsch. staff Nuclear Physics Lab, UIUC, Champaign, Ill., summer 1988, Argonne (Ill.) Nat. Lab., 1989. Contbr. articles to profl. publs. Pastor New Woodville Bapt. Ch., Wyaconda, Mo., 1989—, Cedar Grove Bapt. Ch., Kahoka, Mo., 1991—; entertainer Clark County Nursing Home, Kahoka, 1986—. Ernest Landen fellow, O.M. Stewart fellow, G. Ellsworth fellow, 1990—, NSF, Dept. Def. fellow 1990-92. Fellow Am. Phys. Soc.; mem. Am. Assn. Physics Tchrs., Sigma Pi Sigma. Achievements include development of highest critical currents known in high-temperature superconducting oxides; new circuitry to measure the magnetic hysteresis of materials. Office: U Mo Dept Physics & Astronomy Columbia MO 65211

BOLIE, VICTOR WAYNE, electrical and computer engineering educator; b. Silverton, Oreg., July 23, 1924; m. Earleen Mercia Dale, Mar. 12, 1945. BS in Physics, Iowa State U., 1949, MS in Math., 1950, PhD in Math. and Elec. Engring., 1952; BA in Chemistry, Coe Coll., 1957; MA in Physiology, Stanford U., 1959. Registered profl. engr., Okla., N.Mex.; lic. comml. pilot. Rsch. adminstr. Collins Radio Co., 1952-57; assoc. prof. Iowa State U., 1957-58, prof., chmn. biomed. engring., 1959-63; rsch. adminstr. Rockwell Internat. Corp., 1963-66; prof. elec. engring. U. Ariz., 1966-67; chaired prof. Okla. State U., 1967-71; chmn. dept. elec. and computer engring. U. N.Mex., 1971-76, prof. elec. and computer engring., 1976—; team mem. Engring. Coll. Accreditation Bd. Engring. Tech., 1969-76. Author over 90 publs. in field; mem. editorial bd. Biomed. Engring. Trans. IEEE, 1967-70; dir. 33 MS and PhD theses; 38 patents, 2 copyrights. 1st lt., multi-engine pilot, instr., USAF, 1942-47. NSF sr. postdoctoral fellow, 1958; recipient Gold Ring Highest Acad. Achievement award USAF, 1944, Rsch. Dir. award Morris Animal Found., 1961, Disting Rsch. Svc. award U. N.Mex., 1988, Cert. Recognition Los Alamos Nat. Lab., 1988. Fellow IEEE (nat. chmn. joint com. on engring. in medicine and biology, 1964-65); mem. NSPE, Am. Physiol. Soc., Air Force Assn., Res. Officers Assn. Office: U NMex Dept Elec & Computer Engring Albuquerque NM 87131

BOLIN, EDMUND MIKE, electrical engineer, franchise engineering consultant; b. Bowman, S.C., Sept. 16, 1944; s. Wells Connor and Rebecca May (Dukes) B.; m. Patricia Elmira McGowan, Sept. 1, 1979 (div. Jan. 1990); 1 child, Theresa Michele Lufkin. AA, Brevard (N.C.) Jr. Coll., 1965; BSEE, Clemson U., 1968. Registered profl. engr., S.C. Engr., dir., videographer Communications Ctr., Clemson (S.C.) U., 1965-68; supervisory nuclear engr. Charleston (S.C.) Naval Shipyard, 1968-87; test and evaluation engr. Naval Electronic Systems Engring. Ctr., Charleston, 1987—; quality and tech. cons. Support Systems Internat., Charleston, 1977-79, Exposure 60 Corp., Charleston, 1992—; producer, scriptwriter Trident Prodns., Charleston, 1989; staff writer, photographer Santee (S.C.) Scene and Scene Newspaper Group, 1990-92, North Charleston News, 1992—; lectr. Gibbs Art Mus., Charelston, 1990—. Weekly columnist What's Happening in and Around S.C., 1990—, From the White House, 1991—; inventor automated R-meter, improved satellite receiver, laser alignment system. Mem. Charles County Rep. Com., Rep. Nat. Com. for S.C.; tech. adv. Edn. Svc. Corp. Action 2, 000, S.C., 1991—; photographer and technical adv. S.C. Christian Coalition, 1991—. Scholar Elks Club, 1963. Mem. S.C. Registered Profl. Engrs. Assn., Advt. Fedn. Charleston, Trident C. of C., Charleston Tall Club (media rep. 1990—), Trident Amateur Radio Club. Avocations: computers, inventing, innovative consulting, writing, photography. Home: 7650 Ovaldale Dr Charleston SC 29418-3241 Office: Naval Electronics System Engring Ctr 4600 Marriott Dr Charleston SC 29418-6504

BOLLICH, CHARLES N., agronomist. Rsch. leader USDA Rice Rsch. Unit, Beaumont, Tex. Recipient Agronomic Achievement award Am. Soc. Agronomy, 1991. Office: USDA Rice Rsch Unit RR 7 Box 999 Beaumont TX 77713-8530*

BOLLINGER, JOHN GUSTAVE, engineering educator, college dean; b. Grand Forks, N.D., May 28, 1935; s. Elroy William and Charlotte (Kirchner) B.; m. Heidelore Ladwig, Aug. 16, 1958; children: William, Kristin, Pamela. BSME, U. Wis., 1957, PhDME, 1961; MSME, Cornell U., 1958. Registered profl. engr., Wis. Asst. prof. dept. mech. engring. U. Wis., Madison, 1961-65, assoc. prof., 1965-68, prof., 1968-85, Bascom prof. engring., 1973—, chmn. dept., 1975-79, prof. elec. and computer engring. dept., prof. indsl. engring. dept., 1985—, dean Coll. Engring., 1981—; bd. dirs. Andrew Corp., Orland Park, Ill., Kohler (Wis.) Corp., Nicolet Instrument Corp., Madison, EIT, Inc., Irvine, Calif. Co-author: Introduction to Automatic Controls, 1963, Computer Control of Machines and Processes, 1988; contbr. numerous articles to profl. jours.; founder, editor Jour. Mfg. Systems, 1981-89, assoc. editor, 1990—; holder 11 patents. Bd. dirs. Wis. for Rsch., Madison, 1987—, Meriter Hosp., Madison, 1987-89. Recipient Donald P. Eckman award Am. Automatic Control. Coun. of ASME, IEEE, Am. Inst. Chem. Engring., AIAA, Instrument Soc. Am., 1965, Disting. Univ. Achievement award U. Wis. Alumni Club, 1986, Triple E Educator of Yr. award Marwick Main & Co., Milw., 1989. Fellow ASME (Gold medal 1965, Gustus Larson award 1976, Centennial award 1980), Soc. Mfg. Engrs. (rsch. medal 1978); mem. NAE, Am. Soc. for Engring. Edn., NSPE, Wis. Soc. Profl. Engrs., N.Am. Mfg. Rsch. Confs., Internat. Instn. Prodn. Engring. (pres. 1986-87). Avocations: music, skiing, sailing. Home: 6117 S Highlands Ave Madison WI 53705-1112 Office: U Wis Coll Engring 1513 University Ave Madison WI 53706-1572

BOLON, ALBERT EUGENE, nuclear engineer, educator; b. Montgomery, Ala., July 19, 1939; s. Harry Cloyd and Pearl Juanita (Feen) B.; m. Sherrell Rae Roberts, June 19, 1965; children: Cynthia Pearl, Bruce Thomas. BS in Physics, Mo. Sch. Mines, 1961, MS in Nuclear Engring., 1962; PhD in Nuclear Engring., Iowa State U., 1965. Registered profl. engr., Mo. Grad. instr. Iowa State U., Ames, 1964-65; asst. prof., now assoc. prof. nuclear engring. U. Mo., Rolla, 1965—, reactor dir., 1981—, chmn., 1988-93. Co-author: Starfire: A Commercial Tokamak Power Plant Design, 1981, Wildcat: A Commercial D-D Tokamak Reactor, 1983. Mem. NSPE, Am. Nuclear Soc., Rolla Lions Club (pres. 1988-89), Sigma Xi. Methodist. Home: RR 4 Box 33 Rolla MO 65401-9354 Office: U Mo Nuclear Engring Rolla MO 65401

BOLON, BRAD NEWLAND, veterinary pathologist; b. Kansas City, Mo., Nov. 12, 1961; s. Lucien Milton Jr. and Sammy Louise (Searcy) B.; m. Janine Kaye Dalziel, Mar. 14, 1987. BS, U. Mo., 1983, DVM, 1986, MS, 1986; PhD, Duke U., 1993. Diplomate Am. Coll. Vet. Pathologists. Anatomic pathology resident U. Fla., Gainesville, 1986-88, grad. rsch. assoc., 1988-89; postdoctoral trainee Chem. Industry Inst. Toxicology, Research Triangle Park, N.C., 1989-93; sr. pathologist Pathology Assocs., Inc., Frederick, Md., 1993—. Author: (chpts.) Fish Medicine, 1991, The Experimental Animal in Biomedical Research, vol. 2, 1993. Mem. AAAS, Am. Vet. Med. Assn., Internat. Assn. Aquatic Animal Medicine, Am. Coll. Vet. Pathologists. Office: Pathology Assocs Inc Ste 1 15 Wormans Mill Ct Frederick MD 21701

BOLONCHUK, WILLIAM WALTER, physical educator; b. Winnipeg, Man., Can., Jan. 30, 1934; came to U.S., 1967; s. Walter and Amalia (Deckert) B.; m. Kay Monica Brandhagen, Aug. 25, 1955; children: Thomas, Susan, Gerald, Richard, Robert, Jonathan. BSc in Edn., U. N.D., 1956; MSc, U. Sask., Can., 1966. Tchr. Winnipeg Sch. Div. #1, 1956-61; faculty U. Sask., 1961-67, U. N.D., Grand Forks, 1967-92; rsch. assoc. Grand Forks Human Nutrition Rsch. Ctr., USDA, 1979-91. Author: (with others) Exercise Physiology: Current and Selected Research, 1991; contbr. articles to profl. jours. including Am. Jour. of Phys. Anthropology and Jour. of Phys. Fitness and Sports Medicine. Mem. Mayor's Adv. com., Grand Forks, 1980; chair Neighborhood Assn., Grand Forks, 1982. Rsch. grants USDA, 1979-92. Fellow Human Biology Coun., Am. Alliance for Health Phys. Edn. and Recreation; mem. AAUP, N.D. Acad. Sci., Sigma Xi. Achievements include rsch. on association between bodytype, exercise performance and nutrition. Home: 1038 Boyd Dr Grank Forks ND 58201 Office: U ND Box 8235 Grand Forks ND 58203

BOLONKIN, ALEXANDER ALEXANDROVICH, mathematician; b. Perm, Russia, USSR, Mar. 14, 1933; came to U.S., 1988; s. Olga Dmitrievna (Verevkina) B.; m. Margarita A. Puchnina, 1963 (div. May 1986); 1 child, Vladimir; m. Olga Samuilovna Lyubavina, Feb. 7, 1987. PhD, Moscow Aviation Inst., 1969; postgrad., Leningrad Politech., 1971. Sr. engr. Aircraft Bur. Antonov, C. Kiev, USSR, 1958-60; asst. prof. Moscow Aviation Inst., 1963-65; project dir. Rocket Bur. Glushko, Moscow, 1965-66; assoc. prof. Moscow Aviation Inst., 1966-70; prof. Moscow Tech. U., 1970-72, Tech. Inst., Ulan-Ude, USSR, 1982-88; analyst Shearson Lehman Hutton, N.Y.C., 1990; vis. scientist Courant Inst. Math. Sci., N.Y.C., 1990—; cons. Rsch. Lab. AES, N.Y.C., 1991—; editor-in-chief Math. Notes Moscow, 1963-65. Author: New Methods of Optimization, 1972, Development of Soviet Rocket Engines, 1991; contbr. scientific papers and articles to profl. jours. Pres. Internat. Assn. Former Soviet Polit. Prisoners and Victums of Communist Regime, 1990—. Mem. AIAA, Am. Math. Soc. Achievements include 13 patents. Home: 1001 Ave H #9-C Brooklyn NY 11230

BOLONKIN, KIRILL ANDREW, aeronautical engineer; b. Sydney, Australia, Dec. 5, 1929; s. Andrew Laurence and Maria Nicholas (Morozova) B.; m. Nola Davis, May 21, 1967; children: Nicholas Kirill, Natasha Vera. ASTC, Sydney Tech. Coll., 1953; B in Engring., U. New South Wales, 1963. Apprentice engr. Qantas Empire Airways Pty. Ltd., Sydney, 1948-52, engr., 1953-59; aero. engr. Dept. Civil Aviation, Sydney, Melbourne, Australia, 1959-67; exptl. officer 4 Dept. Supply, Woomera, South Australia, 1968-75, Dept. Def., Canberra, Australia, 1975-83; engr. class 5 Dept. Aviation, Canberra, 1983-92; cons. Canberra, 1992—. Mem. Royal Aero. Soc., Instn. Engrs. Australia, AIAA, Canberra Club. Russian Orthodox. Achievements include development of airworthiness certification requirements for Australian aircraft for commuter and ultralight operations. Home and Office: 27 Brand St, Hughes 2605, Australia

BOLT, BRUCE ALAN, seismologist, educator; b. Largs, Australia, Feb. 15, 1930; came to U.S., 1963; s. Donald Frederick and Arlene (Stitt) B.; m. Beverley Bentley, Feb. 11, 1956; children: Gillian, Robert, Helen, Margaret. BS with honors, New Eng. U. Coll., 1952; MS, U. Sydney, Australia, 1954, PhD, 1959, DSc (hon.), 1972. Math. master Sydney (Australia) Boys' High Sch., 1953; lectr. U. Sydney, 1954-61, sr. lectr., 1961-62; research seismologist Columbia U., 1960; dir. seismographic stas. U. Calif., Berkeley, 1963-89, prof. seismology, 1963-93, prof. emeritus, 1993—, chmn. acad. senate, 1993—; mem. com. on seismology NAS, 1966-72, chmn. nat. earthquake obs. com., 1979-81; mem. earthquake and wind forces com. VA, 1971-75; mem. Calif. Seismic Safety Commn., 1978-93, chmn., 1984-86; earthquake studies adv. panel U.S. Geol. Survey, 1979-83, U.S. Geodynamics Com., 1979-84. Author, editor textbooks on applied math., earthquakes, geol. hazards and detection of underground nuclear explosions. Recipient H.O. Wood award in seismology, 1967, 72; Fulbright scholar, 1960; Churchill Coll. Cambridge overseas fellow, 1980, 91. Fellow Am. Geophys. Union (mem. geophys. monograph bd. 1971-78, chmn. 1976-78), Geol. Soc. Am., Calif. Acad. Scis. (trustee 1981-92, pres. 1982-85, Fellows medal 1989), Royal Astron. Soc. (assoc.); mem. NAE, (IDNDR com. 1992—), Seismol. Soc. Am. (editor bull. 1965-70, bd. dir. 1965-71, 73-76, pres. 1974-75), Internat. Assn. Seismology and Physics Earth's Interior (exec. com. 1964-67, v.p. 1975-79, pres. 1980-83), Earthquake Engring. Research Inst., Calif. Univs. Rsch. Earthquake Engring., (sec. 1988-91), Australian Math. Soc., Sigma Xi, Univ. Club, Chit Chat Club, Bohemian Club. Research on dynamics, elastic waves, earthquakes, reduction geophys. observations; inferences on structure of earth's interior; cons. on seismic hazards. Home: 1491 Greenwood Ter Berkeley CA 94708-1935

BOLT, EUGENE ALBERT, JR., historic natural science museum curator, historian; b. Pitts., Dec. 2, 1963; s. Eugene Albert and Nancy Jane (Double) B. AB, U. Pa., 1988. Asst. to registrar U. Mus., Phila., 1983-86; rsch. historian Clio Group, Inc., Phila., 1984-88; curator Wagner Free Inst. Sci., Phila., 1988—. Author: Apocrypha of Etiquette, 1992; author (play) Appointment in Samarra, 1990; editor (collected essays) Margin to Mainstream, 1992. Bd. dirs. North Phila. Partnership, 1991; mem. Cecil B. Moore Devel. Consortium, Phila., 1991. Fellow Royal Geog. Soc.; mem. Philomathean Soc. (sr., Charles Fine Ludwig award), Explorers Club (Cert. of Merit), Libr. Co. of Phila., Racquet Club Phila., Franklin Inn Club. Lutheran. Achievements include the writing of Nat. Hist. Landmark application for Wagner Free Institute of Science. Home: 2008 Spruce St Philadelphia PA 19103 Office: Wagner Free Inst Sci Montgomery Ave and 17th St Philadelphia PA 19121

BOLT, MICHAEL GERALD, metallurgist; b. Sharon, Pa., Mar. 3, 1953; s. Thomas Bennett and Mary Jane (Lyons) B.; m. Roberta Ann Taylor, Oct. 14, 1972; 1 child, Jennifer Lynn Bolt. BA, Mansfield U., 1975; MS in Student Personnel, Slippery Rock U., 1991. Sci. tchr. Cranford (N.J.) Sch. Dist., 1975-76; metallurgical lab. tech. Wheatland (Pa.) Tube Co., 1976—. Treas. Mercer Crawford County Rails to Trails. Recipient Citizenship award Mercer (Pa.) County Govt., 1974. Mem. Am. Assn. Counseling and Devel., Western Pa. Conservancy, Shenago Conservancy (bd. dirs.), Rails to Trails. Democrat. Roman Catholic. Avocations: bicycling, running, cross country skiing, reading. Home: RD2 Box 252 Patricia Dr W Transfer PA 16154-9305 Office: Wheatland Tube Co Council Ave Wheatland PA 16161

BOLTON, JULIA GOODEN, hospital administrator; b. Wilmington, Del., Nov. 11, 1940; d. Merrill Harvey and Mary Rose (Amoroso) Gooden; m. Roger Edwin Bolton, June 27, 1964; children: Christopher Andrew, Jonathan Hughes. RN with honors, Johns Hopkins Hosp., Balt., 1961; BSN with honors, Case Western Res. U., Cleve., 1964; postgrad., Boston U., 1964-65; MS with honors, Russell Sage Coll., 1986. Lic. nurse, Vt. Staff nurse operating rm., clin. instr. Johns Hopkins Hosp., Balt., 1961-62; instr. practical nursing, acting coord. med. programs Charles H. McCann Vocat. Sch., North Adams, Mass., 1966, clin. instr. manpower devel. tng. act program, 1968, clin. instr. med., surg. and pediatric nursing, 1972-73; staff orientation and tour program for children North Adams Regional Hosp., 1973-74; health edn. cons. Williamstown (Mass.) Pub. Schs., 1978-81, Pine Cobble Sch., 1978-81; clin. cons. patient care stds. project North Adams Regional Hosp., 1985-86; dir. staff edn. and quality assurance Southwestern Vt. Med. Ctr., Bennington, 1986-87, asst. v.p. nursing, 1988, v.p. nursing, 1988-92, interim pres., 1991, sr. v.p., 1992—; mem. client adv. com. Seiler's Corp., 1992. Adv. com. Putnam Meml. Sch. Practical Nursing, 1989—; profl. adv. com. Bennington Home Health Agy., 1988; alt. del. Diocesan Conv. No. Berkshire Deanery, Episcopal Ch., 1987; dir. Vt. div., Bennington County unit, Am. Cancer Soc., 1986-88; mem. Williamstown Betterment Study Com., 1985; adv. com. to plan for declining enrollments Mt. Greylock Reg. High Sch., 1985; bd. dirs. exec. com. Vt. Nursing Initiative Implementation Grant, Pew Charitable Trust Grant, 1992; vestry St. John's Episcopal Ch., Williamstown, 1992; active many other civic and charitable orgns. in past. Recipient Hannah Karp award as outstanding student, Russell Sage Coll., 1985, traineeship, 1983-85, others. Mem. Am. Coll. Healthcare Execs., Am. Orgn. Nurse Execs., Nat. Forum Women Health Care Leaders, Nat. League for Nursing, Vt. Orgn. Nurse Execs., Rotary, Phi Kappa Phi, Sigma Theta Tau.

BOLTON, MERLE RAY, JR., cardiologist; b. Biloxi, Miss., Apr. 8, 1943; m. Ruth Hilden; 2 children. BA in English, U. Kans., 1965; MD, U. Kans., Kansas City, 1969. Diplomate Am. Bd. Internal Medicine, Am. Bd. Cardiology. Intern U. Kans. Med. Ctr., Kansas City, 1969-70, resident, 1970-71, fellow in cardiovascular diseases, 1972-73; pvt. practice, 1971—; sr. attending staff (cardiology, internal medicine) Eisenhower Med. Ctr., Rancho Mirage; chmn. Dept. Medicine, 1978-80, chief cardiology 1980-84, med. dir. CCU, 1980-84, med. dir. cardiac rehab. program, 1980-84, med. edn. com.; asst. clin. prof. medicine U. So. Calif., 1980-91. Contbr. numerous articles, manuscripts to profl. publs. Lt. cmdr. USNR, 1974-76. Named Intern Yr. Dept. Medicine U. Kans. Med. Ctr., 1969-70; recipient Physician's Recognition award AMA, 1974-77, 77-91. Fellow AMA, ACP, Am. Coll. Cardiology, Coun. Clin. Cariology Am. Heart Assn., Am. Coll. Chest Physicians (chmn. bylaws com. 1989-90, scientific program com. 1989-91, awards com. 1991); mem. Calif. Med. Assn., Riverside County Med. Assn., N.Y. Acad. Scis., Am. Inst. of Ultrasound in Medicine, Am. Soc. for Echocardiography, Alpha Omega Alpha. Home: 490 Mariscal Palm Springs CA 92262 Office: 39-000 Bob Hope Dr Rancho Mirage CA 92270

BOLZ, ROGER WILLIAM, mechanical engineer, consultant; b. Cleve., May 17, 1914; s. William and Amelia Anna (Waechter) B.; m. Ruth Elizabeth Wemple, June 19, 1943; children: Charlotte Bolz Talmadge, Hazel J., Lori Bolz Spivey, Woodrow L., Martha Bolz Kusterer. Student, Case Inst., Cleve., 1936-41. Registered profl. engr., Ohio, Calif. Design engr. Weatherhead Co., Cleve., 1934-38, Reliance Electric Co., Cleve., 1938-40, Nat. Carbon Co., Cleve., 1940-41; plant engr. Nat. Carbon Co., Clarksburg, W.Va., 1941-44; editor Penton Pub. Co., Cleve., 1944-68; cons. engr. Automation for Industry, Cleve. and Lewisville, N.C., 1968—; cons. engr. Oak Ridge (Tenn.) Nat. Labs., 1968-78. Author: Production Processes, 1949, 77, Understanding Automation, 1966, How to Automate Your Plant, 1980, Manufacturing Automation Management, 1985; co-author 4 books, 1950-84; editor IEEE Manpower Reports, 1973-79. Recipient 50th Anniversary award USDL, 1961, cert. of appreciation U.S. Dept. Commerce, 1968, Disting. Svc. award Cleve. State U., 1971-74. Fellow ASME (life); mem. IEEE, Soc. Mfg. Engrs., Am. Soc. for Metals Internat. (chmn. 1988-89), Cleve. Engring. Soc. (life). Republican. Methodist. Achievements include design of centrifuge for uranium enrichment. Office: Automation for Industry 203 Wyntfield Dr Lewisville NC 27023-9517

BONA, CHRISTIAN MAXIMILIAN, dentist, psychotherapist; b. Breslau, Schlesien, Germany, Apr. 24, 1937; arrived in Sweden, 1951; s. Humbert Serafin and Ingeborg Jenny (Holmgren) B.; m. Eva Tengblad, June 10, 1962 (div. 1980); children: Christian Georg, Richard Rolf, Henrik Nils; m. Monica Siv Karlsson, Jan. 26, 1982. DDS, Dental High Sch., Malmoe, Sweden, 1962; PhD in Acupuncture, U. Gothenburg, Sweden, 1986; cert. Traditional Chinese Medicine, U. Beijing, 1987. Asst. prof. Dental High Sch., Gothenburg, 1972-76; head dental health dept. Eriksbergs Mer. Verkstads AB, Gothenburg, 1972-78; head dental health and psychotherapy dept. Tandhalsovarden Mariaplan, Gothenburg, 1984—; sec. Swedish Dental Assn., Gothenburg, 1968-72, Swedish Soc. Clin. Hypnosis, Gothenburg, 1979-81, pres., 1981-84; cons. U. Computeraided Adminstrn., Gothenburg, 1972-78, U. Dental Practice Adminstrn., 1976-78. Author: Dental Psychotherapy and Hypnosis, 1986. Mem. German-Sweden Assn. (treas. 1968-71), Partille Sailing Club (founder 1976, pres. 1976-81), Fram Sailing and Yacht Club (bd. dirs. 1981-86), Knights Sovereign Order Hosp. St. John Jerusalem in Denmark. Avocations: sailing, down-hill and cross-country skiing, photography. Office: Tandhalsovarden Mariaplan, Mariagatan 11 B, 41471 Gothenburg Sweden

BONADONNA, GIANNI, oncologist; b. July 28, 1934. MD in Medicine and Surgery, Milan U., 1959; degree in Haematology, Ferrara U., 1973; degree in Oncology, Pavia U., 1973. Resident divsn. pathology Santa Cabrini Hosp., Montreal, 1960-61; postdoctoral rsch. fellow divsn. chemotherapy Sloan-Kettering Cancer Ctr., N.Y., 1963-64; rsch. fellow diagnostics divsn. Nat. Inst. Cancer, Milan, 1964-69; asst. Nat. Inst. Cancer, 1969-76, v.p., 1974-78; pres, Italian Assn. Med. Oncology, 1976—; researcher, head dept. cancer medicine Instituto Nazionale Tumori, Milan, Italy. Recipient ACS Medal of Honor Am. Cancer Soc., 1991; Unrestricted Cancer Research Grantees Program, Bristol-Myers Squibb, 1993; Internat.

Soc. Chemotheraphy award, David A. Karnorsky Meml. award Am. Soc. Clinical Onocology; Josef Steiner Cancer Rsch. prize U. Bern Switzerland. Achievements include the first protocol with adjuvant CMF combination chemotherapy, demonstrating its efficacy in a variety of breast cancers and as primary chemotherapy for operable breast cancer; For Hodgkin's disease developed the ABVD combination therapy showing it to be non-cross resistant and superior to the traditional drug regimen. Office: Istituto Nazionale Tumori, Dept Cancer Medicine, 20120 Milan Italy*

BONAKDAR, MOJTABA, chemistry educator; b. Malayer, Teheran, Iran, Mar. 22, 1959; came to U.S., 1979; s. Morteza and Soghra (Layaghi) B.; m. Norma Ota, Jan. 4, 1992. BS, Southeastern Okla. State U., 1982; PhD, N.Mex. State U., 1987; postgrad., Okla. State U., 1988. Study dir. Analytical Biochemistry Lab., Columbia, Mo., 1988-89; staff scientist Radian Corp., Austin, 1989-90; faculty So. Ark. U., Magnolia, 1990-93; sr. scientist Alcon Lab., Ft. Worth, 1993—. Contbr. articles to profl. jours. including Jour. of Analytical Chemistr, Jour. of Chromatography, Jor. of Electroanalysis and Analytical Chimica Acta. Grantee NSF, 1991; rsch. fellowship Dept. Energy, 1988, Phillips Petroleum, 1987, Am. Heart Assn., 1986. Mem. Am. Chem. Soc. (sec. of treasury 1992-93), Kiwanis, Sigma Xi. Home: 6623 Townlake Cr Arlington TX 76016 Office: Alco Labs Inc 6201 S Fwy Mailcode R1-16 Fort Worth TX 76134-2099

BONATE, PETER LAWRENCE, pharmacologist; b. Elmhurst, Ill., Aug. 27, 1964; s. Peter Paul and Corrine Ann (Bramwell) B.; m. Diana Bonate, May 25, 1990. MS, Wash. State U., 1990, U. Idaho, 1990. Clin. chemist Assoc. Pathologist Labs., Las Vegas, 1982-88; pharmacologist Eli Lilly and Co., Indpls., 1990--. Contbr. articles to profl. jours. Mem. Am. Assn. Pharm. Sci., Am. Statistical Assn. Roman CAtholic. Office: Eli Lilly and Co Wishard Hosp 1001 W 10th St Indianapolis IN 46202

BONAVENTURA, JOSEPH, biochemist, educator, research center director; b. Oakland, Calif., Feb. 15, 1942; s. Filiberto Antonio and Corinne (Fogarty) B.; m. Celia Jean Taylor, Aug. 20, 1960 (div.); children: Marina Celeste, Michelle Celia. BA, San Diego State U., 1964; PhD, U. Tex., 1968. Rsch. assoc. Duke U. Med. Ctr., Beaufort, N.C., 1972-75, asst. med. rsch. prof., 1975-84, dir. Marine Biomed. Ctr., 1978—, assoc. prof., 1984-90, prof., 1990—; Bd. dirs. N.C. Biotech. Ctr., Research Triangle, Marine Environmental Rsch. Inc. Mem. editorial bd. Hemoglobin, 1977—, Jour. Molecular Recognition, 1991—; mem. adv. bd. Molecular Physiology, 1980—; patentee in field, including Hemosponge (oxygen extractor) and human blood substitutes. Established Investigator grantee AMA, 1975-80; Biotech Devel. grantee MacGregor Found., 1989, 91. Mem. AAAS, Am. Soc. Zoologists, Biophysics Soc., Am. Chem. Soc., Sigma Xi. Home: 127 Circle Dr 114 Broad St Beaufort NC 28516-1601 Office: Duke U Marine Lab Marine Biomedical Ctr Beaufort NC 28516

BONCHEV, DANAIL GEORGIEV, chemist, educator; b. Burgas, Bulgaria, Feb. 20, 1937; s. Georgi Nikolov and Penka Danailova (Konstantinova) B.; m. Pravdolyuba Vladimirova, Oct. 31, 1960 (div. 1983); 1 child Adelina Boncheva-Karakoleva; m. Dimitrina Kostova Kostova, June 10, 1984; 1 child, Elina. BS in Chem. Engring., High Inst. Chem. Tech., Sofia, Bulgaria, 1960; PhD in Quantum Chemistry, Acad. Scis., Sofia, Bulgaria, 1970; DSc in Math., Chemistry, State U., Moscow, 1984. Chem. engr. Chem. Komninat, Dimitrovgrad, Bulgaria, 1960-63; asst. prof. chemistry High Inst. Chem. Tech., Burgas, Bulgaria, 1963-72; assoc. prof., head dept phys. chemistry High Inst. Chem. Tech., Burgas, 1973-91, prof. chemistry, 1987—, dean inorganic chemistry faculty, 1987-91; head lab. math. chemistry Bulgarian Acad. Scis., Sofia, 1986-91; rector, founder Free Univ., Burgas, Bulgaria, 1991—; referee internat. jours. in theoretical chemistry. Author: Information-Theoretical Characterization of Chemical Structures, 1983, (textbook) Structure of Matter, 1979; editor: (book series) Mathematical Chemistry; editor bd. mem. Jour. Math. Chemistry, 1987—, MATCH, 1989—; contbr. 160 articles to internat. sci. jours. Decorated Cyril and Methodius order II, State Coun. Bulgaria, Sofia, 1987. Mem. Internat. Soc. Math., Chemistry (officer), Am. Chem. Soc., World Assn. Theoretical Organic Chemistry, Rotary Club. Achievements include contbns. to characterization of molecular topology, molecular branching, cyclicity, centrality; in deriving the properties of chem. elements (transactinids), compounds, polymers and crystals from their structure; in the classification, coding, and complexity of chemical compounds and mechanisms of chemical reactions, in developing chemical information theory, etc. Office: MD Anderson Cancer Ctr Dept Thoracic Surgery 1515 Holcombe Blvd Box 151 Houston TX 77030

BOND, NELSON LEIGHTON, JR., health care executive; b. Glen Ridge, N.J., Apr. 17, 1935; s. Nelson Leighton and Dorothy Louise (Minsch) Hudson B.; m. Susan Priscilla McDonald, June 7, 1958 (div. May 1981); children: Sally Louise, Nelson Leighton III, Trevor Paul, Elizabeth Prescott, Susan Bond Kearney; m. Gwendolen Nash Gorman, July 24, 1982. BA, Lehigh U., 1957; MBA, Harvard U., 1966. Dist. mgr. McGraw Hill, Inc., N.Y.C., 1957-64; assoc. McKinsey and Co., Inc., N.Y.C., 1966-68; fin. analyst Drexel Harriman Ripley, Inc., N.Y.C., 1968-69; instl. salesman Faulkner Dawkins and Sullivan, N.Y.C., 1969-70; v.p. Alex Brown and Sons, Balt., 1970-77; pres., dir. Blood Pressure Testing, Inc., Reisterstown, Md., 1977—; pres. Consumer Micrographics, Inc., Balt., 1980-83; pres. Medscreen, Inc., Balt., 1987-89, also bd. dirs.; mng. dir. Offutt Securities, Inc., 1987—, Bond & Assocs., Reisterstown, Md., 1991—; chmn., pres., chief exec. officer, bd. dirs. Power Source, Inc., Reisterstown, The Green Spring Group, Inc., Reisterstown. Pres. Parents' Club St. Paul's Sch., Brooklandville, Md., 1978-79. 1st lt. USAR, 1958-60. Foote, Cone and Belding fellow Harvard U., 1965. Republican. Episcopalian. Avocations: golf, history, travel. Office: Bond & Assocs PO Box 1053 Reisterstown MD 21136-7053

BOND, RANDALL CLAY, quality assurance engineer; b. Tulsa, Aug. 26, 1953; s. Clay Boyd Bond and Mildred Gaylnn (Christian) Davidson; m. Regina Obutelewicz, Jan. 1975 (div. July 1980); 1 child, Katheryne Marie; m. Kimberll Ann Dingman, Sept. 28, 1984; children: Jamie, Amanda, Andrew. AA, La. Tech. Inst., 1978; AAS, USAF Community Coll., 1980; AS, Salem Community Coll., 1987; BS, Thomas A. Edison Coll., 1989. Cert. mech. and welding insp.; cert. profl. mgr. Pa. State U. With quality control div. Tom Co. Equipment Co., Loganville, Ga., 1979-80; quality engr. Avondale (La.) Shipyards, 1980-81; lead auditor Nuclear Installation Svc., Taft, La., 1981-82; sr. quality engr. Nuclear Installation Svc., Lakeland, Fla., 1983-84; quality engr. ITT-Grinnell, Oswego, N.Y., 1982-83; quality cons. Mich. Quality Systems, Midland, 1983-84; lead engr. Pub. Svc. Electric and Gas, Hancocks Bridge, N.J., 1985—. Author: (slide presentation) Delaware Valley: Our natural trail, 1993; contbr. articles to newspapers. Vol., March of Dimes, Salem County, N.J., 1988, 89, 9, 92, mem. corp. steering com., 1992; youth advisor Our Merciful Savior Ch., Carney's Point, N.J., 1989-92; radiol. officer Emergency Mgmt. Office, Salem County, 1985-86; sect. chmn. United Way, Salem 1988-90; craftsman Tree of Autumn State Fair, 1987; cook barbecue sauce State Fair, 1990; mem. dist. com. So. N.J. coun. Boy Scouts Am., 1991—; b. dirs. Holly Shores Coun. Girl Scouts Am., 1993. With USAF, 1972-79. Mem. Am. Soc. Quality Control (vice chmn. 1990-91, various awrds), PSE and G Speakers Bur (adv. com. 1991—,. Silver Club award 1991, Gold Club award 1992, exceptional speaker award 1993), Tall Cedars of Lebanon, Masons (sr. master of ceremonies 1986-87), Toastmasters Internat. (CTM 1993). Episcopalian. Avocations: golf, fundraising. Office: Pub Svc Electric & Gas PO Box 236 Hancocks Bridge NJ 08038-0236

BOND, WARD C., mathematics and computer educator; b. Sidney, Nebr., June 3, 1961; s. Eugene R. and Clara Kay (Meyer) B. BS, U. Wyo., 1985. Cert. tchr., Wyo., Kans., Nebr., Alaska. Math. tchr. Laramie (Wyo.) High Sch., Decatur Community High Sch., Oberlin, Kans.; math. and computer tchr., coach Carbon County Sch. Dist. 2, Hanna, Wyo.; high sch. math. and sci. tchr. Lower Kuskokwim Sch. Dist., Tuntutuliak, Alaska. Mem. NPA, ACTM, ASCD, NEA, Alaska Edn. Assn., Lower Kuskokwim Edn. Assn., Sigma Nu (Epsilonn Delta chpt. Outstanding Sr. Man). Home: Gen Delivery Tuntutuliak AK 99680

BONDI, ENRICO, engineer; b. Budrio, Bologna, Italy, Jan. 17, 1933; s. Renato and Dina (Andreini) B.; m. Bettina Grassani, Apr. 18, 1960; children: Antonella, Marco, Francesca. Cert., A. Righi, 1952; PhD in Mech. Engring., U. Bologna, 1958, cert., 1959. Registered profl. engr., Italy. Steel shop asst. Thyssen Group, Deusseldorf, Fed. Republic of Germany, 1959-60; rolling mills chief Cogne, S.p.A., Aosta, Italy, 1960-68; rolling mills project mgr. INNSE, S.p.A., Milan, 1968-90; mem. rsch. team Max Planck Inst., Duesseldorf, 1966-67; technology and engring. cons. to various cos., univs., polys., acads.; presenter in field. Patentee in field. Mem. Assn. Italiana di Metallurgia Plastic Deformation Ctr. (v.p. 1986-87, pres. 1988—). Avocations: photography. Home: Alzaia Naviglio Pavese 46, 20143 Milan Italy

BONDURANT, BYRON LEE, agricultural engineering educator; b. Lima, Ohio, Nov. 11, 1925; s. Earl Smith and Joy Koneta (Gesler) B.; m. Lovetta May Alexander, Feb. 28, 1944; children: Connie Jane Bondurant Jaycox, Richard Thayne, Cindy Lynn Bondurant Gardino. Student, Case Inst. Tech., 1943-44, Rensselaer Poly. Inst., 1944; B.S. in Agrl. Engring., Ohio State U., 1949; M.S. in Civil Engring., U. Conn., 1953. Registered profl. engr., Maine, Ohio registered surveyor, Maine. Dist. agrl. engr. western N.Y., N.Y. State Coll. Agr., Cornell U., 1949-50; instr. agrl. engring., extension agrl. engr., dept. agrl. engring. U. Conn., 1950-53; asso. prof. agronomy and agrl. engring., extension agrl. engr. dept. agronomy and agrl. engring. U. Del., 1953-54; prof. agrl. engring., head dept. U. Maine, 1954-64; prof. agrl. engring. Ohio State U., 1964—; also adviser to dean, later dean Coll. Agrl. Engring., Ludhiana, India, 1964-67, 69-71; vis. prof. agrl. engring. U. Nairobi, Kenya, 1974, Fulbright-Hays prof., 1979-80; project mgr. M.U.C.I.A., Mogadiscio, Somalia, 1976-78. Fellow AAAS, Inst. Engrs. (India and Kenya), Indian Soc. Agrl. Engrs. (life), Am. Soc. Agrl. Engrs. (vice chmn. N. Atlantic sect. 1956-57, chmn. Acadia sect. 1961-62, sec.-treas. Ohio and Tri-State sect. 1983-84, dir. internat. div. 1980-82, Kishida Internat. award 1983); mem. AAUP, Soc. Internat. Devel. (life), Am. Soc. Engring. Edn. (chmn. agrl. engring. div.), Nat. Soc. Profl. Engrs., Maine Soc. Profl. Engrs. (pres. 1963), Sigma Xi, Sigma Pi Sigma, Tau Beta Pi, Gamma Sigma Delta, Epsilon Sigma Phi, Alpha Epsilon. Home: 265 Franklin St Dublin OH 43017-1108 Office: Ohio State U 590 Woody Hayes Dr Columbus OH 43210-1057

BONEM, ELLIOTT JEFFREY, psychology educator; b. Chgo., July 2, 1953; s. Stuart Jerome Bonem and Rita Mae (Goldberg) Bear; m. Marilyn K. Krohn, June 19, 1976; children: Megan M., Garrett E. MA, Drake U., 1980; PhD, Utah State U., 1988. Psychology instr. Ea. Mich. U., Ypsilanti, 1987—. Contbr. articles to profl. publs. Mem. Am. Psychol. Soc., Assn. for Behavior Analysis, Behavior Analysis Assn. Mich. (pres. 1993—), Phi Kappa Phi. Achievements include discovery that shortest inter-reinforcement interval controls post-reinforcement latency. Office: Ea Mich U Psychology Dept Ypsilanti MI 48197

BONET, JOSE, mathematics educator, researcher; b. Valencia, Spain, June 18, 1955; s. Jose Bonet and Pilar Solves; m. Encarna Giner, Sept. 26, 1986. B in Math., U. Valencia, 1977, D in Math., 1980. Asst. U. Valencia, 1977-83; asst. prof. Poly. U., Valencia, 1983-86, prof., 1987—; vis. prof. U. Paderborn , Ger., 1989. Author: Espacios Tonelados, 1980, Barrelled Locally Convex Spaces, 1987; also articles; editor: Progress in Functional Analysis, 1992. Recipient 1st Nat. Prize Ministery Edn. Spain, 1978; Real Soc. Math. Spain grantee, 1972-77. Mem. Math. Assn. Am., Am. Math. Soc., Soc. Royale des Scis. de Liege (corr.). Avocations: sci. fiction, soccer. Home: Sequia Rascanya 2-6, 46120 Alboraya, Valencia Spain Office: Poly U, Dept Math, 46022 Valencia Spain

BONEWITZ, ROBERT ALLEN, chemist, manufacturing executive; b. Ottawa, Kans., Dec. 18, 1943; m. Ruth Lasell; children: Paul, Christopher, Cara. BA, Lawrence U., 1965; PhD in Chemistry, U. Fla., 1970. Rsch. engr. Aluminum Co. Am., Alcoa Ctr., Pa., 1970-73, sr. engr., 1973-75, head sect. materials and corrosion, 1975-79, mgr. surface tech., 1979-81, mgr. alloy tech., 1981-82, mgr. product engring., 1982-83, dir. ops. mill products R&D, 1983-84, dir. product design and mfg., 1984—. Contbr. to profl. publs. Mem. Am. Soc. Metals, Indsl. Rsch. Inst. (alt. rep.), Sigma Xi. Office: Alcoa Labs 100 Technical Dr Alcoa Center PA 15069-0001

BONIFACE, CHRISTIAN PIERRE, chemical engineer, oil industry executive; b. Marseille, France, Aug. 1, 1962; s. Jean and Marie-Louise (Truphemes) B. M in Chemistry, U. Marseille, 1984, grad. chem. engring., 1985, PhD in Organic Chemistry, 1987. Postdoctoral rsch. fellow U. Tex., El Paso, 1987-89, U. Fla. Atlantic, Boca Raton, 1989; analytical rsch. chem. engr. Brit. Petroleum, Lavera, France, 1989—; computer sci. tchr. U. Marseille, 1985-88. Contbr. articles to profl. jours. Grantee Robert A. Welch Found., 1987-89, Elsa J. Pardee postdoctoral fellowship grantee, 1989. Avocations: music, hiking, travel, coin and stamp collection, photography. Office: Brit Petroleum, BP1 CRAL, 13117 Lavera France

BONILLA, LUIS LOPEZ, physics and mathematics educator; b. Melilla, Spain, Dec. 27, 1956; s. Luis López Ruiz and María Bonilla Ortega. BS, MS, U. Autónoma, Madrid, 1978; PhD, U. Nat. de Edn. a Distancia, Madrid, 1981. Teaching asst. U. Nat. de Edn. a Distancia, Madrid, 1979-82; postdoctoral fellow math. Stanford (Calif.) U., 1982-85; postdoctoral fellow physics U. Sevilla (Spain), 1985-86, assoc. prof., 1987-90; assoc. prof. physics U. Barcelona (Spain), 1990-92; prof. applied math. U. Carlos III, Madrid, 1992—; vis. asst. prof. math. U. Calif., Irvine, 1986; vis. assoc. prof. math. Duke U., Durham, N.C., 1988-89. Contbr. articles on physics and math. to internat. jours. Fulbright scholar, Stanford, 1983-85. Mem. Soc. Indsl. and Applied Math., Am. Phys. Soc., Am. Math. Soc., European Consortium for Math. in Industry, N.Y. Acad. Scis., Venice Inst. Indsl. and Applied Math. Office: U Carlos III de Madrid Engr, Av Mediterraneo 20, 28913 Leganes Madrid Spain

BONIS, LASZLO JOSEPH, business executive, scientist; b. Budapest, Hungary, May 31, 1931; came to U.S., 1957; s. Joseph and Ilona (Hunvald) B.; m. Eva Markovich, July 31, 1955 (div. 1981); children: Andrea Christine, Peter Anthony Laszlo; m. Cheryl E. Olsen, Dec. 28, 1985. DM Ing. Mech. Engring., U. Tech. Sci., Budapest, 1953; MSc in Metallurgy, MIT, 1959, postgrad., 1959-60. Registered profl. engr., Calif., Mass.; cert. chemist Nat. Cert. Commn. Assoc. dir. material research Electronics, Inc., Budapest, 1953-56; prof. U. Tech. Sci., 1953-56; rsch. asst. MIT, Cambridge, 1957-60; exec. v.p., tech. dir. Ilikon Corp., Natick, Mass., 1960-62, pres., tech. dir., 1962-74; mgmt. cons. Tech. Fin. and Mktg., Inc., Natick, Mass., 1974—; pres., chmn., tech. dir. Composite Container Corp., Medford, Mass., 1977-88; pres. TFM, Inc., Dover, Mass., 1988—. Editor: (4 vols.) Fundamental Phenomena in the Material Science; contbr. articles to profl. jours.; patentee in field. Bd. dirs. The Opera Co., Boston, 1962-85, pres., 1966-85; pres. Boston Arts Coun., 1974—, Boston Opera House, 1991—. Recipient Muse award Pub. Action for the Arts, 1984, George Washington award Am. Hungarian Found., 1984, Golden Door award Internat. Inst., 1980; named One of Outstanding Young Men of Greater Boston C. of C., 1966. Fellow Am. Inst. Chemists; mem. N.Y. Acad. Scis., MIT Club. Office: TFM 52 Haven St Dover MA 02030-2131

BONN, FERDINAND J., geography educator, environmental scientist; b. Erstein, Bas Rhin Alsace, France, Oct. 7, 1943; s. Auguste Joseph and Adele (Andres) Bonn; m. Line Aubert, Sept. 28, 1985; children: Ariane, Etienne, Noemie. Licence es letters, Louis Pasteur U., France, 1968, MSc, 1969, PhD in Physical Geography, 1975. Prof. U. Strasbourg, France, 1969; head of instruction U. Sherbrooke, Canada, 1969-72, adj. prof., 1972-76, prof., 1977-79; vis. prof. U. Calif., Sanata Barbara, Calif., 1979-80; prof. U. Sherbrooke, Canada, 1981—; vis. scientist Inst. Nat. Rsch. Agronony, Avignon, France, 1988-89; prof. U. Sherbrooke, Canada, 1989—; dir. Remote Sensing Rsch. Ctr. (CARTEL), U. Sherbrooke, Canada, 1985—; adv. coun. mem. Remote Sensing Canadian Goverment. Author: Precis de Teledetection, 1992; editor: Remote Sensing Proceedings, 1978-93; Contbr. to profl. jours. Recipient Jacques Rousseau prize French Canadian Assn. for Advancement of Science, 1992. Canadian Remote Sensing Soc. (pres. 1985-86, past chmn.), Assn. Quebecoise de Teledetection (pres. 1979-83, past chmn., special tribute award, 1988), Am. Soc. Photoprocessing. Achievements include founding of the remote sensing center at University of Sherbrooke; conception of Canada's first PhD program in Remote Sensing. Office: Sherbrooke University, 2500 Blvd Universite, Sherbrooke, PQ Canada J1K 2R1

BONNAR, JOHN, obstetrics/gynecology educator, consultant; b. Lanark, Scotland, July 12, 1934; arrived in Ireland, 1975; s. John and Mary (Breen) B.; m. Elizabeth Murray, Sept. 17, 1960; children: John Paul, Christopher Matthew, Clare Elizabeth, James Peter. MB, BCh, U. Glasgow, Scotland, 1958, MD with honors, 1971; fellow, Trinity Coll., Ireland, 1976. Cert. ob-gyn. Sr. registrar ob-gyn. Royal Maternity Hosp. and Victoria Infirmary, Glasgow, Scotland, 1963-69; reader ob-gyn. U. Oxford, Eng., 1969-75; prof., head ob-gyn. Trinity Coll., Dublin, Ireland, 1975—, dean, faculty health scis., 1982-86; coun. mem. Med. Protection Soc., London, 1989—, Royal Coll. Ob-Gyn., London, 1991—. Author: Haemostatic Disorders of the Pregnant Woman and Newborn Infant, 1987; editor: Recent Advances in Obstetrics & Gynecology, 1992. Fellow Royal Coll. Ob-Gyn. (London); mem. Gynecol. Vis. Soc. Great Britain and Ireland, Soc. Pelvic Surgeons USA. Avocations: fishing, hill walking, golf. Home: 58 Deerpark Rd Castleknock, Dublin Ireland Office: Trinity Ctr Dept Ob-Gyn, St James's Hosp, Dublin 8, Ireland

BONNELL, DAVID WILLIAM, research chemist; b. Wichita, Kans., Sept. 29, 1943; s. Gus Travis and Mildred Maxine (CArner) B.; m. Paula Mary Wolfe, June 19, 1968 (div. 1974); m. Patricia McEwan, July 19, 1975; children: Elizabeth Ann and Megan Ruth (twins). BA in Chemistry, Rice U., 1968, PhD in Phys. Chemistry, 1972. Rsch. chemist Rice U., Houston, 1964-68, dept. mass spectrometrist, 1968-69, rsch. assoc., 1972-74; dir. rsch. MCR-Houston, Inc., 1974-75; rsch. chemist Nat. Inst. Stds. and Tech., Gaithersburg, Md., 1975—; mem. adv. com. NASA, 1985—; mem. adv. bd. Jour. Phys. Chem. Ref. Data, 1993—. Contbr. 60 articles to profl. jours. Recipient Dept. of Commerce Bronze medal, 1983, IR-100 award Indsl. Rsch./Devel., 1980. Democrat. Achievements include development of levitation calorimetric technique widely used for high temperature heat content data, liquid metals; co-developer Ideal Mixing of Complex Components thermodynamic model for refractory oxide systems, transpiration mass. spectrometer; research in laser/sputter thin film deposition in situ process diagnostics via MS, optical spectroscopy. Home: 19205 Tilford Way Germantown MD 20874 Office: Nat Inst Stds & Tech B106/223 Gaithersburg MD 20899

BONNER, DAVID CALHOUN, chemical company executive; b. Port Arthur, Tex., Nov. 20, 1946; s. Zora David and Dorothy (Shaw) B.; m. Lillian Yoshiko Hattori, Mar. 31, 1973; children: Marisa, David. BSChemE, U. Tex., 1967, MSChemE, 1969; PhD, U. Calif., Berkeley, 1972. Registered profl. engr., Tex. Asst. prof. chem. engring. Tex. Tech U., Lubbock, 1972-76, assoc. prof., 1976; assoc. prof. chem. engring. Tex. A&M U., College Station, 1976-77; sr. rsch. engr. Shell Devel. Co., Houston, 1977-78, supr. R & D, 1978-82, mgr. tech. support, 1979-82, supr. transp. applications, 1982-86; dir. corp. rsch. B.F. Goodrich Co., Brecksville, Ohio, 1986-88, v.p. R & D, 1988-92; sr. v.p., CTO Premix Inc., Astabula, Ohio, 1992—; bd. dirs. Edison Polymer Innovation Corp., Cleve.; mem. chem. engring. vis. com. U. Tex., Austin, 1987-90; mem. chem. engring. adv. bd. U. Calif. Berkeley, 1989—; mem. governing bd. Coun. for Chem. Rsch., Washigton; mem. nat. bd. Chem. Sci. and Tech. Contbr. numerous articles to profl. jours. Sr. warden Emmanuel Episc. Ch., Houston, 1979-82, Christ Ch. Cathedral, Houston, 1983-86; vestry mem. St. Paul's Episc. Ch., Akron, Ohio, 1989—; pres. music and performing arts Trinity Cathedral Inc. 2d lt. U.S. Army NG, 1970-76. Mem. Am. Inst. Chem. Engrs., Am. Chem. Soc., Ashtabula Country Club, Union Club of Cleve., Tau Beta Pi, Omega Chi Epsilon. Independent. Office: Premix Inc PO Box 886 Ashtabula OH 44004

BONNET, HENRI, engineering manager; b. Paris, France, May 25, 1934; came to U.S., 1962; s. Joseph and Janet (Olzak) B.; m. Ruth, Nov. 3, 1971; children: Sharon, Daniel. BSME, Technion, Haifa, Israel, 1960; BS in Architecture, CCNY, 1970. Air force flight instr. Israeli Air Force, Israel, 1955-61; systems engr. United Parcel Svc., N.Y.C., 1970-80; plant engr. mgr. United Parcel Svc., Atlanta, 1980—; automation engr. mgr. United Parcel Svc., Atlanta, 1980—. Lt. Air Force, 1957-60, Israel. Mem. ASME, IEEE, Optical Soc. Am., Nat. Fire Protection Assn., N.Y. Acad. Sci., Soc. Mfg. Engrs., Soc. Automotive Engrs. Republican. Jewish. Achievements include design and invention of at least 10 new machines for material handling automations (patents applied for). Home: 290 Crosstree Ln Atlant GA 30328 Office: United Parcel Svc Corporate Plant Engr 400 Perimeter Ctr Atlanta GA 30346

BONO, ANTHONY SALVATORE EMANUEL, II, data processing executive; b. N.Y.C., Nov. 24, 1946; s. Anthony S.E. and Lola M. (Riddle) B. BA in Polit. Sci., Hartwick Coll., 1969; cert. in info. systems analysis, UCLA, 1985. Mgmt. trainee Mfrs. Hanover Trust Co., N.Y.C., 1973-74; supr. client services Johnson & Higgins of Calif., Los Angeles, 1974-77, account exec. comml. accounts, 1977-80, coordinator internal systems, 1981-83, mgr. systems devel., 1983-89, v.p., mgr. systems devel., 1989—. Deacon Westwood Presbyn. Ch., 1982-85. Served with USAF, 1969-73. Named Airman of Yr., San Bernadino C. of C., 1970. Mem. Assn. Systems Mgmt. (dir. publicity and awards 1982-84, corr. sec. 1984-85), Channel Island Mensa, Alpha Sigma Phi. Republican. Home: 15010 Reedley St Moorpark CA 93021-2518 Office: Johnson & Higgins 2029 Century Park E Los Angeles CA 90067-2901

BONO, PHILIP, aerospace consultant; b. Bklyn., Jan. 14, 1921; s. Julius and Marianna (Culcasi) B.; m. Gertrude Camille King, Dec. 15, 1950; children: Richard Philip, Patricia Marianna, Kathryn Camille. B.E., U. So. Calif., 1947; postgrad., 1948-49. Research and systems analyst N.Am. Aviation, Inglewood, Calif., 1947; engring. design specialist Douglas Aircraft Co., Long Beach, Calif., 1948-49; preliminary design engr. Boeing Airplane Co., Seattle, 1950-59; dep. program mgr. Douglas Aircraft Co., Santa Monica, Calif., 1960-62; tech. asst. to dir. advanced launch vehicles and space stas. Douglas Aircraft Co., Huntington Beach, Calif., 1963-65; br. mgr. advanced studies, sr. staff engr. advanced tech. McDonnell Douglas Astronautics Co., Huntington Beach, 1966-73; sr. engr.-scientist Douglas Aircraft Co., Long Beach, 1973-83; engring. specialist Northrop Advanced Systems Div., Pico Rivera, Calif., 1984-86; mgr. Cal-Pro Engring. Cons., Costa Mesa, 1986—; lectr. seminars, univs. and insts. including Soviet Acad. Scis., 1965. Author: Destination Mars, 1961, (with K. Gatland) Frontiers of Space, 1969; contbr. articles to profl. jours., chpts. in books. Served with USNR, 1943-46. Recipient Golden Eagle award Council Internat. Events, 1964, A.T. Colwell merit award Soc. Automotive Engrs., 1968, M.N. Golovine award Brit. Interplanetary Soc., 1969, cert. of recognition NASA, 1983, Heritage medallion Project Italia, 1988; named engr. of distinction Engrs. Joint Council, 1971, Knight of Mark Twain, 1979. Fellow AAAS, Royal Aero. Soc. (sr.), Brit. Interplanetary Soc. (editorial adv. bd.), AIAA (assoc.); mem. Am. Astron. Soc. (sr.), N.Y. Acad. Scis., Internat. Acad. Astronautics (academician), ASME, Soc. Automotive Engrs. (chmn. space vehicle com.). Inventor recoverable single-stage space shuttle for NASA. Home: 1951 Sanderling Cir Costa Mesa CA 92626-4799

BONSIGNORE, JOSEPH JOHN, publishing company executive; b. Brooklyn, N.Y., Dec. 9, 1920; s. James Joseph and Isabella (Johnson) B.; m. Madelyn Anne Kleutsch Sept. 27, 1945; children: Mark, Judith, Jay, Andrea, Donna, Regina. BA, Trinity Coll., Hartford, Conn., 1942; MA, U. Chgo., 1945. Mgr. editorial prodn. Time, Inc., Chgo., 1945-69; gen. mgr. Smithsonian Mag., Washington, 1969-78, assoc. pub., 1978-81, pub., 1981-91; pub. Air & Space mag., Smithsonian Inst., 1985-91 (Joseph Henry medal 1991); cons. on formation new mags. Dem. precinct leader Rich Twp., Ill., 1949-69; pres. Rich Twp. Dem. Club, 1965-66; program chmn. Christian Family Movement, 1963-68, pres. Chgo. unit, 1960-62; pres. Civil Rights Orgn. South Suburban Cook County, 1967-68. Recipient Joseph Henry medal, Smithsonian Institution, 1991. Mem. Mag. Pubs. Assn. (bd. dirs. 1984—), Archeol. Inst. Am. (bd. dirs. 1988—), Pi Gamma Mu. Roman Catholic. Office: Smithsonian Arts & Industries Bldg 900 Jefferson Dr SW Washington DC 20560-0001

BONTEMPO, DANIEL, civil engineer; b. Phila., Mar. 10, 1962; s. Dante George and Marie (Carmella) B.; m. Donna Marie Lonegran, Feb. 8, 1985; children: Danielle Eran, Daniel Christopher. BSCE, Spring Garden Coll., Phila., 1984. Civil engr. State of Pa., Norris Town, 1984-85, Conrail, Phila., 1985-86; project engr. Unitech Engrs., Inc., Langhorne, Pa., 1986—; cons. Bristol (Pa.) Twp. Sch. Dist., 1987—. Mem. ASPE. Home: 4209 Remo Cres Bensalem PA 19020-2976

BONTING, SJOERD LIEUWE, biochemist, priest; b. Amsterdam, The Netherlands, Oct. 6, 1924; came to U.S., 1952; s. Sjoerd L. and Johanna H.M. (Hagedoorn) B.; m. Suzanne Maarsen, Jan. 10, 1951 (dec. Jan. 1986); children: Marion S., Paul S., Elizabeth J., Peter J.; m. Erica J.M. Schotman, Feb. 27, 1987. BSc in Chemistry, U. Amsterdam, 1944, MSc in Biochemistry cum laude, 1950, PhD in Biochemistry, 1952; lic. in theology (hon.), St. Mark's Inst. Theology, London, 1975. Ordained priest Episcopal Ch., 1964. Instr. U. Amsterdam, 1947-52; research assoc. State U. Iowa, Iowa City, 1952-55; asst. prof. biochemistry U. Minn., Mpls., 1955-56; asst. prof. U. Ill., Chgo., 1956-60; sect. chief NIH, Bethesda, Md., 1960-65; prof., chmn. dept. biochemistry U. Nymegen, The Netherlands, 1965-85; sci. cons. NASA Ames Research Ctr., Moffett Field, Calif., 1985-93. Editor: Transmitters in the Visual Process, 1976, Evolution and Creation, 1978, Membrane Transport, 1981, Advances in Space Biology and Medicine, 1989—; also articles. Bd. dirs. Multidisciplinary Ctr. for Ch. and Soc., Driebergen, The Netherlands, 1981-85; curate St. Luke's Episcopal Ch., Bethesda, 1963-65; Anglican chaplain Ch. of Eng., various cities, The Netherlands, 1965-85, 93—; asst. priest St. Thomas' Episcopal Ch., Sunnyvale, Calif., 1985-90, St. Mark's Episcopal Ch., Palo Alto, Calif., 1990-93. Postdoctoral fellow USPHS, Iowa City, 1952-54; Rudolf Lehmann Fund scholar, Amsterdam, 1941-46; recipient Fight for Sight citation Nat. Council to Combat Blindness and Assn. for Research in Ophthalmology, N.Y., 1961, 62, Arthur S. Flemming award, Jaycees, Washington, 1964, 1st prize Competition on Enzymology of Leucocytes Karger Found., Basel, Switzerland, 1964. Fellow AAAS; mem. Am. Soc. Cell Biology, Am. Soc. Biol. Chemists, The Netherlands Biochem. Soc. (officer 1973-76, v.p. 1976-79, pres. 1979-81), Sigma Xi. Democrat. Home: 12 Specreyese, 7471 Goor The Netherlands

BONUS, HOLGER, economics educator; b. Berlin, Germany, Feb. 15, 1935; s. Heinz Berthold and Lydia (Petersen) B.; m. Beate Kleindienst, Sept. 17, 1976; children: Bettina, Eva, Tizian. Diploma Volkswirt, U. Bonn, Germany, 1962, D in Polit. Econ., 1967, Dr. Habil Polit. Econs., 1971. Rsch. assoc. U. Bonn, 1962-67; post-doctoral fellow U. Chgo., 1967-69, rsch. assoc., 1969-70; tchr. U. Bonn, 1971-72; prof. U. Dortmund, Fed. Republic of Germany, 1973-78, U. Konstanz, Fed. Republic of Germany, 1978-84, U. MÜnster, Fed. Republic of Germany, 1984—; chmn. bd. dirs. Inst. Genossenschaftswesen, Münster, des Vorstandes Zentrum Umweltforschung, U. Münster, 1991-93; mem. Enquete-Kommission Schutz des Menschen und der Umwelt des Deutschen Bundestages. Author: Konsumgüter, 1975, Umweltschutz, 1984, Geld & Gold, 1990; co-author: Die WA(H)RE Kunst, 1991. Mem. Verein Socialpolitik, Rotary. Office: Inst f Genossenschaftswesen, Am Stadtgraben 9, D-48143 Münster Germany

BOODY, FREDERICK PARKER, JR., nuclear engineer, optical engineer; b. Oak Ridge, Tenn., Feb. 23, 1949; s. Frederick Parker and Ruth (Rich) B. BS, Rensselaer Poly. Inst., 1971, MS, 1973; PhD, U. Mo., 1991. Engr. Knolls Atomic Power Lab., Schenectady, N.Y., 1972-74; sr. devel. engr. Combustion Engring., Inc., Windsor, Conn., 1974-76; rsch. asst. U. Ill., Urbana, 1976-80; diagnostic physicist Princeton (N.J.) Plasma Physics Lab., 1980-86; rsch. scientist U. Mo., Columbia, 1986-91; pres. Ion Light Corp. (formerly Nuclear-Pumped Laser Corp.), Huntsville, Ala., 1981—; program com. Conf. on Physics of Nuclear-Pumped Lasers, Obninsk, Russia, 1991, 92; first western visitor Nuclear Weapons Insts, Arzamas-16 and Chelyabinsk-70, USSR, 1991; vis. prof. Fachhochschule Regensburg, Germany, 1993. Contbr. articles to profl. publs. Recipient Invention award Westinghouse Electric Corp., 1970. Mem. IEEE, Am. Nuclear Soc., Am. Phys. Soc., Optical Soc. Am., Sigma Xi, Tau Beta Pi. Achievements include co-development of photon intermediate direct energy conversion process for converting nuclear energy to light, electricity or useful chemicals. Home: 1202 Kingsway Rd SE Huntsville AL 35802 Office: Ion Light Corp PO Box 12744 Huntsville AL 35815

BOOKSTEIN, ABRAHAM, information science educator; b. N.Y.C., N.Y., Mar. 22, 1940; s. Alex and Doris (Cohen) B.; m. Marguerite Vickers, June 20, 1968. BS, CCNY, 1961; MS, U. Calif., Berkeley, 1966; PhD, Yeshiva U., 1969; MA, U. Chgo., 1970. Asst. prof. U. Chgo., 1971-75, assoc. prof. info. sci., 1975-82, prof. info. sci., 1982—; vis. prof. Royal Inst. Tech., Stockholm, 1982, UCLA, 1985; vis. disting. scholar OCLC, Columbus, Ohio, 1988; bd. dirs. Religion Index, Evanston, Ill., 1984—. Bd. editors Info. Processing and Mgmt., ORSA Jour. Computing; editor: Operations Research, 1972, Prospect for Change in Bibliographic Control, 1977; contbr. articles to profl. jours. NSF grantee, 1985, 87, 93; Fulbright fellow, India, 1992. Mem. IEEE Computer Soc., Am. Soc. Info. Scis. (Rsch. award 1991), Assn. Computing Machinery (gen. chmn. ann. internat. conf. 1991), Phi Beta Kappa. Avocations: personal computing, music, literature. Office: U Chgo/CILS 1100 E 57th St Chicago IL 60637-1596

BOONE, DAVID DANIEL, electrical engineer; b. Kansas City, Mo., Feb. 2, 1960; s. David Thompson and Marilyn (Raupp) B.; m. Jane Lynn Wilson, May 18, 1985. BSEE, U. Mo., Kansas City, 1986, postgrad., 1990—. Dir. engring. Equimed Med. Products, Overland Park, Kans., 1985-88; optical design engr. DIT-MCO Internat., Kansas City, Mo., 1988-90; sr. digital design engr. PPG Biomed. Systems, Lenexa, Kans., 1990—; pres. Blazer Cons., Lee's Summit, Mo., 1987—. Mem. Eta Kapa Nu. Achievements include design of numerous assistive devices for blind diabetics. Home: 1104 NE Long Ridge Rd Lees Summit MO 64064-2400 Office: PPG Biomed Systems 16505 W 113th St Shawnee Mission KS 66219-1383

BOONE, JAMES VIRGIL, engineering executive; b. Little Rock, Sept. 1, 1933; s. Virgil Bennett and Dorothy Bliss (Dorough) B.; m. Gloria Marjorie Gieseler, June 5, 1955; children: Clifford B., Sandra J. Smyser, Steven B. BS in Elec. Engring., Tulane U., 1955; M.S.E.E., Air Force Inst. Tech., Ohio, 1959. Assoc. elec. engr. Martin Co., Balt., 1955; R & D engr. USAF, 1955-62; electronics engr. Nat. Security Agy., Ft. Meade, Md., 1962-77, dep. dir. for rsch. and engring., 1978-81; spl. asst. to gen. mgr. Mil. Electronics div. TRW, Inc., San Diego, 1981-83, asst. gen. mgr., 1983-85, dir. program mgmt. and group devel. TRW Electronic Systems Group, 1985-86, v.p., dir. program mgmt. and group devel, 1986-87; v.p., gen. mgr. Defense Communications div., 1987-91, v.p., gen. mgr., 1991-93; v.p. requirements and group devel. Systems Integration Group, 1993—; assoc. dir. Armed Forces Communications and Electronics Assn., 1991-93; bd. dirs. Entire Corp., E. Rochester, N.Y., 1993—; mem. adv. bd. Tulane U. Coll. Engring., 1991—. Served to capt. USAF, 1955-62. Recipient Nat. Security Agy. Exceptional Civilian Service award, 1975. Mem. IEEE (sr.), AIAA. Republican. Presbyterian (elder). Home: 4905 Oakcrest Dr Fairfax VA 22030 Office: TRW Systems Integration Group 1 Fed Systems Park Dr Fairfax VA 22033

BOONE, THOMAS JOHN, metallurgical engineer; b. Ft. Belvoir, Va., June 1, 1957; s. George Franklin and Carol Louise (Kalk) B. BSChE, Mich. State U., 1980; postgrad., U. Ill., 1992—. Metall. supr. GM Cen. Foundry Div. Saginaw (Mich.) Malleable Iron Plant, 1979-80; supr. die cast quality and prodn. GM Cen. Foundry Div. Bedford (Ind.) Aluminum Plant, 1981-83; melting gen. spr. and advance mfg. project engr. GM Cen. Foundry Div. Massena (N.Y.) Aluminum Plant, 1984-87; advance mfg. project engr. GM Cen. Foundry div. Defiance (Ohio) Iron Plant, 1987-88; metall. gen. supr. GM Cen. Foundry div. Danville (Ill.) Iron Plant, 1988—; mem. foundry ops. seamless task force GM, Powertrain div., 1990—; participant in European Foundry tour, 1990; mem. corp. brake drum sourcing team GM USA and GM of Can., 1991. Mem. ASM, ASTM, Am. Foundrymen's Soc., Soc. Automotive Engrs. Office: GM Powertrain div I-74 at G St Danville IL 61834

BOOR, MYRON VERNON, psychologist, educator; b. Wadena, Minn., Dec. 21, 1942; s. Vernon LeRoy and Rosella Katharine (Eckhoff) B. BS, U. Iowa, 1965; MA, So. Ill. U., 1967, PhD, 1970; MS, U. Pitts., 1981. Lic. psychologist, Kans. Research psychologist Milw. County Mental Health Ctr., 1970-72; asst. prof. clin. psychologist Ft. Hays State U., Hays, Kans., 1972-76; assoc. prof. Ft. Hays State U., Hays, 1976-79; NIMH postdoctoral fellow in psychiat. epidemiology U. Pitts., Western Psychiat. Inst. and Clinic, 1979-81; research psychologist R.I. Hosp. and Butler Hosp., Providence, 1981-84; clin. psychologist Newman Meml. County Hosp., Emporia, Kans., 1985—; clin. psychologist Ft. Hays State U., 1972-79; asst. prof. psychiatry and human behavior Brown U., Providence, 1981-84; adj. faculty Emporia State U., 1985—. Contbr. articles to profl. jours. U.S. Pub. Health Service fellow, 1965-67, NIMH fellow 1979-81. Mem. Am. Psychol. Assn., Soc. for Psychol. Study of Social Issues, Internat. Soc. for Study of Multiple Personalities (charter). Home: 2225 Prairie St Emporia KS 66801-5756 Office: Newman Meml County Hosp 12th & Chestnut Emporia KS 66801

BOORMAN, ROY SLATER, science administrator, geologist; b. Montreal, Que., Can., Apr. 27, 1936; s. Albert George and Roslyn (Young) B.; m. Doreen Margaret Andrews, May 21, 1966; children: Alison, Glen, Lesley. BSc in Engring., Queens U., Kingston, Ont., Can., 1959; AM, Harvard U., 1961; PhD, U. Toronto, Ont., 1966. Mine geologist Eldorado, Can., 1959-60; mineralogist Rsch. and Productivity Coun., Fredericton, N.B., Can., 1966-72, dept. head, 1972-83, exec. dir., 1983—; bd. dirs. Can. Centre Mineral and Energy Tech., mem. bus. com. Ministers Nat. Adv. Coun.; bd. dirs. Fed. Dept. Communications, Ottawa. Contbr. articles to profl. jours. Mem. Can. Inst. Mining and Minerology (Proficiency medal 1985), Profl. Engrs. N.B. Home: RR 6 Box 6, Ste 9, Fredericton, NB Canada E3B 4X7 Office: NB Rsch & Productivity Coun, 921 College Hill Rd, Fredericton, NB Canada E3B 6C2

BOOSE, RICHARD BRADSHAW, biochemist; b. Red Lodge, Mont., June 12, 1928; s. Albert Peter and Frances Ellen (Fogleman) B.; m. Jeanne Marie Richards, July 15, 1951; children: Rebecca, Michelle, Timothy, Theresa, John. Cert., Portland Advent Hosp./U. Oreg., 1953; BS, Walla Walla Coll., 1954; MS, Oreg. State U., 1966. Bioscientist Boeing Co. Aerospace Divsn., Seattle, 1965-68, safety engr., 1968-70; researcher biochemistry dept. Oreg. State U., Corvallis, 1971-78; edn. coord. Hinsdale (Ill.) Hosp., 1978-80; lab. dir. Ctrl. Meml. Hosp., Toppenish, Wash., 1982-85; indsl. hygienist dept. ecology toxics clean-up program State of Wash., Olympia, 1985—; adv. com. occupational medicine and environ. health, U. Wash., 1987—; tech. adv. com. Nat. Inst. Occupational Health and Safety, Washington, 1990; adv. com. Green River Coll., Auburn, Wash., 1987-90; sci., math. instr. C. Cs. Seattle, Auburn, Wash., Albany, Oreg., 1967, 68-69, 76; cons. environ. and health risk assessment. Contbr. articles to profl. jours. Mem. Ctrl. Valley Acad. sch. bd., Wapato, Wash., 1984-85, Olympia Jr. Acad., 1991—; group leader Explorer Scouts, Boy Scouts Am., Toppenish, 1983-84. Recipient Wash. State U. scholarship, 1950-51, U.S. Pub. Health grant for pesticide rsch. Oreg. State U., 1959-65, USAF rsch. grant, 1966, NIH grant for cancer rsch. Oreg. State U., 1971-78. Mem. AAAS, Am. Chem. Soc., Am. Soc. Clin. Pathologists (cert. med. tech.). Seventh Day Adventist. Achievements include publs. on metabolic pathways of nitrosoureas, pesticides, naphthalene and other xenobiotic en vivo effects of carbon monoxide on rat drug metabolism, devel. of methods and tech. for evaluating air purifying respirators. Home: 7720 Loon Ct SE Olympia WA 98513 Office: State of Wash Dept Ecology Woodland Sq Bldg MS PV 13 Olympia WA 98504

BOOTH, KARLA ANN SMITH, biochemist; b. York, Pa., Mar. 8, 1951; d. Karl Ammon and Mary Emma (Black) Smith; m. George Loring Booth, May 19, 1974. BS in Psychology, Ariz. State U., 1973, MA in Ednl. Psychology, 1974; PhD in Biochemistry, Colo. State U., 1992. Rsch. assoc. Mt. St. Mary's Coll., L.A., 1974-77; quality assurance officer Elars Biorsch., Ft. Collins, Colo., 1979-85; rsch. asst. Colo. State U., Ft. Collins, 1985-92; rsch. scientist Paravax, Inc., Ft. Collins, 1992—; qualith assurance officer Colo. Animal Rsch. Ent., Ft. Collins, 1983-85. Author: Tick-Borne Diseases and Their Vectors, 1976, Novel Calcium-Binding Proteins, 1991; contbr. articles to profl. jours. Mem. Biophys. Soc., Sigma Xi. Home: 1025 Turman Dr Fort Collins CO 80525 Office: Paravax Inc 2301 Research Blvd # 110 Fort Collins CO 80526

BOOTHE, RONALD GEORGE, psychology and ophthalmology educator; b. Havre, Mont., May 25, 1947; s. Lynn Norris and Mildred Ethel (Green) B.; m. Marilyn Elizabeth Hondel, July 25, 1982; 1 child, Lyndi Ronica. BA, Concordia Coll., Morehead, Minn., 1968; PhD, U. Wash., 1974. Prof. U. Wash., Seattle, 1977-84; prof. psychology/ophthalmology Emory U., Atlanta, 1984—, chief div. neurobiology and vision Yates Rsch. Ctr., 1993—. Contbr. articles to profl. jours. Mem. AAAS, Am. Psychol. Soc., Soc. for Neurosci., Assn. for Rsch. in Vision and Ophthalmology. Office: Emory U Dept Psychology Atlanta GA 30322

BOOTHROYD, GEOFFREY, industrial and manufacturing engineering educator; b. Radcliffe, Eng., Nov. 18, 1932; came to U.S., 1967; s. Arthur and Annie (Fletcher) B.; m. Shirley Lewis, Apr. 10, 1954; children: Janet Kaye, Lynda Jean. BS in Engring., U. London, 1956, PhD in Engring., 1962, DSc in Engring., 1974. Apprentice Mather & Platt Ltd., Manchester, Eng., 1948-56, designer, 1956-57; designer English Electric Co. Ltd., Leicester, Eng., 1957-58; lectr./reader Salford (Eng.) U., 1958-67; prof. U. Mass., Amherst, 1967-85, U. R.I., Kingston, 1985—; vis. prof. Ga. Inst. Tech., Atlanta, 1964-65; cons. mfg. industries, U.K. and U.S., also various pubs. Author: Fundamentals of Metal Machining, 1965; (with A.H. Redford) Mechanized Assembly, 1968, (in Japanese), 1969; Fundamentals of Metal Machining and Machine Tools, 1975, (in Spanish), 1978, internat. student edit., 1979; (with others) Introduction to Engineering, 1975; (with C.R. Poli) Applied Engineering Mechanics, 1980; (with C.R. Poli and L.E. Murch) Automatic Assembly, 1980, Handbook of Feeding and Orienting Techniques for Small Parts; (with L. Alting) Manufacturing Engineering Processes, 1982; (with P. Dewhurst) Design for Assembly Handbook, Design for Robot Assembly, 1985; also over 100 pub. papers and articles. Recipient Teaching award Western Electric, 1969, Sr. Scholar award U. Mass., 1982, Sci. and Tech. award R.I. Gov., 1989, Nat. medal of Technology, U.S. Dept. Commerce Technology Admin., 1991, Providence Engring. Soc., 1991, UK Mensforth Internat. Gold medal Inst. Elec. Engrs., 1993; grantee NSF, 1967-87, GE, 1967, 69, 81, 83, AMP Inc., 1978, 81-84, IBM, 1983-85, AT&T, 1985, Ford Motor Co., 1984, 86, others. Fellow Soc. Mfg. Engrs.; mem. NAE, Inst. for Prodn. Engring. Rsch. (CIRP). Avocations: squash, tennis, golf, painting. Office: U RI Dept Mfg and Indsl Engring Kingston RI 02881

BORAH, GREGORY LOUIS, plastic and reconstructive surgeon; b. Albuquerque, N.Mex., Apr. 24, 1950; s. Robert Oscar and Martha (Knous) B.; m. Margaret Mary Plantes. BS in Biology, U. N.Mex., 1972; D of Dental Medicine, Harvard U., 1976, MD, 1978. Diplomate Am. Bd. Plastic Surgery. Instr. dept. surgery Harvard Med. Sch., Boston, 1981-83, Yale U. Med. Sch., New Haven, 1983-85; staff surgeon U. Mass. Med. Sch., Worcester, 1985-87, dir. craniofacial ctr., 1987-92; chmn. plastic surgery Robert Wood Johnson Med. Sch. Univ. Medicine and Dentistry of N.J., New Brunswick, N.J., 1992—, cons. Biomedical/Biomaterials, 1989—. Author: Plastic and Reconstructive Surgery, 1987—; editor Reconstructive Surgery News, 1992. Mem. perusing com. Am. Antiquarian Soc., Worcester, 1990-92. Recipient Outstanding Achievement award Intern Coll. Dentists, 1976. Fellow ACS, Am. Soc. Plastic and Reconstructive Surgeons, Am. Soc. Maxillofacial Surgeons (bd. trustees 1988); mem. Materials Rsch. Soc., Beta Beta Beta, Phi Sigma. Achievements include discovery of role of chitosan biopolymer in osteoneogenesis, discovery of anti-imflammatory agents in suppresion of fibroplast proliferation. Office: UMDNJ Robert W Johnson Med Sch 1 RWJ Place CN 19 New Brunswick NJ 08903-0019

BORAH, KRIPANATH, pharmacist; b. Calcutta, India, Mar. 1, 1931; s. Ambicanath and Gunabati (Barooah) B.; married; children: Shambhunath, Arun. BS, Calcutta U., India, 1952, MS, 1956; PhD, Munich U., Germany, 1961. Mgr. R&D Ciba-Geigy, Bombay, India, 1962-76; rsch. assoc. Boston Coll., Boston, 1976-77; group leader W.H. Rorer & Co., Ft. Washington, Pa., 1977-80; dir. pharm. devel. Organon Inc., W. Orange, N.J., 1980-91; assoc. dir. Enzon Inc., S. Plainfield, N.J., 1991-92; sci. dir. G & W Labs., S. Plainfield, N.J., 1992—; adj. prof. Temple U. Sch. Pharm., Phila., 1991—. Fellow Alexander von Humboldt Found., 1959-61. Mem. Am. Assn. Pharm. Sci., ACS, AAiPS, PDA. Home: 34 Overlook Trail Morris Plains NJ 07950 Office: G & W Labs 111 Coolidge St South Plainfield NJ 07080

BORCH, KURT ESBEN, surgeon; b. Ny Osted, Denmark, Nov. 4, 1944; arrived in Sweden, 1978; s. Erik Esben Hansen and Ulla Borch; m. Gerd Gunilla Bjorkblom, Mar. 1, 1985; children: Daniel Visti, Lau Esben, Bjorn Andreas, Eric Olof, Michael Erik. MD, U. Copenhagen, 1977; PhD in Surgery, U. Linkoping, 1985. Intern Cen. Hosp., Eskilstuna, Sweden, 1978-79, resident in surgery, 1980-84, sr. resident surgery, 1984-85; sr. resident surgery U. Hosp., Linkoping, Sweden, 1985-87, asst. prof. surgery, 1987-89, assoc. prof. surgery, 1989—; expert reviewer Med. Poducts Agency Nat. Bd. Health & Welfare, Sweden; mem. study group for endocrine tumors Swedish Med. Rsch. Coun. Reviewer, contbr. articles to med. jours.; mem. internat. editorial bd. Gastroenterology Digest, Hepatology Digest. Mem. Rsch. Animal Wellfare Com., Linkoping, Sweden. Mem. Swedish Assn. Surgery, Swedish Soc. Gastroenterology and Gastrointestinal Endoscopy, Scandinavian Surgical Soc., European Gastro. Club, European Pancreatic Club, Swedish Soc. Medicine, Internat. Assn. Pancreatology, Internat. Soc. Surgery. Achievements include research in diseases of the stomach, pancreas and biliary system, especially hormonal influences and risk factors in tumor development. Home: Fridsbergsgatan 10, S-58247 Linköping Sweden Office: U Hosp, Dept Surgery, S 581 85 Linköping Sweden

BORCH, RICHARD FREDERIC, pharmacology and chemistry educator; b. Cleve., May 22, 1941; s. Fred J. and Martha (Kananen) B.; m. Anne Wright Wilson, Sept. 8, 1962; children: Karen, Eric. BS, Stanford U., 1962; MA, PhD, Columbia U., 1965; MD, U. Minn., 1975. NIH postdoctoral fellow Harvard U., Cambridge, Mass., 1965-66; prof. chemistry U. Minn., Mpls., 1966-82, med. resident, 1975-76; dean's prof. pharmacology and prof. chemistry U. Rochester, N.Y., 1982—; mem. cancer research manpower rev. com. Nat. Cancer Inst., 1982-86, chmn. cancer research manpower rev. com. 1984-86; cons. 3M Pharms., St. Paul, 1972—. Contbr. over 70 articles to profl. jours.; patentee in field. Recipient Coll. Chemistry Tchr. award Minn. sect. Am. Chem. Soc., 1982, Louis P. Hammett award Columbia U., 1965, James P. Wilmot Disting. Professorship, U. Rochester, 1983-86; Alfred P. Sloan Found. fellow, 1970-72. Mem. Am. Chem. Soc., Am. Assn. Cancer Research, Am. Soc. Pharmacology Exptl. Therapeutics, AAAS. Office: U of Rochester Dept Pharmacology 601 Elmwood Ave Rochester NY 14642-9999

BORDEN, ROY HERBERT, JR., civil engineer, geotechnical consultant; b. Quincy, Mass., Sept. 5, 1949; s. Roy H. and Alice P. (Egan) B.; m. Laura Kathryn Horn, Aug. 15, 1981; 1 child, Thomas Andrew. BSCE, Tufts U., 1971; MSCE, Northwestern U., Evanston, Ill., 1975; PhD, Northwestern U., 1980. Registered profl. engr., Ill., N.C. Structural engr. Stone and Webster Engring. Corp., Boston, 1972-75; teaching asst. Northwestern U., Evanston, 1975-80; asst. prof. N.C. State U., Raleigh, N.C., 1980-85; assoc. prof. N.C. State U., Raleigh, 1985—; cons. Hayward Baker Co., Odenton, Md., 1980—, Morrison & Sullivan Engrs., Raleigh, 1991—; adv. bd. mem. Nat. Geotech. Emperimentation Sites Program, 1992—; lectr. Scientist-Tchr. Partnership, Raleigh, 1989—. Editor: Grouting, Soil Improvement and Geosynthetics, 1992; contbr. more than 25 articles to profl. jours. Advisor Christian Bible Study Club, N.C. State U., Raleigh, 1983-86. Mem. ASCE (mem. various coms.), ASTM, Transp. Rsch. Bd. (v.p. soil and rock properties com. 1988-92), Am. Soc. Engring. Edn., Chi Epsilon, Tau Beta Pi. Achievements include devel. of testing procedures for evaluating properties of improved soils, soil improvement techns. with emphasis on grouting, in-situ evaluation of soil properties, specifically residual soil, behavior of founds. subject to static and dynamic loads in residual soils. Office: N C State U Box 7908 Raleigh NC 27695-7908

BORDOGNA, JOSEPH, educator, engineer; b. Scranton, Pa., Mar. 22, 1933; s. Raymond and Rose (Yesu) B. B.S. in Elec. Engring., U. Pa., 1955, Ph.D., 1964; S.M., MIT, 1960. With RCA Corp., 1958-64; asst. prof. U. Pa., Phila., 1964-68; assoc. prof. U. Pa., 1968-72, prof., 1972—, assoc. dean engring. and applied sci., 1973-80, acting dean, 1980-81, dean, 1981-90, dir. Moore Sch. Elec. Engring., 1976-90, Alfred Fitler Moore chair, 1979—; dir. engring. Nat. Sci. Foundation, Washington, 1991—; bd. dirs. AOI Internat., Weston Inc., Univ. City Sci. Ctr.; master Stouffer Coll. House, 1972-76; cons. industry, govt., founds.; mem. Nat. Medal of Sci. com., 1989-91; chair adv. com. for engring. NSF, 1989-91. Author: (with H. Ruston) Electric Networks, 1966, (with others) The Man-Made World, 1971; chmn. editorial bd. Engring. Edn., 1987-90. Served with USN, 1955-58. Recipient commendation for first spacecraft recovery, 1957, Lindback award for disting. teaching U. Pa., 1967, Engr. of Yr. award Phila., 1984, Centennial medal Phila. Coll. Textiles and Sci., 1988; inducted into Engring. Educators Hall of Fame, 1993. Fellow AAAS, IEEE (chmn. Phila. sect. 1987-88, Centennial medal 1984), Am. Soc. for Engring. Edn. (George Westinghouse award 1974, Centennial medal 1993); mem. Sigma Xi, Eta Kappa Nu, Tau Beta Pi, Phi Beta Delta. Home: 1237 Medford Rd Wynnewood PA 19096-2416

BOREL, ARMAND, mathematics educator; b. Chaux-de-Fonds, Switzerland, May 21, 1923; m. Gabrielle Pittet, May 8, 1952; children: Dominique, Anne-Christine. Master Mathematics, Federal Sch. Tech., Zurich, Switzerland., 1947; Dr. Degree, U. Paris, 1952; Ph.D. (hon.), U. Geneva, 1972. Asst. Federal Sch. Tech., Zurich, 1947-49; prof. Federal Sch. Tech., 1955-57, 83-86; attaché de Recherches French Nat. Center Sci. Research, Paris, 1949-50; acting prof. algebra U. Geneva, Switzerland, 1950-52; mem. Inst. Advanced Study, Princeton, 1952-54, prof., 1957-93; prof. emeritus Inst. Advanced Study, Princeton, 1993—; vis. prof. U. Chgo., 1954-55, 76, MIT, 1958, 69, Tata Inst. Fundamental Rsch. Bombay, 1961, 68, 83, 90, U. Paris, 1964, Yale U., 1978. Recipient Brouwer medal Dutch Math. Soc., 1978, Balzan prize, 1992. Mem. NAS, Acad. Arts and Sci., Am. Philos. Soc. (fgn.), Finnish Acad. Scis. and Letters (fgn.), French Acad. Scis. (fgn.), Am. Math. Soc. (Leroy P. Steele prize 1991), Swiss Math. Soc., French Math. Soc. Address: Inst for Advanced Study Olden Ln Sch Mathematics Princeton NJ 08540

BOREL, GEORGES ANTOINE, gastroenterologist, consultant; b. Neuchatel, Switzerland, Oct. 13, 1936; s. Jean and Alice Marie (Perrenoud) B.; m. Beatrice de Przysiecki, Apr. 7, 1964 (div. 1980); children: Pascal, Sibylle, Fabienne. MD, U. Zürich, 1963; Privat Docent (hon.), U. Lausanne, Switzerland, 1977. Rsch. fellow Univ. Hosp., Ann Arbor, Mich., 1968-69; cons. physician Univ. Hosp., Geneva, 1975—; pvt. practice gastroenterologist Lausanne, 1976—. Contbr. articles to profl. jours. Mem. Swiss Soc. Gastroenterology. Avocation: furniture making. Office: Blvd Grancy 7, 1006 Lausanne Vaud, Switzerland

BOREN, KENNETH RAY, endocrinologist; b. Evansville, Ind., Dec. 31, 1945; s. Doyle Clifford and Jeannette (Koerner) B.; m. Rebecca Lane Wallace, Aug. 25, 1967; children: Jennifer, James, Michael, Peter, Nicklas, Benjamin. BS, Ariz. State U., 1967; MD, Ind. U., Indpls., 1972; MA, Ind. U., Bloomington, 1974. Diplomate Am. Bd. Endocrinology, Am. Bd. Nephrology. Intern in pathology Ind. U. Sch. Medicine, Indpls., 1972; intern in medicine Ind. U. Sch. Medicine, 1972-73, resident in medicine, 1975-77, fellow in endocrinology, 1977-79, fellow nephrology, 1979-80, instr., 1980; physician Renal and Endocrine Assocs., Mesa, Ariz., 1980—; chief medicine Mesa Luth Hosp., 1987-89, chief staff, 1990-91. Bd. dirs. Ariz. Kidney Found., Phoenix, 1984—. Lt. USN, 1973-75. Fellow ACP; mem. AMA, Maricopa County Med. Assn., Ariz. Med. Assn., Am. Soc. Nephrology, Internat. Soc. Nephrology, Am. Diabetes Assn. Republican. Latter Day Saints. Home: 4222 E Mclellan #10 Mesa AZ 85205 Office: Renal and Endocrine Assocs 560 W Brown Rd Mesa AZ 85201-3221

BORG, STEFAN LENNART, psychiatrist, educator; b. Stockholm, May 15, 1945; s. Lennart and Inga B.; m. Lena, Feb. 21, 1945; children: Charlotte, Niklas, Jonas. MD, Karolinska Inst., Stockholm, 1971; PhD, Karolinska Inst., 1975. Asst. prof., sr. registrant Karolinska Inst. and Hosp., Stockholm, 1980; assoc. prof. Karolinska Inst., 1985—; program dir. Methadon Maintenance Treatment, Stockholm, 1985—, Treatment Program Hypnotic-Sedative Dependence, Stockholm, 1980—; head dept. psychiatry Karolinska Inst., St. Gorans Hosp., Stockholm, 1991—; cons. Swedish Nat. Bd. Health and Welfare. Assoc. editor Alcohol & Alcoholism jour.; contbr. numerous articles to profl. jours. Grantee Swedish Med. Research Council, Swedish Ministry Health. Mem. Swedish Med. Assn.(sec. alcohol pol. program, 1980), Swedish Psychiatrie Assn., Nat. Swedish Sci. Council, Nat. Bd. Health and Welfare, Swedish Psychiatric Assn. (bd. dirs. 1976-84).

BORGATTI, DOUGLAS RICHARD, environmental engineer; b. Springfield, Mass., Mar. 22, 1952; s. Richard Francis and Patricia Winifred (Hill) B.; m. Patricia Ann Boswell, June 5, 1988. BSCE, Worcester Poly. Inst., 1974; MS in Environ. Engring., Manhattan Coll., 1976; PhD in Environ. Engring., U. Notre Dame, 1981. Cert. profl. engr.; diplomate Am. Acad. Environ. Engrs.; lic. in pub. sewage treatment plant ops. Engr. Lawlea, Matusky & Skelly Engrs., Pearl River, N.Y., 1976-77; chief pollution and treatment control dept. Passaic Valley Sewerage Comm., Newark, 1981-86; supt. Regional Wastewater Treatment Plant, Springfield, 1986-90; exec. v.p. Hydropress N-Viro Svcs., Hatfield, Mass., 1990—; various coms. and tech. revs. Water Environ. Fedn., Washington, 1983—. Author: (tech. manual) Operation of Wastewater Treatment Plants, 1990, Successful Biosolids Processing and Utilization, 1993; contbr. numerous articles to profl. jours. Recipient Linn H. Enslow Meml. award N.Y. Water Pollution Control Assn., 1985. Mem. Sigma Xi. Home: 64 Hadley St Hatfield MA 01075

BORGESE, ELISABETH MANN, political science educator, author; b. Munich, Apr. 24, 1918; arrived in U.S., 1938, naturalized, 1941, became Can. citizen, 1983; d. Thomas and Katia (Pringheim) Mann; m. Giuseppe Antonio

Borgese, Nov. 23, 1939; children: Angelica, Dominica. Diploma, Conservatory of Music, Zurich, 1937; PhD (h.c.), Mt. St. Vincent U., 1986. Research assoc., editor Common Cause, U. Chgo., 1945-51; editor Perspective USA; Diogenes (Intercultural Publs.), 1952-57; exec. sec. bd. editors Ency. Brit., Chgo., 1964-66; sr. fellow Ctr. for Study Dem. Instns., Santa Barbara, Calif., 1964-79; Killam sr. fellow Dalhousie U., Halifax, N.S., Can., 1979-80, prof. dept. polit. sci., 1980—; chmn. planning coun. Internat. Ocean Inst., 1972-92, hon. chmn. life, 1992; chmn. Internat. Ctr. for Ocean Devel., 1986-92; advisor Austrian del. 3d UN Conf. on Law of Sea, 1976-82, Prep. Commn., Jamaica, 1983-86. Author: To Whom It May Concern, 1962, Ascent of Woman, 1963, The Language Barrier, 1965, The Ocean Regime, 1968, The Drama of the Oceans, 1976, Seafarm: The Story of Aquaculture, 1980, The Mines of Neptune, 1985, The Future of the Oceans: A Report to the Club of Rome, 1986, (play) Only the Pyre, 1987, (juvenile) Chairworm an Supershark, 1992; editor: Ocean Yearbook, 10 vols., 1981, Ocean Frontiers, 1992; contbr. short stories and essays to mags. Decorated medal of High Merit, Austria, Order of Merit, Govt. Columbia; recipient Sasakawa Internat. Environ. prize UN, 1987, Order of Can., 1987, Friendship award, Govt. China, 1992, St. Francis of Assisi Internat. Environment prize, 1993. Mem. AAAS, Acad. Polit. Sci., Am. Soc. Internat. Law, World Acad. Arts and Scis., Third-World Acad. Sci., Club of Rome. Office: Dalhousie U Internat Ocean Inst, 1226 LeMerchant St, Halifax, NS Canada B3H 3P7

BORIE, BERNARD SIMON, JR., physicist, educator; b. New Orleans, June 21, 1924; s. Bernard Simon and Ruth (Lastrapes) B.; BS, U. S.W. La., 1944; MS, Tulane U., 1949; PhD, M.I.T., 1956; Fulbright fellow U. Paris, 1956-57; m. Martine Edith Descamps, May 2, 1957 (div. May 1964); children: Kathleen, Fabienne, Marianne. Research physicist metall. div. Oak Ridge Nat. Lab., 1949-53, group leader x-ray diffraction Metals and Ceramics Div., 1957-60, head fundamental research sect., 1960-69, sr. scientist, 1969-85 ; prof., U. Tenn., 1963—; vis. prof. Cornell U., 1971-72, U. Calif., Berkeley, 1980. Served to lt., USNR, 1944-45. Fellow AAAS; mem. AIME, Am. Soc. Metals, Am. Crystallographic Assn., Sci. Research Soc. Am. Research in diffraction effects of thermal motion, x-ray diffraction studies of imperfect solids; order-disorder effects in solid solutions. Home: 13 Brookside Dr Oak Ridge TN 37830-7616 Office: U Tenn Materials Sci & Engring Dept Dougherty Hall Knoxville TN 37996-2200

BORISY, GARY G., molecular biology educator; b. Chgo., Aug. 18, 1942; s. Philip and Mae Borisy; children: Felice, Pippa, Alexis. BS, U. Chgo., 1962, PhD, 1966. Postdoctoral fellow NSF, Cambridge, Eng., 1966-67, NATO, Cambridge, 1967-68; asst. prof. U. Wis., Madison, 1968-72, assoc. prof., 1972-75, prof., 1975-80, Perlman-Bascom prof. life scis., 1980—, chmn. lab. molecular biology, 1981—; mem. numerous panels NIH, other govt. agys., ACS; mem. Marine Biol. Lab. Contbr. over 100 articles to profl. jours.; editor Jour. Biol. Chemistry, 1978-80, Jour. Cell Biology, 1980-82, Internat. Rev. Cytology, 1971-91, Cell Motility and the Cytoskeleton, 1986—, Jour. Cell Sci., 1988—. Recipient Romnes award U. Wis., 1975-80, NIH Merit award, 1989; grantee NIH, NSF, 1975-80. Fellow AAAS; mem. Am. Soc. Cell Biology, Am. Soc. Biochemistry and Molecular Biology, Sigma Xi. Home: 1002 Oak Way Madison WI 53705-1419 Office: Univ Wis Lab Molecular Biology 1525 Linden Dr Madison WI 53703

BORN, ALLEN, mining executive; b. Durango, Colo., July 4, 1933; s. C. S. and Bertha G. (Tausch) B.; m. Patricia Beaubien, Mar. 23, 1953; children: Michael, Scott (dec.), Brett. B.S. in Metallurgy and Geology, U. Tex.-El Paso, 1958; D Engring. (hon.), Colo. Sch. Mines, 1992. Exploration geologist El Paso Natural Gas, Tex., 1958-60; metallurgist Vanadium Corp. Am., 1960-62; gen. foreman Pima Mining, 1962-64; asst. supt. MolyCorp, 1964-67; chief metallurgist and supt., mgr. AMAX Inc., N.Y.C., 1967-76; pres., chief exec. officer Can. Tungsten Mining Corp. Ltd., 1976-81; pres. AMAX of Can. Ltd., 1977-81; pres., chmn., chief exec. officer Placer Devel. Ltd., Vancouver, B.C., Can., 1981-85; pres., chief operating officer AMAX Inc., N.Y.C., 1985—, chief exec. officer, 1986—, chmn., 1988—; chmn. bd. Alumax, Inc., Amax Gold, Inc.; bd. dirs. Can. Tungsten Mining Corp., Aztec Mining Co. Ltd., Australia, Aztec Resources. Contbr. numerous articles to mining jours. Served with U.S. Army, 1952-55. Named Chief Exec. Officer of Yr. in mining industry, silver award winner all industry Financial World mag., 1987; recipient Golden Nugget award Coll. Sci. U. Tex. at El Paso, Daniel C. Jackling award Soc. Mining Engrs., Bus. Leader award Citizens Union of N.Y.C., 1990. Mem. AIME, Am. Mining Congress (chmn., bd. dirs.), Sky Club, Indian Harbor Hacht Club, Vancouver Club. Republican. Office: Amax Inc 200 Park Ave New York NY 10166 also: Alumax Inc Norcross GA

BORN, DAVID OMAR, psychologist, educator; b. Dover, Ohio, June 9, 1944; s. Jesse Woodruff Born and Ellen Rose (Myer) Cass; m. Doris Elaine Musgrave, Sept. 1, 1968; children: Theodore James, Emily Jo, Patricia Elaine. BA, So. Ill. U., 1966, PhD, 1970. From asst. prof. to prof. Schg. Dentistry U. Minn., Mpls., 1970—; cons. to various cos., ednl. instns. and govt. agys. Editorial reviewer Am. Jour. Pub. Health Dentistry, Jour. Dental Edn., Jour. ADA, Jour. Practice Adminstrn.; contbr. articles to profl. publs. Mem. Am. Assn. Dental Schs., Ind. Feature Project. Democrat. Achievements include development of tool for psychological assessment of mid-life crisis in males. Home: 4032 Hunters Hill Way Minnetonka MN 55345 Office: Univ Minn Dental Sch 515 Delaware St SE Minneapolis MN 55455

BORN, GEORGE H., aerospace engineer, educator; b. Westhoff, Tex., Nov. 10, 1939; s. Henry and Lydia (Schulle) B.; m. Carol Ann Leslie, Mar. 21, 1992. BS, U. Tex., 1962, MS, 1965, PhD, 1968. Engr. Ling-Temco-Vought, Dallas, 1962-63; aerospace technologist Johnson Space Ctr., Houston, 1967-70; mem. tech. staff Jet Propulsion Lab., Pasadena, Calif., 1970-83; sr. rsch. engr. U. Tex., Austin, 1983-85; prof. aerospace engring. U. Colo., Boulder, 1985—, dir. Colo. Ctr. for Astrodynamics Rsch., 1985—. Contbr. articles to profl. jours. Recipient Exceptional Svc. medal NASA, 1980. Fellow AIAA, Am. Astronautical Soc.; mem. AAAS, Am. Geophys. Union, Oceanog. Soc., Inst. Navigation, Am. Meteorol. Soc., Am. Soc. Engring. Educators, Tau Beta Pi. Office: U Colo Campus Box 431 Boulder CO 80309

BORN, GUNTHARD KARL, aerospace executive; b. Marienwerder, Fed. Republic Germany, Mar. 31, 1935; s. Karl A. and Elise (Kuczewski) B.; m. Gertraud A. Forstner, Dec. 17, 1963. Diploma in physics, Tech. U. Munich, 1961; DEng, Tech. U. Stuttgart, Fed. Republic of Germany, 1967. Researcher Beckman Instruments, Munich, 1961-63, U.S. Army Electronics Command, Ft. Monmouth, N.J., 1963-69; div. head Messerschmitt-Bolkow-Blohm, Munich, 1969—, tech. dir. high-energy laser rsch. program, 1980—; presenter European, Asian, African, U.S. tours on lasers and musicology Goethe Inst. Fed. Republic of Germany, Munich, 1972—. Author: Mozart's Musiksprache, 1985; inventor in field of linguistics, numerous in fields of lasers and electrooptics; contbr. numerous articles to sci. and tech. publs. Avocation: travel. Home: Donarweg 22, D-81739 Munich Germany

BORNINO-GLUSAC, ANNA MARIA, mathematics educator; b. Naples, Italy, Apr. 2, 1946; came to U.S., 1946; d. Bruno and Anna Maria (De Simone) B.; m. Howard Keith Wolff, July 29, 1966 (div. 1971); 1 child, Francesca Yvonne Wolff; m. Ronald G. Glusac, Sept. 4, 1993. BA in Chemistry, Calif. State U., Dominguez Hills, 1968, MA in Edn. Adminstrv. Svcs., 1986. Cert. standard secondary tchr., Calif., preliminary adminstrv., Calif., Spanish lang. fluency, Calif. Tchr. math. L.A. Unified Sch. Dist., 1968—, dept. chair, 1982-84, 90—. Editor: Accreditation Report, 1983. Mem. Math. Assn. Am., United Tchrs. L.A. Democrat. Roman Catholic. Avocation: needlework, travel, reading, music. Office: Narbonne High Sch 24300 S Western Ave Harbor City CA 90710

BOROVOY, MARC ALLEN, podiatrist; b. Detroit, Oct. 22, 1960; s. Mathew and Joyce Francis (Weisman) B.; m. Michele Lynn Flusty, Oct. 23, 1983; children: Danielle, Brandon. Student, Wayne State U., 1978-81; D. Podiatric Medicine, Ohio Coll. Podiatric Medicine, 1985. Diplomate Am. Bd. Podiatric Sugery, Am. Bd. Quality Assurance and Utilization Rev. Resident Straith Hosp., Southfield, Mich., 1985-86; podiatrist Associated Podiatrists, Oak Park, Mich., 1986—; chief dept. podiatric surgery Providence Hosp., Southfield, Mich. Contbr. articles to profl. jours. Mem. exec. bd. Congregation Bnai Moshe, West Bloomfield, Mich., 1989—. Fellow Am. Coll. Foot Surgeons; mem. APHA, Am. Diabetes Assn., Mich. Podiatric Med. Assn. (exec. sec. 1986-89, chmn. pub. rels. 1988—, v.p. 1990-91, pres.-elect 1992-93, pres. 1993-95). Avocations: swimming, photography. Office: Associated Podiatrists 25725 Coolidge Hwy Oak Park MI 48237-1307 also: Ste B-230 47601 Grand River Ave Novi MI 48374

BORREGO, CARLOS SOARES, environmental engineering educator; b. Malanje, Angola, June 25, 1948; arrived in Portugal, 1966; s. Alberto Soares and Maria Idalina (Diogo) B.; m. Maria Noémia Campos Soares, Sept. 16, 1972; 1 child, Filipa. Cert. mech. engring., Tech. U. Lisbon, 1972; MSc, Free U. Brussels, 1978, PhD, 1981; cert. in habilitation, U. Aveiro, 1991. Head of rsch. U. Aveiro, Portugal, 1981, assoc. prof., 1987-91, head of dept., 1989, prof., 1991—. Author: Camada Limite Atmosférica, 1988; editor: 1a CNQA, 1987, IV ENSB, 1989. Minister Environment, Lisbon, 1991-93. Mem. Am. Chem. Soc., Am. Meteorol. Soc., European Assn. for Sci. of Air Pollution. Office: U Aveiro, Dept Environ, 3800 Aveiro Portugal

BORSON, DANIEL BENJAMIN, physiology educator, inventor, researcher; b. Berkeley, Calif., Mar. 24, 1946; s. Harry J. and Josephine F. (Esterly) B.; m. Margaret Ann Rheinschmidt, May 22, 1974; children: Alexander Nathan, Galen Michael. BA, San Francisco State Coll., 1969; MA, U. Calif., Riverside, 1973; PhD, U. Calif., San Francisco, 1982. Lic. comml. pilot, flight instr. FAA. Musician Composer's Forum, Berkeley, San Francisco, 1961-70; flight instr. Buchanan Flying Club, Concord, Oakland, Calif., 1973-77, pres., 1975-77; lectr. dept. physiology U. Calif., San Francisco, 1984-92, asst. rsch. physiologist Cardiovascular Rsch. Inst., 1988-92; vis. scientist Genentech Inc., South San Francisco, 1990-92, Sch. Law U. San Francisco, 1992—; mem. spl. rev. com. NIH, Washington, 1991—; summer assoc. patent law firm Flehr, Hohbach, Test, Albritton & Herbert, 1993—. Contbr. articles, rev. chpts. and abstracts to profl. jours. and legal periodicals. Fellow NIH, 1976-84, grantee, 1988-93; fellow Cystic Fibrosis Found., 1985, grantee, 1989-91; fellow Parker B. Francis Found., 1985-87; grantee Am. Lung Assn., 1985-87. Mem. Am. Physiol. Soc. (editorial bd. Am. Jour. Physiology 1990—), Am. Soc. Cell Biology, Bay Flute Club (pres. 1978). Avocations: mountain climbing, aviation, music. Home: 146 San Aleso Ave San Francisco CA 94127 Office: 4 Embarcadero Center Ste 3400 San Francisco CA 94127

BORST, LYLE BENJAMIN, physicist, educator; b. Chgo., Nov. 24, 1912; s. George William and Jean Carothers (Beveridge) B.; m. Barbara Mayer, Aug. 19, 1939; children: John Benjamin, Stephen Lyle, Frances Elizabeth. A.B., U. Ill., 1936, A.M., 1937; Ph.D., U. Chgo., 1941. Instr. U. Chgo., 1940-41, rsch. assoc. metall. lab., 1941-43; sr. physicist Clinton Labs., Oak Ridge, 1943-46; (both labs. working on atomic bomb project); asst. prof. dept. chemistry MIT, Cambridge, 1946; chmn. dept. reactor sci. and engring. Brookhaven Nat. Lab., 1946-51; prof. physics U. Utah, 1951; chmn. dept. physics Coll. Engring NYU, 1954-61; prof. physics SUNY, Buffalo, 1961-83, prof. emeritus, 1983—; master Clifford Furnas Coll., 1969-74. Author: Megalithic Software, Part I: England, Part II: Europe and the Near East, Part IIIa: Japan Studies of Prehistoric Science. Fellow Am. Phys. Soc.; mem. AAAS, ACLU (nat. bd. 1958-62, chmn. Niagara Frontier chpt. 1967-69), Phi Beta Kappa, Tau Beta Pi, Sigma Pi Sigma. Achievements include design and operation of Brookhaven reactor; research on sub-critical power reactors. Home: 17 Twin Bridge Ln Buffalo NY 14221-5019

BORTNER, JAMES BRADLEY, wildlife biologist; b. Boston, Jan. 14, 1958; s. James Augustine and Caroline Hinkley (Swaim) B.; m. Sandra Elaine Staples, Apr. 3, 1982; children: Benjamin Dyer, Jefferson Bradley. BS in Wildlife Biology, U. Vt., 1980, BS in Forestry, 1981; MS in Wildlife Biology, U. Md., 1985. Cert. wildlife biologist; registered profl. forester. Ecologist Fed. Energy Regulatory Commn., Washington, 1985-87; woodcock specialist U.S. Fish & Wildlife Svc., Washington, 1987-90, waterfowl specialist, 1990-91, chief population assessment, 1991-92; chief migratory birds and habitat programs U.S. Fish & Wildlife Svc., Portland, Oreg., 1992—. Contbr. articles to profl. jours. Bd. dirs. Gov.'s Task Force on Trees and Forests, Annapolis, Md., 1989-90, Washington County Forest Conservancy Dist., Hagerstown, Md., 1989-92; mem. Md. Farm Game Adv. Com., Annapolis, 1989-91. MacMillan fellow Delta Waterfowl & Wetlands Rsch. Ctr., 1980; New Eng. Outdoor Writers Assn. scholar, 1979. Mem. The Wildlife Soc. (chpt. pres. 1989-91), Soc. Am. Foresters, Am. Ornithologists Union, Cooper's Ornithological Soc. Office: US Fish & Wildlife Svc 911 NE 11th Ave Portland OR 97232-4181

BOS, JOHN ARTHUR, aircraft manufacturing executive; b. Holland, Mich., Nov. 6, 1933; s. John Arthur and Annabelle (Castelli) B.; m. Eileen Tempest, Feb. 15, 1974; children: John, James, William, Tiffany. BS in Acctg., Calif. State Coll., Long Beach, 1971. Officer 1st Nat. Bank, Holland, Mich., 1954-61; gen. mgr. fin. McDonnell Douglas, Long Beach, 1962—. Mem. Inst. Mgmt. Accts. (cert. mgmt. acct. 1979), Nat. Assn. Accts. Avocations: automobile marketing, golf, consulting. Office: McDonnell Douglas Aircraft Co 3855 N Lakewood Blvd Long Beach CA 90846-0001

BOSACCHI, BRUNO, physicist; b. Milan, Italy, Sept. 24, 1938; came to U.S., 1976; s. Camillo and Maria (Tei) B.; m. To-Thi Nguyen, June 15, 1974; children: Massimo, Alessandro. PhD in Physics, U. Milan, 1964. Asst. prof. U. Milan, 1966-69; assoc. prof. U. Parma, Italy, 1971-76; rsch. scientist AT&T Bell Labs., Princeton, N.J., 1978—; tech. asst. to v.p. mktg. AT&T Network Systems, Morristown, N.J., 1989-91; vis. scientist U. Chgo.-Argonne Nat. Lab., 1969-71; vis. prof. U. Ariz., Tucson, 1976-78; vis. fellow Princeton U., 1973; co-chmn. Internat. Conf. Artificial Neural Networks, Orlando, Fla., 1992; chmn. Internat. Conf. Applications of Fuzzy Logic Tech., 1993; presenter at internat. confs.; lectr. in field. Contbr. rsch. papers to sci. publs. Recipient Rsch. and Design award Bell Labs., 1989. Mem. AAAS, IEEE (sr.), Am. Phys. Soc., Optical Soc. Am. Internat. Neural Network Soc, Cath. Theol. Soc. Am. Roman Catholic. Office: AT&T Bell Labs PO Box 900 Princeton NJ 08542-0900

BOSART, LANCE F., meteorology educator; b. N.Y.C., Aug. 24, 1942. BS, MIT, 1964, MS, 1966, PhD, 1969. Resident asst. meteorologist MIT, Cambridge, Mass., 1965-69, asst. prof., 1969-76; assoc. prof. atmos. sci. SUNY, Albany, 1976—; vis. assoc. prof. MIT, 1978-79. NSF grantee, 1970-78. Mem. Am. Meteorol. Soc. (Jule G. Charney award 1992), Royal Meteorol. Soc. Office: SUNY Albany Dept of Atmospheric Sciences 1400 Washington Ave Albany NY 12222

BOSBACH, BRUNO, mathematics educator; b. Marienheide, Oberberg, Germany, Jan. 25, 1932; s. Wilhelm and Maria Anna (Wette); m. Gisela Maria, July 24, 1961; children: Markus, Corinna, Nicola, Johannes. Diploma, Sporthochschule, Cologne, Fed. Republic Germany, 1956; Dr. rer. nat., U. Cologne, 1959, Staatsexamen, 1960; Habilitation, Technische Hochschule, Darmstadt, Fed. Republic Germany, 1970. Diplomsportlehrer Gymnasium, Wipperfürth, Fed. Republic Germany, 1956-58; studienreferendar Gymnasium, Bergneustadt, Bonn, Fed. Republic Germany, 1960-62; studienassessor Gymnasium, Eitorf, Fed. Republic Germany, 1962-64; studienrat Gymnasium, Bensberg, Fed. Republic Germany, 1964-67; oberstudienrat U. Münster, Münster, Fed. Republic Germany, 1967-69; vizedirektor Gymnasium, Neunkirchen, Fed. Republic Germany, 1969-71; lehrbeauftragter Technische Hochschule, Darmstadt, Fed. Republic Germany, 1971—; prof. math. U. Kassel, Fed. Republic of Germany, 1971—. Contbr. articles to profl. jours. in abstract divisibility theory. Mem. Deutscher Hochschulverband, Gesellschaft zur Bekämpfung der Mukoviszidose, Greenpeace. Avocations: hiking, swimming. Home: An Den Vogelwiesen 20, D 3500 Kassel Hessen, Germany Office: GH/U, D 3500 Kassel Germany

BOSCH, ROBERT JOHN, JR., mechanical engineer; b. Bklyn., Oct. 15, 1945; s. Robert John and Virginia Ann (Elsbach) B.; m. Joan Suzzane Dydak, Aug. 31, 1968; children: Jennifer Anne, Jessica Anne. BME, U. Dayton, 1967; MME, Stevens Inst. Tech., 1973, MSM, 1981. Profl. engr., N.J., Utah. Assoc. engr. Gen. Dynamics Corp., Groton, Conn., 1967-69; project engr. Curtiss Wright Corp., Wood Ridge, N.J., 1969-73; project mgr. Foster Wheeler Corp., Livingston, N.J., 1973-80; mgr. bus. devel. F.W. Energy Applications, Inc., Livingston, 1980-84; pres. Planned Mgmt. Systems, Inc., Essex Fells, N.J., 1984-86; regional mgr. Siemens Corp., N.Y.C., 1986-91; comml. dir. Foster Wheeler Energy Corp., Livingston, 1991—. Contbr. to profl. publs. Bd. dirs. Children's Aid and Adoption Soc., Hackensack, N.J., 1989—; pres. Groton Jaycees, 1969. Mem. ASME, Am. Nuclear Soc. Office: Foster Wheeler Energy Corp Perryville Corp Park Clinton NJ 08809

BOSCHI, SRDJAN, radiologist, nephrologist; b. Split, Croatia, Feb. 17, 1927; s. Vladimir and Nila (Covich-Plenkovich) B.; m. Fani Rendich-Miocevich, Jan. 23, 1960; children, Neva, Vladimir. MD, U. Med. Sch., Zagreb, 1957, PhD, 1970, diploma (hon.), 1988; diploma (hon.), Croatian Med. Acad., 1978, Croatian Assn. Nephrology, Zagreb, 1981. Clinician Clin. Hosp. Firule, Split, 1957-63, chief dept., 1963-77, chief inst., 1977—; chair Zagreb U. Sch. Medicine in Split, 1983—; cons. in field. Author: Radiology, 1989; contbr. 91 articles to sci. jours. Mem. AAAS, Croatian Med. Acad., Croatian Assn. for Radiology, Croatian Assn. for Nephrology, Med. Croatian Assn., N.Y. Acad. Scis., M.I.D.I. Acad. (Senator Internat. Parliament Safety and Peace, Knight of Templar Order). Roman Catholic. Home: Karamanova 11, 58000 Split Croatia Office: Inst Radiology Clin Hosp, Spinciceva 1, 58000 Split Croatia

BOSCO, PAUL D., computer scientist; b. Hartford, Conn.; s. Cosimo J. and Edwina T. (Lynch) B.; m. Diane C., June 15, 1991; children: Paul D. Jr. BSEE, Lehigh U., 1980; MS, RPI, 1981, MBA, 1983; MS in Computer Engring., Yale U., 1984. Assoc. engr. IBM, Armonk, N.Y., 1982-84, sr. assoc. engr., 1984-86, staff engr., 1986-87, project mgr., 1987-89, devel. engr., 1989-90, sr. engring. mgr., 1990-91, program mgr., 1991—. Mem. IEEE, AAAS, ACM. Achievements include architected and led the development of advanced networking projects and applications including networked multimedia services and national data super highway initiative. Home: 34 Garden Rd Concord MA 01742 Office: IBM Maildrop 365 1311 Mamaroneck Ave White Plains NY 10605

BOSCO, RONALD F., engineering company executive; b. N.Y.C., May 20, 1950; m. Susan J. Koepper, Jan. 17, 1981; childern: Jennifer, Melissa. BE, Stevens Inst., 1972. Pres., CEO Fed. Engring., Inc., Fairfax, Va.; speaker in field. Contbr. articles to profl. jours. Mem. IEEE, Assn. Computing Machinery, Assn. Pub. Safety Communications Officers,Ind. Computer Cons. Assn., Nat. Assn. State Info. Resource Execs., Nat. Assn. State Telecommmunications Dirs., Nat. Emergency Number Assn., U.S. Distance Learning Assn. Office: Fed Engring Inc 10600 Arrowhead Dr Fairfax VA 22030-7306

BOSE, AJAY KUMAR, chemistry educator; b. Silchar, India, Feb. 12, 1925; s. Abinash C. and Amita Kumari (Chanda) B.; m. Margaret Lois Logan, Sept. 13, 1950; children: Ryan, Ranjan, Indrani, Indira, Krishna, Rajendra. BS, U. Allahabad, India, 1944, MS, 1946; ScD, MIT, 1950; M in Engring. (hon.), Stevens Inst. Tech., Hoboken, N.J., 1963. Rsch. fellow Harvard U., Cambridge, Mass., 1950-51; lectr., then asst. prof. chemistry Indian Inst. Tech., Kharagpur, 1952-56; rsch. assoc. U. Pa., Phila., 1956-57; rsch. chemist Upjohn Co., Kalamazoo, Mich., 1957-59; assoc. prof. Stevens Inst. Tech., 1959-61, prof., 1961-83, George Meade Bond prof. chemistry, 1983—; cons. various chemical cos. Editorial bd. Jour. Heterocyclic Chemistry, 1980-83; contbr. over 200 articles to profl. jours.; patentee in field. Recipient Outstanding Achievement award Nat. Fedn. Indian Am. Assns., 1990; named N.J. Prof. of Yr., Coun. for Advancement and Support of Edn., 1990. Fellow N.Y. Acad. Scis., Indian Nat. Sci. Acad.; mem. Am. Chem. Soc. (councillor 1964-70), Sigma Xi. Avocation: popular sci. writing. Office: Stevens Inst of Tech Dept of Chemistry Castle Point Hoboken NJ 07030

BOSE, AMAR GOPAL, electrical engineering educator; b. Phila., Nov. 2, 1929; s. Noni Gopal and Charlotte (Mechlin) B.; m. Prema Sarathy, Aug. 17, 1960; children: Vanu Gopal, Maya. S.B., S.M., MIT, 1952, Sc.D., 1956. Mem. faculty MIT, Cambridge, 1956—, prof. elec. engring., 1966—; chmn., chief exec. officer Bose Corp., Framingham, Mass. Author: (with Kenneth N. Stevens) Introductory Network Theory, 1965; patentee in acoustics, nonlinear systems and communications. Fulbright fellow India, 1956-57; recipient Baker Teaching award MIT, 1964, Teaching award Am. Soc. Engring. Edn., 1965; named Inventor of Yr., Intellectual Property Owners, 1987. Fellow IEEE; mem. AAAS, Nat. Acad. Engring., Sigma Xi, Tau Beta Pi, Eta Kappa Nu. Office: Bose Corp The Mountain Framingham MA 01701-9168

BOSE, ANIMESH, materials scientist, engineer; b. Calcutta, India, Apr. 8, 1953; came to U.S., 1985; s. Amal Kumar and Bani (Roychoudhury) B.; m. Prarthana Bose, Feb. 24, 1986; 1 child, Pinaki Bose. B. Tech., Indian Inst. Tech., Kharagpur, 1977; PhD, Indian Inst. Tech., 1982. Rsch. assoc. Indian Inst. Tech., Kharagpur, India, 1982-85, Rensselaer Poly. Inst., Troy, N.Y., 1985-89; sr. rsch. engr. S.W. Rsch. Inst., San Antonio, 1989—; cons. Metadyne, Elmira, N.Y., 1986-87, Tex. Instrument, Attleboro, Mass., 1987-88, Ceramic Process Systems, Cambridge, Mass., 1988, Tech. Assoc. Corp., Huntington Beach, Calif., 1990; researcher Calif. Rsch. and Tech., NASA, Def. Advanced Rsch. Project Agy., U.S. Army, Office of Naval Rsch.; mem. program com. 1992 Powder Metallurgy World Congress, San Francisco; presenter in field; co-chmn. Interant. Symposium on Tungsten and Its Alloys. Contbr. more than 60 articles to profl. jours.; inventor, patentee in field. Mem. Am. Powder Mettalurgy Inst., Powder Metallurgy Assn. India (life), Materials Rsch. Soc., Minerals, Metals, Materials Soc. Hindu. Achievements include development of Tungsten and its alloys and composites with applications in defense and commercial industries and the concept and development of a new breed of composites known as micro-in filtrated macro-laminated composites. Office: SW Rsch Inst Div 6 6220 Culebra Rd San Antonio TX 78238-5166

BOSE, ANJAN, electrical engineering educator, researcher, consultant; b. Calcutta, India, June 2, 1946; s. Amal Nath and Anima (Guha) B.; m. Frances Magdelen Pavlas, Oct. 30, 1976; children: Rajesh Paul, Shonali Marie, Jahar Robert. B Tech with honors, Indian Inst. Tech., Kharagpur, 1967; MS, U. Calif., Berkeley, 1968; PhD, Iowa State U., 1974. Systems planning engr. Con Edison Co., N.Y.C., 1968-70; instr., research assoc. Iowa State U., Ames, 1970-74; postdoctoral fellow IBM Sci. Ctr., Palo Alto, Calif., 1974-75; asst. prof. elec. engring. Clarkson U., Potsdam, N.Y., 1975-76; mgr. EMSD, Control Data Corp., Mpls., 1976-81; prof. elec. engring. Ariz. State U., Tempe, 1981-93; Disting. Prof. in Power, dir. Sch. Elec. Engring./Computers Wash. State U., Pullman, 1993—; v.p. Power Math Assocs., Tempe, 1981-84. Contbr. over 50 articles to engring. jours. Fellow IEEE.

BOSHKOV, STEFAN HRISTOV, mining engineer, educator; b. Sofia, Bulgaria, Sept. 29, 1918; came to U.S., 1938, naturalized, 1944; s. Hristo and Karla (Lubich) B.; m. Bianca G. Amaducci, Aug 28, 1943; children—Lynn Karla, Stefan Robert. Diploma, Am. Coll., Sofia, Bulgaria, 1938; B.S., Columbia U., 1941, E.M., 1942. Mem. faculty Columbia, 1946—; prof. Henry Krumb Sch. Mines, 1951—, chmn., 1967-85, 86-88, Henry Krumb prof., 1980-88, Henry Krumb prof. emeritus, 1989—; Disting. prof., sr. scientist (Fulbright program), Yugoslavia, 1969, guest lectr., Taiwan, China, 1972, 76; guest lectr. OAS, Chile, 1972, USSR, 1974, Poland, 1976, Bulgaria, 1976, Bolivia, 1977, People's Republic of China, 1980, 81; cons. engr., 1950—; mem. internat. organizing com. World Mining Congress, 1962—, hon. mem., 1988—; chmn. 4th Internat. Conf. Strata Control and Rock Mechanics, N.Y.C., 1964; tech. chmn. UN Internat. Meeting on New Techs. for Devel. Coal Resources in Developing Countries, Beijing, 1985; mem. adv. com. metal and nonmetallic health and safety standards Dept. Labor, 1978. Mem. editorial bd.: Internat. Jour. Rock Mechanics and Mining Scis, 1964-76. Pres. Benedict Found., 1973—, Harrison (N.Y.) No. 7 Sch. Bd., 1954-63. Served to lst lt. AUS, 1943-46, CBI. Recipient Boleslaw Krupinski medal State Mining Council Poland, 1980, Marin Drinov medal Bulgarian Acad. Scis., 1987. Mem. AIME (Mineral Industry Edn. award 1980, disting. 1984), Am. Arbitration Assn., N.Y. Acad. Scis., Sigma Xi. Presbyterian. (trustee 1965-69). Club: Masons. Home: 119 White Plains Ave White Plains NY 10604-2854 Office: Columbia Univ Mudd Bldg New York NY 10027

BOSIO, ANGELO, pharmacologist, psychiatrist; b. Brescia, Italy, Jan. 18, 1955; s. Giulio and Teresa (Macetti) B.; m. Rosangela Rosola, Nov. 23, 1985; children: Alessandro, Annalisa. MD, Milan U., 1980, degree in pharmacology, 1982, degree in psychiatry, 1988. Intern Milan U. Med. Sch., 1984-88, cons. psychiatrist, 1988—; dir. pharmacological dept. St. Anne Clinic, Brescia, 1987—; cons. Internat. Pharm. Companies, 1983—, World

Health Orgn., 1988, others. Author: Handbook of Reaction Time Evaluation, 1991; editor jour. Neurosciences Collection, 1988—, H & W in Medicine, 1992—; editor videotapes Neurotransmission, 1988, Anxiolytic Drugs: An Up to Date, 1989, The Metamorphosis, 1991, Mioclonus and Piracetam, 1991, The Living Proof, 1991; jouralist Sci. and Med. Press, 1982—. Recipient Nutrition Found. award, Italy, 1982. Fellow AAAS, N.Y. Acad. Scis., Internat. Psychogeriatric Assn., Italian Psychiat. Soc.; mem. Assn. Advancement Neurosci. (pres. Brescia chpt. 1987-92, U.S.A. chpt. 1990—). Roman Catholic. Office: Assn Advancement Neurosci, Via Vivanti 9, Brescia 25100, Italy also: AAN 575 Madison Ave New York NY 10022

BOSLEY, DAVID CALVIN, design engineering executive; b. Cin., Mar. 27, 1959; s. Joseph Jr. and Leotha (Bell) B.; m. Charlene Stewart, Feb. 15, 1986; 1 child, David Anthony. B of Tech., U. Dayton, 1983. Test specialist GE Aircraft Engines Controls and Accessories Test, Fort Wayne, Ind., 1983-86; design engr. GE Aircraft Engnes, Monitoring Systems Engring., Cin., 1986-91; project engr. GE Aircraft Engines, MSE, Cin., 1992—; chmn. Dynamic Ednl. Systems, Inc., Cin., 1989—; pres. Aeroflux Corp., Cin., 1991—. Mem. AIAA, Smithsonian Inst., Phi Beta Sigma. Pentecostal-Apostolic. Acheivements include patent for enclosure ventilation and temperature apparatus. Office: Aeroflux Corp Ste #1 10133 Leacrest Rd Cincinnati OH 45215-1310

BOSLEY, WARREN GUY, pediatrician; b. Palisade, Nebr., Jan. 1, 1922; s. Charles M. and Verna M. (Gruver) B.; m. Alleen Finney, May 20, 1944; children: Michael, Barbara, Matthew, Timothy, David, John. AB, U. Nebr., 1942, MD, 1944. Diplomate Am. Bd. Pediatrics. Intern Johns Hopkins Hosp., Balt., 1944-45; resident in pediatrics Babies Hosp., N.Y.C., 1948-50; pvt. practice Grand Island, Nebr., 1950-91; med. dir. substance abuse unit Hastings (Nebr.) Regional Ctr., 1991—; mem. Am. Bd. Pediatrics, Chapel Hill, N.C., 1981-92, Medicaid Adv. Commn. Nebr., Lincoln, 1988-91, Nebr. State Bd. Health, Lincoln, 1968-71, 77-80. Author: (chpt.) Medical Liability in Pediatrics, 1988, 90. Pres. Bd. Edn., Grand Island, 1959-93, mem., 1993—; mem. Libr. Bd., Grand Island, 1966-80, Friends of Libr., U. Nebr., Lincoln, 1989—, Friends of Librs. U. Nebr. Coll. Medicine, Omaha, 1978—. Capt. med. corps AUS, 1942-47. Mem. AMA, Am. Acad. Pediatrics, Nebr. Med. Assn. (pres. 1975-76), Phi Beta Kappa, Sigma Xi, Alpha Omega Alpha. Home: 1515 W First St Grand Island NE 68801

BOSMA, JAMES FREDERICK, pediatrician; b. Grand Rapids, Mich., Apr. 29, 1916; s. John J. and Gertrude (Jeltes) B.; m. Julia S. Schaafsma, June 16, 1942; children: Ann, John, Peter, Stephen, Elizabeth, Helen, Julia, Martha. BA, Calvin Coll., Grand Rapids, 1937; MD, U. Mich., 1941. Diplomate Am. Bd. Pediatrics. Rotating intern Cleve. City Hosp., 1941-42; intern, then resident in pediatrics West Res. U. Hosps., Cleve., 1942-44; Nat. Found. Infantile Paralysis rsch. fellow U. Minn., Mpls., 1944-47, instr. dept. pediatrics Sch. Medicine, 1947-48; prof. dept. pediatrics Sch. Medicine U. Utah, 1949-59; NIH Rsch. fellow Karolinska Inst., Stockholm, Sweden, 1959-61; chief oral & pharyngeal devel. sect. Nat. Inst. Dental Rsch. NIH, Bethesda, Md., 1961-82; lectr. dept. radiology The Johns Hopkins Sch. Medicine, Balt., 1982-85; rsch. prof. dept. dentistry for children Sch. Dentistry U. Md., Balt., 1984—; rsch. prof. dept. pediatrics Sch. Medicine U. Md., Balt., 1984—, clinical prof. dept. neurology/rehab., 1988—; mem. consulting staff JFK Inst. for Handicapped Children, Balt., 1982-87; symposia condr. in field. Author: Anatomy of the Infant Head, 1986, (with others) The Cranium of the Newborn Infant, 1978, Postnatal Development of the Rat Skull, 1983; contbr. articles to Jour. Neurophysiology, Brain, Physiol. Rev., Pediatrics, Jour. Speech and Hearing, Jour. Applied Physiology, Ann. Otol. Rhinol. and Laryngol., Acta Paediat. Scan, Cleft Palate Jour., Jour. Craniofac. Genet. Develop. Biol., Neurology, Radiology, Jour. AMA, Internat. Jour. Orofacial Myology, Folia Phoniat, Gastrointest. Radiology, Jour. Clin. Nutrition, also chpts. to books. Mem. Am. Pediatric Soc., Am. Soc. for Pediatric Rsch., Am. Acad. Pediatrics (sect. on child devel.), Am. Assn. for Cerebral Palsy and Devel. Medicine. Office: U Md Hosp Dept Pediatrics 22 S Greene St Baltimore MD 21201

BOSNAK, ROBERT J., mechanical engineer, federal agency administrator. BS in Engring., USCG Acad., 1948; MS in Naval Architecture and Marine Engring., degree in Naval Engring., MIT, 1960. Registered profl. engr., Mass. Commd. ensign USCG, 1948, advanced through grades to capt., ret., 1972; tech. reviewer mech. engring. Atomic Energy Commn.; chief mech. engring. br. Office of Nuclear Regulation; dept. dir. divsn. engring. Office Nuclear Regulatory Rsch. U.S. Nuclear Regulatory Commn., Washington, 1987—. Fellow ASME (coun. codes and standards, coun. engring. bd. rsch. adn tech. devel., v.p. codes and standards rsch. planning com., boiler and pressure vessel main com., exec. com., subcom. nuclear power 1964—, Bernard F. Langer Nuclear Codes and Standards award 1982, Codes and Standards medal 1990); mem. Am. Soc. Naval Engrs., Am. Welding Soc., Soc. Naval Architects and Marine Engrs., Tau Beta Pi. Office: Office Nuclear Reg Reseach US Nuclear Regulatory Commn Washington DC 20555*

BOSS, MICHAEL ALAN, biologist; b. London, Jan. 24, 1955. BSc, University Coll., London, 1976; PhD, London U., 1979; MBA, London Bus. Sch., 1986. Sr. scientist Celltech Group PLC, Slough, Eng., 1982-83; product mgr. Celltech Grap PLC, Slough, Eng., 1984-85, regional sales mgr., 1986-87; pres. Oros Instruments Inc., Cambridge, Mass., 1987-90, Alliance Prodns., Cambridge, 1990—; dir. rsch. and devel. Genica Pharms. Corp., Worcester, Mass., 1991—; cons. munerous biotech. cos., Boston and Cambridge, 1990—. Office: Genica Pharms Corp 373 Plantation St Worcester MA 01605

BOSSERT, DAVID EDWARD, electrical engineer; b. Kansas City, Mo., Oct. 11, 1962; s. Edward Francis and Patricia Ann (Aubin) B.; m. Nina Mary Bender, Jan. 19, 1985; children: Katrina Elizabeth, Brett Andrew. BSEE, U. Mo., 1984; MSSM, U. So. Calif., 1988; MSEE, Air Force Inst. of Tech., Dayton, Ohio, 1989. Registered profl. engr., Calif. Commd. 2d lt. USAF, 1985, advanced through grades to capt.; tech. engr. team chief, minuteman/peacekeeper 394th ICBM Test Maintenance Squadron, Vandenberg AFB, Calif., 1985-88; asst. prof. of Aeronautics USAF Acad., Colo., 1988—. Contbr. articles to conf. procs. Mem. AIAA (Rocky Mountain Engr. of Yr. 1992). Republican. Southern Baptist. Achievements include co-developer of prefilter design based upon time response; rsch. on robust control of aircraft systems, robust control of robotics systems, psuedo-continuous time quantitative feedback theory. Home: 5024 Stillwater Rd Colorado Springs CO 80918 Office: USAF Acad HW USAFA/DFAN U S A F Academy CO 80840

BOSTON, LOUIS RUSSELL, packaging engineer; b. Chgo., Sept. 11, 1937; s. Henry William and Alice Jane (Winter) B.; m. Deanna Lee Panebianco, Nov. 30, 1957; children: Terry, Susan, Jeffrey, Colleen. AAS, SUNY, Morrisville, 1957; BS, Syracuse U., 1968. Technician dairy sci. Dairylea, Syracuse, N.Y., 1957-59; technician food sci. Borden Inc., Syracuse, 1959-68, food technologist, 1968-71, project leader product devel., 1971-78, mgr. packaging devel., 1978-87, assoc. dir. packaging, 1987-92, packaging dir., 1992—. Pres. Twp. Community Coun., Town of Sullivan, N.Y., 1976. Mem. TAPPI (assoc.), Inst. Packaging Profls., Ctrl. N.Y. Inst. Food Technologists. Republican. Methodist. Achievements include patents for culture media for starter prodn., method of making and use of resealable package, form, fill, seal, deflation method and apparatus. Office: Borden Inc 1 Gail Borden Dr Syracuse NY 13204

BOSTROM, CARL OTTO, physicist, laboratory director; b. Port Jefferson, N.Y., Aug. 18, 1932; s. Carl Oscar and Dagmar Ester (Anderson) B.; m. Sara A. Herzog, Sept. 6, 1954; children: Robin I. Bostrom Dagan, Jennifer A. Bostrom Simmons, Carl E. BS in Physics, Franklin & Marshall Coll., 1956; MS in Physics, Yale U., 1958, PhD, 1962; ScD (hon.), Franklin & Marshall Coll., 1992. Physicist space dept. Johns Hopkins U. Applied Physics Lab., Laurel, Md., 1960-68, group supr. space dept., 1968-74, chief scientist space dept., 1974-78, assoc. head space dept., 1978-79, dept. head and asst. dir. space dept., 1979, dep. dir., 1979-80, dir., 1980-92; mem.-at-large adv. bd. Def. Intelligence Agy., Washington, 1982-87, 92—, chmn. sci. adv. com., 1988-92; ex officio mem. Def. Sci. Bd., Washington, 1988-92; mem. Pres. Com. on Nat. Medal Sci., 1986-91; mem.-at-large Air Force Sci. Adv. Bd., 1983-87; mem. bd. visitors Naval Surface Weapons Systems Engring. Sta. and Naval Ship Systems Engring. Sta., 1982-92; mem. external adv. bd. Ga. Tech. Rsch. Inst., 1988—. Contbr. articles to profl. jours. including Jour. Geophys. Rsch., Geophys. Rsch. Letters, Sci., Space Sci. Rev., others. Mem. BCC chpt. Izaak Walton League, Poolesville, Md., 1974-79. With U.S. Army, 1950-52. Recipient Air Force medal for exceptional civilian svc./, 1987, DOD Medal for Disting. Pub. Svc., 1992, Def. Intelligency Agy. Medal for exceptional civilian svc., 1992, NASA Disting. Pub. Svc. medal, 1992. Fellow Hudson Inst.; mem. AAAS, Am. Phys. Soc., Am. Def. Preparedness Assn., Am Geophys. Union (com. on govtl. and legis. affairs 1975-79, chmn. 1975-77), Am. Soc. Naval Engrs., Internat. Assn. Geomagnetism and Aeronomy, Cosmos Club, Navy League, Phi Beta Kappa, Sigma Xi, Sigma Pi. Office: Johns Hopkins U Applied Physics Lab Johns Hopkins Rd Laurel MD 20723

BOSWELL, FRED C., retired soil science educator, researcher; b. Monterey, Tenn., Aug. 20, 1930; s. Ferdando Cortez and Julia Ann (Speck) B.; m. Marjorie Sue Brown, Sept. 3, 1954; children: Elaine Joy, Julia Alma Boswell Merry. BS, Tenn. Tech., 1954; MS, U. Tenn., 1956; PhD, Pa. State U., 1960. Asst. agron U. Tenn., Knoxville, 1955-56; asst. soil chemist Ga. Agrl. Experimental Sta., Experimental, Ga., 1956-57; asst. prof. U. Ga., Athens, 1960-82, prof., head agronomy dept., 1982-89; prof. U. Ga., Griffin, 1989-91; prof. emeritus U. Ga., 1991—; adj. prof. U. Tenn., 1993—. Contbr. chpts. to numerous books and sci. jours. on agrl. concerns; mem. editorial com. Fertilizer Technology and Use, 1985. Mem. com. State of Ga. Goals for Ga., 1971; chmn. subcom. Griffin Spalding Co., Ga., 1988. With U.S. Army, 1951-53. Fellow Am. Soc. Agronomy, Soil Sci. Soc. Am. (bd. dirs. 1978-80, so. branch pres. 1983-84), Toastmasters (v.p. 1972). Achievements include research related to certain micronutrients nitrification transformation and nitrogen movement and nitrification inhibitors; published/researched in environmental science, emphasis on land treatment waste materials. Home: 748 Russell Strausse Cookeville TN 38501-4515

BOSWORTH, DOUGLAS LEROY, farm implement company executive; b. Goldfield, Iowa, Oct. 15, 1939; s. Clifford LeRoy and Clara (Lonning) B.; m. Patricia Lee Knock, May 28, 1961; children: Douglas, Dawn. B.S. in Agrl. Engring, Iowa State U., 1962; M.S. in Agrl. Engring, U. Ill., 1964. With Deere & Co., Moline, Ill., 1959—, reliability mgr., 1967-71, div. engr. disk harrows, 1971-76, mgr. mfg. engring., 1976-80, works mgr., 1980-85, mgr. mfg., 1985-89, engring. test mgr., 1989—; mem. Engring. Accreditation Commn., 1985-90; v.p. Skills Inc.; mem. Associated Employers Bd., 1989-91. Active Am. Cancer Soc., Rock Island Unit; bd. dirs. United Med. Ctr., 1984—; mem. exec. com. Quad-City United Way, 1984-89. Mem. Am. Soc. Agrl. Engrs. (chmn. Ill.-Wis. 1973-74, Engring. Achievement Young Designer award 1973, nat. bd. dirs. 1974-76, 79-82, v.p. 1979-82, pres. elect. 1991-92, pres. 1992-93), Sigma Xi, Alpha Epsilon, Gamma Sigma Delta. Lutheran. Lodge: Rotary. Home: 4432 37th Ave Rock Island IL 61201-9214 Office: Deere & Co 1800 158th St N East Moline IL 61244-9532

BOTELHO, ROBERT GILBERT, energy engineer; b. Fall River, Mass., Apr. 20, 1948; s. Louis and Lorraine (Gendreau) B.; m. Arlene Asselin, Oct. 15, 1971 (div. May 1989); children: Lyndsey Dawn, Jennifer Lee; m. Donna Shaw, May 12, 1990. AS, Bristol C.C., 1969; BS in Indsl. Tech., Roger Williams U. Lic. constrn. supt., Mass. Energy conservation mgr. City of Fall River, Mass., 1978-81, energy coord., 1981-85; project mgr. M.P.C. Svc. Corp., St. Louis, 1985-87; tech. dir. Applied Resource Mgmt. Corp., St. Louis, 1987-90; energy cons. Stroh Brewery Corp., Detroit, 1990—; v.p. Assn. of Local Energy Officials, State of Mass., 1982-85; advisor Nat. Community Energy Mgmt. Ctr., Washington, 1983-86. Mem. Jr. High Sch. Site Selection Com., Fall River, 1985; advisor Coalition of N.E. Municipalities, 1982-86; conv. whip John F. Kerry for Senator, Worcester, 1985. Grantee State of Mass., 1984-85, 5 grants HUD, 1981-85. Mem. ASHRAE, Assn. of Energy Engrs., Fall River Lions Evening Club (v.p. 1984-85, Lion of Yr. 1985, 86), Fall River Country Club. Office: RGB Inc 67 Cliffdale Ave Cranston RI 02905

BOTSTEIN, DAVID, geneticist, educator; b. Zurich, Switzerland, Sept. 8, 1942; naturalized, 1954; A.B. cum laude in Biochem. Scis., Harvard U., 1963; Ph.D. in Human Genetics, U. Mich., 1967. Woodrow Wilson fellow, 1963; instr. dept. biology MIT, Cambridge, 1967-69; asst. prof. genetics MIT, 1969-73, assoc. prof. genetics dept. biology, 1973-78, prof., 1978—; mem. sci. adv. bd. Collaborative Research, Inc., 1978-87. Editor in chief Molecular Biology of Cell, 1992—; mem. editorial bd. Virology, 1976-82, Jour. of Virology, 1976-85, Genetics, 1980—, NSF Study Sect. Genetic Biology, 1972-76, ACS Study Sect. Virology and Cell Biology, 1977-81; contbr. over 150 articles to profl. jours. Recipient Career Devel. award NIH, 1972-74; Eli Lilly and Co. award in Microbiology and Immunology, 1978, Rosenstiel award, 1992. Mem. Nat. Acad. Scis., Genetics Soc. Am. (bd. dirs. 1984). Office: Stanford U Dept Genetics Stanford CA 94305 also: Genentech Inc 460 Point San Bruno Blvd South San Francisco CA 94080

BOTT, RAOUL, mathematician, educator; b. Budapest, Hungary, Sept. 24, 1923; s. Rudolf and Margit (Kovach) B.; m. Phyllis Aikman, Aug. 30, 1947; children: Anthony, Jocelyn, Renee, Candace. B Engring., McGill U., Montreal, 1945, M Engring., 1946, DSc (hon.), 1987; DSc, Carnegie Inst. Tech., 1949; DSc (hon.), Notre Dame U., 1979, Carnegie-Mellon U., 1989. Fellow Inst. Advanced Studies, Princeton U., 1949-51, 55-57; instr. math. U. Mich., Ann Arbor, 1951-52, asst. prof., 1952-55, assoc. prof., 1957-59; prof. Harvard U., Cambridge, Mass., 1959—, W. Casper Graustein prof., 1969—. Author books and papers in various branches of math. and its relation to physics. Sloan fellow, 1956-60; hon. fellow St. Catharines Coll., 1985; recipient Nat. Sci. Medal, Pres. of U.S., 1987. Fellow Am. Acad. Arts & Scis., Am. Math. Soc. (Veblen prize 1964, Steel prize 1990); mem. NAS, London Math. Soc. (hon.). Democrat. Roman Catholic. Avocations: music, nature. Home: 1 Richdale Ave # 9 Cambridge MA 02140-2627 Office: Harvard U Dept Math 1 Oxford St Cambridge MA 02138-2901

BOTTENUS, RALPH EDWARD, biochemist; b. Ridgewood, N.J., May 10, 1956; s. Daniel and Louise (Dick) B.; m. Robin Lydell Roberson, June 2, 1984; 1 child, Nicholas Brian. BA in Biology, Wake Forest U., 1978, PhD in Biochemistry, 1986. Sr. fellow U. Washington, Seattle, 1986-90; sr. scientist Ortho Diagnostic Systems, Raritan, N.J., 1990—; mem. analytical instrument subcom. working group Johnson & Johnson, 1991—, drug delivery subcom., 1990—. Contbr. articles to Biochemistry, Jour. Biol. Chemistry, Gene. Mem. AAAS, Sigma Xi. Republican. Lutheran. Achievements include research in gene sequence for human factor XIII b subunit. Office: Ortho Diagnostic Systems Rt 202 Raritan NJ 08869

BOTTERO, PHILIPPE BERNARD, general practitioner; b. Villecresnes, France, May 21, 1940; s. Pierre and Hetty (Saltiel) B. SPCN, Faculty Scis., Paris, 1959; studies of medicine, Faculty Medicine, Paris, 1959-65; MD, 1977; DSc (hon.), London Inst. Applied Rsch., 1991. Intern Montfermeil Hosp., Coulommiers Hosp., Meaux Hosp., 1965-66, Necker Enfants Malades Hosp., Paris, 1970-73; gen. practice medicine Nyons, France, 1978—. Contbr. articles to profl. jours. Mem. MAison Internat. Des Intellectuels Acad. Avocations: travel, hiking, reading, music. Home: Venterol, 26110 Nyons France Office: 10 Ave Henri Rochier, 26110 Nyons France

BOTTJER, DAVID JOHN, geological sciences educator; b. N.Y.C., Oct. 3, 1951; s. John Henry and Marilyn (Winter) B.; m. Sarah Ranney Wright, July 26, 1973. BS, Haverford Coll., 1973; MA, SUNY, Binghamton, 1976; PhD, Ind. U., 1978. NRC postdoctoral rsch. assoc. U.S. Geol. Survey, Washington, 1978-79; asst. prof. dept. geol. scis. U. So. Calif., L.A., 1979-85, assoc. prof. dept. geol. scis., 1985-91, prof. dept. geol. scis., 1991—; rsch. assoc. L.A. County Mus. Natural History, 1979—, U. Calif., 1991—; vis. scientist Field Mus. Natural History, Chgo., 1986; guest prof. Swiss Fed. Inst. Tech., Zurich, 1993. Editor Palaios, 1989—; assoc. editor Cretaceous Rsch., 1988-91; mem. editorial bd. Geology, 1984-89, Hist. Biology, 1988—; co-editor Columbia U. Press Critical Moments in Paleobiology and Earth History (book series), 1990—; chmn. Columbia U. Press Adv. Com. for Paleontology, 1990—. Fellow Geol. Soc. Am., Geol. Soc. London; mem. AAAS, Paleontol. Soc., Soc. Sediment Geology, Internat. Paleontology Assn. Office: U So Calif Dept Geol Sci Los Angeles CA 90089-0740

BOTWIN, MICHAEL DAVID, psychologist, researcher; b. Hamtramck, Mich., Sept. 6, 1957; s. Richard Julius and Joan Loretta (Bruzedwicz) B.; m. Lillian Jarualitis, Aug. 20, 1983; 1 child, David; 1 stepchild, James Veros. BA, Oakland U., 1983; PhD, U. Mich., 1989. Postdoctoral fellow U. Ill., Champaign, 1989-90; asst. prof. Dept. Psychology Calif. State U., Fresno, 1990—. Mem. APA, AAUP, Am. Psychol. Soc., We. Psychol. Assn., Internat. Soc. for Study Individual Differences. Office: Calif State U Dept Psychology 5310 N Campus Dr Fresno CA 93740-0011

BOUCHER, DARRELL A., JR., vocational education educator; b. Conway, Ark., Nov. 3, 1944; s. Darrell A. and Thora Magdaline (Edgeman) B.; m. Zandra Irene Chain, Dec. 26, 1966; children: Michelle, Maegan. A in Aero. Engring. Tech., Okla. State U., 1965, BS in Occupational Adult Edn., 1968, MS in Occupational Adult Edn., 1990. FAA cert. air frame and power plant mechanic. Sr. instrument elec. tech. Tex. Oil and Gas, Canton, Okla., 1976-82; cons./owner/operator Boucher Elec., Oakwood, Okla., 1982-85; maint. supt. Metano Gas, Sayre, Okla., 1985-86; instr. High Plains Area Vo-Tech, Woodward, Okla., 1987—; adv. bd. Okla. Dept. Vocat. Edn., Stillwater, 1991—. Pastor Fellowship Christian Ch., Woodward, 1989—; dist. chmn. Boy Scouts Am., Boiling Springs dist., 1992-93. With U.S. Army, 1969-71. Mem. IEEE (standards com. automobile inverter power systems), Instrument Soc. Am. (sr., com. cert. standards), Nat. Fire Protection Assn., Fluid Power Soc. (cert. fluid power instr.), Phi Delta Kappa, Iota Lambda Sigma. Republican. Office: High Plains Area Vo-Tech 3921 34th St Woodward OK 73801

BOUCHER, THOMAS OWEN, engineering educator, researcher; b. Providence, June 25, 1942; s. Joseph William and Anne Marie (Byrne) B.; m. Unn Gunnerus Jermstad, Mar. 30, 1974. BSEE, U. R.I., 1964; MBA, Northwestern U., 1970; PhD in Indsl. Engring., Columbia U., 1978. Sr. project engr. Continental Can Co., Chgo., 1967-69; sr. staff cons. ABEX Corp., N.Y.C., 1970-72; asst. prof. Cornell U., Ithaca, N.Y., 1978-81; asst. prof. Rutgers U., New Brunswick, N.J., 1981-87, assoc. prof., 1987—. Dept. editor IIE Transactions, 1987-91; area editor Engring. Economist, 1989—; assoc. editor Jour. Productivity Analysis, 1989-91, Internat. Jour. of Flexible Automation and Integrated Mfg., 1992—. 1st lt. U.S. Army, 1965-67, Vietnam. Grantee NSF, Def. Logistics Agy. Mem. IEEE, SME (sr.), Am. Soc. for Engring. Edn. (chmn. engring. econ. dir. 1986-87), N.Y. Acad. Scis., Inst. Indsl. Engrs. (sr.), Sigma Xi. Roman Catholic. Achievements include research in manufacturing automation, computer integrated manufacturing systems, production planning and control, and engineering economics. Home: 65 Douglas Rd Glen Ridge NJ 07028-1227 Office: Rutgers U Coll of Engring PO Box 909 Piscataway NJ 08855-0909

BOUDA, DAVID WILLIAM, insurance medical officer; b. Omaha, Dec. 24, 1945; s. William Paul and Frances Lillian (Kuzela) B.; m. JoAnn Warrell, June 8, 1968; children: William Paul, Tamara Ann. BA in Chemistry, Kans. U., 1968, MD, 1972. Intern St. Luke's Hosp., Kansas City, Mo., 1972-73; resident Wilford Hall USAF Med. Ctr., Lackland AFB, San Antonio, 1975-77; med. oncologist Hutchinson (Kans.) Clinic, 1981-88; healthcare cons. Hutchinson, 1988-89; med. oncologist Internal Medicine Assocs., Omaha, 1989-91; chief med. officer Blue Cross/Blue Shield of Nebr., Omaha, 1992—, sr. v.p. healthcare policy, 1993—; bd. dirs. Nebr. div. Am. Cancer Soc., Omaha, 1992—. Lt. col. USAF/USAFR, 1968-91. Fellow ACP; mem. AMA, Rotary. Democrat. Roman Catholic. Office: Blue Cross/Blue Shield Nebr PO Box 3248 Omaha NE 68180

BOUDART, MICHEL, chemist, chemical engineer, educator; b. Belgium, June 18, 1924; came to U.S., 1947, naturalized, 1957; s. Francois and Marguerite (Swolfs) B.; m. Marina D'Haese, Dec. 27, 1948; children: Mark, Baudouin, Iris, Philip. BS, U. Louvain, Belgium, 1944, MS, 1947; PhD, Princeton U., 1950; D honoris causa, U. Liège, U. Notre Dame, U. Nancy, U. Ghent. Research asso. James Forrestal Research Ctr., Princeton, 1950-54; mem. faculty Princeton U., 1954-61; prof. chem. engring. U. Calif., Berkeley, 1961-64; prof. chem. engring. and chemistry Stanford U., 1964-80, Keck prof. engring., 1980—; cons. to industry, 1955—; co-founder Catalytica, Inc.; Humble Oil Co. lectr., 1958, Am. Inst. Chem. Engrs. lectr., 1961, Sigma Xi nat. lectr., 1965; chmn. Gordon Research Conf. Catalysis, 1962. Author: Kinetics of Chemical Processes, 1968, (with G. Djéga-Mariadassou) Kinetics of Heterogeneous Catalytic Reactions, 1983; editor: (with J.R. Anderson) Catalysis: Science and Technology, 1981, (with Marina Boudart and René Bryssinck) Modern Belgium, 1990; mem. adv. editorial bd. Jour. Internat. Chem. Engring., 1964—, Catalysis Rev., 1968—. Belgium-Am. Ednl. Found. fellow, 1948, Procter fellow, 1949; recipient Curtis-McGraw rsch. award Am. Soc. Engring. Edn., 1962, R.H. Wilhelm award in chem. reaction engring., 1974, Chem. Pioneer award Am. Inst. of Chemists, 1991. Fellow AAAS, Am. Acad. Arts. and Scis.; mem. NAS, NAE, Am. Chem. Soc. (Kendall award 1977, E.V. Murphee award in indsl. and engring. chemistry 1985), Catalysis Soc., Am. Inst. Chem. Engrs., Chem. Soc., Académie Royale de Belgique (fgn. assoc.). Home: 512 Gerona Rd Stanford CA 94305 Office: Stanford U Dept Chem Engring Stanford CA 94305

BOUDINOT, FRANK DOUGLAS, pharmaceutics educator; b. New Brunswick, N.J., Mar. 31, 1956; s. Frank Lins and Dorothy Jean (Libourel) B.; m. Sarah Garrett, Sept. 1992. BS in Biology, Springfield Coll., 1978; PhD in Pharmaceutics, SUNY, Buffalo, 1986. Vet. technician Afton Animal Hosp., Williamsville, N.Y., 1978-79; rsch. technician SUNY-Millard Fillmore Hosp., Buffalo, 1979-80; grad. asst. SUNY, 1980-85; asst. prof. pharmaceutics U. Ga., Athens, 1986-90, assoc. prof., 1990—, head dept. pharm., 1992—; cons. Johnson Matthey, West Chester, Pa., 1991—, Assn. Minority Profl. Health Schs., Drug Devel. Group for AIDS, Tex. So. U. Coll. Pharmacy and Health Scis. Mem. editorial bd. Jour. Pharmacy Teaching, Binghamton, N.Y., 1989—; referee Jour. Pharm. Scis., 1988—, Jour. Pharm. Rsch., 1989—; contbr. more than 50 articles to sci. jours. Bd. dirs. Oconee Animal Shelter, Watkinsville, Ga., 1986-88; vice chair govt. svcs. subcom. Oconee 2000, Waktinsville, 1986-87; del. Ga. State Rep. Conv., Atlanta, 1989, 91, 92. NIH grantee, 1987, 90, U.S. FDA grantee, 1989; named one of Outstanding Young Men of Am., 1987. Mem. AAAS, Am. Assn. Pharm. Scientists, Am. Assn. Colls. of Pharmacy (del. 1989-90, mem. profl. affairs com. 1990-91), Am. Soc. Microbiology, N.Y. Acad. Scis., Soc. Exptl. Biology and Medicine, Rho Chi. Episcopalian. Achievements include research in pharmacokinetics of antiviral drugs, effects of age in drug disposition, veterinary pharmacokinetics, and drug pharmacodynamics. Office: U Ga Coll Pharmacy Coll Pharmacy Brooks Dr Athens GA 30602

BOUDREAU, ROBERT JAMES, nuclear medicine physician, researcher; b. Lethbridge, Alta., Can., Dec. 27, 1950; came to U.S., 1983; s. George Joseph Boudreau and Eleanor Joyce (Dalzell) Hamilton; m. Francine Suzanne Archambault, Jan. 16, 1982. BSc with highest honors, U. Sask., Saskatoon, Can., 1972; PhD, U. B.C., Vancouver, Can., 1975; MD, U. Calgary (Alta.), 1978. Diplomate Am. Bd. Nuclear Medicine. Resident in diagnostic radiology and nuclear medicine McGill U., Montreal, Que., Can., 1978-82; asst. prof. U. Minn., Mpls., 1983-87, assoc. prof., 1987-93, prof., 1993—, dir. nuclear medicine div., dir. grad. studies dept. radiology, 1987-91. Author book chpts.; contbr. articles to profl. jours. Recipient Gold Key award Soc. Chem. Industry, 1972, Soc. Clin. Investigation Young Investigator award, 1978; Can. Heart Found. Med. Scientist fellow, 1976-78. Fellow Royal Coll. Physicians; mem. Am. Heart Assn., Am. Coll. Nuclear Physicians, Assn. Univ. Radiologists, Soc. Chiefs of Acad. Nuclear Medicine Sects. (treas. 1989-93), Soc. Nuclear Medicine (edn. and tng. com. 1983—, bd. govs. cen. chpt. 1989-92, treas. 1992—), Radiol. Soc. N. Am., Am. Coll. Radiology. Avocations: skiing, boating, travel, computers. Office: U Minn PO Box 382 UMHC Minneapolis MN 55455-0382

BOUGAS, JAMES ANDREW, physician, surgeon; b. Bismarck, N.D., Jan. 25, 1924; s. Andrew James and Mary (Psaltiras) B.; m. Tiina Parlin, June 27, 1953; children: Karen I., Tiina Maria. MD, Harvard U., 1948. Diplomate Am. Bd. Surgery, Am. Bd. Thoracic Surgery. Mixed med.-surg. intern Columbia U. Svc., Bellevue Hosp., N.Y.C., 1948-50, chief resident in surgery, 1952-53; resident Presbyn. Hosp., N.Y.C., 1950-52, chief resident surgery, 1953; assoc. Overholt Clinic, Boston, 1955-65; chief thoracic surgery Univ. Hosp., Boston, 1965-70; assoc. prof. surgery Boston U. Sch. Medicine, 1965—; lectr. Tufts U. Sch. Medicine, Boston, 1965-70; chmn. Gordon Rsch. Confs., 1967-68. Contbr. articles to profl. jours. Pres. Heart Assn., Boston, 1967-69; chmn. Mass. Rehab. Commn. Adv. Com.; trustee Boston Tb Assn. With U.S. Army, 1942-44. Fellow: AAAS; mem.: Am. Assn. Thoracic Surgeons, Soc. Thoracic Surgeons, Norfolk Dist. Med. Soc. (pres., tri-state regional planning com.). Achievements include development of combined cardiac catheterization; porous metal fabrication and biology. Office: NE Bapt Hosp 125 Parker Hill Ave Boston MA 02120

BOUKERROU, LAKHDAR, agricultural researcher; b. Constantine, Algeria, Apr. 18, 1952; came to U.S., 1981; s. Ahmed and Aicha (Meniai) B.; m. Cherifa Abbas, Mar. 31, 1979; children: Mohamed, Samir. BSc, ITA Mostaganem, 1974; MSc, U. Minn., 1984, PhD, 1986. Plant breeder IDGC, Guelma, Algeria, 1977-78; sta. dir. IDGC, Sidi Bel Abbes, Algeria, 1978-81; sta. mgr. Pioneer Hi-Bred Internat., Des Moines, 1986-87; postdoctoral fellow U. Minn., St. Paul, 1987-88; sr. scientist ACRES, Waterloo, Iowa, 1988, v.p., 1989-91; pres. CIRTA, Parkersburg, Iowa, 1990—; cons. Experience, Inc., Mpls., 1986, IFAD, Algeria, 1992, Islamic Devel. Bank, Jeddah, Saudi Arabia, 1993; agrl. bus. adv. com. Hawkeye C.C., 1992—. Contbr. numerous articles to profl. publs. Mem. Regional Coord. Coun., Waterloo, 1991—; chair com., v.p., pres. Self Help Found., Waverly, Iowa, 1990—. Recipient Recognition award Minn. Internat. Ctr., 1984; named Hon. Citizen, Elkader, Iowa, 1990—. Mem. Am. Soc. Agronomy, Crop Sci. Soc. Am., North Ctrl. Weed Sci. Soc., Weed Soc. Sci. Am., Coun. on Agrl. Sci. and Tech., European Weed Sci. Rsch. Soc. Muslim. Home: 226 W Park Ln Waterloo IA 50701 Office: CIRTA PO Box 478 Parkersburg IA 50665

BOULIAN, CHARLES JOSEPH, civil-structural engineer; b. Lake Charles, La., Aug. 29, 1955; s. Joseph Henry and Margret (O'Gwynn) B.; m. Sharon Pearson, Dec. 16, 1978; children: Jarrod Scott, Shanna Leann. Student, Auburn U., 1977. Registered profl. engr., Ala., Fla., Miss., Tex., Del., Pa.; cert. asbestos cons. Structural engr. Palmer & Baker Engrs., Mobile, Ala., 1977-78, Strucutural Prestress, Mobile, 1978-79; cons., plant engr. Ideal Basic Industries, Mobile, 1979-83; Gulf Coast mgr. Franki Found. Co., Mobile, 1984-85; structural engr. Palmer & Baker Engrs., Mobile, 1985-86; sect. mgr. BCM Engrs. Inc., Mobile, 1986-91; civil/structural engr. Brown & Root USA Inc., Mobile, 1991—; project mgr. for the asbestos survey Tulane U.; mem. Nt. Coun. Examiners for Engring. and Surveyors. Mem. ASCE. Baptist. Home: 3151 Dog River Rd Theodore AL 36582-2515 Office: Brown & Root Inc PO Box 160689 Mobile AL 36616-1689

BOULOS, ATEF ZEKRY, chemist; b. Cairo, Oct. 6, 1945; came to U.S., 1974; s. Zekry and Aida (Daoud) B.; m. Salwa Youssef, July 12, 1976; children: David, Timothy. BSc, Coll. Sci., Cairo, 1966, MS, 1972; MS, N.J. Inst. of Tech., 1984. Rsch. chemist El-Nasr Chem. Pharm., Cairo, 1967-74, Chase Chem. Co., Newark, N.J., 1974-77, Pharma Tech Co., L.I., N.Y., 1977-78, Beecham Products Inc., Parsippany, N.J., 1978-90; sr. rsch. scientist Smith Kline Beecham, Parsippany, 1990—. Mem. Am. Chem. Soc., Am. Assn. Pharm. Scientists. Democrat. Mem. Coptic Orthodox Ch. Achievements include invention of pharmaceutical over-the-counter products. Office: Smith Kline Beecham 1500 Littleton Rd Parsippany NJ 07054-3884

BOULOS, EDWARD NASHED, transportation specialist; b. Damanhour, Egypt, May 19, 1941; came to U.S., 1979; s. Nashed Boulos and Lila (Habib) Georgy; m. Mervet Saleh, Aug. 31, 1967; children: Nermine E., Yasmine E. BS in Chemistry and Physics, Cairo U., 1963; MS in Solid State Sci., Am. U., Cairo, 1966; PhD in Ceramic Engring., U. Mo., 1970. Supr., cons. Ministry of Industry, Cairo, 1963-79; assoc. prof. Am. U., Cairo, 1972-79; vis. prof. Cath. U. Am., Washington, 1979-81; sr. scientist Anchor Hocking Co., Lancaster, Ohio, 1981-84; team leader Ford Motor Co., Dearborn, Mich., 1984—; cons. USAF, Boston, 1984-89; liaison bd. mem. Alfred (N.Y.) U., 1985—, chmn.-elect, 1992. Co-editor: Advances in the Fusion of Glass, 1988; contbr. articles on glass tech. to prof. jours.; patentee in field. NSF rsch. grantee, 1967-71, 72-79. Fellow Am. Ceramic Soc.; mem. ASTM, Materials Rsch. Soc., Deutsche Glastechnische Gesellschaft, Sigma Xi. Avocations: travel, sports. Office: Ford Motor Co Glass Divsn 15000 N Commerce Dr Dearborn MI 48120-1225

BOULOS, PAUL FARES, civil and environmental engineer; b. Beirut, June 28, 1963; came to U.S., 1983; s. Fares and Marie-Rose (Abou Hadid) B. BS, Beirut U., 1985; BSCE, U. Ky., 1985, MSCE, 1986, PhD, 1989. Asst. prof. U. Ky., Lexington, 1990-91; sr. engr. Montgomery Watson, Pasadena, Calif., 1991—; internat. hydraulic expert Consorcio Nitogoi, Cali, Colombia, 1988-90; cons. in field. Author: Comprehensive Network Analyzer, 1990; contbr. articles to profl. publs. Grantee NSF, 1987, Am. Water Works Rsch. Found., 1992. Mem. ASCE (treas. 1992), Am. Water Works Assn., Sigma Xi, Tau Beta Pi, Chi Epsilon (U.S. delegation to NATO Advanced Study Inst. 1993). Achievements include work on computer-assisted water quality and hydraulic network modeling. Office: Montgomery Watson 301 N Lake Ave Ste 600 Pasadena CA 91101

BOULTER, PATRICK STEWART, surgery educator; b. Annan, Dumfriesshire, Scotland, May 28, 1927; s. Frederick Charles and Flora Victoria (Black) B.; m. Patricia Mary Barlow, Mar. 7, 1946; children: Jennifer Mary Stewart Boulter Bond, Anne Margaret Stewart Boulter Wood. MB, BS with honors, U. London, 1955. House surgeon Guy's Hosp., London, 1955-56, lectr. anatomy, 1956-57, sr. surg. registrar, 1959-62; surg. registrar Middlesex Hosp., London, 1957-59; cons. surgeon Regional Oncological Ctr. Royal Surrey (Eng.) Hosp., Guildford, 1962—; prof. surg. sci. U. Surrey, Guildford, 1985—; examiner U. London, U. Edinburgh, Scotland, U. Newcastle, Eng., U. Nottingham, Eng., U. Glasgow, Eng.; vis. prof. U. Oreg., U. Nebr., U. Bombay, U. Pakistan, U. Queensland, U. Sydney, Australia, U. Auckland, New Zealand, U. Dunedin, New Zealand. Author book chpts. on surgery; contbr. articles to profl. jours. Fellow Royal Coll. Surgeons Edinburgh (pres. 1991), Royal Coll. Surgeons Eng. (Penrose-May tchr. 1985—), Royal Coll. Surgeons Glasgow, Royal Coll. Physicians Edinburgh, Royal Australian Coll. Surgeons (hon.), Royal Soc. Medicine, Assn. Surgeons Gt. Britain and Ireland, Coll. Surgeons of South Africa (hon.), Coll. Surgeons of Sri Lanka (hon.); mem. Brit. Breast Group, Brit. Assn. Surg. Oncology (past v.p.), Surg. Rsch. Soc., Alpine Club, Caledonian Club, New Club, Swiss Alpine Club, Yorkshire Fly Fishers Club. Avocations: mountain climbing, skiing, fly fishing. Home: Cairnsmore Merrow, Guildford Surrey GU1 2xn, England Office: Royal Coll Surgeons, Nicolson St, Edinburgh EH8 9DW, Scotland

BOUNAR, KHALED HOSIE, electrical engineer, consultant; b. Draa Ben Khedda, Algeria, Dec. 10, 1957; came to U.S., 1977; s. Ahmed and Dhaouia (Abderrahmani) B.; married; two children. BSEE, Syracuse U., 1980, MSEE, 1983, PhD in Elec. Engring., 1989. Lectr. Syracuse U., 1981-86; rsch. analyst Radex, Inc., Bedford, Mass., 1987—. Author: Quantized Gaussian Sequences: Fully Observed and Missing Data, 1989. Mem. IEEE, Sigma Xi. Home: 68 Highland St Concord MA 01742-2918 Office: Radex Inc 3 Preston Ct Bedford MA 01730

BOUNOUS, DENISE IDA, veterinary pathology educator; b. Valdese, N.C., Mar. 2, 1952; d. Louis Daniel and Alverta Essie (Fowler) B.; m. John Wesley Kiser, Aug. 1, 1981. BA in English, U. of the South, 1974; DVM, Okla. State U., 1986; PhD in Pathology, La. State U., 1990. Diplomate Am. Coll. Vet. Pathology. Med. technologist U. Va. Med. Ctr., Charlottesville, 1974-77, Phoenix Zoo, 1977-81, Hillcrest Hosp., Oklahoma City, 1981-82; resident in pathology Coll. Vet. Medicine, La. State U., Baton Rouge, 1986-90; asst. prof. dept. pathology Coll. Vet. Medicine, U. Ga., Athens, 1990—. Contbr. articles to Jour. Am. Vet. Med. Assn., Inflammation, Vet. Pediatrics, Avian Diseases, Vet. Immunology and Immunopathology. Mem. Am. Vet. Med. Assn., Am. Soc. Vet. Clin. Pathology, Phi Zeta. Office: U Ga Coll Vet Medicine Dept Pathology Athens GA 30602

BOURDEAU, JAMES EDWARD, nephrologist, researcher; b. Seattle, Feb. 19, 1948; s. Robert Vincent and Beatrice Louella (Oman) B.; m. Susan Gwen Perlman, June 17, 1973 (dec. Dec. 1989); children: Nicole Rochele, Louis Vincent; m. Teri LeAnn Mullikin, May 17, 1991; 1 child, Seth Coran. PhD, Northwestern U., 1973, MD, 1974. Diplomate Nat. Bd. Med. Examiners, Am. Bd. Internal Medicine, Am. Bd. Internal Medicine-Nephrology. Intern, then resident Peter Bent Brigham Hosp., Boston, 1974-76; rsch. assoc. Lab. of Kidney & Electrolyte Metabolism Nat. Heart, Lung and Blood Inst., Bethesda, Md., 1976-79; asst. prof. Northwestern U., Chgo., 1979-84, U. Chgo., 1985-89; assoc. prof. U. Okla. Health Sci. Ctr., Oklahoma City, 1990—; mem. Okla. Health Rsch. Com., Oklahoma City, 1991—. Author: (chpt.) Calcium Metabolism/Clinical Disorders of Fluid and Electrolyte Metabolism, 1993. Surgeon USPHS, 1976-79. Recipient Rsch. grant NIH, 1982-95, Established Investigatorship Am. Heart Assn., 1984-89, fellowship John A. Hartford Found., 1980-83. Mem. Internat. Soc. Nephrology, Am. Soc. Clin. Investigation, Am. Soc. Nephrology, Am. Physiol. Soc. Home: 4509 Val Verde Dr Oklahoma City OK 73142-5152 Office: U Okla Health Sci Ctr Sect of Nephrology 921 NE 13th St Oklahoma City OK 73104-5028

BOURDEAU, PHILIPPE, environmental scientist; b. Rabat, Morocco, Nov. 25, 1926; s. Michel Edgard and Lucienne (Imbrecht) B.; m. Flora E. Gorirossi; 3 children. Cert. in Agrl. Engring., Sch. Agrl. Scis., Gembloux, Belgium, 1949; PhD, Duke U., 1954. Cert. forestry engr. Asst. prof. N.C. State U., Raleigh, 1954-56, Yale U., New Haven, 1956-58, 60-62; prof. U. Belgian Congo, Butare, Rwanda, 1958-60; head biology svc. Euratom Joint Rsch. Ctr., Ispra, Italy, 1962-71; div. head, dir. environ. rsch. Commn. European Communities, Brussels, 1971-91; spl. adviser Commn. European Communities, 1991—; prof. U. Brussels, 1971—; sec. gen. Sci. Problems of Environment, Internat. Coun. Sci. Unions, Paris, 1992. Editor, co-editor books, reports on ecotoxicology; contbr. over 100 articles on plant ecology, radioecology, ecotoxicology and environ. scis. to profl. publs. Named comdr. Order of Couronne, Belgian Govt. Fellow AAAS, Belgian Am. Ednl. Found.; mem. various sci. socs. Achievements include design, implementation and management of research programs regarding the environment. Office: Commn European Communities, 200 rue de La Loi, B 1049 Brussels Belgium

BOURHAM, MOHAMED ABDELHAY, nuclear engineer, educator; b. Mehalla, Gharbeia, Egypt, Apr. 18, 1944; came to U.S., 1987; s. Abdelhay Mohamed Bourham and Badria Ahmed Ghida; m. Laila Gadel Hak, Mar. 22, 1966 (div. 1977); 1 child, Ahmed Mohamed; m. Doria Mahmoud Wafa, Mar. 22, 1987; 1 stepchild, Ahmed Samir Sami. BSc, Alexandria (Egypt) U., 1965; MSc, Cairo U., 1969; PhD, Ain Shams U., Cairo, 1976. Registered profl. engr., N.C. Sr. researcher, asst. prof., then assoc. prof. Nuclear Rsch. Ctr., Cairo, 1965-87; vis. assoc. prof. N.C. State U., Raleigh, 1987-91, rsch. assoc. prof. nuclear engring., 1991—. Contbr. articles on plasma physics, fusion tech., plasma engring., launch tech., nuclear electric propulsion to sci. publs. Maj. arty. Egyptian army, 1968-74. Grantee U.S. Army, 1989, 91, 92, U.S. Dept. Energy, 1992. Mem. IEEE, Am. Phys. Soc., Am. Nuclear Soc., Sigma Xi. Moslim. Achievements include research in plasma microinstabilities and electromagentic emission from core plasmas in magnetically confined fusion devises, development of magnetically collimated electron beams for microelectronics, developed techniques in pulsed power systems for electrothermal launchers, development of diagnostics methodology and techinques for hyper-velocity plasma launchers, developed data bace on plasma-facing components. Office: NC State U Dept Nuclear Engring Raleigh NC 27695-7909

BOURKE, LYLE JAMES, electronics company executive, small business owner; b. San Diego, May 28, 1963; s. Robert Victor and Virginia (Blackburn) B. Cert. in electronics, Southwestern Coll., San Diego, 1984; cert. in microelectronics, Burr Brown, Miramar, Calif., 1985; student, NACS, Scranton, Pa., 1988; AA in Econs., Cuyamaca Coll., 1991, postgrad., 1991-92; student, Wendelstedt Umpire Sch., 1992. Counselor Dept. Parks and Recreation City of Imperial Beach, Calif., 1979-80; warehouse worker Seafood Cannery, Cordova, Alaska, 1981, Nat. Beef Packing, Liberal, Kans., 1983; computer programmer ABC Heating and Air, San Diego, 1985; night mgr. Southland Corp., San Diego, 1983-85; tech. developer Unisys Corp., San Diego, 1985-92; founder Sparrells Ltd., 1992; instr. Harmonium Enrichment Program, 1993. Editor (handbook) College Policies, 1991; contbr. Cleanrooms mag., 1992; inventor Jacuzzi pillow. Vol. United Way, San Diego, 1987—; donor Imperial Beach Boys and Girls Club, 1988-93, Cal Farley's Boys Ranch, 1985-93, Assn. Handicapped Artists, 1988—, San Diego Jr. Theatre, 1992, Cabrillo Elem. Sch. Found., 1992; mem. Save the Earth Com., 1991—. Named Most Valuable Player Mex. Amateur Baseball League, San Diego-Tijuana, 1990. Mem. Am. Assn. Ret. Persons, Am. Mgmt. Assn. (charter), Prognosticators Club. Democrat. Avocation: computer tech., writing, Olympics. Office: Unisys 8011 Fairview Ave La Mesa CA 91941-6416

BOURNE, CAROL ELIZABETH MULLIGAN, biology educator, phycologist; b. Rochester, N.Y., May 4, 1948; d. William Thomas and Ruth Townsend (Stevens) Mulligan; m. Godfrey Roderick Bourne, Dec. 21, 1968. BA in Botany/Bacteriology, Ohio Wesleyan U., 1970; MS in Botany, Miami University, Oxford, Ohio, 1978; PhD in Natural Resources, U. Mich., 1992. Lab. asst. Ohio Wesleyan U., Delaware, 1968-70; biol. lab. tech. USDA-Forest Svc., Delaware, 1970-73; grad. rsch. asst. botany dept. Miami U., Oxford, 1973-75; electron microscopist coll. medicine U. Cin., 1975-76; rsch. asst. sch. pub. health U. Mich., Ann Arbor, 1978-80, rsch. assoc. coll. medicine, 1981-83, grad. rsch. asst. sch. natural resources, 1983-86, grad. teaching asst. dept. biology, 1987; postdoctoral scientist U. Fla., Ft. Lauderdale, 1990-92; adj. instr. ecology Fla. Atlantic U. Coll. Liberal Arts, Davie, 1992-93; adj. asst. prof. dept. biology U. Mo., St. Louis, 1993—. Contbr. articles to scholarly jours. Grantee NSF, 1987-89. Mem. Am. Inst. Biolog. Scis., Am. Soc. Plant Taxonomists, Phycological Soc. Am., Internat. Soc. for Diatom Rsch., Internat. Soc. for Plant Molecular Biology, Brit. Phycological Soc., Soc. for Study of Evolution. Office: U Mo Dept Biology 8001 Natural Bridge Rd Saint Louis MO 63121-4499

BOUWER, HERMAN, laboratory executive; b. Haarlem, Netherlands, July 11, 1927; came to U.S., 1952, naturalized, 1959; s. Eduard and Trinette (Dusschoten) B.; m. Agnes N. Temminck, Mar. 29, 1952; children: Edward John, Herman (Archie) Gerard, Annette Nancy. B.S., Nat. Agr. U., Wageningen, The Netherlands, 1949, M.S., 1952; Ph.D., Cornell U., 1955. Assoc. agr. engr. Auburn U., 1955-59; rsch. hydraulic engr. U.S. Water Conservation Lab., Phoenix, 1959-72, dir., 1972-90; cons. NAS; tech. adv. Nat. Water Rsch. Inst.; lectr. groundwater hydrology Ariz. State U.; cons. in field. Author: Groundwater Hydrology, 1978; Contbr. articles in field to profl. jours. Recipient Superior Svc. awards U.S. Dept., Agr., 1963, 73, 89; named scientist of Yr. 1985; OECD fellow, 1964. Mem. ASCE (Walter Huber Research prize 1966, Royce J. Tipton award 1984), Am. Soc. Agr. Engrs. (Hancor award 1988), Nat. Water Well Assn.(hon. life), Internat. Assn. Water Quality, Water Environ. Fedn., Am. Geophys. Union. Club: Tempe Racquet and Swim. Home: 338 E La Diosa Dr Tempe AZ 85282-2234 Office: US Water Conservation Lab 4331 E Broadway Rd Phoenix AZ 85040-8807

BOVA, MICHAEL ANTHONY, physicist; b. Boston, Oct. 6, 1968; s. Anthony L. and Constance C. (Fama) B. BS in Earth Sci., U. Conn., 1993; BS in Physics, Northeastern U., 1993. Rsch. asst. Ctr. Galactic Astronomy, Danbury, Conn., 1988-90; mgr. physics lab. U. Conn., Danbury, 1988-90; rsch. asst. Ctr. Magnetic Rsch., Boston, 1990-92; rsch. engr. Phillips Lab., Hanscom AFB, Mass., 1992—. Asst. editor Silicon on Diamond Technology, 1992. Mem. Am. Astron. Soc., Am. Phys. Soc., Soc. Physics Students, Planetary Sco. Achievements include design of gravity wave detector and radio telescopes.

BOVEY, FRANK ALDEN, research chemist; b. Mpls., June 4, 1918; s. John Alden and Margaret Eugenia (Jackson) B.; m. Shirley June Elfman, June 19, 1941 (div. 1980); children: Margaret Bovey Glassman, Peter, Victoria A. B.S., Harvard U., 1940; Ph.D., U. Minn., 1948. With 3M Co., 1942, 48-62, Nat. Synthetic Rubber Corp., Louisville, 1942-45; with Bell Telephone Labs., Murray Hill, N.J., 1962—; head polymer chemistry research dept. AT&T Bell Labs., 1967—; v.p., dir. Bodel Corp., Mpls.; adj. prof. Stevens Inst. Tech., 1965-67, Rutgers U., 1971—. Author: Effects of Ionizing Radiation on Polymers, 1958, Nuclear Magnetic Resonance, 1969, Polymer Conformation and Configuration, 1969, High Resolution NMR of Macromolecules, 1972, Chain Structure and Conformation of Macromolecules, 1982, Nuclear Magnetic Resonance Spectroscopy, 2d edit., 1988; also articles.; asso. editor: Macromolecules, 1968—; editorial bd.: Accounts of Chem. Research, 1968-74, Biopolymers, 1972—. Recipient Outstanding Achievement award U. Minn. Fellow N.Y. Acad. Scis.; mem. Nat. Acad. Scis., Am. Chem. Soc. (Union Carbide award 1958, Minn. award 1962, Witco award polymer chemistry 1969, Nichols medal 1978, Phillips award 1983), Am. Phys. Soc. (High Polymer Physics prize 1974), Am. Soc. Biol. Chemists, Sigma Xi, Phi Lambda Upsilon. Home: 9 Lockhaven Ln Bedminster NJ 07921-1728

BOW, SING TZE, engineer, educator; b. Kwangtong, People's Rep. of China, Oct. 3, 1924; s. Shi-yun and Rui-Lien (Chang) B.; m. Xia Fang Wang, July 20, 1957; 1 child, Nai-jun. BSEE, Chiao Tung U., 1947; MS in Elec. Engring., U. Wash., 1952; PhD in Elec. Engring., Northwestern U., 1956; postgrad., Gen. Electric Co., 1949-50. Rsch. engr. Delta Star Electric Co., Chgo., 1955-56; rsch. prof., lab. dir. Academia Sinica, People's Rep. of China, 1956-81, 83-84; prof. elec. engring. Pa. State U., University Park, 1981-83, 84-87; prof. elec. engring., computer engring. Shanghai Jiao Tong U., People's Rep. of China, 1979—; prof. elec. engring, lab. dir. Northern Ill. U., DeKalb, 1987—. Author: Optimal Operation and Control of Power System, 1965, Pattern Recognition-Application to Large Data-Set Problems, 1984, Pattern Recognition and Image Preprocessing, 1992; contbr. articles to profl. jours. Grantee Academia Sinica 1958-81, Digital Equipment Corp. 1984-88, Ideal Industries 1987-90, CTS Knights Div. 1989-90, Dekalb Genetics, Inc. 1990—. Mem. IEEE (sr. mem.), Internat. Soc. Photo-Optical Instrumentation Engrs., Internat. Pattern Recognition Soc., Classification Soc. North Am., Internat. Neural Network Soc., Edit. Bd. Elec. Engring. Electronics Series, Marcel Dekker, Chinese Soc. Space Rsch. (v.p. 1979-86), Chinese Soc. Astronautics (chmn. remote sensing technique com. 1979-86), Shanghai Assn. Automation (exec. bd. 1979—), Eta Kappa Nu, Pi Mu Epsilon, Sigma Xi. Avocations: travel, movie, table tennis. Home: 467 W Hillcrest Dr De Kalb IL 60115-2377 Office: No Ill U Dept Elec Engring De Kalb IL 60115

BOWARD, JOSEPH FRANK, civil engineer; b. Sewickley, Pa., Feb. 15, 1961; s. Joseph Taylor and Esther M. (Hawk) B.; m. Martha Ann Idecker, Jan. 1, 1983; children: Joseph Michael, Sarah Lynn. BSCE, Purdue U., 1984; MSCE, U. Pitts., 1990. Registered profl. engr., Pa. Draftsman Grudovich Land Planning & Surveying, Cannonsburg, Pa., 1978-80; engring. intern Engring. Mechanics, Inc., Pitts., 1981-84, staff engr., 1984-89, project engr., 1989—; lectr. in field. Mem. NSPE, ASCE, PEPP, CLC, Am. Soc. Hwy. Engrs., Am. Concrete Inst., Am. Water Well Assn., Soc. of Environ. Toxicology and Chemistry, Pa. Soc. Profl. Engrs. (dir. 1990—, Young Engr. of the yr. 1992, vice chmn. S.W. region 1992—), Chi Epsilon. Democrat. Methodist. Achievements include research for waste site worker protection, environmental site assessing. Office: Engring Mechanics Inc 4636 Campbells Run Rd Pittsburgh PA 15205

BOWCOCK, ANNE MARY, medical educator; b. Wendover, Great Britian, Feb. 3, 1956; came to U.S., 1984; d. John Brown and Pauline (Dalton) B.; m. Errol C. Friedberg; children: Jonathan Friedberg, Lawrence Friedberg. PhD, U. Witswatersrand, Johannesburg, South Africa, 1984. Postdoctoral Stanford (Calif.) U., 1986, rsch. assoc., 1986-90; asst. prof. U. Tex. Southwestern Med. Ctr., Dallas, 1990—. Office: U Tex Southwestern Med Ctr 5323 Harry Hines Blvd Dallas TX 75235-8591

BOWDEN, DENISE LYNN, civil engineer; b. Salem, Mass., Sept. 29, 1963; d. Richard Corning and Carol Lou (Gustafson) Thompson; m. Bruce Russell Bowden, Oct. 4, 1986; 1 child, Amanda Judith. BSCE, U. Lowell, 1985. Registered profl engr. Draftsperson Thomas K. Dyer, Inc., Lexington, Mass., 1984, engring. tech., 1984-85, track, civil engr., 1985—; mem. Women's Transp. Seminar, 1992—. Assoc. adv. Peabody (Mass.) Police Law Enforcement Explorers, 1986—; mem. explorer com. Boy Scouts Am., 1989—. Named Young Engr. of Yr. Mass. Soc. Profl. Engrs., 1993. Mem. ASCE, Am. Railway Engring. Assn., Mass. Soc. Profl. Engrs. (dir. 1990—, mathcounts coord. 1989-91, asst. treas. met. chpt. 1991-93), Boston Soc. Civil Engrs. (younger mem. com. and transp. group com.). Office: Thomas K Dyer Inc 1762 Massachusetts Ave Lexington MA 02173-5302

BOWDEN, WILLIAM BRECKENRIDGE, natural resources educator; b. Richmond, Va., Sept. 26, 1951; s. William Lukens and Carol Lorraine (Morris) B.; m. Linda Ann Bishop, Aug. 18, 1973; children: Jared Bishop, Seth Elias. BS in Zoology and Chemistry, U. Ga., 1973; MS in Zoology, N.C. State U., 1976, PhD in Zoology, 1982. Rsch. technician Water Resources Rsch. Ctr. N.C. State U., Raleigh, 1974-76; rsch. fellow Ecosystems Ctr. MBL, Woods Hole, Mass., 1976-82; postdoctoral fellow Sch. Forestry Environ. Studies Yale U., New Haven, 1982-84; assoc. researcher Sch. for Environ. Studies Sch. Forestry Environ. Studies, Yale U., New Haven, 1984-86; asst. prof. water resources U. N.H., Durham, 1987-92, assoc. prof. water resources, 1992—. Contbr. articles to profl. publs., chpt. to book. NSF scholar, 1980-82; Jesse Noyes scholar, 1976-78. Mem. Am. Geophys. Union, Ecol. Soc. Am., N.Am. Benthiological Soc., Forestry Honor Soc., Phi Kappa Phi. Office: U NH Dept Natural Resources James Hall Durham NH 03824

BOWEN, JEWELL RAY, academic dean, chemical engineering educator; b. Duck Hill, Miss., Jan. 9, 1934; s. Hugh and Myrtle Louise (Stevens) B.; m. Priscilla Joan Spooner, Feb. 4, 1956; children: Jewell Ray, Sandra L., Susan E. B.S., MIT, 1956, M.S., 1957; Ph.D., U. Calif., Berkeley, 1963. Asst. prof. U. Wis., Madison, 1963-67, assoc. prof., 1967-70, prof. chem. engring., 1970-81, chmn. chem engring. dept., 1971-73, 78-81, assoc. vice chancellor, 1972-76; prof. chem. engring. U. Wash., Seattle, 1981—, dean coll. engring., 1981—; cons. in field; adviser NSF, Dept. Def. Contbr. articles to profl. jours.; editor: 7th-10th Internat. Colloquia on Dynamics of Explosions and Reactive Systems, 1979, 81, 83, 85. Bd. dirs. Wash. Tech. Ctr., 1983—, interim exec. dir., 1989-91; mem. Wash. High Tech. Coordinating Bd., 1983-87. NATO-NSF postdoctoral fellow, 1962-63; sr. postdoctoral fellow, 1968; Deutsche Forschungsgemeinschaft prof., 1976-77. Fellow AAAS; mem. AIAA, Am. Inst. Chem. Engrs., Am. Phys. Soc., Combustion Inst., NSPE, Am. Soc. Engring. Edn. (deans coun. 1985—, chmn. deans coun. 1989-91, bd. dirs. 1989—, 1st v.p. 1991, pres. elect 1992, pres. 1993), Sigma Xi, Tau Beta Pi, Beta Theta Pi. Home: 5324 NE 86th St Seattle WA 98115-3922 Office: Univ Wash Coll of Engring FH-10 Seattle WA 98195

BOWEN, JOHN METCALF, pharmacologist, toxicologist, educator; b. Quincy, Mass., Mar. 23, 1933; s. Loy J. and Marjorie (Metcalf) B.; m. Jean Alma Schmidt, Dec. 26, 1956; children: Mark John, Richard Kelley. DVM, U. Ga., 1957; PhD, Cornell U., 1960. Asst., then assoc. prof. Kans. State U., Manhattan, 1960-63; post-doctoral fellow Emory U., Atlanta, 1963; assoc., then prof., U. Ga., Athens, 1963—, assoc. dean, dir. veterinary med. expt. sta., 1976—. Mem. Am. Veterinary Med. Assn., Soc. Neuroscis., Am. Soc. Pharmacology and Exptl. Therapeutics. Office: U Ga Coll Vet Medicine Athens GA 30602-7371

BOWEN, WILLIAM HENRY, dental researcher, dental educator; b. Enniscorthy, Ireland, Dec. 11, 1933; came to U.S., 1956, naturalized,; s. William H. and Pauline (McGrath) B.; m. Carole Barnes, Aug. 9, 1958 children—William, Deirdre, Kevin, David, Katherine. B.D.S., Nat. U. Ireland, Dublin, 1955; M.Sc., U. Rochester, N.Y., 1959; Ph.D., U. London, 1965; D.Sc., U. Ireland, Dublin, 1974; D Odontologiae (honoris causa), U. Oslo, Norway, 1991, U. Umeå, Sweden, 1993. Diplomate Am. Bd. Dentistry. Assoc. pvt. dental practice private dental practice, London, 1955-56; Quinten Hogg fellow Royal Coll. Surgeons, London, 1956-59, Nuffield Found. fellow, 1962-65, sr. research fellow, 1965-69, Sir Wilfred Fish fellow, 1969-73; acting chief Nat. Inst. Dental Research, NIH, Bethesda, Md., 1973-79; chief Nat. Inst. Dental Research, NIH, Bethesda, MD, 1979-82; chmn. dental research U. Rochester, N.Y., 1982—; dir. Cariology Ctr., Rochester, 1984—. Mem. AAAS (sect. R-Dentistry, chair elect 1989, chair 1990), European Orgn. Caries Rsch., Internat. Assn. Dental Rsch. (treas. 1982-88, v.p. 1988, pres. elect 1989, pres. 1990), Fedn. Dentaire Internationale, Lab. Animal Sci. Assn., Zool. Soc. Roman Catholic. Home: 315 Victor Egypt Rd Victor NY 14564-9710 Office: U Rochester Dental Rsch 601 Elmwood Ave PO Box 611 Rochester NY 14642

BOWER, RUTH LAWTHER, mathematics educator; b. Bellaire, Ohio, Nov. 17, 1917; d. James Hood and Mary Blanche (Studebaker) Lawther; (widowed); 1 child, Bruce Alan. BA, Wooster (Ohio) Coll., 1939; EdS, Fla. Atlantic U., 1974, EdD, 1976. Cert. tchr., Fla. Cost acct. Peasley Constrn., New London, Conn., 1942-43; with Palm Beach County Sch. System, West Palm Beach, Fla., 1964-74, 74-85, chmn. math. dept., 1974-78, maths. cons., 1978-85; prof. maths. Palm Beach Atlantic Coll., West Palm Beach, 1985—; adj. prof. math. Fla. Atlantic U., Boca Raton, 1965-85; prin. summer sch. Palm Beach County Schs., West Palm Beach, 1971, 72; speaker in field. Developer math. games Equivo, NOC, Add-In and others, 1971—; coauthor: Individualizing Mathematics Series, 1970-71. Trustee Admiralty Bank, Juno Beach, Fla. Named Tchr. of the Yr., Fla. Math. Tchrs. Assn., 1977. Mem. Nat. Coun. Tchrs. of Maths., Math. Assn. Am., Fla. Ednl.

Rsch. Assn., Phi Delta Kappa, Delta Kappa Gamma (treas. 1978-83). Home: 525 55th St West Palm Beach FL 33407-2607

BOWERMAN, WILLIAM WESLEY, IV, biologist, researcher; b. Munising, Mich., Feb. 13, 1961; s. William Wesley III and Barbara Loraine (Lindquist) B.; m. Susan Kay Marshall, Nov. 5, 1988; children: Mary Barbara, John Marshall. BA in Biology, Western Mich. U., 1985; MA in Biology, No. Mich. U., 1991; postgrad., Mich. State U., 1989—. Sci. rsch. asst. Mich. Legis. Svcs. Bur., Lansing, 1986-87; biol. technician USDA Forest Svc., Escanaba, Mich., 1987; wildlife technician Mich. Dept. Natural Resources, Shingleton, Lansing, 1988-89; rsch. asst. in fish and wildlife rsch. unit Ohio State Univ., Columbus, 1988; site attendant Nature Conservancy, East Lansing, 1988; program coord. Hiawatha Nat. Forest, Escanaba; consulting ecologist Ecol. Rsch. Svcs., Bay City, Mich., 1989-91; rsch. asst. in dept. fisheries and wildlife Mich. State U., East Lansing, 1989—; cons. on bald eagles; lectr. numerous orgns.; prin. investigator Earthwatch, Watertown, Mass., 1989-91. Contbr. numerous articles to profl. jours. Vol. USDA Forest Svc., Munising, Mich., 1985-88. Recipient numerous grants for bald eagle rsch., 1986—. Mem. Wildlife Soc., Soc. for Environ. Toxicologyand Chemistry, World Working Group on Birds of Prey and Owls, Ecol. Soc. Am., Am. Ornithologists' Union, Assn. Field Ornithologists, Wilson Ornithol. Soc., Cooper Ornithol. Soc., Sigma Xi. Office: Mich State U. 201 Pesticide Rsch Ctr East Lansing MI 48824

BOWERS, CONRAD PAUL, chemist; b. Lower Marion Township, Pa., Jan. 20, 1959; s. Conrad and Inez (Clossey) B. BS, Wheaton Coll., 1981; postgrad., Purdue U., 1981-90. Chemist Indsl. Plating, Lafayette, Ind., 1991—. Mem. Am. Chem. Soc., Phi Lambda Upsilon. Achievements include rsch. in measuring very fast reactions in solution. Home: Apt 1039 3056 Pheasant Run Dr Lafayette IN 47905 Office: Indsl Plating 120 N 36th St Lafayette IN 47905

BOWLBY, RICHARD ERIC, computer systems analyst; b. Detroit, Aug. 17, 1939; s. Garner Milton and Florence Marie (Russell) B.; m. Gwendoline Joyce Coldwell, Apr. 29, 1967. B.A., Wayne State U., 1962. With Ford Motor Co., Detroit, 1962-65, 66—, now computer systems analyst; pres. 1300 Lafayette East-Coop., Inc., 1981-82. Mem. Antiquaries, Friends Detroit Pub. Library, Friends Orch. Hall. Club: Founders Soc. (Detroit). Office: Ford Motor Co PO Box 43314 Detroit MI 48243-0314

BOWMAN, ALBERT W., structural engineer; b. Bronxville, N.Y., June 30, 1965; s. Albert W. and Catherine H. (Zondlo) B. BS in Civil Engring., Clemson U., 1988. Engr. in tng. Field engr. (project I-526 Cooper River Bridge) Howard, Needles, Tammen and Bergendoff, Charleston, S.C., 1988-91; structural engineer HNTB Corp., Atlanta, Ga., 1991—. Mem. ASCE (assoc.), Am. Concrete Inst., Sigma Alpha Epsilon. Home: 634 Akers Ridge Dr Atlanta GA 30339

BOWMAN, DWIGHT DOUGLAS, parasitologist; b. Akron, Ohio, Jan. 26, 1952; s. Lawrence Lincoln and Virginia Lee (Barcus) B.; m. Linda Louise Chavez Bowman, July 20, 1974; children: Anastasia Elizabeth, Vanessa Nicolle, Madeline Rose. BA in Biology with Honors, Hiram (Ohio) Coll., 1974; MS, Tulane U., 1981, PhD, 1983. Predoctoral fellow NIH tng. grant Tulane U., New Orleans, 1975-77, rsch. asst. U.S. EPA grant, 1977-78, rsch. asst. NIH grant, 1978-79, med. rsch. specialist U.S. EPA Grant/Dept. Tropical medicine, 1979-84; lectr., rsch. assoc. Sch. Vet. Medicine U. Wis., Madison, 1984-87; asst. prof. parasitology Coll. Vet. Medicine Cornell U., Ithaca, N.Y., 1987—. Grantee U.S. Dept. Agr.-IR4, 1988, 92, Sears, 1992, Harry M. Zweig Meml. Fund for Equine Rsch., 1990. Mem. AAAS, Am. Assn. Zoo Veterinarians (assoc. editor 1992—), Helminthological Soc. Wash. (editorial bd. 1988-93), Am. Soc. Parasitologists, Am. Assn. Vet. Parasitologists (chair edn. com. 1992—). Achievements include examination of biology of Toxocara canis and Ascaris suum, especially as related to the excretory system, research on immune response of cats to concomitant infections with toxoplasmosis and feline immunodeficiency virus, research on immune response of chickesn to coccidiosis, work on causative agent of equine protozoal myeloencephalitis, examination of compounds for efficacy against helminth, arthopod and protozoan parasites. Office: Cornell U Coll Vet Medicine Microbiology Immunology Parasitology Ithaca NY 14850

BOWMAN, FREDERICK OSCAR, JR., cardiothoracic surgeon, retired; b. Chapel Hill, N.C., May 24, 1928; s. Frederick Oscar and Sallie (Sanders) B.; m. Helen Roberson, 1951 (div. 1956); m. Elizabeth Wallace Schwartz, Jan. 24, 1959; children: Frederick III, John Sanders, Michael Andrew, William Albert. AB, U. N.C., 1948; MD, U. Pa., 1952. Diplomate Am. Bd. Surgery, Am. Bd. Thoracic Surgery. Intern Roosevelt Hosp., N.Y.C., 1952-53, resident in gen. surgery, 1953-54, 56-59; resident in thoracic surgery Bellevue Hosp., N.Y.C., 1959-60, Presbyn. Hosp., N.Y.C., 1960-61; instr. in surgery Coll. Physicians and Surgeons, Columbia U., N.Y.C., 1961-64, asst. prof. clin. surgery, 1964-68, assoc. prof. clin. surgery, 1968-79, prof. clin. surgery, 1979-92, prof. emeritus clin. surgery, 1992—; attending surgeon Presbyn. Hosp., N.Y.C., 1979-91, cons. emeritus, 1991—, sect. chief pediatric cardiac surgery, 1988-91. Contbr. over 100 articles to profl. jours. Mem. U. N.C. Bd. Visitors, Chapel Hill, 1991-94. Capt. U.S. Army, 1954-56. Fellow Am. Coll. Surgeons; mem. AMA, Am. Assn. Thoracic Surgery, Soc. Thoracic Surgeons, N.Y. Soc. Thoracic Surgery (pres. 1985-86), Internat. Cardiovascular Soc., Portugese Soc. Cardiology, N.Y. Soc. Cardiology, N.Y. Heart Assn., N.Y. Surg. Soc., N.Y. Acad. Scis. Home: 12723 Morehead Govs Club Chapel Hill NC 27514

BOWMAN, MARJORIE ANN, physician, academic administrator; b. Grove City, Pa., Aug. 18, 1953; d. Ross David and Freda Louise (Smith) Williamson; m. Robert Choplin; one child, Bridget Williamson Foley. BS, Pa. State U., 1974; MD, Jefferson Med. Coll., 1976; MPA, U. So. Calif., L.A., 1983. Intern, then resident in family practice Duke U., Durham, N.C., 1976-79; med officer USPHS, Hyattsville, Md., 1979-82; clin. instr. uniformed svcs. U. of the Health Scis., Bethesda, Md., 1980-83; dir. family practice residency Sch. Medicine Georgetown U., Washington, 1983-86; prof., chair dept. family and community medicine Wake Forest U., Winston-Salem, N.C., 1986—; Author: Stress and Women Physicians, 1985; contbr. articles to profl. jours. Fellow Am. Acad. Family Physicians; mem. AMA, Soc. Tchrs. Family Medicine (bd. dirs. 1984-88, bd. dirs. Found. 1984—, v.p.1988-91, pres. 1991-92), Am. Pub. Health Assn. Republican. Unitarian. Office: Bowman Gray Sch Medicine Family & Community Medicine Medical Center Blvd Winston Salem NC 27157-1084

BOWMAN, MONROE BENGT, architect; b. Chgo., Aug. 28, 1901; s. Henry William and Ellen Mercedes (Bjork) B.; m. Louise Kohnmann, Nov. 1944; 1 son, Kenneth Monroe; B.Arch., Ill. Inst. Tech., 1924. Registered architect, Ill., Wis., Ind., Ohio, Colo. Asso., Benjamin H. Marshall, Chgo., 1926; exhibited models and photographs of Bowman Bros. comtemporary designs at Mus. Modern Art, N.Y.C., 1931; pvt. practice architecture, Chgo., 1941-44; asso. Monroe Bowman Assos., Chgo., 1945—; cons. Chgo. Dept. City Planning, City of Sparta (Wis.), Alfred Shaw, Architect. Mem. Navy League U.S. Important works include Boeing Aircraft bldgs., Wichita, Kans., Emerson Electric bldgs., St. Louis, Maytag Co., Newton, Iowa, Douglas Aircraft bldgs., Park Ridge, Ill., Shwayder Bros. bldgs., Denver, Clark Equipment Co., Buchannon, Mich., Radio-TV Sta. WHO, Des Moines, Foote, Cone & Belding offices, Chgo., Burridge Devel., Hinsdale, Ill., Yacht Club and recreational facilities, Lake Bemiji, Minn., United Airlines offices downtown Chgo., Automatic Sprinkler Corp., Chgo., King Machine Tool div. Am. Steel Foundries, Cin., Marine Terr. Apts., Chgo., Dorchester Park Apts., Chgo., Manteno (Ill.) State Hosp., No. Ill. Gas Co. bldgs., LaGrange, Joliet, Streator and Morris, 1340 Astor St. Apt. Bldg., Burnham Center, Chgo., NSF, Green Bank, W.Va., Naval Radio Research Sta., Sugar Grove, W.Va., Columbus Boy Choir Sch., Princeton, N.J., office bldg. and hotel, Charleston, W.Va. Home: 422 Davis St Evanston IL 60201

BOWMAN, PATRICIA IMIG, microbiologist; b. Evanston, Ill., Aug. 10, 1951; d. Walter Joseph and Elizabeth Jean (Lutton) Imig; m. Van Martin Bowman, Dec. 16, 1972; B.S., Ga. State U., 1972, M.S., 1975; Ph.D., St. John's U., 1981. Research assoc. Ga. State U., Atlanta, 1975-77; microbiology supr. Personal Products Co., Milltown, N.J., 1981-82, mgr. microbiology, 1982-84; mgr. microbiology Avon Products, Inc., Suffern, N.Y., 1984-86, dir. retail fragrances, 1986-89; dir. tech. L&F Products, Montvale, N.J., 1989-91, v.p. rsch. and devel., 1991-92, sr. v.p. rsch. and devel. household, personal care, and do-it-yourself consumer products, 1993—. Contbr. articles to profl. jours. Mem. Am. Soc. Microbiology, Soc. Indsl. Microbiology, Cosmetics, Toiletries and Fragrances Assn. (microbiology com. 1982—, raw materials subcom. 1982—), Soc. Cosmetic Chemists, Com. Exec. Women, Soap & Det. Assn., Mortar Bd. Soc., Phi Kappa Phi. Republican. Methodist. Office: Eastman Kodak Co L & F Products One Philips Pkwy Montvale NJ 07645

BOWMAN, STEPHEN WAYNE, quality assurance engineer, consultant; b. Charlotte, N.C., Oct. 3, 1949; s. John Wayne and Dagmar Katharine (Hege) B.; m. Patricia Faye Waldron, June 17, 1972 (div. 1988); 1 child, Jennifer Leigh. BS in Physics, Ga. Inst. Tech., 1972, MS in Nuclear Engring., 1974. Registered profl. engr., Tex. Quality assurance engr. GE, Schenectady, N.Y., 1978-81; mgr. IEEE qualification program Stewart and Stevenson Svcs., Houston, 1981-86, mgr. nuclear projects, 1984-86; sr. engr. Pacific Engring. Corp., Portland, Oreg., 1988-89; pres. Bowman and Assocs., Kingwood, Tex., 1986—; project quality assurance mgr. M.W. Kellogg Co., Houston, 1990—; curriculum adv. bd. dept. mfg. technology Houston Community Coll., 1984-87. Contbr. articles to profl. publs. 1st lt. U.S. Army, 1975-78. Mem. Am. Nuclear Soc., Am. Soc. Quality Control, Houston Engring. and Sci. Soc. Home: 1915 Crystal Springs Dr Kingwood TX 77339-3339

BOWMAN, WILLIAM JERRY, air force officer; b. Boise, Idaho, Feb. 9, 1954; s. Bill C. and Phyllis (Erickson) B.; m. Pam Wakefield, June 29, 1977; children: Corey, Eric, Trent. MSE in Mech. Engring., U. Cen. Fla., 1982; PhD in Aero. Engring., Air Force Inst. Tech., 1987. Commd. USAF, 1977—, advanced through grades to maj.; instr. Naval Nuclear Power Sch., Orlando, 1978-82; liaison to NASA Kennedy Space Ctr., Fla., 1982-84; faculty U.S. Air Force Acad., Colorado Springs, 1987-91, Air Force Inst. Tech., Dayton, Ohio, 1991—. Contbr. articles to profl. jours. Scoutmaster Boy Scouts Am., Orlando, 1978-82, team coach, Dayton, 1984-87, 91—, scoutmaster, Colorado Springs, 1987-91. Mem. AIAA (Engr. of the Yr., 5th Region 1990-91, thermophysics com. 1988-91), Sigma Gamma Tau (faculty advisor 1991—), Tau Beta Pi. LDS. Home: 2951 Green Vista Beavercreek OH 45324 Office: Air Force Inst Tech ENY Wright Patterson AFB OH 45433

BOXENHORN, BURTON, aerospace engineer; b. N.Y.C., Apr. 22, 1928; s. Max Boxenhorn and Rose Zvebel; m. Marianne Fedder, Jan. 7, 1962; children; David, Elizabeth. BSME, Clarkson U., 1952; MSME, Carnegie-Mellon, 1956; MSEE, N.Y. U., 1960. Registered profl. engr., Mass. Engr., dept. rsch. United Aircraft, East Hartford, Conn., 1952-54; engr. Hamilton-Standard, Windsor Locks, Conn., 1956-57; instr., dept. mech. engring. N.Y. U., 1957-60; staff engr. MIT Instrn. Lab., Cambridge, Mass., 1962-64; engr. Philco, Willow Grove, Pa., 1965-66; staff engr. C.S. Draper Lab., Cambridge, 1966-93; pvt. practice cons. Chestnut Hill, Mass., 1993—. Author articles, numerous papers in field. Rep. Brookline (Mass.) Town Meeting. Mem. AIAA, IEEE. Achievements include two patents on micromechanical devices, one on a guidance scheme, patents pending.

BOXLEITNER, WARREN JAMES, electrical engineer, researcher; b. Lewiston, Idaho, Jan. 8, 1948; s. Paul Henry and Lois Genelle (Samsel) B.; m. Linda Jane Schraufnagel, Aug. 23, 1969; 1 child, Kirk Lee. BSEE, U. Idaho, 1971. Design engr. Keytronic Corp., Spokane, Wash., 1975-79, project engr., 1981-83, sr. engr., 1983-86, tech. svcs. mgr., 1986-87; internat. sales mgr. Eurokey, Ravensberg, Fed. Republic Germany, 1979-81; dir. engring. Keytek Instrument Corp., Wilmington, Mass., 1987-89, v.p. engring., 1987-92, v.p. tech. and ventures, 1992—; participant symposia in field. Author: ESD and Electronic Equipment, 1989; also articles. 1st lt. USAF, 1971-75. Mem. IEEE (sr., chmn. working group 1989), Elec. Overstress and Electrostatic Discharge Assn. Achievements include research on methods to improve immunity of electronic products to electrostatic discharge and to test for such immunity; developer first statistically accurate test guidelines to be published in national or international test standards. Office: Keytek Instrument Corp 260 Fordham Rd Wilmington MA 01887-2153

BOYARCHUK, ALEXANDER, astronomer; b. Grozny, Russia, June 21, 1931; s. Alexei and Maria (Shiyan) B.; m. Margarita Kropotova, Apr. 27, 1955; 1 child, Cirill. PhD, Pulkovo Obs., Leningrad, USSR, 1958; DSc, Probkovo Obs., Leningrad, USSR, 1967. Minor scientist Crimean Astrophys. Obs., Crimea, USSR, 1956-62, sr. scientist, 1962-70, dep. dir., 1970-87; dir. Inst. Astrolomy, Moscow, 1987—. Editor jour. Sowjet Astronomy, 1987; contbr. over 150 articles to profl. publs. Recipient State award Govt. of USSR, 1983. Mem. Russian Acad. Scis. (full mem. 1987, mem. presidium), Internat. Astron. Union (pres. 1991—, full mem.), Am. Astron. Soc., Am. Phys. Soc., Royal Astron. Soc., Internat. Astron. Acad. Office: Inst Astronomy, 48 Pyatnitskaya St, 109017 Moscow Russia

BOYCE, MEHERWAN PHIROZ, engineering executive, consultant; b. Poona, Maharashtra, India, July 25, 1942; came to U.S., 1960; s. Phiroz Hirjibhoy and Nergesh Phiroz (Colabawala) B.; m. Lola Joan Bennett, June 15, 1963 (div. June 1977); children: Phiroz Meherwan, Anita Meherwan Boyce Bamford; m. Zarine Meherwan, June 1977. BSME, S.D. Sch. Mines & Tech., 1962; MSME, SUNY, Buffalo, 1964; PhD, U. Okla., 1969. Profl. engr., Tex. Project engr. Gen. Turbine Corp., Buffalo, 1962-63; analytical engr. Joy Mfg., Buffalo, 1963-64; project engr. Cummins Engine Corp., Columbus, Ind., 1964-65; group leader comp. tech. Curtiss Wright Corp., Woodridge, N.J., 1965-66; group leader aerodyns. Fairchild Hiller Corp., Bayshore, N.Y., 1966-67; instr. U. Okla., Norman, 1967-69; prof. mech. engring., dir. gas turbine lab. Tex. A&M U., College Station, 1969-79; chmn., chief exec. officer Boyce Engring. Internat. Inc., Houston, 1977—; cons. Enelven, Maracaibo, Venezuela, 1973—, Exxon Chem., Baytown, Tex., 1972-76, Maraven, Lagoven, Meneven, Maracaibo, 1973-86, Wichita Falls (Tex.) Energy Corp., 1985—. Author: Gas Turbine Engineering Handbook, 1985; inventor Surge Controller; contbr. over 100 article to profl. jours. Chmn. 7th N.Am. Zoroastrian Com., Houston, 1990. Recipient Ralph Tertor award Soc. Automotive Engring., 1973, Zoroastrian Bus. award, 1990. Fellow ASME (Herbet Allen award 1974); mem. NSPE, Sigma Xi, Pi Tau Sigma, Phi Kappa Phi, Tau Beta Pi. Office: Boyce Engring Internat Inc 10555 Rockley Rd Houston TX 77099-3581

BOYCE, PETER BRADFORD, astronomer, professional association executive; b. N.Y.C., Nov. 30, 1936; s. Burke and Mabel (Zoeckler) B.; m. Mary Elizabeth Saffell, Nov. 6, 1976; children: Kevin Robert, Colin MacDonald. AB, Harvard U., 1958; MS, U. Mich., 1962, PhD, 1963. Staff astronomer Lowell Obs., Flagstaff, Ariz., 1963-73; program dir. NSF, Washington, 1973-79; sci. cons. to Congressman Morris K. Udall, Congl. fellow Washington, 1977-78; exec. officer Am. Astron. Soc., Washington, 1979—. Contbr. articles to profl. jours. Pres. Sun Found. 1992—, v.p. 1989-92. Dept. Commerce Sci. and Tech. fellow, 1977-78. Mem. AAAS, Am. Astron. Soc., Optical Soc. Am., Am. Phys. Soc., Am. Inst. Physics (bd. dirs. 1979-92, mem. exec. com. 1980-82, 85-92), Internat. Astron. Union, Sigma Xi. Home: 5700 Sherier Pl NW Washington DC 20016-5320 Office: Am Astron Soc 2000 Florida Ave NW Ste 400 Washington DC 20009-1231

BOYCE, RICHARD LEE, forest ecologist, researcher; b. Pittsfield, Mass., Nov. 2, 1959; s. Dick LeRoy and Priscilla Marie Zoraide (Bessette) B.; m. Martha Lynn Viehmann, Sept. 30, 1989. BA in Astrophysics, Williams Coll., 1981; M of Forest Sci., Yale U., 1985, PhD in Ecology, 1990. Rsch. assoc. Dartmouth Coll., Hanover, N.H., 1990—. Contbr. articles to profl. jours. Mem. AAAS, Ecol. Soc. Am., Soc. Am. Foresters, Sigma Xi, Phi Beta Kappa. Democrat. Quaker. Achievements include research in canopy water dynamics of red spruce and balsam fir and winter water relations in red spruce. Office: Dartmouth Coll Environ Stdy 6182 Steele Hall Hanover NH 03755

BOYD, EDWARD LEE, physicist, computer scientist; b. Mexico, Mo., Nov. 27, 1932; s. Lee Moore Boyd and Billy (Richter) Boyd Falk; m. Irene Howe Crossman, July 26, 1969; children—Sloane Victoria, Ashton Lee. B.A. in Physics, Lehigh U., 1954; D.S. in Materials Sci., Kyoto U., Japan, 1967. Engr. Beva Labs., Trenton, 1953-54, Associated Enrs., Poughkeepsie, N.Y., 1954-55; physicist in research IBM, Yorktown Heights, N.Y., 1955-66, physicist in components, Poughkeepsie, N.Y., 1966-76, sr. staff mem. corp. hdqrs., Armonk, N.Y., 1976-79, sr. engr. gen. tech. div., East Fishkill, N.Y., 1979-88, sr. tech. staff mem., 1988-90; cons. in field. Contbr. chpt., articles on physics of magnetism to profl. publs. Patentee in field. Bd. dirs. various cultural orgns. Served as pfc. USAR, 1957-69. Mem. Am. Phys. Soc., Am. Assn. Physics Tchrs., AAAS. Episcopalian.

BOYD, JOHN ADDISON, JR., civil engineer; b. Kansas City, Mo., Dec. 20, 1930; s. John Addison and Sara Frances (Burger) B.; m. Rosemary Kennedy, Jan. 31, 1953; children: Mary Boyd Winter, John K., Thomas K., Christopher K., William K. BSCE, U. Kans., 1952, MSCE, 1960. Registered profl. engr., Mo. Constrn. engr. T.F. Marbut Constrn., Emporia, Kans., 1956-58; engring. instr. U. Kans., Lawrence, 1958-60; engr. Howard, Needles, Tammen & Bergendoff, Kansas City, Mo., 1960-66; pres. Boyd, Brown, Stude & Cambern, Kansas City, 1966—; chmn. Engrs. Joint Documents Com., 1990—, Engring. Adv. Bd., U. Mo., 1986—, U. Kans., 1985—; peer reviewer Am. Cons. Engrs. Coun., Washington, 186—. Bd. dirs. St. Joseph Health Ctr., Kansas City, 1985—; exec. bd. St. Lawrence Ctr., 1984-88; chmn. Mayor's Adv. Bd., Kansas City, 1983-88. Capt. USN, 1952-56. Recipient Disting. Svc. award, Nat. Soc. Profl. Engrs., 1986, Cert. of Appreciation, ASCE, 1989, Claycomb Cup, Phi Delta Theta, 1982. Fellow Cons. Engrs. Coun. Mo. (pres. 1982), Am. Cons. Engr. Coun. (nat. dir. 1983), Soc. Am. Mil. Engrs. (pres. 1972, nat. dir. 1975); mem. Mo. Soc. profl. Engrs., Nat. Soc. Profl. Engrs. (vice chmn. exec. bd. 1984-86), ASCE (pres. 1982, nat. dir. 1986-89), Midwest Concrete Industry Bd. (pres. 1974). Roman Catholic. Avocations: tennis, golf, flying. Home: 8101 El Monte St Shawnee Mission KS 66208-5052 Office: Boyd Brown Stude & Cambern 800 W 47th St Kansas City MO 64112-1251

BOYD, LANDIS LEE, agricultural engineer, educator; b. Orient, Iowa, Dec. 1, 1923; s. Harold Everett and Edith Elizabeth (Lauer) B.; m. Lila Mae Hummel, Sept. 7, 1946; children—Susan Lee, Barbara Edith, Shirley Rae, Carl Steven, Philip Wayne. B.S. in Agrl. Engring, Iowa State U., 1947, M.S., 1948, Ph.D. in Agrl. Engring. and Engring. Mechanics, 1959. Registered profl. engr., N.Y., Minn. Sr. research fellow Iowa State Coll., 1947-48, 54-55; from asst. prof. to prof. Cornell U., Ithaca, 1948-60; coordinator grad. instrn. Cornell U., 1958-64; engring. design analyst Allis-Chalmers Mfg. Co., Milw., 1962-63; mem. faculty U. Minn. at St. Paul, 1964-78, prof. agrl. enging., head dept., 1964-72, asst. dir. Agrl. Exptl. Sta., 1972-78, dir. Coll. Agr. Research Center; asso. dean Coll. Agr., Wash. State U., Pullman, 1978-85; exec. dir. Western Assn. Agrl. Expt. Sta. Dirs., Agrl. Expt. Sta. Colo. State U., Fort Collins, 1985-92, adj. prof. agrl. and chem. engring., 1985—; vis. scholar Ctr. Study Higher Edn.; vis. faculty-in-residence, intern Office Vice Pres. for Research, U. Mich., 1968; (Fed. Exec. Inst.), 1975; Cons. FAO, La Molina, Peru, 1964, 69; part-time cons. in field, 1948—. Supt. farm bldg. project N.Y. State Fair, 1956, 57. Served with USNR, 1943-45. NATO postdoctoral grantee, 1962; recipient Iowa 4-H Alumni Recognition award, 1968; profl. achievement citation in engring. Iowa State U., 1980; Japan Soc. Promotion of Sci. fellow, 1981, U. Tokyo Vis. Faculty fellow, 1993. Fellow Am. Soc. Agrl. Engrs., 1973 (grad. paper award 1949, MBMA award 1969, v.p.-regions 1970-73); mem. Am. Soc. Engring. Edn., Minorities in Agr., Natural Resources, and Related Scis., Sigma Xi, Phi Kappa Phi, Gamma Sigma Delta, Alpha Epsilon, Kappa Sigma. Methodist. Lodge: Rotary (Paul Harris fellow). Home and Office: 1725 Concord Dr Fort Collins CO 80526-1601

BOYD, RICHARD HAYS, chemistry educator; b. Columbus, Ohio, Aug. 12, 1929; s. Robert E. and Charlotte (Hays) B.; m. Patricia A. Scheible, Sept. 5, 1951; children—David Hays, Elizabeth King. B.Sc., Ohio State U., 1951; Ph.D., MIT, 1955. Research chemist E. I. DuPont de Nemours Co., Wilmington, Del., 1955-62; prof. chemistry Utah State U., 1962-67; prof. chem. engring., prof. materials sci. and engring, adj. prof. chemistry U. Utah, Salt Lake City, 1967—, chmn. materials sci. and engring., 1976-82, 88—, disting. prof., 1988—; cons. phys. chemistry and polymer sci.; Swedish NRC vis. fellow Royal Inst. Tech., Stockholm, 1980 Editorial adv. bd. Macromolecules 1986-88. Scoutmaster Great Salt Lake council Boy Scouts Am., 1970-73. Recipient Distinguished Research award U. Utah, 1978. Fellow Am. Phys. Soc. (mem. exec. com. div. high polymer physics 1983-87, chmn. 1985-86, High Polymer Physics prize 1988); mem. Am. Chem. Soc. (Utah award 1986), Am. Phys. Soc., Am. Inst. Chem. Engrs., AAUP, Phi Beta Kappa, Sigma Xi, Phi Eta Sigma, Phi Kappa Phi, Sigma Chi. Achievements include research and publications in physical chemistry, polymer sci. Office: U Utah Dept Materials Sci and Engring 304 EMRO Bldg Salt Lake City UT 84112

BOYD, STEVEN ARMEN, medicinal chemist; b. Inglewood, Calif., Dec. 7, 1956; s. Rodney Ray and Rita Christine (Armen) B. BA, U. Oreg., 1979; PhD, U. Calif., L.A., 1984. Rsch. asst. chemistry dept. U. Oreg., Eugene, 1977-79; rsch. asst., teaching asst. u. Calif., L.A., 1979-84; postdoctoral fellow U. Wis., Madison, 1984-87; rsch. chemist Abbott Labs., Abbott Park, Ill., 1987-89, sr. rsch. chemist, 1990—. Contbr. articles to pubs. Faculty fellow U. Calif., 1983, Nat. Cancer Inst. fellow, 1985. Mem. AAAS, Am. Chem. Soc., Sigma Xi. Democrat. Achievements include patents pending for nonpeptide renin inhibitors, retroviral protease inhibitors. Home: 410 Lakeview Dr Mundelein IL 60060 Office: Abbott Labs Dept 47V AP10 1 Abbott Park Rd Abbott Park IL 60064

BOYD, THOMAS JAMES MORROW, physicist; b. Larne, Antrim, Northern Ireland, June 21, 1932; s. Thomas James and Isobel Cameron (Morrow) B.; m. Marguerite Bridget Snelson, Sept. 5, 1959; children: Rebecca Siobhan, Marguerite Isobel. BSc in Physics, Queens U., Belfast, Ireland, 1953, BA in Math., 1954, PhD in Theoretical Physics, 1957. Rsch. fellow U. Birmingham, Eng., 1957-59; cons. in physics Gen. Dynamics Corp., San Diego, 1959-60; asst. prof. U. Md., College Park, 1960-62; sr. rsch. assoc. Culham Lab. UKAEA, Culham, Eng., 1962-65; sr. lectr. U. St. Andrews, Scotland, 1965-68; prof. applied maths.and computation U. Wales, Bangor, 1968-82, dean of sci., 1981-85, prof. theoretical physics, 1982-90, prof. physics U. Essex, Colchester, Eng., 1990—; mem. U.K.-Austrian Mixed Commn., London and Vienna, 1977-86; vis. prof. U. B.C., Vancouver, Can., 1975, Indian Acad. of Sci., Delhi, 1980, Dartmouth Coll., Hanover, N.H., 1987-88. Author: Plasma Dynamics, 1969; co-author: Electricity, 1979; contbr. numerous articles to profl. jours. Ford Found. fellow Princeton (N.J.) U., 1962, Fulbright Commn. fellow, 1987. Fellow Inst. Physics (chmn. plasma group 1975-77); mem. Am. Phys. Soc., N.Y. Acad. Scis. Avocations: climbing, skiing, traveling, music, gardening. Home: 6 Frog Meadow, Brook St, Dedham CO7 6AD, England Office: U Essex Dept of Physics, Wivenhoe Pk, Colchester CO4 3SQ, England

BOYD, V(IRGINIA) ANN LEWIS, biology educator; b. Shreveport, La., Nov. 15, 1944; d. Fletcher Willard and Bess Juanita (Sherman) Lewis; m. James P. Boyd, June 4, 1965 (div. 1973); children: Kathryn Ann, David Gregory. BS, Northwestern State U., Natchitoches, La., 1965, MS, 1968; PhD, La. State U., 1971. Postdoctoral fellow Baylor Coll. Medicine, Houston, 1971-73; rsch. scientist Frederick (Md.) Cancer Rsch. Ctr., 1973-82; assoc. prof. of biology Hood Coll., Frederick, 1982-88, prof., chair biology dept., 1988—; cons. Frederick Cancer Rsch. Ctr., 1982—; fellowship panelist AAUW, Washington, 1988-92, chmn. fellowship panel, 1992-94; designer, tchr. grad. and undergrad. courses in virology, molecular genetics, bioethics; citizen amb. for biotech. People to People Internat., China, 1988, USSR, 1991. Contbr. articles to sci. jours., chpts. to books. Bd. dirs. Girl Scouts U.S.A., Frederick, 1979-82, Advs. for Homeless, Frederick, 1989-92; vestry officer Episcopal Ch., Frederick, 1980—. Grantee NCI, 1983-86, NSF, 1986-88. Mem. Am. Soc. Virology, Am. Soc. for Microbiology, N.Y. Acad. Scis., AAUW (fellowship panelist 1988-91, chmn. Am. fellowship panel 1992-94), Sigma Xi, Phi Kappa Phi, also others. Achievements include research in microinjection of living cells in cultures; studies are designed to analyze gene expression: oncogenes that cause malignant cell growth, provirus pathology including HIV and other animal retroviruses. Office: Hood Coll Rosemont Ave Frederick MD 21701-8524

BOYER, CHARLES THOMAS, JR., mechanical engineer; b. Hawthorne, Nev., Sept. 1, 1946; s. Charles Thomas B. and Ruth Boyer (Barber) Parker; m. Rebecca Anne Richardson, Mar. 20, 1971; children: Scott A., Matthew D. BS, Va. Tech. U., 1969, MS, 1971, PhD, 1984. Mech. engr. U.S. Marine Engring. Lab., Annapolis, Md., 1969; sr. mech. engr. U.S. Naval Surface Warfare Ctr., Dahlgren, Va., 1974—. Contbr. articles to profl. jours. Bd. dirs. Eden Estates Recreational Club, King George, Va., 1991-93; v.p. Eden Estates Swim Team, King George, 1984-86. Capt. U.S. Army, 1969-73. Named one of Outstanding Young Men Am. Mem. Soc. for Advancement Materials and Process Engring., AIAA. Baptist. Home: 18 Dixie Dr King

George VA 22485 Office: US Naval Surface Warfare Ct Code GH3 Dahlgren VA 22448

BOYER, HERBERT WAYNE, biochemist; b. Pitts., July 10, 1936; m. Grace Boyer, 1959. BA, St. Vincent Coll., Latrobe, Pa., 1958, DSc (hon.), 1981; MS, U. Pitts., 1960, PhD, 1963. Mem. faculty U. Calif., San Francisco, 1966—; prof. biochemistry U. Calif., 1976—; prof. biochemistry U. Calif., San Francisco, 1976-91, prof. emeritus, 1991—; co-founder, dir. Genentech, Inc., South San Francisco, Calif. Recipient V.D. Mattai award Roche Inst., 1977; Albert and Mary Lasker award for basic med. research, 1980, Nat. Tech. medal, 1989, Nat. Sci. medal NSF, 1990. Fellow AAAS; mem. Am. Acad. Arts and Scis., Nat. Acad. Scis. Office: U Calif Dept Biochemistry and Biophysics 513 Parnassus Ave San Francisco CA 94122-2722

BOYER, JOHN FREDERICK, biology educator; b. Evanston, Ill., Oct. 14, 1941; s. Paul Frederick and Betty (Hatton) B.; m. Barbara Ann Conta, Dec. 29, 1968; children: Cynthia Jean, Paul Lewis. BA, Amherst Coll., 1964; PhD, U. Chgo., 1971. Asst. prof. Union Coll., Schenectady, N.Y., 1973-80; assoc. prof. Union Coll., 1980—. Office: Union Coll Dept Biology Schenectady NY 12308-2311

BOYER, NICODEMUS ELIJAH, organic-polymer chemist, consultant; b. Daugavpils, Latgale, Latvia, June 1, 1925; came to U.S., 1949; s. Aloizs and Elvira Adele (Buchholtz) Bojars; m. Annemarie Erika Sielemann, 1952 (div. 1962); children: Arthur John, Evelyne Margarete. Degree in natural scis., U. Göttingen, Germany, 1949; PhD in Chemistry, U. Ill., 1955; postgrad., Princeton U., 1955-56. Rsch. chemist Hooker Chem. Corp., Niagara Falls, N.Y., 1956-61; project leader, lectr. Ill. Inst. Tech., Chgo., 1961-63; rsch. fellow Borg-Warner Chems., Washington, 1964-76; sr. staff mem. Raychem Corp., Menlo Park, Calif., 1976-78; asst. prof. Ind. State U., Terre Haute, 1978-80; sr. rsch. assoc. PPG Industries, Chgo., 1980-88; sr. cons. Delta Sci. Cons., Parkersburg, W.Va., 1988—; Delta Sci. Cons., Three Rivers, Mich.; lectr. evening sch. U. Buffalo, 1958-60. Vol. abstractor Chem. Abstracts Svc., Columbus, Ohio, 1958-71; editor Cosmology Technikas Apskats, Montreal, Que., Can., 1987—; author: Organophosphorus Chemistry, Vol. 1, 1957, Vol. 2, 1959, Radiation Chemistry: Monomers and Polymers, 1977, A New Theory of Cosmology, 1983, The Physics of Creation, 2 vols., 1990, Fire Retardants: A Review and Selected Patents, 1991; 180 chemistry patents; contbr. over 70 articles to profl. jours. Founding mem. Latvian Cath. Students' Assn., Germany, 1946-64; vice chmn. Latvian Acad. Soc. Valdemarija, Ill., Calif., Mich., 1964—; mem. Rep. Presdl. Task Force, 1989-93. With U.S. Army, 1945. Internat. Refugee Orgn. scholar U. Göttingen, 1946-49, Nat. Cath. Welfare Conf. scholar U. Ill., 1949-51; recipient Quality Control & Safety award PPG Industries Inc., 1987. Mem. Am. Chem. Soc., N.Y. Acad. Scis. (life), Phi Lambda Upsilon, Sigma Xi. Republican. Roman Catholic. Achievements include discovery of extremely stable white coatings to heat and ultraviolet radiation for space applications; patent for the first large-scale fire retardant additive for ABS resins; invented a new theory of cosmology. Office: Delta Sci Cons PO Box 312 Three Rivers MI 49093-0312

BOYER, ROBERT ERNST, geologist, educator; b. Palmerton, Pa., Aug. 3, 1929; s. Merritt Ernst and Lizzie Venetta (Reinard) B.; m. Elizabeth Estella Bakos, Sept. 1, 1951; children—Robert M., Janice E., Gary K. B.A., Colgate U., 1951; M.A., Ind. U., 1954; Ph.D., U. Mich., 1959. Instr. geology U. Tex., Austin, 1957-59; asst. prof. U. Tex., 1959-62, asso. prof., 1962-67, prof., 1967—, chmn. dept. geol. scis., 1971-80; dean U. Tex. (Coll. Natural Scis.), 1980—; exec. dir. Natural Scis. Found., 1980—; chmn. exec. com. Geology Found., 1971-80. Author: Activities and Demonstrations for Earth Science, 1970, Geology Fact Book, 1972, Oceanography Fact Book, 1974, The Story of Oceanography, 1975, Solo-Learn in the Earth Sciences, 1975, GEO-Logic, 1976, GEO-VUE, 1978; editor: Tex. Jour. of Sci, 1962-65, Jour. of Geol. Edn., 1965-68. Fellow Geol. Soc. Am., AAAS; mem. Tex. Acad. Sci. (hon. life, pres. 1968), Nat. Assn. Geology Tchrs. (pres. 1974-75), Am. Geol. Inst. (pres. 1983), Am. Assn. Petroleum Geologists, Austin Geol. Soc. (pres. 1975), Gulf Coast Asso. Geol. Soc. (pres. 1977), Soc. Ind. Prof. Earth Sci. Found. (pres. 1992). Home: 7644 Parkview Cir Austin TX 78731-1160

BOYLAN, DAVID RAY, retired chemical engineer, educator; b. Belleville, Kans., July 22, 1922; s. David Ray and Mabel (Jones) B.; m. Juanita R. Sheridan, Mar. 24, 1944; children: Sharon Rae, Gerald Ray, Elizabeth Anne, Lisa Dianne. B.S. in Chem. Engring., U. Kans., 1943; Ph.D., Iowa State U., 1952. Instr. U. Kans., 1942-43; project engr. Gen. Chem. Co., Camden, N.J., 1943-47; sr. engr. Am. Cyanamid Co., Elizabeth, N.J., 1947; plant mgr. Arlin Chem. Co., Elizabeth, 1947-48; faculty Iowa State U., Ames, 1948—; prof. chem. engring. Iowa State U., 1956—; assoc. dir. Iowa State U. (Engring. Expt. Sta.), 1959—; dir. Iowa State U. (Engring. Research Inst.), 1966—; dean Iowa State U. (Coll. Engring.), 1970-88, prof. chem. engring., 1988-92; cons. process engring., 1992—. Fellow AAAS, Am. Inst. Chem. Engrs., Am. Chem. Soc. (Merit award 1987); mem. Nat. Soc. Profl. Engrs. (v.p.), Profl. Engrs. in Edn. (chmn.), Am. Soc. Engring. Edn., Sigma Xi, Phi Lambda Upsilon, Sigma Tau, Phi Kappa Phi, Tau Beta Pi. Research in transient behavior and flow of fluids through porous media, unsteady state and fertilizer tech., devel. fused-phosphate fertilizer processes, theoretical and exptl. correlation of filtration Research in transient behavior and flow of fluids through porous media, unsteady state and fertilizer tech., devel. fused-phosphate fertilizer processes, theoretical and exptl. correlation of filtration. Home: 1516 Stafford Ave Ames IA 50010-5667

BOYLAN, GLENN GERARD, environmental company official, consultant; b. Englewood, N.J., Sept. 29, 1959; s. Matthew Joseph Boylan and Marie (Colagreco) Lawenda; m. Suzanne Lee MacIntyre, June 20, 1981; 2 children: Denise Lee, Patrick Glenn. BME, Ga. Inst. Tech., 1981. Mfg. engr. Torrington Co., Clinton, S.C., 1981-83, indsl. engr., 1983-84, mgr. indsl. engring., 1984-85; mfg. devel. engr. Torrington Co., Norcross, Ga., 1985-86; systems engr. Siemens Energy & Automation Co., Roswell, Ga., 1986-88; systems mgr., project mgr. Rust Environment & Infrastructure, Atlanta, 1988—; founder, Boylan Computer Svcs., Alpharetta, Ga., 1985—. Mem. outreach com. St. David's Ch., Roswell, 1988—, chmn., 1991. Mem. Atlanta Intergraph Users Group (chmn. 1989), North Fulton Ga. Inst. Tech. Club (pres. 1989-91). Episcopalian. Avocations: bicycling, golf, cooking, music. Office: Rust E & I 3980 Dekalb Pky Atlanta GA 30340-2203

BOYLE, FRANCIS WILLIAM, JR., computer company executive, chemistry educator; b. El Paso, Tex., Oct. 16, 1951; s. Francis William and Betty Lou Boyle; m. Sharon Marie McGuire; 1 child, Steven; m. Donna Marie Mader, Mar. 21, 1978; children: Richard, Genevieve, Veronica. BS in Chemistry, N.Mex. State U., 1973, MS in Soil Physics, 1979; PhD in Soil Chemistry, Colo. State U., 1984. Rsch. assoc. N.Mex. State U., Las Cruces, 1971-80, rsch. specialist, asst. prof., 1986-88, vis. prof. chemistry and agr., 1986—; rsch. specialist Colo. State U., Ft. Collins, 1980-84; asst. prof. Calif. State U., Fresno, 1984-85; dir. CATI field rsch. lab. Calif. Agrl. Tech. Inst., Fresno, 1985-86; chief fin. officer Software & Systems Techs., Las Cruces, 1986—; v.p. Internat. Safwater, Las Cruces, 1988-89; environ. cons. Buchanan Cons., Ltd., las Cruces, 1989—; mem. Reseller adv. coun. Everex Systems, Inc., Fremont, Calif., 1991—. Contbr. articles to profl. jours. Computer coord. field days Future Farmers Am., Fresno, 1984-86; judge So.N.Mex. State Sci. Fair, Las Cruces, 1987—. Grantee Kerley Chem. Co., 1985, Hewlett Packard Corp., 1986; N.Mex. eminent scholar N.Mex. Commn. on Higher Edn., 1989-90. Mem. Am. Soc. Agronomy, Soil Sci. Soc. Am., Western Soil Sci. Soc., Sigma Xi, Gamma Sigma Delta. Avocations: silver and gold smithing, software development, model railroading, swimming. Home: 1708 Calle Feliz Las Cruces NM 88001-4332 Office: NMex State U Dept Chemistry Las Cruces NM 88003

BOYLE, KEVIN JOHN, economics educator, consultant; b. Montgomery, Ala., Sept. 15, 1955; s. John Farley and Eliane Ruth (Keaney) B.; m. Nancy Jean Becraft, June 12, 1983; 1 child, Lindsey Jean. BA in Econs., U. Maine, 1978; MS in Econs., Oreg. State U., 1981; PhD in Econs., U. Wis., 1985. Rsch. asst. Oreg. State U., Corvallis, 1979-80; economist U.S. Forest Svc., Corvallis, 1981; rsch. asst. U. Wis., Madison, 1982-85, rsch. assoc., 1985-86; asst. prof. econs. U. Maine, Orono, 1986-91, assoc. prof., 1991—; vis. assoc. scholar N.C. State U., 1992-93; faculty assoc. Ctr. for Econs., Rsch. Triangle Inst., 1992-93; resource economist HBRS, Madison, 1985-86; pres. regional project benefits and costs in natural resource planning USDA, 1988-89. Mem. editorial bd. Jour. Environ. Econs. and Mgmt., 1989-92; contbr. articles to profl. jours. Recipient Cert of Merit, USDA, 1981; Maine Legis.-U.S. Fish and Wildlife Svc. grantee, 1988—, U.S. Forest Svc. grantee, 1989—, Maine Dept. Marine Resources grantee, 1989-91, Oak Ridge Nat. Labs. grantee, 1989-90, Exxon, USA grantee, 1989-92, Maine Dept. Inland Fisheries and Wildlife grantee, 1990-91, Bangor Hydro-Electric Co. grantee, 1991-92, EPA. Mem. Am. Econ. Assn., Am. Agrl. Econs. Assn., Assn. Environ. and Resource Economists, Northeastern Agrl. and Resource Econs. Assn. (bd. dirs. 1993—), Western Region Sci. Assn. Avocations: running, biking, canoeing, reading. Home: 322 Main Rd S Hampden ME 04444-1103 Office: U Maine Econs Dept Winslow Hall Orono ME 04469

BOZANIC, JEFFREY EVAN, marine sciences research center director; b. Miami, Fla., Oct. 25, 1957; s. John and Maxine Iris (Gerstenfeld) B. BA, Humboldt State U., 1979, MA, 1980; MBA, UCLA, 1982. Instr. Humboldt State U., Arcata, Calif., 1979-80; cons. Aquarius Rsch. Internat., Inc., Huntington Beach, Calif., 1980-93; pres. Next Generation Svcs., Inc., Huntington Beach, 1984-93; dive technician Antarctic Svcs., Inc., Paramus, N.J., 1989-90, Antarctic Support Assocs., Inc., Denver, 1991-92; exec. dir. Island Caves Rsch. Ctr., Inc., Melbourne, Fla., 1986—. Editor: The National Association of Underwater Instructors Textbook, 1985; contbg. editor: The NSS-CDS Instructor's Manual, 1985; contbr.: Advanced Diving: Technology and Techniques, 1989, The NSS Cave Diving Manual, 1992, articles to profl. jours. and mags. Recipient Silver Wakulla award Nat. Assn. Cave Diving, 1989. Fellow Explorer's Club; mem. AAAS, Internat. Underwater Found. (chmn. 1991-93), Am. Acad. Underwater Scis., Nat. Assn. Underwater Instrs. (bd. dirs. 1988-95, Outstanding Svc. award 1984, Continuing Svc. award 1986), Nat. Speleol. Soc. (bd. dirs. cave diving sect. 1984-88, chmn. bd. dirs. 1987-88, Abe Davis award 1984, Chmn.'s award 1990). Achievements include development of refinement of techniques to conduct sci. rsch. in submerged cave systems; evaluation of diving equipment during sci. rsch. use in polar diving ops.; discovery of many new species of marine animals, including Agostocaris bozanici and Bahadzia bozanici. Office: Island Caves Rsch Ctr PO Box 3448 Huntington Beach CA 92605-3448

BOZENHARDT, HERMAN FREDRICK, control technologist, automation system designer; b. Bklyn., Dec. 24, 1955; s. Herman Fredrick Sr. and Barbara (Koren) B.; m. Victoria Udovich, Jan. 28, 1978; children: Erich Herman, Heather Victoria. BSChemE, Polytech Inst., 1976, MS in System Engring., 1978. Prodn. supr. Pfizer Pharm., Bklyn., 1976-77; control system engr. Exxon Corp., Linden, N.J., 1977-83; mgr. control systems Fischer and Porter, Warminster, Pa., 1983-88; v.p. engring. Artificial Intelligence Tech., Hawthorne, N.Y., 1988-92; dir. CIM systems Life Scis. Internat., Phila., 1992—; cons. Shell Oil Co., Houston, 1990-91, Mobil Chem. Corp., Macedon, N.Y., 1991-92; tech. advisor U.S. Govt., Washington, 1989—. Contbr. 80 articles to profl. jours. Recipient Polytech Inst. Kulka award, 1976. Mem. AICE, Instrument Soc. Am., Internat. Soc. Pharm. Engrs., Assn. Iron & Steel Engrs., Tau Beta Pi, Omega Chi Epsilon. Republican. Roman Catholic. Home: 10 Saddlebrook Rd Robbinsville NJ 08691 Office: Life Scis Internat 1818 Market St Philadelphia PA 19103

BOZICEVIC, JURAJ, educator; b. Vrbovsko, Croatia, Oct. 7, 1935; s. Emil and Antonija (Crnkovic) B. Diploma in Elec. Engring., U. Zagreb, also Ph.D.; m. Biserka Vitez, Dec. 27, 1961; children—Hrvoje, Zrinka. Research asst. Faculty Elec. Engring., U. Zagreb, 1961-63, lectr., then asst. prof. Faculty Tech., 1964-71, prof. measurement and control, 1972—, head postgrad. study systems sci. and tech. cybernetics, 1971-82, head dept. measurement and control faculty chem. engring., 1991—, dep. minister higher edn. Ministry Culture and Edn., Rep. Croatia, 1993—; research fellow Tech. U. Eindhoven (Netherlands), 1971; vis. prof. U. Trondheim (Norway), 1977; hon. prof. U. Split; founder Internat. Sch. of Measurement, 1980—; observer measurement systems courses, PHoenix, 1986. UNESCO cons. Rizal Technol. Coll., Manila, 1986-89. Recipient Plaque of Appreciation RTC, Philippines, 1988. Mem. IEEE, Yugoslav Assn. Measurement, Control and Automation (chmn. exec. com. 1974-80), Internat. Measurement Confedn. (gen. council 1976-82, chmn. tech. com. meteorol. requirements in devel. countries 1978—, Disting. Service award 1985), Union Internat. Tech. Assn. (chmn. working group metrology 1986-91), Internat. Assn. Cybernetics, Croatian Assn. Inventors, Croation Systems Soc. (founder, 1st pres. 1991—), Croatian Acad. Tech. (sec. gen. 1993), Instrument Soc. Am., UK Systems Soc. Author: Automatic Process Control, 1971; Automatic Control Fundamentals I, 1978, 10th rev. edit, 1987, II, 1991, 9th rev. edit., 1992; also articles; editor books, proc.; patentee Measurement sensors and transducers. Home: 18 Trg kralja Tomislava, Zagreb 41000, Croatia Office: 16 Savska St, Zagreb 41000, Croatia

BOZOZUK, MICHAEL, civil engineer; b. Poland, Nov. 10, 1929; married Marcelle F. M. Daoust, July 20, 1957; children: Lyne, Sylvie, Camille. BSc in Civil Engring., U. Man., Winnipeg, Can., 1952, MSc in Soil Mechanics, 1954; PhD in Geotechnical, Purdue U., 1972. Rsch. officer geotechnical section, divsn. building rsch. Nat. Rsch. Coun. Can., 1953-89, cons., 1989-93; pvt. practice, 1989-93; vis. prof. U. Ottawa, Can., 1976; external examiner MSc, PhD candidates U. Queens, Royal Military Coll., Ottawa, Ecole Polytechique, Sherbrooke; com. soil and rock instrumentation Transp. Rsch. Bd., 1972-81, com. on foundations of bridges and other structures, 1972-81; chmn. adv. com. civil tech. Algonquin Coll., Ottawa, 1972-76; adv. com. Beaufort Sea artificial island Dept. Indian and Northern Affairs, Govt. Can., 1981-84; rsch. com. silo foundations Ont. Silo Assn., 1978-82; chmn. divisional com. safety and health Divsn. Building Rsch., 1980-85; Can. Gen. Standards Bd. Geotextiles, 1980-85; avd. com. environ./geotechnique Sir Sanford Fleming Coll., Lindsay, Ont. 1983-87; tech. com. on foundations Can. Standards Assn., 1983-90; mem. Can. Geoscience Coun., 1985-91; South East China Tour Lectr., 1986; hon. prof. Chengdu U., China, 1986; scientific advisor various orgns. and univs. Contbr. articles to profl. jours. Recipient Hon. award Caisse Populaire St. Genevieve, Ottawa, Can. Engring. Centennial Silver Medal, 1987, Cert. Citizenship City Calgary, 1987. Fellow Engring. Inst. Can. (govt. liaison com. 1989-90, strategic planning and adv. com. 1989-92, hons. and awards com. 1986-91, chmn. engring. liasion com. 1993—, Can. Paper award 1960, John B. Stirling medal 1990); mem. Nat. Rsch. Coun. Can. (assoc. com. geotechnical rsch., tech. advisor 1985-89, exec. sec. 1989-91), Geocontributions (founding mem. 1988, v.p. 1993), Assn. Profl. Engrs. Ontario, Can. Geotechnical Soc. (cross Can. tour lectr. 1979, dir no. and ea. Ont. 1984-85, v.p. tech., past pres. 1988-90, chmn. award com. 1986-90, nominating com. 1987-90, assoc. editor 1982-86, prize for best paper 1973, Svc. award 1988), Ottawa Geotechnical Group (sec. 1957-59, chmn. 1976-78), Internat. Soc. Soil Mechanics and Found., Can. Geotechnical Fund. (treas. 1985-88). Roman Catholic. Home and Office: 691 Sandra Ave, Ottawa, ON Canada K1G 2Z7

BOZZOLA, JOHN JOSEPH, botany educator, researcher; b. Herrin, Ill., Oct. 22, 1946. Ph.D., So. Ill. U., 1977. Instr. Med. Coll. Pa., Phila., 1976-79, asst. prof. microbiology, 1979-83; dir. Electron Microscopy Ctr., So. Ill. U., Carbondale, 1983—, assoc. prof. botany dept., 1985-93, prof., 1993—. Contbr. research articles on electron microscopy to prof. jours. Recipient Young Investigator award Nat. Inst. Dental Research, Washington, 1978. Mem. Electron Microscopy Soc. Am., Am. Soc. Microbiology, Ill. State Acad. Sci., Sigma Xi, Phi Kappa Phi, Kappa Delta Pi. Avocations: photography, bicycling, gardening, painting, computers. Office: So Ill U Ctr for Electron Microscopy Carbondale IL 62901

BRABHAM, DALE EDWIN, product engineer; b. Gainesville, Fla., Dec. 26, 1947; s. George Washington and Aline (Bussard) B.; m. Linda Rose Grant, Apr. 22, 1967; children: Michael, John, David, Christina, Emily, Ellen. BS in Chemistry, U. Fla., 1968; PhD in Molecular Biophysics, Fla. State U., 1973. Rsch. assoc. U. Fla., Gainesville, 1973-74, assoc., 1978-81; NIH fellow U. Ga., Athen, 1974-76; vis. prof. U. Fed. Da paraiba, Joao Pessoa, Brazil, 1976-78; asst. prof. Elmira (N.Y.) Coll., 1981-84; sr. engr. Philips Lighting Co., Bath, N.Y., 1984—. Contbr. articles to jours. Illuminating Engring. Soc., Chem. Physics Letters, Photochemsitry and Photobiology, Jour. Phys. Chemistry. Mem. Am. Chem. Soc., Illuminating Engring. Soc. (sect. pres. 1991-93), Toastmasters Internat. (club pres. 1991). Roman Catholic. Achievements include patents (with other) for High Pressure Series ARC Disharge Lamp, HID Lamp with Multiple Disharge Devices. Office: Philips Lighting Co 7265 Rte 54 Bath NY 14810

BRACEWELL, RONALD NEWBOLD, electrical engineering and computer science educator; b. Sydney, Australia, July 22, 1921; s. Cecil Charles and Valerie Zilla (McGowan) B.; m. Helen Mary Lester Elliott; children: Catherine Wendy, Mark Cecil. BS in Math. and Physics, U. Sydney, 1941, B in Engring., 1943, M. in Engring. with 1st class honors, 1948; PhD, Cambridge (Eng.) U., 1951. Sr. rsch. officer Radiophysics Lab., Commonwealth Sci. and Indsl. Rsch. Orgn., Sydney, 1949-54; vis. asst. prof. radio astronomy U. Calif., Berkeley, 1954-55; mem. elec. engring. faculty Stanford U., 1955—, Lewis M. Terman prof. and fellow in elec. engring., 1974-79, now Terman prof. emeritus elec. engring.; Pollock Meml. lectr. U. Sydney, 1978; Tektronix Disting. Visitor, summer 1981; Christensen fellow St. Catherine's Coll., Oxford, autumn 1987; sr. vis. fellow Inst. Astronomy, Cambridge, U., autumn 1988; mem. adv. panels NSF, Naval Rsch. Lab., Office Naval Rsch., NAS, Nat. Radio Astronomy Obs., Jet Propulsion Lab. Adv. Group on Radio Experiments in Space, Advanced Rsch. Projects Agy. Author: The Fourier Transform and Its Applications, 1965, rev. edit., 1986, The Galactic Club: Intelligent Life in Outer Space, 1974, The Hartley Transform, 1986; co-author: Radio Astronomy, 1955; translator: Radio Astronomy (J.L. Steinberg and J. Lequeux); editor: Paris Symposium on Radio Astronomy, 1959; mem. editorial adv. bd. Planetary and Space Sci.; former mem. editorial adv. bd. Proceedings of the Astron. Soc. Pacific, Cosmic Search, Jour. Computer Assisted Tomography; mem. bd. annual rev. Astronomy and Astrophysics, 1961-68; contbr. articles and revs. to jours., chpts. to books; patentee in field. Recipient Duddell Premium, Instn. Elec. Engrs., London, 1952, Inaugural Alumni award Sydney U., 1992; Fulbright travel grantee, 1954, William Gurling Watson traveling fellow, 1978, 86. Fellow IEEE (life), AAAS, Royal Astron. Soc.; mem. Inst. Medicine of NAS (fgn. assoc), Astron. Soc. Pacific (life), Am. Astron. Soc. (past councilor), Astron. Soc. Australia, Internat. Astron. Union, Internat. Sci. Radio Union, Internat. Acad. Astronautics. Home: 836 Santa Fe Ave Palo Alto CA 94305-1023 Office: Stanford U 329A Durand Bldg Stanford CA 94305

BRACKETT, EDWARD BOONE, III, orthopedic surgeon; b. Fort Worth, Jan. 5, 1936; s. Edward Boone and Bessie Lee (Hudgins) B.; student Tex. Tech. Coll., 1957; M.D., Baylor U., 1961; JD, Ill. Inst. Tech., 1993; m. Jean Elliott, July 11, 1959; children: Bess E., Geoffrey, Elliott Mencken, Edward Boone IV, Anneke Gail; m. Andrea Inman, 1992; 1 child, Amelia. Intern, Cook County Hosp., Chgo., 1961-62; resident Northwestern U., Chgo., 1962-66; practice medicine specializing in orthopedic surgery, Oak Park, Ill., 1966—, Westgate Orthopaedics Ltd., Oak Park, 1969—; mem. staff Loyola U., Oak Park Hosp., Loretto Hosp., Rush Med. Sch.; chmn. dept. orthopedics West Suburban Hosp., pres. med. staff, 1982-84 ; clin. assoc. prof. orthopedics Loyola U.; chmn. bd. Chgo. Loop Mediclinic, 1973-75; cons. orthopedic surgery City Service Oil Co., 1970. Guarantor, Lyric Opera Chgo., 1971-84; guest condr. Chgo. Symphony Orch., 1979 gov. mem. 1992, Chgo. Chamber Orch., 1980; trustee Music of the Baroque; mem. humanities adv. council Triton Coll., 1983-84; charter mem. vis. com. Northwestern U. Sch. Music, 1982—; chmn. Friends of WFMT, Inc. Served as lt. comdr. USNR, 1967-69; Vietnam. Recipient Outstanding Tchr. award Dept. Orthopedic Surgery, West Suburban Hosp., 1978, 79. Diplomate Am. Orthopedic Bd. Surgery, Am. Bd. Neurol. Orthopedic Surgeons. Fellow A.C.S., Am. Acad. Orthopedic Surgeons, Inst. of Medicine of Chgo., Am. Acad. Neurol. and Orthopedic Surgeons, Am. Assn. for Hand Surgery, Internat. Coll. Surgeons; mem. Am. Trauma Soc. (founder), Royal Soc. Medicine, Ill. Orthopedic Soc., Chgo. Orthopedic Soc., AMA, Chgo. Med. Soc. (alt. councilor), Clin. Orthopedic Soc. (chmn. membership com.), Internat. Platform Assn., Civil War Round Table, Friends Chgo. Symphony Orch., Chgo. Chamber Orch. Assn. (dir.), Symphonia Musicale (dir.), Sigma Alpha Epsilon, Phi Eta Sigma, Phi Chi, Alpha Epsilon Delta. Cons. orthopedic editor Jour. Indsl. Medicine, 1966-67. Cert. flight instr. single and multi engine land, single engine sea and airline transport pilot, FAA. Home: 7339 Holly Ct River Forest IL 60305-1915 Office: 1125 Westgate St Oak Park IL 60301-1007

BRADBURY, NORRIS EDWIN, physicist; b. Santa Barbara, Calif., May 30, 1909; s. Edwin Perly and Elvira C. (Norris) B.; m. Lois Platt, Aug. 5, 1933; children—James Norris, John Platt, David Edwin. B.A., Pomona Coll., 1929, D.Sc., 1951; Ph.D., U. Calif., 1932; LL.D., U. N.Mex., 1953; D.Sc., Case Inst. Tech., 1956. NRC fellow in physics M.I.T., 1932-34; asst. prof. physics Stanford U., 1934-37, assoc. prof., 1937-42, prof., 1942-50; prof. physics U. Calif., 1945-70; dir. Los Alamos Sci. Lab., 1945-70. Contbr. tech. articles to phys. revs., jours. Served with USNR, 1941-45; capt. Res. Decorated Legion of Merit. Fellow Am. Phys. Soc. (Enrico Fermi award 1970); mem. Nat. Acad. Sci. Episcopalian. Home: 1451 47th St Los Alamos NM 87544

BRADDOCK, JOHN WILLIAM, safety engineer; b. Phila., May 23, 1947; s. John Thomas and Joan Betty (Faulkner) B; m. Linda Konareski, 1988; 1 child, John Joseph. BS in Commerce and Engring., Drexel U., 1970; MS in Organizational Behavior, U. Hartford, 1976. Cert. safety profl., hazard control mgr. Asst. plant mgr. N.J. Silica Sand Co., Millville, 1972-73; lead safety engr. Pratt & Whitney Aircraft Co., East Hartford, Conn., 1973-78; mgr. occupational safety and indsl. hygiene Sikorsky Aircraft Co., Stratford, Conn., 1978-86; sr. cons. Occupational Safety United Techs. Corp., Hartford, Conn., 1987-89, mgr. loss prevention and regulatory affairs, 1989—; instr. Hartford Grad. Ctr., 1977. Active St. John's Episc. Ch., warden, 1988, vestry, 1991-93; bd. advisors Conn. Safety Coun.; bd. advisors Conn. Occupational Health Clinics. Served to capt. U.S. Army, 1970-84. Mem. APHA, Indsl. Health Found. (trustee 1988), Am. Soc. Safety Engrs., Am. Indsl. Hygiene Assn., Nat. Fire Protection Assn., Res. Officers Assn., Bushnell Park Carousel Soc., Lambda Chi Alpha. Author: (lyrics) A Carousel Ride. Lodge: Knights of Malta, Masons (steward). Home: 60 Beth Ann Cir Meriden CT 06450-7300 Office: United Techs Corp Hartford CT 06101

BRADFORD, PHILLIPS VERNER, engineering research executive; b. Washington, June 15, 1940; s. Henry Knight and Laura Battle (Verner) B.; m. Diane Gnassi, June 15, 1963 (div. 1972); children: Phillip Gnassi Bradford, Anthony Robert Gnassi Bradford; m. Camille Quarrier, Aug. 19, 1972. BS, Johns Hopkins U., 1962; MEE, U. Va., 1964; DSc, Columbia U., 1968. Mem. tech. staff Bell Telephone Labs, Holmdel, N.J., 1967-69; securities analyst Dominick & Dominick, N.Y.C., 1969-73; industry specialist Merrill Lynch, Inc., N.Y.C., 1973-76; sr. scientist AMETEK, Inc., Paoli, Pa., 1976-77; dir. energy product Phelps Dodge Industries Inc., N.Y.C., 1977-82; dir. corp. rsch. Rutgers U., New Brunswick, N.J., 1982-84; exec. dir. Kans. Tech. Enterprise Corp., Topeka, 1984-87; assoc. dir. Ctr. for Advanced Tech., Columbia U., 1988-89; exec. dir. Colo. Advanced Tech. Inst., Denver, 1989—. Co-author: Ota Benga-The Pygmy in the Zoo, 1992, Solar Energy Handbook, 1978 (AMETEK handbook). Mem. Met. Club of N.Y. Achievements include patents on solar absorber plate. Home: 11515 Quivas Way Denver CO 80234 Office: Colo Advanced Tech Inst 1625 Broadway Denver CO 80202

BRADFORD, REAGAN HOWARD, JR., ophthalmology educator; b. Lawton, Okla., July 31, 1954; s. Reagan Howard Sr. and Conita Ann (Hargraves) B.; m. Cynthia Ann McGough, Apr. 22, 1988. BS, U. Okla., 1976; MD, U. Okla., Oklahoma City, 1980. Diplomate Am. Bd. Ophthalmology. Intern Bapt. Med. Ctr., Oklahoma City, 1980-81; resident Dean A. McGee Eye Inst. U. Okla., 1981-84; fellow in vitreo retina Bascom Palmer Eye Inst. U. Miami, Fla., 1984-85; asst. clin. prof. dept. ophthalmology Dean A. McGee Eye Inst., U. Okla., Oklahoma City, 1985—. Author: (with others) Basics of Neurophthalmology; contbr. articles to profl. jours. Fellow Am. Acad. Ophthalmology; mem. Okla. County Med. Soc., Okla. State Med. Assn., AMA, Okla. State Acad. Ophthalmology. Republican. Baptist. Avocations: tennis, softball. Office: Dean A McGee Eye Inst 608 Stanton L Young Blvd Oklahoma City OK 73104-5040

BRADFORD, SUSAN KAY, nutritionist; b. Laramie, Wyo., May 30, 1952; d. Gerald Wesley Butcher and Norma Ellen (Clingenpeel) Appel; m. Dennis Bradford, Aug. 26, 1986. BA, U. Wyo., 1974; cert. plan IV in nutrition, Colo. State U., 1982; MS in Nutrition, NYU, 1983. Registered dietitian, cert. nutrition support dietitian. Dietetic resident NYU Med. Ctr., N.Y.C., 1983-84; staff dietitian Bellevue Hosp. Ctr., N.Y.C., 1984-86; nutrition support coord. L.I. Coll. Hosp., Bklyn., 1986-88, clin. nutrition mgr., 1988-89, asst. dir. clin., 1989-91, assoc. dir. nutrition svc., 1991—; mem. ethics com. L.I. Coll. Hosp., 1991—, mem. subcom. community rels., 1992—, nutrition com., infection control com.; pub. speaker nutrition support topics, med. ethics regarding nutrition and hydration. Editor Hospital Nutrition and

Food Svc. Forms, Checklists, Guidelines, 1992; contbr. articles to Support Line Newsletter, papers and abstracts to jours. Mem. Am. Dietetic Assn. (workshop coord. dietitians in nutrition support group 1987-89, chair pub. rels. 1989-91, chair-elect 1991-92, chair 1992-93, reviewer Handbook Clin. Dietetics 1992). Democrat. Episcopalian. Achievements include devel. of approved pre-profl. practice program of dietetic residency. Home: 161 Remsen St Brooklyn NY 11201 Office: L I Coll Hosp Atlantic Ave and Hick St Brooklyn NY 11201

BRADIC, ZDRAVKO, chemist; b. Ugarci, Croatia, Sept. 3, 1947; came to U.S., 1981; s. Dusan and Marija (Peic) B.; m. Connie Lynn Zothman; children: Daniela L., Ryan J. BS in Pharmacy, Zagreb U., Croatia, Yugoslavia, 1970, MS in Instrumental Methods, 1972, PhD in Chemistry, 1974. Rsch./teaching asst., assoc. Zagreb U., 1970-81; postdoctoral fellow N.Mex. State U., Las Cruces, 1976-77, 82-84, asst. prof. chemistry, 1984-85; postdoctoral fellow U. Minn., Mpls., 1977-78; rsch. scientist Hyland div. Baxter, L.A., 1985-88; sr. scientist Ciba Corning Diagnostics, Irvine, Calif., 1988—. Contbr. over 20 articles to profl. jours. Fellow Am. Inst. Chemists; mem. AAAS, Am. Chem. Soc., Am. Assn. Clin. Chemists. Office: Ciba Corning 17395 Daimler St Irvine CA 92714-5585

BRADLEY, CHARLES MACARTHUR, architect; b. Chgo., Sept. 26, 1918; s. Harold Smith and Helen Frances (MacArthur) B.; B.S. in Architecture, U. Ill., 1940; m. Joan Marie Daane, July 27, 1946; children—Mary Barbara, Nancy Ann, Sally Joan, William Charles. With Holabird & Root, architects, Chgo., 1940-41, Giffels & Vallet, architects and engrs., Detroit, 1941-44; partner, corp. pres. Bradley & Bradley, architects and engrs., Rockford, Ill., 1947—; pres. Bradley Bldg. Corp., 1962—; sec.-treas. Mchts. Police, 1972—; pres. Westshore Plaza Inc., 1979—. Active, Blackhawk council Boy Scouts Am. Served with C.E., U.S. Army, 1945-46. Decorated Bronze Star; recipient Meritorious Service award Ill. Assn. Sch. Bds., 1976. Mem. AIA (pres. No. Ill. chpt. 1962, treas. Ill. council 1973-74), Ill. Soc. Architects (pres. 1974), Edn. Facilities Planners Inst., Ill. Assn. Sch. Bd. Officers. Republican. Congregationalist. Clubs: Rotary, Union League, University, Midday (Chgo.). Lodges: Shriners, Moose. Prin. works include North Sheboygan (Wis.) High Sch. and addition, 1960-68, J.F. Kennedy Middle Sch., Rockford, 1968, Singer Health Clinic, Rockford, 1964, Jacobs High Sch., Algonquin, Ill., 1976, Atwood plant, Rockford, 1977, Admiral Home, Chgo., 1978, Bushnell (Ill.) Jr. High Sch., 1980, Bloom High Schs., 1983, Evenglow Lodge, 1984. Author papers on life cycling old schs., roofing procedures. Office: 3203 Landstrom Rd Rockford IL 61107-1124 also: Bradley & Bradley Inc 924 N Main St Rockford IL 61103

BRADLEY, CHARLES WILLIAM, podiatrist; b. Fife, Tex., July 23, 1923; s. Tom and Mary Ada (Cheatham) B.; m. Marilyn A. Brown, Apr. 3, 1948 (dec. Mar. 1973); children: Steven, Gregory, Jeffrey, Elizabeth, Gerald. Student, Tex. Tech., 1940-42; D. Podiatric Medicine, Calif. Coll. Podiatric Medicine U. San Francisco, 1949, MPA, 1987, D.Sc. (hon.). Pvt. practice podiatry Beaumont, Tex., 1950-51, Brownwood, Tex., 1951-52, San Francisco, San Bruno, Calif., 1952—; assoc. clin. prof. Calif. Coll. Podiatric Medicine, 1992—; chief of staff Calif. Podiatry Hosp., San Francisco; mem. surg. staff Sequoia Hosp., Redwood City, Calif.; mem. med. staff Peninsula Hosp., Burlingame, Calif.; chief podiatry staff St. Luke's Hosp., San Francisco; chmn. bd. Podiatry Ins. Co. Am.; cons. VA. Mem. San Francisco Symphony Found.; mem. adv. com. Health Policy Agenda for the Am. People, AMA; chmn. trustees Calif. Coll. Podiatric Medicine, Calif. Podiatry Coll., Calif. Podiatry Hosp.; mem. San Mateo Grand Jury, 1989. Served with USNR, 1942-45. Mem. Am. Podiatric Med. Assn. (trustee, pres. 1983-84), Calif. Podiatry Assn. (pres. No. div. 1964-66, state bd. dirs., pres. 1975-76, Podiatrist of Yr. award 1983), Nat. Coun. Edn. (vice-chmn.), Nat. Acads. Practice (chmn. podiatric med. sect.), Am. Legion, San Bruno C. of C. (bd. dirs. 1978-91, v.p. 1992, bd. dir. grand jury assoc. 1990), Olympic Club, Commonwealth Club Calif., Elks, Lions. Home: 2965 Trousdale Dr Burlingame CA 94010-5708 Office: 560 Jenevein Ave San Bruno CA 94066-4477

BRADLEY, JAMES ALEXANDER, software engineer, researcher; b. Van Nuys, Calif., May 16, 1965. BA in Math., Computer Sci., U. Colo., 1988, postgrad., 1991—. Software developer Sci. Computer Systems, Inc., Boulder, Colo., 1982-84; teaching asst. Boulder Valley Pub. Schs., Boulder, Colo., 1984-87; software engr. Martin Marietta Aerospace, Littleton, Colo., 1988--. Recipient NASA New Tech. award, Martin Marietta Aerospace, 1990. mem. Am. Math. Soc., Math. Assn. Am., Golden Key Honor Soc. Achievements include design of LASER engraving system, high-speed target tracking and aquisition system. Office: Martin Marietta Aerospace PO Box 179 Mailstop 4372 Denver CO 80201-0179

BRADLEY, RICHARD GORDON, JR., aerospace engineer, engineering director; b. Canton, Ga., June 18, 1932; s. Richard Gordon and Ethel B. (Bishop) B.; m. Margaret Lois Boswell, Sept. 6, 1953; children: Sandra K., Richard G. III. BS in Aero. Engring., Ga. Inst. Tech., 1950-54, MS in Aero. Engring., 1956-57, PhD, 1963-66. Design specialist General Dynamics, Fort Worth, Tex., 1975-77, engring. chief, 1977-81, engring. mgr., 1981-83, engring. dir., 1983—; disting. lectr. AIAA, Washington, 1983-85; fluid dynamics panel mem. AGARD, Paris, 1986-92; aero. and space engring. bd. Nat. Rsch. Coun., Washington, 1987-92; mem. Congressional Aero. Adv. com., 1987-88, NASA aeronautics adv. com., 1982-92. Author: (with others) Progress in Astro & Aero, vols. 81, 125, 1988; editorial bd.: AIAA Progress in Astro. & Aero, 1989—; contbr. articles to profl. jours. First Lt. USAF, 1954-56. Fellow AIAA. Baptist. Office: Lockheed Ft Worth Co P O Box 748 MX 2666 Fort Worth TX 76101

BRADLEY, ROBERT FOSTER, chemical engineer; b. Columbia, S.C., Mar. 26, 1940; s. Robert Foster Sr. and Susan (Tison) B.; m. Doris Peeples McAlpin, Mar. 18, 1969; children: Melrose McAlpin, Susanne Habersham, Robert Foster. BS, U. S.C., 1962, MS, 1964; PhD, Vanderbilt U., 1966. Rsch. engr. E.I. DuPont, Orange, Tex., 1965-66; rsch. engr. Atomic Energy divsn. E.I. DuPont, Aiken, S.C., 1966-71, rsch. supr., 1971-78, chief supr. separation, 1978-81, mgr. hydrogen tech., 1981-86, mgr. non-reactor safety evaluation Atomic Energy div., 1986-89; mgr. facility safety evaluation Westinghouse Savannah River Co., Aiken, 1989—; chmn. DOE NPR Project Office Tech. Working Group on Target Processing, Washington, 1984. Contbr. articles to profl. jours. mem. Am. Inst. Chem. Engrs. (chmn. Savannah river sect. 1968-71), Sigma Xi. Episcopalian. Home: 631 Sandhurst Pl Aiken SC 29801 Office: Westinghouse Savannah Rive Centennial Ave Bldg 2 Aiken SC 29803

BRADLEY, STERLING GAYLEN, microbiology and pharmacology educator; b. Springfield, Mo., Apr. 2, 1932; s. Benn and Lora (Brown) B.; m. Lois Evelyn Lee, May 13, 1951; children—Don, Evelyn, John, Phillip; m. Judith Bond, July 24, 1974; 1 son, Kevin. B.A., B.S., S.W. Mo. State Coll., 1950; M.S., Northwestern U., 1952; Ph.D. (NSF fellow), 1954; Ph.D. certificate med. mycology, Duke, 1957. Grad. teaching asst. Northwestern U., Evanston, Ill., 1950-51; Abbott research asst. Northwestern U., 1951-52, instr. biology, 1954; instr. dept. bacteriology and immunology U. Minn., 1956-57, asst. prof. dept. bacteriology, 1957-59, assoc. prof. dept. microbiology, 1959-63, grad. faculty genetics, 1961-68, prof., 1963-68, chmn. genetics faculty group, 1964; chmn. dept. microbiology Va. Commonwealth U., Richmond, 1968-82; prof. depts. pharmacology and microbiology Va. Commonwealth U., 1979—, dean basic health scis., 1982-93; vis. worker in pharmacology Cambridge (Eng.) U., 1978; mem. bd. sci. counselors NIH, 1968-72, chmn., 1970-72; mem. Internat. Com. Bacteriol. Systematics, 1966-74, exec. bd., 1970-74; mem. U.S. Pharmacopeial Com. of Revision, 1980-85; coord. Project 3 U.S.-USSR Joint Working Group on Microbiology, 1979-82; founder, v.p. Immunotox, 1985-91, pres., 1991—. Mem. editorial bd.: Proc. Soc. Exptl. Biol. Medicine, 1966-72, Conf. on Anti-microbial Agts, 1960, Jour. Indsl. Microbiology, 1985—; editor: Jour. Bacteriology, 1970-78; contbr. articles to profl. jours. Trustee Southeastern U. Rsch. Assn., Inc., 1990-93; bd. dirs. Sci. Mus. Va. Found., 1993—. Recipient Charles Porter award, 1983; named SW Mo. State U. Outstanding Alumnus, 1991; Eli Lilly postdoctoral fellow U. Wis., 1954-55; NSF postdoctoral fellow dept. genetics, 1955-56; NIH Sr. Fogarty internat. fellow, 1978. Fellow AAAS, Va. Acad. Sci. (past mem. council, sec. 1976-77); mem. Am. Assn. Immunology, Am. Acad. Microbiology, Am. Soc. Pharmacology and Exptl. Therapeutics, Am. Soc. Cell Biology, AAUP, Am. Soc. Microbiology (past mem. council, treas. 1985-91), Soc. Protozoologists, Soc. Indsl. Microbiology (past pres.), Am. Inst. Biol. Sci. (past dir., gen. chmn. 41st Meeting, 1989-90), Soc. Toxicology, U.S. Fedn. Culture Collections (pres. 1984-86), Mycol. Soc. Am., Soc. for Exptl. Biology and Medicine (past chmn. Minn. chpt.), Genetics Soc. Am., Torrey Bot. Club (life), N.Y. Acad. Scis. (life), Sigma Xi (life, pres. chpt. 1975-76, fin. com. 1991—). Home: 1324 Brookland Pky Richmond VA 23227-4704 Office: Va Commonwealth U Richmond VA 23298-0678

BRADLEY, SUSAN M., chemistry educator. Prof. chemistry U. New South Wales, Kensington, Australia. Recipient Alice Wilson awd. Royal Soc. of Canada, 1992. Office: U New South Wales Dept Chem, PO Box 1, Kensington 2033, Australia

BRADLEY, WALTER LEE, mechanical engineer, educator, researcher, consultant; b. Corpus Christi, Tex., Dec. 27, 1943; s. Kenneth Carl and Virginia (Fry) B.; m. Carol A. James, Aug. 31, 1965; children: Sharon, Steven. BS in Engring. Sci., U. Tex., 1965, PhD in Materials Sci., 1968. Registered profl. engr. From asst. to assoc. prof. dept. metall. engring. Colo. Sch. of Mines, Golden, 1968-76; from assoc. to full prof. dept. mech. engring. Tex. A&M U., College Station, 1976—; vis. prof. Lawrence Livermore (Calif.) Labs., summer 1973, Fed. U. of Minas Gerias, Belo Horizonte, Brazil, 1974, Fed. Tech. U. of Lausanne, Switzerland, summer 1991. Co-author: Mystery of Life's Orgin: Reassessing Current Theories, 1984, (chpts.) In-Situ Fracture Observation in SEM of Delamination in Composite Materials, (chpt.) Rubber Toughening Plastics. Mem. nat. bd. Christian Leadership Ministries, Richardson, Tex., 1990—. Recipient Best Materials Paper of Yr., Am. Nuclear Soc., 1978. Fellow Am. Soc. for Metals, Tex. Engring. Experiment Sta. (sr.). Republican. Evangelical. Home: 204 Suffolk College Station TX 77840 Office: Tex A&M U Mech Engring Dept College Station TX 77843-3123

BRADLEY, WILLIAM GUY, molecular virologist; b. Naples, Italy, Jan. 29, 1956; came to U.S., 1956; s. William Hearn and Sandra Leta (Jacobs) B.; m. Cindy A. Leach, Apr. 6, 1976 (div. Apr. 1980); children: Paul, Jennifer; m. Teresa Ann D'Orazio, Mar. 2, 1981; children: David, Maria, Sarah. BS with honors, Eckerd Coll., 1984; PhD with honors, U. South Fla., 1990. Supr. cardiopulmonary dept. Lake Seminole Hosp., Seminole, Fla., 1981-82; rsch. asst. Showa U. Rsch. Inst., St. Petersburg, Fla., 1984-85; rsch. asst. med. microbiology and immunology U. South Fla., Tampa, 1985-89; rsch. assoc. Tampa Bay Rsch. Inst., St. Petersburg, Fla., 1989-91, asst. mem., 1990-91; dir. retrovirology lab. All Children's Hosp., St. Petersburg, Fla., 1991—; asst. prof. med microbiology and immunology U. South Fla., Tampa, 1991—. Contbr. articles to profl. jours. Judge Pinellas County Sci. Ctr., St. Petersburg, 1985-90, Internat. Sci. and Engring. Fair, Orlando, Fla., 1991. Recipient Baxter Diagnostics Microscan Young Investigator award, 1992. Mem. AAAS, Am. Assn. for Microbiology, N.Y. Acad. Sci., Sigma Xi (assoc., Outstanding Grad. Rsch. award 1990). Home: 7533 132d St North Seminole FL 34646 Office: All Childrens Hosp Retrovirology Lab 801 6th St S Saint Petersburg FL 33701

BRADLOW, HERBERT LEON, endocrinologist, educator; b. Phila., Mar. 21, 1924; s. Robert and Rose (Rudnick) B.; m. Hattie Gottlieb, Dec. 25, 1947; children: Ellen (dec.), Janet, Alec. BS, U. Pa., 1945; MS, U. Kans., 1948, PhD, 1949. Rsch. assoc. Sloan Kettering Inst., N.Y.C., 1951-57, assoc. prof., 1957-64; sr. rsch. dir. Montefiore Hosp., N.Y.C., 1964-66, prof., 1966-77; prof. Rockefeller U., N.Y.C., 1977-89; pres. Inst. Hormone Rsch., N.Y.C., 1989-92; prof. biochemistry Cornell U. Med. Sch., N.Y.C., 1992—; mem. study sect. NIH, Bethesda, Md., 1967-70. Editorial bd. Endocrinology Jour., 1961-85, Jour. Steriods Biochemistry, 1980-85; editor Steriods, N.Y.C., 1985—; contbr. numerous articles to profl. jours. Mem. Am. Chem. Soc., Am. Asn. Cancer Rsch., Endocrine Soc., Brit. Chem. Soc., Brit. Endocrinology soc., Am. Soc. Biochemistry, Am. Soc. for Preventive Oncology. Jewish. Home: 86-25 Palo Alto Ave Jamaica NY 11423

BRADSHAW, PETER, engineering educator; b. Torquay, Devon, Eng., Dec. 26, 1935; came to U.S., 1988; s. Joseph Newbold and Frances Winifred (Finch) B.; m. Aline Mary Rose, July 18, 1959 (div. 1968); m. Sheila Dorothy Brown, July 20, 1968. BA, Cambridge U., Eng., 1957; DSc (hon.), Exeter U., Eng., 1990. Sci. officer Nat. Phys. Lab., Teddington, Eng., 1957-69; prof. Imperial Coll. Sci. and Tech., London, 1969-88; Thomas V. Jones prof. engring. Stanford U., 1988—; cons. various engring. cos. Author: Introduction to Turbulence, 1971, Momentum Transfer, 1977, Convective Heat Transfer, 1984; author nearly 200 journ. articles on aerodynamics. Recipient Bronze medal Royal Aero. Soc., London, 1971, Busk prize, 1972. Fellow Royal Soc. London. Avocations: cycling, walking. Office: Stanford U Dept of Mech Engring Stanford CA 94305

BRADT, REXFORD HALE, chemical engineer; b. Versailles, Ind., Oct. 17, 1908; s. Fletcher Hale and Mary Elizabeth (Peak) B.; m. Mabel Geraldine Enos, Sept. 16, 1932 (dec. May, 1990); children: Dale Rexford, Constance L., Douglas Hale, Gregory G., Geryce, Camille. BA, Ind. U., 1930; postgrad., U. Wis., 1930-32. Registered profl. engr., Ill. Chemist Nicolet Paper Co., De Pere, Wis., 1935-37; chief chemist Fox River Paper Co., Appleton, Wis., 1937-40; head chemist Oak Ordinance, Illiopolis, Ill., 1941; div. group leader Metallurgical Labs. U. Chgo., Chgo., Oak Ridge, 1942-45; pvt. practice cons. chemist, engr. Chgo., 1946-50; pres., v.p. dir. rsch. Fiberfil Corp., Warsaw, Ind., 1951-60; pres., chief exec. officer Materials Rsch. Inc., Warsaw, 1960—. Recipient fellowship U. Wis., 1930-32. Mem. Am. Chem. Soc., Soc. Plastic Engrs., Phi Lambda Upsilon (hon.). Republican. Presbyterian. Achievements include 4 patents for use of fibrous reinforcement in thermoplastics, developed new and first successful flame resisting plastic molding material for JATO nozzles, first method for separating thin shredded metal wastes based on differential malleability. Currently actively developing and preparing new thermoplastic molding materials for introduction to the industry. Office: Materials Rsch Inc PO Box 1216 Warsaw IN 46581-1216

BRADY, MARY SUE, pediatric dietitian, educator; b. Sedalia, Mo., Mar. 29, 1943; d. H. Wesley and K. Virginia (McGaw) Steele; m. Paul L. Brady, Sept. 2, 1967; 1 child, Chad W. BA, Marian Coll., Indpls., 1968; MS, Ind. U., Indpls., 1970, DMSc, 1987. Registered dietitian. Pediatric dietitian J.W. Riley Hosp. Children, Ind. U. Sch. Medicine, Indpls., 1970-75, acting dir. pediatric nutrition, 1975-78, 80-82, neonatal dietitian, 1978-80, dir. pediatric nutrition, 1982—; asst. prof. Ind. U. Sch. Medicine, Indpls., 1975-78, assoc. prof., 1988—. Contbr. articles to Jour. of Am. Dietetic Assn., Pediatric Pulmonology, Jour. of Pediatrics. Mem. Am. Dietetic Assn. (mem. jour. bd. 1988—, sec. pediatric nutrition practice group 1989-91, Excellence in Practice of Clin. Nutrition award 1991), Sigma Xi. Office: JW Riley Hosp for Children 702 Barnhill Dr Rm 1010 Indianapolis IN 46202-5200

BRADY, M(ELVIN) MICHAEL, engineer, writer; b. San Francisco, Dec. 15, 1933; arrived in Norway, 1962; s. Robert Alexander and Dorothy Elizabeth (Stahl) B.; m. Marianne Hadler, Aug. 11, 1978; children: Thomas, Alexander. BEE, George Washington U., 1956; MS, MIT, 1958; Degree Engr., Stanford U., 1962. Mem. staff Nat. Bur. Standards, Washington and Boulder, Colo., 1952-56; rsch. engr. MIT, 1956-58, Stanford U., Calif., 1959-62, Norwegian Def. Rsch. Establishment, 1958-59, 62-65; project engr. Norconsult, Oslo, 1968-74; cons. telecommunications engr. and editor Oslo, 1975—; guest faculty mem. U. Rochester (N.Y.), 1975, U. Ottawa (Ont., Can.), 1979-80. Author, translator books; contbr. articles to profl. jours. Fulbright fellow, 1958-59; Norwegian Council Sci. and Indsl. Research fellow, 1962-65. Mem. IEEE (sr.). Office: PO Box 8236 Hammersborg, N-0129 Oslo Norway

BRADY, ROSCOE OWEN, neurogeneticist, educator; b. Phila., Oct. 11, 1923; s. Roscoe O. and Martha (Roberts) B.; m. Bennett Carden Manning, 1972; 2 sons. Student, Pa. State U., 1941-43; MD, Harvard U., 1947; postgrad., U. Pa., 1948-49. Intern Hosp. U. Pa., 1947-48; NRC fellow U. Pa., 1948-50, USPHS spl. fellow, 1950-52; sect. chief Nat. Inst. Neurol. Diseases and Blindness, NIH, 1954-67; asst. lab. chief neurochemistry Nat. Inst. Neurol. Diseases and Blindness, NIH, Bethesda, Md., 1967-72; chief developmental and metabolic neurology br. Nat. Inst. Neurol. Disorders and Stroke, 1972—; professorial lectr. George Washington Sch. Medicine, 1963-73; faculty Georgetown U. Sch. Medicine, 1965—; mem. med. staff Children's Hosp. of Washington. Author: (with Donald B. Tower) Neurochemistry of Nucleotides and Amino Acids, 1960, Basic Neurosciences, 1975, (with John A. Barranger) Molecular Basis of Lysosomal Storage Disorders, 1984, also numerous articles. Recipient Gairdner Found. award, 1973, Lasker Found. award, 1982, Passano Found. award, 1982, Warren Alpert Found. prize, 1992, Hadassah Soc. Myrtle Wreath award, 1993. Mem. NAS (J.S. Kovalenko medal 1991), Am. Soc. Biol. Chemists, Am. Acad. Neurology (Kotzias award 1980), Am. Acad. Mental Retardation, Am. Soc. Clin. Investigation, Am. Soc. Human Genetics, Inst. of Medicine. Achievements include first demonstration of enzyme system for fatty acid synthesis; biosynthesis of myelin sheath lipids, nature of metabolic defects in Gaucher's disease, Niemann-Pick disease, Fabry's diseases and Tay-Sachs disease; diagnostic and genetic counseling tests for Gaucher's, Niemann-Pick, Fabry's diseases; enzyme and gene replacement therapy of lipid storage diseases; metabolism of sphingolipids in neoplastic diseases, role of antigenic sphingolipids in neurological diseases; identification of faulty intracellular cholesterol homostasis as a heritable human metabolic disorder. Home: 6026 Valerian Ln Rockville MD 20852-3410 Office: NIH 9000 Rockville Pike Bethesda MD 20892-0001

BRAEUTIGAM, RONALD RAY, economics educator; b. Tulsa, Apr. 30, 1947; s. Raymond Louis Braeutigam and Loys Ann (Johnson) Henneberger; m. Janette Gail Carlyon, July 27, 1975; children: Eric Zachary, Justin Michael, Julie Ann. BS, U. Tulsa, 1969; MSc, Stanford U., 1971, PhD, 1976. Petroleum engr. Standard Oil Ind., Tulsa, 1966-70; staff economist Office of Telecommunications Policy Exec. Office of Pres., Washington, 1972-73; from asst. to prof. econs. Northwestern U., Evanston, Ill., 1975—, Harvey Kapnick prof. Bus. Instns. dept. econs., 1990—; vis. prof. Calif. Inst. Tech., Pasadena, 1978-79. Co-author: The Regulation Game, 1978, Price Level Regulation for Diversified Public Utilities, 1989; assoc. editor Jour. Indsl. Econs., Cambridge, Mass., 1987-90; mem. editorial bd. MIT Press Series on Regulation, Cambridge, 1980—, Jour. Econ. Lit., 1987-91, Rev. Indsl. Orgn., 1991—. Coach Skokie (Ill.) Indians Little League, 1985-91, Evanston (Ill.) Youth Baseball Assn., 1991-93. Grantee Dept. Transp., NSF, Ameritech, Sloan Found., others; Sr. Rsch. fellow Int. Inst. Mgmt., Berlin, Fed. Republic Germany, 1982-83, 91. Mem. Am. Econ. Assn., Econometric Soc., Internat. Telecommunications Soc. (bd. dirs. 1990—), European Econ. Assn., European Assn. for Rsch. in Indsl. Econs. (exec. com. 1992—), Soc. Petroleum Engrs. Avocations: travel, music, German lang., French lang. Home: 731 Monticello St Evanston IL 60201-1745 Office: Northwestern U Dept Econs Evanston IL 60208

BRAGAGNOLO, JULIO ALFREDO, physicist; b. Buenos Aires, Oct. 6, 1941; came to the U.S., 1976; s. Efrain Domingo and Maria Elvira (Bustillo) B.; m. Elsa B. Iturbe, Dec. 15, 1965 (div. 1984); children: Celina Maria, Marcelo Julian; m. Jennifer Anne Ogle, July 11, 1987. Degree in physics, U. Buenos Aires, 1965; PhD, U. Del., 1973. Rsch. assoc. Inst. Energy Conversion U. Del., Newark, 1973-74, assoc. scientist, 1976-81; rsch. scientist Nat. Scientific and Tech. Rsch. Coun. Argentina, Buenos Aires, 1974-76; sr. scientist SES Inc., Newark, 1981-83; mgr. solar cell and module devel. Solarex Corp., Rockville, Md., 1983-85; mgr. amorphous solar cells Spire Corp., Bedford, Mass., 1985-87; dir. advanced product engring. AstroPower, Inc., Newark, 1987—. Contbr. articles to profl. jours. including Jour. Luminescence, Z. Naturforschung, Physica Status Solidi, Solar Cells, Jour. Vacuum Sci. Tech. and various conf. procs. Recipient Excellence in Exporting award Del. Devel. Office, 1989, Indsl. Rsch. 100 R&D Mag., 1979. Mem. IEEE, Am. Vacuum Soc. Achievements include patent for thin-film photovoltaic cell; research in efficiency of polyerstalline thin film solar cells, efficiency of a-Si thin film solar cells, stable, high efficiency thin-film polyerystalline solar cells. Office: AstroPower Inc Solar Park Newark DE 19711

BRAGDON, CLIFFORD RICHARDSON, city planner, educator; b. St. Louis, June 30, 1940; s. Dudley Acton and Ruth (Butler) B.; B.A., Westminster Coll., 1962; M.S., Mich. State U., 1965; Ph.D., U. Pa., 1970. m. Sarah Vaughn, Aug. 21, 1965; children—Katherine, Rachel, Elizabeth. Urban planner West Philadelphia Community Mental Health Consortium, U. Pa., 1967-69; environ. specialist, acting chief bio-acoustics div. U.S. Environ. Hygiene Agy., Edgewood, Md., 1969-72; prof. dept. city planning Ga. Inst. Tech., Atlanta, 1972—, asst. dean, dir. of extension, 1979-82, dir. continuing edn., 1982—, assoc. v.p. 1983-90, special asst. Office of the Pres. 1990-92, head sennsory spatial systems group 1992—; clin. prof. Sch. of Pub. Health, Emory U., Atlanta, 1979—; adj. prof. Auburn U., 1981—; pres. C.R. Bragdon & Assos., environ. planning. Cons. to office noise abatement U.S. EPA, 1972—, also FAA; cons. constrn. research lab. U.S. Army C.E., 1973—; Pres. Carriage Hill Civic Assn., 1977-78, Friends of Redan, 1985—; mem. Atlanta Urban Design Commn., 1979-81; bd. dirs. St. Luke's Tng. and Counseling Center, 1980-83; chmn. DeKalb County Rapid Transit Sta. Planning Task Force, 1976-80, Aviation Consortium for Edn. and Tng., 1992—; mem. adv. bd. Internat. Cultural Center, Inc., 1979; mem. council advs. Nat. Sci. Ctr. for Communications and Electronics, 1984-89; bd. dirs. Network Instrnl. TV, 1983-86; mem. Lincoln Inst. for Land Policy, Harvard U., 1985—. Served to capt. U.S. Army, 1969-72. Fellow Acoustical Soc. Am.; mem. Nat. Acad. Sci., Am. Indsl. Hygiene Assn. (pres. Ga. chpt. 1973-74), Am. Nat. Standards Inst., Am. Planning Assn. (pres. Ga. chpt. 1979-81), Am. Soc. Planning Ofcls., Am. Inst. Cert. Planners, Assn. Energy Engrs. (dir.), ASCE, Transp. Research Bd., Nat. Trust Hist. Preservation, Urban Land Inst., World Future Soc., Air Pollution Control Assn., Ga. Conservancy, Sigma Xi, Omicron Delta Kappa, Kappa Alpha Order. Author: Noise Pollution: The Unquiet Crisis, 1972; Noise Pollution: A Guide to Information Sources, 1979; General Aviation Airport Noise and Land Use Planning, 1979; Municipal Noise Legislation: 1980, 1980; Airport Land Use Planning and Noise Control, 1993. Contbr. chpt. to Environ. Health, 1979, Politics of Neglect, 1974, Airport Noise Planning Transp. Noise Ctrl. Handbook. Contbg. editor Sound and Vibration, 1974—; adv. bd. Airport Noise Reporter. Home: 741 Weatherborn Pl Stone Mountain GA 30083-4735 Office: Ga Inst Tech Prof City Planning 490 10th St Ste # 301 Atlanta GA 30332

BRAGINSKY, STANISLAV IOSIFOVICH, physicist, geophysicist, researcher; b. Moscow, Apr. 15, 1926; s. Iosif Samuilovich Braginsky and Khaya Nutovna Drikker; m. Maya Aronovna Boyarskaya, May 8, 1955; children: Galina, Leonid. Degree in engring. and physics, Moscow Inst. of Mechs., 1948; cand. sci. in physics and math., Inst. Atomic Energy, Moscow, 1953, DSc in Physics and Math., 1966. Sr. scientist I.V. Kurchatov Inst. of Atomic Energy, Moscow, 1948-78, O. Yu. Schmidt Inst. of Physics of the Earth, Moscow, 1978-88; geophysicist, researcher Inst. Geophysics/Planetary Physics UCLA, 1989—. Contbr. over 25 papers on plasma physics to profl. jours., over 60 papers on geophysics to profl. jours. Recipient Lenin prize for rsch. in plasma physics Acad. Sci. USSR, 1958, John Adam Fleming medal for rsch. in geomagnetism Am. Geophys. Union, 1993. Achievements include development of two-temperature equations of plasma dynamics and theory of the pinch-effect in high power electrical discharges in gases; advancement of theory of hydromagnetic dynamo of the Earth and theory of geomagnetic secular variations. Office: U Calif Inst Geophysics & Planetary Physics 405 Hilgard Ave Los Angeles CA 90024

BRAHMBHATT, SUDHIRKUMAR, chemical company executive; b. Dabhoi, Gujarat, India, Apr. 4, 1951; came to U.S., 1973; s. Ramanlal Kalidas and Kamalaben Motilal Barot Brahmbhatt; m. Ashaben Amarsingh, May 22, 1977; children: Tejal Sudhirkumar, Nisha Sudhirkumar. B in Chem. Engring., Nadiad Inst. Tech., India, 1973; M in Chem. Engring., Steven Inst. Tech., 1975; MBA in Internat. Mgmt. and Mktg., Fairleigh Dickinson U., 1982; PhD in Chem. Engring., Kennedy Western U., 1991. Rsch. asst. Stevens Inst. Tech., Hoboken, N.J., 1975-77; chem. engr. Exxon Co. U.S.A., Linden, N.J., 1977-79; sr. process engr. Air Products and Chemicals, Inc., Allentown, Pa., 1979-84; applications engr. MG Industries div. of Hoechst, Valley Forge, Pa., 1984-87, sr. project engr., 1987-89, mgr. chems. group, 1989-92, head R&D dept., 1992—; owner Ashutej Co., Trexlertown, Pa., 1982-84; pres., founder Bal Vihar Sch., St. Louis, 1992—. Patentee in environ. and chem. engring. fields; contbr. articles to profl. jours. Dir., host radio program Music of India, WMUH, Allentown, 1981-91, KDHX, St. Louis, 1992—; pres. Exxon Volleyball League, Linden, 1978-79; pres. Bal Vihar Assn., Hindu Temple Soc., Allentown, Pa., 1989-91; founder, pres. Bal Vihar (Children's Ethnic Sch.) of St. Louis, 1992—. Recipient Merit cert. Poly-Olefins Industries Ltd., Bombay, India, 1972. Mem. AIChE, TAPPI, Am. Powder Metallurgy Inst. (chmn. Phila. sec. 1987-88), Am. Chem. Soc., Am. Ceramic Soc., Am. Soc. Metals. Avocations: cultural programs, radio, overseas travel. Home: 1700 Countryside Ct Glencoe MO 63038-1446 Office: MG Industries #6 Research Park Dr Saint Charles MO 63304

BRAHTZ, JOHN FREDERICK PEEL, civil engineering educator; b. St. Paul, Jan. 29, 1918; s. John Henry August Brahtz and Charlotte Beatrice Peel; m. Lise Vetter, May 11, 1991. BA, Stanford U., 1939, MS, 1948, PhD, 1951. Registered profl. civil and mech. engr., Calif. Various engring. positions Calif., 1939-53; assoc. prof. UCLA, 1953-57; v.p., dir. engring. J.H. Pomeroy & Co. Inc., San Francisco, L.A., 1957-60; mgr. constrn. scis. dvsn. Stanford Rsch. Inst., Menlo Park, Calif., 1960-63; staff cons. U.S. Naval Civil Engring., U.S. Naval Elec. Labs., Port Hueneme, San Diego, Calif., 1963-70; lectr. UCLA, 1963-70; dir. constrn. systems inst. Calif. State U., San Diego, 1970-73; vis. prof. ocean engring. U. Calif., San Diego, 1986-87; cons. rsch. prof. Civil Engring. Stanford U., Palo Alto, Calif., 1987—; cons. various orgns., San Francisco, N.Y., San Diego, Chgo., 1964—. Co-author editor (books) Ocean Engineering: System Planning and Design, 1968, Coastal Zone Management: Multiple Use with Conservation, 1972; editor (book series) Construction Management and Engineering Series, 1986—; patentee in field. Comdr. USN, 1941-46. Fellow ASCE; mem. Am. Soc. for Engring. Edn., Old Capital Club (Monterey, Calif.), Beach and Tennis Club (Pebble Beach, Calif.), La Jolla Beach and Tennis Club. Home: 800 Prospect St La Jolla CA 92037 Office: 2740 16th Ave Carmel CA 93923

BRAID, MALCOLM ROSS, biology educator; b. Balt., June 7, 1947; s. Robert Bruce and Elva Dawn (Outland) B.; m. Linda Lee Grimm, Jan. 22, 1972. BS in Biology, U. Montevallo, 1969; MS in Fisheries Mgmt., Auburn U., 1974, PhD in Fisheries Mgmt., 1977. Biological scis. asst. Neurophysiology sect. Edgewood (Md.) Arsenal, 1970-72, rsch. biologist Ecology sect., 1972; asst. prof. U. Montevallo, Ala., 1977-82; assoc. prof. Auburn U., Montevallo, Ala., 1982-88, chmn., 1990-91, prof., 1989--; liaison Marine Environ. Scis. Consortium, Dauphin Island, Ala., 1990--. Contbr. articles to profl. jours. With U.S. Army, 1970-72. U. Montevallo rsch. grantee 1985, 87, USDA Forest Svc. grantee 1993. Mem. Ala. Acad. Scis., Beta Beta Beta, Gamma Sigma Delta, Sigma Xi. baptist. Home: 340 Comanche St Montevallo AL 35115 Office: U Montevallo Dept Biology Sta 6480 Montevallo AL 35115

BRAIDA, LOUIS BENJAMIN DANIEL, electrical engineering educator. Henry Ellis Warren prof. elec. engring. MIT, Cambridge. Office: MIT Dept Elec Engring & Computer Sci Cambridge MA 02139

BRAIDEK, JOHN GEORGE, agriculturist. With Western Prodr., Saskatoon, Sask., Can. Recipient AIC Fellowship award Agrl Inst. Can., 1992. Office: The Western Producer, PO Box 2500, Saskatoon, SK Canada S7K 2C4*

BRAITHWAITE, CLEANTIS ESEWANU, molecular biologist; b. Freetown, Sierra-Leone; arrived in Can., 1978; d. James Edward and Juliet (Grant) B.; m. John Owusu, Mar. 1, 1985; children: Ronald Denziel. PhD, U. Man., Can., 1987; postdoctoral student, Tex. A&M U., 1989. Instr. biochemistry U. Man., Can., 1984-86; postdoctoral fellow Tex. Agrl. Exptl. Sta., College Station, 1986-88, rsch. assoc., 1988-90; rsch. biochemist Coop. Agrl. Rsch. Ctr., Houston, 1990—; mentor high sch. students Prairie View and Hempstead, Tex. Contbr. articles to Can. Jour. Biochemistry and Cell Biology, other profl. publs. Mem. Am. Soc. Microbiology, Can. Soc. Microbiology, Sigma Xi, Beta Beta Beta. Achievements include construction of diagnostic DNA probe for bovine brucellosis, isolation and characterization of diagnostic markers of corynebacterium, pseudotuberculosis for goat disease, caseous lymphademitis. Office: Prairie View A&M U Prairie View TX 77467

BRAJDER, ANTONIO, electrical engineer, engineering executive; b. Zagreb, Croatia, Sept. 1, 1942; arrived in Germany, 1974; s. Ernest and Mira (Goldoni) B.; m. Miroslava Brozović, Apr. 21, 1973; 1 child, Zrinka. Diploma engring., U. Zagreb, 1967. R&D engr. Inst. Rade Končar, Zagreb, 1967-69, group leader, 1970-74; R&D engr. Siemens AG, Erlangen, Germany, 1974-78, group leader, 1978-84, sr. engr., 1984-90, dep. dir., 1990—. Co-author: Tragedy in the Universe, 1961; contbr. numerous articles to profl. jours., papers to internat. confs.; patentee (20). Mem. Pugwash, Matica Hrvatska, Conf. Internat. Grands Reseaux Electriques. Roman Catholic. Avocations: astronomy, fine arts, diving. Home: Falkenstr 20, 91056 Erlangen Germany

BRAKEL, LINDA A. WIMER, psychoanalyst, researcher; b. N.Y.C., Sept. 6, 1950; d. Walter and Paula (Marcus) Wimer; m. C. Arthur Brakel, May 22, 1983. BA, SUNY, Binghamton, 1972; MD, Tufts U., 1976. Cert. psychoanalyst. Adj. instr. psychiatry U. Mich., Ann Arbor, 1982-88, adj. asst. prof., 1988—; vis. prof. in psychiatry and psychology Mich. State U., East Lansing, 1988-92; program dir. Mich. Psychoanalytic Inst., Southfield, 1991-93. Contbr. articles to Psychoanalytic Quarterly, Jour. the Am. Psychoanalytic Assn. (Jour. prize 1988, 91), Internat. Jour. Psychoanalysis. Mem. AAAS, Am. Psychoanalytic Assn., Am. Psychiat. Assn. Achievements include research in bridging cognitive science and psychoanalysis, converging evidence for unconscious processes, differences between conscious and unconscious category organization. Office: 525 3rd St Ann Arbor MI 48103

BRAKER, WILLIAM PAUL, aquarium executive, ichthyologist; b. Chgo., Nov. 3, 1926; s. William Paul and Minnie (Wassermann) B.; m. Patricia Reese, Sept. 2, 1950. B.S., Northwestern U., 1950; M.S., George Washington U., 1953; student, U. Chgo., 1954-58. Mem. staff John G. Shedd Aquarium, Chgo., 1953—; dir. John G. Shedd Aquarium, 1964—; asst. sec. Shedd Aquarium Soc., 1960-65, sec., 1965—. Served with AUS, 1950-52. Mem. Am. Fisheries Soc., Am. Soc. Icthyologists and Herpetologists, Am. Assn. Zool. Parks and Aquariums, Soc. for Marine Mammalogy. Office: John G Shedd Aquarium 1200 S Lake Shore Dr Chicago IL 60605-2435

BRAMAN, HEATHER RUTH, technical writer, editor, consultant, antiques dealer; b. Wilmington, Ohio, Apr. 27, 1934; d. William Barnett and Violet Ruth (Davis) Hansford; m. Barr Oliver Braman, June 29, 1957 (div.); children: Sean Robert, Heather Paige. BA, Hiram Coll., 1956; postgrad., Sinclair Community Coll., Dayton, Ohio, 1977-85, Wright State U., Dayton, 1986. Pers. clk. USAF, Wright-Patterson AFB, Ohio, 1956, specifications editor, 1956-57, publs. editor, writer, 1957-63; vol. Children's Med. Ctr., 1963-67, Dayton Pubs. Schs., 1969-87; tchr. Gloria Dei Montessori Sch., Dayton, 1973-77; asst. mgr., acctg. mgr., mgr. tennis club USAF, Wright-Patterson AFB, Ohio, 1977-81; tech. writer Miclin, Inc., Alpha, Ohio, 1982, Indsl. Design Concepts, Dayton, 1982-83; tech. writer, cons. Belcan Corp., Cin., 1984—; owner Chimney Sweep Antiques Shoppe, Arcanum, Ohio, 1991—; real estate investor. Founder, bd. dirs. Trotwood (Ohio) Women's Open Tennis Tournament, 1976-81; mem. Harrison Twp. Parks Bd., 1980-82; ballpersons coord. Dayton Pro Tennis Classic, 1977-80; pres. Dayton Tennis Commn., 1978-80; mem. parents exec. com. Hiram (Ohio) Coll., 1985—; ct.-appointed Spl. Advocate/Guardian Ad Litem (CASA GAL), 1988—; tutor English as a second lang. citizenship classes, 1991—. Mem. NOW, NAACP, Dayton Pub. Schs. Orgns., Dayton Tennis Umpires Assn., Mothers Against Drunk Drivers., AARP, WWF, HALT, Sigil of Phi Sigma. Democrat. Mem. Soc. Friends. Avocations: tennis, antiques, reading, property investment. Home: 320 Elm Hill Dr Dayton OH 45415-2943 Office: Belcan Corp 10200 Anderson Way Cincinnati OH 45242-4700

BRAMAN, S. KRISTINE, entomologist; b. Rochester, N.Y., Sept. 28, 1956; d. George W. and Priscilla Rose (Alber) Rich; m. George Delos Braman, Dec. 29, 1979; children: George Rich, Charles Austin. BS in Forest Biology, SUNY, 1978; PhD in Entomology, U. Ky., 1987, posrgrad., 1987-88. Asst. prof. U. Ga., Griffin, 1989—. Contbr. articles to profl. jours. Mem. Am. Inst. Biol. Scis., Entomological Soc. Am., Internat. Orgn. Biol. Control, Ga. Entomological Soc. Achievements include research in ecology of indigenous predators in soybean; ecology and management of adventive azalea lace bugs; ecology of indigenous natural enemies in turf and landscape ornamentals. Office: U Ga Entomology Ga Sta Griffin GA 30223-1797

BRAMBLE, JAMES HENRY, mathematician, educator; b. Annapolis, Md., Dec. 1, 1930; s. Charles Clinton and Edith (Rinker) B.; m. Margaret Hospital Hays, June 25, 1977; children: Margot, Tamara, Mary, James; 1 stepchild, Myron A. Hays. A.B., Brown U., 1953; M.A., U. Md., 1955, Ph.D., 1958; D.Sc. (hon.), Chalmers U. Tech., Göteborg, Sweden, 1985. Mathematician Gen. Electric Co., Cin., 1957-59, Naval Ordnance Lab., White Oak, Md., 1959-60; asst. prof., assoc. prof., prof. U. Md., 1960-68; prof. Cornell U., Ithaca, N.Y., 1968—; dir. Center Applied Math., 1974-80; cons. Brookhaven Nat. Lab., 1976—; vis. prof. Chalmers U. Tech., Göteborg, 1970, 72, 73, 76, 86, U. Rome, 1966-67, Ecole Poly., Paris, 1978, Lausanne, Switzerland, 1979; vis. prof. U. Paris, 1981; lectr. in field. Chmn. editorial bd. Mathematics of Computation, 1975-84; contbr. articles profl. jours. Mem. Am. Math. Soc., Soc. Indsl. and Applied Math. Offices: Cornell U Dept of Math Ithaca NY 14853

BRAMLET, ROLAND CHARLES, radiation physicist; b. Wallowa, Oreg., June 11, 1921; s. Charles David and Edith Pearl (Downard) B.; married; children: Bryan, Steven. MS, NYU, 1961; PhD, St. John's U., 1966. Jr. physicist Queens Gen. Hosp., Jamaica, N.Y., 1959-61, physicist, 1961-69; chief physicist Highland Hosp., Rochester, N.Y., 1969-91; clin. asst. prof. dept. radiation oncology U. Rochester, 1985-91; cons. physicist Finger Lakes Radiotherapy Assocs., Clifton Springs, N.Y., 1991—. Co-author: Basic Nuclear Medicine, 1975; contbr. articles to profl. publs. With U.S. Army, 1942-45, ETO. Sloan-Kettering Inst. grad. fellow, 1958. Mem. Am. Assn. Physicists in Medicine. Achievements include patent for radiation detection and imaging machine. Home: 226 Castleman Rd Rochester NY 14620 Office: Finger Lakes Radiotherapy Assn 7 Ambulance Dr Clifton Springs NY 14432

BRAMON, CHRISTOPHER JOHN, aerospace engineer; b. Marshalltown, Iowa, Oct. 15, 1960; s. Clayton Robert and Judith (Rolston) B.; m. Joy Frances Tomishima, May 4, 1985. BS in Indsl. Engring., Iowa State U., 1984. Mgr. tech. engring. ops. NASA Marshall Space Flight Ctr., Huntsville, Ala., 1985-87, tech. mgr. liquid propulsion system, 1987—. Contbr. article to Aerospace Engring. Recipient Astronaut's Personal Achievement award NASA, 1989, dir.'s commendation NASA Marshall Space Flight Ctr., 1992. Mem. Inst. Indsl. Engrs. Office: NASA Marshall Space Flight Ctr EE31 Huntsville AL 35812

BRAMSON, ROBERT SHERMAN, lawyer; b. N.Y.C., Nov. 11, 1938; s. Oscar David and Gertrude (May) B.; m. Ruth Schaffer, June 27, 1942; children: Jonathan, Jennifer, James, Julia. B.M.E., Rensselaer Poly. Inst., 1959; J.D., Georgetown U., 1963; postgrad., U. Chgo. Sch. Bus., 1963-64. Bar: Ill. 1963, Pa. 1968, N.Y. 1984. Patent examiner U.S. Patent Office, Washington, 1959-60; patent agt. Stevens, Davis, Miller & Mosher, Washington, 1960-63; atty. Abbott Labs., North Chicago, Ill., 1963-66, Scott Paper Co., Phila., 1966-68; ptnr., head computer and tech. law group Schnader, Harrison, Segal & Lewis, Phila., 1968-89; v.p., gen. patent and tech. counsel Unisys Corp., Blue Bell, Pa., 1989-90; founder Robert S. Bramson and Assocs., Bala Cynwyd, Pa., 1991—; pres., CEO InterDigital Technology Corp., King of Prussia, Pa., 1992—; adj. prof. Rutgers U. Law Sch., Camden, N.J., Temple U. Law Sch., Phila. Mem. ABA, Internat. Bar Assn., Am. Law Inst., Am. Patent Law Assn., Phila. Patent Law Assn., Phila. Bar Assn. Club: Racquet (Phila.). Home: 121 Edgehill Rd Bala Cynwyd PA 19004-3148 Office: Ste 105 2200 Renaissance Blvd King Of Prussia PA 19406

BRANCA, ANDREW ANGELO, biochemist, educator; b. New Rochelle, N.Y., Aug. 20, 1950; s. S. Leo and Mary (Lange) B.; m. Gail Mayer, Feb. 2, 1974; children: Erica, Christopher, Adrienne, Angela. BA, N. Adams (Mass.) State Coll., 1975; MS, U. N.H., 1978, PhD, 1980. Rsch. technician Olin Chem. Co., New Haven, Conn., 1973-74; from grad. teaching asst. to grad. rsch. asst. U. N.H., Durham, 1975-80; rsch. assoc. SUNY, Albany, 1980-82; asst. prof. Albany Med. Coll., 1982-89; sr. scientist Procyte Corp., Kirkland, Wash., 1989—; immunobiology grant reviewer Nat. Cancer Inst., Bethesda, Md., 1986; manuscript reviewer Jour. Interferon Rsch., N.Y.C., 1986-90, Proc. Nat. Acad. Scis., Washington, 1990. Contbr. articles to Jour. Biol. Chemistry, Nature, Proceedings NAS, In Vitro Cellular and Devel. Biology, others. Recipient Pub. Health Svc. award Nat. Cancer Inst., 1981, New Investigator award Nat. Inst. Allergies and Infectious Diseases, 1985; Nat. Cancer Inst. fellow, 1981. Mem. AAAS, Am. Soc. Biochemistry and Molecular Biology, Internat. Soc. Interferon Rsch., Tissue Culture Assn., Wound Healing Soc., Sigma Xi. Achievements include pioneering research on human interferon receptors with development of the first radiolabeling techniques; discovery of pathway of interaction of human interferon with its cellular receptor. Office: Procyte Corp 12040 115th Ave NE Kirkland WA 98034-6900

BRANCH, JOHN CURTIS, biology educator, lawyer; b. Buffalo, Okla., Oct. 1, 1934; s. Ernest Samuel and Ethel Imogene (Parsons) B.; m. Jacqueline Joyce Davis, July 20, 1960; children: Kim Renee, Karla Jean, Kay Lynn. BS, Northwestern Okla. State U., 1959; MS, U. Okla., 1963, PhD, 1965; JD, Okla. City U., 1980. Bar: Okla. 1980. Asst. prof. biology dept. Okla. City U., 1964-67, assoc. prof. biology dept., 1967-75, prof. biology dept., 1975—. With U.S. Army, 1955-57. Mem. Okla. County Bar Assn., Okla. Acad. Sci., Okla. Bar Assn., Beta Beta Beta. Methodist. Avocations: hunting, fishing. Home: 2705 Abbey Rd Oklahoma City OK 73120-2702 Office: John C Branch PC 6803 S Western Ave # 300 Oklahoma City OK 73139-1814 Office: Okla City U Dept Biology 2501 N Blackwelder Oklahoma City OK 73106

BRAND, LARRY MILTON, biochemist; b. Mpls., Sept. 18, 1949; s. Arthur Leonard and Raleigh (Murphy) B.; m. Regina Fenster, Jan. 27, 1974; children: Ilana Michele, Sarah Eve, Alicia Leah. BS magna cum laude, U. Minn., 1971; PhD, U. Wis., 1975. NIH rsch. fellow Brandeis U., Waltham, Mass., 1975-77; rsch. biochemist Procter and Gamble Co., Cin., 1977-81, head sect. toxicology, 1981-83, head sect. inflammation, 1983-85, assoc. dir. R&D, 1985-90, assoc. dir. licensing and acquisitions, 1990—. Contbr. articles to Biochemistry, Agents and Actions, other jours. Babcock fellow U. Wis., 1972; recipient Nat. Rsch. Svc. award NIH, 1976. Mem. Am. Coll. Toxicology, Assn. Chemoreception Scis., Inflammation Rsch. Assn., Licensing Execs. Soc., Am. Assn. Pharm. Scientists (charter), Phi Lambda Upsilon. Jewish. Achievements include patent for new type of artificial sweetener, substitute for brominated vegetable oil, analgesic combinations using capsaicin. Home: 4290 Berryhill Ln Blue Ash OH 45242 Office: Procter and Gamble Co Miami Valley Labs PO Box 398707 Cincinnati OH 45239-8707

BRAND, VANCE DEVOE, astronaut, government official; b. Longmont, Colo., May 9, 1931; s. Rudolph William and Donna (DeVoe) B.; m. Joan Virginia Weninger, July 25, 1953; children: Susan Nancy, Stephanie, Patrick Richard, Kevin Stephen; m. Beverly Ann Whitnel, Nov. 3, 1979; children—Erik Ryan, Dane Vance. B.S. in Bus., U. Colo., 1953, B.S. in Aero. Engring., 1960; M.B.A., UCLA, 1964; grad., U.S. Naval Test Pilot Sch., Patuxent River, Md., 1963. With Lockheed-Calif. Co., Burbank, 1960-66; flight test engr. Lockheed-Calif. Co., 1961-62, traveling engr. rep., 1962-63, engring. test pilot, 1963-66; astronaut NASA Johnson Space Ctr., Houston, 1966—; command module pilot Apollo-Soyuz mission NASA Johnson Space Ctr., 1975, comdr. STS-5 Mission,, 1982, comdr. STS 41-B Mission,, 1984, comdr. STS-35 Mission, 1990; chief plans Nat. Aero-Space Plane Joint Program Office, Wright-Patterson AFB, Ohio, 1991—. Served with USMCR, 1953-57. Decorated 2 Disting. Svc. medals NASA, 2 Exceptional Svc. medals, 3 Space medals. Fellow Am. Astron. Soc., Soc. Exptl. Test Pilots, AIAA. Office: ASC/NAX Wright-Patterson AFB Dayton OH 45433

BRANDALISE, SILVIA REGINA, pediatrician; b. Sao Paulo, Brazil, Mar. 26, 1943; d. Antonio Correia and Maria de Lourdes (Vieira) Violante; m. Nelson Ary Brandalise, Sept. 23, 1969; children: André, Fernando, Paola, Marcos. MD, Med. Sch. Sao Paulo, 1967. Resident Med. Sch. Sao Paulo, 1969; prof. pediatrics State U. Campinas, Sao Paulo, 1970—, chief pediatric hematology div., 1978—, coord. rsch. oncology, 1988—; coord. rsch. oncology Brazilian Coop. Group Treatment of Leukemia in Children, 1980—; dir. Centro Info. Boldrini, Campinas, 1978—. Contbr. articles to profl. jours. Recipient honor award Rotary, 1985, honors award Campinas Chamber, 1990, Campinas Med. Soc., 1991, Brazilian Army, 1991, Order of Merit, Bahia League Against Cancer, 1992. Fellow Brazilian Soc. Pediatric Oncology (bd. dirs. 1990—, founder); mem. Am. Soc. Pediatric Hematology and Oncology, Internat. Soc. Imuno Comp. Host, Latin Am. Soc. Pediatric Hematology and Oncology (founder), Internat. Soc. Pediatric Oncology, Am. Soc. Clin. Oncology, Brazilian Soc. Hematology and Hemotherapy, Brazilian Coll. Hematology, Brazilian Soc. Cancer. Home: Av Atilio Martini 454, 13083 Campinas São Paulo, Brazil Office: State U Campinas, 13083 Campinas Sã Paulo, Brazil

BRANDELL, SOL RICHARD, electrical power and control system engineer, research mathematician. Studied piano with Norman Masloff, Juilliard Sch. Music, 1932-41; studied with Sir John Barbirolli, La Follette Sch. Music, N.Y.C., 1939; student, U. Cin., 1943-44, U. Paris, 1945; BEE, CUNY, 1949; postgrad., Poly. Inst. Bklyn., 1952, CUNY, 1954-58. Registered profl. elec. engr. and control systems engr., Calif.; lic. profl. engr., N.Y.; registered profl. engr., N.D. Elec. field engr., designer, estimator Rao Elec. Equipment Co., N.Y.C., 1947-50; elec. designer Edward E. Ashley, P.E., N.Y.C., 1950-51; elec. design engr. Wearn, Vreeland, Carlson, and Sweatt, N.Y.C., 1951-52; sr. elec. engr. Bechtel Assocs., N.Y.C., 1952-57; chief elec. engr. Am. Hydrotherm Corp., N.Y.C., 1957-76; supervising elec. engr. Heyward-Robinson Co., N.Y.C., 1976-77; prin. mem. tech. staff Ralph M. Parsons Co., Pasadena, Calif., 1977-91; pvt. cons., electric power rsch. engr., mathematician Alexandria, Va., 1991—. Co-author: Analysis of Harmonic Pollution on Power Distribution Systems, 1989; contbr. rsch. papers to Am. Hydrotherm Corp. With U.S. Army, 1942-46, ETO. Decorated Bronze Star medal, ETO Campaign medal with 2 bronze battle stars; N.Y. State War Vet. scholar, 1949. Mem. IEEE (life, sr.), VFW, DAV, Combat Infantrymen's Assn., Am. Math. Soc., Sigma Xi. Achievements include patents in electric power applications; research in solid state electronic annunciators, analog to discrete variable conversion system for chemical process temperature control, extremely reliable low-voltage electrical power generating stations for the uninterruptible supply of large-scale air-route traffic control center operations, mathematical modeling and computational harmonic analysis of nonsinusoidal energy flow in electrical power systems, on the decomposition of harmonic distortion power, in various applications of Bessel's equation including experimental work in low frequency eddy-current heating of process liquids in pipes and vessels. Home: 5206 Dover Pl Alexandria VA 22311-1204

BRANDENSTEIN, DANIEL CHARLES, astronaut, naval officer; b. Watertown, Wis., Jan. 17, 1943; s. Walter C. and Agnes (Holzworth) B.; m. Jane A. Wade, Jan. 2, 1966; 1 dau., Adelle. B.S., U. Wis., River Falls, 1965; postgrad., U.S. Naval Text Pilot Sch., Patuxent River, Md., 1971. Commd. officer U.S. Navy, 1965, advanced through grades to capt.; student aviator U.S. Navy, Pensacola, Fla., 1965-67; aviator U.S. Navy, Whidbey Island, Wash., 1967-71; test pilot U.S. Navy, Patuxent River, Md., 1971-74; aviator U.S. Navy, Whidbey Island, Wash., 1974-78; astronaut NASA Johnson Space Ctr., Houston, 1978—; chief astronaut office NASA Johnson Space Ctr., 1987—. Decorated Legion of Honor (France); recipient 26 medals and award USN, 1968-71; recipient Disting. Alumnus award U. Wis., 1982, Space Flight medal NASA, 1982. Mem. AIAA (Haley Space Flight award 1993), Soc. Exptl. Text Pilots, U.S. Naval Inst. Office: NASA Lyndon B Johnson Space Ctr Houston TX 77058

BRANDHORST, BRUCE PETER, biology educator; b. Galveston, Tex., Nov. 14, 1944; s. William Schroeder Brandhorst and Emilie Pontz Pickering; m. Elaine Golds; children: Gregory, Gary. AB, Harvard U., 1966; PhD, U. Calif., San Diego, 1971. Rsch. assoc. U. Colo., Boulder, 1971-73; from asst. to assoc. to prof. McGill U., Montreal, Que., Can., 1973-89; prof., dir. Inst. Molecular Biology and Biochemistry Simon Fraser U., Burnaby, B.C., Can., 1989—; instr. embryology Marine Biol. Lab., Woods Hole, Mass., 1980-82, co-dir. embryology course, 1983-88. Mem. editorial bd. (serials) Molecular and Cellular Biology, 1985-91, Molecular Reproduction and Development, 1988—. Rsch. grantee Natural Scis. and Engring. Rsch. Coun., Can., 1973—, NIH, McGill U., 1984-90. Mem. Am. Soc. for Microbiology, Can. Soc. for Cell and Molecular Biology, Soc. for Developmental Biology, Marine Biology Lab. Corp. Achievements include research in regulation of gene expression in developing embryos. Office: Simon Fraser U, Inst Molecular Biol, Burnaby, BC Canada V5A 1S6

BRANDI, MARIA LUISA, endocrinologist, educator; b. Viterbo, Lazio, Italy, July 31, 1953; d. Domenico and Livia (Guadagni) B.; m. Vittorio de Leonardis, June 23, 1987; 1 child, Brando. MD, U. Florence (Italy), 1977, postgrad. in Endocrinology, 1980, PhD in Cell Biology, 1988. Cert. Italian Nat. Bd. Endocrinology. Rsch. fellow in Endocrinology U. Florence, 1977-84, rsch. assoc. Medicine, 1980-84, 88-92, assoc. prof. clin. pathophysiology dept., 1992—; vis. scientist metabolic diseases br. Nat. Inst. Diabetes Digestive and Kidney Diseases, NIH, Bethesda, Md., 1984-88. Recipient award Premio Roussel Italia, 1988, European Osteoporosis Found., 1989, Premio Schering della Società Italiana di Endocrinologia, 1990, Internat. Gerontol. award Sandoz Found. for Gerontol. Rsch., 1991. Mem. Am. Soc. Cell Biology, Am. Soc. Bone and Mineral Rsch., Endocrine Soc., Italian Endocrine Soc., Italian Osteoporosis Soc. Avocations: travel, gymnastics. Home: Via Chiantigiana 133/C, 50126 Florence Italy Office: Dept Clin Pathophysiology, Viale Pieraccini 6, 50139 Florence Italy

BRANDIN, ALF ELVIN, retired mining and shipping company executive; b. Newton, Kans., July 1, 1912; s. Oscar E. and Agnes (Larsen) B.; m. Marie Eck, June 15, 1936 (dec. 1980); children: Alf R., Jon, Erik, Mark.; m. Pamela J. Brandin, Jan. 28, 1983. A.B., Stanford U., 1936. With Standard Accident of Detroit, 1936-42; bus. mgr. Stanford U., Calif., 1946-52; bus. mgr., exec. officer for land devel. Stanford U., 1952-59, v.p. for bus. affairs, 1959-70; sr. v.p., dir., mem. exec. com. Utah Internat. Inc., San Francisco, from 1970; pres. Richardson-Brandin, 1964-86, also bd. dirs.; bd. dirs. Hershey Oil Co.; vice chmn. bd. dirs. Doric Devel. Inc. Bd. govs. San Francisco Bay Area Council; trustee Reclamation Dist. 2087, Alameda, Calif.; bd. overseers Hoover Instn. on War, Revolution and Peace, Stanford; mem. VIII Olympic Winter Games Organizing com., 1960. Served as comdr. USNR, 1942-46. Mem. Zeta Psi. Clubs: Elk, Stanford Golf, Bohemian, Pauma Valley Country, Silverado Country; Royal Lahaina. Home: 668 Salvatierra St Palo Alto CA 94305-8538 Office: 550 California St San Francisco CA 94104-1006

BRANDON, ROBERT NORTON, zoology and philosophy educator; b. Concord, N.C., Apr. 25, 1952; s. Charles William and Myrtle (Norton) B.; m. Gloria Meares, Aug. 27, 1977; 1 child, Katherine Tierny. BA, U. N.C., 1974; PhD, Harvard U., 1979. Asst. prof. philosophy Duke U., Durham, N.C., 1979-84; assoc. prof. Duke U., 1984-89, assoc. prof. philosophy and zoology, 1989—; vis. prof. U. Pitts., 1984. Author: Adaptation and Environment, 1990; co-editor: Genes, Organisms, Populations: Controversies over the Units of Selection, 1984; contbr. articles to profl. jours. Fellow AAAS; mem. Phi Beta Kappa. Office: Duke U Dept Philosophy Durham NC 27708

BRANDT, I. MARVIN, chemical engineer; b. Shreveport, La., Nov. 26, 1942; s. David and Esta (Epstein) B. BS in Chemistry, Centenary Coll., 1965; postgrad., U. Tex., 1968-70. Gen. mgr. Am. Pipe and Supply, Shreveport, 1970-73; researcher Shell Oil, Houston, 1973-75; rsch. tech. svc. trainer NL Baroid, Houston, 1975-79; researcher, tech. svc. engr. Arco Oil & Gas Co., worldwide, 1979-86; specialist, project mgr. Petrolite Corp., St. Louis, 1986-90; sr. engr., tng. mgr. Marathon Oil Co., Houston, 1990—; cons. Marathon Oil Co. Dallas, South Am., Cen. Am., Calif., Russian, N. Sea, Mid. East, Africa, Alaska; cons. for various drilling and environ. clean-up cos., Tex., Calif. Contbr. articles to profl. jours.; patentee in field. Active Am. Cancer Soc., Houston, Denver, St. Louis, Morris Animal Found., Denver. Recipient Grad. Tching. fellowship, U. Tex., Austin, 1968-70, Robert Welch Rsch. grant, U. Tex., Austin, 1969. Mem. Soc. Petroleum Engrs., Internat. Assn. Drilling Contractors, Am. Chem. Soc., Am. Petroleum Inst. (numerous subcoms. for drilling and environment), Am. Assn. Drilling Engrs. (planning com. for Petro-Safe, com. chmn., chmn. drilling com.), N.Y. Acad. Scis., Internat. Platform Soc. Avocations: tennis, running, bicycling, music, fishing. Home: 2803 Greens Ferry Ct Richmond TX 77469 Office: 5555 San Felipe Houston TX 77253

BRANDT, KATHLEEN WEIL-GARRIS, art history educator; d. Kurt H. and Charlotte (Garris) Weil; m. Werner Brandt (dec. 1983). BA with honors, Vassar Coll., 1952; postgrad., U. Bonn, 1956-57; MA, Radcliffe Coll., 1958; PhD in Art History, Harvard U., 1966. Asst. prof. art history NYU, N.Y.C., 1966-67, assoc. prof. art history, 1967-72, prof. art history, 1973—; vis. prof. Harvard U., Cambridge, Mass., 1980; Renaissance cons. Vatican Mus., Vatican City, 1987—; lectr., tchr. on art history, conservation of art and architecture. Editor-in-chief The Art Bull., 1977-81; author: Leonardo and Central Italian Art, 1974, The Santa Casa di'Loreto, 1977; co-author: The Renaissance Cardinal's Ideal Palace, 1980; contbr. articles to profl. publs. Fellow NEH, 1972, 81-84, J.S. Guggenheim Found., 1976, Henkel fellow Max Plack Inst., 1987; recipient sr. rsch. prize Alexander von Humboldt Found., 1985; named Office of Order of Merit, The Italian Republic, 1993. Fellow N.Y. Acad. Scis.; mem. Met. Soc. History of Sci., Coll. Art Assn. Am. (bd. dirs. 1973-74, 77-81), Renaissance Soc. Am. (bd. editors 1993—). Office: NYU Inst Fine ARts 1 E 78th St New York NY 10021

BRANDT, REINHARD, chemist; b. Konigsberg, Germany, Nov. 14, 1932; s. Melchior Sebastian and Ruthilt (Mannesmann) B.; diplom chemiker Frankfurt U., 1959; Ph.D., U. Calif., Berkeley, 1963; m. Magdalene Brandt-Geller, Nov. 6, 1961; children: Johann-Friedrich, Dorothea Desiree. Postdoctoral fellow CERN, Geneva, 1963-68; mem. faculty Marburg (Germany) U., 1968—, prof. chemistry, 1971—; mem. Adv. Commn. Nuclear Research Fed. Govt., 1976-83; sci. visitor Weizmann Inst., Rehovoth, Joint Inst. Nuclear Research, Dubna, USSR, Los Alamos Nat. Lab., Lawrence Berkeley Lab., CERN. Mem. Am. Chem. Soc., Gesellschaft deutscher Chemiker. Mem. Free Democratic Party. Contbr. articles to profl. jours.; mem. editorial bd. Nuclear Tracks, 1979—, Isotopenpraxis, 1989-92. Mem. Internat. Nuclear Track Soc. (pres. 1988-92). Research in heavy ion nuclear chemistry concentrating on anomalons. Office: Kernchemie Philipps U, 355 Marburg Germany

BRANHAM, RICHARD LACY, JR., astronomer; b. Balt., Mar. 19, 1943; s. Richard Lacy and Mary Magdalene (Mrkonjich) B.; m. Rosa Guadalupe Palma, July 15, 1972; 1 child, Maria Teresita. MA, Harvard U., 1968; PhD, Case Western Res. U., 1977. Astronomer U.S. Naval Obs., Washington, 1968-82; resident dir. Yale-Columbia So. Obs., Barreal, San Juan, Argentina, 1970-72; dir. computing, head math. area, substitute dir. Regional Ctr. for Sci. and Tech. Rsch., Mendoza, Argentina, 1982—; vis. prof. U. La Plata, 1989, U. San Juan, Argentina, 1989, 1990, 1992, U. Cuyo, Argentina, 1992; summer asst. NASA, Washington, 1967. Author: Scientific Data Analysis, 1990; contbr. articles to profl. jours. and pop. mags. Founding mem. Argentine Ctr. for the Investigation and Refutation of Pseudosci., Buenos Aires, 1990. Mem. Internat. Astron. Union, Am. Astron. Soc., Assn. for Computing Machinery, Argentine Astron. Assn., Phi Beta Kappa, Pi Mu Epsilon. Home: Filippini 516, 5501 Godoy Cruz Argentina Office: CRICYT, C C 131, 5500 Mendoza Argentina

BRANKAMP, ROBERT GEORGE, research biochemist; b. Cin., Nov. 24, 1961; s. Robert James and Janice Darlene (Cupp) B. BS, Ky. Wesleyan, 1984. Rsch. asst. U. Cin., 1984-88; rsch. biochemist Marion Merrell Dow Inc., Cin., 1988—. Contbr. articles to Jour. Biol. Chemistry, Jour. of Clin. and Lab. Medicine, Blood Coagulation and Fibrinolysis. Mem. AAAS, Am. Chem. Soc. Office: Marion Merrell Dow 2110 E Galbraith Rd Cincinnati OH 45237-1625

BRANNAN, MICHAEL STEVEN, civil, environmental, automotive engineer; b. Dodge City, Kans., July 16, 1946; s. Joseph Franklin and Fern Adele (Layman) B.; m. Joyce Ann Strode, Feb. 26, 1972; children: Heather Michelle, Kathleen Marie, Megan Elizabeth. BS in Geology, Kans. State U., 1969, BS in Civil Engring. and Environ. Sci., 1976; MS in Civil Engring., U. Okla., 1988. Registered profl. engr., Kans., S.D., Okla. Civil engr. Van Gundy & Assocs., Ellsworth, Kans., 1976-81; geohydrologist Phillips Design Engring., Bartlesville, Okla., 1981-88; sr. automotive engr. Phillips Petroleum Transport, Bartlesville, 1988-93; sr. engr. geohydrologic Phillips Petroleum Corp. Engring., 1993—; task force chmn. Am. Petroleum Inst., Washington, 1990—. Nat. dir. St. Francis Homes for Boys and Girls, Salina, Kans., 1980; chmn. City Traffic Com., Bartlesville, Okla., 1992—. Capt. U.S. Army, 1969-73, Vietnam. Mem. ASCE, Okla. Soc. Profl. Engrs. (membership v.p. 1992-93, Outstanding Engr. 1987), Bartlesville Amateur Radio Club (pres. 1988). Republican. Episcopalian. Achievements include patent for groundwater protection in deep cathodic protection wells; patent pending for deep formation storage of salt water; research in high resolution shallow seismic reflection for groundwater monitoring well location studies. Home: 3313 Wayside Bartlesville OK 74006 Office: Phillips Petroleum Corp Engring Geohydraulics 10 D3 Phillips Bldg Bartlesville OK 74004

BRANSCOMB, LEWIS MCADORY, physicist; b. Asheville, N.C., Aug. 17, 1926; s. Bennett Harvie and Margaret (Vaughan) B.; m. Margaret Anne Wells, Oct. 13, 1951; children—Harvie Hammond, Katharine Capers. AB summa cum laude, Duke U., 1945, DSc (hon.); MS, Harvard U., 1947, PhD, 1949; DSc (hon.), Poly. Inst. N.Y., Clarkson Coll., Rochester U., U. Colo., Western Mich. U., Lycoming Coll., U. Ala., Pratt Inst., Rutgers U., Lehigh U., U. Notre Dame, SUNY, Binghamton; LHD (hon.), Pace U. Instr. physics Harvard U., 1950-51; lectr. physics U. Md., 1952-54; vis. staff mem. Univ. Coll., London, 1957-58; chief atomic physics sect. Nat. Bur. Standards, Washington, 1954-60; chief atomic physics div. Nat. Bur. Standards, 1960-62; chmn. Joint Inst. Lab. Astrophysics, U. Colo., 1962-65, 68-69; chief lab. astrophysics div. Nat. Bur. Standards, Boulder, Colo., 1962-69; prof. physics U. Colo., 1962-69; dir. Nat. Bur. Standards, 1969-72; chief scientist, v.p. IBM, Armonk, N.Y., 1972-86; mem. corporate mgmt. bd. IBM, Armonk, 1983-86; dir. pub. policy program Kennedy Sch. Govt., Harvard U., Cambridge, Mass., 1986—, Albert Pratt pub. service prof., 1988—; mem.-at-large Def. Sci. Bd., 1969-72; mem. high level policy group sci. and tech. info. Orgn. Econ. Coop. and Devel., 1968-70; mem. Pres.'s Sci. Adv. Com., 1965-68, chmn. panel space sci. and tech., 1967-68; mem. Nat. Sci. Bd., 1978-84, chmn., 1980-84; mem. Pres.'s Nat. Productivity Adv. Com., 1981-82; mem. standing com. controlled thermonuclear research AEC, 1966-68; mem. adv. com. on sci. and fgn. affairs Dept. State, 1973-74; mem. U.S.-USSR Joint Commn. on Sci. and Tech., 1977-80; chmn. Com. on Scholarly Communications with the People's Republic of China, 1977-80; mem. tech. assessment adv. coun. Office of Tech. Assessment, U.S. Congress, 1990—; chmn. Carnegie Forum Task Force on Teaching as a Profession, 1985-86; dir. Mobil Corp., Lord Corp., Mitre Corp., Draper Labs., Inc. ; mem. pres.'s bd. visitors U. Okla., 1968-70; mem. astronomy and applied physics vis. coms. Harvard U. 1969-83, bd. overseers, 1984-86; mem. physics vis. com. M.I.T., 1974-79; mem. Pres.'s Com. Nat. Medal Scis., 1970-72; bd. dirs. Am. Nat. Standards Inst., 1969-72; trustee Carnegie Instn., 1973-90, mem. Carnegie Commn. on Sci., Tech. and Govt., 1988-93; trustee Poly. Inst. N.Y., 1974-78, Vanderbilt U., 1980—, Nat. Geog. Soc., 1984—, Woods Hole Oceanographic Instn., 1985-92, 93—. Editor: Rev. Modern Physics, 1968-73. Served to lt. (j.g.) USNR, 1945-46. USPHS fellow, 1948-49; Jr. fellow Harvard Soc. Fellows, 1949-51; recipient Rockefeller Pub. Service award, 1957-58, Gold medal exceptional service Dept. Commerce, 1961, Arthur Flemming award D.C. Jr. C. of C., 1962, Samuel Wesley Stratton award Dept. Commerce, 1966, Career Service award Nat. Civil Service League, 1968, Proctor prize Research Soc. Am., 1972. Fellow Am. Phys. Soc. (chmn. div. electron physics 1961-68, pres. 1979), AAAS (dir. 1969-73), Am. Acad. Arts and Scis.; mem. Nat. Acad. Scis. (council 1972-75), Nat. Acad. Engring. (Arthur Bueche award), Washington Acad. Scis. (Outstanding Sci. Achievement award 1959), Nat. Acad. Public Adminstrn., Am. Philos. Soc., Phi Beta Kappa, Sigma Xi (pres. 1985-86). Office: Harvard U Kennedy Sch Govt 79 J F Kennedy St Cambridge MA 02138

BRAR, GURDARSHAN SINGH, soil scientist, researcher; b. Fazilka, Punjab, India, Dec. 25, 1946; came to U.S., 1983; s. Mall Singh and Gurnam Kaur (Aulakh) B.; m. Kuldeep Kaur Sran; children: Ramandeep, Samrita, Yashmeen. BS, Punjab Agrl. U., Ludhiana, 1969, MS, 1972; PhD, Indian Inst. Tech., Kharagpur, West Bengal, 1986. Soil sci. extension specialist dept. soils Punajb Agrl. U., Ludhiana, India, 1973-77; soil physicist dept. soil sci. Punabj Agrl. U., 1977-83; soil scientist environ. firm. Va., 1985-88; rsch. assoc. Tex. Tech. U., Lubbock, 1988-89; soil scientist agrl. rsch. svc. USDA, Bushland, Tex., 1989-92; rsch. phys. scientist U.S. Army C.E., Hanover, N.H., 1992—. Contbr. articles to profl. jours. Mem. Agronomy Soc. Am., Crop Sci. Soc. Am., Soil Sci. Soc. Am. Office: US Army CRREL 72 Lyme Rd Hanover NH 03755

BRASE, DAVID ARTHUR, neuropharmacologist; b. Orange, Calif., May 9, 1945; s. Arthur Henry and Helen Emma (Rottmann) B. BS, Chapman U., 1967; PhD, U. Va., 1972. Predoctoral fellow NIH, Charlottesville, Va., 1970-72; postdoctoral fellow Univ. Calif., San Francisco, 1972-76, Langley Porter Psychiat. Inst., San Francisco, 1973-74, NIMH, San Francisco, 1974-75, Nat. Inst. Drug Abuse, San Francisco, 1975-76; asst. prof. Ea. Va. Med. Sch., Norfolk, Va., 1976-84; rsch. assoc. Med. Coll. Va., Richmond, 1984-92. Contbr. articles to Jour. Pharmacology Exptl. Therapeutics, Jour. Pharmacy and Pharmacology, Life Scis., Annals of Neurology, Med. Hypotheses. Pharm. Mfg. Assn. Found. rsch. grantee, Norfolk, 1977-78. Mem. AAAS, Am. Soc. for Pharmacology and Exptl. Therapeutics, Soc. for Neurosci., Va. Acad. Sci. Lutheran. Achievements include demonstration that reinitiation of physical dependence occurs by one dose of opiate in post-addicts; demonstration that injection of xanthan gum lowers blood glucose; research in use of an opioid antagonist for prevention of Sudden Infant Death Syndrome; postulated a role for increased intraneuronal sodium in the mechanism of opiate tolerance; published technique with brain slices for screening potential anti-Parkinsonism drugs. Home: 6700 Cabot Dr # C9 Nashville TN 37209

BRASHER, GEORGE WALTER, physician; b. Jackson, Tenn., Dec. 7, 1936; s. George W. and Verla S. Brasher; m. Martha S. Brasher, Dec. 23, 1960; children: Suzanne Cheshier, George Brasher, John Brasher, David Brasher. BA, Lambuth Coll., 1959; MD, U. Tenn., 1961. MD, U. Tenn.; diplomate Am. Bd. Allergy and Immunology, Am. Bd. Pediatrics. Cons. Scott & White Clinic & Hosp., Temple, Tex., 1966—; dir. Allergy and Immunology Scott & White Clinic and Hosp., Temple. Tex., 1975—; prof. Medicine and Pediatrics Tex. A&M U. Coll. of Medicine, Temple, Tex., 1977—. Contbr. articles to profl. jours. Fellow Am. Acad. Allergy and Immunology, Am. Acad. Pediatrics, Am. Coll. Allergy and Immunology; mem. AMA, Tex. Med. Assn., Bell County Med. Soc., Tex. Allergy Soc. Avocations: civil war history, amateur radio. Office: Scott & White Clinic & Hosp 2401 S 31st St Temple TX 76508-0001

BRASHIER, EDWARD MARTIN, environmental consultant; b. New Iberia, La., Sept. 30, 1954; s. Martin Lee and Ann Elizabeth B.; m. Deborah W. Brashier, July 15, 1977 (div. 1987); children: Shannon E., Edward Martin II, Joseph L. II. Student, Jones Jr. Coll., Ellsville, Miss., 1974, U. Miss., 1976, Kensington U., 1990. Chemist Fla. Machine & Fdry., Jacksonville, 1976, Union Carbide, Woodbine, Ga., 1976-77; sr. chemist Nilok Chem., Memphis, 1977-78; tech. mgr. Chem. West Mgmt., Emelle, Ala., 1978-83; dir. Am. Nukem, Rock Hill, S.C., 1983-86; regional environ. mgr. Layne Western, Rock Hill, 1986-87; project mgr. Westinghouse Environ., Pitts., 1987-88; sr. environ. scientist Dames & Moore, Boca Raton, Fla., 1988-91; nat. sales mgr. Mo. Fuel Recycler Inc., Hanible, 1991—. Editor and tech. advisor, editor various manuals. Bd. dirs. Rock Hill (S.C.) Sch. Bd., 1985-86. Fellow Am. Inst. Chemists; mem. Am. Chem. Soc., Am. Water Wks. Assn., Am. Indsl. Hygiene Assn., WHO, Nat. Assn. Environ. Profls., Am. Soc. Safety Engrs. Democrat. Methodist. Avocations: children, golf, karate. Home: 1051 Kennard St Jacksonville FL 52208

BRASIER, STEVEN PAUL, publishing professional; b. Ottawa, Ont., Can., Oct. 24, 1960; s. Charles Steven and Gloria Ethel (Bradley) B.; m. Janice Carolyn Boyle, Aug. 12, 1989; 1 child, Steven Everett. Diploma in print journalism, Loyalist Coll. Arts and Scis., Belleville, Ont., 1981. Staff reporter Morrisburg (Ont.) Leader, 1981; sr. writer., then project mgr. Homeowner's Handbook Consumers' Assn. Can., Ottawa, 1981-84; sr. writer, editor Standards Coun. Can., Ottawa, 1984-87, mgr. publs., 1987—. Editor Consensus mag., 1984—. Office: Standards Coun Can, 1200 45 O'Connor St, Ottawa, ON Canada K1P 6N7

BRASK, GERALD IRVING, sanitary engineer; b. Mpls., Oct. 19, 1927; s. Irving John and Mary Ann (Gerold) B.; m. Lois Beverly Egner, June 3, 1950; children: Gerald Irving Jr., David Andrew, Kenneth Stephen, Jeffrey Peter, Paul Daniel. B in Civil Engring., U. Minn., 1950. Registered profl. engr.; diplomate Am. Acad. Environ. Engrs. Sanitary engr. Consoer, Townsend and Assocs., Chgo., 1950-72, ptnr., 1972-76, v.p., 1976-91, cons., 1992-- . Mem. Water Environ. Fedn. (life), Nat. Soc. Profl. Engrs. Achievements include studies related to and detailed design of wastewater treatment facilities at numerous locations in U.S. Home: 104 S Yale Arlington Heights IL 60005 Office: Consoer Townsend and Assocs 303 E Wacker Dr Chicago IL 60601

BRASSFIELD, PATRICIA ANN, psychologist; b. Lebanon, Oreg., Apr. 22; d. John James and Mabel Dolores (Scott) Smith; children: Byron Scott, Robert Kent, Lisa Michelle Best. Student, U. Oreg.; BS, Oreg. State U.; MEd, U. Hawaii, 1974; PhD, U.S. Internat. U., 1980. Lic. psychologist, Hawaii, Ariz.; cert. substance abuse counselor, Hawaii; cert. hypnotherapist, marriage, family and child counselor, Calif. Family counselor Psychiat. Svcs., Honolulu, 1974-78, San Diego Ctr. for Psychotherapy, 1979-82; unit team mgr. Oahu Community Corrections, Honolulu, 1982-83; sch. counselor Kalaheo & Moanalua High Schs., Oahu, Hawaii, 1983-85; dir. Waipahu (Hawaii) Community Counseling Ctr., 1985-87; forensic psychologist criminal ct. Hawaii Dept. Health, Honolulu, 1987-91; sch. psychologist San Diego City Schs., 1981-82; family counselor Fairlight, Inc., Honolulu, 1983-84; clin. psychologist, pvt. practice, Oahu and Maui, Hawaii, 1982—; cons. United Airlines, Honolulu, 1989—, Sex Abuse Intervention, Wailuku, Maui, 1990—, Family Ct., Wailuku, 1990—. Contbr. articles to profl. jours. Mem. APA, Nat. Assn. Drug and Alcohol Counselors, Mensa. Avocations: dance, weight lifting. Office: 99-209 Moanalua Rd Ste 314 Aiea HI 96701

BRASWELL, J(AMES) RANDALL, retired quality control professional; b. Columbus, Ga., July 7, 1926; s. James Allen and Irma (Pierson) B.; m. Dagmar Enid Santiago, Sept. 4, 1980. AB in Chemistry, Emory U., 1948; BS in Biology summa cum laude, Columbus (Ga.) Coll., 1972; MS in Environ. Engring., U. Fla., 1974. Lab. asst. Chemistry Dept. Emory U., Atlanta, 1948; lab. technician Nehi Corp., Columbus, 1946, 49; mem. tech. svcs. staff Royal Crown Cola Co., Columbus, 1949-65, dir. quality control, 1965-68, mgr. quality control svcs., 1968-70, ret., 1970. Author: Royal Crown Cola Co. Plant Operation Manual, Water and Water Treatment, Quality Control of Carbonated Beverages, Bottle Washing, and Plant Sanitation; contbr. articles to profl. jours. Pres. United Cerebral Palsy of Muscogee County, Ga., 1976-78; bd. dirs. Ga. Jaycees, 1961-62, Cerebral Palsy and Rehab. Ctr., 1971-82; adult advisor Jr. Achievement, Inc., 1956-58; group capt. United Givers Campaign, 1957; trustee Empty Stocking Fund, Inc., 1958-59; bd. dirs. Community Safety Coun., 1959-61, Youth Craft Shop, Inc., 1959-61, Greater Little League Baseball, Columbus, 1962-63; bd. dirs. Goodwill Industries Chattahoochee Valley, 1976-82; bd. dirs. Columbus Devel. Ctr. for Handicapped Children and Adults. With USNR 1944-46. Recipient Key Man award Columbus Jaycees (pres. 1961-62, bd. dirs. 1957-63), named Young Man of Yr., Columbus, 1959. Mem. AARP, Am. Inst. Biol. Scis., Am. Chem. Soc., Profl. Soc. Soft Drink Technologists, Columbus Coll. Alumni Assn. (bd. dirs.), Kiwanis (bd. dirs.), Phi Kappa Phi. Democrat. Baptist. Home: 5321 Emily Dr Columbus GA 31909-5412

BRATCHER, TWILA LANGDON, conchologist, malacologist; b. Smoot, Wyo.; d. Willis G. and Pearl (Graham) Langdon; m. Ford F. Bratcher, Sept. 10, 1942 (dec.). Research assoc. Los Angeles Mus. Natural History, 1965—; mem. Ameripages Sci. Expedition to Galapagos Islands, 1971; author stories for blind children about skin diving, sea shells, creatures of the sea pub. Braille Inst., 1964-72; work with schs. for blind. Mem. Conchological Club So. Calif. (pres. 1966, 88; life hon. mem.), Am. Malacological Union (councilor at large 1971), Western Soc. Malacologists (pres. 1973), Hawaiian Malacological Soc., Conchologists Am. (exec. bd. 1985-88), San Diego Shell Club, Pacific Shell Club (life hon. mem.). Club: So. Calif. Woman's Press (pres. 1977-79). Author: Living Terebras of the World; contbr. articles to sci. jours. Home: 8121 Mulholland Ter Hollywood CA 90046

BRATERMAN, PAUL SYDNEY, chemistry educator; b. London, Eng., Aug. 28, 1938; came to U.S, 1988; BA, Oxford, Eng., 1959, MA, PhD, 1963, DSc, 1985. From lectr. to reader U. Glasgow, Scotland, 1965-88; prof. U. N. Tex., Denton, 1988—. Author: Metal Carbonyl Spectra, 1975, Reactions of Coordinated Ligands, vol. 1, 1987, vol. 2, 1989. Named Gibbs scholar Oxford U., 1958, Dept. Sci. and Industrial Rsch. NATO fellow United Kingdom DSIR, 1962. Office: U No Tex Dept Chemistry Denton TX 70203

BRÄTTER, PETER, chemist, educator; b. Barcelona, Spain, June 23, 1935; arrived in Germany, 1937; s. Fritz and Hede (Gerhardt) B.; m. Virginia Negretti, Dec. 21, 1984; children: Christian, Stefanie, Marena. PhD in Physics, Tech. U. Berlin, 1969, Habilitation in Chemistry, 1978. Researcher Fritz-Haber Inst., Berlin, 1959-64; researcher faculty physics Tech. U. Berlin, 1966-69, assoc. prof. chemistry, 1979-83, prof., 1984—; researcher chemistry Hahn-Meitner Inst., Berlin, 1969-74, dept. head trace element rsch., 1974-78, dept. head trace elements in biomedicine, 1978-84, sci. dir. trace elements in health and nutrition, 1984—; mem. sci. coms. various internat. assns. and confs., Italy, Finland, Venezuela and Fed. Republic of Germany, 1971—; cons. Internat. Atomic Energy Agy., Vienna, Austria, 1976—, Fundacredesa, Caracas, Venezuela, 1981—; mem. sci. coun. Bertelsmann Found., 1991—. Author-editor: (book) Trace Element Analytical Chemistry in Medicine and Biology, vols. 1-5, 1981-89, Mineral and Trace Elements in Nutrition, 1991; editor-in-chief: Jour. Trace Elements and Electrolytes in Health and Disease, 1985—; adv. bd.: Jour. Radioanalytical and Nuclear Chemistry, 1989—. Mem. Soc. Mineral and Trace Elements (chmn. sci. coun. 1985—, v.p. 1992—), Assn. Nat. Rsch. Ctrs. Fed. Republic of Germany (com. Third World Relationships 1988—). Avocations: tennis, music, painter. Office: Hahn-Meitner Inst, Glienicker Str 100, D-1000 Berlin 39, Germany

BRATTON, WILLIAM EDWARD, electronics executive, management consultant; b. Dallas, Oct. 25, 1919; s. William E. and Edna (Walker) B.; m. Betty Thume, May 30, 1942; children: Dale, Janet, Donna. AB in Econs., Stanford U., 1940; MBA, Harvard U., 1945. From v.p. to pres. Librascope, Glendale, Calif., 1947-63; v.p., gen. mgr. Ampex, Culver City, Calif., 1963-66; pres. Guidance Tech., Santa Monica, Calif., 1967-68; v.p. electronics div. Gen. Dynamics, San Diego, 1969-72; pres. Theta Cable T.V., Santa Monica, 1974-82; pres., chief exec. officer Stagecoach Properties, Salado, Tex., 1959—. Served to lt. (j.g.) USNR, 1944-46. Republican. Epicopalian. Club: El Niguel Country (Laguna, Calif.) (pres. 1978-79). Avocations: golf, skindiving.

BRAUER, JOHN ROBERT, electrical engineer; b. Kenosha, Wis., Apr. 18, 1943; s. Robert Charles and Elizabeth Ida (Schloegel) B.; m. Susan Joyce McCord, Oct. 30, 1982. BSEE, Marquette U., 1965; MSEE, U. Wis., 1966, PhD, 1969. Registered profl. engr., Wis. Rsch. asst. dept. elec. engring. U. Wis., Madison, 1966-69; rsch. scientist A.O. Smith Corp., Milw., 1969-74; cons. engr. A.O. Smith Data Systems, Milw., 1974-85; sr. cons. engr. CadComp, Inc., Milw., 1985-87, MacNeal-Schwendler Corp., Milw., 1988—; lectr. finite element analysis at univs. and confs. in U.S., Can., Europe and Japan. Author, editor: What Every Engineer Should Know about Finite Element Analysis, 1988, 2d edit., 1993; contbr. tech. papers to profl. jours. Bd. dirs. Park People, Milw., 1989—. Mem. IEEE (sr., chmn. magnetics soc. 1977-90, Meml. award 1988), Tau Beta Pi. Roman Catholic. Home: Apt 506 929 N Astor St Milwaukee WI 53202 Office: MacNeal-Schwendler Corp 4300 W Brown Deer Rd Milwaukee WI 53223

BRAUER, RIMA LOIS, psychiatrist; b. Bklyn., Feb. 5, 1938; d. Gerald and Freeda (Rubin) Rubenstein; m. Lee David Brauer, Dec. 29, 1959; children: Samuel, Jennifer, Nathan. BA, Goucher Coll., 1959; MD, U. Md., 1964. Biochemistry researcher Sinai Hosp., Balt., 1958-60; med. intern Montefiore Hosp., Bronx, N.Y., 1964-65; psychiatry resident Yale Sch. Medicine, New Haven, Conn., 1966-69, child fellow, 1969-72; psychoanalyst Western New England Inst. for Psychoanalysis, New Haven, 1977-84; pvt. practice Hartford, Conn., 1984—; clin. faculty Yale Sch. Medicine, New Haven, 1973-87, U. Conn. Sch. Medicine, Hartford, 1987—; chair com. on devel. Western New England Inst. Psychoanalysis, 1992—. Mem. Am. Psychoanalytic Assn. (com. on analytic practice 1991—), N.Y. Acad. Sci. Office: 2 Hartford Sq W Hartford CT 06106

BRAUMAN, JOHN I., chemist, educator; b. Pitts., Sept. 7, 1937; s. Milton and Freda E. (Schlitt) B.; m. Sharon Lea Kruse, Aug. 22, 1964; 1 dau., Kate Andrea. B.S., Mass. Inst. Tech., 1959; Ph.D. (NSF fellow), U. Calif., Berkeley, 1963. NSF postdoctoral fellow U. Calif., Los Angeles, 1962-63; asst. prof. chemistry Stanford (Calif.) U., 1963-69, asso. prof., 1969-72, prof., 1972-80, J.G. Jackson-C.J. Wood prof. chemistry, 1980—, chmn. dept., 1979-83; cons. in phys. organic chemistry; adv. panel chemistry div. NSF, 1974-78; adv. panel NASA, AEC, ERDA, Rsch. Corp., Office Chemistry and Chem. Tech., NRC; coun. Gordon Rsch. Confs., 1989—, trustee, 1991—. Mem. editorial adv. bd. Jour. Am. Chem. Soc., 1976-83, Jour. Organic Chemistry, 1974-78, Nouveau Jour. de Chimie, 1977-85, Chem. Revs, 1978-80, Chem. Physics Letters, 1982-85, Jour. Phys. Chemistry, 1985-87, Internat. Jour. Chem. Kinetics, 1987-89; dep. editor for phys. scis. Science, 1985—. Fellow Alfred P. Sloan, 1968-70, Guggenheim, 1978-79; Christensen, Oxford U., 1983-84. Fellow AAAS, Calif. Acad. Scis. (hon.); mem. NAS, Am. Acad. Arts Scis., Am. Chem. Soc. (award in pure chemistry 1973, Harrison Howe award 1976, R.C. Fuson award 1986, James Flack Norris award 1986, Arthur C. Cope scholar 1986, exec. com. phys. chemistry div., com. on sci. 1992—), Brit. Chem. Soc., Sigma Xi, Phi Lambda Upsilon. Home: 849 Tolman Dr Palo Alto CA 94305-1025 Office: Stanford U Dept Chemistry Stanford CA 94305-5080

BRAUN, HANS-BENJAMIN, physicist; b. Basel, Switzerland; s. Benjamin Karl and Heidi Rosemarie (Jundt) B.; m. Annette Elisabeth Birkenmeier, Aug. 8, 1989; children: Emanuel Andreas, Sebastian Johannes. Diploma, Univ. Basel, 1986; PhD, Eth Zurich, 1991. Rsch. and teaching asst. Eth Zurich, 1987-91; postdoctoral researcher U. Calif., San Diego, 1992-93. Contbr. articles to profl. jours. Postdoctoral fellow Swiss Nat. Sci. Found., Bern, 1991. Mem. Am. Phys. Soc. Office: U Calif Dept Physics 0319 9500 Gilman Dr La Jolla CA 92093

BRAUN, JOSEPH CARL, nuclear engineer, scientist; b. Phila., Feb. 12, 1942; m. Aureliz M. Amendola, July 29, 1967. BS in Physics, St. Joseph's U., 1964; MS in Physics, U. Pitts., 1965; PhD in Nuclear Sci., Engring., Carnegie Mellon U., 1971. With Westinghouse Bettis Atomic Power Lab., West Mifflin, Pa., 1965-74, Combustion Engring. Inc., Windsor, Conn., 1974-89; exec. dir. Am. Nuclear Soc., La Grange Park, Ill., 1990-92; mgr. rsch. program Argonne (Ill.) Nat. Lab., 1992—. Office: Argonne Nat Lab 9700 S Cass Ave Argonne IL 60439-4842

BRAUN, MICHAEL WALTER, technology consultant; b. Pullach, Germany, Jan. 3, 1955; s. Guenther W. and Elfriede (Rosenberger) B.; m. Edith Christine Constraint, Sept. 30, 1979; children: Sebastien, Fabienne. MBA, INSEAD, France, 1986; PhD, U. Heidelberg, Germany, 1988. Laser rschr. Max Planck Inst. for Quantum Optics, Garching, 1982-86; cons. Arthur D. Little, Wiesbaden, 1987—. Contbr. articles to profl. jours. Mem. TM Gringisches Institut fur Textil und Kunststoff-Forschung EV. Roman Catholic. Achievements include patents in field of laser research. Office: Arthur D Little, Gustav-Stresemann-Ring 1, Wiesbaden Germany D-65789

BRAUN, TIBOR, chemist; b. Lugos, Romania, Mar. 8, 1932; s. Eugene Braun and Cecilia Friedmann; m. Klara Szepesi, Aug. 27, 1934; children: Robert, András. BSc, U. Cluj, Romania, 1954; PhD, Acad. Budapest, 1968, DSc, 1980. Chemist Med. U., Tirgu Mures, 1955-56; rsch. fellow Inst. Atomic Physics, Bucharest, Romania, 1957-63; asst. prof. L. Eötvös U., Budapest, Hungary, 1963-68, assoc. prof., 1968-80, prof. chemistry, 1980—; dir. gen. Libr. of Acad., Budapest, 1980—. Editor-in-chief Jour Radioanalalytical Nuclear Chemistry, 1968—, Scientometrics, 1978—; editor Fullerene Sci. and Tech., 1992—; author: Polyurethane Foam Sorbents, 1985, Scientometric Indicators, 1985, The Literature of Analytical Chemistry, 1987. Recipient D de Solla Prize internat., 1986. Achievements include discovery and applications of a new class of polyurethane foam sorbents; development of relative scientometric indicators; new results in radioanalytical chemistry. Home: Gvadányi 109, 1144 Budapest Hungary Office: L Eötvös U, PO Box 123, 1443 Budapest Hungary

BRAUNSTEIN, DANIEL NORMAN, management psychologist, educator, consultant; b. N.Y.C., July 29, 1938; s. Samuel and Genia Braunstein; m. Mildred Shaner, June 21, 1980; children: Wayne, Marian, Laura. BA, Cornell U., 1959; MS, Purdue U., 1961, PhD, 1964. Project dir. U.S. Navy Tng. Rsch. Lab., San Diego, 1963-65; asst. prof. U. Rochester, N.Y., 1965-71; assoc. prof. Oakland U., Rochester, Mich., 1971-78, prof. mgmt. psychology, 1978—; vice chair bd. trustees Roeper Sch., Bloomfield Hills, Mich., 1986-92. Co-editor: Decision Making: An Interdisciplinary Inquiry, 1982; contbr. articles to profl. jours. Mem. Sigma Xi, Beta Gamma Sigma. Avocations: travel, mountain hiking, musical theatre. Office: Oakland U Sch Bus Adminstrn Rochester MI 48309

BRAY, ANDREW MALCOLM, chemist; b. Melbourne, Victoria, Australia, Apr. 14, 1962; s. Michael and Christine (Eley) B.; m. Tanya Elliot Lindsey, Jan. 4, 1986; 1 child, Thomas Llewelyn. BSc in Chemistry with honors, U. Melbourne, Parkville, Australia, 1984; PhD, U. Melbourne, Victoria, Australia, 1990. Biochemist Commonwealth Serum Lab., Parkville, Victoria, Australia, 1988; sr. scientist Coselco Mimotopes P/L, Clayton, Victoria, Australia, 1988-91, Chiron Mimotopes P/L, Clayton, 1991—. Contbr. articles to Jour. Immounology Methods, Jour. Organic Chemistry, Tetrahedron Letters. Recipient James Cumming Meml. scholarship U. Melbourne, 1982, Wyselaskie scholarship, U. Melbourne, 1983, Commonwealth Postgrad. Rsch. award, Dept. Edn., 1984. Mem. Am. Chem. Soc., Am. Peptide Soc., Royal Australian Chem. Inst. Achievements include developments in simultaneous multiple peptide synthesis (especially multiple peptide cleavage under mild conditions). Office: Chiron Mimotopes P/L, 11 Duerdin St, Victoria Clayton 3168, Australia

BRAY, WILLIAM HAROLD, naval architect; b. Chgo., May 22, 1958; s. Pierce and Maud Dorothy (Minto) B.; m. Katherine Jane Helsing, Aug. 4, 1990. BS Engring., U. Mich., 1980, MS Engring., 1981; MBA, NYU, 1988. Registered profl. engr., N.Y. Engr. Exxon Internat. Co. R&D, Florham Park, N.J., 1981-83; engr. shipyard field office Esso Internat. Svcs., New Orleans, 1983-84; project engr. marine tech. Exxon Internat. Co., Florham Park, 1984-88; sr. engr. Esso Internat. Shipping-Ops., Rotterdam, The Netherlands, 1988—; engr. aide Naval Ship Engring. Ctr., Washington, 1978; asst. engr. Marine Consultants & Designers, Cleve., 1979. Mem. ASTM, Soc. Naval Architects and Marine Engrs. (scholar 1980), Oil Cos. Internat. Marine Forum (sec. tanker structures coop. forum 1986-88, marine environ. subcom. 1989-91). Avocations: sailing, skiing, other sports. Office: Esso Internat Shipping Bahamas, PO Box 205, 3100 AG Schiedam The Netherlands

BRAYER, ROBERT MARVIN, program manager, engineer; b. Rochester, N.Y., Nov. 11, 1939; s. Henry E. and Jean (Marvin) B.; m. Sandra Lynne Toubman, Jan. 27, 1978; children: Mark, Lisa, Stacey. BS in Engring. Physics, U. Maine, 1961. Reliability mgr. Chrysler Def. Engring., Centerline, Mich., 1976-81, RAM-D mgr., 1981-82; M1 product assurance mgr. Gen. Dynamics Land Systems, Sterling Heights, Mich., 1982-84, M1 chief engr., 1984-85, M1A1 block II program mgr., 1985-86, TMEPS program mgr., 1986-91, armored systems modernization (ASM) program mgr., 1991—. Election com. Rep. Party, Rochester Hills, 1990. Mem. Nat. Mgmt. Assn., Am. Def. Preparidness Assn. (sec. 1971). Office: Gen Dynamics PO Box 2094 Warren MI 48090

BRAYMAN, KENNETH LEWIS, surgeon; b. Boston, Mar. 3, 1955; s. Lawrence Joseph B. and Barbara (Lewis) Franklin; m. Kerrie Ann Lipsky, Oct. 4, 1987; children: Jonathan, Jacqueline, Lawrence. BS in Biochemistry, U. Mass., 1977; MD, U. Pa., 1981, PhD, 1989. Diplomate Nat. Bd. Med. Examiners, Am. Bd. Surgery. Postdoctoral fellow Harrison Dept. Surg. Rsch., Phila., 1984-87; sr. resident Hosp. U. Pa., Phila., 1987-88, chief resident, 1988-89, transplant surgeon, 1992—; fellow transplant surgery U. Minn. Hosp., Mpls., 1989-90; surgery instr. U. Minn. Med. Sch., Mpls., 1990-92; attending surgeon VA Hosp., Mpls., 1991-92. Manuscript reviewer Diabetes Jour., Clin. Transplantation; contbr. articles to Surgery, Transplantation, Archives of Surgery. Measey Found. fellow, 1985-87, Pfizer Postdoctoral fellow, 1985-87, Daland fellow Am. Philos. Soc., 1986-87, 90-91, Sandoz Transplant fellow, 1990-92. Fellow ACS; mem. AMA, Transplantation Soc., Am. Diabetes Assn., Sigma Xi. Office: Hosp U Pa 3400 Spruce St 4 Silverstei Philadelphia PA 19104

BRAYTON, PETER RUSSELL, program director, scientist; b. Glens Falls, N.Y., Aug. 15, 1950; s. Russell Adelbert and Beverly Elizabeth (Gilman) B.;. BS, Union Coll., Schenectady, N.Y., 1972; PhD, U. Mich., 1980. Postdoctoral fellow U. So. Calif., L.A., 1980-83; staff fellow NIH, Bethesda, Md., 1983-87; microbiologist Walter Reed Army Med. Ctr., Washington, 1987-88; program dir. USDA, Washington, 1988—. Contbr. articles to Jour. Gen. Virology, Jour. Virology, Virology. Bd. dirs. Fed. City Performing Arts Assn., Washington, 1986-94. Mem. Am. Soc. for Microbiology, Am. Soc. for Virology, AAAS, N.Y. Acad. Scis. Office: USDA CSRS NRICGP 901 D St SW AG Box 2242 Washington DC 20250-2242

BRAZEL, ANTHONY JAMES, geographer, climatologist; b. Cumberland, Md., Dec. 1, 1941; s. James Anthony and Katherine Ann (Graham) B.; m. Sandra Wardwell, Dec. 30, 1947. BA in Math., Rutgers U., 1963, MA in Geography, 1965; PhD in Geography, U. Mich., 1972. Rsch. scientist U.S. C.E. Lake Survey, Detroit, 1968-70; from instr. to prof. geography Windsor (Ont., Can.) U., 1970-74; asst. prof. geography Ariz. State U., Tempe, 1974—, chair dept. geography, 1991—; state climatologist State of Ariz., Phoenix, 1979—; chair Ariz. Climate Com., Phoenix, 1980—. Contbr. articles to profl. publs. Grantee NSF. Fellow Ariz.-Nev. Acad. Sci., Explorers' Club. Achievements include being one of first geographers to model aspects of Alpine climatology using digital and numerical surface climate models, research on dust storms and desertification of southwest U.S., climate changes in this area. Home: 426 E Geneva Dr Tempe AZ 85282 Office: Ariz State U Dept Geography Tempe AZ 85287-0104

BRAZ-FILHO, RAIMUNDO, organic chemistry educator; b. Pacatuba, Ceara, Brasil, Apr. 19, 1935; s. Raimundo Braz de Souza and Maria Braz Bessa; m. Maria Maroni Monte Braz, May 25, 1963; children: Jamil Monte Braz, Denise Monte Braz. BSc, U. Fed. Ceara, 1962; PhD in Organic Chemistry, U. Fed. Rural Rio de Janeiro, 1971. Tchr. of superior edn. U. Fed. do Ceara, Fortaleza, 1963-65, asst. prof., 1966-71; prof. U. Fed. Rural do Rio de Janeiro, Itaguai, 1971-91, sr. rschr., 1976—. Author: OEA-Editora, 1976, 2d edit., 1983; contbr. articles to profl. jours. including Phytochemistry, Jour. Brazilian Chem. Soc., Jour. Natural Products. Grantee in field, 1976-91. Mem. Brazilian Chem. Soc. (v.p. 1982-84, pres. 1984-86), Acad. Brasileira de Ciencia, Am. Chem. Soc., Am. Soc. Pharmacognosy, Consello Nacional de Desenvoloimento e Tecnologico. Office: UFRRJ Dept De Quimica, Km 47 Antigua Rio Sao Paulo, 23851 Itaguai Brazil

BRDIČKA, MIROSLAV, retired physicist; b. Dvur Králové, Czechoslovkia, Sept. 12, 1913; s. Jan and Antonie (Kočová) B.; m. Blanka Houbová, Mar. 30, 1946; 1 child, Petr. PhD, Charles U., Prague, Czechoslovakia, 1946. Lectr. Charles U., Prague, 1945-54, assoc. prof., 1954-59, prof., 1959-68; prof. Czech Tech. U., Prague, 1968-81. Author: Continuum Mechanics, (Czech) 1959, Theoretical Mechanics (Czech), 1987. Home: Janovskeho 14, 17000 Prague Czech Republic

BREAKFIELD, PAUL THOMAS, III, physicist; b. Nashville, Sept. 19, 1940; s. Paul T. and Katie L. (Hunt) B.; m. Lois Jean Moreland, Sept. 14, 1962; children: Paul Thomas IV, Robert Alan. Student, George Peabody Coll., 1961-62; BA in Math/Physics, David Lipscomb U., 1962. Field engr. C.L. Mahoney, Mech. Contr., Orlando, Fla., 1962; engr. Finfrock Industries, Orlando, Fla., 1963; computer programmer NASA/Kennedy Space Ctr., Fla., 1964-70, systems analyst, 1970-78, shuttle simulation br. chief, 1978-79, launch processing system div. chief, 1979-83, dep. dir. shuttle engring., 1983-86, dir. shuttle payload ops., 1986—. Mem. NASA Kennedy Mgmt. Assn. Ch. of Christ. Office: Kennedy Space Ctr Man Code CS Orlando FL 32899

BREAKSTONE, ROBERT ALBERT, consumer products company executive; b. N.Y.C., Feb. 20, 1938; s. Morris and Minnie E. (Guon) B.; m. Eileen Fogel, Nov. 5, 1966; children: Warren, Ronald, David. B.S. in Math, CCNY, 1960, M.B.A. in Fin, 1964. Systems mgr. IBM, N.Y.C., 1960-64; dir. mgmt. systems Continental Copper & Steel Industries, Inc., N.Y.C., 1964-68; v.p., treas. Systems Audits, Inc., N.Y.C., 1968-70; v.p., group exec. Chase Manhattan Bank, N.Y.C., 1970-74; group v.p., dir. Chesebrough-Pond's, Inc., Greenwich, Conn., 1974-85; chmn. bd., pres., chief exec. officer Health-Tex Inc., 1985-88; exec. v.p., chief oper. officer GTech Corp., West Greenwich, R.I., 1988—; adj. asst. prof. Pace U. and NYU, 1964-71; speaker in field. Bd. dirs. Stamford (Conn.) Mus. and Nature Center. Mem. Soc. Mgmt. Info. Systems, N.Am. Soc. Corporate Planning, Am. Apparel Mfrs. Assn. (dir.), Mu Gamma Tau (pres.). Home: 95 Lynam Rd Stamford CT 06903-4527 Office: GTech Corp 55 Technology Way West Greenwich RI 02816-0498

BREAZEALE, MACK ALFRED, physics educator; b. Leona Mines, Va., Aug. 15, 1930; s. Carl Samuel and Maude Ella (Moore) B.; m. Joanne Morton O'Dell, Oct. 4, 1952 (dec. Nov. 1989); children: Jennifer Lee, David Mark, William Carl; m. Louise Hanna Scott, Nov. 10, 1990. B.A., Berea Coll., 1953; M.S., U. Mo. at Rolla, 1954; Ph.D., Mich. State U., 1957. Asst. rsch. prof. Mich. State U., 1957-62; assoc. prof. U. Tenn., 1962-67, prof. physics and astronomy, 1967—; cons. solid state div. Oak Ridge Nat. Lab., 1962-71, cons. health and safety research div., 1985-87; cons. Naval Rsch. Labs., 1971-75; prin. investigator contracts Office Naval Rsch., AEC, 1963—; disting. rsch. prof. U. Miss., 1988—; prin. scientist Nat. Ctr. for Phys. Acoustics, Miss., 1988—; guest Inst. Basic Tech. Problems, Warsaw, Poland, 1972; vis. prof. Tech. U. of Denmark, 1977; guest U. Paris, 1977; mem. program com. Internat. Symposium on Nonlinear Acoustics, 1975, 76, 78, 81, 84, 87, 90, 93. Contbr. articles to profl. jours. Fulbright travel grantee, 1977-78, NATO rsch. grantee, 1978-81, 92—, NSF U.S.-Italy program grantee, 1982-86, Fulbright rsch. fellow Tech. U., Stuttgart, Fed. Republic Germany, 1958-59. Fellow IEEE (adminstrv. com. ultrasonics, ferroelectronics and frequency control soc. 1987-89, program com. 1979—, named Disting. Lect. 1987-88), Acoustical Soc. Am. (assoc. editor Nonlinear Acoustics 1977—, Silver medal in phys. acoustics 1988), Inst. Acoustics (U.K.); mem. AAUP, Am. Phys. Soc., Sigma Xi, Phi Kappa Phi, Sigma Pi Sigma. Home: 1035 Zilla Avent Dr Oxford MS 38655-2835

BREBBIA, CARLOS ALBERTO, computational mechanic, engineering consultant; b. Rosario, Argentina, Dec. 13, 1948; came to U.S., 1969; s. Carlos Alejandro and Elda (Eiris) B.; m. Carolyn Susan Stones, Oct. 30, 1971; children: Alexander Carlos, Isabel Elena. Diploma in civil engring., U. Litoral, Rosario, 1968; PhD in Civil Engring., U. Southampton, Eng., 1972. Lectr. U. Southampton, 1970-75, reader, 1976-79; asst. prof. Princeton (N.J.) U., 1975-76; prof. U. Calif., Irvine, 1979-81; dir. Wessex Inst. Tech., Southampton, 1981—; pres. Computational Mechanics Inc., Billerica, Mass., 1984—; Mem. several adv. bds. Author 13 books; editor over 100 books; editor 2 profl. jours. Recipient Ville France medal, 1978. Mem. Internat. Soc. Boundary Elements (pres. 1989—). Roman Catholic. Achievements include development of the main concept of the boundary element method, of innovative computational techniques, of an industrial computer aided design code based on boundary element methods; founder of Wessex Institute of Technology. Office: Computational Mechanics Inc 25 Bridge St Billerica MA 01821

BRECHER, EPHRAIM FRED, structural engineering consultant; b. Bklyn., Nov. 22, 1931; s. Morris Abraham and Mary (Wiener) B.; m. Sandra Louise Footer, Aug. 7, 1955 (div. Dec. 1990); children: Leslie, Deanne, Neil, Andrew. BS, MIT, 1953. Registered profl. engr., Pa., N.J., Md., Del., Ky., N.C., Fla., Conn., N.Y., W.Va. Jr. engr. Dorfman-Bloom, Inc., Phila., 1955-59; staff engr. David Bloom, Inc., Phila., 1959-64, assoc., 1964-71; v.p., sr. engr. Caltech & Assocs., Phila., 1971-72; chief structural engr. GBQC, Phila., 1972-74, assoc., chief structural engr., 1974-76, prin., chief structural engr., 1976-90; cons. structural engr. Brecher Assocs., Phila., 1990—; expert witness to legal and ins. communities. Com. mem. for triennial rewrite PCI Design Handbook, 1991. Ednl. counselor MIT Admissions Office, Phila., 1984—. 1st lt. USAF, 1953-55. Fellow ASCE, Am. Conc. Inst., Am. Concrete Inst. (local chpt. pres. 1981-82); mem. Carpenters Co. of the City and County of Phila. (mng. chair 1989, jr. warden 1990, middle warden 1991, sr. warden 1992, sec. 1993), Franklin Inst. Com. for Sci. and Arts. Democrat. Jewish. Avocations: model railroads, fine jewelry, bowling, cooking, gardening. Office: Brecher Assocs 2300 Walnut St Philadelphia PA 19103-5552

BRECHER, KENNETH, astrophysicist; b. N.Y.C., Dec. 7, 1943; s. Irving and Edythe (Grossman) B.; m. Aviva Schwartz, Aug. 18, 1965; children: Karen, Daniel. B.S., MIT, 1964, Ph.D., 1969. Research physicist U. Calif., San Diego, 1969-72; asst. prof. physics MIT, Cambridge, 1972-77; assoc. prof. MIT, 1977-79; assoc. prof. astronomy and physics Boston U., 1979-81, prof., 1981—, dir. Sci. and Math. Edn. Ctr., 1990—. Author, editor: (with G. Setti) High Energy Astrophysics and Its Relation to Elementary Particle Physics, 1974, (with M. Feirtag) Astronomy of the Ancients, 1979; contbr. numerous articles to profl. jours. Mem. Mass. Cultural Coun., 1989-91. Guggenheim fellow, 1979-80; W.K. Kellogg fellow, 1985-88; NRC sr. research assoc., 1983-84. Fellow Am. Phys. Soc. (chmn. astrophysics div. 1990-91); mem. Am. Aston. Soc., Internat. Astron. Union, Sigma Xi. Home: 35 Madison St Belmont MA 02178-3535 Office: Boston U Dept Astronomy 725 Commonwealth Ave Boston MA 02215-1401

BRECK, JAMES EDWARD, fisheries research biologist; b. Lansing, Mich., Mar. 15, 1946; s. John Emil and Dagmar Eugenia (Abrahamson) B.; m. Sandra Kay Leech, July 10, 1971; children: Eric John, Jason Thomas. AB in Biology, Chemistry, Augustana Coll., 1968; MS in Zoology, U. Ill., 1969; PhD in Zoology, Mich. State U., 1978. Project assoc. Ctr. for Biotic Systems U. Wis., Madison, 1978-79, lectr. dept. zoology, summer 1979, project assoc. Marine Studies Ctr., 1979-80, asst. scientist Marine Studies Ctr., 1980-81; rsch. assoc. environ. sci. div. Oak Ridge (Tenn.) Nat. Lab., 1981-86, rsch. staff mem. environ. sci. div., 1987; fisheries rsch. biologist Fisheries Div. Mich. Dept. Nat. Resources, Ann Arbor, 1987—; adj. assoc. prof. of natural resources Sch. Natural Resources and Environ. U. Mich., Ann Arbor, 1991—; participant Inst. for Evaluating Health Risks, Working Conf. on Bioaccumulation of Hydrophobic Organic Chems. in Aquatic Systems, Leesburg, Va., 1992. Co-author: (chpt.) Bioenergetics in Methods for Fish Biology, 1990, Toxic Contaminants and Ecosystem Health: A Great Lakes Focus, 1988; co-editor: Aquatic Plants, Lake Management and Ecosystem Consequences of Lake Harvesting, 1979; editor (tech. publs.) Fisheries Div., Mich. Dept. Nat. Resources, Ann Arbor, 1989-91; assoc. editor Transactions of the Am. Fisheries Soc., 1990-92; contbr. articles to profl. jours. Mem. Thurston Nature Ctr. Com., Ann Arbor, 1990—; cubmaster Pack 160 Wolverine Coun. Boy Scouts of Am., Ann Arbor, 1992-93; actor, singer Thurston Players, Ann Arbor, 1989-93. Recipient Tech. Comm. award East Tenn. chpt. Soc. for Tech. Comm., 1992, Merit award for Scholarly/Profl. Articles Soc. for tech. Comm., 1992. Mem. AAAS, Am. Fisheries Soc., Am. Soc. Naturalists, Am. Inst. Biol. Scis., Sigma Xi, Phi Beta Kappa. Unitarian Universalist. Achievements include research in stunting of bluegill sunfish, bioenergetics models of fish growth, predator-prey interactions in aquatic systems, bioaccumulation of contaminants by fish. Office: Inst for Fisheries Rsch 212 Museums Annex Bldg 1109 N University Ave Ann Arbor MI 48109-1084

BRECK, KATHERINE ANNE, technical writer; b. Atlanta, Feb. 3, 1964; d. Fred Irvin and Katherine (Barton) B. BA in Journalism, Ga. State U., 1986. Editor newspaper DeKalb Community Coll., Clarkston, Ga., 1983-84; prodn. intern Cox Ent., Atlanta, 1986, Jimmy Carter Presdl. Libr., Atlanta, 1987; mktg. coord. Rosser Fabrap, Atlanta, 1987-90; freelance writer MGM Advt, Atlanta, 1990; tech. writer Radian Corp., Atlanta, 1990-93; mktg. coord. PRAD Group, Inc., Atlanta, 1993—. Mem. Am. Mktg. Assn., Soc. for Mktg. Profl. Svcs. (membership com.), Women in Comm. (substitute chmn. nat. profl. conf. com. 1991, media guide chmn. 1991, fin. com. 1991, chpt. del. to regional conv. and nat. profl. conf. 1992, v.p. membership 1992-93, pres. 1993—, judge chpt. media awards 1991, Chpt. Pres.'s award 1992, chpt. del. to regional conv. 1993, regional conf. chair 1993). Pentecostal. Avocations: photography, softball, gourmet cooking, racquetball. Office: PRAD Group Inc 233 Peachtree St Ste 400 Atlanta GA 30303

BRECKE, BARRY JOHN, weed scientist, researcher; b. Milw., Jan. 16, 1947; s. Melvin Albert and Marie Catherine (Goerg) B.; m. Gayle Linda Naggatz, June 14, 1969; children: Darren John, Suzanne Marie. PhD, Cornell U., 1976. Asst. prof. U. Fla., Gainesville, 1976-81, assoc. prof., 1981—. Author: (chpt.) Model Crop Systems: Sorghum, Napiergrass, 1988, Weed Management in Peanuts, 1990; assoc. editor Weed Tech., 1993—; contbr. articles, referee Weed Tech., Peanut Sci., Weed Sci., 1980—. With U.S. Army, 1970-72. Grantee USDA. Mem. Am. Soc. Agronomy, So. Weed Sci. Soc., Am. Peanut Rsch. and Edn. Soc., Weed Sci. Soc. Am., Internat. Weed Sci. Soc., Fla. Weed Sci. Soc. (pres. 1979). Roman Catholic. Achievements include research in weed biology and weed corp interactions. Office: U Fla Agrl Rsch & Edn RR 3 Box 575 Jay FL 32565-9523

BREDE, ANDREW DOUGLAS, research director, plant breeder; b. Pitts., Feb. 4, 1953; s. James Faris and Adele Katherine (Konefal) B.; m. Linda Davis Rudd, Jan. 11, 1992; children from previous marriage: Loralee Elizabeth, Michael Douglas. BS, Pa. State U., 1975, MS, 1978, PhD, 1982. Asst. golf course supt. Valley Brook Country Club, McMurray, Pa., 1975-76; grad. rsch. asst. Pa. State U., University Park, 1976-82; assoc. prof. Okla. State U., Stillwater, 1982-86; dir. rsch. Jacklin Seed Co., Post Falls, Idaho, 1986—; v.p. Turfgrass Breeders Assn., Tangent, Oreg., 1989-91; chmn. variety rev. Lawn Inst., Pleasant Hill, Tenn., 1990—. Assoc. editor Agronomy Jour., 1993—; contbr. articles to Agronomy Jour., 120 articles to mags.; prodr. 15 ednl. videos. Rsch. grantee, 1983-86. Mem. Am. Soc. Agronomy, Am. Radio Relay League, Gamma Sigma Delta, Phi Epsilon Phi, Phi Kappa Phi, Phi Sigma. Republican. Achievements include organization of 1st turfgrass conf. in People's Republic of China; developer 25 plant varieties. Office: Jacklin Seed Co 5300 W Riverbend Rd Post Falls ID 83854-9499

BREEN, JOHN EDWARD, civil engineer, educator; b. Buffalo, May 1, 1932; s. Timothy J. and Alice C. (Keenan) B.; m. Marian T. Killian, June 20, 1953; children: Mary L., Michael T., Dennis P., Sheila A., Sean E., Kerry T., Christopher D. B.C.E., Marquette U., Milw., 1953; M.S. in Civil Engring., U. Mo., 1957; Ph.D., U. Tex., Austin, 1962. Registered profl. engr., Tex., Mo. Structural designer Harnischfeger Corp., Milw., 1952-53; asst. prof. U. Mo., Columbia, 1957-59; mem. faculty U. Tex., Austin, 1959—, prof. civil engring., 1969—, J.J. McKetta prof. engring., 1977-81, Carol Cockrell Curran chair engring., 1981-84, Nasser I. Al-Rashid chair civil engring., 1984—; dir. P.M. Ferguson Structural Engring. Lab., Balcones Research Center, 1967-85; cons. in field. Contbr. articles to profl. jours. Served to lt. USNR, 1953-56. Recipient Teaching Excellence award Gen. Dynamics Corp., 1971, Teaching Excellence award U. Tex. Student Assn., 1963, Teaching Excellence award Std. Oil Found. Ind., 1968, Arthur J. Boase award Reinforced Concrete Rsch. Coun., 1987, Fedn. Internat. Precontrainte medal, 1990. Mem. Am. Concrete Inst. (hon., bd. dirs. 1974-77, Wason medal 1972, 83, Raymond C. Reese Research medal 1972, 79, Kelly medal 1981, Anderson medal 1987, Raymond Davis lectr. 1978, Bloem award 1989), ASCE (T.Y. Lin medal 1985, 89, 91, A.J. Boase Reinforced Concrete Research Council award 1987); mem. Nat. Acad. Engring., Sigma Xi, Chi Epsilon, Tau Beta Pi. Democrat. Roman Catholic. Club: Austin Yacht (commodore 1977). Home: 8603 Azalea Trl Austin TX 78759-7501 Office: Univ Tex Ferguson Lab 10100 Burnet Rd Austin TX 78758-4497

BREESE, JOHN ALLEN, chemical industry executive; b. Chgo, Sept. 21, 1951; s. Leo and Ruth (Pokin) B.; m. Marie Müller, Aug. 2, 1993; children: Sean, Jonathan, Jessica, Andreas, Christine. BA, Knox Coll., Galesburg, IL., 1973; PhD, U. Ill., Urbana, Ill., 1977. Rsch. assoc. Procter & Gamble, Cin., 1977-79; sr. rsch. scientist Kimberly - Clark Corp., Neenah, Wis., 1979-82; exec. v.p., gen. mgr. Paper Chems. divsn. Eka Nobel Inc., Marietta, Ga., 1982—. Author: tech. articles 1974-92; inventor in field. Coach youth sports, Neenah, 1979-81, Eagle River, Wis., 1990, Marietta, 1992—; bd. dirs. Neenah Youth Hockey, 1979-81. Awarded Excellence In Teaching, Kodak, 1976. Mem. TAPPI (secondary fiber com.), Am. Chem. Soc., Nat. Geographic Soc. Avocations: sports. Home: 4781 Taylors Ct Marietta GA 30068 Office: Eka Nobel Inc 2211 Newmarket Pkwy Ste 106 Marietta GA 30067

BREGER, DWAYNE STEVEN, solar energy research engineer; b. Balt., Dec. 28, 1957; s. Eli and Rita (Uzan) B.; m. Leslie Ann Silver, Oct. 13, 1985; children: Alex Eric, Benjamin Silver. BS in Engring., Swarthmore Coll., 1979; MS in Tech. and Policy, MIT, 1983; PhD in Resource Econs., U. Mass., 1993. Researcher Argonne (Ill.) Nat. Lab., 1982-85; cons. CBY Assocs, Inc., Newton, Mass. and, Washington, 1985-88; sr. rsch. assoc. U. Mass., Amherst, 1988--. Co-author: Active Solar Systems--Community Scale Systems, 1993. Link Found. fellow, 1990-91. Mem. Am. Solar Energy Soc., Sigma Xi. Democrat. Office: U Mass Dept Mechanical Engring Amherst MA 01003

BREHM, JACK WILLIAMS, social psychologist, educator; b. Rockwell City, Iowa, Jan. 16, 1928; s. Carl and Charlotte (Williams) B. AB, Harvard Coll., 1952; PhD, U. Minn., 1955. Rsch. asst. Yale U., New Haven, Conn., 1955-57, asst. prof., 1957-58; asst. to prof. Duke U., Durham, N.C., 1958-75; prof. U. Kans., Lawrence, 1975—; mem. exec. com. Soc. Experimental Social Psychology, 1975-79, chair exec. com., 1977-78. Author: A Theory of Psychological Reactance, 1966; co-author Explorations in Cognitive Dissonance, 1962, Perspectives on Cognitive Dissonance, 1976, Psychological Reactance: A Theory of Freedom and Control, 1981. Mem. AAAS, APA, Am. Psychol. Soc., Soc. for Exptl. Social Psychology. Office: U Kans Psychology Dept Lawrence KS 66045

BREIL, DAVID ALLEN, botany educator; b. Brockton, Mass., Mar. 27, 1938; s. Norman David and Helen Mae (Cannon) B.; m. Sandra Durick, Aug. 24, 1963 (div. June 1991); children: Kimberly Ann, Peter David; m. Phyllis Stables Williams, Mar. 20, 1993; stepchildren: Drista Williams, William Williams, Jeremy Williams, Nicholas Williams. BS in Geology, U. Mass., 1960, MA in Botany, 1963; PhD in Botany, Fla. State U., 1968. Instr. Pa. State U., Mont Alto, 1963-65; prof. Mt. Lake Biol. Sta. U. Va., Pembroke, 1974; prof. Longwood Coll., Farmville, Va., 1968—. Mem. Am. Byrological & Lichenological Soc., Assn. Southeastern Biologists, Sigma Xi (pres. 1990-92). Methodist. Office: Longwood Coll Dept Natural Scis Farmville VA 23909

BREINER, SHELDON, geophysics educator, business executive; b. Milw., Oct. 23, 1936; s. James and Fannie (Apple) B.; m. Phyllis Farrington, Feb. 4, 1962; children—David, Michelle. B.S., Stanford U., 1959, M.S., 1962, Ph.D. in Geophysics, 1967. Registered geophysicist. Geologist, Calif. product mgr. Varian Assocs., 1961-68; founder, pres. Geometrics, Sunnyvale, Calif., 1969-83; chmn. EGG Geometrics, 1983; founder, pres. Syntelligence Inc., 1983—; dir. Cadtec Corp., Optical Spltys. Inc., Fracture Techs.; pres., founder Pava Magnetic Logging Inc., 1983 pres. Foothill Assocs.; cons., prof., lectr. geophysics Stanford U. and Grad. Sch. Bus.; cons. archaeol. exploration problems and search for buried objects. Author: Applications Manual for Portable Magnetometers, 1973; contbr. articles to profl. jours.; patentee in oil exploration; inventor gun detector for airports. Trustee Peninsula Open Space Trust; vice chmn. Resource Ctr. for Women; mem. maj. gifts com. Stanford U. Served with U.S. Army, 1960. Honors scholar, 1955-56; NSF grantee for earthquake research, 1965. Fellow Explorers Club; mem. Soc. Exploration Geophysicists (Best Presentation award 1985), Am. Geophys. Union, European Assn. Exploration Geophysicists, Stanford Assocs. Runner Boston Marathon. Office: Geometrics 395 E Java Dr Sunnyvale CA 94089-1142

BREITENBACH, ALLAN JOSEPH, geotechnical engineer; b. Wichita, Kans., Dec. 22, 1949; s. George J. and Cleopha M. (Elpers) B.; married; 2 children. BSCE, Kans. State U., 1972, BS in Geology, 1976; MSCE, Colo. State U., 1976. Registered profl. engr., Colo., Mont., Idaho, N.Mex., Ariz., Nev., Wyo., S.C. Staff engr., project engr. Woodward Clyde Cons., Denver, 1972-83; ind. project engr., cons., 1983-89; project mgr. Westec, Denver, 1989—; Speaker at profl. confs. Contbr. articles to profl. publs. including Geotechnical Testing Jour. and Geofabrics Report. Adult leader Littleton (Colo.) area Boy Scouts Am., 1991-94. Mem. ASCE, Soc. Mining Engrs., Nat. Coun. Engring. Examiners, North Am. Geosynthetics Soc., Denver Mining Club. Roman Catholic. Office: Westec Ste 319 B 5600 S Quebec St Englewood CO 80111

BREITMAN, JOSEPH B., prosthodontist, dental educator; b. Phila., Aug. 4, 1952; s. Abraham A. and Natalie (Ketchurin) B.; m. Barbara Susan Beitman, May 13, 1990; 1 child, Ilana Michell. B.A., LaSalle Coll., 1974; D.M.D., U. Pa., 1977; Cert. of Tng. in Prosthodontics, Temple U., 1979; Cert. of Tng. in Biomaterials, VA Hosp., Elsmere, Del., 1979. Gen. practice dentistry, Lafayette Hill, Pa., 1978-79; prosthodontist Raymond Hancock, Marlton, N.J., 1979-80; pvt. practice prosthodontics, Phila., 1980—; asst. prof. dental materials Temple U., Phila., 1978-84; assoc. in restorative dentistry U. Pa., Phila., 1985—; asst. prof. post-doctoral prosthodontics Temple U., 1990. Oral biology fellow Temple U., Phila., 1978. Mem. ADA, Am. Coll. Prosthodontics, N.E. Dental Soc. (pres. 1988-89), Ea. Dental Soc., Pa. Assn. Dental Surgeons (pres. 1984-85). Jewish. Lodge: Masons. Avocations: play bagpipes, tae kwon do. Office: 8021B Castor Ave Philadelphia PA 19152-2733

BREITSAMETER, FRANK JOHN, safety engineer; b. Chgo., Feb. 9, 1919; s. John and Madeline Breitsameter; m. Irene Marie Mohapp; children: Janice M., John L., George William. BSME, Ill. Inst. Tech., 1952. Registered profl. engr., Ill., Calif. Field engr. Hyatt Bearings divsn. GM, Chgo., 1946-58; sales engr. Rollway Bearing, Chgo., 1958-65; product engr. Am. Steel

Foundrys, Chgo., 1965-70; compliance officer, instr. OSHA U.S. Dept. Labor, Chgo., 1971-75; safety officer U.S. Dept. Def., Chgo., 1976-79; safety engr. GSA, Chgo., 1979-82; cons. forensic safety engr. Chgo., 1982—. Editor numerous profl. articles. Trustee Village of Mt. Prospect, Ill., 1992. With USN, 1944-46, PTO. Mem. ASME, Western Soc. Engrs. (trustee), Ill. Soc. Profl. Engrs., Am. Soc. Safety Engrs. (emeritus). Republican. Roman Catholic. Home: 1005 E Cardinal Ln Mount Prospect IL 60056-2640

BRELAND, HUNTER MANSFIELD, psychologist; b. Mobile, Ala., Aug. 11, 1933; s. Robert Milton and Cora (Peirce); m. Nancy Schact, Aug. 17, 1968; children: Alison, Julia. BS, U. Ala., 1955; MS, U. Tex., 1961; PhD, SUNY, Buffalo, 1972. Propulsion engr. Gen. Dynamics, Ft. Worth, 1955-59; astronautics engr. LTV, Inc., Dallas, 1960-61; mgmt. cons. Harbridge House, Inc., Boston, 1967-68; sr. rsch. scientist Ednl. Testing Svc., Princeton, N.J., 1972—; Mem. editorial bd. Written Communications, 1983-92. Author: Popultion Validity, 1979, Assessing Student Characteristics, 1981, Personal Qualities and College Admission, 1982, Assessing Writing Skill, 1987; consulting editor Revista InterAm. de Psicologia, 1983-86. Fellow Am. Psychol. Assn. (chmn. membership com. 1983-85). Office: Ednl Testing Svcs Rosedale Rd Princeton NJ 08541

BREMER, ALFONSO M., neurosurgeon, neuro-oncologist; b. Mexico City, Aug. 3, 1939; came to U.S., 1966; s. Carlos and Guadalupe (Garcia) B.; m. Maria-Elena Montano, May 25, 1973; children: Maria-Elena, Rosanna Angelica, A. Michael. BS, U. Nacional Autonoma de Mexico, 1958, MD, 1965. Diplomate Am. Bd. Neurol. Surgery. Intern Deaconess Hosp., Buffalo, N.Y., 1966-67; resident Georgetown U. Hosps., Washington, 1967-72; cancer rsch. neurosurgeon I Roswell Park Meml. Inst., Buffalo, 1972-73, cancer rsch. neurosurgeon II, 1973-74, assoc. chief dept. neurosurgery, 1974-79; chief dept. neurosurgery U. Med. Ctr., Jacksonville, Fla., 1979-85; chief div. neurosurgery Meth. Med. Ctr., Jacksonville, 1986-92; asst. rsch. prof. in exptl. pathology SUNY, Buffalo, 1972-79; rsch. prof. in exptl. pathology Niagara U., Niagara Falls, N.Y., 1972-79; prof. neurosurgery U. Fla., Jacksonville, 1979-85. Recipient Stroke Study grant Am. Heart Assn., 1973, 74, Yoshida-Yamagiwa Internat. Meml. Cancer Study grant Internat. Union Against Cancer, Osaka U., Japan, 1975. Fellow ACS, Stroke Coun.-Am. Heart Assn.; mem. Am. Assn. Neurol. Surgeons, Rsch. Soc. Neurol. Surgeons. Achievements include discovery that malignant secondary tumor of brain responded to intra-arterial chemotherapy. Office: Ste 8001 580 W 8th St Jacksonville FL 32209

BREMERMAN, MICHAEL VANCE, reliability engineer; b. Cheverly, Md., Sept. 22, 1954; s. LeRoy Schaeffer and Mary Genevieve (Donovan) B.; m. Connie Diane Hamm, May 14, 1977 (div. 1984); 1 child, Jennifer Marie. BSEE, U. Md., 1978, postgrad., 1993—; MS in Applied Math., Johns Hopkins U., 1984. Registered profl. engr., Tex.; cert. reliability engr. Sr. engr. Westinghouse Electric Corp., Balt., 1979-86; cons. engr. Sperry Corp., Albuquerque, 1986-87, Dynamic Controls Corp., Hartford, Conn., 1987-88, Honeywell Corp., Albuquerque, 1988-89, Westinghouse Oceanic, Annapolis, Md., 1989—; pres. RELTEC, Inc., Columbia, Md., 1986—. Contbr. articles to I.E.S. 1981 Proceedings, Jour. Environ. Scis. Coach Landover Hills (Md.) Boys and Girls Club, 1990. Mem. Md. Soc. Profl. Engrs., Am. Soc. Quality Control, Moose. Republican. Roman Catholic. Home and Office: 7525A Weatherworn Way Columbia MD 21046

BREMNER, JOHN MCCOLL, agronomy and biochemistry educator; b. Dumbarton, Scotland, Jan. 18, 1922; came to U.S., 1959; s. Archibald Donaldson and Sarah Kennedy (McColl) B.; m. Eleanor Mary Williams, Sept. 30, 1950; children: Stuart, Carol. B.S., Glasgow U., 1944, D.Sc., 1987; Ph.D., U. London, 1948, D.Sc., 1959. With chemistry dept. Rothamsted Exptl. Sta., Harpenden, Eng., 1945-59; assoc. prof. Iowa State U., Ames, 1959-61, prof. agronomy and biochemistry, 1961—, C.F. Curtiss disting. prof. agriculture, prof. agronomy, biochemistry, 1975—; tech. expert Internat. Atomic Energy Agy., Austria, 1964-65, Yugoslavia, 1964-65. Author or co-author over 300 publs. including 30 chpts in sci. monographs. Recipient Outstanding Research award First Miss. Corp., 1979, Alexander Von Humboldt award Alexander Von Humboldt Found., Fed. Republic of Germany, 1982, Gov.'s Sci. medal State of Iowa, 1983, Harvey Wiley award U.S. Assn. Ofcl. Analytical Chemists, 1984, Spencer award Am. Chem. Soc., 1987, Burlington No. Found. Faculty Achievement award for Research, Gamma Sigma Delta award of merit for disting. service to agriculture, Regents award for faculty excellence, 1992, Award for Advancement of Agrl. & Food Chemistry, Am. Chem. Soc.; fellow Rockefeller Found., 1957, Guggenheim Found., 1968. Fellow AAAS, Am. Acad. Microbiology, Am. Soc. Agronomy (Agronomic Rsch. award 1985, Environ. Quality Rsch. award 1990), Soil Sci. Soc. Am. (Achievement award 1967, Bouyoucos Disting. Career award 1982, Disting. Svc. award 1993), Iowa Acad. Sci. (disting.); mem. NAS, Am. Soc. Microbiology, Brit. Soc. Soil Sci., Internat. Soil Sci. Soc., Phi Kappa Phi, Sigma Xi, Gamma Sigma Delta. Achievements include patent for nitrification inhibitor; development and evaluation of nitrification and urease inhibitors for control of adverse transformations of fertilizer nitrogen in soils; development of methodology for research on the nitrogen cycle and environmental problems related to agriculture; research on microbial, enzymatic, and chemical processes responsible for nitrogen transformations in soils, such as nitrification, denitrification, chemodenitrification, and urease activity. Home: 2028 Pinehurst Dr Ames IA 50010-4561 Office: Iowa State U Dept Agronomy Ames IA 50011

BRENCHLEY, JEAN ELNORA, microbiologist, researcher; b. Towanda, Pa., Mar. 6, 1944; d. John Edward and Elizabeth (Jefferson) B.; m. Bernard Asbell, July 21, 1990. BS, Mansfield U., 1965; MS, U. Calif., San Diego, 1967; PhD, U. Calif., Davis, 1970; hon. degree, Lycoming Coll., 1992. Rsch. assoc. biology dept. MIT, Cambridge, 1970-71; from asst. prof. to assoc. prof. microbiology Pa. State U., Univ. Pk., 1971-77; head. dept. molecular and cell biology, dir. Biotech. Inst. Pa. State U., University Park, 1984-87, prof. microbiology, dir. Biotech. Inst., 1984-90, prof. microbiology and biotech., 1990—; assoc. prof., then prof. biology Purdue U., West Lafayette, Ind., 1977-81; research dir. Genex Corp., Gaithersburg, Md., 1981-84; mem. Nat. Biotech. Policy Bd., 1990—; trustee Biosis, 1983-88; vis. scholar NIH, 1991. Editor Applied and Environ. Microbiology, 1981-85; mem. editorial bd. Jour. Bacteriology, 1974-84, Butterworth Biotech. Series, 1988—; editor Microbiol. Revs., 1992—. Recipient Outstanding Alumni award Manfield U., 1983; Waksman award Theobald Smith Soc., 1985; named to Pa. Hall of Fame, 1988. Fellow Am. Acad. Microbiology; mem. NAS (bioprocess com.), AAAS (nominating com. 1990—), Am. Soc. Microbiology (pres. 1986-87, ASM Found. lectr. 1975), Assn. Women in Sci., Am. Soc. Biol. Chemists, Am. Chem. Soc., Found. for Microbiology (trustee 1988—), Sigma Delta Epsilon (hon.). Office: Pa State U 209 S Frear University Park PA 16802

BRENDEN, BYRON BYRNE, physicist, optical engineer; b. Bismarck, N.D., Feb. 25, 1927; s. Vernon Rueben and Aurelia Emelia (Ruede) B.; m. Clara Lavelle Haldorson, Dec. 26, 1949; children: Sandra Christine, Susan Cathlene, Michael Andrew. BS in Physics, U. Oreg., 1951; postgrad., Washington State U., Richland, Wash., 1952-54. Physicist GE, Schenectady, N.Y., 1951-52; optical engr. GE, Richland, 1952-65; sr. physicist Battelle N.W., Richland, 1965-69, staff scientist, 1979—; v.p. Holosonics, Richland, 1969-79; mem. adv. bd. dept. elec. engring. Wash. State U., Richland, 1988—. Co-author: An Introduction to Acoustical Holography, 1972; contbg. author: Optical and Acoustical Holography, 1972, Acoustic Imaging, 1976. With U.S. Army, 1944-45. Recipient Gold medal City of Milan, 1972, Disting. Alumni award U. Oreg., 1973; named Inventor of Yr., Assn. for the Advancement of Invention and Innovation, 1973. Achievements include 50 U.S. and foreign patents for 15 different inventions in the field of acoustical holography. Office: Battelle NW PO Box 999 Richland WA 99352

BRENNAN, ANN HERLEVICH, communications professional; b. Wilmington, N.C., Apr. 6, 1956. BS in Engring. Sci. & Mechanics, N.C. State U., Raleigh, 1978. Rsch. engr. Nat. Renewable Energy Lab., Golden, Colo., 1978-88, sr. sci. writer, 1988-92, mgr. tech. comm., 1992—. Contbr. articles to profl. jours., over 60 brochures, reports. Recipient Excellence award Soc. for Tech. Comm., 1992, Achievement award, 1991, 92, Blue Pencil award Nat. Assn. Gov. Communicators, 1992. Mem. ASME, Am. Solar Energy Soc., Internat. Assn. Bus. Communicators. Office: NREL 1617 Cole Blvd Golden CO 80401

BRENNAN, GEORGE GERARD, pediatrician; b. N.Y.C., Nov. 6, 1931; s. George and Bertha (Bradley) Brennan; m. Joan Worfolk, Oct. 31, 1959; children: Jeanne, Gerry, Robert, James, Thomas. AB, Fordham U., 1953; MD, Loyola U., Chgo., 1957. Intern Jersey City Med. Ctr., 1957-58; pediatric resident St. Vincent's Med. Ctr., N.Y.C., 1958-60; pediatrician Old Bridge-Sayneville Med. Group, P.A., 1960—. Fellow Am. Acad. Pediatrics; mem. N.J. State Med. Soc., Middlesex County Med. Soc. Republican. Roman Catholic. Avocations: tennis, literature, music, travel. Office: Old Bridge-Sayneville Med Group PA 26 Throckmorton Ln Old Bridge NJ 08857

BRENNAN, LEONARD ALFRED, research scientist, administrator; b. Westerly, R.I., Aug. 2, 1957; s. Leonard Alfred Brennan Jr. and Louise (Gagne) Ladd; m. Teresa Leigh Pruden, Jan. 1, 1980. BS, Evergreen State Coll., 1981; MS, Humboldt State U., 1984; PhD, U. Calif., Berkeley, 1989. Technician USDA Forest Svc., Arcata, Calif., 1984-85; biologist Calif. Dept. Food & Agr., Ukiah, 1985; rsch. asst. U. Calif., Berkeley, 1986-89; lectr. Humboldt State U., Arcata, 1989-90; rsch. scientist dept. wildlife and fisheries Miss. State U., Mississippi State, 1990-93; dir. rsch. Tall Timbers Rsch. Station, Tallahassee, Fla., 1993—; Habitat ecology cons. The Chukar Found., Boise, 1989—. Author: (chpt.) The Use of Multivariate Statistics for Developing Habitat Suitability Index Models, 1986, The Use of Guilds and Guild-Indicator Species for Assessing Habitat Suitability, 1986, Arthropod Sampling Methods in Ornithology: Goals and Pitfalls, 1989, Influence of Sample Size on Interpretation of Foraging Patterns by Chestnut-backed Chickadee, The Habitat Concept in Ornithology: Theory and Applications; contbr. articles to profl. jours. U. Calif. fellow, 1987; grantee Calif. Dept. Forestry, Internat. Quail Found., USDA Forest Svc.; San Francisco Bay Area chpt. Wildlife Soc. scholar, 1984; judge Mendocino County Pub. Schs. Sci. Fair, Laytonville, Calif., 1988, Miss. Sci. and Engring. Fair, Miss., 1990-91. Mem. AAAS, Am. Ornithologists' Union, Assn. Field Ornithologists, Cooper Ornithological Soc., Ecol. Soc. Am., Wildlife Soc. (faculty advisor Miss. chpt.), Miss. Wildlife Fedn., Pacific N.W. Bird and Mammal Soc., Wilson Ornithological Soc., Ottawa Field Naturalists' Club. Achievements include design of mathematical sex determination model for the Dunlin, of first data-based habitat suitability index models using multivariate statistics; research on contaminant levels in Dunlins in western Washington state, on factors responsible for long-term Northern Bobwhite population decline, on impact of habitat management for the endangered Red-cockaded Woodpecker on terrestrial vertebrates. Office: Tall Timbers Rsch Sta RR 1 Box 678 Tallahassee FL 32312

BRENNAN, SEAN MICHAEL, cancer research scientist, educator; b. Wilmington, Del., Jan. 29, 1954; s. James Patrick and Letitia Rosalie (DePace) B.; married; 4 children. BA, U. Del., 1976; PhD, U. Calif., San Diego, 1982. Rsch. fellow Cancer Rsch. Campaign, Cambridge, Eng., 1983-86; asst. prof. Med. Sch. U. Conn., Farmington, 1986-92; sr. rsch. scientist BioGenex Labs., San Ramon, Calif., 1992—; rsch. assoc. prof. pathology U. So. Calif. Med. Sch., L.A., 1993—; reviewer NSF, Washington, 1987-89; mem. com. on electromagnetic field health effects Conn. Acad. Sci. and Engring., Hartford, 1991-92. Radio show host, producer Sta. WWUH-FM, Hartford, 1987-92. Recipient Nat. Rsch. Svc. award USPHS, 1983-86, Embryonic Gene Activation award A. Cancer Soc., 1988-90, Mechanisms of Teratogenesis award March of Dimes, 1992. Mem. AAAS, Conn. Cancer Inst., Del. Acad. Scis. Democrat. Achievements include co-discovery of first molecular marker for germ layer induction in vertebrate embryo; innovated use of amphibian embryos as a test for substances with potential of causing birth defects. Office: BioGenex Labs 4600 Norris Canyon Rd San Ramon CA 94583

BRENNAN, THOMAS GEORGE, JR., audiologist, speech-language pathologist; b. N.Y.C., Jan. 19, 1953; s. Thomas George Sr. and Gladys Kathleen (Atkinson) B.; m. Brenda Corley, Nov. 28, 1980; 1 child, Matthew Boyd Halfacre. BS in Edn., Stephen F. Austin State U., 1981, MEd, 1984. Cert. clin. competence in audiology and speech-lang. pathology. Piano tuner, technician Brennan's Piano Tuning and Repair, Nacogdoches, Tex., 1972—; pvt. practice audiology assoc. Nacogdoches, 1985-89, pvt. practice speech-lang. pathologist, 1989—, pvt. practice audiologist, 1992—. Author: An Introduction to Biofeedback for Speech-Language Therapists, 1985, CAT: Computer Auditory Test, 1993; author, programmer: (computer program, manual) Fear Survey, 1990. Screener East Tex. Community Health Svcs., Nacogdoches, 1991—; svc. provider Health Fair, Nacogdoches, 1992—, Child Health Jamboree Com., Nacogdoches, 1992—. Mem. Am. Speech-Lang.-Hearing Assn., Am. Tinnitus Assn (referral source), Nat. Stuttering Project. Home: PO Box 4544 SFA Nacogdoches TX 75962

BRENNEN, CHRISTOPHER E., fluid mechanics educator; b. Belfast, No. Ireland. BA in Engring. Sci. with 1st-class honors, Oxford U., 1963, MS in Engring. Sci., 1966, D.Phil. in Engring. Sci., 1966. Postdoctoral fellow Nat. Phys. Lab., London; Fulbright rsch. fellow Calif. Inst. Tech., Pasadena, 1968-69, mem. faculty, 1976—, prof. mech. engring., 1982—, dean students, 1988-92; master student houses Calif. Inst. Tech., 1983-87. Recipient New Tech. award NASA, 1980. Mem. ASME (assoc. editor Jour. Fluids Engring., R.T. Knapp award fluids engring. divsn. 1978, 81, Centennial medallion 1980, Fluids Engring. award 1992). Achievements include research in cavity flows, cavitation inception and noise, turbomachinery and rotordynamics, mechanics of granular material flow. Office: Dept of Mechanical Engineering Calif Inst of Technology Mail Code 104-44 Pasadena CA 91125*

BRENNER, AMY REBECCA, podiatrist; b. N.Y.C., Dec. 31, 1958; d. Frank and Betty (Gitelman) B.; m. Peter Louis Klinge Jr., June 28, 1985. BS, Ithaca Coll., 1980; MA, Columbia U., 1982; D of Podiatric Medicine, N.Y. Coll. Podiatric Medicine, 1988. Cert. athletic trainer, health edn. instr., N.Y. State. Fellow Rusk Inst. Rehab. Medicine, N.Y.C., 1973; podiatrist, athletic trainer Athletic Events, N.Y., 1976—; physiology intern Chase Manhattan Cardiovascular Fitness Lab., N.Y.C., 1981; asst. athletic trainer Barnard Coll., N.Y.C., 1980-81; surg. resident Liberty Med. Ctr., Inc., Balt., 1989; staff podiatrist Bronx Cross County Med. Group, P.C., Bronx, Yonkers, N.Y., 1990—; pvt. practice podiatry N.Y.C. and Westchester County, 1990—; test site adminstrs. com. Nat. Athletic Trainers Assn. Bd. Certification, Greenville, N.C., 1990—. Contbr. articles to profl. jours including Jour. Am. Podiatric Med. Assn. V.p. Bd. Dirs. Vernon Woods Apts., Inc., Mount Vernon, N.Y., 1990—; lectr. Am. Podiatric Med. Assn. Region VIII Meeting, Bethesda, Md., 1989, Liberty Med. Ctr., Inc., Balt., 1989. Mem. Am. Podiatric Med. Assn., Nat. Athletic Trainers Assn., Am. Coll. Sports Medicine, Am. Assn. for Women Podiatrists, Am. Alliance Health, Phys. Edn., Recreation and Dance. Home: 154 Pearsall Dr Apt 4E Mount Vernon NY 10552-3905

BRENNER, BARRY MORTON, physician; b. Bklyn., Oct. 4, 1937; s. Louis and Sally (Lamm) B.; m. Jane P. Deutsch, June 12, 1960; children: Robert, Jennifer. B.S., L.I. U.; M.D., U. Pitts.; MA (hon.), Harvard U.; DSc (hon.), Long Island U.; D.M.Sc. (hon.), U. Paris, (Pierre et Marie Curie). Asst. prof. medicine U. Calif.-San Francisco, 1969-72, asso. prof. medicine and physiology, 1972-75; prof. medicine and physiology U. Calif., San Francisco, 1975-76; Samuel A. Levine prof. medicine Harvard Med. Sch., Boston; with Peter Bent Brigham Hosp., Boston, 1976—; dir. renal div. Brigham and Women's Hosp., Boston, 1979—; dir. physician-scientist program, Harvard Med. Sch., 1984-90, Harvard Ctr. for Study of Kidney Diseases, 1987—; cons. NIH. Co-editor: The Kidney, 2 vols., 1976, 4th edit., 1991, Contemporary Issues in Nephrology, 1978-90, Acute Renal Failure, 1983, 3d edit., 1993, Textbook of Renal Pathology, 2 vols., 1989, Textbook of Hypertension, 2 vols., 1990; founding editor: Current Opinion in Nephrology and Hypertension, 1992—; contbr. numerous articles to publs. Recipient Homer W. Smith award N.Y. Heart Assn., George E. Brown award Am. Heart Assn., Merit award NIH, SKF Disting. Scientist award; rsch. grantee NIH. Mem. Am. Soc. Cell Biology, Am. Physiol. Soc., Assn. Am. Physicians, Am. Soc. Nephrology (councillor, pres.), Am. Soc. Clin. Investigation (councillor, v.p.), Am. Soc. Hypertension (exec. com., pres., Richard Bright award), Western Assn. Physicians, Salt and Water Club, Interurban Clin. Club, Alpha Omega Alpha, Phi Sigma. Office: 75 Francis St Boston MA 02115-6195

BRENNER, BRIAN RAYMOND, structural engineer; b. N.Y.C., Mar. 10, 1960; s. Donald and Marilyn (Scheiner) B.; m. Lauren Rene Hochstat, June 16, 1986; 1 child, Daniel. BSCE, MIT, 1982, MSCE, 1984. Registered profl. engr., Mass. Profl. assoc. Parsons Brinckerhoff Quade and Douglas, Boston, 1984—. Assoc. editor Network mag.; contbr. articles to profl. jours. Recipient William Barclay Parsons fellowship, 1990; named Young Engr. of Yr., Mass. Soc. Profl. Engrs., 1991. Mem. ASCE (reviewer Jour. of Computing in Civil Engring, editorial bd. Civil Engring. Practice), Boston Soc. Civil Engrs. (structural com., computer com.), Mass. Soc. Profl. Engrs., Chi Epsilon. Achievements include structural engineering development of R.R. and hwy. bridges computer-aided design methods, analysis and design for soil-structure interaction; design of cut-and-cover tunnels, rehab. of existing tunnels. Office: Parsons Brinckerhoff Quade and Douglas 1 S Station Boston MA 02110

BRENNER, GEORGE MARVIN, pharmacologist; b. Ottawa, Kans., Sept. 19, 1943; s. Marvin and Olive Lucille (Smith) B.; m. Mary Ann Robinson, Aug. 21, 1966; children: Sharon, John. MS, Baylor U., 1968; PhD, U. Kans., 1971. Registered pharmacist. Asst. prof. S.D. State U., Brookings, 1971-76; assoc. prof. Okla. State U., Tulsa, 1976-82, prof., 1982—, dept. chair, 1989—; adj. prof. Northeastern Okla. State U., Tahlequah, 1981-85; vis. prof. Oral Roberts U., Tulsa, 1980-90. Contbr. articles to profl. jours. Officer Green Country Sierra Club, Tulsa, 1987-88; leader Boy Scouts Am., Jenks, Okla., 1982-90; pres. Band Parents Club, Jenks, 1990. Mem. Am. Soc. Pharmacology and Exptl. Therapeutics, Soc. Exptl. Biology and Medicine, Sigma Xi (treas. 1985-86). Achievements include Nat. Bd. Med. Examiners test constrn.; patent devel. cons. Office: Okla State U 1111 W 17th St Tulsa OK 74107-1898

BRENNER, SYDNEY, biologist, educator; b. Germiston, South Africa, Jan. 13, 1927; s. Morris and Lena (Blacher) B.; m. May Woolf Balkind, 1952; 3 children; 1 stepchild. MSc, U. Witwatersrand, Johannesburg, South Africa, 1947, MB, BCh, 1951; DPhil, Oxford (Eng.) U., 1954; 7 hon. degrees. Mem. sci. staff Med. Rsch. Coun., Cambridge, Eng., 1957-92, dir. lab. molecular biology, 1979-86, dir. molecular genetics unit, 1986-91; fellow King's Coll., Cambridge U., 1959—; hon. fellow Exeter Coll., Oxford U., 1985; rsch. scientist dept. medicine U. Cambridge Sch. Clin. Medicine, 1992—; mem. staff Scripps Rsch. Inst., La Jolla, Calif., 1992—; Carter-Wallace lectr. Princeton U., 1966, 77; Gifford lectr. U. Glasgow, Scotland, 1978-79; Dunham lectr. Harvard U., 1984; hon. prof. genetic medicine U. Cambridge Clin. Sch., 1989—; lectr. in field. Contbr. articles to sci. jours. Recipient Warren Triennial prize, 1968, William Bate Hardy prize Cambridge Philos. Soc., 1969, Albert Lasker Med. Rsch. award, 1971, Royal medal Royal Soc., 1974, Charles-Leopold Mayer prize French Acad., 1975, Gairdner Found. ann. award, 1978, Krebs medal FEBS, 1980, CIBA medal Biochem. Soc., 1981, Feldberg Found. prize, 1983, Rosenstiel award Brandeis U., 1986, Prix Louis Jeantet de Medecine, Switzerland, 1987, medal Genetics Soc. Am., 1987, Harvey prize Technion-Israel Inst. Tech., 1987, Hughlings Jackson medal Royal Soc. Medicine, 1987, Waterford Bio-Med. Sci. award Rsch. Inst. Scripps Clinic, 1988, Kyoto prize Inamori Found., 1990, Gairdner Found. Internat. award, Can., 1991, King Faisal Internat. prize, 1992, Disting. Achievement award Bristol-Myers Squibb, 1992. Fellow Royal Soc. (Croonian lectr. 1986, Royal medal 1974, Copley medal 1991), AAS, IASc (hon.) RSE (hon.), Royal Coll. Physicians (Neil Hamilton Fairley medal 1985) Royal Coll. Pathologists (hon.); mem. Max-Planck Soc., Deutsche Acad. Natural Sci. Leopoldina (Gregor Mendel medal 1970), Am. Philos. Soc. (fgn.), Real Acad. Ciencias (Spain), Am. Acad. Arts and Scis. (fgn. hon.), NAS (U.S., fgn. assoc.), Royal Soc. South Africa (fgn. assoc.), Acad. Europa, Chinese Soc. Genetics (hon.), Assn. Physicians Gt. Brit. and Ireland (hon.); associé étranger, Académie des Scis.; corr. Scientifique Emérite de l'INSERM. Avocation: rumination. Office: U Cambridge Sch Clin Med, Dept Medicine, Hills Rd, Cambridge CB2 2QQ, England

BRENT, ROBERT LEONARD, physician, educator; b. Rochester, N.Y., Oct. 6, 1927; s. Charles and Rose (Katz) B.; m. Lillian H. Hoffman, Aug. 21, 1949; children: David A., James R., Lawrence H., Deborah A. AB, U. Rochester, 1948, MD with honors, 1953, PhD, 1955, DSc (hon.), 1988. Fellow Nat. Found., Strong Meml. Hosp., 1953-54; intern pediatrics Mass. Gen. Hosp., Boston, 1954-55; chief radiation biology Walter Reed Army Inst. Rsch., 1955-57; mem. faculty Jefferson Med. Coll., 1955—, prof. radiology, 1962—, also prof. pediatrics, chmn. dept. and dir., Louis and Bess Stein prof. pediatrics, 1985—; apptd. Disting. prof. Thomas Jefferson U., 1989; hon. prof. Norman Bethume U. Med. Sci., People's Republic of China, 1992, West China U. Med. Scis., Chending, People's Republic of China, 1992; chmn. med. adv. bd. Nat. Found.; chmn. med. adv. bd., mem. fertility and maternal health com. FDA; mem. human embryology study sect. NIH, 1970-74; bd. trustees Health and Environ. Sci. Inst., 1991-94; pres. First Internat. Congress on Birth Defects, People's Republic of China, 1994. Editor in chief Teratology, 1976-93. Served with U.S. Army, 1955-57. Recipient Richie Meml. prize U. Rochester Med. Sch., 1953, Lindback Found. award for disting. teaching, 1968, Burlington Internat. award, 1990; travelling fellow Royal Soc. Medicine, 1971-72, vis. fellow FitzWilliam Coll., Cambridge, 1971-72; Lady Davis scholar Hadassah Med. Ctr., Jerusalem, 1983-84. Mem. Teratology Soc. (pres. 1967-68), AAAS, Radiation Research Soc., Am. Soc. Exptl. Pathology, Soc. Pediatric Research, Am. Pediatrics Soc., Am. Acad. Pediatrics, Soc. Exptl. Biology and Medicine, Phila. Coll. Physicians, Phila. Pediatric Soc., Am. Assn. Immunology, Soc. Developmental Biology, Congenital Malformations Assn. Japan, European Teratology Soc., Sigma Xi, Alpha Omega Alpha.

BRERETON, GILES JOHN, mechanical engineering researcher; b. Iver, England, Sept. 27, 1958; came to U.S., 1981; m. Anne Marie Duncan; children: Robert, Justin. BS, Imperial Coll. U. London, 1980; PhD, Stanford U., 1987. Asst. prof. U. Mich., Ann Arbor, 1987—; cons. in field. Contbr. articles to profl. jours. Imperial Coll. scholar, 1978-80; Fulbright scholar, 1981-82; recipient Silver medal Royal Soc. Arts, 1980. Mem. ASME, Am. Phys. Soc., Fine Particle Soc. Office: U Mich Dept Mech Engring & Applied Mechanics Ann Arbor MI 48109-2121

BRERETON, SANDRA JOY, engineer; b. Toronto, Ont., Can., Nov. 21, 1960, came to U.S. 1983, d. Frank William and M.Joyce (Morton) B.; m. Donald Thomas Blackfield, Sept. 22, 1990. BASc in Chem. Engring., U. Toronto, 1983; SM in Nuclear Engring., MIT, 1985, PhD in Nuclear Engring., 1987. Fusion systems engr. Ont. Hydro Canadian Fusion Fuels Tech. Project, 1987-90; cons. Pleasanton, Calif., 1990-91; safety analysis advisor Lawrence Livermore (Calif.) Lab., 1991—; mem. Safety Analysis Working Group Steering Com. of the Energy Facilities Contractors Group, Livermore, 1992; mem. edit. bd. Jour. of Fusion Energy, N.Y.C., 1992. Contbr. articles to profl. jours. Mem. Sigma Xi. Anglican. Office: Lawrence Livermore Nat Lab L-631 PO Box 808 Livermore CA 94551

BRESLAU, NEIL ART, endocrinologist, researcher, educator; b. Bklyn., Sept. 3, 1947; s. Nathan and Henny (Rabinowitz) B.; m. Sharon Erna Weidman, June 8, 1969; children: Joshua, Jeremy. BS in Biology, CUNY, 1968; MD, Vanderbilt U., 1972. Intern U. Vt. Med. Ctr. Hosp., Burlington, 1972-73; resident Long Island Jewish-Hillside Med. Ctr., New Hyde Park, N.Y., 1973-74; fellow in endocrinology Upstate Med. Ctr., Syracuse, N.Y., 1974-76, from instr. to asst. prof. medicine, 1976-78; asst. prof. medicine U. Tex.-Southwestern Med. Ctr., Dallas, 1978-84, assoc. prof., 1984-92; tenured prof., 1993—; assoc. dir. mineral sect. U. Tex.-Southwestern Med. Ctr., Dallas, 1992—; program dir. Dallas Gen. Clin. Rsch. Ctr., 1989—; attending physician Parkland Meml. Hosp., Dallas, 1978—; cons. in endocrinology and mineral metabolism Dallas VA Hosp., 1980—, Zale Lipsky Univ. Hosp., Dallas, 1990—; mem. instnl. rev. bd. for human rsch. U. Tex.-Southwestern Med. Ctr., 1982—. Contbr. articles to JCEM, Annals Internal Medicine, Am. Jour. Physiology, Am. Jour. Med. Sci.; editorial bd. Jour. Bone and Mineral Rsch., 1989—. Grantee NIH. Mem. Am. Fedn. Clin. Rsch., Endocrine Soc., Am. Soc. for Bone and Mineral Rsch., Assn. Gen. Clin. Rsch. Ctr. Program Dirs. Jewish. Achievements include discovery of new pathogenetic mechanism for hypercalcemia of malignancy production by lymphoma; research in pathophysiology of pseudohypoparathyroidism, interaction of nutritional elementa on Vitamin D metabolism. Home: 529 Tiffany Tr Richardson TX 75081 Office: U Tex Southwestern Med Ctr Dept Medicine 5323 Harry Hines Blvd Dallas TX 75235

BRESLOW, JAN LESLIE, scientist, educator, physician; b. N.Y.C., Feb. 28, 1943; s. Frank and Pearl (Feit) B.; m. Marilyn Ganon, June 25, 1965; children: Noah, Nicholas. AB, Columbia U., 1963, MA, 1964; MD, Harvard Med. Sch., 1968. Diplomate Am. Bd. Pediatrics. Intern in pediatrics, then jr. asst. resident Children's Hosp., Boston, 1968-70; staff assoc. Nat. Heart Lung and Blood Inst., Bethesda, Md., 1970-73; from instr. to assoc. prof. pediatrics Harvard Med. Sch., Boston, 1973-83; prof. Rockefeller U., N.Y.C., 1984-86, Frederick Henry Leonhardt prof., 1986—; dir. Lab. of Biochem. Genetics and Metabolism, Rockefeller U., N.Y.C., 1984—; sr. physician Rockefeller U. Hosp., N.Y.C., 1984—; mem. arteriosclerosis, hypertension and lipid metabolism adv. com. Nat. Heart, Lung and Blood Inst., 1986-90. Mem. editorial bd. Jour. Lipid Research, 1983-85, Arteriosclerosis, 1984—, Genomics, 1987—, Jour. Clin. Investigation, 1988—; cons. editor Jour. Biol. Chemistry, 1989—. Served as surgeon USPHS, 1970-73. Recipient established investigator award Am. Heart Assn., 1981-86, E. Mead Johnson award Am. Acad. Pediatrics, 1984, MERIT award Nat. Heart, Lung and Blood Inst., 1986—, Heinrich Wieland prize, 1991; Eugene Higgins fellow Columbia U., 1963-64. Mem. AAAS, Am. Tissue Culture Soc., Am. Acad. Pediatrics (E. Mead Johnson award 1984), Am. Heart Assn. (fellow Council on Arteriosclerosis 1978—, credentials com. 1984-86, awards com. 1984-87, chmn. awards com. 1987-89, program com. 1986-89, pathology research study com. 1985-87, research council N.Y.C. affiliate 1985-87, chmn. 1987-89, chair policy com., 1989-91, bd. dirs. N.Y.C. affiliate 1987-90; chmn. lipoprotein and lipid metabolism rsch. search, 1982—; mem. exec. com. Council on Arteriosclerosis, 1988—, Nat. Ctr. Rsch. Com., 1989-90, vice chm., 1990—; chmn. Am. Soc., Clin. Nutrition, Soc. Pediatric Research, N.Y. Acad. Scis., Am. Fedn. Clin. Research, Am. Soc. Clin. Investigation (v.p. 1987-88), Internat. Arteriosclerosis Soc., Am. Soc. Biol. Chemists, Assn. Am. Physicians. Avocations: reading, skiing, jogging. Achievements include research in lipid and lipoprotein metabolism, genetic and environ. causes of arteriosclerosis, inborn errors of metabolism. Home: 10 Horseguard Ln Scarsdale NY 10583-2311 Office: Rockefeller U Hosp 1230 York Ave New York NY 10021-6341*

BRESLOW, RONALD CHARLES, chemist, educator; b. Rahway, N.J., Mar. 14, 1931; s. Alexander E. and Gladys (Fellows) B.; m. Esther Greenberg, Sept. 7, 1955; children: Stephanie, Karen. A.B. summa cum laude, Harvard, 1952, M.A., 1953, Ph.D., 1955. NRC fellow Cambridge (Eng.) U., 1955-56; mem. faculty Columbia, 1956—, prof. chemistry, 1962-66, S.L. Mitchell prof., 1966—; univ. prof., 1992—; cons. to industry, 1958—; mem. medicinal chemistry panel NIH, 1964—; mem. adv. panel on chemistry NSF, 1971—; mem. sci. adv. com. Gen. Motors Corp., 1982—; A.R. Todd vis. prof. Cambridge U., 1982. Editor: Benjamin, Inc, 1962—; author: Organic Reaction Mechanisms, 1965, 2d edit., 1969; also articles.; mem. editorial bd. Organic Syntheses, 1964—, Jour. Organic Chemistry, 1969—, Jour. Bioorganic Chemistry, 1972—, Tetrahedron, 1975—, Tetrahedron Letters, 1975—, Proc. of Nat. Acad. Scis., 1984—. Trustee Rockefeller U., 1981—; bd. sci. advisers Alfred P. Sloan Found., 1978-85. Recipient Fresenius award Phi Lambda Upsilon, 1966, Mark Van Doren award Columbia, 1969, Roussel prize, 1978, Gt. Tchr. award Columbia U., 1981, T.W. Richards medal, 1984, A.C. Cope award, 1987, G.W. Kenner award U. Liverpool, Eng., 1988, Paracelsus prize Swiss Chem. Soc., 1990, Arthur Day award, 1990, Nat. Medal of Sci. NSF, 1991, U.S. Nat. medal of sci., 1991; Centenary lectr. London Chem. Soc., 1972. Fellow Am. Acad. Arts and Scis., Indian Acad. Scis.; mem. Am. Philos. Soc. (mem. council 1987—), NAS (chmn. chemistry div. 1974-77, award in chemistry 1989), Am. Chem. Soc. (Pure Chemistry award 1966, Baeheland medal 1969, chmn. div. organic chemistry 1970, Harrison Howe award 1974, Remsen award 1977, J. F. Norris award 1980, N.Y. sect. Nichols medal 1989), Phi Beta Kappa (first marshall 1952). Home: 275 Broad Ave Englewood NJ 07631-4350 Office: Columbia U Dept Chemistry 116th St & Broadway New York NY 10027

BRESSERS, DANIEL JOSEPH, utility executive; b. Kaukauna, Wis., May 3, 1956; s. Harold Henry and Mary Louise (Hanegraaf) B.; m. Nancy Ann Jansen, Mar. 3, 1979; children: Nicolaas, Emma. AAS in Internal Combine Engines, Milw. Sch. of Engring., 1976, BS in Mechanical Engring. Tech., 1978. Registered profl. engr., Wis. Field engr. GE Co., Oak Brook, Ill., 1978-85, svcs. supr., 1985-86; asst. supt. utilities Fort Howard Paper Corp., Green Bay, Wis., 1986-88, Midtec Paper Corp., Kimberly, Wis., 1988-90; utilities supt. Repap Wis. Inc. (formerly Midtec Paper Corp.), Kimberly, 1990—; instr. N.E. Wis. Tech. Coll., Green Bay, 1987—. Roman Catholic. Home: N355 Rogers Ln Appleton WI 54915 Office: Repap Wis Inc 433 N Main St Kimberly WI 54136

BRESSLER, MARCUS N., science administrator. Pres. M.N. Bressler and Assocs., Knoxville, Tenn. Recipient Bernard F. Langer Nuclear Codes and Standards award Am. Soc. Mech. Engrs., 1992. Office: M N Bressler & Assoc 829 Chateaugay Rd Knoxville TN 37923*

BRETSCH, CAREY LANE, civil/architectural engineering administrator; b. Philip, S.D., Feb. 10, 1958; s. Willard Leroy and Joy Ramona (Meseberg) B.; m. Lisa Laine Gilkerson, Mar. 19, 1983; children: Colin, Gayle. BS, S.D. State U., 1981, MS, 1990. Registerd profl. engr., S.D. Constrn. insp. Clark Engring., Aberdeen, S.D., 1981; constrn. engr. Paving Corp., Casper, Wyo., 1981-84; grad. teaching asst. dept. civil engring. S.D. State U., Brookings, 1984, facilities engr. phys. plant, 1984-89, facilities engr., archtl. mgr. phys. plant, 1989—. Mem. ASCE (pres. 1991-93, v.p. 1990-91, Outstanding Svc. award 1986-87, 88, Cert. of Appreciation 1989, Outstanding Young Civil Engr. 1989), S.D. Engring. Soc. (chmn. ethical and employment practices com. 1992—), Assn. Phys. Plant Adminstrs., Brookings Toastmasters (ed. v.p. 1991—, editor newsletter 1991), Brookings Optimists (bd. dirs. 1992). Republican. Lutheran. Home: 1340 5th St Brookings SD 57006 Office: SD State Univ Phys Plant Box 2201 Brookings SD 57007

BRETT, CLIFF, software developer. BS, MIT. Formerly tech. dir. Whitney/Demos Prodns.; then mem. R & D staff Wavefront Techs.; now mem. software R & D team MetroLight Studios, L.A. Office: MetroLight Studios Ste 400 5724 W 3rd St Los Angeles CA 90036-3078

BRETZKE, VIRGINIA LOUISE, civil engineer, environmental engineer; b. St. Louis, Apr. 5, 1963; d. Gerald Herman and Mary Louise (Asbury) Fochtmann; m. Steven M. Bretzke, July 6, 1985. BSCE, U. Mo., Rolla, 1985; MS in Environ. Health Engring., U. Kans., 1992. Registered profl. engr., Mo. Environ. engr. B&V Waste Sci. and Tech. Corp., Kans. City, Mo., 1985—. Mem. ASCE, NSPE, Mo. Soc. Profl. Engrs., Tau Beta Pi, Chi Epsilon. Office: B & V Waste Sci and Tech Corp PO Box 30240 Kansas City MO 64112

BREVIGNON, JEAN-PIERRE, physicist; b. St. Leu la Foret, France, Feb. 18, 1948; s. Andre Louis and Claudie Marie (Lemarchand) B.; m. Beatrice B. Arbis, Jan. 28, 1984; children: Thomas J., Eleonore M. D in Physics, Paris VII U., 1976. Process engr. Companie Internationale pour l'Informatique, Orsay, France, 1975-77, Thomson Sescosem, Orsay, 1977-79; devel. sect. head Thomson Etude et Fabrication de Circuits Integres Speciaux, Grenoble, France, 1979-82; process devel. supr. Fairchild Semiconductor, San Jose, Calif., 1982-83; engring. mgr. Hyundai-Modern Electronics Inc., Santa Clara, Calif., 1983-84, Thomson Semiconducteur, Grenoble, 1984-86; research and devel. mgr. L'Air Liquide, Jouy-en-Josas, France, 1986-88; liaison officer electronics L'Air Liquide, Paris, 1988-91; tech. dir., mgr. field svc. Europe Submicron Systems, Les Ulis, France, 1992—. Patentee semiconductor processing. Mem. IEEE, Internat. Soc. Optical Engring., Inst. Environ. Sci., N.Y. Acad. Sci. Home: Moulin Chamois 4 Rue des Moulins, 60800 Duvy France

BREZOVEC, PAUL JOHN, environmental specialist; b. Johnstown, Pa., Sept. 26, 1957. BS in Chem. Engring., Carnegie Mellon U., 1979; MBA, Pepperdine U., 1987. Registered Profl. engr., Calif., Pa. Engr. I Monsanto Co., Saint Louis, 1979-83; engr. II Monsanto Co., Stonington, Conn., 1983-85; sr. engr. Monsanto Co., Anaheim, Calif., 1985-90; environ. specialist Monsanto Co., Carson, Calif., 1990-92; sr. tech. staff CTC, Johnstown, Pa., 1993—. Home: RD1 Box 60 Johnstown PA 15906-9724

BRICKELL, CHARLES HENNESSEY, JR., marine engineer, retired military officer; b. Memphis, Apr. 13, 1935; s. Charles Hennessey and Mary Ellen (Viau) B.; m. Barbara Virginia Davis, Jan. 4, 1958; children: David Brian, Patricia Ellen, Susan Elizabeth, Timothy Paul, Joel Howard. BS in Marine Engring., U.S. Merchant Marine Acad., 1957; MA in Bus. Mgmt., Cen. Mich. U., 1980. Enlisted USN, 1953, commd. ensign, 1957, advanced through grades to rear adm., 1984; dir. research and devel. Undersea and Strategic Warfare, and Nuclear Energy, 1984-87; dir. USN Strategic Def. Initiative Program, 1984-88; dep. dir. Navy Rsch. Devel., Test and Evaluation, 1987-88; ret. USN, 1988; gen. mgr. advanced technologies Stone & Webster Engring. Corp., Boston, 1988-91; dir. Ops. ea. region N.Am. Energy Svcs., Issaquah, Wash., 1991—; mem. bd. advisors Applied Rsch. Lab. Pa. State U.; cons. NAS. Decorated Def. Superior Service Medal, Legion of Merit with three Gold Stars, Meritorious Service Medal with two Gold Stars. Mem. Sigma Iota Epsilon. Roman Catholic. Avocations: baseball, basketball sports officiating. Home: 5503 Teak Ct Alexandria VA 22309-4218

BRICKEY, JAMES ALLAN, owner environmental testing laboratory, consultant; b. Fayetteville, Ark., Feb. 14, 1965; s. Ralph Allan and Vivian Ann (Cowger) B. BS in Biology, Va. Poly. Inst., 1988. With extraction dept. Commonwealth Lab., Inc., Richmond, Va., 1989-90; lab. mgr. Commonwealth Lab., Inc., Harrisonburg, Va., 1990-93; lab. owner Bent Tree Lab., Manakin-Sabot, Va., 1993—; bd. dirs. Friends of the North River, Bridgewater, Va. Mem. Water Environ. Fedn., Coun. on Soil Testing and Plant Analysis, Am. Soc. Agronomy, Soil Sci. Soc. of Am., Lions. Republican. Home: 1170 Dover Creek Ln Manakin-Sabot VA 23103 Office: Bent Tree Lab 1170 Dover Creek Ln Manakin Sabot VA 23103

BRIDENBAUGH, PETER REESE, industrial research executive; b. Franklin, Pa., July 28, 1940; s. Charles Sumner and Helen Catherine (Reese) B.; m. Mary Ann Ellis, Apr. 17, 1965; children: Matthew B., Gabrielle L. BSME, Lehigh U., 1962, MS in Metallurgy, 1966; PhD in Materials Sci., MIT, 1968. With Alcoa Labs., Alcoa Ctr., Pa., R & D group leader, sect. head, spl. program engr. Warrick Ops., 1968-75, mgr., 1975-78; mgr. quality assurance Alcoa, Tenn., 1978-80; dir. ops. Alcoa Labs., 1980-83, dir., 1983-84, v.p. R & D, 1984-91, exec. v.p., 1991—; mem. vis. com. U. Pitts., 1984—, Carnegie-Mellon U., 1984—, Pa. State U., 1984—, Stanford U., 1987—, U. Va., 1987—, Lehigh U., 1989—, Northwestern U., 1991—; chmn. Fedn. Materials Socs. 10th Biennial Conf., 1988. Patentee in field. Fellow Am. Soc. Metals; mem. AIME, NAE, Indsl. Rsch. Inst., Dirs. Indsl. Rsch., Sigma Xi. Clubs: Duquesne (Pitts.), Fox Chapel Golf (Pitts.). Office: Aluminum Co Am/Alcoa Ctr 100 Technical Dr New Kensington PA 15069-0001

BRIDGE, T(HOMAS) PETER, psychiatrist, researcher; b. Nashville, June 2, 1945; s. Thomas Gale and Hilma Elizabeth (Hartzler) B.; m. Mary L. Matthews, Dec. 15, 1969 (div. Sept. 1974); m. Beth J. Soldo, Sept. 20, 1975. BA, Duke U., 1967; MD, Med. Coll. Va., 1971. Diplomate Am. Bd. Psychiatry and Neurology. Rsch. fellow Duke U., Durham, N.C., 1972-74; clin. staff fellow NIMH, Bethesda, Md., 1977-79; chief unit on geriatrics NIMH, Washington, 1980-83; sci. advisor Alcohol, Drug, and Mental Health Adminstrn., Rockville, Md., 1983-86, AIDS coord., 1986-90; chief clin. trials br. Nat. Inst. on Drug Abuse, Rockville, 1990—. Editor: AIDS Neuropsychiatry, 1989; contbr. more than 75 articles to profl. jours. Named J.D. Lane Outstanding Investigator, USPHS, 1984; recipient New Investigator award Am. Geriatrics Soc., 1985. Fellow Coll. Internat. Neuropsychopharmacology; mem. AAAS, Am. Coll. Neuropsychopharmacology. Achievements include patents for novel pharmacologic treatments for cognitive enhancement, chronic fatigue, and psoriasis. Home: 2910 Brandywine St NW Washington DC 20008 Office: CTB MDD NIDA 11A 55 5600 Fishers Ln Rockville MD 20857

BRIDGEMAN, DAVID MARSH, chemical company executive; b. Bridgeport, Conn., July 10, 1943; s. Paul Loren and Helen Adele (Marsh) B.; m. Hazel Wilma Williams, Oct. 4, 1974; stepchildren: Deborah Diserens, Sheila Floyd. AB in Chemistry, Cornell U., 1965. Tech. sales rep. Union Carbide Corp., Houston, 1970-73, Noury Chem. Co., Houston and L.A., 1973-76, Interplastic Corp., Atlanta, 1979-85; product mgr. Interplastic Corp., San Jacinto, Calif., 1985-87; sr. devel. chemist Dow Chem. Co., Freeport, Tex., 1987-89; account specialist Dow Chem. Co., Atlanta, 1989-92, sr. account specialist, 1992—. Lt. USNR, 1965-69. Mem. Nat. Assn. Corrosion Engrs., Soc. Plastics Industry (com. mem. 1979-87), Alpha Phi Delta (v.p. 1965). Home: 410 Rosewood Ln Cartersville GA 30120 Office: Dow Chem Co Ste 590 115 Perimeter Center Pl Atlanta GA 30346

BRIDGFORTH, ROBERT MOORE, JR., aerospace engineer; b. Lexington, Miss., Oct. 21, 1918; s. Robert Moore and Theresa (Holder) B.; student Miss. State Coll., 1935-37; BS, Iowa State Coll., 1940; MS, MIT, 1948; postgrad. Harvard U., 1949; m. Florence Jarnberg, November 7, 1943; children: Robert Moore, Alice Theresa. Asst. engr. Standard Oil Co., of Ohio, 1940; teaching fellow M.I.T., 1940-41, instr. chemistry, 1941-43, research asst., 1943-44, mem. staff div. indsl. cooperation, 1944-47; asso. prof. physics and chemistry Emory and Henry Coll., 1949-51; rsch. engr. Boeing Airplane Co., Seattle, 1951-54, rsch. specialist 1954-55, sr. group engr., 1955-58, chief propulsion systems sect. Systems Mgmt. Office, 1958-59, chief propulsion rsch. unit, 1959-60; chmn. bd. Rocket Rsch. Corp. (name now Rockcor, Inc.), 1960-69, Explosives Corp. Am., 1966-69. Fellow AIAA (assoc.), Brit. Interplanetary Soc., Am. Inst. Chemists; mem. AAAS, Am. Astronautical Soc. (dir.), Am. Chem. Soc., Am. Rocket Soc. (pres. Pacific NW 1955), Am. Ordnance Assn., Am. Inst. Physics, Am. Assn. Physics Tchrs., Tissue Culture Assn., Soc. for Leukocyte Biology, N.Y. Acad. Scis., Combustion Inst., Sigma Xi. Home: 4325 87th Ave SE Mercer Island WA 98040-4127

BRIESMEISTER, RICHARD ARTHUR, chemist; b. Mansfield, Ohio, Apr. 21, 1942; s. Arthur Conrad and Eileen Patricia (Moore) B.; m. Margaret Knudson (div.); 1 child, Anne Marie Briesmeister Rowlison. BS in Chemistry, U. Wyo., 1965. Technician Los Alamos (N.Mex.) Nat. Lab., 1965-79, staff mem., 1980-85, project leader, 1985-88, dep. group leader, 1988-92, program officer, maintenance mgmt. office, 1992—. Mem. Am. Glovebox Soc. Achievements include research in long pathlength spectroscopic cell. Office: Los Alamos Nat Lab MS M720 PO Box 1663 Los Alamos NM 87545-0001

BRIGGS, DAVID GRIFFITH, mechanical engineer, educator; b. Chgo., Apr. 4, 1932; s. David Reuben and Genevieve (Griffith) B.; m. Beverly Bauer, Sept. 22, 1956; children: David S., Glenn G., Jonathan H. BA, Dartmouth U., 1954, MS, 1955; PhD, U. Minn., 1965. Asst. prof. coll. engring. Rutgers U., New Brunswick, N.J., 1964-68, assoc. prof., 1968-88, prof., 1988—. Lt. j.g. USN, 1955-58. Office: Rutgers Coll Engring Brett & Bowser Rds Piscataway NJ 08855-0909

BRIGGS, GEORGE MADISON, civil engineer; b. Albany, N.Y., May 4, 1927; s. Franklin H. and Emma E. (Briggs) B.; B.S. in C.E., Purdue U., 1952; M.P.A., SUNY, Albany, 1968; m. Jean M. Cully, Oct. 31, 1954; children: George Madison, Barbara Jean. Engr., N.Y. State Dept. Public Works, Albany, 1952-56, asst. planning and location engr. dist. 1, 1956-60, resident engr. Saratoga County, 1961-64; assoc. civil engr. N.Y. State Dept. Transp., Albany, 1965-67, dir. hwy. maintenance, 1968-72, dir. maintenance, 1972-80, asst. commr. ops., 1980-83; pres. Briggs Engring., P.C., Greenwich, N.Y., 1983-92; exec. dir. N.Y. Bituminous Distbrs. Assn., 1983-92; ret. 1992; mem. coms. Transp. Research Bd., Nat. Acad. Scis.; instr. Fed. Hwy. Adminstrn.'s Course in bridge maintenance for state DOT bridge maintenance technicians and suprs. Co-author Am. pub. works assn. Roads and Streets Manual. Chmn. transp. com. 1980 Olympic Winter Games; cons. FHWA , Saudi Arabia, 1983, OAS, Caracas, Venezuela, 1984. Mem. Planning Bd. Town of Easton. Served with AUS, 1945-47. Registered profl. engr., N.Y. Mem. Nat. Soc. Profl. Engrs., N.Y. State Soc. Profl. Engrs. (pres. Tri-County chpt.), Am. Public Works Assn., Nat. Inst. Transp. (exec. council), Nat. Assn. State Hwy. and Transp. Ofcls. (vice chmn. maintenance com., task force leader bridge maintenance), SAR, Soc. Mayflower Descs. Mem. Soc. Friends. Clubs: Masons, Shriners, Elks. Contbr. articles to profl. publs. Home and Office: Burton Rd # 401M Greenwich NY 12834

BRIGGS, JEFFREY LAWRENCE, ecologist; b. Cleve., Jan. 28, 1943; s. Gordon and Dorothy (Callender) B.; m. Jean F. McLean, June 11, 1965; 1 child, Noah W. BS, U. Denver, 1965; MA, Oreg. State U., 1968, PhD, 1970. Asst. prof. U. Wis., Whitewater, 1970-74; acting curator of mammals Milw. Pub. Mus., 1974-75; lectr. U. Wis., Milw., 1975-78, Waukesha, 1978-79; assoc. prof. King Faisal U. Coll. Medicine, Dammam, Saudi Arabia, 1979-84; project mgr. SCS Engrs., Reston, Va., 1985-93, The Earth Technology Corp., 1993—. Mem. Ecol. Soc. Am. (cert. ecologist), Soc. Wetland Scientists, Am. Soc. Ichthyologists and Herpetologists. Office: The Earth Tech Corp 1420 King St Ste 600 Alexandria VA 22314

BRIGGS, RODNEY ARTHUR, agronomist, consultant; b. Madison, Wis., Mar. 18, 1923; s. George McSpadden and Mary Etta (McNelly) B.; m. Helen Kathleen Ryall, June 1, 1944; children: Carolyn, Kathleen, David, Andrew, Amy. Student, Oshkosh (Wis.) State Coll., 1941-42; B.S. in Agronomy, U. Wis., 1948; Ph.D. in Field Crops, Rutgers U., 1953. Extension asso. farm crops Rutgers U., New Brunswick, N.J., 1949-50, 52-53; mem. faculty U. Minn., 1953-73; supt. West Central Sch. and Expt. Sta., Morris, Minn., 1959-60; prof. agronomy, dean U. Minn.; administrv. head, provost U. Minn. (Morris Campus), 1960-69, sec. bd. regents, 1971-72, exec. asst. to pres., 1971-73; on leave of absence Ford Found. as asso. dir., dir. research Internat. Inst. Tropical Agr., Ibadan, Nigeria, 1969-71; pres. Eastern Oreg. State Coll., La Grande, 1973-83; exec. v.p. Am. Soc. Agronomy/Crop Sci. Soc. Am./Soil Sci. Soc. Am., Madison, Wis., 1982-85; ind. cons., 1985—; chmn. Nat. Silage Evaluation Com., 1957; sec. Minn. Corp Improvement Assn., 1954-57; columnist crops and soils Minn. Farmer mag., 1954-59; judge grain and forage Minn. State Fair, 1954-61; mem. ednl. mission to Taiwan, Am. Assn. State Colls. and Univs., 1978, chmn. ednl. mission to Colombia, 1982, state rep., 1974-76, mem. spl. task force of pres.'s on intercollegiate athletics, 1976-77, mem. nat. com. on agr., renewable resources and rural devel., 1978-82, nat. sec.-treas., 1980-82; mem. com. on govt. relations Am. Council Edn. Com., 1981; mem. Gov.'s Commn. on Fgn. Lang. and Internat. Studies, State Oreg., 1980-83; Mem. Gov.'s Commn. Law Enforcement, 1967-69; adv. com. State Planning Agy., 1968-69, Minn. Interinstnl. TV, 1967-69; invited participant The Role of Sci. & Emergency Societies in Devel., African Regional Seminar AAAS, 1984. Com. mem. African sci. jour. AAAS, 1986-88. Bd. dirs. Rural Banking Sch., 1967-69; bd. dirs. Channel 10 ETV, Appleton, Minn., Grande Ronde Hosp., 1980-83; chmn. policy adv. com. Oreg. Dept. Environ. Quality, 1979-81. Served with inf. AUS, 1942-46, 50-52. Recipient Staff award U. Minn., 1959, spl. award U. Minn. at Morris, 1961; commendation Soil Conservation Soc. Am., 1965; Rodney A. Briggs Library named in his honor U. Minn., Morris, 1974. Fellow AAAS, Am. Assn. State Colls. and Univs., Soil Conservation Soc. Am. (pres. emeritus Ea. Oreg. State Coll. chpt. 1988); mem. ACLU, Am. Soc. Agronomy, Am. Inst. Biol. Scis. (bd. dirs. 1982-83), Pres.' Club U. Minn., Sigma Xi, Alpha Gamma Rho. Congregationalist. Home: 1109 Gilbert Rd Madison WI 53711-2504

BRIGGS, WINSLOW RUSSELL, plant biologist, educator; b. St. Paul, Apr. 29, 1928; s. John DeQuedville and Marjorie (Winslow) B.; m. Ann Morrill, June 30, 1955; children: Caroline, Lucia, Marion. B.A., Harvard U., 1951, M.A., 1952, Ph.D., 1956. Instr. biol. scis. Stanford (Calif.) U., 1955-57, asst. prof., 1957-62, asso. prof., 1962-66, prof., 1966-67; prof. biology Harvard U., 1967-73, Stanford U., 1973—; dir. dept. plant biology Carnegie Instn. of Washington, Stanford, 1973-93. Author: (with others) Life on Earth, 1973; Asso. editor: (with others) Annual Review of Plant Physiology, 1961-72; editor (with others), 1972—; Contbr. (with others) articles on plant growth and devel. and photobiology to profl. jours. Recipient Alexander von Humboldt U.S. sr. scientist award, 1984-85; John Simon Guggenheim fellow, 1973-74, Deutsche Akademie der Naturforscher Leopoldina, 1986. Fellow AAAS; mem. Am. Soc. Plant Physiologists (pres. 1975-76), Calif. Bot. Soc. (pres. 1976-77), Nat. Acad. Scis., Am. Acad. Arts and Scis., Am. Inst. Biol. Scis. (pres. 1980-81), Am. Soc. Photbiology, Bot. Soc. Am., Nature Conservancy, Sigma Xi. Home: 480 Hale St Palo Alto CA 94301-2207 Office: Carnegie Inst Washington Dept Plant Biology 290 Panama St Palo Alto CA 94305-4170

BRIGHAM, JOHN CARL, psychology educator; b. Glen Ridge, N.J., Dec. 19, 1942; s. John C. and Jean (Dipman) B.; m. Gayle Bradley (div. 1975); children: Bradley S., Tracy, David; m. Karol A. Solomon, Feb. 21, 1982; 1 child, Jason. BA, Duke U., 1964; MA, U. Colo., 1968, PhD, 1969. Prof. psychology Fla. State U., Tallahassee, 1969—; cons., expert witness, for attys. in U.S. Author: Social Psychology, 1986, 2nd. rev. edition, 1991; contbr. articles to profl. jours., chpt. to book. Mem. Common Cause, Sierra Club; guardian of childs rights, Tallahassee, Fla., 1991—. Grantee NSF, Washington, 1971-73, 78-81, 85-90, Nat. Inst. Justice, Washington, 1982-84. Fellow Am. Psychol. Assn. Democrat. Office: Fla State U Dept Psychology Tallahassee FL 32306

BRIGHTON, JOHN AUSTIN, mechanical engineer, educator; b. Gosport, Ind., July 9, 1934; s. John William and Esther Pauline B.; m. Charlotte L. McCarty, Mar. 20, 1953; children: Jill, Kurt, Eric. B.S., Purdue U., 1959, M.S., 1960, Ph.D., 1963. Draftsman Switzer Corp., Indpls., 1952-56; instr. Purdue U., 1960-63; asst. prof. mech. engring. Carnegie-Mellon U., 1963-65; asst. prof. Pa. State U., 1965-67, asso. prof., 1967-70, prof., 1970-77; prof. Mich. State U., 1977-82, chmn. dept. mech. engring., 1977-82; dir. Sch. Mech. Engring. Ga. Inst. Tech., 1982—; Chmn. Community Sponsors Inc., State College, Pa., 1976-77; chmn. Pre-Trial Alts. Program for First Offenders, State College, 1976-77; bd. dirs. Impression 5. Author: (with Hughes) Fluid Dynamics, 1966. NSF grantee, 1975-77; NIH grantee, 1974-78. Mem. ASME (Engr. of Yr. award for Central Pa. 1977, tech. editor Jour. Biomech. Engring. 1976-79), Am. Soc. Engring. Edn., Am. Soc. Artificial Internal Organs. Home: 525 Kenbrook Dr NW Atlanta GA 30327-4939 Office: Ga Inst Tech Atlanta GA 30332

BRIGMAN, JAMES GEMENY, chemical engineer; b. Fort Worth, Tex., July 9, 1965; s. George Henry and Ada Louise (Gemeny) B. BS in Chem. Engring., Rice U., 1987, M in Chem. Engring., 1988. Registered profl. engr., Tex. Process engr. Solvay Am., Deer Park, Tex., 1988-90, unit mgr., 1990-92, sr. process engr., 1992—. Nat. Merit scholar, 1983, Brown Engring. scholar, 1984. Mem. Am. Inst. Chem. Engrs. (regional coord. 1991--), Am. Chem. Soc., Phi Lambda Upsilon (pres. 1987-88). Methodist. Office: Solvay Interox 1230 Battleground Rd Deer Park TX 77536

BRILES, DAVID E(LWOOD), microbiology educator; b. Hempstead, N.Y., May 26, 1945; s. Worthie E. and Clara R. Briles; m. Marilyn J. Crain; children: Rachel, Travis. BA, U. Tex., 1967; PhD, Rockefeller U., 1973. Grad. fellow Rockefeller U., N.Y.C., 1967-73; USPHS and U.S. NIH rsch. fellow Washington U., St. Louis, 1973-75, rsch. instr., 1975-78; asst. prof. U. Ala., Birmingham, 1978-82, assoc. prof., 1982-88, prof. microbiology, 1989—; bacterial vaccines advisor WHO, Washington, 1992, Pan Am. Health Orgn., Havanna, Cuba, 1992—; lectr. Am. Soc. Microbiology Found. for Microbiology; vis. lectr. U. Edwardsville, Ill., 1977. Recipient Individual Rsch. Fellowship NIH, Nat. Inst. of Allergy and Infectious Diseases, 1973-75, Rsch. and Career Devel. award, 1982-87. Mem. AAAS, Am. Soc. Microbiology, Am. Assn. Immunologists. Office: U Ala 801 SDB University Sta Birmingham AL 35294

BRILES, JOHN CHRISTOPHER, civil engineer; b. Kirkwood, Mo., May 23, 1963; s. John Christopher and Carol Ann Briles; m. Vonda Kay Marner, Aug. 9, 1986; children: Aaron Christopher, Jessica Marie. BS in Civil Engring., U. Mo., Rolla, 1986. Registered profl. engr., Mo. Constrn. inspector Mo. Hwy. & Transp. Dept., St. Louis, 1986-91; field engr., project engr. Massman Constrn. Co., St. Louis, 1991—. Youth leader Washington Park Fellowship, St. Louis, 1989—, ch. elder, 1992—. Mem. ASCE (assoc.), Soc. Am. Mil. Engrs. Republican. Mem. Assembly of God. Home: 7745 New Hampshire Saint Louis MO 63123

BRIMACOMBE, JAMES KEITH, metallurgical engineering educator, researcher, consultant; b. Windsor, N.S., Can., Dec. 7, 1943; s. Geoffrey Alan and Mary Jean (MacDonald) B.; m. Margaret Elaine Rutter, Feb. 6, 1970; children: Kathryn Margaret, Jane Margaret. B of Applied Sci. with honors, U. B.C., 1966; PhD, U. London, 1970, DSc in Engring., 1986. Registered profl. engr., B.C. Asst. prof. metall. engring. U. B.C., Vancouver, Can., 1970-74, assoc. prof., 1974-79, prof., 1979-80, Stelco prof., 1980-85, Stelco/Nat. Scis. and Engring. Rsch. Coun. Can. prof., 1985-91, dir. Ctr. for Metall. Process Engring., 1985—; Alcan chair in materials process engring. U. B.C., Vancouver, 1992—; Arnold Markey lectr. Steel Bar Mill Assn., 1981; retained cons. Hatch Assocs., Toronto, 1984-89; cons. over 60 metall. cos. Author: Continuous Casting, vol. II, 1984, The Mathematical and Physical Modeling of Primary Metals Processing Operations, 1988; contbr. numerous articles to profl. jours.; patentee in field. Capt. Can. Air Force, 1961-70. Recipient B.C. Sci. and Engring. Gold medal Sci. Coun. B.C., 1985, Ernest C. Manning Prin. award The Manning Trust, 1987, Izaak Walton Killam Meml. prize in engring. The Can. Coun., 1989, Corp. Higher End. Forum award, 1989, Commemorative medal for 125th Anniversary of Can. Confedn., numerous awards for publs.; Can. Commonwealth fellow Brit. Coun., 1966, E.W.R. Steacie fellow Nat. Scis. and Engring. Rsch.

Coun. Can., 1980; Officer Order of Can. Fellow Can. Inst. Mining and Metallurgy, Canadian Acad. of Engring., Royal Soc. Can., Minerals, Metals and Materials Soc. (founding chmn. extraction and processing div. 1989-92, extractive metallurgy lectr. 1989, pres. 1993-94); mem. Metall. Soc. of Can. Inst. Mining and Metallurgy (pres. 1985-86, Alcan award 1988), Iron and Steel Soc. (disting. mem., bd. dirs. 1989—, Howe Meml. Lectr. 1993), Inst. Materials, Am. Soc. Metals (now ASM Internat., Can. Coun. lectureship 1986), Iron and Steel Inst. Japan, Sigma Xi, (UBC hon. frat.). Roman Catholic. Avocations: travel, jogging, photography.

BRINCK, KEITH, computer scientist; b. Sydney, NSW, Australia, Mar. 27, 1955; s. Niels Sigvald and Anna-Marie (Koefoed) B. BSc with 1st class honors, U. Sydney, 1977, PhD in Computer Sci., 1982. Sr. tutor U Sydney, 1981-82, sr. programmer, 1982-83; vis. asst. prof. U. Iowa, Iowa City, 1984, asst. prof., 1985-86; systems devel. mgr. Bain & Co., Sydney, 1987—; cons. ASCOMP PTY, Ltd., Sydney, 1978-79, Computer Scis. Australia, Sydney, 1983. Contbr. articles to profl. jours. Recipient U. scholarship Australian Fed. Govt., 1973, Postgrad. Rsch. award, 1977-82. Mem. Assn. for Computing Machinery, Australian Unix Users Group, Revesby heights Squash Club. Avocations: squash, bridge, computing. Home: 11/42-44 St Georges Parade, Hurstville 2220 NSW, Australia Office: Bain & Co, 225 George St, Sydney 2000 NSW, Australia

BRINCKMAN, FREDERICK EDWARD, JR., research chemist, consultant; b. Oakland, Calif., June 24, 1928; s. Frederick Edward and Rose (Mittelman) B.; m. Margaret Jean Hess, Dec. 23, 1954; children: Paul Dent, Brian Edward. BSc, U. Redlands, 1954; AM, Harvard U., 1958, PhD, 1960. Rsch. chemist, head propellant br. U.S. Naval Ordnance Lab., Corona, Calif., 1960-61; sci. staff asst., head gen. rsch. div. U.S. Naval Propellant Plant, Indian Head, Md., 1961-64; rsch. chemist chem. stability and corrosion div. Nat. Bur. Standards (now Nat. Inst. Standards and Tech.), Washington, 1964-78; group leader chem. and biodegradable processes Nat. Bur. Standards (now Nat. Inst. Standards and Tech.), Gaithersburg, Md., 1978-88, rsch. chemist, coord. bioprocessing polymers div., 1988—; rsch. fellow NIH, Washington, 1958-60; teaching fellow Harvard U., Cambridge, Mass., 1956-58; R. Merton vis. prof. geosci. Johannes Gutenberg U., Mainz, Fed. Republic of Germany, 1978-79; adj. prof. inorganic chemistry U. Md., College Park, 1985-86; mem. planning and oversight panel Chesapeake Rsch. Consortium, Solomons, Md., 1990—; Harold A. Iddles lectr. dept. chemistry U. N.H., Durham, 1990; over 150 invited major lectures. Author: Proceedings of the Sixth International Symposium on Controlled Release of Bioactive Materials, 1978, Organometals and Metalloids: Occurrence and Fate in the Environment, 1978, Environmental Speciation and Monitoring Needs for Trace Metal-Containing Substances from Energy-Related Processes, 1981, The Importance of Chemical Speciation in Environmental Processes, 1986; contbr. more than 150 articles to sci. jours. Panel mem. bd. examiners CSC, Washington, 1962-64; scoutmaster troop 1323 Boy Scouts Am., Laytonsville, Md., 1966-78; pres. Upper Rockcreek Civic Assn., Derwood, Md., 1970-74. Recipient silver medal U.S. Dept. Commerce, 1974, gold medal, 1980; scholar Mayr Found., 1953. Fellow Am. Inst. Chemists, Royal Soc. Chemistry (U.K.); mem. Am. Chem. Soc., Internat. Union Pure and Applied Chemistry (affiliate, lectr. 1989), Sigma Xi. Achievements include five patents; research in synthesis and characterization of bioactive organometallic compounds; toxicity, environmental speciation, and monitoring of trace organometals and organometalloids; global biomethylation of the elements and its role in organometallic chemistry and biotechnology; environmental transport and biotransformations of metals; biodegradation of organometallic copolymers. Home: N-8909 560 N St SW Washington DC 20024-4546 Office: Nat Inst Standards and Tech Rm B320 Bldg 220 Gaithersburg MD 20899

BRINE, JOHN ALFRED SEYMOUR, physician, consultant; b. Perth, Australia, Mar. 13, 1926; s. William Lane and Netta Gwendolyn (Wright) B.; m. Robyn Anne Santo; children: Lindesay, Christina, Jennifer. MB BS, Melbourne U., 1949; M.R.C., Physicians, London, 1955. Resident med. officer Royal Perth Hosp., Australia, 1949-52, cons. physician, 1962-91; resident med. officer Royal Postgrad. Med. Sch., London, 1953-55; med. officer Whittington Hosp., London, 1956-57; pvt. practice Perth, 1962—; emeritus cons. Royal Perth Hosp., Australia, 1992—. Author: Looking for Milligan, 1991. Fellow Royal Coll. Physicians; mem. Milligan Soc. (pres. 1980-89). Anglican. Avocations: music, art appreciation, history, bush walking. Office: 7 Richardson St, West Perth 6005, Australia

BRINEGAR, CLAUDE STOUT, oil company executive; b. Rockport, Calif., Dec. 16, 1926; s. Claude Leroy Stout and Lyle (Rawles) B.; m. Elva Jackson, July 1, 1950 (div.); children: Claudia, Meredith, Thomas; m. Mary Katharine Potter Garrity, May 14, 1983. BA, Stanford U., 1950, MS, 1951, PhD, 1954. V.p. econs. and planning Union Oil (now Unocal), L.A., 1965; pres. pure oil div. Union Oil (now Unocal), Palatine, Ill., 1965-69; sr. v.p., pres. refining and mktg. Union Oil (now Unocal), L.A., 1969-73; U.S. sec. of transp. Washington, 1973-75; sr. v.p. adminstr. Unocal Corp., L.A., 1975-85, mem. exec. com., 1975-92, exec. v.p., chief fin. officer, 1985-91, also bd. dirs., vice chmn. bd., also bd. dirs.; founding dir. Consol. Rail Corp., Washington, 1974-75, 90—; bd. dirs. Maxicare Health Plans, Inc.; vis. scholar Stanford U., 1992—. Author: monograph on econs. and price behavior, 1970; contbr. articles to profl. jours. on statistics and econs. Chmn. Calif. Citizens Compensation Commn., 1990—; mem. regional selection panel White House Fellows Program, 1976-83, chmn., 1983. Mem. Am. Petroleum Inst. (bd. dirs. 1976-85, 88-91, hon. life dir. 1992), Calif. Club, Georgetown Club, Internat. Club, Boothbay Harbor Yacht Club, Southport Yahct Club, Phi Beta Kappa, Sigma Xi. Republican. Avocations: collector of first editions of Mark Twain, running. Office: Unocal Corp Unocal Ctr 1201 W 5th St Los Angeles CA 90017-1461

BRINKERHOFF, LORIN C., nuclear engineer, management and safety consultant; b. St. Anthony, Idaho, June 4, 1929; s. James Byron and Bessie Hazel (Miller) B.; m. Donna Lee Lords, Nov. 27, 1951; children: Kathleen Rae, Diane Lee, Sandra Lynne, Bonnie Jo, Dirk Lorin, Michael Lorin. BSChemE, U. Utah, 1955; postgrad. MIT, 1970, Safety and Reliability Directorate (Eng.), 1974, Nuclear Power Devel. Establishment (Scotland), 1981. Rsch. specialist GE, Hanford, Wash., 1952-53; reactor ops. foreman Phillips Petroleum Co., Idaho Falls, Idaho, 1955-58; critical facility mgr. Lawrence Radiation Lab., Nevada Test Site, 1958-62; sr. nuclear engr. Aerojet Gen. Corp., Nevada Test Site, 1962-69; sr. reactor safety specialist AEC, Germantown, Md., 1969-81; chief reactor safety br. U.S. Dept. Energy, Germantown, 1981-86; mgr. tech. safety appraisal team, 1986-89; ind. nuclear safety cons., 1989—. With U.S. Army, 1950-51. Mem. Am. Nuclear Soc. (standards com. 1980-89), Am. Nat. Standards Inst. (standards com. 1978-85). Democrat. Mem. Ch. of Jesus Christ of Latter-day Saints (bishop). Home and Office: 9671 S Countess Way South Jordan UT 84065

BRINKHOUS, KENNETH MERLE, pathologist, educator; b. Clayton County, Iowa, May 29, 1908; s. William and Ida (Voss) B.; m. Frances E. Benton, Sept. 5, 1936; children: William Kenneth, John Robert. Student, U.S. Mil. Acad., 1925; A.B., U. Iowa, 1929, M.D., 1932; D.Sc., U. Chgo., 1967. Asst. in pathology State U. Iowa, 1932-33, instr., 1933-35; asso. in pathology U. Iowa, 1935-37, asst. prof., 1937-45, asso. prof., 1945-46; prof. pathology U. N.C., Chapel Hill, 1946-61; alumni distinguished prof. U. N.C., 1961-80, emeritus, 1980—; Mem. Nat. Adv. Heart and Lung Council, 1969-74; chmn. med. adv. council Nat. Hemophilia Found., 1954-73; sec. gen. Internat. Com. Hemostasis and Thrombosis, 1966-78. Bd. editors Perspectives in Biol. Medicine, 1968—; editor Archives Pathology and Lab. Medicine, 1974-83, Yearbook Pathology Clin. Pathology, 1980-91. Served from capt. to lt. col. M.C. U.S. Army, 1941-46; col. Med. Res. Corps 1946—. Co-recipient Ward Burdick award Am. Soc. Clin. Pathologists, 1941, 1963; O. Max Gardner award, 1961; N.C. award, 1969; Internat. Heart Research award, 1969; Murray Thelin award Nat. Hemophilia Found., 1972; Distinguished Achievement award Modern Medicine, 1973; Maude Abbott award Internat. Acad. Pathology, 1985; Disting. Service award AMA, 1986; H.P. Smith lectr., 1974; recipient 50th Yr. Rsch. award NIH, 1992. Mem. Nat. Acad. Scis. Inst. of Medicine, Am. Acad. Arts and Scis., Assn. Am. Physicians, Internat. Soc. Thrombosis and Haemostasis (pres. 1971, Robert P. Grant award 1985), Am. Assn. Pathologists and Bacteriologists (sec., treas. 1968-71, pres. 1973, Gold-headed Cane award 1981), Am. Soc. Exptl. Pathology (pres. 1965-66), Fedn. Am. Socs. Exptl. Biology (pres. 1966-67), Univs. Assoc. Research and Edn. Pathology (pres. 1964-68), Assn. Pathology Chmn. (Disting. Svc. award 1989), Acad. Clin. Lab. Physicians and Scientists (Cotlove award 1991). Home: 524 Dogwood Dr Chapel Hill NC 27516-2884

BRINLEY, F(LOYD) J(OHN), JR., health science institution executive, physician; b. Battle Creek, Mich., May 19, 1930; s. Floyd John and Neta Fay (Johnson) B.; m. Marlene Schoen, June 12, 1955; children: Floyd John III, Deborah Ann, William Schoen. A.B., Oberlin Coll., 1951; M.D., U. Mich., 1955; Ph.D., Johns Hopkins U., 1961. Diplomate: Nat. Bd. Med. Examiners. Intern Los Angeles County Gen. Hosp., 1955-56; asst. prof. physiology Johns Hopkins U., Balt., 1961-65, assoc. prof., 1965-76; prof. U. Md., Balt, 1976-79; sr. asst. surgeon USPHS, 1957, med. dir., 1979; dir. Neurol. Disorders Program Nat. Inst. Neurol. Disorders and Stroke, 1979-82, dir. div. Convulsive, Developmental and Neuromuscular Disorders, 1982—. Mem. Biophys. Soc., Soc. Neurosci., Am. Neurol. Assn., Am. Acad. Neurology, Am. Physiol. Soc., Am. Biochem. Soc., Soc. Gen. Physiology. Home: PO Box 41021 Bethesda MD 20824 Office: NINDS/NIH Fed Bldg Rm 816 Bethesda MD 20892

BRISCOE, MELBOURNE G., oceanographer, administrator; b. Akron, Ohio, July 17, 1941; s. Melbourne L. and Adelaide B.; m. Vicky Sue Cullen, Dec. 19, 1964 (div. 1975); m. Ellen Dana Straus Wood, Feb. 26, 1983. BSME, Northwestern U., 1963, PhD in Mech. Engring. and Astronautical Sci., 1967. With space power supply divsn. Aerojet-Gen. Corp., Azusa, Calif., 1963; with marine divsn. Westinghouse Corp., Sunnyvale, Calif., 1967; postdoctoral fellow Von Karman Inst. Fluid Dynamics, Rhode-St-Genese, Belgium, 1967-68; postdoctoral fellow NATO SACLANT ASW Rsch. Centre, La Spezia, Italy, 1968-69, with oceanography group, 1969-72; asst. scientist phys. oceanography dept. Woods Hole (Mass.) Oceanographic Inst., 1972-76, assoc. scientist, 1977-89; dir. applied oceanography and acoustics divsn. Office of Naval Rsch., 1989-92; dir. office of ocean and earth scis., Nat. Ocean Svc. Nat. Oceanic and Atmospheric Adminstrn., Silver Spring, Md., 1992—; on intergovernmental personnel assignment Office of Naval Rsch., Arlington, Va., 1987-89; acting head U.S. Global Ocean Observing System, mem. U.S. interagency com.; mem. scientific steering com. U.S. GLOBEC; cons. in field. Assoc. editor Jour. Geophys. Rsch., 1974-77, Jour. Phys. Oceanography, 1977-84, 87-90, editor, 1985-86; contbr. articles to profl. jours. Recipient Bausch and Lomb Sci. award 1959, U.S. Navy Civilian Superior Svc. award and medal 1993; Nat. Merit scholar 1959-63; fellow Nat. Def. Edn. Act, 1963-67, NATO Sci. Com., 1968-69; named to Fed. Exec. Inst. Class 196, 1993. Mem. Acoustical Soc. Am., Am. Geophysical Union (jours. bd. 1982-84, editor Oceanic Internal Waves 1985), Am. Meteorol. Soc., IEEE, Marine Tech. Soc., The Oceanography Soc. (sec.). Achievements include research on oceanography, air-sea interactions, acoustics, telemetry; development of basic-to-applied research teaming techniques, research management and societal applications of oceanography. Office: Nat Ocean and Atmospheric Adminstrn Nat Ocean Svc 1305 East-West Hwy Silver Spring MD 20910

BRISKIN, DONALD PHILLIP, biochemist; b. L.A., Feb. 28, 1955; s. Jerome William and Lynette Helen (Greenbaum) B.; m. Jacqueline Connor, Jan. 6, 1978; children: Brigid Johanna, Alissa Mary, Matthew Connor. BS in Biol. Sci., U. So. Calif., 1977; PhD in Plant Physiology, U. Calif., Riverside, 1981. Postdoctoral fellow biol. dept. McGill U., Montreal, Que., Can., 1982-83; plant physiologist USDA-ARS Utah State U., Logan, 1983-85; asst. prof. dept. agronomy U. Ill., Urbana-Champaign, 1985-88, assoc. prof. dept. agronomy, 1988-92, prof. dept. agronomy, 1992—. Editor: Plant Physiology, 1992—; contbr. articles to Jour. Exptl. Botany, Biochimica et Biophysica Acta, Plant Physiology, Archives of Biochemistry, Biophys, Plant Sci., Phytochemistry, Physiologia Plantarum. Recipient McKnight award, 1986, Young Crop Scientist award Crop Sci. Soc. Am., 1988, Alumni Award of Merit, Gamma Sigma Delta, 1990; faculty scholar U. Ill., 1988; Beckman fellow, 1987. Mem. AAUP, Am. Soc. Plant Physiologists, Sigma Xi. Achievements include research in reaction mechanism of transport ATPases; membrane biochemistry and bioenegetics; signal transduction in plants. Office: U Ill Dept Agronomy 1201 W Gregory Dr Urbana IL 61801

BRISTER, DONALD WAYNE, mechanical engineer; b. Ranger, Tex., Feb. 2, 1949; s. William Murl and Jannie Ruth (Presley) B.; m. Brenda Joyce Pool, Oct. 30, 1970; children: Kimberley Ann, Timothy Murl. BSME, U. Tex., 1972. Registered profl. engr., Tex. Mech. engr. Hewlett Packard, Loveland, Colo., 1972-76; project engr. Frito-Lay, Dallas, 1976-82; sr. design engr. M&M Mars, Cleveland, Tenn., 1982-85; corp. engring. mgr. L.S. Heath & Sons, Robinson, Ill., 1985-87; dir. engring. and svcs. Keebler, Haltom City, Tex., 1987-91; pres. B Assocs., Arlington, Tex., 1991—. Industry mem. Tarrant County Local Emergency Planning Com., Ft. Worth, 1987—; bd. dirs. Planning and Zoning Commn., Colleyville, Tex., 1984-85; cho. chmn. Springtown Ind. Sch. Dist. Tech. Steering Com., 1992. Mem. NSPE, Tex. Soc. Profl. Engrs., Am. Soc. Quality Control. Democrat. Baptist. Achievements include invention of process for producing rippled snack chips and products thereof. Home: Rte 1 Box 350 Springtown TX 76082 Office: B Assocs Inc PO Box 465 Springtown TX 76082

BRISTOL, STANLEY DAVID, mathematics educator; b. Mankato, Minn., Dec. 30, 1948; s. Robert Frederick Bristol and Ruth Charlotte (Buckeye) Bristol Bond; m. Elaine Metzer, Jan. 30, 1970; children: Thomas Alan, Jennifer Elise. BS, Ariz. State U., 1969, MA, 1970. Cert. secondary tchr. with gifted endorsement. Math. tchr. Saguaro High Sch., Scottsdale, Ariz., 1973-74, Poston Jr. High Sch., Mesa, Ariz., 1974-77, Corona del Sol High Sch., Tempe, Ariz., 1977—, Ariz. State U., Tempe, 1989—. Sunday sch. tchr. 1st United Meth. Ch., Tempe, 1983—. With U.S. Army, 1970-73. Named Tchr. of Yr., Tempe Diablos (C. of C.), 1987; recipient Presdl. award for Excellence in Math. Teaching, 1990. Mem. Nat. Coun. Tchrs. Math., Ariz. Assn. Tchrs. Math., Nat. Edn. Assn., Ind. Order of Foresters, Math. Assn. Am. Avocations: photography, reading, bowling, computers. Office: Corona del Sol High Sch 1001 E Knox Rd Tempe AZ 85284-3299

BRISTOW, JULIAN PAUL GREGORY, electrical engineer; b. Barnet, U.K., June 10, 1961; came to U.S., 1986; m. Caleen Mackay, 1985; 1 child, Cara. BSc in Physics, Southampton U., 1982; PhD in Elec. Engring., Glasgow (Scotland) U., 1985. Rsch. engr. Amphenol Fiber Optics, Lisle, Ill., 1986-88; prin. engr. Honeywell S.R.C., Bloomington, Minn., 1988—. Contbr. papers on optics, optoelectronics to profl. publs. Achievements include patent for guided wave polarizers. Office: Honeywell SRC 10701 Lyndale Ave S Bloomington MN 55420

BRISTOW, PRESTON ABNER, JR., civil engineer, environmental engineer; b. Birmingham, Ala., Feb. 14, 1926; s. Preston Abner and Anne Lee (Turner) B.; divorced; children: Kim A., William B., Harris A., Brian C., Li L. B of Gen. Studies, U. Md., 1970. Registered profl. engr., Ala., Calif., Ga., Fla. Land surveyor Trans Arabian Pipe Line Co., Beirut, Lebanon, 1947-49; civil, environ. engr. Hendon & Assocs., Birmingham, 1950-59; environ. sales engr. EIMCO Corp., San Francisco, 1959-66; chief sanitary, civil engr. Adrian Wilson Assocs., Saigon, Vietnam, 1966-70; cons. design and constrn. Saigon Metro. Water Office U.S. Agy. for Internat. Devel., Vietnam, 1970-75; mgr. S.E. office Metcalf & Eddy Engrs., Atlanta, 1975-84; mgr. internat. Lockwood Greene Engrs., Atlanta, 1984-92; v.p., chief engr. Associated Environ., Inc., Atlanta, 1992—; prof. environ. engring. U. of Hue, Vietnam, 1970-75; team staff leader devel. of water supply, wastewater collection and treatment system USAID, Cagayan de Oro, Philippines, 1991;. Lt. col. Ga. State Defense Force (plans and ops. officer, 1987-90). Sgt. USMC, 1942-46; 1st lt. C.E., USAR, 1951-59. Recipient Pub. Works medal, Govt. Vietnam, 1972, Cert. of Satisfaction, Saigon Met. Water Office, 1972. Fellow ASCE, Phi Kappa Phi, Alpha Sigma Lambda, Phi Sigma Alpha. Republican. Episcopalian. Achievements include design and construction of major international water treatment plants, pipelines, air-bases, and other capital projects. Home: 2526 Cardinal Lake Cir Duluth GA 30136-3920 Office: Associated Environ Inc 88 Mansell Ct Roswell GA 30076

BRITO CRUZ, CARLOS HENRIQUE, physicist, science administrator; b. Rio de Janeiro, July 19, 1956; s. Jose Armenio and Helena (Brito) Cruz. Elec. engring. degree, S J Campos, Brazil, Sjcampos, Brazil, 1978; MSc, Unicamp, Campinas, Brazil, 1980; PhD, UNICAMP, Campinas, Brazil, 1983. Prof. Physics Inst. Unicamp, Campinas, 1983-85, 87-91; resident visitor AT&T Bell Labs., Holmdel, N.J., 1986-87; dir. Physics Inst. Unicamp, Campinas, 1991—. Contbr. tech. articles to profl. jours. Home: RAC Lima 385, 13075 Campinas Brazil Office: Cidade Universitaria, Physics Inst Unicamp, 13083-970 Campinas Brazil

BRITT, CHESTER OLEN, electrical engineer, consultant; b. Hughes Springs, Tex., July 20, 1920; s. Bevely Albert and Ida Emma (Martin) B.; m. Patricia Britt, Jan. 4, 1946. BS, U. Tex., 1947, PhD, 1962. Registered profl. engr., Tex. Rsch. engr. Elec. Engring. Rsch. Lab., Austin, 1949-61; rsch. scientist U. Tex., Austin, 1961-84; cons. Austin, 1984—. Home and Office: 2708 Rae Dell Ave Austin TX 78704

BRITT, HAROLD C., physicist; b. Buffalo, Sept. 14, 1934; s. Harold W. and Mary C. B.; m. Donna Louise Cave, Dec. 29, 1956. BS in Physics, Hobart Coll., 1956; MA in Physics, Dartmouth Coll., 1958; PhD in Phyics, Yale Univ., 1961. Mem. staff divsn. physics Los Alamos (N.Mex.) Nat. Lab., 1961-73, group leader P-9 Van de Graaff accelerator, 1974-77, group leader P-7 exptl. nuclear physics, 1978-81, lab. fellow, 1981-86; mem. staff E divsn. physics dept. Lawrence Livermaore Nat. Lab., Washington, 1986-88, leader E divsn., 1988-92, dir. programs nuclear physics, divsn. physics NSF, 1992—; vis. scientist Lawrence Berkeley Lab., 1964; mem. subpanel nuclear instrumentation NAS/Nat. Rsch. Coun., 1970; vis. prof. dept. Physics, nuclear rsch. lab., U. Rochester, 1972-73, U. Munich, 1979-80; mem. super HILAC users exec. com. Lawrence Berkeley Lab., 1972-73, 1977-79, chmn. 1988, program adv. com., 1986-87chmn. Bevalac users exec. com., 1987-89, chmn. 1988; cons. Niels Bohr. Inst. U. Copenhagen, 1976-81; mem. exec. com. HHIRF users Oak Ridge Nat. Lab., 1978-79, 82-83, program adv. com., 1986-87; chmn. Gordon Rsch. Conf. Nuclear Chemistry, 1980; program adv. com. nat. superconducting cyclotron lab., Mich. State U., 1981-83; mem. heavy ion program review U.S. Dept. Energy, 1982, detailee office of nuclear physics, 1983-85, heavy ion facility review panel, 1987, NSF nuclear sci. adv. com., 1988-91; mem. subcom. facility constrn. NSAC, 1986; mem. AGS users exec. com. BNL, 1989-91. Recipient Sr. U.S. Scientist award Alexander von Humbolt-Stiftung, Germany, 1979Sterling fellow, 1960-61. Fellow Am. Phys. Soc. (chmn. programs divsn. nuclear physics 1973-75, exec. com. divsn. nuclear physics 1986-88, selection com. Tom W. Bonner Prize, vice chmn. 1990, chmn. 1991); mem. AAAS, Am. Chem. Soc. (nuclear chem. divsn.). Achievements include research on heavy ion reactions at medium and relativistic energies. Office: National Science Foundation Mathematical & Physical Sciences 1800 G St NW Washington DC 20550

BRITTINGHAM, JAMES CALVIN, nuclear engineer; b. Hamlet, N.C., Apr. 6, 1942; s. James Calvin and Elizabeth (McCanless) B.; m. Margaret Kitchen, Feb. 12, 1978; 1 child, James Robert. BS in Nuclear Engring., N.C. State U., 1964, MS in Nuclear Engring., 1966; PhD in Nuclear Engring., U. Calif., Berkeley, 1975. Registered nuclear engr., Calif. Engr. Rockwell Internat., Canoga Park, Calif., 1975-80; engr. Pacific Gas and Electric, San Francisco, 1981-85; sr. consulting engr. Ariz. Pub. Svc., Phoenix, 1986—; assoc. faculty Ariz. State U. Contbr. articles to profl. jours. Recipient Talent for Svc. scholarship N.C. State U., 1960-63, AEC fellowship, 1964-66, NSF traineeship, 1967-69. Mem. Am. Nuclear Soc. Republican. Achievements include devel. new, improved methodology for ranking new and partially burned fuel assemblies as candidates for reinsertion into a reload core; devel. original rod cluster control assembly inventory model for Diablo Canyon power plant. Home: 3367 W Grandview Rd Phoenix AZ 85023 Office: Ariz Pub Svc 411 N Central Ave Phoenix AZ 85004

BRITTON, LAURENCE GEORGE, research scientist; b. Hampton Court, Eng., Sept. 26, 1951; came to U.S., 1981; s. George and Barbara Mavis (Card) B.; m. Helen Lynn Grass, Apr. 16, 1983 (div. 1987); 1 child, Robert. BS with 1st class honors, U. Leeds, Eng., 1974; PhD in Fuel and Combustion Sci., U. Leeds, 1977. Chartered engr., Eng. Rsch. fellow dept. elec. engring. U. Southhampton, Eng., 1978-81; sr. combustion scientist Union Carbide Corp. Tech. Ctr., South Charleston, W.Va., 1981-84; project scientist process fire and explosion hazards Union Carbide Corp. Tech. Ctr., South Charleston, 1984-89, rsch. scientist, 1990—; guest lectr. Coll. Grad. Studies, Sch. Engring. and Sci., U. W.Va., 1991; mem. fueling systems com. U.K. Ministry Def., 1978-81; speaker at profl. meetings and symposia. Contbr. articles to sci. jours. Fellow Inst. Energy; mem. AICE (William H. Doyle award 1986, 89), Combustion Inst., Ctr. for Chem. Process Safety (engring. design practices com. 1988-93, reactive materials storage and handling com. 1990—, design basis for process safety systems com. 1993—), Nat. Fire Protection Assn. (explosion protection systesm com., hazardous chems. com., static electricity com., properties of flammable liquids com.). Office: Union Carbide Corp Tech Ctr PO Box 8361 South Charleston WV 25303

BROADUS, JAMES MATTHEW, research center administrator; b. Mobile, Ala., Feb. 24, 1947; m. Victoria Anne Gordon; children: Matthew Lee, Victoria Rose, Joseph Gordon. BA, Oberlin Coll., 1969; MA, Yale U., 1972, M.Phil., 1974, PhD, 1976. Economist U.S. Dept. Justice, Washington, 1975-79; vis. asst. prof. econs. U. Ky., Lexington, 1979-81; rsch. fellow Marine Policy Ctr., Woods Hole (Mass.) Oceangraphic Instn., 1981-82; policy assoc. Marine Policy Ctr., 1982-84, social scientist, 1984—, dir., 1986—; mem. marine bd. NRC, Washington, 1989—, Panel on the Law Ocean Uses, N.Y.C., 1988—; bd. govs. Bigelow Lab., W. Boothbay Harbor, Maine, 1983-89; vis. assoc. prof. sci. and soc. Wesleyan U., 1986; adj. prof. marine policy U. Del., 1990—; adv. com. U.S.-Japan Nat. Resources Agreement, 1991—; mem. UN Group of Experts on Sci. Aspects of Marine Pollution (GESAMP), 1986-89; mem. UN Regional Working Groups on Implications of Climatic Change, 1989-91. Editorial bd. Jour. Aquatic Conservation; editorial advisor Oceanus; assoc. editor Jour. Coastal Rsch. Grad. fellow Yale U., 1971 75. Mem. Am. Econ. Assn., Assn. Environ. Resource Economists, Marine Tech. Soc., AAAS, Nat. Man and the Biosphere Program, Marine & Coastal Ecosystems Directorate (chmn.). Office: Marine Policy Ctr Woods Hole Oceanographic Instit Crowell House Woods Hole MA 02543

BROADWELL, CHARLES E., retired agricultural products company executive; b. Kingsville, Ont., Can., May 29, 1931; married; three children. BSA, U. Guelph, 1954. Gen. mgr. Ont. Bean Producers Mktg. Bd., 1968-93; pres. Ont. Inst. Agrologists, 1964; mem. adv. com. Agriculture Can. Rsch. Sta., Harrow, Ont., 1968-93; mem., chmn. Ont. Pulse Com., 1968-93; pres. Agrl. Inst. Can., 1987, pres. and chmn. bd. rsch. found., 1988-89; mem. adv. com. Ont. Agrl. Coll.; past mem. fed. task force n agri-food trade devels. Vice-chmn. Kent County Area Sch. Bd., 1968; mem. senate U. Guelph; vice-chmn. County Terr. Nursing Home, 1992—; bd. dirs. Edgewood Camp, 1993—; chmn. Luth. Ch., Chatham, London. Recipient Disting. Agrologist award Ont. Inst. Agrologists; fellow Agrl. Inst. Can., 1990. Mem. Internat. Pulse Trade and Industry Confederation (pres. 1989—), Can. Bean Coun. (pres. 1991-93), U. Guelph Alumni Assn. (past co-pres., fundraiser), Kiwanis (past pres. Chatham and Forest City, London chpts.), London C. of C. Home: 340 Village Green, London, ON Canada N6J 3Z7 Office: Ontario Bean Producers, 140 Raney Crescent, London, ON Canada N6L 1C3*

BROADWELL, MILTON EDWARD, nurse, anesthetist, educator; b. Plattsburg, N.Y., Sept. 18, 1938; s. Russell Parrott and Carmen (Darrah) B.; m. Elizabeth Longsworth, July 12, 1986; children: Dwayne Scott (dec.), Renee Coleen Demitraszyk, James Russell, Mark Edward. AS in Nursing, Orange Community Community Coll., Middletown, N.Y., 1970; BS in Profl. Studies, Elizabethtown Coll., 1980; MEd, Temple U., 1983. RN anesthetist. Student anesthetist Albany (N.Y.) Med. Ctr., 1971-73; Cert. RN Anesthetist/CPR program coordinator Champlain Valley Physicians Hosp. & Med. Ctr., Plattsburg, 1973; staff CRNA, mem. product standard and evaluation com. Charles S. Wilson Hosp., Johnson City, N.Y., 1973-76; dept. head., chmn. air ambulance team Dakota Midland Hosp., Aberdeen, S.D., 1976-77; instr., trainer in BLS Hershey (Pa.) Med. Ctr., 1980; staff CRNA, instl. inservice programs, chmn. CPR com. Carlisle (Pa.) Hosp., 1977-80; research assoc. in arthroscopic surgery Johns Hopkins U., Balt., 1983; CRNA clin. instr. St. Joseph's Hosp., Sch. Nurse Anesthesia, Lancaster, 1980-87; asst. prof. nurse anesthesiology Med. Coll. Va., Sch. Nurse Anesthesiology, Richmond, 1987-88, asst. prof., dir. clin. edn., 1988-91; owner, operator surg. instrument repair svc. Inventor carbon dioxide intra-articular insufflator; contbr. articles to profl. jours. Treas. Cumberland County E.M.S. Coun., Harrisburg, Pa., 1978-80; emergency preparedness officer

North Middletown Twp., CArlisle, Pa., 1980; instr., trainer BLS Am. Heart Assn., Hershey, 1980; instr. ACLS program Paramedic Tng. Inst., Lancaster, 1983. With USAF, 1956-60, capt. Nurse Corps, U.S. Army Res. Named Hon. Life Mem., USAF Air Defense Team, 1959; recipient Appreciation award Boy Scouts Am., 1979, Commendation Letter, SAC, 1957, Meritorious Achievement award U.S. Army, 1991. Mem. Am. Assn. Nurse Anesthetists, Masons, Shriners, Holy Cross Commandary, Elks. Achievements include research on effect of position of oxygen saturation during transport to the post anesthesia care unit.

BROCA, LAURENT ANTOINE, aerospace scientist; b. Arthez-de-Bearn, France, Nov. 30, 1928; came to U.S., 1957, naturalized, 1963; s. Paul L. and Paule Jeanne (Ferrand) B.; B.S. in Math., U. Bordeaux, France, 1949; Lic. es Scis. in Math. and Physics, U. Toulouse (France), 1957; grad. Inst. Technique Professionnel, France, 1960; Ph.D. in Elec. Engring., Calif. Western U., 1979; postgrad. Boston U., 1958, MIT, 1961, Harvard U., 1961; m. Leticia Garcia Guerra, Dec. 18, 1962; 1 dau., Marie-There Yvonne. Teaching fellow physics dept. Boston U., 1957-58; spl. instr. dept. physics N.J. Inst. Tech., Newark, 1959-60; sr. staff engr. advanced research group ITT, Nutley, N.J., 1959-60; examiner math. and phys. scis. univs. Paris (France) and Caen, Exam. Center, N.Y.C., 1959-69; sr. engr. surface radar div. Raytheon Co., Waltham, Mass., 1960-62, Hughes Aircraft Co., Culver City, Calif., 1962-64; asst. prof. math. Calif. State U., Northridge, 1963-64; prin. engr. astrionics lab. NASA, Huntsville, Ala., 1964-65; fellow engr. Def. and Space Center, Westinghouse Electric Corp., Balt., 1965-69; cons. and sci. adv. electronics, phys. scis. and math. to indsl. firms and broadcasting stations, 1969-80; head engring. dept. Videocraft Mfg. Co., Laredo, Tex., 1974-75; asst. prof. math. Laredo State U., summer, 1975; engring. specialist dept. systems performance analysis ITT Fed. Electric Corp., Vandenberg AFB, Calif., 1980-82; engring. mgr. Ford Aerospace and Communications Corp., Nellis AFB, Nev., 1982-84; engring. mgr. Arcata Assocs., Inc., North Las Vegas, Nev., 1984-85; sr. scientific specialist engring. and devel. EG&G Spl. Projects, Inc., Las Vegas, 1985—. Served with French Army, 1951-52. Recipient Published Paper award Hughes Aircraft Co., 1966; Fulbright scholar, 1957. Mem. IEEE, Am. Nuclear Soc. (vice chmn. Nev. sect. 1982-83, chmn. 1983-84), Am. Def. Preparedness Assn., Armed Forces Communications and Electronics Assn., Air Force Assn. Home: 5040 Lancaster Dr Las Vegas NV 89120-1445 Office: EG&G Spl Projects Inc PO Box 93747 Las Vegas NV 89193-3747

BROCK, JAMES MELMUTH, engineer, venture capitalist; b. Brockton, Mass., Jan. 12, 1944; s. James Melmuth and Ruth Eleanor (Copeland) B.; student U. Hawaii, 1964-65, 1982, Taiwan Normal U., 1969; m. Mary Soong, June 24, 1964; 1 dau., Cynthia. Survey apprentice Malcolm Shaw, Hanson, Mass., 1959-62; with Peace Corps, N. Borneo, 1962-64; engr. Austin, Smith & Assocs., Honolulu, 1964-65, Trans-Asia Engrs., Vietnam, 1965-67; ops. mgr. Teledyne, Bangkok, Thailand, 1967-69; chief surveys Norman Saito Engrs. Hawaii, 1970-73; sr. prin. Brock and Assocs., Maui, Hawaii, 1973-82; pres. Honolulu Cons. Group, Honolulu, 1982-88; dir. Koolau Brewery, Inc., 1985-88; pres., dir. First Pacific Capital, Inc., 1984-88; v.p., dir., ceo, Seaculture Inc., 1988-90; prin. ECM, Inc., 1989-92; v.p., dir., and ceo, USA-China Tech. Corp., 1992—; Del. White House Conf. Small Bus., 1980, 86. Registered land surveyor, Hawaii, registered profl. engr.; mem. NSPE, AWWA. Democrat. Office: Tower 2680 Bishop Square Pacific Honolulu HI 96813 Address: PO Box 4586 Honolulu HI 96812-4586

BROCK, MARY ANNE, research biologist, consultant; b. Aurora, Ill., June 29, 1932; d. Paul Peter and Helen Anna (Mattas) B. BA, Grinnell Coll., 1954; MA, Harvard U., 1956, PhD, 1959. Teaching fellow Harvard U., Cambridge, Mass., 1954-58; rsch. assoc. Harvard Med. Sch., Boston, 1959-60; rsch. biologist Nat. Inst. Aging, NIH, Balt., 1960—; vis. scientist Stanford U., Calif., 1977; cons. NASA, 1981—. Contbr. articles to profl. jours. Bd. dirs. Cross Keys Condo Assn., Balt., 1980-85. Fellow Gerontol. Soc. Am. (fellowship com. 1988-90, pub. policy com. 1991—); mem. AAAS, Cryobiology Soc. (bd. govs. 1973-76, sec. 1971-72, nominating com. 1982-84), Chesapeake Soc. Electron Microscopy (council 1979-82), Am. Soc. Cell Biology (legis. alert com. 1982-92, congl. liason com. 1992—), Internat. Hibernation Soc., Soc. Rsch. Biogl. Rhythms, Internat. Soc. Chronobiology, Sigma Xi, Phi Beta Kappa. Office: Gerontology Rsch Ctr NIA NIH 4940 Eastern Ave Baltimore MD 21224-2780

BROCK, THOMAS GREGORY, plant physiologist; b. Detroit, Nov. 19, 1954; s. Donald Roy and Kay Carolyn (Pollack) B.; m. Carol Martha Fast, July 6, 1984; children: Peter Thomas, Allison Elizabeth. BS, U. Mich., 1976, MS, 1977; PhD, U. Wash., 1985. Rsch. fellow dept. biology U. Mich., Ann Arbor, 1985-90; rsch. assoc. U. Mich. Med. Sch., Ann Arbor, 1990—. Contbr. chpts. to Structural Development of the Oat Plant, 1992, Plant and Animal Cell Bioreactors, other books. Recipient devel. biology award NIH, 1985, rsch. assoc. award NASA, 1986, 87. Mem. AAAS, Am. Soc. Gravitational and Space Biology, Am. Soc. Plant Physiologists. Democrat. Methodist. Achievements include characterization of physiological basis of hormone mediation of plant growth in response to light and gravitational stimulation. Home: 929 Duncan St Ann Arbor MI 48103 Office: U Mich Med Sch 7500 Med Sci Rsch Bldg I Ann Arbor MI 48109

BROCKHOFF, KLAUS K.L., marketing and management educator; b. Koblenz, Germany, Oct. 16, 1939; s. Max and Henny Brockhoff; m. Dagmar Brockhoff, Aug. 26, 1966; children: Claudia M., Alexandra. Diploma, U. Bonn, Germany, 1962, D, 1965; diploma, U. Muenster, Germany, 1966; PhD, U. Bonn, 1965. Rsch. fellow U. Calif., Berkeley, 1967-68; sr. researcher Battelle Inst., Frankfurt, Germany, 1969-71; prof. tech. and innovation mgmt., mktg. U. Kiel, Germany, 1970—; assoc. prof. European Inst. Advanced Studies in Mgmt., Brussels, 1975-77; mem. adv. bd. Datenzentrale, Kiel, 1989—; mem. sci. adv. bd. Prognos AG, Basel, Switzerland, 1990—; mem. sci. coun. German Sci. Coun., Berlin, 1991—. Author 14 books including Forschung und Entwicklung, 3rd edit., 1992; contbr. 170 articles to nat. and internat. jours. Recipient Rsch. award Max Planck Soc., Munich, 1991. Mem. Inst. Mgmt. Scis., Deutsche Ges. Ops. Rsch., Verein Socialpolitik, Jungius Gesellschaft fur Wissenschaften, Verband der Hochschullehrer fur Betriebswirtschaft. Office: U Kiel, Olshausenstr 40, 24098 Kiel 1, Germany

BROCKS, ERIC RANDY, ophthalmologist, surgeon; b. N.Y.C., Apr. 24, 1946; s. William Benjamin and Muriel (Welk) B.; m. Irene Loretta Kraut, Dec. 19, 1970; children: Jason Matthew, Daniel Charles. BA with high honors, U. Rochester, 1968, MD, 1972. Diplomate Am. Bd. Ophthalmology, Nat. Bd. Med. Examiners. Intern medicine NYU Sch. Medicine, N.Y.C., 1973, resident, chief resident ophthalmology, 1973-76; chief resident ophthalmology Bellevue Hosp., NYU Hosp., Manhattan VA Hosp., N.Y.C., 1975-76; attending physician St. Francis Hosp., Beacon, N.Y., 1976-89; asst./assoc. attending physician Vassar Bros. Hosp., Poughkeepsie, N.Y., 1976-80, attending physician, 1980—; clin. asst. ophthalmology Tisch (NYU) Hosp., N.Y.C., 1976—; clin. asst. attending physician Bellevue Hosp. Ctr., N.Y.C., 1976—; eye physician and surgeon Hudson Valley Eye Surgeons, P.C., Fishkill, N.Y., 1976—; cons. ophthalmology Julia Butterfield Hosp., Cold Spring, N.Y., 1981—, West Point (N.Y.) Mil. Acad., Keller Army Hosp., 1989—; chief surgery St. Francis Hosp., beacon, 1988-89, dir. ophthalmology sect., 1981-88, chief of staff, 1979-81; dir. dept. ophthalmology Vassar Bros. Hosp., 1992—; clin. asst. prof. ophthalmology NYU Sch. Medicine, N.Y.C., 1983—, course dir. ophthalmology elective, 1976-91; so. N.Y. coord. Nat. Eye Care Project, San Francisco, 1985—; adj. clin. asst. prof. ophthalmology Mt. Sinai Sch. Medicine, N.Y.C., 1993—. Contbr. articles to profl. jours. Vol. admissions network U. Rochester, 1986—, co-chmn. 25th reunion com., 1993. Fellow ACS, Am. Acad. Ophthalmology (media coord. N.Y. state Nat. Eye Care projects 1978—, pub. info. coun. 1985—); mem. AMA, Am. Acad. Cataract and Refractive Surgery, Med. Soc. State N.Y. (ho. of dels. 1984-89, govt. affairs subcom. 1987, fed. legis. com. 1993—), Dutchess County Med. Soc. (exec. com. 1982—, chmn. legis. liaison com. 1990-92, pres. 1990-91), Boca West Club. Avocations: tennis, golf, reading, family travel. Office: Hudson Valley Eye Surgeons So Dutchess Profl Park 335 Rt 52 Fishkill NY 12524

BRODA, RAFAL JAN, physicist; b. Cieszyn, Poland, Jan. 19, 1944; s. Jan Pawel and Barbara Eryka (Richter) B.; m. Olga Budianska, Sept. 26, 1970; children: Alexander, Joanna. MA, Jagellonian U., Kraków, Poland, 1966, PhD, 1971; Habilitation, Inst. of Nuclear Physics, Kraków, 1981. Asst. Inst. Nuclear Physics, 1966, adiunkt, 1971—, docent, 1981—, prof., 1991—; researcher Joint Inst. Nuclear Rsch., Dubna, U.S.S.R., 1968-71; IAEA Wien fellow Niels Bohr Inst., Copenhagen, Denmark, 1972-74; vis. scitnsits Inst. fuer Kernphysik, Julich, Fed. Republic Germany, 1977-79, 89-91, Purdue U., West Lafayette, Ind., 1982-84; organizer Zakopane internat. schs. on nuclear physics Inst. Nuclear Physics and Inst. of Physics, Jagellonian U., 1985, 86, 87, 92. Contbr. numerous articles to profl. jours. Mem. Polish Phys. Soc., Am. Phys. Soc. Roman Catholic. Avocations: politics, bridge, fishing. Home: Armii Krajowej 77/9, 30-150 Cracow Poland Office: Niewodniczanski, Inst Nuclear Physics, Radzikowskiego 152, 31-342 Kraków Poland

BRODER, CHRISTOPHER CHARLES, microbiologist; b. White Plains, N.Y., June 12, 1961; s. Thomas J. and Jeanne (Comfort) B.; m. Colleen Marie Guay, May 3, 1986. BS, Fla. Inst. Tech., 1983, MS, 1985; PhD, U. Fla., 1989. Grad. rsch. asst. Fla. Inst. Tech., Melbourne, 1983-85; postdoctoral assoc. Coll. Medicine U. Fla., Gainesville, 1989; Nat. Rsch. Coun. assoc. Lab. of Viral Diseases, NIH, Bethesda, Md., 1990-92, IRTA fellow, 1993—. Contbr. articles to science profl. jours. Mem. AAAS, Am. Soc. Microbiology. Achievements include co-discovery of first known bacterial receptor for human plasmin. Office: NIH Lab Viral Diseases Rm 236 9000 Rockville Pike Bldg 4 Bethesda MD 20892

BRODER, IRVIN, pathologist, educator; b. Toronto, Ont., Can., June 27, 1930; married, 1954; 3 children. MD, U. Toronto, 1955, FRCP, 1960. Intern Toronto Gen. Hosp., 1955-56; sr. intern Sunnybrook Hosp., Toronto, 1956-57; asst. resident in medicine Toronto Gen. Hosp., 1958-59, resident physician, 1959-60; clin. instr. allergy U. Mich. Med. Ctr., 1960-62, clin. tchr. medicine, 1963-66, assoc., 1966-68, asst. prof., 1968-71, assoc. prof. medicine, 1971-76; prof. medicine Gage Rsch. Inst., U. Toronto, 1977—; asst. prof. pharmocology U. Mich., 1965-75, asst. prof. pathology, 1965-80; dir. Gage Rsch. Inst., U. Toronto, 1971—, prof. pathology, 1980—, prof. occupational and environ. health, 1982—; resident fellow endocrinology, dept. pathology U. Toronto, 1957-58; resident fellow immunology, dept. pharmacology U. Colo., U. London, 1962-63; resident scholar Med. Rsch. Coun. Can., 1963-66, career investigator, 1966—; mem. Inst. Med. Sci. and Inst. Immunology, U. Toronto, 1971-82. Mem. Can. Med. Assn., Can. Soc. Allergy and Clin. Immunology, Can. Thoracic Soc., Can. Soc. Clin. Investigation, Can. Soc. Immunology. Achievements include research on occupational lung disease, on human obstructive airways disease. Office: Gage Research Institute, 223 College St, Toronto, ON Canada M5T 1R4*

BRODER, SAMUEL, federal agency administrator. Dir. Nat. Cancer Inst., Bethesda, Md. Office: Nat Cancer Inst 9000 Rockville Pike Bldg 31 Bethesda MD 20892-0001*

BRODERICK, DONALD LELAND, electronics engineer; b. Chico, Calif., Jan. 5, 1928; s. Leland Louis and Vera Marguerite (Carey) B.; m. Constance Margaret Lattin, Sept. 29, 1957; children: Craig, Eileen, Lynn. BSEE, U. Calif., Berkeley, 1950; postgrad., Stanford U., 1953-54. Jr. engr. Boeing Co., Seattle, 1950-52; design engr. Hewlett-Packard Co., Palo Alto, Calif., 1952-59; sr. staff engr. Ampex Computer Products, Culver City, Calif., 1959-60; dir. engring. Kauke & Co., Santa Monica, Calif., 1960-61; program mgr. Space Gen. Corp., El Monte, Calif., 1961-68, Aerojet Electronics Div., Azusa, Calif., 1968-89; prin. D.L. Broderick, Arcadia, Calif., 1989—. Contbr. articles to profl. jours. Mem. Jr. C. of C., Woodland Hills, Cailf., 1963-64. With USN, 1945-46. Fellow Inst. for Advancement of Engring; mem. IEEE (chmn. profl. group on audio 1955-59, mem. exec. com. San Francisco sect. 1957-59, chmn. San Gabriel Valley sect. 1964-71, chmn. sections. com. L.A. coun., 1971-72, chmn. L.A. coun. 1972-76, chmn. bd. WESCON conv. 1976-80, bd. dirs. IEEE Electronics Conv. Inc. 1981-84, Centennial medal 1984), AIAA (sec. L.A. sect., 1986-88, sec. nat. tech. com. on command control comm. and intelligence, Washington, 1985-89, chmn. devel. com. L.A. coun. 1986—). Achievements include 2 patents on high frequency communications technology; design of USAF 487-L low frequency communications system; first successful aircraft-ground station communications via satellite; design of first INTELSAT communications station in Africa; design and development of Ground Station computer software program for the USAF Satellite System, which achieved first successful detection and reporting of missile launches. Home: 519 E La Sierra Dr Arcadia CA 91006-4321

BRODKIN, ROGER HARRISON, dermatologist, educator; b. Newark, July 31, 1932. A.B. Lafayette Coll., Easton, Pa., 1954; M.D., Jefferson Med. Coll., 1958; M.M.S. in Dermatology, NYU, 1967. Intern, Lenox Hill Hosp., N.Y.C., 1958-59; resident in dermatology NYU and Bellevue Hosp., N.Y.C., 1959-62; teaching asst. NYU, 1962-64, instr. dermatology, 1964-66; clin. asst. prof. U. N.J. Med. and Dental Sch., Newark, 1966-69, clin. assoc prof., 1969-79, clin. prof., 1979—. Address: 101 Old Short Hills Rd West Orange NJ 07052

BRODL, MARK RAYMOND, biology educator; b. Berwyn, Ill., July 6, 1959; s. Raymond Frank and Ethel Jean (Johnson) B. BA in Biology, Knox Coll., 1981; MS in Plant Biology, U. Ill., 1984; PhD in Plant Biology, Washington U., St. Louis, 1987. Asst. prof. Knox Coll., Galesburg, Ill., 1987-92; assoc. prof. Knox Coll., 1992—; reviewer NRC, Washington, 1991—. Contbr. articles to profl. jours. Recipient Teaching Excellence and Campus Leadership award Sears Roebuck Found., Chgo., 1991; named Presdl. Young Investigator, NSF, Washington, 1991; NSF grantee, 1987, 1988, 1991. Mem. Am. Soc. Plant Physiologists, Midwest Sect. Am. Soc. Plant Physiologists (chair 1992—), Am. Soc. Cell Biologists, Coun. on Undergrad. Rsch. (councilor 1993—), Sigma Xi. Achievements include rsch. in effect of heat shock on the endoplasmic reticulum and secretory protein synthesis in plant secretory cell.s. Office: Knox Coll Dept Biology Campus Box 44 Galesburg IL 61401

BRODNIEWICZ, TERESA MARIA, biochemist, researcher; b. Lodz, Poland, Oct. 1, 1949; arrived in Can., 1981; d. Tadeusz Marian and Zofia Anna (Bielska) B.; m. Zbigniew Aleksander Proba, Dec. 29, 1973; children: Monika-Anna, Joanna-Dorota. MS, Lodz U., Poland, 1972; PhD, Polish Acad. Scis., Warsaw, 1980. Rsch. fellow Polish Acad. Scis., Warsaw, Poland, 1974-81; post-doctoral fellow Armand-Frappier Inst., Laval, Que., Can., 1981; head of unit Blood Fractionating Ctr. Armand Frappier Inst., Laval, 1982-91; head of rsch. Haemacure, Pointe-Claire, Que., Can., 1991-92, v.p. rsch. and regulatory affairs, 1992—; cons. Mich. Dept. Pub. Health, 1988—, Haemacure, Montreal, Quebec, Can., 1990-91. Contbr. articles to profl. jours. Mem. Solidarity, Warsaw, 1980-81. Mem. Polish Scientists and Physicians Assn. in Montreal, Thrombosis and Haemostans Soc., Hound Healing Soc. Office: Haemacure Biotech, 245 Boul Hymus, Pointe Claire, PQ Canada H9R 1G6

BRODRICK, JAMES RAY, mechanical engineer; b. Wellsboro, Pa., Oct. 19, 1948; s. Merrill Shaw and Flora Shara (English) B.; m. Cheryl Anne Crandall, Oct. 22, 1983; children: John Robert, Thomas James. BSME, Pa. State U., 1970; PhD in Mech. Engring., U. Ill., 1979. Devel. engr. Nuclear Devel. Lab. Combustion Engring., Windsor Locks, Conn., 1972-75; rsch. engr. Carrier Corp., Syracuse, N.Y., 1979-83; program mgr. Gas Rsch. Inst., Chgo., 1983-90, Battelle Pacific N.W. Lab., Richland, Wash., 1990-91, U.S. Dept. Energy, Washington, 1991—; speaker, presenter in field. Contbr. articles, papers to profl. publs. Leader Boy Scouts Am., Germantown, Md., 1992. Mem. ASME, Am. Soc. Heating, Refrigerating, and Air-Conditioning Engrs. (program com. 1990—, mem. rev. bd. jour. 1991—), Sigma Xi. Office: US Dept Energy EE-422 1000 Independence Ave Washington DC 20585

BRODSKY, ALLEN, radiological and health physicist, consultant; b. Balt., Nov. 5, 1928; s. Nathan Michael and Gertrude Devera (Silberman) B.; m. Paula Fishman, June 17, 1951 (div. 1983); children: Richard, Karen, Jay; m. Phyllis Levin, Mar. 16, 1984. BS in Engring., Johns Hopkins U., 1949, MA in Physics, 1960; ScD, U. Pitts., 1966. Diplomate Am. Bd. Health Physics, Am. Bd. Indsl. Hygiene, Am. Bd. Radiology. Radiol. physics fellow Oak Ridge (Tenn.) Nat. Lab., 1950; head health physics unit U.S. Naval Rsch. Lab., Washington, 1950-52; physicist region 2 FCDA, Olney, Md., 1956-57; health physicist AEC, Washington, 1957-61; assoc. prof., rsch. assoc. Grad. Sch. Pub. Health U. Pitts., 1961-71; radiation physicist Mercy Hosp., Pitts., 1971-75; sr. health physicist U.S. Nuclear Regulatory Commn., Washington, 1975-86; prin. Allen B. Cons., Inc., Berlin, Md., 1986—; cons. CD, NAS, Washington, 1975; adj. prof. sch. pharmacy Duquesne U., Pitts., 1971-75; radiation sci. fellowship bd. Oak Ridge Associated Univs., 1967-70. Author, editor: Radiation Measurement and Protection, Vol. I, 1979, Vol. II, 1982; contbr. regulatory guides, book chpts., articles in field. Pres. Western Pa. Profs. for Peace in Mid. East, Pitts., 1970-71; witness on radiation effects U.S. Ho. of Reps., Washington, 1978, witness on radiation studies U.S. Senate, Washington, 1978-81, expert witness U.S. Dept. Justice, Washington, 1983-84. Lt. C.E., U.S. Army, 1952-54. Named W.H. Langham lectr., U. Ky., 1979, Failla Meml. lectr., Radiol. and Med. Physics Soc., Health Physics Soc. N.Y., N.Y.C., 1987; recipient Leadership and Sci. Contbns. cert. Conf. on Bioassay, Environ., and Analytical Radiochemistry, 1986. Mem. Am. Nuclear Soc. (radiation sci. and tech. award 1993), Am. Assn. Physicists in Medicine, Am. Indsl. Hygiene Assn., Am. Statis. Assn., Health Physics Soc. (chmn. standards com. 1959-61, 67-70, bd. dirs. 1967-70, pres. Western Pa. chpt. 1967-68, Western Pa. chpt. disting. svc. award 1966, pres. Balt.-Washington chpt. 1982-83, sec.-treas. govt. sect. 1988-92, Founder's award 1986, Fellow award 1992). Avocations: tennis, piano, composing songs, singing, political campaigns. Home and Office: 2765 Ocean Pines Berlin MD 21811-9127

BRODSKY, STANLEY MARTIN, engineering technology educator, researcher; b. Queens, N.Y., Sept. 20, 1924; s. Peter and Hattie Deborah (Weinstein) B.; m. Roslyn Newman, May 5, 1950 (div. June 1979); children: Ellen Amy Nicholson, Janet Terri, Russell David; m. Monica Ann Fuller, July 14, 1982. BME, The City Coll. of N.Y., 1944; MME, Polytechnic Inst. of Brooklyn, 1955; PhD, N.Y.U., 1964. Registered profl. engr., N.Y. Test engr. Mack Mfg. Co., Plainfield, N.J., 1944-45; design-draftsman Lloyd-Rogers Co., N.Y.C., 1945; engr. Huntleigh Co., Schenectady, N.Y., 1945-47; devel. engr. Ralph Cooksley, Inventor, N.Y.C., 1947; jr. instr., instr., asst. prof., assoc. prof., prof. N.Y.C. Tech. Coll. of the CUNY, Bklyn., 1947-91, prof. emeritus, 1991—; div. chair of Engring. Tech., assoc. dean N.Y.C. Tech. Coll. of the City Univ. of N.Y., Bklyn., 1962-79; dir., New Project Devel. CUNY Grad. Sch., Ctr. for Advanced Study in Edn., N.Y.C., 1987—; bd. dirs. Accreditation Bd. for Engring. & Tech., N.Y.C., 1986-90, chmn. tech. accreditation com., 1981-82; mem. engring. manpower com. Am. Assn. Engring. Soc., 1980-85; cons. in field. Co-author: Statics and Strength of Materials, 1960, 4th rev. edit., 1988; contbr. articles to profl. jours. Mem. local Sch. Bd. Dist. 22, Bklyn., 1964-69, Community Planning Bd. 18, Bklyn., 1963-73. Accreditation Bd. for Engring. and Tech. fellow, 1988; recipient citation of svc. bd. dirs. Accreditation Bd. for Engring. and Tech., Am. Soc. of Mechanical Engrs. Mem. ASME (rep., 1st recipient Ben C. Sparks medal 1991), NSPE (Kings County chpt., past bd. dirs., chmn. coms.), Am. Soc. Engring. Edn. (coms. mem.), N.Y. State Engring. Tech. Assn. (past v.p., exec. com.). Democrat. Jewish. Avocations: travel, theatre, sports fan. Office: CASE/CUNY 25 W 43d St Rm 620 New York NY 10036-7406

BRODT, BURTON PARDEE, chemical engineer, researcher; b. Evanston, Ill., June 3, 1931; s. Harry Snowden and Marjorie Florence (Pardee) B.; m. Virginia Faye Futch, June 20, 1954; children: Howard A., Stephen R., Cynthia A., Phillip D. BS in Chem. Engring., U. Fla., 1954, MS in Chem. Engring., 1958. Devel. engr. DuPont Elastomers, Louisville, Ky., 1958-62; sr. rsch. engr. DuPont Elastomers, Wilmington, Del., 1962-66; tech. supr. DuPont Elastomers, Deepwater, N.J., 1966-69; rsch. supr. polymers div. E.I. DuPont de Nemours and Co., LaPlace, La., 1969-83; sr. supr. rsch. E.I. DuPont de Nemours and Co., Wilmington, 1983-89; sr. rsch., assoc. E.I. DuPont de Nemours and Co., LaPorte, Tex., 1989-91; tech. fellow E.I. DuPont de Nemours and Co., LaPlace, 1991—. Contbr. to profl. publs. Chmn. Citizens for Goldwater, Wilmington, 1964; founder, pres. Del. Conservative Union, Wilmington, 1965-69; pres. Homeowners Assn., Chadds Ford, Pa., 1988. Lt. USAF, 1954-56. Mem. AICE. Republican. Achievements include patent for accelerator encapsulation, high-viscosity level control, proprietary processes, powder attrition tester; development of chemical processes now in commercial use, Brodt equation for phase transfer catalysis. Home: 1225 Milan St New Orleans LA 70115 Office: EI duPont de Nemours & Co PO Box 2000 LaPlace LA 70069

BRODY, BARUCH ALTER, medical educator, academic center administrator; b. Bklyn., Apr. 21, 1943; s. Lester and Gussie (Glass) B.; m. Dena Grosser, Aug. 15, 1965; children: Todd, Jeremy, Myles. BA, Bklyn. Coll., 1962; PhD, Princeton U., 1967. Asst. prof. MIT, Cambridge, 1967-75; assoc. prof. Rice U., Houston, 1975-77, prof., 1977—; prof. Baylor Coll. Medicine, Houston, 1982—, dir. ctr. ethics, 1982—; cons. NASA, 1990-91. Author: Abortion and the Sanctity of Human Life, 1975, Identity and Essence, 1981, Life and Death Decision Making, 1988. Chmn. bd. dirs. Hebrew Acad., Houston, 1976—. Recipient Disting. Alumnus award Bklyn. Coll., 1991. Jewish. Avocation: reading. Office: Baylor Coll Medicine Ctr Ethics Medicine Pub Issues Houston TX 77030 Office: Rice U PO Box 1892 6100 South Main Houston TX 77251

BROECKER, WALLACE S., geophysics educator; b. Chicago, IL, Nov. 29, 1931. Attended Wheaton Coll., Wheaton, IL; A.B., Columbia, N.Y.C., 1953, Ph.D., 1958. Asst. prof. Columbia U., N.Y.C., 1959-61, Assoc. prof., 1961-65, prof., 1965—, now Newberry prof. Author 4 books, articles in scholarly jours. Recipient Arthur L. Day Medal, Geol. Soc. Am., 1984, Harold Urey Award, European Geochemical Soc., 1987, Vetlesen Prize, Columbia U., 1987, Priestley award Dickinson Coll., 1990. Mem. Am. Acad. Arts Scis., Nat. Acad. Scis. (Agassiz Medal, 1986); fellow Am. Geophys. Union (Ewing Medal, 1979). Office: Columbia U Ctr Climate Rsch Lamont-Doherty Geol Obsevatory New York NY 10027

BROER, MATTHIJS MENO, physicist; b. Madison, Wis., Aug. 23, 1956; s. Jan Willem and Cornelia Rudolphina (Van Royen) B. BS in Physics, U. Tech., Delft, The Netherlands, 1977; MSc in Physics, U. Wis., 1979, PhD in Physics, 1982. Postdoctoral mem. tech. staff AT&T Bell Labs., Murray Hill, N.J., 1983-84, mem. tech. staff, 1985—; chmn. symposium Materials Rsch. Soc. Meeting, Boston, 1991. Editor: Materials Research Society Proc., Vol. 244, 1992; contbr. articles to Phys. Rev. B, Phys. Rev. Letters, Optics Letters. Mem. Am. Phys. Soc., Optical Soc. Am. (chmn. program com. Conf. on Lasers and Electro Optics 1992, 93). Achievements include patents on Method for Fabricating Devices Including Multicomponent Metal Halide Glasses and the Resulting Devices, Long Wavelength Optical Fiber Communication and Sensing Systems. Office: AT&T Bell Labs 480 Red Hill Rd Middletown NJ 07748

BROG, TOV BINYAMIN, electrical engineer; b. Wiesbaden, Hesse, Germany, Jan. 8, 1960; s. David and Verda Anna (Raney) B.; m. Suzanne Kay Johnson, Mar. 27, 1988. BS in Math., U.S. naval Acad., 1982; MS in Systems Tech., U.S. Naval Postgrad. Sch., 1988; MSEE, Johns Hopkins U., 1992. Commd. 2nd lt. USAF, 1982, advanced through grades to capt., 1986; elec. warefare analyst Air Force Avionics Lab., Dayton, Ohio, 1982-84; program mgr. Aero. Systems Div., Dayton, Ohio, 1984-86; acquisition officer Air Force Systems Command, Washington, 1988-90; staff officer Office of Sec. of Air Force, Washington, 1990; sr. systems engr. Sci. Applications Internat. Corp., San Diego, 1990—. Trustee Barber Tract Homeowners Assn., La Jolla, Calif., 1992—. With Calif. Air N.G., 1992—. Decorated Air Force Commendation medal with one oak leaf cluster. Mem. U.S. Naval Acad. Alumni Assn., Assn. Old Crows. Democrat. Jewish. Office: Sci Applications Internat 10260 Campus Point Dr San Diego CA 92121

BROGUNIER, CLAUDE ROWLAND, environmental engineer; b. Lafayette, Ind., Mar. 17, 1960; s. Joseph and Hope (Anthony) B. BS in Environ. Sci. magna cum laude, U. Mass., 1987, BA in Chemistry cum laude, 1987; MS in Environ. Engring., Ill. Inst. Tech., 1988. Rsch. asst. U. Mass., Amherst, 1984; rsch. chemist Monsanto Chem. Co., Springfield, Mass., 1985, environ. technician, 1986; environ. engr. U.S. EPA, Chgo., 1987-88; environ. engr., cons. DePaul & Assocs., Inc., Chgo., 1988-90; sr. environ. engr. GE Co., Lynn, Mass., 1990-91, hazardous wast mgr., 1991—; chmn. subcom. Assoc. Industries Mass., Boston, 1992—; mem. working group on bus. and environ MIT, Cambridge, 1991—; presenter in field. Ill. Inst. Tech. fellow, 1986. Mem. AAAS, AICE (assoc.), Am. Chem. Soc., Nat. Assn. Environ. Profls. Democrat. Home: 89 Federal St Salem MA 01970 Office: GE Aircraft Engines MD164X9 1000 Western Ave Lynn MA 01910

BROKAW, CHARLES JACOB, educator, cellular biologist; b. Camden, N.J., Sept. 12, 1934; s. Charles Alfred and Doris Evelyn (Moses) B.; m.

Darlene Smith, July 29, 1955; children—Bryce, Tanya. B.S., Calif. Inst. Tech., 1955; Ph.D., King's Coll., Cambridge (Eng.) U., 1958. Research asso. Oak Ridge Nat. Lab., 1958-59; asst. prof. zoology U. Minn., 1959-61; mem. faculty Calif. Inst. Tech., 1961—, prof. biology, 1968—, exec. officer div. biology, 1976-80, 1985-89, assoc. chmn. div. biology, 1980-85. Contbr. articles to profl. jours. Guggenheim fellow, 1970-71. Mem. Biophys. Soc., Am. Soc. Cell Biology, Soc. Exptl. Biology. Home: 940 Oriole Dr Laguna Beach CA 92651-2846 Office: Calif Inst Tech Kerckhoff Marine Lab 101 Dahlia St Corona Del Mar CA 92625

BROLL, NORBERT, physicist, consultant; b. Mulhouse, Haut-Rhin, France, Nov. 10, 1947; s. Marcel and Christiane (Knittel) B.; m. Jeannine Karleskind, Sept. 16, 1972; 1 child, Estelle. MS, U. Haute-Alsace, Mulhouse, France, 1970; Docteur 3d cycle, U. Louis Pasteur, Strasbourg, France, 1976, Dr. Scis. Physics, 1985. Rsch. physicst Centre de Recherches Nucleaires (CRNS), Strasbourg, 1971-74; engring. scientist Siemens AG, Karlsruhe, Germany, 1974-92; founder, dir. FORTEX, France, 1993—; lectr. U. Louis Pasteur, Strasbourg, 1983—, U. Paris VII, 1990—, Ecole Nationale Superieure des Arts et Industries, Strasbourg, 1993—. Contbr. articles to profl. jours. Home: 6 rue de Dingolfing, 67170 Brumath Bas-Rhin France Office: Fortex, 24 Boulevard de la Victoire, 67084 Strasbourg France

BROMERY, RANDOLPH WILSON, geologist, educator; b. Cumberland, Md., Jan. 18, 1926; s. Lawrence Randolph and Edith (Edmondson) B.; m. Cecile Trescott, June 8, 1947; children: Keith, Carol, Dennis, David, Christopher. Student, U. Mich., 1945-46; BS, Howard U., 1956; MS, Am. U., 1962; PhD, Johns Hopkins U., 1968; ScD (hon.), Frostburg State Coll.; EdD (hon.), Western New Eng. Coll.; LLD (hon.), U. Hokkaido, 1976; LHD (hon.), U. Mass., 1979, Bentley Coll., 1993; D Pub. Svc. (hon.), Westfield State Coll., North Adams State Coll. Geophysicist U.S. Geol. Survey, 1948-68, cons., 1968-72; prof., lectr. Howard U., Washington, 1961-65; assoc. prof. geology, dep. chmn. dept. U. Mass., 1967-68, prof. geology, chmn. dept., 1969-70, prof. geology, vice-chancellor, 1970-71, prof. geology, chancellor, 1972-77, Commonwealth prof., 1979-88, 91—, exec. v.p., 1977-79; pres. Geosci. Engring. Corp., 1983—; acting pres. Westfield (Mass.) State Coll., 1988-90; interim chancellor Mass. Bd. of Regents, 1990-91; pres. Springfield Coll., 1992—; pres. Resources Cons. Assocs.; bd. dirs. Chem. Bank, NYNEX Corp., Exxon Corp., John Hancock Mut. Life Ins. Co.; geophys. cons. Kennecott Copper Corp., New Eng. Telephone Co., 1977-81, Northwestern Mut. Life Ins. Co., 1977-79; corp. mem. Woods Hole Oceanographic Instn., Boston Mus. Sci., Boston Mus. Fine Arts. Contbr. numerous articles to sci. and profl. jours. Trustee Babson Coll., 1976-81, Johns Hopkins U., Mount Holyoke; trustee Talladega Coll., 1981-88, Hampshire Coll., 1971-79, Boston Mus. Sci.; mem. Com. Collegiate Edn. Black Students, 1967-72, pres., 1968-72; mem. human devel. com. Cath. Archdiocese, Springfield, 1971-76; vice-chmn. New Coalition for Econ. and Social Change; mem. adv. council NASA; mem. Nat. Acad. Engrs. Com. Minorities in Engring; mem. advisory com. John F. Kennedy Library; mem. bd. overseers vis. com. dept. geol. scis. Harvard U., 1975-79. Served with USAAF, 1943-44. Named hon. pres. Soodo Women's U., Seoul, Korea; Gillman fellow Johns Hopkins, 1964-65. Fellow AAAS, Geol. Soc. Am. (v.p. 1987-88, pres. 1988-89); mem. Am. Geophys. Union, Soc. Exploration Geophysicists, N.Y. Acad. Scis., Coun. on Fgn. Rels., Explorers Club, Cosmos Club, Comml. Club Boston, Sigma Xi, Phi Kappa Phi, Phi Eta Sigma. Office: U Mass Dept of Geophysics Amherst MA 01003

BROMLEY, DAVID ALLAN, physicist, educator; b. Westmeath, Ont., Can., May 4, 1926; s. Milton Escort and Susan Anne (Anderson) B.; m. Patricia Jane Brassor, Aug. 30, 1949 (dec. Oct. 1990); children: David John, Karen Lynn. BS in Engring. Physics, Queen's U., Kingston, Ont., 1948, MS in Physics, 1950; PhD in Nuclear Physics, U. Rochester, 1952; MA (hon.), Yale U., 1961; D of Natural Philosophy (hon.), U. Frankfurt, 1978; Docteur (Physique) (hon.), U. Strasbourg, 1980; DSc (hon.), Queen's U., 1981, U. Notre Dame, 1982, U. Witwatersrand, 1982, Trinity Coll., 1988; LittD (hon.), U. Bridgeport, 1981; Dott. (hon.), U. Padua, 1983; LHD (hon.), U. New Haven, 1987; DSc (hon.), Rensselaer Polytechnic Inst., 1990; LHD (hon.), Ill. Inst. Tech., 1990; DSc (hon.), Lehigh U., 1991, Bklyn. Polytechnic Inst., 1991, U. Guelph, 1991, Fordham U., 1991, Northwestern U., 1991, Coll. of William and Mary, 1991; D Engring. Tech. (hon.), Wentworth Inst., 1991; DSc (hon.), SUNY, U. Mass., Adelphi U., 1993, Chung-Yuan U., Taiwan, 1993; DHL (hon.), Mt. Sinai Med. Ctr., 1993; D. Eng. (hon.), Colo. Sch. Mines, 1992. Oper. engr. Hydro Electric Power Commn. Ont., 1947-48; rsch. officer Nat. Rsch. Coun. Can., 1948; instr., then asst. prof. physics U. Rochester, 1952-55; sr. rsch. officer, sect. head Atomic Energy Can. Ltd., 1955-60; assoc. prof. physics, asso. dir. heavy ion accelerator lab. Yale U., 1960-61, prof. physics, dir. A. W. Wright Nuclear Structure Lab., 1961-89, chmn. physics dept., 1970-77, Henry Ford II prof. physics, 1972-93; Sterling prof. scis. Yale U., New Haven, 1993—; chmn. physics dept. Yale U., 1970-77; asst. to Pres. for sci. and tech. Washington, 1989-93; dir. Office of Sci. and Tech. Policy, Washington, 1989-93; chmn. Pres.'s Coun. Advisers on Sci. and Tech., Washington, 1989-93, Fed. Coordinating Coun. Sci., Engring. and Tech., Washington, 1989-93, Nat. Critical Materials Coun., Washington, 1990-92; cons. Brookhaven, Argonne, Berkley and Oak Ridge Nat. Labs., Bell Telephone Labs., IBM, GTE; mem. panel nuclear physics Nat. Acad. Scis., 1964, chmn. com. on nuclear sci., 1966-74, chmn. physics survey, 1969-74; mem.-at-large, mem. exec. com. div. phys. scis. NRC, 1970-74, mem. exec. com., assembly phys. and math. scis., 1974-78, mem. naval sci. bd., 1974-78; mem. high energy physics adv. panel ERDA, 1974-78; mem. nuclear sci. adv. panel NSF and Dept. Energy, 1980-89; mem. White House Sci. Coun., 1981-89, Nat. Sci. Bd., 1988-89. Editor: Physics in Perspective, 5 vols, 1972, Large Electrostatic Accelerators, 1974, Nuclear Detectors, 1978, Heavy Ion Science, 8 vols, 1981-84; co-editor: Procs. Kingston Internat. Conf. on Nuclear Structure, 1960, Facets of Physics, 1970, Nuclear Science in China, 1979; assoc. editor: Annals of Physics, 1968-89, Am. Scientist, 1969-81, Il Nuovo Cimento, 1970-89, Nuclear Instruments and Methods, 1974-89, Science, Technology and the Humanities, 1978-89, Jour. Physics, 1978-89, Nuclear Science Applications, 1978-89, Technology in Soc, 1981-89; cons. editor: McGraw Hill Series in Fundamentals of Physics, 1967-89, McGraw Hill Ency. Sci. and Tech. Bd. dirs. Oak Ridge Assoc. Univs., 1977-80, U. Bridgeport, 1981-86. Recipient medal Gov. Gen. Can., 1948, U.S. Nat. medal of sci., 1988, Yale medal in sci. and engring., 1991, Disting. Alumnus award U. Rochester, 1986, Presdl. medal N.Y. Acad. Scis., 1989, Disting. Svc. award IEEE, 1991, Louis Pasteur Medal of Sci. U. Strasbourg, 1991, Harvey medal Pierce Found., 1991, Disting. Svc. medal NSF, 1992, Pub. Svc. medal Ctr. Study of Presidency, 1992, Exec. Yr. award R&D Mag., 1992, Disting. Scholar medal U. Rochester, 1993; NRC fellow, 1952; fellow Branford Coll., 1961—; Guggenheim fellow, 1977-78; Humboldt fellow, 1978, 85; Benjamin Franklin fellow Royal Soc. Arts, London, 1979—. Fellow Am. Phys. Soc. (mem. council 1967-71), Am. Acad. Arts and Scis., AAAS (chmn. physics sect. 1977-78, pres.-elect 1980, pres. 1981—, chmn. bd. 1982—, William Carey medal 1993); mem. NAS, Can. Assn. Physicistsopean Phys. Soc., Conn. Acad. Arts and Scis., N.Y. Acad. Arts and Scis., Conn. Acad. Sci. and Engring. (council 1976-78), Internat. Union Pure and Applied Physics (U.S. nat. com. 1969—, chmn. 1975-76, v.p. 1975-81, pres. 1984-87), Southeastern Univ. Research Assn. (bd. dirs. 1984-89), Council on Fgn. Relations, Sigma Xi (pres. Yale 1962-63). Home: 35 Tokeneke Dr North Haven CT 06473 Office: Yale U A W Wright Nuclear Lab 272 Whitney Box 6666 New Haven CT 06511

BRONAUGH, EDWIN LEE, research center administrator; b. Salina, Kans., July 22, 1932; s. Edwin and Violet Mary (Dryden) B.; m. Geraldine Kelley, Dec. 10, 1955: children: Cecilia Ann Bronaugh Snodgrass, Dana Lea Bronaugh Weinberg. BA in Physics and Math., East Tex. State U., Commerce, 1955. Commd. USAF, 1955, advanced through grades to capt., 1961, various communications and ops. assignments, 1955-68; major USAFR, 1968; rsch. scientist Southwest Rsch. Inst., San Antonio, 1968-70, sr. rsch. scientist, 1970-76, rsch. dir., 1976-82; dir. R & D, tech. dir. Electro-Metrics Div. Penril, Amsterdam, N.Y., 1982-89; prin. electromagnetic compatibility scientist Electro-Mechanics Co., Austin, Tex., 1989-91; v.p. engring., 1991—. Author: Electromagnetic Interference Test Methodology and Procedures, 1988; patentee in field; contbr. over 80 articles to profl. jours. Decorated Bronze Star, Air Force Commendation medal. Fellow IEEE; mem. Electromagnetic Compatibility Soc. of IEEE (standards com. 1980—, dir. tech. svcs. 1981-87, v.p. 1988-90, pres. 1990-92, Cert. of Appreciation 1979, Cert. of Achievement 1983, Cert. of Acknowledgement 1985, Richard R. Stoddart award 1985, Standards Medallion 1992, Lawrence G. Cumming Award 1992), SAE (electromagnetic interference standards and test methods tech. com., electromagnetic radiation tech. com., aerospace electromagnetic compatibility com.), Am. Nat. Standards Inst. (vice chmn. accredited standards com. C63 on electromagnetic compatibility 1986—), Nat. Assn. Radio and Telecommunications Engrs. (cert.), Electromagnetic Compatibility Soc. (hon. life). Avocations: music, camping, model railroads. Home: 11007 Crossland Dr Austin TX 87826-1320 Office: Electro-Mechanics 2205 Kramer Ln Austin TX 78758 also: PO Box 1546 Austin TX 78767

BRONZINO, JOSEPH DANIEL, electrical engineer; b. Bklyn., Sept. 29, 1937; s. Joseph Rocco and Antoinette (Saporito) B.; m. Barbara Louise McGrath, Dec. 2, 1961; children: Michael J., Melissa J., Marcella J. BSEE, Worcester Poly. Inst., 1959, PhD in Elec. Engring, 1968; MSEE, U.S. Naval Postgrad. Sch., 1961. Registered profl. engr., Conn. Instr. elec. engring. U. N.H., 1964-66, asst. prof. elec. engring., 1966-67; NSF faculty fellow Worcester Found. for Exptl. Biology, Shrewsbury, Mass., 1967-68; mem. cooperating staff Worcester Found. for Exptl. Biology, 1968—; assoc. prof. engring. Trinity Coll., 1968-75, prof., 1975—, Vernon Roosa prof. applied sci., 1977—, chmn. dept. engring., 1981-91; dir. and chmn. biomed. engring. program Hartford (Conn.) Grad. Center, 1969— ; clin. assoc. dept. surgery U. Conn. Health Center, Farmington, 1971-77; research asso. Inst. for Living, Hartford, 1968—; reviewer NSF; panelist NSF Research Initiation Grants; lectr., speaker in field. Author: Technology for Patient Care, 1977; Computer Application in Patient Care, 1982; Biomedical Engineering Basic Concepts and Instrumentation, 1986, Medical Technology: Economic and Ethical Issues, 1990, Expert Systems: Basic Concepts, 1990, Management of Medical Technology: A Primer for Clinical Engineers, 1992; contbr. articles to profl. publs. Mem. Simsbury (Conn.) Planning Commn., 1977-82. Served to 1st lt. Signal Corps U.S. Army, 1961-63. Fellow IEEE (sr., regional dir. group engring. in medicine and biology 1973-78, v.p. tech. activities 1982-85, pres. 1985-86, chmn. health care engring. policy com. 1986-90, vice chmn. tech. policy coun. 1990-91, chmn. tech. policy coun.), Am. Inst. Med. and Biol. Engrs., Am. Soc. Engring. Edn. (exec. com. div. biomed. engring. 1973-82, vice chmn. career devel. 1974-76, vice chmn. profl. devel. 1976-77, divisional newsletter editor 1977-79, chmn.-elect div. 1979-80, exec. com. 1990-91, chmn. tech. policy coun. 1992—), AAAS, Biol. Psychiatry, Neurosci. Soc., Rotary (pres. Simsbury club 1971-89, 91-93, Hartford club 1989-91), mem. Am. Inst. of Med. and Biol. Engrings., Cosmos Club. Republican. Roman Catholic. Achievements include rsch. in signal analysis concepts and applications, basic neurophysiol. concepts involved in identifying specific neural circuits associated with specific functions of the brain. Home: 12 Brenthaven Avon CT 06001 Office: Trinity Coll Dept Engring Hartford CT 06106

BROOK, JUDITH SUZANNE, psychiatry educator; b. N.Y.C., Dec. 31, 1939; d. Robert and Helen E. (Zimmerman) Muser; m. David W. Brook, Dec. 15, 1962; children: Adam, Jonathan. BA, CUNY, 1961; MA in Psychology, Columbia U., 1962, EdD in Devel. and Ednl. Psychology, 1967. Lic. psychologist, N.Y. Asst. prof. psychology Queens Coll., CUNY, Flushing, 1967-69; rsch. assoc. Columbia U., N.Y.C., 1969-77, sr. rsch. assoc., 1977-80; assoc. prof. psychiatry Mt. Sinai Sch. Medicine, N.Y.C., 1980-90, adj. prof., 1990—; prof. N.Y. Med. Coll., Valhalla, N.Y., 1990—; rsch. scientist devel. Nat. Inst. on Drug Abuse, 1982-90, rsch. scientist, 1992—, ad hoc reviewer, 1989—; ad hoc reviewer NIMH, NSF. Author: The Psychology of Adolescence, 1978; contbr. articles to profl. jours. Recipient 1st ann. Dean's Disting. award N.Y. Med. Coll., 1992; grantee Nat. Inst. on Drug Abuse, 1989—. Mem. Am. Psychol. Soc. (liaison officer 1989—), Assn. for Med. Edn. and Rsch. in Substance Abuse, N.Y. State Psychol. Assn. Home: 350 Central Park W Apt 12H New York NY 10025 Office: Dept Psychiatry-Behavior Sc NY Med Coll Valhalla NY 10595

BROOKE, RALPH IAN, dental educator, vice provost, university dean; b. Leeds, Eng., Apr. 25, 1934; s. Michael and Jeanette (Cohen) B.; m. Lorna Ruth Shields; children: Michael Jeremy Richard, Andrew Timothy. Baccalaureus Chirurgiae Dentium, Licentiate in Dental Surgery, Leeds U., England, 1957. Licentiate Royal Coll. Physicians, 1963. Sr. lectr. Leeds U., 1970-72; prof., chmn. dept. oral medicine U. Western Ont., London, Can., 1972-82, dean dentistry faculty, 1982—, vice provost health scis., 1987—; chief dentistry Univ. Hosp., London, 1973-92. Contbr. articles to profl. jours. Fellow Internat. Acad. Dentistry, Royal Coll. Dentists Can., Royal Coll. Surgeons; mem. Nat. Dental Exam. Bd. (chmn. Can. commn. on dental accreditation), Can. Faculties Dentistry (past pres.), Can. Acad. Oral Medicine (pres.). Avocations: music, cycling.

BROOKER, JEFF ZEIGLER, cardiologist; b. Columbia, S.C., Nov. 1, 1941; s. Jefferson Zeigler and Virginia (Ligon) B.; m. Rhoda Arrowsmith, June 12, 1966; children: Jeff III, John, Rhoda. BS, U. S.C., 1962; MD, Med. U. S.C., 1966. Intern, resident Hosp. U. Pa., Phila., 1966-68; resident internal medicine Stanford U. Med. Ctr., Palo Alto, Calif., 1970-71, rsch. fellow cardiology, 1971-73; staff cardiologist Tex. Heart Inst., Houston, 1973-74; assoc. dir. cardiology Providence Hosp., Columbia, S.C., 1974-81; pvt. practice cardiology Columbia, 1981—; cons. peer rev. Jour. AMA, Chgo., 1976-77; local and regional rsch. com. Am. Heart Assn., Dallas., 1977-86. Mem. editorial bd. Jour. S.C. Med. Assn., Columbia, 1991—; editorial reviewer: Essentials of Echocardiography, 1977. Legis. liaison S.C. Med. Assn., Columbia, 1991-92. Lt. comdr. USN, 1968-70. Recipient Best Sci. Article award Roe Found., Columbia, 1991. Fellow ACP, Am. Coll. Cardiology, Am. Heart Assn. (coun. on clin. cardiology, 1975—); N. Am. Soc. Pacing and Electrophysiology. Achievements include improved method for oral dipyridamole testing for ischemic heart disease; devising a percutaneous method for inserting pacing lead into the internal jugular vein yet still implant and pulse generator on the anterior chest wall. Office: 1625 Brabham Ave Columbia SC 29204

BROOKLER, KENNETH HASKELL, otolaryngologist, educator; b. Winnipeg, Man., Can., Sept. 28, 1938; s. Barney and Mary (Freedman) B.; m. Marcia A. Schuckett, Aug. 24, 1961; children: Richard, Jill, Brent. MD, U. Man., 1962, MS in Otolaryngology, U. Minn., 1968. Intern, Winnipeg Gen. Hosp., 1962-63; fellow Mayo Clinic, Rochester, Minn., 1964-67, fellow in neurotology, 1968; practice medicine specializing in otology and neurotology, N.Y.C., 1968—; pres. Rsch. Inst. for Hearing and Balance Disorders; mem. med. adv. bd. N.Y. State Athletic Commn. Bd. dirs. Hebrew Hosp. Author Dizziness, 1991; editorial bd. Ear, Nose Throat Jour., Jour. Vestibular Rsch.; contbr. chpts. to books, articles to profl. jours. Fellow ACS, Royal Coll. Surgeons (Can.); mem. N.Y. County Med. Soc., Am. Acad. Otolaryngology, N.Y. Otol. Soc., N.Y. Acad. Medicine, N.Y. Acad. Scis., Trilological Soc., Barany Soc., Am. Neurotology Soc., Internat. Vestibulometric Soc., Nat. Hearing Assn. (dir.). Office: Neurotologic Assocs PC 111 E 77th St New York NY 10021-1892

BROOKS, CHARLES IRVING, psychology educator; b. Washington, Mar. 3, 1944; s. Charles M. and Grace (Coakley) B.; m. Joyce Emery, June 11, 1966; children: Audra Ellen, Kelly Lynn. BA, Duke U., 1965; MA, Wake Forest U., 1967; PhD, Syracuse U., 1970. Asst. prof. psychology Wilson Coll., Chambersburg, Pa., 1970-75; prof. psychology, chmn. dept., disting. svc. prof. King's Coll., Wilkes-Barre, Pa., 1975—. Contbr. articles to profl. jours., chpts. to books. Mem. Am. Psychol. Soc., Psychonomic Soc., Sigma Xi, Psi Chi. Home: 93 Dana St Forty Fort PA 18704-4106 Office: Kings Coll Box 1675 133 N River St Wilkes Barre PA 18711-0801

BROOKS, DOUGLAS LEE, retired oceanographic and atmospheric policy analyst, environmental policy consultant; b. New Haven, Conn., Aug. 5, 1916; s. John Lee and Genevieve Marie (Ford) B.; m. Elizabeth Blakey Thatcher, Dec. 27, 1941; children: John Christopher, Alison (dec.), Alan Douglas, Kenneth Thatcher, Donald Lee. BS in Physics, Yale U., 1938; MS, MIT, 1943, ScD, 1948. Analyst Ops. Evaluation Group, Washington, 1948-62; dir. rsch. Naval Warfare Analysis Group, Washington, 1958-62; pres. Travelers Rsch. Ctr., Hartford, Conn., 1963-69; spl. asst. to dir. NSF, Washington, 1970-72; exec. dir. Nat. Adv. Com. Oceans and Atmosphere, Washington, 1972-79. Author: American Looks to the Sea: Ocean Use and the National Interest, 1984; contbr. articles to profl. jours. 1st lt. U.S. Army, 1942-46. Fellow N.Y. Acad. Sci.; mem. Am. Meteorol. Soc., Marine Tech. Soc. Home: 60 Loeffler Rd Apt 303P Bloomfield CT 06002-2275

BROOKS, JAMES ELWOOD, geologist, educator; b. Salem, Ind., May 31, 1925; s. Elwood Edwin and Helen Mary (May) B.; m. Eleanore June Nystrom, June 18, 1949; children: Nancy, Kathryn, Carolyn. A.B., DePauw U., 1948; M.S., Northwestern U., 1950; Ph.D., U. Wash., 1954. Research assoc. Ill. Geol. Survey, 1950; geologist Gulf Oil Corp., Salt Lake City, summers 1951-53; instr. geol. scis. So. Meth. U., Dallas, 1952-55; asst. prof. So. Meth. U., 1955-59, assoc. prof., 1959-62, prof., 1962—, chmn. dept., 1961-70, dean, assoc. provost univ., 1969-72, provost, v.p., 1972-80, interim pres., 1980-81; pres., trustee Inst. for Study Earth and Man, Dallas, 1981—; cons. geologist firm DeGolyer & MacNaughton, Dallas, 1954-59. Contbr. articles to profl. jours. Trustee Hockaday Sch., 1982-88; trustee Dallas Mus. Natural History Assn., 1985—, v.p., 1986-88, pres., 1988-90; mem. exec. bd. Circle Ten coun. Boy Scouts Am., 1982—; bd. visitors DePauw U., 1979-83, chmn., 1983; mem. Mayor's Task Force on Fair Park, 1992; chmn. Coun. Fair Park Instns., 1992—. Fellow AAAS, Geol. Soc. Am., Tex. Acad. Sci., Explorers Club; mem. Am. Assn. Petroleum Geologists, Assn. Petroleum Geochem. Exploration, Dallas Geol. Soc., Sigma Xi, Sigma Gamma Epsilon, Sigma Phi. Home: 7055 Arboreal Dr Dallas TX 75231-7315 Office: Inst Study Earth and Man PO Box 274 Dallas TX 75275-0274

BROOKS, JOHN SAMUEL JOSEPH, pathologist, researcher; b. Phila., Feb. 2, 1948. BS in Biology, St. Joseph's Coll., Phila., 1970; MD, Thomas Jefferson U., 1974. Diplomate Am. Bd. Pathology. Resident in pathology U. Pa., Phila., 1974-78, chief resident, 1978, asst. prof., 1979-84, assoc. prof., 1984-88, prof., 1988-93; chmn. dept. pathology Roswell Pk. Cancer Inst., Buffalo, 1993—; prof., vice chmn. pathologymed. sch. SUNY, Buffalo, 1993—; vis. prof. Royal Marsden Hosp./Inst. Cancer Rsch., London, 1987; expert in immunohistochemistry. Author: Pathology, 1989; contbr. articles to New Eng. Jour. Medicine, Jour. of AMA, Jour. Urology, Internat. Jour. Ob.-Gyn. Pathology, Am. Jour. Pathology; editor Internat. Jour. Surg. Pathology, 1993—; mem. bd. editors: Jour. Modern Pathology, Jour. Surg. Pathology, and reviewer; contbr. over 140 articles to profl. jours. Mem. AAAS, Pathology Soc. Phila. (pres. 1988-90), Eastern Coop. Oncology Group (chmn. sarcoma pathology com. Madison chpt. 1988—), Internat. Acad. Pathology (edn. com. Atlanta chpt. 1989—), Am. Soc. Clin. Pathologists, Arthur Purdy Stout Soc. for Surg. Pathologists (membership com. 1990—), Fedn. Am. Socs. for Exptl. Biology, Nat. Internat. Reputation in Diagnostic Surg. Pathology, Royal Coll. Pathology. Democrat. Roman Catholic. Achievements include research in significance of double phenotypes in sarcomas, growth factors in sarcomas, in immunohistochemistry; posthumous diagnosis of Pres. Cleveland's tumor. Home: 34 Deer Run Orchard Park NY 14127 Office: Roswell Pk Cancer Inst Dept Pathology Elm & Carlton Sts Buffalo NY 14263

BROOKS, LLOYD WILLIAM, JR., osteopath, interventional cardiologist, educator; b. Amarillo, Tex., Nov. 4, 1949; s. Lloyd William and Tina Margaret (Roe) B.; m. Ann Nettleship, Apr. 3, 1987. BS, U. Tex., 1972; DO cum laude, Tex. Coll. Osteo. Medicine, 1985. Diplomate Am. Osteo. Bd. Internal Medicine. Intern Dallas-Ft. Worth Med. Ctr., 1985; resident in internal medicine Ft. Worth Osteo. Med. Ctr., 1986-88; fellow in cardiology Detroit Heart Inst.; fellow in angioplasty and interventional cardiology Riverside Meth. Hosp., Columbus, Ohio; pvt. practice Med. Splty. Assocs., Ft. Worth, 1991—; clin. asst. prof. Tex. Coll. Osteo. Medicine, Ft. Worth, 1991—. Contbr. articles to med. jours. Mem. AMA, Am. Osteo. Assn., Am. Coll. Osteo. Internists (diplomate), Am. Coll. Cardiology. Avocations: photography, music, woodworking, sailplaning, scuba diving. Office: Med Splty Assocs 1002 Montgomery Ste 200 Fort Worth TX 76107-2714

BROOKS, MARC BENJAMIN, civil engineer; b. Denville, N.J., June 25, 1966; s. Jack and Doris (Lohne) B.; m. Gina Pashaian, Aug. 4, 1991. BS in Civil Engring., Rutgers U., 1989. Engr. in tng. Civil engr. Openaka Corp., Denville, N.J., 1989—. Mem. ASTM, Chi Epsilon (treas. 1986-87). Home: 41 Edgewood Ave Springfield NJ 07081 Office: Openaka Corp Inc 565 Openaki Rd Denville NJ 07834

BROOKS, PHILIP RUSSELL, chemistry educator, researcher; b. Chgo., Dec. 31, 1938; s. John Russell and Louise Jane (Seyler) B.; children: Scott, Robin, Christopher, Steven. BS, Calif. Inst. Tech., 1960; PhD, U. Calif., Berkeley, 1964. Rsch. assoc. physics dept. U. Chgo., 1964; from asst. to assoc. prof. chemistry Rice U., Houston, 1964-75, prof., 1975—. Editor: State-to-State Chemistry, 1977. Vol. Boy Scouts Am., Houston, 1970—. Recipient Humboldt prize Alexander von Humboldt Found., 1985; predoctoral fellow NSF, 1960-63, postdoctoral fellow, 1963-64, Alfred P. Sloan fellow, 1970-74, John Simon Guggenheim fellow, 1974-75, Vis. Erskine fellow U. Canterbury, 1991, JSPS fellow Japan Soc. Promotion Sci., 1992. Fellow Am. Phys. Soc.; mem. Am. Chem. Soc. Achievements include research on chemical reaction dynamics. Home: 1026 Glourie Houston TX 77055 Office: Rice U Chemistry Dept PO Box 1892 Houston TX 77251

BROOKS, RICHARD M., mechanical engineer; b. Ridgewood, N.J., July 7, 1962; s. Ronald M. and Patricia (E. Jeynes) B.; m. Sharon Anne Casmer, Nov. 12, 1982. BSME, NYU, 1986, BS in Biomed. Engring., 1989. Med. device designer Osteonics Corp., Allendale, N.J., 1981-88; product devel. engr. Howmedica, Rutherford, N.J., 1988—. Mem. AAAS, Air and Space Smithsonian. Republican. Achievements include patent for orthopaedic device, 3 patents for orthopaedic implants. Office: Howmedica 309 Veterans Blvd Rutherford NJ 07070-2564

BROOKS, ROBERT RAYMOND, pharmacologist; b. Jersey City, Nov. 8, 1944; s. Raymond Edward and Margaret Ann (Keltz) B.; m. Sherry Norris Kimball, July 6, 1968; children: Patrick K., Joshua D., Andrew C., Catherine P. AB in Chemistry, Princeton (N.J.) U., 1966; PhD in Biochemistry, Johns Hopkins U., 1971. Rsch. assoc. dept. biochem. U. Fla., Gainesville, 1971-73; sect. head, group leader, scientist Norwich (N.Y.) Eaton Pharms., Inc., 1973-92; sect. head Procter & Gamble Pharms., Norwich, 1992—; cons. I.L. Richer Co., Inc., 1981-83; peer grant reviewer NIH, Bethesda, Md., 1982. Pres. Peaks n Trails Ski Club, Norwich, 1980; treas. Tri Valley Aviation Assn., Norwich, 1978. Postdoctoral fellowship NIH, 1973. Mem. Am. Soc. for Pharmacology and Exptl. Therapeutics, Am. Chem. Soc. (Norwich sect. chmn. 1977), Am. Soc. for Microbiology, Am. Heart Assn. (basic sci. coun.). Achievements include discovery of antiarrhythmic agent; rsch. on repeated DNA sequences in fungi. Office: Proctor & Gamble Pharms PO Box 191 Norwich NY 13815-0191

BROOKS, SHARON LYNN, dentist, educator; b. Detroit, Oct. 19, 1944; d. Edward Haggit Doubleday and Ila Annabelle (Bobier) Kitamura; m. David Howard Brooks, Aug. 29, 1965. ABEd, U. Mich., 1965, DDS, 1973, MS, 1976, 84, 89. Diplomate Am. Bd. Oral and Maxillofacial Radiology, Am. Bd. Oral Medicine. Pvt. practice Ann Arbor, 1973-76; clin. instr. dentistry U. Mich., Ann Arbor, 1973-76, asst. prof., 1976-80, assoc. prof., 1980-86, prof., 1986—; assoc. prof. radiology Sch. Medicine U. Mich., 1992—; vis. prof. U. Rochester (N.Y.), 1991.; cons. VA Med. Ctr., Ann Arbor, 1981—, Allen Park, 1983—; nat. bd. test constructor dental hygiene Joint Commn. Nat. Dental Examinations, Chgo., 1986-91. Contbr. articles to profl. jours. Rep. U. Mich. Senate Assembly, Ann Arbor, 1981-84, 87-91; bd. dirs. Ann Arbor YWCA, 1985-88; mem. Senate Assembly Com. on U. Affairs, 1989-90. Mem. ADA, Mich. Dental Assn., Washtenaw Dist. Dental Soc. (pres. 1990-91), Orgn. Tchrs. Oral Diagnosis (pres. 1988-89), Am. Assn. Dental Schs. (chair oral diagnosis sect. 1991-92), Internat. Assn. Dental Rsch., Am. Bd. Oral Medicine (bd. dirs. 1990—), Am. Acad. Oral and Maxillofacial Radiology (bd. dirs. continuing edn. 1985-88), Internat. Assn. Oral Maxillofacial Radiology, Health Physics Soc. Avocations: swimming, hiking, nature study, reading, needlework. Office: U Mich Sch Dentistry Dept Oral Medicine Pathology, Surgery Ann Arbor MI 48109-1078

BROOKS, STEVEN DOYLE, environmental engineer; b. St. Joseph, Mo., May 8, 1958; s. Warren Perry and Mary Ann (Stout) B. BS in Chem. Engring., U. Mo., Rolla, 1981; AS in Middle Mgmt., Mo. Western State Coll., St. Joseph, 1990. Cert. solid waste technician. Engr.-in-tng. NOWSCO-Oil Well Svc. Co., Farmington, N.Mex., 1981; environ. engr. St. Joseph Light & Power Co., St. Joseph, Mo., 1981-85; supr. environ. engring. and plant chemistry St. Joseph Light & Power Co., 1985-90, supr. environ. engring., 1990—; chmn. elec. engring. tech. adv. bd. Mo. Western State Coll., St. Joseph, 1990—. Mem. St. Joseph South Side Progressive Assn., 1990—; sec. Benton High Sch. Music Booster Assn., St. Joseph, 1985; vol. Salvation Army, St. Joseph, 1985-90, Spl. Olympics, 1985. Mem. Mo. Waste Control Coalition, Am. Inst. Chem. Engrs., Air Pollution Control Assn., Mo. Elec. Utility Environ. Com., Lions, Masons (33d deg.). Democrat. Methodist. Avocations: weight lifting, running, music. Home: 5705 Osage

Dr Saint Joseph MO 64503-2324 Office: St Joseph Light & Power Co 520 Francis St Saint Joseph MO 64501-1706

BROOME, KENNETH REGINALD, civil engineer, consultant; b. London, May 19, 1925; came to U.S., 1952; s. Reginald Alexander and Violet Frances (Paull) B.; m. Heather Claire Platt, Dec. 27, 1946; children: Claire V., Stephanie A., Rosemary L., H. Elizabeth, J. Cecilia, Martin A. BA, Cambridge (Eng.) U., 1946, MA, 1956. Lic. civil engr., U.K., Calif., Fla., Pa., structural engr., Calif. Civil engr. R.T. James and Ptnrs., London, 1946-52, Austin Co., L.A., 1952-59; project mgr. Aerojet Gen. Co., Covina, Calif., 1959-64; chief civil engr. Bechtel Corp., San Francisco, 1964-72, Gilbert Assocs., Reading, Pa., 1972-82; prin. engr. Williams and Broome, Inc., Reading, Pa., 1982—; com. chmn. long range plan Water Resources Assn. Del. River Basin, Phila., 1980—; chmn., pres. Schuylkill River Greenway, Reading, 1989-91. Contbr. articles to profl. jours. Pres. World Affairs Coun., Berks and Lehigh County, Pa., 1976, com. chmn., 1985—. Recipient grant to develop modular hydro dam U.S. Dept. of Energy, 1978, grant to develop powerbarge concept Pa. Energy Devel. Authority, 1985-86. Mem. ASCE (pres. Reading br.), NSPE (Engr. of Yr. 1985) Inst. Civil Engrs. Democrat. Roman Catholic. Achievements include patents for concept and application technology for powerbarge floating hydroelectric plant; development of floating dry-dock method of launching ships, caisson method of casting solid propellant in large solid space boosters. Office: Williams and Broome Inc 15 Fawn Dr Reading PA 19607

BROSILOW, ROSALIE, engineering editor; b. Montreal, Que., Can., July 7, 1939; came to U.S., 1948; d. Jack and Jennie (Lamdan) Ziegelman; m. Coleman Bernard Brosilow, Feb. 18, 1962; children: Rachelle, Benjamin Jacob. BS in Metall. Engring., U. Mich., 1962; PhD, Case Western Res. U., 1967. Editor Welding Design and Fabrication, Cleve., 1974—. Bd. mem. Solomon Schechter Day Sch., Cleve. Mem. Am. Welding Soc. (B1 com., edn. com.), Am. Soc. Metals Internat. Achievements include research in thermal cutting. Home: 3115 Berkshire Cleveland OH 44118 Office: Welding Design & Fabrication 1100 Superior Ave Cleveland OH 44114

BROTHERTON, VINCE MORGAN, research chemist; b. Seattle, Sept. 21, 1963; s. Daniel Francis and Patricia Ann (Morgan) B. B.Biology/Chemistry, Western Wash. U., 1988; MBA, Seattle Pacific U., Seattle, 1993. Rsch. chemist Abbott Labs., Bothell, Wash., 1988-93; lab. dir. Residential Environ. Analytical Labs., Seattle, 1993—. With USN, 1992. Mem. Am. Chem. Soc. Achievements include patents. Office: R E A L 13346 1st St NE Seattle WA 98125

BROUGHTON, ROBERT STEPHEN, irrigation and drainage engineering educator, consultant; b. Corbetton, Ont., Can., June 23, 1934; s. Arthur Stephen and Luella Margaret (Gray) B.; m. Ruth Mabel Smith, May 11, 1957; children: G. Anne, Sharon Mae, Heather Louise, Stephen Russell. BS in Agr., U. Toronto, Ont., 1956, B in Applied Sci., 1957; MCE, MIT, 1959; PhD in Drainage Engring., McGill U., Montreal, Que., Can., 1972; LLD (hon.), Dalhousie U., Halifax, N.S., Can., 1989. Cert. profl. engr., Ont., Que. Jr. engr. John Deere Plow Co., Welland, Ont., 1956; rsch. asst. MIT, Cambridge, 1957-59; hydraulic engr. conservation br. Ont. Govt., Toronto, 1959-61; lectr. in agrl. engring. McGill U., 1962-63, asst. prof. agrl. engring., 1963-66, assoc. prof., 1966-74, prof., 1974—; from v.p. to pres. Can. Soc. Agrl. Engring., 1968-74, chmn. drainage rsch. com.; speaker farmers' meetings. Author, editor book in field; contbr. over 130 articles to publs. Tchr. Sunday sch. Beaurepaire United Ch., Beaconsfield, Que., 1965-72, clk. of session, 1973-78. Recipient Internat. Achievement award Can. nat. com. Internat. Comsn. Irrigation and Drainage, 1993. Fellow Can. Soc. Agrl. Engring. (Maple Leaf award 1978, James Beamish award 1989); mem. Am. Soc. Agrl. Engring., Ordre des Ingénieurs du Que., Ordre des Agronomes du Que. (cert. profl. agronomist), Assn. Profl. Engrs. Ont., Corrugated Plastic Pipe Assn. (life). Mem. United Ch. Can. Achievements include research on design and construction of subsurface drainage systems for control of waterlogging and salinity of irrigated lands, assisted with irrigation and drainage projects in India, Pakistan, Egypt, Trinidad, El Salvador, Barbados, Canada, etc. Office: McGill U Ctr Drainage Stds, MacDonald Campus, PO Box 950, Sainte Anne de Bellevue, PQ Canada H9X 3V9

BROUSSARD, MALCOLM JOSEPH, pharmacist, consultant; b. Lake Charles, La., Mar. 14, 1953; s. Roy Joseph and Liller Leeova (Tubbs) B. BS in Biology, McNeese State U., Lake Charles, 1975; BS in Pharmacy, Xavier Coll. Pharmacy, 1978. Registered pharmacist. Staff pharmacist Hotel Dieu Hosp., New Orleans, 1978-80, JoEllen Smith Meml. Hosp., New Orleans, 1980-85; asst. dir. pharmacy St. Jude Med. Ctr., Kenner, La., 1985-91; dir pharmacy MacKinnon Ctr., Metairie, La., 1991—; founder, pres. Parenteral Therapy Svcs., Inc., Harvey, La., 1985-90; v.p. Lapalco Pharmacy, Inc., Harvey, 1987-90, South La. Med. Supply Co., Inc., Harvey, 1987-90; pharmacy cons. La. State Bd. Nursing, New Orleans, 1980-93; externship preceptor Xavier Coll. Pharmacy, New Orleans, 1980—; vis. instr., 1982-83, asst. prof., 1991—. Bd. dirs. New Orleans Pharmacy Mus., 1988—; mem., advisor La. Pharmacists Polit. Action Com., 1984—. Mem. Am. Pharm. Assn., Am. Soc. Hosp. Pharmacists (state del. 1991-93), La. Pharmacists Assn. (pres. 1983-84, Pres.'s award 1984, Pharmacist of Yr. 1985), La. Soc. Hosp. Pharmacists (pres. 1988-89), S.E. La. Soc. Hosp. Pharmacists (pres. 1981-82, Pharmacist of Yr. 1984), Greater New Orleans Pharmacists Assn., Rho Chi, Phi Lambda Sigma. Republican. Roman Catholic. Avocation: snow skiing. Home: 1804 Timberlane Estates Dr Harvey LA 70058-5131

BROUSSARD, PETER ALLEN, energy engineer; b. Lake Charles, La., Aug. 23, 1950; s. Roy Joseph and Evelyn Gertrude (Richter) B.; m. Naomi Dominga Lorenzana, Oct. 24, 1973; children: Tessa Marie, Chrysta Laurie. BSME, U. S.W. La., 1972. Registered profl. engr. Ala., Miss., La., Tex. Petroleum engr. Amoco Prodn. Co., Lake Charles, La., 1975-76; environ. engr. Meyer & Assocs., Sulphur, La., 1976-79; indsl. transmission engr. Gulf States Utilities Co., Lake Charles, 1979-85; environ. engr. Associated Cons. Ltd., Meridian, Miss., 1986-89; environ. engr. James River Corp., Pennington, Ala., 1989-91, power plant engr., 1991-92, plant energy chief, 1992—; guest lectr. 12th World Energy Engring. Congress, 1989, Power-Gen Conf., 1989; guest speaker WATTec Conf., 1990. Contbr. chpt.: Strategies for Reducing Natural Gas, Electricity and Oil Costs, 1989. Res. dep. San Bernadino County Search and Rescue, Victorville, Calif., 1972-74; liturgist, catechist St. Patrick Ch., Meridian, 1986—. Major USAF, Miss. ANG, 1972—. Biofuels rsch. grantee U.S. Dept. Energy, 1980; scholar La. Engring. Soc., 1969, T.H. Harris award, 1968, USAF ROTC, 1968. Mem. Assn. Energy Engrs. (sr.), Am. Soc. Am. Mil. Engrs. (sr.). Roman Catholic. Achievements include research in geoenergy potential of S.E. Texas and S.W. Louisiana, potential of wind energy, combustibility of paper mill sludge and tire-derived fuel, integration of alcohol fuels into electric power generation operations. Home: 4252 23d Ave Meridian MS 39305 Office: James River Corp Rt 114 Pennington AL 36916

BROUSSARD, WILLIAM JOSEPH, physician, cattleman, environmentalist; b. Abbeville, La., Feb. 14, 1934; s. Alphe Alcide and Odile Perpetue (Cade) B.; m. Margaret Lucile Reynolds, Apr. 3, 1933; children: Dianne, M. Lynn, Allen, Laura. BS, La. State U., Baton Rouge, 1955; MD, U. Minn., 1959. Diplomate Am. Bd. Ophthalmology. Pvt. practice medicine Melbourne (Fla.) Eye Assocs., 1967—; pres. Med. Staff Holmes Regional Med. Ctr., Melbourne; founder Melbourne Eye Assocs., 1967. Bd. dirs. Landmark Bank, Melbourne, 1976-86, Civic Music Assn., Melbourne, 1970-83. Maj. U.S. Army, 1959-67, Germany. Fellow ACS, Am. Acad. Ophthalmology (ethics com. 1986-88); mem. AMA, Nat. Cattlemen's Assn., Fla. Soc. Ophthalmology (pres. 1983), Fla. Cattleman's Assn., Fla. Med. Assn., Found. Beefmaster Assn. (pres. 1988), Allen Broussard Conservancy, Inc. Avocations: travel, photography, environmental preservation. Home: 3660 N Riverside Dr Melbourne FL 32903-4424 Office: Melbourne Eye Assocs 1355 Hickory St Melbourne FL 32901-3232

BROVOLD, FREDERICK NORMAN, geotechnical and civil engineer; b. N.Y.C., Sept. 7, 1939; s. Peter Brovold and Alice Ann (Becher) Stevens; m. Noriko Inoue, Dec. 20, 1985; children: Ayumi Julia Cox, Mari Kathleen Kunimi. BSCE, U. Fla., 1965, MS in Engring., 1967; MSCE, CE, MIT, 1971. Registered profl. engr., Calif., Oreg., Colo. Wyo., Ga. Group mgr. Law Engring. Testing Co., Atlanta, 1971-73; cons. Geocons., Inc., Atlanta, 1973-77; materials engr. U.S. Forest Svc., Portland, Oreg., 1977-80; sr. engr. CH2M-Hill, Inc., Denver, 1980-81; prin. engr. Rocky Mountain Energy Co., Denver, 1981-87; project mgr. EMCON Assocs., San Jose, Calif., 1987-90; assoc. Roger Foott Assocs., Inc., San Francisco, 1990—. Mem. ASCE, Structural Engrs. Assn. No. Calif. (chmn. founds. subcom. 1992-93). Office: Roger Foott Assocs Inc 3rd Floor 565 Commercial St San Francisco CA 94111

BROWDER, FELIX EARL, mathematician, educator; b. Moscow, July 31, 1927; s. Earl and Raissa (Berkmann) B.; m. Eva Tislowitz, Oct. 5, 1949; children: Thomas, William. SB, MIT, 1946; PhD, Princeton U., 1948; MA (hon.), Yale U., 1962; D (hon.), U. Paris, 1990. C.L.E. Moore instr. math. MIT, 1948-51, vis. assoc. prof., 1961-62, vis. prof., 1977-78; instr. Boston U., 1951-53; asst. prof. Brandeis U., 1955-56; from asst. prof. to prof. Yale U., 1956-63; prof. math. U. Chgo., 1963-72, Louis Block prof. math., 1972-82, Max Mason disting. svc. prof., 1982-87, chmn. dept., 1972-77, 80-85; v.p. rsch. Rutgers, The State U. N.J., 1986-91; univ. prof. Rutgers U., New Brunswick, 1986—; vis. mem. Inst. Advanced Study, Princeton (N.J.) U., 1953-54, 63-64; vis. prof. Princeton U., 1968, Inst. Pure and Applied Math., Rio de Janeiro, 1960, U. Paris, 1973, 75, 78, 81, 83, 85; sr. rsch. fellow U. Sussex, Eng., 1970, 76; Fairchild Disting. visitor Calif. Inst. Tech., Pasadena, 1975-76; invited speaker Internat. Congress of Math., 1970, Sci. Bd. Santa Fe Inst., 1986—. Contbr. theorems to books, including Nonlinear Problems, 1966, Functional Analysis and Related Fields, 1970, Nonlinear Operators and Nonlinear Equations of Evolution in Banach Spaces, 1976, Nonlinear Functional Analysis and Its Applications, 1986. With AUS, 1953-55. Guggenheim fellow, 1953-54, 66-67, Sloan Found. fellow, 1959-63, NSF sr. postdoctoral fellow, 1957-58. Fellow AAAS (chmn. sect. A 1982-83), NAS (coun. mem. 1992-95), Am. Acad. Arts and Scis., Am. Math. Soc. (editor bull. 1959-68, 78-83, mem. coun. 1959-72, 78-83, mng. editor 1964-68, 80, mem. exec. com. coun. 1979-80, colloquium lectr. 1970), Math. Assn. Am., Sigma Xi (pres. chpt. 1985-86). Achievements include development of linear and nonlinear partial differential equations, nonlinear functional analysis and fixed point and mapping theorems.

BROWER, MICHAEL CHADBOURNE, research administrator; b. Washington, Oct. 10, 1960; s. Carleton Chadbourne and Imogene (Cockcroft) B.; m. Julie Scolnik, Oct. 9, 1988; 1 child, Sophia Carolyn Scolnik. BSc, MIT, 1981; Phd, Harvard U., 1986. Analyst Union Concerned Scientists, Cambridge, Mass., 1986-91, rsch. dir., 1991—. Author: Cool Energy: Renewable, 1992. Mem. Sigma Xi, Phi Beta Kappa. Home: 58 Dana St Cambridge MA 02138 Office: Union Concerned Scientists 26 Church St Cambridge MA 02138

BROWN, ALAN CHARLTON, retired aeronautical engineer; b. Whitley Bay, England, Dec. 5, 1929; came to U.S., 1956; s. Stanley and Dorothy (Charlton) B.; m. Gweneth Evelyn Bowler, July 26, 1952; children: Yvonne, Christine, Diane, Maureen. Diploma aeronautics, Hull (Eng.) Tech. Coll., 1950; MS, Cranfield (Eng.) Inst Tech., 1952, Stanford U., 1964. Apprentice Blackburn Aircraft Ltd., Brough, Eng., 1945-50; aerodynamicist Bristol (Eng.) Aeroplane Co., 1952-56; rsch. scientist U. So. Calif., L.A., 1956-58, Wiancko Engring. Co., Pasadena, Calif., 1958-60, Lockheed Missiles & Space Co., Palo Alto, Calif., 1960-66; group leader Lockheed Aeronautical Systems Co., Burbank, Calif., 1966-69; dept. mgr. Lockheed Aerospace Systems Co., Burbank, Calif., 1969-78, chief engr. F-117A, 1978-82, dir. stealth tech., 1982-89; dir. engring. Lockheed Corp., Calabasas, Calif., 1989-92. Fellow AIAA (Aircraft Design award 1990), NAE, Royal Aero. Soc. Democrat. Avocations: music, tennis, chess. Home: 11853 Jellico Ave Granada Hills CA 91344-2113

BROWN, ALAN JOHNSON, chemicals executive; b. Alnwick, Eng., June 18, 1951; came to U.S., 1987; s. George and Margaret Mary (Johnson) B.; m. Cathy Sturlis, May 14, 1988 (div. Feb. 1993). BS in Chemistry, U. East Anglia, Eng., 1974; MS in Enzymology, U. Warwick, Eng., 1975, PhD in Molecular Sci., 1977. Chartered chemist. Group leader ICI Corp. Rsch., Runcorn, Eng., 1977-81; group leader ECC Internat. Cen. Rsch., St. Austell, Eng., 1981-87; rsch. mgr. ECC Internat. Tech. Ctr., Sandersville, Ga., 1987-92; tech. dir. Columbia River Carbonates, Woodland, Wash., 1993—. Patentee in field. Mem. Royal Soc. Chemistry (S.W. Region com. 1986-87), Am. Chem. Soc., Clay Minerals Soc., Tech. Assoc. Pulp and Paper Inst., Chem. Mfrs. Assn. (chmn. analytical methods group 1991, 92). Avocations: chess, home remodeling, gardening, science fiction. Office: Columbia River Carbonates PO Box D Woodland WA 98674

BROWN, ARTHUR EDWARD, physician; b. Trenton, N.J., June 7, 1945; s. Milton Charles and Jeanne Ruth (Swern) B.; m. Jo Frances Meltzer, Nov. 24, 1985. BS, Bucknell U., 1967; MD, Jefferson Med. Coll., 1971. Intern, resident Roosevelt Hosp., N.Y.C., 1971-72, 74-76; fellow infectious diseases Meml. Sloan-Kettering Cancer Ctr., N.Y.C., 1976-78, clin. asst. physician, 1978-82; asst. prof. medicine and pediatrics Cornell U. Med. Coll., N.Y.C., 1979-85, assoc. prof. clin. medicine and pediatrics, 1985—; asst. attending physician Meml. Hosp. for Cancer and Allied Diseases, N.Y.C., 1982-89, assoc. attending physician, 1989—; assoc. attending pediatrician N.Y. Hosp., N.Y.C., 1985—. Editor: Infectious Complications of Neoplastic Diseases Controversies in Mgmt., 1985; mem. editorial bd. European Jour. Clin. Microbiology & Infectious Diseases, 1993—. Surgeon, USPHS, 1972-74. Recipient 2d pl. HeSCA Print Festival, 1985, Bronze Plaque award Film Coun. Columbus, 1985, Bronze medal Internat. Film & TV Festival, N.Y.C., 1985, Semi-Finalist Am. Jour. Nursing Media Festival, 1986. Fellow ACP, Infectious Diseases Soc. Am.; mem. Am. Fedn. Clin. Rsch., N.Y. County Soc. Internal Medicine (pres. elect 1992—), N.Y.C. Soc. Infectious Diseases (sec., treas. 1993—). Achievements include rsch. on AIDS and controversies in the mgmt. of infectious complications of neoplastic diseases. Office: Meml Sloan-Kettering Cancer Ctr 1275 York Ave New York NY 10021

BROWN, BARBARA S., environmental scientist; b. Newark, Aug. 5, 1951; d. Louis and Louise (Mumper) Stein; children: Kristin Leigh, Andrew Hayden. Student Am. U., 1969-71; BS in Biology, U. Miami, 1976, postgrad. Staff scientist Environ. Sci. and Engring., Inc., Miami, 1978-86; dir. environ. crimes unit Dade County Environ. Resource Mgmt., 1986—. Mem. Ecol. Soc. Am., Nat. Assn. Environ. Profls., League Environ. Enforcement and Prosecution (sec. S. Fla. chpt.).

BROWN, CHARLES DURWARD, aerospace engineer; b. Santa Fe, July 23, 1930; s. Charles Edward and Lila Mae (Turk) B.; m. Lawana Jane VanDall, Apr. 13, 1952; children: John, Lawana, Kelsey. BSME, U. Okla., 1952; MSME, So. Meth. U., 1961. With Martin Marietta, Denver, 1958-90; worked on Mariner '71, Viking Orbiter, 1971-77; mgr. Magellan program Martin Marietta, Denver, to 1990; ret., 1990; lectr. in spacecraft design, Colo. U., Boulder, 1981-91. Author: Spacecraft Mission Design, AIAA, 1992; assoc. editor: Rocket Propellant and Pressurization Systems, 1964; editor: Spacecraft Design, 1982. Recipient NASA Pub. Svc. medal for Viking Orbiter, 1977, NASA award of Merit for Magellan, 1987, Jefferson Cup award Martin Marietta, 1971, 90, Silver Snoopy award Astronaut Corps, 1989, Outstanding Engring. Achievement award NSPE, 1989, Dr. Robert H. Goddard Meml. Trophy for leadership of Magellan program, 1992, NASA Pub. Svc. award for Magellan, 1992. Mem. Tau Beta Pi. Achievements include co-design of Magellan, the first planetary spacecraft to fly on shuttle and propulsion for Mariner 9, the first spacecraft to orbit another planet.

BROWN, CRAIG WILLIAM, chemist; b. Denver, Aug. 3, 1953; s. Clarence William and Gail Margaret (Farthing) B. BS in Chemistry, Colo. State U., 1975; MS, Fla. State U., 1977, PhD, 1980. Dep. dir. picosecond and quantum radiation lab. Tex. Tech. U., Lubbock, 1980-83; systems engr. Internat. Marine Systems, Inc., Seattle, 1983-87; freelance cons., 1987-88; staff scientist Heart Interface Corp., Kent, Wash., 1988-91, sr. project engr., 1991-93; cons. scientist Brooks Rand Ltd., Seattle, 1992—; mem. battery charger/inverter project tech. com. Am. Boat and Yacht Coun., Edgewater, Md., 1989—. Contbr. articles to Phys. Rev. Letters and Jour. Chem. Physics; contbr. book revs. to Photochemistry and Photobiology; contbr. to conf. proceedings. Whiteford scholar Colo. State U., 1974-75, Honors scholar U. Denver, 1971-72; Gustavson fellow Colo. State U., 1974-75, Welch postdoctoral fellow Tex. Tech. U., 1980-83. Mem. Am. Chem. Soc., Am. Boat and Yacht Coun. Achievements include patent for switched multi-tapped transformer power conversion method and apparatus; designed power inverters and battery chargers; research in chemical physics, atomic and molecular spectroscopy. Office: Brooks Rand Ltd 3950 6th Ave NW Seattle WA 98107

BROWN, DAVID EDWIN, retired metallurgist, consultant; b. Normal, Ill., Oct. 23, 1923; s. Edwin Stanton and Phyllis Matilda (Bosworth) B.; m. Marcheta Marr, Dec. 23, 1949; children: George Charles, Jeffrey David. BS in Chemistry, Ill. Wesleyan U., 1949; postgrad., U. Wis., 1951-52, Marquette U., 1952-53. Metall. chemist Cutler Hammer Inc., Milw., 1951-56, staff metallurgist, 1956-59, rsch. metallurgist, 1959-77, sr. metallurgist, 1977-78; rsch. supr., materials rsch. Cutler Hammer/Eaton Corp., Milw., 1978-84; prin. engr. metallurgy Eaton Corp., C-H Products, Milw., 1984-85; cons. Eaton Corp., Milw., 1985-88; ind. cons. Milw., 1985—. Author: (with others) Am. Soc. for Metals (Welding), 8th edit., 1971, (Elec. Contacts), 9th edit., 1979; contbr. articles to profl. jours. Adult leader Boy Scouts Am., Brookfield, Wis., 1953-92, Tucson, Ariz., 1992—. With U.S. Army, 1942-46, PTO. Mem. AAAS, Am. Soc. Metals (life), Am. Soc. for Nondestructive Testing, Internat. Metallographic Soc., The Metallurgical Soc., Acoustic Emission Group, Lions. Lutheran. Achievements include patent for electrical terminal seal; research in powder metal electrical contact manufacturing techniques, acoustic emission methods for percussive weld monitor, and ultrasonic methods for electrical contact bond evaluation.

BROWN, DENNIS TAYLOR, molecular biology educator; b. Farmington, Maine, June 6, 1941; s. Frank Ora and Doris Evelyn (Moody) B.; m. Jane French, Feb. 9, 1962 (div. 1977); children—Raymond Peter, Ian Geoffrey; m. Bernadette Maria Brown, Aug. 7, 1978. B.A., U. Pa., 1964, Ph.D., 1967. Asst. prof. Dartmouth Coll., Hanover, N.H., 1968-69; assoc. prof. U. Md., Balt., 1969-73; prof. microbiology U. Tex., Austin, 1978—; dir. cell research Inst. Genetics, U. Cologne, Fed. Republic Germany, 1973-78. Mem. editorial bd. Jour. Virology, 1976-84, Jour. Gen. Virology, 1981—, Virus Research, 1984—; contbr. articles to profl. jours. Mem. Cell Biology Soc., Am. Soc. Virology, Am. Soc. Microbiology, Soc. Gen. Microbiology. Club: Violet Crown. Avocations: bicycle racing, photography. Office: U Tex Austin Cell Rsch Inst Austin TX 78712

BROWN, DONALD DAVID, biology educator; b. Cin., Dec. 30, 1931; s. Albert Louis and Louise (Rauh) B.; m. Linda Jane Weil, July 2, 1957; children: Deborah Lin, Christopher Charles, Sharon Elizabeth. M.S., U. Chgo., 1956, M.D., 1956, D.Sc. (hon.), 1976; D.Sc. (hon.), U. Md., 1983; DSc (hon.), U. Cin., 1992. Staff mem. dept. embryology Carnegie Instn. of Washington, Balt., 1963-76; dir. Carnegie Instn. of Washington, 1976—; prof. dept. biology Johns Hopkins U., 1968—. Pres. Life Scis. Research Found. Served with USPHS, 1957-59. Recipient U.S. Steel Found. award for molecular biology, 1973; V.D. Mattia award Roche Inst., 1975; Boris Pregel award for biology N.Y. Acad. Scis., 1976; Ross G. Harrison award Internat. Soc. Developmental Biology, 1981; Bertner Found. award, 1982; Rosenstiel award for biomed. sci., 1985, Louisa Gross Horwitz award, 1985; Feodor Lynen award U. Miami Winter Symposium, 1987. Fellow Am. Acad. Arts and Scis., AAAS; mem. Nat. Acad. Scis., Soc. Devel. Biology (pres. 1975), Am. Soc. Biol. Chemists, Am. Soc. Cell Biology (pres. 1992), Am. Philos. Soc. Home: 5721 Oakshire Rd Baltimore MD 21209-4291 Office: Carnegie Instn Washington 115 W University Pky Baltimore MD 21210-3399

BROWN, EDWARD JAMES, utility executive; b. Ft. Wayne, Ind., Sept. 30, 1937; s. William Theodore and Jane Elizabeth (Dix) B.; m. Margaret Bessey, June 17, 1989; 1 child, Edward James Jr. BA, Yale U., 1959; MA, Fordham U., 1962. Chartered fin. analyst. Fin. writer E.F. Hutton & Co., N.Y.C., 1970-71; economist N.Y. Power Authority, N.Y.C., 1971-74, prin. economist, 1974-80, mgr., customer svcs., 1980-83, mgr. spl. projects, 1983-86, dir. strategic planning, 1986-93, dir. new bus., 1993—; mem. mgmt. com. Iroquois Gas Transmission System, 1989—. Pres. Park Ave. Meth. Trust, N.Y.C., 1981—; pres. Friends of the Shakers, Inc., Sabbathday Lake, Maine, 1982-84, dir., 1980—; trustee John St. Meth. Episc. Trust Soc., N.Y.C., 1982—; mem. investment com. Meth. Home, Riverdale, N.Y., 1983—; dir. Yorkville Emergency Alliance, N.Y.C., 1982-88; internat. adv. coun. Mus. of Am. Folk Art, N.Y.C., 1988—. Mem. N.Y. Soc. Security Analysts, Assn. Investment Mgmt. and Rsch. Home: 500 E 85th St New York NY 10028-7407 Office: NY Power Authority 1633 Broadway New York NY 10019-6708

BROWN, ELIZABETH ELEANOR, retired librarian; b. Charlotte, Mich., Aug. 29, 1921; d. Delbert Francis and Katherine Eleanor (Griffith) B. AB, Albion Coll., 1943; MLS, Pratt Inst., 1953. Info. specialist Enjay Co., N.Y.C., 1943-50; reports indexer Bakelite Co., Bound Brook, N.J., 1950-52; reference libr. IBM, Poughkeepsie, N.Y., 1953-69, Yorktown Heights, N.Y., 1953-69; info. retrieval specialist, libr. IBM, White Plains, N.Y., 1969-82; ret. IBM, White Plains. Mem. AAAS, ALA, Am. Chem. Soc., Spl. Librs. Assn. (chmn. tech. sci. group N.Y.C. chpt. 1970-71, sec.-treas. engring. div. 1968-70, archivist 1970-72, charter mem., past pres. Hudson Valley chpt.), Am. Soc. for Info. Sci., Libr. Info. Tech. Assn., Soc. Mayflower Descendants, Pilgrim John Howland Soc., New Eng. Historic Geneal. Soc., Colo. Geneal. Soc., Gwynedd Family History Soc., Columbine Geneal. and Hist. Soc., Phi Beta Kappa, Alpha Lambda Delta, Mortar Board. Home: 9875 W Progress Pl Littleton CO 80123

BROWN, ELIZABETH RUTH, neonatologist; b. Washington, Sept. 29, 1946; d. Paul Ambrose and Helene Marie (Kiley) B.; m. William James Pyne, Sept. 28, 1990. BS, Coll. Mt. St. Vincent, Bronx, N.Y., 1968; MD, U. Md., 1972. Diplomate Am. Bd. Med. Examiners, Am. Bd. Pediatrics, sub-board neonatal-perinatal medicine; lic. physician, Mass. Pediatric intern McGill U., Montreal (Que., Can.) Children's Hosp., 1972-73, resident pediatrics, 1973-75; neonatal rsch. fellow dept. pediatrics Harvard Med. Sch. Joint Program in Neonatology, Boston, 1975-78; dir. Infant Follow-Up Program, med. dir. Project Welcome The Children's Hosp., Boston, 1979-85; dir. neonatology Newton (Mass.) Wellesley Hosp., 1984-85; dir. Neonatal Follow Up Clinic Boston City Hosp., 1985-88; co-dir. Boston Perinatal Ctr., 1985—; dir. neonatology Boston City Hosp., 1985—; clin. instr. pediatrics Tufts U. Sch. Medicine, Boston, 1987—; summer fellow tng. program in pub. health N.Y.C. Dept. Pub. Health, Met. Hosp., N.Y.C., 1969; mem. Kellogg Found. Nat. Fellowship Program, Battle Creek, Mich., 1988-91; assoc. prof. pediatrics Boston U. Sch. Medicine, 1987—, asst. prof. ob-gyn., 1985—; instr. pediatrics Harvard Med. Sch., 1979-85, tutor in medicine, 1980-82; lectr. Coll. Pharmacy and Allied Health Professions, Northeastern U., Boston, 1981—; assoc. neonatologist Joint Program in Neonatology, 1979-85. Contbr. articles to profl. jours., chpts. to books. Bd. dirs. Boston Inst. for Devel. of Infants and Parents, Inc., 1983—, NICU Parent Support, Inc., 1980—; mem. adv. bd. Children's Hosp. AIDS Program, 1988; bd. dirs. March of Dimes, Mass. chpt., 1989, Coalition on Addiction, Pregnancy and Parenting, 1990. Named Citizen of Yr. Mass. Assn. Retarded Citizens, 1984; recipient award of excellence Boston Inst. for Devel. Infants and Parents, Inc., 1988. Mem. AMA, Am. Acad. Pediatrics, Am. Med. Women's Assn., Nat. Assn. for Perinatal Addiction Rsch. and Edn., Mass. Perinatal Assn., Mass. Med. Soc., Aircraft Owners and Pilots Assn. Achievements include research in perinatal effects of maternal substance abuse, perinatal transmission of HIV, early childhood growth and development, prevention of prematurity, infant mortality.

BROWN, ELLIOTT ROWE, physicist; b. L.A., Oct. 4, 1955; s. LaMonte Russell and Barbara Lee B.; m. Elayne Beth Reback, Aug. 2, 1981. BS, UCLA, 1979; MS, Calif. Inst. Tech., Pasadena, 1981, DS, 1985. Bachelors fellow Hughes Aircraft Co., El Segundo, Calif., 1977-79, Masters fellow, 1979-81; rsch. staff mem. MIT Lincoln Lab., Lexington, Mass., 1985-92, program mgr., 1992—. Contbg. author: Hot Carriers in Semiconductor Nanostructures, 1992; contbr. over 50 articles to profl. jours. Recipient Achievement awards for coll. scientists ARCS Found., Pasadena, 1983-85, others. Mem. Am. Phys. Soc., IEEE, Optical Soc. Am., Phi Beta Kappa. Achievements include invention of lowest-noise heterodyne receiver above 600 GHz, highest frequency solid-state oscillator of electronic variety, most efficient planar antenna on a high-dielectric substrate; development of highest CW microwave power generated by optoelectronic techniques; patents in field. Office: MIT Lincoln Lab 244 Wood St Lexington MA 02173

BROWN, EUGENE FRANCIS, mechanical engineer, educator; b. Madison, Wis., July 26, 1940; s. Clarence Eugene and Frances Lillian (Mussen) B.; m. Hilda Lutgardis Van Damme, Aug. 9, 1969; children: Ryan Andrew, Eric Todd. BS, U. Wis., 1963; PhD, U. Ill., 1968. Registered profl. engr., Va.

Grad. fellow U. Ill., Urbanna, 1963-69; asst. prof. Va. Tech., Blacksburg, 1969-78; vis. prof. von Karman Inst., Brussels, 1978-79; assoc. prof. Va. Tech., Blacksburg, 1975-85; liaison scientist Office of Naval Rsch., London, 1985-87; prof. Va. Tech., Blacksburg, 1987—; cons. USAF, Dayton, Ohio, 1983-84, McDonnell Douglas Corp., Long Beach, Calif., 1979-84, USN, Washington, 1982-84; hon. com. 1st Joint Europe-U.S. Short Course on Hypersonics, Paris, 1987. Contbr. articles to profl. jours. Recipient Rsch. grant Office of Naval Rsch., 1992—. Fellow AIAA (assoc.), ASME. Achievements include first application of upwind difference methods to internal flows; first implementation of direct numerical simulations of elliptical jets. Office: Va Tech Mech Engring Dept Blacksburg VA 24061-0238

BROWN, FRANCES SUSAN, scientific publication editor; b. London, June 25, 1956; d. Charles Edward and Doreen Frances (Oakley) B. BA in French with honors, Univ. Coll., London, 1978; MA in Internat. Rels., Sussex U., Brighton, Eng., 1986. Head photo libr. Camera Press Ltd., London, 1978-82; editorial asst. Burda Publs., London, 1982-85; editor, group editor Space Policy, other publs. Butterworth-Heinemann, Guildford, Eng., 1987-90; editor Space Policy, Edenbridge, Eng., 1991—. Author: (periodical) Earth Space Rev., 1992, 93. Com. mem. Edenbridge Coun. Footpaths Com., 1992. Mem. AIAA (assoc.), Amnesty Internat., Campaign Nuclear Disarmament.

BROWN, GEORGE EDWARD, JR., congressman; b. Holtville, Calif., Mar. 6, 1920; s. George Edward and Bird Alma (Kilgore) B.; 4 children. B.A., UCLA, 1946; grad. fellow, Fund Adult Edn., 1954. Mgmt. cons. Calif., 1957-61; v.p. Monarch Savs. & Loan Assn., Los Angeles, 1960-68; mem. Calif. Assembly from 45th Dist., 1959-62, 88th-91st congresses from 29th Dist. Calif., 93d Congress from 38th Dist. Calif., 94th-103rd Congresses from 36th (now 42nd) Dist. Calif.; mem. standing com. on agr., chmn. sci. space and tech. com. 94th-101st Congresses from 36th Dist. Calif., 1987; mem. agriculture com., mem. sci., space and tech. com.; chmn. Office of Tech. Assessment; coll. lectr., radio commentator, 1971. Mem. Calif. Gov.'s Adv. Com. on Housing Problems, 1961-62; mem. Mayor Los Angeles Labor-Mgmt. Com., 1961-62, Councilman, Monterey Park, Calif., 1954-58, mayor, 1955-56; candidate for U.S. Senate, 1970. Served to 2d lt., inf. AUS, World War II. Recipient Chairman's award Am. Assn. Engring. Socs., 1993. Mem. Am. Legion, Colton C. of C., Urban League, Internat. Brotherhood Elec. Workers, AFL-CIO, Friends Com. Legislation, Ams. for Dem. Action. Democrat. Methodist. Lodge: Kiwanis. Office: US Ho of Reps 2300 Rayburn House Office Bldg Washington DC 20515-0542

BROWN, GERALD EDWARD, physicist, educator; b. Brookings, S.D., July 22, 1926. BA, U. Wis., 1946; MS, Yale U., 1948; PhD, U. Birmingham, 1950, DSc (hon.), 1957; DSc (hon.), U. Helsinki, 1982, U. Birmingham, 1990. Prof. physics U. Birmingham, 1959-60, Nordic Inst. Theoretic Atomic Physics, 1960-85, Princeton U., 1964-68, SUNY, Stony Brook, 1968-74; leading prof. Nordic Inst. Theoretic Atomic Physics, 1974-88; dist. prof. physics SUNY, Stony Brook, 1988—; lectr. math physics, 1955-58; reader U. Birmingham, 1958-59; dir. nuclear astrophysics Inst. Theoretical Physics NSF, U. Calif., 1960. Recipient Boris Pregel award N.Y. Acad. Sci., 1976, Tom W. Bonner prize Nuclear Physics, 1982, Sr. Dist. Sci. award Alexander von Humboldt Found., 1987, John Price Wetherill medal Franklin Inst., Phila., 1992. Office: SUNY Inst Theoretical Physics Stony Brook NY 11794*

BROWN, GERALDINE, nurse, freelance writer; b. Clemson, S.C., Nov. 1, 1945; d. Isaac and Gladys (Patterson) B. AS in Nursing, U. D.C., Washington, 1973; real estate cert., Long and Foster Inst., College Park, Md., 1984; cert. in TV broadcasting, Columbia Sch., Bailey's Crossroads, Va., 1987; BS in Nursing, Bowie State U., 1989, MA in Communications, 1991; postgrad., Howard U., 1991—. RN, D.C., FCC Third Class License. Supr. staff nurse Walter Reed Hosp., Washington, 1970-76; supr. clin. nurse Dept. Human Svcs., Washington, 1976-78, community health nurse, 1978-84; nursing instr. Phillips Bus. Sch., Alexandria, Va., 1984-85; pvt. nurse Washington, 1973—; dir. pub. affairs Bible Way Chs. Worldwide, Inc., Washington, 1978-91; society columnist As It Happens, Charlotte (N.C.) Post, 1964-66; society editor Washington Cafe. Soc. mag., 1971; contbr. feature stories Capital Spotlight newspaper, 1978—. Asst. organizer DC Mayor's United Nations Day, 1980; vol. Met. Boys and Girls Clubs, Washington, 1980—; vol. Nursing Instr., The Washington Saturday Coll., 1982-84; Co. ARC, 1973—, Big Sisters of the Washington Met. Area, 1988—. Recipient certs. of excellence Govt. of D.C., 1978-84; cert. of appreciation Mayor of D.C., 1980, meritorious pub. svc. award, 1980; svc. trophy Washington Saturday Coll., 1984. Mem. Am. Nurses Assn., Nat. Coun. Negro Women, Smithsonian Inst. (assoc.), NAACP, Nat. Black Nurses Assn., Washington Urban League, Chi Eta Phi. Democrat. Apostolic. Avocations: stamp collecting, traveling, writing poetry.

BROWN, H. WILLIAM, urban economist, private banker; b. L.A., Sept. 6, 1933; s. Homer William Brown and Carol Lee (Thompson) Weaver; m. Verlee Paul, Aug. 1951 (div. 1953); m. Shirley Rom, Jan. 18, 1953 (div. 1962); 1 child, Shirlee Dawn. BA in Pub. Adminstrn., Calif. State U., 1956; MA in Bus. Adminstrn., Western States U., 1983, Phd in Urban Econs., 1984. Pres. Real Estate Econs., Sacramento, 1956-60; dir. spl. projects Resource Agy. Calif., Sacramento, 1960-65; program planning officer U.S. Dept. Housing and Urban Devel., Washington, 1965-66; asst. dir. regional planning U.S. Dept. Commerce, Washington, 1967-69; dir. internat. office Marshall and Stevens, Inc., L.A., 1970-72; vice chmn., CEO Investment Property Econ. Cons., 1972—; chmn., CEO The Northpoint Investment Group, San Francisco, 1986—; chmn. Trade and Devel. Ctr. For UN, N.Y. 1983-88, pres. Ctr. for Habitat and Human Settlements, Washington 1977-90. Author: The Changing World of the Real Estate Market Analyst Appraiser, 1988. Mem. Internat. Real Estate Inst. (fgn. lectr. 1983—), Le Groupe d'Elegance (charge d'affaires, pvt. bankers club). Avocation: worldwide people photography. Office: Northpoint Investment Group 2310 Powell St Ste 205 San Francisco CA 94133-1425

BROWN, HAROLD, corporate director, consultant, former secretary of defense; b. N.Y.C., Sept. 19, 1927; s. A.H. and Gertrude (Cohen) B.; m. Colene Dunning McDowell, Oct. 29, 1953; children: Deborah Ruth (Mrs. Eric Ploumis), Ellen Dunning (Mrs. Ray Merewether). A.B., Columbia U., 1945, A.M., 1946, Ph.D. in Physics (Lydig fellow 1948-49), 1949. Research scientist Columbia U., 1945-50, lectr. physics, 1947-48; lectr. physics Stevens Inst. Tech., 1949-50; div. leader E.O. Lawrence Radiation Lab. U. Calif., Berkeley, 1950-60, staff mem., group leader E.O. Lawrence Radiation Lab., 1952-60; dir. Lawrence Livermore (Calif.) Lab., 1960-61; dir. def. rsch. and engring. Dept. Def., Washington, 1961-65; sec. Dept. Air Force, Washington, 1965-69; pres. Calif. Inst. Tech., Pasadena, 1969-77; sec. def. Washington, 1977-81; disting. vis. prof. Sch. Advanced Internat. Studies Johns Hopkins U., Md., 1981-84, chmn. Fgn. Policy Inst., 1984-92, counselor, Ctr. Strategic & Internat. Studies, 1992—; ptnr. Warburg, Pincus & Co., N.Y.C., 1990—; bd. dirs. AMAX, CBS, IBM, Philip Morris Inc., Cummins Engine Co., Mattel, Inc., Evergreen Holdings, Inc.; mem. Polaris Steering Com., 1956-58; Pres.'s Sci. Adv. Com., 1958-61; sr. sci. advisor Conf. Discontinuance Nuclear Tests, 1958-59; U.S. del. SALT, Helsinki, Vienna and Geneva, 1969-77; chmn. Tech. Assessment Adv. Coun. to U.S. Congress, 1974-77; mem. exec. com. Trilateral Commn., 1973-76, trustee, 1992—; trustee Rand Corp., 1983-92, Calif. Inst. Tech., 1985—, Rockefeller Found., 1983-93. Author: Thinking About National Security: Defense and Foreign Policy in a Dangerous World, 1983. Decorated Medal of Freedom; named One of 10 Outstanding Young Men U.S. Jaycees, 1961; recipient Medal of Excellence Columbia U., 1963; Joseph C. Wilson award in internat. affairs, 1976, Enrico Fermi award U.S. Dept. Energy, 1992. Mem. Nat. Acad. Engring., Am. Phys. Soc., Am. Acad. Arts and Scis., Nat. Acad. Scis., Bohemian Club (San Francisco), Calif. Club (L.A.), Athenaeum (London), Phi Beta Kappa, Sigma Xi. Clubs: Bohemian (San Francisco); California (Los Angeles); Athenaeum (London). Office: Ctr for Strategic & Intl Studies Ste 400 1800 K St NW Washington DC 20006

BROWN, HENRY, chemist; b. Jersey City, Apr. 5, 1907; s. Mayer and Kate (Hearsh) B.; m. Harriet Stone; children: Paula, Dennis. AB, U. Kans., 1928; MS, PhD, U. Mich., 1933. Chemist Udylite Corp., Detroit, 1934-50, dir. rsch., 1950-72; rsch. chemist Manhattan Project Columbia U., N.Y.C., 1943-45. Author: (with others) Modern Electroplating 3rd edit., 1974; contbr. numerous articles to profl. jours., 1932-71. Bd. dirs. Sinai Hosp., Detroit, 1965. Recipient Carl Heussner award Am. Electroplaters Soc., Detroit, 1953, George Hogaboom award, 1963, AES Sci. Achievement award, 1968, Westinghouse prize Inst. of Metal Finishing of Great Britain, 1970; named to Outstanding Inventors List, State of Mich. 1968. Fellow Am. Inst. Chemists; mem. N.Y. Acad. Scis, Am. Chem. Soc. (Midgely Gold Medal 1971), Sigma Xi. Jewish. Achievements include 96 U.S. patents, 250 foreign patents; development of high speed brass electroplating of steel shell casings used during World War II to prevent the sticking of the shell cases inside the cannon after firing, of copper alloy plating for steel pennies during World War II, of high speed electroposition of silver for sleeve linings for Allison airplane engines, of use of perfluoro octane sulfonic acid for the complete suppression of toxic spray during chromium plating; discovery of unsaturated organic addition agents which form the basis of modern brilliant nickel electroplating which underlies the final thin, hard chromium plate. Home: 5270 Gulf Of Mexico Dr Longboat Key FL 34228-2008

BROWN, HERBERT CHARLES, chemistry educator; b. London, England, May 22, 1912; came to U.S., 1914; s. Charles and Pearl (Gorinstein) B.; m. Sarah Baylen, Feb. 6, 1937; 1 son, Charles Allan. AS, Wright Jr. Coll., Chgo., 1935; BS, U. Chgo., 1936, PhD, 1938, DSc (hon.), 1968; hon. doctorate, Wayne State U., 1980, Lebanon Valley Coll., 1980, L.I. U., 1980, Hebrew U. Jerusalem, 1980, Pontificia Universidad de Chile, 1980, Purdue U., 1980; hon. doctorates, U. Wales, 1981, U. Paris, 1982, Butler U., 1982, Ball State U., 1985. Asst. chemistry U. Chgo., 1936-38, Eli Lilly post-doctorate research fellow, 1938-39, instr., 1939-43; asst. prof. chemistry Wayne U., 1943-46, assoc. prof., 1946-47; prof. inorganic chemistry Purdue U., 1947-59, Richard B. Wetherill prof. chemistry, 1959, Richard B. Wetherill research prof., 1960-78, emeritus, 1978—; vis. prof. U. Calif. at L.A., 1951, Ohio State U., 1952, U. Mexico, 1954, U. Calif. at Berkeley, 1957, U. Colo., 1958, U. Heidelberg, 1963, State U. N.Y. at Stonybrook, 1966, U. Calif. at Santa Barbara, 1967, Hebrew U., Jerusalem, 1969, U. Wales, Swansea, 1973, U. Cape Town, S. Africa, 1974, U. Calif., San Diego, 1979; Harrison Howe lectr., 1953, Friend E. Clark lectr., 1953, Freud-McCormack lectr., 1954, Centenary lectr., Eng., 1955, Thomas W. Talley lectr., 1956, Falk-Plaut lectr., 1957, Julius Stieglitz lectr., 1958, Max Tishler lectr., 1958, Kekule-Couper Centenary lectr., 1958, E. C. Franklin lectr., 1960, Ira Remsen lectr., 1961, Edgar Fahs Smith lectr., 1962, Seydel-Wooley lectr., 1966, Baker lectr., 1969, Benjamin Rush lectr., 1971, Chem. Soc. lectr., Australia, 1972, Armes lectr., 1973, Henry Gilman lectr., 1975, others; chem. cons. to indsl. corps; researcher in phys., organic and inorganic chemistry relating chem. behavior to molecular structure, selective reductions, hydroboration and chemistry of organoboranes. Author: Hydroboration, 1962, Boranes in Organic Chemistry, 1972, Organic Synthesis via Boranes, 1975, The Nonclassical Ion Problem, 1977, (with A. W. Pelter and K. Smith) Borane Reagents, 1988; contbr. articles to chem. jours. Bd. govs. Hebrew U., 1969-80; co-dir. war rsch. projects U. Chgo. for U.S. Army, Nat. Def. Rsch. Com., Manhattan Project, 1940-43. Decorated Order of the Rising Sun, Gold and Silver Star (Japan); recipient Purdue Sigma Xi rsch. award, 1951, Nichols medal, 1959, award Am. Chem. Soc., 1960, S.O.C.M.A. medal, 1960, H.N. McCoy award, 1965, Linus Pauling medal, 1968, Nat. Medal of Sci., 1969, Roger Adams medal, 1971, Charles Frederick Chandler medal, 1973, Chem. Pioneer award, 1975, CUNY medal for sci. achievement, 1976, Elliott Cresson medal, 1978, C.K. Ingold medal, 1978, Nobel prize in chemistry, 1979, Priestley medal, 1981, Perkin medal, 1982, Gold medal award Am. Inst. Chemists, 1985, G.M. Kosolapoff medal, 1987, NAS award in chem. scis., 1987, Oesper award Cin. Sect. Am. Chem. Soc. Fellow AAAS, Royal Soc. Chemistry (hon.), Indian Nat. Sci. Acad. (fgn.); mem. NAS, Am. Acad. Arts and Scis., Chem. Soc. Japan (hon.), Pharm. Soc. Japan (hon.), Am. Chem. Soc. (chmn. Purdue sect. 1955-56), Ind. Acad. Sci., Phi Beta Kappa, Sigma Xi, Alpha Chi Sigma, Phi Lambda Upsilon (hon.). Office: Purdue U Dept Chemistry West Lafayette IN 47907

BROWN, IRA BERNARD, data processing executive; b. N.Y.C., July 18, 1927; s. David and Ruth (Ehrlich) B.; m. Myra Dipkin, Oct. 31, 1954; children: Rene, Lori. BS, U.S. Mcht. Marine Acad., 1949; MS, Drexel U., 1959. Data processing mgr. Sperry Rand, Syosset, N.Y., 1962-66; corp. systems mgr. Loral Corp., Scarsdale, N.Y., 1966-68; chmn. bd., CEO Brandon Systems Corp., Secaucus, N.J., 1968—. Bd. dirs. Arthritis Found., N.J. Mem. Assn. for Computing Machinery, Info. Tech. Assn. of Am., Assn. for Systems Mgmt., C. of C., Lake Mohawk Country Club, Sales Exec. Club of N.Y., Compass 1019 Moose. Clubs: Lake Mohawk Country; Sales Exec. of N.Y. Lodges: Compass 1019 F&AM. Home: 2000 Linwood Ave Apt 8V Fort Lee NJ 07024-3038 also: 7 Oakwood Trail Lake Mohawk Sparta NJ 07871 Office: Brandon Systems Corp 1 Harmon Blvd Secaucus NJ 07094-2299

BROWN, J. MARTIN, oncologist, educator; b. Doncaster, Eng., Oct. 15, 1941; married; 2 children. BSc, U. Birmingham, 1963; MSc, U. London, 1965; DPhil in Radiation Biology, Oxford U., 1968. NIH fellow radiation biology Stanford U. Med. Ctr., Calif., 1968-70, rsch. assoc., 1970-71, from asst. prof. to assoc. prof., 1971-84, prof., dir. divsn. radiation biology, 1984—, dir. Cancer Biology Rsch. Lab., 1985—; sr. fellow Am. Cancer Soc. Dernham, 1971-74; mem. adv. com. biol. effects of ionizing radiations NAS, 1971—. Mem. AAAS, Am. Assn. Cancer Rsch., Am. Soc. Therapeutical Radiology & Oncology, Brit. Inst. Radiology, Brit. Assn. Cancer Rsch., Radiation Rsch. Soc. (9th Rsch. award 1980). Achievements include research in mammalian cellular radiobiology, tumor radiobiology, experimental chemotherapy, bioreductive cytotoxic agents, radiation carcinogenesis. Office: Stanford U Med Ctr Cancer Biology Rsch Lab 300 Pasteur Dr Stanford CA 94305*

BROWN, JACK HAROLD UPTON, physiology educator, university official, biomedical engineer; b. Nixon, Tex., Nov. 16, 1918; s. Gilmer W. and Thelma (Patton) B.; m. Jessie Carolyn Schulz, Apr. 14, 1943. B.S., S.W. Tex. State U., 1939; postgrad., U. Tex., 1939-41; Ph.D., Rutgers U., 1948. Lectr. physics Southwest Tex. State U., San Marcos, 1943-44; instr. phys. chemistry Rutgers U., New Brunswick, N.J., 1944-45, rsch. assoc., 1944-48; lectr. U. Pitts., 1948-50; head biol. scis. Mellon Inst., Pitts., 1948-50; asst. prof. physiology U. N.C., Chapel Hill, 1950-52; scientist, prof. biology Oak Ridge Inst. Nuclear Studies, 1952; assoc. prof. physiology Emory U. Med. Sch., Atlanta, 1952-58, prof., 1959-60, acting chmn. dept. physiology, 1958-60; lectr. physiology George Washington U. and Georgetown U. med. schs., Washington, 1960-65; exec. sec. biomed. engring. and physiology tng. coms. Nat. Inst. Gen. Med. Scis., NIH, Bethesda, Md., 1960-62; chief spl. rsch. br. div. Rsch. Facilities and Resources NIH, 1962-63, acting chief gen. clin. rsch. ctrs. br., 1963-64, asst. dir. ops. Div. Research Facilities and Resources, 1964-65; acting program dir. pharmacology/toxicology program Nat. Inst. Gen. Med. Scis., NIH, 1966-70, asst. dir. ops., 1965-66, assoc. dir. sci. programs, 1967-70, acting dir., 1970; spl. asst. to adminstr. Health Services and Mental Health Adminstrn., USPHS, Rockville, Md., 1971-72; assoc. dep. adminstr. for devel. Health Svcs. and Mental Health Adminstrn., USPHS, 1972-73; spl. asst. to adminstr. Health Resources Adminstrn., 1973-78; coord. Southwest Rsch. Consortium, San Antonio, 1974-78; prof. physiology U. Tex. Med. Sch., San Antonio, 1974-78; prof. environ. scis. U. Tex. at San Antonio, 1974-78; adj. prof. health svcs. adminstrn. Trinity U., 1975-78; assoc. provost rsch. and advanced edn. U. Houston, 1978-80, prof. biology, 1980-89, prof. emeritus, 1990—; adj. prof. U. Tex. Sch. Public Health, 1978—; prof. public adminstrn. Tex. Women's U., 1978—; adj. prof. community medicine Baylor Coll. Medicine, Houston, 1986-89; Fulbright lectr. U. Rangoon, 1950; cons. health systems WHO, Oak Ridge Inst. Nuclear Studies, Lockhead Aircraft Co., Drexel Inst. Tech., NASA, Vassar Coll., VA; mem. adv. bd. Ctr. for Cancer Therapy, San Antonio, 1974—; bd. dirs. South Tex. Health Edn. Ctr.; mem. health adv. bd. Tel-Tech Univ. Tex. Health Sci. Ctr., Samitomo, Tokyo. Author: Physiology of Man in Space, 1963, (with S.B. Barker) Basic Endocrinology, 1966, 2d edit., 1970, (with J.F. Dickson) Future Goals of Engineering in Biology and Medicine, 1968, Advances in Biomedical Engineering, vol. II, 1972, vols. III, IV, 1973, vol. V, 1974, vol. VI, 1976, Vol. VII, 1978, (with J.E. Jacobs and L.E. Stark) Biomedical Engineering, 1972, (with D.E. Gann) Engineering Principles in Physiology, vols. I, II, 1973, The Health Care Dilemma, 1977, Integration and Control of Biol. Processes, 1978, Politics and Health Care, 1978, Telecommunications in Health Care, 1981, Management in Health Care Systems, 1983, A Laboratory Manual in Animal Physiology, 1984, 3d edit., 1988, High Cost of Healing, 1985, (with J. Comolo) Productivity in Health Care Systems, 1987, Guide to Collecting Fine Prints, 1989, (with J. Cumolo) Educating for Excellence, 1991, Footsteps in Sci., 1993; editor: (with Ferguson) Blood and Body Functions, 1966, Life Into Space, (Wunder), 1968; contbr. numerous articles on biomed. engring. to sci. jours. Mem. adv. bd. San Antonio Mus. Assn.; mem. spl. effects com. Tex. Sesquicentennial.; bd. dirs. Inst. for Health Policy, U. Tex. Health Sci. Ctr. Served with USNR, 1941. Recipient cert. appreciation NIH, 1969, 1st pl. award Atlanta Internat. Film Festival, 1970, spl. team award NASA, 1978, recognition award Emergency Med. Care, 1980, Best Tchr. award Nat. Mortar Bd., 1986, Most Disting. Alumni award S.W. Tex. State U., 1986; Gerard Swope fellow Gen. Electric Co., 1946-48; Fulbright grantee, 1950; Dept. of Def. grantee, 1950-52; NIH grantee, 1950-60; Cancer Soc. grantee, 1958; Damon Runyon Cancer award grantee, 1959; Dept. Energy grantee, 1980-81; NASA grantee, 1987-89. Fellow AAAS, Nat. Acad. Engring., IEEE (joint com. engring. in medicine and biology 1966—); mem. Am. Chem. Soc. (sr.), Biomed. Engring. Soc. (pres. 1969-70, dir. 1968-69), Inst. Radio Engrs. (nat. sec. profl. group biomed. engring. 1962-64), N.Y. Acad. Scis., Endocrine Soc., Am. Physiol. Soc. (com. mem. 1959-63, nat. com. on animals in research 1985—), Soc. for Exptl. Biology and Medicine, Sigma Xi (research award 1961, founder, pres. Alamo chpt. 1977-78), Council Biology Editors, Soc. Research Adminstrn., Pi Kappa Delta, Phi Lambda Upsilon, Alpha Chi. Club: Cosmos. Inventor capsule manometer, respirator for small animals and basal metabolic apparatus for small animals, dust sampler, apparatus for partitioning human lung volumes, laser credit card patient record system, Warburg apparatus calibrator. Home: 2908 Whisper View St San Antonio TX 78230-3743 Office: U Houston 4800 Calhoun Rd Houston TX 77004-2610

BROWN, JAMES ALLISON, anthropology educator; b. Evanston, Ill., Jan. 16, 1934; s. Richard Paul and Olive (Harris) B.; m. Constance Margaret Kimball, Aug. 5, 1967 (div. 1975); m. Judith Quinn Drick Toland, Oct. 1, 1978 (div. 1981); 1 child, Douglas Alfred Kimball. AB, U. Chgo., 1954, MA, 1958, PhD, 1965. Asst. prof. Anthropology and Computer Inst. Stovall Mus. Okla., 1965-66; asst. prof. dept. anthropology and computer instrn. soc. sci. rsch. Mich. State U., 1966-69, assoc. prof., 1967-71, rsch. assoc., 1967-71; assoc. prof. dept. anthropology Northwestern U., Evanston, Ill., 1971-79, prof., 1979—, chair, 1989—; rsch. assoc. Field Mus. Natural History, Chgo., 1989—; editor Ill. Archaeol. Survey, Urbana, 1966-78, bd. dirs., 1978-85, 88-91, pres., 1991-93; vis. fellow Clare Hall Coll., Cambridge, 1987-88, life fellow, 1989—; advisor dir. registration and edn. State of Ill., 1977, NSF, NEH, Nat. Geographic Soc., AAAS, Time-Life Books, Readers Digest Books, Smithsonian Press, U. Chgo. Press; scientific advisor on redesign Mus. of Ocmulgee Nat. Monument, Macon, Ga., 1978-80. Author: (with others) Pre-Columbian Shell Engravings from Craig Mound at Spiro, Oklahoma, Vols. 1-6, 1975-83, Ancient Art of the American Woodland Indians, 1985; author: Aboriginal Cultural Adaptations in the Midwestern Prairies, 1991; editor: Essays on Archaeological Typology, 1982, Archaic Hunters and Gatherers in the American Midwest, 1983, Prehistoric Hunters and Gatherers: The Emergence of Cultural Complexity, 1985. Sec. Found. for Ill. Archaeology/Ctr. for Am. Archaeology, 1973-83, bd. dirs., 1973—, exec. com., 1984—; commr. Ill. and Mich. Canal Nat. Heritage Corridor Commn., 1985-87; bd. dirs. Ill. State Mus., 1985—, Miss. Valley Archaeol. Ctr., 1986—. With U.S. Army, 1957-59. Grantee NSF, 1970, 72, 74, 77, 87, Nat. Park Svc., 1980, 86, Ill. Dept. Transp., 1978, Ill. Historic Preservation Agy., 1980, 85, 86, Am. Philos. Soc., 1973, Wenner-Gren Found., 1974; fellow NEH. Fellow AAAS, Am. Anthrop. Assn.; mem. Current Anthropology (assoc.). Home: 734 Noyes St Apt L 3 Evanston IL 60201-2865 Office: Northwestern U Dept Anthropology 1810 Hinman Ave Evanston IL 60208-1310

BROWN, JAMES DOUGLAS, materials engineer, researcher; b. Hamilton, Ont., Can., July 17, 1934; s. Samuel William and Hilda Lena (Bennewitz) B.; m. Beverley Joyce Heise, Aug. 5, 1957; children: Douglas George, Robert William, Steven James, Patricia Ann. BS with honors, McMaster U., Hamilton, Ont., 1957; PhD, U. Md., 1966. Registered profl. engr. Rsch. chemist dept. rsch. Imperial Oil Ltd., Sarnia, Ont., 1957-60, College Park (Md.) Rsch. Ctr., U.S. Bur. Mines, 1960-66; vis. rsch. scientist Metaalinstituut, TNO, Apeldoorn, The Netherlands, 1975-76; vis. prof. Inst. for Exptl. Physics, U. Vienna, Austria, 1982-83; asst. prof. faculty of engring. sci. U. Western Ont., London, 1967-71, prof. faculty of engring. sci., 1971—; also chmn. dept. materials engring., 1988—; v.p. Elstat Ltd., London, Can., 1972—. Contbr. articles to profl. jours. Mem. Assn. Profl. Engrs. of Province of Ont., Microbeam Analysis Soc. (charter, pres. 1977, Presdl. award 1988, Corning award 1974, 80), Am. Chem. Soc., X-Ray Spectrometry (editorial adv. bd. 1973—). Achievements include patent for portable electrostatic spraying apparatus; pioneered use of X-ray generation curves as model for quantative electron probe microanalysis. Office: U Western Ont, Dept Materials Engring, London, ON Canada N6A 5B9

BROWN, JAMES EUGENE, III, business executive; b. Greenville, S.C., Oct. 9, 1942; s. James Eugene and Harriett Annette (Coker) B.; m. Sue Keyes, Jan. 31, 1969; children: Angela Marie, Sarah Annette. Student, U. S.C., 1961-66. Sales and svc. rep. Honeywell, 1966-72; pres. Instrumentation Services Inc., Charlotte, N.C., 1972—; chief exec. officer Carolina Comml. Properties, Charlotte, 1989—. Mem. Am. Water Works Assn., N.C. Rural Water Assn. Home: 1128 Greenwood Cliffs Charlotte NC 28204-2821

BROWN, JAMES HARVEY, neuroscientist, government research administrator; b. Yankton, S.D., Sept. 9, 1936; s. Robert Heath and Hildegarde (Grover) B.; m. Betty Jean Pruitt, Aug. 29, 1959; children: Christopher Heath, Karen Elizabeth, Kimberly Frances. B.A., Wesleyan U., 1957; M.S., Purdue U., 1959; Ph.D. (DuPont fellow), U. Va., 1962. Research scientist U.S. Army Med. Research Lab., Ft. Knox, Ky., 1962-66; head vesibular br. U.S. Army Med. Research Lab., 1966-68; asso. program dir. for psychobiology NSF, Washington, 1968-71; program dir. for neurobiology NSF, 1971-76, dep. dir. div. behavioral and neural scis., 1976-77, 79-82, dep. asst. dir., biol., behavioral and social scis. directorate, 1977-79, dir. div. physiology, cellular and molecular biology, 1982-84, dir. divsn. molecular bioscis., 1984-92; dir. divsn. molecular and cellular bioscis., 1992—; adj. prof. U. Ky., Louisville U., 1982-88, George Washington U., 1970-78, v.p. NSF Employees Assn., 1970. Contbr. articles to profl. jours. Active Little League, Potomac, Md. Served to capt. U.S. Army, 1960-65. Mem. Sigma Xi, Phi Sigma, Sigma Chi. Home: 8803 Quiet Stream Ct Potomac MD 20854-4231 Office: 1800 G St NW Washington DC 20550-0002

BROWN, JAMES KENNETH, computer engineer; b. Greensburg, Pa., Jan. 1, 1937; s. James Kenneth and Eleanor Elizabeth (Johnson) B. BSEE, U. Pitts., 1958. Jr. engr. Fairchild Astronautics, Wyandanch, N.Y., 1958-59; engr. Grumman Aerospace and Electronics, Bethpage, N.Y., 1959-63; group leader Grumman Aircraft Systems, Bethpage, N.Y., 1963-82, tech. specialist, 1982—. Mem. AIAA (mem. adv. bd. 1989-92), IEEE. Achievements include design of system architecture and computer architecture for last 3 versions of EA-6B naval aircraft, system architecture for EF-111A aircraft and numerous proposed aircraft, interface, network and processor specifications. Home: 75 Plymouth Blvd Smithtown NY 11987 Office: Grumman Aerospace & Electronics Mail Stop T39-80 Bethpage NY 11714

BROWN, JASON WALTER, neurologist, educator, researcher; b. N.Y.C., Apr. 14, 1938; s. Samuel Robert and Sylvia (Brown) B.; m. Jo-Ann Marie Gelardi, Sept. 21, 1967; children: Jonathan Schilder, Jovana Millay. B.A., U. Calif.-Berkeley, 1959; M.D., U.S.C., 1963. Intern St. Elizabeth's Hosp., Washington, 1963-64; resident in neurology UCLA, 1964-67; practice medicine specializing in neurology N.Y.C., 1970—; instr. Boston U. Med. Sch., 1969-70; asst. clin. prof. Columbia-Presbyn. Hosp., N.Y.C., 1970-75; vis. asst. prof. neurology Albert Einstein Coll. Medicine, N.Y.C., 1972-75; vis. assoc. prof. Rockefeller U., N.Y.C., 1978-79; clin. assoc. prof. neurology NYU, 1975-79, clin. prof., 1979—; pres. Inst. Research in Behavioral Neurosci. Author: Aphasia, Apraxia and Agnosia, 1972, Mind, Brain and Consciousness, 1977, Life of the Mind, 1988; editor: Jargonaphasia, 1982; English Translation of Aphasie by Arnold Pick (Aphasia), 1973, Neuropsychology of Visual Perception, 1989, Classics in Neuropsychology: Apraxia and Agnosia, Self and Process, 1991; contbr. numerous articles on neurology to med. jours.; mem. editorial bd.Jour. Nervous and Mental Disease, Aphasiology, Advances in Neurolinguistics. Grantee NIH; fellow Alexander von Humboldt Found., 1979—, World Rehab. Fund, 1982, Founds. Fund for Research in Psychiatry, 1974-75. Jewish. Office: NYU Med Ctr Dept Neurology 550 1st Ave New York NY 10016-6402 also: Inst Rsch in Behvrl Neuroscience 66 E 79th St New York NY 10021

BROWN, JAY WRIGHT, food manufacturing company executive; b. Seattle, July 26, 1945; s. Warren Wright and Dorothy (Culling) B.; m. Cynthia Anne Berry, Jan. 3, 1981; children: Jessica Morgan, Meghan Lannes; chidren by previous marriage-Susan Kay, Kelly Wright, Jeffrey Wright. B.Sc., Ohio State U., 1966, M.A., 1968. Vice pres. Foodmaker div. Ralston Purina, St. Louis, 1981-83; pres. Van Camp div. Ralston Purina, 1983-84, chief exec. officer Continental Baking subs., 1984—. Office: Ralston Purina Checkerboard Sq Saint Louis MO 63164-0002

BROWN, JERRY MILFORD, medical company executive; b. Anderson, S.C., Apr. 30, 1938; s. James Milford and Jane Elizabeth (McCord) B.; m. Alice Alberta Thompson, July 30, 1960; children—John Milford, Allen Thompson. B.S., Furman U., 1960; M.A. in Biology, Wake Forest U., 1963, Temple U., 1967; Ph.D. in Physiology, Dental Sch., U. Md., 1972. Commd. lt. U.S. Army, 1960, advanced through grades to lt. col., 1980; research instr. Hahnemann Med. Coll., Phila., 1967-68; sect. leader, exptl. medicine div. Biomed. Lab., Edgewood Arsenal, Md., 1967-68; instr. anatomy Med. Sch., U. Md., Balt., 1970-77; sect. leader, exptl. medicine div. U.S. Army Research Inst. Environ. Medicine, Natick, Mass., 1973-76; dep. dir. U.S. Army Med. Intelligence and Info. Agy., Ft. Detrick, Md., 1976-80; dir. internat. health affairs Dept. Def., Washington, 1980-84; editor Mgmt. and Info. Study, Office of Surgeon Gen., 1984; chief plans ops. security, 2nd Gen. Hosp., Federal Republic of Germany, 1984-87; med. co-ordinator, Fed. Emergeny Mgmt. Agy, Washington, 1987-90, nat. disaster med. system staff, bd. govs. Nat. Council Internat. Health, 1980-90; pres., chief operating officer U.S. Neuro Surgical Inc., 1992—; v.p., chief oper. officer M/D Frontiers, Springfield, Va., 1990—; pres. Automated Med. Products, Inc., 1990—; v.p. Automated Systems, 1991—; assoc. dir. rsch. nat. study ctr. for trauma and emergency medicine U. Md.; U.S. mem. Internat. Com. of Mil. Medicine and Pharmacy, 1981-87; U.S. mil. mem. Joint Civil/Mil. Med. Working Group U.S., NATO, 1981—; mem. program planning com. Internat. Assembly on Emergency Med. Services, Balt., 1984; congress lobbyist; cons. in field. Contbr. articles to med. jours; pub. books in field of philately. Commr., Explorer Scouts, Natick, Mass., 1975-76; trustee Cardinal Spellman Philatlic Mus., Weston, Mass., 1980—. Decorated Meritorious Service medal with 3 oak leaf clusters, Legion of Merit; recipient gold medal Res. Officers Assn., 1960. Mem. Electron Microscopy Soc. Am., Am. Stamp Dealers Assn., Central Atlantic Stamp Dealers Assn. (pres. 1977-81), Research and Engring. Soc. Am., Balt. Philatelic Soc., Sigma Alpha Epsilon, Sigma Xi. Republican. Baptist.

BROWN, JOHN DAVID, physicist; b. Shawnee, Okla., Feb. 20, 1957; s. John M. Brown and Nancy (Lee) Smith; m. Stefanie Mari Bilobran, Oct. 5, 1985; children: Andrew M., Michael S. BS, Okla. State U., 1979; PhD, U. Tex., 1985. Postdoctoral asst. Inst. for Theoretische Physik, Vienna, 1985-87; postdoctoral rsch. assoc. U. Tex., Austin, 1987-88, U. N.C., Chapel Hill, 1988-92; asst. prof. N.C. State U., Raleigh, 1992—; dir. Lanczos Internat. Centenary Conf., Raleigh, 1992-93; refree Phys. Rev. D, Phys. Rev. Letters. Author: Lower Dimesional Gravity, 1988; contbr. articles to profl. jours. Mem. Am. Phys. Soc. Achievements include patents for partition function for rotating black holes, quasilocal energy in general relativity and microcanonical functional integral. Office: NC State U Physics Dept Box 8202 Raleigh NC 27695

BROWN, KENNETH GERALD, chemistry educator; b. Syracuse, N.Y., Oct. 19, 1944; s. Kenneth Gerald and Alice Esther (Fish) B.; m. Ellen Jane Bicknell (div. 1985); 1 child, Elizabeth; m. Mary Elizabeth Jarrett, Mar. 26, 1988; children: Melissa, Eric, Wyatt. BA, Syracuse U., 1966; PhD, Brown U., 1971. Asst. prof. U. Detroit, 1976-80, assoc. prof., 1980-81; assoc. prof. chemistry Old Dominion U., Norfolk, Va., 1982-86; prof. chemistry Old Dominion U., Norfolk, 1986—; cons. NASA-Langley, Hampton, Va., 1983—. Contbr. articles to profl. jours. Named Co-Inventor of the Yr. NASA-Langley, 1991. Mem. Am. Chem. Soc. (councilor 1990-92). Achievements include 4 patents for low temperature carbon monoxide oxidation catalysts. Office: Old Dominion U Dept Chemistry Norfolk VA 23529

BROWN, LEO DALE, inorganic chemist; b. Waco, Tex., Apr. 11, 1948; s. Bryce Cardigan and Lillian Marie (Seeliger) B.; m. Paulette Cullen, Sept. 16, 1972; children: Kenneth, Richard, Lisa. BS in Chemistry, Baylor U., 1970; PhD in Chemistry, U. Calif., 1974. Postdoctoral rsch. Northwestern U., Evanston, 1975-77; staff chemist Exxon Rsch. and Engring. Co., Baytown, Tex., 1977-87; rsch. assoc. rsch. and devel. lab. Exxon Co., Baton Rouge, La., 1987--. Contbr. articles to profl. jours. Mem. Am. Chem. Soc., Am. Crystallographic Assn., Sigma Xi. Achievements include patent for cement production from coal conversion residues. Office: Exxon Rsch Devel Lab PO Box 2226 Baton Rouge LA 70821

BROWN, LINDA JOAN, psychotherapist, psychoanalyst; b. Mineola, N.Y., Feb. 18, 1941; d. Charles Harold and Helen (Golbach) B. Student, Smith Coll., Northampton, Mass., 1958-60; BA, Barnard Coll., N.Y.C., 1962; MPS in Art Therapy, Pratt Inst., Bklyn., 1973; MSW, Hunter Coll., N.Y.C., 1976. Cert. social worker, psychoanalyst, N.Y.; lic. clin. social worker, Calif.; diplomate Am. Bd. Examiners Clin. Social Work. Singer, actress Broadway theatres, N.Y.C., 1962-65, pub. rels./community rels. specialist, real estate, publicist/editor, pub., edn. cons., 1965-71; art therapist Bronx (N.Y.) Psychiat. Ctr., 1972-74; clin. social worker North Richmond Community Mental Health Ctr., S.I., N.Y., 1977-79; staff therapist Lincoln Inst. Psychotherapy, N.Y.C., 1978-80; sr. staff therapist Ctr. for Study Anorexia and Bulimia, N.Y.C., 1983-85; staff therapist Inst. Contemporary Psychotherapy, N.Y.C., 1988—; pvt. practice psychotherapy N.Y.C., 1978—; human svc. faculty Tristate Inst. Traditional Chinese Acupuncture, N.Y.C., 1986-89; mem. faculty N.Y. Open Ctr., N.Y.C., 1987—; adj. faculty Health Choices Ctr. for Healing Arts, Princeton, N.J., 1987-90; clin. cons. Personal Performance Cons., EAP, 1988; human resources cons. industry N.Y.C., 1988—; workshop leader seminars on stress mgmt., comm. skills, counseling skills; specialist in creative expression. Mem. Nat. Assn. Social Workers.

BROWN, LOUIS, physicist, researcher; b. San Angelo, Tex., Jan. 7, 1929; s. Metz and Sadie (Johnson) Bishop; m. Lore Elisabeth Frick, July 24, 1952. BS, St. Mary's U., 1950; PhD, U. Tex., 1958. Teaching asst. dept. physics U. Tex. Austin, 1952-58; rsch. asst. dept. physics U. Basle, Switzerland, 1958-61; postdoctoral fellow dept. terrestrial magnetism Carnegie Instn., Washington, 1961-63, staff assoc., 1963-69, staff scientist, 1969—; electronics designer Mil. Physics Lab., Austin, 1955-57; acting dir. dept. terrestrial magnetism Carnegie Instn., 1991-92. Contbr. numerous articles to profl. jours. With U.S. Army, 1950-52, Korea. Recipient Amerbach prize U. Basle, 1963. Fellow Am. Phys. Soc.; mem. AAAS, Am. Geophysical Union. Achievements include collaboration on building of first operating source of polarized ions, on development of techniques for measuring the cosmogenic isotope 10Be in natural materials, on clarifying nature of threshold state in Be-8, and on demonstrating that lavas from island-arc volcanoes have a sedimentary component. Office: Carnegie Inst Washington Dept Terrestrial Magnetism 5241 Broad Branch Rd NW Washington DC 20015

BROWN, MARK STEVEN, medical physicist; b. Denver, July 12, 1955; s. Clarence William and Gail Margaret (Farthing) B.; m. Mary Linda Avery, Oct. 9, 1988. Student, Northwestern U., 1973-74; BS, Colo. State U., 1977; PhD in Phys. Chemistry, U. Utah, 1984. GE postdoctoral fellow Yale U. Sch. Medicine, New Haven, 1984-86, assoc. rsch. scientist, 1986-87; rsch. asst. prof. U. N.Mex. Sch. Medicine, Albuquerque, 1987-89; med. physicist Swedish Med. Ctr. Porter Meml. Hosp., Englewood, Colo., 1989-92; instr. C.C. Denver, Denver, 1990, 91; asst. clin. prof. radiology U. Colo. Sch. Medicine, Denver, 1991-92, asst. prof. radiology, 1992—. Author: (with others) NMR Relaxation in Tissues, 1986; contbr. articles to profl. jours. Mem. Floyd Hill Homeowners Assn., Evergreen, Colo., 1990. Mem. Am. Chem. Soc., Am. Soc. for Magnetic Resonance in Medicine, Soc. for Magnetic Resonanace Imaging. Avocations: lead guitar and vocals for Otis T. & the Hammers, skiing, swimming. Home: 560 Hyland Dr Evergreen CO 80439-4809 Office: Univ Colo Health Scis Ctr Dept Radiology Box A034 4200 E 9th Ave Denver CO 80262

BROWN, MARVIN LEE, retired air force officer; b. Searcy, Ark., June 18, 1926; s. Frank M. and Letha (Litchfield) Brown; m. Doris Emerson, Aug. 5, 1945; children: Carol, Linda, James, Thea. BSCE, U. Ark., 1949; BS in Indsl. Engring., Syracuse U., 1957, MBA, 1958; MPA, George Washington U., 1964. Commd. lt. USAF, 1949, advanced through grades to lt. col., 1966; grad. USAF Command & Staff Coll., Montgomery, Ala., 1962-63; tchr. USAF 1st Postgrad. Sch., Montgomery, 1963-66; mem. staff Comdr. in Chief Pacific, Honolulu, 1967-70, Hdqrs. Air Tng. Command, San Antonio, 1970-73, Hdqrs. Air Tng. Command Tng. Ctr., Biloxi, Miss., 1973-77. Author: Air Force Leadership, 1963. Decorated Bronze Star, Air medal. Mem. VFW (post comdr. 1992-93). Democrat. Achievements include evaluation of programs to make Vietnamese Air Force viable (engring. and fiscal soundness), evaluation of combat rediness taining for foreign Air Forces. Home: 258 Kentucky Blvd New Braunfels TX 78130

BROWN, MICHAEL STUART, geneticist; b. N.Y.C., N.Y., Apr. 13, 1941; s. Harvey and Evelyn (Katz) B.; m. Alice Lapin, June 21, 1964; children: Elizabeth Jane, Sara Ellen. BA, U. Pa., 1962, MD, 1966. Intern, then resident in medicine Mass. Gen. Hosp., Boston, 1966-68; served with USPHS, 1968-70; clin. assoc. NIH, 1968-71; asst. prof. U. Tex. Southwestern Med. Sch., Dallas, 1971-74; Paul J. Thomas prof. genetics, dir. Ctr. Genetic Diseases, 1977—. Recipient Pfizer award Am. Chem. Soc., 1976, Passano award Passano Found., 1978, Lounsbery award U.S. Nat. Acad. Scis., 1979; Lita Annenberg Hazen award, 1982, Albert Lasker Med. Rsch. award, 1985, Horwitz prize, 1985, Nobel Prize in Medicine or Physiology, 1985, Nat. Medal of Sci. U.S., 1988. Mem. Nat. Acad. Scis., Am. Soc. Clin. Investigation, Assn. Am. Physicians, Harvey Soc., Royal Acad. Scis. (fgn. mem.). Office: U Tex Health Sci Ctr Dept Molecular Genetics 5323 Harry Hines Blvd Dallas TX 75235-7200

BROWN, OLEN RAY, medical microbiology research educator; b. Hastings, Okla., Aug. 18, 1935; s. Willis Edward and Rosie Nell (Fulton) B.; m. Pollyana June King, Aug. 30, 1958; children: Barbara Kathryn, Diana Carol, David Gregory. BS in Lab. Tech., Okla. U., 1958, MS in Bacteriology, 1960, PhD in Microbiology, 1964. Diplomate Am. Bd. Toxicology. Instr. Sch. Medicine, U. Mo., Columbia, 1964-65, asst. prof., 1965-70, assoc. prof., 1970-77, prof. dept. molecular microbiology and immunology, 1981—; joint appointments, prof. depts. microbiology and biomed. scis. Coll. Vet. Medicine, U. Mo., 1977—, prof. biomed. scis., 1987—; guest lectr. Ross U., St. Kitts, West Indies, 1984, 88; asst. dir. Dalton Rsch. Ctr. U. Mo., 1974-78, Dalton rsch. investigator grad. sch., 1968—; grant peer reviewer for program projects, SCOR and Superfund grants NIH, 1979, Nat. Inst. Environ. Health Scis., Research Triangle Park, N.C., 1986, 90-92; cons. drug abuse policy office White House, 1982, Immunological Vaccines, Inc., Columbia, 1984—, Lab. Support, Inc., Chgo., 1988-89, Eastern Rsch. Group, Lexington, Mass., 1991—, Teltech, Mpls., 1992—; cons., expert witness tech. adv. svcs. for attys., Blue Bell, Pa., 1986—, Med. Advisors, Phila., 1991—, Tech. Knowledge Svc., Mpls., 1992—, Expert Net, Chgo, 1992—., Teltech, Mpls., 1992—. Author: Laboratory Manual for Veterinary Microbiology, 1973; co-author: elem. and advanced lab. manuals for med. microbiology, 2 vols., 1978, 79; contbr. Progress in Clinical Research, Vol. 21, 1978, 79, Oxygen, 5th Internat. Hyperbaric Conf., Vols I, II, 1974, 79, numerous articles to profl. jours; book and film critic AAAS, Washington, 1986—; item preparer Am. Coll. Test, Med. Coll. Admissions Test, 1981—; mem. editorial staff Biomed. Letters, 1981—; reviewer profl. jours. Track and field ofcl. U. Mo. and Big Eight Conf., Columbia, 1979-86. Investigative rsch. grantee Office Naval Rsch. U.S. Dept. Def., 1968-81, NIH, 1976-88, 88—, U.S. Agy. for Internat. Devel., 1983-86, Nat. Inst. Dental Health Scis., 1989-92. Fellow Am. Inst. Chemists (cert. chemistry and chem engring., profl. program bd. 1989-90, sd com. chemistry and environ. concerns); mem. Top One Percent Soc., Soc. Toxicology, Internat. Soc. Study Xenobiotics, Am. Chem. Soc., Am. Heart Assn., Internat. Soc. Exposure Analysts, Nat. Space Soc., Oxygen Soc., Columbia Track Club (sec.-treas. 1979-82). Avocations: long-distance running, oil painting. Office: U Mo Dalton Rsch Ctr Columbia MO 65211

BROWN, PAUL B., physiologist educator; b. Panama City, Panama, Nov. 29, 1942; s. Harold E. and Nina E. (Wetzler) B.; m. Sally A. Holewinsky, May 25, 1968. BS, MIT, 1964; PhD, U. Chgo., 1968. Rsch. assoc. Cornell U., Ithaca, N.Y., 1968-72; rsch. scientist Boston State Hosp., 1972-74; from asst. to prof. W.Va. Univ., Morgantown, 1974-82, prof., 1982—. Author: Electronics for the Modern Scientist, 1982, (chpt.) Sensory Neurons: Diversity, Development, and Plasticity, 1992; editor: Computer Technology in Neuroscience, 1976, Jour. Electrophysiology Techniques, 1978-87. Exec. dir. W.Va. Assn. Biomed. Rsch., Morgantown, 1990-93; faculty advisor Coalition for Animals and Animal Rsch., Morgantown, 1990-93. Mem. Am. Physiol. Soc., Soc. for Neurosci. Achievements include patents for apparatus and method for long term storage of analog signals, analog neural network and method of implementing same; notable findings include the spinal dorsal horn is organized as a map of the skin sufficiently precise to mediate tactile localization, the morphology of primary afferent projections and dorsal horn cell dendrites is consistent with monosynaptic assembly of dorsal horn cell receptive fields, and with a developmental mechanism in which local map scale is determined by peripheral innervation density. Office: Health Scis Ctr Physiology Dept West Va Univ Morgantown WV 26506

BROWN, PAUL LEIGHTON, electrical engineer; b. Kingston, Jamaica, Jan. 20, 1959; s. Donald and Elorie (Clark) B.; m. Tori Lei Jones, Feb. 15, 1985; children: Taphenese Lanett, Tashelle Lenee. BSEE, MSEE, Drexel U., 1982. Student intern Bell of Pa., Phila., 1977-82; MTS Bell Labs., Phila., 1982-84, Bellcore, Piscataway, N.J., 1984—. Creator computer tng. program for minority youth, Power of Christ Tng. Acad. Nat. youth dir. United Ch. of Jesus Christ; assoc. pastor First Pentecostal Ch. N.J. Pentecostal. Home: 18 Olive St Neptune NJ 07753

BROWN, RANDY LEE, systems engineer; b. Yakima, Wash., Oct. 9, 1963; s. Jack Leroy Brown and Carol Ann (Litchtenburg) Myers. Student, Yakima Valley Vocat. Skills Ctr., 1980-82, Phoenix Inst. Tech., 1982-83. Electronic technician Easy Enterprises Amusements, Yakima, 1983-84; svc. mgr. Cliff Miller's Computers Inc., Yakima, 1984—. Named State Champion radio TV repair Vocat. Industries Clubs Am., 1982. Home: 608 S Yakima Ave Wapato WA 98951-1261 Office: Cliff Millers Computers Inc 22 N 2d St Yakima WA 98901

BROWN, RHONDA ROCHELLE, chemist, health facility administrator; b. Shelbyville, Ky., July 13, 1956; d. Clifton Theophilus and Fannie Mae (Lawson) B. BA in Chemistry, U. Md., 1978; MA, Central Mich. U., 1983; JD, No. Va. Law Sch., 1992. Analytical chemist Dept. Health and Mental Hygiene, Annapolis, Md., 1978-83; epidemiologist Dept. Health and Mental Hygiene, Balt., 1983-88; patent examiner U.S. Patent and Trademark Office, Xtal City, Va., 1989-90; freelance researcher New Carrollton, Md., 1990—; mem. Am. Chem. Soc., Washington, 1978-82; mem., exec. bd. Nat. Lawyers Guild, Washington, 1987—; pres. Voucher Express, 1993—; mediator Superior Ct., Washington, 1993—; legal advt. mgr. Sentinel Newspaper. subcommittee chmn. Anne Arundel County Task Force for Drug and Alcohol Abuse, 1979-80; pres., bd. mem. Md. Ornithological Soc., 1979-82; mem., exec. bd. Md. Condominium and Homeowners Assn., Rockville, Md., 1988-91. Named Outstanding Young Women of Am., 1983. Mem. Nat. Intellectual Propery Law Assn., Anne Arundel County Tennis Assn., Sigma Iota Epsilon.

BROWN, RICHARD MALCOLM, JR., botany educator; b. Pampa, Tex., Jan. 2, 1939; married; 2 children. BS, U. Tex., 1961, PhD in Botany, 1964. Fellow U. Hawaii and U. Tex., 1964-65; asst. prof. botany U. Tex., Austin, 1965-68; assoc. prof. botany U. N.C., Chapel Hill, 1968-73, prof., from 1973, dir. electron microscopy lab., from 1970, now Johnson & Johnson Centennial chair in plant cell biology, prof. botany; NSF fellow U. Freiburg, 1968-69, research grantee, 1970-72. Recipient Anselme Payen awrd 1986, Darbaker prize in phycology, 1978. Fellow Internat. Acad. Wood Scis. Research in airborne algae, algal ecology, ultrastructure of algal viruses, algal ultrastructure, immunochemistry of algae, cytology, Golgi apparatus and cell wall formation, sexual reproduction among algae, cellulose biogenesis. Office: Univ of Tex at Austin Dept of Botany Austin TX 78712

BROWN, ROBERT ARTHUR, chemical engineering educator; b. San Antonio, July 22, 1951; s. Ralph and Lillian (Rilling) B.; m. Beverly Ann Lamb, June 22, 1972; children: Ryan Arthur, Keith Andrew. BS, U. Tex., 1973, MS, 1975; PhD, U. Minn., 1979. Instr. U. Minn., Mpls., 1978; asst. prof. MIT, Cambridge, 1979-82, assoc. prof., 1982-84, prof., 1984—, Warren K. Lewis prof., 1992—, exec. officer dept. chem. engring., 1987-88, head dept. chem. engring., 1989—, co-dir. supercomputer facility, 1989—; cons. Lincoln Labs., Lexington, Mass., 1985-87, Mobil Solar Energy, Waltham, Mass., 1982—. Contbr. over 150 articles to profl. jours. Recipient Outstanding Jr. Faculty award Amoco Oil Co., 1981, Camille and Henry Dreyfus Tchr.-Scholar award 1983; named one of Outstanding Young Texaxs-Execs. U. Tex., 1991. Mem. Am. Isnt. Chem. Engrs. (Allen P. Colburn award 1986), Soc. Indsl. and Applied Math., Am. Assn. Crystal Growth (Young Author award 1985), Am. Phys. Soc., Nat. Acad. Engring. Office: MIT Dept Chem Engring 66-352 Cambridge MA 02139

BROWN, ROBERT GRIFFITH, chemist, chemical engineer, chemicals executive; b. Binghamton, N.Y., Apr. 29, 1941; s. Raymond Nobel and Helen Rebecca (Griffith) B.; m. Sharon Kay Myrick, Apr. 12, 1980; 1 child, Griffith Lauren. B in Chem. Engring., Clarkson U., 1964, PhD in Chemistry, 1972. Rsch. chemist FMC Corp., Balt., 1969-72; sr. rsch. chemist FMC Corp., Princeton, N.J., 1972-75; mgr. rsch. Dixie Chem. Co., Inc., Houston, 1975-79, dir. rsch., 1979-87, v.p. R & D, 1987—. Contbr. articles to profl. jours. Brookhaven Nat. Lab. rsch. fellow, 1965-66, Ayerst Labs. rsch. fellow, 1968-69. Mem. Houston Yacht Club, Sigma Xi. Republican. Achievements include development of high efficiency esterification process for esters of low boiling alcohols, of high yield and purity process for manufacture of guanamines, of process for instantaneous material balancing of gas phase process via on-line mass spectrometry, of process for production of ultra high purity nitriles; research on processing technology for heat sensitive chemicals. Office: Dixie Chem Co Inc 10701 Bay Area Blvd Pasadena TX 77507-1799

BROWN, ROBERT GROVER, engineering educator; b. Shenandoah, Iowa, Apr. 25, 1926; s. Grover Whitney and Irene (Frink) B. BS, Iowa State Coll., 1948, MS, 1951, PhD, 1956. Instr. Iowa State Coll., Ames, 1948-51, 53-55, asst. prof., 1955-56, assoc. prof., 1956-59, prof., 1959-76, Disting. prof., 1976-88; Disting. prof. emeritus Iowa State Coll., 1988—; research engr. N. Am. Aviation, Downey, Calif., 1951-53; cons. various aerospace engring. firms., 1956—. Author: (with R.A. Sharpe, W.L. Hughes) Lines, Waves and Antennas, 1961, (with J.W. Nilsson) Linear Systems Analysis, 1962, (with Patrick Y.C. Hwang) Introduction to Random Signals and Applied Kalman Filtering, 2d edit., 1992. Fellow IEEE; mem. Inst. Navigation (Burka award 1978, 84). Office: Iowa State U Ames IA 50011

BROWN, ROBERT JAMES SIDFORD, physicist; b. Inglewood, Calif., Sept. 7, 1924; s. Andrew Glassell and Ethlyn Bertha (Sidford) B.; m. Phyllis Jeanne White, July 26, 1950 (dec. Dec. 6, 1956); m. Bette Jean Drummond, Aug. 12, 1962 (div. Nov. 22, 1986); children: Erich, Dirk, Eleanor, Sidford, Kurt; m. Margaret Ellwin Steeper Edwards, Feb. 13, 1993. MS, Calif. Inst. Tech., 1948; PhD, U. Minn., 1953. Sr. rsch. assoc. Chevron Oil Field Rsch., La Habra, Calif., 1953-86, Los Alamos (N.Mex.) Sci. Lab., 1944-46; cons. Universita di Bologna, Bologna, Italy, 1987—. Contbr. articles to profl. jours. Sgt. U.S. Army, 1943-46. Mem. Am. Phys. Soc., Am. Geophys. Union, Soc. Exploration Geophysicists, Soc. for Magnetic Resonance in Medicine, Soc. for Magnetic Resonance Imaging, Internat. Soc. for Magnetic Resonance. Achievements include forty patents in field in Nuclear Magnetic Resonance or Nuclear Magnetism Oil Well Logging. Home and Office: 515 W 11th St Claremont CA 91711-3721

BROWN, RONNIE JEFFREY, biologist, researcher; b. Birmingham, Ala., Apr. 27, 1953; s. Eddie and Johnnie Mae (Pitts) B.; m. Fannie M. Garrett, Aug. 11, 1984. BS, U. Ala., 1975; student, Harvard U., 1974-75. Tutor in gross anatomy sch. dentistry U. Ala., Birmingham, 1976-77; biologist VA Med. Ctrs., Dallas, Oklahoma City, 1978-83; supervisory biologist USDA, El Reno, Okla., 1983-84; sr. mfg. supr. UTC Mostex, Carrollton, Tex., 1983-85; prodn. mgr. Kraft Inc., Garland, Tex., 1985-86; dir. surg. rsch. Baylor U. Med. Ctr., Dallas, 1986-90; sr. rsch. assoc. dept. surgery U. Ala., 1990—; coord. metro edn. Am. Assn. Lab. Animal Sci. Tex. Br., Dallas, 1987-90. Contbr. articles to Jour. Ala. Jr. Acad. Sci., So. Med. Jour., and others. Named One of Outstanding Young Men of Am., 1979, 92. Mem. APHA, AAAS, Am. Assn. Lab. Animal Sci. (southeastern br.), Urban League, Young Men Rep. Club, Acad. Surg. Rsch., Masons, Alpha Phi Alpha, Beta Beta Beta. Methodist. Achievements include development of continuing education program in laboratory animal management. Hom: 1820 Forestdale Blvd Birmingham AL 38214 Office: U Ala Dept Surgery Div Gen Surger Univ Sta Birmingham AL 35205

BROWN, SAMUEL JOSEPH, JR., mechanical engineer; b. New Orleans, May 6, 1941; s. Samuel Joseph and Camille (Trumbatory) B.; m. Josephine Monistere; children: Troy Joseph, Tricia Maria Brown Kenworthy, Kamryn Leigh. BS in Mech. Engring., U. Southwestern La., 1966; MS, U. Fla., 1968; PhD, U. Akron, 1982. Registered profl. engr., Ohio, Tex., La., Okla., Pa., Ala., Miss. New constrn. inspector New Orleans Port Authority, 1964; project mech. engr. Mid South Utilities, New Orleans, 1966; R&D cons. U. Fla., Gainesville, 1969-70; with design and devel. of prototype equipment Babcock & Wilcox McDermott Co., Akron, Ohio, 1970-78; cons. Sci. Mgt. Corp./O'Donnell & Assocs., Pitts., 1979-80; pres./cons. Quest Engring. Devel. Corp., Humble, Tex., 1980—; bd. dirs. Intertech Svcs. Inc., Houston, 1984—; univ. faculty, vis. lectr., profl. devel. instr. in courses on computer simulation, failure analysis, fluid structure dynamics, component design and analysis, explosions and hazardous release protection. Author: (handbook) Pressure Systems Energy Release Protection, 1986; co-author: Am. Soc. Metals's Handbook of Engineering Mathematics, 1983, Handbook of Case Histories in Failure Analysis, Vol. 1, 1993, Non-Linear Analysis of Light Water Reactor Components: Areas of Investigation/Benefits/Recommendations, 1980; author, editor vols. for ASME: A Decade of Progress, 1976, 85, Failure Data and Failure Analysis, 1977, Dynamics of Fluid Structure Systems, 1979, Metallic Bellows and Expansion Joints, Part I, 1981, Part II, 1984, Impact, Fragmentation, Blast, 1984, Proc. Pressure Vessel and Piping Conf., 1985, The Product Liability Handbook, 1991, Hazardous Release Protection, 1993, others; co-editor Jour. Process Mech. Engring. (UK), 1990-92; author 350 tech. reports; contbr. 90 publications to profl. jours. Sponsor U. Southwestern La. Alumni Assn., 1990, U. Akron Alumni Assn., U. Fla. Alumni Assn., 1991. NASA fellow, 1981, NDEA fellow, Wisdom Soc. fellow, 1989; Personalities in Am. award ABI, 1990. Fellow ASME (vice chmn. tech. com. 1979-83, pressure vessel and piping tech. divsn. and codes 1974—, chmn. subcom. on hazardous release, chmn. pressure vessel and piping divsn., edn., honors and awards, chmn. OAC com. 1982-85, editor newsletter ASME PVP 1982-83, mem. high pressure com. 1982—, tech. divsn. 1989—, chmn. conf. tech. program com. 1985, Outstanding Tech. Paper award 1984, Bd. Govs. svc. to codes and standards award, 1992), Am. Soc. Chem. Engrs. (tech. divsn. 1989—), Am. Soc. Civil Engrs. (tech. divsn. 1979—); mem. Am. Soc. for Metals (tech. divsn. 1980—), Am. Soc. Exptl. Mechanics (tech. divsn. 1978—), Nat. Assn. Profl. Accident Reconstruction Specialists, Post Tng. Inst., Houston C. of C., Sigma Xi. Achievements include design of PWR, LWR, breeder, naval nuclear and geothermal power systems, and new mechanical-civil-aeronautical equipment and structural concepts. Office: Quest Engring Devel Corp 7500 N Sam Houston Pky E Humble TX 77396-2625

BROWN, SCOTT WILSON, educational psychology educator; b. Greenwich, Conn., Dec. 1, 1952; s. Vernon Watson and Elizabeth (Rounds) B.; m. Mary Margret pearson, July 28, 1974; children: Matthew, Melissa, Steven. AB in Psychology, Boston U., 1974; MS in Psychology, Montana State U., 1975; PhD in Psychology, Syracuse U., 1980. Prof. ednl. psychology U. Conn., Storrs, 1980—. Mem. editorial bd. Computers in Schs. jour., 1985—; contbr. articles to profl. jours. Recipient grant IBM, 1991, U.S. Ctrs. for Disease Control, 1991, 92, Conn. Dept. Health Svcs., 1992. Mem. APA, Am. Psychol. Soc., Am. Ednl. Rsch. Assn., Northeastern Ednl. Rsch. Assn. (pres. 1991-92). Congregational. Home: 26 Cowles Rd Willington CT 06279 Office: U Conn 249 Glenbrook Rd Storrs Mansfield CT 06269

BROWN, SERENA MARIE, mathematics and home economics educator; b. Stockton, Calif., June 30, 1951; d. Harold Petersen Burkett and Margarite Della (Claeys) Hatfield; m. Richard Frank Brown, July 21, 1973; children: Jeremy, Eric, Christopher, Alexander. BS in Textiles and Clothing, U. Calif., Davis, 1973; MA in Home Econs. Edn., U. Minn., St. Paul, 1983; tchr. cert., Cen. Mo. State U., 1987, Mt. Mary Coll., Milw., 1988. Cert. home economist; cert. secondary math. and home econs. tchr. Dept. mgr., interior designer Yardage Shop, Sacramento, 1977; interior designer Weinstocks, Sacramento, 1977-78; asst. home economist U. Pa. Extension, Gettysburg, 1979-80; teaching asst. home econs. edn. dept. U. Minn., St. Paul, 1981-82, rsch. asst., 1982-83; substitute tchr. Lee's Summit (Mo.) Sch. Dist., 1983-85; high sch. tchr. Lee's Summit High Sch., 1985-86; freelance arts and crafts tchr. Milw., 1987—; Audio-visual chairperson Minn. Home Econs. Assn. Conv. Task Force, St. Paul, 1983; curriculum/evaluation chairperson Coop. Neighborhood Early Edn. Ctr., White Bear Lake, Minn., 1980-83. Designer line of toddler clothing, 1990-91; author latchkey curriculum, 1988-90; artist crewel, flower arranging, ceramics. Coun. sch. night for scouting coord. Potawatami coun. Boy Scouts Am., Waukesha, Wis., 1990-92, scouting for food Southbrook dist., 1989-90, com. chair, 1988-91. Mem. Am. Home Econs. Assn. (vol. project home safe 1988-89), Nat. Coun. Tchrs. Math., Am. Ednl. Rsch. Assn., Wis. Home Econs. Assn., Proud Parents of Twins (2d v.p. 1992-93), Wis. Orgn. ot Mothers of Twins Clubs (2d v.p.). Democrat. Methodist. Avocations: art, ceramics, needlework, soccer, crafts. Home and Office: S104 W20942 Cindy Dr Muskego WI 53150-9592

BROWN, STEPHEN LAWRENCE, environmental consultant; b. San Francisco, Feb. 16, 1937; s. Bonnar and Martha (Clendenin) B.; m. Ann Goldsberry, Aug. 13, 1961; children: Lisa, Travis, Meredith. BS in Engring. Sci., Stanford U., 1958, MS in Physics, 1961; PhD in Physics, Purdue U., 1963. Ops. analyst Stanford Rsch. Inst., Menlo Park, Calif., 1963-74, program mgr., 1974-77, dir. Ctr. Resource and Environ. Systems Studies, 1977-80, dir. Ctr. Health and Environ. Rsch., 1980-83; assoc. dir. Commn. on Life Scis. NAS, Washington, 1983-86; prin. Environ Corp., Arlington, Va., 1986-91; mgr. risk assessment ENSR Cons. and Engring., Alameda, Calif., 1992—; mem. sci. adv. bd. EPA, Washington, 1991—; mem. coms. SAB, NAS, 1980-87. Contbr. over 20 articles to profl. jours., chpts. to books; author over 100 reports in field. Mem. AAAS, Internat. Soc. Exposure Assessment, Soc. for Risk Analysis, Phi Beta Kappa, Sigma Xi, Sigma Pi Sigma, Tau Beta Pi. Office: ENSR Cons and Engring Ste 210 1320 Harbor Bay Pkwy Alameda CA 94501

BROWN, THEODORE LAWRENCE, chemistry educator; b. Green Bay, Wis., Oct. 15, 1928; s. Lawrence A. and Martha E. (Kedinger) B.; m. Audrey Catherine Brockman, Jan. 6, 1951; children: Mary Margaret, Karen Anne, Jennifer Gerarda, Philip Matthew, Andrew Lawrence. BS in Chemistry, Ill. Inst. Tech., 1950; PhD, Mich. State U., 1956. Faculty U. Ill., Urbana, 1956—, prof. chemistry, 1965—; vice chancellor for rsch., dean Grad. Coll., 1980-86; dir. Beckman Inst. for Advanced Sci. and Tech., 1987—; vis. scientist Internat. Meteorol. Inst., Stockholm, 1972; Boomer lectr. U. Alta. (Can.), Edmonton, 1975; Firth vis. prof. U. Sheffield (Eng.), 1977; bd. dirs. Champaign County Opportunities Industrialization Ctr., 1970-79, chmn., 1975-78; bd. govs. Argonne Nat. Lab., 1982-88, Mercy Hosp., Urbana, Ill., 1985-89, Chem. Abstracts Svc., 1991—. Author: (with R.S. Drago) Experiments in General Chemistry, 3d edit., 1970, General Chemistry, 2d edit., 1968, Energy and the Environment, 1971, (with H.E. LeMay) Chemistry: The Central Science, 1977, 2d edit., 1981, 3d edit., 1985, 4th edit., 1988, 5th edit., 1991; assoc. editor: Inorganic Chemistry, 1969-78; contbr. articles to profl. publs. Mem. Govt.-Univ.-Industry Roundtable Coun., 1989—. With USN, 1950-53. Sloan research fellow, 1962-66; NSF sr. postdoctoral fellow, 1964-65; Guggenheim fellow, 1979. Fellow AAAS; mem. Am. Chem. Soc. (award in inorganic chemistry 1972, award for disting. svc. in advancement of inorganic chemistry 1993), Sigma Xi, Alpha Chi Sigma. Home: 309 Yankee Ridge Ln Urbana IL 61801-7115 Office: U Ill Noyes Lab Box 31 505 S Mathew St Urbana IL 61801

BROWN, THOMAS EDWARD, chemist; b. Athens, Ga., May 29, 1967; s. Emory Thomas and Lila Eugenia (Mitchell) B.; m. Randi Kay Petzke, May 9, 1992. BSA, U. Ga., 1989. Chief chemist Kraft Gen. Foods, Decatur, Ga., 1990-91; quality assurance mgr. Savannah (Ga.) Cocoa Inc., 1991—. Contbr. articles to profl. jours. Vol. Spl. Olympics, Cobb County, Ga., 1984-92. Mem. AOAC, Pa. Mfg. Chocolate Assn., Internat. Food Tech. Republican. Pentecostal. Achievements include continuous processing of cocoa beans. Office: Savannah Cocoa Inc PO Box 7824 102 Eli Whitney Blvd Savannah GA 31418

BROWN, TREVOR ERNEST, environment risk consultant; b. Cheviot, Canterbury, New Zealand, Oct. 31, 1939; s. Ernest Carl and Helen Farely (Dickinson) B.; m. Joyce Burnett, Mar. 10, 1964 (div. Apr. 1984); children: Ross Trevor, Lenaire; m. Kerrie Therese Conant, Jan. 7, 1985; 1 child, Brett. Diploma in med. tech., U. Otago, New Zealand, 1962; B Applied Sci., U. Otago, 1964. Chief technologist Med. Labs., Napier, New Zealand, 1964-75; sect. leader Cockburn Sound Study, Perth, Australia, 1975-78; environ. supt. Pancon Mining Ltd., Jabiluka, Australia, 1978-80, Rundle Oil Shale Project, Gladstone, Australia, 1980-81, CSR Ltd., Sydney, Australia, 1981-83; tech. mgr. Brown Coal Liquefaction, Morwell, Victoria, Australia, 1983-85; dir. environment Tasmania Dept. Environ., Hobart, Australia, 1985-89; prin. Dames & Moore, Sydney, 1989-91, AGC Woodward-Clyde, 1991—. Contbr. articles to profl. jours. Bd. dirs. Marineland New Zealand, Napier, 1968-72; chmn. Litter Control Assn., Hobart, 1985-89, State Oil Pollution Com., Hobart, 1985-89, Australian Environ. Coun., 1988-89. Mem. Royal Australian Chem. Inst., Australian Soc. Microbiology (assoc.), Australia Med. Lab. Scientists Inst., Environ. Inst. Australia (assoc.), Am. Chem. Soc. Presbyterian. Avocations: golf, computing, bush walking. Office: AGC Woodward-Clyde, Park Rd, Milton Qld 4064, Australia Mailing address: PO Box 377, Albion Qld 4010, Australia

BROWN, VICKI LEE, civil engineering educator; b. Johnstown, Pa., June 14, 1954; d. Paul Harshberger and Janet LaVerne (Ellenberger) Saylor; m. Larry Bruce Brown, Feb. 19, 1977; children: Nathaniel Paul, Katherine Lee. BS in Civil Engring. Tech., U. Pitts., Johnstown, Pa., 1976; PhD in Civil Engring., U. Del., 1988. Registered profl. engr., Pa. Structural engr. Bechtel Power Corp., Gaithersburg, Pa., 1976; engr. Pa. Electric Co., Johnstown, 1976-79; teaching asst. Pa. State U., University Park, 1980-81; lectr. Widener U., Chester, Pa., 1981-88, asst. prof., 1988-92, assoc. prof., 1992—. Bd. mem., sec. Pennwood Park Civic Assn., West Chester, Pa., 1989-93. Mem. ASCE, NSPE, Am. Concrete Inst. (assoc., com. 440 1992—), Pa. Soc. Profl. Engrs. (edn. com. 1990—), Am. Soc. Engring. Edn., Soc. Women Engrs. Achievements include research in engring. mechanics. Home: 1540 Pennsbury Dr West Chester PA 19382 Office: Widener U 1 University Pl Chester PA 19013

BROWN, WALTER CREIGHTON, biologist; b. Butte, Mont., Aug. 18, 1913; s. D. Frank and Isabella (Creighton) B.; m. Jeanette Snyder, Aug. 20, 1950; children: Pamela Hawley, James Creighton, Julia Elizabeth. AB, Coll. Puget Sound, 1935, MA, 1938; PhD, Stanford U., 1950. Chmn. dept. Clover Park High Sch., Tacoma, Wash., 1938-42; acting. instr. Stanford U., Calif., 1949-50; instr. Northwestern U., Evanston, Ill., 1950-53; dean sci. Menlo Coll., Menlo Park, Calif., 1955-66, dean instrn., 1966-75; rsch. assoc., fellow Calif. Acad. Sci., San Francisco, 1978—; lectr. Sillman U., Philippines, 1954-55, dir. rsch. Program on Ecology and Systematics of Philippine Amphibians and Reptiles, 1958-74; ; vis. prof. biology Stanford U., 1962, 64, 66, 68, Harvard U., Cambridge, Mass., 1969, 72. Author: Philippine Lizards of the Family Gekkonidae, 1978, Philippine Lizards of the Family Scincidae, 1980, Lizards of the genus Emoia (Scincidae) with Observations of Their Evolution and Biogeography, 1991; contbr. 73 articles to profl. jours. Served with U.S. Army, 1942-46. Fellow AAAS; mem. Am. Soc. Ichthyologists and Herpetologists, Am. Inst. Biol. Scis., Sigma Xi. Office: Calif Acad Scis Dept Herpetology Golden Gate Park San Francisco CA 94118-4501

BROWN, WALTER REDVERS JOHN, physicist; b. Toronto, Ont., Can., Aug. 22, 1925; s. Ernest Redvers and Rita Mary (Brooks) B.; m. Anita Catherine Goggio, June 5, 1948 (div. 1972); children: Paul, Susan, Patricia, Judith; m. Beth Susan Southard, Oct. 12, 1974; 1 child, Amy. BA, U. Toronto, 1947; MS, U. Rochester, 1949. Sr. physicist Eastman Kodak Co., Rochester, N.Y., 1947-55; rsch. assoc. Boston U., 1955-57; asst. to dir. rsch. Itek Corp., Lexington, Mass., 1957-62; v.p. R & D United Carr Inc., Boston, 1962-69; exec. v.p. Ealing Corp., Cambridge, Mass., 1969-71; pres. Daedalon Corp., Salem, Mass., 1971—. Patentee in field. Fellow Optical Soc. Am. (Adolph Lomb medal 1956); mem. Eastern Yacht Club, St. Botolph Club. Roman Catholic. Home: 120 Atlantic Ave Marblehead MA 01945-3049 Office: Daedalon Corp PO Box 2028 Salem MA 01970-6228

BROWN, WARREN JOSEPH, physician; b. Bklyn., July 17, 1924; s. Benjamin Oscar and Angela Marie (Cahill) B.; m. Greet Roos, July 3, 1970; children—Warren James, Robert E., Suzanne J., Annemarie, Eric Jan. Student, Ursinus Coll., 1942-43; B.S., Bethany Coll., 1945; M.D., Ohio State U., 1949. Diplomate Am. Bd. Family Practice. Intern U.S. Naval Hosp., Long Beach and Oceanside, Calif.; resident Pottstown Hosp., Pa., 1950-51; assoc. Roos Loos Med. Group, Alhambra, Calif., 1951; practice medicine specializing in family practice Largo, Fla., 1953—; sr. civilian flight surgeon FAA, 1964—; pres. Aero-Med. Consultants, Inc., Largo, 1969—. Author: Florida's Aviation History, 1980, 2d edit., 1993, Child Yank Over the Rainbow, 1977, Patients' Guide to Medicine, 10th edit., 1987, The World's First Airline: The St. Petersburg-Tampa Airboat Line, 1914, 1981, 2d edit., 1984. Historian Fla. Aviation Hist. Soc., 1978—, St. Petersburg-Clearwater-Tampa Hangar Order of Quiet Birdmen, 1969—. With USN, 1943-45, 49-50, 51-53. Fellow Am. Acad. Family Physicians; mem. Pinellas County Med. Assn., Fla. Med. Assn., Aircraft Owners and Pilots Assn., Am. Radio Relay League. Home: 14607 Brewster Dr Largo FL 34644-4822 Office: 10912 Hamlin Blvd Largo FL 34644-5099

BROWN, WILLIAM SAMUEL, JR., communication processes and disorders educator; b. Pottstown, Penn., Apr. 25, 1940; s. William Samuel and Elizabeth (Gallager) B.; m. Elaine Kay Whitehouse, Aug. 18, 1962; children: William Samuel III, Allen Reed. MA, SUNY, Buffalo, 1967, PhD, 1969. Speech therapist Crawford Cty. Schools, Meadville, Penn., 1962-65; rsch. asst. SUNY, Buffalo, N.Y., 1965-68; prof. U. Fla., Gainesville, Fla., 1970—; Contrib. numerous publications to scientific jours. Post-doctoral fellow U. Fla, Gainsville, 1968-70. Fellow Internat. Soc. Phonetic Sci. (coun. rep. 1980—), Am. Speech-Lang.-Hearing Assn., Acoustical Soc. Am.; mem. Am. Assn. Phonetic Sci. (exec. sec. 1980—). Republican. Presbyterian. Office: U Fla IASCP Dauer 63 Gainesville FL 32611

BROWNE, JAMES CLAYTON, computer science educator; b. Conway, Ark., Jan. 16, 1935; s. Walter E. and Louise (James) B.; m. Gayle Moseley; children: Clayton Carleton, Duncan James, Valerie Siobhan. BA, Hendrix Coll., 1956; PhD, U. Tex., 1960. Asst. prof. physics U. Tex., Austin, 1960-64; prof. Computer Sci. Dept. U. Tex., Austin, 1968—, acting chmn., 1968-69, chmn., 1971-75, 85-87, regents centennial chmn. #2, 1987—; postdoctoral fellow Queen's U., Belfast, Ireland, 1964-65, prof. computer sci., dir. computer lab., 1965-68; chmn. bd. Info. Research Assocs., Austin. Assoc. editor Computer Physics Communications, 1987—; author over 150 papers in computer sci. and physics. Fellow Am. Phys. Soc., Brit. Computer Soc., Assn. Computing Machinery (chmn. spl. interest group on operating systems 1975-77); mem. Soc. Indsl. and Applied Math. Avocations: skiing, jogging. Office: Univ of Tex Dept of Computer Scis & Physics/Taylor Hall 5 126 Austin TX 78712

BROWNE, RICHARD HAROLD, statistician, consultant; b. St. Louis, Sept. 24, 1946; s. Basil Campbell and Evelyn Beatrice (Biver) B.; m. Dennise Marie Richardson, Aug. 10, 1970. B.S., U. Mo.-Rolla, 1968; M.S., Okla. State U., 1970, Ph.D., 1973. Statistician M.D. Anderson Hosp., Houston, 1971-72; asst. prof. U. Tex. Health Sci. Ctr., Dallas, 1973-79; statistician Criterion Inc., Dallas, 1979-81; sr. mgmt. analyst Sun Co., Dallas, 1981-83; sr. biostatistician Teams, Inc., Dallas, 1983-85; sr. cons. RHB Cons. Svcs., Dallas, 1979—; adminstrv. dir. rsch. Tex. Scottish Rite Hosp., Dallas, 1988—; asst. prof. So. Meth. U., Dallas, 1974-77, Health Sci. Ctr., U. Tex.-Dallas, 1979-82; adj. assoc. prof. Tex. Women's U., Dallas, 1984—. Contbr. articles to profl. jours. Mem. Am. Statis. Assn., Biometric Soc. Republican. Club: Dallas Camera. Avocation: photography. Home: 12045 Inwood Rd Dallas TX 75244

BROWNER, CAROL, federal official; d. Michael Browner and Isabella Harty-Hugues; m. Michael Podhorzer; 1 child, Zachary. Grad., U. Fla., 1977, JD, 1979. Gen. counsel govt. ops. com. Fla. Ho. of Reps.; with Citizen Action Com., Washington; chief legis. aide environ. issues to Sen. Lawton Chiles, legis. dir. to Sen. Al Gore, Jr., 1988-91; sec. Dept. Environ. Regulation, Fla., 1991-93; adminstr. EPA, Washington, 1993—. Office: Environmental Protection Agency Office of the Administrator 401 M St SW Washington DC 20460*

BROWNING, BURT OLIVER, mechanical engineer, consultant; b. Marion, N.C., May 5, 1957; s. Harley Ray and Alice Dean (Gray) B.; m. Leslie Carol Lee, May 24, 1980; 1 child, Elliot Montgomery. BSME, N.C. State U., 1979. Registered profl. engr., N.C. Tooling engr. Stanley Tools, Cheraw, S.C., 1979-81; mfg. engr. Schrader-Bellows, Wake Forest, N.C., 1981-82; mech. engr. Duracell, Lancaster, S.C., 1982-85; sr. mech. engr. Gas Springs, Gastonia, N.C., 1985-89, Day Engring., Charlotte, N.C., 1989-90, Wix Filters, Gastonia, 1990—; pres. Browning's Bee Svc., 1989—, Browning Engring., 1989—. Program chmn. Mecklenburg County Beekeepers, Charlotte, 1990-91, v.p., 1990-91, pres., 1992. Mem. NSPE, Soc. Mfg. Engrs., N.C. Profl. Engrs. Home: 6244 Netherwood Dr Charlotte NC 28210

BROWNING, CHARLES BENTON, university dean, agricultural educator; b. Houston, Sept. 16, 1931; s. Earl William and Emma (Summerlin) B.; m. Magda Luest, Jan. 14, 1956; children: Susan Elaine Browning Kreps, Charles Benton Jr., Steven Randolph, Karen Diane Browning Bassetti, Heidi Charlene, Gary Thomas. B.S. in Animal Sci., Tex. Tech., 1955; M.S. in Dairy Sci., Kans. State U., 1956, Ph.D. in Animal Nutrition, 1958. Asst. prof. Miss. State U., State Coll., 1958-60, assoc. prof., 1960-61, prof., 1961-66; head dept. dairy sci. U. Fla., Gainesville, 1966-69; dean resident instrn. U. Fla., Gainsville, 1969-79; dean and dir. div. agr. Okla. State U., Stillwater, 1979—; mem. numerous state agrl. coms.; head team to rev. agrl. problems and programs of Jamaica for AID, 1978, team to rev. agrl. edn. programs in Honduras, 1983, MIAC team to rev. agrl. project in Morocco, 1985, 86, 89, in Tunisia, 1988, in Kenya, 1989, Gov.'s Reverse Trade Mission to Japan, 1986; mem. USDA, SEA Joint Council on Food and Agrl. Scis., 1977; cons. Dept. State Internat. Communication Agy., Venezuela, 1978; del. 6th Working Conf. of Repas, Paris, 1978. Active United Way, Boy Scouts Am., PTA. Named Outstanding Prof. Alpha Zeta, 1966; recipient Z.W. Craine Research award Nat. Silo Assn., 1967, Disting. Agriculture Alumnus award Tex. Tech U., 1985. Mem. Am. Dairy Sci. Assn., Am. Soc. Animal Sci., Am. Grassland Council, AAAS, Sigma Xi, Phi Kappa Phi, Omicron Delta Kappa, Gamma Sigma Delta, Alpha Zeta. Episcopalian. Club: Kiwanis. Home: 6505 W Coventry Dr Stillwater OK 74074-1024 Office: Okla State U 139 Agricultural Hall Stillwater OK 74078

BROWNING, DAVID GUNTER, physicist; b. Wakefield, R.I., May 3, 1937; s. Harold William and Mary Emma (Williams) B. BS, U. R.I., 1958; MS, Mich. State U., 1961. Physicist Naval Underwater Warfare Ctr., New London, Conn., 1965—; rsch. scientist U. Conn.-Avery Pt., Groton, 1989—; exchange scientist Defence Sci. Establishment, Auckland, New Zealand, 1973, Defence Rsch. Establishment-Pacific, Victoria, Can., 1983; vis. scientist Naval Postgrad. Sch., Monterey, Calif., 1979. Contbr. numerous articles to profl. jours. Fellow Acoustical Soc. Am.; mem. Am. Phys. Soc., Can. Acoustical Soc., New Zealand Assn. Scientists. Mem. Soc. of Friends. Achievements include research in sound absorption in the sea, underwater acoustics (ambient noise, sound propagation). Home: 139 Old North Rd Kingston RI 02881 Office: NUWC-New London Detach Code 31 New London CT 06320

BROWNING, RONALD KENNETH, aerospace engineer; b. Evansville, Ind., Aug. 2, 1934; s. Joseph Kenneth and Esther (Scheessele) B.; m. Patricia Brown, Aug. 27, 1960; children: Juliet, Timothy, Jeanne, Theodore. BS in Indsl. Engring., U. Evansville, 1957. Rsch. engr. U.S. Naval Ordinance Lab., White Oak, Md., 1959; aerospace engr. Nat. Aerospace Adminstrn. Goddard, Greenbelt, Md., 1959-67; project mgr. Nat. Aerospace Adminstrn. Goddard, Greenbelt, 1967-83, dep. program dir., 1983-89; program mgr. Ford Aerospace, Seabrook, Md., 1989-90; v.p. Loral AeroSys, Seabrook, 1990—. Bd. dirs. Coll. Heights (Md.) Estates Assn., 1992. Recipient Disting. Svc. medal NASA, 1984. Mem. AIAA. Home: 7017 Partridge Pl Hyattsville MD 20782 Office: Loral Aerosystems 7375 Executive Pl Seabrook MD 20706

BROWNING, WILLIAM ELGAR, JR., chemistry consultant; b. Warren, Ohio, Oct. 1, 1923; s. William Elgar and Selina (Clark) B.; m. Alice Cornuelle, June 6, 1944 (dec. July 1982); children: Arthur William, Susan Carol, Carl Douglas, Kenneth Robert; m. Ellen Mary Nash, May 18, 1985. BA in Chemistry, Ohio State U., 1943. Chemist U-235 purification group U.S. Army Manhattan Engring. Dist., Los Alamos, 1943-46; group leader chem. effects radiation Fairchild Engine and Aircraft Corp., Oak Ridge, Tenn., 1947-51; group leader Calif. Rsch. and Devel. Corp., Livermore, 1951-53; group leader Oak Ridge Nat. Lab., 1953-68, 71-73, sci. coord. nuclear safety rsch. program, 1966-69; guest scientist OECD High Temperature Gas-Cooled Reactor Project OECD U.K. Atomic Energy Authority, Winfrith, Dorset, Eng., 1969-70; corp. mgr. environ. impact analysis Cabot Corp., Boston, 1973-86, corp. mgr. toxic substances control, 1976-86; cons. on health effects and environ. effects indsl. chems., Boston, 1986—; pres. High Nickel Alloys Health and Safety Group, Boston, 1975-84; referee Nuclear Tech. Jour., 1973—; judge New Eng. Regional Sci. Fair, Somerville, Mass., 1975—. Contbr. articles to sci. jours., chpts. to books. Chmn., sec. Kingston (Tenn.) Regional Planning Commn., 1959-64. Fellow Am. Inst. Chemists; mem. AAAS, Am. Chem. Soc., Am. Nuclear Soc., Sigma Xi. Achievements include research on nuclear fuel meltdown and fission product behavior, preparation of ultra-pure materials, effects of radiation on materials, thermometry in radiation fields, measurement and trapping of gas-borne radioactivity, nuclear safety, environmental impact analysis of nuclear power reactors and other energy sources, health effects and environmental effects of industrial chemicals. Office: 79 Iroquois St Boston MA 02120-2831

BROXMEYER, HAL EDWARD, medical educator; b. Bklyn., Nov. 27, 1944; s. David and Anna (Gurman) B.; m. C. Beth Biller, 1969; children: Eric Jay, Jeffrey Daniel. BS, Bklyn. Coll., 1966; MS, L.I. U., 1969; PhD, NYU, 1973. Postdoctoral student Queens U., Kingston, Ont., Can., 1973-75; assoc. researcher, rsch. assoc. Meml. Sloan Kettering Cancer Ctr., N.Y.C., 1975-78, assoc., 1978-83, assoc. mem., 1983; asst. prof. Cornell U. Grad. Sch., N.Y.C., 1980-83; assoc. prof. Ind. U. Sch. Medicine, Indpls., 1983-86, prof. medicine, microbiology and immunology, 1986—; sci. dir. Walther Oncology Ctr., Indpls., 1988—; mem. hematology II study sect. NIH, Bethesda, Md., 1981-86; adv. com. NHLBI, NIH, Bethesda, 1991—. Assoc. editor Exptl. Hematology, 1981-90, Jour. Immunology, 1987-92; mem. editorial bd. Blood, 1983-87, Jour. Exptl. Medicine, 1992; contbr. articles to profl. jours. Mem. ednl. com. Leukemia Soc. Am., Indpls., 1983—, nat. study sect., N.Y., 1991—. Recipient Merit award Nat. Cancer Inst., 1987—, Scholar award Leukemia Soc. Am. Mem. AAAS, N.Y. Acad. Scis., Soc. for Leukocyte Biology, Am. Assn. Cancer Rsch., Am. Assn. Immunologists, Internat. Soc. Exptl. Hematology (pres. 1990-91), Am. Soc. Hematology. Avocations: competitive Olympic-style weightlifting, running. Home: 1210 Chessington Rd Indianapolis IN 46260-1630 Office: Ind U Sch Medicine 975 W Walnut St # 501ib Indianapolis IN 46202-5121

BROYLES, MICHAEL LEE, geophysics and physics educator; b. Corpus Christi, Tex., Apr. 3, 1942; s. Ned Lee and Marion (Richardson) B.; m. Laura Ruth Ferguson, July 30, 1983; 1 child, William Matthew. BA in Phys. Sci., San Francisco State U., 1965; MST in Physics, U. Wis., Superior, 1972; MS in Geophysics, U. Hawaii, 1977. Cert. sec. sch. tchr. Tchr. sci. and math. Upper Lake and Sonoma (Calif.) Pub. Schs., 1966-72; geophys. rschr. Hawaii Inst. Geophysics, Honolulu, 1973-79; rsch. geophysicist Amoco Prodn. Co., Tulsa, 1979-80; exploration geophysicist Mobil Oil Corp., Dallas, 1980-86; prof., chmn. dept. physics and astronomy Collin County Community Coll., Plano, Tex., 1986—. Editorial bd. UFO Phenomena Mag., Bologna, Italy, 1976-79; Contbr. Sci. Ency., 1990-92; contbr. articles to profl. jours. Grantee NSF, 1991—. Mem. Am. Assn. Physics Tchrs., Tex. Jr. Coll. Tchrs. Assn., Mutual UFO Network (state dir. Hawaii 1975-79, state sect. dir. Dallas County 1987-92). Methodist. Achievements include geothermal research and developing teaching methodologies for elementary particle physics. Office: Collin County Community Col 2800 E Spring Creek Pkwy Plano TX 75074

BRUCE, ROBERT DOUGLAS, acoustics, noise and vibration control consultant; b. Livingston, Tex., Jan. 13, 1941; s. Vivian Eugene and Edna Lee (St. Clair) B.; m. Lydia Marcelyn Meynig, May 31, 1963; children: Robert Douglas Jr., James Elliott. BSEE, Lamar U., 1963; SMEE, EE, MIT, 1966. Registered profl. engr., Tex. Assoc. engr. Mobil Oil Co., Beaumont, Tex., 1963-64; teaching asst. MIT, Cambridge, 1964-66; cons., dir. Bolt Beranek and Newman Inc., Cambridge, 1966-81; prin. cons. Hoover Keith & Bruce Inc., Houston, 1981-87, Collaboration in Sci. and Tech. Inc., Houston, 1987—; cons. to fed. agys., 1970—. Contbr. 9 chpts. to books; contbr. over 50 technical papers and reports; lectr. in field. Recipient telephone counseling award 1st Bapt. Ch., Houston, 1988. Fellow Acoustical Soc. Am. (chmn. Boston chpt. 1970, Houston chpt. 1980); mem. Inst. Noise Control Engring. (cert., bd. dirs. 1982-84, 86-88, pres. 1986), MIT Enterprise Forum Tex. (bd. dirs. 1984—). Home: 5019 Chantry Houston TX 77084-2307 Office: Collaboration Sci and Tech Inc 15835 Park Ten Pl Ste 105 Houston TX 77084-5131

BRUCK, JAMES ALVIN, telecommunications professional; b. Bedford, Pa., Jan. 19, 1947; s. Herbert Alvin and Julia Christine (Lockard) B.; m. Patricia Marie McDermott, Oct. 19, 1968 (div. Nov. 1983); children: Jeffrey, Jonathan, Gabrielle. AS in Bus. Adminstrn., Goldey Beacom Coll., 1985; postgrad., Wilkes U., 1990-93. FCC radiotelephone lic. gen. class. Maintenance engr. Cablevision of Del., Wilmington, 1969; comms. technician AT&T, Salem, N.J., 1969-70, Diamond State Telephone, Wilmington, 1970-84; comms. technician AT&T, Hawley, Pa., 1984-87, instr. Satellite Tng. Ctr., 1987-90, supr. Satellite Control Facility, 1990—. Author: (tng. manuals) Satellite Technology, 1987, Vitalink Earth Stations, 1987, STS-SCAMP, 1988, Globe McKay Earth Station-Manilla, 1989, JDI Earth Station-Montgomery, 1988. Staff sgt. USAF, 1964-68. Mem. Nat. Assn. Bus. and Ednl. Radio, NE Coun. Telephone Pioneers (publicity chmn. 1988—), Chpt. 65 Telephone Pioneers (publicity chmn. 1991—). Democrat. Lutheran. Office: AT&T RD # 1 Box 672 Hawley PA 18428

BRUENING, ROBERT JOHN, physicist, researcher; b. Balt., May 20, 1939; s. John S. Jr. and Bessie C. (Miller) B.; m. Elizabeth M. Fiskerbeck. BA, Johns Hopkins U., 1960. Rsch. physicist Nat. Bur. Standards (now Nat. Inst. Standards and Tech.), Gaithersburg, Md., 1960—, 1988—; mem. Lamp Testing Egnrs. Conf., Cortland, N.Y., 1978—. Contbr. articles to profl. publs. Vice-pres. White Flint Park Garrett Park Estates Civic Assn., Kensington, Md., 1974-80; corp. trustee Augustana Luth. Ch., Washington, 1980-82. Mem. Coun. Optical Radiation Measurements, Optical Soc. Am. (all offices nat. capital sect. 1963-87). Office: Nat Inst Standards/Tech Rm B208 Physics Bldg Gaithersburg MD 20899

BRUES, ALICE MOSSIE, physical anthropologist, educator; b. Boston, Oct. 9, 1913; d. Charles Thomas and Beirne (Barrett) B. A.B., Bryn Mawr Coll., 1933; Ph.D., Radcliffe Coll., 1940. Faculty U. Okla. Sch. Medicine, 1946-65, prof., 1960-65; vis. prof. anthropology U. Colo., Boulder, 1965-66; prof. U. Colo., 1966—, chmn. dept. anthropology, 1969-71. Asso. editor: Am. Jour. Phys. Anthropology, 1962-66; Author: People and Races, 1977, contbr. articles to profl. jours. Mem. Am. Assn. Phys. Anthropologists (v.p. 1966-68, pres. 1971-73), Soc. Study Evolution, Am. Acad. Forensic Scis., Soc. Naturalists, Sigma Xi. Home: 4325 Prado Dr Boulder CO 80303-9629

BRUHN, HJALMAR DIEHL, retired agricultural engineer, educator; b. near Spring Green, Wis., Aug. 5, 1907; s. Aksel Theodor and Emma Bertha (Diehl) B.; m. Janet Helen Weber, Aug. 7, 1938; 1 child, Janet Margaret Bruhn Jeffcott. BS in Agrl. Engring., U. Wis., 1931, BSME, 1933; MSME, MIT, 1937. Registered profl. engr., Wis. Mem. faculty dept. agrl. engring. U. Wis., Madison, 1933—, chmn. dept., 1962-66, prof. emeritus, 1978—; cons. H.D. Bruhn and Assocs., Madison, 1978—; external examiner U. Ibadan, Nigeria, Africa, 1976, 77. Contbr. over 150 articles to sci. jours. Past bd. dirs. Highlands-Mendota Beach Sch., Madison Twp.; pres., bd. dirs. Middleton (Wis.) Sportsmen's Club, 1962—. Fellow Am. Soc. Agrl. Engrs. (McCormick Case Gold medal 1992, Engr. of Yr. Wis. sect. 1971, Outstanding Paper award 1955, 74); mem. AAAS, NSPE, AAUP, Am. Forage and Grassland Coun., Wis. Soc. Profl. Engrs., Wis. Acad. Scis., Arts and Letters,m N.Y. Acad. Sci., Soc. Green Vegetation Rsch., others. Achievements include development of machinery for pelletting alfalfa, mechanical cherry picking, irrigation and well jetting procedures, mechanical tree-acreage planting, plant protein extraction to feed the world's malnourished. Office: U Wis 460 Henry Mall Madison WI 53706

BRUMBACK, GARY BRUCE, industrial and organizational psychologist; b. New Castle, Ind., July 23, 1935; s. Donald Clair and Doris Lydia (Utterberg) B.; m. Doris Anne Ast, June 15, 1958; children: Babette Anne, Lyndia Claire. BA, Ind. U., 1958; MA, Ohio State U., 1960, PhD, 1963. Rsch. asst. Ohio State U., Columbus, 1958-62; rsch. scientist N.Am. Aviation, Columbus, 1962-63; rsch. psychologist U.S. Dept. HEW, Washington, 1964-73; sr. rsch. scientist Am. Inst. for Rsch., Washington, 1973-79; psychologist pers. mgr. U.S. Dept. Health and Human Svcs., Washington, 1979—. Editorial bd. Pub. Pers. Mgmt. Jour., 1986, 91, 93; contbr. articles to profl. jours., chpts. to books. Sec. Banana Beach Assn., Ocean City, Md., 1978-84. Fellow APA, Am. Psychol. Soc. (charter); mem. Am. Soc. Pub. Adminstrn., Acad. Mgmt., Internat. Pers. Mgmt. Assn., Sigma Xi, Phi Beta Kappa. Achievements include creation of MBR, a model approach for managing performance-showcased at request of government around the country; developed model personnel procedures for nine occupations in the government. Home: 724 Wilson Ave Rockville MD 20850

BRUMBAUGH, PHILIP S., minerals manager, quality control consultant; b. St. Louis, Nov. 14, 1932; s. Richard I. and Grace (Lischer) B.; m. Bettina A. Viviano, Feb. 25, 1978. AB, Washington U., St. Louis, 1954, PhD, 1963. Cert. quality engr. Ops. analyst Humble Oil and Refining Co., Houston, 1961-63; indsl. engr. Falstaff Brewing Co., St. Louis, 1963; asst. prof. Washington U., St. Louis, 1964-70; assoc. prof. U. Mo., St. Louis, 1970-74; pres. Quality Assurance, Inc., St. Louis, 1974-85; quality control cons. pvt. practice, St. Louis, 1985—; trustee Tex.-St. Louis Land Trust, St. Louis, 1982—. Contbr. chpts. to books: Mech. Engrs. Handbook, 1992, Handbook Indsl. Engring., 1992; contbr. articles to various tech. jours. Mem. adv. com. St. Louis C.C., Florissant, Mo., 1975—; bd. dirs. Lighthouse for the Blind, St. Louis, 1985—. Mem. Nat. Assn. Royalty Owners (bd. dirs., v.p. 1987—, Disting. Svc. award 1989), Am. Soc. for Quality Control (chmn. St. Louis sect. 1977-78, Vezeau award 1979), Inst. Indsl. Engrs. (pres. St. Louis chpt. 1973-74), Tex. Ind. Producers and Royalty Owners, Nev. Petroleum Soc. Office: Tex St Louis Land Trust PO Box 11499 Saint Louis MO 63105

BRUMER, PAUL WILLIAM, chemical physicist, educator; b. Bklyn., June 8, 1945; s. Abraham and Barbara (Feldman) B.; m. Abbey Pohrille, Aug. 18, 1968; children: Debbie, Jeremy, Eric, Sharon. BS in Chemistry with honors, Bklyn. Coll., 1966; PhD in Chem. Physics, Harvard U., 1972. Postdoctoral fellow dept. chem. physics Weizmann Inst. Sci., 1972-73; postdoctoral fellow Harvard Coll. Obs., 1973-74; lectr. dept. astronomy Harvard U., 1974-75; from asst. to assoc. prof. dept. chemistry U. Toronto, Ont., Can., 1975-83, prof. chem. physics theory group, 1983—; project leader Ont. Laser and Lightwave Rsch. Ctr.; Hudnall Disting. lectr. U. Chgo., 1983; Michael vis. prof. Weizmann Inst., 1985, 88, Einstein vis. prof., 1992. Recipient Noranda award Chem. Inst. Can., 1985, CIC Palladium medal, 1993, Chem. Inst. Can. medal, 1993; Woodrow Wilson fellow, 1966-67, predoctoral fellow NSF, 1966-71, inst. fellow Weizmann Inst. Sci., 1972-73, fellow Alfred P. Sloan Found., 1977-81, Chem. Inst. Can., 1993—, Killam Rsch. fellow, 1981-83. Mem. Phi Beta Kappa, Sigma Xi. Achievements include research on quantum interference and laser control of atomic and molecular processes, non-linear classical and quantum mechanics, conservative classical and quantum chaos, classical/quantum correspondence, intramolecular dynamics, dynamical stability and statistical behavior. Office: U Toronto Chem Physics Theory Group, 80 St George St, Toronto, ON Canada M5S 1A1

BRUNALE, VITO JOHN, aerospace engineer; b. Mt. Vernon, N.Y., July 2, 1925; s. Donato and Antoinette (Wool) B.; m. Joan Florence Montuori, Apr. 23, 1949; 1 child, Stephen. AAS, Stewart Aero. Inst., 1948; BSAE, Tri-State U., 1958; MSME, U. Bridgeport, 1966; DSc, Nev. Inst. Tech., 1973; PhD (hon.), Internat. U., Spain, 1987; DSc, Pacific Western U., 1984. Rsch. engr. Norden Labs., White Plains, N.Y., 1948-55; instr. Tri-State U., Angola, Ind., 1955-58; engring. cons. Norden Div. United Aircraft, Norwalk, Conn., 1958-67; chief engring. cons. Singer-Kearfott Corp., Pleasantville, N.Y., 1967-73; chief engr. Diagnostic/Retrieval Systems, Mt. Vernon, N.Y., 1973-76; tech. problem mgr. Fairchild Republic Co., Farmingdale, N.Y., 1977-87; sr. tech. expert Sikorsky Aircraft, 1987—; cons. in field; engring. tutor to coll. students; v.p. Lithoway, Inc., 1969-73; lectr. in field; tech. guest speaker numerous tech. soc. meetings.; participant engring. exchange program, USSR, People's Republic China. Contbr. articles to profl. jours. including Product Engring., Aviation Week, Environ. Scis. Participant U.S.A. Citizen Amb. Program. Served with USAAF, 1943-45. Decorated Purple Heart (3), Air medals, D.F.C. Tri-State U. teaching fellow, 1955-58; NSF grantee; recipient Aircraft Design award, 1948, Inst. Aero. Sci. Lecture award, 1948, Norden Rsch. award, 1963, Cost Reduction award, 1965, Singer Engring. award, 1970, 72, Fairchild outstanding achievement award, Fairchild award of excellence, 1984, Am. Biographical Inst. and Research Assn. Outstanding Performance award, 1989, Aircraft Recognition award, 1986, citation N.Y. State Assembly, 1988, Conspicuous Service Cross N.Y. State, 1988, Prisoner of War medal, 1988, others. Mem. AIAA, U.S. Naval Inst., Air Force Assn., Am. Ordnance Assn., Inst. Environ. Sci., Nat. Space Inst., K.C., VFW, DAV, Newman Club, Internat. Students Assn., Internat. Platform Assn., World Inst. of Achievement. Roman Catholic. Achievements include patent (with others) for Bearing Spin Rail Test; development of method of discriminate displacement for equilibrium of structures, of the position point vibration isolation technique, of the vapress vibration system, of advanced techniques for structural and vibration analyses, of the Doppler-Inertial-Loran system, of state of the art mathematical and structural analyses techniques, of Mars Doppler Lander system, computer time studies, anti-corrosion methods; resolution of 140 technical problems on the Fairchild A-10 aircraft, of more than 30 technical problems with the Saab-Fairchild 340; solution of Grumman A-6A radar tracking problem in Vietnam; elimination of technical problems on LEM inertial guidance; rsch. in mfg. productivity, co-planer structural analyses. Home: 459 Bronxville Rd Bronxville NY 10708-1102 Office: Main St Bridgeport CT 06604-5706

BRUNDAGE, GERTRUDE BARNES, pediatrician; b. Neptune, N.J., May 13, 1941; d. John Holt and Mary Downey (Chatham) B. BS in Chemistry, Marietta Coll., 1964; MD, Jefferson Med. Coll., 1971. Diplomate Am. Bd. Pediatrics. Chemist Lederle Labs., Pearl River, N.Y., 1964-67; intern pediatrics Harrisburg (Pa.) Polyclinic Hosp., 1971-72; resident pediatrics Wilmington (Del.) Med. cTr., 1972-74; pediatrician Orange, N.J., 1974—; chief dept. pediatrics Hosp. Ctr. at Orange, 1990—. Mem. AMA, N.J. Med. Women's Assn., Am. Med. Women's Assn., Essex County Med. Soc., Med. Soc. N.J., Alpha Gamma Delta, 1st Presbyn. Ch. (elder, trustee 1982-87, 89-92). Republican. Presbyterian. Avocations: choral singing, needlework, gardening. Home: 18 Farrington St West Caldwell NJ 07006-7716 Office: Gertrude B Brundage MD 572 Park Ave East Orange NJ 07017-1998

BRUNEAU, ANGUS A., engineer; b. Toronto, ON, Can., Dec. 12, 1935; s. Earl Angus and Lois McClung (Gordon) B. BA, U. Toronto, 1958; DIC, Imp. Coll. London, 1962; PhD, U. London, 1962. Chmn., pres., CEO Fortis, Inc., 1988—; chmn. Newfoundland Power, 1990—, Air Nova, 1988—, Unitel-Newfoundland, Ltd., Fortis Trust, Fortis Properties; dir. Maritime Electric Co. Ltd., 1990—, The SNC Group, Inc., 1991—; lectr. in Math. U. Waterloo, 1958-59; lectr. in Materials Sci. Queen Mary's Coll. and Imp. Coll. London, 1960-62; lectr. and research materials egring. Material Sci. Group Leader U. Waterloo, 1962-68, dir. gen. engring.. 1966-68. Recipient Julian C. Smith medal Engring. Inst. Can., 1992. Fellow Canadian Acad. of Engring. (recipient Gold medal); mem. Order of Can. (officer), Assn. Profl. Engrs. Nfld. (recipient award of merit). *

BRUNELLE, EUGENE JOHN, JR., mechanical engineering educator; b. Montpelier, Vt., Mar. 17, 1932; s. Eugene John Sr. Brunelle and Maxine Gertrude (Chatfield) Corson; m. Raylene Julia Clark, June 12, 1955 (div. Sept. 1967); children: Steven, Alison, Holly. BS in Engring., U. Mich., 1953, MS in Engring., 1955; ScD, MIT, 1962. Asst. prof. Princeton (N.J.) U., 1960-64; assoc. prof. Rensselaer Poly. Inst., Troy, N.Y., 1964—; vis. prof. Air Force Inst. Tech., Dayton, Ohio, 1983-85; cons. aerospace firms, 1958—. Contbr. chpt. Principles of Aeroelasticity, 1962; contbr. over 70 papers to profl. jours., also reviewer. Patron Challenger Space Ctrs., 1988—. Fellow Boeing Aerospace Co., 1959; grantee NSF, 1967, NASA, 1975-83. Mem. ASME, AIAA, Am. Acad. Mechs. (founding), Soc. Indsl. and Applied Maths., Am. Math. Soc. Achievements include discovery of similarity rules for composite structures, of affine transformations for all the field equations of classical physics and mechs. Home: 66 Thimbleberry Rd Ballston Spa NY 12020 Office: Rensselaer Poly Inst 4006 Jonsson Engring Ctr Troy NY 12180-3590

BRUNER, JANET M., neuropathologist; b. East Liverpool, Ohio, Oct. 2, 1949; d. Russell Edward and Hazel Isabel (Pride) Roof; m. Charles Thomas Bruner, July 8, 1972. BS in Pharmacy, U. Toledo, 1972, MS in Pharm. Sci., 1974; MD, Med. Coll. of Ohio, 1979. Resident in pathology Med. Coll. of Ohio, Toledo, 1979-82; fellow neuropathology Baylor Coll. Medicine, Houston, 1982-84; chief neuropathologist M.D. Anderson Cancer Ctr., Houston, 1984—; test com. neuropathology Am. Bd. Pathology, Tampa, Fla., 1988—. Contbr. articles to profl. jours. Grantee NIH, 1992—. Mem. AMA, Houston Soc. of Clin. Pathologists (pres. 1990-91), Am. Assn. of Neuropathologists, Am. Soc. of Clin. Pathologists, U.S./Can. Acad. of Pathology, Tex. Med. Assn. Achievements include research on brain tumor biologic behavior, glial cell transformation, tumor suppressor genes, cell proliferation kinetics, and gene therapy of brain tumors. Home: 5115 S Braeswood Houston TX 77096 Office: MD Anderson Cancer Ctr Pathology 1515 Holcombe Houston TX 77030

BRUNETTO, FRANK, electrical engineer; b. Graniti, Messina, Italy, July 19, 1921; came to U.S., 1930; s. Rosario and Maria (Mannino) B.; m. Bella Forman, Apr. 13, 1942; children: Russ L., Theodore. BEE, Bklyn. Poly. Inst., 1953. Electronics engr. U.S. Naval Applied Sci. Lab., Bklyn., 1953-69; revenue officer IRS, N.Y.C., 1971-72; gen. engr. HUD, N.Y.C., 1972—; participant testing of hydrogen and atom bombs, off Johnston Island, Pacific Ocean, 1962. Author: Reforming Selenium Rectifiers, 1966, Introducing the Gate Turn-Off Switch, 1963. With U.S. Army, 1942-45, ETO. Mem. AIAA, Sci. Rsch. Soc. Roman Catholic. Achievements include patents on dynamic testing of silicon control rectifiers and of gate turn-off switches. Home: 4817 Avenue L Brooklyn NY 11234-3105 Office: US HUD 26 Federal Pla New York NY 10278

BRUNGARD, MARTIN ALAN, civil engineer. BCE, U. Fla., 1984, MCE, 1985. Registered profl. engr., Fla. Staff engr. Ardaman & Assocs., Orlando, Fla., 1983; geotech. engr. CH2M Hill, Gainesville, Fla., 1986—. Contbr. article to Geotech. Fabrics Report. Mem. ASCE (assoc.), NSPE, Fla. Engring. Soc. (mathcounts coord. 1991-92). Office: CH2M Hill 7201 NW 11 Pl POB 147009 Gainesville FL 32614

BRÜNGER, AXEL THOMAS, biophysicist, researcher, educator; b. Leipzig, Germany, Nov. 25, 1956; came to U.S., 1982; s. Hans and Hildegard (Müller) B. Diploma, U. Hamburg (Germany), 1980; PhD, Tech. U. Munich, 1982. Postdoctoral fellow Max-Planck Inst., Martinsried, Germany, 1984; rsch. assoc. Harvard U., Cambridge, Mass., 1982-83, 85-87; asst. investigator Howard Hughes Med. Inst., New Haven, 1987-92, assoc. investigator, 1992—; asst. prof. Yale U., New Haven, 1987-91, assoc. prof., 1991-93, prof., 1993—. NATO postdoctoral fellow Deutscher Akademischer Austauschdienst, Bonn, Germany, 1982-83. Mem. AAAS, Am. Crystallographic Assn., Am. Chem. Soc. Achievements include studies of protein structure and function, developments in macromolecular x-ray crystallography and solution NMR spectroscopy. Office: Yale U 266 Whitney Ave New Haven CT 06511-3748

BRUNGRABER, LOUIS EDWARD, retired mechanical engineer; b. Stanton, Mich., July 30, 1926; s. Louis Rudolph and Beatrice Imogene (Crawford) B.; m. Irene Phylis Meyer, June 25, 1949; children: Sue Ann, Carol Lynn, Laurie Kay, Kurt Edward. BSME, Iowa State U., 1948. Sales engr. trainee Bigelow-Liptak Corp., Detroit, 1948-52; sales engr. Bigelow-Liptak Corp., N.Y.C. and Phila., 1952-58; refractory product engr. Gen. Refractories Co., Chgo., 1958-61; refractory sales engr. Gen. Refractories Co., Cleve., 1961-62; KA-Weld sales engr. Bloom Engring. Co., Chgo., 1962-67; mgr. rolling mill products Bloom Engring. Co., Pitts., 1967-92. Lt. (j.g.) USNR, 1944-48. Republican. Presbyterian. Achievements include patents in field. Home: 390 Parkway Dr Pittsburgh PA 15228

BRUNK, SAMUEL FREDERICK, oncologist; b. Harrisonburg, Va., Dec. 21, 1932; s. Harry Anthony and Lena Gertrude (Burkholder) B.; m. Mary Priscilla Bauman, June 24, 1976; children: Samuel, Jill, Geoffrey, Heather, Kirsten, Peter, Christopher, Andrew, Paul, Barbara. BS, Ea. Mennonite Coll., 1955; MD, U. Va., 1959; MS in Pharmacology, U. Iowa, 1967. Diplomate Am. Bd. Internal Medicine, Am. Bd. Internal Medicine in Med. Oncology. Straight med. intern U. Va., Charlottesville, 1959-60; resident in chest diseases Blue Ridge Sanatorium, Charlottesville, 1960-61; resident in internal medicine U. Iowa, Iowa City, 1962-64, fellow in clin. pharmacology (oncology), 1964-65, 66-67, asst. prof. internal medicine, 1967-72; assoc. prof. internal medicine, 1972-76; fellow in medicine (oncology) Johns Hopkins U., Balt., 1965-66; vis. physician bone marrow transplantation unit Fred Hutchinson Cancer Treatment Ctr., U. Wash., Seattle, 1975; practice medicine specializing in med. oncology Des Moines, 1976—; attending physician Iowa Luth. Hosp., 1976—, Iowa Meth. Med. Ctr., 1976—, Charter Hosp., 1976—, Mercy Hosp. Med. Ctr., 1976—; chief of staff Iowa Luth. Hosp., 1990, chmn. dept. internal medicine, 1988; cons. physician Des Moines Gen. Osteo. Hosp., 1976—; prin. investigator Iowa Oncology Rsch. Assn. in assn. with N. Cen. Cancer Treatment Group and Ea. Coop. Oncology Group, 1978-83; prin. investigator Iowa Oncology Rsch. Assn. Community Clin. Oncology Program, 1983-84. Contbr. articles to profl. jours. Bd. dirs. Iowa div. Am. Cancer Soc., 1971-89, Johnson County chpt., 1968-72. Mosby scholar, U. Va., 1959. Fellow ACP, Am. Coll. Clin. Pharmacology; mem. AMA, Iowa Med. Soc., Polk County Med. Soc., Iowa Thoracic Soc., Am. Thoracic Soc., Iowa Clin. Med. Soc., Am. Fedn. Clin. Rsch., Iowa Heart Assn., Am. Assn. Cancer Edn., Am. Soc. Hematology, Am. Soc. Clin. Pharmacology and Therapeutics, Cen. Soc. Clin. Rsch., Raven Soc., Alpha Omega Alpha. Roman Catholic. Home: 3940 Grand Ave West Des Moines IA 50265-5730

BRUNO, MICHAEL STEPHEN, ocean engineering educator, researcher; b. Nutley, N.J., Apr. 16, 1958; s. Frank Joseph and Annie Marie (Golden) B.; m. Lise Simard, Aug. 6, 1988. BS, N.J. Inst. Tech., 1980; MS, U. Calif., Berkeley, 1981; PhD, MIT, 1986. Registered profl. engr., N.J. Prin. engr. N.J. Dept. Environ. Protection, Toms River, 1981-82; rsch. asst. MIT, Cambridge, 1982-86; asst. prof. N.J. Inst. Tech., Newark, 1986-89; assoc. prof., dir. Davidson Lab. Stevens Inst. Tech., Hoboken, N.J., 1989—; young investigator Office of Naval Rsch., Washington, 1991. Editor-in-chief Jour. Marine Environ. Engring., 1991—; contbr. articles to profl. jours. Mem. Planning Bd., West Caldwell, N.J., 1990—. Mem. ASCE (Outstanding Svc. award 1987), ASME, Am. Geophys. Union, Pan Am. Fedn. Ocean Engrs. (sec.-gen. 1990—), Soc. Naval Architects and Marine Engrs., Tau Beta Pi. Office: Stevens Institute of Technology Castle Point Sta Hoboken NJ 07030-5907

BRUNS, BILLY LEE, consulting electrical engineer; b. St. Louis, Nov. 21, 1925; s. Henry Lee and Violet Jean (Williams) B.; B.A., Washington U., St. Louis, 1949, postgrad. Sch. Engring., 1959-62; EE, ICS, Scranton, Pa., 1954; m. Lillian Colleen Mobley, Sept. 6, 1947; children—Holly Rene, Kerry Alan, Barry Lee, Terrence William. Supt., engr., estimator Schneider Electric Co., St. Louis, 1950-54, Ledbetter Electric Co., 1954-57; tchr. indsl. electricity St. Louis Bd. Edn., 1957-71; pres. B.L. Bruns & Assos., cons. engrs., St. Louis, 1963-72; v.p., chief engr. Hosp. Bldg. & Equipment Co., St. Louis, 1972-76; pres., prin. B.L. Bruns & Assos. cons. engrs., St. Louis, 1976—; tchr. elec. engring. U. Mo. St. Louis extension, 1975-76. Mem. Mo. Adv. Council on Vocat. Edn., 1969-76, chmn., 1975-76; leader Explorer post Boy Scouts Am., 1950-57. Served with AUS, 1944-46: PTO, Okinawa. Decorated Purple Heart. Registered profl. engr., Mo., Ill., Wash., Fla., La., Wis., Minn., N.Y., N.C., Iowa, Pa., Miss., Ind., Ala., Ga., Va., R.I. Mem. Nat. Soc. Profl. Engrs., Mo.' Soc. Profl. Engrs., Profl. Engrs. in Pvt. Practice, Am. Soc. Heating, Refrigeration and Air Conditioning Engrs., Illuminating Engrs. Soc., Am. Mgmt. Assn., Nat. Fire Protection Assn. (health care div., archtl./engr. div.), Masons. Baptist. Tech. editor The National Electrical Code and Blueprint Reading, Am. Tech. Soc., 1959-65. Home: 1243 Hobson Dr Saint Louis MO 63135-1422 Office: 10 Adams St Ste 111 Saint Louis MO 63135-2751

BRUNS, MICHAEL WILLI ERICH, virologist, veterinarian; b. Bückeburg, Fed. Republic of Germany, Nov. 15, 1945; s. Heinz and Kathrin (Niere) B.; m. Rosemarie Goltz, June 14, 1974; children: Evelyn, Alexander. Diploma for veterinarian, Justus-Liebig-U., Giessen, Fed. Republic Germany, 1973, grad., 1976, habilitation, 1990. Sci. co-worker Bundesforschungsanstalt Viruskrankheiten der Tiere, Tübingen, Fed. Republic of Germany, 1974-77; sci. rsch. asst. Heinrich-Pette-Inst. Exptl. Virology & Immunology, Hamburg, Fed. Republic of Germany, 1978—; guest researcher Ctr. Biotechnology, Imperial Coll. of Sci. and Tech., London, 1987-88. Contbr. articles to profl. jours. Grantee Deutsche Forschungsgemeinschaft, 1985, 89, 91. Avocations: sports, reading, painting, travel, photography. Home: Quedlinburger Weg 15, Hamburg 22455, Germany Office: Heinrich-Pette-Inst, Martinistr 52, Hamburg 2025, Germany

BRUNSON, JOEL GARRETT, pathologist, educator; b. Greenville, S.C., Apr. 22, 1923; s. James Edwin and Leila (Ballenger) B. MD, SUNY, Buffalo, 1950. Diplomate Am. Bd. Med. Examiners, Am. Bd. Pathology. Internship U. Ala. Teaching Hosp., Birmingham, 1950-51; residency U. Minn. Hosp., Mpls., 1951-53; asst. prof. U. Minn. Med. Sch., Mpls., 1955-59; prof., chmn. dept. U. Miss. Med. Ctr., Jackson, 1959-77, Morehouse Med. Sch., Atlanta, 1978-80, U. Ife, Nigeria, 1980-81; prof. pathology St. George's (Grenada) U. Med. Sch., 1982-89, prof., chmn. dept., 1989—; asst. prof. pathology U. Minn., 1953-55; vis. prof. Am. U. Beirut, 1977-78; cons. rsch. grants award com. NIH, 1963-67, chmn. com., 1965-67, mem. internat. fellowship rev. panel, 1980; mem. instl. rsch. programs VA, 1965-70. Author, editor: (text) Concepts of Disease, 1971; mem. editorial bd. Am. Jour. Pathology, 1966-80; contbr. numerous rsch. articles to sci. publs. Am. Cancer Soc. fellow U. Minn., 1953-55, NIH sr. rsch. fellow, 1955-57. Mem. AAAS, AMA, Coll. Am. Pathologists, Am. Assn. Pathology Chmn. (v.p. 19709-71, pres. 1972-72), Am. Soc. for Exptl. Pathology, Internat. Acad. Pathology, Am. Assn. Pathologists (editor symposia series 1973-76), others. Avocations: photography, swimming. Home: 4152 Markin Dr W Jacksonville FL 32211-1526 Office: St Georges U Med Sch, PO Box 7, Saint George's Grenada

BRUNSTROM, GERALD RAY, engineering executive, consultant; b. Hoquiam, Wash., Apr. 21, 1929; s. Alfred Ferdinand and Ingrid Helen (Ranta) B.; m. Betty Grace Elgin, Sept. 8, 1951 (div. 1980); children: Dianne Marie Brunstrom Denney, Janice Elaine, Gerald Alan, Eric Elgin; m. Mary Elizabeth Reid, Sept. 12, 1980; 1 child, Michael Reid. BS in Civil Engring., Wash. State U., 1951; MS in Civil Engring., U. Wis., 1952. Ptnr., prin. Tracey & Brunstrom Cos., Seattle, 1955-79; pres., prin. Olympic Engring. Corp., Seattle, 1971-79; v.p. constrn. St. Joe Internat. Corp., N.Y.C., 1979-85; dir. engring. St. Joe Minerals Corp., St. Louis, 1985-88; exec. v.p. React Environ. Engrs., St. Louis, 1988-89; cons. St. Louis, 1989-90; dir. engring. svcs. Big River Minerals Corp., St. Louis, 1990—; v.p. Big River Relcamation Corp., St. Louis, 1992—. Trustee Highline C.C., Federal Way, Wash., 1976-78; mem. Lafayette Sq. Restoration Com., St. Louis, 1985—. 1st lt. USAF, 1952-54. Mem. ASCE, NSPE, Air and Waste Mgmt. Assn., Engrs. Club St. Louis. Presbyterian. Achievements include direction of engineering and construction of process facilities and infrastructure for new remote area mines in Australia, Chile, Brazil and Canada; leader in corporation's recent entrance into field of environmental restoration and remediation and recycling heavy metals utilizing innovative technologies. Office: Big River Minerals Corp 150 N Meramec Ave Ste 400 Saint Louis MO 63105-3753

BRUSA, DOUGLAS PETER, purchasing executive; b. Oceanside, N.Y., Sept. 28, 1958; m. Nancy Anne Taratko; 1 child, Daniel Harry. BS, Adelphi U., 1980; MPA, Columbia U., 1991. Cert. purchasing mgr. Park ranger U.S. Dept. Interior, N.Y.C., 1980-83; asst. dir. Metro Area Ednl. Svcs., N.Y.C., 1983-85; buyer Joint Purchasing Corp., N.Y.C., 1985-87; purchasing mgr. Lamont-Doherty Earth Obs. of Columbia U., Palisades, N.Y., 1987—; chmn. Purchasing Mgrs.'s Adv. Com., N.Y.C., 1989. Contbg. author: Purchasing Management, 1990, The International Service Quality Handbook, 1993. Mem. Nat. Assn. Purchasing Mgmt. (com. chair 1986-88), Nat. Assn. Ednl. Buyers (pres. 1991-92). Office: Lamont-Doherty Earth Obs Rt 9W Palisades NY 10964

BRUSCA, RICHARD C., zoologist, museum curator, researcher, educator; b. Los Angeles, Jan. 25, 1945; s. Finny John and Ellenora C. (McDonald) B.; m. Caren Irene Spencer, 1964 (div. 1971); m. Anna Mary Mackey, 1980 (div. 1987); children: Alec Matthew, Carlene Anne. BS, Calif. Poly. State U., 1967; MS, Calif. State U., 1970; PhD, U. Ariz., 1975. Curator and researcher Aquatic Insects Lab., Calif. State U., Los Angeles, 1969-70; resident dir. U. Ariz. and U. Sonora (Mex.) Cooperative Marine Lab., Sonora, 1970-72; prof. biology U. So. Calif., Los Angeles, 1975-86; head Invertebrate Zoology sect. Los Angeles County Mus. Natural Hist., 1984-87; Joshua L. Baily curator, chmn. dept. invertebrate zoology San Diego Natural History Mus., 1987—; dir. acad. programs Catalina Marine Sci. Ctr., U. So. Calif.; field researcher North, Cen. and South Americas, Polynesia, Australasia, Antarctica, Europe; bd. dirs. Orgn. for Tropical Studies; mem. panels NAS/NSF; chair adv. com. Smithsonian Instn., Systematics Agenda 2000. Author: Common Intertidal Invertebrates of the Gulf of California, 1980; co-author: A Naturalist's Seashore Guide, 1978, Invertebrates, 1990; contbr. over 70 sci. articles to profl. jours. Recipient U.S. Antarctic Svc. medal, 1965, numerous rsch. awards and grants NSF, Nat. Geog. Soc., Charles Lindberg Fund, also others. Mem. AAAS, NAS (panel), Crustacean Soc. (pres.), Soc. for Systematic Biology, Willi Hennig Soc., U. Edinburgh Biol. Study Group, S.Am. Explorers Club, Assn. Sea of Cortez Rschrs. (hon. life), Sigma Xi. Avocations: photography, Mesoamerican indigenous art and culture, Latin American politics. Office: Natural History Mus PO Box 1390 San Diego CA 92112-1390

BRUSCH, JOHN LYNCH, physician; b. Boston, Nov. 9, 1943; s. Charles and Margaret Agnes (Lynch) B.; m. Patricia Gahan, May 12, 1972; children: Amy Claire, Meaghan, Patrick. BS, Tufts U., 1965, MD, 1969. Intern New England Med. Ctr., Boston, 1969-71, resident in infectious disease, 1971-74; asst. chief medicine Brighton Meml. Hosp., Boston, 1974-76; pvt. practice physician Cambridge, Mass., 1976—; chief medicine Youville Hosp., Cambridge, 1991—. Assoc. editor Infectious Disease Practice, 1984—; contbr. articles to med. jours. Dir. North Cambridge Coop Bank, 1980—. With USPHS, 1974-79. Fellow ACP; mem. Am. Soc. Microbiology. Home: 52 Radcliffe Rd Belmont MA 02178 Office: Cambridge Hosp 1493 Cambridge St Cambridge MA 02139

BRUSEWITZ, GERALD HENRY, agricultural engineering educator, researcher; b. Green Bay, Wis., June 1, 1942; s. Henry Jackson and Wardeen Mae (Thiel) B.; m. Glenna Sue Williams, May 12, 1990; children: Kelly K., Nicole J. BS in Agr., U. Wis., 1964, BSME, 1965, MSAE, 1966; PhDAE, Mich. State U., 1969. Registered profl. engr., Okla. Rsch. asst. U. Wis., Madison, 1965-66; asst. prof. Okla. State U., Stillwater, 1969-75, assoc. prof., 1975-80, prof., 1980-92, interim dept. head, 1985; regents prof., 1992—; sabbatical leave U. Calif., Davis, 1979, Cornell U., 1988; vis. engr. solar energy dept. Kuwait Inst. for Sci. Rsch., 1980; cons. Cen. Machine & Tool Co., Enid, Okla., 1986, Clements Food Co., Oklahoma City, 1986-87, Omnidata Internat. of Logan, 1990. Recipient Dist. Svc. to Students award Alpha Epsilon, Outstanding Faculty award Halliburton, 1991. Fellow Am. Soc. Agrl. Engrs. (Paper award 1990); mem. Am. Assn. Cereal Chemists, Inst. Food Technologists, Alpha Zeta. Methodist. Avocations: racquetball, cycling, hiking. Office: Okla State U Agrl Engring Dept 216 Ag Hall Stillwater OK 74078

BRUSSAARD, GERRIT, telecommunications engineering educator; b. Zegveld, Utrecht, Netherlands, Oct. 5, 1942; s. Anne Pieter and Janna Willempje (Vermaas) B.; m. Pieternella J. de Vries, Feb. 24, 1967 (div. 1991); children: Anne Pieter, Gerrit J.H., Francois M. Ingenieur, Tech. U. Delft, Netherlands, 1966; Dr.Applied Sci., U. Catholic de Louvain, Belgium, 1985. Head section PTT Rsch. Lab., Leidschendam, Netherlands, 1966-74, European Space Tech. Ctr., Noordwijk, Netherlands, 1974-88; prof. telecommunication engring. Eindhoven U. Tech., 1988—; chmn. ITU-CCIR Working Party 5A, Geneva, 1990—, URSI Commn. F, Brussels, 1990-93; cons. European Space Agy., Noordwijk, 1988—. Contbr. articles to profl. jours. 1st lt. Royal Netherlands Air force, 1966-68. Mem. IEEE (sr.), Netherlands Royal Inst. Engrs., Rotary. Achievements include research on wave propagation through the atmosphere including theories on crosspolarization and radiative transfer of microwave emissions. Office: Eindhoven U Tech, PO Box 513, 5600MB Eindhoven The Netherlands

BRUSSEAU, MARK LEWIS, environmental educator, researcher; b. Pontiac, Mich., June 17, 1958. PhD, U. Fla., 1989. Asst. prof. U. Ariz., Tucson, 1989-93, assoc. prof., 1993—; vis. scientist on groundwater rsch.

dept. environ. engring, Tech. U. of Denmark, 1989; mem. groundwater contamination com. USDA, 1990—; mem. exploratory rsch. proposal rev. panel EPA, 1990—; mem. subsurface sci. program rsch. proposal rev. panel U.S. Dept. of Energy, 1990—; cons. EPA, Office of Sci. Rsch., USAF, Ariz. Dept. Environ. Quality. Assoc. editor Jour. Contaminant Hydrology, 1990—; contbr. over 40 articles to profl. jours. Rsch. fellow Air Force Office Sci., 1990, U.S. Dept. Energy, 1992; recipient Emil Truog award for Outstanding PhD, Soil Sci. Soc. Am., 1990, Young Scientist postdoctoral rsch. Am. Chem. Soc., 1990, Young Faculty award U.S. Dept. Energy, 1992, Young Investigator award NAS, 1993. Mem. Assn. Groundwater Scientists and Engrs., Am. Geophys. Union, Am. Chem. Soc., Internat. Assn. Hydrogeologists, Soil Sci. Soc. Am., Phi Kappa Phi, Sigma Xi. Achievements include design of mathematical models for transport of chemicals in heterogeneous porous media; research in effect of solvent, sorbent, and solute properties on rate-limited sorption and nonequilibrium transport of organic chemicals in porous media. Office: U Ariz Soil and Water Sci Dept 429 Shantz Tucson AZ 85721

BRUST, JOHN CALVIN MORRISON, neurology educator; b. Syracuse, N.Y., Aug. 20, 1936; s. John C.M. and Constance (Cook) B.; m. Mary Duncan, Oct. 23, 1965; chldren: Mary Duncan, Frederick Eliot Noyes, James Charles Morrison. AB, Harvard U., 1958; MD, Columbia U., 1962. Diplomate Am. Bd. Psychiatry and Neurology. Intern Presbytn. Hosp., N.Y.C., 1962-63, resident in neurology, 1966-69, attending in neurology, 1969—; attending in neurology Harlem Hosp. Ctr., N.Y.C., 1969-75; dir. dept. neurology Harleur Hosp. Ctr., N.Y.C., 1975—; prof. clin. neurology Columbia U., N.Y.C., 1975—. Author: Neurological Aspects of Substance Abuse, 1993; editor: Neurological Complications of Substance Abuse, 1993; contbr. over 140 articles to profl. jours. and textbooks. Lt. USNR, 1962-65. Fellow Am. Acad. Neurology; mem. Am. Neurol. Assn., Am. Clin. and Climatological Assn., Century Assn., N.Y. Practitioners Soc., Alpha Omega Alpha. Office: Harlem Hosp Ctr Dept Neurology 506 Lenox Ave New York NY 10037

BRY, PIERRE FRANÇOIS, engineering manager; b. Tarare, France, Aug. 12, 1950; s. Rene V. and Yolande Caroline (Bassetti) B.; m. Lesly Patricia Rose, July 1, 1972; children: Madeleine, Emilie Rose. Diplom Ingenieur, ETH, Zurich, 1974. Rsch. engr. Office Nat. d'Etudes et de Recherches Aerospatiales, Paris, 1975-81; supr. SNECMA, Villaroche, France, 1981-88, dep., head turbine dept., 1988-90; engring. mgr. SNECMA, Cin., 1990–; assoc. prof. ENSICA, Toulouse, France, 1986-90, U. Orsay, Paris, 1990–. Author: Engineering Turbulence, 1990, ICIDES III, 1991. Mem. Am. Soc. Mechanical Engrs., Am. Inst. Aeronautics and Astronautics. Achievements include design of low pressure turbines for highly successful CFM56-5A and -5C commercial engines, of turbines for M88 military engine. Office: SNECMA Inc PO Box 15667 MD Y7 Cincinnati OH 45215

BRYAN, HAYES RICHARD, aerospace engineer; b. Bozeman, Mont., Apr. 2, 1937; s. Hayes Glenn and Anna Edith (Simonson) B.; m. Patsy Jean Kinshella, Aug. 14, 1960; children: Ann Marie, Colleen Diane, Daniel Hayes. BS in Mech. Engring., Mont. State U., 1958; MS in Astronautics, USAF Inst. Tech., 1967; PhD, U. Tex., 1976. Registered profl. engr., Colo.; licensed airline transport pilot. Commd. 2d lt. USAF, 1958, advanced through grades to col., 1979, ret., 1983; mgr. tech. devel. Gen. Electric Mgmt. and Data Systems Ops., Springfield, Va., 1984-88, sr. test and evaluation engr., 1988-93. Dir. choir Black Forest Luth. Ch., Colorado Springs, 1974-79; dir. Black Forest Pine Beetle Assn., Colorado Springs, 1976-79; dir. folk choir St. Paul Luth. Ch., Stafford, Va., 1981-85; active New Life Community Ch. Decorated Legion of Merit. Mem. AIAA, Aircraft Owners and Pilots Assn., Seaplane Pilots Assn. Home: 1006 Plymouth Dr Stafford VA 22554 Office: Martin Marietta Mgmt Data System 8080 Grainger Ct Springfield VA 22153

BRYAN, HENRY COLLIER, private school mathematics educator; b. Atlanta, Apr. 10, 1941; s. Thomas Harper and Rubye (Collier) B. Student, Temple U., 1959-63, 64, 70; BEd, Cheyney U., 1962; postgrad., Va. Union U., 1965-66; MDiv, Ea. Bapt. Theol. Sem., 1968; postgrad., Howard Law Sch., 1962-63, U. Alaska, Juneau, 1990. Cert. math. tchr., Phila.; ordained to ministry Am. Bapt. Ch., 1968. Tchr. math. Masterman Demonstration Sch., Phila., 1968-71, Phila. High Sch. for Girls, 1971—; chaplain Alpha Phi Alpha Fraternity, Phila, 1968—. Charter mem. North br. Y's Men Assn., Phila., 1972—; bd. dirs. Cherry Hill (N.J.) Civic Assn., 1992—. Recipient Outstanding Young Men Am. award Wynnefield Presbyn. Ch., Phila., 1971. Mem. ASCD, NAACP (life), Assn. Tchrs. Math. Phila. (life), Nat. Coun. Tchrs. Math. (life), Phila. Fedn. Tchrs., Math. Assn. Am., Nat. Coun. Suprs. of Math., Pa. Coun. Suprs. of Math., Pa. Coun. Tchrs. Math., Nat. Sci. Tchrs. Assn. (life), Am. Bapt. Mins. Coun. (life), Phila. Health Users Group (life), Phi Delta Kappa (life), Alpha Phi Alpha (life). Avocations: computers, electronics, sports, chess, world travel. Office: Phila High Sch for Girls Broad St and Olney Ave Philadelphia PA 19141

BRYAN, JOHN LELAND, retired engineering educator; b. Washington, Nov. 15, 1926; s. George W. and Buena (Youe) B.; m. Sarah Emily Barton, June 7, 1950; children—Joan Marie, Steven Leland. A.A., Okla. State U., 1950, B.S., 1953, M.S., 1954; D.Ed., Am. U., 1965. Field rep. Grain Dealers Mut. Ins. Co., Indpls., 1950, Jackson, Miss., 1950-52; sr. instr. U. Md., 1954-56, prof., 1956-93; ret., 1993; fire prevention engr., civil engring. div. U.S. Coast Guard, Washington, summers 1960-64. Author: Fire Detection and Suppression Systems, 1973, 3d edit., 1993, Automatic Sprinkler and Standpipe Systems, 1976, 2d rev. edit., 1990. Mem. Am. Soc. for Engring. Edn., Soc. Fire Protection Engrs., Nat. Fire Protection Assn., ASTM, Iota Lambda Sigma, Psi Chi, Kappa Delta Pi, Phi Kappa Phi. Home: 2399 Bear Den Rd Frederick MD 21701-3201

BRYAN, RAYMOND GUY, mechanical engineer, consultant; b. Inglewood, Calif., July 22, 1961; s. Douglas Robert and Linda Lou (Slater) B.; m. Jennifer Louise Queyrel, June 30, 1984; 1 child, Elaina Katharine. BSME, U. Nev., Reno, 1983, MSME, 1993. Registered profl. engr., Nev. Engring. intern, mech. engr., then project engr. Internat. Game Tech., Reno, 1981-87; project engr., then leader engring. group Linear Instruments/Spectra-Physics, Reno, 1987-92; pres. Synergy Tech., Inc., Reno, 1992—; instr. Truckee Meadows C.C., Reno, 1988. Mem. Rep. Cen. Com., Washoe County, 1986—; mem. Nat. Ski Patrol, 1979—, region dir., 1990—. Named Outstanding Patroller Nat. Ski Patrol, 1986-87. Mem. ASME, NSPE. Achievements include patents for gimbal mounting of optical elements, collet mounting of optical elements, off-axis rotation of diffraction gratings, seals for high-pressure and small volume sample cells. Office: Synergy Tech Inc PO Box 10662 Reno NV 89510

BRYAN, SPENCER MAURICE, cost engineer; b. Covington, Ga., July 26, 1962; s. William Monroe and Cornelia (French) B.; m. Michelle Roy, Feb. 18, 1989; children: Brittany Elizabeth, Chelsea Rae. BSEET, Ga. So. U., 1985. Cert. cost. cons.; engr.-in-tng. Engring. asst. Ga. Pacific Corp., Monticello, Ga., 1982-84; staff engr. AT&T, Sandy Springs, Ga., 1984-85; cost engr. Project Time & Cost, Inc., Atlanta, 1985—. Mem. NSPE, Am. Assn. Cost Engrs. (sec. 1986-88, membership dir. 1988-89), Am. Soc. Profl. Estimators. Methodist. Achievements include research in assembly pricing techniques. Office: Project Time & Cost Inc Ste 1600 3390 Peachtree Rd NE Atlanta GA 30326

BRYANT, CLIFTON DOW, sociologist, educator; b. Jackson, Miss., Dec. 25, 1932; s. Clifton Edward and Helen (Dow) B.; m. Nancy Ann Arrington, Sept. 13, 1953; m. Patty Maurine Watts, Feb. 1, 1957; children: Melinda Dow, Deborah Carol, Karen Diane, Clifton Dow II. Student, U. Miss., 1950-53, B.A., 1956, M.A., 1957; postgrad., U. N.C., Chapel Hill, 1957-58, La. State U., 1958-60; Ph.D., La. State U., 1964. Vis. instr. dept. sociology and anthropology Pa. State U., summer, 1958; instr., research asso. dept. sociology and anthropology U. Ga., 1960-63; asst. prof., asso. prof., chmn. dept. sociology and anthropology Millsaps Coll., Jackson, Miss., 1963-67; summer research participant, tng. and tech. project Oak Ridge Asso. Univs., summer 1967; prof., head dept. sociology and anthropology Western Ky. U., Bowling Green, Ky., 1967-72; prof. sociology Va. Poly. Inst. and State U., Blacksburg, 1972—; head dept. Va Poly. Inst. and State U., Blacksburg, 1972-82; vis. prof. Xavier U., Philippines, 1984-85; vis. prof., vis. research scholar Miss. Alcohol Safety Edn. Program, Miss. State U., (summer), 1985; vis. Fulbright prof. dept. grad. inst. sociology Nat. Taiwan U., Taipei, Republic of China, 1987-88; vis. scientist U.S. Army summer faculty rsch. and engring. program, 1993; participant Fulbright-Hays Seminar Abroad program, Hungary, 1993. Author: Khaki-Collar Crime: Deviant Behavior in Military Context, 1979, Sexual Deviancy and Social Proscription, 1982; editor and contbr.: Deviant Behavior: Occupational and Organizational Bases, 1974, The Social Dimensions of Work, 1972, Sexual Deviancy in Social Context, 1977, Deviant Behavior: Readings in the Sociology of Norm Violations, 1990; co-editor, contbr.: Deviancy and the Family, 1973, The Rural Work Force: Nonagricultural Occupations in America, 1985; compiler: Handbook of Audio-Visual Resources to Accompany Social Problems Today, 1971; editor: Social Problems Today: Dilemmas and Dissensus, 1971; co-editor: Introductory Sociology: Selected Readings for the College Scene, 1970; editor in chief Deviant Behavior: An Interdisciplinary Jour., 1978-91; editor So. Sociologist, 1970-74; mem. editorial bd. Criminology: An Interdisciplinary Jour, 1978-91; chmn. editorial policy bd., founding editor-in-Chief Deviant Behavior: An Interdisciplinary Journal, 1992—; chmn. editorial bd. Sociol. Symposium, 1968-80; assoc. editor Sociol. Forum, 1979-80, Sociol. Spectrum, 1981-85; mem. bd. adv. editors Sociol. Inquiry, 1981-85; assoc. editor spl. issue Marriage and Family Relations, fall 1982; contbr. chpts. to books, articles, book reviews to profl. publs. Served to 1st lt., M.P. U.S. Army, 1953-55. Recipient E. Gordon Ericksen Outstanding Grad. Faculty award sociology dept. Va. Poly. Inst. and State U., 1992, 93, spl. award for continuing contbn. to undergrad. teaching enterprise, 1992. Mem. Am. Sociol. Assn., Am. Soc. Criminology, So. Sociol. Soc. (pres. 1978-79), Mid-South Sociol. Assn. (pres. 1981-82, Disting. Career award 1991), Rural Sociol. Soc., Soc. Anthropology of Work, Internat. Sociol. Assn., Inter-Univ. Seminar on Armed Forces and Society, So. Assn. Agr. Scientists (rural sociology sect.), Phi Kappa Phi, Pi Kappa Alpha, Alpha Phi Omega, Alpha Kappa Delta. Presbyterian. Home: 1724 E Ridge Dr Blacksburg VA 24060-8568 Office: Va Poly Inst State U Dept Sociology Blacksburg VA 24061

BRYANT, DONALD ASHLEY, molecular biologist; b. LaGrange, Ky., Mar. 12, 1950; s. Roger William Jr. and Wanda Lillian (Partin) B.; m. Rita Jean Corio, July 20, 1974 (div. Aug. 1988); m. Veronica Lynne Stirewalt, Aug. 24, 1991. BSc in Chemistry, MIT, 1972; PhD in Molecular Biology, UCLA, 1977. Postdoctoral assoc. Pasteur Inst., Paris, 1977-79, Cornell U., Ithaca, N.Y., 1979-81; prof. molecular biology Pa. State U., University Park, 1981—, Ernest C. Pollard prof. biotech., 1992; vis. prof. Swiss Fed. Tech. U., Zurich, 1989-90. Assoc. editor Archives of Microbiology, 1988—, Photosynthesis Rsch., 1988—; mem. editorial bd. Jour. Bacteriology, 1986-91; editor 3 books in field; contbr over 85 articles to profl. jours., chpts. to books. Mem. AAAS, Am. Soc. Microbiology, Am. Soc. Plant Physiology, Am. soc. Biochemistry and Molecular Biology, Internat. Soc. Plant Molecular Biology. Democrat.

BRYANT, FRED BOYD, psychology educator; b. Princeton, N.J., Nov. 26, 1952; s. George Macon and Merrilee (Miles) B.; m. Linda Sue Perloff, July 12, 1980; children: Hilary Jacyln, Erica Lindsay. BA, Duke U., 1974; MA, Northwestern U., Evanston, 1977, PhD, 1980. Postdoctoral fellow Inst. for Social Rsch., U. Mich., Ann Arbor, 1979-82; asst. prof. Loyola U., Chgo., 1982-85, assoc. prof., 1985-90, prof., 1990—; rsch. cons. in field, 1982—; legal cons., N.Y., Ill., 1985—. Editor: Methodological Issues in Applied Social Psychology, 1992; contbr. numerous articles to profl. jours. Mem. APA, Am. Evaluation Rsch. Assn., Midwestern Psychol. Assn. Office: Loyola U Dept Psychology 6525 N Sheridan Rd Chicago IL 60626

BRYANT, LARRY MICHAEL, structural engineer, researcher; b. Newnan, Ga., Dec. 24, 1946; s. John Wilson Bryant and Eloise (Bradford) Roberts; m. Brenda Elaine Coker, Aug. 2, 1968; children: Christopher Michael, Alison Michelle. B in Bldg. Constrn., Auburn U., 1968; MS in Civil Engring., U. Tex., 1971, PhD in Civil Engring., 1977. Asst. instr. U. Tex., Austin, 1969-76, rsch. asst., 1972-77; sr. engr. Petro-Marine Engring., Gretna, La., 1977-79, supr. engring. rsch. and devel., 1979-86; sr. staff engr. JAYCOR Structures Divsn., Vicksburg, Miss., 1986—; mem., advisor several Am. Petroleum Inst. coms., Gretna, 1976-86; presented papers Offshore Tech. Conf., 1977, 78, 79, 82, ASCE Specialty Confs., 1978, 82, Shock and Vibration Symposium, 1988, 89. Mem. ASCE. Achievements include development of numerous state-of-the-art computer programs for structural analysis and design. Office: JAYCOR Structures Divsn 1201 Cherry St Vicksburg MS 39181

BRYANT, PETER JAMES, biologist, educator; b. Bristol, Eng., Mar. 2, 1944; came to U.S., 1967; s. Sydney Arthur and Marjorie Violet (Virgurs) B.; m. Toni Boettger, 1980; children: Katherine Emily, Sarah Grace. B.Sc. (Special) in Zoology with 1st class honors, Kings Coll., London, Eng., 1964; M.Sc. in Biochemistry, Univ. Coll., London, Eng., 1965; D.Phil. in Genetics, U. Sussex, Falmer, Eng., 1967. Postdoctoral research fellow Devel. Biology Ctr., Case Western Res. U., Cleve., 1967-70; postdoctoral research fellow Devel. Biology Lab., U. Calif.-Irvine, 1967-70, lectr. dept. devel. and cell biology, 1970-71, asst. prof., 1971-74, assoc. prof., 1974-77, prof. dept. devel. and cell biology, 1977—, vice chmn. dept. devel. and cell biology, 1978-79, dir. Devel. Biology Ctr., 1979—; nat. and internat. lectr.; reviewer NSF, NIH. Contbr. numerous articles, book revs., abstracts to Cell, Proc. Nat. Acad. Sci., Jour. Cell Sci., Devel., Nature Sci., Jour. Insect Physiology, Developmental Biology, other profl. publs.; reviewer Cell, Devel. Biology, Genetics, Sci., Can. Jour. zoology, numerous other profl. jours.; assoc. editor: Developmental Biology, 1976-85, editor-in-chief, 1985—. Bd. dirs. Orange County Natural History Mus., 1987-92. Recipient numerous grants NIH, NSF, 1974—, Am. Cetacean Soc. grantee, 1980-81, Nat. Marine Fisheries Svc. grantee, 1981; recipient Disneyland Community Svc. award (on behalf of Am. Cetacean Soc., Orange County chpt.), 1980. Mem. Soc. Devel. Biology, Genetics Soc. Am., Internat. Soc. Devel. Biologists, Soc. Exptl. Biology, Am. Cetacean Soc. (pres. 1982, founding pres. Orange County chpt. 1977-80). Avocations: insect photography, sailing, hiking, model building, conservation. Office: Univ California Devel Biology Ctr Irvine CA 92717

BRYANT, ROBERT LEAMON, mathematics educator; b. Harnett County, N.C., Aug. 30, 1953; s. James Ray and Josephine (Strickland) B. BS, N.C. State U., 1974; PhD, U. N.C., 1979. Asst. prof. Rice U., Houston, 1979-81, assoc. prof., 1981-82, prof., 1982-85, Noah Harding prof., 1985-87; Arts and Scis. prof. Duke U., Durham, N.C., 1987-88, J.M. Kreps prof., 1988—. A.P. Sloan fellow, 1983; recipient Presdl. Young Investigator award NSF, 1984. Democrat. Home: 6310 Turkey Farm Rd Chapel Hill NC 27514-9784 Office: Duke U Dept Math Durham NC 27706

BRYANT, WENDY SIMS, medical physicist; b. Hartsville, S.C., June 20, 1967; d. Mitchell Carlyle and Dora Jean (McElveen) Simms; m. Richard James Jr., Aug. 5, 1989. BS, Francis Marion Coll., 1989; MS, U. N.C., 1991. Med. physicist JOF, Inc., Charleston, S.C., 1990-92, McLeod Regional Med. Ctr., Florence, S.C., 1992-93, JOF, Inc., Charleston, 1993—. Mem. Am. Assn. of Physicists in Medicine. Office: Tuomey Regional Med Ctr 130 N Washington St Sumter SC 29150

BRYCHEL, RUDOLPH MYRON, engineer; b. Milw., Dec. 4, 1934; s. Stanley Charles and Jean Ann (Weiland) B.; m. Rose Mary Simmons, Sept. 3, 1955; children: Denise, Rita, Rudolph Myron Jr., Patrick, Bradford, Matthew. Student, U. Wis., Stevens Point, 1953, U.S. Naval Acad., 1954-55, U. Del., 1957, Colo. State U., 1969, North Park Coll., Chgo., 1973, Regis U., Denver, 1990-91. Lab. and quality tech. Thiokol Chem. Co., Elkton, Md., 1956; final test insp. Martin Aircraft Co., Middle River, Md., 1956-57; system final insp. Delco Electronics Co., Oak Creek, Wis., 1957-58; test equipment design engr. Martin Marietta Co., Littleton, Colo., 1958-64; prodn. supr. Gates Rubber Co., Denver, Colo., 1964-65; freelance mfr., quality and project engr. Denver and Boulder, Colo., Raton, N.Mex., 1965-67; quality engr. IBM, Gaithersburg (Md.), Boulder (Colo.), 1967-73; sr. quality engr. Abbott Labs., North Chicago, Ill., 1973-74; instrumentation and control engr. Stearns Roger Co., Glendale, Colo., 1974-81; staff quality engr. Storage Tech., Louisville, Colo., 1981-83; sr. quality engr. Johnson & Johnson Co., Englewood, Colo., 1983-84; quality engr., cons. Staodynamics Co., Longmont, Colo., 1984-85; sr. engr. for configuration and data mgmt. Martin Marietta Astronautics Group, Denver, 1985-91; freelance cons. Littleton, Colo., 1991—. With USN, 1953-56. Mem. Am. Soc. Quality Control (cert. quality engr.), Regulatory Affairs Profl. Soc., Soc. for Tech. Communications (regional chpt. chmn. 1970), KC. Democrat. Roman Catholic. Avocations: berry and fruit gardening. Home and Office: 203 W Rafferty Gardens Ave Littleton CO 80120-1710

BRYDON, HAROLD WESLEY, entomologist, writer; b. Hayward, Calif., Dec. 6, 1923; s. Thomas Wesley and Hermione (McHenry) B.; m. Ruth Bacon Vickery, Mar. 28, 1951 (div.); children: Carol Ruth, Marilyn Jeanette, Kenneth Wesley. AB, San Jose State Coll., 1948; MA, Stanford U., 1950. Insecticide sales Calif. Spray Chem. Corp., San Jose, 1951-52; entomologist, fieldman, buyer Beech-Nut Packing Co., 1952-53; mgr., entomologist Lake County Mosquito Abatement Dist., Lakeport, Calif., 1954-58; entomologist, adviser Malaria Eradication Programs AID, Kathmandu, Nepal, 1958-61, Washington, 1961-62, Port-au-Prince, Haiti, 1962-63; dir. fly control research Orange County Health Dept. Santa Ana, Calif., 1963-66; free-lance writer in field, 1966—; research entomologist U. N.D. Sch. Medicine, 1968; developer, owner Casierra Resort, Lake Almanor, Calif., 1975-79; owner Westwood (Calif.) Sport Shop, 1979-84; instr. Lassen Community Coll., Susanville, Calif., 1975—; bio control cons., 1980—. Mem. entomology and plant pathology del. People to People Citizen Ambassador Program, China, 1986; citizen ambassador 30th Anniversary Caravan to Soviet Union, 1991, Vietnam Initiative Del., 1992. Contbr. profl. jours. and conducted research in field. Served with USNR, 1943-46. Recipient Meritorious Honor award for work in Nepal, AID, U.S. Dept. State, 1972. Mem. Entomol. Soc. Am., Am. Mosquito Control Assn., Pacific Coast Entomol. Soc., Am. Legion. Republican. Methodist. Club: Commonwealth of California. Lodges: Masons, Rotary. Home: PO Box 312 Westwood CA 96137-0312

BRYHAN, ANTHONY JAMES, metallurgist; b. Detroit, Nov. 17, 1945; s. Donald and Caroline Bryhan. BS, Mich. State U., 1974, MS, 1976. Metallurgist Mich. Hwy. Dept., Lansing, 1970-76; devel. engr. Linde div. Union Carbide, Ashtabula, Ohio, 1976-77; sr. rsch. metallurgist Amax Corp., Ann Arbor, Mich., 1977-85; rsch. specialist Exxon Rsch., Houston, 1985-86; tech. staff mem. TRW Space and Def., Redondo Beach, Calif., 1987-88; mgr. fabrication devel. Martin Marietta (formerly GE Aerospace), San Jose, Calif., 1988—; chmn. field welding com. Welding Rsch. Coun., N.Y.C., 1981-85; mem. peer rev. com. Am. Welding Soc., Miami, Fla., 1981-85. Author (booklet): Joining of Molybdenum Base Metals and Factors Which Influence Ductility, 1986; contbr. articles to profl. publs.; presented papers in field. Active local politics; pres. local condo. assn. Mem. USN, 1966-70. Mem. AAAS, Am. Soc. Metals (chmn., contbr. to jour. 1983-84), Mensa. Democrat. Achievements include patent in field. Home: 6093 Calle de Amor San Jose CA 95124-6537

BRYSON, KEITH, obstetrician/gynecologist; b. Spruce Pine, N.C., Nov. 12, 1951; s. Lloyd C. and Mildren (Pittman) B.; Vicki Bryson, June 12, 1990. BA, Appalacian State U., 1974; MD, Wake Forest U., 1979. Rotating intern Erlanger Med. Ctr., Chattanooga, 1979-80, resident in ob-gyn., 1980-83; mem. staff Med. Ctr. at Bowling Green, Ky., HCA Greenville Hosp., Bowling Green, T. J. Samson Hosp., Glasgow, Ky.; aviation med. examiner FAA; clin. instr. Vanderbilt U., Nashville; clin. advisor We. Ky. U., Bowling Green; clin cons. Surgical Laser Techs., Malvern, Pa., Laserscope (KTP), San Jose, Calif., Surgilase Corp., Warwick, R.I., Advanced Laser Svcs., Grove City, Ohio, Edn. Design, Inc., Malvern, Trace Med. Eqipment, Burr Ridge, Ill., Minn. Laser Corp., Roseville, Mich., TAP Pharms., Deerfield, Ill; mem. faculty various presentations in Ohio, Pa., Md., Tenn., Ga., Wis., La., Calif.; speaker in field, including St. Charles Hosp., Toledo, 1988, Ob-gyn. Med. Soc., Milw., 1988, Lexington County Hosp., Columbia, S.C., 1989, Mercy Med. Ctr., Des Moines, 1989, Meridia Hosps., Cleve., 1989, Delray Med. Ctr., Delray Beach, Fla., 1990, St. Francis Hosp., Pitts., 1990, Deaconess Hosp., Cin., 1990, Laparoscopic Appendectomy and Cholecystectomy for Gen. Surgeons, Ga., Fla., Tenn., 1990, GYN Operative Ednoscopy, Milw., N.Y., Ill., Ind., N.J., Pa., Fla., S.C., 1990. Contbr. articles to med. jours.; several TV appearances. Fellow Am. Coll. Ob-Gyn. (jr.); mem. AMA, Am. Soc. for Laser Medicine and Surgery, Am. Fertility Soc., Am. Assn. Gynecol. Laparoscopists, Ky. Med. Assn., Tenn. State Med. Soc., N.C. State Med. Soc., Warren County Med. Assn., Nashville Acad. Medicine. Home: 1306 Euclid Ave Bowling Green KY 42101 also: 236 Yellow Buckeye Rd Glasgow KY 42141 Office: Commonwealth Med Plz 720 Second St Ste 305 Bowling Green KY 42101

BRYSON, VERN ELRICK, nuclear engineer; b. Woodruff, Utah, May 28, 1920; s. David Hyrum and Luella May (Eastman) B.; m. Esther Sybil de St Jeor, Oct. 14, 1942; children: Britt William, Forrest Lee, Craig Lewis, Nadine, Elaine. Commd. 2d lt. USAAF, 1941; advanced through grades to lt. col. USAF, 1960, ret., 1961; pilot, safety engr., civil engr., electronic engr., nuclear engr., chief Aeronaut. Systems div., Aircraft Nuclear Propulsion Program, Wright-Patterson AFB, Ohio, 1960-61; chiefRadiation Effects Lab., also chief Radiation Effects Group Boeing Airplane Co., Seattle, 1961-65; nuclear engr. Aerospace Corp., San Bernardino, Calif., 1965-68; service engr., also head instrumentation lab., Sacramento Air Logistic Ctr. USAF, McClellan AFB, Calif., 1968-77; owner, mgr. Sylvern Valley Ranch, Calif., 1977—; Mem. panel Transient Radiation Effects on Electronics, Weapons and Effects Bd., 1959-61. Contbr. research articles on radiation problems to profl. pubs. Decorated D.F.C. with oak leaf cluster, Air medal with 12 oak leaf clusters. Mem. IEEE. Mem. Ch. Jesus Christ of Latter-day Saints. Home: 1426 Caperton Ct Penryn CA 95663-9515

BRYT, ALBERT, psychiatrist; b. Marburg, Germany, Mar. 8, 1913; came to U.S., 1944; s. David Naftula and Rajzla (Malc) B.; m. Meta Sebag, June 17, 1935 (div. 1943); m. Natalie Lewy, April 19, 1957; children: Marguerite Maude, Allison Bartley. BS, Oberrealschule, Butzbach, Germany, 1932; PCN in Natural Sciences, Ecole de Plein Exercise, Tours, France, 1934; MD, U. de la Sorbonne, Paris, 1939; cert. in psychoanalysis, William Alanson White Inst., N.Y.C., 1950. Diplomate Am. Bd. Psychiatry and Neurology. Pvt. practice Tunisia, 1941-44; Fellow, intern, resident, jr. psychiatrist Bellevue Hosp., N.Y.C., 1945-47; pvt. practice N.Y.C., 1947--; psychiatrist in charge Adolescent Girls Ward Bellevue Hosp., 1949; chief Adolescent Out patient Univ. Hosp., N.Y.C., 1951-57; supr., clin. dir. Northside Ctr. Child Devel., N.Y.C., 1957-60; psychiatrist in charge Adolescent Outpatient Bellevue Hosp., 1964-70; resident tng. Adolescent Psychiatry NYU-Bellevue Med Ctr., 1970-72; cons. psychiatrist Human Resource Adminstrn. City of N.Y., 1978-89; team psychiatrist Project Assist Manhattan Children's Psychiat. Ctr., N.Y.C., 1989; mem. attending staff Univ. Hosp., N.Y.C., 1950; vis. neuro-psychiatrist Bellevue Hosp. Ctr., 1950; mem. exec. com., dir. of curriculum William Alanson White Inst., 1960, 62, 64-67, 69-71; cons. Luth. Community Svcs., 1954—; cons. The Salvation Army Social Svcs. for Children, 1975-90; cons. Dept. Mental Health, State of N.Y. Co-author: Facial Disfigurement and Plastic Surgery, 1953; contbr. to other books, scientific jours. With French Army Med. Corps, 1940. Named team psychiatrist USPH Rsch. Grant, NYU-Bellevue, 1948-51. Fellow Am. Psychiat. Assn. (life), N.Y. Soc. for Adolescent Psychiatry; mem. N.Y. Coun. on Child Psychiatry, William Alanson White Psychoanalytic Soc. (treas. 1954-56, com. on ethics 1989). Democrat. Jewish. Avocations: photography (movies), woodworking, landscaping. Home: 130 E 75th St New York NY 10021-3277 Office: 4 E 89th St New York NY 10128-0636

BRZOSKO, JAN STEFAN, physicist; b. Grodzisk, Warsaw, Poland, Sept. 12, 1939; s. Jan and Helena (Usakiewicz) B.; m. Ewa Maria Schmid, June 24, 1962; children: Maria M., Jan Ryszard. PhD in Nuclear Physics, Warsaw U., 1968, Habilitated Dr. of Physics, 1971. Docent Warsaw U., 1971-81; sr. scientist, project leader J.I.N.R., Dubna, USSR, 1972-73; head of phys. dept. Warsaw U., Bialstok Div., Bialstok, Poland, 1974-81; dean of sci. faculty Warsaw U., Bialstok, Poland, 1975-78, v.p., 1978-81; sci. advisor ENEA Frascati Rsch. Ctr., Rome, 1981-85; rsch. prof. of physics Stevens Inst. Tech., Hoboken, N.J., 1985—; mem. sci. coun. High Voltage Tech., Ministry of Tech. Sci., Warsaw, 1976-81, Neutron Physics, JINR, Dubna, 1971-73. Editorial bd. Mala Delta jour., Poland, 1979-81. Recipient award Polish Phys. Soc., 1962, Polish Atomic Energy Com., 1963, Minister of Tech. and Sci., Warsaw, 1972, 78, 80. Mem. IEEE (sr.), Am. Phys. Soc., Polish Phys. Soc. (v.p. Bialystok div. 1977-81), European Phys. Soc., N.Y. Acad. Sci. Republican. Roman Catholic. Achievements include observation and interpretation of EXIT states of compound nucleus produced in nuclear reactions; first heavy-light ion fusion (with others) in focused discharges; first integral experiment (with other) on radioactivity accumulation from the nuclear explosion field in body of mammals; development of statistical model of surface discharges; research in nuclear reactions, surface discharges, nuclear fusion technology, plasma physics. Office: Stevens Inst Tech Physics Dept Castle Point Hoboken NJ 07030

BRZUSTOWICZ, RICHARD JOHN, neurosurgeon, educator; b. Bklyn., Dec. 19, 1917; s. John B. and Victoria Eleanor (Szutarska) B.; m. Alice Lorraine Cinq-Mars, May 30, 1945; children: Richard John, Thaddeus P., Victoria, Barbara, John, Teresa, Krystyna, Mary. B.S., CCNY, 1938; M.D., SUNY, 1942; M.S. in Neurol. Surgery, U. Minn., 1951. Diplomate: Am. Bd. Psychiatry and Neurology, Am. Bd. Neurol. Surgery. Intern Bklyn. Hosp., 1942-43, asst. resident in surgery, 1947-48; resident in pathology Kings County Hosp., Bklyn., 1946-47; fellow in neurol. surgery Mayo Found., U. Minn.-Rochester, 1948-51; practice medicine specializing in neurol. surgery Rochester, N.Y., 1951—; asst. prof. anatomy U. Rochester Sch. Medicine and Dentistry, 1961-68, clin. sr. instr. neurol. surgery, 1962-71, clin. asst. prof., 1971—; chmn. dept. neuroscis., div. neurol. surgery St. Mary's Hosp., Rochester, 1951-83, pres. med. adv. bd., 1972-75; cons. in neurosurgery to various hosps., 1951—; vol. physician St. Mary's Hosp., Rochester, 1987—; mem. adv. bd. mil. pers. supplied com. on helmets, NRC, 1972-78. Assoc. editor Bull. Polish Med. Sci. and History, 1963-67; mem. editorial bd. Bull. of the Monroe County Med. Soc., 1957-79, chmn., 1967-73, editor, 1976-79; contbr. articles on neurology to med. jours. Mem. parents adv. coun. U. Rochester, 1966-69, adv. bd. Alliance Coll., Cambridge Springs, Pa., 1967-70; sec. med. adv. bd. St. Mary's Hosp., Rochester, 1970-72, pres. med. adv. bd. and med. and dental staff, 1972-75, med. adv. bd., 1975-82; served to capt. U.S. Army, 1943-46. Recipient Vincentian award St. Mary's Hosp., 1989. Mem. N.Y. State Med. Soc., Monroe County Med. Soc. (editor bull. 1967-73, 76-79), Am. Assn. Neurol. Surgeons, N.Y. State Neurosurg. Soc., Rochester Acad. Medicine (Paine prize 1952, award of merit 1988), Am. Acad. Neurology, Am. Assn. Med. Systems and Informatics, Congress Neurol. Surgeons, Catholic Physicians Guild (pres. 1956-57, 86-87), Nat. Med. and Dental Assn., Polish Inst. Arts and Scis. (med. sect.), Am. Med. Writers Assn., AMA, N.Y. Acad. Scis., AAAS, Am. Heritage Soc., Polish Med. Alliance, Polish Am. Hist. Assn., Am. Physicians Art Assn., Alumni Assn. Mayo Found., Am. Philatelic Soc., Polonus Philatelic Soc., Sigma Xi. Roman Catholic. Home: 366 Oakdale Dr Rochester NY 14618-1130 Office: 909 Main St W Ste 108 Rochester NY 14611-2635

BUB, ALEXANDER DAVID, acoustical engineer; b. Milw., Oct. 19, 1949; s. Alex Robert and Rose (Monafo) B.; m. Kay Lynn Johannes, Jan. 5, 1982; 1 child, David. AAS in Electronic Communications, Milw. Sch. Engring., 1969; BA in Econs., History and Anthropology, U. Wis., Milw., 1976. Nuclear weapons specialist USAF, 1969-73; with Harley Davidson, Inc., Milw., 1977—, project coord., 1986—, with powertrain devel. group, 1993—. U.S. nat. champion 410 Superbike, 1979, Mexican champion 750 Prodn. and Open Superbike, 1980, Midwest champion Supertwins and Formula Twins, 1985, 86, 87. Mem. Acoustical Soc. Am. (guest speaker conf. 1990, 92), Soc. Automotive Engrs., Western/Eastern Roadracing Assn. Avocations: professional cycle racing, mountain bikes, skiing, amateur radio (WA9OLH). Home: 6435 Western Rd Cedarburg WI 53012-1821 Office: Harley Davidson Inc 3700 W Juneau Ave Milwaukee WI 53208-2865

BUBASH, JAMES EDWARD, engineering executive, entrepreneur, inventor; b. St. Louis, Oct. 26, 1945; s. Joseph C. and Helen M. (Grab) B.; m. Patricia F. Mann, Nov. 27, 1988. BSME, U. Mo., 1968. Design engr. McDonnel Aircraft Co., St. Louis, 1968, 70-74; mfg. engr. Unidynamics, St. Louis, 1974-76, Broderick and Bascom Rope Co., St. Louis, 1976-80; project engr. Bussmann Mfg., St. Louis, 1980-88; pres., chief exec. officer C.R. Magnetics, Inc., Fenton, Mo., 1986—, also bd. dirs. With U.S. Army, 1968-70. Recipient Best 20 Designs of 80s award Design News mag., 1990, Readers Choice award Control Engring. mag., 1987. Mem. Instrumentation Soc. Am. (sr.). Achievements include patents for Measure Master and Current Ring. Home: 9600 Carrimae Ct Saint Louis MO 63126-2000 Office: CR Magnetics Inc 308 Axminister Dr Fenton MO 63026-2902

BUCCINI, FRANK JOHN, molecular geneticist; b. Jersey City, N.J., Mar. 28, 1959; s. Frank George and Catherine Mary (Kamrowski) B.; m. Judith Patricia Frazier, May 26, 1984; 1 child, Michelle Catherine. BS, Seton Hall U., 1981, MS, 1985. Cert. qualifications lab. tech. and supr., N.Y. Med. rsch. technician VA Med. Ctr., East Orange, N.J., 1983-87; microbiology lab. mgr. Gen. Care Biomed. Rsch. Corp., Mountainside, N.J., 1987-88, supr. cancer immunology lab., 1988—; supr. molecular genetics lab. Gen. Care Biomed. Rsch. Corp., Mountainside, 1987—. Contbr. articles to Jour. of Infection, Antimicrobial Agts. and Chemotherapy, Jour. of Antimicrobial Chemotherapy, Am. Assn. of Clin. Chemistry. Mem. Theobald Smith Soc., Am. Soc. for Microbiology, Sigma Xi. Office: Gen Care Biomed Rsch Corp 271 Sheffield St Mountainside NJ 07092

BUCHAN, RONALD FORBES, preventive medicine physician; b. Concord, N.H., Sept. 24, 1915; s. Robert and Mary Jean (Forbes) B.; m. Maureen O'Regan, June 17, 1940; children: Robert Bruce, Joan Dallas (Mrs. Fleming), Ian Forbes Morgan. A.B., U. N.H., 1936; M.D., C.M., McGill U., 1942; postgrad., Princeton U., 1958. Diplomate Nat. Bd. Med. Examiners, Am. Bd. Preventive Medicine. Reporter Concord Daily Monitor, 1936; asst. exec. sec. Unemployment Compensation Commn., N.H. Dept. Labor, 1937; sanitarian City of Concord and Eastern Health Dist. N.H., 1938; chief, med. unit Bur. Indsl. Hygiene, Conn. Dept. Health, 1943-46; dir. Hartford Small Plant Indsl. Med. Svcs., 1946; clin. dir., asst. prof. indsl. medicine Yale U. Inst. Occupational Medicine and Hygiene, 1946-48; assoc. clin. prof. indsl. medicine N.Y.U. Bellevue Post Grad. Med. Sch., 1948-57; assoc. med. dir. Prudential Ins. Co. Am., 1948-49, dir. employee health, 1949-57; med. dir., v.p. med. svcs. Prudential Ins. Co. Am., Boston, 1957-74, cons. occupational medicine, environ. medicine, toxicology, 1974—; chief med. dir., v.p. Mediscreen, 1974-87; propr. Portsmouth (N.H.) Athenaeum; assoc. clin. prof. preventive medicine Tufts U. Sch. Medicine, 1958-74; vis. lectr. numerous med. schs., 1948-89. Narrator (audio hist. tour) The Freedom Trail, Boston, (audio visual hist. survey) Shipbuilding on the Kennebec-Maine Maritime Mus.; author: Industrial Toxicology; contbr. Oxford Medicine, Current Therapy, Occupational Medicine, Encyclopedia-Medico-Chirurgicale (Paris); also numerous articles to profl. and lit. jours. Chmn. rsch. adv. com. Brattleboro (Vt.) Retreat, 1960-70; mem. sci. adv. bd. Office Chief Staff USAF, chmn. life scis. human factors facilities, 1960-65, protocol rank, lt. gen.; cons. R.I. Group Health Assn., 1973-75, Harvard Community Health Plan, 1972-75; bd. dirs. Met. Boston chpt. ARC, 1971-73, chmn. com. on safety, 1972-74; founding mem. Challenger Space Ctr., 1987; trustee Miles Meml. Hosp., Damariscotta, Maine, 1988-91. Sr. asst. surg., USPHS, 1943-46; surgeon-lt. York (Maine) Militia-Gov.'s Footguard, 1971—. Recipient Honor award Wisdom Soc., 1970. Fellow Am. Coll. Occupational and Env. Medicine (past pres.), Am. Coll. Preventive Medicine (chmn. com. on clin. procedures 1972-74), Am. Coll. Occupational Med. (past pres.), Acad. Medicine N.J. (past pres.); mem. AAAS, Am. Indsl. Hygiene Assn., Am. Acad. Ins. Medicine, AMA (assoc. editor Archives environmental Health), Assn. Internationale Pour La Medicine Du Travail (permanent commn. 1965-74), Mass. Med. Soc., Ramazzini Soc., Academie Europeene des Arts, Sciences et des Lettres, Am. Assn. Sr. Physicians, N.Y. Acad. Scis., Nat. Trust Historic Preservation, Soc. for Preservation of New Eng. Antiquities, John Buchan Soc. (Edinburgh), Osler Libr. (patron Montreal), Soc. for Protection of N.H. Forests, North Country Authors and Scientists League (past pres.), Newcomen Soc. N.Am., St. Andrew's Soc. of Maine, Clan Buchan U.S.A., Clan Forbes U.S.A., U. N.H. Alumni Assn. (gen. awards com. 1987-90, sec. U. N.H. class of '36, 1981—), McGill U. Alumni Assn., Friends of Bowdoin Coll. Home: Hill Winds Millay Hill RR 3 Box 8400 Union ME 04862-9500

BUCHANAN, DIANE KAY, pharmacist; b. Connersville, Ind., July 31, 1960; d. Everett and Wanda Lee (James) B. BS in Pharmacy, Purdue U., 1984. Registered pharmacist. Pharmacist, store mgr. Walgreen Co., Houston, 1984-86; staff pharmacist Randall's, Houston, 1986-88; staff pharmacist Osco Drug, Evansville, Ind., 1988-89, pharmacy mgr., 1989-92; cons. pharmacist Insta-Care, Decatur, Ga., 1992; asst. pharmacy mgr. Publix Super Markets, Norcross, Ga., 1992—; community cons. Osco Drug, Evansville, Ind., 1988-92. Mem., assoc. dir. Tri State Christian Singles, Evansville, 1988-91; mem. Evansville Mus. Contemporaries, 1989-92. Mem. Ind. Pharm. Assn. (D Pharmacy), Southwestern Ind. Pharmacist's Assn., Purdue Alumni Assn. (life), Sigma Kappa (life). Avocations: physical fitness, photography. Home: 2533 Marcia Dr Lawrenceville GA 30244 Office: Publix Super Markets 1250 Tech Dr Norcross GA 30093

BUCHANAN, JOHN GRANT, chemistry educator; b. Dumbarton, Scotland, Sept. 26, 1926; s. Robert Downie and Mary Hobson (Wilson) B.; m. Sheila Elena Lugg, July 14, 1956; children: Andrew, John, Neil. BA, Cambridge (Eng.) U., 1947, MA, 1951, PhD, 1951, ScD, 1966. Rsch. fellow U. Calif., Berkeley, 1951-52; rsch. asst. Lister Inst. Preventive Medicine, London, 1952-54; lectr. King's Coll., U. Durham, Newcastle upon Tyne, Eng., 1955-62; sr. lectr. U. Newcastle upon Tyne, 1962-65, reader, 1965-69; prof. organic chemistry Heriot-Watt U., Edinburgh, Scotland, 1969-91; professorial fellow U. Bath, U.K., 1991—; mem. editorial adv. bd. Nucleosides and Nucleotides, Dekker Jour., 1982-91. Mem. editorial bd. Carbohydrate Rsch., Elsevier Jour., 1966-91, editor, 1992—; contbr. articles to profl. publs. Fellow Royal Soc. Edinburgh (mem. coun. 1980-82), Royal Soc. Chemistry (mem. coun. 1982-85); mem. European Carbohydrate Orgn. (pres. 1989-93). Avocation: golf. Office: U Bath Sch Chemistry, Claverton Down, Bath BA2 7AY, England

BUCHANAN, JOHN MACHLIN, biochemistry educator; b. Winamac, Ind., Sept. 29, 1917; s. Harry James and Eunice Blanche (Miller) B.; m. Elsa Nilsby, Dec. 11, 1948; children—Claire Louise, Stephen James, Lisa Renne, Peter Nilsson. A.B., De Pauw U., 1938, D.Sc., 1975; M.S., U. Mich., 1939, D.Sc., 1961; Ph.D., Harvard, 1943. Instr. dept. physiol. chemistry Sch. Medicine U. Pa., 1943-46, asst. prof., 1946-49, asso. prof., 1949-50, prof., 1950-53; NRC fellow Med. Nobel Inst., Stockholm, 1946-48; prof., head div. biochemistry dept. biology Mass. Inst. Tech., 1953-67, Wilson prof. biochemistry, 1967-88, Wilson prof. emeritus, 1988—; lectr. Harvey Soc., 1958. Mem. editorial bd.: Jour. Biol. Chemistry, 1961-67, Jour. Am. Chemistry Soc, 1961-72, Physiol. Revs, 1957-60, 65-71. Civilian with Nat. Def. Research Com., 1943; mem. subcom. blood and related substances NRC, 1951-55, mem. med. fellowship bd., 1954—; mem. sci. adv. bd. Boston Biomed. Rsch. Inst., 1975-93, Papanicoulaou Cancer Research Inst., 1975-81. Fellow Guggenheim Meml. Found., 1964-65; leave of absence to Salk Inst. Biol. Studies LaJolla, Calif. Mem. Am. Soc. Biol. Chemists (sec. 1969-72), Am. Chem. Soc. (Eli Lilly award in biol. chemistry 1951), Am. Acad. Sci., Internat. Union Biochemists (mem. nat. com.), Nat. Acad. Scis., Am. Acad. Arts and Scis., Sigma Xi. Home: 56 Meriam St Lexington MA 02173-3622

BUCHELE, WESLEY FISHER, agricultural engineering educator, consultant; b. Cedar Vale, Kans., Mar. 18, 1920; s. Charles John and Bessie (Fisher) B.; m. Mary Wanda Jagger, June 12, 1945; children: Rod, Marybeth, Sheron, Steven. BS, Kans. State U., 1943; MS, U. Ark., 1951; PhD, Iowa State U., 1954. Registered profl. engr., Iowa, Calif. Jr. engr. John Deere Tractor Works, Waterloo, Iowa, 1946-48; asst. prof. U. Ark., Fayetteville, 1948-51; agrl. engr. USDA, Ames, Iowa, 1954-56; assoc. prof. Mich. State U., East Lansing, 1956-63; prof. Iowa State U., Ames, 1963-89; vis. prof. U. Ghana, Legon, 1968-69, Beijing Agrl. Engring. U., 1983-84; vis. scientist Commonwealth Sci. and Indsl. Rsch. Orgn., Australia, Internat. Inst., Tropical Agr., Abadan, Nigeria, 1979-80, Internat. Rice Rsch. Inst., Manila, The Philippines, 1991-92; cons. engr. GM, Detroit, 1974-76; bd. dirs. Farm Safety 4 Just Kids, Earlham, Iowa, Self-Help, Inc., Waverly, Iowa, JAC Tractor Co. Author 18 books; inventor 23 patents. Mem. Ames Energy Com., 1974-75; advisor Living History Farm, Urbandale, Iowa, 1965—, bd. govs., 1984—. Maj. U.S. Army, 1943-46, PTO; maj. Ordnance Corps, USAR, 1946-69, ret. Named Eminent Engr., Iowa Engring. Soc., 1989. Fellow Am. Soc. Agrl. Engrs. (bd. dirs. 1978-80, McCormick-Case award 1988), Nat. Inst. Agrl. Engrs.; mem. AAAS, Society Automotive Engrs., Am. Soc. Agronomy (mem. com. 1961-65), Steel Ring, Internat. Assn. Mechanization of Field Experiments (v.p. 1964—), Internat. Platform Assn., Osborne Club, Toastmasters. Avocations: photography, travel, golf, inventing, writing. Home and Office: 239 Parkridge Cir Ames IA 50010-3695

BUCHER, BERNARD JEAN-MARIE, immunopathologist, researcher, consultant; b. Fort-de-France, Martinique, May 1, 1962; came to U.S., 1984; s. Camille and Marcelle Irlande (Mathurin) B. BS, SUNY, N.Y.C., 1986; PhD in Immunopathology and Pub. Health, Union Inst., Cin., 1990. Predoctoral fellow psychoneuroimmunology Med. Coll. Pa., Phila., 1986-87; intern lab. medicine, 1988; doctoral trainee Wistar Inst. Anatomy and Biology, Phila., 1988-89; rsch. fellow Universite des Antilles-Guyane Med. Sch., Fort-de-France, Martinique, 1989-90; lectr. Sch. Nursing, Univ. Hosp., Fort-de-France, Martinique, 1989-90; vis. rsch. attaché dept. neurology Univ. Hosp., Fort-de-France, Martinique, 1991, sr. rschr. dept biochemistry, 1992—; ad hoc cons. Ministry of Health, Castrie, St. Lucia, W.I., 1988-90; advisor Inserm-Martinique Rsch. Coun., Fort-de-France, 1990—; v.p., session chair Internat. Symposium on HTLV-I, Castrie, 1990; sr. rschr. depts. neurology & biochemistry Univ. Hosp., Fort-de-France, 1992; guest rschr. neuroepidemiology br. CNP, NINDS, NIH, Bethesda, Md., 1992. Author: Neurology and Neurobiology, 1989, (with others) Manifestations Neurologiques et Infections Retrovirales, 1991; contbr. articles to profl. jours. Recipient Vol. Svc. award United Hosp. Fund of N.Y., 1985; Conseil Regional and Conseil Gen. fellow and ednl. grantee, Fort-de-France, Martinique, 1988-90. Mem. N.Y. Acad. Scis., Am. Soc. for Microbiology, Am. Pub. Health Assn., Am. Soc. Tropical Medicine & Hygiene, Royal Soc. Tropical Medicine & Hygiene (U.K.). Achievements include development of a permanent surveillance system for the detection of communicable diseases in the island nation of Saint Lucia, W.I. Office: Univ Hosp Dept Biochemistry, La Meynard BP 632, Fort-de-France 97200, Martinique

BUCHER, GAIL PHILLIPS, cosmetic chemist; b. Natick, Mass., June 29, 1941; d. Ellsworth Samuel and Emma Florence (Tupper) Phillips; m. Edward Andrew Bucher, Apr. 20, 1974; children: Sabina Lee, Eric Andrew. BS, Mass. Coll. Pharmacy, 1963; MS, Northeastern U., 1975. Pharm. chemist Colgate Palmolive Co., Piscataway, N.J., 1963-67; rsch. pharm. chemist Colgate Palmolive Co., Pascataway, N.J., 1967; devel. chemist Gillette Co., Boston, 1967-71, corp. quality analyst, 1971-88, clearance officer, database adminstr., 1988—; mem. exec. com. Mass. Coll. Pharmacy and Allied Health Scis., Boston, 1985—, trustee, chmn. edn. com., 1987-92, auditor, 1992—. Contbr. articles to New Eng. Nucleus-Am. Chem. Soc., Cosmetic and Toiletries Mfg. Mem. Faith and Sci. Exch., Concord/Newton, Mass., 1989—. Mem. Internat. Fedn. Socs. Cosmetic Chemists (pres. 1990-91), Soc. Cosmetic Chemists (chair instl. grant com. 1987-91, pres. 1984, chair NE chpt. 1972, 92), Mass. Coll. Pharmacy Alumni Assn. (exec. dir. 1987-93, treas. 1993—), Belmont Aquatic Team (treas. 1991—). Lutheran. Home: 91 Spring Valley Rd Belmont MA 02178-1717 Office: Gillette Co Gillette Park # 2U-1 Boston MA 02127-1096

BUCHHOLZ, RICHARD, ethologist; b. Manhasset, N.Y., Nov. 5, 1964; s. Eduard and Elisabeth Buchholz. BS in Biology, SUNY, Binghamton, 1986; MS in Zoology, U. Fla., 1989. Peer acad. advisor SUNY, Binghamton, 1984-86; grad. teaching asst. U. Fla., Gainesville, 1986—; mem. organizing com. 2d Internat. Cracid Symposium, Caracas, Venezuela, 1987-88. Co-editor Cracid Newsletter, 1991-92; contbr. articles to profl. publs. Sigma Xi grantee, 1988, 91; recipient Chapman Fund award Am. Mus. Natural History. Mem. Am. Behavior Soc., Am. Ornithologists Union, Am. Soc. Naturalists, Soc. for Conservation Biology. Achievements include documentation of the reproductive behavior of the rare and elusive yellow-knobbed currassow in Venezuela. Office: U Fla Dept Zoology Gainesville FL 32611

BUCHI, GEORGE HERMANN, chemistry educator; b. Baden, Aargau, Switzerland, Aug. 1, 1921; came to U.S., 1948; s. George Jakob and Martha (Müller) B.; m. Anne Barkmann. Diploma in Chem. Engring, Swiss Fed. Inst. Tech., Zurich, 1945, DSci, 1947, D Natural Wissenschaft honoris causa, 1987; D honoris causa, U. Heidelberg, Federal Rep. of Germany, 1983. Firestone postdoctoral fellow U. Chgo., 1948-51; asst. prof. chemistry MIT, Cambridge, 1951-56, assoc. prof., 1956-58, prof., 1958-71, Camille Dreyfus prof., 1971—; cons. Firmenich SA, Geneva, Switzerland, 1954—, Hoffman-LaRoche, Nutley, N.J., 1963—. Decorated Order of the Rising Sun with Gold Rays and Neck Ribbon, Govt. of Japan, 1986. Mem. Am. Chem. Soc. (Ernest Guenther award 1958, award for creative work organic chem. 1973), Swiss Chem. Soc. (Ruzicka award 1957), Brit. Chem. Soc., German Chem. Soc., Japanese Chem. Soc., Pharm. Soc. Japan (hon. fgn. mem.), Nat. Acad. Sci. Republican. Office: MIT Dept of Chemistry 77 Massachusetts Ave Cambridge MA 02139-4307

BUCHWALTER, STEPHEN L., chemist, researcher; b. Nyack, N.Y., July 9, 1947; s. Omar Ranck and Dorothy Marie B.; m. Katherine Elaine Kent, June 7, 1969 (div. 1981); children: Elena Lee, Rachel Anne; m. Leena Paivikki Forsstrom, June 19, 1982; children: Abigail Lynn, Jonathan Omar. BA, Coll. of Wooster, 1969; PhD, Harvard U., 1974. Rsch. assoc. dept. chemistry U. Chgo., 1974-76; sr. rsch. chemist Coatings and Resins Rsch. Labs., Allison Park, Pa., 1976-82; group leader Beacon (N.Y.) Rsch. Labs., Texaco, Inc., 1982-85; mem. rsch. staff Thomas J. Watson Rsch. Ctr., IBM Corp., Yorktown Heights, N.Y., 1985—; mgr. Very Large Scale Integration Packaging Materials, 1987-92; invited lectr. Clarkson U., Potsdam, N.Y., 1984, SUNY-Brockport, 1986, Rensellaer Poly. Inst., Troy, N.Y., 1990, CUNY, 1991, Gordon Rsch. Conf., New Hampton, N.H., 1991. Reviewer Jour. Organic Chemistry, Chemistry of Materials; contbr. articles to profl. jours. Elder 1st Presbyn. Ch. Wappingers Falls, N.Y., 1983-91; mem. Reformed Ch. Am., Hopewell Junction, N.Y., 1991—. Mem. AAAS, Am. Chem. Soc. (polymer div.). Achievements include discovery of 1,3 Cyclopentadiyl, discovery of rapidly epimerizing methylenecyclopropane, discovery of method of electrophoretic deposition of polyimide, 11 U.S. patents. Office: IBM Corp TJ Watson Rsch Ctr PO Box 218 Yorktown Heights NY 10598

BUCINELL, RONALD BLAISE, mechanical engineer; b. Johnson City, N.Y., Feb. 3, 1958; s. Felix James and Irene Mary (Novak) B.; m. Jill Bucinell, Aug. 24; children: Ryan Michael, Benjamin David. AAS, Rochester Inst. Tech., 1978, BS, 1981; MS, Drexel U., 1983, PhD, 1987. Engr. Boeing Aerospace, Seattle, 1979-80; rsch. asst. Dyna East Corp., Wynnewood, Pa., 1983-85; rsch. analyst Hercules Aerospace, Magna, Utah, 1987-89; rsch. engr. Materials Scis. Corp., Blue Bell, Pa., 1989-93; v.p. Innotech, Schenectady, N.Y., 1991—; adj. asst. prof. Temple U., Phila., 1991—, U. Utah, Salt Lake City, 1988-90; asst. prof. Union Coll., Schenectady, 1993—. Publ. reviewer: Jour. Composite Materials, 1988—; contbr. articles to profl. jours. Mem. ASME, Nat. Soc. Profl. Engrs., ASTM, K.C. (4th degree). Achievements include the development of methodology for altering failure modes in composite overwrapped pressure vessels, scaling methodology for response of composite materials, stochastic damage progression model for composite materials. Home: 1022 Maryland Ave Schenectady NY 12308 Office: Innotech PO Box 9561 Schenectady NY 12309

BUCK, CHRISTIAN BREVOORT ZABRISKIE, independent oil operator; b. San Francisco, Oct. 18, 1914; s. Frank Henry and Zayda Justine (Zabriskie) B.; student U. Calif., Berkeley, 1931-33; m. Natalie Leontine Smith, Sept. 12, 1948; children—Warren Zabriskie, Barbara Anne. Mem. engring. dept. U.S. Potash Co., Carlsbad, N.Mex., 1933-39; ind. oil operator, producer, Calif., 1939-79, N.Mex., 1939—; owner, operator farm, ranch, Eddy County, N.Mex., 1951-79; dir. Belridge Oil Co. until 1979; dir. Buck Ranch Co. (Calif.). Served with RAF, 1942-45. Democrat. Episcopalian. Club: Riverside Country (Carlsbad). Home: PO Box 5368 599 Lariat Circle # 2 Incline Village NV 89450 Office: PO Box 2183 Santa Fe NM 87504-2183

BUCK, DOUGLAS EARL, chemist; b. Idaho Falls, Idaho, Apr. 26, 1936; s. Orland C. and Erma (Dingman) B.; m. Lorna Jean Seavey, Aug. 7, 1965; children: Andrew Douglas, Eric Stephen. BS in Chemistry, Coll. Idaho, 1958; PhD in Phys. Chemistry, MIT, 1965. Rsch. chemist U.S. Borax Rsch. Corp., Anaheim, Calif., 1965-83; mgr. pilot plant U.S. Borax Rsch. Corp., Boron, Calif., 1983-87; mgr. process rsch. U.S. Borax Rsch. Corp., Anaheim, 1987-90, mgr. product rsch., 1990-91, dir. rsch., 1991—; mem. sci. adv. com. Chapman Coll., Orange, Calif., 1978-83. Contbg. author: Mellors Comprehensive Treatise on Inorganic Chemistry, 1981. Adult leader Boy Scouts Am., Apple Valley, Calif., 1983-87. Mem. Am. Chem. Soc. Home: 18782 Deville Dr Yorba Linda CA 92686-2620 Office: US Borax Rsch Corp 26877 Tourney Rd Santa Clarita CA 91355

BUCK, GREGORY ALLEN, molecular biology educator; b. Burlington, Wis., Mar. 28, 1951; s. E. Dale and Juanita Irene (Keske) B. BS, U. Wis., 1973; PhD, U. Wash., 1981. Postdoctoral asst. prof. Inst. Pasteur, Paris, 1981-84; asst. prof. microbiology, immunology Va. Commonwealth U., Richmond, 1984-91, assoc. prof., 1991—; mem. NIH biol. scis. III study sect., Bethesda, Md., 1991—, charter mem. nuclear acids com. Assn. for Biomolecular Resource Facilities, Phila., 1991-93; vis. scientist U. Sao Paulo, Brazil, summers 1989, 91, 92. Contbr. articles to profl. jours. Recipient grants NIH, 1986-94, 87-93, Am. Heart Assn., 1991-94, NSF, 1991-94, Jeffress Meml. Trust, 1986-88. Mem. AAAS, Am. Soc. for Microbiology, Am. Heart Assn. (basic scis. coun.). Home: 1620 Buford Rd Richmond VA 23235 Office: Va Commonwealth U Med Coll of Va Box 678 MCV Sta Richmond VA 23298

BUCK, M. SCOTT, electrical engineer; b. Highland Park, Mich., July 24, 1966; s. Gordon Scott and Rosemary (Paul) B. BSEE, Lawrence Tech. U., 1989; BS in Computer Sci., Wayne State U., 1992. Registered profl. elec. engring. technician. Elec. engr. U.S. Army, Warren, Mich., 1989—. Mem. NSPE, IEEE, Soc. Automotive Engrs., Mich. Soc. Profl. Engrs. Office: Software Engring Divsn Amsta-OS Warren MI 48397-5000

BUCK, RICHARD FORDE, physicist; b. Enterprise, Kans., Dec. 8, 1921; s. Charles Fay and Ruth (Scott) N.; m. Harriet J. Ojers, June 4, 1944; children: David R., Janet H., Paul S., Bryan T., Neal A., Daniel C. BS in Physics, U. Kans., 1943; MS in Physics, Okla. State U., 1955. Registered profl. engr., Okla. Cereal chemist Flour Mills Am., Kansas City, Mo., 1940-41; application engr. Tung-Sol Lamp Works, Newark, 1946-48; from instr. to assoc. prof. physics Okla. State U., Stillwater, 1948-76, assoc. prof. tech., 1976-84, prof. emeritus, 1984—; from engr. to lab. dir. Rsch. Found. Electronics Lab., Stillwater, 1948-76; dir. electronics lab. div. engring., tech. and architecture, Stillwater, 1976-84; cons. to USAF, NASA, various comml. mfrs. in aerospace industry, 1954-84. Founder, life bd. mem. Sheltered Workshop for Payne County, Okla., 1969—; past pres., life bd. mem. Stillwater Group Homes, 1980—; past chmn. bd. 1st Christian Ch., Stillwater, 1950—. 1st lt. U.S. Army, 1942-46, ETO. Republican. Home: 1301 Westwood Dr Stillwater OK 74074

BUCKELEW, ROBIN BROWNE, aerospace engineer, manager; b. York, Pa., Mar. 14, 1947; d. Grant Hugh and Frances (Coleman) Browne; m. William Paul Buckelew, June 5, 1971; children: Leon, Christina. BS in Aerospace Engring., U. Ala., 1970; MS in Engring., U. Ala., Huntsville, 1977. Registered profl. engr., Ala. Aerospace engr. U.S. Army Missile Command, Redstone Arsenal, Ala., 1970-74; systems engr. U.S. Army Missile Intelligence Agy., Huntsville, Ala., 1974-81; group leader air vehicle Sentry U.S. Army Ballistics Missile Def. System Command, Huntsville, 1981-83, interceptor engr. High Endoatmospheric Def. Interceptor, 1983-85; chief air vehicle div. HEDI project U.S. Army Strategic Def. Command, Huntsville, 1985-88, chief Ground Based Interceptor Experiment officer, 1988-91, chief engr. HEDI project, 1991-92; dir. systems directorate U.S. Army Space and Strategic Def. Command, Huntsville, 1993—. Contbr. articles to AIAA conf. proceedings. Bd. dirs. Trinity Personal Growth Ctr., Huntsville, 1990-92. Named Strategic Def. Engr. of Yr., NSPE, 1990, Disting. Engring. fellow U. Ala., 1993; recipient Superior Civilian Svc. award U.S. Army, 1991. Mem. AIAA (sr. mem.), Capstone Engring. Soc. (dir. dist. 4). Methodist. Home: 117 Bel Air Rd SE Huntsville AL 35802-3107 Office: US Army Space-Strat Def Cmd Huntsville AL 35807-3801

BUCKINGHAM, MICHAEL JOHN, oceanography educator; b. Oxford, Eng., Oct. 9, 1943; s. Sidney George and Mary Agnes (Walsh) B.; m. Margaret Penelope Rose Barrowcliff, July 15, 1967. BSc with hons., U. Reading (Eng.), 1967, PhD, 1971. Postdoctoral rsch. fellow U. Reading, 1971-74; sr. sci. officer Royal Aircraft Establishment, Farnborough, Eng., 1974-76; prin. sci. officer Royal Aircraft Establishment, 1976-82; exchange scientist Naval Rsch. Lab., Washington, 1982-84; vis. prof. MIT, Cambridge, 1986-87; sr. prin. sci. officer Royal Aircraft Establishment, 1983-86, 1987-90; prof. oceanography Scripps Instn. of Oceanography, La Jolla, Calif., 1990—; vis. prof. Inst. Sound and Vibration rsch., Southampton, Eng., 1990—; cons. Commn. of European Communities, Brussels, Belgium, 1989—; dir. Arctic rsch. Royal Aerospace Establishment, Farnborough, 1990—. Editor-in-chief Jour. Computational Acoustics; author: Noise in Electronic Devices and Systems, 1983; patentee in field; contbr. articles to profl. jours. Recipient Clerk Maxwell Premium, Inst. Electronic and Radio Engrs. London, 1972, A.B. Wood Medal, Inst. Acoustics, Bath, Eng., 1982, Alan Burman Pub. award, Naval Rsch. Lab., 1988, Commendation for Disting. Contbns. to ocean acoustics Naval Rsch. Lab., 1986. Fellow Inst. Acoustics (U.K.), Inst. Elec. Engrs. (U.K.); mem. Acous. Soc. Am. (chmn. acoustical oceanography tech. com. 1991—), Sigma Xi. Avocations: photography, squash, flying gliders. Home: 7921 Caminito Del Cid La Jolla CA 92037-3404 Office: Scripps Inst Oceanography Marine Phys Lab La Jolla CA 92093

BUCKIUS, RICHARD O., mechanical and industrial engineering educator; b. Sacramento, Calif., July 17, 1950; s. Orland Edwin and Holley (Lynip) B.; m. Kathleen Marie Mariani Buckius, Aug. 21, 1972; children: Sarah Jane, Emily Ann. BS, U. Calif., Berkeley, 1972, MS, 1973, PhD, 1975. Asst. prof. U. Ill., Urbana, 1975-80, assoc. prof., 1980-84, prof., 1984--, assoc. head mechanical and indsl. engring dept., 1985-87, assoc. vice chancellor for rsch., 1988-91, Richard W. Kritzer prof., 1992--; program dir. Nat. Sci. Found., Washington, 1987-88; assoc. tech. editor, Am. Soc. Mechanical Engrs. Jour of Heat Transfer, 1987-93. Co-author: Fundamentals of Engineering Thermodynamics, SI version, 1987, English/SI version, 1987, 2d edit. 1992; contbr. articles to profl. jours. Recipient Stanley H. Pierce Faculty award U Ill., 1979, Everitt award, 1980, Campus award, 1980, Halliburton Engring. Edn. Leadership award, 1987, CIC Academic Leadership fellow, 1989. Fellow Am. Soc. Mechanical Engrs. (exec. com. heat transfer div. 1989--); mem. AIAA, The Combustion Inst., Am. Soc. for Engring. Edn. Office: U Ill Dept Mech and Indsl 1206 W Green St Urbana IL 61801

BUCKLAND, PETER GRAHAM, structural engineer; b. London, Sept. 7, 1938; arrived in Can., 1963; s. Walter Basil and Norah Elaine Buckland; m. Aralee Buckland, July 21, 1974; 3 children. BA, Cambridge U., 1960, MA, 1972; LLD (hon.), U. B.C., 1992. Profl. engr., B.C. Engr. Freeman Fox & Ptnrs., London, 1960-63; engr., project mgr. Canron Inc., Vancouver, B.C., 1963-68; engr. Swan Wooster-CBA, Vancouver, 1968-70; pres. Buckland & Taylor Ltd., North Vancouver, B.C., 1970—. Contbr. 22 articles to profl. jours. Recipient Meritorious Achievement award Assn. Profl. Engrs. B.C., 1981. Fellow Can. Acad. Engring., Can. Soc. for Civil Engring. (recipient Le Prix P. L. Pratley award, 1991, 1992); mem. Inst. Civil Engrs. (U.K.), Am. Soc. Civil Engrs. Achievements include derivation of loading for long span bridges; contributed to seismic upgrading of Golden Gate Bridge in San Francisco; independent engineer for 13 km-long Northumberland Strait crossing bridge. Home: 1591 Bowser Ave, North Vancouver, BC Canada V7P 2Y4

BUCKLAND, ROGER BASIL, university dean, educator, vice principal; b. Lower Jemseg, N.B., Can., May 18, 1942; s. Basil John and Nancy (Coates) B.; m. Vicki Nealson, Aug. 7, 1965; children—Kenneth Roger, Adrienne Elise. Diploma, Nova Scotia Agrl. Coll., Truro, 1961; B.Sc., McGill U., 1963, M.Sc., 1965; Ph.D., U. Md., 1968. Assoc. prof. Macdonald Coll., McGill U., Ste. Anne de Bellevue, Que., Can., 1973-80, chmn. prof., 1979-85, vice prin., dean, 1985—; sabbatical leave Sta. de Recherche Avicole and Poultry Research Ctr., Nouzilly, France, 1977-78. Bd. dirs. Morgan Arboretum Assn., 1985—, Louis G. Johnson Found., 1989—; mem. hon. bd. West Island Community Radio, 1985-88; mem. bd. govs. RCS-Netherwood, Rothesay, N.B., 1987—. Recipient Scroll award Poultry Sci. Research Assn., 1972; Une Mission France-Quebec, 1977. Mem. Confederation Can. Faculties Agr. and Vet. Medicine, Poultry Sci. Assn., Centre de Recherche en Zootechnie, Can.-Cuba Joint Poultry Working Group (chmn. 1978—). Office: McGill U Macdonald Campus, 21111 Lakeshore Rd, Sainte Anne de Bellevue, PQ Canada H9X 3V9

BUCKLEY, JOHN LEO, retired environmental biologist; b. Binghamton, N.Y., Sept. 22, 1920; s. Leo J. and Anna (Rounds) B.; m. Claire Bennett, May 24, 1947; children: Susan, John, James, David. BS, SUNY, 1942, MS, 1948, PhD, 1951. Instr. U. Alaska, Fairbanks, 1950-51, adj. prof., 1958; leader coop. wildlife rsch. unit U.S. Fish and Wildlife Svc., Fairbanks, 1951-58; dir. Patuxent Wildlife Rsch. Ctr. U.S. Fish and Wildlife Svc., Laurel, Md., 1959-64; staff mem. office of sci. and tech. Office of the Pres., Washington, 1962-65, 68-71; environ. quality advisor sci. advisors office Dept. Interior, Washington, 1965-68; dep. dir. rsch. EPA, Washington, 1971-75, ret., 1975, cons. sci. adv. bd., 1975-81; cons. Nat. Inst. Environ. Health & Sci., Rsch. Triangle Park, N.C., 1975-76; cons. Broome County Health Dept., Binghamton, 1981—; mem. U.S. delegations on environ. pollution to Sweden, Japan, Federal Republic of Germany, USSR, Egypt, 1962-76. Author: (Nat. Acad. Press publ.) Ecological Knowledge and Environmental Problem Solving, 1986. City councilman Laurel, 1970-72; mem. Broome County Environ. Mgmt. Coun., 1978—, Broome County Resource Recovery Agy., 1984-88. Capt. USMC, 1942-46, PTO. Fellow AAAS; mem. The Wildlife Soc., Sigma Xi. Democrat. Home: 3 Avon Rd Binghamton NY 13905

BUCKLEY, REBECCA HATCHER, physician; b. Hamlet, N.C., Apr. 1, 1933; d. Martin Armstead and Nora (Langston) Hatcher; m. Charles Edward Buckley, III, July 9, 1955; children—Charles Edward, IV, Elizabeth Ann, Rebecca Kathryn, Sarah Margaret. B.A., Duke U., 1954; M.D., U. N.C., 1958. Intern Duke U. Med. Ctr., Durham, N.C., 1958-59, resident, 1959-61, practice medicine, specializing in pediatric allergy and immunology, 1961—; dir. Am. Bd. Allergy and Immunology, Phila., 1971-73, chmn. exam. com., 1971-73, co-chmn. bd. dirs., 1982-84; chmn. Diagnostic Lab. Immunology, 1984-88; mem. staff Duke U. Med. Ctr.; asst. prof. pediatrics and immunology, 1968-72, assoc. prof. pediatrics, 1972-76, assoc. prof. immunology, 1972-79, prof. immunology, 1979—, J. Buren Sidbury prof. pediatrics, 1979—. Contbr. numerous articles to med. publs. Recipient Allergic Diseases Acad. award Nat. Inst. Allergy and Infectious Diseases, 1974-79, Merit rsch. award NIH, 1987—, Gen. Clin. Rsch. Ctrs. award NIH, 1990, Nat. Bd. award Med. Coll. Pa., 1991, Clemens von Parquet award Georgetown, 1993, Disting. Tchr. award Duke U. Med. Alumni Assn., 1993. Fellow Am. Acad. Allergy (mem. exec. com. 1975-82, pres. 1979-80); mem. Am. Assn. Immunologists, Soc. Pediatric Rsch., Am. Acad. Pediatrics (Bret Ratner award 1992), Southeastern Allergy Assn. (pres. 1978-79), Am. Pediatric Soc. (coun. mem. 1991-97). Republican. Episcopalian. Home: 3621 Westover Rd Durham NC 27707-5032 Office: Duke U Med Ctr PO Box 2898 Durham NC 27710-0001

BUCKMORE, ALVAH CLARENCE, JR., computer scientist, ballistician; b. Lewiston, Maine, Sept. 11, 1944; s. Alvah Clarence and Mary (Begin) B. Student, Holyoke Community Coll., Nat. Radio Inst., Famous Writers Sch., U. Mass. Cert. firearms instr.; lic. amateur radio operator. Chief exec. officer, chief scientist Buckmore Enterprises, Westfield, Mass., 1974—; developer math./engring. software database for microcomputer Calculated Solutions (formerly SC Applied Tech. Inc.), Columbia, S.C.; mgmt. cons. firearms industry; instr. Mass. Mil. NCO Acad., 1976; mem. Mass. State Rifle and Pistol Team, 1976. Contbr. Collier's Ency., articles to profl. jours. Mem. Mass. Rep. Com., Rep. Presdl. Task Force, Mass. Rep. Senate Com. With U.S. Army, 1974, former prisoner of war, Vietnam; with Mass. Army N.G., 1975-78. Recipient Internat. Recognition award, 1979; NSF fellow, 1978—. Mem. AAAS, Computer Soc. of IEEE, Am. Def. Preparedness Assn., NRA (life), DAV (life), Vietnam Vets. Am. (mem. vets. coun. Liberty chpt. 219), Assn. for Computer Tng. and Support, Math. Assn. Am., Am. Radio Relay League, N.Am. Hunting Club, Am. Fedn. Police, Am. Legion. Achievements include development of amateur radio satellite communications, of parallel processing techniques and code for ballistic applications; over 38 major discoveries made in ballistics, including the discovery of 3 new sciences: time physics, the study of the physical properties of time; force-fields, the study of the absorption, displacement, projection, or reflection of kinetic energy; and ballistic signatures, the study of the physical characteristics of a bullet in terminal flight. Address: 18 Tannery Rd Westfield MA 01085-4822

BUCKNER, HARRY BENJAMIN, civil engineer; b. Asheville, N.C., June 25, 1970; s. Harry and Dorothy Ann (Black) B.; m. Virginia Gail Dillingham, Oct. 3, 1992. BSCE, N.C. State U., 1992. Rsch. asst. N.C. State U., Raleigh, 1991-92; assoc. McGill Assocs., Asheville, N.C., 1992—. Mem. ASCE (assoc.). Baptist.

BUCKNER, JAMES LEE, forester, biologist; b. East Flat Rock, N.C., Oct. 20, 1940; s. Guy Vernon Buckner and Clara (Cairnes) Corn; m. Marian Arlene Sexton, June 8, 1962; children: John Forrest, Mitchell Andrew. BS in Forestry, U. Ga., 1964, MS in Wildlife, 1969. Registered forester, Ga., Ala., N.C. Asst. ops. forester Internat. Paper Co., Bainbridge, Ga., 1964-66; regional wildlife specialist Internat. Paper Co., Panama City, Fla., 1966-69; project leader wildlife rsch. Internat. Paper Co., Bainbridge, 1969-71, sect. leader wildlife rsch., 1971-83; mgr. wildlife ecology Internat. Paper Co., Dallas, 1983-85; owner, pres. Forestry and Wildlife Mgmt. Svcs., Thomasville, Ga., 1985—. Author, coauthor 32 publs. on wildlife, forestry and fisheries. Mem. adv. com. Abarham Baldwin Coll. Sch. of Forestry and Wildlife Mgmt.. Ga. Forest Conservationist of Yr., Ga. Wildlife Fedn., 1981. Mem. Soc. Am. Forsters, The Wildlife Soc. (cert. wildlife biologist 1979), Xi Sigma Pi. Avocation: family activities. Home and Office: PO Box 589 Thomasville GA 31799-0589

BUCKNER, JOHN KENDRICK, aerospace engineer; b. Indpls., June 13, 1936; s. Roland Kendrick and Lucille (Cave) B.; m. Nancy Ann Smith, June 13, 1974; children: James Kendrick, Bari Kay, Kendrick Ann. BA in Math., DePauw U., 1958; MS in Aero-Engring., Stanford U., 1960. Aerodynamics sr. engr. Gen. Dynamics, Ft. Worth, 1960-69, supr. aerodyns., 1969-75, aircraft project engr., 1975-77, mgr. flight controls, 1977-80, dir. advanced programs, 1980-89, v.p. spl. programs, 1989—; com. mem. Nat. Rsch. Coun./Naval Studies Bd., Washington, 1990, Aeronautics and Space Engring. Bd., 1992—; mem. aerospace rsch. and tech. subcom. aeronautics adv. com. NASA, Washington, 1988-93. Bd. dirs. Am. Heart Assn., Ft. Worth, 1990—. Assoc. fellow AIAA (chmn. aircraft design tech. com. 1990-92, pub. policy com. 1988—). Achievements include aerodynamic design of high performance jet fighters F-111 and F-16. Home: 5401 Benbridge Dr Fort Worth TX 76107-3209 Office: Lockheed Ft Worth Co PO Box 748 Fort Worth TX 76101-0748

BUCKSBAUM, PHILIP HOWARD, physicist; b. Grinnell, Iowa, Jan. 14, 1953; s. Arnold M. and Corinne P. (Schlass) B.; m. Roberta J. Morris, June 15, 1985. AB, Harvard Coll., 1975; PhD, U. Calif., Berkeley, 1980. Postdoctoralfellow Lawrence Berkeley Lab., Berkeley, Calif., 1980-81, Bell Labs., Holmdel, N.J., 1981-82; mem. tech. staff AT&T Bell Labs., Murray Hill, N.J., 1982-90; prof. physics U. Mich., Ann Arbor, 1990—. Fellow Am. Phys. Soc. Office: Physics Dept U of Michigan Ann Arbor MI 48109

BUDA, ALEKS, science administrator, history researcher, educator; b. Elbasan, Albania, Sept. 7, 1910; s. Dhimiter and Elena (Kasapi) B.; m. Vasilika Stratoberdha, Nov. 24, 1941; 1 child, Tatjana. Student, U. Vienna, Austria, 1938, U. Tirana, Albania, 1958, Acad. Sci., Tirana, Albania, 1972, Bulgarian Acad. Sci., Sofia, 1979. History educator Gymnasium, Tirana, Albania, 1938-40, Lycium, Korca, Albania, 1940-41; dir. Nat. Libr., Tirana, Albania, 1945-46; history rsch. Inst. of Scis., Tirana, Albania, 1947-55, Inst. Linguistics, History, Tirana, Albania, 1955-57; prof. history U. Tirana, 1957-72; pres. Acad. Sci., Tiranë, 1972—. Author: Historical Notes, vol. 1, 2, 1986; co author: (chief ed.) History of Albania , vol. 1, 1959, vol. 2, 1965; contbr. articles to profl. jours. Mem. Gen. Coun. Dem. Front, Tirana, 1985; deputy People's Coun., 1950-91. Recipient Tchr. of People award Presidium of People's Coun., 1977, Great Golden medal Pres. Austria, 1981. Mem. Internat. Soc. for South-East Europe Studies (vice-chmn. 1066, medal 1973), German Soc. for South-East Europe (hon., diploma 1981). Avocations: music, literature. Home: Asim Vokshi 91, Tiranë Albania Office: Acad of Scis, Tiranë Albania*

BUDALUR, THYAGARAJAN SUBBANARAYAN, chemistry educator; b. India, July 14, 1929; came to U.S., 1969, naturalized, 1977; s. Subbanarayan Subbuswamy and Parvatham (Gopalakrishnan) B.; children: Chitra, Poorna, Kartik. M.A., U. Madras, 1951, M.Sc., 1954, Ph.D., 1956. Reader organic chemistry U. Madras, 1960-68; prof. chemistry U. Idaho, Moscow, 1968-74; prof. chemistry, dir. div. earth phys. sci. U. Tex., San Antonio, 1974—; lectr. in field. Author: Mechanisms of Molecular Migrations; Selective Organic Transformations; Editorial bd. chem. jours.; contbr. articles to profl. jours.; 3 patents in field. Recipient Intra Sci. Research award, 1966. Fellow Am. Chem. Soc.; mem. Chem. Soc. London, Soc. Cosmetic Chemistry N.Y. Acad. Sci., Am. Inst. Chemists, Sigma Xi, Phi Kappa Phi. Club: Lions. Home: 12200 Interstate Apt 1902 San Antonio TX 78230 Office: U Tex Loop 1604 NW San Antonio TX 78249

BUDDINGTON, PATRICIA ARRINGTON, engineer; b. Takoma Park, Md., Dec. 25, 1950; d. Warren and Elsie (Miller) B. BS, Northrop Inst. Tech., 1973; MS, Fla. Inst. Tech., 1986. With Air Force Systems Command, Edwards AFB, Calif., 1973-78; various positions Boeing Def. & Space Group, Huntsville, Ala., 1978-81, test engr. reaction control system inertial upper stage, 1981-86, lead engr. microgravity material processing facility, 1986-88, task leader advanced civil space systems, 1988—. Mem. AIAA (sr. mem.). Office: Boeing Advanced Civil Space PO Box 240002 JW-21 499 Boeing Blvd Huntsville AL 35824-6402

BUDELMANN, BERND ULRICH, zoologist, educator; b. Hamburg, Germany, Apr. 1, 1942; came to the U.S., 1987; s. Gunther and Minna (Siemssen) B. PhD, U. Munich, 1970; degree, U. Regensburg, 1975. Asst. prof. U. Regensburg, Germany, 1973-78, assoc. prof., 1978-87, Heisenberg fellow, 1979-84; assoc. prof. U. Tex., Galveston, 1987-93, prof., 1993—; mem. scientific adv. bd. Stazione Zoologica Anton Dohrn, Naples, Italy, 1992—. Contbr. articles to Nature, Philos. Transactions of Royal Soc., Jour. Comparative Physiology. Grantee Deutsche Forschungsgemeinschaft, 1979-85, NIH, 1989—, Wellcome Trust, 1991. Mem. Am. Soc. for Gravitational and Space Biology, Am. Soc. Zoologists, Assn. for Rsch. in Otolaryngology, Barany Soc., Deutsche Zoologische Gesellschaft, Gesellschaft Deutscher Naturforscher und Arzte, J.B. Johnson Club, Neurootological and Equilibriometric Soc., Soc. for Exptl. Biology, Soc. for Neurosci., Verband Deutscher Biologen, Sigma Xi (sec. chpt. 1988—). Lutheran. Home: 415 E Beach Dr # 901A Galveston TX 77550 Office: U Tex Med Br Marine Biomedical Institute Galveston TX 77555-0863

BUDEN, DAVID, aerospace power engineer, nuclear power researcher; b. Phila., July 10, 1930; s. Samuel and Stella Bessie (Friend) Budenstein; m. Pennie Miller, Aug. 23, 1953 (dec. June 1978); children: Susan, Janice, Barry; m. Rosemary Jacobson Vidale, Apr. 13, 1984. BS, Pa. State U., 1952, MS, 1954; M of Bus. Econs., Claremont (Calif.) Grad. Sch., 1966. From staff mem. to mgr. Gen. Electric Co., Cin., 1955-61; mgr. Aerojet Nuclear Systems Co., Sacramento, 1961-72; from staff mem. to mgr. Los Alamos (N.Mex.) Nat. Lab., 1972-84; program mgr. Sci. Application Internat., Albuquerque, 1984-87; prin. scientist Idaho Nat. Engring. Lab., Idaho Falls, 1987—; mem. Lunar Power Working Group, Washington, 1992; mem. Stafford Synthesis Group, Washington, 1990-91; mem. space power com. NRC, Washington, 1987; tech. co-chair Space Nuclear Power Symposium, 1986; tech. organizer Intersoc. Energy Conversion Energy Conf., 1982, 85-90, 92. Co-author: Space Nuclear Power, 1985, Aerospace Nuclear Safety, 1993; contbr. articles on space nuclear power and propulsion to profl. jours. With U.S. Army, 1953-55. Mem. AIAA (tech. com. 1983-86, 91—), Am. Nuclear Soc. (tech. co-chair 1992). Achievements include patents on nuclear reactor refuelable in space, survivable pulse power space radiator, turbojet speed control. Office: Idaho Nat Engring Lab PO Box 1625 Idaho Falls ID 83415

BUDENHOLZER, ROLAND ANTHONY, mechanical engineering educator; b. St. Charles, Mo., Nov. 24, 1912; s. Joseph P. and Mary (Willey) B.; m. Florence C. Christiansen, Nov. 28, 1941; children: Francis Edward, John Christopher, Robert Joseph. B.S. in Mech. Engring, N.Mex. State U., 1935; M.S. in Mech. Engring, Calif. Inst. Tech., 1937, Ph.D., 1939. Grad. asst. Calif. Inst. Tech., 1935-39; rsch. fellow Am. Petroleum Inst., 1939-40; faculty Ill. Inst. Tech., 1940—, prof. mech. engring., 1947-78, prof. emeritus, 1978—; resident rsch. assoc. Argonne Nat. Lab., summer 1961; cons. IIT Rsch. Inst., 1946-84; dir. Midwest Power Conf., 1949-52; dir. Am. Power Conf., 1952-78, chmn., 1978-90, chmn. emeritus, 1990—; rep. Am. Power Conf. to World Energy Conf., 1972-78; bd. dirs. U.S. nat. com. World Energy Conf., 1972-78, mem. exec. com., 1973-78; bd. dirs. Triodyne, Inc. Author handbooks; contbr. to encys., profl. jours. Recipient George Westinghouse gold medal Am. Soc. M.E., 1968, award Chgo. Tech. Socs. Council, 1975, Disting. Alumni award N.Mex. State U., 1981. Mem. AAUP (pres. Ill. Inst. Tech. chpt. 1963-64), ASME (hon., sec., exec. com. power div. 1967-68, chmn. 1970-71), Am. Soc. Engring. Edn., Western Soc. Engrs. (trustee 1969-72), Armour Faculty Club, Sigma Xi (pres. Ill. Inst. Tech. chpt. 1948-49), Triangle, Tau Beta Pi, Pi Tau Sigma, Tau Kappa Epsilon. Home: 306 Harris Ave Clarendon Hills IL 60514 Office: Ill Inst Tech Chicago IL 60616

BUDINGER, FREDERICK CHARLES, geotechnical engineer; b. Chgo., June 13, 1936; m. Shirley M. Watson, Aug. 15, 1961 (div. 1985); children: Teri, Beth, Vince, Bernie, Steve, Clare, Martin; m. Jo Ann Packey Bender, May 25, 1986. BA, Ariz. State U., 1959. Registered profl. engr., Calif., Wash., Idaho, Mont., Oreg., Alaska. Tchr. math. Scottsdale (Ariz.) High Sch., Spokane, Wash., 1956-60; tchr. physics/sci. Ctrl. Cath. High Sch., Portland, Oreg., 1960-61; field tech. Phoenix, 1961-67; engring. tech. S/G Test Labs., Lampoc, Calif., 1967-71; br. mgr., engr. Braun Skaggs & Kevorkian Engring., Bakersfield, Calif., 1971-76; pres., geotechnical engr. Budinger and Assocs., Spokane, Wash., 1976—. Author: Soil Compaction, 6th edit., 1992. Mem. ASCE (Engr. of Yr. 1993), ASTM (com. on soil and rock), Wash. Soc. Profl. Engrs. (sponsor high sch. bridge bldg. contest 1990-93), Rotary. Republican. Episcopalian. Home: 1729 E 18th St Spokane WA 99203 Office: Budinger and Assocs 3820 E Broadway Spokane WA 99202

BUDINGER, THOMAS FRANCIS, radiologist, educator; b. Evanston, Ill., Oct. 25, 1932; married, 1965; 3 children. BS, Regis Coll., 1954; MS, U. Wash., 1957; MD, U. Calif. Berkeley, 1964, PhD, 1971. Asst. chemist Regis Coll., Colo., 1953-54; analytical chemist Indsl. Labs., 1954; sr. oceanographer U. Wash., 1961-66; physicist Lawrence Livermore Lab., U. Calif., 1966-67; resident physician Donner Lab. and Lawrence Berkeley Lab., 1967-76; H. Miller Prof. med. rsch. and group leader rsch. medicine Donner lab., prof. elec. engring. and computer sci. Donner Lab., U. Calif. Berkeley, 1976—; with Peter Bent Brigham Hosp., Boston, 1964; dir. med. svc. Lawrence Berkeley Lab., 1968-76, sr. staff scientist, 1980—; chmn. study sect. NIH, 1981-84; prof. radiology U. Calif. San Francisco, 1984—. Recipient Special award Am. Nuclear Soc., 1984. Mem. AAAS, Am. Geophysical Union, N.Y. Acad. Sci., Soc. Nuclear Medicine, Soc. Magnetic Rsch. Medicine (pres. 1984-85). Achievements include research in imaging body functions, electrical, magnetic, sound and photon radiation fields, electron microscopy, polar oceanography, nuclear magnetic resonance, reconstruction tomography and instrument development, cardiology. Office: U Calif Lawrence Berkeley Lab Mail Stop 55-121 Berkeley CA 94720*

BUDINSKI, KENNETH GERARD, metallurgist; b. Rochester, N.Y., June 29, 1939; s. Anthony C. and Loretta C. (Hunt) B.; m. Marilyn A., Aug. 26, 1961; children: Michael K., Mark D., Steven T. BS in Mech. Engring., GM Inst., 1961; MS in Metallurgical Engring., Mich. Tech. U., 1963. Devel. engr. Rochester Products div. GM, N.Y., 1957-62; metallurgist Eastman Kodak Kodak Park, Rochester, 1964-75, sr. metallurgist, 1975-85, tech. assoc., 1985--. Author: Engineering Materials: Properties and Selection, 4th edit. 1992, Surface Engineering for Wear Resistance, 1989. Pres. Greece Civic Orgn., Rochester, 1965-70; com. mem. Greece Rep. Com., Rochester, 1965--. Fellow Am. Soc. for Metals (chpt. advisor 1986--, chmn. 1971-72), Am. Soc. for Testing Materials (subcom. chmn. 1987--, chmn. G2 com. 1983-85); mem. Welding Rsch. Coun., Surface Engring. Soc. Roman Catholic.

BUDMAN, CHARLES AVROM, aerospace engineer; b. Washington, Feb. 18, 1963; s. Jack Franklin Budman and Janet Helene Goldfarb Wineland; m. Phaedra Lucara, Aug. 24, 1986; 1 child, Emily Tovah. BS in Aero. Engring. magna cum laude, U. Md., 1986, MS in Aero. Engring., 1988. Co-op student Applied Physics Lab., Johns Hopkins U., Laurel, Md., 1983-86, assoc. engr., 1986-92, sr. engr., 1992—. Mem. AIAA. Democrat. Jewish. Achievements include rsch. and devel. in field; flight test and reliability evaluation of USN missile systems. Office: Johns Hopkins Univ Applied Physics Lab Johns Hopkins Rd Laurel MD 20723

BUDNICK, THOMAS PETER, social worker; b. Ludlow, Mass., Feb. 16, 1947; s. Henry F. and Mildred Mary (Killian) B. BS, Am. Internat. Coll., 1972, MA, 1975. Lic. cert. profl. social worker. Mailhandler U.S. Postal Svc., Springfield, Mass., 1970-72; substitute tchr. Pub. Schs. Dept., Ludlow, Mass., 1973-74; social worker Mass. Dept. Pub. Welfare, Springfield, 1975—; pres. Am.'s Manifest Destiny Soc., Inc., West Harwich, Mass., 1979—; bd. dirs. Mass. Astronomy Club, Boston, 1988—. Contbr. numerous articles to jours. V.p. Local 509, Boston, 1989. Democrat. Home: 3 Harding Ave Ludlow MA 01056-2327

BUDWANI, RAMESH NEBHANDAS, consultant; b. Shikarpur, Sind, Pakistan, July 30, 1940; came to U.S., 1969; s. Nebhandas J. and Jaidevi Budwani; m. Meena Rajpal, Mar. 28, 1977; children: Rocky, Deepak. Degree in chem. engring., Laximinarayan Inst. Tech., Nagpur, India, 1965; degree in nuclear engring., Indian Atomic Energy Sch., Bombay, 1966. Cert. profl. estimator; cert. cost estimator/analyst; apptd. to panel arbitrators Am. Arbitration Assn. Sci. officer Indian Atomic Energy Commn., Bombay, 1966-69; planning and scheduling engr. Parsons Jurden Corp., N.Y.C., 1970-71; constrn. scheduler Gibbs & Hill, Inc., N.Y.C., 1971-72; supr. planning and scheduling United Engrs. and Constrn., Phila., 1972-76; mgr. cost analysis Burns & Roe Inc., Oradell, N.J., 1976-92; prin. cost engr. GE Nuclear Energy, San Jose, Calif., 1993—; expert witness/analysis investigations regarding power plants, N.Y., Tex., Ill., Conn., Ohio, Ga., Calif. N.C., Ariz., Pa.; developer privately-held power plant database; lectr. in field. Contbr. articles to profl. jours. Recipient Jesse H. Neal Cert. of Merit, Assn. Bus. Pubs., 1985. Mem. Am. Assn. Cost Engrs. (pres. Rampo Valley sect.), Project Mgmt. Inst., Inst. Cost Analysis (cert.), Nat. Estimating Soc. (cert.), Performance Mgmt. Assn., Soc. Cost Estimating and Analysis (cert.). Hindu. Avocations: stamp collecting, photography, dancing, sports, reading. Home: 131 Rt 46W Rules Ct W # 12W Lodi NJ 07644

BUERK, DONALD GENE, medical educator; b. St. Louis, Jan. 28, 1946; s. Charles Albert and Virginia (Kirkpatrick) B.; m. Steffie Greif, July 7, 1968; children: Jesse Nathaniel, Daniel Joshua. BS, Case Western Res. U., 1969, MS, 1976; PhD, Northwestern U., 1980. Biomed. engr. St. Vincent Charity Hosp., Cleve., 1969-75; rsch. asst. Case Western Res. U., Cleve., 1976; rsch. fellow Evanston (Ill.) Hosp., Northwestern U., 1980-82; prof. La. Tech. U., Ruston, La., 1982-87; rsch. prof. Drexel U., Phila., 1987-90; rsch. prof. dept. ophthalmology Sch. Medicine, U. Pa., Phila., 1990—; vis. scientist Cath. U., Nijmegen, The Netherlands, 1988; vis. prof. Johns Hopkins U., Balt., 1986—. Author: Biosensors--Theory and Applications, 1993; contbr. over 40 articles to profl. jours. Recipient Travel award Internat. Soc. Oxygen Transport, Eng., 1978, Outstanding Rschr. award La. Tech. U., Ruston, 1984, New Investigator award NIH, 1986, Vis. Scientist award Dutch Govt., 1988, Ind. Investigator award NSF, 1990. Mem. IEEE, Am. Heart Assn., Am. Physiol. Soc., Assn. for Rsch. in Vision and Ophthalmology, Biomed. Engring. Soc., Microcirculatory Soc., Sigma Xi. Avocations: jazz saxophonist, film making. Office: U Pa IFEM-1 John Morgan Bldg Philadelphia PA 19104-6068

BUESCHEN, ANTON JOSLYN, physician, educator; b. Toledo, June 7, 1940; s. Robert F. and Mary J. (Joslyn) B.; m. Norma Jean McClanahan, Sept. 5, 1964; children—Anton, Elaine. Student, Va. Mil. Inst., 1958-61; M.D., U. Va., 1965. Diplomate: Am. Bd. Urology. Intern in surgery Vanderbilt U., 1965-66, asst. resident in surgery, 1966-67; resident in urology Ind. U., Indpls., 1969-72; practice medicine specializing in urology Birmingham, Ala., 1973—; instr. urology Tulane U. Sch. Medicine, 1972-73; asst. prof. div. urology dept. surgery U. Ala., Birmingham, 1973-75; assoc. prof. U. Ala., 1975-79, prof., 1979—, dir. div. urology, 1975—; chief urology sect. Children's Hosp., Birmingham, 1978-86. Contbr. numerous articles on urology to profl. jours. Served with M.C. U.S. Army, 1967-69. Mem. ACS, AMA (Billings Gold medal 1978), AAUP, Am. Urol. Assn., Am. Assn. Clin. Urologists, Soc. Univ. Urologists, Birmingham Urology Club, Jefferson County Med. Soc., Soc. for Pediatric Urology, Soc. Urologic Oncology, So. Med. Assn. (chmn. urology sect. 1987), Soc. Nuclear Medicine, Med. Assn. Ala. Office: U Ala Div Urology University Sta Birmingham AL 35294

BUESCHER, ADOLPH ERNST (DOLPH BUESCHER), aerospace company executive; b. St. Louis, Mo., Oct. 6, 1922; s. Adloph E. Sr. and Eugenie K. (Stroh) B.; m. Ruth L. Fleming, Aug. 21, 1948; children: Timothy Wayne, Philip Clay. BS in Mech. Engring., U. Mo., 1946; MS in Mech. Engring., Stanford U., 1950; postgrad., UCLA, 1950-52, Harvard U., 1958. Registered profl. engr., N.Y., Calif., Mo. Devel. engr. Eastman Kodak Co., Rochester, N.Y., 1946-49; supr. flight test, rsch. engr. Northrop Aircraft Inc., Hawthorne, Calif., 1952-53; mgr. controls and instruments Sverdrup & Parcel, Inc., St. Louis, 1953-56; program mgr. ATLAS, ICBM GE, Valley Forge, Pa., 1956-59, various positions, mktg. mgr., engring. mgr., gen. mgr., 1959-63; mgr. strategic planning, chief staff GE Aerospace Group, Valley

Forge, 1963-88; lectr. Franklin Inst., 1965-68. Author: Loran-C, 1990, Radar, 1991; contbr. articles to profl. jours. V.p. Citizens Coun., Greater Phila., 1958-60; chmn. Planning Commn., Whitemarsh Twp., 1959-68; chmn. Zoning Hearing Bd., Whitemarsh Twp., 1968-74. With USAF/AUS, 1942-45, ETO. Recipient Fellowship Std. Oil of Calif., 1950. Fellow ASME, AIAA (assoc.), The Explorer's Club, Am. Rocket Soc. (chpt. pres. 1942), Inst. Aero. Sci. (assoc.); mem. NSPE, Air Force Assn., U.S. Power Squadron (dist. exec. officer, nat. com.), Am. Def. Preparedness Assn. (bd. dirs. 1964-85), Greater Phila. C. of C. (mem. aviation com. 1963-73, Engr. of Yr. 1966), Sassafras River Yacht Club (commodore 1981), Chesapeake Bay Yacht Clubs Assn. (bd. govs. 1990, vice commodore 1991, commodore 1992), Pi Tau Sigma, Tau Beta Pi. Republican. Lutheran. Achievements include patent for automatic celestial navigation, others in field of aircraft/spacecraft instrumentation; first use of automatic control theory applied to hypersonic fluid flow, and incompressible liquid flow, early, high altitude test flights, in unpressured aircraft. Home: 6044 Cannon Hill Rd Fort Washington PA 19034

BUFFET, PIERRE, computer scientist; b. Sens, France, Apr. 2, 1946; s. Etienne and Marie-Magdeleine (Guinand) B.; m. Sylvie Chabauty, Sept. 10, 1977; children: Adeline, Noemie. Dipl.Engr., IDN, Lille, France, 1969. Researcher French Navy, Toulon, 1969-70; head computer dept. CNRS-CDST, Paris, 1971-78; database mgr. Questel, Paris, 1978-82, sci. dir., 1983—; chmn. bd. Questel, Inc., Arlington, Va., 1989—; dir. Hampden Data Svcs. Ltd., Abingdon, Eng. Assoc. editor Jour. Chem. Info. and Computer Scis., 1989; contbr. articles to profl. jours. Named European Info. Personality of Yr. European Assn. Information Scis., 1987. Mem. Internat. Coun. Sci. and Tech. Info., Eusidic (chmn. 1992—). Avocations: cooking, music, gardening. Office: Questel Le Capitole, 55 Ave des Champs-Pierreux, Nanterre France 92000

BUFFINGTON, DENNIS ELVIN, agricultural engineering educator; b. Elizabethville, Pa., Aug. 11, 1944; s. Edwin E. and Sally M. (Schwalm) B.; m. Anna Carol Stafford, June 30, 1968; children: Kristin, Melanie. BS, Pa. State U., 1966, MS, 1970; PhD, U. Minn., 1971. Registered profl. engr., Fla., Pa. Instr., grad. asst. Pa. State U., Univ. Pk., 1966-67; grad. asst. U. Minn., Mpls., 1968-71; asst. prof., then assoc. prof. Agrl. Engring. U. Fla., Gainesville, 1971-82, prof., 1982-85; prof., head agrl. engring. dept. Pa. State U., University Park, 1985—. Author or co-author 110 tech. articles. Recipient Outstanding Teaching awards Coll. Engring., 1977 and Agrl. Engring. Dept., 1982, U. Fla. Mem. Am. Soc. Agrl. Engrs (Young Educator award 1980), Am. Soc. Engring. Edn., ASHRAE, Alpha Epsilon (nat. pres. 1984-85). Republican. Lutheran. Avocations: woodworking, gardening, antique restoration. Office: Pa State Univ 250 Agrl Engring Bldg University Park PA 16802

BUFFINGTON, GARY LEE ROY, safety standards engineer, construction executive; b. Custer, S.D., Dec. 6, 1946; s. Donald L. B. and Madge Irene (Selby) Lampert; m. Kathleen R. Treloar, Aug. 3, 1965; children: Katherine, Lowell, Gary Jr. BS in Bus. Edn., Black Hill State Coll., 1971; AA in Criminal Justice, U. S.D., 1972, MS, 1974. Cert. safety profl., EMT, law enforcement officer, mine safety and health adminstrn. instr., OSHA instr.; Canadian registered safety profl.; lic. pvt. investigator. Contract miner Homestake Mining Co., Lead, S.D., 1966-72; dep. sheriff, criminal investigator Pennington County Sheriff's Dept., Rapid City, S.D., 1972-77; fed. mine inspector U.S. Dept. of Labor, Mine Safety and Health Adminstrn., Birmingham, Ala., 1977-79; supr., spl. investigator U.S. Dept. of Labor, Mine Safety and Health Adminstrn., Birmingham, 1979-81; supr., mine inspector U.S. Dept. of Labor, Mine Safety and Health Adminstrn., Grand Junction, Colo., 1981-83; safety and security mgr. Black & Veatch Engrs. Stanton Energy Ctr., Orlando, Fla., 1983-87; loss control mgr. Black & Veatch Engrs. AES Thames Cogeneration Plant, Uncasville, Conn., 1987-90; loss control mgr. Trans-Mo. River Tunnel project Black & Veatch, Engrs.-Architects, Kansas City, Mo., 1990-92; mgr. safety and security. metro rail constrn. mgr. Parsons-Dillingham, L.A., 1992—; mem. ANSI A-10 Accredited Standards Com., Washington, 1984—, Mine Safety and Health Adminstrn. Standards Com., Arlington, Va., 1981-83. Named Police Officer of the Year, Sundown Optimist Club, Rapid City, 1975; recipient Meritorious Achievement award, U.S. Dept. of Labor, Arlington, 1979, Monetary Spl. Achievement award, U.S. Dept. Labor, Arlington, 1980. Mem. Am. Soc. Safety Engrs., World Safety Orgn., Am. Indsl. Hygiene Assn., Am. Soc. for Indsl. Security, Nat. Safety Council, Moose Lodge. Republican. Lutheran. Avocations: photography, sports. Home: 505 N Kenwood St #1 Glendale CA 91206 Office: Parsons-Dillingham 523 W 6th St Ste 400 Los Angeles CA 90014

BUGG, CHARLES EDWARD, biochemistry educator, scientist; b. Durham, N.C., June 5, 1941; s. Everett I. and Annie Laurie (Newsom) B.; m. Barbara Bradshaw Bugg, Dec. 23, 1962; children: Barbara Jean, Elizabeth Anne, Charles Edward Jr. AB in Chemistry, Duke U., 1962; PhD in Physical Chemistry, Rice U., Houston, 1965. Rsch. chemist E.I. duPont de Nemours & Co., Wilmington, Del., 1966-67; postdoctoral fellow Calif. Inst. Tech., Pasadena, 1965-68; prof. biochemistry, assoc. dir., sr. scientist Comprehensive Cancer Ctr., sr. scientist Rsch. Ctr. in Oral Biol. U. Ala., Birmingham, 1968—, dir. Ctr. for Macrmolecular Crystallography, 1985—; cons. Schering Plough Rsch., Kenilworth, N.J., 1986—, BioCryst Pharms., Inc., Birmingham, Ala., 1986—, Eastman Kodak Co., Rochester, N.Y., 1988—; chmn. commn. on jours. Internat. Union of Crystallography, Chester, Eng., 1989—; vice chmn. U.S. Nat. Com. Crystallography, Washington, 1989-92. Contbr. articles to profl. jours., 1962—; editor jour. Acta Crystallographica, 1987—; co-editor: Crystallographic & Modeling Methods in Molecular Design, 1990; patentee in crystallography. Mem. Am. Crystallographic Assn. (pres. 1987-88), Am. Chem. Soc., Am. Assn. for Advancement of Sci., Fedn. Am. Socs. for Experimental Biology, Ala. Acad. Scis., Sigma Xi, Rotary (v.p. 1990—), Mountain Brook Club, Redstone Club. Avocations: hunting, fishing, traveling. Office: U Ala-Ctr Macromolecular Crystallography U Sta Box 79 THT Birmingham AL 35294-0005

BUGGIE, STEPHEN EDWARD, psychology educator; b. Mpls., Aug. 27, 1946; s. Gregory L. Lawrence and Elizabeth Eleanor (Stone) B.; m. Shauna Dee Galloway, Sept. 2, 1972; children: Kevin David, Brian G. Student, Foothill Coll., Stanford U., 1966; BA with distinction and honors, San Jose State Coll., 1968; MA in Psychology, U. Oreg., 1970, PhD in Psychology, 1974. Grad. rsch. asst. cognitive lab. psychology dept. U. Oreg., 1968-69, 70-71, grad. teaching fellow psychology dept., 1969-70, 71-73, instr., 1972, 73, rsch. asst. Ctr. for Ednl. Policy and Mgmt., 1974; psychology rsch. assoc. Inst. for Study of Cognitive Systems Tex. Christian U., 1970; asst. prof. psychology U. Ams., Puebla, Mexico, 1975-76; lectr. psychology U. Zambia, Lusaka, Africa, 1977-80; lectr., founding head psychology dept. U. Malawi, Zomba, Africa, 1980-89; assoc. prof. psychology Presbyn. Coll., Clinton, S.C., 1990—; psychology external examiner U. Swaziland, 1989; presenter 41st Ann. Nebr. Symposium on Motivation, 1992; English lang. examiner Malawi Nat. Parliament, 1983, 87. Assoc. editor Jour. Psychology in Africa, 1992; contbr. articles to Am. Cage-Bird, JSAS Catalogue of Selected Documents, Memory and Cognition, Psychonomic Sci., Jour. Rehab. Natural Scis. scholar, 1965; fellow Calif. State, 1968-69, Tex. Christian U., 1970, Nat. Coun. U.S.-Arab Rels. 1993; Belk Faculty grantee Presbyn. Coll., 1990-93; Fulbright fellow, 1992. Fellow Nat. Coun. U.S.-Arab Rels.; mem. APA, Am. Psychol. Soc. (presenter conv. 1991), African Studies Assn. (presenter conv. 1991, 92), Southeastern Psychol. Assn. (presenter conv. 1991, 92,93), Tau Delta Phi, Psi Chi, Epsilon Eta Sigma. Republican. Presbyterian. Achievements include rsch. in placebo effect in African traditional vs. western medication, psychology of oral memory performance in Malawi, developmental psychology of bird song, bibliotherapy with chronic schizophrenics. Home: 410 Chestnut Clinton SC 29325 Office: Presbyn Coll Phychology Dept Clinton SC 29325

BUGNO, WALTER THOMAS, civil engineer; b. Scranton, Pa., Oct. 11, 1942; s. Walter A. and Elizabeth (Phillips) B.; m. Helen N. Steltzner, Nov. 30, 1963; children:Melia A., Dawn A., Janelle A., Timothy A., Peter W. BSCE, U. Minn., 1967, MSCE, 1967. Registered profl. engr., Calif., Alaska. Sr. engr. Chevron Rsch. and Tech. Co., San Francisco, 1968—; adviser Pres.'s Arctic Rsch. Commn., Washington, 1992. Contbr. articles on developing petroleum res. using tunnelling, other petroleum drilling topics to profl. publs. Bd. dirs. Calif. Christian Coll., Stockton, 1982-84; corp. chmn. Ch. of Christ, Pleasant Hill, 1986—. Fellow ASCE. Republican. Office: Chevron Rsch and Tech Co 2400 Camino Ramon San Ramon CA 94583

BUHAC, H(RVOJE) JOSEPH, geotechnical engineer; b. Brijesta, Dalmatia, Croatia, Aug. 8, 1938; came to U.S., 1966; s. Jozo Ivan and Marija (Božovic) B.; divorced; 1 child, Ella Joane. MS, Columbia U., 1970. Registered profl. engr., Ohio, N.Y., Va. Asst. to prof. U. Sarajevo, Bosnia, 1963-66; project mgr. Seeley, Stevenson, Value & Knecht, N.Y.C., 1966-75; soil engr. Stone & Webster, N.Y.C., 1975-78; sr. geotech. engr. AEP Svc. Corp., Columbus, Ohio, 1978—; cons. in civil engring. Translator Karst Hydrogeology, 1981; contbr. articles to tech. mags. Scholar Columbia U., 1968. Fellow ASCE. Achievements include rsch. in stability of concrete dam and new role of drainage systems in dam stability, safety of dams. Home: 1400 B Lake Shore Dr Columbus OH 43204 Office: AEPS Co 1 Riverside Plz Columbus OH 43215

BUHKS, EPHRAIM, electronics educator, researcher, consultant; b. Kishinev, U.S.S.R., Apr. 30, 1949; came to U.S., 1980; BS in Physics, Kishinev U., 1971; PhD in Chemistry, Tel-Aviv (Israel) U., 1980. Rsch. fellow U. Del., Newark, 1980-81; project leader Solavolt Internat. (Shell), Newark, 1981-83; mgr. R&D B.F. Goodrich R&D Ctr., Brecksville, Ohio, 1983-87; tech. dir. Sunstone Inc., Dayton, N.J., 1987-89; asst. dir. supervisory acad. svcs. ORT Tech. U.S.A., N.Y.C., 1990—; cons. Johnson Rsch. Found., U. Pa., Phila., 1981-83, Kingston Tech., Inc., Dayton, 1989, Energia, Inc., Princeton, N.J., 1989-90. Editor: Protein Structure & Electronic Reactivity, 1987; contbr. over 42 articles to profl. jours. Solar Energy Rsch. Inst. fellow, 1980, Von Humboldt Found. fellow, 1980. Mem. Am. Chem. Soc., Am. Phys. Soc., Optical Soc. Am., Soc. Photo-Electric Engrs. Achievements include patent for Electrodeless Heterogeneous Polypyrrole Composites; patent pending for Method and Device for Optical Storage of Information; invention of Fiber-Optic Viewer, Application of IR Stimulation Phosphors in IR Detectors, IR Imaging System, X-Ray Imaging with Fluorescence Dyes and Memory Phosphors, PVC/Copper Sulfide Electrical Composites, Electrochromic Displays, Solar Cells, Sensors, Optical Disc Replication Process, Resistance Heating Device Bond on Polypyrrole, Electronic and Optical Ice Sensors, High Technology Education in the Areas of Electronics, Computers, Lasers and Fiber-Optics. Home: 33 Thoreau Dr Plainsboro NJ 08536-3018 Office: ORT Ops USA 200 Park Ave S New York NY 10003-1503

BÜHRER, HEINER GEORG, chemist, educator; b. Basel, Switzerland, Aug. 2, 1943; s. Eugen W. and Agnes (Reichwein) B.; m. Anna Regula, Sept. 8, 1973; children: Christian G., Adrian T., Elaine R. Diploma in Chem. Engring., Swiss Fed. Inst. Tech., Zurich, 1967, ScD, 1970. Postdoctoral fellow Swiss Fed. Inst. Tech., Zurich, 1970-71, Mich. Molecular Inst., Midland, 1971-72; head R&D Schüpbach AG, Burgdorf, Switzerland, 1972-74; lectr. indsl. chemistry Technikum, Winterthur, Switzerland, 1974-81, head dept. chemistry, 1983-89, prof. indsl. chemistry, 1981—; cons. to various cos. in plastics industry, 1974—. Contbr. to profl. publs. Mem. Neue Schweizerische Chemische Gesellschaft, Chimia (adv. bd.), Schweizerischer Verband Dozenten HTL (v.p. 1989-91), Lions Club Tösstal (charter pres. 1991-92). Achievements include patents in flexible packaging, contributions to stereoselective polymerization, thermal methods. Office: Technikum, Postfach 805, CH 8401 Winterthur Switzerland

BU-HULAIGA, MOHAMMED-IHSAN ALI, information systems educator; b. Damascus, Syria, May 12, 1956; s. Ali Hussain and Najah Deeb (Kuwaifatie) Bu-H.; m. Enaya Mousa, July 17, 1975; children: Mervat, Raula, Deema, Safa, Marwa. BS, U. Petroleum, Dhahran, Saudi Arabia, 1979, MBA, 1981; PhD, U. Wis., Milw., 1987. Fin. analyst Nat. Comml. Bank, Dammam, Saudi Arabia, 1979-80; fgn. banks officer Arab Nat. Bank, Khobar, Saudia Arabia, 1980-81; lectr. King Fahd U. Petroleum & Minerals, Dhahran, 1981-87, asst. prof., 1987-90; dir. indsl. data bank, sr. expert info. systems Gulf Orgn. for Indsl. Cons., Doha, Qatar, 1990—. Author: (with others) Recent Development in Production Research, 1987; commentary on tech. Al-Eqtisadiah newspaper, 1991—; mem. editorial bd. Indsl. Cooperation in the Arabian Gulf; contbr. articles to profl. jours. Sim doctoral fellow Soc. Info. Mgmt., 1986; dissertation grantee Wis. Gas, 1986. Mem. Am. Assn. Artificial Intelligence, Inst. Mgmt. Sci., Internat. Assn. for Decision Support, Beta Gamma Sigma. Home: PO Box 5114, Doha Qatar Office: Gulf Orgn Indsl Cons, PO Box 5114, Doha Qatar

BUI, JAMES, defense industry researcher; b. Can Tho, South Vietnam, June 26, 1961; came to U.S., 1975; s. Anh Van and Mai (Dinh) B.; m. Tina Dang, Aug. 26, 1984; 1 child, Amanda. BS, Va. Tech., 1983, M in Engring Adminstrn., 1987. Systems engr. DCS Corp., Alexandria, Va., 1983-84; O.R. analyst Delex Systems Inc., Vienna, Va., 1984-85; fin. specialist Martin Marietta Corp., Washington, 1985-88; rsch. staff Inst. for Def. Analyses, Alexandria, Va., 1988—. Office: Inst for Def Analyses 1801 N Beauregard Alexandria VA 22311

BUICK, FRED J.R., physiologist, researcher; b. July 16, 1951; married. BS in Kinesiology, York U., Toronto, Ont., 1974, MS in Physiology, 1979; PhD in Med. Scis., McMaster U., Hamilton, Ont., 1985. Dir. fitness and assessment YMCA, Toronto, Ont., 1974-76; part-time physiology lab. instr. York U., Toronto, Ont., 1976-78, McMaster U., Hamilton, Ont., 1979-82; def. scientist in aerospace physiology Def. & Civil Inst. Environ. Medicine, North York, Ont., 1983—; project officer aerospace medicine, life support and aircrew systems Air Standardization Coordinating Com., Washington, 1984—; mem. adv. group aerospace r & d NATO, Paris, 1988-91; adj. mem. faculty grad. program excercise & health scis. York U., 1988—, grad. program community health U. Toronto, 1992—. Co-author: High Gz Physiological Protection Training, 1990; contbr. articles to profl. jours.; pub. papers, reports. Grad. scholar McMaster U., 1979-82. Fellow Aerospace Med. Assn. (assoc., Eric Liljencrantz award 1993). Achievements include research in physiology of hypoxia and altitude life support equipment, chemical defense protection, physiological responses to high G-forces, human factors in CF-18 pilot environment. Office: Def Civil Inst Environ Medicine, 1133 Sheppard Ave W PO Box 2000, North York, ON Canada M3M 3B9*

BUKONDA, NGOYI K. ZACHARIE, public health administrator; b. Lubumbashi, Shaba, Zaire, Feb. 14, 1951; came to U.S., 1987; s. Munyuka Kalambayi and Tumba (Tshileo) Marie; m. Muyumba Kapinga Agnes, Aug. 29, 1975; children: Munyuka Ngoyi, Muyumba Ngoyi, Kalambayi Ngoyi, Tshileo Ngoyi, Kashala Ngoyi. BS in Health Systems Mgmt., U. Kinshasa, Zaire, 1981; Diploma in Teaching, U. Zaire, 1983; MPH, U. Minn. Sch. Pub. Health, 1989; postgrad., U. Minn. Hosp. adminstr. Gen. Hosp., Bukavu, Zaire, 1975-76; chief of bur. Ministry of Health Zaire, Kinshasa, 1981-83, chief of div., 1983-87; health planner Sanru B.P. 3355 Kinshasa, Kinshasa, 1987; asst. prof. Inst. Superieur de Techniques Medicales, Kinshasa, 1981-87; grad. fellow African Am. Inst., N.Y.C., 1987—; grad. teaching asst. Grad. Program in Social & Adminstrv. Pharmacy, Mpls., 1991; acad. sec. Inst. Superieur de Techniques Medicales, Kinshasa, 1983-86. Recipient Afgrad fellowship African Am. Inst., 1987, Melendy Grad. fellowship Coll. of Pharmacy, 1991; grantee Mac Arthur Interdisciplinary Program on Peace Internat., 1991; named Hon. Citizen of Louisville, 1986. Mem. Am. Pub. Health Assn., Am. Pharmacy Assn., Assn. des Adminstrs. Gestionnaires (pres. 1981-87). Roman Catholic. Home: 1504 2nd St NE Minneapolis MN 55413-1136 Office: Grad Program Adminstrv Pharmacy 388 Harvard St SE Rm 7-174 Minneapolis MN 55455-0361

BUKRY, JOHN DAVID, geologist; b. Balt., May 17, 1941; s. Howard Leroy and Irene Evelyn (Davis) Snyder. Student, Colo. Sch. Mines, 1959-60; BA, Johns Hopkins U., 1963; MA, Princeton U., 1965, PhD, 1967; postgrad., U. Ill., 1965-66. Geologist U.S. Army Corp Engrs., Balt., 1963; research asst. Mobil Oil Co., Dallas, 1965; geologist U.S. Geol. Survey, La Jolla, Calif., 1967-84, U.S. Minerals Mgmt. Service, La Jolla, 1984-86, U.S. Geol. Survey, Menlo Park, Calif., 1986—; research assoc. dept. geol. research div. U. Calif.-San Diego, 1970—; cons. Deep Sea Drilling Project, La Jolla, 1967-87; lectr. Vetlesen Symposium, Columbia U., N.Y.C., 1968, 3d Internat. Planktonic Conf., Kiel, Fed. Republic Germany, 1974; shipboard micropaleontologist on D/V Glomar Challenger, 5 Deep Sea Drilling Project cruises, 1968-78; mem. stratigraphic correlations bd. NSF/Joint Oceanographic Instns. for Deep Earth Sampling, 1976-79. Author: Leg I of the Cruises of the Drilling Vessel Glomar Challenger, 1969, Coccoliths from Texas and Europe, 1969, Leg LXIII of the Cruises of the Drilling Vessel Glomar Challenger, 1981; editor: Marine Micropaleontology, 1976-83, mem. editorial bd. Micropaleontology, 1985-90. Mobil Oil, Princeton U. fellow, 1965-67; Am. Chem. Soc., Princeton U. fellow, 1966-67. Fellow AAAS, Geol. Soc. Am., Explorers Club; mem. Hawaiian Malacological Soc., Paleontol. Rsch. Inst., Am. Assn. Petroleum Geologists, Soc. Econ. Paleontologists and Mineralogists, Internat. Nannoplankton Assn., European Union Geoscis., The Oceanography Soc., Nat. Sci. Tchrs. Assn., U. Calif. at San Diego Ida and Cecil Green Faculty Club, San Diego Shell Club, Princeton Club No. Calif., Sigma Xi. Avocations: basketball, photography, shell and mineral collecting. Achievements include research in stratigraphy, paleoecology and taxonomy for 300 new species of marine nannoplankton used in ocean history studies. Office: US Geol Survey MS-915 345 Middlefield Rd Menlo Park CA 94025-3591

BULGAK, VLADIMIR BORISOVICH, telecommunications engineer; b. Moscow, May 9, 1941; s. Boris Victorovich and Maria Mikhailovna (Markova) B.; m. Galina Aleksandrovna Rudaya, Apr. 30, 1970; 1 child, Maria Vladimirovna. Diploma in engring., Moscow Electrotec. Inst. Comm., 1963, mgr., 1972, candidate of tech. scis., 1972. Leader youth orgns. various comm. enterprises, Moscow, 1963-68; tech. dir. several comm. orgns., Moscow, 1968-72; head Moscow Radio Networks, 1972-83; head several depts. Ministry of Comms. of USSR, Moscow, 1983-90, min., 1990—; chmn. State Com. on Radio Frequencies, Moscow, 1991—, State Com. on Telecomm., Moscow, 1991—, Regional Commonwealth on Comm., Moscow, 1991—. Author, editor: Cable Broadcasting, 1985; author over 70 publs.; mem. editorial bd. News of Comm., 1983—, Tele Vestnik, 1992—. Mem. organizing com. Kremlin Cup Internat. Tennis Tournament, Moscow, 1991, 92, 93. Recipient Spl. award/medal USSR Govt., 1985. Mem. Russian Scientific-Tech. Soc. on Radiotechnics, Electronics & Comm., Internat. Acad. Info. Tech. (Gold medal 1992), Acad. Tech. Scis. Mem. Orthodox Ch. Office: Min of Comm,Info,Tech & Space, Delegatskaya ul 5, 103091 Moscow Russia

BULKIN, BERNARD JOSEPH, oil industry executive; b. Trenton, N.J., Mar. 9, 1942; s. Jacob and Beatrice Bulkin; m. Susan H. Lees, Dec. 31, 1975; children: Anna, Noah, David. BS, Poly Inst. Bklyn., 1962; PhD, Purdue U., 1966. Postdoctoral fellow Eidg. Tech. Hochschule, Zurich, Switzerland, 1966-67; prof. chemistry CUNY, 1967-75, chmn. chemistry dept., 1973-75; dean arts and sci. Poly. Inst. N.Y., Bklyn., 1975-82, v.p., 1982-85; dir. downstream oil R&D Sohio Corp., Cleve., 1985-88; mgr. products div. Brit. Petroleum, London, 1988-90, head oil rsch., 1991-92; head R&D BP Oil Internat., London, 1992—; bd. dirs. Viscon Internat., N.Y.C.; cons. numerous cos., 1970-85. Author: Chemical Application of Raman Spectroscopy, 1983, Raman Spectroscopy, 1991; contbr. articles to profl. jours. Bd. dirs. Cleve. Edn. Fund, 1986-88. Recipient medal N.Y. Soc. Applied Spectroscopy, 1978. Fellow Royal Soc. Chemistry; mem. Soc. Applied Spectroscopy (numerous coms.), Am. Chem. Soc., Coblentz Soc. (Coblentz award 1975), Sigma Xi (Rsch. award 1982). Office: BP Oil Tech Ctr, Chertsey Rd Sunbury on Thames, Middlesex TW16 7LN, England

BULLARD, ROGER DALE, aerospace engineer; b. Mt. Sterling, Ill., Feb. 2, 1942; s. Gilbert Ray and Bessie Lois (Colwell) B.; m. Joyce Ann Elbus, Nov. 21, 1971 (div. 1980); children: Larissa, Aimee, Jennifer. BS in Engring., Calif. State U., 1976. Engring. specialist F/A-18A divsn. Northrop Aircraft, Hawthorne, Calif.; tech. mgr. B-2 divsn. Northrop Aircraft, Pico Rivera, Calif., 1981-83; tech. mgr. F-20A divsn. Northrop Aircraft, Hawthorne, 1983-86, tech. mgr. YF-23A divsn., 1986-89, tech. dir. F/A-18 E/F divsn., 1989—. Capt. USMC, 1967-74, Vietnam. Mem. AIAA (chmn. system engring. and safety tech. com. 1985-89), Tau Beta Pi. Office: Northrop Aircraft F/A-18 E/F Vlnrblty 3932/W3 1 Northrop Ave Hawthorne CA 90250-3277

BULLEN, DANIEL BERNARD, nuclear engineering educator; b. Iowa City, July 20, 1956; s. John Bernard and Helen May (Ferguson) B.; m. Elizabeth Ann Clark, Aug. 17, 1979; children: Katherine Andrea, Mark Bernard, Sarah Elizabeth, Rachel Suzanne. BS in Engring. Sci., Iowa State U., 1978; MS in Nuclear Engring., U. Wis., 1979, MS in Material Sci., 1981, PhD in Nuclear Engring., 1984. Registered profl. engr., Calif., N.C., Ga., Iowa. Engr. Lawrence Livermore (Calif.) Nat. Lab., 1984-86; sr. engr. Sci. and Engring. Assocs., Inc., Pleasanton, Calif., 1986-88; pres. DG Engring., Inc., Livermore, 1988-89; asst. prof. nuclear engr. N.C. State U., Raleigh, 1989-90, Ga. Inst. Tech., Atlanta, 1990-92; assoc. prof. nuclear engring. Iowa State U., Ames, 1992—, coord. nuclear engring. program, 1993—; cons. Lawrence Livermore Nat. Lab., 1988-91, Electric Power Rsch. Inst., Palo Alto, Calif., 1989—, Internat. Lead Zinc Rsch. Orgn., Research Triangle Park, N.C., 1990—, HDR Engring., Inc., Omaha, 1991—. Contbr. 25 articles to profl. jours. Mem. NSPE, ASME, ASM Internat., Mineral, Metals and Materials Soc. AIME, Am. Nuclear Soc., Am. Ceramic Soc. (tech. reviewer 1986—), Materials Rsch. Soc., Am. Soc. Engring. Edn. Roman Catholic. Home: PO Box 1768 Ames IA 50010-1768 Office: Iowa State U Nuclear Engring Program 103 Nuclear Engring Lab Ames IA 50011-2241

BULLIS, W(ILLIAM) MURRAY, physicist, consultant; b. Cin., Aug. 29, 1930; s. Ralph M(argetts) and Edith M(urray) (Bryant) B.; m. Kathleen Page Andree, Sept. 5, 1953; children: Keith, Bruce, Claire. AB magna cum laude, Miami U., Oxford, Ohio, 1951; PhD, MIT, 1956. Staff mem. Los Alamos (N.Mex.) Sci. Lab., 1956-57; sr. engr. Farnsworth Electronics Co., Ft. Wayne, Ind., 1957-59; mem. tech. staff Tex. Instruments, Dallas, 1959-65; mem. tech. staff Nat. Bur. Standards, Gaithersburg, Md., 1965-78, div. chief, 1978-81; mgr. semiconductor materials Fairchild Adv. R & D Lab., Palo Alto, Calif., 1981-83; v.p. R & D Siltec Silicon Co., Menlo Park, Calif., 1983-91; founder, pres. Materials & Metrology, Sunnyvale, Calif., 1991—; lectr. elec. engring. U. Md., College Park, 1966-69. Author, editor tech. publs. Recipient Silver medal U.S. Dept. Commerce, 1979. Mem. IEEE, ASTM, Am. Phys. Soc., Electrochem. Soc. (div. chair 1987-89). Office: Materials & Metrology 1477 Enderby Way Sunnyvale CA 94087

BULLOCK, DANIEL HUGH, computational neuroscience educator, psychologist; b. Peoria, Ill., July 20, 1952; s. Roy and Jane (Morris) B.; m. Laura Lee Merrill, May 31, 1980; children: Evan Merrill, Ian Merrill. BA, Reed Coll., 1974; PhD, Stanford U., 1979. Postdoctoral fellow U. Denver, 1979-81, instr., 1981-82, asst. prof., 1982-86; sr. rsch. assoc. Boston U., 1985-88, asst. prof. computational neurosci., 1988-90, assoc. prof., 1990—. Mem. editorial bd. Neural Networks, 1987—; contbr. over 40 articles to Child Devel., Psychol. Rev., Human Movement Sci., Neural Networks, Jour. Cognitive Neurosci., also chpts. to books. Fellow NSF, 1974; rsch. grantee NSF, 1974, 87, 90, Office Naval Rsch., 1992. Mem. Soc. for Neurosci., Soc. for Math. Psychology, Internat. Neural Network Soc., Am. Psychol. Soc. Democrat. Achievements include research in cognitive psychology and computational neuroscience, mathematical studies of neural networks that make voluntary movement possible, including known spinal neural networks and central neural networks distributed across the cerebral cortex, the basal ganglia, and the cerebellum; proposed convergence rate hierarchy theory of intellectual development. Home: 26 Greenough St Newton MA 02165 Office: Boston U CNS Dept 111 Cummington St Boston MA 02215

BULLOCK, THEODORE HOLMES, biologist, educator; b. Nanking, China, May 16, 1915; s. Amasa Archibald and Ruth (Beckwith) B.; m. Martha Runquist, May 30, 1937; children—Elsie Christine, Stephen Holmes. Student, Pasadena Jr. Coll., 1932-34; A.B., U. Calif. at Berkeley, 1936, Ph.D., 1940; Sterling fellow zoology, Yale U., 1940-41, Rockefeller fellow exptl. neurology, 1941-42. Research assoc. Yale U. Sch. Medicine, 1942-43, instr. neuroanatomy, 1943-44; instr. Marine Biol. Lab., Woods Hole, Mass., 1944-46; head invertebrate zoology Marine Biol. Lab., 1955-57, trustee, 1955-57; asst. prof. anatomy U. Mo., 1944-46; asst. prof. zoology U. Calif. at Los Angeles, 1946, assoc. prof., 1948, prof., 1955-66; Brain Research Inst., U. Calif. at Los Angeles, 1960-66; prof. neuroscis. Med. Sch., U. Calif. at San Diego, 1966-82, prof. emeritus, 1982—; mem. AEC 2d Resurvey of Bikini Expdn., 1948. Author: (with G.A. Horridge) Structure and Function in the Nervous Systems of Invertebrates, 2 vols., 1965; (with others) Introduction to Nervous Systems, 1977; (with W. Heiligenberg) Electroreception, 1986, (with E. Basar) Brain Dynamics, 1989, (with E. Basar) Induced Rhythms in the Brain, 1992. Fulbright scholar Stazione Zooologica, Naples, 1950-51; fellow Center Advanced Study in Behavioral Scis., Palo Alto, 1959-60. Fellow AAAS; mem. Am. Soc. Zoologists (chmn.

comparative physiology div. 1961, pres. 1965), Soc. Neurosci. (pres. 1973-74), Internat. Soc. Neuroethology (pres. 1984-86), Am. Physiol. Soc., Soc. Gen. Physiologists, Am. Acad. Arts and Scis., Nat. Acad. Scis., Am. Philos. Soc., Internat. Brain Research Orgn., Phi Beta Kappa, Sigma Xi.

BULMAN, WILLIAM PATRICK, data processing executive; b. Corona, N.Y., Jan. 11, 1925; s. William T. and Bridget A. (Gibbons) B.; m. Jane G. Jones, June 30, 1952. BS, U. Upper N.Y., 1947; BBA, Syracuse (N.Y.) U., 1949, MBA, 1977. In systems/programming Mohawk Airlines, Utica, N.Y., 1951-55; data processing mgr. Gold Medal Packing, Utica, 1956-59, West End Brewing, Utica, 1960-73; coord. on-line data processing systems Sperry-Univac, Utica, 1973-76, data processing mgr., 1976-77; programmer/analyst MDS, Herkimer, N.Y., 1977-86; sr. programmer, analyst, Momentum Techs., Herkimer, N.Y., 1986-89; ret., 1989; cons. Bilb-Tech, 1989—. With USN, 1941-46. Mem. Data Processing Mgmt. Assn. (v.p., treas.), Assn. Systems Mgmt. Address: 35 Ashwood Ave Whitesboro NY 13492

BULMAN PAGE, PHILIP CHARLES, chemistry researcher; b. Erith, Kent, Eng., Dec. 28, 1955; s. Henry Charles and Doris (Bulman) Page; m. Patricia Rosina Howard, June 19, 1980. BSc, Imperial Coll., London, 1978, PhD, 1981. Lectr. U. Liverpool, 1983-90, sr. lectr., 1990—. Contbr. over 70 articles to profl. jours. Fellow Royal Soc. of Chemistry; mem. Am. Chem. Soc., Soc. of Chem. Industry. Office: Dept Chemistry, U Liverpool, Oxford St, Liverpool L69 3BXP, England

BUNCH, JEFFREY OMER, mechanical engineer; b. Amarillo, Tex., Mar. 14, 1958; s. James Earl and Etola Katherine (Parrish) B.; m. Candace Louise Cook; 1 child, Justin Charles. BSME, U. Houston, 1980; PhD, U. Conn., 1986. Mem. tech. staff McDonnell Douglas Helicopter Co., Mesa, Ariz., 1986-88; engring. specialist Northrop B-2 Div., Pico Rivera, Calif., 1988—. Mem. ASTM, Am. Soc. Metals. Office: Northrop B2 Div 8900 E Washington Blvd Pico Rivera CA 90660

BUNCH, MICHAEL BRANNEN, psychologist, educator; b. Miami, Fla., Oct. 19, 1949; s. Edwin Bunch and Janet (Morgan) Bradley; m. Kathryn Ann Campbell, Jan. 17, 1970; children: Melissa Anne, Amy Kathryn. BS, U. Ga., 1972, MS, 1974, PhD, 1976. Tests and measurement specialist Mountain Plains Corp., Glasgow, Mont., 1975-76; rsch. psychologist Am. Coll. Testing Program, Iowa City, 1976-78; sr. profl. NTS Rsch. Corp., Durham, N.C., 1978-82; v.p. Measurement Inc., Durham, N.C., 1982—; mem. Durham Pub. Edn. Task Force, Durham, 1983-90; chmn. Durham Math. Coun., 1985-90; adj. faculty N.C. Ctrl. U., Durham, 1988—. Mem. Am. Psychol. Soc., Am. Ednl. Rsch. Assn., Nat. Coun. Measurement in Edn., Ga. Ednl. Rsch. Assn., Sigma Xi. Home: 1000 Sedwick W Durham NC 27713 Office: Measurement Inc 2408 Reichard St Durham NC 27705

BUNCHER, CHARLES RALPH, epidemiologist, educator; b. Dover, N.J., Jan. 9, 1938. BS, MIT, 1960; MS, Harvard U., 1964, ScD, 1967. Statistician Atomic Bomb Casualty Comsn., NAS, 1967-70; chief biostatistician Merrell-Nat. Labs., 1970-73, asst. prof. stats., 1970-73; prof. and dir. divsn. epidemiology and biostatistics Med. Coll., U. Cin., 1973—. Mem. AAAS, APHA, Am. Stats. Assn., Biometrical Assn., Soc. Epidemiol. Rsch. Achievements include research in cancer epidemiology; screening, diagnosis and treatment, as well as occupational and environmental epidemiology; statistical research; clinical trials; design of experiments; pharmaceutical research; biostatistical analysis, pharmaceutical statistics, ALS epidemiology, risk analysis. Office: University of Cincinnati Div of Epidemiology & Biostatist ML 183 Cincinnati OH 45267*

BUNCHMAN, HERBERT HARRY, II, plastic surgeon; b. Washington, Feb. 23, 1942; s. Herbert H. and Mary (Halleran) B.; m. Marguerite Fransioli, Mar. 21, 1963 (div. Jan. 1987); children: Herbert H. III., Angela K., Christopher. BA, Vanderbilt U., 1964; MD, U. Tenn., 1967. Diplomate Am. Bd. Surgery, Am. bd. Plastic Surgery. Resident in surgery U. Tex., Galveston, 1967-72, resident in plastic surgery, 1972-75; practice medicine specializing in plastic surgery Mesa, Ariz., 1975—; chief surgery Desert Samaritan Hosp., 1978-80. Contbr. articles to profl. jours. Eaton Clin. fellow, 1975. Mem. AMA, Am. Soc. Plastic and Reconstructive Surgery, Am. Soc. Aesthetic Plastic Surgery, Singleton Surgical Soc., Tex. Med. Assn., So. Med. Assn. (grantee 1974), Ariz. Med. Assn. Office: Plastic Surgery Cons PC 1520 S Dobson Rd Ste 314 Mesa AZ 85202-4783

BUNDGAARD, NILS, sound and acoustics professional; b. Vejlby, Aarhus, Denmark, Dec. 30, 1952; s. Knud and Rigmor B.; children: Svante, Kristoffer. BS in Math., U. Aarhus, 1991, BS in Computer Sci., 1992. Sound engr. AB Musik Sound Hire, Aarhus, 1975-79; mgr., Hire div. AB Musik, Aarhus, 1979-84; founder, owner EAR Electro Acoustic Rsch., Aarhus, 1984—; cons. Danish Music Coun., State of Denmark/Copenhagen, 1989—, Roskilde Festival, 1985—, Tonder Festival, Denmark, 1983—, Mitjyn Festival, 1993—. Author: To Build a Rehearsal Room, 1987, PA-Systems and Acoustic Problems in Music Clubs, 1990. Mem. Audio Engring. Soc. Avocations: sailing, philosophy, quantum mechanics. Office: Ear Electro Acoustic Rsch, Klostergade 68 1, 8000C Aarhus Denmark

BUNDY, KIRK JON, biomaterials educator, researcher, consultant; b. Highland Park, Mich., May 21, 1947; s. LaForde Edison and Eunice Bundy; m. Pia Kriistina Ilmalahti, Dec. 12, 1974; children: Erik, Jennifer. BS, Mich. State U., 1968; MS, Stanford U., 1970, PhD, 1975. Teaching asst. materials sci. and engring. dept. Stanford U., Palo Alto, 1968; metallurgist Lawrence Radiation Lab., Livermore, Calif., 1969; rsch. asst. aeros. and astronautics dept. Stanford U., 1971, sci. assoc. Biomed. Engring. Inst. Swiss Fed. Inst. Tech., Zurich, 1971-75; postdoctoral fellow metallurgy program sch. chem. engring. Ga. Inst. Tech., Atlanta, 1975-78; asst. prof. materials sci. and engring. dept. sch. engring. Johns Hopkins U., Balt., 1978-83; asst. prof. biomed. engring. dept. sch. medicine Johns Hopkins U., 1980-83; assoc. prof. biomed. engring. dept. Tulane U., New Orleans, 1983-92, prof., 1992—, cons. USF&G Corp., Balt., 1981, Nat. Bur. Standards, Gaithersberg, Md., 1981-83, DePuy, Inc., Warsaw, Ind., 1992, Ceracon, Inc., 1993, various attys., 1983—; reviewer various archival jours., 1982—. Contbr. articles to profl. jours. T.J. Watson Meml. Nat. Merit scholar IBM Corp., 1965; Internat. Nickel Co. fellow Stanford U., 1970, Fogerty Sr. Internat. fellow NIH, 1990. Mem. Nat. Assn. Corrosion Engrs., ASTM, Am. Soc. for Metals, AAAS, Soc. for Biomaterials, Biomed. Engring. Soc., Acad. Dental Materials, Tau Beta Pi, Alpha Eta Mu Beta, Sigma Xi. Office: Tulane U Biomed Engring Dept New Orleans LA 70118

BUNGE, RICHARD PAUL, cell biologist, educator; b. Madison, S.D., Apr. 15, 1932; married, 1956; 2 children. BA, U. Wis., 1954, MS, 1956, MD, 1960. Asst. anatomist U. Wis., 1954-57, instr., 1957-58; from asst. prof. to assoc. prof. anatomy, coll. physicians and surgeons Columbia U., 1962-70; prof. anatomy sch. medicine Wash. U., 1970-89; sci. dir. Miami project/paralysis divsn., prof. neurosurg. cell biology anatomy Miami (Fla.) U., 1989—; vis. Auburn U. prof. Harvard Med. Sch., 1968-69; Nat. Multiple Sclerosis Soc. fellow surgeon Coll. Physicians and Surgeons, Columbia U., 1960-62. Mem. Am. Assn. Anatomy, Am. Soc. Cell Biology, Am. Assn. Neuropathology, Tissue Culture Assn., Soc. Neuroscience. Achievements include research in biology of cells of the nervous system in vivo and in vitro. Office: University of Miami Miami Project to Cure Paralysis 1600 NW 10th Ave R-48 Miami FL 33136*

BUNGER, ROLF, physiology educator; b. Hamburg, Germany, Oct. 19, 1941; came to U.S., 1979; s. Heinz Johannes Albert and Helga (Franz) B.; m. Margriet Akkerman, Dec. 14, 1973; children: Nils, Frank. MD, U. Hamburg, 1969; Dr. med., U. Heidelberg, Germany, 1970; PhD, U. Munich, 1979. Med. intern Heidberg Infirmary, Hamburg, 1970; asst. of physiology U. Aachen, Germany, 1970-79, U. Munich, 1979; asst. prof. dept. physiology F. E. Hebert Med. Sch., USUHS, Bethesda, Md., 1979-82, assoc. prof., 1983-92; prof. USUHS, Bethesda, Md., 1992—; cons. U. Buffalo, 1983, U. Ala., 1986-89, U. Tex., Fort Worth, 1990—; referee, editorial reviewer domestic and fin. sci. jours. NIH, 1984-89; vis. prof. Erasmus U., Rotterdam; lectr. in field. Mem. editorial bd.: Internat. Jour. Purine & Pyrimidine Rsch., 1989-93, Internat. Jour. Angiology, 1991—. Webelo leader Boy Scouts Am., McLean, Va., 1986-87, packmaster, 1987-89. Capt. German Air Force. Grantee NIH, 1982—, USUHS, 1979—. Fellow Am. Heart Assn.; mem. Internat. Study Group for Heart Rsch., Am. Physiol. Soc., Deutsche Physiol. Gesellschaft. Achievements include clarification of adenylate compartments in myocardium; demonstration of energy-linked and work-dependance of myocardial pyruvate dehydrogenase flux, of interstitial free AMP in myocardium; research in substrate enhancement of isolated preischemic and postischemic heart preparations; metabolic protection of cytosolic phosphorylation potential by pyruvate and adenosine during myocardial reperfusion. Home: 1922 Kenbar Ct Mc Lean VA 22101-5321 Office: USUHS Dept Physiology 4301 Jones Bridge Rd Bethesda MD 20814-4799

BUNN, JOE MILLARD, agricultural engineering educator; b. Wayne County, N.C., Jan. 20, 1932; s. Clarence S. and Zora S. (Woodall) B.; m. F. Marie Baker, June 26, 1955; children: Ronnie Joe, Kenneth Bruce. BS Agrl. Engr., N.C. State Coll., 1955, MS Agrl. Engr., 1957; PhD Agrl. Engr., Math., Iowa State U., 1960. Registered Agrl. Engr., Ky. 1963. From asst. prof. to assoc. prof. U. Ky., Lexington, 1960-78; engr. AID-Ky. Team, Khon Kean, Thailand, 1968-70; prof. Clemson (S.C.) U., 1978—. Contbr. tech. papers and chpts. in books in field. mem. Meth. Ch. (Sunday sch. tchr. and bd. deacons), Lexington, Ky., 1965-78., Presbyn. Ch. (Sunday sch. tchr. and elder), Sandy Springs, S.C., 1979—. Grantee various pub. and pvt. agcys. $1.3m for rsch. Fellow Am. Soc. Agrl. Engrs. (sec., vice chmn. various nat. state coms.); mem. Coun. Agr. Scis. Tech. Democrat. Presbyterian. Avocations: gardening, bowling. Office: Clemson U Agrl Engring Dept 116 McAdams Hall Clemson SC 29634-0357

BUNNETT, JOSEPH FREDERICK, chemist, educator; b. Portland, Oreg., Nov. 26, 1921; s. Joseph and Louise Helen (Boulan) B.; m. Sara Anne Telfer, Aug. 22, 1942; children—Alfred Boulan, David Telfer, Peter Sylvester (dec. Sept. 1972). B.A., Reed Coll., 1942; Ph.D., U. Rochester, 1945. Mem. faculty Reed Coll., 1946-52, U. N.C., 1952-58; mem. faculty Brown U., 1958-66, prof. chemistry, 1959-66, chmn. dept., 1961-64; prof. chemistry U. Calif. at Santa Cruz, 1966-91; prof. emeritus; Erskine vis. fellow U. Canterbury, N.Z., 1967; vis. prof. U. Wash., 1956, U. Würzburg, Fed. Republic Germany, 1974, U. Bologna, Italy, 1988; rsch. fellow Japan Soc. for Promotion of Sci., 1979; Lady Davis vis. prof. Hebrew U., Jerusalem, Israel, 1981; mem. adv. council chemistry dept. Princeton U., 1985-89; mem. Nat. Rsch. Coun. com. on alternative chemical demilitarization techs., 1992—. Contbr. articles to profl. jours. Trustee Reed Coll., Societá Chimica Italiana (hon.). Fulbright scholar Univ. Coll., London, Eng., 1949-50; Guggenheim fellow, Fulbright scholar U. Munich, Germany, 1960-61; recipient James Flack Norris award in Physical Organic Chemistry, Am. Chemical Soc., 1992. Fellow AAAS; mem. Am. Acad. Arts and Scis., Am. Chem. Soc. (editor jour. Accounts of Chem. Research 1966-86), Royal Soc. of Chemistry (London), Internat. Union Pure and Applied Chemistry (chmn. commn. on phys. organic chemistry 1978-83, sec. organic chemistry div. 1981-83, v.p. 1983-85, pres. 1985-87), Pharm. Soc. Japan (hon.), Acad. Gioenia (U. Catania, Italy) (hon.), Sociedad Argentina de Investigaciones en Quimica Organica (hon.). Home: 608 Arroyo Seco Santa Cruz CA 95060-3148 Office: U Calif Thimann Labs Santa Cruz CA 95064

BUNNEY, BENJAMIN STEPHENSON, psychiatrist; b. Lansing, Mich., Sept. 27, 1938; s. William E. and Nora Orpha (Null) B.; m. Marjorie Bunney, Oct. 6, 1984; children: Edward Bradshaw, Katherine Stephenson, Elizabeth Janice. BA, NYU, 1960, MD, 1964. Resident in internal medicine Bellevue Hosp. NYU, 1964-66; resident in psychiatry Yale U., New Haven, 1968-71; asst. prof. psychiatry Yale U., 1971-74, asst. prof. pharmacology, 1974-75, assoc. prof. psychiatry, 1975-84, assoc. prof. pharmacology, 1976-84, prof. psychiatry, 1984—, prof. pharmacology, 1984, vice chair dept. psychiatry, 1986-87, acting chair dept. psychiatry, 1987-88, chair dept. psychiatry, 1988—; mem. bd. sci. counselors NIMH; mem. Inst. Medicine NAS, 1993. Contbr. articles to profl. jours. Capt. USAF, 1966-68. Recipient Daniel H. Efron award for rsch. Am. Coll. Neuropsychopharmacology, 1983, Lieber prize for rsch. Nat. Alliance for Rsch. in Schizophrenia and Depression, 1987, MERIT award NIMH, 1990. Fellow Am. Coll. Neuropsychopharmacology; mem. AAAS, Am. Psychiatric Assn. Office: Yale U Dept Psychiatry 25 Park St New Haven CT 06519-1189

BUNNI, NAEL GEORGES, engineering consultant, international arbitrator, conciliator; b. Kirkuk, Iraq, Apr. 23, 1939; s. Georges Azeez and Adeeba (Fathalla) B.; m. Anne Carroll, Apr. 7, 1962; children: Nadia, Layth, Siobháin, Lara, Layla, Lydia. BS with 1st honors, Baghdad U., Iraq, 1959; MS, Victoria U., Manchester, Eng., 1962; PhD, Queen Mary Coll., London, 1964. Chartered civil engr. Part time lectr. N.E. London Poly., 1962-64; sr. lectr., asst. prof., constrn. cons. Baghdad Coll. Tech., 1964-69; agt. Iraq Nat. Ins. Co., Baghdad, 1964-69; cons. T.J. O'Connor & Assocs., Dublin, Ireland, 1969-72, sr. dir., 1978—; sr. dir. T.J. O'Connor Internat. Ltd., Dublin, Ireland, 1978—. Author: Construction Insurance & the Irish Conditions of Contract, 1984, Construction Insurance, 1986, The FIDIC Form of Contract, 1991. Recipient scholarship, Manchester U., 1960, London U., 1962. Fellow Inst. Engrs. of Ireland (Mullins Silver medal 1987, 91, Smith Testimonial award 1988), Inst. Civil Engrs., Inst. Structural Engrs., Chartered Inst. Arbitrators (chmn. br. 1986-88), Inst. Dirs.; mem. Assn. Cons. Engrs. (pres. 1986-87). Avocations: painting, travel, tennis, music. Home: Bearna Thormanby Rd, Howth Ireland Office: TJ O'Connor & Assocs, Corrig House Corrig Rd, Dublin 18, Ireland

BUNT, RANDOLPH CEDRIC, mechanical engineer; b. Pascagoula, Miss., Dec. 3, 1958; s. Cedric and Linda Lou (McGuire) B.; m. Raechel Amy Ellis, May 15, 1982; children: Ashley Michele, Ryan Christian. BME, Auburn U., 1979, MS, 1982. Registered profl. engr., Ala., Ga. Asst. engr. So. Co. Svcs., Birmingham, Ala., 1982-84, engr. II, 1984-86, engr. I, 1986-87, sr. engr., 1987-88; sr. engr. Ga. Power Co., Birmingham, 1988-89; project engr. So. Nuclear Oper. Co., Birmingham, 1989—; project mgr. Ga. Power Co., Atlanta, Birmingham, 1987-89. Capt. Birmingham Amateur Hocker Assn., 1985. Mem. ASME, NSPE, Am. Nuclear Soc., Ala. Soc. Profl. Engrs. (chmn. student engring. yr. com. 1987-92, chmn. Math. Counts program 1990—, sec. 1992-93, v.p. 1993-94, co-chairperson 1993 Conf. Young Engr. of Yr. 1991), So. Nuclear Nat. Mgmt. Assn. (treas. 1991-92), Terry Turbine Users Group (vice chmn. 1993-94). Republican. Baptist. Home: 1005 Muscadine Cir Leeds AL 35094-1027 Office: So Nuclear Oper Co 42 Inverness Center Pky Birmingham AL 35242-4817

BUNTING, GARY GLENN, operations research analyst, educator; b. Toledo, Ohio, Mar. 19, 1947; s. Glenn Rose and Maxine (Hunt) B.; m. Glenda Marlene Mechum, Aug. 23, 1974; children: Wendy Daniele, Bradley Glenn, Max Alan. BS, Auburn U., 1969; MS, Troy State U., 1977. Research analyst City of Jacksonville, Fla., 1972-76; ops. research analyst U.S. Army Aviation Ctr., Ft. Rucker, Ala., 1976-78, U.S. Army Tng. Support Ctr., Ft. Eustis, Va., 1978-80, USAF Tactical Air Warfare Ctr., Eglin AFB, Fla., 1980-85, Office Sec. Def. Tng. and Performance Data Ctr., Orlando, Fla., 1985-92; chief ops. analysis Spl/ Missions Operational Test and Evaluation Ctr.; adj. assoc. prof. U. West Fla., Ft. Walton Beach, 1982, Troy State U., Ft. Walton Beach, 1983—. Contbr. articles to profl. jours. Served with U.S. Army, 1969-71. Recipient Civilian Excellence award, USAF Tactical Air Warfare Ctr. and Air Force Assn., 1983. Mem. Air Force Assn. Republican. Lodge: Elks. Avocations: reading, photography. Home: 52 Sweetwater Creek Cir Oviedo FL 32765-6469 Office: 3280 Progress Dr Orlando FL 32826

BUNTROCK, GERHARD FRIEDRICH RICHARD, mathematician; b. Berlin, Aug. 4, 1954; s. Werner and Else (Poppe) B.; m. Christine Gonser, Mar. 29, 1985; children: Lydia Rosmarie Else Christine, Raphael Werner Hansjörg Gerhard. Diploma math., Tech. U., Berlin, 1978, Tech. U., 1984; promotion for natural scis., Tech. U., 1989. Scientist hilfskraft Tech. U., Berlin, 1981-84, scientist mitarbeiter, 1984-89; scientist asst. Tech. U., Wurzburg, Germany, 1989—. Contbr. articles to profl. jours. Mem. European Assn. Theoretical Computer Sci., Assn. Math. Okonometrie and Ops. Rsch., Assn. for Computing Machinery. Home: Hessenstr 78, D 97078 Wuerzburg Bayern, Germany Office: U Wurzburg Inst Info, Am Exerzierplatz 3, D 97072 Wuerzburg Bavaria, Germany

BUNYARD, ALAN DONALD, designer, inventor; b. London, July 31, 1931; s. Harold Stanley and Florence Elizabeth (Thompson) B.; m. Mavis Mary Moss, Dec. 23, 1959; children: Martin Noel, Adam Guy. D-raughtsman Crosby Valve Co., Middlesex, Eng., 1946-48, Royal Elec. and Mech. Engrs., Eng., 1949-51; chief draughtsman Electroflo Meters Co., Park Royal, London, 1952-54; ptnr. Preston Designs Ltd., Peacehaven, Sussex, Eng., 1955-57; prodn. mgr. Norris Bros. Ltd., Haywards Heath, Sussex, 1958-64; exec. dir. Norbro Engring. Ltd., Haywards Heath, Sussex, 1965-76; mng. dir. Norbro Ltd., Burgess Hill, Sussex, England, 1977-79; ptnr. Bunyard and Co., Brighton, Sussex, 1980—; mng. dir. Forac Ltd., Hove, Sussex, 1986—. Designer, patentee speec contr., rotary actuator, air cylinder, 4-rack actuator (award Brit. Design Coun. 1973). Mem. Soc. Genealogists. Avocations: genealogy, tropical marine fish. Office: Forac Ltd, Brunswick St E, Hove, Sussex BN3 1AU, England

BUONANNI, BRIAN FRANCIS, health care facility administrator, consultant; b. Pawtucket, R.I., Sept. 2, 1945; s. James and Roselle B.; m. Lynne Buonanni (div. 1982); children: Donna, Karen, Jamie; m. Diane Manenty, Feb. 23, 1985. BA, Providence Coll., 1967; EdM, Boston Coll., 1968; M in Health Adminstrn., St. Louis U., 1973. Lic. nursing home adminstr., N.J. Rehab. counselor, tchr. R.I. Assn. for Blind, Providence, 1968-71; adminstrv. resident Carney Hosp., Boston, 1972; asst. adminstr. Alton (Ill.) Meml. Hosp., 1973-77, Gnaden Huetten Meml. Hosp., Lehighton, Pa., 1977-80; v.p. ops. Burdette Tomlin Hosp., Cape May Ct. House, N.J., 1980-85; chief oper. officer St. Elizabeth's Hosp., Elizabeth, N.J., 1985—, exec. v.p., 1989—; pres. Health Care Practice Mgmt., Jenkinstown, Pa., 1984—; chmn., mem. adv. bd. Shifa, McFaul & Lyons, Morristown, N.J., 1987—; mem. rev. com. N.J. Health Council, Trenton, 1987—. Fellow Am. Coll. Healthcare Execs.; mem. NAACP, Nat. Assn. Purchasing Agts., Rotary (pres.). Home: 12 Coldevin Rd Clark NJ 07066-1237 Office: St Elizabeth Hosp 225 Williamson St Elizabeth NJ 07207

BURBANK, ROBINSON DERRY, crystallographer; b. Berlin, N.H., Oct. 3, 1921; s. Paul William and Hazel Louise (Robinson) B.; m. Jeannette Murielle Bisson, July 14, 1945 (div. 1975); children: Paul Robinson, Claudia Olive. BA, Colby Coll., 1942; PhD, MIT, 1950. Rsch. asst. Manhattan Project, MIT, Cambridge, 1942-45, Lab. Insulation Rsch., MIT, 1945-50; sr. physicist Gaseous Diffusion Plant, Oak Ridge, Tenn., 1950-53; group leader, crystallography Olin Industries, New Haven, Conn., 1953-55; tech. staff Bell Telephone Labs., Murray Hill, N.J., 1955-86; U.S. del. Internat. Union Crystallography, Stony Brook, L.I., N.Y., 1969, Amsterdam, 1975; mem. U.S.A. Nat. Com. Crystallography, 1968-76. Contbr. technical papers to profl. jours. Bd. dirs. Chester Twp. Taxpayers Assn., N.J., 1961-65, 70-74, pres. 1973. Mem. Am. Crystallographic Assn. (treas. 1965-68, v.p. 1974, pres. 1975), AAAS, Phi Beta Kappa, Sigma Xi. Achievements include X-ray crystallography of inorganic compounds, interhalogen compounds, noble gas compounds, phase transformations, thin films. Home: 45 Woodland Ave Summit NJ 07901-2152

BURBEA, JACOB N., mathematics educator; b. Livorno, Italy, Dec. 13, 1942; came to U.S., 1974; s. Amos Clemente and Dvora (Jonas) B.; m. Claire M. Moss; children: John, Michelle. BS, Hebrew U., Jerusalem, 1966; MS, Weizmann Inst., Rehovoth, Israel, 1968; PhD, Stanford (Calif.) U., 1971. Rsch. assoc. Stanford U., 1970-71; lectr. Tel Aviv (Israel) U., 1971-74; vis. prof. Pa. State U., University Park, 1974-76; prof. U. Pitts., 1976—; vis. prof. U. Pisa, Italy, 1972-73; vis. scholar Swedish Acads. Sci.-Inst., Mittag Leffler Djursholm, Sweden, 1981-82, 91, IBM T.J. Watson Rsch. Ctr., Yorktown Heights, N.Y., 1982-83, Korea Inst. of Tech., Taejon, 1987, U. Barcelona, Spain, 1988-89, 91, U. Provence, Marseilles, France, 1989-90; sr. scientist IBM Corp., Yorktown, 1982-83, Westinghouse Corp., Pitts., 1985-86, PPG Industries, Pitts., 1986—. Author: Banach & Hilbert Spaces, 1986; contbr. articles in areas of math. analysis, applied math., stats. and fluid mechanics to profl. jours. U.S. Army grantee, 1973, NSF grantee, 1976, 78, French Acad. Sci. grantee, 1984. Fellow Rend Circle Math. Palermo; mem. Am. Math. Soc., London Math. Soc., Sci. and Tech. Agy., Japan Assn. Advancement Rsch. Coop. (Sci. and Tech. Agy. 1993). Avocation: stamps. Home: 409 S Dallas Ave Pittsburgh PA 15208-2818 Office: U Pitts Dept Math Pittsburgh PA 15260

BURBIDGE, GEOFFREY, astrophysicist, educator; b. Chipping Norton, Oxon, Eng., Sept. 24, 1925; s. Leslie and Eveline Burbidge; m. Margaret Peachey, 1948; 1 dau. B.Sc. with spl. honors in Physics, Bristol U., 1946; Ph.D., U. Coll., London, 1951. Asst. lectr. U. Coll., London, 1950-51; Agassiz fellow Harvard, 1951-52; research fellow U. Chgo., 1952-53, Cavendish Lab., Cambridge, Eng., 1953-55; Carnegie fellow Mt. Wilson and Palomar Obs., Calif. Inst. Tech., 1955-57; asst. prof. dept. astronomy U. Chgo., 1957-58, assoc. prof., 1958-62; assoc. prof. U. Calif. San Diego, La Jolla, 1962-63; prof. physics U. Calif. San Diego, 1963-83, 88—; dir. Kitt Peak Nat. Obs., Tucson, 1978-84; Phillips vis. prof. Harvard U., 1968; bd. dirs. Associated Univs. Research in Astronomy, 1971-74; trustee Associated Univs., Inc., 1973-82. Author: (with Margaret Burbidge) Quasi-Stellar Objects, 1967; editor Ann. Rev. Astronomy and Astrophysics, 1973—; contbr. articles to sci. jours. Fellow Royal Soc. London, Am. Acad. Arts and Scis., Royal Astron. Soc., Am. Phys. Soc.; mem. Am. Astron. Soc., Internat. Astron. Union, Astron. Soc. of Pacific (pres. 1974-76). Office: U Calif-San Diego Ctr for Astrophysics Space Scis C-011 La Jolla CA 92093

BURCH, JOHN WALTER, mining equipment company executive; b. Balt., July 14, 1925; s. Louis Claude and Constance (Boucher) B. m. Robin Neely Sinkler, Apr. 19, 1952; children—John C., Robert L., Charles C., Anne N. BS in Commerce, U. Va., 1951; postgrad., U.S. Coast Guard Acad., 1951. With Procter & Gamble Co., Phila., 1953-65, sales mgr., 1960-65; v.p. Warner Co., Phila., 1965-73; chmn. bd., chief exec. officer S.S. Keely Co., Phila., 1973-75; pres., chmn. bd., chief exec. officer Burch Materials Co., Inc., Wayne, Pa., 1975—; dir. Eagle's Eye, Inc., Wayne. Bd. dirs. Nat. Multiple Sclerosis Soc., 1970-81, v.p., mem. exec. com., 1974-77; bd. dirs. Pa. Sports Hall of Fame, 1974—, v.p., mem. exec. com., 1974-79; chmn. Am. Legion Tennis Tournaments for State of Pa., 1975-82; mem. U.S. Congl. Adv. Bd., 1982—; bd. dirs. Eagle's Eye Lacrosse Club, 1982—. With USN, 1943-46, USCG, 1951-53. Named All Am. in lacrosse, 1949. Mem. Am. Mgmt. Assn., Soc. Advancement of Mgmt., Internat. Platform Assn., Merion Cricket Club, Merion Golf Club. Republican. Roman Catholic. Home: 412 Conestoga Rd Wayne PA 19087-4812 Office: Burch Materials Co Inc 685 Kromer Ave Berwyn PA 19312-1317

BURCHAM, JEFFREY ANTHONY, nuclear engineer; b. Cin., Feb. 7, 1964; s. Jess Edmund and Claire Francis (Beulter) B. BS in Aerospace Engring., U. Notre Dame, 1986; cert., Bettis Reactor Engring. Sch., 1987. Fluid systems engr. Naval Sea System Command, Naval Reactors, Washington, 1986; asst. program mgr. for surface ships Naval Sea System Command, Naval Reactors, Washington, 1987-92, asst. program mgr. shipyards, 1992—. Lt. USN, 1986—. Mem. AIAA. Roman Catholic. Home: 2409 Culpeper Rd Alexandria VA 22308 Office: Naval Sea System Command Naval Reactors Washington DC 20362-5160

BURCHFIEL, BURRELL CLARK, geology educator; b. Stockton, Calif., Mar. 21, 1934; s. Beryl Edward and Agnes (Clark) B.; m. Leigh H. Royden; children: Brian Edward, Brook Evans, Benjamin Clark. B.S., Stanford U., 1957, M.S., 1958; Ph.D., Yale U., 1961. Prof. geology Rice U., 1961-76; Prof. geology MIT, 1977-84, Schlumberger prof. geology, 1984—. Served with U.S. Army, 1958-59. Fellow Geol. Soc. Am., Am. Acad. Arts and Scis., Nat. Acad. Scis., Am. Geophys. Union, European Union Geoscis. (hon. fgn.); mem. Geol. Soc. Australia, Am. Assn. Petroleum Geologists. Home: 9 Robinson Pk Winchester MA 01890-3717 Office: MIT 54-1010 77 Massachusetts Ave Cambridge MA 02139

BURCIAGA, JUAN RAMON, physics educator; b. Ft. Worth, Tex., June 24, 1953; s. Ramon Medellin and Aurora (Vega) B. BS in Physics, U. Tex., 1975, MA in Physics, 1977; PhD, Tex. A&M U., 1986. Asst. prof. Austin Coll., Sherman, Tex., 1986-93, Colo. Coll., Colo. Springs, 1993—. Contbr. articles to profl. jours. including Jour. of Molecular Spectroscopy, Proc. of the Workshop on Comp. Physics, Phys. Rev. A, Jour. Physics B. Advisor Grayson C.C., Sherman, 1987-93, Sherman Pub. Libr., 1991-93. Mem. Am. Astrophys. Soc., Am. Assn. of Physics Tchrs., Am. Phys. Soc., Soc. of Physics Students, Sigma Pi Sigma. Office: Colo Springs Coll Dept Physics 14 E Cache la Poudre Colorado Springs CO 80903

BURD, ROBERT MEYER, hematologist, educator; b. N.Y.C., Aug. 25, 1937; s. David and Anne (Popkin) B.; m. Alice Stoller, May 30, 1964; children: Russell J., Stephen J. AB, Columbia U., 1959, MD, 1963. Diplomate Am. Bd. Internal Medicine, Am. Bd. Hematology and Oncology. Intern Albert Einstein Med. Sch., N.Y.C., 1963-64, resident in internal medicine, 1964-66; hematology fellow Montefiore Hosp., N.Y.C., 1966-67;

pvt. practice medicine, specializing in hematology and oncology, Fairfield, Conn., 1969—; assoc. prof. medicine Yale U., New Haven, 1975; assoc. clin. prof. of medicine, 1975—; chief of hematology/oncology St. Vincent's Med. Ctr., 1980—, chmn. oncology practice com.; attending physician Yale Hosp., New Haven; mem. staff Park City, Yale-New Haven hosps., Bridgeport Hosp. Editorial bd. Conn. Medicine, 1974-78; med. cons. U.S. News and World Report, 1990; dir. oncology fellowship Yale-St. Vincent, 1991-93. Active Leukemia Soc. Am., Hemophilia Found.; chmn. profl. edn. com. Am. Cancer Soc. Lt. comdr. USN, 1967-69. Ettinger Meml. fellow Am. Cancer Soc., 1982. Fellow ACP; mem. AMA, Am. Soc. Hematology, Am. Soc. Internal Medicine, Am. Soc. Clin. Oncology, N.Y. Acad. of Scis., Soc. Columbia Grads., Columbia U. Alumni Fedn. Coun., Columbia U. Alumni Club (pres. Fairfield County 1983-85, editor newsletter 1982—), Bridgeport Medical Soc. (Physician of Yr. 1993). Office: 425 Post Rd Fairfield CT 06430-6016

BURDETT, BARBRA ELAINE, biology educator; b. Lincoln, Ill., Mar. 18, 1947; d. Robert Marlin and Klaaska Johanna Baker; m. Gary Albert Burdett, Sept. 27, 1968; children: Bryan Robert, Heather Lea, Amanda Rose. AA, Lincoln Coll., 1981; BS, Millikin U., Ill. State U. Edn. Core, 1985. Cert. tchr., Ill. Tchr. biology, botany and human physiology Brown County High Sch., Mt. Sterling, Ill., 1985—; dir. Drama Club, Brown County High Sch., 1988-90. Author: Misty White, 1991. Sponsor Children, Inc., Richmond, Va., 1985—, Internat. Wildlife Coalition, North Falmouth, Mass., 1991—; vol. Vets. Hosp., St. Louis, 1988—. Mem. Nat. Assn. Biology Tchrs., Ill. Sci. Tchrs. Assn., Phi Delta Kappa (editor newsletter 1990), Phi Theta Kappa. Episcopalian. Avocation: classical guitar.

BURDICK, DAVID MAALOE, marine ecological researcher; b. Flushing, N.Y., Dec. 30, 1954; s. Edward and Joanne Fuller (Maaloe) B.; m. Funi Burdick, Dec. 26, 1982; 1 child, Benjamin. BS, Hobart Coll., 1977; PhD, La. State U., 1988. Postdoctoral fellow Woods Hole (Mass.) Oceanographic Instn., 1988-90; rsch. scientist Jackson Estuarine Lab., Durham, N.H., 1990-92; asst. rsch. prof. dept. natural resources U. N.H., Durham, 1992—. Contbr. articles to profl. jours. Mem. Am. Inst. Biol. Scis., Estuarine Rsch. Fedn., Bot. Soc. Am., Sigma Xi. Achievements include relation of anatomical adaptations to metabolic responses during flooding of Spartina; demonstration of top-down control of plant composition during eutrophication of eelgrass habitats. Office: Jackson Estuarine Lab 85 Adams Point Rd Durham NH 03824

BURDICK, WILLIAM MACDONALD, biomedical engineer; b. Providence, R.I., Apr. 24, 1952; s. Franklin Pierce and Lola Alice (Cook) B. BS, Ind. U. Pa., 1975; M of Engring., Tex. A&M U., 1981; postgrad., U. Tex., 1982-86. Engring. analyst FDA, Winchester, Mass., 1988-90; reviewer neurological devices FDA, Rockville, Md., 1990—. Inventor in field; contbr. articles to profl. jours. Active Native Am. Rights Fund. With USAF, 1976-78. Mem. Internat. Platform Assn., Biomed. Engring. Soc., Environ. Def. Fund, Nature Conservancy, Humane Soc. of U.S., Nat. Multiple Sclerosis Soc. Congregationalist. Avocations: reading, writing (poetry, songs, fiction), gardening, sports. Office: HHS/PHS/FDA/ODE/NEDB 1390 Piccard Dr Rockville MD 20850

BURFORD, ALEXANDER MITCHELL, JR., physician; b. Memphis, Mar. 21, 1929; s. Alexander Mitchell and Mary Young (Tittle) B.; BS, Florence (Ala.) State Coll., 1951; MD, U. Tenn., Memphis, 1957. Intern, U. Tenn., Knoxville, 1957-58, resident in pathology, Memphis, 1958-62; asso. pathologist Eliza Coffee Meml. Hosp., Florence, Ala., 1962-73, dir. lab., chief pathology, 1973—; practice medicine specializing in pathology, 1958—, Florence Pathologists P.C., 1983—. Active Florence Tree Commn., 1987-92. Mem. Ala. Assn. Pathologists (pres. 1974-75), Coll. Am. Pathologists (del. 1972-90), Am. Soc. Clin. Pathologists, Am. Assn. Blood Banks, Am. Forestry Assn., Am. Rifleman Assn., Nat. Wildlife Fedn., Shoals Symphony Assn., Florence C. of C., Florence Tree Commn., Friends of Florence-Lauderdale Pub. Library (pres. 1993—), U. Tenn. Coll. Med. Alumni Coun., Alpha Kappa Kappa, Kappa Mu Epsilon, Alpha Psi Omega. Home: 652 Howell St Florence AL 35630-3537 Office: Eliza Coffee Meml Hosp PO Box 818 Florence AL 35631-0818

BURG, MAURICE BENJAMIN, renal physiologist, physician; b. Boston, Apr. 9, 1931; s. Charles and Augusta (Green) B.; m. Judith Anne Braverman (dec.); m. Ruth Cooper, Dec. 30, 1967; children: Elizabeth, Laurence, Joan, Robert. AB, Harvard U., 1952, MD, 1955. Investigator Lab. Kidney/Electrolyte Metabolism Nat. Heart Lung and Blood Inst., NIH, Bethesda, Md., 1956—, chief Lab. Kidney/Electrolyte Metabolism, 1975—. Contbr. over 145 articles to profl. jours. Mem. NAS. Office: Nat Heart Lung Blood Inst Bldg 10 Rm 6N307 Bethesda MD 20892

BURGARELLA, JOHN PAUL, electronics engineer, consultant; b. Gloucester, Mass., Feb. 1, 1928; s. Joseph James and Mary Frances (Alves) B.; m. Claire T. Courchene, Oct. 20, 1956; children: Paul, Jane, Carol, Steven. BSEE with distinction, Worcester Poly. Inst., 1950, MSEE, 1952. Jr. engr. to sr. project engr. Honeywell, Newton and Boston, 1952-60; sr. engr. to sr. engring. fellow Polaroid Corp., Cambridge and Waltham, Mass., 1960-86; mem. dean's adv. com. U. Mass., 1983-86; mem. EE adv. com. Worcester (Mass.) Poly. Inst., 1986-93; occasional cons. and expert witness in photog. patent litigation. Recipient Master Design award Product Engring. Mag., 1964, Robert H. Goddard award Outstanding Profl. Achievement, Worcester Poly. Inst., 1984; honoree for one of the seven modern discoveries among Worcester Poly. Inst. Aumni, 1992. Mem. Sigma Xi, Tau Beta Pi, Eta Kappa Nu. Achievements include 25 patents in electronics, magnetics and mechanics; research related to automatic exposure control and camera systems for amateur photographic cameras. Home: 111 Pokonoket Ave Sudbury MA 01776

BURGARINO, ANTHONY EMANUEL, environmental engineer, consultant; b. Milw., July 20, 1948; s. Joseph Francis Burgarino and Mardelle (Hoeffler) T.; m. Gail Fay DiMatteo, Mar. 13, 1982; children: Paul Anthony, Joanna Lynn. BS, U. Wis., 1970; MS, Ill. Inst. Tech., 1974, PhD, 1980. Registered profl. engr., Ariz. Sales engr. Leeds & Northrup, Phila., 1970-72; rsch. asst. Ill. Inst. Tech., Chgo., 1972-75; chemist City of Chgo., 1975-79; instr. Joliet (Ill.) Jr. Coll., 1978-79; project engr. John Carollo Engrs., Walnut Creek, Calif., 1980—; cons. City of Clovis, Calif., 1981-83, City of Fresno, Calif., 1983—, City of Phoenix, 1981-90, City of Yuma, Ariz., 1989—, City of Santa Maria, Calif., 1991—. Contbr. articles to profl. jours. EPA grantee, 1970-72; NSF fellow, 1973, Ill. Inst. Tech. Rsch. Found. fellow, 1974. Mem. Am. Water Works Assn. Roman Catholic. Avocations: mechanical and electronics projects building, stock and real estate investments. Home: 4355 Oakdale Pl Pittsburg CA 94565-6258 Office: John Carollo Engrs 450 N Wiget Ln Walnut Creek CA 94598

BURGER, GEORGE VANDERKARR, wildlife ecologist, researcher; b. Woodstock, Ill., Jan. 22, 1927; s. Irwin Louis and Nettie Ann (Vanderkarr) B.; m. Jeannine Ingram Willis, June 23, 1949; children: Suzanne Linda Burger Campbell, Christine Melissa Burger Rice, Nancy Willis Burger Smith. BS, Beloit Coll., 1948; MS, U. Calif., Berkeley, 1950; PhD, U. Wis., 1958. Cert. wildlife biologist. Instr. Contra Costa (Calif.) Jr. Coll., 1952-54; field rep. Sportsmen's Svc. Bur., La Crosse, Wis., 1958-62; mgr. wildlife mgmt. Remington Arms Co., Chestertown, Md., 1962-66; gen. mgr. Max McGraw Wildlife Found., Dundee, Ill., 1966-92; commr. Ill. Nature Preserves Commn., 1985-89; mem. Ill. Surface Mining Adv. Coun., 1990—, Kane County Solid Waste Management Adv. Com., Geneva, Ill. Author: Practical Wildlife Management, 1975; editor: Pheasants: Symptoms of Wildlife Problems, 1988, (proc.) N.Am. Wood Duck Symposium, 1988. Mem. Kane County Regional Planning Commn., Ill., 1980—, Elgin (Ill.) Parks and Reclamation Commn., 1986—. Sgt. U.S. Army, 1945-46. Wis. Alumni Rsch. Found. fellow, 1954-56; Green Trees Club grantee, 1958; recipient nat. award Nature Conservancy, 1954. Mem. Am. Fisheries Soc., Izaak Walton League of Am., Outdoor Writers Assn. Am., Soil and Water Conservation Soc. (nat. adv. com. 1989—), Wildlife Soc. (hon., editor bull. 1972-75). Office: Max McGraw Wildlife Found PO Box 9 Dundee IL 60118-0009

BURGER, HENRY G., anthropologist, vocabulary scientist, publisher; b. N.Y.C., June 27, 1923; s. B. William and Terese R. (Felleman) B.; m. Barbara G. Smith, Nov. 29, 1991. B.A. with honors (Pulitzer scholar), Columbia Coll., 1947; M.A., Columbia U., 1965, Ph.D. in Cultural Anthropology (State Doctoral fellow), 1967. Indsl. engr. various orgns., 1947-51, Midwest mfrs. rep., 1952-55; social sci. cons. Chgo. and N.Y.C., 1956-67; anthropologist Southwestern Coop. Ednl. Lab., Albuquerque, 1967-69; assoc. prof. anthropology and edn. U. Mo., Kansas City, 1969-73, prof., 1973—, founding mem. univ.wide doctoral faculty, 1974—; lectr. CUNY, 1957-65; adj. prof. ednl. anthropology U. N.Mex., 1969; anthrop. cons. U.S. VA Hosp., Kansas City, 1971-72; speaker at numerous confs. Author: Ethno-Pedagogy, 1968, 2d edit., 1968; compiler, pub.: The Wordtree, a Transitive Cladistic for Solving Physical and Social Problems, 1984, selected for exhibit at 3 insts.; author linguistic-periodical column New Times, New Verbs, 1988—; contbr. to anthologies, articles to profl. jours., cassettes to tape librs. Mem. editorial bd. Council on Anthropology and Edn., 1975-80. Served to capt. AUS, 1943-46. NSF Instl. grantee, 1970. Fellow AAAS, World Acad. Art and Sci., Am. Anthrop. Assn. (life), Royal Anthrop. Inst. Gt. Britain (life), European Assn. for Lexicography; mem. Nat. Assn. for Practice of Anthropology, Internat. Soc. for Knowledge Orgn., Dictionary Soc. N.Am. (life, terminology com.), Assn. Internationale de Terminologie, Academie Europeene de scis., arts et lettres (corr.), Soc. Conceptual and Content Analysis by Computer, Columbia U. Club, Phi Beta Kappa. Achievements include discovery of the branchability of processes (corresponding, for materials, to the periodic table of elements); research on computerized causality and reasoning. Office: The Wordtree 10876 Bradshaw St Overland Park KS 66210-1148

BURGESS, KATHRYN HOY, biologist. AB in Biology, Dartmouth Coll., 1983; AM in Biology, Harvard U., 1985, PhD in Biology, 1991. Rsch. assoc. Ctr. Insect Sci., U. Ariz., Tucson, 1991—. Dir. Nat. Youth Sci. Camp, Charleston, W.Va., 1990-92. Office: Univ Ariz Dept Ecology/Evolution Biol Bio Sciences West Tucson AZ 85721

BURGESS, LARRY LEE, aerospace executive; b. Phoenix, May 13, 1942; s. Byron Howard and Betty Eileen (Schook) B.; m. Sylvia Wynnell, Sept. 30, 1964 (div. Dec. 1984); children: Byron, Damian; m. Mary Jane Ruble, Mar. 10, 1985. BSEE, MSEE, Naval Postgrad. Sch. Officer USN, Washington, 1964-85; corp. exec. Martin Marietta, Denver, 1985—; pres. L & M Investments, Denver, 1987—. Coach Youth Activities, Corpus Christi, Tex., 1976-78; speaker in local schs., Littleton, Colo., 1987-90. Inducted into the Kans. Basketball Hall of Fame, 1993. Mem. AIAA (dir.), SASA, Armed Forces Comm. and Electronic Agy. Republican. Home: 3 Red Fox Ln Littleton CO 80127 Office: Martin Marietta PO Box 179 DC 4001 Denver CO 80201

BURGESS, ROBERT LEWIS, educator, ecologist; b. Kalamazoo, Mich., Sept. 12, 1931; s. James Lewis and Hazel Lira Mae (Warren) B.; BS, U. Wis.-Milw., 1957; MS, U. Wis.-Madison, 1959, PhD, 1961; m. Vera Ballegoin, July 30, 1955; children: Karen, Steven, Susan, Ellen, Jonathan. Teaching asst. U. Wis., 1957-58, rsch. asst., 1958-60; asst. prof. Ariz. State U., 1960-63; dir. Summer Inst. in Desert Biology, 1963; asst. to assoc. prof. N.D. State U., 1963-70; dep. dir. Eastern Deciduous Forest Biome, U.S. Internat. Biol. Program, Oak Ridge Nat. Lab., 1970-77, program mgr., 1972-77, sect. head, 1975-79, sr. rsch. staff, 1980-81; prof. ecology U. Tenn., 1974-81; prof., chmn. dept. environ. and forest biology Coll. Environ. Sci. Forestry, SUNY, Syracuse, 1981—; vis. prof. Pahlavi U., Shiraz, Iran, 1965-66; research collaborator Nat. Park Svc., 1961-64; traveling lectr. Ariz. Acad. Sci., 1962-63, Am. Inst. Biol. Scis., 1968-70, Oak Ridge Associated Univs., 1973-75; mem. N.D. Wildlife Adv. Com., 1967-72; mem. adv. panel RANN program NSF, 1974-75; co-chmn. IV Internat. Congress Ecology, 1986. Served with AUS, 1953-55; Korea. Recipient N.D. Conservationist of Year award Nat. Wildlife Fedn., 1969; N.D. Cons. of Year award Safari Club Internat., 1970; award of distinction Soc. for Tech. Communication, 1978, Disting. Svc. citation Ecological Soc. Am., 1988. Fellow AAAS; mem. Am. Inst. Biol. Scis. (governing bd. 1981-84, membership chmn. 1981-82, meetings com. 1984-86, bd. dirs. 1990-92), Ecol. Soc. Am. (membership com. 1965-73, com. on professionalism 1971-81, governing coun. 1971-80, 83-88, com. on hist. records 1976—, chmn., 1982-88, program chmn. 1977-80, awards com. 1990—), Internat. Assn. for Ecology, Internat. Soc. for Tropical Ecology, N.D. Natural Sci. Soc. (pres. 1967-68), Forest History Soc., Botanical Soc. of Am., N.D. Acad. Sci. (pres. 1970-71), S.W. Assn. Naturalists, Nature Conservancy (bd. dirs. Tenn. chpt. 1975-81, chmn. Tenn. chpt. 1976-77, bd. dirs. N.Y. state 1987—), Wilderness Soc., Sigma Xi. Democrat. Methodist. Contbr. numerous articles to profl. jours: rev. editor Ecology, 1971-78; mem. edit. bd. Arid Lands Abstracts, 1979-84; bd. editors Ecology and Ecological Monographs, 1971-78, Forest Ecology and Mgmt., 1985-88. Home: 4049 Lafayette Rd Jamesville NY 13078-9771 Office: SUNY Dept Environ & Forest Biology Coll Environ Sci & Forestry Syracuse NY 13210

BURGHDUFF, JOHN BRIAN, mathematics educator; b. Augusta, Ga., July 16, 1958; s. Richard Dean and Betty Kay (Hebeler) B. BS in Applied Maths., Tex. A&M U., 1980; MS in Maths., Ohio State U., 1982. Teaching asst. Tex. A&M U., College Station, 1978-80, Ohio State U., Columbus, 1980-82; instr. San Jacinto Coll., Houston, 1982-88, U. Houston, 1988-92, Kingwood Coll., 1992—. Vol. youth dir. League City (Tex.) Ch. of Christ, 1982-86; faculty sponsor San Jacinto Coll. Bapt. Student Union, Houston, 1982-86; vol. Magnificat House Homeless Shelter, Houston, 1989—. Mem. Math. Assn. Am., Am. Math. Soc., Inst. for Combinatorics and its Applications. Democrat. Baptist. Achievements include research in spectra of graphs and permanents of matrices. Home: 2600 Westridge St Apt 292 Houston TX 77054-1543 Office: Kingwood Coll Dept Math Kingwood TX 77339

BURHANS, FRANK MALCOLM, mechanical engineer; b. Hagerstown, Md., Dec. 11, 1920; s. William Humphrey Sr. and Ethel Adella (Forthman) B.; m. Jean Maria Dermott, Oct. 10, 1943; children—Stephen William, Douglas Allan, Jeffrey Malcolm. B.E. in Mech. Engring., Johns Hopkins U., 1942; postgrad. U. Conn., 1942-43. Registered profl. engr., Wash. Design engr. Pratt & Whitney, East Hartford, Conn., 1942-55, Ford Motor Co., Dearborn, Mich., 1955-58; sr. design engr. Fairchild Engine Div., Deer Park, N.Y., 1958-59; sr. specialist engr. Turbine Div. Boeing Co., Seattle, 1959-66, prin. engr. Boeing Aircraft Engine Installations, 1967-86. Active Boy Scouts Am. Served with AC, U.S. Army, 1945-47. Recipient Silver Beaver award Boy Scouts Am. Mem. AIAA, ASME. Presbyterian (elder). Club: Masons (past master) (Bellevue). Pioneering designer gas turbines and gas turbine installations.

BURHENNE, HANS JOACHIM, physician, radiology educator; b. Hannover, Germany, Dec. 27, 1925; emigrated to U.S., 1955, naturalized, 1959; s. Adolph and Clara (Ditges) B.; m. Linda Jean Warren, Oct. 20, 1978; children by previous marriage: Mark, Antonia, Yvonne. Matura, Gymnasium, Salzburg, Austria, 1944; M.D. magna cum laude, Maximilian Med. Sch., Munich, 1951. Intern Monmouth Med. Center, Long Branch, N.J.; resident in radiology Peter Bent Brigham Hosp., Boston, 1955-59; instr. Harvard U., 1958-59; chmn. dept. radiology Children's Hosp., San Francisco, 1960-78; clin. prof. radiology U. Calif., San Francisco, 1960-78; prof. radiology U. B.C., 1978—, head dept. radiology, 1978-91. Author: Sierra Spring Ski Touring, 1971, (with A.R. Margulis) Alimentary Tract Roentgenology, 4th edit., 1989, Biliary Lithotripsy, 1990, Practical Alimentary-Tract Radiology, 1993; editor: Mammography, 1969; editorial bd. Radiologica Clinica, 1964-90, Oncology, 1973-77, Gastrointestinal Radiology, 1976—, Western Jour. of Medicine, 1975-79, Radiology, 1983-91, Lithotripsy and Stone Disease, 1988—. Chmn. bd. dirs. Cathedral Sch., San Francisco, 1976-77; bd. dirs. Sterling-Winthrop Imaging Rsch. Inst., 1989-92. NIH fellow, 1959; recipient Walter B. Cannon medal, 1982, Forsell Lectr. and medal Swedish Acad. Medicine, Stockholm, 1990; named Disting. Lectr., U. B.C., 1987. Fellow Am. Coll. Radiology (counselor 1973-77), Royal Coll. Physicians Can., Royal Coll. Surgeons Ireland (hon. faculty radiology); mem. Calif. Radiol. Soc. (pres. 1977-78), Internat. Soc. Radiology (exec. com. 1985—, chmn. diagnostic radiology 1990), Soc. Gastrointestinal Radiologists (pres. 1977), Internat. Soc. Biliary Radiology (pres. 1989-91). Home: 1063 W 7th Ave #1, Vancouver, BC Canada V6H 1B2 Office: 10th Ave and Heather St, Vancouver, BC Canada V5Z 1M9

BURK, ROBERT DAVID, physician, medical educator; b. Washington, Aug. 20, 1951; s. Meyer and Elaine Ruth Burk; m. Esther Platovsky, Oct. 21, 1984; children: Elana Ruth, Meir Solomon. Student, George Washington U., 1969-71, MD, 1976; student, McGill U., Montreal, Que., Can., 1971-72. Diplomate Nat. Bd. Med. Examiners, Am. Bd. Pediatrics, Am. Bd. Med. Genetics; lic. physician, Md., Calif., N.Y., Pa. Biologist molecular biology sect. Lab. Biochem. Genetics, Nat. Heart and Lung Inst., NIH, Bethesda, Md., 1973-74, biol. lab. technician sect. on somatic cell genetics, 1974-75; surg. intern U. Calif., San Francisco, 1976-77, pediatric resident, 1978-80; genetics fellow Johns Hopkins U. Sch. Medicine, Balt., 1980-83; rsch. assoc. molecular hepatology Albert Einstein Coll. Medicine, Bronx, N.Y., 1983-84, asst. prof. pediatrics, 1984-89, assoc. prof. pediatrics, asst. prof. ob-gyn., 1989—; asst. prof./assoc. prof. microbiology and immunology Albert Einstein Coll. Medicine, Bronx, 1987—; attending physician Montefiore Med. Ctr., Einstein div., Bronx, 1984—, Hosp. Albert Einstein Coll. Medicine, Bronx, 1984—; asst. attending North Cen. Bronx Hosp., Bronx Mcpl. Hosp. Ctr., 1984—, assoc. attending, 1989—; investigator Cancer Rsch. Ctr., 1986. Recipient NIH clin. investigator award, 1984-87, First Prize, Henry L. Moses award for outstanding clin. rsch. Montefiore Hosp., 1987, Jr. Faculty Rsch. award Am. Cancer Soc., 1987-90, Rsch. award Sinsheimer Found., 1987-90, 2nd prize Henry L. Moses award for outstanding basic rsch. Montefiore Hosp., 1989, Faculty Rsch. award Am. Cancer Soc., 1991—, Ramapo Trust Rsch. award, 1985-87, Sinsheimer Scholar award, 1987-90; grantee NIH, 1984-91, 92—, Ctr. Disease Control, 1987-92. Mem. AAAS, Am. Soc. Human Genetics, Am. Fedn. Clin. Rsch., Am. Soc. Microbiology, Soc. Pediatric Rsch. Achievements include the cloning of human DNA from the Y chromosome; description of the use of cervicovaginal lavage for the detection of human papillomavirus; important observations on the edidemology, natural history and role of cervical human papilloma virus infection and cervical dysplasia; produced one of the first trasngenic mice containing hepatitis B virus. Home: 255 W 84th St Apt 1E New York NY 10024 Office: Albert Einstein Coll Medicine 1300 Morris Park Ave Bronx NY 10461

BURKA, MARIA KARPATI, chemical engineer; b. Ujpest, Hungary, June 24, 1948; came to U.S., 1958; d. Jozsef and Katalin (Szentirmai) Karpati; m. Robert Alan Burka, Dec. 22, 1968; children: Jacqueline, Michael, Jennifer. BS, MIT, 1969, MS, 1970; MA, Princeton U., 1972, PhD, 1978. Process design engr. Scientific Design Co., N.Y.C., 1970-71; asst. prof. U. Md., College Park, 1978-81; environ. scientist EPA, Washington, 1981-82; program dir. NSF, Washington, 1984—. mem. AAUW, Am. Inst. Chem. Engrs. (sec., treas. computing and systems tech. div. 1988-93, bd. dirs. 1993—), Sigma Xi. Home: 5056 Macomb St NW Washington DC 20016-2673 Office: NSF 1800 G St NW Washington DC 20550-0002

BURKE, BERNARD FLOOD, physicist, educator; b. Boston, June 7, 1928; s. Vincent Paul and Clare (Brine) B.; m. Jane Chapin Pann, May 30, 1953; children—Geoffrey Damian, Elizabeth Chapin, Mark Vincent, Matthew Brine. S.B., M.I.T., 1950, Ph.D., 1953. Staff mem. terrestrial magnetism Carnegie Instn. of Washington, 1953-65, chmn. radio astronomy sect., 1962-65; prof. physics, Burden prof. astrophysics Mass. Inst. Tech., 1965—; vis. prof. U. Leiden, Netherlands, 1971-72, U. Manchester, Eng., 1992-93; trustee N.E. Radio Obs. Corp., 1973—, vice chmn., 1975-82, chmn., 1982—; cons. NSF, NASA, Dept. Transp.; Oort lectr. U. Leiden, 1993. Trustee Associated Univs., Inc., 1972-90; mem. Nat. Sci. Bd., 1990—. Recipient Helen Warner prize Am. Astron. Soc., 1963; Rumford prize Am. Acad. Arts and Scis., 1971; Sherman Fairchild scholar Calif. Inst. Tech., 1984, Smithsonian Regents fellow, 1985. Fellow AAAS; mem. Nat. Acad. Scis., Am. Acad. Arts and Scis., Am. Phys. Soc., Am., Royal astron. socs., Internat. Astron. Union, Internat. Sci. Radio Union. Research on microwave spectroscopy, radio astronomy, galactic structure, antenna design, comsmology. Office: Mass Inst Tech Dept Physics Cambridge MA 02139

BURKE, JERRY ALAN, retired chemist; b. Elkins, W.Va., June 30, 1937; s. John Albert and Iyone Dell (Robinson) B.; m. Janet Helen Berg, Apr. 15, 1961; children: John, James. BS in Chemistry, W.Va. Wesleyan Coll., 1959. Rsch. chemist FDA, Washington, 1959-65, head halogenated cpds. sect., 1965-71, chief analytical chemistry and phys. br., 1971-78, assoc. dir. rsch. mgmt. div. chemistry and physics, 1978-80, dir. div. chem. tech., 1980-87, acting dir. Office Phys. Scis., 1987-90, dir. Office Phys. scis., 1990-93. Editorial bd. Jour. Food Safety, 1984—, Jour. Assn. Ofcl. Analytical Chemists, 1986-92; contbr. over 50 articles to profl. jours. Bd. trustees Immanuel Luth. Ch., Alexandria, Va., 1981—. Recipient Award of Merit FDA, 1965, 76, 86, Group Recognition award, 1992. Fellow Assn. Ofcl. Analytical Chemists (various tech. coms.); mem. Nat. Wild Turkey Fedn., Nat. Rifle Assn. (life), Trout Unltd. Lutheran. Achievements include research on analytical methods for pesticide residues, contribution to development, extension and validation for uniform use among laboratories, multiresidue analytical methods for pesticide residues. Office: FDA 200 C St SW Washington DC 20204

BURKE, JOHN FRANCIS, surgeon, educator, researcher; b. Chgo., July 22, 1922; s. Frank A. and Mary V. Burke; m. Agnes Redfearn Goldman, June 24, 1950; children: John Selden, Peter Ashley, Ann Campbell, Andrew Thomas. B.S., U. Ill., 1947; M.D., Harvard U., 1951. Intern Mass. Gen. Hosp., Boston, 1951-52; resident in surgery Mass. Gen. Hosp., 1952-54, 56-57; rsch. fellow Lister Inst., London, 1955; vis. surgeon Mass. Gen. Hosp., Boston, 1968—; chief trauma services Mass. Gen. Hosp., 1980—; program dir. Burn Trauma Research Center, 1973—; assoc. prof. surgery Harvard Med. Sch., 1969-75, prof. surgery, 1975-76, Helen Andrus Benedict prof. surgery, 1976—; chief of staff Shriners Burns Inst., Boston, 1969-80; vis. prof. MIT, 1977—; pioneer in development of artificial skin; developer concept antibiotic use to prevent post-operative infection; vis. fellow Baliol Coll. Oxford U., 1990; program dir. New Eng. Burn Demonstration Program, 1977-80; chmn. bd. dirs. Boston Med. Flight. Co-editor 12 books in field; contbr. articles to profl. jours. Served with USAAF, 1942-45. Moseley Traveling fellow, 1955. Mem. Am. Burn Assn. (pres. 1982-83), Boston Surg. Soc. (pres. 1983), N.Y. Acad. Scis., AMA, Am. Thoracic Soc., Mass. Med. Soc., A.C.S., Soc. Univ. Surgeons, Infectious Disease Soc. Am., Am. Surg. Assn., New Eng. Surg. Soc. (pres. 1989), Am. Assn. Surgery of Trauma, Surg. Infection Soc. (founding mem., pres. 1985), Internat. Soc. Burn Injuries, Am. Trauma Soc. (founding mem.). Home: 216 Prospect St Belmont MA 02178-2616 Office: Mass Gen Hosp Harvard Med Sch Trauma Svc Trauma Svc Boston MA 02114

BURKE, J(OHN) MICHAEL, environmental company executive; b. Takoma Park, Md., Apr. 27, 1946; s. John Richard and Doris Jean (Waltman) B.; m. Mary Jane Elenewski, May 24, 1975; children: Alexander, Mairead. AB, Thomas More Coll., 1966; PhD, Case Western Res. U., 1971. Staff scientist U.S. Army Missile Command, Redstone Arsenal, Ala., 1971; presdl. intern Nat. Bur. Standards, Gaithersburg, Md., 1972-73; rsch. assoc. Princeton (N.J.) U., 1973-77; project leader Procter and Gamble Co., Cin., 1977-87; v.p. Roslon Internat. Corp., Point Pleasant, N.J., 1987-91; pres. Pyrogenics of N.J., Ringoes, 1991-92; dir. environ. projects Tiger Constrn. Svc. Corp., Spring Lake Heights, N.J., 1992—; chmn. Environ. Commn., Spring Lake, N.J., 1991—; mem. Spring Lake Planning Bd., 1991—; adj. instr. Brookdale Community Coll., Lincroft, N.J., 1992—. Contbr. articles to environ. jours. Pres. Spring Lake/Spring Lake Heights Soccer, 1992. NSF fellow, 1965; NASA trainee Case Western Res. U., 1968; NIH postdoctoral fellow, 1976. Mem. Am. Chem. Soc., N.J. Fedn. Planning Ofcls., Nassau Club. Avocations: tennis, photography, bridge. Home: 309 Jersey Ave Spring Lake NJ 07762

BURKE, LAURENCE DECLAN, chemistry educator; b. Cork, Ireland, July 5, 1939; s. Lawrence and Elizabeth (O'Connell) B.; m. Susan Elizabeth Allen, June 11, 1966; children: Vivienne, John, Alan. BSc, Univ. Coll., Cork, Ireland, 1959, MSc, 1961; PhD, Queen's U., Belfast, Ireland, 1964. Asst. lectr. chemistry U. Coll., Cork, 1965-80, lectr., 1980-83, assoc. prof., 1983—. Contbr. numerous articles to profl. jours. Alexander von Humboldt Stiftung fellow, Fed. Republic Germany, 1966-67. Mem. Internat. Soc. Electrochemistry, The Electrochem. Soc. Roman Catholic. Avocations: travel, gardening, reading, history. Home: Templehill, Carrigrohane, Cork Ireland Office: U Coll, Dept Chemistry, Cork Ireland

BURKE, MARGARET ANN, computer and communications company specialist; b. N.Y.C., Feb. 25, 1961; d. David Joseph and Eileen Theresa (Falvey) B. BS in Computer Sci., St. John's U., Jamaica, N.Y., 1982. Cert. data processor. Software specialist Bell Atlantic Corp., Washington, 1983—. Commr. C&P Telephone Softball League, 1986-90; mem. Corcoran Gallery Art, 1989—, Smithsonian Resident Assoc. Program, Premier, Chevy Chase Women's Rep. Club. Mem. NAFE, Alliance Francaise, Nat. Fedn. Rep. Women, Am. Film Inst., Data Processing Mgmt. Assn., Internat. Platform

Assn. Roman Catholic. Home: 6652 Hillandale Rd # A Bethesda MD 20815-6406 Office: Bell Atlantic 13101 Columbia Pike Silver Spring MD 20904-5248

BURKE, RICHARD JAMES, optical physicist, consultant; b. Barberton, Ohio, Apr. 19, 1917; s. Edward Richard Joseph and Mary Margaret (Hildum) B.; m. Louise Morgan, Nov. 9, 1940 (div. 1961); children: Richard Lewis, Pamela Jean Zwehl-Burke; m. Polly Pring Deveau, Dec. 21, 1985. BS in Engring. Physics, U. Ill., 1940; MS in Solid State Physics, U. Md., 1950, PhD in Optical Physics, 1954. Devel. engr. Eastman Kodak Co., Rochester, N.Y., 1937-40; devel. physicist Naval Ordnance Lab., Silver Spring, Md., 1940-56; dept. mgr. Lockheed Missiles and Space Co., Palo Alto, Calif., 1956-60; pres., founder Applied Systems Corp., Palo Alto, 1960-64, h nu Systems, Inc., Menlo Park, Calif., 1964-68; dept. mgr. Coherent Inc., Palo Alto, 1972-80; pres., founder Burke Concepts Unltd., San Mateo, Calif., 1980—. Author: Alternatives to Economic Disaster, 1972, The Fifth Force, 1988, Boundaries of Knowledge, 1991. Mem. Tau Beta Pi, Phi Kappa Phi, Sigma Xi, Phi Eta Sigma, Phi Lambda Upsilon. Achievements include invention of many technical devices. Home and Office: 741 Cuesta Ave San Mateo CA 94403-1203

BURKE, ROBERT HARRY, surgeon, educator; b. Cambridge, Mass., Dec. 22, 1945; s. Harry Clearfield and Joan Rosalyn (Spire) B.; m. Margaret Cauldwell Fisher, May 4, 1968; children: Christopher David, Catherine Cauldwell. Student, U. Mich. Coll. Pharmacy, 1964-67; DDS, U. Mich., 1971, MS, 1976; MD, Mich. State U., 1980. Diplomate Am. Bd. Oral and Maxillofacial Surgery, Am. Bd. Cosmetic Surgery. Pvt. practice cosmetic and reconstructive surgery Ann Arbor, Mich., 1976—; house officer oral and maxillofacial surgery U. Mich. Sch. Dentistry, U. Mich. Hosp., Ann Arbor, 1973-76; clin. asst. prof. dept. oral surgery U. Detroit Sch. Dentistry, 1976-77; adj. asst. rsch. scientist Ctr. Human Growth and Devel. U. Mich., 1976-77, adj. rsch. investigator, 1982-85; clin. asst. prof. Mich. State U., East Lansing, 1978-80, 1987—; house officer surg. emphasis St. Joseph Mercy Hosp., Ann Arbor, 1980-81; adj. rsch. investigator dept. anatomy U. Mich. Med. Sch., 1982-85; clin. asst. prof. oral and maxillofacial surgery U. Mich., 1984-86; lectr. U. Detroit Sch. Dentistry, 1986, assoc. clin. prof. oral and maxillofacial surgery, 1987-90; cons., lectr. dept. occlusion U. Mich. Sch. Dentistry, 1986; head sect. dentistry and oral surgery dept. gen. surgery St. Joseph Mercy Hosp., 1982-87, mem. exec. com. dept. gen. surgery, 1984-87; chmn. com. emergency care rev. Beyer Meml. Hosp., Ypsilanti, Mich., 1986, also active, 1987, 1990—; active staff St. Joseph Meml. Hosp.; courtesy staff Saline (Mich.) Community Hosp., 1978-88; Chelsea (Mich.) Med. Ctr., 1978-88, 90-92, McPherson Community Hosp., Howell, Mich., 1984-87. Mem. editorial bd. Topics in Pain Mgmt., 1985—; contbg. editor Am. Jour. Cosmetic Surgery, 1990-91; section editor Internat. Jour. of Aesthetic and Restorative Surgery. Campaign chmn. med. and dental sects. United Way Washtenaw County, Ann Arbor, 1982, dental sect. 1983; profl. adv. com. March of Dimes Genesee County Valley Chpt., Flint, 1979; pres. Huron Pkwy. Pla. Condominium, 1984—. Fellow Internat. Coll. Surgeons, Am. Coll. Oral and Maxillofacial Surgeons (v.p. 1987-88, pres.-elect 1989-90, pres. 1991-93), Am. Acad. Aesthetic and Restorative Surgery; mem. AMA, Am. Assn. Craniomaxillofacial Surgeons (pres. 1992—), Internat. Soc. Cosmetic Laser Surgeons (trustee 1992-93, sec. 1992), British Soc. for Oral and Maxillofacial Surgeons (assoc.), European Soc. Aesthetic Surgery and Liposuction, Chalmers Lyons Acad. Oral Surgery, European Assn. for Cranio-Maxillofacial Surgery (assoc.), Washtenaw County Med. Soc. (exec.com. sec. 1987-88, pres. 1990), Inst. Study Profl. Risk (bd. dirs. 1985-90), Victor's Club, Pres.'s Club, Omicron Kappa Upsilon. Congregationalist. Avocations: triathlon, chung do kwan, tae kwon do. Home: 720 Watershed Dr Ann Arbor MI 48105-9412 Office: 2260 Huron Pky Ann Arbor MI 48104-5126

BURKE, SHAWN EDMUND, mechanical engineer; b. Waterville, Maine, Oct. 28, 1959; s. Frances Mary (Bernier) B.; m. Monica Arlene Schnitger, Sept. 7, 1985. BS in Mech. and Aero. Engring., Princeton U., 1981; MS in Mech. Engring., MIT, 1983, PhD, 1989. Assoc. engr. Bolt, Beranek and Newman, Inc., Cambridge, Mass., 1980-81; sr. cons. engr. Chase Inc., Boston, 1984-86; cons. scientist Tech. Integration, Inc., Billerica, Mass., 1985; sr. tech. staff C.S. Draper Lab., Inc., Cambridge, Mass., 1989—; dept. instr. MIT, Cambridge, 1984-85. Vol. Climb Against Cancer, Brookline, Mass., 1990—; reader Recording for Blind, Cambridge, 1991. Recipient Donald J. Dike prize Princeton U., 1981, John Marshall II Meml. award, 1980. Mem. Acoustical Soc. Am., ASME, Sigma Xi. Achievements include patent for pressure distribution characterization system. Home: 39 Kathleen Dr Andover MA 01810

BURKE, THOMAS JOSEPH, civil engineer; b. Grosse Pointe Park, Mich., Sept. 1, 1927; s. Cyril Joseph and Marie Estelle (Sullivan) B.; BCE, Villanova U., 1949; m. Elaine Kiefer, Nov. 10, 1951; children: Judy Lee Burke Brooks, Kathleen Marie Harness, Maureen Elaine Beck, Thomas P. Chmn., Burke Rental Service, Sterling Heights, Mich., 1949—, Cyril J. Burke, Inc., Sterling Heights, Mich., 1949—. Trustee Villanova U., 1980—. Served to lt. USAF, Korea. Mem. ASCE, Detroit Builders Exchange (v.p. 1976-78, dir. 1975-78), Associated Equipment Distbrs. (dir. 1955-58, 75-78), Associated Underground Contractors (dir. 1965-68), Mich. Ready Mix Concrete Assn. (dir. 1960-65), Villanova U. Alumni Assn. (nat. v.p. 1978-79, nat. pres. 1980), Detroit Engring. Soc. Roman Catholic. Clubs: Grosse Pointe Yacht, Otsego Ski, Ocean Reef, Detroit Athletic, Villanova U. of Detroit (pres. 1955-65). Home: 578 Shelden Rd Grosse Pointe MI 48236-2640 also: 688 N Lakeshore Rd Port Sanilac MI 48469-9713 Office: 36000 Mound Rd Sterling Heights MI 48311

BURKETT, EUGENE JOHN, chemical engineer; b. Cin., Nov. 13, 1937; s. James E. and Amelia (Kues) B.; married, Apr. 15, 1977; 1 child, Matthew. BSChemE, U. Cin., 1962. Mgr. chem. plants engring. Goodyear Tire and Rubber Co., Akron, Ohio, 1973-75, mgr. corp. environ. engring., 1975-84, engring. mgr., 1985-88; tech. supt. Shell Chem. Co. (formerly Goodyear Tire and Rubber Co.), Apple Grove, W.Va., 1988—. Mem. AICE (sect. pres. 1978). Office: Shell Chem Co Ste Rt 2 Apple Grove WV 25502

BURKETT, MARJORIE THERESA, nursing educator, gerontology nurse; b. Jamaica, West Indies, Mar. 21, 1931; d. David Cameron and Mabel Louise (McKenzie) Espeut; m. Leo A. Burkett, Apr. 4, 1962; 1 child, Catherine Ann. Diploma in Midwifery and Nursing, Kingston Sch. of Nursing, Kingston, Jamaica, 1953; diploma in Nursing Edn., U. Edinburgh (Scotland UK), 1963; diploma in Psychiat. Nursing, U. Edinburgh, Scotland, 1970; BA, U. West Indies, 1975, MSN Edn., 1977, PhD, 1990; adult health nurse practitioner, Fla. Internat. U., 1992. RN, Fla., Tenn., Eng. and Wales UK, Jamaica; cert. midwife, Jamaica. Asst. prof. Fla. Internat. U., North Miami, 1988—, coord. adult med. and surg. nursing, 1990, coord. Childbearing Nursing, 1992—; faculty mem. numerous community colls. and profl. coms. Contbr. articles to profl. jours. Mem. ANA, Nat. League Nursing, Fla. League Nursing, Fla. Nurses Assn., Transcultural Nursing Soc., Gerontol. Soc. Fla., Nat. Coun. on Aging, Sigma Theta Tau, Phi Lambda Pi, Phi Delta Kappa. Office: Fla Internat U Sch Nursing 3000 NE 145 St N Miami Campus Miami FL 33186

BURKHOLDER, DONALD LYMAN, mathematician, educator; b. Octavia, Nebr., Jan. 19, 1927; s. Elmer and Susie (Rothrock) B.; m. Jean Annette Fox, June 17, 1950; children: Kathleen, Peter, William. BA, Earlham Coll., 1950; MS, U. Wis., 1953; PhD, U. N.C., 1955. Asst. prof. U. Ill. at Urbana, 1955-60, assoc. prof., 1960-64, prof. math., 1964—; prof. Center for Advanced Study, 1978—; sabbatical leaves U. Calif., Berkeley, 1961-62, Westfield Coll., U. London, 1969-70; vis. prof. Rutgers U., 1972-73; researcher Stanford U., 1961, Hebrew U., 1969, Mittag-Leffler Inst., Sweden, 1971, 82, U. Paris, 1975, Institut des Hautes Études Scientifiques, 1986; Mordell lectr. Cambridge U., 1986; Zygmund lectr. U. Chgo., 1988; trustee Math. Scis. Rsch. Inst., 1979-84; bd. govs. Inst. Math. and Its Applications, 1983-85, chmn., 1985. Editor: Annals Math. Statistics, 1964-67. Fellow Inst. Math. Statistics (Wald lectr. 1971, pres. 1975-76); mem. NAS, Am. Math. Soc. (mem. editorial bd. Trans. 1983-85), London Math. Soc., Am. Acad. Arts and Scis. Achievements include research in probability theory and its applications to other branches of analysis. Home: 506 W Oregon St Urbana IL 61801-4044

BURKHOLDER, WENDELL EUGENE, entomologist; b. Octavia, Nebr., June 24, 1928; s. Elmer and Susie (Rothrock) B.; m. Leona Rose Flory, Aug. 18, 1951; children: Paul Charles, Anne Carolyn, Joseph Kern, Stephen James. A.B., McPherson Coll., 1950; M.Sc., U. Nebr., 1956; Ph.D., U. Wis., 1967. Research entomologist U.S. Dept. Agr., 1956—, Madison, Wis., 1965—; asst. prof. U. Wis.-Madison, 1967-70, asso. prof., 1970-75, prof. entomology, 1975—. Mem. editorial bd.: Jour. Chem. Ecology, 1980—, Jour. Stored Products Rsch., 1992—; contbr. chpts. to books and articles to profl. jours. Served with U.S. Army, 1951-53. NSF grantee, 1972-75, 79; Rockefeller Found. grantee, 1974-77; Nat. Inst. Occupational Safety and Health grantee, 1977-79. Mem. Entomol. Soc. Am., Soc. Invertebrate Pathology, AAAS, Wis. Entomol. Soc. (pres. 1980), Wis. Acad. Sci. Arts, and Letters, Internat. Soc. Chem. Ecology, Sigma Xi. Patentee in field. Home: 1726 Chadbourne Ave Madison WI 53705-4108 Office: U Wis 537 Russell Lab Madison WI 53706-1598

BURL, JEFFREY BRIAN, electrical engineer, educator; b. Lansing, Mich., Apr. 19, 1956; s. Louis Lester and Florence (Rosenow) B.; m. Suzanne Williams, May 29, 1992. BS, U. Mich., 1978; MS, U. Calif., Irvine, 1982, PhD, 1987. Engring. specialist in radar Ford Aero. and Communications Corp., Newport Beach, Calif., 1979-87; assoc. prof. elec. engring. Naval Postgrad. Sch., Monterey, Calif., 1987—. Contbr. articles to Solar Physics, IEEE publs., Internat. Jour. Modelling and Simulation, other profl. jours. Basketball coach Monterey County Spl. Olympics, 1989-92. Am. Bentley rsch. fellow, 1984; McDonnell Douglas rsch. fellow, 1986. Mem. AIAA (sr.), IEEE (sr., sect. chmn. 1988-91, treas. 1991-92). Achievements include development of novel algorithm for estimating motion from images, novel algorithm for system identification with a plethora of sensor data, development of adaptive controller for crew and equipment retrieval for space station. Home: 1223 Funston Ave Pacific Grove CA 93950 Office: Naval Postgrad Sch Code EC/BL Monterey CA 93943-5000

BURLANT, WILLIAM JACK, chemical company executive; b. Chgo., Oct. 20, 1928; m. Arlene Rosenberg, July 31, 1955; children: Diana, Michael. BS, CCNY, 1949; MA, Bklyn. Coll., 1953; PhD in Chemistry, Poly. Inst. N.Y. Asst. dir. chem. scis. Ford Motor Co., Dearborn, Mich., 1955-79; tech. mgr. Gen. Electric Corp., Columbus, Ohio, 1979-81; dir. Lexington Lab. Kendall Co., Boston, 1981-84; v.p. research GAF Chems. Corp., Wayne, N.J., 1984-90; prin. scientist ISP, Inc., Wayne, 1990—; mem. adv. bd. NSF. Author: (book) Block and Graft Polymers, 1959; contbr. articles to profl. jours.; patentee in field. Mem. Am. Chem. Soc. Office: GAF Chems Corp 1361 Alps Rd Wayne NJ 07470-3700

BURLESKI, JOSEPH ANTHONY, JR., information services professional; b. Poughkeepsie, N.Y., June 30, 1960; s. Joseph Anthony Burleski Sr. and Fredeline Cyr; m. Judith Ann Lezon, June 10, 1989. BSBA, Marist Coll., 1982; MBA Mktg., U. Phoenix, 1992; grad. in human rels. and effective speaking, Dale Carnegie, 1990. Computer operator IBM Corp., Poughkeepsie, 1982-83, lead/sr. computer operator, 1983-84, systems programmer, 1984-85, assoc. systems programmer, 1985-86, mgr. offshift computer ops., 1986-87; mgr. info. processing IBM Corp., Boulder, Colo., 1987-88, mgr. MVS systems programming, 1988-91; mgr. location and field svcs. devel. Integrated Systems Solutions Corp. (subs. IBM), Boulder, 1991-93, mgr. location and field svc. devel. ind. test, 1992-93; mgr. VM/VSE svcs. Integrated Systems Solutions Corp. (subs. IBM Corp.), Boulder, 1993—; mem. IBM Data Processing Ops. Coun., Poughkeepsie, 1983—; grad. asst. Dale Carnegie Inst., Boulder, 1990—. Coach Spl. Olympics, 1987—; mem. Order of the Arrow Hon. Soc., sec., editor, 1976-77, pres. 1977-78, treas. 1980-81; patrol leader, store dir., asst. camp dir. Boy Scouts Am., Cub Scouts Summer Camp, 1985-87. Mem. Marist Coll. Alumni Assn. (contbr.), Vigil Nat. Honor Soc., IBM Runners' Club. Roman Catholic. Avocations: running, reading, camping, hiking, raising tropical fish. Home: 1826 Lashley St Longmont CO 80501-2061 Office: ISCC Corp 7R6A/024W 5600 N 63rd St Boulder CO 80314

BURLINGAME, ALMA LYMAN, chemist, educator; b. Cranston, R.I., Apr. 29, 1937; s. Herman Follett and Rose Irene (Kohler) B.; children: Mark, Walter. BS, U. R.I., 1959; PhD, MIT, 1962. Asst. prof. U. Calif., Berkeley, 1963-68, assoc. chemist, 1968-72, rsch. chemist, 1972-78; prof. U. Calif., San Francisco, 1978—. Editor: Topics in Organic Mass Spectrometry, 1970, Mass Spectrometry in Health and Life Science, 1985, Biological Mass Spectrometry, 1990; contbr. articles to profl. jours. With USAR, 1954-62. Guggenheim Found. fellow, 1970. Fellow AAAS. Office: U Calif Dept of Pharmaceutical Chemistry San Francisco CA 94143-0446

BURMASTER, MARK JOSEPH, software engineer; b. Caledonia, Minn., July 13, 1961; s. Myron George and Delores Marian (Hosch) B. BS, U. Wis., La Crosse, 1984. Software engr. NMT Corp., La Crosse, 1986—. Home: 2809 Robinsdale Ave La Crosse WI 54601 Office: NMT Corp 2004 Kramer St La Crosse WI 54603

BURNER, ALPHEUS WILSON, JR., optical engineer; b. Jacksonville, Fla., Dec. 4, 1947; s. Alpheus Wilson and Dorothy Mae (Howard) B.; m. Judy Irene Driggers, Sept. 1, 1968; children: Alpheus III, Katherine Maria. BS in Physics, U. S.C., 1969; MS in Optics, U. Rochester, 1976. Physicist NASA Langley Rsch. Ctr., Hampton, Va., 1969—. Contbg. author: Non-Topographic Photogrammetry, 2d edit., 1989; contbr. articles to profl. publs. With U.S. Army, 1970-72. Mem. Optical Soc. Am. Methodist. Achievements include application of holographic techniques for wind tunnel flow visualization, development of calibration schemes for video cameras, application of video photogrammetric techniques to measure wind tunnel model deformation. Office: NASA Langley Rsch Ctr Mail Stop 236 Hampton VA 23681

BURNETT, JAMES RAY, physicist, consultant; b. Detroit, Feb. 24, 1933; s. James Elliot and Hazel Mary (Johns) B.; m. Danielle Garland, Apr. 15, 1981; children: Anne Catherine, Susan Elizabeth. BS in Aeronautics, U. Mich., 1955, MS in Nuclear Physics and Math., 1957; PhD in Physics, Century U., 1983. Registered profl. nuclear engr., Calif. Rsch. asst. U. Mich., Ann Arbor, 1953-57; sr. engr. Bendix Aerospace Corp., Ann Arbor, 1957-61; assoc. dept. mgr. EG & G, Santa Barbara, Calif., 1961-64; project mgr. Electro-Optical Systems, Pasadena, Calif., 1965-72; chief engr. Xerox Word Processing Products, Pasadena, 1972-77; mgr. new mfg. plants Intel Corp., Santa Clara, Calif., 1978-82; pres., founder Burnett Tech., Los Gatos, Calif., 1982—; instr. microelectronics mfg. U. Calif., U. Ariz., 1985—. Author: Operating in Clean Room, Class 1 to Class 100,000 Practice, 1987; columnist World Class Practice mag., 1982-83; contbr. over 50 articles to tech. jours. Mem. Am. Assn. Scientists, Inst. Environ. Sci. Achievements include patents for nuclear instrumentation, reactor control, fiber-optics and space power systems. Office: Burnett Tech 208 Vista De Sierra Los Gatos CA 95032

BURNFIELD, DANIEL LEE, engineering program manager; b. Warren, Ohio, Jan. 19, 1952; s. Robert Lee and Virginia Ruth (Dicks) B.; m. Joyce Lynn Buxton, June 17, 1972; children: Jennifer, Deborah. BS in Nuclear Engring., Purdue U., 1979; cert., Bettis Reactor Engring. Sch., West Mifflin, Pa., 1981. Enlisted USN, 1970, advanced through grades to lt., 1984; quality assurance engr. Naval Reactors USN, Arlington, Va., 1981-86, prodn. engr. Naval Reactors, 1986-87, radiol. engr. Naval Reactors, 1987-90; nuclear engr. Def. Nuclear Facilities Safety Bd., Washington, 1990-91, Pantex program mgr., 1991—; pres. Springfield (Va.) Computer Assocs., 1987-91. Bd. dirs. Franklin Commons Assn., Springfield, 1987-89. Mem. Am. Nuclear Soc., Health Physics Soc. (plenary). Republican. Baptist. Home: 6973 Villa Del Rey Ct Springfield VA 22150-3067 Office: Def Nuclear Facilities Safety Bd 625 Indiana Ave NW Washington DC 20004-2901

BURNHAM, DUANE LEE, pharmaceutical company executive; b. Excelsior, Minn., Jan. 22, 1942; s. Harold Lee and Hazel Evelyn (Johnson) B.; m. Susan Elizabeth Klinner, June 22, 1963; children—David Lee, Matthew Beckwith. B.S., U. Minn., 1963, M.B.A., 1972. C.P.A., Wis. Fin. and mgmt. Maremont Corp., Chgo., 1969-75; v.p. fin., chief fin. officer Bunker Ramo Corp., Oak Brook, Ill., 1975-79, exec. v.p., dir., 1979-80, pres., chief exec. officer, dir., 1980-82; sr. v.p. fin., chief financial officer Abbott Labs., North Chgo., 1982-84, exec. v.p., chief financial officer, dir., 1985-87, vice chmn., chief fin. officer, 1987-89, chmn., chief exec. officer, 1990—; bd. dirs. Evanston Hosp. Corp., Sara Lee Corp., Fed. Res. Bank Chgo. Bd. dirs. Mus. Sci. and Industry, Chgo., Lyric Opera, Healthcare Leadership Coun.; trustee Northwestern U.; mem. adv. bd. J.L. Kellogg Grad. Sch. Mgmt.; chmn. Emergency Com. for Am. Trade. Office: Abbott Labs 1 Abbott Park Rd North Chicago IL 60064-3500*

BURNHAM, VIRGINIA SCHROEDER, medical writer; b. Savannah, Ga., Dec. 9, 1908; d. Henry Alfred and Natalie Morris (Munde) Schroeder; children: Douglass L., Peter B., Gilliat S. (dec.), William W., Virginia L., Daniel B. Student, Smith Coll., Barnard Coll. V.p., sales mgr., bd. dirs. Conn. Mfg. Co. Inc., Waterbury, 1952-61, pres., chief exec. officer, bd. dirs., 1961-73; pres. Burnham Industries, Watertown, Conn., 1956-59; pres., bd. dirs. NuTip Corp., Waterbury, 1962-64, Maretta Inc., N.Y.C., 1963-65; pres., treas., bd. dirs. Tech., Inc., Waterbury, 1969-73; sec. The Gaylord Hosp., Wallingford, Conn., 1970-81; pres. Tech. Internat. Corp., 1973-75, Tech. Interaction, Med. Cons., Greenwich, Conn., 1975—, The Paper Mill, Inc., 1989-92; dir. Community Mental Health Ctr., Inc., Stamford, Conn., 1974-78; mem. nat. adv. food and drug com. FDA, Dept. Health, Edn., and Welfare, Washington, 1973-76; mem. health rsch. facilities coun. NIH, Washington, 1959-61, nat. adv. heart coun., 1957-60; mem. ad-hoc com. cons. on med. rsch. Subcom. Labor, Health and Welfare, U.S. Senate Appropriations Com., Washington, 1959-60. Co-author: Knowing Yourself, 1992, The Two-Edged Sword, 1990, The Lake With Two Dams, 1993; contbr. articles to Am. Health Found. Newsletter, Jour. Sch. Health, Conn. Med. Jour. Mem. adv. coun. steering com. The Episc. Ch. Found., 1970-76, Commn. to Reform the Ct. System, Gen. Assembly, State of Conn., 1974-78, nat. adv. coun. SBA, Washington 1976-77, chmn. dist. adv. coun., Hartford, Conn., 1976-77; pres. Conn. Citizens for Judicial Modernization, Inc., Hartford, 1976-78; mem. Presdl. Task Force on Rehab. of Prisoners, Washington, 1969-70; bd. dirs. Assn. to Unite the Democracies, Washington, 1982—; chmn. dist. adv. coun. Small Bus. Adminstrn., Hartford, 1976-77; pres. Conn. chpt. Am. Health Found., N.Y.C., 1971-73; dir. exec. com. Greater N.Y. Safety Coun., N.Y.C., 1971-79, dir., 1964-79; vice-chmn., dir. Conn. Coun. of Nat. Coun. Crime and Delinquency, Hartford, 1965-73; mem. adv. com. Conn. Regional Med. Program, 1973-76, FDA, Washington and many more civic and health orgns. Decorated Knighthood of Honour and Merit, Sovereign Hospitaler Order of St. John of Jerusalem, Knights of Malta, 1985; recipient Ira V. Hiscock award Conn. Pub. Health Assn., 1986; Silver Key award The Gaylord Hosp., 1970, Disting. Svc. award Conn. Heart Assn., 1960, Conn. Mother of Yr. award Am. Mothers Com., 1951, Cert. of Honor, Conn. Cancer Soc., 1950, Merit award Am. Heart Assn., 1960. Mem. AAAS, APHA, Am. Cancer Soc., Am. Heart Assn., Am. Holistic Med. Found., Am. Women in Sci., Conn. Bus. & Industry Assn., Conn. Pub. Health Assn., Nat. Assn. Mfrs., N.Y. Acad. Sci. Republican. Episcopalian. Home and Office: 41 Duncan Dr Greenwich CT 06831-3616

BURNO, JOHN GORDON, JR., microbiologist; b. Pitts., Oct. 4, 1963; s. John Gordon and Adrain Diane (Lotz) B.; m. Rosanna Rocha, Sept. 9, 1989; children: Raven, Nicole. BS, U. Ariz., 1985, PhD, 1991. Rsch. asst. Dept. Microbiology, U. Ariz., Tucson, 1986-91; postdoctoral rsch. assoc. Armstrong Lab. Brooks, AFB, San Antonio, 1991—. Contbr.a rticles to profl. jours. Achievements include patent application filed for construction of composite redox system in mammdlian cells. Office: AL/OEDR Brooks AFB Bldg 175 E San Antonio TX 78235

BURNS, BRIAN PATRICK, electrical engineer; b. Camden, N.J., Mar. 17, 1962; s. James Francis and Mary Margaret (Grogan) B.; m. Alison Carol Barwis, May 16, 1987. BSEE, Drexel U., 1985; MA in Psychology for Human Factors Engring., George Mason U., 1992, postgrad., 1992—. Team leader Coop. Employment Tn. Act, Stratford, N.J., 1980; estimator E.C. Ernst, Phila., 1981-82; coop. student jr. engr. Harris Govt. Systems Group, Melbourn, Fla., 1982; coop. student fed. systems div. IBM, Manassas, Va., 1983; coop. student systems product div. IBM, Boca Raton, Fla., 1984; assoc. engr./scientist fed. systems div. IBM, Gaithersburg, Md., 1985-87, sr. assoc. engr./scientist systems integration div., 1987-88, staff engr./scientist systems integration div., 1988-90; adv. engr./scientist U.S. mktg. and svcs. group IBM; lead systems engr./architect systems integration div. IBM, Gaithersburg, Md., 1988-89, exec. area tech. lab. coord., 1989, bus. solutions cons. project mgr./tech. lead U.S. mktg. and svcs. group, 1989-91, adv. engr./scientist U.S. mktg. and svcs. group, 1990-91; adv. human factors engr. Gen. Sector div. IBM, Bethesda, Md., 1992—; lab. coord., IBM, 1986-89, adv. devel. prototype investigator, 1987-88; resident tutor Drexel U., Phila., 1982-85. Background vocalist for Christian folk music album I Lift Up My Eyes, 1987. Enumerator 1980 Census Bur., Westmont, N.J., 1980; election poll judge and clk. Camden County, N.J. Bd. of Elections, 1981-85; assessment commr. Twp. of Cherry Hill, N.J., 1982; vocalist Queen of Heaven Cath. Ch., Cherry Hill, 1985—. Recipient Dean J. Peterson Ryder award Drexel U., 1985. Mem. AIAA, IEEE, Computer Soc. of IEEE (student v.p. 1984-85), Human Factors Soc. (presenter 35th ann. meeting 1991), Assn. Computing Machinery. Democrat. Roman Catholic. Avocations: music, volleyball, karate, tennis. Home: 24701 Kings Valley Rd Damascus MD 20872-2230 Office: IBM Mail Drop 3061 10401 Fernwood Rd Bethesda MD 20817

BURNS, DONALD RAYMOND, structural and mechanical design engineer; b. St. Louis, May 12, 1961; s. John Francis and Mildred Sue (Johnson) B.; m. Dawn Michelle Nash, Feb. 12, 1988. BSCE, U. Mo., Rolla, 1983. Sr. specialist/design engring. McDonnell Douglas, St. Louis, 1984—. Contbr. articles to profl. jours. Lt. USNR, 1987—. Mem. AIAA, U.S. Naval Inst., Naval Res. Assn. Achievements include research in nondestructive method for measuring coating thickness; design and testing of hot-enriched ion source for deuterium simulation. Home: 1902 Spring Beauty Dr Florissant MO 63031

BURNS, JAMES KENT (JASPER), science illustrator; b. Charlottesville, Va., Mar. 1, 1952; s. James Richard and Jaquelin Lee (Caskie) B. BS, George Mason U., 1981. Self-employed author/illustrator, 1981—. Illustrator: P.B.'s Quick Index to Bird Nesting, 1983, Fishing in the Chesapeake Bay, 1993; author/illustrator: Selected Lives, 1986, Fossil Collecting in the Mid Atlantic States, 1991. Recipient Design and Effectiveness award The Washington Book Pubs., 1991. Home: 726 Hornet Dr Gardnerville NV 89410-8322

BURNS, MARCELLINE, psychologist, researcher. BA in Psychology, San Diego State U., 1955; MA, Calif. State U., L.A., 1969; PhD, U. Calif., Irvine, 1972. Co-founder Calif. Rsch. Inst., L.A., 1973—; cons., expert witness alcohol and drug effects on performance, FSTs, HGN, and drug recognition; lectr. in field. Contbr. articles to profl. jours. Recipient Public Svc. award U.S. Dept. Trans., 1993. Achievements include the development of software for a databacs of Drug Evaluations; research of alcohol and drug effects on boating skills; field sobreity tests for the boating environment. Office: Southern California Rsch Instit 11912 W Washington Blvd Los Angeles CA 90066

BURNS, MARSHALL, physicist; b. Toronto, Ont., Can., Aug. 24, 1954; s. Perry Birenbaum and Evelyn Weinrib. BS in Physics, MIT, 1979; PhD in Physics, U. Tex., 1991. Pres. Ennex Corp., Toronto, 1975—, Ennex Tech. Mktg., Inc., Austin, Tex., 1982—; rsch. cons. Ennex Fabrication Techs., L.A., 1992—. Author: Automated Fabrication-Improving Productivity in Manufacturing, 1993; co-author, editor: Rapid Prototyping: System Selection and Implementation Guide. Office: Ennex Fabrication Techs 549 Landfair Ave Los Angeles CA 90024

BURNS, NED HAMILTON, civil engineering educator; b. Magnolia, Ark., Nov. 25, 1932; s. Andrew Louis and Ila Mae (Martin) B.; m. Martha Ann Fontaine, June 11, 1955; children: Kathryn Jane, Stephanie Ann, Michael Everett. B.S., U. Tex., 1954, M.S., 1958; Ph.D., U. Ill., 1962. Registered profl. engr., Tex. Instr. U. Tex., Austin, 1957-59, asst. prof., 1962-65, assoc. prof., 1965-70, prof. civil engring., 1970-83, Zarrow Centennial prof. engring., 1983—; assoc. dean engring. for acad. affairs, 1989-93; research asst. U. Ill., Urbana, 1959-62. Author: (with T. Y. Lin) Design of Prestressed Concrete Structures, 1981 (McGraw Hill Book of Month 1982), S.I. Version—Design of Prestressed Concrete Structures, 1982; contbr. tech. papers, reports on structural engring. to profl. publs. Served with U.S. Army, 1955-57. Recipient Gen. Dynamics Teaching award U. Tex. Coll. Engring., 1965; recipient AMOCO Teaching award, 1983. Mem. Am. Concrete Inst. (bd. dirs. 1983—, Joe Kelley award for contbns. to engring. edn. 1990), Post-Tensioning Inst. (dir. 1975—), ASCE (com. chmn. 1975—), Prestressed Concrete Inst. (com. mem. 1968—), Nat. Soc. Profl. Engrs. (chpt. pres.

1970), Tex. Soc. Profl. Engring. (Young Engr. of Yr. award 1970, Travis chpt. Engr. of Yr. award 1987). Democrat. Baptist. Home: 3917 Rockledge Dr Austin TX 78731-2921 Office: U Tex Dept Civil Engring Austin TX 78712

BURNS, PADRAIC, physician, psychiatrist, psychoanalyst, educator; b. Des Moines, Aug. 31, 1929; s. Charles and Ethel P. (Bentz) B.; m. Ikuko Kawai, Oct. 19, 1959; children—Kenneth, Amelia, Margaret. B.A., U. Chgo., 1948; postgrad. NYU, 1949-51; M.D., Yale U., 1955. Diplomate Am. Bd. Psychiatry, Sub-Bd. Child Psychiatry. Asst. prof. psychiatry Boston U., 1969-72, assoc. prof. psychiatry, 1972—. Capt. U.S. Army, 1959. Fellow Am. Acad. Child Psychiatry; mem. Am. Psychiat. Assn., Mass. Psychiat. Soc., Boston Psychoanalytic Soc. and Inst. Home: 9 Downing Rd Brookline MA 02146-2114 Office: Boston U Med Ctr P-905 720 Harrison Ave Boston MA 02118

BURNS, PAT ACKERMAN GONIA, infosytems specialist, software engineer; b. Birmingham, Ala., July 16, 1938; d. Richard Lee and Hattie Eugenia (Bragg) Ackerman; m. Robert Edward Gonia, June 4, 1957 (div. Jan. 1973); children: Deborah Hayes, Junita Grantham, Ronald Gonia; m. James Clayton Burns, June 23, 1984 (dec. Dec. 1989). BS in Math., U. Ala., 1970, postgrad., 1971-77. Cert. secondary tchr., Ala. Missionary United Meth. Bd. of Missions, Sumatra, Indonesia, 1961-64; homebound tchr. Huntsville (Ala.) City Schs., 1970-75; mem. tech. staff Gen. Rsch. Corp., Huntsville, 1975-79; rsch. scientist Nichols Rsch. Corp., Huntsville, 1979-84, mgr. personnel div., 1984-87, mgmt. info. systems dept. head, 1987-90, dir. info. systems div., 1990—; program mgr. MIS and tech. MIS U.S. Army Space and Strategic Defense Systems, 1990—; mem. adv. com. Drake Tech. Sch., Huntsville, 1988—; program mgr. USASDC MIS/TMIS, 1990—. Mem. PTA, Huntsville, Ch. Women United, Huntsville, Community Chorus, Huntsville. Mem. NAFE, Data Processing Mgmt. Assn., Assn. Personnel Adminstrs., IEEE, Am. Computer Soc., Huntsville C. of C. (speaker 1986—). Democrat. Methodist. Avocations: music, old movies. Office: Nichols Rsch Corp 4040 S Memorial Pky Huntsville AL 35802-1396

BURNS, PETER DAVID, imaging scientist; b. Perth, Australia, May 6, 1952; came to U.S., 1967; s. R. David and Barbara Mary (Franks) B.; m. Gloriela Eleida Olivares, Jan. 31, 1976; children: Kevin, Courtney. BSEE, Clarkson U., 1974, M of Elec. Engring., 1977; postgrad., Rochester Inst. Tech., 1991—. Engring.-tech. specialist Xerox Corp., Webster, N.Y., 1975-82; sr. rsch. scientist Eastman Kodak, Rochester, N.Y., 1983—; adj. faculty Rochester Inst. Tech., 1986-89. Contbr. articles to Jour. Imaging Sci., Proc. SPIE, Applied Optics. Mem. IEEE, Soc. Imaging Sci. and Tech. (tech. program com. of ann. conf. 1990). Achievements include patent pending (with other) for image scanner and method for improved microfilm image quality. Office: Eastman Kodak Co Rsch Labs Rochester NY 14650-1925

BURNS, ROBERT, JR., architect, freelance writer, artist; b. Jackson, Miss., Jan. 29, 1936; s. Robert Sr. and Grace Hortense (Inmon) B. BS in Architecture, Ga. Inst. Tech., 1959, BArch, 1960. Registered architect, Miss. Architect Overstreet, Ware, Ware & Lewis, Jackson, 1961-70, Ware, Lewis & Eaton, Jackson, 1970-71, Jones & Haas, Jackson, 1971-74, Leon Burton & Assocs., Jackson, 1974-75, Glenn Albritton Designer, Jackson, 1975-83, Breland & Farmer, Jackson, 1983-84, The Plan House, Jackson, 1984-86; part-time tchr. art dept. Miss. Coll., 1987; freelance writer and painter Jackson, 1987—; tenor soloist 1st Christian Ch., Jackson, Miss., 1968-71, Covenant Presbyn. Ch., Jackson, 1971-74. Mem. Friends of the Gallery, Mcpl. Art Gallery, Jackson; mem. rev. panel Arts Alliance of Miss., 1989. Sgt. USAR, 1961-67. Mem. Miss. Poetry Soc., Handel Soc. Jackson, Miss., 1991-93, Am. Hemerocallis Soc., Inc. Republican. Methodist. Avocations: art, writing, music. Home: 609 Broadway Ave Jackson MS 39216-3206

BURNS, ROBERT WAYNE, computer scientist; b. Austin, Tex., July 10, 1953; s. Wayne Adrian and Laura Mae (Moeller) B. BSEE, U. Tex., 1975. Cert. netware engr. Sr. analyst U. T. Computation Ctr., Austin, Tex., 1973-78; v.p. engr. Balcones Computer Corp., Austin, Tex., 1978-87; owner Robert Burns Consulting, Austin, Tex., 1985—; chief scientist U.S. Microlabs, Austin, Tex., 1988-89; v.p. engr. Mesa Software, Inc., Austin, Tex., 1989—; cons. Citi Corp Bank, London, U.K., 1986-89, Bankers Trust, London, 1987-88, Fidelity Internat., London, 1989-91, Sun Life Plc., U.K., 1992—. Author: Xerox 820-II, 1982, GEMS, 1986. Del. State Dem. Conv., Houston, 1992. Democrat. Roman Catholic. Home: 2801 Twin Oaks Dr Austin TX 78757-2761 Office: Mesa Software Inc 3435 Greystone Dr #106 Austin TX 78731-2363

BURNS, ROGER GEORGE, mineralogist, educator; b. Wellington, N.Z., Dec. 28, 1937; s. Alexander Parker and Jean Gertrude (Rodgers) B.; m. Virginia Anne Mee, Sept. 7, 1963; children: Kirk George, Jonathan Roger. BSc (Sir George Grey scholar 1958, Emily Lilias Johnson scholar 1959), Victoria U. of Wellington, 1959, MSc, 1961; PhD (Sci. fellow), U. Calif., Berkeley, 1965; MA in Geology (Brit. Council scholar 1965-66, Natural Environ. Research Council, Eng. fellow 1966), Oxford U., 1968, DSc, 1984. Demonstrator chemistry dept. Victoria U. of, Wellington, 1959-60; sr. lectr. geochemistry Victoria U. of, 1967; sci. officer Dept. Sci. and Indsl. Research, Wellington, 1961; research asso. dept. engring. scis. U. Calif., Berkeley, 1965; sr. research visitor, dept. mineralogy and petrology Cambridge U., Eng., 1966; lectr. geochemistry Oxford U., 1968-70; asso. prof. geochemistry MIT, Cambridge, 1970-72; prof. mineralogy and geochemistry MIT, 1972—; vis. prof. Scripps Instn. Oceanography, La Jolla, Calif., 1976; UNESCO prof. Jadavpur U., Calcutta, India, 1981; Hallimond lectr. Mineral. Soc. Eng., 1987; Guggenheim prof. Manchester U., England, 1991; prin. investigator lunar sample analysis team Apollo program NASA, 1970—, mem. lunar and planetary proposal rev. panel, 1978-81; mem. exec. com. Manganese Nodule Project Seabed Assessment Program Internat. Decade Ocean Exploration NSF, 1974-80; mem. adv. panel Marine Minerals Office, 1976; mem. rev. panel Nat. Scis. and Engring. Rsch. Can., 1985; mem. steering com. NASA MSATT project, 1990—. Author: Mineralogical Applications of Crystal Field Thepry, 2d edit., 1993; editor: Chem. Geology, 1968-85, Canadian Mineralogist, 1988-90; assoc. editor: Geochimica et Cosmochimica Acta, 1978—; contbr. articles to profl. publs. Fulbright travel grantee U.S. Govt., 1961; Sci. Research fellow Com. for Exhbn. of 1851, London, 1961-63; Pacific scholar English Speaking Union, San Francisco, 1961-63; fellow Wolfson Coll. Oxford U., 1970, Guggenheim fellow, 1990-91. Fellow Mineral. Soc. Am. (life; award 1976, councillor 1978-82, rep. for Geol. Soc. Am. abstracts rev. com. 1984-88); mem. Mineral. Soc. Gt. Britain; Mem. Am. Geophys. Union (mineral physics com. 1984-86), Geochem. Soc., N.Z. Geochem. Group. Presbyterian. Home: 7 Humboldt St Cambridge MA 02140-2804 Office: MIT 54-816 Dept Earth Atmos & Plant Cambridge MA 02139

BURNS, SALLY ANN, medical association administrator; b. Findlay, Ohio, Dec. 13, 1959; d. Van Larson and Marian (Delia) B. Student, Findlay Coll., 1980-82, Bowling Green State U., 1982-83; AAS, Houston C.C., 1985. Lic. physical therapist asst., Tex. Intern in clin. studies various Hosps., Houston, 1984-85; patient care Spring Br. Meml. Hosp., Houston, 1985-86; pres. Burns Phys. Therapy Clinic, Inc., Houston, 1986—; pres. Phys. Therapy Plus, Inc., Houston, 1988—, also bd. dirs.; pres. FYI Med. Suppliers, Inc., 1991—; pres. FYI Med., Inc., 1991. Author: Physical Therapy for Multiple Sclerosis. Mem. Inst. for Profl. Health Svc. Adminstrs. (charter mem.), Am. Judicature Soc., Am. Phys. Therapy Assn., Tex. Phys. Therapy Assn., Community Health Adminstrn. Avocations: breeding emu's, Tennessee walking horses and long horn cows, photography. Home: 1914 Potomac Houston TX 77057 Office: Phys Therapy Plus Inc 3303 Audley St Houston TX 77098-1921

BURNS, WILLIAM EDGAR, chemical engineer; b. Altus, Okla., Oct. 23, 1924; s. Edgar Willimer and Thelma Mollie (Kelly) B.; m. Geraldine Era McCarty, Aug. 30, 1946; children: David W, Edgar M., Molly I., Patricia Gail, James H. BS in Chem. Engring., U. Okla., 1945. Engring. mgr. Phillips Petroleum Co., Bartlesville, Okla., 1946-69; group v.p. Iowa Beef Processors, Dakota City, Nebr., 1969-75; mgr. process engring. Crest Engring., Tulsa, 1976-85; cons. William E. Burns, Inc., Tulsa, 1985-89; mgr. process and tech. John Brown E&C, Tulsa, 1989—. Lt. (j.g.) USN, 1943-46, PTO. Mem. Am. Inst. Chem. Engrs. Republican. Mem. Ch. of Christ. Achievements include 3 patents in petrochems.

BURNSIDE, EDWARD BLAIR, geneticist, educator, administrator; b. Madoc, Ont., Can., Mar. 16, 1937; s. Maxwell Blair and Vera Marion Burnside; m. Laurene Ruth Mayen, Sept. 5, 1959; children: Margaret Ruth, Stephen Alexander Blair, Shelley Anne Burnside Barnes, Joanne Lea. BSA, U. Toronto, Can., MSA; PhD, N.C. State U. Asst. prof. Ont. Coll., 1964-68; assoc. prof. U. Guelph, Ont., 1968-72, prof., 1972—, dir. Ctr. for. Improvement Livestock, 1985—; cons. CIDA, Ottowa, 1986-88, FAO, Rome, 1989-90, Semex Can., Guelph, 1988—. Co-author Canadian Animal Agriculture, 1993; contbr. over 150 tech. papers to sci. jours. Rsch. grantee NSERC, 1988, Agriculture Can., 1990, Ont. Ministry Agriculture and Food, 1991, CAAB, 1993. Mem. Agrl. Inst. Can. (rsch. scholar 1961-64), Ont. Inst. Agrologists (pres. Guelph chpt. 1968-69). Home: RR # 2, Elora, ON Canada N0B 1S0 Office: Univ of Guelph, Dept of Animal Sci, Guelph, ON Canada N1G 2W1

BURNSIDE, OTIS HALBERT, mechanical engineer; b. Montgomery, Ala., May 26, 1943; s. Dorothy (Benbow) B.; m. Rita Freitag, July 4, 1969. BSEM, Ga. Tech. Inst., 1965, MSEM, 1967; PhD in Applied Mechanics, U. Mich., 1975. Registered profl. engr., Tex. Sr. rsch. engr. S.W. Rsch. Inst., San Antonio, 1973-79, group leader, 1979-86, mgr., 1986-92, dir., 1992—. Mem. Zoning Commn., Leon Valley, Tex., 1984-89. Capt. U.S. Army, 1967-70. Fellow ASME (tech. editor Applied Mechanics Reviews 1979—); mem. AIAA (assoc. fellow, Engr. of Yr. 1991). Office: Southwest Rsch Inst 6220 Culebra Rd San Antonio TX 78228

BURR, BROOKS MILO, zoology educator; b. Toledo, Aug. 15, 1949; s. Lawrence E. and Beverly Joy (Herald) B.; m. Patti Ann Grubb, Mar. 5, 1977 (div. July 1987); 1 child, Jordan Brooks. BA, Greenville Coll., 1971; MS, U. Ill., 1974, PhD, 1977. Cert. scuba diver, Nat. Assn. Underwater Instrs. Lab. instr. dept. biology Greenville (Ill.) Coll., 1971-72; rsch. asst. Ill. Natural History Survey, Urbana, 1972-77, affiliate scientist Ctr. for Biodiversity, 1989—; from asst. prof. to prof. dept. zoology So. Ill. U., Carbondale, 1977—; mem. adv. panel U.S. Fish and Wildlife Svc., 1990—; adj. prof. dept. biology U. N.Mex., Albuquerque, 1991—. Co-author: A Distributional Atlas of Kentucky Fishes, 1986, A Field Guide to Fishes, North America North of Mexico, 1991; contbr. more than 80 articles to profl. jours. Recipient Paper of Yr. award Ohio Jour. Sci., 1986. Mem. AAAS, Am. Soc. Ichthyologists and Herpetologists (sec. and mem. exec. com. 1990—), Soc. Systematic Zoology, Biol. Soc. Washington, Sigma Xi (Leo M. Kaplan Meml. award 1990). Achievements include the discovery and description of 6 species of fish new to science from North American fresh waters. Home: RR 2 Box 338 Murphysboro IL 62966 Office: So Ill Univ Dept Zoology Carbondale IL 62901-6501

BURRELL, VICTOR GREGORY, JR., marine scientist; b. Wilmington, N.C., Sept. 12, 1925; s. Victor Gregory and Agnes Mildred (Townsend) B.; m. Katherine Stackley; Jan. 7, 1956; children: Cheri, Cathey, Charlene, Sarah. BS, Coll. Charleston, 1949; MA, Coll. William and Mary, 1968, PhD, 1972. Rsch. assoc. Va. Inst. Marine Sci., Gloucester Point, 1966-68, asst. marine scientist, 1968-70, assoc. marine scientist, 1970-72; assoc. marine scientist S.C. Marine Resources Rsch. Inst., Charleston, 1972-73, assoc. marine scientist, asst. dir., 1973-74, sr. marine scientist, dir., 1974-91, dir. emeritus, 1991—. Contbr. numerous articles to profl. jours. With USN, 1943-46, PTO. Mem. Nat. Shell Fisheries Assn. (pres. 1982-83), Estuarine Research Fedn. (sec. 1975-77), Gulf and Caribbean Fisheries Inst., Southeastern Estuarine Research Soc. (pres. 1986-88). Episcopalian. Office: SC Marine Resources Rsch Inst PO Box 12559 Charleston SC 29422-2559

BURRI, PETER HERMANN, anatomy, histology and embryology educator; b. Pfaeffikon, Zurich, Switzerland, July 18, 1938; s. Hermann Oskar and Clara (Gautschi) B.; m. Laurette Ida Barbey, Nov. 20, 1964; children: Christine, Anne-Françoise, Isabelle. Fed. diploma in medicine, U. Zurich and U. Paris, 1964; MD, U. Bern, Switzerland, 1968. Asst. Inst. Pathology, U. Zurich, 1965-66; asst. Inst. Anatomy U. Bern, 1966-68, grad. asst., 1968-73, lectr., 1969-73; rsch. fellow dept. pathology Royal Postgrad. Med. Sch.-Hammersmith Hosp., London, 1973-74; asst. prof. Inst. Anatomy U. Bern, 1974-78, prof., 1978—, head dept. devel. biology, 1974—, chmn. Inst. Anatomy, 1993—; vis. prof. medicine Boston U., 1987; Sam Stein Meml. lectr. L.I. Jewish Med. Sch., New Hyde Park, N.Y., 1988. Contbr. numerous articles and revs. on lung devel., growth and regeneration and microvascular devel. and growth to med. jours., also chpts. to books. Recipient Degussa award, 1987. Mem. Swiss Soc. Anatomy, Histology and Embryology (bd. dirs. 1983-89, pres. 1986-89), Anat. Soc., Union Swiss Socs. Exptl. Biology (bd. dirs., v.p. 1987-90), Swiss Acad. Scis. (pres. biology sect. 11987-90), Am. Assn. Anatomists, Swiss Soc. Cell and Molecular Biology (bd. dirs. 1978-81), Swiss Soc. Optics and Electron Microscopy (bd. dirs. 1981-87). Avocations: photography, chess, model helicopter flying. Home: Gurnigelstrasse 20, CH-3132 Riggisberg Switzerland Office: U Bern Inst Anatomy, Buehlstr 26, CH-3000 Bern Switzerland

BURRIDGE, MICHAEL JOHN, veterinarian, educator, academic administrator; b. St. Albans, Eng., Apr. 27, 1942; came to U.S., 1973; s. Arthur Wilfred Bailey and Georgina Augusta (Davis) Burridge; m. Desree Margaret Wiggins, Aug. 13, 1973 (div. Sept. 1981); m. Karen Maureen Bengtsson, Jan. 1, 1983; 1 child, Christina Michelle. BVM&S, U. Edinburgh, Scotland, 1966; MPVM, U. Calif., Davis, 1974, PhD, 1976. Rsch. asst. East African Trypanosomiasis, Toporo, Uganda, 1966; vet. practitioner Grant and Arnold, Woking, Eng., 1967-68; animal health officer Food & Agr. Orgn., Kabete, Kenya, 1968-73; grad. rsch. asst. U. Calif., Davis, 1973-76; assoc. prof. U. Fla., Gainesville, 1976-82, prof., 1982—, chmn. dept., 1984—; mem. com. on animal health NAS, Washington, 1980-83; cons. vet. medicine Williams & Wilkins, Balt., 1982—; cons. U.S. AID, India, 1987, 91; bd. dirs. Internat. Laveran Found., Annecy, France, 1991—. Editor: Impact of Diseases on Livestock Production in the Tropics, 1984. Grantee U.S. AID, 1985—. Mem. Royal Coll. Vet. Surgeons (U.K.), Am. Vet. Med. Assn. Achievements include co-invention of attractant decoy for tick control. Home: 10021 SW 67th Dr Gainesville FL 32608 Office: U Fla Ctr Tropical Agr Mowry Rd Bldg 471 Gainesville FL 32611

BURRIER, GAIL WARREN, physician; b. Newark, Ohio, Apr. 6, 1927; s. Harold I. and Esther M. (Simpson) B.; m. Mary Lou Miller, June 12, 1948 (dec. 1982); children: Dale, Marie. BS, Ohio State U., 1950, MS, 1952, MD, 1956. Diplomate Am. Bd. Family Practice; cert. geriatrist. Intern Grant Hosp., Columbus, Ohio, 1956-57; pvt. practice Canal Winchester, Ohio, 1957-73; dir. family practice Grant Med. Ctr., Columbus, 1973-88; med. dir. Alum Crest Nursing Home, Columbus, 1988—; clin. instr. Ohio State U., Columbus, 1969-74, clin. asst. prof., 1974-81, clin. assoc. prof., 1981—; med. dir. Winchester Place Nursing Home, Canal Winchester, 1983—; med. tech. asst. State of Ohio, Columbus, 1988—. Trustee Columbus Area Mental Health, 1974-81; bd. dirs. Meth. Ch., Canal Winchester, 1959-63; team physician high sch. tournaments, Columbus, 1957—. Fellow Am. Acad. Family Physicians (local pres. 1973); mem. Am. Coll. Physician Execs., AMA, Am. Geriatric Soc., Am. Soc. on Aging, Ohio State U. Alumni Club (pres. Columbus chpt. 1985-86). Republican. Home and Office: 45 Trine St Canal Winchester OH 43110-1229

BURRIGHT, RICHARD GEORGE, psychology educator; b. Freeport, Ill., Aug. 10, 1934; s. Willard Clarence and Emma Charlotte (Waller) B.; m. Shirley Louise Rock, Sept. 1, 1957; children: Lisa Ann, Scott Richard. BS, U. Ill., 1959, MA, 1962, PhD, 1966. Acting asst. prof. SUNY, Binghamton, 1963-66, asst. prof., 1966-69, assoc. prof., 1969-85, prof., 1985—. Co-author chpt. in Genetics of the Brain, 1982, The Behavioral Biology of Early Brain Damage, 1984, Preoperative Events: Their Effects on Behavior Following Brain Damage, 1989, Rehabilitation of the Brain Injured: A Neuropsychological Perception, 1990; co-author cpt. in: The Vulnerable Brain: Nutrition and Toxins, 1992. Coach Vestal (N.Y.) Jr. Baseball League, 1981, Vestal (N.Y.) Jr. Soccer League, 1983; webelo leader cub scouts Boy Scouts of Am., Vestal, 1980; mem. troop com. Boy Scouts of Am., Vestal, 1981-86. With U.S. Army, 1954-56. Grantee NIMH, 1969-72, NSF, 1974-76, 79-83, SUNY Rsch. Found., 1974-75, Nat. Inst. Childhood Devel., 1974-77. Mem. AAAS, Am. Psychol. Soc. (charter), N.Y. Acad. Sci., Behavior Genetics Assn., Sigma Xi. Home: 1201 Front St Vestal NY 13850-1221 Office: State Univ New York Vestal Pkwy Binghamton NY 13902-6000

BURRIS, HARRISON ROBERT, computer and software developer; b. Phila., July 13, 1945; s. Harrison Roosevelt and Mabel Eynon (Bosler) B. BS in Elec. Engring., Pa. State U., 1967, MS in Computer Sci., 1969; MBA in Mgmt., Fairleigh Dickinson U., 1973. Chief technologist U.S. Army Office Project Mgr. for Army Tactical Data Systems, Fort Monmouth, N.J., 1970-74; mgr. advanced computing Systems Engring. & Integration Divsn TRW, Redondo Beach, Calif., 1974-84, mgr. engring. operation Electron Systems Divsn., 1984-86, mgr. NASA Spaceborne Computing, Electronic Systems Group, 1986-90; pres. Neotechnic Industries Inc., Redondo Beach, 1980—; lectr. computer sci. Grad. Sch. Calif. State U., Fullerton, 1977-82. Editor microprogramming Simulation mag., 1982—. Recipient Am. Spirit Honor medal Assn. U.S. Army, 1970. Mem. IEEE, Assn. Computing Machinery, Res. Officers Assn. (v.p. Army Dept. Calif. 1993). Republican. Baptist. Office: 620 Via Monte Doro Redondo Beach CA 90277

BURRIS, JAMES FREDERICK, academic dean, educator; b. Mauston, Wis., Apr. 15, 1947; s. James Duane and Margaret Katherine (Jones) B.; m. Christine Tuve, July 3, 1971; 1 child, Cameron William Tuve. AB, Brown U., 1970, ScB, 1970; MD, Columbia U., 1974. Diplomate Am. Bd. Internal Medicine, Subspecialty Bd. Geriatrics, Am. Bd. Clin. Pharmacology. Intern Roosevelt Hosp., N.Y.C., 1974-75; resident in internal medicine Georgetown U. Med. Ctr., Washington, 1977-79; fellow in hypertension VA Med. Ctr., Washington, 1979-81; asst. prof. Sch. Medicine, Georgetown U., Washington, 1981-86, assoc. prof., 1986-91, coord. MD/PhD program, 1988—, prof., 1991—, asst. dean, 1987-90; assoc. dean Sch. Medicine, Georgetown U., 1990—; bd. dirs. Inst. for Clin. Rsch., Washington, 1989-92; mem. bd. regents Am. Coll. Clin. Pharmacology 1990—, Am. Bd. Clin. Pharmacology, 1992—, rsch. administr. cert. coun.; rsch. assoc. Hypertension Unit VA Med. Ctr., 1981—; vis. investigator Centre Hospitalier U. Vaudois, Lausanne, Switzerland, 1982-83; dir. clin. rsch. Cardiovascular Ctr. No. Va., Falls Church, 1988-92. Contbr. over 195 articles to profl. jours. Bd. dirs. Nation's Capital affiliate Am. Heart Assn, fellow couns. on high blood pressure rsch., circulation, epidemiology, clin. cardiology. Lt. comdr. USPHS, 1975-77. Recipient Svc. award ARC, 1970, Outstanding Svc. citation DAV, 1987; Commd. Officer Student Tng. and Extern Program scholar USPHS, 1973-74; rsch. fellow Found. for Rsch. of Cardiovascular Illness, Lausanne, 1983. Fellow ACP, Am. Geriatrics Soc., Am. Coll. Preventive Medicine, Am. Coll. Clin. Pharmacology (mem. bd. regents 1990—, Disting. Svc. award 1992), Am. Coll. Cardiology; mem. AMA (Physician's Recognition award 1982, 85, 88, 91), Sigma Xi. Democrat. Achievements include education and research in hypertension, hyperlipidemia, preventive cardiology and clinical pharmacology; grants and contracts management and regulatory affairs administration. Office: Georgetown U Sch Medicine Rm NE 120 Med Dent Bldg 3900 Reservoir Rd NW Washington DC 20007

BURRIS, JOHN EDWARD, biologist, educator, administrator; b. Madison, Wis., Feb. 1, 1949; s. Robert Harza and Katherine (Brusse) B.; m. Sally Ann Sandermann, Dec. 21, 1974; children: Jennifer, Margaret, Mary. AB, Harvard U., 1971; postgrad. U. Wis., 1971-72; PhD, U. Calif.-San Diego, 1976. Asst. prof. biology Pa. State U., University Park, 1976-83, assoc. prof., 1983-85, adj. assoc. prof., 1985—; dir. bd. basic biology Nat. Research Council, Washington, 1984-92, dir. Marine Biological Laboratory, Woods Hole, Mass. 1992— Home: 2332 N Oak St Falls Church VA 22046-2326 Office: Marine Biological Laboratory Woods Hole MA 02543

BURROWS, ADAM SETH, physicist; b. Salt Lake City, Nov. 11, 1953; s. Reynold Zachary and Diane Marian (Axelrad) B. AB, Princeton U., 1975; PhD, MIT, 1979. Postdoctoral fellow Mich. U., Ann Arbor, 1979-80; postdoctoral fellow SUNY, Stony Brook, 1980-83, asst. prof., 1983-86; assoc. prof. U. Ariz., Tucson, 1986-89, prof., 1989—. Contbr. articles to profl. jours. Recipient Dudley award Fullam Found., N.Y., 1984, Sloan fellowship, 1983. Fellow Am. Phys. Soc.; mem. Am. Astron. Soc., Phi Beta Kappa. Office: Physics and Astronomy Depts Univ of Ariz Tucson AZ 85721

BURROWS, BARBARA ANN, veterinarian; b. Columbia, S.C., Dec. 15, 1947; d. Robert Beck and Betty Elizabeth (Rabon) Burrows; m. Richard M. Duemmler, Aug. 31, 1968 (div. Aug. 1975); 1 child, Sandra Lynn. BA, Hartwick Coll., 1969; VMD, U. Pa., 1983. Bacteriologist Johnson & Johnson, North Brunswick, N.J., 1969-70; microbiologist Ciba-Geigy, Summit, N.J., 1973-79; veterinarian Amboy Ave Vet. Hosp., Metuchen, N.J., 1983-84, Black Horse Pike Animal Hosp., Turnersville, N.J., 1984-93, San Juan Animal Hosp., San Juan Capistrano, Calif., 1993—. Capt. USAF. Mem. Am. Vet. Med. Assn., Assn. of Women Vets., So. Calif. Vet. Med. Assn., Am. Assn. Feline Practitioners. Avocations: dancing, racquetball. Home: 25582 Breezewood Dana Point CA 92629 Office: San Juan Animal Hosp 32391 San Juan Creek Rd San Juan Capistrano CA 92675

BURROWS, BENJAMIN, physician, educator; b. N.Y.C., Dec. 16, 1927; s. Samuel and Theresa Helen (Handelsman) B.; m. Nancy Kreiter, June 14, 1949; children—Jan C., Susan K., Lynn A., Steven M. M.D., Johns Hopkins, 1949. Intern Johns Hopkins Hosp., 1949-50; resident King County Hosp., Seattle, 1950-51; resident U. Chgo., 1953-55, instr. to asso. prof. medicine, 1955-68; prof. internal medicine U. Ariz. Coll. Medicine, Tucson, 1968—, head section pulmonary diseases, 1968-87, Chalfant-Moore prof. of medicine, 1987—; cons. Tucson VA Hosp.; dir. Respiratory Scis. Ctr., Nat. Heart Lung and Blood Inst. Specialized Ctr. Research in Pulmonary Diseases, U. Ariz. Coll. Medicine, 1971—. Mem. editorial bd. Am. Rev. Respiratory Diseases, 1967-71, 74-80, Chest, 1971-76, Annals Internal Medicine, 1973-76, Archives of Environ. Health, 1976-93, Jour. of Allergy and Clinical Immunology, 1992—; contbr. articles to profl. jours., chpts. to books. Served to capt. USAF, 1951-53. Research grantee USPHS, 1958—. Fellow Am. Coll. Chest Physicians (regent dist. 11 1970-75), A.C.P.; mem. Am. Thoracic Soc. (counsilor), Ariz. Thoracic Soc. (pres.), Assn. Am. Physicians, Am. Soc. Clin. Investigation (emeritus), Am. Physiol. Soc. Home: 6840 N Table Mountain Rd Tucson AZ 85718-1329 Office: U Ariz Health Scis Ctr Tucson AZ 85724

BURROWS, BRIAN WILLIAM, research and development manufacturing executive; b. Burnie, Tasmania, Australia, Nov. 15, 1939; came to U.S., 1966; s. William Henry and Jean Elizabeth (Ling) B.; m. Inger Elisabeth Forsmark; 1 child, Karin. BSc, U. Tasmania, 1960, BSc with honors, 1962; PhD, Southampton U., 1966. Staff scientist Tyco Labs., Inc., Waltham, Mass., 1966-68; lectr. Macquarie U., Sydney, Australia, 1969-71; chef de sect. Battelle-Geneva, Switzerland, 1971-75; group leader Inco, Ltd., Mississauga, Ont., Can., 1976-77; mgr. program, lab. dir. Gould, Inc., Rolling Meadows, Ill., 1977-86; v.p. rsch. and devel. USG Corp., Chgo., 1986—; bd. dirs. U.S. Gypsum Co., Chgo., Ibacos Inc., Pitts., Indsl. Rsch. Inst., Washington, also. co. rep. Contbr. articles to tech. jours.; patentee in field. Mem. IEEE, AAAS, Am. Chem. Soc., Materials Rsch. Soc., Meadow Club (Rolling Meadows, Ill.). Home: 835 Fairfax Ct Barrington IL 60010-3153 Office: USG Rsch Ctr 700 N Us Hwy 45 Libertyville IL 60048-1296

BURROWS, JAMES H., computer scientist; b. Williamsport, Pa.. BS in Engring., MIT, 1949; MS in Math., U. Chgo., 1951. Dir. computer systems lab. NIST, 1979—; past chmn. info. systems standards bd. Am. Nat. Standards Inst.; mem. adv. bd. depr. computer sci. U. Md.; mem. com. on computer rsch. and applicationd Fed. Coord. Coun. on Sci., Engring., and Tech.; mem. Fed. Info. Resources Mgmt. Policy Coun.; bd. dirs. Coop. for Open Systems. Recipient Annual Excellence award Govt. Computer News, 1988, Disting. Presdl. Rank award 1989, Exec. Excellence award Interagency Com. on ADP, Silver medal U.S. Dept. Commerce, IRM Hall of Fame award, 1989, Fed. IRM 100 award Fed. Computer Week, 1990, 91, IRM Leadership award Assn. for Fed. Info. Resources Mgmt, 1991, Fed. Office Systems Exposition award, 1991; named IRM Exec. of Yr., Fedn. Govt. Info. Processing Couns., 1993. Mem. IEEE, AAAS, Assn. for Computing Machinery, Data Processing Mgrs. Assn. Office: Dept Commerce Nat Computer Systems Lab Tech Bldg Rte 270 Gaithersburg MD 20899*

BURROWS, RICHARD STEVEN, chemist; b. Parma, Ohio, Mar. 15, 1966; s. Charles Alexander and Catherine Ann (Bost) B.; m. Karen Scipione, June 23, 1990. BS in Chemistry, Muskingum Coll., 1988; MS in Chemistry, U. Cin., 1991, PhD, 1993. Technician Cleve. Electric Illuminating Co., 1985, 86, Glidden Co., Strongsville, Ohio, 1987, 88; rsch. asst. dept. chemistry U. Cin., 1988-93. Contbr. articles to profl. publs. Mem. Am. Chem. Soc., Am. Physical Soc., Sigma Xi. Achievements include discovery of copper-based ferro-magnetic materials containing no mid-transition metals with Curie points of several hundred degrees centigrade; discovery of direct evidence for superconductivity at room temperature. Office: Univ Cincinnati Dept Chemistry Mail Location 172 Cincinnati OH 45221

BURSAL, FARUK HALIL, mechanical engineer, educator; b. Boston, Nov. 4, 1963; s. Nasuhi Ismail and Nazan Ayse (Akyol) B. BSME, MIT, 1985; PhD in Mech. Engring., U. Calif., Berkeley, 1991. Grad. researcher U. Calif., Berkeley, 1989-91, instr., 1991-92; asst. prof. mech. engring. Va. Poly. Inst. and State U., Blacksburg, 1992—. NSF grad. fellow, 1985. Mem. ASME, Sigma Xi, Pi Tau Sigma, Tau Beta Pi. Democrat. Achievements include research on vibrations of mechanical systems, mathematical modeling of systems and analysis of their dynamical behavior. Office: Va Poly Inst and State U Dept Mech Engring Blacksburg VA 24061-0238

BURSIEK, RALPH DAVID, information systems company executive; b. Cin., Dec. 7, 1937; s. Ralph Carl and Marjorie (deCamp) B.; m. Judith Ann Bauer, July 27, 1963; children: Brian, Suzanne, Elizabeth. BCE, U. Cin., 1961; MBA, Harvard U., 1963. Mgr. large account mktg. IBM, Chgo., 1970-72; br. mgr. IBM, Green Bay, Wis., 1972-74; dir. fed. mktg. ops. IBM, Bethesda, 1974-76, dir. pub. sector, 1976-77; dir. product mgmt. IBM, White Plains, N.Y., 1977-81, div. dir., 1981-83; v.p. systems and line of bus. Burroughs Corp., Detroit, 1983-84, v.p. worldwide product mktg., 1984-85; v.p., regional mgr. UNISYS, Berkeley Heights, N.J., 1985-89; v.p. sales and mktg. Syncsort, Inc. Syncsort, Inc., Woodcliff Lake, N.J., 1989-92; pres. Sapiens USA, Cary, N.C., 1992—. Guides program Darien (Conn.) YMCA, 1977-81; soccer coach Town of Darien and Grosse Pointe (Mich.) 1979-85, Millburn (N.J.) Twp., 1986-91. Avocations: skiing, tennis, golf, gardening. Office: Sapiens USA 4001 Weston Pkwy Cary NC 27513

BURSIK, DAVID JAMES, software engineer; b. Chgo., Oct. 22, 1952; s. George and Doris Lorraine (Klecka) B. BS, Mich. State U., 1975, MS, 1980. Engr. II ITT Telecom/North Electric Co., Delaware, Ohio, 1975-78; mem. tech. staff AT&T Bell Labs., Columbus, Ohio, 1980-88; sr. mem. tech. staff GTE Labs./Contel Tech. Ctr., Chantilly, Va., 1988-92; sr. sytems analyst LCC Inc., Arlington, Va., 1992—; grad. rep. engring. computer adv. com. Mich. State U., East Lansing, 1979-80. Treas. Wooded Glen II Homeowners Assn., Burke, Va., 1990. Mem. IEEE, Assn. Computing Machinery, Tau Beta Pi, Eta Kappa Nu. Achievements include devising and implementing a computer software algorithm for validating traffic data collected from the 5ESS telephone switching system. Home: 9360 McCarthy Woods Ct Burke VA 22015

BURSTEIN, STEPHEN DAVID, neurosurgeon; b. Bklyn., Apr. 10, 1934; s. Moe and Anna (Bloch) B.; m. Ronnie Sue Deutsch, Oct. 8, 1972; 1 dau., Alissa Aimee. B.A. with distinction, U. Mich., 1954; M.D., SUNY-Bklyn., 1958; M.S. in Neurosurgery, U. Minn.-Rochester, 1965. Diplomate Am. Bd. Neurol. Surgery Surg. intern Johns Hopkins Hosp., Balt., 1958-59; neurosurgery fellow Mayo Clinic, Rochester, 1961-65; chief dept. neurosurgery South Nassau Community Hosp., Oceanside, N.Y., 1980—, pres. med. staff, 1980-82; chief dept. neurosurgery Franklin Gen. Hosp., Valley Stream, N.Y., 1980—; prin. Neurol. Surgery & Neurology, P.C., Freeport, N.Y., 1965—. Contbr. articles to med. jours. Bd. dirs. South Nassau Community Hosp., 1978—. Served to lt. USNR, 1959-61. Recipient Neurosurg. Travel award Mayo Found., 1966. Fellow ACS; mem. L.I. Hearing and Speech Soc. (bd. dirs.), N.Y. State Neurosurgeons Soc. (bd. dirs.), N.Y. State Neurosurg. Soc. (pres. 1981-82), Sigma Xi, Alpha Omega Alpha. Hebrew. Avocations: theatre; travel. Home: 19 Bridle Path Roslyn NY 11576-3115 Office: Neurol Surgery & Neurology 88 S Bergen Pl Freeport NY 11520-3510

BURT, ROBERT NORCROSS, diversified manufacturing company executive; b. Lakewood, Ohio, May 24, 1937; s. Vernon Robert and Mary (Norcross) B.; m. Lynn Chilton, Apr. 19, 1969; children: Tracy, Randy, Charlie. BSChemE, Princeton U., 1959; MBA, Harvard U., 1964. With Mobil Oil Corp., N.Y.C. and Tokyo, 1964-68; dir. corp. planning and acquisitions Chemetron Corp., Chgo., 1968-70, mgr. internat. div., 1970-73; dir. corp. planning FMC Corp., Chgo., 1973-76; v.p. agrl. chems. group FMC Corp., Phila., 1976-83; v.p. def. group FMC Corp., San Jose, Calif., 1983-88; exec. v.p. FMC Corp., Chgo., 1988-90, pres., 1990-91, chmn., pres., CEO, 1991—; chmn., CEO FMC Gold Co., 1989-91; mem. vis. com. Coll. Arts and Scis., Northwestern U., 1990—. Mem. Def. Policy Adv. Com. on Trade, 1992—; chmn. leadership coun. Princeton U. Sch. Engring. and Applied Sci., 1991—; bd. dirs. Rehab. Inst. Chgo., 1991—; governing mem. Orchestral Assn. of Chgo. Symphony Orch., 1992—. Lt. USMC, 1959-62. Mem. Mfrs. Alliance Productivity and Innovation (trustee 1990—), Chem. Mfrs. Assn. (bd. dirs. 1992—), Comml. Club Chgo., Econ. Club Chgo. (bd. dirs. 1992—), Chgo. Coun. Fgn. Rels. (mem. bd. 1993—), Execs. Club Chgo. (dir. 1993—), Bus. Roundtable (policy com.), Ill. Bus. Roundtable, U.S.-Japan Bus. Coun. Avocations: reading, golfing, spectator sports. Home: 1171 Oakley Ave Winnetka IL 60093-1437 Office: FMC Corp 200 E Randolph Dr Chicago IL 60601-6401

BURTIS, CARL A., JR., chemist; b. Flagstaff, Ariz., July 3, 1937. BS, Colo. State U., 1959; MS, Purdue U., 1964, PhD in Biochemistry, 1967. Chemist Ind. State Ctrl. Labs., 1963-66; fellow bioanal. Oak Ridge Nat. Lab., 1966-67, rsch. assoc. molecular anat., 1967-69; sr. chemist Varian Aerograph, 1969-70; group leader gemsaec fast analyzer proj. molecular anat, program Oak Ridge Nat. Lab., 1970-76, coord. biothechnol. program, 1973-77; chief anal. biochem. br. Ctr. Disease Control, 1976-79; chief clin. chemist Oak Ridge Nat. Lab., 1979—. Recipient Internat. fellowship Am. Assn. for Clin. Chemistry, 1992. Achievements include research in chemical inducement of ovine ketosis, separation and quantitation of body fluid constituents by liquid chromatography, separation of nucleic acid constituents by liquid chromatography, rapid clinical analysis, reference methods and materials. Office: Oak Ridge Nat Laboratory Oak Ridge TN 37830*

BURTLE, JAMES LINDLEY, economist, educator; b. Bremerton, Wash., July 18, 1919; s. James Andrew and Hazel (Lindley) B.; m. Vasanti Erulkar, Sept. 1952 (div. 1961); children: Anthea, Meriel. BA, U. Chgo., 1942, MA, 1948. Mem. staff Internat. Labour Office, Geneva, 1949-58; v.p. W.R. Grace & Co., N.Y.C., 1958-80; mng. editor Internat. Country Risk Guide Internat. Reports, N.Y.C., 1980-82; prof. econs. Iona Coll., New Rochelle, N.Y., 1982-90; dir. fgn. exch. svc. Wharton Econometric Forecasting Assocs., Bala Cynwyd, Pa., 1990—. Author: (with Sydney Rolfe) The Great Wheel, 1974. Staff sgt. USAAF, 1942-46, PTO. Mem. Econometric Soc. of Am. Econ. Assn. Home: 25 W 13th St New York NY 10011-7926 Office: 401 City Ave Bala Cynwyd PA 19004

BURTOFT, JOHN NELSON, JR., cardiovascular physician assistant; b. Pitts., Jan. 29, 1944; s. John Nelson Sr. and Elizabeth Louise (Lyons) B.; m. Jo Ann Stewart, Aug. 16, 1963 (div. Apr. 1966); 1 child, John Nelson III; m. Artie Ann Spilman, Sept. 19, 1966 (div. Dec. 1969); 1 child, Bonnie Beth; m. Sandra Ellen Bishop, Jan. 28, 1971; 1 child, Tracy Lynne. Grad., Sch. Health Care Sci., Sheppard AFB, 1975; grad. clin. preceptorship, Naval Hosp., 1976, BS, U. Nebr., Omaha, 1977. Physician asst. USN, 1961-80; retired chief warrant officer, 1980; surg. intern, resident Montefiore Hosp., Bronx, N.Y., 1980-82; cardiovascular physician asst. Cen. Ill. Cardiac Surgery Assocs., Peoria, Ill., 1982-83, Sanger Clinic, Charlotte, N.C., 1983-89, Dr. R. Carlton, Hickory, N.C., 1989-92, Mid-Atlantic Cardiothoracic Surgeons, Ltd., Norfolk, Va., 1992—. Fellow Am. Acad. Physician Assts., Assn. Physician Assts. in Cardiovascular Surgery, N.C. Acad. Physician Assts. Avocations: model railroads, Bonsai trees. Office: Mid-Atlantic Cardiothoracic Surgeons Ltd 400 W Brambleton Ave Ste 200 Norfolk VA 23501

BURTON, EDWARD LEWIS, industrial procedures and training consultant; b. Colfax, Iowa, Dec. 8, 1935; s. Lewis Harrison and Mary Burton; m. Janet Jean Allan, July 29, 1956; children: Mary, Cynthia, Katherine, Daniel. BA in Indsl. Edn., U. No. Iowa, 1958; MS in Indsl. Edn., U. Wis.-Stout, 1969; postgrad., Ariz. State U., 1971-76. Tchr. apprentice program S.E. Iowa Community Coll., Burlington, 1965-68; tchr. indsl. edn. Keokuc (Iowa) Sr. High Sch., 1965-68, Oak Park (Ill.)-River Forest High Sch., 1968-70; tchr. Rio Salado Community Coll., Phoenix, 1972-82; tchr. indsl. edn. Buckeye (Ariz.) Union High Sch., 1970-72; cons. curriculum Westside Area Career Opportunities Program - Ariz. Dept. Edn.; instr. vocat. automotive Dysart High Sch., Peoria, Ariz., 1979-81; tng. adminstr. Ariz. Pub. Service Co., Phoenix, 1981-90; tng. devel. cons. NUS Corp., 1991—; mem. dispatcher tng. com. Western Systems Coord. Coun., Salt Lake City, 1986-90; owner Aptitude Analysis Co., 1987—; mem. IEEE Dispatcher Tng. Work Group, 1988-91. Editor: Bright Ideas for Career Education, 1974, More Bright Ideas for Career Education, 1975. Mem. Citizens Planning Com., Buckeye, 1987-90, Town Governing Coun., Buckeye, 1990-91. NDEA grantee, 1967. Mem. NEA (life), Ariz. Indsl. Edn. Assn. (life), Personnel Testing Council of Ariz., NRA (life, endowment), Cactus Combat League, Mensa (test proctor 1987—), Masons. Republican. Methodist. Avocations: combat shooting, camping, travel. Home: 19845 W Van Buren St Buckeye AZ 85326-9134

BURTON, LAWRENCE DEVERE, agriculturist, educator; b. Afton, Wyo., May 27, 1943; s. Lawrence VanOrden and Maybell (Hoopes) B.; m. Arva Merrill, Nov. 20, 1967; children: LauraLee, Paul, Shawn, Renee, Kaylyn, Kelly, Brett. BS, Utah State U., 1968; MS, Brigham Young U., 1972; PhD, Iowa State U., 1987. Agr. tchr. Box Elder County Sch. Dist., Brigham City, Utah, 1967-68, Morgan County Sch. Dist., Morgan, Utah, 1968-70, Minidoka County Sch. Dist., Rupert, Idaho, 1972-79, Cassia County Sch. Dist., Declo, Idaho, 1979-84; instr. Iowa State U., Ames, 1984-87; area vocat. edn. coord. Idaho State Div. Vocat. Edn., Pocatello, 1987-88; state supr. agrl. sci. and tech. Idaho State Div. Vocat. Edn., Boise, 1988—; bio/chem. cons. rep. Ctr. for Occupational Rsch. and Devel., Waco, Tex., 1989—. Author: Agriscience and Technology, 1991; contbr. articles to profl. jours. Vice chmn. Minidoka County Fair Bd., Rupert, Idaho, 1977-80. Mem. Am. Vocat. Assn., Am. Assn. Agrl. Edn., Idaho Vocat. Agrl. Tchrs. Assn. (pres. 1981-82, Adminstr. of the Yr. 1989), Nat. Assn. Suprs. Agrl. Edn. (Western v.p. 1990-91, nat. pres.-elect 1992-93, nat. pres. 1993—), Gamma Sigma Delta, Alpha Zeta. Mem. Ch. of Jesus Christ of Latter Day Saints. Home: 10966 W Highlander Rd Boise ID 83709-6307 Office: State Div Vocat Edn 650 W State St Boise ID 83720-0001

BURTON, RODNEY LANE, engineering educator, researcher; b. Evanston, Ill., June 20, 1940; s. Joseph Roy and Jean (Brandon) B.; m. Linda Margaret Barnes, July 22, 1978; children: Carl Chapin, Rodney Lane Jr., Alice Julia. BS, Princeton U., 1962, MA, 1964, PhD, 1966. Profl. engr., Calif. Asst. prof. U. Calif., San Diego, 1966-71; rsch. scientist Dutcher Industries, San Diego, 1971-77; sr. scientist Jaycor, Alexandria, Va., 1977-79; rsch. engr. Princeton (N.J.) U., 1979-81; sr. scientist G.T. Devices, Inc., Alexandria, 1981-89; assoc. prof. U. Ill., Urbana, 1989—; cons. G.T. Devices, Inc., Alexandria, 1989—; gen. chmn. Internat. Electric Propulsion Conf., Orlando, Fla., 1990. Contbr. 24 articles to profl. jours., mostly on electric propulsion of rockets and projectiles. Trustee South Kent (Conn.) Sch., 1985—. Recipient Ralph R. Teetor award Soc. Automotive Engrs., 1971. Mem. AIAA (assoc. fellow), IEEE (sr.). Achievements include creation and development of field of pulsed electrothermal rocket propulsion for satellites and spacecraft; 5 patents, including Pulsed Electrothermal Thruster. Office: U Ill Dept Aero Astronautical Engring 104 S Wright St Urbana IL 61801

BURTON, WILLIAM JOSEPH, engineering executive; b. Gaffney, S.C., Mar. 22, 1931; s. Emory Goss and Olivia (Copeland) B.; m. Joan Holland Burton, Sept. 26, 1987. BSME, U. S.C., 1957, MSME, 1964; PhDME, Tex. A&M U., 1970. Registered profl. engr., Tenn. Sr. dynamics engr. Lockheed-Ga. Co., Marietta, 1957-62; sr. project engr. Allison div. GM Corp., Indpls., 1964-67; asst. prof., researcher Tex. A&M U., College Station, 1968-70; asst. prof. U. Tenn., Knoxville, 1970-74; projects mgr. Tenn. Valley Authority, Chattanooga, 1974-79; program mgr. Dept. Navy, Washington, 1979—; chmn. equal employment opportunity com. Chesapeake div. Naval Facilities Engring. Command, Washington, 1982-83. Author: On the Heating Surface Effects of Nucleate Boiling Data Correlation, 1964, The Effects of Surface Roughness on the Wave Forces on a Circular Cylindrical Pile, 1970; contbr. articles to profl. jours. Secretary, mem. hospitality com. Exch. Club, Knoxville, 1975, bd. dirs., 1976; coord. campaign Naval Facilities Engring. Com., Washington, 1982. With U.S. Army, 1951-53. Recipient Antarctic Svc. medal U.S. Dept. of Navy, 1962. Fellow ASME (chmn. exec. com. ocean engring. divsn. 1985, com. on tech. planning coun. on engring. 1992—, Golden Cert. 1989), Va. Soc. Profl. Engrs. (no. Va. regional coun. 1988); mem. AAAS, NSPE (pres.-elect Fairfax chpt. 1988), Soc. Mfg. Engrs., Soc. Naval Architects and Marine Engrs. Baptist. Avocations: travel, bicycling, classic guitar, golf, tennis. Home: 307 Miramar Rd Lakeland FL 33803-2633 Office: Naval Facilities Engring Command Chesapeake Divsn FPO-1 901 M St SE Bldg 212 Washington DC 20374

BURTT, EDWARD HOWLAND, JR., ornithologist, natural history educator; b. Waltham, Mass., Apr. 22, 1948; s. Edward Howland and Barbara Louise (Pride) B.; m. Pam Young, June 2, 1972; children: Michelle Erian, Jeremy Bredon. AB, Bowdoin Coll., 1970; MS, U. Wis., 1973, PhD, 1977. Vis. instr. U. Tenn., Knoxville, 1976-77; asst. prof. Ohio Wesleyan U., Delaware, 1977-83; sr. rsch. fellow Ohio State U., Columbus, 1982-83; assoc. prof. Ohio Wesleyan U., Delaware, 1983-87, prof., 1987—, chmn. zoology, 1991-93. Author: Evolution of Color in Wood-Warblers, 1986, Photographic Guide to Birds of the World, 1991; editor, author: Behavioral Significance of Color, 1979; contbr. articles to profl. jours.; Editor: Jour. Field Ornithol 1985-90 (v.p. 1989-91, pres. 1991-93). Mem. Assn. Field Ornithologists (pres. 1991—), Alliance for the environment (pres.-elect 1992-93, pres. 1993-94). Democrat. Episcopalian. Office: Ohio Wesleyan Univ Dept Zoology Delaware OH 43015

BUSBY, KENNETH OWEN, applied physicist; b. Livingston, Mont., Oct. 17, 1950; s. Alvah Kenneth and Mary (Hoiland) B. BS in Math., Mont. State U., 1972, BS in Physics, 1973; MA in Physics, U. Tex., 1976; PhD in Physics, Dartmouth Coll., 1980. Postdoctoral assoc. Lab. Plasma Studies, Cornell U., Ithaca, N.Y., 1980-82; staff scientist Naval Rsch. Lab., Washington, 1982-84, Mission Rsch. Corp., Albuquerque, 1984-87; cons. Albuquerque, 1987-89; sr. design engr. AAI Corp., Hunt Valley, Md., 1989-91; sr. scientist B-K Dynamics, Inc., Rockville, Md., 1991—; cons. Rockville, Md., 1993—. Contbr. chpt. to book, articles to profl. jours. Mem. IEEE, Am. Phys. Soc. Achievements include patent for fast rise time pulse power system; patent for conformed ground referenced self-integrating electric field sensor. Home: Apt 1123 9314 Cherry Hill Rd College Park MD 20740

BUSCAGLIA, ADOLFO EDGARDO, economist, educator; b. Buenos Aires, Jan. 10, 1930; s. Adolfo Angel and Francisca (Summa) B.; m. Maria Teresa Reyna Vila, Apr. 10, 1955; children: Edgardo Adolfo, Mariano Jorge, Teresita Sofia. BA in Econs., Nat. U. La Plata, Argentina, 1956; MA in Econs, Stanford U., 1965, PhD, 1966. Diplomate economist and academic. From economist to dir. econ. rsch. div. Cen. Bank Argentina, Buenos Aires, 1956-69; gen. dir. Analysis and Programming Inc., Buenos Aires, 1970-81, 83—; pres., chmn. Bank of Province of Buenos Aires, 1982; prof. econs. Cath. U. Buenos Aires, 1967-76, U. Buenos Aires, 1977—; bd. dirs. Bagley SA, 1979-81, Salto Grande's Hydroelectric Multipurposes Project, Buenos Aires, 1969-71, Cen. Market, Buenos Aires, 1972-73, Martin Garcia's Island Devel. Plan, Buenos Aires, 1971-72; hon. prof. monetary theory U. Buenos Aires, 1991. Author: Hacia una Argentina Posible, 1983; author, editor: Dinero, Inflación e Incertidumbre, 1985; contbr. articles to profl. jours. Founder Club of Argentina's Nat. Constn., Buenos Aires, 1988. With Argentina infantry, 1951. Fellow Argentine Assn. Fiscal Legis.; mem. Nat. Acad. Econ. Sci. Argentina (hon.), Am. Econ. Assn., Argentine Econ. Assn., Info. and Ops. Rsch. Assn., Buenos Aires Stock Exch., Jockey Club Buenos Aires. Roman Catholic. Avocations: golf, fishing. Office: Edificio Torre Tucuman 540, Ste 25 D, 1049 Buenos Aires Argentina

BUSCHE, ROBERT MARION, chemical engineer, consultant; b. St. Louis, June 14, 1926; s. Ferdinand and Irma (Seim) B.; m. Norma Jean Nickles, Sept. 17, 1950 (div. Mar. 1978); children: Robert Eric, David Clay, Kristin Anne, Amy Ellen; m. Emma Elizabeth Ruch, June 21, 1980. BSChE, Washington U., St. Louis, 1948, MSChE, 1949, DSc in Chem. Engring., 1952. Project engr. coal-to-oil demonstration br. U.S. Bur. Mines, Louisiana, Mo., 1950-53; rsch. supr. plastics dept. rsch. div. E.I. DuPont de Nemours & Co. Inc., 1953-62; tech. svcs. supr. plastics dept. mktg. div. E.I. DuPont de Nemours & Co. Inc., Wilmington, Del., 1962-64, tech. mgr. devel. dept. heat transfer products div., 1964-68, staff cons. devel. dept. mgmt. svcs. div., 1968-74, planning cons. cen. rsch. dept. life scis. div., 1974-85; pres. Bio En-Gene-Er Assocs. Inc., Wilmington, 1985—; adj. prof. chem. engring. U. Pa., Phila., 1983—; mem. adv. bd. Nat. Solar Energy Plan, Mitre Corp., Washington, 1980-85; mem. adv. bd. liaison com. Forest Products Lab., Madison, Wis.; mem. adv. bd. NSF cellulose hydrolysis program N.C. State U., Raleigh, 1980-82. Contbr. numerous articles to profl. jours. Deacon, elder Presbyn. Ch., Wilmington and Orange, Tex., 1957—; dist. and coun. com. Delmarva coun. Boy Scouts Am., Wilmington, 1959—; bd. dirs. News Castle Cotillion, Wilmington, 1989-91. Heerman's fellow Washington U., 1948, Honor scholar, 1943; recipient Silver Beaver award Delmarva coun. Boy Scouts Am., 1974, Wood Badge award, 1974. Mem. Am. Chem. Soc., Am. Inst. Chem. Engrs., Tau Beta Pi, Tau Kappa Epsilon, Alpha Chi Sigma, Alpha Phi Omega. Republican. Home and Office: 533 Rothbury Rd Wilmington DE 19803-2439

BUSH, DAVID FREDERIC, psychologist, educator; b. Watertown, N.Y., July 12, 1942; s. Frederic Ralph and Charlotte Mary (Ellingworth) B.; m. Joanne Arena; 1 child, Lara A. BA, U. South Fla., 1965; MA, U. Wyo., 1968; PhD, Purdue U., 1972. Instr. psychology Hiram Scott Coll., Scottsbluff, Nebr., 1967-69, Purdue U., West Lafayette, Ind., 1971-72; asso. prof. psychology West Chester (Pa.) State Coll., 1972-73; asst. prof., chmn. grad. program psychology Villanova (Pa.) U., 1972-77, assoc. prof., 1978-84, prof., 1984—, assoc. dir. human resource devel., grad. program in human orgn. sci.; instr. seminar Am. Coll., Bryn Mawr, Pa., 1976-78; ptnr. Quality Mgmt. Group; cons. in field. Mem. coun. Unitarian fellowship Lafayette, Ind., 1970-72; bd. dirs. Ars Moriendi, Dennis Burton Day Care Ctr., 1971-72, Life Guidance Services, Inc., 1977-82; pres. Bush Assocs. NDEA fellow 1969-71, David Ross summer fellow, 1972; Villanova U. grantee, 1974, 77, 83. Mem. Am., Ea. psychol. assns., Soc. Psychol. Study Social Issues, Am. Soc. Quality Control, Assn. of Mgmt., Internat. Human Resource Mgmt. Assn., Human Resources Profl. Assn. (pres. 1991-93), Acad. Mgmt., Soc. for Human Resource Mgmt., Assn. for Quality and Participation, Sigma Xi, Psi Chi. Author: Human Development: The Psychology of the Life-Span, 1974; Canterbury Press Memory Improvement Course. Editor: Researcher decision making and communication, orgn. behavior, human resource mgmt., quality improvement programs. Home: 16 Pennswood Dr Glenmoore PA 19343 Office: Villanova U Dept Psychology Villanova PA 19085

BUSH, EUGENE NYLE, pharmacologist, research scientist; b. McKeesport, Pa., Apr. 14, 1952; s. Nyle E. and Rosalia M. (Merlino) B.; m. Janet Rosemary Ruscitto, May 7, 1977; children: Stephen Michael, Rebecca Renee, Timothy George. BS in Pharmacy, U. Pitts., 1977, PhD in Pharmacology, 1981. RPh. Teaching asst. U. Pitts., Pitts., 1978-81; staff pharmacist Western Pa. Hosp., Pitts., 1977-81; pharmacologist II Abbott Labs., 1981-84, pharmacologist I, 1984-87; sr. rsch. sci. Abbott Labs., Abbott Park, Ill., 1986-88, rsch. investigator, 1988-89, group leader, endocrine pharmacol., 1989-91; sr. group leader endocrine pharmacol. Abbott Labs., 1991—. Co-author of numerous publications; contbr. articles to profl. jours. Cubmaster Cub Scout Pack 60, St. Joseph Ch., Libertyville, Ill., 1990, 91, 92. Mem. Endocrine Soc., Nat. Eagle Scout Assn., Sigma Xi, Am. Pharm. Assn. Republican, Roman Catholic. Avocations: piano, photography, bicycling, scouting. Home: 816 Bedford Ct Libertyville IL 60048-3002 Office: Abbott Labs Dept 46R 1 Abbott Park Rd North Chicago IL 60064-3500

BUSH, JOHN BURCHARD, JR., consumer products company executive; b. Lake Charles, La., Sept. 3, 1933; s. John Burchard and Mary Eleanor (Goldsworthy) B.; m. Earline Stowe, June 25, 1955; children: Catherine, Carolyn, John. BS, U. Calif., Berkeley, 1957, PhD, 1963. Asst. prof. U. Oreg., Eugene, 1960-63; chemist Gen. Electric Co., Schenectady, N.Y., 1963-68, mgr. polymer unit, 1968-70, mgr. tech. adminstrn., 1970-73, mgr. electrochem. br., 1973-76, mgr. energy programs, 1977-78, research and devel. mgr. materials science and engring., 1978-79; v.p. corp. research and devel. Gillette Co., Boston, 1980—. Patentee in field; contbr. articles to profl. jours. Served to 1st lt. U.S. Army, 1954-56. Mem. Am. Chem. Soc., Electrochemistry Soc., AAAS, N.Y. Acad. Sci., Sigma Xi. Republican. Episcopalian. Office: Gillette Co Prudential Twr Bldg Boston MA 02199*

BUSH, JOHN WILLIAM, business executive, federal official; b. Columbus, Ohio, Sept. 17, 1909; s. William Hayden and Esther (Brushart) B.; m. Mary Elizabeth Van Doren, June 4, 1932 (dec. 1958); children: Jan Hayden (Mrs. Richard L. Jennings), Emily Van Doren Bush; m. 2d Dorothy Vredenburgh, Jan. 13, 1962. BS in Bus. Adminstrn., Va. Poly. Inst., 1931. With Standard Oil Co. La., 1932-37, T.K. Brushart Oil Co., Portsmouth, Ohio, 1937-49; pres. Ohio System Inc., 1946—, dir. purchasing Ohio, 1949-57; dir. commerce, 1959-61; commr. ICC, 1961-71, chmn., 1966; spl. transp. adviser Senate Commerce Com., 1971—; v.p. Coastal Petroleum Co., 1972-89; bd. dirs. Can. So. Petroleum, Ltd., 1972-89, R.C. Williams & Co. Inc., N.Y.C.; chmn. bd. Old Judge Foods Corp., St. Louis, 1957-59. Mem. U.S. Nat. Commn. Pan Am. Ry. Congress Assn., 1962, Govs. Adv. Coun., Fla., 1985-86; councilman, Portsmouth, 1941-44; Dem. nominee for Congress, 1974; dir. Ohio Vietnam Vets. Bonus Commn., 1973-74. Mem. Nat. Assn. State Purchasing Ofcls. (pres. 1954). Address: Box C104 106 Moorings Park Dr Naples FL 33942

BUSH, KIRK BOWEN, electrical engineering educator, consultant; b. Idaho Falls, Idaho, Oct. 11, 1955; s. Wayne B. and E. Jeanine (Anderson) B.; m. Debra A. Lindstrom, Apr. 22, 1977; children: Daniel, Ian, Sarah, Travis, Peter, Nicholas, Caitlin, Morgan. BSEE, Brigham Young U., 1979; MSEE, Syracuse U., 1985; PhD, SUNY, Binghamton, 1990. Registered profl. engr., Pa.; cert. Tempest profl., Nat. Security Adminstrn. Rsch. asst. Brigham Young U., Provo, Utah, 1978-79; R&D engr. IBM Corp., Endicott, N.Y., 1979-90; assoc. prof. Wilkes U., Wilkes-Barre, Pa., 1990—; cons. various orgns., Owego, N.Y., 1990—; mem. computer bus. architecture task force IBM Corp., Endicott, 1984-89. Contbr. articles to profl. jours. Mem. com., scoutmaster Boy Scouts of Am., Owego, 1981—; various lay clergy positions LDS Ch., Owego, 1979—. Recipient Lab. Improvement grant NSF, 1991. Mem. IEEE, Armed Forces Communications Electronics Assn., Nat. Rowing Assn., Eta Kappa Nu (faculty advisor 1991—), Sigma Xi. Achievements include development of a printed circuit bd. extender for insitu measurements in a "shoe box" package; of a method for transformation of arbitrary fields from near-field measurements; of a fully automated method of printed circuit network checking, of a statistical delay calculator for very large scale integrated circuits; of a method of switching large numbers of integrated circuit drivers simultaneously. Home: 410 Ford Rd Owego NY 13827 Office: Wilkes Univ ECE Dept Wilkes Barre PA 18766

BUSH, MARK BENNETT, ecologist, educator; b. Croydon, Surrey, Eng., May 23, 1958; came to U.S., 1987; s. Dennis James Bennett and Avisa Jeanne Mary (Morley) B.; m. Katherine Marie Casola, July 8, 1991. BSc in Botany, Geography with honors, Hull (Eng.) U., 1979, PhD, 1986. Postdoctoral rsch fellow Ohio State U., Columbus, 1987-91; Mellon fellow Smithsonian Tropical Rsch. Inst., Panama City, Panama, 1991-92; vis. asst. prof. Duke U., Durham, N.C., 1992—; regional officer Brit. Trust Conservation Vols., Hull, 1980-82. Contbr. articles on biogeography and paleoecology to sci. publs. Mem. Internat. Assn. Vegetation Sci., Sigma Xi. Office: Dept Botany Duke Univ Durham NC 27713

BUSH, NORMAN, research and development executive; b. N.Y.C., Dec. 10, 1929; s. Louis and Ida (Trembola) B.; m. Audrey Faith Blumberg, Dec. 28, 1952; children: Stewart Alan, I. Jeffrey, Ellen Gail Dash. BBA, CUNY, 1951, MBA, 1952; PhD, N.C. State U., 1962. Statistician Army Chem. Ctr., Edgewood, Md., 1952-56, RCA Svc. Co., Patrick AFB, Fla., 1956-58, DBA and ICF, Melbourne, Fla., 1962-64, Pan Am Airlines, Patrick AFB, Fla., 1964-72; div. mgr. ENSCO Inc., Melbourne, Fla., 1972-83; pres., chief oper. officer ENSCO Inc., Springfield, Va., 1983—, chmn. bd., 1989—. Contbr. articles to statis. jours. With U.S. Army, 1952-54. Mem. Am. Statis. Assn., IEEE. Republican. Avocation: travel. Home: 1300 Crystal Dr # 1509 Arlington VA 22202-3234 Office: ENSCO Inc 5400 Port Royal Rd Springfield VA 22151-2312

BUSHMAN, DAVID MARK, aerospace engineer; b. Provo, Utah, Mar. 8, 1962; s. Jess Richard Bushman and Mildred (Price) Bushman Shultz; m. Noriko Nakaoka, Oct. 5, 1984; children: Emi Janae, Ana Michel, Elina Mari. BSME, Brigham Young U., 1988. Computer lab. expert humanities dept. Brigham Young U., Provo, 1982-87, video lab. mgr. Coll. Engring., 1986-88, software project leader Coll. Humanities, 1988-89; expdn. leader Ezra Taft Benson Inst., Oaxaca, Mex., 1988; pub. affairs asst. Advanced Combustion Engring. Rsch. Ctr., Provo, 1987-88; space sta. tech. leader Marshall Space Flight Ctr., NASA, Hunstville, Ala., 1989-90, mem. Fisher-Price study team, 1990, sta. calibration leader, propulsion components leader, 1990—; communicaitons cons. Ezra Taft Benson Inst., Provo, 1988—; selected by NASA to attend Russian Aerospace sch. Moscow Aviation Inst., 1992; delegation to Russia as expert on Russian docking system on assured crew return vehicle, 1992. Author planetarium presentation Fire from the Sky, 1986; software Japanese Dialogs, 1988, Intro to Humanities, 1988; writer, producer video Combustion Engring., 1989. Scholar Internat. Space U., 1993. Mem. AIAA (chmn. 1987-88), ASME, Space Rsch. Assn. (pres. 1987-89), Huntsville Area Tech. Socs. (communicaitons cons. 1991), Huntsville Amateur Radio Assn. (coord. earthquake drill 1990). Mem. LDS Ch. Avocations: spelunking, scuba diving, Soviet space program, foreign langauges. Home: 919 Speake Rd NW Huntsville AL 35816-3533 Office: NASA Marshall Space Flight Ctr Mail Code EP63 Huntsville AL 35812

BUSHNELL, DENNIS MEYER, mechanical engineer, researcher; b. New Haven, Conn., May 10, 1941; s. Jordan Lawrence and Anna Marie Bushnell; m. Judith Anne Simoni, June 8, 1963; 1 child, Matthew Gregory. BSME with distinction, U. Conn., 1963; MSME, U. Va., 1967. Rsch. engr. NASA Langley Rsch. Ctr., Hampton, Va., 1963-68, sect. head, 1969-72, br. chief viscous flows, 1973-88, assoc. chief fluid mechanics div., 1989—; con. Office of Sci. Rsch., USAF, Washington, 1980—, Nat. Aerospace Plane, Dayton, Ohio, 1986—, Def. Advanced Rsch. Projects Agy., Washington, 1988—. Author: Mixing Augmentation Technique for Hypervelocity Scramjets, 1989, Turbulence Control in Wall Flows, 1989, Supersonic Aircraft Drag Reduction, 1990, Synopsis o f Drag Reduction in Nature, 1991; editor: Viscous Drag Reduction in Boundary Layers, 1990. External v.p. Jaycees, Gloucester County, Va., 1967-69, emergency med. technician Abingdon Rescue Squad, Gloucester County, 1969-75; scoutmaster Boy Scouts Am., Gloucester County, 1976-80. Recipient Sperry award AIAA, 1976, Fluid and Plasma Dynamics award AIAA, 1991; Exceptional Sci. Achievement medal NASA, 1981. Fellow AIAA; mem. ASME, Pi Tau Sigma, Phi Kappa Phi, Tau Beta Pi, Sigma Xi. Achievements include patent on transonic wing passive boundary layer control device, on submicron high pressure particle generator, also riblets technique to reduce turbulent drag, high speed quiet tunnel device, and transition estimation approach. Home: Harbour Hill Hayes VA 23072 Office: NASA Langley Rsch Ctr M/S 197 Hampton VA 23665

BUSHUK, WALTER, agricultural studies educator; b. Pruzana, Poland, Jan. 2, 1929. BSc, U. Manitoba, 1952; MSc, McGill U., 1953, PhD in Physical Chemistry, 1956; DAgr (hon.), Agrl. Acad., Poznan. Chemist Grain Rsch. Lab., Manitoba, Can., 1953-61; sect. head Grain Rsch. Lab., 1961-62; dir. Ogilvie Flour Mills Co., Ltd., 1962-64; rsch. chemist Grian Rsch. Lab., 1964, head wheat sect., 1964-66, provost, 1980-84; prof. plant sci. U. Manitoba, 1966—, prof. food sci., 1981—; fellow Nat. Rsch. Coun. Can., Macromolecule Rsch. Ctr., France, 1957-58. Recipient Osborne medal, Neumann medal; AIC fellowship award Agrl. Inst. Can., 1990. Address: 26 Millikin Rd, Winnipeg, MB Canada R3T 2N2*

BUSHWAY, RODNEY JOHN, food science educator; b. Milo, Maine, June 23, 1949; s. Alfred Joseph and Dorothy Elizabeth (Landers) B.; m. Sandra Marie Carlson, Mar. 2, 1984; children: Nicholas David, Meghan Elizabeth. BS, U. Maine, 1971; MS, Tex. A&M U., 1973, PhD, 1977. Rsch. analytical chemist Phillips Petroleum, Bartlesville, Okla., 1977-78; prof. food sci. U. Maine, Orono, 1978—; sci. adv. bd. Immuno Systems, Scarborough, Maine, 1991—; cons. for food and chem. industries; lectr. in field. Author: (with others) HPLC Analysis of Pesticide Residues in Food, 1991; contbr. articles to refereed scientific jours. including Food Chemistry Bull. Environ. Contamination and Toxicology; mem. editoral bd. Jour. Food Protection, 1991. Judge sch. sci. fairs, Maine, 1980—; mem. sch. bd. USDA grantee, 1983-84, 86-91. Mem. Inst. Food Technologists, Am. Potato Assn., Am. Chem. Soc., Assn. Official Analytical Chemists (assoc. referee 1980—). Roman Catholic. Achievements include development of an official analytical method for analysis of Rotenone in pesticide formulations; development of several analytical methods using HPLC and immunoassay techniques for pesticide residues in food and water. Office: U Maine Dept Food Sci 5736 Holmes Hall Orono ME 04469-5736

BUSICK, ROBERT JAMES, computer scientist; b. Lima, Ohio, Jan. 4, 1950; s. Paul and Helen (Klay) B.; m. Judith Santacross, May 5, 1975 (div. 1978); m. Denise Tierney, July 2, 1978 (div. 1992); children: Katrina, John-Paul. Student, Oral Roberts U., Tulsa, 1968-71; diploma in computer sci., Ohio Inst. of Tech., Columbus, 1982; student, Ohio State U., 1990. Tchr. Hampshire Country Sch., Rindge, N.H., 1975-76; social worker Syntaxis Inc., Columbus, 1976-81; programmer Ohio Interactive Computer Sys., Columbus, 1981-82; programmer, analyst OHIONET, Columbus, 1982-85, asst. dir. computer svcs., 1985—, interim exec. dir., 1992; cons. Digitec Inc., Dayton, 1984-86. Contbr. articles to mag. Mem. Nat. Data Gen. Users Group, Nat. Bus. BASIC Users Group (libr. 1987-89), Columbus Info. Ctr. Assn., Internat. Platform Assn. Avocations: movies, motorcycling, raquetball. Home: 498 S Hague Ave Columbus OH 43204 Office: OHIONET 1500 W Lane Ave Columbus OH 43221-3975

BUSKIRK, ELSWORTH ROBERT, physiologist, educator; b. Beloit, Wis., Aug. 11, 1925; s. Ellsworth Fred and Laura Ellen (Parman) B.; m. Mable Heen, Aug. 28, 1948; children: Laurel Ann Buskirk Wiegand, Kristine Janet Buskirk Hallett. Student, U. Wis., 1943; BA, St. Olaf Coll., Northfield, Minn., 1950; MA, U. Minn., 1951, PhD, 1954. Lab. and teaching asst. Lab. Physiol. Hygiene, U. Minn., 1951-53; rsch. fellow Life Inst. Med. Rsch. Fund, 1953-54; physiologist Environ. Rsch. Ctr., Natick, Mass., 1954-57, Nat. Inst. for Arthritis, Metabolic and Digestive Diseases, NIH, Bethesda, Md., 1957-63; prof. applied physiology Pa. State U., University Park, 1963-92, dir. Lab. Human Performance Rsch., 1963-92; Marie Underhill Noll prof. Human Performance Pa. State U., 1988-92, emeritus, 1992—; mem. sci. adv. com. Pres.' Coun. on Phys. Fitness, 1959-61; mem. applied physiology study sect. divsn. rsch. grants NIH, 1964-68, 76-80; mem. com. on interplay of engring. with biology and medicine NAS-NAE, 1968-74, 82-88; mem. rsch. com. Pa. Heart Assn., 1970-73, 82-86, 87-89, 90—; mem. Pa. Gov.'s Coun. on Phys. Fitness and Sports, 1978-82; mem. com. on mil. nutrition rsch. NAS/NRC, 1982-90; mem. clin. scis. study sect. divsn. rsch. grants NIHh, 1989-92. Sect. editor Jour. Applied Physiology, 1974-78, assoc. editor, 1978-84; co-editor Sci. and Medicine in Sports and Exercise, 1974, editor, 1973-75, editor-in-chief, 1984-88, cons. editor, 1989—; mem. editorial bd. Physician and Sports Medicine, 1974-85, Jour. Cardiopulmonary Rehab., 1980—, Am. Jour. Clin. Nutrition, 1982-92, Jour. Gerontology, 1982-92, Exptl. Gerontology, 1989; also over 225 articles on physiology, revs. to sci. jours. Bd. visitors Sargent Coll., Boston U., 1976—; bd. dirs. Center Community Hosp., Pa., 1966-70, sec., 1971-72, v.p., 1973, pres., 1974-75. Served with U.S. Army, 1943-46, ETO. Recipient Disting. Alumni award St. Olaf Coll., 1969, Healthy Am. Fitness Leaders award 1992; rsch. grantee NIH, 1963-92, U.S. Olympic Com., 1965-68, USAF, 1965-69, Pa. Dept. Health, 1966-67, Pa. Heart Assn., 1966, 76-80, NSF, 1968-70, Nat. Inst. Occupational Safety and Health, 1969-74; NATO sr. fellow in sci., 1977. Mem. AAAS, AAHPERD, ASHRAE, Aerospace Med. Assn., Am. Acad. Phys. Edn., Am. Coll. Sports Medicine (citations 1973, 75, Honor award 1984, editorial award 1989, 93, Mid-Atlantic regional chpt. Svc. award 1991), Am. Inst. Nutrition, Am. Physiol. Soc. (pres. environ. and exercise sect. 1987-91, com. on coms. 1988-92, Honor award environ. exercise physiology sect. 1993), Am. Heart Assn. (coun. on epidemiology), N.Y. Acad. Scis., NIH Alumni Assn., Pa. Heart Assn. (rsch. com. 1988—), Am. Diabetes Assn., Coun. Biology Editors (Healthy Am. Fitness Leaders award 1992). Lutheran. Club: Centre Hills Country. Home: 216 Hunter Ave State College PA 16801-6947 Office: Pa State U 119 Noll Lab University Park PA 16802-6900

BUSMANN, THOMAS GARY, chemical engineer; b. Twin Falls, Idaho, Mar. 11, 1958; s. Phil Reed and Betty Jane (Hyde) B.; m. Rebecca Louise Atchley, June 12, 1982; 1 child, Katelyn Elizabeth. BS in Chem. Engring., U. Idaho, 1980. Registered profl. engr., Tenn. Project engr. I.T. Corp., Knoxville, Tenn., 1980-89, sr. project engr., 1989-92, sr. project mgr., 1992—. Mem. Am. Inst. Chem. Engrs., Inst. Chem. Engrs. Office: IT Corp 312 Directors Dr Knoxville TN 37923

BUSS, EDWARD GEORGE, geneticist; b. Concordia, Kans., Aug. 28, 1921; s. George E. and Kathryn (Luginsland) B.; m. Dorothy Ruth Arvidson, May 7, 1949; children: Ellen, Norman. BS, Kans. State Coll., 1943; MS, Purdue U., 1949, PhD, 1956. Grad. rsch. teaching asst. Purdue U., West Lafayette, Ind., 1946-49; asst. prof. Colo. A&M Coll., Ft. Collins, 1949-55; instr. Purdue U., 1955-56; assoc. prof., prof. Pa. State U., University Park, 1956-86; prof. emeritus Pa. State U., 1987—; cons. P.T. Anputraco Ltd., Surabaya Indonesia, 1987—; sr. scientist Biopore Inc., State College, Pa., 1987—. Co-author: Meat Production in Turkeys, 1990; contbr. articles to profl. jours. Capt. U.S. Army, 1943-46. Mem. Am. Genetic Assn. (coun. mem.), Am. Inst. Biological Scis. (gov. bd., exec. com.), AAAS (fellow 1962); Poultry Sci. Assn. (fellow 1988), World's Poultry Sci. Assn. Democrat. Home: 1420 S Garner St State College PA 16801-6330 Office: Pa State U Dept Poultry Sci 213 Henning Bldg University Park PA 16802

BUSS, LEO WILLIAM, biologist, educator; b. Alexandria, Va., Sept. 27, 1953; s. Leo Alfred and Margaret (Nyhan) B.; m. Jane Moore, June 12, 1982; children: Evan Daniel, Blake William. BA, Johns Hopkins U., 1975, MA, 1977, PhD, 1979; MA (hon.), Yale U., 1990. Asst. prof. dept. biology Yale U., New Haven, 1979-84, assoc. prof., 1984-90, assoc. prof. dept. geology and geophysics, 1988-90, prof., 1990—, chair Program in Organismal Biology, dir. Inst. Biospheric Studies, 1991—; curator Peabody Mus. Nat. History, New Haven, 1979—; dir. Sears Found. Marine Rsch., 1992—. Author: The Evolution of Individuality, 1987; co-editor: Population Biology and Evolution of Clonal Organisms, 1985. Bd. dirs. Sears Found. Marine Rsch., 1992—. John Simon Guggenheim fellow, 1984, Prize fellow John D. and Catherine MacArthur Found., 1989. Office: Yale U Dept Biology New Haven CT 06511

BUSSONE, DAVID EBEN, hospital administrator; b. Portland, Maine, Mar. 21, 1947; s. Patrick and Elizabeth (Frost) B.; m. Doreen Elizabeth Spears, Aug. 23, 1969; children: Kristin Rebecca, Marc David. BS, U. Mass., 1971; MBA, Boston U., 1978; postgrad., Nova U., 1988—. Dept. mgr. Beverly (Mass.) Hosp., 1971-78; cons. Mass. Hosp. Assn., Burlington, 1978-79, Hosp. Corp. Am., Nashville, 1979-80, 87—; asst. adminstr. West Paces Ferry Hosp., Atlanta, 1980-81, assoc. adminstr., 1981-84; assoc. hosp. dir. Univ. Hosp., Jackson, Miss., 1984-85, hosp. dir., 1985-91; pres., chief exec. officer Tampa (Fla.) Gen. Hosp., 1991—. Contbr. articles to profl. jours. Active Infant Mortality Task Force, Miss., 1985—, Gov.'s Com. on Indigent Care, Miss., 1988; chmn. Addie McBryde Ctr. for Blind, Jackson, 1986-87; bd. dirs. Ronald McDonald House, Jackson, 1986—; pres. Jackson-Vicksburg Hosp. Coun., Jackson, 1987-88. Fellow Am. Coll. Healthcare Execs.; mem. Healthcare Fin. Mgmt. Assn. (advanced mem.), Assn. Univ. Programs in Healthcare Adminstrn., Miss. Hosp. Assn. (chmn. elect 1990-91, chmn. 1991—), Coun. Teaching Hosps., Rotary. Republican. Baptist. Avocations: soccer referee, jogging. Home: 1150 Shipwatch Circle Tampa FL 33602 Office: Tampa Gen Hospital Davis Island PO Box 1289 Tampa FL 33601

BUSTARD, THOMAS STRATTON, engineering company executive; b. Balt., Feb. 18, 1934; s. Thomas Harrison and Helen Isabella (Slough) B.; m. Dorothy Lee Harren, Aug. 27, 1957 (div. Sept. 1979); children: Richard Todd, Cara Lynn Bustard Clark; m. Dubra Doris Schoen, Jan. 7, 1983. BSChemE, Johns Hopkins U., 1955; MSME, Drexel U., 1962; PhDChemE, U. Md., 1965; cert. in bus. mgmt., George Washington U., 1969. Registered profl. engr., Md. Chemist E.I. Dupont de Nemours & Co., Inc., Phila., 1955-56; process engr. Davison Chem. Co., Curtis Bay, Md., 1956-57; heat transfer/thermodynamic specialist The Martin Co./Nuclear Div., Middle River, Md., 1957-65; from sr. engr. to chief engr. Hittman Corp., Columbia, Md., 1965-77; on-site cons. U.S. Dept. Energy, Washington, 1977-79; founder, pres. Energetics, Inc., Columbia, 1979—; past instr. Essex C.C., Loyola Coll., Johns Hopkins U. Contbr. articles to profl. jours.; patentee in field. With USCG, 1957-65. Recipient four Purple Martin awards Martin-Marietta, Outstanding Grad. Student award U. Md., 1965. Mem. Sigma Xi. Democrat. Lutheran. Avocations: bridge, gardening, cycling, golf. Office: Energetics Inc 7164 Columbia Gateway Dr Columbia MD 21046-2101

BUSTER, JOHN EDMOND, gynecologist, medical researcher; b. Oxnard, Calif., July 18, 1941; s. Edmund B. and Beatrice (Keller) B.; m. Frances Bunn (dec.). Student, Stanford U., 1959-62; M.D., UCLA, 1966. Diplomate Am. Bd. Gynecology. Intern., Harbor UCLA Med. Ctr., Torrance, 1966-67, resident, 1967-71, research fellow, 1971-73, faculty, 1975—; prof. obstetrics and gynecology UCLA Sch. Medicine, 1983, dir. divsn. reproductive endocrinology; prof. obstetrics and gynecology U. Tenn., Memphis, 1987—; dir. research group human embryo transplants UCLA; examiner Am. Bd. Obstetrics and Gynecology. Contbr. articles to profl. jours. Served to lt., U.S. Army, 1973-75. Mem. Am. Gynecologic & Obstetrics Soc., Soc. for Gynecologic Investigation. Presbyterian. Office: University of Tennessee Div of Reproductive & Endo 956 Ct Ave Rm D328 Memphis TN 38163

BUTCH, JAMES NICHOLAS, electronics company executive; b. Elkins, W.Va., Nov. 6, 1951; s. Gus and Mary (Pizzoferrato) B.; m. Frances Maria Rasi, May 31, 1975; children: Maria, Nikki, Natalie. BSEE with honors, W.Va. Inst. Tech., 1974. Electronic engr. Preiser Scientific, Charleston, W.Va., 1975-76; co-founder, v.p. The Computer Store, Inc. W.Va., Charleston, 1976-77, Eagle Rsch. Corp., Charleston, 1976-86; pres., treas. Eagle Rsch. Corp., Scott DePot, W.Va., 1986—; co-founder, sec., bd. dirs. The Computer Store, Inc. W.Va., 1986—; chpt. pres., bd. dirs. Software Valley Corp., Morgantown, W.Va., 1989—. Mem. IEEE, NRA, Am. Radio Relay League, Pacific Coast Gas Assn., Planetary Soc., Charleston Regional C. of C., W.Va. State C. of C., Kanawha Amateur Radio Club, Tau Beta Pi, Eta Kappa Nu. Avocations: amateur radio, gun collecting, hunting, amateur astronomy, antique radio collecting. Home: 1502 Lyndale Dr Charleston WV 25314-2140 Office: Eagle Rsch Corp 107 Erskine Ln Scott Depot WV 25560-9752

BUTCHER, BRIAN RONALD, software company executive; b. Watford, Herts, U.K., Apr. 20, 1947; s. Ronald Stephen and Lizzy Maud (Haddrell) B.; m. Lesley Elaine Greenslade, Oct. 14, 1972 (div. Dec. 1989); children: Timothy Brian, Samantha Emily; m. Gayle Jozwiak, July 17, 1990; 1 child, Victoria. B in Tech. with honors, Brunel U., London, 1972. Engring. apprentice Pressed Steel Fisher, Oxford, Eng., 1966-72; project engr. Pressed Steel Fisher, Oxford, Eng., 1972-75; rsch. asst. Brunel U., 1975-77; applications mgr. SDRC, London, 1977-79; project leader GKN Tech. Ltd., Wolverhampton, Eng., 1979-81; project mgr. SDRC Ltd., Hitchin, Herts, Eng., 1981-83; team leader GM, Detroit, 1983-85; support mgr. PDA Engring. Internat. Ltd., Basingstoke, 1985-86, country mgr., 1986—; presenter papers at seminars and symposia in field, Europe and U.S. Mem. Soc. Automotive Engrs., Inst. Mech. Engrs. Achievements include applications of CAE methods to manufacturing engineering. Office: PDA Engring Internat Ltd, Rowan Woodlands Bus Village, Basingstoke RG212SX, England

BUTCHER, FRED R., biochemistry educator, university administrator; b. Rochester, Pa., Aug. 11, 1943; s. Goble S. and Monnie (Gibson) B.; m. Letty Jean Lytton, June 19, 1965; children: Allen Ray, Amy Jo. B.S., Ohio State U., 1965, Ph.D., 1969. Postdoctoral fellow U. Wis., Madison, 1969-71; asst. prof. Brown U., Providence, 1971-76; assoc. prof. Brown U., 1976-78; prof. W.Va. U., Morgantown, 1978—; chmn. dept. biochemistry W.Va. U., 1981-84, assoc. dean Sch. Medicine, 1984-89, dir. MBR Cancer Ctr., 1989—, sr. assoc. v.p., 1993—; cons. NIH, Bethesda, Md., 1976—, Cystic Fibrosis Found., Bethesda, 1977—. Contbr. articles to profl. jours. NIH grantee, 1975—. Mem. Am. Soc. Biol. Chemists. Home: RD 1 Box 242 Independence WV 26374 Office: WVa U Sch Medicine Health Sciences Ct Morgantown WV 26506

BUTENANDT, ADOLF FRIEDRICH JOHANN, physiological chemist; b. Bremerhaven, Germany, Mar. 24, 1903; s. Otto and Wilhelmine (Thomfohrde) B.; m. Erika von Ziegner, 1931; 2 sons, 5 daus. PhD; student, Oberrealschule Bremerhaven, U. Marburg, U. Göttingen. Sci. asst. Chem. Inst. Göttingen U., 1927-30, docent in organic and biol. chemistry, 1931; prof. chemistry, dir. Organic Chemistry Inst. Danzig (Germany) Inst. Tech., 1933-36; dir. Kaiser Wilhelm Inst. Biochemistry, Berlin; dir. Max Planck Inst. Biochemistry, Tubingen and Munich, Munich, 1936-72; prof. physiol. chemistry Munich U., 1956-71, prof. emeritus, 1971—. Author: Biochemie der Wirkstoffe; also articles. Recipient Nobel prize for chemistry, 1939; Adolf von Harnack medal Max Planck Soc., 1973; decorated Orden pour le Mérite, 1962, comdr. Légion d'Honneur, Ordr Palmes Académiques; Österreichisches Ehrenzeichen für Wissenschaft und Kunst. Mem. Max Planck Soc. (pres. 1960-72, hon. pres. 1972), Acad. Scis. Paris (fgn.).

BUTLER, ALISON, chemist, educator; b. Chgo., Nov. 19, 1954; d. Warren Lee and Lila Storrs (Bowen) B. BA, Reed Coll., 1977; PhD, U. Calif., San Diego, 1982. Postdoctoral fellow UCLA, 1982-84, CalTech, Pasadena, 1984-85; prof. chemistry U. Calif., Santa Barbara, 1986—. Contbr. articles to Jour. Am. Chem. Soc., Jour. Biol. Chemistry, other profl. publs. Named NIH postdoctoral fellow, 1983-85, Alfred P. Sloan Found. fellow, 1992—; recipient Jr. Faculty Rsch. award Am. Cancer Soc., 1988-91. Office: Univ Calif Dept Chemistry Santa Barbara CA 93106

BUTLER, BYRON CLINTON, physician, cosmologist, gemologist, scientist; b. Carroll, Iowa, Aug. 10, 1918; s. Clinton John and Blance (Prall) B.; m. Jo Ann Nicolls; children: Marilyn, John Byron, Barbara, Denise; 1 stepdau., Marrianne. MD, Columbia Coll. Physicians and Surgeons, 1943; ScD, Columbia U., 1952; G.G. grad. gemologist, Gemol. Inst. Am., 1986. Intern Columbia Presbyn. Med. Ctr.; resident Sloane Hosp. for Women; instr. Columbia Coll. Physicians and Surgeons, 1950-53; dir. Butler Rsch. Found., Phoenix, 1953-86, pres., 1970—; pres. World Gems/G.S.G., Scottsdale, Ariz., 1979—, World Gems Software, 1988, World Gems Jewelry, 1990—; cosmologist, jewelry designer Extra-Terrestrial-Alien Jewelry & Powerful Personal Talismans, 1992—. Featured in life; patentee in field; discovery of cause of acute fibrinolysis in humans; research on use hypnosis for relief of pain in cancer patients, use of tPA (tissue plaminogen activator) in acute coronary occlusion treatment. Bd. dirs. Heard Mus., Phoenix, 1965-74; founder Dr. Byron C. Butler, G.G., Fund for Inclusion Research, Gemol. Inst. Am., Santa Monica, Calif., 1987. Served to capt. M.C. AUS, 1944-46. Grantee Am. Cancer Soc., 1946-50, NIH, 1946-50. Fellow AAAS; mem. Am. Gemstones Trade Assn., Ariz. Jewelers Assn., Ariz. Soc. Astologers, Mufon, Mutual UFO Networks. Home: 6302 N 38th St Town Paradise Valley AZ 85253-3825

BUTLER, CLARK MICHAEL, aerospace engineer; b. Tulsa, Apr. 28, 1946; s. Clark Maurice and Doris Louise (Wynn) B.; m. Carolyn Sue Ray, Aug. 16, 1975; children: Kimberly Ann, Christina Louise, Kelli Annette. BS in Aerospace Engring., U. Okla., 1968, MS in Aerospace Engring., 1971; MBA, Cen. State U., 1976. Registered profl. engr., Okla., Tex. Aero. engr. DOD, USAF-AFLC, Oklahoma City, 1968-70; propulsion R & D Engr. Beech Aircraft Corp., Wichita, Kans., 1971; propulsion supr. Rockwell Internat. Corp., Oklahoma City, 1971-79; chief engr. Fairchild Industries, Inc., San Antonio, 1979-84; pres. Butler Aero., San Antonio, 1984-86; v.p. engring. Swearingen Aircraft Corp., San Antonio, 1986-87; pres. Bulter Aerospace, Inc., San Antonio, 1987—, Aircraft Design Svcs., Inc., 1993—; bus. cons. Butler & Assocs., Oklahoma City, 1976-79. Mem. NSPE, Soc. Automotive Engrs.-Aerospace., FAA (designated engring. rep. powerplants, systems, equipment and spl. adminstrv.). Republican. Home: 2515 Hidden Glen St San Antonio TX 78232-4231 Office: Butler Aerospace Inc 16414 San Pedro Ave Ste 675 San Antonio TX 78232-2246

BUTLER, DONALD PHILIP, electrical engineer, educator; b. Toronto, Ont., Can., Nov. 13, 1957; s. Clifton Aubrey and Helen Eunice (Roy) B.; m. Zeynep Celik, Aug. 23, 1986; 1 child, Melissa. BA in Sci., U. Toronto, 1980; MS, U. Rochester, 1981, PhD, 1986. Fellow U. Rochester, N.Y., 1980-83, rsch. asst., 1983-85, rsch. assoc., 1985-86; asst. prof. elec. engring. So. Meth. U., Dallas, 1987—. Contbr. articles to Applied Physics Letters, Jour. Applied Physics, others. Grantee Tex. Higher Edn. Coord. Bd., 1989-92. Mem. IEEE (chmn. Dallas chpt. electron device soc. 1988—), Am. Phys. Soc. Achievements include investigation of nonequilibrium properties of superconductors observing dynamic intermediate state, transient magnetic superheating and phase-slip induced anomalous relaxation, high T superconductor microbridge mixers. Office: So Meth Univ Dept Elec Engring Dallas TX 72575-0335

BUTLER, GEOFFREY SCOTT, systems engineer, educator, consultant; b. Jacksonville, Fla., July 19, 1958; s. George Lauritzen and Mary Elizabeth (Cox) B.; m. Diana Lynn Martin, Aug. 29, 1987. BS in Aerospace Engring., U. Fla., 1981; MS in Aerospace Engring., San Diego State U., 1986; MS in Aerospace Systems, West Coast U., 1988. Engr. Lockheed Missiles & Space Co., Sunnyvale, Calif., 1981-83; engring. specialist Convair div. Gen. Dynamics, San Diego, 1983-92; project mgr. Horizons Tech., Inc., San Diego, 1992—; cons. WEB Engring., San Diego, 1992—. Speaker Scott's Valley Homeowners Assn., Encinitas, Calif., 1992. Mem. AIAA (sr. mem., tech. com. mem. 1992). Republican. Roman Catholic. Achievements include conception and direction of first tests of store separation at hypersonic speeds within a ballistic range, specialized software development activities. Office: Horizons Tech Inc 3990 Ruffin Rd San Diego CA 92123-1826

BUTLER, JAMES HALL, oceanographer, atmospheric chemist; b. San Antonio, June 25, 1948; s. Franklin Hall and Audrey (Chaffin) B.; m. Sherie Jean Kittrell, Dec. 30, 1972 (dec. 1982); m. Kathleen Ann Hawes, Aug. 4, 1984; children: Stephanie, Michael. BA in Biology, U. Calif., Santa Barbara, 1970; MS in Natural Resources, Humboldt State U., 1975; PhD in Chem. Oceanography, Oreg. State U., 1986. Regional mgr., lab. dir. Environ. Rsch. Cons., Arcata, Calif., 1975-79; instr. dept. oceanography Humboldt State U., Arcata, 1979-82; rsch. assoc. U. Colo., Boulder, 1986-89; rsch. chemist NOAA, Boulder, 1989—; cons. on coastal planning, wastewater use cities of Arcata, Trinidad and Watsonville, County of Humboldt, Calif., 1977-81. Contbr. articles to Nature, Marine Chemistry, other profl. publs. Sec. Homeowners' Assn., Boulder, 1990—; mem. Supt. Schs. Adv. Com., Longmont, Colo., 1991—. Mem. AAAS, Am. Geophys. Union, Oceanography Soc. (charter), Sigma Xi. Democrat. Achievements include research in ocean regulation of trace-gas composition of atmosphere, sources and fate of ozone-depleting and radiative trace gases in atmosphere. Office: NOAA 325 Broadway Boulder CO 80303

BUTLER, JAMES LEE, agricultural engineer, researcher; b. Sevierville, Tenn., Jan. 8, 1927; s. James Lawson and Dora Mae (Fox) B.; m. Jane Isabell Hollis, Nov. 20, 1948; children: Kathryn Jo, Nancy Lee, Benjamin Hollis. BS, U. Tenn., 1950, MS, 1951; PhD, Mich. State U., 1958. Asst. agrl. engr. U. Ga., Experiment, 1951-56, assoc. agrl. engr., 1958-60; grad. asst. Mich. State U., East Lansing, 1956-58; agrl. engr. Agrl. Research Service, USDA, Tifton, Ga., 1960-62, research investigations leader, 1962-80, research leader, 1985-89; mgr. So. Agrl. Energy Ctr., Tifton, 1980-85. Assoc. editor Peanut Sci., 1974-80; mem. editorial bd. Energy in Agriculture, 1982-87; contbr. numerous articles to profl. publs. Served to cpl. AC, U.S. Army, 1944-46. Recipient Golden Peanut Research award Nat. Peanut Council, 1979, Ga. Research and Edn. award. 1988; USDA Merit cert. 1980, 82, 86. Fellow Am. Soc. Agrl. Engrs. (bd. dirs. 1980-82, 87-89, best paper award 1960, Ga. sect. Engr. of Yr. 1981, trustee ASAE Found. 1989—); mem. Am. Peanut Research and Edn. Soc. (pres. 1982-83, Bailey award 1978), NSPE, Council Agrl. Sci. and Tech. Baptist. Home: 2823 Rainwater Rd Tifton GA 31794-2530 Office: USDA-ARS PO Box 748 Tifton GA 31793-0748

BUTLER, JOHN BEN, III, physician, computer specialist; b. Newark, May 24, 1948; s. John Ben and Regina (Woody) B.; m. Susan Margaret Lesser, Sept. 4, 1977; children: Margaret Rhodes, John Ben IV. MD, NYU, 1975; MS in Computer Sci., U. Wash., 1982. Resident in internal medicine U. Pitts. Health Ctr., 1975-78; fellow in cardiology Med. Coll. of Wis., Milw., 1978-79; fellow in bioengring. U. Wash. Ctr. for Bioengring., Seattle, 1979-82; software designer Microsoft, Bellevue, Wash., 1982-86, sr. trainer, 1986-91, pen computing specialist, 1991-92; mem. X3H3 Am. Nat. Std.Inst., Washington, 1983-86, chmn. X3H36 subcom. on windowing stds., 1985-86. Author: (video courseware) Introduction to Microsoft Windows Programming, 1988. Fellow AAAS; mem. Am. Med. Informatics Assn., Fedn. of Am. Scientists. Achievements include development of ANSI X3H3.6 standards subcommittee to develop a standard for windowing computer software; designed programmers interface to drawing functions in Microsoft Windows. Home: 5811 Princeton Ave NE Seattle WA 98105

BUTLER, KEVIN CORNELL, physicist; b. Queens, N.Y., Sept. 10, 1963; s. Richard Cornell and Wanda Lee (Roberts) B. BS, N.C. A&T State U., 1989. Staff technologist Bellcore, Redbank, N.J., summer 1988; commd. 1st lt. U.S. Army; platoon leader 10th Mountain Div. U.S. Army, Fort Drum, N.Y., 1990-92; asst. S2/3 10th Mountain Div. U.S. Army, Fort Drum, 1992—. Recipient ROTC scholarship U.S. Army Rev., 1986, Ron McNair scholarship N.C. A&T Physics Dept., 1986-87. Home: 161 Poplar Ave Staten Island NY 10309-1220

BUTLER, PAUL CLYDE, psychologist; b. N.Y.C., May 8, 1950; s. Thomas William and Catherine Christine (Finney) B. BA, Hunter Coll., 1975; MA, Nat. U., San Diego, 1987; PhD, U.S. Internat. U., San Diego, 1991. Mental health profl. Rancho Park Hosp., El Cajon, Calif., 1988—; asst. dir. Adolescent Long Term Treatment Unit Scripps McDonald's Ctr., LaJolla, Calif., 1988—. Mem. Am. Psychol. Soc., Calif. Psychol. Assn., Assn. Black Psychologists. Democrat. Achievements include research on effects of drugs and alcohol on youth. Home: 5700 Baltimore Dr #513 La Mesa CA 91942

BUTLER, ROBERT RUSSELL, systems engineer; b. Glasgow, Ky., Aug. 30, 1954; s. Robert Allen and Dorothy Jean (McGlasson) B.; m. Margaret Belue, June 25. BSEE, U. Ky., 1976; MSEE, U. S.C., 1984. Engr. R.R. Donnelley & Sons Co., Glasgow, Ky., 1978-80; project engr. R.R. Donnelley & Sons Co., Spartanburg, S.C., 1981-84; sr. engr. R.R. Donnelley & Sons Co., Glasgow, 1985-91, sr. systems engr., 1991—; mem. adv. bd. computer sci. dept. Western Ky. U., Bowling Green, 1990-92. Achievements include invention of binding line book tracking system and method patent.

BUTRY, PAUL JOHN, engineer; b. Niagara Falls, N.Y., Mar. 12, 1946; s. John Steven and Sophie (Zaczek) B.; m. Sharon Lee Wall, Oct. 14, 1968 (div. Nov. 1976); 1 child, Paul John II; m. Deborah Lynn D'Agostino, Aug. 16, 1980; children: Taylor Bethany, Piers Alexander. AS in Math. and Sci., Niagara Community Coll., 1981; BS in Indsl. Tech., Buffalo State U., 1988, MS in Indsl. Tech., 1993. Chem. technician SKW Alloys, Niagara Falls, 1968-84; chemist Bell Aerospace Textron, Wheatfield, N.Y., 1986-89; tech. svc. engr. Sigri Great Lakes Carbon Corp., Niagara Falls, 1989—; cons., owner Butry Analytical Svcs., Niagara Falls, 1982—. Mem. Geraldine Mann Sch. Parents Edn. Group, 1991—, Cayuga Island Homeowners Assn., 1991—; mem. Pres. Coun. on Phys. Fitness. With USN, 1966-67. Recipient Running award Pres. Coun. on Phys. Fitness, 1991. Mem. Am. Soc. Quality Control, Soc. Mfg. Engrs., Am. Running and Fitness Assn., Skip Barber Racing Assn. (racing cert. 1984). Avocations: physical fitness, antique auto restoration. Home: 8630 Hennepin Ave Niagara Falls NY 14304-4430 Office: Sigri Great Lakes Carbon Corp 6200 Niagara Falls Blvd Niagara Falls NY 14304-1534

BUTT, JIMMY LEE, retired association executive; b. Tippo, Miss., Oct. 13, 1921; s. H.W. and Jimmie O. (Davis) B.; m. Jane F. Williams, June 23, 1943; children—Janie Lake, Melanie Maryanne, Jimmy Lee. BS, Auburn U., 1943, MS, 1949. Grad. asst. agrl. engring. dept. Auburn U., 1947-48, asst., 1948-50, asso. agrl. engr., 1950-56; exec. v.p. Am. Soc. Agrl. Engrs., 1956-86. Mem. rsch. adv. coun. Auburn U., 1985-88. Capt. F.A. AUS, 1943-46. Decorated Bronze Star; officer Ordre du Merite Agricole (France), 1979; recipient Industry Leader citation Farm and Indsl. Equipment Inst., 1986, Disting. Svc. award Agrl. Engring. Dept. Mich. State U., 1975, Outstanding Alumni award Agrl. Engring. Dept. Auburn U., 1988. Fellow Am. Soc. Agrl. Engrs. (pres. 1987-88); mem. Coun. Engring. and Sci. Soc. Execs. (pres. 1977-78), Am. Assn. Engring. Socs. (past dir.), Sigma Xi, Tau Beta Pi, Phi Kappa Phi, Gamma Sigma Delta, Alpha Zeta, Omicron Delta Kappa, Spades. Club: Economic. Lodge: Lions. Home: 2572 Stratford Dr Saint Joseph MI 49085-2714

BUTTERFIELD, D. ALLAN, chemistry educator, university administrator; b. Milo, Maine, Jan. 14, 1946; s. Francis H. and Gwendolyn (Fish) B.; m. Marcia Tuthill, June 9, 1968; 1 child, Nyasha E. BA, Univ. Maine, 1968; PhD, Duke Univ., 1974. High sch. tchr. United Meth. Ch., Mrewa, Zimbabwe, 1968-71; rsch. asst. Duke Univ., Durham, N.C., 1971-74, NIH postdoctoral fellow, 1974-75; asst. prof. chemistry Univ. Ky., Lexington, 1975-78, assoc. prof. chemistry, 1978-83, prof. chemistry, 1983—, dir. ctr. membrane sci., 1987—; Editorial bd. Journal Membrane Sci., 1989—; contbr. articles to profl. jours. Active Habitat for Humanity, Lexington, 1991—. Recipient numerous grants for rsch. in field. Mem. Am. Chemical Soc., North Am. Membrane Sco., Am. Sci. Biochemist & Molecular Biology. Episcopalian. Achievements include insight into molecular basis of neurological disorders; molecular studies of membrane structure and function. Office: Univ Ky Dept Chemistry Lexington KY 40506

BUTTERWECK, HANS JUERGEN, electrical engineering educator; b. Gevelsberg, Germany, June 23, 1932; arrived in Netherlands, 1964; s. Hans and Anny (Schaefer) B.; m. Hannelore Bachmann, Feb. 11, 1960; children: Ute, Christoph. Diploma in engring., Aachen Inst. Tech., Fed. Republic Germany, 1956; D of Engring., Aachen Inst. Tech., 1959. Lectr. Aachen Inst. Tech., 1958-64; rsch. fellow Philips Rsch., Eindhoven, Netherlands, 1964-67; prof. elec. engring. Eindhoven U. Tech., 1967—. Author: Mikrowellenband-filter, 1959, Elektrische Netwerken, 1974. Mem. IEEE, Verein Deutscher Elektrotechniker, European Assn. Signal Processing. Avocations: music, walking, history. Home: Akert 148, 5664 RL Geldrop The Netherlands Office: Eindhoven U, Den Dolech, Eindhoven The Netherlands

BUTTERWORTH, MICHAEL, computer programmer; b. Hartford, Conn., July 4, 1942; s. Oliver and Miriam (Brooks) B.; m. Carol Ann Hastings, Aug. 6, 1966; 1 child, Beth. BS, Dartmouth Coll., 1963; MS, Stanford U., 1968; postgrad., Rensselaer Poly. Inst., 1992—. Statistician Univ. Calif. Med. Ctr., San Francisco, 1968-70; statistician, demographer Greater Hartford Process, 1970-75; mktg. specialist ADVO-System, Inc., Hartford, 1975-77, programmer, 1978-79; sr. software engr. Gerber Systems Tech., South Windsor, Conn., 1979-91. Contbr. articles to profl. jours. Sec. Arts of Tolland (Conn.), Inc., 1979-84, pres., 1985—. Mem. Assn. Computing Machinery. Home: 106 Cedar Swamp Rd Tolland CT 06084

BUTTKE, THOMAS FREDERICK, mathematics educator; b. Wausau, Wis., Sept. 25, 1956; s. William Frank and Ethel Loretta (Krohn) B.; m. Joni Lynn Steiner, Sept. 12, 1981 (div. Aug. 1990); children: Patricia Christine, Alexandra Elizabeth; m. Judith Lynn Waters, Apr. 21, 1991; 1 child, Carl Thomas. BS in Applied Math., Engring. and Physics, U. Wis., 1978; MA, U. Calif., Berkeley, 1980; PhD, U. Calif., 1986. Postdoctoral fellow Livermore (Calif.) Nat. Lab., 1986-87; vis. rschr. Princeton (N.J.) U., 1987-89; prof. math. NYU, N.Y.C., 1989—; mem. Inst. for Advanced Study, Princeton, 1991-92. Author: Vortex Methods and Vortex Motion, 1991; contbr. articles to Comm. in Pure and Applied Math., Jour. Computational Physics. NSF postdoctoral fellow Princeton U., 1987-89. Mem. Am. Math. Soc., Am. Phys. Soc., Soc. for Indsl. and Applied Math. Achievements include discovery of Hamiltonian formulation of incompressible fluid flow. Home: 38 Marion Rd W Princeton NJ 08540 Office: NYU Courant Inst 251 Mercer St New York NY 10012

BUTTS, DAVID PHILLIP, science educator; b. Rochester, N.Y., May 9, 1932; s. George Albert and Susie Bertha (Hicks) B.; m. Velma M. Walton, Aug. 2, 1958; children—Carol Sue, Douglas Paul. B.S., Butler U., 1954; M.S., U. Ill., 1960, Ph.D., 1962. Asst. prof. Olivet Nazarene Coll., Kankakee, Ill., 1961-62; prof. U. Tex., Austin, 1962-74; prof., chmn. dept. sci. edn. U. Ga., 1974-85, Aderhold Disting. prof., 1985-92; ednl. cons., writer AAAS. Author: (with A. Lee) Vanilla, 1964, Chocolate, 1965, Watermellon, 1966, The Teaching of Science A Self Directed Guide, 1973, Teaching Science in the Elementary School, 1973, (with Hall) Science and Children, 1976; Editor: Designs for Progress in Science Education, 1970, (with others) Science and Society, 1985, Research-Development in Science Education, 1971, Jour. of Research in Sci. Teaching, 1974-79. Served to capt. USAF, 1954-57. Fellow A.A.A.S., Tex. Acad. Sci.; mem. Assn. for Edn. Tchrs. Sci. (regional v.p. 1966-68, pres. 1973-75), Nat. Sci. Tchrs. Assn. (dir. 1970-72), Council Elementary Sci. Internat., Nat. Assn. Research in Sci. Teaching (pres. 1984-87), Am. Ednl. Research Assn., Am. Sci. Affiliation. Home: 145 Deerfield Rd Bogart GA 30622-1737

BUTTS, EDWARD P., civil engineer, consultant; b. Ukiah, Calif., July 29, 1958; s. Edward Oren Butts and Orvilla June (Daily) Hutcheson; m. JoAnne Catherine Zellner, Aug. 14, 1978; children: Brooke C., Adam E. Cert. continuing studies in Irrigation Theory and Practices, U. Nebr., 1980. Registered profl. engr., Oreg., Wash.; cert. water rights examiner, Oreg. Technician Ace Pump Sales, Salem, Oreg., 1976; technician Stettler Supply Co., Salem, 1976-78, assoc. engr., 1978-86, chief engr., 1986-90, v.p. engring., 1990—; profl. engr. exam. question reviewer Nat. Coun. Engring. Examiners, Clemson, S.C., 1989—; profl. engr. exam. supr. Oreg. State Bd. Engring. Examiners, Salem, 1986—; lectr. various water works profl. groups; mem. Marion County Water Mgmt. Coun., 1993—. Contbr. articles to Jour. Pub. Works Mag., AWWA Opflow, Water Well Jour. Coach Little League Cascade Basketball Leage, Turner, Oreg., 1990—. Recipient Merit award Am. City and County Mag., 1990. Mem. Profl. Engrs. Oreg. (mid-Willamette chpt. v.p. 1990-91, pres. 1992-93, state v.p. 1993—); Am. Water Works Assn. Republican. Achievements include devel. of system used to install multiple pumps in water wells, cert. sprinkler irrigation designer. Office: Stettler Supply Co 1810 Lana Ave NE Salem OR 97303

BUTTS, WILLIAM RANDOLPH, nuclear medicine technologist; b. Landstuhl, Germany, May 12, 1966; s. William T. and Blanche E. (Hughes) B.; m. Patti Miller, July 29, 1989; 1 child, Heather. BA in Biology, Furman U., 1987. Cert. nuclear medicine tech. Nuclear chemistry technician Ga. Power Co., Waynesboro, 1988-89; nuclear medicine tech. Univ. Hosp., Augusta, Ga., 1990—. Mem. Am. Assn. Physicists in Medicine, Health Physics Soc., Soc. Nuclear Medicine. Baptist. Home: 3963 Carson Cutoff Martinez GA 30907

BÜTZ, MICHAEL RAY, psychologist; b. Glendale, Ariz., Aug. 1, 1962; s. Bobby Ray and Joan Rose (Thies) B.; m. Kelly Ann Decker, Jan. 8, 1983 (div. Oct. 1984); 1 child, Lindsay Ann. BA in Clin. Psychology, San Francisco State U., 1987, MS in Clin. Psychology, 1989; PhD in Clin. Psychology, Wright Inst., 1992. Banker Lloyds Bank of Calif., Long Beach, Calif., 1983, Continental Bank, Scottsdale, Ariz., 1984; student intern San Francisco Gen. Hosp., 1986-87, San Francisco State U. Psychology Clinic, 1987-88, Kaiser Permanente Dept. of Psychiatry, Vallejo, Calif., 1988-89, Pacific Grad. Sch.-Psychology Clinic, Palo Alto, Calif., 1989-90; predoctoral intern Nat. Asian Am. Psychology Tng. Ctr., San Francisco, 1991; asst. prof. Eastern Mont. Coll., Billings, 1992—; dir. clin. svcs. Rivendell Psychiat. Ctr., Billings; coord. clin. tng. Eastern Mont.State Coll. Psychology Dept., Billings, 1992-93; family therapist Rivendell Psychiatric Ctr., Billings, Mont., 1992-93; post-doctoral internship, Aspen Counseling, Billings, Mont., Dr. Richard D. Recor, 1992-93. Contbr. articles to profl. jours. Named to Ariz. Select Team (Soccer), 1979, Phoenix Inferno Farm Team (Soccer), 1980; recipient letter in Varsity Soccer, U. La Verne, Calif., 1981-82, San Francisco State U., 1986. Mem. APA, Soc. for Chaos Theory in Psychology, San Francisco State U. Alumni Assn., Sierra Club, Western Psychol. Assn., Mont. Psychol. Assn., Yellowstone Valley Psychol. Assn. (bd. mem.). Achievements include theoretical research and presentations on the integration of chaos theory for models of development and therapy in psychology. Office: Rivendell Psychiat Ctr 701 S 27th St Billings MT 59101-4597

BUTZER, KARL W., archaeology and geography educator; b. Mülheim-Ruhr, Germany, Aug. 19, 1934; s. Paul A. and Wilhelmine (Hansen) B.; m. Elisabeth Schlösser, May 12, 1959. B.Sc. honors in Math., McGill U., 1954, M.Sc., 1955; D.Sc., U. Bonn, Germany, 1957. Asst. prof., then assoc. prof. geography U. Wis., 1959-66; prof. anthropology and geography U. Chgo., 1966-80, Henry Schultz prof. environ. archaeology, 1980-84; Raymond Dickson centennial prof. liberal arts U. Tex.-Austin, 1984—; chair, prof. human geography Swiss Fed. Inst. Tech., 1981-82. Author: Environment and Archeology, 1964, rev., 1971, Desert and River in Nubia, 1968, History of an Ethiopian Delta, 1971, Geomorphology from the Earth, 1976, Early Hydraulic Civilization in Egypt, 1976, Archeology as Human Ecology, 1982; editor: After the Australopithecines, 1975, Dimensions of Human Geography, 1978, The Americas Before and After 1492, 1992, Jour. Archeol. Sci., 1978—. Recipient Busk medal Royal Geog. Soc., 1979; Fryxell medal Soc. Am. Archeology, 1981; Stopes medal Geologists Assn. of London, 1982; Archeol. Geology award Geol. Soc. Am., 1985; G.K. Gilbert award Assn. Am. Geographers, 1986; Guggenheim fellow, 1977. Fellow Am. Acad. Arts and Scis, Am. Geog. Soc. (hon.). Office: U Tex Dept Geography Austin TX 78712

BUYER, LINDA SUSAN, psychologist, educator; b. Orange, N.J., June 29, 1956; d. James Robert and Shera Helaine (Feldman) Buyer; m. Herbert Halsten Stenson, May 14, 1988. BA, U. Ill., Chgo., 1979; PhD, U. Ill., 1989. Asst. prof. psychology Roosevelt U., Chgo., 1989-90, U. Notre Dame du Lac, Notre Dame, Ind., 1990—; cons. Equity Group, Chgo., 1990, 91. Contbr. articles to profl. jours. Bd. dirs. LaPorte County Mental Health Orgn., Ind., 1993—; master gardener Duneland Master Gardeners, Valparaiso, 1992—. Inst. for Scholarship in the Liberal Arts grantee, 1992. Mem. Am. Psychol. Soc., Midwest Psychol. Assn., Psychonomic Soc. (assoc.). Office: Univ of Notre Dame 111 Haggar Hall Notre Dame IN 46556

BUZARD, KURT ANDRE, ophthalmologist; b. Lakewood, Colo., Apr. 9, 1953; s. Donald Keith and Sonja Marie (Vik) B.; m. Carol Ann Moss, Aug. 4, 1989. BA in Math. and Physics, Northwestern U., 1975; MA in Applied Physics, Stanford, U., 1976; MD, Northwestern U., 1980. Diplomate Am. Bd. Ophthalmology, Nat. Bd. Med. Examiners. Intern medicine L.A. County-U. So. Calif. Med. Ctr., 1980-81; resident Jules Stein Eye Inst. UCLA, 1982-85; fellow cornea/refractive surgery Richard C. Troutman, MD, 1985-86; ophthalmologist, corneal specialist Las Vegas, Nev., 1986—; staff physician Rancho Los Amigos Hosp., 1981-82; clin. asst. prof. div. ophthalmology dept. surgery U. Nev. Sch. Medicine, 1988—; clin. asst. prof. dept. ophthalmol. medicine Tulane U. Med. Ctr., New Orleans, 1991; med. dir. S.W. Eye Procurement Ctr., Las Vegas, 1989—; affiliate Humana Hosp.-Sunrise, 1989—, Las Vegas Surg. Ctr., 1989—, Las Vegas Surg. Ctr., Med. Ctr. So. Nev., 1989—; assoc. staff Valley Hosp., Las Vegas, 1986—; mem. med. adv. bd. Donor Orgn. Referral Svc. Author: (with Richard Troutman) Corneal Astigmatism: Etiology, Prevention and Management, 1992; contbr. articles to profl. jours. Mem. Las Vegas C. of C., 1989. Recipient Rsch. award Jules Stein Inst., L.A., 1985. Fellow Am. Acad. Ophthalmology, Am. Coll. Surgeons; mem. Am. Soc. Cataract and Refractive Surgery, AMA, Assn. for Rsch. in Vision and Ophthalmology, Castroviejo Soc., Colombian Soc. Ophthalmology (corr.), Eye Bank Assn. of Am.-Paton Soc., Internat. Soc. for Eye Rsch., Internat. Soc. Refractive Keratoplasty (long-range planning com., alternative rep. to Am. Acad. Ophthalmology, bd. dirs. 1992-94), Pan Am. Assn. Ophthalmology, Pan Am. Implant Assn., Phi Eta Sigma, Phi Dela Kappa. Avocations: computers, photography. Office: 2575 Lindell Rd Las Vegas NV 89102-5409

BY, ANDRE BERNARD, engineering executive, research scientist; b. Detroit, May 19, 1955; s. Bernard Joseph and Margaret (Voytish) B. BS in Aerospace Engring., U. Mich., 1977, BS in Mech. Engring., 1977; MS in Mech. Engring., MIT, 1979, postgrad., 1985. Mech./chem. engr. Motor Vehicle Emissions Lab. EPA, Ann Arbor, Mich., 1977; teaching asst., rsch. asst. MIT, Cambridge, 1977-79; engr., sr. engr., sr. project engr. Computer Aided Engring. Group, No. Rsch. and Engring. Corp., Woburn, Mass., 1979-84; mgr. automated sytems group No. Rsch. and Engring. Corp., Woburn, Mass., 1984-90; rsch. assoc., lectr. mech. engring. dept. Tufts U., Medford, Mass., 1990—; pres., tech. dir. Automation Engring. Inc., Wakefield, Mass., 1990—; v.p. rsch. and devel. Productivity Tech. Inc., Sunnyvale, Calif., 1992—; seminar lectr., panel mem. Cell Contr. Seminar, Soc. Mfg. Engrs., Detroit, 1989. Contbr. articles to profl. jours. Mem. AIAA, ASME, Soc. Mfg. Engrs., Soc. Automotive Engrs. Avocations: music, contemporary literature, impressionist art, skiing, electronics. Office: Automation Engring Inc PO Box 350 Wakefield MA 01880-0950

BYCZKOWSKI, JANUSZ ZBIGNIEW, toxicologist; b. Gdansk, Poland, May 29, 1947; came to U.S., 1979; s. Stanislaw and Halina (Osterczy) B.; m. Janina K. Slosarska, Aug. 6, 1977; children: Jan S., L. Peter. MSc in Toxicology, Acad. Medicine, Gdansk, 1970, PhD in Pharmacology, 1975, DSc in Biochem. Pharmacology, 1979. Cancer rsch. scientist dept. exptl. therapeutics Roswell Park Meml. Inst., Buffalo, 1979-80, 1985-87; adj. ast. prof. pharmacology Acad. Medicine Gdansk, 1980-83; pharmacologist and dir. of pharmacy Internat. Red Cross and Red Crescent, Tobruk, Libya, 1983-84; asst. prof. and rsch. scientist Coll. Pub. Health U. South Fla., Tampa, 1987-91; project scientist and study dir. ManTech. Environ. Tech., Inc., Dayton, Ohio, 1991—; editorial reviewer Bull. Environ. Contamination and Toxicology, Reno, Nev., 1989—, Free Radical Biology and Medicine, Baton Rouge, 1989—. Contbr. articles to profl. publs., chpts. to books. Active mem. Solidarity, Poland, 1980-83. Recipient Rsch. award 1st degree Sci. Soc. Gdansk, 1975, Polish Pharmacol. Soc., 1977, Ministry Health and Social Welfare of Poland, 1977. Mem. AAAS, Soc. for Rsch. on Polyunsaturated Fatty Acids (pres. 6th sci. meeting 1992—, travel grantee 1992), N.Y. Acad. Scis., Oxygen Soc., Soc. Toxicology. Achievements include finding mechanism of action of DDT on mitochondrial respiration; discovery of NAD-Dependent mode of action of vanadium, co-oxygenation of benzopyrene by lipoxygenase; developing physiologically-based pharmacokinetic model for lactational transfer. Home: 212 N Central Ave Fairborn OH 45324 Office: ManTech Environ Tech Inc PO Box 31009 Dayton OH 45437

BYDZOVSKY, VIKTOR, surgeon; b. Prague, Czechoslovakia, May 24, 1944; s. Viktor and Milada (Raus) B.; m. Anne Marie Berset, Feb. 26, 1977; children—Patricia, Pierre. Dr.Med., U. Prague, 1967. Hosp. asst., Kolin, Czechoslovakia, 1968-70, Montreux, Switzerland, 1970-71, Fribourg, Switzerland, 1971-73, Geneva, 1973-78; practice medicine specializing in gen. surgery, Fribourg, 1978—. Lt. M.C., Czechoslovakian Army. 1967-68, 1st lt. M.C., Swiss Army, 1990—. Mem. Internat. Soc. Surgery, Internat. Coll. Surgeons, Swiss Surg. Soc., Swiss Soc. Sports Medicine, Soc. Suisse Senologie. Avocations: sports, music. Home: Chemin de Primeveres 35, CH-1700 Fribourg Switzerland Office: Locarno 3, CH-1700 Fribourg Switzerland

BYERLY, RADFORD, JR., science policy official; b. Houston, May 22, 1936; s. Radford and Garvis N. (Cook) B.; m. Kathryn Jester, May 13, 1960 (div. 1980), children: Laura, Hamilton, Charles; m. Carol Ann Ries, Apr. 10, 1987. BA, Williams Coll., 1958, MA, 1960; PhD, Rice U., 1967. Sr. engr. No. Rsch. & Engring. Co., Cambridge, Mass., 1961-63; postdoctoral fellow U. Colo., Boulder, 1967-69, dir. Ctr. for Space and Geoscis. Policy, 1987-91; physicist, mgr. Nat. Bur. Standards, Washington, 1969-75; mem. profl. staff com. on sci. and tech. U.S. Ho. of Reps., Washington, 1975-87, chief of staff, com. on sci. and tech., 1991-93; v.p. pub. policy Union Corp. for Atmosphere Rsch., Boulder, 1993—; dir. Roberts Inst., Boulder, 1993—; mem. space sta. adv. com., space sci. adv. com. NASA, 1988-91; hon. lectr. Mid-Am. State Univs. Assn., 1988-89. Contbr. articles to profl. jours. NSF fellow, 1963-67. Mem. AAAS, ASTM, AIAA (chmn. civil space subcom. 1988-89), Am. Phys. Soc., Am. Astronautical Soc., Williams Club (N.Y.C.), Phi Beta Kappa, Sigma Xi. Avocations: skiing, hiking, squash. Home: 3870 Birchwood Dr Boulder CO 80304 Office: UCAR Sci Space & Tech PO Box 3000 Boulder CO 80307

BYERS, JIM DON, research chemist; b. Levelland, Tex., May 26, 1954; s. Z.T. and Carol Jim (Williams) B.; m. Ruth Ann Bailey, Dec. 27, 1974; children: Angela Michelle, John Michael. BS, Ouachita Bapt. U., Arkadelphia, Ark., 1974; MS, N.E. La. U., 1977; PhD, Duke U., 1980. Rsch. chemist Phillips Petroleum Co., Bartlesville, Okla., 1980-86; sr. rsch. chemist Phillips Petroleum Co., Bartlesville, 1986-91, rsch. assoc., 1991—. Author: Behavior Modifying Chemicals for Insect Management, 1990, Advances in Organic Geochemistry, 1981; contbr. articles to profl. jours. Mem. Am. Chem. Soc. (chmn. N.E. Okla. sect. 1990-91). Republican. Baptist. Achievements include 14 patents, extensive work in field of organic peroxides. Home: 1457 Oakdale Dr Bartlesville OK 74006 Office: Phillips Petroleum Co 92G PRC Bartlesville OK 74004

BYERS, NINA, physics educator; b. Los Angeles, Jan. 19, 1930; d. Irving M. and Eva (Gertzoff) B.; m. Arthur A. Milhaupt, Jr., Sept. 8, 1974. BA in Physics with highest honors, U. Calif., Berkeley, 1950; M.S. in Physics, U. Chgo., 1953, Ph.D., 1956; M.A., U. Oxford, Eng., 1967. Research fellow dept. math. physics U. Birmingham, Eng., 1956-58; research assoc., asst. prof. Inst. Theoretical Physics and dept. physics Stanford, 1958-61; asst. then assoc. prof. physics UCLA, 1961-67, prof. physics, 1967—; mem. Sch. Math., Inst. Advanced Studies, Princeton, N.J., 1964-65; fellow Somerville Coll., Oxford, 1967-68, Janet Watson vis. fellow, 1968-75; faculty lectr. Oxford U., 1967-68, sr. vis. scientist, 1973-74. John Simon Guggenheim Meml. fellow, 1963-64, Sci. Rsch. Coun. fellow Oxford U., 1978, 85. Fellow AAAS (mem.-at-large physics sect. 1983-86, com. on ethics and sci. responsibility), Am. Phys. Soc. (councillor-at-large 1977-81, panel pub. affairs 1980-83, vice chmn. Forum on Physics and Soc. 1981-82, chmn. 1982-83); mem. Fedn. Am. Scientist (nat. coun. 1972-76, 78-80, exec. com. 1974-76, 78-80). Achievements include work in research theory particle physics and superconductivity. Office: U Calif Dept Physics Los Angeles CA 90024

BYLSMA, FREDERICK WILBURN, neuropsychologist; b. Kingston, Ont., Can., Jan. 29, 1957; came to U.S., 1987; s. Hinne and Weipkje (Vander Meulen) B.; m. Neerja Raj Narain, Oct. 15, 1982; 1 child, Tara. BA, U. Ottawa, Ont., 1980, PhD, 1987. Lic. psychologist, Md. Postdoctoral fellow Johns Hopkins U., Balt., 1987-89; instr. The Johns Hopkins U., Balt., 1989-92, asst. prof., 1992—. Contbr. articles to profl. jours. Mem. APA, Internat. Neuropsychol. Soc. Office: Johns Hopkins U Meyer 218 600 N Wolfe St Baltimore MD 21287-7218

BYNES, FRANK HOWARD, JR., physician; b. Savannah, Ga., Dec. 3, 1950; s. Frank Howard and Frenchye (Mason) B.; m. Janice Ratta, July 24, 1987; children: Patricia, Frenchye. BS, Savannah State Coll., 1972; MD, Meharry Med. Coll. Resident gen. surgery Staten Island (N.Y.) Hosp., 1978-82; resident internal medicine N.Y. infirmary Beekam Downtown Hosp., N.Y.C., 1983-86; dir. medicine USAF Sheppard Regional Hosp., Sheppard AFB, Tex., 1986-87; pvt. practice internal medicine N.Y.C., 1987-90; attending physician Bronx (N.Y.) Lebanon Hosp., 1990—. Maj. USAF, 1986-87. Mem. AMA, AAAS, ACP, N.Y. Acad. Scis., Assn. Mil. Surgeons of U.S., Alpha Phi Alpha.

BYNUM, BARBARA STEWART, federal health institute administrator; b. Washington, June 13, 1936; d. Oliver Walton and Mabel (Easton) Stewart; m. Elward Bynum, Apr. 4, 1959; 1 son, Christian. B.A. in Chemistry, U. Pa., 1957; postgrad. in biochemistry, Georgetown U., 1958-60. Chemist Nat. Cancer Inst. NIH, Bethesda, Md., 1958-71; adminstrv. asst., office assoc. dir. for adminstrn. NIH, Bethesda, 1971-72, sci. grants program specialist div. rsch. grants, 1972-75, health scientist adminstr. div. rsch. grants, 1975-78, asst. chief for spl. programs, sci. rev. br. div. rsch. grants, 1978-81, dir. div. extramural activities Nat. Cancer Inst., 1981—; reviewer, cons. AAAS, Washington, 1974—; chmn. HHS Task Force Working Group on Cancer in Minorities, 1986—. Recipient Dirs. award NIH, 1980; recipient Sr. Exec. Service Superior Performance award HHS, 1982, 1987, 91. Mem. Am. Assn. Cancer Research, Am. Assn. Pathologists, AAAS, Biophys. Soc. Democrat. Roman Catholic. Office: Nat Cancer Inst Extramural 9000 Rockville Pike Bethesda MD 20892-0001

BYRD, MARY LAAGER, life science researcher; b. N.Y.C., Apr. 7, 1935; d. Creston Frederick Laager and Mary King (Poteat) Lindgren; m. Mark Willard Byrd, Dec. 17, 1954 (div. Apr. 1967); children: Carolyn Byrd Hill, Christopher M., Cynthia Byrd Becker. Student, U. Pa., 1952-55, U. Madrid, Spain, 1964-65, Mitchell Coll., 1965-66. Researcher Zool. Soc., San Diego, Calif., 1974-87; researcher Found. for Endangered Animals, La Jolla, Calif., 1985—, also bd. dirs., sec./treas., 1985—; cons. South Fla. Aguaculture, Florida City, 1989—; cons. Save the Manatee, Orlando, Fla., 1990. Editor: One Medicine, 1984; contbr. articles to profl. jours. Active Fla. Keys Land Trust, Marathon, 1989—. Recipient grant Zool. Soc. of San Diego, Chaco, Paraguay, 1985-90. Mem. The Nature Conservancy, Zool. Soc. of San Diego, Found. Moises Bertoni de Paraguay, Nat. Audubon Soc., Worldlife Fund, NOW. Republican. Avocations: sailing, diving, skiing, travel, crafts. Home: 213 Wildwood Cir Key Largo FL 33037-4220 Office: South Fla Aquaculture 40801 SW 232nd Ave Homestead FL 33034-6703

BYRD, WILLIAM GARLEN, clinical pharmacist, medical researcher; b. Hazard, Ky., Oct. 14, 1947; s. William and Nancy (Baker) B.; m. Vickie J. Wilford, Dec. 1, 1979. BS in Pharmacy, U. Ky., 1973, Dr. Pharmacy, 1979. Lic. pharmacist, Ky. Pharmacist various firms, 1973-76; resident in clin. pharmacy Chandler Med. Ctr., Lexington, Ky., 1976-79; asst. prof. U. Ky., Lexington, 1979-87; assoc. mgr. clin. rsch. Miles, Inc., Elkhart, Ind., 1987-88, mgr. clin. rsch., 1988-91, assoc. dir. med. dept., 1991—; cons. in field. Rev. editor Miles Sci. Jour., 1990-91; contbr. articles to profl. publs. Capt. U.S. Army Nat. Guard, 1984-92. Mem. Am. Soc. Clin. Pharmacology and Therapeutics, Am. Coll. Nutrition, Am. Coll. Clin. Pharmacy, Ky. Soc. Hosp. Pharmacists (pres.-elect 1986-87). Republican. Lutheran. Office: Miles Inc 1127 Myrtle St Elkhart IN 46514-2282

BYRNE, DANIEL WILLIAM, computer specialist, medical researcher, biostatistician, educator; b. Bklyn., Jan. 21, 1958; s. Thomas Edward and L.M. (Collins) B.; m. Loretta Marie May, June 22, 1985; children: Michael, Virginia. BA in Biology, SUNY, Albany, 1983; MS in Biostatistics, N.Y. Med. Coll., 1991. Programmer, med. software Dept. Surgery N.Y. Med.

Coll., Valhalla, 1983-84; computer/data analyst N.Y. Med. Coll., Westchester County Med. Ctr. and affiliate hosps., 1984-87; rsch. asst. prof. med. software Dept. Surgery N.Y. Med. Coll., Valhalla, 1988—; founder, computer cons. Byrne Research, Ridgefield, Conn., 1989—. Author: (tng. manual) Guidelines for Analyzing Clinical Research on a Microcomputer, 1986; author/programmer various software including Trauma Management System, 1990, Occupational Stress Database, 1990, Nuclear Disaster Evacuation Plan Database, 1990; contbr. numerous articles to med. jours. Grantee Centers Disease Control 1987-90, 89-92. Mem. Am. Statis. Assn., Inst. Math. Stats., Biometric Soc. Roman Catholic. Home and Office: 17 Dogwood Dr Ridgefield CT 06877-2707

BYRNE, GEORGE MELVIN, physician; b. San Francisco, Aug. 1, 1933; s. Carlton and Esther (Smith) B.; BA, Occidental Coll., 1958; MD, U. So. Calif., 1962; m. Joan Stecher, July 14, 1956; children: Kathryne, Michael, David; m. Margaret C. Smith, Dec. 18, 1982. Diplomate Am. Bd. Family Practice, 1971-84. Intern, Huntington Meml. Hosp., Pasadena, Calif., 1962-63, resident, 1963-64; family practice So. Calif. Permanente Med. Group, 1964-81, physician-in-charge Pasadena Clinic, 1966-81; asst. dir. Family Practice residency Kaiser Found. Hosp., L.A., 1971-73; clin. instr. emergency medicine Sch. Medicine, U. So. Calif., 1973-80; v.p. East Ridge Co., 1983-84, sec., 1984; dir. Alan Johnson Porsche Audi, Inc., 1974-82, sec., 1974-77, v.p., 1978-82. Bd. dirs. Kaiser-Permante Mgmt. Assn., 1976-77; mem. regional mgmt. com. So. Calif. Lung Assn., 1976-77; mem. pres.'s circle Occidental Coll., L.A. Drs. Symphony Orch, 1975-80; mem. profl. sect. Am. Diabetes Assn. Fellow Am. Acad. Family Physicians (charter); mem. Am., Calif., L.A. County Med. Assns., Calif. Acad. Family Physicians, Internat. Horn Soc., Am. Radio Relay League (Pub. Service award), Sierra (life). Home and Office: 528 Meadowview Dr La Canada Flintridge CA 91011-2816

BYRNE, JOHN VINCENT, academic administrator; b. Hempstead, N.Y., May 9, 1928; s. Frank E. and Kathleen (Barry) B.; m. Shirley O'Connor, Nov. 26, 1954; children: Donna, Lisa, Karen, Steven. AB, Hamilton Coll., 1951; MA, Columbia U., 1953; PhD, U. So. Calif., 1957. Research geologist Humble Oil & Refinery Co., Houston, 1957-60; assoc. prof. Oreg. State U., Corvallis, 1960-66, prof. oceanography, 1966—, chmn. dept., 1968-72, dean Sch. Oceanography, 1972-76, acting dean research, 1976-77, dean research, 1977-80, v.p. for research and grad. studies, 1980-81, pres., 1984—; adminstr. NOAA, Washington, 1981-84; Program dir. oceanography NSF, 1966-67. Recipient Carter teaching award Oreg. State U., 1964. Fellow AAAS, Geol. Soc. Am., Am. Meteorol. Soc.; mem. Am. Assn. Petroleum Geologists, Am. Geophys. Union, Sigma Xi, Chi Psi. Club: Arlington (Portland, Oreg.). Home: 3520 NW Hayes Ave Corvallis OR 97330-1746 Office: Oreg State U Office of Pres Corvallis OR 97331

BYRNES, CHRISTOPHER IAN, academic dean, researcher; b. N.Y.C., June 28, 1949; s. Richard Francis and Jeanne (Orchard) B.; m. Catherine Morris, June 24, 1984; children: Kathleen, Alison, Christopher. BS, Manhattan Coll., 1971; MS, U. Mass., 1973, PhD, 1975. Instr. U. Utah, Salt Lake City, 1975-78; asst. prof. Harvard U., Cambridge, Mass., 1978-81, assoc. prof., 1981-85; rsch. prof. Ariz. State U., Tempe, 1985-89; prof., chmn. dept. systems sci. and math. Washington U., St. Louis, 1989-91, dean engring. and applied sci., 1991—; adj. prof. Royal Inst. Tech., Stockholm, 1985-90; cons. Sci. Systems, Inc., Cambridge, 1980-84, Systems Engring., Inc., Greenbelt, Md., 1986; mem. NRC. Editor: (book series) Progress in Systems Control, 1988—, Foundations of Systems and Control, 1989—; Nonlinear Synthesis, 1991, 10 other books; contbr. numerous articles to profl. jours., book revs. Recipient Best Paper award, IFAC, 1993. Fellow IEEE (George Axelby award 1991), Japan Soc. for Promotion Sci.; mem. AAAS, Soc. for Indsl. Applied Math. (program com. 1986-89), Am. Math. Soc., Sigma Xi., Tau Beta Pi. Avocations: cooking, fishing, travel. Office: Washington U Sch Engring and Applied Sci 1 Brookings Dr Saint Louis MO 63130

BYRNES, JOHN JOSEPH, medical educator; b. Waterbury, Conn., Oct. 27, 1942; s. Thomas Joseph and Anne Elizabeth (Shanahan) B.; m. Sixta Herazo, Apr. 2, 1986; children: John, Michael, Matthew, Ann, Diana, Stephen, Daniel. BS, Holy Cross Coll., 1964; MD, Tufts U., 1968. Diplomate Am. Bd. Internal Medicine and the subspecialties of hematology, med. oncology and diagnostic lab. immunology. Med. resident SUNY, Buffalo, 1968-70; hematology fellow U. Miami, 1970-72, chief, hematology, 1989-92, asst. prof., 1974-77, assoc. prof., 1977-85, prof., 1985—, dir. Ctr. for Blood Diseases, 1989-92; chief, hematology/oncology VA Med. Ctr., 1992—; bd. dirs. Leukemia Soc. S. Fla., Miami; vis. prof. John Hopkins U., Balt., 1985-86. Contbr. articles to profl. jours. Bd. dirs. Leukemia Soc. S. Fla., Miami, 1990—, IPARC Internat., Rio de Janeiro, 1980—. Lt. comdr. USN, 1969-75. Rsch. assoc. VA, Miami, 1974-77, clin. investigator, 1978-81, merit rev. grantee, 1974-90; grantee Nat. Inst. Health, Miami, 1986—; Harrington Professorship, U. Miami, 1986—. Mem. Am. Soc. Clin. Investigation, Am. Soc. Hematology, Am. Soc. Biochemistry and Molecular Biology, Am. Coll. Physicians. Achievements include discovery of DNA polymerase delta, treatment for thrombotic/thrombocytopenic purpura; description of selective inhibition of 3' to 5' exonuclease associated with DNA polymerases, action of cancer and antiviral chemotherapy drugs, nucleotides and purine analogs upon DNA synthesis, innovative targeted drug treatment of idiopathic thrombocytopenic purpura and autoimmune hemolytic anemias using vinblastine loaded platelets. Home: PO Box 3321 Key Largo FL 33037 Office: Miami Vets Med Ctr 1201 NW 16ST Miami FL 33136

BYRNES, MICHAEL FRANCIS, podiatrist; b. Chgo., Aug. 11, 1957; s. Edward and Dorothy Franchi; m. Debra Michelle Moody, July, 31, 1982. BA, Loyola U., Chgo., 1979; D in Podiatry Medicine, Ill. Coll. Podiatry Med., 1984. Diplomate Am. Bd. Podiatric Surgery. Practice medicine specializing in podiatrics Ridgeland Foot Clinic., Chgo., 1984—; surgeon Mercy Surg. Ctr., Justice, Ill., 1984—, Holy Cross Hosp., 1987—; assoc. prof. Dr. Scholl Coll. Podiatric Medicine; mem. sci. and med. staff Mercy Hosp. and Med. Ctr. Contbr. case reports to Jour. Foot Surgery, 1985. Precinct capt. 49th Dem. Ward., Chgo., 1976-80. Winner state skating championship, 1980. Fellow Am. Coll. Foot Surgeons; mem. Am. Podiatric Med. Assn., Ill. Podiatric Med. Assn. (co-chmn. legis. com. 1985-86, del.), Am. Acad. Podiatric Sports Medicine. Roman Catholic. Avocation: skating. Home: 218 Janet Ave Westmont IL 60559-4282 Office: 10021 SW Hwy Oak Lawn IL 60453-3725

BYRON, JOSEPH WINSTON, pharmacologist; b. N.Y.C., Apr. 23, 1930; s. Joseph Adolphos and Florence Augusta (Coull) B.; 1 child, Annette. BS, Fordham U., 1952; MS, Phila. Coll. Pharmacy and Sci., 1955; PhD, U. Buffalo, 1960. Sr. rsch. scientist Paterson Cancer Rsch. Inst., Manchester, U.K., 1962-65; prin. rsch. scientist Paterson Cancer Rsch. Inst., Manchester, 1965-73; assoc. prof. U. Md. Med. Sch., Balt., 1973-76, prof. pharmacology, 1976-82; prof., chmn. pharmacology Meharry Med. Coll., Nashville, 1982—; mem. bd. sci. counselors NIOSH, CDC, Atlanta, 1991—, Div. Cancer Treatment, Nat. Cancer Inst., 1980-82; mem. com. Nat. Bd. Med. Examiners, Phila., 1989-90; nat. rsch. coun. fellowship panelist, Washington, 1986-90. Editorial bd. mem.: Exptl. Hematology, 1975-78, Internat. Jour. of Cell Cloning, 1983-88; contbr. over 60 articles to profl. jours. and books. Recipient Postdoctoral fellowship NSF, 1959-60, Am. Cancer Soc., 1961-62. Mem. Internat. Soc. for Exptl. Hematology (mem. coun. 1978-81), Am. Soc. for Pharmacology and Exptl. Therapeutics. Achievements include significant contbns. to approaches for treatment of radiation injury; first to assoc. drug-receptors, histamine H2 receptors with bone marrow stem cells. Office: Meharry Med Coll 1005 D B Todd Jr Blvd Nashville TN 37208

BZIK, DAVID JOHN, parasitologist, researcher; b. N.Y.C., July 1, 1955; s. John and Mary F. (Frank) B.; m. Barbara A. Fox, July 16, 1983. MS, Pa. State U., 1980, PhD, 1983. Postdoctoral fellow Pa. State U., State College, 1983-84, Glasgow (Scotland) U., 1984-85, Dartmouth Coll., Hanover, N.H., 1985-87; asst. prof. Dartmouth Med. Sch., Hanover, N.H., 1987—. Recipient fellowship European Molecular Biology Assn., 1984-85. Mem. AAAS, Am. Soc. Tropical Med. Hygiene. Achievements include discovery of enlarged malaria parasites RNA polymerases. Home: PO Box 1064 Grantham NH 03753 Office: Dartmouth Med Sch Dept Microbiology Hanover NH 03755-3842

CABALLES, ROMEO LOPEZ, pathologist, bone tumor researcher; b. Pagsanjan, Laguna, Philippines, Dec. 10, 1925; came to U.S., 1955; s. Eladio Luna and Ceferina (Lopez) C.; m. Lucia Rose Mercadante, June 29, 1958; children: Romeo Jr., Rose, Theresa, Nancy, James. AA, U. Santo Tomas, 1946-48, MD, 1948-54. Diplomate Am. Bd. Pathology. Asst. pathologist St. Michael's Med. Ctr., Newark, N.J., 1960-68; assoc. pathologist United Hosp. Med. Ctr., Newark, 1968-74, attending pathologist, 1974-85, interim dir. labs., 1978-79; pathologist United Hosp. Orthopedic Ctr., Newark, 1968-85; clin. asst. prof. N.J. Med. Sch., Newark, 1971-81, clin. assoc. prof., 1981—; pro adv. panelist Med. Lab. Observer, Fairlawn, N.J., 1971-72; course dir. United Hosp. Orthopedic Ctr. Skeletal Symposium, Newark, 1977. Contbr. articles to profl. jours. Fellow Coll. Am. Pathologists, N.Y. Acad. Medicine; mem. N.Y. Acad. Scis., N.J. Soc. Pathologists, Philippine-Am. Med. Assn. (N.J. founding mem.), Practitioners Club N.J. Republican. Roman Catholic. Home: 2 Keystone Dr Livingston NJ 07039

CABANISS, GERRY HENDERSON, geophysicist, analyst; b. Winter Haven, Fla., Apr. 22, 1935; s. William Fredrick and Adelaide (Henderson) C.; m. Judith Ann McAuliffe, June 17, 1962; children: Jennifer, Paul, William. BA in Geology, Dartmouth Coll., 1957; MS in Geophysics, Boston Coll., 1968; PhD in Geology, Boston U., 1975. Staff scientist Air Force Geophysics Lab., Bedford, Mass., 1958-82; prin. staff mem. BDM Internat. Inc., Albuquerque, 1982—; proposal/review com. NASA, Washington, 1970, 76. Compiler Sci. Studies on U.S. Arctic Ocean Drift Stations, 1968.]em. AAAS, Am. Geophys. Union (geodynamics com. 1978). Achievements include research in deployed arrays of shallow and deep borehole tiltmeters to measure tidal and long-period earth deformations, significant advances in techniques, design and performance. Office: BDM Federal 1801 Randolph Rd SE Albuquerque NM 87106

CABARET, JOSEPH RONALD, electronics company executive; b. Astoria, N.Y., Dec. 26, 1934; s. Joseph Henry and Henrietta (Nevejans) C.; m. Giovanna Longhitano, Dec. 26, 1960; children: Grace, Joseph, Corinne. BSME, Stevens Inst. Tech., 1957; MBA, Pepperdine U., 1980. Chief engr. Space Ordnance Systems, Canyon Country, Calif., 1962-67, sales rep., 1968-76, program mgr., 1977-79, asst. gen. mgr., 1980, gen. mgr., 1981, pres., 1982-85; v.p. corp. devel. Transtech., Sherman Oaks, Calif., 1985-89; pres. Unidynamics/Phoenix, Goodyear, Ariz., 1989—. Office: Unidynamics/Phoenix 102 S Litchfield Rd Goodyear AZ 85338-1295

CABASSO, ISRAEL, polymer science educator; b. Jerusalem, Israel, Nov. 17, 1942; came to U.S., 1973; s. Victor and Judith (Zimmerman) C. BS in Chemistry and Physics, Hebrew U., Jerusalem, 1966, MS in Polymer Chemistry with distinction, 1968; PhD in Polymer Chemistry, Weizmann Inst. Sci., Rehovot, Israel, 1973. Scientist Weizmann Inst. Sci., Rehovot, 1973-76, Meyerhoff prof., 1987-88; head ploymer dept. Gulf South Rsch. Inst., New Orleans, 1976-80; clin. assoc. prof. biomaterials La. State U., New Orleans, 1978-80; prof. polymer chemistry Coll. Environ. Sci. and Forestry SUNY, Syracuse, 1980—, dir. polymer rsch. inst. Coll. Environ. Sci. and Forestry, 1982—; assoc. dir. ctr. memb. engring. and sci. Syracuse U., 1987—; cons. to govt. and industry in field; vis. prof. Japan, Korea, Taiwan, China, Israel. Contbr. over 100 articles to sci. jours., encyclopedias. Capt. Armored Cavalry, 1961-63, Israel. Recipient Syracuse Sect. award Am. Chem. Soc., 1987, over 30 indsl. grants and rsch. awards. Achievements include over 15 patents in field. Home: 131 Buckingham Ave Syracuse NY 13210 Office: SUNY Coll Environ Sci and Forestry 1 Forestry Dr Syracuse NY 13210

CABAUP, JOSEPH JOHN, geology educator; b. Bronx, N.Y., Nov. 5, 1940; s. Joseph Christopher and Angelina (DeVenuta) C.; m. Barbara Louise Mellor, June 26, 1965 (div. Dec. 1987); children: Joseph E., Jean M. BA in Physics, Hunter Coll., 1962; MS, U. N.C., 1969. Cert. geologist, Maine. Asst. prof. chemistry and physics Winthrop Coll., Rock Hill, S.C., 1967-70; tchr. Franconia (N.H.) Coll., 1970-71, Dartmouth High Sch., North Dartmouth, Mass., 1971-72; asst. prof. physics R.I. Jr. Coll., 1972-74; prof. geology N.H. Tech. Coll., Berlin, 1974—; environ. cons. Environ. Survey and Analysis, Bethlehem, N.H., 1971—. Mem., past pres. Bethlehem Conservation Commn., 1984—. Mem. AAAS, Geol. Soc. Am., Nat. Assn., Geology Tchrs. Home: 24 Church St PO Box 760 Bethlehem NH 03574-0760 Office: NH Tech Coll 2020 Riverside Dr Berlin NH 03570

CABBAGE, WILLIAM AUSTIN, chemical engineer; b. Liberty Hill, Tenn., Jan. 21, 1934; s. Fred Austin and Lillieth Thelma (Beeler) C.; m. Julia Ann Kidd, Jan. 11, 1963; children: Michael Christian, Brian William. BS, U. Tenn., 1956. Engr. Cramet, Inc., Chattanooga, 1956; process engr. Hercules Inc., Radford, Va., 1958-88; sr. tech. engr. Hercules Inc., Radford, 1988—. Author: Hazards Analysis of Nitroglycerin Manufacturing Plant, 1990; contbr. papers to seminars. With U.S. Army, 1956-58. Mem. Am. Radio Relay League, Southeastern Repeater Assn., Toastmasters (sec. 1992). Baptist. Achievements include rsch. in effect of environ. influences on ultraviolet detector response. Home: 905 Custis St Radford VA 24141 Office: Hercules Inc Radford Army Ammunition Plant Radford VA 24141

CABE, JERRY LYNN, mechanical engineer, researcher; b. Oakhill, W.Va., Oct. 12, 1953; s. James Earl Sr. and Christine Mary (Jones) C. BSME, W.Va. U., 1979. Registered profl. engr., Ohio. Design engr. Pratt and Whitney Aircraft, West Palm Beach, Fla., 1979-84; program mgr. GE Aircraft Engines, Cin., 1984—; chief engr. Engineered Designs, Inc., Cin., 1989—. Mem. AIAA, NSPE, Soc. Automotive Engrs., Ohio Soc. Profl. Engrs. Office: Engineered Designs Inc 4733 Devitt Dr Cincinnati OH 45246

CABLE, CHARLES ALLEN, mathematician; b. Akeley, Pa., Jan. 15, 1932; s. Elton Thomas and Margaret (Fox) C.; m. Mabel Elizabeth Yeck, Dec. 19, 1955; children: Christopher A., Carolyn E. B.S., Edinboro State Coll., 1954; M.Ed., U. N.C., 1959; Ph.D. in Math., Pa. State U., 1969. Instr. math. Interlaken High Sch., N.Y., 1954-55, Tidioute High Sch., Pa., 1957-58; asst. prof. math Juniata Coll., Huntingdon, Pa., 1959-67; assoc. prof. dept. math. Allegheny Coll., Meadville, Pa., 1969-75, prof. dept. math., 1975—, chmn. dept., 1970-90. Editorial reviewer: Math. Mag., 1975-80; assoc. editor: Focus, 1981-85. Served with AUS, 1955-57. Gen. Elec. fellow, 1958; NSF fellow, 1959, 61, 68, 73; NDEA fellow, 1969. Mem. Am. Math. Soc., Math. Assn. Am. (chmn. Allegheny Mountain chpt. 1973-75, bd. govs. 1981-84, mem. newsletter editorial com. 1981-85, com. on student chpts. 1987-93, publs. co. 1983-86), AAUP. Republican. Presbyterian. Home: 199 Jefferson St Meadville PA 16335-1108 Office: Allegheny Coll N Main St Meadville PA 16335-1111

CABRERA, ALEJANDRO LEOPOLDO, physics researcher and educator; b. Santiago, Chile, May 26, 1950; came to U.S., 1974; s. Guillermo Raul Cabrera-Guzman and Eliana Catalina (Oyarzun-Fontaine) de Cabrera; m. Lilly Ann Suarez, Oct. 4, 1975; 1 child, Guerau Bernat. Lic., U. Chile, Santiago, 1974; MS in Physics, U. Calif., San Diego, 1976, PhD in Physics, 1980. Rsch. asst. U. Chile, 1971-74, U. Calif., San Diego, 1974-80; rsch. assoc. U. Calif., Berkeley, 1980-82; sr. physicist, prin. Air Products & Chems., Allentown, Pa., 1982-90; prof. physics Pontificia U. Catolica de Chile, Santiago, 1990—; creator ctrs. for excellence in physics, Chile; active efforts to improve Am.-Chilean sci. exch. Patentee in field; contbr. articles to sci. publs. Ind. activist to stop waste contamination. Grantee Fundacion Andes, Santiago, 1990, NSF for Latin Am., 1993. Mem. Am. Phys. Soc., Chilean Phys. Soc. Roman Catholic. Avocations: swimming, photography, modeling. Office: Pontificia U Catolica Chile, Casilla 306, 22 Santiago Chile

CACOURIS, ELIAS MICHAEL, economist, consultant; b. Cairo, Oct. 25, 1929; s. Michael Dimitrios and Evangelia (Karaiskou) C.; m. Mary Maravelli, Nov. 24, 1957 (dec. Apr. 1983); children: Michael-Alkis, Theodore; m. Karin Ghar, May 31, 1985; stepchildren: Ariane, Irene. BA in Lit., U. London, 1957. Fin. officer U.S. Med. Rsch. Unit, Cairo, 1950-61; fin. asst., adminstrn. officer, chief pers. officer UN Operation in the Congo, Zaire, 1961-65; sr. pers. officer, chief geog. divsns., resident rep./coord. UN Devel. Programme, N.Y.C., Geneva and devel. countries, 1966-89; rep. sec. gen. UN; cons. missions Poland, Chad, Mali, Angola, other countries in Africa and Asia, Kyrghyzstan, 1990-93. Mem. Hellenic Assn. for the UN. Avocations: tennis, sailing, writing. Home: 8 Markou Botsari, 17455 Kalamaki Athens, Greece

CADUTO, RALPH, nuclear biochemist; b. Providence, Aug. 20, 1927; s. Ralph and Anna (Durante) C.; m. Esther Martone, Apr. 11, 1952; children: Linda, Michael, Nancy, Mary. BSc, U. R.I., 1952; postgrad., R. I. Coll., 1976—. Chief chemist C. V. Chapin Hosp., Providence, 1952-57; analytical chemist, supr., mgr. pharm. production CIBA-Geigy, Cranston, R.I., 1957-78; nuclear biochemist, supr. radio-immunoassay lab. Va. Med. Ctr., Providence, 1978—; presenter in field. Pres. PTA, Warwick, R.I., 1956-60, CIBA-Geigy Mgr. Assn., 1962-70, West Cranston (R.I.) Garden Club, 1964-68, West Cranston (R.I.) Square Dancer, 1977-80. Mem. Am. Assn. Chemists, Am. Soc. Chemists (sr. mem.), Am. Assn. Clin. Chemists, N.Y. Acad. Sci., Ligand Soc., Phi Sigma, Sigma Chi. Roman Catholic. Achievements include patents in field. Home: 69 Indian Run Trail Esmund RI 02917

CADWALDER, HUGH MAURICE, psychology educator; b. Mt. Ayer, Iowa, July 1, 1924; s. Hugh M. and Mary (Crouch) C.; m. Melba Atwood, May 22, 1944 (div. 1975); children: Mark M., Mindy M.; m. Dianna Renfro-Reeves, May 15, 1980. MA, Baylor U., Waco, Tex., 1955, PhD, 1962; DD, Houston Bible Coll., 1963; ArtsD (hon.), Inst. Applied Rsch., London, 1970. Cert. social worker, Tex., cert. chem. dependency specialist, Tex., cert. clin. assoc. Am. Bd. Med. Psychotherapists; ordained to ministry Baptist Ch., 1944. Instr. psychology Baylor U., Waco, Tex., 1955-62; acad. dean Southwestern Agrl. Coll., Waxahachie, Tex., 1962-64; sr. minister Christian Life Community Ch., Dallas, 1964-69, 1st Assembly of God Ch., Corpus Christi, Tex., 1969-74; chmn. psychology dept. San Jacinto Coll., Pasadena, Tex., 1974—; seminar dir. Cadwalder Behavioral Ctr., Houston, 1982—; staff psychologist Charter Hosp., Sugar Land, Tex., 1990—; seminar dir. Sharpstown Christian Singles, Houston, 1975-92; sec. 1st Colony Mcpl. Utility Dist. 6, Sugar Land, 1988—; staff mem. West Oaks Psychiat. Hosp.; host radio talk show KTEK. Author: Some Psychological Determinants Involved in Religious Attitudes, 1955, The Spiritual Dimension of Recovery From Addiction, 1988, Emotional Adhesions and Their Cure, 1989. Pres. bd. Community Ednl. TV Network, 1989-92; bd. dirs. Cadwalder Behavioral Ctr., Christendom Interdominational Ministerial Ch.; mem. Rep. Presdl. Task Force. Mem. NASW, Am. Assn. Coll. Profs., Am. Pub. Health Assn., Community Coll. Humanities Assn., Am. Assn. Retired Persons, Tex. Jr. Coll. Tchrs. Assn., Am. Assn. Christian Counselors, Am. Bd. Med. Psychotherapy, Am. Coun. on Alcoholism, Am. Assn. Coll. Profs., Community Coll. Humanities Assn., Harris County Sheriff's Deputies Assn. (cert. hon. membership 1985), Nat. Mental Health Assn., N.Y. Acad. Sci., Tex. Assn. Social Workers, Nat. Assns. Ind. Bus., Forum Club Houston, Phi Delta Kappa. Office: Cadwalder Counseling 7324 SW Freeway Ste 850 Houston TX 77074

CADY, ELWYN LOOMIS, JR., medicolegal consultant, educator; b. Ames, Iowa, Feb. 21, 1926; s. Elwyn Loomis Sr. and Annabel (Lacey) C.; m. Jane Carolyn Elliott, Jan. 27, 1964 (dec. Dec. 1989); children: James Anson, Kathryn Anne; stepchildren: Martin Norman Jensen III, Paul Elliott Jensen. JD, Tulane U., 1951; BS in Medicine, U. Mo., 1955. Bar: Mo. 1951, U.S. Supreme Ct., 1965. Sci. tchr. Vermillion (Kans.) Rural High Sch., 1948-49; pvt. practice Kansas City, St. Louis, Independence, Mo., 1951—; dir. law-medicine program U. Kansas City, 1951-56; asst. dir. Law-Sci. Inst. U. Tex., Austin, 1956-57, sec. Law-Sci. Acad. Am., 1956-57; of counsel Koenig & Dietz, St. Louis, 1959-74; gen. counsel Elliott Oil, Inc., Independence, 1966—; mem. com. on mgmt. Ea. Jackson County Planned Parenthood Clinics, Independence, 1970-75. Author: (book) Law and Contemporary Nursing, 1961, 1st. rev. edit., 1963; Author: (with others) Immediate Care of the Acutely Ill and Injured, 1974, Cardiac Arrest and Resuscitation, 1958, 3rd rev. edit., 1974, West's Federal Practice Manual, 1960, 3rd rev. edit., 1989, Gradwohl's Legal Medicine, 1954; book reviewer: sci. books and films. Legal counsel Friends of the Truman Campus (engring.), Independence, 1987—, Community Assn. for Arts, Independence, 1991—; charter mem. Friends of Nat. Frontier Trials Ctr., Independence, 1990—, Independence Historic Trails City Com., 1991—. With U.S. Army, 1944-45, ETO. Fellow Harry S. Truman Libr. Inst. for Nat. and Internat. Affairs (hon.); mem. AAAS (life), Nat. Geog. Soc. (life), Am. Legion (past comdr., judge adv., chaplain, chmn. state blood donor program, chmn. dist. oratorical contest), Phi Alpha Delta (life), Phi Beta Pi, Tau Kappa Epsilon. Home and Office: 1919 Drumm Ave Independence MO 64055-1836

CAFFARELLI, LUIS ANGEL, mathematician, educator; b. Buenos Aires, Argentina, Dec. 8, 1948; came to U.S., 1973; s. Luis and Hilda Delia (Cespi) C.; m. Irene Andrea Martinez-Gamba; children: Alejandro, Nicolas, Mauro. Masters degree, U. Buenos Aires, 1969, Ph.D., 1972. Postdoctoral asst., asst. prof. U. Buenos Aires, 1972-73; asst. prof. to prof. U. Minn., 1973-83; prof. math. U. Chgo., 1983-86; faculty Inst. for Advanced Study, Princeton, N.J., 1986—; prof. NYU, 1980-82. Contbr. articles to profl. jours. Recipient Stampacchia prize Scuola Normale de Pisa, 1983, Bocher prize Am. Math. Soc., 1984; Pius XI medal, 1988; Guggenheim grantee. Mem. NAS, AAAS, Accademia dei XL, Academia Argentina de Ciencias. Office: Inst Advanced Study Sch Mathematics Olden Ln Princeton NJ 08540-4920*

CAFFEE, MARCUS PAT, computer consulting executive; b. Tulsa, Feb. 23, 1948; s. Malcolm Wesley and Martha Marjorie (Deming) C.; m. Virginia Maureen Gladden, May 31, 1975; 1 child, Katheryn Elizabeth. Student, Tulsa U., 1965-66, Okla. State U., 1966-67, 77-78. Pres. Computer Sales & Svc., Tulsa, 1972-75; owner Data Mgmt. Systems, Tulsa and Houston, 1975-77; staff analyst Okla. State U., Stillwater, 1977-78; project leader Ranger Ins. Cos., Houston, 1979-80; group mgr. corp. and fin. svcs. Am. Gen. Life Ins., Houston, 1980-82; mgr. systems devel. U. Tex. Health Sci. Ctr., Houston, 1982-84; owner Marcus Caffee, Cons., Conroe, Tex., 1984-89; pres., chief exec. officer Emcee Systems, Inc., 1989-90; dir. ops. I.C. Svcs., St. Petersburg, Fla., 1989-91; pvt. practice computer consulting Largo, Fla., 1991—. Author: Time Scaled Real Time Simulators, U.S. Navy, 1970, Satelite Data Communication Criteria Between Mainframe Computer Sites, Am. General 1981, Evaluation, Justification and Purchasing Guidelines for MicroComputers and Word Processors, U. Texas Health Science Ctr. at Houston 1982; copyright computer operaing system IBOL, 1972, integrated bus. software Office Master!, 1988, 89; author, editor bus. newsletter Read.Me, 1986, 89, 91. Mem. Montgomery County Econ. Devel. Team, 1987; mem. adminstrv. bd. First United Meth. Ch., Conroe, 1987, instr. computer literacy, 1985-87. Served with USN, 1967-71. Mem. Airman's Aero Club, Rotary (Conroe chpt., guest speaker 1988). Republican. Avocations: chess (Okla. regional champion), sailing (Master's lic.), bridge, flying. Office: 13985 105th Terr N Largo FL 34644

CAFFEY, BENJAMIN FRANKLIN, civil engineer; b. Jacksonville, Fla., Nov. 18, 1927; s. Eugene Mead and Catherine (Howell) C.; m. Laura Marlowe, Oct. 2, 1949 (dec. Jan. 1991); children: Benjamin, John, Lochlin; m. Suzanne Morris, Aug. 10, 1991; stepchildren: Jay, Julie, Kelly. BCE, Ga. Inst. Tech., 1949; MSCE, U. So. Calif., L.A., 1964. Registered profl. engr., Calif.; cert. project mgmt. profl. Resident engr. Sch. Dist. of Glendale, Calif., 1950-51; constrn. engr. Fluor Corp., L.A., 1951-56; v.p. Petroleum Combustion & Engring. Co., L.A., 1956-58, Sesler & Caffey, Inc., Gardena, Calif., 1958-69, Fluor Corp., L.A., 1969-76; pres. Fluor Arabia, Ltd., Saudi Arabia, 1976-81; sr. v.p. Fluor Daniel, Inc., Irvine, Calif., 1981—; adv. bd. Ga. Tech. Sch. Civil Engring., Atlanta, 1992—. Lt. U.S. Army, 1946-48; lt. col. USAR (ret.). Fellow ASCE; mem. Project Mgmt. Inst. (trustee 1991), Soc. Am. Mil. Engrs., Chi Epsilon. Republican. Episcopalian. Office: Fluor Daniel Inc 3333 Michelson Dr Irvine CA 92730

CAFFEY, JAMES ENOCH, civil engineer; b. Rockdale, Tex., May 5, 1934; s. Enoch Arden and Leevicy Viola (Stephens) C.; m. Patricia Louise Latham, June 4, 1960; children: Jeffrey E., Jeanne Erin, Jerald E. BS in Civil Enging., Tex. A&M U., 1955, MS in Civil Engring., 1956; PhD, Colo. State U., 1965. Registered profl. engr., Tex. Prof. U. Tex., Arlington, 1959-74; dept. head Turner, Collier & Braden, Inc., Houston, 1974-76; assoc. Rady and Assocs., Inc., Ft. Worth, 1976-85; pres. Caffey Engring., Inc., Arlington, 1985—; asst. city engr. City of Arlington, 1991—; adv. bd. Cancer Rsch. Found. North Tex., Arlington, 1990-91. Capt. USAF, 1956-68. Fellow ASCE, Am. Water Resources Assn. (charter, bd. dirs. 1973-75); mem. NSPE, Am. Geophys. Union, Am. Inst. Hydrology (profl.), Kiwanis, Phi Eta Sigma, Tau Beta Pi, Phi Kappa Phi, Chi Epsilon. Baptist. Office: Caffey Engring Inc 1506 Wagonwheel Trl Arlington TX 76013-3140

CAFFIN, ROGER NEIL, research scientist, consultant; b. Canberra, ACT, Australia, July 30, 1945; s. Neil Rupert and Bessie Gwendolyn (Jones) C.; m. Roberta Susan Wilcox, Jan. 17, 1967; children: Fiona Susan, Elanor Emma. BS with honors, Melbourne (Australia) U., 1967, MS, 1968; PhD, City Univ., London, 1972; diploma, Inst. Corp. Mgrs., 1991. Exptl. scientist Commonwealth Sci. and Indsl. Rsch. Orgn. div. Wool Tech., Sydney, NSW, Australia, 1968-73, rsch. scientist, 1973-79, sr. rsch. scientist, 1979-89, prin. rsch. scientist, 1989—. Author: Hitch Hikers Guide to Digital, 1991; contbr. articles to profl. jours. Mem. IEEE, Assn. for Computing Machinery, Digital Equipment Computer Users Soc. (bd. dirs. 1983-90, Paul Raynor Meml. award 1991). Avocations: computing, bushwalking, mountain climbing, ski touring. Office: CSIRO Div Wool Tech, PO Box 7, Ryde New South Wales 2112, Australia

CAGAN, JONATHAN, mechanical engineering educator; b. Worcester, Mass., Sept. 25, 1961; s. George and Beverly (Feinberg) C. BS, U. Rochester, 1983, MS, 1985; PhD, Univ. Calif., 1990. Registered profl. engr., Calif., Pa. Cooperative intern Eastman Kodak Co., Rochester, N.Y., 1981-84, applied rsch. engr., 1984-86; rsch. asst. Univ. Calif., Berkeley, 1986-90; asst. prof. Carnegie Mellon Univ., Pitts., 1990—; program com. mem. AAAI Symposium on Design, 1992, internat. adv. bd., Second Internat. Round-Table Conf. on Model-Based Creativity, 1992. Contbr. articles to profl. jours. Recipient NYI Initiation award NSF. Mem. ASME, Am. Assn. Artificial Intell., Am. Soc. Engring. Edn., Tau Beta Pi, Phi, Beta Kappa, Sigma Xi. Office: Carnegie Mellon Univ Dept Mech Engring Pittsburgh PA 15213

CAGIN, TAHIR, physicist; b. Izmir, Turkey, Nov. 27, 1956; came to U.S. 1984; s. Mehmet and Bahriye (Teksoz) C.; m. Gul Yenici, Sept. 18, 1980; children: Elif, Kerem. BS, Middle East Tech. U., Ankara, Turkey, 1981, MSc in Physics, 1983; PhD in Physics, Clemson (S.C.) U., 1988. Rsch. assoc. chemistry U. Houston, 1988-89; rsch. scientist Systran Corp., Dayton, Ohio, 1989-90; sr. scientist, project mgr. Molecular Simulations, Inc., Pasadena, 1990—; vis. scientist Matls. Lab., Wright Paterson AFB, Dayton, 1989-90; vis. assoc. chemistry Calif. Tech., 1991—. Contbr. articles to profl. jours. Mem. AAAS, Am. Phys. Soc., Am. Chem. Soc., Matls. Rsch. Soc. Achievements include co-development of fluctuation expressions to obtain higher order elastic properties of materials from computer simulations; co-formulated new computer simulation methods for phase equilibria problems, new algorithms for simulating/ modeling icosahedral virions. Office: Molecular Simulations Inc 199 S Los Robles Ave Ste 540 Pasadena CA 91101-2458

CAGLE, THOMAS MARQUIS, electronics engineer; b. Chillicothe, Tex., Apr. 26, 1927; s. William Robert and B. Clyde (White) C.; m. Jane E. DeBute, May 16, 1964; children: Kent Mark, Thomas DeBute. BS, U. So. Calif., 1968. Engr. North Am. Rockwell, L.A., 1950-71; cons. Scottsdale, Ariz., 1971-77; electronics engr. Dept. of Def., L.A., 1977-89; engr. Butler Engring. Co., Orange, Calif., 1989-90, Rockwell Internat. Corp., Downey, Calif., 1990-92; cons. Santa Ana, Calif., 1992—. Contbr. numerous publs. to profl. jours. Pres., dir. Inglewood (Calif.) Jaycees; pres. Inglewood Youth Counseling YMCA. With USN, 1945-46. Home: 10461 Greenbrier Rd Santa Ana CA 92705

CAHAY, MARC MICHEL, electrical engineer, educator; b. Liege, Belgium, Oct. 11, 1959; came to U.S., 1983; s. Oscar and Mary (Bony) C. BS in Physics, U. Liege, 1981; MS in Physics, Purdue U., 1986, PhD in Elec. Engring., 1987. Rsch. scientist dept. nuclear physics U. Liege, 1981-82; substitute tchr. physics various high schs., Belgium, Germany, 1983; teaching asst. physics Purdue U., West Lafayette, Ind., 1983-84, rsch. asst. dept. elec. engring., 1984-87; rsch. scientist Sci. Rsch. Assocs., Glastonbury, Conn., 1987-89; asst. prof. elec. and computer engring. U. Cin., 1989—; presenter at profl. confs. Contbr. articles to profl. publs. Grantee NSF, 1991—. Mem. IEEE, Internat. Soc. Optical Engring., Am. Phys. Soc. (session chmn. 1991), Electro Chem. Soc. (membership chmn. 1992—), Sigma Xi, Eta Kappa Nu. Office: Univ Cin Elec Computer Engring Dept 832 Rhodes Hall Cincinnati OH 45221

CAHEN, DAVID, materials chemist; b. Vught, Holland, Aug. 14, 1947; arrived in Israel, 1966; s. Max and Henriette (Elion) C.; m. Geula-Pimentel-Meruk, Apr. 5, 1985; children: Anat Meruk, Maya Meruk, Yair, Yael. BS, Hebrew U., 1969; PhD, Northwestern U., 1973. Weizmann fellow Weizmann Inst. of Sci., Rehovot, Israel, 1973-74; Bogen fellow Hebrew U. of Jerusalem, Israel, 1974-75; scientist Weizmann Inst. of Sci., Rehovot, 1975-78, sr. scientist, 1978-93, assoc. prof., 1993—; mem. editorial adv. bd. Marcel Dekker Pub. Co., 1991—, Solar Cells, 1984-91, Applied Physics Communications, 1990—, Solar Energy Material Solar Cells, 1992—. Contbr. over 180 scientific publs. to profl. jours. Recipient E.D. Bergmann prize Weizmann Inst., 1980. Mem. IEEE (sr.), Materials Rsch. Soc., Am. Phys. Soc. Achievements include patents in materials, semiconductors and energy conversion. Office: Weizmann Inst of Sci, Rehovot 76100, Israel

CAHILL, LAWRENCE BERNARD, environmental engineer; b. Medford, Mass., Nov. 10, 1947; s. Alfred L. and Virginia (Goggin) C.; m. Claire J. Chaill, Nov. 15, 1980; children: Brendon, Bryan. BSME, Northeastern U., 1970; MS in Environ. Engring., Northwestern U., 1974; MPA, U. Pa., 1976. Project engr. Exxon Rsch. & Engring., Florham Park, N.J., 1970-73; prin. Booz, Allen & Hamilton, Inc., Washington, 1976-85; v.p. R.F. Weston, Inc., West Chester, Pa., 1985-87; COO Hart Environ. Mgmt. Corp., Phila., 1987-90; v.p. McLaren/Hart Environ. Engring. Corp., Phila., 1990—; conductor seminars in field; lectr. in field. Editor, prin. author: Environmental Audits, 2d edit. 1983, 6th edit. 1989; contbr. numerous articles to profl. jours. Environ. commr. City of Camden, N.J., 1976. Named Disting. Instr. Govt. Inst., Rockville, Md., 1990. Mem. Am. Mgmt. Assn., Environ. Auditing Roundtable, Pi Tau Sigma. Achievements include design of noise assessment computer program used worldwide by Exxon affiliates. Home: 326 Brookwood Dr Downingtown PA 19335 Office: McLaren/Hart Environ Engr 300 Stevens Dr Lester PA 19113

CAHILL, PAUL AUGUSTINE, chemistry researcher; b. Akron, Ohio, May 21, 1959; s. Walter H. and Geraldine (Crano) C.; m. Elizabeth Clair Lind, May 10, 1981; 1 child, Diana Nicole. BA, Rice U., 1981; PhD, U. Ill., 1986. Sr. mem. tech. staff Sandia Nat. Labs., Albuquerque, 1986—. Mem. Am. Chem. Soc., Materials Rsch. Soc. Office: Sandia Nat Labs Dept 1811 Albuquerque NM 87185

CAHILL, THOMAS ANDREW, physicist, educator; b. Paterson, N.J., Mar. 4, 1937; s. Thomas Vincent and Margery (Groesbeck) C.; m. Virginia Ann Arnoldy, June 26, 1965; children: Catherine Frances, Thomas Michael. B.A., Holy Cross Coll., Worcester, Mass., 1959; Ph.D. in Physics; NDEA fellow, U. Calif., Los Angeles, 1965. Asst. prof. in residence U. Calif., Los Angeles, 1965-66; NATO fellow, research physicist Centre d'Etudes Nucleaires de Saclay, France, 1966-67; prof. physics U. Calif., Davis, 1967—; acting dir. Crocker Nuclear Lab., 1972, dir., 1980-89; dir. Inst. of Ecology, 1972-75; cons. NRC of Can., Louvre Mus.; mem. Internat. Com. on PIXE and Its Application, Calif. Atty. Gen., Nat. Audubon Soc., Mono Lake Com. Author: (with J. McCray) Electronic Circuit Analysis for Scientists, 1973; editor Internat. Jour. Pixe, 1989—; contbr. articles to profl. jours. on physics, applied physics, hist. analyses and air pollution. Prin. investigator IMPROVE Nat. Air Pollution Network.; co-dir. Crocker Hist. and Archeol. Projects.; mem. internat. com. Ion Beam Analysis. OAS fellow, 1968. Mem. Am. Phys. Soc., Air Pollution Control Assn., Am. Chem. Soc., Am. Assn. Aerosol Rsch., Sigma Xi. Democrat. Roman Catholic. Club: Sierra. Home: 1813 Amador Ave Davis CA 95616-3104 Office: U Calif Dept Physics Crocker Nuclear Laboratories Davis CA 95616

CAHN, GLENN EVAN, psychologist; b. Washington, Jan. 10, 1953; s. Julius Norman and Ann (Foote) C.; m. Emily Zofnass, Sept. 2, 1990 (dec. June 1993). BA, Washington U., St. Louis, 1975; MA, Calif. Sch. Profl. Psychology, San Diego, 1978, PhD, 1980. Psychologist The Arbour, Boston, 1980-82, Stoughton Counseling Ctr., Mass., 1982, Mass. Rehab., Boston, 1982—, Fuller Meml. Hosp., S. Attleboro, Mass., 1985-90, Hurst Assocs., Boston, 1990—; pvt. practice psychology Sharon, Mass., 1988—. Contbr. articles to profl. jours. Avocations: gardening, woodworking. Home: 373 Massapoag Ave Sharon MA 02067-2716

CAHN, JOHN WERNER, metallurgist, educator; b. Germany, Jan. 9, 1928; came to U.S., 1939, naturalized, 1945; s. Felix H. and Lucie (Schwarz) C.; m. Anne Hessing, Aug. 20, 1950; children: Martin Charles, Andrew Blender, Lorie Selma. BS, U. Mich., 1949; PhD, U. Calif. at Berkeley, 1953; DSc (hon.), Northwestern U., 1990. Instr. U. Chgo., 1952-54; with research lab. Gen. Electric Co., 1954-64; prof. metallurgy MIT, 1964-78; ctr. scientist Nat. Inst. Standards and Tech. (formerly Nat. Bur. Standards), 1978—, sr. fellow, 1984—; vis. prof. Israeli Inst. Tech., Haifa, 1971-72, 80; cons. in field, 1963—; chmn. Gordon conf. Phys. Metallurgy, 1964; vis. scientist Nat. Bur. Standards, Gaithersburg, Md., 1977; affiliate prof. physics U. Wash., Seattle, 1984—. Rsch. and articles on surfaces and interfaces, thermodynamics, phase changes, quasicrystals. Guggenheim fellow, 1960; research fellow Japan Soc. for Promotion of Sci., 1981-82; recipient Dickson prize Carnegie Mellon U., 1981, Gold medal U.S. Dept. Commerce, 1982, Von Hippel award Materials Research Soc., 1985, Stratton award Nat. Bur. Standards, 1986, Michelson-Morley prize Case Western Res. U., 1991, Hume-Rothery award, 1993, William Hume-Rothery award Minerals, Metals and Materials Soc., 1993. Fellow Am. Acad. Arts and Scis., Am. Inst. Metall. Engrs., Am. Soc. Metals Internat. (Saveur award 1989); mem. NAS. Home: 6610 Pyle Rd Bethesda MD 20817-5454 Office: Nat Inst Standards and Tech Gaithersburg MD 20899

CAHN, ROBERT NATHAN, physicist; b. N.Y.C., Dec. 20, 1944; s. Alan L. and Beatrice (Geballe) C.; m. Frances C. Miller, Aug. 22, 1965; children: Deborah, Sarah. BA, Harvard U., 1966; PhD, U. Calif., Berkeley, 1972. Rsch. assoc. Stanford (Calif.) Linear Accelerator Ctr., 1972-73; rsch. asst. prof. U. Wash., Seattle, 1973-76; asst. prof. U. Mich., Ann Arbor, 1976-78; assoc. rsch. prof. U. Calif., Davis, 1978-79; sr. staff physicist Lawrence Berkeley Lab., 1979-91, div. dir., 1991—. Author: Semi Simple Lie Algebras and Their Representations, 1984; co-author: Experimental Foundations of Particle Physics, 1989. Fellow Am. Phys. Soc. (sec.-treas. div. particles and fields 1992—).

CAI, ZHENGWEI, biomedical researcher; b. Shanghai, China, Aug. 19, 1946; came to U.S., 1985; s. Chifu and Ruying (Wang) C.; m. Yuguang Song, Jan. 7, 1975; 1 child, Cheng. MS, Fudan U., Shanghai, 1981; PhD, Oreg. State U., 1988. Lectr. Shanghai Fisheries U., 1982-85; rsch. asst. Oreg. State U., Corvallis, 1987-88; rsch. assoc. dept. pharmacology and toxicology U. Miss. Med. Ctr., Jackson, 1988-92, sr. rsch. assoc. dept. pediatrics and newborn, 1992—. Contbr. articles to Toxicology Applied Pharmacology, Brian Rsch., European Jour. Pharmacology, Biochem. Pharmacology, Pediatric Rsch. Recipient Young Scientist Travel award Am. Soc. for Pharmacology and Exptl. Therapeutics, 1992. Mem. Soc. Toxicology (assoc.), Am. Fisheries Soc., Sigma Xi. Office: U Miss Med Ctr Dept Pediatrics/Newborn 2500 N State St Jackson MS 39216

CAILLOUET, CHARLES WAX, JR., fisheries scientist; b. Baton Rouge, Dec. 15, 1937; s. Charles Wax and Elida Pulcherie (Millet) C.; m. Jeanne Mae Engels, Aug. 1959 (div. 1976); children: Suzanne Renee, Theresa Elaine, Michelle Marie, Charles Christopher; m. Nancy Louise Laird, July 30, 1977. BS in Forestry, La. State U., 1959, MS in Game Mgmt., 1960; PhD in Fishery Biology, Iowa State U., 1964. Asst. prof. biology U. Southwestern La., Lafayette, 1964-67; assoc. prof. fisheries and applied estuarine ecology Rosenstiel Sch. Marine and Atmospheric Scis., U. Miami, Fla., 1967-72; supervisory fishery biologist Nat. Marine Fisheries Svc., Galveston (Tex.) Lab., 1972—; mem. Tex. Coastal Clean-up Steering Com., Austin, 1986-89. Co-editor symposium procs.; contbr. articles to numerous sci. jours. Mem. Baton Rouge Geneal. and Hist. Soc., United We Stand America. Fellow Am. Inst. Fishery Rsch. Biologists; mem. AAAS, Tex. Orgn. Endangered Species, Sigma Xi. Republican. Office: Nat Marine Fisheries Svc Galveston Lab 4700 Ave U Galveston TX 77551-5997

CAIN, B(URTON) EDWARD, chemistry educator; b. Batavia, N.Y., Sept. 11, 1942; s. Burton Leo and Bettie S. (Williams) C. BA, SUNY, Binghamton, 1964; PhD, Syracuse U., 1971. Biochemist Onondaga County (N.Y.) Pub. Health Labs., Syracuse, N.Y., 1971-72; chemist O'Brien & Gere Cons. Engrs., Inc., Syracuse, N.Y., 1972-74; asst. prof. chemistry Nat. Tech. Inst. Deaf, Rochester (N.Y.) Inst. Tech., N.Y., 1974-80, assoc. prof. dept. chemistry, 1980-84, prof., 1984—; asst. chemistry dept. head Rochester Inst. Tech., 1988—; reader Advanced Placement chemistry exams. Ednl. Testing Svc., June 1987, 88, 89, 90, 91, 92. Author: The Basics of Technical Communicating, 1988; contbr. articles to profl. jours. Reviewer grant proposals coll. sci. instrument program NSF, 1987, instrumentation and lab. improvement program NSF, 1992. Recipient Eisenhart Outstanding Tchr. award, 1980. Mem. Am. Chem. Soc., AAAS, AAUP, Nat. Sci. Tchrs. Assn., Nat. Assn. Deaf, Conf. Am. Instrs. for Deaf, Registry of Interpreters for Deaf, Sigma Xi, Phi Lambda Upsilon, Gamma Epsilon Tau (Tchr. of Yr. award 1983). Home: PO Box 40257 Rochester NY 14604-0257 Office: 1 Lomb Memorial Dr Rochester NY 14623-0887

CAIN, MICHAEL DEAN, research forester; b. Pascagoula, Miss., Nov. 9, 1946; s. Thomas R. and Bennie (Gleghorn) C. AS, Perkinston (Miss.) Jr. Coll., 1966; BS, Miss. State U., 1969; MS, La. State U., 1973. Registered forester, Miss. Rsch. forester So. Forest Experiment Sta., Pineville, La., 1975-78, Crossett, Ark., 1978-87, Monticello, Ark., 1987—. Contbr. articles to Forest Sci., So. Jour. Applied Forestry, Forest Ecology and Mgmt., Internat. Jour. Wildland Fire, New Forests, Proceedings of the So. Weed Sci. Soc., Univ. rsch. publs., USDA Forest Svc. rsch. publs. With U.S. Army, 1969-71. Mem. Soc. Am. Foresters, Weed Sci. Soc. Am., So. Weed Sci. Soc., Ecol. Soc. Am., Internat. Assn. Wildland Fire, Soc. for Conservation Biology. Office: So Forest Experiment Sta PO Box 3516 Monticello AR 71655-3516

CAIN, STEVEN LYLE, aerospace engineer; b. Landstuhl, Germany, Feb. 10, 1960; s. Vern Leon and Estiline (Robertson) C.; m. Lisa Annette Dunn, Nov. 10, 1990; 1 child, Aaron Benjamin. BS in Aero. Engring., U. Colo., 1984. Engring. draftsman Nat. Bur. Standards, Boulder, Colo., 1983-84; aerospace engr. F100 Engine Maintenance, Kelly AFB, Tex., 1984-88, aerosapce engr. fgn. mil. sales, 1988-93; program mgr. Logistics Studies, Gunter AFB, Ala., 1993—. Mem. Gideons Internat. Mem. Assembly of God. Home: 1200 Hilmar Ct Montgomery AL 36117

CAINE, STEPHEN HOWARD, data processing executive; b. Washington, Feb. 11, 1941; s. Walter E. and Jeanette (Wenborne) C. Student Calif. Inst. Tech., 1958-62. Sr. programmer Calif. Inst. Tech., Pasadena, 1962-65, mgr. systems programming, 1965-69, mgr. programming, 1969-70; pres. Caine, Farber & Gordon, Inc., Pasadena, 1970—; lectr. applied sci. Calif. Inst. Tech., Pasadena, 1965-71, vis. assoc. elec. engring., 1976, vis. assoc. computer sci., 1976-84. Dir. San Gabriel Valley Learning Ctrs., 1992—. Mem. Pasadena Tournament of Roses Assn., 1976—. Mem. AAAS, Nat. Assn. Corrosion Engrs., Am. Ordnance Assn., Assn. Computing Machinery, Athanaeum Club (Pasadena), Houston Club. Home: 77 Patrician Way Pasadena CA 91105-1039

CAIRNS, JOHN, JR., environmental science educator, researcher; b. Conshohocken, Pa., May 8, 1923; s. John and Eunice S. (Fesmire) C.; m. Jean Ogden, Aug. 5, 1944; children: Karen Jean, Stefan Hugh, Duncan Jay, Heather. A.B., Swarthmore Coll., 1947; M.S., U. Pa., 1949, Ph.D., 1953. Curator limnology Acad. Natural Scis., Phila., 1948-66; prof. zoology U. Kans., Lawrence, 1966-68; univ. disting. prof. Va. Poly. Inst. and State U., Blacksburg, 1968—, dir. Univ. Ctr. Environ. and Hazardous Materials Studies. Author: Testing for Effects of Chemicals on Ecosystems, 1981, Artificial Substrates, 1982, Biological Monitoring, 1982, Modeling the Fate of Chemicals in the Aquatic Environment, 1982, Multispecies Toxicant Testing, 1985, EcoAccidents, 1985, Community Toxicity Testing, 1986, Environmental Regeneration II: Managing Water Resources, 1986, Rehabilitating Damaged Ecosystems, 1988, Functional Testing for Hazard Estimation, 1989, Integrated Environmental Management, 1990, Estimating Ecosystem Risk, 1992, Restoring Aquatic Ecosystems, 1992; contbr. chpts. to books, articles to profl. jours. Served with USN, 1942-46. Recipient Presdl. Commendation, 1971, Dudley award for Outstanding Publ. ASTM, 1978, Superior Achievement award U.S. EPA, 1980, Founders award Soc. Environ. Toxicology and Chemistry, 1981, Morrison medal for Outstanding Accomplishments in the Environ. Scis., 1984, Environ. Program medal UN, 1988, Excellence award Am. Fisheries Soc., 1989, Life Achievement award in Sci. Commonwealth Va. and Sci. Mus. Va., 1991. Fellow AAAS, Am. Acad. Arts and Scis., Linnean Soc. of London; mem. NAS, Am. Microscopiacal Soc. (pres. 1980), Am. Water Resources Assn. (editorial bd. 1975-81, Icko Iben award 1984), Inst. Ecology (founder), Acad. Natural Scis. (rsch. assoc.). Unitarian. Office: Va Poly Inst and State Univ Ctr Environ and Hazardous 1020 Derring Hall Blacksburg VA 24061-0415

CAIRNS, THEODORE LESUEUR, chemist; b. Edmonton, Alta., Can., July 20, 1914; came to U.S., 1936, naturalized, 1945; s. Albert William and Theodora (MacNaughton) C.; m. Margaret Jean McDonald, Aug. 17, 1940; children: John Albert, Margaret Eleanor (Mrs. William L. Etter), Elizabeth Theodora (Mrs. Ernest I. Reveal III), James Richard. B.S., U. Alta., 1936, LL.D., 1970; Ph.D., U. Ill., 1939. Instr. organic chemistry U. Rochester, 1939-41; research chemist central research dept. E.I. duPont de Nemours & Co., Wilmington, Del., 1941-45; research supr. E.I. duPont de Nemours & Co., 1945-51, lab. dir., 1951-63, dir. basic scis., 1963-66, dir. research, 1966-67, asst. dir. central research and devel. dept., 1967-71, dir., 1971-79; Regents prof. UCLA, 1965-66; mem. adv. bd. Organic Syntheses, from 1958; mem. Pres.'s Sci. Adv. Com., 1970-73, Pres.'s Com. Nat. Medal Sci., 1974-75; chmn. Office of Chemistry and Chem. Tech., NRC, 1979-81. Editorial bd.: Organic Reactions, from 1959, Jour. Organic Chemistry, 1965-69. Recipient award for creative work in synthetic organic chemistry Am. Chem. Soc., 1968; Perkin medal, 1973; Cresson medal Franklin Inst., 1974. Mem. Nat. Acad. Scis., Am. Chem. Soc. (chmn. organic div. 1964-65), AAAS, Sigma Xi, Phi Lambda Upsilon, Alpha Chi Sigma, Phi Lambda Upsilon (hon.).

CAIRO, JIMMY MICHAEL, physiologist; b. New Orleans, Jan. 8, 1952; s. John August and Mary Evelyn (Peterson) C.; m. Rhonda Lynn Philmon, Jan. 24, 1986; children: Brooke Cambre, Allyson Marie. BS, U. Southwestern La., 1974; MS, Tulane U., 1977; PhD, La. State U., 1986. Instr. physiology La. State U., New Orleans, 1978-85, asst. prof., 1985-88, assoc. prof., 1988—, head dept. cardiopulmonary sci., 1989—; adj. assoc. prof. Tulane U., New Orleans, 1991—; cons. pulmonary Children's Hosp., New Orleans, 1988—. Author: Introduction to Respiratory Care, 1990; contbr. articles to profl. jours. Mem. Orgn. for Establishment of Sci. Magnate Sch. for New Orleans, 1991—. Mem. AAAS, Am. Thoracic Soc., Am. Assn. Respiratory Care, N.Y. Acad. Sci. Office: La State U Med Ctr 1900 Gravier St New Orleans LA 70112-2262

CALABI, EUGENIO, mathematician, educator; b. Milan, May 11, 1923; Naturalized, U.S.; married, 1952; 1 child. BS, MIT, 1946; AM, U. Ill., 1947; PhD in Math., Princeton U., 1950. Asst. and instr. Princeton (N.J.) U., 1947-51; asst. prof. math. La. State U., 1951-55; from asst. prof. to prof. U. Minn., 1955-64, prof., 1964-69, chmn. dept., 1973-76; Thomas A. Scott Prof. math. U. Pa., Phila., 1969—; With Inst. Advanced Study, 1958-59, fellow, 1962-63. Mem. NAS, Am. Math. Soc. (Leroy P. Steele prizes 1991). Achievements include research on differential geometry of complex manifolds. Office: U Pa Dept Math 209 S 3d-33 St Philadelphia PA 19104*

CALABRO, JOSEPH JOHN, III, physician; b. Carbondale, Pa., Sept. 4, 1955; s. Joseph J. and Judith A. (Fidati) C.; m. Anne Wroblewski Calabro, Jan 25, 1985; children: Lia Jude, J. John. IV. Secondary cert., Scranton Prep. Sch., 1973; BS in Biology cum laude, U. Scranton, 1977; DO, Phila. Coll. Osteo. Medicine, 1981. Commd. 2d lt. U.S. Army, 1977, advanced through grades to maj., 1989; intern Tripler Army Med. Ctr., Honolulu, 1981-82; resident in emergency medicine Madigan Army Med. ctr., Tacoma, Wash., 1984-86; chief dept. ambulatory care and emergency med. svcs. Letterman Army Med. Ctr., San Francisco, 1986-90; asst. clin. prof. Sch. Nursing U. Calif., San Francisco, 1987-89; attending physician San Francisco Gen. Hosp., 1987-91; asst. clin. prof. U. Calif. Sch. Medicine, San Francisco, 1987—; chmn. Dept. Emergency Medicine Jersey Shore Med. Ctr., Neptune, N.J., 1990-92; chmn. dept. emergency medicine Beth Israel Med. Ctr., Newark, 1992—; residency dir. emergency medicine Beth Israel Med. Ctr., San Francisco, 1992—; asst. clin. prof. U. Medicine and Dentistry of N.J., 1991—; chmn. San Francisco City and County Emergency Med. Care Com., 1988-90; chmn. emergency med. care com. San Francisco chpt. Am. Heart Assn., 1989-90. Reviewer Jour. AMA, Jour. EMS, Annals of Emergency Medicine, Rescue. Mem. AMA, Am. Coll. Emergency Physicians (N.J. chpt. 1990—, bd. dirs., chmn. EMS/trauma com., chmn. emergency medicine residency com.), nat. ACEP, EMS com. 1988—, Am. Coll. Osteo. Emergency Physicians (nat. com. chmn., com. mem.), Am. Osteo. Assn., Assn. Mil. Osteo. Physicians and Surgeons, Assn. Mil Surgeons of U.S., Soc. Acad. Emergency Medicine, Nat. Assn. EMS Physicians, Phi Lambda Upsilon. Roman Catholic. Home: 18 Spier Ave Allenhurst NJ 07711

CALDERON, ALBERTO P., mathematician, educator; b. Mendoza, Argentina, Sept. 14, 1920; m. Mabel E. Molinelli Wells, 1950 (dec. 1985); m. Alexandra Bagdasar, 1989; 2 children. Grad. in civil engring., U. Buenos Aires, 1947; PhD in Math., U. Chgo., 1950; D. (hon.), U. Buenos Aires, 1969, Technion, Israel, 1989. Assoc. prof. Ohio State U., 1950-53; vis. mem. Inst. Advanced Study, Princeton, N.J., 1953-55; assoc. prof. to prof. math. MIT, 1955-75; prof. math. U. Chgo., 1959-68, Louis Block prof., 1968-72, Univ. prof., 1975-85, prof. emeritus, 1985—, chmn. dept. math., 1970-72; hon. prof. U. Buenos Aires, from 1975. Recipient Bocher prize, 1978, Wolf Found. prize in math., 1989, Steele prize, 1989, Consagracion Nacional prize Arg., 1989, Nat. Medal of Sci. NSF, 1991. Fellow Am. Acad. Arts and Scis.; mem. NAS, Acad. Sci. Latin Am., Acad. Nacional Ciencias Exactas, Arg., Third World Acad. Sci., Royal Acad. of Scis. (Spain), French Acad. of Sci. Office: U Chgo Dept Math 5801 S Ellis Ave Chicago IL 60637

CALDERON, EDUARDO, general practice physician, researcher; b. Tacna, Peru, Mar. 19, 1953; s. Roberto and Flora (Lade) C.; m. Martha I. Arrieta, Aug. 31, 1984; children: Lucia, Diego. MD, San Agustin Nat. Univ., Arequipa, Peru, 1987. Resident internal medicine British Am. Hosp., Lima, Peru, 1986-88; visiting fellow Univ. South Fla. Coll. Medicine, Tampa, 1988-90, rsch. assoc., 1990-92; asst. prof. allergy & immunology div. Univ. South Fla., Tampa, 1992—. Editor, founder Actualidades en Medicina; contbr. articles to profl. jours. Founder Task Force Against Tuberculosis of Arequipa, 1987, Edmundo Escomel Scientific Assn., 1981. Mem. ACP, Am. Acad. Allergy and Immunology, Peruvian Coll. Physicians. Home: 14623 Pine Glen Cir Lutz FL 33549 Office: USF Coll of Medicine Box 19 12901 Bruce B Downs Blvd Tampa FL 33612-4799

CALDERON, NISSIM, tire and rubber company executive; b. Jerusalem, Apr. 1, 1933; s. Jacob and Rina (Behar) C.; m. Rivka Rapoport, July 26, 1961; children: Rina, Meir. MS in Organic Chemistry, Hebrew U., 1958; PhD in Polymer Chemistry, U. Akron, 1961, postgrad., 1962. Sr. research chemist The Goodyear Tire & Rubber Co., Akron, Ohio, 1962-67, sect. head, 1967-75, mgr., 1975-86, v.p. research, 1986—; mem. adv. bd. Consortium on Polymer Simulation. Numerous patents in field; contbr. articles to profl. jours. Mem. Am. Chem. Soc. Office: Goodyear Tire & Rubber Co Tech Ctr PO Box 3531 Akron OH 44309-3531*

CALDERONE, MARLENE ELIZABETH, toxicology technician; b. DuBois, Pa., Apr. 10, 1940; d. James Joseph and Elizabeth Madge (Marando) C. Student, Canisius Coll., 1960. Rsch. chemist SUNY, Buffalo, 1960-64, UCLA, 1965-67; toxicology technician Damon Reference Labs., Newbury Park, Calif., 1967—; lectr. in toxicology Calif. Luth. U., Thousand Oaks, 1985-86. Contbr. abstracts to profl. jours., presenter papers to sci. meetings. Mem. Am. Assn. Clin. Chemistry, Calif. Assn. Toxicologists. Office: Damon Reference Lab 1011 Rancho Conejo Blvd Newbury Park CA 91320

CALDERWOOD, WILLIAM ARTHUR, physician; b. Wichita, Kans., Feb. 3, 1941; s. Ralph Bailey and Janet Denise (Christ) C.; m. Nancy Jo Crawford, Mar. 31, 1979; children: Lisa Beth, William Arthur II. MD, U. Kans., 1968. Diplomate Am. Bd. Family Practice. Intern Wesley Med. Ctr., Wichita, 1968-69; gen. practice family medicine Salina, Kans., 1972-80, Peoria, Ariz., 1980—; pres. staff St. John's Hosp., Salina, 1976; 28th jud. dist. coroner State of Kans., Salina, 1973-80; clin. instr. U. Kans., Wichita, 1978-80; cons. in addiction medicine VA Hosp., 1989—. Inventor, patentee lighter-than-air-furniture. Lt., M.C., USN, 1969-70. Fellow Am. Acad. Family Physicians; mem. AMA, Ariz. Med. Soc. (physicians med. health com., exec. com. 1988-92), Maricopa County Med. Soc., Ariz. Acad. Family Practice (med. dir. N.W. Orgn. Vol. alternatives 1988-91), Am. Med. Soc. on Alcoholism and Other Drug Dependencies (cert.), Shriners. Home: 7015 W Calavar Peoria AZ 85381 Office: 14300 W Granite Valley Dr Sun City West AZ 85375-5783

CALDWELL, ALLAN BLAIR, health services company executive; b. Independence, Iowa, June 13, 1929; s. Thomas James and Lola (Ensminger) C.; B.A., Maryville Coll., 1952; B.S., N.Y.U., 1955; M.S., Columbia U., 1957; M.D., Stanford U., 1964; m. Elizabeth Jane Steinmetz, June 13, 1955; 1 child, Kim Allistair; m. Susan A. Koss, Feb. 12, 1984. Med. intern Henry Ford Hosp., Detroit, 1964-65; resident Jackson Meml. Hosp., Miami, Fla., 1956-57; admstr. Albert Schweitzer Hosp., Haiti, 1957-58; asst. admstr. Palo Alto-Stanford Hosp. Center, 1958-59; asso. dir. program in hosp. adminstrn. U. Calif. at Los Angeles, 1965-67; dir. bur. profl. service Am. Hosp. Assn., Chgo., 1967-69; v.p. Beverly Enterprises, Pasadena, Calif., 1969-71; exec. v.p., med. dir. Nat. Med. Enterprises, Beverly Hills, Calif., 1971-73; pres., chmn. bd. Emergency Physicians Internat., 1973—, Allan B. Caldwell, M.D., Inc., 1973—; dir. indsl. medicine Greater El Monte Community Hosp., South El Monte, Calif., 1973-83; pres. Am. Indsl. Med. Services, 1978—; chmn. bd. dirs. Technicraft Internat., Inc., San Mateo, Calif., 1970-76; pres., med. dir. Shelton-Livingston Med. Group, 1984—; dir. Career Aids, Inc., Glendale, Calif., 1969-75; cons. TRW Corp., Redondo Beach, Calif., 1966-71; lectr. UCLA, 1965—, Calif. Inst. Tech., 1971—, Calif. State U., Northridge, 1980-85; examiner Civil Service Commn., Los Angeles, 1966; adviser Western Center for Continuing Edn. in Hosps. and Related Health Facilities, 1965—; cons. Los Angeles Hosp. and Nursing and Pub. Health Dept., 1965—; adv. council Calif. Hosp. Commn., 1972-78; commr. Emergency Med. Services Commn. Bd. dirs. Comprehensive Health Planning Assn. Los Angeles County, 1972-76; vice chmn. Emergency Med. Care Commn., Los Angeles County, 1977-78, chmn., 1978-79. Recipient Geri award Los Angeles Nursing Home Assn., 1966, Outstanding Achievement award Health Care Educators, 1978; USPHS scholar, 1961-63. Diplomate Am. Bd. Med. Examiners. Mem. Am., United (pres. 1971-72) hosp. assns., Am. Coll. Hosp. Adminstrs., Am., Calif. med. assns., Los Angeles County Med. Assn., Am. Coll. Emergency Physicians (v.p. 1975-77 dir. continuing med. edn. for Western U.S., Hawaii, Australia, N.Z., 1976-84), Hosp. Fin. Mgmt. Assn., Am. Indsl. Hygiene Assn., Rolls Royce Owners' Club (dir. 1982—, vice chmn. 1982, chmn. 1983), Classic Car Club of Am. (life 1983). Home: 4405 Medley Pl Encino CA 91316-4344 Office: 1414 S Grand Ave Ste 123 Los Angeles CA 90015

CALDWELL, ANDREW BRIAN, quality control engineer; b. Salt Lake City, June 4, 1958; s. Calvin and Virginia C.; m. Doreen, June 25, 1987; children: Leslie, Leann, Elizabeth, Evan. BA in Design, Calif. Poly., 1985; MS in Quality Engring., Kennedy Western, Boise, Idaho, 1992. V.p. ops. Aubrecht Duran Optical Systems, LaVerne, Calif., 1981-87; lead engr. Northrop Electro-Mech. Div., Anaheim, Calif., 1986-88; sr. engr./quality group mgr. Pent div. Group Dekko Inc., Kendallville, Ind., 1989—; mem. ind. adv. group Underwriters Labs., Northbrook, Ill., 1991—. Mem. Am. Soc. Quality Control. Achievements include patent for new design of fluorescent lampholder. Office: Pent div Group Dekko Inc Box 246 6928 N 400 E Kendallville IN 46755

CALDWELL, BILLY RAY, geologist; b. Newellton, La., Apr. 20, 1932; s. Leslie Richardson and Helen Merle (Clark) C.; m. Carolyn Marie Heath; children: Caryn, Jeana, Craig. BA, Tex. Christian U., 1954, MA, 1970. Cert. petroleum geologist, cert. profl. geologist. Geologist, Geol. Engring. Svc. Co., Ft. Worth, Tex., 1954-60; sci. tchr. Ft. Worth and Lake Worth Sch. Dists., 1960-63; mgr. Outdoor Living, 1963-71; instr. geology Tarrant County Jr. Coll., Ft. Worth, 1971—; petroleum and environ. geologist cons., Ft. Worth, 1971—. Bd. dirs. Ft. Worth and Tarrant County Homebuilders Assn., 1973. Named Dir. of Yr., Ft. Worth Jaycees, 1966-67. Mem. Am. Inst. Profl. Geologists, Am. Assn. Petroleum Geologists, Geol. Soc. Am., Ft. Worth Geol. Soc., Soc. Profl. Well Log Analysts, Ft. Worth C. of C. (environ. com). Republican. Baptist. Avocations: traveling, ch. work. Home: 305 Bodart Ln Fort Worth TX 76108-2032 Office: Ste 104 101 Jim Wright Fwy S Fort Worth TX 76108-2032

CALDWELL, CURTIS IRVIN, acoustical engineer; b. Columbus, Ohio, Mar. 4, 1947; s. Elmer Irvin and May Alice (Wing) C.; m. Susan Marion Belcher, Dec. 22, 1972; 1 child, Joshua Benjamin Lee. BS, U. S.C., 1972; MA, U. North Fla., 1983; postgrad., Pa. State U., 1984-93. Sr. analyst Analysis and Tech., Inc., North Stonington, Conn., 1973-81, cons., 1981-84; instr. U. North Fla., Jacksonville, 1984; grad. rsch. asst. Pa. State U., State College, 1984-91; v.p., editor Pa. State U. Engring. Grad. Student Coun.; mem. NOISECON 87 com. Inst. Noise Control Engring.; mem. East Coast Multi-REDCOM Tech. Tng. Meeting, 1993; chair Session Info. Mgmt. Issues; lectr. on stats. of complex variables with applications to sonar signal processing. Vol. Clinger for Congress campaign, State College, 1988, 90, 92; vol. in Clinger's Thornburg for Senate campaign, State College, 1991. With USN, 1968-70, 73. Mem. Acoustical Soc. Am. (conf. com. 1989-90, sec. local chpt. 1985-86), Inst. Math. Statistics, Am. Math. Soc., Am. Statis. Assn., Math. Assn. Am., Sons of Am. Revolution. Republican. Episcopalian. Achievements include research in development of general multivariate exponential family maximum likelihood detector as an estimator-subtractor, adaptive beamforming, multivariate statistics of complex variables, trigonometry of complex matrices and zonal polynomials of two complex matrix arguments. Home: 2030 Highland Dr State College PA 16803-1309

CALDWELL, DOUGLAS RAY, oceanographer, educator; b. Lansing, Mich., Feb. 16, 1936; s. Ray Thornton and Pearl Elizabeth (Brown) C.; m. Joan Hannauer, Sept. 9, 1961; children: Michael G., Elizabeth C., Katherine L. AB, U. Chgo., 1955, BS, 1957, MS, 1958, PhD, 1964. Research assoc. U. Chgo., 1963; post doctoral study Cambridge U., Eng., 1963-64; research asst. Inst. Geophysics, La Jolla, Calif., 1964-68; asst. prof. Oreg. State U., Corvallis, 1968-73, assoc. prof., 1973-78, prof., 1978—, assoc. dean, 1983-84, dean, 1984—. Office: Oreg State U Coll of Oceanography Oceanography Admin Bldg 104 Corvallis OR 97331

CALDWELL, ELWOOD FLEMING, food science educator, researcher, editor; b. Gladstone, Man., Can., Apr. 3, 1923; s. Charles Fleming and Frances Marion (Ridd) C.; m. Irene Margaret Sebille, June 13, 1949; children: John Fleming, Keith Allan; m. Florence Annette Zar, June 23, 1979. BS, U. Man., 1943; MA in Food Chemistry, U. Toronto, 1949, PhD in Nutrition, 1953; MBA, U. Chgo., 1956. Chemist Lake of the Woods Milling Co., Can., 1943-47; research chemist Can. Breweries Ltd., Toronto, Ont., 1948-49; chief chemist Christie, Brown & Co. div. Nabisco, Toronto, 1949-51; research assoc. in nutrition U. Toronto, 1951-53; with Quaker Oats Co., Barrington, Ill., 1953-72, dir. research and devel., 1969-72; prof., head dept. food sci. and nutrition U. Minn., St. Paul, 1972-86, exec. assoc. to dean Coll. Agr., 1986-88; dir. sci. svcs. Am. Assn. Cereal Chemists, 1988—, exec. editor Cereal Foods World, 1986-91; chmn. bd. Dairy Quality Control Inst., Inc., St. Paul, 1972-88, R. & D. Assocs. for Mil. Food & Packaging, Inc., San Antonio, 1970-71; chmn. evening program in food sci. Ill. Inst. Tech., Chgo., 1965-69. Contbr. articles to sci. jours. Chmn. North Barrington (Ill.) Bd. Appeals, 1966-69, mayor, 1969-72; vice-chmn. Barrington Area Council Govts., 1972; bd. dirs. Family Guidance Barrington, 1971-72. Recipient cert. of appreciation for civilian service U.S. Army Materiel Command, 1970. Fellow Am. Assn. Cereal Chemists, Inst. Food Technologists (Chmn.'s Svc. award Chgo. sect. 1975, Chmn.'s award Minn. sect. 1988, Calvert L. Willey Disting. Svc. award 1991); mem. Am. Home Econs. Assn., Phi Tau Sigma, Gamma Sigma Delta (award of merit 1988), Phi Upsilon Omicron. Republican. Lutheran. Office: Am Assn Cereal Chemists 3340 Pilot Knob Rd Saint Paul MN 55121-2055

CALDWELL, GARY WAYNE, chemist, researcher; b. Berea, Ky., Dec. 5, 1953; s. Reed and Susie (Saylor) C.; m. Patricia Ann Thomas, Aug. 11, 1979; children: Erin Thomas, Steven Micheal. BS in Chemistry, Ill. State U., 1978; PhD in Chemistry, Ind. U., 1982. Postdoctoral rschr. U. Alberta/Edmonton, Can., 1982-84; rsch. chemist Celanese Chem. Co., Corpus Christi, Tex., 1984-85; rsch. mgr. R. W. Johnson Pharm. Rsch. Inst., Spring House, Pa., 1985—. Contbr. articles to Jour. Am. Chem. Soc., Jour. Phys. Chemistry, Jour. Chem. Physics, Can. Jour. Chemistry, Tetrahedron Letters. Mem. Am. Chem. Soc., Am. Soc. Mass Spectrometry. Achievements include research in negative gas-phase ion-molecule reactions. Home: 1855 Meredith Ln Blue Bell PA 19422 Office: R W Johnson Pharm Rsch Inst Welsh & Mckean Rds Spring House PA 19477

CALDWELL, LYNTON KEITH, social scientist, educator; b. Montezuma, Iowa, Nov. 21, 1913; s. Lee Lynton and Alberta (Mace) C.; m. Helen A. Walcher, Dec. 21, 1940; children: Edwin Lee, Elaine Lynette. Ph.B., U. Chgo., 1935, Ph.D., 1943; M.A., Harvard U., 1938; LL.D. (hon.), Western Mich. U., 1977. Asst. prof. govt. Ind. U., Bloomington, 1939-44; dir. advanced studies in sci., tech. and public policy Ind. U., 1965—, Arthur F. Bentley prof. polit. sci., 1971-84, prof. pub. and environ. affairs, 1970—; dir. research and publs. Council of State Govts., 1944-47; faculty U. Chgo., 1945-47; prof. polit. sci. Syracuse U., 1947-54; dir. Pub. Adminstrn. Inst. for Turkey and Middle East, UN, Ankara, 1954-55; prof. polit. sci. U. Calif.-Berkeley, 1955-56; mem. environmental adv. bd. C.E., 1970—; mem. sea grant adv. panel NOAA, 1971—; panel mem. Office Tech. Assessment, 1977—; cons. U.S. Senate Com. on Interior and Insular Affairs, 1969—, UN, 1973-74, UNESCO, 1975—, Army Environ. Policy Inst., 1991—, Nat. Com. on New Direction for Nat. Wildlife Refuge System, 1990—; mem. Nat. Commn. on Materials Policy, 1971—, Nat. Acad. Scis. Com. on Internat. Environ. Programs, 1970—; chmn. com. internat. law, policy and adminstrn. IUCN, 1969-77; mem. sci. adv. bd. Internat. Joint Commn., 1984-91; Franklin lectr. Auburn U., 1972; Disting. Profl. lectr. U. Ala., 1981, William and Mary, 1991, U. Houston, 1992. Author: Administrative Theories of Hamilton and Jefferson, 1944, 2d edit. 1988, Environment: A Challenge to Modern Society, 1970, In Defense of Earth, 1972, Environmental Policy and Administration, 1975, Citizens and the Environment, 1976, Science and the National Environmental Policy Act, 1982, International Environmental Policy: Emergence and Dimensions, 1984, 90, Biocracy: Public Policy and the Life Sciences, 1987, Perspectives on Ecosystem Management for the Great Lakes, 1988, Between Two Worlds: Science, The Environmental Movement and Policy Choice, 1990, (with K. Schrader) Policy for Land: Law and Ethics, 1993; bd. editors Environ. Conservation jour., 1973—, Natural Resources Jour., 1973—, Sci., Tech. and Soc., 1979-91, Environ. Profl. Jour., 1981-89, Politics and the Life Scis., 1982—, Colo. Jour. Internat. Environ. Law and Policy, 1990—, Ambiente y Recursos Naturales (Argentina), 1985—, Environmental Awareness (India), 1989—, Duke U. Law and Policy Forum, 1991—, Environ., 1993—. Bd. govs. The Nature Conservancy, 1959-65, Shirley Heinze Environ. Fund. Recipient Sagamore of Wabash award State of Ind., 1980, H. and M. Sprout award Internat. Studies Assn., 1985, Global 500 award UN Environ. Programme, 1991; grantee Conservation Found., 1968-69, NSF, 1963—, Conservation and Research Found., 1969-70, U.S. Office Edn., 1973; guest fellow Woodrow Wilson Internat. Ctr. for Scholars Smithsonian Instn., 1971-72, East-West Ctr. fellow, 1981; named to Royal Order of Crown Thailand. Fellow AAAS; mem. Am. Soc. Pub. Adminstrn. (William Mosher award 1966, Laverne Burchfield award 1972, Marshall E. Dimock award 1981), Nat. Acad. Pub. Adminstrn., Royal Soc. Arts, Nat. Acad. Law and Social Scis. (hon. Cordoba, Argentina chpt.), Internat. Assn. for Impact Assessment (Rose Hulman Inst. Tech. award for outstanding achievement 1989), Am. Polit. Sci. Assn., Assn. Integrative Studies, N.Am. Assn. for Environ. Edn., Internat. Soc. Naturalists. Home: 4898 E Heritage Woods Rd Bloomington IN 47401-9158

CALDWELL, MICHAEL DEFOIX, surgeon, educator; b. Spartanburg, S.C., Aug. 29, 1943. BS in Chemistry, U. S.C., 1964; MD, Med. U. S.C., 1968; PhDin Physiology, Vanderbilt U., 1980. Diplomate Am. Bd. Surgery, Am. Bd. Nutrition. Intern Med. U. S.C., Charleston, 1968-69, resident in surgery, 1969-71, fellow neonatal div., 1971-72; post-doctoral fellow Vanderbilt U., Nashville, 1972-76; resident in surgery Hosp. of U. Pa., Phila., 1980-81, chief resident in gen. surgery, 1981-82, fellow clin. nutrition ctr., 1980-82, instr. dept. surgery, 1981-82; assoc. prof. surgery Brown U., Providence, 1982-87, prof. surgery, 1987-90; acting chief dept. surgery Rhode Island Hosp., 1988-90; prof. surgery U. Minn. Hosps., Mpls., 1990-91, prof. surgery and biochemistry, 1991—; lectr. dept. nutritional sci. U. Calif., Berkeley, 1979-80; clin. coordinator Vanderbilt Clin. Nutrition Service, 1972-76; chief div. surg. metabolism Letterman Army Inst. of Research, Presidio of San Francisco, Calif., 1976-78, chief nutritional support service, 1977-80; dir. nutrional support service and surg. metabolism lab. R.I. Hosp., Providence, 1982-90, exec. com. dept. surgery, 1982—, pediatric parenteral nutrition com., 1984—, dir. med. dietetics com., 1984—, mem. dietetic internship adv. com., 1984—, trustee, mem. pension com. Surgery Found., Inc., 1984—, home nutrition com., 1985—, search coms. for cellular immunologist, 1985—, vice chmn. dept. surgery, 1985—, mem. digestive tract and liver market focus group, 1986, mem. spl. task force recruitment for chmn. dept. med. 1986, chmn. dept. medicine, 1986, surgeon-in-charge div. surg. research, 1986—; mem. full-time faculty com. dept. surgery, Brown U., 1982—, MD curriculum com., 1983—, course dir., 1983—, long-range planning subcom., 1983-86, MD/PhD curriculum com., 1984-86, pre-clin. subcom., 1984-86, grad. council, 1987—; mem. nutrition com. Miriam Hosp., 1983—; dir. wound healing program U. Minn. Hosp. and Clinics, 1990—; lectr. in field. Co-editor: Clinical Nutrition I: Enteral and Tube Feeding, 1984. Clinical Nutrition II: Parenteral Nutrition, 1985, Growth Factors and Other Aspects of Wound HEaling: Biological and Clinical Implications-Proceedings of 2d International Symposium on Tissue Repair, Biological and Clinical Aspects, 1987; founder, editor-in-chief Jour. Parenteral and Enteral Nutrition, 1977-81, assoc. editor, 1981-86; edit. bd. Am. Jour. Clin. Nutrition, 1986; contbr. articles to profl. jours., chpts. to books. Bd. dirs. Oley Found. Served to lt col. U.S. Army, 1976-80. Fellow Am. Coll. Nutrition; mem. Am Bd. Nutrition (sec./treas. 1988—), Soc. Univ. Surgeons, Assn. Acad. Surgery, Soc. for Surgery of Alimentary Tract, New Eng. Surg. Soc., Am. Inst. Nutrition, Am. Soc. Clin. Nutrition (chmn. subcom. on JCAH standards, com. on clin. practice issues 1986-87), Shock Soc., Am. Fedn. Clin. Research, Ravdin-Rhoads Surg. Soc., Am. Soc. for Parenteral and Enteral Nutrition (organizing com. 1975-76, exec. council 1976-81, nat. bd. nutritional support cert. 1985-86, chmn. standards com. 1985-86), Internat. Soc. Parenteral Nutrition, AMA, R.I. Med. Soc. (del. 1983-85), Providence Med. Assn., Diabetes and Endocrine Soc. R.I. Inc., AAAS, N.Y. Acad. Scis., Nat. Insts. Health (site visit and spl. rev. com. 1985—), Am. Bd. Nutrition (bd. dirs. 1986—, credentials com. 1986—, fin. and fundraising com. 1986—), Wound Healing Soc. (bd. dirs. 1990). Office: U Minn PWB-B150 516 Delaware St SE Minneapolis MN 55455-0374

CALDWELL, PETER DEREK, pediatrician, pediatric cardiologist; b. Schenectady, N.Y., Apr. 16, 1940; s. Philip Graham and Mary Elizabeth (Glockler) C.; m. Olga Hoang Hai Miller, May 31, 1969. BA, Pomona Coll., 1961; MD, UCLA, 1965. Intern King County Hosp., Seattle, 1965-66; resident in pediatrics U. Wash., 1969-71, fellow in pediatric cardiology, 1971-73; pvt. practice Hawaii Permanente Med. Group, Honolulu, 1973—. Author: Bac-Si: A Doctor Remembers Vietnam, 1990, Adventurer's Hawaii, 1992; contbr. articles to profl. jours. Lt. comdr. USNR, 1966-69. Fellow Am. Acad. Pediatrics; mem. Am. Coll. Sports Medicine, Wilderness Med. Soc., Internat. Soc. Mountain Medicine. Avocations: ocean sports, hiking, backpacking, cycling, photography. Office: Kaiser Punawai Clinic 94-235 Leoku St Waipahu HI 96797

CALE, WILLIAM GRAHAM, JR., environmental sciences educator, university administrator, researcher; b. Phila., Dec. 10, 1947; s. William Graham and Kathryn (Rowland) C.; m. Betty Jean Byrd, June 8, 1974. B.S., Pa. State U., 1969; Ph.D. in Zoology, U. Ga., 1975. Asst. prof. ecology and environ. scis. U. Tex.-Dallas, Richardson, 1975-80, assoc. prof. environ. scis., 1980-87, full prof. 1987-89, assoc. dean Sch. Natural Scis. and Math., 1983-85, 87-89, chmn. dept. environ. scis., 1984-89; dean Coll Natural Scis. and Math. Ind. Univ. Pa., 1989—; vis. scientist Oak Ridge Nat. Lab., 1981, 84, 85. Mem. NSF grant adv. panel, 1985-88—, Dept. Energy grant rev. panel, 1989-90; contbr. articles to profl. jours. NSF grantee, 1978, 81, 83, 85. Mem. Ecol. Soc. Am., Am. Inst. Biol. Scis., Internat. Assn. for Ecology, Internat. Soc. for Ecol. Modelling, Sigma Xi. Democrat. Avocations: tournament bridge, golf. Home: 1051 Mansfield Ave Indiana PA 15701-2415 Office: Ind U of Pa Coll of Natural Scis & Math 305 Weyandt Hall Indiana PA 15705-1087

CALHOUN, JESSE, JR., electrical engineer; b. Chgo., July 21, 1958; s. Jesse Sr. and Katherine (Collins) C.; m. Judy Ann Richardson, June 6, 1987; 1 child, Jesse Irving. BS, So. Ill. U., 1983. Profl. engr. Ky. Electrical engr. Martin Marietta, Paducah, Ky., 1983-88, telecommunication mgr., 1988--. Mem. Nat. Orgn. Black Chem. and Chem. Engrs. (v.p. Western Ky. Area chpt.). Democrat. Baptist. Home: 175 Circle Dr Paducah KY 42001 Office: Martin Marietta 5600 Hobbs Rd Paducah KY 42001-4080

CALIE, PATRICK JOSEPH, molecular biologist; b. Paterson, N.J., July 8, 1953; s. Marcel Joseph and Dorothy Rose (Ineman) C.; m. Mary Reeves, Jan. 31, 1981; 1 child, Martha Edens. MS, U. Tenn., 1980, PhD, 1986. Postdoctoral fellow U. Mich., Ann Arbor, 1986-90, sr. rsch. fellow, 1990-92; asst. prof. Ea. Ky. U., Richmond, 1992—. Contbr. articles to Jour. Bacteriology, Procs. Nat. Acad. Sci., other profl. publs. Recipient Nat. Rsch. Svc. award NIH, 1987. Mem. Internat. Plant Molecular Biology Assn., Sigma Xi, Phi Sigma. Home: 130 Dogwood Heights Berea KY 40403 Office: Ea Ky Univ 235 Moore Sci Bldg Richmond KY 40475

CALISE, ANTHONY JOHN, aerospace engineering educator; b. Chester, Pa., Feb. 27, 1943; m. Alice A. Boyle; children: John V., Linda M., Jean K., Anthony W. BSEE, Villanova U., 1964; MSEE, U. Pa., 1966, PhD, 1968. Asst. prof. Widner U., Chester, 1967-68; rsch. engr. Raytheon Missile Systems, Lexington, Mass., 1968-69; sr. rsch. engr. Dynamic Rsch. Corp., Wilmington, Mass., 1969-78; assoc. prof. Mech. Engring. Drexel U., Phila., 1978-85, prof. Mech. Engring., 1985-86; prof. Aerospace Engring. Ga. Inst. Tech., Atlanta, 1986—; cons., 1978—. Assoc. editor: AIAA, Washington, 1988-90, IEEE, Washington, 1987—; over 100 pub. articles and rsch. reports. Recipient Tech. Achievement award USAF Systems Command, 1973. Fellow AIAA (Mechanics and Control of Flight award, 1992); mem. IEEE (sr.). Achievements include theory and application of singular perturbation methods for near optimal real time guidance of aerospace vehicles. Office: Ga Inst Tech Aerospace Engring Atlanta GA 30332

CALKINS, EVAN, physician, educator; b. Newton, Mass., July 15, 1920; s. Grosvenor and Patty (Phillips) C.; m. Virginia McC. Brady, Sept. 9, 1946; children: Sarah Calkins Oxnard, Stephen, Lucy McCormick, Joan Calkins Bender, Benjamin, Hugh, Ellen Rountree, Geoffrey, Timothy. Grad., Milton Acad., 1939; A.B., Harvard U., 1942, M.D., 1945. Intern, asst. resident medicine Johns Hopkins, 1946-47, 48-50; chief resident physician Mass. Gen. Hosp., 1951-52, mem. arthritis unit, 1952-61; NRC fellow med. scis. Harvard, 1950-51, instr., asst. prof. medicine, 1952-61; practice medicine, specializing in rheumatology Boston, 1951-61, Buffalo, 1961—; prof. medicine SUNY, Buffalo, 1961—; chmn. dept. SUNY, 1965-77; head dept. medicine Buffalo Gen. Hosp., 1961-68; dir. medicine E.J. Meyer Meml. Hosp., 1968-78; head gerontology sect. Buffalo VA Med. Ctr., 1978-90; head div. geriatrics/gerontology SUNY-Buffalo, 1978-90; founder, pres. Network in Aging of Western N.Y., Inc., 1980-83; cons. Nat. Inst. Arthritis and Metabolic Diseases Tng. Grants Com., 1958-62, Program Project Com., 1964-68, Nat. Insts. Spl. Study Sect. for Health Manpower, 1969-77, for Behavioral Medicine, 1978-79; mem. acad. awards com. Nat. Inst. on Aging, 1979-80, mem. nat. adv. coun., 1985-88; dir. Western N.Y. Geriatric Edn. Ctr., 1983-88, co-dir., 1988-90; dir. Multidisciplinary Ctr. on Aging SUNY-Buffalo, 1989-90, prof. family medicine, 1987—; sr. physician and coord. geriatric program Health Care Plan, 1990—. Editor: Handbook of Medical Emergencies, 1945, Geriatric Medicine, 1983, Practice of Geriatrics, 1986, 2nd edition 1991; contbr. articles to profl. jours. Pres. Nat. Assn. Geriatric Edn. Ctrs., 1992—. Served to capt., M.C. AUS, 1943-45, 46-48. Recipient Presdl. citation for Community Service, 1983. Fellow ACP (master 1989), Am. Coll. Rheumatism (founder, pres. 1967-68, master 1986), Gerontol. Soc. Am. (chair clin. med. sect. 1989, Freeman award 1991), Am. Geriatrics Soc. (Milo D. Leavitt award 1986); mem. Am. Clin. and Climatological Assn. (v.p. 1987), Am. Soc. Clin. Investigation, Assn. Am. Physicians, Cen. Soc. for Clin. Rsch., Soc. Medicine Argentina (hon.), Soc. of Fellows John Hopkins Univ., Alpha Omega Alpha. Home: 3799 Windover Dr Hamburg NY 14075-6338 Office: SUNY Beck Hall 3435 Main St Buffalo NY 14215

CALKINS, ROBERT BRUCE, aerospace engineer; b. Pasadena, Calif., Apr. 10, 1942; s. Bruce and Florence May (Bennit) C.; m. Dana B. Ericson. BS in Aerospace Engring., Calif. State Polytech., 1965; BA in Applied Math., San Diego State U., 1970; MS in Computer Sci., Wright State U., 1984. Project engr. U.S. Air Force Flight Test Ctr., El Centro, Calif., 1965-75; sr. engr. U.S. Air Force Aero. Systems Div., Dayton, Ohio, 1975-85; prin. engr. Douglas Aircraft Co., Long Beach, Calif., 1985-90; project engr. McDonnell Douglas Missile Systems Co., Long Beach, 1990—. Recipient U.S. Presidential citation, Fed. Govt., 1967. Fellow AIAA (Disting. Svc. award 1992, assoc., chmn. tech. standards com.); mem. SAFE Assn. (sec. chpt. 1 1991, pres. chpt. 1 1992). Achievements include 2 patents. Home: 7901 Southwind Cir Huntington Beach CA 92648-5458 Office: McDonnell Douglas 3855 N Lakewood Blvd Long Beach CA 90846-0001

CALL, KATHERINE MARY, biologist; b. Chgo., May 2, 1956; d. Leonard Mooney and Dolores Jean (Stull) C.; m. Howard Lewis Liber, Dec. 28, 1985. BA with highest honors, U. Calif., 1979; PhD, MIT, 1985. Postdoctoral fellow Ctr. for Cancer Rsch., MIT, Cambridge, 1985-90; asst. prof. Dept. Molecular/Cell Toxicology, Harvard Sch. Pub. Health, Boston, 1990—. Contbr. articles to profl. jours. including Cell, Sci. and Somat. Cell. Molecular Genetics. Scientific writing vol. Sierra Club, Santa Cruz, 1977-79. Postdoctoral fellowship NIH, 1986-89. Mem. AAAS, Am. Soc. Human Genetics. Achievements include patent applications for localization and characterization of Wilm's Tumor Gene; research includes isolation and characterization of a candidate 11p13 Wilms' Tumor Gene. Home: 99 Woodland Rd Malden MA 02148 Office: Harvard Sch Pub Health Dept Molec & Cell Toxicol 665 Huntington Ave Boston MA 02115

CALLAHAN GRAHAM, PIA LAASTER, medical researcher, virology researcher; b. Chapelle-lez-Herlaimont, Belgium, Sept. 21, 1955; came to U.S., 1956; d. Heino and Helga (Sepp) Laaster; m. Lynn T. Callahan III, June 26, 1981 (div. 1992); m. Donald J. Graham, Oct. 24, 1992. BS in Microbiology, Cornell U., 1977; M in Clin. Microbiology, Hahnemann U., 1979. Registered microbiologist. Research asst. Temple U. Med. Coll., Phila., 1979-80; microbiologist Thomas Jefferson Hosp., Phila., 1980-81; staff virologist Merck Sharp & Dohme Rsch. Labs., West Point, Pa., 1981-84, research virologist, 1984-90, rsch. assoc., 1990—. Contbr. articles to profl. jours. Mem. NAFE, Am. Soc. Microbiology. Republican. Lutheran. Avocations: golf, reading, gardening. Home: 432 Sterners Rd Green Lane PA 18054 Office: Merck Sharp and Dohme Rsch Labs Sumneytown Pike West Point PA 19486

CALLAWAY, JOSEPH, physics educator; b. Hackensack, N.J., July 1, 1931; s. Joseph and Sybil Leigh (Mock) C ; m. Mary Louise Morrison, July 30, 1949; children: Joseph, Paul, Sybil. BS, Coll. William & Mary, 1951; PhD, Princeton U., 1956. Asst. prof. U. Miami, Coral Gables, Fla., 1954-60; assoc. prof. U. Calif., Riverside, 1960-64, prof., 1964-67; prof. La. State U., Baton Rouge, 1967—. Author: Quantum Theory of the Solid State, 1974, 91, Energy Band Theory, 1964; contbr. over 200 articles to profl. jours. Mem. AAAS, Am. Phys. Soc., Inst. Phys. Achievements include theoretical studies of electron scattering by atoms; theoretical studies of electronic structure and related properties of solids. Office: La State U Dept Physics Baton Rouge LA 70803

CALLEJO, GERALD RODRIGUEZ, aerospace engineer; b. Honolulu, Mar. 28, 1963; s. Aurelio Bactista and Zenaida Rodriguez (Yamauchi) C. BS in Aerospace Engring., U. Kans., 1986. Computer operator Synthes, Ltd., Monument, Colo., 1986-87; flight ops. engr. Continental Airlines, Inc., Denver, 1987—. Mem. AIAA. Democrat. Roman Catholic. Home: 5975 S Taft Way Littleton CO 80127 Office: Continental Airlines Inc 8250 E Smith Rd Rm 305 Denver CO 80207

CALLENDER, CLIVE ORVILLE, surgeon; b. N.Y.C., Nov. 16, 1936; s. Joseph and Ida (Burke) C.; m. Fern Irene Marshall, May 25, 1968; children—Joseph, Ealena, Arianne. A.B., Hunter Coll., 1959; M.D., Meharry Med. Coll., 1963. Diplomate Am. Bd. Surgery. Intern U. Cin., 1963-64; asst. resident Harlem Hosp., N.Y.C., 1964-65, Howard U. and Freedmans Hosp., Washington, 1965-66, 67-68; chief resident Howard U. and Freedmans Hosp., 1968-69, instr. dept. surgery, 1969-71; asst. resident Meml. Hosp. for Cancer and Allied Diseases, N.Y.C., 1966-67; cons. surgery Port Harcourt Gen. Hosp., Nigeria, 1970, 71; med. officer D.C. Gen. Hosp., 1970-71; postdoctoral research and clin. transplant fellow U. Minn., 1971-73; asst. prof. surgery Howard U. Med. Coll., Washington, 1973-76; assoc. prof. Howard U. Med. Coll., 1976-81, prof. surgery, 1981—, vice-chmn. dept. surgery, 1980—, dir. transplant ctr., 1973—; transplantation cons. Bermuda, 1977, V.I., 1978, 82-86; cons. Ethiopian Surgical, Amenity Med. Sch., 1984; G.P.A. Ford Meml. lectr., 1978; mem. task force on organ procurement and transplantation, HEW, 1984; testifier com. on labor and human resources, U.S. Senate, 1983; mem. end stage renal disease study com. Inst. Medicine, 1989—, to the Sec. Health, 1990-94; fellowship in liver transplantation Pitts. U., 1986-87. Mem. editorial adv. bd. New Directions, 1974-91; contbr.

articles to med. jours. Recipient Hoffman LaRoche award, 1961, Charles Nelson Gold medal, 1963, Hudson Meadows award, 1963, Charles R. Drew research award, 1969, Daniel Hale Williams award, 1969, William Alonzo Warfield award, 1977, Howard U. Faculty Outstanding Unit award, 1982, 1st Humanitarian award Community of Caring Ctr., 1990, Disting. Svc. award Surg. Sect. Nat. Med. Assn., 1990; appreciation plaque for 1st renal transplant in V.I. Gov. of St. Thomas, 1983, plaque for outstanding contbns. V.I. Legislature, 1984; Howard U. Health Affairs Disting. Service award, 1984; named to Hunter Coll. Hall of Fame, 1989, Practitioner of Yr. Nat. Med. Assn., 1989. Fellow ACS, Am. Cancer Soc.; mem. D.C. Med. Soc., Soc. Acad. Surgeons Transplantation Soc., Am. Soc. Transplant Surgeons (chmn. membership com. 1986, organ placement com. 1991), N.Y. Acad. Medicine, Nat. Kidney Found. (bd. dirs. nat. 1991—, nat. capital area 1977-90), Am. Surg. Assn., Am. Coun. on Transplantation (bd. dirs.), Alpha Omega Alpha, Alpha Phi Omega, Alpha Phi Alpha. Home: 509 Kimblewick Dr Silver Spring MD 20904-6341 Office: 2041 Georgia Ave NW Washington DC 20060-0002

CALLENDER, ROBERT HOWARD, biophysics educator, research scientist; b. St. Paul, July 8, 1942; s. Howard Charles and Esther Marie (Peterson) C.; m. Ann Brooks Weathers, Sept. 7, 1974; children: David, Sean, James. BA in Math., U. Minn., 1963; PhD in Solid State Physics, Harvard U., 1969; postgrad., U. Paris, 1969-70. From physics faculty to disting. prof. biophysics CUNY, 1970—; vis. prof. Hebrew U., Jerusalem, 1977, Columbia U., 1986; mem. program com. Internat. Laser Sci. Conf., 1985-86; panel mem. biophysics program NSF, 1987-90, mem. numerous ad hoc rev. panels, 1979—; mem. Am. Phys. Soc. divsn. Biol. Physic's Exec. Com., 1988-91, vice-chmn.-elect, vice-chmn. chmn. divsn. biol. physics, 1991-93. Mem. editorial bd. Biophysics Jour., 1985-91. Fellow Am. Phys. Soc.; mem. Am. Chem. Soc., Am. Soc. Photobiology, Protein Soc., Biophys. Soc., N.Y. Acad. Sci. (biophysics adv. coun. 1984—). Achievements include rsch. in understanding how biologically important molecules function on a molecular level using state-of-the-art spectroscopic techniques to obtain relevant structural, molecular and kinetic information. Office: CUNY Dept Physics 160 Convent Ave New York NY 10031-9101

CALLERAME, JOSEPH, physicist; b. N.Y.C., Feb. 9, 1950; s. Salvatore and Josephine (Salanitro) C.; m. Patricia Holland, June 18, 1972; children: Nicholas, Stephen. BA in Chem. Physics, Columbia Coll., 1970; MA, Harvard U., 1971, PhD in Physics, 1975. Postdoctoral fellow MIT, Cambridge, Mass., 1975-77; sr. scientist rsch. divsn. Raytheon Co., Lexington, Mass., 1977-80, mgr. stable sources lab. rsch. divsn., 1980-89, mgr. infrared detectors rsch. divsn., 1982-89, asst. gen. mgr. rsch. divsn., 1989—. Mem. IEEE (sr.). Office: Raytheon Co Rsch divsn 131 Spring St Lexington MA 02173

CALLISTER, JOHN RICHARD, mechanical engineer; b. Virginia, Minn., Dec. 18, 1961; s. Richard Mead and Paulina Anne (Erickson) C. BSME, U. Minn., 1984; MSME, Cornell U., 1992; MBA, U. Chgo., 1993. Project engr. GM, Milford, Mich., 1985—. Vol. Dukakis campaign for Pres., Warren, Mich., 1988. Minn. Higher Edn. Bd. scholar, 1980; GM fellow, 1989. Mem. AIAA, ASME, Acoustical Soc. Am., Soc. Automotive Engrs. Democrat. Roman Catholic. Achievements include development of method of analytical prediction of wind noise more advanced and more accurate than recent techniques; research in automobile wind noise. Home: 131 Whitetail Dr Ithaca NY 14850

CALMES, JOHN WINTLE, architect; b. Topeka, Sept. 5, 1942; s. Baxter D. and Kathleen Edith (Wintle) C.; m. Virginia Irene McLemore, Dec. 6, 1966 (div. 1980); children: John Joseph, Angela Jo. BArch., Kans. State U., 1968; postgrad., Jones Bus. Coll., 1974-76, U. Wis., 1983. Assoc. architect Gerald D. Ervin, Architect, Junction City, Kans., 1967-71; v.p. Kencraft, Inc., Topeka, 1971-72; v.p., Vista Properties Vero Beach (Fla.), Inc., 1972-76; pres. John W. Calmes, Architect, P.A., Vero Beach, 1976-89; chmn. bd. Calmes & Pierce, Vero Beach, 1989—. Chair Bd. Adjustments and Appeals, Indian River County, Fla., 1978—, Bd. Bldg. Appeals, Vero Beach, 1978-83; vice chair Constrn. Bd., Sebastian, Fla., 1981-83, Planning and Zoning Bd., Vero Beach, 1983—; mem. Bldg. and Dept. Study Com., Indian River County, 1982-83. Sgt. U.S. Army, 1964-67, Viet Nam. Recipient Jaycee of Month award Jaycee, Junction City, Kans., 1971, cert. honor City Coun. City of Sebastian, 1983, Appreciation award City Coun. City of Vero Beach, 1990. Mem. AIA (bd. dirs. Fla. chpt. 1982-88, pres. Indian River chpt. 1981-82, award excellance 1985), C. of C., Hundred Club (bd. dirs., treas. 1986-90), Vero Beach Country Club. Republican. Episcopalian. Avocations: golf, boating, art, coin collecting. Office: Calmes & Pierce Inc 1125 12th St Ste B Vero Beach FL 32960-3791

CALVAER, ANDRE J., electrical science educator, consultant; b. Ixelles, Belgium, July 4, 1921; s. John B. and Flore M. (Romain) C.; m. Cecile G. Mahaux, Sept. 2, 1953; children: Myriam M., Yves F., Sophie M. MA in Civil and Mining Engring., State U. Liege, Belgium, 1946; PhD, State U. Liege, 1957. Lic. civil and elec. engr., Belgium. Chief engr. Nat. Elec. Coordn. Co. CPTE, Brussels, 1947-66; adviser to bd. Nat. Elec. Coordn. Co., Brussels, 1966-86; assoc. prof. U. Liege, 1961-63, full prof., 1964-86, head dept. elec. engring., 1968-74, dean faculty of engring., 1972-75, bd. dirs., 1975-79; cons. Laborelec and Tractebel, Brussels, 1986—; mem. Internat. Conf. on Large Elec. Systems (CIGRE), Paris, 1954-90; mem. Internat. Electrotech. Commn., Geneva, 1962-88. Author: Electricite Theorique (5 vols.), 1963; editor: Power Systems, 1989; contbr. numerous papers to sci. and tech. publs. Decorated comdr. Order of Crown, grand officer Order of Leopold II (Belgium), Officer Order Palmes Academiques (France); recipient internat. Montefiore prize, 1960. Fellow N.Y. Acad. Scis., IEEE; mem. Assn. Engrs. U. Liege (chmn. sci. com. 1978-82; gold medal 1982). Avocations: abstract painting, modern art history, electronics. Home: Ave Blonden 23/091, B 4000 Liège Belgium

CALVERT, DAVID VICTOR, soil science educator; b. Chaplin, Ky., Feb. 26, 1934; s. Stanford Byron and Willia Neal Calvert; m. Joyce Faye LeMay, July 27, 1957; children: Victor Neal, Yvonne Carole Calvert Lee. BS, U. Ky., 1956, MS, 1958; PhD, Iowa State U., 1962. Cert. profl. soil scientist, Am. Registry of Cert Profls. in Agronomy, Crops and Soils, Ltd. Grad. rsch. asst. U. Ky., Lexington, 1956-58, Iowa State U., Ames, 1958-62; asst. prof. soil sci. U. Fla., Ft. Pierce, 1962-68, assoc. prof., 1968-76, prof., 1976—, acting dir. Agrl. Rsch. & Edn. Ctr., 1978-79, dir. Agrl. Rsch. & Edn. Ctr., 1979—; ofcl. collaborator S.E. region USDA, Athens, Ga., 1965—; cons. World Bank, Jamaican Sch. Agr., Kingston, 1970-71; cons. soil sci. Coun. for Agrl. Sci. and Tech., St. Louis. Contbr. over 100 articles to profl. jours. including Soil Sci. Soc. Am. Proceedings, Jour. Agrl., Food Chem., Jour. Environ. Quality, Soil Sci., Proceedings Internat. Soc. Citriculture. Recipient Soil-Water-Air-Plant grant USDA Agrl. Rsch. Svc., Fla., 1968-80; U.S. EPA grant, 1970-73, Water Quality Rsch. grant City of Okeechobee, Fla., 1990-93, Rsch. Achievement award Fla. Fruit and Vegetable Assn., 1979; U. Ky. fellow; named Outstanding Conservationist of Yr., Soil Conservation Svc. USDA, Fla., 1983. Mem. Soil Sci. Soc Am., Internat. Soc. Soil Sci., Am. Soc. for Hort. Sci., Coun. of Agrl. Sci. and Tech., Soil and Crop Sci. Soc. Fla., Fla. State Hort. Soc., Internat. Soc. Citriculture, Rsch. Ctr. Adminstrs. Soc., Am. Soc. Agronomy, Farmhouse Fraternity, Scovell Soc. U. Ky. (charter mem.), Sigma Xi, Gamma Sigma Delta, Alpha Zeta. Achievements include quantification of the adverse effects of intensive dairy land-use and inadequate waste management on phosphorus and nitrogen nutrient concentrations, and eventual loads of receiving streams and lake for southern Fla. watershed. Home: 1007 Grandview Blvd Fort Pierce FL 34982-4323 Office: U Fla IFAS Agrl Rsch & Edn Ctr PO Box 248 Fort Pierce FL 34954-0248

CALVERT, GEORGE DAVID, consumer products company executive; b. Washington, Feb. 9, 1964; s. Robert and Janice Cornell (Mills) C.; m. LeVerne Kirkland, Aug. 19, 1989; 1 child, Robert David. BS in Chemistry, Coll. William and Mary, 1986; PhD of Analytical Chemistry, U. S.C., 1989. Mgr. Spectroscopy Lab. U. S.C., Columbia, 1987-89; group leader Analytical Svcs. Amway Corp., Ada, Mich., 1989-91; group leader Nutritional Foods Products Div. Amway Corp., Ada, 1991—. Contbr. articles to profl. jours. Tech. advisor Rockford (Mich.) Pub. Schs., 1992—. Mem. Am. Chem. Soc., Soc. for Applied Spectroscopy, Inst. Food Technologists, Sigma Xi. Home: 2265 Old Dominion Ct Kentwood MI 49508 Office: Amway Corp 7575 E Fulton Rd Ada MI 49355

CALVERT, WILLIAM PRESTON, radiologist; b. Warrensburg, Mo., July 2, 1934; s. William Geery and Elizabeth (Spaulding) C.; m. Mary Kay Kersh, Apr. 4, 1976. BS, MIT, 1956; MD, U. Pa., 1960. Diplomate Am. Bd. Nuclear Medicine, Am. Bd. Radiology. Intern Pa. Hosp., Phila., 1960-61, resident in medicine, 1961-62, 64-66, chief med. resident, chief resident physician, 1965-66; resident in gastroenterology U. Miami, 1966-67, NIH fellow in gastroenterology, 1967-68, resident in radiology, 1968-71; radiologist Meml. Hosp., Hollywood, Fla., 1971-72; chief dept. radiology Larkin Gen. Hosp., South Miami, Fla., 1972-80, radiologist, 1980-89; radiologist Jackson Meml. Hosp., U. Miami, 1989-93, Univ. Hosp., Tammarac, Fla., 1993—; clin. instr. radiology U. Miami Sch. Medicine, 1971-76, clin. asst. prof. radiology, 1984-88, clin. assoc. prof. radiology, 1988—. Bd. dirs. Wediko Farms Children's Svcs., Carbondale, Ill. Served with M.C., USAF, 1962-64. Mem. AMA, Fla. Med. Assn., Fla., Greater Miami radiol. socs., Soc. Nuclear Medicine, Radiol. Soc. N.Am., Explorers Club.

CALVIN, ALLEN DAVID, psychologist, educator; b. St. Paul, Feb. 17, 1928; s. Carl and Zelda (Engelson) C.; m. Dorothy VerStrate, Oct. 5, 1953; children—Jamie, Kris, David, Scott. B.A. in Psychology cum laude, U. Minn., 1950; M.A. in Psychology, U. Tex., 1951, Ph.D. in Exptl. Psychology, 1953. Instr. Mich. State U., East Lansing, 1953-55; asst. prof. Hollins Coll., 1955-59, assoc. prof., 1959-61; dir. Britannica Center for Studies in Learning and Motivation, Menlo Park, Calif., 1961; prin. investigator grant for automated teaching fgn. langs. Carnegie Found., 1960; USPHS grantee, 1960; pres. Behavioral Research Labs., 1962-74; prof., dean Sch. Edn., U. San Francisco, 1974-78; Henry Clay Hall prof. Orgn. and leadership, 1978—; pres. Pacific Grad. Sch. Psychology, 1984—. Author textbooks. Served with USNR, 1946-47. Mem. Am. Psychol. Assn., AAAS, Sigma Xi, Psi Chi. Home: 1645 15th Ave San Francisco CA 94122-3523 Office: U San Francisco San Francisco CA 94117

CALVIN, MELVIN, chemist, educator; b. St. Paul, Minn., Apr. 8, 1911; s. Elias and Rose I. (Hervitz) C.; m. Genevieve Jemtegaard, 1942; children: Elin, Karole, Noel. BS, Mich. Coll. Mining and Tech., 1931, DSc, 1955; PhD, U. Minn., 1935, DSc, 1969; hon. rsch. fellow, U. Manchester, Eng., 1935-37; Guggenheim fellow, 1967; DSc, Nottingham U., 1958, Oxford (Eng.) U., 1959, Northwestern U., 1961, Wayne State U., 1962, Gustavus Adolphus Coll., 1963, Poly. Inst. Bklyn., 1962, U. Notre Dame, 1965, U. Gent, Belgium, 1970, Whittier Coll., 1971, Clarkson Coll., 1976, U. Paris Val-de-Marne, 1977, Columbia U., 1979, Grand Valley U., 1986. With U. Calif., Berkeley, 1937—; successively instr. chemistry, asst. prof., prof., Univ. prof., dir. Lab. Chem. Biodynamics U. Calif., 1963-80, assoc. dir. Lawrence Berkeley Lab., 1967-80; Peter Reilly lectr. U. Notre Dame, 1949; Harvey lectr. N.Y. Acad. Medicine, 1951; Harrison Howe lectr. Rochester sect. Am. Chem. Soc., 1954; Falk-Plaut lectr. Columbia U., 1954; Edgar Fahs Smith Meml. lectr. U. Pa. and Phila. sect. Am. Chem. Soc., 1955; Donegani Found. lectr. Italian Nat. Acad. Sci., 1955; Max Tishler lectr. Harvard U., 1956; Karl Folkers lectr. U. Wis., 1956; Baker lectr. Cornell U., 1958; London lectr., 1961, Willard lectr., 1982; Vanuxem lectr. Princeton U., 1969; Disting. lectr. Mich. State U., 1977; Prather lectr. Harvard U., 1980; Dreyfus lectr. Grinnell Coll., 1981, Berea Coll., 1982; Barnes lectr. Colo. Coll., 1982; Nobel lectr. U. Md., 1982; Abbott lectr. U. N.D., 1983; Gunning lectr. U. Alta., 1983; O'Leary disting. lectr. Gonzaga U., 1984; Danforth lectr. Dartmouth Coll., 1984, Grinnell Coll., 1984; R.P. Scherer lectr. U. S. Fla., 1984; Imperial Oil lectr. U. Western Ont., Can., 1985; disting. lectr. dept. chemistry U. Calgary, Can., 1986; Melvin Calvin lectr. Mich. Tech. U., 1986; Eastman prof. Oxford (Eng.) U., 1967-68. Author: (with G. E. K. Branch) The Theory of Organic Chemistry, 1940, Isotopic Carbon, (with others), 1949, Chemistry of Metal Chelate Compounds, (with Martell), 1952, Path of Carbon in Photosynthesis, (with Bassham), 1957, (with Bassham) Photosynthesis of Carbon Compounds, 1962, Chemical Evolution, 1969, Following the Trail of Light: A Scientific Odyssey, 1992; contbr. articles to chem. and sci. jours. Recipient prize Sugar Research Found., 1950, Flintoff medal prize Brit. Chem. Soc., 1953, Stephen Hales award Am. Soc. Plant Physiologists, 1956, Nobel prize in chemistry, 1961, Davy medal Royal Soc., 1964; Virtanen medal, 1975, Priestley medal, 1978, Am. Inst. Chemists medal, 1979, Feodor Lynen medal, 1983, Sterling B. Hendricks medal, 1983, Melvin Calvin Medal of Distinction Mich. Tech. U., 1985, Nat. Medal of Sci., 1989, John Ericsson Renewable Energy award U.S. Dept. Energy, 1991. Mem. Britain's Royal Soc. London (fgn. mem.), Am. Chem. Soc. (Richards medal N.E. chpt. 1956, Nichols medal N.Y. chpt. 1958, award for nuclear applications in chemistry, pres. 1971, Gibbs medal Chgo. chpt. 1977, Priestley medal 1978, Desper award Cin. chpt. 1981), Am. Acad. Arts and Scis., Nat. Acad. Scis., Royal Dutch Acad. Scis., Japan Acad., Am. Philos. Soc., Sigma Xi, Tau Beta Pi, Phi Lambda Upsilon. Office: U Calif Dept Chemistry Berkeley CA 94720

CALVIN, WILLIAM HOWARD, neurophysiologist; b. Kansas City, Mo., Apr. 30, 1939; s. Fred Howard and Agnes (Leebrick) C.; m. Katherine Graubard, Sept. 1, 1966. BA, Northwestern U., 1961; PhD, U. Wash., 1966. Affiliate prof. psychiatry U. Wash., Seattle, 1966—. Author: Inside the Brain, 1980, The Throwing Madonna, 1983, The River That Flows Uphill, 1986, The Cerebral Symphony, 1989, The Ascent of Mind, 1990, How the Shaman Stole the Moon, 1991. Pres. ACLU of Wash., 1973-74. NIH rsch. grantee, 1967-84; recipient Gov.'s Writing prize State of Wash., 1984, 88. Mem. Soc. for Neurosci., Internat. Brain Rsch. Orgn., N.Y. Acad. Sci., Internat. Soc. for Human Ethology, Lang. Origins Soc., Am. Assn. Phys. Anthropologists, Internat. Astronom. Union, Am. Psychol. Soc. Achievements include research in biophysics and anthropology. Office: Univ of Wash NJ-15 Seattle WA 98195

ÇALVINO, PHILIPPE ANDRE MARIE, engineer, manufacturing executive; b. Commercy, Lorraine, France, May 11, 1939; s. Louis Antoine and Jeanne Simone (Taule) C.; m. Anne Marguerite Schneider, Aug. 6, 1965; children: Nathalie, Jean-Charles, Laurent. BAC/Math., Henry IV, Paris, 1959; degree in engring., Violet, Paris, 1963. Mfg. engr. Railways Vehicles Co./Cimt, Valenciennes, France, 1966-69; product planning mgr. Bearing Co./SKF, Avallon, France, 1969-72; plant mgr. automotive equipment Tecalemit, Paris, 1972-75; mng. dir. controls Robertshaw Controls, Reims, France, 1975-80; div. mgr. appliances Moulinex, Mayenne, France, 1980-83; internat. engring. ops. mgr. Moulinex, Caen, France, 1984—; chmn. elec. components GMX, Ireland, Mouli-Misr, Egypt, 1990. Lt. French Navy, 1963-65; Lt. comdr. French Navy Res., 1985-89. Decorated Vol. Mil. Svc. medal, 1990. Mem. Employers Assn. (treas. 1975-80), Am. C. of C. (Paris), France-USA Assn., Rotary. Home: 18 rue des Carrieres, St Julien, 14000 Caen France

CAMBALOURIS, MICHAEL DIMITRIOS, aircraft engineer; b. Nikea, Greece, Sept. 8, 1959; s. Dimitrios M. and Maria J. C. B in Engring., Helenic Air Force Acad., Tatoi, Greece, 1982. Flight line officer Helenic Air Force, Crete, Greece, 1982-85, quality assurance officer, 1985-87; quality assurance officer Helenic Air Force, Greece, 1987-89; flight engr. Helenic Air Force, Koridallos, Greece, 1988—, chief engr., 1989-92; chief engr. SAR SQN Helenic Air Force, Holargos, Greece, 1992—. Mem. AIAA, Am. Helicopter Soc. Home: Samou 143, 18121 Koridallos Piraeus, Greece

CAMBEL, ALI BULENT, engineering educator; b. Merano, Italy, Apr. 9, 1923; came to U.S., 1943, naturalized, 1951; s. H. Cemil and Remziye (Hakki) C.; m. Marion dePaar, Dec. 20, 1946; children—Metin, Emel, Leyla, Sarah. BS, Robert Coll., Istanbul, Turkey, 1942; postgrad., U. Istanbul, 1942-43, MIT, 1943-45; MS, Calif. Inst. Tech., 1946; PhD, U. Iowa, 1950. Registered profl. engr. Instr. State U. Iowa, 1947-50, asst. prof., 1950-53; assoc. prof. mech. engring. Northwestern U., 1953-56, prof. mech. engring., 1956-61, Walter P. Murphy disting. prof., 1961-68, dir. gas dynamics lab., 1955-66, chmn. dept. mech. engring. and astronautical scis., 1957-66; dir. research and engring. support div. IDA, 1966-67, v.p. for research, 1967-68; dean Coll. Engring., Wayne State U., Detroit, 1968-70; exec. v.p. for acad. affairs Wayne State U., 1970-72; v.p., dir. System Research div. Gen. Research Corp., 1972-74; dep. asst. dir. for sci. and tech. NSF, 1974-75; prof. engring. and applied sci. George Washington U., Washington, 1975-88, prof. emeritus, 1988—; chmn. dept. civil, mech. and environ. engring. George Washington U., 1978-80, dir. energy programs, 1976-88; tech. cons. govt. agys., various firms; staff dir. Pres.'s Interdept. Energy Study, 1963; engring. scis. adv. com. USAF Office Sci. Research, 1961-63; mem. Commn. Engring. Edn., 1966-68, Army Sci. Advisory Panel, 1966-72; nat. lectr. Sigma Xi, 1961-62. Author: Plasma Physics and Magnetofluidmechanics, 1963; co-author: Gas Dynamics, 1958, Real Gases, 1963, Plasma Physics, 1965, Applied Chaos Theory: A Paradigm for Complexity, 1993; co-editor: Transport Properties in Gases, 1958, The Dynamics of Conducting Gases, 1960, Magnetohydrodynamics, 1962, Second Law Analysis of Energy Devices and Processes, 1980, Dissipative Structures in Integrated Systems, 1989; co-editor AIAA Jour., Jet Propulsion, 1955-60, Energy, The Internat. Jour., 1975—; mem. editorial bd. Energy, Environment, Economics, 1991—; contbr. numerous papers in the field. Bd. dirs. YMCA. Recipient leadership award YMCA, 1953; citation for solar satellite power system evaluation Dept. Energy/NASA, 1981; cert. for patriotic service Sec. of Army; award for excellence NSF/RANN; award for contbns. to sci. and edn. U.S. Immigrants League.; Washburn scholar, 1938. Fellow AIAA (J. Edward Pendray award 1959, nat. dir.); mem. Am. Soc. Engring. Edn. (Curtis McGraw award 1960, George Washington award 1966, chmn. engring. and pub. policy div. 1986-87), ASME (founding chmn. energy systems analysis tech. com. 1980-82), Cosmos Club (Washington), Sigma Xi, Pi Tau Sigma, Tau Beta Pi. Mem. Soc. of Friends. Home: 6155 Kellogg Dr Mc Lean VA 22101-3120 Office: George Washington U Sch Engring and Applied Sci Washington DC 20052

CAMBONI, SILVANA MARIA, environmental sociologist; b. Columbus, Ohio; d. Louis and Margherita (DeFeo) C. BA, Ohio State U., 1970, MA, 1971, PhD, 1984. Rsch. assoc. II Ohio State U., Columbus, 1982-84; asst. dir. rsch. devel. The Ohio State U. Rsch. Found., Columbus, 1985-88; assoc. dir. Ohio State U. Rsch. Found., Columbus, 1988—; adj. asst. prof. Ohio State U., 1987—; faculty assoc. The Mershon Ctr., Columbus, 1990-92; dir., founder Human and Natural Resources Preservation Ctr., Columbus, 1988—. Contbr. articles to profl. jours., chpts. to books. Founding com. Hunger & Devel. Coalition of Ohio, Columbus, 1984; dir. pub. rels. UNICEF, Franklin County, 1972. USDA rsch. grantee, 1991, Mershon Ctr. rsch. grantee, 1990. Mem. AAAS, AIAA, World Assn. Soil and Water Conservation, Soil and Water Conservation Soc. Am., Nat. Coun. Univ. Rsch. Adminstrs., Gamma Sigma Delta. Achievements include conduction of first-known national social science policy study in which quantitative data were collected on the topic of orbital space debris. Office: Ohio State Univ Rsch Found 1960 Kenny Rd Columbus OH 43210

CAMERON, CHARLES BRUCE, naval officer, electrical engineer; b. Toronto, Ontario, Can., May 3, 1954; came to U.S., 1980; s. William Walter and Wilda Mae Cameron; m. Nancy Diane Cooper, Apr. 4, 1981; children: Erin Elizabeth, Heather Diane. BS, U. Toronto, 1977; MSEE, Naval Postgrad. Sch., 1989, PhD, 1991. Mktg. rep. IBM Can. Ltd., Toronto, 1977-79; sr. EDP analyst No. Telecom, Toronto, 1979; commd. ensign USN, 1981, advanced through grades to lt. comdr., 1992; asst. combat dir. ctr. officer USS John F. Kennedy, Norfolk, Va., 1991-92; adminstrv. officer Carrier Airborne Early Warning Squadron 112, FPO AP 96601-6400, AP, 1992—. Mem. IEEE, AIAA, Mensa. Achievements include patent pending for demodulators for optical fiber interferometers with (3x3) outputs. Home: 13393 Gabilan Rd San Diego CA 92128-4079

CAMERON, MARYELLEN, science association administrator, geologist, educator; b. New Orleans, Nov. 15, 1943; d. James T. and Olga M. Jones; m. Kevin D. Crowley. BS in Geology, U. Houston, 1965, MS, 1969; PhD, Va. Poly. Inst. and State U., 1972. Rsch. and teaching asst. Va. Poly. Inst. and State U., Blacksburg, 1968-70; pre-doctoral fellow geophysical lab. Carnegie Inst. Washington, 1971; postdoctoral rschr. SUNY, Stony Brook, 1971-73; asst. rsch. earth scientist U. Calif., Santa Cruz, 1973-81; assoc. prof. U. Okla., Norman, 1981-87; prof., chair Miami U., Oxford, Ohio, 1987-92; program. dir. NSF, Washington, 1992—; mem. proposal review panel NSF, 1985-88, mem. adv. com. for earth scis., 1992; disting. vis. scientist facility for high resolution electron microscopy Ariz. State U., Tempe, 1990; vis. scientist dept. earth and planetary scis. U. N.Mex., Albuquerque, 1991; Donnay lectr. Carnegie Inst. Washington, 1993. Contbr. over 80 articles and abstracts to sci. jours. Grantee NSF, 1977-92. Feloow Mineralogical Soc. Am. (lectr. 1980, assoc. editor Am. Mineralogist 1980-83, sec. 1987-91, Innaugural lectr. 1989-90); mem. AAAS, Am. Geophysical Union, Geochemical Soc., Materials Rsch. Soc., Geol. Soc. Washington, Sigma Xi. Achievements include research on structural variations in silicate and phosphate minerals as a function of composition and tempurature, thermal annealing systematics and radiation damage of apatites, igneous petrology and geochemistry of Tertiary igneous rocks of the southwestern U.S. and Mexico. Home: 6408 Recreation Ln Falls Church VA 22041 Office: NSF 1800 G St NW Washington DC 20550

CAMERON, ROBERT H., water association administrator; b. Spokane, Wash., Apr. 4, 1953. AA, Spokane C.C., 1974. Water svc. oper. Union Pacific R.R., Hermiston, Oreg., 1974-86; cir. rider Nast. Rural Water Assn., Duncan, Okla., 1988-90; program mgr. Wash. Rural Water Assn., Spokane, 1990-91, CEO, 1991—; local chmn. Sheet Metal Workers Union, Hermiston, 1983-86; emm. needs assessment policy com. Seattle Dept. Health, 1991-92; tech. adv. com. Nat. Rural Water Assn., Duncan, 1990-91; mem. Wash. State Tng. Coalition, Seattle, 1992, Intergovtl. Pub. Facilities Fin. Com., 1989-92; mem. Wash. State Wellhead Protectin Com., 1992. Mem. Internat. Assn. Continuing Edn. and Tng., Spokane C. of C., Am. Backflow Prevention Assn., Am. Soc. Assn. Execs. Office: Wash Rural Water Assn Ste G E9211 Mission St Spokane WA 99206

CAMERON, WILLIAM DUNCAN, plastic company executive; b. Harrell, N.C., June 14, 1925; s. Paul Archiebald and Atwood (Herring) C.; m. Betty Gibson, Oct. 3, 1953; children: Phillip MacDonald, Colleen Kay. Student Duke U., 1945-49. Chmn. Reef Industries Inc., Houston, 1958—. Pres. bd. trustees Trinity Episcopal Sch., Galveston, Tex., 1981-82; trustee William Temple Found., 1987-90. Served with U.S. Army, 1943-45. Mem. World Bus. Coun., Houston C. of C. (chmn. mfg. com. 1967), Rotary, Galveston Artillery, Bob Smith Yacht. Home: PO Box 310 Smith Ranch Rd Cuero TX 77954-0310 Office: Reef Industries Inc 9209 Almeda Genoa Rd Houston TX 77075-2339

CAMMARATA, ANGELO, surgical oncologist; b. Italy, 1936; s. Giuseppe and Giuseppina (Ruggiero) C.; m. Diane M. Donner, Apr. 25, 1965; children: Joseph, Marisa, Michael, Christina. BA, Upsala Coll., 1958; MD, N.Y. Med. Coll., 1962. Diplomate Am. Bd. Surgery. Intern N.Y. Polyclin. Hosp., N.Y.C., 1962; resident, chief resident Met. Hosp. N.Y.C., 1963-67, asst. surgeon, 1968—; resident in surgery Meml. Hosp. Cancer and Allied Diseases, N.Y.C., 1967-68; assoc. surgeon, attending surgeon Cabrini Med. Ctr., N.Y.C.; attending surgeon Beth Israel North Hosp., N.Y.C.; instr. surgery N.Y. Med. Coll., N.Y.C., 1968-74, clin. asst. prof. surgery, 1974—; vis. attending surgeon Met. Hosp. Ctr., N.Y.C. Contbr. articles to profl. jours. Fellow ACS, Internat. Coll. Surgeons; mem. AMA, N.Y. Cancer Soc., N.Y. Met. Breast Cancer Group, Meml. Alumni Soc., Alpha Club. Office: 55 E 87th St New York NY 10128

CAMP, FRANCES SPENCER, retired nurse; b. Lake Charles, La., Feb. 8, 1924; d. Henry Wesley and Annie Erle (Allen) S.; m. John Clayton Camp, Nov. 3, 1944; children: John C., Elizabeth C., Martha L., Charles H. Student, So. Meth. U., Dallas, 1943-44; BSRN, McNeese State U., Lake Charles, La., 1957. Surg. nurse St. Patrick's Hosp., Lake Charles, La., 1963-65; founder, dir. Home Health Svcs. Inc., Lake Charles, 1969-80; pres., founding com. mem. Hosp. Aux. for Charity Hosp., Lake Charles; mem. task group; on ethical dilemmas caused by med. tech. George Washington U. Med. Ctr., Washington, 1991-92, mem. adv. coun. to Cancer Ctr., 1993—. Artist commd. oil paintings. Active sustainer Jr. League of Am., 1957—; mem. Hospitality and Info. Svcs. of Meridian House, Washington, 1980—, steering com., 1992-93. Recipient Woman of Yr. award, Lake Charles, La., 1962. Mem. Nat. Presbyn. Ch. Women's Assn. (1st v.p. 1990-91, exec. bd., by-laws chmn., parliamentarian), Congl. Country Club, City Club of Washington, Sigma Theta Tau. Avocations: reading, travel, photography, bridge. Home: 5450 Whitley Park Ter Apt 510 Bethesda MD 20814-2059

CAMP, THOMAS HARLEY, economist; b. Charlotte, N.C., Aug. 13, 1929; s. Thomas Franklin and Agnes Mae (Davis) C.; m. Frances Ann Rogers, Mar. 20, 1953; children: Thomas Harley Jr., Landon G. BSc, U. N.C., 1956; postgrad, Am. U., 1965-67. Industry econ. USDA, Washington, 1959-70; location leader USDA, Austin, Tex., 1970-74; rsch. leader USDA, College Station, Tex., 1974-86; program leader USDA, Weslaco, Tex., 1986-88; agrl. mktg. specialist USDA, Lane, Okla., 1988-90; cons. Georgetown, Tex., 1990—. Author, co-author 44 sci. publs.; contbr. articles to profl. jours.

Cubmaster Boy Scouts Am., Springfield, Va., 1965-69, asst. scoutmaster, 1966-70, scoutmaster, Round Rock, Tex., 1970-72, asst. scoutmaster, Austin, 1972-74. With USN, 1946-51, Korea. Mem. Am. Soc. Agrl. Engrs., Animal Air Transp. Assn., Food Distbn. Rsch. Soc., Transp. Rsch. Forum, Masons. Methodist. Avocations: photography, boating. Home and Office: 230 Mesa Dr Georgetown TX 78628-1529

CAMPBELL, ASHLEY SAWYER, retired mechanical engineering educator; b. Montclair, NJ, Dec. 24, 1918; s. George A. and Caroline G. (Sawyer) C.; m. Mary Letitia Fishler, July 18, 1942; children: Ashley Sawyer Jr., Christopher, Martha, Gordon, Philip, Ben. BS, Harvard U., 1940, ScD, 1948. Test engr. Wright Aero. Corp., Paterson, N.J., 1940-45; asst. prof. Harvard U., Cambridge, Mass., 1948-50; dean Tufts U., Medford, Mass., 1957-68; dean U. Maine, Orono, 1950-57, prof. mech. engring., 1968-80; ret., 1980. Author: Thermodynamic Analysis of Combustion Engines, 1979. Selectman Town of Randolph, N.H., 1981-88. Democrat. Home: 118 Rim Rd Santa Fe NM 87501

CAMPBELL, CHARLES, geologist. Recipient Francis J. Pettijohn medal Soc. for Sedimentary Geology, 1993. Home: POB 27 Florence MT 59833*

CAMPBELL, CHARLES LEE, plant pathologist; b. Denver, July 5, 1953; s. Charles Ernest and Mary Jane (Camp) C.; m. Karen Kay Korfhage, Mar. 3, 1979. BS, Colo. State U., 1974, MS, 1976; PhD, Pa. State U., 1979. Rsch. asst. dept. plant pathology Colo. State U., Ft. Collins, 1974-76; grad. asst., then grad. fellow dept. plant pathology Pa. State U., University Park, 1976-79; asst. prof. dept. plant pathology N.C. State U., Raleigh, 1974-85, assoc. prof., 1985-91, prof., 1991—; cons., instr. Colegio de Postgraduados, Montecillo, Mex., 1987—; tech. dir. environ. monitoring and assessment program agroecosystem component EPA/USDA Agr. Rsch. Svc., Raleigh, 1992—. Co-author: Introduction to Plant Disease Epidemiology, 1990, Introduction to Plant Diseases: Identification and Management, 1992; editor: The Fischer-Smith Controversy, 1981; co-exec. producer film Healthy Plants: Our Future, 1987; contbr. to profl. publs. Recipient Media award of excellence Nat. Assn. Colls. and Tchrs. Agr., 1990; grantee USDA Agrl. Rsch. Svc., 1992—. Mem. AAAS, So. Assn. Agrl. Scientists (bd. dirs. 1989-91), Am. Phytopathological Soc. (publs. bd., newsletter editor 1991—, pres. So. div. 1990-91), Sigma Xi, Phi Beta Kappa, Phi Kappa Phi. Achievements include first implementation of spatial analysis research for root disease epidemics in U.S., identification of magnitude of role of viruses in yield losses in white clover, study of effect of acidity in rainfall on major crop diseases. Office: NC State U Dept Plant Pathology Box 7616 Raleigh NC 27695-7616

CAMPBELL, DAVID BRUCE, biology educator, researcher; b. Newport, R.I., May 15, 1953; s. G. David and Barbara (Nuttall) C.; m. C. Shannon Briscoe, June 6, 1987; children: Cassandra, Liam. BS, U. R.I., 1975; MS, Fla. Inst. Tech., 1978; PhD, U. R.I., 1983. Instr. Sch. for Field Studies, St. John, V.I., 1983-84; asst. prof. dept. zoology U. N.H., Durham, 1983-86; adj. assoc. prof. Shoals Marine Lab. Cornell U., Appledore Island, Maine, 1985-92; assoc. prof. dept. biology Rider Coll., Lawrenceville, N.J., 1986—, chair of biol. dept., 1993—; cons. Ednl. Testing Svc., Lawrenceville, 1992. Author: (chpt.) Foraging Behavior, 1987. EMT Durham Ambulance Corps, 1984-86, hyperbaric chamber U. N.H., Durham, 1984-86. Recipient Fulbright award Coun. for Internat. Exchange of Scholars, 1990, Oceanography fellowship NSF, 1991, Rsch. grant N.J. Marine Scis. Consortium, 1988, 90, Rsch. fellowship Rider Coll., 1990. Mem. Nat. Assn. Underwater Instrs. (asst. instr. 1976—), Am. Soc. Zoologists, Sigma Xi (pres. Rider Coll. club 1987—). Office: Rider Coll 2083 Lawrenceville Rd Lawrenceville NJ 08648

CAMPBELL, DAVID GEORGE, ecologist; b. Decatur, Ill., Jan. 28, 1949; s. George Robert and Jean Blossom (Weilepp) C.; m. Karen S. Lowell; 1 child, Tatiana Claire. BA, Kalamazoo Coll., 1971; MS, U. Mich., 1973; PhD, Johns Hopkins U., 1984. Exec. dir. Bahamas Nat. Trust, Nassau, 1974-77; ecologist N.Y. Bot. Garden, Bronx, 1984-88, leader Amazon Expdns., 1974-92, research fellow, 1989—; Henry R. Luce prof. in Nations and the Global Environment Grinnell (Iowa) Coll., 1991—; adj. prof. U. Nanjing, People's Republic of China; cons. Internat. Union for Conservation of Nature, 1978-79; biologist and lectr. M.V. World Discoverer in Amazon and Artarctic, 1981-87; biologist Brazilian Antarctic Expdn., 1987-88. Author: The Ephemeral Islands, 1978, The Crystal Desert, 1992; editor: Floristic Inventory of Tropical Countries, 1989; contbr. articles to profl. jours. Fellow John Simon Guggenheim Found., 1989-90; recipient Fulling award Soc. Econ. Botany, 1987, Houghton Mifflin Lit. Fellowship, 1992, Pen/Martha Albraud award for nonfiction, 1993. Fellow Linnean Soc. London. Home: 4069 Coquina Dr Sanibel FL 33957-5205 Office: Grinnell Coll Dept Biology Grinnell IA 50112

CAMPBELL, DONALD ACHESON, nuclear engineer, consultant; b. Glendale, Ohio, Jan. 7, 1919; s. Acheson Meacham and Helen Gertrude (Boyle) C.; m. Ruth Marian Cory, Sept. 15, 1945 (div. June 1966); children: Bruce Hawick, Kathleen Cory; m. Tomiko Sugahara, June 1, 1968. Degree in mech. engring., U. Cin., 1940. Machine-design engr. Alvey Ferguson Co., Cin., 1940-41; heating and air-conditioning engr. Wright Field, Ohio, 1941-44; pres. Airco Inc., St. Bernard, Ohio, 1945-56; application engr. Westinghouse Electric, Pitts., 1956-66; nuclear-plant dir. Westinghouse of Japan, 1966-67; prin. engr. Westinghouse, 1967-84; nuclear-standards cons. Internat. Atomic Energy Agy., Vienna, Austria, 1988; standards coord. Westinghouse Electric Corp., Pitts., 1969-84. Contbr. to many nat. standards on nuclear power plants. Chmn. youth program com. YMCA, Dayton, Ohio, 1944-45, duplicate bridge club, 1944-45, gen.-program com., Cin., 1948-49, leader tng. series, 1948-49, bldg.-addn. com. Monroeville Sr. Citizen Ctr., 1991—; bd. dirs. Suburban YMCA, Wyoming, Ohio, 1946-49; pres. Monroe Heights Civic Assn., Monroeville, Pa., 1961-62; 1st pres. Gateway Heights Recreation Club, Monroeville, 1963-64; mgr. Gateway Heights Softball Team, Monroeville, 1985-86; pres. Circle Eight Sq. Dance Club, North Versailles, Pa., 1985-86; pres. Prime Timers Club, 1987-78;. Ohio Dept. Edn. scholar, 1935. Mem. Am. Nuclear Soc., Pi Tau Sigma, Tau Beta Pi. Achievements include patents for methodology to provide accurately ongoing measures of total amounts of leakage of all water and steam from pressure retaining systems within the reactor-containment enclosure of a nuclear-power plant and provide for discrimination of the proportion of leakage attributable to systems with radioactivity in contained fluids. Home: 2439 Saunders Station Rd Monroeville PA 15146-4451

CAMPBELL, DORIS KLEIN, retired psychology educator; b. Tazewell County, Ill.; d. Emil L. and Cora May (Osterdock) Klein. AB, Augustana Coll., 1930; MA, U. Ill., 1931; EdD, U. Fla., 1962. Instr. Arlington Hall, Washington, 1931-33, Cen. Coll., Mc Pherson, Kans., 1933-37; supr. student teaching Seattle (Wash.) Pacific Coll., 1937-39; tchr. The Harris Schs., Chgo., 1939-41; instr. to full prof. East Tenn. State Univ., Johnson City, 1960-77; Fulbright prof. psychology Silliman Univ., Dumaguete, Philippines, 1977-80; cons. Philippine-Am. Ednl. Found., Manila, 1968-69. Recipient Fulbright grant Coun. for Internat. Exchange of Scholars, Washington, 1968-69. Mem. APA, Am. Assn. Retired Persons, Am. Coun. for the Blind, Fulbright Alumni Assn., Phi Kappa Phi, Delta Kappa Gamma, Tau Kappa Alpha, Psi Chi. Methodist. Avocations: music, travel, reading, writing for children. Home: Lime Plaza #903 400 S Florida Ave Lakeland FL 33801-5262

CAMPBELL, HARRY L., systems analyst; b. Bklyn., Mar. 5, 1966; s. Alfred Agustus and Dorcas (Belnavis) C.; m. Sophier McElveen, May 18, 1986; children: Daria Allanna, Harry Jarrell Lewis. AS in Computer Sci., Westchester Bus. Inst., 1985. With shipping and recieving Am. Brotec Corp., Ossining, N.Y., 1985-87, quality control engr., 1987-89, systems analyst, 1989—. Usher, choir mem., jr. deacon The True Ch. of God, 1984—. Home: 27 William St Ossining NY 10562 Office: Am Biotec Corp 24 Browning Dr Ossining NY 10562

CAMPBELL, HENRY J., JR., mechanical engineering consultant; b. Queens, N.Y., Jan. 3, 1922; s. Harry J. and Mary Ann (Henry) C.; m. Lillian E. Gallimore; children: Barbara Ann Campbell Yonai, Nancy E. Campbell Chiesa. Degree in mech. engring., Stevens Inst. of Tech., 1942. Registered profl. engr. Prin. Henry J. Campbell Jr. Cons. Engr., Albertson, N.Y., 1953—. Fellow ASHRAE (life); mem. IEEE (life), NSPE (life), ASME (life). Office: 1125 Willis Ave Albertson NY 11507

CAMPBELL, IAIN LESLIE, biomedical scientist; b. Sydney, Australia, Aug. 25, 1955; came to U.S., 1989; s. John Leslie and Anne Gai (Nott) C.; m. Julia Ruth Quigley, Dec. 4, 1976; children: Hamish Iain, Belinda Ruth. BSc with honors, U. Sydney, 1978, PhD, 1982. Royal Soc. exch. fellow Biomedicum, U. Uppsala, Sweden, 1981; D.W. Keir rsch. fellow Royal Melbourne (Australia) Hosp., 1982-86; sr. rsch. officer Walter & Eliza Hall Inst. Med. Rsch., Australia, 1986-91; assoc. prof. Scripps Rsch. Inst., La Jolla, Calif., 1991—; dir. Australian Soc. Med. Rsch., 1985-89; mem. organizing com. satellite symposium Internat. Diabetes Found., Melbourne, 1988. Contbr. articles to sci. jours. Recipient Roslyn Flora Goulston award U. Sydney, 1978, Hoechst R&D award Hoechst Pty. Ltd., Melbourne, 1987, Rsch. award Juvenile Diabetes Found. Internat., 1988; postdoctoral fellow Nat. Multiple Sclerosis Soc. Am., 1990. Mem. AAAS, Am. Diabetes Soc., Australian Soc. Med. Rsch. Achievements include research in cytokine biology and pathogenesis of autoimmune and viral diseases. Office: Scripps Rsch Inst CVN9 10666 N Torrey Pines Rd La Jolla CA 92037

CAMPBELL, JAMES EDWARD, physicist, consultant; b. Kingsport, Tenn., Jan. 13, 1943; s. Edward Montroe and Iola (Church) C.; m. Judy Priscilla Cameron, June 12, 1966; children: Jennifer Marie, James Kyle. BA in Math., Physics, Catawba Coll., 1965; PhD in Physics, Va. Tech. Inst., 1969. Nuclear radiation expert Naval Weapons Evaluation Facility, Albuquerque, 1970-76; mem. tech. staff Sandia Nat. Labs., Albuquerque, 1976-80; v.p. Intera Techs., Inc., Denver, 1980-88; disting. mem. tech. staff Sandia Nat. Labs., Albuquerque, 1988—; bd. dirs. TechLaw, Inc., Denver, 1983-88. Contbr. articles to profl. jours. Recipient IEEE Outstanding Paper award, 1991, NDEA fellowship, 1965-68, Acad. Honors scholarship, Catawba Coll., 1961-65. Mem. AAAS, Am. Phys. Soc. Republican. Lutheran. Achievements include design and development of reliability analysis software for SEMATECH consortium companies, equipment reliability guidelines task force, computer automated litigation support for EPA hazardous waste site program; direction of program to develop risk methodology for radioactive waste disposal. Office: Sandia Nat Labs Divsn 6613 Albuquerque NM 87185

CAMPBELL, JOHN, engineering educator; b. Leicester, Eng., Dec. 2, 1938; s. Clarence Preston and Catherine Mary (Crossley) C.; m. Sheila Margaret Bacon, Sept. 11, 1982. MA, Cambridge U., Eng., 1962; MMet, Sheffield, Eng., 1964; PhD, Birmingham U., Eng., 1967, DEng, 1988. Cert. FEng, FIM, FIBF, MAFS. Rsch. mgr. Brit. Iron and Steel Rsch. Assoc., Sheffield, 1967-70; mgr. Fulmer Rsch. Inst., Slough, Eng., 1970-78; dir. Cosworth R&D Ltd., Worcester, Eng., 1978-85, Triplex Alloys, West Midlanders, Eng., 1985-90; vis. prof. U. Birmingham, Eng., 1989-92, Baxi prof. casting tech., 1992—; expert in casting UN Indsl. Devel. Orgn., Vienna, Austria, 1973, 76; bd. dirs. Light Metals Founders Assoc. R&D Ltd., Casting Metals Devel. Ltd. Author: Castings, 1991. Named Officer Order Brit. Empire, 1993.Recipient Acta/Scripta Metallurgica Lecture, Acta Metallurgica, 1992. Achievements include the invention of Cosworth Casting process. Office: Univ of Birmingham, IRC in Metals, Birmingham B15 2TT, England

CAMPBELL, JOHN ROY, animal scientist educator, academic administrator; b. Goodman, Mo., June 14, 1933; s. Carl J. and Helen (Nicoletti) C.; m. Eunice Vieten, Aug. 7, 1954; children: Karen L., Kathy L., Keith L. B.S., U. Mo., Columbia, 1955, M.S., 1956, Ph.D., 1960. Instr. dairy sci. U. Mo., Columbia, 1960-61, asst. prof., 1961-65, assoc. prof., 1965-68, prof., from 1968; assoc. dean, dir. resident instrn. Coll. Agr., U. Ill., Urbana-Champaign, 1977-83; dean Coll. Agr. U. Ill., Urbana, 1983-88; pres. Okla. State U., Stillwater, 1988—. Author: (with J.F. Lasley) The Science of Animals That Serve Humanity 1969, 2d edit., 1975, 3d edit., 1985, In Touch with Students, 1972, (with R.T. Marshall) The Science of Providing Milk for Man, 1975. Recipient Gamma Sigma Delta Superior Teaching award, 1967, Gamma Sigma Delta Internat. award for disting. svc. to agr., 1985, Coll. Osteo. Medicine Okla. State U. Disting. Svc. award, 1992. Mem. Am. Dairy Sci. Assn. (dir. 1975-78, 80-86, pres. 1980-81, Ralston Purina Disting. Teaching award 1973, Award of Honor 1987), Nat. Assn. Coll. Tchrs. Agr. (Ensminger Interstate Disting. Tchr. award 1973, Teaching fellow 1973, Disting. Educator award 1990, Nat. Assn. State and Univ. and Land-Grant Colls. (commns. on home econs. and vet. medicine, com. on water resources, coun. of presidents), Okla. Futures, Gamma Sigma Delta. Office: Okla State U 107 Whitehurst Hall Stillwater OK 74078

CAMPBELL, JONATHAN WESLEY, astrophysicist, aerospace engineer; b. Alexander City, Ala., Sept. 1, 1950; s. Harry Underwood and Sarah Ruth Campbell; m. Mary Magdalene Sanders, Dec. 11, 1974; children: Jason Jonathan, Christopher Sanders, Benjamin Robert. BS distinguished grad., Auburn U., 1972, MS, 1974; MS, U. Ala., 1988, PhD, 1992. Cert. flight instr. Coop. engr. Pratt & Whitney Aircraft, West Palm Beach, Fla., 1968-70; instr. physics Auburn U., 1972-74; astrophysicist, aerospace engr. Missile and Space Intelligence Ctr., Huntsville, Ala., 1978-80; space scientist, supervisory aerospace engr. propulsion, exec. asst. to dir., lead engr. space telescope fine guidance sensor NASA/Marshall Space Flight Ctr., Huntsville, Ala., 1980—; pres. Redstone Aerospace Inc.; cons. Starflight Assocs. Served to capt. AUS, 1975-78; col. USAFR. Recipient Eagle Scout award. Mem. AIAA, Air Force Assn., Res. Officer Assn., Aircraft Owners and Pilots Assn., Scabbard and Blade, Tau Beta Pi, Sigma Gamma Tau, Sigma Pi Sigma. Methodist. Home: PO Box 37 Harvest AL 35749-0037 Office: NASA E51 Marshall Space Flight Ctr Huntsville AL 35812

CAMPBELL, JUDITH LOWE, child psychiatrist; b. Indpls., Jan. 21, 1946; d. Albert St. Clair and Adele V. (Lobraico) Lowe; m. Robert Frank Campbell, Nov. 30, 1968; children: Christiaan Robert, Kevin Lowe, Geoffrey Ford. B.S. in Zoology, Butler U., 1967; M.D., Ind. U., 1971. Resident in psychiatry Ind. U. Sch. Medicine, 1971-73, fellow in child psychiatry 1973-75; asst. dir. Riley Child Guidance Clinic, Indpls., 1975-79, dir. child psychiatry consultation, liaison service to pediatrics, 1975-79; dir. child psychiatry services Riley Hosp. for Children, 1979-85; pvt. practice child psychiatry, Indpls., 1985—; child psychiatry cons. Center for Mental Health of Madison County (Ind.), Anderson, 1975-77, Lutheran Child Welfare Assn., Indpls., 1974—, Lutherwood Children's Home, Indpls., 1974—, Jewish Family and Children's Services, 1983-84, child and adolescent div. Midtown Community Mental Health Center, 1983-85; assoc. med. dir. child and adolescent psychiat. svcs. Community Hosps. of Indpls., Inc., 1989-90; med. dir. outreach svcs. Arbor Hosp. of Greater Indpls., 1990, med. dir. children's unit, 1990-92, pres. med. staff, 1990-92; med. dir. Arbor Hosp., 1992—. instr. Ind. U. Sch. Medicine, Indpls., 1974-75, asst. prof. dept psychiatry, 1975-89, clin. assoc. prof., 1989—. Vice-precinct committeeman Rep. Party, 1990—; mem. parent's adv. coun. Butler U., 1989—, pres., 1990—. Recipient Physician's Recognition award in Continuing Edn. AMA, 1974, 77; Helen McQuiston award in sci., 1967. Fellow Am. Psychiat. Assn., Ind. Psychiat. Soc. (councilor 1978-80, 90-91, sec. 1981-83, editor newsletter 1981-83, chmn. com. women 1983-92, mem. ethics com. 1992—), Am. Acad. Pediatrics (Ind. br.), Am. Acad. Child and Adolescent Psychiatry, Ind. Coun. Child and Adolescent Psychiatry (sec. 1986-87, pres.-elect 1987-88, pres. 1988-89, Smithsonian Assocs., Indpls. Mus. Art, Indpls. Zool. Soc., Pi Beta Phi. Clubs: Eastern Star, Woodland Country. Contbr. articles on child psychiatry to profl. jours. Research on emotional aspects of burns in children, craniofacial anomalies in children, also sex differences in child and adolescent population groups. Office: 11075 N Pennsylvania St Indianapolis IN 46280-1093

CAMPBELL, KATHLEEN CHARLOTTE MURPHEY, audiology educator; b. Sioux Falls, S.D., Mar. 20, 1952; d. Chester Humphrey and Ruth Maxine (Thompson) Murphey; m. Craig Anthony Campbell, Nov. 15, 1975. BA, S.D. State U., 1973; MA, U. S.D., 1977; PhD, U. Iowa, 1989. Cert. audiologist. Clin. grad. asst. dept. communication U. S.D., Vermillion, 1976-77; regional audiologist II British Columbia Ministry Health, Cranbrook, 1977-82; audiologist II dept. otolaryngology head and neck U. Iowa, Iowa City, 1983-88, rsch. asst. dept. speech, pathology and audiology, 1985; doctoral fellow Health Svcs. R&D, VA, Iowa City, 1987-88; assoc. prof. div. otolaryngology dept. surgery So. Ill. U. Sch. Medicine, 1989—; cons. Packer Engring., Naperville, Ill., 1992—. Editorial cons. Am. Jour. Audiology, 1992; reviewer Annals of Otolaryngology, 1992; contbr. articles to profl. jours. Mem. Midamerica Playwrights Theatre, Springfield, Ill., 1989—, Sierra Club, Springfield, 1989—. Recipient Clin. Investigator Devel. Award grant NIH, 1990, Small Bus. Innovative Rsch. grant NIH, 1990, Ctrl. Rsch. Coun. grant So. Ill. U., 1991, Children's Miracle Network award So. Ill. U., 1991, 92, Alzheimer Disease Ctr. grant So. Ill. U. Sch. Medicine, 1992. Mem. Am. Speech-Lang.-Hearing Assn., Am. Acad. Audiology, Assn. Rsch. in Otolaryngology, Am. Acad. Otolaryngology-Head/Neck Surgery (assoc.), Mensa. Achievements include development of a device for treatment of Meniere's disease; research in electrocochleography and perilymphatic fistual; in ototoxicity. Office: SIU Sch Medicine 301 N 8th St PAV 5B Springfield IL 62794-9230

CAMPBELL, LINZY LEON, microbiologist, educator; b. Panhandle, Tex., Feb. 10, 1927; s. Linzy Leon and Eula Irene (McSpadden) C.; m. Alice P. Dauksa, Feb. 7, 1953. B.A. in Bacteriology and Chemistry, U. Tex., 1949, M.A., 1950, Ph.D., 1952. Research scientist U. Tex., 1947-51; predoctoral research fellow NIH, 1951-52; postdoctoral research fellow Nat. Microbiol. Inst., U. Calif. at Berkeley, 1952-54; asst. prof., then asso. prof. Wash. State U., 1954-59; asso. prof. Western Res. U. Sch. Medicine, 1959-62; sr. research fellow USPHS, 1959-62; prof. microbiology U. Ill. at Urbana, 1962-72, head dept., 1963-71, dir. Sch. Life Scis., 1971-72; prof. microbiology, provost and v.p. acad. affairs U. Del., Newark, 1972-88, univ. rsch. prof. molecular bioscis., 1988-89, Hugh M. Morris rsch. prof. molecular bioscis., 1989—. Editorial bd.: Jour. Bacteriology, 1961-65; editor, 1964-65, editor-in-chief, 1965-77; Contbr. articles to profl. jours. Served with USNR, 1944-46. Fellow AAAS; mem. Am. Soc. Microbiology (chmn. publ. bd. 1965-80, councilor at large 1962-64, v.p. 1972-73, pres. 1973-74), Am. Soc. Biochemistry and Molecular Biology. Office: U Del 400 Morris Newark DE 19717

CAMPBELL, NAOMI FLOWERS, biochemist; b. Collins, Miss., Oct. 6, 1960; d. Jerome and Archie Lee Flowers; m. Edwin Glenn Campbell, Feb. 18, 1990. BS in Chemistry, Tougaloo Coll., 1982; PhD in Chemistry, U. So. Miss., 1989. Biochemist Olin E. Teague VA, Temple, Ariz., 1988; chemist Ft. Polk, La., 1989-90; biochemist USDA, New Orleans, La., 1990-92, Phila., 1992-93; asst. prof. U. South Ala., Mobile, 1993—. Co-author: Nucleic Acids Hybridization Detection, 1988, Erythrocyte Aging, 1989, Enzymatic Modification of Soy, 1992; contbr. articles to profl. jours. Mem. Am. Chem. Soc., Am. Assn. for the Advancement Sci., Sigma Xi. Democrat. Baptist. Achievements include biosensor development for detecting nucleic acids hybridization using piezoelectric resonance and development of bioaffinity purification method for aminoglycosides in biological fluids. Office: U South Alabama Dept Chem 600 E Mermaid Ln Mobile AL 36688

CAMPBELL, ROBERT P., information scientist. BA in Bus. Adminstrn., St. Lawrence U.; MBA, Am. U. CEO Advanced Info. Mgmt., Woodbridge, Va. Author: Am. Bankers Assn. Contingency Planning Manual; columnist Govt. Computer News; mem. editorial adv. bd. Inf. Systems Security, Privacy and Am. Bus., Infosecurity News, Computer Security Jour., Computer Fraud and Security Bull., Corporate Crime and Security Bull., IFIP Jour. Computers and Security, Info. Security Monitor; contbr. articles to profl. jours. Recipient Disting. Info. Scis. award Data Processing Mgmt. Assn., 1991. Office: Advanced Info Mgmt 1515 Davis Ford Rd Ste 6 Woodbridge VA 22191*

CAMPBELL, SHARON LYNN, communications company executive; b. Denver, Sept. 23, 1955; d. Frank Belanger and June Jewell (Wallace) Sauter; m. Gerald Ray Campbell, Jan. 7, 1978. BS in Safety and Health, U. Colo., 1977; MS in Occupational Health, NYU, 1980. Cert. safety profl. Environ. safety and health coord. Columbia U., N.Y.C., 1980-83; pres. S.L.C. Comm., N.Y.C., 1983-88, St. Louis, 1988-92, Fairless Hills, Pa., 1992—. Mem. AAAS, Am. Soc. Safety Profls. Achievements include education of disabled people in safety precautions and in educating safety profls. about the needs of disabled people. Home and Office: SLC Comm 509 Barbara Circle Fairless Hills PA 19030-3603

CAMPBELL, WILBUR HAROLD, research plant biochemist, educator; b. Santa Ana, Calif., Apr. 23, 1945; s. Russell Carton and Vivian (Yates) C.; m. Ellen Roth, June 6, 1981. AA, Santa Ana Coll., 1965; BA, Pomona Coll., 1967; PhD, U. Wis., 1972. Postdoctoral U. Ga., Athens, 1972-73, Mayo Clinic, Rochester, Minn., 1973-74, Mich. State U., East Lansing, 1974-75; asst. prof. Coll. Environ. Sci. and Forestry. SUNY, Syracuse, 1975-80, assoc. prof., 1980-85; assoc. prof. Mich. Technol. U., Houghton, 1985-86, prof., 1986—, adj. prof. Dept. Chemistry and Chem. Engring, 1990—, coord. Pytotech. Rsch. Ctr., 1990—; pres. & CEO The Nitrate Elimination Co., Inc.; guest prof. Botanisches Inst., U. Bayreuth, Fed. Republic Germany, 1982; vis. scientist Molecular Biology Computer Rsch. Resource Harvard U., 1987; vis. prof. U. Stockholm, 1991-92. Mem. editorial bd. Plant Physiology, 1982-92; contbr. numerous articles to profl. jours. Recipient Excellence in Rsch. award Mich. Tech. U., 1988, Disting. Faculty award Mich. U. Bd. Govs., 1989. Mem. AAAS, Am. Chem. Soc., Am. Soc. of Plant Physiologists, Am. Soc. Agronomy/Crop Sci. Soc. Am., Japanese Soc. Plant Physiologists, Plant Growth Regulator Soc. Am., N.Y. Acad. Sci., Internat. Soc. for Plant Molecular Biology, Scandinavian Soc. Plant Physiology, France Soc. Plant Physiology, Sigma Xi. Home: 334 Hecla St Lake Linden MI 49945-1323

CAMPBELL, WILLIAM BUFORD, JR., materials engineer, chemist, forensic consultant; b. Clarksdale, Miss., Nov. 23, 1935; s. William Buford and Bertha Lucille (Atkins) C.; m. Joan E. Stakem, June 29, 1963 (div. 1983); children: Lisa Anne, William Buford II, Heather Katherine, Matthew Rush. B in Ceramic Engring., Ga. Inst. Tech., 1958, MS in Engring., 1960; AM in Mineralogy, Inorganic Chemistry, Harvard U., 1962; PhD in Materials Engring., Ohio State U., 1967; postgrad., MIT, 1963-65, Ohio State U., 1969-72, NYU, 1980. Registered profl. engr., Ga. Asst. prof. dept. ceramic engring. Ohio State U., Columbus, 1967-69, assoc. prof., 1969-74; chief biomed. engring. dept. Doctor's Hosp., Morristown, Tenn., 1974-76; assoc. prof. engring. sci. and mechanics U. Tenn., Knoxville, 1974-77; sr. ptnr. Campbell, Churchill, Zimmerman & Assocs., Cons., Knoxville, 1977-84; cons., pvt. practice Biomed. Engring. and Forensic Sci., Knoxville, 1984—; ptnr. Brae Arden Farms Ltd., Phila., Tenn., 1979-83, tech. dir., chief exec. officer Southeastern Mobility Co., Inc., Phila., Tenn., 1977-84; founder Biomed. Systems Inc., Knoxville, 1986; project engr. TVA, 1987-89; dir. EPRI Ctr. for Materials Fabrication, Battelle Meml. Inst., Columbus, Ohio, 1989-90, dir., program mgr. Innovation and Tech. Transfer, Battelle Meml. Inst., Columbus, 1990. Patentee: Holds 3 U.S. Patents; contbr. articles to profl. jours. Mem. Adminstrv. Bd. Bearden United Meth. Ch., 1982-85; chmn. exec. com. The Ch. of the Redeemer, Concord, Tenn., 1985-87; mem. fin. com. Americans for Responsible Govt, 1984-85, fin. com. 50th Am. Presidential Inaugural, 1984-85, Nat. Adv. Bd. on Tech. and the Disabled; trustee Lakeshore Mental Health Insts., Dept. Mental Health, State of Tenn., 1986-93; bd. dirs. Vols. of Am., Knoxville, 1987-90; adv. bd. Knoxville Mus. Art, 1987-88; mem. adv. coun. Tenn. Sci. and Tech. Coun., State Tenn., 1993—. Recipient Cert. Recognition, NASA, 1973, Freeman award Am. Coun. for the Blind, 1976. Fellow Am. Inst. Chemists (cert. chemist), Am. Acad. Forensic Sci., Am. Ceramic Soc. (abstracter and reviewer 1962—, nat. programs and meetings com. 1968-72, div. chmn. 1969-70, other offices, life mem.); mem. ASTM (liaison mem. for ceramics and med., surg. materials and devices 1975—), Nat. Soc. Profl. Engrs., Am. Soc. for Engring. Edn., Am. Soc. for Nondestructive Testing, Nat. Inst. Ceramic Engrs. (life), Phi Lambda Upsilon, Tau Beta Pi (Eminent Engr. 1974), Sigma Xi (rsch. awards 1958, 60, 71), KERAMOS, others. Republican. Office: Performance Cons PO Box 51825 Knoxville TN 37950-1825

CAMPER, JEFFREY DOUGLAS, biology educator; b. Aurora, Ill., Feb. 5, 1960; s. Wayne Douglas and Gayle Joan (Sims) C.; m. Hope Elizabeth Armstrong, May 21, 1989. MA, Drake U., Des Moines, 1985; PhD, Tex. A&M U., 1990. Field scientist Harris County Mosquito Control Dist., Houston, 1989-90; rsch. assoc. Tex. A&M U., College Station, 1990-91; vis. asst. prof. biology Austin Coll., Sherman, Tex., 1991—; cons. Morrison-Knudsen Co., Inc., San Antonio, 1985-86, Horizons Environ. Svcs., Inc., Austin, Tex., 1991; guest speaker College Station (Tex.) Ind. Sch. Dist., 1986, Navasota (Tex.) Ind. Sch. Dist., 1990, Sherman (Tex.) Ind. Sch. Dist., 1991. Mem. Am. Soc. Ichthyologists and Herpetologists (grad. student com.), Soc. Sytematic Biology, Herpetologist's League, Sigma Xi.

CAMPIGOTTO, CORRADO MARCO, physicist; b. Olten, Solothurn, Switzerland, Sept. 20, 1962; s. Giuseppe and Maria Teresa (Facen) C. BSc in Physics, U. Basle, Switzerland, 1988; PhD, Inst. Nuclear Physics, Lyon, France, 1993. Researcher Inst. Nuclear Physics, Moscow, 1988-90; cons.

Solar platform, Almeria, Spain, 1987. Grantee Govt. of USSR, 1988-90, Govt. of France, 1991, Govt. of Switzerland, 1991. Avocations: jazz, astronomy, photography. Office: Inst Nuclear Physics, Theory 1 Lyon 1, Villeurbanne France 69622

CAMPO, J. M., mechanical engineer; b. Oct. 14, 1943; m. Vicky R. Herran, Dec. 2, 1971; children: Vanessa, Gunnar, Ingrid. MS, U. Barcelona, Spain; PhD in Mech. Engring., ETSIIB, 1967; MBA, IESE, Barcelona, 1968. Sr. instrument engr. Davy-McKee, Cleve., 1968-74; prin. instrument engr. Parsons Co., Pasadena, Calif., 1974-76; resident engr. Alcan, 1976-80; project mgr. Aramco Overseas Co., The Hague, Holland, 1980-81; gen. mgr. Enertec Engring., Houston and Mexico City, 1981-83; sr. project mgr. Intecsa Industrial, 1983—; bd. dir. TTX Corp. Patentee in field. Fellow Assn. Indsl. Engrs. Home: Ebro 33, Mayflower, Penascales, 28291 Torrelodones Madrid Spain

CAMPOS, JOAQUIN PAUL, III, chemical physicist, regulatory affairs specialist; b. L.A., Feb. 16, 1962; s. Joaquin Reyna and Maria Luz (Chavez) C.; m. Barbara Ann Esquivel, Oct. 31, 1987; children: Courtney Luz, Nathaniel Alexander. Student, U. Calif., Santa Cruz, 1980-85, UCLA, 1985-86. Tutor U. Calif., Santa Cruz, 1980-82, admissions liaison, 1982-84; chem. teaching assoc. L.A. Unified Sch. Dist., 1985-87; pvt. tutor Santa Clara, L.A., 1987-89; tech. specialist Alpha Therapeutics Corp., L.A., 1989—; cons. L.A. Unified Sch. Dist., 1985-87. Docent in tng. L.A. Mus. of Sci. and Industry, 1989—. Scholar, grantee So. Calif. Gas Co., L.A., 1980-84, Sloan Rsch. fellow, 1981-82. Mem. Am. Chem. Soc., N.Y. Acad. Sci., Am. Inst. Chemists, Am. Assn. Physics Tchrs., AAAS, Fedn. Am. Scientists, Internat. Union of Pure and Applied Chemistry, So. Calif. Paradox User Group, Math. Assn. Am. Avocations: reading, playing chess, computer programming, family. Office: Alpha Therapeutic Corp 5555 Valley Blvd Los Angeles CA 90032-3548

CAMPOS, LUÍS MANUEL BRAGA DA COSTA, mathematics, physics, acoustics and aeronautics educator; b. Lisbon, Portugal, Mar. 28, 1950; s. Elmano Neves and Francelina (dos Reis Braga) da Costa Campos; m. Maria Isabel Carreira de Vila-Santa, Aug. 8, 1978; children: Nuno Luis, Ana Isabel. Diploma Mech. Engring., Inst. Superior Tecnico, Lisbon Tech. U., 1972, Sc.D., 1982; Ph.D., Cambridge U., 1977. Lectr. applied mechanics and math. Inst. Superior Tecnico, Lisbon Tech. U., 1972-78, aux. prof., 1978-80, assoc. prof., 1980-85, prof., 1985—. Counsellor Nat. Inst. Sci. Rsch., 1985—. Author: Funcoes Complexas e Campos Potenciais; contbr. articles to profl. jours. Fellow Cambridge Philos. Soc., Royal Astron. Soc., Royal Aero. Soc.; mem. ASME, AIAA, Am. Math. Soc., Am. Astron. Soc., European Astron. Soc., European Math. Soc., London Math. Soc., Soc. Indsl. and Applied Math., Internat. Astron. Union, Adv. Group for Aerospace Rsch. and Devel. (chmn. Flight mechanics panel), Acoustic Soc. Am., European Sci. Found. (mem. space sci. com.), NSF (liaison mem. space sci. bd.), Societe Francaise d'Acoustique, Internat. Coun. for Aero. Scis., Aero. Rsch. and Tech. (v.p., mgmt. com. European Community Aero. program). Avocations: classical music, plastic arts, photography, swimming. Office: Inst Superior Tech, Av Rovisco Pais, 1096 Lisbon Portugal

CAMRAS, MARVIN, electrical and computer engineering educator, inventor; b. Chicago, Ill., Jan. 1, 1916; s. Samuel and Ida (Horwich) C.; m. Isabelle Pollack, 1952; children: Robert, Carl, Ruth, Michael, Louis. BSEE, Armour Inst. Tech. (now Ill. Inst. Tech.), 1940; MSEE, Ill. Inst. Tech., 1942, LLD (hon.), 1968. Registered profl. engr., Ill. Mem. staff Armour Rsch. Found. now Ill. Inst. Tech. Rsch. Inst., Chgo., 1940—, asst. physicist, 1940-45, assoc. physicist, 1945-46, physicist, 1946-49, sr. physicist, 1949-58, sr. engr., 1958-65, sci. adviser electronics div., 1965-69, sr. sci. adviser, 1969—; rsch. prof. elec. and computer engring. Ill. Inst. Tech., 1986—; mem. Am. Nat. Standards Inst., 1966—, chmn. s-4 com., 1966;. Author: Magnetic Tape Recording, 1985, Magnetic Recording Handbook, 1988, others; editor: Inst. Radio Engrs. Trans. on Audio, 1958-63. 500 patents in devel. wire and tape recorders, stereo sound reproduction, motion picture sound, video recorders, others in field. Recipient Disting. Svc. award Ill. Inst. Tech. Alumni Assn., 1948, Achievement award for outstanding contbn. motion picture photography U.S. Camera mag., 1949, John Scott medal Franklin Inst. Phila., 1955, citation Ind. Tech. Coll., 1958, Achievement award I.R.E., 1958, product award Indsl. Rsch. mag., 1966, Merit award Chgo. Tech. Socs., 1973, Alumni medal Ill. Inst. Tech., 1978, Inventor of Yr. award Patent Law Assn., Chgo., 1979, Nat. medal tech. Pres. of U.S., 1990, Am. Ingenuity award Coors Co./Nat. Assn. Mfrs., 1992; named to Ill. Inst. Tech. Hall of Fame, 1981, Am. Ingenuity Hall of Fame, 1992, Acad. Chgo. Assn. Tech. Socs., 1981, Pioneers of Electronics Foothills Electronics Mus., 1981, Nat. Inventors Hall of Fame Nat. Coun. Patent Law Assns., Patent and Trademark Office, U.S. Dept. Commerce, 1985. Fellow IEEE (sec.-treas. 1951-53, Consumer Electronics award 1964, nat. chmn. profl. group on audio I.R.E. 1953-54, Info. Storage award Magnetics Soc. 1990, other awards), Acoustical Soc. Am. (patent rev. bd.), AAAS, Soc. Motion Picture and TV Engrs. (bd. mgrs. Chgo. 1986-88, Hon. Mem. award 1990); mem. NAE, Western Soc. Engrs. (Washington award 1979), Physics Club Chgo. (bd. dirs. 1969—, pres. 1973-74), Radio Engrs. Club Chgo., Chgo. Acoustic and Audio Group (bd. dirs. 1967-68), Audio Engring. Soc. (cen. v.p. 1972-73, hon. gov. 1970—, John S. Potts Meml. Gold medal 1969), Midwest Acoustics Conf. (bd. dirs. 1969—), Sigma Xi (chpt. pres. 1959-60), Tau Beta Pi, Eta Kappa Nu. Home: 560 Lincoln Ave Glencoe IL 60022-1420 Office: Ill Inst Tech Dept Elec & Computer Engring Chicago IL 60616

CANALAS, ROBERT ANTHONY, nuclear engineer; b. Indpls., Mar. 14, 1940; s. Anton and Maria (Gabrovec) C.; m. Constance Turk, Aug. 7, 1965; children: Maria, Robert II, Christina, Anne. BS in Chemistry, Marian Coll., 1964; MS, U. Mo., Rolla, 1967, MS in Nuclear Engring., 1971. Registered profl. engr., Calif.; cert. sr. reactor operator. Chief chemist Link-Belt Co., Indpls., 1964-65; rsch. trainee Gen. Elec. Co., Schenectady, N.Y., 1967-69; nuclear engr. Ge. Electric Co., San Jose, Calif., 1974-86; chemist/nuclear engr. Commonwealth Edison Co., Chgo., 1971-74, Boston Edison Co., 1986—. Contbr. articles to Jour. Geophys. Rsch. Mem. K.C. (grand knight). Roman Catholic. Office: Boston Edison Co 25 Braintree Hill Park Boston MA 02184-8786

CANBY, CRAIG ALLEN, anatomy educator; b. Mt. Pleasant, Iowa, Oct. 23, 1959; s. Allen Rex and Betty Lou (Ross) C.; m. Lisa Ann Coen, June 23, 1990. BS, Iowa Wesleyan Coll., 1982; PhD, U. Iowa, 1988. Asst. prof. Grand View Coll., Des Moines, 1988-91, assoc. prof., 1991-92; adj. asst. prof. U. Osteo. Medicine and Health Scis., Des Moines, 1991-92, asst. prof., 1992—; lectr. Mercy Med. Ctr., Des Moines, 1992—; clin. anatomy cons. VA Med. Ctr., Des Moines, 1992—. Co-author (book chpt.) Subcellular Growth of Cardiocytes During Hypertrophy, 1987; contbr. articles to profl. jours. Judge South Ctrl. Iowa Sci. & Engring. Fair, Indianola, 1990-92, Hawkeye Sci. Fair, Des Moines, 1991; mem. sci. concepts com., Des Moines, 1992. Lectureship program grantee GTE, 1991. Mem. Sigma Xi. Office: U Osteo Medicine Health Sci 3200 Grand Ave Des Moines IA 50312

CANCRO, ROBERT, psychiatrist; b. N.Y.C., Feb. 23, 1932; s. Joseph and Marie E. (Cicchetti) C.; m. Gloria Costanzo, Dec. 8, 1956; children: Robert, Carol. Student, Fordham U., 1948-51; M.D., SUNY, 1955. Intern Kings County Hosp., Bklyn., 1955-56; resident in psychiatry Kings County Hosp., 1956-59; attending staff Gracie Sq. Hosp., N.Y.C., 1959-66; clin. instr. SUNY Downstate Med. Ctr., Bklyn., 1959-66; staff psychiatrist Menninger Found., Topeka, Kans., 1966-69; cons. Topeka State and VA Hosps., 1967-69; prof. dept. psychiatry U. Conn. Health Ctr., Farmington, 1970-76; prof., chmn. dept. psychiatry NYU Med. Ctr., 1976—; dir. N.S. Kline Inst. Psychiat. Research, 1982—; cons. psychiat. edn. br. NIMH; biol. scis. sect. NIMH. Editor 10 books.; Contbr. articles on schizophrenia to profl. jours. Recipient rieda Fromm-Reichmann award, 1975, Strecker award, 1978, Dean award, 1981, Lehmann award, 1992. Fellow A.C.P., Am. Coll. Psychiatrists, Am. Psychiat. Assn.; mem. Am. Psychol. Assn., Assn. Am. Med. Colls., Am. Assn. Social Psychiatry (pres.-elect 1982-84), N.Y. Acad. Scis., AAAS, AMA. Home: 118 Mclain St Mount Kisco NY 10549 Office: NYU Med Ctr 550 1st Ave New York NY 10016-6402

CANDIA, OSCAR A., ophthalmologist, physiology educator; b. Buenos Aires, Argentina, Apr. 30, 1935; came to U.S., 1964; s. Jose F. and Luisa P. (Mitri) C.; m. Blanca E. Fernandez, Jan. 8, 1960; children: Roberto, Leticia, Silvina. BS, Belgrano, Curuzu, Argentina, 1952; MD, U. Buenos Aires, 1959. Instr. U. Buenos Aires, 1960-63; asst. prof. U. Louisville, 1964-68; assoc. prof. Mt. Sinai Sch. Medicine, N.Y.C., 1968-77, prof. ophthalmology and physiology, 1977—; mem. research com. Nat. Eye Inst., Bethesda, Md., 1979-83, study sect. NIH, Bethesda, 1984-88, adv. bd. Eye-Bank for Sight Restoration, Inc., 1985—. Contbr. articles to profl. jours. Recipient research awards Am. Heart Assn., 1972, Nat. Eye Inst., Fight for Sight, Inc., 1975, recognition award Alcon Research Inst., 1985, Research to Prevent Blindness Sr. Investigator award, 1988. Mem. Am. Physiol. Soc., Assn. for Research in Vision and Ophthalmology (chmn. physiology 1976), Internat. Soc. for Eye Research. Avocations: tennis; auto mechanics. Office: Mt Sinai Sch Medicine 1 Gustave Levy Pl New York NY 10029

CANE, GUY, engineering test pilot, aviation consultant; b. N.Y.C., Aug. 3, 1929; s. Peter John and Claire (German) C.; m. Simone Desiree Rosier, June 6, 1954; children: Peter Guy, John Hendrik. BSEE, U.S. Naval Acad., 1954; MS in Internat. Affairs, George Washington U., 1975. Commd. ensign USN, 1954, advanced through grades to capt., 1974, served in various locations including Vietnam, ret., 1977; dir. engring. OmniJet, Inc., Rockville, Md., 1977-79; v.p. engring. Washington Jet, Inc., Rockville, 1979-80; pres. Cane Assocs., Inc., Annapolis, Md., 1980—. Contbr. articles to profl. publs. Trustee U.S. Naval Acad. Found., Inc., Annapolis, 1984—. Decorated Silver Star, DFC, Air medal. Office: Cane Assocs Inc PO Box 765 Annapolis MD 21404

CANE, MARK ALAN, oceanography and climate researcher; b. Bklyn., Oct. 20, 1944; s. Philip and Ida Deborah C.; m. Barbara Jane Haak, Oct. 28, 1968; children: Laura, Jacob. BA, Harvard U., 1965, MS, 1968; PhD, MIT, 1975. Earth scientist NASA, 1975-79; assoc. prof. oceanography MIT, Cambridge, 1979-84; Doherty sr. scientist Lamont Doherty Earth Obs. of Columbia, Palisades, N.Y., 1984—; cons. pvt. practice, Nyack, N.Y., 1984—. Contbr. articles to profl. jours. Organizer Student Non-Violent Coord. Conv., Ala., 1965. Mem. Am. Meteorol. Soc. (Harald Ulrick Sverdup Gold medal 1992), Am. Geophys. Union, Oceanography Soc., N.Y. Acad. Scis. Democrat. Jewish. Achievements include prediction of climate variations known as El Nino and the Southern Oscillation. Office: Lamont Doherty Earth Obs Palisades NY 10964

CANEBA, GERARD TABLADA, chemical engineering educator; b. Manila, Philippines, June 2, 1958; came to U.S., 1980; s. Doroteo Naval and Saturnina Manegdeg (Tablada) C.; m. Mary Ann Naret, Aug. 26, 1980; children: Christine Ann, Richard Noel, Benjamin Edwin, Katherine Lee. MS, U. Calif., Berkeley, 1982, PhD, 1985. Registered profl. engr., Philippines. Instr. U. of the Phlippines, Diliman, Quezon City, 1978-80; grad. student, rsch. asst. Lawrence Berkeley Lab., Berkeley, Calif., 1984-85; vis. asst. prof. Mich. Tech. U., Houghton, Mich., 1985-90, asst. prof., 1990—; assoc. rsch. engr. U. of the Philippines, 1978-80; adj. faculty Mich. Molecular Inst., Midland, 1985—; mem. tech. com. in polymer recycling Mich. Materials Processing Inst., 1990—. Contbr. articles to profl. jours. Recipient fellowship UNESCO, 1980-82, Faculty Devel. Fund award Dow Corning Corp., 1985-90, fellowship Filtration Soc., 1983-84, Rsch. grant NSF, 1989-92, Rsch. Excellence Fund award State of Mich., 1989-94. Mem. AICE, Am. Chem. Soc., Soc. Plastics Engrs., Sigma Xi. Achievements include patent on formation of anisotropic polymer membranes; devel. free-radical retrograde-precipitation polymerization process for environmentally friendly polymer systems; devel. phenomenological theory in polymer phase transition. Office: Mich Tech Univ 1400 Townsend Dr Houghton MI 49931-1295

CANER, MARC, physicist; b. Paris, Dec. 26, 1936; s. Marcel Moshe and Ophelia (Dub) C.; m. Hava Volokita, July 11, 1967; children: Roy, Yuval, Niv. BA, U. Buffalo, 1960; MS, Rensselaer Poly. Inst., 1963; DSc, Technion, Haifa, Israel, 1976. Vis. scholar Ohio State U., Columbus, 1984-86; rschr. Soreq Nuclear Rsch. Ctr., Yavne, Israel, 1964—; cons. Ben Gurion U., Beer-Sheva, Israel, 1979-82. Contbr. articles Jour. Chem. Physics, Nuclear Sci. and Engring., Annals Nuclear Energy, Phys. Rev. Mem. AIAA, Am. Physical Soc., Israel Nuclear Soc., Israel Physics Soc., Info. Processing Assn. Israel. Home: Herzog 23, Rehovot 76310, Israel Office: Soreq Nuclear Rsch Ctr, Yavne 70600, Israel

CANFIELD, GLENN, JR., metallurgical engineer; b. Springfield, Ill., Sept. 20, 1935; s. Glenn Sr. and Ruth (Kestel) C.; m. Beverly Joyce Brown, June 18, 1955 (div. 1981); children: Cheryl Lee, Charisse Marie, Glenn III (dec.), Derek Ethan; m. Virginia Nell Davis, June 28, 1982. Assoc. in Sci., Springfield Jr. Coll., 1957; BS in Metall. Engring., U. Ill., 1959; MBA, U. Tex., Tyler, 1987. Registered profl. engr., Tex. Sr. mfg. engr. Westinghouse Electric Co., Cheswick, Pa., 1961-63; supr. quality control Latrobe (Pa.) Steel Co., 1963-69; mgr. spl. products Crucible Specialty Metals, Syracuse, N.Y., 1969-73; mgr. product constrn. Lone Star (Tex.) Steel, 1974-76, supt. melting, 1976-86; pres. Thermo-Tech Co., Longview, Tex., 1986—; mng. dir. The Plum Group, Longview, 1986—; pres. Canfield & Assocs., Longview, 1986—; chmn. Canfield Engring. Inc., Longview, 1987—; bd. dirs. Dickson, Inc., Longview. Editor: The Americanist Newsletter, 1962-66, The Plum Index of Steel, 1986—; pub.: (newspaper) The Plum Crier, 1961-64, the Melt Shop Messenger; author tech. papers on deoxidation. Chmn. Plum Twp. (Pa.) Rep. Com., 1962-65; chmn. Gregg County Rep. Com., 1991—, chmn. ballot security, 1988; sec. Am. Party, N.Y., 1968-72; chmn. Tax Reform Immediately Com., Longview, 1980-84. With USN, 1952-56, Korea. Recipient Thomas Jefferson award Constl. Party Pa., 1966. Mem. NSPE, Am. Soc. for Engring. Mgmt., Am. Computer Mgrs. Assn., Am. Soc. for Metals, Am. Inst. Metall. Engrs. (Silver award 1980). Baptist. Achievements include patents on Deoxidation Products. Home: 303 Ramblewood Pl Longview TX 75601-3059 Office: Canfield Cos PO Box 9955 Longview TX 75608-9955

CANFIELD, JUDY OHLBAUM, psychologist; b. N.Y.C., May 15, 1947; d. Arthur and Ada (Werner) Ohlbaum; m. John T. Canfield (div.); children: Oran David, Kyle Danya. BA, Grinnell Coll., 1963; MA, New Sch. Social Rsch., 1967; PhD, U.S. Internat. U., 1970. Psychologist Mendocino State Hosp., Talmage, Calif., 1968-69, Douglas Coll., New Westminster, BC, Can., 1971-72, Family & Childrens Clinic, Burnaby, BC, Can., 1971-72; psychologist, trainer, cons. VA Hosp., Northampton, Mass., 1972-75; dir. New England Ctr., Amherst, Mass., 1972-76; dir., psychologist Gateways, Lansdale, Pa., 1977-78; asst. prof., psychologist Hahnemann Med. Ctr., Phila., 1978-84; pres., dir. Inst. Holistic Health, Phila., 1978-85; psychologist, cons. Berkeley, Calif., 1986—. Mem. task force, tng. com. Berkeley Dispute Resolution Svc., 1986-89; mem. measure H com. Berkeley United Sch. Dist., 1987-88. Mem. APA, Nat. Register Health Svc. Providers in Psychology, Nat. Assn. Advancement Gestalt Therapy (steering com. 1990), Calif. Psychol. Assn., Alameda County Psychol. Assn. (info.-referral svc. 1989—), Assn. Humanistic Psychology. Avocations: piano, horseback riding, ice skating. Office: 2031 Delaware St Berkeley CA 94709-2121

CANFIELD, PHILIP CHARLES, electrical engineer; b. La Jolla, Calif., Sept. 17, 1956; s. Charles Clinton and Marilyn Loree (Miller) C.; m. Lingzhou Li, July 25, 1986; 1 child, Jason. BS in Physics, Engring. Physics, Oreg. State U., 1984, MSEE, 1986, PhD, 1990. Grad. rsch. asst. Oreg. State U., Corvallis, 1984-90, instr., 1989; summer intern Tri-Quint Semiconductor, Inc., Beaverton, Oreg., 1985, 89; engr. semiconductor mfg. devel. Hewlett-Packard Co., Santa Rosa, Calif., 1990—. Contbr. articles to profl. publs. Mem. IEEE (sr.), Sigma Xi. Achievements include solution of longstanding problems in GaAs MESFET technology creating opprotunity for new circuit applications in GaAs MESFET technology. Office: Hewlett Packard Co 1412 Fountaingrove Pky Santa Rosa CA 95403

CANIZARES, CLAUDE ROGER, astrophysicist, educator; b. Tucson, June 14, 1945; s. Orlando and Stephanie (Bolan) C.; m. Jennifer Wilder, Aug. 31, 1968; children: Kristen, Alexander. BA, Harvard U., 1967, MA, 1968, PhD, 1972. Rsch. staff MIT, Cambridge, 1971-74, asst. prof. physics, 1974-78, assoc. prof. physics, 1978-84, prof. physics, 1984—, deputy dir. Ctr. for Space Rsch., 1989-90, dir. Ctr. for Space Rsch., 1990—; assoc. dir. NASA-AXAF Sci. Ctr.; chair Space Sci. Applicatin adv. com. NASA, 1992—, mem. Space Earth Sci. adv. com., Washington, 1986-88, mem. adv. coun. NASA, 1992—; mem. Astron. Astrophysics Survey Com. NRC, Washington, 1989-91. Contbr. over 100 articles to profl. jours. Royal Soc. vis. fellow, Cambridge, Eng., 1981-82, Alfred P. Sloan Found. fellow, 1980-84; NASA grantee, 1975--. Fellow Am. Phys. Soc.; mem. NAS, AAAS, Am. Astron. Soc., Internat. Astron. Union, Phi Beta Kappa, Sigma Xi. Achievements include first implementation of studies in x-ray spectroscopy and plasma diagnostics of supernova remnants, clusters of galaxies. Office: MIT Rm 37-241 Cambridge MA 02139

CANNADY, EDWARD WYATT, JR., retired physician; b. East St. Louis, Ill., June 20, 1906; s. Edward Wyatt and Ida Bertha (Rose) C.; m. Helen Freeborn, Oct. 20, 1984; children by previous marriage: Edward Wyatt III, Jane Marie Starr. AB, Washington U., St. Louis, 1927, MD, 1931. Intern in internal medicine Barnes Hosp., St. Louis, 1931-33, resident physician, 1934-35, asst. physician, 1935-74, emeritus, 1974—; asst. resident Peter Bent Brigham Hosp., Boston, 1933-34; fellow in gastroenterology Washington U. Sch. Medicine, 1935-36, instr. internal medicine 1935-74, emeritus, 1974—; cons. internal medicine Washington U. Clinics, 1942-74; physician St. Mary's Hosp., East St. Louis, 1935-77, pres. staff, 1947-49, chmn. med. dept., 1945-47; physician Christian Welfare Hosp., 1935-77, chmn. med. dept., 1939-53, dir. electrocardiography, 1936-77; dir. electrocardiography Centreville Twp. Hosp., East St. Louis; mem. staff Meml. Hosp., Belleville, Ill., St. Elizabeth Hosp., Belleville; pres. C.I.F. Dir. health service East St. Louis pub. schs., 1936-37; chmn. med. adv. bd. Selective Svc., 1941-45; pres. St. Clair County Coun. Aging, 1961-62; chmn. St. Clair County Home Care Program, 1961-68, St. Clair County Med. Soc. Com. Aging, 1960-70; del. White House Conf. Aging, 1961, 71, 81; mem. Adv. Coun. Improvement Econ. and Social Status Older People, 1959-66; bd. dirs., exec. com. Nat. Council Homemaker Svcs., 1966-73, chmn. profl. adv. com. 1971-73; bd. dirs. St. Louis Met. Hosp. Planning Commn., 1966-70; mem. Ill. Coun. Aging, 1966-74; mem. Gov.'s Council on Aging, 1974-76; mem. Ill. Regional Heart Disease, Cancer and Stroke Com.; mem. exec. com. Bi-State Regional Com. on Heart Disease, Cancer and Stroke; pres. Ill. Joint Council to Improve Health Care Aged, 1959-61; dir. Ill. Coun. Continuing Med. Edn., 1972-77, v.p., 1974-75. Trustee McKendree Coll., 1971-79; adv. bd. Belleville Jr. Coll. Sch. Nursing, 1970-78; bd. dirs. United Fund Greater East St. Louis, 1953-58. Recipient Disting. Service Award Am. Heart Assn., 1957, Disting. Achievement award, 1957; award Ill. Public Health Assn., 1971; Greater Met. St. Louis award in geriatrics, 1976. Diplomate Am. Bd. Internal Medicine. Fellow Am. Coll. Cardiology, Am. Geriatrics Soc., ACP (gov. 1964-70); mem. AMA (ho. dels. 1961-71, mem. aging com.; editorial adv. bd. Chronic Illness News Letter 1962-70, chmn. Ill. delegation 1964-66, mem. council vol. health agys.), Am. (dir. 1956-62, personnel and personnel tng. com. 1956-60), Ill. (pres. 1950-51) heart assns., St. Clair County (pres. 1952, bd. censors 1953-57), Ill. (sec. cardiovascular sect. 1957, chmn. sect., 1958-59; chmn. com. on aging, 1959-69, speaker Ho. Dels. 1964-68, pres. 1969-70) med. socs., Mason, St. Clair Country Club, Palmbrook Country Club (Sun City, Ariz.), Sun Cities Physicians Club, Palmbrook Country Club, Beta Theta Pi, Nu Sigma Nu, Alpha Omega Alpha. Presbyterian. Contbr. articles to med. jours. Address: 14406 Bolivar Sun City AZ 85351

CANNON, ALBERT EARL, JR., electrical engineer; b. New Bern, N.C., Dec. 1, 1951; s. Albert Earl and Peggy O'Neil (Boyd) C.; m. Vicki Lynn Potter, Oct. 29, 1972; children: Susan Michelle, Albert Earl III, Thomas Christofer Robert. BSEE, N.C. State U., 1973. Registered profl. engr., Fla. Mem. field engr. program GE Installation and Svc. Engring., Richmond, Va., 1973-74; field engr. GE Installation and Svc. Engring., Wilmington, N.C., 1974-79; area engr. GE Installation and Svc. Engring., Orlando, Fla., 1979-85, Wilmington, 1986—; area engr. GE-Industry Svcs. Engring., Nashville, 1985-86. Mem. IEEEĄ Nat. Soc. Profl. Engrs., Fla. Engring. Soc., Jaycees (pres. Altamonte-South Seminole chpt. 1982-83, dist. dir. Fla. 1984-85, N.C. state chmn. 1990-91, Outstanding Fla. pres. award 1982-83, Bill Roleson Meml. award 1984-85, Keith Upson Meml. award 1984-85, Seth L. Crapps Meml. award 1989-90). Republican. Methodist. Home: 6259 Turtle Hall Dr Wilmington NC 28409-2132 Ofifce: GE - IS&E PO Box 3916 Wilmington NC 28406

CANNON, GRACE BERT, immunologist; b. Chambersburg, Pa., Jan. 29, 1937; d. Charles Wesley and Gladys (Raff) Bert; m. W. Dilworth Cannon, June 3, 1961 (div. 1972); children: Michael Quayle, Susan Radcliffe, Peter Bert Cannon. AB, Goucher Coll., 1958; PhD, Washington U., St. Louis, 1962. Fellow Columbia U., N.Y.C., 1962-64, Columbia U. Coll. Physicians and Surgeons, N.Y.C., 1964-65; staff fellow NIH Nat. Cancer Inst., Bethesda, Md., 1966-67; cell biologist Litton Bionetics, Inc., Kensington, Md., 1972-80, head immunology sect., 1980-85; dir. sci. ImmuQuest Labs., Inc., Rockville, Md., 1985-88; pres. Biomedical Analytics, Inc., Rockville, Md., 1988—; mgr. ATLIS Fed. Svcs., Inc., Rockville, Md., 1991—; Mem. contract rev. coms. Nat. Cancer Inst., 1983-87. Contbr. articles to profl. jours. Mem. Pub. Svc. Health Club, Bethesda, Md., 1984—, sec., 1990—. Grantee USPHS, 1959-65, NSF, 1959. Mem. AAAS, Am. Assn. for Cancer Research, N.Y. Acad. Sci., Sigma Xi. Republican. Presbyterian. Achievements include dissemination of cancer clinical trials information. Home and Office: 4905 Ertter Dr Rockville MD 20852-2203

CANNON, PAUL JUDE, physician, educator; b. N.Y.C., Feb. 20, 1933; s. John W. and Rita K. Cannon; m. Chantal de Cannart d'Hammale, Aug. 5, 1959; children: Christopher P., Peter J., Anne B., Karen C. B.A., Holy Cross Coll., 1954, D.Sc. (hon.), 1977; M.D., Harvard U., 1958. Asst. prof. medicine Columbia U. Coll. Physicians and Surgeons, N.Y.C., 1965-70, assoc. prof., 1970-75, prof. medicine, 1975—; assoc. attending physician Presbyn. Hosp., N.Y.C., 1970-75, attending physician, 1975—; mem. cardiovascular and renal study sect. Nat. Heart, Lung and Blood Inst., 1974-77, mem. diagnostic radiology study sect., 1983—. Contbr. articles to profl. jours.; mem. editorial bd. Circulation Jour., 1976—, Am. Jour. Medicine, 1980—, Nephron, 1976—. Recipient Career Devel. award NIH, 1965-75; grantee Nat. Heart, Lung and Blood Inst., 1965—. Fellow Am. Coll. Cardiology, ACP; mem. Am. Assn. Physicians, Am. Soc. Clin. Investigation, Am. Physiol. Soc., Am. Heart Assn., Practicioners' Soc. N.Y. (sec. 1983-85). Democrat. Roman Catholic. Avocations: hiking; tennis; skiing; violin. Home: 96 Avondale Rd Ridgewood NJ 07450-1302 Office: Columbia Coll of Physicians and Surgeons 630 W 168th St New York NY 10032-3702

CANNON, RANDY RAY, civil engineer; b. Newberry, S.C., Oct. 14, 1953; s. Edgar Ray and Delene Mavis (Moates) C. BS, U. S.C., 1975, MEng, 1980. Registered profl. engr., S.C. Civil engr. I S.C. Dept. Hwys., Edgefield, 1975-78; civil engr. II S.C. Dept. Hwys., Columbia, 1978-83, asst. sr. bridge design engr., 1983-86, sr. br. design engr., 1986-90, bridge geotech. engr., 1990—. Producer: Geotechnical Guidelines for Structures. Councilman Colony Luth. Ch., Newberry, S.C., 1985-88, 89-92, treas., 1984-92. Mem. ASCE (br. treas. 1992-93), S.C. Soc. Engrs. (bd. dirs. 1989-92). Home: Rt 2 Box 240 Newberry SC 29108 Office: SC Dept Hwys 955 Park St Columbia SC 29202

CANNON, RICHARD ALAN, chemical process engineer; b. Spartanburg, S.C., June 10, 1944; s. Harold Alan and Mattie Cora (Settle) C.; m. Cheri Lynn Bracken, June 12, 1971; children: Kimberly Ann, Richard Alan, Michael Andrew, Courtney Lynn. B of Chem. Engring., Vanderbilt U., 1967. Registered profl. engr., Tex., S.C., Ala. Shift supr. Deering Milliken, Inc., Spartanburg, S.C., 1971-72; process engr. J. E. Sirrine Co., Greenville, S.C., 1972-73; sr. process engr. J. E. Sirrine Co./CRS Sirrine, Houston, 1973-85, CRS Sirrine, Inc., Greenville, 1985-88, Rust Internat., Birmingham, Ala., 1988—. Referee Timberline Youth Soccer Assn., Spring, Tex., 1980-85, Greenville Recreational Swimming League, 1986-88; asst. scoutmaster troop 266 Boy Scouts Am., Greenville, 1986-88. Lt. USN, 1967-71. Named Eagle Scout Boy Scouts Am., 1958. Mem. AICE (sec. ctrl. Ala. sect. 1991—), KC. Roman Catholic. Office: Rust Internat 100 Corporate Pkwy Birmingham AL 35242

CANTERBURY, RONALD A., biologist; b. Cleve., Nov. 20, 1966; s. Mac Canterbury and Dollie Mae (Smithson) Stover; m.Beverly Dawn Hamilton, June 15, 1991. BS, Concord Coll., 1989; MS, Marshall U., 1991. Gypsy Moth trapper USDA, Ripley, W.Va., 1986-88; grad. asst., field researcher biology dept. Marshall U., Huntington, W.Va., 1989-91; grad. asst. biology dept. Cleve. State U., 1991—; faculty part-time biology dept. Cuyahoga Community Coll., Parma, Ohio, 1992—. Contbr. articles to profl. pubs. Grad. sch. grantee Marshall U., 1990; recipient Hewey A. Wells Sr. Biologist award, 1989. Mem. AAAS, Am. Soc. Zoologists, Am. Ornithologists Union, Assn. Field Ornithologists, W.Va. Acad. Sci., Wilson Ornithologcial Soc., Brooks Bird Club (grantee 1993), Cooper Ornithological Soc. Home: Apt

205 1395 Cleveland Hts Rd Cleveland Heights OH 44121 Office: Cleve State U Biology Dept 1983 E 24th St Cleveland OH 44115

CANTILLI, EDMUND JOSEPH, safety engineer educator, author; b. Yonkers, N.Y., Feb. 12, 1927; s. Ettore and Maria (deRubeis) C.; m. Nella Franco, May 15, 1948; children: Robert, John, Teresa. AB, Columbia U., 1954, BS, 1955; cert., Yale Bur. Hwy. Traffic, 1957; PhD in Transp. Planning, Poly. Inst. Bklyn., 1972; postgrad. in urban planning and pub. safety, NYU, 1968-71. Registered profl. engr., N.Y., N.J., Calif.; profl. planner, N.J.; cert. safety profl. (BCSP); cert. planner (AICP). Supervising engr. safety rsch. and studies Port Authority of N.Y. & N.J., 1955-69; prof. transp. and safety engring. Poly. U., N.Y.C., 1969-90; pres. Urbitran Assocs., 1973-81; exec. dir., chmn. bd. Internat. Inst. for Safety Trans., Inc., 1977—; pres. EJC Safety Assocs., Inc., 1989—; tchr. Italian, algebra, traffic engring., urban planning, transp. planning, urban and transp. geography, land use planning, aesthetics, environment, indsl., traffic and transp. safety engring., human factors engring., ethics for engrs.; cons. transp. and traffic safety engring., community planning, traffic engring., transp. planning, accident reconstrn., environ. impacts, 1969—; vis. prof. transp. safety engring. Inst. Superior Técnico, Lisbon, 1987—; advisor to doctorate students Poly. U., CUNY, 1969—, Politecnico di Milano, U. Trieste, Italy, 1980—. Author: Programming Environmental Improvements in Public Transportation, 1974, Transportation and the Disadvantaged, 1974, Transportation System Safety, 1979; editor: Transportation and Aging, 1971, Pedestrian Planning and Design, 1971; editor, contbr.: Traffic Engineering Theory and Control, 1973; editor and calligrapher There Is No Death That Is Not Ennobled by So Great A Cause, 1976; contbr. over 200 articles to profl. jours. and trade jours.; developer methods of severity evaluation of accidents, identification, priority-setting and treatment of roadside hazards, transp. system safety, methodology; expert systems for improving traffic safety; introduced diagrammatic traffic signs, collision energy-absorption devices. With U.S. Army, 1945-49, 50-51. Fellow ASCE, Inst. Transp. Engr.; mem. NSPE, Am. Planning Assn. (charter), Am. Inst. Cert. Planners (cert.), Am. Soc. Safety Engrs., N.Y. Acad. Scis., Nat. Assn. Profl. Accident Reconstrn. Specialists, Internat. Assn. for Accidents and Traffic Medicine, Human Factors Soc., System Safety Soc., Sigma Xi. Home: 134 Euston Rd West Hempstead NY 11552-1024 Office: PO Box 63 Franklin Sq New York NY 11010

CANTLIFFE, DANIEL JAMES, horticulture educator; b. N.Y.C., Oct. 31, 1943; s. Sarah Lucretia Keesler C.; m. Elizabeth F. Lapetina, June 5, 1965; children: Christine, Deanna, Danielle, Cheri. MS, Purdue U., 1967, PhD, 1971. Asst. prof. horticulture U. Fla., Gainesville, 1974-76, assoc. prof., 1976-81, prof., 1981—, asst. chair dept., 1983-84, acting chair dept., 1984-85, chmn. dept., 1985-92, acting chair dept. fruit crops, 1991-92, chair dept. hort. scis., 1992—; vis. prof. U. Hawaii, Honolulu, 1979-80; sci. cons. Sun Seeds Genetics, Hollister, Calif., 1987, Pillsbury Co., 1987—, Teltech Inc., Bloomington, Minn., 1988—. Author: (with others) Tissue Culture Propagation Develpment, 1985, Development of Artificial Seeds, 1987, Somatic Embryos as a Tool for Synthetics, 1990, Automated Evaluation of Somatic, 1991, Micropropagation of Sweet Potato, 1991. Recipient Rsch. award Fla. Fruit and Vegetable Assn., Orlando, 1986, Alumni Achievement award Delaware Valley Coll., Doylestown, 1990. Fellow Am. Soc. Hort. Sci. (v.p. rsch. 1991—, pres.-elect 1993-94, Outstanding Grad. Educator award 1991); mem. Fla. State Hort. Soc. (v.p. vegetable sect. 1984-85, pres. 1991-92, chair exec. com. 1992-93), Internat. Soc. Hort., Am. Soc. Plant Physiologists, Sigma Xi, Delta Tau Alpha, Phi Kappa Phi, Gamma Sigma Delta, Phi Beta Delta. Office: U of Fla Hort Scis Dept 1255 Fifield Hall Gainesville FL 32611-0514

CANTONI, GIULIO LEONARDO, biochemist, government official; b. Milan, Italy, Sept. 29, 1915; s. Umerto L. and Nella (Pesaro) C.; m. Gabriella S. Sobrero, May 29, 1965; children: Allegra, Serena. M.D., U. Milan, 1939. Research asst. dept. pharmacology Oxford U., Eng., 1940; instr. NYU, N.Y.C., 1943-45; asst. prof. L.I. Coll. Medicine, N.Y., 1945-48; sr. fellow Am. Cancer Soc., N.Y.C., 1948-50; assoc. prof. Western Res. U., Cleve., 1950-54; chief lab. cellular pharmacology NIMH, NIH, Bethesda, Md., 1954-56, chief lab. gen. and comparative biochemistry, 1956—. Patentee 3-Deazaaristeromycin and uses. Mem. Nat. Acad. Scis., Am. Acad. Arts and Scis., Biochem. Soc. Eng., Am. Soc. Biol. Chemists. Home: 6938 Blaisdell Rd Bethesda MD 20817-3039 Office: NIMH Bldg 36 Room 3D-06 9000 Rockville Pike Bethesda MD 20892-0001

CANTONI, LOUIS JOSEPH, psychologist, poet, sculptor; b. Detroit, May 22, 1919; s. Pietro and Stella (Puricelli) C.; m. Lucile Eudora Moses, Aug. 7, 1948; children: Christopher Louis, Sylvia Therese. AB, U. Calif., Berkeley, 1946; MSW, U. Mich., 1949, Ph.D., 1953. Personnel mgr. Johns-Manville Corp., Pittsburg, Calif., 1944-46; social caseworker Detroit Dept. Pub. Welfare, 1946-49; counselor Mich. Div. Vocat. Rehab., Detroit, 1949-50; conf. leader, tchr. psychology, coordinator family and community relations program Gen. Motors Inst., Flint, Mich., 1951-56; from assoc. prof. to prof., dir. rehab. counseling Wayne State U., Detroit, 1956-89, prof. emeritus, 1989—. Author books including: The 1939-1943 Flint Michigan Guidance Demonstration, 1953, Marriage and Community Relations, 1954, (with Mrs. Cantoni) Counseling Your Friends, 1961, Supervised Practice in Rehabilitation Counseling, 1978, Writings of Louis J. Cantoni, 1981, Essays, Theses and Projects in Rehabilitation Counseling, 1989; (poetry) With Joy I Called to You, 1969, Gradually The Dreams Change, 1979. Editor: Placement of the Handicapped in Competitive Employment, 1957; co-editor: Preparation of Vocational Rehabilitation Counselors through Field Instruction, 1958; prin. editor: (poetry) Golden Song Anthology, 1985. Editor jours.: Mich. Rehab. Assn. Digest, 1961-63, Grad. Comment, 1963-64; poetry editor Cathedral Digest, 1973-75. Contbr. articles, revs., poems and illustrations to jours. Judge Mich. regional and nat. essay and poetry contests, 1965-77; bd. dirs. Mich. Rehab. Assn., 1962-64, 78-79, Mich. Rehab. Conseling Assn., 1985-87. Served to 2d lt. AUS, 1942-44. Recipient award for leadership and service Mich. Rehab. Assn., 1964, Mich. Rehab. Counseling Assn., 1985, 87, 88, Outstanding Service award Mich. State Bd. Edn., 1989; South and West ann. poetry award, 1970; Award for Meritorious Service Wayne State U., 1971, 81, 86, 87, 89; Outstanding Service award Poetry Soc. Mich., 1984. Fellow AAAS; mem. Coun. of Rehab. Counselor Educators (sec. 1957-58, chmn. 1965-66), Am. Assn. Univ. Profs., Am. Psychol. Assn., Nat. Rehab. Assn., Nat. Assn. Rehab. Profls. in Pvt. Sector, Nat. Congress of Orgns. of the Physically Handicapped, Nat. Assn. of the Physically Handicapped, Nat. Alliance for the Mentally Ill, Mich. Rehab. Assn. (pres. 1963-64), Detroit Rehab. Assn. (pres. 1958), Mich. Assn. for Counseling and Devel., Mich. Career Devel. Assn., Internat. Inst. Met. Detroit, World Poetry Soc. (Edwin A. Falkowski Meml. award 1990), Acad. Am. Poets, Detroit Inst. Arts, Friends of Detroit Pub. Libr., Soc. For Study of Midwestern Lit., U.S. Hist. Soc., Italic Studies Inst., USN Meml., Internat. Sculpture Ctr., Nat. Sculpture Soc., Sculptors Guild Mich., Lladro Collectors Soc., Birmingham-Bloomfield Art Assn., Poetry Soc. Mich., Univ. Club, Scarab Club (Detroit), Phi Kappa Phi, Phi Delta Kappa. Democrat. Episcopalian. Achievements include research in theory and practice of counseling and psychotherapy, psychosocial aspects of disabling conditions, therapeutic and vocational counseling with disabled persons, workplace accommodation for the disabled, vocational rehabilitation of the severely disabled. Home: 2591 Woodstock Dr Detroit MI 48203

CANTRALL, IRVING J(AMES), entomologist, educator; b. Springfield, Ill., Oct. 6, 1909; s. Ula J. and Elsie M. (LaRue) C.; m. Dorothy Louise Ransom, Dec. 24, 1932; children: Marion Louise, James Bruce. AB, U. Mich., 1935, PhD, 1940. Asst. Mus. Zoology U. Mich., Ann Arbor, 1935-37, tech. asst. Mus. Zoology, 1937-42, from asst. to prof. Zoology, 1949-77; curator Edwin S. George Res., Ann Arbor, 1949-59; curator insects Mus. Zoology U. Mich., Ann Arbor, 1959-77, curator insects emeritus, 1978—, prof. emeritus, 1978—; jr. aquatic biologist TVA, 1942, asst. aquatic biologist, 1942-43; asst. prof. biology U. Fla., Gainesville, 1946-49. Editor Great Lakes Entomologist, 1971-76; contbr. numerous articles on zoology to profl. jours. 1st lt. USAAF, 1943-46. Fellow AAAS; mem. Am. Entomol. Soc., Mich. Entomol. Soc. (pres. 1957-58, editor 1971-75), Soc. for Study of Evolution, Ecol. Soc. Am., Soc. Systematic Zoology, The Orthopterists Soc., Mich. Acad. Sci., Art and Letters (sec. 1963-65), N.Y. Acad. Sci., Ind. Acad. Sci., Fla. Acad. Sci., Sigma Xi, Phi Kappa Phi, Phi Sigma, Gamma Alpha. Home: 1531 Las Vegas Dr Ann Arbor MI 48103-5765

CANTRELL, CHRISTOPHER ALLEN, atmospheric chemist; b. Aurora, Colo., Dec. 31, 1954; s. Clifford Allen Cantrell and Martha Almeda Steakley Cansler; m. Terri Ann Matlin, Jan. 30, 1982; children: Kathryn Ann, Ryan Allen. BA, Kans. Wesleyan U., 1977; MS, PhD, U. Mich., 1983. Postdoctoral fellow Nat. Ctr. for Atmospheric Rsch., Boulder, Colo., 1983-85, scientist in atmospheric chemistry, 1985—. Contbr. articles to profl. jours. Mem. Am. Geophys. Union, Universal Round Dance Coun. (dir.). Achievements include development of instruments to measure gas phase NO2 by luminol chemiluminescence, and to measure gas phase peroxy radicals by chemical amplification. Office: NCAR 1850 Table Mesa Dr Boulder CO 80303

CANTWELL, THOMAS, geophysicist, electrical engineer; b. Buffalo, June 25, 1927; s. Thomas and Helen (Robinson) C.; children: Elizabeth Raye, Thomas III, Douglas. BSChemE, MIT, 1948, MSChemE, 1949; MBA, Harvard, Boston, 1951; PhD in Earth Sci., MIT, 1960. Registered profl. engr., Tex.; lic. geologist, Calif. Project engr. nuclear engring. dept. MIT, Cambridge, 1951-58, mem. faculty, 1958-65; pres. Mandrel Industries, Houston, 1966-70, Petroleum Holdings, Inc., Houston, 1970-78, Ind. Exploration, Inc., Houston, 1978-84, Tech. Computer Graphics, Inc., Houston, 1984—. With U.S. Army, 1946-47. Fellow Royal Geographic Soc.; mem. IEEE, Soc. Exploration Geophysicists, Am. Assn. Petroleum Geologists. Achievements include patent for Helium Leak Detector. Home: 3949 Ann Arbor Dr Houston TX 77063-6396 Office: Tech Computer Graphics Inc 3949 Ann Arbor Dr Houston TX 77063-6396

CAO, JIA DING, mathematics educator; b. Shanghai, People's Republic China, Sept. 14, 1940; s. Cao An Shi and Shen Xi Fen. Grad. in math., Fudan U., Shanghai, 1962. Asst. dept. math. Fudan U., Shanghai, 1962-85, lectr. dept. math., 1985-88, prof. dept. math., 1988—. Reviewer editorial office Math. Revs., U.S., 1991—; contbr. articles and abstracts to numerous profl. and internat. jours. in U.S., Germany, Italy, Portugal, Hungary, and China. Mem. Am. Math. Soc., Am. Planetary Soc. Home: Feng Yang Rd 376 No 13, Shanghai 200003, China Office: Fudan U Dept Math, Han Dan Rd No 220, Shanghai 200433, China

CAO, YOU SHENG, materials science and engineering researcher; b. Shanghai, China, July 31, 1955; came to U.S., 1983; s. Weiming Cao and Man Yu; m. Weihua Bi, Feb. 1983; children: Hannah, Hanson. BS, Shanghai U. Sci. and Tech.; MS, Oreg. State U.; PhD, U. Pitts. Rsch. asst. U. Pitts., Pa., 1985-91; rsch. assoc. U. Pitts., 1991—; rsch. cons. Al Co. of Am., Pitts., 1992—. Contbr. articles to Jour. Phys. Chemistry. Recipient Chinese Outstanding Coll. Students awards Shanghai, 1980, 81, 82; named Teaching fellow U. Pitts., Pa., 1987. Mem. Am. Physics Soc., Sigma Xi. Home: 3333 Juliet St Pittsburgh PA 15213 Office: Univ Pitts Dept Materials Sci/Engring Pittsburgh PA 15261

CAO, ZHIJUN, physicist; b. Wuxi, Jiangsu, People's Republic of China, Nov. 1, 1963; came to U.S., 1991; s. Shunlin Cao and Mingxia He. MS in Physics, Chinese Acad. Scis., 1989. Rsch. asst. Inst. High Energy Physics, Chinese Acad. Scis., Beijing, 1986-89; rsch. asst., teaching asst. Iowa State U., Ames, 1991—. Mem. Am. Phys. Soc. Achievements include research on quark degrees of freedom in nuclei, epsecially on the EMC effect. Home: 133 B University Village Ames IA 50010 Office: Iowa State Univ Dept Physics Ames IA 50011

CAPE, RONALD ELLIOT, biotechnology company executive; b. Montreal, Que., Can., Oct. 11, 1932; came to U.S., 1967, naturalized, 1972; s. Victor and Fan C.; m. Lillian Judith Pollock, Oct. 21, 1956; children: Jacqueline R., Julie A. AB in Chemistry, Princeton U., 1953; MBA, Harvard U., 1955; PhD in Biochemistry, McGill U., Montreal, 1967; postgrad., U. Calif., Berkeley, 1967-70. Customs, purchasing and advt. clk. Merck and Co., Ltd., Montreal, 1955-56; pres. Profl. Pharm. Corp., Montreal, 1960-67; chmn. bd. Profl. Pharm. Corp., 1967-73; pres. Cetus Corp., Emeryville, Calif., 1972-78; chmn. bd. Cetus Corp., 1978-91; chmn. Darwin Molecular Corp., 1992—; mem. adv. coun. dept. molecular biology Princeton U.; adj. prof. bus. adminstrn. U. Pitts.; vis. prof. biochemistry Queen Mary Coll., U. London; bd. dirs. Neutrogena Corp., The Found. Nat. Medals of Sci. & Tech., 1992—; mem. bus. affairs com. Ann. Revs., Inc., 1975-80; mem. impacts of applied genetics adv. panel to Office Tech. Assessment; mem. adv. com. on life scis. Natural Scis. and Engring. Rsch. Coun. Can. Mem. Rockefeller U. Coun.; bd. dirs. U. Calif. Art Mus. Coun., Berkeley, 1974-76; trustee Head-Royce Schs., Oakland, Calif., 1975-80, Rockefeller U., 1986-90, San Francisco Conservatory Music, The Keystone Ctr., 1987-93; bd. dirs. San Francisco Opera Assn., mem. budget and fin. com., 1992—, U. Waterloo Inst. for Biotech. Rsch.; mem. bus. adv. com. U. Calif., Berkeley; mem. Berkeley Roundtable on Internat. Economy; mem. sci. adv. bd. Bio-Technology Mag., 1987—; trustee Princeton U., 1989-93; mem. bd. regents Nat. Libr. of Medicine, 1989-92. Fellow AAAS; mem. Am. Soc. Microbiology (Found. for Microbiology lectr. 1978-79), Can. Biochem. Soc., Fedn. Am. Scientists, Royal Soc. Health, Soc. Indsl. Microbiology, Indsl. Biotech. Assn. (founding mem., pres. 1983-85, dir.), N.Y. Acad. Scis., Princeton Club of N.Y., Commonwealth Club of Calif., Sigma Xi. Jewish. Office: 220 Montgomery St Ste # 1010 San Francisco CA 94104

CAPECCHI, MARIO RENATO, geneticist, educator; b. Verona, Italy, Oct. 6, 1937; m. 1963. BS, Antioch Coll., 1961; PhD in Biophysics, Harvard U., 1967. Soc. fellows, jr. fellow biophysics Harvard U., 1966-68, from asst. prof. to assoc. prof. biochemistry med. sch., 1968-73; prof. human genetics U. Utah Sch. Medicine, Salt Lake City, 1973—; established investigator Am. Heart Assn., 1969-72. Recipient Biochemistry award Am. Chem. Soc., 1969, Career Devel. award NIH, 1972-74, Am. Cancer Faculty Rsch. award 1974-79, Gairdner Found. Internat. award Gairdner Found. (Can.), 1993. Mem. NAS, Am. Biochemical Soc., Am. Soc. Biol. Chemistry. Achievements include research in gaining an understanding of how the information encoded in the gene is translated by the cell, on expression in Eucaryotic and Procaryotic cells, in somatic cell genetics. Office: U Utah Sch Medicine Dept Human Genetics Salt Lake City UT 84132*

CAPELLOS, CHRIS SPIRIDON, chemist; b. Athens, Greece, Oct. 22, 1934; came to U.S., 1966, naturalized, 1976; s. Spiridon Em. and Melpo Christou (Christidou) C.; m. Helen Nicholaou Sakkoulas, Dec. 3, 1959; children: Melina, Maria. B.S. in Chemistry, Athens, 1959; D.I.C. in Nuclear Tech., Imperial Coll., London U., 1962, Ph.D., London U., 1965. Rsch. assoc. Brookhaven (N.Y.) Nat. Lab., 1966-68, assoc. chemist, 1968, vis. assoc. chemist, 1968-72; sr. rsch. chemist Energetic Materials Div., Armament Rsch. and Devel. Command, Dover, N.J., 1968—, sr. scientist, 1972—; vis. scientist Davy Faraday Lab., Royal Inst., 1970-71; NRC rsch. advisor; bd. dirs. NATO Advanced Study Inst., 1980, 85; vice-chmn. Gordon Rsch. Conf. on Energetic Materials for 1990, chmn., 1992. Editor NATO Conf. Proceedings, 1986. Served with Greek Army, 1965-66. Air Force Office Sci. Rsch. fellow, 1962-65; NATO awardee, 1979-80, 83. Mem. Am. Chem. Soc., Radiation Rsch., N.Y. Acad. Scis., Sigma Xi (pres. Picatinny chpt.). Author: Kinetic Systems, 1972, Japanese transl., 1978; editor profl. conf. procs., 1980; contbr. writings to sci. jours. Home: 11 Cambridge Rd Morris Plains NJ 07950-1529 Office: USA ARRADCOM Attn DRDAR-LCE-D Dover NJ 07801

CAPEN, CHARLES CHABERT, veterinary pathology educator, researcher; b. Tacoma, Sept. 3, 1936; s. Charles Kenneth and Ruth (Chabert) C.; m. Sharron Lee Martin, June 27, 1968. DVM, Wash. State U., 1960; MS, Ohio State U., 1961, PhD, 1965. Diplomate Am. Coll. Vet. Pathologists (pres. 1978-79, coun. 1975-81). Instr. dept. vet. pathology Ohio State U., Columbus, 1962-65, asst. prof. dept. vet. pathology, 1965-67, assoc. prof., 1967-70, prof., 1970—, prof. endocrinology Coll. Medicine, 1972—, acting chmn. dept. vet. pathobiology, 1981-82, chmn., 1982—; cons. clinician Ohio State U. Vet. Teaching Hosp., 1973—; Israel Doniach Meml. lectr. Brit. Endocrine Soc. meeting, Manchester, 1989. Editor: (series) Animal Models of Human Disease, 1979—; mem. editorial bd. Lab. Investigation, 1988—, Vet. Pathology, 1986-87. Mem. Opera Columbus, 1982—, Columbus Symphony Assn., 1972—. Recipient Nat. Borden Rsch. award Am. Vet. Med. Assn., 1975, Small Animal Rsch. award, 1984, Gaines Rsch. award, 1987, Disting. scholar award, Ohio State U., 1993, Dean;'s Teaching Excellence award Edn. Coll. Veterinary Medicine, Ohio State U., 1993. Mem. NAS, Inst. Medicine, Internat. Acad. Pathology (coun. U.S.-Can. div. 1989-92). Avocations: wildlife and nature photography, travel. Office: Ohio State U Dept Vet Pathobiology 1925 Coffey Rd Rm 207A Goss Lab Columbus OH 43210-1093

CAPLAN, ARNOLD I., biology educator; b. Chgo., Jan. 5, 1942; s. David and Lillian (Diskin) C.; m. Bonita Wright, July 4, 1965; children: Aaron M., Rachel L. BS, Ill. Inst. Tech., 1963; PhD, Johns Hopkins U., 1966. Asst. prof. Case Western Res. U., Cleve., 1969-75, assoc. prof., 1975-81, prof. devel. genetics, anatomy, 1981-88, dir. cell molecule basis aging tng. program, 1981—, dir. skeletal rsch. ctr., 1986—, prof. biophysics, physiology, 1989—; vis. prof. U. Calif., San Francisco, 1973, Inst. de Chimie Biologique, Strasbourg, France, 1976-77; Erna and Jakob Michael vis. prof. Weizmann Inst. Sci., Rehovot, Israel, 1984-85. Contbr. articles to profl. jours. Recipient Career Devel. award NIH, 1971-76; Am. Cancer Soc. fellow, 1967-69; Josiah Macy Faculty scholar Case Western Res. U., 1976-77. Mem. Am. Assn. Orthopaedics Surgery (Elizabeth Winston Lanier Kappa Delta award 1990), Orthopaedics Rsch. Soc., Soc. Devel. Biology, AAAS, Am. Soc. Cell Biology. Jewish. Office: Case Western Res U Dept Biology 2080 Adelbert Rd Cleveland OH 44106-2623

CAPLAN, DAVID NORMAN, neurology educator; b. Montreal, Que., Can., Feb. 4, 1947; came to U.S., 1988; s. Hyman and Sonia (Roskies) C.; 1 child, Hilary. PhD, MIT, 1971; MD, McGill U., Montreal, 1975. Asst. prof. medicine U. Ottawa, Ont., Can., 1980-81; assoc. prof. neurology Temple U., Phila., 1981-82, McGill U., 1982-88, Harvard U., Boston, 1988—; dir. neuropsychology Mass. Gen. Hosp., Boston, 1988—; mem. CMS study section NIH, Bethesda, Md., 1992—. Author: Neurolinguistics and Linguistic Aphasiology, 1987, Language: Structure, Processing and Disorders, 1992; co-author: Disorders of Syntactic Comprehension, 1988. NIH grantee, 1989—. Achievements include discovery of various specific disorders affecting sentence comprehension in aphasic patients. Office: Neuropsychol Lab MGH Fruit St Boston MA 02114

CAPLAN, YALE HOWARD, toxicologist, consultant; b. Balt., Dec. 27, 1941; s. Louis and Yetta (Edelson) C.; m. Suzanne Joan Libowitz, Feb. 4, 1965; children: Ilene, Lee, Erica. BS in Pharmacy, U. Md., 1963, PhD, 1968. Registered pharmacist, Md; diplomate Am. Bd. Forensic Toxicology. Rsch. assoc., supr. dept. rsch., oncology, cell biology Sinai Hosp., Balt., 1968-70; asst. toxicologist office of chief med. examiner State of Md., Balt., 1970-74, chief toxicologist, 1974-91; dir. Nat. Ctr. Forensic Sci., Md. Med. Lab., Inc., Balt., 1988—, dir. forensic and spl. toxicology, 1992—; adj. prof. dept. pharmacology and toxicology U. Md., 1986—, dept. biomed. chemistry, 1986—; clin. prof. dept. pathology, 1985—; professorial lectr. dept. forensic scis. George Washington U., Washington, 1975-76; asst. in surgery Johns Hopkins U., Balt., 1969-70, lectr. div. forensic pathology, 1973-84; faculty Nat. Inst. Drug Abuse, SAMSHA, 1992—, drug testing adv. bd. 1990—; cons. Abbott Labs., Chgo., 1988-90, Coll. Am. Pathologists, 1987—; sr. assoc. Bensinger, DuPont and Assocs., Chgo., 1987-90; cons. in field. Author: Review Questions in Analytical Toxicology, 1982; editorial adv. bd. Jour. Forensic Scis., 1984—, Jour. Analytical Toxicology, 1976—; contbg. editor Employee Testing and the Law, 1986—; editor Forensic Scis. Rev., 1989—; assoc. editor Jour. Forensic Scis., 1993—; contbr. chpts. to books, articles to refereed jours. Fellow Am. Inst. Chemists, Am. Acad. Forensic Scis. (bd. dirs. 1986-89, exec. com. 1986-88, pres. 1987-88); mem. AAAS, Am. Assn. Clin. Chemistry, Am. Bd. Forensic Toxicology (bd. dirs. 1984—, pres. 1988—), Am. Chem. Soc., Am. Soc. Crime Lab. Dirs., Calif. Assn. Toxicologists, Soc. Forensic Toxicologists (pres. 1981), Forensic Scis. Found. (trustee 1989-92), Internat. Assn. Forensic Toxicologists, Mid-Atlantic Assn. Forensic Scientists, Nat. Registry Clin. Chemistry (bd. dirs. 1992—), Nat. Saftey Coun., others. Home: 3411 Philips Dr Baltimore MD 21208 Office: Nat Ctr Forensic Sci 1901 Sulphur Spring Rd Baltimore MD 21227

CAPLIN, JERROLD LEON, health physicist; b. Phila., Jan. 25, 1930; s. Samuel Harry and Katherine (Socloff) C.; children: Sally C. Daniels, Patricia Graham Reed. AB, Temple U., 1951, postgrad., 1952-53; postgrad. Vanderbilt U., 1951-52. Supervisory health physicist U.S. Army C.E., Fort Belvoir, Va., 1959-61; health physicist AEC, U.S. Nuclear Regulatory Commn., Washington, 1961-81, ret., 1981; cons., 1981—; guest lectr. radiation sci. Georgetown U. Grad Sch., 1987—; sr. scientist Advanced Sys. Tech., Inc., 1993—. Photographer, newspaper editor, sci. writer, 1983—. Coauthor, editor Manual Respiratory Protection Against Airborne Radioactive Materials, 1976. Active Nat. Mus. of Women in Arts, Friends of the Nat. Zoo, Friends of the Kennedy Ctr. Lt. USNR, 1953-58. AEC radiol. physics fellow Vanderbilt U., 1951-52; mem. Am. Nat. Standards Inst., Health Physics Soc., Am. Conf. Gov. Indsl. Hygienists, (chmn. com. 1977-83), Internat. Radiation Protection Assn., Am. Assn. Physics Tchrs., AAAS, ASTM, U.S. Naval Inst., Am. Film Inst., Nat. Wildlife Fedn., Nat. Geog. Soc., Smithsonian Instn. (resident assoc. 1970—). Home and Office: 9 Goodport Ln Gaithersburg MD 20878-1001

CAPONE, ANTONIO, psychiatrist; b. Afragola, Naples, Italy, Feb. 18, 1926; came to U.S., 1954; s. Giulio and Giovanna (Fico) C.; m. Maria Morello, Mar. 21, 1957; children: Antonio Jr., John, Walter. MD, U. Naples, 1953. Diplomate Am. Bd. Med. Psychotherapists, Am. Bd. Psychiatry and Neurology, Internat. Acad. Behavioral Medicine. Intern Ospedale Incurabili, Naples, 1953-54, St. Francis Hosp., Jersey City, 1954-55; resident physician St. Clare's Hosp., Denville, N.J., 1955-56; courtesy staff Butler Hosp., Providence, 1959—; chief psychiatry John E. Fogarty Meml. Hosp., North Smithfield, R.I., 1969-79, Pawtucket (R.I.) Meml. Hosp., 1971-80, St. Joseph Hosp., Providence, 1971—; dir. Psychiat. Svcs., Inc., Providence, 1970—; clin. asst. prof. psychiatry and human behavior Brown U. Med. Sch., Providence, 1980—; med. dir. St. Joseph Ctr. for Psychiat. Svcs., Providence, 1987—; cons. psychiatrist Pawtucket Meml. Hosp., 1980—, John E. Fogarty Meml. Hosp., North Smithfield, 1979; chief psychiat. cons. R.I. Div. Vocat. Rehab., Providence, 1967-72; med. advisor Dept. HEW, 1967—; clin. elective course leader Brown Med. Sch., Providence, 1982—. Contbr. articles to profl. jours. Various presentations on mental health and alcoholism, Lions Club, Kiwanis, TV, and radio. Fellow Am. Psychiat. Assn. (pres. R.I. dist. br. 1968-70, mem. peer rev. com. 1982—, mem. fellowship com. 1984—, mem. ad hoc on referral svc. 1987-88); mem. AMA, R.I. Med. Soc., Providence Med. Assn., Psychiatry and Neurology, Pan Am. Med. Assn., Am. Soc. Vienna. Roman Catholic. Avocations: piano playing, gardening, jogging, walking, traveling. Office: Psychiat Svcs Inc 911 Smith St Providence RI 02908-2789

CAPORALI, RENSO L., aerospace executive; b. 1933. BSCE, Clarkson Coll., 1954, MS, 1960; MAE, Princeton U., 1962. With Grumman Corp., 1959—, sr. v.p. tech. ops., 1983-85, pres. aircraft systems divsn., 1985-88, vice chmn. tech., 1988-90, chmn. bd. dirs., pres., CEO, 1990-91, chmn. bd. dirs., CEO, 1991—; chmn. bd. dirs., pres., CEO Grumman Aerospace Corp., Bethpage, N.Y. With USN, 1954-58. Office: Grumman Corp 1111 Stewart Ave Bethpage NY 11714-3533

CAPPITELLA, MAURO JOHN, architect; b. N.Y.C., N.Y., July 11, 1934; s. Gaetano and Maria (D'Errico) C.; m. Christine Wilhelmine Otte, Oct. 11, 1964; children: Mark, Christina, Nicole. BS in Architecture, CCNY, 1956; postgrad., Columbia U., 1960-62; M in Urban Planning, NYU, 1967. Registered architect, N.Y., N.J.; lic. Nat. Coun. Archtl. Registration Bds.; profl. planner, N.J. Designer Garfinkel & Marenberg, N.Y.C., 1956-57; architect Western Electric Co., Inc., N.Y.C., 1957-68; cons. architect Norwood, N.J., 1968-76, Upper Saddle River, N.J., 1976—. Served to 1st lt. U.S. Army, 1957-59. Recipient Anton Vegliante award 1993. Mem. AIA, N.Y. Soc. Architects, N.J. Soc. Architects (bd. dirs. 1983-84, 87-89), Architects League No. N.J. (bd. dirs. 1980-83, 89-91, sec. 1984-85, v.p. 1985, 1st v.p. 1986, pres.-elect 1987, pres. 1988, 93, Dir. of Yr. award 1981, 81), Soc. 3d U.S. Inf. Div. U.S. Army, Saddle River Tennis Club, Windham Ridge Resorts Country Club, Rotary. Republican. Roman Catholic. Office: 332 E Saddle River Rd Saddle River NJ 07458-2108

CAPRIOLI, RICHARD MICHAEL, biochemist, educator; b. N.Y.C., Apr. 12, 1943. BS, Columbia U., 1965, PhD in Biochemistry, 1969. Rsch. assoc. chemistry Purdue U., 1969-70, from asst. to assoc. prof., 1970-75; assoc. prof. biochemistry med. sch. U. Tex., Houston, 1975-80, prof., 1980—. Mem. Am. Soc. Biol. Chemists, Am. Soc. Mass Spectrometry. Achievements include research in peptide sequencing by mass spectrometry, intermediary metabolism using stable isotopes, mechanisms of enzyme action. Office:

University of Texas Hlth Sc Ctr Analytical Chemistry Ctr PO Box 20708 Houston TX 77225*

CAPUTO, WAYNE JAMES, surgeon, podiatrist; b. Newark, Feb. 18, 1956; s. James Vincent and Jennie (DeMaio) C.; m. Phyllis A. Grillo, Nov. 20, 1984; children: Karla, Stefanie. BS in Biology, Syracuse (N.Y.) U., 1978; DPM, N.Y. Coll. Podiatric Medicine, 1982. Diplomate Am. Bd. Podiatric Surgery. Chief dept. podiatric surgery Clara Maass Med. Ctr., Belleville, N.J., 1987—; dir. residency in podiatric surgery Union (N.J.) Hosp., 1990—. Contbr. articles to profl. jours. Fellow ACS. Office: Clara Maass Profl Med Ctr 5 Franklin Ave Belleville NJ 07109-3532

CÁRABE LÓPEZ, JULIO, physicist; b. Tangier, Morocco, Spain, Oct. 14, 1959; s. Julio and Ana (López) C.; m. Delia Muñoz, July 28, 1989. MS, U. Complutense, Madrid, 1983, PhD in Physics, 1990. Researcher Instituto de Energia Solar, Madrid, 1981-83, European Space Agy., Noordwyk, Netherland, 1986-87, Inst. de Energias Renovables, Madrid, 1987—. 2nd lt. Spanish Arty., 1983-84. Achievements include development of novel method for measuring non-linear solar cell spectral responses, new approaches for the deposition of high-quality amorphous silicon thin films. Office: CIEMATIER, Avda Complutense 22, E-28040 Madrid Spain

CARAHER, MICHAEL EDWARD, systems analyst; b. Indpls., Dec. 22, 1953; s. Gregory Thomas and Mary Margaret (Shevlin) C.; m. Debra Sue Sedam, Jan. 6, 1979 (div. 1986); children: Joseph Michael, Erin Michelle. BA, Butler U., Indpls., 1976. Systems mgr. Alexander Typesetting, Indpls., 1984-88; systems specialist Shepard Poorman Comm. Corp., Indpls., 1988—. Vice precinct committeeman Ind. Dem. Party, Indpls., 1972-78. Mem. Ancient Order of Hibernians, Saenger Chor. Democrat. Roman Catholic. Home: 5205 E Washington St Indianapolis IN 46219 Office: Weimer Graphics 111 E McCarty St Indianapolis IN 46206

CARASSO, ALFRED SAM, mathematician; b. Alexandria, Egypt, Apr. 9, 1939; came to U.S., 1962; s. Samuel and Renee (Ades) C.; m. Beatrice Kozak, June 12, 1964; children: Adam Leonard, Rachel Lisa. BSc in Physics, U. Adelaide, Australia, 1960; PhD in Math., U. Wis., 1968. Meteorologist Bur. Meteorology, Adelaide, 1960-62; rsch. asst. grad. sch. U. Wis., Madison, 1962-68; asst. prof. of math. Mich. State U., East Lansing, 1968-69; asst. prof. of math. U. N.Mex., Albuquerque, 1969-72, assoc. prof., 1972-76, prof., 1976-81; mathematician Nat. Inst. Standards and Tech., Gaithersburg, Md., 1982—; cons. Los Alamos (N.Mex.) Nat. Lab., 1972-81. Contbr. articles to profl. publs. Mem. Am. Math. Soc., Soc. for Indsl. and Applied Math. Jewish. Achievements include research in mathematical and computational analysis of ill-posed inverse problems in partial differential and integral equations; applications in heat transfer, system identification, image processing, ultrasonics, and electromagnetics; patent pending for new approach to image deblurring. Office: Nat inst Standards & Tech Applied Computational Math Divsn Gaithersburg MD 20899

CARBERRY, EDWARD ANDREW, chemistry educator; b. Milw., Nov. 20, 1941; s. Edward Andrew Carberry Sr. and Sophie Teresa (Hologa) Ryall; m. Linda Lee Querry, July 22, 1967; children: Daniel Edward, Cristin Lee. BSChemE, Marquette U., 1963; PhD in Inorganic Chemistry, U. Wis., Madison, 1968. Prof. chemistry S.W. State U., Marshall, Minn., 1968—; lectr. Higher Coll. of Chemistry Russian Acad. of Scis., 1992. Author: Glassblowing: An Introduction to Scientific and Artistic Frameworking, 1991, Everyday Chemicals and Food for Thought, 1989; contbr. articles to profl. jours. Mem. Am. Chem. Soc., Am. Scientific Glassblowers' Soc., Mendeleev Chem. Soc. (Moscow), Assn. for Advancement of Chem. Edn. (Moscow). Home: 700 S 1st St Marshall MN 56258 Office: SW State Univ Dept Chemistry Marshall MN 56258

CARBERRY, JAMES JOHN, chemical engineer, educator; b. Bklyn., Sept. 13, 1925; s. James Thomas and Alice (McConnin) C.; BS, U. Notre Dame, 1950, MS, 1951; DEng, Yale U., 1957; m. Judith Ann Bower, Sept. 12, 1959 (div.); children: Alison Ann, Maura O'Malley; m. Margaret V. Bruggner, Sept. 24, 1974. Process engr. E.I. duPont de Nemours & Co., Gibbstown, N.J., 1951-53, sr. rsch. engr., Wilmington, Del., 1957-61; teaching and research fellow Yale U., 1953-57; prof. chem. engring. U. Notre Dame (Ind.), 1961—; mem. U.S.-Soviet working com. chem. catalysis; cons. in field. Mem. adv. coun. chem. engring. dept. Princeton U., 1980-88. With USNR, 1944-46. Recipient award for advancement pure and applied sci. Yale Engring. Assn., 1968. NSF fellow Cambridge U. (Eng.), 1965-66; Fulbright-Hays sr. scholar, Italy, 1973-74; Kelley lectr. Purdue U., 1978; Richard King Mellon fellow, Sir Winston Churchill fellow Cambridge U., 1979-82; vis. fellow Clare Hall Coll., Cambridge U., 1987; vis. chaired prof. Politecnico di Milano, Italy, 1987; life fellow Clare Hall, Cambridge U., 1988; Gruppo Attivitá Verdiane, Roncole Verdi, 1987—. Fellow Royal Soc. Arts (London), N.Y. Acad. Scis., Am. Inst. Chemists; mem. NAE, Am. Chem. Soc. (Murphree award indsl. and engring. chemistry 1993), Am. Inst. Chem. Engrs. (R.H. Wilhelm award in chem. reaction engring. 1976, Autoclave Engrs. award 1988, William H. Walker award 1989), AICHE, Nat. Acad. Engring., Yale Alumni Assn. (rep.), Yale Engring. Assn., Lucrezia Borgia Soc., Sigma Xi, Fellowship of Cath. Scholars, Yale Club (N.Y.C.). Author: Chemical and Catalytic Reaction Engineering, 1976; editor: Catalysis Revs., 1974-90. Office: U Notre Dame Dept Chem Engring Notre Dame IN 46556

CARBO, RAMON, chemistry educator, researcher; b. Girona, Catalonia, Spain, Oct. 19, 1940; s. Marcos and Clara (Carre) C.; m. Caterina Arnau, June 24, 1964 (div. 1986); 1 child, Joan-Marc; m. Mima Masip; children: Martina, Blai, Laia. PhDChemE, Inst. Quimic de Sarriá, Barcelona, Spain, 1968; PhD in Chemistry, U. Autónoma de Barcelona, 1974. Prof. numerari Inst. Quimic de Sarriá, Barcelona, 1964-86; prof. titular U. Autónoma de Barcelona, 1970-87; catedràtic Universitat de Girona, 1988—. Author: A General SCF Theory, 1978, Algebra Matricial y Lineal, 1987, Current Aspects of Quatum Chemistry, 1982; editor: Quantum Chemistry: Basic Aspects, 1989, SCF Theory, 1990; contbr. more than 130 papers to sci. jours. Mem. Am. Phys. Soc. (life), Am. Chem. Soc., N.Y. Acad. Scis., Real Soc. Española de Química, Soc. Catalana de Quimica, Grup de Química Quàntica de Catalunya, Internat. Soc. Quantum Biology (life). Avocations: gardening, Oriental langs., Oriental religions, sci. fiction. Home: Francesc Romaguera 27, 17003 Girona Spain Office: Universitat de Girona, Albereda, 5, 17071 Girona Spain

CARBO-FITE, RAFAEL, physicist, researcher; b. Valladolid, Spain, July 13, 1942; s. Modesto Miguel and Concepcion (Fite) C.; m. Marisol Rubiera, July 30, 1970; children: Javier Ignacio, Carolina. D Physics in Signal Processing, U. Grenoble, France, 1969; D Physics in Underwater Acoustics, U. Madrid, 1980. Engr. mgr. Femsa-Bosch Mfg., Madrid, 1969-75; researcher Consejo Superior Investigation Sci., Madrid, 1975—; mem. sci. com. European Conf., Luxemburg, 1991—. Author, editor: Acoustics and Ocean Bottom, 1987; contbr. articles to profl. jours.; patentee (6). French Govt. scholar, 1967-69. Mem. Acoustical Soc. Spain, Acoustical Soc. Am., Acoustical Soc. France, Inst. de Acustica (sec. 1979-82, vice dir. 1991—). Roman Catholic. Avocations: classic music, reading, basketball. Office: Inst de Acustica, 144 Serrano, 28006 Madrid Spain

CARBON, MAX WILLIAM, nuclear engineering educator; b. Monon, Ind., Jan. 19, 1922; s. Joseph William and Mary Olive (Goble) C.; m. Phyllis Camille Myers, Apr. 13, 1944; children—Ronald Allen, Jean Ann, Susan Jane, David William, Janet Elaine. B.S. in Mech. Engring, Purdue U., 1943, M.S., 1947, Ph.D., 1949. With Hanford Works div. Gen. Electric Co., 1949-55, head heat transfer unit, 1951-55; with research and advanced devel. div. Avco Mfg. Corp., 1955-58, chief thermodynamics sect., 1956-58; prof., chmn. nuclear engring. and engring. physics dept. U. Wis. Coll. Engring., 1958-92, emeritus prof., collateral faculty, 1992—; group leader Ford Found. program, Singapore, 1967-68; mem. Adv. Com. on Reactor Safeguards, 1975-87; chmn. U. Chgo. Spl. Com. for Integral Fast Reactor, 1984—; mem. INPO Nat. Nuclear Accrediting Bd., 1990—; mem. nuclear safety rev. and audit com. Kewaunee Nuclear Power Plant, 1993—. Served to capt. ordnance dept. AUS, 1943-46. Named Disting. Engring. Alumnus, Purdue U. Fellow Am. Nuclear Soc.; mem. AAAS, Am. Nuclear Soc., Sigma Xi, Tau Beta Pi. Office: U Wis Engring Rsch Bldg Madison WI 53706

CARBONE, PAUL PETER, oncologist, educator, administrator; b. White Plains, N.Y., May 2, 1931; s. Antonio and Grace (Cappelieri) C.; m. Mary Iamurri, Aug. 20, 1954; children—David, Kathryn, Karen, Kim, Paul J., Mary Beth, Matthew. Student, Union Coll., Schenectady, 1949-52; M.D., Albany (N.Y.) Med. Coll., 1956. Diplomate: Am. Bd. Internal Medicine; cert. medical oncology. Joined USPHS, 1956; intern USPHS Hosp., Balt., 1956-57; resident in internal medicine USPHS Hosp., San Francisco, 1958-60; mem. staff Nat. Cancer Inst., NIH, Bethesda, Md., 1960-76; chief medicine br. Nat. Cancer Inst., NIH, 1968-72, asso. dir. for med. oncology, div. cancer treatment, 1972-76, dep. clin. dir., 1972-76; clin. prof. Georgetown U. Med. Sch., 1971-76; lectr. hematology Walter Reed Army Inst. Research, 1962-76; prof. medicine and human oncology U. Wis., Madison, 1976—, dir. div. clin. oncology, 1976—, chmn. dept. human oncology, 1977-87; dir. Wis. Clin. Cancer Center, 1978—. Contbr. profl. jours. Decorated USPHS Commendation medal; recipient Trimble Lecture award Md. Chirurgical Faculty, 1968; Lasker award clin. cancer chemotherapy, 1972; Rosenthal award for improvement in clin. cancer care, 1977, Medal of Honor Am. Cancer Soc., 1987, Jeffrey A. Gottlieb Meml. award M.D. Anderson Cancer Ctr., U. Tex., 1990; NIMMO vis. prof. Royal Adelaide Hosp., Adelaide, Australia, 1989. Mem. Am. Soc. Clin. Oncology (pres. 1972-73), Am. Soc. Clin. Investigation, Assn. Am. Physicians, ACP (bd. govs. Wis. chpt. 1986-90), Am. Assn. Cancer Research (pres. 1978-79), Am. Fedn. Clin. Research, AMA, Alpha Omega Alpha. Home: 600 Highland Ave Rm K4/614-6164 Madison WI 53792-0001 Office: U Wis Comprehensive Cancer Ctr 600 Highland Ave Madison WI 53792-0001*

CARBONELL, RUBEN GUILLERMO, chemical engineering educator; b. Havanna, Cuba, Dec. 27, 1947; came to U.S., 1958; s. Ruben and Guillermina (Lopez-Silvero) C.; m. Augustina Rafaela Rodriguez, June 8, 1969; children: Tomas, David, Rebecca. BSChemE, Manhattan Coll., 1969; MA in Chem. Engring., Princeton U., 1971, PhD in Chem. Engring., 1973. Asst. prof. chem. engring. U. Calif., Davis, 1973-80, assoc. prof. chem. engring., 1980-83, prof. chem. engring., 1983; prof. chem. engring. N.C. State U., Raleigh, 1984—, chmn. bioprocessing/bioanalytical interest group, 1988—, interim chmn. biotechnology faculty, 1990; rsch. fellow Slovenian Rsch. Coun., Dept. Chem. Engring., U. Guanajuato, Mexico, 1983; vis. prof. NATO, 1983; lectr. dept. chem. engring. U. Bologna, Italy, 1985, 87, 88; cons. Aerojet Corp., Sacramento, 1977, Lawrence Livermore (Calif.) Lab., 1978-81, Lockheed Missiles and Space Co., Inc., Palo Alto, Calif., 1983-85, Chevron Rsch. and Devel., Richmond, Calif., 1983-85, Owens-Corning Fiberglass, Granville, Ohio, 1984-86, Hoechst Celanese Corp., Charlotte, N.C., 1991-92, others. Contbr. articles to Internat. Jour. Quantum Chemistry, Jour. Chem. Physics, Jour. Statis. Physics, Biotechnology and Bioengineering, The Chem. Engring. Jour., Chem. Phys. Lett., Jour. Chromatography, Phys. of Fluids, Chem. Physics, Physics Letters, Indsl. and Engring. Chemistry Fundamentals, and many others. Recipient N.C. State U. Alumni Assn. Outstanding Rsch. award N.C. State U., 1989, Maurice Simpson Tech. Editors award for excellence in th field of contamination control Inst. Environ. Scis., 1992; recipient grants NSF, 1985-88, 86-88, 87, 89-91, 92—, Gas Rsch. Inst., 1986-87, N.C. Biotechnology Ctr., 1986-87, 92—, others. Mem. AAAS, AICE, Am. Chem. Soc., Am. Phys. Soc., N.Y. Acad. Scis., Soc. Rheology, Sigma Xi, Tau Beta Pi. Roman Catholic. Achievements include patents in Purification by Affinity Binding to Liposomes, Chromatography Apparatus and Method and Material for Making the Same, Affinity Precipitation of Proteins Using Biospecific Surfactants, others. Home: 6105 Godfrey Dr Raleigh NC 27612 Office: NC State Univ Dept Chem Engring PO Box 7905 223 Riddick Raleigh NC 27695-7905

CARBONNEAU, COME, mining company executive; b. Saint-Jean-des-Piles, Que., Can., Nov. 24, 1923; s. Omer and Edith (Bordeleau) C.; m. Francoise Pettigrew, Sept. 15, 1951; children: Helene, Marie, Jean, Pierre, Lise, Alaine. B.A., Laval U., 1943, B.A. in Sci. Geol. Engring., 1948; M.A. in Sci., U. B.C., 1959; Ph.D. in Geology, McGill U., 1953. Assoc. prof. geology Ecole Polytechnique, Montreal, Que., 1951-63; exec. v.p. SOQUEM, Sainte Foy, Que., 1963-65, founding pres., chief exec. officer, 1965-77; prof. Laval U., Quebec City, Que., 1977-81, chmn. dept. geology, 1979-81; pres., chief exec. officer Corp. Falconbridge Copper, Toronto, Ont., Can., 1981—; cons. geologist St. Lawrence Columbium, Oka, Que., 1953-59; chmn., dir. Bachelor Lake Gold Mines Inc., Montreal, 1980—; dir. Les Releves Geophysiques Inc. Decorated officer Order of Can. Fellow Geol. Assn. of Can.; mem. Can. Inst. Mining and Metallurgy (A.O. Dufresne award 1978, Selwyn G. Blaylock medal 1992), Sigma Xi. Clubs: Cercle Universitaire (Quebec); Engineers (Toronto). Home: 2540 Rue Keable, Sanite Foy, PQ Canada G1W 1L3 Office: Xerox Tower Ste 1210, 3400 de Maisonneuve Blvd W, Montreal, PQ Canada H3Z 3B8

CARDÉ, RING RICHARD TOMLINSON, entomologist, educator; b. Hartford, Conn., Sept. 18, 1943. BS, Tufts U., 1966; MS, Cornell U., 1968, PhD, 1971. Postdoctoral fellow N.Y. State Agriculture Exptl. Sta. Cornell U., Geneva, 1971-75; asst. prof. Mich. State U., East Lansing, 1975-78, assoc. prof., 1978-81; assoc. prof., dept. head U. Mass., Amherst, 1981-83, prof., dept. head, 1983-87, prof., 1987-89, disting. prof., 1989—. Contbr. over 145 articles to profl. jours. and chpts. to books; co-editor: Chemical Ecology of Insects, 1984. Mem. Entomol. Soc. Am. (mem. governing bd. 1988-91, Outstanding Contbn. to Agriculture 1980). Office: U Mass Dept Entomology Amherst MA 01003

CARDENAS, DIANA DELIA, physician, educator; b. San Antonio, Tex., Apr. 10, 1947; d. Ralph Roman and Rosa (Garza) C.; m. Thomas McKenzie Hooton, Aug. 20, 1971; children: Angela, Jessica. BA with highest honors, U. Tex., 1969; MD, U. Tex., Dallas, 1973; MS, U. Wash., 1976. Diplomate Nat. Bd. Med. Examiners, Am. Bd. Phys. Medicine & Rehab., Am. Bd. Electrodiagnostic Medicine. Asst. prof. dept. rehab. medicine Emory U., Atlanta, 1976-81; instr. dept. rehab. medicine U. Wash., Seattle, 1981-82, asst. prof. dept. rehab. medicine, 1982-86, assoc. prof. dept. rehab. medicine, 1986-92, prof. rehab. medicine, 1992—; med. dir. rehab. medicine clinic U. Wash. Med. Ctr., Seattle, 1982; project dir. N.W. Regional Spinal Cord Injury System, Seattle, 1990—. Editor: Rehabilitation & The Chronic Renal Disease Patient, 1985, Maximizing Rehabilitation in Chronic Renal Disease, 1989; contbr. articles to profl. jours. Co-chairperson Lakeside Sch. Auction Student Vols., Seattle, 1991. Mem. Am. Spinal Injury Assn. (chairperson rsch. com. 1991), Am. Acad. Phys. Medicine and Rehab., Am. Congress of Rehab. Medicine (chairperson rehab. practice com. 1980-84, Ann. Essay Contest winner 1976), Am. Assn. Electrodiagnostic Medicine. Avocations: art collecting, sewing, painting. Office: Univ Wash RJ-30 Dept Rehab Medicine 1959 NE Pacific St Seattle WA 98195-0001

CARDENAS, JOHN I., electrical engineer, consultant; b. San Antonio, May 24, 1924; s. Indalecio and Claudia (Hernandez) C.; m. Maedean Hance, May 31, 1951; children: John D., Rita D., Karen D. BSEE, Tex. A&I U., 1950. Lic. profl. engr., Tex. Design engr. Lodal and Assocs., San Antonio, 1950-53, Am. Engring., San Antonio, 1953-55; sr. engr. Nelson Miner Engrs., San Antonio, 1957-59, U.S. Dept. Def., Europe, 1962-75; supr. engr. U.S. Dept. Def., U.S., Europe, Asia, 1975-89; cons. Groves and Assocs., Engrs., San Antonio, 1989—. Author, pub. John's Book of Acronyms, Etc., 1990. With U.S. Army, 1953-54. Mem. Am. Legion, Masons, Scottish Rite, York Rite. Roman Catholic. Home: 4215 Timberhill St San Antonio TX 78238 Office: JC Consulting Engrs PO Box 681986 San Antonio TX 78268-1986

CARDIS, THOMAS MICHAEL, chemist; b. Indpls., June 29, 1945; s. Frank and Eleanor A. (Koers) C.; m. Sheila Mary Olohan, Nov. 28, 1970; children: Thomas, Michael, Daniel, William. BS in Chemistry, Marian Coll., Indpls., 1967; MS in Chemistry, Roosevelt U., Chgo., 1977. Chemist Naval Avionics Facility, Indpls., 1967-72; analytical chemist Nalco Chem., Chgo., 1972-75, applications chemist, 1975-78; applications chemist Foxboro Co., Norwalk, Conn., 1978-79, lab. supr., 1979-82; staff engr. ABB Process Analytics, Lewisburg, W.Va., 1982-86, lab. mgr., 1986—. Mem. Instrument Soc. Am. (Best Paper award 1986). Roman Catholic. Avocations: poetry, tennis, volleyball, sports cars. Home: 213 Church St Lewisburg WV 24901-1305

CARDNER, DAVID VICTOR, chemical engineer; b. Wilmington, Del., June 15, 1935; s. Roland Carl and Aileen Templeton (Shaw) C.; m. Mary Linn Pond, Aug. 5, 1961; children: Roland Russell, John Charles, William Merrill. BA, Rice Inst., 1957; PhD, Rice U., 1963. Rsch. engr. E.I. DuPont De Nemours & Co., Inc., Orange, Tex., 1963-65, asst. div. supt. rsch., 1965-68, asst. div. supt. tech., 1968-74, sr. engr., 1974-83, tech. assoc., 1983-93, rsch. assoc., 1993—. Cons. United Way of Orange County, 1989—; mem. State Rep. Exec. Com., Austin, 1976-80; commr. Sabine River Compact Commn., Orange, 1981-86, 89—. Mem. Am. Inst. Chem. Engrs., Instrument Soc. of Am. Republican. Methodist. Achievements include patents for system and method for improving model product property estimates; system and method for improved flow data reconciliation. Home: 4616 Emerson Rd Orange TX 77630 Office: E I DuPont de Nemours PO Box 1089 Orange TX 77630

CARDOZA, JAMES ERNEST, wildlife biologist; b. Falmouth, Mass., Aug. 10, 1944; s. Ernest Jr. and Dorothy Parker (Bailey) C. BS, U. Mass., 1966, MS, 1976. Cert. wildlife biologist. Cons. helper Mass. div. Fisheries and Wildlife, Westboro, 1969; asst. game biologist Mass. div. Fisheries and Wildlife, Westbond, 1969-74, game biologist, 1974-92, game biologist III, 1992—. Author: (booklet) Black Bear in Massachusetts, 1976; assoc. editor Trans. N.E. Sec. Wildlife Soc., 1989-90, N.E. Wildlife, 1992—. 1st lt. U.S. Army, 1967-69. Named Sportsman of Yr., Worcester County Sportsmens League, 1992, Non-Mem. Sportsman of Yr., New Eng. Outdoor Writers Assn., 1992; recipient Pride in Performance award Gov. of Mass., 1992. Mem. Nat. Wild Turkey Fedn. (tech. com. mem. 1980—), Am. Ornithologists Union, Am. Soc. Mammalogists, Internat. Assn. for Bear Rsch. and Mgmt., Soc. for Conservation Biology, Wildlife Disease Assn., Wildlife Soc. (mem. many coms.), Wilson Ornithol. Soc. Methodist. Achievements include direction of program to restore wild turkeys to Mass., of management and research of black bear to Mass. Office: Mass Divsn Fisheries & Wildlife 1 Rabbit Hill Rd Westborough MA 01581

CARDOZO, MIGUEL ANGEL, telecommunications engineering educator; b. Buenos Aires, Jan. 7, 1932; s. Joaquin and Aurelia (Gomez) C.; m. Silvia Capelli, May 12, 1967; children: Magali, Lisandro, Agustin, Griselda. Degree in telecommunications engr., Buenos Aires U., 1958. Design engr. Standard Electric Argentina, San Isidro, 1958-60; enlisted Argentine Navy, 1960, advanced through grades to lt. comdr., 1972, served as design engr., 1960-62; lt. engr. Argentine Navy, Puerto Belgrano Base, 1963-82; prof. electronics U. Nat. Sur, Bahía Blanca, Argentina, 1965—. Contbr. articles to profl. jours. Roman Catholic. Avocation: ham radio. Home: San Juan 718, 8000 Bahía Blanca Buenos Aires, Argentina Office: U Nacional del Sur, Ave Alem 1253, 8000 Bahía Blanca Buenos Aires, Argentina

CAREK, GERALD ALLEN, mechanical engineer; b. Lorain, Ohio, Oct. 25, 1960; s. Jerry Louis and Loretta Ann (Imbrogono) C.; m. Laura Kay Pitts, Sept. 30, 1989. BSME, U. Toledo, 1983, MSME, 1987. Mech. engr. NASA Lewis Rsch. Ctr., Cleve., 1983—; pvt. cons. Corsa Performance, Berea, Ohio, 1989—. Mem. AIAA. Republican. Achievements include management of facility used for fundamental research and technology development in multistage turbomachinery. Office: NASA Lewis Rsch Ctr 21000 Brookpark Rd Cleveland OH 44135

CAREY, GERALD JOHN, JR., former air force officer, research institute director; b. Bklyn., Oct. 1, 1930; s. Gerald John and Madeline (McNamara) C.; m. Joan Bennett, Apr. 24, 1954; children: Gerald John, III, Cathleen, John Kevin, Daniel. B.S., U.S. Mil. Acad., 1952; M.S. in Aero. Engring, Tex. A&M U., 1961. Commd. 2d lt. USAF, 1952, advanced through grades to maj. gen., 1978; pilot trainee Victoria, Tex., 1953; flight instr. Laredo, Tex., 1954-56; asst. air attache Tokyo, 1958-61; aero. engr. Air Force Systems Command, Andrews AFB, Md., 1963-66; flight comdr. Seymour Johnson AFB, 1967; ops. officer Udorn, Thailand, 1969-70; wing comdr. 1st and 56th Tactical Fighter Wings, Tampa, Fla., 1973-75; asst. dep. chief of staff ops. Tactical Air Command Hdqrs., Langley AFB, Va., 1975-78; comdr. USAF Tactical Air Warfare Center, Eglin AFB, Fla., 1978-81; ret., 1981; assoc. dir. Rsch. Inst. Ga. Inst. Tech., Atlanta, 1981—. Mem. USAF Sci. Adv. Bd., 1991. Decorated Legion of Merit, D.S.M., D.F.C. with 2 oak leaf clusters. Mem. Air Forces Assn., Daedalians, Tau Beta Pi, Sigma Gamma Tau. Office: Ga Inst Tech Rsch Inst Atlanta GA 30332

CAREY, MARTIN CONRAD, gastroenterologist, molecular biophysicist, educator; b. Clonmel, County Tipperary, Ireland, June 18, 1939; came to U.S., 1967; s. John Joseph and Alice (Broderick) C.; m. Gracia Antonieta Fernández, July 1, 1972 (div. 1987); children: Julian Albert, Dermot Martin. MB, BCh, BAO with 1st class honors, Nat. U. Ireland, 1962, MD, 1981, DSc, 1984; AM (hon.) Harvard U., 1989; LLD (hon.) Nat. U., Ireland, 1992. Intern, St. Vincent's Hosp., Dublin, Ireland, 1962-63, resident, 1965-67; resident Nat. Maternity Hosp., Dublin, 1963, St. Luke's Hosp., Dublin, 1964, Queen Charlotte's Hosp., London, 1964; postdoctoral fellow, rsch. assoc. Boston U. Sch. Medicine, 1968-73, asst. prof. medicine, 1973-75; asst. prof. medicine Harvard U. Med. Sch., Boston, 1975-79, assoc. prof., 1979-88, Lawrence J. Henderson assoc. prof. health sci. and tech., 1979-88; faculty mem. Grad. Sch. of Arts and Scis., assoc. mem. Dept. of Cellular and Molecular Physiology, Harvard U. Med. Sch., Boston, 1983—, prof. medicine 1988—, Lawrence J. Henderson prof. health sci. and tech., 1988-91, prof. health sci. and tech., 1991—; mem. staff Brigham and Women's Hosp., Boston, 1975—; cons. West Roxbury VA Hosp. and Dana-Faber Cancer Inst., Boston, 1975—, Calif. Biotech. Inc., Palo Alto, 1983-89, Gipharmex S.P.A., Milan, 1984-87, Dow Chem. Co., Midland, Mich., 1984-87, Merix Inc., Needham, 1986—, Oculon, Cambridge, 1987—, Ciba-Geigy, Summit, N.J., 1988-93, Labs. Fournier, Dijon-Daix, 1992-93, Hoechst AG, Frankfurt, 1993—. Author: Bile Salts and Gallstones, 1974, Hepatic Excretory Function, 1975; contbr. numerous articles to med. and sci. jours.; assoc. editor Jour. Lipid Rsch., 1978-81; mem. editorial bds. Am. Jour. Physiology, 1976-81, Gastroenterology, 1983-88, Hepatology, 1981-84. Recipient Acad. Career Devel. award NIH, 1976, also MERIT award, 1986, Adolf Windaus prize Falk Found., 1984, Fitzgerald medal Univ. Coll., 1993; Guggenheim Found. fellow, 1974; Fogarty internat. fellow NIH, 1968; McIlrath guest prof., Royal Prince Alfred Hosp., U. Sydney, Australia, 1987. Fellow Royal Coll. Physicians Ireland; mem. Gastroenterology Rsch. Group (vice chmn., steering coms.), Am. Soc. Clin. Investigation, Am. Gastroent. Assn. (disting. achievement award 1990), Am. Oil Chemists Soc., Biophys. Soc., Interurban Clin. Club, Am. Assn. Physicians, Babson Club (Wellesley, Mass.). Democrat. Roman Catholic.

CAREY, PAUL RICHARD, biophysicist, scientific administrator; b. Dartford, Kent, Eng., June 17, 1945; arrived in Can., 1969; s. Charles Richard and Winifred Margaret (Knight) C.; m. Julia Smith, Sept. 4, 1966 (div. May 1991); children: Emma, Sarah, Matthew; m. Marianne Pusztai, Mar. 7, 1992. BS in Chemistry with honors, U. Sussex, Eng., 1966, PhD, 1969. Postdoctoral fellow Nat. Rsch. Coun., Ottawa, Ont., Can., 1969-71, rsch. officer, 1971—, mgr. Ctr. for Protein Structure Design, head protein lab. Inst. for Biol. Scis., 1987—; mem. internat. adminstrv. com. Internat. Conf. on Lasers and Biol. Molecules, 1987—, adj. prof. Dept. Biochemistry, U. Ottawa, 1987—. Author: Biochemical Applications of Raman and Resonance Raman Spectroscopies, 1982; contbr. over 140 articles to profl. jours.; patentee in field. Mem., past pres. Ottawa br. Amnesty Internat., 1980—. Fellow Chem. Inst. Can.; mem. Can. Biophysics Soc., Am. Chem. Soc., Can. Protein Engring. Network (Adminstrv. body 1990—), Internat. Network Protein Engring. Ctrs. Achievements include first demonstration of resonance Raman spectroscopy providing vibrational spectrum of a substrate or drug in active site of an enzyme; generation of first quantitative relationship between active site bond lengths and reactivity by combining resonance Raman spectroscopy, enzyme kinetics and x-ray crystallography; elucidation of mechanism of sunlight degradation of biological insecticide from B. thuringiensis; research on use of lasers in fingerprint detection. Avocations: literature, music, birding. Office: Nat Rsch Coun Inst Biol Scis, Montreal Rd Bldg M-54, Ottawa, ON Canada K1A 0R6

CAREY, ROBERT MUNSON, university dean, physician; b. Lexington, Ky., Aug. 13, 1940; s. Henry Ames and Eleanor Day (Munson) C.; m. Theodora Vann Hereford, Aug. 24, 1963; children: Adonice Ames, Alicia Vann, Robert Josiah Hereford. BS, U. Ky., 1962; MD, Vanderbilt U., 1965. Diplomate Am. Bd. Internal Medicine, Am. Bd. Endocrinology and Metabolism, Nat. Bd. Med. Examiners. Intern in medicine U. Va. Hosp., Charlottesville, 1966; jr. asst. resident in medicine N.Y. Hosp.-Cornell Med. Ctr., N.Y.C., 1968-69, sr. asst. resident, 1969-70; instr. endocrinology, dept. medicine Vanderbilt U. Sch. Medicine, Nashville, 1970-72; postdoctoral fellow in medicine St. Mary's Hosp. Med. Sch., London, 1972-73; asst. prof.

internal medicine, endocrinology and metabolism U. Va. Sch. Medicine, Charlottesville, 1973-76, assoc. prof., 1976-80, prof., 1980—, James Carroll Flippin prof. medical sci. and dean, 1986—, assoc. dir. Clin. Rsch. Ctr., 1975-86, head. div. endocrinology and metabolism, dept. internal medicine, 1978-86, chmn. gen. faculty, chmn. med. adv. com., chmn. exec. com., 1986—; attending staff U. Va. Hosp., Charlottesville, 1973—, pres. clin. staff, 1977-79, vice chmn. med. policy com., 1986—, adv. bd. 1986—; mem. study sect. on exptl. cardiovascular scis. NIH, 1982-85; mem. cardiovascular and renal adv. com. USDA, 1988—; vis. prof. div. nephrology, U. Miami Med. Sch., Fla., 1979, 83, 84, Hosp. das Clinicas da Univ., Fed. do Ceara, Forteleza, Brazil, 1981, hypertension div. Mt. Sinai Sch. Medicine, N.Y.C., 1981, div. pediatric endocrinology N.Y. Hosp.-Cornell Med. Ctr., 1981, dept. endocrinology St. Vincent's Hosp., Univ. Coll., Dublin, Ireland, 1982, depts. physiology and endocrinology Mayo Grad. Sch. Medicine, Rochester, Minn., 1984, div. rsch. Cleve. Clinic Found., 1984, Genentech, Inc., San Francisco, 1984, divs. endocrinology and metabolism U. Mass., U. Pa. Sch. Medicine, Boston U. Med. Sch., 1984, U. N.C. Sch. Medicine, 1985, Harvard Med. Sch., Boston, 1987, Jefferson Med. Coll., 1988; Bley Stein vis. prof. endocrinology U. So. Calif., 1987; Pfizer vis. prof. in pharmocology U. Chgo., 1988; co-organizer 3d Internat. Meeting on Peripheral Actions of Dopamine, Charlottesville, 1989; v.p. Va. Ambulatory Surgery, Inc., 1986—; speaker, presenter numerous nat. and internat. profl. meetings and congresses. Author: (with E.D. Vaughn) Adrenal Disorders, 1988; co-editor: Hypertension: An Endocrine Disease, 1985; mem. editorial bd. Jour. Clin. Endocronology and Metabolism, 1981-84, Hypertension jour., 1983-84, Am. Jour. Physiology: Heart and Circulatory Physiology, 1987-89, Am. Jour. Hypertension, 1987—; author over 150 articles, revs., papers for profl. jours., contbr. 19 chpts. to books. Mem. exec. com. and fin. com. U. Va. Health Services Found., 1986—; bd. dirs. Va. Kidney Stone Found., Inc., 1986—, The Harrison Found., Inc. U. Va., 1986—, Dyslexia Ctr., Charlottesville, 1986—. Surgeon (lt. comdr.) USPHS, 1966-68, res., 1968—. Recipient Attending Physician of Yr. award dept. internal medicine, U. Va. Med. Ctr., 1983-84; USPH fellow Vanderbilt U., 1970-72; recipient numerous NIH grants as co-prin. and prin. investigator, 1972—. Fellow Am. Coll. Physicians (program com. regional meeting 1987), Coun. for High Blood Pressure Rsch. AMA (program com. 1984-86, exec. and long rang planning coms. 1992—); mem. Inst. Medicine of NAS, Am. Heart Assn. (established investigator 1975-80), Va. affiliate Am. Heart Assn. (bd. dirs. 1977-83, pres. 1979-80, Disting. Service award), The Endocrine Soc. (fin. com. 1988—, chair devel. com. 1991-92), Am. Fedn. Clin. Rsch. (so. sect. councilor 1978-81, nominating com. 1982), So. Soc. Clin. Investigation (nominating com. 1982, sec.-treas. 1985-86), Inter-Am. Soc. for Hypertension, Am. Soc. Clin. Investigation, Am. Clin. and Climatol. Assn., Am. Soc. Hypertension (intersocietal affairs com. 1986—), Internat. Soc. Hypertension, Assn. Am. Physicians, AMA, Albemarle County Med. Soc., Med. Soc. Va., Assn. Am. Med. Col, Inst. of Medicine, Nat. Acad. of Scis., The Raven Soc., Alpha Omega Alpha. Home: Pavilion VI East Lawn Charlottesville VA 22903 Office: U of Va Sch Medicine Med Ctr Box 395 MaKim Hall Charlottesville VA 22908

CARFORA, JOHN MICHAEL, economics and political science educator; b. New Haven, Conn., July 24, 1950; s. John Michael and Rose Mary (Mitro) C.; m. Linda Louise Palmer, July 22, 1972; 1 child, Rachel Ellen. BS, U. New Haven (Conn.), 1973, M in Pub. Adminstrn., 1975; MS in Econs. and Polit. Sci., London Sch. Econs. and Polit. Sci., 1978; AM, Dartmouth Coll., 1985, EdM Harvard U., 1993. Vis. asst. prof. dept. def., 1979-80; vis. sr. lectr. Poly. of Central London, 1980; research asst. London Sch. Econs. and Polit. Sci., 1980-81; vis. asst. prof. internat. relations So. Conn. State U., New Haven, 1982; lectr. dept. polit. sci. Albertus Magnus Coll., New Haven, 1982-83; lectr. dept. econs. and quantitative analysis U. New Haven, 1982-83; program cons. Dartmouth Coll., 1984-85; asst. prof. internat. econ. Sch. Internat. Tng., 1985-90; v.p. rsch. and acad. affairs, dir. Soviet-Am. projects Global-Genesis; Internat. Cons. (formerly Global Consultancy Group), 1989-91, dir. east and west projects, 1992—; ednl. cons. USSR Acad. Mgmt., Moscow, 1991-92; cons. Commonwealth Acad. Mgmt., Moscow, 1992—; lectr. in field. Author book reviews; contbr. articles to profl. jours. Served with USAR, 1970-76. Recipient Roy E. Jenkins award, 1972; fellow Radio Free Europe-Radio Liberty, 1979, Internat. Research and Exchanges Bd., 1981-84. Mem. Am. Assn. Advancement Slavic Studies, Nat. Assn Fgn. Student Advisors (internat. educators), Am. Acad. Polit. Sci., Am. Econ. Assn., Am. Polit. Sci. Assn., Acad. Polit. Sci., N.E. Slavic Assn., Royal Acad. Pub. Adminstrn. (Eng.), Atlantic Econ. Soc., Am. Friends of the London Sch. Econs. (Conn. program chmn. 1981-85, N.H.-Vt. program chmn. 1985-87, alumni bd. dirs. 1983-92). Home: PO Box 964 Northampton MA 01061-0964 Office: Global-Genesis Internat Cons PO Box 964 Northampton MA 01061-0964

CARIDAD, JOSE MARIA, computer scientist, educator; b. Seville, Spain, Mar. 29, 1949; s. Jose Maria and Maria Teresa (de Ocerin) C.; m. Rosa Maria Lopez del Rio, Oct. 26, 1972; children: Daniel, Lorena. M in Math., Seville U., Spain, 1971, PhD Math., 1973; M in Econs., Malaga U., Spain, 1975. Asst. Seville U., 1971-73; asst. prof. Bilbao (Spain) U., 1973-77; prof. Dept. Statistics and Econometrics Cordoba (Spain) U., 1977—; cons. Banco de Bilbao, 1973-77, Junta de Andalucia, Seville, 1984-90; vice decan Engring. Agronomical Sch., Cordoba, 1978-81; dir. Computer Ctr., Cordoba, 1978-90; vice-rector U. Cordoba, 1984-90. Author: Informe Economico, 1990; contbr. books and articles to profl. jours. Mem. IEEE, Assn. Computing Machinery, Econometric Soc., Biometric Soc., Inst. Andaluz de Prospectiva, Club Pineda Sevilla, Club Los Villares. Roman Catholic. Avocations: tennis, golf. Home: Ronda de Los Tejares, 19-3-6-6 Aptdo 961, 14008 Cordoba Andalucia, Spain Office: U De Cordoba, Aptdo 3048 Etsia, 14071 Cordoba Spain

CARIGNAN, CLAUDE, astronomer, educator; b. Montreal, Que., Can., Dec. 20, 1950; s. Philippe and Gilberte (Frenette) C.; m. Lucie Houde, Aug. 1972 (div. Oct. 1985); life ptrnr. Suzanne Cardinal; children: Stephane Carignan, Veronik Carignan, Marilis Cardinal. MSc, U. Montreal, 1978; PhD, Australian Nat. U., Canberra, 1983. Postdoctoral fellow Kapteyn Lab., Groningen, Holland, 1983-85; univ. rsch. fellow U. Montreal, 1985-90, asst. prof., 1990-91, assoc. prof., 1991—; mem. adv. bd. observatory user's com. Dominion Radio-Astronomy, Penticton, B.C., Can., 1991. Contbr. sci. papers to Astrophysical Jour. and Astron. Jour. Australian Nat. U. scholar, 1978. Mem. Can. Astron. Soc. (adv. bd. mem. radio-astronomy com. 1990). Achievements include research in neutral hydrogen in galaxies from radio synthesis observations, detailed kinematics, mass distribution and properties of dark matter in spiral and dwarf galaxies. Home: 6730 Place La Bataille, Laprairie, PQ Canada J5R 3X8 Office: U Montreal Dept Physics, CP 6128 Succ A, Montreal, PQ Canada H3C 3J7

CARIM, ALTAF HYDER, materials science and engineering educator; b. Bombay, India, Feb. 17, 1961; came to U.S., 1962; s. Hyder M. and Mahera H. (Vasi) C.; m. Merrilea Joyce Mayo, Oct. 7, 1989. SB in Materials Sci. and Engring., MIT, 1982; MS in Materials Sci. and Engring., Stanford U., 1984, PhD in Materials Sci. and Engring., 1989. Postdoctoral rsch. scientist Philips Rsch. Labs., Eindhoven, The Netherlands, 1987-88; rsch. asst. prof. U. N.Mex., Albuquerque, 1988-90; asst. prof. materials sci. and engring. Pa. State U., University Park, 1990—. Contbr. over 40 articles to profl. jours., chpt. to book. SEM Inc. presdl. scholar, 1985; recipient Young Investigator award U.S. Office Naval Rsch., 1991. Mem. Am. Ceramic Soc., Materials Rsch. Soc., Electrochem. Soc., Microscopy Soc. Am., Sigma Xi, Tau Beta Pi. Achievements include transmission electron microscopy of thin films; development and characterization of dilute ceramics. Office: Pa State U 118 Steidle Bldg University Park PA 16802

CARIOTI, BRUNO MARIO, civil engineer; b. St. Andrea, Catanzaro, Italy, Nov. 3, 1929; came to U.S., 1935; s. Vincenzo and Caterina (Stillo) C.; m. Ida Annamaria Schmutz, July 16, 1955; children: Daniela, Laura. BCE cum laude, CCNY, 1951; MS in Engring., Princeton U., 1961. Registered profl. engr., N.Y. Commd. ensign Civil Engrs. Corps USN, 1953, advanced through grades to comdr., 1965; project mgr. USN Contracts for Missile and Space Prog. USN, Cape Canaveral/Missile, Range, 1961-63; cons. U.S. Naval Mission to Chile, Valparaiso, 1963-68; engring. dir. U.S. Naval Forces, Danang, South Vietnam, 1968-69, Naval Facilities Engring. Command, Washington, 1969-70; dep. pub. works dir. U.S. Naval Dist., Washington, 1970-74; constrn. dir., cons. Iranian Navy, 1974-79; retired USN, 1974; sr. cons. Royal Commn. for Jubail and Yanbu, Riyadh, Saudi Arabia, 1979—; conf. participant Internat. Environ. Protection, Washington, 1990, World Conf. on Desalination and Water Reuse, Washington, 1991. Contbr. tech. articles on earthquake engring. and ocean engring. to profl. jours.; author Master Plan for Devel. Chilean Navy Shipyards, 1967. Project mgr. Alliance for Progress Program, Chile and South America, 1963-68; head damage assessment and disaster relief Govts. of Chile and Iran for major earthquakes, 1965, 75. Decorated Bronze Star, Navy Commendation medal, Chilean Navy Diploma of Merit. Fellow ASCE; mem. Nat. Soc. Profl. Engrs., Soc. Am. Mil. Engrs., U.S. Naval Inst, Marine Tech. Soc., Jacque Cousteau Soc. Roman Catholic. Avocations: scuba diving, underwater photography, tennis, bridge. Home: 4238 Embassy Park Dr NW Washington DC 20016 Office: Psc 1203 PO Box # 1374R APO AE 09803-1374

CARLIER, JEAN JOACHIM, cardiologist, educator, administrator; b. Jemeppe-sur-Sambre, Belgium, July 31, 1926; s. Edouard-Georges and Hermance (Doucet) C.; m. Marie-Louise Barbier, Aug. 17, 1954; children: Pierre, Philippe. Specialist in internal medicine, U. Liège (Belgium), 1958, specialist in cardiology, 1961, Agrégé de l'Enseignement Supérieur, 1967, specialist in cardiac rehab., 1981. Asst., U. Liège, 1953-60, chef de travaux, 1960-67, agrégé de Faculté, 1967-68, chargé de cours associé, 1968-73, associated prof., 1973—, founder Unities of Cardiac Pharmacology and Cardiac Rehab. Contbr. numerous articles to med. jours. Recipient prix Masius U. Liège, 1958, prix Amis U. Lg, 1968; pres.'s Comm. Agreation Cardiologistes de Langue Française, Ministry of Pub. Health, 1972—. Fellow Internat. Coll. Angiology; mem. Belgian, European, French Socs. Cardiology, Belgian and European Socs. Hypertension, N.Y. Acad. Sci. Home: Rue Boulanger-Duhayon 23, B-5190 Jemeppe-Sur-Sambre Belgium

CARLILE, LYNNE, private school educator; b. Hereford, Tex., June 22, 1947; d. V.H. and Norene (Vines) Poarch; m. Jeff R. Carlile, Sept. 26, 1974; children: Marta, Robbin, Chad, Lyndi. BS in Edn., West Tex. State U., 1987. Cert. tchr., Tex. Sci. tchr. Nazarene Christian Acad., Hereford, 1988—. Home: Rt 5 PO Box 32 Hereford TX 79045

CARLIN, HERBERT J., electrical engineering educator, researcher; b. N.Y.C., May 1, 1917; s. Louis Aaron and Shirley (Salzman) C.; children—Seth Andrew, Elliot Michael; m. Mariann J. Hartmann, June 29, 1975. B.E.E., Columbia Coll., 1938, M.E.E., 1950; Ph.D. in Elec. Engring., Poly. Inst. N.Y., 1947. Engr. Westinghouse Corp., Newark, 1940-45; from asst. to assoc. prof. Poly. Inst. Bklyn., 1945-60, prof., head electrophysics, 1960-66; J. Preston Levis prof. engring. Cornell U., Ithaca, N.Y., 1966—, dir. elec. engring., 1966-75; mem. adv. panel Nat. Bur. Standards, Boulder, Colo., 1967-70; mem. rev. com. Lehigh U., Bethlehem, Pa., 1966-74, U. Pa., Phila., 1979-82; vis. prof. Ecole Normale Superieure, Paris, 1964-67, MIT, Boston, 1973-74; vis. scientist Nat. Ctr. for Telecommunications, Issy Les Moulineaux, France, 1979-80; vis. lectr. U. Genoa, Italy, summer 1973, U. London, Dec. 1979, The Technion, Haifa, Israel, Mar. 1980, Tianjin U., China, summer 1982, Univ. Coll., Dublin, Ireland, summer 1983, Polytech. of Turin, Italy, summer 1985, 91, Fed. Polytech., Lausanne, Switzerland, summer 1992. Fellow NSF, 1964; recipient Outstanding Achievement award U.S. Air Force, 1965. Fellow IEEE (chmn. profl. group on circuit theory 1955-56, Centennial medal 1985). Avocations: flute; photography. Home: 8 Highland Park Ln Ithaca NY 14850-1452 Office: Cornell U 201 Phillips Hall Ithaca NY 14853

CARLISLE, ALAN ROBERT, aeronautical design engineer; b. Salt Lake City, Aug. 2; s. Robert P. and Norma (Dean) C.; m. Alice Brubaker, Aug. 29, 1986; children: Thomas Brubaker, Laura Elizabeth, Earl Alan. BS, U. Utah, 1986. Summer tech. Hanford ops. Rockwell, Inc., Richland, Wash., 1985; rsch. tech. Mech. Engring. Dept., U. Utah, Salt Lake City, 1986; design engr. Thiokol Corp., Brigham City, Utah, 1986—; mem. shuttle booster postflight assessment team Thiokol Corp., 1988-91. Mem. AIAA. LDS. Home: 455 Springcreek Rd Providence UT 84332 Office: Thiokol Corp Hwy 83 Thiokol UT 84302-0707

CARLISLE, MARK ROSS, naval aviator; b. Madisonville, Ky., Apr. 27, 1966; s. Melvin Ross and Sherry Ann (DeBow) C.; m. Shelia Andréa Middleton, Dec. 31, 1988. BSCE, Memphis State U., 1989. Commd. ensign USN, 1989, advanced through grades to Lt., 1992; naval aviator USN, Corpus Christi, Tex., 1990—; plane commdr. USN, Barbers Point, Hawaii, 1992-93, VT-27 primary squadron instr. pilot, 1993—. Office: VP-4 FPO AP 96601

CARLO, GEORGE LOUIS, epidemiologist; b. Jamestown, N.Y., Aug. 24, 1953; s. Louis Samuel and Josephine Lenora (Butera) C.; m. Patricia Herrgott; children: Jody, Doug, Matthew, Michael, Aaron, David, Kristen. BA, SUNY, Buffalo, 1975, MS, 1977, PhD, 1979; JD, George Washington U., 1988. Cert. Am. Coll. Epidemiology. Asst. prof. U. Ark. Med. Sci., Little Rock, 1979-81; rsch. leader Dow Chem. Co., Midland, Mich., 1981-84; chmn. Health and Environ. Scis. Group, Washington, 1984—; adviser Ark. Hazard Waste Adv. Com., Little Rock, 1979-81, Office Tech. Assesment adv. panel U.S. Congress, Washington, 1982—; sci. adviser various industry and govt. groups. Author 75 articles, reports, chpts. in books and commentaries. Named one of Outstanding Young Men Am., 1979; recipient Silver Anvil award Pub. Rels. Soc. Am., 1991, Toth award Pub. Rels. Soc. Am., 1991. Mem. Soc. for Epidemiology Rsch., AAAS, N.Y. Acad. Scis., Soc. for Risk Analysis. Achievements include research in the health sciences and environmental health policy in U.S., Europe, Australia, and New Zealand. Office: HES Group Ltd 1513 16th St NW Washington DC 20036

CARLOMAGNO, GIOVANNI MARIA, engineering educator; b. Gardone V.T., Brescia, Italy, Nov. 27, 1940; s. Giuseppe and Giacinta (Menaldi) C.; m. Rosa Amabile Guerci, Sept. 6, 1967; children: Francesca, Giuseppe. Diploma liceo scientifico, Liceo Statale Mercalli, Naples, Italy, 1958; D Ingegneria Meccanica, U. Naples, 1965. Olivetti fellow faculty engring. U. Naples, 1964-65; IBM fellow IBM Edn. Ctr., Rivoltella, Italy, 1964; Consiglio Nazionale delle Ricerche fellow Istituto di Aerodinamica, U. Naples, 1966-67; rsch. asst. Princeton (N.J.) U., 1967-68; asst. prof. U. Naples, 1967-69, assoc. prof. physics, 1969-75, assoc. prof. gasdynamics, 1975-86, prof. gasdynamics, 1986—; Adv. Group for Aerospace Rsch. and Devel. fellow Heat Transfer Labs., 1968; dir. Univ. Librs. Automation Svc., Naples, 1986-90; dean Mech. Engring. Sch., Naples, 1989—. Editor: Computers and Experiments in Fluid Flow, 1989, Computers and Experiments in Stress Analysis, 1989, Computational Methods and Experimental Measurements V, 1991, Quantitative Infrared Thermography, 1992; patentee in field; author more than 140 sci. papers in field of thermo-fluid dynamics; editorial bd. Expts. in Fluid, Stability and Analysis of Continuous Media; assoc. editor Aerotecnica, Missili e Spazio; bd. jour. Flow Visualization Image Processing; mem. adv. bd. Meccanica, Jour. Italian Assn. Theoretical and Applied Mechanics. Trustee U. Naples, 1991—. Mem. Internat. Soc. Flow Visualization (gen. bd.), Internat. Coun. Aeronautical Sci. (internat. bd.), Associazione Italiana Aeronautica Astronautica (bd. dirs. 1987—). Home: Via Cintia 38, Parco San Paolo, 80126 Naples Italy Office: Faculty Engring, Piazzale Tecchio 80, 80125 Naples Italy

CARLOTTO, MARK JOSEPH, electrical and computer engineer, researcher; b. New Haven, May 16, 1954; s. Andrew J. and Gloria M. (Bigelli) C.; m. Jeanne Marie Skettino, July 20, 1979 (div. Nov. 1989); m. Mary Katherine Brennan, Nov. 24, 1990; children: Katherine Anne, Hayley Elizabeth. BSEE, Carnegie-Mellon U., 1977, MSEE, 1979, PhDEE, 1981. Div. staff analyst TASC, Reading, Mass., 1981, chief project engr., dept. mgr., sect. mgr.; asst. adj. prof. Boston U., 1981-83; chmn. Digital Image Processing and Visual Comm. Techs. in Earth and Atmospheric Scis. Conf., Soc. Photo-Optical Instrumentation Engrs., 1992. Author: The Martian Enigmas, 1992; also over 50 publs. Mem. IEEE (sr., best paper award Conf. Applied Imagery and Pattern Recognition 1985). Democrat. Achievements include development of phase-coding technique for matching multiple patterns in parallel; analysis of possible extraterrestrial artifacts on Mars. Office: TASC 55 Walkers Brook Dr Reading MA 01867

CARLOW, JOHN SYDNEY, research physicist, consultant; b. Kingston-upon-Thames, Surrey, Eng., May 8, 1943; s. Sidney George and Gwendoline (Craymer) C.; m. Carole Evelyn Harbidge, May 9, 1973; children: Anne-Marie Evelyn, Helen-Louise Gwendoline. BSc in Physics, Exeter U., Eng., 1964; PhD in Physics, Exeter (Eng.) U., 1971. Scientist GE (G.B.), Wembley, Eng., 1968-71; rsch. fellow chemistry dept. Dalhousie U., Halifax, N.S., Can., 1972-74; rsch. fellow Southampton U., Eng., 1974-77; sr. fellow CERN, Geneva, 1977-80; cons. Southampton, 1981-87; rsch. assoc. AEA Tech., Abingdon, Eng., 1988—. Contbr. articles to sci. jours. Killam postdoctoral scholar Dalhousie U., 1972. Mem. Coun. European Engring. Assns., Engring. Coun. (U.K.) (chartered), Inst. Physics (U.K.) (chartered), Am. Phys. Soc., Inst. Materials, Brit. Computer Soc., Royal Soc. Chemistry (U.K.) (assoc.). Avocations: family history, astronomy, pollution control, home computer electronics and programming, home design. Home: 139 Drayton Rd, Sutton Courtenay Abingdom, Oxford OX14 4HA, England Office: AEA Tech, Culham Lab, Abingdon, Oxford OX14 3DB, England

CARLSMITH, ROGER SNEDDEN, chemistry and energy conservation researcher; b. N.Y.C., Oct. 2, 1925; s. Leonard Eldon and Hope (Snedden) C.; m. Thelma Kathleen Sutton, July 31, 1954; children: David, Nancy Lynn. AB in Chemistry cum laude, Harvard, 1948; MSCE, MIT, 1950. Rsch. engr. Oak Ridge (Tenn.) Nat. Lab., 1950-62, group leader, 1962-70, sect. mgr., 1970-78, prog. dir. conservation and renewable energy, 1978—; mem. Gov.'s Energy Task Force, Tenn., 1972-74, adv. com. Fed. Power Commn., Washington, 1973; bd. dirs. Am. Coun. Energy Efficient Economy., Washington. Author: (book with others) World Energy Conference Survey of Energy Resourses, 1974. Sgt. USAF, 1943-46. Mem. AAAS, Sierra Club, The Wilderness Soc. Achievements include research on development of advanced technology for improved energy efficiency, alternative energy sources, environmental impacts of energy, energy and the economy. Home: 1052 W Outer Dr Oak Ridge TN 37830-8641 Office: Oak Ridge Nat Lab PO Box 2008 Oak Ridge TN 37831-2008

CARLSON, CURTIS EUGENE, orthodontist, periodontist; b. Mar. 30, 1942; m. Dona M. Seely; children: Jennifer Ann, Gina Christine, Erik Alan. BA in Divisional Scis., Augustana Coll., 1965; BDS, DDS, U. Ill., 1969; cert. in periodontics, U. Wash., 1974, cert. in orthodontists, 1976. Dental intern Oak Knoll Navy Hosp., Oakland, Calif., 1969-70; dental officer USN, 1970-72; part-time dentist VA Hosp., Seattle, 1972-73; part-time periodontist Group Health Dental Coop., Seattle, 1973-76, part-time orthodontist, 1976-78; clin. instr. U. Wash., 1976; prin. Bellevue (Wash.) Orthodontic and Periodontic Clinic, 1976—; clin. instr., trainer Luxar Laser Corp., Bothell, Wash., 1992—; presenter in field. Master of ceremonies Auctioneer Friendship Fair, Augustana Coll., 1965, orientation group leader, 1965, mem. field svcs. com. for high sch. recruitment, 1965. Mem. ADA, Am. Acad. Periodontology, Am. Assn. Orthodontics, Western Soc. Periodontology (bd. dirs. 1984-85, 86, program chmn. 1986, v.p. 1988, pres. elect 1989, pres. 1990), Seattle King County Dental Soc. (grievance, ethics and pub. info. coms.), Wash. State Dental Assn., Wash. State Soc. Periodontists (program chmn., pres. elect 1987, pres. 1988, 89), Wash. Assn. Dental Specialists (com. rep. 1987, 88, 89), Omicron Kappa Upsilon (dental hon. fraternity), Pi Upsilon Gamma (social chmn. 1964, pres. 1965). Home: 16730 Shore Dr NE Seattle WA 98155-5634 Office: Bellevue Orthodontic Periodontic Clinic 1248 112th Ave NE Bellevue WA 98004-3712 also: Luxar Corp 19204 N Creek Pkwy Ste 100 Bothell WA 98011-8009

CARLSON, ELOF AXEL, genetics educator; b. Bklyn., July 15, 1931; s. Axel Elof and Ida Charlotte (Vogel) C.; m. Helen Zuckerman, Sept. 1, 1954 (div. 1958); 1 child, Claudia; m. Nedra Ann Miller, Mar. 28, 1959; children: Christina, Erica, John, Anders. BA, NYU, 1953; PhD, Ind. U., 1958. Lectr. Ind. U., Bloomington, 1958, Queen's U., Kingston, Ont., Can., 1958-60; asst. prof. UCLA, 1960-64, assoc. prof., 1964-68; prof. SUNY, Stony Brook, 1968-74, Disting. Teaching prof., 1974—; disting. vis. prof. San Diego State U., 1969, U. Minn., 1974, U. Utah, 1984, Tougaloo (Miss.) Coll., 1987; master Honors Coll., SUNY, 1988—. Author: The Gene: A Critical History, 1966, 89, Genes, Radiation and Society: The Life and Work of H.J. Muller, 1981, Human Genetics, 1984. Mem. staff Lilly Endowment Workshop in the Liberal Arts, Colorado Springs, Color., 1979—; mem. adv. coun. Danforth Found., St. Louis, 1974-79. Recipient Harbison award Danforth Found., 1972; fellow Inst. Advanced Study, Ind. U., 1986; scholar-at-large United Negro Coll. Fund, Tougaloo, Miss., 1987. Fellow AAAS; mem. Am. Soc. Human Genetics, Genetics Soc. Am., History of Sci. Soc. Democrat. Unitarian. Home: 19 Mud Rd Setauket NY 11733-1409 Office: SUNY Dept Biochemistry Stony Brook NY 11794

CARLSON, LAWRENCE EVAN, mechanical engineering educator; b. Milw., Dec. 22, 1944; s. John Walfred and Louise Marie (Altseimer) C.; m. Elizabeth M. Studley, Jan. 28, 1967 (div. 1979); 1 child, Jeremy L.; m. Poppy Carlson Copeland, June 15, 1985. BS, U. Wis., 1967; MS, U. Calif., Berkeley, 1968, DEng, 1971. Asst. prof. U. Ill., Chgo., 1971-74; asst. prof. U. Colo., Boulder, 1974-78, assoc. prof., 1978—; cons. Ponderosa Assn., Lafayette, Colo., 1982—. Contbr. over 40 articles to profl. jours. Disting. rsch. fellow Nat. Inst. on Disability and Rehab. Rsch., 1990-91; recipient Bronze award Lincoln Arc Welding, 1981, Ralph R. Teetor award Soc. Automotive Engrs., 1976. Mem. ASME, Internat. Soc. Prosthetics/Orthotics. Achievements include patent in upper-limb prosthetics. Office: Univ Colo Campus Box 427 Boulder CO 20309

CARLSON, MARVIN, analytical chemist; b. Harmony Twp., Wis., Nov. 19, 1937; s. Edgar John and Marit (Donali) C.; m. Marilyn Joan Hill, Nov. 10, 1970; children: Kristin Marie, Stephen Jan. BS, Platteville State U., 1961; postgrad. in Analytical Chemistry, Georgetown U., 1966. Analytical chemist FDA, Mpls., 1961—. Contbr. articles to profl. jours. Sgt. USAR, 1962-68. Mem. Am. Chem. Soc., Assn. Ofcl. Analytical Chemists Internat., Minn. Chromatography Forum, Minn. Astron. Soc., Sons of Norway. Office: FDA 240 Hennepin Ave Minneapolis MN 55401

CARLSON, PER J., physics educator, academic administrator; b. Stockholm, May 27, 1938; s. Fritz D. and Marie Louise (Ljungberger) C.; m. Birgitta Gunilla Lundin, Jan. 3, 1963; children: Ulf, Mans, Emma. PhD in Physics, U. Stockholm, 1969. Rsch. staff CERN, Geneva, 1975-79; sr. physicist Swedish Nat. Sci. Rsch. Coun., Stockholm, 1980-86; prof. physics, dir. Manne Siegbahn Inst. Physics, Stockholm, 1987—. Mem. Royal Swedish Acad. Scis. Office: Manne Siegbahn Inst Physics, Frescativägen 24, S-10405 Stockholm 50, Sweden*

CARLSON, RICHARD MERRILL, aeronautical engineer, research executive; b. Preston, Idaho, Feb. 4, 1925; s. Carl and Oretta C.; m. Venis Johnson, Nov. 26, 1946; children: Judith, Jennifer, Richard. BS in Aero. Engring., U. Wash., Seattle, MS; PhD in Engring. Mechanics, Stanford U. Registered profl. engr., Calif., chartered engr., U.K. Chief aeros., structures engr. Hiller Aircraft, Menlo Pk., Calif., 1949-64; engr. rotary wing div. Lockheed Calif. Co., Burbank, 1964-72; chief advanced systems rsch. Air Mobility R&D Lab. U.S. Army, Moffett Field, Calif., 1972-76, dir. Rsch. & Tech. Labs., Ames Rsch. Ctr., 1976—; lectr. Stanford U.; designated engring. rep. FAA. Contbr. articles to profl. jours. Lt. (j.g.) AC, USN, 1943-46. Recipient Meritorious Civilian Svc. awards U.S. Army, 1975, 77, 83, Presdl. Rank Meritorious Exec., 1987; Consol.-Vultee fellow, 1947. Fellow AIAA, Royal Aero. Soc., Am. Helicopter Soc. (hon.); mem. NAE, Swedish Soc. Aeros. and Astronauts, Sigma Xi, Elks. Mem. LDS Ch. Office: US Army Rsch & Tech Lab Activity Ames Rsch Ctr Moffett Field CA 94035

CARLSON, RICHARD RAYMOND, statistician, consultant; b. Chgo., Aug. 5, 1957; s. Gustaf Raymond and Nancy Jane (Teske) C.; m. Victoria Lynn Sausser, June 23, 1979. BS in Math., Physics summa cum laude, Butler U., 1979; MS in Stats., U. Ill., 1982. Teaching asst. U. Ill., Urbana, 1979-83; assoc. quality engr. Unisys Corp., St. Paul, 1983-85, quality engr., 1985-87, sr. statistician, 1987-90, prin. statistician, 1990-91; community faculty Metro. State U., St. Paul, 1990—; specialist, statistician Northwest Airlines, Inc., St. Paul, 1991-92; statis. cons. Group Health, Inc., Mpls., 1992—; examiner, sr. examiner Minn. Quality Award Minn. Coun. for Quality, Mpls., 1991-93; tech. adv. bd. St. Paul Tech. Coll., 1991—; examiner Unisys Total Quality Award, St. Paul, 1991; book reviewer Technometrics Jour., 1987, 90, 92, 93, Am. Statis. Jour., 1993. Vol. Community Meals, Mpls., 1987—. Mem. Am. Soc. Quality Control (sr.), Am. Stat. Assn. (coun. chpts. 1989—), North Cen. Deming Mgmt. Forum (annual conf. task chmn. 1990-92, sec., bd. dirs. 1987-90), Phi Kappa Phi. Home: 4517 Portland Ave S Minneapolis MN 55407

CARLSON, ROLF STANLEY, psychologist; b. Chgo., Feb. 2, 1943; s. Rolf J.H. and Evelyn M. (Hakanson) C.; children: Erik, Kirsten, Chris. BA, U. Wis., 1965; PhD, U. N.C., 1973. Clin. psychologist HRC, Chapel Hill,

N.C., 1973-77, Peninsula Psychol. Hosp., Knoxville, Tenn., 1975-82, SWMHC, Worthington, Minn., 1982-83, Worthington Med. Ctr., 1983—; mem. ethics com. TPA, Knoxville, 1976-82. Pres. Prairieland Film Soc., Worthington, 1983-88. Mem. Am. Psychol. Assn. Office: Worthington Med Ctr 508 10th St Worthington MN 56189

CARLSON, WILLIAM SCOTT, structural engineer; b. Columbia, S.C., Feb. 13, 1963; s. William Paul and Nancy Oleta (Long) C.; m. Teresa Lynn Bullock, July 21, 1990. BS in Civil Engring., Clemson U., 1985, ME in Civil Engring., 1988. Structural engr. Boyle Engring. Corp., Orlando, Fla., 1987-88, Fluor Daniel, Inc., Greenville, S.C., 1988-92; knowledge engr. Fluor Daniel, Inc., 5, S.C., 1992--. Sec. housing corp. Sigma Chi Housing Corp., Clemson, 1990--; bd. stewards Mauldin (S.C.) United Meth. Ch., 1993--. Mem. Am. Soc. Civil Engrs. Republican. Home: 319 Bethel Way Simpsonville SC 29681 Office: Fluor Daniel Inc 100 Fluor Daniel Dr Greenville SC 29607-2762

CARLSSON, ANDERS EINAR, physicist; b. Vätervik, Sweden, Dec. 14, 1953; s. Sven Erik Carlsson and Anna-Maria (Gunnarsdotter) Westbeck; m. Christiane Steinbach, Aug. 11, 1988; children: Nils, Anna. BA, Harvard U., 1975, PhD, 1981. Asst. prof. Physics Washington U., St. Louis, 1983-87, assoc. prof., 1987-91, prof. physics, 1991--; cons. McDonnell Douglas Labs., St. Louis, 1988-90. Contbr. articles to profl. jours. Mem. Am. Phys. Soc. Achievements include development of quantum mechanically based real space pictures of bonding in methods; ab-mitio calculation of transition metal alloy phase diagrams; atamistic analysis of ductile vs. brittle behavior of materials. Office: Washington U Dept Physics 1 Brookings Dr Saint Louis MO 63130

CARLTON, ROBERT L., clinical psychologist; b. Murray, Ky., June 1, 1918; s. Albert B. and Ophelia (Hughes) C.; m. Frances Suiter, July 10, 1946; children: Glenn R., Keith H. BS, Murray State U., 1948; MA, Ohio State U., 1950, PhD, 1953. Lic. psychologist, Ohio. Staff psychologist Children's Mental Health Ctr., Columbus, Ohio, 1952-55, chief psychologist, 1955-63, assoc. dir., 1963-82; pvt. practice Columbus, 1982—; clin. asst. prof. dept. psychiatry Ohio State U., Columbus, 1966-77, adj. assoc. prof. dept. psychology, 1972-81. 1st lt. USAF, 1942-46, PTO. Mem. Am. Psychol. Assn., Ohio Psychol. Assn. (spl. recognition award 1972), Cen. Ohio Psychol. Assn. Avocations: environment and wildlife conservation, organic gardening, flying. Home: 928 Meadowview Dr Columbus OH 43224-1930

CARMAN, CHARLES JERRY, chemical company executive; b. Tucumcari, N.Mex., Nov. 14, 1938; s. Charles Brents and Geraldine (Dodgen) C.; m. Patsy June Turner, Jan. 18, 1961; children: Davis, Carol, Darin, Gregg. BS in Chemistry and Math., Ea. N.Mex. U., 1961; MS in Phys. Chemistry, U. Calif., Davis, 1963. Scientist corp. rsch. div. B.F. Goodrich Co., Brecksville, Ohio, 1963-71, rsch. assoc., 1971-77, sr. rsch. assoc., 1977-79, mgr. corp. rsch., 1981-85; mgr. tech. resources engineered products B.F. Goodrich Co., Akron, Ohio, 1979-81; sr. mktg. devel. mgr. Geon vinyl div. B.F. Goodrich Co., Cleve., 1985-87, dir. polymer composites, 1987-88, dir. med. market, 1988-90, project dir. mktg., 1990-93, ret., 1993; internat. lectr. on blood product safety and sufficiency; internat. leader to 70 country hemophilia socs. Contbr. over 60 articles to profl. jours. Pres. Nat. Hemophilia Found., N.Y.C., 1979-87, bd. dirs. 1985-88, chmn., 1982-84; pres., chief exec. officer World Fedn. Hemophilia, Montreal, Que., Can., 1988—; cons. Internat. Global Blood Safety Initiative, WHO, Geneva, 1989-90; cons. ARC, Wash., 1987-90, HHS, 1980-90. Sci. Rsch. Coun. sr. vis. fellow, Eng., 1976; named Vol. of Yr., Nat. Hemophilia Found., 1982. Mem. Am. Chem. Soc., Franklin Club Akron (pres. 1985-86). Republican. Mem. Ch. of Christ. Achievements include first use of carbon 13 NMR for characterizing polymers; patent for thermoplastic polymer blends of an EPDM polymer having a high ethylene length index with polyethylene; development of 10 year global strategic plan for medical care for persons affected by hemophilia. Home: 184 Brentwood Dr Hudson OH 44236-1664

CARMAN, ROBERT LINCOLN, JR., physicist; b. Bay Shore, N.Y., Jan. 15, 1941; s. Robert Lincoln and Helen Mae (McCabe) C.; m. Deedre Florence Norris, June 23, 1964 (div. 1975); 1 child, Yvonne, 1 stepchild, Jennifer Lynn Mundinger; m. Constance Joan Welty, May 3, 1975. BA in Physics, Adelphi Coll., 1962; PhD in Physics, Harvard U., 1968. Staff MIT Lincoln Lab., Lexington, 1964-70; group leader Lawrence Livermore (Calif.) Nat. Lab., 1970-74; spl. project leader/program mgr. Los Alamos (N.Mex.) Nat. Lab., 1974-84; program mgr. Rockwell Internat. Corp./Rocketdyne, Canoga Park, Calif., 1984—; co-founder, CEO, pres. KDC Tech., Livermore, 1982-88; cons. in field. Contbr. articles to profl. jours., chpts. to books; contbg. author: Heritage of Copernicus, 1974. NSF fellow, 1961-62; Harvard U. postdoctoral fellow, 1968-70; N.Y. State Regents scholar, 1958-62, Adelphi Coll. gen. scholar, 1959-62. Mem. Am. Phys. Soc., Nat. Mgmt. Assn. Achievements include discovery of underlying physics limiting scale up of solid state lasers; generated first terrawatt sub picosecond pulse lasers and performed fundamental studies on self focusing phenomena using them; others in weapons and propulsion systems; 1 patent, 3 pending. Home: 334 W Siesta Ave Thousand Oaks CA 91360 Office: Rocketdyne 6633 Canoga Ave IB47 Canoga Park CA 91304

CARMICHAEL, CHARLES WESLEY, industrial engineer; b. Marshall, Ind., Jan 18, 1919; s. Charles Wesley and Clella Ann (Grubb) C.; B.S., Purdue U., 1941; m. Eleanor Lee Johnson, July 2, 1948 (dec. 1984); 1 dau., Ann Bromley Carmichael Biada; m. Bernadine P. Carlson, Dec. 21, 1985. Owner, operator retail stores, West Lafayette, Ind., 1946-48, Franklin, Ind., 1950-53; mem. staff time study Chevrolet Co., Indpls., 1953-55; indsl. engr. Mallory Capacitor Co., Indpls., 1955-60, Greencastle, Ind., 1960-70, plant engr., 1970-81; contract cons. Northwood Assocs., 1981—; lectr. in field. Chmn. Greencastle br. ARC, 1962-63; bd. dirs. United Way Greencastle, 1976-79, 84-86. Served to capt., F.A., U.S. Army, 1941-46; ETO. Decorated Bronze Star, Purple Heart with oak leaf cluster. Mem. Greencastle C. of C. (dir. 1962-64), Am. Inst. Plant Engrs., Ind. Bd. Realtors (dir. 1983-85), Putnam County Bd. Realtors (pres. 1983-84), Ind. Hist. Soc, Am. Legion. Republican. Methodist. Clubs: John Purdue, Soc. Ind. Pioneers, Beacon Club. Lodges: Masons, Shriners, Kiwanis (past pres.). Home and Office: 3628 Woodcliff Dr Kalamazoo MI 49008-2513

CARMICHAEL, IAN STUART EDWARD, geologist, educator; b. London, England, Mar. 29, 1930; came to U.S., 1964; s. Edward Arnold and Jeanette (Montgomerie) C.; m. Kathleen Elizabeth O'Brien; children by previous marriages—Deborah, Graham, Alistair, Anthea. B.A., Cambridge U., Eng., 1954; Ph.D., Imperial Coll. Sci., London U., 1958. Lectr. geology Imperial Coll. Sci. and Tech., 1958-63; NSF sr. fgn. sci. fellow U. Chgo., 1964; mem. faculty U. Calif.-Berkeley, 1964—, prof. geology, 1967—, chmn. dept., 1972-76, 80-82, assoc. dean, 1976-78, 85—, assoc. provost, 1986—. Author: Igneous Petrology, 1974; editor-in-chief Contbns. to Mineralogy and Petrology, 1973-90. Geol. Soc. Am. Guggenheim fellow, 1992; recipient Arthur L. Day medal Geol. Soc. Am., 1991. Fellow Mineral Soc. Am., Mineral Soc. Gt. Britain (Schlumberger medal 1992), Am. Geophys. Union (Bowen award 1986). Office: U Calif Berkeley Provost for Rsch Berkeley CA 94720*

CARMICHAEL, STEPHEN WEBB, anatomist, educator; b. Detroit, July 17, 1945; s. Lucien Webb and Sue (Peil) C.; m. Susan Lee Stoddard, May 16, 1992; 1 child, Allen St. Pierre. AB with honors, Kenyon Coll., 1967; PhD, Tulane U., 1971; DSc, Kenyon Coll., 1989. Instr. to assoc. prof. W.Va. U., Morgantown, 1971-82; assoc. prof. to prof. Mayo Clinic, Rochester, Minn., 1982—, chair anatomy, 1991—. Author: 6 books on adrenal medulla; contbr. over 80 articles to profl. jours. Men's sponsor Women's Shelter, Rochester, 1984—; ski instr. Courage Alpine Skiiers, Golden Valley, Minn., 1989—. Recipient over 20 awards for teaching and rsch. accomplishments. Office: Mayo clinic Medical Scis 3 Rochester MN 55905

CARMICHAEL, VIRGIL WESLY, mining, civil and geological engineer, former coal company executive; b. Pickering, Mo., Apr. 26, 1919; s. Ava Abraham and Rosevelt (Murphy) C.; m. Emma Margaret Freeman, Apr. 1, 1939 (dec.); m. Colleen Fern Wadsworth, Oct. 29, 1951; children: Bonnie Rae, Peggy Ellen, Jacki Ann. BS, U. Idaho, 1951, MS, 1956; PhD, Columbia Pacific U., San Rafael, Calif., 1980. Registered profl. geol., mining and civil engr., geologist, land surveyor. Asst. geologist Day Mines, Wallace, Idaho, 1950; mining engr. De Anza Engring. Co., Troy, Idaho, Santa Fe, 1950-52; hwy. engring. asst. N.Mex. Hwy. Dept., Santa Fe, 1952-53; asst. engr. U. Idaho, 1953-56; minerals analyst Idaho Bur. Mines, 1953-56; mining engr. No. Pacific Ry. Co., St. Paul, 1956-67; geologist N.Am. Coal Corp., Cleve., 1967-69, asst. v.p. engring., 1969-74, v.p., head exploration dept., 1974-84; travel host Satrom Travel and Tour, Bismarck, N.D., 1988-92; mem., advisor (photogeology) for People to People "Hard Rock" Minerals Del. to China, 1981; mem., leader People to People Coal Mechanization Del. to China, 1982; advisor (photogeology) to Carbocol, Colombia, S.A., 1984-85; mem. Bismarck Scottish Rite Children's Hearing Impairment Bd., 1991—. Asst. chief distbr. DC Emergency Mgmt. Fuel Resources of N.D., 1968-92; bd. dirs., chmn. fund drive Bismarck-Mandan Orch. Assn., 1979-83; 1st v.p., bd. dirs., chmn. fund drive Bismarck Arts and Galleries Assn., 1982-86; mem. and spl. advisor (Minerals) Nat. Def. Exec. Res., 1983—; mem. Fed. Emergency Mgmt. Agy, 1983—; mem. adv. bd. Bismarck Salvation Army, 1988—, chmn. 1993—, sci. rsch. bd. N.D. Acad. Sci. Found., 1986—, award chmn. 1989-91. Recipient A award for sci. writing Sigma Gamma Epsilon. Mem. Am. Inst. Profl. Geologists (past pres. local chpt.), N.D. Geol. Soc. (past pres.), Breezy Shores Resort and Beach Club (bd. dirs. 1987—), Kiwanis (past pres., dist. lt. gov., dist. chmn. internat. found. 1991—), Masons (past master, trustee 1987-92, chmn. 1991-92, chmn. N.D. found. 1990-92, named Mason of Yr. Bismarck lodge 1992), Elks, Scottish Rite (award), York Rite (award, named KTCH 1993), Shriners, Sigma Xi. Republican. Home: 1013 N Anderson St Bismarck ND 58501-3446

CARMICIANO, MARIO, electrical engineer; b. N.Y.C., Nov. 29, 1946; s. Gaetano and Palma (Dovile) C.; m. Josephine Ferro; children: Thomas, Michelle, Nicole. B.E.E., Pratt Inst., 1970. Registered profl. engr., N.Y., N.J. Pa., Calif., Mass., Conn. Engr. Syska & Hennessy, N.Y.C., 1967-78; engr. Edward A. Sears Assocs., N.Y.C., 1978-88, ptnr., 1988—; adj. lectr. Coll. S.I., N.Y.C., 1974-75. Mem. code com. N.Y. Assn. Consulting Engrs., 1989-91. Mem. NSPE, IEEE, Illuminating Engr. Socs. Office: Edward A Sears Assocs 48 W 37 St New York NY 10018

CARNAHAN, BRICE, chemical engineer, educator; b. New Philadelphia, Ohio, Oct. 13, 1933; s. Paul Tracy and Amelia Christina (Gray) C. BS, Case Western Res. U., 1955, MS, 1957; PhD, U. Mich., 1965. Lectr. in engring. biostats. U. Mich., Ann Arbor, 1959-64; asst. prof. chem. engring. and biostatics U. Mich., 1965-68, assoc. prof., 1968-70, prof. chem. engring., 1970—; vis. lect. Imperial Coll., London, England, 1971-72; vis. prof. U. Pa., 1970, U. Calif.-San Diego, 1986-87; mem., chmn. Curriculum Aids for Chem. Engring. Edn. com. Nat. Acad. Engring., 1974-75. Author: (with H.A. Luther and J.O. Wilkes) Applied Numerical Methods, 1969, (with J.O. Wilkes) Digital Computing and Numerical Methods, 1973; Editorial bd.: Jour. Computers and Fluids, 1971—, Computers and Chemical Engineering, 1974—. Mem. communications com. Mich. Council for Arts, 1977—. Recipient Chem. Engr. of Yr. award Detroit Engring. Soc., 1987, 3M award Am. Soc. for Engring. Edn., 1990. Mem. AAAS, Am. Inst. Chem. Engrs. (Computers in Chem. Engring. award 1980, chmn. CAST div. 1981), Assn. for Computing Machinery, Soc. for Computer Simulation, Sigma Xi, Sigma Nu. Home: 1605 Kearney Rd Ann Arbor MI 48104-4065

CARNES, ROBERT MANN, aerospace company executive; b. De Valls Bluff, Ark., May 1, 1942; s. Samuel Abda and Elizabeth Thompson (Mann) C.; m. Merilyn Elizabeth Latour, Aut. 7, 1977; 1 child, Karne. BS, N.C. state U., 1966; MA, Cen. Mich. U., 1975. Researcher Cen. Rsch. Sta., Raleigh, N.C., 1961-63; psychiat. aide Dorothea Dix Hosp., Raleigh, 1963-65; commd. 2d lt. USAF, 1966, advanced through grades to col., 1982, served in various locations worldwide, 1966-87, ret., 1987; v.p. Plasti-Tech, Inc., Sumter, S.C., 1987-89; pres. Striptech Internat., Inc., Mt. Pleasant, S.C., 1989—; cons. numerous govt. and indsl. orgns., 1987—. Contbr. to profl. publs. Mem. Soc. Mfg. engrs., Robotic Internat., Soc. Advancement Mfg. Process Engrs., Masons, Shriners. Achievements include development of non-chemical methods for removal of paint from airplanes, design of world's largest non-chemical dry-strip facility for commercial aircraft, research in animal nutrition, genetics. Home: 6 Park Ave Sumter SC 29150 Office: Striptech Internat Inc 300 W Coleman Blvd Ste 105 Mount Pleasant SC 29464

CARNEY, JEAN KATHRYN, psychologist; b. Ft. Dodge, Iowa, Nov. 10, 1948; d. Eugene James and Lucy (Devlin) C.; m. Mark Krupnick, Jan. 1, 1977; 1 child, Joseph Carney Krupnick. BA, Marquette U., Milw., 1970; MA, U. Chgo., Chgo., 1984; PhD, U. Chgo., 1986. Registered Clin. Psychologist, Ill. Reporter Milw. Jour., 1971-76, editorial writer, 1976-79; asst. prof. psychology St. Xavier Coll., Chgo., 1985-86; dir. Lincoln Park Clinic, Chgo., 1986-87; pvt. practice psychotherapist Chgo., 1987—; mem. sci. staff Michael Reese Hosp. Med. Ctr., Chgo., 1987—; instr. Northwestern U. Med. Sch., 1991—. Recipient Best Series Articles, 1975, Best Editorial, 1978, Milw. Press Club, William Allen White Nat. Award for Editorial Writing, 1978, Robert Kahn Meml. Award for Research on Aging, Univ. Chgo., 1985. Mem. APA, Ill. Psychol. Assn., Chgo. Assn. Psychoanalytic Psychology. Home: 915 Burns Ave Flossmoor IL 60422-1107 Office: 55 E Washington St Ste 1219 Chicago IL 60602

CARNEY, RONALD EUGENE, chemist; b. Waukegan, Ill., Mar. 15, 1943; s. Eugene George and Letty Jean (Church) C.; m. Mary Lenore Healy, Nov. 15, 1969; 1 child, Lori Jean. BS in Chemistry, Northeastern Ill. U., 1976. Lab. tech. Abbott Labs., North Chicago, Ill., 1962-77; rsch. chemist Abbott Labs., 1978-82, rsch. chemist I, 1983-87, sr. scientist, 1987—; speaker in field. patentee in field. Mem. Am. Chem. Soc., AAAS. Avocations: camping, gardening, music, boating. Office: Abbott Labs 14th & Sheridan North Chicago IL 60064

CAROLAN, DONALD BARTLEY ABRAHAM, JR., electrical engineer; b. Chgo., July 20, 1960; s. Donald Bartley and Marilyn Ann (Concannon) C.; m. Debra Beth Frankel, July 6, 1985; 1 child, Leah Beth. BA in Music, Berklee Coll. Music, 1985; BSEE, Northeastern U., 1991. Sr. technician GTE Sylvania, Danvers, Mass., 1987-89; staff engr. Northeastern U., Boston, 1989-91; Fourier Transform InfraRed engr. Bruker Instruments, Billerica, Mass., 1991-93; biomedical engr. Boston U., 1993—. Contbr. to publs. Analog Computation Circuits, Electronic Noise and Random Processes. Active Atlantic Salmon Soc., Boston, 1990—, Hop Brook Protection Soc., Sudbury, Mass., 1992—. Mem. IEEE. Home: 66 Robbins Rd Sudbury MA 01776 Office: Boston Univ Biomedical Dept Billerica MA 02215

CAROLINE, LEONA RUTH, retired microbiologist; b. Bklyn., June 17, 1912; d. Hyman and Clara (Baron) C. BA, Bklyn. Coll., 1933; MS, Wagner Coll., 1973. Lic. supr. Dept. Health of N.Y.C. Rsch. asst. Mt. Sinai Hosp., N.Y.C., 1933-35, 41-42, Overly Biochem. Rsch. Lab., N.Y.C., 1942-47, Coll. Medicine, NYU, N.Y.C., 1948-53, 56-58; asst. microbiologist North Shore Hosp., Manhasset, N.Y., 1953-56; rsch. assoc. Maimonides Med. Ctr., Bklyn., 1961-69, chief microbiologist, 1969-77, ret., 1977. Co-contbr. articles to Sci., Jour. Investigative Dermatology, Blood. Newsletter editor East 87th St. 400 Block Assn. Recipient Pres.'s award Am. Coll. Obstetricians and Gynecologists, 1968, Cert. of Merit, Citizens Commn. for N.Y.C., 1984-85. Mem. Internat. Soc. for Human and Animal Mycology, Am. Soc. Microbiology (Lifetime Achievement award N.Y.C. Br., 1988), N.Y. Acad. Scis. (life), Med. Mycology Soc. of Ams. Achievements include discovery of iron-binding activity in egg white (conalbumen), of iron-binding ability in Cohn plasma fraction IV3.4 (transferrin), that untreated acute myelogenous leukemia patients had 100% blood plasma saturation with iron; demonstration of reversal of growth inhibition of C. albicans, that serum fungistasis is caused by unsaturated iron-binding capacity of their plasma, reversed by iron addition (transferrin). Home: 448 E 87th St Apt 2C New York NY 10128-6542

CARPENTER, ALLAN LEE, civil engineer; b. Huntington, W.Va., Mar. 10, 1931; s. Cecil Clayton and Leola Belle (Klopp) C.; m. Bess Mann Spoonamore, Feb. 20, 1954 (dec.); m. Eudell Denton Jones, May 23, 1961; 1 child, Lee Ella Carpenter, 1stepchild, Fred Carl Curtis II. BS, U. Ill., 1952. Registered profl. engr. and land surveyor, Ky., Va. Apprentice supr. So. Rlwy. System, Cin., 1952-56; track supr. So. Rlwy. System, New Orleans, Lexington, Ky., 1956-59; instrument man C.J. Fuller Cons. Engrs., Inc., Lexington, 1960; jr. engr. Chesapeake & Ohio Rlwy. Co., Richmond, Va., 1961; asst. engr. and chief engr. Norfolk & Portsmouth Beltline R.R. Co., Norfolk, Va., 1962-69, 78-91; traffic engr. City of Chesapeake (Va.), 1969-78; owner All Trans Engring., Norfolk, 1991—; cons. pvt. practice, Norfolk, 1962—; instr. Tidewater C.C., Portsmouth, Va., 1971-73. mem. and officer The Exchange Club of Norfolk, 1963-84; trustee River Forest Shores, Wayside Manor, Easton Pl. Civic League, Norfolk, 1991-92, pres. 1993—. Lt. U.S. Army, 1954-56. Mem. ASCE, Inst. Transp. Engrs., Am. Rlwy. Engring. Assn., Va. Assn. of Surveyors (sec.-treas. Tidewater chpt. 1971, v.p. 1972). Republican. Home and Office: All Trans Engring 336 E McGinnis Cir Norfolk VA 23502-5336

CARPENTER, F. LYNN, biology educator; b. Oklahoma City, Feb. 14, 1944; d. Clarence Coe and Beth (Woodman) C. BA, U. Calif., Riverside, 1966; PhD, U. Calif., Berkeley, 1972. Asst. prof. U. Calif., Irvine, 1972-78, assoc. prof., 1978-84, prof., 1984—. Author: Ecology and Evolution in an Andean Hummingbird, 1976; bd. editors The Am. Zoologist, 1984-89; contbr. articles to profl. jours. Chair ecology Am. Soc. Zoologists, 1983-84, symposium organizer, 1984. NSF grantee, 1978-80, 81-84, 84-86. Fellow AAAS; mem. Ecol. Soc. Am., Cooper Ornithol. Soc., Am. Orinthol. Union. Achievements include research in the behavior of territorial birds. Office: Univ of Calif Dept Eco Evol Biol Irvine CA 92717

CARPENTER, FRANK CHARLES, JR., retired electronics engineer; b. L.A., June 1, 1917; s. Frank Charles and Isobel (Crump) C.; A.A., Pasadena City Coll., 1961; B.S. in Elec. Engring. cum laude, Calif. State U.-Long Beach, 1975, M.S. in Elec. Engring., 1981; m. Beatrice Josephine Jolly, Nov. 3, 1951; children—Robert Douglas, Gail Susan, Carol Ann. Self-employed design and mfgr. aircraft test equipment, Los Angeles, 1946-51; engr. Hoffman Electronics Corp., Los Angeles, 1951-56, sr. engr., 1956-59, project mgr., 1959-63; engr.-scientist McDonnell-Douglas Astronautics Corp., Huntington Beach, Calif., 1963-69, spacecraft telemetry, 1963-67, biomed. electronics, 1967-69, flight test instrumentation, 1969-76; lab. test engr. Northrop Corp., Hawthorne, Calif., 1976-82, spl. engr., 1982-83; mgr. transducer calibration lab. Northrop Corp., Pico-Rivera, Calif., 1983-86. Served with USNR, 1941-47. Mem. IEEE (sr.), Amateur Radio Relay League. Contbr. articles to profl. jours. Patentee transistor squelch circuit; helicaland whip antenna. Home: 2037 Balearic Dr Costa Mesa CA 92626-3514

CARPENTER, KENNETH RUSSELL, international trading executive; b. Chgo., May 22, 1955; s. Kenneth and Margaret (Lucas) C.; m. Holly Lee Nelson, July 20, 1985; 1 child, Matthew. AS in Aviation, Prairie State Coll., Chicago Heights, Ill., 1979. Respiratory therapist Harvey, Ill., 1980-83; dir., owner, ptnr. Pulmonary Therapy Inc., Harvey, 1983-91; v.p. Home Air Joliet Ltd., Harvey, 1984—; dir., owner Air Systems, Chicago Heights, 1981—, Ft. Lauderdale, 1991—; dir., owner Home Ortho Ltd., Harvey, 1985—; pres., CEO Profl. Yacht Svcs., Inc., Chicago Heights, 1987—; owner CL2 Exporting Inc., Chicago Heights, 1993—; dir. pub. rels. Lansing (Ill.) Med. Group, 1990; dir. pulmonary rehab. Cardio-Pulmonary Assocs., Munster, Ind., 1990, CLZ Exporting, 1992—; major importer/exporter in the Middle East. Pilot CAP, 1979-86. With USN, 1973-77. Mem. Am. Assn. Respiratory Therapy (cert.), Nat. Assn. Med. Equipment Suppliers, Internat. Trade Assn., World Trade Assn. Avocations: flying, boating, computer programming. Home and Office: 23030 Miller Rd Chicago Heights IL 60411-5932

CARPENTER, MALCOLM SCOTT, astronaut, oceanographer; b. Boulder, Colo., May 1, 1925; s. Marion Scott and Florence Kelso (Noxon) C.; m. Rene Louise Price, Sept. 9, 1948 (div.); children—Marc Scott, Robyn Jay, Kristen Elaine, Candace Noxon; m. Maria Roach, 1972; children—Matthew Scott, Nicholas André; m. Barbara Curtin, 1988; 1 child, Zachary Scott. B.S. in Aero. Engring., U. Colo., 1962. Commd. ensign U.S. Navy, 1949, advanced through grades to comdr., 1959; assigned various flight tng. schs., 1949-51, (Patrol Squadron 6), Barbers Point, Hawaii, 1951; also in (Patrol Squadron 6), Korea, 1951-52; grad. (Navy Test Pilot Sch.), 1954; assigned electronics test div. (Naval Air Test Center), 1954-57, (Naval Gen. Line Sch.), 1957, (Naval Air Intelligence Sch.), 1957-58; air intelligence officer (U.S.S. Hornet), 1958-59; joined Project Mercury, man-in-space project NASA, 1959; completed 3 orbit space flight mission in spacecraft (Aurora 7), May 1962; mem. (U.S. Navy Aquanaut Project), 1965-67; retired from U.S. Navy, 1969; now engaged in pvt. oceanographic and energy research business. Fellow Inst. Environ. Scis. (hon.); mem. Assn. Space Explorers, Explorers Club, Delta Tau Delta.

CARPENTER, RAY WARREN, materials scientist and engineer, educator; b. Berkeley, Calif., 1934; s. Fritz Josh and Ethel Thordis (Davisson) C.; m. Ann Louise Leavitt, July 10, 1955; children: Shannon R., Sheila A., Matthew L. BS in Engring., U. Calif., Berkeley, 1958, MS in Metallurgy, 1959, PhD in Metallurgy, 1966. Registered profl. engr., Calif. Sr. engr. Aerojet-Gen. Nucleonics, San Ramon, Calif., 1959-64; sr. metallurgist Stanford Rsch Inst., Menlo Park, Calif., 1966-67; mem. sr. rsch. staff Oak Ridge (Tenn.) Nat. Lab., 1967-80; prof. Solid State Sci. & Engring. Ariz. State U., Tempe, 1980—, dir. Facility for High Resolution Electron Microscopy, 1980-83, dir. Ctr. for Solid State Sci., 1985-91, also bd. dirs. Ctr. for Solid State Sci.; chmn. doctoral program on Sci. and Engring. of Materials Ariz. State U., 1987-90; vis. prof. U. Tenn., 1976-78; adj. prof. Vanderbilt U., Nashville, 1979-81. Contbg. author books; contbr. articles to profl. rsch. jours. and symposia. Recipient awards, Internat. Metallographic Soc. and Am. Soc. for Metals competition, 1976, 77, 79; Faculty Disting. Achievement award Ariz. State U. Alumni Assn., 1990. Mem. Electron Microscopy Soc. Am. (pres. 1989, dir. phys. sci. 1980-83), Metall. Soc. of AIME, Materials Rsch. Soc., Am. Phys. Soc., Am. Ceramic Soc., Sigma Xi. Office: Ariz State U Ctr Solid State Sci Tempe AZ 85287-1704

CARPENTER, ROBERT VAN, JR., laser and engineering technician; b. Gainesville, Fla., June 22, 1965; s. Robert Van and Betsy (Parrish) C. Degree in Laser Electro-optics Tech., Orlando Vocat. Tech. Ctr., 1990. With shipping and receiving dept. Laser Photonics, Inc., Orlando, Fla., 1984-86, ops. assembly staff, 1986-87, laser technician, 1987-89, supr. fin. test dept., 1990-92, sr. laser technician, engr., 1992—; sr. optical technician, engr. Advanced Lasersystems Tech., Orlando, 1989-90. Home: PO Box 678156 Orlando FL 32867 Office: Laser Photonics Inc 12351 Research Pkwy Orlando FL 32826

CARPER, KENNETH LYNN, architect, educator; b. Colfax, Wash., Nov. 2, 1948; s. Emery C. and Marjorie A. (Hain) C.; m. Tanya J. Corcoran, June 9, 1970; children: Brent, Corin, Daren. BArch, Wash. State U., 1972, MSCE, 1977. Registered architect, Wash. Architect, designer Sylvester Assocs., Spokane, Wash., 1968-74; prof. arch. Wash. State U., Pullman, 1974—. Author: Forensic Engineering, 1989; editor: Forensic Engineering: Learning from Failure, 1986; founding editor Jour. of Performance of Constructed Facilities, 1986—; contbr. articles to profl. jours. Facilities adv. com. pub. schs. Pullman, 1985—. Named Outstanding Prof. Coll. Engring. & Arch., Wash. State U., 1985, Sch. Arch., 1983, 84, 85, 92; recipient Outstanding Faculty award N.W. Coop. Edn. Assn., 1992. Mem. ASCE (Richard R. Torrens award 1991, Daniel Mead award 1983, chair tech. coun. on forensic engring. 1990, chair pubs. com. 1986—, chair com. on dissemination of failure info. 1985), Am. Soc. Engring. Edn. Democrat. Office: Wash State Univ Coll Engring & Arch Sch of Arch Pullman WA 99164-2220

CARPER, STEPHEN WILLIAM, biochemist, researcher; b. Boise, Idaho, Mar. 25, 1958; s. William Joseph and Florence Stephanie (Cooper) C.; m. Barbara Ann Worcester, June 7, 1980; children: Sarah Katheryn, JoeAnn Stephanie, Elizabeth Stacey. BS in Chemistry and Biology, Ea. Oreg. State Coll., 1981; PhD in Biochemistry, Utah State U., 1986. Postdoctoral fellow U. Ariz., Tucson, 1986-88; asst. scientist U. Wis., Madison, 1988-91; asst. prof. U. Nev., Las Vegas, 1991—. Reviewer: Cancer Rsch., Radiation Rsch.; contbr. articles to profl. jours. and chpts. to books. Named Young Investigator Nat. Cancer Inst., 1992. Mem. Am. Soc. for Biochemistry and Molecular Biology, Sigma Pi Sigma. Achievements include research in function and regulation of heat shock protein 27, polyamine metabolism. Office: U Nev Las Vegas Dept Chem 4505 Maryland Pkwy Las Vegas NV 89154

CARR, ALBERT ANTHONY, organic chemsit; b. Covington, Ky., Dec. 20, 1930; s. Albert Anthony and Virginia Charlotte (Wendel) C.; m. Irene Eleanor Mauntel (div.); children: Virginia I., Michael P., Gregory J., Jerome R. BS, Xavier Univ., 1953, MS, 1955; PhD, Univ. Fla., 1958. Rsch. chemist Wm. S. Merrell Co., Cin., 1958-65; section head Merrell Nat. Labs., Cin., 1965-76; sr. section head Richardson-Merrell Inc., Cin., 1976-85; assoc. scientist Merrell Dow Rsch. Inst., Cin., 1985-88, sr. assco. scientist, 1988-92; disting. scientist Marion Merrell Dow Rsch. Inst., Cin., 1992-93; lectr.

Xavier Univ., 1962-68. Contbr. articles to profl. jours.; patentee in field. Named Chemist of Yr., Cin. Am. Chem. Soc., 1987; recipient Disting. Scientist award Tech. Soc. Coun. Engrs., 1988, Am. Chem. Soc. award Creative Invention, 1993. Mem. Am. Chem. Soc. (Creative Invention award 1993), N.Y. Acad. Scis. Office: Marion Merrell Dow 2110 E Galbraith Rd Cincinnati OH 45215

CARR, DANIEL BARRY, anesthesiologist, endocrinologist, medical researcher; b. N.Y.C., Apr. 6, 1948; s. Andrew Joseph and Florence (Glassman) C.; m. Justine M. Meehan, Nov. 11, 1978; children: Nora, Rebecca, Andrew. BA, Columbia U., 1968, MA, 1970, MD, 1976. Diplomate Am. Bd. Internal Medicine (subsplty. bds. Endocrinology and Metabolism, Anesthesiology). Intern Columbia-Presbyn. Med. Ctr., N.Y.C., 1976-78; resident med. svc. Mass. Gen. Hosp., Boston, 1978-79, endocrine fellow, 1979-82, staff physician endocrine unit, 1982—, clin. assoc. physician, clin. rsch. ctr., 1982-84, fellow in anesthesiology, 1984-86; dir. analgesic peptide research unit, 1986—, staff physician anesthesia svc. and co-dir. anesthesia pain unit, 1986-91, dir. div. pain mgmt., 1991—; clin. asst. in medicine, 1983—, anesthetist, 1992—; cons. internal medicine Mass. Eye and Ear Infirmary, 1980-82; instr. medicine Harvard U. Med. Sch., 1982-84, asst. prof., 1984-88; assoc. prof., 1988—; rsch. staff Shriners Burns Inst., Boston, 1986—; co-chair pain mgmt. guidelines panel, Agy. for Health Care Policy and Rsch., U.S. Dept. Health and Human Svcs., 1990—; mem. Gov. Mass. spl. commn. pain mgmt. Contbr. articles, rsch. reports, essays, revs. to profl. lit. Daland fellow Am. Philos. Soc., 1980-83. Mem. Am. Pain Soc., Am. Soc. Anesthesiologists, Am. Burn Assn., Internat. Assn for Study Pain (editor-in-chief Pain Clin. Updates), Endocrine Soc., Soc. for Neurosci., Internat. Anesthesia Rsch. Soc., Assn. Univ. Anesthetists, Alpha Omega Alpha. Research on pain, analgesic peptides and stress responses, relationship between analgesia and clinical outcome; identification of links between endogenous opioid (endorphin) secretion on infection and exercise; development of guidelines and regulatory policies for improved pain treatment in hospital, hospice, and home care settings. Office: Mass Gen Hosp Dept Anesthesia Fruit St Boston MA 02114-2620

CARR, DOLEEN PELLETT, computer service and environmental specialist, consultant; b. Alameda, Calif., Sept. 23, 1950; d. Charles Joseph Ziegler and Dola Faye (Cushing) Peterson; m. Glen Allwin Pellett, June 26, 1971 (div. 1986); children: Mark D., Michael J.; m. Danny Lynn Carr, Dec. 29, 1986. BA, U. Wis., Madison, 1973. Notary Pub., Mich. Budget analyst Ednl. Testing Svc., Princeton, N.J., 1979-80; tech. recruiter Uniforce Svcs., Inc., Rock Hill, S.C., 1983-84; mgr. tng. and documentation Electronic Data Systems Corp., Troy, Mich., 1985-87; tech. writer, trainer, analyst cons. CES, Inc., Troy, 1989-92; pres. D'Carr Co., Inc., Roseville, Mich., 1988—; tech. writer, trainer, cons. Eaton Corp., Southfield, Mich., 1992—; installer Gt. Plains Acctg., Fargo, N.D., 1990—; cons. Hazardous Materials Info. Exch., Washington, 1989—; cons., tech. writer Saturn Corp., 1991-92, Blue Cross/Blue Shield, Southfield, Mich., 1992-93; tech. writer FANUC Robotics, N.A., Inc., Auburn Hills, Mich., 1993—. Co-author: CIW-Weld Monitor, 1990, 93. Mem. ASTD, NAFE, Internat. Platform Assn., Greater Trenton Musicians Union, Profl. Bus. Women Assn., Macintosh Users' Club, Cen. Macomb County C. of C. Democrat. Roman Catholic. Avocations: piano, swimming, computers. Office: 16020 Thirteen Mile Rd Roseville MI 48066

CARR, FLOYD EUGENE, sales engineer; b. Indpls., Mar. 1, 1955; s. Nelson Carr and Louetta (Teets) Cole; m. Terry Rickard, Oct. 1980 (div. 1985); children: Eric Eugene, Evan Trent; m. Cathy L. Clark, Dec. 26, 1986; 1 stepchild, Chris A. Hines. Grad., Ind. U., Indpls., 1985. Sales engr. Valley-Todeco, Inc., Sylmar, Calif., 1980-86, Shur-Lok Corp., Irvine, Calif., 1986-91; co-founder BFS Techs., Lancaster, Calif., 1986—; sales mgr. Flexalloy, Inc., Cleve., 1993—. Author tng. manuals, tech. and sales guides. Mem. Warren Twp. Sch. Bd., Indpls., 1990-91; exec. bd. Warren Fine Arts Found., Indpls., 1990-91. Mem. Chgo. Nut and Bolt Assn., Masons. Achievements include design of aerospace fasteners currently used in military and commercial gas turbine engines. Home: 10036 Park Glen Ct Indianapolis IN 46229

CARR, GERALD PAUL, former astronaut, business executive, former marine officer; b. Denver, Aug. 22, 1932; s. Thomas Ernest and Freda (Wright) C.; divorced; children: Jennifer, Jamee, Jeffrey, John, Jessica, Joshua; m. Patricia Musick, Sept. 14, 1979. BS in Mech. Engring., U. So. Calif., 1954; BS in Aero. Engring., U.S. Naval Postgrad. Sch., 1961; MS in Aero. Engring., Princeton U., 1962; DSc (hon.), St. Louis U., 1976. Registered profl. engr., Tex. Commd. 2d lt. USMC, 1954, advanced through grades to col., 1974, ret., 1975; jet fighter pilot U.S., Mediterranean, Far East, 1956-65; astronaut NASA, Houston, 1966-77; comdr. 3d Skylab Manned Mission, 1973-74; pres. CAMUS, Inc., Huntsville, Ark.; project dir. Ark. Aerospace Edn. Ctr.; adv. bd. Nat. Space Soc., Space Dermatology Found. Recipient Group Achievement award NASA, 1971, Distinguished Service medal, 1974; Gold medal City of Chgo., 1974; Gold medal City of N.Y., 1974; Alumni Merit award U. So. Calif., 1974; Distinguished Eagle Scout award Boy Scouts Am., 1974; Robert J. Collier Trophy, 1974; Robert H. Goddard Meml. trophy, 1975; FAI Gold Space medal; others. Fellow Am. Astronautical Soc. (Flight Achievement award 1975); mem. NSPE, Marine Corps Assn., Marine Corps Aviation Assn., Soc. Exptl. Test Pilots, Tex. Soc. Profl. Engrs., U. So. Calif. Alumni Assn., Tau Kappa Epsilon. Presbyterian. Home and Office: PO Box 919 Huntsville AR 72740-0919

CARR, JACQUELYN B., psychologist, educator; b. Oakland, Calif., Feb. 22, 1923; d. Frank G. and Betty (Kreiss) Corker; children: Terry, John, Richard, Linda, Michael, David. BA, U. Calif., Berkeley, 1958; MA, Stanford U., 1961; PhD, U. So. Calif., 1973. Lic. psychologist, Calif; lic. secondary tchr., Calif. Tchr. Hillsdale High Sch., San Mateo, Calif., 1958-69, Foothill Coll., Los Altos Hills, Calif., 1969—; cons. Silicon Valley Companies, U.S. Air Force, Interpersonal Support Network, Santa Clara County Child Abuse Council, San Mateo County Suicide Prevention Inc.,Parental Stress Hotline, Hotel/Motel Owners Assn.; co-dir. Individual Study Ctr.; supr. Tchr. Edn.; adminstr. Peer Counseling Ctr.; led numerous workshops and confs. in field. Author: Learning is Living, 1970, Equal Partners: The Art of Creative Marriage, 1986, The Crisis in Intimacy, 1988, Communicating and Relating, 1984, 3d edit., 1991, Communicating with Myself: A Journal, 1984, 3d edit., 1991; contbr. articles to profl. jours. Mem. Mensa. Club: Commonwealth. Home: 837 Miller Ave Cupertino CA 95014 Office: Foothill College 12345 El Monte Ave Los Altos CA 94022-4504

CARR, MICHAEL, secondary education educator; b. Stockton, Calif., Sept. 20, 1951; s. Frank Edward and Eleanor (Adair) C. AA, San Joaquin Delta Community, 1971; BA, Calif. State U., Long Beach, 1975; MA, Pacific Oaks Coll., 1977; postgrad., U. Calif., Irvine, 1981. Cert. tchr./trainer, Calif. Tchr. Carlsbad (Calif.) Unified Schs.; presenter workshops on writing process, accelerated learning and learning styles. Contb. author: Thinking/Writing, Practical Ideas for Teaching Writing as a Process, CBEST Preparation Guide, Guiding Young Children's Learning: An Activities Handbook, (audio tape series) Success Through Writing; cons. film series. Mem. NEA, Calif. Tchrs. Assn., Internat. Reading Assn., SCTE. Office: US Geological Survey 345 Middlefield Rd Menlo Park CA 94025 *Died July 27, 1993.*

CARR, MICHAEL HAROLD, geologist; b. Leeds, Eng., May 26, 1935; came to U.S., 1956, naturalized, 1965; s. Harry and Monica Mary (Burn) C.; m. Rachel F. Harvey, Apr. 14, 1961; son, Ian M. B.Sc., London U., 1956; M.S., Yale U., 1957, Ph.D., 1960. Rsch. assoc. U. Western Ont., 1960-62; with U.S. Geol. Survey, 1962—; chief astrogeologic studies br. U. Geol. Survey, Menlo Park, Calif., 1973-79; mem. Mariner Mars Imaging Team, 1969-73; leader Viking Mars Orbiter Imaging Team, 1969-80; mem. Voyager and Galileo Jupiter Imaging Teams, 1978—; Interdisciplinary scientist, Mars observer. Author: The Surface of Mars, The Geology of the Terrestrial Planets. Recipient Exceptional Sci. Achievement medal NASA, 1977, Disting. Meritorious Svc. award Dept. Interior, 1988. Mem. Geol. Soc. Am., AAAS, Am. Geophys. Union. Home: 1389 Canada Rd Redwood City CA 94062-2452 Office: US Geol Survey Menlo Park CA 94025

CARR, THOMAS MICHAEL, analytical chemist; b. Shamokin, Pa., Mar. 5, 1953; s. Thomas Edward and Mary Irene (Novosel) C. BS in Chemistry, Elizabethtown (Pa.) Coll., 1975; MS in Chemistry, Case Western Res. U., 1977, PhD in Chemistry, 1980. Rsch. assoc. Cath. U., Nijmegen, Netherlands, 1979-80; project chemist Dow Corning Corp., Midland, Mich., 1980-83, analytical specialist, 1983-86; sr. scientist Owens Corning Fiberglas, Granville, Ohio, 1986—; adj. instr. Saginaw (Mich.) State Coll., 1984-86. Mem. various offices Midland Jr. C. of C., 1981-86, Newark (Ohio) Jr. C. of C., 1986-90. Mem. AAAS, Am. Chem. Soc. (program chmn. 1989—), Soc. for Applied Spectroscopy, U.S. Jr. C. of C. (cert. instr., various offices, Seiji Horiuchi award 1990), Columbus (Ohio) Jr. C. of C. (pres.-elect 1990-91, pres. 1991-92), Ohio Jr. C. of C. (v.p. mgmt. devel. 1992—). Republican. Roman Catholic. Avocations: travel, education and training, community service. Home: 8357 Morningdew Dr Reynoldsburg OH 43068-9642 Office: Owens Corning Fiberglas Tech Ctr 2790 Columbus Rd Granville OH 43023-1200

CARRADINI, LAWRENCE, comparative biologist, science administrator; b. Astoria, N.Y., Apr. 18, 1953; s. George John and Florence (Camuti) C.; m. Susan Marie Peterson, Sept. 23, 1972; 1 child, Daniel Lawrence. BS in Zoology, Columbia Pacific U., 1989, MS in Vertebrate Reproductive Physiology and Physiol. Ecology, 1992. Technician Charles River Labs., Wilmington, Mass., 1978, coord., tech. advisor, 1979, area supr., 1979, rsch. animal surgeon, 1980-87, coord., 1987-89, lab. supr., 1989-91, researcher, 1991; sr. scientist, mgr. Biol. Labs., Mass. Health Rsch. Inst., Boston, 1992—; mem. Lake Survey, U. N.H., Salem, 1982-83. Mem. adv. bd. Internat. Jour. Advances in Contraceptive Delivery Systems; contbr. articles to Jour. Lab. Animal Sci., Jour. Am. Vet. Med. Assn. Officer, selectman apptd. mem. 208 Water Quality Study Com., Salem, N.H., 1981-82, chmn., selectman apptd. mem., 1982-83; mem. adv. bd. Internat. M.C. Chang Meml. Festschrift. Recipient N.Y. State Regents Scholarship award, 1971. Mem. Soc. for Cryobiology, Nat. Am. Assn. Lab. Animal Sci., Am. Assn. Lab. Animal Sci. (New England chpt., cert. nat. lab. animal technologist), Lab. Animal Mgmt. Assn., Internat. Platform Assn., Atlantic Salmon Fedn. (assoc.), Am. Mus. Natural History (assoc.), Internat. Soc. for Pharm. Engring. Democrat. Achievements include development of reliable method to cycle estrus in syrian hamsters; co-development of commercially available cryopreserved 1-cell mouse embryos for use in media assays; co-application of 1-cell technology toward development of commercially available cryopreserved, fertilized, pronuclear-staged mouse oocytes for DNA microinjection. Home: 10 N Shore Rd Derry NH 03038-5111 Office: 305 South St Boston MA 02130

CARRASQUILLO, RAMON LUIS, civil engineering educator, consultant; b. San Juan, P.R., July 28, 1953; s. Ramon L. and Abigail Carrasquillo; m. Gladys Mateu (div. 1983); children: Ramon L. Jr., Jessica Marie; m. Peggy Musser, 1985; 1 child, Travis Andrew. MSCE, Cornell U., 1978, PhD, 1980. Registered profl. engr., Tex. Grad. rsch. asst. Cornell U., Ithaca, N.Y., 1976-79; asst. prof. civil engring. U. Tex., Austin, 1980-84, assoc. prof. civil engring., 1984-89, prof. civil engring., 1980—, researcher, 1980—; ptnr. Carrasquillo Assocs., Marble Falls, Tex., 1980—; assoc. dir. Internat. Ctr. for Aggregates Rsch., Austin, 1992—. Co-author: Production of High Strength Concrete, 1986, also papers in field. Austin Industries fellow, 1991. Fellow Am. Concrete Inst. (T.Y. Lin award 1990); mem. ASTM, ASCE, Tex. Aggregates and Concrete Assn. (Outstanding Speaker award 1980). Home: Rt 4 Box 528 Cow Creek Rd Marble Falls TX 78654 Office: U Tex 10100 Burnet Rd Bldg 18B Austin TX 78758-4497

CARREKER, JOHN RUSSELL, retired agricultural engineer; b. Cook Springs, Ala., Aug. 15, 1908; s. John Robert and Cora Selina (Polk) C.; m. Helen Mackey Garrett, Feb. 10, 1934; children: Joan Louise, James Russell. BS, Ala. Poly. Inst., 1930, MS, 1933. Agrl. engr. Civilian Conservation Corps, Dadeville, Ala., 1933-34, USDA-SCS, Anniston, Ala., 1935-38; rsch. agrl. engr. USDA-SCS & Agrl. Rsch. Svc., Watkinsville, Ga., 1938-58; rsch. lia. rep. USDA-SCS & Agrl. Rsch. Svc., Athens, Ga., 1958-61; rsch. leader USDA-Agrl. Rsch. Svc., Athens, 1961-73; ret., 1973. Author over 80 tech. papers on erosion control, irrigation and water mgmt. on farm land. Chmn. PTA, Athens, 1950; pres. Civitan Club, Athens, 1948, Residents Assn., Atlanta, 1986. Fellow Am. Soc. Agrl. Engrs. (Soil & Water Engring. award 1970), Soil & Water Conservation Soc.; mem. SAR.

CARREL, JAMES ELLIOTT, biologist; b. San Pedro, Calif., June 10, 1944; s. Leonard Willis and Louise Hanna (Schaffer) C.; m. Jan Carol Weaver, Jan. 14, 1978; children: Margaret Alice, Gary Lee Thomas. AB, Harvard U., 1966; PhD, Cornell U., 1971. From asst. prof. to prof. biology U. Mo., Columbia, 1971—; vis. prof. U. Manchester, Eng., 1979-80, SUNY-Stony Brook, 1990; vis. assoc. prof. Fla. State U., Tallahassee, 1981-82; vis. biologist Archbold Biol. Sta., Lake Placid, Fla., 1967-93; chmn. sci. talent search Mo. Acad. Sci., Kirksville, 1991—. Editorial bd. Jour. Arachnology, 1989—; contbr. articles to Sci., Environ. Entomology, others. Chmn. Citizens for Ecol. Action, Ithaca, N.Y., 1969-70, East Campus Neighborhood Orgn., Columbia, 1978. NIH grantee, 1982-90; Kemper fellow for teaching excellence, 1993. Mem. AAAS, Am. Arachnol. Soc., Entomol. Soc. Am., Brit. Arachnol. Soc., Internat. Soc. Chem. Ecology, Sigma Xi, Phi Kappa Phi, Beta Beta Beta. Democrat. Unitarian Universalist. Achievements include discovery of first natural sedative for spiders, first beetle species to feed on pine pollen, massive biosynthesis of cantharidin in male blister beetles. Office: Mo Univ 209 Tucker Hall Columbia MO 65211

CARRERA, MARTIN ENRIQUE, research scientist; b. Almirante, Panama, Aug. 16, 1960; s. Manuel and Isenith Ester (Patiño) C.; m. Elizabeth Anne Lott, May 24, 1985; children: William Martin, Robert Martin. SB in Chem. Engring., MIT, 1982, SB in Math., SB in Chemistry, 1983; MS in Phys. Chemistry, U. Chgo., 1987, PhD in Phys. Chemistry, 1988. Rsch. scientist Amoco Chem. Co., Naperville, Ill., 1989—. Contbr. articles to Phys. Rev., Jour. Chem. Physics, others. Mem. indsl. adv. coun. on minority edn. MIT, Cambridge, 1990-93. Mem. AIChE, Am. Phys. Soc., Am. Chem. Soc., Sigma Xi. Office: Amoco Rsch Ctr PO Box 3011 150 W Warrenville Rd Naperville IL 60566

CARRERA, RODOLFO, nuclear engineer; b. Barcelona, Spain, Dec. 18, 1953; came to U.S., 1976, naturalized, 1990; s. Rodolfo and Rosario (Zabaleta) C.; m. Elena Montalvo, Jan. 7, 1978. BS, Poly. U., Madrid, 1975; MS, U. Wis., 1979, PhD, 1983. Rsch. engr. Spanish Nuclear Agy., Madrid, 1975-77; rsch. asst. nuclear engring. dept. U. Wis., Madison, 1978-83; rsch. fellow physics dept. Inst. Fusion Studies U. Tex., Austin, 1983-86, project mgr. Ctr. Fusion, 1986-88, chief scientist Ctr. Fusion, 1988-91, lectr. mech. engring. dept., 1987-90; sr. rsch. scientist, founder Valley Rsch. Corp., Austin, 1991—; cons. Nat. Commn. on Sci. and Tech., Spain, 1991—; rsch. asst. Sci. Applications, Inc., Boulder, Colo., summer, 1982; lectr. in major plasma and nuclear rsch. labs. and univs., U.S. and overseas; referee to major plasma and nuclear sci. and engring. tech. jours. Contbr. articles, reports to Plasma and Nuclear Sci. and Engring.. Mem. AAAS, AIAA, IEEE, Am. Vacuum Soc., Am. Nuclear Soc., Am. Phys. Soc., Am. Chem. Soc., Material Rsch. Soc., Instruments Soc. Am., Internat. Soc. Optical Engrs. Achievements include development of a kinetic theory of thermal barrier in tandem mirrors, theory for non linear dynamics of magnetic islands in high-temperature tokamak plasmas; design and analysis of fusion ignition experiment.

CARRICO, JOHN PAUL, physicist; b. Detroit, June 26, 1938; s. John Francis and Jean Reinette (Kennedy) C.; m. Clairellen Collom, Aug. 20, 1960 (div. Aug. 1975); children: John Paul, Timothy Kermit, Kevin Charles, Laura Ellen; m. Anita Gilden, May 6, 1990. BS with honors, U. Windsor, Ont., Can., 1961; PhD, Brandeis U., 1966. Instr. Brandeis U., Waltham, Mass., 1966-67; sr. scientist, engring. mgr. Bendix Corp., 1967-83; div. chief, assoc. dir. U.S. Army R&D Ctr., Edgewood, Md., 1983-88; dir. SRI Internat., Menlo Park, Calif., 1988-90; spl. asst. Office of Sec. of Def., Washington, 1990—. Contbr. more than 100 articles to profl. jours. and confs. Recipient R&D Achievement award U.S. Army, 1986, Meritorious Civilian award U.S. Army, 1976, award of excellence Sec. of Def., 1993; Woodrow Wilson Found. fellow, 1961. Roman Catholic. Achievements include 10 patents; developement of concepts in mass spectrometry, remote sensing, ion mobility; establishment of new lower limits to existence of electric dipole moment of atoms and electrons. Home: 220 Warren Ave Baltimore MD 21230 Office: DATSDCAE Rm 3C124 The Pentagon Washington DC 20301-3050

CARRIER, GEORGE FRANCIS, applied mathematics educator; b. Millinocket, Maine, May 4, 1918; s. Charles Mosher and Mary (Marcaux) C.; m. Mary Casey, June 30, 1946; children: Kenneth, Robert, Mark. Degree in Mech. Engr., Cornell U., 1939, PhD, 1944. From asst. prof. to prof. Brown U., 1946-52; Gordon McKay prof. mech. engring. Harvard U., 1952-72, T. Jefferson Coolidge prof. applied math., 1972-88, emeritus, 1988—; mem. coun. emeritus Cornell U. Engring. Coll. Co-author: Functions of a Complex Variable, 1966, Ordinary Differential Equations, 1968, Partial Differential Equations, 1976; assoc. editor: Quar. Applied Math. Former trustee Rensselaer Poly. Inst., Troy, N.Y. Recipient Von Karman prize ASCE, 1977, Pres.'s Nat. medal sci. NSF, 1990. Fellow Am. Acad. Arts and Scis., Brit. Inst. Math. and Its Applications (hon.); mem. ASME (hon., Timoshenko medal 1978, Centennial medal 1980), NAS (Applied Math. and Numerical Analysis award 1980), AIAA (Dryden medal 1989), Soc. Indsl. and Applied Math. (Von Karman prize 1979), Nat. Acad. Engring., Am. Philos. Soc., Am. Phys. Soc. (Fluid Dynamics prize 1984), Internat. Soc. Interaction Mechs. and Math., Sigma Xi. Office: Harvard U Div Applied Sci Pierce 311 Cambridge MA 02138

CARRIERE, SERGE, physiologist, physician, educator; b. Montreal, Que., Can., July 21, 1934; s. Virgile and Angelina (Malouin) C.; m. Irene Lafond, Dec., 1976; children: Sylvie, Brigitte, Alain, Francois. B.A., U. Montreal, 1954, M.D., 1959. Intern Notre Dame Hosp., Montreal, 1958-59; resident in internal medicine Notre Dame Hosp., 1959-62; practice medicine specializing in nephrology Montreal, 1964—; instr. physiology Harvard Med. Sch., Boston, 1962-64; asst. prof. dept. medicine U. Montreal, 1964-70, asso. prof., 1970-74, prof., 1974-80, prof., head dept. physiology, 1980-86, prof., head dept. medicine, 1986-88, dean med. sch., 1989—; mem. staff Maisonneuve-Rosemont Hosp. Contbr. numerous articles on research in physiology and nephrology to sci. and med. jours. Med. Rsch. Coun. Can. fellow, 1962-64, grantee, 1971—, career investigator, 1971-80. Fellow Royal Coll. Physicians; mem. Can. Soc. Physiology, Am. Soc. Physiology, Internat. Soc. Physiology, Am. Soc. Nephrology, Can. Soc. Nephrology, Am. Soc. Clin. Investigation, Can. Soc. Clin. Investigation. Home: 40 Du Chene Vandreuil, Sur Le Lac, PQ Canada J7V 8T3 Office: University of Montreal, CP 6128 Succursale A, Montreal, PQ Canada H3C 3J7

CARR-LOCKE, DAVID LESLIE, gastroenterologist; b. Trowbridge, Wiltshire, Eng., Aug. 19, 1948; s. Dennis Charlton and Ruby Marjorie (Gibbs) Carr-Locke; children: Alexander, Antonia. MA, Gonville & Caius, Cambridge, Eng., 1972; MD, London and Cambridge U., 1972. House officer in medicine/surgery Kettering (Eng.) Gen. Hosp., 1972-73; house officer obgyn. Orsett (Essex, Eng.) Hosp., 1973-74; house officer med. specialties Leicester (Eng.) Hosps., 1974-75; lectr. medicine/gastroenterology Leicester Royal Infirmary, 1973-83, cons. gastroenterologist, 1983-89; dir. endoscopy Brigham and Women's Hosp., Boston, 1989—. Editor: Endoscopy, 1991. Fellow Royal Coll. Physicians (London), Am. Coll. Gastroenterology; mem. Internat. Hepato-Biliary Soc. (sec.-treas. 1989—). Achievements include research on biliary diseases and the pancreas. Office: Brigham Womens Hosp Endoscopy Ctr 75 Francis St Boston MA 02115

CARROLL, CHARLES LEMUEL, JR., mathematician; b. Whitsett, N.C., Sept. 16, 1916; s. Charles Lemuel and Erma Ruth (Greeson) C.; m. Geraldine Budd, June 8, 1938; children: Jill, Charlda Sizemore, Charles Lemuel III. BS, Guilford Coll., 1936; AM, U. N.C., 1937, PhD, 1945. Instr. math. Ga. Inst. Tech., Atlanta, 1939-42; assoc. prof. math. N.C. State U., Raleigh, 1946-55; research adminstr. Air Force Office Sci. Research, Balt., 1955-56; mgr. systems analysis RCA Service Co., Patrick AFB, Fla., 1956-61; asst. dir. Aerospace Corp., Atlantic Missile Range, Fla., 1961-62; mgr. tech. staff Pan Am World Airways, Patrick AFB, 1962-72; mgr. Japanese Tech. Assistance project Pan Am, 1972-82; mgr. info. systems Pan Am World Services, Cocoa Beach, Fla., 1972-86, cons. Indialantic, Fla., 1986—. Elder, Eastminster Presbyn. Ch., Indialantic, 1962—; bd. dirs. South Brevard YMCA, Melbourne, Fla., 1959-65, v.p., 1962-63. Served to lt. comdr. USNR, 1943-57. Assoc. fellow AIAA; mem. Am. Math. Soc., Inst. Math. Stats. Sigma Xi. Democrat. Home: 109 Michigan Ave PO Box 033343 Indialantic FL 32903

CARROLL, DAVID TODD, computer engineer; b. West Palm Beach, Fla., Apr. 8, 1959; s. David Irwin and Lois Ellen (Spriggs) C. Student, U. Houston, 1978-81. Lab. technician Inst. for Lipid Rsch., Baylor Coll. Medicine, Houston, 1978-81; software specialist Digital Equipment Corp., Colorado Springs, Colo., 1982-86, systems engr., 1986-91, systems support cons., 1991—. Mem. AAAS, Digital Equipment Corp. Users Soc. Home: 7332 Aspen Glen Ln Colorado Springs CO 80919-3024 Office: Digital Equipment Corp 305 S Rockrimmon Blvd Colorado Springs CO 80919-2303

CARROLL, EDWARD WILLIAM, anatomist, educator; b. Chgo., Sept. 20, 1942; s. Edward Charles and Lorraine Marion (Nagel) C.; m. Jeanne Ann Hebbring, Sept. 19, 1981; children: Doniella Christina, Kimberly Jeanne; stepchildren: John Scott Lesak, Richard Alan Lesak. BS, U. Wis., Milw., 1967, MS, 1973; PhD, Med. Coll. Wis., 1982. Instr. U. Wis., Waukesha, 1972-74, 76; rsch. assoc. Med. Coll. Wis., Milw., 1981-84; assoc. prof. anatomy Sch. Dentistry Marquette U., Milw., 1985—. Contbr. articles to sci. jours. Dep. coroner County of Washington, Wis., 1989—. Sgt. U.S. Army, 1963-69. Recipient postdoctoral rsch. award NIH, 1984. Mem. Am. Assn. Anatomists, Cajal Club, Am. Legion, Sigma Xi. Achievements include research in gene-induced degeneration in the visual system, metabolic changes in the primate visual system using cytochrome oxidase histochemical methods, neural pathways in the visual system. Office: Marquette Univ Sch Dentistry 604 N 16th St Milwaukee WI 53233

CARROLL, J. RAYMOND, engineering educator; b. Maywood, Ill., July 8, 1922; m. Darlene Brown, Aug. 29, 1942; children: John, Virginia, Andrew. BSME, U. Ill., 1943, MS in Engring., 1947. Registered profl. engr., Ill., Wis. Prof. engring. U. Ill., Urbana, 1947-75; cons. engr. Carroll-Henneman & Assoc., Champaign, Ill., 1961-75; dir. engr. Office of the Architect Capitol, Washington, 1975—; engr. rep. adv. bd. Sch. Safety, Ill., 1964-73; mem. profl. engrs. exam bd., 1968-74; chmn. Rockville, Md. Energy Commn., 1982-84. Co-author: texts Winter Air Conditioning, 1955, Summer Air Conditioning, 1955; contrmr. articles in field. Capt. USArmy, 1942-46, ETO. Named Disting. Alumni U. Ill., 1968, 83. Fellow NSPE (v.p. 1973-75, asst. treas. 1975—), Am. Soc. Heating, Refrigerating and Air Conditioning Engrs. (life, disting. svc. award); mem. Am. Arbitration Assn. (Nat. Const. panel), Ill. Soc. Profl. Engrs. (pres.), Am. Soc. Engring. Edn., Assn. Energy Engrs., Inst. Noise Control Engrs., Assn. Commerce (v.p. 1972-74), U. Ill. Alumni Assn. Washington (pres. 1977-85, Loyalty award 1986), Pi Tau Sigma, Tau Beta Pi, Lamda Alpha. Home: 15 Hawthorn Ct Rockville MD 20850 Office: Office of Architect of Capitol US Capitol Washington DC 20515

CARROLL, JOHN MOORE, retired electrical engineer; b. Butte, Mont., Oct. 27, 1911; s. William Craig and Harriet Lane (McKay) C.; m. Kathleen McClintock, Apr. 18, 1938 (div. 1964); children: Susan, William; m. Virginia Lee Roberts, Dec. 31, 1966; stepchildren: Thomas Gilbert, Rodney Gilbert, Gary Gilbert. Student, Oreg. State U., 1931; BA in Engring., N.W. Schs., 1958. Elec. supt. Harvey Aluminum Co., The Dalles, Oreg., 1957-62; sr. elec. engr. Aetron-Blume-Atkinson, Stanford, Calif., 1962-66, Bechtel Power Corp., San Francisco, 1968-75; sr. cost engr., contract adminstr. Burns & Roe, Inc., Paramus, N.J., 1975-79; sr. elec. engr. J.A. Jones, Inc., Hanford, Wash., 1979-88; cons., elec. estimator Kennewick, Wash., 1988-90; ret., 1990; elec. estimator Tempest Co., Omaha, 1985; elec. engr. Metcalf & Eddy, Inc., Long Beach, Calif., 1987. Capt. U.S. Army, 1942-46, PTO. Decorated Bronze Star. Mem. Nat. Soc. Profl. Engrs., IEEE, Am. Assn. Cost Engrs., Am. Arbitration Assn., Masons, Shriners. Republican. Presbyterian. Avocation: amateur radio. Address: 2917 W 19th # 126 Kennewick WA 99337-2310

CARROLL, ROGER CLINTON, medical biology educator; b. Mt. Clemens, Mich., Sept. 28, 1947; s. Lee Stanley and Evelyn Marie (Badgett) C.; m. Andrea Kristine Skrec, Sept. 13, 1969; children: Brian Roger, Alicia Helene. BS, Cornell U., 1969, PhD, 1977. Rsch. assoc. U. Calif. San Diego, LaJolla, 1976-78; rsch. assoc. U. Okla. Health Sci. Ctr., Oklahoma City, 1978-79, asst. prof. dept. pathology, 1979-80, adj. asst. prof. dept. biochemistry, dept. physiology, 1980-84; assoc. prof. dept. med. biology U. Tenn. Med. Ctr., Knoxville, 1984-90, prof. dept. med. biology, 1990—; asst. mem. Okla. Med. Rsch. Found. (Merrick award 1984), Oklahoma City, 1982-84; cons. Nat. Heart Lung and Blood Inst., 1985— (grantee 1980-90).

Contbr. articles to profl. jours., chpts. to books. Mem. Am. Heart Assn. (thrombosis coun., rsch. com. chmn., peer rev. com. chmn., Tenn. affiliate, grantee 1981-84, 1987—), Internat. Soc. on Thrombosis and Haemostasis, Sigma Xi. Democrat. Roman Catholic. Avocations: swimming, tennis, gourmet cooking. Home: 706 Ala Dr Knoxville TN 37920-6364 Office: U Tenn Med Ctr Grad Sch Medicine 1924 Alcoa Hwy Knoxville TN 37920-1511

CARROLL, STEPHEN DOUGLAS, chemist, research specialist; b. Clarendon, Ark., Nov. 2, 1943; s. Albert Genson and Wilma Mae (Hill) C.; m. Nonnie Lee Dyer, June 8, 1991; children: Geoffrey Genson, Raymond Loyd. BA, Hendrix Coll., 1965; MS, U. Ark., 1970. Del. chemist Chicopee Mfg. Co., North Little Rock, Ark., 1969-73, Mgr. Quality Assurance, 1973-80; cons. self employed, Clarendon, Ark., 1980-82; rsch. asst. U. Ark., Marianna, 1982-87, rsch. specialist, 1987—. Mem. Am. Chem. Soc. Democrat. Methodist. Avocations: photography, writing, painting. Office: U Ark Hwy 1 Bypass Marianna AR 72360-2100

CARRUTHERS, PETER AMBLER, physicist, educator; b. Lafayette, Ind., Oct. 7, 1935; s. Maurice Earl and Nila (Ambler) C.; m. Jean Ann Breitenbecher, Feb. 26, 1955; children: Peter, Debra, Kathryn; m. Lucy J. Marston, July 10, 1969; m. Cornelia B. Dobrovolsky, June 20, 1981; m. Lucy Marston Carruthers, Mar. 3, 1990. B.S., Carnegie Inst. Tech., 1957, M.S., 1957; Ph.D., Cornell U., 1960. Asst. prof. Cornell U., N.Y., 1961-63, assoc. prof., 1963-67, prof. physics, atomic and solid state physics, nuclear studies, 1967-73; div. leader, theoretical div. Los Alamos (N.Mex.) Sci. Lab., 1973-80, group leader of elem. particles and field theory, 1980-85, sr. fellow, 1980-86; prof., dept. head physics U. Ariz., Tucson, 1986-93, dir. Ctr. for Study Complex Systems, 1987—; vis. asso. prof. Calif. Inst. Tech., 1965, vis. prof., 1969-70, 77-78; mem. physics adv. panel NSF, 1975-80, chmn., 1978-80; trustee Aspen Center for Physics, 1976-82, chmn. exec. com., 1977-79, chmn. bd. trustees, 1979-82, advisor, 1982-89, hon. trustee, 1989—; mem. High Energy Physics Adv. Panel, 1978-82, com. on U.S.-USSR cooperation in physics NAS, 1978-82; cons. SRI Internat., 1976-81, MacArthur Found., 1981-82, 84-88, Inst. for Def. Analysis, 1985-89; chmn. Ariz. Superconducting Super Collider Tech. Com., 1986-89; editor Multiparticle Prodn. Dynamics, 1988. Author: (with R. Brout) Lectures on the Many-Electron Problem, 1963, Introduction to Unitary Symnetry, 1966, Spin and Isospin in Particle Physics, 1971; editor: (with D. Strottman) Hadronic Matter in Collision, 1986, Hadronic Multiparticle Dynamics, 1988, (with J. Rafelski) Hadronic Matter in Collision, 1988; cons. editor Harwood Soviet Physics Series. Trustee Santa Fe Inst., 1984-86, v.p., 1985-86, mem. sci. bd. 1986-93. Recipient Merit award Carnegie Mellon U., 1980; Alfred P. Sloan research fellow, 1963-65; NSF sr. postdoctoral fellow U. Rome, 1967-68; Alexander von Humboldt sr. fellow, 1987—. Fellow AAAS, Am. Phys. Soc. (panel on pub. affairs 1984-86), Univs. Rsch. Assn. (Superconducting Super Collider bd. overseers 1990-93. Home: 2220 E Camino Miraval Tucson AZ 85718-4939 Office: Univ Ariz Dept Physics PAS Bldg 81 Tucson AZ 85721

CARSON, GORDON BLOOM, engineering executive; b. High Bridge, N.J., Aug. 1, 1911; s. Whitfield R. and Emily (Bloom) C.; m. Beth Lacy, June 19, 1937; children—Richard Whitfield, Emily Elizabeth (Mrs. Lee A. Duffus), Alice Lacy (Mrs. William P. Allman), Jeanne Helen (Mrs. Michael J. Gable). BSMechE, Case Inst. Tech., 1931, D Engring., 1957; MS, Yale U., 1932, ME, 1938; LLD, Rio Grande Coll., 1973. With Western Electric Co., 1930; instr. mech. engring. Case Inst. Tech., 1932-37, asst. prof., 1937-40, asso. prof. indsl. engring. charge indsl. div., 1940-44; with Am. Shipbldg. Co., 1936; patent litigation, 1937; research engr., dir. research Cleve. Automatic Machine Co., 1939-44; asst. to gen. mgr. Selby Shoe Co., 1944, mgr. engring., 1945-49, sec. of corp., 1949-53; sec., dir. Pyrrole Products Co., 1948-53; dean engring. Ohio State U., Columbus, 1953-58; v.p. bus. and finance, treas. Ohio State U., 1958-71; dir. Engring. Exptl. Sta., 1953-58, Accuray Corp., 1960-82, Cardinal Funds, Inc., 1962—; exec. v.p. Albion (Mich.) Coll., 1971-76, exec. cons., 1976-77; asst. to chancellor, dir. fin. Northwood Inst., 1977-82; v.p. Mich. Molecular Inst., 1982-88; prin. Whitfield Robert Assocs., 1988—. Editor: The Production Handbook, 1958; cons. editor, 1972—; Author of tech. papers engring. subjects. Trustee White Cross Hosp. Assn., 1960-71; bd. dirs. Cardinal Funds, 1966—; bd. dirs. Goodwill Industries, 1959-67, 1st v.p., 1963-64; bd. dirs. Orton Found., 1953-58; v.p. Ohio State U. Research Found., 1958-71; v.p., chmn. adv. council Center for Automation and Soc., U. Ga., 1969-71; Chmn. tool and die com. 5th Regional War Labor Bd., 1943-45; chmn. Ohio State adv. com. for sci., tech. and specialized personnel SSS, 1965-70. Fellow ASME, AAAS, Am. Inst. Indsl. Engrs. (pres. 1957-58); mem. Columbus Soc. Fin. Analysts (pres. 1964-65), Fin. Analysts Fedn. (bd. dirs. 1964-65), C. of C. (bd. dirs., treas. 1952-53), Am. Soc. Engring. Edn., Assn. Univs. for Rsch. in Astronomy (bd. dirs. 1958-71), Midwestern Univs. Rsch. Assn. (bd. dirs. 1958-71), U.S. Naval Inst., Nat. Soc. Profl. Engrs. (life), Romophos, Sphinx, Sigma Xi (fin. com. 1975-89, nat. treas. 1979-89), Masons (32 deg.), Tau Beta Pi, Zeta Psi, Phi Eta Sigma, Alpha Pi Mu, Omicron Delta Epsilon. Home: 5413 Gardenbrook Dr Midland MI 48642-3236 Office: Whitfield Robert Assocs 220 W Ellsworth St Rm 004 Midland MI 48640

CARSON, REGINA EDWARDS, pharmacy administrator, educator; b. Washington; d. Reginald Billy and Arcola (Gold) Edwards; m. Marcus T. Carson; children: Marcus Reginald, Ellis K., Imani R. BS in Pharmacy, Howard U., 1973; MBA in Mktg., Health Care Adminstrn., Loyola Coll., Balt., 1987. Asst. prof., asst. dir. pharmacy U. Md., Balt., 1986-88; asst. prof., coord. profl. practice Howard U., Washington, 1988—; exec. v.p. Marrell Inc., Randallstown, Md., 1985—; drug utilization rev. cons. Md. Pharmacy Assn., Balt., 1986—; pharmacist, cons. Balt. County Adv. Coun. Drug Abuse, Towson, 1984-86; adv. com. longterm care com. Nat. Assn. Retail Druggists. Bd. dirs. Balt. County Hosp. Aux., Randallstown, Joshua Johnson Coun. Balt. Mus. Art. Fellow Am. Soc. Cons. Pharmacists; mem. Nat. Assn. Retail Druggists, Am. Assn. Colls. Pharmacy, Nat. Pharm. Assn. (life, Outstanding Women Pharmacy 1984). Avocations: pharmacognosy, Windsor chairs, American art. Office: Howard U Coll Pharmacy 2300 4th St NW Washington DC 20059-0001

CARSON, RICHARD MCKEE, chemical engineer; b. Dayton, Ohio, June 6, 1912; s. George E. and Gertrude (Barthelemy) C.; children: Joan Roderer, Linda McCartan. BS in Chem. Engring., U. Dayton (Ohio), 1934. Registered profl. engr., Ohio. Rsch. chemist Dayton Mall Iron Co., 1934-45; pres. Carson-Saeks, Inc., Dayton, 1945-80, Carson & Saeks Cons. Assocs. Inc., Dayton, 1980—. Mem. AAAS, Am. Chem. Soc. Achievements include 6 patents for clinical test procedures, reagents, and closet accessories. Home: 2310 Kershner Rd Dayton OH 45414-1214

CARTA, FRANKLIN OLIVER, retired aeronautical engineer; b. Middletown, Conn., July 16, 1930; s. Salvatore and Anna (DeMauro) C.; m. Ann J. DiMauro, Sept. 25, 1954; children: Lisa Ann, Christopher Pace, Maura Ferragut. BS, MIT, 1952, MS, 1953. Rsch. asst. MIT Aeroelastics Lab., Cambridge, 1952-53; rsch engr. United Aircraft Rsch. Labs., East Hartford, Conn., 1953-60; aeroelastics engr. Cornell Aero. Lab., Buffalo, 1960; sr. engr./supr. aeromechanics United Tech. Rsch. Ctr., East Hartford, 1960-93; lectr. aero von Karman Inst., Rhode-St-Genese, Belgium, 1970, 77, Iowa State U., 1975, 77, 80—; panel mem. Adv. Group for Aerospace Rsch. and Devel. of NATO, 1979-84. Assoc. editor Jour. of Fluids and Structures, 1985-87; contbr. articles to profl. jours., chpts. to books. Assoc. fellow AIAA; fellow ASME (v.p. 1992—, bd. dirs. Internat. Gas Turbine Inst. of ASME 1985-90, chmn. IGTI bd. dirs. 1988-89, Gas Turbine Power award 1967); mem. Sigma Xi, Tau Beta Pi, Gamma Alpha Rho. Achievements include patents in field; development of research on unsteady deep stall aerodynamics of wings and rotors, on turbomachinery coupled flutter.

CARTA, GIORGIO, chemical engineering educator; b. Cagliari, Italy, Apr. 23, 1957; arrived in U.S. 1980.; s. Mario and Miranda (Altieri) C.; m. Beth A. Naumann, Aug. 3, 1984; children: Julian, Anna Marie. Laurea ChE, U. Cagliari, 1980; PhD in Chem. Engring., U. Del, 1984. Asst. prof. U. Va., 1984-89, assoc. prof., 1989--. Contbr. articles to profl jours. Mem. Am. Inst. Chem. Engrs. (com. mem.), Am. Chem. Soc. Achievements include one patent. Office: U Va Dept Chem Engring Charlottesville VA 22903-2442

CARTER, CECIL NEAL, environmental engineer; b. Savannah, Ga., July 1, 1939; s. H. Cecil and Gertrude B. (Brown) C.; m. Gloria A. Fitzkee, Dec. 24, 1961; children: Wendy Sue, Rebecca Jo, Gregory Scott, Jennifer Lynn. BS, N.C. State U., 1961; PhD, Inst. of Paper Chemistry, Appleton, Wis., 1966. Process engr. P.H. Glatfelter Co., Spring Grove, Pa., 1965-66, schedule coord., 1966-70, mgr. storeroom, 1970-72, tech. mgr., 1972-81, tech. dir., 1981-89, corp. environ. mgr., 1989—; mem. operating com. NCASI, N.Y.C., 1981-92. Pres. West York (Pa.) Sch. Bd., 1988-89. Mem. TAPPI, NSPE, Am. Assn. Environ. Engrs. Republican. Baptist. Home: 4740 Darlington Rd York PA 17404 Office: P H Glatfelter Co 228 S Main St Spring Grove PA 17362

CARTER, CHARLEATA A., cancer researcher, developmental biologist, cell biologist, toxicologist; b. Asheville, N.C., Dec. 6, 1960; d. Charles E. and Oleata J. Carter. BS in Biology, Mars Hill Coll., 1981; MA, Appalachian State U., Boone, N.C., 1983; PhD, Clemson U., 1988. Nat. Rsch. Svc. Award fellow U. N.C., Chapel Hill, 1988-91; Intramural Rsch. Tng. Award fellow Nat. Inst. Environ. Health Sci., Research Triangle Park, N.C., 1991-93; assoc. scientist Lovelace Biomed. & Environ. Rsch. Inst., Albuquerque, 1993—. Contbr. articles to profl. jours. Mem. Am. Soc. for Cell Biology, Am. Assn. Cancer Rsch., Metastasis Rsch. Soc., Sigma Xi (grantee 1986). Achievements include research in extracellular matrix and cytoskeletal structural alterations induced by oncogenes, chemicals and growth factor pathways and involvement of these protein alterations in tumorigenesis and metastasis, developmental biology, toxicology and tumorigenesis in fish. Office: Inhalation Toxicology Rsch Inst Lovelace Biomed & Environ Rsch Inst P O Box 5890 Albuquerque NM 87185

CARTER, CRAIG NASH, veterinary epidemiologist, educator, researcher, software developer; b. Gary, Ind., Mar. 20, 1949; s. Frank Lynn and Harriet May (Nash) C. AA Computer Sci., Riverside (Calif.) City Coll., 1972; BS in Vet. Sci. magna cum, Tex. A&M U., 1980, DVM cum laude, 1981, MS in Epidemiology, 1985, PhD in Pub. Health, 1993. Diplomate Am. Coll. Vet. Preventive Medicine. Computer systems engr. U.S. Navy, Civilian, Mechanicsburg, Pa., 1976-77; computer specialist USAF, Civilian, Universal City, Tex., 1977-78; vet. clinical assoc. Tex. Vet. Med. Diagonstic Labs, College Station, Tex., 1981-83; vet. epidemiologists, adj. prof. Tex. A&M U., College Station, 1983-88, adj. prof., 1988—; head dept. epidemiology and informatics Tex. Vet. Med. Diagnostic Labs, College Station, 1988—; pres. Am. Vet. Acad. Disaster Medicine, 1988—; pres. Carter-Melloy Corp., 1989—; presenter more than 50 nat. and internat. sci. presentations. Contbr. 35 articles to profl. jours.; inventor in field. Capt. USAFR, Vietnam, Desert Storm. Mem. Am. Coll. Vet. Informatics (pres. 1990—), Am. Vet. Computer Soc. (pres. 1987—), Am. Vet. Med. Assn., Tex. Vet. Med. Assn., Am. Assn. Vet. Labs. Diagnostic Med., Internat. Vet. Acad. Disaster Med., U.S. Animal Health Assn. Democrat. Congregationalist. Avocations: music, physical fitness, reading, writing, travel. Home: 3107 Manorwood Dr Bryan TX 77801-4204 Office: Tex Vet Med Diagnostic Lab PO Box 3040 College Station TX 77841-3040

CARTER, DALE WILLIAM, psychologist; b. Woodbury, N.J., Jan. 27, 1949; s. Charles Elmer and Dorothy Adele (Seibold) C. BS, Wake Forest U., 1971; MS, Radford U., 1976; PhD, U. Ga., 1982. Tchr. Gaston Day Sch., Gastonia, N.C., 1971-73, Charlotte (N.C.) Country Day Sch., 1973-74; psychologist Roanoke County Schs., Salem, Va., 1976-83, Gwinnett County Schs., Lawrenceville, Ga., 1983—; pvt. practice Lilburn, Ga., 1985—; adj. prof. Mercer U., Atlanta, 1984-85; cons. N.E. Counseling Ctr., Lawrenceville, 1985—; intern supervision Gwinnett County Schs., Lawrenceville, 1985—. Mem. APA (div. sch. psychology), Ga. Assn. Sch. Psychologists, Nat. Assn. Sch. Psychologists, Beta Beta Beta, Kappa Delta Pi, Phi Kappa Phi. Home: 1004 Camp Creek Dr SW Lilburn GA 30247-5460 Office: Gwinnett County Schs 52 Gwinnett Dr SW Lawrenceville GA 30245-5624

CARTER, DAVID LAVERE, soil scientist, researcher, consultant; b. Tremonton, Utah, June 10, 1933; s. Gordon Ray and Mary Eldora (Hirschi) C.; m. Virginia Beutler, June 1, 1953; children: Allen David, Roger Gordon, Brent Ryan. BS, Utah State U., 1955, MS, 1957; PhD, Oreg. State U., 1961. Soil scientist USDA Agrl. Research Service, Corvallis, Oreg., 1956-60; research soil scientist, line project leader USDA Agrl. Research Service, Weslaco, Tex., 1960-65; rsch. soil scientist USDA Agrl. Rsch. Svc., Kimberly, Idaho, 1965-68, supervisory soil scientist, rsch. leader, 1968-86, supervisory soil scientist, rsch. leader, dir., 1986—; cons., adviser to many projects and orgns. Contbr. articles to profl. jours.; author, co-author books. Recipient Emmett J. Culligan award World Water Soc. Fellow Am. Soc. Agronomy (cert.), Soil Sci. Soc. Am. (cert.); mem. Soil Conservation Soc. Am. (Soil Conservation award 1985), AAAS, Internat. Soc. Soil Sci., Western Soc. Soil Sci., CAST, Internat. Soc. Soil Sci., OPEDA. Mormon. Office: Siol & Water Mgmt Rsch USDA-ARS 3793 N 3600 E Kimberly ID 83341-9801

CARTER, DAVID MARTIN, dermatologist; b. Doniphan, Mo., June 10, 1936; s. Joseph and Elizabeth (Estes) C.; m. Anne Babson; children: Anna, Christopher, Elizabeth. AB, Dartmouth Coll., 1955-58; MD, Harvard U. Med. Sch., 1961; PhD, Yale U., 1971. Diplomate Am. Bd. Dermatology. Intern U. Rochester, 1961-62, asst. resident, 1962-63; teaching fellow USPHS Ctr. for Disease Control, Atlanta, 1963-65; dermatology resident U. Pa., Phila., 1965-67; postdoctoral fellow Yale U. Sch. Medicine, New Haven, Conn., 1967-70; attending physician Yale-New Haven Hosp., New Haven, 1970-81; prof. dermatology Yale U. Sch. Medicine, New Haven, 1977-81; co-head div. dermatology N.Y. Hosp.-Cornell Med. Ctr., N.Y.C., 1981—; prof., sr. physician The Rockefeller U., N.Y.C., 1981—; bd. dirs. Soc. Investigative Dermatology, N.Y.C., pres. 1985-86; mem. adv. council Nat. Inst. for Arthritis, Musculoskeletal and Skin Diseases, 1988-91; mem. Nat. Commn. Orphan Diseases, 1987-89; lectureships and vis. professorships include Am. Physicians Fellowship, Inc. for Medicine, Israel, Coll. Physicians and Surgeons of Columbia U., N.Y.C., Ind. U., Indpls., U. Ala., Birmingham, Barney Usher vis. prof. McGill U., Montreal, Washington U., St. Louis , Kyushu U., Japan, 1984, Kobe U., Japan, 1984, Kitasato U., Japan, 1984, U. Pitts. Sch. Medicine, 1984, U. Ariz., Tucson, 1985, U. Calif., San Francisco, 1985, British Soc. Investigative Dermatology Oxford U., 1985, M.H. Samitz lectr. U. Pa., 1986, U. Conn., 1986, All India Inst. Medicine, 1986, Columbia U., 1986, Robert N. Buchanan vis. prof. Vanderbilt U., 1988, Taiwan, New Delhi. Author numerous books and articles in field; assoc. editor Yale Jour. Biology and Medicine, 1977-81, Jour. Investigative Dermatology, 1977-82, Jour. of Am. Acad. Dermatology, 1979-84. Daniel Webster Nat. scholar Dartmouth Coll., 1954-58; Howard Hughes med. investigator Yale U., 1971-78. Fellow AAAS, Coll. Physicians of Phila.; mem. Am. Acad. Dermatology, Am. Dermatol. Assn. (edn. com.), Soc. Investigative Dermatology (pres. 1985-86, bd. dirs. 1975-80, 84-87), Nat. Program Dermatology (task force on genetics), New England Dermatol. Soc., Dermatology Found. (med. and sci. com. for grants and fellowships), Am. Fedn. Clin. Research, Internat. Pigment Cell Soc., N.Y. Acad. Sci., N.Y. Acad. Medicine (pres. sect. on dermatology and syphilology, 1986-87, sec. 1985-86), Assn. Profs. Dermatology (genetics com., bd. dirs.), NIH (adv. com. on formation of nat. inst. arthitis, musculoskeletal, and skin diseases), Japanese Soc. for Investigative Dermatology (hon.), French Soc. Dermatology and Syphilology (hon.). Avocations: music, opera. Office: Rockefeller U 1230 York Ave New York NY 10021-6399

CARTER, EDWARD FENTON, III, pathologist, medical examiner; b. Tampa, Fla., July 27, 1948; s. Edward Fenton Jr. and Ruth Louise (Chastain) C.; m. Debra Ann Pittman (div. 1986); 1 child, Jaquelin Heather; m. Sandra Jean Turner, Apr. 10, 1987. BS, Tulane U., 1971, MD, 1975. Diplomate Am. Bd. Pathology. Resident pathologist Charity Hosp. of La., New Orleans, 1975-79; assoc. pathologist Suburban Pathology Assocs., Atlanta, 1979-81; chief of pathology Washington County Hosp., Chipley, Fla., 1981-84; assoc. pathologist Sumter Regional Hosp., Americas, Ga., 1984-86; chief of pathology Spalding Regional Hosp., Griffin, Ga., 1986—; lab. dir. Spalding Regional Hosp., Griffin, 1986—; med. examiner Spalding and Pike Counties, Ga., 1986—. Recipient Cancer Rsch. grant Cancer Assn. of Greater New Orleans, 1970. Fellow Coll. Am. Pathologists, Am. Soc. Clin. Pathologists, Internat. Acad. Pathology. Avocations: vintage sports car racing, snow skiing, boating. Home: 509 N Pinehill Rd Griffin GA 30223 Office: Spalding Regional Hosp S 8th St Griffin GA 30223-3001

CARTER, HERBERT JACQUE, biologist, educator; b. Kankakee, Ill., Mar. 18, 1953; s. Herbert John and Doris Mae (Hasemeyer) C.; m. Judy Ann Nishimura, Aug. 18; children: Jared Herbert, Megan Doris. BS in Biology, No. Ill. U., 1975, MS in Biology, 1978; PhD in Marine Sci., Coll. of William and Mary, 1984. Rsch. fellow, conservation biologist Wildlife Conservation Internat. N.Y. Zool. Soc., Bronx, 1984-87; asst. prof. Bucknell U., Lewisburg, Pa., 1987-89, U. New Eng., Biddeford, Maine, 1989—; assoc. faculty Sch. Marine Sci. Va. Inst. Marine Sci., 1986—,; vis. scientist inresidence, 1986—; tech. advisor, mem. Belize govt. commn. to protect whales, 1986; instr. tropical conservation biology, Belize, 1988—; mem. project document planning team Belize Conservation Strategy Plan Environ. Facility-UN Devel. Program, 1992; co-author, prin. investigator rsch. component Global Environ. Facility, 1992—; prin. investigator Ctrl. Maine Power, 1990—; coral reef specialist, cons. UN Devel. Program. Contbr. articles and papers to Nat. History Mag., Bull. Marine Sci., Deep-Sea Rsch., Copeia, Gulf and Caribbean Fisheries Inst. Mem. environ. awareness study com. U. New Eng., 1990-91, mem. global awareness study group com., 1992—, mem. budget and fin. com. faculty senate appointment, 1990-92, coord. faculty colloquium series, 1992—, faculty advisor student senate, 1992—, co-chmn. new med. sci. facility com., 1992—, chmn. budget and fin. com., 1992—. Grantee So. Regional Edn. Bd., 1978. Mem. AAAS, Am. Fisheries Soc., Am. Soc. Icthyologists and Herpetologists (Raney award 1980), Internat. Conservation Orgn. (bd. sci. advisors 1988—), Nat. Biol. Rsch. Soc., Belize Audubon Soc., Belize Zool. Soc., Gulf and Caribbean Fisheries Inst., Soc. for Conservation Biology, Sigma Xi, Phi Sigma. Lutheran. Achievements include planning, design and establishment of the Hol Chan Marine Reserve, the 1st coral reef park protected by govt. legislation in Belize. Home: 3 Settlers Way Kennebunk MN 04043 Office: U New Eng Hills Beach Rd Biddeford MN 04005

CARTER, JAMES SUMTER, oil company executive, tree farmer; b. Rock Hill, S.C., June 3, 1948; s. James Roy Jr. and Sumter Inez (McWatters) C.; m. Melinda Ruth Roberts, Mar. 25, 1972; children: James Sumter Jr., Stephanie Jane, Lauren Elizabeth. BSME, Clemson U., 1970; MBA, Tulane U., 1974. Various mktg. positions Exxon Co. USA, Houston, 1974-79; dist. mgr. Exxon Co. USA, Linden, N.J., 1980-81; adv. Exxon Corp., N.Y.C., 1982-83; analysis mgr. Exxon Internat., Florham Pk, N.J., 1984-85; coord. mgr. Exxon Co. USA, Houston, 1986, exec. asst. to pres., 1987, distbn. mgr., 1988, downstream planning mgr., 1989, fuel products mgr., 1990—. Lt. U.S. Army, 1971-72. John Jay assoc. Columbia U. Mem. Am. Petroleum Inst., Petroleum Marketers Edn. Found. (bd. dirs. 1990-92), Forest Farmers Assn., Ben Tilman Soc., Petroleum Club Houston, Columbia U. Club of N.Y., Tau Beta Pi, Beta Gamma Sigma. Republican. Baptist. Avocation: restoration of S.C. plantation home. Office: Exxon USA 800 Bell St Houston TX 77002-7426

CARTER, JOHN ANGUS, geologist, geochemist, environmental engineer; b. Winsted, Conn., May 2, 1935; s. John Angus and Frances Wilhelmina (Hanauer) C.; m. Anne Marie Morin, May 16, 1981. BS, SUNY, Albany, 1977. Registered environ. analyst, Conn. Asst. engr. Winsted Bearing Corp., Winsted, 1950-58; plant mgr. MacArthur Photo Processing Svc., 1958-64; engr. Orbit Tool and Mfg. Co., Winsted, 1968-77; prin. Iemco/Gen., Winsted, 1977—, Geoteknika, Winchester, Conn., 1977—. Contbr. articles to sci. publs. Office: Geoteknika PO Box 193 Winchester CT 06094

CARTER, JOHN PHILLIP, civil engineering educator, consultant, researcher; b. Sydney, NSW, Australia, June 12, 1950; s. George Leslie and Dorothy Amy (Nettle) C.; m. Heather Jean Kells, Jan. 11, 1975; children: Anna Jane, Sophie Elizabeth. BE, U. Sydney, 1973, PhD in Civil Engring., 1977. Trainee engr. Electricity Commn. NSW, Sydney, 1969-73; rsch. asst. Cambridge (Eng.) U., 1977-79; lectr. U. Queensland, Brisbane, Australia, 1979-82; lectr. civil engring. U. Sydney, 1982-90, prof., 1990—; vis. scholar King's Coll., U. London, 1974; cons. various engring. firms and cos., Eng., Australia, U.S.A., 1977—. Editor internat. conf. proc., 1991; contbr. articles to profl. jours. and conf. proc. Fellow Instn. Engrs. Australia; mem. ASCE, Internat. Soc. for Soil Mechanics and Found. Engring., Internat. Soc. for Rock Mechanics. Avocations: music, reading. Office: U Sydney Civil Engring Dept, Sydney NSW 2006, Australia

CARTER, L. PHILIP, neurosurgeon, consultant; b. St. Louis, Mo., Feb. 26, 1939; s. Russell G. and Dorothy Ruth (Zerwick) C.; m. Marcia L. Carlson, Aug. 26, 1960 (div. Apr., 1989); children: Kristin, Melinda, Chad Philip; m. Colleen L. Harrington, Oct. 20, 1990. MD, Wash. U., 1964. Active staff Barrow Neurol. Inst., Phoenix, 1976-88, dir. microsurg. lab., 1978-88, chief cerebral vascular surgery, 1983-88; prof. neurosurgery, chief neurosurg. svcs. Coll. Medicine U. Ariz., Tucson, 1988—; med. cons. Flowtronics, Inc., Phoenix, 1980—; vis. prof. Japan Neurosurg. Soc., Kyoto, 1983. Co-editor: Cerebral Revascularzation for Stroke, 1985; contbr. articles to profl. jours. Cons. Ariz. Head Injury Found., 1988—, Ariz. Epilepsy Found., 1973-75. Capt. USAF, 1965-67. Recipient Internat. Coll. Surgeons fellowship, 1973, Ariz. Disease Control for the Study of Treatment of Stroke grant, 1986. Fellow Am. Heart Assn., Am. Coll. Surgeons; mem. Ariz. Neurol. Soc. (sec., treas. 1985-91), Western Neurosurg. Soc. (program chmn. 1990), Am. Assn. Neurol. Surgeons, Rocky Mtn. Neurosurg. Soc. Republican. Achievements include patents that describe quantification of surface techniques measuring cortical blood flow; design of new instruments for osseus disection and microsurgery. Home: 2701 E Camino Pablo Tucson AZ 85718-6625 Office: Univ Ariz Med Ctr 1501 N Campbell Ave Tucson AZ 85724-0001

CARTER, OLICE CLEVELAND, JR., civil engineer; b. Columbia, Ala., Aug. 30, 1955; s. Olice Cleveland and Burma Louise (Hasty) C. BS in Biology, U. Ala., Tuscaloosa, BSCE, 1985, MS in Environ. Engring., 1987. Registered profl. engr., Ala. Environ. engr. U.S. Bur. Mines, Tuscaloosa, 1985—. Contbr. articles to profl. jours. Mem. ASCE, NSPE, Ala. Soc. Profl. Engrs., Nat. Eagle Scout Assn., Chi Epsilon. Republican. Episcopalian. Achievements include patents in the development of the liquid anhydrous ammonia leaching process used on activated carbon. Home: 1015 17th Ave Tuscaloosa AL 35401-3027 Office: US Bur Mines TURC PO Box L Tuscaloosa AL 35486-6119

CARTER, PAUL R., agronomist consultant; b. Park Rapids, Minn., May 29, 1955. AA, Golden Valley Luth. Coll., 1975; BS in Agronomy, N.D. State U., 1978; MS in Agronomy, U. Minn., 1980, PhD in Agronomy, 1982. Asst. prof, ext. agronomist U. Wis., Madison, 1982-87, assoc. prof., ext. agronomist, 1987-91, prof., ext. agronomist, 1991-93; agronomy support mgr. No. Am. Seed Divsn, Pioneer Hi-Bred Internat. Inc., Iowa; sterring com. mem. Nat. Corn Handbook. 1986-92; program chair, Ext. Edn. Am. Soc. Agronomy; drought task force Gov. Wis., 1988; bd. dirs., sec, treas., Wis. Corn Growers Assn., 1982—. Assoc. Editor: Journal of Production Agriculture, 1992—; reviewer for crop. sci.: Agronomy Journal, Journal of Production Agriculture. Recipient Second Mile award Wis. County Agents Assoc., 1989, CIBA-GEIGY Agronomy award Am. Soc. Agronomy, 1990. Office: U Wisconsin 1575 Linden Dr Madison WI 53706*

CARTER, REBECCA DAVILENE, surgical oncology educator; b. Alma, Ga., Nov. 24, 1932. BS, Valdosta State Coll., 1959; MD, Med. Coll. Ga., 1965. Intern Meml. Hosp., Savannah, Ga., 1965-66, resident, 1966-70; NIH postdoctoral fellow Tulane U. Sch. Medicine, New Orleans, 1970-72, instr. dept. surgery, 1970-74, asst. prof., 1974-78, assoc. prof., 1978-83, prof. surgery, 1983—, acting chief sect. surg. oncology, 1987-91, dir. surg. oncology rsch. and edn., dept. surgery, 1991—; adj. asst. prof. Tulane U. Sch. Pub. Health and Tropical Medicine, New Orleans, 1977-79, co-dir. oncology RN program, 1977-79, adj. prof. dept. applied health sci., 1983—; clin. dir. surg. svcs. Charity Hosp. New Orleans, 1981—. Chmn. bd. La. Cancer and Lung Trust Fun, 1990-92. Mem. Am. Assn. for Cancer Edn. (pres. 1992-93), Am. Cancer Soc. (chmn. bd. dirs. dirs. La. div. 1993). Home: 2314 Metairie Ct Metairie LA 70001-2167 Office: Tulane U Sch of Medicine Dept of Surgery 1430 Tulane Ave New Orleans LA 70112-2699*

CARTER, REGINA ROBERTS, physicist; b. Huntsville, Ala., Sept. 28, 1962; d. Thomas George and Alice Anne (Harbin) Roberts; m. Kelly Kealoha Carter, July 6, 1985; 1 child, Matthew Thomas. BS, Auburn U., 1985; MS, Ga. Inst. Tech., 1986. Staff engr. Phys. Dynamics, Inc., Huntsville, Ala., 1986-89; staff scientist TECHNOCO, Huntsville, 1990—. Mem. Am. Phys. Soc., Huntsville Electro-Optical Soc. Episcopalian. Home: 1206 Wilmington Rd Huntsville AL 35803 Office: TECHNOCO PO Box 4723 Huntsville AL 35815-4723

CARTER, ROSCOE OWEN, III, chemist; b. Cin., Apr. 10, 1946; s. Roscoe Owen and Janet (Weigand) C.; m. Margaret Anne Barnes, June 16, 1973; m. Matthew Owen, Paul Joseph, Janet Alford. BA, Wittenberg U., 1968; MS, Miami U., 1971; PhD, U. S.C., 1973. Asst. prof. Old Dominion U., Norfolk, Va., 1976-81; sr. prin. scientist, assoc. Ford Rsch., Dearborn, Mich., 1981-88, staff scientist, 1988--. Contbr. articles to profl. jours. Mem. Detroit Round Table, !990-92. Mem. Am. Chem. Soc., Soc. for Applied Spectcospy (mem. edit. bd. 1991-92), Sigma Xi. Office: Ford Rsch Lab PO Box 2053 MS 3061 Dearborn MI 48121

CARTER, THOMAS ALLEN, engineering executive, consultant; b. Cin., July 12, 1935; s. Fernando Albert and Mary Gladys (Gover) C.; m. Janet Tucker, Oct. 14, 1956; children: Barry Everett, Duane Allen, Sarita Anne. AB, Jones Coll., 1980, BBA, 1982. Cert. constrn. insp. Enlisted USN, 1954, advanced through grades to master chief, ret., 1976; contract adminstr. Red Lobster Restaurants, Orlando, Fla., 1976-78; pvt. practice Orlando, 1978-80; sec. Blacando Devel. Corp., Orlando, 1980-84; chief engr. D.A.M.S., Inc., Orlando, 1991, SA Williams Inc, Orlando, 1991—. Mem. Fleet Res. Assn., Am. Bowling Congress. Democrat. Methodist. Avocations: bowling, tennis. Office: SA Williams Inc 5750 Major Blvd Orlando FL 32819-7921

CARTMELL, JAMES V., research and development executive; b. Dayton, Ohio, Aug. 8, 1938; s. Vernon Louis and Mary (Shoemaker) C.; m. m. Helen Claire Ruble, Sept. 16, 1960 (div. Aug. 1980); m. Barbara Louise McLean, May 21, 1981; children: James Jeffery, Cynthia Ann. A in Chem. Tech., U. Dayton, 1966, B in Chemsitry, 1976. Rsch. technician Monsanto Chem. Co., Dayton, 1956-61; sr. rsch. technician NCR Corp., Dayton, 1961-72; scientest NDM Corp., Dayton, 1972-80, dir. R & D, 1980-89, v.p. R & D, 1989—. Contbr. articles to Jour. Colloid, Interface Sci. Republican. Achievements include patents for Method of Making Medical Electrode, Method of Manufacutirng Medical Electode Pads, X-Ray Tranparent Medical Electrodes and Leadwires and Assemblies, Medical Elctrode: Skin Conducting, Medical Electrode, and others. Home: 2046 Winding Brook Way Xenia OH 45385 Office: NDM Corp 3040 E River Rd Dayton OH 45401

CARTWRIGHT, KEROS, hydrogeologist, researcher; b. L.A., July 25, 1934; s. Eugene Ewing and Charlotte Lucy (Searle) C.; m. Jenifer Elizabeth Moberley, Mar. 9, 1962 (div. Sept. 1988); children: Sylvia, Jennifer, David, Bridget; m. Madalene Rose Tierney, Feb. 16, 1990. AB in Geology, U. Calif., 1959; MS in Geology, U. Nev., 1961; PhD in Geology, U. Ill., 1973. Cert. profl. geologist, profl. hydrologist. Hydrogeologist Humboldt River Rsch. Project, Winnemucca, Nev., 1959-61; hydrogeologist Ill. State Geol. Survey, Champaign, 1961 , head hydrogeology and geophysics section, 1975-84, prin. scientist and head gen. and environ. geology group, 1984-88, prin. rsch. scientist, 1988—; adj. prof. geology No. Ill. U., DeKalb, 1979—, U. Ill., Urbana, 1985—; cons. pvt. practice in hydrogeology, N.Am. and Europe, 1968—, U.S. Environ. Protection Agy. Sci. Adv. Bd., Washington, 1983—, Savannah River Site Environ. Adv., Aiken, S.C., 1988—. Mem. editorial bd. Elsevier Sci. Publ. Jour. of Hydrology, 1982-85; contbr. articles to profl. jours. Named Disting. Lectr. Assn. Groundwater-Water Scientists and Engrs., 1987; recipient Cert. Appreciation U.S. Environ. Protection Agy., 1988. Fellow Geol. Soc. Am. (officer hydrogeology sect. 1975-78, chmn. 1978-79, editorial bd. Jour. Water Resources Rsch. 1975-81, Bull. 1981-83, Birdsall disting. lectr. 1987-88, George B. Maxey Disting. Svc. award 1981), Explorers Club; mem. ASTM (vice chmn. subcom. D-14 1984-88), Am. Inst. Hydrology (editorial bd. Jour. Hydrological Sci. and Tech. 1985—), Am. Geophys. Union (assoc. editor 1975-81), Am. Water Resources Assn., Internat. Assn. Hydrogeologists (U.S. com. 1985-89). Avocations: farming. Office: Ill Geol Survey 615 E Peabody Dr Champaign IL 61820-6964

CARTY, JOHN BROOKS, civil engineer; b. New Brunswick, N.J., June 4, 1961; s. John Robert and Catharine Wilhelmina (Brooks) C.; m. Alicia Jean Malinowski, Sept. 14, 1985; children: Erin Alicia, Emily Analise, Matthew Brooks. BSCE, Drexel U., Phila., 1984; MSCE, Drexel U., 1987. Profl. engr, N.J., Pa., profl. planner, N.J.; cert. sewage enforcement officer, Pa., underground storage tank, N.J. Staff engr. Pennoni Assocs., Inc., Phila., 1984-86; civil engr. Whitesell Constrn. Co., Delran, N.J., 1986-87; project engr. NTH Cons., Exton, Pa., 1987-89; project mgr. Lippincott Engring. Assocs., Riverside, N.J., 1989—; sec., v.p., pres.-elect ASCE N.J. Section, S. Jersey Branch, 1990-92, 1992-93, 93—. Pianist Hainesport (N.J.) Community Bapt. Ch. Named Dean's List Drexel U., Phila., 1984. Mem. ASCE (assoc. N.J. sect. South Jersey br. sec. 1990-92, v.p. 1992-93, pres.-elect 1993—). Republican. Baptist. Home: 931 Oriental Ave Collingswood NJ 08108

CARUSO, NANCY JEAN, chemist; b. East Orange, N.J., Mar. 1, 1957; d. Anthony John and Sandra Mary (Mucherino) C. BS in Chemistry, Rutgers Coll., 1980; MBA, U. Rochester, 1987. Devel. engr. Westinghouse Elec. Corp., Bloomfield, N.J., 1980-84, N.A. Philips Lighting Corp., Bath, N.Y., 1984-88; sr. devel. engr. Philips Lighting N.V., Turnhout, Belgium, 1988—. Mem. Restoration and Integration Mechelen, Belgium, 1991—. Mem. Am. Chem. Soc., Illuminating Engr. Soc. (assoc.). Achievements include 3 U.S. patents covering inventions in HID metal halide lamps. Home: Gierlesteenweg 38, 2300 Turnhout Belgium Office: Philips Lighting, Gierlesteenweg 417, 2300 Turnhout Belgium

CARVALHO, JULIE ANN, psychologist; b. Washington, Apr. 11, 1940; d. Daniel H. and Elizabeth Cecilia (Gardiner) Schmidt; BA with high honors, U. Md., 1962, postgrad., 1973; MA, George Washington U., 1966; postgrad. Va. Poly. Inst., 1979—; children: Alan R., Dennis M., Melanie D., Celeste A., Joshua E. Social sci. rsch. analyst Mental Health Study Ctr., NIMH, 1963-67; edn. and tng. analyst Computer Applications, Inc., 1967-68; edn. program specialist Nat. Ctr. for Ednl. Rsch. and Devel., U.S. Office of Edn., Washington, 1969-70, program analyst, 1970-73; equal opportunity specialist Office of Sec., HEW, Washington, 1973-77; legis. program, civil rights analyst Office for Civil Rights Dept. Health and Human Svcs., 1977-85; ind. cons.; adj. lectr., No. Va. Community Coll., George Mason U., Montgomery Coll., 1986—. Mem. steering com. Alliance for Child Care, 1975-80; bd. dirs. Child Care Centers, 1970-76, HEW Employees Assn., 1973-78. Mem. Am. Psychol. Assn. (panel conductor 1969—, editor Bulletin of Peace Psychology 1991—, div. 48), Am. Soc. Public Adminstrn., (condr. panels), Capitol Area Social Psychologists Assn. (conf. chmn. 1985). Federally Employed Women (nat. editor 1975-79), Fairfax County Assn. for the Gifted (pres. 1980), Psi Chi, Phi Alpha Theta. Contbr. articles on ednl. programs to profl. publs. Home and Office: 16614 Fern Pl Manassas VA 22191

CARVER, RON G(EORGE), chemical engineer; b. Pryor, Okla., Nov. 25, 1952; s. Mitchell and Clauidene (Westenhaver) C.; m. Melinda Davis, May 30, 1975; children: Carrie, Christine, Molly, Jeff. BS, Okla. State U., 1975, MChemE, 1976, MS in Environ. Engr., 1992. Registered profl. engr., Ark., Okla., Tex. Mgr. spl. projects Occidential Oil and Gas, Tulsa, 1976-89; prin. environ. engring. and sr. process engr. John Brown Engring., Houston, 1989-90; mgr. environ. affairs Ark. Western Gas Co., Fayetteville, 1990—. Contbr. articles to profl. jours. Vol. Day Start for Homeless, Tulsa, 1989-90, Little League, Tulsa, 1989. Mem. AICE, Am. Petroleum Inst., Gas Processors Assn., Am. Air and Waste Mgmt. Assn.

CARVILLE, THOMAS EDWARD, environmental engineer; b. New Orleans, Feb. 8, 1949; s. Jules A. Jr. and Doris C. (Lasseigne) C.; m. Jill Thena Sellers, Apr. 16, 1975; children: Christopher, Alexis. BSME, La. State U., 1971, MSCE, 1973. Registered profl. engr., La. Environ. engr., mgr. energy and environ. affairs CF Industries, Inc., Donaldsonville, La., 1973—. Mem. NSPE, La. Assn. Bus. and Industry (past chmn. environ. coun., past chmn. energy coun.), Fertilizer Inst. (mem., past chmn. mfg. environ. com.), La. Chem. Assn. (mem. environ. com.), Lower Miss. Environ. Control Assn. (founding mem.), Air and Waste Mgmt. Assn. (past program. chmn., past sec.-treas. La. sect.), La. Engr. Soc., Sigma Xi. Home: 1402 Main St La Place LA 70068 Office: CF Industries PO Box 468 Donaldsonville LA 70346

CASAGRANDA, ROBERT CHARLES, industrial engineer; b. Iron River, Mich., Oct. 8, 1939; s. Charles Casagranda and Lillian Otto Seppi; m. Sheila Adele Mikkola, Nov. 24, 1961; children: Gregory Charles, Wendy Jean, Jodi Marie, Renee Lynn. AA, Cerritos Coll., 1974. Sr. planner McDonnell Douglas, Long Beach, Calif., 1966-72; parts planner N. Am. Rockwell, El Segundo, Calif., 1972; quality analyst White Sunstrand, Belvidere, Ill., 1972-78; supr. Ares Inc., Port Clinton, Ohio, 1978-81; mgr. MFG. Systems-Ex-Cell-O, Rockford, Ill., 1981-84; mfg. con. Ingersoll Engrs., Rockford, 1984-88; project engr. Ingersoll Milling Machine Co., Rockford, 1988-90; sr. ptnr., owner, cons. The Mfg. Cons. Group, Inc., Rockford, 1990—. With U.S. Army, 1958-60. Mem. Soc. Mfg. Engrs., Soc. Mfg. Technologists. Home: 5450 Tam Oshanter Dr Rockford IL 61107-3764 Office: The Mfg Cons Group Inc 5450 Tam Oshanter Dr Rockford IL 61107-3764

CASALE, JOSEPH WILBERT, environmental organic chemist, researcher; b. Hartford, Conn., Dec. 7, 1961; s. Albert Joseph and Lois Hellen (Johns) C. BS, Fairfield U., 1984. Chemist trainee dept. health State of Conn., Hartford, 1985-86, chemist dept. health, 1986-89, sr. chemist dept. health, 1989—. Actor: appeared in Is Conn.'s Ground Water Safe, Conn. Pub. TV, 1985. Mem. Jaycees (com. chair Greater Hartford chpt. 1990—). Achievements include research in gas chromatography, mass spectrometry and electronic design and development of transducers and circuits employing them. Home: 112 Orchard Rd West Hartford CT 06117-2913 Office: Bur Labs 10 Clinton St Hartford CT 06106-1684

CASALS, JUAN FEDERICO, economist, consultant; b. Guayaquil, Guayas, Ecuador, June 27, 1930; s. Juan Feliciano and Maria Amada (Martinez) C.; m. Alicia Torres Egü; children: Mariangeles, Juan Francisco, Maria de los Milagros, Maria Soledad. Grad. in Econ. Scis., Guayaquil U.; diploma in Econ. and Fin. Stats., CIEF-Centro Interamericano de Estadistacas Financieras, Santiago, Chile; diploma in Agrl. Econ., IICA-Andean Zone, Lima, Peru. Researcher economist Cen. Bank of Ecuador, Guayaquil and Quito, 1960-69; deputy dir. Nat. Colonization Inst., Quito, 1957-60; tech. adviser Econ. Planning Bd., Quito, 1961-62; exec. dir. Nat. Colonization Inst., Quito, 1963-64; gen. dir. Land Reform Inst., Quito, 1964-67; tech. adviser Interamerican Devel. Bank, Washington, 1967-69; fin. funds mgr., tech. mgr. Cen. Bank Ecuador, Quito, 1970-86; min. counsellor, rep. Embassy of Ecuador and Aladi, Montevideo, Uraguay, 1986-88; tech. cons. Quito, 1989—; tech. cons. Food & Agrl. Orgn., Quito, 1982, Agy. Internat. Devel.-U.S.A., Quito, 1971-72, Internat. Devel. Bank, Internat. Inst. Agrl. Coop., Quito, 1989-90, Andean Group, Lima, 1989. Author: La Estructure Agraria, 1964, Fondos Fiduciarios para el Desarrollo, 1971, La Division Tecnica de Banco Central del Ecuador, 1985, Politica Monetaria y Funciones del Banco Central, 1985; contbr. articles to profl. jours. Mem. Assn. Economistas y Egresad, Quito, 1960-71; pres. Assn. Filatélica Ecuatoriana, Quito, 1989-90; tchr. Economy Faculty of Cen. Univ., Quito, 1962-67. Recipient medals Sociedad Filantropica del Guayas, 1943-49, Gold medal Assn. Ex-Alumnos Lasallanos, 1949. Mem. Colegio de Economistas, Assn. Empleados Beo. Cen., Assn. de Jubilados del Banco Cen. del Ecuador, Assn. Ex-Alumnos de LaSalle. Roman Catholic. Avocations: philately, numismatist, classical music, swimming. Home: Martinez Mera # 653, Quito Ecuador Office: PO Box 17.17.37, Quito Ecuador

CASANI, JOHN RICHARD, electrical engineer; b. Phila., Sept. 17, 1932; s. John Charles and Julia Jean (Bateman) C.; divorced; 1 son, John Charles; m. J. Lynn Seitz, Dec. 13, 1969; children: Jason, Josh, Drew. BSEE, U. Pa., 1955, DSc, 1992. Spacecraft mgr. Jet Propulsion Lab., Pasadena, Calif., 1966-70, project mgr. Mariner Mars 69, 1970-71, spacecraft mgr. Mariner Venus Mercury, 1971-73, div. mgr. guidance and control, 1973-75, Voyager project mgr., 1975-77, Galileo project mgr., 1977-88, asst. lab. dir., 1988—. Recipient Exceptional Svc. medal, NASA, 1965, Outstanding Leadership medal, 1974, 81. Fellow AIAA (Space Systems award 1979, Astronautics Engr. award 1981, Von Karmen Lectr. Astronautics 1991); mem. NAE, Internat. Acad. Astronautics. Republican. Roman Catholic. Home: 281 S Orange Grove Blvd Pasadena CA 91105-1748 Office: Jet Propulsion Lab 4800 Oak Grove Dr Pasadena CA 91109-8099

CASANOVA-LUCENA, MARIA ANTONIA, computer engineer; b. Cienfuegos, Las Villas, Cuba, Jan. 1, 1954; came to U.S., 1979; d. Manuel José and Loida Eugenia (Ojeda) Casanova; m. Angel de Jesus Lucena, Aug. 12, 1978; 1 child, Ingrid. BSEE cum laude, U. Miami, 1985. Software engr. Martin Marietta Corp., Orlando, Fla., 1986-89; computer engr., mgr. software acquisition Naval Tng. Systems Ctr., Orlando, 1989—. Mem. IEEE, Golden Key, Sigma Xi, Tau Beta Pi, Eta Kappa Nu, Phi Kappa Phi. Achievements include co-development of weapons system for Desert Storm. Home: 3212 Lake George Cove Dr Orlando FL 32812 Office: Naval Tng Systems Ctr Code 242 12350 Research Pky Orlando FL 32826

CASAÑ-PASTOR, NIEVES, chemist, researcher; b. Valencia, Spain, Oct. 31, 1959; d. Jose Casañ-Beut and Nieves Pastor-Villalonga; m. Pedro Gomez-Romero, Dec. 26, 1982; 1 child, Daniel Gomez-Casañ. B, U. Valencia, Spain, 1981; M, U. Valencia, 1982; PhD of Chemistry, Georgetown U., 1988. Fellow Ministry of Edn., Valencia, 1988-90; staff rschr. Inst. Ciencia Materials, Barcelona, Spain, 1990—. Author: Polyoxometalates from Platonic Solids, 1993; contbr. articles to profl. jours. Grantee Midas Program 1992, 93—. Mem. Am. Chem. Soc., Sigma Xi. Achievements include research on the influence of electron delocalization on magnetic properties, on possible artifacts using squid magnetometers, on charge reorganization in La2cu04 d sinthetized electrochemically superconducting wires. Office: Inst Ciencia Materials, Campus U A B, 08193 Barcelona Spain

CASE, ELDON DARREL, materials science educator; b. Logan, Kans., Aug. 23, 1949; s. Eldon George and Ila Marie (Lewis) C.; m. Linda Lee Lubken, Aug. 29, 1975 (div. Mar. 1993); 1 child, Carl Allen. BA in Physics and Math., U. Colo., 1971; PhD in Materials Sci., Iowa State U., 1980. Rsch. asst. dept. materials sci. Iowa State U., Ames, 1976-80; NRC postdoctoral assoc. Nat. Bur. Standards, Gaithersburg, Md., 1980-82; rsch. engr. in materials sci. and mining engring. U. Calif., Berkeley, 1982-85; asst. prof. metallurgy, mechanics and materials sci. Mich. State U., East Lansing, 1985-88, assoc. prof., 1988—; cons. Indsl. Tech. Inst., Ann Arbor, Mich., 1990, Westinghouse, West Mifflin, Pa., 1991-92. Contbr. articles to Jour. Materials Sci., Materials Sci. Engring., Applied Physics Letters. Speaker to sch. groups Okemos (Mich.) Pub. Schs., 1986-90; asst. with middle-sch. activities Episcopal Ch., East Lansing, 1988-92. Regents scholar U. Colo., 1967-71; NRC postdoctoral assoc., 1980-82; grantee NASA, 1987, NSF, 1987-90, Mich. State U., 1989. Mem. The Metall. Soc. (sec. structural materials div. 1988-91, chair non-metall. com. 1988-91), Am. Ceramic Soc., Sigma Xi. Democrat. Achievements include first neutron scattering study from microcracks in a polycrystalline ceramic; statistical analysis of water drop impact damage cracks in infrared windows; adhesion studies of diamond thin-films on brittle substrates; thermal-shock and thermal fatigue studies on ceramics and ceramic composites. Home: 4469 Fairlane Dr Okemos MI 48864 Office: Materials Sci and Mechanics Sci Dept Rm A403 Engring Bldg East Lansing MI 48824

CASE, GERARD RAMON, drafting technician; b. Bklyn., Dec. 22, 1931; s. James Sanford and Adele Elizabeth (Harris) C. Student BFA program, Pratt Inst., Bklyn., 1955-59. Cert. coml. artist and draftsman. Pvt. practice advt. and art N.Y.C., 1955-72, Ultra Cooling Corp., N.Y.C., 1972-85; drafting technician Engring. div. Dept. of Pub. Works, Hackensack, N.J., 1986—. Author: Fossil Shark-Fish Remains of North America, 1967, Fossils Illustrated, 1968, Handbook of Fossil Collecting, 1972, Fossil Sharks: A Pictorial Review, 1973, Pictorial Guide to Fossils, 1982; contbr. 81 articles to profl. jours. With USN, 1951-55. Recipient Harrell L. Strimple award Paleontol. Soc., 1992; Rsch. grantee Griffis Found./Am. Littoral Soc., 1976-86. Mem. AAAS, Am. Littoral Soc., Soc. Vertebrate Palaeontology, Paleontological Soc., Am. Legion, Soc. Col. Wars, Hereditary Order First Families Mass. (life), Sons of the Revolution (life). Republican. Baptist. Achievements include research in genera and species of fossil remains; discovery of insects in amber, new genera and species of fossil fish, and a new order of fossil fishes, the Iniopterygians. Office: Bergen County Dept Pub Works Engring Div 21 Main St Ste 201E Hackensack NJ 07601

CASE, HADLEY, oil company executive; b. N.Y.C., Mar. 28, 1909; s. Walter Summerhayes and Mary Soule (Hadley) C.; m. Julie Marguerite Ill, June 8, 1935 (dec. Mar. 1975); children: Mary C. Durham, Julie Anne, Rosalie C. Clark, Deborah Joan; m. Elizabeth M. McCabe, Nov., 1975. Student, Kent (Conn.) Sch., 1924-29, Antioch Coll., 1929-33; DSc (hon.), Antioch U., 1991, DS (hon.), 1991. Geol. field work Australia, 1933-34, Tex., 1935-36; with geol. dept. Case, Pomeroy & Co., Inc., 1936-39, v.p., 1939-41, pres., chief exec. officer, dir., 1941-83, chmn. bd., chief exec. officer, 1983—; pres. chief exec. officer Felmont Oil Corp., 1952-72; chmn. bd., chief exec. officer Felmont Oil Corp. (merger Felmont and Homestake Mining Co.), 1972-84; dir. Homestake Mining Co., 1984—, Brown Bros. Harriman Trust Co. Fla., 1986-93; bd. dirs. N.W. Airlines, Inc., 1957-78, Copper Range Co., 1966-77, Nashua Corp., 1965-81, Numac Oil & Gas Ltd., 1963-88. Trustee Kent Sch., 1959-75, Brewster Acad., 1956-63, Boys' and Girls' Camps, Inc., Boston, 1971-76, Hosp. St. Barnabas, Newark, 1942-59, pres. bd. trustees, 1949-52; bd. dirs. Greenwich Boys CLub Assn., 1957-73, hon mem., 1974—; trustee Naples (Fla.) Community Hosp., 1985-91, Antioch U., 1987-93; bd. dirs. Naples Philharmonic Ctr. for Arts, 1988—; dir. of The Conservancy, Naples, 1985-91; chancellor Kent Sch., 1985—, trustee, 1986—. Mem. Am. Inst. Mining and Metall. Engrs., Am. Petroleum Inst., Ind. Petroleum Assn. Am. (past v.p., dir.). Office: Case Pomeroy & Co Inc 529 5th Ave New York NY 10017-4608

CASE, MARK EDWARD, geotechnical engineer, consultant; b. Phila., Jan. 25, 1964; s. Albert Charles and Catherine Julia (Castorina) C.; m. Stephanie Marie Forgione, Feb. 21, 1987. BSCE, Drexel U., 1986, MSCE, 1993. Registered profl. engr., N.J. Engring. technician Phila. Dept. Commerce, 1982, Geotech, Inc., Maple Shade, N.J., 1984-85; engring. aide Joseph Gaudet & Assocs., Phila., 1983; engr., geotechnician Geotech. Bur., N.J. Dept. Transp., Trenton, 1986-88; engr., project geotech. engr. Golder Assocs. Inc., Mt. Laurel, N.J., 1988-93, sr. geotech. engr., 1993—. Mem. ASCE, N.Am. Geosynthetics Soc. Roman Catholic. Achievements include research on strain behavior of geosynthetic clay liners. Office: Golder Assocs Inc 305 Fellowship Rd Ste 200 Mount Laurel NJ 08054

CASELLA, PETER F(IORE), patent and licensing executive; b. N.Y.C., June 5, 1922; s. Fiore Peter and Lucy (Grimaldi) C.; m. Marjorie Eloise Enos, Mar. 9, 1946 (dec. Aug. 1989); children: William Peter, Susan Elaine, Richard Mark. BChE, Poly. Inst. Bklyn. (Now Poly. U.), 1943; student in chemistry St. John's U., N.Y.C., 1940. Registered to practice by U.S. Patent and Trademark Office, Can. Patent and Trade Mark Offices. Head patent sect. Hooker Electrochem. Co., Niagara Falls, N.Y., 1943-54; mgr. patent dept. Hooker Chem. Corp. (named changed to Occidental Chem. Corp. 1981), Niagara Falls, 1954-64, dir. patents and licensing, 1964-81, asst. sec., 1966-81, ret., 1981; pres. Intra Gene Internat., Inc., Lewiston, N.Y., 1981-92; chmn. bd. In Vitro Internat., Inc., Linthicum, Md., 1983-86; cons. patents and licensing, Lewiston, N.Y., 1981—; Dept. Commerce del. on patents and licensing exchange, USSR, 1973, 90, Poland and German Dem. Republic, 1976. Editor: Drafting the Patent Application, 1957. Mem. Lewiston Bd. Edn., 1968-70. With AUS, 1944-46; MTO. Recipient Centennial citation Poly. Inst. Bklyn., 1955, Golden Jubliee Soc., 1993. Mem. Assn. Corp. Patent Counsel (emeritus, mem. exec. com. 1974-77, charter mem.), N.Y. Patent Law Assn., Niagara Frontier Patent Law Assn. (pres. 1973-74, Founder award 1974), Licensing Execs. Soc. (v.p. 1976-77, Trustees award 1977), Chartered Inst. Patent Agts. Gt. Britain, Patent and Trademark Inst. Can., Internat. Patent and Trademark Assn., U.S. Trademark Assn., Nat. Assn. Mfrs. (patent com.), Mfg. Chemists Assn., Pacific Indsl. Property Assn., U.S. Patent Office Soc. (assoc.), U.S. Trade Mark Office Soc. (assoc.), Am. Chem. Soc., Am. Inst. Chem. Engrs. Clubs: Chemists (N.Y.C.); Niagara (pres. 1973-74) (Niagara Falls).

CASELLA, RUSSELL CARL, physicist; b. Framingham, Mass., Nov. 6, 1929; s. Rosario and Lena Mary (D'Amico) C.; m. Marilyn Smith, Jan. 27, 1952; children: Sheryl M., Cynthia L. Conturie. BS in Physics, MIT, 1951, MS in Physics, 1953; PhD in Physics, U. Ill., 1956. Physicist Cambridge (Mass.) AF Rsch. Ctr., 1951-52; teaching and rsch. asst. physics dept. U. Ill., Urbana, 1953-55, rsch. fellow physics dept., 1955-56, rsch. assoc. physics dept., 1956-58; theoretical physicist IBM T.J. Watson Rsch. Ctr., Yorktown Heights, N.Y., 1958-65, Nat. Inst. Standards and Tech., Gaithersburg, Md., 1965—. Contbr. over 50 articles to profl. jours. Recipient Silver medal U.S. Dept. Commerce, 1973. Mem. Am. Physical Soc., Sigma Xi. Achievements include development of theory of condensed-matter and of elementary-particle physics; research in (broken) symmetries; neutron scattering; Bose condensation of excitons; tests of time reversal and CPT symmetries in Kaon physics; neutrino scattering; topology in neutron interferometry; high-temperature superconductivity; hydrogen in metals; quark-parton-sea content of the nucleon in deep-inelastic electroweak scattering. Home: 1485 Dunster Ln Potomac MD 20854-6107 Office: Nat Inst Standards and Tech Gaithersburg MD 20899

CASEM, CONRADO SIBAYAN, civil/structural engineer; b. Luna, La Union, Philippines, Feb. 19, 1945; came to U.S., 1984; s. Pedro Nuesca and Francisca (Sibayan) C.; m. Corazon Nieveras Noble, May 7, 1972; children: Christopher, Conrado Jr. BSc in Civil Engring. cum laude, Feati U., Manila, 1967; M Engring. in Structural Engring., Asian Inst. Tech., Bangkok, 1970. Lic. profl. engr., N.Y., N.J., Ohio, Ill., Pa., Calif., Mass., Minn., Ont. Civil engr. Philippine Engring. & Constrn. Corp., Rizal, Philippines, 1970-73; structural engr. Morrison Knudsen Internat. Co., Singapore, 1973-75; design engr. Monenco (Asia) Ltd., Singapore, 1975-78; sr. engr. Montreal (Ont.) Engring. Co., 1978-83; sr. engr. Flour City Archtl. Metals, Glen Cove, N.Y., 1984-87, chief engr., 1987—. Recipient Scholarship Grant Asian Inst. Tech., Bangkok, 1968-70. Mem. ASCE, Assn. Profl. Engrs. Ont. Democrat. Roman Catholic. Achievements include structural engring. svcs. to exterior wall systems of bldgs. such as United Airlines Terminal at O'Hare Airport, Soc. Bank Tower, others; rsch. in the elastic flexural-torsional buckling of beam-columns by discrete elements tech. Office: Flour City Archtl Metals 175 Sea Cliff Ave Glen Cove NY 11542

CASERO, JOSEPH MANUEL, civil engineer; b. Tineo, Asturias, Spain, Dec. 23, 1936; came to the U.S., 1947; s. Manuel Garcia and Oliva (Bolano) C.; m. Karina Mittermaier, July 22, 1981. BS, Rutgers U., 1959, MS, 1961. Design engr., soil cons. Kellogg Internat. (London), Malaga, Spain, 1963-64; tech. dir. Soil Testing Espanola, S.A., Madrid, 1965-66; chief engr. Soil Testing Svcs. Iowa, Davenport, 1966-69; v.p. Caribbean Soil Engr., St. Croix, Virgin Islands, 1969-73, pres., owner, 1973-78; internat. cons. Greenwood, Va., 1979—. Fellow ASCE. Republican. Roman Catholic. Home and Office: PO Box 304 Greenwood VA 22943

CASEY, MURRAY JOSEPH, obstetrician/gynecologist, educator; b. Armour, S.D., May 1, 1936; s. Meryl Joseph and Gladice (Murray) C.; m. Virginia Anne Fletcher; children: Murray Joseph Jr., Theresa Marie, Anne Franklin, Francis Xavier, Peter Colum, Matthew Padraic. Student, Chanute Jr. Coll., 1954-55, Rockhurst Coll., 1955-56; AB, U. Kans., 1958; MD, Georgetown U., 1962; postgrad., Suffolk U. Law Sch., 1963-64, Howard U., 1965, U. Conn., 1977; MS in Mgmt., Cardinal Stritch Coll., 1984; MBA, Marquette U., 1988. Diplomate Am. Bd. Med. Examiners, Am. Bd. Ob-Gyn. Intern USPHS Hosp.-Univ. Hosp., Balt., 1962-63; staff physician USPHS Hosp., Boston, 1963-64; staff assoc. Lab Infectious Diseases, Nat. Inst. Allergy and Infectious Diseases, NIH, Bethesda, Md., 1964-66; virologist, resident physician Columbia-Presbyn. Med. Ctr. also Francis Delafield Hosp., N.Y.C., 1966-69, USPHS sr. clin. trainee, 1969-70; fellow gynecol. oncology, resident dept. surgery Meml. Hosp. Cancer and Allied Diseases, Meml. Sloan-Kettering Cancer Ctr., N.Y.C., 1969-71; Am. Cancer Soc. fellow, 1969-71; ofcl. observer in radiotherapy U. Tex. M.D. Anderson Hosp. and Tumor Inst., Houston, 1971; vis. scientist Radiumhemmet Karolinska Sjukhuset and Inst., Stockholm, 1971; asst. prof. ob-gyn U. Conn. Sch. Medicine, 1971-75, asso. prof., 1975-80, dir. gynecologic oncology, 1971-80, also mem. med. bd.; prof., assoc. chmn. dept. ob-gyn U. Wis. Med. Sch., 1980-89; prof., chmn. dept. ob-gyn. Creighton U., Omaha, 1989—; chief ob-gyn and dir. gynecologic oncology St. Joseph Hosp., Creighton U. Med. Ctr., Omaha, 1989—; chief ob-gyn Mt. Sinai Med. Ctr., Milw., 1980-82, dir. gynecologic oncology, 1980-89, also mem. med. exec. com.; chmn. research adv. com., mem. council Conn. Cancer Epidemiology Unit. Editor, contbr. articles in sports medicine to profl. jours., chpts. to books; rsch. in oncogenesis and tumor immunology. Bd. dirs., mem. exec. com., chmn. profl. edn. com. Hartford unit Am. Cancer Soc., dir. Milw. div., exec. com. 1985-87, v.p., 1985-86, pres.-elect, 1986-87, 1st v.p. exec. com. Wis. div. 1987-89, bd. dirs., chmn. profl. edn. com., 1987-89, bd. dirs., exec. com. Nebr. div., 1989—, pub. edn. and communications com., profl. edn. com. vice chair, 2nd v.p., 1990-91, 1st v.p., pres. elect, 1991-92, pres., 1992—; mem. med. svcs. 1980 Winter Olympic Games, Lake Placid, N.Y.; mem. med. supervisory team U.S. Nordic Ski Team. Lt. (j.g.) USPHS, 1962-64, lt. comdr., 1964-66; col. USAR, 1988—. Fellow ACS, Am. Coll. Obstetricians and Gynecologists; mem. AAAS, Soc. of Gynecologic Surgeons, Cen. Assn. Ob-Gyns., Am. Coll. Sports Medicine, N.Y. Acad. Scis., Am. Soc. Colposcopy, Am. Assn. Gynecologic Laporoscopists, Am. Fertility Soc., Soc. Gynecologic Oncologists, European Soc. Gynecologic Oncologists, New Eng. Assn. Gynecologic Oncologists (pres. 1980-81), Internat. Gynecological Cancer Soc., Am. Radium Soc., Am. Soc. Clin. Oncology, Internat. Menopause Soc., N.Am. Menopause Soc., Internat. Assn. for Advancement of Humanistic Studies in Gynecology, Soc. Meml. Gynecologic Oncologists (exec. bd. 1979-84; pres. 1982-83), Lake Placid Sports Medicine Soc. (v.p. 1981-84, pres. 1984-86), Am. Urogynecologic Soc., Am. Assn. Gynecologic Laporoscopists, Internat. Gynecologic Soc., Assn. Mil. Surgeons, Soc. of Gynecologic Surgeons, Cedarburg C. of C. (Ambassadors com. 1983—, dir. 1983-85, chmn. bus. indsl. program com. 1985, 87-89), St. George Soc. Office: Creighton U Sch Medicine Dept Ob-Gyn 601 N 30th St # 4810 Omaha NE 68131-2100

CASEY, STEVEN MICHAEL, ergonomist, human factors engineer; b. Downey, Calif., Aug. 19, 1952; s. Reuben L. and Mary Lou C.; m. Mary C. Morgan, Dec. 28, 1974; children: Erin, Stephanie. MS, N.C. State U., 1977, PhD, 1979. Cert. profl. ergonomist, Calif. Engr., psychologist Naval Air Devel. Ctr., Warminster, Pa., 1977; prin. scientist Anacada Scis., Inc., Santa Barbara, Calif., 1978-87; pres. Ergonomic Systems Design, Santa Barbara, 1987—; bd. dirs. Bd. Cert. Profl. Ergonomics, Bellingham, Wash., 1990—. Author: Developing Effective User Documentation, 1988, Set Phasers on Stun and Other True Tales of Design, Technology, and Human Error, 1993; contbr. articles to profl. publs. Fellow Human Factos Soc. (exec. coun., Alexander C. Williams award 1987); mem. APA, Soc. Indsl. and Orgnl. Psychologists, Soc. Engring. Psychologists. Achievements include founding of Board of Certification in Professional Ergonomics. Office: Ergonomics Systems Design 5290 Overpass Rd Ste 105 Santa Barbara CA 93111-2042

CASH, (CYNTHIA) LAVERNE, physicist; b. Statesville, N.C., Oct. 7, 1956; d. William J. and Martha Lee (Stroud) C. BS, Appalachian State U., 1979; MS, Clemson U., 1982; AA, Mitchell Community Coll., 1976; postgrad., Johns Hopkins U. Physicist U.S. Army Material Systems Analysis Activity, Aberdeen Proving Ground, Md., 1984-88; rsch. physicist U.S. Army Chem. Rsch., Devel. and Engring. Ctr., Aberdeen Proving Ground, 1988—. Contbr. articles to profl. publs. Mem. Oak Grove Bapt. Ch, Bel Air, Md., singer in choir, sound engr., numerous others. Mem. Am. Phys. Soc., Sigma Phi Sigma, Pi Mu Epsilon, Phi Theta Kappa, Gamma Beta Phi. Baptist. Home: 100 Drexel Dr Bel Air MD 21014-2002

CASILLAS, ROBERT PATRICK, research toxicologist; b. San Bernardino, Calif., Mar. 7, 1962; s. Friedrich Karl and Eleanor (Guzman) Tietz; m. Christy Scales Casillas, Oct. 13, 1991. BS, Ga. State U., 1986, PhD, 1992. Rsch. scientist USPHS, Jefferson, Ark., 1990-91; rsch. fellow Vanderbilt U. Sch. Medicine, Nashville, 1991—; adj. faculty David Lipscomb U., Nashville, 1991—; cons. Lab. Design Svcs., Little Rock, 1991—. V.p. Alternative Colors, Inc., Little Rock, 1991-92. Ensign USPHS, 1990-91. Predoctoral fellow Nat. Inst. Environ. Health Scis., Nat. Ctr. for Toxicol. Rsch., 1991. Mem. AAAS, Am. Soc. Microbiology. Mem. Ch. of Christ. Achievements include research in transformation of selected biologically active compounds by fungi. Office: Vanderbilt U Dept Biochemistry 23rd and Pierce 607 LH Nashville TN 37232

CASPAR, DONALD LOUIS DVORAK, physics and structural biology educator; b. Ithaca, N.Y., Jan. 8, 1927; s. Caspar V. and Blanche (Dvorak) C.; m. Gwladys Williams, Dec. 20, 1962; children: Emma, David. BA in Physics, Cornell U., 1950; PhD in Biophysics, Yale U., 1955. Postdoctoral fellow Calif. Inst. Tech., 1954-55, MRC Lab. Molecular Biology, Cambridge, Eng., 1955-56; instr. in biophysics Yale U., New Haven, 1956-58, asst. prof., 1958-59; rsch. assoc. Harvard U., Cambridge, Mass., 1962-63; lectr. Harvard Med. Sch., Boston, 1963-73; rsch. assoc. in pathology Children's Hosp. Med. Ctr., Boston, 1959-73; prof. of physics, rsch. prof. of structural biology Rosenstiel Basic Med. Scis. Rsch. Ctr., Brandeis U., Waltham, Mass., 1972—, acting dir., 1987-88; mem. biophysics and biophys. chem. study sect. NIH, 1969-73; guest rsch. assoc. in biology Brookhaven Nat. Lab., 1973—, mem. chmn. biology dept. vis. com., 1974-77; mem. nat. laser users facility steering com. Lab. for Laser Energetics, U. Rochester, N.Y., 1981-84; mem. sci. adv. com. European Molecular Biology Lab., Heidelberg, Fed. Republic Germany, 1976-81; mem. editorial com. Ann. Revs. Biophysics and Bioengring., 1970-73. Contbr. articles, rsch. papers to profl. publs. Grantee NIH, 1969-88, 88—, NSF, 1983-86. Fellow Am. Acad. Arts & Scis.; mem. Biophys. Soc. (pres. 1991—, nat. lectr. 1985), Am. Crystallographic Soc. (Fankuchen Memorial Award in X-Ray Crystallography 1992). Achievements include research in structural biology of viruses, membranes and protein interactions. Home: 9 Hyslop Rd Brookline MA 02146-5712 Office: Structural Biology Lab Rosenstiel Med Sci Rsch Ctr Brandeis U 415 South St Waltham MA 02254-9110

CASPER, PATRICIA A., human factors research engineer; b. Ann Arbor, Mich., Oct. 12, 1959; d. James Earl and Jean Ann (Johnson) Lakin; m. Daniel Paul Casper, July 23, 1983; 1 child, Gerhard Daniel. BS in Psychology, U. Iowa, 1983; MS in Psychology, Purdue U., 1985, PhD in Psychology, 1988. Sr. human factors rsch. engr. Sikorsky Aircraft, Stratford, Conn., 1988—; reviewer lecture proposals Human Factors Soc. Annual Meeting, Sant Monica, Calif., 1990. Co-author: Human Factors in Modern Aviation, 1988. Mem. Seymour Oxford Newcomer's Club, Oxford, Conn., 1992 Digital Equip. Corp. Bus. AI fellow, 1990. Mem. Human Factors Soc., Assn. Aviation Psychologists, Am. Helicopter Soc., Army Aviation Assn. Am. Office: Sikorsky Aircraft 6900 Main St MS 5326A Stratford CT 06601-1381

CASSADY, PHILIP EARL, mechanical engineer; b. Springfield, Mass., Sept. 8, 1940; s. Julian Willard and Vivian Celema (Guindon) C.; m. Katherine Leslie Palmer, Sept. 8, 1963; children: Sean Erik, Edward Julian. BS and MS, MIT, 1963; PhD, Calif. Inst. Tech., 1970. Rsch. scientist Lockheed Rsch. Lab., Palo Alto, Calif., 1970-73; prin. scientist Math. Sci. N.W., Bellevue, Wash., 1973-83; assoc. tech. fellow Boeing, Seattle, 1983—; adj. prof. U. Wash., Seattle, 1978, 80; invited lectr. symposia in Russia, Belgium, Israel; program chmn. Nat. Tech. Conf. on Laser Tech.; tech. session chmn. numerous nat. tech. confs.; invited guest Russian Acad. Scis. Contbr. articles to profl. jours. Coun. mem. MIT, Cambridge, 1984—; chmn. adminstrv. bd. United Meth. Ch., Snoqualmie, Wash., 1980. Boeing Co. Assoc. Tech. fellow, 1991; recipient Von Karman award Von Karman Inst., 1965. Assoc. fellow AIAA (com. chmn. 1981—, Survey Paper Citation 1980, 85). Office: Boeing Co PO Box 3999 M/S 8H-18 Seattle WA 98124

CASSEDAY, JOHN HERBERT, neurobiologist; b. Pasadena, Calif., Aug. 11, 1934; s. John Herbert and Marjorie (Hoban) C.; m. Margrid Krueger; children: Patrick, Tara; m. Ellen Covey, Feb. 15, 1981. BA, U. Calif., Riverside, 1960; PhD, Ind. U., 1970. Asst. prof. Duke U. Med. Ctr., Durham, N.C., 1972-79, assoc. med. rsch. prof., 1979-88, assoc. prof. dept. neurobiology, 1988—. Author: Neural Basis of Echolocation in Bats, 1989; contbr. articles to profl. jours., chpts. to books. Rsch. grantee NIH, 1981, 84, 87, 92, NSF, 1982, 85. Mem. AAAS, Internat. Soc. Neuroethology, Soc. Neurosci. Home: 1021 Dacian Ave Durham NC 27701 Office: Duke U Med Ctr Dept Neurobiology Box 3209 Durham NC 27710

CASSEDY, EDWARD SPENCER, JR., electrical engineering educator; b. Washington, Aug. 19, 1927; s. Edward S. and Edna Marjorie (Muddle) C.; m. Sylvia Levine (dec.); m. Bernice Herman Cassedy. BSEE, Union Coll., 1949; SM, Harvard U., 1950; D of Engring., Johns Hopkins U., 1959. Jr. engr. Potomac Elec. Power Co., Washington, 1950-51; electronics. engr. US Naval Ordnance Lab., White Oak, Md., 1951-54; rsch. scientist The Johns Hopkins U., Balt., 1954-60; rsch. asst. prof. Microwave Rsch. Inst. Polytech. Inst., Bklyn., 1960-63, assoc. prof., 1963-68, prof., 1968—. Author: Introduction to Energy-Resources, Technology and Society, 1990. Mem. IEEE, Am. Phys. Soc., Internat. Assn. Energy Economists. Democrat. Office: Poly U 6 Metro Tech Ctr New York NY 11201

CASSEL, CHRISTINE KAREN, physician; b. Mpls., Sept. 14, 1945; d. Charles Moore and Virginia Julia (Anderson) C.; AB U. Chgo., 1967; MD U. Mass., 1976. Intern, resident in internal medicine Children's Hosp., San Francisco, 1976-78; fellow in bioethics, Inst. Health Policy Studies, U. Calif., San Francisco, 1978-79; fellow geriatrics Portland (Oreg.) VA Hosp., 1979-81; asst. prof. medicine and public health U. Oreg. Health Scis. U., 1981-83; asst. prof. geriatrics and medicine Mt. Sinai Med. Ctr., N.Y.C., 1983-85; assoc. prof. medicine, prof. pub. policy, chief gen. internal medicine, U. Chgo., 1985—. Bd. dirs. Lutheran Gen. Health Care Systems, Greenwall Found. Woodrow Wilson fellow, 1967; Henry J. Kaiser Family Found. faculty scholar, 1982-85; Hastings Ctr. fellow; Ctr. Advanced Study in Behavioral Sci. fellow, 1991-92; diplomate Am. Bd. Internal Medicine. Fellow Am. Geriatrics Soc., ACP (vice chair bd. regents 1991); mem. Physicians for Social Responsibility (dir. 1983—, pres. 1988—), Soc. Health and Human Values (pres. 1986), Inst. of Medicine, Nat. Acad. Sci., Am. Bd. Internal Medicine, Am. Soc. Law & Medicine (bd. dirs.), Inst. of Medicine. Author: Ethical Dimensions in the Health Professions, 1981; Geriatric Medicine: Principles and Practice, 1984, 2d edit., 1990; Nuclear Weapons and Nuclear War: A Sourcebook for Health Professionals, 1984. Office: U Chgo Pritzker Sch Medicine Sect Gen Internal Medicine MC6098 Chicago IL 60690-0012

CASSETTA, SEBASTIAN ERNEST, industry executive; b. July 30, 1948; m. Linn Miller; children: Christopher, Sebastian III, Kathryn. BSBA, U. Denver, 1971; MBA, U. Pa. Spl. asst. to gov./v.p. Nelson A. Rockefeller Albany, N.Y. and Washington, 1971-75; pres. Transec Inc., Dallas, 1975-80; v.p. internat. devel. Brinks, Inc., Darien, Conn., 1980-85; pres., chief exec. officer, mem. bd. dirs. Burns and Roe Securacom, Inc., Oradell, N.J., 1985—; bd. dirs. Brinks Europe, Paris; mem. Industry Sector Adv. Coun., 1983, U.S. Export Coun., 1984, V.P.'s Task Force on Terrorism, Washington, 1986, Nat. Def. Exec. Res., U.S. Dept. Commerce. Author: Winning America's Biggest Security Contract. Trustee Worcester (Mass.) Acad., 1983—; dir. Christian edn. St. Michael's Episc. Ch., Dallas, 1976-80. Mem. Am. Soc. Indsl. Security, Am. Nuclear Soc., Am. Def. Preparedness Assn., Nat. Security Indsl. Assn. (bd. dirs.), Young Pres. Orgn. (bd. dirs.), U.S. Rowing Assn. Office: Burns and Roe Securacom 7 Morningside Ln Westport CT 06880 also: 477 Devlin Rd Napa CA 94558

CASSIDY, JAMES EDWARD, chemistry consultant; b. Springfield, Mass., Aug. 17, 1928; s. Robert Adam and Ethel Louise (Shaw) C.; m. Barbara J. Odorkiewicz, June 6, 1959; children: Sean C., Ian A. BS, U. Mass., 1949; MS, U. Vt., 1954; PhD, Rensselaer Poly. Inst., 1958. Chemist engring. expt. sta. U. N.H., Durham, 1949-50; rsch. analysis chemist Am. Cyanamid Co., Stamford, Conn., 1958-64; metabolism chemist Geigy, Ardsley, N.Y., 1964-73; sr. group leader Ciba-Geigy, Greensboro, N.C., 1973-86; pvt. practice cons. Greensboro, N.C., 1987-88; sr. chemist, environ. cons. Jellnek, Schwartz and Connolly, Inc., Washington, 1988—. Contbr. articles to profl. jours. With U.S. Army, 1951-52. Mem. Internat. Soc. for Study of Xenobiotics, Am. Chem. Soc., N.Y. Acad. Scis., Sigma Xi (vice chmn. Greensboro chpt. 1986), Phi Lambda Upsilon. Republican. Congregationalist. Achievements include work in pesticide metabolism in plants and animals. Home: 7846 Somerset Ct Greenbelt MD 20770-3023 Office: JSCF Inc 1015th St NW Washington DC 20005

CASSIDY, JOHN FRANCIS, JR., research center executive; b. Troy, N.Y., Nov. 26, 1943; s. John F. Sr. and Beverly A. (Blowers) C.; m. Paulina C. DiBacco, July 24, 1965; children: Rachel, Sean. BEE, Rensselaer Poly. Inst., 1965, MEE, 1967, PhD, 1969. Various R & D mgmt. positions GM, 1969-81; with control systems R & D GE Corp. R & D Labs., 1981-89; corp. dir. tech. mgmt. United Techs. Corp., 1989-90, chief tech. officer, 1990-92; dir. United Techs. Rsch. Ctr., 1992-93, v.p., 1993—. Bd. dirs. Convergence Electronics Transp. Assn., Convergence Ednl. Found., Detroit. Sr. mem. IEEE, Soc. Automotive Engrs.; mem. Conn. Acad. Sci. & Engring. Office: United Techs Rsch Ctr MS 129-04 411 Silver Ln East Hartford CT 06108

CASSON, ANDREW J., mathematician educator. Prof. math. U. Calif., Berkeley. Recipient Oswald Veblen Geometry prize Am. Math. Soc., 1991. Office: Univ of Calif Berkeley Dept of Math Berkeley CA 94720*

CASTAGNEDE, BERNARD ROGER, physicist, educator; b. Pau, France, Apr. 20, 1956; s. Etienne Marcel and Simone Joséphine (Delhoste) C.; m. Dalia Lafuente, Apr. 20, 1986. BS, U. Bordeaux I, France, 1978, MS, 1980, DSc, 1984; Habilitation, U. Bordeaux I, 1991; PhD, Cornell U., 1987. Tchr. Lycee de la Sauque, Labrede, France, 1979-82; rsch. asst. U. Bordeaux I, 1982-84; postdoctoral assoc. SUNY, Syracuse, 1985-87; rsch. assoc. Cornell U., Ithaca, N.Y., 1987-89; asst. prof. physics U. Bordeaux I, 1989-91, assoc. prof. physics, 1991-92; full prof. physics U Maine, France, 1992—; sci. expert Saint-Gobain Paper-Wood divsn., Talence, France, 1990—. Contbr. articles to profl. jours. Mem. Am. Phys. Soc., European Phys. Soc., French Phys. Soc., N.Y. Acad. Scis., Acoustical Soc. Am., Assn. Universitaire de Mecanique, Internat. Physics Group, French Acoustical Soc. Home: 15 rue de Normandie, 72000 Le Mans France Office: Lab Acoustique U Maine, Av Olivier Messiaen B P 535, 72017 Le Mans France

CASTAIN, RALPH HENRI, physicist; b. L.A., Nov. 23, 1954; s. Henry Ulrich and Anni (Springmann) C.; m. Cynthia Ellen Nicholson, Dec. 28, 1976; children: Kelson, Alaric. BS in Physics, Harvey Mudd Coll., 1976; MS in Physics, Purdue U., 1978, MSEE in Robotics, PhD in Nuclear Physics, 1983. Sr. engr. Harris Semiconductor, Palm Bay, Fla., 1978-79; mem. staff Jet Propulsion Lab., Pasadena, Calif., 1983-84; mem. staff Los Alamos (N.Mex.) Nat. Lab., 1984-92, chief scientist nonproliferation and arms control, 1992—; cons. Jet Propulsion Lab., Pasadena, 1984-91. Editor, contbr. to Dept. Energy Office Arms Control publs. Recipient Maths. award Bank of Am., 1972, Gold Seal, State of Calif., 1972; Calif. State scholar, 1972-76. Mem. IEEE, Am. Phys. Soc., Internat. Neural Network Soc., VLSI Spl. Interest Group (chmn. electronics com.). Achievements include patents pending for technique incorporating neural networks and genetic algorithms into cryptography, for neural network-based face verification technique for advanced security system; devel. of new CCD camera system plus high speed data reduction hardware for beam sensing on neutral particle beam program, nuclear proliferation detection techs. Home: 431 Bryce Ave Los Alamos NM 87546-3605 Office: Los Alamos Nat Lab NAC MS-F650 Los Alamos NM 87545

CASTANEDA, MARIO, chemical engineer, educator; b. Mexicali, BC, Mex., Oct. 16, 1954; came to U.S., 1979; s. J. Guadalupe and Plasida (Olivares) C.; m. Rosalva Gutierrez, Apr. 21, 1989; children: Mario Alberto, Mariela. BSChemE, U. Sonora, Hermosillo, Mex., 1977; MS in Petroleum Engring., Stanford U., 1981; postgrad., U. Kans., 1988—. Cert. ground water profl.; registered environ. assessor, Calif. Lab. mgr. Inst. de Investigaciones Electricas., Mexicali, 1978-79, project engr., 1981-85; quality assurance mgr. LN Safety Glass, Mexicali, 1986-87; rsch, engr. Inst. Engring., Mexicali, 1987; project geologist Alton Geoscience, Irvine, Calif., 1989-90; project engr. Mittelhauser Corp., Laguna Hills, Calif., 1990-92; hydrologist Ariz. Dept. Environ. Quality, Phoenix, 1992—; asst. prof. U. Sonora, 1977, U. Baja California, Mexicali, 1992. Contbr. articles to sci. jours. Fellow Consejo Nacional de Ciencia Y Tech., Mexico, 1976, Calif., 1979, Bank of Mex., Stanford, Calif., 1979, UN, 1981, Am. States Orgn., 1988. Mem. Soc. Petroleum Engrs., Assn. Ground Water Scientists and Engrs., Chem. Engrs. Mex. Inst. (v.p. 1984-85). Republican. Roman Catholic. Achievements include development of method to determine feed zones in geothermal wells using pressure gradient logging. Home: 6608 N 90th Ave Glendale AZ 85305 Office: Ariz Dept Environ Quality 3033 N Central Ave Phoenix AZ 85012

CASTANES, JAMES CHRISTOPHER, architect; b. Vallejo, Calif., Jan. 12, 1951; s. James Christopher and Helen C.; m. Diane Allenbach, June 22, 1991. BArch, U. Ark., Fayetteville, 1975. Apprentice Schmidt, Garden, Erickson, Chgo., 1969-71; staff architect Ibsen Nelsen & Assocs., Seattle, 1975-76, Jouce, Copeland, Vaughn & Nordfors, Seattle, 1976-80; project architect Jean Fraley & Assocs., Seattle, 1980-85; pvt. practice architect Seattle, 1985-87; ptnr. Castanes/Gibson Architects, Seattle, 1987—. Recipient Home of Yr. award Seattle Am. Inst. Architects/Seattle Times Newspaper, 1987. mem. N.F.G. 1969—. Office: Castanes/Gibson Architects 1932 First Ave Suite 928 Seattle WA 98101

CASTEEL, MARK ALLEN, psychology educator; b. Cedar Rapids, Iowa, Feb. 10, 1960; s. Gerald Robert and Kathryn Lucille (Purdy) C.; m. Stephanie Sue Clark, Dec. 27, 1986 (div. Aug. 1989); m. Judith Lynn Rouiller, Mar. 7, 1992; 1 stepchild, Laura Beth Miller. BA, Coe Coll., 1982; MA, U. Nebr., Omaha, 1985; PhD, U. Nebr., Lincoln, 1988. Asst. prof. psychology Pa. State U., York Campus, 1988—. Contbr. articles to Devel. Psychology, Jour. of Rsch. in Reading, Jour. Exptl. Psychology: Learning, Memory and Cognition. Mem. rsch. and evaluation com. Ctrl. Pa. Health Edn. Ctr., York, 1991—. Mem. APS, Soc. for Rsch. in Child Devel., Teaching of Psychology, The Psychonomic Soc. (assoc.). Office: Pa State U York Campus 1031 Edgecomb Ave York PA 17403

CASTEL, GÉRARD JOSEPH, physician; b. Gardanne, France, Nov. 16, 1934; s. Roger Alphonse and Marguerite Henriette (Bossy) C.; m. Maryse Tartaise (div. 1965); 1 child, Gilles; m. Charlotte Elisabeth Gaglio; 1 child, David. MD, U. Marseille, France, 1961; D of History, U. Aix en Provence, France, 1978. Extern, then resident Conception Hosp., Marseille, 1956-57, Timone Hosp., Marseille, 1957-58, Salvator Hosp., Marseille, 1958-59, Sainte-Marguerite Hosp., Marseille, 1959-60; gen. practice medicine Rognac, France, 1961—; historian, conferenceer history Faculty Letters, Aix en Provence, 1978—, associated researcher, 1988; physician French Soc. Rys., 1966—; instr. physician French Red Cross, 1966-90. Author: Contribution a l'Etude Historique de Rognac, 1969, Rognac Depuis 3000 Ans, 1976, Raymond des Baux, Premier Seigneur de Berre, 1983, Histoire de la Paroisse de Rognac du XIo au XXo siècle, 1984, Histoire de Berre, 10 vols., 1983-91, Histoire de la Grande-Bastide de Rognac, 1986, Dictionnaire Archeologique de la Commune de Rognac, 1987, Musical Memories, 1987, Rognac un Antique Terroir, 1987, Souvenirs Musicaux, 1987, Réflexions sur l'Histoire, 1988; contbr. articles to profl. jours.; composer romantic piano works, also several records by internat. pianists. Recipient French League Instrn. medal, 1978. Mem. History Soc. Rognac (chmn. 1965—), Sci. and Culture Assn. Berre, Authors, Composers and Editors Music, Chopin Soc. Paris. Office: 4 Rue Lamartine, 13340 Rognac France

CASTELAZ, PATRICK FRANK, computer scientist; b. Milw., Sept. 25, 1952; s. Harvey William and Dorothy Marie (Seilenbinder) C.; m. Denise Lenore Passante, Nov. 4, 1952; 1 child, Julia Marie. BEE, Marquette U., 1974, MEE, 1975, PhD, 1978. Mem. tech. staff Rockwell Internat., Anaheim, Calif., 1978-79; staff engr. Hughes Aircraft Co., Fullerton, Calif., 1979-81; sr. scientist Hughest Aircraft Co., Fullerton, Calif., 1985-91; mgr. signal processing Aerojet Electrosystems, Azusa, Calif., 1981-85; dir. advanced tech. ctr., prof. neurology Loma Linda (Calif.) U., 1991—; pres. PFC Enterprises, Yorba Linda, Calif., 1986—. Contbr. articles to profl. jours. Recipient Crozier prize Am. Def. Preparedness Assn., 1988. Mem. IEEE, Internat. Neural Network Soc., UCLA Cognitive Sci. Coun., Mensa, Sigma Xi. Achievements include over 25 patents in advanced computing architectures, signal processing, image and sensor processing, neural networks, parallel computing tehniques and technologies. Office: Loma Linda U Office of Dean Sch Medicine Loma Linda CA 92350

CASTELLA, XAVIER, critical care physician, researcher; b. Manresa, Catalonia, Spain, May 16, 1958; s. Josep Castella and Palmira Picas. MD, Autonomous U. Barcelona, Spain, 1982, Master in Statis. Methods Pub. Health, 1991. Resident in critical care medicine Hosp. Sant Pau, Barcelona, 1984-88; staff mem. ICU Ctr. Hosp. Manresa, Spain, 1988-90, Hosp. Sabadell, Spain, 1990-92; head critical care dept. Hosp. Gen. de Manresa, Spain, 1992—. Contbr. articles to profl. jours. Fellow Am. Coll. Chest Physicians; mem. European Soc. Intensive Care Medicine, European Diploma Intensive Care Medicine, European Med. Assn. Avocations: swimming, computers. Home: San Juan Bta La Salle 28-5, Manresa 08240, Spain Office: Hosp Gen de Manresa, La Culla Sin, 08240 Manresa Spain

CASTELLI, WILLIAM, cardiovascular epidemiologist, educator; b. N.Y.C., Nov. 21, 1931; s. Rudolph Edward and Alma Veronica (McNeil) C.; m. Marjorie Irene Fish, July 15, 1961; children: Laurence Edward, William Alton, Allyson Irene. BS Zoology, Yale Coll., 1953; MD, Catholic U. Louvain, Belgium, 1959; D honoris causa, Framingham State Coll., 1993. Intern Kings County Hosp. Ctr., Bklyn., 1958-59; asst. resident medicine Lemuel Shattuck Hosp., Jamaica Plain, Mass., 1959-61, sr. resident medicine, 1961-62; fellow Rheumatology Mass. Gen. Hosp., Boston, 1960-62; rsch. fellow Preventive Medicine Harvard Med. Sch., Boston, 1962-65; dir. labs. Framingham (Mass.) Heart Study, 1965-79, dir., 1979—; lectr. preventive medicine Harvard Med. Sch., Boston, 1962—; lectr. medicine U. Mass. Med. Sch., Worster, 1982—; adj. prof. medicine Boston U. Med Sch., 1979—; chmn. Coun. Epidemiology Am. Heart Assn., Dallas, 1988-89; pres. New England Hypertension Coun., 1982—. Author: Good Fat, Bad Fat. Pres. Marlborough Friends of the Libr., 1965; trustee Marlborough Pub. Libr., 1966-83; med. dir. USPHS, Framingham, 1965—. Recipient Meritorius Svc. medal USPHS, Paul Dudley White award Mass. Affiliate Am. Heart Assn., Disting. faculty award Fla. Acad. Family Physicians, Drake award Maine affiliate Am. Heart Assn. Fellow Coun. Epidemiology, Am. Coll. Epidemiology; mem. Société Médicale des Hôpitaux de Paris. Roman Catholic. Home: 74 White Terr Marlborough MA 01752 Office: Framingham Heart Study 5 Thurber St Framingham MA 01701*

CASTELLON, CHRISTINE NEW, information systems manager; b. Pittsfield, Mass., June 22, 1957; d. Edward Francis Jr. and Helen Patricia (Cordes) New; m. John Arthur Castellon, Oct. 1, 1988. BS in Elec. and Computer Engring., U. Mass., 1979; MBA, Northeastern U., 1986. Engr. microwave radio system design New Eng. Telephone Co., Framingham, Mass., 1979-82; mgr. minicomputer support group New Eng. Telephone Co., Dorchester, Mass., 1982-85; mgr. current systems planning/network svcs. NYNEX Svc. Co., Boston, 1985-87; mem. tech. staff computing environments Bellcore, Piscataway, N.J., 1987-90; assoc. dir. info. systems provisioning NYNEX Telesector Resources Group, N.Y.C., 1990-93; speaker New Eng. Telephone Careers-In-Engring. Program, 1980-92. Leader 2d violin sect. Cen. Jersey Symphony Orch., Raritan Valley Community Coll., N.J., 1988—; prin. 1st violinist New Eng. Conservatory Extension Div., Boston, 1979-87; violinist Civic Symphony Orch., Boston, 1982-87. Named Monument Mountain High Sch. valedictorian, 1975; recipient Arion Music award, 1975, cert. Applied Music and Theory Pittsfield Community Music Sch., 1975. Mem. IEEE, U. Mass. Alumni Assn. (coll./industry adv. com. for women). Roman Catholic. Home: 622 Old York Rd Neshanic Station NJ 08853-3601

CASTENSCHIOLD, RENE, engineering company executive, author, consultant; b. Mt. Kisco, N.Y., Feb. 7, 1923; s. Tage and Juno (Hagemeister) C.; m. Martha Naomi Stinson, Dec. 14, 1947; children—Gail F., Frederick T., Lynn Castenschiold Jones. BEE, Pratt Inst., 1944. Registered profl. engr., N.Y., N.J., Pa.; registered profl. planner, N.J. Test engr. (Manhattan Project) GE, Schenectady, N.Y., 1944-45; design engr. GE, Schenectady, 1946-47; sr. product engr. Am. Transformer Co., Newark, 1947-50; design engr. Automatic Switch Co., Florham Park, N.J., 1951-57; chief customer engr. Automatic Switch Co., Florham Park, 1957-74, exec. engring. mgr., 1974-85; pres. LCR Cons. Engrs. P.A., Green Village, N.J., 1986—; lectr. N.J. Inst. Tech., Newark, 1967-79; adviser Underwriter Labs. Inc., Melville, N.Y., 1973-85; chmn. U.S. Tech. Adv. Group and U.S. del. Internat. Electrotech. Commn., Geneva, 1981-90. Contbr. articles to profl. jours., chpts. to books, promulgation of numerous nat. and internat. elec. standards; patentee in fields transformer design, relays, automatic transfer switches and controls. Chmn. Bd. of Adjustment, Harding Twp., 1975-77, chmn. Planning Bd., 1982-85, dir. Civil Def., 1966-70; trustee Wash. Assn. of N.J., Morristown, 1984—, sec., 1985-88, v.p. 1989-92, pres., 1992—. Named to Disting. Alumni Bd. Visitors Pratt Inst., 1979; recipient The James H. McGraw award, 1986, Disting. Svc. award Morristown Nat. Hist. Park, 1993. Fellow IEEE (standards bd. 1983-85, recipient IEEE Achievement award, 1988, Richard Harold Kaufmann award 1990), NSPE, Instrument Soc. Am., Nat. Elec. Mfrs. Assn. (chmn. automatic transfer switch com. 1982-88), Internat. Assn. Elec. Insps., Nat. Acad. Forensic Engrs., Coun. Engring. Splty. Bds. (cert. diplomate), Am. Cons. Engrs. Coun., Nat. Fire Protection Assn., N.J. Christmas Tree Growers' Assn., Can. Standards Assn., Nat. Forensic Ctr. Republican. Episcopalian (vestryman 1991—). Clubs: Skytop (Pa.); The Morristown (N.J.). Avocations: swimming, photography, hiking. Home: Lees Hill Rd New Vernon NJ 07976 Office: LCR Cons Engrs PA PO Box 2 Green Village NJ 07935-0002

CASTENSON, ROGER R., agricultural engineer, association executive; b. Galveston, Tex., Aug. 6, 1943. BSAE, Tex. A&M U., 1972; MBA, Mich. State U., 1977. Rsch. assoc. Blackland Rsch. Ctr., Temple, Tex., 1972-73; mgr. pub. rels., mem. activities Am. Soc. Agrl. Engrs., St. Joseph, Mich., 1973-80, exec. v.p., 1987—; ops. mgr. Soc. Petroleum Engrs., 1980-81, bus. mgr., 1981-86. Mem. Am. Soc. Agrl. Engrs., Soc. Petroleum Engrs., Sigma Xi. Office: Am Soc Agrl Engrs 2950 Niles Rd Saint Joseph MI 49085-9659*

CASTERLINE, JAMES LARKIN, JR., research biochemist; b. Lexington, S.C., Dec. 25, 1931; s. James L. and Hattie (Steppe) C.; m. Dorothy C. Sueoka, July 15, 1961; children: Rex L., Jonathan P. BA, Gallaudet U., 1959; MS, U. Md., 1979, PhD, 1982. Rsch. chemist U.S. FDA, Washington, 1962—; mem. adv. bd. FDA, 1964—. Contbr. articles to profl. jours. Mem. Am. Chem. Soc., Sigma Xi. Lutheran. Achievements include discovery of role of level and quality of proteins in diet in body's defenses against pesticides, stimulation of metabolizing enzymes in liver by pretreating rats with chlorinated hydrocarbon compounds; method developments include a sensitive in-vitro biological test to detect the presence of carcinogenic contaminants, such as dioxins and furans in foods. Office: US FDA 200 C St SW Washington DC 20204

CASTIGNETTI, DOMENIC, microbiologist, biology educator; b. Boston, Sept. 22, 1951; s. Andrew and Anna (Guarino) C.; m. Dorothy Ann Papalia, June 22, 1975; children: Nancy, Lisa, Michael. BA, Merrimack Coll., 1973; MS, Colo. State U., 1977; PhD, U. Mass., 1980. Post-doctoral fellow Brandeis U., Waltham, Mass., 1980-82; asst. prof. Loyola U., Chgo., 1982-88, assoc. prof., 1988—. Lectr. St. Isaac Jogues Parish, Niles, Ill., 1989—. Recipient grant Cystic Fibrosis Found., Bethesda, Md., 1989. Mem. AAAS, Am. Soc. for Microbiology, Sigma Xi (pres. Loyola U. chpt. 1991-92). Roman Catholic. Achievements include first publ. of rsch. establishing that microbial iron chelators, siderophores, were present during a clinical infection of humans/cystic fibrosis patients. Office: Loyola Univ Chgo Biology Dept 6525 N Sheridan Rd Chicago IL 60626

CASTLE, RAYMOND NIELSON, chemist, educator; b. Boise, Idaho, June 24, 1916; s. Ray Newell and Lula (Nielson) C.; m. Ada Necia Van Orden, June 16, 1937; children: Raymond Norman, Dean Lowell, David Elliott, George Leonard, Elizabeth Anne, Edith Eilene, Christian Daniel, Lyle William. Student, Boise Jr. Coll., 1934-35; BS, Idaho State U., 1939; MA., U. Colo., 1941, PhD, 1944. Instr. chemistry U. Idaho, 1942-43, U. Colo., 1943-44; research chemist Battelle Meml. Inst., Columbus, Ohio, 1944-46; faculty U. N.M., 1946-70, prof. chemistry, 1956-70, chmn. dept., 1963-70; prof. chemistry Brigham Young U., Provo, Utah, 1970-81; disting. rsch. prof. chemistry U. South Fla., 1981—; rsch. fellow U. Va., 1952-53; vis. rsch. scholar Tech. U. Denmark, Copenhagen, 1962; pres. 1st Internat. Congress Heterocyclic Chemistry, N.M., 1967, sec. 2d congress, France 1969, v.p. 3d congress, Japan, 1971, 4th congress, Utah, 1973, 5th congress, Yugoslavia, 1975, 6th congress, Iran, 1977, adv. com., 1992-93; mem. organizing com. 2d Internat. Symposium Chemistry and Pharmacology of Pyridazines, Vienna, 1990, 3d Internat. Symposium, Como, Italy, 1992; Plenary lectr. 11th Symposium on chemistry of heterocyclic components, Prague, 1993. Editor, co-author: Chemistry of Heterocyclic Compounds, vols. 27, 28, 1973; editor Jour. Heterocyclic Chemistry, 1964—, Topics in Hetrocyclic Chemistry, 1979, Lectures in Heterocyclic Chemistry, vol. I, 1972, vol. II, 1974, vol. III, 1975, vol. IV, 1977, vol. V, 1980, vol. VI, 1982, vol. VII, 1984, vol. VIII, 1985, vol. IX, 1987, vol. X, 1990, vol. XI, 1992; adv. editor: English transl. Russian Jour. Heterocyclic Compounds; contbr. rsch. articles to profl. jours. Recipient Idaho State U. Profl. Achievement award, 1992. Fellow Chem. Soc. London (Eng.); mem. Am. Chem. Soc., Internat. Soc. Heterocyclic Chemistry (pres. 1973-75, past pres. 1976-77, Biennial award, 2d award 1983, adv. com. 1992—), Sigma Xi. Mem. Ch. of Jesus Christ of Latter-day Saints (bishop 1957-61). Home: 2775 Kipps Colony Dr Apt 303 Saint Petersburg FL 33707-3946

CASTLE, WILLIAM EUGENE, academic administrator; b. Thomas, S.D., Sept. 5, 1929; s. Eugene Albert and Kathryn (Barkley) C.; m. Diane Lee Sklar, Aug. 8, 1963. B.S., No. State Tchrs. Coll., 1951; M.A., U. Iowa, 1958; Ph.D., Stanford U., 1963. Tchr. Faulkton (S.D) High Sch., 1951; instr. St. Cloud (Minn.) Tchrs. Coll., 1958-60, Central Wash. Tchrs. Coll., Ellensburg, 1961; asst. prof. U. Va., 1963-65; asso. sec. for research and sci. affairs Am. Speech, Lang. and Hearing Assn., Washington, 1965-68; dean Nat. Tech. Inst. for Deaf, Rochester Inst. Tech., N.Y., 1968-79; v.p. Nat. Tech. Inst. for Deaf, Rochester Inst. Tech., 1979—, dir., 1977—. Author: The Effect of Narrow Band Filtering on the Perception of Certain English Vowels, 1964. Served with USAF, 1952-56. Named Outstanding Alumnus, No. State Coll., 1984. Mem. Nat. Assn. Deaf., Am. Speech Lang. and Hearing Assn., Assn. Am. Sch. Adminstrs., Conf. Ednl. Adminstrs. Serving the Deaf, Conv. Am. Instrs. of Deaf, Assn. of Deaf, Am. Deafness and Rehab. Assn., Am. Assn. Higher Edn., Alexander Graham Bell Assn. for Deaf (pres. 1982-84, 90-92). Home: 4272 Clover St Honeoye Falls NY 14472-9323 Office: 1 Lomb Memorial Dr Rochester NY 14623-5603

CASTRACANE, JAMES, physicist; b. Buffalo, July 29, 1954; s. Nick Bennett and Elizabeth (Berger) C.; m. Kathleen Mary McCarthy; 1 child, Caitlin. BS in Physics, Canisius Coll., Buffalo, 1976; MA in Physics, Johns Hopkins U., 1978, PhD in Physics, 1982. Teaching asst. Johns Hopkins U., Balt., 1976-79, rsch. asst., 1979-82; sr. rsch. physicist MPB Tech., Inc., Montreal, 1982-88, Intersci., Inc., Troy, N.Y., 1988—. Contbr. articles to profl. jours. Mem. Am. Phys. Soc., Biomed. Optics Soc.

CASTRO, GONZALO, ecologist; b. Lima, Peru, Nov. 21, 1961; came to U.S., 1985; s. Ramiro and Elsa (Valdivia) C.; m. Jodi Shull, Aug. 29, 1987; children: Bradford, Brook. BS, Cayetano Heredia, Lima, 1983, MS, 1984; PhD, U. Pa., 1988. Lectr. U. Cayetano Heredia, Lima, 1983-84; rsch. asst. Acad. Natural Scis., Phila., 1985-88; teaching asst. U. Pa., Phila., 1985-88; postdoctoral fellow Colo. State U., Ft. Collins, 1988-90; program mgr. Western Hemisphere Shorebird Res. Network, Manomet, Mass., 1990-92; exec. dir. Wetlands for the Ams., Manomet, 1992—; sci. reviewer auk, condor functional ecology NATO, 1986-91; chair organizing com. Shorebird Symposium, Quito, Ecuador, 1990-91; co-chair 58th N.Am. Wildlife and Natural Resources Conf., 1993; cons. Troy Ecol. Rsch. Assocs., Anchorage, Nature Conservancy, W. Alton Jones Found. Sci. reviewer Can. Jour. of Zoology, Earthwath, Jour. of Field Ornithology, Nat. Fish and Wildlife Found., Southwestern Naturalist, Wildlife Conservation Internat., Wilson Bull.; contbr. articles to Ecology, Auk, Condor, Am. Birds Physiol. Zoology, Jour. Field Ornithology, Vista, others. Grad. fellow OAS, 1985-87; recipient Kathleen Anderson award Manomet Bird Obs., 1986, Alexander Wilson prize Wilson Ornithol. Soc., 1988. Mem. AAAS, Am. Inst. Biol. Scis., Am. Ornithologists Union (chair wetlands conservation subcommittee 1992, conservation com. 1993), Cooper Ornithol. Soc., Ecol. Soc. Am., Friends of Peruvian Rain Forest, Soc. Conservation Biology, Internat. Soc. Ecol. Econs., Internat. Coun. Bird Preservation (bd. dirs. 1991), Internat. Coun. Bird Preservation Pan-Am. (vice-chmn. 1991-93), Internat. Waterfowl and Wetlands Rsch. Bur. (standing com. 1992), Peruvian Assn. Conservation (co-founder 1983), Peruvian Group Study of Waterbirds (hon. 1992). Office: Wetland for the Americas PO Box 1770 Manomet MA 02345-1770

CASTRO, JOSEPH RONALD, physician, oncology researcher, educator; b. Chgo., Apr. 9, 1934; m. Barbara Ann Kauth, Oct. 12, 1957. B.S. in Natural Sci., Loyola U., Chgo., 1956, M.D., 1958. Diplomate: Am. Bd. Radiology, 1964. Intern Rockford (Ill.) Meml. Hosp.; resident U.S. Naval Hosp., San Diego; assoc. radiotherapist and assoc. prof. U. Tex.-M.D. Anderson Hosp. and Tumor Inst., 1967-71; prof. radiology/radiation oncology U. Calif. Sch. Medicine, San Francisco, 1971—; vice-chmn. dept. radiation oncology U. Calif. Sch. Medicine, 1980—; dir. particle radiotherapy Lawrence Berkeley Lab., 1975—, faculty sr. scientist, 1991—; mem. program project rev. com. NIH/Nat. Cancer Inst. Cancer Program, 1982-85. Author sci. articles. Past pres., chmn. bd. trustees No. Calif. Cancer Program, 1980-83. Served to lt. comdr., M.C. USN, 1956-66. Recipient Teaching award Mt. Zion Hosp. and Med. Center, San Francisco, 1972. Fellow Am. Coll. Radiology; mem. Rocky Mountain Radiol. Soc. (hon.), Am. Soc. Therapeutic Radiology, Gilbert H. Fletcher Oncologic Soc. (past pres. 1988). Office: Lawrence Berkeley Lab Bldg 55 Berkeley CA 94720

CASTRO-BLANCO, DAVID RAPHAEL, clinical psychologist, researcher; b. N.Y.C., June 7, 1958; s. David Rafael and Frances Agnes (Salerno) Castro-Blanco. BA, St. John's Coll., Jamaica, N.Y., 1980; PhD, St. John's U., Jamaica, N.Y., 1990. Lic. psychologist, N.Y. Applied behavior specialist Nassau Ctr. for Developmentally Disabled, Woodbury, N.Y., 1986-91; rsch. scientist N.Y. State Psychiat. Inst., N.Y.C., 1991—; instr. clinical psychology Columbia U., Coll. Physicians and Surgeons, N.Y.C., 1991—; asst. profl. psychologist Columbia Presbyn. Med. Ctr., N.Y.C., 1992—; adj. asst. prof. L.I. U., C. W. Post Coll., Old Brookville, N.Y., 1991—. Co-author: Treating Depressed and Disturbed Adolescents, 1991, Successful Negotiation-Acting Positively, 1991; contbr. articles to profl. jours. Mem. Amnesty Internat., 1988, People to People Citizen Amb. Program, Seattle, 1990. Named Most Influential Person in a Career, Westinghouse Sci. Talent Search Awards Com., 1984. Mem. APA, Psychologists for Social Responsibility, N.Y. State Psychol. Assn., Ea. Psychol. Assn. Democrat. Roman Catholic. Achievements include development of multi-method matrix to aid research design in studies of timeout. Office: NY State Psychiat Inst Box 29, 722 W 168th St New York NY 10032

CASTROGIOVANNI, ANTHONY G., aerospace engineer, researcher; b. Bklyn., Sept. 26, 1965; s. Angelo and Vita (Catanzaro) C. BSME, Poly. U., 1987, MS in Aeronautics & Astronautics, 1991. Sr. scientist Gen. Applied Sci. Labs., Inc., Ronkonkoma, N.Y., 1987—. Mem. ASME, AIAA. Achievements include research in modelling of two phase flows and boiling heat transfer. Office: Gen Applied Sci Labs Inc 77 Raynor Ave Ronkonkoma NY 11779

CASWELL, HAL, mathematical ecologist; b. L.A., Apr. 27, 1949; s. Herbert Hall and Ethel (Preble) C. BS with high honor, Mich. State U., 1971, PhD, 1974. Asst. to assoc. prof. U. Conn., Storrs, 1975-81; assoc. scientist Woods Hole (Mass.) Oceanographic Inst., 1981-88, sr. scientist, 1988—; disting. vis. prof. U. Miami, 1989. Editorial bd. Ecol. Soc. Am., 1987-90; author: Matrix Population Models, 1989; contbr. articles to profl. jours. John Simon Guggenheim Found. fellow, 1989. Fellow AAAS; mem. Ecol. Soc. Am., Brit. Ecol. Soc., Am. Soc. Naturalists, Soc. of Conservation Biology. Achievements include development and application of matrix models for population dynamics of plants and animals; analysis of effects of environmental variation in ecological systems; mathematical models in conservation biology. Office: Woods Hole Oceanographic In Dept Biology Woods Hole MA 02543

CASWELL, ROBERT DOUGLAS, aerospace engineer; b. Markdale, Ont., Can., Nov. 8, 1946; s. Robert G. and Helen (Hawken) C.; children: Laura, Andrea. BASc, U. Toronto, 1969; MASc, Inst. for Aerospace Studies, Toronto, 1970. Thermal engr. SPAR Aerospace, Toronto, 1970-73; systems space engr. Govt. of Can., Ottawa, Ont., 1973-76, shuttle arm test mgr., 1976-80, MSAT platform mgr., 1981-83, RADARSAT spacecraft mgr., 1983-86; space sta. robotics mgr. Nat. Rsch. Coun., Ottawa, Ont., 1986-88; Olympus platform mgr. ESA/ESTEC, Noordwijk, The Netherlands, 1988—. Tech. dir. film: Arm in Space, 1981; contbr. articles to profl. jours. Recipient J.A.D. McCurdy award U. Toronto, 1969, NASA Group Achievement award, 1981, Estec Performance award, 1990, Estec metal for Olympus Spacecraft Recovery, 1991,. Mem. AIAA, Assn. Profl. Engrs. Ont., Can. Aeronautics and Space Inst. Home: Seinpostduin 74, Den Haag 2586 EC, The Netherlands Office: Estec/CP, PO Box 299, 2200AG Noordwijk The Netherlands

CATALANO, LOUIS WILLIAM, JR., neurologist; b. Bklyn., Apr. 20, 1942; s. Louis William and Aileen (Bobb) C.; m. Kathleen Jamea Ferrari, Oct. 1, 1966; children: Louis W. III, Jamea E. BS cum laude, U. Pitts., 1963, MD, 1967. Diplomate Am. Bd. Neurology, Am. Bd. Electroencephalography, Am. Bd. Medical Examiners. Intern Presbyn.-St. Luke's Hosp., Chgo., 1967-68; rsch. assoc. NIH, Bethesda, Md., 1968-70; fellow neurology The Neurol. Inst., N.Y.C., 1970-73; clin. asst., prof. neurology U. Pitts. Sch. Med., 1973—; pvt. practice Greensburg, Pa., 1973—; staff Latrobe (Pa.) Area Hosp., 1973—, Westmoreland Hosp., Greensburg, 1973—, Torrance (Pa.) State Hosp., 1978—, Indiana (Pa.) Hosp., 1983—; cons. Jeannette (Pa.) Dist. Meml. Hosp., 1984—, Frick Community Health Ctr., Mt. Pleasant, Pa., 1991—; lectr. in field. Contbr. articles to profl. jours. With USPHS, 1968-70. Spl. fellow Columbia U., NIH, 1970-73, epilepsy minifellow, Bowman Gray Sch. Medicine, Winston-Salem, N.C., 1988. Fellow Royal Soc. Medicine, Am. Acad. Neurology; mem. AMA, Am. Med. Electroencephalographic Assn., Am. Soc. Neuroimaging, Am. Acad. Clin. Neurophysiology, Am. Sleep Disorders Assn., Pa. Med. Soc., Westmoreland County Med. Soc., Latrobe Acad. Medicine, Pittsburgh Neurosci. Soc., Sigma Xi, Alpha Omega Alpha. Avocations: sport fishing, scuba diving, skiing, travel. Office: Cen Med Arts RD 7 Old Rte 30 Greensburg PA 15601

CATALANO, ROBERT ANTHONY, ophthalmologist, physician, hospital administrator, writer; b. Albany, N.Y., Nov. 24, 1956; s. Anthony Joseph and Ida Santa (Muscolino) C.; m. Madeline Faye Kalmer, Aug. 6, 1978; children: Christopher, Ruth, Thomas, Matthew. BS, Union Coll., Schenectady, 1978; MD, U. Va., 1982; MBA, Rensselaer Poly. Inst., 1992. Resident in ophthalmology Albany Med. Coll., 1983-86, vice-chmn. dept. ophthalmology, 1989-90, acting chmn., 1990-91; fellow in pediatric ophthalmology Wills Eye Hosp., Phila., 1986-87; v.p. med. affairs Olean (N.Y.) Gen. Hosp., 1991—. Author: Atlas of Ocular Motility, 1989, Ocular Emergencies, 1992; contbr. articles to profl. jours. Recipient Nat. Found. award March of Dimes Found., 1978, Robert D. Reinecke award Albany Med. Coll., 1985, Shannon award U. Va., 1982; Heed Found. fellow, 1986. Fellow Am. Acad. Ophthalmology; mem. Am. Coll. Physician Execs., Am. Coll. Healthcare Execs., Western N.Y. Hosp. Assn. (bd. dirs. 1992—), Alpha Omega Alpha. Roman Catholic. Office: Olean Gen Hosp 515 Main St Olean NY 14760

CATALFO, BETTY MARIE, health service executive, nutritionist; b. N.Y.C., Nov. 2, 1942; d. Lawrence Santo and Gemma (Patrone) Lorefice; children—Anthony, Lawrence, Donna Marie. Grad. Newtown High Sch., Elmhurst, N.Y., 1958. Sec., clk. ABC-TV, N.Y.C., 1957-60; lectr., nutritionist Weight Watchers, Manhasset, N.Y., 1976-75; founder, pres. Every-Bodys Diet, Inc. dba Stay Slim, Bronx, N.Y., 1976—; dir. in-home program N.Y. State Dept. Health, N.Y.C., 1985—; founder, pres. Delitegul Diet Foods, Inc., 1988—; lectr. in field. Author: 101 Stay-Slim Recipes, 1983, Get Slim and Stay Slim Diet Cook Book, rev. ed., 1987. Author, dir., producer: (video) Dancersize for Overweight, 1986, Get Slim and Stary Slim Diet Cook Book, Eating Right for Your Life, Hello It's Me and I'm Slim; author, editor: (video) Eating Right For Life, 1985, Isometric Techniques for Weight Reduction, Dance Your Calories A-Weigh; author, producer: (video) Eating Habits, 1986—, (video) Isometric Techniques for Weight Reduction, 1986, (video) Patience Is a Virtue When Weight Loss is the Goal, 1986; producer, dir.: (video) Positive and Negative Diet Forces, 1987, (video) Hello It's Me and I'm Thin, 1987, (video) Dance Your Calories A-Weigh, 1987, (video) Positive and Negative Diet Forces, 1987. Sponsor, lectr. St. Pauls Ctr., Bklyn., 1981—, Throgs Neck Assn. Retarded Children, Bronx, 1985—; active ARC, LWV, Am. Italian Assn., United Way Greenwich, Council Chs. and Synagogues, Heart Assn., N.Y. Meals on Wheels, 1985—, Health Assn. Fairfield County. Named Woman of Yr., Bayside Womens Club, N.Y., 1983, O, PK Woman of Yr., 1986—, Woman of Yr. Richmond Boys Club, 1987, Woman of Yr. Bronx Press Club Assn., 1987; recipient Merit award for Service Catholic Archdiocese of Bklyn., 1985, Community Service award Sr. Citizens Sacred Heart League Bklyn./Queens Archdiocese. N.Y. State Nutritional Guidance for Children Nat. Assn. Scis. Mem. Nat. C. of C. for Women (Woman of Yr. 1987, 90), Pres.'s Coun. on Nutrition, Roundtable for Women in Food Service, Bus. and Profl. Women's Club, Pres. Council for Phys. Fitness, Nat. Assn. Female Execs., Assn. for Fitness in Bus. Inc., Nat. Assn. Female Bus. Owners. Democrat. Roman Catholic. Club: Mothers Sacred Heart Sch. (chairperson 1979-82). Avocations: reading; travel, golf, family. Home: 214-22 27th Ave Flushing NY 11360 also: 58 Riverside Ave Greenwich CT 06878 Office: 100-05 101st Ave Ozone Park NY

CATE, RODNEY MICHAEL, university dean; b. Sudan, Tex., May 9, 1942; s. Tommy A. and Elsie P. (Cherry) C.; m. Patricia Cate, June 11, 1941; children: Brandi, Shani. BS in Pharmacy, U. Tex., 1965; MS in Family Studies, Tex. Tech. U., 1975; PhD in Human Devel. and Family Studies, Pa. State U., 1979. Asst. prof. Tex. Tech. U., Lubbock, 1978-79; asst. prof. Oreg. State U., Corvallis, 1979-83, assoc. prof., 1983-85; prof., dept. chmn. Washington State U., Pullman, 1985-90; assoc. dean Iowa State U., Ames, 1990—; bd. dirs. Internat. Soc. for the Study Presonal Relationships, 1992--. Co-author: Courtship, 1992; editor: Home Econs. Rsch. Jour., 1992; contbr. articles to profl. jours. Lt. USN, 1966-69. Named Researcher of Yr. Washington Home Econs. Assn., 1990. Mem. Am. Home Econ. Assn. (James D. Moran Meml. Rsch. award 1991), Am. Psychological Assn., Nat. Coun. on Family Relations, Internat. Soc. for the Study Personal Relationships (bd. dirs.). Democrat. Office: Iowa State U Family & Consumer Sci Rsch Inst 126 Mackay Hall Ames IA 50011

CATES, HAROLD THOMAS, aircraft and electronics company executive; b. Key West, Fla., Aug. 24, 1941; s. Joseph Livingston and Dorothy Louise (Whitehead) C.; m. Margaret Anne Browne, Nov. 9, 1978; children: Harold Joseph, Kimberly Lisa, Andrew Franklin. BSEE, Valparaiso U., 1963; MSEE, U. N.Mex., 1965, PhDEE, 1967. Registered profl. engr., Fla. Rsch. assoc. and asst. Bur. Engring. Rsch., U. N.Mex., Albuquerque, 1963-67, instr., asst. prof. elec. engring., 1967-68; rsch. asst. Kirtland AFB and Sandia Labs., Albuquerque, 1963-67; program dir., chief engr. Martin-Marietta Co., Orlando, Fla., 1968-87; v.p., asst. gen. mgr. E-Systems, Inc., Greenville, Tex., 1987—; pres. Spectrum Engring., Testing & Cons., Key West, Fla., 1973—; prof. Rollins Coll., Winter Park, Fla.; vis. prof. Fla. Inst. Tech., Melbourne; assoc. Carriage Realty, Orlando. Patentee process for preparing nuclear hardened semicondrs. and microelectronic devices. Mem. IEEE (sr.), Armed Forces Communications and Electronics Assn., Air Force Assn., Theta Chi, Sigma Chi. Avocations: baseball, sailing, coins, scuba diving, tennis. Home: RR 5 Box 261B Greenville TX 75401-9805 Office: Martin Marietta Electronics and Missiles Group PO Box 555837 MP 1549 Orlando FL 32855-5837*

CATEY, LAURIE LYNN, mechanical engineer; b. Muskegon, Mich., Mar. 2, 1962; d. Joe Strange and Jane Louise (Morton) C. BSME, Mich. Tech. U., 1984. Registered profl. engr., Mich. Sr. mech. engr. and assoc. Smith, Hinchman & Grylls, Detroit, 1984—. Mem. ASHRAE, Soc. Am. Mil. Engrs. Office: SH & G 150 W Jefferson Ste 100 Detroit MI 48226

CATH, STANLEY HOWARD, psychiatrist, psychoanalyst; b. June 11, 1921; m. Claire Muriel Cohen, Mar. 14, 1943; children: Phyllis, Alan, Sandra Mahon. BS, Providence Coll., 1942; MD, Boston U., 1946; postgrad., Phila. Psychoanalytic Inst., 1949-52; grad., Boston Psychoanalytic Inst., 1958. Diplomate Am. Bd. Psychiatry and Neurology; cert. physician Nat. Bd. Med. Examiners; registered physician Mass., R.I. Intern Pawtucket (R.I.) Meml. Hosp., 1946-47; resident Taunton (Mass.) State Hosp., 1947-48, Norristown (Pa.) State Hosp., 1949-52; asst. to dir. Eugenia Meml. Hosp., Norristown, 1950-52; staff physician VA Hosp., Bedford, Mass., 1952-54; physician in psychiatry Boston Dispensary, 1952-56; psychiatrist Community Health Project Harvard Sch. Pub. Health, 1954-57; founder, med. dir. The Family Adv. Svc. and Treatment Ctr., 1978-89; pvt. practice Arlington, Mass., 1952—; asst. psychiatrist Children's Hosp. Med. Ctr., Boston, 1953-54, jr. assoc. med. psychiatry Peter Bent Brigham Hosp., 1953-54; clin. instr. Tufts U. Med. Sch., 1953-59, sr. clin. instr., 1959-60, asst. clin. instr., 1960-86; cons. Symmes Hosp., Arlington, 1953—, Mary McArthur Respiratory Unit Boston Children's Hosp., Wellesley, 1954-55, Worcester State Hosp., 1962-65, Health and Human Svcs. Social Security Adminstrn., 1963—, VA Hosp., Bedford, Mass., 1956—, Life Studies Found., 1984-89; rsch. assoc. in mental health Sch. Pub. Health Harvard U., 1954-56; courtesy staff psychiatrist Bournewood Hosp., Brookline, Mass., 1981—, Santa Maria Hosp., Cambridge, 1981-89, McLean Hosp., Belmont, Mass., 1977—. Author: (with others) The Patient and the Mental Hospital: Hospital Image and Post-hospital Experience, 1957, Explaining Divorce to Children, 1969, Research, Planning and Action for the Elderly, 1972, The Five Minute Hour, 1976, Psychological Issues for Mental Health Staff, 1977, Love and Hate on the Tennis Court, 1977, Mental Health in the Nursing Home: An Educational Approach for Staff, 1979, The Age of Aging, 1979, Geriatric Psychopharmacology, 1979, Father and Child, 1982, Death of a Father, 1982, Cults, 1983, Late Life Awakening, 1984, Psychotherapy of Depressive States in Last Half of LIfe, 1985, Fathering, 1986, Caesar, 1988, Lincoln and the Fathers, 1989, Fathers and Their Families, 1989, Senescent Cell Antigen, 1990. Capt. Med. Corps, U.S. Army, 1948-49. Named honored lectr. Lewin Meml. Symposium, Phila., Erich Lindemann Mental Health Ctr., Boston, honored lectr. and given meritorious recognition Mil. and Hospitaller Order of St. Lazarus of Jerusalem. Fellow Am. Geriatric Soc. (life), Am. Gerontol. Soc., Am. Psychiat. Assn. (life, cons. task force treatments psychiat. disorders), Boston Med. Libr.; mem. Am. Acad. Psychiatry and Law, Am. Psychoanalytic Assn. (chmn. Issues in Paternity Study group 1978—), Israel Psychoanalytic Assn., Nat. Found. for Grandparents (adv. bd. 1982—), Flying Physicians Assn., Mass. Gerontology Assn., Mass. Med. Soc., Mass. Psychiat. Soc., Bay State Health Care Found., Middlesex South County Med. Assn., Boston Med. Libr., Boston Psychoanalytic Soc. and Inst., Boston Soc. for Gerontologic Psychiatry. Avocations: flying, tennis. Office: Agricultural Research Inst 9650 Rockville Pike Bethesda MD 20814

CATHEY, WADE THOMAS, electrical engineering educator; b. Greer, S.C., Nov. 26, 1937; s. Wade Thomas Sr. and Ruby Evelyn (Waters) C.; children: Susan Elaine, Cheryl Ann. BS, U. S.C., 1959, MS, 1961; PhD, Yale U., 1963. Group scientist Rockwell Internat., Anaheim, Calif., 1962-68; from assoc. prof. to prof. elec. engring. U. Colo., Denver, 1968-85, chmn. dept. elec. engring. and c.s., 1984-85; prof. U. Colo., Boulder, 1985—; chmn. faculty senate U. Colo., Denver, 1982-83; cons. in field, 1968—; dir. ctr.NSF Optoelectronic Computing Systems, Boulder, 1985—. Author: Optical Information Processing and Holography, 1978; contbr. articles to profl. jours.; inventor in field. Fellow Croft, U. Colo., 1982, Faculty, U. Colo., 1972-73. Fellow Optical Soc. Am. (topical editor 1977-79, 87-90; mem. IEEE (sr.), Soc. Photo-optic Instrumentation Engrs. Avocations: skiing, hiking, reading, writing. Home: 228 Alpine Way Boulder CO 80304-0406 Office: U Colo Dept Elec Engring Boulder CO 80309-0425

CATHOU, RENATA EGONE, chemist, consultant; b. Milan, Italy, June 21, 1935; d. Egon and Stella Mary Egone; m. Pierre-Yves Cathou, June 21, 1959. BS, MIT, 1957, PhD, 1963. Postdoctoral fellow, research assoc. in chemistry MIT, Cambridge, 1962-65; research assoc. Harvard U. Med. Sch., Cambridge, 1965-69, instr., 1969-70; research assoc. Mass. Gen. Hosp., 1965-69, instr., 1969-70; asst. prof. dept. biochemistry Sch. Medicine, Tufts U., 1970-73, assoc. prof., 1973-78, prof., 1978-81; pres. Tech. Evaluations, Lexington, Mass., 1983—; sr. cons. SRC Assocs., Park Ridge, N.J., 1984—; sr. investigator Arthritis Found., 1970-75; vis. prof. dept. chemistry UCLA, 1976-77; mem. adv. panel NSF, 1974-75; mem. bd. sci. counselors Nat. Cancer Inst., 1979-83; ind. cons. and writer. Mem. editorial bd. Immunochemistry, 1972-75; contbr. chpts. to books and articles to profl. jours. MIT Company Founders citation, 1989; NIH predoctoral fellow, 1958-62; grantee Am. Heart Assn., 1969-81, USPHS, 1970-81. Fellow Am. Inst. Chemists; mem. AAAS, Am. Soc. for Biochemistry and Molecular Biology, Am. Assn. Immunologists, U.S. Power Squadron (comdr.), Charles River Squadron (comdr.), MIT Enterprise Forum (treas., exec. bd.), Clin. Ligand Assay Soc. (exec. bd. New Eng. chpt. 1987-91). Avocations: photography, sailing, fine arts. Office: Tech Evaluations PO Box 23 Lexington MA 02173

ČATIC, IGOR JULIO, mechanical engineering educator; b. Zagreb, Croatia, Mar. 14, 1936; s. Julio Ohran and Darinka (Vrus) Č; m. Ranka Mijo Brcic, June 23, 1963. Dipl.Ing., Faculty Mech. Engring., Zagreb, 1960, MSc, 1970, Dr.Ing., 1972. Mould designer Štanca, Zagreb, 1954-60; devel. engr. Me-Ga, Zagreb, 1960-63, TOZ, Zagreb, 1963-65; asst. Faculty Mech. Engring., Zagreb, 1965-74, asst. prof., 1974-80, assoc. prof., 1980-86, prof., 1986—; cons. TOZ, 1970—, Tvornica strojeva Belisce, 1978—. Author: Heat Exchange in Mould for Injection Moulding, 1985, System Analysis of Injection Moulding of Polymers, 1991; editor proc. in field; contbr. articles to profl. publs. Recipient Sci. award Nikola Tesla, Parliament of Croatia, 1977, Golden Wreath, Pres. of Yugoslavia, 1983; grantee CEMP, 1964, Deutsche Acad. Austauschdienst, 1968, Alexander von Humboldt Found., 1970-72, 86. Fellow Inst. of Materials; mem. Soc. Plastics and Rubber Engring. (v.p. 1991—, founding mem., award 1983, Sci. award 1987), Gesellschaft Kunststofftechnik, Club Croatians Humboldt (sec. 1992—, founding mem.). Roman Catholic. Home: Fancevljev prilaz 9, Zagreb 41010, Croatia Office: Fakultet strojarstva, Salajeva 1, Zagreb 41000, Croatia

CATOE, BETTE LORRINA, physician, health educator; b. Washington, Apr. 7, 1926; d. John Booker and Laura Beola (Adams) C.; B.S. cum laude,

Howard U., 1948, M.D., 1951; m. Warren J. Strudwick, Sept. 17, 1949; children—Laura Christina, Warren J., William J. Intern, Freedmen's Hosp., Washington, 1951-52; pediatric resident Howard U. Freedman's Hosp., 1952-55; practice medicine specializing in pediatrics, Washington, 1956—; instr. bacteriology Howard U., 1955-57; mem. staff Providence Hosp., Columbia Hosp., Howard U. Hosp., Washington Hosp. Center; sch. health officer Dept. Health, Washington, 1960-64; clin. instr. Howard U., 1956-58. Mem. D.C. Health Planning Adv. Council, 1967-77, chmn., 1973-77; chmn. D.C. Devel. Disabilities Adv. Council, 1970-74; mem. D.C. Mayor's Commn. on Food and Nutrition, 1971-72, Mayor's Commn. on Maternal and Child Health. 1978-84; mem. D.C. Commn. Jud. Tenure and Disabilities, 1977—, chair, 1984—; bd. dirs. United Way of Nat. Capital Area, 1974-76, chmn. social planning com., 1974-75; bd. govs. St. Alban's Sch., 1978-84; bd. dirs. D.C. Health and Welfare Council, 1968-73, pres., 1973-74; del. Democratic Nat. Conv., 1976; bd. dirs. Met. Washington Health and Welfare Council, 1970-72, Parent Council of Washington, 1974-75, Met. Med. Founds., Inc., Silver Spring YMCA, 1977-80. Mem. AMA, D.C. Chirurg. Soc., D.C. Med. Soc., Nat. Med. Assn. (chmn. pediatric sect. 1981-83), Am. Med. Women's Assn., NAACP, Urban League, Assn. Comprehensive Health Planners (dir. 1975-77), Women's Aux. Medico-Chirurg. Soc., Jack and Jill Am., Century Club of Nat. Assn. Negro Bus. and Profl. Women's Clubs (pres. 1985-89), Alpha Kappa Alpha. Baptist. Clubs: Links, Carrousels (nat. v.p. 1986-88, nat. pres. 1988-90), Women's Nat. Dem. Home: 1748 Sycamore St NW Washington DC 20012-1031 Office: 5505 5th St NW Washington DC 20011-6513

CATRAN, JACK, aerospace and physiology scientist, writer; b. N.Y.C., Jan. 22, 1918; s. Joseph and Rachel (Ohana) C.; m. Gladys Louise Goldfin, Apr. 12, 1945; children—Diane Lee, Steven Jay, Rachel Ann, Lisa Beth. Ph.D. in Exptl. Psychology, Sussex Coll. Tech., Eng., 1962. Cons. human factors, man-machine systems, life support, behavioral sci. Apollo Space Program Hughes N.Am., Calif., 1941—; cons. Hughes Aerospace, Culver City, Calif., 1962-70. Author: Is There Intelligent Life on Earth, 1981, Walden Three, 1985; How to Speak English Without a Foreign Accent, 15 lang. edits.; newspaper columnist; contbr. articles to profl. jours. Decorated USAF Commendation medal. Mem. Aviation and Space Writers Assn. Club: Mensa. Home: 8758 Sophia Ave Sepulveda CA 91343-4718 Office: Sci Features Syndicate PO Box 5567 Sherman Oaks CA 91413-5567

CATTELL, HEATHER BIRKETT, psychologist; b. Carlisle, eng., Dec. 16, 1936; came to U.S., 1955; d. Wilfred B. and Anne Birkett; m. Russel B. Shields, June 10, 1953 (div. 1963); children: Vaughn, Gary, Heather Luanne; m. Raymond B. Cattell, May 9, 1981. BA, U. Hawaii, 1974, MA, 1977, PhD, 1979. Lic. clin. psychologist, Hawaii. Dir. rsch. Salvation Army, Honolulu, 1979-81; pvt. practice Honolulu, 1981—; lectr., workshop leader, U.S., Australia, Can., and United Kingdom, 1989—. Author: The 16PF: Personality in Depth, 1989. Mem. Phi Beta Kappa. Office: 1188 Bishop St Ste 1702 Honolulu HI 96813-1551

CAUGHLIN, DONALD JOSEPH, JR., engineering educator, research scientist, experimental test pilot; b. San Pedro, Calif., Dec. 17, 1946; s. Donald Joseph Sr. and Eileen Teresa (Mangan) C.; m. Barbara Bell Schultz, Aug. 18, 1973; children: Amy Marie, Jon Andrew. MBA, U. Utah, 1975; PhD, U. Fla., 1988. Instr. pilot 3525th Pilot Tng. Wing, Williams AFB, Ariz., 1969-71; A-1 skyraider pilot 1st Spl. Ops. Squadron, Thailand, 1972; air ops. officer 3rd Tactical Fighter Wing, Kunson AB, Korea, 1973; program mgr. USAF Tactical Air Warfare Ctr., Eglin AFB, Fla., 1976-78; aerospace rsch. test pilot Test Ops. 3246th Test Wing, Eglin AFB, Fla., 1979-80, chief flight mgmt. sect., 1980-81; dir. LANTIRN programs 3246th Test Wing, Eglin AFB, Fla., 1979-82; chief resource div. USAF Test Pilot Sch., Edwards AFB, Calif., 1983-85; F-16E(XL) combined test force dir. 6510th Test Wing, Edwards AFB, 1985-86; asst. for sr. officer mgmt. Hdqrs. Air Force systems Command, Andrews AFB, Md., 1986-87; chief test and evaluation br. Directorate Pentagon Advanced Programs Directorate, Washington, 1987-88; dir. ops. Strategic, SOF, and Airlift Programs Directorate, Eglin AFB, 1988-90; assoc. dean Sch. Engring. Air Force Inst. Tech., Wright Patterson AFB, Ohio, 1991-92; sr. scientist Mission Rsch. Corp., Colorado Springs, Colo., 1991—; F-16 avionics review team mem. 3246th Test Wing, Eglin AFB, 1979-80, night attack quick look flight test team, 1980-81, night attack workload stering group, 1979-82; variable inflight stability test aircraft steering group USAF Test Pilot Sch., Edwards AFB, 1983-85; electronic combat devel. test review panel The Pentagon, Washington, 1988. Col. USAF, 1964—. AEC Nuclear Sci. and Engring. fellow Atomic Energy Commn., 1968; Disting. grad. USAF Test Pilot Sch., Edwards AFB, Calif., 1978. Mem. AIAA (sr.), IEEE (sr.), Soc. Exptl. Test Pilots, KC. Roman Catholic. Home: 235 Cobblestone Way Monument CO 80132 Office: Mission Rsch Corp 4445 N 30th St Colorado Springs CO 80919

CAULDER, JERRY DALE, weed scientist; b. Gideon, Mo., Nov. 7, 1942. BA & BS, Southeast Mo. State U., 1964; MS, U. Mo., 1966, PhD in Agronomics, 1969. Rsch. asst. weed sci. U. Mo., 1966-70; mktg. devel. specialist Monsanto Co., 1969-71; mgr. Monsanto Co., Columbia, 1971-73; devel. assoc. Monsanto Co., 1973, tech. mgr. herbicides, 1973-74, mew product mgr., 1974-84; pres. Mycogen Corp., San Diego, 1984—. Achievements include research in coordination of the discovery, development and manufacture of herbicides and plant growth regulators. Office: Mycogen Corp 5451 Oberlin Dr San Diego CA 92121*

CAULFIELD, HENRY JOHN, physics educator. BA in Physics, Rice U., 1958; PhD in Physics, Iowa State U., 1962. Staff scientist Cen. Rsch. Labs. Tex. Instrument Inc., Dallas, 1962-67; tech. dirs. night vision dept. Raytheon Co., Melville, N.Y., 1968; prin. scientist Sperry Rand Rsch. Ctr., 1968-72; mgr. laser tech. dept. Block Engring. Inc., 1972-77; pres. Innovative Optics, Caulfield Cons., 1981-85; dir Ctr. for Applied Optics U. Ala., Huntsville, 1985-91; University Eminent Scholar, prof. physics Ala. Agrl. and Mech. U. Normal, 1991—; adj. prof. U. Louis Pasteur, France, Degli Studi di Cagliari, Italy, U. Ala. Huntsville; bd. dirs. Helen Keller Eye Rsch. Inst., Innovision, Phys. Optics. Corp., AL Cryogenic Engring., Nodal Systems Corp., AIA Optical Info. Processing Team; mem. patent bd. Applied Optics, evaluation bd. NAS, NBS, adv. com. Mus. Holography, N.Y.C., hon. exec. bd. Nat. Inst. for Emerging Tech., Frederick, Md.; chmn. bd. Loki; curator Holography Works, N.Y. Mus. Holography, 1984; tchr. course Optical Info. Processing, Munich, Paris, Longon, Tel Aviv, 1986; tchr. course Strategic Infrared Systems, U. Ala. Huntsville, 1986. Author: (with Sun Lu) The Applications of Holography, 1972, (with Gregory Gheen) Optical Computing, 1990; editor: (with Jean Robillard) Industrial Applications of Holography, 1990, editor Optical Memory and Neural Networks; former editor Optical Engring.; me. editorial bd. Fiber and Integrated Optics (former), Applied Optics (info. processing editoral adv. bd., Laser Focus, Holography News, Microwave and Optical Tech. Letters, Jour. Neural network Computing, Internat. Jour. Optical Computint, Optical Computingand Processing, Jour. Math. Imaging; series editor PWS Kent Pub. Co., 1988; publ. bd. Acad. Press; contbr. articles to profl. jours., chpts. to books, presentations to confs. and seminars; patentee in field. Fellow Optical Soc. Am. (Eastman lectr. 1982, 92), Soc. Photooptical-Instrumentational Engrs. (Svc. award, Gov.'s award, pres.'s award; mem. IEEE (sr.). Address: Ala A&M U PO Box 1268 Normal AL 35762

CAUNA, NIKOLAJS, physician, medical educator; b. Riga, Latvia, Apr. 4, 1914; came to U.S., 1961; s. Nikolajs and Marija (Manika) C.; m. Dzidra Priede, June 23, 1942. M.D., U. Latvia, 1942; M.Sc., U. Durham (Eng.), 1954, D.Sc., 1961. Lectr. anatomy U. Latvia, Riga, 1942-44; gen. practice medicine Sarsted and Eschershausen, West Germany, 1944-46; acting chmn. anatomy dept. Baltic U., Hamburg, Germany, 1946-48; lectr. anatomy Med. Sch. U. Durham (Eng.), 1948-57, reader, 1958-61; prof. anatomy Sch. Medicine U. Pitts., 1961-84, chmn., 1975-83, prof. emeritus, 1984—. Mem. editorial bd. Anat. Record, 1969-91, Histology and Histopathology, 1985-90; contbr. articles to profl. jours. Recipient Golden Apple award (tchr. of year) U. Pitts., 1964, 67, 73; research grantee Royal Soc. Eng., 1958-60; USPHS grantee, 1962-82; Am. Cancer Inst. grantee, 1961. Mem. AAAS, Anat. Soc. Gt. Britain and Ireland, Am. Assn. Anatomists, Royal Micros. Soc., Anatomische Gesellschaft, Histochem. Soc., Am. Soc. Cell Biology, Internat. Assn. for Study Pain. Achievements include research in normal and pathol. sensory receptor organs, in autonomic control mechanism, in devel. and evolution of sense organs and limbs. Home: 5850 Meridian Rd Apt 311C Gibsonia PA 15044-9605

CAVALET, JAMES ROGER, engineering executive, consultant; b. Dean, Pa., Jan. 5, 1942; s. Irvin Gordon and Elizabeth Ann (Nevling) C.; m. Margaret Joan Burkey, June 17, 1961; children: Peggy Ann, James Jr., Beth Ann, Deborah. Assoc. Mech. Engring., Pa. State U., 1964; MCE, U. Pitts., 1981. Registered profl. engr., Pa., Ohio, W.Va., Minn., La., Tex., Va., WIs., Ala., N.J. Chief engr. Acme Design Co., Pitts., 1966-67; asst. mgr. civil engring. Auburn Engring., Inc., Pitts., 1967-69; asst. chief civil engr. Dravo Corp., Pitts., 1969-76; chief engr. Emp Projetos div. Dravo Corp., Belo Horizonte, Brazil, 1976-77; asst. chief facility engring. Dravo Corp., Pitts., 1977-80, chief design engr., 1980-82; div. mgr. Sci. Applications Internat., San Diego, 1982-88; v.p., gen. mgr. SEI Engrs. & Cons., Inc., Pitts., 1988-90, pres., 1990—; bd. dirs. SEI Engrs. & Cons., Inc. Contbr. articles to profl. jours. Mem. ASCE, Soc. Mining Engrs., Nat. Assn. Investment Clubs, Nat. Rifle Assn. Avocations: hunting, swimming, archery, reading. Home: 1438 Swede Hill Rd Greensburg PA 15601-4748 Office: SEI Engrs & Cons Inc 300 6th Ave Pittsburgh PA 15222-2511

CAVALLARO, JOSEPH JOHN, microbiologist; b. Lawrence, Mass., Mar. 18, 1932; s. John and Salvatrice (Zappala) C.; m. Kathleen Frances Kraus, Dec. 2, 1972; children: Theresa Margaret, Sandra Marie, Elizabeth Camille, Danielle Kay, Gina Kathleen. BS, Tufts U., 1952; MS, U. Mass., 1954; PhD, U. Mich., 1966. Pub. health sanitarian Hartford (Conn.) Health Dept., 1954-55, 57-61; teaching assoc. dept. microbiology U. Mass., Amherst, 1961-62; rsch. virologist Med. Rsch. Labs., Charles Pfizer & Co., Groton, Conn., 1966-67; rsch. assoc. dept. epidemiology Sch. Pub. Health, U. Mich., Ann Arbor, 1967-70; microbiologist, diagnostic immunology tng. br. Ctrs. for Disease Control, Atlanta, 1971-86, research microbiologist anaerobic bacteria br., Ctrs. for Disease Control, 1986—; lectr. resident pathologists Grady Meml. Hosp., Atlanta, 1975; asst. prof. pathology Morehouse Sch. Medicine, 1982-85, clin. assoc. prof., 1986—; adj. asst. prof. pathology and lab. medicine Emory U. Sch. of Medicine, 1985—; cons. Pan Am. Health Orgn., Colombia and Brazil, 1976, 77. Served with M.C., AUS, 1955-57. Registered specialist microbiologist Nat. Registry Microbiologist, Am. Acad. Microbiology. Fellow Am. Acad. Microbiology; mem. Am. Soc. Microbiology, Am. Assn. Immunologists, N.Y. Acad. Sci., KC, Sigma Xi. Democrat. Roman Catholic. Prin. author/co-author over 11 lab. manuals; contbr. articles to profl. jours., chpts. to books. Home: 1325 Balsam Dr Decatur GA 30033-2905 Office: 1600 Clifton Rd Atlanta GA 30333

CAVE, DAVID RALPH, gastroenterologist, educator; b. Tunbridge Wells, Kent, Eng., Dec. 15, 1946; came to U.S., 1976; m. Anne Edwina Jones. MB BS, U. London, 1970, PhD, 1976, MRCP, 1976. Intern Hayday Hosp., Croydon, England, 1970-71, St. Georges Hosp., London, 1971; resident U. Chgo., 1976-77; asst. prof. medicine Boston U. Med. Ctr., 1983-90, assoc. prof. medicine, 1990-92; chief divsn. of gastroenterology St. Elizabeth's Hosp., Boston, 1992—. Author papers in field. Fellow ACP; mem. New Eng. Endoscopy Soc. (sec.-treas. 1987—), Royal Coll. Physicians, Am. Gastroenterol. Assn., Am. Soc. Gastrointestinal Endoscopy. Office: St. Elizabeths Hosp 736 Cambridge St Boston MA 02135

CAVERLY, ROBERT H., electrical engineer, educator; b. Cin., June 29, 1954. BS, N.C. State U., 1976, MS, 1978; PhD, Johns Hopkins U., 1983. Assoc. prof. U. Mass., North Dartmouth, 1983-88, 1988—; cons. M/A Com, Burlington, Mass., 1983—. Contbr. articles to profl. jours. Mem. IEEE (mem. editorial bd. Microwave Transactions 1990—, sr. mem. 1992), Am. Soc. Engring. Edn. (Outstanding Young Faculty 1987). Office: U Mass North Dartmouth MA 02747

CAVICCHI, LESLIE SCOTT, health facility administrator; b. Plymouth, Mass., May 22, 1954; s. Alphonso John Jr. and Mary Louise (Brookings) C.; m. Christine Anne Lafayette, Apr. 9, 1977; children: Douglas Clifton Cushing, Jarrod Scott. BS, Stonehill Coll., 1976; MPA, Suffolk U., 1988. Lic. nursing home administrator, Mass. Respiratory therapist Mass., 1976-77; supr. South Shore Hosp., Weymouth, Mass., 1978; mgr. Brockton (Mass.) Hosp., 1978-83, dir. purchasing, 1984; dir. Start program Lakeville (Mass.) Hosp., 1984-86, assoc. dir., 1986-89; sr. mgr. Health Care Svcs. of N.E., Braintree, Mass., 1989-91; material adminstr. HMO Blue Cross/Blue Shield, Framingham, 1991-92; v.p. Tech. Aid Corp., Newton, Mass., 1992—; cons., distbr. Success Motivation Inst., Waco, Tex., 1987-90. Contbr. articles to profl. jours. Mem. exec. bd. Am. Cancer Soc., 1984—, chmn. pub. issues com., 1992, Mass. unit pres., 1986-88. Mem. Hosp. Purchase Assn. Mass., Health Care Material Mgmt. Soc. (pres. 1991). Avocations: computer scis., travel, art appreciation, music. Home: 60 Indian Pond Rd Kingston MA 02364 Office: Tech Aid Corp Med Divsn 60 Wells Ave Newton MA 02159

CAVIGLI, HENRY JAMES, petroleum engineer; b. Colfax, Calif., Mar. 14, 1914; s. Giovanni and Angelina (Giachi) C.; m. Ruth Loree Denton, June 11, 1942; children: Henry James Jr., Robert D., Paul R., Loree Ann McIntire. BS in Petroleum Engring., U. Calif., Berkeley, 1937, MS in Mech. Engring., 1947. Sr. engr. Chevron Corp., Rio Vista, Calif., 1954-57, supt. No. Calif., 1958-69; mgr. non operated joint ventures Chevron Corp., LaHabra, Calif., 1970-76; cons. Cavigli & Mee, petroleum cons., Sacramento, Calif., 1976—. Mem. sch. bd. Rio Vista High Sch., 1962-67. Maj. USAF, 1942-47. Decorated Bronze Star with 4 oak leaf clusters. Mem. Soc. Petroleum Engrs., Petroleum Prodn. Pioneers, Calif. Conservation Commn. Oil Producers (chmn. 1971-72), Sutter Club, C. of C., Lion, Sigma Xi, Theta Tau Epsilon. Republican. Roman Catholic. Achievements include research in mech. sampling-field oil tanks, determination of minimum chem., productivity index of pumping wells, rotating piston pressure recorder. Home: 6271 Eichler Sacramento CA 95831 Office: Cavigli & Mee PO Box 22815 Sacramento CA 95822

CAWLEY, CHARLES NASH, enviromental scientist; b. Shreveport, La., Aug. 21, 1937; s. Charles Preston and Carnall (Nash) C. BA, U. Okla., 1960, MA, 1970; MS, U. Tex.-Dallas, Richardson, 1976, PhD, 1978. Registered environ. mgr., environ. profl. Project leader Tex. Woman's U. Research Inst., Denton, 1964-73, gen. ptnr. Southwest Textile Lab., Denton, 1973-77; research assoc. U. Tex.-Dallas, 1978; asst. prof. Cornell U., Ithaca, N.Y., 1979-83; prin. scientist Hanford Ops. div. Rockwell Internat., Richland, Wash., 1983-87, Westinghouse Corp., Richland, 1987-88; licensing supr. Bechtel Nat. Inc., Oak Ridge, 1988—; cons. Saint St. Cons., Oak Ridge, Tenn., 1984—. Contbr. articles in field to profl. jours. Served with U.S. Army, 1962-64. Mem. AAAS, Soc. Risk Analysis, N.Y. Acad. Sci., Am. Nuclear Soc., Health Physics Soc., Nat. Assn. Environ. Profls. (registered environ. mgr.), Nat. Registry Environ. Profls. (cert. environ. profl.), Am. Statis. Assn., Sigma Xi, Delta Upsilon. Home: 130 Brandeis Ln Oak Ridge TN 37830-7601 Office: Bechtel Environmental Inc Oak Ridge TN 37831-0350

CAWNS, ALBERT EDWARD, computer systems consultant; b. Houston, Apr. 3, 1937; s. Harry William and Blanche Ophelia (Bays) C.; m. Sheila Mathie Climie, June 24, 1961; children: Elizabeth Carrick, Jennifer Kathryn. AB in Math., Drury Coll., Springfield, Mo., 1958; BS in Mech. Engring., U. Mo., Rolla, 1959, MS in Computer Sci., 1984; M Engring. Adminstrn., Washington U., St. Louis, 1965. Engr. White Rodgers Co., St. Louis, 1959-62, McDonnell Aircraft Co., St. Louis, 1962-64; v.p. Thomas Inc., St. Louis, 1964-82; pres. Talos Co., St. Louis, 1982—; adj. faculty Webster U., St. Louis, 1986-91, asst. prof. math. and computer sci., 1991—. Moderator Southeast Mo. Presbytery, 1971; mem. Gen. Assembly Mission Bd., Presbyn. Ch. U.S.A., 1973; trustee Westminster Presbyn. Ch., St. Louis, 1976; pres. alumni adv. coun. Sch. Engring., Washington U., St. Louis, 1990-92. Cpl. USMCR, 1954-62. Home: 7391 Stratford Ave Saint Louis MO 63130-4138 Office: Talos Co PO Box 3069 Saint Louis MO 63130-0469

CAYWOOD, JAMES ALEXANDER, III, transportation engineering company executive, civil engineer; b. Kona, Ky., Jan. 28, 1923; s. James Alexander and Mary Viola (Crawford) C.; m. Carol Ann Fries, Mar. 20, 1959; children: Daniel, Malinda, Elizabeth; children from previous marriage: Beverly, James. B.S.C.E., U.Ky., 1944. Registered profl. engr., 50 states. Asst. engr., sr. instrumentman Louisville & Nashville R.R., Ky., 1946-47; chief engr., gen. mgr. C & O-B R.R., 1961-64; pres., dir. Royce Kershaw Co., 1964-65; v.p., then exec. v.p. De Leuw, Cather & Co., Washington, 1965-78, pres., 1978—. Served to lt. USN, 1944-46. Inducted into Hall of Disting. Alumni U. Ky., Lexington, 1985. Mem. Am. Pub. Transit Assn., Am. Ry. Bridge and Bldg. Assn., Am. Ry. Engring. Assn., Am. Rd. and Transp. Builders Assn. (Guy Kelcey award 1978, chmn., Washington, 1982), ASCE (John I. Parcel - Leif J. Sverdrup Civil Engr. Mgmt. award 1978, Laurie Prize, 1990), D.C. Soc. Profl. Engrs., Inst. Transp. Engrs., Nat. Soc. Profl. Engrs., Roadmasters and Maintenance of Way Assn., Soc. Am. Mil. Engrs., The Moles. Clubs: University, Burning Tree (Washington). Office: De Leuw Cather & Co 1133 15th St NW Washington DC 20005-2710

CAZES, JACK, chemist, marketing consultant, editor; b. N.Y.C., Feb. 2, 1934; s. Angel and Esther (Calderon) C.; m. Eleanor Harriet Schwartz, Mar. 25, 1961; children: Elliot Evan, Larry Alan. BS in Chemistry, CCNY, 1955; MS in Organic Chemistry, NYU, 1962, PhD in Organic Chemistry, 1963. Sr. rsch. chemist Mobil Chem. Co., Edison, N.J., 1963-69; supervising chemist Mobil Oil Corp., Paulsboro, N.J., 1970-74; with Waters Assocs., Inc., Milford, Mass., 1974-81; v.p. Marcel Dekker, Inc., N.Y.C., 1981-83, Elf Aquitaine, S.A., N.Y.C., 1983-84, Varex Corp., Rockville, Md., 1984-87, Sanki Labs., Inc., Mt. Laurel, N.J., 1987—; instr. Rutgers U., New Brunswick, N.J., 1963-69; lectr. Queens Coll., Flushing, N.Y., 1961-62; cons. various cos., 1959—. Editor-in-chief Jour. Liquid Chromatography, 1973—, numerous chromatographic sci. books; contbr. articles to profl. jours. With U.S. Army, 1955-57. Mem. Am. Chem. Soc. (prof. short course 1967-71), Sigma Xi. Jewish. Achievements include research on chromatography theory and applications. Avocations: electronics, computers. Home: 107 Society Hill Blvd Cherry Hill NJ 08003-2402 Office: Sanki Labs Inc 520 Fellowship Rd Ste 406 Mount Laurel NJ 08054-3410

CEASOR, AUGUSTA CASEY, medical technologist, microbiologist; b. Birmingham, Ala., Feb. 22, 1943; d. Augustus and Willie Mae (Stubbs) C. AS, SUNY, 1981; BS, So. Ill. U., 1981. Cert. med. technologist, Washington. Lab. asst. Mt. Sinai Hosp., Miami Beach, Fla., 1967-68; lab. technician Coordinated Lab. Svcs., Jamaica, N.Y., 1969-71; med. technician Andrew Radar U.S. Army Health Clinic, Ft. Myer, Va., 1972-76; med. technologist Armed Forces Inst. Pathology, Washington, 1976-91, Dept. Army, Mil. Dist. Wash., Washington, 1991—; sci. fair judge Am. Soc. Microbiology, Washington, 1988—; high sch. sci. mentor Minority Women in Sci., 1989—; speaker to profl. groups. Mem. editorial bd. Metroscope Newsletter, 1985—, editor, 1989—; tech. asst. Mycobacteriology Rsch., 1985-90. Active minority alumni scholarship com. So. Ill. U., Carbondale, Ill., 1981—; mem. Montgomery Knolls Community Assn., Silver Spring, Md., 1983—, v.p., chairperson safety & environ. com., 1984-85. Recipient Cert. of Meritorious Svc., Dept. of Army, 1991. Fellow Alpha Mu Tau; mem. Am. Soc. Med. Tech. (mem. Region II Coun. 1986-93, Region II microbiology chair 1988-89, Region II membership chair 1990-93, Cert. of Recognition 1990), D.C. Soc. Med. Tech. (profl. and pub. rels. chair 1985-86, 92-93, program com. chair 1986-87, pres. 1987-88, microbiology chair 1988-89, awards chair 1988—, Svc. award 1989, Mem. of Yr. 1989-90, Past Pres. award 1988, Disting. Svc. award 1991), Omicron Sigma (award 1988-93). Democrat. Roman Catholic. Achievements include research in unique toxin of mycobacterium ulcerans. Home: 9114 September Ln Silver Spring MD 20901-3705

CEBALLOS, GERARDO, biology educator, researcher; b. Toluca, Mex., Oct. 3, 1958; s. Oscar Ceballos and Leonor Gonozalez; m. Guadalupe Mondragón, Dec. 15, 1983; 1 child, Pablo. MSc, U. Wales, Bangor, 1982; PhD, U. Ariz., 1989. Asst. prof. Inst. Ecology, Mexico City, 1979-80, U. Ariz., Tucson, 1984-87; vis. prof. Nat. U. Mex., Mexico City, 1988, prof. biology, 1990—; vis. prof. U. N.Mex., Albuquerque, 1989—; mem. adv. bd. Agrupacion Sierra Madre, Mexico City, 1990—; cons. Mexican Ministry Ecology, Mexico City, 1990—; mem. Fundacion Ecologica Cuixmala, Mexico City, 1988—; advisor U.S. Fish and Wildlife Svc., 1990—. Co-author: Mammals of the Mexico Basin, 1984, Mammals from Chamela, Jalisco, 1988, Wildlife Conservation in Mexico, 1991; contbr. articles to profl. jours. Mem. Com. for Protection Mexican Pacific, Mexico City, 1990—. Rsch. grantee Nat. Coun. Sci. and Tech., 1980-82, World Wildlife Fund, 1987, Wildlife Preservation Trust Internat., 1988; study grantee Nat. U. Mexico, 1984-89. Mem. Mexican Assn. Mammalogists (founder, pres. 1988-89, editor 1990-93), Internat. Union for Conservation (group coord. 1989—). Achievements include creation of Chamela-Cuixmala Biosphere Reserve, Jalisco; research on role of seasonality of climate in dry tropical forest for ecology and conservation of mammals; founder program to protect endangered mammals from extinction in Mexico. Home: Manuel Sotelo 421, 50120 Toluca Mexico Office: Nat U Mex Ctr Ecology, Apartado Postal 70-275, 04510 Mexico City Mexico

CECH, THOMAS ROBERT, chemistry and biochemistry educator; b. Chicago, Ill., Dec. 8, 1947; m. Carol Lynn Martinson; children: Allison E., Jennifer N. BA in Chemistry, Grinnell Coll., 1970, DSc (hon.), 1987; PhD in Chem., U. Calif., Berkeley, 1975; DSc (hon.), U. Chgo., 1991. Postdoctoral fellow dept. biology MIT, Cambridge, Mass., 1975-77; from asst. prof. to assoc. prof. chemistry U. Colo., Boulder, 1978-83, prof. chemistry and biochemistry also molecular cellular and devel. biology, 1983—, disting. prof., 1990—; research prof. Am. Cancer Soc., 1987—; investigator Howard Hughes Med. Inst., 1988—; co-chmn. Nucleic Acids Gordon Conf., 1984; Phillips disting. visitor Haverford Coll., 1984; Vivian Ernst meml. lectr. Brandeis U., 1984, Cynthia Chan meml. lectr. U. Calif., Berkeley; mem. Welch Found. Symposium, 1985; Danforth lectr. Grinnell Coll, 1986; Pfizer lectr. Harvard U., 1986; Verna and Marrs McLean lectr. Baylor Coll. Medicine, 1987; Harvey lectr., 1987; Mayer lectr. MIT, 1987; Martin D. Kamen disting. lectureship, U. Calif., San Diego, 1988; Alfred Burger lectr. U. Va., 1988; Berzelius lectr. Karolinska Inst., 1988; Osamu Hayaishi lectr. Internat. Union Biochemistry, Prague, 1988; Beckman lectr. U. Utah, 1989, HHMI lectr. MIT, 1989; Max Tishler lectr. Merck, 1989; Abbott vis. scholar U. Chgo., 1989; Herriott lectr. Johns Hopkins U., 1990; J.T. Baker lectr., 1990; G.N. Lewis lectr. U. Calif., Berkeley, 1990; Sonneborn lectr. Ind. U., 1991; Sternbach lectr. Yale U., 1991; W. Pauli lectr., Zürich, 1992; Carter-Wallace lectr. Princeton U., 1992; Hastings lectr. Harvard U., 1992; Stetten lectr. NIH, 1992; Dauben lectr. U. Wash., 1992; Marker lectr. U. Md., 1993; Hirschmann lectr. Oberlin Coll., 1993; Beach lectr. Purdue U., 1993. Assoc. editor Cell, 1986-87; mem. editorial bd. Genes and Development; dep. editor Science mag. NSF fellow, 1970-75, Pub. Health Service research fellow Nat. Cancer Inst. 1975-77, Guggenheim fellow, 1985-86; recipient medal Am. Inst. Chemists, 1970, Research Career Devel. award Nat. Cancer Inst., 1980-85, Young Sci. award Passano Found., 1984, Harrison Howe award, 1984, Pfizer award, 1985, U.S. Steel award, 1987, V.D. Mattia award, 1987, Louisa Gross Horowitz prize, 1988, Newcombe-Cleveland award AAAS, 1988, Heineken prize Royal Netherlands Acad. Arts and Scis., 1988, Gairdner Found. Internat. award, 1988, Lasker Basic Med. Rsch. award, 1988, Rosenstiel award, 1989, Warren Triennial prize, 1989, Nobel prize in Chemistry, 1989, Hopkins medal Brit. Biochemical Soc., 1992; named to Esquire Mag. Register, 1985, Westerner of Yr. Denver Post, 1986. Mem. AAAS, Am. Soc. Biochem. Molecular Biology, NAS, Am. Acad. Arts and Scis., European Molecular Biology Orgn. Office: U Colo Dept Chemistry & Biochemistry Boulder CO 80309

CEDARBAUM, JESSE MICHAEL, neurologist; b. Chgo., July 6, 1951; s. David I. and Sophia M. (Nahamkin) C.; m. Shari Koldony, Nov. 18, 1979; children: Zachary, Derek. BA, MA, Stanford U., 1973; MD, Yale U., 1978. Asst. prof. neurology Cornell U. Med. Coll., N.Y.C., 1983-89, assoc. prof. neurology, 1989-90, clin. assoc. prof. neurology, 1990—; dir. clin. rsch. Regeneron Pharm., Tarrytown, N.Y., 1990-93, v.p. clin. affairs, 1993—. Contbr. articles to profl. jours. Office: Regeneron Pharm 777 Old Saw Mill River Rd Tarrytown NY 10538

CELLURA, ANGELE RAYMOND, psychologist; b. Rochester, N.Y., Dec. 22, 1932; s. Raymond Anthony and Helen (Balistrere) C.; children: Jon, Jane, Todd. BA, St. Francis Coll., 1957; MS, L.I. U., 1960, SUNY, New Paltz, 1960; D of Edn., U. Rochester, 1965. Lic. psychologist, Mass., Ga. Psychologist City Sch. Dist., Rochester, 1961-63; sr. clin. psychologist Dept. Mental Hygeine, State of N.Y., 1964-65; asst. dir. Community Mental Health Rsch. Tng. Program, Wash. U., St. Louis, 1964-65; asst. prof. Grad. Inst. Edn. Wash. U., St. Louis, 1964-65; head Dept. of Human Devel., U. Mass., Amherst, 1965-68; asso. prof. psychology R.I. Coll., Providence, 1968-70; pres. EDPSI, Inc., Sharon, Mass., 1970-89; psychologist IV S.C. Dept. Mental Health, Columbia, 1989-91; cons. Behavior Consults, Abbeville, S.C., 1991-92; med. cons. Disability Determination S.C. Divsn. Voc. Rehab., 1992—. Contbr. articles to profl. jours. including Am. Edn. Rsch. Jour. and Am. Jour. Mental Deficiency. Mem. AAAS, Am. Psychol. Assn., N.Y. Acad. Scis. Achievements include introduction and development of concept of anterior hypothalamic type II schizophrenia; development of need potential for acad. achievement questionnaire. Office: Behavior Consults RR 3 Box 67aa Abbeville SC 29620-9803

CELNIKER, SUSAN ELIZABETH, geneticist, molecular biologist; b. Culver City, Calif., Mar. 13, 1954; d. Leo and Phyllis Mae Ann (Finneran) C.; m. Richard Francis Galle, Jan. 4, 1976; children: Samuel Ellsworth Galle, Justin Edward Galle. BA, Pitzer Coll., 1975; PhD, U. N.C., 1983. Rsch. fellow Calif. Inst. Tech., Pasadena, 1983-86, sr. rsch. fellow, 1986-92, sr. rsch. assoc., 1992—; divsn. RAD safety officer Caltech, Pasadena, 1992—. Contbr. articles to profl. jours. Recipient Nat. Rsch. Svc. award NIH, 1983-86; fellow Calbiochem, 1974-75. Mem. AAAS, Genetics Soc. Am. Achievements include cloning, sequencing and characterization of products of Drosophilia Abdominal-B gene of the bithorax complex; identification of first homeotic product found to be distributed in a graded fashion. Home: 1826 Diamond Ave South Pasadena CA 91030 Office: Calif Inst Tech Divsn Biology Pasadena CA 91125

CEPLUCH, ROBERT J., retired mechanical engineer. Registered profl. engr., Calif. Former supr. Hartford Steam Boiler Inspection and Inst. Co.; former dir. welding engring. and quality assurance ABB/Combustion Engring., Inc., Pawleys Island, S.C., contract cons.; tech. cons. Pressure Vessel Mfrs. Assn. Mem. ASME (organizer divsn. pressure vessels and piping, exec. com. 1966-71, chmn. 1970-71, boiler and pressure com., chmn. main com., vice chmn. exec. com., chair subcom. pressure vessels, bd. pressure tech. codes and standards, dept. profl. devel., Pressure Vessel and Piping award, 1990). Achievements include research in manufacturing quality assurance and control, inspection and the development of codes and standards for the pressure vessel and piping industry. Office: Lichfield Country Club PO Box 1079 223 Cypress Dr Pawleys Island SC 29585*

CERAMI, ANTHONY, biochemistry educator; b. Newark, Oct. 3, 1940; s. Anthony and Hazel (Kirk) C.; m. Helen Vlassara, May 1, 1981; children: Carla, Ethan. B.S., Rutgers U., 1962; Ph.D., Rockefeller U., 1967. Asst. prof. biochemistry Rockefeller U., N.Y.C., 1969-72, assoc. prof., 1972-78, prof., 1978-91, head lab. med. biochemistry, 1972-91, dean grad. and postgrad. studies, 1986-91; pres. Picower Inst. for Med. Rsch., 1991—. Editor Jour. Exptl. Medicine, 1981-93. Mem. NAS, AAAS, Am. Soc Biol. Chemists, Am. Soc. Pharmacology and Exptl. Therapeutics, Am. Soc. Hematology, Am. Soc. Biochemistry and Molecular Biology, Am. Diabetes Assn., Internat. Diabetes Fedn., N.Y. Acad. Sci., Clin. Immunology Soc., N.Y. Soc. Tropical Medicine, Gerontol. Soc. Am., Am. Assn. for Cancer Rsch. Home: Ram Island Dr Shelter Island NY 11964 Office: Picower Inst for Med Rsch 350 Community Dr Manhasset NY 11030

CERNUSCHI-FRIAS, BRUNO, electrical engineer; b. Montevideo, Uruguay, Apr. 7, 1952; became citizen of Argentina, 1978; s. Felix and Zulema (Frias) Cernuschi. B.E.E., U. Buenos Aires, 1976; M.E.E., Brown U., Providence, R.I., 1983, Ph. D. in Elec. Engring., 1984. Asst. faculty engring. U. Buenos Aires, Argentina, 1976-78, chief asst., 1979-83, asst. prof., 1984-87, sec. for rsch. and PhD degree studies, 1987-89, prof. dept. elec. engring., 1990—; rschr. Consejo Nacional de Investigaciones Cientificas y Tecnicas, 1984—. Contbr. articles to profl. publs. Recipient 1st Prize for ship engine control design, Secretaria de Estado de Intereses Maritimos, Buenos Aires, 1978, Bernardo Houssay prize Consejo Nacional de Investigaciones Cientificas y Tecnicas, 1987. Mem. IEEE, Assn. Computing Machinery, AAAS, Am. Math. Soc., Sigma Xi. Roman Catholic. Home: Las Heras 2269 (4A), 1127 Buenos Aires Argentina Office: Cassilla 8, Sucursal 12 B, 1412 Buenos Aires Argentina

CERNY, JOSEPH CHARLES, urologist, educator; b. Oak Park, Ill., Apr. 20, 1930; s. Joseph James and Mary (Turek) C.; m. Patti Bobette Pickens, Nov. 10, 1962; children: Joseph Charles, Rebecca Anne. BA, Knox Coll., 1952; MD, Yale U., 1956. Diplomate Am. Bd. Urology. Intern U. Mich. Hosp., Ann Arbor, 1956-57, resident, 1957-62; practice medicine specializing in urology, Ann Arbor, and Detroit since 1962—; inst. surgery (urology) U. Mich., Ann Arbor, 1962-64, asst. prof., 1964-66, assoc. prof., 1966-71, clin. prof., 1971—; chmn. dept. urology Henry Ford Hosp., Detroit, 1971—; pres. Resistors, Inc., Chgo., 1960—; cons. St. Joseph Hosp., Ann Arbor, 1973—. Mem. editorial bd. Am. Jour. Kidney Diseases, 1988—; contbr. articles to profl. jours., chpts. in books. Bd. dirs., trustee Nat. Kidney Found. Mich., Ann Arbor, 1988—, chmn. urology council 1987—, exec. com. 1987—, pres. 1988—; bd. dirs. Ann Arbor Amateur Hockey Assn., 1980-83; pres. PTO, Ann Arbor Pub. Schs., 1980. Served to lt. USNR, 1956-76. Recipient Disting. Service award Transplantation Soc. Mich., 1982. Fellow ACS (pres.-elect Mich. br. 1984-85, pres. 1985—); mem. Am. Acad. Med. Dirs., Am. Coll. Physician Execs., Internat. Soc. Urology, Am. Urol. Assn. (pres. Mich. br. 1980-81, pres. North Cen. sec. 1985-86, Manpower com. 1987-88, Jud. Rev. com. 1987-91, tech. exhibits 1987-88, fiscal affairs rev. commn. 1985-89, manpower commn. 1990-92, audit commn. 1992-93, exec. commn. 1993—, Best Sci. Exhibit award 1978, Best Sci. Films award 1980, 82), Transplantation Soc. Mich. (pres. 1983-84), ACS (pres. Mich. chpt. 1985-86), Am. Assn. Transplant Surgeons, Endocrine Surgeons, Soc. Univ. Urologists, Am. Assn. Urologic Oncology, Am. Fertility Soc, Am. Coll. Physician Execs., Am. Acad. Med. Dirs., S.W. Oncology Group. Republican. Methodist. Clubs: Barton Hills Country; Ann Arbor Raquet (Ann Arbor). Avocations: tennis, fishing, Civil War. Home: 2800 Fairlane St Ann Arbor MI 48104-4110 Office: Henry Ford Hosp Dept Urology 2799 W Grand Blvd Detroit MI 48202-2689

CEROFOLINI, GIANFRANCO, laboratory administrator; b. Milan, June 14, 1946; s. Ernesto and Angela (Senna) C.; m. Elena Lonati, Dec. 7, 1973. D Physics, U. Milan, 1970. Sr. scientist Telettra, Vimercate, Italy, 1970-77; staff scientist SGS-Thomson, Agrate, Italy, 1977-88; head lab. Enichem, Milan, 1988—; lectr. U. Pisa, Milan, Modena, Italy, 1983-91; vis. scientist Boston U., 1991—. Author: Physical Chemistry Of, In and On Silicon, 1989, Chemistry for Innovative Materials, 1991. Office: Instituto Guido Donegani, EniChem, 28100 Novara Italy

CERROLAZA, MIGUEL ENRIQUE, civil engineer, educator; b. Caracas, Venezuela, Dec. 3, 1957. BCE, Cen. U. Venezuela, 1979; MSc, Fed. U., Rio de Janeiro, 1981; PhD, Polytech. U., Madrid, 1988. Rsch. asst. sch. civil engring. Cen. U. Venezuela, 1977-79; vis. rsch. fellow U. Calif., Irvine, 1980, Fed. U., Rio de Janeiro, 1980-81; researcher Inst. for Materials and Structural Models Cen. U. Venezuela, 1981—; assoc. prof. civil engring. Cen. U. Venezuela, 1981-83, prof., 1984—; vis. prof. Poly. U., Madrid, 1985-88; advisor in non-conventional structures. Author: Structural Problems, 1981; editor: Métodos Numéricos Aplicados en Ingeniería, 1993, Contribuciones Recientes a la Ing. Estructural y Sismoresistente; contbr. chpts. to books., articles to profl. jours. Mem. Internat. Soc. Computational Mechanics in Engring., Spanish Soc. Numerical Methods in Engring., Venezuelan Soc. Engrs., Assn. for Rsch. Progress Venezuela, Venezuelan Soc. Numerical Methods in Engring. (pres.). Office: Cen U Venezuela Inst Materials of Models, PO Box 50361, Caracas 1050A, Venezuela

ČERVENY, LIBOR, chemistry educator; b. Prague, Czech Republic, July 2, 1942; s. Josef and Ludmila (Kubešová) Č; m. Milada Smitková, Dec. 2, 1971. Ing., Inst. Chem. Tech., 1964, Dr., 1969. Rschr. Rsch. Inst. of Organ. Syntheses, Pardubice, Czech Republic, 1969-70; scientific worker Inst. Chem. Tech., Prague, 1970-90, prof., 1990—; cons. Aroma, Prague, 1975. Editor: Catalytic Hydrogenation, 1986; contbr. over 200 articles to profl. jours. Recipient Nat. Prize Czech Parliament, 1987. Mem. Am. Chem. Soc. Achievements include more than 50 patents in field; research in catalysis, technology, syntheses and technologies of chemical specialties such as flavors and fragrances. Home: U smaltovny 32, 170 00 Prague 7 Czech Republic Office: Inst Chem Tech, Technická 5, 166 28 Prague Czech Republic

CERVERO, JOSE MARIA, physics educator; b. Madrid, Spain, Nov. 22, 1946; s. Jose M. and Ana (Santiago) C.; m. Pilar G. Estevez de Cervero, July 19, 1980; children: Jose M., Ana, Jorge, Elvira. Licenciado en Fisica, U. Madrid, Spain, 1968, PhD, 1973. Rsch. fellow Harvard U., Cambridge, Mass., 1976-78; assoc. prof. U. Salamanca, Spain, 1978-82; vis. fellow CERN, Geneva, Switzerland, 1982-83; prof. U. Salamanca, Spain, 1983—; vis. fellow Bristol U., U.K., 1986-88. Contbr. over 40 papers and articles to profl. jours. Fulbright fellow, 1976-78, Brit. Coun. fellow, 1986-88. Avocations: history, fgn. langs., sports. Office: Dept de Fisica Teorica, Facultad de Ciencias, 37008 Salamanca Spain

CESKA, MIROSLAV, biochemist, researcher; b. Budimerice, Nymburk, Czechoslovakia, July 23, 1932; naturalized U.S. citizen; s. Antonin and Marie (Valtrová) C.; m. Dagmar Rama, May 18, 1957 (div. Nov. 1982); children: Mirko Rene, Michele Diane, Patrick Daniel; m. Krassimira Jordan, Apr. 27, 1984 (div. Aug. 1990); 1 child, Edward Anthony. Student, Vysoka Skola Chemicko-Tech., Prague, Czechoslovakia, 1952-53, Tech. U., Berlin, 1953-55; MS, Fla. State U., 1957, PhD, 1960. Rsch. chemist Fla. State U., Tallahassee, 1960; postdoctoral fellow Free U. Brussels, Belgium, 1961-62, Max Planck Inst. for Virus Rsch., Tübingen, Fed. Republic Germany, 1962; head dept. biochemistry Pfizer Ltd., Sandwich, Eng., 1962-65; head dept. diagnostics Pharmacia AB, Uppsala, Sweden, 1966-72; rsch. asst. prof. biochemistry U. Miami (Fla.), 1965-66; scientist Sandoz Rsch. Inst., Vienna, Austria, 1972—; chem. cons. Swedish Royal Acad. Arts, Stockholm, 1966-69; chmn. 2d Internat. Congress Immunology, Brighton, Eng., 1974; co-chmn. VI Internat. Conf. on Immunofluorescence, Vienna, 1978. Author: Biological and Biomechanical Applications of Bioelectrofocusing, 1977, Methods in Enzymology, 1981, more than 100 scientific publications. Coach Swedish gymnastics Olympic team, 1967-69; internat. judge Mini-World Championship in Gymnastics, 1979, 80. Recipient citation classics Current Contents, Phila., 1986, Austrian State Prize for Rsch. in Rheumatology, 1990. Mem. AAAS, N.Y. Acad. Scis. Avocations: tennis, swimming, hiking. Home: Brabbeegasse 5, A-1220 Vienna Austria Office: 59 Brunnerstrasse, A-1235 Vienna Austria

CETEGEN, BAKI M., mechanical engineering educator; b. Istanbul, Turkey, Nov. 3, 1956; came to U.S., 1978; s. Fahri and Selma Cetegen. MS, U. Calif., Berkeley, 1979; PhD, Calif. Tech., Pasadena, 1982. Group leader Energy & Environ. Rsch. Corp., Irvine, Calif., 1982-86; rsch. fellow U. Calif., Irvine, 1986-87; asst. prof. U. Conn., Storrs, 1987-92, assoc. prof. mech. engring., 1992—. Office: U Conn Mech Engring Dept U-139ME Storrs Mansfield CT 06269-3139

CEVENINI, ROBERTO MAURO, gas and oil industry executive, entrepreneur; b. Bologna, Emilia, Italy, Oct. 28, 1957; came to U.S., 1980; s. Romano and Camilla Cevenini; m. Carol Jean Porter, Aug. 9, 1985; children: Dino, Marco. BSME, Cath. U., Caracas, 1979; BS in Indsl. Engring. cum laude, U. Miami, Fla., 1983, MS in Indsl. Engring., 1985, MBA, 1986, MS in Mgmt. Sci., 1992. Registered profl. engr., Fla. Prodn. engr. Metalmaster Prodenca, Caracas, 1978, quality control engr., 1979; mfg. mktg. engr. Metal Master Prodenca, Caracas, 1980-83; mfg. engr. Rolls-Royce, Inc., Miami, 1983-84; grad. rsch. tchr. asst. U. Miami, 1984-85; corp. mgmt. cons. Fla. Power & Light Co., Miami, 1986, nuclear plant ops. coord., 1987, prin. analytical engr., 1988-89, power plant supt., 1989-90; project mgr. and internat. cons. Qualtec Quality Svcs., Palm Beach, 1991; exploration and prodn. dir. total quality mgmt. systems Texaco, Inc., Denver, 1992—; pres. Dynamic Technologies Internat., Denver, 1993—; cons. Goodwill Industries, Miami, 1983; adj. faculty Broward C.C., Miami, 1987, Fla. Internat. U., Miami, 1987-92, U. Miami, 1990-92, Denver U., 1993—. Author: (tech. manuals) Statistical Team Leader Training, Statistical Team Member Training, Reliability Engineering, 1 & 2, Statistical Process Control, 1 & 2, Process Management, Team Facilitator, Applications Expert Engineering, Benchmarking in the Technical Fields, tech. manuals on integration of engring. reliability and indsl. stats. applications, also 16 tech. papers. Mem. Fla. Power & Light Co. Track Team, Miami, 1986, Fla. Power & Light Co. Golf League, 1986; coach Denver Little League Soccer, 1993—. Recipient Gold medal Swimming Fedn., 1971. Mem. ASME, IEEE, Soc. Petroleum Engrs., Am. Inst. Indsl. Engrs., Am. Nuclear Soc., Inst. Mgmt. Scis., Am. Soc. for Quality Control, Internat. Platform Assn., Texaco Athletic Corp. Team, Phi Kappa Phi, Tau Beta Pi. Roman Catholic. Avocations: marathon running, triathlete, reading, writing-publishing, fgn. langs. Office: Texaco Inc 4601 DTC Blvd Denver CO 80237

CHA, CHUAN-SIN, chemistry educator; b. Nanjing, Jiansu, China, Apr. 11, 1925; s. Chian Cha and Mense Hwang; m. Wan-hui Zhang, Feb. 12, 1954; children: Leping Zha, Xiao-hui Zha, Lenian Zha. BSc, Wuhan (China) U., 1950; PhD, Moscow State U., 1959. Asst. chemistry dept. Wuhan U., 1950-56, lectr., 1956-62, asst. prof., 1962-78, prof., 1978—; vis. scientist chemistry dept. Moscow State U., 1957-59; vis. scholar Electronic Design Ctr./Case Western Res. U., Cleve., 1989, chemistry dept. U. Ill., Urbana-Champaign, 1990. Mem. adv. bd. Jour. Applied Electrochemistry, 1983—, Accounts of Chem. Rsch., 1991—, Elektrohimiya, 1993—; author: (monograph) Introduction to Kinetics of Electrode Processes, 1976, 2d edit., 1987; author or co-author over 200 sci. papers. Rsch. grantee Chinese NSF, Beijing, 1982-90; recipient Achievement of Nature Sci. award Nat. Com. Scis., Beijing, 1987. Mem. Chinese Acad. Scis., Chinese Chem. Soc., Am. Chem. Soc., Internat. Electrochemistry Soc. Achievements include fundamental studies of adsorption on electrode surfaces, electrocatalysis theory and applications, high energy density lithium batteries, biomedical electrochemical sensors. Office: Wuhan U, Dept Chemistry, Wuhan 430072, China

CHA, DONG SE, economist, research institute administrator; b. Haman, Korea, Mar. 3, 1943; s. Jin-Soo and Moo-Nyun (Shin) C.; m. Young-Sook Je, Mar. 16, 1969; children: Seung-Eun, Jung-Eun, Young-Lan. BA in Econs., Seoul (Korea) Nat. U., 1965; MA in Econs., Vanderbilt U., Nashville, 1973; PhD in Econs., Vanderbilt U., 1978. Staff Bank of Korea, Seoul, 1965-67; economist Korea Exchange Bank, Seoul, 1967-75; sr. rsch. fellow Korea Inst. Econs. and Tech., Seoul, 1978-86; v.p. Korea Inst. Econs. and Tech., 1986-87; sr. rsch. fellow Sejong Inst., Seoul, 1987-88; pres. Lucky Gold Star Econ. Rsch. Inst., Seoul, 1989-93, Korea Inst. Econs. and Trade, Seoul, 1993—; lectr., Sung Kyoon Kwan U., Seoul, 1979; cons., Korean Ministry Trade and Industry, 1983, Korea Exchange Bank, Seoul, 1979-80. Author: Foreign Capital and Korean Economy, 1983, A Study of Competitiveness among Asian Countries, 1984, How To Expand Exports in Japanese Markets, 1985, International Financial Markets, 1987. Avocation: golf. Home: 22-803 Woosung Apt, Seocho-Dong Seocho-Ku, Seoul Republic of Korea Office: Korea Inst Econs and Trade, Cheongryangri-Dong, Seoul Republic of Korea

CHA, PHILIP DAO, engineering educator; b. Manila, Philippines, July 10, 1962; came to U.S., 1975; s. Hsien-Lin and Jean (Wan) C. BS, Cornell U., 1984; MS, U. Mich., 1985, PhD, 1989. Sr. rsch. engr. Ford Motor Co., Dearborn, Mich., 1989-91; asst. prof. Harvey Mudd Coll., Claremont, Calif., 1991—. Contbr. articles to profl. jours. including AIAA Jour., Jour. of Sound and Vibration, Jour. of Applied Mechanics, IEEE Transactions of Edn. Mem. ASME, IEEE (transaction on edn.). Office: Harvey Mudd Coll Claremont CA 91711

CHABOT, CHRISTOPHER CLEAVES, biology educator; b. Bangor, Maine, Mar. 21, 1961; s. Maurice Joseph and Meredyth (Cleaves) C.; m. Heather Leigh Frasier, July 7, 1990. BA, Colby Coll., 1983; PhD, U. Va., 1990. Grad. teaching fellow U. Oreg., Eugene, 1983-85; rsch. assoc. U. Va., Charlottesville, 1987-88; postdoctoral scholar Miami U., Oxford, Ohio, 1990-92; asst. prof. biology Plymouth (N.H.) State Coll., 1992—; vis. instr. Sweet Briar (Va.) Coll., 1989-90. Contbr. articles to Jour. Comparative Physiology, Physiology and Behavior, Behavioral Neurosci., Jour. Biol. Rhythms. Office: Plymouth State Coll Dept Natural Sci Plymouth NH 03264

CHACHOLIADES, MILTIADES, economics educator; b. Omodos, Limassol, Cyprus, June 22, 1937; came to U.S. 1962; s. Panagis Themistokli and Hariclea (Miltiadou) C.; m. Mary Modenos, Dec. 30, 1962; children: Lea, Marina, Linda. BA, Sch. Bus. & Econs., Athens, 1961; PhD, MIT, 1965. Asst. prof. NYU, 1965-68; vis. assoc. prof. UCLA, 1970; assoc. prof. econs. Ga. State U., Atlanta, 1968-71; prof. econs. Ga. State U., 1971-73, rsch. prof. econs., 1973-87, chmn. dept. econs., 1986-89, prof. econs., 1989-93; prof. econs. U. Cyprus, Nicosia, 1993—. Author: The Pure Theory of International Trade, 1973, Brit. edit. 1974, Internat. Monetary Theory and Policy, 1978, internat. student edit., 1978, International Trade Theory and Policy, 1978, internat. student edit., 1978, Principles of International Economics, 1981, Spanish edit., 1982, Malaysian edit. 1988, Microeconomics, 1986, Greek edit., 1989, Microeconomics: Instructors Manual, 1986, internat. econs. edit., 1990, International Economics: Instructors Manual, 1990; contbr. articles to profl. jours.; editorial advisor Greek Econ. Rev.; editorial bd. Cyprus Jour. Econs. Athens Sch. fellow, Am. Hellenic Ednl. and Welfare Fund fellow, 1962-64, MIT Sloan rsch. assistantship, 1962-63, econs. fellow, 1963-64, others. Mem. Royal Econ. Soc., Ea. Econ. Assn., Greek Econ. Assn., So. Econ. Assn., Am. Econ. Assn. Office: Univ of Cyprus, Dept Econs, Nicosia TT134, Cyprus

CHACKO, GEORGE KUTTICKAL, systems science educator, consultant; b. Trivandrum, India, July 1, 1930; came to U.S., 1953.; s. Geevarghese Kuttickal and Thankamma (Mathew) C.; m. Yo Yee, Aug. 10, 1957; children: Rajah Yee, Ashia Yo Chacko Lance. MA in Econs. and Polit. Philosophy, Madras (India) U., 1950; postgrad., St. Xavier's Coll., Calcutta, India, 1950-52; B in Commerce, Calcutta U., 1952; cert. postgrad. tng., Indian Stat. Inst., Calcutta, 1951; postgrad., Princeton U., 1953-54; PhD in Econometrics, New Sch. for Social Rsch., N.Y.C., 1959; postdoctoral, UCLA, 1961. Asst. editor Indian Fin., Calcutta, 1951-53; comml. corr. Times of India, 1953; dir. mktg. and mgmt. rsch. Royal Metal Mfg. Co., N.Y.C., 1958-60; mgr. dept. ops. rsch. Hughes Semicondr. div., Newport Beach, Calif., 1960-61; cons., 1961-62; ops. research staff cons. Union Carbide Corp., N.Y.C., 1962-63; mem. tech. staff Research Analysis Corp., McLean, Va., 1963-65, MITRE Corp., Arlington, Va., 1965-67; sr. staff scientist TRW Systems Group, Washington, 1967-70; cons. def. systems, computer, space, tech. systems and internat. devel. systems, assoc. in math. test devel. Ednl. Testing Service, Princeton, N.J., 1955-57; asst. prof. bus. adminstrn. UCLA, 1961-62; lectr. Dept. Agr. Grad. Sch., 1965-67; asst. professorial lectr. George Washington U., 1965-68; professorial lectr. Am. U., 1967-70, adj. prof., 1970; vis. prof. def. systems Mgmt. Coll., Ft. Belvoir, Va., 1972-73; vis. prof. U. So. Calif., 1970-71, prof. systems mgmt., 1971-83, prof. systems sci., 1983-93, prof. emeritus, 1993—; sr. Fulbright prof. Nat. Chengchi U., Taipei, 1983-84, sr. Fulbright rsch. prof., 1984-85; prin. investigator and program dir. Tech. Transfer Project, Taiwan Nat. Sci. Coun., 1984-85; disting. fgn. expert lectr. Taiwan Ministry Econ. Affairs, 1986; sr. vis. rsch. prof. Taiwan Nat. Sci. Coun. Nat. Chengchi U., Taipei, 1988-89; sr. vis. rsch. prof. Dah-Yeh Inst. Tech., Dah-Tsuen, Chang-Hwa, Taiwan, 1993—; v.p. program devel. Systems and Telecom. Corp., Potomac, Md., 1987-90; chief sci. cons. RJO Enterprises, Lanham, Md., 1988-89; cons. Med. Svcs. Corp. Internat., vector biology and control project U.S. Agy. for Internat. Devel., 1991; guest lectr. Tech. Univs. Tokyo, Taipei, Singapore, Dubai, Cairo, Warsaw, Budapest, Prague, Bergen, Stockholm, Helsinki, Berlin, Madras, Bombay, London, 1992, Yokohama, Taipei, Hong Kong, Kualalampore, Bangkok, Alexandria, Jerusalem, Paris, 1993—; USIA sponsored U.S. sci. emissary to Egypt, Burma, India, Singapore, 1987; USIA sponsored U.S. expert on tech. transfer and military conversion 1st Internat. Conf., Hannover, Germany, 1992; keynote speaker 2d annual conf. on mgmt. edn. in China, Taipei, Taiwan, 1989, world conf. on transition to advanced market economies, Warsaw, Poland, 1992, annual conv. Indian Inst. Indsl. Engring., Hyderabad, India, 1993. Author: 22 books in field, including Applied Statistics in Decision-Making, 1971, Computer-aided Decision-Making, 1972, Systems Approach to Public and Private Sector Problems, 1976, Operations Research Approach to Problem Formulation and Solution, 1976, Management Information Systems, 1979, Trade Drain Imperative of Technology Transfer-U.S. Taiwan Concomitant Coalitions, 1985, Robotics/Artificial Intelligence/Productivity-U.S.-Japan Concomitant Coalitions, 1986, Technology Management Applications to Corporate Markets and Military Missions, 1988, The Systems Approach to Problem-Solving from Corporate Markets to National Missions, 1989, Toward Expanding Exports Through Technology Transfer-IBM-Taiwan Concomitant Coalitions, 1989, Dynamic Program Management: From Defense Experience to Commercial Application, 1989, Decision-Making under Uncertainty: An Applied Statistics Approach, 1991, Operations Research/Management Science: Case Studies in Decision-Making under Structured Uncertainty, 1993; contbr. articles to profl. publs.; editor contbr. 20 books including The Recognition of Systems in Health Services, 1969, Reducing the Cost Space Transportation, 1969, Systems Approach to Environmental Pollution, 1972, National Organization of Health Services-U.S., USSR, China, Europe, 1979, Educational Innovation in Health Services-U.S., Europe, Middle East, Africa, 1979, Management Education in the Republic of China: Second Annual Conference, 1989, Expert Systems World Congress Proceedings, 1991, Transition to Advanced Market Economies Internat. Conf. Proceedings, 1992; guest editor Jour. Rsch. Comm. Studies, 1978-79; assoc. editor Internat. Jour. Forecasting, 1982-85. Active Nat. Presbyn. Ch., Washington, 1967-84, mem. ch. coun., 1969-71, mem. chancel choir, 1967-84, co-dean ch. family camp, 1977, coord. life abundant discovery groups, 1979; chmn. worship com. Taipei Internat. Ch., 1984, chmn. membership com., 1985, chmn. Stewardship com., 1985, chmn. com. Christian edn., 1989, Sunday Sch. supt., 1989; adult Sunday Sch. leader 4th Presbyn. Ch., Bethesda, Md., 1986-87, mem. sanctuary choir, 1985—; participant 9th Internat. Ch. Mus. Festival, Coventry Cathedral, 1992; mem. Men's Ensemble, 1986—; mem. Men's Ministry Com., 1990—; founder-dir. Prayer Power Partnership, 1990—. Recipient Gold medal Inter-Collegiate Extempore Debate in Malayalam U. Travancore, Trivandrum, India, 1945, 1st pl. Yogic Exercises Competition U. Travancore, 1946, Jr. Lectureship prize Physics Soc. U. Coll., 1946, 1st prize Inter-Varsity Debating Team Madras, 1949, NSF internat. sci. lectures award, 1982, USIA citation for invaluable contbr. to America's pub. diplomacy, 1992; Coll. scholar St. Xavier's Coll., 1950-52; Inst. fellow Indian Stat. Inst., 1951, S.E. Asia Club fellow Princeton U., 1953-54, Univ. fellow UCLA, 1961. Fellow AAAS (mem. nat. coun. 1968-73, chmn. or co-chmn. symposia 1971, 72, 74, 76, 77, 78), Am. Astronautical Soc. (v.p. publs. 1969-71, editor Tech. Newsletter 1968-72, mng. editor Jour. Astronautical Scis. 1969-75); mem. Ops. Rsch. Soc. Am. (vice chmn. com. of representation on AAAS 1972-78, mem. nat. coun. tech. sect. on health 1966-68, editor Tech. Newsletter on Health 1966-73), Washington Ops. Rsch. Coun. (trustee 1967-69, chmn. tech. colloquia 1967-68, editor Tech. Newsletter 1967-68, Banquet chmn. 1992-93), Inst. Mgmt. Scis. (rep. to Internat. Inst. for Applied Systems Analysis in Vienna, Austria 1976-77, session chmn. Athens, Greece 1977, Atlanta 1977), World Future Soc. (editorial bd. publs. 1970-71), N.Y. Acad. Scis. Democrat. Club: Kiwanis (charter 1st v.p. Costa Mesa North Club, charter pres. Friendship Heights Club, charter dir. Taipei Yang-Min Club, Capital Dist. Div. One Disting. Service award 1968, 70, capital dist. chmn. 1967, 69-70, 71-72, inter-div. chmn. Green Candle of Hope Dinner 1965-82, Friendship Heights Club Outstanding Service award 1972-73, Life Mem. award Capital Dist. Found. 1982, disting. dir. Taipei-Keystone Club 1978, spl. rep. of internat. pres. and counselor to dist. of Republic of China 1983-86, Pioneer Premier Project award Asia-Pacific Conf. 1986, Legion of Honor 1985, dir. Bethesda Club 1967-69, chmn. numerous coms., 1966—). Office: U So Calif Inst Safety and Systems Mgmt Los Angeles CA 90089-0021

CHACRON, JOSEPH, mathematics educator; b. Alexandria, Egypt, June 28, 1936; s. Elie and Israel Victorine Chacron; m. Chanel Josette, Oct. 1958; children: Eric, Valérie, Sophie. BA in Elem. Math., Alexandria French Sch.; lic. math., U. Marseilles, France; postgrad., U. Paris; PhD in Math., U. Amiens, France, 1970. Asst. U. Amiens, 1967-70, maître asst., 1970—, cours algebra; conf. speaker U. Paris. Contbr. algebra articles to profl. jours. Home: 52 Rue Cozette, 80000 Amiens France Office: Univ Amiens, 33 Rue St Leu, Amiens France

CHADHA, NAVNEET, chemical engineer, environmental consultant; b. New Delhi, India, Feb. 22, 1955; came to U.S., 1977; s. O.P. and Usha Chadha; m. Kavita Chadha; 1 child, Tanya. BSChemE, Indian Inst. Tech., 1977; MSChemE, U. Ky., 1978. Registered profl. engr., Fla., Ga., Ohio, Tenn. Process engr. Davy McKee Corp., Cleve., 1979-83; project engr. Sherex Chem. Corp., Mapleton, Ill., 1983-87; mgr. air engring. IT Corp., Knoxville, Tenn., 1987—. Contbr. articles to profl. jours. Mem. AICE (environ. divsn.). Office: IT Corp 312 Directors Dr Knoxville TN 37923

CHADIMA, SARAH ANNE, geologist; b. Newport, R.I., June 27, 1956; d. Robert Shanor and Charlotte Kay (Wilson) C.; m. Richard Horace Hammond, Mar. 16, 1985; 1 child, Paul Chadima Hammond. BS, Iowa State U., 1979, MS, 1982. Geologist S.D. Geol. Survey, Vermillion, 1983—. Mem. Clay Minerals Soc., Geol. Soc. Am., Nat. Assn. Geology Tchrs., Nat. Earth Sci. Tchrs. Assn. Office: SD Geol Survey U SD Sci Ctr Vermillion SD 57069

CHADWICK, ROBERT WILLIAM, toxicologist; b. Buffalo, Mar. 16, 1930; s. Elihu Clare and Helen Evelyn (Murray) C.; m. Claire Jeannette Crisp, Aug. 20, 1966; 1 child, Natanya Laurel. MS, Western Res. U., 1962; PhD, Utah State U., 1966. Diplomate Am. Bd. Toxicology. Chemist Republic Steel Rsch. Ctr., Independence, Ohio, 1958-62; pharmacologist PHS Pesticide Rsch. Lab., Perrine, Fla., 1966-68, FDA Perrine Primate Rsch. Lab.,

1968-69, EPA Perrine Primate Lab., 1969-73; chief of metabolic pathways sect. EPA Health Effect Rsch. Lab., Research Triangle Park, N.C., 1973—. Contbr. articles to profl. jours. including Jour. of Pesticide Toxicology and Physiology, Food, Cosmetics Toxicology, Jour. of Agricultural and Food Chemistry, Jour. of Pesticide Biochemistry and Physiology, Drug Metabolism Revs., others. With U.S. Army, 1951-53. NIH fellowship in toxicology Utah State U., 1964. Mem. Soc. of Toxicology, Am. Chem. Soc., Internat. Soc. for the Study of Xenobiotics, Assn. of Govt. Toxicologists, Fla. Chess Assn. (pres. 1972-73, plaque 1973), Sigma Xi, Pi Delta Epsilon. Democrat. Achievements include rsch. on knowledge of the organochlorine insecticide, lindane and its metabolism, interactions between environmental toxicants, dietary factors and microbial flora, pesticide metabolites and metabolic pathways. Home: 6937 Wade Dr Apex NC 27502 Office: US EPA Environ Rsch Ctr HERL/GTD/MD-68A Research Triangle Park NC 27711

CHAFFEE, FREDERIC H., JR., astronomer; b. West Point, N.Y., Apr. 4, 1941; s. Frederic H. and Winnifred (Waddell) C.; m. Holly Lowry, Nov. 19, 1969 (div. 1977). AB, Dartmouth Coll., 1963; PhD, U. Ariz., 1968. Astronomer Smithsonian Astrophys. Obs., Cambridge, Mass., 1968-70; resident astronomer Mt. Hopkins Obs., Amado, Ariz., 1970-81, resident dir., 1981-84; dir. Multiple Mirror Telescope Obs., Tucson, 1984—. Contbr. articles to sci. jours. V.p. Ariz. Friends of Chamber Music, Tucson, 1984—. With U.S. Army, 1968-1970. Mem. Am. Astron. Soc., Internat. Astron. Union, Astron. Soc. of the Pacific. Democrat. Avocations: golf, classical piano. Home: 2256 E Prince Rd Tucson AZ 85719-2002 Office: U Ariz Multiple Mirror Telescope Obs Tempe AZ 85721

CHAFFIN, DONALD B., industrial engineer, researcher; b. Sandusky, Ohio, Apr. 17, 1939; married, 1966; 1 child. B of Indsl. Engring., Gen. Motors Inst., 1962; MS in Indsl. Engring., U. Toledo, 1964; PhD in Engring., U. Mich., 1967. Jr. draftsman Mack Iron Steel Co., Ohio, 1955-57; quality ctrl. engr. New Departure Divsn. GM Corp., Ohio, 1960-62, inspection foreman, 1962-63; project engr. Micrometrical Divsn. Bendix Corp., Mich., 1963-64; asst. prof. phys. medicine U. Kans., 1967-68, asst. prof. indsl. engring., 1968-70, assoc. prof. indsl. engring. and bioengineering, 1970-77; prof. indsl. and ops. engring. U. Mich., Ann Arbor, 1977—; cons. Bendix Corp., Mich., 1964. Bioengineering grantee Western Electric Co., 1967-71, NASA, 1970-71, Aerospace Med. Rsch. Labs., 1970-71, Nat. Inst. Occupational Safety and Health, 1971-72. Mem. NSPE, Am. Inst. Indsl. Engrs. (Baker Disting. Rschr. award 1991), Brit. Ergonomics Rsch. Soc., Human Factors Soc., Biomedical Engring. Soc., Sigma Xi. Achievements include research on the effects and applications of electromyography for bettering human performance, on relation of concepts of mechanics to the study of the skeletal-muscle system; expanding the teaching of physiological, neurological, and anatomical concepts as related to the bettering of man-machine systems. Office: U Mich Ctr Ergonomics 2308 EECS Ann Arbor MI 48109-2117*

CHAHID-NOURAI, BEHROUZ J.P., economist, corporate executive; b. Teheran, Iran, Dec. 12, 1938; s. M. Hassan and Eliane Y. (Peron) Chahid-N.; m. Michele E.M. Poirel, July 6, 1962; children: Cecile, Alexis. Diplome, Institut D'Etudes Politiques, Paris, 1959; MBA, Columbia U., 1960; PhD, London Sch. Econs., 1962. Head rsch. dept. Lazard Freres & Cie, Paris, 1962-70; sr. cons. McKinsey & Co., Inc., Paris, 1970-71; dir. Pricel-Banque Morin Pons, Paris, 1972-77; chief fin. officer Michelin & Cie, Clermont-Ferrand, France, 1977-90; mng. ptnr. Financiere Indosuez S.C.A., 1990-91; adminstr. dir. gen. S.G. Warburg France; dir. S.G. Warburg & Co. Ltd., 1991-92; pres., CEO Finanval Conseil SA, 1992—; prof. École Supérieure de Sciences Economiques et Commerciales, Paris, 1966-72; affiliate prof. Institut Européen D'Administration des Affaires, Fontainebleau, France, 1972-77; prof. CEDEP, Fontainebleau, 1992—. Decorated knight Palmes Academiques. Mem. Cercle Interallie. Roman Catholic. Home: 4 Rue Alfred-Bruneau, 75016 Paris France Office: Finanval Conseil SA, 18 ave Matignon, 75008 Paris France

CHAHINE, MOUSTAFA TOUFIC, atmospheric scientist; b. Beirut, Lebanon, Jan. 1, 1935; s. Toufic M. and Hind S. (Tabbara) C.; m. Marina Bandak, Dec. 9, 1960; children: Tony T., Steve S. B.S., U. Wash., 1956, M.S., 1957; Ph.D., U. Calif., Berkeley, 1960. With Jet Propulsion Lab., Calif. Inst. Tech., Pasadena, 1960—; mgr. planetary atmospheres sect. Jet Propulsion Lab., Calif. Inst. Tech., 1975-78, sr. research scientist, mgr. earth and space scis. div., 1978-84, chief scientist, 1984—; vis. scientist MIT, 1969-70; vis. prof. Am. U., Beirut, 1971-72; regent's lectr. UCLA, 1989-90; mem. NASA Space and Earth Sci. Adv. Com., 1982-85; mem. climate rsch. com. Nat. Acad. Scis., 1985-88, bd. dirs. atmospheric scis. and climate, 1988—; chmn. sci. steering group Global Energy and Water Cycle Experiment World Meteorol. Orgn., 1988—; cons. U.S. Navy, 1972-76. Contbr. articles on atmospheric scis. to profl. jours. Recipient medal for exceptional sci. achievements NASA, 1969, NASA Outstanding Leadership medal, 1984, William T. Pecora award, 1989, Jule G. Charney award, 1991, Losey Atmospheric Scis. award AIAA, 1993. Fellow AAAS, Am. Phys. Soc., Royal Soc.; mem. Am. Meteorol. Soc., Internat. Acad. of Astronautics, Sigma Xi. Office: 4800 Oak Grove Dr Pasadena CA 91109-8099

CHAI, WINBERG, political science educator, foundation chair; b. Shanghai, China, Oct. 16, 1932; came to U.S., 1951, naturalized, 1973; s. Ch'u and Mei-en (Tsao) C.; m. Carolyn Everett, Mar. 17, 1966; children: Maria May-lee, Jeffrey Tien-yu. Student, Hartwick Coll., 1951-53; BA, Wittenberg U., 1955; MA, New Sch. Social Rsch., 1958; PhD, NYU, 1968. Lectr. New Sch. Social Rsch., 1957-61; vis. asst. prof. Drew U., 1961-62; asst. prof. Fairleigh Dickinson U., 1962-65; asst. prof. U. Redlands, 1965-68, assoc. prof., 1969-73, chmn. dept., 1970-73; prof., chmn. Asian studies CCNY, 1973-79; disting. prof. polit. sci., v.p. acad. affairs, spl. asst. to pres. U. S.D., Vermillion, 1979-82; prof. polit. sci., dir. internat. programs U. Wyo., Laramie, 1988—; chmn. Third World Conf. Found., Inc., Chgo., 1982—; pres. Wang Yu-fa Found., Taiwan, 1989—. Author: (with Ch'u Chai) The Story of Chinese Philosophy, 1961, The Changing Society of China, 1962, rev. edit., 1969, The New Politics of Communist China, 1972, The Search for a New China, 1975; editor: Essential Works of Chinese Communism, 1969, (with James C. Hsiung) Asia in the U.S. Foreign Policy, 1981, (with James C. Hsiung) U.S. Asian Relations: The National Security Paradox, 1983; (with Carolyn Chai) Beyond China's Crisis, 1989, In Search of Peace in the Middle East, 1991, Chinese Human Rights, 1993; (with Cal Clark) Political Stability and Economic Growth, 1991; co-translator: (with Ch'u Chai) A Treasury of Chinese Literature, 1965. Haynes Found. fellow, 1967, 68; Ford Found. humanities grantee, 1968, 69, Pacific Cultural Found. grantee, 1978, 86, NSF grantee, 1970, Hubert Eaton Meml. Fund grantee, 1972-73, Field Found. grantee, 1973, 75, Henry Luce Found. grantee, 1978, 80, S.D. Humanities Com. grantee, 1980, Pacific Culture Fund grantee, 1987, 90-91. Mem. Am. Assn. Chinese Studies (pres. 1978-80), AAAS, AAUP, Am. Polit. Sci. Assn., N.Y. Acad. Scis., Internat. Studies Assn., NAACP. Democrat. Home: 1071 Granito Dr Laramie WY 82070-5045 Office: PO Box 4098 Laramie WY 82071-4098

CHAIFETZ, L. JOHN, industrial engineer; b. Detroit, Aug. 10, 1950; s. Edward and Sally (Marco) C.; m. Joanne Testa, Aug. 10, 1975; children: Jason, Jordan. B of Indsl. Engring., Ga. Tech U., 1972; MBA, Ga. State U., 1981. Cert. cost engr. Project engr. Systems/Project Mgmt., Atlanta, 1973-76, DDR Internat., Atlanta, 1976-79; sr. cons., v.p. Constrn. Consulting Group, Atlanta, 1980-88; pres. Chaifetz/Brooks, Inc., Atlanta, 1989—; edn. com. chair Associated Builders and Contractors, Atlanta, 1986-87; sec., treas. Project Mgmt. Inst., Atlanta, 1987-90. Guest lectr. So. Coll. of Tech., Marietta, Ga., 1991, 92. Mem. Project Mgmt. Inst., Associated Builders and Contractors, Am. Subcontractors Assn., Am. Arbitration Assn. Republican. Jewish. Office: Chaifetz/Brooks Inc 400 Interstate N Pkwy NW Ste 880 Atlanta GA 30339

CHAIKOF, ELLIOT LORNE, vascular surgeon; b. Toronto, Ont., Can., Apr. 9, 1957; s. Leo and Bayla (Appel) C.; m. Melissa Kershman, Aug. 7, 1983; children: Rachel, Adam. BA, Johns Hopkins U., 1979, MD, 1982; PhD, MIT, 1989. Diplomate Am. Bd. Surgery. Intern Mass. Gen. Hosp./Harvard Med. Sch., Boston, 1982-83, resident in gen. surgery, 1983-85, 89-91; asst. prof. Emory U. Sch. Medicine, Atlanta, 1992—, fellow in vascular surgery, 1991-92; adj. prof. chem. engring. Ga. Inst. Tech. Contbr. to profl. publs. Mem. Am. Inst. Chem. Engrs., Am. Chem. Soc., Materials Rsch. Soc., Soc. Biomaterials, Phi Beta Kappa, Sigma Xi, Alpha Omega Alpha. Achievements include patent on antithrombogenic devices containing polysiloxanes.

CHAIYABHAT, WIN, engineering manager; b. Bangkok, Jan. 9, 1950; came to U.S., 1967; s. Thavorn and Boonchia (Khomsiri) C.; m. Nancy Benzie, June 1, 1975; children: Whit, Shaun. BS in Engring. Physics, U. Maine, 1973, MEd, 1974; EdD, Boston U., 1981. Teaching asst. physics dept. U. Maine, Orono, 1972-73; teaching fellow Boston U., 1975-76; sect. head data compilation EDP, Petroleum Authority Thailand, Bangkok, 1980-82, asst. dir. corp. planning, adminstrv. mgr. pipeline project, 1982-83; dep. mgr. Delta Engring. & Constrn. Co., Bangkok, 1983-84; tech. rep. highly protected risk dept. Kemper Group, North Quincy, Mass., 1976-79, tech. rep., 1984-86; edn. coord. highly protected risk dept. Kemper Group, Long Grove, Ill., 1986-88; engring. mgr. internat. dept. Kemper Internat. Corp., Chgo., 1988—. Violinist, concert master Boston Light Opera, 1977-79; founder, chmn. bd. Bangkok Symphony Orch., 1982-84; violinist Woodstock (Ill.) Opera Theatre, 1986-88, Fox Valley Symphony Orch., Aurora, Ill., 1986-88. Recipient Cultural award Govt. of Australia, Bangkok, 1983; scholar U. Maine Coll. Engring., 1967-72. Mem. Soc. Fire Protection Engrs., Nat. Fire Protecetion Assn. (tech. coms. 72E, 72H, 75 and 16 1988—), Inst. Indsle Engrs. (sr. mem.). Avocations: personal computing, violin, concerts, museums, tennis. Home: 589 Parkside Ct Crystal Lake IL 60012-3366 Office: Kemper Group 500 W Madison St Chicago IL 60661-2511

CHAKOIAN, GEORGE, aerospace engineer; b. Providence, June 14, 1924; s. Daniel and Margaret (Derderian) C.; m. Marion Mahdesian, Aug. 29, 1948; children: Janis, Cynthia, Laura. BS in Machine Design, R.I. Sch. Design, 1949; BSME, Tri-State U., 1950. Registered profl. engr., R.I., Mass. Engr. B.I.F. Industries, Inc., Providence, 1950-55; mech. engr., R. M. Hallam Consulting Engrs. U.S. Naval Underwater Ordnance Sta., Newport, R.I., 1955-56; mech. engr. U.S. Naval Air Sta., Quonset Point, R.I., 1956-58, asst. tech. dir., supr. mech. engr., 1958-66; aerospace engr. U.S. Army Natick Rsch., Devel. and Engring. Ctr., Natick, Mass., 1966-81, supr. aerospace engr., 1981-90. Contbr. articles to profl. jours.; patentee in field. Chmn. Sch. Bldg. Com., R.I., 1968; mem. parish coun. St. Sahag & St. Mesrob Armenian Apostolic Ch., 1967-74, diocesan del., 1975-83, 87—, chmn. ch. bldg. com. 1987—; Oud player Leader of New Eng. Ararat Orch. Tech. sgt. 1943-45, PTO. Decorated air medal with four oak leaf clusters and five battle stars. Mem. AIAA (gen. chmn. aerodynamic deceleration internat. conf. 1989), Armenian Students Assn. (bd. dirs. 1989—, Sci. award 1975), Knights of Vartan (Past Achievement award 1989), Armenian Gen. Benevolent Union. Avocations: music, housebuilding. Home: 11 Southwick Dr Lincoln RI 02865-4819

CHAKRABARTTY, SUNIL KUMAR, mathematician; b. Jirrah, India, Nov. 19, 1946; s. Khudiram and Menoka (Goswami) C.; m. Shikha Biruni, Mar. 7, 1976; children: Subrata, Shreya. MS, Jadavpur U., Calcutta, India, 1970, PhD, 1976. Jr. rsch. fellow Jadavpur U., Calcutta, 1972-75; sr. rsch. fellow Indian Inst. Sci., Bangalore, India, 1975-77; rsch. assoc. Indian Inst. Tech., Kharagpur, India, 1977-78; scientist Nat. Aero. Lab., Bangalore, 1978—. Contbr. articles to AIAA Jour., Acta Mechanica. Mem. AIAA, Aero. Soc. India. Office: CTFD Divsn, Nat Aero Lab, Bangalore 560017, India

CHAKRAVARTHY, SREENATHAN RAMANNA, physicist; b. Bangalore, Karnatak, India, Jan. 23, 1937; arrived in Can., 1966; s. Bangalore Sunderachar Ramanna and Kollupalli V. Rukmini; m. Sudha Sharda, Feb. 17, 1964; children: Vandana, Archana, Jivas. Cert. in Chemistry, Karnatak U., Dharwar, India, 1960, Punjab U., Chandibarh, India, 1965; cert. in Health Physics, U. Western Ont., London, Can., 1972. Med. physicist St. Joseph Hosp., Chgo., 1991-92, U. Ill. Hosp., Chgo., 1992—. Bd. dirs. Third World Found. Conf., Chgo., 1978—. Mem. Therpeutic Radiol. Physics. Office: Dept Radiation Oncology 1740 W Taylor St Ste C200 Chicago IL 60612

CHAKRAVARTY, DIPTO, computer-performance engineer; b. Calcutta, India, Oct. 29, 1963; came to U.S., 1984; s. Dhirendra Nath and Rina (Banerjee) C.; m. Aloka Gangopadhyay, Jan. 15, 1990. BS, Bhopal (India) U., 1984; BS, MS, U. Md., 1987. System programmer U. Md., Balt., 1985-88; sr. software engr. A&T Systems, Silver Spring, Md., 1988-90, product devel. mgr., 1990-92; performance engr. IBM Corp., Rockville, Md., 1992—; dir. adv. bd. AIX Users Group, Washington, 1991—; Unix cons. Bell, DEC, IBM, Bull, AT&T, Uniprime, N.H., Pa., Tex., Mass., N.C., Md., Dept. of Justice, Washington, 1989—; cons. U.S. Army, Ft. Lee, Va., 1992. Author: Computer Automation in Clinical Biochemistry and Lab, 1988, POWER RISC System/6000: Architecture and Internals, 1993. Mem. IEEE, Computer Measurements Group (reviewer 1991—, conf. chair 1993, internat. publ. award 1992), Digital Equipment Corp. User Soc., Washington Area Unix Users Group. Achievements include patent pending for computer response time analysis methods for interactive workloads; research in scalability of wide area computer networks, performance engineering and tuning techniques, enhancement of time measurement resolution under Unix oper. system, simulation of RISC CPU-s using transputers. Home: 8216 Langport Ter Gaithersburg MD 20877-1199 Office: IBM Corp 9201 Corporate Blvd Rockville MD 20850

CHAKRAVORTY, KRISHNA PADA, chemist, spectroscopist; b. Calcutta, India; came to U.S., 1965; s. Ramani and Kamala C.; m. Chandana Banerjee. MTech., Calcutta U., 1964; MS in Forestry, Syracuse U., 1968; MS, SUNY, Syracuse, 1968; PhD, Temple U., 1982. Mem. faculty Calcutta U., 1971-77, Ohio State U., Columbus, 1982-84, U. Ill., Chgo., 1986-87; researcher Iowa Laser Facility, U. Iowa, Iowa City, 1984-86; chemist Def. Logistics Agy, Def. Pers. Support Ctr., Phila., 1987-90, Def. Logistics Agy., Def. Gen. Supply Ctr., Richmond, Va., 1990—. Contbr. articles to sci. jours. Mem. Relief, Welfare, Ambulance Corps, India, 1964. Fulbright Found. scholar, 1966, NSF scholar, 1966. Mem. ASTM, U.S. Ednl. Found. in India. Home: 2513 Noel St Richmond VA 23237 Office: Def Gen Supply Ctr DGSC-SSC 8000 Jefferson-Davis Hwy Richmond VA 23297-5000

CHAKROUN, WALID, mechanical engineering educator; b. Beirut, Lebanon, Sept. 7, 1963; came to U.S., 1982; s. Mohamed and Amne (Mansour) C.; m. May Mansour, Nov. 31, 1990; 1 child, Nadim. MS, Va. Tech. U., 1987; PhD, Miss. State U., 1992. Researcher Va. Tech. U., Blacksburg, 1986-87, 88-89; instr. Am. U. Beirut, 1987-88, Miss. State U., Starkville, 1989—. Contbr. articles to profl. jours. Mem. AIAA, ASME, Sigma Xi, Pi Tau Sigma. Office: Miss State Univ PO Drawer ME Mississippi State MS 39762

CHAKU, PRAN NATH, metallurgist; b. Srinagar, India, May 22, 1942; s. Gopi Nath an Leelawati (Dhar) C.; m. Asha Ganju, Sept. 25, 1966; children: Ashish, Maneesh. BS in Metallurgy Engring. with honors, Benares Hindu U., Varanasi, India, 1964; MS in Metallury and Material Sci., U. Pa., 1972. Registered profl. engr., Tex. Staff engr. Tata Iron & Steel Co., Jamshed Pur-Bihar, India, 1964-69; fellow U. Pa., Phila., 1969-72; chief process metallurgy Midvale Heppenstall, Phila., 1972-75; chief materials applications Midvale Heppenstall, Pitts., 1975-78; mgr. metallurgy Gulf Forge, Houston, 1978-80; sr. engr. Fluor Daniel, Houston, 1980-81, prin. engr., 1981-87, tech. dir., 1987-90; chief metallurgist ABB Lummus Crest Inc., Houston, 1990—. Mem. Nat. Assn. Corrosion Engrs. (com. Houston chpt. 1988—), ASME (com. 1988—), Materials Properties Coun. (tech. adv. com. 1988—). Home: 16225 St Helier St Houston TX 77040-2061 Office: ABB Lummus Crest Inc 12141 Wickchester Ln Houston TX 77079-1207

CHALFOUN, NADER VICTOR, architect, educator; b. Alexandria, Egypt, July 14, 1949; came to U.S. 1983; s. Victor George and Leila (Roman) C.; m. Marie Albert Toutounji, Jan. 22, 1983; children: Debora N., Fadi N. MArch, U. Ariz., 1985, PhD, 1989. Tchr. U. Zajarie, Cairo, 1972; owner consulting office arch. and energy svcs., 1972—; assoc. prof. Arch. U. Ariz., 1989; rsch. assoc. Environ. Rsch. Lab., Tucson, 1989-91; assoc. prof. arch. U. Ariz., Tucson, 1991—; owner consulting office of arch. and energy svcs., 1972—; cons. Am. U., Cairo, 1989. Contbr. articles to profl. jours. Roman Catholic. Achievements include invention of Azimuth Protractor: a device to measure Azimuth angles of existing buildings using solar time; development of house energy doctor programs. Home: 4230 E Whittier St Tucson AZ 85711 Office: Univ of Ariz Coll of Arch Tucson AZ 85721

CHALLINOR, DAVID, scientific institute administrator; b. N.Y.C., July 11, 1920; s. David and Mercedes (Crimmins) C.; m. Joan Ridder, Nov. 22, 1952; children: Julia M., Mary E., Sarah L., D. Thompson. B.A., Harvard U., 1943; M.F., Yale U., 1959, Ph.D., 1966. With Offerman-Anderson, Clayton & Co., Houston, 1947-51; cotton farmer Culberson County, Tex., 1951-53; asst. sec. First Mortgage Co., Houston, 1953-57; research asst. Conn. Agr. Expt. Sta., New Haven, 1959-60; dep. dir. Yale Peabody Mus., New Haven, 1960-65; acting dir. Yale Peabody Mus., 1965-66; spl. asst. in tropical biology Smithsonian Instn., Washington, 1966-67, dep. dir. office internat. activities, 1967-68, dir. office internat. activities, 1968-70, asst. sec. sci., 1971-87, sci. advisor to sec., 1988—; v.p. for No. Am. Charles Darwin Found., 1971-92. Contbr. articles to sci. jours. Trustee Manhattanville Coll., 1964-70, Environ. Law Inst., 1975-84, 86-92; bd. dirs. Environ Def. Fund, 1982—, African Wildlife Found., 1980—, chmn. bd. Fixing Tree Assn., 1988—, Ctr. for Marine Conservation, 1992—. Served with USNR, 1943-46. Fellow AAAS; mem. Sigma Xi. Home: 3117 Hawthorne St NW Washington DC 20008-3588 Office: Smithsonian Inst Nat Zoo 3000 Connecticut Ave NW Washington DC 20008-2509

CHALMERS, FRANKLIN STEVENS, JR., engineering consultant; b. Atlanta, Mar. 21, 1928; s. Franklin Stevens Sr. and Martha (Bratton) C.; m. Anne Upshaw, June 11, 1955; children: F. Steven, Martha Chalmers Hamilton, James MacAllen, S. Elizabeth Chalmers Smith. BChemE, Ga. Inst. Tech., 1949; postgrad., Harvard U., 1987. With plant ops. Stauffer Chem. Co., Houston and Baytown, Tex., 1952-56; with project engring. and mgmt. CF Braun & Co., Alhambra, Calif., 1956-67; with bus. devel. Jacobs Engring. Co., Pasadena, Calif., 1967-70; asst. dir. alt. fuels devel. and cons. devel. Arthur G. McKee & Co., Cleve., 1970-76; dir. govt. rels. Davy McKee Corp., Washington, 1976-83; cons. Chalmers & Co., Washington, 1983-86, 89—; dir. policy OSHA/U.S. Dept. Labor, Washington, 1986-89; presenter seminars in People's Republic of China. Contbr. articles to profl. jours.; patentor tar sands oil recovery, apparatus for oil recovery, fluegas desulfurization. Deacon San Marino (Calif.) Community Ch., 1969; pres. YMCA-YMen's Club, South Pasadena, Calif., 1967; trustee Plymouth Ch. of Shaker Heights, Ohio, 1973-76. With Chem. Corps, U.S. Army, 1950-52. Mem. Am. Inst. Chem. Engrs., Am. Conf. Govtl. Indsl. Hygienists, Wash. Export Coun., Nat. Constructors Assn. (chair govt. affairs com. 1979), Nat. Bldg. Mus. (docent), Congl. Country Club, Cosmos Club, Rotary. Republican. Presbyterian. Achievements include research on technical, economic and political feasibility of methods for generation of electric power at remote locations; hosting of delegations from People's Republic of China to U.S. regarding recovery of sulfur for industrial use by flue gas desulfurization. Home and Office: 2022 Columbia Rd NW # 401/403 Washington DC 20009

CHAMBERLAIN, JOSEPH MILES, astronomer, educator; b. Peoria, Ill., July 26, 1923; s. Maurice Silloway and Roberta (Miles) C.; m. Paula Bruninga, Dec. 12, 1945; children: Janet Ann, Susan Louise, Barbara Jean. BS, U.S. Mcht. Marine Acad., 1944; BA, Bradley U., 1947; AM, Tchrs. Coll. Columbia, 1950, EdD, 1962. Instr. Columbia Jr. High Sch., Peoria, 1943; instr. nav. War Shipping Adminstrn., 1944-45; boys sec. YMCA, Peoria, 1946-47; instr. U.S. Mcht. Marine Acad., Kings Point, N.Y., 1947-50; asst. prof. U.S. Mcht. Marine Acad., 1950-52; asst. curator Am. Museum-Hayden Planetarium, N.Y.C., 1952-53; gen. mgr., chief astronomer Am. Museum-Hayden Planetarium, 1953-56, chmn., 1956-64; asst. dir. Am. Mus. Natural History, 1964-68; dir. Adler Planetarium, Chgo., 1968-91, pres., 1977-91, ret., 1991; prof. astronomy Northwestern U., 1968-78; professorial lectr. U. Chgo., 1968-71; led eclipse expdns. to Can., 1954, 79, Ceylon, 1955, Pacific Ocean, 1977, 91, astro-geodetic expdns. to Can., 1956, 57, Greenland, 1958; dean coun. of sci. staff Am. Mus. Nat. History, 1960-62. Co-author: Planets, Stars and Space, 1957; author: Time and the Stars, 1964; also articles on popular astronomy. Active Boy Scouts Am., Met. Chgo. YMCA. Served to lt. USNR, 1945-46; staff Naval Res. Officers Sch. 1953-54, N.Y.C. Mem. Am. Astron. Soc., Internat. Astron. Union, Internat. Planetarium Dirs. Conf. (vice chmn. 1968-77, chmn. 1977-87), Am. Polar Soc., Am. Assn. Museums (mem. council 1965-77, v.p. 1971-74, pres. 1974-75), Phi Delta Kappa, Phi Kappa Phi, Kappa Delta Pi. Republican. Presbyn. (elder). Clubs: University (Chgo.), Tavern (Chgo.), Dutch Treat. Home: 208 W Wolf Rd Peoria IL 61614

CHAMBERLAIN, JOSEPH WYAN, astronomer, educator; b. Boonville, Mo., Aug. 24, 1928; s. Gilbert Lee and Jessie (Wyan) C.; m. Marilyn Jean Roesler, Sept. 10, 1949; children: Joy Anne, David Wyan, Jeffrey Scott. A.B., U. Mo., 1948, A.M., 1949; M.S., U. Mich., 1951, Ph.D., 1952. Project sci. aurora and airglow USAF Cambridge Research Center, 1951-53; rsch. assoc. Yerkes Obs., Chgo., 1953-55; asst. prof. Yerkes Obs., 1955-59, assoc. prof., 1959-60, prof., 1961-62, assoc. dir., 1960-62; assoc. dir. planetary scis. div. of Kitt Peak Nat. Obs., 1962-70, astronomer, planetary scis. div., 1970-71; dir. Lunar Sci. Inst., Houston, 1971-73; prof. dept. space physics and astronomy Rice U., Houston, 1971-90, prof. emeritus, 1990—. Author: Physics of the Aurora and Airglow, 1961, Theory of Planetary Atmospheres, 1978, 2d edit., 1987, Global Change and American Culture, 1993; editor: Revs. of Geophysics and Space Physics, 1974-80; mem. editorial bd.: Planetary Space Sci.; assoc. editor Astrophys. jour., 1960-62,. Recipient Warner prize Am. Astron. Soc., 1961; Alfred P. Sloan research fellow, 1961-63. Fellow Royal Astron. Soc. (fgn.), AAAS, Am. Geophys. Union (councilor 1968-70); mem. Am. Astron. Soc. (councilor 1961-64, chmn. div. planetary scis. 1969-71), Am. Phys. Soc., Am. Meteorol. Soc., Internat. Astron. Union, Internat. Union Geodesy Geophysics, Internat. Sci. Radio Union, NRC Assembly Math. and Phys. Scis. (exec. com. 1973-78), Nat. Acad. Sci. (chmn. geophysics sect. 1972-75), Soc. Sci. Exploration. Office: Rice U Dept Space Physics and Astronomy PO Box 1892 Houston TX 77251-1892

CHAMBERLAIN, OWEN, nuclear physicist; b. San Francisco, July 10, 1920; divorced 1978; 4 children; m. June Steingart, 1980 (dec.). AB (Cramer fellow), Dartmouth Coll., 1941, PhD, U. Chgo., 1949. Instr. physics U. Calif., Berkeley, 1948-50, asst. prof., 1950-54, assoc. prof., 1954-58, prof., 1958-89, prof. emeritus, 1989—; civilian physicist Manhattan Dist., Berkeley, Los Alamos, 1942-46. Recipient Nobel prize (with Emilio Segré) for physics, for discovery anti-proton, 1959, The Berkeley citation U. Calif., 1989; Guggenheim fellow, 1957-58; Loeb lectr. at Harvard U., 1959. Fellow Am. Phys. Soc., Am. Acad. Arts and Scis.; mem. Nat. Acad. Scis., Berkeley Fellows. Office: U Calif Physics Dept Berkeley CA 94720

CHAMBERLIN, PAUL DAVIS, metallurgical engineer; b. Saginaw, Mich., Jan. 28, 1937; s. Marshall Vernon and Ruth (Davis) C.; m. Dixie Ann Coughran, Aug. 9, 1958; children: Michael, Julie. BS in Metall. Engring., Mich. Tech. U., 1959; MBA, Ariz. State U., 1968. Registered profl. engr., Colo. Mill metallurgist Anaconda Co., Butte, Mont., 1962-64; smelter metallurgist Inspiration (Ariz.) Consolidated Copper Co., 1964-67; process engr. Hazen Rsch., Golden, Colo., 1968-73; sr. metallurgist Conoco Minerals, Denver, 1973-78; mgr. metallurgy Occidental Minerals Co., Denver, 1978-83, Nerco Minerals Co., Fairbanks, Alaska, 1983-85; owner, cons. Chamberlin and Assocs., Denver, 1985—. Contbr. articles to profl. jours. 1st lt. U.S. Army, 1960-62. Mem. Soc. Mining Engrs. (chmn. govt., edn. and mining com. 1991-92), Can. Inst. Mining and Metallurgy, Am. Assn. Cost Engrs. (bd. dirs. 1978-83, cert. bd. 1980-81, Recognition award 1981, Cost Engr. of Yr. 1992), Colo. Mining Assn. Achievements include design of static mixer/settler for copper solvent extraction systems; gold heap leach building and operating techniques. Home and Office: 7463 W Otero Pl Littleton CO 80123

CHAMBERS, DONALD ARTHUR, biochemistry educator; b. N.Y.C., Sept. 24, 1936. AB, Columbia U., 1959, PhD, 1972. Rsch. biochemist dept. surgery Harvard Med. Sch./Mass. Gen. Hosp., Boston, 1961-66; rsch. fellow in hematology dept. surgery Harvard Med. Sch./Beth Israel Hosp., Boston, 1967-68; faculty fellow in chem. biology Columbia U., N.Y.C., 1969-71; asst. rsch. biochemist Ctr. for Med. Genetics Dept. Medicine U. Calif. Med. Ctr., San Francisco, 1972-74, lectr. in biochemistry and biophysics Dept. Biochemistry, 1972-74, asst. prof. molecular biology and biochemistry, 1974-75; asst. prof. biol. chemistry and dermatology U. Mich., Ann Arbor, 1975-79, assoc. prof. biol. chemistry, 1979; prof. molecular biology U. Ill., Chgo., 1979—, prof. biol. chemistry, 1980—, rsch. prof. dermatology, 1981—; assoc. mem. Dental Rsch. Inst. U. Mich., 1978-79, adj. rsch. investigator Dept. Biol. Chemistry, 1979—; dir. Ctr. for Molecular Biology of Oral Disease, U.

Ill., Chgo., 1979—, interim head dept. biochemistry, 1985, head dept. biochemistry, 1986—, fellow Honors Coll., 1985—, Phi Kappa Phi lectr., 1991, mem. rsch. adv. bd. Coll. Dentistry, 1979—, co-dir. MD/PhD program Coll. Medicine, 1988—, mem. exec. com. Coll. Medicine, 1990—, others; vis. prof. Oxford U., 1989—; mem. nat. action com. Am. Assn. dental Rsch., 1981—;. Editorial reviewer Analytical Biochemistry, Archives Biochemistry and Biophysics, Archives Dermatology, Biochemica Acta, Biophysica Acta, Biochem. Jour., Blood, Cell, Endocrinology, Jour. Investigative Dermatology, Jour. Dental Rsch., Jour. Clin. Investigation, Jour. Biol. Chemistry, Molecular and Cell Biology, Nature, New Eng. Jour. Medicine, Procs. NAS, Sci., Thrombosis Rsch.; contbr. numerous articles to profl. jours. including Archives Oral Biology, Jour. Periodontal Rsch., Jour. Investigative Dermatology, Procs. NAS, Jour. Experimental Medicine, Cell, Archives Biochemistry and Biophysics, Nature. Recipient James Howard McGregor prize Columbia U., 1971; named Inventor of Yr., U. Ill., 1990; fellow in hematology NIH, 1967-68; fellow in chem. biology, 1969-71; rsch. grantee NIH-NIAMDD, 1975-79, 75-85, 79-80, 84—, NIH-NIDR, 1975-79, 78-79, Mich.-Phoenix Project, 1975-76, Am. Cancer Soc., 1977-78, 78, The Upjohn Co., 1982-87, Xytronyx, Inc., 1984—, Office of Naval Rsch., 1986—, Helene Curtis, Inc., 1988—; tng. grantee NIH-NIGMS, 1975-79, NIH-NIAMDD, 1976-79, 77-80, NIH-NIDR-NIAMDD, 1980—, NIH-NCI, 1982-88. Mem. AAAS, Am. Assn. Med. Colls., Am. Chem. Soc., Am. Fedn. Clin. Rsch., Am. Soc. Biol. Chemistry, Am. Soc. Cell Biology, Am. Soc. Microbiology, Internat. Assn. Dental Rsch. (com. on rsch. progress 1982-85, chmn. 1984-85, chmn. grad. tng. forum com. exptl. pathology sect. 1983), Assn. Dept. Chmn. Biol. Chemistry, Chgo. Assn. Immunologists, Chgo. Cancer Assn., N.Y. Acad. Scis. (organizer meeting The Double Helix, 40 Yrs. 1993), Royal Soc. Medicine, Soc. Investigative Dermatology, Phi Kappa Phi, Sigma Xi. Achievements include patents (U.S., Can.) for Method of Determining Periodontal Disease, (with other) Method of Quantifying Aspartate Amino Transferase in Periodontal Disease; research in role of cyclic nucleotides, prostaglandins, hormones and other regulatory factors in the regulation of cell function, proliferation and differentiation, in the molecular biology of disease, and in the regulatory mechanisms of host-microbial interactions. Office: U Ill Dept Biochemistry Coll Medicine 1853 W Polk St Chicago IL 60612-4345 Office: Ctr Molecular Biol Oral Diseases 801 S Paulina Chicago IL 60612

CHAMBERS, EDWARD ALLEN, aerophysical systems designer, consultant engineer; b. Arnold, Pa., Sept. 28, 1933; s. Buford Allen and Elva Viola (Trautman) C.; m. Virginia Ann Cole, Sept. 4, 1955 (dec.); children: Mark Allen, Brian Cole, Kurt Trautman; m. Bianca Marie Montafla, Mar. 20, 1965. BS in Aero. Engring., Tri State U., Angola, Ind., 1953; engring. mgmt. cert., Harvard U., 1974, bus. mgmt. cert., 1975. Aero. engr. N.Am. Aviation, Columbus, Ohio, 1954-56; aero. engr. Missile Systems div. Raytheon, Bedford, Mass., 1956-59, aero. lead engr. advanced project, 1960-64; project aero. engr. Goodyear Aircraft, Akron, Ohio, 1959-60; lead engr. systems design Missile Systems div. Raytheon, Bedford, Mass., 1964-92; missile program developer Lockheed Marietta, Ga., 1964; v.p. tech. devel. Analytical Systems Inc., Sandwich, Mass., 1992—. Mem., chmn. Acton (Mass.) Planning Bd., 1966-72, Acton Long Range Planning Com., 1972-76, Acton Conservation Commn., 1978-82. Recipient Disting. Citizen award Acton Lions Club, 1973. Mem. AIAA (sr., mem., chmn. solid rocket tech. com. 1984-91, tech. co-chmn. joint propulsion conf. 1989, bd. dirs. New Eng. region 1988-92, Disting. Mem. award 1989). Achievements include patent for guided anti-armor mortar projectile, patent pending for trimavector-3D TVC flight control from a single nozzle, air slew aerodynamics and controllability of guided missiles-munitions minimum response time flight controls for Smart munitions; system design for super-penetration in guided missles-munitions. Home: 407 N Park Ave Indianapolis IN 46202 Office: Analytical Systems Inc PO Box 1046 11 School St Sandwich MA 02563

CHAMBERS, HENRY GEORGE, surgeon; b. Portsmouth, Va., June 22, 1956; s. Walter Charles and Teresa Frances (Fernandez) C.; m. Jill Annette Swanson, June 10, 1978; children: Sean Michael, Reid Christopher. BA summa cum laude in Biochemistry, U. Colo., 1978; MD, Tulane U. Sch. Medicine, 1982. Diplomate Am. Bd. Orthopaedic Surgery. Commd. 2d lt. U.S. Army, 1978; advanced through grades to maj.; intern Fitzsimmons Army Med. Ctr., Aurora, Colo., 1982-83; orthopaedic surgery resident Brooke Army Med. Ctr., Ft. Sam Houston, Tex., 1983-87, chief resident, 1986-87, staff orthopaedic surgeon to asst. residency program dir., 1987-89, asst. chief surgeon orthopaedic surgery svc., 1990-92; staff orthopaedic surgeon DeWitt Army Hosp., Ft. Belvoir, Va., 1987; pediatric orthopaedic fellow San Diego Children's Hosp., 1989-90; adj. prof. natural scis. Incarnate Word Coll., San Antonio, 1986—; asst. prof. surgery Uniformed Svcs. U. Health Scis., Bethesda, Md., 1987—; asst. program dir. Brooke Army Med. Ctr. Orthopaedic Surgery, 1987-92; asst. prof. U. Calif. San Diego Med. Ctr., 1989—. Co-author: Long Distance Runner's Guide to Training, 1983; contbr. various articles to profl.jours. Physician St. Vincent de Paul Clinic for Homeless, San Diego, 1989-92; bd. dirs. United Cerebral Palsy. Recipient Comdrs. award for oustanding rsch. Brooke Army Med. Ctr., 1987. Fellow Acad. Cerebral Palsy Devel. Medicine; mem. Am. Acad. Orthopaedic Surgeons, Am. Coll. Sports Medicine, Greenpeace, Earth Island Inst., World Wildlife Fedn., Friends of Earth, Handgun Control, Phi Beta Kappa. Democrat. Roman Catholic. Avocations: bicycling, weight lifting, golf, tennis, raquetball. Home: 5458 Sandburg Ave San Diego CA 92122

CHAMBERS, JAMES PATRICK, mechanical engineer; b. Silver Spring, Md., June 23, 1968; s. Edward Francis and Mary Suzanne (Dwyer) C.; m. Julie Ann Morris, Aug. 31, 1991. BME with highest honors, Ga. Inst. Tech., 1990, postgrad., 1990—. Engring. intern E.I. duPont de Nemours & Co., Newark, 1990; Hertz fellow mech. engring. Ga. Inst. Tech., Atlanta, 1990—. Mem. South Atlantans for Neighborhood Devel., Atlanta, 1991—. Mem. ASME, Ga. Acoustical Soc. of Am., Acoustical Soc. Am., Tau Beta Pi, Pi Tau Sigma. Roman Catholic. Home: 878 Gilbert St SE Atlanta GA 30316

CHAMBERS, KENNETH CARTER, astronomer; b. Los Alamos, N.Mex., Sept. 27, 1956; s. William Hyland and Marjorie (Bell) C.; m. Jeanne Marie Hamilton, June 28, 1986; children: Signe Hamilton, William Hamilton. BA in Physics, U. Colo., 1979, MS in Physics, 1982; MA in Physics and Astronomy, Johns Hopkins U., 1985, PhD in Physics and Astronomy, 1990. Rsch. asst. dept. physics U. Colo., Boulder, 1982-83; rsch. asst. dept. physics and astronomy Johns Hopkins U., Balt., 1983-86; mem. instrument team Hopkins Ultraviolet Telescope, Balt., 1983-86; rsch. asst. Space Telescope Sci. Inst., Balt., 1986-90; postdoctoral fellow Leiden (The Netherlands) Obs. Leiden U., 1990-91; asst. prof. Inst. Astronomy U. Hawaii, Honolulu, 1991—; Contbr. articles to Astrophys. Jour., Nature mag., Phys. Rev.; contbr. conf. procs. in field. Mem. Am. Astron. Soc. (Chretein award 1989), Am. Phys. Soc. Achievements include discovery of most distant known galaxy (4C41.17), of alignment effect in high redshift radio galaxies; research on observational cosmology, galaxy formation and evolution, active galaxies, observing techniques, spacecraft observations. Office: Inst Astronomy U Hawaii 2680 Woodlawn Dr Honolulu HI 96822

CHAMBERS, KENTON LEE, botany educator; b. Los Angeles, Sept. 27, 1929; s. Maynard Macy and Edna Georgia (Miller) C.; m. Henrietta Laing, June 21, 1958; children—Elaine Patricia, David Macy. A.B. with highest honors, Whittier Coll., 1950; Ph.D. (NSF fellow), Stanford U., 1955. Instr. biol. scis. Stanford U., 1954-55; instr. botany, Yale U. 1956-58, asst. prof., 1958-60; assoc. prof. botany Oreg. State U., Corvallis, 1960-65, prof., 1965-90, prof. emeritus, 1991—, curator Herbarium, 1960-90; program dir. systematic biology NSF, Washington, 1967-68. NSF fellow, 1955-56. Mem. Bot. Soc. Am. (BSA Merit award 1990), Am. Soc. Plant Taxonomists, AAAS, Am. Inst. Biol. Scis., Calif. Bot. Soc., Soc. Systematic Biology. Democrat. Presbyterian. Clubs: Triad, Oreg. State U. Contbr. articles in field to profl. jours. Home: 3220 NW Lynwood Cir Corvallis OR 97330-1134 Office: Oreg State U Herbarium Botany Dept Corvallis OR 97331

CHAMBOLLE, THIERRY JEAN-FRANCOIS, environmental scientist; b. Beychac et Caillon, Gironde, France, June 12, 1939; s. Jean Michel and Renèe Marie (Minchini) C.; m. Claude Elisabeth DeLord, Aug. 29, 1963; children: Damien, David, Antonin, Etienne, Thomas. Buccalaurēat in Math and Philosophy, Lycée Montesquieu, Bordeaux, France; diploma of engineering, Ecole Polytechnique, Paris, 1959-61, Ecole Nationale des Ponts et Chaussées, Paris, 1963-64. Dir. de l Eau et la Prevention des Pollutions, 1978-88; dé légué aux risques majeurs Ministerè de l Environment France; directeur de la recherche Lyonnaise Des Eaux, Dumez, France, 1989-91, directeur general adj., 1991—; pres. CRMAGREF, France, 1989. Chevalier du merite agricole, 1979, Chevalier de la legion d' honneur, 1986, officier du mérite national, 1992. Mem. Acad. Scis. (assoc. mem. com. applications). Office: Lyonnaise des Eaux Dumez, 72 Av de la Liberte, 92000 Nanterre France

CHAMIS, CHRISTOS CONSTANTINOS, aerospace scientist, educator; b. Sotira, Greece, May 16, 1930; came to U.S., 1948; s. Constantinos and Anastasia (Kyriakos) C.; m. Alice Yanosko, Aug. 20, 1966; children: Chrysanthie, Anna-Lisa, Constantinos. BS in Civil Engring., Cleve. State U., 1960; MS, Case Western Res. U., 1962, PhD, 1967. Draftsman, designer Cons. Engring., Cleve., 1955-60; research asst. Case Western Res. U., Cleve., 1960-62, rsch. assoc., 1964-68; rsch. mathematician B. F. Goodrich, Brecksville, Ohio, 1962-64; aerospace engr. Lewis Rsch. Ctr., NASA, Cleve., 1968-78, sr. rsch. engr., 1978-86, sr. aerospace scientist, 1986—; cons. Lawrence Livermore Labs., Calif., 1974-79; adj. prof. Cleve. State U., 1968—, Akron U., 1980—, Case Western Reserve U., 1984—. Editor: Composites Analysis/Design, 1975, Test Methods and Design Allowables for Composites, 1979, 89; mem. editorial bd. Jour. Composites Rsch. and Tech., Reinforced Plastics and Composites. Contbr. numerous articles to sci. jours. Patentee in field for Intraply Hybrid Composites; researcher in hygrothermal composite micromechanics, computational composite mechanics-computer codes, high-temperature composite structures, structural tailoring of engine structures, computational simulation of progressive fracture, engine structures computational simulations, computational simulation/tailoring of coupled multi-discipline problems, and probabilistic structural analysis. Served with USMCR, 1953-54. Fellow ASME, AIAA (assoc. fellow, assoc. editor 1986-88); mem. ASCE, ASTM, Soc. Exptl. Mechanics, Soc. Aerospace Processing and Materials Engring., Soc. Plastics Industry, Sigma Xi. Clubs: Dodoni, Hellenic U. Home: 24534 Framingham Dr Cleveland OH 44145-4902

CHAMKHA, ALI JAWAD, research engineer; b. Beirut, Lebanon, June 1, 1964; came to U.S., 1983; s. Jawad Mohammed and Samira Mohammed (Ketiesh) C. PhD, Tenn. Tech. U., 1989. Grad. asst. Tenn. Tech. U., Cookeville, 1986-87, grad. instr., 1987-89, asst. prof., 1990—; staff rsch. engr. Fleetguard Inc., Cookeville, 1991—. Assoc. editor Fluid/Particle Separation Jour., 1993; contbr. over 45 articles to profl. jours. and confs. Recipient Students Choice award Best Mech. Engring. Prof., 1991. Mem. AIAA, ASME, Am. Filtration Soc. (mem. edn. com. 1992, reviewer 1992, sr. co-chmn. meeting 1992), Nat. Soc. Profl. Engrs., Tenn. Acad. Sci., Scientific Rsch. Soc., Soc. Engring. Sci., Phi Kappa Phi, Tau Beta Pi. Moslem. Office: Fleetguard Inc 1200 Fleetguard Cookeville TN 38502

CHAMP, STANLEY GORDON, scientific company executive; b. Hoquiam, Wash., Feb. 15, 1919; s. Clifford Harvey and Edna Winnifred (Johnson) C.; m. Anita Knapp Wegener, Sept. 6, 1941; children: Suzanne Winnifred Whalen, Colleen Louise Szurszewski. BS, U. Puget Sound, 1941; MS, U. Wash., 1950; postgrad., MIT, 1955, 57, UCLA, 1959. Cert. tchr., adminstr., Wash. Tchr. Lake Washington Sch. Dist., Kirkland, Wash., 1942-48; prof. math. U. Puget Sound, Tacoma, 1948-51; supr. mathematician Puget Sound Naval Shipyard, Bremerton, Wash., 1951-55; rsch. specialist Boeing Co., Seattle, 1955-68; v.p. R.M. Towne & Assocs., Seattle, 1968-75; founder, pres. Dynac Scis., Tacoma, 1975—; cons. R.M. Towne Assocs., Seattle, Yantis Assocs., Bellevue, Wash. Contbr. articles to profl. jours.; patent method and apparatus determination soil dynamics insitu. Mem. N.Y. Acad. Sci., Phi Delta Kappa. Presbyterian. Avocation: model building. Home: 1540 S Fairview Dr Tacoma WA 98465-1314

CHAMPLIN, WILLIAM GLEN, clinical microbiologist-immunologist; b. Rogers, Ark., Sept. 10, 1923; s. Glen and Anna Champlin; m. Helen Elizabeth Garner, Feb. 2, 1951; 1 child, Steven. BS, N.E. Okla. State U., 1948; MS, U. Ark., 1965, PhD, 1971. Lab. dir. VA Med. Ctr., Fayetteville, Ark., 1955-65, clin. microbiologist, lab. dir., 1965-80; cons. ANL Med. Lab. Wash. Regional Med. Ctr. VA Med. Ctr., 1965-90; edn. coord. Antaeus Inst. Sch. Med. Tech., 1980-90; vis. prof. microbiology U. Ark., 1978-85. With U.S. Army, 1943-45. Mem. Am. Acad. Microbiology (specialist), Am. Soc. Clin. Pathologists (specialist), Sigma Xi.

CHAN, AH WING EDITH, chemist; b. Hong Kong, June 11, 1963; d. Siu-Tong and Kit-Ching (Law) C. BS in Chemistry cum laude, U. Utah, 1986; PhD, Cornell U., 1991. Vis. scientist U. Barcelona, 1989, Acad. Sci. USSR, Moscow, 1990; with drug design dept. Italfarmaco, Milan, Italy, 1991—. Contbr. articles to Jour. Chem. Physics, Jour. Vacuum Sci. and Tech., Langumir, Jour. Inorganic Chemistry. Active Chinese Hong Kong Student for Chinese Affairs, Ithaca, N.Y., 1989—. Mem. Am. Chem. Soc. Achievements include research in protein folding, interaction of enzymes and inhibitors, and drug design. Office: Italfarmaco SpA, Via Dei Lavoratori 54, 20092 Milan Italy

CHAN, ANDREW MANCHEONG, engineering executive; b. Hong Kong, Jan. 21, 1957; s. Ing Kang and Mei Yin (Cheung) C.; m. Trudy Suk-Yin Ko, Sept. 13, 1976; children: Jennifer, Kathleen, Ellena. BS, Queens U., 1979; MBA, Chinese U. Hong Kong, 1987. Project engr. Fook Lee Constrn. Co. Ltd., Hong Kong, 1979-82; project mgr. Leader Civil Engring. Corp. Ltd., Hong Kong, 1982-87; dir. Keystone Bus. and Project Cons., Hong Kong, 1987-88; sr. mgr. Dynamic Holdings Ltd., Hong Kong, 1988—; dir., gen. mgr. Kenworth Eng Ltd. subs. of Dynamic Holdings Ltd., Hong Kong, 1988—; bd. dirs. Nutec Demolition Ltd., Chiu Woo Cheong Land Investment Co. Ltd., Chiu Woo Cheong Porcelainware Ltd.. Mem. New Hong Kong Alliance, 1989. Avocations: hiking, chess. Home: 38 Cloudview Rd Flat K5, North Point Hong Kong Office: Kenworth Eng Ltd, 41 Yau Tong Marine 7/F Chevalier Comm Cr, 8 Wang Hoi Rd Kowloon Bay Hong Kong

CHAN, CLARA SUET-PHANG, physician; b. Swatow, Guandong, People's Republic of China, Sept. 23, 1949; came to U.S., 1969; d. Hon-Kwong and Suet-Hing (Wong) C. BS, Mary Manse Coll., 1972; MD, George Washington U., 1976. Diplomate Am. Bd. Internal Medicine, Am. Bd. Hematology, Am. Bd. Oncology. Intern U. Miami (Fla.) Hosp., 1976-77, med. resident, 1977-79; fellow hematology, oncology George Washington U., 1979-81; fellow oncoloy research City of Hope Med. Ctr., Duarte, Calif., 1981-83, instr. medicine, 1982-83; asst. chief hematology VA Med. Ctr., Washington, 1983-91; with Hematology/Oncology Cons., Greenbelt, Md., 1991—; asst. prof. medicine George Washington U., 1983-88, assoc. prof., 1988—; prin. investigator stem cell lab. VA Med. Ctr., Washington, 1983-91; physician Hematology Oncology Cons., Greenbelt, Md., 1991—; project chmn. S.E. Cancer Study Group, Birmingham, Ala., 1982-85; mem. med. staff George Washington U., 1983—. Contbr. articles to profl. jours. Del. cancer update Citizen Ambassador program People to People Internat., 1986. Recipient Internat. Peace scholarship George Washington U. 1972-76; Med. Student Research grantee Pan Am. Health Orgn., 1974; Reader's Digest Internat. fellow United Christian Hosp., Hong Kong, 1976; research fellow VA Career Devel. program, Washington, 1981. Fellow ACP; mem. Am. Soc. Clin. Oncology, Am. Soc. Hematology, N.Y. Acad. Sci., William Beaumort Med. Soc. Home: 7001 Bybrook Ln Bethesda MD 20815-3166 Office: Hematology Oncology Cons 7525 Greenway Center Dr Ste 205 Greenbelt MD 20770-3532

CHAN, DAVID CHUK, software engineer; b. Hainan, Canton, China, July 16, 1962; came to U.S., 1986; s. Kwok Keung and So Yu (Lam) Yuen. BS in Engring., U. Guelph, 1986; MS in Computer Sic., Fla. Inst. Tech., 1988. Instr. U. Minn., Morris, 1988-89; sr. software engr. Firearms Tng. Systems, Norcross, Ga., 1989—. Recipient Top Achievement award Firearms Tng. Systems, Norcross, Ga., 1992. Mem. Assn. Computing Machinery. Achievements include leading a small team that developed an engineering software product from concept and won two international contracts worth 80 million over similar products from ten industry giants such as the British Aerospace. Office: Firearms Tng Systems 110 Technology Pkwy Norcross GA 30092

CHAN, DONALD PIN-KWAN, orthopaedic surgeon, educator; b. Rangoon, Burma, Jan. 21, 1937; s. Charles Y.C. and Josephine (Golamco) C.; m. Dorothy Chan, July 31, 1966; children: Joanne, Elaine. BS, U. Rangoon, Burma, 1955, MD, 1960. Intern medicine U. Hong Kong, 1960-61, residnet surgery and orthopaedics, 1961-68; resident orthopaedic surgery U. Vt., 1968-71; assoc. orthopaedist Strong Meml. Hosp., Rochester, N.Y., 1972-80, sr. assoc. orthopaedist, 1980-86, attending orthopaedist, 1986—; asst. prof. U. Rochester, 1972-80, assoc. prof., 1980-87, prof., 1987—; dir. Goldstein Fellowship, Rochester, Orthopaedic Clin. Svcs., Rochester; chief sect. spine surgery dept. orthopaedics, Rochester. Contbr. articles on clin. rsch. related to the spine. Bd. dirs. Rochester Chinese Assn., 1991. Traveling fellow Scoliosis Rsch. Soc. Fellow ACS, Am. Acad. Orthopaedic Surgeons, Am. Orthopaedic Assn., Scoliosis Rsch. Soc.; mem. AMA, N.Am. Spine Soc., Am. Spinal Injury Assn., Ea. Orthopaedic Assn. Avocations: tennis, fishing. Office: U Rochester Sch Medicine 601 Elmwood Ave Rochester NY 14642

CHAN, ERIC PING-PANG, industrial designer; b. Canton, China, Sept. 1, 1952; came to U.S., 1978; s. Chung-Po and Kit-Jon (So) C.; m. Juliana Po-Ying Young, Sept. 18, 1985; 1 child, Kevin Yu-Hinn. Higher diploma in indsl. design, Hong Kong Poly., 1976; MFA, Cranbrook Acad. Art, 1980. Sr. designer Henry Dreyfuss Assocs., N.Y.C., 1980-84; design assoc. Emilio Ambasz Assocs., N.Y.C., 1984-87; prin. designer Wang Lab, Lowell, Mass., 1987; pres., design dir. Chan & Dolan Design, Inc., N.Y.C., 1987-90; pres. Ecco Design, Inc., N.Y.C., 1990—; guest lectr. Parson Sch. of Design, N.Y.C., 1990—; cons. Chinese History Mus., N.Y.C., 1988-91. Exhibited in group shows, including Arango Internat. Design Exhbn., 1987, Milw. Mus. Art, 1988; represented in permanent collections Denver Mus. Art, London Design Mus., Israel Mus.; contbr. articles to profl. jours.; patentee in field. Recipient Forma Finlandia Internat. Design award, Finland, 1989, award Internat. Furniture Design Competition, Progressive Architecture, 1982, Best Product Design award Accent on Design, N.Y., 1987, Best Product award in Good Office, Arango Design Exhbn., Mus. of Art, Fla., 1989, Best Product Design of 1989 Bus. Week, Best Product of 1990 Design News, Design Plus award Frankfurt (Germany) Fair, 1990, Highest award Design Innovation 90, Germany, others. Mem. Indsl. Design Soc. Am. (Bronze award 1989, 90). Office: Ecco Design 89 Fifth Ave Ste 600 New York NY 10003

CHAN, JACK-KANG, anti-submarine warfare engineer, mathematician; b. Toyshan, China, Oct. 20, 1950; came to U.S., 1975; s. David En-Shek and Yip-Ching (Yuen) C.; m. Suet-Fong Ng, June 3, 1982; children: Me-Fun, Kang-Ray. PhD in Elec. Engring., Poly. U., 1982, PhD in Math., 1990. Microwave engr. Sedco Systems div. Raytheon, Melville, N.Y., 1979-80; sr. mem. tech. staff Norden Systems div. United Tech., Melville, 1980—. Author papers in field. Mem. IEEE (reviewer signal processing 1989-90, vice chmn. L.I. signal processing chpt.), Am. Math. Soc., Math. Assn. Am., Soc. Indsl. and Applied Math. Office: Norden Systems United Tech 75 Maxess Rd Melville NY 11747-3182

CHAN, JOHN DODDSON, aerospace engineer; b. Urbana, Ill., Oct. 3, 1962; s. Carl Wah and Sou (Lin) C. BS in Aero. and Astron. Engring., U. Ill., 1984; MS in Astron. Engring., Air Force Inst. Tech., 1986. Commd. 2d lt. USAF, 1985, advanced through grades to capt., 1990; test mgr. system program office Consolidated Space Ops. Ctr., L.A. AFB, 1987-90; chief test div. comms. site activation task force USAF, Falcon AFB, Colo., 1990—. Singles dir. Christian Ctr. Colo. Springs, 1991. Mem. AIAA, Air Force Assn. Republican. Mem. Christian Ch. (Disciples of Christ). Home: 940 Chapman Dr Apt 3 Colorado Springs CO 80916 Office: Detachment Space System Div MS 82 Falcon AFB CO 80912

CHAN, KWAI SHING, materials engineer, researcher; b. Hong Kong, Feb. 2, 1955; came to U.S., 1973; s. Mau F. and Sau L. (Siu) C.; m. Ing M. Oei, June 5, 1982; children: Candace Kay, Kara Lynn, Stacey Karman. BS in Metall. Engring., Mich. Tech. U., 1976, MS in Metall. Engring., 1978, PhD in Metall. Engring., 1980. Jr. rsch. engr. Olin Material Rsch. Lab., New Haven, 1976; postdoctoral affiliate Stanford (Calif.) U., 1980-82; rsch. engr. S.W. Rsch. Inst., San Antonio, 1982-85, sr. engr., 1985-88, prin. engr., 1988-92, staff engr., 1992—; vis. scholar Harvard U., Cambridge, Mass., 1991; bd. rev. Metall. Transactions A, Pitts., 1992—. Co-editor: Forming Limit Diagrams: Concepts, Methods and Applications, 1989; contbr. articles to profl. jours. Mem. ASCE (Alfred Nobel prize 1991), ASME (assoc. editor Jour. Engring. Materials and Tech., 1993—), ASM Internat. (Marcus A. Grossmann award 1986), Am. Inst. Mining & Petroleum Engrs. (Rossiter M. Raymond Meml. award 1990), Minerals, Metals & Materials Soc., Am. Ceramic Soc., Sigma Xi, Alpha Sigma Mu. Achievements include contributions to the understanding of the mechanical behavior (deformation, fracture and fatigue) of engineering metals (metals, ceramics and composites) at the microstructure level through integrated micromechanical modeling. Office: SW Rsch Inst PO Drawer 28510 6220 Culebra Rd San Antonio TX 78228-0510

CHAN, MICHAEL CHIU-HON, chiropractor; b. Hong Kong, Aug. 31, 1961; came to U.S., 1979; s. Fuk Yum and Chun Wai (Ma) C. D of Chiropractic, Western States Chiropractic Coll., 1985; fellow, Internat. Acad. Clin. Acupuncture, 1986. Assoc. doctor Widoff Chiropractic Clinic, Phoenix, 1986, Horizon Chiropractic Clinic, Glendale, Ariz., 1986-88; dir. North Ranch Chiropractic Assoc., Scottsdale, Ariz., 1988-91; pvt. practice Phoenix, 1991—; dir. Neighborhood Chiropractic, Phoenix, 1988-89. Contbr. articles to profl. jours. Mem. Am. Chiropractic Assoc., Coun. on Diagnostic Imaging, Paradise Valley Toastmaster Club. Avocations: golf, reading, traveling. Office: 6544 W Thomas Rd Ste 37 Phoenix AZ 85033

CHAN, RAYMOND HONFU, mathematics educator; b. Hong Kong, Apr. 28, 1958; s. Kwok Kang and Yuk Chee (Lou) C.; m. Katharine Siuyuen Yu, Dec. 17, 1981; 1 child, Sung-hin. BSc, Chinese U. of Hong Kong, 1980; MSc, NYU, 1984, PhD, 1985. Teaching asst. Chinese U. of Hong Kong, 1980-81; asst. prof. U. Mass., Amherst, 1985-86; lectr. dept. of math. U. Hong Kong, 1986-92, U. Hong Kong of Sci. and Tech., 1993; sr. lectr. Chinese U. of Hong Kong, 1993—. Contbr. articles to profl. jours. NSF grantee, 1986; recipient Leslie Fox prize Inst. Math. and Its Applications, Cambridge, Eng., 1989. Office: Chinese U of Hong Kong, Shatin, Hong Kong Hong Kong

CHAN, TAK HANG, chemist, educator; b. Hong Kong, June 28, 1941; s. Ka King and Ling Yee (Yick) C.; m. Christina W.Y. Hui, Sept. 6, 1969; children—Juanita Y., Cynthia S. B.A., U. Toronto, 1962; M.A., Princeton U., 1963, Ph.D., 1965. Rsch. assoc. Harvard U., 1965-66; asst. prof. McGill U., Montreal, Que., Can., 1966-71, assoc. prof., 1971-77, prof. chemistry, 1977—, chmn. dept., 1985-91, dean sci., 1991—. Contbr. articles to profl. jours. Killam fellow, 1983-85; recipient R.U. Lemieux award Can. Soc. Chem., 1993, Merck, Sharp andDohme award, 1982. Fellow Royal Soc. Can.; mem. Chem. Inst. Can., Am. Chem. Soc., Royal Soc. Chemistry. Office: 801 Sherbrooke St, Montreal West, Montreal, PQ Canada H3A 2K6

CHAN, WAI-YEE, molecular geneticist; b. Canton, China, Apr. 28, 1950; came to U.S., 1974; s. Kui and Fung-Hing (Wong) C.; BSc with first class honors, Chinese U. of Hong Kong, 1974; PhD, U. Fla., 1977; m. May-Fong Sheung, Sept. 3, 1976; children: Connie Hai-Yee, Joanne Hai-Wei, Victor Hai-Yue, Amanda Hai-Pui, Bessie Hai-Lui. Teaching asst. dept. biochemistry and molecular biology U. Fla., Gainesville, 1974-77; rsch. assoc. U. Okla., Oklahoma City, 1978-79, asst. prof. dept. pediatrics, 1979-82, assoc. prof., 1982-89, asst. prof. dept. biochemistry and molecular biology, 1979-82, assoc. prof., 1982-89; staff affiliate Pediatric Endocrine Metabolism & Genetic Svc., Okla. Children's Meml. Hosp., Oklahoma City, 1979-89, dir. Clin. Trace Metal Diagnostic Lab., 1979-85, asst. sci. dir. Biochem. Genetics and Metabolic Screening Lab., 1980-87; cons. VA Med. Center, Oklahoma City, 1981-87; co-sci. dir. State of Okla. Teaching Hosp., 1982-87; prof. dept. pediatrics, biochemistry and molecular biology, anatomy and cell biology, mem. Vincent T. Lombardi Cancer Rsch. Ctr. Georgetown U., Washington, 1989—. Assoc. mem. Okla. Med. Res. Found., Oklahoma City, 1987-89, affiliate assoc. mem., 1989—. Chinese U. Hong Kong scholar, 1972-74, 73-74; NATO fellow, 1979; recipient Okla. Med. Rsch. Found. Merrick award 1988. Mem. Am. Inst. Nutrition, Am. Soc. Biochem. Molecular Biology, Nutrition Soc. (U.K.), Am. Soc. Human Genetics, Am. Fedn. Clin. Rsch., Am. Chem. Soc., Biochem. Soc. (U.K.), Internat. Assn. BioInorganic Scientists, N.Y. Acad. Sci., Soc. Exptl. Biology and Medicine Am. Soc. Cell Biology Soc. Pediatric Rsch., Am. Assn. Immunology, Endocrinology Soc., Tissue Culture Assn., Am. Genetic Assn., Am. Coll. Nutrition, Sigma Xi. Editor 2 books and monograph; editor Jour. of Am. Coll. Nutrition; contbr. articles to profl. jours. Achievements include patent on application of pregnancy specific glycoproteins andevelopment of in vitro diagnostic method for Wilson's disease. Home: 10708 Butterfly Ct North Potomac MD 20878 Office: Georgetown U Childrens Med Ctr Dept Pediatrics 3800 Reservoir Rd NW Washington DC 20007

CHAN, WAN CHOON, mining company executive; b. Batu Gajah, Perak, Malaysia, Aug. 6, 1937; s. Tong Thye and Seow Ying (Ng) C.; m. Nguk Lan Wong, June 5, 1965; children: Mayee, Mayin. ACSM, Camborne Sch. of Mines, 1960. Mining engr. Malayan Tin Dredging Ltd., Batu Gajah, Perak, Malaysia, 1960-64; chief planning dept. Anglo Oriental (M) Ltd., Batu Gajah, 1964-66; asst. supt. Selangor Dredging Bhd., Kuala Lumpur, Malaysia, 1966-67, supt., 1967-79, gen. mgr., 1979—. Chmn. mining Standards and Indsl. Research Inst., Malaysia, 1980—; mem. employers panel Indsl. Ct., 1984-92. Recipient Pingat Jasa Kebaktian, His Royal Highness the Sultan of Selangor, 1983. Fellow Instn. Mining and Metallurgy (overseas council mem. 1977-88, chmn. Malaysian sect. 1975-76, 88-90), Inst. Mineral Engring. (council mem. 1973-74, 90—); mem. Instn. Engrs. Malaysia, Council Engring. Instns., Malayan Mining Employers Assn. (council 1973—, pres. 1977-78, 80-82), Malaysian Employers Fedn. (council 1986-92). Home: 88 Lorong Buluh Perindu 1, Damansara Heights, Kuala Lumpur 59000, Malaysia Office: Selangor Dredging, 142-C Jalan Ampang, Kuala Lumpur 50450, Malaysia

CHAN, WING-CHUNG, pathologist, educator; b. Hong Kong, Oct. 11, 1947; came to U.S., 1975; s. Kwok-Ping and Yuet-Wah (Ching) C.; m. Angelina H. Li, May 16, 1981; children: Eric J., Jason E. MBBS, U. Hong Kong, 1973, MD, 1988. Diplomate in anat. pathology, clin. pathology and hematology Am. Bd. Pathology. Resident in pathology U. Chgo., 1975-79, rsch. assoc. in immunology, 1979-80; asst. prof. pathology Emory U. Sch. Medicine, Atlanta, 1980-86, assoc. prof., 1986-91; prof. pathology U. Nebr. Med. Ctr., Omaha, 1991—. Mem. editorial bd. Am. Jour. Clin. Pathology, 1990—; contbr. chpts. to books, numerous articles to profl. jours. Mem. U.S. and Can. Acad. Pathology, Hematopathology Soc. (charter), Am. Soc. Hematology, Am. Assn. Immunologists. Achievements include research in the lymphoproliferative disorder involving large granular lymphocytes; study of myeloperoxidase gene expression in health and in leukemic conditions; study of retroviral gene sequences in lymphomas and leukemias. Home: 10617 Castelar St Omaha NE 68124 Office: U Nebr Med Ctr 600 S 42nd St Omaha NE 68198-3135

CHAN, YUPO, engineer; b. Canton, Kwantung, China, Aug. 12, 1945; came to U.S. 1964; s. Wui and Yun-yin (Yeung) C.; m. Susan Johnson, Apr. 20, 1991. BS, MIT, 1967, MS, 1969, PhD, 1972. Registered profl. engr., Pa. Cons. Peat Marwick Mitchell, Washington, 1972-74; asst. prof. Pa. State U., University Park, 1974-79; Congl. fellow Office of Tech. Assessment, Washington, 1979-80; assoc. prof. SUNY, Stony Brook, 1980-83, Wash. State U., Pullman, 1983-87; prof., dep. head operational sci. Air Force Inst. Tech., Wright-Patterson AFB, Ohio, 1987—. Author: Facility Location and Land Use, 1994. Harvard-MIT Joint Ctr. Urban Studies fellow, 1971. Fellow ASCE (chmn UPDD exec. com. 1992-93), AIAA, Mil. Ops. Rsch. Soc. (edn. com. 1990—), Ops. Rsch. Soc. Am. (Koopman prize 1991). Office: Air Force Inst Tech Wright Patterson AFB Dayton OH 45433

CHANCE, BRITTON, biophysics and physical chemistry educator emeritus; b. Wilkes Barre, Pa., July 24, 1913; s. Edwin M. and Eleanor (Kent) C.; m. Jane Earle, Mar. 4, 1938 (div.); children: Eleanor, Britton, Jan, Peter; m. Lilian Streeter Lucas, Nov. 1956 (div.); children: Margaret, Lilian, Benjamin, Samuel; stepchildren—Ann Lucas, Gerald B. Lucas, A. Brooke Lucas, William C. Lucas. B.S. and M.S., U. Pa., 1936, Ph.D. (E.R. Johnson Found. fellow), 1940; Ph.D., U. Cambridge, 1942, D.Sc., 1952; M.D. (hon.), Karolinska Inst., Stockholm, 1962; D.Sc. (hon.), Med. Coll. Ohio, 1974, Semmel Weis U., Budapest, 1976; M.D. (hon.), Hahnemann Coll. and Hosp., 1977; D.Sc. (hon.), U. Pa., 1985, U. Helsinki, 1990; M.D. (hon.), U. Dusseldorf, Fed. Republic Germany. Asst. prof. biophysics U. Pa., 1940-48, prof., chmn., 1949—, acting dir. Johnson Found.., 1940-41, dir. Johnson Found.., 1949-83, Eldridge Reeves Johnson prof. biophysics and phys. biochemistry, 1949, 77-83, emeritus prof. biophysics and phys. chemistry, also univ. prof. emeritus, 1983—; dir. Inst. Structural and Functional Studies, 1982-90, Inst. for Biophys. and Biomed. Rsch., 1990—; staff MIT, 1941-46; Cons. NSF, 1952-55; mem. Pres.'s Sci. Adv. Com., 1959-60; mem. adv. council Nat. Inst. Alcohol Abuse and Alcoholism, 1971-75; mem. molecular control working group Nat. Cancer Inst., 1973—. Author: (with F.C. Williams, V. Hughes, E.F. McNichol, David Sayre) Waveforms, 1949, (with R.I. Hulsizer, E.F. McNichol, F.C. Williams) Electronic Time Measurements, 1949, Energy-linked Functions of Mitochondria, 1964, (with Q.H. Gibson, R. Eisenhardt, K.K. Lonberg-Holm) Rapid Mixing and Sampling Techniques in Biochemistry, 1964, (with R.W. Estabrook, J.R. Williamson) Control of Energy Metabolism, 1965, (with R.W. Estabrook, T. Yonetani) Hemes and Hemoproteins, 1966, (with others) Probes of Structure and Function of Macromolecules and Enzymes, 1971, Alcohol and Aldehyde, Vol. I, 1974, II, III, 1977, Tunneling in Biological System, 1979; rev. articles Advances in Enzymology, Vo. 12, 1951, Vol. 17, 1956, Ann. Rev. of Biochemistry, 1952, 70, 76, The Enzymes, Vol. II. Part 1, 1952, Vol. XIII, 1976, Ann. Rev. Plant Physiology, 1958, 68; Bd. editors: Physiol. Revs, 1951-54, FEBS Letters, 1973-75, BBA Reviews, 1972—, Photobiochemistry and Photobiophysics, 1979—; Contbr.: articles to Am., Brit., Swedish, German and Japanese Jours. Presdl. lectr. U. Pa., 1975; Julius L. Jackson Meml. lectr. Wayne State U., 1976; Da Costa oration Phila. County Med. Coll., 1976; Recipient Paul Lewis award for enzyme chemistry, 1950; Pres.'s Certificate of Merit for services, 1941-45, as staff mem. Radiation Lab. of M.I.T., 1950; Guggenheim fellow Stockholm, 1946-48; Harvey lectr., 1954; Phillips lectr., 1955, 65; Pepper lectr., 1957; Exchange scholar to USSR, 1963; Genootschapps medal Dutch Acad. Scis., 1965; Heineken medal, 1970; Keilin medal Brit. Biochem. Soc., 1966; Harrison Howe award, 1966; Franklin medal, 1966; Overseas fellow Churchill Coll., 1966; Herter lectr. NYU, 1968; Pa. award for excellence in life scis., 1968; Nichols award N.Y. sect. Am. Chem. Soc., 1970; Phila. sect. award, 1969; Redfearn lectr., 1970; Gairdner award, 1972; Post-Congress Festschrift Stockholm, 1974; Semmelweis medal, 1974; Nat. medal Sci., 1974, Benjamin Franklin medal Am. Philos. Soc., 1990; Troy C. Daniels lectr. U. Calif.-San Francisco, 1984; Pendergrass lectr. U. Pa., 1991, Christopher Columbus Discovery award in biomedical rsch. NIH, 1992. Fellow Am. Phys. Soc., IEEE (Morlock award 1961, Phila. sect. award 1984), AAAS, Am. Inst. Chemists; mem. NAS, Internat. Union Pure and Applied Biophysics (pres. 1972-75), Chem. Soc., Royal Soc. Arts, Biochem. Soc. Eng., Am. Soc. Biol. Chemists (Sober lectr. 1984), Am. Philos. Soc. (v.p. 1984-90), Am. Acad. Arts and Sci., Am. Physiol. Soc., Am. Acad. Orthopaedic Surgeons (Elizabeth Winston Lanier award 1981, Kappa Delta award 1986), Soc. Magnetic Resonance in Medicine (Gold medal 1988), Soc. Gen. Physiologists (council 1957-60), Am. Inst. Physics, Soc. for Neurosci., Biophys. Soc. (council 1959-62), Swedish Biochem. Soc., Royal Swedish Acad. Scis., Royal Acad. Arts and Scis., Sweden, Bavarian Acad. Scis., Acad. Leopoldina DDR, Max-Planck Gesellschaft für Forerung der Wissenschaften (fgn.), Argentine Nat. Acad. Sci., Royal Soc. London (fgn.), Harvey Soc., Sigma Xi, Tau Beta Pi. Clubs: Corinthian Yacht (Phila.); St. Anthony. Holder numerous patents on automatic steering devices, ship spectrophotometers, radar and bombing devices, nuclear magnetic resonance and optical imaging. Gold medal winner (yachting) 1952 Olympics. Home: 4014 Pine St Philadelphia PA 19104-4027 Office: Univ of PA School of Medicine Dept Biochemistry and Biophysics 36th & Hamilton Walk Philadelphia PA 19104

CHANCE, HUGH NICHOLAS, mechanical engineer; b. Bromsgrove, Eng., Mar. 6, 1940; came to U.S., 1981; s. Sir Hugh and Lady Cynthia (Baker-Creswell) C.; m. Rose Elenor Heft Baldwin, Mar. 18, 1990; children: Timothy W.H., Henry C.H., Lucy E. B.Eng.(Mech), McGill U., Montreal, Quebec, Can., 1964. Owner Harcourt Farm, Brignorth, U.K., 1966-75; project design leader F.W. McConnel Ltd., Ludlow, U.K., 1974-80; pres. Hugh Chance Ltd., Santa Cruz, Calif., 1981—. Mem. Woodmen of Arden. Achievements include patents in field; original work on transportation of particulate material. Home and Office: Hugh Chance Ltd 107 Chrystal Ter Santa Cruz CA 95060-3623

CHANCELLOR, WILLIAM JOSEPH, agricultural engineering educator; b. Alexandria, Va., Aug. 25, 1931; s. John Miller and Caroline (Sedlacek) C.; m. Nongkarn Bodhiprasart, Dec. 13, 1960; 1 child, Marisa Kuakul. BS in Agr., BSME, U. Wis., 1954; MS in Agrl. Engring., Cornell U., 1956, PhD, 1957. Registered profl. agrl. engr., Calif. Prof. agrl. engring. U. California.-Davis, 1957—; vis. prof. agrl. engring. U. Malaya, Kuala Lumpur, Malaysia, 1962-63; UNESCO cons. Punjab Agrl. U., 1976. Contbr. articles to profl. jours.; patentee transmission, planters, dryer, 1961-73. East/West Ctr. sr. Fellow, Honolulu, 1976. Fellow Am. Soc. Agrl. Engrs. (Kishida Internat. award 1984); mem. Soc. Automotive Engrs., Sigma Xi. Office: U Calif Dept Agrl Engring Davis CA 95616

CHANCO, AMADO GARCIA, surgeon; b. Manila, Nov. 8, 1936; came to U.S., 1962; s. Amado L. and Mercedes S. (Garcia) C.; m. Ruby S. Chanco, Jan. 12, 1961; children: Arlene, Anne, Alma, Adele. Student, U. Philippines, Quezon City, 1953-56; MD, Far Ea. U. Inst. Medicine, Manila, 1961. Rotating intern Wilkes-Barre (Pa.) Gen. Hosp., 1961-62; gen. surg. resident Riverside Meth. Hosp., Columbus, Ohio, 1962-65; gen. practice N.D., 1965-69; gen. surgery resident U. Iowa Hosp., Iowa City, 1969-73; gen. surgeon Ind. Med.-Surg. Group, Mason City, Iowa, 1973-89, Mason City Clinic, 1989—; pres. med. staff St. Joseph Mercy Hosp., Mason City, 1985. Mem. Iowa State Med. Soc. (chmn. grievance com. 1990-92), Cerro Gordo County Med. Soc. (pres. 1984), Christian Med. Soc. Republican. Methodist. Achievements include research in mesenteric fibromatosis. Office: Mason City Clinic 300 Eisenhower Mason City IA 50401

CHANDLER, BRUCE FREDERICK, internist; b. Bohemia, Pa., Mar. 26, 1926; s. Frederick Arthur and Minnie Flora (Burkhardt) C.; m. Janice Evelyn Piper, Aug. 14, 1954; children: Barbara, Betty, Karen, Paul, June. Student, Pa. State U., 1942-44; MD, Temple U., 1948. Diplomate Am. Bd. Internal Medicine. Commd. med. officer U.S. Army, 1948, advanced through grades to col., 1967; intern Temple U. Hosp., Phila., 1948-49; chief psychiatry 7th Field Hosp., Trieste, Italy, 1950; resident Walter Reed Gen. Hosp., Washington, 1949-53; battalion surgeon 2d Div. Artillery, Korea, 1953-54; chief renal dialysis unit 45th Evacuation Hosp. and Tokyo Army Hosp., Korea, Japan, 1954-55; various assignments Walter Reed Gen. Hosp., Fitzsimons Gen. Hosp., Letterman Gen. Hosp., 1955-70; comdg. officer 45th Field Hosp., Vicenza, Italy, 1958-62; pvt. practice internist Ridgecrest (Calif.) Med. Clinic, 1970-76; chief med. svc. and out-patients VA Hosps., Walla Walla, Spokane, Wash., 1976-82; med. cons. Social Security Adminstrn., Spokane, Wash., 1983-87; ret. Panel mem. TV shows, 1964-70; lectr.; contbr. numerous articles to med. profl. jours. Decorated Legion of Merit. Fellow ACP, Am. Coll. Chest Physicians; mem. AMA, Am. Thoracic Soc., N.Y. Acad. Scis., So. European Task Force U.S. Army Med. Dental Soc. (pres., founder 1958-62). Republican. Methodist. Avocations: photography, travel, fishing, collecting books. Home: 6496 N Callisch Ave Fresno CA 93710-3902

CHANDLER, HENRY WILLIAM, environmental engineer; b. Greenville, S.C., May 31, 1961; s. Henry William Bauer Jr. and JoAnne (Wauchope) Chandler; m. Pamela Renee Anthony, June 22, 1985; 1 child, Brandon Anthony. BSCE, The Citadel Mil. Coll., 1983; MA, Clemson U., 1991. Registered profl. engr., S.C., N.C., Ga., Md., N.H., Del., Conn. Commd. 2d lt. U.S. Army, Washington, 1983; advance through grades to capt. U.S. Army, 1991; with USAR, 1983—; project mgr. LAW Engring., Greenville, 1984-88; sr. engr. Westinghouse Environ., Spartanburg, S.C., 1988-91; dist. mgr., sr. engr. ATEC Assocs., Inc., Greenville, 1991—; adv. bd. mem. Greenville Tech. Coll., 1986-87. Mem. NSPE, ASCE, Soc. Am. Mil. Engrs. Home: 100 Brigham Creek Ct Greer SC 29650 Office: ATEC Assocs Inc 1200 Woodruff Rd Ste C23 Greenville SC 29607

CHANDLER, JOHN W., biochemistry educator, ophthalmology educator; b. Richland Center, Wis., June 10, 1940. BS in Med. Sci., U. Wis., 1962, MD, 1965. Diplomate Am. Bd. Ophthalmology; lic. in Ill., Wash., Wis. Intern sch. medicine U. Wash., Seattle, 1965-66, rsch. fellow depts. ophthalmology and microbiology, 1968-69, resident in ophthalmology, 1969-72, from asst. to assoc. prof. dept. ophthalmology, 1973-77, clin. assoc. prof., 1978-86; Corneal fellow dept. ophthalmology coll. medicine U. Fla., Gainesville, 1972-73; prof., chmn. dept. ophthalmology U. Wis., Madison, 1986-90; prof., head dept. ophthalmology UIC Eye Ctr. U. Ill., Chgo., 1990—, prof. biochemistry, 1992—, chief ophthalmology eye and ear infirmary, 1990—; chief ophthalmology Michael Reese Hosp. Med. Ctr., Chgo., 1990—; surg. dir. Lions Eye Bank, U. Wash., 1973-77; attending physician Harborview Med. Ctr., Seattle, VA Hosp., Univ. Hosp., 1973-77, Swedish Hosp. Med. Ctr., Seattle, 1978-86, dir. eye inst., 1985-86; dir. ophthalmic rsch. lab. Swedish Hosp., 1978-86; attending physician, chief svc. U. Wis. Hosp. and Clinics, 1986-90; mem. data and safety monitoring com., Herpes eye disease study NIH, 1988—; mem. internat. adv. com. Tissue Banks Internat., 1988—, mem. nat. adv. com., 1990—; mem. adv. bd. Saudi Eye Found. Rsch. and Prevention Blindness, 1989—; mem. task force rev. office of vice chancellors rsch./dean grad. coll. U. Ill., 1991—, chmn. search com. vice chancellor rsch./dean grad. coll., 1992—, mem. health scis. strategic planning task force, 1992—, mem. various coms.; mem. 25th anniversary organizing com. Nat. Eye Inst., 1992—, chmn. bd. counselors, 1993—; cons. in field. Mem. editorial bd. Archives Ophthalmology, 1984—; mem. adv. bd. Saudi Bull. Ophthalmology, 1989—. Bd. dirs. North Fla. Eye Bank, 1972-73. Fellow ACS (mem. program com. 1988—, adv. coun. ophthalmology surgery 1988—); mem. AAAS, AMA, Am. Acad. Ophthalmology and Otolaryngology (instr. 1976—, Honor award 1982, Sr. Honor award 1992), Am. Assn. Immunologists, Am. Soc. Microbiology, Eye Bank Assn. Am., Nat. Soc. Prevent Blindness (chmn. sci. adv. com. 1990—), Ill. State Med. Soc., Chgo. Ophthalmology Soc., Assn. Rsch. Vision and Ophthalmology, Assn. Univ. Profs. Ophthalmology (mem. rsch. com. 1990—, patient care com. 1990—), Clin. Immunology Soc., Internat. Soc. Eye Rsch., Ocular Microbiology and Immunology Soc., Pacific Coast Oto-Ophthal. Soc., Rsch. Prevent Blindness, Castroviejo Soc. Office: Univ of Illinois at Chicago Lions Ill Eye Rsch Inst 1905 W Taylor St Chicago IL 60612

CHANDLER, PAUL ANDERSON, physicist, researcher; b. Clinton, S.C., Sept. 13, 1933; s. Carl Irby and Maggie (Lynch) C.; m. Mary Jo Wasson, Feb. 2, 1959; children: Margaret, Paula. BS cum laude, Presbyterian Coll., 1959; MS, Inst. Textile Tech., Charlottesville, Va., 1961. Physicist Union Carbide Corp., South Charleston, W.Va., 1960-61; sr. scientist, group leader Allied-Signal Corp., Petersburg, Va., 1961—. Staff Sgt. USAF, 1951-55. Inst. Textile Tech. fellow, 1960. Mem. ASTM, Physics Club of Richmond (charter, treas. 1968-69). Republican. Presbyterian. Avocations: reading, motorcycling, travel. Home: 14011 Howlett Line Dr Colonial Heights VA 23834-5864 Office: Allied Signal Corp PO Box 31 Petersburg VA 23804-0031

CHANDLER, THOMAS EUGENE, industrial engineer; b. Florence, Ala., Aug. 7, 1958; s. Willie Eugene and Rita Renee (Rivas) C.; m. Robin Jeanice Giles, June 6, 1980; children: Kathryn Michelle, Tyler Giles. BS Indsl. Engring. with honors, Auburn U., 1980. Assoc. engr. maintenance Internat. Paper, Mobile, Ala., 1980-81; project engr. constrn. Internat. Paper, Moss Point, Miss., 1981-82; sr. project engr. plant engring. Internat. Paper, Mobile, 1982-83, design engr. plant engring., 1983-84, sr. design engr. tech. svcs., 1985-87, dept. engr. constrn., 1988-90, sr. dept. engr. corp. process tech., 1991—. Mem. Corp. Citizenship Team, Mobile, 1991—; bd. advisors Crown Ministries, Mobile, 1989; bd. dirs. Christian Bus. Men's Com., Mobile, 1992; mem. fin. com. St. John Ch., Mobile, 1987—. Named Eagle Scout, Boy Scouts Am., 1972. Mem. NSPE, Ala. Soc. Profl. Engrs. Home: 5854 Saint Gallen Ave S Mobile AL 36608

CHANDLEY, ANN CHESTER, research scientist, cytogeneticist; b. Gatley, Eng., Mar. 30, 1936; d. Samuel and Freda (Chester) C. BSc, Manchester (Eng.) U., 1957, MSc, 1961, PhD, 1968, DSc, 1978. Scientist various orgns. Manchester, 1957-69; scientist with spl. responsibility MRC Human Genetics Unit Western Gen. Hosp., Edinburgh, Scotland, 1969—. Co-author: Aneuploidy, 1983; mem. editorial bd. Cytogenetics and Cell Genetics, 1981-86, Jour. Med. Genetics, 1986-93, Jour. Reproduction and Fertility, 1993—. Fellow Inst. of Biology, Royal Soc. Edinburgh; mem. Assn. Clin. Cytogeneticists, European Soc. Human Genetics, Genetical Soc. Great Britain (sec. 1975-81, v.p. 1981-84). Avocations: music, cycling, walking, gardening. Home: 20 Comely Bank, Edinburgh EH4 1AL, Scotland Office: MRC Human Genetics Unit, Western Gen Hosp, Crewe Rd, Edinburgh EH4 2XU, Scotland

CHANDRA, AJEY, chemical engineer; b. India, June 17, 1964; came to U.S., 1968; s. Suresh and Krishna (Agarwal) C.; m. Ann Piccolo, July 30, 1988. BS in Chem. Engring., Tex. A&M U., 1986; MBA, U. Houston, 1991. Registered profl. engr., Tex. Engr. Amoco Prodn. Co., Houston, 1986—. Chartered orgn. rep. Boy Scouts Am., Houston, 1986—; treas. Westlake Asian Am. Assn. Recipient Gold Congl. award U.S. Congress, 1987, United Way Project Blueprint, 1993. Mem. NSPE, Am. Inst. Chem. Engrs., Instrument Soc. Am., Soc. Petroleum Engrs, Gas Processors Assn. Office: Amoco Prodn Co 501 Westlake Park Blvd Houston TX 77079

CHANDRA, KAVITHA, electrical engineer, educator; b. Bangalore, India, Aug. 1, 1961; came to the U.S., 1985; d. Sajjan and Uma Devi (Chandrashekariah) C. MS in Computer Engring., U. Mass., Lowell, 1987, D in Engring., 1992. Teaching asst. U. Mass., 1986-90, rsch. asst., 1990-92, rsch. asst. prof., 1992—. Contbr. articles to Jour. Acoustical Soc. Am., Frontiers of Nonlinear Acoustics, Jour. Nat. Tech. Assn. Advisor to minority undergrads. U. Mass. Lab. for Advanced Computation, 1990. Named one of Outstanding Young Women Am., 1991. Mem. Acoustical Soc. Am., IEEE, Soc. Women Engrs., Sigma Xi. Achievements include first introduction of Padé approximant theory in two dimensional ultrasonic scattering problems; research in characterization of bone tissue and diagnosis of osteoporotic conditions using scattered ultrasonic waves. Home: 1009 Westford St # 21 Lowell MA 01851 Office: U Mass Dept Elec Engring 1 University Ave Lowell MA 01854

CHANDRA, RAMESH, medical physics educator; b. Nakur, India, June 9, 1942; s. Brij Bhushan and Prakash (Vati) Gupta; m. Mithilesh Aron, Aug. 19, 1966; children: Anurag, Ritu. MSc in Physics, Allahabad (India) U., 1961; PhD in Physics, Boston U., 1968. Asst. prof. NYU Med. Ctr., N.Y.C., 1970-74, assoc. prof., 1974-78, prof., 1978—. Author: Introductory Physics of Nuclear Medicine, 1976, 4th edit., 1992; co-editor: Physics of Nuclear Medicine: Recent Advances, 1984. Mem. Am. Assn. Phusicists in Medicine, Soc. Nuclear Medicine, Soc. Magnetic Resonance in Medicine, Soc., Magnetic Resonance and Imaging, Soc. Photo-optical Instrumentation Engrs. Hindu. Office: NYU Med Ctr Dept Radiology 560 1st Ave New York NY 10016

CHANDRASEKARAN, RENGASWAMI, biochemist, educator; b. Bangalore, India, June 26, 1939; came to U.S., 1968; s. Panchanathan and Savithri R.; m. Reva Ramachandran, May 12, 1967; 1 child, Sujatha. MS in Physics, U. Madras, India, 1960, PhD in X-Ray Crystallography, 1966. Rsch. assoc. U. Chgo., 1968-71; asst. prof. Indian Inst. Sci., Bangalore, 1974-77; rsch. assoc. Purdue U., Lafayette, Ind., 1971-74, rsch. scientist, 1977-86, assoc. prof., 1986-92, prof. structural biochemistry, 1992—; referee various jours., 1980—. Editor: Frontiers in Carbohydrate Research 1, 1989, Frontiers in Carbohydrate Research 2, 1992; contbr. articles on structures of polypeptides, polysaccharides and nucleic acids to internat. jours. Mem. Am. Crystallographic Assn., Am. Chem. Soc., Am. Soc. Biochemistry and Molecular Biology, Am. Assn. Cereal Chemists, Inst. Food Technologists. Achievements include research in structure-function relationships in proteins, nucleic acids and polysaccharieds, computer modeling of macromolecules, molecular basis of functionality of food materials. Office: Purdue Univ 1160 Smith Hall Lafayette IN 47907

CHANDRASEKHAR, SUBRAHMANYAN, theoretical astrophysicist, educator; b. Lahore, India, Oct. 19, 1910; came to U.S., 1936, naturalized, 1953; m. Lalitha Doraiswamy, Sept. 1936. M.A., Presidency Coll., Madras, 1930; Ph.D., Trinity Coll., Cambridge, 1933, Sc.D., 1942; Sc.D., U. Mysore, India, 1961, Northwestern U., 1962, U. Newcastle Upon Tyne, Eng., 1965, Ind. Inst. Tech., 1966, U. Mich., 1967, U. Liege, Belgium, 1967, Oxford (Eng.) U., 1972, U. Delhi, 1973, Carleton U., Can., 1978, Harvard U., 1979. Govt. India scholar in theoretical physics Cambridge, 1930-34; fellow Trinity Coll., Cambridge, 1933-37; rsch. assoc. Yerkes Obs., Williams Bay and U. Chgo., 1937, asst. prof., 1938-41, assoc. prof., 1942-43, prof., 1944-47, Disting. Service prof., 1947-52, Morton D. Hull Disting. Service prof., 1952-86, prof. emeritus, 1986—; Nehru Meml. lectr., Padma Vibhushan, India, 1968. Author: An Introduction to the Study of Stellar Structure, 1939, Principles of Stellar Dynamics, 1942, Radiative Transfer, 1950, Hydrodynamic and Hydromagnetic Stability, 1961, Ellipsoidal Figures of Equilibrium, 1969, The Mathematical Theory of Black Holes, 1983, Eddington: The Most Distinguished Astrophysicist of His Time, 1983, Truth and Beauty: Aesthetics and Motivations in Science, 1987, Selected Papers, 6 Vols., 1989-90; mng. editor: The Astrophysical Jour., 1952-71; contbr. various sci. periodicals and articles to profl. jours. Recipient Bruce medal Astron. Soc. Pacific, 1952, gold medal Royal Astron. Soc., London, 1953; Rumford medal Am. Acad. Arts and Scis., 1957; Nat. Medal of Sci., 1966; Nobel prize in physics, 1983; Dr. Tomalla prize Eidgenössiches Technische Hochschule, Zurich, 1984. Fellow Royal Soc. (London) (Royal medal 1962, Copley medal 1984); mem. Nat. Acad. Scis. (Henry Draper medal 1971), Am. Phys. Soc. (Dannie Heineman prize 1974), Am. Philos. Soc., Cambridge Philos. Soc., Am. Astron. Soc., Royal Astron. Soc. Club: Quadrangle (U. Chgo.). Office: U of Chicago Astrophysics & Space Rsch Lab 933 E 56th St Chicago IL 60637-1460

CHANDRA SEKHARAN, PAKKIRISAMY, forensic scientist; b. Nagapattinam, India, June 15, 1934; s. Pakkirisamy Sattayappan and Meenakshisundaram; m. Evelyn Swamy, Dec. 28, 1968; children: Meena, Anand. MSc, Annamalai U., Tamil Nadu State, 1958; PhD, U. Madras, Tamil Nadu State, 1986; BL, Madras U., 1986. Head of sect. in physics and chemistry Tamil Nadu Poly., Madurai, 1958-59; second physicist Govt. Erskine Hosp., Madurai, 1959-65; asst. dir. physics Forensic Sci. Lab., Madras, 1965-71, dir., 1971-80; dir., prof. Forensic Scis. Dept., Madras, 1980—; mem. World Body of Cranial Indentification Group, Kiel, West Germany, 1990—. Editor: Frontiers of Forensics, 1989, A Treatise on Bank Crime, Prevention and Detection, 1988, Forensic Science As is What Is?, 1987; chief-editor Indian Jour. of Forensci Scis., 1986 ; contbr. articles to profl. jours. Pres. Madras Young Men's Christian Assn., 1982; chmn. Denaturant Com., Madras; convener Working Group for 8th Plan Recommendation, Madras, 1988. Recipient J.L. Eustace Meml. award U. Adelaide, South Australia, 1988, Scroll of Honour Soc. of Toxicology of India, 1984, For the Sake of Honour Rotary Club of Madras Metro, 1991, Bharatiya Udyog Jyoti award, 1992. Fellow Inst. Physics, Indian Acad. Forensic Scis., Forensic Sci. Soc. of India (pres. 1979—), Indian Soc. Criminology; mem. Y's Men Internat. (no. region Madras Meritorious award 1988, dist. gov. no. region 1988), Presidency Club, Madras Gymkhana Club. Roman Catholic. Achievements include design of revised skull face superimposition technique; rsch. in propounder of skull suture theory and identification technique via skull suture pattern, propounder of scientific methods of indexing bronzes, propounder of classification system for check writing in Indian currency; fabricated electronic skull identification device. Home: 25 Thomas Nagar, Little Mount Saidapet, Madras 600 015, India Office: Forensic Scis Dept, Govt of Tamilpadu, 20 A Kamarajar Salai, Mylopore, Madras 600 004, India

CHANDRASHEKAR, VARADARAJ, endocrinologist; b. Karnataka, India, Aug. 12, 1942; came to U.S., 1968; s. Varadaraj and Logambal Pillai; m. Padmalatha Puddukotai, June 21, 1972; 1 child, Rupa. PhD, Rutgers U., 1975. Rsch. fellow U. Mysore, India, 1964-68; rsch. asst. Rutgers U., New Brunswick, N.J., 1968-74; postdoctoral/rsch. assoc. U. Wis., Madison, 1974-80; vis. asst. prof. depts. Physiology & Zoology So. Ill. U., Carbondale, 1980-84, assoc. scientist, 1984—; rsch. fellow Ford Found., U. Mysore, 1964-68. Contbr. articles to profl. jours., chpts. to books. Mem. Endocrine Soc., Soc. for Study of Reproduction. Achievements include research in transgenic mice expressing the human growth hormone gene are hypoprolactinemic and their neuroendocrine functions are alerted. Office: So Ill U Sch Medicine Dept Physiology Carbondale IL 62901

CHANDROSS, EDWIN A., chemist, polymer researcher; b. N.Y.C., Oct. 13, 1934. BS, MIT, 1955; MA, Harvard U., 1957, PhD, 1960. Mem. tech. staff AT&T Bell Labs., Murray Hill, N.J., 1959—, head dept. organic chemistry R&D, 1980—; adv. bd. Chem. Revs., 1978—, MIT program in polymer sci. and tech., Cambridge, 1990—. Achievements include over 40 U.S. patents in optical properties of polymers, and photosensitive materials. Office: AT&T Bell Labs 1D246 New Providence NJ 07974

CHANENCHUK, CLAIRE ANN, chemical engineer, market developer; b. Huntingdon, Pa., Apr. 4, 1965; d. David Lee Battistella and Dixie Darlene (Krouse) Norris; m. Bruce Francis Chanenchuk, Aug. 15, 1987; 1 child, Bruce Clement. BS in Chem. Engring., Johns Hopkins U., 1987; MS in Chem. Engring. Practice, MIT, 1989, PhD in Chem. Engring., 1991. Tech. cons. GE, Selkirk, N.Y., 1988, Dow Chem. Co., Midland, Mich., 1988; dir.

tech. ops. Molten Metal Tech., Waltham, Mass., 1991—; presenter at profl. confs. Contbr. articles on energy, fuels and pollution control to profl. publs. Mem. Am. Inst. Chem. Engrs., Am. Chem. Soc., Sigma Xi, Tau Beta Pi, Phi Mu. Achievements include research in reaction engineering and heterogeneous catalysis, iron and cobalt catalyzed Fischer-Tropsch synthesis, catalytic extraction processing, an environmental technology using molten metal. Home: 14 Penobscot Rd Natick MA 01760 Office: Molten Metal Tech 2 Univ Park 510 Sawyer Rd Waltham MA 02154

CHANEY, RONALD CLAIRE, environmental engineering educator, consultant; b. Tulsa, Okla., Mar. 26, 1944; s. Clarence Emerson and Virginia Margaret (Klinger) C.; m. Barbara Marquardt, Jan. 18, 1969 (div. Sept. 1974); m. Patricia Jane Robinson, Aug. 11, 1984. BS, Calif. State U., Long Beach, 1969; MS, Calif. State U., 1970; PhD, UCLA, 1978. Prof. engr., Calif.; geotech. engr., Calif. Structural engr. Fluor Corp. Ltd., L.A., 1968-70; rsch. engr. UCLA, 1972-74; lab. mgr. Fugro Inc., Long Beach, 1974-79; assoc. prof. Lehigh U., Bethlehem, Pa., 1979-81; dir. Telonicher Marine Lab. Humboldt State U., Trinidad, Calif., 1984-90; prof. Humboldt State U., Arcata, Calif., 1981—; geotech. engr. LACO Assocs., Eureka, Calif., 1988-93; panel mem. Humboldt County Solid Waste Appeals, Eureka, 1992-93; mem. shipboard measurement panel Joint Oceanog. Instn., Washington, 1991-93. Co-editor: Marine Geotechnology and Nearshore/Offshore Structures, 1986, Symposium on Geotechnical Aspects of Waste Disposal in the Marine Environment, 1990; editor Marine Geotech. jour., 1981-92; co-editor Marine Georesources and Geotech. jour., 1992-93. Mem. ASCE, ASTM (Hogentogler award 1988, Std. Devel. award 1991, Outstanding Achievement award 1992), Seismological Soc. Am., Earthquake Engring. Rsch. Inst., Sigma Xi, Phi Kappa Phi. Office: Humboldt State U Dept Environ Engring Arcata CA 95521

CHANG, ANDREW C., engineer, educator; b. Jan. 28, 1940; s. C. Y. and Josephine (Tu) C.; m. Linda L. Chao, Sept. 2, 1967; children: Edie, Joseph. MS, Va. Poly. Inst. and State U., 1965; PhD, Purdue U., 1971. Registered profl. engr., Calif. Asst., assoc., prof. U. Calif., Riverside, 1991—; dir. Kearney Found. of Soil Sci., U. Calif., Riverside, 1991—. Recipient Disting. Svc. award USDA, 1991, 1st Pl. award for Rsch. on Beneficial Use of Mcpl. Sludge, U.S. EPA, 1991. Mem. Soil Sci. Soc. of Am., Internat. Assn. Water Quality, Calif. Water Pollution Control Assn. Office: U of California Riverside CA 92521

CHANG, CHARLES SHING, biology educator, researcher; b. Tamshui, Taipei, Taiwan, Nov. 18, 1940; came to U.S., 1972; s. Ang Pong and Tau (Lin) C.; m. Celia Chien-Huey Shiao, Dec. 28, 1969; children: Alice, Betty, Christina. BS, Nat. Taiwan Normal U., Taipei, 1965; MS in Pharmacy, U. Chinese Culture, Taipei, 1969; MS in Biology, Midwestern State U., Wichita Falls, Tex., 1976; PhD in Microbiology and Biochemistry, U. North Tex., 1980. Rsch. assoc. Chinese Pharm. Rsch. Inst. U. Chinese Culture, Taipei, 1970; rsch. assoc. Tex. Coll. Osteo. Medicine U. North Tex., Denton, 1980; rsch. assoc. U. Tex. and VA Hosp. at Dallas, 1982; vis. prof. Fu-Jen Cath. U., Taipei, 1984; dir. Xin-Ichi Rsch. Inst., Arcadia, Calif., 1990—; researcher Republic of China Nat. Sci. Ctr., Taipei, 1984; cons. Pen T'sao Health Food Co., L.A., 1990. Author: Molecular Biology I, Molecular Biology II, 1985-87, Modern Virology, 1989; contbr. articles to profl. jours. 2d lt. ROTC, 1966-67, Taipei. Sci. and Tech. grantee Republic of China Nat. Sci. Ctr., 1983-84; World-wide Famous Books Translation grantee Republic of China Compilation Com., 1985-87. Mem. AAAS, N.Y. Acad. Scis., Taiwan Pharm. Assn., Am. Soc. Microbiology, Am. Official Analytical Chemistry.

CHANG, CHENG-HUI (KAREN), medical physicist; b. Tainan, Taiwan, Nov. 18, 1948; came to U.S., 1973; d. Wei-Chang Chang and H.C. Yeh. PhD in Physics, U. Ark., 1978. Diplomate Am. Bd. Radiology. Rsch. fellow in med. physics U. Tex. Health Sci. Ctr. at Dallas, 1983-85, med. physicist, 1985-86; asst. prof. W.Va. U., Morgantown, 1986-87; med. physicist North Shore U. Hosp., Manhasset, N.Y., 1988-89; asst. prof. Baylor Coll. of Medicine, Houston, 1989-90; med. physicist Moncrief Radiation Ctr., Ft. Worth, 1990-92; asst. prof. U. Tex. S.W. Med. Ctr., Dallas, 1992—. Contbr. articles to profl. jours. Mem. Am. Assn. Physicists in Medicine, Am. Physics Soc., Am. Soc. Therapeutic Radiology and Oncology, Sigma Xi. Office: U Tex SW Med Ctr 5323 Harry Hines Blvd Dallas TX 75235-9071

CHANG, CHIA-WUN, chemist, researcher; b. Taipei, Taiwan, Feb. 22, 1957; arrived in Japan, 1987; s. Tsung-Ming M and Su-Chen (Lai) C. BSc, U. Chicago, 1979, PhD in Chemistry, 1987. Postdoctoral fellow Suntory Inst. for Biorganic Rsch., Osaka, Japan, 1987-90; researcher Kanagawa Acad. Sci. and Tech., Kawasaki, Kanagawa, Japan, 1990-93, Jasco Rsch. Ctr., Hachioji, Japan, 1990-93. Contbr. articles to Jour. Am. Chem., Proceedings Nat. Acad. Sci., FEBS Let. Office: Rohm & Haas Japan, Kaisei Bldg 1-8-10, Azabudai Minato-Ku Tokyo 213, Japan

CHANG, CHONG EUN, chemical engineer; b. Seoul, Korea, Dec. 4, 1938; came to U.S., 1968; BS in Chem. Engring., Seoul Nat. U., 1964; PhD in Chem. Engring., U. So. Calif., 1973, MS in Mech. Engring., 1977. Asst. chief of lab. Hyundai Co., Seoul, 1964-68; postdoctoral rsch. assoc. U. So. Calif., L.A., 1973-75; sr. process engr. Allis-Chalmers Corp., Stansteel Products, L.A., 1976-78; mgr. process devel. Alpha Therapeutics Corp., L.A., 1979-81, sr. prin. scientist, 1981-92; v.p. RAAS, Inc., Agoura Hills, Calif., 1992—. Contbr. articles to Jour. Crystal Growth, Internat. Jour. Heat and Mass Trans. With Korean Army, 1959-60. Archimedes Cir. scholar. 1971. Mem. AICE. Achievements include patents for fractionation of blood plasma, albumin purification. Home: 5833 Briartree Dr La Canada Flintridge CA 91011 Office: RAAS Inc 30423 Canwood St Agoura Hills CA 91301

CHANG, CHUAN CHUNG, physicist; b. Tainan, Taiwan, Nov. 28, 1938; came to U.S., 1958; s. Shiang-Hua and Ching-Hua Chang; m. Merry Chang, Aug. 17, 1963; children: Eileen, Sue-Lynn. BS in Physics, Rensselaer Poly. Inst., 1962; PhD in Physics, Cornell U., 1967. Mem. tech. staff Bell Labs., Murray Hill, N.J., 1967-83; disting. mem. tech. staff Bellcore, Red Bank, N.J., 1983—. Contbr. over 200 articles to profl. jours. Achievements include patent on GaAs MOS on low energy electron diffraction, auger electron spectroscopy, semiconductor device processing, electronics materials science, and superconductivity, image storage, cleaning compound semiconductor surfaces ohmic contacts, cleaning probe tips, a new FET structure, use of nitridel oxides in Si ICs, also others. Over 100 internal reparts on telecommunications equiptment. Office: Bellcore 331 Newman Springs Rd Red Bank NJ 07701

CHANG, CHUN-HSING, psychologist, educator; b. Changlo, Shantung, China, Nov. 18, 1927; s. Hwa-Yun Chang and Yu (Yun) Kau; m. Chou Hwei-Chiang, Dec. 27, 1958; children: Hsiu-Jan, Jieh-Jan, Y-Jan. BEd, Taiwan Normal U., Taipei, 1955, MEd, 1959; MA, U. Hawaii, 1962; PhD, U. Oregon, 1972. Prof. psychology Taiwan Normal U., Taipei, 1966-72, chmn. psychology dept., 1972-77; vis. prof. Purdue U., West Lafayette, Ind., 1977-78, Taiwan Normal U., Taipei, 1978—; cons. Ministry Edn., Taipei, 1987—. Author: Principles of Psychology, 1977, Educational Psychology, 1980, Series in Today's Youth and Education (6 vols.), 1985-89, Modern Psychology, 1990; editor: Chang's Dictionary of Psychology, 1988, Perspectives in Educaitonal Psychology: Pedagogical, Holistic, and Cultural, 1993. Mem. Chinese Psychol. Assn. (pres. 1983-85), Chinese Ednl. Soc., Chinese Mental Health Assn., Chinese Guidance Assn. Office: Taiwan Normal U Ednl Psycho, 162 Hoping Rd E Sec 2, Taipei Taiwan

CHANG, DANIEL HSING-NAN, aerospace engineer; b. Taipei, Taiwan, Mar. 22, 1967; s. Chen-Che and Meng-Lan (Mi) C.; m. Leann Marie Abbate, Feb. 28, 1992. BS in Aero./Astronautical, MIT, 1989, MS in Aero./Astronautical, 1991. Engr. Ford Motor Co., Dearborn, Mich., 1989; mem. tech. staff Jet Propulsion Lab., Pasadena, Calif., 1991—. Mem. AIAA, Sigma Xi. Office: Jet Propulsion Lab M/S 198-326 4800 Oak Grove Dr Pasadena CA 91214

CHANG, DONALD CHOY, biophysicist; b. Shunzhen, Guangdong, People's Republic China, Aug. 28, 1942; came to U.S., 1965; BS, Nat. Taiwan U., 1965; MA, Rice U., 1967, PhD, 1970. Rsch. assoc. Rice U., Houston, 1970-74; asst. prof. dept. pediatrics Baylor Coll. Medicine, Houston, 1974-77, asst. prof. dept. physiology, 1978-85, assoc. prof. dept. molecular physiology and biophysics, 1985-91; prof. Hong Kong U. Sci. and Tech., 1989—; conf. chmn. Internat. Conf. on Structure and Function in Excitable Cells, Woods Hole, Mass., 1981, Internat. Conf. on Electroporation and Electrofusion, Woods Hole, 1990. Editor: Structure and Function in Excitable Cells, 1983, Guide to Electroporation and Electrofusion, 1991. Pres. Nat. Assn. Chinese Am., Houston, 1985-86, Chinese in Neurosci. Club, N.Am., 1984-85. Welch fellow Welch Found., 1970-73; grantee NIH, 1976-92, NSF, 1984-88, 91—, Advanced Tech. Program, 1988-90, 91-93. Mem. Marine Biol. Lab. at Woods Hole (corp.), Soc. Chinese Bioscientists in Am. (life, sec. Houston chpt. 1988-90), Biophys. Soc., Soc. for Cell Biology. Achievements include patents for electroporation method using RF field; first to visualize electric field-induced membrane pores; application of nuclear magnetic resonance for detection of cancer; research on the physicochemical basis of potential generation in nerve cells, which has lead to the modification of the Goldman-Hodgkin-Katz equation. Office: Hong Kong U Sci and Tech, Clear Water Bay, Kowloon Hong Kong

CHANG, FRANCIS, medical physicist; b. China, Oct. 2, 1945; s. Chao-Chien and Yin-Lien (Wu) C.; m. Hsuan Hung, Mar. 20, 1976; children: Christine, Justin, Oliver. MS, U. Tex. Arlington, 1973; PhD, U. Ala., Birmingham, 1977. Diplomate Am. Bd. Radiology, Am. Bd. Med. Physics. Tech. dir. St. Frnacis Hosp. and Med. Ctr., Topeka, 1977-83; dir. med. physics North Shore Med. Ctr., Miami, Fla., 1983—; bd. dirs., dep. radiation safety officer North Shore Med. Ctr., Miami, 1983—. Contbr. articles to profl. jours. Bd. dirs., v.p. Lake Elizabeth Home Assn., Miami, 1990. Mem. Am. Assn. Physicists in Medicine, Health Physics Soc., Am. Coll. Radiology. Achievements include design and execution of the quality assurance program for radiation therapy dept.; set up and supervision of the radiation safety program for community hosp. and med. ctr. Office: North Shore Med Ctr Griffith Cancer Ctr 1100 NW 95th St Miami FL 33150

CHANG, HSUAN HUNG, dosimetrist; b. China, July 31, 1950; came to U.S., 1972; d. Yi and Hui-Ying Hung; m. Francis Chang, Mar. 20, 1976; children: Christine, Justin, Oliver. MS, U. Tex. at Dallas, 1974; PhD, U. Ala. at Birmingham, 1977. Cert. med. dosimetrist. Rsch. asst. prof. U. Kans., Lawrence, 1978-83; dosimetrist North Shore Med. Ctr., Miami, Fla., 1983—. Contbr. articles to profl. jours. Bd. dirs. Miami Chinese Lang. Sch., U. Miami, 1992. Office: North Shore Med Ctr Cancer Ctr 1100 NW95th St Miami FL 33150

CHANG, HUAI TED, environmental engineer, educator, researcher; b. Chaochou, Pingtong, Taiwan, Apr. 12, 1955; came to U.S., 1980; s. Tsang Shun and Jui Hsia (Fang) C.; m. Jane Liao, Aug. 13, 1984; children: Kelly Chiryurn, Alice. MS in Environ. Engring., Nat. Cheng Kung U., Tainan, Taiwan, 1976; PhD in Environ. Engring., U. Ill., 1985. Registered profl. engr., Taiwan. Rsch. assoc. U. Ill., Urbana, 1985-88; sr. environ. engr. Twin City Testing Corp., St. Paul, 1988-90; rsch. engr. Amoco Rsch. Ctr., Naperville, Ill., 1990-92, Ill. Inst. Tech., Chgo., 1992—. V.p. Minn. Taiwanese Assn., St. Paul, 1989-90, treas., 1988-89. Lt. Chinese Army, 1978-80. Mem. ASCE, Assn. Environ. Engring. Profs., Taiwanese Assn. Am. (sec. Chgo. chpt. 1990-92). Office: Pritzker Dept Environ Engr 3200 S State St Chicago IL 60616-3793

CHANG, JAE CHAN, physician, educator; b. Chong An, Korea, Aug. 29, 1941; s. Tae Whan and Kap Hee (Lee) C.; came to U.S., 1965, naturalized, 1976; M.D., Seoul (Korea) Nat. U., 1965; m. Sue Young Chung, Dec. 4, 1965; children: Sung-Jin, Sung-Ju, Sung-Hoon. Intern, Ellis Hosp., Schenectady, 1965-66; resident in medicine Harrisburg (Pa.) Hosp., 1966-69, fellow in nuclear medicine, 1969-70; fellow in hematology and oncology, instr. in medicine U. Rochester, N.Y., 1970-72; chief hematology sect. VA Hosp., Dayton, Ohio, 1972-75; hematopathologist Good Samaritan Hosp., Dayton, 1975—, dir. oncology unit, 1976—, coord. of med. edn., 1976-77, chief oncology-hematology sect., 1976—; co-dir. hematology lab., 1988—; asst. clin. prof. medicine Ohio State U., Columbus, 1972-75; assoc. clin. prof. medicine Wright State U., Dayton, 1975-80, clin. prof., 1980—, co-dir. hematology and med. oncology fellowship program Wright State U. Sch. Medicine, 1993—; staff Kettering Med. Ctr., St. Elizabeth Med. Ctr., Dayton, Miami Valley Hosp., Dayton; cons. in hematology VA Hosp. Mem. med. adv. com. Greater Dayton Area chpt. Leukemia Soc. Am., 1977; trustee Montgomery County Soc. for Cancer Control, Dayton, 1976-85, Dayton Area Cancer Assn., 1985-88, Community Blood Ctr., 1982-86; bd. dirs. Samaritan Health Found., 1992. Recipient Wright State U. Acad. Medicine award, 1985, Med. Econ. Essay Competition award, 1990. Nat. Cancer Inst. fellow in hematology and oncology, 1970-72; diplomate Am. Bd. Internal Medicine, Am. Bd. Pathology. Fellow ACP, Am. Soc. Clin. Pathologists; mem. AAAS, Am. Soc. Hematology, Am. Fedn. Clin. Research, Am. Soc. Clin. Oncologists, Am. Assn. Cancer Research, Dayton Oncology Club, Dayton Soc. Internal Medicine (pres. 1989), Montgomery County Med. Soc. (dir. 1990-93). Contbr. articles to profl. med. jours., essays to newspaper columns, mags. and periodicals. Home: 1905 Kresswood Cir Kettering OH 45429 Office: Good Samaritan Hosp & Health Ctr 2222 Philadelphia Dr Dayton OH 45406-1891 also: 2661 Salem Ave Ste 232 Dayton OH 45406 also: 7083 Corporate Way Dayton OH 45459

CHANG, JERJANG, veterinarian, educator; b. Taichung, Taiwan, June 11, 1940; came to U.S., 1966; s. Sung-Chiang and Yueh-Sing (Tsai) C.; m. Renee Cheng Huang, Dec. 21, 1969; 1 child, Ricahrd, Jacqueline. DVM, Nat. Taiwan U., 1964; PhD, U. Mo., Columbia, 1971. Rsch. asst. NAMRU-2, Taipei, Taiwan, 1965-66, U. Mo., Columbia, 1966-71; sr. microbiologist SmithKline Beecham, West Chester, Pa., 1971-73; asst. pathologist Emory U., Yerkes Primate Ctr., Atlanta, 1973-76; chief veterinarian Duke U., Durham, N.C., 1976-81; assoc. prof. pathology U. N.C., Chapel Hill, 1981—; cons. Durham (N.C.) Vets. Hosp., 1979-83; mem. comparative medicine rev. com. NIH, Bethesda, 1989-92. Contbr. articles to Lab. Animal Scis., Jour. Medicinal Chemistry, Pesticide Biochemistry and Physiology, Cancer Letter. Mem. AAAS, Am. Coll. Lab. Animal Medicine, Am. Veterinary Med. Assn., Am. Assn. Lab. Animal Scis. Office: Univ NC CB #7115 300 Bynam Hall Chapel Hill NC 27599

CHANG, JIAN CHERNG, research analyst; b. Malacca, Malaysia, Feb. 28, 1939; came to U.S., 1962.; s. Wen Mei and Fay Ching (Chan) Chang. BS, Nanyang U., 1962; PhD, U. Mo., 1970. Rscher. Cancer Rsch. Ctr., Columbia, Mo., 1970-77; rsch. analyst Mo. Dept. Health, Columbia, 1977—. Contbr. articles to profl. jours. Mem. AAAS, Internat. Assn. Math. Modelling, Am. Phys. Soc., N.Y. Acad. Scis. Achievements include development in mathematical modeling for the survival of cancer patients. Home: 414 S William St Columbia MO 65201-5837 Office: Mo Cancer Registry Mo Dept Health 201 Business Loop 70 W Columbia MO 65203

CHANG, KAI SIUNG, medical physicist; b. Teluk-Betung, Lampung, Indonesia, Aug. 29, 1939; arrived in Can., 1967; s. Tien Lim and Pat Moy (Fuk) Tjong; div.; children: Peggy S.F., Allison H.F., Jamie S.L., Angela S.F. BSc, Nat. Taiwan U., Taipei, 1965; MSc, Meml. U. of Newfoundland, Can., 1971, PhD, 1974. Med. rsch. scientist Faculty of Medicine, Meml. U. of Newfoundland, St. Johns, Can., 1974-85; sr. phys. scientist Centre for Cold Ocean Resources Engring., St. Johns, 1982-83; med. physicist Ont. Cancer Treament and Rsch. Found., Ont., Can., 1985-88, Newfoundland Cancer Treament and Rsch. Found., St. John's, 1988-89; asst. prof. U. Mich., Ann Arbor, 1989-91; chief physicist Oakwood Hosp., Dearborn, Mich., 1989-91, The Meth. Hosp., Merrillville/Gary, Ind., 1991—. Contbr. articles to profl. jours. Councillor YWCA, Taipei, 1964-67; coord. Friends of Refugee, St. John's, 1980-85. 2d lt. Army, 1965-66. Mem. Am. Assn. of Physicists in Medicine, Am. Coll. Radiology, Internat. Bioelectric Repair and Growth Soc, Chinese Assn. Physicists, Can. Assn. Physicists. Office: Meth Hosp 8701 Broadway Blvd Merrillville IN 46410

CHANG, KEH-CHIN, aerospace engineer, educator; b. Taitung, Taiwan, Republic of China, Jan. 14, 1952; s. Wen-Chen and Pin-Mei (Tu) C.; m. Jane Y. Chen, Feb. 16, 1989; children: Jim B., Judith Y. BS, Nat. Taiwan U., 1975; PhD, U. Ill., 1984. Engr. China Steel Corp., Koashiung, Taiwan, 1977-79; postdoctorate Argonne (Ill.) Nat. Lab., 1984-85; prof. Nat. Cheng-Kung U., Tainan, Taiwan, 1986—, dep. dir. aerospace sci. & tech. rsch. ctr., 1991-93; cons. China Steel Corp., 1987-89, Energy Rsch. Lab., Shinchu, Taiwan, 1986-88, Chun-Shun Inst. Sci. & Tech., Chung-Li, Taiwan, 1987-90. Contbr. articles to profl. jours. Exec. sec. 1st and 2d Nat. Conf. on Combustion, Taiwan, 1991, 92. 2d lt. Free China Army, 1975-77. Recipient rsch. award Ministry Econ. Affairs, Taiwan, 1978, 81, univ. fellowship, U. Ill., 1983-84. Mem. ASME, AIAA., Chinese Soc. Mech. Engrs. Achievements include development of hybrid turbulence model to improve prediction accuracy of flow field using conventional two-equation models; exploration of composition change effects on non-intrusive temperature measurement using holographic interferometry. Office: Nat Cheng Kung Univ, Inst Aero & Astronautics, Tainan 70101, Taiwan

CHANG, KER-CHI, environmental engineer, civil engineer; b. Taipei, Taiwan, China, Feb. 1, 1950; came to U.S., 1975; s. Chen-Yen and Chu-Fan (King) C.; m. Yee-Wen Liang, July 18, 1976; children: Arthur, Daniel, Allison. BS, Chung Yuan U., Taiwan, 1973; MS, Northwestern U., 1978; PhD, Ga. Inst. Tech., 1982. Environ. engr. Metcalf & Eddy, Inc., Des Plaines, Ill., 1977-78, U.S. Environ. Protection Agy., RTP, N.C., 1985-88; sr. project engr. Lufkin/Cooper Ind., Apex, N.C., 1983-85; environ. coord. ICI Ams., Inc., Dighton, Mass., 1988-89; dir. CPCT, ITRI, Taiwan, 1989-90; site mgr. CDM FPC, Boston, 1990-91; program mgr. U.S. Dept. Energy, Washington, 1991—. Recipient Spl. Achievement award EPA, 1988. Office: US Dept Energy EM-352 Washington DC 20585

CHANG, LEROY L., physicist; b. Kaifung, China, Jan. 20, 1936; came to U.S., 1959; s. Hsin-Fu and Hsien-Hen (Lee) C.; m. Helen H. Chang, 1962; children: Justin, Leslie. BS, Taiwan U., 1957; MS, U. S.C., 1961; PhD, Stanford U., 1963. Mem. rsch. staff IBM T.J. Watson Rsch. Ctr., Yorktown Heights, N.Y., 1963-68, 69-75, mgr. molecular beam epitaxy, 1975-84, tech. planning staff, 1984-85, mgr. quantum structures, 1985-92; dean of sci. Hong Kong U. of Sci. Tech., 1993—; assoc. prof. MIT, Cambridge, 1968-69; adj. prof. Brown U., Providence, 1989—. Fellow IEEE (David Sarnoff award 1990), Am. Phys. Soc. (Internat. Prize New Materials 1985); mem. NAE, Am. Vacuum Soc., Materials Rsch. Soc., Franklin Inst. (Stuart Ballantine award 1993). Office: Sch of Sci, Hong Kong U of Sci & Tech, Kowloon Hong Kong

CHANG, LOUIS WAI-WAH, medical educator, researcher; b. Hong Kong, July 1, 1944; s. Ernest Y.P. and Jeanne (Ma) C.; m. Jane Chi-Chih Wang, Sept. 26, 1989; children: Michael, Jennifer-Michelle. BA, U. Mass., 1966; MS, Tufts U., 1969; PhD, U. Wis., 1972. Asst. prof. pathology U. Wis. Med. Sch., Madison, 1972-76; assoc. prof. pathology U. Ark. Med. Sch., Little Rock, 1977-80, prof. pathology, 1980—, dir. exptl. pathology, 1980—, prof. pharmacology and toxicology, 1984—, dir. toxicology, 1989-91; pres. Am. Chinese Toxicologist Soc., 1988-90; leader, organizer Asian Toxicology Conf. Dels., 1989—; organizer toxicology/environ. health workshop, Japan, Korea, China, Taiwan, 1989—; mem. U.S. Congress Adv. Panel on Neurosci., 1988. Assoc. editor Neurotoxicology and Teratology, 1984-90; mem. editorial bd. Fundamentals and Applied Toxicology, 1986-91, Exptl. Biology and Medicine, 1990—; mng. editor Biomed. and Environ. Sci., 1992—; editor-in-chief: Handbooks of Neurotoxicology, I, II, and III, 1993, Toxicology Risk Assessment I & II, 1993, Toxicology of Metals I & II, 1994. Recipient Gold Forceps award Nat. Histotechnol. Soc., 1975. Mem. Am. Assn. Neuropathology, Am. Assn. Pathologists, Soc. Toxicology, Soc. Neurosci. Office: U Ark Med Scis 4301 W Markham Slot 517 Little Rock AR 72205

CHANG, R. P. H., materials science educator; b. Chung King, Peoples Republic China, Dec. 22, 1941; s. Joseph K. Cho; m. Bennie Chang; children: Vivian, Samuel. BS in Physics, MIT, 1965; PhD in Plasma Physics, Princeton U., 1970. Postdoctoral fellowship Princeton Plasma Physics Lab., 1970-71; mem. tech. staff AT&T Bell Labs., Murray Hill, N.J., 1971-86; prof. Material Sci. & Engring. Northwestern U., 1986—; dir. Materials Rsch. Ctr., 1989—. 7 original inventions 1977—; author over 170 sci. publs.; co-author chpts. in Plasma Diagnostics and Material Sci. & Engring.; co-editor: Plasma Synthesis & Etching of Electronic Materials, 1985. Fellow Am. Vacuum Soc.; mem. Am. Physics Soc., Materials Rsch. Soc. (pres. 1990), Internat. Union of Materials Rsch. Socs. Home: 2330 Iroquois Dr Glenview IL 60025-1034 Office: Northwestern U Dept Tech Materials Sci Engring 2145 Sheridan Rd Evanston IL 60208-0002

CHANG, RUEY-JANG, life science researcher; b. Huwei, Taiwan, Dec. 1, 1957; s. Wan-Lian and Ming (Wu) C.; m. Wen-Ling Wang, Dec. 27, 1986; 1 child, Sanford. MS, Nat. Chung Hsing U., Taichung, Taiwan, 1982; PhD, U. Ill., 1991. Rsch. asst. Nat. Chung Hsing U., Taichung, Taiwan, 1980-82, teaching asst., 1984-87; rsch. asst. U. Ill., Urbana-Champaign, Ill., 1988-91, rsch. assoc., 1991—. 2nd lt. Taiwan Army, 1982-84. Mem. Am. Phytopathological Soc., Sigma Xi. Office: U of Illinois N-519 1102 S Goodwin Ave Urbana IL 61801

CHANG, SHI-KUO, electrical engineering and computer science educator, novelist; b. Szechuan, China, July 17, 1944; m. Judy Pan, Mar. 15, 1969; children—Emily, Cybele. B.S.E.E., Nat. Taiwan U., 1965; M.S., U. Calif.-Berkeley, 1967, Ph.D. in Elec. Engring., 1969. Research mem. T.J. Watson Research Ctr., IBM, Yorktown Heights, N.Y., 1969-70, 71-72, 73-75; asst. prof. elec. engring. Cornell U., Ithaca, N.Y., 1970-71; vis. prof. dept. computer sci. Nat. Chiao-Tung U., Taiwan, 1972-73; dir. computation lab. Inst. Math., Academia Sinica, Taiwan, 1972-73; assoc. prof. dept. info. engring. U. Ill.-Chgo., 1975-78, prof., 1978-82; prof., chmn. dept. elec. engring., dir. info. systems lab. Ill. Inst. Tech., Chgo., 1982-86; prof., chmn. dept. computer sci. U. Pitts., 1986—; founder grad. sch. of computer info. scis. Knowledge Systems Inst., Skokie, Ill.; cons. Standard Oil Co. (Ind.), Bell Telephone Labs., IBM, Nat. Electric Data Processing Ctr., Republic of China, others. Author: Intelligent Database Systems, 1978, Management and Office Information Systems, 1982, Visual Languages, 1986, Principles of Pictorial Information Systems, 1989, Visual Programming Systems, 1990; (novel) Chess King, English edit., 1986, musical, 1988, movie, 1991, also 8 other novels, 7 collections short stories, numerous articles; co-editor: Pictorial Information Systems, 1980, Fuzzy Sets: Theory and Applications to Policy Analysis and Information Systems, 1980; editor in chief Internat. Jour. Policy Analysis and Info. Systems, 1977-82, Internat. Jour. Software Engring. and Knowledge Engring., 1991—; editor Jour. Visual Langs. and Computing, 1990—. Grantee NSF, YOSIN Found., Office Naval Research, U.S. Army; recipient Best Short Stories of Yr. award United Daily News, 1982; IEEE Disting. lectr., 1979-80. Fellow IEEE; mem. IEEE Computer Soc. (standing steering com. 1984 symposium), Assn. computing Machinery, Chinese Inst. Elec. Engrs., Chinese Lang. Computer Soc. (council). Home: 410 Greenleaf Ave Glencoe IL 60022-1908 Office: Knowledge Systems Inst 3420 Main St Skokie IL 60076-2453

CHANG, STEPHEN S., food scientist, educator, researcher, inventor; b. Beijing, China, Aug. 15, 1918; came to U.S., 1947, naturalized, 1962; s. Zie K. and Hui F. (Yuang) C.; m. Lucy Ding, June 2, 1952. B.S., Nat. Ji-nan U., 1941; M.S., Kans. State U., 1949; Ph.D., U. Ill., 1952. Research chemist Swift & Co., Chgo., 1955-57; sr. research chemist A.E. Staley Co., Decatur, Ill., 1957-60; assoc. prof. Rutgers U., New Brunswick, N.J., 1960-62, prof. food chemistry, 1962-88, prof. emeritus, 1988—, chmn. food sci. dept., 1977-86; cons. to industry; convenor adv. com. to Taiwan Food Industry; sci. advisor Hong Kong Inst. Biotech.; mem. tech. bd. Inst. Internat. Devel. and Edn. in Agr. and Life Scis., Inc.; chmn. bd. Cathay Food Cons. Co., Inc. Contbr. articles to profl. jours. Patentee in field. Named hon. prof. Wuxi Inst. Light Industry, Jinan U., Peoples Republic of China, 1987; recipient award for excellence in rsch., bd. trustees Rutgers U., 1984, Rutgers U. award and medal, 1988, Citation Prime Min. of Republic of China, 1989, cert. of appreciation Japan Oil Chemists Soc., 1993. Mem. Am. Oil Chemists Soc. (pres. 1970-71; Lipid Chemistry award 1979, A.E. Bailey award 1974), Inst. Food Technologists (Disting. Food Scientist award 1970, Nicholas Appert award 1983, Internat. award 1989, fellow 1974). Methodist. Office: Cathay Food Cons Co Inc 29 Gloucester Ct East Brunswick NJ 08816-3319

CHANG, SYLVIA TAN, health facility administrator, educator; b. Bandung, Indonesia, Dec. 18, 1940; came to U.S., 1963.; d. Philip Harry and Lydia Shui-Yu (Ou) Tan; m. Belden Shiu-Wah Chang, Aug. 30, 1964; children: Donald Steven, Janice May. Diploma in nursing, Rumah Sakit Advent, Indonesia, 1960; BS, Philippine Union Coll., 1962; MS, Loma Linda (Calif.) U., 1967; PhD, Columbia Pacific U., 1987. Cert. RN, PHN, ACLS, BLS Instr., IV, TPN, Blood Withdrawal. Head nurse Rumah Sakit Advent, Bandung, Indonesia, 1960-61; critical care, spl. duty and medicine nurse, team leader White Meml. Med. Ctr., L.A., 1963-64; nursing coord. Loma Linda U. Med. Ctr., 1964-66; team leader, critical care nurse, relief head nurse Pomona (Calif.) Valley Hosp. Med. Ctr., 1966-67; evening supr. Loma Linda U. Med. Ctr., 1967-69, night supr., 1969-79, adminstrv. supr., 1979—; sr. faculty Columbia Pacific U., San Rafael, Calif., 1986—; dir. health svc. La Sierra U., Riverside, Calif., 1988—; site coord. Health Fair Expo La Sierra U., 1988-89; adv. coun. Family Planning Clinic, Riverside, 1988—; blood drive coord. La Sierra U., 1988—. Counselor Pathfinder Club Campus Hill Ch., Loma Linda, 1979-85, crafts instr., 1979-85, music dir., 1979-85; asst. organist U. Ch., 1982-88. Named one of Women of Achievement, YWCA, Greater Riverside Cs. of C., The Press Enterprise, 1991. Mem. ASCD, AACN, Am. Coll. Health Assn., Assn. Seventh-day Adventist Nurses, Pacific Coast Coll. Health Assn., Adventist Student Pers. Assn., Sigma Theta Tau Internat. Republican. Seventh-day Adventist. Avocations: music, travel, collecting coins, shells and jade carvings. Home: 11466 Richmont Rd Loma Linda CA 92354 Office: La Sierra U Health Svc 4700 Pierce St Riverside CA 92515

CHANG, THOMAS MING SWI, biotechnologist, medical scientist; b. Swatow, Kwantang, China, Apr. 8, 1933; m. Lancy Yuk Lan, June 21, 1958; children: Harvey, Victor, Christine, Sandra. B.Sc., McGill U., 1957, M.D., C.M., 1961, Ph.D., 1965, F.R.C.P.(C), 1972. Intern Montreal (Que.) Gen. Hosp., 1961-62; rsch. fellow depts. physiology and chemistry McGill U., Montreal, 1962-65, lectr. dept. physiology, 1965, asst. prof., 1966-69, assoc. prof., 1969-72, prof. physiology, 1972—, prof. medicine, 1975—, dir. artificial organs rsch. unit, 1975-79, dir. artificial cells and organs rsch. ctr., 1979—, assoc. dept. chem. engring., 1985—, assoc. dept. chemistry, 1986—, prof. biomed. engring., 1990—; lab. and clin. researcher med. scis., biotech. and biomed. engring. Montreal, 1972—; staff Royal Victoria Hosp.; hon. Montreal Chinese Hosp., 1970—; cons. Montreal Children's Hosp., 1979-93, Med. Rsch. Coun. fellow, 1962-65, scholar, 1965-68, career investigator, 1968—; hon. prof. Nankai U. Inventor artificial cells; author: Artificial Cells, 1972, Biomedical Application of Immobilized Enzymes and Protiens, Vols. I and II, 1977, Artificial Kidney, Artificial Liver and Artificial Cells, 1978, Hemoperfusion-Kidney and Liver Supports and Detoxification, 1980, Hemoperfusion, 1981, Past Present and Future of Artificial Organs, 1983, Microencapsulation and Artificial Cells, 1984, Hemoperfusion and Artificial Organs, 1985, Blood Substitutes, 1988, Blood Substitutes and Oxygen Carriers, 1993; editor-in-chief Jour. Artificial Cells, Blood Substitutes and Immobilization Biotech.; sect. editor: Internat. Jour. Artificial Organs, 1977—, Trans. Am. Soc. Artificial Organs, 1987—; assoc. editor: Jour. Artificial Organs, 1977-92; editorial bd.: Jour. Biomaterial Med. Devel. and Orgn. 1972-87, Jour. Membrane Sci., 1975-92, Jour. Bioengring, 1975-79, Jour. Enzyme and Microbial Tech, 1978-86. Decorated officer Order of Can., 1991. Fellow Royal Coll. Physicians Can.; mem. Internat. Soc. Artificial Organs (trustee 1982-87, 89-92, congress pres. 1991, pres. elect 1992, pres. 1993-95), Can. Soc. Artificial Organs (pres. 1980-82), Internat. Soc. Artificial Cells, Blood Substitutes and Immobilization Biotech. (hon. pres. 1990—). Office: Artificial Cells & Organs Rsch Ctr, 3655 Drummond St Rm 1005, Montreal, PQ Canada H3G 1Y6

CHANG, WEI, medical physicist; b. Chongching, Szechuan, China, Oct. 20, 1945; came to U.S., 1969; s. Cheng Jen and Hui Ying (Chueh) C.; m. Ching Lin, Jan. 5, 1974; children: Brian Lin, Amanda Yiping. BS, Nat. Taiwan U., 1968; PhD, SUNY, Buffalo, 1976. Med. physicist Loyola U. Med. Ctr., Maywood, Ill., 1977-83, U. Iowa Hosp., Iowa City, 1984-91, Rush Presbyn. St. Luke Med. Ctr., Chgo., 1991—; assoc. prof. Loyola U. Med. Ctr., 1981-83, U. Iowa Med. Sch., 1984-91; prof. Rush U. and Med. Sch., Chgo., 1991—. Contbr. articles to Jour. Nuclear Medicine, other profl. publs. Mem. IEEE, Soc. Nuclear Medicine, Am. Assn. Physicists in Medicine. Achievements include two patents on emission tomography system designs. Office: Rush Presbn St Luke Med Ctr 1653 W Congress Pky Chicago IL 60612

CHANG, WON SOON, research scientist, mechanical engineer; b. Suwon, Korea, July 30, 1949; s. Seok Yeon and Yoen Sook (Choi) C.; m. Jill Kilja Lee, May 17, 1975; children: Catherine, Andrew. MSME, Okla. State U., 1977; PhD, Ga. Inst. Tech., 1981. Asst. prof. No. Ill. U., DeKalb, Ill., 1980-83; assoc. prof. W.Va. Inst. Tech., Montgomery, 1984-87; rsch. scientist Wright Lab., Wright-Patterson AFB, Ohio, 1987—. Contbr. articles to profl. jours. Recipient Heron award Aero Propulsion and Power Directorate, 1992. Mem. ASME, AIAA. Roman Catholic. Home: 376 Twelve Oaks Trail Beavercreek OH 45434

CHANG, Y(UAN)-F(ENG), mathematics and computer science educator; b. Tientsin, Republic of China, Oct. 7, 1928; came to U.S., 1944, naturalized, 1962; s. Peng-Chung and Sieu-Tsu C.; m. Hanping Chu, Nov. 21, 1955; children: Claire, Leon. MSEE, Purdue U., 1950; PhD in Applied Physics, Harvard U., 1959. Asst. and assoc. prof. Purdue U., Lafayette, Ind., 1958-66; research fellow Battelle Meml. Inst., Columbus, Ohio, 1966-71; vis. assoc. prof. Ind. U., Bloomington, 1972-74; prof. computer sci. U. Nebr., Lincoln, 1974-84; W.M. Keck prof. applied math. and computer sci. Claremont (Calif.) McKenna Coll., 1984—; vis. assoc. prof. UCLA, 1964-65; vis. prof. San Diego State U., 1981-82; assoc. La Jolla (Calif.) Inst. Contbr. articles to profl. jours. Research grantee NSF, Naval Air System Command, Office Naval Research, NASA. Mem. Am. Math. Soc., Assn. Computing Machinery. 24 yrs. research in numerical analysis and application of Taylor series to sci. problems; indsl. research in solid state physics. Home: 4258 N Piedmont Mesa Dr Claremont CA 91711-2332

CHANG, ZHAO HUA, biomedical engineer; b. Zibo, Shandong, People's Republic of China, July 3, 1963; s. Xuzhong and Sumai (Wang) C.; m. Xiaojing Yang, Aug. 5, 1987; 1 child, Brian Yale. MS, Shanghai Inst. Mech. Engring., 1985; PhD, SUNY, Binghamton, 1992. Rsch. assoc. SUNY, Binghamton, N.Y., 1987-90; cons. Cryomedical Sci., Inc., Bethesda, Md., 1990; sr. engr. Cryomedical Sci., Inc., Rockville, Md., 1991, dir. R & D, 1993; cons. Life Cell Inc., Houston, 1988-89. Author: Cryobiological Engineering, 1985; contbr. articles to profl. jours. Mem. Cryobiology Soc. Achievements include invention of and patent for new cryosurgical systems that have simplified many surgeries; provision of the first calometric evidence of nuclei formation below the glass transition temperature. Home: 19125 Hempstone Ave Poolesville MD 20837

CHANG-HASNAIN, CONSTANCE JUI-HUA, educator; b. Taipei, Taiwan, Oct. 1, 1960; came to the U.S., 1978; d. Ping-Jen and Chia-Fu (Wan) Chang; m. Ghulam Hasnain, June 22, 1984; 1 child, Katherine Mei. BS, U. Calif., Davis, 1982; MS, U. Calif., Berkeley, 1984, PhD, 1987. Mem. tech. staff Bellcore, Red Bank, N.J., 19878-92; asst. prof. dept. elec. engring. Stanford (Calif.) U., 1992—. Contbr. articles to IEEE Jour. Quantum Electronics, Applied Phys. Letters, Photonics Tech. Letters, Jour. Lightwave Tech. Quantum fellow Am. Electronics Assn., 1984-87, David Sakrison prize, 1989; best paper Soviet-Am. Interacad. Workshop on Semiconductor Laser Physics, 1991; named Outstanding Young Elec. Engr. Eta Kappa Nu, 1991, Nat. Young Investigator NSF, 1992; Packard fellow David and Lucille Packard Found., 1992, Reid and Polly Anderson faculty scholar, 1992. Mem. IEEE (sr., assoc. editor Circuits and Devices mag. 1991—, chairperson semiconductor laser workshop 1991), Conf. Lasers and Electro-Optics (subcom. chair 1993), Optical Soc. Am., Am. Phys. Soc. Achievements include patents for cross-coupled quantum well stripe laser array, for multiple wavelength laser array and method of making. Home: 837 Allardice Way Stanford CA 94305 Office: Stanford Univ Edward L Ginzton Lab Stanford CA 94305-4085

CHANGHO, CASTO ONG, power plant construction executive; b. Calauan, Laguna, The Philippines, Mar. 28, 1939; came to U.S., 1967; s. San Ong and Su (Ching) C.; m. Aurora D. Diego, Sept. 12, 1970; children: Christine, Emelyn, Kathleen, Alexander. BSChemE, Mapua Inst. Tech., Manila, 1966; MS in Nuclear Engring., U. Mo., 1970. Registered profl. engr., Kans. Nuclear engr. Black & Veatch, Kansas City, Mo., 1970-82, chem. engr., 1982-90; project mgr. Black & Veatch, Raleigh, N.C., 1990-92; field project mgr. fgn. power plant constrn. Black & Veatch Internat., Kansas City, Mo., 1992—; mem. adv. bd. nat. environ. studies projects Atomic Indsl. Forum, Washington, 1980-81. Mem. Am. Nuclear Soc., Tau Beta Pi. Republican. Roman Catholic. Achievements include work in construction and process design management, environmental radiological monitoring, radwaste management facilities design, radiological engineering, chemical process engineering, general and ultra-pure water systems design and power plant construction management. Home: 103 Coventry Ln Cary NC 27511 Office: Black & Veatch PO Box 8405 Kansas City MO 64114

CHANG-MOTA, ROBERTO, electrical engineer; b. Caracas, Venezuela, Dec. 28, 1935; s. Roberto W. and Mary C. (Mota) Chang; m. Alicia Santamaria-Gonzales, May 4, 1968; children: Roberto Ignacio, Roxana Ivette, Ricardo Ignacio. DEE, U. Cen. Venezuela, 1960; MS, U. Ill., 1962; AR, Harvard U., 1970; PhD, U. Calif., 1983. Dir. Sch. Elec. Engring., also prof. Central U., 1964-69; prof., dean Schs. Engring., Architecture and Sci., Simon Bolivar U., 1971-77; pres. Colegio de Ingenieros de Venezuela, 1974-79; dir. Venezuelan Power Co., 1974-79; pres. Latin Am. Orgn. Engring., 1977-79, Corporoil, 1981-85, Audio Interface Corp., 1983—; v.p. ESCA Corp., 1991—; cons. in field. Spl. cons. Venezuelan Navy and Army, 1971-75, Venezuelan Congress, 1989—; mem. tech. com. Venezuelan Supreme Election Coun., 1971-81, exec. dir., 1981-82, gen. dir., 1982-89; cons. Ministry of Interior, 1990; bd. dirs. Edelca, 1989—; v.p. Electronic Cir. Corp., 1991; trustee Simon Bolivar U., 1985—. Mem. IEEE, Am. Soc. Engring. Edn., Venezuelan Soc. Elec. and Mech. Engring. (pres. 1972-73), Instn. Elec. Engrs., Puerto Azul Club, Playa Pintada Club, Caracas Racquet Club. Roman Catholic. Home: Quinta Cumana Calle Colon, Prados de Este Estado, Miranda Venezuela Office: Torres Centro Simon Bolivar, Consejo Supremo Electoral, Esq Pajarito, Caracas Venezuela

CHANUSSOT, GUY, physics educator; b. Dijon, France, Nov. 12, 1943; s. Martial and Eugenie (Perron) C.; children: Caroline, Anne. Licence, U. Dijon, 1964, Doctorate, 1967, State Doctorate, 1970. Asst. U. Dijon, 1964-70; maitre de conf. U. Mohamed V, Morocco, 1970-72; engr. Bell Labs., N.J., 1976; physicst Acad. Sci., Moscow, 1977; dean U. Dijon, 1979; sci. attache French Embassy, Brasilia, Brazil, 1979-83, Mexico City, 1983-87; sci. tech. advisor ONU, N.Y.C., 1988; prof. physics U. Dijon, 1989—. Editor Publitech, 1991—; contbr. numerous articles to profl. jours. Citoyen d'honneur Etat of Amazonas, Brazil, 1983; mem. of honor Inst. d Neurologia and Neuropsychiatria (Mex.), 1987. Mem. Am. Phys. Soc., AVRIST, Internat. Sci. and TEch. Soc. Avocations: skiing, clarionet. Home: 3A Cours Gl de Gaulle, Dijon France 21000 Office: Lab Solid State, LPS BP 138, Dijon France 21004

CHANY, CHARLES, microbiology educator; b. Budapest, Sept. 1, 1920; s. Nicolas and Irene (Lobl) Csanyi; m. Francoise Fournier, Dec. 18, 1975. MD, U. Paris, 1957. Rockefeller fellow Rockefeller Inst., N.Y.C., 1957-58; rsch. asst. NIH Inst. Pasteur, Paris and Bethesda, Md., 1954-60; assoc. idr. Inserm, Paris, 1960-66; dir. U-43 Inserm, Paris, 1966-86; prof. Univ. Paris V, 1966-88; dir. Virus Diagnosis Unit, Hosp. St. Vincent de Paul, Paris, 1954-86. Discoverer anti-tumor function of interferons, 1958, Reversion of Cancer Cells to Non-Malignancy, 1960, Sarcolectin, 1982, Measles Virus Hemagglutination, 1960; contbr. to first recombinant Hepatitis-B Vaccine, 1980. Officer Order of Nat. Merite, France, 1980. Mem. Internat. Soc. Interferon Rsch. (hon.), Rapid Viral Diagnosis Assn. (hon.), Rsch. Assn. A.D.B.E.A. (pres. Paris br. 1986—). Home: 34 Rue Du Docteur Blanche, 75016 Paris France

CHAO, BEI TSE, mechanical engineering educator; b. Soochow, China, Dec. 18, 1918; came to U.S., 1948, naturalized, 1962; s. Tse Yu and Yin T. (Yao) C.; m. May Kiang, Feb. 7, 1948; children: Clara, Fred Roberto. B.S. in Elec. Engring. with highest honor, Nat. Chiao-Tung U., China, 1939; Ph.D. (Boxer Indemnity scholar), Victoria U., Manchester, Eng., 1947. Asst. engr. tool and gage div. Central Machine Works, Kunming, China, 1939-41; asso. engr. Central Machine Works, 1941-43, mgr. tool and gage div., 1943-45; research asst. U. Ill., Urbana, 1948-50; asst. prof. dept. mech. engring. U. Ill., 1951-53, assoc. prof., 1953-55, prof., 1955-87, prof. emeritus, 1987—, head thermal sci. div., 1971-75, head dept. mech. and indsl. engring., 1975-87; assoc. mem. U. Ill. (Center for Advanced Study), 1963-64; cons. to industry and govtl. agys., 1950—; Russell S. Springer prof. mech. engring. U. Calif., Berkeley, 1973; mem. reviewing staff Zentralblatt für Mathematik, Berlin, 1970-82; mem. U.S. Engring. Edn. Del. to Visit People's Republic of China, 1978; mem. adv. screening com. in engring. Fulbright-Hays Awards Program, 1979-81, chmn., 1980, 81; mem. com. U.S. Army basic sci. research NRC, 1980-83; Prince disting. lectr. Ariz. State U., 1984; bd. dirs. Aircraft Gear Corp., 1989—. Author: Advanced Heat Transfer, 1969; contbr. numerous articles on mech. engring. to profl. jours.; tech. editor: Jour. Heat Transfer, 1975-81; mem. adv. editorial bd.: Numerical Heat Transfer, 1977—; mem. hon. editorial bd. Internat. Jour. Heat and Mass Transfer, 1987—, Internat. Communications in Heat and Mass Transfer, 1987—. Recipient Outstanding Tchr. award Ill. Mech. Engring. Alumni, 1978, Max Jakob Meml. award ASME/Am. Inst. Chem. Engring., 1983; Tau Beta Pi Daniel C. Drucker eminent faculty award, 1985; Univ. scholar, 1985. Fellow AAAS, ASME (Blackall award 1957, Heat Transfer award 1971, William T. Ennor Mfg. Tech. award 1992), Am. Soc. Engring. Edn. (Outstanding Tchr. award 1975, Western Electric Fund award 1973, Ralph Coats Roe award 1975, Benjamin Garver Lamme award 1984, Centennial Medallion 1993); mem. Nat. Acad. Engring., Academia Sinica, Soc. Engring. Sci., Chiao-Tung U. Alumni Assn. (pres. Mid-West sect. 1975-76), Tau Beta Pi, Pi Tau Sigma (hon.). Home: 704 Brighton Dr Urbana IL 61801-6315 Office: Univ Ill 264 Mech Engring Bldg 1206 W Green St Urbana IL 61801

CHAO, CONRAD RUSSELL, obstetrician; b. Elmhurst, Ill., Apr. 21, 1957; s. William Kuan-Hua and Dorothy Regina (Yee) C.; m. Ruth Ann Nelson, May 22, 1982. BA, Ill. Wesleyan, 1978; MD, U. Ill., Chgo., 1982. Cert. gen.. and maternal-fetal medicine subspecialty Am. Bd. Ob-Gyn. Intern L.A. County Harbor-UCLA Med. Ctr., Torrance, Calif., 1982-83; resident L.A. County Harbor-UCLA Med. Ctr., Torrance, 1983-86, fellow maternal-fetal medicine and physiology Loma Linda (Calif.) U., Loma Linda, 1986, Oreg. Health Scis. U., Portland, 1987-88; asst. prof. and Mortimer G. Rosen scholar Columbia U., N.Y.C., 1988-93; asst. prof. UCLA, 1993—. Contbr. articles to Jour. Cerebral Blood Flow and Metabolism, Am. Jour. Ob-Gyn., Devel. Brain Rsch. Mem. com. on sci. and theology Episcopal Diocese of Newark, 1989-93. Rsch. tng. grantee NIH, 1987, Am. Heart Assn., 1988, rsch. grantee NIH Nat. Inst. of Child Health and Human Devel., 1990—, Nat. Inst. on Drug Abuse, 1991—. Fellow Am. Coll. Ob-Gyn.; mem. N.Y. Acad. Scis. (conf. com. 1991-93), Soc. of Perinatal Obstets. Achievements include research on the effects of ischemia, hyperglycemia, behavioral state, sound stimulation, and cocaine on cerebral blood flow, metabolism, and behavior in the fetus. Office: Rsch and End Inst RB-1 1124 W Carson Torrance CA 90502

CHAO, JAMES LEE, chemist; b. Lafayette, Ind., Sept. 4, 1954; s. Tai Siang and Hsiang Lin (Lee) C.; m. Juliana Meimei Ma, Apr. 4, 1992. BS in Chemistry, U. Ill., 1975, MS in Chemistry, 1976; PhD in Chemistry, U. Calif., Berkeley, 1980. Applications scientist IBM Instruments, Inc., Danbury, Conn., 1980-87; vis. assoc. prof. dept. chemistry Duke U., Durham, N.C., 1986-87, adj. asst. prof. dept. chemistry, 1987-91, adj. assoc. prof., 1992—; advisory scientist IBM Corp., Research Triangle Park, N.C., 1987—; cons. Lab. for Laser Energetics U. Rochester, N.Y., 1979-80; postdoctoral fellow lab. for chem. biodymanics Lawrence Berkeley Lab., 1980; referee Applied Spectroscopy, 1982—, Applied Physics Letters, Jour. Applied Physics, 1989—; grant referr N.C. Biotech. Ctr., 1991. Contbr. articles to profl. jours. Edmund James scholar, 1972-75, Dow Chem. scholar, 1977. Fellow N.Y. Acad. Scis.; mem. Am. Inst. Chemists, Am. Chem. Soc. (chmn. N.C. sect. 1991, alt. councilor 1992, councilor 93—), Soc. for Applied Spectroscopy, Coblentz Soc., Am. Inst. Chemists, Sigma Xi. Achievements include development of step-scan implementation for FT-IR spectrometers to study photothermal and time-resolved spectroscopies; first reported 2-D FT-IR correlation spectroscopy using step-scan; project leader for IBM gaseous corrosion testing. Home: 7424 Ridgefield Dr Durham NC 27713-9503 Office: IBM Corp Dept E81/061 PO Box 12195 Research Triangle Park NC 27709

CHAO, JAMES MIN-TZU, architect; b. Dairen, China, Feb. 27, 1940; s. T. C. and Lin Fan (Wong) C.; came to U.S., 1949, naturalized, 1962; m. Kirsti Helena Lehtonen, May 15, 1968. BArch, U. Calif., Berkeley, 1965. Registered architect, Calif.; cert. instr. real estate, Calif. Intermediate draftsman Spencer, Lee & Busse, Architects, San Francisco, 1966-67; asst. to pres. Import Plus Inc., Santa Clara, Calif., 1967-69; job capt. Hammaberg and Herman, Architects, Oakland, Calif., 1969-71; project mgr. B A Premises Corp., San Francisco, 1971-79; constrn. mgr. The Straw Hat Restaurant Corp., 1979-81, mem. sr. mgmt., dir. real estate and constrn., 1981-87; mem. mktg. com. Straw Hat Coop. Corp., 1988-91; pvt. practice architect, Berkeley, Calif., 1987—; pres. Food Svc. Cons. Inc., 1987-89; pres., chief exec. officer Stratsac, Inc., 1987-92; prin. architect Alpha Cons. Group Inc., 1991—; v.p. Intersyn Industries Calif., 1993—; lectr. comml. real estate site analysis and selection for profl. real estate seminars; coord. minority vending program, solar application program Bank of Am.; guest faculty mem. N.W. Ctr. for Profl. Edn.; bd. dirs Ambrosia Best Corp., 1992—. Patentee tidal electric generating system; author first comprehensive consumer orientated performance specification for remote banking transaction. Recipient honorable mention Future Scientists Am., 1955. Mem. AIA, Encinal Yacht Club (bd. dir. 1977-78). Republican.

CHAO, KOUNG-AN, physics educator; b. Chungking, Szechuan, People's Republic China, Feb. 19, 1940; arrived in Norway, 1989; s. Hsing-Yuan and Daisy (Huang) C.; m. Lindy Ling-Ying Lin,June 21, 1966 (dec. May 1979). PhD in Physics, Rensselaer Poly. Inst., 1969. Rsch. assoc. IBM, Yorktown Heights, N.Y., 1969-71; rsch. fellow Chalmers U. Tech., Gothenborg, Sweden, 1971-73, U. Warwick, Coventry, Eng., 1973-74; vis. prof. Fed. U. Pernambuco, Recife, Brazil, 1975; rsch. prof. U. Linköping, Sweden, 1976-89; chair prof. U. Trondheim Norwegian Inst. Tech., 1989—. Mem. editorial bd. mem. World Sci. Pub. Co., Singapore, 1987—, adv. bd. mem., 1986—; contbr. articles to profl. jours. Mem. Am. Phys. Soc. Avocation: carpentry. Home: Smestuveien 11B, N 7040 Trondheim Norway Office: U Trondheim, Norwegian Inst Tech Physics, N 7034 Trondheim Norway

CHAO, KWANG-CHU, chemical engineer, educator; b. Chongqing, China, June 7, 1925; came to U.S., 1954, naturalized, 1969; s. Chung-Pu and Jui-Pu (Chou) C.; m. Jiun-Ying Su, May 2, 1953; children: Howard Honshuen, Albert Honchi, Bernard Honwei. B.S., Nat. Chekiang U., 1948; M.S., U. Wis., 1952, Ph.D., 1956. Chem. engr. Taiwan Alkali Co., 1948-51, 52-54; research engr. Chevron Research Co., Richmond, Calif., 1957-63; asso. prof. Ill. Inst. Tech., Chgo., 1963-64, Okla. State U., 1964-68; prof. Purdue U., West Lafayette, Ind., 1968—, Harry C. Peffer Disting. prof. chem. engring., 1989—; cons. to industry, 1964—. Author: (with R.A. Greenkorn) Thermodynamics of Fluids, 1975; Editor: Applied Thermodynamics, 1968, Equations of State in Engineering and Research, 1979; Equations of State-Theories and Applications, 1986. Fellow Am. Inst. Chem. Engrs. (editorial bd. jour., also Ind. Engring. Chem. Ann. Revs.); mem. Am. Chem. Soc., AAUP, Sigma Xi, Omega Chi Epsilon. Home: 2909 Henderson Ave West Lafayette IN 47906-1542

CHAO, YEI-CHIN, aerospace engineering educator; b. Taiwan, Republic of China, Jan. 22, 1955; s. Pin-Tu and Yen-Kuay (Chen) C.; m. Show-Learn Kuo, Sept. 16, 1981; children: Sheen, Vivia. MSME, Nat. Cheng Kung U., 1979; PhD in Aerospace, Ga. Inst. Tech., 1984. Lectr. Nat. Cheng Kung U., Tainan, 1979-80, assoc. prof., 1984-91, prof., 1991—, assoc. dir., 1991-93; rsch. asst. Ga. Inst. Tech., Atlanta, 1980-83; cons. Chung-Sun Inst. of Sci. and Tech., Tai Chung, 1985—, Indsl. Tech. Rsch. Inst., Hsinchu, 1990—; founding mem. Inst. of Aeronautics and Astronautics of Nat. Cheng-Kung U., 1983. Editorial mem. The Chinese Dictionary of Mechanics, 1989—; contbr. more than 60 articles to profl. jours. Mem. Organizing com. The Combusion Inst. of R.O.C., Tainan, 1991-92, bd. dirs., 1992—; mem. The Ko-Min Party, Taiwan, 1984—. Soldier Army, 1977. Recipient Ann. Rsch. award Nat. Sci. Coun., 1986-92, rsch. grants, 1985-92, Def. Rsch. Com., 1986-92. Mem. AIAA, Chinese Soc. of Mech. Engring., Chinese Sco. of Theoretical and Applied Mechanics, The Chinese Combustion Inst. (bd. dirs. 1992—, organizing mem. 1992). Achievements include patent pending for control of the jet flames by acoustic excitation. Office: Nat Cheng Kung U, Inst Aero/ Astro #1 Ta-Hsuey Rd, Tainan 701, Taiwan

CHAPANIS, ALPHONSE, human factors engineer, ergonomist; b. Meriden, Conn., Mar. 17, 1917; s. Anicatas and Mary (Barkevich) C.; m. Marion Amelia Rowe, Aug. 23, 1941 (div. 1960); children: Roger, Linda Chapanis Fox; m. Natalia Potanin, Mar. 25, 1960 (div. 1987). BA, U. Conn., 1937; MA, Yale U., 1942, PhD, 1943. Cert. Human Factors Profl. Prof. psychology The Johns Hopkins U., Balt., 1946-82; pres. Alphonse Chapanis P.A., Balt., 1974—; mem. tech. staff Bell Labs., Murray Hill, N.J., 1953-54, adv. panel USAF Office of Sci. Rsch., Washington, 1956-59, com. on human factors, NRC, Washington, 1980-85; liaison scientist Office of Naval Rsch., U.S. Embassy, London, 1960-61; cons. IBM, Yorktown Heights, N.Y. and Bethesda, Md., 1960—. Author: (book) Applied Experimental Psychology, 1949, Research Techniques in Human Engineering, 1959, Man-Machine Engineering, 1965, Ethnic Variables in Human Factors Engineering, 1975; co-editor: Human Engineering Guide to Equipment Design, 1963; contbr. over 175 articles to profl. jours. Capt. USAAF, 1943-46. Recipient Disting. Contbn. for Applications in Psychology award APA, 1978, Outstanding Sci. Contbn. to Psychology, Md. Psychol. Assn., 1981. Fellow AAAS, Soc. Engring. Psychologists (Franklin V. Taylor award 1963), Human Factors Soc. (Paul M. Fitts award 1973, Pres.'s Disting. Svc. award 1987), Ergonomics Soc. (hon.); mem. Internat. Ergonomics Assn. (Outstanding Contbn. award 1982). Achievements include patent (with others) on Correlation of Seismic Signals. Office: 8415 Bellona Ln Ste 210 Baltimore MD 21204-2055

CHAPEL, THERON THEODORE, quality assurance engineer; b. Jackson County, Mich., Jan. 31, 1918; s. Theron Eugene and Monica Iris (Paton) C.; m. Lucy Chapel (dec.); m. Sue Chapel; children: Robert, James, Katie. BA, Albion Coll., Mich., 1938; BSE, U. Mich., 1940. Analytical chemist Minn. Mining & Mfg. Co., Detroit, 1940-41; rubber compounder, 1941-43; research/devel. chemist The Simonitz Co., Chgo., 1946-51, quality control supr. The Simoniz Co., 1951-53; process devel. engr. Armour Pharm. Labs., Bradley, Ill., 1954-56; product control supr. Acheson Colloids Co., Port Huron, Mich., 1956-61; matl. trtmt. and processes quality control rep. Inspector of Naval Material, Chgo., 1962-63; shipbuilding quality control rep., supr. shipbuilding USN, Bay City, Mich., 1963-70; quality assurance specialist supr. Shipbuilding, Conversion & Repair USN, Pascagoula, Miss., 1970-72, quality assurance engr. supr. Shipbuilding, Conversion & Repair, 1972-93; ret. Shipbuilding, Conversion & Repair, Pascagoula, Miss., 1993. With U.S. Army, 1943-46. Mem. ACS, Am. Soc. for Quality Control, ASTM, Am. Soc. for Nondestructive Testing, Phi Lambda Upsilon. Mem. LDS Ch. Avocations: genealogy, photography. Home: PO Box 1708 Escatawpa MS 39552-9999

CHAPLINE, GEORGE FREDERICK, JR., theoretical physicist; b. Teaneck, N.J., May 6, 1942; s. George Frederick Chapline and Ferne Louise (Copeland) C.; m. Marie Jeanne Hjort, Mar. 24, 1968; 1 child, Michael. B.A., UCLA, 1961. Teaching asst. Calif. Inst. Tech., 1962-64; asst. prof. physics U. Calif.-Santa Cruz, 1967-69; physicist Lawrence Livermore Lab., Livermore, Calif., 1969—. Patentee x-ray laser. Recipient E.O. Lawrence award Dept. Energy, 1983. Office: Dept Physics Lawrence Livermore Nat Lab Livermore CA 94550

CHAPMAN, DARIK RAY, chemical process engineer; b. Midland, Mich., June 16, 1970; s. Dwain R. and Twila Virginia (Humphrey) C.; m. Sharon Ruth Meltzer, Apr. 14, 1990; 1 child, Calee Joanna. Grad. high sch., Adrian, Mich. Cert. hazardous waste mgr. Lab. technician Sil-Tech Corp., Tecumseh, Mich., 1988-90, v.p. ops., 1990—, also bd. trustees, 1990—. Mem. Ch. LDS. Home: 710 LeLand Ct Adrian MI 49221 Office: Sil-Tech Corp 810 S Maumee St Tecumseh MI 49286

CHAPMAN, GARRY VON, mechanical engineer; b. Granite City, Ill., Aug. 29, 1960; s. Franklin Delano Sr. and Lucy Marie (Dodson) C.; m. Carol Ann Schumacher, Sept. 3, 1983. BSME, U. Mo., 1983. Registered profl. engr., Mo. Engring. intern Burns & McDonnell, Kansas City, Mo., 1979-82; engr. Union Electric Co., Fulton, Mo., 1983—; mem. task group Electric Power Rsch. Inst., Palo Alto, Calif., 1991-92. Vol. earthquake damage insp. State Emergency Mgmt. Agy., Jefferson City, Mo., 1991-92. With Naval ROTC, 1979-80. Mem. NSPE, ASME, Mo. Soc. Profl. Engrs., Nat. Inst. Engring. Ethics, U.S. Golf Assn., Columbia Country Club, Tau Beta Pi, Pi Tau Sigma. Roman Catholic. Achievements include development of training program for Callaway Nuclear Plant for seismic technical evaluations. Home: 5500 Arrowwood Dr Columbia MO 65202 Office: Union Electric Nuclear Engring PO Box 620 Fulton MO 65251

CHAPMAN, GILBERT BRYANT, physicist; b. Uniontown, Ala., July 8, 1935; s. Gilbert Bryant and Annie Lillie (Stallworth) C.; m. Loretta Woodward, June 5, 1960; children: Annie L., Bernice M., Cedric N., David O., Ernest P., Frances Q.H., Gilbert Bryant III. BS in Math. and Chemistry, Baldwin Wallace Coll., Berea, Ohio, 1968; MS in Physics, Cleve. State U., 1973; MBA, Mich. State U., 1990; postgrad, Kent State U., Ohio, 1974-76. Phys. sci. technician NASA-Lewis Rsch. Ctr., Cleve., 1953-68, emission spectroscopist, 1968-75, materials engr., 1975-77; sr. rsch. engr. Ford Motor Co., Redford Twp., Mich., 1977-83, project engr., 1983-86; adv. materials testing specialist Chrysler Corp., Highland Park, Mich., 1986-89; adv. materials specialist Chrysler Corp., Madison Heights, Mich., 1989-91, advanced materials and product specialist, 1991—; chmn. auto. com. '87 Soc. Mfg. Engrs. Composites Group, Dearborn, 1987; chmn. ind. adv. bd. NDE/Ctr., Iowa State U., Ames, 1989, 90; mem. indsl. adv. bd. Inst. for Mfg. Rsch., Wayne State U. Contbr. articles to profl. jours., chpts. to books. Lay leader, elder SDA Ch. of Southfield, Mich., 1983-93; bd. trustees Mt. Vernon Acad., Ohio, 1972-76; lay adv. coun. Ohio Conf. SDA, 1974-77. With USAF, 1959-61. Recipient Group Achievement award, NASA Lewis Rsch. Ctr., 1970, Apollo Achievement award, 1968. Fellow Am Soc. Nondestructive Testing; mem. AAAS, ASM (polymer composites program com. paper 1986), ASTM, IEEE, SAE (Award for Excellence in Oral Presentation), Am Chem. Soc., Am. Phys. Soc., Am. Soc. for Composites, Engring. Soc. Detroit (sci. com. mem.), Fedn. of Analytical Chemists, Nat. Tech. Assn. (Cleve. program com.), Soc. for Applied Spectroscopy (Cleve. vice chair, sec.), Soc. for Mfg. Engrs. (sr. mem.), Soc. Physics Students (pres.), Sigma Pi Sigma. Achievements include patent for infrared inspection method for friction welds in thermoplastics and advanced vehicle concepts; development of low-frequency ultrasonic inspection methods for adhesive bond joints; co-development of D.C. arc method of determining work functions of refractory alloys, spectrochemical analysis of microgram-size samples. Home: 17860 Bonstelle Ave Southfield MI 48075-3452 Office: Chrysler Corp 30900 Stephenson Hwy Madison Heights MI 48071-1617

CHAPMAN, JOHN ARTHUR, agricultural engineering executive; b. White Hall, Ill., Sept. 26, 1933; s. Ralph Burwell and Floy May (Kinser) C.; m. Mary R. Cunningham, Dec. 14, 1952; children: Julie A., Andrew D. BS in Agrl. Engring., U. Ill., 1959; MS in Agrl. Engring., U. Nebr., 1972. Registered profl. engr., Nebr., S.D. Asst. mgr. Harrison Cropsaver Co., Bondville, Ill., 1959-60; sr. project devel. engr. Oliver Corp., Shelbyville, Ill., 1960-65; engr. Valmont Industries Inc., Valley, Nebr., 1965-72, dir. product mgmt., 1972-79, v.p. engring. N.Am. Irrigation div., 1979-91, dir. rsch., 1991—. Patentee in irrigation improvement. Bd. dirs. dean's coun. Coll. Engring. U. Nebr., Lindoln, 1981—. Cpl. U.S. Army, 1953-55. Fellow Am. Soc. Agrl. Engrs. Republican. Presbyterian. Home: 1306 Birch St Wahoo NE 68066-1128 Office: Valmont Industries Inc Hwy 275 Valley NE 68064-9678

CHAPMAN, MARK ST. JOHN, chemist; b. Birkenhead, U.K., Apr. 30, 1960; s. Maurice Thomas and Elizabeth Mary Beryl (Coogan) C.; m. Laura Jane Barrat, Aug. 29, 1987; children: Antony Maurice, Adam John. Degree in chemistry, math., U. Newcastle-upon-Tyne, Eng., 1981. Sr. technologist Berger Paints U.K., 1981-84; tech. mgr. Berger Paints Overseas, 1984-87; export tech. svc. mgr. Mebon U.K., 1987-89; regional tech. dir. Berger Paints Singapore, 1989—; chmn. tech. sub com. Singapore Paint Mfrs. Assn., 1989-92. Mem. Oil and Colour Chemists Assn., Steel Structures Painting Coun., Nat. Assn. Corrosion Engrs., Inst. of Corrosion. Roman Catholic. Home: 7 Faber Park, Singapore 0512, Singapore

CHAPMAN, MARTIN DUDLEY, immunochemist; b. Dudley, Eng., Dec. 14, 1954; came to U.S., 1983; s. Raymond and Ethel and (Lake) C.; m. Madeleine Cicely Watkins, June 25, 1982; children: Ellen, Alice, Emma. BSc with honors, North East London Poly., 1976; PhD, U. London, 1981. Postdoctoral fellow London Sch. Hygiene and Tropical Medicine, 1981-83, UCLA Sch. Medicine, L.A., 1983-85; asst. prof. medicine and microbiology U. Va., Charlottesville, 1985-91, assoc. prof. medicine and microbiology, 1992-93; assoc. dir. U. Va. Asthma and Allergic Disorders Ctr., 1993—. Med. editorial bd. Jour. Allergy and Clin. Immunology, 1989-93, Allergy: The European Jour. Allergy and Clin. Immunology, 1992, Jour. Immunol. Methods; contbr. articles to Jour. Immunology, Jour. Allergy, Clin. Immunology, Jour. Exptl. Medicine, PNAS USA, Clin. Exptl. Allergy. Recipient Pharmacia Allergy Rsch. Found. award Pharmacia AB, Sweden, 1987. Fellow Am. Acad. Allergy and Immunology; mem. Brit. Soc. for Immunology, Am. Assn. Immunologists, Brit. Soc. Allergy and Clin. Immunology, Collegicum Internationale Allergologicum. Achievements include research on purification of dust mite, cat, cockroach and fungal allergens, on analysis of immune response to allergens, on the role of allergens in the etiology of asthma; development of monoclonal antibody based tests for measuring environmental allergen exposure. Office: U Va Divsn of Allergy Box 225 Charlottesville VA 22908

CHAPMAN, ORVILLE LAMAR, chemist, educator; b. New London, Conn., June 26, 1932; s. Orville Carmen and Mabel Elnora (Tyree) C.; m. Faye Newton Morrow, Aug. 20, 1955 (div. 1980); children: Kenneth, Kevin; m. Susan Elizabeth Parker, June 15, 1981. B.S., Va. Poly. Inst., 1954; Ph.D., Cornell U., 1957. Prof. chemistry Iowa State U., 1957-74; prof. chemistry UCLA, 1974—; Cons. Mobil Chem. Co. Recipient John Wilkinson Teaching award Iowa State U., 1968, award Nat. Acad. Scis., 1974; Founders prize Tex. Instruments; George and Freda Halpern award in photochemistry N.Y. Acad. Scis., 1978. Mem. Am. Chem. Soc. (award in pure chemistry 1968, Arthur C. Cope award 1978, Midwest award 1978, Havinga medal 1982, McCoy award UCLA, 1985). Home: 1213 Roscomare Rd Los Angeles CA 90077-2202 Office: UCLA Dept Chemistry 405 Hilgard Ave Los Angeles CA 90024

CHAPMAN, RUSSELL LEONARD, botany educator; b. Bklyn., May 30, 1946; s. Russell Hood and Helen C.; m. Melanie Anne Chapman, June 28, 1969; children: Christopher John, Timothy Sean. BA, Dartmouth Coll., 1968; MS, U. Calif., Davis, 1970, PhD, 1973. NSF grad. fellow dept. botany U. Calif., Davis, 1971-73; asst. prof. dept. botany La. State U., Baton Rouge, 1973-77, assoc. prof. dept. botany, 1977-83, prof. dept. botany, 1983—, assoc. dean Coll. of Arts and Scis., 1979-83, assoc. dean Coll. of Basic Scis., 1983-84, chmn. dept. botany, 1988—; vis. prof. dept. molecular, cellular and devel. biology, U. Colo., Boulder, 1984. Editorial bd.: Jour. of Phycology, Algologia, Molecular Phylogenetics and Evolution; author book chpts. in field; contbr. articles to profl. jours. Bd. dirs. Baton Rouge Earth Day, Inc., 1990-92; mem. Found. for Hist. La., Baton Rouge, 1973—. Recipient Outstanding Undergrad. Teaching award Amoco Found., Inc., 1978, Disting. Faculty award La. State U. Alumni Fedn., Baton Rouge, 1981. Fellow Linnaen Soc. London; mem. Phycol. Soc. Am. (sec., v.p., pres. 1985-90), Botanical Soc. Am. (chmn. phycol. sect. 1983-85), British Phycol. Soc., Internat. Phycol. Soc., Internat. Soc. for Evolutionary Protistology, Willie Hennig Soc., La. Soc. Electron Microscopy (treas., pres. 1976-80), Phi Kappa Phi, Sigma Xi. Episcopalian. Home: 6920 Bayou Paul Rd Saint Gabriel LA 70776 Office: Dept of Botany 502 Life Scis Bldg La State Univ Baton Rouge LA 70803-1705

CHAPMAN, SANDRA BOND, neurolinguist, researcher; b. Houston, June 5, 1949; d. Paul Ray and Maurine (Parrish) Little; m. Carroll Bond, Aug. 23, 1969 (dec. Oct. 1982); m. Donald Mason Chapman; 1 child, Noah Mason. MA in Communication Disorders, U. North Tex., 1974; PhD in Communication Disorders, U. Tex., Dallas, 1986. Cert. speech-lang. pathologist. Mem. faculty U. North Tex., Denton, 1977-80; rsch. scientist U. Tex., Dallas, 1986—, adj. instr. Southwestern Med. Ctr., 1989—. Contbg. author: Aphasia and Aging, 1991, Neurolinguistics and Aging, 1991; contbr. articles to profl. jours. Grantee Nat. Inst. Aging, 1991—, Nat. Inst. Neurological Disorders and Stroke, 1991—, Tex. Advanced Rsch. Program/Advanced Tech. Program, 1992-93; named one of Top 100 Grads. in 100 Yrs., U. North Tex., 1990. Mem. AAAS, Am. Gerontol. Assn., Am. Speech-Hearing Assn., Tex. Speech-Hearing Assn., Dallas Aphasiology Assn. (coord. 1990-92), Acad. Aphasia. Achievements include research in patterns of discourse disruption in children with frontal lobe injury, in deficits in depth of information processing in aphasia, in discourse processing in dementia and old-elderly. Office: U Tex Callier Ctr 1966 Inwood Rd Dallas TX 75235

CHAPMAN, WILLIAM EDWARD, pathologist; b. Miami, Fla., Aug. 23, 1945; s. Ralph Edward Chapman and Emily (Del Mar) Muller; m. Eva Louise Morgenstern, Aug. 15, 1975; children: Claire, Andrew. BA, U. Conn., 1972; MD, U. Conn., Farmington, 1976. Cert. Am. Bd. Pathology in Anatomic and Clin. Pathology. Resident anatomic pathology U. Conn. Sch. Medicine, 1976-78; rsch. trainee U. Conn., 1979-80; intern straight med. New Britain (Conn.) Gen. Hosp., 1979-80; resident in medicine Norwalk (Conn.) Hosp.; resident clin. pathology Westchester County Med. Ctr., Valhalla, N.Y., 1981-82; fellow anatomic pathology Meml. Sloan-Kettering Cancer Ctr., N.Y., 1982-83; asst. attending pathologist Beth Israel Med. Ctr., N.Y., 1983-86, White Plains (N.Y.) Med. Ctr., 1986-87; staff pathologist Vets. Hosp., Oteen, N.C., 1987—; presenter 1st Internat. Congress on AIDS, 1985, 2d Internat. Congress on AIDS, Paris, 1986. Contbr. articles to profl. jours. including Jour. Immunology, Sci., Am. Jour. Pathology, Am. Jour. Medicine, Jour. Clin. Endo and Metab., Chest, Surg. Rounds. With U.S. Army, 1966-70, Vietnam. Fellow Coll. Am. Clin. Pathologists. Office: Vets Hosp Oteen NC 28805

CHAPPELL, CHARLES RICHARD, space scientist; b. Greenville,, S.C., June 2, 1943; s. Gordon Thomas and Mabel Winn (Ownbey) Chappell; m. Barbara Lynne Harris, May 15, 1968; 1 child, Christopher Richard. B.A., Vanderbilt U., 1965; Ph.D. in Space Sci., Rice U., 1968. Assoc. research scientist Lockheed Palo Alto (Calif.) Research Lab., 1968-70, research scientist, 1970-72, staff scientist, 1972-74; chief magnetospheric physics br. NASA-Marshall Space Flight Ctr., Huntsville, Ala., 1974-80, chief solar terrestrial physics div., 1980-87, assoc. dir. for sci., 1987—; selected as Alternate Payload Specialist for the ATLAS-1 mission of the Space Shuttle, 1985. Author: (ency.) Plasmasphere, 1970, Spacelab Mission, 1985; contbr. numerous articles to profl. jours. Recipient medal for Exceptional Sci. Achievement NASA, 1981, 84; NASA trainee, 1966-68. Mem. Am. Geophys. Union, Congress of Space Research, Phi Beta Kappa, Phi Eta Sigma. Methodist. Avocations: distance running; windsurfing. Home: 2803 Downing Ct SE Huntsville AL 35801-2246 Office: NASA- Marshall Space Flight Ctr Huntsville AL 35812

CHARACKLIS, WILLIAM GREGORY, research center director, engineering educator; b. Annapolis, Md., Aug. 21, 1941; s. Gregory A. and Artemis Characklis; m. Nancy Crowley; children: Gregory William, Erin Elizabeth. BS, Johns Hopkins U., 1964, PhD, 1970; MS, U. Toledo, 1967. Prof. Rice U., Houston, 1970-80; dir. Inst. Process Analysis, Bozeman, Mont., 1980-90; dir. Ctr. for Interfacial Microbial Process Engring. Montana State U., Bozeman, 1990—; pres. CCE Inc. Author: (with P.A. Wilderer and John Wiley) Structure and Functions of Biofilms, 1989, (with K.C. Marshall and J. Wiley) Biofilms, 1990; contbr. articles to profl. jours. Fellow NSF. Mem. Am. Inst. Chem. Engrs., Assn. Environ. Engring. Profs., Am. Soc. Microbiology, Water Pollution Control Fedn., Cooling Tower Inst., Nat. Assn. Corrosion Engrs. Greek Orthodox. Office: Ctr for Interfacial Microbial Process Engring 407 Cobleigh Hall Bozeman MT 59717

CHARBENEAU, RANDALL J., environmental and civil engineer; b. Ann Arbor, Mich., Oct. 5, 1950; s. Gerald T. and Margaret (Dohm) C.; m. Nancy Jean Haskell, Sept. 7, 1974; children: Cynthia Catherine, Robert James. BS, U. Mich., 1973; MS, Oreg. State U., 1975; PhD, Stanford U., 1978. Prof. U. Tex., Austin, 1978—, dir. Ctr. for Rsch. in Water Resources, 1989—. Office: U Tex Dept Civil Engring Austin TX 78712

CHARBONNEAU, LARRY FRANCIS, research chemist; b. Faribault, Minn., Aug. 14, 1939; s. Francis Leroy and Lois Margaret (Pleschourt) C.; m. Kathryn Elizabeth Anderson, Apr. 16, 1966; children: Angela K., Mark F. BS, Mankato U., 1964; PhD, U. Ill., Urbana, 1972. Chemist 3M Co., St. Paul, 1964-67, GM, Warren, Mich., 1972-77; sr. rsch. chemist Celanese Corp., Summit, N.J., 1977-83, staff scientist, 1983-85; rsch. assoc. Hoechst Celanese Corp., Summit, 1985-89; rsch. supr. Hoechst Celasene, Celanese Corp., Summit, 1989—. Contbr. articles to profl. jours. Mem. K.C. (treas. N.J. chpt. 1989—, chpt. chmn. 1987-88), Am. Chem. soc. (councilor 1987-90), Am. Chem. Soc. Polymeric (materials sci. and engring. div.). Republican. Roman Catholic. Achievements include patents in the field of liquid crystal polymers and polyimides. Home: 64 Mountainside Rd Mendham NJ 07945 Office: Hoechst Celanese Corp 86 Morris Ave Summit NJ 07901

CHARGAFF, ERWIN, biochemistry educator emeritus, writer; b. Austria, Aug. 11, 1905; came to U.S., 1928, naturalized, 1940; s. Hermann and Rosa C.; m. Vera Broido; 1 son, Thomas. Dr. phil., U. Vienna, 1928; Dr. phil. h.c, U. Basel, 1976; Sc.D. (hon.), Columbia U., 1976. Research fellow Yale U., 1928-30; asst. U. Berlin, Germany, 1930-33; research assoc. Inst. Pasteur, Paris, France, 1933-34; faculty Columbia U., 1935—, prof. biochemistry, 1952-74, prof. emeritus, 1974—, chmn. dept. biochemistry, 1970-74; vis. prof., Sweden, 1949, Japan, 1958, Brazil, 1959; vis. prof. Coll. de France, 1965, Naples, Palermo, Cornell, 1966, Stazione biologica, Naples, 1969. Author: Essays on Nucleic Acids, 1963, Voices in the Labyrinth, 1977, Heraclitean Fire, 1978, Das Feuer des Heraklit, 1979, Unbegreifliches Geheimnis, 1980, Bemerkungen, 1981, Warnungstafeln, 1982, Kritik der Zukunft, 1983, Zeugenschaft, 1985, Serious Questions, 1986, Abscheu vor der Weltgeschichte, 1988, Alphabetische Anschlaege, 1989, Vorlaeufiges Ende, 1990, Vermächtnis, 1992, Ueber das Lebendige, 1993; numerous articles in field, other lit. work in English and German; Editor: The Nucleic Acids, 3 vols, 1955, 60. Guggenheim fellow, 1949, 58; recipient Pasteur medal, 1949, Soc. Biol. Chemistry, Paris, 1961, Neuberg medal Am. Soc. European Chemists, 1958; Bertner Found. award Houston, 1965; C.L. Mayer prize French Acad. Scis., 1963, Dr. H.P. Heineken prize Netherlands Acad. Scis., 1964; Gregor Mendel medal German Acad. Scis. Leopoldina, 1973; Nat. Medal of Sci., 1975; medal N.Y. Acad. Medicine, 1980; Disting. Svc. award Columbia U., 1982, Johann-Heinrich Merck prize German Acad. Lang. and Lit., 1984. Fellow Am. Acad. Arts and Scis.; mem. Nat. Acad. Scis., Am. Philos. Soc.; fgn. mem. Royal Swedish Physiographic Soc., German Acad. Scis. Leopoldina. Home: 350 Central Park W Apt 13G New York NY 10025-6547

CHARKIEWICZ, MITCHELL MICHAEL, JR., economics and finance educator; b. Springfield, Mass., Sept. 29, 1946; s. Mitchell Michael Sr. and Helen (Nycz) C.; m. Sandra Isabel Miranda Amaral, June 30, 1990. BSBA, Am. Internat. Coll., 1969, MBA, 1984; postgrad., U. Conn., 1986—. Prof. econs. Am. Internat. Coll., Springfield, 1985—, acting chmn. fin. dept., chmn. econs. dept., 1988-91; prof. econs. Central Conn. State U., New Britain, 1989—, Tunxis Community Coll., Farmington, Conn., 1986—; bus. tchr. Mt. Greylock Regional High Sch., Williamstown, Mass., 1969-71, Montachusett Regional Tech. High Sch., Fitchburg, Mass., 1971-74; unit mgr. Sports Prodns., Cleve., 1974; with sales and promotions San Antonio Spurs NBA, 1974-76; mgr. Bristol (Conn.) Racquet Club, 1976-80; sports dir. Sta. WBIS Radio, Bristol, 1980-85; cons. Tecnomed, J. Trapp, CA, Caracas, Venezuela, 1989, De-Yang Devel. Corp. of Sci. and Tech., China, 1990. Author test bank for Principles of Economics (by Colander), 1993; editor: AIC Sch. of Bus. Jour., 1988-91; referee jour. Iranian Econ. Rev., 1990. Chmn. softball marathon Easter Seals, New Britain, 1985; celebrity chmn. Conn. Pub. TV, Hartford, 1980-84; mem. World Affairs Coun., Springfield, 1985-91; judge Internat. Chili Soc., L.A., 1980—. Recipient Appreciation award Optimist Club, Woodland Hills, Tex., 1976, San Antonio, 1976, Conn. Pub. TV, Hartford, 1982-84, Easter Seals, Conn., 1985, U. Kuwait, 1988. Mem. AAUP, Am. Econ. Assn., Ea. Econ. Assn., N.E. Bus. and Econs. Assn. (bd. dirs. 1992—), We. Mass. Econ. Devel. Assn., Conn. Internat. Trade Assn. (chmn. scholarship com.), New Eng. Coun. Latin Am. Studies, Omicron Delta Epsilon (dir. 1988-91). Republican. Roman Catholic. Avocations: judging food contests, world travel. Home: 900 Jerome Ave Bristol CT 06010-2407 Office: Cen Conn State U Dept Econs 1615 Stanley St New Britain CT 06050-9999

CHARLES, JOEL, audio and video tape analyst, voice identification consultant; b. Phila., Jan. 12, 1914; s. Samuel William and Minnie (Fink) Blumenstein; m. Lillian DuBowe, May 31, 1938 (div. 1964); children: Mark Blumenstein, Richard Blumenstein; m. Nancy Sher, Oct. 24, 1988. BSChemE, Drexel U., 1938. Pres. The Charles Agy., Phila., 1938-42, 45-64; physicist Naval Air Exptl. Station, Phila., 1942-45; pres. The Dento-Med. Tapes, Upper Darby, Pa., 1957-73, Associated TV Prodns., Inc., Phila., 1948-52, Computerized Electronic Edn., Upper Darby, Pa., 1969-73; dir. continuing edn., media instructional methodology Pa. Coll. of Podiatric Medicine, Phila., 1973-77; pvt. practice, audio and video tape analysis and voice identification cons. Plantation, Fla., 1977—; expert witness on tape recordings in over 200 trials. Contbr. articles to profl. jours. Mem. Nat. Assn. Criminal Def. Lawyers (assoc.), Nat. Forensic Ctr., Am. Fedn. Musicians, Am. Dialect Soc., Audio Engring. Soc. (chmn. forensic tape com.). Achievements include development of only rapid form of computerized voice identification, designed first high-speed portable audio cassette duplicator. Home: 9951 NW 5th Pl Plantation FL 33324-7040

CHARLES, M. ARTHUR, endocrinologist, educator; b. Alhambra, Calif., Mar. 23, 1941. Student, L.A. State Coll.; MD, U. Calif., Irvine, 1967; PhD, U. Calif., San Francisco, 1975. Diplomate Am. Bd. Internal Medicine, Endocrine Subspecialty Bd.; lic. Calif. Commd. U.S. Army, 1966; intern Walter Reed Army Med. Ctr., Washington, 1967-68, resident, 1968-70, chief med. resident, 1970-71; endocrine fellow U. Calif., San Francisco, 1971-73, Metabolic Rsch. Unit, 1975-76; lectr. dept. biochemistry U. Calif. Med. Ctr., San Francisco, 1973-76; rsch. assoc. Metabolic Rsch. Unit U. Calif., 1971-75, asst. rsch. physician Metabolic Rsch. Unit, 1975-76; clin. attending physician, rsch. physician Letterman Army Med. Ctr. Presidio, San Francisco, 1975-76; asst. clin. prof. dept. medicine U. Colo. Med. Ctr., Denver, 1976-79; rsch. physician Clin. Investigation Svc. Fitzsimons Army Med. Ctr., 1976-79; resigned U.S. Army, 1979; assoc. prof. depts. medicine and physiology U. Calif., Irvine, 1979-85, prof. depts. medicine and physiology, 1985—, dir. Focused Rsch. Program in Diabetes, 1985—, founder, dir. Diabetes Rsch. Ctr., 1986—; founder ann. Diabetes Program's External Review Retreat U. Calif., Irvine, 1984—; founder rsch. seminar series U. Calif., Irvine, Med. Ctr., Long Beach VA and CHOC, 1991; site visit team mem. NIH Diabetes Endocrine Rsch. Ctr., U. Wash., Seattle, 1981, 86, U. Mass., Worcester, 1988; site visit team mem. NIH Diabetes Rsch. and Tng. Ctr. Albert Einstein Med. Ctr., N.Y., 1981, 87, Wash. U., St. Louis, 1987; mem. tng. grant study sect. NIH, Washington, 1987-90; mem. editorial bd. Metabolism, 1992—; reviewer various profl. jours., 1978—; mem. evaluation site unit team We. Assn. Schs. and Colls., 1989, 93; mem. ad hoc review com. UCLA, 1991; USA cons. GETREM study group of Europe for intensive therapy for New Onset Diabetic Children; speaker numerous seminars in field. Contbr. numerous papers and abstracts to profl. jours. With USAR, 1979—. Grantee NIH, 1980-85, 85-88, 88-91, FDA 1986-87; recipient J. Gordon Hatfield award, 1967, V.P. Carroll award, 1967, Ben J. Carey Meml. award, 1967, Disting. Alumnus award Nat. Assn. State Univs. and Land-grant Colls., 1988, Dick Roberts award, 1984-85, Clin. Rsch. award NOVO Industries, 1980-81, 81-82, 82-83, Parker-Hannifin Pacesetter Corps. award, 1981-82, Rsch. award NOVO Industries, 1983-84, Parker-Hannifin Corp. award, 1983-84, Pacesetter Systems award, 1983, Nordisk award, 1984-84, Squibb-NOVO, Inc. award, 1984-85, 87-88, Roerig-Pfizer award, 1988-89, 89-90, 85-91, Stuart ICI award, 1985-88, 87-88, Pacesetter Infusion award, 1985-86, 86-88, Hoechst Pharms. award, 1986-88, 81-91, Lily award, 1987-88, Trental Neuropathy award Hoechst, 1990-92, Infusaid Implantable Pump award, 1990-93, Minimed Implantable Pump award, 1991-92, Fujisawa award, 1991—, Wyeth-Ayerst award, 1991-93, Hoechst award, 1991-92. Mem. Am. Fedn. Clin. Rsch., Am. Diabetes Assn. (bd. dirs. Orange County chpt. 1989-91, 93, pres. elect 1990, pres. 1991-92, Calif. affiliate profl. sci. com. 1992-93, profl. edn. com. 1990—, sci. rsch. com. So. Calif. affiliate 1982-84, prof. edn. com. 1983-85, rsch. com. 1990—, abstract reviewer ann. sci. meeting 1984-87, profl. edn. subcom. Orange County chpt. 1988—, rsch. grantee 1979-81, 83-84, 84-85), We. Soc. Clin. Rsch., Orange County Endocrine Soc. (seminar series coord., seminar dir. 1986—, pres. 1988-89), Endocrine Soc., Juvenile Diabetes Assn. (bd. dirs. Orange County chpt. 1989-91), U. Calif. Alumni Assn. (bd. dirs. 1989—), Alpha Omega Alpha. Achievements include research in highly purified porcine insulins, portable infusion pump, semisynthetic human insulin, immunogenicity of human insulin, semisynthetic human long acting insulin, oral agents and new onset type I diabetes, type II diabetes, aldose reductase inhibitor, implantable infusion system, biosynthetic human insulin, captopril, glipizide as an immunosuppressant, aldose reductase inhibitor trial, use of glipizide to prevent dts, HOE490 prevention of diabetes in BB rats. Office: U Calif Irvine Diabetes Rsch Program Dept Med Sci I C264 Irvine CA 92717*

CHARLESWORTH, ERNEST NEAL, immunologist, educator; b. Denver, June 11, 1945; s. Albert Ernest and Wilma Nadine (Wright) C.; m. Margaret Louise Gay, July 12, 1969; children: Richard Neil, Mark Edward. BS, U. Houston, 1967; MD, U. Tex., 1971. Diplomate Am. Bd. Allergy and Immunology, Am. Bd. Dermatology, Am. Bd. Internal Medicine, Diagnostic Lab. Immunology Bd. Commd. capt. USAF, 1971, advanced through grades to col., 1987; intern in medicine Wilford Hall USAF Med. Ctr., Lackland AFB, Tex., 1971-72, resident in dermatology, 1973-76, resident in internal medicine, 1984-86; staff dermatologist USAF Med. Ctr., Keesler AFB, Miss., 1976-78; pvt. practice in dermatology Jackson, Miss., 1978-81; clin. faculty dept. medicine U. Miss. Med. Sch., Jackson, 1979-81; chief dermatology svc. USAF Regional Med. Ctr., Clark Air Base, The Philippines, 1981-84; fellow in allergy and immunology Johns Hopkins U. Sch. Medicine, Balt., 1986-89, clin. faculty divsn. allergy and clin. immunology, 1988-89; asst. chief allergy-immunology svc. Wilford Hall USAF Med. Ctr., Lackland AFB, 1989—; assoc. prof. medicine Uniformed Svcs. Univ. of The Health Scis., Bethesda, Md., 1990-92; clin. assoc. prof. U. Tex. Health Sci. Ctr., San Antonio, 1990-92; cons. to surgeon gen. for dermatology PACAF Med. Command, Hickam AFB, Hawaii, 1981-84; presenter Harold S. Nelson Allergy-Immunology Symposium, Fitsimons Army Med. Ctr., Aurora, Colo., Johns Hopkins Asthma and Allergy Ctr., Balt., 16th Hawaii Dermatology Seminar, Maui. Contbr. articles to Arch. Dermatology, Jour. Mil. Assn. Dermatology, Jour. Clin. Investigation, Internat. Arch. Allergy Immunology, Insights in Allergy, Jour. Pediatrics, Jour. Investigative Dermatology. Recipient Clemens Von Pirquet Rsch. award Am. Coll. Allergy and Immunology, 1987. Fellow ACP (presenter), Am. Acad. Dermatology, Am. Coll. Allergy (presenter); mem. Am. Acad. Allergy and Immunology (dermatologic disease interest sect., presenter, Young Investigator award 1989), Assn. Air Force Allergists, Soc. Air Force Physicians, Soc. Investigative Dermatology, San Antonio Allergy Soc. (sec.). Episcopalian. Achievements include rsch. in late-phase allergic reaction sites, intractable sneezing, cutaneous late-phase response to allergen, decline in pulmonary function tests in extrinsic asthmatics immediately following immunotherapy. Home: 12811 Kings Forest San Antonio TX 78230 Office: Wilford Hall Med Ctr SGHMA 2200 Bergquist Dr Ste 1 Lackland AFB TX 78236-5300

CHARLEY, PHILIP JAMES, testing laboratory executive; b. Melbourne, Australia, Aug. 18, 1921; came to U.S., 1940, naturalized; 1948; s. Walter George and Constance Mary (Macdonald) C.; BS, U. Wis., 1943; MS in Mech. Engring., U. So. Calif., 1947, PhD in Biochemistry, 1960; m. Katherine Truesdail, Jan. 31, 1948; children: James Alan, Linda Kay, William John. Test engr. Gen. Electric Co., Schenectady, 1943-44; lectr. in engring. U. So. Calif., L.A., 1947-49; project engr. Standard Oil of Calif., El Segundo, 1948-55; v.p. Truesdail Labs., L.A., 1955-70, pres., 1970—. Served to lt. Royal Can. Elec. and Mech. Engrs., 1943-45. Recipient Dueul award U. So. Calif., 1960, rsch. assoc., 1960-65; registered profl. engr., Calif.. Ariz., Nev. Mem. AAAS, ASTM, ASME, Am. Soc. Metals, Am. Soc. Safety Engrs., Am. Chem. Soc., Sigma Xi, Tau Beta Pi, Beta Theta Pi. Republican. Club: Rotary. Home: 1906 Calle De Los Alamos San Clemente CA 92672-4309 Office: Truesdail Labs Inc 14201 Franklin Ave Tustin CA 92680-7094

CHARLTON, CLIVEL GEORGE, neuroscientist, educator; b. Manchester, Jamaica; came to U.S., 1969; s. Ferdinand Nehemiah and Mazy Mildred (Davy) C.; m. Pauline Hyacinth Harrison. BS, Tuskegee U., 1972, MS, 1974; MS, U. Calif., San Francisco, 1976; PhD, Howard U., 1983. Instr. Fla. A & M Univ., Tallahassee, Fla., 1974; rsch. assoc. Univ. Calif., San Francisco, 1976-78, Howard Univ., Washington, 1979-83; post-doctoral scholar Uniformed Svcs. Univ., Bethesda, Md., 1984-86; student scientist NIH, 1979-83; asst. prof. Meharry Med. Coll., Nashville, 1986-93, assoc. prof., 1993—; reviewer NSF, Washington, 1987, 89, 91, NIH, Bethesda, 1989, 92, Nat. Inst. Drug Abuse, Bethesda, 1990; mem. study section NIH, 1992—. Contbr. chpt. to Drug Addiction: New Aspect of Analytical and Clinical Toxicology, 1974, Basic Clinical and Therapeutic Aspects of Alzheimer's and Parkinson's Disease, 1990; contbr. articles to Jour. of Neuroscience and others. Recipient Travel award Am. Soc. for Pharmacology and Experimental Therapeutics, 1973; grantee NIH. Mem. Soc. for Neuroscience (travel fellowship 1984), Internat. Brain Rsch. Orgn., World Fedn. Neuroscientists. Achievements include discovery that the Parkinson's disease casuing agent MPP may serve as a methyl donor, that S-adensylmethionine can cause Parkinsonian type of effects: indentification of the hormone secretin in brain. Home: 529 Janice Dr Antioch TN 37013-

4157 Office: Meharry Medical Coll 1005 D B Todd Blvd Nashville TN 37208

CHARLTON, GORDON RANDOLPH, physicist; b. Newport News, Va., Aug. 30, 1937; s. George Randolph and Sarah Louise (Harper) C.; children: George Thomas, Anne Louise. BSc, Ohio State U., 1957; MSc in Physics, W.Va. U., 1960; PhD in Physics, U. Md., 1966. Charge de recherches Ecole Poly. Lab., Paris, 1966-69; asst. physicist high energy physics div. Argonne (Ill.) Nat. Lab., 1969-72; rsch. assoc. Stanford (Calif.) Linear Accelerator Ctr., 1972-73; systems mgr. physics dept. U. Toronto (Ont., Can.), 1973-75; sr. physicist div. high energy physics Office of Energy Rsch., U.S. Dept. Energy, Germantown, Md., 1975—. Contbr. articles to Phys. Rev., Phys. Rev. Letters, Physics Letters. Mem. AAAS, Am. Phys. Soc., U.S. Croquet Assn. Office: US Dept Energy Divsn High Energy Physics 19901 Germantown Rd Germantown MD 20874-1290

CHARLTON, JOHN KIPP, pediatrician; b. Omaha, Jan. 26, 1937; s. George Paul and Mildred (Kipp) C. A.B., Amherst Coll., 1958; M.D., Cornell U., 1962; m. Susan S. Young, Aug. 15, 1959; children: Paul, Cynthia, Daphne, Gregory. Intern, Ohio State U. Hosp., Columbus, 1962-63; resident in pediatrics Children's Hosp., Dallas, 1966-68, chief pediatric resident, 1968-69; nephrology fellow U. Tex. Southwestern Med. Sch., Dallas, 1969-70; pvt. practice medicine specializing in pediatrics, Phoenix, 1970; chmn. dept. pediatrics Maricopa Med. Ctr., Phoenix, 1971-78, 84-93, assoc. chmn. dept. pediatrics, 1979-84, med. staff pres., 1991; med. dir., bd. dirs. Crisis Nursery, Inc., 1977—; sr. clin. lectr. dept. pediatrics U. Ariz. Pres. Maricopa County Child Abuse Coun., 1977-81; bd. dirs. Florence Critenton Svcs., 1980-83, Ariz. Children's Fund, 1987-91; mem. Gov.'s Coun. on Children, Youth and Families, 1984-86. Officer M.C., USAF, 1963-65. Recipient Hon Kachina award for volunteerism, 1980, Jefferson award for volunteerism, 1980, Horace Steel Child Advocacy award, 1993. Mem. Am. Acad. Pediatrics, Ariz. Pediatric Soc., Maricopa County Pediatric Soc. (past pres.). Author articles, book rev. in field. Home: 6230 E Exeter Blvd Scottsdale AZ 85251-3060 Office: Maricopa Med Ctr 2601 E Roosevelt St Phoenix AZ 85008-4973

CHARLTON, KEVIN MICHAEL, aerospace engineer, support contractor; b. Detroit, Jan. 24, 1966; s. James Martin and Patricia (Burke) C. BS in Engring. cum laude, Princeton U., 1988; MS in Engring., U. Mich., 1990. Engring. aide Naval Rsch. Lab., Washington, 1986-87; intern Office of Sen. Albert Gore, Washington, 1987-88; mgmt. intern Consolidated Edison of N.Y., N.Y.C., 1988-89; staff mem. BDM Internat., Inc., Washington, 1990—. Tutor Washington Pub. Schs., 1992; vol. Washington Lupus Found., Arlington, 1992. Recipient Benton grad. fellowship U. Mich., 1990. Mem. AIAA. Democrat. Roman Catholic. Office: BDM Internat 409 3d St SW Ste 340 Washington DC 20024

CHARNESS, MICHAEL EDWARD, neurologist, neuroscientist; b. Montreal, Quebec, Can., Nov. 27, 1950; came to U.S., 1972; BSc, McGill U., Montreal, 1972; MD, Johns Hopkins U., Balt., 1976. Diplomate Am. Bd. Internal Medicine, Am. Bd. Psychiatry and Neurology. Instr. neurology U. Calif., San Francisco, 1981-83, asst. prof. neurology, 1983-89; assoc. prof. neurology and neurosci. Harvard Med. Sch., Boston, 1989—. Contbr. articles to Sci., Annals of Neurology, Jour. Biol. Chemistry, New Eng. Jour. Medicine, Proceedings Nat. Acad. Sci. Recipient The Sandoz award, Sandoz Corp., San Francisco, 1981, Rsch. Scientist Devel. award. Nat. Inst. Alcohol Abuse & Alcoholism, Washington, 1984-89, Basil O'Connor award March of Dimes, 1985-87. Fellow Am. Acad. Neurology; mem. Rsch. Soc. Alcoholism (program com. chmn. 1991-92), Soc. for Neurosci., Performing Arts Medicine Assn. Office: Harvard Med Sch VA Med Ctr 1400 VFW Pkwy West Roxbury MA 02132-4927

CHARNEY, PHILIP, dermatologist; b. N.Y.C., Dec. 15, 1939; s. Louis and Rose (Shay) C. BA cum laude, CUNY, Bklyn., 1960; MD, SUNY, Bklyn., 1964. Diplomate Am. Bd. Dermatology; lic. physician, N.Y., D.C., N.J., Calif., Nev., Ariz. Intern Interfaith Med. Ctr., Bklyn., 1964-65; resident dermatology USPHS Hosp., S.I., N.Y., 1965-67; vis. fellow dermatology Columbia-Presbyn. Med. Ctr., N.Y.C., 1967-68; asst. chief medicine dermatology outpatient clinic USPHS, Washington, 1968-70; asst. dir. clin. rsch. Schering Corp., Bloomfield, N.J., 1970-73, assoc. dir. clin. rsch., 1973; pvt. practice dermatology South Lake Tahoe, Calif., 1973-76, Lake Tahoe, Carson City, Nev., 1976-84, Phoenix, 1984-87, Castro Valley, Calif., 1987—; attending staff dept. dermatology Howard U., Washington, 1968-70; clinic asst. St. Luke's Hosp. Ctr., N.Y.C., 1971-73; attending staff Barton Meml. Hosp., South Lake Tahoe, 1973-84; courtesy staff Carson Tahoe Hosp., Carson City, 1978-84; active staff Chandler (Ariz.) Regional Hosp., 1985-87, Eden Hosp. Ctr., Castro Valley, Calif. Contbr. articles to profl. jours. Mem. AMA, Am. Acad. Dermatology, Am. Soc. Dermatologic Surgery, Internat. Soc. Dermatology, Pacific Dermatological Assn., San Francisco Dermatological Soc., Sacramento Valley Dermatological Soc., Alameda/Contra Costa County Med. Assn., Calif. Med. Assn., N.Y. Acad. Sci., ACLU, Sierra Club. Democrat. Jewish. Avocations: piano, photography, tennis, sailing, skiing. Office: 2457 Grove Way Ste 105D Castro Valley CA 94546-7183

CHARPAK, GEORGES, physicist; b. Dabrovica, Poland, Aug. 1, 1924; arrived in France, 1946.; s. Maurice and Anna (Szapiro) C.; m. Dominique Vidal, 1953; children: Yves, Nathalie, Serge. Student, Ecole des Mines de Paris, 1945-47; D of Physics, Coll. of France, 1954; hon. doctorate, U. Geneva. Lic. civil mining engr. Prof. Centre Nation de la Recherche Scientifique, 1948-59, Centre Européen pour la Recherche Nucléaire, Geneva, 1959—; researcher Cern Lab. for Particle Physics, Geneva; Joliot-Curie prof. Ecole Supérieure de Physique et Chimie de la Ville de Paris, 1984—. Contbr. articles to profl. jours. With French Army, prisoner of war, Dachau. Decorated Croix de Guerre (France); recipient Ricard prize European Physics Soc., 1980, High Energy and Particle Physics prize, 1989, Nobel prize for physics Nobel Found., Stockholm, 1992. Mem. NAS (fgn. assoc.), French Acad. Scis. (Commissariat prize of Atomic Energy 1984), Austrian Acad. Scis. (hon.). Achievements include invention of multiwire proportional chambers, drift chambers, diverse types of flash chambers without photography; development of particle detectors in high energy physics, installations for biological research using B wave imagery; research in nuclear structure by reactions. Home: 37 rue de la Plaine, 75020 Paris France Office: Cern Lab for Particle Physics, CH 1211 Geneva Switzerland

CHARRY, JONATHAN M., psychologist; b. N.Y.C., May 13, 1948; s. H. Paul and June B. Charry; A.B., Tufts U., 1970; Ph.D., NYU, 1976, M.A., 1973; cert. Alfred Adler Inst., 1977. Teaching fellow applied social psychology N.Y. U., N.Y.C., 1972-74, assoc. research scientist community psychology, 1973-75, Warburg scholar, 1974-75, postdoctoral fellow applied social psychology, 1975-77, adj. asst. prof., 1976-78; Rockefeller Found. postdoctoral fellow, guest investigator environ. scis. and physiol. psychology Rockefeller U., N.Y.C., 1977-79, asst. prof., 1979-83; sr. research scientist Inst. Basic Research, 1983-85; pres., chief exec. officer Environ. Research Info., Inc., N.Y.C.; pres., chief exec. officer Ion Technologies Inc., 1983; mem. clin. faculty Alfred Adler Inst., N.Y.C., 1977-83; adj. prof. psychology N.Y. U., 1979-83. Mem. Am. Psychol. Assn., Eastern Psychol. Assn., Am. Inst. Med. Climatology (dir. 1979-83, pres. 1981-82), N.Y. Acad. Scis., Bioelectromagnetics Soc., Sigma Xi, Phi Delta Kappa, AAAS. Research in biol. and behavioral effects of phys. stressors, toxic agts., electric fields, ionization, environ. risk assessment and mgmt. Home: 5220 Arlington Ave Bronx NY 10471-2822

CHARYK, JOSEPH VINCENT, retired satellite telecommunications executive; b. Canmore, Alta., Can., Sept. 9, 1920; came to U.S., 1942, naturalized, 1948; s. John and Anna (Dorosh) C.; m. Edwina Elizabeth Rhodes, Aug. 18, 1945; children: William R., J. John, Christopher E., Diane E. B.Sc., U. Alta., 1942, LL.D., 1964; M.S., Calif. Inst. Tech., 1943, Ph.D., 1946; D.Engring. (hon.), U. Bologna, 1974. Sect. chief Jet Propulsion Lab., Calif. Inst. Tech., 1945-46, instr. aeros., 1945-46; asst. prof. aeros. Princeton (N.J.) U., 1946-49, assoc. prof., 1949-55; dir. aerophysics and chemistry lab., missile systems div. Lockheed Aircraft Corp., 1955-56; dir. aero. lab. Aeronutronic Systems, Inc. subs. Ford Motor Co., 1956-58, gen. mgr. space tech. div., 1958-59; asst. sec. for research and devel. USAF, 1959, under sec., 1960-63; pres. Communications Satellite Corp., 1963-79, chief exec. officer, 1979-85, chmn., 1983-85; chmn. Draper Labs., 1987-90; bd. dirs. Comm. Satellite Corp. Recipient Lloyd V. Berkner Space Utilization award, 1967, Disting. Aviation Aerospace Svc. award, 1973, Gugliemo Marconi Internat. award, 1974, TV Arts and Scis. Directorate award, 1974, Theodore Von Karman award, 1977, Goddard Astronautics award, 1978, award Computer and Comm. Found., 1985, Nat. Medal of Tech., 1987, Arthur C. Clarke award, 1992, Disting. Alumni award U. Alta., 1993. Fellow AIAA, IEEE; mem. Nat. Acad. Engring., Internat. Acad. Astronautics, Nat. Space Club, Chevy Chase Country Club, Burning Tree Country Club, Gulf Stream Golf Club, Gulf Stream Bath and Tennis Club, Met. Club, Sigma Xi. Home: A-302 790 Andrews Ave Delray Beach FL 33483

CHASE, MERRILL WALLACE, immunologist, educator; b. Providence, Sept. 17, 1905; s. John Whitman and Bertha H. (Wallace) C.; m. Edith Steele Bowen, Sept. 5, 1931 (dec. 1961); children: Nancy Steele (Mrs. William W. Cowles), John Wallace, Susan Elizabeth (dec. 1985); m. Cynthia Hambury Pierce, July 8, 1961. AB, Brown U., 1927, ScM, 1929, PhD, 1931, ScD honoris causa, 1977; ScD honoris causa, Rockefeller U., 1988; MD honoris causa, U. Münster, Fed. Republic Germany, 1974. Instr. biology Brown U., 1931-32; staff mem. Rockefeller Inst. Med. Research, 1932-79; prof. immunology and microbiology, head lab. immunology and hypersensitivity Rockefeller U., 1956-79; med. adv. council Profl. Ednl. and Research Task Force, Asthma and Allergy Found. Am., 1955-83. Editor: (with C.A. Williams) Methods in Immunology and Immunochemistry, Vol. 1, 1967, Vol. 11, 1968, Vol. III, 1970, vols. IV, 1977, and V, 1976. Fellow Am. Acad. Allergy (hon., Disting. Svc. award 1969), Am. Coll. Allergists (hon.), Am. Acad. Arts and Scis.; mem. AAAS, Am. Assn. Immunologists (pres. 1956-57), Am. Soc. Microbiology (program chmn. 1959-61), Harvey Soc., N.Y. Acad. Scis., N.Y. Allergy Soc. (hon.), Nat. Acad. Sci. Republican. Universalist-Unitarian. Spl. research hypersensitivity to simple chem. allergens, studies Kveim antigen in sarcoidosis, studies tuberculins and mycobacterial antigens. Office: Rockefeller U 1230 York Ave New York NY 10021-6399

CHASE, PHILIP NOYES, psychologist; b. Ipswich, Mass., June 29, 1954; s. John Garvey and June Eileen (Savi) C.; m. Karen Anne Kenny, May 23, 1981; children: Nina, Monica. BA, U. Mass., 1977, PhD, 1982. Asst. prof. St. Cloud (Minn.) State U., 1981-82; Asst. prof. W.Va. U., Morgantown, 1982-86, assoc. prof., 1986-92, prof., 1992—; chair W.Va. U., 1993—; tng. cons. Ctr. Entrepreneurial Studies, Morgantown, 1982—, Enabling Techs., Inc., Chgo., 1985-88. Co-author: Performance Analysis, 1993; co-editor: Dialogues on Verbal Behavior, 1991, Psychological Aspects of Language, 1986, others; contbr. articles to profl. jours. Coach Youth Hockey, Morgantown, 1982-87. Fulbright scholar, Rome, 1990. Mem. Am. Psychol. Soc. (charter mem.), Southeastern Assn. Behavior Analysis (bd. dirs. 1991—), Assn. Behavior Analysis, Soc. Experimental Analysis of Behavior. Home: 709 Park St Morgantown WV 26505 Office: WVa U Dept Psychology PO Box 6040 Morgantown WV 26506-6040

CHASE, RAMON L., aerospace engineer; b. Mich., June 19, 1933; m. Arlene Torsch; children: Kirt, Diane Rodney. AA in Liberal Arts, Graceland Coll., 1953; BSE in Aero. Engring., U. Mich., 1956, MSME, 1961; MA in Adminstrn., U. Calif., 1975. Group supr. performance Chrysler Missile Div., 1956-63; group supr. aerodynamics Ling Temco Vought, 1963-65; MTS Aerospace Corp., 1965-71; registrar Riverside City Coll., 1971-73; mem. tech. staff Calif. Inst. Tech., JPL, 1973-76; MTS Gen. Rsch. Corp., 1976-77; div. mgr. space tech., prin. ANSER, Arlington, Va., 1977; participant Joint NASA/Dept. of Defense Space Transp. Architecture Study, 1986; tech. advisor ad hoc com. Air Force Sci. Adv. Bd., 1988-92. Contbr. articles to profl. jours. Mem. AIAA, Planetary Soc. Achievements include formulation of first Air Force Space Architecture and Weapons Master Plan. Home: 4575 Lawnvale Dr Gainesville VA 22065-1228 Office: ANSER Ste 800 1215 Jefferson Davis Hwy Arlington VA 22202

CHASE, RICHARD LIONEL ST. LUCIAN, geology and oceanography educator; b. Perth, Australia, Dec. 25, 1933; s. Conrad Lucien Doughty and Vera Mabel (Saw) C.; m. Mary Malcolm Nafe, Aug. 28, 1965; children: Sarah, Samuel, Elijah. B.Sc. with honors, U. Western Australia, 1956; Ph.D., Princeton U., 1963. Geologist Geosurveys of Australia, Adelaide, 1956-57; sr. asst. geologist Ministere des Mines, Que., Can., 1959; geologist Ministerio de Minas, Venezuela, 1960-61; postdoctoral fellow Woods Hole (Mass.) Oceanographic Instn., 1963-64; asst. scientist, 1964-68; asst. prof. U. B.C., Vancouver, 1968-73; assoc. prof. U. B.C., 1973-78, prof. dept. geol. scis., 1978-80, prof. depts. geol. scis. and oceanography, 1980—, acting head dept. geol. scis., 1990-91. Contbr. articles to profl. jours. Mem. Am. Geophys. Union, Geol. Assn. Can., Geol. Soc. Am. Club: Faculty U. B.C. Home: 4178 W 12th Ave, Vancouver, BC Canada V6R 2P6 Office: U BC, 6339 Stores Rd, Vancouver, BC Canada V6T 1Z4

CHASE, ROBERT WILLIAM, petroleum engineering educator, consultant; b. Scranton, Pa., Oct. 24, 1950; s. Elmer Frank and Doris Lynette (Shafer) C.; m. Carol Ann Leh, Oct. 11, 1975; children: Christopher Robert, Thomas William. BS in Petroleum Engring., Pa. State U., 1972, MS in Petroleum Engring., 1974, PhD in Petroleum Engring., 1980. Engr.-in-tng. Halliburton Svcs., Elkview, W.Va., 1972; rsch. engr. Gulf R & D Corp., Harmarville, Pa., 1974; asst. prof. W.Va. U., Morgantown, 1976-78; prof., dir. petroleum engring. program Marietta (Ohio) Coll., 1978—; cons. Gulf Sci. & Devel. Corp., Harmarville, 1978, Sci. Applications Inc., Morgantown, 1978-79, Columbia Gas, Charleston, W.Va., 1984—. Bd. dirs. Greater Marietta United Way, 1989—, Marietta Country Club, 1986-89. Rsch. grantee U.S. Dept. Energy, Morgantown, 1977-78. Mem. Soc. Petroleum Engrs. (bd. dirs. 1991-94), Assn. Petroleum Engring. Dept. Heads (pres. 1990-91), S.E. Ohio Oil & Gas Assn. (adv. com. 1989-91). Republican. Methodist. Achievements include patent for Explosive Earth Fracturing Process; developed dimensionless inflow performance relationship curves for fractured gas wells. Office: Marietta Coll 215 5th St Marietta OH 45750

CHASIN, MARSHALL LEWIS, audiologist, educator; b. Toronto, Ont., Can., Nov. 24, 1954; s. Philip and Lillian (Parker) C.; m. Joanne DeLuzio, Oct. 9, 1992; children: Courtney, Meredith, Shaun. BSc, U. Toronto, 1978; MSc, U. B.C., Vancouver, Can., 1981. Audiologist Can. Hearing Soc., Toronto, 1981-83, sr. audiologist, 1983-85; pvt. practice audiology Marshall Chasin & Assocs., Toronto, 1985—; lectr. U. Toronto, 1990—; dir. rsch. Can. Hearing Soc., Toronto, 1992—; mem. numerous standards and adv. coms. in field, Ont., 1983—. Contbr. numerous articles to profl. jours.; chpt. to book. Mem. Ont. Assn. Speech Pathology and Audiology (chair consumer advocacy com. 1983-85, Founder's award 1985), Can. Acoustical Assn., Acoustical Soc. Am., Can. Assn. Speech-Lang. Pathology/Audiology (Eve Kassirir award 1991). Home: 34 Banstock, North York, ON Canada M2K 2H6 Office: Can Hearing Soc, 271 Spadina Rd, Toronto, ON Canada M5R 2V3

CHASMAN, DANIEL BENZION, aerospace engineer; b. Haifa, Israel, Mar. 21, 1949; came to U.S., 1978; s. David and Haddassah (Hasanoff) C.; m. Judith Mandel, Aug. 14, 1974 (div. Mar. 1982); children: Mike Moses, Inbal Jill; m. Edit Orna Zaphir, Apr. 13, 1989; 1 child, Roey David. BSME, Tel Aviv U., 1976; MSCE, Colo. State U., 1982; PhD in Aerospace Engring., U. Colo., 1989. Design engr., project mgr. Israeli Mil. Industries, Tel Aviv, 1976-78; rschr. Colo. State U., Ft. Collins, 1979-82; wind tunnel test engr. Israeli Aircraft Industries, Lod, 1982-83; ind. cons. micro computers Ft. Collins, 1983-84; rschr. U. Colo., Boulder, 1984-90; sr. analyst Space Systems divsn. Rockwell Internat., Huntsville, Ala., 1990—. Contbr. articles to profl. jours. Sgt. Israeli Mil., 1968-71. Mem. AIAA. Home: 8221 Madison Pike Madison AL 35758 Office: Rockwell Internat Space Systems Div 555 Discovery Dr Huntsville AL 35806

CHASSAY, ROGER PAUL, JR., engineering executive, project manager; b. Chgo., Aug. 30, 1938; s. Roger Paul Sr. and Ruth Ruby (Taylor) C.; m. Katheryne Faye Roper, Jan. 1961 (div. Mar. 1988); children: Cynthia, Terri, Donald, Dean, Paul, Brett; m. Judith Marie Armstrong, Mar. 1990; stepchildren: Angela, Dana. BS, La. State U., 1961; postgrad., Ohio State U., 1962, Calhoun Community Coll., Huntsville, Ala., 1975-76, U. Ala., Huntsville, 1978-79. F8U-3 fighter wing designer Dallas, 1958; aerospace engr. Saturn & Skylab Program Offices NASA/Marshall Space Flight Ctr., Huntsville, Ala., 1964-74; SPAR project mgr. Spl. Projects Office NASA/Marshall Space Flight Ctr., Huntsville, 1974-77, mgr. integration/ test office Microgravity Projects Office, 1977-82; mgr. experiment carriers office Microgravity Projects Office, NASA/Marshall Space Flight Ctr., Huntsville, 1982-86, dep. mgr. Microgravity Projects Office, 1986-88; mgr. space sta. and advanced projects office Microgravity Projects Office NASA/Marshall Space Flight Ctr., Huntsville, 1988—; chmn. orbiter motion subcom. NASA, 1984-86, ctr. rep. flight assignments working group, Huntsville and Washington, 1985-86, chmn. space sta. microgravity requirements integration group, Huntsville, 1990—. Author: Application of Mathematics to the XB70 Bomber, 1963, Low-g Measurements by NASA, 1986, Processing Materials in Space: History and Future, 1987, (chpt.) Low Gravity Materials Experiments in Space Sta., 1989, Cooperation Betwen NASA and ESA for the First Microgravity Materials Science Glovebox, 1992; author, editor (NASA movie) Space Processing Applications Rocket Project, 1979. Sr. arbitrator Better Bus. Bur., Huntsville, 1987—; pres. Holy Spirit Ch. Coun., Huntsville, 1982, N.E. Ala. Ch. Coun., Huntsville, 1984-85; scoutmaster Boy Scouts Am., Dayton, Ohio, 1963; clarinetist Huntsville Concert Band, 1987-91. Capt. USAF, 1961-64. Assoc. fellow AIAA (gen. chmn. sounding rocket conf. 1977, space processing com. 1981-83). Roman Catholic. Achievements include management of test program for world's largest turbojet engine, management of first successful space levitation experiments, management of first materials science glovebox in space, management of first commercial product made in space (monodisperse latex spheres), mgmt. of first nuclear detector and beryllium expts. in space. Office: NASA/Marshall Space Flt Ctr JA82 Huntsville AL 35812

CHASTAIN, DENISE JEAN, process engineer, consultant; b. Casper, Wyo., Dec. 12, 1961; d. Jerry and Nancy Gayle (Stewart) C. BAChemE, Ga. Inst. Tech., 1986. Registered profl. engr., Ga. Product devel. engr. Ga. Pacific, Atlanta, 1986-89; process engr. Lockwood Greene Engrs., Atlanta, 1989—. Mem. AICE (chmn. 1991-93, vice chmn. 1990-91, sec. 1989-90), Engrs. for Edn., Nat. Profl. Devel. Com. and Subcom. for Profl. Standards. Achievements include patent pending for silicone as water proofing agent in gypsum products.

CHATELIER, PAUL RICHARD, aviation psychologist, training company officer; b. St. Petersburg, Fla., Oct. 1, 1938; s. Paul Andrew and Mary (Knecht) C.; m. Mary Lu Moss, Sept. 26, 1964; children: Michael Andrew, Suzanne Margaret. BS in Biology, Chemistry, Psychology, U. Fla., 1960; MA in Psychology, U. Miss., 1962; postgrad., U. N.Mex., 1967-69. Commd. ensign USN, 1962, advanced through grades to capt., 1986; sr. v.p. strategic planning Perceptronics, Inc., Washington, 1986—; U.S. rep. on human factors NATO, Brussels, 1978-86; mem. task force tng. and wargaming Def. Sci. Bd., 1986-88; U.S. rep. on tng. The Tech. Cooperation Panel, Washington, 1987-86; mem. indsl. adv. com. U. Cen. Fla. Inst. for Simulation and Tng.; edn. and tng. cons. Office Sci. and Tech. White Ho., 1993—; workshop dir. internat. tng. and human factors. Author: (chpt.) Psychology of Reality, 1985; editor: Manprint & System Integ, 1988, International Human Factors, 1991, Advanced Technology for Training Design, NATO, 1993. Career advisor Fairfax County Pub. Sch., 1982-88. Mem. Nat. Human Factors Soc. (exec. coun. 1982-85), Va. Human Factors Soc. (pres. 1982-83), Nat. Security Indsl. Assn. (chmn. manpower pers. tng. 1986-89). Avocations: tennis, community activities. Home: 8021 W Point Dr Springfield VA 22153-3023 Office: Perceptronics Inc #510 1901 N Beauregard St Alexandria VA 22311

CHATO, JOHN CLARK, mechanical engineering educator; b. Budapest, Hungary, Dec. 28, 1929; s. Joseph Alexander and Elsie (Wasserman) C.; m. Elizabeth Janet Owens, Aug. 1954; children: Christine B., David J., Susan E. ME, U. Cin., 1954; MS, U. Ill., 1955; PhD, MIT, 1960. Co-op student, trainee Frigidaire div. GMC, Dayton, Ohio, 1950-54; grad. fellow U. Ill., Urbana, 1954-55; grad. fellow, inst. MIT, Cambridge, 1955-58, asst. prof., 1958-64; assoc. prof. U. Ill., Urbana, 1964-69, prof., 1969—, chmn. exec. com. bioengring. faculty, 1972-78, 82-83, 84-85; cons. Industry and Govt., 1958—; dir., founder Biomed. Engring. Systems Team, Urbana, Ill, 1974-78; assoc. editor Jour. Biomech. Engring., 1976-82. Patentee in field; contbr. articles to profl. jours., chpts. to books. Trustee, elder 1st Presbyn. Ch., Urbana, 1976-78, 82-85; bd. dirs. Univ. YMCA, Champaign, Ill., 1976-78, 87-90; com. mem. troop 6 Boy Scouts Am., Urbana, 1984-86; com. mem. Urbana Planning Commn., 1973-78; mem. adv. com. Urbana Park Dist., 1981-84; 2d v.p. Champaign County Izaak Walton League, 1986, 1st v.p., 1987, pres., 1988-92, Tobin award, 1992. Named Disting. Engring. Alumnus, U. Cin., 1972; NSF fellow, 1961, Fogarty sr. internat. fellow, 1978-79. Fellow ASME (Charles Russ Richards Meml. award 1978, H.R. Lissner award 1992, bioengring. div. exec. com. 1992--); mem. ASHRAE (treas. East Cen. Ill. chpt. 1984, sec. 1985, 87, 1st v.p. 1988, pres. 1989), IEEE, Am. Soc. for Engring. Edn., Internat. Inst. Refrigeration (assoc.), Exch. Club (bd. dirs. 1989-91), Audubon Soc. (bd. dirs. 1988-89, v.p. 1990, treas., 1991-92). Avocations: tennis, photography, bird watching, hiking. Office: U Ill Dept Mech Indsl Engring 1206 W Green St Urbana IL 61801-2906

CHATRY, FREDERIC METZINGER, civil engineer; b. New Orleans, Aug. 14, 1923; s. Lloyd Chatry and Bertha (Metzinger) Pearce; m. Leah Fernon Park, Nov. 5, 1945; children: Martha, Rebecca, Deborah, Mark, John. B of Elec. Engring., Tulane U., 1945. Registered profl. engr., La. Hydraulics engr. U.S. Army Engr. Dist., New Orleans, 1946-63, chief planning sect., 1963-67, chief basin planning br., 1967-72, chief planning divsn., 1972-75, chief engr., 1975-90; pvt. practice as cons. New Orleans, 1990—; dir. rehabilitation of Old River control structure, 1975-85. Author tech. papers. Commdr. USN, 1943-46, 50-52. Fellow ASCE, Soc. Am. Mil. Engrs. (pres. La. post 1990-92); mem. La. Engring. Soc. (life). Roman Catholic. Achievements include development, design and direction of plan which protected water supply of met. New Orleans area from drought. Home: 8905 Ormond Pl River Ridge LA 70123

CHATT, ALLEN DARRETT, psychologist, neuroscientist; b. Phoenix, Ariz., July 17, 1949; s. Arthur Henry Ellis Jr. and Helen Berta (Scheidt) Chatt; m. Gail Nancy Anguish, Aug. 21, 1971. BS in Psychology with honors, SUNY, Buffalo, 1971; MS in Psychology, Fla. State U., 1974, PhD in Neuroscience, 1978. Rsch. asst. Fla. State U., Tallahassee, 1971-76; postdoctoral fellow U. Tex. Med. Br., Galveston, 1977; postdoctoral fellow sch. medicine Yale U., New Haven, Conn., 1979-81; rsch. psychologist VA Med. Ctr., West Haven, Conn., 1978-84, sr. rsch. psychologist, 1985-90; rsch. asst. prof. sch. medicine Yale U., New Haven, Conn., 1982-87; rsch. assoc. prof. Yale U., New Haven, 1988-90; founder, dir., consulting psychologist Phoenix Fund for Neurologically Challenged, Guilford, 1991—; vis. prof. neurosci. Beijing Normal U., 1990; grant reviewer NSF, NIH, VA, 1982—; sci.-by-mail scientist Mus. Sci., Boston, 1991—; psychol. cons. for neurologically impaired. Contbr. chpts. to books, articles to profl. jours.; mem. editorial bd. Brain Rsch., 1983-86, Exptl. Neurology, 1982-86, Quar. Jour. Exptl. Physiology, 1986, Exptl. Brain Rsch. Mem. Rep. Senatorial Inner Circle, Washington, 1985, Rep. Town Com., Guilford, 1992; life mem. Rep. Nat. Com., 1993—, Eisenhower Commn., 1993—. Merit Review Rsch grantee VA Med. Ctr., 1982-90; RO-1 Rsch. grantee NIH, 1982-87. Fellow Am. Epilepsy Soc., N.Y. Acad. Scis.; mem. AAAS, Am. Psychol. Soc., Epilepsy Found. Am., Soc. for Neurosci., Am. Epilepsy Soc. Republican. Methodist. Achievements include discovery of neuronal circuits involved in focal and secondarily generalized seizure activity in neocortial animal model of epilepsy, mid. brain neuronal circuits modulating pain in an animal model, Thermal evoked potential in humans and the localization of cortical cells responsive to pain in an animal model. Home: PO Box 1449 699 Goose Ln Guilford CT 06437-2114 Office: Phoenix Fund for Neurologically Challenged Concept Park Profl Ctr 741 Boston Post Rd Guilford CT 06437

CHATTERJEE, AMIT, structural engineer; b. Dacca, Bangladesh, Jan. 16, 1943; came to U.S., 1967; s. Santosh and Nilima (Chakraborty) C.; m. Koely Sanyal, Nov. 19, 1973; children: Swarna, Piyali. MS in Engring., Brigham Young Y., 1969; MBA, NYU, 1973. Registered profl. engr. N.Y., N.J. Engr. Steinman, Boynton, Gronquist, N.Y.C., 1969-70; sr. engr. Edwards & Kelcey Inc., Livingston, N.J., 1973-77, Willis & Paul Corp., Denville, N.J., 1977-80, CF Braun, Murray Hill, N.J., 1980-83, Sverdrup and Parcel, N.Y.C., 1970-73, 83-84; prin. engr. Port Authority of N.Y. and N.J., N.Y.C., 1984—. Mem. ASCE, Am. Inst. Steel Constrn., Bengal Engring. Coll. Alumni Assn. (gen. sec. 1992—). Office: Port Authority NY & NJ 241 Erie St Jersey City NJ 07310

CHATTERJEE, AMITAVA, electronics engineer; b. Calcutta, India, July 11, 1956; came to U.S., 1978; s. Jyoti Prasad and Chinmoyee (Mukherjee) C.; m. Radharani Goswami, Dec. 15, 1986; 1 child, Korok. MS, La. State U., 1980; PhD, Rensselaer Poly. Inst., 1985. Mem. tech. staff Tex. Instruments,

Dallas, 1985-91, sr. mem. tech. staff, 1991—. Contbr. articles to profl. jours. Nat. Coun Ednl. Rsch. and Tng. scholar, India, 1972. Mem. IEEE (sr.). Achievements include patents and patents pending in areas of electrostatic discharge protection and programmable interconnects, articles in journals in areas of semiconductor device modeling and reliability of very large scale integrated circuits. Office: Tex Instruments PO Box 655012 MS461 Dallas TX 75265

CHATTERJEE, HEM CHANDRA, electrical engineer; b. Hirapur, W. Bengal, India, Jan. 3, 1940; came to U.S., 1969; s. Kishory Mohon and Katayani (Mukherjee) C.; m. Kamal Renu Mukherjee, Feb. 27, 1967; children: Madhumita, Biswajit. Diploma E.E., Indsl. Tng. Inst., India, 1960; diploma engring., Brit. Inst. Engring. Tech., 1965; MSEE, U. Pa., Phila., 1972; PhD in E.E., City U., L.A., 1979. Registered profl. engr., Pa., N.J., Md., S.C., Va., Ga., D.C., N.H., Conn., W.Va., Mass., Maine, N.C.; cert. profl. mgr.; chartered engr., India; lic. nat. elec. supr. Govt. of West Bengal; cert. cogeneration profl. Elec. chargehand Rallis India Ltd., Calcutta, West Begal, 1961-62; elec. engr. Schindler Aufzuge G.m.b.H., Berlin, 1965-67; sr. elec. engr. Simco, Lessard, Thomson, Dixon Assocs., Windsor, Ont., Can., 1967-69; chief elec. engr. Vinokur-Pace Engring. Svcs., Inc., Jenkintown, Pa., 1969-80; mgr. elec. dept. Walter F. Spiegel, Inc., Jenkintown, 1980-82; sole proprietor, prin. Chatterjee Internatl Engrs., Jenkintown, 1982-85; dir. elec. engring. GSGSB Architects, Engrs. & Planners, Clarks Summit, Pa., 1986-87; chief elec. engr. Robert G. Werden & Assocs., cons. engrs., Jenkintown, 1987-88; v.p. engring. Marvin Waxman Cons. Engrs., P.C., Wyncote, Pa., 1988-90; sole-proprietor, prin. Unique Engrs., Willow Grove, Pa., 1991—. Contbr. more than 20 engring. articles to profl. jours. Founder, mem. Pragati-Bengali Cultural Assn., Phila., 1969-92. Fellow India Soc. Engrs. (pres. U.S. chpt. 1970-82), Inst. Engrs. (India), Inst. Elec. & Telecom. Engrs. (New Delhi); mem. IEEE (sr.). Hindu. Achievements include interior and exterior power distbn. systems, interior and exterior lighting systems, interior and exterior comm. systems, generating plants, substas. and switchgears, motor controls and ctrs., remote monitoring and controls. Completion of several thousand projects with industrial, commercial, institutional, and residential applications. Home and Office: Unique Engrs 1703 Alba Rd Willow Grove PA 19090

CHATTERJEE, MONISH RANJAN, electrical engineering educator; b. Calcutta, India, Jan. 2, 1959; came to U.S., 1980; s. Suranjan and Lilabati (Bhattacharya) C.; m. Joyoti Chowdhury, Jan. 15, 1990. BTech with honors, Indian Inst. Tech., Kharagpur, 1979; MS, PhD in Elec. Engring., U. Iowa, 1985. Grad. rsch. asst. dept. ECE U. Iowa, Iowa City, 1980-85, grad. teaching asst. dept ECE, 1980-84, vis. lectr. dept. ECE, 1985-86; vis. asst. prof. elec. engring. SUNY, Binghamton, 1986-88, asst. prof. elec. engring., 1988-92, assoc. prof. elec. engring., 1992—, MSEE advisor, 1991—; presenter in field of optics and wave propagation. Translator: Kamalakanta, 1992; contbr. articles to profl. jours. Panelist N.Am. Bengali Assn., Lowell, Mass., 1991; organizer, narrator Rabindranath Tagore commemoration SUNY, Binghamton, 1991, Indian cultural celebration, 1992. NSF rsch. grantee, 1992. Mem. IEEE (sr., vice chair tech. seminars, mem. exec. bd. Binghamton chpt. 1989—), Optical Soc. Am., Am. Soc. Engring. Educators, Sigma Xi. Hindu. Achievements include establishment of an accurate mathematical model for electronic holography and nonlinear echoes. Office: SUNY Watson Sch Dept Elec Engring P12/EB Binghamton NY 13902-6000

CHATTERJEE, TAPAN KUMAR, astrophysics researcher; b. Calcutta, Bengal, India, Jan. 15, 1952; arrived in Mex., 1986; s. Santosh Kumar and kanika (Mukerjee) C. MS in Astronomy, Osmania U., Hyderabad, India, 1977, PhD in Astronomy, 1984. Rsch. assoc. Osmania U., 1984-86; scientist Nat. Inst. Astrophysics, Puebla, Mex., 1986-88; assoc. prof. physics U. Puebla, 1988—. Contbr. articles to jours. Astrophysics and Space Sci., I.A.U. Colloquium, I.A.U. Symposium, Rev. Mexicana Astron. Astrophys. Named Disting. Fgn. Citizen, Municipality of Puebla, 1990. Mem. Internat. Astron. Union, Am. Astron. Soc., Nat. System Investigators (scholarship 1989—), Nat. System Profs. (scholarship 1990—), Astron. Soc. India. Hindu. Achievements include determination of the expected frequencies of interacting galaxies (including merging galaxies) and an analysis of their dynamical implications; discovery that during galactic collisions leading to ring formation a relationship between density maximum of the ring and collision parameters is found, that fractional change in internal energy corresponding to merger of galaxies as a funciton of the impact parameter defines a fundamental unit for measuring energy; research in evolutionary trends in elliptical galaxies due to interactions leading finally to mergers and their possible secular character in the light of the existence of the fundamental parameter plane on which the global properties of such galaxies lie; research in induced nuclear activity in interacting galaxies. Office: U Puebla Fis-Mat, Apartado Postal 1152, Puebla 72000, Mexico

CHATTOPADHYAY, ADITI, aerospace engineering educator; b. Krishnagar, India, Mar. 6, 1958; came to U.S., 1980; d. Narayan Prosad and Samita (Mitra) C.; m. Johnny Richardson Narayan, Apr. 6, 1985. B of Tech. with honors, Indian Inst. of Tech., Kharagpur, India, 1980; MS, Ga. Inst. Tech., 1981, PhD in Aerospace, 1984. Rsch. asst. Ga. Inst. of Tech., Atlanta, 1980-84, postdoctoral fellow, 1985-86; sr. scientist Nat. Aero. Lab., Bangalore, India, 1984-85; rsch. scientist NASA Langley Rsch. Ctr., Hampton, Va., 1986-90; asst. prof. Ariz. State U., Tempe, 1990—; vis. faculty NASA Ames Rsch. Ctr., Moffett Field, Calif., 1991—. Author: Optimization of Structural Systems and Industrial Applications, 1991, (with others) Structural Optimization: Status & Promises, 1992; contbr. articles to profl. jours. Rsch. grant Allied Signal Co., 1991, NASA AMes Rsch. Ctr., 1992, 93, Nat. Ctr. for Supercomputing, 1992, 93, Nat. Sci. Found. 1992, Army Reserve Office, 1993. Mem. AIAA, ASME, Am. Helicopter Soc., Sigma Gamma Tau. Achievements include devel. of optimization procedure with multidisciplinary couplings for rotary and fixed wing aircraft design, monlinear sensitivity analysis, large deformation and energy absorption in composites, smart structures. Home: 4294 E Agave Rd Phoenix AZ 85044 Office: Ariz State U Dept Mech and Aerospace Engring Tempe AZ 85287-6106

CHATTREE, MAYANK, mechanical engineering educator, researcher; b. Ajmer, India, Jan. 11, 1957; came to U.S., 1979; s. Amar Singh and Sunder Devi (Rohatgi) C.; m. Neelu Sharma, Sept. 16, 1986; 1 child, Gaurav Mohit. B of Tech. Mechanical Engring., Indian Inst. of Tech., Kanpur, 1979; MSME, U. Miami, 1982, PhDME, 1986. Grad. rsch. asst. U. Miami, Coral Gables, Fla., 1979-86; rsch. asst. prof. W.Va. U., Morgantown, 1986-89; asst. prof. U. New Orleans, 1989—; cons. Tech. Internat. Inc., Laplace, La., 1979—; reviewer West Pub., San Francisco, 1989—. Contbr. articles to profl. jours. Sec. India Students' Assn., U. Miami, 1981, v.p. Internat. House, 1979. Rsch. grantee Martin Marietta, 1990. Mem. ASME (step award 1984), AIAA, Am. Soc. of Engring. Edn., Phi Kappa Phi, Sigma Xi. Hindu. Avocations: tennis, walking, traveling, music, movies. Office: U New Orleans Dept Mech Engring New Orleans LA 70148

CHAU, LAI-KWAN, research chemist; b. Hong Kong, Sept. 3, 1956; came to U.S., 1984; s. Yuen-Ming and Wing-Ngar (Leung) C. BS, Chinese U. Hong Kong, 1980; MS, U. Houston, 1986; PhD, Iowa State U., 1990. Jr. quality engr. Kalex Circuit Bd. Co. Ltd., Hong Kong, 1981; chemist Wing Hang Trading Co. Ltd., Hong Kong, 1982; prodn. supr. Elcap Elecs. Ltd., Hong Kong, 1982-83; asst. rsch. specialist U. Ariz., Tucson, 1990—. Contbr. articles to profl. jours. Pres. The Ninety-Seven Soc., Ames, Iowa, 1989-90. Iowa State U. Phillips Petroleum Grad. fellow, 1989; recipient Materials Rsch. Soc. Grad. Student award, 1989, Ames Lab. FY Inventor Incentive award, 1991, Iowa State U. Rsch. Excellence award, 1991. Achievements include optical calcium sensor patent. Office: U Ariz Dept CHemistry Tucson AZ 85721

CHAUDHARY, SHAUKAT ALI, plant taxonomist, ecologist; b. Sialkot, Punjab, Pakistan, Mar. 1, 1931; s. Allah-Rakha and Raisham Bibi (Din) C.; m. Zahida Sarwar, Oct. 22, 1967; children: Naveed, Naila, Ayesha, Samir. MSc, U. Punjab, Lahore, 1953; PhD, Wash. State U., 1965. Lectr. Gordon Coll., Rawalpindi, Punjab, 1953-54; asst. prof. Agrl. U., Faisalbad, Punjab, 1954-62, reader, head dept., 1965-70; sr. lectr. Am. U. Beirut (Lebanon), 1970-76; assoc. prof. Sana'a (Yemen) U., 1976-78; sr. scientist Am. U. Beirut on secondment to USDA Team in Saudi Arabia, Riyadh, 1978-89, UN FAO Team in Saudi Arabia, Riyadh, 1989—; dean students Agr. U., Faisalabad, 1966-70; curator Nat. Herbarium Saudi Arabia, Riyadh, 1979—. Author: Weeds of Yemen, 1983, Weeds of Saudi Arabia and Arabian Peninsula, 1987, Weed Control Handbook for Saudi Arabia, 1985, Grasses of Saudi Arabia, 1989, Natural History of Saudi Arabia, 1992. Mem. Aril Soc. Internat. (dir. at large 1974-78), Pakistan Bot. Soc. (founding sec.-treas. 1968-69), Sigma Xi. Avocations: study of deserts, photography. Home and Office: Nat Agr and Water Rsch Ctr, PO Box 17285, Riyadh 11484, Saudi Arabia

CHAUDHRY, G. RASUL, molecular biologist, educator; b. Multan, Pakistan, June 6, 1948; came to U.S., 1980; s. Abdul Majid and Sakeena B. C.; m. Zeenat Farzana, Dec. 30, 1982; children: Sadia, Saad, Sarah, Sophia. BSc with honors, U. Agr., Pakistan, 1969, MSc with honors, 1971; PhD, U. Man., Winnipeg, Can., 1980. Rsch. officer Agrl. Rsch. Inst., Pakistan, 1972-77; teaching asst. U. Man., 1978-80; rsch. assoc. Georgetown U., Washington, 1980-82; vis. fellow, then sr. staff fellow NIH, Bethesda, Md., 1982-85; asst. prof. U. Fla., Gainesville, 1985-89; assoc. prof. molecular biology Oakland U., Rochester, Mich., 1989—. Editor: Biological Degradation of Hazardous Chemicals, 1994; contbr. chpts. to books, articles to sci. jours.; author rsch. reports. Grantee USDA, 1986-88, NSF, 1991-92, 91-93, 92-93, Fla. Dept. Environ. Regulation, 1987-92. Mem. AAAS, Am. Soc. Biochemistry and Molecular Biology, Am. Soc. Microbiology, Am. Pub. Health Assn., NIH Alumni Assn., Sigma Xi. Achievements include rsch. in presence of HIV in wastewater, hazardous waste-degrading recombinant Microorganisms for Bioremediation; microbial metabolism of pesticides; DNA repair genes; varied consulting experience worldwide in agr., biotech., more. Office: Oakland Univ Dept Biol Scis Rochester MI 48309

CHAUHAN, VED PAL SINGH, biochemist, researcher; b. Meerut, India, Aug. 10, 1953; came to U.S. 1981; s. Charan Singh and Sona Devi (Tanwar) C.; m. Abha Jain, Nov. 23, 1980; children: Deepti, Neha. PhD, Postgrad. Inst. Med. Edn./Rsch, Chandigarh, India, 1980. Rsch. assoc. U. So. Calif., L.A., 1981-82; rsch. scientist I Inst. for Basic Rsch. in Devel. Disabilities, S.I., N.Y., 1983-86, rsch. scientist II, 1986-87, rsch. scientist III, 1987-90, rsch. scientist IV, head cellular neurochem. lab., 1990—. Contbr. articles to profl. jours. NIH grantee, 1992—. Mem. AAAS, N.Y. Acad. Sci., Am. Soc. Neurochemistry, Am. Soc. Biochemistry and Molecular Biology. Office: Inst for Basic Rsch 1050 Forest Hill Rd Staten Island NY 10314

CHAWLA, AMRIK SINGH, chemist, educator; b. Derajara, Punjab, Pakistan, July 28, 1941; s. Ajit Singh and Jasmeet Kaur Chawla; m. Inderjit Kaur, May 27, 1973; children: Pavneet, Manreet. MPharm., B.H.U., 1963; PhD, Panjab U., 1970. Mfg. chemist Fairdeal Corp., Bombay, 1964-65; sr. rsch. fellow UGC/CSIR, Chandigarh, India, 1965-71; lectr. Panjab U., Chandigarh, 1971-79, reader, 1979-89, prof., 1989—. Author: Medicinal Chemistry Research in India, 1985, Organic Chemistry: An Introduction, 1985, 2d edit., 1989; contbr. articles and revs. to profl. jours. and chpts. to books. Mem. Assn. Pharm. Tchrs. India (sec. 1989-92), Indian Pharm. Assn., Am. Chem. Soc. (assoc. medicinal chemistry div.). Achievements include research in Indian medicinal plants for chemical constituents and biological potentials, isolation of new compounds by spectral studies chemical transformations. Home: E-1/7 Sector 14, 160014 Chandigarh India Office: Panjab U, Dept Pharm Scis, 160014 Chandigarh India

CHAWLA, LAL MUHAMMAD, mathematics educator; b. Mahatpur, India, Nov. 1, 1917; s. Haji Umardin and Fatimah (Bibi) C.; m. Sakina Begum, Mar. 26, 1930 (dec. De. 1988); childen: Surriaya, Ehsan-ul-Haq, Rehana, Ikram-ul-Haq. BA in Math., U. Punjab, 1937, MA in Math., 1939; DPhil, U. Oxford, 1955. Lectr. Islamia Coll., Lahore, 1939-47; sr. lectr. Govt. Coll., Lahore, 1948-57, sr. prof., 1957-69; prin. Cen. Tng. Coll., Lahore, 1969; chmn. Bd. Edn., Sargodha, Punjab, 1970; prof. math. Kans. State U., Manhattan, 1970-82, King A Univ., Jeddah, Saudi Arabia, 1982-86; vis. prof. U. Ill., Urbana, 1965-66, U. Fla., Gainesville, 1966-67; presenter Internat. Congress Mathematicians, Stockholm, 1962; former mem. Am. Math. Assn., London Math. Soc. Author four books on study of Holy Quran, 1993; editor math. sect. Jour. Natural Scis. and Math., 1960-83; reviewer Am. Math. Revs., 1962—; contbr. articles to profl. jours. Mem. Am. Math. Soc. Achievements include discovery of Chawla's Arithmatic Function, 1968; guiding research in Number Theory, Group Theory, Lattice Theory and Boolean Algebras, Isotopy Theory of Linear Systems of Algebras, Invariants of Multilinear Forms and Abstract Algebras. Home: 41-J Model Town, Lahore Pakistan

CHAWLA, MANMOHAN SINGH (MONTE CHAWLA), armor, weapon system and arms control technology specialist; b. Bikaner, Rajasthan, India, July 12, 1940; s. Gian Singh and Laj (Rani) C.; children: Andrew, Julie. BSc, U. Rajasthan, Bikaner, 1960; MSc, U. Rajasthan, Pilani, India, 1963; A. Inst. N.P., Calcutta U., 1965; PhD, La. State U., 1970. Asst. prof. math. Fla. Inst. Tech., Melbourne, 1970-72; mem. rsch. staff S-Cubed, La Jolla, Calif., 1972-75; rsch. scientist U.S. Army Ballistic Rsch. Lab., Aberdeen, Md., 1975-77; sr. scientist Physics Internat., San Leandro, Calif., 1977-78; mem. rsch. staff Rockwell Internat., Columbus, Ohio, 1978-80; sr. scientist Dresser Industries, Houston, 1980-85; rsch. dir. Baker Perforating Svcs., Houston, 1985-86; project mgr. Ferranti ISC Techs., Lancaster, Pa., 1986-87; prin. scientist System Planning Corp., Arlington, Va., 1987—; cons. Intersea Rsch. Corp., La Jolla, 1975, Ship Systems, La Jolla, 1975-76, Ops. Rsch. Assocs., Silver Spring, Md., 1980-81, Tracor MBA, San Ramon, Calif., 1984, Nuclear Metals, Concord, Mass., 1984-87, LTV Vought, Dallas, 1983-84, System Planning Corp., Arlington, 1981-87, Western Atlas, Houston, 1990—, Los Alamos Nat. Lab., 1990—. Patentee shaped charge perforating apparatus, shaped charge perforating apparatus, bullet perforating apparatus gun assembly, high temperature shaped charge, implosion shaped charge perforators. Highest merit scholar Govt. of India, 1956-63; rsch. fellow Coun. Sci. & Indsl. Rsch., India, 1963-64, Atomic Energy Commn., India, 1964-65, teaching and rsch. fellow La Stte U., 1965-70. Mem. Am. Def. Preparedness Assn., Soc. Exploration Geophysicist, Pi Mu Epsilon, Phi Kappa Phi. Avocation: inventing. Home: 2614 Osage St Hyattsville MD 20783-1740 Office: System Planning Corp 1500 Wilson Blvd Arlington VA 22209-2404

CHAZELL, RUSSELL EARL, environmental chemist; b. Baxter Springs, Kans., Nov. 21, 1964; s. Danny Earl and Connie Sue (Tippit) C.; m. Meredith Jane Thompson, Aug. 18, 1984 (div. Mar. 1990); 1 child, Dwight Sutherland; m. Julie Diane Kalous, Aug. 23, 1991; 1 child, Jarod Glenn Crawford. Student, U.S. Mil. Acad., 1983; BS, SUNY, Albany, 1990. Phys. sci. technician Tooele Army Depot U.S. Dept. Army, Tooele, Utah, 1989-93; chief metals chemist SEG Labs., Lansing, Mich., 1993—. Capt. Civil Air Patrol, Utah, 1978-. Mem. Aircraft Owner's and Pilots Assn., Nat. Rifle Assn., Am. Chem. Soc. (nat. affiliate 1991-). Republican. Mormon. Home: 214A Falls Ct Lansing MI 48917 Office: SEG Labs 1120 May St Lansing MI 48906

CHAZEN, MELVIN LEONARD, chemical engineer; b. St. Louis, Sept. 26, 1933; s. Saul and Tillie (Kramer) C.; m. Dorothea Glazer, June 29, 1958; children: Jamie Lynn, Avery Glazer. BS in Chem. Engring., Washington U., St. Louis, 1955. Registered profl. engr., Mo. Thermodynamics engr. Bell Aerospace Textron, Buffalo, 1958-59; devel. engr. Bell Aerospace Textron, 1959-62, project engr., 1962-65, chief sec. rocket engines, 1965-72, prog. mgr., tech. dir., 1972-74, project engr., 1974-84, chief engr. rocket devel., 1984-87; sr. staff engr. TRW-Applied Tech. div., Redondo Beach, Calif., 1987—; bd. dirs. Unimed Corp., Rochester. Contbr. articles to profl. jours. Mem. Alpha Chi Sigma. Avocations: photography, travel, sports. Home: 12522 Inglenook Ln Cerritos CA 90701-7837 Office: TRW Applied Tech Divsn One Space Park Redondo Beach CA 90278

CHE, CHUN-TAO, chemistry educator; b. Hong Kong, Apr. 16, 1953; s. Yue-kee and Wing-tung (Ling) C.; m. Julie K. Yee, Oct. 17, 1987; 1 child, Kelsey V. BS, Chinese U., Hong Kong, 1975, M of Philosophy, 1977; PhD, U. Ill., 1982. Rsch. chemist Ludwig Inst. for Cancer Rsch., Toronto, Ont., Can., 1982-83; rsch. assoc. U. Ill., Chgo., 1983-85, rsch. asst. prof., 1985-91; lectr. Hong Kong U. of Sci. and Tech., 1991—. Mem. adv. bd. Internat. Jour. Pharmacognosy, Jour. of Ethnopharmacology; contbr. articles to profl. jours. Mem. Am. Chem. Soc., Am. Soc. Pharmacognosy, Am. Assn. Pharm. Scientists, Hong Kong Chem. Soc. Home: Hong Kong U Sci and Tech, Sr Staff Qtr Tower 3, 6-B, Clear Water Bay Hong Kong Office: Hong Kong U Sci and Tech, Dept Chemistry, Clear Water Bay Hong Kong

CHEATHAM, THOMAS EDWARD, JR., computer scientist, educator; b. Robinson, Ill., Sept. 20, 1929; s. Thomas Edward and Eva Irene (McGuigan) C.; m. Mary Elizabeth Johnson, Oct. 25, 1951; children: Daniel, Stephen, Thomas Edward III, Benjamin. BS, Purdue U., 1951, MS, 1953; AM (hon.), Harvard U., 1969. Chief computer tech. group Nat. Security Agy., Washington, 1955-56, dep. dir. Project Omega, 1956-60; pres., tech. dir. Mass. Computer Assocs., Inc., Wakefield, 1961-69; Gordon McKay prof. computer sci. Harvard U., Cambridge, Mass., 1969—, dir. Ctr. Rsch. Computing Tech., 1970—; chmn. Software Options, Inc., 1983—; cons. sci. adv. bd. USAF, 1970—. Editor various jours., trans. With U.S. Army, 1954-56. Fellow Am. Acad. Arts and Scis.; mem. Assn. Computing Machinery, AAAS, Soc. Indsl. and Applied Math., Sigma Xi. Home: 40 Hancock St Lexington MA 02173-3432 Office: Harvard U Ctr Rsch Computing Tech 33 Oxford St Cambridge MA 02138-2901

CHEDDAR, DONVILLE GLEN, chemist, educator; b. St. Mary, Jamaica, June 8, 1946; came to U.S., 1968; s. Joshua A. and Delmaria Adassa (Wilson) C.; children: Michael Glen, Angela Rose. BS in Chemistry, U. Oreg., 1973; PhD in Chemistry, U. Ariz., 1980. Lectr. U. West Indies, Kingston, Jamaica, 1981-83; rsch. biochemist U. Ariz., Tucson, 1984—. Contbr.: (textbook) Flash Photolysis and Pulse Radiolysis, 1983; contbr. articles to jours. Photochemistry and Photobiology, Biochemistry, Archives of Biochemistry and Biophysics. Group leader Youth Summer Program, Tucson, 1978. Achievements include construction of energy storage and solar energy conversion system based on photosynthesis; elucidation of the factors which determine reaction rate constants and biological specificity for electron transfer proteins. Office: Univ Ariz Bioscience W Tucson AZ 85721-0001

CHEEVER, ALLEN WILLIAMS, pathologist; b. Brookings, S.D., June 4, 1932; s. Herbert E. and Margaret Haynes (Williams) C.; m. Jane Ellen Gilkerson, Aug., 1953; children: Carol, Erik, Laura, Angela. BS, Carleton Coll., 1954; MD, Harvard U., 1958. Diplomate Am. Bd. Pathology. Commd. 2d lt. USPHS, 1960, advanced through grades to capt., 1968; researcher NIH, Bethesda, Md., 1960—. Contbr. articles to profl. jours. Capt. USPHS, 1960—. Home: 4507 Conifer Ln Bethesda MD 20814-4009 Office: NIH Bldg 4 Rm 126 Bethesda MD 20892

CHEGINI, NASSER, cell biology educator, endocrinologist; b. Malayer, Iran, Sept. 11, 1949; came to U.S., 1981; s. Mohammad and Fakhrol (Montazm) C.; m. Cherry Ruth Middleton, Dec. 30, 1981; 1 child, Claudine Jasmine. BSc in Cell Biology with honors, Nat. U. Iran, Tehran, 1973; PhD in Cell and Molecular Biology, U. Southampton, Eng., 1980. Postdoctoral fellow dept. biology U. Southampton, 1980-81; postdoctoral rsch. asso. Reproductive Endocrinology Labs., U. Louisville Sch. Medicine, 1981-87, instr., 1987-88, asst. prof., 1988-89; asst. prof. reproductive endocrinology dept. ob-gyn U. Fla. Coll. Medicine, Gainesville, 1989-93, assoc. prof. ob-gyn, 1993—; speaker in field, 1981—; reviewer Endocrinology, Ob-Gyn, Adolescent and Pediatric Gynecology, Regulatory Peptides, Biology Reprodn., March of Dimes. Contbr. over 70 articles to sci. jours., including Jour. Cellular Biochemistry, Cancer Letters, Exptl. Cell Rsch., Molecular Cellular Endocrinology, Jour. Clin. Endocrinology Metabolism, Jour. Biol. Chemistry, Jour. Cell Sci., Cell and Tissue Rsch., Exptl. Cell Biology, Biomaterials, Endocrinology, Jour. Reproductive Medicine, Biol. Reprodn., Jour. Gynecol. Surgery, also chpts. to books. 2d lt. Iranian Army, 1973-75. Recipient award for best sci. presentation ACOG, 1988; award for best rsch. dept. ob-gyn U. Fla., 1992, resident rsch. adv. award, 1992; grantee NIH, 1985-88, 91—, U. Louisville, 1988-89, U. Fla., 1989-91, Regent Hosp. Products, Ltd., 1991-94. Mem. Brit. Inst. Biology (chartered), Am. Soc. for Cell Biology, Endocrine Soc., Soc. for Study Reproduction, N.Y. Acad. Scis. Achievements include scientific contribution toward understanding the expression, production and the presence of growth factor and their receptors in human reproductive tissue.

CHEN, CHAOZONG, chemistry educator; b. Shanghang, Fujian, China, Nov. 28, 1935; came to U.S., 1993; s. Xinshan Chen and Julian Xie; m. Meizhen Xu, Apr. 6, 1967; children: Xingdeng, Xinghong. BS in Chemistry, Peking U., China, 1964, MS in Chemistry, 1964. Asst. prof. chemistry Peking U., 1964-78, lectr., 1978-89, assoc. prof., 1989-92; sr. rsch. chemist Xiankun Hu Crustal Stress Rsch. Peking, 1992—; tech. cons. Rare Earth Elements Factory Peking U., 1970-78, Phosphors Factory Yingkou, Liaoning Province, China, 1981-86, Tricolor Phosphors Factory Peking U., 1991-92. Co-author: Experiments of Advanced Inorganic Chemistry, 1987; contbr. articles to profl. jours. Recipient Sci. and Tech. award Nat. Ednl. Com., Beijing, 1986, Sci. and Tech. Improvement award Nat. Sci. and Tech. Com., Beijing, 1987. Mem. Am. Chem. Soc., Chinese Physics Soc. Achievements include research in production of high performance and energy saving phosphors used in various lamps, in purification of chemical reagents that successfully apply to industrial manufacture.

CHEN, CHAUR-FONG, civil engineer, educator; b. Tainan, Taiwan, Oct. 22, 1956; came to U.S 1982.; s. Jung-Chang and Ju-Yuan (Kuo) C.; m. Jang Luh, June 30, 1983; children: Lotos, Elton. MS, Oreg. State U., 1988, PhD, 1992. Rsch. asst. nat. Taiwan U., Taipei, 1980-81; rsch. engr. Fu-Shen Energy Resource, Inc., Taipei, 1980-81; engring. rsch. assoc. N.W. Econ. Assocs., Battle Groud, Wash., 1984-86; rsch. engr. CROPIX, Inc., Irrigon, Oreg., 1986-89; dir. systems devel. Environ. Aeroscientific, Corvallis, 1989-91; dir. application devel. Pen Metrics, Inc., Corvallis, 1991--; asst. prof. Oreg. State U., Corvallis; cons. CROPIX, Inc., 1989--, Environ. Aeroscientific, 1991--; adv. bd. mem. OSU Remote Sensory/G25/Super Computing Com., Corvallis, 1991--. Co-author: Advances in Image Analysis, 1992; contbr. articles to profl. jours. Lt. Taiwan Army, 1978-90. Wade Rain scholr, 1988. Mem. AM. Soc. Photogrammethy and Remote Sensing, Am. Soc. Civil Engrs., Nat. Soc. Profl. Engrs., Alpha Epsilon, Gamma Sigma Delta, Phi Kappa Phi. Achievements include the developement of near real time satellite remote sensing procedures for agricultural production management; established the first nat. ctr. of excellence for light rsch. aircraft. Office: Oreg State U Dept Bioresource Engring Corvallis OR 97331

CHEN, CHEN-TUNG ARTHUR, chemistry and oceanography educator; b. Changhua, Taiwan, China, Apr. 22, 1949; s. Wei-Fu and Shaw-Ming (Wei) C.; m. Joanne Hwa Chao, Mar. 5, 1976; children: Rana Wan-Hui, Diane Wan-Jia. BS, Nat. Taiwan U., Taipei, 1970; MS, U. Miami, 1974, PhD, 1977. Asst. researcher U. Miami, Fla., 1972-77; asst. prof., assoc. prof. Oreg. State U., Corvallis, 1977-81, 81-84; vis. researcher Oak Ridge (Tenn.) Nat. Lab., 1980; prof.. Sun Yat-Sen U., Kaohsiung, Taiwan, China, 1984—; vis. prof. U. Paris VI, France, 1985; dir. Inst. Marine Geology, Kaohsiung, Taiwan, China, 1985-89, Ctr. for Marine Sci. Rsch., Kaohsiung, Taiwan, China, 1989—; dean Coll. Marine Scis., Kaohsiung, Taiwan, China, 1989—; mem. program planning com. World Ocean Circulation Experiment, 1992—; exec. mem. Internat. Ocean Technol. Congress, Honolulu, 1987—; reseracher Sci. and Technol. Adv. Group, Taipei, 1988—; chmn. Nat. Global Change Com., Taipei, 1990—; chmn. Joint Global Ocean Flux Studies Marginal Seas Task Force; mem. internat. program Asian Workshop in IGBP; mem. Sci. Com. on Oceanographic Rsch. Editor: Solubility of Gases, 1980-82; contbr. articles to profl. jours. Dir. Fisheries Extension Svc., Kaohsiung, 1989—, Underwater Soc., Taipei, 1991—. Recipient Outstanding Young Man of Am. award U.S. Jaycees, 1978, Sun Yat-Sen Rsch. award, 1984, Outstanding Rsch. award NSF, 1990. Mem. Internat. OTEC Assn., Soc. Ocean Sci. and Tech. (bd. dirs. 1990—). Office: Inst Marine Geology, Nat Sun Yat-Sen U, Kaohsiung 804, Taiwan

CHEN, CHIH-YING, mathematics and computer science educator, researcher; b. Taichung, Taiwan, June 26, 1951; s. Tain-Sheng and Yueh-Er (Wang) C.; m. Shwu-Jyy Lee, Mar. 29, 1981; children: Yi-Jen, yi-Hwan. B, Tamkang U., Taipei, Taiwan, 1974; M, Cen. U., Chungli, Taiwan, 1976. Instr. Feng Chia U., Taichung, Taiwan, 1976-82; from assoc. prof. to prof. math. and computer sci. Feng Chia U., Taichung, 1982-93; cons. Consulting Ctr. of Statistics, Feng Chia U., 1979-82. Contbr. numerous articles to profl. jours. Recipient Rsch. award Nat. Sci. Coun., Taiwan, 1987-89. Mem. Math. Soc. Republic of China, Chinese Statistics Assn., Computer Soc. Republic of China, Yang Ann Tennis Club (capt. 1980—), Phi Tau Phi. Home: 11-4 Ln 359 Minchen Rd, Taichung 40404, Taiwan Office: Feng Chia U, 100 Wenhwa Rd Seatwen, Taichung 40724, Taiwan

CHEN, CHING-HONG, medical biochemist, researcher; b. Pingtung, Taiwan, July 15, 1935; came to U.S., 1963; s. Ching-Da and Jen-Mei (Yang)

C.; m. Su-Wan Yang, Aug. 29, 1964; 1 child, Sung-Wei. BS in Biology, Taiwan Normal U., Taipei, 1959; MS in Biology, U. Wash., 1966, MS in Chemistry, 1972; PhD in Biochemistry, U. N.D., 1978. Instr. Taiwan Chung-Hsing U., Taichung, 1959-62; biologist U. Wash., Seattle, 1966-72, rsch. scientist dept. biology, 1972-75, rsch. assoc. dept. medicine, 1978-80, mem. rsch. faculty dept. medicine, 1980-88; pres., CEO Alpha Biomed. Labs., Bellevue, Wash., 1988—; vis. scientist Biophys. Inst. Boston U., 1982; cons. Bainbridge Lab., Bainbridge Island, Wash., 1984-87, Solomon Pk. Rsch. Inst., Kirkland, Wash., 1985-88, Medix Biotech., Inc., Foster City, Calif., 1988—. Contbr. articles to profl. jours. Recipient NSF award (Taiwan), 1957; NIH fellow, 1975-78. Fellow Am. Heart Assn. (scientific coun.); mem. AAAS, Am. Assn. for Clin. Chemistry, Wash. State Biotech. Assn. Achievements include discovery of new apolipoprotein activators and regulators for lecithin-cholesterol acyltransferase; development of new purification and assay methods for lecithin-cholesterol acyltransferase; co-development of first purification and assay methods for lipid transfer proteins. Home: 7248 29th Ave NE Seattle WA 98115 Office: Alpha Biomed Labs 920 180th Ave NE # 1 Bellevue WA 98004

CHEN, CHU-CHIN, food chemist; b. Hsin-Chu, Taiwan, Republic of China, July 2, 1954; s. Wei-Bin and So-Han (Chang) C.; m. Sue-Hsia Chen, Dec. 25, 1977; children: Henry, Wendy, Hank. MS in Biochem. Sci., Nat. Taiwan U., 1978; PhD in Food Sci., Rutgers U., 1987. Assoc. food scientist Food Industry R&D Inst., Hsinchu, Taiwan, 1980-86, food scientist, 1986—, group/project leader, 1989—; grad. adv. bd. Nat. Taiwan U., Taipei, 1989-91, Nat. Ocean U., Keelung, Taiwan, 1991. Author: Enzymic Formation of Volatile Compounds in Shiitake, 1986, Volatile Compounds in Ginger Oil Generated by Thermal, 1989, Interaction of Orange Juice and Inner Packaging Material; contbr. articles to profl. jours. Rsch. grantee Pres. Enterprise, 1991. Mem. Am. Chem. Soc., Inst. of Food Technologists (profl.), Chinese Inst. of Food Tech. (Young Scientist 1990). Achievements include patent on manufacturing of volatile oil from edible mushroom (Republic of China). Home: #23 Lane 26 Jin-Hwa St, 300 Hsinchu Taiwan, Republic of China Office: Food Industry R&D Inst, PO Box 246, 30099 Hsinchu Taiwan

CHEN, CHUN-FAN, biology educator; b. Taipei, Taiwan, Republic of China, Mar. 20, 1937; came to U.S., 1973; s. Min-mei C.; married, June 20, 1969; children: Edith, Emma. MS, U. Mich., 1967, PhD, 1971. Asst. physiologist U. Mich., Ann Arbor, 1971; asst. scientist Dept. Physiology, U. Mainz, Fed. Republic of Germany, 1971-73; asst. prof. biology Fla. Internat. U., Miami, 1973-77, assoc. prof. biology, 1977—; chmn. state course numbering system Dept. Edn., Tallahassee, 1990—. Author: Oxygen Transport to Tissue, 1988; contbr. articles to Advanced Exptl. Medicine and Biology, Pfluger Archives, Nature. Recipient Excellence in Advising award Fla. Internat U., 1990, Excellence in Svc. award, 1992, Excellence in Teaching award, 1993. Mem. Am. Physiol. Soc., Internat. Soc. on Oxygen Transp., N.Am. Taiwanese Prof. Assn. Office: Fla Internat U Biology Dept University Pk Miami FL 33199

CHEN, CHUNG LONG, software engineer; b. Taipei, Taiwan, Jan. 31, 1958; s. Ten Jen and Chou Mei Chou; m. Hui Mei Tai, June 30, 1989; 1 child, Yen Ming. BSc, Tsing Hua U., Taiwan, 1980; MSc, London U., 1984, PhD, 1988. Chem. engr. Brit. Petroleum Co., London, 1989—. Mem. Soc. Indsl. and Applied Math., Am. Inst. Chem. Engrs. Avocation: swimming.

CHEN, CHUNG-HSUAN, research physicist; b. Amoy, Fu-Gen, Republic of China, Jan. 4, 1948; came to U.S., 1970; s. King-Shu and Tsuan-Hwei (Lin) C.; m. Shan-Lan Kung, July 20, 1974; children: Nelson, Franklin. MS, U. Chgo., 1971, PhD, 1974. Group leader Oak Ridge (Tenn.) Nat. Lab., 1974—. Contbr. over 100 articles to sci. jours. Grantee Indsl. Rsch., 1984, 89, 92. Mem. Am. Phys. Soc., Am. Chem. Soc., Am. Optical Soc. Achievements include patents for rare gas atom counter, crystal laser beam monitor, freon ratiometer; rsch. in demonstration of rare gas atom counting with isotopic selectivity, fast DNA analysis by laser mass spectrometry, resonance ionization spectoscropy. Home: 1812 Plumb Branch Rd Knoxville TN 37932 Office: Oak Ridge Nat Lab Bldg 5500 MS-6378 PO Box 2008 Oak Ridge TN 37831

CHEN, CONCORDIA CHAO, mathematician; b. Peiping, China; came to U.S., 1955, naturalized, 1969; d. Chun-fu and Kwie Hwa (Wong) Chao; BA in Bus. Adminstrn., Nat. Taiwan U., 1954; MS in Math., Marquette U., 1958; postgrad. Purdue U., 1958-60, M.I.T., 1961-62; m. Chin Chen, July 2, 1960; children: Marie Hui-mei, Albert Chao. Teaching asst. Purdue U., Lafayette, Ind., 1958-60; system analysis engr. electronic data processing div. Mpls.-Honeywell, Newton Highlands, Mass., 1960-63; mgmt. planning asst. Lederle Labs., Am. Cyanamid Co., Pearl River, N.Y., 1964, computer applications specialist, 1967, ops. analyst, 1967; staff programmer IBM, Sterling Forest, N.Y., 1968-73, adv. programmer Data Processing Mktg. Group, Poughkeepsie, 1973-80, mgr. systems programming and systems architecture, Princeton, N.J., 1980-82, sr. systems analyst, 1982-83, data processing mktg. cons., Beijing, 1983-88 ; sr. planner IBM DSD, Poughkeepsie, 1988-92; program mgr. Chiang Indsl. Charity Found Ltd., 1993—. Chmn. ednl. council Hudson region MIT. Mem. Am. Math. Soc., Soc. Indsl. and Applied Maths., MIT Club Hudson Valley (pres.). Home: 12 Mountain Pass Rd Hopewell Junction NY 12533-5331 Office: Chiang Indsl Charity Found Ltd, 7/F Chinaweal Ctr 414-424 Jaffe Rd, Wanchai Hong Kong

CHEN, FANG CHU, mechanical engineer; b. Chekiang, China; came to U.S., 1968; m. Teresa C. Hsuean. BS, Nat. Taiwan U., Taipei, 1967; PhD, Harvard U., 1973. Registered profl. engr., Md. Mech. engr. NUS Corp., Rockville, Md., 1973-77; mem. tech. staff TRW Energy Systems Group, McLean, Va., 1977; project mgr. Oak Ridge (Tenn.) Nat. Lab., 1977-83, group leader, 1984—; mech. engring. evaluator Accreditation Bd. on Engring. and Tech., N.Y.C., 1992—. Co-editor symposium publs. Founder, pres. Orgn. of Chinese Ams. East Tenn. chpt., 1982; v.p. Park Village Homeowners Assn., Knoxville, 1980. Recipient R&D 100 Award R&D Mag., 1992. Mem. ASME (founder heat pump tech. com. 1988), ASHRAE (com. chmn. 1991-92). Achievements include patent for charged particle mobility refrigerant analyzer, (patent pending) electron attachment refrigerant leak detector, liquid overfeeding air conditioning systems and methods. Home: 12114 Butternut Circle Knoxville TN 37922

CHEN, FRANCIS F., physics and engineering educator; b. Canton, Kwangtung, Republic of China, Nov. 18, 1929; s. M. Conrad and Evelyn (Chu) C.; m. Edna Lau Chen, Mar. 31, 1956; children: Sheryl F., Patricia A., Robert F. AB, Harvard U., 1950, MA, 1951, PhD, 1954. Research staff mem. Princeton (N.J.) Plasma Physics Lab., 1954-69; prof. elec. engring. UCLA, 1969—; chmn. plasma physics div. Am. Phys. Soc., N.Y.C., 1983. Author: Introduction to Plasma Physics and Controlled Fusion, 1974, 2d edit., 1984; contbr. over 100 articles to sci. jours. Fellow IEEE, Am. Phys. Soc.; mem. N.Y. Acad. Scis., Fusion Power Assocs., Phi Beta Kappa. Avocations: tennis, running, photography, backpacking, woodworking. Office: U Calif L A 56-125B Engr IV Mail code 159410 Los Angeles CA 90024

CHEN, GONG NING, mathematics educator; b. Fujian, People's Republic of China, Nov. 25, 1939; s. Qin Zhuo Chen and Zhi Fang Wang; m. Shu Yun Tian, July 22, 1962; 1 child, Zhong Chen. Diploma, Beijing Normal U., Beijing, 1960, grad. diploma, 1962; cert. of postgrad. study, U. Calif., Santa Barbara, 1981-82. Rsch. scholar U. Calif., Santa Barbara, 1980-82; lectr. Beijing Normal U., 1978-80, 82-85, assoc. prof. math., 1985-88, prof., 1988-91, 91—; vis. prof. Moscow State U., 1991; vice chmn. Inst. of Math. and Math. Edn. Beijing Normal U., 1988-92, dir. doctoral candidates, 1993—. Author (books) Introduction to Computational Mathematics, 1988, Matrix Theory with Applications, 1990; editor Jour. of Beijing Normal U., 1988—; contbr. articles to profl. jours. Mem. Chinese Math. Soc., Beijing Math. Soc. (bd. dirs. 1985—), Am. Math. Soc. Avocations: collecting stamps, photographing, classic music, volleyball. Home: Beijing Normal U, 10291 Bldg Leyu-8, Beijing 100875, China Office: Beijing Normal U, Dept Math, Beijing 100875, China

CHEN, GUANRONG (RON), electrical engineer, educator, applied mathematician; b. Canton, China, July 5, 1948; came to U.S., 1982; m. Helen Qiyun Xian, Sept. 16, 1976; children: Julie Yuxiu, Leslie Yuli. MS, Sun Yatsen U., Canton, 1981; PhD, Tex. A&M U., 1987. Post-doctoral fellow Tex. A&M U., College Station, 1987; vis. asst. prof. Rice U., Houston, 1987-90; asst. prof. U. Houston, 1990—. Co-author: Kalman Filtering with Real-Time Applications, 1987, 2d edit., 1991, Linear Systems and Optimal Control, 1989, Signal Processing and Systems Theory, 1992, Nonlinear Feedback Control Systems, 1993, Approximate Kalman Filtering, 1993. Mem. IEEE (sr., assoc. editor IEEE publ.), AIAA. Office: Univ Houston Dept Elec Engring Houston TX 77204

CHEN, GUI-QIANG, mathematician, educator, researcher; b. Ningbao, Zhejiang, People's Republic of China, May 25, 1962; came to U.S., 1987; parents Zhi-Biao and Jin-Er (Hu) C. BS, Fudan U., Shanghai, People's Republic China, 1982; PhD, Acad. Sinica, Beijing, 1987. Asst. prof. Inst. Systems Sci., Acad. Sinica, 1987; vis. scientist Courant Inst. Math. Scis., N.Y.C., 1987; asst. prof. math. U. Chgo., 1989—; cons. Argonne Nat. Lab., Chgo., 1989—. Recipient Young Investigator award NSF, Beijing, China, 1987, Nat. Medal of Sci., People's Republic of China, 1989; Alfred P. Sloan Rsch. fellow, 1991; named Excellent Young Scientist, Beijing Soc. for Sci. and Tech., 1988. Mem. Am. Math. Soc., Soc. for Indsl. and Applied Math. Office: U Chgo Dept Math 5734 S University Ave Chicago IL 60637-1546

CHEN, GWO-LIANG, physicist, researcher; b. Ping Jong, Taiwan, Apr. 5, 1948; s. Wen-Shan and Jin-Show (Tsai) C.; m. Su-Line Su, June 5, 1984. MA, U. Louisville, 1976; PhD, U. Tex., 1981. Mem. rsch. staff Oak Ridge (Tenn.) Nat. Lab., 1981—. Mem. Am. Phys. Soc. Office: Oak Ridge Nat Lab Oak Ridge TN 37831

CHEN, HAI-CHIN See FUKUZUMI, NAOYOSHI

CHEN, HENRY, systems analyst; b. Tainan, China, Feb. 27, 1959; came to U.S., 1971; s. Liang-Foo and Dorothy (Wong) C.; m. Anne Lee Chen, Dec. 23, 1986; 1 child, Lauren. BA, NYU, 1982; MS, Poly. U., 1985. Programmer IBM, White Plains, N.Y., 1982-84, assoc. programmer, 1984-85, sr. assoc. programmer, 1985-86; staff programmer IBM, North Tarrytown, N.Y., 1987-89, adv. systems analyst, 1989-91, mem. steering com., 1989—, cons., 1990—, program mgr., 1991—. Treas. Woodside (N.Y.) Townhouse Condo Bd., 1989-92; bd. dirs. Chinese N.Y. Jaycees, N.Y.C., 1990-91, Univoice Chorus, 1989-90. Mem. Computer Soc. IEEE, Assn. Computing Machinery, 1991, N.Y. Acad. Scis., Chinese Mei Soc. (pres. 1981-82).

CHEN, HO-HONG H., industrial engineering executive, educator; b. Taiwan, Apr. 11, 1933; s. Shui-Cheng and Mei (Lin) C.; m. Yuki-Lihua Jenny, Mar. 10, 1959; children: Benjamin Kuen-Tsai, Carl Joseph Chao-Kuang, Charles Chao-Yu, Eric Chao-Ying, Charmine Tsuey-Ling, Dolly Hsiao-Ying, Edith Yi-Wen, Yvonne Yi-Fang, Grace Yi-Sing, Julia Yi-Jiun. Owner Tai Chang Indsl. Supplies Co., Ltd., 1967—; pres. Pan Pacific Indsl. Supplies, Inc., Ont., Can., 1975—, Maker Group Inc., Md., 1986—, Wako Internat. Co., Ltd., Md., 1986—; prof. First Econ. U., Japan. Author: 500 Creative Designs for Future Business, 1961; A Summary of Suggestions for the Economic Development in Central America Countries, 1979; Access and Utilize the Potential Fund in Asia, 1980. Mem. Internat. Club (Washington), Kenwood Golf & Country Club (Bethesda, Md.). Office: PO Box 5674 Friendship Sta Washington DC 20016

CHEN, IRVIN SHAO YU, microbiologist, educator; b. Toms River, N.J., Sept. 29, 1955; s. Tseh-An and Cheh-Chen (Chang) C.; m. Diven Sun, June 21, 1981; children: Katrina Nai Ching, Kevin Nai Hong. BA, Cornell U., 1977; PhD, U. Wis., 1981. Asst. prof. UCLA Sch. Medicine, 1984-86, assoc. prof., 1986-90, prof., 1990—; dir. AIDS Inst. UCLA, 1992—, Core BSL3 SCID-hu Mouse Lab., 1989—, Core Human Retrovirus Lab., 1989—, AIDS Ctr. Virology Lab., 1986—. Mem. editorial adv. bd. Oncogene, 1986; mem. editorial bd. AIDS Rsch. and Human Retrovirus, 1990, Jour. of Virology, 1991; contbr. articles to Sci., Nature, Cell; contbr. chpt.: HTLV-1 and HTLV-II in Virology, 1990. Grantee NIH, 1982—, U. Calif. U. Task Force on AIDS, 1986—; recipient Jr. Faculty award Am. Cancer Soc., 1984, Scholar award Leukemia Soc. Am., 1989, Stohlman Scholar award, 1992. Mem. AAAS, Am. Soc. Microbiology, Jonsson Comprehensive Cancer Ctr. Achievements include patent for retroviral polypeptides associated with human transformation; first to achieve molecular cloning of human T-cell leukemia virus type II, discovery of trans-activation gene as essential gene for HTLV-II, molecular basis for HIV-1 tropism for macrophages. Office: UCLA Sch Medicine Dept Medicine and Immunology 11-934 Factor Los Angeles CA 90024-1678

CHEN, JAMES JEN-CHUAN, electrical engineer; b. Hsinchu, Taiwan, July 30, 1964; came to U.S. 1968; s. Yung C. and Lily L. (Lin) C. BA, U. Calif., Berkeley, 1985; MS, U. Ill., 1987, PhD, 1991. Rsch. asst. U. Ill., Urbana, 1985-91; device engr. Intel Corp., Hillsboro, Oreg., 1991—. Contbr. articles to profl. jours. Mem. IEEE, Am. Phys. Soc. Achievements include discovery of photon emission from avalanche breakdown in the GaAs/Al-GaAs HBT collector; first report of ballistic transport in an InGaAs/InAlAs hot electron transistor. Home: 222 SW Harrison St #16A Portland OR 97201 Office: Intel Corp 5200 NE Elam Young Pkwy Hillsboro OR 97124

CHEN, JAMES PAI-FUN, biology educator, researcher; b. Fengyuan, Taichung, Republic of China, May 1, 1929; came to U.S., 1952; s. Chuan and Su-wuo (Lin) C.; m. Metis Hsiu-chun Lin, Dec. 19, 1964; children: Mark Hsin-tzu, Eunice Hsin-yi, Jeremy Hsin-tao. BS, Houghton (N.Y.) Coll., 1955; MS, St. Lawrence U., 1957; PhD, Pa. State U., 1961. From instr. to assoc. prof. Houghton Coll., 1960-64; rsch. assoc. Coll. of Medicine U. Vt., Burlington, 1964-65; rsch. assoc. Sch. of Medicine SUNY, Buffalo, 1965-68; asst. prof. U. Tex. Med. Br., Galveston, 1968-75; sr. rsch. assoc. NASA/Johnson Space Ctr., Houston, 1973-76, rsch. assoc. prof. U. Tenn. Meml. Rsch. Ctr., Knoxville, 1976-78; assoc. prof. Coll. of Medicine U. Tenn., Knoxville, 1978-84, prof. Coll. of Medicine, 1984—; mem. rsch. rev. com. Tex. affiliate Am. Heart Assn., Austin, 1974-76; co-investigator Spacelab I project, Johnson Space Ctr., Houston, 1976-83; vis. prof. Trnovo Hosp. Internal Medicine, Ljubljana, Yugoslavia, 1985. Contbr. articles to Thromb. Haemosts, over 41 articles to profl. jours. Grantee Robert Welch Found., 1970-74, Ortho Rsch. Found., 1971-75, NIH, 1975-82, Am. Heart Assn. Tex. affiliate, 1969-72, 74-75, Am. Heart Assn. Tenn. affiliate, 1984-85, 89-90, U.S. Army Med. Rsch., 1988-91. Fellow Internat. Soc. Hematology; mem. Am. Assn. Immunologists, Am. Soc. Biochemistry and Molecular Biology, Internat. Soc. Thrombosis and Haemostasis, Internat. Fibrinogen Rsch. Soc., Am. Bd. Bioanalysis (clin. lab. dir.). Achievements include research in thrombosis and hemostasis; discovery of additional proteolytic fragmentation in the high temperature trypsin cleavage of human IgM; development of a radioimmunoassay for fragment E-neoantigen and applied it to the clinical assay of hypercoagulable state; discovered evidence of the coagulopathy in Pichinde virus-infected guinea pigs. Office: U Tenn Med Ctr 1924 Alcoa Hwy Knoxville TN 37920-6999

CHEN, JIAN NING, mathematics educator; b. Baotou, Mongolia, China, Jan. 21, 1959; s. Ren-Jie and Yu-Lin (Zhou) C.; m. Chuan-Jin Yu, Jan. 10, 1988. BS, Inner Mongolia U., Huhot, Peoples Republic China, 1982, MS, 1985, PhD, 1988. Lectr. dept. math. Inner Mongolia U., Huhot, 1988-90, assoc. prof., 1990—; vis. rsch. fellow U. Bristol, Eng., 1990-91, U. Manchester, Eng., 1991-92; reviewer Math. Revs., Ann Arbor, Mich., 1990—. Contbr. articles to profl. jours. Mem. Chinese Dem. League, Beijing, 1988, All-China Fedn. of Youth, 1993—. Mem. Am. Math. Soc. Avocations: sports, travel, reading. Home: Flat 403, 1 Da Xue Rd Bldg 41, Huhot Inner Mongolia China Office: Inner Mongolia U, Dept Math, Huhot China

CHEN, LIH-JUANN, materials science educator; b. Honan, People's Republic China, Aug. 13, 1946; s. Zhi-Li Chen and Yann Tang; m. Hsiang Wu, Aug. 18, 1984; children: Hsueih-An, Hsueih-Shih. BS in Physics, Nat. Taiwan U., Taipei, Republic of China, 1968; PhD in Physics, U. Calif., Berkeley, 1974. Rsch. assoc. UCLA, 1974-77; assoc. prof. Nat. Tsing Hua U., Hsinchu, Republic of China, 1977-79, prof., 1979—, chmn. dept. materials sci. and engring., 1982-84, assoc. dir. materials sci. ctr., 1984-85; vis. assoc. prof. Cornell U., Ithaca, N.Y., 1980-81; vis. Scientist Xerox Palo Alto (Calif.) Rsch. Ctr., 1986-87; cons. Materials Rsch. Labs., Hsinchu, 1981—, Electronics rsch. and Svc. Orgn., 1990—; mem. acad. com. Republic of China Ministry Edn., Taipei, 1987-91. Author: Electronic Microscopy of Materials, 1990, Electronic Materials, 1990; editor Materials Chemistry and Physics Jour., 1992—; editor proc. 1984 ROC Internat. Electronic Devices and Materials Symposium, 1984; contbr. over 150 articles to internat. jours. and 170 articles to conf. proc. Recipient Disting. Rsch. in Engring. award Republic of China Ministry Edn., 1986, Outstanding Rsch. award Nat. Sci. Coun., Republic of China, 1985-93, Dr. Sun Yat-Sen award for excellence in rsch., 1991. Mem. Chinese Soc. for Materials Sci. (councilor 1983-93, Lu Chi-Hung medal 1993), Materials Rsch. Soc.-Taiwan (v.p. 1991—), Materials Rsch. Soc. USA, Bohnische Phys. Soc. (hon. soc.). Avocations: badminton, music, travel. Office: Nat Tsing Hua U, Dept Materials Sci & Engring, Hsinchu 300, Taiwan

CHEN, MICHAEL MING, mechanical engineering educator; b. Hankow, China, Mar. 10, 1933; came to U.S., 1953, naturalized, 1965; s. Kwang Tzu and Hwei Chuing (Deng) C.; m. Ruth Hsu, Oct. 15, 1961; children: Brigitte (dec.), Derek, Melinda. BS, U. Ill., 1955; SM, MIT, 1957, PhD, 1961. Sr. staff scientist research and devel. Avco Corp., Wilmington, Mass., 1960-63; asst. prof. engring. and applied sci. Yale U., 1963-69; asso. prof. mech. engring. N.Y. U., 1969-73; prof. mech. engring. and bioengring., dept. mech. and indsl. engring. U. Ill., Urbana-Champaign, 1973-91; prof. dept. mech. engring. and applied mechanics U. Mich., Ann Arbor, 1991—; bd. dirs. thermal systems program NSF, 1991—; cons. A.D. Little Co., NIH, Argonne Nat. Lab., Bell Labs. Asso. editor: Applied Mechanics Rev.; contbr. to profl. publs. Fellow ASME (Heat Transfer Meml. award 1990); mem. Am. Phys. Soc., Sigma Xi, Phi Kappa Phi, Tau Beta Pi, Pi Tau Sigma. Office: U Mich G G Brown Bldg Ann Arbor MI 48109-2125 also: NSF 1800 G St NW Washington DC 20550

CHEN, MIN-CHU, mechanical engineer; b. Hsiang-Hsiang, Hu-Nan, China, June 30, 1949; came to U.S., 1975; s. Pai-Hsun and Kua-Fan (Lee) C.; m. Yuh-Mei Chung, Aug. 1, 1975; children: Willis Fan, Thomas Li. BS, Nat. Ocean & Marine U., 1971; MS, Nat. Taiwan U., 1975; PhD, Oreg. State U., 1979. Sr. engring. Brown & Root, Inc., Houston, 1978-80; sr. rsch. engr. Exxon Prodn. Rsch. Co., Houston, 1980-81; sr. engr. Sonat Offshore Drilling Inc., Houston, 1981-85; v.p. Act Engring. Inc., Houston, 1988—; chief exec. officer C & C Internat. Svcs., Inc., Houston, 1985—; chmn. bd. C & C Internat. Svcs., Ltd., Hong Kong, 1991—. Contbr. tech. articles to profl. jours. Mem. ASME, Sigma Xi. Home: 11427 Wickersham Ln Houston TX 77077-6827

CHEN, NAIXING, thermal science educator, researcher; b. Shanghai, China, Sept. 29, 1933; s. De-Pai and Zhoushi C.; m. Zhichu Yan, Mar. 12, 1958; children: Xiaohong, Xiaodong. BSME, Moscow Baumann St. U., 1958. Dep. div. head Harbin (China) Turbine Factory, 1958-61; div. head Marine Turbine and Boiler Rsch. Inst., Harbin, 1961-73, System Engring. Rsch. Inst., Beijing, 1973-78; dep. div. head inst. mechanics Chinese Acad. Scis., Beijing, 1978-80, vice dir. inst. engring thermophysics, 1981-86, prof. and dir. inst. engring. thermophysics, 1986—; com. mem. Internat. Congress of Heat and Mass Transfer, Minsk, Byelorussia, 1989—, Internat. Symposium on Internal Aerothermodynamics, 1990—; guest prof. Tsingua U. and U. Sci. and Tech., China, 1988—. Author: Aerothermodynamics of Turbomachinery, 1981; editor-in-chief: Journal of thermal Science, 1992—; editor: Experimental and Computational Aerothermodynamics of Internal Flows, 1990; contbr. articles to profl. jours. Recipient award China Sci. Congress, Beijing, 1978, praises Ministry of Shipbuilding, Beijing, 1978. Mem. AIAA (sr.), Chinese Soc. Engring. Thermophysics. Office: Inst Engring Thermophysics, Beijing 100080, China

CHEN, PETER, chemistry educator. Prof. dept. chemistry Harvard U. Recipient Arthur C. Cope Scholar award Am. Chemical Soc., 1993. Office: Harvard U Dept Chemistry Cambridge MA 02138*

CHEN, PETER PIN-SHAN, electrical engineering and computer science educator, data processing executive; b. Taishan, Kwangtung, China, Jan. 3, 1947; came to U.S., 1969; s. Man-See and T.T. Chen; m. Li-Chuang Ho; children: Victoria, Angela, Gloria Lily. BSEE, Nat. Taiwan U., Republic of China, 1968; MS, Harvard U., 1970, PhD, 1973. Student assoc. IBM, Yorktown Heights, N.Y., 1970; teaching fellow Harvard U., Cambridge, Mass., 1970-71; prin. engr. Honeywell, Waltham, Mass., 1973-74; vis. researcher Digital Equipment Corp., Maynard, Mass., 1974; asst. prof. MIT, Cambridge, Mass., 1974-78; assoc. prof. UCLA, 1978-82; Sinclair vis. prof. MIT, 1986-87; Foster Disting. Chair prof. La. State U., Baton Rouge, 1983—; vis. prof. Harvard U., Cambridge, 1990, MIT, Cambridge, 1990-92; chmn. Chen & Assocs. Inc., Baton Rouge, 1978—; pres. ER Inst., Baton Rouge, 1980—. Author: ER to Logical DB Design, 1978, ER to Systems Analysis, 1980, ER to Information Modeling, 1983; patentee in field. Tech. officer with Republic of China mil. svcs., 1968-69. Recipient faculty career award UCLA, 1979, Info. Tech. award Data Adminstrn. Mgmt. Assn., 1990; NSF rsch. grantee, 1978. Fellow IEEE; mem. Assn. for Computing Machinery. Office: La State Univ Computer Sci Dept Baton Rouge LA 70803

CHEN, ROGER KO-CHUNG, electronics executive; b. Taiwan, Apr. 16, 1951; arrived in U.S., 1980; s. Ja-bing and Chi-ping (Tsui) C.; m. Shu-chen Kao, May 25, 1980; 1 child, Wickham. BS, Nat. Taiwan Normal U., Taipei, Rep. China, 1974; MS, U. Tex., 1982. Tchr. Tainan 1st High Sch., 1976-80; teaching asst. U. Tex., Dallas, 1980-82; jr. applications engr. Internat. Power Machines, Dallas, 1982-83, sr. applications engr., 1983-85, regional mgr., 1985, Pacific dir., 1985-86; gen. mgr. Liebert Hong Kong Ops., 1987-88, mng. dir., 1989—. Served to lt., infantry Republic China armed forces, 1974-76. Mem. IEEE. Lodge: Masons. Avocation: sports. Office: Liebert Internat 1050 Dearborn Dr Columbus OH 43085-4709 also: Liebert Hong Kong Ltd, 489 Hennessy Rd 19/F, Causeway Bay Pla I Hong Kong Hong Kong

CHEN, SHIH-HSIUNG, aerospace engineer; b. Huwei, Taiwan, China, Aug. 6, 1955; came to U.S., 1980; s. Chung-Soong and Ing-Shei (Lin) C.; m. Li-Ping Wang, July 19, 1980; children: Jessie, Theodore. MS, Stanford U., 1981; PhD, Purdue U., 1987. Instr. Nat. Cheng Kung U., Tainan, Taiwan, China, 1982-83; tech. staff Rockwell Internat. Rocketdyne Div., Canoga Park, Calif., 1987—; pres. Jetpro Tech., Thousand Oaks, Calif., 1991—. Contbr. articles to profl. jours. With Chinses Air Force, 1978-80. Mem. AIAA, ASME (Arch Cowell Merit award 1986), Chinese-Am. Engrs. & Scientist Assn., Reaction Rsch. Soc., Soc. Automotive Engrs., Experimental Aircraft Assn. Achievements include development of numerical code CRT for advanced counter rotating propfan gas turbine engine performance, code CASUN for accurate and efficient rocket turbo-engine life assessment and failure prevention, three-dimensional multi-stage turbomachinery flow code MST3D to improve turbo-engine design procedure and optimize configuration. Home: 3970 Calle Del Sol Thousand Oaks CA 91360

CHEN, SHILU, flight mechanics educator; b. Dongyang, Zhejiang, People's Republic of China, Sept. 24, 1920; came to U.S., 1987; s. Yanneng and Fayun C.; m. Xiaosu Gong, Jan. 30, 1948; children: Zhaoyi, Qingyi, Liyi. BS in Aeronautics, Tsinghua U., Kunming, 1945; PhD in Aeronautics, Moscow Aero. Inst., 1958. Asst. Tsinghua U., Peking, 1945-48; instr. Chao-Tung U., Shanghai, 1948-52, East-China Aero. Inst., Nanking, 1952-56; prof. Northwestern Poly. U., Xian, 1958—, chmn., hon. chmn. dept. astronautics, 1960-87, hon. dean Coll. of Astronautics, 1988—; vis. scholar Moscow Aero. Inst., 1956-58; vis. prof. Tech. Univ. Braunschweig, Fed. Republic Germany, 1986; cons. Aerodynamic Rsch. Inst., Beijing, 1964-66; mem. session aeronautics and astronautics, Nat. Adv. Com. for Acad. Degree, Beijing, 1981-85, chmn. sessions, 1985-92. Editor-in-chief: Modern Vehicle Flight Dynamics, 1987; contbr. articles to profl. jours., publs. Recipient First Class award for Progress in Sci. and Tech., China Nat. Edn. Com., Beijing, 1990. Mem. AIAA, Chinese Soc. Astronautics (bd. dirs. 1979-93), Chinese Soc. Aeronautics (bd. dirs. 1964-88), Shaanxi Provincial Soc. Astronautics (v.p. 1984—). Achievements include research on dynamic stability and control of elastic vehicles, establishment of first Dept. of Astronautics in China, 1960—. Home: Northwestern Polytech U, North Apt 2-4-4, Xian Shaanxi 710072, People's Republic of China Office: Northwestern Poly U, Xi'an Shaanxi 710072, China

CHEN, TIAN-JIE, physicist, educator; b. Shanghai, China, Aug. 9, 1939; came to U.S., 1991; s. Zhongken and Jianhuan (Sun) C.; m. Lier Song, Jan. 14, 1966; 1 child, Xuhui. BS, Peking U., 1962; PhD, Columbia U., 1983. Assoc. prof. Peking U., Beijing, China, 1984-93; vis. prof. So. Ill. U., Carbondale, 1991—; mem. faculty senate Peking U., 1991—. Author:

Fundamentals of Laser, 1986, Laser Spectroscopy, 1987; contbr. articles to profl. jours. Grantee Chinese Govt. 1985, Third World Acad., Italy, 1988. Mem. Am. Phys. Soc., Optical Soc. Am., N.Y. Acad. Sci. Achievements include invention of two-photon tri-level echo. Home: 126-18 Southern Hills Carbondale IL 62901

CHEN, WALTER YI-CHEN, electrical engineer; b. Shanghai, Sept. 8, 1956; came to the U.S., 1980; s. Frank L. and Sally X. (Wang) C.; m. Nancy Ran Xu, May 21, 1986; children: Aaron W., Brian R. MEE, Calif. Inst. Tech., Pasadena, 1983; PhD in Elec. Engring., Poly. U., Bklyn., 1989. Mem. tech. staff AT&T Bell Labs., Holmdel, N.J., 1982-87, NYNEX Sci. and Tech., White Plains, N.Y., 1987-89, Bellcore, Morristown, N.J., 1989—. Contbr. articles to profl. jours. Mem. IEEE. Republican. Achievements include patents for dual mode LMS nonlinear data echo canceller. Office: Bellcore 445 South St Morristown NJ 07960-6454

CHEN, WAYNE H., electrical engineer, educator; b. Soochow, China, Dec. 13, 1922; came to U.S., 1947, naturalized, 1957; s. Ting Li and Yung-Chin (Hu) C.; m. Dorothy Teh Hou, June 7, 1957; children: Avis Shirley and Benjamin Timothy (twins). BSEE, Nat. Chiao Tung U., China, 1944; M.S., U. Wash., 1949; Ph.D., 1952. Registered profl. engr., Fla. Electronic engr. cyclotron project Applied Physics Lab. U. Wash., 1949-50, assoc. in math., 1950-52; mem. faculty U. Fla., Gainesville, 1952—, prof. elec. engring., 1957—, chmn. dept., 1965-73, dean Coll. Engring., dir. Engring. and Indsl. Expt. Sta., 1973-88; vis. prof. Nat. Chiao Tung U., Nat. Taiwan U., spring 1964; vis. scientist Nat. Acad. Scis. to USSR, 1967; mem. tech. staff Bell Tel. Labs., summers 1953, 54, cons., 1955-60; mem. tech. staff Hughes Aircraft Co., summer 1962; vis. prof. U. Carabobo, Venezuela, summer 1972; pres. College CAD/CAM Consortium Inc. (not-for-profit corp.), 1985-87. Author: The Analayis of Linear Systems, 1963, Linear Network Design and Synthesis, 1964, The Robotosyncrasies (pseudonym Wayne Hawaii), 1976, The Year of the Robot, 1981. Recipient Fla. Blue Key Outstanding Faculty award, 1960, Outstanding Publs. award Chia Hsin Cement Co. Cultural Fund, Taiwan, 1964, Tchr.-Scholar award U. Fla., 1971. Fellow IEEE; mem. AAUP, Am. Soc. Engring. Edn., Fla. Engring Soc., Nat. Soc. Profl. Engrs., Blue Key, Sigma Xi, Sigma Tau, Eta Kappa Nu, Tau Beta Pi, Epsilon Lambda Chi, Omicron Delta Kappa, Phi Tau Phi., Phi Kappa Phi. Lodge: Rotary. Patentee in field. Office: U Fla Coll Engring Gainesville FL 32611

CHEN, WEI R., physics educator; b. Shanghai, People's Republic of China, July 5, 1958; came to U.S., 1982; m. Chinyun Lu, June 28, 1986; 1 child, Jason Yunti. M Physics, U. Oreg., 1984, PhD in Physics, 1988. Lectr. in physics Parks Coll. St. Louis U., Cahokia, Ill., 1988-89; rsch. assoc. U. Oreg., Eugene, 1989; physics instr. Okla. Sch. Sci. and Math., Oklahoma City, 1989—; vis. rsch. assoc. U. Okla., Norman, 1991, adj. asst. prof., 1990—. Contbr. publs. to Phys. Rev. Recipient Minority Sci. Tchr. Rsch. award NIH, 1991, 93. Mem. Am. Phys. Soc. Office: Okla Sch Sci and Math 1141 N Lincoln Blvd Oklahoma City OK 73014

CHEN, WEN FU, otolaryngologist; b. Taiwan, Apr. 25, 1942; came to U.S., 1969, naturalized, 1977; s. Wainan and Wangchien C.; m. Huiying Wu, Sept. 13, 1973; children: David W., Jeffrey W., Justin W. MD, Kaohsiung Med. Coll., Taiwan, 1968. Diplomate Am. Bd. Otolaryngology. Intern, Augustana Hosp., Chgo., 1969-70; resident in surgery VA Hosp., Dayton, Ohio, 1970-72; resident in otolaryngology Homer Phillip Hosp., St. Louis, 1972-75; asst. chief otolaryngology VA Hosp., Kansas City, Mo., 1975-77; mem. staff Kansas Med. Center, Kansas City, 1975-77; practice medicine specializing in otolaryngology, Chillicothe, Ohio, 1977—. Fellow Am. Acad. Otolaryngology, Am. Cosmetic Surgery; mem. Am. Soc. Liposuction Surgery, Ohio State Med. Assn. Home: 22 Oakwood Dr Chillicothe OH 45601-1923 Office: 3 Medical Center Dr Chillicothe OH 45601

CHEN, WEN H., laboratory executive, educator; b. Hsin-chu, Taiwan, Dec. 8, 1938; came to U.S., 1964; s. Toh-huang Chen and Mei-mei Tzen; m. Lung-yu Hong; children: Caren H., Christina H., Colleen Y. BS, Nat. Taiwan U., 1962; MS, Kans. State U., 1966; PhD, U. So. Calif., 1973. Elec. engr. Allis-Chalmers Co., Havey, Ill., 1966-68; instr. U. So. Calif., L.A., 1968-70, rsch. assoc., 1970-73; sr. engring. specialist Fox Aerospace & Comm. Corp., Palo Alto, Calif., 1973-77; sr. v.p., chief scientist Compression Labs., Inc., San Jose, Calif., 1977—; prof. San Jose State U., 1982—; chiar prof. Beijing U. Posts & Telecomm., China, 1983—. Recipient Best Algorithm award Picture Coding Symposium, 1973, Pitcure Coding Achievement awards, 1975, 77, Best Speech award Bay Area Profl. Assn., 1992. Mem. IEEE, Chinese Info. Networking Assn. (bd. dirs.), Tech. Adv. Com. of U.S., Taiwanese Engrs. Assn. (life). Office: Compression Labs Inc 2860 Junction Ave San Jose CA 95134-1900

CHEN, WEN-YIH, chemical engineer; b. Kadushing, Republic of China, Jan. 12, 1958; s. Shea-Long and Yu-Sha Chen; m. Hui-Fen Deng, Feb. 26, 1983; 1 child, Ann. PhD, Okla. State U., 1989. Assoc. researcher Fluid Engring. Svc., Inc., Stillwater, Okla., 1988-89; postdoctoral fellow Stanford Rsch. Inst. Internat., Menlo Park, Calif., 1989; assoc. prof. Nat. Cen. U., Chung-Li, Taiwan, 1989—. Contbr. articles to Biotech. Progress, Chem. Engring. Sci., Bioindustry. Recipient Rsch. award Nat. Sci. Coun., 1989-91. Office: Nat Cen U, Dept Chem Engring, Chung-Li Taiwan

CHEN, YANG-FANG, physicist, educator; b. Tainan, Taiwan, Peoples Republic of China, July 2, 1953; s. Shang and I-chu (Lin) C.; m. Ih-Ling Chen, Jan. 26, 1974; children: I-Ying, Ji-Pei. BS, Nat. Tsing-Hua U., Hsin-Chu, Taiwan, 1976; PhD, Purdue U., 1984. Rsch. asst. Purdue U., West Lafayette, Ind., 1981-84; rsch. assoc. Harvard U., Cambridge, Mass., 1985-86; assoc. prof. Nat. Taiwan U., Taipei, 1986-91, prof., 1991—. Referee Material Chemistry and Physics; contbr. articles to profl. jours. Cpl. Chinese Air Force, 1976-78. Mem. Phys. Soc. Republic China (editor in chief 1989-90, exec. councillors 1991—, editor 1991—), Am. Phys. Soc., Nat. Sci. Coun. Republic of China (editor newsletter 1991—, project coord. 1992—, Outstanding Rsch. award 1990-93). Avocations: ping pong, bridge. Home: E Nankin Rd Sect 5 Lane 59, Alley 31 No 5 Fl 2, Taipei 10764, People's Republic of China Office: Nat Taiwan U Dept Physics, 1 Sect 4 Roosevelt Rd, Taipei 10764, Taiwan

CHEN, YUAN JAMES, chemical company executive; b. Keelung, Taiwan, China, June 18, 1949; came to U.S., 1975; s. Hong and Shu-chen (Cheng) C.;m. Ruey-chi Shuai, July 8, 1983; children: Eric Yen-Fu, Albert Hsin-Fu. BS in Mech. Engring., Chung-Hsing U., Taichung, Taiwan, 1971; MS in Mech. Engring., Ga. Inst. Tech., 1976, postgrad., 1976-78; postgrad., U. Houston, 1981-83. Registered profl. engr., Ala., Tex. Mech. engr. China Tech. Cons. Inc., Taipei, Taiwan, 1973-75; sr. engr. Monsanto Chems. Co., Guntersville, Ala., 1978-81; engring. specialist Monsanto Chems. Co., Texas City, Tex., 1981-86; sr. engring. specialist Sterling Chems. Inc., Texas City, 1986—; bd. dirs., tech. adv. com. Heat Transfer Rsch., Inc., College Station, Tex., 1986—. Mem. ASME. Avocations: tennis, table tennis. Home: 14218 Ridgewood Lake Ct Houston TX 77062 Office: Sterling Chem Inc 201 Bay St S Texas City TX 77592-1311

CHEN, ZHAN, physicist; b. Nantong, Jiangsu, People's Republic of China, Dec. 7, 1945; m. Ning Yuan; 1 child, Yijia Chen. MS, Fudan U., Shanghai, People's Republic of China, 1980, PhD, 1985. Elec. engr., mem. rsch. staff Guayang (People's Republic China) Bearing Mfg. Factory, 1968-79; mem. rsch. staff Jiangsu (People's Republic China) Laser Inst., 1982-83; postdoctoral fellow U. Toledo, 1986-88, U. Toronto, Ont., Can., 1988-89; prof. physics Fudan U., 1989-91; sr. R&D scientist Briteview Techs., Toledo, 1991—; mem. faculty Chinese Ctr. Advanced Sci. and Tech., 1989—. Author, editor: Handbook of Modern Science, 1989. Fellow Soc. Physics Students (hon.); mem. Optical Soc. China, Physics Soc. China. Achievements include patent in collimated back light system for LCD. Office: Wande Inc 4088 Lindberg Dr Dallas TX 75244

CHEN, ZHEN, research engineer; b. Shanghai, China, June 27, 1958. Cert. in Chem. Engring., Shanghai Inst. Chem. Engring., 1977; BS in Precision Mech. Engring., Shanghai U. Sci. Tech., 1982; MS in Solid and Computational Mech., U. N.Mex., 1985, PhD in Solid and Computational Mech., 1989. Registered engr., N.Mex. Technician water-treatment divisn. Changjiang Transistors Co., 1977-78; asst. engr. design divsn. Shanghai Med. Electronics Co., 1982-83; rsch. asst. U. N.Mex., Albuquerque, 1983-86, teaching asst., cons. CAD-CAM lab., 1988, rsch. assoc. applied mech. divisn., 1986-90, rsch. engr. applied mech. divisn., 1990-92, sr. rsch. engr. applied mech. divisn., 1992—. Contbr. papers to Jour. Applied Mechs., Jour. Engring. Mechs., Computers and Structures, others; author conf. procs., abstracts, reports in field. SNL grantee, DOE/WERC grantee. Mem. ASCE, Am. Acad. Mech., U.S. Assn. Computational Mech. Achievements include rsch. in theoretical and computational aspects of engring. mech., constitutive modeling using continuous and discontinuous approaches based on macro- and micro-experiments, efficient numerical schemes for failure prediction of single and composite structures. Office: U NMex Engring Rsch Inst 901 University Blvd SE Albuquerque NM 87106-4339

CHENG, ALEXANDER LIHDAR, computer scientist, researcher; b. Taichung, Taiwan, Aug. 1, 1956; came to U.S., 1980; s. Pei-Kao and Kuang-Kun (Shiong) C.; m. Wei-Hong Mao, Feb. 16, 1988; children: Alexander Raymond, Bernard King. BS, Nat. Taiwan U., 1978; MS, U. Ky., 1982; PhD, Poly. U., Bklyn., 1992. Rsch. asst. Taiwan Hydraulic Bur., Taichung, 1978; teaching asst. U. Ky., Lexington, 1981-82; sci. programmer Megadata Corp., Bohemia, N.Y., 1982-83; sr. software engr. Siemens Data Switching, Inc., Hauppauge, N.Y., 1983-87; tech. staff NYNEX S&T, Inc., White Plains, N.Y., 1987—; bd. dirs. C&M First Svcs., Inc., N.Y.C.; adj. prof. Pace U.; advisor to ministry of post and telecom. UN Devel. Programme, People's Republic of China. V.p. Woodcrest Hts. Assn., White Plains, 1991-92. Mem. IEEE, IEEE Computer Soc., Assn. Computing Machinery, Upsilon Pi Epsilon. Avocations: music, stereo, skiing, bicycling, travel. Home: 11 Springdale Ave White Plains NY 10604

CHENG, ANSHENG, mathematics educator, researcher; b. Xinghua, Jiangsu, People's Republic of China, May 28, 1938; s. Yingzhen and Yufang (Cao) C.; m. Peiming Shen, Sept. 25, 1972 (div. 1987); 1 child, Yijia. Diploma in sci., Peking U., Beijing, 1962. Rsch. asst. East China Inst. Computing Tech., Shanghai, 1962-69; asst. prof. Shanghai Inst. Computing Tech., 1969-80, assoc. prof., 1980-88, prof., 1988—; part-time prof. East China Normal U., Shanghai, 1981—; mem. staff Dist. Consultation and Svc. Ctr. for Sci. and Tech., 1982—; reviewer Mathematical Reviews, Am. Math. Soc., 1988—. Contbr. articles to sci. jours and chpts. to books. Dep. Dist. People's Congress, Shanghai, 1987—; mem. Dist. Com. on Edn., Sci., Culture and Health, Shanghai, 1989—. Mem. Ocean and Limnology Soc. China, Shanghai Computation Math. Soc. (trustee 1986—). Avocations: basketball, swimming, music, bridge, photography. Home: 9 Tianlin Estate Apt 1 601, Shanghai 200233, China Office: Shanghai Inst Computing Tec, 546 Yuyuan Rd, Shanghai 200040, China

CHENG, CHUEN YAN, biochemist; b. Hong Kong, June 18, 1954; came to the U.S., 1981; s. C. Yin and Tak Ying (Ho) C.; m. Po Lee, Mar. 17, 1978; children; Yan Ho, Chin Ho. BS with honors, Chinese U. Hong Kong, 1978; PhD, U. Newcastle, 1982. Postdoctoral fellow Population Coun., N.Y.C., 1981-82, rsch. investigator, 1983-84, staff scientist, 1985-87, scientist, 1988-90, sr. scientist, 1991—; asst. prof. Rockefeller U., N.Y.C., 1986-90; prof. U. Rome, 1990—; cons. Angelini Pharms., Inc., River Edge, N.J., 1985-91. Contbr. articles to Jour. Biol. Chemistry, Biochemistry. Mem. Endocrine Soc. (Richard E. Weitzman Meml. award 1988). Achievements include patent for abnormally glycosylated variants of d2 macroglobulin. Office: Population Coun 1230 York Ave New York NY 10021

CHENG, CHUNG P., chemical engineer; b. Canton, China, Aug. 30, 1954; came to U.S., 1972; s. Man-Kee and Yuet-Wah (Wong) C.; m. Jennifer Chung, May 27, 1982. BS, U. Wis., 1976; PhD, U. Del., 1981. Sr. rsch. engr. Akzo (Stauffer Chem.), Dobbs Ferry, N.Y., 1981-88; sect. leader Quantum Chem., Morris, Ill., 1988-91; tech. supt. Catalyst Resources, Inc., Pasadena, Tex., 1991—. Contbr. articles to profl. jours.; patantee in field. Mem. Am. Inst. Chem. Engrs. (chmn. Tappan Zee sect. 1987, Jiolet sect. 1990). Office: Catalyst Resources Inc 10001 Chemical Rd Pasadena TX 77507

CHENG, DAVID KEUN, educator; b. Kiangsu, China, Jan. 10, 1918; came to U.S., 1943, naturalized, 1955; s. Han J. and Ying H.C.; m. Enid Kwok, Mar. 27, 1948; 1 child, Eugene. B.S. in Elec. Engring., Nat. Chiao Tung U., 1938; S.M., Harvard U., 1944, Sc.D., 1946; D. Engr. (hon.), Nat. Chiao Tung U., 1985. Electronics and project engr. research labs. U.S. Air Force, Cambridge, Mass., 1946-48; asst. prof. elec. and computer engring. Syracuse U., N.Y., 1948-51, assoc. prof., 1951-55, prof., 1955—, Centennial prof., 1970—; hon. prof. Beijing Inst. Posts and telecomm., 1982—, N.W. Inst. Telecomm. Engring., 1982—, Shanghai Jiao Tong U., 1985—, People's Republic of China; exch. scientist NAS, Hungary, 1972, Yugoslavia, 1974, Poland and Romania, 1978; liaison scientist Office of Naval Rsch., London, 1975-76; disting. European lectr. IEEE, 1975-76; pres., chmn., bd. trustees Li Instn. Sci. & Tech., 1992—. Author: Analysis of Linear Systems, 1959, Field and Wave Electromagnetics, 1983, 2d edit., 1989, Fundamentals of Engineering Electromagnetics, 1993; cons. editor elec. sci. Addison-Wesley, 1961-78, elec. engring. monographs Intext Edn. Pubs., 1969-72; mem. editorial bd. Jour. Electromagnetic Waves and Applications, 1987—; mem. internat. adv. bd. book series on Progress in Electromagnetic Rsch., 1989—; contbr. numerous articles to profl. jours. Recipient Disting. Achievement award Chinese Inst. Engrs., 1962, Disting. Engr. award Li Inst. Sci. and Tech., 1979; Guggenheim fellow, 1960-61; Chancellor's citation, 1981. Fellow IEEE, AAAS, Inst. Elec. Engrs. (U.K.); mem. AAUP, Am. Soc. Engring. Edn., N.Y. Acad. Scis., Sigma Xi, Eta Kappa Nu, Phi Tau Phi (Disting. Svc. award 1975). Home: 4620 N Park Ave Apt 104E Bethesda MD 20815-4550 Office: Syracuse U Link Hall Syracuse NY 13244

CHENG, HERBERT SU-YUEN, mechanical engineering educator; b. Shanghai, China, Jan. 15, 1929; came to U.S., 1949; s. Chung-Mei and Jing-Ming (Xu) C.; m. Lily D. Hsiung, Apr. 11, 1953; children: Elaine, Elise, Edward, Earl. BSME, U. Mich., 1962; MSME, Ill. Inst. Tech., 1956; PhD, U. Pa., 1961. Jr. mech. engr. Internat. Harvester Co., Chgo., 1952-53; project engr. Machine Engring. co., Chgo., 1953-56; instr. Ill. Inst. Tech., Chgo., 1956-57, U. Pa., Phila., 1957-61; asst. prof. Syracuse (N.Y.) U., 1961-62; rsch. engr. Mech. Tech. Inc., Latham, N.Y., 1962-68; assoc. prof. Northwestern U., Evanston, Ill., 1968-74, prof., 1974—, Walter P. Murphy prof., 1987—, dir. Ctr. for Engring. Tribology, 1984-88, 92—; v.p. Gear Rsch. Inst., Naperville, Ill., 1985-90; cons. GM, Chrysler Corp., Deere Co., Nissan, E.T.C., 1970—. Contbr. articles to profl. jours. Deacon South Presbyn. Ch., Syracuse, 1961-62, 1st Presbyn. Ch. Schenectady, N.Y., 1962-68. Named a hon. prof. Nat. Zhejiang (People's Republic of China) U., 1985. Fellow ASME (hon., Mayo D. Hersey award 1990); mem. NAE, Soc. Tribologists & Lubrication Engrs. (hon., Nat. award 1987), Inst. Mech. Engrs. (U.K., Tribology gold medal 1992), Am. Gear Mfrs. Assn. (acad. mem.), Japan Soc. Lubrication Engrs. Avocations: Peking opera, tennis. Office: Northwestern U 219 Catalysis Bldg 2145 Sheridan Rd Evanston IL 60208-0002

CHENG, H(WEI) H(SIEN), agriculture and environmental science educator; b. Shanghai, China, Aug. 13, 1932; came to U.S., 1951, naturalized, 1961; s. Chi-Pao and Anna (Lan) C.; m. Jo Yuan, Dec. 15, 1962; children: Edwin, Antony. BA, Berea Coll., 1956; MS, U. Ill., 1958, PhD, 1961. Rsch. assoc. Iowa State U., Ames, 1962-64, asst. prof. agronomy, 1964-65; asst. prof. dept. agronomy and soils Wash. State U., Pullman, 1965-71, assoc. prof., 1971-77, prof., 1977-89, interim chmn., 1986-87, chmn. program environ. sci. and regional planning, 1977-79, 88-89, assoc. dean Grad. Sch., 1982-86; prof., head dept. soil sci. U. Minn., St. Paul, 1989—; vis. scientist Julich Nuclear Rsch. Ctr., Fed. Republic Germany, 1971-73, 79-80, Academia Sinica, Taipei, Republic of China, 1978, Fed. Agrl. Rsch. Ctr., Braunschweig, Fed. Republic Germany, 1980; mem. acad. adv. coun. Inst. Soil Sci., Academia Sinica, Nanjing, People's Republic China, 1987—; mem. adv. bd. Inst. Botany, Academia Sinica, Taipei, 1991—; mem. first sci. adv. bd. Dept Ecology State of Wash., 1988-89. Editor: Pesticides in the Soil Environment: Processes, Impacts, and Modeling, 1990; assoc. editor Jour. Environ. Quality, 1983-89; mem. editorial bd. Bot. bull. Academia Sinica, 1988—, cons. editor: Pedosphere, 1991—; contbr. articles to profl. jours. Fulbright rsch. scholar State Agrl. U., Ghent, Belgium, 1963-64. Fellow AAAS, Am. Soc. Agronomy (bd. dirs. 1990—), Soil Sci. Soc. Am. (div. chair 1985-86, bd. dirs. 1990—); mem. Am. Chem. Soc., Soc. Environ. Toxicology and Chemistry, Internat. Soc. Chem. Ecology, Internat. Soc. Soil Sci., Coun. for Agrl. Sci. and Tech., Sigma Xi, Phi Kappa Phi, Gamma Sigma Delta. Methodist. Office: U Minn Dept of Soil Sci 1991 Upper Buford Cir Saint Paul MN 55108-6028

CHENG, J. S. (ZHONG-ZHI ZHENG), nuclear astrophysical chemist, educator; b. Luqiao, Huanyan, China, Mar. 19, 1914; s. Yuan-You and Xiao-yong (Cai) C.; m. Gui-Hua Jia, Oct. 6, 1933; children: Xing-Fei, Wen-Xiu, Tong-Jie, Chao-Mei, He-Zhong, Xiao-Lu. BS in Chemistry, Nat. Jinan U., Shanghai, China, 1936. Asst. prof. analysis chemistry dept. Nat. Jinan U., 1936-37; lectr. chemistry and physics Provincial Fujian Med. Inst., Shaxian, China, 1941-42; engr. Zhejiang Paper Factory and Cheng's Soda Lab., Bihu, Lugiao, China, 1942-45; assoc. prof. phys. chemistry Nat. Ying-shi U., Jinhua, China, 1946; ministerial assoc. prof. Nat. Yingshi U., Nat. Edn. Ministry, Jinhua, 1947-49; assoc. prof. phys. chemistry East North Normal U., Changchun, 1950-54; assoc. prof., then prof. East China Normal U., Shanghai, 1955—; dir., head Irreversible Thermodynamics and Reaction Kinetics Rsch. Group, Shanghai, 1982—; head, advisor Laser-Catalyzed Chain Rsch. Group, Nat. Sci. Found. China, Shanghai, 1988-90; referee Acta Physico-Chimica Sinica, Beijing, 1990—. Author: Potash Soda Process, A Practical Treatise, 1946, Thermonuclear Reaction Kinetics and H-He Cycle, 1980; editor-in-chief: Physical Chemistry, 2 vols., 1986, 86, Irreversible Thermodynamics and Modern Reaction Kinetics, 2 vols., 1987, 90; contbr. articles to Acta Phys. Acad. Sci., Hungaricae, Chem. Abstracts, Phys. Abstracts, Nuclear Sci. Abstracts, Bull. Am. Astron. Soc.; inventor in fields of astrophysics, high energy astrophysics, nuclear astrophys. chemistry, and soda process. Recipient 2d prize for invention nat. class Nat. Edn. Ministry, 1948, Honor Cert. of Achievement, Nat. Edn. Com., 1990, Cert. Govt. Spl. Allowance, State Coun., 1992. Mem. Am. Astron. Soc., Am. Inst. Physics, Astron. Soc. Pacific. Achievements include patent (with other) in soda process (China); discovery of role of Strongly Endoergic Thermonuclear Reaction in the kinetic research of thermonuclear reactions, of empirical equality relations between mean gravitational pressure Pm and central radiation pressure Pc for nondegenerate stars (and obtained Quick Method for determining central temperatures of these stars), of the principle of Irreversible Cyclic Evolution and its applications on origins and evolutions of elements, stars and life, on ways to health, teaching, learning and researching, on agricultural and industrial productions, etc.; derivation of Rate Formula of Strongly Endoergic Thermonuclear Reactions; elucidation of the Neutronization Mechanism of Elements in the core of massive stars in last evolutionary stage; prediction of the extensive existence of Neutron Stars which was confirmed by the discovery of pulsars, of hydrogen and neutrinos production in type II supernova explosion confirmed by observations of SN1987A, various rates confirmed by experimental cross section determination; proposition of the H-He Cycle in the evolution of massive stars, also confirmed. Home: East China Normal U, No 476, 1-Chun, Shanghai 200062, China Office: East China Normal U, Chemistry Dept, Shanghai 200062, China

CHENG, KWONG MAN, structural/bridge engineer; b. Guang Dong, China, Feb. 15, 1952; came to U.S. 1970; s. Kao Chiu and Miu Chun (Koo) C.; m. King-Yu Yeou, Mar. 31, 1984; 1 child, Natalie. BSCE, MIT, 1976; MEng. Civil Engring., Rensselaer Poly. Inst., 1978. Registered profl. engr. 15 states and U.K. Rsch. asst. Rensselaer Poly. Inst., Troy, N.Y., 1976-78; staff engr. T.Y. Lin Internat., San Francisco, 1978-84, sr. bridge engr., 1984-87, dep. chief bridge engr., 1987-89, prin. and chief bridge engr., 1989-90, v.p., chief bridge engr., 1990-91; pres., prin. engr. OPAC Cons. Engrs., San Francisco, 1992—; dir. T.Y. Lin Internat., San Francisco; lectr. in field. Contbr. articles to profl. jours. Recipient Engring. Design award James F. Lincoln Arc Welding Found., Cleve., 1983. Mem. ASCE, Instn. of Structural Engrs. U.K., Internat. Assn. of Bridge and Structural Engring., Chi Epsilon. Achievements include development of complex construction procedures for long span bridges; complex prestressing details in design of long span concrete bridges. Home: 152 Lombard St #208 San Francisco CA 94111 Office: OPAC Cons Engrs 315 Bay St San Francisco CA 94133

CHENG, LI, mechanical engineer, educator; b. Baoding, China, Feb. 1, 1963; arrived in Can., 1989; s. Mantang Cheng and Cuiyan Liu; m. Yuping Lu, June 18, 1988; 1 child, Jassy. BSc, Xi'an Jiaotong U., China, 1984; PhD, Inst Nat. Scis. Appliques, Lyon, France, 1989. Postdoctoral researcher Sherbrooke (Que.) U., Can., 1989-91, rsch. assoc., 1991-92; assoc. prof. U. Laval, Quebec, Que., 1992—. Contbr. articles to profl. publs. Grantee NSERC, 1992—. Mem. Can. Acoustical Assn. (E.M. Shaw prize 1990, 91), Acoustical Soc. Am. Achievements include research in study of damping mechanism in vibroacoustic systems, room acoustics, new technique of model scaling, simulation of vibroacoustic structures. Office: Laval Univ Dept Mech Eng, Cite Universitaire, Quebec, PQ Canada G1K 7P4

CHENG, MINQUAN, chemical engineer; b. Anhui, China, June 10, 1965; came to U.S., 1989; s. Binwen and Guomei (Su) C.; m. Lei Zheng, May 9, 1990. BS, East China Inst. of Chem. Tech., Shanghai, China, 1984; MS, 1987. Rsch. assoc. East China Inst. of Chem. Tech., Shanghai, China, 1987-89, U. Notre Dame, Ind., 1989—; cons. WuJin Chem. Corp., Shanghai, 1988-89. Contbr. articles to profl. jours. Recipient Francois N. Frenkiel award for Fluid Mechanics, Am. Phys. Soc., 1991, Alumni Assn. award for Excellence in Rsch., U. Notre Dame, 1992. Mem. AIChE, Am. Phys. Soc., Tech. Assn. of Pulp and Paper Industry. Achievements include rsch. interests in hydrodynamic instability and transition to turbulence, pattern formation of chem. reaction system. Home: PO Box 259 Univ Village Notre Dame IN 46556 Office: U Notre Dame Dept Chem Engring Notre Dame IN 46556

CHENG, SHENG-SAN, chemistry research scientist, consultant; b. Taipei, Taiwan, Apr. 14, 1952; s. Lin-Hai Cheng and Chon-Wu Chu; m. Jaw-Hua Chiao-Cheng, Mar. 25, 1978; 1 child, Kane. BA, Nat. Tsing-Hua U., 1974; MA, Kansas State U., 1979, Doctor, 1984. Sr. rsch. scientist Chung-Shan Inst. Sci. and Tech., Taipei, 1984—; assoc. prof. Nat. Tsing-Hua U., Hsin-Chu, Taiwan, 1988-90; cons. Taiwan SM Corp., Taipai, 1988—, Chinese Petroleum Corp., Chia-Yi, Taiwan, 1988-90; exec. sec. Synthesis and Combustion Rsch. in High Energy Materials, Hsin-Chu, 1989-91. Contbr. articles to profl. jours. Mem. Am. Chem. Soc., Am. Inst. Aeornautics and Astronautics, Chinese Chem. Soc. Roman Catholic. Achievements include 8 patents on high energy fuel and 1 patent on energetic binder. Office: Chung-Shan Inst Sci & Tech, PO Box 90008-17, Lungtan Taiwan

CHENG, STEPHEN ZHENG DI, chemistry educator, polymeric material researcher; b. Shanghai, China, Aug. 3, 1949; came to U.S., 1981, naturalization, 1992; s. Luzhong and Jingzhi (Zhang) C.; m. Susan L.Z. Xue, June 28, 1978; 1 child, Wendy D.W. PhD, Rensselaer Poly. Inst., 1985. Postdoctoral and rsch. assoc. Rensselaer Poly. Inst., Troy, N.Y., 1985-87; asst. prof. polymer sci. U. Akron, Ohio, 1987-91, assoc. prof. polymer sci., 1991—; mem. adv. bd. Polymer Internat. Jour., 1990—, Jour. Macromolecular Sci., Rev. Macromolecular Physics and Chemistry, Thermochemia Acta, 1992—, Trends Polymer Sic., 1992—. Contbr. over 100 articles to sci. jours., also chpts. to books. Recipient Presdl. Young Investigator award NSF and White House, 1991, cert. of appreciation U. Akron, 1992. Fellow N.Am. Thermal Analysis Soc.; mem. Am. Chem. Soc., Am. Phys. Soc., N.Am. Thermal Analysis Soc., Soc. Plastics Engrs., Materials Rsch. Soc., Soc. Advancement Material and Process Engring. Achievements include research on solid state of polymeric materials including phase transition thermodynamics, kinetics, molecular motion, crystal structure and morphology, liquid crystal polymers, surface and interface structures, high-performance polymer fibers, films for microelectronic and optical applications, high temperature composites, computer simulation of molecular dynamics and modeling. Office: U Akron Inst and Dept Polymer Sci Akron OH 44325-3909

CHENG, THOMAS CLEMENT, parasitologist, immunologist, educator, author; b. Nov. 5, 1930; came to U.S., 1946; s. James Tsu-Mook and Dorothy (Lee) C.; m. Barbara Ann Schimmel, May 31, 1957 (div. 1982); children—Thomas C., J. Bradford, Allison E.; m. Anne Foos Whitelaw, June 19, 1982 (div. 1985). A.B., Wayne State U., 1952; M.S., U. Va., 1956, Ph.D., 1958. Asst. prof. U. Md. Med. Sch., Balt., 1958-59; from asst. prof. to assoc. prof. Lafayette Coll., Easton, Pa., 1959-64; chief immunology and parasitology Northeast Marine Health Sci. Lab USPHS, Narragansett, R.I., 1964-65; from assoc. prof. to prof. U. Hawaii, Honolulu, 1965-69; prof., dir. Ctr. for Health Sci., Lehigh U., Bethlehem, Pa., 1969-80; dir. Marine Biomed. Research Med. U. S.C., Charleston, S.C., 1980-91; sr. prof. Med. U. S.C., Charleston, 1991-92, sr. prof. cell biology, 1980—; sr. sci. cons.

Atlantic Clam Farms, Folly Beach, S.C., 1993—; cons. Acad. Press, San Diego, Calif., 1969—, Univ. Park Press, Balt., 1969— , Div. Microbiology FDA, Washington, 1968-72, biomed. research Internat. Copper Research Assn., Inc., N.Y.C., 1970-92, Sandoz Pharm. Co., Winter Park, Fla., 1979—, Xytronyx, Inc., San Diego, 1984—, Atlantic Littleneck Clamfarms, Folly Beach, South Carolina, 1993—; mem. environ. biology and chemistry study sect. NIH, Bethesda, Md., 1969-71, spl. study sect. Marine Environ. Health, Bethesda, 1973-80, spl. com. on Use of Marine Invertebrates in biomed. research, Bethesda, 1973-75, planning com. FDA-HHS, Washington, 1969-71; mem. adv. bd. Ctr. for Pathobiology U. Calif.-Irvine, 1969—; mem. rev. panel Div. Ocean Scis., Office of Internat. Decade of Ocean Exploration NSF, Washington, 1977-78, div. cell physiology, 1980-83; chmn. molecular biology Office Naval Research, Arlington, Va., 1983-92; mem. USPHS Commn. on Food Protection, Washington, 1965-66; dir. WHO Collaborative Lab. of Vector Biology, Charleston, S.C., 1982-92, research Centre Nationale de la Recherche, France, 1986-87; prin. lectr. Internat. Soc. Comparative Physiology, Switzerland, 1989; co-chmn. Internat. Congress Parasitology, Paris, 1990; mem. sci. directorate Internat. Orgn. Pathology Marine Aquaculture. Author: The Biology of Animal Parasites (1st prize Phila. Book Show), 1964, 65, Marine Molluscs as Hosts for Symbiosis, 1967, Symbiosis: Organisms Living Together, 1970, General Parasitology, 1973, 2d edit. 1986, Human Parasitology, 1990; editor, contbr. Some Biochemical and Immunological Aspects of Host-Parasite Relationships, 1963, Aspects of the Biology of Symbiosis, 1971, Molluscicides in Schistosomiasis Control, 1974, Invertebrate Immune Responses, 1977, Invertebrate Models for Biomedical Research, 1978, Structure of Membranes and Receptors, 1984, Invertebrate Blood: Functions of Serum Factors and Cells, 1984; editor: Pathogens of Invertebrates: Application in Biological Control and Transmission Mechanisms, 1984; co-author: Medical and Economic Malacology, 1974, Biology of Microsporidia, 1976, Systematics of the Microsporidia, 1977, Pathology in Marine Science, 1990; editor Jour. Invertebrate Pathology, 1969-91, Exptl. Parasitology jour., 1969-88, Comparative Pathobiology jour., 1975-84; contbr. numerous articles to profl. jours. and revs. Mem. Mayor's Marine Mus. Project, Charleston, 1984—; regional coordinator Nat. Disaster Med. System, Charleston, 1987—. Served to capt. USPHS, U.S. Army, 1952-54, 64-65; Korea. Recipient George C. Wheeler Disting. Lectureship award U. N.D., Grand Forks, 1973; Disting. Lectr. Southwestern Assn. Parasitologists, 1973; Disting. Lectureship award Tulane Med. Sch., New Orleans, 1977; Parasitology prin. lectr. award Coll. of Physicians of Phila., 1980; Andrew Fleming Research award U. Va., Charlottesville, 1958; Roy and Ira Jones Superior Teaching award Lafayette Coll., Easton, Pa., 1962; Disting. Alumnus award Wayne State U., Detroit, 1975, Medal of Honor U. Montpellier, 1992; Fulbright research scholar, France, 1986-87; fellow U. Va., 1955-57, NSF, 1958, 63; grantee 1958-59, 60-62, NSF, 1959-61, 78-83, 82-85, NIH, 1961-64, 71-76, 75-78, 77-79, Am. Cancer Soc., 1966-69, FDA, 1973-75, Internat. Copper Research Assn., 1971-76, 75-77, 81-84, Nat. Marine Fisheries, NOAA, 1991-93, Dept. Commerce, Dept. Energy, WHO, USDA, Am. Cyanamid Co.; Disting. Stoll lectr. Rutgers U., 1988; named Hon. Citizen City of Montpellier, France, 1992. Fellow Royal Soc. Tropical Medicine and Hygiene, AAAS; mem. Am. Microscopical Soc. (pres. 1980-81), Am. Physiol. Soc., Jefferson Soc. U. Va., Am. Soc. Zoologists (rep. pub. affairs 1982-88), Am. Soc. Parasitologists (exec. council 1974-78), Soc. for Exptl. Biology and Medicine, Soc. Protozoologists, Reticuloendothelial Soc., N.Y. Acad. Scis., Helminthological Soc. Washington, Soc. for Invertebrate Pathology (chmn. publs. bd. 1969-84), N.Y. Soc. Tropical Medicine (sec.-treas. 1975-76), Council of Biology Editors, N.J. Soc. Parasitology (pres. 1977-78), Sigma Xi, Tau Kappa Epsilon. Democrat. Episcopalian. Home and Office: 8 Queen St Charleston SC 29401-2111 also: Atlantic Clam Farms PO Box 12139 Charleston SC 29422 Office: Atlantic Clam Farms Box 12139 Charleston SC 29422

CHENG, WING-TAI SAVIO, medical technology consultant; b. Hong Kong, Apr. 22, 1955; came to U.S., 1974; s. Huen and Chu (Ou) C. BSc, U. Wis., Green Bay, 1979; PhD, Dartmouth Coll., 1986. Postdoctoral scholar U. Ala., Birmingham, 1986-88; rsch. assoc. Dartmouth Med. Sch., Hanover, N.H., 1988-89, instr., 1990-92; med. tech. cons. Aetna Life and Casualty, Hartford, Conn., 1992—. Contbr. articles to profl. jours. Mem. Am. Physiol. Soc., Internat. Soc. Tech. Assessment in Health Care, Endocrine Soc. Home: 3213 Town Brooke Middletown CT 06457 Office: Aetna Life & Casualty 151 Farmington Ave MCA3 Hartford CT 06156

CHEREMISINOFF, PAUL NICHOLAS, environmental engineer, educator; b. N.Y.C., Feb. 20, 1929; s. Nicholas P. and Luba Cheremisinoff; m. Louise Nappi, Aug. 25, 1951 (dec. Mar. 1987); children: Nicholas, Peter; m. Louise Ferrante, July 3, 1991. BChemE, Pratt Inst., 1949; MS, Stevens Inst. Tech., 1952. Lic. profl. engr., N.J. Chem. engr. Joseph Turner Co., Ridgefield, N.J., 1949-56; plant mgr. Am. Polyglas-Alsynite Co., Clifton, N.J., 1956-60; sr. chem. engr. Celanese Corp., Newark, N.J., 1960-63, Tenneco Corp., Garfield, N.J., 1963-67; chief environ. engr. Engelhard Minerals & Chems. Co., Newark, 1967-73; prof. civil and environ. engring. N.J. Inst. Tech., Newark, 1973—; cons. engr. various orgns. and govt. agys., 1973—. Author, editor more than 300 books; editor: Pollution Engineering, 1973-90, Nat. Environ. Jours., 1991-92; contbr. articles to profl. jours. Fellow N.Y. Acad. Sci. (chmn. engring. 1973-76); mem. Sigma Xi, Tau Beta Pi. Achievements include patents on pollution control devices and chemical processes. Home: 230 Terrace Ave Hasbrouck Heights NJ 07604 Office: NJ Inst Tech King Blvd Newark NJ 07102

CHERENKOV, PAVEL ALEXEYEVICH, physicist; b. Novaya Chigla, Voronezh, Nov. 28, 1904. Ed., Voronezh U. Prof. Moscow Engring. Phys. Inst.; mem. Inst. Physics USSR Acad. Scis.; mem. CPSU, 1946; corr. mem. Acad. Scis., 1964-70, academician, 1970—. Recipient Nobel prize for physics (with Tamm and Frank), 1958, State Prize (3), Order of Lenin (2), Order of Red Banner of Labor (2), Badge of Honor, other decorations. Discoverer of Cherenkov Effect. Mem. NAS (fgn. assoc.). Office: USSR Acad Scis Lebedev Physics Inst, Leninsky Prospekt 53, Moscow Russia

CHEREPAKHOV, GALINA, metallurgical and chemical engineer; b. Moscow, July 10, 1936; came to U.S., 1979; d. Lev and Khanna Degtyarov; m. Naum Cherepakhov, Jan. 31, 1964; 1 child, Alexander. MS, Steel and Alloy Inst., Moscow, 1959; PhD, Chem. Engring. Inst., Moscow, 1966. Rsch. metallurgist Chromaloy Metal Tectonics, West Nyack, N.Y., 1979-80; metall. engr. The M.W. Kellogg Co., Hackensack, N.J., 1980-82; metall. and chem. engr. Consol. Edison Co., N.Y.C., 1982—, sr. engr., 1989—; cons. Assn. Engrs. and Scientists for New Ams., N.Y., 1989—. Author: Corrosion and Materials Protection, 1974; contbr. articles to profl. jours. Mem. Nat. Assn. Corrosion Engrs. Achievements include research and corrosion evaluation of materials for petrochemical and chemical industry, project management of research and development projects on decontamination and chemical cleaning in the nuclear industry. Office: Consol Edison Co 4 Irving Pl Rm 1250S New York NY 10003-3502

CHEREPANOV, GENADY PETROVICH, mathematician, mechanical engineer; b. Krutaia, Melenki Dist., Vladimir, Vladimir Region, USSR, Jan. 8, 1937; came to U.S., 1990; s. Petr Vasilievich and Alexandra Petrovna (Gorkina) C.; m. Galina P. Lebed (div. 1965); 1 child, Andrew; m. Alexandra V. Dvoichenkova (div. 1971); 1 child, Yury; m. Elena F. Odintsova (div. 1980); 1 child, Daria; m. Larisa Yakovlevna Beyleen, July 4, 1985; 1 child, Peter. BS in Engring. and Physics, Moscow Inst. Physics and Tech., 1960; PhD in Applied Mechanics, Moscow State U., 1962; ScD in Applied Math. & Theoret Mechanics, USSR Acad. Scis., Moscow, 1965. Cert. math. engring. Sr. scientist Inst. Mechanics, USSR Acad. Scis., Moscow, 1962-69; mgr. math. modeling lab. Pacific Oceanology Inst., USSR Acad. Scis.; prof. applied math. Moscow Mining Inst., 1969-78; sr. scientist Moscow Rsch. Inst. Drilling Tech., 1978-87; disting. rsch. assoc. solid mechanics Harvard U., Cambridge, Mass., 1990-91; prof. mech. engring. Fla. Internat. U., Miami, 1991—; vis. prof. Moscow State U., 1967-72; advisor Spacecraft Tech. Rsch. Inst., Moscow, 1967-72; head coord. Spl. Math. Modeling Bd., Baku, 1978-80. Author: Mechanics of Brittle Fracture, 1979, Fracture Mechanics of Composite Materials, 1983, Rock Fracture Mechanics in Drilling, 1987, Elastic-Plastic Problems, 1988, Decline of the Evil Empire, and others; mem. editorial bd. and referee Internat. Jour. Engring. Fracture Mechanics, Jour. Applied Math. and Mechanics, and others; contbr. over 250 articles to profl. jours. Mem. coord. com. Christian-Dem. Union, Moscow, 1989. Recipient Lenin Komsomol prize, 1971. Fellow Internat. Congress Fracture (hon.); mem. ASME, Am. Math. Soc., N.Y. Acad. Scis. (hon. life). Orthodox Christian. Achievements include founding of contemporary fracture mechanics based on invariant or path-independent integrals called Eshelby-Cherepanov-Rice integrals and founding of the mechanics of quantum fracture. Office: Fla Internat Univ 465 ECS Miami FL 33199

CHERN, JENG-SHING, aerospace engineer; b. Tainan, Taiwan, Republic of China, May 30, 1947; s. Shin-Tay and Yeou Huang Chern; m. Hsueh-Chin Lin, Sept. 28, 1973; children: Leland L., Eolande Y. BSME, Cheng Kung U., Tainan, 1969; PhD in Aerospace Engring., U. Mich., 1979. Instr. Cheng Kung U., Tainan, 1972-73; asst. scientist Chung Shan Inst. Sci. and Tech., Lungtan, Taiwan, 1973-79, assoc. scientist, 1979-89, sr. scientist, 1989—; assoc. prof. Cheng Kung U., Tainan, 1983-87; prof. Nat. Ctrl. U., Chungli, Taiwan, 1989—. Contbr. articles to Acta Astro. Jour., Jour. Guidance, Control and Dynamics. Recipient Excellence in Acad. Study award Ministry Def., Taipei, 1979, Excellence in Govt. Svc. award Ministry Def., 1987, Excellence in Rsch. award Ministry Def., 1987. Fellow Aero. and Astronautical Soc. (Republic of China), Chinese Soc. Mech. Engrs.; mem. AIAA (com. standards 1991—), Republic of China Neygong Assn. (bd. dirs. 1991—), Tau Beta Pi. Achievements include discovery of optimal reentry trajectory, of complete footprint of shuttle-type vehicle, of methodology for low earth orbit keeping using low thrust, of optimal trajectory for vertical ascent to geosynchronous earth orbit. Home: 6th Flr # 13, Yang Ming 10th St, Taoyuan Taiwan 330 Office: Chung Shan Inst Sci & Tech, PO Box 90008-6-7, Lungtan Taiwan 325

CHERN, JENN-CHUAN, civil engineering educator; b. Kin-Men, Fuken, Republic of China, July 28, 1954; s. Juh-Tang Chern and Wu-Teng Lu; m. Huey-Jen Lee, May 19, 1979. BSCE, Nat. Taiwan U., Taipei, Republic of China, 1976; MCE, Rice U., 1980; PhDCE, Northwestern U., 1984. Structural engr. Fu-Tai Engring. Co., Taipei, 1978-79; rsch. engr. Argonne (Ill.) Nat. Lab., 1984-85; assoc. prof. dept. civil engring. Nat. Taiwan U., Taipei, 1985-89, prof. dept. civil engring., 1989—; chmn. dept. civil engring., 1992—. Chief cons. City Mayor of Keelung, Taiwan, 1989; com. mem. Urban Planning Com., Keelung, 1989; com. mem. Nat. Mus. of Prehistory Planning Bur., China, 1989. Recipient Excellent Rsch. award Nat. Sci. Coun., China, 1987, 88, Outstanding Rsch. award, 1989, 91, Outstanding Indsl. Waste Reduction award Ministry of Economy, Environ. Protection Bur., China, 1991. Mem. ASTM, RILEM Paris, Chinese Inst. Civil Hydraulic Engring. (com. chmn. 1990—), Am. Concrete Inst., Soc. Exptl. Mechanics, Chinese Inst. Engrs., Chinese Soc. Structural Engring. (com. chmn. 1993—). Avocations: basketball, volleyball, swimming, country music. Office: Nat Taiwan U, Dept Civil Engring, Taipei 10617, Taiwan

CHERN, JIUN-DER, medical physicist; b. Taipei, Taiwan, July 14, 1962; came to U.S., 1989; s. Ben-San and (Chu) Chen. BS, Nat. Cen. U., 1984; MS, U. Colo., Denver, 1992. Pvt. tchr. Tai-Ann Dist., Taipei, 1986-87; med. physicist Jen-Ai Hosp., Taipei, 1987-89; resident in med. physics Meth. Hosp. Ind., Indpls., 1992—. Mem. Am. Assn. Physicists in Medicine. Home: 8415 Rothbury Dr Apt G Indianapolis IN 46260 Office: Meth Hosp Ind PO Box 1367 1700 N Senate Blvd Indianapolis IN 46202

CHERN, JI-WANG, pharmacy educator; b. Tainan County, Taiwan, Republic of China, Sept. 6, 1953; s. Yuan-Chun and Shih-Shia (Zhau) C.; m. Mei-Ying Kuo, Sept. 4, 1978; children: Ting-Rong, Ting-Zhao. BS in Pharmacy, Nat. Def. Med. Ctr., Taipei, Taiwan, 1977; MS in Medicinal Chemistry, U. Mich., 1982, PhD in Medicinal Chemistry, 1985. Registered pharmacist, Taiwan. Teaching asst. Nat. Def. Med. Ctr., Taipei, 1977-80; postdoctoral fellow Coll. Pharmacy U. Mich., Ann Arbor, 1985; assoc. prof. Nat. Def. Med. Ctr., Taipei, 1985-89, prof. Inst. Pharmacy, 1989—; adj. prof. sch. pharmacy Nat. Taiwan U., 1993—. Mem. editorial bd. Chinese Pharm. Jour., 1988—; asst. editor Jour. Med. Scis., 1988-89. Patentee in field. Lt. col. Taiwan mil., 1990. Recipient Best Rch. Scientist award Nat. Sci. Coun., Taipei, 1988, 89, Outstanding in Rsch. Scientist award, 1992; named One of 10 Outstanding Young Persons in Republic of China, 1991. Mem. AAAS, Am. Chem. Soc., Internat. Soc. Heterocyclic Chemistry, Am. Cancer Rsch. Assn., N.Y. Acad. Scis., Chinese Chem. Soc. Mem. Nationalist Party. Avocations: running, softball, baseball, golf. Home: 4F 30, 5 Alley, 24 Ln, Sect 3, Ting-Zou Rd, Taipei 100, Taiwan Office: Nat Def Med Ctr, Inst Pharmacy PO Box 90048-512, Taipei 100, Taiwan

CHERN, SHIING-SHEN, mathematics educator; b. Kashing, Chekiang, China, Oct. 26, 1911; s. Lien Ching and Mei (Han) C.; m. Shih-ning Chern, July 28, 1939; children—Paul, May. B.S., Nankai U., Tientsin, China, 1930; M.S., Tsing Hua U., Peiping, 1934; D.Sc., U. Hamburg, Germany, 1936, D.Sc. (hon.), 1972; D.Sc. (hon.), U. Chgo., 1969, SUNY-Stony Brook, 1985; LL.D. honoris causa, Chinese U., Hong Kong, 1969; Dr. Math (hon.), Eidgenossische Technische Hochschule, Zurich, Switzerland, 1982. Prof. math. Nat. Tsing Hua U., China, 1937-43; mem. Inst. Advanced Study, Princeton, N.J., 1943-45; acting dir. Inst. Mathematics, Academia Sinica, China, 1946-48; prof. math. U. Chgo., 1949-60, U. Calif., Berkeley, 1960-79; prof. emeritus U. Calif., 1979—; dir. Math. Scis. Rsch. Inst., 1981-84, dir. emeritus, 1984—; dir. Inst. Mathematics, Tianjin, P.R., China. hon. prof. various fgn. univs.; Recipient Chauvenet prize Math. Assn. Am., 1970, Nat. Medal of Sci., 1975, Wolf prize Israel, 1983-84. Fellow Third World Acad. Sci. (founding mem. 1985); mem. NAS, Am. Math. Soc. (Steele prize 1983), Am. Acad. Arts and Scis., N.Y. Acad. Scis. (hon. life), Am. Philos. Soc., Indian Math. Soc. (hon.), Brazilian Acad. Scis. (corr.), Academia Sinica, Royal Soc. London (fgn.), Academia Peloritana (corr. mem. 1986), London Math. Soc. (hon.)., Acad. des sciences Paris (fgn. mem.), Acad. der Lincei Rome (stranieri). Home: 8336 Kent Ct El Cerrito CA 94530-2548 Office: Univ Calif Berkeley Dept of Mathematics Berkeley CA 94720

CHERNIAVSKY, JOHN CHARLES, computer scientist; b. Boston, Feb. 15, 1947; m. Ellen Huage Abelson, June 22, 1968; children: John Philip, Neva Anne. BS, Stanford U., 1969; MS, Cornell U., 1971, PhD, 1972. Asst. prof. SUNY, Stony Brook, 1972-80, assoc. prof., 1980-81; program dir. NSF, Washington, 1980-84; prof., chmn. dept. Georgetown U., Washington, 1984-90; program dir. NSF, Washington, 1990-91, office head, 1991—. Office: Natl Science Foundation Computer Science & Engineering 1800 G St NW Washington DC 20550

CHERNICK, CEDRIC L., research grants administrator; b. Manchester, Lancashire, Eng., Nov. 2, 1931; came to U.S., 1957; s. Joseph and Frieda (Steinberg) C.; m. Judith Ellen Davis, June 14, 1959 (div. 1972); children: Devra Naomi, Sarah Elizabeth; m. Judy Marks, Aug. 2, 1972; 1 stepchild, Deborah Jeanne. BS with honors, Manchester U., Eng., 1953, PhD, 1956. Postdoctoral fellow Ind. U., Bloomington, 1957-59; rsch. chemist Argonne (Ill.) Nat. Lab., 1959-66, asst. to dir., 1966-69; asst. v.p. U. Chgo., 1969-76, assoc. v.p., 1976-78, v.p., 1978-80; dir. The Searle Scholars Program, Chgo., 1980—; alt. mem. Nat. Commn. on Productivity, Washington, 1971-74; mem. Com. on Govtl. Rels., Washington, 1978-80. Contbr. over 40 articles to profl. jours. Pres. Jackson Towers Condo Assn., Chgo., 1987-90; sec. Lake Mich. Performance Handicap Racing Fleet, Oshkosh, Wis., 1990—. Recipient Disting. Contbn. to Rsch. Adminstrn. award Soc. of Rsch. Adminstrs., 1992. Mem. AAAS, Am. Chem. Soc., Nat. Acad. Univ. Rsch. Adminstrs., Jackson Park Yacht Club (sec. 1976-79), Sigma Xi, Phi Lambda Upsilon. Jewish. Office: Searle Scholars Program 222 N La Salle St Ste 1444 Chicago IL 60601-1009

CHERNISH, STANLEY MICHAEL, physician; b. N.Y.C., Jan. 27, 1924; s. Michael B. and Veronica (Hodon) C.; m. Lelia M. Higgins, June 19, 1949; 1 child, Dwight. BA, U. N.C., 1945; MD, Georgetown U., 1949. Diplomate Nat. Bd. Med. Examiners, Am. Bd. Internal Medicine. Intern Washington Gen. Hosp., 1949-51; resident Marion County Gen. Hosp., Indpls., 1953-55; with clin. rsch. div. Eli Lilly & Co., Indpls., 1954-85; from asst. to assoc. in medicine Sch. Medicine, Ind. U., 1957-66, asst. prof., 1967-76, clin. assoc. prof., 1977-80, assoc. prof., 1981—; rsch. cons. Meth. Hosp., Indpls., 1986—; vis. staff Marion County Gen. Hosp., 1965—. Contbr. more than 100 articles to profl. jours., chpts. to books. Served with USNR, 1943-45, 50-53, ret. comdr. 1984. Fellow ACP, Am. Coll. Gastroenterology; mem. Am. Coll. Clin. Pharmacology and Therapeutics, AMA (Physicians Recognition award in continuing med. edn. 1972—), Ind. State Med. Soc. (mem. subcommn. on accreditation), Marion County Med. Soc., Am. Pancreatic Study Group, Assn. Am. Physicians and Surgeons, Am. Fedn. Clin. Rsch., Am. Gastroent. Assn., Sci. Rsch. Soc., Sigma Xi. Office: Meth Hosp Ind Dept Med Rsch PO Box 1367 1701 N Senate Blvd Indianapolis IN 46204

CHERNYAK, BORIS VICTOR, biochemist, researcher; b. Moscow, Nov. 21, 1954; came to U.S., 1990; s. Victor Ya Chernyak and Agness Heft; m. Marine Gasparian, Apr. 2, 1988. Masters, Moscow State U., 1976, PhD, 1980. Rsch. scientist Moscow State U., 1980-88, sr. rsch. scientist, 1988-90; vis. scientist SUNY, Syracuse, 1990-91, U. Cin., 1991-92; sr. rsch. scientist Moscow State U., 1992—. Office: Moscow State Univ AN Belozersky Inst, Lieninskie Gory Lab Bldg A, Moscow 119899, Russia

CHERON, JAMES CLINTON, mechanical engineer; b. New Orleans, Dec. 24, 1943; s. Claude Edward and Edna (Moran) C.; m. Patricia Ann Brossette, May 9, 1975; children: Jeremy Clinton, Patrick Robert. BS in Engring. Sci., U. New Orleans, 1970; M of Engring., Tulane U., 1973. Plant engr. Mid-La. Gas Co., New Orleans, 1972-73; engr. ICS McDermott Internat., New Orleans, 1973-80, sr. supr. ICS, 1981-84; sr. instn. supt. Asamera S. Sumatra Ltd., Sakarta, Indonesia, 1984-89, maintenance mgr., 1990—. With USN, 1962-66. Mem. ASME, Instrument Soc. of Am., Hash House Harriers, Indonesia Petroleum Assn. Roman Catholic. Achievements include being named ICS engr. on 3 of 5 tallest offshore structures ever set. Home: 4608 Rebecca Blvd Metairie LA 70003 Office: Asamera South Sumatra Ltd, JL-G ATOT Subroto PO Box 2858, Jakarta Indonesia

CHERRY, ANDREW L., JR., social work educator, researcher; b. Dothan, Ala., Nov. 11, 1943; s. Andrew L. Cherry and Wyalene Cain; m. Mary Elizabeth Dillon, July 16, 1988. MSW, U. Ala., Tuscaloosa, 1974; D Social Work, Columbia U., 1986. Child welfare worker Escambia County Dept. Pensions and Securities, Brewton, Ala., 1968-72; psychiat. social worker Bryce State Hosp., Tuscaloosa, 1974-79; instr. Salisbury (Md.) State Coll., 1981-85; asst. prof. Marywood Coll. Sch. Social Work, Scranton, Pa., 1986-87; prof. Barry U. Sch. Social Work, Miami, Fla., 1987—; conf. Informed Families Dade County, Miami, 1990—, Miami Coalition for Care to Homeless, 1991—. Contbr. articles to profl. jours. Scholar NIMH, 1979. Fellow Am. Orthopsychiat. Assn.; mem. NASW, Conf. Social Work Edn., N.Y. Acad. Scis. Democrat. Achievements include research and development of the social bond theory; extensive work and research among the homeless. Office: Barry U Sch Social Work 11300 NE 2d Ave Miami Shores FL 33161

CHESTER, ARTHUR NOBLE, physicist; b. Seattle, Aug. 5, 1940; s. Arthur Malbridge and Marjorie (Stenberg) C.; m. Cynthia Anne Ashford, Sept. 6, 1961 (div. June 1968); m. Catherine Rogers Buchanan, Aug. 10, 1969. BS in Physics, U. Tex., 1961; PhD in Theoretical Physics, Calif. Inst. Tech., 1965. Mem. tech. staff Bell Labs., Murray Hill, N.J., 1965-69; mem. tech. staff Hughes Research Labs., Malibu, Calif., 1969-73, mgr. laser dept., 1973-75, assoc. dir., 1975-80; program mgr. very high speed integrated circuits Hughes Aircraft Co., El Segundo, Calif., 1980-83, mgr. tactical engring. div., 1984-85; group v.p., mgr. space and strategic systems div. Hughes Aircraft Co., 1985-88; v.p. and dir. research labs. Hughes Aircraft Co., Malibu, Calif., 1988-93, sr. v.p. rsch. and technology, 1993—; cons. U.S. Dept. Def., Washington, 1975-79; co-dir. Internat. Sch. Quantum Electronics, Erice, Sicily, Italy, 1980—. Co-editor: Integrated Optics: Physics and Applications, 1983, Free Electron Laster, 1983, Analytical Laser Spectroscopy, 1985, Laser Photobiology and Photomedicine, 1985, Optica Fiber Sensors, 1987, Laser Science and Technology, 1988, Progressin Microemulsions, 1989, Nonlinear Optics and Optical Computing, 1990, Laser Systems for Photobiology and Photomedicine, 1991, Phase Transitios in Liquid Crystals, 1992, Industrial Laser Applications, 1993, Integrated Optics, 1993; contbr. articles to jours. Pres. Masterwork Chorus, Morristown, N.J., 1968-69; bd. dirs. Fellows Contemporary Art, Los Angeles. Recipient A.A. Bennett Calculus prize U. Tex., 1959; recipient Nat. Merit scholar, 1957; NSF fellow, 1961; Howard Hughes doctoral fellow, 1963. Fellow Optical Soc. Am., IEEE (chmn. com. 1982—, Centennial medal 1984); mem. AAAS, IEEE Lasers and Electro-Optics Soc. (pres. 1980), Am. Phys. Soc., Sigma Xi. Office: Hughes Aircraft Co Rsch Labs 3011 Malibu Canyon Rd Malibu CA 90265-4797

CHESTNUT, HAROLD, foundation executive, electrical engineer; b. Albany, N.Y., Nov. 25, 1917; s. Harry and Dorothy (Schulman) C.; m. Erma Ruth Callaway, Aug. 24, 1944; children: Peter Callaway, H. Thomas, Andrew T. BS in Elec. Engring., Mass. Inst. Tech., 1939; MS, MIT, 1940; DE (hon.), Case Western Res. U., 1966, Villanova U., 1972. With Gen. Electric Co., 1940-83; cons. systems engr., aeros. and ordnance dept. Advanced Tech. Lab., Schenectady, 1956-66; mgr. Research and Devel. Center, 1966-71; cons. systems engr., 1972-83; pres. SWIIS Found., Inc., 1983—. Editor: Systems Engring. and Analysis, 1965-83, Contributions of Technology to International Conflict Resolution, 1987; author: Servomechanisms and Regulating Systems Design, Vol. I, 1951, Vol. II, 1955, Systems Engineering Tools, 1965, Systems Engineering Methods, 1967; editor: Jour. Automatica, 1961-67. Mem. commn. sociotech. systems NRC, 1975-78. Case Western Res. U. Centennial scholar, 1980. Fellow IEEE (v.p. tech. activities 1970-71, v.p. regional activities 1972, pres. 1973, exec. com. 1967-75, Centennial medal 1984, Richard M. Emberson award 1990), AAAS, Instrument Soc. Am.; mem. ASME (Rufus Oldenburger award 1990), Nat. Acad. Engring., Internat. Fedn. Automatic Control (pres. 1957-58), World Federalists Assn. (bd. dirs. 1980—, exec. com. 1984—), Am. Automatic Control Council (pres. 1962-63, Honda prize 1981, Bellman Control Heritage award 1985), Nat. Soc. Profl. Engrs., First Unitarian Soc. (pres. Schenectady ch. 1983-84). Home and Office: 1226 Waverly Pl Schenectady NY 12308-2627

CHETELAT, ROGER TOPPING, geneticist; b. Greenbrae, Calif., Sept. 22, 1957; s. Guy Felix Joseph and Mary Jane (Rogers) C.; m. Catherine Jeane-Marie Samora, June 30, 1982; children: Alan, Thomas, Laurynne. BS in Biology, Santa Clara U., 1979; MS in Plant Physiology, U. Calif., Davis, 1983. Programmer, analyst Eurosoft, Paris, 1984; rsch. assoc. Campbell Soup Co., Davis, Calif., 1985-90; staff rsch. assoc. U. Calif., Davis, 1990—; mem. Tomato Crop Adv. Com., 1992—. Contbr. articles to profl. jours. Recipient Staff Devel award U. Calif. Davis, 1991, Rsch. grant USDA, 1991. USDA rsch. grantee, 1991; Jastro-Shields rsch. grantee U. Calif., Davis, 1992. Democrat. Office: U Calif Davis Dept Vegetable Crops Davis CA 95616

CHEUNG, KWOK-WAI, optical network researcher; b. Hong Kong, May 14, 1956; came to U.S., 1980; s. Kwun-hung and Pik-wan (Lee) C. BSEE with 1st honors, U. Hong Kong, 1978; MS in Physics, Yale U., 1981; PhD in Physics, Calif. Inst. Tech., 1987. Mem. tech. staff Bell Comm. Rsch., Morristown, N.J., 1987-90, Red Bank, N.J., 1990-92; assoc. prof. dept. engring. Chinese U., Hong Kong, 1993—; sr. rsch. assoc. Columbia U., N.Y.C., 1990; vis. rsch. assoc. Chinese Acad. Scis., 1992—; referee to various jours. Contbr. articles to IEEE Jour. of Selected Areas in Communications, IEE Globecom, IEEE Electronic Letters. Founder, trustee Olive Tree Ministries Internat., N.J., 1989—. Recipient Electronics Letters Premium award 1992. Mem. IEEE (sr.), Optical Soc. Am., Soc. Photo-Optical Instrumentation Engrs., Photonic Soc. Chinese-Americans. Achievements include patents for Switch for Selectively Switching Optical Wavelenghts, (with others) Integrated Acousto-Optic Filters and Switches, (with other) Polarization-Dependent and Polarization-Diversified Opto-Electronic Devices Using a Strained Quantum Well, (with other) Wavelength Division Multiplexing Using a Tuanble Acoustic Optic Filter. Office: Chinese Univ of Hong Kong, Dept Info Engring, Sha Tin NT Hong Kong

CHEUNG, LIM HUNG, physicist; b. Hong Kong, Feb. 18, 1953. BS, Calif. Inst. Tech., 1975; PhD, U. Md., 1980. Astrophysicist Harvard Smithsonian Ctr. for Astrophysics, Cambridge, Mass., 1980-81; sr. rsch. geophysicist Gulf Oil, Pitts., 1981-84, Standard Oil, Dallas, 1984-86; head optical physics lab. Grumman Corp. Rsch. Ctr., Bethpage, N.Y., 1986—. Contbr. articles to Astrophys. Jour. Achievements include one patent. Office: Grumman Aerospace MS A01-26 5 Oyster Bay Rd Bethpage NY 11714

CHEUNG, WILKIN WAI-KUEN, entomologist, educator; b. Hong Kong, May 12, 1941. BS, U. Hong Kong, 1965, BS Spl., 1966, PhD, 1971. Lectr. U. Singapore, 1972-75; lectr., pest cons. Chinese U. of Hong Kong, 1975—. Contbr. over 40 sci. papers to profl. publs. Recipient Hong Kong Govt. studentship, 1966-68, Commonwealth scholarship Hull U., 1968-69, La Trobe U., 1969-71. Mem. Entomol. Soc. Am., Electron Microscopy Soc.

Am., AAAS. Avocations: jogging, travel, swimming, computer. Office: Chinese U Hong Kong, Biology Dept, Sha Tin NT Hong Kong

CHEUNG, WILSON D., electrical engineer; b. Wilmington, Del., Sept. 16, 1966; s. Peter and Evangeling (Soh) C. BSEE, Villanova U., 1987, MSEE, 1989. Neural network engr. E. I. Du Pont de Nemours & Co., Wilmington, 1989-92; design engr. E. I. Du Pont de Nemours & Co., Newark, Del., 1992—. Home: 112 Hunter Ct Wilmington DE 19808 Office: E I DuPont de Nemours & Co Process Engring Imaging Systems Eagle Run Bellvue Bldg Box 6 Newark DE 19714

CHEVALIER, JEAN, physics educator; b. Moutier, Jura, Switzerland, Mar. 6, 1936; s. René and Renée (Joray) C. BSc, Ecole Cantonale, Porrentruy, Switzerland, 1955; diploma in physics, Swiss Fed. Inst. Tech., Zürich, 1961; PhD in Physics, U. Berne, Switzerland, 1970. Prof. math. Ecole Cantonale, Porrentruy, 1961-66; sci. collaborator Inst. Theoretical Physics U. Geneva, 1970-76, lectr. Inst. Theoretical Physics, 1974-78; prof. physics Lycée Porrentruy, 1976—; researcher in gen. relativity. Contbr. articles to profl. jours. Mem. Swiss Phys. Soc. Avocations: choir, chess. Home: Rue des Vignes 15, CH-2822 Courroux, Jura Switzerland Office: Lycée Cantonal, Place Blarer-de-Wartensee, CH-2900 Porrentruy Switzerland

CHEVALIER, ROGER ALAN, astronomy educator, consultant; b. Rome, Sept. 26, 1949; came to U.S., 1962; s. Frank Charles and Marion Helen (Janhke) C.; m. Margaret Mary With, July 27, 1974.; children: Chase Arthur, Max Toussaint. B.S. in Astronomy, Calif. Inst. Tech., 1970; Ph.D. in Astronomy (Woodrow Wilson and NSF fellow), Princeton U., 1973. Asst. astronomer Kitt Peak Nat. Obs., Tucson, 1973-76, assoc. astronomer, 1976-79; assoc. prof. astronomy U. Va., Charlottesville, 1979-85, prof. astronomy, chmn. dept., 1985-92, W.H. Vanderbilt prof. astronomy, 1990—; dir. Leander McCormick Obs., 1985-92; cons. Lawrence Livermore Nat. Lab., Livermore, Calif., 1981-90. Contbr. numerous research articles to Astrophys. Jour., other astronomy and physics jours. Named Va. Outstanding Scientist, Sci. Mus. Va., 1991; Woodrow Wilson Found. fellow Princeton U., 1970-71, NSF fellow Princeton U., 1970-73. Mem. Am. Astron. Soc. (councilor 1988-91), Internat. Astron. Union, Ill. Sci. Lectr. Assn. (v.p. 1975-85), Sigma Xi. Home: 1891 Westview Rd Charlottesville VA 22903-1632 Office: U Va Leander McCormick Obs PO Box 3818 Charlottesville VA 22903-0818

CHI, HSIN, entomology and ecology educator; b. Hong Kong, China, Dec. 4, 1949; s. Chih and Hsueh-Der (Ma) C.; children: San-Hue, Kan Chi. MS, Chung-Hsing U., Taichung, Taiwan, 1975; PhD, U. Goettingen, Germany, 1980. Assoc. prof. Chung-Hsing U., Taichung, Taiwan, 1980-89; prof. Chung-Hsing U., Taichung, 1990—; rsch. assoc. U. Calif., Berkeley, 1985-86. Contbr. articles to Environ. Entomology, Jour. Econ. Entomology, Bulletin Inst. Zool. Academia Sinica. Lt. Army, 1975-77, Taiwan. Mem. Ecol. Soc. Am., Entomol. Soc. Am., Am. Inst. Biol. Scis., Entomol. Soc. China. Achievements include rsch. in age-stage, two sex life table theory; mass rearing and harvesting theory; timing of control. Office: Chung-Hsing U, PO Box 17-25, Taichung 40098, Taiwan

CHI, VERNON L., computer science educator, administrator; b. Tieujin, China, Mar. 18, 1940; came to U.S., 1941; s. Hilary Shou-Yû Chi and Emily Green Exner; married; 1 child. BS in Physics, Antioch Coll., 1964. Staff and project engr. Electro Sci. Industries, Portland, Oreg., 1979-81; lectr. computer sci., dir. Microelectronic Systems Lab. dept. computer sci. U. N.C., Chapel Hill, 1981—; advisor, cons. Microelectronics Ctr. N.C., Research Triangle Park, 1981-91. Research includes salphasic distribution of timing signals for the synchronization of physically separated entities, biologic sequence comparative analysis mode, general purpose mech-connected SIMD engine, fuzzy logic inference engine, head mounted displays, 3D graphics supercomputers, salphasic clock planes, tracking systems for head mounted displays, testbeds for Gbit/sec networking; contbr. articles to profl. jours. Recipient Young Investigator award NSF, 1993. Mem. IEEE (sr., computer, circuits and systems soc.), ACM. Office: U NC Computer Sci Dept Microelectronic Systems Lab Sitterson Hall CB3175 Chapel Hill NC 27599-3175

CHIALVO, ARIEL AUGUSTO, chemical engineering research scientist; b. Rafaela, Argentina, Apr. 9, 1955; came to U.S., 1984; s. Hoevel Mireillt and Judith Emilia (Milessi) C.; m. Maria Elida Sosa, Jan. 6, 1984; children: Sebastian, Pablo. BSChemE, U. Nat. Littoral, Santa Fe, 1978; MSChemE, U. Nat. South, Bahia Blanca, Argentina, 1983; DSChemE, Clemson U., 1988. Rsch. assoc. Dept. Chem. Engring., Princeton U., 1988-91; rsch. assoc. Dept. Chem. Engring., U. Va., Charlottesville, 1991-92, rsch. scientist, 1992—. Contbr. articles to profl. jours. including Jour. Chem. Physics, Jour. of Phys. Chemistry, Molecular Physics, Indsl. and Engring. Chemistry, Phys. Rev. A, Fluid Phase Equilibrium, 1992. Fellowship Conicet, 1980-86, NSF, 1986, 92; grantee Pitts. Supercomputer Ctr., 1991—. Mem. AICE, Am. Chem. Soc., Sigma Xi. Office: U Va Dept Chem Engring McCormick Rd Charlottesville VA 22903-2442

CHIANG, ALBERT CHINFA, polymer chemist; b. Pai-ho, Tainan, Taiwan, Jan. 3, 1946; came to U.S., 1973; s. Long and Ping (Su) C.; m. Geraldine Chin, June 4, 1978; 1 child, Scott Jinlong. BS, Nat. Chung-Hsing U., Taichung, Taiwan, 1970; MS, Georgetown U., 1977; PhD, Am. U., 1980. Teaching asst. Georgetown U., Washington, 1974-77, Am. U., Washington, 1977-80; assoc. chemist Pitney Bowes, Stamford, Conn., 1980-81, chemist, 1982-83, staff chemist, 1984-86, sr. chemist, 1987-89, tech. advisor, 1989-92; dir. Mearthane Products, Cranston, R.I., 1992—; mem. Chinese Oversea Scholar, Taipei, Taiwan, 1980—. Mem. adv. bd. Am. Security Coun., Washington, 1984. Dissertation fellow Am. U., 1979. Mem. Am. Chem. Soc. (rubber div. 1987—), Soc. Plastics Engring. (sr. mem.), Photography of Sci. and Engring. Achievements include 5 patents and 5 patents pending; development of processes for preparation of polyphenylacetylene and desulfurization of coal; invention of materials for electrophotographic toners, high solid content emulsion formation, flourescent thermal transfer ribbon formation, new dual-step thermal transfer printing; research in rubber, photopolymers, thermal printing, conducting polymer and high temperature ceramic superconducting material formation, non-impact printing technology and printing materials for postage meter and other mailing system machines. Home: 11 Heath St Mystic CT 06355

CHIANG, CHENG-WEN, internal medicine educator, physician; b. I-Lan, Taiwan, Oct. 24, 1943; s. Are-Jee (Yuh) C.; m. Yang Mei-Yu Chiang, Nov. 8, 1972; children: Chiang Yih-Shien, Chiang Yih-Tsung. MD, Nat. Taiwan U., Taipei, 1971. Intern Nat. Taiwan U. Hosp., 1970-71, resident dept. internal medicine, 1972-75; lectr. Chang Gung Meml. Hosp., Taiwan, 1979-82, assoc. prof., 1982-88, dir. CCU, 1987-92; prof. internal medicine Chang Gung Med. Coll., Taiwan, 1989—, dir. 1st cardiovascular divsn., 1992—; vis. physician Case Western Reserve U., Cleve., 1981. Editor: Jour. of Ultrasound in Medicine of the Republic of China, 1987—, Acta Cardiologica Sinica, 1987—; chief editor: Chang Gung Medical Jour., 1990—. Recipient Cheng-Hsing Med. award Med. Assn. Taiwan, 1979; Internat. scholar Cleve. Clinic, 1982; rsch. fellow Johns Hopkins Hosp., Balt., 1982. Fellow Sci. Coun. Internat. Coll. Angiology, Am. Coll. Chest Physicians; mem. The Soc. Ultrasound in Medicine Republic of China (standing supr. 1992—, chmn. med. consultation com. 1992—), Western Pacific Assn. Critical Care Medicine Taiwan (bd. dirs., chmn. edn. com. 1992—), Republic of China Soc. Cardiology (bd. dirs. chmn. Preventive Cardiology com. 1993—). Avocations: music, bicycling, boating. Office: Chang Gung Med Coll, 199 Tung Hwa N Rd, Taipei 10591, Taiwan

CHIANG, KIN SENG, optical physicist, engineer; b. Zhongshan, Guangdong, People's Republic China, Aug. 18, 1957; d. Arthur and Lai Kam (Lei) Jan; m. Yuan Li Chiang, July 3, 1984; children: Cordelia, Shannon. BEE, U. NSW, Sydney, Australia, 1982, PhD, 1986. Rsch. officer Australian Def. Force Acad., Canberra, Australia, 1986; from rsch. scientist to sr. rsch. scientist div. applied physics Commonwealth Sci. and Indsl. Rsch. Orgn., Sydney, 1986-92; vis. scientist Electrotech. Lab., Tsukuba City, Japan, 1987-88; vis. fellow City Polytechnic Hong Kong, 1992. Contbr. articles to profl. jours., chpt. to book; inventor, patentee optical fiber ultrasonic sensors. Recipient rsch. award for fgn. specialists Govt. of Japan, 1987; rsch. scholar Govt. of Australia, 1982. Mem. Optical Soc. Am., Internat. Soc. for Optical Engring., Australian Optical Soc. Avocations: Chinese literature, translation. Office: CSIRO Div Applied Physics, Bradfield Rd, Lindfield NSW 2070, Australia

CHIANG, RICHARD YI-NING, aerospace engineer; b. Taipei, Taiwan, Jan. 15, 1953; s. Chung Hsin Chang and Yen (Shu) Cheng; m. Carol M. Chiang, July 7, 1981; children: Brian, Evette. BS, Nat. Cheng Kung U., Tainam, Taiwan, 1976; MS, San Diego State U., 1979; PhD, U. So. Calif., 1988. Engr. Sargent Industries, City of Industry, Calif., 1979-80; control analyst Garrett AiResearch, Torrance, Calif., 1980-82, HTL Advanced Tech., Azusa, Calif., 1982-83, Lear Siegler, Santa Monica, Calif., 1983-85; rsch. asst. U. So. Calif., L.A., 1985-88; flight control engr. Northrop Aircraft, El Segundo, Calif., 1989-91; spacecraft control engr. Jet Propulsion Lab., Pasadena, Calif., 1991—; vis. scientist, instr. U. So. Calif., 1989—. Author software packages Robust Control Toolbox, 1988—, Modeling and Control Synthesis Toolbox, 1993—; also numerous papers, reports in field. TRW Rsch. grantee, 1987-88. Mem. AIAA, IEEE. Office: Jet Propulsion Lab MS 198-326 4800 Oak Grove Dr Pasadena CA 91109

CHIANG, TAI-CHANG, physics educator; b. Taipei, Taiwan, Republic of China, Aug. 28, 1949; came to U.S., 1973.; BS, Nat. Taiwan U., Taipei, 1971; PhD, U. Calif., Berkeley, 1978. Postdoctoral assoc. IBM, Yorktown Heights, N.Y., 1978-80; asst. prof. U. Ill. at Urbana-Champaign, Urbana, 1980-84, assoc. prof., 1984-88, prof. physics, 1988—. Contbr. over 100 articles to profl. jours. Named Presdl. Young Investigator NSF, 1984-89; recipient Faculty Devel. award IBM, 1984-85, Xerox Faculty award U. Ill., 1985. Fellow Am. Phys. Soc. Office: U Ill Physics Dept 1110 W Green St Urbana IL 61801

CHIANG, TOM CHUAN-HSIEN, biochemist; b. Apr. 1, 1944. MS, Ill. State U., 1971; PhD, U. Tex. Dallas, 1976. Grad. asst. U. Tex., Dallas, 1971-75, postdoctoral fellow, 1975-78; asst. prof. med. sch. Tex. A&M U., College Station, 1978-81; sect. chief Tex. Agrl. Experiment Sta., College Station, 1981-85; supr., staff scientist Lockheed Engring. & Scis. Co., Las Vegas, 1985-91; project mgr. Lockheed Environ. Svcs. and Techs., Las Vegas, 1992—. Contbr. articles to profl. jours. Grantee NIH, EPA, U.S. Army. Mem. Assn. Ofcl. Analytical Chemists (assoc. referee 1981—). Office: Lockheed Environ Svcs Techs 980 Kelly Johnson Las Vegas NV 89119

CHIBA, KIYOSHI, chemist; b. Muroran, Hokkaido, Japan, Nov. 6, 1946; s. Jiro and Kaoru Chiba; m. Chikako Tajima, May 27, 1978; children: Yu, Atuko. BS, U. Tokyo, 1969, MS, 1971, DEng, 1988. Registered engr. Sr. chemist Cen. Research Labs., Teijin Ltd., Tokyo and Hino, Japan, 1971—; research assoc., MIT, Boston, 1974-75, U. Tokyo, 1979-80. Contbr. articles to profl. jours.; patentee in field. Mem. Optical Soc. Am., Soc. Photo-Optical Instrumentation Engrs. Home: C-204, 2-8-3 Somechi, Chyofu, Tokyo 182, Japan Office: Teijin Ltd, 4-3-2 Asahigaoka, Hino Tokyo 191, Japan

CHIBA, YOSHIHIKO, biology educator; b. Pyongyang, North Korea, Dec. 1, 1931; arrived in Japan, 1945; s. Goro and Ichi (Kikuchi) C.; m. Kazuko Hayasaka, Mar. 27, 1959; children: Akiko, Masako. B, Tohoku U., Sendai, Japan, 1954, M, 1956, D, 1967. Asst. faculty sci. Tohoku U., Sendai, 1963-70; assoc. prof. faculty lit. and sci. Yamaguchi (Japan) U., 1970-75, assoc. prof. faculty sci., 1975-79, prof., 1979—, dir. lib., 1992—; rsch. fellow U. Minn., St. Paul, Mpls., 1969-70; vis. prof. U. Ga., Athens, 1980; invited lectr. faculty sci. Kyoto (Japan) U., 1988, Grad. Sch. Okayama (Japan) U., 1988-93. Author: The Biological Clock, 1975 (prize 1975), Organisms and Time, 1982; editor: Chronobiology, 1978, Handbook of Chronobiology, 1991; mem. editorial bd. Japanese Jour. Biometeorology, 1986—; mem. editorial and adv. bd. Internat. Jour. Chronobiology, 1972-84, Chronobiologia, 1979-87, Chronobiology Internat., 1984-87, Jour. Insect Physiology, 1992—. Mem. univ. coun. Yamaguchi U., 1990—, libr. curator, 1992—. Recipient Prize, Japanese Soc. Zoology, Tokyo, 1987. Mem. Zool. Soc. Japan (councilor 1983-87, 93—), Japanese Soc. Biometeorology, Biol. Rhythms Soc. of Japan (pres. 1992—), Japanese Soc. Comparative Biochemistry and Physioloogy (councilor 1990), Internat. Soc. Chronobiology (bd. dirs.). Office: Biol Inst Faculty Sci, Yamaguchi U, Yamaguchi 753, Japan

CHIBBARO, ANTHONY JOSEPH, environmental and occupational health and safety professional; b. N.Y.C., Sept. 16, 1946; s. Biagio and Gaetana (Miceli) C.; m. Patricia Daly, June 8, 1985; children: Marcella, Gabriella. BS, CCNY, 1974; MS, Hunter Coll., 1985. Rsch. asst. Meml. Sloan-Kettering, N.Y.C., 1974-85; assoc. radiophysicist N.Y. State Dept. Labor, Bklyn., 1985-90; safety dir. Albert Einstein Coll. Medicine, Bronx, N.Y., 1990—. Mem. Am. Conf. Govtl. Indsl. Hygienists, Pub. Health Assn., N.Y.C. Acad. Sci. Achievements include co-development of lectin separation of bone marrow cells for human transplantation. Office: Albert Einstein Coll Medici Chanin 203 1300 Morris Park Ave Bronx NY 10461

CHICZ-DEMET, ALEKSANDRA, science educator, consultant; b. Toronto, Ont., Can., Nov. 21, 1951; came to U.S., 1955.; d. Oleksa and Ksenia (Fersonowicz) Chicz; m. Edward Michael DeMet, Oct. 22, 1983. BA, U. Chgo., 1969-73; PhD, Ill. Inst. Tech., 1977-85. Lab. technician dept. mental health State of Ill., Chgo., 1971, lab. technician dangerous drugs commn., 1980-81; rsch. technician U. Chgo., 1973-76; rsch. assoc. U. Calif., Irvine, 1985-89, asst. adj. prof., 1989-92, asst. prof. in residence, 1992—; rsch. scientist State Devel Rsch. Insts., Costa Mesa, Calif., 1985—; cons. Silliker Labs., Chgo. Heights, Ill., 1982, Abbott Labs., Abbott Park, Ill., 1987, VA Med. Ctr., West L.A., Calif., 1992—. Contbr. articles to profl. jours. Scientific Exchange fellow NSF, Budapest, Hungary, 1977. Mem. AAAS, Soc. for Neuroscience, N.Y. Acad. Sci., West Coast Coll. Biol. Psychiatry. Office: Psychiatry and Human Behavior Irvine CA 92717

CHIEN, LARRY See CHIEN, LUNG-SIAEN

CHIEN, LUNG-SIAEN (LARRY CHIEN), mechanical engineer; b. Keelung, Taiwan, May 8, 1955; came to U.S., 1979; s. Ching-Yuan and Chu-Chu (Chen) C.; m. Jenny C.C. Lu, June 15, 1991. MS, Stanford U., 1980; PhD, Purdue U., 1989. Registered profl. engr., Calif. Mech. engr. ITT, San Jose, Calif., 1983-85; instr. Purdue U., West Lafayette, Ind., 1989; rsch. scientist NASA Langley Rsch. Ctr., Hampton, Va., 1990, Air Force Inst. Tech., Wright-Patterson AFB, Ohio, 1990-91; rsch. assoc. U.S. Army Ballistic Rsch. Lab., Aberdeen Proving Ground, Md., 1992—; cons. Armstrong Air Force Med. Rsch. Lab., Wright-Patterson AFB, 1991—. Reviewer AIAA Jour., 1990—, Composites Engring. Jour., 1991—; contbr. chpt. to book and articles to AIAA Jour., Internat. Jour. Nonlinear Mechanics, Composites Engring., Jour. Applied Mechanics. Recipient David Ross fellowship Purdue U., West Lafayette, 1986, Army Post Doctoral fellowship U.S. Army Ballistic Rsch. Lab., Aberdeen Proving Ground, 1992. Mem. AIAA (sr. mem.), ASME, ASCE (tech. com. 1991—), Am. Acad. Mechanics, Sigma Gamma Tau, Tau Beta Pi. Home: 345 Sheridan Ave Apt # 216 Palo Alto CA 94306-2035 Office: SLCBR-IB-M US Army Ballistic Rsch Lab Aberdeen Proving Ground MD 21005-5066

CHIEN, SHU, physiology and bioengineering educator; b. Beijing, June 23, 1931; came to U.S., 1954; s. Shih-liang and Wan-tu (Chang) C.; m. Kuang-Chung Hu, Apr. 7, 1957; children: May Chien Busch, Ann Chien Guider-a. MB, Nat. Taiwan U., Taipei, Republic of China, 1953; PhD, Columbia U., 1957. Instr. physiology Columbia U. Coll. Physicians & Surgeons, N.Y.C., 1956-58, asst. prof. physiology, 1958-64, assoc. prof. physiology, 1964-69, prof. physiology, 1969-88, dir. div. circulatory physiology and biophysics, 1973-88; dir. Inst. Biomed. Scis. Academia Sinica, Taipei, 1987-88; prof. bioengring and medicine U. Calif.-San Diego, La Jolla, 1988—, bioengring. group coord., 1989—, dir. Inst. Biomed. Engring., 1991—; chmn. adv. com. Am. Bur. for Med. Advancement in China, N.Y.C., 1991—, Inst. Biomed. Scis., Academia Sinica, Taipei, 1991—, Nat. Health Rsch. Inst., Taipei, 1991—. Editor: Vascular Endothelium in Health and Disease, 1988, Molecular Biology in Physiology, 1989, Molecular Biology of Cardiovascular System, 1990; co-editor: Nuclear Magnetic Resonance in Biology and Medicine, 1986, Handbook of Bioengineering, 1986, Clinical Hemorheology, Applications in Cardiovascular and Hematological Disease, Diabetes, Surgery and Gynecology, 1987, Fibrinogen, Thrombosis, Coagulation and Fibrinolysis, 1990, Biochemical and Structural Dynamics of the Cell Nucleus, 1990, others; contbr. more than 300 sci. articles on physiology, bioengring. and related biomed. rsch. to profl. jours. Recipient Fahraeus award European Soc. for Clin. Haemorheology, London, 1981, Melville award ASME, 1990, Zweifach award World Congress of Microcirculation, Louisville, 1991, Spl. Creativity Grant award NSF, 1985-88, Merit Grant award NIH, 1989-99. Mem. Am. Physiol. Soc. (pres. 1990-91), Biomed. Engring. Soc. (sr.), Internat. Soc. Biorheology (v.p. 1983-89), Microcirculatory Soc. (pres. 1990-91, Landis award 1983), N.Am. Soc. Biorheology (chmn. steering com. 1985-86), Fedn. Am. Socs. for Exptl. Biology (pres. 1992-93). Achievements include elucidation of the mechanism of red cell aggregation in terms of energy balance at cell surface; demonstration of the role of endothelial cell turnover in the transport of protein molecules into the artery wall; research on the molecular basis and physiological implications of blood cell deformability. Office: U Calif San Diego Inst Biomed Engring 9500 Gilman Dr La Jolla CA 92093-0412

CHIEN, YEW-HU, aquaculture educator; b. Taipei, Taiwan, Republic of China, Oct. 22, 1951; s. In-Leih and Chin (Lin) C. BS, Nat. Taiwan U., Taipei, 1973; MS, La. State U., 1978, M. Ap. Stats., 1980, PhD, 1980. Rsch. asst. La. State U., Baton Rouge, 1975-80, fisheries statistician, 1980-82; data processing mgr. Louis J. Capzolli and Assoc., Co., Baton Rouge, 1982-83; vis;. assoc. prof. Nat. Taiwan Ocean U., Keelung, Republic of China, 1983-85, assoc. prof., 1985-88, prof., 1988—, chmn., 1992—. Assoc. editor Jour. World Aquaculture Soc., Asian Marine Biology, China Fisheries Monthly. Mem. Asian Fishers Soc. (sec. Taipei br. 1986), World Aquaculture Soc. (bd. dirs.), Soc. Environ. Toxicology & Chemistry, Fisheries Soc. Taiwan (assoc. editor jour.), China Fisheries Assn., Soc. of Stream. Home: No 8 Ln 23 Chien-Kuo N Rd, Sect 1, Taipei Taiwan Office: Nat Taiwan Ocean U Dept Aquaculture, No 2 Pei-Ning Rd, Keelung Taiwan

CHIEN, YIE WEN, pharmaceutics educator; b. Keelung, Taiwan, Oct. 20, 1938; came to U.S., 1967; s. Chou-lin and Ai-wen (Chen) C.; m. Margaret C. Chuang, Apr. 23, 1964; children: Steven, Linda. BSc in Pharmacy, Kaohsiung Med. Sch., Taiwan, 1963; PhD in Pharmaceutics, Ohio State U., 1972. Group leader, scientist G.D. Searle and Co., Skokie, Ill., 1972-78; sect. head Endo Lab. The Dupont Co., Garden City, N.Y., 1978-81; prof. pharmaceutics Coll. Pharmacy Rutgers U., Piscataway, N.J., 1981-86, prof. II, 1986-89, dpet. chmn., 1982-88, Parke-Davis chair, 1989—; dir., founder Controlled Drug-Delivery Rsch. Ctr., Piscataway, 1982—; cons. WHO, UN, 1988—; mem. editorial bd. several sci. jours., U.S., Spain, France, 1983—. Author: Novel Drug Delivery Systems, 1982, 2d rev. edit., 1992, Nasal Systemic Drug Delivery, 1989; editor: Transdermal Controlled Systemic Medications, 1987; contbr. 248 articles to profl. jours. Recipient Sci. and Tech. Achievement award, Bd. of Trustees award for rsch. excellence, Disting. lectureship Parke-Davis Endowed chair. Fellow Acad. Pharm. Sci./Am. Pharm. Assn., Am. Assn. Pharm. Scientists, Am. Inst. Chemists; mem. AAAS, Controlled Release Soc. (bd. dirs. 1984-87), Acad. Pharm. Rsch. and Sci., Parenteral Drug Assn., Fedn. Internationale Pharmaceutique, Am. Chem. Soc. (polymeric materials sci. and engring. div.), Am. Assn. Coll. of Pharmacy, Am. Found. for Pharm. Edn., N.Y. Acad. Sci., Sigma Xi, Rho Chi. Achievements include patents for microsealed pharm. delivery devices; for injectable metronidazole compositions; for transdermal fertility control system and process; for transdermal absorption dosage unit for estradiol and other estrogenic steroids; for transdermal estrogen/progestin dosage unit, system and process; transdermal iontotherapeutic system, dosage unit and process. Office: Rutgers U Controlled Drug Delivery Ct 41 Gordon Rd Ste D Piscataway NJ 08854

CHIERI, PERICLE ADRIANO CARLO, educator, consulting mechanical and aeronautical engineer, naval architect; b. Mokanshan, Chekiang, China, Sept. 6, 1905; came to U.S., 1938, naturalized, 1952; s. Virginio and Luisa (Fabbri) C.; m. Helen Etheredge, Aug. 1, 1938. Dr Naval Engring., U. Genoa, Italy, 1927; ME, U. Naples, Italy, 1927; Dr Aero. Engring., U. Rome, 1928. Registered profl. engr., Italy, N.J., La., S.C. chartered engr., U.K. Naval architect. mech. engr. research and exptl. divs., submarines and internal combustion engines Italian Navy, Spezia, 1929-31; naval architect, marine supt. Navigazione Libera Triestina Shipping Corp., Libera Lines, Trieste, Italy, 1931-32, Genoa, 1933-35; aero. engr., tech. adviser Chinese Govt. commn. aero. affairs Nat. Govt. Republic of China, Nanchang and Loyang, 1935-37; engring. exec., dir. aircraft materials test lab., supt. factory's tech. vocational instrn. SINAW Nat. Aircraft Works, Nanchang, Kiangsi, China, 1937-39; aero. engr. FIAT aircraft factory, Turin, Italy, 1939; aero. engr. and tech. sec. Office: Air Attache, Italian Embassy, Washington, 1939-41; prof. aero. engring. Tri-State Coll., Angola, Ind., 1942; aero. engr., helicopter design Aero. Products, Inc., Detroit, 1943-44; sr. aero. engr. ERCO Engring. & Research Corp., Riverdale, Md., 1944-46; assoc. prof. mech. engring. U. Toledo, 1946-47; assoc. prof. mech. engring., faculty grad div. Newark (N.J.) Coll. Engring., 1947-52; prof., head dept. mech. engring. U. Southwestern La., Lafayette, La., 1952-72; cons. engr. Lafayette, 1972—; research engr., adv. devel. sect., aviation gas turbine div. Westinghouse Electric Corp., South Philadelphia, Pa., 1953; exec. dir. Council on Environment, Lafayette, 1975—. Instr. water safety ARC Nat. Aquatic Schs., summers 1958-67; Bd. dirs. Lafayette Parish chpt. ARC. Fellow Royal Instn. Naval Architects London (life); assoc. fellow Am. Inst. Aeronautics and Astronautics; mem. Soc. Naval Architects and Marine Engrs. (life mem.), AAAS, AAUP (emeritus), Am. Soc. Engring. Edn. (life), Am. Soc. M.E., Soc. Automotive Engrs., Instrument Soc. Am., Soc. Exptl. Stress Analysis, Nat. Soc. Profl. Engrs., N.Y. Acad. Scis., La. Engring. Soc., La. Tchrs. Assn., AAHPER, La. Acad. Scis., Commodore Longfellow Soc., Cons. Engrs. Council La., Phi Kappa Phi., Pi Tau Sigma (hon.). Home: 142 Oakcrest Dr Lafayette LA 70503-2726 Office: Oil Ctr PO Box 52923 Oil Ctr Sta Lafayette LA 70505

CHIERICI, GIAN LUIGI, petroleum engineer, educator; b. Parma, Italy, Dec. 1, 1926; s. Giuseppe and Anna (Ferrari) C.; D.Chemistry, U. Parma, 1949; D.Chem. Engring., U. Padua, 1955; D.Physics, U. Parma, 1965; Ph.D. in Petroleum Engring., U. Bologna, 1963; m. Graziella Antonioli, Jan. 7, 1956; 1 son, Marcello. Research chemist Carlo Erba, 1949-51; head thermodynamics lab. AGIP SpA, 1951-55, head reservoir physics dept., 1955-70, head phys. chemistry dept., 1970-78, v.p. fields devel. and prodn., 1978-82, v.p. petroleum engring., 1982-87, chmn. research and devel. com., 1984-91; chmn. working group on exploration and exploitation of petroleum reservoirs ENI Coll. for Rsch., 1991—; assoc. prof. petroleum engring. U. Bologna, 1961-85, prof. petroleum reservoir engring., 1987—; hon. prof. U. Patagonia (Argentina), 1962—; mem. adv. com. mgmt. projects on geothermal energy EEC, Brussels, 1976-83, cons. to dirs. gen., 1987—; mem. Italian nat. com. for World Petroleum Congresses, 1980-91; mem. permanent council World Petroleum Congresses, 1982-91, mem. sci. program com., 1983-91; chmn. com. for tech. and sci. cooperation Ministry of Oil Industry of USSR and Agip S.p.A., 1980-91; chmn. steering com. for research on petroleum recovery Norsk Agip/Statoil/IKU, 1980-85; mem. sci. program com. European Symposium on Enhanced Oil Recovery 1982—; mem. sci. com. Rogalandsforskning U., Stavanger, Norway, 1987-91; mem. sci. com. Osservatorio Geofisico Sperimentale, Trieste, Italy, 1993—. Mem. Soc. Petroleum Engrs., AIME. Author: Volumetric and Phase Behaviour of Hydrocarbon Reservoir Fluids, 1962; Enhanced Oil Recovery-A State-of-the-Art Review, 1980, Principles of Petroleum Reservoir Engineering, 1989; tech. editor SPE Reservoir Engineering; assoc. editor Jour. Petroleum Science and Engineering; contbr. articles to internat. tech. mags. Home: Via Triulziana 36/A, San Donato Milanese, I-20097 Milan Italy Office: Inst Earth Scis, Faculty Engring, Viale del Risorgimento 2, Bologna I-40136, Italy

CHIGNELL, COLIN FRANCIS, pharmacologist; b. London, Eng., Apr. 7, 1938; came to U.S., 1962, naturalized, 1969; s. Francis George and Elsie Mary (Lee) C.; B. Pharmacy with honors, U. London, 1959, Ph.D., 1962; m. Anke K. Chignell, Nov. 19, 1966. Vis. fellow Lab. of Chemistry, Nat. Inst. Arthritis and Metabolic Diseases, NIH, Bethesda, Md., 1962-64; vis. asso. Lab. of Chem. Pharmacology, Nat. Heart, Lung and Blood Inst., NIH, Bethesda, 1964-69, research pharmacologist 1969-74, research pharmacologist Pulmonary Br., 1974-77; chief Lab. of Molecular Biophysics, Nat. Inst. Environ. Health Scis., NIH, Research Triangle Park, N.C., 1977—; adj. prof. pharmacology U. N.C., Chapel Hill, 1978—. Recipient J. J. Abel award Am. Soc. Pharmacology and Exptl. Therapeutics, 1973. Mem. Am. Soc. Pharmacology and Exptl. Therapeutics, Am. Soc. Biol. Chemists, Biophys. Soc., Am. Soc. Photobiology, AAAS. Lutheran. Editor: Methods in Pharmacology, Vol. 2, 1972; mng. editor Jour. of Biochem. and Biophys. Methods, 1979—. Contbr. articles to various sci. and tech. jours. Home: 128 Bruce Dr Cary NC 27511-6304

CHIH, CHUNG-YING, physicist, consultant; b. Yuki, Fukien, China, Dec. 11, 1916; s. Lai Sui and Sung-Yee (Lin) C.; BSc, Nat. Tsing Hua U., Peking, China, 1937; PhD, U. Calif., Berkeley, 1954; m. Alice Yuen, Aug. 15, 1955; came to U.S., 1948, naturalized, 1962. Instr. physics Fukien Med. Coll., 1937-40; instr., then asso. prof. Fukien Tchrs. Coll., 1940-44; assoc. prof., then prof. physics Nat. Chi-Nan U., 1944-45; prof. physics Kiang-su Coll., 1945-48; physicist Radiation Lab., U. Calif., Berkeley, 1948-54, summer 1956; mem. faculty Middlebury (Vt.) Coll., 1954-68, prof. physics, 1966-68; sci. cons., Bridgeport, Conn., 1968—. NSF grantee, 1957-60. Mem. Am. Phys. Soc. Address: PO Box 2556 Noble Sta Bridgeport CT 06608

CHILD, FRANK MALCOLM, biologist educator; b. Jersey City, Nov. 30, 1931; s. F. Malcolm and Florence Marie (Lilienthal) C.; m. Julia Carroll Swope, June 19, 1960; children: Malcolm Swope, Alice Hamilton, Rachel Hayward. AB, Amherst Coll., 1953; PhD, U. Calif., Berkeley, 1957. Lectr. U. Calif., Berkeley, 1957; instr., asst. prof. U. Chgo., 1957-65; from assoc. to prof. Trinity Coll., Hartford, Conn., 1965—. Contbr. articles to profl. jours. Fellow AAAS; mem. Corp. Marine Biol. Lab., Phi Beta Kappa, Sigma Xi. Democrat. Home: 187 Griswold Rd Wethersfield CT 06109 Office: Dept Biology Trinity Coll Hartford CT 06106

CHILDERS, NORMAN FRANKLIN, horticulture educator; b. Moscow, Idaho, Oct. 29, 1910; s. Lucius Franklin and Frances M. (Norman) C.; 4 children. BS in Horticulture, U. Mo., 1933, MS in Horticulture (Gregory scholar 1933-34), 1934; PhD in Pomology, Cornell U., 1937. Grad. asst. pomology. Cornell U., 1934-37; asst. prof. horticulture, asst. research specialist Ohio State U. and Ohio Agr. Expt. Sta., 1937-39; asso. in research Ohio Agr. Expt. Sta., 1939-44; asst. dir., sr. plant physiologist fed. expt. sta. U.S. Dept. Agr., Mayaguez, P.R., 1944-47; prof. horticulture, research specialist Rutgers U., New Brunswick, N.J., 1948-81; chmn. dept. Rutgers U., 1948-66, Maurice A. Blake distinguished prof., 1966-81, prof. emeritus, 1981—; adj. prof. U. Fla., Gainesville, 1981—. Author: Arthritis-Childers' Diet to Stop It, 4th edit., 1993; pub., co-author 10 horticulture books; rsch. author in Classic Papers in Horticultural Science, 1989; co-editor: The Peach-World Cultivars to Marketing, 1988; contbr. numerous articles to profl. jours. Councilman, Milltown, N.J., 1953-56. Recipient Best Tchr. award Alpha Zeta, Rutgers U., 1980, award Nat. Peach Coun., 1981, Disting. Svc. to Agr. award N.J. State Bd. Agr., 1982, Internat. Dwarf Fruit Tree award, 1982, Disting. Svc. award to U.S. Agr., Chevron Chem. Co., N.Am. Strawberry Growers award, 1990. Fellow Am. Soc. Hort. Sci. (recipient L.M. Ware Disting. Teaching award 1968), Internat. Acad. Preventive Medicine (Spokesman of Yr. 1979—), Am. Acad. Neur. and Orthopaedic Surgeons; mem. N.J. Hort. Soc. (pres. 1976), Columbus Hort. Soc. (pres. 1941-43), Am. Pomology Soc. (Wilder award 1991), Pa. State Hort. Assn., Fla. State Hort. Soc. (editor Procs. 1988—), N.J. Garden Club (life, Hon. Disting. award 1977). Achievements include discovery of relationship between nightshades (tobacco, tomato, potato, eggplant and peppers) and arthritis, other key nutritional diseases. Office: U Fla Hort Scis Dept Gainesville FL 32611

CHILDERS, RICHARD HERBERT, JR., chemist; b. Indpls., Aug. 27, 1954; s. Richard Herbert Sr. and Mary Louise (Norman) C.; m. Cynthia Ann Dunn. Degree in auto truck tech., Lincoln Tech. Inst., 1975; BS in Chemistry, Marian Coll., 1984. Chemist intern Ind. State Bd. Health, Indpls., 1981-82, chemist, 1985-87; grad. asst. Wayne State U., Detroit, 1987; chemist EMS Labs., Inc., Indpls., 1988, Ind. Dept. Environ. Mgmt., Indpls., 1988—. Active Friends of the Libr. Named Outstanding Chemistry student Am. Chem. Soc., 1984. Mem. Modern Poetry Assn., Acad. Am. Poets. Democrat. Roman Catholic. Office: Ind Dept Environ Mgmt 105 S Meridian St Indianapolis IN 46206

CHILDS, SADIE L., mathematician, chemist, patent agent; b. Winston Salem, N.C., May 18, 1952; d. Robert Hubert and Pollie (James) C. BS, Howard U., 1974; MS, U. D.C., 1993. Patent examiner U.S. Patent and Trademark Office, Washington, 1974-90; cons. Washington, 1990—. Trustee Sargent Meml. Presbyn. Ch., Washington, 1978-84, deacon, 1984—; vol. Children's Hosp., Washington. Mem. Nat. Coun. Negro Women (svc. award 1983, educator 1982—), Phi Delta Kappa (educator 1989—), Delta Sigma Theta Sorority Inc. Avocations: piano, singing, modern jazz dancing, travel, tutoring in math. Home and Office: 1814 Bryant St NE Washington DC 20018

CHILINGARIAN, GEORGE VAROS, petroleum and civil engineering educator; b. Tbilisi, Ga., July 22, 1929; s. Varos and Klavdia (Gorchakova) C.; m. Yelba Maria, June 12, 1953; children: Modesto George, Mark Steven, Eleanore Elizabeth. B.E. in Petroleum Engring., U. So. Calif., 1949, M.S., 1950, Ph.D. in Geology and Petroleum Engring., 1954; hon. degree, Pepperdine U., 1976, Clayton U., Pacific States U., Kensintgon U., Pacific Western U. Profl. geologist, Calif.; cert. Am. Assn. Petroleum Geologists. Chief Petroleum and Chems. Labs. Wright-Patterson AFB, Dayton, Ohio, 1954-56; prof. petroleum engring. U. So. Calif., L.A., 1956-90, prof. civil and petroleum engring., 1990—. Author 35 books in field; contbr. over 300 articles to profl. jours. Recipient numerous awards and medals; named to Sci. Hall of Fame. Fellow Geol. Soc. Am., Am. Chem. Soc.; mem. Soc. Petroleum Engrs. of AIME, Am. Assn. Petroleum Geologists, Soc. Econ. Paleontologists and Mineralogists, AAUP, Am. Soc. Engring. Edn., Calif. Acad. Sci., N.Y. Acad. Sci., Russian Acad. Scis., Sigma Xi, Phi Kappa Phi, Tau Beta Pi, Pi Epsilon Tau. Office: Univ So Calif Dept Civil Engring Los Angeles CA 90089-1211

CHILTON, MARY-DELL MATCHETT, chemical company executive; b. Indpls., Feb. 2, 1939; d. William Elliot and Mary Dell (Hayes) Matchett; m. William Scott Chilton, July 9, 1966; children—Andrew Scott, Mark Hayes. B.S. in Chemistry, U. Ill., 1960, Ph.D. in Chemistry, 1967; Dr. honoris causa, U. Louvain, Belgium, 1983. Research asst. prof. U. Wash., Seattle, 1972-77, research assoc. prof., 1977-79; assoc. prof. Washington U., St. Louis, 1979-83; exec. dir. agrl. biotech CIBA-Geigy Corp., Research Triangle Park, N.C., 1983-91, v.p. agrl. biotech, 1991—; adj. prof. genetics N.C. State U., Raleigh, 1983—; adj. prof. biology Washington U., 1983—; Gov. appointee N.C. Bd. Sci. and Tech. Mem. editorial bd. Proceedings of the NAS; contbr. articles to profl. jours. Recipient of Rank Prize for Nutrition, 1987. Mem. NAS (coun. 1988—). Office: CIBA-Geigy Corp Biotech Facility PO Box 12257 Durham NC 27709-2257*

CHILTON, WILLIAM DAVID, architect; b. Tulsa, Jan. 4, 1954; s. Horace Thomas Jr. and Betty Jane (Gray) C.; m. Laura Ann Johnson, Aug. 22, 1981. BA in Architecture, Iowa State U., 1976; MArch, U. Minn., 1980. Registered architect, Minn. Designer CDG, Tulsa, 1976; assoc. architect Olson & Coffey Architects, Tulsa, 1977-78, Leonard Parker Assocs., Mpls., 1980-81; sr. architect Conoco, Inc., Ponca City, Okla., 1981-89; v.p., project dir. Ellerbe Becket, Inc., Mpls., 1989—. Prin. works include Milne Point (Alaska) Ops. Complex (award Best of Engring. News Record 1986, Excellence in Architecture award North Cen. Okla. chpt. AIA 1987, Honorable Mention Builder mag. 1985), Conoco Corp. Offices, Wilimington, Del. (Excellence in Architecture award North Cen. Okla. chpt. AIA 1987). Mem. Minn. Soc. AIA (sec. North Ctrl. Okla. chpt. 1986, v.p. 1987, pres. 1988, bd. dirs. 1986-88, bd. dirs. Okla. Coun. 1987-88), Minn. Internat. Ctr., Leadership Mpls., Interlachen Country Club (Edina, Minn.). Lutheran. Avocations: fly fishing, golf, tennis, reading, music. Home: 101 Maple Hill Rd Hopkins MN 55343-8544 Office: Ellerbe Becket Inc 800 LaSalle Ave Minneapolis MN 55402-2014

CHIN, ALAND KWANG-YU, physicist; b. Canton, Republic of China, May 7, 1950; came to U.S., 1952; s. George See-Ng and Liawah (Gee) C.; m. Virginia Sook-Ping Wong, Sept. 3, 1978; children: Victoria Hao-Yun, Richard Hee-Yang. BA, Brandeis U., 1972; PhD, Cornell U., 1977. Sr. engr. Honeywell Electro-Optics Ctr., Lexington, Mass., 1977-78; mem. tech. staff AT & T Bell Labs., Murray Hill, N.J., 1978-84; supr. laser fabrication and process devel. AT&T Bell Labs., Murray Hill, N.J., 1984-85; dir. opto-electronics Polaroid Micro-Electronics Lab., Cambridge, Mass., 1986—. Contbr. more than 70 articles to profl. jours. Mem. IEEE, AAAS, Am. Phys. Soc., Electrochem. Soc., Phi Beta Kappa. Achievements include 12 patents in field. Home: 45 Manomet Rd Sharon MA 02067-2967 Office: Polaroid Micro-Electronics 21 Osborne St Cambridge MA 02139-3500

CHIN, ALEXANDER FOSTER, electronics educator; b. Moneague, St. Ann, Jamaica, Dec. 12, 1937; s. Humphrey and Betty (Chen) C.; m. Barbara Kittner, May 18, 1974; children: Micah, Michelle. BS, Okla. State U., 1969, MS, 1974, EdD, 1983. Self-employed grocer Jamaica, W.I., 1955-62; salesman Singer Sewing Machine Co., Jamaica, 1962-64; with Bell Fibre Co., Marion, Ind., 1964-68; electronics engr. Electronic Engring. Co. of Calif., Santa Ana, 1970-71; assoc. prof. electronics Tulsa Jr. Coll., 1971—; adj. prof. Okla. State U., Tulsa, 1988-89. Author: Electronic Instrument & Measurements, 1983; author numerous tech. lab. manuals. Vol. pub. sch. tchr. Monroe Middle Sch., Tulsa, 1990—. NSF grantee, 1971; Am. Inst. Fgn. Study ednl. awardee, 1981, 86, 91, Work award, Tulsa Jr. Coll., 1984. Mem. Okla. Tech. Soc. (bd. dirs. 1971—), Am. Tech. Edn. Assn., Phi Delta Kappa. Plymouth Brethren Ch. Avocations: reading, music, Bible study. Office: Tulsa Jr Coll NE 3727 E Apache St Tulsa OK 74115-3151

CHIN, DER-TAU, chemical engineer, educator; b. Zhejiang, China, Sept. 14, 1939; came to U.S., 1963, naturalized, 1977; s. Tsu-Kang and Shou-Chen (Chen) C.; B.S. in Chem. Engring., Chungyuan Coll. Sci. and Engring., 1962; M.S. in Chem. Engring., Tufts U., 1965; Ph.D. in Chem. Engring., U. Pa., 1969; m. Lorna Fe Gencianeo, July 17, 1971; children—Janet G., Lynn G. Plant engr. Lungyen Sugar Factory, 1962-63; sci. programmer U.S. Air Force Cambridge (Mass.) Research Lab., Lexington, Mass., 1965; sr. research engr. research labs. Gen. Motors Corp., Warren, Mich., 1969-75; prof. Clarkson U., Potsdam, N.Y., 1975—; vis. scientist Brookhaven Nat. Lab., Upton, N.Y., summers 1977, 80, U.S. Army Belvoir Research Devel. Ctr., Ft. Belvoir, Va., summer 1985, U.S. Army Electronics Tech. and Devices Lab., Ft. Mammouth, N.J., summer, 1986; vis. prof. U. Calif., Berkeley, 1981, Swiss Fed. Inst. Tech., Zurich, 1981, Nat. U. Singapore, 1982, 87, Nat. Tsing Hua UNI, 1989—; cons. Centro de Pesquisas do Energia Electrica, Rio de Janiero, Brazil, summer 1979, Los Alamos Nat. Lab., 1981—, Hooker Chem. Devel. Center, Niagara Falls, N.Y., 1981—, Inst. Hydrogen Studies, U. Toronto, 1983—, St. Joe Minerals Corp., Monaca, Pa., 1983. Mem. Electrochem. Soc. (Young Authors award 1971), Am. Inst. Chem. Engrs., Am. Electroplaters Soc., Am. Chem. Soc., Inst. Colloid and Surface Sci. Office: Clarkson U Box 5705 Potsdam NY 13699-5705

CHIN, HONG WOO, oncologist, educator, researcher; b. Seoul, Korea, May 14, 1935; came to U.S., 1974; s. Jik H. and Woon K. (Park) C.; m. Soo J. Chung, Dec. 27, 1965; children: Richard Y., Helen H., KiSik. MD, Seoul Nat. U., 1962, PhD, 1974. Diplomate Am. Bd. Radiology; cert. Korean bd. internal medicine. Resident in radiation oncology Royal Victoria Hosp., Montreal (Que., Can.) Gen. Hosp., 1975-79; asst. prof. U. Ky., Lexington, 1979-86; assoc. dir. Radiarium Found., Overland Park, Kans., 1987-88; clin. prof. radiology U. Mo., Kansas City, 1987-91; chief radiation oncology Va. Med. Ctr., Shreveport, La., 1988; assoc. prof. La. State U., Shreveport, 1988; prof. and dir. radiation oncology Creighton U. Sch. Medicine, Omaha, 1988-90; dir. dept. radiation oncology Creighton U. Cancer Ctr., Omaha, 1988-90; chief radiation oncology Overton Brooks VA Med. Ctr., Shreveport, La., 1990—; prof. La. State U. Med. Ctr., Shreveport. Author monographs. Lt. comdr. USN, 1967-70. Mem. Pan Am. Med. Assn. (mem. coun. 1984—), AMA, Am. Coll. Radiology, Am. Soc. Therapeutic Radiology and Oncology, Radiation Rsch. Soc., Am. Biograph Assn. (rsch. bd. advisors 1988), Internat. Platform Assn. Roman Catholic.

CHIN, JAMES KEE-HONG, surgeon; b. Rangoon, Burma, Sept. 2, 1934; arrived in Hong Kong, 1962; s. Swe-Htyan and Soo-Sin (Cho) C.; m. Julia Yu Siu Yang, May 5, 1964; children: Victor, Sammy, Angellina. ISc, U. Rangoon, 1959; MB, BS, Rangoon Med. Coll., 1961. Cert. London Med. Coun. House physician and surgeon Rangoon Gen. Hosp., 1961-62; med. officer Tung Wah Groups of hosp., Hong Kong, 1962-64; rsch. fellow U. Southern Calif., L.A., 1964-65; prin. James K.H. Chin Clins., Hong Kong, 1965—. Durham fellow Am. Cancer Soc., 1965. Mem. Hong Kong Med. Assn. (life), Hong Kong Coll. Gen. Practitioners, Automobile Assn., Asia Gun Club (life mem.), Soc. Arts and Antiques. Avocations: swimming, gerneral sports, shooting sports, travelling. Home: Flat 3 Grand Ct No 135, Kadoorie Ave, Kowloon Hong Kong Office: 301 Nathan Rd, Champion Bldg, Ste 907, Kowloon Hong Kong

CHIN, NEEOO WONG, reproductive endocrinologist; b. Hong Kong, Nov. 27, 1955; came to U.S., 1958; s. Bing Leong and Din Sui (Gee) C.; m. Shelly Loraine Crumrine, June 25, 1977; children: Jason Lei, Taryn Mae. BA, U. Cin., 1977; MD, Ohio State U., 1981. Diplomate Am. Bd. Ob-Gyn. Resident Duke U. Med. Ctr., Durham, N.C., 1981-84; chief resident Duke U. Med. Ctr., Durham, 1984-85; fellow Ohio State U. Coll. Medicine, Columbus, Ohio, 1985-87; teaching staff Good Samaritan Hosp., Cin., 1987—; clin. asst. prof. U. Cin. Med. Ctr., 1987—; dir. assisted reproductive techs. The Christ Hosp., Cin., 1992—; mem. High Sch. for the Health Profl. subcom., Cin., 1989—. Author: (with others) Current Therapy in Obstetrics, 1988; contbr. articles to profl. jours. Named to Honorable Order of Ky. Cols., Gov. Martha Collins of Ky., 1987. Fellow Am. Coll. Ob-Gyn.; mem. AAAS, Am. Fertility Soc., Soc. Assisted Reproductive Tech., Soc. for Immunology Repro., Cin. Ob-Gyn. Soc. (med. malpractice com. 1989—), Acad. Medicine Cin. Avocations: tennis, karate. Office: The Christ Hosp 2123 Auburn Ave Ste 044 Cincinnati OH 45219

CHIN, ROBERT ALLEN, engineering graphics educator; b. San Francisco, Oct. 3, 1950; s. Suey Hey and Stella (Yee) C.; m. Susan Curtis Fleming, June 18, 1976. AAS, Community Coll. Air Force, 1982; BA, U. No. Colo., 1974; MA in Edn., Ball State U., 1975; PhD, U. Md., 1986. Cert. sr. indstrl. technologist. Grad. teaching asst. Ball State U., Muncie, Ind., 1974-75; instr. Sioux Falls Sch. Dist., S.D., 1975-79; instr. mech. drawing U. Md., College Park, 1979-86, asst. prof. engring. graphics East Carolina U., Greenville, N.C., 1986-92; assoc. prof. engring. graphics dept. indsl. tech., 1992—; acting chmn. Dept. Constrn. Mgmt., 1989-90; aircraft maintenance officer 113th Logistics Group DCANG, Andrews AFB, Md., 1983—. Jour. reviewer; contbr. articles to profl. jours. Faculty advisor Disabled Student Alliance, U. Md., 1982-86. Served with USAF, 1968-72, Air N.G., 1977—. Decorated Master Aircraft Munitions and Maintenance badge USAF., Air Res. Forces Meritorious Svc. medal, Nat. Def. Svc. medal with Bronze star, Armed Forces Res. medal. Mem. Air Force Assn. (chpt. sec. 1988-92, chpt. pres. 1992—), N.G. Assn. U.S., N.G. Assn. D.C., Nat. Assn. Indsl. Tech., Internat. Tech. Edn. Assn., Am. Soc. Engring Edn., Nat. Assn. Indsl. and Tech. Tchr. Educators, Coun. Tech. Tchr. Edn., Phi Kappa Phi, Epsilon Pi Tau, Iota Lambda Sigma (v.p. Nu chpt. 1984-85, pres. Nu chpt. 1985-86). Republican. Presbyterian.

CHING, CHAUNCEY TAI KIN, agricultural economics educator; b. Honolulu, July 25, 1940; m. Theodora Lam, July 7, 1962; children: Donna, Cory. AB in Econs., U. Calif., Berkeley, 1962; MS in Agrl. Econs., U. Calif., Davis, 1965, PhD in Agrl. Econ., 1967. Asst. prof. U. N.H., Durham, 1968-72; assoc. prof. U. Nev., Reno, 1972-77, prof., head div. agrl. and resource econs., 1977-80; prof., chmn. dept. agrl. and resource econs. U. Hawaii, Honolulu, 1980-84, prof. agrl. econs., 1992—, dir. Hawaii Inst. Tropical Agr. and Human Resources, 1984-92. Recipient Charles H. Seurferle award U. Nev., Reno, 1977. Office: Hawaii Inst Tropical Agr 3050 Maile Way # 202 Honolulu HI 96822

CHING, DANIEL GERALD, civil engineer; b. Honolulu, Dec. 17, 1948; s. Daniel K.F. and Ethel (Lo) C.; m. Mei Yung Chen, Mar. 15, 1976; 1 child, Karen Denise. BSCE, U. Denver, 1973; MCE, Cornell U., 1974. Registered profl. engr. Hawaii, Guam. Estimator Swinerton and Walberg Co., Honolulu, 1974-76; project engr. EE Black, Ltd., Honolulu, 1976; structural engr. Alfred A. Yee and Assocs., Honolulu, 1976-80; project mgr. Blackfield Hawaii Corp., Honolulu, 1980-82; pres. Daniel G. Ching and Assocs., Honolulu, 1982—. 1st lt. U.S. Army, 1968-71, Vietnam. Mem. ASCE, Am. Concrete Inst., Bldg. Industry Assn. Episcopalian. Home: 906 Kahena St Honolulu HI 96825 Office: Daniel G Ching & Assocs Ste 2309 1188 Bishop St Honolulu HI 96813

CHING, WAI YIM, physics educator, researcher; b. Shaoshing, China, Oct. 18, 1945; came to U.S., 1969; s. Di-Son and Hung-Wong (Sung) C.; m. Mon Yin Lung, Dec. 27, 1975; children: Tianyu, Kunyu. BSc, U. Hong Kong, 1969; MS, La. State U., 1971, PhD, 1974. Rsch. assoc., lectr. U. Wis., Madison, 1974-78; asst. prof. U. Mo., Kansas City, 1978-81, assoc. prof., 1981-84, prof. physics, 1984-88, curators' prof., 1988—, chmn. physics dept., 1990—; cons. Argonne (Ill.) Nat. Lab., 1978-82, vis. scientist, 1985-86; vis. prof. U. Sci. and Tech., Hefei, China, 1983. Contbr. articles to profl. jours. Recipient N.T. Veatch award for disting. rsch., 1985; Trustee fellow U. Mo., 1984, 90. Mem. AAAS, Am. Phys. Soc., Am. Ceramic Soc., Am. Vacuum Soc., Materials Rsch. Soc., Sigma Xi. Achievements include the study of theoretical dondensed matter physics and materials sciences; electronic, magnetic, optical, dynamical structural and superconducting properties of ordered and disordered solids. Home: 2809 W 119 St Leawood KS 66209 Office: U Mo Physics Dept 1110 E 48th St Kansas City MO 64110

CHINN, REX ARLYN, chemist; b. Bosworth, Mo., Apr. 5, 1935; s. Loren Herbert and Lima (Stanton) C.; m. Wanda June Williams, May 31, 1959; children: Timothy Michael, Sharon Rose Chinn-Heritch, Jonathan Daniel. BS in Chemistry, S.W. Mo. State Coll., 1961; grad., Cleve. Inst. Electronics. Lic. Bapt. minister. Rsch. asst. U. Mo. Med. Ctr., Columbia, 1961-65, William S. Merrell Co., Cin., 1965-67; lab. supr. U.S. Indsl. Chem. Co., Rsch. div., Cin., 1967-72; mgr. quality assurance Cloudsley Co., Cin., 1972-74; dir. tech. affairs Woodson Tenent Labs., Memphis, 1974-77; quality engr. Nat. Ind. for the Blind, Earth City, Mo., 1977—; owner/mgr. The Master's Image, Maryland Hts., Mo., 1987—; free lance field prodns. KNLC, Channel 24, St. Louis, 1987—; video cons. Contbr. articles to profl. jours; producer/dir.: More Than a Fighting Chance, 1989. With U.S. Army, 1954-56. Mem. Internat. Platform Assn. Republican. Avocations: art, photography, electronics, motorcycling, guitar. Home and Office: The Masters Image 12032 Wesford Dr Maryland Heights MO 63043

CHINNASWAMY, RANGAN, cereal chemist; b. Coimbatore, Tamil Nadu, India, May 25, 1957; came to U.S., 1986; s. Rangaswamy and Valliammal Gounder; m. Chitra Kuppuswamy, Feb. 8, 1988; 1 child, Meera. BS, Madras (India) U., 1978, MS, 1980; PhD, UN U., Mysore, India, 1985. Asst. prof. Gandhigram Internat. Inst., India, 1980; jr. rsch. fellow CFTRI program UN U., 1980-82, sr. rsch. fellow CFTRI program, 1982-85; rsch. scientist CFTRI, Mysore, 1985-86; postdoctoral fellow U. Nebr., Lincoln, 1986-89, asst. prof. chemistry, 1989-93, asst. prof. food sci. and tech. and biol. systems engring., 1989-93; vis. scientist Inst. Nat. de la Recherche Agronomique, France, 1992-93; prin. scientist Midwest Grain Products, Inc. Contbr. articles to profl. publs., chpts. to books; editor Food Structure. Grantee USDA, United Soybean Bd., State Commodity Bds., 1988-92; Scanning Microscopy Internat. Presdl. scholar, 1992; recipient Gardners Best Rsch. Paper award Indian Assn. Food Scientists, 1987. Mem. Am. Assn. Cereal Chemists, Inst. Food Technologists, Am. Soc. Agrl. Engrs. (assoc.), Am. Chem. Soc. Achievements include design and study of macromolecular changes in starch, protein polymers doing thermal processing and extrusion cooking; patent pending on biodegradable polymers. Office: Midwest Grain Products Inc 1300 Main St Atchinson KS 66002

CHIOU, GEORGE CHUNG-YIH, pharmacologist, educator; b. Taoyuan, Taiwan, July 11, 1934; came to U.S., 1964, naturalized, 1973; s. Chang and Mei (Wei) C.; m. Tricia Ten-Sian Cheng, Sept. 23, 1961; children: Linda Y., Faye Y. BS in Pharmacy, Nat. Taiwan U., 1957, MS in Pharmacology, 1960; PhD, Vanderbilt U., 1967. Instr. pharmacology China Med. Coll., Taiwan, 1962-64; rsch. assoc. Vanderbilt U., 1967-68; research assoc. U. Iowa, 1968-69; from asst. prof. to prof. pharmacology and therapeutics U. Fla., Gainesville, 1969-78; prof. pharmacology, head dept. med. pharmacology and toxicology Tex. A&M U., College Station, 1978—, co-dir. Inst. Molecular Pathogenesis and Therapeutics, 1984-93, dir. Inst. Ocular Pharmacology, 1984—, asst. dean medicine, 1985, assoc. dean medicine, 1987-90, prof. ophthalmology, 1987—; adj. prof. biotech. Baylor Coll. Medicine, Houston, 1986—, adj. prof. ophthalmology, 1987—; sr. cons. Houston Biotech. Inc., Woodlands, 1986-89; chmn. sci. adv. bd. Orbon Corp., Palo Alto, Calif., 1989—; vis. scientist NIH, 1975; cons. in field; hon. prof. pharmacology Nanjing (China) Med. Coll., 1987—, China Pharm. U., 1992—, Nanjing Inst. Materia Medica, 1992—. Chief editor Jour. Ocular Pharmacology, 1985—; mem. editorial bd. Internat. Jour. Oriental Medicine, 1987—; editor: Ophthalmic Toxicology, 1992; contbr. articles to revs. in field. 2nd lt. Chinese Air Force, 1961-62. Recipient Health Scis. Achievement award NIH, 1967, Disting. Achievement award Nat. Taiwan U., 1989; named Frank Duckworth eminent scholar U. Fla., 1993; Mead Johnson & Co. fellow, 1964-67; Pfizer scholar, 1955; grantee NIH, Am. Cancer Soc., Cooper Vision Labs., Merck, Sharp & Dohme, Merrell-Dow Pharm. Inc., Houston Biotech., Inc., Barnes-Hind, Inc., The Retina Rsch. Found., Am. Cyanamid/Storz, Orbon Corp. Mem. Am. Soc. Pharmacology and Exptl. Therapeutics, Soc. Exptl. Biology and Medicine, N.Y. Acad. Scis., Assn. Rsch. Vision and Ophthalmology, Assn. Med. Sch. Pharmacology, Sigma Xi. Republican. Buddhist. Office: Tex A&M U Coll Medicine Dept Med Pharmacology College of Medicine College Station TX 77843

CHIOU, WIN LOUNG, pharmacokinetics educator, director; b. Hsinchu, Taiwan, China, Aug. 29, 1938; came to U.S., 1964; s. Shing T. and Lee C. (Lee) C.; m. Ming H. Lin, Feb. 21, 1963 (dec. June 1991); children: David, Lena; m. Linda Liaw, Dec. 5, 1992. BS, Nat. Taiwan U., Taipei, 1961; PhD, U. Calif., San Francisco, 1969. Asst. prof. pharmacy Coll. Pharmacy, Wash. State U., Chgo., 1969-71; asst. prof. pharmacy Coll. Pharmacy, U. Ill., Chgo., 1971-73, assoc. prof. pharmacy and occupational and environ. medicine, 1973-76, dir. clin. pharmacokinetics lab., 1975-79, prof. pharmacy, dept. pharmacy, 1979-82, prof. pharmacokinetics, dept. pharmacodynamics, 1982—; prof. occupational and environ. medicine Sch. Pub. Health, U. Chgo., 1976-77; mem. generic drugs adv. com. FDA, 1990—; mem. pharmacology study sect. NIH, 1981; mem. adv. subcom. Nat. Health Rsch. Inst., Taiwan, 1992-93. Mem. editorial bd. Jour. Pharmacokinetic Biopharm., Jour. Pharm. Scis., Jour. of Chromatography-Biomed. Applications, Internat. Jour. Clin. Pharmacology, Toxicology and Therapy, Jour. Clin. & Hosp. Pharmacy, Biopharm. Drug Disposal; contbr. over 200 articles to profl. jours. FDA grantee, 1973-76, U.S. Dept. Edn. and Pub. Welfare grantee, 1975-78, NIH, 1981-84. Fellow Am. Coll. Clin. Pharmacology, Am. Pharm. Assn., Acad. Pharm. Scis., Am. Assn. Pharm. Scientists; mem. Nat. Taiwan U. Alumni Assn. (Disting. Achievement award N.Am. chpt. 1987). Achievements include patent for Gris-PEG. Office: U Ill (M/C 865) 833 S Wood St Chicago IL 60612-4324

CHIPMAN, JOHN SOMERSET, economist, educator; b. Montreal, P.Q., Can., June 28, 1926; s. Warwick Fielding and Mary Somerset (Aikins) C.; m. Margaret Ann Ellefson, June 24, 1960; children: Thomas Noel, Timothy Warwick. Student, Universidad de Chile, Santiago, 1943-44; B.A., McGill U., Montreal, 1947, M.A., 1948; Ph.D., Johns Hopkins U., 1951; postdoctoral, U. Chgo., 1950-51; Doctor rerum politicarum honoris causa, U. Konstanz, Germany, 1991. Asst. prof. econs. Harvard U., Cambridge, Mass., 1951-55; assoc. prof. econs. U. Minn., Mpls., 1955-60; prof. U. Minn., 1961-81, Regents' prof., 1981—; fellow Ctr. for Adv. Study in Behavioral Scis., Stanford, Calif., 1972-73; Guggenheim fellow 1980-81; vis. prof. econs. various univs.; permanent guest prof. U. Konstanz, 1985-91. Author: The Theory of Intersectoral Money Flows and Income Formation, 1951; editor: (with others) Preferences, Utility, and Demand, 1971, (with C.P. Kindleberger) Flexible Exchange Rates and the Balance of Payments, 1980, (with others) Preferences, Uncertainty, and Optimality, 1990; co-editor: Jour. of Internat. Econs., 1970-76; assoc. editor: Econometrica, 1956-69, Can. Jour. Stats., 1980-82; mem. adv. bd. Jour. Multivariate Analysis, 1988-92. Recipient Humboldt Rsch. award for Sr. U.S. Scientists 1992. Fellow AAAS, Econometric Soc. (coun. 1971-76, 81-83), Am. Statis. Assn., Am. Acad. Arts and Scis.; mem. NAS, Am. Econ. Assn., Inst. Math. Stats., Can. Econ. Assn., Royal Econ. Soc., Soc. for Advancement of Econ. Theory, Hist. of Econs. Soc. Home: 2121 W 49th St Minneapolis MN 55409 Office: U Minn Dept Econs 1122 Mgmt and Econs Bldg 217 19th Ave S Minneapolis MN 55455

CHIRLIAN, PAUL MICHAEL, electrical engineering educator; b. N.Y.C., Apr. 29, 1930; S. Gustave and Leonora (Morrison) C.; m. Barbara Ellen Schein, Aug. 27, 1961; children: Lisa Emily, Peter Jonathan. BEE, NYU, 1950, MEE, 1952, DSc in Engring., 1986. Instr. NYU, N.Y.C., 1951-57, asst. prof. elec. engring., 1957-60; assoc. prof. elec. engring. Stevens Inst. Tech., Hoboken, N.J., 1960-65, prof., 1965-85, Anson Wood Burchard prof. elec. engring., 1985—; cons. in field. Author: Analysis and Design of Integrated Electronic Circuits, 1986; author 30 textbooks and contbr. numerous articles to profl. jours.; developer of effective bandwidth of signals. Recipient Great Tchr. award Stevens Inst. Tech. Fellow IEEE; mem. Am. Soc. Engring. Edn., Sigma Xi, Tau Beta Pi, Eta Kappa Nu. Achievements include development of effective bondworth of a signal. Office: Stevens Inst of Tech Castle Pt Hoboken NJ 07030

CHIRMULE, NARENDRA BHALCHANDRA, immunologist, educator; b. India, Jan. 7, 1961; came to U.S., 1987; s. Bhalchandra Chirmule and Sharayu Jeurkar; m. Preeti Lokhande, Feb. 22, 1987; 1 child, Anisha. MS, U. Bombay, 1982, PhD, 1986. Rsch. officer Cancer Rsch. Inst., Bombay, 1986-87; rsch. assoc. Northshore U. Hosp, Cornell U. Med. Coll., Manhasset, N.Y., 1987-89, instr., 1989-91, asst. prof. immunology, 1991—, dir. clin. lab. pediatric immunology, 1992—; adj. prof. C.W. Post U., Brookville, N.Y., 1991—. Achievements include development of techniques for diagnosis of HIV infection in newborn infants; involvement in studies of mechanism of immune suppression early in HIV infection. Home: 717 Willis Ave Williston Park NY 11596 Office: North Shore U Hosp 350 Community Dr Manhasset NY 11030

CHISHOLM, MALCOLM HAROLD, chemistry educator; b. Bombay, India, Oct. 15, 1945; came to U.S., 1972; s. Angus MacPhail and Gweneth (Robey) C.; m. Cynthia Ann Truax, May 1, 1982; children: Calum R.I., Selby Scott, Derek Adrian. BS in Chemistry, Queen Mary Coll., London, 1966, PhD in Chemistry, 1969; DSc (hon.), London U., 1981. Postdoctoral fellow U. Western Ont., London, 1969-72; asst. prof. Princeton (N.J.) U., 1972-78; assoc. prof. chemistry Ind. U., Bloomington, 1978-80, prof., 1980-85, Disting. prof. chemistry, 1985—; cons. in field. Editor: Polyhedron; mem. editorial bd. Inorganic Chemistry, Organometallics, Inorganic Chimica Acta; contbr. over 350 rsch. articles to profl. jours. Fellow AAAS, Ind. Acad. Scis., Royal Soc. (London), Royal Soc. for Chemistry (Corday Morgan medal 1981, award for Transition Metal Chemistry, mem. editorial bd. Chem. Communications), Am. Chem. Soc. (Akron sect. award 1982, Buck-Whitney award 1987, Inorganic Chemistry award, mem. editorial bd. ACS Books). Home: 515 S Hawthorne Dr Bloomington IN 47401-5023 Office: Ind Univ Dept of Chemistry Bloomington IN 47401

CHISHOLM, TOM SHEPHERD, environmental engineer; b. Morristown, N.J., Nov. 28, 1941; s. Charles Fillmore and Eileen Mary (Fenderson) C.; m. Mary Virginia Carrillo, Nov. 7, 1964; children: Mark Fillmore, Elaine Chisholm. Student, Northeastern U., Boston, 1959-61; BS in Agrl. Engring., N.Mex. State U., 1964; MS in Agrl. Engring., S.D. State U., 1967; PhD in Agrl. Engring., Okla. State U., 1970. Registered profl. engr., Ariz., La.; cert. Class A indsl. wastewater operator. Agrl. engr. U.S. Bur. Land Mgmt., St. George, Utah, 1964-65; asst. prof. U. P.R., Mayaguez, 1970-74, La. State U., 1974-77; assoc. prof. S.D. State U., 1977-81; environ. engr. Atlantic Richfield Subsidiary, Sahuarita, Ariz., 1981-86, Ariz. Dept. Environ. Quality, Phoenix, 1986-88; environ. mgr. Galactic Resources, Del Norte, Colo., 1988-91; v.p. M&E Cons., Inc., Phoenix, 1991—; cons. various mfrs., Calif., Tex., Ill., Mex., 1980-91. Contbr. articles to profl. jours. NSF fellow, 1965-66, 68-69. Mem. Am. Soc. Agrl. Engrs. (faculty advisor student chpt. 1978-79), Phi Kappa Phi, Sigma Xi, Alpha Epsilon, Beta Gamma Epsilon. Avocations: hiking, running, investing, solar energy. Home: 2323 E Paradise Dr Phoenix AZ 85028-1018 Office: M&E Cons Inc 2338 W Royal Palm Rd Ste E Phoenix AZ 85021-9339

CHISHOLM, TOMMY, lawyer, utility company executive; b. Baldwyn, Miss., Apr. 14, 1941; s. Thomas Vaniver and Rubel (Duncan) C.; m. Janice McClanahan, June 20, 1964; children: Mark Alan (dec.), Andrea, Stephen Thomas, Patrick Ervin. B.S.C.E., Tenn. Tech. U., 1963; J.D., Samford U., 1969; M.B.A., Ga. State U., 1984. Registered profl. engr., Ala., Ark., Del., Ga., Fla., Ky., La., N.H., Miss., N.C., Pa., Tenn., S.C., Va., W.Va. Civil engr. TVA, Knoxville, Tenn., 1963-64; design engr. So. Co. Svcs., Birmingham, Ala., 1964-69; coord. spl. projects So. Co. Svcs., Atlanta, 1969-73; sec., house counsel So. Co. Svcs., 1977-82, v.p., sec., house counsel, 1982—; asst. to pres. So. Co., Atlanta, 1973-75; sec., asst. treas. So. Co., 1977—; mgr. adminstrv. svcs. Gulf Power Co., Pensacola, Fla., 1975-77; sec. So. Electric Internat., Atlanta, 1981-82, v.p., sec., 1982—; sec. So. Devel. and Investment Group, 1985—, So. Electric R.R. Co., 1993—. Mem. Am. Bar Assn., State Bar Ala., Am. Soc. Corp. Secs., Am. Corp. Counsel Assn., Phi Alpha Delta, Beta Gamma Sigma. Home: 1611 Bryn Mawr Cir Marietta GA 30068-1607 Office: So Co Svcs Inc 64 Perimeter Ctr E Atlanta GA 30346-2205

CHISHTI, ATHAR H., biochemist; b. Aligarh, India, Dec. 25, 1957; s. M. Tahir and Naim (Akhtar) Husain; m. Yasmin Qadiri, July 25, 1987; 1 child, Imran H. MSc, A.M.U., Aligarh, India, 1980, MPhil, 1980; PHD, U. Melbourne, Victoria, Australia, 1984. Postdoctoral fellow, rsch. lab. Biol. Labs. Harvard U., Cambridge, Mass., 1984-88, teaching fellow, 1986-88; asst. prof., assoc. investigator St. Elizabeth's Hosp./Tufts U. Sch. Medicine, Boston, 1988—; sci. advisor Immunetics, Inc., Cambridge, 1991—. Contbr. articles to profl. publs. Am. Heart Assn. fellow, 1987; grantee Am. Cancer Soc., 1991, NIH, 1992. Achievements include patent pending for human erythroid p55 and methods of use. Office: St Elizabeth's Hosp 736 Cambridge St Boston MA 02135

CHISM, JAMES ARTHUR, information systems executive; b. Oak Park, Ill., Mar. 6, 1933; s. William Benjamin Thompson and Arema Eloise (Chadwick) C. AB, DePauw U., 1957; MBA, Ind. U., 1959; postgrad. advanced mgmt. program U. Pa., 1984; postgrad. sr. exec. devel. program U. Notre Dame, 1988. Mgmt. engr. consumer and indsl. products div. Uniroyal, Inc., Mishawaka, Ind. and N.Y., 1959-61, sr. mgmt. engr., office mgr., 1961-63; systems analyst Miles, Inc., Elkhart, Ind., 1963-64, sr. systems analyst, 1965-69, project mgr. distbn./logistics systems, 1969-71, mgr. systems and programming for corp. fin. and adminstrv. depts., 1971-73, mgr. adminstrv. systems and corp. staff svcs., 1973-75, group mgr. consumer products group systems and programming, 1975-79; dir. corp. orgnl. analysis, adminstrn. and staff svcs. Berkeley, Calif., Elkhart, Ind., London, Toronto, Can., Cutter/Miles, 1979-81, dir. advanced office systems and corp. adminstrn. 1982-84; dir. advanced office systems Internat. MIS and Adminstrn., 1984-85, dir. advanced office systems, tng. and adminstrn., 1985-87; exec. dir. Office Info Systems Tng., Fin. and Adminstrn., 1987-90, exec. dir. fin. and adminstrv. svcs., 1991-92. CFO, N.Am. Information Systems and Tech. Bd. dirs. United Way Elkhart County, 1974-78, Ind. Colls. of Ind. Found., 1988— (devel. comm., 1992—), Snite Mus. Art, U. Notre Dame, 1990—, vice chmn., bd. dirs. Sorin Soc.; sustaining fellows Art Inst. Chgo., 1970—; mem. adv. coun. sch. bus. Ind. U., Coun. of Sagamores of Wabash, 1993—; bd. visitors Depauw U., 1990-93; mem. Mich. Arts and Scis. Coun., 1983—. With AUS, 1954-56. Mem. Guide and Common Dataprocessing Assn., Assn. Systems Mgmt. (chpt. pres. 1969-70, div. dir. 1972-77, Merit award 1975, Achievement award 1977, cert. systems profl. 1984, Disting. Svc. award 1986, 25 Yrs. Leadership award 1988), Disting. Dean's Assocs. of Ind. U. Sch. Bus. Bloomington (mem. adv. coun.), Ind. U. Alumni Assn. (life mem.), Well House Soc., Assn. Internal Mgmt. Cons. (exec. com., bd. dirs., v.p.), Fin. Execs. Inst., Econ. Club Chgo., Office Automation Soc. Internat., Nat. Assn. Bus. Economists, Assn. Mgmt. Orgn. Design, DePauw U. Alumni Assn. (Pres.'s Cir., regents program 1989, nat. ann. fund exec. com. 1990—), DePauw Deke Realty Assn., Delta Kappa Epsilon, Deke Club, (pres. bd. dirs. Deke Realty), Sigma Delta Chi, Sigma Iota Epsilon, Beta Gamma Sigma. Republican. Episcopalian. Clubs: Morris Park Country (South Bend, Ind.), Univ. Club (Notre Dame, Ind.), Yale of N.Y.C.; Skyline Club (Indpls.), Ind. Soc. of Chgo., Ind. Soc. Washington, Deke Club of N.Y.C. Home: 504 Cedar Crest Ln Mishawaka IN 46545-5772 Office: Miles Inc 1127 Myrtle St PO Box 40 Elkhart IN 46515-0040

CHIU, ARTHUR NANG LICK, engineering educator; b. Singapore, Mar. 9, 1929; came to U.S., 1948; s. S.J. and Y.N. (Wong) C.; m. Katherine N. Chang, June 12, 1952; children: Vicky, Gregory. BSCE, BA, Oreg. State U., 1952; MSCE, MIT, 1953; PhD in Structural Engring., U. Fla., 1961. Instr. U. Hawaii, Honolulu, 1953-54, asst. prof., 1954-59, assoc. prof., 1959-64, chmn. dept. civil engring., 1963-66; prof. structural engring. Colo. State U. (on assignment to Asian Inst. Tech., Bangkok, Thailand), 1966-68; acting assoc. dean research, tng. and fellowships grad. div. U. Hawaii, Monoa, 1968; assoc. dean research, tng. and fellowships grad. div. U. Hawaii, Manoa, 1972-76, prof. civil engring., 1964—; rsch. specialist Space and Info. Systems div. N.Am. Aviation, Inc. (now Rockwell Internat.), Downey, Calif; vis. scholar UCLA and vis. assoc. Calif. Inst. Tech., Pasadena, 1970; vis. rsch. scientist Naval Civil Engring. Lab., Port Hueneme, Calif., 1976-77; mem. several univ. coms., U. Hawaii; co-chmn. Indo-US Workshop on Wind Disaster Mitigation, 1985; chmn. US-Asia Conf. on Engring. for Mitigating Natural Hazards Damage, 1987, 92. Contbr. articles to profl. jours. NSF research grantee 1970—. Mem. (hon.) ASCE (wind effects com. 1992—, Kaoiki earthquake damage assessment team 1983, past pres. Hawaii sect.), NSPE, NRC (leader Hurricane Iwa damage assessment team 1982, com. on natural disasters 1985—), Am. Concrete Inst., Structural Engrs. Assn. of Hawaii, Am. Soc. Engring. Edn., Earthquake Engring. Rsch. Coun. Inc., Internat. Conf. on Engring. for Protection Against Natural Disasters, Pan-Pacific Tall Bldgs. Conf., Sigma Xi, Chi Epsilon, Kappa Mu Epsilon, Phi Eta Sigma, Tau Beta Pi. Home: 1654 Paula Dr Honolulu HI 96816-4316 Office: U Hawaii Manoa Dept Civil Engring 2540 Dole St Honolulu HI 96822-2382

CHIU, TAK-MING, magnetic resonance physicist; b. Hong Kong, Aug. 26, 1949; came to the U.S., 1984; s. Shiu-Hay and Yuet-Shim (Leung) C.; m. Ming Fung, July 7, 1978; 1 child, Michael. BS in Chemistry cum laude, McMaster U., 1974; PhD in Phys. Chemistry, U. Western Ontario, 1984. Postdoctoral appointee chemistry div. Argonne (Ill.) Nat. Lab., 1984-87; rsch. assoc. dept. neurology Henry Ford Hosp., Detroit, 1987-88; instr. in biophysics dept. neurology Harvard Med. Sch., Mass. Gen. Hosp., Boston, 1988—; magnetic resonance physicist dept. neurology and brain imaging ctr. McLean Hosp., Belmont, Mass., 1988—. Contbr. articles to Chem. Phys. Letters, Jour. Phys. Chemistry, Annals of Neurology. Mem. Am. Chem. Soc., AAAS, Soc. Magnetic Resonance in Medicine. Achievements include research in biomedical applications of magnetic resonance, particularly in vivo localized nuclear magnetic resonance of cerebro-disorders in humans, spectroscopic imaging and functional magnetic resonance imaging. Office: McLean Hosp Brain Imaging Ctr Belmont MA 02178

CHLOUPEK, FRANK RAY, physicist; b. South Holland, Ill., June 25, 1968; s. Frank James and Trilby Rae (Chvatal) C. BS in Physics and Math., Mich. State U., 1990. Undergrad. teaching asst. dept. math. Mich. State U., East Lansing, 1988-90; fellow Ohio State U., Columbus, 1990-91, grad. teaching asst. dept. physics, 1991-93; del. univ. senate Ohio State U., 1991-93, del. coun. grad. students, 1991-93, pres. 1993—, com. on acad. misconduct, 1990—. Mich. State U. Disting. Freshman scholar, 1986. Mem. Phi Beta Kappa, Phi Kappa Phi, Sigma Pi Sigma. Office: Ohio State Univ Dept Physics 174 W 18th St Columbus OH 43210

CHMELEV, SANDRA D'ARCANGELO, laboratory administrator; b. Sacramento, May 2, 1959; d. A.M. and Helene E. (Prohaske) D'Arcangelo; m. Alexander Chmelev, June 1, 1985. BA in Biology, SUNY, Oswego, 1981. Rsch. technician Syracuse (N.Y.) U., 1981-82; sci. buyer Cold Spring Harbor (N.Y.) Lab., 1982-85, supr. purchasing, 1986-89, mgr. purchasing, 1990—; chem. buyer Waters Div. Millipore Corp., Milford, Mass., 1985-86; chair Instl. Rev. Bd. for the Protection of Human Subjects in Rsch. Mem. Nat. Assn. Purchasing Mgmt. Office: Cold Spring Harbor Lab 1 Bungtown Rd Cold Spring Harbor NY 11724-2209

CHMELL, SAMUEL JAY, orthopedic surgeon; b. Chgo., Aug. 21, 1952; s. Samuel and Elsie (Wauterlek) C.; m. Nancy Jean Aumiller, June 22, 1974; children: Jessica, Carson, Alexis, Lesley, Samuel Jayson. BS, U. Notre Dame, 1974; MD, Loyola U., 1977. Diplomate Am. Bd. Orthopedic Surgery. Intern Loyola U. Med. Ctr., Maywood, Ill., 1977-78, resident in orthopedic surgery, 1980-84; emergency room physician USPHS Indian Health Service, Chinle, Ariz., 1978-80; attending surgeon Hines (Ill.) VA Hosp., 1984-88, Shriners Hosp. for Crippled Children, Chgo., 1985-89, Gallup (N.Mex.) Indian Hosp., 1988-89, Humana-Michael Reese Hosp. and Health Plan, Chgo., 1989—; chmn. sect. orthopaedic surgery Humana-Michael Reese HMO, Chgo., 1991—; asst. prof. dept. orthopaedic surgery U. Ill., Chgo., 1991—; clin. instr. orthopedic surgery Loyola U. Med. Ctr., Maywood, 1985-88; attending orthopaedic surgeon Gallup Indian Med. Ctr., 1988-89, Humana/Michael Reese Hosp. and Health Plan, Chgo., 1989—; asst. prof. dept. orthopaedic surgery U. Ill., Chgo. Contbr. articles on orthopedic surgery to profl. jours. Sofield Travelling Fellow Orthopedic Research Soc. Great Britain, 1985. Fellow ACS, Am. Acad. Orthopaedic Surgeons; mem. AMA, Ill. State Med. Soc., Ill. Orthopaedic Soc., Chgo. Med. Soc., Olmsted Hist. Soc. Riverside (Ill.), Am. Acad. Orthopaedic Surgeons, Founders' Cir. of Sorin Soc. U. Notre Dame, Alpha Omega Alpha. Office: Humana/Michael Reese Health Plan 2545 S King Dr Chicago IL 60616-2498

CHMIEL, CHESTER T., adhesive chemist, consultant; b. Lackawanna, N.Y., Feb. 26, 1926; m. Margaret Fox, Apr. 14, 1956; children: Stephanie, Catherine, Carolyn, Geraldine, Gregory. BS, Canisius Coll., 1949, MS, 1951; PhD, Cornell U., 1956. Chemist Monsanto, Springfield, Mass., 1956-60, Uniroyal, Inc., Wayne, N.J., 1960-67; sect. mgr. Uniroyal, Inc., Mishawaka, Ind., 1967-87; v.p. R & D Uniroyal Adhesives and Sealants Co., Mishawaka, 1987—; pres. Chester T. Chmiel, Cons., Granger, Ind., 1990—. Contbr. articles to profl. jours. Republican. Roman Catholic. Achievements include 8 patents for adhesives. Home: 16674 Kent Dr Granger IN 46530-9583

CHO, ALFRED YI, electrical engineer; b. Beijing, China, July 10, 1937; came to U.S., 1955, naturalized, 1962; s. Edward I-Lai and Mildred (Chen) C.; m. Mona Lee Willoughby, June 16, 1968; children: Derek Ming, Deidre Lin, Brynna Ying, Wendy Li. BSEE, U. Ill., 1960, MS, 1961, PhD, 1968. Rsch. physicist Ion Physics Corp., Burlington, Mass., 1961-62; mem. tech. staff TRW-Space Tech. Labs., Redondo Beach, Calif., 1962-65; mem. tech. staff Bell Labs., Murray Hill, N.J., 1968-84, dept. head, 1984-87; dir. Materials Processing Rsch. Lab. AT&T Bell Labs. Murray Hill, 1987-90; dir. semiconductor rsch. lab. AT&T Bell Labs., Murray Hill, N.J., 1990—; rsch. U. Ill., Urbana, 1965-68, vis. prof. dept. elec. engring., vis. rsch. prof. coordinated sci. lab., 1977-78, adj. prof. dept. elec. engring. adj. rsch. prof. coordinated sci. lab., 1978—; bd. dirs. Instruments S.A., Edison, N.J., 1984—. Contbr. over 300 articles to profl. jours.; developer molecular beam epitaxy. Recipient Disting. Tech. Staff award AT&T Bell Labs., 1982, Elec. and Computer Engring. Disting. Alumnus award U. Ill., 1985, Disting. Achievement award Chinese Inst. Engrs. U.S.A., 1985, Internat. Gallium Arsenide Symposium award, 1986, Heinrich Welker Gold medal, 1986, The Coll. Engring. Alumni Honor award U. Ill., 1988, World Materials Congress award ASM Internat., 1988, Achievement award Indsl. Rsch. Inst., Inc., 1988, Thomas Alva Edison Sci. award N.J. Gov., 1990, Internat. Crystal Growth award Am. Assn. for Crystal Growth, 1990, Asian Am. Corp. Achievement award, 1992, Chinese Am. Engrs. and Scientists Assn. Achievement award, 1993, Nat. Medal of Sci., Nat. Sci. Found., 1993. Fellow IEEE (Morris N. Liebman award 1982), Am. Phys. Soc. (Internat. prize for new materials 1982); mem. NAS, Am. Vacuum Soc. (Gaede-Langmuir award 1988), Electrochem. Soc. (electronic divsn. award 1977, Solid State Sci. and Tech. medal 1987), Materials Rsch. Soc., Academia Sinica (Taiwan), Am. Acad. Arts and Scis., Nat. Acad. Sci., Nat. Acad. Engring., Sigma Xi, Tau Beta Pi, Eta Kappa Nu, Sigma Tau. Office: AT&T Bell Labs 600 Mountain Ave PO Box 636 Murray Hill NJ 07974-0636

CHO, CHANG GI, polymer scientist; b. CheongJu, Korea, Mar. 12, 1955; s. Haangtai and Youngwha (Park) C.; m. Aehwa Kwon, Aug. 27, 1983; 1 child, Kyunghee. BS in Chem. Tech., Seoul Nat. U., Korea, 1978; PhD, Va. Poly. Inst. and State U., 1988. Rsch. scientist Korea Rsch. Inst. Chem. Tech., Daejon, Korea, 1980-83; vis. scientist DuPont Exptl. Sta., Wilmington, Del., 1988-89; rsch. scientist Case Western Res. U., Cleve., 1989-90; mgr. Lucky Ctrl. Rsch. Inst., Daejon, 1991—; mem. editorial bd. Korean Polymer Soc., Seoul, 1991. Contbr. articles to profl. jours. Mem. Am. Chem. Soc., Korean Polymer Soc., Korean Chem. Soc., Korean Soc. Indsl. and Engring. Chemistry. Achievements include research in a new concept development in polymerization, polymer blends and compatibilizers, and biodegradable materials. Office: Lucky Ctrl Rsch Inst, Yoosung-Gu Jang-Dong 84, Daejon 305-343, Republic of Korea

CHO, JAI HANG, internist, hematologist, educator; b. Busan, Republic of Korea, May 1, 1942; came to U.S., 1972; s. Neung Whan and Heo Jai (Min) C.; m. Jawon Nam, Oct. 8, 1971; children: Karen, Austin. M.D., Catholic Med. Coll., Seoul, Republic of Korea, 1968. Diplomate Am. Bd. Internal Medicine. Intern, White Plains Hosp., N.Y., 1972-73; resident in internal medicine Nassau Hosp., Mineola, N.Y., 1973-76, fellow in hematology, 1976-77; fellow in hematology and oncology U. South Fla. Med. Coll., Tampa, 1977-79; practice medicine specializing in internal medicine and hematology, Tampa, 1979—; mem. staff Univ. Community Hosp.; clin. asst. prof. medicine U. South Fla. Med. Coll., 1985—. Served to capt. Korean Army, 1968-71. Mem. AMA, ACP, Fla. Med. Assn., Hillsborough County Med. Assn. Avocation: art, antiques. Home: 16114 Ancroft Ct Tampa FL 33647-1040 Office: 13701 Bruce B Downs Blvd Ste 105 Tampa FL 33613

CHO, MICHAEL YONGKOOK, biochemical engineer, educator, consultant; b. Seoul, Republic of Korea, Nov. 13, 1945; came to U.S., 1968; s. Jae Kyu and Mal K. (Kim) C.; m. Grace Y. Paik, Feb. 1, 1969; children: John M., Jerome I. BS, Seoul Nat. U., 1963; MS, U. Houston, 1975, PhD, 1977. Hosp. adminstr. Korea Air Force Acad. Hosp., Seoul, 1963-66; rsch. asst. U. Houston, 1972-77; rsch. engr. Ill. Water Treatment Co., Rockford, 1977-78; process engr. Am. Cyanamid Co., Bound Crook, N.J., 1978-80; rsch. assoc. La. State U., Baton Rouge, 1980-84; asst. prof. Fla. State U., Tallahassee, 1984-88; assoc. prof. N.Mex. State U., Las Cruces, 1988—; cons. Shepherd Oil Co., La., 1980-82, Unifaith Co., Houston, 1987—. Dir. steering com. Korean Bapt. Ch., Baton Rouge, 1981-84, Korean Ch., Tallahassee, 1984-88. 1st lt. Korean AF, 1963-66. Grantee NSF, 1988, Dept. Energy, 1990, 91. Mem. Am. Inst. Chem. Engrs., Am. Chem. Soc., Soc. Indsl. Microbiology. Achievements include patents in high loading of immobilized enzymes, high pressure fermentation for enhanced enzyme production. Home: 1800 Regal Ridge Las Cruces NM 88001-4924 Office: N Mex State U Dept Box 3805 Las Cruces NM 88003

CHO, NAM MIN, physicist; b. Mokpo, South Korea, Mar. 1, 1945; came to U.S., 1980; s. Bok Tae and Nang Ja (Park) C.; m. Eun Hee Nah, Nov. 24, 1975; children: Dooyon, Nayon, Brian. BS in Physics, Seoul Nat. Univ., 1967; PhD, Univ. New Mex., 1985. Physics instr. Korean Air Force Acad., Seoul, 1968-72; sr. researcher Korean Agy. Defense Devel., Seoul, 1973-80; rsch. asst. Univ. Calif., La Jolla, 1980-81, Univ. New Mex., Albuquerque, 1981-85; sr. engr. Rocky Mountain Instrument Co., Longmont, Colo., 1986-88, Polaroid Corp., Cambridge, Mass., 1988-89; mem. tech. staff NYNEX Sci. and Tech. Inc., White Plains, N.Y., 1989—; scientist, cons. Litton Laser System Rsch. Ctr., Albuquerque, 1985-86; exchange scientist Frankford Arsenal, Phila., 1974, U.S. Army Elec. Command, Fort Monmouth, N.J., 1975. Contbr. articles to profl. jours. Lt., 1968-72. Named Excellent Instr. of Yr., Korean Air Force Acad., 1971; recipient grad. scholarship Korean Traders Scholarship Found., 1977. Mem. Internat. Soc. Optical Engring, Optical Soc. Am. Achievements include initiation of patents on PZT path-length modulators and optical components for wavelength-division multiplexed optical network. Designed dichroic filters and deposition on fiber facets for high-efficiency neodymium-doped fiber lasers. Home: 1 Cedar Ln Chappaqua NY 10514 Office: NYNEX Sci & Tech Inc 500 Westchester Ave White Plains NY 10604

CHO, SECHIN, medical geneticist; b. Soowon, Korea, Jan. 18, 1947; came to U.S. 1972; s. Che Yong and Bok Nam (Shin) C.; m. Young Lee, May 20, 1978; 1 child, Dennis Hoyle. MD, Seoul Nat. U., 1971. Cert. in pediatrics, clin. genetics, clinical cyto genetics. Asst. prof. U. Kans., Wichita, 1978-81, assoc. prof., 1981-86, prof., 1986-, chmn., 1991-; mem. com. on Genetics AAP, Chgo., 1992-; pres. Great Plains Genetic Soc., Omaha, Nebr., 1988-89. Author: Progress in Medical Genetics, 1977, Cellular and Molecular Regulation of Hemoglobin Switching, 1978. Mem. profl. adv. bd. March Dimes, Wichita, 1979-, Midwest Cancer Found., Wichita, 1990-. Recipient Chancellor's award U. Kans., 1983; Genetic fellow Johns Hopkins Sch. Medicine, 1978. Mem. Am. Human Genetic Soc., Am. Acad. Pediatrics, Great Plains Genetic Soc. (pres. 1988-89). Presbyterian. Office: Genetic Svcs 550 N Hillside Wichita KS 67214

CHO, YOUNG IL, mechanical engineering educator; b. Seoul, Nov. 26, 1949; came to U.S., 1976; s. Sungshik and Keunsook (Oh) C.; m. Sunyoung Uhm, Oct. 6, 1973; children: Joseph, Daniel. BS, Seoul Nat. U., 1972; MBA, Korea U., 1975; MS, U. Ill., Chgo., 1977, PhD, 1980. Rsch. fellow Energy Resources Ctr., U. Ill., Chgo., 1980-81; mem. tech. staff Jet Propulsion Lab., Calif. Inst. Tech., Pasadena, 1981-85; prof. dept. mech. engring. and mechanics Drexel U., Phila., 1985-91, prof., 1992—. Author: Advances in Heat Transfer, 1982, Handbook of Heat Transfer, 1985; editor Advances in Heat Transfer, 1991—; contbr. articles toprofl. jours. Recipient award NASA, 1988, Lindback award of Excellence in Teaching; grantee NSF, 1987, NASA, 1988, Dept. Energy, 1989-92. Mem. ASME, Am. Electrophoresis Soc., Electrochem. Soc. Am., Am. Physics Soc., Korean Scientists and Engrs. in Am. Home: 8 Niamoa Dr Cherry Hill NJ 08003-1219 Office: Drexel U Dept Mech Engring 32d and Chestnut Sts Philadelphia PA 19104

CHOAIN, JEAN GEORGES, physician, acupuncturist; b. West Cappel, France, Dec. 8, 1917; s. Felix Auguste and Marguerite Marie (Van Der Schooten) C.; m. Christiane Marie Maurin, Apr. 21, 1954; 1 child, Françoise-Marie. MD, Faculty of Medicine, Lille, France, 1945. Cons. physician in acupuncture Lille, 1953—; tchr. acupuncture Faculty of Medicine, Lille, 1979—; physician Centre Hospitalier Regional, Lille, 1979—. Author: La Voie Rationelle de la Medecine Chinoise, 1957, Introduction au Yi-king, 1983; contbr. articles to profl. jours. Mem. Assn. Scientifique des Medecins Acuponcteurs France (v.p. 1980—), Soc. Gens de Lettres. Home: 83 Rue Reine Astrid, Marcq en Baroeul 59700, France

CHOBANIAN, ARAM VAN, physician; b. Pawtucket, R.I., Aug. 10, 1929; s. Van and Marina (Arsenian) C.; m. Jasmine Goorigian, June 5, 1955; children: Karin, Lisa, Aram. B.A., Brown U., 1951; M.D., Harvard U., 1955. Intern, resident Univ. Hosp., Boston, 1955-59, cardiovascular research fellow, 1959-62; asst. prof. Boston U. Sch. Medicine, 1964-67, assoc. prof., 1967-70, prof. medicine, 1970—, prof. pharmacology, 1975—, dir. U.A. Whitaker Labs. for Blood Vessel Rsch., 1973-88, dir. Hypertension Specialized Ctr. Rsch., 1975—, dir. Cardiovascular Inst., 1975-93, dean, 1988—; dir. Nat. Rsch. and Demonstration Ctr. in Hypertension, 1985-90; chmn. FDA Cardiovascular and Renal Adv. Com., 1978-80, NIH Hypertension and Arteriosclerosis adv. com., 1977-78; chmn. Cardiovascular Study Sect. B. NIH, 1982-84; chmn. 4th Joint Nat. Com. on Hypertension NIH, 1990-91; Sandoz lectr. Royal Coll. Physicians and Surgeons Can., 1989; Sandson distinguished prof. health scis., 1993—; mem. NIH Nat. Heart, Lung and Blood Adv. Coun., 1993—. Author: Heart Risk Book, 1982; mem. editorial bd. New Eng. Jour. Medicine, Hypertension, Jour. Hypertension, Jour. Vascular Biology, Hypertension Rsch., Cardiovascular Pharmacology. Pres. Am. Heart Assn., Boston, 1974-75; trustee Armenian Library of Am., 1975-85; bd. dirs. Armenian Culture Soc., 1976—; fellow trustee Armenian Assembly of Am. Served to capt. USAF, 1956-57. Recipient Community Edn. and Disting. Svc. Am. Heart Assn., Boston, 1975, 78, Eastman Kodak award Nat. Acad. Clin. Biochemistry, 1987, Abbott award Am. Soc. Hypertension, 1993. Fellow ACP, Am. Heart Assn. (chmn. coun. high blood pressure rsch. 1984-86, Corcoran lectr. 1989, award of merit 1990, Modern Medicine award 1990, Lifetime Achievement award in hypertension Bristol-Myers Squibb), Am. Soc. of Hypertension (Abbott award 1993), Soc. Clin. Investigation, Assn. Am. Physicians, Am. Physiol. Soc., New Eng. Cardiovascular Soc. (pres. 1985-86), Phi Beta Kappa, Sigma Xi, Alpha Omega Alpha. Home: 5 Rathbun Rd Natick MA 07160 Office: Boston U Sch Medicine 80 E Concord St Boston MA 02118-2394

CHOBOTOV, VLADIMIR ALEXANDER, mechanical engineer; b. Zagreb, Yugoslavia, Apr. 2, 1929; came to U.S., 1946; s. Alexander M. and Eugenia I. (Scherbak) C.; m. Lydia M. Kazanovich, June 22, 1957; children: Alexander, Michael. BSME, Pratt Inst., 1951; MSME, Bklyn. Poly. Inst., 1956; PhD, U. So. Calif., 1963. Dynamics engr. Sikorsky Aircraft, Bridgeport, Conn., 1951-53, Republic Aviation, Farmingdale, N.Y., 1953-57, Ramo-Wooldridge, Redondo Beach, Calif., 1957-62; mgr. The Aerospace Corp., El Segundo, Calif., 1962—; adj. prof. Northrop U., L.A., 1982-91; instr. UCLA, 1984—; cons. Univ. Space Rsch. Assn., Washington, 1984-85; ad hoc advisor USAF Sci. Adv. Bd., Washington, 1985-87; cons. NASA Space Sta. Adv. Com., Washington, 1990-91; course leader Space Debris, Washington, 1990-91. Author: Spacecraft Attitude Dynamics and Control, 1991; author, editor: Orbital Mechanics, 1991; contbg. author: Space Based Radar Handbook, 1989, Earth, Sea and Solar System, 1987; contbr. numerous articles and reports to profl. publs. Assoc. fellow AIAA; mem. Internat. Acad. of Astronautics. Achievements include pioneering in the analysis and modeling of space debris. Office: The Aerospace Corp PO Box 92957 Los Angeles CA 90009-2957

CHOCK, CLIFFORD YET-CHONG, family practice physician; b. Chgo., Oct. 15, 1951; s. Wah Tim and Leatrice (Wong) C. BS in Biology, Purdue

U., 1973; MD, U. Hawaii, 1978. Intern in internal medicine Loma Linda (Calif.) Med. Ctr., 1978-79, resident in internal medicine, 1979; resident in internal medicine U. So. Calif.-L.A. County Med. Ctr., L.A., 1980; physician Pettis VA Clinic, Loma Linda, Calif., 1980; pvt. practice Honolulu, 1981—; physician reviewer St. Francis Med. Ctr., Honolulu, 1985—, chmn. family practice care, 1990—, chmn. utilization rev., 1990-91, acting chmn. credentials com., 1992, chmn. care evaluation com., 1990—; physician reviewer Peer Rev. Orgn. Hawaii, Honolulu, 1987—. Mem. Am. Acad. Family Physicians. Avocations: photography, model construction, soccer, piano, Bible studies. Office: 321 N Kuakini St Ste 513 Honolulu HI 96817-2361

CHOE, WON-HO (WAYNE), plasma physicist; b. Jinhae, Korea, Mar. 30, 1955; came to U.S., 1978; s. Yong-Sok and Choon-Young (O) C.; m. Jong-Suk Park, May 20, 1978; children: Susan, Daniel, Michael. MS in Nuclear Engring., U. Wash., 1980; PhD in Nuclear Engring., MIT, 1985. Asst. prof. U. Ill., Urbana, 1985-92; dir. Inst. Advanced Tech., Urbana, 1992—; adj. asst. prof. U. Ill., 1992—; spl. faculty cons. Argonne (Ill.) Nat. Lab., 1988—. Contbr. articles to profl. jours. NSF Rsch. grantee, 1990-92, Expedited Rsch. award, 1986-88; recipient Profl. awards Dept. Energy, 1987. Mem. Am. Phys. Soc. (com. mem. internat. freedom of scientists 1989—), Am. Nuclear Soc. Achievements include first analytic study of the second stability regime in tokamak plasmas; discovery of self-similar solutions for imploding screw-pinch plasmas; first computer simulation study of conversion of laser energy to strong magnetic field. Home: 2310 Shurts Cir Urbana IL 61801 Office: Inst Advanced Tech 808 W Oregon Urbana IL 61801

CHOI, DAE HYUN, precision company executive; b. Yundong, Choobuk, Republic of Korea, Apr. 25, 1932; s. Dosul Choi and Yong Ak Kim; m. Young Sook, May 8, 1959; children: Soo Young, Gun Young, Jin Young. BSEE, Seoul Nat. U., 1956. Registered profl. engr., Korea. Rsch. mem. Sci. Rsch. Inst., Seoul, Republic of Korea, 1956-60; tech. dir. Internat. Electric Co., Ltd., Seoul, 1960-69; pres. Orient Electric Co., Ltd., Changwon, Republic of Korea, 1973-77; pres. Tae San Precision Co., Ltd., Seoul, 1981-87, chmn., chief exec. officer, 1987—. Avocation: golf. Office: Tae San Precision Co Ltd, 1024-4 Daechi-dong, Gangnam-ku, 135-280 Seoul Republic of Korea

CHOI, JUNHO, electronic engineer; b. Jungnam, Whasung, Korea, Apr. 19, 1941; s. Kwon-Sik and Changnyer (Yoo) C.; m. Seungyun Ham, Aug. 20, 1970; children: Charles D., John D. MSEE, SUNY, Buffalo, 1972; PhD in Elec. Engring., Duke U., 1978. Field engr., electrician Fedders Corp., Buffalo, 1972-74; electronic engr. Planning Rsch. Corp., Kennedy Space Ctr., Cannaveral, Fla., 1978-79; asst. prof. Fla. Inst. Tech., Melbourne, 1979-83; electronic engr. Naval Coastal Systems Ctr., Panama City, Fla., 1983-85, U.S. Info. Agy./VOA, Washington, 1985-86, Naval Rsch. Lab., Washington, 1986—; evaluator IEEE ABET/TAC, 1987—; mem. alumni admissions adv. bd. Duke U., 1982—. Contbr. articles to profl. jours. 2nd lt. Republic of Korea Army, 1964-66. Mem. IEEE (sect. chmn. 1981-83, Centennial Medal 1984), Armed Forces Comm. Electronic Assn., Korean Scientist and Engrs. Assn. (sec.-treas. 1974-91), Assn. Old Crows, Sigma Xi. Republican. Roman Catholic. Home: 7209 Hadlow Dr Springfield VA 22152-3529 Office: US Naval Rsch Lab 4555 Overlook Ave SW Washington DC 20375-5354

CHOI, KWING-SO, physics and engineering educator; b. Toyohashi, Aichi, Japan, Feb. 9, 1953; s. Jyun-Hyun and Chung-Ja (Park) C.; m. Tomoko Nakagawa, Dec. 23, 1985; children: Seneka, Tsubasa. BS, Tottori (Japan) U., 1976, MS, 1978; PhD, Cornell U., 1983; postgrad., Henley Mgmt. Coll., Oxfordshire, Eng., 1987—. Chartered engr., Eng. Rsch. asst. Tottori U., 1975-78, Va. Poly. Inst. and State U., Blacksburg, 1978-79; rsch. asst. Cornell U., Ithaca, N.Y., 1979-83, rsch. fellow, 1983-84; sr. rsch. engr. BMT Fluid Mechanics Ltd., Teddington, Eng., 1984-86, group leader, 1986-90, program mgr., 1987-90; lectr. U. Nottingham, Eng., 1991—. Editor: Recent Developments in Turbulence Management, 1991; inventor improvements on or related to reduction of drag; convened special interest group "drag reduction" European Rsch. Community on Flow Turbulence & Combustion, Brussels, Belgium, 1991—. Rsch. grantee U. Nottingham, 1991; travel grantee Royal Soc., London, 1991, 93, rsch. grantee, 1992. Mem. AIAA, AAAS, Japan Soc. Mech. Engrs., Am. Phys. Soc., Royal Aero. Soc., N.Y. Acad. Scis., Sigma Xi. Home: 20 Ridgway Rd, Leicester LE2 3LH, England Office: U Nottingham Dept Mech Engring, University Park, Nottingham NG7 2RD, England

CHOI, SIU-TONG, aeronautical engineering educator; b. Macao, May 6, 1960; s. Kai Un and Sou Iong (Chao) Choi; m. Huei-Hsin Joyce Hsu, May 23, 1987; children: May Kay, May Yi. MS, Vanderbilt U., 1985; PhD, Columbia U., 1988. Teaching asst. Nat. Taiwan U., Taipei, 1982-83; structural engr. Weidlinger Assocs. Cons. Engr., N.Y.C., 1989; assoc. prof. aero. engring. Nat. Cheng Kung U., Tainan, Taiwan, 1989—. Mem. AIAA, Soc. Theoretical and Applied Mechanics of Republic of China, Aero. and Astronautical Soc. of Republic of China. Office: Inst Aero and Astronautics, Nat Cheng Kung U, Tainan 701, Taiwan

CHOOK, EDWARD KONGYEN, disaster medicine educator; b. Shanghai, China, Apr. 15, 1937; s. Shiu-heng and Shuiking (Shek) C.; m. Ping Ping Chew, Oct. 30, 1973; children by previous marriage: Miranda, Bradman. MD, Nat. Def. Med. Ctr., Taiwan, 1959; MPH, U. Calif., Berkeley, 1964, PhD, 1969; ScD, Phila. Coll. Pharmacy & Sci., 1971. Assoc. prof. U. Calif., Berkeley, 1966-68; dir. higher edn. Bay Area Bilingual Edn. League, Berkeley, 1970-75; prof., chancellor United U. Am., Oakland and Berkeley, Calif., 1975-84; regional adminstr. U. So. Calif., L.A., 1984-90; vis. prof. Nat. Def. Med. Ctr., Taiwan Armed Forces U., 1982—, Tongju U., Shanghai, 1992—, Foshan U., People's Republic of China, 1992—; founder, pres. United Svc. Coun., Inc., 1971—; pres. Pan Internat. Acad., Changchun, China, San Francisco, 1992—; pres. U.S.-China Gen. Devel. Corp., 1992—; pub. Power News, San Francisco, 1979—; mem. Nat. Acad. Scis./NRC, Washington, 1968-71. Trustee Rep. Presdl. Task Force, Washington, 1978—; mem. World Affairs Coun., San Francisco, 1989—; deacon Am. Bapt. Ch.; sr. advisor U.S. Congl. Adv. Bd.; mem. Presdl. Adv. Commn., 1991—. Mem. Rotary (chmn. cm. 1971—). Achievements include research on hearing conservation program in U.S. Army, criteria for return to work. Office: 555 Pierce St # 1338-9 Albany CA 94706

CHOPKO, BOHDAN WOLODYMYR, neuroscientist; b. Berea, Ohio, Apr. 13, 1966; s. Bohdan and Jean (Kerick) C; m. Rhonda Stoll. PhD, Kent State U., 1991; MD, Northeast Ohio U., 1993. Rsch. instr. Northeast Ohio U. Coll. Medicine, Rootstown, 1991-93; resident in neurological surgery San Diego Med. Ctr. U. Calif., 1993—. Recipient Med. Student Rsch. award Am. Fedn. for Clin. Rsch., 1991, 92; Rsch. grantee Aultman Found., 1992. Mem. Soc. for Neuroscience, Neurotrauma Soc., Am. Fedn. for Clin. Rsch., Phi Beta Kappa. Achievements include conduction of basic sci. rsch. relating to the transplantation of fetal brain tissue into the adult brain. Office: NEOUCOM 4209 SR 44 Box 95 Rootstown OH 44272

CHOPPIN, GREGORY ROBERT, chemistry educator; b. Eagle Lake, Tex., Nov. 9, 1927; s. Gilbert P. and Nellie M. (Guidroz) C.; m. Ann M. Warner; children—Denise, Suzanne, Paul, Nadine. B.S. in Chemistry, Loyola U., New Orleans, 1949, D.Sc. (hon.), 1969; Ph.D. in Chemistry, U. Tex, 1953; D.Sc. Tech. (hon.), Chalmers U., Goteborg, Sweden, 1985. Rsch. scientist Lawrence Radiation Lab., Berkeley, Calif., 1953-56; faculty Fla. State U., Tallahassee, 1956—, now R.O. Lawton Disting. prof. chemistry; vis. scientist Centre d'Etude Nucleaire Mol, Belgium, 1962-63; vis. prof. Sci. U. Tokyo, 1978; vis. scientist European Transuranium Inst., Karlsruhe, W.Ger., 1979-80; cons. Los Alamos Nat. Lab., N.Mex., Lawrence Livermore Nat. Lab., Calif., Sandia Nat. Lab., N.Mex., Westinghouse Hanford Co., Mallinckrodt Med. Co. Co-author: Nuclear Chemistry: Theory and Applications, 1980, editor: Plutonium Chemistry, 1983, Actinide-Lanthanide Separations, 1985, Lanthanide Probes in Life, Chemical and Earth Sciences, 1989, Principles and Practice of Solvent Extraction, 1992; mem. editorial bd. sci. jours. including Handbook on Physics and Chemistry of Rare Earths; contbr. numerous articles to sci. jours. Served to cpl. U.S. Army, 1946-48. Recipient Alexander von Humboldt Stiftung award, 1979, Chem. Mfrs. Assn. Edn. award, 1979, Presdl. citation Am. Nuclear Soc., 1991, Scientist of Yr. award Fla. Acad. of Sci., 1992, Seaborg Actinide Separations Sci. award, 1989. Mem. Am. Chem. Soc. (So. Chemist award 1971, Fla. sect. award 1973, Nuclear Chemistry award 1985), Rare Earth Rsch. Conf. (pres. bd. 1981-83, chmn. 16th conf. 1983). Avocations: sailing; racquetball. Office: Fla State U Dept Chemistry Tallahassee FL 32306

CHOPPIN, PURNELL WHITTINGTON, research administrator, virology researcher, educator; b. Baton Rouge, July 4, 1929; s. Authur Richard and Eunice Dolores (Bolin) C.; m. Joan Harriet Macdonald, Oct. 17, 1959; 1 dau., Kathleen Marie. MD, La. State U., 1953; DSc (hon.), Emory U., 1988, La. State U., 1988, Tulane U., 1989, Washington U., 1991; D Medicine (hon.), U. Cologne, 1988. Diplomate: Am. Bd. Internal Medicine. Intern Barnes Hosp., St. Louis, 1953-54, asst. resident, 1956-57; postdoctoral fellow, rsch. assoc. Rockefeller U., N.Y.C., 1957-60, asst. prof., 1960-64, assoc. prof., 1957-60, prof., sr. physician, 1970-85, Leon Hess prof. virology, 1980-85, v.p. acad. programs, 1983-85; dean grad. studies Rockefeller U., 1985; v.p., chief sci. officer Howard Hughes Med. Inst., 1985-87, pres., 1987—; chmn. sect. 43 microbiology and immunology NAS, 1989-92, chmn. class IV med. scis., 1983-86, mem. com. on reorganization structure, 1985-86; coun. Inst. Medicine, 1987-92, exec. com., 1988-91; mem. virology study sect. NIH, 1968-72, chmn. virology study sect., 1975-78; bd. dirs. Royal Soc. Medicine Found. Inc., N.Y.C., 1978—; mem. adv. com. fundamental rsch. Nat. Multiple Sclerosis Soc., 1979-84; chmn. adv. com. fundamental rsch., 1983-84; mem. adv. coun. Nat. Inst. Allergy and Infectious Diseases, 1980-83; mem. bd. scis., cons. Meml. Sloan-Kettering Cancer Ctr., N.Y.C., 1981-86; chmn. bd. scis., 1983-84; mem. commn. on life scis. NRC, Washington, 1982-87; mem. sci. rev. com. Scripps Clinic and Rsch. Found., La Jolla, Calif., 1983-85, chmn. sci. rev. com., La Jolla, Calif., 1984; mem. coun. for rsch. and clin. investigation Am. Cancer Soc., N.Y.C., 1983-85; mem. com. priorities for vaccine devel. Inst. Medicine, Washington; mem. governing bd. NRC, 1990-92. Contbr. numerous articles to profl. pubs., chpts. on virology, cell biology, infectious diseases to profl. publs., 1958—; editor: Procs. Soc. Exptl. Biology and Medicine, 1966-69; assoc. editor Virology, 1969-72, editor, 1973-86; assoc. editor: Jour. Immunology, 1968-72, Jour. Supramolecular Structure, 1972-75; mem. editorial bd. Jour. Virology, 1972-85, Comprehensive Virology, 1972; mem. overseas adv. panel Biochem. Jour., 1973-77. Served as capt. USAF, 1954-56, Japan. Recipient Howard Taylor Ricketts award U. Chgo., 1978; Waksman award for excellence in microbiology Nat. Acad. Scis., 1984; named to alumni Hall of Distinction La. State U., Baton Rouge, 1983. Fellow AAAS; mem. NAS, Am. Acad. Arts and Scis., Am. Philos. Soc., Assn. Am. Physicians, Am. Soc. Clin. Investigation, Am. Soc. Microbiology (chmn. virology div. 1977-79, div. group councilor 1983-85), Harvey Soc., Am. Assn. Immunologists, Soc. Cell Biology, Infectious Diseases Soc. Am., Practitioners Soc. N.Y., Am. Clin. and Climatological Assn., Am. Soc. Virology (pres. 1985-86), Sigma Xi (chpt. pres. 1980-81), Alpha Omega Alpha. Office: Howard Hughes Med Inst 4000 Jones Bridge Rd Chevy Chase MD 20815-6789

CHORNENKY, VICTOR IVAN, laser scientist, researcher; b. Kiev, USSR, Sept. 8, 1941; came to U.S., 1989; s. Ivan Andrew and Vera (Nechiporenko) C.; m. Larissa Kim, July 2, 1968; children: Alice, Diana, Dennis. MS in Physics, Moscow Inst., 1965, PhD in Physics, 1974. Leading engr. Inst. Quantum Electronic "Polus", Moscow, 1968-70, sr. scientist, 1970-75; prof. Physics Moscow Inst. for Automobiles and Hwys., 1975-89; sr. scientist Inst. Applied Geophysics, Moscow, 1975-89; scientist Harvard Smithsonian Ctr. for Astrophysics, Cambridge, Mass., 1989-90; vis. scientist MIT, Cambridge, 1991-92; sr. scientist GV Med. (SpectraScience), Mpls., 1992—. Contbr. articles to profl. jours. Achievements include 11 patents on different laser applications in navigation, interferometry, and air pollution monitoring. Home: 5525 Mayview Rd Minnetonka MN 55345 Office: SpectraScience 3750 Annapolis Ln Minneapolis MN 55447

CHORPENNING, FRANK WINSLOW, immunology educator, researcher; b. Marietta, Ohio, Aug. 17, 1913; s. Roy Albert and Laura Leola (Klintworth) C.; m. Annie Laurie Kay; children: Anne Kay, Jonathan Edward, Kathleen, Janie Cecelia. AB, Marietta Coll., 1939; MSc, Ohio State U., 1950, PhD, 1963. Immunologist USAREUR Med. Lab./U.S. Army, Germany, 1952-55; chief clin. pathology Brooke Gen. Hosp., Ft. Sam Houston, Tex., 1955-61; cons. Nationalist Chinese Army, Taiwan, 1960; from lectr. to prof. Ohio State U., Columbus, 1961-81, prof. emeritus, 1981—; me. coop. study group WHO, 1953-55. Div. editor Ohio Jour. Sci., 1974-83; editor: Clinical Pathology Procedures, 1959; author: (chpt.) Regulation of Immune Response Dynamics, 1982, Immunology of Bacterial Cell Envelope, 1983; author: The Man from Somerset, 1993; contbr. articles to profl. jours. Mem. Epidemiol. Com., San Antonio, 1949, Rep. nat. Com., Delaware, Ohio, 1979-88, Rep. presdl. Task Force, Delaware, 1983. Lt. col. U.S. Army, 1941-61. Recipient Commendation, Chinese Surgeon Gen., 1960, C.G. Brooke Gen. Hosp., 1960. Fellow Am. Acad. Microbiology, Ohio Acad. Sci.; mem. Am. Assn. Immunologists, Ohio Acad. Sci., Assn. for Gnotobiotics. Roman Catholic.

CHOTIROS, NICHOLAS PORNCHAI, research engineer; b. Bangkok, Thailand, Jan. 5, 1951; s. Hsien-Shu Cho and Prachumporn Chotiros; m. Ashra Anantavara, Sept. 15, 1975; children: Kathleya, Carissa, Joseph. BA in Engring., MA in Engring., Cambridge (Eng.) U., 1973; PhD in Engring., U. Birmingham (Eng.), 1981. Lectr. U. Birmingham, 1974-81; prin. investigator applied rsch. labs. U. Tex., Austin, 1981—; cons. DB Instrumentation Ltd., Aldershot, U.K., 1980-82. Contbr. articles to Jour. Acoustical Soc. Am., Jour. Sound and Vibration, Soc. Photo-optical Instrument Engrs. Procs. Mem. IEEE, Acoustical Soc. Am. (tec. com. 1991), Soc. Photo-optical Instrument Engrs., Instn. Elec. Engrs. (U.K.). Achievements include patents for Doppler Scanning Correlator, for Tracking of Spatial Patterns. Office: U Tex Applied Rsch Labs Austin TX 78713-8029

CHOU, CHUNG LIM, electrical engineer; b. Canton, People's Republic China, July 18, 1943; arrived in Hong Kong, 1953; s. Loo Sheung and Wai Ching Chou; m. Tammy Tak-Mui, Nov. 18, 1970; children: Kwan Pang, Kwan Yao. BScEE, Taiwan U., Taipei, Republic of China, 1965. Quality control engr. Fairchild Semiconductor Ltd., Hong Kong, 1965-69, quality assurance engr., 1969-74, prodn. mgr., 1974-76, mfg. mgr. Century Electronic Ltd., Hong Kong, 1976-77; operation dir. Clover Display Ltd., Hong Kong, 1977—. Mem. Soc. for Info. Display. Office: Clover Display Ltd, 166 Wai Yip St 5/F, Kwun Tong Hong Kong

CHOU, CLIFFORD CHI FONG, research engineering executive; b. Taipei, Taiwan, Dec. 19, 1940; came to U.S., 1966, naturalized, 1978; s. Ching piao and Yueh li (Huang) C.; m. Chu hwei Lee, Mar. 23, 1968; children: Kelvin Lin yu, Renee Lincy. Ph.D., Mich. State U., 1972. Research asst. Mich. State U., East Lansing, 1967-70; research asst. Wayne State U., Detroit, 1970-72, research assoc., 1972-76; research engr. Ford Motor Co., Dearborn, 1976-81, sr. research engr., 1981-82, prin. research engr. assoc., 1982-89, prin. staff engr., 1989—; lectr. to China under U.N. Devel. Program, 1987, lectr. to Taiwan under Automotive Rsch. and Test Ctr., 1991; tchr. in field. Contbr. articles to profl. jours. Recipient Safety Engring. Excellence award Nat. Hwy. Traffic Safety Adminstrn., 1980, Innovation award Engring. and Mfg. Staff Ford Motor Co., 1986, Tech. Accomplishment awards, 1989, 91, 92, 93; coord. Detroit Auto. Tech. Conf., 1993; grantee Soc. Automotive Engrs. Mem. ASME, AIAA, Soc. Automotive Engrs., Ford Chinese Club (pres. 1991-92), Mich. Chinese Acad. Profl. Assn. (bd. dirs. 1992—), Sigma Xi. Club: Detroit Chinese Am. Assn. Home: 28970 Forest Hill Dr Farmington MI 48331-2439 Office: Ford Motor Co Body Engring Dept 800 Village Plz 23400 Michigan Ave Dearborn MI 48124-1915

CHOU, DAVID CHIH-KUANG, analytical chemist; b. Taipei, Taiwan, Oct. 26, 1955; came to U.S. 1978.; s. Chi-Shiang, and Mei-Chuang (Tan) C.; m. Tammy Yueh-Kwei, June 25, 1978; children: Christopher David, Cynthia Tammy. BS, Fu-Jen Cath. U., 1978; PhD, CUNY, 1985. Rsch. assoc. Am. Health Found., Valhalla, N.Y., 1984-85; sr. scientist Hoffmann-LaRoche, Inc., Nutley, N.J., 1985-89; assoc. rsch. investigator Hoffman-LaRoche, Inc., Nutley, N.J., 1989-91, rsch. investigator, 1991--; dir. N.Y. Chromatography Topical Group Am. Chemistry Soc., N.Y., 1985. Bd. dirs. No. Westchester Chinese Sch., N.Y., 1991-93. Mem. Am. Chem. Soc., Am. Assn. Pharm. Scientist. Achievements include development of chiral seperation on high performance liquid chromatography, lab. automation tech., supercritical fluid extraction and capillary electrophoresis.

CHOU, LIH-HSIN, materials science and engineering educator; b. Taoyuan, Taiwan, Republic of China, Oct. 28, 1957; parents Der-Ming Chou and Chin-Bih Liang. BS, Nat. Tsing Hua U., Hsinchu, Republic of China, 1979; MS, U. Ill., 1981, PhD, 1985. Rsch. assoc. U. Ill., Urbana, 1985-87; assoc. prof. Nat. Tsing Hua U., Hsinchu, 1987-93, prof., 1993—. Contbr. articles to profl. jours. Rsch. grantee Nat. Sci. Coun., Republic of China, 1988, 89, 90, 91. Mem. Am. Soc. for Metals, Materials Rsch. Soc., Am. Vacuum Soc. Avocations: tennis, volleyball, picnic, reading, music. Home: 5F No 8 Alley 7 Lane 81, Yune-Yeh Rd, Hsin Tien Taipei, Taiwan Office: Nat Tsing Hua U Materials Sci, 101 Sec 2 Kuang Fu Rd, Hsinchu 300, Taiwan

CHOU, MIE-YIN, physicist; b. Taiwan, Feb. 17, 1958. BS, Nat. Taiwan U., Taipei, 1980; PhD, U. Calif., Berkeley, 1986. Postdoctoral fellow Exxon Rsch. and Engring., Annandale, N.J., 1986-88; asst. prof. Physics Ga. Inst. Tech., Atlanta, 1989-93, assoc. prof. physics, 1993—. Contbr. over 40 articles to profl. jours. Packard fellow David and Lucile Packard Found., 1990-95; recipient Pres. Young Investigator award NSF, 1991-96. Mem. Am. Phys. Soc., Materials Rsch. Soc. Office: Ga Inst Tech Sch Physics Atlanta GA 30332

CHOU, TEIN-CHEN, economics educator; b. Taipei, Taiwan, Republic of China, Jan. 22, 1953; s. Chrong-rong and Shuer (Wu) C.; m. Jin-I Ching, Jan. 14, 1979; 1 child, Chin. BA, Nat. Chung Hsing U., Taipei, 1975; MA, Nat. Taiwan U., Taipei, 1975-78; PhD, U. Cath. de Louvain, Belgium, 1985. Rsch. and teaching asst. Nat. Chung Hsing U., Taipei, 1975-78; journalist Cen. Daily News, Taipei, 1978-81; assoc. prof. Nat. Chung Hsing U., Taipei, 1985-89, prof., 1988—; researcher Chung-Hua Inst. for Econs. Rsch., Taipei, 1985-86; pres. Taipei Foun. Fin., 1991—; cons. Far-East Economist Rsch. Consortium, Taipei, 1990—. Author: Industrial Organization in the Process of Economic Development: The Case of Taiwan, 1985, Essays on Taiwan's Industrial Organization, 1991; contbr. articles to profl. jours. Scholar Le Conseil du Tiers-Monde, UCL, Belgium, 1981-85; grantee Chi-Yu Found., 1977, Nat. Sci. Coun., Taipei, 1987-93; recipient Acad. Publ. award Sun Yat-Sen Acad. Cultural Found., 1991. Mem. Chinese Econ. Assn. (bd. dirs. 1989—), European Assn. Rsch. for Indsl. Econs. (Belgium), Atlantic Econ. Assn. (U.S.), Ctr. for European Policy Studies (Belgium). Avocation: travel. Office: Nat Chung Hsing U Inst Econs, 68 Ho-Kiang St, Taipei 104, Taiwan

CHOUDARY, SHAUKAT HUSSAIN, science administrator, chemist; b. Rawalpindi, Punjab, Pakistan, June 20, 1958; s. Mohammad Yaqub and Iqbal Jan Choudary; m. Sajida Aslam, June 3, 1988; children: Sundus Shaukat, Osama Shaukat. F.sc, Sirsyed Coll., Rawalpindi, 1975; BSc, Gordon Coll., Rawalpindi, 1979; MSc in Inorganic/Analytical Chemistry, Quaid-I-Azam U., Islamabad, Pakistan, 1982. Asst. chemist lab. Attock Refinery Ltd., Rawalpindi, 1983-86; chemist lab. Hydro-Carbon Devel. Inst. Pakistan, Karachi, 1986-87, Petro Lube (Petromin), Jubail, Saudi Arbia, 1987-89; chief chemist lab., asst. mgr. Arabian Gulf Oil Co. Ltd., Yanbu, Saudi Arbia, 1989—. Mem. Am. Soc. Quality Control. Avocation: radio, reading, cricket, travel. Home: Vill Kotha Kalan, P/O Attock Oil Co, Rawalpindi Pakistan Office: Arabian Gulf Oil Co, PO Box 30439, Yanbu Al Sin Saudi Arabia

CHOUDHARY, MANOJ KUMAR, chemical engineer, researcher; b. Panchobh, Bihar, India, May 22, 1952; came to U.S., 1974; s. Madan and Mahamaya (Thakur) C.; m. Saraswati Jha, June 24, 1974; children: Niharika, Vikas A. MS, SUNY, Buffalo, 1976; ScD, MIT, 1980. Registered profl. engr., Ohio. Rsch. assoc. Dept. Materials Sci. and Engring., Cambridge, Mass., 1980-82; advanced engr. Owens Corning Fiberglas, Granville, Ohio, 1982-85, sr. engr., 1985-90, mem. sr. tech. staff, team leader, 1990—; mem. bd. review Metall. Transactions, Pitts., 1985-92, cons. Prof. J. Szekely, MIT, Cambridge, 1980-82; mem. rsch. staff SUNY, Buffalo, 1976. Contbr. over 25 articles to profl. jours. Recipient Best Paper award IEEE Glass Industry Com., 1985, Prof. S.K. Nardi Gold medal Indian Inst. Tech., Kharagpur, India, 1974. Mem. AIChE (Outstanding Young Chem. Engr. 1985), Am. Ceramic Soc., Minerals, Metals and Materials Soc., Sigma Xi. Achievements include patent in method and aparatus for electrically heating molten glass; development of math. and physical models for transport phenomena in glass mfg. and processing; of novel tech. approaches for the design of glass furnaces by combining math. modeling of transport phenomena with statis. process control; of math. tools for analysis of electromagnetically driven flow and heat transfer. Office: Owens Corning Fiberglas 2790 Columbus Rd Rte 16 Granville OH 43023

CHOUDHURY, DEO CHAND, physicist, educator; b. Darbhanga, India, Feb. 1, 1926; came to U.S., 1955; s. Kapleshwar and Gutainya Choudhury; BSc, U. Calcutta, 1944, MS., 1946; PhD, UCLA, 1959; m. Annette Patricia DuBois, Aug. 3, 1963; 1 son, Raj. Rsch. fellow Niels Bohr Inst., Cophenhagen, 1952-55; rsch. asst. physics U. Rochester, N.Y., 1955-56; rsch. and teaching asst. physics UCLA, 1956-59; asst. prof. physics U. Conn., Storrs, 1959-62; assoc. prof. Poly. Inst. of N.Y., Bklyn., 1962-67, prof. physics, 1967—; vis. asst. physicist Brookhaven Nat. Lab., summer 1960; vis. physicist Oak Ridge Nat. Lab., summer 1962, Niels Bohr Inst., 1978-79. Govt. India Coun. Sci. and Indsl. Rsch. scholar U. Calcutta Coll. Sci., 1947-52. Mem. Am. Phys. Soc., N.Y. Acad. Scis., AAAS, Indian Phys. Soc., Sigma Xi, Sigma Pi Sigma. Contbr. chpt. to book, numerous articles on high energy nuclear scattering, nuclear models, structure and reactions to profl. publs. Home: 90 Gold St New York NY 10038-1833 Office: Poly U Dept Physics 6 Metrotech Ctr Brooklyn NY 11201-2990

CHOUINARD, WARREN E., engineer; b. San Diego, Aug. 13, 1955; s. Ronald Eli and Caroline Dixie (Young) C.; m. Marjorie Diane Fortney, May 1, 1986 (div.); children: Kristina Marie, Diana Sue. Student, Mesa City Coll., 1973-74, S.D. City Coll., 1974-76. Mechanic, shipmate Bahia & Catamaran Hotels, San Diego, 1973-79, estimator, foreman P.V.T. Insulation Co., San Diego, 1979-81; asst. chief engr. Sheraton Harbor Island East, San Diego, 1981-90; chief engr. U.S. Grant Hotel, San Diego, 1990—. Contbr. articles to profl. jours. Activist, leader Rep. Com., San Diego, 1992. Mem. Hotel Engrs. Assn., Assn. of Energy Engrs., Environ. Mgrs. Inst. Roman Catholic. Achievements include patent pending for controller for cooling towers to eliminate bacteria; rsch. in radon problems in San Diego area. Office: US Grant Hotel 326 Broadway San Diego CA 92101

CHOVAN, JOHN DAVID, biomedical engineer; b. Canton, Ohio, Sept. 14, 1958; s. John Jr. and Esther Lee (Baker) C. BS, Ohio State U., 1980, BS in Audio Recording, 1980, BSEE, 1982, MS, 1984, PhD, 1990. Grad. rsch. assoc. Ohio State U., Columbus, 1982-84, lead programmer-analyst, 1984-85, grad. rsch. assoc., 1987-90, postdoctoral researcher, 1990-91; evaluation programs assoc. Nat. Bd. Med. Examiners, Phila., 1985-87; rsch. scientist Battelle Meml. Inst., Columbus, 1991—. Author: Educom Selected Academic Software, 1990; editor: Preprints of the 1991 IFIP Working Group on Intelligent CAD, 1991; author conf. papers, tech. reports. Mem. Columbus AIDS Task Force, 1985; mem. Ohio State U. AIDS Edn. and Rsch. Com., Columbus, 1987-90, Am. Rose Soc. Mem. IEEE (Engring. in Medicine and Biology Soc.), Biomed. Engring. Soc., Internat. Neural Networks Soc., Mensa, Sigma Xi. Home: 135 Arden Rd Columbus OH 43214 Office: Battelle Meml Inst 505 King Ave Columbus OH 43201

CHOW, ANTHONY WEI-CHIK, physician; b. Hong Kong, May 9, 1941; s. Bernard Shao-Ta and Julia Chen (Fan) C.; m. Katherine Cue, May 20, 1967; children: Calvin Anthony, Byron Calbert. Student, Brandon (Man., Can.) Coll., 1961-63; M.D., U. Man., 1967. Intern Calgary (Atla., Can.) Gen. Hosp., 1967-68; resident in internal medicine Winnipeg (Man.) Gen. Hosp., 1968-70; fellow in infectious disease UCLA Harbor Gen. Hosp., 1970-72, from asst. prof. to assoc. prof., assoc. head div. infectious disease, 1972-78; practice medicine specializing in infectious disease; prof. medicine, head div. infectious disease U. B.C., Vancouver Gen. Hosp., 1979—; mem. Can. Bacterial Disease Network, 1989—, MRC, NIH, FDA coms.; councilor Can. Soc. Clin. Invest., Western Soc. Clin. Invest.; apptd. Can. Inst. Acad. Medicine, 1993. Contbr. articles to profl. jours. Med. Research Council Can. grantee, 1979—. Mem. Am. Soc. Microbiology, Am. Fedn. Clin. Rsch., Western Assn. Physicians, Infectious Disease Soc. Am., Western Soc. Clin. Investigation, Can. Soc. Clin. Investigation, Can. Infectious Disease Soc., Can. Inst. of Acad. Medicine, Can. Bacterial Diseases Network, Nat. Ctr. of Excellence. Roman Catholic. Achievements include rsch. in microbial pathogenesis, cellular and molecular immunology, staphylococcal toxins. Home: 1119 Gilston Rd, West Vancouver, BC Canada V7S 2E7 Office: Vancouver Gen Hosp, 2733 Heather St Div Infectious Diseases, Vancouver, BC Canada V5Z 1M9

CHOW, GAN-MOOG, physicist; b. Hong Kong, Sept. 24, 1957; came to U.S., 1978; s. Young Jew and Fung-Lin (Leung) C.; m. Deborah K. Nadolny, July 28, 1984; children: Timothy, Benjamin. BS, SUNY, Stony Brook, 1983; MS, U. Conn., 1985, PhD, 1988. Rsch. assoc. Nat. Rsch. Coun., Washington, 1989-90; sr. scientist Geo-Ctrs., Washington, 1990-91; rsch. physicist Naval Rsch. Lab., Washington, 1991—. Contbr. articles to Applied Physics Letters, Science, Jour. Materials Rsch., others. Recipient Performance award Naval rsch. Lab., 1992, Associateship award Nat. Rsch. Coun., 1989. Mem. Am. Phys. Soc., Materials Rsch. Soc., Sigma Xi. Democrat. Achievements include discovery of self arrangement of molybdenum cubes formed in vapor; patent for novel technique to form phase separated materials. Office: Naval Rsch Lab Code 6900 Washington DC 20375

CHOW, JOHN LAP HONG, biomedical engineer; b. Bangkok, Thailand, Oct. 20, 1960; came to U.S., 1975; naturalized; s. Pius C. S. and Veronica S. Y. Chow. BS in Cell and Molecular Biology, U. Wash., 1983, BA in Chemistry, 1983; MS in Biomedical Engring., U. Calif., Davis, 1988; post-grad., Med. Coll. Wis. Cert. tchr., Calif. Rsch. assoc. U. Wash. Sch. Medicine, Seattle, 1980-85; rsch. asst. Ctr. for Bioengring., U. Wash., Seattle, 1985; rsch. asst. biomed. engring. U. Calif., Davis, 1985-86, assoc. teaching asst., dept. elec. computer engring., 1986-88; programmer, analyst Genentech, Inc., South San Francisco, 1988-90; summer rsch. fellow neonatology U. So. Calif. Sch. Medicine, L.A., 1991; elected com. mem. admissions com. Med. Coll. Wis., Milw., 1991—; biostats. cons. Cyanotech, Corp., Seattle, 1985; com. mem. human subjects review com. U. Calif., Davis, 1986-87. Contbr. articles to profl. jours. including Annals N.Y. Acad. Scis., Annals Biomed. Engring. Lang. interpreter ARC Lang. Bank (King County chpt), Seattle, 1984-85; med. vol. hematology/oncology Milw. County Hosp., 1991; team capt. 1991 fund raising Med. Coll. Wis., Milw., 1991. Scholar Med. Coll. Wis., 1992-93. Mem. AAAS (1st place award for excellence 1989), AMA, Biomed. Engring. Soc., N.Y. Acad. Scis., Sigma Xi. Roman Catholic. Achievements include development of an artificial intelligence expert system for quantification of cardiac metabolites from 31-phosphorus NMR spectroscopy, software systems for biomedical laboratory automations, research in non-invasive hemodynamic monitoring of critical ill patients. Home: PO Box 26041 Milwaukee WI 53226

CHOW, STEPHEN HEUNG WING, physician; b. Hong Kong, Nov. 26, 1954; s. Hon Wing Chow and Kam Wah Choi; m. Clare Chau Fung Sin, Mar. 10, 1986; children: Tin Yee, Tin Yan, Chow Tin Bo. MBBS, Hong Kong U., 1979. House officer Queen Elizabeth Hosp., Hong Kong, 1979, Nethersole Hosp., Hong Kong, 1980; gen. practice medicine Hong Kong, 1981—; founder Human Bioenergy Devel. Ltd., Hong Kong, 1990—. U.S. Bicentennial Scholar, 1978. Mem. Hong Kong Med. Assn. Avocation: devel. of human psychic energy. Home: Bl 240 Prince Edward Rd, Kowloon Hong Kong Office: 235 Ground Fl, Nam Cheong St, Kowloon Hong Kong

CHOWBEY, SANJAY KUMAR, mechanical engineer; b. Ranchi, Bihar, India, Jan. 8, 1968; came to U.S., 1991; s. Ram Dayal and Jayanti (Devi) C. MS in Aero. Engring., Birla Inst. Tech., Ranchi, 1991; MS in Mechanical Engring., Tenn. Tech. U., 1993. Mgmt. trainee Steel Authority of India Ltd., 1990-91. Sec. Gandhi Constructive Com., Dhanbad, India, 1985-89; pres. Indian Assn. Cookeville, Tenn., 1992-93. Mem. ASME, Acoustical Soc. Am. (fellowship 1992), Phi Kappa Phi. Office: Tenn Tech U TTU Box 13052 Cookeville TN 38505

CHOWDHURI, PRITINDRA, electrical engineer, educator; b. Calcutta, July 12, 1927; came to U.S., 1949, naturalized, 1962; s. Ahindra and Sudhira (Mitra) C.; m. Sharon Elsie Hackebeil, Dec. 28, 1962; children: Naomi, Leslie, Robindro, Rajendro. B.Sc. in Physics with honors, Calcutta U., 1945, M.Sc., 1948; M.S., Ill. Inst. Tech., 1951; D.Eng., Rensselaer Poly. Inst., 1966. Jr. engr. lightning arresters sect. Westinghouse Electric Corp., East Pittsburgh, Pa., 1951-52; elec. engr. high voltage lab. Maschinenfabrik Oerlikon, Zurich, 1952-53; research engr. High Voltage Rsch. Commn., Daeniken, Switzerland, 1953-56; devel. engr. high voltage lab. GE, Pittsfield, Mass., 1956-59; elec. engr. research and devel. ctr. GE, Schenectady, N.Y., 1959-62; engr. elec. investigations transp. systems div. GE, Erie, Pa., 1962-75; staff mem. Los Alamos (N.Mex.) Nat. Lab., 1975-86; prof. elec. engring. Ctr. Elec. Power Tenn. Technol. U., Cookeville, 1986—; lectr. Pa. State U. Behrend Grad. Ctr., Erie, 1969-75. Patentee in field. Fellow AAAS, Instn. Elec. Engrs. (Eng.), N.Y. Acad. Scis.; mem. IEEE (sr.). Democrat. Unitarian. Home: 690 Valley Forge Rd Cookeville TN 38501-1574 Office: Tenn Technol U Ctr Elec Power PO Box 5032 Cookeville TN 38505

CHOY, CLEMENT KIN-MAN, research scientist; b. Fukien, China, Aug. 4, 1947; came to U.S., 1970.; s. Yick-Chu and Hui-Keng (Sy) C.; m. Anna K. Chan, Oct. 4, 1975; 1 child, Jennifer. Diploma, Hong Kong Baptist Coll., 1970; MS, Cleve. State U., 1972; PhD, Case Western Reserve U., 1976. Technician Univ. Hosps., Cleve., 1974-76; asst. dir. Gen. Med.Labs, Warrensville, Ohio, 1974-76; tech. staff Procter and Gamble, Cin., 1976-80; scientist Clorox, Pleasanton, Calif., 1980-81, sr. scientist, 1981-82, project leader, 1982-89, sr. rsch. assoc., 1989-93. Pres. Chinese Assn. of Greater Cleve., 1972-74. Mem. Am. Chem. Soc., Am. Soc. Oil Chemists, Am. Assn. Clin. Chemists. Achievements include U.S. patents in Surfactant Cake Composition, Passive Dosing Dispenser Exhibiting Improved Resistance to Clogging, Poly (ethylene oxide) Compositions with Controlled Solubility Characteristics, Thickened Aqueous Abrasive Scouring Cleanser, Thixotropic Acid-Abrasive Cleaner, Thickened Aqueous Abrasive Scouring Cleanser, Thickened Aqueous Cleanser, Timed-Release Bleach Coated with an Inorganic Salt and an Amine with Reduced Dye Damage, Timed-Release Hypochlorite Bleach Compositions, Hard Surface Acid Cleaner, Polymer Film Composition for Rinse Release of Wash Additives, Aqueous Based Acidic Hard Surface Cleaner, Thickened Liquid Improved Stability Abrasive Cleanser, Isotropic Fabric Softener Composition Containing Fabric Mildewstat, Timed-Release Bleach Coated with an Amine with Reduced Dye Damage, Rinse Release Laundry Additive and Dispenser, Aqueous Based Acidic Hard Surface Cleaner, Rinse Release Laundry Additive and Dispenser; foreign patents in field; patents pending; research in determination of the Si-o-Si Bond Angle common to the shift reagent compounds (CH) SiO (PcSiO) Si(CH) where x=1-5, iron and ruthenium phthalocyanines shift reagents. Home: 1345 Sugarloaf Dr Alamo CA 94507 Office: Clorox Tech Ctr 7200 Johnson Dr Pleasanton CA 94566

CHOY, PATRICK C., biochemistry educator; b. China, June 16, 1944; 1 child. BSc, McGill U., 1969; MSc, U. N.D., 1972, PhD, 1975. Teaching fellow biochemistry U. B.C., 1975-79; from asst. prof. to assoc. prof. U. Man., Can., 1979-86, prof. biochemistry, 1986—; sci. Med. Rsch. Coun. Can., 1984—. Mem. Am. Soc. Biol. Chemists, Can. Biochem. Soc., Can. Cardiovascular Soc. Achievements include rsch. in lipid metabolism in mammalian hearts; pathogenesis of heart failure. Office: Can Biochem Soc, Dept of Biochem/U Western Ont, London, ON Canada N6A 5C1 also: U Man, Dept Biochem Fac Med, Winnipeg, MB Canada R3E 0W3*

CHOYCE, DAVID PETER, ophthalmologist; b. London, Mar. 1, 1919; s. Charles Coley and Gwendolen Alice (Dobbing) C.; m. Diana Graham, Sept. 3, 1949; children: Jonathan, David Gregory, Matthew Quentin. BSc, U. London, 1939, MB, BS, 1943, MS, 1962. Cons. Southend Hosp., Essex, Eng., 1954-84; ophthalmologist Hosp. for Tropical Diseases, London, 1953-88; overseas cons., ophthalmologist Henry Ford Hosp., Detroit, 1980—; pvt. practice London, 1955—; cons. London Centre for Refractive Surgery, 1987—; mem. Brit. Acad. Experts, 1991—. Author: Intraocular Lenses and Implants, 1964; contbr. articles to profl. jours. Med. Officer, Brit. Navy, 1943-46. Recipient Disting. Achievement award Am. Soc. Contemporary Ophthalmology, 1986, Mericos award Mericos Eye Inst., La Jolla, Calif., 1986, Binkhorst medal Am. Implant Soc., 1981, Internat. award of excellence in ophthalmology Hawaiian Eye Found., 1991, Innovators Lecture, ASCRS, 1993. Fellow Royal Coll. Surgeons (Eng.; Hunterian prof.), Japanese Implant Soc.; mem. Yugoslav Implant Soc. (hon.), Kerato-Refractive Soc. (pres. 1986-89, Palaeologus medal 1984), Am. Soc. Cataract and Refractive Soc. (40th Anniversary Pioneer award 1989), Internat. Intraocular Implant Club (pres. 1977-80), U.K. Intraocular Implant Soc. (pres. 1980-82), So. African Implant Soc., Brit. Acad. of Experts, Moor Park Golf Club. Mem. Conservative Party. Avocation: golf. Home: 9 Drake Rd, Westcliff-on-Sea SS0 8LR, England Office: 45 Wimpole St, London W1M 7DG, England

CHRETIEN, MICHEL, physician, educator, administrator; b. Shawinigan, Que., Can., Mar. 26, 1936; s. Willie and Marie (Boisvert) C.; m. Micheline Ruel, July 9, 1960; children—Marie, Lyne. M.D., U. Montreal, Que., 1960; M.Sc. in Exptl. Medicine, McGill U., Montreal, 1962; D.Sc. (hon.), U. Liege, Belgium, 1981, U. Reneé Descartes, Paris, 1992. Intern Montreal Hosps. and McGill U., 1959-60, research fellow, 1960-62; resident in medicine Peter Bent Brigham Hosp. and Harvard U., 1962-64; asst. Biochem. Hormone Research Lab. U. Calif.-Berkeley, 1964-67; dir. Lab. Molecular Neuroendocrinology Clin. Research Inst. Montreal, 1967—, sci. dir., chief exec. officer, 1984—; physician dept. medicine Hotel-Dieu Hosp., Montreal, 1967—, head endocrinology sect., 1978-84; prof. exptl. medicine McGill U., 1969—; titular of medicine U. Montreal, 1975—; vis. scientist Cambridge U., Eng., 1979, Lab. Neuroendocrinology, Salk Inst. of San Diego, 1980; mem. grant com. Biochem. Med. Rsch. Coun. Can.; frequent nat. and internat. guest speaker, lectr.; Sandoz speaker Can. Soc. Endocrinology and Metabolism. Author, co-author numerous publs.; mem. various editorial bds. Recipient Clarke Inst. Psychiatry Toronto award, 1977, Archambault medal Assn. Canadienne-Francaise pour l'Avancement des scis., 1978; faculty scholar Josiah Macy Jr. Found., 1979-80; named. Hon. Prof. Chinese Acad. Med. Scis. and Peking Union Med. Coll., 1986; decorated Order of Can., 1986; Fuller Albright lectr. Peripatetic Club, 1992, McLaughlin medal Royal Soc. Can., 1993, Boehringer-Mannheim award Can. Biochem. Soc., 1993, 150th Anniversary medal U. Montreal, 1993. Fellow ACP, Royal Coll. Physicians (Can.); mem. Am. Fedn. Clin. Research, Can. Soc. Clin. Investigation (pres. 1977), AAAS, Club de Recherches Cliniques du Que. (Michel Sarrazin award 1977), Can. Biochemical Soc. (pres. 1983, Boehringer-Mannheim Can. prize 1993), Endocrine Soc., N.Y. Acad. Scis., Can. Soc. Endocrinology and Metabolism (pres. 1986-88), Am. Soc. Clin. Investigation, Am. Clin. and Climatol. Assn. (Jeremiah Metzger lectr. 1990), Assn. des medecins de langue francaise du Can. (gen. council, pres. research com., past bd. dirs., Fundamental Research award 1971), Assn. Am. Physicians, Sigma Xi. Home: Apt 1404, 1 Côte Ste Catherine, Montreal, PQ Canada H2V 1Z8 Office: Clin Rsch Inst Montreal, 110 Pine Ave W, Montreal, PQ Canada H2W 1R7

CHRISMAN, VINCE DARRELL, information scientist; b. Highland Park, Mich., June 4, 1956; s. Buford Wayne and Geraldine Norita (Evans) C.; m. Marilyn Sue Bruck, May 15, 1981; 1 child, Alyssa Nicole. BBA, Ea. Ky. U., 1981; MSA, Ctrl. Mich. U., Mount Pleasant, 1988. Cert. data processing. Programmer II Nat. Lead of Ohio, Fernald, 1981; programmer analyst II 1st Nat. Bank Cin., 1981-84; programmer analyst K Mart Corp., Troy, Mich., 1984; systems programmer Hygrade Food Products, Southfield, Mich., 1984-86; supr. tech. support Borg Warner Automotive, Sterling Heights, Mich., 1986-89, mgr. info. systems, 1989-93; dir. mgmt. info. svcs. Sandy Corp., Troy, Mich., 1993—; bd. mem. Windows World/Comdex, Mass., 1991—. Republican. Roman Catholic. Home: 2906 Saratoga Troy MI 48083-2650 Office: Sandy Corp 1500 W Big Beaver Rd Troy MI 48084

CHRISTENSEN, CLYDE MARTIN, plant pathology educator; b. Sturgeon Bay, Wis., Aug. 8, 1905; s. Peter Karl and Christine Ann (Christensen) C.; m. Katherine Wallace Barry, Sept. 27, 1935; children—Sarah Ellen Christensen Nelson, Melanie Barry Christensen Behrendt, Jane Martin Christensen Vance. B.S., U. Minn., 1929, M.S., 1930, Ph.D., 1937; postgrad., U. Halle, Halle an der Saale, Germany, 1932-33. Instr. U. Minn., 1929-37, asst. prof., 1937-46, asso. prof., 1946-48, prof. plant pathology, 1948—, Regents' prof., 1973—; Cons. various grain storage and processing firms. Author: Common Edible Mushrooms, 1943, Common Fleshy Fungi, 1946, The Molds and Man, 1953; Spanish edit. Los Hongos y El Hombre, 1964, (with H.H. Kaufmann) Grain Storage: The Role of Fungi in Quality Loss, 1969, Molds, Mushrooms and Mycotoxins, 1975, E.C. Stakman, Statesman of Science, 1984, (with R.A. Meronuck), Quality Maintenance in Stored Grains and Seeds, 1986; also articles. Fellow Am. Phytopath. Soc., Am. Coll. Allergists (hon.); mem. Am. Soc. Microbiology, Mycol. Soc. Am., Sigma Xi. Home: 1666 Coffman St # 315 Saint Paul MN 55108-1326

CHRISTENSEN, DAVID WILLIAM, mathematician, engineer; b. San Francisco, Jan. 19, 1937; s. Christopher Drost and Wilma (Hallowell) C.; m. Felicity Ann Bush, Nov. 2, 1963; children: Karen Ann, Paul Thomas. Student, MIT, 1954-58; BA, BS in Math., U. Calif., Santa Barbara, 1960; MIM in Internat. Mgmt. with honors, Am. Grad. Sch., Glendale, Ariz., 1973. Registered profl. engr., Calif. Project engr. North Am. Rockwell, Anaheim, Calif., 1963-70, Litton Ingalls Ships, Pascagoula, Miss., 1970-73; coord. of fin. Sonatrach Oil Co., Algiers, Algeria, 1975-78; revenue cons. Saudi Arabian Bechtel, Jubail, Saudi Arabia, 1978-80; sr. planner Arabian Am. Oil Co., Dhahran, Saudi Arabia, 1980-86; strategic planning and project controls Bechtel Power Corp., San Francisco, 1987—; mem. (from Jubail) Saudi Royal Commn. Com. on Indsl. Devel., Riyadh, Saudi Arabia, 1978-80; lectr. U. Calif., Santa Barbara, 1964. Contbr. articles to profl. publs. Mem. Charcot-Marie-Toothe Assn., Balt., 1987—, Gertrude Herbert Art Inst., Augusta, Ga., 1990—, Jr. C. of C., Santa Barbara, Calif., 1960-64. Recipient Boit prize, 1957. Mem. NSPE (power group 1990), Soc. Am. Mil. Engrs. Republican. Episcopalian. Achievements include development with others of hydrocarbon and industrial resources in the Middle East (Algeria, Egypt, Saudi Arabia). Office: Bechtel Inc 50 Beale St San Francisco CA 94119

CHRISTENSEN, JULIEN MARTIN, psychologist, educator; b. Capron, Ill., Sept. 3, 1918; s. Peter Martin and Lucile Marie (Edson) C.; m. Imogene E. Willis, Mar. 27, 1954; children: Kim, Kyle, Karen. BS, U. Ill., 1940; MA, Ohio State U., 1952, PhD, 1959; ScD (hon.), U. Dayton, 1989. Diplomate Am. Bd. Forensic Psychology; lic. psychologist, Ohio, lic. scuba diver, pilot. Statis. clk., personnel technician U.S. Air Force Tng. Command, Ft. Worth, 1941-43; research scientist Aerospace Med. Research Lab., Wright-Patterson AFB, Ohio, 1946-56; dir. human engring. div. Aerospace Med. Research Lab., 1956-74; prof. dept. indsl. engring. and ops. research Coll. Engring. Wayne State U., Detroit, 1974-78; dept. chmn. Wayne State U., 1974-77; dir. human factors div. Stevens, Scheidler, Stevens, Vossler Inc., Dayton, 1978-80; chief scientist human factors Gen. Physics Corp., Dayton, 1980-83; chief scientist Ergonomics Div. Universal Energy Systems, Inc., 1983—, UES, Inc. (formerly Universal Energy Systems, Inc.), 1983—; vis. lectr. U. Mich., Ann Arbor, 1959—; chmn. NASA Behavior/Tech. Com., 1976; vis. lectr. USPHS Sch. Medicine, 1977-81; adj. prof. Wright State U., U. Dayton, Wittenberg U., Sinclair Coll., Ohio State U.; vis. lectr. Air Force Inst. Tech,; cons. and lectr. in field; one of six experts on govt./industry adv. group for NASA man-systems integration standards, 1986-87. Mem. editorial bd. Jour. Systems Engring., 1969, Jour. Safety Research, 1969. Contbr. chpts. to books and articles to profl. jours. Cons. and mem. of numerous editorial bds. of profl. jours. Chmn. human performance sci. adv. com. for Manned Orbiting Lab. Developer of methods for gathering in-flight activity data. Served with USAF, 1943-46. NSF fellow, 1957; named Outstanding Scientist Engring. and Sci.Found., Dayton, Ohio, 1986; recipient AF Assn. Citation of Honor, 1966, AF Decoration of Exceptional Civilian Svc., 1966, Pres. disting. svs. award Human Factors Soc., 1992. Fellow Human Factors Soc. (pres. 1964-65, 86-87), Am. Psychol. Assn. (Franklin V. Taylor award 1969); mem. NAS/NRC (naval studies bd. future carrier study 1990), Explorers Club, Am. Soc. Safety Engrs. (acad. accreditation council 1979-85), Am. Soc. Safety Research (bd. govs. 1975-79), Soc. Automotive Engrs. (chmn. human factors com. 1975-77, exec. com.), Soc. Engring. Psychologists (past pres., past chmn. exec. council), Can. Human Factors Assn. (Julien M. Christensen award), Internat. Ergonomics Research Soc., Miami Valley Psychol. Assn. (past pres.), Soc. Logistics Engrs., S.E. Mich. Human Factors Soc. (past pres.), Systems Safety Soc., Inst. Nuclear Power Ops. (adv. coun. 1989—), Alpha Kappa Psi, Pi Mu Epsilon, Psi Chi, Tau Beta Pi. Club: Pole Vaulters. Home: 5950 Little Sugar Creek Rd Dayton OH 45440 Office: UES, Inc. 4401 Dayton-Xenia Rd Dayton OH 45432

CHRISTENSEN, KARL REED, aerospace engineer; b. Moses Lake, Wash., Mar. 20, 1962; s. Robert Reed and Joanne L. (McMahan) C.; m. Lisa Kay Human, July 1, 1983; children: Todd, Rachael, Scott. BS in Design Tech., Brigham Young U., 1986; MSME, Ariz. State U., 1989. Engr. Garrett Engine div. Allied-Signal Aerospace Co., Phoenix, 1989—. Mem. ASME. Home: 2215 E Cathedral Rock Dr Phoenix AZ 85044

CHRISTENSEN, OLE, general practice physician; b. Copenhagen, May 12, 1944; s. Gunnar and Rigmor (Sørensen) C.; m. Mariann Drasbek Jensen, Nov. 5, 1945; children: Mette, Søren. Cand. med., U. Copenhagen, 1971. Mem staff dept. surgery Amtssygehuset Nakskov, Denmark, 1971-73, mem. staff dept. medicine, 1973-74; pvt. practice Nakskov, 1974—; cons. Egeborg addict treatment ctr., Copenhagen ministry of health, 1990—, Nakskov alcohol abuse clinic, 1974—, County Coun., Storstrøm, 1986-92. Contbr. articles to profl. jours. Recipient Town of Yr. prize of Initiative, 1993. Mem. Danish Med. Assn., Danish Assn. Gen. Practitioners, Rotary. Office: Krøyers Gaard 1, DK 4900 Nakskov Denmark

CHRISTENSEN, RALPH J., chemistry educator; b. Payson, Utah, Aug. 10, 1953; s. Jerry Blaine and Wilda (Youd) C.; m. Amylyn Melanie Hansen, Aug. 4, 1978; children: Jay Ralph, Aimee Rae, Mark Jerry, sherman Kent. ME, U. Utah, 1979; PhD, Tex. A&M U., 1987. Assoc. prof. chemistry and physics Howard Coll., Big Spring, Tex., 1983-88; assoc. prof. chemistry Pikeville (Ky.) Coll., 1988—, div. head, 1989—. Scout leader Boy Scouts Am., Big Spring, 1984-88, Pikeville, 1990—; vol. coach YMCA, Pikeville, 1989—. Mem. Am. Chem. Soc., Nat. Sci. Tchrs. Assn. LDS. Achievements include research in general, organic, allied health, physical, instrumental, environmental, radiation, inogranic and anlytical chemistry; general and engineering physics. Office: Pikeville Coll 214 Sycamore Pikeville KY 41501

CHRISTENSEN, RICHARD MONSON, mechanical engineer, materials engineer; b. Idaho Falls, Idaho, July 3, 1932; married, 1958; 2 children. BSc, U. Utah, 1955; ME, Yale U., 1956, DEng, 1961. Structural engr. Convair Divsn., Gen. Dynamics, 1956-58; with technical staff TRW Systems, 1961-64; asst. prof. mech. engring. U. Calif. Berkeley, 1964-67; staff rsch. engr. Shell Devel. Co., 1967-74; prof. mech. engring. Washington U., 1974-76; sr. scientist technical staff Lawrence Livermore (Calif.) Nat. Lab., 1976—; lectr. U. So. Calif., 1962-64, U. Calif. Berkeley, 1969-70, 78, 80, U. Houston, 1973; mem. U.S. Nat. Com. Theoretical and Applied Mechanics, 1980—. Fellow ASME (chmn. applied mechanics divsn. 1980-81, hon. mem. 1992); mem. Nat. Acad. Engring., Am. Chem. Soc., Soc. Rheology. Achievements include research in properties of polymers, in wave propagation, in failure theories, in crack kinetics, in composite materials. Office: Lawrence Livermore Nat Lab PO Box 808 L-355 Livermore CA 94551-0808*

CHRISTENSEN, SØREN BRØGGER, medicinal chemist; b. Tønder, Denmark, Aug. 19, 1947; s. Hans and Inger (Bang) C.; m. Helle Flyger, July 1, 1972; children: Elise Flyger, Cecilie Brøgger. PharmM, Royal Danish Sch. of Pharmacy, Copenhagen, 1971, PhD, 1975. Vis. prof. UCLA, 1974; from assoc. prof. to asst. prof. Royal Danish Sch. Pharmacy, Copenhagen, 1975-90, readership, 1990—, chmn. dept. organic chemistry, 1993—; hon. rsch. fellow Univ. Coll., London, 1984; ofcl. apptd. censorship Technical U., Copenhagen, 1990—. Editor: Natural Products and Drug Development, 1984, New Leads and Targets in Drug Research, 1992; contbr. articles to profl. jours; patentee in field. Mem. Am. Chem. Soc., Danish Chem. Soc., Danish Acad. Natural Scis. Avocation: rowing. Home: Aatoften 187, DK-2990 Nivaa Denmark Office: Royal Danish Sch Pharmacy, Universitetsparken 2, DK-2100 Copenhagen Denmark

CHRISTHILF, DAVID MICHAEL, aerospace engineer; b. Springfield, Ohio, Dec. 7, 1959; s. Franklin Donald and Doris Lucille (Overholt) C.; m. Corinne Fay Barnes, Feb. 15, 1986; children: Nathaniel David, Reneé Elizabeth. BS in Aerospace and Ocean Engring., Va. Tech U., 1984. Assoc. engr. Planning Rsch. Corp., Hampton, Va., 1986-89; engr. Lockheed Engring. and Scis. Co., Hampton, Va., 1989—. Contbr. articles and tech. memos to profl. publs. Recipient Group Achievement award NASA, 1990, Army Achievement medal Dept. of Army, 1991. Mem. AIAA. Republican. Baptist. Achievements include co-design and testing of an eight sensor four-control surface flutter suppression control system for the full-span, free-to-roll Active Flexible Wing Wind-tunnel model. Office: Lockheed Engring & Scis Co Attn: MS 489 144 Research Dr Hampton VA 23666

CHRISTIAN, CARL FRANZ, electromechanical engineer; b. Naugatuck, Conn., May 19, 1929; s. Joseph Francis and Isabel Rose (Lizdas) C.; m. Florence (Bates) Christian, May 15, 1955 (div. 1977); children: Peter Bates, Matthew Bates; m. Cynthia Belle Suntup, Apr. 15, 1977. BSEE, U. Bridgeport, 1963; MS in Ocean Engring., U. R.I., 1967; MSEE, Pacific Western U., 1977, PhD in Elec. Engring., 1978. Registered profl. engr., Conn. Engr. Lewis Engring. Co., Naugatuck, 1952-64; sr. engr. electric boat div. Gen. Dynamics Co., various locations, 1964-77; systems engr. Martin Marietta Co., Cape Canaveral, Fla., 1978-87; owner, cons. C2 Engring., New London, Conn., 1987-92; ret., 1992. Mem. Elks. Achievements include patent on arc welding machine with memory of last heat achieved. Home: 2 Cove View Rd New London CT 06320

CHRISTIAN, FREDERICK ADE, entomologist, physiologist, biology educator; b. Lagos, Nigeria, July 8, 1937. BS, Allen U., 1962; MS, Wayne State U., 1964; PhD in Parasitology, Ohio State U., 1969. Teaching assoc. biology Wayne State U., 1962-64; teaching asst., instr. zoology Ohio State U., 1965-69; from asst. to assoc. prof. So. U., Baton Rouge, 1969-74, prof. biology, dir. health rsch. ctr., 1975—, dir. rsch. coll. sci., 1976—; prin. investigator USDA, 1972-77, NIH, 1972—, NSF, 1980—; coun. mem. marine consortium La. U., 1978—; extra-mural assoc. NIH, 1979; chmn. u. rsch. coun. So. U., 1979—, mem. rsch. incentive com., 1981—. Mem. Am. Soc. Parasitologists, Am. Microscopic Soc., Nat. Minority Health Affairs Assn., Am. Inst. Biol. Sci. Achievements include research in physiology of host-parasite relationships of Fasciola hepatica liver fluke of cattle and man, the effects of environmental pollution pesticides on parasites physiology and occurrences. Office: So U Agrl & Mech Coll Health Rsch Ctr Leehold Rm 128 Ste F Baton Rouge LA 70813*

CHRISTIAN, JOE CLARK, medical genetics researcher, educator; b. Marshall, Okla., Sept. 12, 1934; s. Roy John and Katherine Elizabeth (Beeby) C.; m. Shirley Ann Yancey, June 5, 1960; children: Roy Clark, Charles David. BS, Okla. State U., 1956; MS, U. Ky., 1959, PhD, 1960, MD, 1964. Cert. clin. geneticist, Am. Bd. Med. Genetics. Resident internal medicine Vanderbilt U., Nashville, 1964-66; from asst. prof. to full prof. med. genetics Ind. U., Indpls., 1966-87, prof., chmn. med. genetics, 1978—. Served with USAR, 1953-60. Mem. AMA, Am. Soc. Human Genetics. Democrat. Methodist. Avocations: bicycling, farming. Office: Ind U Dept Med/Molecular Genetics 975 W Walnut St Indianapolis IN 46202-5251

CHRISTIAN, THOMAS FRANKLIN, JR., aerospace engineer, educator; b. Macon, Ga., Mar. 2, 1946; s. Thomas Franklin and Lucille Vanessa (Solomon) C.; B.A.E., Ga. Inst. Tech., 1968, M.S.A.E., 1970, Ph.D., 1974; MS in Engring. Adminstrn., U. Tenn., 1976. Registered profl. engr.; cert. profl. logistician; m. Jan McGarity, Apr. 30, 1983; children: Ellen Caroline, Thomas Franklin III. Sr. design engr. nuclear analytical engring. Combustion Engring., Inc., Chattanooga, 1973-77; team mgr. IF-1 pilot plant Procter & Gamble, Macon, 1977-80; program mgr. durability and damage tolerance assessment, Warner Robins Air Logistics Center, Robins AFB, Ga., 1980-85, chief engr., 1985-87, sect. chief, 1987-90, br. chief, 1990, div. chief sof fixed wing div., 1990—; adj. prof. math. Cleveland State Community Coll., 1976-77; adj. prof. engring. Mercer U., 1986—; continuing engr. edn., DTA short course coord., instr. George Washington U. Registered profl. engr., Ga., Tenn. Adminstrn. br. mem., sewardship chmn. Trinity United Meth. Ch., 1991—. Assoc. fellow AIAA (mem., nat. tech com. on aerospace maintenance 1982-86, structures 1985-87, system safety and effectiveness 1987-89); mem. N.Y. Acad. Scis., ASME, Soc. for History Tech., Soc. for Exptl. Stress Analysis, AAAS, Soc. Logistics Engrs. (Ga. state dir. 1983-89, Schoenberg award 1985, Joyner award, 1992), ASTM, Air Force Assn., Am. Acad. Mechanics, Ga. Inst. Tech. Alumni Assn., Univ. Tenn. Alumni Assn., Order of Engr., Sigma Xi, Pi Tau Chi. Home: 101 Chadwick Dr Warner Robins GA 31088-6401 Office: LUF Warner Robbins Air Logistics Ctr Robins AFB GA 31098

CHRISTIANSEN, DENNIS LEE, transportation engineer; b. Madison, Wis., Feb. 10, 1948; s. Herman L. and Faith H. (Haase) C.; m. Gayla Ann Eppright, Feb. 27, 1971. BS, Northwestern U., 1970; MS, Tex. A&M U., 1972, PhD, 1977. Registered profl. engr., Tex. Hwy. engr. Howard Needles Tammon & Bergendoff, Kansas City, Mo., 1970-71; program mgr. Tex. Transportation Inst., College Station, 1972-83, div. head, 1983-92, assoc. dir., 1992—. Contbr. over 100 rsch. reports and papers to profl. jours. Fellow Inst. Transp. Engrs. (Engr. of Yr. Tex. sect. 1989, tech. Paper award

1984, Tech. Coun. award 1988); mem. Transp. Rsch. Bd. (Fred Burgraff award 1979), NSPE, Internat. Inst. Transp. Engrs. (bd. dirs. 1991—). Office: Tex A&M U Tex Transp Inst College Station TX 77843

CHRISTIANSEN, TIM ALAN, ecologist; b. Port Clinton, Ohio, Nov. 2, 1951; s. August and Betty Lou (Raven) C. BS, Ohio State U., 1975; PhD, U. Wyo., 1988. Rsch. asst. Savannah River Ecology Lab., Aiken, S.C., 1976-80; researcher U. Okla., Norman, 1980-82; cons. pvt. archaeologist firms Laramie, Wyo., 1982-85; grad. asst. U. Wyo., Laramie, 1985-88, rsch. asst., 1988-89, state extension entomologist, asst. prof., 1989-91; post-doctoral fellow W.Va. U., Morgantown, 1991—; advisor, researcher Nat. Park Svc., Laramie, 1989-91. Contbr. articles to profl. jours. Vol. Natural Heritage Found., Okla., 1980-82, Nature Conservancy, Wyo., 1988-91, Nature Conservancy, W.Va., 1992. Office: Wst Virginia U Div Forestry PO Box 6125 Morgantown WV 26506

CHRISTIE, LAURENCE GLENN, JR., surgeon; b. Houston, May 13, 1930; s. Laurence Glenn and Tommie Katherine (Myers) C.; m. Constance Graham Kelsey, Sept. 15, 1973; 1 child, Susan Elizabeth. BS, Washington and Lee U., 1953; MD, Med. Coll. Va., 1957. Diplomate Am. Bd. Surgery. Intern surgery Med. Coll. Va., Richmond, 1957-58, resident surgery, 1957-62, clin. instr., 1963—; practice medicine specializing in gen. and vascular surgery, Ft. Smith, Ark., 1962-63, Richmond, Va., 1963—; mem. active staff Henrico Doctors Hosp.; mem. courtesy staff Johnston-Willis Hosp., Stuart Circle Hosp., St. Mary's Hosp., Richmond Meml. Hosp., St. Luke's Hosp., Retreat Hosp.; chmn. dept. surgery chmn. med. exec. com., med. dir. Henrico Doctors Hosp., also vice chmn. bd. trustees, 1981—, chief staff, 1982—; courtesy staff Richmond Met. Hosp., Johnston-Willis Hosp.; pres. Med. Planning Corp.; mem. sci. adv. bd. Richmond chpt. Nat. Found. for Ileitis and Colitis, Inc. Contbr. articles to profl. jours. Fellow ACS; mem. Southeastern Surg. Congress, So. Med. Assn., Richmond Acad. Medicine, Richmond Surg. and Gynecol. Soc., Med. Soc. Va., AMA, Humera Soc. Episcopalian. Clubs: Bull and Bear, Irish Setter of Greater Richmond, Irish Setter of Am. Home: 1224 The Forest Crozier VA 23039-2419 Office: 7605 Forest Ave Ste 402 Richmond VA 23229-4936

CHRISTIE, RICHARD G., plant pathologist; b. Dunedin, Fla., 1934. Bs, U. Fla. Lab. tech. plant virus lab. U. Fla., Gainesville, prof. dept. agronomy. Recipient Ruth Allen award Am. Phytopathological Soc., 1993. Achievements include the development of innovative techniques for detecting and identifying plant virus inclusions with the light microscope. Office: Univ of Florida Dept Agronomy Gainesville FL 32611*

CHRISTIE, STEVEN LEE, aerospace engineering researcher; b. Tacoma Park, Md., June 7, 1960; s. Tom Edward and Nancy (Knight) C. BS in Applied Mechanics, U. Calif., San Diego, 1983; MS in Aerospace Engring., U. So. Calif., 1988, PhD in Aerospace Engring., 1993. Cert. engr.-in-tng., Calif. Mem. tech. staff Rocketdyne divsn. Rockwell Internat., Canoga Park, Calif., 1983-88; rsch. asst. dept. aerospace engring. U. So. Calif., L.A., 1988-92, teaching asst., 1992—. Mem. Am. Phys. Soc., Am. Geophys. Union. Achievements include discovery that statistical distributions and heat transfer relations used to characterize "hard" and "soft" turbulence (two regimes of thermal turbulence) in Rayleigh-Benard connection depend on cell geometry and are not universal; that exponential distributions indicate small scales of motion and Gaussian distributions indicate large scales, rather than hard and soft regimes as previously claimed. Home: 2105 Via Visalia Palos Verdes Estates CA 90274 Office: U So Calif Dept Aerospace Engring University Pk Los Angeles CA 90089-1191

CHRISTMAN, EDWARD ARTHUR, physicist; b. Lakewood, Ohio, Aug. 3, 1943; s. John N.H. and Mary Elizabeth (Fuller) C.; m. Florence T. Cua, July 21, 1979. MS, Rutgers U., 1975, PhD, 1977. Mech. engr. missile systems div. AVCO Corp., Wilmington, Mass., 1966-72; instr. Rutgers U., New Brunswick, N.J., 1975-77, radiol. physicist, 1977-89, assoc. dir., 1989-91; dir. environ. health and safety Columbia U., N.Y.C., 1991—; cons. in field, North Brunswick, N.J., 1977—; assoc. faculty Rutgers U., 1978—; faculty Columbia U., 1991—. Mem. Health Physics Soc. N.J. (pres. 1989-90), Health Physics Soc. Home: 1353 Seneca Rd North Brunswick NJ 08902 Office: Columbia U Health Scis 630 W 168th St New York NY 10032

CHRISTMAN, LUTHER PARMALEE, university dean emeritus, consultant; b. Summit Hill, Pa., Feb. 26, 1915; s. Elmer and Elizabeth (Barnicoat) C.; m. Dorothy Mary Black, Dec. 5, 1939; children: Gary, Judith, Lillian. Grad., Pa. Hosp. Sch. Nursing for Men, 1939; BS, Temple U., 1948, EdM, 1952; PhD, Mich. State U., 1965; LHD (hon.), Thomas Jefferson U., 1980. Cons. Mich. Dept. Mental Health, Lansing, 1956-63; assoc. prof. psychiat. nursing U. Mich., 1963-67; rsch. assoc. Inst. Social Rsch., U. Mich., 1963-67; prof. nursing and sociology, dean nursing Vanderbilt U., 1967-72; DON Vanderbilt U. Med. Ctr. Hosp.; prof. sociology Rush Coll. Health Scis., Chgo.; sr. scientist Rush-Presbyn.-St. Luke's Med. Center; prof. nursing, v.p. nursing affairs Coll. Nursing Rush U., 1972-87; dean Coll. Nursing Rush U., 1972-87; dean emeritus Coll. Nursing, Rush U., 1987—; sr. advisor to pres. Ctr. of Nursing, Am. Hosp. Assn., 1989; pvt. cons., 1989—; pres. Christman-Cornesky & Assocs., 1990—; cons. community svcs. and rsch. br. NIMH, 1963-66; psychiat. rsch. project So. Regional Edn. Bd., 1964-67; Chmn. planning com. 1st Midwest Conf. Psychiat. Nursing, Mpls., 1956; mem. team to survey mental health facilities of Colo. NIMH, 1962, of Ga., 1964; mem., workshop leader White House Conf. on Children, 1970; mem. nursing panel Nat. Commn. for Study Nursing and Nursing Edn., 1968-70; mem. regional med. programs rev. com. Health Svcs. and Mental Health Adminstrn., HEW, 1968-72; cons. dept. medicine and surgery VA Cen. Office, 1968-71, 74-77; mem. panel nurse cons. to com. on nursing AMA, 1968-71; mem. health svcs. adv. com. Am. Assn. Med. Colls., 1968-71; mem. action com. pub. health Am. Health Found., 1970-72; mem. membership com. Inst. Medicine, Nat. Acad. Sci., 1972-76, mem. com. on edn. in health professions, 1973-75; participant numerous confs. in field; mem. S.D. Bd. Nursing, Tenn. Bd. Nursing. Contbr. numerous articles to profl. jours. Recipient Disting. Practitioner award Nat. Acads. Practice, 1985, Old Master Purdue U., 1985, Coun. of Specialists in Psychiat. and Mental Health Nursing award, 1980, Hon. Recognition award Ill. Nurses Assn., 1987, Edith Copeland Founders award for Creativity, 1981, Lifetime Achievement award Sigma Theta Law Internat., 1991, History Makers in Nursing award Ctr. for Advancement of Nursing Practice, Beth Israel Hosp./Mass. Gen. Hosp., 1992, Mem. of Yr. award Am. Assembly for Men in Nursing, 1992, Dist. Alumnus award Temple U., 1992; named dean emeritus Am. Assn. Colls. of Nursing, 1988. Fellow AAAS, Am. Acad. Nursing, Inst. Medicine Chgo., Soc. Applied Anthropology; mem. ANA (3d v.p., Jesse M. Scott award 1985), Mich. Nurses' Assn. (pres. 1961-65), Am. Sociol. Assn., Soc. Gen. Systems Rsch., Inst. Medicine, N.Y. Acad. Scis., Biomed. Engring. Soc., Nat. Acads. of Practice (chmn. acad. nursing 1985-92), Alpha Omega Alpha (hon.), Alpha Kappa Delta. Home: 5535 Nashville Hwy Chapel Hill TN 37034-2074

CHRISTODOULOU, DEMETRIOS, mathematics educator; b. Athens, Oct. 19, 1951; s. Lambros Christodoulo and Maria Georgiades; m. Kathleen Kelly, Mar. 8, 1973; children: Penelope, Alexandra. MA in Physics, Princeton U., 1970, PhD in Physics, 1971. Rsch. fellow Calif. Inst. Tech., 1971-72; prof. U. Athens, 1972-73; vis. scientist CERN, 1973-74, Internat. Ctr. for Theoretical Physics, 1974-76; Humboldt fellow Max Planck Inst., 1976-81; vis. mem. Courant Inst., 1981-83; from assoc. prof. to prof. Syracuse U., 1983-87; prof. math. Courant Inst. NYU, 1988-92; prof. math. Princeton U., 1992—. Author: (with S. Klainerman) The Global Nonlinear Stability of the Minkowski Space, 1992; contbr. articles on gravitational collapse and formation of black holes and singularities to Comms. in Math. Physics, Comms. on Pure and Applied Math., 1984-93. Recipient Otto Hahn medal for math. physics Max Planck Soc., 1980, Basilis Xanthopoulos award for gen. relativity, 1991, MacArthur Fellowship award MacArthur Found., 1993. Achievements include discovery of nonlinear memory effect of gravitational waves; research in nonlinear partial differential equations, general relativity, fluid dynamics. Home: Princeton Univ 270 Carriage Way Princeton NJ 08540 Office: Princeton U Dept Math Princeton NJ 08544-1000

CHRISTOFORIDIS, A. JOHN, radiologist, educator; b. Greece, Dec. 24, 1924; s. John P. and Ada A. C.; m. Ann Dimitriadis, Nov. 11, 1961; children: John, Gregory, Alex, Jimmy. M.D. summa cum laude, Nat. U. Athens, Greece, 1949; M.M.Sc., Ohio State U., 1957; Ph.D., Aristotelian U., Greece, 1969. Instr. to prof. Ohio State U., Columbus, 1956-74; clin. prof. Ohio State U., 1974—; chmn. dept. radiology Aristotelian U., Salonika, Greece, 1971; prof., chmn. dept. radiology Med. Coll. Ohio, Toledo, until 1982; prof., chmn. dept. Ohio State U., Columbus, 1982—; researcher in chest and gastrointestinal radiology; cons. Greek Ministry Health, Batelle Meml. Inst., Columbus. Contbr. to textbook Atlas of Axial Sagittal and Coronal Anatomy with Computed Tomography and Magnetic Resonance; author: Radiology for Medical Students, 4th edit., 1988, Diagnostic Radiology-Thorax, 1989; contbr. several chpts. to books, over 100 articles to med. jours. Served to lt. M.C. Greek Army, 1950-52. Recipient Silver award Ohio Med. Assn., 1969, awards Heart Assn., 1960, awards Batelle Meml. Inst., 1965, awards Astra Co., 1967, awards Lung Assn., 1970-71; named Hon. Citizen City of Thessalonike, 1973; Ohio Geriatrics Med. grantee, 1980; NSF grantee, 1980. Fellow Am. Coll. Chest Physicians, Am. Coll. Radiology; mem. AAA, AMA, AAUP, Ohio Radiol. Soc., Assn. Univ. Radiologists, Radiol. Soc. N. Am., Soc. Chmn. Acad. Radiology Depts., Fleishner Soc. (charter), Am. Hellenic Ednl. Progressive Assn., Greek-Am. Progressive Assn., Acad. of Athens (corr. mem.). Greek Orthodox. Office: Ohio State U 410 W 10th Ave Columbus OH 43210-1236

CHRISTON, MARK ALLEN, mechanical engineer; b. Denver, Nov. 1, 1958; s. Robert H. and Ethel L. (Matt) C.; m. Elisa Lucille Capron, June 7, 1980; children: Jennifer Marie, Adam Jordan. BSME, Colo. State U., 1982, MSME, 1986, PhD in Mech. Engring., 1990. Mech. design engr. Tex. Instruments, Dallas, 1982-83; rsch. assoc. advanced tech. computing staff Colo. State U., Ft. Collins, 1986-90; engr. Lawrence Livermore (Calif.) Nat. Lab., 1990—. Author: NATO ASI Series Proceedings on Seasonal Snowcovers, 1986; contbr. articles to profl. jours. Mem. ASME (Roy Rothermel scholar 1985), AIAA, Assn. Computing Machinery (program com. 1991, overall best student paper 1990), Sigma Xi, Tau Beta Pi, Pi Tau Sigma. Achievements include demonstration that the diffusion enhancement in snow undergoing temperature gradient metamorphism is greater than unity and bounded. Home: 1820 Bristlecone Dr Tracy CA 95376

CHRONISTER, RICHARD DAVIS, physicist; b. Birmingham, Ala., Aug. 17, 1943; s. Richard D. and Mary Anne (Bealmear) C.; m. Vickie A. Bacon, Apr. 10, 1965; children: Susan K., Karen J. BS in Physics, U. Okla., 1965; MS in Nuclear Engring., Ohio State U., 1968. Cert. electromagnetic compatibility engr.; registered environ. profl. Commd. 2d lt. USAF, 1965; advanced through grades to maj., 1977; Project mgr. USAF Aeropropulsion Lab., Dayton, Ohio, 1965-69; electronics survivability officer Field Command Def. Nuclear Agy., Livermore, Calif., 1969-72; grad. student U. Okla., Norman, 1972-75; mgr., transient radiation effects on electronics USAF Weapons Lab., Albuquerque, 1975-78; chief, radiation analysis lab. USAF Tech. Applications Ctr., Sacramento, Calif., 1979-83; chief, aircraft space systems USAF Nuclear Criteria Group Secretariat, Albuquerque, 1983-86; sr. engr./physicist BDM Internat., Albuquerque, 1986—. Author, co-author 32 tech. reports. Sr. mem. Am. Inst. Aeronautics and Astronautics; mem. AAAS, Am. Phys. Soc., Nat. Assn. Radio and Telecommunications Engrs. Methodist. Achievements include support of development of Army, USN and USAF systems in the areas of integrated electromagnetics, nuclear and natural environmentals, test and evaluation, and technological assessments. Home: 13005 Rebonito Rd NE Albuquerque NM 87112-4819 Office: BDM Internat 1801 Randolph Rd SE Albuquerque NM 87106-4295

CHRYSIKOPOULOS, CONSTANTINOS VASSILIOS, environmental engineering educator; b. Corfu, Greece, Mar. 4, 1960; s. Vassilios and Stavroula Chrysikopoulos. BS in Chem. Engring., U. Calif., San Diego, 1982; MS in Chem. Engring., Stanford U., 1984, Engr. degree in Civil Engring., 1986, PhD in Civil Engring., 1990. Postdoctoral fellow Stanford U., 1990-91; asst. prof. U. Calif., Irvine, 1991—. Contbr. articles to profl. jours. Grantee NSF. Home: 80 Whitman Ct Irvine CA 92715 Office: Univ Calif Civil Engring Dept Irvine CA 92717

CHRYSOCHOOS, JOHN, chemistry educator; b. Icaria, Greece, Feb. 27, 1934; came to U.S., 1964; s. Michal P. and Irene (Glaros) C.; m. Alexandra Kratsas, May 17, 1964; children: Michael J., Constantine J., Irene J. Diploma chemistry, U. Athens, Greece, 1957; MSc, U. B.C., Can., 1962, PhD, 1964. Rsch. fellow chemistry Harvard U., Boston, 1964-65; rsch. assoc. Michael Reese Hosp. & Med. Ctr., Chgo., 1965-67; asst. prof., assoc. prof., prof. chemistry U. Toledo, Ohio, 1967—; sr. vis. fellow U. Western Ont., London, 1980; adj. prof. Ctr. for Photochem. Scis., Bowling Green (Ohio) State U., 1986—; invited prof. U. Crete, Greece, 1989; cons. Owens Ill. Inc., Toledo, 1977-82; internat. panel Centre Interdisciplinary Studies in Chem. Physics, London, 1981-83; sect. chair nat. and internat. confs. Contbr. 77 articles to profl. jours. Chmn. nominating com. Hellenic Profl. Soc., Toledo, 1989-91. 1st lt. Greek Royal Navy, 1958-60. Rsch. grantee Owens-Ill. Inc., Toledo, 1969-75, NSF, Washington, 1984. Mem. AAAS, Am. Chem. Soc. (councilor 1990—), Soc. Applied Spectroscopy, Inter-Am. Photochem. Soc., N.Y. Acad. Scis., Sigma Xi. Greek Orthodox. Achievements include research in optical generation and reactivity of hydrated electron, spectroscopic properties of trivalent lanthanides in liquid laser solvents, glass-ceramic lasers, solar concentrators, spectroscopic and catalytic properties of semiconductor clusters. Home: 4708 Elm Pl Toledo OH 43613 Office: U Toledo Dept Chemistry 2801 W Bancroft Toledo OH 43606

CHRYSS, GEORGE, chemical company executive, consultant; b. Sacramento, Oct. 23, 1941; s. John and Anna (Lesko) Chryssikos; m. Joan Christie; children: Lauri, Mark, Kelly. BA in Lang., Syracuse U., 1961; BA in Econs., Rutgers U., 1971, MA in Econs., 1974. CPA, Fla. Corp. controller Burmah Castrol Corp., N.Y., 1966-69, corp. treas., 1969-71; div. v.p. Burmah Castrol Corp., Jacksonville, Fla., 1971-74; corp. v.p. mfg. Burmah Castrol Corp., Hackensack, N.J., 1974-80; exec. v.p. Burmah Castrol Corp., Irvine, Calif., 1980-84; exec. v.p., chief operating officer Hatco Chem. Corp., Fords, N.J., 1984—; bd. dirs. Bel-Ray Inc., Bel-Ray Internat.; instr. econs. Rutgers U., New Brunswick, N.J., 1975-76. Lt. USAF, 1959-63. Avocation: golf. Home: 21 Burnside Dr Short Hills NJ 07078-2105 Office: Hatco Corp King Georges Rd Fords NJ 08863-1821

CHU, ALEXANDER HANG-TORNG, chemical engineer; b. Taiwan, Republic of China, Oct. 30, 1955; came to the U.S., 1979; s. Wu-Lung and Su-Chin (Cheng) C.; m. Wei Jeng-Chu, June 22, 1981; children: Albert P., Jocelyn C. BS, Nat. Taiwan U., 1977; PhD, U. Wis., 1984. Cert. chem. engr. Rsch. asst. U. Wis., Madison, 1979-84, teaching asst., 1980-84; rsch. engr. Internat. Minerals & Chem., Terre Haute, Ind., 1984-87, project leader, 1987-88; prin. devel. investigator Abbott Labs., North Chicago, Ill., 1988—. Contbr. articles to Jour. Chromatography, Analytical Chemistry, Jour. Chem. Rsch., British Chem. Engring. Jour., Internat. Conf. on Separation Tech. 2d lt. Taiwanese army, 1977-79. Recipient Internat. Edison prize Thomas A. Edison Found., 1973, Book Coupon award, Nat. Taiwan U., 1977, Meritorious Svc. award China Youth Corps, 1975-76, Engr. award Chinese-Am. Inst., 1977. Mem. Am. Inst. Chem. Engrs. (vice chmn. 1986-88), Chinese Inst. Engrs. (pres. 1975-77), Sigma Xi. Achievements include patents for polyether antibiotics recovery and purification processes, patents for process for making silica gel for chromatography and glycopolypeptide antibiotics recovery and purification. Office: Abbott Labs 14th & Sheridan North Chicago IL 60064

CHU, ALLEN YUM-CHING, automation company executive, systems consultant; b. Hong Kong, June 19, 1951; arrived in Can., 1972; s. Luke King-Sang and Kim Kam (Lee) C.; m. Janny Chu-Jen Tu, Feb. 27, 1993. BSc in Computer Sci., U. B.C., Vancouver, Can., 1977; BA in Econs., U. Alta., Edmonton, Can., 1986. Rsch. asst. dept. neuropsychology and rsch. Alta. Hosp., Edmonton, 1977-78; systems analyst dept. agr. Govt. of Alta., Edmonton, 1978-81; systems analyst for computing resources City of Edmonton, 1981-86; pres. ANO Automation Inc., Vancouver, 1986—. Mem. IEEE Computer Soc., N.Y. Acad. Sci. Office: ANO Automation Inc, 380 W 2d Ave 2d Flr, Vancouver, BC Canada V5Y 1C8

CHU, BENJAMIN THOMAS PENG-NIEN, chemistry educator; b. Shanghai, China, Mar. 3, 1932; came to U.S., 1953; s. Charles C. and Gladys (Chen) C.; m. Louisa King, Mar. 30, 1959; children: Peter, Joanne, Laurence. BS magna cum laude, St. Norbert Coll., 1955; PhD, Cornell U., 1959. Research assoc. Cornell U., Ithaca, N.Y., 1958-62; asst. prof. U. Kans., Lawrence, 1962-65, assoc. prof., 1965-68; prof. chemistry SUNY, Stony Brook, 1968-88, Leading prof. chemistry, 1988-92, Disting. prof., 1992—, chmn. chemistry dept., 1978-85, prof. materials sci. and engring., 1982—; vis. prof. U. New South Wales, Australia, 1974, Australian Nat. U., 1974, Wayne State U., Hokkaido U., 1975; vis. scientist Inst. for Theoretical Physics, U. Calif., Santa Barbara; cons. Calgon, Pitts., 1978-80, E.I. DuPont de Nemours, Wilmington, Del., 1979—, Brookhaven (N.Y.) Instruments, 1981, USRA, Microgravity Sci. and Applications div. NASA, 1988, Bristol-Myers Squibb Co., 1990-92; hon. prof. Academia Sinica, People's Republic of China, 1992—. Author: Molecular Forces, 1967, Problems in Chemical Thermodynamics, 1967, Laser Light Scatterings, 1974; editor: NATO ASI series B: Physics, Vol. 73, 1981, SPIE Milestone series: Selected Papers on Quasielastic Light Scattering by Macromolecular, Supramolecular, and Fluid Systems, Vol. MS 12, 1990, Laser Light Scattering: Basic Principles and Practice, 2d edit., 1991; patentee prism light scattering cells, method and apparatus for determining viscosity, light scattering and spectroscopic detector, magnetic needle rheometer. Sloan Research fellow, 1966-68, John Simon Guggenheim fellow, 1968-69, Japan Soc. Promotion Sci. fellow, 1975-76, 92-93; recipient Humboldt award 1976-77, 92-93, Disting. Achievement award St. Norbert Coll., 1981. Mem. Am. Phys. Soc. (High Polymer Physics prize 1993), Am. Inst. Chemists; mem. Am. Crystallographic Assn., Am. Chem. Soc.

CHU, DAVID YUK, chemical engineer; b. Canton, People's Republic China, May 12, 1945; came to U.S., 1962; s. Kwok Tsing and Yuet Moi (Ma) C.; m. Margaret Po Yee, June 29, 1969; children: David Yue, William, Randolph. BSChemE, Lowell Tech. Inst., 1969, BS in Paper Engring., 1969. Tech. svc. engr. Westvaco, Mechan, N.Y., 1969-7l; lab. supr. Boston Insulated Wire and Cables Co., Boston, 1971-76, materials specialist, 1977-84; mgr. materials tech. svc. BIW Cable Systems, Inc., Plymouth, Mass., 1985-86; mgr. materials devel. BIW Cable Systems, Inc., North Dighton, Mass., 1987—. Fellow Am. Chem. Soc.; mem. TAPPI, AICE, Soc. Plastics Engrs., Boston Rubber Group, Jackson Park Club (pres. 1987—). Office: BIW Cable Systems Inc 20 Joseph E Warner Blvd N Dighton MA 02764-1345

CHU, JAMES CHIEN-HUA, medical physicist, educator; b. Nanking, China, July 6, 1948; came to U.S., 1972; s. Tao-tsun and Yu (Auyang) C.; m. Sherry Yuan; 1 child, Michael. MS, U. Tex., 1973, PhD, 1978. Diplomate Am. Bd. Radiology, Am. Bd. Med. Physics. Asst. prof. U. Pa. Med. Sch., Phila., 1978-87, assoc. prof., 1987-90; prof., chmn. Rush-Presbyn.-St. Luke's Med. Ctr., Chgo., 1991—; chief physicist Fox Chase Cancer Ctr., Phila., 1985-90. Contbr. articles to profl. jours. Mem. Am. Assn. Physicists in Medicine, Am. Coll. Radiology, Am. Coll. Med. Physics, Am. Soc. Therapeutic Radiology and Oncology. Office: Rush Presbyn St Lukes Med Ctr Dept Med Physics 1653 W Congress Pkwy Chicago IL 60612

CHU, JOHNSON CHIN SHENG, physician; b. Peiping, China, Sept. 25, 1918; came to U.S., 1948, naturalized, 1957; s. Harry S.P. and Florence (Young) C.; m. Sylvia Cheng, June 11, 1949; children—Stephen, Timothy. M.D., St. John's U., 1945. Intern Univ. Hosp., Shanghai, 1944-45; resident, research fellow NYU Hosp., 1948-50; resident physician in charge State Hosp. and Med. Ctr., Weston, W.Va., 1951-56; chief services, clin. dir. State Hosp., Logansport, Ind., 1957-84; active mem. Meml. Hosp., Logansport, Ind., 1968—. Research in cardiology and pharmacology; contbr. articles to profl. jours. Fellow Am. Psychiat. Assn., Am. Coll. Chest Physicians; mem. AMA, Ind. Med. Assn., Cass County Med. Soc., AAAS. Home: E 36 Lake Shafer Monticello IN 47960 Office: Southeastern Med Ctr Walton IN 46994

CHU, MON-LI HSIUNG, molecular biologist; b. Kwangtung, China, July 27, 1948; came to U.S. 1970; d. Tsun-Shiang and Ah-Wha (Yang) Hsiung; m. Shaw-Chang Chu, Nov. 10, 1972; children: Emily, Andy. BS, Nat. Taiwan U., 1970; PhD, U. Fla., 1975. Adj. asst. prof. U. Med./Dentistry N.J.-Rutgers Med. Sch., Piscataway, N.J., 1979-84, adj. assoc. prof., 1984-86; assoc. prof. Thomas Jefferson U., Phila., 1986-90, prof. molecular biology, 1990—. Contbr. over 70 articles to profl. jours. NIH grantee, 1986—. Mem. AAAS, Am. Soc. Biochemistry and Molecular Biology. Achievements include isolation and characterization of cDNAs and genomic DNAs for many human collagens, including Type I, III, VI, XVI collagens; definition of the first deletion mutation in type I collagen in a patient with Osteogenesis Imperfecta. Office: Thomas Jefferson U 233 S 10th St Philadelphia PA 19107

CHU, SHIRLEY SHAN-CHI, retired educator; b. Beijing, Feb. 16, 1929; came to U.S., 1952; d. Ching Tao and Chi Chun (Yao) Yu; m. Ting Li CHu, Sept. 4, 1954; children: Dennis, Dora, Daniel. BS, Nat. Taiwan U., 1951; MS, Duquesne U., 1954; Phd, U. Pitts., 1961. Rsch. assoc. U. Pitts., 1961-67; asst. prof. So. Meth. U., Dallas, 1968-73, assoc. prof., 1973-81, prof., 1981-88; prof. U. South Fla., Tampa, 1988-91, prof. emeritus, 1992—; cons. Poly Solar Inc., Dallas, 1978-88, Nat. Renewable Energy Lab., 1992—. Contbr. numerous articles to profl. jours. Panelist NRC, Washington, 1982—; mem. com. Coun. Internat. Exch. of Scholars, Washington, 1986-89. U. Pitts. scholar, 1960. Mem. Materials Rsch. Soc., Am. Crystallographic Assn. (publs. com. 1979-82). Avocations: knitting, needlepoint, cooking.

CHU, STEVEN, physics educator; b. St. Louis, Feb. 28, 1948; s. Ju Chin and Ching Chen (Li) C.; children: Geoffrey, Michael. BS in Physics, AB in Math., U. Rochester, 1970; PhD in Physics, U. Calif., Berkeley, 1976. Post doctoral fellow U. Calif., Berkeley, 1976-78; mem. tech. staff Bell Labs., Murray Hill, N.J., 1978-83; head quantum electronics rsch. dept. AT&T Bell Labs., Holmdel, N.J., 1983-87; prof. physics and applied physics Stanford (Calif.) U., 1987—, Frances and Theodore Geballe prof. physics and applied physics, 1990—, chmn. physics dept., 1990-93, Morris Loeb lectr. Harvard U., Cambridge, Mass., 1987-88; vis. prof. Coll. de France, fall 1990. Contbr. papers in laser spectroscopy and atomic physics, especially laser cooling and trapping, and precision spectroscopy of leptonic atoms. Co-recipient King Faisal Prize for Sci. award, 1993; Woodrow Wilson fellow, 1970, NSF doctoral fellow, 1970-74, NSF postdoctoral fellow, 1977-78; Richtmeyer Meml. Prize lectr., 1990. Fellow Am. Phys. Soc. (Herbert P. Broida prize for laser spectroscopy 1987, chair laser sci. topical group 1989), Optical Soc. Am., Am. Acad. Arts and Scis.; mem. NAS, Phi Beta Kappa. Office: Stanford Univ Dept Physics Stanford CA 94305

CHU, TING LI, electrical engineering educator, consultant; b. Beijing, Dec. 26, 1924; came to U.S., 1948; m. Shirley S. Yu, Sept. 4, 1954; children: Dennis, Dora, Daniel. BS, Catholic U. Peking, 1945, MS, 1948; PhD, Washington Univ., 1952. Asst. prof., assoc. prof. Duquesne Univ., Pitts., 1952-56; fellow rsch. scientist, mgr. elec. materials Westinghouse Rsch. Ctr., Pitts., 1956-67; prof., dir. elec. scis. ctr. Southern Methodist Univ., Dallas, 1967-88; grad. rsch. prof. elec. engring. Univ. South Fla., Tampa, 1988-91; emeritus, 1992—; cons. Tex. Instruments, Dallas, 1968-74, Nat. Renewable Energy Lab., Golden, Colo., 1992—, Golden Photon, Inc., El Paso, Tex., 1992—. Author book chpts.; contbr. articles to profl. jours. Recipient Outstanding Prof. award Southern Methodist Univ., 1973; Cert. Recognition, 1972-76. Mem. IEEE, Electrochemical Soc. Home: 12 Duncannon Ct Dallas TX 75225-1809

CHU, WAYNE SHU-WING, food industry entrepreneur, researcher; b. Canton, China, Sept. 12, 1949; came to U.S., 1970; s. Chung-Hay and Mei-Siu (Hung) C.; m. Julie Merilyn Stohler, Feb. 5, 1980; children: Nathan, Jane. BS, Rutgers U., 1974; MS, Cornell U., 1975. Application engr. Kearney Industries, South Plainfield, N.J., 1975-77; biochem. engr. Milbrew, Inc., Juneau, Wis., 1977-80; sr. rsch. engr. Van Camp Seafood Co., San Diego, 1980-84; rsch. mgr. Ralston Purina Co., St. Louis, 1984-88; tech. dir. Campbell Soup Far East, Camden, N.J., 1989-90; gen. mgr. Heinz-China, Pitts., 1990-91; pres. A. P. Technotrade Internat., Hong Kong, 1991—; guest lectr. U. Calif., San Diego, 1982, U. Hong Kong, 1991-92. Mem. Zool. Soc., San Diego, 1980-89. Recipient Procter & Gamble scholarship, 1974-75. Mem. AICE, Inst. Food Technologists (profl. mem.). Achievements include patent for fish skinning process. Home: 17H Seabird Ln, Discovery Bay Hong Kong Office: A P Technotrade Internat, Witty Comml Bldg Ste 1712D, Kowloon Hong Kong

CHUANG, FRANK SHIUNN-JEA, engineering executive, consultant; b. Taiwan, China, Sept. 5, 1942; came to U.S., 1966, naturalized, 1974; s. Swiss S. and Chin-May C.; m. Lily L. Chuang, Aug. 14, 1971; 1 child, Eugene. BS, Nat. Taiwan U., 1964; MS, U. Mass., 1968, PhD, 1971. Instr.

engring. U. Conn., 1971-72; dept. mgr. C.E. Maguire; cons. engrs. New Britain, Conn., 1972-78; v.p., cons. engrs. Hayden, Harding & Buchanan, Inc., East Hartford, Conn., 1978-82; pres., cons. engrs. L-C Assocs., Inc., Rocky Hills, Conn., 1982—; bd. dirs. Equity Bank, Wethersfield, Conn. Chmn. Wethersfield Flood Encroachment Control Bd. U. Mass. Water Resource Rsch. Ctr. grantee, 1966-71. Mem. ASCE, Nat. Soc. Profl. Engrs., Water Pollution Control Fedn., Wethersfield Country Club. Home: 38 Stonegate Dr Wethersfield CT 06109-3652 Office: L-C Assocs Inc 1960 Silas Deane Hwy Rocky Hill CT 06067-1310

CHUBB, SCOTT ROBINSON, research physicist; b. N.Y.C., Jan. 30, 1953; s. Charles Frisbie and Lydia Atherton (Robinson) C.; m. Anne Lauren Pond, June 5, 1982; 1 child, Scott Robinson Jr. BA, Princeton U., 1975; MA, SUNY, Stony Brook, 1978, PhD, 1982. Rsch. assoc. Northwestern U., Evanston, Ill., 1982-85, NRC/Naval Rsch. Lab., Washington, 1985-88; rsch. physicist Sachs Freeman Assocs., Inc., Landover, Md., 1988-89, Naval Rsch. Lab., Washington, 1989—. Contbr. articles to profl. jours. Mem. Am. Phys. Soc., Am. Geophys. Union, Sigma Xi. Presbyterian. Home: 9822 Pebble Weigh Ct Burke VA 22015 Office: Naval Rsch Lab Remote Sensing Div Code 7234 Washington DC 20375-5320

CHUBUKOV, ANDREY VADIM, physicist; b. Moscow, USSR, Feb. 24, 1959; came to U.S., 1990; s. Vadim F. and Natalia M. Chubukov; m. Katya A. Gulko, Oct. 16, 1980; children: Victor, Boris. MSc, Moscow State U., 1982, PhD, 1985. Rschr. P.L. Kapitza Inst. for Phys. Problems, Moscow, 1985—. Contbr. 55 articles to sci. jours. Postdoctoral fellow U. Ill., Urbana, 1990-92, Yale U., New Haven, Conn., 1992—. Office: Yale U Dept Physics 217 Prospect St New Haven CT 06511-8167

CHUCK, LEON, materials scientist; b. Balt., Mar. 7, 1955; s. Billy and Yuk Yin C. BSME, U. Md., 1978, MSME, 1984. Ceramic engr. Naval Rsch. Lab., Washington, 1976-79; rsch. engr. Nat. Bur. Stds., Gaithersburg, Md., 1979-86; sr. rsch. engr. Norton Co. High Performance Ceramics Div., Northboro, Mass., 1986-88; owner Advanced Structural Materials Consulting, Auburn, Mass., 1988-89; assoc. materials scientist U. Dayton (Ohio) Rsch. Inst., 1989—. Faculty advisor, coach U. Dayton Men's Volleyball, 1991, 92, 93; chmn. St. Paul's Spares and Pairs Group, Oakwood, Ohio, 1991. Mem. ASME, Am. Ceramic Soc., Am. Soc. Testing and Materials (task group leader 1991—), Am. Soc. Materials, Soc. Automotive Engrs., Soc. Exptl. Mechanics, Soc. Application and Materials Processing Engrs., Nat. Inst. Ceramic Engrs. Achievements include invention of high temperature ceramic instruments; research of failure and degradation mechanisms of structural ceramics for high temperature applications (turbine engine blades & combustion chambers)for long term reliability. Home: 560 Oaknoll Dr Springboro OH 45066-9676 Office: U Dayton Rsch Inst 300 College Park Dayton OH 45469-0172

CHUDZIK, DOUGLAS WALTER, internist; b. Newark, Dec. 18, 1946; s. Stanley Anselm and Irene Victoria (Winkler) C.; m. Jeanmarie Murphy, Jan. 18, 1975; children: Douglas, Jeanmarie, Gregory. BS in Biology, Xavier U., 1968; MD, U. Autonoma de Guadalajara, Mex., 1972. Intern St. Barnabas Med. Ctr., Livingston, N.J., 1974, resident in internal medicine, 1975-77; attending physician dept. medicine Bayshore Community Hosp., Holmdel, N.J., 1977—, dir. medicine, 1988-89; assoc. attending physician Riverview Med. Ctr., Red Bank, N.J., 1977—. Fellow Internat. Coll. Physicians, Am. Coll. Internat. Physicians. Democrat. Roman Catholic. Home: 88 Mallard Rd Middletown NJ 07748-2950 Office: 31 Village Ct Hazlet NJ 07730-1533

CHUE, SECK HONG, mechanical engineer; b. Singapore, Singapore, Aug. 9, 1942; arrived in Malaysia, 1946; s. Poh Fun Chu and Siew Mui Chiu; m. Lee Na Lim, 1972. B in Engring., U. Malaya, Kuala Lumpur, Malaysia, 1967; MSME, Purdue U., 1969, PhD, 1970. Registered profl. engr., Malaysia. Rsch. asst. Purdue U., Lafayette, Ind., 1968-70; lectr. U. Singapore, 1970-73, Sci. U. of Malaysia, Penang, 1973-75; cons. engr. R. J. Brown & Assocs. (Far East) Pvt. Ltd., Singapore, 1975-1976; lectr. U. Malaya, Kuala Lumpur, 1976-80; cons. engr. Jurutera Konsultant (S.E. Asia) Sendirian Berhad, Kuala Lumpur, 1980-84; cons. engr., assoc. dir. Perunding HHYC, Kuala Lumpur, 1985-87; cons. engr. Sepakat Setia Perunding Sendirian Berhad, Kuala Lumpur, 1987—; assoc. Sepakat Setia Perunding (E & M) Sdn, Kuala Lumpur, 1991—. Author: (tech. report) Analytical and Experimental Study of Heat Transfer in a Simulated Martian Atmosphere, 1970; (book) Thermodynamics--A Rigorous Postulatory Approach, 1977; (monograph) Pressure Probes for Fluid Measurement, 1975. Fulbright scholar U.S. Govt., 1967, Pres. scholar Purdue U., 1968-69. Mem. Bd. Engrs. Malaysia. Achievements include research in fluid mechanics, offshore and coastal engineering. Home: 2 Jalan Dedap, 53000 Kuala Lumpur Malaysia

CHUGH, YOGINDER PAUL, mining engineering educator; came to U.S., 1965, naturalized, 1975; s. Atma Ram and Dharam (Devi) C.; m. Evangeline Negron, July 18, 1970; children: Anjeli K., Shimilee M., Pauline E. BS, Banaras Hundu U., 1961; MS, Pa. State U., 1968, PhD, 1970. Cert. 1st class mine mgr., India. Instr. Banaras Hindu U., India, 1961-64; asst. mgr. Andrew Yule Coal Co., 1961-64; research asst. Pa. State U., University Park, 1965-70; research assoc. Henry Krumb Sch. Mines, 1971, Columbia U., N.Y.C., 1971; research engr. Ill. Inst. Tech. Research Inst., Chgo., 1971-74; planning engr. Amax Coal Co., Indpsl., 1974-76; assoc. prof. Dept. Mining Engring. So. Ill. U., Carbondale, 1977-81, prof. Dept. Mining Engring., 1981—; acting chmn., Dept. Mining Engring., So. Ill. U., 1981-82, chmn., 1984—; chmn. PhD. com., 1983-86, active numerous other univ. coms; cons. to nat. and internat. coal cos., state and fed. mining and mineral agys.; dir. Coal Combustion Residues Mgmt. Program, 1990—; bd. dirs. Accreditation Bd. for Engring. and Tech., 1989-92. Author: (with K.V.K. Prasad) Workshop on Design of Coal Pillars in Room-and-Pillar Mining, Workshop on Design of Mine Openings in Room-and-Pillar Mining, 1984; editor (with others) Proceedings of the First Conference on Ground Control Problems in the Illinois Coal Basin, 1980, Proceedings First International Conference on Ground Control in Longwall Mining and Mining Subsidence, 1982, Proceedings of the Polish-American Conference on Ground Control in Room-and-Pillar Mining, 1983; editor Ground Control Room and Pillar Mining, 1983 (Soc. Mining Engrs. award 1983), Longwall Mining Subsidence, 1983 (Soc. Mining Engrs. award 1984), Proceedings of the Second Conference on Ground Control Problems in the Illinois Coal Basin, 1985, Proceedings of the Third and Fourth Conference on Ground Control Problems in the Midwestern U.S., 1990, 92; contbr. over 50 articles to profl. jours., also many research reports; inventor roof truss, 1990. V.p. India Assn., Indpls., 1975. Recipient numerous research grants state and fed. agys., pvt. coal cos.; named Disting. Alumnus Banaras Hindu U., 1985. Mem. AIME (active rock mechanics unit com. 1978-82, various pubs. coms.1979-85, geomechanics com. 1984-85), ASTM, ASCE, Internat. Soc. Rock Mechanics (coordinator 1986—), Internat. Bur. Strata Mechanics, Ill. Mining Inst. (bd. dirs.), Ind. Mining Inst., Soc. Geologists and Mining Engrs. (faculty advisor 1977-78, 80-84), Soc. Exptl. Stress Analysis, Am. Soc. Higher Edn., Sigma Xi. Lodge: Rotary. Avocations: tennis, boating, badminton, computers. Office: Southern Illinois U-Carbondale Mining & Mineral Rscs Rsrch Inst Dept Mining Engring Carbondale IL 62901

CHUN, LOWELL KOON WA, architect; b. Honolulu, Sept. 2, 1944; s. Kwai Wood and Sara Lau C. BA in Eng., U. Hawaii, 1967; BArch, Cornell U., 1971. Registered profl. architect, Hawaii. Archtl. designer Wilson, Okamoto & Assocs., Honolulu, 1972-74; architect, planner Aotani & Assocs., Inc., Honolulu, 1974-82; design planner Daniel, Mann, Johnson & Mendenhall, Manila and Honolulu, 1982-84; architect, planner Alfred A. Yee div. Leo A. Daly Co., Honolulu, 1984-87; prin. Lowell Chun Planning & Design, Honolulu, 1987-89; dir. planning Daniel, Mann, Johnson & Mendenhall, Honolulu, 1989; assoc. AM Ptnrs., Inc., Honolulu, 1989-92; planning svcs. officer Hawaii Community Devel. Authority, Honolulu, 1992—; pres. LPC Internat., 1993—; rsch. bd. advisors Am. Biog. Inst., Raleigh, S.C., 1990—; dep. dir. gen. The America's, 1990—; mem. IBC adv. coun. Internat. Biog. Centre, Cambridge, Eng., 1990—. Prin. author: Kauai Parks and Recreation Master Plan, 1978, Hawaii State Recreation Plan (Maximum Fed. Eligibility award 1980), Maui Community Plans, 1981, Pauahi Redevel. Project, Honolulu, 1974, State Tourism Plan, Physical Resources Element, 1977, City & County of Honolulu Urban & Regional Design Plans, 1979, Hilo Civic Center Master Development Plan, 1989, New Communities, U. Petroleum and Minerals, Dhahran, Saudi Arabia, 1982, Lake Pluitt Resdl./Comml. District, Jakarta, Indonesia, 1982, Destination Resorts: Key Biscayne, Fla., Sint Maarten, Netherlands Antilles, St. Croix, U.S. Virgin Islands, Palm Springs, Calif., 1988, The Imperial Plaza Residential Commercial Complex, 1989, Kailua Elderly Housing Community Master Plan, 1990, Waimano Ridge Master Development Plan, 1991, and others. Advisor, locations officer Maitreya Inst., Honolulu, 1983-84; v.p., treas. Kagyu Theg Chen Ling Tibetan Ctr., Honolulu, 1982, 84; rep. Environ. Coalition to Hawaii State Legislature, 1974; mem. Waikiki Improvement Assn., Phys. Improvement Task Force, 1990, Hawaii Soc. Corp. Planners, 1991. Recipient Master Plan award Nat. Assn. Counties, 1975. Mem. AIA, Am. Planning Assn. (local exec. com. mem.-at-large 1987-88), Sierra Club (local vice-chmn. 1974-76), Cornell Club of Hawaii (Honolulu), Kiwanis. Buddhist. Avocations: creative writing, photography, hiking. Home: 456 N Judd St Honolulu HI 96817-1754 Office: Hawaii Community Devel Authority Ste 1001 677 Ala Moana Blvd Honolulu HI 96813

CHUNG, CHO MAN, psychiatrist; b. Hong Kong, Mar. 3, 1918; s. Tze Chun and Sau Fong (Lai) C.; m. Ping Fai Ho, Jan. 18, 1947 (dec. Mar. 1966); children: John Tze Nang, Mary; m. Lillian Auyang, Sept. 27, 1968; 1 child, Grace Yuet Chee. MD, Lingnan U., Canton, People's Republic of China, 1943; DPM, U. London, 1956. Intern, 1942-43; tech. expert Kwongtung Provincial Health Adminstrn., Canton, 1945-48; med. supt. Canton Mcpl. Hosp., 1948-49; resident psychiatry, 1954-55; psychiat. cons. Hong Kong Govt., 1957-58; lectr. psychiatry U. Hong Kong, 1957-58; pvt. practice psychiat. cons. Hong Kong, 1961—; bd. dirs. Hong Kong Cen. Hosp., 1961—; researcher in field. Lt. col. Med. Corps, 1943-45, People's Republic of China. Fellow WHO, 1955-57; corr. fellow Am. Psychiat. Assn., 1982. Fellow Royal Australian & New Zealand Coll. Psychiatrists, Royal Coll. Psychiatrists, Am. Assn. Social Psychiatrist; mem. World Psychiat. Assn. (hon., individual), Hong Kong Psychiat. Assn. (pres. 1980-82, 88-90), The Soc. Physicians of Hong Kong (pres. 1980), Hong Kong Golf Club, World Trade Ctr. Club Hong Kong. Avocation: carpentry. Office: 611 Melbourne Pla, Queens Rd Cen, Hong Kong Hong Kong

CHUNG, DOUGLAS CHU, pharmacist, consultant; b. N.Y.C., Jan. 20, 1951; s. Gook Wah and Yot Woy (Chin) C. BS, NYU, 1974; BS in Pharmacy, L.I. U., 1979; PharmD, Mercer U., 1981; postgrad., Harvard U., 1984. Lic. pharmacist, N.Y., Fla., Ala., Mass., Conn.; cert. in CPR and emergency cardiac care. Design engr. Jaros, Baum and Bolles, Cons. Engrs., N.Y.C., 1972-75; staff amd emergency disaster medications pharmacist Hosp for Joint Disease North Gen., N.Y.C., 1979, 80; postdoctoral splty. clin. resident in parenteral and enteral nutrition/metabolic care Bapt. Med. Ctr., Gadsden, Ala., 1981-83, Mercer U., Atlanta, 1981-83; clin. dir. Metabolic Support Pharmacy, Suffield, Conn., 1983-86; founder, pres., chmn., chief exec. officer Metabolic Homecare, Inc., Suffield, 1986-92; pres. Metabolic Support Inc., Suffield, 1993—; clin. cons. Caremark, Inc., Deerfield, Ill., 1985, 86; adj. prof. clin. pharmacy Mercer U., 1984-86; mem. med. staff Park City Hosp., Bridgeport, Conn., 1984—, Noble Hosp., Westfield, Mass., 1986—; lectr. in field, 1980—. Contbr. articles to Jour. Toxicology: Clin. Toxicology. Vol. health screener Asian Ams. for Community Health, N.Y.C., 1977-79. Fellow Am. Coll. Nutrition (cert.); mem. Am. Soc. for Parenteral and Enteral Nutrition, Am. Soc. Nutrition Support Svcs. (bd. dirs. 1984-86), Am. Pharm. Assn., Am. Soc. Hosp. Pharmacists, Atlanta Soc. Instnl. Pharmacists, Ala. Soc. Hosp. Pharmacists, N.Y. Acad. Scis., N.Y. State Coun. Hosp. Pharmacists, N.Y.C. Soc. Hosp. Pharmacists, European Soc. Parenteral and Enteral Nutrition, Conn. Soc. Parenteral and Enteral Nutrition (toxicology pres. 1993—), Conn. Soc. Hosp. Pharmacists, Arnold and Marie Schwartz Coll. Pharmacy Alumni Assn., Mercer U. Pharmacy Alumni Assn., NYU Alumni Assn. Achievements include research in in-vivo comparison of adsorption capacity of superactive charcoal and fructose with activated charcoal and fructose. Home: 15 Shilling Rd Englishtown NJ 07726-4310 Office: Metabolic Homecare Inc 133 Mountain Rd Suffield CT 06078-2084

CHUNG, FUNG-LUNG, cancer research scientist; b. Keelung, Taiwan, Republic of China, Nov. 10, 1949; came to U.S., 1973; s. Tse-Yung and Carol (Cheng) C.; m. Judy Chu, Aug. 2, 1975; children: Christine, Christopher, Clifford. BS in Chemistry, Chung-Yuan U., Chung-Li, Taiwan, 1971; PhD, U. Utah, 1978. Postdoctoral fellow Columbia U., N.Y.C., 1979-80; rsch. assoc. Am. Health Found., Valhalla, N.Y., 1980-85, sect. head, 1985-89, assoc. chief, 1990—; study sect. mem. Nat. Cancer Inst., Washington, 1990—; mem. sr. adv. com. Am. Health Found., 1991—. Contbr. articles to profl. jours. Recipient Young Investigator award Nat. Cancer Inst., 1982, grantee, 1982—. Mem. Am. Assn. Cancer Rsch., Am. Chem. Soc., Am. Soc. Preventive Oncology, The Oxygen Soc. Office: Am Health Found 1 Dana Rd Valhalla NY 10595-1549

CHUNG, HWAN YUNG, neurosurgeon; b. Seoul, Korea, June 16, 1927; s. Yoon Sik and Bok Hyun (Bak) C. MD, Junnam U., 1949; PhD, Korea U., 1966. Diplomate Korean Neurosurgery Specialty, Korean Gen. Surgery Specialty. Bd.; m. Jong Sun Kim; children: Hyo Min, Hyo Sook, Hyo Sun, Chun Kee, Hyo Gyung, Soon Gi. Commd. lt. Republic of Korea Army, 1951, advanced through grades to col., 1965, discharged, 1965; intern Junnam Univ. Hosp., Gwangju, Korea, 1949-50; gen. surg. resident Gwangju Mil. Gen. Hosp., 1952-56; neurosurg. resident Korea Univ. Hosp., Seoul, 1956-60; neurosurgeon Korea U. Hosp., Seoul, 1956-60, 121st Evacuation Hosp., U.S. Army in Korea, 1960-61, Letterman Gen. Hosp., San Francisco, 1961-62; chief neurosurgeon 1st Korean Army Hosp., Daegu, Chung-Ang Gil Hosp., Inchun, 1993—; clin. asst. prof. Gyungbook U., Daegu, 1963-65; asst. prof. Korea U., 1965-66; asst. prof. Yonsei U., Seoul, 1966-69, assoc. prof., 1969-72, prof., 1972; prof. and chmn. neurosurgery Hanyang U., Seoul, 1972-92, emeritus prof. 1992—, hosp. dir. Joong-Ang Gen. Hosp., 1992-93, Hanyang U. Hosp., 1986-87. Decorated Bronze Star (U.S.A.), Hwarang Medal of Hon., Korea, 1952; Recipient Citation of Merit Ministry Def., Republic of Korea, 1964, Citation of Merit, Ministry Health and Welfare, 1987. Mem. Korean Neurosurg. Soc. (pres. 1978-79), Korean Microsurg. Soc. (pres. 1984-85), Korean Vascular Surg. Soc. (adviser 1984—), Pan-Pacific Surg. Assn. (pres. Korean chpt. 1984—, v.p. hdqrs. 1984—), Spinal Neurosurgery Rsch. Soc. (pres. 1987-91, hon. pres. 1991—). Home: 80-102 Hyundai-Apt, Abgoojung Gangnam, Seoul 135-110, Republic of Korea

CHUNG, KING-THOM, microbiologist, educator; b. Tou Fen, Taiwan, Apr. 25, 1943; came to U.S., 1966; s. Aa-Yuan and Yi-Ing (Buu) C.; m. Lan-Seng Fang, Oct. 27, 1973; children: Theodore, Serena. MA, U. Calif., Santa Cruz, 1967; PhD, U. Calif., Davis, 1972. Scientist Frederick (Md.) Cancer Rsch. Ctr., 1972-77; vis. asst. prof. Food Sci. Inst. Purdue U., West Lafayette, Ind., 1977-78; assoc. prof. Tunghai U., Taichung, Taiwan, 1978-80; prof., chmn. dept. Soochow U., Taipei, Taiwan, 1980-87, dean, 1983-87; vis. scientist U.S. Meat Animal Rsch. Ctr., Clay Center, Nebr., 1987-88; assoc. prof. biology Memphis State U., 1988-93, prof., 1993—; mem. adv. bd. Dept. Agr. and Forestry, Taiwan Provincial Govt., Taichung, 1982-87; exec. sec. Internat. Symposium on Biogas, Microalgae and Livestock Wastes, Taipei, 1980. Editor Jour. Biomass Energy Soc. China, Taipei, 1984—; author (in Chinese): Environment and Pollution, 1987, Intellectuals and Academic Education, 1987; contbr. articles to profl. publs. Grantee Am. Inst. Cancer Rsch., 1992. Mem. Am. Soc. Microbiology, Inst. Food Technologists, Sigma Xi. Achievements include discovery that azo reducation is the initial step of azo dye mutagenesis and carcinogenesis. Office: Memphis State U Dept Biology Memphis TN 38152

CHUNG, PAUL MYUNGHA, mechanical engineer, educator; b. Seoul, Korea, Dec. 1, 1929; came to U.S., 1947, naturalized, 1956; s. Robert N. and Kyungsook (Kim) C.; m. E. Jean Judy, Mar. 8, 1952; children: Maurice W., Tamara P. B.S. in Mech. Engring, U. Ky., 1952, M.S., 1954; Ph.D., U. Minn., 1957. Asst. prof. mech. engring. U. Minn., 1957-58; aero. research scientist Ames Research Center, NASA, Calif., 1958-61; head fluid physics dept. Aerospace Corp., San Bernardino, Calif., 1961-66; prof. fluid mechanics U. Ill.-Chgo., 1966—, head dept. energy engring., 1974-79, dean engring., 1979—; mem. nat. tech. com. AIAA on Plasmadynamics, 1972-74, com. on propellants and combustion, 1976-80; mem. tech. adv. com. Ill. Inst. Environ. Quality, 1975-77; corp. mem. Underwriters Lab., 1983—; cons. to industry, 1966—. Author numerous papers in field; author: Electric Probes in Stationary and Flowing Plasmas, 1975, Russian edit., 1978; contbr. chpt. to Advances in Heat Transfer, 1965, to Dynamics of Ionized Gases, 1973. Bd. govs. Redlands (Calif.) YMCA, 1965-67. Fellow AIAA; mem. Am. Soc. Engring. Edn. (exec. bd. engring. dean's coun. 1983-84), AIChE (nat. com. on internat. activities 1992—), Sigma Xi, Tau Beta Pi, Pi Tau Sigma, Phi Kappa Phi. Home: 2003 E Lillian Ln Arlington Heights IL 60004-4215 Office: U Ill Chicago IL 60680

CHUNG, SUNG-KEE, chemistry educator; b. Andong, Korea, Dec. 15, 1945; came to U.S., 1968; m. Saehyang Park, June 8, 1974. BSc, Yonsei U., Seoul, Republic of Korea, 1968; PhD, U. Ill., 1972. Rsch. fellow Yale U., New Haven, 1972-75, univ. staff, 1975-77; prof. Tex. A&M U., College Station, 1977-83; group leader Smith Kline & French Lab., Phila., 1983-87; prof., head dept. chemistry Pohang (Republic of Korea) Inst. Sci. & Tech., 1987—, dean acad. affairs, 1990-92. Mem. editorial bd. Korean Chem. Soc. Grantee NSF, 1983, NIH, 1980, Robert Welch Found., 1979, Rsch. Corp., 1977. Fellow Am. Inst. Chemists; mem. Am. Chem. Soc., N.Y. Acad. Sci. Achievements include research on structure elucidation of hedamycin; vitamin B12 model; glycopeptide antibiotics. Office: Pohang Inst Sci and Tech, Pohang 790-600, Republic of Korea

CHUNG, YOUNG CHU, chemist; b. Seoul, Korea, July 19, 1956; came to U.S., 1974; s. Kyuho and Heeja Chung. BA, U. Chgo., 1978; PhD, Mich. State U., 1985. Postdoctoral rsch. assoc. U. Calif., Irvine, 1986; lectr. Northeastern U., Boston, 1987-88; rsch. scientist Naval Rsch. Labs., Washington, 1988-89; dir. devel. tech. Molecular Displays Inc., Lexington, Mass., 1989—. Author: Advances in Raman & IR, 1989; contbr. articles to Jour. of Chem. Physics, Jour. of Electrochem. Soc. Yates Meml. fellow Mich. State U., 1985; NRC grantee, 1988. Mem. Am. Chem. Soc., Am. Phys. Soc. Achievements include patents pending for electrochromic materials. Home and Office: 2754 Moorhead Ave #208 Boulder CO 80303

CHUNG-WELCH, NANCY YUEN MING, biologist; b. N.Y.C., July 28, 1960; d. Thomas Richard and Jennie Kan Fee (Lew) Semler; m. James Michael Welch, June 29, 1985. BS, Northeastern U., Boston, 1982; PhD, Boston U., 1990. Rsch. technician Dept. Biology, Boston U., 1983-85, teaching fellow, 1987-89; rsch. fellow surgery Mass. Gen. Hosp., Harvard Med. Sch., Boston, 1989—. Contbr. articles to profl. jours. including Jour. Cellular Physiology, Differentiation, Analytical Biochemistry, Surg. Forum. Boston U. Grad. Sch. grad. rsch. award, 1987, Biology Dept. grad. travel award, 1988, 89, Grega-Zacharkow Young Investigator award Microcirculatory Soc., 1988; named Outstanding Young Woman of Mass., 1988; Repligen Corp fellow, 1993-95. Mem. AAAS, Am. Soc. Cell Biology, N.Y. Acad. Scis., Tissue Culture Assn. Achievements include development of tissue culture technique for the isolation and culture of pulmonary microvascular endothelial cells and mesothelial cells in vitro; demonstrated presence of simple epithelial keratins in endothelial cells; research on the phenotypic properties between endothelial and mesothelial cells using histochemical and biochemical criteria and in vitro assays of angiogenic potential. Office: Mass Gen Hosp 149 13th St Charlestown MA 02129-2060

CHUPP, TIMOTHY E., physicist, educator, nuclear scientist, academic administrator; b. Berkeley, Calif., Nov. 30, 1954. AB, Princeton U., 1979; PhD in Physics, U. Washington, 1983. Postdoctoral in physics Princeton U., 1983-85; from asst. prof. to assoc. prof. physics Harvard U., 1989-91; assoc. prof. physics U. Mich., Ann Arbor, 1991—; fellow Alfred P. Sloan Found., 1987. Recipient Presdl. Young Investor award NSF, 1987. Mem. Am. Phys. Soc. (I. I. Rabi prize 1993). Achievements include research in low energy particle physics particularly by study of symmetries accessible with polarization; tests of time reversal; local Lorentz invariance and linearity in quantum mechanics; structure of the neutron. Office: U Mich Dept Physics Ann Arbor MI 48109*

CHURCH, KERN EVERIDGE, engineer, consultant; b. North Wilkesboro, N.C., July 22, 1926; s. Wilford Albert and Rosa Bell (Everidge) C.; m. Agnes Elouise Pardue, Dec. 25, 1948; children: Ronald Kern, David Albert, Deborah Jean, Stephen Sherwood, Anne Michelle. BS in Gen. Engring., N.C. State U., Raleigh, 1949. Registered profl. engr., N.C. Plan rev. engr. State Bldg. Codes Div., Raleigh, N.C., 1949-67; dir. bldg. codes State of N.C., Raleigh, 1967-82; cons. engr. W.H. Gardner and Assocs., PA, Durham, N.C., 1983-85; consulting engr. pvt. practice, Raleigh, 1985—; mem. constrn. panel Am. Arbitration Assn., N.Y.C., 1984—; mem. safety to life com. Nat. Fire Protection Assn., Boston, 1954-68; mem. bldg. code com. So. Bldg Code Cogress, Birmingham, 1965-80; bd. dirs. Nat. Conf. of States on Bldg. Codes, Herndon, Va., 1967-82, pres. 1971-72; mem. fire coun. Underwriters Labs., Chgo., 1970-82. Editor N.C. State Building Code, 1967. Recipient Frank Turner award, N.C. chpt. AIA, Profl. Engrs. and Associated Contractors, 1983. Mem. N.C. chpt. AIA (hon.), Nat. Soc. of Profl. Engrs., Nat. Soc. Fire Protection Engrs. Democrat. Home and Office: 1217 Trailwood Dr Raleigh NC 27606

CHURCH, RICHARD DWIGHT, electrical engineer; b. Ogdensburg, N.Y., June 27, 1936; s. Dwight Perry and Carmeta Elizabeth (Walters) C.; m. Vernice Naomi Ives, Aug. 26, 1961; children: Joel, Benjamin. BEE, Clarkson Coll. Tech., 1963. Electronic design engr. IBM, Owego, N.Y., 1963-69; prin. engr., pres. ASL Systems, Inc., Afton, N.Y., 1969—, chmn. bd. dir.; sr. electronic design engr. Magnetic Labs., Inc., Apalachin, N.Y., 1980-82; power supply engring. cons., 1982—. Patentee in field. Treas., trustee Candor Congregational Ch., 1972-84; vice chmn. Town Planning Bd. Candor, 1975-82; rep., mem. Candor Fire Co., 1972-87; bd. dirs., treas. Candor Community Club, 1970-72. With USAF, 1955-59. Recipient Dr. Carl Michel award Clarkson Coll. Tech., 1960. Mem. IEEE, N.Y. Assn. Fire Chiefs, Assn. Energy Engrs. (sr.), Afton Bd. Fire Commrs., Candor Coin Club (pres. 1978-81). Republican. Home: RD 1 Box 702 Long Hill Rd Afton NY 13730 Office: PO Box 110 Afton NY 13730-0110

CHURCH, ROBERT MAX, JR., sales executive; b. Bowling Green, Ohio, Nov. 14, 1949; s. Robert Max and Dorrise Pennock (Cromwell) C.; m. Janet Lewellen, May 4, 1977; children: Benjamin Alexander, Deanna Leigh. BS, Bowling Green State U., 1971. Paint line foreman Ajusto Equipment Co., Bowling Green, 1974-75, asst. prodn. mgr., 1975-76, prodn. mgr., 1976-87, sales mgr., 1989—; prodn. mgr. Mill Bus. Furniture, Toledo, Ohio, 1987-89. Asst. scoutmaster troop 358 Boy Scouts Am., Bowling Green, 1992. Mem. Lions (pres., sec., treas. 1974—). Office: Ajusto Equipment Co 20163 Haskins Rd PO Box 348 Bowling Green OH 43402

CHURCHILL, RALPH JOHN, environmental chemist; b. Pitts., July 16, 1944. BS, U. Ky., 1966; MS, U. Houston, 1970; PhD in Civil Engring., U. Calif., Berkeley, 1973. Engr. pollution control Shell Oil Co., 1966-71; cons. Engr.-Sci., Inc., 1973-75; group leader water rsch. Tretolite Divsn. Petrolite Corp., 1975-81, cons., 1981-89, v.p. tech. dept., 1989—. Mem. AICE, Am. Water Works Assn., Water Pollution Control Fedn. Achievements include research in water and wastewater investigation, oil-water separation, water and wastewater treatment technology, mineral scale deposition-inhibition, municipal and industrial wastewater management. Office: Petrolite Corp 369 Marshall Ave Saint Louis MO 63119

CHURCHILL, SHARON ANNE-KERNICKY, research engineer, consultant; b. Detroit, May 16, 1957; d. Sylvester and Irene (Scott-Hutton) Kernicky; m. Perry Forrest Churchill, Sept. 8, 1979. BS in Chemistry and Math, Mid. Tenn. State U., 1982; BSChemE, U. Ala., Tuscaloosa, 1988, PhD in Civil Engring., 1993. Rsch. asst. Wayne State U., Detroit, 1975-77; rsch. engring. technician steering and gear divsn. TRW, Inc., Sterling Heights, Mich., 1976-79; sr. rsch. technician Vanderbilt U., Nashville, 1983-86; grad. rsch. engr. U. Ala., 1989—; environ. cons. State of Ala., Tuscaloosa, 1990. Contbr. articles to profl. jours. Vol. reader for blind, Nashville, 1983-85, Tuscaloosa, 1987-91. Recipient Creativity award NSF, 1991—; ACS-PRF scholar Am. Chem. Soc., 1980-82, scholar grad. sch. Nat. Alumni Assn., U. Ala., 1992. Mem. AICE, Am. Assn. Water Pollution Control (pres. univ. chpt. 1991, grad. award AL-MS chpt. 1992), Am. Soc. Microbiology, Am. Chem. Soc. (Petroleum Rsch. Fund scholar), Soc. Women Engrs., Air and Waste Mgmt. Assn., Hazardous Materials Rsch. Control Inst. Achievements include research on use of natural surfactants as bioremediation enhancers, production of microbial polyesters and surfactants, structure-function of membrane proteins, study of exoenzymes (microbial) on recalitrant natural compounds and relative activities on xenobiotic compounds in the environment. Home: Box 870344 Tuscaloosa AL 35487-0344 Office: U Ala Dept Civil Engring Box 870205 Tuscaloosa AL 35487-0205

CHURCHWELL, EDWARD BRUCE, astronomer, educator; b. Sylva, N.C., July 9, 1940; s. Doris L. Churchwell; m. Dorothy S. Churchwell, June 24, 1964; children: Steven T., Beth M. BS, Earlham Coll., 1963; PhD, Ind. U., 1970. NASA fellow Ind. U., Bloomington, 1963; postdoctoral fellow Nat. Radio Astronomy Obs., Charlottesville, Va., 1970; Heinrich Hertz postdoctoral fellow Max Planck Inst. Radioastronomie, Bonn, Fed. Republic Germany, 1970-72, staff scientist, 1972-77; asst. prof. U. Wis., Madison, 1977-79, assoc. prof., 1979-83, prof. of astronomy, 1983—. Fellow NASA, 1985, Fulbright rsch. fellow, 1988-89. Mem. Am. Astron. Soc., Internat. Astron. Union, Union Concerned Scientists. Office: U Wis Washburn Observatory 475 N Charter St Madison WI 53706-1582

CHURG, JACOB, pathologist; b. Dolhinow, Poland, July 16, 1910; came to U.S., 1936, naturalized, 1943; s. Wolf and Gita (Ravich) C.; m. Vivian Gelb, Oct. 18, 1942; children: Andrew Marc, Warren Bernard. M.D., U. Wilno, Poland, 1933; M.D. in pathology, 1936. Diplomate: Am. Bd. Pathology. Intern City Hosp., Wilno and State Hosp., Wilejka, Poland, 1933-34; asst. in gen. and exptl. pathology U. Wilno, 1934-36; asst. in bacteriology Mt. Sinai Hosp., N.Y.C., 1938; fellow in pathology Mt. Sinai Hosp., 1941-43, research asso., 1946—, attending physician, 1962-81, cons., 1982—; resident in pathology Beth Israel Hosp., Newark, 1939-40; pathologist Barnert Meml. Hosp., Paterson, N.J., 1946—; prof. pathology and community med. Mt. Sinai Sch. Med., N.Y.C., 1966-81, prof. emeritus, 1982—; cons. pathologist VA Hosp., Bronx, N.Y., Nassau County Med. Ctr., East Meadow, N.Y., St. Barnabas Med. Ctr., Livingston, N.J., Valley Hosp., Ridgewood, N.J., St. Joseph's Hosp., Paterson, Englewood Hosp.; chmn. mesothelioma reference panel Internat. Union Against Cancer, 1965-81, mem., 1982—; chmn. com. for histologic classification renal diseases WHO, 1975—; Lady Davis vis. prof. pathology, Jerusalem, 1975; past mem. sci. adv. group NIH, Bethesda, Md.; clin. prof. pathology U. Medicine and Dentistry N.J. Author: Histological Classification of Renal Diseases, Renal Disease—Present Status, Glomerular Diseases, Tubulo-Interstitial Diseases, Tumors of Serosal Surfaces, Vascular Diseases of the Kidney, Developmental and Hereditary Diseases of the Kidney, Infections and Tropical Diseases of the Kidney, Systemic Vasculitides, Urinary Tract Pathology, Kidney in Collagen-Vascular Diseases; also numerous articles in sci. jours.; discoverer Churg-Strauss syndrome, 1951. Served to capt., M.C. AUS, 1943-46. Mem. Am. Assn. Pathologists, Am. Soc. Nephrology (John P. Peters award 1987), N.Y. Acad. Medicine, Internat. Acad. Pathology, Harvey Soc., Internat. Soc. Nephrology, Alpha Omega Alpha. Achievements include research in vascular diseases and renal structure, in pneumokonioses. Address: 711 Ogden Ave Teaneck NJ 07666

CHUSID, MICHAEL THOMAS, architect; b. Chgo., Dec. 8, 1952; s. Fred M. and Ruth E. (Sacks) C.; m. Charlotte Dunn, June 16, 1976 (div. 1992); children: Aaron, Andy, Abra. BA in Design, So. Ill. U., 1975; MArch, U. Ill., 1978. Cert. architect, Okla., Wis.; cert. constrn. specifier. Designer Inryco, Inc., Milw., 1977-79; architect Stubenrauch Assocs., Inc., Sheboygan, Wis., 1979-82; chief specifier HTB, Inc., Oklahoma City, 1982-84; v.p. Design Drafting, Yukon, Okla., 1984; pres. Chusid Assocs., Oklahoma City, 1985—; conducted various seminars and presentations. Mem. editorial adv. bd. Constrn. Bus. Rev.; contbg. editor Archtl. & Engring. Systems, 1988, 89, Progressive Architecture, 1988-93; columnist Constrn. Mktg. Today, 1991, 92, 93; contbr. articles to profl. jours. Mem. AIA (Commendation 1985, Honor award 1985), Constrn. Specifications Inst. (Regional Ednl. award 1985, 86, Specification Proficiency award 1985, Commendation 1989), Am. Arbitration Assn. (panel of arbitrators), Specification Cons. in Ind. Practice. Office: Chusid Assocs 14951 Califa St Van Nuys CA 91411-3002

CHYNOWETH, ALAN GERALD, telecommunications research executive; b. Harrow, Eng., Nov. 18, 1927; came to U.S. 1952; s. James Charles and Marjorie (Fairhurst) C.; m. Betty Freda Edith Boyce, Sept. 22, 1950; children: Trevor Alan, Kevin Ray. BS in physics, U. London Kings Coll., 1948, PhD, 1950. Post doctoral fellow Nat. Research Council, Ottawa, Can., 1950-52; mem. tech. staff Bell Labs., Murray Hill, N.J., 1953-60, dept. head, 1960-65, dir., 1965-76, exec. dir., 1976-83; v.p. applied rsch. Bellcore, Morristown, N.J., 1984-92; R & D cons., 1993—; cons. advanced study inst. and rsch. workshops com. NATO, Brussels, 1982-90; lectr. Electrochem. Soc., 1983; alt. dir. Microelectronics and Computer Tech. Corp., Austin, Tex., 1984-92; dir. Optoelectronic Industry Devel. Assn., 1991-92; mem. adv. bd. dept. elec. engring. and computer sci. U. Calif., Berkeley, 1987-93; mem. natural scis. adv. bd. U. Pa., 1988—; mem. adv. bd. dept. elec. engring. U. So. Calif., 1988—; dir. Indsl. Rsch. Inst., 1990-92. Assoc. editor Solid State Communications, 1975-83; co-editor: Optical Fiber Telecommunications, 1979; contbr. articles to profl. jours.; patentee in field. Mem. vis. com. Cornell U. Materials Sci. Ctr., 1973-76, Am. Mgmt. Assn. R/D Coun., 1989-93. Fellow IEEE (mem. com. on U.S. competetiveness 1988-89, mem. bd. adv. task force on new initiatives 1989-90, chmn. Marconi Award com. 1987, mem. Alexander Graham Bell Prize com. 1990—, chmn. 1992—, W.R.G. Baker prize 1967, Frederik Philips award 1992), Am. Phys. Soc. (mem. indsl. affiliates com. 1984-87, editorial bd. Physics Today 1985-88, George E. Pake prize 1992), Inst. Physics and Phys. Soc. (London); mem. The Metall. Soc. of AIME, N.Y. Acad. Scis., Materials Rsch. Soc. NRC (survey dir. com. on survey of materials sci. and engring. 1970-74, panel chmn. com. on mineral resources and the environment 1973-75, panel chmn. materials sci. engring. study com. 1986-88, nat. materials adv. bd. 1976-80, advisor to panel on high performance computing and comm. Office of Sci. and Tech. Policy, The White House, 1991-92). Avocations: travel, boating. Home: 6 Londonderry Way Summit NJ 07901-2914 Office: Bellcore Applied Rsch Box 7040 331 Newman Springs Rd Red Bank NJ 07701-7040

CHYU, JIH-JIANG, structural engineer, consultant; b. Wu-Tai, Shan-Xi, China, Oct. 24, 1935; s. Yueng Cheng and Jian Yin (Hu) C.; m. Chi-Oy Wei, 1966. BSCE, Cheng Kung U., 1959; MA in Math., U. Wash., 1969; MSCE, Columbia U., 1973, PhD/DSc in Civil Engring & Engring Mechanics 1981. Registered profl. engr., N.Y., Wash. Grad. teaching and rsch. asst. Columbia U., N.Y.C., 1972-76; engring. specialist ITT Fed. Electric Corp., Paramus, N.J., 1976-77; cons. on engring. rsch. and designs, 1978-82; project engr. DRC Cons., Inc., N.Y.C., 1983-85; project mgr. Baker Engrs., Elmsford, N.Y., 1987-89; pvt. practice Forest Hills, N.Y., 1985-87, 89—; disting. speaker Tamkang U., Taiwan, 1982. Contbr. articles to profl. jours. Mem. AAAS, ASCE, Am. Concrete Inst., N.Y. Acad. Scis. Achievements include development of plane strain analogy for a class of elasticity problems, pulse propagation in an elastic hollow circular cylinder of semi infinite length, flow law of ice under combined action of torsion and compression; design guidelines for cable-stayed bridges; research on effects of bridge inspection on bridge designs.

CIACCIO, LEONARD LOUIS, chemist researcher, science administrator; b. N.Y.C., June 21, 1924; s. Baldassare J. and Accursia (Bacchi) C.; m. Eva Imelda Agostini, July 1, 1946 (div. Nov. 1982); children: Gloria Maria, Luke Joseph, Dominic Joseph, Imelda Liana, Rita Assunta; m. Mae Margaret Searles, June 18, 1983. BS in Chemistry, St. Peter's Coll., 1951; MS in Chemistry, Poly. U. of N.Y., 1956, PhD in Chemistry, 1962. Rsch. chemist Allied Chem. Corp., Queens, N.Y., 1951-53; co-dir. analytical rsch. Charles Pfizer & Co. Inc., Groton, Conn., 1953-60; mgr. tech. svc. Thomas J. Lipton, Inc., Englewood Cliffs, N.J., 1961-65; head water and waste rsch. Wallace & Tiernan Inc. (PennWalt), Cedar Knolls, N.J., 1965-69; sect. head environ. sci. GTE Rsch. Lab. Inc., Bayside, N.Y., 1969-72; prof. chemistry and environ. sci. Ramapo Coll. of N.J., Mahwah, 1972-88, prof. emeritus, 1990; v.p. R&D TCY Applied Technologies, Inc., Pomona, N.Y., 1990—. Mem. editorial bd. Standard Methods for the Exam. of Water and Waste Water, 1970—, Environ. Letters, 1970—; editor: (4 vol. book) Water and Water Pollution Handbook, 1971-73; author: (with others) Encyclopedia of Environmental Science and Engineering, 3d revised edit., 1992; contbr. articles to profl. jours. including Water Sci. Tech., Environ. Sci. and Tech., Analytical Chemistry, Jour. Am. Pharm. Assn. Chmn., treas. Glen Rock (N.J.) Religious Communities, 1974-80; mem. Bergen-Passaic Lung Assn., Paramus, N.J., 1977-80, Bergen County (N.J.) on Religion and Race, 1962-66. Mem. AAAS, Am. Chem. Soc. (chair Hudson Bergen N.J. sect. 1986-87), Internat. Assn. on Water Quality, Water Environ. Fedn. Achievements include patent for BOD Measuring Instrument, for Method (Enzymatic) for Cleaning Dairy Equipment. Home: 124 Kiel Ave Butler NJ 07405 Office: TCY Applied Technologies 45 White Birch Dr Pomona NY 10970

CIAPPENELLI, DONALD JOHN, chemist, academic administrator; b. Worcester, Mass., Dec. 4, 1943; s. John Steven and Domenica (Ruffo) C.; m. Gloria Rossi, Aug. 21, 1965; children: Robert Donald, Leah Denise. BS, U. Mass., 1966; AM, Brandeis U., 1967, PhD, 1970. NDEA fellow, 1966-70; research fellow M.I.T. Cambridge, 1970-72, NIH postdoctoral fellow, 1970-72; dir. chemistry Brandeis U., Waltham, Mass., 1972-77, dir. summer programs, 1974-77; dir. chem. labs. Harvard U., Cambridge, Mass., 1977-92; pres. Galliea, Inc., 1992—; chmn. Cambridge Lab. Cons. Inc., 1980—; bd. dirs. Chem. Design Corp., Cambridge Molecular Structures, Inc.; treas., dir. Cambridge Lab. Technologies Inc., 1982—; bd. dirs. Chem. Design Corp. Mem. AAAS, Am. Chem. Soc. (chmn.-elect N.E. sect. 1984, chmn. 1986). Office: Harvard U 12 Oxford St Cambridge MA 02138-2900

CIERESZKO, LEON STANLEY, chemistry educator; b. Holyoke, Mass., July 31, 1917; s. Albert Wojciech and Valerie Ann (Keller) C.; m. Esther Wynona Martin, May 1, 1943; 1 child, Leon Stanley. B.S. in Chemistry magna cum laude, Mass. State Coll., 1939; Ph.D. in Physiol. Chemistry, Yale, 1942. Research biochemist med.-research div. Sharp & Dohme, 1942-45; instr. biol. chemistry U. Utah Sch. Medicine, 1945-46; instr. chemistry U. Ill., 1946-48; mem. faculty U. Okla., Norman, 1948—; prof. chemistry U. Okla., 1956-91, spl. prof. chemistry, 1991—, chmn. dept. chemistry, 1969-70; research participant Oak Ridge Inst. for Nuclear Studies, 1951-52; vis. research assoc. Brookhaven Nat. Lab., summer 1953; ofcl. participant U.S. Program in Biology, Internat. Indian Ocean Expdn., 1963; cons. prof. U. Okla. Health Scis. Center, 1967-79; vis. research prof. Coll. V.I., 1971; prin. investigator Sea Grant Project, 1971—; vis. prof. marine scis. Port Aransas Marine Lab. U. Tex., summer 1978; mem. Nat. Acad. Scis. exchange program with Council of Acads. Yugoslavia, 1978; vis. investigator various marine labs. Lalor fellow Marine Biol. Lab., Woods Hole, Mass., 1951, 52; Fulbright fellow Stazione Zoologica, Naples, Italy, 1955-56; recipient Regents' award U. Okla., 1967. Fellow AAAS, Explorers Club; mem. Am. Chem. Soc., Geochem. Soc., Sigma Xi, Phi Kappa Phi, Phi Sigma, Lambda Tau, Phi Lambda Upsilon. Research and publs. in comparative biochemistry of marine invertebrates, chemistry of coelenterates, biogeochemistry of coral reefs. Home: 639 S Lahoma Ave Norman OK 73069-4506

CIFUENTES, ARTURO OVALLE, civil and mechanical engineer; b. Santiago, Chile, July 3, 1953; came to U.S. 1980; s. Arturo and Lucia (Ovalle) C.; m. Ventura L. Charlin, Jan. 26, 1979. Ingeniero Civil Degree, U. Chile, 1977; MS in Civil Engring., Calif. Inst. Tech., 1981, PhD, 1984. Asst. prof. U. Chile, Santiago, 1977-80; mem. part time faculty Calif. State U., L.A., 1984, U. So. Calif., L.A., 1986-89; project engr. MacNeal Schwendler, L.A., 1984-89; adv. engr. IBM East Fishkill, N.Y., 1989; rsch. engr. IBM Watson Rsch. Ctr., Yorktown, N.Y., 1990--. Author: Using MSC/Nastran, 1989; co-author: Handbook of Earthquake Engineering, 1992. Mem. IEEE, Am. Soc. Civil Engrs., Am. Soc. Mechanical Engrs. Office: IBM Rsch Rt 134 Yorktown Heights NY 10598

CILEK, JAMES EDWIN, medical and veterinary entomologist; b. Crown Point, Ind., July 1, 1952; s. Edwin James and Bernice Ruth (Burkley) C. BS, Purdue U., 1974; MS, La. State U., 1981; PhD, U. Ky., 1989. Cert. Am. Registry Profl. Entomologists. Dir. East Flagler Mosquito Control Dist., Flagler Beach, Fla., 1975-78; grad. rsch. asst. La. State U., Baton Rouge, 1979-81; pest control inspector East Baton Rouge Mosquito/Rodent Control Dist., 1981-82; grad. rsch. asst. U. Ky., Lexington, 1983-89, postdoctoral scholar, 1989-92; rsch. assoc. Southwest Kans. Rsch. Extension Ctr., Garden City, Kans., 1992-93; rsch. assoc. J.A. Mulrennan Sr. Rsch. Lab. Fla. A&M U., Panama City, 1993—; manuscript reviewer Am. Mosquito Control Assn.; cons. Quest Engring., Inc., Lexington, 1991—. Contbr. to profl. publs. Grantee Nat. Agrl. Pesticide Impact Assessment Program, 1988, 89, 91, 92. Mem. Entomol. Soc. Am., Am. Mosquito Control Assn., Ky. Mosquito/Vector Control Assn., Fla. Mosquito Control Assn. Democrat. Achievements include characterization of negative cross-resistance of organophosphate insecticides to pyrethroid resistant horn flies. Home: Apt A202 2175 Frankford Ave Panama City FL 32405 Office: JA Mulrennan Sr Rsch Lab Extension Ctr 4000 Frankford Ave Panama City FL 32405

CIMATTI, GIOVANNI ERMANNO, rational mechanics educator; b. Ferrara, Emilia, Italy, Aug. 17, 1945; s. Gaetano and Fernanda (Bagolini) C.; m. Luciana Pardi, Jan. 5, 1975; 1 child, Marcello. Grad., Inst. Math., Bologna, Italy, 1970. Researcher NRC, Pisa, Italy, 1970-75; asst. prof. U. Pisa, 1975-80, prof. rational mechanics, 1980—. Author: Research in Applied Mathematics, 1970-93. Office: U Pisa Dept Math, Via Buonarroti 2, 56127 Pisa Italy

CIMBLERIS, BORISAS, engineering educator, writer; b. Kaunas, Lithuania, Feb. 2, 1923; arrived in Brazil, 1935; s. Arkadijus and Rima (Felmanas) C.; m. Maria Honorina Prates, Dec. 23, 1953; children: Claudia Regina, André Carlos, Antonio Cesar. Mining, civil, metall. enring. degree, ENMM/UB, Ouro Preto, Brazil, 1948; MS in Nuclear Engring., N.C. State U., 1956; ScD in Thermodynamics, Fed. U. Minas Gerais, Belo Horizonte, Brazil, 1964. Instr. Brazilian Ctr. Phys. Rsch., Rio de Janeiro, 1950-52; project engr. Engring. Hidrotermica Instaltech, São Paulo, Brazil, 1952-53; assoc. physicist Armour Rsch. Found., Chgo., 1955-56; teaching fellow U. Mich., Ann Arbor, 1956-57; head tech.-sci. div. Nat. Nuclear Engring. Commn., Rio de Janeiro, 1957-60; various teaching and adminstrv. positions EE/UFMG Sch. of Engring., Belo Horizonte, Brazil, 1958-64, EE/UFMG Sch. Mines, Ouro Preto, Brazil, 1961-63; chief engr. Inst. for Radioactive Rsch., Belo Horizonte, 1969-72; prof. emeritus Fed. Univ. Minas Gerais, Belo Horizonte, 1984—; guest prof. MIT, Cambridge, 1967; participant tech. missions CEA, Saclay, France, 1964, reactor engring., San Juan, P.R., 1970-72. Author 6 books, 150 tech. and didactic papers, 8 book translations. Mem. Am. Phys. Soc., Am. Nuclear Soc., Soc. for History of Sci., Acad. Scis. Minas Gerais, Soc. for Advancement Sci., N.Y. Acad. Scis., Brazilian Soc. History of Sci. (founding mem.), Brazilian Soc. Mech. Scis. Home: São João do Paraiso 105, 30315-450 Minas Gerais Brazil

CINK, JAMES HENRY, chemical safety consultant, educator; b. Wichita, Kans., June 9, 1959; s. George Andrew and Twilladean Ruth (Milner) C.; m. Cecilia Jo Proft, Nov. 28, 1987. BS, Okla. State U., 1982, MS, 1986. Sr. agriculturist Okla. State U., Stillwater, 1982-87; asst. prof. U. Minn., St. Paul, 1987-89; grad. rsch. asst. Iowa State U., Ames, 1990-91, predoctoral rsch. assoc., 1991-93; v.p. Ozark-Prairie chpt. Soc. of Environ. Toxicology & Chemistry, 1993—; adv. bd. cert. tng. Minn. Dept. Agr., St. Paul, 1987-89. Author, editor: (tng. manuals) Stored Grain Pest Management, 1989, 1st rev. edit. 1991, Food Processing Pest Management, 1988, Structural Pest Management, 1988. Mem. Am. chem. Soc., Agronomy Soc. Am., Weed Sci. Soc. Am., Entomol. Soc. Am., Am. Registry of Profls. in Agronomy, Crops and Soils (cert. profl. agronomist). Republican. Methodist. Home: 4114 Aplin Rd Ames IA 50014-7604

CIPALE, JOSEPH MICHAEL, software engineer; b. Des Moines, Feb. 20, 1958; s. Joseph Adam and Doris Irene (Hargis) C.; m. Peggy Jane Schmid, July 14, 1979. BSEE, Portland (Oreg.) State U., 1989, MSEE, 1993. With Weyerhauser, Aberdeen, Wash., 1978-86; grader/tutor Portland State U., 1986-89; tech. writer Mentor Graphics, Wilsonville, Oreg., 1990-91; quality assurance, software engr. intern Analogy, Inc., Beaverton, Oreg., 1991-92; software engr. Cypress Semiconductor, Beaverton, 1992—. Contbr. articles to profl. jours. Mem. IEEE (student mem. of yr. 1989). Home: 1000 W 23rd St Vancouver WA 98660-2333 Office: 8196 SW Hall Blvd Ste 100 Beaverton OR 97005

CIRAULO, STEPHEN JOSEPH, nurse, anesthetist; b. Danville, Pa., Feb. 25, 1960; s. Leonard Joseph and Mary Louise (Purpuri) C. Diploma, Geisinger Med. Ctr. Sch. Nursing, Danville, 1980; cert., Sch of Anesthesia for Nurses Univ. Health Ctr. Pitts., 1983; student, Ottawa U., 1986—. Nursing asst. Geisinger Med. Ctr., Danville, 1978-80; staff RN, part time charge RN cardiac care unit Williamsport (Pa.) Hosp., 1980-81; asst. gastroenterology research group Presbyn. Univ. Hosp., Pitts., 1982-83; staff nurse anesthetist dept. anesthesia Duke U. Med. Ctr., Durham, N.C., 1983-90; with Anesthesia Anytime, Winston-Salem, N.C., 1990, Nash Gen. Hosp., Rocky Mount, N.C., 1991-92; staff nurse anesthetist, mem. epidural analgesia svc. Nash Gen. Hosp., 1991-92; staff nurse anesthetist Wake Anesthesiology Assocs., Inc., Raleigh, N.C., 1992—; mem. coun. for nurse anesthetists dept. anesthesia Duke U. Med. Ctr., Durham, 1985-89; staff nurse anesthetist Epidural Analgesia Svc., Rocky Mount, 1991-92, Wake Anesthesiology Assocs., Inc., 1992—. Charter mem. Outstanding Young Ams., 1988. Mem. Am. Assn. Nurse Anesthetists, N.C. Assn. Nurse Anesthetists (bylaws com. 1984-86, chmn. fin. com. 1986-88, mem. fin. com. 1988—, fall program com. 1988, spring program speaker 1990, treas. 1991-93, pres.-elect 1993—), Triangle Transplant Recipient Internat. Orgn. (charter), Music Theatre Assocs. Democrat. Avocations: music, art, weight lifting, traveling, gardening. Home and Office: 1710 Falls Church Rd Raleigh NC 27609 also: Anesthesia Anytime Inc PO Box 15500 Winston-Salem NC 27113

CISKO, GEORGE JOSEPH, JR., applied mechanics engineer, research lab manager; b. Chgo., Nov. 21, 1958; s. George Joseph and Isabelle Christine (Dziekan) C.; m. Susan Carol Loitfellner, July 18, 1987; children: Ryan Joseph, Christine Carol. BS in Engring. Mechanics, U. Ill., 1985; MS in Applied Mechanics, Calif. Inst. Tech., 1987. Mem. tech. staff Aerospace Corp., L.A., 1985-88; design engr. Wilson Sporting Goods Co., River Grove, Ill., 1988-89; mgr. racquet sports test and prototype labs. Wilson Sporting Goods Co., Chgo., 1989—. Mem. Soc. Exptl. Mechanics (Best Paper award 1987). Office: Wilson Sporting Goods Co 5551 N Milton Pky Rosemont IL 60018

CITA, MARIA, geology educator; b. Milan, Sept. 12, 1922. Degree in Geology, U. Milan, 1946. Prof. of Geology U. Milan, U. Degli Studi, Parma, Italy. Recipient Francis P. Shepard medal Soc. Sedimentary Geology, 1994. Office: U Degli Studi Area della Sci, Via Universita 12, Parma 43100, Italy*

CITARDI, MATTIO H., chemist, researcher, system manager; b. N.Y.C., Jan. 20, 1966; s. Mattio and Timotea G. Citardi; m. Ann Marie Delli Pizzi, June 27, 1993. BS in Biology, Manhattan Coll., 1989; MS in Computer Sci., Pace U., 1992. Chemist Pepsico Inc., Valhalla, N.Y., 1985-88, Gen. Foods Corp., Tarrytown, N.Y., 1988-89; analytical chemist, system mgr., lab. automation specialist Am. Cyanamid Co., Pearl River, N.Y., 1989—. Co-contbr. article to Pace U. Tech. Notes, 1991. Mem. Am. Assn. Pharm. Scientists, Assn. Computing Machinery, Am. Chem. Soc., Hewlett-Packard LDS-Users Group. Republican. Roman Catholic. Achievements include research in new technologies, networking and analytical method development; co-development of the algorithm to find longest up sequence (LUP) in a series of data points. Office: Am Cyanamid Co North Middletown Rd Pearl River NY 10965

CITROEN, CHARLES LOUIS, information scientist, consultant; b. Amsterdam, Netherlands, May 18, 1939; s. I. Jozef and Sophia (Speyer) C.; m. Elisabeth Schakel, Dec. 22, 1965; children: Micky, Paul. D Chemistry, U. Amsterdam, 1966. Head dept. info. Gist-Brocades N.V., Delft, Netherlands, 1967-68; info. scientist Royal Netherlands Chem. Soc., The Hague, 1968-72; dir. Netherlands Orgn. Chem. Info., The Hague, 1973-77; dep. dir. ctr. info and documentation Netherlands Orgn. Applied Sci. Rsch. TNO, Delft, 1978-86, dir., 1987—; organizer, intern chem. structures conf., Netherlands, 1987, 90, 93; chmn. Computers in Chemistry div., The Hague, 1988-91. Editor ann. book: Perspective in Information Management, 1987—; editorial bd. Jour. Chem. Info. and Computer Scis., 1991—. Fellow Inst. Info. Scientists (London). Office: TNO, PO Box 6042, 2600JA Delft The Netherlands

CIVISH, GAYLE ANN, psychologist; b. Lynnwood, Calif., Sept. 29, 1948; d. Leland and Arline (Frazer) Civish; children: Nathan Morrow, Shane Morrow. BA, U. Nev., Reno, 1970; MA, U. Colo., 1973, PhD, 1983. Lic. psychologist, Colo.; cert. sch. psychologist, Colo. Sch. psychologist Jefferson County (Colo.) Schs., 1983-89; psychologist in pvt. practice Lakewood, Colo., 1983—. Contbr. articles to profl. jours. Mem. Colo. Psychol. Assn. (bd. dirs. 1990—), Am. Psychol. Assn., Colo. Women Psychologists (external liaison), Am. Soc. Clin. Hypnosis, Phi Kappa Phi, Phi Delta Kappa. Democrat. Office: 3000 Youngfield St Ste 376 Lakewood CO 80215-6552

CIZEK, JOHN GARY, safety and fire engineer; b. St. Louis, Sept. 16, 1948; s. John Ernst and Ann Margaret (Seith) C.; m. Carolyn MArie Haas, Dec. 4, 1971; children: Laura Suzanne, John David. BSCE, U. Mo., 1971. Registered profl. engr.; cert. safety profl. Loss prevention engr. Factory Mutual Engring. Assn., St. Louis, 1971-76; safety engr. Diamond Shamrock Corp., Cleve., 1977-80; from corp. safety specialist to mgr. safety Diamond Shamrock Corp., Dallas, 1980-87; cons. safety and fire protection, asst. v.p. M&M Protection Cons., Houston, 1987-90, v.p., 1990—. Mem. AICE, Am. Soc. Safety Engrs., Soc. Fire Protection Engr. Lutheran. Office: M&M Protection Cons 1100 Milam Bldg Ste 4500 Houston TX 77002

CIZEWSKI, JOLIE ANTONIA, physics educator, researcher; b. Frankfurt a Main, Fed. Republic Germany, Aug. 24, 1951; came to U.S., 1951; d. Stanley Joseph and Ludmilla (Kotaczka) C. BA, U. Pa., 1973; MA, SUNY, Stony Brook, 1975, PhD, 1978. Rsch. asst. Brookhaven Nat. Lab., Upton, N.Y., 1975-78; postdoctoral staff Los Alamos Nat. Lab., N.Mex., 1978-80; asst. prof. physics Yale U., New Haven, 1980-85, assoc. prof., 1985-86, Rutgers U., 1986-92, prof. 1992—; cons. Livermore Nat. Lab., Calif., 1982—; mem. vis. staff Los Alamos Nat. Lab., 1980—. Contbg. author: Interacting Bose-Fermi Systems, 1981, Bosons in Nuclei, 1984. Recipient Women in Sci. Faculty award NSF, 1991. Mem. Nat. Abortion Rights Action League, NOW. Jr. faculty fellow Yale U., New Haven, 1984-85, A. P. Sloan Found. fellow, N.Y.C., 1983-88. Fellow Am. Phys. Soc.; mem. Assn. Women in Sci., Sigma Xi. Avocations: photography, gourmet cooking. Home: 1 Amur Rd Martinsville NJ 08836 Office: Rutgers U Serin Physics Lab New Brunswick NJ 08903

CLAES, DANIEL JOHN, physician; b. Glendale, Calif., Dec. 3, 1931; s. John Vernon and Claribel (Fleming) C.; AB magna cum laude, Harvard U., 1953, MD cum laude, 1957; m. Gayla Christine Blasdel, Jan. 19, 1974. Intern, UCLA, 1957-58; Bowyer Found. fellow for rsch. in medicine, L.A., 1958-61; pvt. practice specializing in diabetes, L.A., 1962—; v.p. Am. Eye Bank Found., 1978-83, pres., 1983—, dir. rsch., 1980—; pres. Heuristic Corp., 1981—. Mem. L.A. Mus. Art, 1960—. Mem. AMA, Calif. Med. Assn., L.A. County Med. Assn., Am. Diabetes Assn., Internat. Diabetes Fedn. Clubs: Harvard and Harvard Med. Sch. of So. Calif.; Royal Commonwealth (London). Contbr. papers on diabetes mellitus, computers in medicine to profl. lit. Office: Am Eyebank Found 15327 W Sunset Blvd Ste 236 Pacific Palisades CA 90272-3674

CLAESSENS, PIERRE, chemist, researcher; b. Brussels, Sept. 5, 1939; married, 1968; 2 children. Lic., U. Louvain, 1963, D in Electrochemistry, 1967. Asst. prof. chemistry U. Montreal, 1967-68; rsch. chemist Noranda Rsch. Ctr., Pointe Claire, Can., 1968-70, group leader, 1970-73, head dept., 1973-81, chief divsn. scientist, 1981-89, prin. scientist, 1989—. Recipient Sherrit Hydrometallurgy award 1991. Mem. Am. Inst. Mining and Petroleum Engrs., Can. Inst. Mining and Metallurgy (Sherritt Hydrometallurgy award 1991), Nat. Assn. Corrosion Engrs., Electrochem. Soc. Achievements include research in electrodeposition of metals, cathodic process, study of the physical properties of solutions. *

CLAGETT, ARTHUR F(RANK), JR., psychologist, sociologist, qualitative research writer, retired sociology educator; b. Little Rock, Dec. 3, 1916; s. A.F. and Mary Gertrude (Bell) C.; m. Dorothy Ruth Pinckard, Dec. 23, 1954. BA in Chemistry, Baylor U., 1943; MA in Psychology, U. Ark., 1957; PhD in Sociology, La. State U., 1968. Shift chemist Celanese Corp., Cumberland, Md., 1942-44; shift supr. penicillin prodn. Comml. Solvents Corp., Terre Haute, Ind., 1944-45; rsch. supr. streptomycin pilot plant Schenley Labs., Lawrenceburg, Ind., 1945-48; asst. mgr. Clagett's Feed and Seed Store, Donna, Tex., 1948-50; med. svc. rep. Blue Line Chem. Co., St. Louis, 1952-56; prison classification officer La. State Penitentiary, 1956-59, classification supr. new admissions, 1959-60; counseling psychologist, Baker, La., 1960-64; asst. prof. sociology Lamar State Coll. Tech., Beaumont, Tex., 1964-66; assoc. prof. sociology Stephen F. Austin State U., 1968-83, prof., 1983-85, prof. emeritus, 1986—; consulting sociologist, social psychologist, criminologist, Nacogdoches, Tex., 1986-91; qualitative rsch. writer, 1992—. Recipient Twentieth Century award for achievement IBC, 1993; named Outstanding Educator in Am., 1974-75, Internat. Man of Yr., 1991-92, World Intellectual 1993. Mem. univ. rsch. coun., 1973-75, Sch. Liberal Arts coun., 1970-71, 78-79. Mem. editorial bd. Quar. Jour. Ideology, 1982-93; contbr. numerous articles to profl. jours. including Jour. of Offender Counseling, Internat. Re. Mod. Sociol., Jour. Offender Rehab. Mem. AAUP,

Internat. Platform Assn., So. Sociol. Soc., Am. Soc. Criminology, Am. Acad. Criminal Justice Scis., Am. Sociol. Assn., Inst. Criminal Justice Ethics (chaired annual meetings, presented 28 papers). Methodist. Avocations: reading, fishing, classical music. Home and Office: 609 Egret Dr Nacogdoches TX 75961-6579

CLAIBORNE, JIMMY DAVID, electrical engineer; b. Lebanon, Tenn., Nov. 22, 1961; s. Earl and Anna Lois (West) C.; m. Lisa Lynn Hackett, Dec. 7, 1985; 1 child, Spenser David. BSEE, Tenn. Tech. U., 1984. Engr. Pulsar Control Corp., Hendersonville, Tenn., 1985-91; project engr. Fruit of the Loom, Bowling Green, Ky., 1991—. Mem. IEEE, Scottsville Country Club. Baptist. Home: 4253 Lafayette Rd Scottsville KY 42164

CLAMPITT, OTIS CLINTON, JR., health agency executive; b. Burlington, N.C., Nov. 17, 1947; s. Otis Clinton and Audrey Mae (Brafford) C.; m. Martha Jane Redding, Apr. 3, 1971. BA in English, Guilford Coll., 1971. Unit exec. dir. N.C. div. Am. Cancer Soc., Winston-Salem, 1972-73, area rep., 1973-74, met. area dir., 1974-75; dir. pub. edn./info. S.C. div. Am. Cancer Soc., Columbia, 1976-78, dir. devel., 1978-79, dep. exec. v.p., 1979-81; exec. v.p. Miss. div. Am. Cancer Soc., Jackson, 1981-89; nat. v.p. Am. Cancer Soc., Washington, 1989—; faculty, cons. Am. Cancer Soc. Acad., Atlanta, 1989—. Co-founder, pres. Forsyth County Interagy. Health Coun., Winston-Salem, 1975; chmn. Miss. Combined Fed. Campaign, Jackson, 1987-89, Miss. Com. on Indigent Patient Care, Jackson, 1989; appointed Govs. Task Force on Agy. Registration, Jackson, 1989; bd. dirs. Miss. Seatbelt Coalition, Jackson, 1986-90. Mem. Miss. Soc. Assn. Execs., Nat. Soc. of Fund Raising Execs. Avocations: sailing, snow skiing, water skiing, traveling. Home: 1712 Woodlore Rd Annapolis MD 21401-6568 Office: Am Cancer Soc PO Box 6640 Annapolis MD 21401-0640

CLANCY, THOMAS JOSEPH, civil engineer; b. N.Y.C., Nov. 29, 1946; s. Joseph Edward and Rita (Weiss) C.; m. Lana Schettino, June 5, 1970. BSCE, Poly. Inst. Bklyn., 1972, MSCE, 1976. Registered profl. engr., N.Y. Project engr. Hardesty and Hanover, N.Y.C., 1972-78; assoc., mgr. br. office A.G. Lichtenstein and Assocs., N.Y.C., 1978-90; v.p., dir. engring. URS Cons., Inc., N.Y.C., 1990—. Sgt. U.S. Army, 1966-68, Vietnam. Mem. ASCE, Am. Ry. Engring. assn., Soc. Am. Mil. Engrs. Home: 106 Center St Ramsey NJ 07446 Office: URS Cons Inc 1 Penn Plz New York NY 10119

CLAPHAM, WILLIAM MONTGOMERY, plant physiologist; b. N.Y.C., Sept. 23, 1948; s. Wentworth Beggs and Mittie McGaw (Boardman) C.; m. Jayne Brewer, May 18, 1984 (div. Feb. 1987); 1 child, Katherine Clapham; m. Sarah Barnes, June 2, 1990; 1 child, Abigale Clapham. BA, Ill. Wesleyan U., 1970; PhD, U. Mass., 1981. Postdoctoral fellow U. Maine, Orono, 1982-85; plant physiologist USDA Agrl. Rsch. Svc., Orono, 1985-87, acting rsch. leader, 1987-89, rsch. leader, 1989—; advisor Maine Potato Bd., Presque Isle, 1989-91. Contbr. articles to profl. jours. Mem. AAAS, Am. Assn. Agronomy, Internat. Lupin Assn. Office: USDA Agrl Rsch Svc NE Plant Soil and Water Lab Univ of Maine Orono ME 04469

CLAPP, BEVERLY BOOKER, university administrator, accountant; b. Savannah, Ga., Oct. 26, 1954; d. Herschel Ray and Ida Marie (Bove) Beville; m. William L. Clapp III; 1 child, Matthew Anthony. BS in Med. Tech., Med. Coll. Ga., 1976; MS in Clin. Lab. Sci., U. Ala., Birmingham, 1977, BS in Acctg., 1989. CPA, Ala. Blood bank technologist U. Ala. Hosp., Birmingham, 1976-77; asst. supr. physiology Bapt. Med. Ctr., Montclair, Birmingham, 1977-79; rsch. chemist Nephrology Rsch. and Tng. Ctr. U. Ala., Birmingham, 1979-91; med. technologist VA Med. Ctr., Gainesville, Fla., 1991-92; CPA, mgr. J.J. Lucky & Co., Gainesville, 1992; sr. grants specialist U. Fla., Gainesville, 1992—; acct. Bambino Minor League Allstars; presenter sci. meetings. Treas. Pack 82 Boy Scouts Am. Mem. AICPA, Ala. Soc. CPAs, Am. Soc. Clin. Pathology, Fla. Inst. CPAs, Fla. Soc. Med. Tech., Alpha Aeta, Phi Kappa Phi. Roman Catholic.

CLARE, GEORGE, safety engineer, system safety consultant; b. N.Y.C., Apr. 8, 1930; s. George Washington and Hildegard Marie (Sommer) C.; student U. So. Calif., 1961, U. Tex., Arlington, 1963-71, U. Wash., 1980; m. Catherine Saidee Hamel, Jan. 12, 1956; children: George Christopher, Kristine René. Enlisted man U.S. Navy, 1948, advanced through grades to comdr., 1968; naval aviator, 1951-70; served in Korea; comdr. Res., 1963-70; ret., 1970; mgr. system safety LTV Missiles and Electronics Group, Missiles div., Dallas, 1963-90. Mem. Nat. Republican Com., Rep. Senatorial Com., Rep. Congl. Com., Tex. Rep. Com., Citizens for Republic. Decorated Air medal with gold star, others; cert. product safety mgr. Mem. AIAA, Am. Security Council, Internat. Soc. Air Safety Investigators, System Safety Soc., Am. Def. Preparedness Assn., Assn. Naval Aviation, Ret. Officers Assn., Air Group 7 Assn. (pres.). Roman Catholic. Home and Office: 825 Bayshore Dr Ste 500 Pensacola FL 32507-3463

CLARIDGE, RICHARD ALLEN, structural engineer; b. Chgo., Feb. 22, 1932; s. Dalbert Otis and Lucille Alma (Lindquist) C.; m. Joan Elaine Powell, June 12, 1952; children: Cathy L. Jansen, Richard Allen Jr., Jaylynn P. Cook. BSBA, Fla. State U., Tallahassee, 1953; BCE, U. Fla., 1959; postgrad., U. Cen. Fla., 1972-75. Registered profl. engr. Fla., S.C. With McDonnell Douglas Astro, 1963-89; ground supt. edn., design engr. McDonnell Douglas Astro, Cape Canaveral, Fla., 1974-78; stress analyst McDonnell Douglas Astro, Titusville, Fla., 1982-89; structural cons., analyst Atlantic Cons., Titusville, 1989—; group engr. McDonnell Douglas Astro, Kennedy Space Center, 1963-74, sect. chief stress McDonnell Douglas Missile Systems, Titusville, 1983-89; structural analyst, designer Gen. Dynamics Space Systems Div., Cape Canaveral, Fla., 1989—. Mem. Titusville Shoreline Authority, 1965. Lt. (j.g.) USNR, 1953-57. Mem. ASCE (computer practices reviewer), Nat. Soc. Profl. Engring., Fla. Engring. Soc., Am. Inst. Aero. and Astron. Engrs., U.S. Naval Inst. (life), Internat. Soc. Allied Weight Engrs. Avocations: woodworking, photography, volleyball, surfing. Office: Atlantic Cons PO Box 443 Titusville FL 32781-0443

CLARK, BENJAMIN CATES, JR., flavor chemist, organic research chemist; b. Knoxville, Tenn.; s. Benjamin Cates and Frances (Foute) C. BA, Duke U., 1963; MS, PhD in Organic Chemistry, Emory U., 1967. Postdoctoral fellow U. Ga., Athens, 1967-69; prin. staff chemist Coca-Cola Co., Atlanta, 1969-75, sr. rsch. scientist, 1975-86, prin. investigator, 1986—; Grad. Sch. rep. Emory U. Alumni Assembly, Atlanta, 1991—. Contbr. chpts. to books, articles to sci. jours. Mem. Druid Hills Civic Assn., Atlanta, 1981—. NASA grad. fellow Emory U., 1964-67. Mem. Am. Chem. Soc. (exec. com. Ga. sect. 1984—), Inst. Food Technology, Royal Soc. Chemistry, Phi Beta Kappa, Sigma Xi. Achievements include patent in flavor chemistry, research in flavor chemistry of beverages, citrus flavors, terpene chemistry, reactions of essential oils in dilute aqueous systems, micelle chemistry, photochemistry, chemistry of plant gums and citrus emulsions, analytical flavor chemistry. Office: Coca-Cola Co 1 Coca-Cola Plz Atlanta GA 30313

CLARK, BILLY PAT, physicist; b. Bartlesville, Okla., May 15, 1939; s. Lloyd A. and Ruby Laura (Holcomb) C. BS, Okla. State U., 1961, MS, 1964, PhD, 1968. Grad. asst. dept. physics Okla. State U., 1961-68; postdoctoral rsch. fellow dept. theoretical physics U. Warwick, Coventry, Eng., 1968-69; sr. mem. tech. staff Booz-Allen Applied Rsch., 1969-70; sr. mem. tech. staff field svcs. div. Computer Scis. Corp., Leavenworth, Kans., 1970-73, sr. mem. tech. staff field svcs. div., Hampton, Va., 1973-76, head quality assurance engring. Landsat project Goddard Space Flight Center, NASA, Greenbelt, Md., 1976-77, quality assurance sect. mgr., 1977-79, sr. staff scientist engring. dept., 1979-80, sr. staff scientist image processing ops., 1980-82, sr. prin. engr./scientist GSFC sci. and application operation, system scis. div., 1982-83, sr. adv. staff CSC/NOAA Landsat Operation, 1983-91; sr. scientist Computer Scis. Corp. Ctr. Excellence in Geographic Info., 1991—; tech. rep. internat. Landsat Tech. Working Group (representing USA Landsat operation). Author tech. publs. Recipient undergrad. scholarships Phillips Petroleum Co., 1957-61, Am. Legion, 1957-58, Okla. State U., 1957-58. Mem. AAAS, IEEE, Am. Acad. Polit. and Social Sci., Internat. Platform Assn., Am. Phys. Soc., N.Y. Acad. Scis., Soc. Photo Optical Instrumentation Engrs., Internat. Soc. for Photogrammetry and Remote Sensing (organizer and dir. plenary sessions XVII congress meeting 1992), Am. Soc. for Photogrammetry and Remote Sensing, Victory Hills Golf and Country Club, Crofton Country Club, Pi Mu Epsilon, Sigma Pi Sigma. Home: 5811 Barnwood Pl Columbia MD 21044-2811

CLARK, BRIAN THOMAS, mathematical statistician, operations research analyst; b. Rockford, Ill., Apr. 7, 1951; s. Paul Herbert and Martha Lou (Schlensker) C.; m. Suzanne Drake, Nov. 21, 1992. B.S. cum laude, No. Ariz. U., 1973; postgrad. Ariz. State U., 1980-82. Math. aide Center for Disease Control, Phoenix, 1973-74, math. statistician, 1979-83; math. Statistician Ctrs. for Disease Control, Atlanta, 1983-84 ops. research analyst U.S. Army Info. Systems Command, Ft. Huachuca, Ariz., 1984—; math. statistician U.S. Navy Metrology Engring. Center, Pomona, Calif., 1974-79. Mem. Am. Statis. Assn., Am. Soc. Mil. Comptrs. Republican. Mormon. Office: US Army Info Systems Command Dep Chief Staff Resource Mgmt Chargeback Test Divsn Fort Huachuca AZ 85613

CLARK, CARL ARTHUR, retired psychology educator, researcher; b. Oak Park, Ill., Sept. 20, 1911; s. Alfred Houghton and Mary (Geist) C.; m. Janet Picquet; 1 child, Peter Picquet. BA cum laude, Colo. Coll., 1948, MA, 1951; PhD, State U. Iowa, 1954. Mem. faculty Colorado Springs High Sch., 1948-50; rsch. asst. State U. Iowa, Iowa City, 1951-53; rsch. assoc. U. Chgo., 1953-54; prof. psychology Chgo. State U., 1954-76, chmn. psycol. dept., prof. emeritus, 1976—; adj. prof. U. Mo.-Rolla, 1976-77; rsch. evaluator Ford Found. projects, Chgo., 1963-66. Mem. editorial bd. Ill. Schs. Jour., 1966-76. Contbr. articles and revs. to profl. jours. Served with U.S. Army, 1942-45. Recipient Outstanding Achievement award Black Students Psychology Assn., 1973. Mem. AAAS, N.Y. Acad. Scis., Am. Psychol. Assn., Sigma Xi. Achievements include invented portable theft and intrusion alarm; improvements in electromagnetic transmission; developed electromagnetic ground transmission method. Home: 616 W Washington Ave Saint Louis MO 63122-3835

CLARK, CAROLYN ARCHER, technologist, scientist; b. Leon County, Tex., Feb. 16, 1944; d. Ray Brooks and Dena Mae (Green) Archer; m. Frank Ray Clark, Nov. 20, 1960 (div. Oct. 1979); children: Frank Ray, Valerie Lynn, Bruce Layne. BA, Sam Houston State U., 1961; MS, Tex. A&M U., 1973, PhD, 1977. Supr., bookkeeper Rep. Sewing Machine Distbrs., Dallas, 1961-65; door-to-door sales Avon Products, Inc., Bryan, Tex., 1965-72; lectr. Tex. A&M U., College Station, Tex., 1977, rsch. assoc., 1977-79; sr. sci. Lockheed Emsco., Houston, 1979-82, prin. scientist, 1983-85; aerospace technologist, phys. scientist NASA Stennis Space Ctr., Miss., 1985-87; staff scientist Lockheed EMSCO, Houston, 1987-88; sr. project mgr., office mgr. Ctr. for Space and Advanced Tech., Houston, 1988-91; staff scientist Lockheed Engring. and Scis. Co., Houston, 1991—; cons. in field. Contbr. articles to profl. publs. Recipient Commendation for Outstanding Contbns. Lockheed, 1979-80, 91, Commendation for Excellence, 1984; Cert. of Merit U.S. Dept. Agr. 1980; Grad. Rsch. fellow Tex. A&M, 1975-76; NSF cograntee Tex. A&M, 1976-77. Mem. Am. Soc. Plant Taxonomists, Bot. Soc. Am., Nat. Mgmt. Assn., Sigma Xi, Phi Sigma, Alpha Chi, Kappa Delta Pi. Republican. Avocations: sailing, scuba diving, tennis, piano. Office: Lockheed Engring/Scis Co 2400 Nasa Rd 1 Houston TX 77058-3799

CLARK, DAVID LEIGH, marine geologist, educator; b. Albuquerque, N.Mex., June 15, 1931; s. Leigh William and Sadie (Ollerton) C.; m. Louise Boley, Aug. 31, 1951; children: Steven, Douglas, Julee, Linda. B.S., Brigham Young U., 1953, M.S., 1954; Ph.D., U. Iowa, 1957. Geologist Standard Oil Calif., Albuquerque, 1954; asst. prof. So. Meth. U., Dallas, 1957-59; assoc. prof. Brigham Young U., Provo, Utah, 1959-63; prof. geology and geophysics U. Wis.-Madison, 1963—, chmn. dept. geology and geophysics, 1971-74, assoc. dean natural scis., 1986-91. Author Fossils, Paleontology, Evolution, 1968,72; author and coordinator: Treatise on Invertebrate Paleontology-Conodonts, 1981. Recipient Fulbright award Bonn, W.Ger., 1965-66; Disting. Professorship U. Wis., 1974. Fellow Geol. Soc. Am.; mem. Paleontol. Soc., Am. Assn. Petroleum Geologists, Soc. Econ. Paleontologists and Mineralogists, Am. Geophys. Union, Pander Soc., Paleontol. Assn., N.Am. Micropaleontology Soc., AAAS. Mem. LDS Ch. Home: 2812 Oxford Rd Madison WI 53705-2218 Office: U Wis Dept Geology and Geophysics Madison WI 53706

CLARK, GORDON HOSTETTER, JR., physician; b. New Haven, Aug. 5, 1947; s. Gordon Hostetter and Elizabeth Master (Mapes) C.; m. Gail Marie Theroux, July 23, 1988; children from previous marriage: Emily Blakeslee Clark, Christopher Robert. BA, Yale U., 1970; MDiv, Pacific Sch. Religion, 1973; MD, George Washington U., 1977. Diplomate Am. Bd. Psychiatry and Neurology; cert. in adminstrv. psychiatry, APA, 1992. Intern, then resident, then fellow Dartmouth-Hitchcock Med. Ctr., Hanover, N.H., 1977-81; staff psychiatrist Lakes Region Med. Health Ctr., Laconia, N.H., 1981-82, med. dir., 1982-86; dir. psychiat. unit Lakes Region Gen. Hosp., Laconia, 1986-89; med. dir. behavioral svcs. St. Vincent Health Ctr., Erie, Pa., 1990—; dir. med./profl. adminstrn. Deerfield Mgmt. Group, 1991—; pres. Deerfield Profl. Assocs., 1992—; adj. asst. prof. clin. psychiatry Dartmouth Med. Sch., Hanover, 1983-90; clin. asst. prof. psychiatry U. Pitts. Sch. Medicine, 1990—; chmn. com. psychiatrists in N.H. Community Mental Health Ctrs., Concord, 1982-86; med. liason to Pa. office of Mental Health and Mental Retardation and Erie County Office of Mental Health and Mental Retardation, 1991—. Exec. v.p. Erie Phiharm., 1991-92. Recipient Exemplary Psychiatrist award Nat. Alliance for Mentally Ill, 1992; recipient Benjamin Manchester award George Washington U., 1977. Fellow Am. Psychiat. Assn. (task force to develop guidelines for psychiat. practice in community mental health ctrs., com. on state and community psychiatry systems, com. chronically mentally ill, Falk fellow 1979-81), Am. Coll. Mental Health Adminstrn., Am. Assn. Social Psychiatry (mem. coun.); mem. AMA, Am. Assn. Community Psychiatrists (com. to develop guidelines for psychiat. practice in community mental health ctrs., founding pres. 1984-90, Disting. Svc. award 1990), Am. Assn. Psychiat. Adminstrs., Am. Assn. Gen. Hosp. Psychiatrists, Am. Coll. Physician Execs., Am. Coll. Psychiatrists, Pa. Med. Soc. (coun. on med. practice, physician execs. liaison com.), Nat. Psychiatric Alliance (chmn. med. staff com. 1992—, exec. com. 1992—), Psychiat. Physicians Pa. (coun., govt. rels. com., fed. legis. rep.), Western Pa. Psychiat. Soc. (pres. elect. 1992—). Avocations: hockey, tennis, biking, hiking, camping. Home: 203 Hunter Willis Rd Erie PA 16509-3701 Office: St Vincent Health Ctr 232 W 25th St Erie PA 16502-2701

CLARK, GREGORY ALTON, research chemist; b. Warren, Ohio, Oct. 5, 1947; s. Elmer Edwin and Marie Laurette (Wantz) C.; m. Mary Ann Margaret Kapp, Sept. 8, 1973; children: Hilary Ann, Melissa Marie. BS, Youngstown State U., 1971; MBA, Kent State U., 1985. Analytical chemist The Youngstown (Ohio) Sheet & Tube Co., 1973-79; rsch. chemist Babcock & Wilcox Co., Alliance, Ohio, 1979—. Contbr. to books and articles to profl. jours. Mem. Humane Soc. U.S., Washington. 1st lt. U.S. Army, 1971. Mem. ASTM (com. D-5), Am. Chem. Soc., Air and Waste Mgmt. Assn. Roman Catholic. Office: Babcock & Wilcox 1562 Beeson St Alliance OH 44601-2196

CLARK, JACK I., civil engineer, researcher. Grad., Acadia U., Tech. U. Nova Scotia; DEng (hon.), Tech. U. Nova Scotia; grad., U. Alta., Can. With major civil engring. projects, 1957—; dir. Ctr. for Cold. Ocean Resources Engring. Meml. U. Newfoundland, St. John's, 1984-91, 1st pres., CEO Ctr. for Cold OCean Resources Engring., 1991—. Past editor Can. Geotechnical Jour. Recipient R. F. Legget award, 1983., Xerox Can.-Forum award, 1991. Fellow Engring. Inst. Can. (Julian C. Smith medal 1987), Can. Soc. Civil Engrs.; mem. Can. Acad. Engring., Nat. Scis. and Engring. Rsch. Coun. (v.p., exec. com. for coun., policy review com., allocations task force), Can. Geotechnical Rsch. Bd. (chmn.), Founds. for Offshore Structures (chmn. Can. Standards Assn. Com. S472). Office: Meml U Newfoundland, Ctr for Cold Ocean Rsch Engring, Saint John's, NF Canada A1B 3X5

CLARK, JEFF RAY, economist; b. Waynesboro, Va., Nov. 6, 1947; s. Jefferson Davis and Mildred (Cameron) C.; m. Arlene Donowitz, Dec. 17, 1989. BS, Va. Commonwealth U., 1970; MA, Va. Tech. U., Blacksburg, 1972, PhD, 1974. Assoc. dir. Joint Coun. Econ. Edn., N.Y.C., 1974-78, dir., 1978-80; chmn. econ./fin. Fairleigh Dickinson U., Madison, N.J., 1980-87; rsch. fellow Princeton (N.J.) U., 1987; Hendrix chair econs. U. Tenn., Martin, 1987-92; Probasco chair econs. U. Tenn., Chattanooga, 1992—; cons. Pew Charitable Trusts, Phila., 1987—, IT&T, Nutley, N.J., 1985, Fed. Res. Bank N.Y., N.Y.C., 1980, The Johns Hopkins U., Balt., 1984-86; disting. teaching fellow NSF, Washington, 1977, 78. Author: The Science of Cost, Benefit and Choice, 1988, Essential of Economics, 1982, 86, Economics Cost and Choice, 1987, Macroeconomics for Managers, 1990. Bd. dirs. The William B. Cockroft Found., The Plamer Chitester Fund. Mem. Assn. for Pvt. Enterprise (v.p. 1991, pres. 1992), Am. Econ. Assn., Ea. Econ. Assn. (bd. dirs. 1980-85), Western Econ. Assn., So. Econ. Assn., Midwest Econ. Assn. Avocations: aviation, skiing, boating. Home: 166 Glenwood Dr Martin TN 38237-2316 Office: U Tenn 615 Mc Callie Ave Chattanooga TN 37403

CLARK, JEFFREY ALAN, mechanical engineer; b. Johnson AFB, Japan, July 6, 1959; s. Jimmie Ray and Elaine Mary (Bautista) C.; m. Laura Jean Elmes, Feb. 25, 1989. BS, Clemson (S.C.) U., 1981; MS, Fla. Inst. Tech., 1985; grad. cert. of completion, U. Fla., 1984. Sr. quality engr. Harris Corp., Melbourne, Fla., 1981-82, head prodn. engr., 1982-85; lead design engr. Tex. Instruments, Dallas, 1985-87; dir. bus. devel. Keith & Schnars, Ft. Lauderdale, Fla., 1987-92; project dir. Keith & Schnars, Ft. Lauderdale, 1992—. Pres. Broward County Young Reps., Ft. Lauderdale, 1991, 2nd Century Broward, 1991; co-chmn. Mainstream Broward, 1990; exec. bd. Young Profl. for Covenant House; co-chmn. Rep. Exec. Com., 1992, South Broward Regional Bush re-election campaign, 1992; treas. Broward County Rep. Exec. Com. Robert McDonald scholar, 1977, Robert Earle awardee, Clemson Coll. Engring., 1981; named Top Ten Fla. Young Reps., 1992. Mem. Fla. Atlantic Bldrs. Assn., Nat. Assn. Indsl. and Office Parks, Pompano Econ. Group (chmn. govt. rels. com. 1990-91, chmn. construction devel. com. 1989-90), Congl. Club, Mensa, Tau Beta Pi, Phi Kappa Phi. Republican. Methodist. Avocations: running, biking, scubadiving. Home: 2701 N Ocean Blvd Apt 15A Fort Lauderdale FL 33308-7541

CLARK, JOHN F., aerospace research and engineering educator; b. Reading, Pa., Dec. 12, 1920; s. John F. Clark and Edith Dix (Long) Guenther; m. June Teubner Schweiger, July 14, 1974; children from previous marriage: Linda J. Marks, James C. BSEE with honors, Lehigh U., 1942, EE, 1947; MS in Math., George Washington U., 1946; PhD in Physics, U. Md., 1956. Registered profl. engr., N.J. Electronic engr. Naval Rsch. Lab., 1942-47, physicist, atmospheric electricity br. head, 1948-58; asst. prof. elec. engring. Lehigh U., 1947-48; dir. physics and astronomy programs NASA, 1958-63, dep. assoc. adminstr. space sci. and applications (scis.), 1963-65, chmn. space sci. steering com., 1963-65; dir. Goddard Space Flight Center, 1965-76; dir. space applications and tech. RCA Corp., Princeton, N.J., 1976-86; past-time cons. Gen. Electric Astro Space Div., 1987-88; NAVSPACE rsch. prof. U.S. Naval Acad. aerospace engring. dept., Annapolis, Md., 1988-90; acad. chmn., prof. space systems Fla. Inst. Tech., Titusville, 1990—; part-time lectr. math. George Washington U., 1956-58; part-time cons. rsch. Grad. Coun., 1960-66; part-time lectr. physics U. Md., 1958; mem. indsl. and profl. adv. coun. Pa. State U., 1963-65; mem. vis. com. physics Lehigh U., 1966-74; mem. Com. on Fed. Labs., 1971-75, Md. Gov.'s Sci. Adv. Coun., 1972-76, N.J. Gov.'s Sci. Adv. Com., 1980-86, Am. Geophys. Union-URSI Bd. Radio Sci., 1974-78; mem. study panel Office Telecommunications, Nat. Assembly Engring., 1976-77; chmn. adv. com. FCC, 1981-83; mem. U.S. del. to Internat. Telecommunication Union Conf., Regional Adminstrv. Radio Conf., 1983, World Adminstrv. Radio Conf., 1985; chmn. Direct Broadcast Satellite Assn., 1986; mem. spectrum planning adv. com. U.S. Dept. Commerce, 1986-92; bd. dirs. ECON Inc.; mem. Calif. Inst. Tech. Jet Propulsion Lab.'s Mars Observer Program Rev. Bd., 1986—. Contbr. numerous articles to profl. jours.; cons. editor space tech. McGraw-Hill Ency. Sci. and Tech, 1977—. Recipient NASA medals for Disting. Service, Outstanding Leadership, Exceptional Service, Collier trophy Nat. Aero. Assn. Fellow Am. Astron. Soc., AIAA (gen. chmn. Communications Satellite System Conf. 1984, v.p. pub. policy 1986-90), IEEE, Explorers Club; mem. Am. Geophys. Union, Satellite Broadcasting and Communications Assn. (chmn. 1987, chmn.'s coun. 1989-90, 1st Pres.'s award 1993), Internat. Soc. Satellite Profls. (bd. dirs. 1985-89), Internat. Acad. Astronautics, Phi Beta Kappa, Sigma Xi, Pi Mu Epsilon, Tau Beta Pi, Sigma Phi Epsilon, Sigma Pi Sigma. Patentee electronic circuits and systems. Home: 555 Fillmore Ave #406 Cape Canaveral FL 32920

CLARK, JOSEPH DANIEL, ecologist; b. Atlanta, Mar. 22, 1957; s. Bobby O'Neill and Winifred Jessica (Jones) C.; m. Sherri Lynn Addis, Aug. 22, 1982; 1 child, Daniel Ethan. BS in Forest Resources, U. Ga., 1980, MS, 1982; PhD, U. Ark., 1991. Biol. technician U.S. Fish and Wildlife Svc., Athens, Ga., 1979-80; rsch. asst. Sch. Forest Resources, Athens, 1980-82, rsch. technician II, 1982-83; bear/furbearer project leader Ark. Game and Fish Commn., Little Rock, 1983-90, asst. chief rsch., 1990-92; rsch. ecologist Nat. Park Svc., Knoxville, 1992—. Named Ark. Wildlife Artist of Yr., Ark. Wildlife Fedn., 1988. Mem. Wildlife Soc. (pres. Ark. chpt. 1988-89, cert. wildlife biologist), Am. Soc. Mammalogists, Internat. Assn. for Bear Rsch. and Mgmt. Home: 4711 Lakeview Rd Louisville TN 37777 Office: U Tenn Nat Park Svc CPSU 274 Ellington Plant Sci Bld Knoxville TN 37901-1071

CLARK, KENNETH WILLIAM, mechanical engineer; b. Royal Oak, Mich., July 6, 1960; s. Ralph Waldo and Shirley Anne (Cutright) C. BS in Mech. Engring., Mich. Tech. U., 1983. Engr.-in-tng. status, Mich. Design engr. Troy (Mich.) Design Svcs., 1983-84; application engr. NOK, Inc., Bloomfield Hills, Mich., 1984-86, supr. application engring., 1986-88; asst. mgr. application engring. Freudenberg-NOK, Plymouth, Mich., 1988-92; mgr. application engring., 1992—. Mem. Soc. Automotive Engrs. Avocations: comml. aircraft pilot, auto racing, fishing, hunting, backpacking. Office: Freudenberg-NOK 47690 E Anchor Ct Plymouth MI 48170-2460

CLARK, LYNNE WILSON, speech and language pathology educator; b. Bridgeport, Conn., Apr. 29, 1947; d. John Lucas and Lillian (Stekl) Wilson; m. Arthur Edward Clark, Sept. 5, 1970; children: Meggan, Matthew. MA, Columbia U., 1972; PhD, CUNY, 1984. Speech pathologist I.C.D. Rehab. Ctr., N.Y.C., 1973-75, NYU Med. Ctr.-Goldwater Hosp., N.Y.C., 1975; clin. supr. Burke Rehab. Ctr., White Plains, N.Y., 1975-77; instr. Queens Coll., CUNY, Flushing, 1977-79; prof. speech-lang. path., chmn. dept. Hunter Coll., CUNY, 1984—; editorial cons. Williams & Wilkens, Balt.,1986; mem. Hunter-Mt. Sinai Geriatric Edn. Ctr., N.Y.C., 1986—; mem. faculty Brookdale Ctr. on Aging, N.Y.C., 1987-88; cons. New Medico, Darien, Conn., 1982-84; frequent presenter to nat., state and regional gerontol. and aphasia confs. Contbr. articles to profl. jours., chpts. to books; contbr. two books. Co-leader Girl Scouts U.S.A., Valley Cottage, N.Y., 1988—. Fellow Brookdale Ctr. on Aging, 1988; grantee CUNY, 1989, Adminstrn. on Aging, 1989. Mem. Am. Speech-Lang. and Hearing Assn. (cert., peer reviewer 1987, legis. counselor 1990-93), N.Y. State Speech-Lang. and Hearing Assn. (sec. 1986-87, treas. 1987-89), N.Y. Acad. Scis. Office: Hunter Coll CUNY 425 E 25th St New York NY 10010-2590

CLARK, MARY DIANE, psychologist, educator; b. Trenton, N.J., Feb. 20, 1953; d. John Joseph and Elizabeth (Bellington) C. BA, Shippensburg State Coll., 1975; MA, Marshall U., 1977; PhD, U. N.C., Greensboro, 1985. Postdoctoral fellow Gallaudet U., Washington, 1985-86; asst. prof. Mt. St. Mary's Coll., Emmittsburg, Md., 1986-88; assoc. prof. dept. psychology Shippensburg (Pa.) U., 1988—; bd. dirs. Shippensburg's Alliance, 1992—; chair bd. dirs. Cumberland Valley Links, Mont Alto, Pa., 1993—. Co-editor: Psychological Perspectives on Deafness, 1993. Mem. APA, Soc. Rsch. Child Devel., Southeastern Psychol. Assn. (chair com. equality of profl. opportunity 1992—). Home: 130 E King St Shippensburg PA 17257 Office: Shippensburg Univ Dept Psychology Shippensburg PA 17257

CLARK, MICHAEL EARL, psychologist; b. Berea, Ohio, July 20, 1951; s. William Gray and Marguerite Jane (Charles) C.; m. Laura Lynn Putt, June 19, 1976 (div. Nov. 1987); 1 child, Brian Gray. BA, Kent State U., 1974, PhD, 1978. Asst. dir., chem. dependency unit N.D. State Hosp., Jamestown, N.D., 1978-79; staff psychologist VA Med. Ctr., Chillicothe, Ohio, 1979-84, Bay Pines, Fla., 1984-89; clin. dir., pain program James A. Haley Vets. Hosp., Tampa, Fla., 1989—; assoc. prof. dept. psychology U. South Fla., Tampa, 1986—, clin. asst. prof. dept. neurology Sch. Medicine, Tampa, 1991—; pvt. practice program cons., Tampa, 1987—. Contbr. chpts. to Innovations in Clinical Practice, 1991, Social Psychology: A Sourcebook, 1983; contbr. articles to Biofeedback and Self-Regulation, Jour. Personality Assessment, Jour. Clin. Psychology, Jour. Dental Rsch. Vice chmn. Paint Valley Mental Health Bd., Chillicothe, 1980-84. Mem. APA, Am. Psychol. Soc. (charter), Am. Pain Soc., N.Y. Acad. Scis. Democrat. Home: 9645 Fox Hearst Rd Tampa FL 33647 Office: Psychology Svc (116B) VAMC 13000 Bruce B Downs Blvd Tampa FL 33612-4798

CLARK, PATRICIA, molecular biologist; b. Lake Village, Ark., Mar. 21, 1928; d. Cleburn Clem and Helen Miller (Baker) C. BA, Washington U., St. Louis, 1950, MA, 1955; PhD, Purdue U., 1962. Microanalyst Washington U., St. Louis, 1950-51; rsch. chemist The Chemstrand Corp., Decatur, Ala., 1953-55; assoc. chemist So. Rsch. Inst., Birmingham, Ala., 1955-56; grad. asst. Purdue U., West Lafayette, Ind., 1956-61; biochemist Gerontology Rsch. Ctr. Child Health and Human Devel. NIH, Balt., 1961-78, Nat. Inst. on Aging NIH, Balt., 1978—. Fellow Am. Inst. Chemists; mem. AAAS, Am. Chem. Soc., Gerontol. Soc., N.Y. Acad. Scis., Sigma Xi, Pi Mu Epsilon, Alpha Lambda Delta. Home: Dulaney Towers 1 Smeton Pl Unit 1206 Baltimore MD 21204-2734 Office: Gerontology Rsch Ctr Nat Inst on Aging 4940 Eastern Ave Baltimore MD 21224-2780

CLARK, PHILIP RAYMOND, nuclear utility executive, engineer; b. Bklyn., June 16, 1930; s. Daniel Joseph and Freda (Rogerson) C.; m. Jeanne Marie Cushing, Aug. 22, 1953; children: Philip, Margaret, Andrew, Mary, Michael, Jeanne, Robert. B in Civil Engring., Polytech. Inst., 1951; postgrad., Oak Ridge Sch. of Reactor Tech., 1954. Naval architect N.Y. Naval Shipyard, Bklyn., 1951-53; nuclear engr. U.S. Dept. of Navy, Washington, 1954-64; dir. reactor engring. U.S. AEC/DOE, Washington, 1964-79; exec. v.p. GPU Nuclear, Parsippany, N.J., 1980-83, pres., chief exec. officer, 1983—; dir. INPO, Atlanta, 1987—, Am. Nuclear Energy Coun., Washington, 1993—, World Assn. Nuclear Operators, Atlanta, 1988—, Advanced Reactor Corp., 1991—, NUMARC, 1987-92; mem., com. chmn. Assn. Edison Illuminating Cos., N.Y.C., 1984—. Recipient Disting. Svc. award U.S. Navy, 1972, U.S. Energy R&D Adminstrn. Spl. Achievement award, 1976. Fellow Am. Nuclear Soc.; mem. NAE. Roman Catholic. Office: GPU Nuclear Corp 1 Upper Pond Rd Parsippany NJ 07054-1050

CLARK, RAYMOND JOHN, mechanical engineer; b. Gary, Ind., Dec. 22, 1951; s. John and Anne (Kenders) C.; m. Angelika Karna, Sept. 12, 1982; children: Kristine, Jonathan. BSME, Purdue U., 1973, MSME, 1975. Asst. project engr. Skidmore, Owings and Merrill, Chgo., 1975-78, assoc. project engr., 1978-82, assoc. ptnr., 1982-86, ptnr., 1986—; bd. dirs. Intelligent Bldgs. Inst., Washington, Indsl. and Profl. Adv. Coun. Pa. State U., University Park. Contbr. articles to profl. jours. Mem. ASHRAE, Chartered Inst. of Bldg. Svcs. Engrs. Office: Skidmore Owings and Merrill 224 S Michigan Ave Ste 1000 Chicago IL 60604

CLARK, STEVEN JOSEPH, energy engineer, business owner, inventor, author; b. Great Falls, Mont., Nov. 1, 1957; s. Maurice P. and Patricia G. (Myers) C.; m. Susan K. Falk, May 10, 1981; children: Amber, Adam, Aaron, Austin. BSME, Mont. State U., 1980; postgrad., Trane Grad. Engr. Tng. Ctr., 1982. Registered profl. engr.; cert. energy mgr. Devel. engr. Trane Co., LaCrosse, Wis., 1981-82; applications engr. Trane Co., Honolulu, 1982-84; project engr. Kohloss & Assocs., Wheatridge, Colo., 1984-85; project mgr. Flack and Kartz, Denver, 1985-86; v.p. Energy Conservation Cons., Billings, Mont., 1986-89; dir. energy dept. Assn. Constrn. Engrs., Belgrade, Mont., 1989-91; owner Energy Engring., Whitehall, Mont., 1991—, HVQC Systems, Whitehall, Mont., 1992—. Author: HVAC for the 21st Century, 1993. Co-founder, vice chmn. Mont. Libertarian Party, 1980. Mem. Assn. Energy Engrs. (sr.), Am. Soc. Heating, Refrigeration and Air Conditioning Engrs., Mont. Soc. Hosp. Engrs. Mem. LDS Ch. Achievements include patents for ocean wave electric power generator and for integrated piping system-heating, ventilating and air conditioning system. Office: Energy Engring 255 Franich Ln Whitehall MT 59759

CLARK, STUART ALAN, mechanical engineer; b. Nov. 26, 1962. BS in Engring., Calif. Poly. Inst., 1987. Cert. profl. engr. Mech. engr. Pacific Gas & Electric Co., San Francisco, 1991, Life Scis., Inc., Pleasanton, Calif., 1991-92, Gannett Fleming, Inc., San Francisco, 1992—. Mem. ASHRAE (assoc.). Office: GFI 685 Market St Ste 860 San Francisco CA 94105

CLARK, SUSAN MATTHEWS, psychologist; b. Newton, Kans., Aug. 5, 1950; d. Glenn Wesley Matthews and Jane Buckles; m. S. Bruce Clark, Aug. 14, 1971; children: Casandra Jane, Ryan Matthews. BME, Wichita State U., 1971, MME, 1975, MA, 1982; PhD, North Tex. State U., 1985. Elem. tchr. Derby (Kans.) Pub. Schs., 1972-74; profl. musician Amarillo (Tex.) Symphony, 1974-77; psychol. cons. Achenbach Ctr., Hardtner, Kans., 1983-85; psychologist VA Med. Ctr., Wichita, Kans., 1984-85, St. Francis Acad., Inc., Salina, Kans., 1986-89, Psychiat. Clinic Wichita, 1989-93; psychol. cons. Affiliated Psychiat. Svcs., Wichita, 1993—; bd. dirs. Salina Coalition for the Prevention of Child Abuse, 1986-87. Author: Grant, 1987. Deacon Plymouth Congl. Ch., Wichita, 1989-92, bd. deacons, mem. bd. Christian Edn., 1993—. Recipient: Phi Kappa Phi, Mu Phi Epsilon, Psi Chi. Mem. APA, Nat. Acad. Neuropsychology, Southwestern Psychol. Assn., Kans. Psychol. Assn., Wichita Area Psychol. Assn., Kans. Assn. Profl. Psychologists, Menninger Found., Beta Sigma Phi. Republican. Congregationalist. Avocations: stained glass, photography, needlepoint, tennis. Office: Affiliated Psychiat Svcs Ste 104 1148 S Hillside Wichita KS 67211-4005

CLARK, SYDNEY PROCTER, geophysics educator; b. Phila., July 26, 1929; s. Sydney P. and Isabella L. (Mumford) C.; m. Elizabeth Frey, 1963; 4 children. AB, Harvard U., 1951, MA, 1953, PhD in Geology, 1955. Research fellow in geophysics Harvard U., Cambridge, Mass., 1955-57; geophysicist Carnegie Inst., 1957-62; Weinberg prof. geophysics Yale U., New Haven, 1962—. Fulbright scholar Australian Nat. U., 1963. Office: Yale U Dept Geology PO Box 6666 New Haven CT 06511-8101

CLARK, TERESA WATKINS, psychotherapist, mental health counselor; b. Hobart, Okla., Dec. 18, 1953; d. Aaron Jack Watkins and Patricia Ann (Flurry) Greer and Ralph Gordon Greer; m. Philip Winston Clark, Dec. 29, 1979; children: Philip Aaron, Alisa Lauren. BA in Psychology, U. N.Mex., 1979, MA in Counseling and Family Studies, 1989. Nat. cert. counselor. Child care worker social svcs. div. Family Resource Ctr., Albuquerque, 1978-79; head tchr., asst. dir. Kinder Care Learning Ctr., Albuquerque, 1979-80; psychiat. asst. Vista Sandia Psychiat. Hosp., Albuquerque, 1980-87; psychotherapist outpatient clinic Bernalillo County Mental Health Ctr.-Heights, 1989-91; therapist adolescent program Heights Psychiat. Hosp., Albuquerque, 1991—. Mem. ACA, Am. Assn. Multicultural Counseling and Devel., Phi Kappa Phi. Democrat. Avocations: music, camping, horseback riding, reading. Office: Heights Psychiat Hosp 103 Hospital Loop Albuquerque NM 87110

CLARK, TERRENCE PATRICK, veterinarian, researcher; b. Green Bay, Wis., Dec. 28, 1960; s. Henry Lee and Marlene Ann (Fenlon) C. BS, U. Wis., 1983, DVM, 1987. Intern Okla. State U., Stillwater, 1987-88; assoc. vet. Moon Valley Animal Hosp., Phoenix, 1988-89, Catalina Pet Hosp., Tucson, 1989-90; rschr. Auburn (Ala.) U., 1990—; cons. Endocrine Diagnostic Svc., Auburn, 1990—. Contbr. articles to profl. jours. Advisor Omega Tau Sigma, Auburn U., 1992. Alice Boise scholar U. Wis., 1984-87. Mem. AVMA, Am. Physiol. Soc., N.Y. Acad. Scis., Endocrine Soc. Democrat. Roman Catholic. Home: 336 W Drake Ave Auburn AL 36830 Office: Auburn U Coll Vet Medicine Auburn AL 36849

CLARK, WILLIAM ANTHONY, neuropharmacologist; b. Teaneck, N.J., Sept. 18, 1961; s. Edward Albert and Kathleen Francis (Mulqueen) C.;m. Janet Elizabeth Aylward, Nov. 3, 1984. BS in Chemistry, Boston Coll., 1983; MPhil in Pharmacology, Yale U., 1989, PhD in Pharmacology, 1993. Chemist Lehn & Fink Products Co., Montvale, N.J., 1983-86; postdoctoral assoc. Dept. Pharmacology, Yale U., New Haven, 1993; postdoctoral fellow cell biology NIH, Bethesda, Md., 1993—; lectr. pediatric pharmacology Sch. Nursing, Yale U., 1990; founder Pharmacology Rsch. Colloquium, Yale U., 1988. Contbr. articles to profl. jours. Scientist, educator Sci.-By-Mail Program, Boston Mus. of Sci., 1991-92. Nat. Inst. Gen. Med. Scis. postdoctoral fellow, 1993; recipient Nat. Rsch. Svc. award Pub. Health Svc., 1991-92, Predoctoral award, 1986-91, John Thomas Coll Meml. award Boston Coll., 1983. Mem. N.Y. Acad. Sci., Soc. for Neurosci., SPEBSQSA. Achievements include patent in field. Office: Lab Cell Biology NIH/NIMH Bldg 36 Rm 2D23 9000 Rockville Pike Rockville MD 20892

CLARK, WILLIAM BURTON, IV, dentist, educator; b. Madison, Fla., Nov. 19, 1947; s. William Burton III and Eunice (Priest) C.; m. Mae Merchant, Dec. 27, 1969; children: William Burton V, Corrie Elizabeth. BA in Biology, U. of the South, 1969; DDS, Emory U., 1973; DMS in Biology and Microbiology, Harvard Sch. Dental Medicine, 1979. Cert. in periodontology; cert. in oral medicine. Lectr. health sci. Northeastern U., 1976-77; asst. prof. basic dental scis., immunology and med. microbiology U. Fla., Gainesville, 1977-82, assoc. prof. oral biology, immunology and med. microbiology, 1982-87, prof., 1987—, mem. grad. faculty div. grad. studies, 1978—; mem. attending staff Shands Hosp., 1977—; mem. staff VA Hosp., Gainesville, 1978—; adj. asst. prof. microbiology and cell sci., U. Fla., 1980-83, adj. assoc. prof., 1983-87, acting dir. Periodontal Disease Rsch. Ctr., 1980-82, dir. Periodontal Disease Rsch. Ctr., 1982—, mem./past mem. various univ. coms., mem. univ. faculty senate, 1980-82, mem. faculty group practice Coll. Dentistry, 1977-85; vis. scientist Tokyo Dental Coll., 1982, Nat. Inst. Dental Rsch., 1985; vis. prof. microbiology, U. B.C., Vancouver, Can., 1990. Contbr. numerous articles to profl. jours., 1972—. Mem. Alachua County dem. exec. com., Gainesville, 1978-83. NIH postdoctoral fellow, 1975-77; recipient NIH Rsch. Career Devel. award, 1982-87, Robert Wood Johnson Health Policy fellowship, 1992-93. Mem. ADA, AAAS, Internat. Assn. Dental Rsch. (treas. periodontal rsch. group 1991—), Am. Assn. Dental Rsch. (pres. microbiology/immunology sect. 1985-86, mem. nat. affairs com. 1986-89, mem. nominations com. 1989-92), Am. Soc. Microbiology, Am. Acad. Periodontology (mem. Orban com. 1984-85), Am. Assn. Oral Biologists (founding), Fla. Dental Assn. (Cen. Dist. Dental Soc.), Fla. Soc. Periodontists, N.Y. Acad. Sci., Rotary Internat., Sigma Xi, Xi Psi Phi (chpt. pres. 1973). Baptist. Achievements include research in mechanisms of bacterial attachment to teeth and development of vaccines made from bacterial adhesions to prevent or reduce colonization by periodontal disease-associated bacteria, clinical trials for antiplaque and anti-gingivitis agents, antibiotic therapy in treatment of peridontal disease. Home: 2811 NW 23rd Ter Gainesville FL 32605-2861 Office: Univ Fla Coll Dentistry PO Box 424J Gainesville FL 32602-0424

CLARKE, KIT HANSEN, radiologist; b. Louisville, May 24, 1944; d. Hans Peter and Katie (Jones) Hansen; A.B., Randolph-Macon Woman's Coll., 1966; M.D., U. Louisville, 1969; m. Dr. John M. Clarke, Feb. 14, 1976; children: Brett Bonnett, Blair Hansen, Brandon Chamberlain; stepchildren: Gray Campbell, Jeffrey William John M. Intern, Louisville Gen. Hosp., 1969-70; resident in internal medicine and radiology U. Tenn., Knoxville, 1970-73; resident in radiology U.S. Fla., Tampa, 1973-74; staff radiologist, chief spl. procedures Palms of Pasadena, St. Petersburg, Fla., 1974—. Active Fla. Competitive Swim Assn. of AAU. Diplomate Am. Bd. Radiology. Fellow Am. Coll. Radiology; mem. Fla. West Coast Radiology Soc., Radiol. Soc. N.Am., AMA, Fla. Med. Assn., Pinellas County Med. Soc., Fla. Radiology Soc. Episcopalian. Home: 7171 9th St S Saint Petersburg FL 33705-6218 Office: 1609 Pasadena Ave S Saint Petersburg FL 33707

CLARKE, NICHOLAS CHARLES, metallurgical executive, mineral technologist; b. Ipswich, Suffolk, Eng., Feb. 2, 1948; arrived in Malaysia, 1988; s. Stanley Charles and Edna May (Howes) C.; m. Lynda Power, Mar. 24, 1974. BS in Mineral Tech. with honors, Imperial Coll., London, 1969; Assoc., Royal Sch. of Mines, London, 1969; PhD, Leeds (Eng.) U., 1975. Registered profl. engr. Operator Falconbridge Nickel Mine, Sudbury, Ont., Can., 1968; mineral dressing engr. Sierra Leone Devel. Co. Ltd., Marampa, 1969-71; lectr. U. Tech., Lae, Papua New Guinea, 1975-77; sr. metallurgist Bougainville Copper Ltd., Panguna, Papua New Guinea, 1977-86; mgr. metallurgy Agnew (West Australia) Mining Co. Proprietary Ltd., 1986; resident mgr. Golden Spec Mine, Nullagine, West Australia, 1986-87; prin. cons. Normet Proprietary Ltd., Perth, West Australia, 1987-89; mng. dir., prin. cons. Normet (Malaysia) Sdn Bhd, Kuala Lumpur, 1989-92; prin. Imtech P/L, Applecross, W.A., Australia, 1992—; cons. Kalgoorlie (West Australia) Consolidated Gold Mines, 1991. Fellow Instn. Mining and Metallurgy; mem. Australasian Instn. Mining and Metallurgy. Avocations: sailing, stage lighting, carpentry, theatre and cinema.

CLARKE, ROY, physicist, educator; b. Bury, Lancashire, England, May 9, 1947. BSc in Physics, U. London, PhD, 1973. Rsch. assoc. Cavendish Lab., Cambridge, U.K., 1973-78; James Frank fello U. Chgo., 1978-79; prof. U. Mich., Ann Arbor, 1979—. Editor: Synchrotron Radiation in Materials Research, 1989. Fellow Am. Phys. Soc. Achievements include development of novel methods for real-time x-ray studies; patent for quasiperiodic optical coatings. Office: U Mich Randall Lab Ann Arbor MI 28109

CLARKE, THOMAS EDWARD, research and development management educator; b. Vancouver, B.C., Can., Nov. 1, 1942; s. Thomas Edward and Lavona (Burton) C.; m. Jean Reavley, Dec. 13, 1969; children: Thomas E., James R. BS, U. B.C., 1964, MS in Physics, 1967, MBA in Rsch. & Devel. Mgmt., 1971. Sci. polit. advisor Ministry of State Sci. & Tech., Ottawa, Ont., Can., 1973-74; program mgr. Dept. Industry Trade & Commerce, Ottawa, Ont., Can., 1974-80; pres. Stargate Cons. Ltd., Ottawa, Ont., Can., 1980—; exec. dir. Innovation Mgmt. Inst. Can., Ottawa, 1980-91. Author: Science and Technology Management Bibliography 1993; contbr. articles to profl. jours. Govt. B.C. Univ. scholar, 1960, 61, 63. Mem. Rsch. & Devel. Mgmt. Assn., The Tech. Transfer Soc. Achievements include the establishment of Canada's only federal industrial innovation centres. Office: Stargate Cons Ltd, PO Box 8235, Cold Lake, ON Canada T0A 0V0

CLARKSON, JOCELYN ADRENE, medical technologist; b. Bennettsville, S.C., July 9, 1952; d. Henry Louis and Frankie Allene (Carter) C. BA in Biology, Columbia (S.C.) Coll., 1973; cert. med. tech., Presbyn. Hosp., Charlotte, N.C., 1975. Coll. tutor of Germanic language Columbia Coll., 1970-73, switchboard operator, 1972-73; lab aide Richland Meml. Hosp., Columbia, 1974, now, med. technologist; profl. model. Appeared (TV commls.) Back Porch Restaurant and Meat Market, 1992, (film) The Chasers, author: poems, compilation, short stories, Messages from Hijae, 1989. Mem. Am. Soc. Clin. Pathologists (assoc.), Assn. for Studies of Classical African Civilization, African Am. Resource Inst. Roman Catholic. Avocations: world traveling, reading, forming teen music orgns., performing and writing music, keyboard. Home: 201 H L Clarkson Rd Hopkins SC 29061-9723

CLARKSON, JOHN J., dentist, dental association administrator; b. Mullingar, Ireland, Oct. 23, 1941; m. Marie Bannon; children; Ruth, Robert, Alan. B Dental Surgery, Nat. U. Ireland, Dublin, 1964; PhD, Nat. U. Ireland, Cork, 1987. Asst. in gen. dental practice Dartford, Kent, Eng., 1964, London, Eng., 1964-65; dental surgeon in gen. dental practice Armagh, No. Ireland, 1965-67; dental surgeon Dublin Health Auth./Eastern Health Bd., Ireland, 1968-79; acting. sr. dental surgeon Crumlin Area, Dublin, 1979-80; dental officer Dept. Health, Dublin, 1980-82, dep. chief dental officer, 1982-90; exec. dir. Internat. Assn. for Dental Rsch./Am. Assn. for Dental Rsch., Washington, 1990—; hon. lectr. in community dental health and preventive dentistry Trinity Coll., Dublin, 1983-90; mem. WHO Collaborating Ctr. for Health Svcs. Rsch., U. Coll., Cork, Ireland, 1987-90; mem. steering coms. Nat. Survey Children's Dental Health, Ireland, 1984, Nat. Survey Adult Dental Health, Ireland, 1988; sci. advisor Dental Health Found., Ireland; mem. radiation adv. com. Nuclear Energy Bd., Ireland, 1987-90; cons. European Community, France, Britain, Ireland. 1989-90, WHO, 1989—. Contbr. articles to Jour. Irish Dentistry, Jour. Dental Rsch., others; presenter papers in field. Sec. Oatlands Coll. Parents Coun., 1987-90. Coun. Europe fellow, 1981. Fellow Royal Acad. Medicine (Ireland); mem. ADA, AAAS, Am. Soc. Assn. Execs., Greater Washington Soc. Assn. Execs., Internat. Assn. for Dental Rsch., European Orgn. for Caries Rsch., Irish Dental Assn. (pub. dental officers com., sci. com., rep. staff negotiations bd. on conciliation and arbitration), Irish Soc. Dentistry for Children (pres. 1989), Brit. Assn. for Study Community Dentistry, Brit. Dental Assn., Fedn. Dentaire Internat. (cons. 1988-90), Local Govt. and Pub. Svcs. Union (sec. dental br.), Blainroe Golf Club, St. Mary's Lawn Tennis Club, Carderok Swim and Tennis Club. Home: 12400 Bobink Ct Potomac MD 20854 Office: American Assoc for Dental Rsch 1111 14th St NW Ste 1000 Washington DC 20005

CLARKSON, KENNETH WRIGHT, economics educator; b. Downey, Calif., June 30, 1942; s. William Wright and Constance (Patch) C.; m. Mary Jane Purdy, June 20, 1965; children: Steven Wright, Thomas David. A.B., Calif. State U., 1964; M.A., UCLA, 1966, Ph.D., 1971. Economist Office Mgmt. and Budget, Washington, 1971-72, assoc. dir., 1982-83; asst. prof. econs. U. Va., 1969-75; prof. econs. U. Miami, Coral Gables, Fla., 1975—, dir. Law & Econs. Ctr., 1981—; cons. in field; mem. Pres.'s Task Force on Food Assistance, 1983-84; mem. governing bd. Credit Rsch. Ctr. Purdue U., 1981—; rsch. com. Fla. C. of C. Found., 1985-91; nat. adv. bd. Nat. Ctr. for Privatization, 1986—, Washington Legal Found., 1985—, Capital Rsch. Ctr., 1990—; mem. Fla. adv. com. U.S. Commn. on Civil Rights, 1985—; mem. Gov.'s Coun. Econ. Advisors, Fla., 1988-90; corp. strategic planning task force United Way, 1986-89; adv. bd. James Madison Inst., 1990—. Author: Food Stamps and Nutrition, 1975, Intangible Capital and Rates of Return, 1975; co-author: Correcting Taxes for Inflation, 1975, Distortions in Official Unemployment Statistics, 1979, Industrial Organization: Theory, Evidence and Public Policy, 1982, West's Business Law, 1980, 5th edit., 1992, The Federal Trade Commission Since 1970, 1981, Economics Sourcebook of Government Statistics, 1983, The Role of Privatization in Florida's Growth, 1987, Using Private Management to Foster Florida's Growth: Initial Steps, 1987, A Proposal for Medical Malpractice Insurance in Florida, 1987, Alternative Service Delivery Project: An Analysis of the City and County of Los Angeles, 1988, Products Liability at a Glance, 1990, 92; contbr. numerous articles to profl. jours. Mem. Reagan-Bush Transition Team, Washington, 1980; mem. econ. adv. panel Fla. State Comprehensive Plan Com., 1986-87; mem. Met. Dade County (Fla.) Budget Reform Commn., 1990-91. NSF grantee, 1972-74; Heritage Found. adj. scholar, 1977—. Mem. Am. Econ. Assn., Am. Bus. Law Assn., Western Econ. Assn., Mont Pelerin Soc., Phil. Soc., Sigma Xi. Home: 6101 W Suburban Dr Miami FL 33156-1922 Office: U Miami 5801 S Red Rd Miami FL 33124-3801

CLARKSON, THOMAS BOSTON, comparative medicine educator; b. Decatur, Ga., June 13, 1931. DVM, U. Ga., 1954; Diploma, Am. Coll. Lab. Animal Medicine, 1963. Rsch. assoc. pharmacology and exptl. therapeutics sect. S. E. Massengill Co., 1954-57; from asst. to assoc. prof. exptl. medicine, dir. vivarium Wake Forest U., Winston-Salem, N.C., 1957-64, assoc. prof. lab. animal medicine, head dept., 1964-65, prof., chmn. dept. Bowman Gray sch. medicine, 1965—, dir. arteriosclerosis rsch. ctr., 1971—, dir. comparative medicine clin. rsch. ctr., 1989—; mem. sci. adv. com. regional primate rsch. ctr. U. Wash., 1971—; mem. adv. com. Cerbrovascular Rsch. Ctr., 1973—; mem. com. vet. med. sci. NAS-Nat. Rsch. Coun., 1975—; chmn. arteriosclerosis, hypertension and lipid metabolism adv. com. Nat. Heart Lung & Blood Inst., 1983-85. Recipient Griffin award Am. Assn. Lab. Animal Sci., 1977. Fellow Am. Soc. Primatology, Acad. Behavioral Med. Rsch., Soc. Behavioral Medicine; mem. NAS (mem. clin. sci. panel study nat. needs biomedical and behavioral rsch. pers. com. and task force animal models atherosclerosis 1976—), Am. Heart Assn. (mem. com. coronary artery lesions and myocardial infarctions 1970—, chmn. task force rsch. animal use, vice-chmn. coun. arteriosclerosis 1979-81, chmn. 1981-83, G. Lyman Duff Meml. lectr. 1985, Award of Merit 1987), Am. Assn. Advancement Lab. Animal Sci., Am. Assn. Pathologists, Am. Soc. Exptl. Pathology, Am. Vet. Medicine Assn. (Charles River prize 1978), Sigma Xi. Achievements include research in comparative and experimental atherosclerosis, particularly factors affecting susceptibility and resistance to the disease and the mechanisms by which risk factors affect the pathogenesis. Office: Wake Forest U Dept Comparative Medicine 300 S Hawthorne Rd Winston Salem NC 27103*

CLARREN, STERLING KEITH, pediatrician; b. Mpls., Mar. 12, 1947; s. David Bernard and Lila (Reifel) C.; m. Sandra Gayle Bernstein, June 8, 1970; children: Rebecca Pia, Jonathan Seth. BA, Yale U., 1969; MD, U. Minn., 1973. Pediatric intern U. Wash. Sch. Medicine, Seattle, 1973-74, resident in pediatrics, 1974-77, asst. prof. dept. pediatrics, 1979-83, assoc. prof., 1983-88, prof., 1988, Robert A. Aldrich chair in pediatrics, 1989—; head divsn. congenital defects U. Wash. Sch. Medicine, 1987—; dir. dept. congenital defects Children's Hosp. and Med. Ctr., Seattle, 1987—. Contbr. articles to profl. jours.; patentee for orthosis to alter cranial shape. Cons. pediatrician Maxillofacial Rev. Bd., State of Wash., Seattle, 1984—, chmn. Health-Birth Defects Adv. Com., Olympia, 1980—; mem. fetal alcohol adv. com. Children's Trust Found., Seattle, 1988—; mem. adv. bd. Nat. Orgn. on Fetal Alcohol Syndrome. Rsch. grantee Nat. Inst. Alcohol Abuse & Alcoholism, 1982—. Mem. AAAS, Am. Acad. Pediatrics, Soc. for Pediatric Rsch., Teratology Soc., Rsch. Soc. on Alcoholism (pres. fetal alcohol study group 1993), Am. Cleft Palate Assn., N.Y. Acad. Scis. Avocations: cross-country skiing, fishing, hiking. Home: 8515 Paisley Dr NE Seattle WA 98115-3944 Office: Children's Hosp and Med Ctr Divsn Congenital Defects PO Box C-5371 Seattle WA 98105

CLAUSEN, JERRY LEE, psychiatrist; b. Wausau, Wis., Nov. 5, 1939; s. Douglas William and Florence Jean (Amidon) C.; m. Nancy Eileen Longdon, Aug. 3, 1962; children: Keith Russell, Pamela Dawn. BA, Wesleyan U., Middletown, Conn., 1961; MD, Albany Med. Coll., N.Y., 1965. Diplomate Am. Bd. Psychiatry and Neurology; cert. Am. Soc. Addiction Medicine, N.Y. State Alcoholism Counselor. Psychiatry intern Upstate Med. Ctr., Syracuse, N.Y., 1965-66; psychiatric resident Upstate Med. Ctr., 1966-67, 69-71, asst. attending, 1971-72, attending, 1972-80; staff psychiatrist Onondaga Mental Health Clinic, Syracuse, 1971-72; courtesy staff Benjamin Rush Psychiatric Ctr., Syracuse, 1971-84, active staff, 1984—; pvt. practice psychiatry Syracuse, 1971—; clin. asst. prof., 1972—; staff psychiatrist Onondaga Pastoral Counseling Ctr., Syracuse, 1971-73, 81—, psychiatric dir., 1973-81; cons. psychiatrist Loretto Rest Geriatric Ctr., Syracuse, 1972-74. Tchr. First Universalist Ch., Syracuse, 1966—. Lt. comdr. USN, 1967-69. Fellow Am. Psychiat. Assn. (chmn. ins. mktg. com. 1979-88); mem. Onondaga County Med. Soc., N.Y. State Med. Soc. Universalist-Unitarian. Avocations: walking, tennis, cross-country skiing. Office: 1014 State Tower Bldg Syracuse NY 13202

CLAUSEN, THOMAS HANS WILHELM, chemist; b. Luebeck, Fed. Republic Germany, Apr. 20, 1950; s. Hans August and Lotte Henny Adele (Stubbendorf) C.; m. Monika Maria Epe, June 21, 1980; children: Lars Christopher, Lisa Katharina Joseline. Diploma in Chemistry, U. Kiel, Fed. Republic Germany, 1974, PhD in Chemistry, 1977. Asst. Christian Albrechts U., Kiel, 1974-78; hair dye rschr. Wella Ag, Darmstadt, Fed. Republic Germany, 1978-83, head dept. rsch. hair dyes, organic chemistry, 1983-87, head dept. chem. rsch., 1987-89, R & D hair care, 1989—. Co-author: Ullmann Encyclopedia, 1989; speaker Chitosan Congress, 1988; contbr. articles to profl. jours.; patentee in field of hair dyes, hair care and organic chemistry. Mem. Gesellschaft Deutscher Chemiker, Am. Chem. Soc. Avocations: house and garden activities, photography, video, electronics. Home: Ernst Pasque Strasse 35A, D 64665 Alsbach Germany Office: Wella Ag, Berliner Allee 65, D 64274 Darmstadt Germany

CLAVELLI, LOUIS JOHN, physicist; b. Chgo., Oct. 7, 1939. BS, Georgetown U., 1961; PhD, U. Chgo., 1967. Rsch. assoc. Yale U., New Haven, Conn., 1967-69; physicist Fermilab, Batavia, Ill., 1969-71; vis. asst. prof. U. Md., College Park, 1973-75, 77-78; assoc. prof. U. Bordeaux, France, 1975-77; vis. scientist U. Bonn, Germany, 1978-81; vis. assoc. prof. Ind. U., Bloomington, 1983-85; prof. U. Ala., Tuscaloosa, 1985--. Editor: Lewes String Theory Workshop, 1985, Superstrings and Particle Theory, 1989; contbr. articles to profl.jours. Office: U Ala Dept Physics and Astronomy Tuscaloosa AL 35487

CLAY, AMBROSE WHITLOCK WINSTON, telecommunication company executive; b. Marion, Ohio, Dec. 20, 1941; s. Ambrose Whitlock Winston and Ann Bernadette (Robinson) C.; m. Sharon Lee Boyd, June 25, 1966; children: Susan Rose, Allison Win. BEE, MIT, 1964; MBA, U. Chgo., 1972. Sr. supr. GTE Labs, Northlake, Ill., 1972-74, mgr. project plan, 1974-79, mgr. software devel., Phoenix, 1979-81; dir. systems devel. GTE Communication Systems R & D, Phoenix, 1981-84, dir. tech. devel. and applications, 1985-87, v.p. R & D AG Communication Systems (formerly GTE Communication Systems), 1987—; corp. contact to U. Ariz., Phoenix, 1984—; chmn. tech. presentation, 1979; mem. Ariz. Math., Engring., Sci. Achievement Industry Adv. Bd., chmn., 1985-86; mem. industry adv. bd. U. Ariz. Coll. Engring. and Mines, 1987—, vice chmn., 1988, chmn., 1990; mem. adv. coun. for engring. Ariz. State U., 1987—, adv. coun. Western Communication Forum, 1989—; bd. A Patentee telephone switching, 1975. Active MIT Ednl. Council, Chgo., 1972-76; mem. sch. bd. Glen Ellyn Sch. Dist. 89, 1976-79, pres. 1979; mem. computer sci. edn. com. Ariz. State U., Phoenix, 1981-87; mem. governing bd. Ariz. Math. Coalition, 1990—. Mem. IEEE, Communication Soc. (vice-chmn. Phoenix chpt. 1982-83, chmn. 1983-84), MIT Club. Republican. Roman Catholic. Home: 215 E Acapulco Ln Phoenix AZ 85022-3618 Office: AG Communication Systems 2500 W Utopia Rd Phoenix AZ 85027-4199

CLAY, HARRIS AUBREY, chemical engineer; b. Hartley, Tex., Dec. 28, 1911; s. John David and Alberta (Harris) C.; m. Violette Frances Mills, June 19, 1948 (dec. 1972); m. 2d Garvice Stuart Shotwell, Apr. 28, 1973 (dec. 1989); m. Leona G. Steele, May 25, 1991. BS, U. Tulsa, 1933; ChE, Columbia U., 1939. Pilot plant operator Phillips Petroleum Co., Burbank, Okla., 1939-42, resident supr. Burbank pilot plants, 1942-44; process design engr. Phillips Petroleum Co., Bartlesville, Okla., 1944-45; process engring. supr. Philtex Plant, Phillips, Tex., 1946-56; tech. adviser to pilot plant mgr. Bartlesville, 1957-61, chem. engring. assoc., 1961-73; ind. cons. engr., 1974—; chmn. tech. com. Fractionation Research, Inc., 1966-71, mem. tech. com., 1972-73. Contbr. articles to profl. jours.; patentee in field. Mem. dist. commn. Boy Scouts Am. Fellow Am. Inst. Chem. Engrs.; mem. Nat. Soc. Profl. Engr., Okla. Soc. Profl. Engrs., Am. Chem. Soc., Electrochem. Soc., Elks, Lions. Presbyterian. Home: 1723 Church Ct Bartlesville OK 74006-6401

CLAY, KELLI SUZANNE, mechanical engineer; b. Columbus, Ga., Aug. 27, 1964; d. Billy Jerrell and Martha Jane (Kirkland) C.; m. Joseph Brett Newsom, Mar. 28, 1992. BME, Ga. Inst. Tech., 1987. Engr. Shumate, Inc., Tucker, Ga., 1987-88; energy systems engr. Atlanta Gas Light Co., 1988—. Vol. United Way, Athens, Ga., 1989-90; youth motivator Merit Employment Assn., 1990. Mem. ASHRAE, Soc. Automotive Engrs., Ga. Soc. Profl. Engrs., Ga. Soc. Profl. Engrs. Buckhead chpt. (outstanding svc. award 1991, 92). Home: 3901 Sentry Walk Marietta GA 30068 Office: Atlanta Gas Light Co PO Box 4569 235 Peachtree St NE Atlanta GA 30302

CLAY, STANTON TOWER, technical consultant; b. Denver, Oct. 19, 1932; s. Claude S. and Avo May (Russell) C.; m. Mildred R. Morrison; children: Paul, Kathleen, Diana, Barbara. BA in Math., U. Denver, 1962. Mgr. quality control and tech. svc. GAF Corp., Binghamton, N.Y., 1962-72; mgr. tech. svc. Ilford Inc., Paramus, N.J., 1972-80; v.p. R&D Nimslo Corp., Atlanta, 1980-88; v.p. tech. svc. Delphi Tech., Atlanta, 1988—; cons. Clay Assocs., Atlanta, 1985—. Chmn. Ga. Environ. Silver Coun., Atlanta, 1992. With U.S. Army, 1956-58. Mem. Imaging Sci. and Tech. (Svc. award 1984, pres. Atlanta chpt. 1982-92). Achievements include patent for tamperproof security card. Home and Office: 8605 S Mount Dr Alpharetta GA 30202

CLAYCAMP, HENRY GREGG, radiobiologist educator; b. Wa Keeney, Kans., Oct. 15, 1952; s. Henry John and Oakie Joanne (Hillman) C.; m. Leslie Suzanne Hodge, June 21, 1973 (div. Jan., 1980); m. Rebecca Dawn Tuttle, Dec. 29, 1983; children: Kathryn T., Benjamin J. BA, Stanford U., 1974; MS, Northwestern U., 1977, PhD, 1982. Indsl. hygienist US OSHA, Chgo., 1975-76; asst. prof. U. Kans., Lawrence, 1981-85, U. Iowa Coll. Medicine, Iowa City, 1985-90; assoc. prof. U. Pitts., 1990—. Contbr. articles to Internat. Jour. Radiation Biology, Radiation Rsch., Mutation Rsch., Carcinogenesis. Grantee NIH, Nat. Cancer Inst., 1988, '89; Nuclear Regulatory Commn., Pitts., 1991—. Mem. Health Physics Soc. (plenary), Radiation Rsch. Soc. (plenary), Soc. for Risk Analysis, Sigma Xi. Office: U Pitts CEOHT 260 Kappa Dr Pittsburgh PA 15238

CLAYTON, DONALD DELBERT, astrophysicist, nuclear physicist, educator; b. Shenandoah, Iowa, Mar. 18, 1935; s. Delbert Homer and Avis (Kembery) C.; children: Donald, Devon, Alia, Andrew; m. Nancy McBride. B.S., So. Meth. U., 1956; Ph.D., Calif. Inst. Tech., 1962. Research fellow in physics Calif. Inst. Tech., 1961-63; staff scientist Aerospace Corp., El Segundo, Calif., 1961-63; mem. faculty Rice U., Houston, 1963-89; assoc. prof. physics and space sci. Rice U., 1965-69; prof. physics and space sci. and faculty assoc. Wiess Coll., 1969-77, Andrew Hays Buchanan prof. astrophysics, 1975-89; prof. physics and astronomy Clemson U., S.C., 1989—; vis. assoc. physics Calif. Inst. Tech., 1966-67; vis. fellow Inst. Theoretical Astronomy, Cambridge, summers 1967-72. Author: Principles of Stellar Evolution and Nucleosynthesis, 1969, The Dark Night Sky, 1975, The Joshua Factor, 1986; also numerous rsch. articles. Sloan fellow, 1966-70; Humboldt awardee Max Planck Inst., Heidelberg, 1977, 82; Fulbright fellow, Heidelberg, 1979-80; recipient NASA Exceptional Scientific Achievement medal, 1992, So. Meth. U. Disting. Alumni award, 1993. Fellow Am. Phys. Soc., Meteoritical Soc. (Leonard medal 1991); mem. AAAS, Am. Astron. Soc., Royal Astron. Soc. (G.H. Darwin lectr. 1981), Cosmos Club (Washington), Phi Beta Kappa, Sigma Xi. Office: Clemson U Dept Physics & Astronomy Clemson SC 29634-1911

CLAYTON, DWIGHT ALAN, electrical engineer; b. Carthage, Tenn., Dec. 9, 1958; s. Joseph Harold and Barbara Ann (Dodd) C.; m. Pamela Anne Ausmus, Oct. 20, 1990; 1 child, Joseph Alan. BSEE, Tenn. Tech. U., 1981, MSEE, 1983. Summer engring. intern Arnold Enging. Devel. Ctr., Tullahoma, Tenn., 1980; rsch. asst. Tenn. Tech. U., Cookeville, 1981-82, rsch. assoc., 1982-83; devel. engr. Oak Ridge (Tenn.) Nat. Lab./Martin Marietta Energy Systems, 1983—; tech. forum coord. Oak Ridge Nat. Lab., 1991—. Mem. IEEE, IEEE Computer Soc. (newsletter editor 1988-90), IEEE Signal Processing Soc. Republican. Baptist. Home: 832 Sailview Rd Farragut TN 37922 Office: Oak Ridge Nat Lab PO Box 2008 Bldg 3500 Oak Ridge TN 37831

CLAYTON, JOHN, engineering executive; b. Marlboro, Mass., Aug. 16, 1930; s. John and Esther Elizabeth (Gray) C.; m. Carol Ann Kopp, Feb. 19, 1954; children: Susan A., Dianne G., Jacqueline. Miscellaneous positions, 1948-52; with Tech. Instrument Corp., Acton, Mass., 1952-56; engring. mgr. R&D Waters Mfg. Inc., Wayland, Mass., 1956—. Past chmn. Stow Bd. Selectmen; chmn. Stow Zoning Bd. Appeals; pres. Stow Community Housing Corp., Stow Elderly Housing Corp. Mem. IEEE, Computer Soc. of IEEE, Amateur Radio Relay League. Republican. Achievements include 9 patents in electro-mechanical transducers and conductive plastics. Home: 15 Walnut Ridge Rd Stow MA 01775-1109 Office: Waters Mfg Inc Longfellow Ctr Wayland MA 01778-1890

CLAYTON, PAULA JEAN, psychiatry educator; b. St. Louis, Dec. 1, 1934; 3 children. B.S., U. Mich., 1956; M.D., Washington U., St. Louis, 1960. Intern St. Luke's Hosp., St. Louis, 1960-61; asst. resident and chief resident psychiatry Barnes and Renard Hosp., St. Louis, 1961-65; from instr. to assoc. prof. psychiatry Sch. Medicine Washington U., 1965-74, prof., 1974—; prof., head dept. psychiatry U. Minn. Med. Sch., Mpls., 1980—; dir. tng. and rsch., 1975; dir. psychiat. inpatient svc. Barnes and Renard Hosp., 1975-81. Author 3 books; contbr. numerous articles to profl. pubs. Fellow Am. Psychiat. Assn.; mem. Psychiat. Rsch. Soc., Assn. Rsch. in Nervous and Mental Diseases, Am. Psychopath. Assn., Soc. Biol. Psychiatry, Am. Coll. Neuropsypharm. Office: U Minn Hosps & Clinic Harvard St & E River Rd Minneapolis MN 55455 also: U Minn Med Sch 420 Delaware St SE Minneapolis MN 55455

CLEARY, JAMES W., retired university administrator; b. Milw., Apr. 16, 1927; married, 1950. PhB, Marquette U., 1950, MA, 1951; PhD, U. Wis., 1956. Instr., dir. forensics high sch. Wis., 1949-51; instr. speech, head coach debate Marquette U., 1951-53; from instr. to prof. speech U. Wis., 1956-63, vice chancellor academic affairs, 1966-69; pres. Calif. State U., Northridge, 1969—; mem. Pres.'s Commn. NCAA. Author: books in field including Robert's Rules of Order Newly Revised, 1970, 80; editor: books in field including John Bulwer's Chirologia . . . Chironomia, 1644, 1974; co-editor: books in field including Bibliography of Rhetoric and Public Address, 1964. Served to 2d lt. AUS, 1945-47. Recipient Disting. Alumni award U. Wis., 1990; named one of 100 Most Effective Coll. Pres. in U.S., Exxon Edn. Found., 1986; U. Wis. fellow, 1954-55. Mem. Speech Assn. Am., Am. Assn. State Colls. and Univs. (chmn. 1983), NCAA (pres.' commn. 1984—, chmn. div. II com.). Address: Calif State U Office of the Pres 18111 Nordhoff St Northridge CA 91330

CLEARY, MICHAEL, pathologist, educator. Prof. dept. pathology Stanford (Calif.) U. Sch. Medicine. Recipient Warner-Lambert/Parke-Davis award Am. Assn. Pathologists, 1993. Office: Stanford U Sch Med Dept Pathology Stanford CA 94305*

CLEARY, ROBERT EMMET, gynecologist, infertility specialist; b. Evanston, Ill., July 17, 1937; s. John J. and Brigid (O'Grady) C.; M.D., U. Ill., 1962; m. June 10, 1961; children—William Joseph, Theresa Marie, John Thomas. Intern, St. Francis Hosp., Evanston, 1962-63, resident, 1963-66; practice medicine specializing in gynecology and infertility, Indpls., 1970—; head Sect. of Reproductive Endocrinology and Infertility, Chgo. Lying-In Hosp., U. Chgo., 1968-70; head Sect. of Reproductive Endocrinology and Infertility, Ind. U. Med. Center, Indpls., 1970-80; prof. ob-gyn Ind. U., Indpls., 1976-80, clin. prof. ob-gyn, 1980—. Recipient Meml. award Pacific Coast Obstetrical and Gynecol. Soc., 1968; diplomate Am. Bd. Ob-Gyn, Am. Bd. Reproductive Endocrinology and Infertility. Fellow Am. Coll. Ob-Gyn, Am. Fertility Soc.; mem. Endocrine Soc.,Soc. Gynecol. Investigation, Pacific Coast Fertility Soc., Soc. Reproductive Endocrinologists, Soc. Reproductive Surgeons, N.Y. Acad. Scis., Sigma Xi. Roman Catholic. Contbr. articles in field to med. jours. Home: 7036 Dubonnet Ct Indianapolis IN 46278-1541 Office: 8091 Township Line Rd Indianapolis IN 46260-2495

CLEAVE, MARY L., environmental engineer, former astronaut; b. Southampton, N.Y., Feb. 5, 1947. BS in Biol. Scis., Colo. State U., 1969; MS in Microbiol. Ecology, Utah State U., 1975, PhD in Civil and Environ. Engring., 1979. Mem. rsch. staff Utah State U., 1971-80; astronaut NASA, Lyndon B. Johnson Space Ctr., Houston, 1980-90, mission specialist STS 61-B, 1985, mission specialist STS-30, 1989; now dep. project mgr. NASA Ocean Color Satellite Program, Greenbelt, Md. Mem. Tex. Soc. Profl. Engrs., Water Pollution Control Fedn., Sigma Xi, Tau Beta Pi. Office: Goddard Space Flight Ctr Washington DC 20071

CLELAND, NED MURRAY, civil engineer; b. Lynchburg, Va., Apr. 20, 1951; s. James Murray and Jean (Collier) C.; m. Nancy Elizabeth Sforza (div. 1977); m. Betty Jo Flagler Cleland, July 11, 1981; children: John, Jennifer. BS, Rensselaer Poly. Inst., 1973, M Engring., 1974; PhD in Civil Engring., U. Va., 1984. Project engr. Thomas a. Hanson & Assocs., Richmond, Va., 1974-78; chief engr. Concrete Structures, inc., Richmond, 1979; project engr. Shockey Bros., Inc., Winchester, Va., 1980-81, chief engr., 1982-86; pres. Blue Ridge Design, Inc., Winchester, 1986—; v.p. engring. Shockey Industries, Inc., Winchester, 1990—; bd. dirs. Frederick County Sanitation Authority, Winchester, 1988-93, Frederick-Winchester Svc. Authority, Winchester, 1989—; mem. state bd. geology Dept. Commerce, Richmond, 1987—. Capt. U.S. Army Corps Engrs., 1974-81. Named Young Engr. of Yr. Va. Soc. Profl. Engrs., 1985, NSPE, 1986. Mem. ASCE, Am. Concrete Inst., Prestressed Concrete Inst. (chmn. software com. 1986-90), Va. Soc. Profl. Engrs. (chpt. pres. 1985-86), Coalition Am. Structural Engrs. (sec.-treas. Va. sect. 1990-92, vice chmn. 1992-93, chmn. 1993—), Tau Beta Pi, Chi Epsilon. Achievements include research into lateral stability of precast concrete ledger berms; discovery of primary mechanism of lateral displacement of this class of beams. Home: 151 Jane's Way Winchester VA 22602 Office: Shockey Industries Inc 1057 Martinsburg Pike Winchester VA 22603

CLELAND, W(ILLIAM) WALLACE, biochemistry educator; b. Balt., Jan. 6, 1930; s. Ralph E. and Elizabeth P. (Shoyer) C.; m. Joan K. Hookanson, June 18, 1967; children: Elsa Eleanor, Erica Elizabeth. A.B. summa cum laude, Oberlin Coll., 1950; M.S., U. Wis., 1953, Ph.D., 1955. Postdoctoral fellow U. Chgo., 1957-59; asst. prof. U. Wis., Madison, 1959-62, assoc. prof., 1962-66, prof., 1966—, M.J. Johnson prof. biochemistry, 1978—, Steenbock prof. chem. sci., 1982—. Contbr. articles to profl. biochem. and chem. jours. Served with U.S. Army, 1957-59. NIH grantee, 1960—; NSF grantee, 1960—. Mem. Am. Acad. Arts and Scis., Am. Soc. Biochemistry and Molecular Biology (Merck award 1990), Am. Chem. Soc. (Alfred R. Bader Bioinorganic or Bioorganic Chem. award 1993), NAS. Achievements include development of dithiothreitol (Cleland's Reagent) as reducing agent for thiol groups; development of application of kinetic methods for determining enzyme mechanism. Office: Enzyme Inst 1710 University Ave Madison WI 53705-4098

CLELLAND, MICHAEL DARR, electrical engineer; b. Marfa, Tex., Oct. 23, 1947; s. L. Maurice and Frances B. (McBrayer) C.; m. Becki Sue Ripley, May 24, 1969; children: Douglas, Andrew, Timothy, Darra. BSEE, U. Tex., El Paso, 1976. Elec. engr. technician U.S. Bur. Reclamation, Loveland, Colo., 1973-74; civil engr. technician U.S. Bur. Reclamation, El Paso, 1974-76; elec. engr. U.S. Bur. Reclamation, Truth or Consequences, N.Mex., 1976-77, Sacramento, 1977-79; supr. elec. engring. U.S. Bur. Reclamation, Loveland, 1988-90; supr. gen. engr. U.S. Army, White Sands MR, N.Mex., 1988-90, elec. engr., 1990—. Bd. dirs. Longs Peak Fed. Credit Union, Loveland, 1986-88, chmn. credit com., 1983-86. Mem. IEEE (mem. section exec. com. 1977-79). Democrat. Methodist. Office: US Army Stews-EH-PD White Sands Missle Range NM 88002-5076

CLEM, JOHN RICHARD, physicist, educator; b. Waukegan, Ill., Apr. 24, 1938; s. Gilbert D. and Bernelda May (Moyer) C.; m. Judith Ann Paulsen, Aug. 27, 1960; children—Paul Gilbert, Jean Ann. B.S., U. Ill., 1960, M.S., 1962, Ph.D., 1965. Rsch. assoc. U. Md., College Park, 1965-66; vis. rsch. fellow Tech. U., Munich, Ger., 1966-67; asst. prof. physics Iowa State U., Ames, 1967-70, assoc. prof., 1970-75, prof. physics, 1975—, chmn. dept. physics, 1982-85, disting. prof. in liberal arts and scis., 1989—; vis. staff mem. Los Alamos Nat. Lab., 1971-83; cons. Argonne Nat. Lab., Ill., 1971-76, Oak Ridge (Tenn.) Nat. Lab., 1981, Brookhaven Nat. Lab., Upton, N.Y., 1980-81, Allied-Signal, Torrance, Calif., 1990-92; guest prof. U. Tuebingen, W. Germany, 1978; cons. IBM Watson Rsch. Ctr., Yorktown Hts., N.Y., 1982-86, vis. scientist, 1985-86; vis. scientist Electric Power Rsch. Inst., Palo Alto, Calif., 1992-93, vis. prof. applied physics Stanford U., 1992-93. Sci. editor newsletter High-Tc Update; contbr. articles to profl. jours.; patentee in field. Recipient award for sustained outstanding rsch. in solid state physics, U.S. Dept. Energy; Fulbright sr. rsch. fellow, 1974-75; NATO grantee, 1979-82. Fellow Am. Phys. Soc. (chair-elect divsn. condensed matter physics 1993-94); mem. AAUP, AAAS, Iowa Acad. Sci., Sigma Xi, Tau Beta Pi, Phi Kappa Phi. Democrat. Presbyterian. Avocation: singing. Home: 2307 Timberland Rd Ames IA 50010-8251 Office: Iowa State Univ A517 Physics Ames IA 50010

CLEMENDOR, ANTHONY ARNOLD, obstetrician, gynecologist, educator; b. Port-of-Spain, Trinidad, West Indies, Nov. 8, 1933; s. Anthony Arnold and Beatrice Helen (Stewart) C.; came to U.S., 1954, naturalized, 1959; A.B., NYU, 1959; M.D., Howard U., 1963; m. Elaine Browne, May 31, 1958 (dec. May, 1991); children—Anthony Arnold, David Alan. Intern, USPHS, S.I., N.Y., 1963-64; resident Met. Hosp. Ctr., N.Y.C., 1964-68, chief outpatient dept. ob-gyn, 1969-73; med. dir. family planning Human Resources Adminstrn., N.Y.C., 1973-74; assoc. dean student affairs, dir. office of minority affairs N.Y. Med. Coll., Valhalla, N.Y., 1974—, assoc. clin. prof. dept. ob-gyn., 1978-90, prof. clin. ob-gyn, 1990—. Bd. dirs. Elmcor, Caribbean-Am. Ctr. N.Y.C.; mem. Nat. Urban League, N.Y. Urban League; life mem. NAACP. Diplomate Am. Bd. Ob-Gyn. Fellow Am. Coll. Ob-Gyn, Am. Pub. Health Assn.; mem. Royal Soc. Medicine, Nat. Med. Assn., N.Y. State Med. Soc., N.Y. County Med. Soc. (sec. 1989, v.p. 1990, pres. elect 1991, pres. 1992—), N.Y. Gynecol. Soc. (v.p. 1986, pres. 1988).

CLEMENS, ROGER ALLYN, medical and scientific affairs manager; b. L.A., Sept. 7, 1946; s. Marvin Leo and Betty Jane (Kent) C.; m. Catherine Christine Sprecher, June 19, 1970; children: Stephanie, Janna, Brittany. AB, UCLA, 1972, MPH, 1973, DrPH, 1978. Rsch. nutritionist Carnation Rsch. Labs. div. Calreco, Inc., 1978-83, sr. rsch. nutritionist, 1983-84; mgr. nutrition rsch. Westreco div. Nestle S.A., Van Nuys, Calif., 1984-91; mgr. med. and sci. affairs Carnation, Glendale, Calif., 1991—; adj. asst. prof. food sci. and nutrition Chapman Coll., 1980—; vis. lectr. UCLA, 1981-88, adj. asst. prof. nutritional scis., 1988—; nutrition del. Citizen Amb. Program, Soviet Union, 1990; chmn. adv. and devel. coun. Calif. State U., Loma Linda U. Contbr. articles to Food Tech., Am. Jour. Clin. Nutrition, Jour. Oral Medicine, Am. Jour. Physiology, others. Mem. dean's coun. UCLA, 1980—; mem. adv. and devel. coun. Loma Linda U., 1986-87, Calif. State U., Long Beach, 1985—, chmn., 1987—; active Calif. Mus. Found., 1984-88; bd. dirs. and mem. exec. bd. L.A. chpt. ARC, 1981-82, 84-85, bd. dirs. San Fernando Valley dist., 1978-85, 86-88, chmn. 1981-82, 84-85, mem. exec. com., 1979-85. With U.S. Army, 1967-69. Recipient Spotlight award, 1985, 25-Yr. Vol. Svc. award, 1992, Meritorious Svc. award Calif. Dietetic Assn., 1987; Marilyn Magaram Ctr. for Food Sci., Nutrition and Dietetics fellow, 1993. Mem. Am. Dietetic Assn. (assoc.), Am. Inst. Nutrition, Am. Soc. Parenteral and Enteral Nutrition, Calif. Nutrition Coun. (pres. 1988-89), Greater L.A. Nutrition Coun. (bd. dirs. 1982-87, pres. 1985-86), Inst. Food Technologists, Internat. Life Sci. Inst.-Nutrition Found. (several subcoms.), Nat. Coun. Against Health Fraud, Nat. Food Processors Assn., N.Y. Acad. Sci., Soc. for Nutrition Edn., Calif. Inst. Food Technologists (councilor 1989—, chmn. univ. rsch. awards com. 1989—). Republican. Baptist. Achievements include research on iron availability and chemistry in processed products; stero isomerization of vitamin A in thermal processing; bioavailability of Pb particulate in evaporated milk; effects of dietary fiber on GI morphology, iodine. Office: Carnation Nutritional Products Div 800 N Brand Blvd Glendale CA 91203-1244

CLEMENT, DOUGLAS BRUCE, medical educator; b. Montreal, Que., Can., July 15, 1933. BSc, U. Oreg., 1955; MD, U. B.C., Vancouver, Can., 1959. Intern St. Mary's Hosp., San Francisco, 1960; adj. assoc. prof. Simon Fraser U., 1976-79; from asst. to assoc. prof. U. B.C., Vancouver, 1979-90, prof., 1990—; vis. assoc. prof. Simon Fraser U., 1979-80; coach track and field team U. B.C., 1981-87, mem. family practice promotion and tenure com., 1986-93, mem. search com. dept. family practice, 1988-89, co-founder clin. fellowship sports medicine, Allan McGavin sports medicine ctr.; cons. in sports medicine; radio, TV presenter, speaker in field. Regional editor Sports Tng., Medicine and Rehab., 1987-88; mem. editorial bd. Clin. Sports Medicine, 1989-90, Clin. Jour. Sports Medicine, 1991—; sect. editor Can. Jour. Sport Scis., 1990-93; contbr. articles to profl. jours. and chpts. to books. Recipient Vanier award, 1965, Centennial medal, 1967, Op. Lifestyle medal, 1980, Longines/Wittnauer Coaching Excellence award, 1989, Order of Can. medal, 1991, Can. 125 medal, 1992, Lifetime Achievement award Sports Medicine Coun. Can., 1992; named Coach of Yr., Sport B.C., 1988; grantee Sport Can., 1984, 86, Beecham Labs., 1984, Ciba Geigy, 1985, Sport Medicine Coun. Can., 1985-86, Internat. Olympic Com., 1987-89, CFLRI, 1990; Town Club scholar U. Oreg., 1954. Fellow Am. Coll. Sports Med., Can. Acad. Sport Medicine (travelling, past pres., mem. accreditation com. 1987—); mem. Can. Med. Assn., Can. Assn. Sport Sci. (past sec.), Can. Ctr. Drug Free Sport (mem. sci. adv. com. 1992), Can. Olympic Assn. (gen. mgr. Can. team Pan Am. Games 1975, mem. com. doping control, com. accreditation sports sci. facilities, bd. dirs. 1985-93), Athletics Can. (chmn. med. com. 1981-87, mem. sports sci. com. 1983-87, nat. coach various games), B.C. Athletics (master coach 1988—, Coach of Yr. 1987-92), B.C. Med. Assn. (mem. com. athletics and recreation health planning coun. 1971—), Sports Medicine Coun. B.C. (chmn. high performance sports sci. unit 1985-89, com. anti-doping 1989—), Achilles Internat. Athletics Soc. (chmn. 1964-72, 86—), Kajak Track and Field Club (co-founder, coach 1962—). Achievements include research on effect of exercise on the human body, stressor effects of running on local tissue as well as central effects. Office: U BC Allan McGavin Sports Medicine Ctr, 3055 Wesbrook Mall, Vancouver, BC Canada V6T 1Z3

CLEMENTE, CARMINE DOMENIC, anatomist, educator; b. Penns Grove, N.J., Apr. 29, 1928; s. Ermanno and Caroline (Friozzi) C.; m. Juliette Vance, Sept. 19, 1968. A.B., U. Pa., 1948, M.S., 1950, Ph.D., 1952; postdoctoral fellow, U. London, 1953-54. Asst. instr. anatomy U. Pa., 1950-52; mem. faculty UCLA, 1952—, prof., 1963—, chmn. dept. anatomy, 1963-73, dir. brain research inst., 1976-87; prof. surg. anatomy Charles R. Drew Postgrad. Med. Sch., L.A., 1974—; hon. rsch. assoc. Univ. Coll., U. London, 1953-54; vis. scientist Nat. Inst. Med. Rsch., Mill Hill, London, 1988-89, 91; cons. Sepulveda VA Hosp., NIH; mem. med. adv. panel Bank Am.-Giannini Found.; chmn. sci. adv. com., bd. dirs. Nat. Paraplegia Found.; bd. dirs. Charles R. Drew U., 1985—. Author: Aggression and Defense: Neurol Mechanisms and Social Patterns, 1967, Physiological Correlates of Dreaming, 1967, Sleep and the Maturing Nervous System, 1972, Anatomy, An Atlas of the Human Body, 1975, 3d edit., 1987; editor: Gray's Anatomy, 1973—, 30th Am. edit., 1985, also Exptl. Neurology; asso. editor: Neurol. Research; contbr. articles to sci. jours. Recipient award for merit in sci. Nat. Paraplegia Found., 1973; 23d Ann. Rehfuss Lectr. and recipient Rehfuss medal Jefferson Med. Coll., 1986; John Simon Guggenheim Meml. Found. fellow, 1988-89. Mem. Pavlovian Soc. N.Am. (Ann. award 1968, pres. 1972), Brain Research Inst. (dir. 1976-87), Am. Physiol. Soc., Am. Assn. Anatomists (v.p. 1970-72, pres. 1976-77, Henry Gray award 1993), Am. Acad. Neurology, Am. Assn. Clin. Anatomists (Honored Mem. of Yr. 1993), Am. Acad. Cerebral Palsy (hon.), Am. Neurol. Assn., Assn. Am. Med. Colls. (exec. com. 1978-81, disting. service mem. 1982), Council Acad. Socs. (adminstrv. bd. 1973-81, chmn. 1979-80), Assn. Anatomy Chairmen (pres. 1972), Biol. Stain Commn., Inst. Medicine of Nat. Acad. Scis. (sci. adv. bd.), Internat. Brain Research Orgn., AMA-Assn. Am. Med. Colls. (liaison com. on med. edn. 1981-87), Med. Research Assn. Calif. (dir. 1976—), N.Y. Acad. Sci., Nat. Bd. Med. Examiners, Nat. Acad. Sci. (mem. com. neuropathology, BEAR coms.), Japan Soc. Promotion of Sci. (Research award 1978), Soc. for Neurosci., Sigma Xi. Democrat. Home: 11737 Bellagio Rd Los Angeles CA 90049-2158 Office: UCLA Sch Medicine Dept Anatomy and Cell Biology Los Angeles CA 90024

CLEMENTS, BRIAN MATTHEW, computer consultant, educator; b. Glens Falls, N.Y., May 4, 1946; s. Robert Edward and Lois Jennie (Gubitz) C.; m. Rene Ann Hammond, Apr. 18, 1970; children: Chad Aaron, Andrew Hammond. BA in Edn. and Math., Potsdam Coll., 1969; postgrad., Adirondack Community Coll., Queensbury, N.Y., 1971-82, SUNY, Plattsburgh, 1971-82; MS in Edn. and Microcomputers, SUNY, Albany, 1984. Cert. ednl. specialist IBM; cert. nursery, kindergarten, elem. and higher edn. educator, N.Y. Instr. elem./mid. sch. Glens Falls (N.Y.) City Sch. Dist., 1969-84; instr. high sch. Southern Adirondack Ednl. Ctr., Hudson Falls, N.Y., 1984-87; under-grad. instr. Saratoga extension The New Sch. for Social Rsch., N.Y.C., 1987—; pres. computer divsn. Foothills Computer, Inc., Queensbury, N.Y., 1982-93, pres. Xerox divsn., 1984-93; ptnr. Clements Assocs., Glens Falls, 1988—; pres., CEO Foothills Computer, Inc., Glens Falls, 1989-93; grad. instr. Wilton extension The New Sch. for Social Rsch., N.Y.C., 1990—; founder, dir., instr. Adirondack Computer Camp, Glens Falls, 1982-85; dir., presenter microcomputers in banking N.Y. State Bankers Assn., Syracuse, 1983; cons. computer systems Bd. Coop. Edn. Svcs., Hudson Falls, N.Y., 1984-90, Lower Adirondack Regional Arts Coun., Glens Falls, 1989-91; v.p. Queensbury All-Sports Booster Club, 1990-93, pres., 1993-94. Author, editor: (manual/software) Basic Basic Plus, 1983, software WWSC Election Coverage, 1983-89. Presenter Instrument Soc. Am. (Adirondack), Research Triangle Park, N.C., 1983; sponsor computer svc. Am. Diabetes Assn., Queensbury, 1983, Am. Heart Assn., Queensbury, 1984—; supporter, presenter internship program Adirondack C.C., Queensbury, 1984-93, mem. adv. bd., 1986—. Named Tchr. of the Yr., Glens Falls Schs. and Tchr. Assn., 1981, Hon. Life Mem., N.Y. State Congress of Parents and Tchrs., 1982, Best Spot News Election Coverage award, UPI NY/NJ in conjunction with Normandy Broadcasting, 1985. Mem. Vocat. Indsl. Clubs Am. (chmn. N.Y. state skills olympics data processing 1986, advisor 1985-87), N.Y. State United Tchrs. (local chpt. bldg. rep., treas., negotiating team, grievance chmn. 1969-87), N.Y. State Assn. Ednl. Data Systems, N.Y. State Assn. for Computers and Techs. in Edn., Queensbury Parent-Tchr.-Student Assn., ABCD: The Microcomputer Industry Assn., Adirondack Regional C. of C. (edn. com.), Rotary, Elks. Republican. Home: 70 Cronin Rd Queensbury NY 12804-1419 Office: Foothills Computer Inc 21 Ridge St Glens Falls NY 12801-3608

CLEMENTS, JAMES DAVID, psychiatry educator, physician; b. Pineview, Ga., May 7, 1931; s. Marcus Monroe and Dewey Thelma (Gammage) C.; m. Janet Collier Swan, Aug. 25, 1952; children—Leiliar Ann, David Marcus. B.A., Emory U., 1952; M.D., Med. Coll. Ga., 1956. Intern Temple U., Phila., 1956-57; resident in pediatrics Temple U., 1957-59; fellow mental retardation Sch. Medicine, Yale U., 1959-60; med. dir. Gracewood (Ga.) State Sch. Hosp., 1960-62, asst. supt., 1963-64; dir. planning mental retardation Ga. Dept. Pub. Health, Atlanta, 1964-65; dir. Ga. Retardation Center, Atlanta, 1964-79; med. cons. mental retardation Ga. Dept. Human Resources, 1979-81; resident in psychiatry Emory U., Atlanta, 1983-86; clin. asst. prof. pediatrics and psychiatry Emory U. Sch. Medicine, Atlanta, from 1964, asst. prof. psychiatry, 1985—; assoc. clin. prof. neurology, asst. clin. prof. pediatrics Med. Coll. Ga., Augusta, 1970—; spl. cons. neurology mental retardation dept. pediatrics Ga. Bapt. Hosp., 1965—; mem. adv. com. program exceptional children Ga. Dept. Edn., 1968-70; mem. adv. bd. Sch. Allied Health Sci., Ga. State U., 1971-76; mem. accreditation council mental retardation council Joint Commn. on Accreditation Hosps., Chgo., 1975-79; del. White House conf. Ga. com. children youth, 1970; mem. Pres.'s Com. on Mental Retardation, 1975-78; chmn. Willowbrook rev. panel Fed. Ct. Eastern Dist. N.Y.; reviewer NSF; cons. Inst. Society, Ethics and Life Scis., Hastings Center; commr. Am. Bar Assn., 1976—. Contbr. articles to profl. jours., anthologies, seminars. Mem. adv. bd. Arbor Acad., DeKalb County (Ga.) Dept. Edn., 1973-75; mem. bd. founders, adv. com. Ashdun Hall, 1965-70; trustee Gatchell Sch., Mental Health Law Project; adv. com. Kennedy Center, Johns Hopkins U. Recipient Leadership award Am. Assn. Mental Deficiency, 1980. Fellow Am. Acad. Pediatrics (cons. head start med. cons. service), Am. Assn. Mental Deficiency (pres. 1974-75), Pan Am. Med. Assn., Am. Geriatrics Soc.; mem. Ga. Pediatric Soc., Nat. Assn. Supts. Pub. Residential Facilities Mentally Retarded, Nat. Assn. Retarded Citizens (legal advocacy adv. com. 1975), Internat. Assn. Sci. Study Mental Deficiency (chmn. local organizing com. 4th internat. congress, mem. council 1976—), Am. Psychiatric Assocs. Home: 475 Grant St SE Atlanta GA 30312-3154

CLEMENTS, LUTHER DAVIS, JR., chemical engineer, educator; b. Miami Beach, Fla., Nov. 2, 1944; s. Luther Davis and Phyllis Ann (Morton) C.; m. Helen Lee Peeler, May 22, 1966 (div. May 1983); 1 child, Edward Davis; m. Emilia Gonzalez, June 11, 1983; stepchildren: Samantha Lynn Cohen, Rebecca Ann Cohen. BSChemE, Okla. State U., 1966; MSChemE, U. Ill., 1968; PhD, U. Okla., 1973. Engr. Ethyl Corp., Baton Rouge, summer 1966; rsch. engr. Chevron Rsch. Corp., Richmond, Calif., 1973-75; from asst. to assoc. prof. dept. chem. engring. Tex. Tech. U., Lubbock, 1976-84; prof., chmn. dept. chem. engring. U. Nebr., Lincoln, 1984-92, prof. dept. chem. engring., 1984—; assoc. dean coll. engring. and tech. U. Nebr., 1993—; dir. office of agrl. materials USDA-Coop. State Rsch. Svc., Washington, 1992-93; cons. various co. and agencies, 1976—; founder, dir. Indsl. Agrl. Products Ctr., U. Nebr., 1987-90. Author (computer program) Chem. Engring. Libr. for TI74/TI95, 1986; contbr. over 50 articles to profl. jours. Mem. steering com. Nebr. Futures-Agrl. Uses, Lincoln, 1988; pres. Lincoln Unitarian Ch., 1989-91; regional co-dir. Unitarian Universalist Svc. Com., Prairie Star Dist., 1985-91. 1st lt. U.S. Army, 1971-73. Fulbright fellow, 1983-84; recipient Nat. Grad. fellowship NSF, 1966-68, Rsch. Excellence award Halliburton Found., 1983, over $2.8 million funded rsch., 1972—. Mem. AICE (vice chmn./cnmn. Nebr. sect. 1985-89), Am. Soc. Engring. Edn. (chmn. chem. engring. div. 1993-94), Am. Chem. Soc. Achievements include 3 patents in field; development of first solar-thermal-electric commercial power system. Home: 8010 Lillibridge Lincoln NE 68510 Office: Univ Nebr 116 Engring 116 Engring Omaha NE 68182-0176

CLEMENTS, MICHAEL CRAIG, health services consulting executive, retired renal dialysis technician; b. Cin., Sept. 17, 1945; s. Marvin Hubert and Mildred Helen (Rabe) C.; m. Minnie Faye Pospisil, Dec. 1, 1972; children: Melissa Ayn, Michael Aaron. Student, U. Cin., 1968-70; EMT/paramedic, Good Samaritan Health Ctr., 1980. Cert. renal dialysis technician. Hemodialysis technician Christ Hosp., Cin., 1968-79; tech. svcs. dir. Dialysis Clinic, Inc., Cin., 1980-91; pres. Critical Care Svcs., Inc., Mason, Ohio, 1987—; firefighter/paramedic Mason Vol. Fire Co., 1978-85, EMS tng. officer, 1984, EMS capt., 1985. Contbr. articles to profl. jours. Mem. Mason Environ. Adv. Commn., 1990—, vice chmn., 1992-93, bus. and parent curriculum review com. Mason City Schs., 1992; employer advisor coop. program Cin. Tech. Coll. Biomed. Engring. Tech., 1986-91. With USN, 1964-70. Mem. AAMS, Assn. for Advancement of Med. Instrumentation, Am. Water Works Assn., Ohio Acad. Sci. Mem. Ch. of Christ.

CLERKE, WILLIAM HENRY III, forester; b. Easton, Pa., May 28, 1939; s. William Henry Jr. and Olivia Amelia (Messinger) C.; m. Carol Helen Mobley, july 23, 1975; children: Kathryn Gail, William Henry IV, Suzanne Carol. AAS in Forestry, Paul Smith Coll., 1964; BS in Forestry, U. Mich., 1965, postgrad., 1965-67. Entomologist USDA Forest Svc., Asheville, N.C., 1967-74; survey improvement specialist so. region USDA Forest Svc., Atlanta, 1974-80, remote sensing specialist so. region, 1980—. Co-Author: Evaluation of Spruce and Fir Mortality in the Southern Appalachain Mountains, (So. Forest Insect Work Conf. award 1989); pub. numerous papers and reports. Scoutmaster Boy Scouts Am., 1958-64; guest speaker on conservation at local schs., 1977-92. With USCG, 1958-62. Recipient Contbn. award Southeastern Forest Experiment Station, 1988. Mem. Am. Soc. Photogrammerty and Remote Sensoring (Pres.'s award 1984), Nature Conservancy. Democrat. Presbyterian. Achievements include development (with others) of integrated earth resources data analysis system, multispectral camera system and sampling intervelometer, radiographic systems and techniques for estimating southern pine populations as part of biological evaluations and research studies, color infared aerial photography for pest damage surveys; first to evaluate and implement time sharing computer acess to support field units; research in LORAN C navigation systems to aerial photo aircraft navigation. Home: 761 E Morningside Dr Atlanta GA 30324 Office: USDA Forest Svc So Region 1720 Peachtree Rd Atlanta GA 30367

CLEVE, HARTWIG KARL, medical educator; b. Braunschweig, Fed. Republic Germany, June 9, 1928; s. Hartwig and Elisabeth (Schomburg) C.; m. Annelore Bohnig, Feb. 12, 1955 (div. Feb. 1991); children: Hartwig, Cornelia; m. Burgis Garmann, Mar. 22, 1991; 1 child, Christian. MD, U. Goettingen, Fed. Republic Germany, 1953. Intern, resident U. Hosp., Goettingen, Marburg, 1954-59; guest investigator Pasteur Inst., Paris, 1959-60; rsch. assoc. Rockefeller Inst., N.Y.C., 1960-63; docent U. Marburg, 1963-67; assoc. prof. Cornell U. Med. Coll., N.Y.C., 1967-72, prof., 1972-73; prof., chmn. Inst. Anthropology and Human Genetics U. Munich, 1973—. Contbr. articles to profl. jours. Mem. Am. Soc. Human Genetics, Am. Assn. Immunologists, Genetics Soc. Am., Soc. Clin. Investigators, Soc. Anthropology and Human Genetics (officer 1975-77, 81-87), German Soc. Human Genetics (founding pres. 1987-91). Lutheran. Home: Paul-Gerhardt-Allee 25, D-81245 Munich Germany Office: Inst Anthropology and Human Genetics, Richard-Wagner Str 10, D-80333 Munich Germany

CLEVENGER, LARRY ALFRED, electronics company researcher; b. Oakland, Calif., July 19, 1961; s. Jesse Floyd and Maybelle (Nielsen) C.; m. Leigh Anne Hodges, June 30, 1984; 1 child, Jill Deanne. BS, UCLA, 1984; PhD in Electronic Materials, MIT, 1989. Engr. IBM, Tucson, 1981-82; asst. tech. staff MIT Lincoln Lab., Lexington, 1984; rsch. asst. MIT, Boston, 1984-89; mem. rsch. staff IBM Rsch., Yorktown Heights, N.Y., 1989—. Contbr. articles to profl. jours. Fin. sec. Fishkill (N.Y.) United Meth. Ch., 1990—,. Indsl. processing fellow MIT, 1981, Am. Ceramic Soc. fellow UCLA, 1983. Mem. Materials Rsch. Soc., Sigma Xi, Phi Kappa Tau (domain dir. 1987—, Stennis award 1992), Tau Beta Pi. Achievements include patent pending for a process to improve the thermal stability of cobalt silicide by the addition of selected elements; first scientist to discover exptl. evidence for nucleation controlled thin film phase formation. Home: 377 Andrews Rd LaGrongeville NY 12540 Office: IBM T J Watson Rsch Ctr PO Box 218 Yorktown Heights NY 12540

CLEVIDENCE, DEREK EDWARD, biologist; b. Janesville, Wis., Dec. 8, 1966; s. Darrell Fay and Faye Christine (Vos) C.; m. Brenda Elaine Stewart, Aug. 10, 1991. AB in Biology, Washington U., St. Louis, 1989. Rsch. asst., med. scientist tng. program fellow U. Ill., Chgo., 1989—. Co-chair Student Adv. Com., Ill. State Bd. Edn., Springfield, 1985. Mem. AAAS. Office: Univ Ill Dept Biochemistry A-312 Coll of Medicine West 1853 W Polk St Chicago IL 60612

CLIFF, JOHNNIE MARIE, mathematics and chemistry educator; b. Lamkin, Miss., May 10, 1935; d. John and Modest Alma (Lewis) Walton; m. William Henry Cliff, Apr. 1, 1961 (dec. 1983); 1 child, Karen Marie. BA in Chemistry, Math., U. Indpls., 1956; postgrad., NSF Inst., Butler U., 1960; MA in Chemistry, Ind. U., 1964; MS in Math., U. Notre Dame, 1980. Cert. tchr., Ind. Rsch. chemist Ind. U. Med. Ctr., Indpls., 1956-59; tchr. sci. and math. Indpls. Pub. Schs., 1960-88; tchr. chemistry, math. Martin U., Indpls., 1989—, chmn. math. dept., 1990—, divsn. chmn. depts. sci. and math., 1993—; adj. instr. math. U. Indpls., 1991. Contbr. rsch. papers to sci. jours. Grantee NSF, 1961-64, 73-76, 78-79, Woodrow Wilson Found., 1987-88; scholarship U. Indpls., 1952-56, NSF Inst. Reed Coll., 1961, C. of C., 1963. Mem. AAUW, NAACP, NEA, Assn. Women in Sci., Urban League, N.Y. Acad. Scis., Am. Chem. Soc., Nat. Coun. Math. Tchrs., Am. Assn. Physics Tchrs., Nat. Sci. Tchrs. Assn., Am. Statis. Assn., Am. Assn. Ret. Persons, Neal-Marshall-Ind. U. Alumni Assn., U. Indpls. Alumni Assn., U. Notre Dame Alumni Assn., Ind. U. Chemist Assn., Notre Dame Club Indpls., Kappa Delta Pi, Delta Sigma Theta. Democrat. Baptist. Avocations: gardening, sewing. Home: 405 Golf Ln Indianapolis IN 46260-4108 Office: Martin U 2171 Avondale Pl Indianapolis IN 46218-3878

CLIFF, STEVEN BURRIS, engineer; b. Knoxville, Tenn., Mar. 30, 1952; s. Edgar Burris and Otella (Patterson) C.; m. Sharon Grace Davis, Sept. 11, 1971 ; children: Sarah Elizabeth, Susan Rebecca, Steven John. BS in Engring. Sci., U. Tenn., 1974, MS in Engring. Sci., 1976; postgrad., So. Sem., 1974-75. Rsch. asst. U. Tenn., Knoxville, 1972-75, asst. rsch. prof., 1975-76; program analyst Oak Ridge (Tenn.) Nat. Lab., 1976-77, rsch. engr., 1977-79; chief tech. officer Computer Concepts Corp., Knoxville, 1979-81; pres. Productive Programming Inc., Knoxville, 1981-82; v.p. R&D Control Tech. Inc., Knoxville, 1982—, corp. sec., 1991—; ptnr. Middlebrook Indsl. Properties, 1985—, Cliff Bros. Investments, 1988—. Contbr. articles to profl. jours. Mem. exec. bd. Rocky Hill Parent-Tchr. Orgn., Knoxville, 1987, 91—; deacon West Knoxville Bapt. Ch., 1984-87, Loveland Bapt. Ch., Knoxville, 1976-82. U. Tenn. scholar, 1970. Mem. Soc. Mfg. Engrs. (sr.), Nat. Electronic Mfg. Assn. (chmn. com. 1987—, seminar speaker 1988—), Am. Assn. for Artificial Intelligence. Avocations: photography, bicycling, home improvement projects. Home: 8210 Northshore Dr SW Knoxville TN 37919-8711

CLIFFORD, MAURICE CECIL, physician, former college president, foundation executive; b. Washington, Aug. 9, 1920; s. Maurice C. and Rosa P. (Linberry) C.; m. Patricia Marie Johnson, June 15, 1945; children: Maurice Cecil III, Jay P.L., Rosemary Clifford McDaniel. AB, Hamilton Coll., 1941, ScD, 1982; AM, U. Chgo., 1942; MD, Meharry Med. Coll., 1947; LHD, LaSalle Coll., 1981, Hamilton Coll., 1982, Hahnemann U., 1985, Meharry Med. Coll., 1992; LLD, Med. Coll. Pa., 1986. Diplomate Am. Bd. Ob-Gyn. Intern Phila. Gen. Hosp., 1947-48, resident in ob-gyn, 1948-51, asst. chief service ob-gyn, 1951-60; mem. faculty Med. Coll. Pa., Phila., 1955—, prof. ob-gyn., 1975-91, prof. emeritus, 1992—, v.p. for med. affairs, 1978-80, pres., 1980-86, bd. dirs., 1980—, pres. emeritus, 1992—; commr. pub. health City of Phila., 1986-92; chmn. HMA Found., Phila., 1991—. Contbr. articles to profl. jours. Trustee Phila. Award, Phila. Art Mus.; trustee, vice chmn. Phila. Coll. Textiles and Sci.; trustee emeritus Phila. Acad. Natural Scis.; life trustee Meharry Med. Coll.; former alumnus trustee Hamilton Coll.; mem. nat. med. com. Planned Parenthood, 1975-78; mem. adv. com. on arts John F. Kennedy Ctr. for Performing Arts, 1978-80. Capt. M.C., U.S. Army, 1952-54. Recipient Dr. Martin Luther King, Jr. award PUSH, 1981, Dr. William H. Gray, Jr. award Educators Roundtable Assn., 1981, Ann. award Phila. Tribune Charities, 1981, Disting. Am. award Edn. and Rsch. Fund Am. Found. for Negro Affairs, 1980; Outstanding Svc. award Phila. br. NAACP, 1965, others. Fellow Am. Coll. Obstetricians and Gynecologists (life); mem. Nat. Med. Assn., Pa. Med. Soc., Med. Soc. Eastern Pa., Philadelphia County Med. Soc., Phi Beta Kappa, Alpha Omega Alpha. Office: 1314 Chestnut St 15th Fl Philadelphia PA 19107

CLIFFORD, STEVEN FRANCIS, science research director; b. Boston, Jan. 4, 1943; s. Joseph Nelson and Margaret Dorothy (Savage) C.; children from previous marriage: Cheryl Ann, Michelle Lynn, David Arthur. BSEE, Northeastern U., Boston, 1965; PhD, Dartmouth Coll., 1969. Postdoctoral fellow NRC, Boulder, Colo., 1969-70; physicist Wave Propagation Lab., NOAA, Boulder, 1970-82, program chief, 1982-87, dir. lab., 1987—; mem. electromagnetic propagation panel, NATO, 1989-93; vis. sci. closed acad. city Tomsk, Siberia, USSR. Author: (with others) Remote Sensing of the Troposphere, 1978; contbr. 120 articles to profl. jours.; patentee in acoustic scintillation liquid flow measurement, single-ended optical spatial filter. Recipient 4 Outstanding publs. awards Dept. Commerce, 1972, 75, 89. Fellow Optical Soc. Am. (editor atmospheric optics 1978-84, advisor atmospheric optics 1982-84), Acoustical Soc. Am.; mem. IEEE (sr.), Internat. Radio Sci. Union, Am. Geophys. Union. Avocations: running, cross country skiing. Office: NOAA Wave Propagation Lab 325 Broadway St Boulder CO 80303-3328

CLIFTON, DAVID SAMUEL, JR., research executive, economist; b. Raleigh, N.C., Nov. 15, 1943; s. David Samuel and Ruth Centelle (Paker) C.; m. Karen Lisette Buhrer (div. June 1980); children: Derek Scott, Mark David; m. Eileen Lois Cooley, July 30, 1983;children: Dana Cooley, Michael Cooley. B in Indsl. Engring., Ga. Inst. Tech., 1966; MBA in Econs., Ga. State U., 1970, PhD in Econs., 1980. Customer facilities engr. Lockheed Ga. Co., Marietta, 1966-70; prin. rsch. scientist Ga. Tech. Rsch. Inst., Atlanta, 1970-93, dir. econ. devel. lab., 1979-90, dir. econ. devel. and tech. transfer, 1990-93, dir. Ctr. for Internat. Standards and Quality, 1991-92; exec. assoc. dir. Ga. Tech. Econ. Devel. Inst., Atlanta, 1993—; mem. So. Tech. Coun. Ga. Tech. Rsch. Inst., Atlanta, 1992—; bd. dirs. Sea Adventure Unltd., Inc., Atlanta; cons. UN Indsl. Devel. Orgn., Vienna, 1982, Inst. de Adminstn. Cientifica de los Empreos, Mexico City, 1978. Co-author: Project Feasibility Analysis, 1977; contbr. articles to profl. jours. Mem. Am. Econs. Assn., Atlanta Power Squadron Club, Sigma Xi. Avocation: sailing, navigating. Home: 1362 Briarcliff Rd NE Atlanta GA 30306-2214 Office: Georgia Institute Of Tech Economic Devel & Tech Transfer Georgia Tech Rsch Institute Atlanta GA 30332-0001

CLIFTON, MARCELLA DAWN, dentist; b. Galax, Va., Feb. 28, 1956; d. Hugh Clayton and Yoshiko (Sano) C. Grad. dental hygienist with honors, Ctrl. Piedmont C.C., 1977; DDS, U. N.C., Chapel Hill, 1989. Assoc. Forest Irons & Assocs., Biscoe, N.C., 1989-91; pvt. practice Raleigh, N.C., 1991—; asst. clin. prof. dept. operative dentistry Sch. of Dentistry U. N.C., Chapel Hill, 1991—; Mem. task force Acad. Gen. Dentistry Task Force for Young Dentists. Bd. dirs. N.C. State U. Theatre Endowment. Recipient Outstanding Achievement and Contbn. award Southeastern Acad. Prosthodontics. Mem. ADA, Am. Assn. of Women Dentists (Wake County Branch sec./treas. 1992, v.p. 1993), N.C. Dental Soc., Wake County Dental Soc., Acad. of Gen. Dentistry, C. of C. (task force for exec. womwn), Psi Omega. Avocations: aerobics, weight lifting, sailing, biking, stained glass. Office: 4601 Western Blvd Raleigh NC 27606

CLINE, CAROLYN JOAN, plastic and reconstructive surgeon; b. Boston; d. Paul S. and Elizabeth (Flom) Cline. BA, Wellesley Coll., 1962; MA, U. Cin., 1966; PhD, Washington U., 1970; diploma Washington Sch. Psychiatry, 1972; MD, U. Miami (Fla.) 1975. Diplomate Am. Bd. Plastic and Reconstructive Surgery. Rsch. asst. Harvard Dental Sch., Boston, 1962-64; rsch. asst. physiology Laser Lab., Children's Hosp. Research Found., Cin., 1964, psychology dept. U. Cin., 1964-65; intern in clin. psychology St. Elizabeth's Hosp., Washington, 1966-67; psychologist Alexandria (Va.) Community Mental Health Ctr., 1967-68; research fellow NIH, Washington, 1968-69; chief psychologist Kingsbury Ctr. for Children, Washington, 1969-73; sole practice clin. psychology, Washington, 1970-73; intern internal medicine U. Wis. Hosps., Ctr. for Health Sci., Madison, 1975-76; resident in surgery Stanford U. Med. Ctr., 1976-78; fellow microvascular surgery dept. surgery U. Calif.-San Francisco, 1978-79; resident in plastic surgery St. Francis Hosp., San Francisco, 1979-82; practice medicine, specializing in plastic and reconstructive surgery, San Francisco, 1982—. Contbr. chpt. to plastic surgery textbook, articles to profl. jours. Mem. Am. Bd. Plastic and Reconstructive Surgeons (cert. 1986), Royal Soc. Medicine, Calif. Medicine Assn., Calif. Soc. Plastic and Reconstructive Surgeons, San Francisco Med. Soc. Address: 490 Post St Ste 735 San Francisco CA 94102

CLINE, DAVID BRUCE, physicist, educator; b. Kansas City, Kansas, July 12, 1933; s. Andrew B. Cline and Ella M. Jacks; married; children: Heather, Bruce, Richard, Yasmin. BS, MS, Kansas State Univ., 1960; PhD, Univ. Wis., 1965. Asst. prof. physics Univ. Wis., Madison, 1965-66, assoc. prof. physics, 1966-68, prof. physics, 1969; prof. physics and astronomy UCLA, 1969—; vis. appts. U. Hawaii, Lawrence Berkeley (Calif.) Lab., Fermilab, CERN; mem. various high energy physics adv. panels and program coms., theory and lab. astrophysics panel, panel on particles NRC Astronomy & Astrophysics Survey Com.; past co-dir. Instit. for Accelerator Physics at U. Wis.; founder Ctr. for Advanced Accelerators, UCLA, 1987. Editor numerous books. With U.S. Army, 1956-58. Recipient Sloan fellow A.P. Sloan Found., 1967. Fellow N.Y. Acad. Scis; mem. Phi Beta Kappa. Democrat. Achievements include first search for weak neutral currents that charge flavor; co-discovery of Weak Neutral current at FNAL (HPWF exp) early evidence for charm particle; devise of the antiproton-proton collider; co-discovery of the W and Z intermediate boson at CERN, Geneva; discovery of B 0 - Bo mixing; patentee for PET medical imagery technique. Office: UCLA Dept Physics 405 Hilgard Ave Los Angeles CA 90024

CLINE, DAVID CHRISTOPHER, geologist; b. Muncie, Ind., Feb. 6, 1955; s. Lowell Edward and Mary Jane (Langdon) C.; m. Sally Sue Love, Dec. 28, 1977; children: Rachel, Amanda, Elizabeth Anne. BS in Geology and Anthropology, Ball State U., 1977; postgrad., Dresser Ctr., 1978. Registered profl. geologist, Ind., Tex. Field engr. Dresser Atlas Ind. Inc., Houston, 1978-81; sales engr., asst. mgr. Gearhart, Ind. Inc., Fort Worth, 1981-86; geologist, owner San Angelo, 1986-88; sales engr. Weaver Svcs., Inc., Snyder, Tex., 1982-88; chief environ. sect. S.K. Geo-Sci., 1989-91; hydrogeologist, v.p. field ops. White Buffalo Environ. Svc., Inc., San Angelo, 1991—. Bd. dirs. Disciples, Inc., San Angelo, 1992; mem. Local Emergency Prepardness Com., San Angelo, 1992. Mem. Am. Assn. Petroleum Geologist, Am. Ins. of Profl. Geologists, Assn. of Engring. Geologists, San Angelo Geologic Soc. (West Tex. Geologic sect.), Tex. Bioremediation Coun., Tex. Environ. Health Assn., Nat. Water Well Assn., Coun. of Tex. Archaeologists, Concho Valley Archaeol. Soc. Mem. Disciples of Christ. Home: 5302 Oriole Dr San Angelo TX 76903 Office: White Buffalo Environ Svcs 1809 W Avenue N San Angelo TX 76904

CLINE, DOUGLAS, physicist, educator; b. York, Eng., Aug. 28, 1934; came to U.S., 1963; s. William Patrick and Annie Rita C.; m. Lorraine Van Meter, Dec. 28, 1975; children: Julia Van Meter, Geoffrey Karl. B.Sc. with 1st class honours, Manchester U., 1957, Ph.D., 1963. Mem. faculty U. Rochester, N.Y., 1963—; prof. physics U. Rochester, 1977—; dir. N.S.R.L., 1988. Author papers in field. Office: U Rochester Nuclear Structure Rsch Lab Rochester NY 14627

CLINE, JAMES MICHAEL, physicist; b. Bamberg, Fed. Republic Germany, Nov. 7, 1960; came to U.S., 1960; BS, Harvey Mudd Coll., 1982; MS, Calif. Inst. Tech., 1985, PhD, 1988. Postdoctoral fellow Ohio State U., Columbus, 1988-91, U. Minn., Mpls., 1992—; rsch. fellow McGill U., Montreal, Que., Can., 1991-92. Contbr. articles to profl. jours. NSF fellow, 1982-85, NSERC internat. fellow, 1991-92. Office: U Minn TPI 116 Church St SE Minneapolis MN 55455

CLINE, THOMAS WARREN, molecular biologist, educator; b. Oakland, Calif., May 6, 1946; married, 1986. AB, U. Calif. Berkeley, 1968; PhD in Biochemistry, Harvard U., 1973. Fellow devel. genetics Helen Hay Whitney Found., U. Calif. Irvine, 1973-76; from asst. to prof. biology Princeton U., 1976-90; prof. genetics U. Calif. Berkeley, 1990—. Mem. AAAS (NAS award in Molecular Biology 1992), Genetics Soc. Am., Soc. Devel. Biology. Achievements include developmental regulation of gene expression and pattern formation in Drosophila melanogaster with emphasis on oogenesis, sex determination, and X-chromosome dosage compensation. Office: U Calif Dept Molecular and Cell Biology Berkeley CA 94720*

CLINTON, LAWRENCE PAUL, psychiatrist; b. Lubbock, Tex., Apr. 27, 1945; s. Lewis Paul Clinton and Dorothy E. (Higgins) Clinton-Billingslea; m. Bonnie Gail Orenstein, June 22, 1969; children: Kerry Elizabeth, Andrew James, Alexander Geoffrey, Kaylin Lee. BA with honors, So. Conn. State Coll., 1966; postgrad., Ohio State U., 1966-68; MD, Hahnemann U., 1972. Diplomate Am. Bd. Psychiatry and Neurology. Teaching asst. Ohio State U., Columbus, 1966-68, research fellow, 1966-68; clin. instr. psychiatry Hahnemann U., Phila., 1975-82, asst. clin. prof., 1982—; chief exec. officer Bldg. Mgmt. Group, Vineland, N.J., 1986—; psychiat. dir. James Guiffre Med. Ctr., Phila., 1976-79; cons. Superior Ct. N.J., 1975—, Ranch Hope, Alloway, N.J., 1989-92. Contbr. articles to profl. jours. Mem. Am. Security Coun., 1975—, Rep. Senatorial Com., 1978—, Rep. Nat. Com., 1978, The Pres.'s Club, 1990—. Recipient awards Am. Security Coun., 1982, Buena Regional Sch. Dist., N.J., 1983, Vineland Parent Support and Adv. Group, 1990, Rep. Presdl. Legion of Merit medal, 1992; decorated Chevalier Comdr. Ordre Souverain et Militaire de la Milice du Saint Sepulcre, 1990—. Mem. AMA, Am. Psychiat. Assn., Internat. Assn. Group Psychotherapy, N.J. Psychiat. Soc., Phila. Coll. Physicians and Surgeons, Internat. Platform Soc., Med. Club Phila., World Fedn. Mental Health, InterAm. Coll. Physicians and Surgeons, Hahnemann Undergrad. Rsch. Soc. (treas. 1971-72), Confedn. of Chivalry, Am. Chem. Soc., Phi Lambda Kappa (v.p. 1972), Societe d'Chemie (pres. 1965-66), SPQR Club (pres. 1961-62) (Milford, Conn.). Avocations: gardening, art collecting, book collecting, historical biography. Office: 1138 E Chestnut Ave Bldg 6 Vineland NJ 08360-5053

CLOES, ROGER ARTHUR JOSEF, economist, consultant; b. Schwalmstadt, Fed. Republic Germany, May 22, 1956; s. Arthur Karl and Felicitas (Wundersee) C.; m. Charon Regina Gerta König, Sept. 25, 1992. Intermediate exams. in econs., Phillips-U. Marburg, Marburg, Germany, 1979; diploma in internat. banking and fin., City of London Poly., 1980; cert. internat. mgmt. in banking and fin, Sch. Higher Comml. Studies (HEC), Paris, 1982; diploma in econs. and banking, U. Cologne, Cologne, Germany, 1983; cert. internat. mgmt. program in banking/fin., NYU, 1985; D in Internat. Econs., U. Cologne, Fed. Republic Germany, 1988. Stagiaire Kreissparkasse Marburg/Lahn, 1979; stagiaire econ. dept. Res. Bank Australia, Melbourne, 1979; stagiaire Commerzbank AG Succursale de Paris, 1982; asst. western Africa project dept. indsl. devel.-fin. div. World Bank, Washington, 1984; reseacher dept. internat. econ. and social affairs Office Devel. Rsch. and Policy Analysis, UN Secretariat, N.Y.C., 1985; research Bd. Govs. FRS, Washington, 1985; asst. prof. econs. U. Cologne, 1987-88; cons. sect. budget and fin. reference and rsch. svcs. Deutscher Bundestag, Bonn, Germany, 1988-90, chief secretariat subcom. Treuhandanstalt budget com., 1990-93; chief secretariat com. Treuhandanstalt Deutscher Bundestag, 1993—; del. Conf. on Atlantic Community, Washington, 1983. Author: The Problem of Country Risk Associated with International Capital Movements—Methods of Analysis and Risk Management, 1988, The European Monetary Union: Problem Analysis from a German Point of View, 1989, The Harmonization of Banking, Bank Supervision and Investment Law with Regard to the European Single Market, 1990, German-German Monetary Union—Consequences for Larger Germany and Europe, 1990, The Parliamentary Control of the Treuhandanstalt, 1992, Critical Analysis of the DM-Opening-Balance Sheet of the Treuhandanstalt, 1993; also articles. Scholar Carl Duisberg Soc., 1979, German Acad. Exch. Svc., 1980, German French Youth Office, 1981, Acad. Exch. Svc. U. Cologne, 1982, 84. Mem. World Assn. Former UN Interns and Fellows (N.Y.C. cons. on internat., fin., econ. and devel. issues 1985—), Assn. Reserve Officers ts Deutsche Bundeswehr, Alumni Assn. Sch. Higher Comml. Studies (HEC), Banking and Fin. Group Paris, PIM-CEMS Club Cologne. Avocations: travel, swimming, tae kwon do, shooting. Home: Fritz-Hönig-Strasse 5, 50935 Cologne Germany Office: Deutscher Bundestag, Bundeshaus, NH 2131, 53113 Bonn Germany

CLOGG, CLIFFORD COLLIER, statistician, educator; b. Oberlin, Ohio, Oct. 16, 1949; s. Richard Gould and Margaret Narise (Puder) C.; m. Vicki Lynne Bowman, Dec. 20, 1971 (div. 1974); m. Judy Marie Ellenberger, Sept. 24, 1977; children: Katye, Edna, Roberta, Edith. BA, Ohio U., 1971; MA, MS, U. Chgo., 1974, PhD, 1977. From asst. prof. to prof. Pa. State U., University Park, 1976-86, disting. prof., 1990—. Author: Measuring Underemployment, 1979; editor J. Am. Statis. Assn., 1989-91; contbr. over 50 articles to profl. jours. Recipient Spl. Creativity award NSF, 1982; Fellow Ctr. for Advanced Study, Stanford (Calif.) U., 1983-84. Fellow AAAS; Am. Statis. Assn.; mem. Population Assn. Am. (bd. dirs. 1992—), Am. Sociol. Assn. (chair methodology sect. 1990-91, Lazarsfeld award 1987), Biometric Soc. Republican.

CLOSE, EDWARD ROY, hydrogeologist, environmental engineer, physicist; b. Pilot Knob, Mo., Oct. 7, 1936; s. Edward Theodore and Bernice Marie (Tyndall) C.; m. Roberta Jane Lamb, June 20, 1959 (div. 1978); children: Edward M., Christiana J.; m. Jacquelyn Ann Hill, July 7, 1979; 1 child, Joshua J. BA in Math. and Physics, Cen. Meth. Coll., Fayette, Mo., 1962; postgrad., U. Calif., Davis, 1968, Johns Hopkins U., 1972; PhD in Environ. Engring., Pacific Western U., L.A., 1988. Registered environ. assessor, Calif. Hydrologist U.S. Geol. Survey, Iowa City and Rolla, Mo., 1974-67; rsch. hydrologist U.S. Geol. Survey, Arlington, Va., 1967-72; project hydrologist U.S. Geol. Survey, San Juan, P.R., 1972-74; supr. hydrologist U.S. Geol. Survey, Tampa, Fla., 1974-78; environ. engr. Ralph M. Parsons/S.A.P. Ltd., Pasadena and Yanbu, Saudi Arabia, 1979-85; sr. environ. scientist Amartech, Ltd., Jeddah, Saudi Arabia, 1985-87; hydrology project mgr. W.W. Irwin, Inc., Long Beach, Calif., 1987-90; sr. hydrologist, mgr. R.L. Stollar & Assocs., Palo Alto, Calif., 1990-92; assoc. engr. hydrogeologist Dames & Moore Environ. Svcs., Cape Girardeau, Mo., 1992—; sons. sr. scientist Tetratech, Pasadena, Calif., 1987-88. Author: The Book of ATMA, 1978; contbr. articles to profl. jours. Asst. scoutmaster Boy Scouts Am., North Liberty, Iowa, 1966. Recipient 10-yr. svc. award U.S. Geol. Survey, 1975, outstanding performance award, 1976. Mem. Am. Math. Soc., Am. Inst. Hydrology (cert. profl. hydrologist), Nat. Water Well Assn. Groundwater Scientists and Engrs., Mensa N.Am., Kappa Mu Epsilon. Mem. Self-Realization Fellowship. Achievements include development of calculus of distinctions with applications to theoretical physics, of theory of infinite continuity unified field concept, of method for indirect determination of aquifer characteristics in shallow alluvial aquifers. Office: Dames & Moore 2 Spanish St Ct Cape Girardeau MO 63701

CLOUGH, RAY WILLIAM, JR., civil engineering educator; b. Seattle, July 23, 1920; s. Ray William and Mildred (Nelson) C.; m. Shirley Claire Potter, Oct. 30, 1942; children—Douglas Potter, Allison Justine, Meredith Anne. B.S. in Civil Engring, U. Wash., 1942; M.S., Calif. Inst. Tech., 1943; S.M., MIT, 1947, Sc.D., 1949; D.Tech. (hon.), Chalmers U., Goteborg, Sweden, 1979, U. Trondheim, (Norway), 1982. Registered profl. engr., Wash. Faculty U. Calif.-Berkeley, 1949—, prof. civil engring., 1959—, chmn. div. structural engring. and structural mechanics, 1967-70, dir. Earthquake Engring. Research Ctr., 1973-76, Nishkian prof. structural engring., 1983-87, emeritus, 1987—; cons. in field, 1953—; mem. NAS-NAE adv. com. Environ. Sci. Svcs. Adminstrn., 1967-70; mem. dynamics panel NAS adv. bd. on hardened electric power system, 1964-70; mem. U.S. C.E. Structural Design Adv. Bd., 1967—. Served to capt. USAAF, 1942-46. Recipient Sr. Rsch. award Am. Soc. for Engring. Edn., 1986, Congress medal Internat. Assn. Computer Mechanics, 1986, citation U. Calif. at Berkeley, 1987, A.C. Eringen medal Soc. of Engring. Sci., 1992; Fulbright fellow Ship Rsch. Inst., Trondheim, Norway, 1956-57, Tech. U. Norway, 1972-73, Overseas fellow Churchill Coll., Cambridge (Eng.) U., 1963-64; hon. researcher Lab. Nacional De Engenharia Civil, Lisbon, Portugal, 1972. Fellow ASCE (chmn. engring. mechanics divsn. 1964-65, Rsch. award 1960, Howard award 1970, Newmark medal 1979, Moissieff medal 1980, hon. mem. 1989), Inst. Water Conservation and Hydroelectric Power Rsch. (hon. mem. People's Republic of China 1992); mem. NAS, NAE, Structural Engrs. Assn. No. Calif. (dir. 1967-70), Earthquake Engring. Rsch. Inst. (dir. 1957-60, 70-73), Seismol. Soc. Am. (dir. 1970-73). Home: PO Box 4625 Sunriver OR 97707

CLOUTIER, STEPHEN EDWARD, chemical engineer; b. Laurium, Mich., Sept. 25, 1949; s. Edward Joseph and Barbara Ruth (Howe) C.; m. Connie Sue Wong, June 12, 1971; 1 child, David. MDiv, Trinity Evang. Div. Sch., Deerfield, Ill., 1980; M Chem. Engring., Ill. Inst. Tech., 1987. Registered profl. engr., Ill. Devel. engr. Universal Oil Prods., Des Plaines, Ill., 1972-74, design engr., 1974-81, process design coord., 1981-89, sr. process design coord., 1989—; mem. Ctr. for Chem. Process Safety, N.Y.C., 1990-92. Mem. Am. Inst. Chem. Engrs., Am. Petroleum Inst. (chmn. 1987-92), Am. Soc. Safety Engrs. (assoc.), Am. Sci. Affiliation. Achievements include research in sizing, selection, installation of pressure relieving devices; guide for pressure relieving and depressuring systems; industry practice guidelines for engineer design for process safety. Home: 1175 Ivy Hall Ln Buffalo Grove IL 60089 Office: UOP 25 E Algonquin Rd Des Plaines IL 60017

CLOWES, GARTH ANTHONY, electronics executive, consultant; b. Didsbury, Eng., Aug. 30, 1926; came to U.S., 1957; s. Eric and Doris Gladys (Worthington) C.; m. Katharine Allman Crewdson, July 29, 1950; children: John Howard Brett, Peter Miles, Vicki Anne. BSc Higher Nat. Cert., Stockport Coll., Cheshire, Eng., 1953; postgrad., UCLA, 1965-66, Birmingham (Eng.) Coll. Tech., 1955-56. Gen. mgr., v.p., dir. Eldon Industries, Inc., El Segundo, Calif., 1962-69; chief exec. officer, founder Entex Industries, Inc., Compton, Calif., 1969-83; pres., founder Entex Electronics, Inc., Valley Ford, Calif., 1983—; pres., founder TTC, Inc., Carson, Calif., 1984-86; pres. Universal Telesis Electronics, Inc., Carson, 1986-87; gen. mgr. Matchbox Toys (U.S.A.) Ltd., Moonachie, N.J., 1987-88; dir. gen. Matchbox Spain, S.A., Valencia, 1988-89; cons. Matchbox Internat. Ltd., worldwide, 1986-89. Inventor electronic voice recognition devices, numerous others. Mem. pres.'s com. UNICEF, N.Y., 1972-74, Senate Adv. Bd., Washington, 1982-83; cons. Interracial Coun., L.A., 1967-69. Mem. Knights of Malta. Avocations: antiques, gardening, art. Home: 13950 Coast Hwy 1 Valley Ford CA 94972

CLUFF, LEIGHTON EGGERTSEN, physician; b. Salt Lake City, June 10, 1923; s. Lehi Eggertsen and Lottie (Brain) C.; m. Beth Allen, Aug. 19, 1944; children: Claudia Beth, Patricia Leigh. BS, U. Utah, 1944, ScD (hon.), 1989; MD with distinction, George Washington U., 1949; ScD (hon.), Hahnemann Med. Sch., 1979, L.I. U., 1988, St. Louis U., George Washington U., 1990. Intern Johns Hopkins Hosp., Balt., 1949-50, asst. resident, 1951-52; asst. resident physician Duke Hosp., Durham, N.C., 1950-51; vis. investigator, asst. physician Rockefeller Inst. Med. Research, 1952-54; fellow Nat. Found. Infantile Paralysis, 1952-54; mem. faculty Johns Hopkins Sch. Medicine, Balt.; staff Johns Hopkins Hosp., Balt., 1954-66, prof. medicine, 1964-66, physician, head div. clin. immunology, allergy and infectious diseases, 1958-66; prof., chmn. dept. medicine U. Fla., Gainesville, 1966-76; VA disting. physician U. Fla., 1990, prof. dept. medicine, 1990—; exec. v.p. Robert Wood Johnson Found., Princeton, N.J., 1976-86, pres., 1986-90, trustee emeritus, 1990—; U.S. del. U.S.-Japan Coop. Med. Sci. Program, 1972-81; mem. council drugs AMA, 1965-67; mem. NRC-NAS Drug Research Bd., 1965-71; mem. expert adv. panel bacterial diseases (coccal infection) WHO; mem. council Nat. Inst. Allergy and Infectious Diseases, 1968-72; cons. FDA; mem. tng. grant com. NIH, 1964-68. Author, editor books on internal medicine, infectious diseases, clin. pharmacology; contbr. articles to profl. jours. Recipient Career Research award NIH, 1962, Edward Jill award Acad. of Medicine N.J., 1990; Markle scholar med. scis., 1955-62, Johns Hopkins U. scholar, 1988. Mem. Inst. Medicine of NAS, Am. Soc. Clin. Investigation, Assembly Life Scis. of NAS, Assn. Am. Physicians, Soc. Exptl. Biology and Medicine, Am. Assn. Immunologists, Am. Fedn. Clin. Research, Harvey Soc., N.Y. Acad. Sci., Infectious Disease Soc. Am. (pres. 1973), So. Soc. Clin. Investigation, ACP (Fla. gov. 1975-76, Mead-Johnson postgrad. scholar 1954-55, Ordronaux award med. scholarship 1949), Am. Clin. and Climatol. Assn., Alpha Omega Alpha. Office: U Fla Dept of Veteran Affairs 1601 SW Archer Rd Gainesville FL 32608-1197

CLUNIE, THOMAS JOHN, chemical engineer; b. Racine, Wis., Mar. 8, 1940; s. Milton Orville and Geraldine Mae (Koehler) C.; m. Deborah Kathleen Rutherford, May 2, 1970; children: Sarah Kathleen, Amy Elizabeth. BSc, Northwestern U., Evanston, 1963; PhD, U. Notre Dame, 1968. Engr. Exxon Rsch. and Engring. Co., Florham Park, N.J., 1967-70, group leader, 1970-76, sect. head, 1976-78; sr. sect. head Carter Oil Co./Exxon Coal, U.S.A., Houston, 1978-82; coordination assoc. Exxon Co., U.S.A., Houston, 1982-84, energy advisor, 1984—; bus. cons. Jr. Achievement of S.E. Tex., Houston, 1990—. Mem. planning bd. Clear Creek Ind. Sch. Dist., Houston, 1991; exec. bd. E.A. Smith YMCA, Houston, 1992—. Recipient Parent/Child Vol. of Yr. E.A. Smith YMCA, 1991, 92. Mem. AICE, Am. Chem. Soc., Sigma Xi. Office: Exxon Co USA 4025D EB PO Box 2180 Houston TX 77252-2180

CLUTTER, MARY ELIZABETH, federal official; b. Charleroi, Pa.; BS, Allegheny Coll., 1953, DSc., 1986, MS, U. Pitts., 1957, PhD in Botany, 1960; Rsch. assoc. Yale U., 1961-73, lectr. biology, 1965-78, sr. rsch. assoc., 1973-78; program dir. NSF, Washington, 1976-81, sect. head, 1981-84, div. dir., 1984-85, 87-88, sr. sci. advisor, 1985-87, asst. dir., 1989—. Mem. AAAS (bd. dirs. 1986-90), HFSP (trustee), Internat. Soc. Plant Molecular Biology, Am. Soc. Cell Biology, Am. Soc. Plant Physiologists, Soc. Devel. Biology, Assn. Women in Sci. Office: Nat Sci Found 1800 G St NW Washington DC 20550-0002

CLYMER, BRADLEY DEAN, electrical engineering educator; b. Bluffton, Ohio, Dec. 27, 1957; s. Thomas Lynn and Donna Jean (Ferrell) C. BSEE, Ohio State U., 1981, MS, 1982; PhD, Stanford (Calif.) U., 1987. Mem. tech. staff Space and Comms. Group, Hughes Aircraft Co., El Segundo, Calif., 1981-83; asst. prof. dept. elec. engring. Ohio State U., Columbus, 1987-93; assoc. prof., 1993—; cons. Honeywell, Inc., Bloomington, Minn., 1988. Contbr. articles to profl. jours. Summer Faculty fellowship Am. Soc. for Engring. Educators, 1990-91; fellowship Hughes Aircraft Co., 1981-82. Mem. IEEE (sr.), Optical Soc. of Am., Soc. of Photo-Optical Instrumentation Engrs., Sigma Xi. Mem. Religious Soc. of Friends. Achievements include patent (with Stuart A. Collins Jr.) on computer optical switching system. Office: Ohio State U Dept Elec Engring 2015 Neil Ave Columbus OH 43210

COAKER, JAMES WHITFIELD, mechanical engineer; b. Boston, Nov. 12, 1946; s. George W. and Margaret N. Coaker; m. Ruth Johnson, May 17, 1969; children—James W., John A., Stephen D. BSME, Lafayette Coll., 1968; MSB, Va. Commonwealth U., 1976. Registered profl. engr., Va. Application engr. pump and condenser div. Ingersoll-Rand Co., Richmond, Va., 1972-76; project mgr. Reco Industries, Inc., Richmond, Va., 1976-77; asst. mgr. engring. Reco Industries, Inc., Richmond, 1977-79, mgr. engring., 1979-83; systems engr., program mgr. Advanced Tech., Inc., Arlington, Va., 1983-87; program mgr. Boiler and Elevator Safety U.S. Postal Svc., Washington, 1987—; lectr. and educator in field. With USN, 1969-72; capt. USNR. Mem. ASME (past nat. chmn. plant engr. and maintenance div., chmn. elevator insps. manual com., mem. bd. profl. devel., bd. on safety codes and standards), Nat Coun. Examiners for Engring. and Surveying (affiliate), Naval Res. Assn. (life). Home: 11675 Captain Rhett Ln Fairfax Station VA 22039

COALSON, JAMES A., grain company executive, researcher; b. Winters, Tex., Sept. 6, 1942; m. Clara Coalson; children: Stacy Michelle, J. Dee. BS in Animal Sci., Abilene Christian U.; MS in Animal Sci., Okla. State U., PhD in Animal Nutrition. Asst. prof. N.C. State U., 1970-73; swine nutritionist Ctrl. Soya Feed Co., Inc., Decatur, Ind., 1973-74, sr. swine nutritionist, 1974-75, dir. swine feeds, 1975-83, dir. internat. feed rsch., 1983-87, asst. dir. feed rsch., 1986-87, dir. feed. rsch., 1987—. Mem. Am. Soc. Animal Sci., Am. Feed Industry Assn. (chmn. quality assurance group nutrition coun.). Office: Central Soya Feed Co Inc Feed Rsch Dept 1200 N 2nd St Decatur IN 46733-1175

COAST, MORGAN K., civil engineer, consultant; b. Wheeling, W.Va., July 3, 1953; s. Harold Campbell and Kathleen Regina (Silber) C.; m. Janis Ann Biega, Nov. 29, 1974; children: Kara, Erin, Kristen. BSCE, W.Va. U., 1975, MBA, 1984. Registered profl. engr., W.Va., Ohio. Engr.-in-tng. Cerrone & Vaughn, Inc., Wheeling, W.Va., 1976-80, project engr., 1980-84, project mgr., 1984-86; project mgr. Vaughn Consultants, Inc., St. Clairsville, Ohio, 1986-89, prin., owner, 1989—; chmn. Wheeling Bldg. Code Bd. of Appeals, 1986—. Author: (tech. paper) W.Va. Am. Waterworks Assn. Ann. Conf., 1983. Mem. NSPE (pres. Wheeling chpt. 1980), W.Va. Water Pollution Control Assn. (pres. 1991-92), Am. Waterworks Assn., Am. Concrete Inst. Home: 36 Maser Ave Wheeling WV 26003-7247

COATE, DAVID EDWARD, acoustician consultant; b. Kansas City, Mo., Sept. 10, 1955; s. Arthur Dale and Martha (Goodrich) C.; m. Sheryl Marie Luebbert, Aug. 8, 1981; children: Allison Marie, Brian Joseph, Michelle Grace. BA in Math/Chemistry/Physics cum laude, Westminster Coll., 1978; MS, MIT, 1980. Rsch. staff scientist MIT Energy Lab., Cambridge, 1980-81; energy auditor Volt Energy Systems, Boston, 1981-82; owner Turning Point Records, Boston, 1982-85; sr. scientist Bolt, Beranek, and Newman, Cambridge, 1986-89; sr. cons. Acentech Inc., Cambridge, 1989—. Albums include Time Keeps on Running, 1980, State of the Heart, 1985, Still Small Voice, 1990. Mem. Acoustical Soc. Am., Inst. Noise Control Engring. (assoc.), Transp. Rsch. Bd. Presbyterian. Avocations: stamp collecting, weight lifting, jogging, water skiing. Office: Acentech Inc 125 Cambridge Park Dr Cambridge MA 02140-2327

COATES, JOHN PETER, marketing professional; b. Coventry, Eng., Apr. 4, 1946; came to U.S., 1978; s. Harry and Barbara Joan (Snape) C.; m. Laura Francis Curran, July 28, 1979; children: Jonathan Edmund, Kristen Elizabeth, Ross James. BS/MS in Chemistry, Slough Coll. of Tech. now Thames Valley Univ., Eng., 1972; PhD in Chemistry, Brunel U., London, 1987. Analytical chemist Castrol Oil Co., Bracknell, Eng., 1964-73; sr. chromatographer Burmah Oil, Bromboro, Eng., 1973-74; sr., chief chemist Perkin-Elmer Ltd., Beaconsfield, Eng., 1974-78; sr. staff scientist Perkin-Elmer Corp., Norwalk, Conn., 1978-85; dir. mktg. Spectra-Tech Inc., Stamford, Conn., 1985-88; dir. analyzer div. Nicolet Instrument Corp., Madison, Wis., 1988-92; dir. mktg. real time systems divsn. (PAI) Perkin-Elmer, Norwalk, Conn., 1992—. Co-author: (with L.C. Setti) Oils, Lubricants and Petroleum Products--Characterization by Infrared Spectra, 1985; patentee in field; contbr. chpts. to books and articles to profl. jours. Fellow Royal Soc. Chemistry; mem. ASTM, Am. Chem. Soc., Soc. Automotive Engrs., Soc. Tribologists and Lubricant Engrs. Avocations: writing, photography, music, computers. Office: Perkin-Elmer PAI 761 Main Ave M/S 201 Norwalk CT 06859-0201

COBB, JAMES TEMPLE, JR., chemical engineer, engineering educator; b. Cin., Mar. 9, 1938; s. James Temple and Norma Mary (Wellman) C.; m. Lana Jo Lane, June 14, 1964; children: Christopher James, Kendall Lane. SB, MIT, 1960; MS, Purdue U., 1963, PhD, 1966. Registered profl. engr., Pa. Engr. Esso Rsch. Labs., Baton Rouge, 1967-70; asst. prof. U. Pitts., 1970-75, assoc. prof., 1975—; sr. engr. K&M Engring. Cons. Corp., Washington, 1991—; sr. cons. Goodyear Tire and Rubber Co., Akron, Ohio, 1980-85; advisor Pa. Energy Devel. Authority, Harrisburg, 1985—; bd. dirs. Cannon Boiler Works, Inc., New Kensington, Pa.; dir. energy resources program U. Pitts., 1979—, undergrad. coor. chem. and petroleum engring. dept., 1979-86; co-chmn. program com. Pitts. Coal Conf., 1983—. Contbr. articles to Jour. Catalysis, Biotech. and Bioengring., Chem. Engring. Progress, IEC Process Design Devel., Coal Processing Tech., ACS Symposium Series. Mem. Penn Hills (Pa.) Sch. Authority, 1975; dist. committeeman Penn Hills Republican Party, 1970-79. Capt. U.S. Army, 1965-67. Recipient Cert. of Merit, Am. Soc. for Engr. Edn., 1977. Fellow AICE (chmn. Pitts. sect. 1977, 89, chmn. continuing edn. com. 1981-82, speakers bur. 1974—); mem. NSPE (pres. Pitts. chpt. 1993), Sigma Xi, Phi Lambda Upsilon. Republican. Christian Scientist. Home: 141 Deerfield Dr Pittsburgh PA 15235 Office: U Pitts 1137 Benedum Hall Pittsburgh PA 15261

COBB, JOHN WINSTON, plasma physicist; b. Nashville, Mar. 10, 1963; s. Lewis Latané and Eulalia Ann-Bell (Fuson) C.; m. Lynda Joy Hawk, July 6, 1985. BS, Mich. State U., 1985, BA, 1985; PhD, U. Tex., 1993. Instr. Computer Libr., Metairie, La., 1983; undergrad. rsch. asst. Argonne Nat. Lab., Lemont, Ill., 1984; contract programmer Schlumberger Well Svcs., Austin, Tex., 1985-88; grad. rsch. asst. Inst. for Fusion Studies, U. Tex., Austin, 1986-91, rsch. engring. scientist asst., 1991—; vis. scientist Los Alamos (N.Mex.) Nat. Lab., 1990. Contbr. articles to profl. pubs. Vol. reader Recording for the Blind of Tex., Austin, 1991—. Nat. Merit scholar, 1981, Alumni Disting. Scholar, 1981-85. Mem. Am. Phys. Soc. Home: 3903 Duval St Austin TX 78751 Office: U Tex Inst Fusion Studies RLM 11.222 Austin TX 78712

COBB, RICHARD E., mechanical engineer; b. Jessup, Md., Feb. 11, 1947; s. Ralph B. and Gloria (Kurze) C.; m. Patricia A. Nickinson, Aug. 29, 1981. BSME, Va. Poly. Inst. and State U., 1979, MSME, 1981, PhD in Mech. Engring., 1988. Registered engr.-in-tng., Va. Rsch. technician Atlantic Rsch., Gainesville, Va., 1974-75; engr. E.I. du Pont de Nemours & Co., Brevard, N.C., 1981-83; instr. mech. engring. Va. Poly. Inst. and State U., Blacksburg, 1983-88; contract engr., 1988-90; sr. engr. NKF Engring., Arlington, Va., 1990—; reviewer Internat. Jour. Analytical and Exptl. Modal Analysis, 1988—. Contbr. articles to profl. jours., conf. procs. and popular pubs. With U.S. Army, 1965-68, Vietnam. Decorated Purple Heart; recipient cert. of teaching excellence Va. Poly. Inst. and State U., 1981; grad. fellow NSF, 1979-81. Mem. ASME, Sigma Xi, Phi Kappa Phi, Tau Beta Pi, Pi Tau Sigma. Achievements include discovery of improved method for modeling of highly non-linear behavior of rubber isolators under shock inputs; research on methods for estimating noise levels and statistical confidence bands to experimental frequency response function measurements. Home: 623 W Foster Ave State College PA 16801 Office: NKF Engring 4000 Wilson Blvd Ste 900 Arlington VA 22203

COBB, WILLIAM THOMPSON, environmental consultant; b. Spokane, Wash., Nov. 10, 1942; s. Elmer Jean and Martha Ella (Napier) C.; m. Sandra L. Hodgson, Aug. 29, 1964 (div. 1988); children: Mike, Melanie, Megan, Bill II. BA, Ea. Wash. U., 1964; PhD, Oreg. State U., 1973. Cert. profl. agronomist. Mgr., agronomist Sun Royal Co., Royal City, Wash., 1970-74; sr. scientist Lilly Rsch. Labs., Kennewick, Wash., 1974-87; environ. cons. Cobb Cons. Svcs., Kennewick, Wash., 1988—; bd. dirs. Bentech Labs., Portland, Oreg., 1989; dir. spl. projects Bioremediation, Inc., Lake Oswego, Oreg., 1990-92; adv. bd. Adv. Coun. Tri-Cities, Wash., 1991. Contbr. articles to profl. jours. 1st lt. U.S. Army, 1964-67. Mem. Am. Phytopathol. Soc., Weed Sci. Soc. Am., Am. Soc. Agronomy, Nat. Assn. Environ. Profls., Hazardous Matls. Control Rsch. Inst., Sigma Xi. Republican. Home and Office: Cobb Cons Svcs 815 S Kellogg St Kennewick WA 99336-9369

COBBAN, WILLIAM AUBREY, paleontologist; b. Anaconda, Mont., Dec. 31, 1916; s. Ray Aubrey and Anastacia (McNulty) C.; m. Ruth Georgina Loucks, Apr. 15, 1942; children—Georgina, William, Robert. B.A., U. Mont., 1940; Ph.D., Johns Hopkins U., 1949. Geologist Carter Oil Co., Tulsa, 1940-46; paleontologist U.S. Geol. Survey, Washington, 1948-92, emeritus scientist, 1992—. Contbr. numerous articles to profl. jours. Recipient Meritorious Service award Dept. of Interior, Washington, 1974, Paleontol. medal Paleontol. Soc. Am., 1985, Disting. Svc. award. U.S. Dept. Interior, 1986. Fellow Geol. Soc. Am., AAAS; mem. Rocky Mountain Assn. Geologists (hon., Disting. Pioneer Geologist award 1985), Mont. Geol. Soc. (hon.), Wyo. Geol. Assn. (hon.); mem. Paleontol. Soc., Soc. Econ. Paleontologists and Mineralogists (Raymond C. Moore Paleontolgoy medal 1990), Am. Assn. Petroleum Geologists, Phi Beta Kappa, Sigma Xi. Republican. Mem. United Ch. of Christ. Office: U S Geol Survey Federal Ctr Box 25046 MS 919 Denver CO 80225

COBBS, JAMES HAROLD, engineer, consultant; b. Bristow, Okla., Aug. 25, 1928; s. Harold Martin and Ella A. (Rountree) C.; m. Charlotte Marie Fisher, Aug. 16, 1953 (dec. June 1990); children: James Harold, David Charles, Gregory Lee, Matthew Louis. BS in Petroleum Engring., U. Okla., 1949, postgrad., 1949-51; postgrad. U. Tulsa, 1955-68. Assoc. engr. Tidewater Oil Co., Midland, Tex., 1951-52; reservoir engr., Houston, 1952-55, div. reservoir engr., Tulsa, 1955-59; pvt. practice cons. engr., 1959-63; sr. engr. Fenix & Scisson Inc., Tulsa, 1963-69; pres. Cobbs Engring., Inc., cons. engrs., Tulsa, 1969—; faculty U. Wis. Extension. Various positions including scoutmaster Indian Nations coun. Boy Scouts Am., 1962-81; instr. first aid ARC, 1969-81; active Vols. in Tech. Assistance, 1978—. Registered profl. engr., 8 states; cert. of qualification Nat. Council Engring. Examiners. Mem. Petroleum Engrs., Nat. Soc. Profl. Engrs., Okla. Soc. Profl. Engrs., Inst. Shaft Drilling Tech., Nat. Acad. Forensic Engrs., World Rock Boring Assn. Republican. Mem. Christian Ch. (elder, chmn. bd. elders 1971, 79). Contbr. articles to profl. jours.; patentee in field. Home: 5144 S New Haven Ave Tulsa OK 74135-3963 Office: 5350 E 46th St Tulsa OK 74135-6601

COBEL, GEORGE BASSETT, chemical engineer; b. Lincoln, Nebr., June 6, 1928; s. Paul Samuel and Helen (Bassett) C.; m. Betty Ann Gartner, June 22, 1946; children: Scott D., Sandra R. BS in Chem. Engring., U. Nebr., 1952; MS in Chem. Engring., U. Mich., 1958. Mgr. primary metals Dow Chem. Co., Midland, Mich., 1952-85; gen. mgr. Kelsey Hayes Corp., Traverse City, Mich., 1985-87; cons. Omni Tech Ltd., Midland, Mich., 1987—. With USN, 1946-53, Cuba. Mem. AIME (chmn. light metal com. 1973-75), Internat. Magnesium Assn. (pres. 1979-81), Am. Soc. Materials, Sigma Xi. Achievements include 11 patents in field including purification of calcium, calcium by electrolysis and prepartios of titanium. Home: 74-711 Dillon Rd Desert Hot Springs CA 92240 Office: Omni Tech Ltd 2715 Ashman St Ste 100 Midland MI 46840

COBURN, RONALD MURRAY, ophthalmic surgeon, researcher; b. Detroit, Aug. 25, 1943; s. Sidney and Jean (Goldberg) C.; m. Barbara Joan Levy, Feb. 21, 1969; children: Nicholas Scott, Lauren Joy. BS, Wayne State U., 1965, MD, 1969. Diplomate Am. Bd. Eye Surgery, Am. Bd. Ophthalmology. Dir. The Coburn Clinic, Dearborn, Mich., 1976—; chief ophthalmology Straith Hosp. for Spl. Surgery, Southfield, Mich., 1985—; cons. CooperVision, Inc., Bellevue, Wash., 1985-88, Alcon Surg., Inc., Ft. Worth, 1988—. Co-author: Lens-Stat Intraocular Lens Modeling System; editorial advisor Phaco and Foldables, 1990. Trustee Straith Hosp. for Spl. Surgery, 1986—. Capt. Mich. N.G., 1969-76. Fellow Am. Acad. Ophthalmology, Am. Coll. Eye Surgeons, Soc. Eye Surgeons, Royal Soc. Medicine (London); mem. AAAS, Am. Soc. Cataract and Refractive Surgery, Am. Diabetes Assn., Mich. Ophthalmol. Soc., Wayne County Med. Soc., Rsch. to Prevent Blindness, N.Y. Acad. Sci., Internat. Assn. Ocular Surgeons, Internat. Eye Found., Soc. Geriatric Ophthalmology, Soc. for Excellence in Eye Care, Am. Bd. Eye Surgery (surg. examiner), Internat. Glaucome Cong., Phi Beta Kappa. Achievements include design of Am. Med. Optics PC19LB intraocular lens, CILCO CPLU CP20 intraocular lenses, CooperVision CP10BG posterior chamber intraocular lens, Alcon CZ20BD intraocular lens. Home: 1490 W Long Lake Rd Bloomfield Hills MI 48302-1340 Office: The Coburn Clinic 19855 Outer Dr # 12L Dearborn MI 48124-2027

COBURN, THEODORE JAMES, retired physicist; b. Newton, Mass., June 11, 1926; s. Charles Arthur and Viola Mabel (Hunter) C.; student U. Louisville, 1944-46; B.Sc. in Physics, Ohio State U., 1947, Ph.D. in Physics, 1957; m. Edith Marshall Banta, June 16, 1949; children—Sue Ellen, Charles Edwin, Joanne Edith. Sr. design engr. Eastman Kodak Co., Rochester, N.Y., 1957-60, sr. research physicist research lab., 1960-68, research assoc. in solid state physics, sr. lab. staff, 1968-86; pres. Coburn Cons. to Microelectronic Industry, 1986—. Pres. elementary PTA West Irondequoit Sch. Dist., Rochester, 1970-71, life mem., 1974—, pres. High Sch. Parent Tchr. Orgn., 1975-76; chmn. Citizens Com. for Selection Sch. Bd. Candidates West Irondequoit Sch. Dist., 1976-84; mem. West Irondepuoit Sch. Dist. Civil Service Grievance Rev. Panel, 1971-84; ruling elder, deacon, chmn. coms. Brighton Presbyn. Ch., Rochester. Served with USN, 1944-46, 53-55. U.S. Army research fellow, 1951-53; USN research fellow, 1955-57. Mem. IEEE, Am. Phys. Soc., Electrochem. Soc. Author numerous co., Dept. def. classified reports; contbr. articles to sci. jours.; patentee in field. Home: 156 Pinecrest Dr Rochester NY 14617-2227 Office: Eastman Kodak Co Research Labs B-81 Kodak Park Rochester NY 14650-0001

COCA-PRADOS, MIGUEL, molecular biologist; b. Salamanca, Spain, Dec. 10, 1948; came to U.S., 1976; s. Manuel Coca Rebollero and Ines Maria Prados Martin; m. Silvia D. Ducach, May 24, 1980; children: Daniel, Natalie. MSc, U. Salamanca, 1972, PhD cum laude, 1975. Postdoctoral fellow Ohio State U., Columbus, 1976-77, Rockefeller U., N.Y.C., 1977-80; rsch. assoc. Yale U., New Haven, 1981-83, assoc. rsch. scientist, 1983-87, rsch. scientist, 1988-92, sr. rsch. scientist, 1993—; prin. investigator Nat. Eye Inst., NIH, 1984—. Contbr. articles to profl. publs. Recipient Outstanding Contbns. in Field of Vision Rsch. award Alcon Rsch. Inst., 1989. Mem. Am. Soc. for Microbiology, Assn. for Rsch. in Vision. Democrat. Roman Catholic. Achievements include establishment of cell lines from human ocular tissues from donor patients by transformation with Simian Virus 40, construction cDNA libraries from mRNA of ocular tissues of donor patients with glaucoma. Home: 70 Horseshoe Rd Guilford CT 06437 Office: Yale U Sch Medicine 330 Cedar St New Haven CT 06510

COCHRAN, ADA, data specialist, writer; b. Lost Creek, Ky., Dec. 21, 1933; d. Shade And Doshie (Combs) Fugate; m. Alan Cochran, Oct. 1, 1956; children: Debra, Evangeline, William. Student, Am. U, Kent, Ill., 1957-62; AA, C.C. La Plata, 1972. Dental office mgr. Dr. Arthur F. Furman, Brandywine, Md., 1968-77; data mgr. FCC, Washington, 1978—. Contbr. articles to profl. jours. Mem. Bapt. Womens Group, (pres. 1964-65). Served with USA 1956-57. Mem. So. Md. Writer's Vineyard (founder, dir. 1986—), Writer's Inst., Freelance Writer's Assn. Democrat. Clubs: Toastmaster (pres. 1986-87), Writer's (dir. 1985—). Home: 5819 Castle Brook San Antonio TX 78218 Office: Cochran's Corner PO Box 2036 Waldorf MD 20604-2036

COCHRAN, GEORGE VAN BRUNT, physician, surgery educator, researcher; b. N.Y.C., Jan. 20, 1932; s. George Gilfillan and Mary Lott (Van Brunt) C.; m. Caroline Winston, June 13, 1970; children: Linsay, Alexander, John. AB magna cum laude, Dartmouth Coll., 1953; MD, Columbia U., 1956, ScD, 1967. Diplomate Am. Bd. Orthopaedic Surgery. Intern in surgery Presbyn. Hosp., N.Y.C., 1956-57; asst. resident, then resident N.Y. Orthopaedic Hosp., N.Y.C., 1961-63; NIH rsch. fellow dept. orthopaedics Columbia U., N.Y.C., 1964-69, prof. clin. orthopaedic surgery, 1981—; dir. orthopaedic rsch. St. Lukes Hosp., N.Y.C., 1970-80; dir., founder biomechs. rsch. unit Helen Hayes Hosp., West Haverstraw, N.Y., 1972-80, dir. and founder orthopaedic engring. and rsch. ctr., 1980—; attending orthopaedic surgeon rsch. svc. VA Med. Ctr., Castle Point, N.Y., 1975—; adj. prof. biomed. engring. Rensselaer Poly. Inst., Troy, N.Y., 1981—; dir. Charles A. Lindbergh Fund, Mpls., 1983-86, Health Rsch. Inst., Albany, N.Y., 1981-85. Author: A Primer of Orthopaedic Biomechanics, 1982; contbr. numerous articles to profl. jours. Capt. USAF, 1957-59, ETO. HEW/Social and Rehab. Svcs. fellow, 1968-69, A.O. Lab. for Exptl. Surgery/Arctic Inst. N.Am. fellow, 1967, USPHS fellow and grantee, 1964-66; NIH/VA grantee. Fellow N.Y. Acad. Scis., Am. Acad. Orthopaedic Surgeons, The Explorers

Club (pres. 1981-85); mem. Bioelec. Repair and Growth Soc. (pres. 1990), Internat. Soc. Orthopaedic Surgery and Traumatology, Orthopaedic Rsch. Soc., Soc. for Exptl. Stress Analysis, Am. Congress Rehab. Medicine, Am. Orthopaedic Assn., Phi Beta Kappa, Kappa Delta (Orthopaedic Rsch. award 1968, Nicholas Andry co-award 1975, Spl. Achievement award 1987). Republican. Avocations; exploring, mountaineering, Arctic glaciology. Office: Helen Hayes Hosp RR 9 West Haverstraw NY 10993

COCHRAN, HENRY DOUGLAS, chemical engineer; b. Plainfield, N.J., Sept. 13, 1943; m. Elizabeth Penland, Aug. 4, 1973; children: Sara E., M. John. BS, Princeton U., 1965; MS, MIT, 1967, PhD, 1973. Devel. engr. Oak Ridge (Tenn.) Nat. Lab., 1968-69, 73-76, group leader, 1976-78, program leader, 1978-83, mem. sr. rsch. staff, 1983--; adj. assoc. prof. U. Tenn., Knoxville, 1985--. Mem. Am. Inst. Chem. Engrs., Am. Chem. Soc., Am. Phys. Soc., Sigma Xi. Achievements include development of molecular theory of solutions in supercritical solvents and establishment of coal liquefaction program. Office: Oak Ridge Nat Lab PO Box 2008 Oak Ridge TN 37831-6224

COCHRANE, JAMES LOUIS, economist; b. Nyack, N.Y., Aug. 31, 1942; s. Thomas and Anna (Yaroscak) C.; m. Katherine Prince Schirmer, Mar. 24, 1984; 1 child, Katherine Anne. BA, Wittenberg U., 1964; PhD, Tulane U., 1968. Instr. Tulane U., New Orleans, 1967-68; asst. prof. U. S.C., Columbia, 1968-70, assoc. prof., 1970-72, prof., 1972-77; sr. staff mem. NSC, Washington, 1978-79; directorate of intelligence CIA, Washington, 1980-83; sr. v.p., chief economist Tex. Commerce Bancshares Inc., Houston, 1984-88, N.Y. Stock Exch., 1988—; assoc. staff mem. Brookings Instn., Washington, 1972-74, 76-78; mem. editorial bd. History of Polit. Economy Duke U., 1974-80, So. Econ. Jour. U. N.C., 1976-79; 1st v.p. So. Econ. Assn., U. N.C., 1976-77; vis. prof. U. Melbourne, Australia, 1972, U. Tex., Austin, 1973-74; adv. bd. U. Pa. White Ctr. Fin. Rsch., Vanderbilt U. Fin. Markets Rsch. Ctr., Columbia U. Ctr. Law and Econ. Studies, N.Y. Assembly Bd. of Advisors; mem. dean's adv. bd. Sch. Bus. Hofstra U., Pace U. Study of Equities Markets; mem. internat. adv. com. Harvard U. Ctr. for Internat. Affairs, Pace U. for study equities mrkts., U.S. Nat. Com. for Pacific Econ. Cooperation. Author: Macroeconomics Before Keynes, 1970, Macroeconomics Analysis and Policy, 1974, Industrialism and Industrial Man in Retrospect, 1977; editor: Multiple Criteria Decision Making, 1975. Mem. History of Econs. Soc. (treas. 1974-80), Asia Soc. (adv. dir. 1986), Am. Econ. Assn., Western Fin. Assn. Avocations: tennis, singing, writing. Home: 25 Burnside Dr Short Hills NJ 07078-2105 Office: NY Stock Exch 11 Wall St New York NY 10005-1916

COCHRANE, THOMAS THURSTON, tropical soil scientist, agronomist; b. Gisborne, N.Z., Mar. 18, 1936; came to U.S., 1990; s. Thomas Nicholson and Muriel Hope (Morrison) C.; m. Rosa Elena Fajardo de las Muñecas, Mar. 18, 1970; 1 child, Thomas Arey. B Agrl. Sci., U. N.Z., 1960; Assoc., Imperial Coll. Tropical Agr., Trinidad, B.W.I., 1962; PhD, U. W.I., Trinidad, W.I., 1969. Agronomist U. W.I., Trinidad, 1962; soil scientist Ministry Overseas Devel., U.K., Bolivia, 1963-74; cons. Tate and Lyle, U.K., Bolivia and Mex., 1974-77; land resource specialist Centro Internacional de Agricultura Tropical, Colombia, 1977-81; soil scientist S.E. Consortium for Internat. Devel., Washington, 1981-82; land resource specialist Inter-Am. Inst. for Agrl. Cooperation, Brazil, 1982-85; cons. West Lafayette, Ind., 1986—. Author: The Land Use Potential of Bolivia, 1973, Land in Tropical America, 1985; contbr. over 50 articles to sci. publs. Decorated El Condor de los Andes, Bolivia, 1973. Mem. Internat. Soil Sci. Soc., Soil Sci. Soc. Am., Am. Soc. Agronomy. Achievements include development of an equation for liming acid soils to compensate aluminum toxicity, a differential equation for predicting crop fertilizer response, an equation for calculating the contribution of the separate ions and molecules of water solutions to osmotic pressure; research showing that the Amazonian forests depend on both climate and soils. Home: 623 N Salisbury St West Lafayette IN 47906-2711

COCK, JAMES HEYWOOD, agricultural scientist; b. Norwich, Norfolk, Eng., May 1, 1944; s. John H. and Stella M. C.; m. Angela Maria Misas, Aug. 22, 1975; children: Juan Camilo, Andrew Philip. BA in Agr., Cambridge U., Eng., 1966; PhD in Agrl. Botany, Reading U., Eng., 1969. Rsch. fellow Internat. Rice Rsch. Inst., Los Banos, Philippines, 1969-71; cassava program leader Internat. Ctr. for Tropical Agr., Cali, Colombia, 1971-89; sr. agriculturist World Bank, Washington, 1989-91; dir. gen. Cenicaña, Cali, Colombia, 1991—; cons. World Bank, 1991—; editor Internat. Soc. Sugar Cane Technologists, Baton Rouge, La., 1991—. Author: Cassava: New Potential for a Neglected Crop, 1985; contbr. articles to profl. jours. Recipient Don Plucknet award Internat. Soc. Tropical Root Crops, Bangkok, 1988. Mem. Crop Sci. Soc. Am. (Internat. Achievement award 1989), Latin Am. Soc. Plant Physiology (pres. 1980-82). Achievements include development of integrated Cassava projects to assist small farmers in tropics; description of ideal plant type for cassava to assist breeders in developing better varieties; promoted nat. and internat. rsch. on cassava; dir. of institute that has provided technology to make the Colombian sugar industry among the must productive in the world. Home: Calle 16 # 107 71, Cali Colombia Office: Cenicana, AA 91-38, Cali Colombia

COCKE, JOHN, computer scientist; b. Charlotte, N.C., 1925. BS, Duke U., 1946, PhD in Math., 1956. Engr. Air Engring. Co., 1946-49, Gen. Electric Co., 1949-50; with rsch. staff IBM Rsch. Divsn., Thomas J. Watson Rsch. Ctr., Yorktown Heights, N.Y., 1956—; vis. prof. elec. engring. dept. MIT, 1962, Courant Inst. math. and Sci., 1968-69; mem. corp. tech. com. IBM, 1977-78. Contbr. numerous articles to profl. jours. Recipient Eckert-Maunchy award Am. Computer Soc. and Inst. Elec. and Electronics Engrs., 1985, Am. Turing award Assn. Computer Machinery, 1987, Nat. medal Tech., U.S. Dept. Commerce, Tech. Adminstrn., 1991. Fellow AAAS; mem. NAE. Achievements include invention of reduced instruction set computer (RISC). Office: IBM T J Watson Rsch Ctr PO Box 704 Yorktown Heights NY 10598-0704

COCKE, WILLIAM MARVIN, JR., plastic surgeon, educator; b. Balt., Aug. 2, 1934; s. William M. and Clara E. (Bosley) C.; m. Sue Ann Harris, Apr. 25, 1981; children: Gregory William, Laura Marie, Ann Julie; children by previous marriage: William Marvin III, Catherine Lynn, Deborah Kay, Brian Thomas. B.S. with honors in Biology, Tex. A&M U., 1956; M.D., Baylor U., 1960. Diplomate: Am. Bd. Plastic Surgery (guest examiner 1978). Intern surgery Vanderbilt U. Hosp., Nashville, 1960-61; fellow gen. surgery Ochsner Clinic and Found. Hosp., New Orleans, 1961-64; chief resident surgery Monroe (La.) Charity Hosp., 1963-64; resident reconstructive surgery Roswell Park Meml. Inst., Buffalo, 1965-66; chief resident plastic surgery VA Hosp., Bronx, N.Y., 1966; practice medicine specializing in plastic surgery Nashville, 1968-75, Sacramento, 1976-79; pvt. practice medicine specializing in plastic surgery Bryan, Tex., 1980-92; prof. surgery, head div. plastic/reconstructive surgery Marshall U. Sch. of Medicine, Huntington, W.Va., 1993—; mem. staff St. Joseph's Hosp., Bryan, Humana Hosp., Brazos Valley; asst. prof. plastic surgery Vanderbilt U. Sch. Medicine, Nashville, 1968-69, asst. clin. prof. plastic surgery, 1969-75; assoc. prof. plastic surgery Ind. U. Sch. Medicine, Indpls., 1975-76; chief plastic surgery service Wishard Meml. Hosp., Ind. U., 1975-76; assoc. prof. surgery U. Calif. Sch. Medicine, Davis, 1976-79, chmn. dept. plastic surgery, 1976-79; prof. surgery, chief div. plastic surgery Tex. Tech. U. Sch. Medicine, Lubbock, 1979-80, dir. Microsurg. Research Lab., 1979-80; clin. prof. surgery Tex. A&M U. Sch. Medicine, 1983-92; prof. plastic surgery, 1986-89; chief plastic surgery svc., dept. surgery, Olin Teague VA Med. Ctr., Temple, Tex., 1986-92; prof. head surgery divsn. plastic and reconstruction Marshall U. Sch. Medicine, 1992—. Author textbooks on plastic surgery; contbr. articles to profl. jours. Served with M.C. USAF, 1966-68. Recipient Dean Echols award Ochsner Hosp. Found., 1963. Mem. ACS, Am. Assn. Plastic Surgeons, Soc. Head and Neck Surgeons, Assn. for Acad. Surgery, Alton Ochsner Surg. Soc. Episcopalian. Home: 45 Olde Farm Rd Ona WV 25545 Office: Marshall U Sch Medicine Dept Surgery 1801 6th Ave Huntington WV 25755-0001

CODDING, PENELOPE WIXSON, chemistry educator; b. Emporia, Kans., Sept. 18, 1946; d. Robert Mernis and Maxine Elliot (Clark) Wixson; children: Dana Stephanie Codding, Nathaniel James Codding. BSc, Mich. State U., 1968, PhD, 1971. Heritage med. scholar Alta. Heritage Found. for Med. Rsch., Edmonton, 1981-91; asst. prof. chemistry U. Calgary, 1981-85, assoc. prof. chemistry, 1985-89, prof. chemistry, 1989—, head of dept., 1993—; mem. cellular and molecular biophysics study sect. NIH, Washington, 1987-91; lectr. in field. Contbr. articles to profl. jours.; co-editor Acta Crystallographica, Chester, Eng., 1990—. Com. mem. Ministers Com. on High Sch. Sci. Curriculum, Province of Alta., 1989-90. Fellow Chem. Inst. Can.; mem. Am. Chem. Soc., Am. Crystallographic Assn. (SIG chair 1988-89), Internat. Union of Crystallography (Exec. com. 1990—), Can. Biochem. Soc. Office: Univ of Calgary Dept Chemistry, 2500 University Dr NW, Calgary, AB Canada T2N 1N4

CODY, JOHN THOMAS, forensic toxicologist, biological chemist researcher; b. Susquehanna, Pa., Oct. 17, 1949. BA, Iowa Wesleyan Coll., 1971; MS, U. Iowa, 1974, PhD, 1979. Clin. biochemistry instr. Sch. Health Care Scis., Sheppard AFB, Tex., 1980-85; chief quality control Air Force Drug Testing Lab., Brooks AFB, Tex., 1985-87, chief analytical scis., 1987-89, dep. dir., 1989-91; clin. investigations fellow Wilford Hall Med. Ctr., Lackland AFB, Tex., 1991-92, chief lab. scis., 1992, dep. dir., 1992—; nat. lab. cert. program inspector Dept. Health and Human Svcs., 1989—, com. mem., 1989; cons. USAF Surgeon Gen., 1991—. Referee: Journal of Analytical Toxicology, 1990—, Biological Mass Spectrometry, 1991; editorial bd. mem.: The Chemist, 1988; contbr. articles to profl. jours. Bd. mem. Chrosome 18 Registry and Rsch. Soc., San Antonio, 1990—. Lt. Col. USAF, 1980—. Recipient Physician, Scientist of Yr. award Air Force Assn., Tex., 1988, Sci. Achievement award USAF, 1989. Fellow Am. Acad. Forensic Scis.; mem. Am. Soc. Mass Spectrometry, Am. Chem. Soc., Soc. Forensic Toxicologists, Southwestern Assn. Toxicologists. Achievements include clinical investigations fellowship program; research in forensic toxicology, drug metabolism, analytical methods development, computer analysis of complex instrumantal data. Office: Wilford Hall Med Ctr 1255 Wilford Hall Loop Lackland AFB TX 78236-5319

COE, EDWARD HAROLD, JR., agronomist, educator, geneticist; b. San Antonio, Dec. 7, 1926; married, 1949; 2 children. BS, U. Minn., 1949, MS, 1951; PhD in Botany, U. Ill., 1954. Rsch. fellow genetics Calif. Inst. Tech., 1954-55, rsch. assoc. field crops, 1955-58, assoc. prof., 1959-63; prof. agronomy U. Mo., Columbia, 1964—; rsch. leader plant genetics U. Mo., 1977—. Editor Maize Genetics Coop Newslett, 1975—. Mem. AAAS, Genetics Soc. Am. (Thomas Hunt Morgan medal 1992), Am. Genetic Assn., Crop Sci. Soc. Am., Am. Soc. Naturalists. Achievements include research in genetics of maize, in genome mapping, in extramendelian inheritance, in fertilization and development. Office: Univ of Missouri Curtis Hall Agronomy Dept & USDA Columbia MO 65211 Address: 206 Heather Ln Columbia MO 65203*

COELHO, ANTHONY MENDES, JR., health science administrator; b. Danbury, Conn., May 26, 1947; s. Anthony Mendes and Angela (Fernandes) C.; m. Linda Straw, Jan. 12, 1974. BS in Social Scis., Western Conn. State U., 1970; MA in Phys.-Biol. Anthropology, U. Tex., 1973, PhD in Phys.-Biol. Anthropology, 1975. Cert. social scis. secondary tchr., Conn. Asst. prof. anthropology Tex. Tech U., Lubbock, 1974-75; instr. social-cultural anthropology U. Tex., Austin, 1971-72, teaching asst. phys.-biol. anthropology, 1972-74; asst. scientist S.W. Found. for Biomed. Rsch., San Antonio, 1975-76, assoc. scientist, 1976-86, scientist, 1986-92, head Behavioral Medicine Lab., 1975-92; health sci. adminstr. Nat. Heart Lung & Blood Inst., NIH, Bethesda, Md., 1992—; adj. asst. prof. pediatrics U. Tex. Health Scis. Ctr., San Antonio, 1976-84, adj. assoc. prof., 1984-92, adj. assoc. prof. dental diagnostic scis., 1984-90, adj. prof. surgery and neurosurgery, 1989-92; lectr. social and behavioral scis. U. Tex., San Antonio, 1977-82; mem. rsch. manpower rev. com. NIH, Bethesda, Md., 1988-92; grant reviewer NSF, Nat. Geog. Soc., NIMH, Alcohol Drug Abuse and Mental Health Adminstrn., Nat. Scis. and Engring. Rsch. Coun., Can. Wenner-Gren Found. for Anthrop. Rsch.; reviewer various sci. jours. Contbr. articles to sci. jours.. Active Am. Heart Assn. Scholar Command Security Corp., 1970, U. Tex., 1972-74; grantee NIH, 1983, 85, 89. Mem. Am. Soc. Primatologists (exec. sec., bd. dirs. 1982-84, cons. editor Am. Jour. Primatol., 1986—, editor book reviews 1989-91), Am. Assn. Phys. Anthropologists, Soc. Behavioral Medicine, Nat. Coun. U. Rsch. Adminstrs., Animal Behavior Soc., Human Biology Coun., Inst. for Advancement Health, Internat. Primatol. Soc., Latin Am. Soc. Primatology, Soc. Rsch. Adminstrs., Sigma Xi, Phi Kappa Phi, Delta Tau Kappa. Home: 8427 Timber Crest Dr San Antonio TX 78250-4406 Office: NIH-NHLBI 5333 Westbard Ave Bethesda MD 20892

COELHO, HÉLIO TEIXEIRA, physics researcher, consultant; b. Bonito, Brazil, Feb. 16, 1941; s. Fructuoso and Nadir (Teixeira) C. Degree in elec. engring., U. Federal de Pernambuco, Recife, Brazil, 1964; degree in nuclear sci., U. de São Paulo (Brazil), 1966; MS, U. Pa., 1968, PhD, 1971. Asst. prof. U. Federal de Pernambuco, Recife, Brazil, 1965-72, full prof., 1973—, chmn. dept. physics, coun. mem., 1982-84; researcher U. Frankfurt (Fed. Republic of Germany), 1972-73; vis. prof. U. Strassbourg (France), 1972, U. Manchester (Eng.), 1975, McMaster U., Hamilton, Ont., Can., 1976, 85, 91, Kyoto (Japan) U., 1977, U. Lisbon (Portugal), 1978, Ruhr U., Bochum, Fed. Republic of Germany, 1981, Drexel U., Phila., 1985, U. São Paulo, 1990; cons. State Sec. Sci. and Tech., Recife, 1990; chmn. nuclear forces session Few-Body Internat. Conf., Nanking, China, 1985; program com. Internat. Nuclear Physics Conf., São Paulo, 1989; organizing com. Internat. Gauss Symposium. Contbr. sci. articles to profl. jours. Rsch. grantee Rsch. Found. Pernambuco, Recife, 1990, Nat. Rsch. Found., Brasilia, Brazil, 1991. Mem. Am. Phys. Soc., Brazilian Phys. Soc. (sec. 1978-79), Brazilian Assn. for Advancement Sci. (sec. 1988-89), Brazilian Acad. Sci. Roman Catholic. Avocations: reading, music, conversation, swimming, walking. Office: U Fed de Pernambuco Dept Fisica, 50739 Recife Brazil

COFFEY, HOWARD THOMAS, physicist; b. Bristol, Va., Sept. 23, 1934; s. Charles Walker and Verna Lee (Pardue) C.; m. Rose Marie DePasse, Nov. 10, 1962 (div. 1978); children: Nancy Lee, Sandra Marie. AB, King Coll., 1956; PhD, U. N.C., 1964. Sr. scientist Westinghouse S&T Ctr., Pitts., 1962-67; mgr. cryogenic applications Stanford Rsch. Inst., Menlo Park, Calif., 1967-76; systems mgr. Am. Magnetics, Oak Ridge, Tenn., 1976-82; dir. mktg. Biomagnetic Techs., San Diego, 1983-85; systems mgr. Sub-Sea Systems, Escondido, Calif., 1985-88; dep. mgr. Maglev Argonne (Ill.) Nat. Lab., 1988—; bd. dirs., corp. sec. Applied Superconductivity Conf., Inc., Batavia, Ill.; bd. dirs. Am. Magnetics, Inc., 1976-82; mem. Maglev tech. adv. com. U.S. Sen., 1989. Contbr. articles to Phys. Rev., IEEE Publ., Jour. Applied Physics, Advances in Cryogenic Engring. Achievements include development of first 10 Tesla superconducting magnet using Nb-Ti alloy, of magnetic levitation for transportation; design and construction of Maglev test sled. Home: 1109 Bristlecone Ct Darien IL 60561 Office: Argonne Nat Lab 9700 S Cass Ave Argonne IL 60439

COFFEY, KEVIN ROBERT, materials scientist, researcher; b. Merced, Calif., Apr. 2, 1954; s. John Brindley and Valerie Althea (Kendall) C.; m. Randi Payne (div. July 1980); 1 child, Alexandra Lynn; m. Shelly Alane Martin. BA in Physics, New Coll., Sarasota, Fla., 1975; PhD in Materials Sci., MIT, 1989. Physicist Saxon Copystatics, Miami, Fla., 1977-80, Nashua (N.H.) Corp., 1980-85; adv. scientist/engr. IBM, San Jose, Calif., 1989—. Contbr. articles to profl. jours. Mem. IEEE, Am. Vacuum Soc., Materials Rsch. Soc., Sigma Xi. Democrat.

COFFEY, TIMOTHY, physicist; b. Washington, June 27, 1941; s. Timothy and Helen (Stevens) C.; m. Paula Marie Smith, Aug. 24, 1963; children: Timothy, Donna, Marie. B.S. in Elec. Engring. (Cambridge scholar 1958), MIT, 1962; M.S. in Physics, U. Mich., 1963, Evening News Assn. fellow, 1964, Ph.D., 1967. Rsch. physicist Air Force Cambridge Rsch. Lab., 1964; theoretical physicist EGG, Inc., Boston, 1966-71; head plasma dynamics br., then supt. plasma physics div. Naval Rsch. Lab., Washington, 1971-80; assoc. dir. rsch. for gen. sci. and tech. Naval Rsch. Lab., 1980-83; dir. rsch. Naval Research Lab., 1983—. Recipient award Naval Rsch. Lab., 1974, 75, Disting. Civilian award Dept. Defense, 1991. Fellow Am. Phys. Soc., Washington Acad. Scis.; mem. AAAS, Franklin Inst. (Delmar S. Fahrney medal 1991 com. for sci. and arts), Am. Inst. Physics. Office: Naval Research Lab Code 1001 4555 Overlook Ave SW Washington DC 20375-0001

COFFIELD, RONALD DALE, mechanical engineer; b. Monongahela, Pa., Feb. 27, 1943; s. Ronald Dale and Dorthy Lourene (Wood) C.; m. Patricia Earle Lightholder, June 22, 1968; children: Erin Elizabeth, Grant Edward. BSME, U. Pitts., 1965, MSME, 1966, PhD in Mech. Engring., 1969. Adv. engr. Westinghouse Electric, Pitts., 1969—; lectr. U. Pitts., Pa. State U., 1987-92. Co-author: Thermal Analysis of Liquid Metal Fast Breeder Reactors, 1978; contbr. over 90 articles to profl. publs. Adviser Jr. Achievement, Greensburg, Pa., 1981-84; mem. adminstrv. bd. Charter Oak Meth. Ch., 1992. Mem. ASME, Sigma Xi. Home: Hilton Rd RD 8 Box 155F Greensburg PA 15601

COFFILL, CHARLES FREDERICK, JR., aerospace engineer, educator; b. Rockville Centre, N.Y., Feb. 6, 1946; s. Anna Erma (Lemke) Windsor; m. Madeline Lee Evans, Jan. 2, 1970; children: Victoria, Lorraine. BS in Aeronautics & Astronautics, NYU, 1968; M in Engring., U. Redlands, 1972. Commd. 2d lt. USAF, 1968, advanced through grades to capt., 1971; reentry systems engr. Advanced Ballistic Reentry Vehicles Office, Norton AFB, Calif., 1968-73; liquid rocket engine mgr. USAF Launch Vehicle Office, L.A., 1973-76; fgn. mil. sales program mgr. USAF, Norton AFB, 1976-80; resigned USAF, 1980; staff engr. Hercules Aerospace Divsn., Magna, Utah, 1980-92; math. & sci. tchr. Utah Pub. Sch. System, Salt Lake City, 1992—; substitute tchr. Jordan Sch. Dist., Sandy, Utah. Mem. AIAA. Methodist. Home: 2060 Wildwood Dr Salt Lake City UT 84121-1428

COFFIN, DEBRA PETERS, ecologist, researcher; b. Waverly, Iowa, Nov. 19, 1958; d. Erwald Herman and Lavon Florence (Wendt) Peters; m. Paul Clinton Coffin, July 18, 1981; children: Stacey Lea, Allie Marie. BS, Iowa State U., 1981; MS, San Diego State U., 1983; PhD, Colo. State U., 1988. Postdoctoral assoc. Colo. State U., Ft. Collins, 1988-89, rsch. assoc., 1989-92, rsch. scientist, 1992—; mem. Cen. Plains Exptl. Range Sci. Adv. Com., Ft. Collins, 1991—. Contbr. articles to sci. jours. Mem. AAAS, Ecol. Soc. Am., Soc. Range Mgmt., Phi Beta Kappa, Sigma Xi. Office: Colo State U Range Sci Dept Fort Collins CO 80523

COFRANCESCO, DONALD GEORGE, health facility administrator; b. New Haven, May 29, 1953; s. George William and Marie Teresa (Marra) C. BS with distinction in Chemistry and Life Scis., Worcester Poly. Inst., 1975; MA in Gerontology, U. New Haven, 1979; MPH, Yale U., 1992. Lic. nursing home adminstr., Conn. Dir. biostats. and health planning Dept. of Health, New Haven, 1980; adminstr. Golden Manor Convalescent Home, New Haven, 1980-81, West Haven (Conn.) Nursing Ctr., 1981-85, Independence Manor, Meriden, Conn., 1986-87, Hillside Manor, Hartford, Conn., 1987-88; asst. in rsch. Sch. Medicine, Yale U., New Haven, 1975-77, asst. adminstr., fin. analyst, 1990—; cons. Hospice: Project Care, Inc., Watertown, Conn., 1989-90; v.p., CFO Environ. Health Corp., Hamden, Conn., 1991—. Bd. dirs. Partnerships Ctr. for Adult Day Care, Inc., Hamden, 1991—; mem. Health Systems Agy. South Cen. Conn., Inc., Woodbridge, 1982-87; mem. Peabody Mus., Yale U. Named one of Outstanding Young Men of Am., 1983. Mem. APHA, NRA (cert. firearms instr., light rifle expert, air pistol sharpshooter), Am. Chem. Soc., Conn. Pub. Health Assn., Planetary Soc. Roman Catholic. Home: 104 Hillfield Rd Hamden CT 06518-1810 Office: Yale U Sch Medicine PO Box 208041 333 Cedar St New Haven CT 06520-8041

COGAN, KENNETH GEORGE, quality assurance engineer; b. Neptune, N.J., Sept. 10, 1960; s. Kenneth Leonard Cogan and Mary Anne (Stevens) Taylor; m. Maureen Patricia Flanagan, Oct. 6, 1990. BS in Indsl. Mgmt., U. Lowell, 1988. Cert. quality engr., cert. quality auditor. Mech. insp. M/A-COM Semicondr. Products, Burlington, Mass., 1979-87, quality assurance engr., 1988-90; quality assurance engr. Paramax Systems, Lanham, Md., 1990-93; quality assurance assessor Lloyd's Register Quality Assurance, Ltd., Hoboken, N.J., 1993—. Editor newsletter Quality Nexus, 1991-92; mem. editorial review bd. Quality Progress. Mem. Am. Soc. for Quality Control (sr., treas. 1991, chair-elect 1991-92, chair 1992-93, mixed media rev. bd.). Avocations: amateur radio, music, golf. Home: 10283 Wayover Way Columbia MD 21046 Office: Lloyd's Register Quality Assurance Ltd 33-41 Newark St Hoboken NJ 07030

COGGESHALL, NORMAN DAVID, former oil company executive, investment executive; b. Ridge Farm, Ill., May 15, 1916; s. Lester B. and Grace (Blaisdell) C.; m. Margaret Josephine Danner, Aug. 22, 1940; children: Nancy Ellen Von der Ohe, David M., M. Gwen Calabretta, Phillip A. BA, U. Ill., 1937, MS, 1938, PhD, 1942. Tchr. physics U. Ill., 1942-43; scientist Gulf Oil Rsch., Pitts., 1943-50, asst. dir. physics div., 1950-55, dir. analytical sci. div., 1955-61, dir. phys. scis. div., 1961-67, v.p. process scis., 1967-70, v.p. exploration and prodn., 1970-76, v.p. tech. govt. coodination, 1976-81; pvt. investor and pvt. cons., Lynn Haven, Fla., 1981—. Contbg. author: Colloid Chemistry, 1946, Physical Chemistry of Hydrocarbons, 1950, Organic Analysis, 1953, Advances in Mass Spectrometry, 1963; contbr. articles to tech. jours.; patentee in field. Recipient Resolution of Appreciation, Am. Petroleum Inst., 1970. Fellow Am. Phys. Soc.; mem. Am. Chem. Soc. (award in chem. instrumentation 1970), Spectroscopy Soc. Pitts., Mass. Spectrometry Soc., Bay County C. of C. (mil. affairs com.), Rotary. Republican. Home and Office: 701 Driftwood Dr Lynn Haven FL 32444-3434

COGLEY, ALLEN C., mechanical engineering educator, administrator. BS in Aero. Engring., Iowa State U., 1962; MS in Aero. Engring., U. Va., 1964; PhD in Aero. and Astron. Engring., Stanford U., 1968. Rsch. engr. Naval Ordinance Lab, White Oak, Md., 1961, NASA-Langley Rsch. Ctr., Hampton, Va., 1962-63; ASEE-NASA postdoctoral fellow NASA-Ames Rsch. Ctr., Moffett Field, Calif., 1969, 70, vis. scientist, 1974-75; asst. prof. energy engring. U. Ill., Chgo., 1968-71, assoc. prof., 1971-79, prof., 1979-83; prof., chmn. mech. engring. U. Ala., Huntsville, 1983-87; prof., head mech. engring. Kans. State U., 1987—. Contbr. numerous articles to sci. jours. Grantee NSF, 1969-71, 75-77, 78-80, Water Resource Ctr., Dept. Interior, 1972-74, NASA-Ames Rsch. Ctr., 1974-75, 76-78, 78-80, Air Force Contract, 1979-83, IBM, 1985, 89, U.S. Army, 1985, 86 (twice), Univ. Advanced Design Program, 1989, NASA, 1991, NSF/IBM, 1991. Mem. AIAA (com. atmosphereic environ. 1978-80, com. thermophysics 1981-83, bd. dirs. S.E. sect. 1986-87, faculty advisor), AAUP (pres. local chpt. 1978-80), ASME, Kans. U. Senate, Sigma Gamma Tau, Sigma Xi, Tau Beta Pi, ASEE, AMS. Achievements include research in radiative transfer, fluid mechanics, atmosphereic sciences, and heat transfer. Home: 3118 Harahey Ridge Manhattan KS 66502 Office: Kans State U Coll Engring Dept Mech Engring Durland Hall Manhattan KS 66506

COGUT, THEODORE LOUIS, environmental specialist, meteorologist; b. Royal Oak Twp., Mich., Jan. 3, 1928; s. Louis and Mary Agnes (Evanish) C.; m. Martha Marie Nordstrom, Nov. 1, 1945; children: Leta Marie Cogut Mach, Willa Lynette Cogut Swartz, Pamela Anne Cogut Bryant. Grad. several meteorol. schs., USAF and U.S. Army; BA with honors, U. Md., 1965; MA in Teaching, Wayne State U., 1970. Weather forecaster USAF, 1948-53, 56-62; environ. analyst Climatology Ctr. USAF, Washington, 1962-64; instr. U.S. Army Arty. Meteorology Sch., Ft. Sill, Okla., 1965; grad. rsch. asst. in meteorology U. Okla., 1972-73; chief meteorologist Phelps Dodge Corp., Morenci, Ariz., 1974-79; environ. svcs. supr. Phelps Dodge Morenci, Inc., 1979-93; environ. cons. Tucson, Ariz., 1993—; mem. SKYWARN Spotter Network of Nat. Weather Svc., 1991-93; Citizen Amb. Environ. Del. to China, 1988; editor-in-chief Morenci Copper Rev., 1985—. Author: (programmed text) Ballistic Wind Plotting, 1968; author meteorol. newsletters, 1968-69; inventor computerized air quality and weather prediction system MCAPS, 1975. Pres. Greenlee County Hist. Soc., Clifton, Ariz., 1990. Chief warrant officer arty. U.S. Army, 1965-69, Vietnam. Decorated Bronze Star, Legion of Merit; Gallantry Cross (Vietnam). Mem. AIME (chmn. Morenci chpt. 1989), Am. Meteorol. Soc. (pres. So. Ariz. chpt. 1980), Nat. Weather Assn. (mem. indsl. meteorology com. 1980), Air and Waste Mgmt. Assn., Greenlee County C. of C. (pres. 1992-93), Rotary Internat. (sec. Clifton-Morenci chpt. 1979), Phi Kappa Phi (U. Md. chpt.). Avocations: writing, painting, gardening, designing brochures. Home and Office: 5810 Paseo San Valentine Tucson AZ 85715

COHEN, AARON, aerospace engineer; b. Tex., Jan. 31, 1931; s. Charles and Ida (Moloff) C.; m. Ruth Carolyn Goldberg, Feb. 7, 1953; children—Nancy Ann Santana, David Blair, Daniel Louis. BS, Tex. A&M U., 1952; MS in Applied Math., Stevens Inst. Tech, 1958, D Engring. (hon.), 1982. Microwave tube design engr. RCA, Camden, N.J., 1954-58; sr. research engr. Gen. Dynamics, San Diego, 1958-62; mgr. Apollo command and service module lunar module guidance nav. and control NASA, Houston, 1962-70, mgr. command and service module project, 1970-72, mgr. shuttle orbiter project, 1972-82, dir. research and engring., 1982-86, dir. Johnson Space

Ctr., 1986—. Editor Astronautics sect. Marks Mechanical Engineer's Handbook, 9th edit.; contbr. articles to profl. jours. Vice chmn. engring. task force Target 2000 Tex. A&M U., College Station, 1981-83. Served to lt. C.E., U.S. Army, 1952-54, Korea. Recipient Exceptional Service medal NASA, Houston, 1969, Disting. Service medal, 1973, 81, 88, Goddard Meml. trophy, 1988; Presdl. Rank of Meritorious Exec., U.S. Govt., Washington, 1981, Presdl. Rank of Disting. Exec., 1982, 88; Named NASA Engr. of Yr., Washington, 1982. Fellow Am. Astron. Soc. (W. Randolph Lovelace II award 1982), AIAA (Von Karman lectureship 1984), ASME (medal 1984), Tau Beta Pi. Jewish. Avocation: tennis. Home: 406 Shadow Creek Dr Seabrook TX 77586-6016 Office: NASA Lyndon B Johnson Space Ctr 2101 NASA Rd 1 Houston TX 77058*

COHEN, ALAN, civil engineer; b. Phila., Apr. 15, 1932; s. Max E. and Shirley R. (Chipin) C.; m. Patricia Smith, June 24, 1956 (div. 1986); children: Jeffrey S., Susan J. BCE, Cornell U., 1955; MSCE, MIT, 1956; Cert. of Adv. Engring. Study, Cornell U., 1988. Registered profl. engr., Pa. Md., N.J., Fla., Del., Mich.; accredited OSHA constrn. health and safety insp. Installation engr., 1st lt. USAF, Osaka, Japan, 1956-57; project engr. Moran, Proctor, Mueser & Rutledge, N.Y.C., 1957-60; v.p. Joseph S. Ward and Assoc., Phila., 1960-61; pres. Site engrs., Inc., Moorestown, N.J., 1961-71; v.p. Day & Zimmermann, Inc., Phila., 1971-79; cons. Alan Cohen, P.E., Phila., 1979-86; pres. Alan Cohen Assoc. Inc., Conshohocken, Pa., 1986—; mem. Montgomery County Bd. of View, Norristown, Pa., 1988—, State Bd. of Pvt. Licensed Schs., Harrisburg, Pa., 1986-88. Bd. dirs. Jr. Achievement, Phila., 1985; adv. bd. mem. Fedn. of Allied Jewish Appeal, Phila., 1989. Mem. ASCE, Am. Arbitration Assn., Am. Nat. Standards Inst., Nat. Soc. Profl. Engrs., Cornell Soc. Engrs. (nat. pres. 1980-81), Pa. Soc. Profl. Engrs. (v.p. PEPP 1990—), Consulting Engring. Coun. (state treas. 1968-69), Nat. Safety Coun. Home: Unit 16 The Court 681 Meetinghouse Rd Elkins Park PA 19117 Office: Alan Cohen Assocs Inc 1100 E Hector St Ste 330 Conshohocken PA 19428

COHEN, ALAN SEYMOUR, internist; b. Boston, Apr. 9, 1926; s. George I. and Jennie (Laskin) C.; m. Joan Elizabeth Prince, Sept. 12, 1954; children: Evan Bruce, Andrew Hollis, Robert Adam. A.B. magna cum laude, Harvard U., 1947; M.D. magna cum laude, Boston U., 1952. Intern Harvard Med. Service, Boston City Hosp., 1952-53, resident, 1953-55; exchange registrar in medicine Dundee Royal Infirmary and U. St. Andrews, Scotland, 1955-56; rsch. and clin. fellow in medicine (rheumatology) Mass. Gen. Hosp., Boston, 1956-58; instr. Harvard Med. Sch. and Mass. Gen. Hosp., 1958-60; head arthritis and connective tissue disease sect. Evans dept. clin. rsch. Mass. Univ. Hosp., Boston, 1960-72; Conrad Wesselhoeft prof. medicine Boston U. Sch. Medicine, 1972-93, disting. prof. of medicine in rheumatology, 1993—, prof. pharmacology, 1974—; dir. Arthritis Center, 1977—; dir. div. medicine Boston City Hosp., 1973-93; dir. Thorndike meml. lab., 1973-93. Editor: Laboratory Diagnostic Procedures in the Rheumatic Diseases, 1967, rev. edit., 1975, 3d edit., 1985, (with others) Symposium on Amyloidosis, 1968, (with R. Friedin and M. Samuels) Medical Emergencies: Diagnostic and Management Procedures from Boston City Hospital, 1977, (with J. Combes and H. Koh) 2d edit., 1983, Rheumatology and Immunology, 1979, (with J.C. Bennett) 2d edit., 1986, Progress in Clinical Rheumatology, 1984, (with D. Goldenberg) Drugs in the Rheumatic Diseases, 1986, Amyloidosis, 1986, Clinical Problems in Acute Care Medicine (J.J. Heffernan, R.A. Witzburg, A.S. Cohen), 1989; editor-in-chief Amyloid: An Internat. Jour. of Exptl. and Clin. Rsch., 1993—; contbr. over 600 articles to profl. jours. Trustee Arthritis Found., Atlanta, 1976-82, trustee Mass. chpt., 1966-85, vice chmn., 1971-84, pres., 1981-94; vice sec. for N.Am., mem. exec. com. Pan Am. League Against Rheumatism, 1982-85; chmn. Boston City Hosp. Physician Alumni Reunion Com., 1992; pres. Boston City Hosp. Fund for Excellence, 1992. Served to surg. USPHS, 1953-55. Recipient Outstanding Alumnus award Boston U. Sch. Medicine, 1975, Purdue Frederic Arthritis award, 1979, James H. Fairclough Jr. award for disting. svc. to Mass. chpt. Arthritis Found., 1981, Alumni award for spl. distinction Boston U., 1981, Jan Van Bremeen Gold medal Dutch Rheumatism Soc., 1990, Commrs. Disting. Physician award Boston City Hosp., 1991. Master Am. Coll. Rheumatology; fellow ACP; mem. Am. Rheumatism Assn. (pres. 1978-79), Am. Soc. Clin. Investigation, Assn. Am. Physicians, Am. Fedn. Clin. Research, Am. Soc. Exptl. Pathology, Interurban Clin. Club, Soc. Exptl. Biology and Medicine, Electron Microscopy Soc. Am., New Eng. Soc. for Electron Microscopy, Am. Soc. Cell Biology, N.Y. Acad. Sci., AMA, Mass. Med. Soc., New Eng. Rheumatism Assn. (past pres.), Italian Rheumatism Soc. (hon.), Spanish Rheumatism Soc. (hon.), Finnish Rheumatism Soc. (hon.), Brazilian Rheumatism Soc. (hon.), Irish Soc. Rheumatism and Rehab. (hon.), Boston U. Sch. Medicine Alumni Assn. (past pres.), Phi Beta Kappa, Alpha Omega Alpha. Jewish. Clubs: Harvard (Boston); Wightman Tennis Center (Weston, Mass.). Office: Arthritis Ctr Conte 5 Boston U Sch Med Boston MA 02118-2999

COHEN, ALBERTO, cardiologist; b. Rio de Janeiro, Aug. 12, 1932; came to U.S., 1959; s. Nessim and Anneta (Rabischoffsky) C.; m. Bertha Kalichztein, Dec. 27, 1958; children: Deborah Cohen Stein, Annabel, Miriam Cohen Disner. BS, Edn. Rui Barboza, Rio de Janero, 1952; MD, Brazil U., Rio de Janero, 1958. Diplomate Am. Bd. Internal Medicine, Am. Bd. Cardiovascular Disease; cert. bd. med. examiners, Ariz., Fla., Tex., Calif. Intern Middlesex Gen. Hosp., New Brunswick, N.J., 1959-60; resident Detroit Meml. Hosp., 1960-61; fellow cardiology medicine Harper Hosp., Detroit, 1961, Detroit Receiving Hosp., Wayne State U., 1963-65; practice medicine specializing in cardiology and internal medicine Mt. Clements, Mich., 1966—; attending staff St. Joseph's Mercy Hosp., Clemton Twp./Mt. Clemens, Mich., 1966; staff Detroit Gen. Hosp., 1966—; instr. medicine Wayne State U., Detroit, 1964-65, asst. prof. medicine, 1965-67, clin. asst. prof. medicine, 1967-87. Contbr. numerous articles to profl. jours. Pres. Macomb County Heart Unit, 1974-77. Fellow Am. Coll. Chest Physicians, Am. Coll. Angiology, Am. Coll. Internat. Physicians, Am. Coll. Cardiology, Am. Coll. Internal Angiology Internat. Coll. Angiology; mem. AMA, Brazilian Soc. Medicine, Brazilian Soc. Cardiology, Am. Soc. Internal Medicine, Mich. Soc. Internal Medicine (trustee 1974-77), Mich. State Med. Soc., Wayne County Med. Soc., Macomb County Med. Soc., Am. Heart Assn. (fellow scientific council clin. cardiology), Detroit Heart Club (pres. 1989-90). Jewish. Avocations: music, art, reading. Home: 1477 Lochridge Rd Bloomfield Hills MI 48302-0734

COHEN, ALLEN BARRY, health science administrator, biochemist; b. Ft. Wayne, Ind., July 14, 1939; s. Samuel and Dorothy (Weisse) C.; m. Geraldine Stein, June 12, 1960; children: Rachel, Deborah. BA, George Washington U., 1960, MD, 1963; PhD, U. Calif., San Francisco, 1972. Diplomate Am. Bd. Internal Medicine, Am. Bd. Pulmonary Diseases. Intern U. Rochester/Strong Meml. Hosp., Rochester, N.Y., 1963-64; resident Barnes Hosp., St. Louis, 1964-66; rsch. fellow Cardiovascular Inst., U. Calif., 1969-71, asst. prof.-in-residence, 1971-75, assoc. mem., 1973-75; asst. prof. medicine San Francisco Gen. Hosp., U. Calif., 1973-75; assoc. prof. medicine Temple U., Phila., 1975-83, prof. physiology and medicine, 1978-83; exec. assoc. dir. U. Tex. Health Ctr., Tyler, 1983—; prof. medicine and biochemistry, 1983—; prof. medicine with tenure U. Tex. Southwestern Med. Sch., Dallas, 1983—; mem. pulmonary disease adv. com. Nat. Heart, Lung and Blood Inst., Bethesda, Md., 1977-82. Assoc. editor Am. Jour. Physiology, Lung Cellular Molecular Physiology, 1989—; contbr. articles to Jour. Clin. Investigation, Am. Rev. Respiratory Disease, Am. Jour. Physiology. Surgeon USPHS, 1966-68. Rsch. grantee NIH, 1990. Mem. Am. Soc. Clin. Investigation, Am. Lung Assn. (rsch. policy com. 1988—), Am. Lung Assn. Tex. (bd. dirs. 1988-90), Tex. Thoracic Soc. (pres. 1988-90), Phi Beta Kappa, Alpha Omega Alpha. Jewish. Achievements include discovery that neutrophil elastase may be an important cause of the adult respiratory distress syndrome, interleuken 8 is the major neutrophil chemotaxin in the lungs of patients with ARDS, pulmonary edema in patients with severe heart failure may be related to platelet constituents. Office: U Tex Health Ctr PO Box 2003 Tyler TX 75710-2003

COHEN, DONALD JAY, pediatrics, psychiatry and psychology educator, administrator; b. Chgo., Sept. 5, 1940; m. Phyllis Cohen, 1964; children—Matthew, Rebecca, Rachel, Joseph. B.A. in Philosophy and Psychology summa cum laude, Brandeis U., 1961; Student in philosophy and psychology, U. Cambridge, 1961-62; M.D., Yale U., 1966. Diplomate Am. Bd. Psychiatry and Neurology, Am. Bd. Child Psychiatry. Intern in pediatric medicine Children's Hosp. Med. Ctr., Boston, 1966-67; resident in child psychiatry Judge Baker Guidance Ctr., Children's Hosp. Med. Ctr., Boston, 1969-70; resident in psychiatry Mass. Mental Health Ctr., Boston, 1967-69; fellow in child psychiatry Hillcrest Children's Ctr. and Children's Hosp., Washington, 1970-72; asst. in medicine Children's Hosp., Boston, 1967-69; asst. to dir. child devel. Dept. Health, Washington, 1970-72; assoc. prof. pediatrics, psychiatry, and psychology Yale U., New Haven, Conn., 1972-79; prof. pediatrics, psychiatry, psychology Yale U., New Haven, 1979—; Irving B. Harris prof. child psychiatry, pediatrics and psychology Yale U., New Haven, Conn., 1987—, dir. Child Study Ctr., 1983—; clin. assoc. adult psychiatry bd. NIMH Sect. on Twin and Sibling Studies, 1970-72; vis. prof. Hebrew U., Hadassah Med. Ctr., summer 1982; mem. Nat. Commn. on Children, 1988; tng. and supervising analyst Western New Eng. Psychoanalytic Inst., 1992—; bd. trustees Anna Freud Ctr., London, 1992. Contbr. numerous articles to profl. jours., chpts. to books; author monographs: Serving Sch. Age Children, 1972, Serving Presch. Children, 1974; editor monographs: Schizophrenia Bull., Vol. 8, No. 2, 1982, Jour. Autism and Devel. Disorders, 1982; co-editor monographs: (with A. Donnellan) Handbook of Autism and Disorders of Atypical Development, 1985; (with A.J. Solnit, J.E. Schowalter) Psychiatry, 1985; author of book revs.; mem. editorial bd. Jour. Am. Acad. of Child Psychiatry, 1972-76, 80—, Israel Jour. Psychiatry, 1983—; mem. adv. bd. Jour. Child Psychology and Psychiatry, 1977. Chmn. profl. adv. bd. Nat. Soc. for Autistic Children, 1981—; mem. med. adv. bd. Tourette Syndrome Assn., 1980—, mem. profl. adv. bd. Benhaven, New Haven, Conn., 1972—; bd. dirs. NIMH Treatment Devel. and Assessment Study Sect., 1979-82, Psychoanalytic Research and Devel. Fund, 1982—, Found.'s Fund for Research in Psychiatry, 1977-81, Spl. Citizens, Futures Unlimited, Inc., 1983—, Ounce of Prevention Fund Nat. Adv. Com., 1983—, B'nai B'rith Hillel Found., Yale U., 1984—; trustee Brandeis U., 1982—, Western New Eng. Inst. for Psychoanalysis, 1984—. Served with USPHS, 1970-72. Recipient Ann. Pub. Svc. award Nat. Soc. for Autistic Children, 1972, Spl. Recognition, Hofheimer prize Am. Psychiatric Assn., 1977, Ittleson award Am. Psychiat. Assn., 1981, Strecker award Inst. of Pa. Hosp., U. Pa., 1990; Woodrow Wilson fellow, 1961, Falk fellow Am. Psychiat. Assn., 1970-71; Fulbright scholar Trinity Coll., U. Cambridge, 1961-62. Fellow Am. Acad. Child Psychiatry (chmn. com. on rsch. 1975-81), Am. Pediatric Soc., Am. Acad. Pediatrics; mem. Inst. Medicine of NAS, Soc. for Rsch. in Child Devel., Internat. Assn. Child and Adolescent Psychiatry and Allied Professions (pres. 1992—), Internat. Psychoanalytic Soc., Am. Psychoanalytic Assn., Israel Psychoanalytic Soc. (corr.), Western New Eng. Psychoanalytic Soc., Phi Beta Kappa, Sigma Xi, Alpha Omega Alpha. Office: Yale Child Study Ctr PO Box 207900 New Haven CT 06510-7900

COHEN, EDWARD, private consultant; b. N.Y.C., Jan. 21, 1930; s. Joseph and Marilyn Cohen; m. Elizabeth Susan Birtwell, Dec. 31, 1952 (div. July 1979); m. G. Patricia Firth, Sept. 3, 1991; children from previous marriage: John, Ann Katherine, Susan, Alicia, Pamela. BSS, CCNY, 1951; MS, Columbia U., 1955. Writer USIA, Washington, 1955-57; press attaché Am. Embassy, Belgrade, Yugoslavia, 1957-61; ednl. attaché Am. Embassy, Caracas, Venezuela, 1961-63; v.p. Internat. Devel. Found., N.Y.C., 1963-67; dir. 2-yr. colls. N.J. Dept. Higher Edn., Trenton, 1967-70, asst. chancellor, 1970-85; exec. dir. N.J. Com. Sci. and Tech., Trenton, 1982-91; pvt. sci. and tech.-based econ. devel. cons. Lambertville, N.J., 1991—; mem. N.J. Bd. Med. Examiners, Newark, 1972-75, N.J. Bd. Dentistry, Trenton, 1977-82. Trustee, pres., chmn. bd. Newark Comprehensive Health Svcs. Plan, 1972-80; exec. dir. N.J. Gov.'s Commn. on Profl. Health Svcs., Trenton, 1977-79. Sgt. inf. U.S. Army, 1952-54, Korea. Recipient Meritorious Svc. award USIA, 1960, award of med. achievement Seton Hall U., 1979, cert. of recognition N.J. Health Svcs. Corp., 1980; fellow CCNY, 1951-52. Mem. Internat. Assn. Polit. Cons. (founding). Home: 45 York St Lambertville NJ 08530-2017 Office: Box 365 Lambertville NJ 08530

COHEN, EDWARD, civil engineer; b. Glastonbury, Conn., Jan. 6, 1921; s. Samuel and Ida (Tanewitz) C.; m. Elizabeth Belle Cohen, Dec. 19, 1948 (dec. June 1979); children: Samuel, Libby M. Wallace, James; m. Carol Simon Kalb, Jan. 11, 1981; stepchildren: Anne Kalb Bronner, Paul Kalb. BS in Engring., Columbia U., 1945, MS in Civil Engring., 1954. Registered profl. engr., N.Y., Conn., Fla., Ga., Md., N.J., La., Mass., Mich., Pa., D.C., Okla., Va., Wis., Del., Nat. Council Engring. Examiners; chartered civil engr., Gt. Britain; cert. Eur ING (FEANI Europe); lic. land surveyor, N.Y., Conn., Mass., N.J. Engring. aide Conn. Hwy. Dept., 1940-42; asst. engr. East Hartford Dept. Pub. Works, 1942-44; structural engr. Hardesty & Hanover, N.Y.C., 1945-47, Sanderson & Porter, N.Y.C., 1947-49; lectr. architecture Columbia U., 1948-51; with Ammann & Whitney, N.Y.C., 1949—; ptnr. Ammann & Whitney, 1963-74, sr. ptnr., 1974-77, mng. ptnr., 1977—, dir. co. work as engrs. of record restoration of Statue of Liberty, West Face and Olmsted Ters. of U.S. Capitol Bldg. and Roebling Del. Canal Bridge; exec. v.p. Ammann & Whitney, Inc., 1974-77, in charge bldg., transp., communications, mil. and hist. preservation projects, chmn., chief exec. officer, 1977—; v.p. Ammann & Whitney Internat. Ltd., 1963-73; pres. Safeguard Constrn. Mgmt. Corp., 1973-77, chmn., chief exec. officer, 1977—; cons. RAND Corp., Santa Monica, Calif., 1958-72, Dept. Def., 1962-63, Hudson Inst., Croton-on-Hudson, N.Y., 1967-71, World Bank, 1984, TVA, 1987, Nat. Trust for Hist. Preservation, Drayton Hall Restoration, 1990; Stanton Walker lectr. U. Md., 1973, Henry M. Shaw lectr. N. Carolina State U., 1987; deptl. adv. com. Urban and Civil Engring. U. Pa., 1978-84, Rutgers U., 1984—; mem. engring. coun. Columbia U., 1975—, vice chmn., 1985-86; chmn. Building Rsch. Bd. Com. on Fed. Constrn. Stds. to control building life-cycle costs, 1989-91; mem. Planning Group Nat. Consortium for infrastructure, rsch. and tech. transfer, 1987—, Nat. Rsch. Coun. Com. for infrastructure and rsch. agenda, 1992—; commr. Bklyn. Bridge Centennial Commn., 1981-83; spl. adv. centennial commn. N.Y. State Statue of Liberty, 1985; chmn. engring. com. NEA first U.S. Presdl. Awards, 1985, mem., 1991. Mem. adv. bd. Jour. Resource Mgmt. and Tech., 1981-91; co-editor: Handbook of Structural Concrete, 1983; contbr. more than 100 articles to profl. jours. and govt. manuals on structural, siesmic, hardened design, wind forces, dynamic analysis, ultimate strength and plastic design, guyed towers and shell structures. Bd. dirs. Cejwin Youth Camps, 1972—; mem. com. of 100 Trailblazer Summer Camp for Underpriviledged Children, 1985-89; trustee Hall of Sci., N.Y.C., 1976—; mem. exec. com. March of Dimes Transp. Award Luncheon, 1983—; mem. exec. com. Architects/Engrs. div. United Jewish Appeal-Fedn., 1985— (Ann. honoree 1991); mentor in engring. N.Y. Alliance for Pub. Schs., 1986-91; N.Y. area chmn. engring. sect. Orgn. for Rehab. Through Tng., 1983—, nat. dir., 1989— (Sci. and Tech. award 1987). Recipient Illig medal in Applied Sci. Columbia U., 1946, Patriotic Civilian Svc. award Dept. of Army, 1973, Egleston medal Columbia U., 1981, Goethals medal for Engring. Achievement Soc. Am. Mil. Engrs., 1985, Mayor's Award of Honor for Sci. and Tech., N.Y., 1988, U.S. Presdl. Design Excellence award for Roebling Del. Aqueduct Bridge Restoration, NEA, 1988, Prize Bridge award Am. Inst. of Steel Contrn. for Engring. Trinity Ch. Pedestrian Bridge, 1989; Best of Program award for Achievement in Arc Welded Design Engring. and Fabrication for Trinity Ch. Pedestrian Bridge, Bronze award Roebling (Del.) Aqueduct Bridge James F. Lincoln Arc Welding Found., 1988, Nat. Historic Preservation award for engring. U.S. Capitol restoration U.S. Dept. Interior and Adv. Coun. Historic Preservation, 1988. Fellow Am. Cons. Engr. Council (Grand award for Engring. Excellence 1986), Inst. Civil Engrs. (Gt. Britain); mem. NAE (elected, 1975), ASCE (hon com. design loads for bldgs. and other structures A58, 1968-88, chmn. reinforced concrete rsch. coun. 1980-89, Civil Engring. State-of-the-Art award 1974, Raymond Reese award 1976, Ernest Howard Gold Medal 1983, Svc. to People award 1987, met. sect. v.p., 1978-79, pres., 1980, Ridgeway award, 1946, Met. Civil Engr. of Yr. award 1986), N.Y. Assn. Cons. Engrs. (bldg. code adv. com., bd. dirs. 1981-82, 85-89), N.Y. Acad. Scis. (hon. life, bd. govs., v.p. 1991—, Laskowitz Aerospace rsch. gold medal 1970, chmn. engring. sect. 1977-79, N.Y. Acad. Scis. award 1989, mem. bd. govs. 1991—, v.p. 1991—, Charles Darwin Assocs. inaugural mem., 1992), Am. Concrete Inst. (hon. mem., dir. 1966-76, v.p. 1970-72, pres. 1972-73, chmn. com. bldg. code requirements for reinforced concrete 1963-71, Wason medal 1956, Delmar Bloem award 1973), N.Y. Concrete Industry Bd. (dir. 1976—, pres. 1978-79), Columbia U. Sch. Engring. Alumni Assn. (bd. dirs. 1985-86), N.Y. Concrete Constrn. Design Inst. (pres. tall bldgs. coun. 1975-80), NSPE (Outstanding Engring. Achievement award 1987), N.Y. State Soc. Profl. Engrs. (Engr. of Yr. 1987, Nassau chpt. Engr. of Yr. 1987), Soc. Am. Mil. Engrs., Internat. Bridge and Turnpike Assn., Internat. Assn. Bridge and Structural Engrs., Am. Welding Soc., Comite European de Beton (specialist mem.), Moles, Century Assn., Sigma Xi, Chi Epsilon, Tau Beta Pi. Clubs: Engrs. N.Y.C. (dir. 1974-75), Wings, Club at World Trade Ctr. Lodge: B'nai Brith. Avocations: tennis, golf. Home: 56 Chestnut Hl Roslyn NY 11576-2824 Office: Ammann & Whitney 96 Morton St New York NY 10014-3326

COHEN, HARLEY, civil engineer, science educator; b. Winnipeg, Man., Can., May 12, 1933; s. Joseph and Ettie (Gilman) C.; m. Estelle Brodsky, Dec. 25, 1956; children: Brent, Murray, Carla. B.Sc. hons., U. Man., 1956; Sc.M., Brown U., 1958; Ph.D., U. Minn., 1964. Registered profl. engr., Man. Research engr. Boeing Co., Seattle, 1958-60; sr. research scientist Honeywell, Inc., Mpls., 1960-64; asst. prof. aero. and engring. mechanics U. Minn., Mpls., 1965-66; assoc. prof. civil engring. U. Man., Winnipeg, 1966—, prof., 1968-89, disting. prof., 1983—, head dept., 1984-89, prof. applied math., 1989—, dean faculty of sci., 1989—; J.L. Record prof. U. Minn.; invited vis. prof. U. Pisa, Italian Rsch. Coun., 1987; bd. dirs. Man. Rsch. Coun., Tri-Univ.-Meson Facility, U. B.C., Premier's Econ. Innovation and Tech. Coun. Co-author: Theory of Psuedo-Rigid Bodies, 1988; contbr. over 90 articles to profl. jours. Killam scholar, 1982; Brit. sci. fellow, 1985. Fellow Am. Acad. Mechanics (bd. dirs. 1988-91); mem. Soc. Natural Philosophy, Soc. Engring. Sci. Home: 55 Tanoak Park Dr, Winnipeg, MB Canada R2V 2W6 Office: U Man Deans Office, Faculty of Sci, Winnipeg, MB Canada R3T 2N2

COHEN, HARVEY JAY, physician, educator; b. Bklyn., Oct. 21, 1940; s. Joseph and Anne (Margolin) C.; m. Sandra Helen Levine, June 1964; children: Ian Mitchell, Pamela Robin. BS, Bklyn. Coll., 1961; MD, Downstate Med. Coll., Bklyn., 1965. Diplomate Am. Bd. Internal Medicine, Am. Bd. Hematology. Intern, then resident internal medicine Duke U. Med. Ctr., Durham, N.C., 1965-67, fellow hematology and oncology, 1969-71; chief hematology-oncology VA Med. Ctr., Durham, N.C., 1975-76, chief med. service, 1976-82, assoc. chief of staff-edn., 1982-84, now dir. geriatric research, edn. and clin. ctr.; assoc. prof. medicine Duke U. Med. Ctr., Durham, 1976-80, now prof. medicine, chief geriatric div., also dir. Ctr. for Study of Aging. Author: Medical Immunology, 1977; editor: Cancer I and II, 1987; editor Jour. Gerontology: Med. Scis., 1988-92; contbr. numerous articles to profl. jours. Served as surgeon USPHS, 1967-69. Fellow ACP, Am. Geriatrics Soc. (bd. dirs. 1987—, sec. 1991—, pres.-elect 1993—), Gerontology Soc. Am. (clin. sect., rsch. com. 1987—, publs. com. 1992—); mem. Am. Soc. Clin. Oncology, Am. Soc. Hematology, Am. Assn. Cancer Rsch. Home: 2811 Friendship Cir Durham NC 27705-5521 Office: Duke U Med Ctr for Study Aging & Human Devel PO Box 3003 Durham NC 27710-3003

COHEN, IRA MYRON, aeronautical and mechanical engineering educator; b. Chgo., July 18, 1937; s. Harry Nathan and Esther (Lenchner) C.; m. Linda Barbara Einstein, June 12, 1960; children: Susan Ellen Bolstad, Nancy Beth. B. in Aero. Engring., Poly. Univ., Bklyn., 1958; M.A., Princeton U., 1961, Ph.D. in Aero. Engring, 1963; M.A. (hon.), U. Pa., 1971. Mem. tech. staff Sandia Labs., Albuquerque, summers 1971, 74, 77; asst. prof. engring. Brown U., Providence, 1963-66; asst. prof. mech. engring. U. Pa., Phila., 1966-67, assoc. prof., 1967-76, prof., 1976-92, chmn. dept., 1992—; guest prof. Technische Hochschule Aachen, W. Ger., 1966; cons. fluid mechanics related problems to industry, 1966—, attys., 1966—; Mem. bd. The Sch. in Rose Valley, Moylan, Pa., 1969-74. Contbr. articles to various publs. Recipient Fulbright Travel grant, 1966. Assoc. Fellow AIAA (sect. sec. 1977-80, 85—); mem. AAUP, ASME, Am. Phys. Soc., Internat. Soc. for Hybrid Microelectronics, Sigma Xi. Office: U Pa Dept Mech Engring and Applied Mechanics 297 Towne Bldg Philadelphia PA 19104-6315

COHEN, IRWIN, economist; b. Bronx, N.Y., Feb. 29, 1936; s. Samuel and Gertrude (Levy) C.; B.S. in Accounting, N.Y. U., 1956, M.B.A. in Finance, 1964, M.A. in Econs., 1969; B.S. in Math., CCNY, 1970. Financial analyst U.S. SEC, N.Y.C., 1965-67, Fed. Res. Bank N.Y., N.Y.C., 1967-72, Prudential Ins. Co. Am., 1973-74, SEC, N.Y.C., 1974—. Life Fellow Internat. Biog. Assn., Am. Biog. Inst. Research Assn. (dep. gov.), World Acad. Scholars, World Literary Acad., World Inst. Achievement; mem. Internat. Biographical Ctr. (dep. dir. gen.), Internat. Platform Assn (life), Math. Assn. Am., Am. Finance Assn., Econ. History Assn. Home: 372 Central Ave Apt 2K Scarsdale NY 10583-1303

COHEN, ISAAC LOUIS (IKE COHEN), data processing executive; b. N.Y.C., Sept. 15, 1948; s. Louis and Dora (Dostis) C.; m. Lucille Competello, June 3, 1982 (div.); children: Janice, Matthew. AAS in Bus. Adminstrn., Kingsborough Community Coll., 1977. Asst. v.p. Mfrs. Hanover Trust Co., N.Y.C., 1966-87; data processing ops. mgr. First Boston Corp., Princeton, N.J., 1987—; owner ILC Liquidators Co., South Bound Brook, N.J., 1990—, ILC Vending Co., South Bound Brook, N.J., 1991—, ILC 900 Co., South Bound Brook, N.J., 1991—, ILC Fin. Mgmt. Co., 1991—. Editor booklet: Annadale Memorial Day Parade, 1985, 86, 87. Asst. v.p. Annadale Community Assn., 1982-89; umpire S.I. High Sch. League & Semi-Pro Baseball, 1985—; assoc. scout Kansas City Royals Profl. Baseball Team, 1988—. With USNR, 1967-74. Mem. U.S. Submarine Vets., Am. Legion, Jewish War Vets., Vietnam Vets. Avocations: baseball, motorcycle riding, skiing, roller skating, stamp and coin collecting. Home: 5 Koehler Dr South Bound Brook NJ 08880-1304 Office: First Boston Corp 700 College Rd E Princeton NJ 08540-6689

COHEN, JEROME BERNARD, materials science educator; b. Bklyn., July 16, 1932; s. David I. and Shirley Anne C.; m. Lois Nesson, Sept. 15, 1957; children: Elissa Diane, Andrew Neil. BS, MIT, 1954, ScD, 1957; DSc (hon.), Linköping U., Sweden, 1991. Sr. scientist materials AVCO Corp., Wilmington, Mass., 1958-59; mem. faculty Northwestern U., 1959—, prof. materials sci. and engring., 1965—, chmn. dept. materials sci. and engring., 1973-78, Frank C. Engelhart prof., 1974—, fellow Center Teaching Professions, 1971—, prof. McCormick Sch. Engring., 1983—, dean McCormick Sch. Engring., 1986—; sci. liaison officer Office Naval Research, London, 1966-67; cons. to govt. and industry. Author: Diffraction Methods in Materials Science, 1966; co-author: Diffraction from Materials, 1978, 2d edit., 1987; co-author Residual Stress: Measurement by Diffraction and Interpretation, 1987; co-editor: Local Atomic Arrangements Studied by X-Ray Diffraction, 1967, Jour. Applied Crystallography, Modulated Structures, 1979; patentee in field. All-Star coach Glencoe (Ill.) Hockey Assn., 1974-77. Served as 1st lt. AUS, 1959. Fulbright fellow U. Paris, 1957-58; recipient Tech. Inst. Teaching award Northwestern U., 1976, C.S. Barrett Diffraction award, 1989, Gold medal Acta Metallurgica, 1992. Fellow AIME (Hardy gold medal 1960, R.F. Mehl Mehl and Inst. of Metals lectr. 1992), Am. Soc. Metals (Henry Marion Howe medal 1981); mem. AAUP, NAE, Am. Soc. for Engring. Edn. (George C. Westinghouse award 1976), Am. Ceramic Soc., Am. Crystallographic Assn., Metall. Soc. (Franklin H. Mehl award 1992), Royal Instn. Gt. Britain, Econ. Club Chgo., Sigma Xi (nat. lectr. 1989-90), Tau Beta Pi, Alpha Sigma Mu, Phi Lambda Upsilon. Jewish. Home: 574 Woodlawn Ave Glencoe IL 60022-2040 Office: Nothwestern U McCormick Sch Engring 2145 Sheridan Rd Evanston IL 60208-3100

COHEN, JOHATHAN, orthopedic surgery educator, researcher; b. N.Y.C., Feb. 26, 1915; s. Paul and Clara (Lobel) C.; m. Louise A. Horwood, July 14, 1972. BS, NYU, 1933; MD, St. Louis U., 1938. Diplomate Am. Bd. Orthopaedic Surgery. Intern Jewish Hosp., St. Louis, 1938-39; resident Cen. Ga. Rwy. Hosp., 1939-40; resident Children's Hosp., Montreal, Can., 1940-41, Boston, 1945-48; resident Children's and Mass. Gen. Hosp., 1948-50; instr. Harvard Med. Sch., Boston, 1950-60, prof., 1960-70; prof. orthopaedic surgery Tufts U. Med. Sch., Boston, 1970—. Author: Pathophysiology of Bone Encyclopedia, 1990; dep. editor Jour. Bone and Joint Surgery, 1950—; also over 100 articles. Maj. M.C., U.S. Army, 1941-46, PTO. Milton fellow Harvard U., 1946-50. Fellow Am. Orthopaedic Assn., Am. Acad. Orthopaedic Surgery. Achievements include patent for pressure bandage; researched interactions between metals and tissues, characteristics of variety of pathological lesions in bone including those caused by growth disturbances, bone seeking radio-isotopes, and tumors. Home: 24 Grozier Rd Cambridge MA 02138 Office: Franciscan Children's Hosp 30 Warren St Brighton MA 02135

COHEN, KENNETH ALLAN, chemist; b. Bklyn., Mar. 30, 1953; s. Philip and Ruth C.; m. Dory A. Cohen, Oct. 26, 1980; 1 child, Bryan M. BS, L.I. U., 1976. Chemist Coty div. Pfizer, N.Y.C., 1976, A.R. Winarick, Carlstadt, N.J., 1976-78, Cosmair Inc., Clark, N.J., 1978-80, CCM, Islip, N.Y., 1980-83, Vanity Cosmetic Labs., 1983-86, Revlon Rsch. Ctr., Edison, N.J., 1986-91; with Proctor & Gamble, Hunt Valley, Md., 1991—. Mem. Soc. Cosmetic Chemists. Jewish. Achievements include patent on gel mascara, on solid powder preparation; patent pending on liquid powder foundation. Office: Proctor & Gamble 11050 York Rd Cockeysville Hunt Valley MD 21030-2098

COHEN, LOIS RUTH KUSHNER, government research institute official; b. Phila., May 31, 1938; d. Joseph George and Doris (Bronstein) Kushner. Tchr.'s diploma, Gratz Coll., Phila., 1957; BA, U. Pa., 1960; MS, Purdue U., 1961, PhD, 1963, LittD (hon.), 1989. Rsch. coord. dept. sociology Purdue U., 1963-64; social sci. analyst div. dental health USPHS, Washington, 1964-70; chief applied behavioral studies div. dental health USPHS, NIH, Bethesda, Md., 1970-71; chief Office Social and Behavioral Analysis, 1971-74; spl. asst. to dir. Div. Dentistry, 1974-76, Nat. Inst. Dental Rsch., 1976—; vis. lectr. Howard U., spring 1964, health policy and social medicine Harvard U., 1981-88; Percy T. Phillips vis. prof. Columbia U. Sch. Dental and Oral Surgery, N.Y.C., 1988; cons. WHO, 1970, 74, 75, dental health unit WHO, 1970—, Inst. Medicine Nat. Acad. Sci., 1977-80; co-dir. Internat. Collaborative Study Dental Manpower Systems in Relation to Oral Health Status, 1970-84. Co-editor: Toward a Sociology of Dentistry, 1971, Social Sciences and Dentistry, Vol. I, 1971, Vol. II, 1984; editorial reviewer Social Sci. and Medicine, 1975—, Jour. Preventive Dentistry, 1973—, Scandinavian Jour. Dental Rsch., 1973—; contbr. numerous articles to profl. jours., books. Recipient Phila. High Sch. for Girls Rowen stipend, 1956, Superior Svc. award Pub. Health Svc., 1988; Senatorial scholar U. Pa., 1960; David Ross fellow NSF, 1963. Fellow AAAS (gov. coun. 1971), Am. Coll. Dentists (hon.), Internat. Coll. Dentists (hon.); mem. Am. Pub. Health Assn, Am. Dental Assn. (hon.), Am. Sociol. Assn., Behavioral Scientists in Dental Rsch. (founder, pres. 1971-72), Federation Dentaire Internationale (cons.), Internat. Assn. Dental Rsch. (dir. 1976-77, chmn. internat. rels. com. 1979-83, disting. sr. scientist award 1987), Am. Assn. Dental Rsch. (dir. 1980-81). Office: DHHS USPHS NIH NIDR Westwood Bldg 5333 Wesbard Ave Bethesda MD 20892

COHEN, MALCOLM MARTIN, psychologist, researcher; b. New Brunswick, N.J., May 13, 1937; s. Nathan and Esther (Greenhaus) C.; m. Marilyn Jerrow, Jan. 2, 1959 (dec. 1967); m. Eleanor Johnson, June 30, 1969 (div. 1988); m. Suzana Gal, Feb. 14, 1988. BA, Brandeis U., 1959; MA, U. Pa., 1961, PhD, 1965. Lic. psychologist, Pa. Asst. instr. U. Pa., Phila., 1961-63; rsch. psychologist Naval Air Engring. Ctr., Phila., 1963-67; supervisory rsch. psychologist Naval Air Devel. Ctr., Warminster, Pa., 1967-82; asst. chief biomed. rsch. div. NASA-Ames Rsch Ctr., Moffett Field, Calif., 1982-85, chief neuroscis. br., 1985-88, rsch. scientist, 1988—; lectr. dept. aeros. and astronautics Stanford (Calif.) U., 1982-92; mem. aerospace med. adv. panel Am. Inst. Biol. Scis., Washington, 1984-92; v.p. Nat. Hand Rehab. Fund, Washington, 1975—. Contbr. numerous articles to profl. jours. Patentee light bar to monitor human acceleration tolerance. Founding mem. Common Cause of Phila., 1973. Fellow Aerospace Med. Assn. (editorial bd. aviation, space and environ. medicine 1985—, Environ. Sci. award 1985, William F. Longacre award 1989), Aerospace Human Factors Assn. (pres. 1992); mem. AAAS, AIAA, N.Y. Acad. Sci., Psychonomics Soc., Sigma Xi. Jewish. Avocation: scuba diving. Office: NASA Ames Rsch Ctr Mail Stop 239-11 Moffett Field CA 94035

COHEN, MARTIN GILBERT, physicist; b. Bklyn., Jan. 13, 1938; s. Benjamin and Betty (Frishman) C.; m. Marcia Dimond, Mar. 26, 1961 (div. 1977); children: Bruce, Benjamin, Lisa; m. Rita Norma Rover, June 1, 1986. AB, Columbia Coll., 1957; MA, Harvard U., 1958, PhD, 1964. Mem. tech. staff Bell Telephone Labs., Murray Hill, N.J., 1964-69; dept. head Quantronix Corp., Smithtown, N.Y., 1969-75, dir. rsch., 1975-84, v.p. rsch., 1984-90, v.p. tech., 1990—. Contbr. articles to profl. jours. Mem. adv. bd. Laser & Fiber Optics Tech. Queensborough Community Coll., Queens, N.Y., 1986—. Mem. Laser and Electro-optics Soc. of IEEE (chmn. engrs. com. 1990-91), Am. Phys. Soc., Optical Soc. Am. (conf. on lasers and electro-optics steering com. 1982-86). Achievements include 6 patents in laser technology. Office: Quantronix Corp 45 Adams Ave Hauppauge NY 11788-3914

COHEN, MELVIN R., physician, educator; b. Chgo., May 24, 1911; s. Louis M. and Anna S. (Friedman) C.; m. Miriam, May 19, 1946; children—Nancy, Alan. B.S., U. Ill., 1931, M.S. in Pathology, 1933, MD, 1934. Diplomate: Am. Bd. Ob-Gyn. Practice medicine specializing in infertility Chgo.; sr. attending physician Michael Reese Med. Ctr., Chgo., Northwestern Meml. Hosp., Chgo.; founder, dir. Fertility Inst. Ltd., Chgo.; prof. Northwestern U. Med. Sch., Chgo., prof. emeritus; guest vis. prof. first Martin Clyman postgrad. course in infertility Mount Sinai Hosp., N.Y.C., 1982. Author: Laparoscopy, Culdoscopy and Gynecography: Technique and Atlas, 1970; contbr. numerous chpts. in med. books and articles to med. jours. on infertility, endometriosis and Spinnbarkeit. Dir., producer: 8 teaching films on infertility; video films during surgery; ektochrome slides established world-wide technique. Pioneer use of Pergonal for stimulating ovulation. Served with MC, AUS, 1942-45. Co-recipient Gold Medal for Infertility exhibit, AMA, 1951; recipient award for film on endometriosis 10th World Congress of Fertility and Sterility, Madrid, Spain, 1980. Fellow Chgo. Gynecol. Soc. (life); mem. AMA, Am. Fertility Soc., Am. Coll. Ob-Gyn., Am. Assn. Gynecol. Laparoscopists, Internat. Fertility Assn., Internat. Family Planning Research Assn., Ill. State Med. Soc., Chgo. Gynecol. Soc., Kansas City Gynecol. Soc. (hon.), Los Angeles Gynecol. Soc., Inst. Medicine Chgo., Midwest Bio-Laser Inst., Indian Assn. Gynecol. Endoscopists (hon.), Soc. Reproductive Surgeons, Chgo. Assn. Reproductive Endocrinologists (pres. 1984-85), Sigma Xi, Alpha Omega Alpha. Named Father of Modern Am. Laparoscopy, 1974. Address: 990 N Lake Shore Dr Chicago IL 60611

COHEN, MICHAEL LEE, statistics educator; b. Chgo., June 20, 1953; s. Sol William and Beverly Mary (Thierhouse) C.; m. Janice Kay Teece, Dec. 28, 1980 (div. Nov. 1986); m. Sharman Elaine Brown, Oct. 5, 1990. BS in Math., U. Mich., 1975; MS in Statistics, Stanford U., 1977, PhD in Statistics, 1981. Statistician Energy Info. Adminstrn. Dept. Energy, Washington, 1980-81; vis. lectr. Dept. Stats. Princeton U., 1981-82; consulting statistician Applied Math. Scis., Silver Spring, Md., 1982-83; rsch. assoc. Com. on Nat. Stats. Nat. Rsch. Coun., Washington, 1983-85; asst. prof. Sch. Pub. Affairs U. Md., College Park, 1985—; cons. Rand Corp., Santa Monica, Calif., 1985; vis. scholar Brookings Inst., Washington, 1990; cons. Dept. Energy, 1992—, Com. on Nat. Stats., 1985—. Co-editor: The Bicentennial Census: New Directions for M..., 1985; contbr. articles to profl. jours. Recipient NSF/Am. Statis. Assn. census rsch. fellowship, 1988. Mem. Am. Statis. Assn., Inst. Math. Stats., Wash. Statis. Assn. (pres. 1992-93). Jewish. Home: 9005 Walden Rd Silver Spring MD 20901 Office: U Md Sch Pub Affairs Morrill Hall College Park MD 20742

COHEN, MONTAGUE, medical physics educator; b. London, July 24, 1925; arrived in Can., 1975; s. Nathan Cohen and Annie Chissik; m. Dora Weinles, Sept. 7, 1947; children: Laurence John, Robert Anthony, Andrew Michael. BS, Imperial Coll. of Sci., London, 1946; PhD, London Hosp. Med. Coll., 1958. Physicist Royal Aircraft Establishment, Farnborough, Hampshire, Eng., 1946-47, GE, Wembley, Eng., 1947-48; sr. profl. officer Internat. Atomic Energy Agy., Vienna, Austria, 1961-66; asst. physicist The London Hosp., 1948-61, chief physicist, 1966-75; chief physicist Mont. Gen. Hosp., P.Q., Can., 1975-79; prof., dir. med. physics med. physics unit McGill U., Mont., 1979-91, prof., curator Rutherford Mus. physics dept., 1984—; cons., mem. Internat. Commn. Radiation, Washington, 1961-90; cons. radiation physics WHO, Geneva, 1975-84; mem. task forces Units and Measurements; chmn. physics adv. com. Healing Arts Radiation Protection (HARP) Commn., Ont. Govt., Toronto, 1983-89; vis. prof., India, 1976. Author, compiler: Atlas of Radiation Dose Distributions, 1966, Cobalt-60 Teletherapy: A Compendium of International Practice, 1984; author, compiler, editor: Central Axis Depth Dose Data for Use in Radiotherapy, 1972; contbr. articles to Fontanus. Recipient Landauer award Met. Assn. of Radiotherapists, 1979. Fellow Can. Coll. of Physicists in Medicine, Inst. Physics (U.K.); mem. Am. Inst. Physics, Am. Assn. Physicists in Medicine, Brit. Inst. Radiology (sec. 1974-75, Stanley Melville Meml. award 1960, Roentgen prize 1973). Reform Judaism. Achievements include first application of wedge filters in radiotherapy, depth-dose and isodose data for use in radiotherapy. Office: McGill U, 3600 University St, Montreal, PQ Canada H3A 2T8

COHEN, NOEL LEE, otolaryngologist, educator; b. N.Y.C., Sept. 20, 1930; s. Victor Max and Esther Lily (Schonfeld) C.; m. Baukje Philippina Boersma, June 1, 1957; 1 child, Mark Bennett. AB, NYU, 1951; MD, U. Utrecht, The Netherlands, 1957. Intern Stads-en Aacademish Ziekenhuis, Utrecht, 1955-57; resident in otolargyngology Bellevue Med. Ctr., NYU, 1959-62; instr. NYU Sch. Medicine, 1962-64, asst. prof., 1964-69, assoc. prof., 1969-73, clin. prof., 1973-80, prof. otolargyngology, 1980—, chmn. dept., 1980—; bd. dirs. N.Y. League for Hard of Hearing, 1985—; mem. rsch. adv. bd. EAR Found., Nashville, 1987—. Mem. editorial bd. Am. Jour. Otology, 1986—; reviewer articles and books for profl. jours.; contbr. numerous articles to profl. jours., author chpts. in books. Lt. USNR, 1957-59. Fellow ACS, Am. Acad. Otolaryngology-Head-Neck Surgery (Honor award 1985), Am. Laryngol., Rhinol. and Otol. Soc., Am. Soc. Head and Neck Surgery, Am. Bronchoesophagol. Assn., Am. Otol. Soc., N.Y. Acad. Medicine, N.Y. State Soc. Otolaryngology-Head and Neck Surgery (pres. 1988-89), N.Y. Head and Neck Soc. (charter mem. Pres.' award 1984), Soc. Univ. Otolaryngologists, Soc. Acad. Depts. Otolaryngology, N.Y. Acad. Scis., Acoustic Neuroma Soc., Alexander Graham Bell Assn. (med. adv. bd.). Democrat. Jewish. Avocations: tennis, skiing, gardening, carpentry. Office: NYU Med Ctr 530 1st Ave New York NY 10016-6402

COHEN, PATRICIA ANN, biochemist; b. Buffalo, N.Y., June 20, 1944; d. Lawrence Lionel and Jean (Rosenbloom) C.; m. Douglas Dix, July 7, 1970; children: Sam, Joe, Ben, Aaron, Naomi, Joel. BA, Conn. Coll., 1966; MS, So. Conn. State U., 1975; PhD, SUNY, Buffalo, 1973. Postdoctoral fellow Yale U., New Haven, Conn., 1972-73; dir. Haynes Med. Lab., Manchester, Conn., 1976-80; adj. faculty U. Hartford, West Hartford, Conn., 1980-84; co-dir. U. Hartford, 1990—; asst. prof. St. Joseph Coll., West Hartford, 1984-90. Contbr. articles to European Jour. Cancer Prevention, Sci. Scope, Jour. Chem. Edn., European Jour. Cancer. Grantee NSF, 1990, Conn. Higher Edn., 1991, Dept. Energy, 1991, City of Hartford, 1992. Mem. AAAS, Nat. Assn. Sci. Tchrs. Home: 6 Cobblestone Rd Bloomfield CT 06002

COHEN, PHILIP, hydrogeologist; b. N.Y.C., Dec. 13, 1931; s. Isadore and Anna (Katz) C.; m. Barbara Sandler, Dec. 26, 1954; 1 son, Jeffery. B.S. cum laude, CCNY, 1954; M.S., U. Rochester, 1956. Cert. profl. geologist, Va. With U.S. Geol. Survey, 1956—, chief Long Island program, 1968-72; assoc. chief land info. and analysis office U.S. Geol. Survey, Reston, Va., 1975-78, asst. chief hydrologist water resources div., 1978-79, chief hydrologist water resources div., 1979—. Contbr. numerous articles on geology and hydrology to profl. jours. Recipient Ward medal Coll. City, N.Y., 1954; Meritorious Ser. award Dept. Interior, 1975, Disting. Ser. award, 1979, Presdl. Meritorious Exec. Rank award, 1986, Presdl. Disting. Exec. Rank award, 1988. Fellow Geol. Soc. Am.; mem. Am. Water Resources Assn., Am. Geophys. Union, Am. Inst. Hydrology (C.V. Theis award 1993), Sigma Xi. Office: US Geol Survey 409 Geol Survey Reston VA 22092

COHEN, RANDY WADE, neuroscientist; b. L.A., Sept. 1, 1954; s. Julian and Lauretta Claire (Levin) C.; m. Susan Marie Duke, Jan. 9, 1982; children: Rachel Erin, Sarah Ashley, Joshua Evan. BS, U. So. Calif., 1976; PhD, U. Ill., 1987. Instr. Calif. State U., Northridge, 1990—; rschr. U. Calif., L.A., 1987—. Contbr. articles to profl. jours. including Jour. of Insect Physiology and Molecular Brain Rsch. Mem. AAAS, Entomological Soc., Soc. for Neuroscience. Achievements include research on insect diet mixing. Office: UCLA Sch Medicine Mental Retardation Rsch Ctr 760 Westwood Plz Los Angeles CA 90024-8300

COHEN, RAYMOND, mechanical engineer, educator; b. St. Louis, Nov. 30, 1923; s. Benjamin and Leah (Lewis) C.; m. Katherine Elise Silverman, Feb. 1, 1948 (dec. May 1985); children—Richard Samuel, Deborah, Barbara Beth; m. Lila Lakin Cagen, Nov. 30, 1986. B.S., Purdue U., 1947, M.S., 1950, Ph.D., 1955. Instr. mech. engring. Purdue U., 1948-55, asst. prof., 1955-58, assoc. prof., 1958-60, prof., 1960—, asst. dir. Ray W. Herrick Labs., 1970-71, dir., 1971—, acting head Sch. Mech. Engring., 1988-89; cons. to industry. Departmental editor of: Ency. Brit, 1957-62; editorial bd. Jour. Sound and Vibration, 1971-87. Served as sgt. inf. AUS, 1943-46. NATO sr. fellow in sci., 1971. Fellow ASME, ASHRAE; mem. Am. Soc. Engring. Edn., Soc. Exptl. Mechanics, Nat. Soc. Profl. Engrs., Internat. Inst. Refrigeration (chmn. U.S. nat. com. 1992—), Acoustical Soc. Am., Inst. Noise Control Engring. (pres. 1990), Sigma Xi, Pi Tau Sigma, Tau Beta Pi. Home: 316 Leslie Ave West Lafayette IN 47906-2412 Office: Purdue U Ray W Herrick Labs Sch Mechanical Engring West Lafayette IN 47907-1077

COHEN, RAYMOND JAMES, radio astronomy educator; b. Brisbane, Queensland, Australia, June 25, 1948; arrived in Eng., 1970; s. Raymond Murdoch and Elsie Beatrice (Waldie) C.; m. Anne Merryl Treverton, Apr. 5, 1975; children: Helen Mary, Elizabeth Anne. BSc in Math. with 1st class honors, U. Queensland, Brisbane, 1970, MSc in Applied Math., 1970; PhD in Radio Astronomy, U. Manchester, Eng., 1974. Rsch. fellow in radio astronomy U. Manchester, 1974-82; lectr. radio astronomy Nuffield Radio Astronomy Labs., U. Manchester, 1982-89, sr. lectr., 1989—; mem. U.K. study group 7, del. meetings Internat. Radio Consultative Com., 1986—; mem. radio astronomy and space rsch. frequency com., radio astronomy rep. to radiocommunications agy. working groups U.K. Dept. Trade and Industry, 1987—; U.K. rep. com. on radio astronomy frequency European Sci. Found., 1988—; mem. U.K. panel Internat. Union Radio Sci. (formerly Brit. Nat. Com. Radio Sci.), 1988—; U.K. corr. Inter-Union Commn. Frequency Allocations for Radio Astronomy and Space Sci., 1988—. Contbr. numerous articles to sci. publs. Mem. Royal Astron. Soc., Internat. Astron. Union. Avocations: music, playing flute, reading. Office: Nuffield Radio Astron Labs, Jodrell Bank, Macclesfield Cheshire SK11 9DL, England

COHEN, RICHARD LAWRENCE, electrical engineer, mechanical engineer; b. N.Y.C., Aug. 25, 1937; s. Maxwell and Pearl Cohen; m. Paula Malamud, Nov. 17, 1962; children: David Robert, Steven Peter. BSME, CCNY, 1960; MSME, CUNY, 1963; MSEE, N.J. Inst. Tech., 1968. Supervising engr. Bendix Corp., Teterboro, N.J., 1960-89; dir. engring. Cox & Co., N.Y.C., 1989—. Named Divisional Patent Leader, Allied Signal Corp., 1987. Mem. AIAA, SAMPE. Achievements include 15 patents; 3 patents-pending. Office: Cox & Co 200 Varick St New York NY 10014

COHEN, ROBERT FADIAN, research engineer; b. N.Y.C., Mar. 21, 1969; s. Barry and Marianne (Clark) C. BS, U. Md., 1991. Rsch. engr. U. Md., College Park, 1991—. Mem. AIAA.

COHEN, ROBERT SONNÉ, physicist, philosopher, educator; b. N.Y.C., Feb. 18, 1923; m. Robin Gertrude Hirshhorn, June 18, 1944; children: Michael, Daniel, Deborah. BA, Wesleyan U., Middletown, Conn., 1943, LHD, 1986; MS, Yale U., 1943, PhD (NRC fellow), 1948. Instr. physics Yale U., 1943-44, instr. philosophy, 1949-51; sci. staff, war research div. Columbia U. and Communications Bd., U.S. Joint Chiefs Staff, 1944-46; asst. prof. physics and philosophy Wesleyan U., 1949-57; assoc. prof. physics Boston U., 1957-59, prof. physics and philosophy, 1959-93, chmn. dept. physics, 1959-73, chmn. dept. philosophy, 1986-88, prof. emeritus, 1993—; acting dean Coll. Liberal Arts, 1971-72; chmn. Boston U. Center for Philosophy and History Sci., 1970-93, chmn. emeritus, 1993—; vis. lectr. humanities and philosophy of sci. Mass. Inst. Tech., 1958-59, 61-62; vis. prof. history of ideas Brandeis U., 1959-60; lectr. history and philosophy of sci. Am. U., Washington, summers 1958-68; vis. fellow Polish and Yugoslav Acad. Sci., 1963, Hungarian Acad. Sci., 1964; vis. prof. philosophy U. Calif., San Diego, 1969, Yale U., 1973; rsch. fellow history of sci. Harvard U., 1974; mem., chmn. U.S. Nat. Com. for Internat. Union History and Philosophy of Sci., 1969-75; trustee Wesleyan U., 1968-84, emeritus, 1984—, Tufts U., 1984-93, emeritus, 1993—. Author, editor articles, books and jours. in field.; Editor: Boston Studies in Philosophy of Sci., Vienna Circle Collection, Sci. in Context. Trustee Bill of Rights Found. Am. Council Learned Soc. fellow philosophy and sci., 1948-49; Ford faculty fellow Cambridge, Eng., 1955-56; fellow Wissenschaftskolleg zu Berlin, 1983-84. Fellow AAAS (chmn. sect. L history and philosophy of sci. 1978-79); mem. AAUP, Am. Phys. Soc., Am. Assn. Physics Tchrs., Am. Philos. Assn. (exec. com. 1988-91), History Sci. Soc., Philosophy Sci. Assn. (v.p. 1972-75, pres. 1982-84), Nat. Emergency Civil Liberties Com. (mem. nat. coun.), Am. Inst. Marxist Studies (chmn. 1964-82), Fedn. Am. Scientists (nat. coun. 1967-70), Inst. for Unity of Sci. (exec. com. 1960-74). Home: 44 Adams Ave Watertown MA 02172-1391 Office: Boston U Dept Philosophy Boston MA 02215

COHEN, SAUL MARK, chemist, retired; b. Springfield, Mass., Oct. 6, 1924; s. Samuel and Celia (Drapkin) C.; m. Betty Clarice Fineberg, Mar. 8, 1953; 1 child, Bruce Alan. BS, U. Mass., 1948; MS, U. Ill., 1949, PhD, 1952. Rsch. chemist Eastman Kodak Co., Rochester, N.Y., 1952-54; rsch. chemist Shawinigan Resins Corp., Springfield, Mass., 1954-60, rsch. group leader, 1960-65; rsch. specialist Monsanto Co., Springfield, 1965-68, sr. rsch. specialist, 1968-85. Contbr. articles to profl. jours. 1st lt. U.S. Army Air Force, 1943-45, ETO. Fellow Am. Inst. Chemists; mem. AAAS, N.Y. Acad. Sci., Am. Chem. Soc. (Arthur K. Doolittle award Div. Organic Coatings and Plastic Chemicals 1972), Sigma Xi. Jewish. Achievements include patents covering synthesis and applications of novel monomers, polymers and cross linkers. Home: 15 Lindsay Rd Springfield MA 01128

COHEN, STANLEY, biochemistry educator; b. Brooklyn, N.Y., Nov. 17, 1922; s. Louis and Fannie (Feitel) C.; m. Olivia Larson, 1951 (div.); children: Burt Bishop, Kenneth Larson, Cary; m. Jan Elizabeth Jordan, 1981. BA, Bklyn. Coll., 1943; MA, Oberlin Coll., 1945, PhD, 1989; PhD in Biochemistry, U. Mich., 1948; PhD, U. Chgo., 1985, Washington U., 1993. Instr. dept. biochemistry and pediatrics U. Colo., Denver, 1948-52; Am. Cancer Soc. fellow in radiology Washington U., St. Louis, 1952-53, assoc. prof. dept. zoology, 1953-59; asst. prof. biochemistry, sch. medicine Vanderbilt U., Nashville, 1959-62, assoc. prof., 1962-67, prof. biochemistry, 1967-86, disting. prof., 1986—; rsch. prof. biochemistry Am. Cancer Soc., Nashville, 1976—; Charles B. Smith vis. rsch. prof. Sloan Kettering, 1984; Feodor Lynen lectr. U. Miami, 1986, Steenbock lectr. U. Wis., 1986. Mem. editorial bd. Abstracts of Human Developmental Biology, Jour. of Cellular Physiology. Cons. Minority Rsch. Ctr. for Excellence. Recipient Research Career Devel. award NIH, 1959-69, William Thomson Wakeman award Nat. Paraplegia Found., Earl Sutherland Research Prize Vanderbilt U., 1977, Albion O. Bernstein MD award Med. Soc. State N.Y., 1978, H.P. Robertson Meml. award Nat. Acad. Sci., 1981, Lewis S. Rosentiel award Brandeis U., 1982, Alfred P. Sloan award Gen. Motors Cancer Research Found., 1982, Louisa Gross Horwitz prize Columbia U., 1983, Disting. Achievement award UCLA Lab. Biomed. and Environ. Scis., 1983, Lila Gruber Meml. Cancer Research award Am. Acad. Dermatology, 1983, Bertner award MD Anderson Hosp. U. Tex., 1983, Gairdner Found. Internat. award, 1985, Fred Conrad Koch award Endocrine Soc., 1986, Nat. Medal Sci., 1986, 89, Albert and Mary Lasker Found. Basic Med. Research award, 1986, Nobel Prize in physiology or medicine, 1986, Tennessean of Yr. award Tenn. Sports Hall of Fame, 1987, Franklin Medal, 1987, Albert A. Michaelson award Mus. Sci. and Industry, 1987. Fellow Jewish Acad. Arts and Sci.; mem. Nat. Acad. Sci., Am. Soc. Biol. Chemists, Am. Chem. Soc., AAAS, Internat. Inst. Embryology, Internat. Acad. Sci. (hon. internat. coun. for sci. devel.). Office: Vanderbilt U Sch Medicine Dept Biochemistry Nashville TN 37232-0146

COHEN, STEPHEN DOUGLAS, mathematician, educator; b. London, Jan. 7, 1944; s. Eli and Olive Mary (McKay) C.; m. Yvonne Joy Roulet, July 25, 1967; children: Shirley Joy, Brenda Grace. BSc, U. Glasgow, Scotland, 1966, PhD, 1969. Lectr. U. Glasgow, 1968-88, sr. lectr., 1988-93, reader, 1993—; vis. lectr. U. Witwatersrand, Johannesburg, South Africa, 1987; reviewer Math. Revs., Ann Arbor, Mich., 1971—; referee Acta Arithmetica, Warsaw, Poland, 1975—; rsch. visits to U.S., Portugal, Russia, Australia. Editor Glasgow Math. Jour., 1988—, Proc. Edinburgh Math. Soc., 1991—; founding editor Finite Fields and Their Applications, 1993—; contbr. articles to profl. jours. Mem. London Math. Soc., Edinburgh Math. Soc. Avocations: music appreciation, walking, travel. Home: 40 Earlspark Ave, Glasgow G43 2HW, Scotland Office: U Glasgow Dept Math, University Gardens, Glasgow G12 8QW, Scotland

COHEN-TANNOUDJI, CLAUDE NESSIM, physics educator; b. Constantine, Algerie, France, Apr. 1, 1933; s. Abraham and Sarah (Sebba) Cohen-T.; m. Jacqueline Veyrat, Nov. 24, 1958; children: Alain, Joelle, Michel. Student, Ecole Normale Superieure, Paris, 1953-57; PhD in Physics, U. Paris, 1962. Researcher Centre Nat. La Recherche Scientifique, Paris, 1960-64; prof. U. Paris, 1964-73, Coll. de France, Paris, 1973—. Author 4 books. Recipient Julius Edgar Lilienfeld prize Am. Physical Soc., 1992. Mem. Académie des Sciences, Am. Acad. Arts and Scis. Home: 38 Rue Des Cordelieres, 75013 Paris France Office: Lab Physique Ens, 24 Rue Lhomond, 75005 Paris France

COHN, MILDRED, biochemist, educator; b. N.Y.C., July 12, 1913; d. Isidore M. and Bertha (Klein) Cohn; m. Henry Primakoff, May 31, 1938; children: Nina, Paul, Laura. BA, Hunter Coll., 1931, ScD (hon.), 1984; MA, Columbia U., 1932, PhD, 1938; ScD (hon.), Women's Med. Coll., 1966, Radcliffe Coll., 1978, Washington U., St. Louis, 1981, Brandeis U., 1984, U. Pa., Phila., 1984, U. N.C., 1985; PhD (hon.), Weizmann Inst. Sci., Israel, 1988; ScD (hon.), U. Miami, 1990. Rsch. asst. biochemistry George Washington U. Sch. Medicine, 1937-38; rsch. assoc. Cornell U., 1938-46; research assoc. Washington U., 1946-50, 51-58, assoc. prof. biol. chemistry, 1958-60; assoc. prof. biophysics and phys. biochemistry U. Pa. Med. Sch., 1960-61, prof., 1961-78, emeritus, 1982—, Benjamin Rush prof. physiol. chemistry, 1978-82; sr. mem. Inst. Cancer Research, Phila., 1982-85; Chancellor's disting. prof. biophysics U. Calif., Berkeley, spring 1981; vis. prof. biol. chemistry Johns Hopkins U. Med. Sch., 1985—; research assoc. Harvard U., 1950-51; established investigator Am. Heart Assn., 1953-59, career investigator, 1964-78. Editorial bd. jour. Biol. Chemistry, 1958-63, 67-72. Recipient Cresson medal Franklin Inst., 1976, award Internat. Assn. Women Biochemists, 1979, Nat. Medal Sci., 1982, Chandler medal Columbia U., 1986, Disting. Svc. award Coll. Physicians, Phila., 1987, Gov.'s award for Excellence in Sci., Pa., 1993. Mem. Am. Philos. Soc., NAS, Am. Chem. Soc. (Garvan medal 1963, Remsen award Md. sect. 1988), Harvey Soc., Am. Soc. Biol. Chemists (pres. 1978-79), Am. Biophys. Soc., Am. Acad. Arts and Scis., Phi Beta Kappa, Sigma Xi, Iota Sigma Pi (hon. nat. mem. 1988). Office: U Pa Med Sch Dept Biochemistry and Biophysics Philadelphia PA 19104-6089

COIRO, DOMENICO PIETRO, research scientist, educator; b. S. Rufo, Salerno, Italy, July 4, 1960; s. Mario and Giuseppina (Spinelli) C. Degree U. Naples, 1985. Researcher Gasdynamic Inst., Naples, 1985-86; teaching asst. Pa. State U., State Coll., 1986-87; head of transonic group Italian Ctr. for Aerospace Rsch., Capua, Italy, 1987-90; rsch. scientist U. Naples, 1990—. Contbr. articles to profl. jours. Mem. AIAA. Office: U Naples Istituto Procetto Velivoli, Via Claudio 21, Naples 80125, Italy

COLARDYN, FRANCIS ACHILLE, physician; b. Kortrijk, Belgium, Dec. 3, 1944; s. Evariste and Carolina (Tarkanyi) C.; m. Michelle Fauconnier, July 10, 1967 (div. 1989); children: Gregory, Anouk; m. Marie-Antoinette Van Den Berghe, Sept. 10, 1989. MD, State U., Ghent, Belgium, 1970, postgrad., 1973-76, specialist in internal medicine degree, 1975. Asst. in internal medicine Univ. Hosp., Ghent, 1970-75, asst. prof. intensive care unit and emergency room, 1976-86, head dept. intensive care, 1981—, prof. intensive care, 1986—; mem. Med. Bd., 1979-80, Univ. Hosp., Ghent, 1988—; mem. directory bd. Univ. Hosp., Ghent, 1991—, mem. hosp. com. on medication, 1984—, mem. hosp. com. on hosp. hygiene, 1983—, mem. nutrition com., 1984—. Mem. med. adv. bd. Internal Medicine Digest, 1983-87. Mem. Commn. to the Ministry of Health for med. urgencies, 1976-85. Mem. Belgium Soc. Ultrasonography, Belgian Soc. Med. Informatics, Belgian Soc. Intensive Care (bur., teaching commn.), N.Y. Acad. Scis., Belgian Soc. Endoscopy, Belgian Assn. Burn Injuries, European Soc. Intensive Care Medicine, Belgian Soc. Oxygen Metabolisms, Toxicological Soc. Belgium and Luxembourg, Belgian Working Group on Cardiac Pacing and Electrophysiology, Am. Soc. for Microbiology, Dutch-Belgian Soc. for A.R.D.S. (bd. dirs. 1990—), Soc. Critical Care Medicine. Office: Univ Hosp, De Pintelaan 185, B 9000 Ghent Belgium

COLARELLI, STEPHEN MICHAEL, psychology educator, organizational psychologist; b. Denver, Apr. 26, 1951; s. Michael Nicholas and Margie Aileen (Ferguson) C.; m. Margaret Cary, June 30, 1979; children: Catherine Lydia, Julia Rose. BA, Northwestern U., 1973; MA, U. Chgo., 1979; PhD, NYU, 1982. Vol. Peace Corps, Animation Rurale, Senegal, 1973-75; asst. prof. psychology Ctrl. Mich. U., Mt. Pleasant, 1985-90, assoc. prof., 1990—, dir. grad. programs in indsl. and orgnl. psychology, 1990—. Contbr. chpts. to books, articles to profl. jours. Dissertation fellow Social Sci. Rsch. Coun., 1981-82. Mem. Human Behavior and Evolution Soc., Acad. of Mgmt., Soc. Indsl. and Orgnl. Psychology, Am. Psychol. Soc. Republican. Episcopalian. Achievements include organizational influences on the use of human resource technologies, social ecology of personnel programs. Office: Ctrl Mich U Dept Psychology Mount Pleasant MI 48859

COLBAUGH, RICHARD DONALD, mechanical engineer, educator, researcher; b. Pitts., Oct. 31, 1958; s. Richard Donald and Anne Marie (McCue); m. Kristin Lea Glass, July 18, 1987; 1 child, Allison Collette. BS in Mechanical Engring., Pa. State U., 1980, PhD in Mechanical Engring., 1986. Mechanical engr. McDonnell Douglas Corp., Long Beach, Calif., 1980-81; instr. mechanical engring. Pa. State U., State College, 1981-86; asst. prof. mechanical engring. N.Mex. State U., Las Cruces, 1986-90, assoc. prof. mechanical engring., 1990--; cons. Dept. Energy, Albuquerque, 1987-- Jet Propulsion Lab., Pasadena, Calif., 1988--. Assoc. editor: Internat. Jour. of Robotics and Auto., 1991—, Internat. Jour. Environ. Conscious Mfg., 1992—; editor: (jours.) Redundant Robots, 1993, Redundant and Flexible Robots, 1992; co-author: Robotics and Remote Systems in Hazardous Environ., 1992; contbr. articles to profl. jours. Recipient NASA Space Act Tech Brief award, 1990, 91, 92, Best Paper award Am. Automatic Control Coun., 1990; NASA/ASEE Summer Faculty fellow 1991, 92. Mem. IEEE, Am. Soc. Mechanical Engrs., Sigma Xi. Achievements include patent for Obstacle Avoidance for Redundant Robots Using Configuration Control; design of one of first adaptive inverse kinematics algorithms for robots; development of first real time control algorithm for robots possessing any combination of kinematic or actuator redundancy, one of first globally stable direct adaptive robot compliant motion control system. Office: NMex State U Dept Mechanical Engring Box 30001 Dept 3450 Las Cruces NM 88003

COLBERG-POLEY, ANAMARIS MARTHA, virologist; b. San Juan, PR, Dec. 17, 1954; d. Amado and Pura (Ortiz) Colberg; m. Gerald E. Poley Jr., July 22. 1978; children: Celeste RC Poley, Marian FC Poley. BS, U. PR, Mayagüez, 1976; PhD, Penn State U., 1980. Post doctoral fellow Nat. Inst. Health, Bethesda, Md., 1980-83; rsch. assoc. Zentrum für Molecular Biology, Heidelberg, Germany, 1983-86; sr. rsch. scientist DuPont, Wilmington, Del., 1986-92; assoc. prof. Children's Nat. Med. Ctr., Washington, 1992—; Assoc. editor Virology, 1989—. Author: J. Virology 64: 2033-2040, 1990, Virology 182: 199-210, 1991, J. Virology 65: 6724-6734, 1991, Biotechniques 11: 739-743, 1991, J. Virology 66: 95-105, 1992. Recipient Charles Darwin award 1976, Faculty of Arts and Scis. award, 1976 Luis Stefani Raffucci award, 1976, U. PR; grantee Am. Cancer Inst. Pa. State U., 1977; fellow Post Doctoral Nat. Rsch. Svc., Bethesda, Md., 1980-83, Alexander von Humboldt AVH Stiftung, Bonn, Germany, 1983-85. Mem. Am. Soc. Microbiology, Am. Soc. Virology, Sigma Xi, Phi Kappa Phi. Office: Ctr I Research R-213 Childrens Nat Med Ctr 111 Michigan Ave NW Washington DC 20010-2916

COLBERT, EDWIN HARRIS, paleontologist, museum curator; b. Clarinda, Iowa, Sept. 28, 1905; s. George Harris and Mary (Adamson) C.; m. Margaret Mary Matthew, July 8, 1933; children: George Matthew, David William, Philip Valentine, Daniel Lee, Charles Diller. Student, N.W. Mo. State Tchrs. Coll., 1923-26; BA, U. Nebr., 1928, ScD, 1973; AM, Columbia U., 1930, PhD, 1935; ScD, U. Ariz., 1976, Wilmington Coll., 1984. Student asst. Univ. Mus. U. Nebr., 1926-29; univ. fellow Columbia U., 1929-30, lectr. dept. zoology, 1938-39, prof. vertebrate paleontology, 1945-69, prof. emeritus, 1969—; research asst. Am. Museum Natural History, 1930-32, asst. curator, 1933-42, acting curator, 1942, curator, 1943, chmn. dept. amphibians and reptiles, 1943-44, curator of fossil reptiles and amphibians, 1945-70, chmn. dept. geology and paleontology, 1958-60, chmn. dept. vertebrate paleontology, 1960-66, curator emeritus, 1970—; hon. curator vertebrate paleontology Mus. No. Ariz., Flagstaff, 1970—. Author: Evolution of the Vertebrates, 1955, 69, 80, 91, Millions of Years Ago, 1958, Dinosaurs, 1961, (with M. Kay) Stratigraphy and Life History, 1965, The Age of Reptiles, 1965, Men and Dinosaurs, 1968, Wandering Lands and Animals, 1973, The Year of the Dinosaur, 1977, A Fossil Hunter's Notebook, 1980, Dinosaurs: An Illustrated History, 1983, Digging into the Past, 1989, William Diller Matthew, Paleontologist, 1992; also sci. papers and monographs. Recipient John Strong Newberry prize Columbia U., 1931, Am. Mus. Natural History medal, 1970, Nat. Ghost Ranch Found. Spl. award, 1986, Disting. Alumni award Dept. Geology, U. Nebr., 1986, George W. Clinton award The Buffalo Soc. Natural Scis., 1987. Fellow AAAS, Geol. Soc. Am., Paleontol. Soc. (v.p. 1963), N.Y. Zool. Soc.; mem. Soc. Vertebrate Paleontology (sec.-treas. 1944-46, pres. 1946-47, Romer-Simpson medal 1989), Soc. Mammalogy, Soc. Ichthyology and Herpetology, Soc. for Study Evolution (editor 1950-52, v.p. 1957, pres. 1958), Nat. Acad. Sci. (Daniel Giraud Elliot medal 1935), Sigma Xi. Office: Mus of No Ariz RR 4 Box 720 Flagstaff AZ 86001-9302

COLBERT, ROBERT FLOYD, aerospace engineer; b. Atlanta, Apr. 6, 1960; s. Floyd Columbus and Marjorie (Meacham) C.; m. Frances Lynn Price, Dec. 28, 1984; 1 child, Jennifer Renee Colbert. BS in Aerospace Engring., U. Ala., 1983. Engr. Morton Thiokol, Inc., Huntsville, Ala., 1986-89; engr. Sverdrup Technology, Inc., Huntsville, 1989—. Author: (NASA tech. memorandum) Thermal Analysis Workbook, 1992; contbr. articles to AIAA Jour., Soc. Automotive Engrs. Jour. Dir. Profl. Devel. and Resource Ctr. Emmanuel Ch. of Christ, Huntsville, 1991—. Mem. AIAA. Republican. Pentecostal. Home: 104 Waterbury Dr Harvest AL 35749

COLBORN, GENE LOUIS, anatomy educator, researcher; b. Springfield, Ill., Nov. 23, 1935; s. Adin Levi and Grace Downey (Tucker) C.; divorced; children: Robert Mark, Adrian Thomas, Lara Lee Colborn Russell; m. Sarah Ellen Crockett, Aug. 14, 1976; children: Jason Matthew, Nathan Tucker. BA with honors, Ky. Christian Coll., 1957; BS with honors, Milligan Coll., 1962; MS in Anatomy, Wake Forest U., 1964, PhD in Anatomy, 1967. Postdoctoral fellow U. N.Mex. Sch. Medicine, Albuquerque, 1967-68; asst. prof. U. Tex. Health Sci. Ctr., San Antonio, 1968-72, assoc. prof., 1972-75; assoc. prof. anatomy Med. Coll. Ga., Augusta, 1975-88, prof. anatomy, 1988—, prof. surgery, 1993—, dir. Ctr. for Clin. Anatomy, 1987—, dir. med. gross anatomy, 1975—, cons. dept. surgery, 1977—, prof. surgery, 1993—; pres. Ga. State Anatomical Bd., 1983-93; cons. Eisenhower Hosp., 1990—. Author: Practical Gross Anatomy, 1982, Surgical Anatomy, 1987, Hernias, 1988, Musculoskeletal Anatomy, 1989, Workbook of Surgical Anatomy, 1990, Clinical Gross Anatomy, 1993; contbr. numerous articles on cardiac conduction, nervous system, primate anatomy, cell culture and clin. and surg. anatomy to profl. jours. Mem. San Antonio Symphony Mastersingers, 1970-75, Augusta Opera, 1975—, Augusta Choral Soc., 1975—; judge Regional Sci. Fairs, Augusta, 1978—. Recipient Golden Apple award U. Tex. Health Sci. Ctr., 1975; Outstanding Med. Educator award Med. Coll. Ga., 1976, 77, 78, 82, 87, 88, 90, 91, Disting. Faculty award, 1978. Mem. AAUP, Am. Assn. Clin. Anatomists (membership chmn. 1982-86, councillor 1992—), Am. Assn. Anatomists, K.C. (4th degree). Republican. Avocations: opera, chorales, church soloist, tennis, camping. Office: Med Coll Ga 15th at Laney-Walker Blvd Augusta GA 30912

COLBY, GEORGE VINCENT, JR., electrical engineer; b. Montpelier, Vt., Sept. 4, 1931; s. George Vincent and Clara Pauline (Tebbetts) C.; m. Barbara Ann Gardner, Sept. 5, 1955; children: George V. III, Ann G. Colby Cummings, Catherine M. Colby Fielding. BEE, MEE, MIT, 1954. Tech. staff engr. Gen. Electric Co., West Lynn, Mass., 1954-55; mgr. radar lab. L.F.E. Corp., Waltham, Mass., 1957-70; staff mem. MIT Lincoln Lab., Lexington, Mass., 1970-80; dep. dir. Textron Def. Systems, Wilmington, Mass., 1980—. Treas. troop 159 Boy Scouts Am., Lexington, 1970-76; dir. teen mass St. Brigid's Ch., Lexington, 1974-81. 1st lt. U.S. Army, 1955-57. Mem. IEEE (sr.). Roman Catholic. Achievements include patents for Doppler radar system, phase shifter, radar altimeter, Doppler radar altimeter, data acquisition and analysis system. Avocation: oil painting. Home: 7 Hawthorne Rd Lexington MA 02173-1731 Office: Textron Def Systems 210 Lowell St Wilmington MA 01887

COLBY, RALPH HAYES, chemical engineer; b. Providence, Feb. 27, 1958; s. Edwin F. and Janice (Abel) C.; m. Patricia Cook, July 15, 1982; children: Melissa S., Edwin F., Graham R. BS in Engring., Cornell U., 1979; PhD in Chem. Engring., Northwestern U., Evanston, Ill., 1985. Rsch. scientist Eastman Kodak Co., Rochester, N.Y., 1985—; adj. asst. prof. dept. physics and astronomy, U. Rochester, 1990—. Contbr. articles to profl. jours. Recipient Walter P. Murphy fellowship Northwestern U., 1981. Mem. Am. Phys. Soc., Soc. Rheology, British Soc. Rheology. Achievements include theory and experiment on complex fluid dynamics; polymers, polymer blends, block copolymers, liquid crystal polymers, branched polymers and networks. Office: Eastman Kodak Co Corporate Rsch Labs Rochester NY 14650-2110

COLBY, S(TANLEY) BRENT, medical physicist; b. Williston, N.D., Nov. 22, 1964; s. John S. and Verna Jean (Simle) C. BA, Concordia Coll., 1988; MS, U. Colo., 1991. Radiation physicist Gundersen Clinic, LaCrosse, Wis., 1991-93, Roger Maris Cancer Ctr., Fargo, N.D., 1993—; cons. regional clinics, hosps., LaCrosse, 1992—. Contbr. articles to Health Physics Jour., Med. Physics. Mem. Am. Assn. Physicists in Medicine (North Ctrl. chpt.), Health Physics Soc. (North Ctrl. chpt.), Soc. Nuclear Medicine. Achievements include research on photon contribution to total dose in radioimmunotherapy. Home: 511 40th St SW # 316 Fargo ND 58103 Office: Roger Maris Cancer Ctr 820 N 4th St Fargo ND 58122

COLE, CHARLES R., hydrologist, researcher; b. Canfield, Ohio, Aug. 31, 1942; s. John Orville and Ann Elizebeth (Vogel) C.; m. Karen Lee Decker, June 20, 1964; children: Heidi Lee Cole Fisher, Russell John. BS in Physics, Kent State U., 1964; MS in Computer Sci., Wash. State U., 1975. Rsch. scientist Monsanto Rsch. Corp., Miamisburg, Ohio, 1964-66; staff scientist Battelle Pacific Northwest Labs., Richland, Wash., 1966—. Contbg. author: Dynamics of Fluids in Hierarchical Porous Media, 1990. Mem. AAAS, Am. Geophysical Union, Nat. Water Well Assn. Home: 602 Riverside Dr West Richland WA 99352

COLE, DEAN ALLEN, biomedical researcher; b. Clarkfield, Minn., Jan. 22, 1952; s. Virgil Howard and Alvina Selora (Alness) C.; m. Monica Lea Eckstrom, May 10, 1975; children: Kenan Lindsay, Cole. Assocs. in Nuclear Medicine, U. Iowa, 1975, BS, 1977, MS, 1981, PhD, 1985. Head Clin. Nuclear Medicine Eugene (Oreg.) Hosp. and Clinic, 1975-78; grad. rsch. asst. U. Iowa, Iowa City, 1978-80, rsch. staff scientist, 1980-85; postdoctoral research fellow Los Alamos (N.Mex.) Nat. Lab., 1985-87, staff mem., dir. Lung Cancer program, 1987—; clin. assoc. prof. Radiopharmacy U. N.Mex., Albuquerque, 1990-92; adj. prof. Medicine Johns Hopkins U., Balt., 1990—; dir. med. rsch. U.S. Dept. Energy, Washington, 1992—; mem adv. coun. Nat. Cancer Insts., 1993—. Author: (book sect.) Water Chlorination—Chemistry, Environmental Impacts and Health Effects, Vol. 5, 1984; contbr. articles to profl. jours. Recipient Internat. Travel award, 1983, Disting. Performance award Los Alamos Nat. Lab., 1992; rsch. grantee Iowa Acad. of Scis., 1979; Los Alamos Nat. Lab. Dir. fellowship, 1985. Mem. Radiation Rsch. Soc., Am. Soc. for Photobiology, Soc. Nuclear Medicine, Iowa Acad. of Scis., Soc. Clin. Radiology, Soc. Clin. Pathology. Avocations: sailing, skiing, hiking, camping. Home: 11739 Othello Terr Germantown MD 20876 Office: Office Health & Environ Rsch US Dept Energy ER-73 E225 GTN Washington DC 20585

COLE, JACK HOWARD, mechanical engineer; b. Tulsa, Okla., June 7, 1934; s. Louie Howard and Lillie Almeda (Reamey) C.; m. Carol Sue Smith, June 10, 1955; 1 child, Suzanne Elaine. BSME, Okla. State U., 1958, MSME, 1963, PhD, 1968. Registered profl. engr., Okla. Mem. tech. staff Rockwell Corp., Tulsa, Okla., 1964-68; asst. prof. mech. enging. U. Ark., Fayetteville, 1968-70, assoc. prof. mech. engring., 1970-74, prof. mech. engring., 1974-81; rsch. assoc. Conoco, Inc., Ponca City, Okla., 1982-85, engring. dir., 1985-92, rsch. assoc., 1992—; cons. in field. Contbr. articles to profl. jours. Cpl. U.S. Army, 1953-55. Rockwell fellow, 1965. Mem. ASME, Soc. Automotive Engrs., Soc. Exploration Geophysicists, Soc. Petroleum Engrs., Phi Kappa Phi, Sigma Xi, Pi Tau Sigma. Achievements include patents for Mass Flow Metering and Geophysics Methods and Apparatus, Elliptically Polarized Shear Waves in Seismic Exploration; development of powerful new techniques for synthesizing asynchronous switching networks. Office: Conoco Inc 1000 S Pine Ponca City OK 74601-7501

COLE, JONATHAN JAY, aquatic scientist, researcher; b. N.Y.C., Jan. 14, 1953; s. Leonard and Selma Ruth (Greenblatt) C.; m. Nina F. Caraco, Nov. 25, 1980; children: Aaska H. Puccoon, Zak LeH Puccoon. BA, Amherst Coll., 1976; PhD, Cornell U., 1982. Post-doctoral fellow Woods Hole (Mass.) Oceanographic Inst., 1981-82, Marine Biol. Lab., Woods Hole, 1982-83; asst. scientist Inst. Ecosystems Studies, Millbrook, N.Y., 1983-89, assoc. scientist, 1989—. Editor: Comparative Analysis of Ecosystems, 1991; contbr. articles to profl. jours. Mem. Am. Soc. Microbiology, Am. Soc. Limnology and Oceanography (editorial bd. 1987-90), Internat. Soc. Limnology. Achievements include elucidation of the role of microorganisms in aquatic ecosystems. Office: Inst Ecosystem Studies Cary Arboretum Sharon Tpke Millbrook NY 12543

COLE, JULIAN D., mathematician, educator; b. Bklyn., Apr. 2, 1925; m. 1949; children. BME, Cornell U., 1944, MB, AE, 1946, PhD in Aeronautics and Math., 1949. Asst. prof. aeronautics and applied mechanics Calif. Inst. Tech., 1951-59, prof., 1959-69; chmn. dept. applied sci. and engring. UCLA, 1969-76, prof., 1969-82; prof. applied math. Rensselaer Poly. Inst., Troy, N.Y., 1982—; prof. math. UCLA, 1976-82. Recipient Von Karman prize Soc. Indsl. and Applied Math., 1984. Mem. AIAA (Fluid and Plasmadynamics award 1992), NAS, NAE, AAAS. Achievements include research in aeronautics and applied mathematics. Office: Rensselaer Poly Inst Dept Applied Math 110 8th St Troy NY 12180-3522*

COLE, JULIAN WAYNE (PERRY COLE), computer educator, consultant, programmer, analyst; b. LaFayette, Ala., Dec. 16, 1937; s. William Walter and Hattie Lucille (Berry) C.; m. Judith Elaine Riley, June 27, 1959; children—Jeffrey Paul, Jarrett David. B.S. in Bus. Adminstrn., Ariz., State U., 1967; M. in Computer Sci., Texas A&M U., 1969. Joined U.S. Air Force, 1956, advanced through grades to capt., 1970, ret., 1978; programmer/analyst Hewlett Packard Corp., Colorado Springs, Colo., 1978-79, Digital Equipment Corp., Colorado Springs, 1979-85; lectr. U. Colo., Colorado Springs, 1978-92; tng. dir. System Devel. Corp., Colorado Springs, 1979-85; tng. dir. Unisys Corp., Colorado Springs, 1986-91; computer cons., Colorado Springs, 1980—; sr. systems analyst, tech. trainer MCI, Inc., Colorado Springs, 1992—; pres. Advanced Info. Methodology and Systems, Colorado Springs, 1982—. Author: ANSI Fortran IV, 1978; ANSI Fortran IV with Fortran 77 Extensions, 1983, ANSI Fortran 77: Structured Problem Solving Approach, 1987. Mem. Assn. Computing Machinery (chmn. 1981-82), Data Processing Mgmt. Assn., Beta Gamma Sigma, Phi Kappa Phi, Upsilon Pi Epsilon. Republican. Baptist. Club: Business. Home: 735 Big Valley Dr Colorado Springs CO 80919-1004 Office: MCI 2424 Garden of the Gods Colorado Springs CO 80919

COLE, MALVIN, neurologist, educator; b. N.Y.C., Mar. 21, 1933; s. Harry and Sylvia (Firman) C.; A.B. cum laude, Amherst Coll., 1953; M.D. cum laude, Georgetown U. Med. Sch., 1957; m. Susan Kugel, June 20, 1954; children: Andrew James, Douglas Gowers. Intern, Seton Hall Coll. Medicine, Jersey City Med. Ctr., 1957-58; resident Boston City Hosps., 1958-60; practice medicine specializing in neurology, Montclair and Glen Ridge, N.J., Montville, N.J., 1963-72, Casper, Wyo., 1972—; teaching fellow Harvard Med. Sch., 1958-60; Research fellow Nat. Hosp. for Nervous Diseases, St. Thomas Hosp., London, Eng., 1960-61; instr. Georgetown U. Med. Sch., 1961-63; clin. assoc. prof. neurology N.J. Coll. Medicine, Newark, 1963-72, acting dir. neurology, 1965-72; assoc. prof. clin. neurology U. Colo. Med. Sch., 1973-88, prof., 1988—; mem. staff Wyo. Med. Ctr., Casper, U. Hosp., Denver. Served to capt. M.C., AUS, 1961-63. Licensed physician, Mass., N.Y., Calif., N.J., Colo., Wyo.; diplomate Am. Bd. Psychiatry and Neurology, Nat. Bd. Med. Examiners. Fellow ACP, Am. Acad. Neurology, Royal Soc. Medicine; mem. Assn. Research Nervous and Mental Disease, Acad. Aphasia, Am. Soc. Neuroimaging, Internat. Soc. Neuropsychology, Harveian Soc. London, Epilepsy Found. Am., Am. Epilepsy Soc., Am. EEG Soc., N.Y. Acad. Sci., Osler Soc. London, Alpha Omega Alpha. Contbr. articles to profl. jours. Office: 246 S Washington St Casper WY 82601-2921

COLE, SAMUEL JOSEPH, computational chemist; b. Sandoval, N.Mex., Aug. 21, 1950; s. Samuel Eugene and Ethel Modena (Reynolds) C.; m. Shirley Ann Nightingale, June 1, 1970 (div. 1989). BA, Colo. U., 1971; PhD, U. Calif., Santa Barbara, 1977. Prof. Antillian Coll., Mayagüez, P.R., 1977-83; rsch. assoc. U. Fla., Gainesville, 1983-86; mgr. computer ops. U. Utah, Salt Lake City, 1986-90; sr. chemist Cache Scientific, Beaverton, Oreg., 1990—; sec. Array-FPS User Group, Beaverton, 1987-88, v.p., 1988-89. Contbr. articles to profl. jours. including Jour. Chem. Physics and Internat. Jour. of Quantum Chemistry. Regents fellowship U. Calif., 1972, U. of Calif. fellowship, 1973. Mem. AAAS, Am. Chem. Soc., Am. Phys. Soc. Office: Cache Scientific MS 13-400 PO Box 500 Beaverton OR 97077

COLE, THEODORE JOHN, osteopathic physician; b. Covington, Ky., May 30, 1953; s. John N. and Florence R. (Bruener) C.; children: Joren, Emily. BA, Centre Coll., Danville, Ky., 1975; MA, Western Ky. U., 1978; DO, Ohio U., 1986. Diplomate Am. Osteo. Bd. Gen. Practice, Nat. Bd. Osteo. Examiners. Psychologist Comprehensive Mental Health Svcs., St. Petersburg, Fla., 1978-82; intern Detroit Osteo. Hosp., 1986-87; resident Doctors Hosp., Columbus, Ohio, 1987-88; pvt. practice, West Chester, Ohio, 1989—; preceptor Ohio U. Coll. Osteo. Medicine, Athens, 1990—, U. Cin. Med. Sch., 1990—; dir. So. Ohio Coll. Nursing. Coach, Soccer Assn. for Youth, West Chester, 1989, 90, Liberty Sports Orgn., West Chester, 1990. Mem. Am. Osteo. Assn., Am. Assn. Osteopathy, Am. Coll. Gen. Practitioners, Nat. Inst. Electromed. Info., Ohio Osteo. Assn., Am. Acad. Environ. Medicine, Cranial Acad., Am. Acad. Advancement of Medicine. Avocations: soccer, collecting art, hunting, camping, raising exotic fish. Office: West Chester Family Practice 9678 Cincinnati-Columbus Rd Cincinnati OH 45241

COLE, TIMOTHY DAVID, chemist; b. Troy, N.Y., Oct. 15, 1955; s. Warren James and Naomi Ruth (Beadleston) C.; m. Chari Lee Knights, Apr. 7, 1988; children: Melanie, Jessica, Rebecca. AA in Ind. Studies, Hudson Valley Community Coll., 1989; BS in Chemistry, Rensselaer Poly. Inst., 1992. Med. technician Albany (N.Y.) Med. Ctr., 1992; lab. technician, researcher N.Y. State Dept. of Health, Albany, 1992-93. Recipient Class of 1902 Rsch. prize Rensselaer Poly. Inst., 1992, William Pitt Mason prize, 1992, John and Mary Cloke prize, 1992. Mem. Phi Lambda Upsilon. Democrat. Home: 5534 Pennsylvania Blvd Concord CA 94521

COLEF, MICHAEL, engineering educator; b. Cimpina, Prahova, Romania, Apr. 18, 1956; came to U.S., 1981; s. Ivan Mihailof and Jana Ioana (Rucareanu) C.; m. Doina Gabriela Iancu, Feb. 15, 1990; children: Diana-Ioana, Robert Gabriel. BEEE, Polytech. Inst., Bucharest, 1979; MEEE, CCNY, 1983, PhD in Engring., 1987. Rsch. asst. Polytech. Inst. Bucharest, 1977-79; rsch. asst. engring. dept. CCNY, 1982-87; cons. SCS Telecom, Inc., Port Washington, N.Y., 1987-88; asst. prof. engring. N.Y. Inst. Tech., Old Westbury, 1989—; adj. prof. engring. Manhattan Coll., N.Y.C., 1988; adj. lectr. engring. CCNY, 1982-87. Contbr. articles to profl. jours. Treas. Industry Acad. Conf., L.I., 1990-92. Recipient Pope, Evans and Robbins scholarship Soc. Am. Mil. Engrs., CCNY, 1985. Mem. IEEE (facilities chmn. L.I. sect. 1990-93), Eta Kappa Nu, Tau Beta Pi. Avocations: music, tennis, skiing, computers. Home: 82 Dogwood Ln Staten Island NY 10305-2813 Office: NY Inst Tech Old Westbury NY 11568

COLEMAN, BRIAN FITZGERALD, electrical engineer; b. Inglewood, Calif., Dec. 16, 1960; s. John Joseph and Esther Hope (Fitzgerald) C. BS summa cum laude, Rensselaer Poly. Inst., 1982; MS, MIT, 1987. Elec. engr. Alphatech Inc, Burlington, Mass., 1985-86, GE, San Jose, Calif., 1988—. Contbr. to profl. publs. Sybron Corp. scholar, 1981; fellow Am. Electronics Assn.-Hewlett-Packard, 1983. Mem. Sigma Xi, Tau Beta Pi. Democrat. Roman Catholic. Home: 1231 Minnesota Ave San Jose CA 95125 Office: GE Nuclear Energy 175 Curtner Ave San Jose CA 95125

COLEMAN, DALE LYNN, electronics engineer, aviator, educator; b. Topeka, June 17, 1958; s. Dale R. Coleman and Linda C. (Parks) Meiergerd; m. Patricia Bermödez, Nov. 20, 1982; 1 child, Athena C. AS in Electronic Engring. Tech. with honors, Cleve. Inst. Electronics, 1987, postgrad.; B summa cum laude in Elec. Engring. Tech., World Coll., 1993. Electronics technician Litton G & CS, L.A., 1979-82; sr. electronics technician Cedars-Sinai Med. Ctr., L.A., 1982-85; svc. engr. Litton AMS, San Diego, 1985-86; assoc. electrical engr. IMED Corp. R & D, San Diego, 1987—; communications officer USCG Aux., San Diego, 1987—; participant Space Life Scis. (SLS) mission Space Sta. Freedom, NASA. Co-author: The Art of Hsin Hsing Yee Ti Kenpo Kung Fu, 1991; contbr. articles to various publs. Mem. UN Assn., 1979—, sr. officer USCG Aux., 1980—; USCG liaison U.S. Naval Sea Cadet Corps, NAS Miramar, 1985—; liaison NASA, 1992—. With USN, 1976-79. Named Outstanding Citizen, Exch. Club, San Diego, 1989, Outstanding Grad., Cleveland Inst. Electronics, 1990. Mem. IEEE, Planetary Soc., Nat. Space Soc., Star Trek Assn. for Revival (chpt. pres. 1973-79, legion of honor 1975). Achievements include contributions to patent for improved switching power supply; research in osmotic migration of water through PVC, in H2 generation of reversed biased capacitors, and in lead-acid battery life prolongation; patent pending for microprocessor controlled battery mgmt. system. Office: IMED Corp R & D 9775 Businesspark Ave San Diego CA 92131-1699

COLEMAN, EDWIN DEWITT, III, electronics engineer; b. Phoenix, Apr. 2, 1954; s. Edwin DeWitt and Joanne Edwina (Moore) C.; m. Nancy Robin Broome, Oct. 8, 1977; children: Kristen Erica, Jillian Ashley. BSEET, Devry Inst. Tech., Phoenix, 1979. Electronic engr. III Advanced Products R&D Lab./Motorola, Inc., Mesa, Ariz., 1979-83; mfg. engr. II S.W. Acquisition Ctr., Digital Equip. Corp., Phoenix, 1983; product specialist Schlumberger Ltd./Technologies, Dallas, 1983-85; electronics engr. A.T.E. Div., Schlumberger Tech., Tempe, Ariz., 1985—. Mem. IEEE, N.Y. Acad. Sci. Home: 1363 N Valencia Dr Chandler AZ 85226 Office: Schlumberger Tech 7855 S River Pkwy Tempe AZ 85284-1825

COLEMAN, HOWARD S., engineer, physicist; b. Everett, Pa., Jan. 10, 1917; s. Howard Solomon and Amy (Ritchey) C.; children: Michael Howard, Madeline Frances, Thomas Robert, Carl William, Stephen Mitchell Rosenberg; m. Jeannette Eve Dresher, Dec. 27, 1969. BS, Pa. State U., 1938, MS in Physics, 1939, PhD, 1942. Registered profl. engr., Va., Ariz., Tex. Mem. faculty Pa. State U., 1934-47, dir. optical inspection lab., 1941-47; dir. optical rsch. lab., assoc. prof. physics U. Tex., 1947-51; with Bausch & Lomb, Inc., 1951-62, mgr., v.p. rsch. and engring., 1954-62; head physics rsch. dept., tech. asst. to v.p. charge rsch. Melpar, Inc., Falls Church, Va., 1962-64; dean U. Ariz. Coll. Engring., prof. elec. engring., 1964-68; dir. Spl. Projects Ctr., Schellenger Rsch. Labs., U. Tex., El Paso, 1968-75, Howard S. Coleman and Assos., El Paso, 1975—; dep. dir. solar energy div. ERDA, 1976-77; dep. dir. div. solar energy tech. U.S. Dept. Energy, 1977-78, dir. central solar tech. div., 1978-80, dir. tech. and utilization alcohol fuels, 1980-81, prin. dep. asst. sec. for conservation and renewable energy, 1981-84, dir. Div. Solar Thermal Tech., 1984-90, dir. Office Grants Mgmt., 1990—; cons. to industry, govt., 1941—; spl. rsch. optical inspection devices; mem. Ariz. Bd. Tech. Registration.; Mem. adv. vis. com. electronics U. Rochester, 1952; chmn. vis. com. math. Clarkson Coll. Tech., 1953-63. Recipient Joint Svc. award, 1942. Fellow Optical Soc. Am.; mem. Am. Phys. Soc., Meteorol. Soc., Inst. Aero. Scis., Am. Assn. Physics Tchrs., Am. Soc. Metals, Internat. Commn. Optics, Am. Geophys. Union, Am. Inst. Physics, Am. Soc. Engring. Edn., Nat. Soc. Profl. Engrs., N.Y. Acad. Scis., Illuminating Engring. Soc., Soc. Photo-Optical Instrumentation Engrs. Patentee in field. Home: PO Box 26368 El Paso TX 79926-6368

COLEMAN, JERRY TODD, chemist; b. Humbolt, Tenn., Dec. 29, 1957; s. Jerry Leon and Eddie Dean (Richardson) C.; m. Louise Strickland Long, Mar. 4, 1989; 1 child, Joshua Todd. BS, Memphis State U., 1980; MS, U. Tenn., 1984. Sr. chemist DePont de Nemours, Aiken, S.C., 1984-88, area supr., 1988-89; mgr. Westinghouse, Aiken, S.C., 1989-91, prin. scientist, 1991—. Mission pilot U.S. CAP, S.C., 1988—. Mem. ASTM (vice chmn. 1988-92). Republican. Presbyterian. Achievements include patent on apparatus and method for in-vivo measurement of chemical concentrations. Home: 142 Moss Creek Ct Martinez GA 30907 Office: Westinghouse Bldg 773-41A Aiken SC 29802

COLEMAN, JOHN HOWARD, physicist; b. Danville, Va., Aug. 21, 1925; m. Virginia M. Robertson, 1964; 4 children. BEE, U.Va., 1946; postgrad., Princeton U., 1946-52. Rsch. scientist RCA, N.J., 1946-52; pres. Radiation Rsch. Corp., N.Y., 1952-67, Plasma Physics Corp., Locust Valley, N.Y., 1967—, Solar Physics Corp., Locust Valley, N.Y., 1990—. Achievements include numerous patents. Office: Solar Physics Corp PO Box 548 Locust Valley NY 11560 Home: 40 Overlook Rd Locust Valley NY 11560

COLEMAN, JOHN MORLEY, transportation research director; b. Ottawa, Ont., Can., Dec. 24, 1948; s. Morley Hillis and Marion Sloan (McKelvie) C.;

m. Rebecca J. Truxal, June 1, 1974; 1 child, Adam J. BEng, Carleton U., Ottawa, 1971; MBA, U. Western Ont., London, 1973. Registered profl. engr., Ont. Micrographics cons. tech. divsn. Pub. Archives of Can., 1973-77; policy analyst industry br. Ministry State for Sci. and Tech., 1977-81; contracts coord. contract svcs. office Nat. Rsch. Coun. Can., 1981, spl. projects program svcs. secretariat, 1982-85, exec. asst. to pres., 1986-87, coord. transp. program, 1987-98, head indsl. liaison office Inst. Mech. Engring., 1985-89, head ground transp. tech. program Inst. Mech. Engring., 1989-93, dir. gen. Ctr. for Surface Transp. Tech., 1993—; mem. Can. Railway Rsch. adv. bd., 1993—; mem., co-founder Tech. Transfer Round Tabel Ministère des Transports du Québec; mem. steering com. Future Logging Truck Project, Transport Can., and Forest Rsch. Engring. Inst. Can., 1992—; lectr. in field. Mem. Transp. Assn. Can. (R & D coun. 1988—, chmn. heavy vehicle rsch. coordination com. 1989—, conf. session planning coms., R & D coun. 1992, 93, lectr.), Assn. Am. Railroads (implementation officers com. 1991—), Assn. Profl. Engrs. of Ont. Home: 20 Shannondoe Cres, Kanata, ON Canada K2M 2H1 Office: National Rsch Council of Can, U-89 Montreal Rd, Ottawa, ON Canada K1A 0R6

COLEMAN, LESTER LAUDY, otolaryngologist; b. N.Y.C., Mar. 17, 1911; s. Avron and Ann (Blum) C.; m. Felicia Slatkin, Sept. 30, 1945 (dec. 1981); 1 child, Lisa; m. Elizabeth Smith Pantano, Mar. 9, 1986; 1 child, Lynn Ann Dale. B.S., Johns Hopkins U.; M.D., L.I., Coll. Medicine, 1932. Diplomate: Am. Bd. Otolaryngology. Asst. resident in otolaryngology Johns Hopkins Hosp., 1936-38; practice medicine specializing in otolaryngology N.Y.C., 1940—; med. dir. Morton Prince Center Psychotherapy, N.Y.C.; attending surgeon Manhattan Eye, Ear and Throat Hosp., N.Y. Hosp. Cornell Med. Center; asst. clin. prof. Albert Einstein Sch. Medicine; cofounder Internat. Grad. U. Med. columnist: Speaking of Your Health, King Features, Inc.; producer show, NBC-TV, 1953-56. Served to maj. M.C. AUS, 1942. Fellow Am. Trilogical Soc.; mem. Am. Acad. Psychosomatic Medicine (past v.p.). Home: 1000 Park Ave New York NY 10028-0934 Office: 114 E 72d St New York NY 10021

COLEMAN, MARK DAVID, mechanical engineer; b. Santa Monica, Calif., May 11, 1960; s. Gilbert and Mary Patricia (Daly) C.; 1 child, David Alan. BS in Civil Engring., U. Portland, 1978; BSME, U. Del., 1990; MSME, LaSalle U., 1993. Designer Westinghouse, Richland, Wash., 1979-80; design engr. Ebasco Svcs., Inc., N.Y.C., 1980-83; mech. engr. United Engrs. and Constructors, Seabrook, N.H., 1983-84, Bechtel Corp., Gaithersburg, Md., 1984—. Contbr. articles to profl. jours. Office: Bechtel Corp 9801 Washington Blvd Gaithersburg MD 20878

COLEMAN, MICHAEL DORTCH, nephrologist; b. Jackson, Tenn., June 19, 1944; s. Ivery R. and Kathleen (Campbell) C.; children by previous marriage: Michael Dortch, Christopher Mathew; m. Stephanie Sherean Summers; 1 child, Cassandra Sherean. BA in Chemistry, U. Ark., 1966; MD, Duke U., 1970. Diplomate Am. Bd. Internal Medicine. Intern, Duke U. Med. Sch., Durham, N.C., 1970-71, resident internal medicine, 1971-72, nephrology fellow, 1972-74; practice medicine specializing in nephrology, Durham, 1972-74, Kannapolis, N.C., 1973-74, Ft. Smith, Ark., 1974—; nephrology cons. Cabarrus County Hosp., Kannapolis, 1973; chief dept. nephrology Holt Krock Clinic, Ft. Smith, 1974—, dir. dialysis Holt Krock Dialysis Ctr., 1974—, Sparks Regional Med. Center, Ft. Smith, 1974—, St. Edward's Mercy Med. Ctr., Ft. Smith, 1980—; assoc. prof. medicine U. Ark., Ft. Smith, 1976—; mem. med. rev. bd. Ark. Kidney Disease Commn., 1974—; nephrology cons., 1974—; mem. exec. com. and med. rev. bd. Ark.-Okla. Endstage Renal Disease Coun., 1977—. Bd. dirs. Ark. Tennis Assn., Jr. Tennis Coun., Holt Krock Clinic, Ft. Smith, Ark.; bd. dirs., mem. fin. com. Holt Krock Clinic. Mem. Internat. Soc. Nephrology, Renal Physician Assn., Am. Soc. Nephrology, Am. Heart Assn., AMA, Ark. Med. Assn., Sebastian County Med. Assn., Ft. Smith Racquet Club (bd. dirs., pres.), Town Club of Ft. Smith, Hardscrabble Country Club, Alpha Omega Alpha. Contbr. articles to med. jours. Office: 1500 Dodson St Fort Smith AR 72901

COLEMAN, PAUL JEROME, JR., physicist, educator; b. Evanston, Ill., Mar. 7, 1932; s. Paul Jerome and Eunice Cecile (Weissenberg) C.; m. Doris Ann Fields, Oct. 3, 1964; children: Derrick, Craig. BS in Engring. Math., U. Mich., 1954, BS in Engring. Physics, 1954, MS in Physics, 1958; PhD in Space Physics, UCLA, 1966. Rsch. scientist Ramo-Wooldridge Corp. (name now TRW Systems), El Segundo, Calif., 1958-61; instr. math. U. So. Calif., L.A., 1958-61; mgr. interplanetary scis. program NASA, Washington, 1961-62; rsch. scientist UCLA, 1962-66, prof. geophysics, space physics, 1966—, dir. Inst. Geophysics and Planetary Physics, 1989-92; pres. Univs. Space Rsch. Assn., Columbia, Md., 1981—; bd. dirs. Lasertechnics Inc., Albuquerque, Caci Internat., Washington, Fairchild Space and Def. Co., Germantown, Md., others; mem. adv. bd. San Diego Supercomputer Ctr., 1986—, chmn., 1987-88, others; trustee Univs. Space Rsch. Assn., Columbia, Md., 1981—, Am. Tech. Initiative, 1990—, Internat. Small Satellite Org., 1992—; vis. scholar U. Paris, 1975-76; vis. scientist Lab. for Aeronomy Ctr. Nat. Rsch. Sci., Verrières le Buisson, France, 1975-76; asst. lab. dir., mgr. Earth and Space Scis. div., chmn. Inst. Geophysics and Planetary Physics Los Alamos (N.Mex.) Nat. Lab., 1981-86; com. mem. numerous scientific and ednl. orgns., cons. numerous fin. and indsl. cos. Co-editor: Solar Wind, 1972; co-author: Pioneering the Space Frontier, 1986; mem. editorial bd. Geophysics and Astrophysics Monographs, 1970—; assoc. editor Cosmic Electrodynamics, 1968-72; contbr. revs. to numerous profl. jours. Apptd. to Nat. Commn. on Space, Pres. of U.S., 1985, apptd. to Space Policy Adv. Bd., Nat. Space Coun., V.p. of U.S., 1991; bd. dirs. St. Matthew's Sch., Pacific Palisades, Calif., 1979-82, v.p., 1981-82. 1st lt. USAF, 1954-56, Korea. Recipient Exceptional Sci. Achievement Medal NASA, 1970, 1972, spl. recognition for contributions to the Apollo Program, 1979; Guggenheim fellow 1975-76, Fulbright scholar, 1975-76, Rsch. grantee NASA, NSF, Office Naval Research, Calif. Space Inst., Air Force Office Sci. Research, U.S. Geol. Survey. Mem. AAAS, AIAA, Am. Geophys. Union, Am. Phys. Soc., Internat. Acad. Astronautics, Bel Air Bay Club (L.A.), Birnam Wood Golf Club (Santa Barbara, Calif.), Cosmos Club (Washington), Explorers Club (N.Y.C.), Eldorado Country Club (Indian Wells, Calif.), Tau Beta Pi, Phi Eta Sigma. Avocations: flying, skiing, racquetball, tennis, golf. Home: 1323 Monaco Dr Pacific Palisades CA 90272-4007 Office: UCLA Inst Geophysics & Planetary Physics 405 Hilgard Ave Los Angeles CA 90024-1301

COLEMAN, RICHARD WALTER, biology educator; b. San Francisco, Sept 10, 1922; s. John Crisp and Reta (Walter) C.; m. Mildred Bradley, Aug. 10, 1949 (dec.); 1 child, Persis C. BA, U. Calif., Berkeley, 1945, PhD, 1951. Rsch. asst. div. entomology and parasitology U. Calif., Berkeley, 1946-47, 49-50; ind. rsch., 1951-61; prof. biology, chmn. dept. Curry Coll., Milton, Mass., 1961-63; chmn. div. scis. and math. Monticello Coll., Godfrey, Ill., 1963-64; vis. prof. biology Wilberforce U., Ohio; 1964-65; prof. sci. Upper Iowa U., Fayette, 1965-89; prof. emeritus sci., 1989—; collaborator natural history div. Nat. Park Svc., 1952; spl. cons. Arctic Health Rsch. Ctr., USPHS, Alaska, 1954-62; apptd. explorer Commr. N.W. Ty., Yellowknife N.W. Ty., Can., 1966. Contbr. articles to profl. reports. Mem. AAAS, Nat. Health Fedn., Iowa Acad. Sci., Geol. Soc. Iowa (affiliate), Am. Inst. Biol. Scis., Nat. Sci. Tchrs. Assn., Ecol. Soc. Am., Am. Soc. Limnology and Oceanography, Am. Bryological and Lichenological Soc., Arctic Inst. N.Am., N.Am. Benthological Soc., Am. Malacological Union, Assn. Midwestern Coll. Biology Tchrs., The Nature Conservancy, Nat. Assn. Biology Tchrs., Sigma Xi. Methodist. Home: PO Box 13321 Baltimore MD 21203-3321

COLEMAN, ROBERT GRIFFIN, geology educator; b. Twin Falls, Idaho, Jan. 5, 1923; s. Lloyd Wilbur and Frances (Brown) C.; m. Cathryn J. Hirschberger, Aug. 7, 1948; children: Robert Griffin Jr., Derrick Job, Mark Dana. B.S., Oreg. State U., 1948, M.S., 1950; Ph.D., Stanford U., 1957. Mineralogist AEC, N.Y.C., 1952-54; geologist U.S. Geol. Survey, Washington, 1954-57, Menlo Park, Calif., 1958-80; prof. geology Stanford U., Calif., 1981-93, prof. emeritus, 1993; vis. petrographer New Zealand Geol. Survey, 1962-63; br. chief isotope geology U.S. Geol. Survey, Menlo Park, 1964-68, regional geologist, Saudi Arabia, 1970-71, br. chief field geochemistry and petrology, Menlo Park, 1977-79; vis. scholar Woods Hole Oceanographic Inst., Mass., 1975; vis. prof. geology Sultan Qaboos U., Oman, 1987, 89; cons. geologist, 1993—. Author: Ophiolites, 1977; contbr. articles to profl. jours., monographs. Named Outstanding Scientist, Oreg. Acad. Sci., 1977; Fairchild scholar Calif. Inst. Tech., Pasadena, 1980; recipient Meritorious award U.S. Dept. Interior, 1981. Fellow Geol. Soc. Am. (coun.), Am. Mineral Soc. (coun., editor), Am. Geophys. Union; mem. Nat. Acad. Scis. Republican. Avocations: wood carving; art. Home: 2025 Camino Al Lago Menlo Park CA 94027-5938 Office: Stanford Univ Geology Dept Stanford CA 94305

COLEMAN, ROBERT SAMUEL, chemistry researcher, educator; b. Sioux City, Iowa, Apr. 26, 1959; s. Robert and Elizabeth Emily (Meier) C. BS with honors in chemistry, U. Iowa, 1981; MS in Medicinal Chemistry, U. Kans., 1984; PhD in Chemistry, Purdue U., 1987. Grad. rsch. asst. Dept. Chemistry, Purdue U., West Lafayette, Ind., 1985-87; NIH postdoctoral fellow Dept. Chemistry, Yale U., New Haven, 1988-89; asst. prof. chemistry U. S.C., Columbia, 1989—; lectr. in field; conductor seminars in field. Contbr. articles to profl. jours. Lilly Teaching fellow U. S.C., 1992-93, NIH postdoctoral fellow, Yale U., 1988-89, David Ross Grad. fellow, Purdue U., 1986-87; NIH predoctoral trainee U. Kans., 1984-85; recipient Jr. Faculty Rsch. award Am. Cancer Soc., 1991-93, Disting. New Faculty award Camille and Henry Dreyfus Found., 1989-94, Genentech Investigator in Biomolecular Chemistry award, 1990-92, Am. Cyanamid Faculty award, 1992-93, A.W. Davidson award Phi Lambda Upsilon, 1984. Achievements include research on first total synthesis of the potent antitumor agent CC-1065; achieved with others the first synthesis of the core structure of the antitumor agent calicheamicin; other research includes natural products total synthesis and mechanism of action, synthetic organic methods development, antisense oligonucleotides. Office: Univ of SC Dept Chemistry/Biochemistry Columbia SC 29208

COLEMAN, SAMUEL EBOW, chemist, engineer; b. Kumasi, Ghana, Feb. 13, 1945; came to U.S., 1978; s. Samuel Atta and Lucy (Sam) C.; m. Akosua Adufa Mintah, Aug. 31, 1974; 1 child, Kwesi. BSc in Chemistry, U. Sci. and Tech., Kumasi, 1972; MSc in Cement Prodn. Tech., Inst. Chem. Tech., Prague, Czechoslovakia, 1977; PhD, Purdue U., 1981. Gen. sec. All African Students Union, Accra, 1972-73; rsch. asst. Bldg. and Rd. Rsch. Inst., Kumasi, 1974; sr. rsch. engr. Dowell Schlumberger, Tulsa, 1982-86; project engr. Law Engring., Houston, 1986-88; tech. mgr. Southwestern Labs., Houston, 1989-91; prin. cons. S.E. Coleman & Assocs., Houston, 1991—; presenter at profl. confs. Exec. com. Nat. Union Ghana Students, 1970-71; pres. Union Ghanaian Students in Czechoslovakia, 1976, African Students Union Purdue U., 1979. Fellow Am. Inst. Chemists; mem. NSPE, Am. Soc. Testing and Materials, Internat. Concrete Repair Inst. (pres. Houston chpt.), UN Orgn. Ea. Okla., Ghana U. Sci. and Tech. Alumni of Houston (pres.). Achievements include shared patents for CO2-Enhanced Hydrocarbon with Corrosion-Resistant Cement and Method for Underground Support and Removal of Hazardous Ions in Ground Waters. Home: 12866 Westleigh Dr Houston TX 77077-3739 Office: S E Coleman & Assocs PO Box 820608 Houston TX 77282-0608

COLEMAN, SIDNEY RICHARD, physicist, educator; b. Chgo., Mar. 7, 1937; s. Harold Albert and Sadie (Shanas) C. B.S., Ill. Inst. Tech., 1957; Ph.D., Calif. Inst. Tech., 1962. Research fellow dept. physics Harvard U., 1961-63, asst. prof., 1963-66, assoc. prof., 1966-69, prof., 1969—; vis. prof. U. Rome, Italy, 1968, U. Calif., Berkeley, 1989, Princeton U., 1973, Stanford U., 1979-80; partner Advent Pubs. Author: Aspects of Symmetry, 1985. Recipient prize for physics lectures Ettore Majorana Centre Sci. Culture, Boris Pregel award N.Y. Acad. Sci., Disting. Alumnus award Calif. Inst. Tech., Dirac medal Internat. Centre for Theoretical Physics 1990. Fellow NAS (J. Murray Lack award for sci. revs.), Am. Acad. Arts and Sci., Am. Phys. Soc.; mem. Lilapa. Home: Unit 12 1 Richdale Ave Cambridge MA 02140 Office: Harvard U Lyman Lab Cambridge MA 02138

COLEMAN, TOMMY LEE, soil science educator, researcher, laboratory director; b. Baxley, Ga., Nov. 8, 1952; s. E.C. and Lucille (Fussell) C.; m. Mildred Cross, Dec. 22, 1974 (div. 1977); m. Edna Thompson, Mar. 6, 1982; children: Sherri, Thomas, Brian. BS in Agronomy, Fort Valley State Coll., 1974; MS, U. Ga., 1977; PhD, Iowa State U., 1980. Soil scientist USDA/Soil Conservation Svc., Statesboro, Ga., 1974-77; rsch. assoc. Iowa State U., Ames, 1977-80; postdoctoral fellow in rsch. Ala. A&M U., Normal, 1981-83, asst. prof., 1983-89, assoc. prof., 1989-93, prof. soil sci. and remote sensing, 1993—, dir. remote sensing lab, 1990—; rsch. phys. scientist USGS, Reston, Va., 1992—; cons. Abiola Farms Ltd., Lagos, Nigeria, 1988, U.S. AID-Botswana, Gaborone, 1989-91, INRAN-DRE and U.S. AID-Niger, Niamey, 1990. Contbr. articles to profl. jours. Mem. Am. Soc. Agronomy, Soil Sci. Soc. Am., Am. Soc. Photogrammetry and Remote Sensing, NAACP, North Ala. Golf Club (pres. Huntsville chpt. 1987-92), Omega Psi Phi (Xi Omicron chpt., chair scholarship com. Huntsville chpt. 1986-90). Democrat. Baptist. Office: Ala A&M U Dept Plant & Soil Sci PO Box 1208 Normal AL 35762-1208

COLEMAN, WILLIAM ELIAH, psychologist; b. Utica, N.Y., May 30, 1921; s. Michael and Mildred (Hoffman) C.; m. June Juster, Oct. 3, 1943; children: Nancy, Lawrence, Karen. BA, Ohio State U., 1942, MA, 1946, PhD, 1949. Teaching asst. Ohio State U., Columbus, 1946-47; instr. U. Tenn., Knoxville, 1947-49, asst. prof., 1949-51, dir. testing and guidance, 1951-56; human factors scientist RAND/SDC, Santa Monica, Calif., 1956-60; dir. psychological svcs. Ward J. Jenssen, Inc., L.A., 1960-61; pres. Coleman and Assocs., Santa Monica, Calif., 1961--; cons. Dept. of Labor, 1963-71; counseling and testing cons. Houghton-Mifflin, 1957-59; lectr. ULCA, 1960-85. Author: Personnel Selection and Recruitment, 1985; co-author: (test) Life Goals, 1965, Clerical Skills Inventory, 1966; contbr. articles to profl. jours. Vice chmn. TFED, Torrance, Calif, 1988--; pres. New Deal Dem. Club, Santa Monica, 1989--; mem. RSVP Bd., Santa Monica, 1992--. Lt. U.S Army, 1942-45, ETO. Fellow APA, Am. Psychol. Soc.; mem. Am Counseling Assn (life). Jewish. Office: Coleman and Assocs 1238 7th St Santa Monica CA 90401

COLES, RICHARD W(ARREN), biology educator, research administrator; b. Phila., Sept. 16, 1939; s. Henry Braid and Katherine Warren (Baker) C.; m. Mary Sargent, June 10, 1962, children: Christopher Sargent, Deborah Coles Ryan. BA with highest honors, Swarthmore Coll., 1961; MA, Harvard U., 1967, PhD, 1967. Teaching fellow Harvard U., 1961-65; asst. prof. biology Claremont Coll., Calif., 1966-70; dir. Tyson Rsch. Ctr., prof. biology Washington U., St. Louis, 1970—; vis. instr. Nature Place, Florissant, Colo., 1981—; participant seven ornithol. field trips, Venezuela, 1986-93, leader Costa Rica, Ecuador, Galapagos and U.S., 1990, 91; cons., sec.-treas. Orgn. Biol. Field Stas., 1976—; cons. NSF, Nat. Geogrpahic Soc., book pubs., park planners. Contbr. articles to profl. jours. Grantee in field. Mem. AAAS, Am. Inst. Biol. Scis., Am. Soc. Zoologists, Am. Soc. Mammalogists, Am. Ornithologists Union, Soc. Conservation Biology, Ecol. Soc. Am., Wilderness Soc., Animal Behavior Soc., Webster Groves Nature Study Soc., World Wildlife Fund, Phi Beta Kappa, Sigma Xi. Quaker. Clubs: Explorers (N.Y.C.); Naturalists (St. Louis). Home: 11 Hickory Ln Eureka MO 63025-3104 Office: Washington U Tyson Rsch Ctr PO Box 258 Eureka MO 63025-0258

COLGAN, EVAN GEORGE, materials scientist; b. Oakland, Calif., May 16, 1960; s. William George and Helen Hope (Evans) C. BS, Calif. Inst. Tech., 1982; PhD, Cornell U., 1987. Staff engr. IBM Tech. Products, East Fishkill, N.Y., 1987-91; adv. engr. IBM Rsch., Yorktown Heights, N.Y., 1991—. Contbr. articles to profl. jours. Mem. Bohmische Phys. Soc., Sigma Xi. Office: IBM Watson Rsch Ctr 12-258 PO Box 218 Yorktown Heights NY 10598

COLGATE, SAMUEL ORAN, chemistry educator; b. Amarillo, Tex., Oct. 5, 1933; s. Cleon Edward C.; m. Betty Joyce Hart, Sept. 3, 1955; children: Rebecca Ann, James Edward. MS, Okla. State U., 1956; PhD, MIT, 1959. Prof. chemistry U. Fla., Gainesville, 1959—; rschr. Acoustic Resonance Spectroscopy. Contbr. articles to profl. jours. Mem. Am. Chem. Soc., Am. Phys. Soc., Am. Vacuum Soc. Achievements include patents in fluid idspensing, electric switching, hermetic seals; patents pending in acoustic flow monitoring, area selective chemical vapor deposition. Home: 4132 NW 38th St Gainesville FL 32606 Office: Univ of Fla Dept Chemistry Gainesville FL 32611-2046

COLGATE, STIRLING AUCHINCLOSS, physicist; b. N.Y.C., Nov. 14, 1925; s. Henry A. and Jeannette (Pruyn) C.; m. Rosemary B. Williamson, July 12, 1947; children; Henry A., Sarah, Arthur S. BA, Cornell U., 1948, PhD in Physics, 1952. Physicist Radiation Lab., Univ. Calif., Berkeley, 1951-52, Lawrence Livermore (Calif.) Lab., 1952-64; pres. N.Mex. Inst. Mining and Tech., Socorro, N.Mex., 1964-74; physicist Los Alamos (N.Mex.) Nat. Lab., 1976—. Contbr. over 200 articles to profl. jours. Served with Merchant Marines, 1943-46. Fellow Am. Phys. Soc.; mem. Am. Astron. Soc., Nat. Acad. Home: 422 Estante Way Los Alamos NM 87544-3812 Office: MS B275 Los Alamos Nat Lab Los Alamos NM 87545

COLICE, GENE LESLIE, physician; b. N.Y.C., May 1, 1950; s. Joseph V. and Matilda (Finkel) C.; m. Elizabeth O'Hare, June 3, 1973; children: C. Max, Benjamin, Anne. AB, Brown U., 1972; MD, NYU, 1976. Diplomate internal medicine, pulmonary and criticalcare medicine. Asst. prof. medicine U. South Fla., Tampa, 1982-85; Asst. prof. medicine Dartmouth Coll., Hanover, N.H., 1985-91, assoc. prof., 1991—; chief pulmonary sect., dir. med. ICU VA Hosp., White River Junction, Vt., 1985—; bd. dirs. Vt. Lung Assn., Burlington. Contbr. articles to profl. jours. Fellow Am. Coll. Chest Physicians; mem. Am. Thoracic Soc. (sec. coun. chpt. reps. 1991-92, chmn. com. health care 1992—). Home: RR 2 Box 17 Norwich VT 05055 Office: VA Hosp White River Junction VT 05001

COLLAN, YRJÖ URHO, pathologist, medical educator, physician, toxicopathology consultant; b. Helsinki, June 23, 1941; s. Yrjö Johannes and Toini Lahja (Sederholm) C.; m. Eira Kyllikki Lehto, Oct. 9, 1971; children: Yrjö William, Lauri Urho, Anni Toini. Candidate in medicine, U. Helsinki, 1963, lic. in medicine, 1968, DMS, 1972. Bd. cert. pathologist. Registrar dept. pathology U. Helsinki, 1968-73; registrar dept. surgery U. Cen. Hosp., Helsinki, 1973-74; assoc. prof. dept. pathology U. Helsinki, 1976-77, cons. dept. pathology, 1977-78; lab. supr. Inst. Occupational Health, Helsinki, 1978-80; prof., chmn. pathology U. Kuopio, Finland, 1980-88; dir. histopathology U. Cen. Hosp., Turku, Finland, 1988-89; prof. U. Turku, 1989—; chmn. com. for postgrad. edn. U. Kuopio, 1982-88; dir. pathology service U. Cen. Hosp., Kuopio, 1980-88; contract prof. U. Ancona, Italy, 1985-87; expert cons. Nat. Bd. Health, Helsinki, 1987—. Author: Medical English, 1975; editor: Morphometry in Morphological Diagnosis, 1982, Stereology and Morphometry in Pathology, 1984, Annals of Clinical Reserach, Annales Chirugiae et Gynaecologiae, 1969-74; editorial bd. Acta Sterologica, 1981—, Analytical and Cellular Pathology, 1987, Forma, 1989—. Mgr. Soc. Young Friends of Nature in Finland, Helsinki, 1961-62; bd. dirs. student union Helsinki U., 1966; v.p. North Savo Cancer Soc., Kuopio, 1982-88. Lt. Finnish mil., 1964. Fogarty internat. fellow, 1974-75, NSF/Acad. Finland grantee, 1988-89; recipient Class I medal Order of White Rose of Finland, 1982, Symbol of Accademia Medico Chirurgica del Piceno, Italy, 1986, Medal City Milan, 1986, Ancone, 1991. Mem. Internat. Acad. Pathology (internat. councillor 1982-88, pres. Finnish sect. 1986-88), Internat. Soc. Stereology (Scandinavian rep. 1983-87), European Soc. Pathology (chmn. com. for diagnostic quantitative pathology 1988-92), European Soc. Analytical Cellular Pathology (coun. 1986-93), Soc. for Cytometry and Morphometry in Finland (chmn. 1982-87, 90—), Finnish Cancer Soc. (bd. dirs. 1992—), Turku/Soc. for Cancer Rsch. (chmn. 1993—). Lutheran. Avocations: ornithology, languages. Office: U Turku Dept Pathology, Kiinamyllynkatu 10, SF20520 Turku Finland

COLLARD, RANDLE SCOTT, chemist; b. San Antonio, July 31, 1951; s. George C. and Dorcas Elizabeth (McLaughlin) C.; m. Vicki Jean Rosenbaum, Aug. 17, 1973; children: Megan Ruth, Matthew Scott. BS, Trinity U., 1973; PhD, La. State U., 1978. Project leader R&D Dow, USA, Plaquemine, La., 1978-84, supr. instrument lab., 1984-88, group leader analyzer devel. and environ. rsch., 1988-90, rsch. mgr. engring. scis., 1990-92; mgr. computation and simulation Dow, USA, Midland, Mich., 1992—; mem. La. Gov.'s Task Force, 1987; chmn. On-Line Monitoring Task Force, Baton Rouge, 1989-92; chmn. Dow Computational Fluid Dynamics Task Force, Palquemine, 1991—; chmn. chem. adv. bd. Trinity U., San Antonio, 1990—; presenter at profl. confs. Author tech. reports. Bd. dirs. Cerebral Palsy Ctr., Baton Rouge, 1989-90, Chapel Trafton Sch., Baton Rouge, 1991—. Mem. Am. Chem. Soc. Achievements include development of method for determining PCBs in complex chlorinated matrices. Office: Dow USA Central Rsch 1776 Midland MI 48764

COLLAZOS GONZALEZ, JULIO, internist, researcher; b. Tordehumos, Valladolid, Spain, Mar. 11, 1955; s. Pablo Collazos and Concepción González. MD with honors, U. Complutense, Madrid, 1978, internal medicine specialist, 1984; PhD, U. Autonoma, Madrid, 1990. Resident Found. Jimenez Diaz, Madrid, 1980-84; attending physician Hosp. Provincial, Alicante, Spain, 1984-87; assoc. prof. U. Alicante, 1984-87; attending physician Hosp. de Galdácano, Vizcaya, Spain, 1987-93, chief infectious diseases sect., 1993—. Author, editor: The Tumor Markers in Benign Liver Diseases, 1990; also numerous articles. Avocations: sports, photography, computers. Office: Hosp de Galdácano, Bo Labeaga s/n, 48960 Galdácano Vizcaya, Spain

COLLEN, MORRIS FRANK, physician; b. St. Paul, Nov. 12, 1913; s. Frank Morris and Rose (Finkelstein) C.; m. Frances B. Diner, Sept. 24, 1937; children: Arnold Roy, Barry Joel, Roberta Joy, Randal Harry. BEE, U. Minn., 1934, MB with distinction, 1938, MD, 1939. Diplomate Am. Bd. Internal Medicine. Intern Michael Reese Hosp., Chgo., 1939-40; resident Los Angeles County Hosp., 1940-42; chief med. service Kaiser Found. Hosp., Oakland, Calif., 1942-43; chief of staff Kaiser Found. Hosp., San Francisco, 1953-61; med. dir. Permanente Med. Group, West Bay Div., 1953-79, dir. med. methods research, 1962-79, dir. tech. assessment, 1979-83, cons. div. research, 1983—; chmn. exec. com. Permanente Med. Group, Oakland, 1953-73; dir. Permanente Services, Inc., Oakland, 1958-73; lectr. Sch. Pub. Health, U. Calif., Berkeley, 1966-78; lectr. info. sci. U. Calif., San Francisco, 1970-85; lectr. U. London, 1972, Stanford U. Med. Ctr., 1973, 75, 84-86, Harvard U., 1974, Johns Hopkins U., 1976, also others; cons. Bur. Health Services, USPHS, 1965-68, chmn. health care systems study sect., 1968-72, mem. adv. com. demonstration grants, 1967; advisor VA, 1968; cons. European region WHO, 1968-72; cons. med. fitness program U.S. Air Force, 1968; cons. Pres.'s Biomed. Research Panel, 1975; mem. adv. com. automated Multiphasic Health Testing, 1971; discussant Nat. Conf. Preventive Medicine, Bethesda, Md., 1975; mem. com. on tech. in health care NAS, 1976; mem. adv. group Nat. Commn. on Digestive Diseases, U.S. Congress, 1978; mem. adv. panel to U.S. Congress Office of Tech. Assessment, 1980-85; mem. peer rev. adv. group TRIMIS program Dept. Def., 1978-90; program chmn. 3d Internat. Conf. Med. Informatics, Tokyo, 1980; chmn. bd. sci. counselors Nat. Library Medicine, 1985-87. Author: Treatment of Pneumococcic Pneumonia, 1948, Hospital Computer Systems, 1974, Multiphasic Health Testing Systems, 1977, Informatics: A Historical Review, 1991; editor: Permanente Med. Bull., 1943-53; mem. editorial bd. Preventive Medicine, 1970-80, Jour. Med. Systems, Methods Information Medicine, 1980-93, Diagnostic Medicine, 1980-84, Computers in Biomed. Rsch., 1987-93; contbr. articles to med. jours., chpts. to med. books. Johns Hopkins Centennial scholar, 1976; fellow Ctr. Adv. Studies in Behavioral Scis., Stanford, 1985-86; scholar-in-residence Nat. Libr. Medicine, 1987-93; recipient Pioneer award Computers in Health Care Jour., 1992. Fellow ACP, Am. Coll. Cardiology, Am. Coll. Chest Physicians, Am. Inst. Med. & Biol. Engring.; mem. AMA, Inst. Medicine of NAS (chmn. tech. subcom. for improving patient records 1990, chmn. workshop on informatics in clin. preventive medicine 1991), Am. Fedn. Clin. Rsch., Salutis Unitas (v.p. 1972), Soc. Adv. Med. Systems (pres. 1973), Nat. Acad. Practice in Medicine (chmn. 1983-88, co-chmn. 1989-91), Am. Coll. Med. Informatics (pres. 1987-88), Am. Med. Informatics Assn. (bd. dirs. 1989-91), Internat. Med. Informatics Assn. (hon.), Internat. Health Evaluation Assn. (bd. dirs. 1985-93, Lifetime Achievement award 1992, Computers in Healthcare Pioneer award 1992), Alpha Omega Alpha, Tau Beta Pi. Home: 4155 Walnut Blvd Walnut Creek CA 94596-5834 Office: 3451 Piedmont Ave Piedmont CA 94611-5463

COLLEY, ROGER J., environmental biotechnology research company executive; b. Phila., Feb. 10, 1938; s. Emil William and Josephine Marie Colley; m. Janice Marie Schnell, June 29, 1963; children: Kenneth, Carolyn, Deborah, Alexander. BS in Econs., U. Pa., 1960. From mem. staff to pres. Betz Labs. Inc., Trevose, Pa., 1966-85; pres., chief exec. officer Envirogen Inc., Lawrenceville, N.J., 1988—. Author: (manual) Positive Cash Flow, 1988. Mem. township planning com., Lower Moreland, Pa., 1988-90; mem. Ancillae Assumpta Acad. Bus. Bd., 1983-91; trustee Children's Hosp. Pa., Phila., 1984—. Mem. AICPA, Am. Mgmt. Assn., Huntingdon Valley Country Club. Office: Envirogen Inc 4100 Quakerbridge Rd Trenton NJ 08648-4702

COLLEY, THOMAS ELBERT, JR., psychologist; b. Washington, Feb. 6, 1928; s. Thomas Elbert Sr. and Clara Ellen (Miller) C.; m. May 20, 1952 (div.); children: Glenn, Wendy, Jennifer. BA, Am. U., 1952, MA, 1954; PhD, Syracuse (N.Y.) U., 1965. Lic. psychologist, Ky., Ind. Psychologist Cen. State Hosp., Lakeland, Ky., 1958-65; chief psychologist U. Louisville Child Evaluation Ctr., 1965-73; pvt. practice clin. and child psychology Louisville, 1973—; cons. Ky. Disability Determination Svcs., Jefferson County Pub. Defender's Office, Multichannel Cochlear Implant Program, Whitney Young Job Corps Ctr., Washington County Schs. Spl. Edn., U. Louisville Med. Sch., Surrogate Parenting Assocs., Inc., Family and Children's Agy., Ky. Easter Seal Hearing and Speech Ctr., Louisville Sch. for Autistic Children, Tri-County Spl. Edn. Coop., Ky. Dept. Vocat. Rehab., Louisville Deaf-Oral Sch., Shelby County Schs. Spl. Edn., St. Matthews Area Ministries Day Care. Mem. Ky. Bd. Psychology Examiners, Gov.'s Commn. on Mental Health; chair adv. bd. Sch. for Autistic Children; bd. dirs. Ky. Assn. for Mental Health. Sgt. U.S. Army, 1950-52, Korea. Mem. APA, Am. Assn. Mental Deficiency, Ky. Psychol. Assn., Ky. Psychol. Soc., Ind. Psychol. Assn., Louisville Psychol. Soc. (pres.), Louisville Pediatric Soc., Southeastern Psychol. Assn. Avocations: canoeing, singing, guitar, camping. Office: 804 Medical Towers North Louisville KY 50202

COLLIER, JOHN GORDON, nuclear scientist; b. London, Jan. 22, 1935; s. John Collier and Edith Georgina née de Ville; m. Ellen Alice Mary Mitchell; 2 children. BSE, University Coll., London; DSc (hon.), Cranfield, 1988. Apprentice in mech. and chem. engring. AERE, Harwell, 1951-56, SO, then SSO chem. engring. divsn., 1957-62; sect. head, then br. head, chem. engring., adviser reactor engring. divsn. Atomic Energy Can., Ltd., 1962-64; SSO, then PSO chem. engring. divsn. AERE, Harwell, 1964-66; head engring. divsn. Atomic Power Constrns., Ltd. R&D Lab., Heston, 1966-70; head engring. scis. group, later engring. scis. br. UKAEA, Heston, 1970-75, head chem. engring. divsn., 1975-77, mem. atomic energy tech. unit, 1977-79, head atomic energy tech. unit, 1979-81; dir. tech. studies UKAEA, Harwell, 1981-82; dir. safety and reliability directorate UKAEA, Culcheth, 1982-83; dir. gen., generation devel., constrn. divsn. CEGB, Barnwood, Gloucester, 1983-86; dept. chmn. CEGB, Barnwood, Glos., 1986, chmn., 1987-90, chmn. nationalised industries chairmen's group, 1990—. Author: Convective Boiling and Condensation, 1972, 2d edit., 1981; co-author: Introduction to Nuclear Power, 1987. Office: Nuclear Electric PLC, Barnett Way, Barnwood Gloucester GL4 7RS, England*

COLLIER, STEVEN EDWARD, utilities executive, consultant; b. Alamogordo, N.Mex., July 19, 1952; s. Homer Edward and Wanda JoAnn (Harred) C.; m. Trella Jean Wallace, May 26, 1973; children: Rachel, Joel, Lori, Ginna. BSEE, U. Houston, 1976; MSEE, Purdue U., 1977. Registered profl. engr., Okla., Tex. Planning engr. Houston Lighting & Power Co., Houston, 1970-76; researcher Purdue Electric Power Ctr., W. Lafayette, Ind., 1976-77; analytical cons. Power Technologies, Inc., Schenectady, N.Y., 1977-79; mem. tech. staff Sandia Nat. Lab., Albuquerque, 1979-80; sr. v.p. C.H. Guernsey & Co., Oklahoma City, 1980-89; v.p. energy resources and govt. affairs Cap Rock Electric, Stanton, Tex., 1989—; pres. New West Resources, Austin, Tex., 1991—; exec. v.p. New West Fuels, 1993—, New West Fuels, L.C., 1993—; bd. dirs. Trident Cogeneration, Inc., Houston; instr. NRECA MIP and Internat. Mgmt. Devel. Ctr.; speaker at confs. and convs., 1978—. Mem. editorial adv. bd. NRECA Rural Electrification Mag., 1988—, featured cover story, 1991; contbr. articles to profl. jours. Elder, chmn. bd. dirs. Community Christian Ch., Round Rock, Tex., 1990—. Grad. fellow in engring. NSF, 1976. Mem. IEEE, IEEE Industry Applications Soc. (bd. dirs. 1987-90, power com. chmn. 1986-87). Republican. Avocations: bicycling, running, guitar. Office: Cap Rock Electric 8140 Burnet Rd Austin TX 78758-7799

COLLING, DAVID ALLEN, industrial engineer, educator; b. Lancaster, N.Y., Apr. 21, 1935; s. Edward and Mabel Harriet (Clifford) C.; m. Genevieve DeMeo, Apr. 19, 1958 (div. 1985); children: Jeanne Maider, Gregory, Mark, Elizabeth, Suzanne. BS, MIT, 1957, MS, 1961, DSc, 1965. Registered profl. engr., Mass.; cert. safety profl. Metallurgist Army Materials Mechanics Rsch., Watertown, Mass., 1958-61, 71-74; sr. scientist Avco Corp., Wilmington, Mass., 1961-64, Westinghouse Rsch., Pitts., 1966-71; materials dir. Magnetic Corp. Am., Waltham, Mass., 1974-77; prof. U. Mass., Lowell, 1977—; cons. in field, Cambridge, Mass., 1977—. Author: Industrial Safety, 1990; contbr. articles to profl. jours. Bd. dirs. Greater Boston Assn. Retarded Citizens, 1990—. Mem. Nat. Safety Coun., Am. Soc. Safety Engrs., Nat. Assn. Indsl. Tech. Achievements include 2 patents in field. Home: 159 Concord Ave Cambridge MA 02138 Office: U Mass 1 University Ave Lowell MA 01854

COLLINS, ANGELO, science educator; b. Chgo., June 15, 1944; d. James Joseph and Mary (Burke) C. BS, Marian Coll., 1966; MS, Mich. State U., 1973; PhD, U. Wis., 1986. High sch. biology tchr. various schs., Wis., 1966-81; rsch. asst. U. Wis., Madison, 1981-86; asst. prof. Kans. State U., Manhattan, 1986-87, Stanford (Calif.) U., 1988-90, Rutgers U., New Brunswick, N.J., 1990-91; assoc. prof. Fla. State U., Tallahassee, 1991—; mem. Working Group on Sci. Stds., Washington, 1992, coord. 1993—, sci. com. Nat. Bd. Profl. Teaching Stds., Washington, 1991—; chmn. adv. bd. Bio-Quest, Beloit, Wis., 1988—; bd. dirs. Jour. for Rsch. in Sci. Teaching. Editor Tchr. Edn. Quarterly, 1991; reviewer several books; contbr. articles to profl. jours. Recipient Henry Rutgers fellowship, Rutgers U., 1990, grant NSF, 1987-90, 92, grant Carnegie, 1989, 90, grant Holmes Group N.E., 1990. Mem. AAAS, Nat. Assn. Biology Tchrs. (Outstanding Biology Tchr. Wis. 1977), Nat. Assn. Rsch. Sci. Teaching, Assn. Edn. Tchrs. Sci., Am. Ednl. Rsch. Assn., Sch. Sci. and Math., Assn. Tchr. Educators, Phi Delta Kappa. Office: Fla State Univ 203 Carothers Tallahassee FL 32306

COLLINS, ANITA MARGUERITE, research geneticist; b. Allentown, Pa., Nov. 8, 1947; d. Edmund III and Virginia (Hunsicker) C. BSc in Zoology, Pa. State U., 1969; MSc in Genetics, Ohio State U., 1972, PhD in Genetics, 1976. Instr. biology Mercyhurst Coll., Erie, Pa., 1975-76; rsch. geneticist Honey Bee Breeding Lab. Agrl. Rsch. Svc., USDA, Baton Rouge, 1976-88; rsch. leader Honey Bee Rsch. Lab. Agrl. Rsch. Svc., USDA, Weslaco, Tex., 1988—. Co-author: Bee Genetics & Breeding, 1986; contbr. articles to profl. jours. Mem. Entomol. Soc. Am. (chair subsect. Cb 1991), Assn. for Women in Sci. (pres. Baton Rouge chpt. 1982), Am. Beekeeping Fedn. (rsch. com. 1990, 92), Am. Genetics Assn., Animal Behavior Soc., Internat. Union for Study Social Insects, Sigma Xi. Office: USDA ARS Honey Bee Rsch 2413 E US Hwy 83 Weslaco TX 78596-8344

COLLINS, CHARLES CURTIS, pharmacist, educator; b. Hazard, Ky., Sept. 18, 1954; s. James Edward and Pecola (Logan) C.; m. Frances Allene Halcomb, Dec. 21, 1973; children: Charles Curtis II, Sarah Elizabeth, Candis Ashley. BS in Pharmacy, W.Va. U., 1977, PhD, 1984. Registered pharmacist, W.Va., Pa. Pharmacist Rite Aid of W.Va., Madison, 1977-79, Weston, 1979-81; pharmacist Monongahela Gen. Hosp., Morgantown, W.Va., 1981-83, Children's Hosp. Pitts., 1985—; asst. prof. Duquesne U., Pitts., 1983—. Coach Mt. Lebanon (Pa.) Girls Softball Assn., 1990—. Grantee Pa. State Matching Funds, 1986, Internat and Van Kel Industries, 1987, 88, 91; recipient Univ. Presdl. award, 1988. Mem. Am. Assn. Pharma. Scientists, Am. Pharma. Assn., Controlled Release Soc. Pa., Mason Lodge 684 (worshipful master 1993). Democrat. Baptist. Achievements include evaluation of new dissolution device which led to modifications and is now included as an official device in the U.S. Pharmacopiea, of new diffusion cell, now modified and undergoing further examination. Can be used with existing equipment. Office: Sch of Pharmacy Duquesne Univ Pittsburgh PA 15282

COLLINS, DENNIS GLENN, mathematics educator; b. Gary, Ind., June 26, 1944; s. Glenn and Irene Martha (Richman) C.; m. Barbara Jean Hamilton, July 14, 1979; 1 child, Glenn H. BA, Valparaiso U., 1966; MS, Ill. Inst. Tech., 1970, PhD, 1975. Temp. instr. Mich. State U., East Lansing, 1975-76; instr. U. New Orleans, 1976-79; asst. prof. Valparaiso (Ind.) U., 1979-82; from asst. prof. to prof. math. U. P.R., Mayaguez, 1982—; vis. assoc. prof. dept. math. Mich. State U., 1988-89; judge computer sci. 38th Internat. Sci. and Engring Fair, San Juan, P.R., 1987. Created copyrighted set postcards of mathematicians and physicists, 1983. NSF fellow, 1966-67; vis. scholar Mich. State U., 1988-89. Mem. Soc. Photo-optical Instrumentation Engrs., Internat. Soc. for Optical Engring., Am. Math. Soc. (presenter ann. meetings 1985-87, invited address 5th Internat. Conf. on info. rsch., informatics and cybernetics 1990), Sigma Xi, SIAM. Lutheran. Home: 7108 Grand Blvd Hobart IN 46342-6628 Office: U PR Dept Math Mayaguez PR 00681

COLLINS, EILEEN MARIE, astronaut; b. Elmira, N.Y., Nov. 19, 1956; d. James Edward and Rose Marie (O'Hara) C.; m. James Patrick Youngs, Aug. 1, 1987. AS in Math., Sci., Corning C.C., 1976; BA in Math., Econs., Syracuse U., 1978; grad., USAF Undergrad. Pilot Tng., Vance AFB, Okla., 1979, USAF Test Pilot Sch., Edwards AFB, Calif., 1990; MS in Ops. Rsch., Stanford U., 1986; student, USAF Inst. Tech., 1986; MA in Space Systems Mgmt., Webster U., 1989. Commd. 2d lt. USAF, 1978, advanced through grades to lt. col., 1993; instr. pilot 71st flight tng. wing USAF, Vance AFB, 1979-82; aircraft comdr. 86th mil. airlift squadron USAF, Travis AFB, Calif., 1983-85; asst. prof. math. USAF Acad., Colorado Springs, Colo., 1986-89; astronaut Johnson Space Ctr. NASA, Houston, 1990—. Decorated Air Force Commendation medal with one oak leaf cluster, Meritorious svc. medal with one oak leaf cluster, Air Force Expeditionary medal. Mem. U.S. Space Found., Am. Inst. Aeronautics and Astronautics, Air Force Assn., Women Mil. Aviators, Order Daedalians.

COLLINS, EMMANUEL GYE, aerospace engineer; b. Monrovia, Liberia, July 15, 1959; s. Emmanuel Gye and Esther Charlene (Hardman) C.; m. Bonita Lorraine Caldwell, Sept. 12, 1992; children: Nakia, Caleb. MSME, Purdue U., 1982, PhD in Aero. Engring., 1987. Sr. tech. assoc. AT&T Bell Labs., Holmdel, N.J., 1980, mem. tech. staff, 1981; staff engr. govt. aerospace systems divsn. Harris Corp., Melbourne, Fla., 1987—; adj. faculty Fla. Inst. Tech., Melbourne, 1992—. Author: (chpts.) Recent Advances in Robust Control, 1990, Control and Dynamic Systems: Advances in Theory and Applications, 1992. Mentor Brevard County Pub. Schs., Melbourne, 1991, 92. Recipient Superior Accomplishment award NASA Langley Rsch. Ctr., 1991. Mem. AIAA, IEEE (tech. assoc. editor 1992—). Achievements include research in the active control of flexible spacecraft. Office: Harris Corp GASD PO Box 94000 MS 22/4849 Melbourne FL 32902

COLLINS, ERIK, psychologist, researcher; b. Grand Rapids, Mich., May 17, 1938; s. Kreigh Taylor and Theresa (Van Der Laan) C.; divorced; children: Brett and Brian; m. Janice Louise Lloyd, Dec. 19, 1987; children: Nicole, Ben, Toby. BBA, U. Mich., 1959, MA, 1963, PhD, 1969. Lic. psychologist, Pa., Del. Tchr. Plymouth (Mich.) Community Schs., 1963-67; vis. lectr. edn. Eastern Mich. U., Ypsilanti, 1967-69; asst. prof., rsch. project dir. SUNY, Fredonia, 1969-74; rsch. assoc., project dir. U. Md., Princess Anne, 1974-76; dir. rsch. Geneva Acad., Phila., 1976-77; ptnr. Greely, Collins & Assocs., N.Y.C., 1977-78; psychologist Embreeville Ctr., Coatesville, Pa., 1978-87, Haverford (Pa.) State Hosp., 1987-89, Del. State Hosp., New Castle, 1989—; cons. test devel. SUNY, Brockport, 1971-72, evaluation N.Y. Bd. Edn., 1974-78, rsch. projects and proposal devel. N.Y.C. Community Sch. Dists. Stoney Pointe, N. Y., U. Md., 1976-78; consulting psychologist Pinehill Rehab. Ctr., Phila., 1986-84; pvt. practice West Chester, Pa., Wilmington, Del., 1980—; consulting psychologist, Phila., Elwyn, 1991—. Author poetry and tech. reports on assessment of emotional devel. and screening for learning disabilities; contbr. articles to profl. jours. Evaluation grantee N.Y. State Ctr. for Migrant Studies, Geneseo, 1971-72, Innovative Project grantee ESEA Title I and VII, Appalachian Regional Commn., 1972-78, rsch. grantee Coop. State Rsch. Svcs., 1974-76, Innovative Project grantee Dept. of Spl. Edn., Harrisburg, Pa., 1977. Mem. Am. Psychol. Assn., Lower Shore Assn. for Children Learning Disabilities (pres. Eastern Shore chpt. 1975-76), Dunkirk Yacht Club (fleet capt. 1971-72).. Democrat. Episcopalian. Avocations: building boats, sailing.

COLLINS, EUGENE BOYD, chemist, molecular pathologist, consultant; b. Los Angeles, May 28, 1917; s. Harold Porter and Mina Rosannah (Eversoll) C.; m. Frances Louise File, Aug. 4, 1946 (div. May 1962); children: Dana, Diane, Eric; m. Helen Lucille Schultz, Oct. 16, 1966; 1 child, Dane. BS in Chemistry, UCLA, 1951, diploma in edn., 1962; DSc (hon.), De Landas U., 1952; cert. advanced med. tech., Calif. State U., Dominguez Hills, 1977; MD, U. Cen. del Este, San Pedro, Dominican Republic, 1982. Lic. clin. lab. technologist, Calif.; cert. tchr., Calif. Assoc. dir. spectroscopy Union Oil Co. (Unocal), Wilmington, Calif., 1951-57; prof. chemistry Los Angeles Harbor Coll., Wilmington, 1957-74; cons. chemist Collins and Assocs., Carson, Calif., 1974-79, 83—; pres. Boyd Collins Co., South Gate, Calif., 1960-70; cons. Holley Carburetor Co. Research Lab., San Pedro, Calif., 1957-60; lectr. biology and clin. science Calif. State U., Dominquez Hills, 1988-92. Contbr. articles to profl. jours. Commr. Boy Scouts Am., Long Beach, Calif., 1958-59. Served as sgt. U.S. Army, 1944-46, ETO. Fellow Royal Soc. Arts; mem. AAAS, Am. Inst. Chemists, Am. Chem. Soc., Internat. Union of Pure and Applied Chemistry (affiliate), Am. Pharm. Assn., Acad. Pharm. Scis., N.Y. Acad. Scis., Am. Assn. Clin. Chemsitry. Avocations: history, chess, internat. affairs. Office: Collins and Assoc 470 Deep Woods Dr Selma AL 36701

COLLINS, FRANCIS S., medical research scientist. MD, PhD in physical chemistry. Former staff mem. Howard Hughes Med. Inst., U. Mich. Med. Ctr., Ann Arbor; now dir. Nat. Ctr. for Human Genome Rsch. NIH, Bethesda, Md. Co-recipient Gairdner Found. Internat. award for work on cystic fibrosis, 1990. Mem. NAS. Office: Nat Ctr Human Genome Rsch NIH, Bldg 38A Rm 605 9000 Rockville Pike Bethesda MD 20892

COLLINS, GEORGE BRIGGS, retired physicist; b. Washington, Jan. 3, 1906; s. Guy N. and Christine (Schmidt) C.; m. Elsa Leser, July 10, 1937 (dec. Oct. 1969); children: Peter, Lucy, Robert; m. Emily Ambler, June 22, 1971. PhD, Johns Hopkins U., 1932. Instr. Notre Dame (Ind.) U., 1935-39, Radiation Lab., Cambridge, Mass., 1940-46; prof. U. Rochester, N.Y., 1946-50; scientist Brookhaven Nat. Lab., Upton, N.Y., 1950-71; prof. Va. Tech., Blacksburg, 1971-76, retired, 1976. Editor: Microwave Microtions, 1947. Fellow Am. Phys. Soc.; mem. Phi Beta Kappa. Achievements include development of microwave magnetions; first to manage proton accelerator above one billion electron volts. Home: 1380 Locust Ave Blacksburg VA 24060

COLLINS, GEORGE JOSEPH, chemist; b. Pitts., May 15, 1948; s. Homer McDonald and Gladys Pauline (Bienko) C.; m. Susan Elisabeth Schwehm, Dec. 28, 1971; children: Jenny Marie, Erin Rebecca, Katharine Elisabeth. BS, Thiel Coll., 1970, PhD, Rutgers U., 1979, MBA, 1990. Asst. prof. Rutgers U., New Brunswick, N.J., 1979-80; sr. rsch. chemist Stauffer Chem. Co., Dobbs Ferry, N.Y., 1980-84; product mgr. Finnigan MAT, San Jose, Calif., 1984-87; regional mgr. Finnigan MAT, Livingston, N.J., 1987-91, TopoMetrix, Bedminster, N.J., 1991—. Contbr. articles to profl. publs. Chair Summerdale Sch. Site Coun., San Jose, 1985-87; com. mem. San Jose City Parks Commn., 1986. With USN, 1970-74. Mem. Am. Chem. Soc. (chair North Jersey mass spectrometry group 1983-84), Am. Soc. Mass Spectrometry, Microscopy Soc. Am., Sigma Xi. Office: TopoMetrix Ste 18 1 Robertson Dr Bedminster NJ 07921

COLLINS, HARRY DAVID, forensic engineering specialist, mechanical and nuclear engineer, retired army officer; b. Brownsville, Pa., Nov. 18, 1931; s. Harry Alonzo and Cecelia Victoria (Morris) C.; BS in Mech. Engring., Carnegie Mellon U., 1954; MS in Physics, U.S. Naval Postgrad. Sch., 1961; postgrad., U.S. Army Command and Gen. Staff Coll., 1970; postgrad. in exptl. physics, George Washington U., 1971-72; m. Suzanne Dylong, May 11, 1956; children: Cynthia L., Gerard P. Commd. 2d lt. C.E., U.S. Army, 1954, advanced through grades to lt. col. 1969; comdr. 802d Heavy Engr. Constrn. Bn., Korea, 1972-73; dep. dist. engr. and acting dist. engr. Army Engr. Dist., New Orleans, 1973-75; v.p. deLaureal Engrs., Inc., New Orleans, 1975-78; v.p. Near East mktg. and project mgmt. Kidde Cons., Inc., 1978-82; dir. new bus. devel. and project mgmt. for Middle East, Am. Middle East Co., Inc., 1982-84; sr. cons. Wagner, Hohns, Inglis, Inc., 1984-91; chief engr. bd. commrs. Orleans Levee Dist. State of La., 1991-92; pres. Harry D. Collins and Assocs., 1992—. Contbr. articles to profl. jours. Decorated Legion of Merit, Bronze Star with oak leaf cluster, Meritorious Service medal with oak leaf cluster, Joint Svc. Commendation medal, Armed Forces Expeditionary medal, Vietnam Svc. medal, Vietnam Nat. Commendation medal, Vietnam Tng. Svc. medal; registered profl. engr., Miss., La. Mem. ASME, Am. Soc. Mil. Engrs. (past pres. La Post), La. Engring. Soc., N.Y. Acad. Sics., NSPE, Am. Nuclear Soc., Am. Arbitration Assn. (panel of arbitrators and mediators), Nat. Acad. Forensic Engrs. (diplomate, cert.), Constrn. Specifications Inst. (cert.), Sigma Xi. Home: 2024 Audubon St New Orleans LA 70118-5518

COLLINS, HEATHER LYNNE, government official; b. Oak Park, Ill., Mar. 5, 1951; d. James Richard and Carolyn (Baumgartner) C. BA in Geology, DePauw U., Greencastle, Ind., 1973; MA in Geology, Washington U., St. Louis, 1975; MBA in Mgmt., U. Utah, 1989. Cartographer Def. Mapping Agy., St. Louis, 1976-77; petroleum geologist Phillips Petroleum Co., Denver and Oklahoma City, 1978-82, Aminoil U.S.A., Oklahoma City, 1982-83, Tex. Oil & Gas, Wichita, 1984-85, Applied Geophysics Inc., Salt Lake City, 1985-86; mgmt. intern U.S. Dept. Energy, San Francisco, Washington, 1989-91; ops. specialist U.S. Dept. Energy, Golden, Colo., 1991—. Mem. AAAS. Office: US Dept Energy 1617 Cole Blvd Golden CO 80401

COLLINS, JOHN, molecular genetics educator, researcher; b. Bournemouth, Eng., Apr. 7, 1945; s. Alfred Ernest and Pearl (Levy) C.; m. Marie-Christiane Martin, Apr. 12, 1969; children: Simon, Pascal. BSc in Microbiology, U. London, 1967; PhD, U. Leicester (Eng.), 1971. Postdoctoral fellow U. Calif., San Diego, 1971-74, U. Copenhagen, 1974-75; rsch. asst. Nat. Ctr. for Biotech., Inst. for Biotech. Rsch., Braunschweig, Fed. Republic Germany, 1975-81, head genetics dept., 1981-87, head sect. cell biology and genetics, 1987—; prof. genetics Tech. U. Braunschweig, 1987—; assoc. com. applications Acad. Scis., Paris, 1988—; counselor Biofutur, 1982—; media appearances and pub. debater on gene tech. Mem. editorial bd. Methods in Molecular and Cellular Biology, 1988—, Clin. Biotech., 1989—; contbr. numerous articles to profl. jours.; patentee process for prodn. hybrid bacteria. Advisor, grant reviewer Ministry Rsch. and Tech., Bonn, Fed. Republic Germany, 1984—, German Sci. Found., Bonn, 1983—; mem. com. on ethics of gene tech. investigation Evang. Ch., Hannover, Fed. Republic Germany, 1983-88. Jane Coffin Childs Meml. Fund fellow, 1971-74. Mem. European Molecular Biology Orgn. (fellow 1974-75), Human Genome Orgn. (founding com., fin. com. 1988-89), German Biol. Chem. Soc., German Genetics Soc. Avocations: magic, cuisine, problem solving, history and philosophy of science, boules. Office: Nat Ctr Biotech Rsch GBF, Mascheroder Weg 1, D-3300 Braunschweig Germany

COLLINS, MALCOLM FRANK, physicist, educator; b. Crewe, Eng., Dec. 15, 1935; s. Bernard and Ethel Collins; m. Eileen Ray, Apr. 22, 1961; children: Adrian Bernard, Andrew Malcolm, Gillian Olive. BA, Cambridge U., 1957, PhD, 1962. Staff scientist Atomic Energy Rsch. Establishment, Harwell, U.K., 1961-69; assoc. prof. McMaster U., Hamilton, Ont., Can., 1969-73, prof. physics, 1973—, chmn. dept. physics, 1976-82, dir. McMaster Nuclear Reactor, 1987—; sec. Can. Inst. for Neutron Scattering, 1986-92; mem. bd. editors Solid State Comm., 1970—. Author: Magnetic Critical Scattering, 1989; contbr. over 100 articles to profl. jours. Alfred P. Sloan Found. fellow, 1980-82. Mem. Can. Assn. Physicists, Am. Phys. Soc., Can. Nuclear Soc. Office: McMaster University, Nuclear Reactor, Hamilton, ON Canada L8S 4K1

COLLINS, OLIVER MICHAEL, electrical engineer, educator, consultant, researcher; b. Washington, Feb. 24, 1964; s. Lester Albertson and Petronella Dorothea (leRoux) C. BS, Calif. Inst. Tech., 1986, MS, 1987, PhD, 1989. Asst. prof. Johns Hopkins U., Balt., 1989—; cons. Jet Propulsion Lab., Pasadena, Calif., 1989—. Contbr. articles to profl. jours. Recipient rsch. grant NSF, 1990—. Quaker. Achievements include patent for next generation coding system for Deep Space Communications. Office: Johns Hopkins Univ Elec Engring Barton Hall Baltimore MD 21218

COLLINS, PAUL ANDREW, industrial designer; b. Munich, Germany, Dec. 28, 1955; s. Paul and Anna Elisabeth (Lauterbach) C. BSME, Columbia U., 1978; MS in Indsl. Design, Domus Acad., Milan, Italy, 1988. Registered profl. engr., N.Y., N.J. Rschr. Steven Winter Assocs., N.Y.C., 1978-80; project engr. Langer Cons., N.Y.C., 1981-82; sole proprietor P.A. Collins P.E., N.Y.C., 1983-87; designer Milan, 1988-90, Technologias Modulares S.A., Barcelona, Spain, 1991—. Contbr. articles to profl. jours. Home: Calle Valencia 287, Barcelona Spain 08009 Office: Technologias Modulares, Av Sant Julia 100, Barcelona Spain 08400

COLLINS, RICHARD EDWARD, physicist; b. Apr. 23, 1940; s. T.E. and N.M. Collins; m. Marilyn Martin; 3 children. BSc with hons., Univ. Sydney, Australia; PhD, NYU, New York. Rsch. physicist Amalgamated Wireless Australasia Ltd., 1961-65; instr. NYU, 1965-68; sr. rsch. physicist Amalgamated Wireless Australasia Ltd., 1968-72, chief physicist, 1972-78; prin. lectr. N.S.W. Inst. Tech., 1978-80; chmn. Lucas Heights Rsch. Lab.; chmn. solar energy adv. com. Energy Authority N.S.W., 1982-86; bd. dir. Austech Ventures Ltd., 1984-88. review com. Australia Atomic Energy Commn., 1986. Office: Lucas Heights Research Lab, New Illawarra Rd, Lucas Heights NSW 2234, Australia*

COLLINS, VINCENT PETER, pathologist; b. Dublin, Ireland, Dec. 3, 1947; s. James Vincent and Mary Ann (Blanche) C. MB, BCh, BAO, Nat. U. Ireland, 1971; MD, Karolinska Inst., Stockholm, 1978. Registered med. practitioner, Ireland, Eng., Sweden; specialist in pathology, Eng., Ireland, Sweden, specialist in clin cytology, Sweden. Assoc. prof. pathology Karolinska Inst., Stockholm, 1979-93, dir. studies Inst. Pathology, 1983-84; cons. pathologist and cytologist Karolinska Hosp., Stockholm, 1982-86, sr. cons. pathologist, 1986-90; chief for clin. research Ludwig Inst. for Cancer Research, Stockholm, 1986—; prof. pathology U. Gothenburg, Sweden, 1990-93; sr. cons. pathologist Salgrenska Hosp., Gothenburg, 1990-93; prof. tumor pathology Karolinska Inst., Stockholm, 1993—; sr. cons. pathologist Karolinska Hosp., Stockholm, 1993—; mem. sci. bd., chmn. subcom. A, Swedish Cancer Assn., Stockholm, 1988-93; mem. sci. rev. bd. Swedish Med. Rsch. Coun., 1991—; William O. Russell lectr. in anatomical pathology U. Tex. M D Anderson Cancer Ctr., Houston, 1990. Mem. editorial bd. Cancer Letters, 1985—, Brain Pathology, 1990—; mem. internat. editorial bd. Excerpta Medica, 1985—; contbr. articles to profl. jours. Recipient Minerva Found. prize, 1987, Joanne Vandenberg Hill award U. Tex. M D Anderson Cancer Ctr., 1990; grantee NIH, 1991-95, Swedish Cancer Soc., 1980-86, 90—, Stockholm Cancer Soc., 1974-86; Karolinska Inst. rsch. fellow, 1982, 86. Fellow Royal Coll. Physicians Ireland; mem. Royal Coll. Pathologists (London), Brit. Neuropath. Soc., Am. Cell. Biology Assn., N.Y. Acad. Scis., Swedish Med. Soc., Stockholm Cancer Soc. Avocations: sailing, skiing, music, bonsai. Home: Skillingrannd 9, S-11220 Stockholm Sweden Office: Karolinska Hosp Dept Pathology, Box 100, S-17176 Stockholm Sweden

COLLINS, WILLIAM HENRY, environmental scientist; b. Balt., Mar. 1, 1930; s. John Edward and Evelyn Agnes (Himmel) C.; m. Marie Grzechowiak, Nov. 11, 1951; children: William Patrick, Michael Shawn. BS, Loyola Coll., 1952; cert. in bus., Alex Hamilton Inst., N.Y.C., 1966. Rsch. chemist Catalyst Rsch. Corp., Balt., 1952-60; chief engr. Miller Rsch. Corp., Balt., 1960-67; prin. engr., program mgr. AAI Corp., Cockeysville, Md., 1967-70; dir. engring. Franklin Inst. Rsch. Lab., Phila., 1970-74; div. chief U.S. Army Chem. RD&E Ctr., Edgewood, Md., 1974-88, dep. dir. for rsch., 1988-93; dir. Ctr. to Enhance Capabilities of Handicapped Franklin Inst. Rsch. Labs, 1972-74; lectr. European Confs. on Environ. Pollution, 1985-91; tech. project officer Data Exch. Agreement U.S. and Germany, 1985-93. Editor proceedings U.S.-Germany Conf. on Environ. Tech., 1992; contbr. over 75 articles to profl. publs. Mem. Am. Def. Preparedness Assn., Sigma Xi (pres. 1972-74). Republican. Roman Catholic. Home: 1913 Forest Guard Ct Jarrettsville MD 21084 Office: Chem RD&E Ctr Attn SMCCR-RS Edgewood MD 21010

COLLINS, WILLIAM LEROY, telecommunications engineer; b. Laurel, Miss., June 17, 1942; s. Henry L. and Christene E. (Finnegan) C. Student, La Salle U., 1969; BS in Computer Sci., U. Beverly Hills, 1984. Sr. computer operator Dept. Pub. Safety, Phoenix, 1975-78, data communications specialist, 1978-79, supr. computer ops., 1981-82; mgr. network control Valley Nat. Bank, Phoenix, 1979-81; mgr. data communications Ariz. Lottery, Phoenix, 1982-85; mgr. telecommunications Calif. Lottery, Sacramento, 1985—; Mem. Telecomm. Study Mission to Russia, Oct. 1991. Served as sgt. USAF, 1964-68. Mem. IEEE, Nat. Systems Programmers Assn., Centrex Users Group, Accunet Digital Svcs. User Group, Telecommunications Assn. (v.p. edn. Sacramento Valley chpt. 1990-93), Assn. Data Communications Users, Soc. Mfg. Engrs., Data Processing Mgmt. Assn., Am. Mgmt. Assn., Assn. Computing Machinery, K.C. Roman Catholic. Home: 116 Valley Oak Dr Roseville CA 95678-4378 Office: Calif State Lottery 600 N 10th St Sacramento CA 95814-0393

COLLIPP, BRUCE GARFIELD, ocean engineer, consultant; b. Niagara Falls, N.Y., Nov. 7, 1929; s. Planton G. Collipp and Audrey O. Collipp; m. Priscilla Jane Collipp; children: Gary, Richard. BS, MIT, 1952, MS, 1954. Registered profl. engr., Tex.; lic. marine engr. Engring. offficer Lykes Bros., New Orleans, 1951, 53; teaching asst. MIT, Cambridge, Mass., 1954; designer, contractor Shell Oil, 1954-56, mgr. design constrn. and operation semisubmersible rig, 1956-61, prin. lectr. floating drilling and subsea completions, 1961; divsn. engr. Shell Oil, L.A., 1962-65, Lafayette, La., 1965-70; sr. staff rschr. Shell Oil, 1970-74, offshore designer Gulf of Mex., 1974-78, constrn. engr., 1978-80, project mgr., 1980-83, cheif naval architect, 1984, designer, contractor, 1985-87; cons. Shell Offshor Inc., Shell Oil Co., Exxon, World Bank, Reading and Bates Drilling C., U. of Texas, Austin, Noble Denton, PMB Systems Engring., British Petroleum, Elf Acquitaine Petroleum, Homestake Mining Inc., CBS Engring., Shell Pecten Internat., ARCO Internat., Lemle and Kelleher, 1987—; vis. prof. U. Tex., Austin, 1976—. Author: Buoyancy and Stability, 1976; contbr. over 50 tech. papers to profl. jours. Mayor Hidden Coves, Tex., 1976-86; dir. Spring Branch Meml. Sports Assn., Houston, 1972-76; mem. marine bd. Nat. Rsch. Coun., Washington, 1993—. Lt. USNR. Recipient Holley medal Am. Soc. Mech. Engrs., 1979, Gibbs Bros. medal Nat. Acad. Scis., 1991; Fulbright scholar. Mem. Nat. Acad. Engring., Soc. Naval Arch. and Marine Engring. (Blakley Smith medal 1993), Sigma Xi. Republican. Presbyterian. Achievements include patents for drill barge anchor system, floating drilling platform, pitch period reduction apparatus for tension leg platforms, curved conductor well template, tension leg platform anchoring method and apparatus; invention of semisubmersible drilling rig. Home: 511 Kickerillo Houston TX 77079

COLLURA, THOMAS FRANCIS, biomedical engineer; b. Cleve., Jan. 5, 1952; s. Howard John and Jean Anna (Fitzgerald) C.; m. Wendy Lois Herman, July 11, 1974; children: Jessica, Elisabeth, Joseph, Benjamin, David. AB, BS, Brown U., 1973; MS, Case Western Res. U., 1977, PhD, 1978. Registered profl. engr., Ohio, Ill. Engr. Eaton Corp., Cleve., 1973-74; staff engr. Bell Labs., Allentown, Pa., 1978-80; supr. Bell Labs., Chgo., 1980-88; staff engr. Cleve. Clinic, 1988—. Contbr. articles to IEEE Computer, Brain Topography, IEEE Biomed. Engring.; contbr. chpts. to books. Scoutmaster pack 150 Cub Scouts, Chagrin Falls, Ohio, 1990-92. NIH grantee, 1974-78. Mem. IEEE, Am. Electroencephalographic Soc., Human Factors Soc. Achievements include first application of steady-state EEG evoked potentials to study of human attention; development of new DC EEG system, EPILOG system for long-term EEG monitoring. Office: Cleve Clinic Found 9500 Euclid Ave Cleveland OH 44195

COLLVER, KEITH RUSSELL, agricultural products exective; b. Simcoe, Ont., Can., May 20, 1925. BScA in Horticulture, Ont. Agrl. Coll., 1945-49. Mgr. western prodn. procurement Birds Eye Foods (Can.) Ltd., 1949-52; gen. mgr., treas. Pacific Co-Operative Union, 1952-56; asst. mgr. The Norfolk Fruit Growers Assn., 1956-63, Norfolk Berry Growers Assn., 1956-63; gen. mgr., treas. Norfolk Fruit Growers Assn., 1963-86, exec. v.p., treas., 1986-91, exec. asst., sec., 1986—; ptnr. Kech Enterprises Inc., 1992—; dir. Apple Mktg. Commn., 1969-91, Internat. Apple Inst., Washington, D.C., 1973-79. Recipient Merit award The Ont. Fruit and Vegetable Growers Assn., Spl. Recognition award The Can. Soc. Horticulture Sci., Uniroyal Golden Apple award; Agrl. Inst. of Can. fellow, 1992. Office: Norfolk Fruit Growers, 99 Queensway East Box 279, Simcoe, ON Canada N3Y 4M5*

COLMAN, ROBERT WOLF, physician, medical educator; b. N.Y.C., June 7, 1935; s. Jack K. and Miriam (Greenblatt) C.; m. Roberta Fishman, June 16, 1957; children: Sharon, David. AB summa cum laude, Harvard U., 1956, MD cum laude, 1960. Intern Boston City Hosp., 1960-61; resident Beth Israel, Brookline, Mass., 1961-62; clin. assoc. USPHS, NIH, 1962-64; resident Barnes Hosp., St. Louis, 1964-65, fellow in hematology, 1965-67; assoc. in medicine Harvard Med. Sch., Cambridge, Mass., 1967-69, asst. prof., 1969-73, assoc. prof., 1973; assoc. prof. U. Pa., Phila., 1973-77, prof. medicine, 1977-78; prof. medicine Temple U. Sch. Medicine, Phila., 1978—, prof. thrombosis rsch., 1981—, prof. physiology, 1992—, dir. Sol Sherry Thrombosis Rsch. Ctr., 1979—, Sol Sherry prof. of medicine, 1989—; mem. hematology study sect. NIH, Bethesda, Md., 1977-81; invited lectr. Gordon confs., Internat. Congress Hemostasis and Thrombosis, Fedn. Am. Socs. Exptl. Biology meetings, Internat. Soc. Kallikreins and Kinins, others. Contbr. numerous papers to profl. publs., 1959—; editor: Hemostasis and Thrombosis, 1982, 2d edit., 1987; editor Platelet Jour., 1989—; mem. editorial bds. Jour. Clin. Investigation, Blood, Proc. Soc. Exptl. Biology, Thrombosis Rsch. Platelets, 1980-88. Surgeon USPHS, 1962-64. Recipient Leon Resnick prize Harvard U., Career Devel. award NIH, Sr. Investigator award S.E. Pa. chpt. Am. Heart Assn., Disting. Career award Internat. Soc. Thrombosis and Hemostasis. Fellow ACP; mem. Assn. Am. Physicians, Am. Soc. Clin. Investigation, Am. Soc. Biochemistry and Molecular Biology, Internat. Soc. Hemostasis and Thrombosis (councillor 1989—, Disting. Career award to hemostatis), Peripatetic Club, Interurban Clin. Club, Phi Beta Kappa, Sigma Xi, Alpha Omega Alpha. Office: Temple U Sch Medicine Sol Sherry Thrombosis Rsch Ctr 3400 N Broad St Philadelphia PA 19140-5196

COLMENARES, JORGE ELIECER, mechanical engineer; b. Caracas, Venezuela, Mar. 16, 1966; came to U.S., 1990; s. Jacinto and Maria Cristina (Matheus) C.; m. Mariela Ch. Miralles, June 3, 1992. Ingeniero Mecanico summa cum laude, U. Simon Bolivar, Caracas, Venezuela, 1988; MSME, MIT, 1992. Registered profl. engr., Calif. Mech. engring. asst. Maraven, Cardon, Venezuela, 1987; mech. engr. Tecnoconsult, Caracas, 1988-90; rsch. asst. MIT, Cambridge, 1990-92; mech. engr. I Fluor Daniel, Inc., Irvine, Calif., 1992—. Named London Internat. Youth Scientist, British Consulate, 1988. Mem. ASME, MIT Alumni, Sigma Xi. Roman Catholic. Home: 148 Pergola Irvine CA 92715 Office: Fluor Daniel Inc 3333 Michelson Dr Irvine CA 92730

COLMENARES, NARSES JOSE, electrical engineer; b. Caracas, Venezuela, Apr. 29, 1945; came to U.S., 1977, naturalized, 1992; s. Jose and Isabel (Guevara) C.; m. Linda Burns, July 23, 1988. Communications Engr., Escoelfa, Caracas, 1974; Elec. Engr., Metropolitana U., Caracas, 1976; MS in Engring, Princeton U., 1980. Asst. prof. Simon Bolivar U., Caracas, 1975-77; systems engr. CSEE, Paris and Caracas, 1981-82, Ram Broadcasting Corp., Avenel, N.J., 1983-85, Metromedia Telecommunications, Englewood City, N.J., 1985-86; telecom. engr. CE Caracas Telecom. Group, Bloomfield, N.J., 1982-83; communications engr. N.Y. Power Authority, White Plains, N.Y., 1986-87; mem. wireless strategic planning staff AT&T Bell Labs., Holmdel, N.J., 1987—; asst. researcher Princeton U., N.J., 1978-79; cons. Thevenin S.A., Caracas, 1974-77. Contbr. articles to profl. jours. Lt. Venezuelan Navy, 1965-75. Scholar GMA Found., 1977, Venezuela Sec. of Def., 1972. Mem. IEEE, Venezuela Assn. Elec. and Mech. Engrs. (former exec. dir.), Princeton Club. Avocations: tennis, skiing, jogging. Home: 182 Bristol Ct Old Bridge NJ 08857-3240 Office: AT&T Bell Labs 1K-220 Crawfords Corner Rd Holmdel NJ 07733

COLODNY, EDWIN IRVING, lawyer, retired airline executive; b. Burlington, Vt., June 7, 1926; s. Myer and Lena (Yett) C.; m. Nancy Dessoff, Dec. 11, 1965; children: Elizabeth, Mark, David. AB with distinction, U. Rochester, 1948; LLB, Harvard U., 1951; D in Comml. Sci. (hon.), Robert Morris Coll., 1985; LLD (hon.), Middlebury Coll., 1986; HHD (hon.), Kings Coll., 1988. Bar: N.Y. 1951, D.C. 1958. With Office Gen. Counsel, GSA, 1951-52, CAB, 1954-57; with USAir Inc. (formerly Allegheny Airlines Inc.), 1957-91, exec. v.p. mktg. and legal affairs, 1969-75, pres., 1975-90, chief exec. officer, 1975-91, chmn. bd. dirs., 1978-92, ret., 1992; also chmn. USAIR Group Inc., 1978-92; of counsel Paul, Hastings, Janofsky and Walker, Washington, 1991—; bd. dirs. COMSAT Corp., Martin Marietta Corp., Esterline Techs., USAIR Group, Inc., USAIR, Inc. Trustee U. Rochester, Shelburne (Vt.) Mus.; commr. Nat. Mus. Am. Art of Smithsonian Instn. Served to 1st lt. AUS, 1952-54. Recipient James D. McGill Meml. award U. Rochester, Wright Bros. Lectr. in Aeronautics AIAA, 1990, Tony Jannus award 1990. Mem. ABA, U. Rochester (bd. trustees).

COLOMBO, ANTONIO, cardiologist; b. Busto Arsizio, Varese, Italy, June 17, 1950; s. Luigi and Lidia (Filippini) C.; m. Antonia Busetto, Mar. 30, 1978; children: Andrea, Paola. MD summa cum laude, U. Milan, 1975; degree in cardiology, U. Parma, Italy, 1978. Resident Centro Gasperis, Milan, 1975-78; intern N.Y. Med. Coll., N.Y.C., 1978-79, resident in cardiology, 1979-80, chief resident, 1980-82; fellow in cardiology VA Med. Ctr., Long Beach, Calif., 1982-84, dir. cardiology wards, 1985-86, asst. dir. cath. lab., 1985-86; fellow in cardiology St. Joseph's Hosp., Syracuse, N.Y., 1984-85; asst. prof. medicine U. Calif., Irvine, 1985-89; dir. Catheterization Lab. Columbus Hosp., Milan, 1986—; cons. for interventional cardiology Clinica Villa Bianca, Bari, 1991—; chmn. cardiopulmonary resuscitation VA Med. Ctr., Long Beach, Calif., 1985-86. Recipient Best Teaching award Cabrini Med. Ctr., N.Y., 1981. Fellow Am. Coll. Cardiology, Am. Heart Assn. Office: Columbus Hospital, Via M Buonarroti N 48, Milan 20145, Italy

COLOMBO, MICHAEL PATRICK, mechanical engineer; b. Edina, Minn., Dec. 13, 1966; s. Arthur and Susan (Woods) C. BSME, U. Nev., Reno, 1990. Registered profl. engr., Nev. Mech. engr. Cubix Corp., Carson City, Nev., 1990—. Mem. ASME, NSPE. Home: Apt 137B 8200 Offenhauser Dr Reno NV 89511 Office: Cubix Corp 2800 Lockheed Way Carson City NV 89706

COLOMBO, UMBERTO PAOLO, Italian government official; b. Livorno, Italy, Dec. 20, 1927; s. Eugenio and Maria (Eminente) C.; ScDr in Phys. Chemistry, U. Pavia, 1950; Prof. Indsl. Chemistry, U. Genoa, 1964; m. Milena Piperno, July 5, 1951; children: Carla, Claudia. Scientist in phys. chemistry Montecatini's G. Donegani Rsch. Inst., Novara, Italy, 1951-66, dir., 1967-78; Fulbright fellow MIT, Cambridge, 1953; gen. mgr. R & D div. Montedison's, Milan, Italy, 1973-78; chmn. Italian Atomic Energy Commn., Rome, 1979-82, ENEA Italian Nat. Agy. New Tech., Energy and the Environ., Rome, 1982-93; minister for Scientific and Technological Rsch. & Univ., 1993—. Chmn. European Econ. Commn. Com. for Sci. and Tech., 1983-88, UN Com. Sci. and Tech. for Devel., 1984-86; mem. internat. coun. UN U.; hon. trustee Aspen Inst. for Humanistic Studies; gov. Internat. Devel. Rsch. Centre, Can.; pres. European Sci. Found., Aurelio Peccei Found. Mem. AAAS, Italian Chem. Soc., Am. Chem. Soc., Swiss, Swedish, U.S.-Japanese Acads. Engring. (fgn.), N.Y. Acad. Scis., Sci. Policy Found. London (mem. internat. adv. bd. 1975—), Am. Acad. Arts and Scis. (fgn. hon.). Club: Chemist (N.Y.C.). Co-author: WAES Report-Italy, 1977; Beyond the Age of Waste, 1978, Il Secondo Pianeta, 1983, Scienza e Tecnologia verso XXI Secolo, 1988, Le Frontiere della Tecnologia, 1990; contbr. articles to profl. jours. Home: Via San Martino ai Monti 26 BIS, 00184 Rome Italy

COLONEY, WAYNE HERNDON, civil engineer; b. Bradenton, Fla., Mar. 15, 1925; s. Herndon Percival and Mary Adore (Cramer) C.; m. Anne Elizabeth Benedict, June 21, 1950; 1 child, Mary Adore. B.C.E. summa cum laude, Ga. Inst. Tech., 1950. Registered profl. engr. and surveyor, Fla., Ga., Ala., N.C. Project engr. Constructora Gen. S.A., Venezuela, 1948-49, Fla. Rd. Dept., 1950-55; hwy. engr. Gibbs & Hill, Inc., Guatemala, 1955-57; project mgr. Gibbs & Hill, Inc., Tampa, Fla., 1957-59; project engr., then assoc. J.E. Greiner Co., Tampa, 1959-63; ptnr. Barrett, Daffin & Coloney, Tallahassee, 1963-70; pres. Wayne H. Coloney Co., Inc., Tallahassee, 1970-78, chmn., bd. chief exec. officer, 1978-85; pres., sec. Tesseract Corp., 1975-85; dep. chmn. Howden Airdynamics Am., Tallahassee, 1985-90; pres. Coloney Co. Cons. Engrs., Inc., 1978—; v.p., dir. Howden Coloney Inc., Tallahassee, 1985-90; prin. Coloney-Von Soosten & Assocs. Inc., Tallahassee, 1990—; chmn. adv. com. Area Vocat. Tech. Sch., 1965-78; pres. Retro Tech. Corp., 1983-93, Profl. Mgmt. Con. Group, 1983-87, pres., bd. dirs. Internat. Enterprises Inc, 1967-73. Patentee roof framing system, dense packing external aircraft fuel tank, tile mounting structure, curler rotating device, bracket system for roof framing; contbr. articles to profl. jours. Pres. United Fund Leon County, 1971-72; bd. dirs. Springtime Tallahassee, 1970-72, pres., 1981-82; bd. dirs. Heritage Found., 1965-71, pres., 1967; mem. Pres.'s Adv. Council on Indsl. Innovation, 1978-79; bd. dirs. LeMoyne Art Found., 1973, v.p., 1974-75; bd. dirs. Goodwill Industries, 1972-73, Tallahassee-Popoyan Friendship Commn., 1968-73; mem. Adv. Com. for Hist. and Cultural Preservation, 1969-71, Govs. Commn. for Purchase from the Blind, 1980—. Served with AUS, 1943-46. Fellow ASCE, Nat. Acad. Forensic Engrs.; mem. Am. Def. Preparedness Assn., NSPE, Fla. Engring. Soc. (sr.), Fla. Inst. Cons. Engrs., Fla. Soc. Profl. Land Surveyors, Tallahassee C. of C., Anak, Koseme Soc., Am. Arbitration Assn., Fla. Small Bus. Assn. (pres. 1981), Am. Mktg. Assn., Phi Kappa Phi, Omicron Delta Kappa, Sigma Alpha Epsilon, Tau Beta Pi. Episcopalian. Clubs: Governor's, Killearn Golf and Country. Home: 503 McDaniel St Tallahassee FL 32303 Office: Coloney Co Cons Engrs Inc PO Box 668 1014 N Adams St Tallahassee FL 32303

COLON-OTERO, GERARDO, hematologist, oncologist; b. San Juan, Feb. 21, 1956; s. Antonio and Gladys (Otero) C.; m. Nelly Mauras; children: Daniel Gerardo, Christina Marie. BS magna cum laude, U. Puerto Rico, 1975, MD, 1979. Intern, resident Mayo Grad. Sch. Medicine, Rochester, Minn., 1979-82, fellow in hematology, 1982-84; fellow in oncology U. Va., Charlottesville, 1984-86; staff physician Mayo Clinic Jacksonville, Fla., 1986—; chmn. sect. hematology Mayo Clinic Jacksonville, 1986-91, St. Lukes Hosp., Jacksonville, 1988—; mem. adv. bd. Hospice N.E., Jacksonville, 1992—. Contbr. chpts. to books and articles to profl. jours. Fellow ACP; mem. Alpha Omega Alpha. Office: Mayo Clinic Jacksonville 4500 San Pablo Rd Jacksonville FL 32224

COLOSKE, STEVEN ROBERT, mechanical engineer; b. Livonia, Mich., Sept. 22, 1960; s. Robert Henry and Helen (Hartmeyer) C.; m. Jill Speck, May 16, 1987; 1 child, Taylor Carr. BSME, U. Mich., 1983; MSIE, U. Tenn., 1993. Maintenance engr. Sverdrup Tech., Inc., Tullahoma, Tenn., 1983-88, project mgr., 1988-90, mech. svcs. engr., 1990-92; mech. svcs. engr. Sverdrup Tech., Inc., Livonia, Mich., 1992-93; sr. engr. AVL N.Am., Inc., Novi, Mich., 1993—. Mem. ASME (com. mem. 1992-97), Soc. Automotive Engrs. Office: AVL NAm Inc 41169 Vincenti Ct Novi MI 48375

COLTON, CLARK KENNETH, chemical engineering educator; b. N.Y.C., July 20, 1941; s. Sidney and Goldie (Chases) C.; m. Ellen Ruth Brandner, June 20, 1965; children: Jill Erin, Jason Adam, Michael Ross, Brian Scott. B of Chem. Engring., Cornell U., 1964, PhD, MIT, 1969. Asst. prof. chem. engring. MIT, Cambridge, 1969-73, assoc. prof., 1973-76, prof., 1976—, Bayer prof. chem. engring., 1980-85, dep. head dept. chem. engring., 1977-78, chmn. centennial chem. engring. edn., 1988; cons. to NIH, FDA, various indsl. orgns.; mem. adv. bd. mil. personnel supplies NRC, 1971-75. Mem. editorial bd. Jour. Membrane Sci., 1975-81, Jour. Bioengring., 1976-79, Preparative Chromatography, ASAIO Jour., Cell Transplantation; contbr. articles to sci. jours. Ford found. fellow, 1969-70; recipient Tchr./Scholar award Camille and Henry Dreyfus Found., 1972. Mem. AAAS, N.Y. Acad. Scis., Am. Inst. Chem. Engrs. (dir. food, pharm. and bioengring. div. 1978-81, Allan P. Colburn award 1977), Am. Soc. Artificial Internal Organs (editorial bd. 1978-84), Am. Diabetes Assn., Am. Soc. for Apheresis, Am. Soc. for Engring. Edn. (Curtis W. McGraw rsch. award 1980), North Am. Membrane Soc., Am. Heart Assn., Cell Transplantation Soc., Internat. Soc. on Oxygen Transport to Tissue, Am. Chem. Soc., Am. Inst. Med. and Biological Engring., Internat. Soc. Articificial Organs, Internat. Soc. Blood Purification (Gambro award 1986), Biomed. Engring. Soc., Cornell Club, Sigma Xi, Tau Beta Pi, Phi Lambda Upsilon. Home: 279 Commonwealth Ave Chestnut Hill MA 02167-1012 Office: MIT Dept Chem Engring Cambridge MA 02139

COLTON, FRANK BENJAMIN, retired chemist; b. Bialystok, Poland, Mar. 3, 1923; came to U.S., 1934, naturalized, 1934; s. Rubin and Fanny (Rosenblat) C.; m. Adele Heller, Mar. 24, 1950; children—Francine, Sharon, Laura, Sandra. B.S., Northwestern U., 1945, M.A., 1946; Ph.D., U. Chgo., 1949. Research fellow Mayo Clinic, Rochester, Minn., 1949-51; with G.D. Searle & Co., Chgo., 1951-86, asst. dir. chem. research, 1961-70, research advisor, 1970-86. Contbr. articles to profl. jours. Pioneer in organic and steroid chemistries. Patentee first oral contraceptive. Recipient Discovery medal for first oral contraceptive Nat. Assn. Mfrs., 1965, Profl. Achievement award U. Chgo., 1978, Achievement award Indsl. Research Inst., 1978; inducted in Nat. Inventors Hall of Fame, 1988. Mem. Am. Chem. Soc., Chgo. Chemists Club. Home: 3901 Lyons St Evanston IL 60203-1324

COLUCCI, ROBERT DOMINICK, cardiovascular pharmacologist; b. Stamford, Conn., Apr. 15, 1961; s. Dominick and Mira (Rex) C.; m. Kathryn DiCristofaro, May 4, 1990. BS in Pharmacy, Mass. Coll. Pharmacy, 1984, PharmD, 1987. Lic. pharmacist, Conn., Mass. Critical care pharmacologist Bronx (N.Y.) VA Med. Ctr., 1986-87; postdoctoral fellow cardiovascular pharmacology/therapeutics Hartford Hosp., 1987-89; sr. med. rsch. assoc., cardiovascular clin. rsch. Schering Plough Corp., Kenilworth, N.J., 1989-93, asst. dir. clin. pharmacology, 1993—; vis. scientist Pharm. Manufacture Assn., Washington, 1991—; clin. instr. Mt. Sinai Sch. Medicine, N.Y.C., 1990—; pharmacology/rsch. cons. Bronx VA Med. Ctr., 1987—. Mem. editorial bd. Jour. Clin. Pharmacology, 1991—; contbr. articles to Jour. Electrophysiology, Jour. Clin. Pharmacology, Am. Jour. Gastroenterology, Critical Care Medicine, DICP The Annals of Pharmacotherapy. Recipient Community Pharmacy Internship award Pfizer Pharm. Corp., 1984. Fellow Am. Coll. of Critical Care Medicine, Am. Coll. Clin. Pharmacology; mem. Soc. Critical Care Medicine (pharmacology bd.), Am. Coll. Clin. Pharmacy, N.Y. Acad. Scis., Am. Fedn. Clin. Rsch., KC (4th degree), Sigma Xi. Office: Schering Plough Corp 2015 Galloping Hill Rd Kenilworth NJ 07033-1310

COLUCCIO, LYNNE M., biochemistry educator; b. Albany, N.Y., June 25, 1956. PhD, Rensselaer Poly. Inst., 1982. Postdoctoral fellow in biology U. Pa., Phila., 1982-84; postdoctoral fellow in cell biology Baylor Coll. Medicine, Houston, 1984-86; rsch. assoc. in biochemistry Cornell U., Ithaca, N.Y., 1986-89; asst. prof. biochemistry Emory U. Sch. Medicine, Atlanta, 1989—. Recipient Jr. Faculty Rsh. award Am. Cancer Soc., 1991-94; rsch. grantee NIH, 1990-95. Mem. Am. Soc. for Biochemistry and Molecular Biology, Am. Soc. for Cell Biology. Office: Emory U Sch Medicine Dept Biochemistry Atlanta GA 30322

COLVIN, JOHN TREVOR, physicist; b. L.A., June 26, 1957; s. John Trevor Roseboom and Sue Carolyn (Colvin) Colvin; m. Gail Suzanne Vockeroth, Jan. 12, 1986; children: Jacqueline Rae, John Trevor, Melanie Suzanne. BS, BA in Physics and Math., U. Chgo., 1978, MBA, 1988; MS, PhD in Physics, U. Ill., 1984. Staff physicist Hughes Aircraft, Malibu, Irvine, Calif., 1984-86; prin. Smith Breeden Assocs., Chapel Hill, N.C., 1988—. Contbr. articles to Phys. Rev., Jour. Chem. Physics; co-author chpt.: Ajustable Rate Mortgages. Achievements include development of model using fractals to explain vibrational density of states in amorphous materials. Home: 909 Pinehurst Dr Chapel Hill NC 27514-3427 Office: Smith Breeden Assocs 727 Eastowne Dr 200A Chapel Hill NC 27514

COLVIN, THOMAS STUART, agricultural engineer, farmer; b. Columbia, Mo., July 17, 1947; s. Charles Darwin and Miriam Elizabeth (Kimball) C.; m. Sonya Marie Peterson, Sept. 11, 1982; children: Christopher, Kristel. BS, Iowa State U., 1970, MS, 1974, PhD, 1977. Registered profl. engr., Iowa. Farmer Hawkeye and Cambridge, Iowa, 1970—; rsch. assoc. Iowa State U., Ames, 1972-77; agrl. engr. USDA/Agrl. Rsch. Svc., Ames, 1977—; cons. WillowCreek Cons., Manning, Iowa, 1978-85. Sgt. USAF, 1970-72, Vietnam. Recipient Air Force Commendation medal USAF, 1971. Mem. Am. Soc. Agrl. Engrs. (power machinery stds. com. St. Joseph, Mich. 1989—, Young Engr. of Yr. 1986), Am. Soc. Agronomy, Soil and Water Conservation Soc., Iowa Acad. Sci. (chair agrl. scis. sect. 1991-92), Sigma Xi, Alpha Epsilon (pres. 1978), Gamma Sigma Delta, Phi Mu Alpha. Achievements include design and development of first computer program to help farmers manage tillage and residue cover for erosion control. Office: Nat Soil Tilth Lab 2150 Pammel Dr Ames IA 50011-4420

COLVIS, JOHN PARIS, aerospace engineer, mathematician, scientist; b. St. Louis, June 30, 1946; s. Louis Jack and Jacqueline Betty (Beers) C.; m. Nancy Ellen Fritz, Mar. 15, 1969 (div. Sept. 1974); 1 child, Michael Scott; m. Barbara Carol Davis, Sept. 3, 1976; 1 child, Rebecca Jo; stepchildren: Bruce William John Zimmerly, Belinda Jo Zimmerly Little. Student, Meramec Community Coll., St. Louis, 1964-65, U. Mo., 1966, 72-75, Palomar Coll., San Marcos, Calif., 1968, U. Mo., Rolla, 1968-69; BS in Math., Washington U., 1977. Assoc. system safety engr. McDonnell Douglas Astronautics Co., St. Louis, 1978-81; sr. system safety engr. Martin Marietta Astronautics Group-Strategic Systems Co., Denver, 1981-87; sr. engr. Martin Marietta Astronautics Group-Space Launch Systems Co., Denver, 1987—; researcher in field. Precinct del., precinct committeeman, congl. dist. del., state del. Rep. Party; mem. state adv. bd. Colo. Christian Coalition. Lance cpl. USMC, 1966-68, Vietnam. Mem. AAAS, Math. Assn. Am., Colo. Home Educators' Assn. (pres. 1989), Colo. Christian Coalition (state bd. dirs.). Evangelical. Avocations: camping, hiking, swimming. Home: 4978 S Hoyt St Littleton CO 80123-1988 Office: Martin Marietta Astronautics Group-SLS PO Box 179 Denver CO 80201-0179

COLWELL, RITA ROSSI, microbiologist, molecular biologist; b. May 31, 1956; m. Jack H. Colwell, May 31, 1956; children: Alison E.L., Stacie A. BS in Bacteriology with distinction, Purdue U., 1956, MS in Genetics, 1958; PhD, U. Wash., 1961; DSc, Heriot-Watt U., Edinburgh, Scotland, 1987; DSc (hon.), Hood Coll., 1991; DSc, Purdue U., 1993. Rsch. asst. genetics lab. Purdue U., West Lafayette, Ind., 1956-57; rsch. asst. U. Wash., Seattle, 1957-58, predoctoral assoc., 1959-60, asst. rsch. prof., 1961-64; asst. prof. biology Georgetown U., Washington, 1964-66, assoc. prof. biology, 1966-72; prof. microbiology U. Md., 1972—, v.p. for acad. affairs, 1983-87; dir. Ctr. Marine Biotech., 1987-91; pres. Md. Biotech. Inst. U. Md., 1991—; hon. prof. U. Queensland, Brisbane, Australia, 1988; cons., advisor Washington area comms. media, congressman, legislators, 1978—; external examiner various univs. abroad, 1964—; mem. coastal resources adv. com. dept. natural resources State of Md., 1979; NAS ocean scis. bd., 1977-80, vice-chair polar rsch. bd., 1990-94; mem. Nat. Sci. Bd., 1984-90, sci. adv. bd. Oak Ridge Nat. Labs., 1988-90, 93-96, adv. com. FDA, 1991-92, food adv. com., 1993-96. Author 14 books including (manual numerical taxonomy) Collecting the Data, 1970, (with M. Zambruski) Rodina-Methods in Aquatic Microbiology, 1972, (with L. H. Stevenson) Estuarine Microbial Ecology, 1973, (with R. Y. Morita) Effect of the Ocean Environment on Microbial Activities, 1974, (with A. Sinsky and N. Pariser) Marine Biotechnology, 1983, Vibrios in the Environment, 1985, Nucleic Acid Sequence Data, 1988; mem. editorial bd. Microbial Ecology, 1972-91, Applied and Environ. Microbiology, 1969-81, Oil and Petrochemical Pollution, 1980-91, Jour. Washington Acad. Scis., 1981-87, Johns Hopkins U. Oceanographic Series, 1981-84, Revue de la Fondation Oceanographique Ricard, 1981-92, Estuaries, 1983-89, Zentralblatt fur Bacteriologie, 1985—, Jour. Aquatic Living Resources, 1987—, System. Applied Microbiology, 1985—; contbr. articles and revs. to profl. jours. including Can. Jour. Fisheries and Aquatic Scis., Soc. Gen. Microbiology, Jour. Bacteriology, others. Recipient Gold medal Internat. Biotech. Inst., 1990, Purkinje Gold medal Achievement In Scis. Czechoslavakian Acad. Scis., 1991; Civic award Gov. Md., 1990, Cert. Recognition NASA, 1984, Alice Evans award Amer. Soc. Microbiol. Com. on Status of Women, 1988; named Phi Kappa Phi Scholar of Yr., 1992, Outstanding Women on Campus U. Md., 1979. Fellow AAAS, Grad. Women Sci., Can. Coll. Microbiologists, Am. Acad. Microbiology (chmn. bd. govs. 1989-94), Washington Acad. Scis. (bd. mgrs. 1976-79), Marine Tech. Soc., (exec. com. 1982-88), Sigma Delta Epsilon; mem. Am. Soc. Microbiology (various sci. coms. 1961—, pres. 1985, chmn. program com. REGEM-1 1988, Fisher award 1985), World Fedn. Culture Collections, Internat. Union Microbiol. Soc. (v.p. 1986-90, pres. 1990—), Am. Inst. Biol. Scis. (bd. govs. 1976-82), Am. Soc. Limnology and Oceanography, Internat. Coun. Sci. Unions (gen com.), U.S. Fedn. Culture Collections (governing bd. 1978-88), Soc. Indsl. Microbiology (bd. govs. 1976-79), Classification Rsch. Group Eng. (charter), Soc. Gen. Microbiology, Phi Beta Kappa, Sigma Xi (Ann. Achievement award 1981, Rsch. award 1984, nat. pres. 1991—), Omicron Delta Kappa, Delta Gamma. Achievements include research in marine biotechnology, marine and estuarine microbial ecology, survival of pathogens in aquatic environment, ecology of Vibrio cholerae and related organisms, microbial systematics, marine microbiology, antibiotic resistance, indexing of E. coli to identify sources of fecal contamination in water, environmental aspects of Vibrio cholerae in transmission of cholera. Office: U Md System Biotech Inst Office of Pres 4321 Hartwick Rd Ste 500 College Park MD 20740

COMA-CANELLA, ISABEL, cardiologist; b. Vigo, Pontevedra, Spain, Sept. 12, 1948; d. Fernando and Rosina (Canella) C. License in medicine, Navarra U., Pamplona, Spain, 1971. Resident Cantabria Hosp., Santander, Spain, 1972-73; resident in internal medicine Puerta de Hierro Hosp., Madrid, 1973-74, resident in cardiology, 1974-76; cardiology asst. coronary care unit La Paz Hosp., Madrid, 1977—. Author: Ischemic Heart Disease, 1981, 88, Cholesterol and Infarction, 1991; contbr. articles to profl. jours.; editorial bd. Spanish Jour. Cardiology, 1982; reviewer European Jour. Cardiology, 1990. Founder Verin Cultural Orgn., Madrid, 1991. Recipient award Knoll Labs., 1982, Rsch. award, 1992. Fellow European Soc. Cardiology; mem. Spanish Soc. Cardiology. Roman Catholic. Avocations: music, tennis. Office: La Paz Hosp Coronary Care Unit, Paseo de la Castellana 261, 28046 Madrid Spain

COMBERG, HANS-ULRICH, physician; b. Iden, Altmark, Germany, Feb. 20, 1948; s. Gustav-Wilhelm and Erika (Scherer) C.; m. Ingrid Rettenmeier, Oct. 11, 1991; 1 child, Hans-Christian. MD, U. Heidelberg, 1974. Med. lic., Germany, 1976, Bd. Family Practice, Hamburg, 1984. Intern U. Heidelberg, 1975-76; researcher Claxton/Georgia and N.Y., 1976-77, FLEX, San Francisco, 1977; resident U. Heidelberg, 1978-85; physician, gen. and family medicine pvt. practice, Hamburg, 1986—; lectr. gen. and family medicine U. Hamburg, 1988—. Contbr. articles to profl. jours. Avocations: photography. Home and Office: Juergensallee 42, D-22609 Hamburg Germany

COMBS, LINDA JONES, business administration educator, researcher; b. Jonesboro, Ark., Apr. 12, 1948; d. Dale Jones and Neva Mae (Craig) Green; m. Nathan Lewis Combs, Jan. 13, 1968; 1 child, Nathan Isaac. BSBA, U. Ark., 1971, MBA, 1972, PhD in Bus. Adminstrn., 1983. Assoc. economist Bur. Bus. and Econ. Rsch., Fayetteville, Ark., 1973-76; pres. Combs Mgmt. Co., Springdale, Ark., 1976-83; asst. prof. fin. U. Ark., Fayetteville, 1983-87; asst. prof. fin. and mktg. Western Ill. U., Macomb, 1987-88; asst. prof. bus. adminstrn. Cen. Mo. State U., Warrensburg, 1988-89; assoc. prof. bus. adminstrn. N.E. State U., Tahlequah, Okla., 1989-93; cons. Sears, Roebuck, Chgo., 1973-74, Fayetteville Adv. Coun., 1975-76, King Pizza, Fayetteville, 1985-87; cons. in fin. and banking, Fayetteville, 1973-76. Contbr. articles to profl. jours. Mem. gov.'s inaugural com. State of Ark., Little Rock, 1985; county co-chmn. Clinton for Gov., Washington County, Ark., 1984, 86, 90; bd. dirs. Shiloh Mus., Am. Cancer Soc., North Ark., Symphony Soc.; bd. dirs. Ark State Hosp., Little Rock, sec., chmn., 1993; active many polit. campaigns for candidates and issues. Mem. Am. Mktg. Assn. (health care mktg.), Transp. Rsch. Forum. Avocations: gardening, quilting. Office: Combs Mgmt Co PO Box 1452 Fayetteville AR 72702-1452

COMER, JAMES PIERPONT, psychiatrist; b. East Chicago, Ind., Sept. 25, 1934; s. Hugh and Maggie (Nichols) C.; m. Shirley Ann Arnold, June 20, 1959; children: Brian Jay, Dawn Renee. AB, Ind. U., 1956; MD, Howard U., 1960; MPH, U. Mich., 1964; DSc (hon.), U. New Haven, 1977; LittD (hon.), Calumet Coll., 1978; LHD (hon.), Bank St. Coll., N.Y.C., 1987, Albertus Magnus Coll., 1989, Quinnipiac Coll., 1990, DePauw U., 1990; DSc (hon.), Ind U., 1991, Wabash Coll., 1991; EdD (hon.), Wheelock Coll., 1991; LLD, U. Conn., 1991; LHD (hon.), SUNY Buffalo, 1991, New Sch. for Social Rsch., 1991; D Pedagogy (hon.), R.I. Coll., 1991; DSc (hon.), Amherst Coll., 1991; LHD (hon.), John Jay Coll. Criminal Justice, 1991, Wesleyan U., 1991; DH (hon.), Princeton U., 1991; DSc (hon.), Northwestern U., 1991, Worcester Poly. Inst., 1991; LHD (hon.), U. Pa., 1992; DPD (hon.), Niagara U., 1992; LHD, Hamilton Coll., 1992, LHD (hon.), 1991; DSc (hon.), Brown U., 1992; LHD (hon.), U. Mass. at Lowell, 1992, Med. Coll. Ohio, 1992; DSc (hon.), Howard U., 1993, W.Va. U., 1993; LLD (hon.), Lawrence U., 1993. Served with USPHS, Washington and Chevy Chase, Md., 1961-68; intern St. Catherine's Hosp., East Chicago, 1960-61; resident Yale Sch. Medicine, 1964-67; asst. prof. psychiatry Yale Child Study Center and dept. psychiatry, 1968-70, assoc. prof., 1970-75, prof., 1975-76, Maurice Falk prof. psychiatry, 1976—; assoc. dean Yale Med. Sch., New Haven, 1969—; dir. pupil svcs. Baldwin-King Sch. Project, New Haven; dir. sch. devel. program Yale Child Study Ctr.; dir. Conn. Energy Corp.; co-dir. Black Family Roundtable Greater New Haven, 1986—; cons. Joint Commn. on Mental Health of Children, Nat. Commn. on Causes and Prevention of Violence, NIMH; mem. nat. adv. mental health coun. HEW; Henry J. Kaiser Sr. fellow Center for Advanced Study in the Behavioral Scis., Stanford, 1976-77. Author: Beyond Black and White, 1972, Black Child Care, 1975, 2d edit., 1992, School Power, 1980, Maggie's American Dream, 1988; mem. editorial bd. Am. Jour. Orthopsychiatry, 1970-76, Youth and Adolescence, 1971-87, Jour. Negro Edn., 1978-83; guest editor Jour. Am. Acad. Child Psychiatry, 1985; columnist Parents mag.; contbr. articles to profl. jours. Bd. dirs. Dixwell Soul Sta. and Yale Afro-Am. House; trustee Carnegie Corp. of N.Y., 1990, Albertus Magnus Coll., 1989—, Conn. State U., 1991—; Nat. Coun. for Effective Schs., 1985—; bd. dirs., mem. profl. adv. bd. Children's TV Workshop, 1972-88; trustee Wesleyan U., 1978-84, Hazen Found., 1975-79, Field Found., 1981-88, Carnegie Corp., 1990; mem. profl. adv. coun. Nat. Assn. Mental Health; mem. ad hoc adv. com. Conn. Rsch. Commn.; mem. adv. coun. Nat. Com. for Citizens in Edn.; mem. nat. adv. coun. Hogg Found. for Mental Health, 1983-86; mem. adv. com. adolescent pregnancy prevention Children's Def. Fund. Recipient Child Study Assn.-Wel-Met Family Life book award, 1975, Howard U. Disting. Alumni award, 1976, John and Mary Markle Found. scholar, 1969—, Rockefeller Public Service award, 1980, Media award NCCJ, 1981, Community Leadership award Greater New Haven C. of C., 1983, Disting. Fellow award Conn. chpt. Phi Delta Kappa, 1984, Elm and Ivy award New Haven Found., 1985, Disting. Service award Conn. Assn. Psychologists, 1985, Disting. Educator award Conn. Coalition of 100 Black Women, 1985, Outstanding Leadership award Children's Def. Fund, 1987, Whitney M. Young Jr. Svc. award Boy Scouts Am., 1989, Prudential Leadership award Prudential Found., 1990, Harold W. McGraw Jr. prize in Edn., 1990, James Bryant Conant award Edn. Commn. States, 1991, Charles A. Dana prize in Edn., 1991, Disting. Svc. award Coun. Chief State Sch. Officers, 1991; James Comer NIMH Minority Fellowship established in his honor, 1991. Mem. APA (Disting. Svc. award 1993), Am. Acad. Child Adolescent Psychiatry, Nat. Med. Assn., Nat. Mental Health Assn. (Lela Rowland Prevention award 1989), Am. Psychiat. Assn. (Agnes Purcell McGavin award 1990, Solomon Carter Fuller award 1990, Spl. Presdendation 1990, Disting. Svc. award 1993), Am. Orthopsychiat. Assn. (Vera S. Paster award 1990), Am. Acad. Child Psychiatry, Black Psychiatrists of Am., NAACP, Black Coalition of New Haven, Greater New Haven Black Family Roundtable (co-dir. 1986—), Alpha Omega Alpha, Alpha Phi Alpha. Avocations: photography, travel, sports fan. Office: Yale U Child Study Ctr 230 S Frontage Rd New Haven CT 06519-1112

COMER, NATHAN LAWRENCE, psychiatrist, educator; b. Phila., Nov. 10, 1923; s. Rubin L. and Fannie (Cassover) C.; m. Rita Ellis, June 19, 1949 (dec. Mar. 1978); children: Robert, Susan Comer Kitei, Debra R., Marc J. BA, U. Pa., 1944; MD, Hahnemann Med. Coll., 1949. Diplomate Am. Bd. Psychiatry and Neurology, Am. Bd. Profl. Disability Cons. Intern Hahnemann Med. Coll., Phila., 1949-50; NIMH fellow Inst. of Pa. Hosp., Phila., 1951-53, sr. attending psychiatrist, 1968—; chief of psychiatry Thomas Jefferson Univ. Hosp.-Ford Road Campus, Phila., 1978; clin. assoc. prof. Med. Coll. Pa., Phila., 1978—; emeritus sr. attending Phila. Psychiat. Ctr., 1988—, pres. med. staff, 1975-77; pres. med. staff Inst. of Pa. Hosp., 1983-85. Contbr. articles to profl. jours. Bd. dirs. Temple Adath Israel of Main Line, Merion, Pa., 1958-78. Fellow Am. Psychiat. Assn. (life); mem. AMA, Am. Soc. for Adolescent Psychiatry, Hahnemann Med. Coll. Alumni Assn. (pres. 1973-74), B'nai Brith. Republican. Jewish. Home: 1100 Hillcrest Rd Narberth PA 19072-1224 Office: Inst Pa Hosp 111 N 49th St Philadelphia PA 19139-2718

COMETTO-MUÑIZ, JORGE ENRIQUE, biochemist; b. Buenos Aires, May 31, 1954; came to U.S., 1988; s. Jorge Raul and Teresa (Muñiz) Cometto; m. Andriana Nora Merlo, Nov. 5, 1979; children: Carolina S., Lucas M., Tomás P. Biochemist, U. Buenos Aires, 1977, D in Biochemistry, 1986. Lic. in clin. analysis. From rsch. fellow to asst investigator Nat. Coun. for Sci. & Tech., Buenos Aires, 1978-88, assoc. investigator, 1988—; asst. prof. biolog. chemistry U. Buenos Aires, 1986—; assoc. rsch. scientist Yale U., New Haven, 1988—; asst. fellow John B. Pierce Lab., New Haven, 1991—; vis. asst. fellow John B. Pierce Lab., New Haven, 1988-91. Contbr. articles to Annals the N.Y. Acad. Sci., Pharmacology, Biochemistry & Behavior, Physiology & Behavior, Perception & Psychophysics. Mem. Soc. Sci. Argentina, Assn. for Chemoreception Sci., European Chemoreception Rsch. Orgn., Internat. Soc. Indoor Air Quality and Climate. Office: John B Pierce Laboratory 290 Congress Ave New Haven CT 06519

COMIS, ROBERT LEO, oncologist, educator; b. Troy, N.Y., July 16, 1945; s. James Carl and Mary (Casile) C.; m. Virginia Martin; children: Larissa, Robert Leo Jr., Anthony. BS, Fordham U., 1967; MD, SUNY, Syracuse, 1971. Diplomate Am. Bd. Internal Medicine, Am. Bd. Internal Medicine-Med. Oncology. Intern SUNY Upstate Med. Ctr., Syracuse, 1971-72, resident, 1974-75, chief med. oncology, 1978-84; staff assoc. cancer evaluation program Nat. Cancer Inst., NIH, Bethesda, Md., 1972-74; dir. Cen. N.Y. Regional Oncology Ctr., N.Y.C., 1984-88; chmn. med. oncology Fox Chase Cancer Ctr., Phila., 1984-88, med. dir., 1984—, v.p. med. sci., 1987—; prof. medicine Temple U. Hosp., Phila., 1987—; staff assoc. cancer therapy evaluation program; asst. prof. medicine, chief solid tumor oncology svc., sect. hematology/oncology SUNY Upstate Med. Ctr., 1976-78, assoc. prof., 1978-83, prof. medicine, 1983-84, chief sect. med. oncology, 1978-84; chmn. respiratory disease com. Cancer and Leukemia Group, 1979-84; assoc. chmn. lab. sci. Eastern Coop. Oncology Group, 1991—; mem. cancer award selection com. Bristol Myers-Squibb, 1985-91; asst. in medicine Peter Pent Brigham Hosp., Boston, 1975-76. Contbr. articles to profl. publs. V.p. Am. Cancer Soc., Onodaga County, N.Y., 1981-82, pres., 1982, mem. med. affairs com. N.Y. State div. With USPHS, 1972-74. Grantee, Ea. Coop. Oncology Group, 1985—, Bristol-Myers, 1985-89, Small Instrumentation Program, 1989-90, Biomed. Rsch. Support, 1989-90, Clin. Oncology Rsch. Career DEvel. Program, 1992—. Mem. AAAS, Am. Soc. Clin. Oncology (program chair 1985-86, editorial bd. jour. 1985, chmn. program com. 1985, clin. trial com.), Am. Radium Soc. (resident awards com. 1990—, sci. program com. 1990-91), Am. Assn. Cancer Rsch., Inc., Internat. Assn. for Study of Lung Cancer, N.Y. State Cancer Programs Assn., Pa. Soc. Hematology/Oncology, Pa. Med. Soc., Phila. County Med. Soc., Soc. Biol. Therapy. Office: Jefferson Cancer Ctr Bluemle Life Scis Bldg 233 S 10th St Philadelphia PA 19107-5541

COMISO, JOSEFINO CACAS, physical scientist; b. Narvacan, The Philippines, Sept. 21, 1940; came to U.S., 1964; s. Severino Cacho and Silvestra (Cacas) C.; m. Diana Pareñas Jimenez, June 27, 1970; children: Glen Arnold, David Arnel, Melissa Jane. BS in Physics, U. The Philippines, Quezon City, 1962; MS in Physics, Fla. State U., 1966; PhD in Physics, UCLA, 1972. Scientist Philippine Atomic Rsch. Ctr., Quezon City, 1962-63; instr. U. The Philippines, Quezon City, 1963-64; asst. rsch. physicist UCLA, 1972-73; rsch. assoc. U. Va., Charlottesville, 1973-77; sr. mem. tech. staff Computer Scis. Corp., Greenbelt, Md., 1977-79; phys. scientist Goddard Space Flight Ctr. NASA, Greenbelt, 1979—. Co-author: Arctic & Antarctic Sea Ice, 1992; contbr. articles to profl. jours. Pres. Philippine-Am. Acad. Scientists and Engrs., Washington, 1987. Mem. Am. Geophys. Union, Am. Phys. Soc., Internat. Glaciol. Soc., Com. on Polar Meteorology and Oceanography, Electromagnetics Acad. Achievements include discovery of a new technique for measuring thickness distributions of Arctic sea ice using Lidar, a comprehensive study of ocean color pigment distributions in the entire southern ocean, and discovery of a recent large decrease (highly correlated with surface temperature increase) of sea ice cover in the Bellingshausen Sea. Home: 11013 Elon Dr Bowie MD 20720-3509 Office: Labr for Hydrospheric Processes NASA/GSFC Code 971 Greenbelt MD 20771

COMMITO, RICHARD WILLIAM, podiatrist; b. Chgo., May 2, 1951; s. Mario Fiore and Aileen Margaret (Stang) C. BS, U. Ill., Chgo., 1972; DPM, Ill. Coll. Podiatric Medicine, 1976. Diplomate Nat. Bd. Podiatry Examiners, Am. Bd. Podiatric Surgery, Am. Coun. Cert. Podiatric Physicians and Surgeons, Am. Acad. Pain Mgmt. Podiatrist Chgo., 1976—; dir. podiatry svcs. Community Hosp., Evanston, Ill., 1978-80; cons. staff podiatry Ridgeway Hosp., Chgo., 1981—; owner Conserv Environ. Products, FootDoc Products, NewFoot Pharmical; dir. podiatry svc. Lawndale Pla. Surgicenter, Chgo.; ind. med. examiner of counsel R. S. Connors Assocs., Chgo. Bd. dirs. Little Village unit Chgo. Boys' Clubs, 1981—, mem. One Hundred Club, 1982, mem. 400 Club, Marshall Sq. unit, 1981-82; mem. Art Inst. Chgo., 1980-89, Lincoln Park Zool. Soc., Chgo., 1980-89; mem. Sailboat Bend Civic Assoc. Fellow Acad. Ambulatory Foot Surgery, Am. Inst. Foot Medicine (cert.); mem. Soaring Soc. Am., Ill. Podiatry Edn. Group, Am. Podiatric Med. Assn., Ill. Podiatry Soc., Am. Med. Soc. of Vienna (life), Nat. Assn. Professions, Am. Soc. Podiatric Legal Medicine, Am. Podiatric Circulatory Soc., Am. Acad. Pain Mgmt., Am. Soc. Podiatric Physicians and Surgeons, Am. Coun. Cert. Podiatric Physicians and Surgeons (cert.), Internat. Inst. Reflexology (cert.), Am. Pain Soc., Nat. Assn. of the Self-Employed, Internat. Inst. Continuing Med. Edn. Roman Catholic.

COMMONER, BARRY, biologist, educator; b. Bklyn., May 28, 1917; s. Isidore and Goldie (Yarmolinsky) C.; m. Lisa Feiner, 1980; children by previous marriage: Lucy Alison, Frederic Gordon. A.B. with honors, Columbia U., 1937; M.A., Harvard U., 1938, Ph.D., 1941; D.Sc. (hon.), Hahnemann Med. Coll., 1963, Clark U., 1967, Grinnell Coll., 1968, Lehigh U., 1969, Williams Coll., 1970, Ripon Coll., 1971, Colgate U., 1972, Cleve. State U., 1980; LL.D. (hon.), U. Calif., 1974, Grinnell Coll., 1981; D.Sc. (hon.), St. Lawrence U., 1988; D.H.L. (hon.), Lowell U., 1990; DSc (hon.), Conn. Coll., 1992. Asst. biology Harvard, 1938-40; instr. biology Queens Coll., 1940-42; asso. editor Sci. Illus., 1946-47; asso. prof. plant physiology Washington U., St. Louis, 1947-53; prof. Washington U., 1953-76, chmn. dept. botany, 1965-69; dir. Washington U. (Center for the Biology of Natural Systems), 1965-81, Univ. prof. environ. sci., 1976-81; prof. dept. geology Queens Coll., Flushing, N.Y., 1981-87, prof. emeritus, 1987—, dir. Center for the Biology of Natural Systems, 1981—; vis. prof. community health Albert Einstein Coll. of Medicine, N.Y.C., 1981-87; disting. univ. prof. indsl. policy, U. Mass., Lowell, 1992—; pres. St. Louis Com. for Nuclear Info., 1965-66, bd. dirs. 1966; mem. Nat. Tb Commn. on Air Conservation, 1966-68; bd. dirs. Scientists Inst. Pub. Info., 1963—, co-chmn., 1967-69, chmn., 1969-78, chmn. exec. com., 1978—; chmn. spl. cons. group sonic boom Dept. Interior, 1967-68; adv. council on environ. edn. Office Edn. HEW, 1971; internat. sponsoring com. Chaim Weizmann Centenary Celebration, 1974-75; adv. com. Coalition Health Communities, 1975; mem. sec.'s adv. council U.S. Dept. Commerce, 1976; sci. adv. council on dioxin Vietnam Vets. Am. Found., 1985—; sci. adv. N.Y. State Com. on Sci. and Tech., 1981—; adv. bd. Com. for Responsible Genetics, 1983—. Author: Science and Survival, 1966, The Closing Circle, 1971 (Phi Beta Kappa award), (Internat. prize City of Cervia, Italy), La Technologia del Profitto, 1973, The Poverty of Power, 1976 (Premio Iglesias award, Sardinia, Italy 1978), Ecologia e Lotte Sociali, 1976, l'energia alternativa, 1978, The Politics of Energy, 1979 (Premio Iglesias award 1982), Se Scoppia La Bomba, 1984, Il Cerchio Da Chiudere, 1986, Making Peace With the Planet, 1990; editorial bd. World Book Ency., 1968-73, Environment mag., 1977; mem. adv. bd. Science Year, 1967-72; editorial adv. bd. Hon. Chemosphere, from 1972; bd. sponsors In These Times, 1976—. Bd. cons. experts Rachel Carson Trust for Living Environment, 1967—; adv. com. Center for Devel. Policy, 1978; mem. bd. Univs. Nat. Anti-War Fund; adv. bd. Fund for Peace, 1978, Citizens Party candidate for pres. of U.S., 1980. Served to lt. USNR, 1942-46. Recipient Newcomb Cleveland prize AAAS, 1953; 1st Humanist award Internat. Humanist and Ethical Union, 1970; medal AIA, 1979; decorated comdr. Order of Merit Italy, 1977. Fellow AAAS (chmn. com. sci. in promotion of human welfare 1958-65, dir. 1967-74, chmn. com. on environ. alterations 1969-72), Am. Sch. Health Assn. (hon.); mem. Soc. Biol. Chemists, Soc. Gen. Physiologists, Am. Soc. Plant Physiologists, Sierra Club, Nat. Parks Assn. (trustee 1968-70), Soil Assn. Eng. (hon. life v.p.), Am. Chem. Soc., Am. Soc. Biol. Chemists, Fedn. Am. Scientists, Ecol. Soc. Am., Inst. Environmental Edn. (trustee), Phi Beta Kappa, Sigma Xi. Office: Queens Coll Ctr for Biology Natural Systems Flushing NY 11367

COMO, FRANCIS W., plastics manufacturing executive; b. Fall River, Mass., 1931. Degree, Bryant Coll., 1951. V.p., gen. mgr. Owens-Corning Fiberglass Corp., Toledo, Ohio. Recipient Clare E. Bacon Person of the Yr. award Soc. Plastics Industry, 1992. Office: Owens-Corning Fiberglas Corp Fiberglas Tower Toledo OH 43659*

COMOGLIO, PAOLO MARIA, cell and molecular biologist, educator; b. Torino, Piedmont, Italy, May 29, 1945; s. Pier Giovanni and Ada (Brunetti) C.; m. Adriana Gillio, July 26, 1969; children: Francesca, Pier Andrea. D in Medicine, U. Torino, 1969. Rsch. fellow Washington U., St. Louis, 1969-71; rsch. assoc. U. Pa., Phila., 1973; asst. prof. cell biology U. Trieste, Italy, 1974-80, prof., 1980-83; prof., chmn. U. Torino, 1983—; mem. editorial bd. European Jour. Cell Biology, 1987—. Author: (book) The Cell: Molecules Anatomy and Biology, 1983, rev. edit., 1987, 92, 93; contbr. articles to Nature. Recipient Prodi award for Cancer Rsch. IX Centennial of U. Bologna, Italy, 1989. Mem. European Molecular Biology Orgn., European Assn. Cancer Rsch. (exec. coun. 1983-88), Am. Assn. Immunologists. Achievements include discovery of action of moyoclonal antibodies against phospho-tyrosine; elucidation of the stucture and function of the C-Met oncogene and the encoded HGF receptor. Office: U Torino Med Sch Dept Biomed Sci & Oncology, C.so D'Azeglio 52 10126, Torino Piedmonte, Italy

COMPADRE, CESAR MANUEL, chemistry educator; b. Mexico City, July 1, 1953; came to U.S., 1980; s. Cesar and Maria (Reyes) C.; m. R. Lilia Lopez, Dec. 16, 1978; children: Irene, Amanda. BS, Nat. U. Mexico, Mexico City, 1977, MS, 1980; PhD, U. Ill., 1985. Rsch. assoc. U. Ill., Chgo., 1985-86, Pomona Coll., Claremont, Calif., 1986-88; assoc. prof. U. Ark., Little Rock, 1988—. Contbr. articles to Science, NAS Procs., Experientia, Biochem. Biophys. Alta, Jour. Pesticide Biochem. Grantee Ark. Sci. and Rsch. Found., 1990, Am. Cancer Soc., 1991, NSF, 1992; recipient Mr. and Mrs. William Walker award, U. Ark. Med. Endowment, Little Rock, 1990. Mem. Am. Chem. Soc., Sigma Xi. Achievements include patent on sugar substitute Hernandulcin; co-patent (pending) for process for poultry decontamination. Office: Univ Ark Dept Med Sci 4301 N Markham Little Rock AR 72205

COMPOSTO, RUSSELL JOHN, materials science engineering educator, researcher; b. Hoboken, N.J., Aug. 8, 1960; m. Karen I. Winey, July 11, 1987. BA, Gettysburg Coll., 1982; PhD, Cornell U., 1987. Postdoctoral U. Mass., Amherst, 1987-90; asst. prof. U. Pa., Phila., 1990—. Contbr. articles to profl. jours. Recipient Presdl. Young Investigator, NSF, 1991. Mem. Am. Phys. Soc., Am. Chem. Soc., Materials Rsch. Soc. Achievements include first rigorous study of tracer and mutual diffusion in binary polymer blends; observation of surface segregation in thin films consisting of binary polymer blends; using neutron reflectivity, first measurement of the volume fraction profile of end-fuctionalized polymers absorbed to a silicon oxide surface. Office: Univ Pa 3231 Walnut Philadelphia PA 19105-6272

COMPTON, DALE LEONARD, space agency administrator; b. Pasadena, Calif., June 18, 1935; s. John Leonard and Gladys Imnachuck (Foster) C.; m. Marilyn Doris Garland, June 21, 1959; children: David, Debora. BSME, Stanford U., 1957, MS in Aero. Engring., 1958, PhD, 1969; MMS in Mgmt. Sci., MIT, 1975. Rsch. scientist NASA-Ames Reseach Ctr., Moffett Field, Calif., 1957-72; Tech. asst. to dir. NASA-Ames Research Ctr., Moffett Field, Calif., 1972-73; dep. dir. astronautics NASA-Ames Research Ctr., 1973-74; Sloan fellow MIT, Cambridge, Mass., 1974-75; chief space sci. div. NASA-Ames Research Ctr., Moffett Field, 1974-80, mgr. IRAS Project, 1980-81, dep. dir. astronautics, 1981-82, dir. engring. & computer systems, 1982-85, dep. dir., 1985-88, dir., 1988—. Recipient NASA's Outstanding Leadership medal, SES Presdl. Ranks of Meritorious and Disting. Exec. Fellow AIAA (named Outstanding Engr./Astro., 1983-84); mem. AAAS, Internat. Acad. Astronautics, Tau Beta Pi, Sigma Xi. Avocations: woodworking, reading, sailing, bird watching. Office: NASA Ames Rsch Ctr Mail Stop 200-1 Moffett Field CA 94035

COMPTON, JAMES ALLAN, chemical engineer; b. Ottawa, Kans., Feb. 27, 1951; s. Robert Otis and Madge Arlene (Wait) C. BS in Chem. Engring., U. Kans., 1975. Registered profl. engr., Wash. Chem. engr. Westinghouse Hanford Co., Richland, Wash., 1975—. Mem. Am. Inst. Chem. Engrs. Republican. Presbyterian. Office: Westinghouse Hanford Co PO Box 1970 M/S T5-12 Richland WA 99352

COMPTON, MARK MELVILLE, poultry science educator; b. Richmond, Va., Nov. 21, 1953; s. Archie M. and Betty I. (Sparkhall) C.; m. Kathy Y. Yoho, June 23, 1979; 1 child, Aurora L. MS, Va. Polytech. Inst. & State U., 1978, PhD, 1978. Postdoctoral rsch. assoc. Med. Coll. Va., Richmond, 1979-83, U. N.C., Chapel Hill, 1983-87; asst. prof. Poultry Sci. U. Ga., Athens, 1987—. Contbr. articles to profl. jours. Am. Cancer Soc. rsch. grantee, 1982, Nat. Insts. of Health rsch. grantee, 1986, USDA rsch. grantee, 1988. Mem. Poultry Sci. Assn., Endocrine Soc., Sigma Xi. Avocations: woodworking, photography, silk screening, stained glass, motorcycles. Office: U Ga Dept Poultry Sci Athens GA 30602

COMPTON, W. DALE, physicist; b. Chrisman, Ill., Jan. 7, 1929; s. Roy L. and Marcia (Wood) D.; m. Jeanne C. Parker, Oct. 14, 1951; children: Gayle Corinne, Donald Leonard, Duane Arthur. B.A., Wabash Coll., 1949; M.S., U. Okla., 1951; Ph.D., U. Ill., 1955; D.Eng. (hon.), Mich. Technol. U., 1976. Physicist U.S. Naval Ordnance Test Sta., China Lake, Calif., 1951-52, U.S. Naval Research Lab., Washington, 1955-61; prof. physics U. Ill. at Urbana, 1961-70, dir. coordinated sci. lab., 1965-70; dir. chem. and phys. scis., exec. dir. sci. research staff, v.p. research Ford Motor Co., Dearborn, Mich., 1970-86; sr. fellow Nat. Acad. Engring., 1986-88; disting. prof. indsl. engring. Purdue U., West Lafayette, Ind., 1988—; mem. Presdl. Commn. for Award of Medal of Sci., 1978-80; mem. vis. com. Nat. Bur. Standards, 1975-79, chmn. vis. com., 1979; bd. govs. NRC, 1991—. Author: (with J.H. Schulman) Color Centers in Solids, 1962; editor: Interaction of Science and Technology, 1969, Design and Analysis of Integrated Manufacturing Systems, 1988; co-editor (with J. Heim): Manufacturing Systems, Foundations of World Class Practice, 1992. Bd. dirs. Mich. Cancer Found., 1975-86, Coordinating Rsch. Coun., 1983-85; adv. com. Combustion Rsch. Facility, Sandia Nat. Lab., 1983-86; mem. energy rsch. adv. bd. Dept. Energy, 1979-80; bd. govs. Argonne Nat. Lab., 1983-86. Fellow AAAS, NAE (coun. 1990—), Am. Phys. Soc., Washington Acad. Scis., Soc. Automotive Engrs., Engring. Soc. Detroit; mem. Rsch. Soc. Am., Nat. Rsch. Coun. (bd. govs. 1991—).

COMSTOCK, M(ARY) JOAN, publishing executive, chemist; b. Norfolk, Va., Dec. 1, 1946; d. Franklin Leonard and June (Bennett) C.; m. Leroy D. Resnick, May 20, 1978; children: Lis J. Resnick, Laura J. Resnick. BS in Chemistry, Alvernia Coll., 1969; MBA, George Washington U., 1991. Tchr. chemistry Bishop Conwell High Sch., Levittown, Pa., 1969-70; tchr. sci. LeReine High Sch., Suitland, Md., 1971-74; sales-svc. rep. Drew Chem. Corp., Parsippany, N.J., 1974-75; copy editor Am. Chem. Soc., Washington, 1975-77, acquisitions editor, 1977-79, head dept. books, 1979—. Mem. AAAS, Am. Chem. Soc., Soc. for Scholarly Pub. (chair ann. meeting 1993). Office: Am Chem Soc 1155 16th St NW Washington DC 20036

CONANT, DAVID ARTHUR, architectural acoustician, educator, consultant; b. Biloxi, Miss., Dec. 22, 1945; s. Roger and Lillian Rose May (Lovell) C.; m. Nancy Hayes, June 17, 1972; children: Christopher, Tyler. BS in Physics, Union Coll., 1968; MA in Geology, Columbia U., 1972; BArch, Rensselaer Poly. Inst., 1975, MArch, 1975. Faculty fellow Lamont-Doherty Geol. Obs., Palisades, N.Y., 1970-72; teaching asst. Rensselaer Poly. Inst., Troy, N.Y., 1973-76; asst. prof. dept. architecture Calif. State Poly. U., Pomona, 1976-78; sr. cons. Bolt Beranek Newman Inc., Canoga Park, Calif., 1977-87; prin. McKay Conant Brook, Inc., Westlake Village, Calif., 1987—; cons. IBM Bldg. Energy Rsch. Group, Marina Del Rey, Calif., 1976-77, Expo '93, Taejon, Korea; prin. acoustical cons. Euro Disneyland, Paris; lectr. acoustics UCLA. Co-author: (textbook) Fundamentals and Abatement of Highway Traffic Noise, 1980; author computer software. Instr., vol. Upward Bound, Schenectady, N.Y., 1967-68; overseas vol. Am. Friends Svc. Com., Yugoslavia and Denmark, 1967. With U.S. Army, 1968-70. Recipient Honor award Am. Inst. Architects, 1991. Mem. ASHRAE, Acoustical Soc. Am., Nat. Coun. Acoustical Cons., Constrn. Specifications Inst., Sigma Xi (univ. chpts. lectr. physics). Republican. Presbyterian. Achievements include development of binaural/analog computer and psychoacoustic model to assess and predict degree of subjective "immersion" in auditorium sound fields. Home: 1504 Grissom St Thousand Oaks CA 91362 Office: McKay Conant Brook Inc 5655 Lindero Canyon Rd Ste 325 Thousand Oaks CA 91362

CONCEICAO, JOSIE, chemistry researcher; b. Macao, Mar. 24, 1961; came to U.S., 1980; s. Joao Haracio and Beatriz (Batalha) C. MS, Yale U., 1986; PhD, Rice U., 1992. Rsch. assoc. U. Utah, Salt Lake City, 1992—; peer review (referee) Jour. Chem. Physics, 1992, Jour. Phys. Chemistry, 1993. Contbr. articles to profl. jours. Chmn. Svc. Com. Rice U. Cath. Student Union, 1989-90; founder, pres. Rice U. Badminton Club, Houston, 1991-92. Named Outstanding Student Am. Inst. Chemistry, 1984, Sigma Xi, 1984, Phi Lambda Upsilon, 1987. Mem. Am. Chem. Soc., Phi Beta Kappa. Office: U Utah Henry Eyring Bldg Salt Lake City UT 84112

CONDER, GEORGE ANTHONY, parasitologist; b. Albuquerque, Oct. 7, 1950; s. George L. and Theresa Angelina (Naccarato) C.; m. Kathleen Brooke Cunningham, Aug. 27, 1977; children: Brie Anne, Ty Anthony, Lucie Brooke. BA, Pomona Coll., 1972; MS, U. N.Mex., 1975; PhD, Brigham Young U., 1979. Teaching/rsch. asst. U. N.Mex., Albuquerque, 1973-75; teaching asst., rsch. assoc. Brigham Young U., Provo, Utah, 1975-79; postdoctoral tng. grantee Mich. State U., East Lansing, 1979-81; scientist I-IV Upjohn Labs., Kalamazoo, Mich., 1981—; adj. assoc. prof. U. Tex., El

Paso, 1988—; temp. advisor onchocerciasis chemotherapy project steering com. WHO, 1985-87; external reviewer Helminth Disease Lab., Animal Parasitology Inst., USDA, Beltsville, Md., 1987. Co-editor: Onchocerciasis/Filariasis Procs., 1987; contbr. articles to profl. publs. Sci. fair judge Gull Lake Schs., Richland, Mich., 1976; coach Youth Basketball, Richland, 1989-93. Grantee Argonne Nat. Lab., 1975, WHO, 1985-87, Upjohn/Nat. Lab. Collaborative Rsch. Program, 1988-90. Mem. Am. Assn. Vet. Parasitologists (mem. newsletter editorial bd. 1990—, v.p., program chmn. 1991-92, pres. 1993-94), Am. Soc. Parasitologists, Am. Soc. Tropical Medicine and Hygiene, Helminthological Soc. Wash., N.Y. Acad. Scis., Wildlife Disease Assn., Phi Kappa Phi. Achievements include 4 patents in field. Office: Upjohn Labs 7923-25-5 Kalamazoo MI 49001

CONDON, THOMAS BRIAN (BRIAN CONDON), hospital executive; b. Beverly, Mass., June 1, 1942; s. Thomas William and Marguerite Mary (Welch) C.; m. Carol Therese Sicilliano, Apr. 29, 1969; children: Therese Beth, Tara Bridget, Colleen Marguerite, Caroline Susan. BA in English, Boston Coll., 1964; MPA, U. New Haven, 1973, MA in Community Psychology, 1975, MA in Indsl. and Organizational Psychology, 1977. Dir. unit mgmt. Yale New Haven Hosp., 1971—, asst. adminstr., 1975, v.p., 1980; pres., dir. Shirley Frank Found., New Haven, 1980—; chmn., dir. South Cen. Community Coll. Found., New Haven, 1988—; dir. Nat. Inst. Community Health Edn., Hamden, Conn., 1990—; bd. dirs. Hill Devel. Corp., New Haven; mem. adv. coun. New England Organ Bank, 1990—. Elected mem. Cheshire (Conn.) Planning & Zoning Commn., 1976-87; mem. Gov.'s Task Force on Student Aid, Hartford, Conn., 1986-87; chmn., dir. Conn. Student Loan Found., Rocky Hill, 1976—. Capt. U.S. Army, 1964-70. Recipient Community Svc. award Bd. Trustees Conn. State Coll., 1991. Roman Catholic. Avocations: antiques, netsuke, film collector. Home: 150 Hotchkiss Ridge Cheshire CT 06410 Office: Yale New Haven Hosp 20 York St New Haven CT 06504

CONDY, SYLVIA ROBBINS, psychologist; b. Vallejo, Calif., Sept. 16, 1931; d. Charles Richard and Stella Marie (Stannard) C.; m. Ramon Romain Tappero, Mar. 22, 1950 (div. Feb. 1959); children: Susan, Lucia, Rhys; m. Vernon Oliver Robbins, May 1, 1960 (div. Feb. 1979); 1 child, Vernon Oliver Jr. MA, U. Calif., Chico, 1976; PhD, Calif. Sch. Profl. Psychology, 1985. Lic. clin. psychologist, Alaska, Calif. Clinician Anchorage (Alaska) Com. Mental Health, 1977-82, supr. testing, 1985-86, clin. deputy dir., 1986-90; pvt. practice clin. psychology, 1985—. Bd. dirs. Hospice, Anchorage, 1985-86, STAR Rape Crisis, Anchorage, 1986. Mem. APA, Assn. Humanist Psychology, Child Advocacy Network (pres. 1989), Statewide Child Advocacy, Rotary (East Anchorage). Democrat. Mem. Ch. of Religious Sci. Avocations: gardening, reading, ocean kayaking, traveling. Home: 3300 Hiland Dr Anchorage AK 99504-4078 Office: 2550 Denali St Ste 1610 Anchorage AK 99503-2737

CONE, RICHARD ALLEN, biophysics educator; b. St. Paul, May 23, 1936; s. E. Richard and Martha (Egloff) C.; married; children: Jenny Ahern, Ariel Martin-Cone. SB, MIT, 1958; PhD, U. Chgo., 1963. Asst. prof. biology Harvard U., Cambridge, Mass., 1965-69; assoc. prof. biophysics Johns Hopkins U., Balt., 1969-73, prof. biophysics, 1973—, prof. biology, 1985—; Visual scis. study sect. NIH, Washington, 1977-81; program planning panel Nat. Adv. Eye Coun., Washington, 1981-83; cons. Prophylactic Contraceptive Devel. Recipient Cole award, 1979. Mem. Soc. for Study Reproduction. Office: Johns Hopkins U 3400 N Charles St Baltimore MD 21218

CONES, VAN BUREN, electronics engineer, consultant; b. Indpls., July 4, 1918; s. Ben and Fanette (Miller) C.; m. Eloise Winifred Knoll, Sept. 8, 1951; children: Diane Lee Cones Serban, Anita Sue Cones Cohee. BSEE, Purdue U., 1940; postgrad., Mass. Inst. Tech., 1942, Harvard U., 1942. Registered profl. engr., Ind. Field engr. Powers Regulator Co., Indpls., 1940-42, 46-52; electronics engr. Naval Avionics Ctr., Indpls., 1952-84; pvt. practice Indpls., 1984—. Post comdr. Am. Legion, Indpls., 1974. Capt. USAF, 1942-46. Mem. IEEE, NSPE. Methodist. Achievements include patents for self-triggered sawtooth wave generator, self-triggered sawtooth voltage wave generator, headlamp for both day and night driving, direct measurement probe for RF energy, temperature stabilized transistor amplifier, appliance theft control alarm system; co-inventor RF interference supression control. Home: 5503 Skyridge Dr Indianapolis IN 46250

CONFORTI, RONALD ANTHONY, JR., communications and electronics consultant; b. Pitts., May 6, 1960; s. Ronald Anthony and Blanche (Thiem) C. Diploma in Info. Systems, Robert Morris Coll., 1982. Sr. systems designer Applied Sci. Assocs., Inc., Butler, Pa., 1982-93. Author: Electronic Component Training, Electronic Troubleshooting and Maintenance, 1988. Mem. Soc. Applied Learning Tech. Roman Catholic. Office: Applied Sci Assocs Inc 292 Three Degree Rd Butler PA 16001

CONGER, JEFFREY SCOTT, electrical engineer; b. Mapleton, Iowa, Mar. 31, 1961; s. Walter Lee and Betty Elaine (Sanford) C. BS, Iowa State U., 1983; MS, U. Minn., 1988, PhD, 1992. Product engr. Tex. Instruments, Inc., Sherman, 1983-84; teaching asst. U. Minn., Mpls., 1984-85; student aide Honeywell Systems & Rsch. Ctr., Bloomington, Minn., 1985-92; sr. circuit design engr. Cray Rsch., Inc., Chippewa Falls, Wis., 1992—. Contbr. articles to profl. jours. Recipient fellowship Am. Electronics Assn., 1985-89. Mem. IEEE, Eta Kappa Nu. Achievements include patent for hotclock complex logic. Home: 3411 Miller St Eau Claire WI 54701 Office: Cray Rsch Inc 850 Industrial Blvd Chippewa WI 54729

CONKLIN, HAROLD COLYER, anthropologist, educator; b. Easton, Pa., Apr. 27, 1926; s. Howard S. and May W. (Colyer) C.; m. Jean M. Morisuye, June 11, 1954; children: Bruce Robert, Mark William. A.B., U. Calif.-Berkeley, 1950; Ph.D., Yale U., 1955. From instr. to assoc. prof. anthropology Columbia U., 1954-62; lectr. anthropology Rockefeller Inst., 1961-62; prof. anthropology Yale U., 1962—, chmn. dept., 1964-68; curator of anthropology Yale Peabody Mus. Natural History, 1974—; Franklin Mussy Crosby prof. human environment, 1990—; mem. Inst. for Advanced Study, Princeton, N.J., 1972; fellow Ctr. for Advanced Study in Behavioral Scis., Stanford, Calif., 1978-79; field research in Philippines, 1945-47, 52-54, 55, 57-58, 61, 62-65, 68-69, 70, 73, 80-81, 82-85, 90-91, Malaya and Indonesia, 1948, 57, 83, Melanesia, 1987, Calif. and N.Y., 1951, 52, Guatemala, 1959, Peru, 1987; dir., com. problems and policy Social Sci. Research Council, 1963-70; bd. dirs. Survival Internat. USA; spl. cons. Internat. Rice Research Inst., Los Baños, Philippines, 1962—; book rev. editor Am. Anthropologist, 1960-62; mem. Pacific sci. bd. Nat. Acad. Scis.-NRC, 1962-66. Author: Hanunóo Agriculture, 1957, Folk Classification, 1972, Ethnographic Atlas of Ifugao, 1980; other publs. on ethnol., linguistic and ecol. topics. Served with AUS, 1944-46. Guggenheim fellow, 1973; recipient Internat. Sci. prize Fyssen Foundation, 1983. Mem. NAS; Fellow Am. Acad. Arts and Scis., Am. Anthrop. Assn. (exec. bd. 1965-68), Royal Anthrop. Inst., N.Y. Acad. Scis. (sec. sect. anthropology 1956); mem. Am. Ethnol. Soc. (councilor 1960-62, pres. 1978-79), Koninklijk Inst. voor Taal- Land- en Volkenkunde, Conn. Acad. Arts and Scis., Linguistic Soc. Am., Kroeber Anthrop. Soc., Phila. Anthrop. Soc., Am. Geog. Soc., Am. Oriental Soc., Assn. Asian Studies, Classification Soc., Linguistic Soc. Philippines, Indo-Pacific Prehistory Assn., Soc. Econ. Botany, Internat. Assn. Plant Taxonomy, AAAS, Phi Beta Kappa, Sigma Xi. Home: 106 York Sq New Haven CT 06511-3625 Office: Yale Univ Dept Anthropology New Haven CT 06520

CONKLIN, JOHN ROGER, electronics company executive; b. Poughkeepsie, N.Y., Dec. 20, 1933; s. Leland Thomas and Eleanor (Warren) C.; m. Catharine Becker, Dec. 28, 1956 (div. Apr. 1976); children: Thomas Stephen, Todd Roger; m. Nancy Plank, July 16, 1983. BS in Mil. Sci., U.S. Mil. Acad., 1956; postgrad., Xavier U., Cin., 1961-62, Northeastern U., 1974. Engr. Procter & Gamble Co., Cin., 1960-64; sales engr. Orville Simpson Co., Cin., 1964-67; various sales positions DeLaval Separator Co., Poughkeepsie, 1967-74, pres., 1974-78; pres. Standard Gage Co., Poughkeepsie, 1979-86; pres., owner Chief Electronics Inc., Poughkeepsie, 1985—, Discount Data Products, Inc., Poughkeepsie, 1988—; pres. Chief Indsl. Controls, Poughkeepsie, 1987—; bd. dirs. Fargo Mfg. Co., Poughkeepsie; adv. bd. Dutchess divsn. Bank of N.Y., Poughkeepsie, 1974—. Contbr. articles to profl. jours.; patentee basket centrifuge. Bd. dirs. Area Fund-Dutchess County, Poughkeepsie, 1975, Poughkeepsie C. of C., 1981-86, YMCA, Poughkeepsie, 1982; campaign chair United Way, Dutchess County, N.Y., 1986. 1st lt. U.S. Army, 1956-60. Mem. D.C. Mycological Soc., D.C. Hist. Soc. Avocations: skiing, trout fishing, gardening, mycology. Home: 4 Dutchess Ter Rhinecliff NY 12574-9999

CONKLIN, ROBERT EUGENE, electronics engineer; b. Loveland, Ohio, Apr. 21, 1925; s. Charles and Alberta (Reynolds) C.; m. Virginia E. McCann, June 14, 1952; children—Carl Lynn, Jill Elaine Conklin Bradford. B.S. in Edn., Wilmington Coll., 1949, B.S. in Sci., 1949. Electronic scientist Electronic Technol. Lab., Wright-Patterson AFB, Ohio, 1951-55; electronic engr. AF Avionics Lab., Wright-Patterson AFB, 1956-60 supervisory elec. engr., 1960-72, cons. electronic engr., 1972-78, supervisory electronic engr., 1978-82, electronic engr. (VHSIC), 1982-84; cons. engr. REC Electronics, Fairborn, Ohio, 1984—; mem. Inst. Nav., 1968-72. Mgr. Babe Ruth Boys' Baseball, 1969-74; mgr. and pres. Little League, Fairborn, 1965-68. Served with USAAC, 1943-46. Mem. IEEE. Republican. Quaker. Lodge: Lions (Fairborn) Home: 114 Wayne Dr Fairborn OH 45324-5228 Office: 47 N Broad St Fairborn OH 45324

CONLEY, EDWARD VINCENT, JR., research engineer; b. McKeesport, Pa., Apr. 14, 1940; s. Edward Vincent Sr. and Antoinette Marie (Schettino) C.; m. Kathleen Elizabeth Carr, June 11, 1966; children: Shannon Michelle, Sean Michael. Student, Carnegie Mellon U., 1960, 1966-68; BS in Chemistry, U. Pitts., 1976. Rsch. engr. U. S. Steel Corp., Monroeville, Pa., 1976-80, Kennametal, Inc., Greensburg, Pa., 1980-83; corp. quality assurance engr. Kennametal, Inc., Latrobe, Pa., 1983-85; sr. engr. Kennametal, Inc., Greensburg, Pa., 1985-89; staff engr. Kennametal, Inc., Latrobe, Pa., 1989—; presenter Materials Week 93 Conf. Past contbr. to ASM Handbook of Metals, also articles to profl. publs. Advisor Jr. Achievement, Duquesne-Pitts., 1973-76; tchr. Explorer Scout Program, Monroeville, Pa., 1977-79. With U.S. Army, 1963-65. Mem. Am. Soc. Metals, Am. Powder Metal Inst., Soc. Carbide and Tool Engrs., Internat. Platform Assn. Democrat. Roman Catholic. Achievements include patents on cermet cutting tool, earth working tool having a working element fabricated from WC compositions with enhanced properties; development of several commercial metal cutting tool products and a number of commercial coal, construction and rock drilling products. Home: 10187 Lavonne Dr Irwin PA 15642-2640 Office: Kennametal Inc Rte 981 S Latrobe PA 15650

CONLEY, JOHN JOSEPH, otolaryngologist; b. Carnegie, Pa., 1912. MD, U. Pitts., 1937. Intern Mercy Hosp., Pitts., 1937-38; resident Kings County Hosp., Bklyn., 1938-41. Recipient Disting. Svc. award AMA, 1992. Home: 211 Central Park W New York NY 10024-6020*

CONN, REX BOLAND, JR., physician, educator; b. Marengo, Iowa, Aug. 3, 1927; s. Rex Boland and Helena Dorothea (Schoenfelder) C.; m. Victoria Grace Sellens, Dec. 28, 1950; children: Elizabeth Marian, Victoria Anne, Mary Catherine. BS, Iowa State U., 1949; MD, Yale U., 1953; BSc, U. Oxford, Eng., 1955; MS, U. Minn., 1960. Prof. pathology, dir. clin. labs. W.Va. Med. Center, Morgantown, 1960-68; prof. lab. medicine, dir. dept. Johns Hopkins Med. Instns., Balt., 1968-77; prof. pathology and lab. medicine, dir. clin labs. Emory U., Atlanta, 1977-87; prof. and vice chmn. dept. pathology and cell biology, dir. clin. labs. Thomas Jefferson Med. Coll., Phila., 1987—; mem. pathology tng. com. NIH, 1972-73, mem. pathology A study sect., 1968-72; cons. Walter Reed Army Med. Center, 1972-77; cons. Armed Forces Inst. of Pathology, 1984-88. Editor: Current Diagnosis, 1991, Yearbook of Pathology and Clinical Pathology, 1980. Served with USNR, 1945-46. Mem. Coll. Am. Pathologists, Am. Soc. Clin. Pathologists (dir. 1975-81, pres. 1993-94), Acad. Clin. Lab. Physicians and Scientists (pres. 1972). Office: Thomas Jefferson Univ Hosp 204 Pavilion Philadelphia PA 19107

CONN, ROBERT WILLIAM, applied physics educator; b. N.Y.C., Dec. 1, 1942; s. William Conrad and Rose Marie (Albanese) C.; m. Gloria Trovato, Sept. 21, 1963; children: Carole, William. BChemE, Pratt Inst., 1964; MS in Mech. Engring., Calif. Inst. Tech., 1965; Ph.D. in Engring. Sci., 1968. NSF postdoctoral fellow Euratom Community Research Center, Ispra, Italy, 1968-69; research assoc. Brookhaven Nat. Lab., Upton, N.Y., 1969-70; vis. assoc. prof. U. Wis., Madison, 1970-72; assoc. prof. U. Wis., 1972-75, prof., 1975-80, dir. fusion tech. program, 1974-79, prof., 1975-80, Romnes faculty prof., 1977-80; prof. engring. and applied sci. UCLA, 1980—, dir. Inst. Plasma and Fusion Research, 1987—; chmn. bd. Plasma & Matls. Tech., Inc., L.A., 1986—; chair, sec. Energy's Fusion Energy Adv. Com., 1991—; cons. to govt. and industry. Author papers, chpts. in books. Recipient Curtis McGraw Research award Am. Assn. Engring. Edn., 1982; Outstanding Service cert. U.S. Dept. Energy, also E.O. Lawrence Meml. award, 1984, Fusion Power Assocs. Leadership award, 1992. Fellow Am. Nuclear Soc. (Outstanding Achievement award for excellence in research fusion div. 1979), Am. Phys. Soc.; mem. NAE. Office: UCLA 44-139 Engineering IV Inst Plasma & Fusion Rsch Los Angeles CA 90024

CONNEALLY, P. MICHAEL, medical educator. BS, Univ. Coll., Nat. U. Ireland, 1954; MS, U. Wis., 1960, PhD, 1962. With Ind. U.-Purdue U., Indpls., 1964—, now Disting. prof. med. genetics and neurology. Office: Ind U Dept Med Genetics Bloomington IN 47405

CONNELL, DESLEY WILLIAM, chemist, educator, administrator; b. Monto, Australia, July 31, 1938; s. William Eugene and Alva Caroline (Dowse) C.; m. Patricia Ann Wright, Jan. 17, 1960; children: Luke Daulton, Melissa Tace. BS, U. Queensland, Brisbane, Australia, 1961, MS, 1964, PhD, 1968; DSc, Griffith U., Brisbane, Queensland, Australia, 1991. Dir. Gippsland Lakes Environ. Study Ministry for Conservation, Melbourne, Victoria, Australia, 1973-76; dean, prof. Sch. of Australian Environ. Studies Griffith U., Brisbane, Queensland, 1976-89; dir. Govt. Chem. Lab., Brisbane, 1989—; vis. prof. SUNY, Stonybrook, N.Y., 1981, 87; mem. Great Barrier Reef Consultative Com., Townsville, Qld Australia, 1976-82. Author: Experiments in Environ. Chemistry, 1980, Water Pollution: Causes and Effects in Australia and New Zealand, 1981, Chemistry and Ecotoxicology of Pollution, 1984, Bioaccumulation of Xenobiotic Compounds, 1990. Mem. Water Quality Coun. of Queensland, Brisbane, Qld Australia, 1979-89, Australian Conservation Found. Coun., 1978-81. Recipient Commonwealth scholarship, Australian Govt., Brisbane, 1957-58, Commonwealth Postgrad. award, Australian Govt., Brisbane, 1964-68, Australian Inst. Food Sci. and Tech. award, 1969, Churchill fellowship, Churchill Found., 1975. Fellow Royal Australian Chem. Inst. (inaugural chmn. environ. div. 1987-90). Achievements include a major contribution to the devel. of an understanding of the behavior of chemicals in the environment. Home: 82 Roderick St Cornubia, Queensland 4130, Australia Office: Govt Chem Lab, Kessels Rd Coopers Plains, OLD Queensland 4108, Australia

CONNELL, JAMES ROGER, atmospheric turbulence researcher, educator; b. Rolla, Kans., Oct. 25, 1929; s. James Edward and Alys Louella (Boman) C.; div.; children: Heather Joanna, Jonathan Edward. BS, U. Ill., 1953; PhD, Colo. State U., 1973. Physicist U.S. Nat. Bur. Standards, Boulder, Colo., 1959-63; instr., grad. rsch. asst. Colo. State U., Ft. Collins, 1963-69, rsch. engr., 1973-74; asst. prof. atmospheric sci. St. Louis U., 1969-70; asst. prof. U. Wyo., Laramie, 1970-73; assoc. prof. U. Tenn., Tullahoma, 1974-78; staff scientist Battelle Northwest Lab., Richland, Wash., 1978-89; vis. prof. and cons. Colo. State U., Ft. Collins, 1989—. Contbr. articles to Jour. Applied Meteorology, Jour. Solar Energy Engring. other profl. publs. Asst. scoutmaster Boy Scouts Am., Tenn., Wash., Colo., 1974—; coach ch. league United Presbyn. Ch., Richland, Wash., 1986. With U.S. Army, 1954-56. Grantee NSF, NASA, Dept. Energy, Dept. Def., 1974—; recipient Disting. Assoc. award U.S. Dept. Energy, 1984. Mem. Am. Meteorol. Soc., Am. Solar Energy Soc., Am. Inst. Physics Tchrs. Unitarian Universalist. Achievements include development of a turbulence model that explains observed large stresses on rotor blades and propeller anemometers, measurements that explain character of frictional dead band in propeller anemometers and improve use of anemometers in dynamic wind conditions. Office: Colo State Univ B225 Engring Rsch Ctr Fort Collins CO 80523

CONNELLEE-CLAY, BARBARA, quality assurance auditor, laboratory administrator; b. Hereford, Tex., Dec. 4, 1929; d. Herman and Audrey Stella (Carroll) Galbraith; m. Rodger Sadosa Connellee, 1950 (dec.); m. Edward Lee Clay, 1983; children: Alison Elaine Stephens, Rebecca Diane Connellee Crabtree, Calvin Clay, Larry Clay, Becky Clay. BS, U. N.Mex., 1976, MBA, 1981; Cert. advanced facilitator, Quality Circle Inst.; insp. Occupational Safety and Health Adminstrn. Mem. adminstrv. staff U. Calif. Los Alamos Nat. Lab., 1976—. Editor nuclear materials tech. div. Safety and News bulletin. Pres., Wesleyan Service Guild, 1958. Recipient Women at Work award region 8 Dept. Labor Council on Working Women, 1983, N.Mex. Women at Work Spl. award Coun. Working Women Inst. Women and Minority Affairs, 1985. Mem. NAFE, Women in Sci., Assn. for Quality and Participation (cert. facilitator). United Methodist (past dir. edn.). Address: PO Box 1663 MS E583 Los Alamos NM 87544

CONNELLY, ALAN B., career officer, engineer. Diploma, Royal Military Coll. Can., 1931; student, Nova Scotia Tec. Coll., 1932; B in Engring., McGill U., 1933; postgrad., Sch. Military Engring., England, 1935-36. Commd. lt. R.C.E., 1931, advanced through the grades to hon. col., 1990—; works officer, asst. Dack Mountain Forestry Hdqs., R.C.E., 1933-35; attached Royal Engrs., R.C.E., England, 1935-36; comdr. First Field Co., R.C.E., 1937; acting comdr. Sch. Military Engring., Halifax, N.S., 1937; apptd. dist. engr. officer R.C.E., 1937-38; camp. engr. R.C.E., Petawawa, 1937-39; apptd. distr. engr. officer R.C.E., Kingston, 1939; apptd. adj. commdr. R.C.E., 1939; apptd. adjutant C.R.E., R.C.E., 1940; engr., defense bldg. Belgian border C.R.E., R.C.E., Belgian, 1940; maj. First Can. Infantry Divsn., R.C.E., 1940; from staff officer to chief engrs. VII corps. R.C.E., 1940; apptd. engr. instr. First Army Staff Course, R.C.E., England, 1940; promoted lt. col., appointed comdr. First Can. Infantry Divsn., R.C.E., 1941; apptd. commdr. Fourth Can. Armes Divsn., R.C.E., 1942; promoted brigadier, apptd. chief engr. First Can. Corps., R.C.E., 1943; engr. Italian campaign R.C.E., 1943-44, apptd. comdr. reinforcement B group, 1944-45; apptd. comdr. Can. Repatriations Units, R.C.E., 1946; comdr. Can. Troops N.W. Europe, R.C.E., 1946; dep. chief staff C.M. Hdgs., R.C.E., England, 1946; apptd. dep. chief adj. gen. Army Hdgs, R.C.E., Ottawa, Can., 1946; apptd. comdr., chief engr. N.W. Hwy. System., R.C.E., 1948-50; apptd. comdr. R.C.E., Saskatshewan, 1950; apptd. G.O.C., R.C.E., Winnipeg, 1951; apptd. comdr. Can. Commn. for East, R.C.E., Tokyo, Japan, 1951-52; served supplementary reserve R.C.E., 1953-68; hon. lt. col. Third Field Engr. Regiment, R.C.E., 1979-89, hon. col., 1990—; v.p. Mannix Construction, 1955; adv. Ministry of Fin. and Planning, Burma, 1974-75. Mem. Engring. Inst. Can. (exec. chmn., 1980-81; John B. Stirling Medal, 1991), Assn. Profl. Engrs of Ont. (chmn. Ottawa chpt. 1961-62). Office: care Engring Inst of Canada, 202 280 Albert St, Ottawa, ON Canada K1P 5G8*

CONNERY, STEVEN CHARLES, computer scientist; b. North Kingstown, R.I., Jan. 22, 1950; s. Robert Charles and Marjorie Viola (Cook) C.; m. Sandra Perrotti, Apr. 10, 1980; children: Luke Stevenson, Michael Christopher, Alyssa Marie. AS in Bus. Mgmt., Johnson and Wales U., Providence, 1977, BS in Computer Sci., 1992. Computer operator Systems Cons. Corp., Newport, R.I., 1981-84; computer operator Naval Underwater Systems Ctr., Newport, 1984-86, operator, programmer, 1986-92; lead operator Naval Undersea Warfare Ctr., Newport, 1992—; mem. Naval Undersea Warfare Ctr. PC Group, 1990—; tech. adminstr. for installation guide Reference in Johnson and Wales U. Libr., 1990. Scout master Boy Scouts Am., 1993. Sgt. USAF, 1971-72. Mem. Air Force Sgts. Assn. (life), Nat. Assn. of Govt. Employees. Democrat. Roman Catholic. Office: Naval Undersea Warfare Ctr Div Ops and Graphics Br Newport RI 02841-5047

CONNICK, ROBERT ELWELL, chemistry educator; b. Eureka, Calif., July 29, 1917; s. Arthur Elwell and Florence (Robertson) C.; m. Frances Spieth, Dec. 19, 1952; children—Mary Catherine, Elizabeth, Arthur, Megan, Sarah, William Beach. B.S., U. Calif. at Berkeley, 1939, Ph.D., 1942. Mem. faculty U. Calif., Berkeley, 1942-88, researcher Manhattan project, 1943-46, asst. prof. then assoc. prof. chemistry, 1945-52, prof., 1952-88, chmn. dept. chemistry, 1958-60, dean Coll. Chemistry, 1960-65, vice chancellor acad. affairs, 1965-67, vice chancellor, 1969-71, acting dean Coll. Chemistry, 1987-88. Contbr. articles profl. jours. Guggenheim fellow, 1949, 59. Mem. Am. Chem. Soc., Nat. Acad. Scis., Phi Beta Kappa, Sigma Xi, Pi Mu Epsilon. Home: 50 Marguerita Rd Kensington CA 94707-1020

CONNOLLY, CONNIE CHRISTINE, nurse anesthetist; b. Ft. Lauderdale, Fla., July 19, 1947; d. Cyrus Robert and Betty Jane (Burcham) Christensen; m. Mark T. Connolly, Nov. 2, 1990; 1 child, Sarah. Nursing diploma, Petersburg (Va.) Gen. Hosp. Sch. Nursing, 1968; diploma, Richland Meml. Hosp. Sch. Nurse Anesthesia, Columbia, S.C., 1975; postgrad., Stephens Coll., 1975-78. Lic. parachutist/jumpmaster, instr. rating. Enlisted student nurse program U.S. Army Nursing Corps, 1967, advanced through grades to lt. col., 1987; operating rm. nurse U.S. Army Nurse Corps, various locations, 1968-72, South Vietnam, 1970-71; nurse USAR, 3270th Hosp. AUG, Ft. Jackson, S.C., 1973-76, 324th Gen. Hosp., Perrine, Fla., 1976—; nurse anesthetist, clin. instr. Richland Meml. Hosp., Columbia, S.C., 1975-76; asst. chief nurse res. Hialeah (Fla.) Hosp., 1978-82; office and clinic anesthetist, 1980-85, chief nurse anesthetist, asst. officer in charge operating rm. svcs., 1985—; sec.-treas. Office Anesthesia Specialists, P.A., Hollywood, Fla., 1986-89. Decorated Gallantry Cross with palm (Vietnam). Mem. Am. Assn. Nurse Anesthetists, Fla. Assn. Nurse Anesthetists (sec.-treas. edn. dist. II 1976-77, nominating com. 1976-77, chmn. edn. workshop program 1981, trustee 1980-83), Res. Officers Assn (life, health armed svcs. com. 1981), DAV (life), Am. Soc. Mil. Surgeons U.S., VFW (life), Am. Legion, Vietnam Vets Am. (life), Vets. Vietnam War, Harley Owners Group, U.S. Mil. Vets. M/C. Republican. Presbyterian. Avocations: body building, motorcycle riding, writing, community service, public speaking. Home: 1141 SW 8th Terr Fort Lauderdale FL 33315-1262 Office: 4651 Sheridan St Ste 400 Hollywood FL 33021-3634

CONNOLLY, JOHN EARLE, surgeon, educator; b. Omaha, May 21, 1923; s. Earl A. and Gertrude (Eckerman) C.; m. Virginia Hartman, Aug. 12, 1967; children: Peter Hart, John Earle, Sarah. A.B., Harvard U., 1945, M.D., 1948. Diplomate: Am. Bd. Surgery (bd. dirs. 1976-82), Am. Bd. Thoracic and Cardiovascular Surgery, Am. Bd. Vascular Surgery. Intern. in surgery Stanford U. Hosps., San Francisco, 1948-49, surg. research fellow, 1949-50, asst. resident surgeon, 1950-52, chief resident surgeon, 1953-54, surg. pathology fellow, 1961, instr. surgery, 1957-60, John and Mary Markle Scholar in med. scis., 1957-62; surg. registrar professional unit St. Bartholomew's Hosp., London, 1952-53; resident in thoracic surgery Bellevue Hosp., N.Y.C., 1955; resident in thoracic and cardiovascular surgery Columbia-Presbyn. Med. Ctr., N.Y.C., 1956; from instr. to assoc. prof. surgery Stanford U., 1957-65; prof. U. Calif., Irvine, 1965—, chmn. dept. surgery, 1965-78; attending surgeon Stanford Med. Ctr., Palo Alto, Calif., 1959-65; chmn. cardiovascular and thoracic surgery Irvine Med. Ctr. U. Calif., 1968—; attending surgeon Children's Hosp., Orange, Calif., 1968—, Anaheim (Calif.) Meml. Hosp., 1970—; vis. prof. Beijing Heart, Lung, Blood Vessel Inst., 1990, A.H. Duncan vis. prof. U. Edinburgh, 1984; Hunterian prof. Royal Coll. Surgeons Eng., 1985-86; Kinmonth lectr. Royal Coll. Surgeons, Eng., 1987; mem. adv. coun. Nat. Heart, Lung, and Blood Inst.-NIH, 1981-85; cons. Long Beach VA Hosp., Calif., 1965—. Contbr. articles to profl. jours.; editorial bd.: Jour. Cardiovascular Surgery, 1974—, chief editor, 1985—; editorial bd. Western Jour. Medicine, 1975—, Jour. Stroke, 1979—, Jour. Vascular Surgery, 1983—. Bd. dirs. Audio-Digest Found., 1974—; bd. dirs. Franklin Martin Found., 1975-80; regent Uniformed Svcs. U. of Health Scis., Bethesda, 1992—. Served with AUS, 1943-44. Recipient Cert. of Merit, Japanese Surg. Soc., 1979, 90. Fellow ACS (gov. 1964-70, regent 1973-82, vice chmn. bd. regents 1980-82, v.p. 1984-85), Royal Coll. Surgeons Eng. (hon.), Royal Coll. Surgeons Ireland (hon.), Royal Coll. Surgeons Edinburgh (hon.); mem. Am. Surg. Assn., Soc. Univ. Surgeons, Am. Assn. Thoracic Surgery (coun. 1974-78), Pacific Coast Surg. Assn. (pres. 1985-86), San Francisco Surg. Soc., L.A. Surg. Soc., Soc. Vascular Surgery, Western Surg. Assn., Internat. Cardiovascular Soc. (pres. 1977), Soc. Internat. Chirurgie, Soc. Thoracic Surgeons, Western Thoracic Surg. Soc. (pres. 1978), Orange County Surg. Soc. (pres. 1984-85), James IV Assn. Surgeons (councillor 1983—). Clubs: California (Los Angeles); San Francisco Golf, Pacific Union, Bohemian (San Francisco); Cypress Point (Pebble Beach, Calif.); Harvard (N.Y.C.); Big Canyon (Newport Beach). Home: 7 Deerwood Ln Newport Beach CA 92660-5108 Office: U Calif Dept Surgery Irvine CA 92717

CONNOLLY, PATRICIA ANN STACY, physical education educator; b. Sewickley, Pa., May 14, 1939; d. Clyde Grant Stacy and Dorothy Rose (Rupnik) Zupancic; m. Thomas R. Connolly, Dec. 2, 1963 (div. Sept. 1971). BS, Slippery Rock (Pa.) U., 1961; MEd, U. Pitts., 1967. Cert. health and phys. edn. tchr., Pa. Tchr. South Hills High Sch., Pitts., 1961-76; tchr. John A. Brashear High Sch., Pitts., 1976—, instructional chairperson, 1978-

87; swim coach Brashear High Sch., 1971—. Sponsor, coord. Am. Cance Soc.'s Swim-a-Thon, Pitts., Am. Heart Assn. Jump-Rope-for-Heart, community collector, Mt. Lebanon, Pa.; sec. Western chpt. Sports Hall of Fame, Pitts., 1979—. Recipient Profl. Honor award, Pa., 1976, named to Slippery Rock U. Hall of Fame. Mem. Pa. Assn. Health, Phys. Edn., Recreation and Dance (profl., pres. 1985-86, Elmer B. Cottrell award 1991), Allegheny County Assn. for Health, Phys. Edn., Recreation and Dance (pres. 1974-75), Order Eastern Star (worthy matron 1983-84, 89-90, dist. dep. 1987-88), Alpha Delta Kappa (pres.-elect 1992—). Republican. Baptist. Avocations: golf, swimming, reading, puzzles, gardening. Home: 245 Orchard Spring Rd Pittsburgh PA 15220-1713 Office: Brashear High Sch 590 Crane Ave Pittsburgh PA 15216-3999

CONNOLLY, WILLIAM MICHAEL, civil and structural engineer; b. Lakewood, Ohio, May 12, 1952; s. Thomas and Mary (Boylan) C.; m. Sharon Rose Zarbo, Sept. 24, 1971 (div. Apr. 1993); children: Jennifer, Megan. Assoc. in Architecture and Bldg., Cuyahoga C.C., 1975; BS in Civil Engring. Tech., Cleve. State U., 1979. Registered profl. engr., Ohio, N.C. Plant engr. Goodyear Aerospace, Akron, Ohio, 1982-85; engring. mgr. Osborn Engring. Co., Cleve., 1985-88; asst. constrn. engr. Cuyahoga County Engrs. Office, Cleve., 1988-90; engring. mgr. Gerber Products Co., Asheville, N.C., 1990—; mem. A/B Water Authority Conservation Task Team, Asheville, 1990—; mem. St. Lawrence Capital Improvements Com., Asheville, 1992—. With U.S. Army, 1970-74, Vietnam. Decorated Purple Heart, Combat Infantry Badge, Vietnamese Cross of Gallantry, 1970. Mem. NSPE, Disabled Am. Vets. (life), Vietnam Vets. Am., County Engrs. Assn. Ohio, Order of Engr. Roman Catholic. Achievements include installation of baby food prodn. line. Home: 14 Park Ave Asheville NC 28803 Office: Gerber Products Co PO Box 950 Skyland NC 28776-6000

CONNOR, CHARLES WILLIAM, airline pilot; b. Miami, Fla., Aug. 2, 1935; s. Robert Hugh and Mary (Cauthen) C.; m. Retha Moeller, Mar. 12, 1988; childen: Charles W. Jr., Christine Wendy, Elizabeth Tammy. MAS with distinction, Embry Riddle Aero. U., 1970, MBA with distinction, 1970; PhD in Behavioral Psychology/Aero. Sci., Columbia Pacific U., 1982. Exptl. test pilot Boeing Vertol Co., 1962-65; C-5A program contract adminstr. Lockheed Ga. Aircraft Corp., 1965-66, Delta Air Lines, Inc., Atlanta, 1966—; v.p. Aviation Systems Concepts, Inc., 1983—; adj. prof., grad. curriculum advisor Embry Riddle Aero. U., 1979-82; pres. aerospace Behavioral Engring. Tech., 1982—, chmn. SAE HBT, G-10 Internat. Aviation Human Factor Standards; planning chmn. over 100 tech. sessions on aerospace behavioral engring. tech. Contbr. articles, reports to profl. publs., confs. With USMC, 1957-62. Recipient Chinese Air Force Wings, Chinese Govt., 1959. Mem. AIAA, Soc. Automotive Engrs. (mem. gen. com. aerospace div. 1985—, div. chmn. engring. and activity bd. 1989—, proc. editor for confs., Forest R. McFarland award 1986, 93, Tech. Bd. award 1988), Inst. Navigation, Human Factors Soc., Air Line Pilots Assn. (bd. dirs. 1968-74, nat. spokesman 1980—) Soc. Exptl. Test Pilots, Nat. Aviation Club. Achievements include development of line oriented flight training scenarios for air carrier crew members, progressive cognitive branching concept for advanced adaptive air crew training. Home: 206 Elm Ave Melbourne Beach FL 32951 Office: Delta Airlines Inc Flight Ops Atlanta Hartsfield Internat Atlanta GA 30320

CONNOR, JOHN ARTHUR, neuroscientist; b. Kansas City, Mo., June 18, 1940; s. John William and Dorothy Marie (Biebel) C.; m. Julia M. Philips; children: Julia Marie, Bridget Ann, John Hazard, Ellen Katrina. BSEE, U. Mo., Columbia, 1963; PhD in Elec. Engring., Biomed. Engring., Northwestern U., 1967. Postdoctoral fellow physiology and biophysics U. Wash., Seattle, 1967-69; from asst. prof. to prof. physiology, biophysics U. Ill., Urbana, 1969-81; disting. mem. tech. staff AT&T Bell Labs., Murray Hill, N.J., 1981-89; head div. neuroscis. Roche Inst., Nutley, N.J., 1989—. Editorial bd. Cell. and Molecular Neurobiology, 1979—, Neuroimage, 1992—; contbr. articles to Jour. Physiology, Nature, other profl. publs. Recipient Camillo Golgi award FIDIA Rsch. Found., 1991. Mem. Sigma Xi, Tau Beta Pi, Eta Kappa Nu. Achievements include research on physiological role of neuronal A current, discovery of cyclic nucleotide-gated sodium-calcium current in neurons, pioneering in imaging studies of calcium ion dynamics in living neurons. Office: Roche Inst Molecular Biology 340 Kingsland St Nutley NJ 07110

CONNOR, JOHN MURRAY, agricultural economics educator; b. Attleboro, Mass., July 7, 1943; s. John Murray Sr. and Victoria Rose (Moro) C.; m. Ulla Maija Niemelä, Apr. 3, 1972; 1 child, Timo. BA cum laude, Boston Coll., 1965; MA, U. Fla., 1974; MS, U. Wis., 1974, PhD, 1976. Vol. U.S. Peace Corps, Nigeria, Uganda, 1966-68; agrl. economist Econ. Rsch. Svc. USDA, Madison, 1976-79; head food mfg. rsch. Econ. Rsch. Svc. USDA, Washington, 1979-83; assoc. prof. agrl. econs. Purdue U., West Lafayette, Ind., 1983-89; prof. Purdue U., West Lafayette, 1989—, asst. dept. head, 1985-88; adj. prof. Catholic U. Sacred Heart, Piacenza, Italy, 1991—; cons. subcom. on multinats. U.S. Senate, Washington, 1974-76, select com. on nutrition, 1977-78; cons. UN Ctr. on Transnats., N.Y.C., 1981-82; chair Orgn. and Performance World Food Systems, 1988-93. Author: Market Power of Multinationals, 1977, Food Processing Industries, 1988, (with others) Food Manufacturing Industries, 1985; contbr. articles to profl. jours. chpts. to books. Grantee U.S. Office Tech. Assessment, 1984-85, Inst. Food Technologists, 1986-88, Ind. Dept. Commerce, 1987-91, Econ. Rsch. Svc., USDA, 1988-89, Coop. State Rsch. Svc., USDA, 1989—. Mem. AAUP (pres. Purdue U. chpt. 1988-90, exec. bd. Ind. conf. 1990—, nat. coun. 1991-92), Am. Agrl. Econs. Assn. (policy award 1980, comm. award 1985, Disting. Extension Program award 1993), Indsl. Orgn. Soc., Am. Econ. Assn., Royal Econ. Soc., ACLU. Avocations: tennis, sailing, classical music, food and wine. Home: 180 Sumac Dr West Lafayette IN 47906-2157 Office: Purdue U Dept Agrl Econs West Lafayette IN 47907-1145

CONOBY, JOSEPH FRANCIS, chemist; b. Albany, June 12, 1930; s. Joseph Francis and Helen Emma (Brucker) C.; B.S., Union Coll., 1952; m. Mary Joan A. Ryan, June 21, 1958; children—James Francis, Mark Joseph. Sr. tech. service engr. Allied Chem. Corp., Syracuse, N.Y., 1956-66; research chemist Conversion Chem. Corp., Rockville, Conn., 1966-69; environ. engr., indsl. hygienist, mgr. environ. and health engring. Honeywell Bull, Billerica, Mass., 1969-87, mgr. environ. engring. Bull HN Worldwide Info. Systems, 1987—; mem. adv. bd. Mass. Water Resources Authority Sewer Use (rules and regulations, policy and procedures, and facilities planning task forces); cons. exptl. project course Mass. Inst. Tech., 1977-78. Served to lt. USN, 1952-56. Mem. Am. Electroplaters Soc. (chmn. project com.), Am. Electroplaters Soc. (pres. Merrimack br.), Am. Indsl. Hygiene Assn. Patentee in field, U.S., Germany. Contbr. articles to profl. jours. Home: 5 Samuel Parlin Dr Acton MA 01720-3206 Office: Bull H N Worldwide Info Systems Billerica MA 01821

CONOVER, LLOYD HILLYARD, retired pharmaceutical research scientist and executive; b. Orange, N.J., June 13, 1923; s. John Howard and Marguerite Anna (Cameron) C.; m. Virginia Rogers Kirk, Aug. 24, 1944 (dec. Dec. 1988); children: Kirk Howard, Roger Lloyd, Heather Cameron, Craig Scott; m. Marie Strauss Solomons, Oct. 18, 1990. BA, Amherst Coll., 1947; PhD, U. Rochester, 1950. Rsch. chemist Chas. Pfizer & Co., Bklyn. and Groton, Conn., 1950-58; supr., mgr. Chas. Pfizer & Co., Groton, 1958-68; dir. chem. rsch. and chemotherapy Pfizer Cen. Rsch., Groton, 1968-71; rsch. dir. Europe, Pfizer Cen. Rsch., Sandwich, Eng., 1971-74; v.p. agrl. R & D Pfizer Cen. Rsch., Groton and Sandwich, 1975-84; drug rsch. cons. Sandwich, Conn., 1984-86. Contbr. articles to sci. jours.; patentee tetracycline. Chmn. Waterford Planning, 1961-63. Lt. (j.g.) USNR, 1943-46, PTO. Recipient Eli Whitney award Conn. Patent Law Assn., 1983, Third Century award Found. Creative Am., 1990; inductee Nat. Inventors Hall of Fame, 1992. Fellow Royal Soc. Chemistry (chartered chemist), Royal Soc. Arts; mem. Am. Chem. Soc., Phi Beta Kappa, Sigma Xi. Republican. Avocations: travel, gardening, theatre, biographical research. Home: 27 Old Barry Rd Quaker Hill CT 06375-1019

CONRAD, BRUCE R., earth scientist. Recipient Sherritt Hydrometallurgy award Can. Inst. Mining and Metallurgy, 1992. Office: care Xerox Tower Ste 1210, 3400 de Maisonneuve Blvd W, Montreal, PQ Canada H3Z 3B8*

CONRAD, HAROLD THEODORE, psychiatrist; b. Milw., Jan. 25, 1934; s. Theodore Herman and Alyce Barbara (Kolb) C.; A.B., U. Chgo., 1954, B.S., 1955, M.D., 1958; m. Elaine Marie Blaine, Sept. 1, 1962; children—Blaine, Carl, David, Erich, Rachel. Intern USPHS Hosp., San Francisco, 1958-59; commd. sr. asst. surgeon USPHS, 1958, advanced through grades to med. dir., 1967; resident in psychiatry USPHS Hosp., Lexington, Ky., 1959-61, Charity Hosp., New Orleans, 1961-62; chief of psychiatry USPHS Hosp., New Orleans, 1962-67, clin. dir., 1967; dep. dir. div. field investigations NIMH, Chevy Chase, Md., 1968; chief NIMH Clin. Research Center, Lexington, 1969-73; cons. psychiatry, region IX, USPHS, HEW, San Francisco, 1973-79; dir. adolescent unit Alaska Psychiat. Inst., Anchorage, 1979-81, supt., 1981-85 ; clin. assoc. prof. psychiatry U. Wash. Med. Sch., 1981-85; med. dir. Bayou Oaks Hosp., Houma, La., 1985—. Decorated Commendation medal; recipient various community awards for contbns. in field of drug abuse and equal employment opportunity for minorities. Diplomate Am. Bd. Psychiatry. Fellow Royal Soc. Health, Royal Soc. Medicine, Am. Psychiat. Assn.; mem. AMA, Alpha Omega Alpha, Alpha Delta Phi. Contbr. to publs. in field. Office: 855 Belanger St Houma LA 70360-4463

CONRAD, JEFFREY PHILIP, mechanical engineer, manager; b. Toledo, May 11, 1950; s. Keith and Elizabeth Mary (Gottron) C. BSME, U. Notre Dame, 1972. Mfg. engr. Nat. Semiconductor Corp., Santa Clara, Calif., 1973-75; design engr. Peterbilt Motors Co., Newark, Calif., 1976-78; sr. facilities engr. Synertek, Inc., Santa Clara, 1978-83; engring. mgr. Semi-Gas Systems Inc., San Jose, Calif., 1984-88, v.p. engring., 1988—. Mem. ASTM, Am. Vacuum Soc. Office: Semi Gas Systems Inc 625 Wool Creek Dr San Jose CA 95112

CONRADI, MARK STEPHEN, physicist, educator; b. St. Louis, Jan. 25, 1952; s. Joseph John and Mary Ann (Knezevich) C.; m. Debra Jean Mutchler, Aug. 17, 1985; children: Gabriel, Nicholas. BS, Washington U., St. Louis, 1973; PhD, Washington U., 1977. Eugene P. Wigner fellow Oak Ridge (Tenn.) Nat. Lab., 1977-79; asst. prof., then assoc. prof. physics Coll. of William and Mary, Williamsburg, 1979-85; assoc. prof. Washington U., St. Louis, 1985-91, prof. physics, 1991—; chmn. Gordon conf. on magnetic resonance, Wolfesboro, N.H., 1993. Contbr. articles to Phys. Rev. Letters, Jour. Magnetic Resonance, others. NSF grad. fellow, 1974-77; Alfred P. Sloan Found. fellow, 1983-85. Mem. Am. Phys. Soc., Am. Chem. Soc. Achievements include patent on ultrasonic continuous wave particle monitor. Office: Washington Univ Dept Physics 1105 One Brookings Dr Saint Louis MO 63130

CONRON, JOHN PHELAN, architect; b. Brookline, Mass., Dec. 4, 1921; s. Carl Edward and Katherine (Phelan) C. Student, U. So. Calif., 1940-41; B.Arch., Yale U., 1948. Draftsmn. Whelan & Westman, Boston, 1948-52; owner, prin. John P. Conron (Architect), Santa Fe, N.Mex., 1952-61; ptnr. Conron-Lent Architects, Santa Fe, 1961-86, Conron-Muths (restoration architects), Santa Fe and Jackson Hole, Wyo., 1975-88, Conron-Woods Architects, Santa Fe, 1986—; pres. The Centerline, Inc., Santa Fe., 1952-86. Prin. works include Centerline, Inc., Santa Fe, KB Ranch, nr. Santa Fe, Henry R. Singlton residence, Lamy, N.Mex., Amtech Tech. Corp., Santa Fe, restorations Stephen W. Dorsey Mansion State Monument, Colfax County, N.Mex., Palace of Govs., Santa Fe, Pipe Spring Nat. Monument, Ariz.; editor La Cronica de Nuevo Mexico, 1976—; co-editor N.Mex. Architecture mag., 1960-66, editor, 1966—. Vice Chmn. N.Mex. Cultural Properties Com., 1968-80; founder v.p. Las Trampas Found., 1967-80; trustee Internat. Inst. Iberian Colonial Art, Santa Fe, pres., 1978—; bd. dirs. Preservation Action., 1976-80; bd. dirs. Hist. Soc. N.Mex., 1976—, pres, 1982-86. Served with USAAF, 1941-45. Recipient Merit award AIA, 1962, Spl. Commendation award, 1970. Fellow AIA, Am. Soc. Interior Designers (pres. N.Mex. chpt. 1966-68, 74-75, regional v.p. 1970-73, Historic Preservation award for restoration Palace of Govs. 1980); mem. N.Mex. Soc. Architects (past pres.), Am. Soc. Man Environ. Relations (dir. 1973-86), Hist. Soc. N.Mex. chpt. Am. Soc. interior (1st v.p. 1979-83, pres. 1983-87, bd. dirs. 1987—). Office: Conron & Woods Architects 1807 2d St Ste 44 Santa Fe NM 87501

CONSEJO, EDUARDO, aeronautics engineer; b. San Sebastian, Spain, Jan. 6, 1957; s. Claudio and María Rosario (Prieto) C.; m. María Montserrat Iglesias, Feb. 11, 1991; 1 child, Alejandro. Degree in aeronautical engring., Poly. U., Madrid, 1985. Registered profl. engr., Spain. Design engr. Eurotren Monoviga, Madrid, 1985; sales engr. Danfoss, Madrid, 1985-88; sales mgr. MBB Helicopters, Madrid, 1988-90, gen. mgr., 1990-93; gen. mgr. Helicopteros Eurocopter, Madrid, 1993—; cons. in field, Madrid, 1988—. Mem. Spanish Aeronautical Engrs. Assn., Ofcl. Coll. Aeronautical Engrs. Spain. Achievements include design of heliports. Office: Helicoptros Eurocopter, Santa Cruz de Marcenado 33, Madrid 28015, Spain

CONSTANCE, MERVYN, utility executive; b. Trinidad, May 9, 1950; s. Julius and Winifred (William) C.; m. Kathleen McCleta Browne, Dec. 26, 1975; children: Mark Kevin, Marlon Jason, Davida Alexandre. Trade man asst. Water Sewage Authority, 1973, pipe fitter, 1979-83, water works foreman I, 1983—. Avocations: reading, cycling, basketball, travel, engring. drawing. Home: 21 Oleander Dr, Pleasant Ville, San Fernando Trinidad and Tobago

CONSTANT, CLINTON, chemical engineer, consultant; b. Nelson, B.C., Can., Mar. 20, 1912; came to U.S., 1936, naturalized, 1942; s. Vasile and Annie (Hunt) C.; m. Margie Robbel, Dec. 5, 1965. BSc with honors, U. Alta., 1935, postgrad., 1935-36; PhD, Western Res. U., 1939. Registered profl. engr. Devel. engr. Harshaw Chem. Co., Cleve., 1936-38, mfg. foreman, 1938-43, sr. engr. semi-works dept., 1948-50; supt. hydrofluoric acid dept. Nyotex Chems., Inc., Houston, 1943-47, chief devel. engr., 1947-48; mgr. engring. Ferro Chem. Co., Bedford, Ohio, 1950-52; tech. asst. mfg. dept. Armour Agrl. Chem. Co. (name formerly Armour Fertilizer Works), Bartow, Fla., 1952-61, mfg. research and devel. div., 1961-63; mgr. spl. projects Research div. (co. name changed to USS Agri-Chems 1968), Bartow, Fla., 1963-65, project mgr., 1965-70; chem. adviser Robert & Co. Assocs., Atlanta, 1970-79; chief engr. Almon & Assocs., Inc., Atlanta, 1979-80; project mgr. Engring. Service Assocs., Atlanta, 1980-81; v.p. engring. ACI Inc., Hesperia, Calif., 1981-83; sr. v.p., chief engr. MTI (acquisition of ACI), Hesperia, 1983-86; engring. cons. San Bernardino County APCD, Victorville, Calif., 1986-90; instr. environ. chemistry Victor Valley C.C., 1990; pvt. cons. Victorville, Calif., 1991—; cons. in engring., 1992—. Author tech. reports, sci. fiction; patentee in field. Fellow AAAS, Am. Inst. Chemists, Am. Inst. Chem. Engrs., N.Y. Acad. Scis., AIAA (assoc.); mem. Am. Chem. Soc., Am. Astron. Soc., Astron. Soc. Pacific, Royal Astron. Soc. Can., NSPE, Am. Water Works Assn., Calif. Water and Pollution Control Assn., Air Pollution Control Assn., Soc. Mfg. Engrs., Calif. Soc. Profl. Engrs.

CONSTANT, WILLIAM DAVID, chemical engineer, educator; b. Bunkie, La., May 15, 1954; s. Warren LeRoy and Montez Henning (Haas) C.; m. Donna Gail Hall, Nov. 14, 1987; 1 child, Justin Glen Germany. BS in Chem. Engring., La. State U., 1977, MS in Chem. Engring., 1980, PhD in Chem. Engring., 1984. Registered profl. engr., La. Lab. tech. U.S. Forest Svc., Pineville, La., 1973, 74; chem. engr. Ethyl Corp., Baton Rouge, La., 1977-78; Exxon fellow dept. chem. engring. La. State U., Baton Rouge, 1978-84, asst. prof. petroleum engring., 1984-88, asst. dir. hazardous waste rsch. ctr., 1988-91; dir. hazardous waste rsch. ctr. La. State U., 1991—, asst. dir. hazardous substance rsch. ctr. south and S.W., 1991—; affiliate faculty member dept. chem. engring. La. State U., Baton Rouge, 1989—, dir. La. water resources rsch. inst., 1989—; cons. in field, Baton Rouge, 1989—. Contbr. numerous articles to profl. jours. Recipient numerous rsch. grants, 1985—. Fellow Am. Inst. Chemists; mem. Am. Chem. Soc., Am. Inst. Chem. Engrs., Soc. Petroleum Engrs., Air and Waste Mgmt. Assn., U. Coun. on Water Resources, Am. Water Resources Assn., Gamma Beta Phi, Tau Beta Pi, Pi Epsilon Tau, Phi Lambda Epsilon (chpt. pres. 1980-81). Republican. Methodist. Avocations: golf, hunting. Office: La State U Hazardous Waste Rsch Ctr 3418 CEBA Baton Rouge LA 70803

CONSTANTINO-BANA, ROSE EVA, nursing educator, researcher; b. Labangan Zamboanga del Sur, Philippines, Dec. 25, 1940; came to U.S., 1964; naturalized, 1982; d. Norberto C. and Rosalia (Torres) Bana; m. Abraham Antonio Constantino, Jr., Dec. 13, 1964; children: Charles Edward, Kenneth Richard, Abraham Anthony III. B.S. in Nursing, Philippine Union Coll., Manila, 1962; M.Nursing, U. Pitts., 1971, Ph.D., 1979; J.D., Duquesne U., Pitts., 1984. Lic. clin. specialist in psychiatric-mental health nursing; registered nurse. Instr. Philippine Union Co., 1963-65, Spring Grove State Hosp., Balt., 1965-67, Montefiore Sch. Nursing, Pitts., 1967-70; instr. U. Pitts., 1971-74, asst. prof., 1974-83, assoc. prof., 1983—, chmn. Senate Athletic Com., 1985-86, 89-90, univ. senate sec., 1991-92, univ. senate v.p., 1993—; project dir. grant div. of nursing HHS, Washington, 1983-85; prin. investigator NIH NCNR, 1991—; bd. dirs. Internat. Council on Women's Health Issues, 1986—. Author: (with others) Principles and Practice of Psychiatric Nursing, 1982; contbr. chpts. to books and articles to profl. jours. Mem. Republican Presdl. Task Force, Washington, 1980, Rep. Senatorial Com., Washington, 1980. Mem. ABA, Pa. Bar Assn., Women's Bar Assn., Assn. Trial Lawyers Am., Am. Assn. Nurse Attys., Am. Nurses Assn., Pa. Nurses Assn., Nat. League Nursing, Pa. League Nursing (chairperson area 6), U. Pitts. Sch. Nursing Alumni Assn., U. Duquesne Law Alumni Assn., Sigma Theta Tau, Phi Alpha Delta. Seventh-Day Adventist. Avocations: cooking, playing the piano. Home: 6 Carmel Ct Pittsburgh PA 15221-3618 Office: U Pitts Sch Nursing 467 Victoria St Pittsburgh PA 15261-0001

CONTAG, CHRISTOPHER HEINZ, biomedical researcher; b. New Ulm, Minn., Sept. 8, 1959; s. Carlos Heinz and Ann Louise (Schwermann) C.; m. Pamela Mary Reilly, Oct. 10, 1987; children: Caitlin Ann, Ashlyn Grace. BS, U. Minn., 1982, PhD, 1988. Postdoctoral fellow dept. microbiology Dept. Microbiology U. Minn., Mpls., 1988-89, Stanford (Calif.) U., 1989—. Author: (book chpts.) Simian Immunodeficiency Virus, 1991, Age-dependent Poliomyelitis Virol. and Immunol. Facotrs, 1992. Named Scholar Am. Found for Aids Rsch., 1990-93; recipient Nat. Rsch. Svc. award, NIH, 1986-90. Mem. AAAS, Am. Soc. for Virology, Am. Soc. for Microbiology, U.S. Tang Soo Do Fedn. Achievements include elucidation of two virus mechanisms of pathogenisis for fatal paralytic disease of mice. Office: Stanford U Dept Micro Immunology Fairchild Bldg Stanford CA 94305-5402

CONTAG, PAMELA REILLY, biologist, researcher; b. St. Paul, Apr. 27, 1957; d. John William and Juanita Marie (O'Connell) Reilly; m. Christopher Heinz Contag, Oct. 10, 1987; children: Caitlin Ann, Ashlyn Grace. MS, U. Minn., 1985, PhD, 1989. Postdoctoral fellow Stanford U., Palo Alto, Calif., 1990—; course coord. Marine Biol. Lab., Woods Hole, Mass., 1990-92. Contbr. articles to Jour. Indsl. Microbiology, Jour. Applied and Environ. Microbiology. Mem. Am. Soc. Microbiology, Sigma Xi (Young Investigators award 1982, Grad. Rsch. award 1989). Office: Stanford U Fairchild Science Bldg D309 Stanford CA 94305-5402

CONTE, JEAN JACQUES, medical educator, nephrologist; b. Tours, France, Nov. 6, 1938; s. Rene G. and Anne J. (Grangier) C.; m. Anne Y. Viatge, June 4, 1970; 1 dau., Stephanie C. M.D., Faculte de Medecine, Toulouse, France, 1966. Cert. nephrologist Comité Consultatif des Universités. Prof. medicine Universite P. Sabatier, Toulouse, France, 1971—, cons. Mission de la Recherche, 1983—; chief nephrology dept. Centre Hospitalo-Universitaire Purpan, Toulouse, 1974—; vice doyen Faculte de Medecine Purpan, Toulouse, 1980—; dir. Research Unit in Renal Immunopathology and Immunopharmacology, 1984—; expert in medicine Toulouse Ct. Appeals, 1984—; pres. U. Paul Sabatier, Toulouse III, 1986—; v.p. de la Conf. des Pres. Univ., 1987, Assn. des Univs. Partiellement ou Entièrement de Langue Française, 1987, Comité Economique et Social de la Région Midi-Pyrénées, 1989—; v.p. de l'Agence de Coopération Midi-Pyrénées; conseiller Mcpl. de Portet Sur Garonne, Tech. de la Found. HARIRI, mem. Chargé de Mission au Ministére de l'Edn. Nationale, 1991; conseiller technique de la Chambre de Commerce et d'Industrie de Toulouse, 1992; pres. du Groupement Interprofessionnel Régional pour la Promotion de l'Emploi des Personnes Handicappés, 1992. Co-author: Glomerulonephritis, 1973; Advanced in Nephrology, 1974, 76; Plasmapheresis, 1983. Mem. European Soc. for Clin. Investigation, Internat. Soc. Nephrology, French Soc. Immunology, French Soc. of Plasma Exchange (founding mem.), Chevalier dans l'Ordre Nat. de Mérite (médaille d'Or de la Ville de Toulouse), de l'Ordre du Lion du Sénégal, chevalier des Palmes Académiques, Compagnie des Mousquetaires d'Armagnac (Auch, France), Confrérie des Vins de Cahors, Golf of Vieille-Toulouse (Toulouse), Golf of Pals (Costa Brava). Roman Catholic. Office: U Purpan Centre Hosp, Dept Néphrologie-Hémodialyse, 31059 Toulouse France

CONTOIS, DAVID FRANCIS, biochemist; b. Schenectady, N.Y., Oct. 31, 1963; s. Francis Xavier and Helen Agnes (Supranowicz) C. BS, Lynchburg Coll., 1985. Lab. technician Health Rsch. Inc., Albany, N.Y., 1985-87; biochem. researcher Uniformed Svcs. Univ. of Health and Scis., Bethesda, Md., 1987—. Contbr. articles to Prostaglandins, Prostaglandins, Leukotrienes and Lipoxins. Vol. ARC, Montgomery County, 1990—; assoc. mem. Kensington (Md.) Vol. Fire Co., 1990—. Recipient Cert. Outstanding Performance, 1988-93, Outstanding Contbn. in Svc. award Kensington Vol. Fire Dept., 1992. Mem. Nat. Audubon Naturalist Soc., Am. Horse Protection Assn. Democrat. Roman Catholic. Office: Uniformed Svcs Univ Dept Ob/Gyn 4301 Jones Bridge Rd Bethesda MD 20814

CONTRADA, RICHARD J(UDE), psychologist; b. Bklyn., Sept. 11, 1954; s. Pasquale Richard and Theresa (Lubrano) C. BA, L.I. U., 1976; PhD, CUNY, 1985. Asst. prof. Rutgers U., New Brunswick, N.J., 1986-92, assoc. prof., 1992—. Mem. edit. bd. Jour. Personality & Social Psychology, Health Psychology, Jour. Applied Social Psychology; contbr. articles to profl. jours. Postdoctoral fellow Uniformed Svcs. U. Health Scis., Bethesda, Md., 1985-86. Mem. AAAS, Am. Psychol. Assn., Am. Psychol. Soc., Soc. Psychol. Rsch. Office: Rutgers U Dept Psychology New Brunswick NJ 08903

CONVERSANO, GUY JOHN, civil engineer; b. Jersey City, May 3, 1924; s. John and Angela Conversano; m. Pauline Marie Zuchowski, Jan. 6, 1951; children: Carol Ann, Guy Allen, Donna Jean, Sandra Rose, Lisa Ellen. B Maritime Sci., N.Y. State Maritime Coll., Bronx, 1950; BSCE, Humboldt State Coll., 1968; lifetime teaching credential, Coll. of Redwoods, 1970. Lin. civil engr. Calif., Oreg., Wash., Nev., traffic engr. Calif. Jr. civil engr. Madigan/Hyland, Newark, 1952-54; assoc. hwy. engr. Calif. Div. Hwys., Eureka, 1954-66; dir. pub. works, city engr. City of Arcata, Calif., 1966-74; pres. Laco Assocs. Cons. Engrs., Eureka, 1974—. Mem. Bldg. Bd. Appeals City of Eureka, 1982-88, Humboldt County, Eureka, 1985—; mem. Local Agy. Formation Commn., Humboldt County, 1985—. With USN, 1941-45, 50-52, ret. comdr., USNR. Fellow ASCE (life); mem. Cons. Engrs. and Land Surveyors Calif., Calif. Geotechnical Engr. Assn., Eureka C. of C. Home: PO Box 1276 Willow Creek CA 95573 Office: Laco Assocs PO Box 1023 Eureka CA 95501

CONWAY, JOHN THOMAS, computational fluid dynamacist, educator; b. Ballymoney, No. Ireland, Sept. 17, 1956; arrived in Can., 1985; s. Malachy John and Irene Steele (Mulholland) C. BA in Physics, King's Coll., Cambridge, Eng., 1978; MA in Physics, Coll. of Aeronautics, Cranfield, Eng., 1981, MSc in Aerodynamics, 1982. Registered profl. engr., Eng., Europe. Aero. engr. Brit. Aerospace (Dynamics), Eng., 1982-85; staff specialist Canadair Aerospace Group, Mont., Que., Can., 1985-92; assoc. prof. to prof. aero. engring. Agder Coll. Engring., Grimstad, Norway, 1992—; course instr. U. Mont., 1986-92, external examiner MSc thesis, 1990-91. Mem. AIAA, Instn. of Mech. Engrs., Royal Aero. Soc., Inst. of Physics, Can. Aeronautics and Space Inst. Achievements include development of an unsteady 3-D panel method with a vortex sheets wake model and wake convection; co-demonstration that cross-flow separation is crucial in determining the forces experienced by a turning submarine. Home: Apt 2 Grooseveien 70, N-4890 Grimstad Norway Office: Agder Coll Engring, Mech & Aero Engring Groosevn 36, N-4890 Grimstad Norway

CONWAY, LYNN ANN, computer scientist, educator; b. Mount Vernon, N.Y., Jan. 2, 1938. B.S., Columbia U., 1962, M.S. in Elec. Engring., 1963. Mem. research staff IBM Corp., Yorktown Heights, N.Y., 1964-68; sr. staff engr. Memorex Corp., Santa Clara, Calif., 1969-73; mem. research staff Xerox Corp., Palo Alto, Calif., 1973-78, research fellow, mgr. VLSI systems area, 1978-82, research fellow, mgr. knowledge systems area, 1982-83; asst. dir. for strategic computing Def. Advanced Research Projects Agy., Arlington, Va., 1983-85; prof. elec. engring. and computer sci., assoc. dean Coll. Engring. U. Mich., Ann Arbor, Mich., 1985—; vis. assoc. prof. elec. engring. and computer sci. MIT, Cambridge, Mass., 1978-79; mem. sci. adv. bd. USAF, 1987-90. Co-author: textbook Introduction to VLSI Systems, 1980. Mem. coun. Govt.-Univ.-Industry Rsch. Roundtable, 1993—. Recipient Ann. Achievement award Electronics mag., 1981, Harold Pender award U. Pa., 1984, Wetherill Medal Franklin Inst., 1985, Sec. of Def. Meritorious Civilian Svc. award, 1985; sr. fellow U. Mich. Soc. Fellows, 1987-91. Fellow IEEE; mem. AAAS, NAE, Am. Assn. for Artificial Intelligence, Soc. Women Engrs. (Ann. Achievement award 1990). Office: U Mich 170 ATL Bldg Ann Arbor MI 48109

COOEY, WILLIAM RANDOLPH, economics educator; b. Wheeling, W.Va., Feb. 23, 1942; s. William Earl and Marguerite Ruth (Potts) C.; m. Linda Faye Whiteman, Aug. 11, 1973; children: William Justin, Crissa Kaye. BA, Bethany Coll., 1964; MS, W.Va. U., 1966; postgrad., Miss. State U., 1973-74. Prof. Bethany (W.Va.) Coll., 1966—; v.p., bd. dirs. Cooey-Bentz Co., Wheeling, 1986-90; part-time assoc. prof. Ohio U., St. Clairsville, 1967-86, W.Va. U., West Liberty, 1976-84; pvt. practice legal cons., Bethany, 1975—. Contbr. articles to profl. publs. Advisor Boy Scouts Am., Bethany, 1986-90; asst. coach Little League Baseball, Bethany, 1986-90. Mem. Ea. Econs. Assn., Beta Beta Beta, Omicron Delta Epsilon, Gamma Sigma Kappa. Avocations: woodworking, making videos, computers. Home: 102 Pt Breeze Dr Bethany WV 26032 Office: Bethany Coll Harlan Hall Bethany WV 26032

COOHILL, THOMAS PATRICK, biophysicist, photobiologist; b. N.Y.C., Aug. 25, 1941; s. Francis John and Mary (Donelley) C.; m. Patricia Ann Trutty, Sept. 8, 1962; children: Joseph, Thomas, Matthew. BSc, U. Toronto, Ont., Can., 1962; PhD, Pa. State U., 1968. Rsch. scientist U.S. VA Hosp., Pitts., 1968-72; asst. rsch. prof. U. Pitts. Med. Sch., 1968-72; prof. biophysics Western Ky. U., Bowling Green, 1972-92; pres. Ultraviolet Cons., Bowling Green, 1992—; advisor Scope, Paris, 1990—; cons. Advanced Interventional Systems, Irvine, Calif., 1989-91; NAS/NRC sr. fellow Calif. Tech. Jet Propulsion Lab. Grantee FDA, 1974-80, NIH, 1982-83, NASA-Ky. Space Grant Consortium, 1992. Mem. Am. Soc. Photobiology (pres. 1988-89), Biophys. Soc., Sigma Xi (nat. lectr. 1991-93), Sigma Phi Sigma. Democrat. Roman Catholic. Achievements include discovery of capacity enhancement and the large plaque effect in herpes virus; research on effects of ultraviolet radiation of living systems especially as it related to stratospheric ozone depletion.

COOK, ADDISON GILBERT, chemistry educator; b. Caracas, Venezuela, Apr. 1, 1933; s. Harold Reed and Florence (Sloan) C.; m. Nancy Lois Spriggs, Aug. 18, 1956; children—Virginia Lynn, Shirley June, Diane Joyce. B.S., Wheaton Coll., 1955; Ph.D., U. Ill., 1959. Research assoc. Cornell U., 1959-60; from asst. prof. to prof. chemistry Valparaiso U., 1960—, chmn. dept., 1970—; cons. chemistry divsn. Argonne (Ill.) Nat. Lab., 1961-69; rsch. assoc. Amoco, Whting, Ind., 1960. Editor, contbr.: Enamines: Synthesis, Structure, and Reactions, 1969, 2d edit., 1988; Contbr. articles profl. jours. Recipient Research Corp. grant, 1960-61; Petroleum Research Fund grant, 1963-69. Mem. Am. Chem. Sco., Chem. Soc. (London), Ind. Acad. Sci., Sigma Xi, Phi Lambda Upsilon, Pi Mu Epsilon. Mem. Evangel. Free Ch. Am. Home: 2308 Shannon Dr Valparaiso IN 46383-2427

COOK, ALFRED ALDEN, chemical engineer; b. St. Louis, May 2, 1930; s. Jesse Alfred and Catherine Weston (Fuller) C.; m. Leath Lorraine Koester, June 14, 1952; children: Steven Alfred, Gregory Alden, David Lee. B-SChemE, Washington U., St. Louis, 1952. Registered profl. engr. Mo., Ill. Chem. project engr. Mallinckrodt, St. Louis, 1952-69; process engr. Miles Labs. Union Div., Granite City, Ill., 1969-71; sr. process engr. Armour Pharm. Co., Kankakee, Ill., 1971-73, pilot plant supt., 1973; profl. engr. Sverdrup & Parcel and Assocs., St. Louis, 1973-83; process engr. Interglobal Tech. Svcs., St. Louis, 1983-84, Butler Svcs. Group, St. Louis, 1984-85; process and chem. engr. TBG Inc., St. Louis, 1986-91; process engr. HL Yoh Inc., Sauget, Ill., 1991; retired engr., 1992. Co-author: (with C.A. Carter) EPA Document Phosphorus Industry Waste Water Guidelines, 1977; contbr. articles to profl. jours. With U.S. Army Corp Engrs., 1953-54. Recipient Jauncy Meml. prize Washington U., 1950. Mem. AICE, Am. Chem. Soc., Instrument Soc. Am., Air and Waste Mgmt. Assn., Tau Beta Pi, Pi Mu Epsilon, Alpha Chi Sigma. Independent. Home: 375 Paul Dr Florissant MO 63031

COOK, ANDREW ROBERT, experimental chemical physicist; b. Easton, Pa., Dec. 31, 1965; s. Robert C. Cook and Eileen N. (Woundy) Reifsnyder; m. Pamela Kaye Parr, July 9, 1988. BS with honors, U. Calif., 1988; PhD, MIT, 1993. Programmer Sandia Nat. Labs., Livermore, Calif., 1983-84; chem. technician Lawrence Livermore Nat. Labs., 1985-88; teaching asst. MIT, Cambridge, 1988-89, rsch. asst., 1989—. Author: (with others) Picosecond Laser Spectroscopy, 1993; contbr. articles to profl. jours. Recipient Undergrad. award in Analytical Chemistry Am. Chem. Soc., 1987; fellowship Associated Western U., 1985, Clorox Undergrad. scholarship Clorox Co., 1987. Office: MIT Rm 2-052 77 Massachusetts Ave Cambridge MA 02139

COOK, ANTHONY MALCOLM, aerospace engineer; b. Liverpool, Eng., Jan. 10, 1936; came to U.S., 1948; s. Edward George Cook; m. Dianne Lorene Combest, July 22, 1956; children: Robert L., Barbara L. BS in Engring., San Jose State U., 1958. Aero. rsch. engr. NASA Ames Rsch. Ctr., Moffett Field, Calif., 1962-68, aero. project engr. advanced aircraft office, 1968-70, tech. asst. to dir. aeronautics, 1970-74, asst. chief Flight Systems & Simulation Rsch. div., 1974—; chmn. flight simulation lab. adv. com. Calif. Poly. State U., San Luis Obispo, 1983—. Adviser, Los Gatos (Calif.) High Sch. Dist., 1978-79. Capt. USAF, 1958-60. Recipient Exceptional Svc. medal NASA, 1985, Silver Snoopy Astronaut award NASA, 1988. Mem. AIAA (chmn. working group on r&d flight simulation 1982-83), Am. Arbitration Assn. Avocations: tennis, skiing, scuba diving. Office: NASA Ames Rsch Ctr Flight Systems & Simulation Rsch Divsn Bldg 243-1 Moffett Field CA 94035

COOK, HAROLD RODNEY, military officer, medical facility administrator; b. Sterling, Colo., Feb. 13, 1944; s. Harold E. Cook and Adelaide Cook; 1 child, Dawn. BS in Bus. and Psychology, Kearney State Coll., 1973; MA in Human Rels., Webster U., 1983; MHA, Baylor U., 1985. Commd. 2d lt. U.S. Army, 1974, advanced through grades to maj., 1986; med. adminstr. U.S. Gen. Hosp., Nürnberg, Germany, 1975-78; comdr. 560th Ambulance Co., Korea; chief ops. med./surg. div. Acad. of Health Sci., Ft. Sam Houston, Tex., 1980-83; with health care adminstrn. Baylor U., Waco, Tex., 1983-85; surgery adminstr. Fitzsimons Army Med. Ctr., Aurora, Colo., 1985—; dep. comdr. USAHC, St. Louis, 1988-90; with administr. directorate ancillary svcs. Brook Army Med. Ctr., Ft. Sam Houston, 1991-92; with Team Am. Consulting, San Antonio, 1992—; exec. dir., pres. Colo. Petrolon Inc., 1986-88; regional v.p. Petrolon Inc., 1988—; pres. Team Am. Cons. Mem. Am. Coll. Hosp. Adminstrs., Am. Soc. Mgmt., Nat. Assn. Collegiate Vets. (exec. bd.), Fitz Alpine Club (pres. 1985-87), Colo. Pantera Club. Office: Team Am Consulting 7704 Painted Ridge Dr San Antonio TX 78239

COOK, JOHN BELL, chemist; b. Wellington, New Zealand, Aug. 9, 1933; s. Alfred Bell and Isobel Louise (Taylor) C.; m. Dianne Scott, Nov. 7, 1970 (div. 1981); children: Fiona, Rachel. BA in Physiology, U. New Eng., Armidale, 1988, MSc in Rural Sci., 1991, PhD in Argl. Chemistry, 1993. Tech. officer Merck Sharpe & Dohme, Sydney, NSW, Australia, 1971-74, Vet. Rsch. Sta., NSW Agr., Sydney, 1974-80; chemist Cotton Rsch. Ctr., NSW Agr., Narrabri, 1980-82; lab. mgr. analitical chemistry Agr. Rsch. Ctr. Tamworth, NSW, 1982—. With Australian Army, New Zealand Army 1952-70. Mem. Am. Chem. Soc., Australian Inst. Agrl. Scientists, Returned Svc. League Australia (life), U. New Eng. Alumni Assn. (life). Home: 9/46 Church St, Tamworth 2340 NSW, Australia Office: Agr Rsch Ctr, RMB 944, Tamworth 2340 NSW, Australia

COOK, JULIA LEA, geneticist; b. Athens, May 28, 1958; came to U.S., 1961; d. K.L. and V.M. Cook; m. Jawed Alam, Sept. 19, 1987. PhD in Genetics, N.C. State U., 1986. Postdoctoral fellow La. State U. Med. Ctr., New Orleans, 1986-89, rsch. asst. prof. dept. ophthalmology, 1991—; rsch. asst. prof. dept. biochem. and molecular biology, 1992—; staff scientist Ochsner Med. Found., New Orleans, 1989—. Mem. editorial bd.; Analytical Biochemistry, 1991—; contbr. articles to Biology and Medicine, Analytical Biochemistry-Reporter Genes. Mem. AAAS, N.Y. Acad. Scis., Genetics Soc. Am. Home: 4909 Rebecca Blvd New Orleans LA 70065 Office: Ochsner Med Found Div Rsch 1516 Jefferson Hwy New Orleans LA 70121-2484

COOK, RAYMOND DOUGLAS, biomedical engineer; b. Fairfax, Va., Mar. 13, 1966; s. Douglas Roger and Marianne (Danek) C.; m. Doreen Ann Gagnon, Aug. 4, 1989. BS in Biomed. Engring., Rensselaer Polytech. Inst., 1988, PhD in Biomed. Engring., 1992. Tech. aid Mitre Corp., McLean, Va., 1982-86; rsch. asst. Rensselaer Polytechnic Inst., Troy, N.Y., 1988-92; engring. InAir Ltd., Delmar, N.Y., 1992—; rsch. cons. InAir Ltd., Delmar, 1992—. Contbr. articles to profl. jours.; patentee for high speed, high precision elec. impedance tomograph. Recipient Rensselaer fellowship Rensselaer Polytechnic Inst., Troy, 1989. Mem. IEEE, Sigma Xi, Tau Beta Pi. Office: JEC Rm 7049/Biomed Engring Rensselaer Polytechnic Inst Troy NY 12180-3590

COOK, ROBERT EDWARD, educator, plant ecology researcher; b. Providence, R.I., Sept. 26, 1946; s. John Edward and Suzanne Marie (Boisvert) C. A.B., Harvard U., 1968; Ph.D., Yale U., 1973. Instr. Yale U., New Haven, 1973-74; postdoctoral fellow Harvard U., Cambridge, Mass., 1974-75; asst. prof. Harvard U., Cambridge, 1975-80, assoc. prof., 1980-83, Arnold prof., 1989—; program dir. NSF, Washington, 1982-83; assoc. prof. Cornell U., Ithaca, N.Y., 1983-88, dir. Cornell Plantations, 1983-88; dir. Arnold Arboretum, Boston, 1989—; vis. prof. Cornell U., Ithaca, 1981-82. Contbr. articles to profl. jours. Capt. U.S. Army, 1968-74. NSF grantee. Office: Harvard U Arnold Arboretum 125 Arborway Jamaica Plain MA 02130-3500

COOK, ROBERT EUGENE, SR., engineer; b. Mpls., Feb. 13, 1921; s. George Henry and Florence Elizabeth (King/Foote) C.; m. Mary Ann Anderson, Sept. 26, 1942; children: Diane Cook Knudson, Robert Eugene Jr., James Lee, Jaclyn Cook Michlin. BS in Engring., U. Minn., 1969; MA, Govs. State U., 1973. Pres. Baco Engring., Inc., Mpls., 1948-51; mgr. HVAC (heating, ventilation and air conditioning) divsn. Standard Plumbing Supply, Inc., Mpls., 1951-56; dist. mgr. A.O. Smith Corp., CPD, Mpls., 1956-70; mgr. product engring. A.O. Smith Corp., CPD, Kankakee, Ill., 1970-76, dir. engring., 1976-80, dir. internat., 1980-82; cons. R.E. Cook, Cons., Seminole, Fla., 1982—; instr. Kankakee C.C., 1972-74; community prof. Govs. State U., Park Forest South now University Park), Ill., 1976-78. Contbr. articles to profl. jours. State rep. Minn. State Legis., St. Paul, 1967-70; bd. dirs. Govs. State U. Found., 1980-85; pres., bd. dirs. Kankakee County ARC, 1972-86. With U.S. Army, 1942-45, to lt. col. USAR, 1945-70. Fellow ASHRAE (life, Award of Merit 1975, Cert. of Appreciation 1983, 92, Disting Svc. award 1987). Achievements include 5 patents for solar water heating system, cold water inlet tube, corrosion protection system, system for heating and storing a liquid, improved heating and cooling system. Home: 7701 Starkey Rd #747 Largo FL 34647-4325

COOK, RONALD LEE, petroleum engineer; b. Lima, Ohio, June 22, 1936; s. Lee Benjamin and Evalyn Goldie (Dirth) C.; m. Sandra Lou Sexten, June 14, 1958; children: Ronald William, Andrew Wayne, Jill Suzanne. BS, Ohio State U., 1960. Registered profl. engr., Ohio, Kans. Petroleum engr. Panhandle Eastern Pipeline Co., Liberal, Kans., 1960-78, Kansas City, Mo., 1978-82; cons. Petroleum Cons., Inc., Overland Park, Kans., 1982—; cons. Kans. Corp. Commn., Wichita, 1982-87, 92. Contbr. articles to Jour. Petroleum Tech. Pres. Liberal Kids, Inc., 1961. Mem. Soc. Petroleum Engrs. (chmn. 1968-69), Am. Petroleum Inst. (chmn. 1990-91), Kans. Engring. Soc., Nat. Soc. Profl. Engrs. Republican. Methodist. Office: Petroleum Cons Inc Ste 440 8717 W 110th St Overland Park KS 66210

COOK, STEPHEN PATTERSON, physical sciences and astronomy educator, researcher; b. Burbank, Calif., May 5, 1951; s. Edgar N. and Ruth E. (Raplinger) C.; m. Laurie A. Moore, Dec. 28, 1976 (div. 1990); children: Casey, Keziah. BA in astronomy, UCLA, 1973, MS in physics, 1976. Instr. North Ark. Community Coll., Harrison, 1977-83; cons., author CompuSOLAR, Jasper, Ark., 1983-85; asst. prof. physical sci. Ark. Tech. Univ., Russellville, 1985—. Author: Achieving Self Reliance-Backyard Energy Lessons, 1984, Coming of Age in the Global Village, 1990, Physicial Science in the Laboratory, 1992. Recipient Tech. grant Ark. Dept. Energy, 1979, 1980, project grant, 1981. Mem. Am. Assn. Variable Star Observers, Internat. Dark Sky Assn., Nat. Assn. Sci., Tech. Soc. Achievements include first residential use of photovoltaics solar energy, software design package; development of new findings about various eclipsing binary stars. Home: Rt 2 Box 492B Russellville AR 72801 Office: Ark Tech Univ Dept Physical Scis Russellville AR 72801

COOK, WILLIAM R., JR., chemist; b. Boston, Nov. 28, 1927; s. William R. Sr. and Ramona (Graham) C.; m. Anne D. Johnson, Sept. 23, 1950; children: William III, Elizabeth, Barbara, Susan. BA, Oberlin U., 1949; MA, Columbia U., 1950; PhD, Case Western Res., 1971. Scientist Brush Devel. Co./Clevite Corp., Cleve., 1951-54; head crystallography sect. Clevite Corp./Gould Inc., Cleve., 1954-74; corp. sec. Cleve. Crystals, Inc., 1973—. Author: (book chpts.) Piezoelectrics and Ferroelectrics on Digest of Literature on Dielectrics, 1959-69, Piezoelectrics and Electro-optics in Landolt-Bornstein, 1979, 84, 93; joint author Piezoelectric Ceramics. Fellow Am. Ceramic Soc.; mem. Am. Chem. Soc., Am. Crystallographic Assn., Mineralogical Soc. Am., Mineralogical Soc. Can. Achievements include patents in Piezoelectric ceramics and Harmonic generation. Office: Cleveland Crystals Inc 19306 Redwood Ave Cleveland OH 44110

COOKE, DAVID LAWRENCE, chemical engineer; b. Billinge, Lancashire, Eng., Mar. 24, 1954; came to Can., 1982; s. Cedric Lawrence and Joan (Ince) C.;m. Liliane Valensi, July 5, 1976; children: Graham, Steven. BSc with honors, U. Birmingham, Eng., 1975; diploma in mgmt. studies, Poly. of Wales, Pontypridd, 1981. Chartered eng., U.K. Chem. engr. Courtaulds, Ltd., Coventry, Eng., 1975-77, BP Chems. Ltd., Baglan Bay, Wales, 1977-82; sr. bus. devel. engr. Novacor Chems. Ltd., Calgary, Alta., Can., 1982-87, tech. svc. mgr., 1987—. Mem. TAPPI, Instn. Chem. Engrs., Soc. Plastics Engrs. Achievements include rsch. on linear low density polyethylene polymers and addition of branched molecules, high molecule weight molecules to improve optical properties of linear low density polyethylene film and future applications for recycled plastics. Office: Novacor Chems Ltd, 3620 32 St NE, Calgary, AB Canada T1Y 6G7

COOKE, G. DENNIS, biological science educator; b. Ravenna, Ohio, June 29, 1937; B. George H. and Mabel E. (Brown) C.; m. Marian Moore; children: Kathryn, Eric, Tom; m. Angela B. Martin, Oct. 14, 1989. BS, Kent State U., 1959; MS, U. Iowa, 1963, PhD, 1965. Instr. U. Iowa, Iowa City, 1965; post-doctoral fellow Inst. Ecology U. Ga., Athens, 1965-67; from asst. prof. to prof. biol. sci. Kent (Ohio) State U., 1967—. Author: Lake and Reservoir Restoration, 1986, Reservoir Management for Water Quality, 1989, In-Reservoir Water Quality Management Techniques, 1989, Restoration and Management of Lakes and Reservoir, 1993. Mem. N.Am. Lake Mgmt. Soc. (pres. 1980-81, bd. dirs. 1986-89, Tech. Excellence award 1990), Ohio Lake Mgmt. Soc. (bd. dirs. 1985-88, Disting. Svc. award 1989). Office: Kent State U Biol Scis Cunningham Hall Biological Sciences Kent OH 44242

COOKE, ROBERT EDMOND, physician, educator, former college president; b. Attleboro, Mass., Nov. 13, 1920; s. Ronald Melbourne and Renee Jeanne (Wuillumier) C.; m. Sharon Riley, Nov. 20, 1978; children—Susan R., Anne R.; children by previous marriage—Robyn (dec.), Christopher, Wendy, W. Robert, Kim. B.S., Sheffield Sci. Sch., Yale U., 1941, M.D., 1944; postgrad. (NIH postdoctorate fellow), Sch. Medicine, 1948-50, John and Mary R. Markle scholar, 1951-55. Intern, asst. resident dept. pediatrics New Haven Hosp., 1944-46; instr. pediatrics Yale, 1950-51, asst. prof. pediatrics, physiology, 1951-54, asso. prof., 1954-56; resident to asso. pediatrician Grace-New Haven Community Hosp., 1951-56; pediatrician-in-chief Johns Hopkins Hosp., 1956-73; chmn. dept. Johns Hopkins Sch. Medicine, 1956-73; Grover Powers prof. pediatrics Nat. Assn. Retarded Children, 1957-59, Given Found. prof. pediatrics, 1962-73; vis. prof. Harvard U., 1972-73; vice chancellor for health scis., prof. pediatrics U. Wis., 1973-77; pres. Med. Coll. Pa., 1977-80; A. Conger Goodyear prof. pediatrics, med. dir. Children's Rehab. Center SUNY, Buffalo, 1982-88, prof. emeritus, 1988—, chmn. dept. pediatrics, 1985-88; pediatrician-in-chief Children's Hosp. of Buffalo, 1985-88; chief med. officer Spl. Olympics Internat.; with Mass. Dept. Mental Health, 1980-82; chmn. med. adv. bd. Kennedy Found.; mem. adv. bd. Nat. Ctr. Rehab. Rsch., Nat. Inst. Child Health and Human Devel., 1991—. Editor, contbr. to pediatric textbooks, profl. jours. Trustee Children's Rehab. Inst. Served from lt. to capt. M.C. AUS, 1946-48. Recipient Mead Johnson award in pediatrics, 1954; Kennedy Internat. award for disting. svc. in field of mental retardation, 1968; Howland medal, 1992; medallion of the Surgeon Gen., 1993. Fellow Am. Acad. Pediatrics; Distinguished fellow Am. Psychiat. Assn.; mem. Am. Pediatric Soc. (John Howland award 1991), Soc. for Pediatric Research (pres. 1965-66), Am. Soc. for Clin. Investigation, Md. Med. Soc., Am. Pub. Health Assn., Am. Fedn. Clin. Research, Inst. of Medicine, Aurelian Hon. Soc., Phi Beta Kappa, Sigma Xi, Alpha Omega Alpha. Home: 865 Painted Bunting Ln Vero Beach FL 32963-2026

COOKE, STEVEN JOHN, scientist, chemical engineer, consultant; b. Grand Rapids, Mich., Oct. 1, 1954; s. Edward G. and Annette M. (Minnema) C.; m. Marguerite K. Oldenburger, June 18, 1977; children: Allison, Jonathan. BS in Chemistry, Calvin Coll., 1977; M in Chem. Engring., Ill. Inst. Tech., 1987. Registered profl. engr., Ill.; cert. profl. chemist, quality engr. Chemist, lab. supr. Matheson Gas Products, Joliet, Ill., 1977-80; chief chemist Cardox, Countryside, Ill., 1980-85; scientist Am. Air Liquide, Countryside, 1985-92; asst. quality mgr. Alphagaz Divsn. of Liquid Air, Countryside, 1992-93; quality assurance/quality control mgr. Am. Air Liquide, Countryside, 1993—. Group leader Hazardous Materials Emergency Response Team; treas. Christian Reformed Ch. Mission, Western Springs, Ill., 1982—, Chicagoland Diaconal Task Force Bd., Palos Heights, Ill., 1989-92. Fellow Am. Inst. Chemists; mem. Fed. Analytic Chem. and Specialty Soc., Am. Soc. Quality Control, Am. Inst. Chemists, Am. Chem. Soc. (publicity chairperson ind. and engring. chemists divsn. 1989—). Achievements include patent for portable gas analyzer. Office: Liquid Air 5230 East Ave La Grange IL 60525-3133

COOKSON, ALBERT ERNEST, telephone and telegraph company executive; b. Needham, Mass., Oct. 30, 1921; s. Willard B. and Sarah Jane (Jack) C.; m. Constance J. Buckley, Sept. 10, 1949; children—Constance J., William B. B.E.E., Northeastern U., 1943; M.E.E., Mass. Inst. Tech., 1951; Sc.D., Gordon Coll., 1974. Group leader Research Lab. Electronics, Mass. Inst. Tech., 1947-51; lab. dir. ITT Fed. Labs., Nutley, N.J., 1951-59; v.p., dir. operations ITT Fed. Labs. (Internat. Elec. Corp. div.), Paramus, N.J., 1959-62; pres. ITT Intelcom, Falls Church, Va., 1962-65; dep. gen. tech. dir. Internat. Tel. & Tel. Corp., N.Y.C., 1965-66, v.p., tech. dir., 1966-68, sr. v.p., gen. tech. dir., 1968-84, ret., 1984; pres., chief exec. officer Richmond Properties, 1982—; chmn. bd. ITT Interplan; pres., chmn. Comtexco Industries, 1980—; chmn. tech. adv. bd. U.S. Postal Svc., 1983-91; bd. dirs. Internat. Standard Electric, ITT Industries; mem. Def. Communications Satellite Panel; adviser research and engring. on def. communications satellite systems Dept. Def.; mem. indsl. panel sci. and tech. NSF; mem. Fairfax County Econ. and Indsl. Devel. Com., 1962-65; mem. nat. coun. Northeastern U.; mem. pride com. U. Hartford, 1973-76; elec. engring./computer adv. bd. MIT, 1977-82. Bd. dirs. Fundacion Chile, 1983-89. Served with USNR, 1943-46. Fellow IEEE; mem. Armed Forces Communications and Electronics Assn., Am. Mgmt. Assn., Am. Inst. Aeros. and Astronautics, Electronic Industries Assn., Sigma Xi, Tau Beta Pi. Patentee frequency search and track system. Office: 320 Park Ave New York NY 10022-6815

COOLES, PHILIP EDWARD, physician; b. London, Oct. 6, 1953; arrived in Dominica, 1982; s. James Austen and Hilda Alice (Prangnell) C.; m. Sandra Newton Hogarth; children: Abigail, Holly, Anisa. BSc 1st class honors, U. London, 1975, MB, BS, 1978; Diploma of Tropical Medicine and Hygiene, U. Liverpool (Eng.), 1987. Intern St. George's Hosp., London, 1978-79; resident Aberdeen (Scotland) Royal Infirmary, 1979-82, registrar (medicine), 1981-82; specialist physician, head of dept. Princess Margaret Hosp., Roseau, Dominica, 1982-88; prof. medicine Ross U. Med. Sch., Portsmouth, Dominica, 1988—, dean basic med. scis., 1992, exec. dean, 1992—; cons. physician, Roseau, 1988—; translator in field. Contbr. articles to profl. jours. Mem. Nat. Spiritual Assembly of Baha'is of Dominica. Mem. Royal Coll. Physicians of U.K., Royal Soc. Tropical Medicine (London), Dominica Club. Avocations: painting, squash, carpentry. Home: Castle Comfort, PO Box 335, Roseau Dominica Office: Ross U Med Sch, Portsmouth Campus, Portsmouth Dominica

COOLEY, DENTON ARTHUR, surgeon, educator; b. Houston, Aug. 22, 1920; s. Ralph C. and Mary (Fraley) C.; m. Louise Goldsborough Thomas, Jan. 15, 1949; children: Mary, Susan, Louise, Florence, Helen. B.A., U. Tex., 1941; M.D., Johns Hopkins U., 1944; Doctorem Medicinae (hon.), U. Turin, Italy, 1969; H.H.D. (hon.), Hellenic Coll., 1984, Holy Cross Greek Orthodox Sch. of Theology, 1984; DSc honoris causa, Coll. of William and Mary, 1987. Diplomate: Am. Bd. Surgery, Am. Bd. Thoracic Surgery. Intern Johns Hopkins Sch. Medicine, Balt., 1944-45; resident surgery Johns Hopkins Sch. Medicine, 1945-50; sr. surg. registrar thoracic surgery Brompton Hosp. for Chest Diseases, London, Eng., 1950-51; assoc. prof. surgery Baylor U. Coll. Medicine, Houston, 1954-62; prof. surgery Baylor U. Coll. Medicine, 1962-69; clin. prof. surgery U. Tex. Med. Sch., Houston, 1975—; founder, surgeon-in-chief Tex. Heart Inst., 1962—. Served as capt., M.C., 1946-48. Named one of ten Outstanding Young Men in U.S., U.S. C. of C., 1955; Man of the Yr. award Kappa Sigma, 1964; Rene Leriche prize Internat. Surg. Soc., 1967; Billings Gold medal Am. Surg. Soc., 1967; Vishnevsky medal Vishnevsky Inst., USSR, 1971; Theodore Roosevelt Award, 1980; Presdl. Medal of Freedom, presented by Pres. Reagan, 1984; Gifted Tchr. award Am. Coll. Cardiology, 1987. Hon. fellow Royal Coll. Physicians and Surgeons of Glasgow, Royal Coll. Surgeons of Ireland, Royal Australasian Coll. Surgeons, Royal Coll. Surgeons of Eng.; mem. ACS, Am. Surg. Assn., Internat. Cardiovascular Soc., Am. Assn. Thoracic Surgery, Soc. Thoracic Surgery, Soc. Univ. Surgeons, Am. Coll. Cardiology, Am. Coll. Chest Physicians, Soc. Clin. Surgery, Soc. Vascular Surgery, Western Surg. Assn., Tex. Surg. Soc., Halsted Soc. Performed numerous heart transplants; implanted 1st artificial heart, 1969. Office: Tex Heart Inst PO Box 20345 Houston TX 77225-0345

COOLEY, PHILIP CHESTER, computer modeller; b. Summit, N.J., Apr. 30, 1941; s. Emerson Frisbie and Helen (White) C.; divorced; children: Dagmar, Jacob. BA, Northwestern U., 1964; MS, U. N.C., 1976. Analyst Rsch. Triangle Inst., Research Triangle Park, N.C., 1969-79, sr. mathematician, 1980-85, asst. dir., 1985—; cons. CDC, Atlanta, 1989-90, Family Health Internat., Research Triangle Park, 1991-92. Reviewer Sci. mag., 1990—, JAMA, 1989—; contbr. articles to profl. jours. Grantee Nat. Inst. on Drug Abuse, 1987, 90. Mem. AAAS, Math. Programming Soc. Achievements include patents for General Purpose Queuing Simulator, for Forms Management System for Distributed Data Entry Applications, for a Statistical Set of Procedures Not Supported by Most Conventional Packages. Home: 134 W Margaret Ln Hillsborough NC 27278

COOLEY, RICK EUGENE, chemist; b. Indpls., Mar. 18, 1953; s. Harry Arthur and Wilma Mae (Ashley) C.; m. Karen Jo Porter, Jan. 28, 1972; children: Jason Matthew, Jennifer Lynn. BS in Chemistry, Purdue U., 1979. Lab. technician Eli Lilly and Co., Indpls., 1972-79, assoc. chemist, 1979-82, chemist, 1982-85, asst. sr. chemist, 1985-90, assoc. sr. chemist, 1990—; indsl. advisor Ctr. for Process Analytical Chemistry, U. Wash., Seattle, 1985-87; presenter Pitts. Conf. Analytical Chemistry and Applied Spectroscopy, 1979, Fedn. Analytical Chemistry and Spectroscopy Socs., 1979, Internat. Forum on Process Analytical Chemistry, 1991, Spring nat. meeting Am. Chem. Soc., 1993. Contbr. articles to Analytical Chemistry, Jour. Automatic Chemistry, Process Control and Quality, chpt. to Analyzers in Instrumentation and Control. Coach Ctr. Grove T-ball and Football leagues, Greenwood, Ind., 1977-81; EMT, fire fighter White River Twp. Vol Fire Dept., Greenwood, 1980-82, treas., 1982—; pres. Ctr. Grove Band Boosters, Greenwood, 1988-89. Achievements include development of automated dissolution sample collection device, automated polarograph for solution labile compounds, automated nitrosamine analyzer, on-line liquid chromatograph analyzer. Home: 2597 Lake Crossing Greenwood IN 46143 Office: Eli Lilly and Co Corporate Center Indianapolis IN 46285

COOLEY, SHEILA LEANNE, psychologist, consultant; b. Oakland, Calif., July 25, 1956; d. Philips Theadore and Helen Ellene (Newbill) C. BA, St. Leo Coll., 1979; MS, U. So. Miss., 1986; PhD, Miss. State U., 1990. Lic. psychologist, Ky. Counselor Charter Counseling Ctr., Jackson, Miss., 1988-89; staff psychologist Rivendell Psychiat. Ctr., Bowling Green, Ky., 1989-90; program dir. MidSouth Hosp., Memphis, 1990-91; resource ctr. dir. Mid-South Resource Ctr., Ridgeland, Miss., 1991-92; partial hosp. dir. Pathways Partial Hospitalization, Ridgeland, 1991-92; edn. specialist, sr. position Miss. Dept. of Edn., Bur. Spl. Svcs., Jackson, 1993—. Campaign organizer for Dem. mayor, Jackson, 1992. Mem. APA, Miss. Psychol. Assn., Phi Delta

Kappa, Psi Chi, Theta Phi Sigma. Baptist. Home: 816 Harbor Bend Dr Brandon MS 39042

COOLING, THOMAS LEE, civil engineer, consultant; b. Peoria, Ill., Mar. 13, 1950; s. Lawrence John and Eloise Mardell (Meister) C.; m. Annette Wilson, Dec. 30, 1972 (div. June 1990); 1 child, Nathan. BSCE, U. Ill., 1972; MSCE, U. Calif., Berkeley, 1975. Registered profl. engr., Ill., Tenn., Ga., Mo., Calif.; cert. geotech. engr., Calif. Project engr. Chevron Oil Co., San Francisco, 1972-74; geotech. engr. Dames and Moore, San Francisco, 1974-76, McClelland Engrs., St. Louis, 1976-78; sr. assoc. Woodward-Clyde Cons., St. Louis, 1978—. Mem. ASCE (head geotech. subcom. 1987-88), NSPE, Earthquake Engring. Rsch. Inst. (mem. adv. bd. 1991—), Am. Concrete Inst. Home: 420 W Adams Apt E Kirkwood MO 63122 Office: Woodward Clyde Cons 2318 Millpark Dr Saint Louis MO 63043

COOLMAN, FIEPKO, retired agricultural engineer, researcher, consultant; b. 'T Zandt, Croníncen, The Netherlands, May 22, 1918; s. Egge and Afina (Offeringa) C.; m. Anna Maatje Derect, Apr. 29, 1943. Agrl. degree, U. Wageningen, 1942. Cert. agrl. engr. mgr. Prof. Agrl. High Sch., Groningen, 1942-44; cons. Royal Dutch Shell, The Hague, 1947-53; rsch. officer Min. Agriculture, The Hague, 1953-58, leader rsch. dept., 1958-64, dir. rsch. engring. inst., 1964-89; cons. various projects numerous developing countries, 1978-89. Author: Technical Data Farm Tractors, 1948, Small Farm Mechnization, 1956, Who is Who in Agricultural Engineering, 1975; contbr. intro. to The Literature of Agricultural Engineering, 1983; contbr. articles to profl. jours. Recipient Officer Orange Nassau award Dutch Royal Family, 1974, Officer Crown Order award Belgian Govt., 1978, Officer Merite Agricole award French Min. Agriculture, 1974, Gold medal Am. Soc. Agrl. Engrs., 1979. Mem. Internat. Com. Agrl. Engring., Dutch Assn. Agrl. Engring. Office: Thijsselaan 62, 6705 AS Wageningen The Netherlands

COON, FRED ALBERT, III, mechanical engineer; b. Monroe, La., Feb. 28, 1949; s. Fred Albert Jr. and Margaret Estell (Stuntz) C.; m. Harryette DeLaRue Brock, Apr. 14, 1973; children: Angela D., Fred A. IV. BS in Mech. Engring., La. Tech. U., 1971, MS in Biomed. Engring., 1973. Registered profl. engr., La., W.Va. Instr. La. Tech. U., Ruston, 1973-74; design engr. Union Carbide Corp., Charleston, W.Va., 1974-80; sr. control syems engr. FB&D Industries, Inc., Monroe, 1980—. Mem. Instrument Soc. Am. (sr., pres. Monroe sect. 1983-85, 92-93), ASME. Lutheran. Office: FB&D Industries 4001 Jackson St Monroe LA 71202

COON, HILARY HUNTINGTON, psychiatric genetics educator, researcher; b. Los Alamos, N.Mex., June 6, 1961; d. James H. and Joan (Newman) C.; m. James R. Yehle. PhD, U. Colo., 1991. Software engr. CyberMedic, Louisville, Colo., 1984-85; grad. rsch. assoc. U. Colo., 1986-91; rsch. assoc. U. Utah Med. Sch., Salt Lake City, 1991-92, rsch. asst. prof., 1992—. Contbr. articles to profl. jours. Nat. Inst. of Child Health and Devel. predoctoral fellow, 1987-91; recipient NIMH Nat. Rsch. Svc. award, 1992—. Mem. AAAS, APA, Am. Soc. Human Genetics, Behavior Genetics Assn., Phi Beta Kappa. Democrat. Home: 1585 E Woodland Ave Salt Lake City UT 84106 Office: U Utah Med Sch Psychiatry Dept 50 N Medical Dr Salt Lake City UT 84132

COONEY, NED LYHNE, psychologist; b. Cleve., Jan. 3, 1955; s. Ned Joseph and Jacqueline Marie (Lyhne) C.; Judith Lynn Lifshitz, Aug. 21, 1983; children: Sarah, Daniel. BA in Psychology, SUNY, Stony Brook, 1978; PhD in Clin. Psychology, Rutgers U., 1981. Asst. prof. U. Conn., Farmington, 1985-89; dir. substance abuse treatment program VA Med. Ctr., West Haven, Conn., 1989—; assoc. prof. Dept. Psychiatry Yale U. Sch. Medicine, New Haven, Conn., 1989—. Co-author: Treating Alcohol Dependence, 1989. Named prin. investigator Nat. Inst. on Alcohol Abuse and Alcoholism, 1987—. Mem. APA, Assn. for Advancement of Behavior Therapy. Office: VA Med Ctr 116B 950 Campbell Ave West Haven CT 06516

COOPER, ALLEN DAVID, research scientist, educator; b. N.Y.C., Oct. 18, 1942; s. Samuel and Fay (Sussman) C.; m. Phyllis Butler, June 22, 1968; children: Ian, Todd. BA, NYU, 1963; MD, SUNY Downstate Med. Ctr., N.Y.C., 1967. Intern then resident Boston City Hosp., 1967-69; resident fellow in gastroenterology U. Calif., San Francisco, 1969-72; clin. asst. prof. medicine U. Tex. Med. Sch., San Antonio, 1972-74; asst. prof. medicine Stanford (Calif.) U., 1974-80, assoc. prof. medicine, 1980-89, courtesy assoc. prof. physiology, 1987-90, prof. medicine, 1990; dir. Palo Alto (Calif.) Med. Found. Rsch. Inst., 1986—. Recipient Scholastic Achievement award Am Inst. Chem., 1963; Univ. fellow Stanford U., 1981-83, Andrew W. Mellon Found. fellow, 1977-79. Fellow ACP; mem. Am. Soc. Clin. Investigation, Am. Soc. Biochemistry and Molecular Biology, Western Soc. Clin. Investigation (sec.-treas. 1988, pres. 1992), Am. Fedn. Clin. Rsch. (pres. 1974), Coyote Point Yacht Club (San Mateo, Calif.), Pi Lambda Xi, Alpha Omega Alpha. Avocation: sailing. Office: Rsch Inst Palo Alto Med Found 860 Bryant St Palo Alto CA 94301-2799*

COOPER, BENITA ANN, federal agency administrator; b. Sante Fe, May 4, 1944; d. William Warder and Sylvia Ann Abeyta; m. Robert S. Cooper, Oct. 5, 1985. BA in Polit. Sci., Earlham Coll., 1965; MA in Govt., Ind. U., 1968; MS in Mgmt., MIT, 1985. Dir. pers. Goddard Space Flight Ctr. NASA, Greenbelt, Md., 1975-77; exec. asst. to dep. adminstr. NASA, Washington, 1977-79; dep. dir. adminstrn. and mgmt. NASA, Greenbelt, Md., 1979-80, dir. mgmt. ops., 1981-88; asst. adminstr. for HQ ops. NASA, Washington, 1988-91, assoc. adminstr. for mgmt. systems and facilities, 1991—. Recipient Meritorious Exec. award Pres. of U.S., 1989. Avocation: house renovation. Office: NASA Code J 300 E St SW Washington DC 20546

COOPER, CHESTER LAWRENCE, research administrator; b. Boston, Jan. 13, 1917; s. Israel and Hannah (Levenson) C.; m. Orah Pomerance, July 23; children: Joan Laurence Gould, Susan Louise. BS, NYU, 1939, MBA, 1941; PhD, Am. U., Washington, 1960. Asst. dep. dir. CIA, Washington, 1947-62; sr. staff White House/NSC, Washington, 1962-66, U.S. Dept. State, Washington, 1966-70; dir. internat. div. Inst. Def. Analysis, Arlington, Va., 1970-72; fellow Woodrow Wilson Internat. Ctr. Scholars, Washington, 1972-75; dep. dir. Inst. Energy Analysis, Oak Ridge, Tenn., 1975-83; dep. dir., acting dir. Internat. Inst. Applied Systems Analysis, Laxenburg, Austria, 1983-85; coord. internat programs Resources for the Future, Washington, 1985-92; assoc. dir. Battelle Pacific N.W. Labs., Washington, 1992—; mem. adv. bd. Greyhaven Inst., Carmel, Calif.; cons. Aspen Inst., Sci. Policy Assocs., Washington, Screenscope Films, Washington. Author: The Lost Crusade, 1971 (award 1971), The Lion's Last Roar, 1977; editor: Growth in America, 1976, Science for Public Policy, 1987. Bd. dirs. D.C. Tb Soc., Washington, 1960-62, D.C. Common Cause, Washington, 1970, Chevy Chase (Md.) Fire Dept., 1986-91, Lester Cook Found., Washington, 1987-89. Nat. War Coll. scholar, Washington, 1952-53, Internat. Inst. Applied Systems Analyses hon. scholar, Laxemburg, Austria, 1986. Mem. Am. Com. U.S.-Soviet Rels., Coun. Fgn. Rels., Poets, Essayists, Novelists, Cosmos Club. Avocations: fishing, gardening, sculpting, 18th Century Furniture and Silver. Home: 7514 Vale St Bethesda MD 20815-4004 Office: Battelle Pacific NW Labs 1616 P St NW Washington DC 20036-1434

COOPER, DAVID ROBERT, geotechnical testing professional; b. San Francisco, July 16, 1956; s. Robert Stephen Cooper and Phyllis Ann (Frazer) Thompson; m. Elisa Ann Biagiotti, May 25, 1985; 1 child, Katelyn Ann. Cert. in advanced geotech. testing, U. Mo., Rolla, 1983. Lab. mgr. Cooper Engrs., Redwood City, Calif., 1976-87; owner, mgr. Cooper Testing Labs., Inc., Mountain View, Calif., 1987—. Mem. ASTM, Am. Coun. Ind. Labs. Office: Cooper Testing Labs Inc 1951 Colony St Mountain View CA 94043-1752

COOPER, DOUGLAS WINSLOW, research physicist; b. N.Y.C., Dec. 21, 1942; s. Michael J. and Priscilla (Taylor) C.; m. Carol Imbt, July 29, 1972 (div. June 1982); m. Tina Su Chiang, June 2, 1984; children: Theodore Chiang, Philip Chiang. AB, Cornell U., 1964; MS, Pa. State U., 1969; PhD, Harvard U., 1974. Physicist Pa. State U., State College, 1966-67, GCA/Tech., Bedford, Mass., 1973-76; asst. prof., assoc. prof. environ. physics Harvard U., Boston, 1976-83; jr. physicist IBM, Kingston, N.Y., 1964; engr. IBM, Yorktown Heights, N.Y., 1983-93; dir. contamination control Texwipe, Upper Saddle River, N.J., 1993—; mem. sci. adv. bd. Adv. Environ. Ctrl. Tech. Rsch. Ctr., U. Ill., Urbana, 1980-83; mem. exec. coun. Fine Particle Soc., Tulsa, 1985-89; plenary speaker European Aerosol Conf., Karlsruhe, Germany, 1991. Contbg. author: Handbook of Multiphase Systems, Handbook of Power Science and Technology, 1984, Handbook of Electronic Manufacturing, 1992, Aerosol Measurement, 1992; co-editor, contbg. author: Control and Dispersion of Air Pollutants, 1978; also over 100 articles. With U.S. Army, 1964-66. Mem. AAAS, Inst. Environ. Scis. (sr., working groups 1984—, W.J. Whitfield award 1990), Am. Assn. for Aerosol Rsch. (bd. dirs. 1990-93). Achievements include patents in field; research on environmental science, aerosols, applied statistics, optimization. Home: 91 Jean St Ramsey NJ 07446 Office: Texwipe 650 E Crescent Ave Upper Saddle River NJ 07458

COOPER, EUGENE BRUCE, speech-language pathologist, educator; b. Utica, N.Y., Dec. 20, 1933; s. Clements Everett and Beulah (Wetzel) C.; m. Crystal Silverman, Sept. 12, 1965; children: Philip Adam, Ivan Bruce. BS, SUNY, Geneseo, 1955; MEd, Pa. State U., 1957, DEd, 1962. Pathologist speech and lang. Franklin County Schs., Chambersburg, Pa., 1957-59; asst. prof. Ohio U., 1962-64, Pa. State U., 1964-66; program specialist U.S. Office Edn., 1966; exec. sec. sensory study sect., research and demonstrations Rehab. Services Adminstrn., HEW, Washington, 1966-67; mem. faculty U. Ala., Tuscaloosa, 1967—, prof. speech-lang. pathology, 1969—, chmn. dept. communicative disorders, dir. Speech and Hearing Ctr., 1967—; chmn. Ala. Bd. Examiners Speech Pathology and Audiology, 1979; cons.-at-large Nat. Student Speech-Lang.-Hearing Assn., 1983-88. Author: Personalized Fluency Control Therapy, 1976, Understanding Stuttering: Information for Parents, 1979, rev. 1990; (with Crystal Cooper) The Cooper Personalized Fluency Control Therapy Program, 1985; also articles. Fellow Am. Speech, Lang. and Hearing Assn. (legis. council 1971-72, 85—), Am. Fluency Assn. (mem. steering com. 1993—), Am. Speech, Lang. and Hearing Found. (chmn. adv. and devel. bd. 1988-89, trustee 1989—); mem. Council Exceptional Children (pres. div. children communication disorders 1975-76), Nat. Council Grad. Programs in Speech, Lang. Pathology and Audiology (pres. 1978-80), Nat. Council State Bds. Examiners Speech-Lang. Pathology and Audiology (pres. 1980, mem. exec. bd. 1988-91, pres. 1991), Nat. Council Communication Disorders (chmn. 1982), Nat. Alliance Prevention and Treatment Stuttering (chmn. 1984), Nat. Alliance on Stuttering (pres. 1985-86), Internat. Fluency Assn. (bd. dirs. 1991—), Nat. Rehab. Assn., Sigma Xi. Office: University of Alabama Speech & Hearing Ctr PO Box 870242 Tuscaloosa AL 35487-0242

COOPER, GERALD RICE, clinical pathologist; b. Scranton, S.C., Nov. 19, 1914; s. Robert McFadden and Viola Lavender Cooper; m. Lois Corrina Painter, Mar. 9, 1946; children: Annetta, Gerald Jr., Rodney. AB, Duke U., 1936, MA, 1938, PhD, 1939, MD, 1950. Cert. Am. Bd. Clin. Chemistry. Intern Atlanta VA Hosp., 1950-51, resident, 1951-52; rsch. assoc. Duke U. Sch. Medicine, Durham, N.C., 1939-46; chief chemistry, hematology and pathology Ctrs. for Disease Control, Atlanta, 1952-72; rsch. med. officer Ctrs. for Disease Control, Nat. Ctr. Environ. Health, Atlanta, 1973—. Author (with others) books; contbr. articles to profl. jours. Col. USPHS. Decorated commendation medal, Superior Svc. award, Disting. Svc. medal, Asst. Sec. for Health award for exceptional achievement; recipient Hektoen Silver medal AMA, 1954, Fulton County Med. Achievement award, 1954, Billings Silver medal, 1956. Mem. Am. Assn. for Clin. Chemistry (pres. 1984, bd. dirs. 1975-77, chmn. bd. editors of selected methods 1967-80, bd. editors Clin. Chemistry jour. 1970-76, Fischer award 1975, Dade Internat. award 1975, N.J. Gerulat award 1979, SE Sect. Meritorious Svc. award 1989, Outstanding Contbn. Clin. Chemistry award 1992), Internat. Fedn. Clin. Chemistry (apolipoprotein expert panel 1985), Am. Soc. Clin. Pathologists (chmn. clin. chemistry coun. 1974, Continuing Edn. award 1967, 77). Methodist. Home: 2165 Bonnevit Ct NE Atlanta GA 30345-4126 Office: Ctrs for Disease Control Chamblee 17/1103 F20 1600 Clifton Rd NE Atlanta GA 30329-4046*

COOPER, JANELLE LUNETTE, neurologist; b. Ann Arbor, Mich., Dec. 11, 1955; d. Robert Marion and Madelyn (Leonard) C.; children: Lena Christine, Nicholas Dominic. BA in Chemistry, Reed Coll., 1978; MD, Vanderbilt U., 1986. Diplomate Nat. Bd. Med. Examiners; diplomate in neurology Am. Bd. Psychiatry and Neurology. Med. technologist Swedish Hosp. Med. Ctr., Seattle, 1978-80, U. Wash. Clin. Chemistry, Seattle, 1980-82, Vanderbilt U. Hosp., Nashville, 1983-84; intern medicine Vanderbilt U. Med. Ctr., Nashville, 1986-87, resident neurology, 1987-90; instr. neurology Med. Coll. Pa., Phila., 1990-91, asst. prof., clerkship dir., 1991—, mem. curriculum com., 1990-91, vis. asst. prof., 1991—; neurologist Greater Ann Arbor Neurology Assocs., 1991-93; dir. neurological svcs. St. Francis Hosp., Escanaba, Mich., 1993—; physician MCP Neurology Assocs., Phila., 1990-91; emergency rm. physician Tenn. Christian Med. Ctr., 1989-90. Contbr. articles to Annals of Ophthalmology, Ophthalmic Surgery. Vol. Rape & Sexual Abuse Ctr., Nashville, 1988-90; mem. adminstrv. bd. Edgehill United Meth. Ch., Nashville, 1989-90; editorial bd. mem. Nashville Women's Alliance, Nashville, 1989-90. Recipient Svc. award for outstanding contbns. Rape & Sexual Abuse Ctr., 1990. Mem. AMA (Physician's Recognition award 1989-92), NOW, Am. Acad. Neurology, Am. Med. Women's Assn. (del. nat. meeting 1990), N.Y. Acad. Scis. Methodist. Achievements include first synthesis of Difluoromethanedisulfonic Acid; research on neurobehavioral disorders; on neuroendocrinology of sexual development, identity and orientation; on the history of women in medicine. Home: 519 South Eighth St Escanaba MI 49829 Office: St Francis Hosp 3401 Ludington Ave Escanaba MI 49829

COOPER, JOHN EDGAR, SR., research technician; b. Chillicothe, Ohio, Nov. 16, 1947; s. George Warren and Phyllis Ilene (McFadden) C.; m. Shirley Mae Wilson, June 11, 1967; children: John II, Kimberly, Jennifer. Grad., Paint Valley High Sch., Bainbridge, Ohio, 1966. Mfg. operator DuPont, Circleville, Ohio, 1974-81, quality control insp., 1981-86, lab. technician, 1986-89, tech. asst., 1989—; pay/progression com. DuPont, Circleville, 1990—, job evaluation team, 1988—. Min. Churches of Christ in Christian Union, Richmond Dale, Ohio, 1988—; para-profl. counselor Paint Valley-Scioto Mental Health Ctr., Chillicothe, Ohio, 1991. Mem. Electron Microscopy Soc. Am. Home: 5634 State Rt 772 Chillicothe OH 45601 Office: DuPont PO Box 89 Circleville OH 43113-0089

COOPER, JOHN EDWIN, fisheries biologist; b. Ann Arbor, Nov. 17, 1948; s. Edwin Lavern and Margaret (Simmons) C. BA, Pa. State U., 1970; MS, Frostburg (Md.) State U., 1979; AAS, Pitt Community Coll., Greenville, N.C., 1982. Biol. aide US Fish and Wildlife Svc., Laconia, N.H., 1970, Rosemont, N.J., 1971; mem. faculty, rsch. asst. U. Md., Solomons, 1971-74; field supr. Aquatic Ecology Assocs., Erie, Pa., 1974-76; assoc. scientist Ecol. Analysts Inc., Sparks, Md., 1978-80; sci. illustrator Pa. State U., University Park, 1981-82; fisheries biologist East Carolina U., Greenville, N.C., 1983-93, SUNY, Syracuse, N.Y., 1993—. Co-editor: Common Strategies of Anadromous and Catadromous Fishes, 1987; contbr. articles to profl. jours. Mem. Am. Fisheries Soc., Am. Inst. Fishery Rsch. Biologists, Estuarine Rsch. Fedn., Am. Soc. Ichthyologists and Herpetologists, Sigma Xi. Office: SUNY-ESF 1 Forestry Dr Syracuse NY 13210

COOPER, KHERSHED PESSIE, metallurgist; b. Bombay, Sept. 26, 1952; came to U.S., 1975; s. Pessie Khershedji and Coomie Pessie (Nicholson) C.; m. Michiko Dara Bhagwagar, Jan. 14, 1988; children: Vivian K., Johan K. MS, U. Wis., 1977, PhD, 1982. Rsch. scientist Olin Metals Rsch. Lab., New Haven, 1982-84; group supr. Geo-Ctrs., Inc., Fort Washington, Md., 1984-90; metallurgist Naval Rsch. Lab., Washington, 1990—; reviewer panel NASA, Washington, 1992. Author: (with others) Metals Handbook, vol. 18, 1992; contbr. articles to profl. jours. Mem. Am. Soc. for Metals (chmn. DC chpt. 1993-94, treas. 1991-93, arrangements com. 1987-91, Outstanding Young Mem. 1988-89), Am. Soc. for Metals Internat., The Metall. Soc., Zorastrian Assn. of Washington (v.p. 1991-93), Sigma Xi, Alpha Sigma Mu. Zorastrian. Achievements include discovery (with others) of smallest glass-encased metal volume in scanning electron microscope; design of novel liquid metal atomizing process. Office: Naval Rsch Lab Code 6321 Washington DC 20375

COOPER, LEON N., physicist, educator; b. N.Y.C., Feb. 28, 1930; s. Irving and Anna (Zola) C.; m. Kay Anne Allard, May 18, 1969; children: Kathleen Ann, Coralie Lauren. AB, Columbia U., 1951, AM, 1953, PhD, 1954, DSc, 1973; DSc hon. degrees; DSc, U. Sussex, Eng., 1973, U. Ill., 1974, Brown U., 1974, Gustavus Adolphus Coll., 1975, Ohio State U., 1976, U. Pierre et Marie Curie, Paris, 1977. NSF postdoctoral fellow, mem. Inst. for Advanced Study, 1954-55; rsch. assoc. U. Ill., 1955-57; asst. prof. Ohio State U., 1957-58; assoc. prof. Brown U., Providence, 1958-62, prof., 1962-66, Henry Ledyard Goddard U. prof., 1966-74, Thomas J. Watson Sr. prof. sci., 1974—; dir. Ctr. for Neural Sci., 1978-91, Inst. Brain and Neural Systems, 1991—; lectr. pub. lectures, internat. conf. and symposia; vis. prof. various univs. and summer schs.; cons. indsl., ednl. orgns.; sponser Fedn. Am. Scientists; mem. Def. Sci. Bd.; co-chair Nester Inc., assoc. Neuoscience Rsch. Program. Author: Introduction to The Meaning and Structure of Physics, 1968; Contbr. articles to profl. jours. Alfred P. Sloan Found. rsch. fellow, 1959-66, John Simon Guggenheim Meml. Found. fellow, 1965-66; recipient Nobel prize (with J. Bardeen and J.R. Schrieffer), 1972; award of Excellence, Grad. Facilities Alumni of Columbia, U., 1974; Descartes medal Acad. de Paris, U. Rene Descartes, 1977, John Jay award Columbia Coll., 1985, award for Disting. Achievement Columbia U., 1990. Fellow Am. Phys. Soc., Am. Acad. Arts and Scis.; mem. AAAS, Am. Philos. Soc., Nat. Acad. Scis. (Comstock prize with J.R. Schrieffer 1968), Soc. Neurosci, Internat. Neural Network Soc. (governing bd.), Phi Beta Kappa, Sigma Xi. Office: Brown U Dept Physics Providence RI 02912

COOPER, LEROY GORDON, JR., former astronaut, business consultant; b. Shawnee, Okla., Mar. 6, 1927; s. Leroy Gordon and Hattie Lee (Herd) C.; m. Trudy B. Olson, Aug. 29, 1947 (div.); children—Camala Keoki, Janita Lee; m. Susan Taylor, May 6, 1972; children—Elizabeth Jo, Colleen Taylor. Student, U. Hawaii, 1946-49, European extension U. Md., 1951-53; B.S. in Aero. Engring., Air Force Inst. Tech., 1956; grad., Exptl. Test Pilot Sch., USAF, 1957. Commd. lt. USAF, 1949, advanced through grades to col., 1965; jet fighter pilot, 1950-54, pilot exptl. flight test engring., 1957-59; astronaut with Project Mercury, NASA, from 1959; piloted Mercury-Atlas 9 mission (last Mercury flight), 1963; command pilot Gemini V, 1965; worked with Apollo lunar program, until 1970; v.p. R & D W.E.D. Enterprises, 1974-80; pres. XL, Inc., Beverly Hills, Calif., from 1980, Galaxy Group Inc., 1988-93. Recipient Harmon trophy; Collier trophy. Mem. AIAA, Soc. Exptl. Test Pilots, Am. Astronautical Soc. Lodge: Masons (Shriner, Jester). Address: 7135 Haven Hurst Van Nuys CA 91406

COOPER, MARY CAMPBELL, information services executive; b. Meadville, Pa., Aug. 14, 1940; d. Paul F. and Margaret (Webb) Campbell; m. James Nicoll Cooper, June 8, 1963; children: Alix, Jenny. BA, Mt. Holyoke Coll., 1961; MLS, Simmons Coll., 1963; MEd, Harvard U., 1965. Cert. museum adminstrn. With Harvard U. Libr., Cambridge, Mass., 1961-63, Carleton U. Libr., Ottawa, Can., 1965-85; archive cons. U.S., Can., 1985-86; info. mgr. Haley & Aldrich Inc., Cambridge, 1986-88, Tsoi/Kobus & Assocs., Cambridge, 1988-90; pres., founder Cooper Info., Cambridge, 1990—; pres. Mass. Com. for Preservation of Archtl. Records, Boston, 1991—, past bd. dirs. Author: Records In Architectural Offices, 1992. Mem. Berkshire County Hist. Commn., Pittsfield, Mass., 1990—. Travel grantee Nat. Hist. Pub. Records Commn., 1991. Mem. Spl. Librs. Assn., Am. Museums Assn., Assn. of Ind. Info. Profls., Nat. Govt. Archivists and Records Adminstrs., Am. Soc. Info. Sci., Assn. Records Mgrs. and Adminstrs. (nat. com. 1991—). Avocations: travel, tennis, swimming. Home and office: 5 Ellery Pl Cambridge MA 02138

COOPER, MAX DALE, physician, medical educator, researcher; b. Hazelhurst, Miss., Aug. 31, 1933; s. Ottis Noah and Lily (Carpenter) C.; m. Rosalie Lazzara, Feb. 6, 1960; children: Owen Bernard, Melinda Lee Cooper Holladay, Michael Kane, Christopher Byron. Student, Holmes Jr. Coll., 1951-52, U. Miss., 1952-54; postgrad., U. Miss. Med. Sch., 1954-55; MD, Tulane U., 1957. Diplomate Am. Bd. Pediatrics. Intern Saginaw (Mich.) Gen. Hosp., 1957-58; resident dept. pediatrics Tulane Med. Sch., New Orleans, 1958-60; house officer Hosp. for Sick Children, London, 1960, rsch. asst. dept. neurophysiology, 1961; allergy fellow dept. pediatrics U. Calif. Med. Ctr., San Francisco, 1961-62; instr. Tulane Med. Sch., New Orleans, 1962-63; med. fellow specialist U. Minn., Mpls., 1963-64, instr., 1964-66; asst. prof. dept. pediatrics U. Ala., Birmingham, 1967—, prof. dept. pediatrics, 1967—, assoc. prof. dept. microbiology, 1967-71, dir. rsch. Rehab. Rsch. and Tng. Ctr., 1968-70, prof. dept. microbiology, 1971—, dir. Cell. Identification Lab., 1987—, dir. Ctr. Interdisciplinary Rsch. in Immunological Diseases, 1987—, dir. Div. Devel./Clin. Immunology, 1987—, prof. dept. medicine, 1987—, investigator Howard Hughes Med. Inst., 1988—; sr. scientist Comprehensive Cancer Ctr. U. Ala., Birmingham, 1971—, Multipurpose Arthritis Ctr., 1979—, Cystic Fibrosis Rsch. Ctr., 1981—; dir. Cellular Immunobiology Unit of Tumor Inst. U. Ala., Birmingham, 1976-87; vis. scientist tumor immunology unit dept. zoology, U. Coll. London, 1973-74, Inst. D'Embryologie, Nogent-Sur-Marne and Inst. Pasteur, Paris, 1984-85. Co-author: Acute Hemiplegia in Childhood, 1962, Ontogeny of Immunity, 1967, Immunologic Incompetence, 1971, Immunodeficiency in Man and Animals, 1975, numerous others; mem. editorial bds. Immunology Today, 1986, Immunodeficiency Revs., 1987—, Clin. Immunology and Immunopathology, 1987-90, Internat. Immunology, 1988—; assoc. editor Jour. Immunology, 1972-76, Arthritis and Rheumatism, 1985-90, Jour. Clin. Immunology, 1979—; co-editor Seminars in Immunopathology, 1988-91; editor Current Topics in Microbiology and Immunology, 1981—; contbr. over 450 articles to profl. jours. Faculty rsch. assoc. Am. Cancer Soc., 1966-71; mem. bd. sci. advisors St. Jude Hosp., Memphis, 1981-84, 91—, Becton-Dickinson Monoclonal Antibody Ctr., 1980-90; mem. med. adv. com. Immune Deficiency Found., 1981—; mem. bd. sci. counselors Nat. Cancer Inst., Bethesda, Md., 1982-86, Nat. Inst. Allergy and Infectious Diseases, 1978-82, 91—, Inst. Merieux, Lyons, France, 1985-90, Med. Biology Inst., La Jolla, Calif., 1986; mem. internat. sci. adv. bd. Basel (Switzerland) Inst. Immunology, 1987-91; mem. U.S.-Japan Panel in Immunology, 1980-91, NIH Immunobiology Study Sect., 1974-78; trustee Leukemia Soc. Am., 1983-88. Special Postdoctoral Rsch. fellow USPHS, 1964-66; recipient Teaching Traineeship award Nat. TB Assn., 1962-63, Samuel J. Meltzer Founder's award Soc Exptl. Biology and Medicine, 1966, Life Scis. award 3M, 1990, Sandoz Prize for Immunology, 1990. Mem. NAS, AAAS, AAUP, Am. Assn. Immunologists (pres. 1988-89, councilor 1983-86, chmn. mem. com. 1974-77), Am. Soc. Exptl. Pathology, Am. Soc. Clin. Investigation, Am. Assn. Cancer Rsch., Am. Acad. Pediatrics, Am. Pediatric Soc., Fedn. Am. Scientists, Med. Assn. State Ala., Internat. Soc. Devel. and Comparative Immunology, Soc. Francaise d'Immunologie (life Membre d'Honneur), Soc. Pediatric Rsch. (v.p. 1978), So. Soc. Pediatric Rsch. (pres. 1975), Cen. Soc. Clin. Rsch., Jefferson County Med. Assn., Clin. Immunology Soc., Am. Acad. Scis., Inst. Medicine, Am. Acad. Arts and Scis., Soc. Mucosal Immunology, Alpha Omega Alpha, Sigma Xi. Achievements include research in developmental immunobiology with emphasis on B cell and T cell differentiation, in clinical immunology with emphasis on immunodeficiency diseases and lyhmphoid malignancies. Office: U Ala Sch Medicine Dept Pediatrics 1717 7th Ave S Birmingham AL 35294 also: Howard Hughes Medixth Ave S #378 Birmingham AL 35294*

COOPER, PETER SEMLER, physicist; b. Paterson, N.J., Nov. 7, 1949; s. Harry and Ruth (Semler) C.; m. Irene Gail Racker, Aug. 16, 1970. BA, Rutgers U., 1970; PhD in Physics, Yale U., 1975. Assoc. prof. physics Yale U., 1979-88; asst. head, computing div. Fermi Nat. Accelerator Lab., Batavia, Ill., 1988-93. Contbr. articles to profl. jours. Mem. Am. Phys. Soc. Office: Fermi Nat Accelerator Lab PO 500 MS 122 Batavia IL 60510

COOPER, REGINALD RUDYARD, orthopaedic surgeon, educator; b. Elkins, W.Va., Jan. 6, 1932; s. Eston H. and Kathryn (Wyatt) C.; m. Jacqueline Smith, Aug. 22, 1954; children—Pamela Ann, Douglas Mark, Christopher Scott, Jeffrey Michael. B.A. with honors, W.Va. U., 1952, B.S., 1953; M.D., Med. Coll. Va., 1955; M.S., U. Iowa, 1960. Diplomate Am. Bd. Orthopedic Surgeons (examiner 1968—). Orthopedic surgeon U.S. Naval Hosp., Pensacola, Fla., 1960-62; assoc. in orthopedics U. Iowa Coll. Medicine, Iowa City, 1962-65; asst. prof. orthopaedics U. Iowa Coll. Medicine, 1965-68, assoc. prof. orthopedics, 1968-71, prof. orthopedics, 1971—, chmn. orthopedics, 1973—; research fellow orthopedic surgery Johns Hopkins Hosp., Balt., 1964-65; exchange fellow to Britain for Am. Orthopedic Assn., 1969. Trustee Nat. Easter Seals Research Found., 1977-81, chmn., 1979-81. Served to lt. comdr. USNR, 1960-62. Mem. Iowa, Johnson County med. socs., Orthopedic Research Soc. (sec.-treas. 1970-73, pres. 1974-75), Am. Acad. Orthopedic Surgeons (Kappa Delta award for outstanding research in orthopedics 1971), Canadian, Am. Orthopedic assns., N.Y. Acad. Sci., Assn. Bone and Joint Surgeons, AMA, Am. Rheumatism Assn., Am. Acad. Cerebral Palsy, Am. Acad. Orthopedic Surgeons (chmn.

exams. com. 1978-82, sec. 1982, 2d v.p. 1985-86, 1st v.p. 1986-87, pres. 1987-88). Home: 201 Ridgeview Ave Iowa City IA 52246-1625 Office: U Iowa Hosps & Clinics 450 Newton Rd Iowa City IA 52242

COOPER, RICHARD ARTHUR, oceanographer; b. Ann Arbor, Mich., July 3, 1936; s. Gerald Paul and Alma Elizabeth (Hosner) C.; B.S., U. Mich., 1958; M.S., U. R.I., 1960, Ph.D., 1965; m. Patricia W. Davis, Aug. 13, 1960; children—Christopher, Jeffrey, Wendelyn, Catherine. Program leader lobster biology and ecology Nat. Marine Fisheries Service, Boothbay Harbor, Maine, 1965-73, oceanographer, Woods Hole, Mass., 1973-85, div. chief manned undersea research and tech., 1973-85, diving officer, 1965-85; prof. marine scis. U. Conn.-Groton, 1985—, dir. nat. undersea research program, 1985—; dir. marine scis. inst. U. Conn.-Avery Point. Bd. dirs., instr. scuba diving YMCA, Boothbay Harbor. Cert. aquanaut, U.S. Navy, NOAA; recipient various awards and grants in field. Mem. Am. Fisheries Soc., Am. Inst. Fishery Biologists. Republican. Methodist. Author textbooks in field; contbr. articles to profl. jours. Home: PO Box 89 North Stonington CT 06359-0089 Office: Univ of Conn at Avery Point Marine Sciences Institute Groton CT 06340*

COOPER, ROBERT MICHAEL, nuclear energy industry specialist; b. Little Rock, Nov. 27, 1948; s. John William and Rachel Lou Ann (Merritt) C.; m. Beverly Jean Wiles, Aug. 25, 1972; 1 child, Kristen Amanda. AA in Gen. Studies, Ark. Tech. U., 1984; BS in Indsl. Tech., So. Ill. U., 1990. Electrician helper Ark. Power and Light Co., Pine Bluff, 1971-75, journeyman electrician Ark. Nuclear One, 1975-77, relay engr. Ark. Nuclear One, 1977-79; sr. quality assurance engr. Ark. Nuclear One, Russellville, 1979-90, nuclear safety and licensing specialist, 1990—. Author: Quality Surveillance Handbook, 1991; contbr. articles to Am. Nuclear Soc., Am. Soc. Quality Control. Lt. London (Ark.) Rural Vol. Fire Dept., 1985—. With USN, 1967-71. Mem. Am. Nuclear Soc., Am. Soc. Quality Control (cert. quality engr. 1982, vice chair quality verification subcom. 1988-90, chair bylaws com. 1990). Baptist. Home: RR 2 Box 250 Russellville AR 72801-9553

COOPER, STEPHEN RANDOLPH, chemistry educator and industrial research manager; b. Bethesda, Md., Jan. 9, 1950; s. Milton Bernard and Eva (Korolishin) C.; m. Trilby Addison, Sept. 30, 1989. BA, U. Calif. San Diego, La Jolla, 1971; PhD, U. Calif., Berkeley, 1976. Postdoctoral researcher U. Calif., Berkeley, 1976-79; asst. prof. Harvard U., Cambridge, Mass., 1979-84, assoc. prof., 1984; univ. lectr. Oxford (Eng.) U., 1984-92; dir. discovery rsch. Mallinckrodt Med.,Inc., Hazelwood, Mo., 1992—. Editor: Crown Compounds: Towards Future Applications: contbr. articles to profl. jours.; patentee in field. Mem. Am. Chem. Soc., Chem. Soc. Japan, Royal Soc. Chemistry, Soc. Nuclear Medicine. Office: Mallinckrodt Med Inc 675 McDonnell Blvd Saint Louis MO 63134

COOPER, STUART LEONARD, chemical engineering educator, researcher, consultant; b. N.Y.C., Aug. 28, 1941; s. Jacob and Anne (Bloom) C.; m. Marilyn Portnoy, Aug. 29, 1966; children: Gary, Stacey. B.S., MIT, 1963; Ph.D., Princeton U., 1967. Asst. prof. chem. engring. U. Wis., Madison, 1967-71, assoc. prof., 1971-74, prof., 1974, chmn. dept., 1983-89, 92, Paul A. Elfers prof., 1989-93; dean, H. Rodney Sharp prof. Coll of Engring. U. Del., Newark, 1993—; vis. assoc. prof. U. Calif.-Berkeley, 1974; vis. prof. Technion, Haifa, Israel, 1977; cons. in field; trustee Argonne Univs. Assn., Argonne Nat. Lab., 1975-81. Editor: Multiphase Polymers, 1979, Biomaterials: Interfacial Phenomena and Applications, 1982; author: Polyurethanes in Medicine, 1986; contbr. numerous articles in field to profl. jours. Lady Davis fellow, 1977. Fellow AIChE (Materials Engring. and Sci. div. award 1987), Am. Phys. Soc., Am. Inst. Med. and Biol. Engrs. (founding); mem. Am. Chem. Soc. (Best Paper award 1976), Am. Soc. Artificial Internal Organs, Soc. Biomaterials (Clemson award for basic rsch. 1987). Office: U Del Coll of Engring Newark DE 19716

COOPER, WILLIAM ALLEN, JR., audiologist; b. Detroit, Aug. 16, 1932; s. William Allen and Ida Louise (Ford) C.; m. Auguste Ingrid Schneider, Oct. 5, 1957; children: Ingrid Louise, Robert William, James Allen. BS, Wayne State U., 1957; PhD, Okla. U., 1964. Chief audiology VA Hosp., Oklahoma City, 1963-71; asst. prof. U. Okla., Oklahoma City, 1964-71; assoc. prof. Purdue U., West Lafayette, Ind., 1971-81; prof. U. S.C., Columbia, 1981—; vis. fellow Inst. Sound & Vibration Rsch., Southampton, U.K., 1979. Contbr. articles to profl. jours. Chmn. bd. dirs. Unitarian Fellowship, Lafayette, Ind., 1975, 88-89; bd. dirs. Unitarian Ch., Oklahoma City, 1969-70. With U.S. Army, 1954-56, Germany. Am. Speech and Hearing Assn. fellow, 1978; recipient Honors of Assn. award S.C. Speech and Hearing Assn., 1989. Mem. AAAS, Internat. Soc. Audiology, Am. Acad. Audiology, Am. Auditory Soc. Unitarian. Home: 6503 Christie Rd Columbia SC 29209

COOPER, WILLIAM EUGENE, consulting engineering executive; b. Erie, Pa., Jan. 11, 1924; s. William Hall and Ruth E. (Dunn) C.; m. Louise I. Ferguson, June 23, 1946; children: Margaret, Glenn, Keith, Joyce, Carol. Student, Stevens Inst. Tech., 1941-43; B.S., Oreg. State Coll., 1947, M.S., 1948; Ph.D., Purdue U., 1951. Instr. Purdue U., 1948-52; cons. engr. Knolls Atomic Power Lab., GE, 1952-63; engring. mgr. Lessells and Assos., Waltham, Mass., 1963-68; sr. v.p., tech. dir. Teledyne Materials Rsch., Waltham, Mass., 1968-76; cons. engr. Teledyne Engring. Svcs., Woburn, 1976—. Contbr. articles to tech. jours. Served with AUS, 1943-46. Named Distinguished Engring. Alumnus Purdue U., 1974. Mem. ASME (hon., B.F. Langer Nuclear Codes and Standards award 1978, hon. mem. boiler and pressure vessel com. 1980, v.p. codes and standards and mem. exec. com. of council 1980-81, sr. v.p. codes and standards 1981-84, Pressure Vessel and Piping medal 1983, codes and standards medal 1986), NAE, Soc. Exptl. Mechanics (Murray lectr. 1977), Am. Nat. Standards Inst. (dir. 1981-84), Sigma Xi, Pi Tau Sigma, Sigma Pi Sigma. Achievements include development of comprehensive design criteria for pressure vessels and piping in critical services. Home and Office: 1010 Waltham St # 352C Lexington MA 02173-8044

COOPER, WILLIAM JAMES, chemist; b. Rochester, N.Y., Dec. 1, 1945; s. Raymond Burton and Florence Catherine (Bush) C.; children: Jonathan Edwards, Derek William. BS, Allegheny Coll., 1968; MS, The Pa. State U., 1971; PhD, U. Miami, 1987. Research chemist U.S. Army Med. Bioengineering Research & Devel. Lab., Frederick, Md., 1972-80; assoc. prof. Fla. Internat. U., Miami, 1980-90, prof., 1990—. Office: Fla Internat U Drinking Water Rsch Ctr Univ Park Miami FL 33199

COOPER, WILLIAM THOMAS, natural history artist; b. Adamstown, NSW, Australia, Apr. 6, 1934; s. William and Coral (Bird) C.; m. Wendy Elizabeth Price, June 25, 1979. Exhibited in group shows at Artarmon Galleries, Sydney, 1973, City of Newcastle Art Gallery, 1973, Artarmon Galleries, Sydney, 1980, Leigh Yawkey Woodson Art Mus., Wis., 1986; represented in permanent collections Woodhall Art Found., Australian Nat. Libr., Papua New Guinea Govt., Newcastle Art Gallery, Rockhampton City Art Gallery; work represented in, A Portfolio of Australian Birds, 1968, The Birds of Paradise and Bowerbirds, 1977, Kingishers & Related Birds, vol. I, 1983, vol. II, 1985, vol. III, 1987, Parrots of the World, 1988, Visions of a Rainforest, 1992; illustrator, Fierce Encounter, 1970; designer stamps Papua New Guinea, 1972, 73. Recipient Gold medal Distinction Natural History Art Acad. Natural Scis. Phila., 1992. Office: Malanda 4885, PO Box 314, Queensland Australia

COOPERRIDER, TOM SMITH, botanist; b. Newark, Ohio, Apr. 15, 1927; s. Oscar Harold and Ruth Evelyn (Smith) C.; m. Miwako Kunimura, June 13, 1953; children: Julie Ann, John Andrew. B.A., Denison U., 1950; M.S., U. Iowa, 1955, Ph.D. (NSF fellow), 1958. Instr. biol. scis. Kent State U., Ohio, 1958-61; asst. prof. Kent State U., 1961-65, asso. prof., 1965-69, prof., 1969-93, dir. exptl. programs, 1972-73, emeritus prof., 1993—, curator herbarium, 1968—; dir. Bot. Gardens and Arboretum, 1972-93; mem. editorial bd. Univ. Press, 1976-79; on leave as asst. prof. botany U. Hawaii, 1962-63; NSF researcher Mountain Lake Biol. Sta., U. Va., summer 1958; faculty mem. Iowa Lakeside Lab., U. Iowa, summer 1965; cons. endangered and threatened species U.S. Fish and Wildlife Service, Dept. Interior, 1976-83; cons. Davey Tree Expert Co., 1979-85, Ohio Natural Areas Council, 1983. Author: Ferns and Other Pteridophytes of Iowa, 1959, Vascular Plants of Clinton, Jackson and Jones Counties, Iowa, 1962; editor: Endangered and Threatened Plants of Ohio, 1983. Active YMCA-YWCA Students in Govt., Washington, 1950; personnel placement U.S. Census Bur., Washington, 1950-51; Quaker Internat. Vol., Fed. Republic Germany, 1951. Served with U.S. Army, 1945-46. NSF predoctoral fellow, 1957-58; NSF research grantee, 1965-72. Fellow AAAS, Ohio Acad. Sci. (v.p. 1967), Explorers Club; mem. Am. Soc. Plant Taxonomists, Internat. Assn. Plant Taxonomists, Bot. Soc. Am., Nature Conservancy, Wilderness Soc., So. Appalachian Bot. Soc., Blue Key, Sigma Xi. Home: 548 Bowman Dr Kent OH 44240-4512

COOPERSMITH, BERNARD IRA, obstetrician/gynecologist, educator; b. Chgo., Oct. 19, 1914; s. Morris and Anna (Shulder) C.; m. Beatrice Klass, May 26, 1940; children: Carol, Cathie. BS cum laude, U. Ill., 1936, MD cum laude, 1938. Diplomate Am. Bd. Ob-Gyn. Intern Michael Reese Hosp., Chgo., 1938-39, resident in ob-gyn, 1939-42; practice medicine specializing in ob-gyn Chgo., 1942—; mem. staff Prentice Women's Hosp. of Northwestern Meml. Hosp., Michael Reese Hosp., Mt. Sinai Hosp., Chgo. Maternity Ctr.; asst. prof. ob-gyn Northwestern U. Med. Sch., Chgo., 1948—. Contbr. articles to profl jours. Pres. Barren Found. Chgo., 1971-73. Fellow ACS; mem. AMA, Ill. Med. Soc., Chgo. Med. Soc., Chgo. Gynecol. Soc., Cen. Assn. Ob-Gyn, Am. Coll. Ob-Gyn, Alpha Omega Alpha. Jewish. Clubs: Bryn Mawr Country, Carleton. Home: 1110 N Lake Shore Dr Chicago IL 60611-1054 Office: 680 N Lake Shore Dr Ste 1030 Chicago IL 60611-4402

COORS, WILLIAM K., brewery executive; b. Golden, Colo., Aug. 11, 1916. BSChemE, Princeton U., 1938, MSChemE, 1939. Pres. Adolph Coors Co., Golden, Colo., from 1956, Chmn. bd., 1970—, also corp. pres. Office: Adolph Coors Co BC350 Golden CO 80401

COPES, JOHN CARSON, III, mechanical engineer; b. Baton Rouge, Sept. 6, 1923; s. John Carson Jr. and Beatrix (Lyons) C.; m. Edith Estelle Givens, Mar. 4, 1944; 1 child, John Carson IV. BSME, La. State U., 1947. Mech. design engr. Esso Rsch. Labs., Baton Rouge, 1947-50; design engr. Ethyl Corp., Baton Rouge, 1950-56; head mech. sect. F.R. Harris Engrs., New Orleans, 1956-61; chief engr. A.G. Keller Inc., Baton Rouge, 1961-65; cons. engr. pvt. practice, Baton Rouge, 1965—. State rep. SAR, La., 1991; nat. pres. 14th Armed Div. Assn., 1989; comdr. Baton Rouge Chpt. Mil. Order World Wars, 1991-92. 2d lt. U.S. Army, 1942-45. Fellow ASME; mem. SAR (pres. 1991-92, Patriot medal 1991), Sons of Rep. Tex., Fost Landry Post No. 2 Amvets (comdr. 1987-88, Outstanding Amvet 1992). Democrat. Epsicopalian. Achievements include 4 patents on split mechanical seals. Home: 2750 McConnell Dr Baton Rouge LA 70809-1113 Office: 1956 Wooddale Ct Baton Rouge LA 70806-1526

COPES, PARZIVAL, economist, researcher; b. Nakusp, B.C., Can., Jan. 22, 1924; s. Jan Coops and Elisabeth Catharina Coops-van Olst; m. Dina Gussekloo, May 1, 1946; children: Raymond Alden, Michael Ian, Terence Franklin. BA, U. B.C., 1949, MA, 1950; PhD, London Sch. Econs., 1956; D of Mil. Sci., Royal Roads Mil. Coll., Victoria, B.C., Can., 1991; D (honoris causa), U. Tromsö, Norway, 1993. Economist, statistician Dominion Bur. of Stats., Ottawa, Can., 1953-57; from assoc. prof. to prof. Meml. U. Nfld., St. John's, Can., 1957-64; prof. Simon Fraser U., Burnaby, B.C., Can., 1964-91, head dept. econs. and commerce, 1964-69, chmn. dept. econs. and commerce, 1972-75, dir. Ctr. for Can. Studies, 1978-85, founding dir. Inst. of Fisheries Analysis, 1980—, prof. emeritus, 1991—; governor Inst. Can. Bankers, Montreal, Que., 1967-71; dir. "Can.-Fgn. Arrangements Project," Can. Govt. Dept. Environment, 1976; pres., chmn. Pacific Regional Sci. Conf. Orgn., 1977-85; assoc. Oceans Inst. Can., Halifax, N.S., 1992—. Author: St. John's and Newfoundland: An Economic Survey, 1961, The Resettlement of Fishing Communities in Newfoundland, 1972. Lt. Can. Army, 1945-46, 50-51. Fgn. fellow Acad. Natural Scis. of Russia, Moscow, 1992. Mem. Internat. Inst. Fisheries Econs. and Trade (exec. com. 1982-86), Internat. Assn. for Study of Common Property, Can. Regional Sci. Assn. (pres. 1983-85), Can. Econs. Assn. (v.p. 1972-73), Assn. for Can. Studies, Western Regional Sci. Assn. (pres. 1977-78), Social Sci. Fedn. Can. (dir., v.p. 1979-83), Can. Assn. Univ. Tchrs. Achievements include one of earliest research contributions to establish sub-discipline of fisheries economics; research and international consulting in fisheries resource management. Home: Simon Fraser University, 2341 Lawson Ave, West Vancouver, BC Canada V7V 2E5 Office: Simon Fraser U, Inst of Fisheries Analysis, Burnaby, BC Canada V5A 1S6

COPPER, CHRISTINE LEIGH, chemist; b. Camden, N.J., Nov. 18, 1969; d. Walter Logan Jr. and Barbara Dena (VanGulik) C. BS, Mary Washington Coll., 1991. Grad. teaching asst. U. Tenn., Knoxville, 1991—. Sec. Knox Area Literary Coun., Knoxville, 1992-93. Mem. Am. Chem. Soc., Alpha Phi Sigma (pres. 1990-91), Chi Beta Phi. Home: 3636 Taliluna Ave # 212 Knoxville TN 37919 Office: Univ Tenn Dept Chemistry 575 Buehler Hall Knoxville TN 37996

CORAN, ARNOLD GERALD, pediatric surgeon, educator; b. Boston, Apr. 16, 1938; s. Charles and Anne (Cohen) C.; m. Susan Williams, Nov. 17, 1960; children: Michael, David, Randi Beth. B.A. cum laude, Harvard U., 1959, M.D. cum laude, 1963. Diplomate Am. Bd. Surgery, Am. Bd. Thoracic Surgery, Am. Bd. Pediatric Surgery, Am. Bd. Surg. Critical Care. Intern Peter Bent Brigham Hosp., Boston, 1963-64; resident in surgery Peter Bent Brigham Hosp., 1964-68, chief surg. resident, 1969; resident in surgery Children's Hosp. Med. Center, Boston, 1965-66; sr. surg. resident Children's Hosp. Med. Center, 1966, chief surg. resident, 1968; instr. surgery Harvard, Cambridge, Mass., 1967-69; asst. clin. prof. surgery George Washington U., 1970-72; head physician pediatric surgery Los Angeles County-U. So. Calif. Med. Center, 1972-74; asst. prof. surgery U. So. Calif., 1972-73, assoc. prof., 1973-74; prof. surgery U. Mich., Ann Arbor, 1974—; head sect. pediatric surgery U. Mich. Hosp., 1974—; Surgeon-in-chief Mott Children's Hosp. Editor: Pediatric Surgery Internat.; contbr. over 130 articles to profl. jours. Bd. dirs. U. Fla. Found., 1989-91. Served to lt. comdr. MC AUS. Fellow ACS; mem. Am. Acad. Pediatrics, Am. Surg. Assn., Soc. Univ. Surgeons, Am. Pediatric Surg. Assn., Western, Central surg. assns. Achievements include early sci. contbns. to reentry aerodynamics, contbns. to modelling blood flow, early studies using time dependent blood rheology expts. for input to constitutive modelling, original contbns. to fluid dynamics of artificial hearts, principle of end diastolic washing to allow use of smooth blood contacting surgaces, prosthetic valve fluid dynamics and initial laser Doppler studies of unsteady biofluid dynamics. Home: 505 E Huron St Apt 802 Ann Arbor MI 48104-1541 Office: Pediatric Surgery Assocs Pediatric Surgery Rsch Lab L2110 Maternal-Child Health Ctr Ann Arbor MI 48109-0245

CORBATO, FERNANDO JOSE, electrical engineer and computer science educator; b. Oakland, Calif., July 1, 1926; s. Hermenegildo and Charlotte (Jensen) C.; student UCLA, 1943-44; B.S., Calif. Inst. Tech., 1950; Ph.D., MIT, 1956; m. Isabel Blandford, Nov. 24, 1962 (dec. July 7, 1973); children: Carolyn Suzanne, Nancy Patricia; m. Emily Susan Gish, Dec. 6, 1975; stepchildren: David Lawrence Gish, Jason Charles Gish. With Computation Center, MIT, Cambridge, 1955-66, dep. dir., 1963-66, head computer systems research group of project MAC, 1963-72, co-head systems research div., 1972-73, co-head automatic programming div., 1972-73, faculty mem., 1962—, prof. elec. engring. and computer sci., 1965—, assoc. dept. head computer sci. and engring., 1974-78, 83-93, Cecil H. Green prof. computer sci. and engring., 1978-80, dir. computing and telecommunication resources, 1980-83, Ford prof. engring, 1993—; computer sci. and engring. bd. Nat. Acad. Sci., 1971-73. Served with USNR, 1944-46. Recipient Harry Goode Meml. award Am. Fedn. Info. Processing Socs., 1980. Fellow IEEE (W.W. McDowell award 1966; Computer Pioneer award IEEE Computer Soc. 1982), AAAS; mem. Nat. Acad. Engring., Am. Acad. Arts and Scis., Assn. Computing Machinery (council 1964-66, A.M. Turing award 1990), Am. Phys. Soc., Sierra Club, Sigma Xi. Co-author: The Compatible Time Sharing System, 1963; Advanced Computer Programming, 1963. Home: 88 Temple St Newton MA 02165-2307 Office: Room 524 545 Technology Sq Cambridge MA 02139

CORBEN, HERBERT CHARLES, physicist, educator; b. Portland, Dorset, Eng., Apr. 18, 1914; came to U.S., 1946, naturalized, 1950; s. Harold Frederick and Margaret (Hart) C.; m. Beverly Balkum, Oct. 25, 1957; children: Deirdre Corben McGowan, Sharon, Gregory. B.A., U. Melbourne, 1933, B.Sc., 1934, M.A., 1936, M.Sc., 1936; Ph.D., Cambridge U., 1939. Lectr. math. and physics New Eng. U. Coll., Armidale, Australia, 1941; lectr. math., physics U. Melbourne, Australia, 1942-46; acting dean Trinity Coll., Melbourne, 1942-46; asso. prof. Carnegie Inst. Tech., 1946-51, prof., 1951-56; part-time lectr. physics U. Pitts., 1947; Fulbright vis. prof. U. Genoa, Milan, and Bologna, 1951-53; part-time lectr. physics U. So. Calif., 1957-58; asso. dir. Research Lab. Ramo-Wooldridge Corp. and Space Tech. Labs, Inc., Los Angeles, 1956-60; dir. Quantum Physics Lab., 1961-68; chief scientist Phys. Research Center, 1966-68; distinguished vis. prof. physics Queens Coll., 1968; acting dean faculties Cleve. State U., 1968-69, dean faculties, 1969-70, v.p. acad. affairs, 1970-72, dean Coll. Grad. Studies, prof. physics, 1968-72; prof. physics Scarborough Coll., U. Toronto, 1972-78, vis. prof., 1980-82, chmn. phys. scis. group, 1972-76; faculty Harvey Mudd Coll., Claremont, Calif., 1978-80, scholar-in-residence, 1982-85, sr. prof., 1985-88; Commonwealth Fund fellow U. Calif. and; Princeton U., 1939-41. Author: Classical and Quantum Theories of Spinning Particles, 1968, The Struggle to Understand, A History of Human Wonder and Discovery, 1992; co-author: Classical Mechanics, 1950, 2d edit., 1960, internat. edit., 1974; contbr. to International Dictionary of Physics and Electronics, 1956. Mem. Am. Phys. Soc., Am. Soc. Physics Tchrs. Home: 4304 Oleary Ave Pascagoula MS 39581-2352

CORBETT, GERARD FRANCIS, electronics executive; b. Phila., Apr. 6, 1950; s. Eugene Charles and Dolores Marie (Hoffmann) C.; m. Marcia Jean Serafin, July 9, 1983; 1 child, Daniel Gerard. AA, Community Coll. of Phila., 1974; BA in Pub. Rels., San Jose State U., 1977. Sci. programmer Sverdrup Inc., NASA Ames Rsch. Ctr., Moffett Field, Calif., 1970-77; sr. writer Four-Phase Systems, Inc., Cupertino, Calif., 1977-78; with Nat. Semicondr. Corp., Santa Clara, Calif., 1978-79; sr. account exec. Creamer Dickson Basford, Providence, 1979-81; mgr. tech. and exec. communications Internat. Harvester Co., Chgo., 1981-82; mgr. corp. tech. communications Gould Inc., Rolling Meadows, Ill., 1982-83, dir. corp. pub. rels., 1983-86, bd. dir. corp. communications 1986-89; sr. corp. communications exec., ASARCO Inc., N.Y.C., 1989—; pub. rels. and communications cons. on high tech. Recipient Vice Presdl. award of honor Calif. Jaycees, 1977. Mem. AIAA, Nat. Investor Rels. Inst., Pub. Rels. Soc. Am. (accredited, Pres.'s citation 1981), Am. Mining Congress (communications com.), LEAD Industries Assn. (chmn. pub. affairs com.), Nat. Assn. Sci. Writers, Internat. Coun. on Metals and the Environ. (comm. adv. panel), Advanced Lead Acid Battery Consortium (public affairs and mktg. com.), Kappa Tau Alpha. Republican. Roman Catholic. Clubs: Capital Hill, Commonwealth of Calif., Meadow (Rolling Meadows, Ill.). Home: 48 Woodland Ave Glen Ridge NJ 07028-1232 Office: ASARCO Inc 180 Maiden Ln New York NY 10038-4925

CORBETT, JULES JOHN, microbiology educator; b. Natrona, Pa., Apr. 12, 1919; s. Anthony and Theodosia (Kuczynski) C.; m. Gabrielle Ann Wengel, June 24, 1950; children: Brian Lee, Alan Jeffrey, Christine Marie. AA, North Park Coll., 1941; cert. med. tech. Franklin Sch. Sci. and Arts, 1946; SB, U. Chgo., 1950; MS, Ill. Inst. Tech., 1957. Med. technologist St. Bernard Hosp., Chgo., 1947-49; microbiologist Englewood Hosp., Chgo., 1949-52; dir. labs. Beverley Arts Bldg., Chgo., 1954; microbiologist Borden Co., Hammond, Ind., 1955-64; instr. Roosevelt U., Chgo., 1956-59, asst. prof., 1959-64, assoc. prof., 1964-72, prof. biology, 1972-87, prof. emeritus, 1987—, chmn. biology dept., 1974-79; cons. Metro Labs., Chgo. Served with U.S. Army, 1937-39, with USNR, 1941-45. HEW grantee, 1969-72, Ill. Bd. Higher Edn. grantee, 1971-72. Mem. Am. Soc. Microbiology, Ill. Soc. for Microbiology, AAAS, N.Y. Acad. Scis., Ill. State Acad. Sci., Am. Legion (4th dist. comdr. 1975, dep. comdr. 1988), La Societe des 40 Hommes et 8 Chevaux Chef de Gare, Navy League, VFW, Sigma Xi. Republican. Roman Catholic. Clubs: Immaculate Conception Men's, Clown Unit, Voiture 220 (Chgo.). Home: 8318 S Komensky Ave Chicago IL 60652-3114 Office: Roosevelt U 430 S Michigan Ave Chicago IL 60605-1301

CORCORAN, MICHAEL JOHN, orthopedic surgeon; b. Chgo., Nov. 10, 1960; s. Thomas and Carole (Juric) C.; m. Linda M. Nieders, Nov. 2, 1991. BS, No. Ill. U., 1983, MS, 1987; MD, So. Ill. U., 1991. Unit coord. emergency rm., trauma dept. U. Chgo. Hosps., 1983-87; rsch. asst. dept. biology No. Ill. U., DeKalb, 1983-87; rsch. asst. dept. internal medicine So. Ill. U., Springfield, 1989-90; intern gen. surgery Butterworth Hosp., Grand Rapids, Mich., 1991-92; resident orthopedic surgery U. Mo.-Columbia Hosp., 1992—. Co-author: One Hundred Orthopaedic Conditions Every Doctor Should Understand, 1991. Mem. AMA, Sigma Xi, Phi Sigma (v.p. 1986). Office: Univ Mo Columbia Hosp 1 Hospital Dr Columbia MO 65212

CORDAIN, LOREN, physical education educator, graduate program director, researcher; b. Oct. 24, 1950. BS in Health Scis., Pacific U., 1974; MS in Phys. Edn., U. Nev., 1978; PhD in Phys. Edn., U. Utah, 1981. Lectr. dept. exercise and sport sci. Colo. State U., 1981-82, from asst. prof. to assoc. prof., 1982-90, prof., 1990—, dir. Human Performance Lab., 1981—, dir. grad. ecercise sci. program, 1983—. Author: Laboratory Manual for Essentials of Physical Activity, 1982, Body Composition Determination by Underwater Weighing: A User's Manual for the Chatillon Model 1309H Scale, 1982; author: (with others) Computers in Cardiology, 1978, Microcomputing in Sport and Physical Education, 1983; contbr. articles to profl. jours. Recipient Faculty Svc. award Coll. Profl. Studies, 1982, 84, Outstanding Rschr. award Phi Delta Kappa, 1983. Achievements include research on the interface of a Benedict Roth type spirometer to a microcompter via a rotary potentiometer, effects of carbonic anhydrase inhibition by acetazolamide on exercise tolerance, procurement of spirometer calibration syringe and D.O.S. for IBM PC, effects of aerobic training on ventilatory muscle strength, effects of aerobic training on residual lung volume relative to pulmonary pressures, acquisition of digital manometer, biomechanics of successful downhill skiing, response of plasma pyridoxal and pyridoxal 5 phosphate concentrations to different intensities of exercise. Office: Human Performance Lab Dept of Exercise & Sport Science Colorado State University CO 80523

CORDELL, FRANCIS MERRITT, instrument engineer, consultant; b. South Pittsburg, Tenn., Sept. 11, 1932; s. Lucien Hall and Sara Frances (Taliaferro) C.; m. Olivia Elizabeth West, June 17, 1950; 1 child, Francis Merritt Jr. LittB, Hamilton Coll., 1966; PhD in Physics, U. Del., 1973. Low speed code operator Dept. Army, Ft. Devens, Mass., 1949-52; materials tester TVA, Stevenson, Ala., 1952-53; instrument mechanic TVA, Stevenson, 1953-57, sr. instrument mechanic, 1957-80, instrumentation supr., 1980-86; prin. restorer, telescope and observatory project U. of the South, Sewanee, Tenn., 1982—; info. cons. South Pittsburg, 1980—; mem. Tenn. Vis. Scientists Program, Associated Univs. for the Tenn. Acad. Sci., Oak Ridge, 1989—. contbng. writer Barnard Astronomical Soc. Jour., 1973—. Recipient Llewellyn Evans award Barnard Astronomical Soc., Chattanooga, 1983. Mem. AAAS, Barnard Astronomical Soc., Instrument Soc. Am., Tenn. Acad. Sci., Astronomical Soc. Pacific. Achievements include restoration of Alvan Clark & Sons refractor and observatory. Home: Medius Lodge Dogwood Trail South Pittsburg TN 37380 Office: Info Consulting 1018 Holly Ave South Pittsburg TN 37380-1432

CORDER, LOREN DAVID (ZEKE CORDER), quality assurance engineer; b. Nampa, Idaho, Sept. 9, 1949; s. H. Burton and E. Eldora (Crill) C.; m. Sheryl Ann Kinney, 1968 (div. 1971); m. Charlene Louise Brown, 1972; children: Jesse A., Nathaniel A., Quintan A., Dana A. (dec.). Cert., USAF Electronics, Chanute AFB, Ill., 1970, Ctr. For Employment Tng., Nampa, 1980. Fabrication process engr. Micron Semiconductor Inc., Boise, Idaho, 1981-83; assembly engr. Micron Tech., Inc., Boise, Idaho, 1983-85, quality assurance engr., 1986—; trustee Idaho Migrant Coun., Boise, 1980-82; mem. adv. bd. Ctr. For Employment Tng., Nampa, 1982-90. Contbr. articles to Surface Mt. Tech., Test and Measurement World and Semiconductor Internat. With USAF, 1969-72. Mem. AAAS, ASME (affiliate), EOS/ESD Assn., Am. Soc. Metals Internat., Inst. Environ. Scis. (sr.), Soc. Mfg. Engrs., Planetary Soc., Foresters, Treasure Valley Antique Power Assn., Early Day Gas Engine and Tractor Assn., Inc., Boise Blues Soc. Achievements include patent for a new specimen mount for scanning electron microscope. Office: Micron Semiconductor Inc 2805 E Columbia Rd # 502 Boise ID 83706-9698

CORDES, EUGENE HAROLD, biochemist; b. York, Nebr., Apr. 7, 1936; s. Elmer Henry and Ruby Mae (Hofeldt) C.; m. Shirley Ann Morton, Nov. 9, 1957; children: Jennifer Eve, Matthew Henry James. B.S., Calif. Inst.

Tech., 1958; Ph.D., Brandeis U., 1962. Instr. chemistry Ind. U., Bloomington, 1962-64; asst. prof. Ind. U., 1964-66, asso. prof., 1966-68, prof., 1968-79, chmn., 1972-78; exec. dir. biochemistry Merck, Sharp and Dohme Research Labs., Rahway, N.J., 1979-84, v.p. biochemistry, 1984-87; v.p. research and devel. Eastman Pharmaceuticals, Malvern, Pa., 1987-88; pres. Sterling Winthrop Pharms. Rsch. Divsn. Sterling Winthrop Inc., Collegeville, 1988—. Author: (with Henry Mahler) Biological Chemistry, 1966, 2d. edit., 1971, Basic Biological Chemistry, 1969, (with Riley Schaeffer) Chemistry, 1973; also articles. NIH Career Devel. award, 1966; Alfred P. Sloan Found. fellow, 1968. Mem. AAAS, Am. Soc. Biol. Chemists. Home: 867 Lesley Rd Villanova PA 19085-1117 Office: Sterling Winthrop PO Box 5000 1250 S Collegeville Rd Collegeville PA 19426-0900

CORDINGLEY, JAMES JOHN, systems engineer; b. Providence, Jan. 19, 1960; s. Harvey James and Joan A. (Walsh) C.; m. Patricia Lawson, Sept. 28, 1985. BSEE, U. R.I., 1984. Laser engr. Teradyne Inc., Boston, 1984-90; laser systems engr. Gen. Scanning Inc., Watertown, Mass., 1990—. Achievements include patent pending for method and apparatus for severing integrated circuit connection paths by means of a phase plate adjusted laser beam. Home: 22 Tower Hill Rd Cumberland RI 02864 Office: Gen Scanning TLSI div 32 Cobble Hill Rd Somerville MA 02143

CÓRDOVA, FRANCE ANNE-DOMINIC, astrophysics educator; b. Paris, Aug. 5, 1947; came to U.S., 1953; d. Frederick Ben Jr. and Joan Francis (McGuinness) C.; m. Christian John Foster, Jan. 4, 1985; children: Anne-Catherine Cordova Foster, Stephen Cordova Foster. BA in English with distinction, Stanford U., 1969; PhD in Physics, Calif. Inst. Tech., 1979. Staff scientist earth and space sci. div. Los Alamos Nat. Lab., 1979-89, dep. group leader space astronomy and astrophysics group, 1989; prof., head dept. astronomy and astrophysics Pa. State U., University Park, 1989—; mem. Nat. Com. on Medal of Sci., 1991—; mem. adv. com. for astron. scis. NSF, 1990-93, external adv. com. Particle Astrophysics Ctr., 1989—; bd. dirs. Assn. Univs. for Rsch. in Astronomy, 1989—; mem. Space Telescope Inst. Coun., 1990-93; mem. com. space astronomy and astrophysics Space Sci. Bd., 1987-90, internat. users com. Roentgen X-ray Obs., 1985-90, extreme ultraviolet explorer guest ovserver working group NASA, 1988-92, com. Space Sci. and Applications Group, NASA, 1991-93; mem. Hubble Telescope Adv. Camera Team, 1993; chair Hubble Fellow Selection Com., 1992. Author: The Women of Santo Domingo, 1969; guest editor Mademoiselle mag., 1969; editor: Multiwavelength Astrophysics, 1988, The Spectroscopic Survey Telescope, 1990; contbr. numerous articles, abstracts and revs. to Astrophysics Jour., Nature, Astrophysics and Space Scis., Advanced Space Rsch., Astron. Astrophysics, Mon. Nat. Royal Astron. Soc., chpts. to books. Named One of Am.'s 100 Brightest Scientists under 40, Sci. Digest, 1986; numerous grants NASA, 1979—, recipient group achievement award, NASA, 1991. Mem. Internat. Astron. Union (U.S. nat. com. 1990—), Am. Astron. Soc. (v.p. 1993—,. chair high energy astrophysics div. 1990, vice chair 1989), Sigma Xi. Achievements include analysis of ultra-soft x-ray emission from active galactic nuclei; observations and modeling of the winds from accretion disks; studies of the interstellar medium using ultraviolet spectroscopy of nearby hot binary stars; observations and modeling of extended x-ray emitting regions in close binary systems; understanding the accretion geometry of magnetic binaries with accreting white dwarfs; coordinating radio and x-ray observations of x-ray binaries in an effort to find a unified model for correlated behavior; search for evidence of galactic magnetic monopoles by identifying a class of ultrasoft x-ray emitters; studying the thermal emission from neutron stars; making observations of an x-ray emitting pulsar and its associated supernova remnant in the radio and infrared; conceiving space instruments and data systems for imaging detectors (U.S. principal investigator for optical/UV Telescope to fly 1999 on ESA's X-Ray Multi-Mirror mission); making multifrequency observations of high-energy sources. Office: Pa State U Dept Astronomy and Astrophysics 525 Davey Lab University Park PA 16802-6305

CORELL, ROBERT WALDEN, science administration educator; b. Detroit, Nov. 4, 1934; s. George W. and Grace (Hagland) C.; m. Billie Jo Proctor, June 16, 1956; children: Robert Walden, David Richard, Beth Anne. BSME, Case Inst. Tech., 1956; MS, MIT, 1959, PhD, 1964. Engr. GE, Cleve., 1955, program engr., Lynn, Mass., 1956-57; instr. U. N.H., 1957-58, asst. prof., 1959-60, assoc. prof., 1964-66, prof., 1966-90, chmn. dept. mech. enging., 1964-72, dir. marine program, 1975-87; now asst. dir. for geoscis. NSF, Washington, 1987—; rsch. engr. Huggins Hosp., Wolfeboro, N.H., 1957-60, Highland View Hosp., Cleve., 1960-64; vis. investigator Woods Hole Oceanographic Inst., 1965; rsch. assoc., vis. prof. Scripps Instn. Oceanography, 1971-72; vis. prof. U. Wash., 1985; chair U.S. Global Change Rsch. Com. of U.S. Govt., 1987—; numerous positions as chair of interagy. sci. coms. and internat. bodies. Contbr. articles to profl. jours. Founding chair Internat. Group of Funding Agencies for Global Change Rsch., 1988-90; chair Implementation Com. for Inter-Am. Inst. for Global Change Rsch., 1992—; dir. White House Conf. on Sci. and Econs. to Global Change Rsch., 1990. Mem. AAAS, Sigma Xi, Tau Beta Pi, Sigma Alpha Epsilon. Mem. AAAS, Am. Soc. Engring. Edn., IEEE, Marine Tech. Soc., Sigma Xi, Tau Beta Pi, Sigma Alpha Epsilon. Achievements include research in global change, climate and environmental research, medicine, medical engineering, ocean science and technology. Home: 1B2 Spa Creek Landing Annapolis MD 21403 Office: Nat Science Foundation Asst Dir for Geoscis 1800 G St NW Rm 510 Washington DC 20550-0002

COREY, ELIAS JAMES, chemistry educator; b. Methuen, Mass., July 12, 1928; s. Elias and Tina (Hashem) C.; m. Claire Higham, Sept. 14, 1961; children: David, John, Susan. BS, MIT, 1948, PhD, 1951; AM (hon.), Harvard U., 1959; DSc (hon.), U. Chgo., 1968, Hofstra U., 1974, Oxford U., 1982, U. Liege, 1985, U. Ill., 1985. From instr. to asst. prof. U. Ill., Champaign-Urbana, 1951-55, prof., 1955-59; prof. chemistry Harvard U., Cambridge, Mass., 1959—, Sheldon Emory prof., 1968—. Contbr. articles to profl. jours. Bd. dirs. phys. sci. Alfred P. Sloan Found., 1967-72; mem. sci. adv. bd. dirs. Robert A. Welch Found. Recipient Intrasci. Found. award, 1968, Ernest Guenther award in chemistry of essentials oils and related products, 1968, Harrison Howe award, 1971, Ciba Found. medal, 1972, Evans award Ohio State U., 1972, Linus Pauling award, 1973, Dickson prize in sci. Carnegie Mellon U., 1973, George Ledlie prize in sci. Harvard U., 1973, Nichols medal, 1977, Buchman award Calif. Inst. Tech., 1978, Franklin medal in sci. Franklin Inst., 1978, Sci. Achievement award CCNY, 1979, J.G. Kirkwood award, Yale U., 1980, Chem. Pioneer award, Am. Inst. Chemists, 1981, Wolf prize (chem.), Wolf Found., 1986, Japan prize, 1989, Nat. Med. Sci., 1988, Nobel prize in chemistry, 1990, Gold Medal Award, AIC, 1990, numerous others; fellow Swiss-Am. Exch., 1957, Guggenheim Found., 1957-58, 68-69, Alfred P. Sloan Found., 1956-59. Mem. Am. Acad. Arts and Scis., AAAS, Am. Chem. Soc. (award in synthetic chemistry 1971, Pure Chemistry award 1960, Fritzche award 1968, Md. sect. Remsen award 1974, Arthur C. Cope award 1976, Roger Adams award organic chemistry 1993), Nat. Acad. Sci., Sigma Xi. Office: Harvard U Dept Chemistry 12 Oxford St Cambridge MA 02138

CORINTHIOS, MICHAEL JEAN GEORGE, electrical engineering educator; b. Cairo, Jan. 19, 1941; arrived in Can., 1965; s. Jean George and Gisele Michel (Cabbabe) C.; m. Maria Scigalski, Nov. 18, 1967; children: Angela, Gisele, John. Diploma in art, Leonardo DaVinci Art Sch., Cairo, 1956; BSc, Ain Shams U., Cairo, 1962; MASc, U. Toronto, 1968, PhD, 1971. Engr. Radio Transmission, Abu Zaabal, Cairo, Egypt, 1962-65; engr. Bell Can., Toronto, Ont., 1965-66, Litton Systems Canada, Toronto, 1968-69; asst. prof. Ecole Poly., Montreal, Que., Can., 1971-74; prof. Ecole Poly., Can., 1977—; vis. scientist U. Nice, France, 1979-80; engring. cons. Huntec Exploration Corp., Toronto, 1970-71; prin. investigator Def. Research Lab., Victoria, B.C., Canada, 1977-81; acad. visitor Imperial Coll. Sci., Tech. and Medicine, London, Eng., 1992-93; pres. Corinthian Games Ltd., Montreal, 1980—; chmn. 2d Internat. Spectral Workshop, Montreal, 1986. Author: How to Patent Your Invention, 1971; Analyse des Signaux Ecole Polytech., 1982; contbr. chpts. to books, numerous papers to sci. and profl. jours. Mary H. Beatty fellow U.Toronto, 1969. Mem. IEEE (1st prize art competition Montreal chpt. 1971), Art Group 80 (Montreal). Achievements include patents for processors for high speed vector transformation, 1973, for trademarked symmetric chess game Miniatures, 1977. Home: # 1204, 5999 Monkland Ave, Montreal, PQ Canada H4A 1H1 Office: Ecole Poly Montreal, CP 6079 Succ A, Montreal, PQ Canada H3C 3A7

CORK, RANDALL CHARLES, anesthesiology educator; b. St. Paul, Oct. 6, 1948; s. Willis Lawson and Ruth (Plotz) C.; m. Patricia Bee, June 6, 1969 (dec. Sept. 1990); 1 child, Malcolm; m. Dixie Holmes, Oct. 19, 1991. BS in Engring., Ariz. State U., 1970, MS in Engring., 1972, PhD, 1974; MD, U. Ariz., 1977. Systems engr. Honeywell Info. Systems, Phoenix, 1970-74; asst. prof. U. Ariz. Dept. Anesthesiology, Tucson, 1980-86, assoc. prof., 1986-93; prof. dept. anesthesiology La. State U., New Orleans, 1993—; lectr. dept. anesthesiology U. Zimbabwe, Harare, 1983-84; sect. head cardiac anesthesia Tucson VA Med. Ctr., 1991-93. Contbr. chtps. to books and articles to profl. jours. Cons. physician Tucson Spl. Olympics, 1991-93. Rsch. grantee Am. Cancer Soc., Janssen Pharm., 1984-85, Bristol Labs., 1984-85, Organin, 1984-85, Am. Heart Assn., 1985-86, Stuart Pharms., 1987-88, Hoffman-LaRoche, 1987-88, DuPont Pharms., 1989-90, Glaxo Inc., 1990-91, 91-92, Maruishi Pharms. Co., 1990-91, Anaquest, 1991-92, Astra Pharms., 1991-92. Fellow Am. Coll. Angiology; mem. AMA, Am. Soc. Anesthesiologists, Internat. Anesthesia Rsch. Soc., Ariz. State Soc. Anesthesiologists, N.Y. Acad. Scis., Soc. Cardiovascular Anesthesiologists, Soc. Critical Care Medicine, Soc. Ambulatory Anesthesia, U. Oxford Anesthesic Rsch. Soc. Avocations: motorcycling, tennis.

CORL, WILLIAM EDWARD, environmental chemist; b. Liveoak, Fla., Apr. 3, 1964; s. William Edward Jr. Corl and Mary Jo (Bain) Scott;m. Lynn Ann Miller, Sept. 30, 1989; 1 child, Heather Renee. Assoc. in Sci., Tidewater Community Coll., Portsmouth, Va., 1985; B in Chemistry, Old Dominion U., 1988. Field svc. engr. Ecolochem, Inc., Norfolk, Va., 1988-89; sr. environ. chemist PWC Lab. Norfolk Naval Base, 1989--. Home: 3609 King St Portsmouth VA 23707 Office: Environ Lab Code 931 Bldg z 140 Norfolk VA 23511

CORLEY, DANIEL MARTIN, physicist; b. Columbus, Ohio, July 2, 1940; s. John Leo and Florence (Dietlin) C.; m. Mary M. Ryan, Sept. 1, 1962; children: Cynthia, Lyn, Stephanie. BA in Physics, Cath. U., 1962; PhD in Physics, U. Md., 1968. Rsch. physicist NIST Commerce Dept., Gaithersburg, Md., 1961-93; prof. physics Montgomery Coll., Rockville, Md., 1969—. Mem. Am. Physical Soc. Achievements include first research into the technology of the scanning tunnelling microscope. Office: Montgomery Coll 51 Mannakee St Rockville MD 20850-1195

CORLEY, JEAN ARNETTE LEISTER, infosystems executive; b. Charleston, S.C., June 16, 1944; d. William Audley and Arnette (Mason) Leister; widowed; children: Arnette Elizabeth, Daniel Lee. BS, Med. Coll. Ga., 1970; MBA, M of Pub. Adminstrn., Southeastern U., 1980. Various positions health care orgns., Augusta, Ga., 1960-70; office mgr., counselor Info. Ctr. for Alcohol and Drug Abuse, Augusta, 1970-71; planner health care systems Nat. Med. Assn. Found., Washington, 1971-72; research assoc., systems analyst GEOMET, Inc., Gaithersburg, Md., 1972-74; dir. med. records Georgetown U. Hosp., Washington, 1974-80; dir. med. info. systems Lahey Clinic Med. Ctr., Burlington, Mass., 1980-84; sales mgr. 3M Health Info. Systems, Boston and Atlanta, 1984-91; mktg. mgr. Code 3 Health Info. Systems/3M, Atlanta, 1992—. Contbr. articles to profl. jours. Mem. adv. bd. various colls., 1973-92. Mem. CPRI, Am. Health Info. Mgmt. Assn. (program com. 1977-80, chmn. 1981-82, fed. health program adv. com. 1978-80, computerized health info. task force 1983-87, subcom. on edn. 1990-93, WEDI 1992—), Women in Info. Processing, New Eng. Med. Records Conf. (exec. dir. 1984-88), LWV, Common Cause. Democrat. Presbyterian. Avocations: sailing, reading, restoring old houses. Home: 545 N DeSoto St Salt Lake City UT 84103

CORLEY, WILLIAM GENE, engineering research executive; b. Shelbyville, Ill., Dec. 19, 1935; s. Clarence William and Mary Winifred (Douthit) C.; m. Jenny Lynd Wertheim, Aug. 9, 1959; children: Anne Lynd, Robert William, Scott Elson. BS, U. Ill., 1958, MS, 1960, PhD, 1961. Registered profl. engr., Ill., Va., Wash., Calif., Miss., Fla., La.; lic. structural engr., Ill. Devel. engr. Portland Cement Assn., Skokie, Ill., 1964-66, mgr. structural devel. sect., 1966-74, dir. engring. devel. div., 1974-86; v.p. Constrn. Tech. Labs., Inc. (formerly Portland Cement Assn.), Skokie, 1986—; mem. adv. panels NSF. Contbr. aritcles to tech. and profl. jours. Pres. caucus Glenview (Ill.) Sch. Bd., 1971-72; elder United Presbyn. Ch., 1975-79; sec. bd. dirs. Assn. Ho., Chgo., 1976, treas., 1977, pres., 1978-79; chmn. bd. dirs. North Cook dist. ARC, bd. dirs. Mid-Am. chpt., chmn. North Region Coun., 1988-92; mem. Gov.'s (Ill.) Earthquake Preparedness Task Force. Recipient Wason medal for Research, 1970; Martin Korn award Prestressed Concrete Inst., 1978; Reinforced Concrete Research Council Arthur J. Boase award, 1986; Outstanding Paper award SEAOI, 1993. Fellow ASCE (T.Y. Lin award 1979), Am. Concrete Inst. (Bloem award 1978, Reese Structural Research award 1986, Henry C. Turner award 1988, Ferguson lectr. 1991); mem. NSPE, Réunion Internationale des Laboratoires d'Essais et Recherchas sur Matériaux Construction, Earthquake Engring. Research Inst. (chpt. sec., treas. 1980-82, chmn. 1984-86), Internat. Assn. Bridge and Structural Engring., Structural Engrs. Assn. Ill. (pres. 1986-87, Meritorious Publ. award 1993), Post-Tensioning Inst., Chgo. Com. High-Rise Bldgs. (vice-chmn. 1978-82, chmn. 1982-84, Bldg. Seismic Safety Council (vice-chmn. 1983-85, sec. 1985-87). Presbyterian. Office: Construction Tech Labs Inc 5420 Old Orchard Rd Skokie IL 60077 Home: 744 Glenayre Dr Glenview IL 60025-4411

CORMA, AVELINO, chemist, researcher; b. Moncofar, Castellon, Spain, Dec. 15, 1951; s. Avelino and Dolores (Canos) C.; m. Brisa Gomez, Feb. 5, 1982; 1 child, Anais. BS in Chemistry, U. Valencia, Spain, 1973, PhD in Chemistry, 1976. Assoc. researcher Queen's U., Kingston, Ontario, Can., 1977-79; assoc. research Inst. Catalisis Consejo Superior Investigaciones Cientificas, Madrid, 1979-86; rsch. prof. Inst Catálisis CSIC, Madrid, 1986-90; dir. Inst. Tech. Chemistry Valencia Poly. U.-Consejo Superior Investigaciones Sci., Valencia, 1991—; cons. CEPSA, Madrid, Crosfield, Manchester, Eng., BP Am., Cleve., Shell, Amsterdam, Catlytica, San Diego; adv. bd. Jour. Applied Catalysis, Elsevier, Catalysis Rev., Marcel Dekker, Zeolites Butterworths. Author book, 1986, over 200 sci. papers; contbr. articles to Jour. Catalysis, Applied Catalysis, Catalysis Rev. Sci., others. Cons. UNIDO, Vienna, 1986. Recipient Nat. Bd. for Sci. award, Minister Edn., Madrid, 1991, Sci. Com. award, Pres. CSIC, 1990. Mem. Royal Soc. Chemistry, Am. Chem. Soc. (petroleum div.). Achievements include 20 patents in catalysis, 5 commercially used; discovery of active sites and mechanism for catalytic cracking ion zeolites; development first synthesis of large pore zeolite with Ti, active for oxidations under mild conditions. Home: Street Yecla 41, 46022 Valencia Spain Office: CSIC Univ Politécnica, Camino de Vera s/n, 46071 Valencia Spain

CORMACK, ALLAN MACLEOD, physicist, educator; b. Johannesburg, South Africa, Feb. 23, 1924; came to U.S., 1957, naturalized, 1966; s. George and Amelia (MacLeod) C.; m. Barbara Jeanne Seavey, Jan. 6, 1950; children: Margaret, Jean, Robert. BS, U. Cape Town, South Africa, 1944, M.Sc., 1945; research student, Cambridge (Eng.) U., 1947-50. Lectr. U. Cape Town, 1946-47, 1950-56; research fellow Harvard U., 1956-57; asst. prof. physics Tufts U., Medford, Mass., 1957-60; assoc. prof. Tufts U., 1960-64, prof., 1964-80, University prof., 1980—; researcher nuclear and particle physics, computed tomography and related math. topics. Recipient Ballou medal Tufts U., 1978, Nobel prize in medicine and physiology, 1979, Medal of Merit U. Cape Town, 1980, Nat. Medal Sci. NSF, 1990. Fellow AAAS, Am. Phys. Soc., Am. Acad. Arts and Sci., Royal Soc. South Africa (fgn.); mem. South African Phys. Soc., Nat. Acad. Sci., Am. Assn. Physicists in Medicine (hon.), Inst. Medicine, Sigma Xi. Office: Tufts U Physics Dept Medford MA 02155

CORMIA, FRANK HOWARD, industrial engineering administrator; b. Montreal, Que., Can., Nov. 17, 1936; s. Frank Edward Cormia and Elizabeth Kulp (Hall) St. Louis; m. Mary Irene Porter, Aug. 29, 1959; children: John Howard, Carl William, Ross Michael, Judith Anne. BS in Engring., Calif. Inst. Tech., 1960. Indsl. engr. Alcoa, Vernon, Calif., 1960-64; chief indsl. engr. Warrick ops. Alcoa, Evansville, Ind., 1968-76; mgr. indsl. engring. Tenn. ops. Alcoa, 1976-1991; chief indsl. engr. Wear-ever Aluminum, Chillicothe, Ohio, 1964-68. Past chmn. bd. dirs. Sch. Indsl. Systems Engring., Ga. Inst. Tech., Atlant; bd. dirs., past chmn. dept. indsl. systems engring. Va. Poly. Inst. and State U., Blacksburg; chmn. long range planning com. Smoky Mountain coun. Boy Scouts Am., 1988-89. 1st lt. USAR, N.G., 1960-66. Mem. Inst. Indsl. Engrs. (sr.). Episcopalian. Office: Aluminium Co Am PO Box 9128 Alcoa TN 37701-9128

CORNELIS, ERIC RENE, mathematician; b. Namur, Belgium, July 24, 1960; s. Christian and Gisele (Deville) C. Lic. Math., U. Notre Dame de la Paix, Namur, Belgium, 1982, PhD in Sci. Math., 1992. Rsch. asst. Nat. Fund Sci. Rsch., Brussels, 1983-87; asst. dept. math. Facultes U. Notre Dame de la Paix, Namur, 1987-91, rsch. assoc. dept. math.-transp. rsch. group, 1984—; reviewer Math. Reviews, Providence, 1985—; reviewer, referee Revue des Questions Scientifiques, Brussels, 1987—; reviewer Tugboat, Tex User Group, Providence, 1991—. Author: Premiers pas avec Latex, 1991; contbr. articles on math. and computer scis. to profl. jours. Recipient Laureate of Univ. Competition, Communauté francaise de Belgique, Brussels, 1983; mem. d'honneur Groupe des Utilisateurs de Tex francophones, 1991, Gutenberg. Mem. Assn. des Anciens de Math. de Namur (sec. 1984-92, pres. 1992—), Sci. Soc. Brussels, GUTenberg (hon.), Am. Math. Soc., European Math. Soc., Belgium Soc. Math., Belgium Soc. Profs. Math. d'expression francaise, European Assn. Theoretical Computer Sci., Soc. d'Informatique Fondamentale, Tex Users Group, Soc. Indsl. and Applied Math. Avocations: reading, collecting comics, gardening. Home: Ave Gen Gracia 18, B-5170 Profondeville, Province de Namur Belgium Office: U Notre Dame de la Paix, Dept Math, Rempart de la Vierge 8, B-5000 Namur Belgium

CORNELISSEN-GUILLAUME, GERMAINE GABRIELLE, chronobiologist, physicist; b. Brussels, Nov. 22, 1949; came to U.S., 1976; d. Alphonse and Helene A. (Minne) Cornelissen; m. Francis M.G. Guillaume, Nov. 22, 1975. MS, MEd, U. Brussels, 1971, PhD, 1976. Asst. prof. Chronobiology Lab., U. Minn., Mpls., 1987—, sr. rsch. assoc., 1991—, dir. biometry, 1992—; sec., mem. N.Am. br. sci. coun. Internat. Soc. Rsch. on Civilization Diseases and the Environment; mem. adv. bd. internat. meetings on chronobiology, 1989—. Co-editor Chronobiologia; mem. editorial bd. Il Policlinico; referee several sci. jours. Recipient grant NIH, 1981-84, 88-91, Chronobiology award Hoechst, Italy, 1983. Mem. AAAS, Internat. Soc. Chronobiology (bd. dirs. 1985—), Am. Phys. Soc., Am. Stat. Assn. (biometric sect. 1984—), Sigma Xi. Home: 511 Ryan Ave W Roseville MN 55113 Office: Univ Minn 5-187 Lyon Labs 420 Washington Ave SE Minneapolis MN 55455

CORNELIUS, WAYNE ANDERSON, engineering technology educator, consultant; b. Russellville, Ky., Nov. 8, 1923; s. Eldon and Mabel Ruth (Gentle) C.; m. Elizabeth Grider (dec. Aug. 1946); children: Johanna Vastola, Keith, John; m. Linda Brady, Apr. 27, 1985; stepchildren: Pam Gondzur, Mark, Todd, Allison Smith. BS, U. Ky., 1953, EE, 1966; MS, U. Louisville, 1962; postgrad., U. Cin., 1969-72. Elec. engr. U.S. Naval Ordnance Sta., Louisville, 1953-66, dir. engring. electronics lab., 1973-85; rsch. assoc. Pa. State U., State College, 1966-67; prof. engring. tech. Miami U, Oxford, Ohio, 1967-72; elec. engr. System Devel. Corp., Dayton, Ohio, 1972-73; chair dept. electronics tech. Ivy Tech. Coll., Sellersburg, Ind., 1985-90; adj. prof. elec. engring. tech. Purdue U., New Albany, 1992—, U. Louisville, 1976-84; adj. prof. math. Bellarmine Coll., Louisville, 1964-66, Ind. U., New Albany, 1990-91. With USN, 1942-45. Named to Honorable Order of Ky. Cols., 1963. Mem. NSPE, Am. Soc. for Engring. Edn., Phi Delta Kappa. Democrat. Presbyterian. Office: 9005 Lethborough Dr Louisville KY 40299-1437

CORNELL, CHARLES ALFRED, engineering executive; b. Bronx, N.Y., May 11, 1946; s. Charles Alfred and Helen Marie (O'Mara) C.; m. Elizabeth Anne Lopano, July 27, 1968; children: Kristin, Justin, Matthew, Rebecca. BS in Physics, Fordham U., Bronx, N.Y., 1969; BSEE, Columbia U., N.Y.C., 1969, MSEE and Computer Sci., 1970. Mem. tech. staff Bell Telephone Labs., Holmdel, N.J., 1969-73; systems programmer Interdata-Perkin Elmer Data Systems, Oceanport, N.J., 1973-75; dir. advanced devel. IPL Systems, Waltham, Mass., 1975-80, dir. systems engr., 1980-84; v.p. engring. VMARK Software, Inc., Framingham, Mass., 1984-93, v.p. advanced tech., 1993—. Contbr. articles to profl. jours. Mem. IEEE Computer Soc., Assn. Computing Machinery. Roman Catholic. Achievements include development of first successful implementation of plug compatible micro code assists for IBM operating systems; built (with others) OS/32 operating system for first commercially successful 32 bit micro computer from Interdata. Home: 206 Hudson Rd Stow MA 01775-1207 Office: VMARK Software Inc 30 Speen St Framingham MA 01701

CORNELL, JAMES FRASER, JR., biologist, educator; b. Charlotte, N.C., Dec. 19, 1940; s. James Fraser and Catherine Odom (Parker) C.; m. Sandra Johnson, June 13, 1965; children: James F. III, Thomas Alexander Duncan. BS in Geology, U. N.C., 1963; MS in Entomology, N.C. State U., 1965; PhD, Oreg. State U., 1971. Asst. prof. biology Appalachian State U., Boone, N.C., 1968-71; prof. N.C. State U., Raleigh, 1971-72; prof. biology Lees McRae U., Banner Elk, N.C., 1972-73; pres. Scientific Equipment Co., Inc., Charlotte, N.C., 1973-85; with internat. studies program Charlotte-Mecklenburg Sch. System, 1986-91, with internat. baccalaureate program, 1991—; cons. forensic entomology Charlotte Mecklenburg Police Dept., 1991—. Contbr. articles to profl. jours. Fulbright grantee, 1990, 92. Mem. Pacific Entomol. Soc., Washington Entomol. Soc., Am. Topical Assn. (pres. biology unit 1988-92), N.C. Entomol. Soc. Achievements include research in anophthalmic edaphic and nidicolous Coleoptera, and rare and endangered insects, particularly beetles. Home: 1616 Euclid Ave Charlotte NC 28203 Office: Charlotte Mecklenburg Sch 1967 Patriot Dr Charlotte NC 28222

CORNELL, JAMES S., critical care and pulmonary physician; b. Harrisberg, Pa., Apr. 9, 1947; s. Don LaRue and Esther Louise (Johnson) C.; m. Agnes M. Collis, June 23, 1985; children: Kristin Rene, Emily Elizabeth, Amanda Dawn. PhD, UCLA, 1973; MD, Cornell U., 1988. Diplomate Am. Bd. Internal Medicine. Critical care and pulmonary physician Cornell U. Med. Coll., N.Y. Hosp., N.Y.C., 1973—; assoc. prof. Cornell U. Med. Coll., N.Y.C., 1973—; asst. dir. lab. for clin. biochemistry Cornell U. Med. Coll., N.Y. Hosp., N.Y.C.; cons. Amsterdam Nursing Home, N.Y.C., 1992—. Contbr. articles to profl. jours. Mem. Begen County Dem. Com., Hackensack, 1970's. Recipient Tchr./Scientist award Andrew W. Mellon Found., 1977-78, fellow Meml. Sloan Kettering Cancer Ctr., 1991—, Burke Rehab. Ctr., 1991—; Martin Newman fellow Will Rogers Inst. Mem. Am. Thoracic Soc., Am. Coll. Chest Physicians, Am. Chem. Soc., Endocrine Soc. Achievements include research in obstructive mass lesion in pregnancy, bronchoscopy role in diagnosis of TB, PCP, oxytocinase from term human placenta, biochemistry in modern medical curruculam, structure and function of pituitary TSH/LH. Office: Cornell U Med Coll NY Hosp Starr 505 1300 York Ave New York NY 07652

CORNELL, ROBERT WITHERSPOON, engineering consultant; b. Orange, N.J., Aug. 16, 1925; s. Edward Shelton and Helen Lauretta (Lawrence) C.; m. Patricia Delight Plummer, June 24, 1950; children: Richard W., Delight W. Cornell Dobby, Elizabeth Cornell Wilkin, Roberta Shelton. B. Mech. Engring., Yale U., 1945, M. Mech. Engring., 1947, D. Engring., 1950. Registered profl. engr., Conn., N.Y. Instr. math. New Haven Jr. Coll., 1947-48; analytical engr. Pratt & Whitney Aircraft, East Hartford, Conn., 1947; with Hamilton Standard, Windsor Locks, Conn., 1948-87, chief applied mechanics and aerodynamics, 1961-87; instr. enging. Hillyer Coll., Hartford, 1955; pres. Cornell Cons., 1973—, Cornell Enterprises, West Hartford, 1984—; adj. prof. Yale U., 1985, 90. Contbr. articles to profl. jours. Patentee in field. Bd. dirs., treas. Yale Sci. and Engring. Assn., 1969—; Conn. State Taxpayers Assn., Stratford, 1984-86; past pres., bd. dirs. West Hartford Taxpayers Assn., 1972—; rep. state senatorial candidate 5th dist. State of Conn., 1988, state rep. candidate 18th dist., 1990; mem. SCORE, 1990—. With USN, 1943-46. Fellow ASME mem. Sigma Xi, Tau Beta Pi. Clubs: Yale (Hartford and N.Y.C.); Hartford Golf. Avocations: tennis, squash, jogging, swimming, gardening. Home and Office: 40 Belknap Rd West Hartford CT 06117-2819

CORNFORTH, SIR JOHN WARCUP, chemist; b. Sydney, Australia, Sept. 7, 1917; s. John William and Hilda (Eipper) C.; m. Rita H. Harradence, Sept. 27, 1941; children: Brenda (Mrs. David Osborne), John, Philippa (Mrs. William Horder). BSc, U. Sydney, 1937, MSc, 1938; D.Phil., Oxford U., 1941, D.Sc. (hon.), 1976; D.Sc. (hon.), E.T.H. Zurich, 1975, Trinity Coll., Dublin, Univs. Liverpool, Warwick, Aberdeen, Hull, Sussex and Sydney. Mem. sci. staff Med. Research Council, London, 1946-62; dir. Milstead Lab. Chem. Enzymology, Shell Research Ltd., Sittingbourne, Kent, Eng., 1962-75; Royal Soc. research prof. Sch. Molecular Scis. U. Sussex, Brighton, Eng., 1975-82. Contbr. articles on chemistry of penicillin, total synthesis of steroids and other biologically active natural products, chemistry of heterocyclic compounds, biosynthesis of steroids, enzyme chemistry to profl. jours. Decorated comdr. Brit. Empire; knighted, 1977; apptd. Companion of the Order of Australia, 1991; recipient Stouffer prize, 1967, Prix Roussel, 1972, Nobel Prize in Chemistry, 1975. Fellow Royal Soc., 1953 (Davy medal 1968, Royal medal 1976, Copley medal 1982), Royal Soc. Chemistry (Corday-Morgan medal 1953, Flintoff medal 1966), Am. Chem. Soc. (Ernest Guenther award 1969); mem. Biochem. Soc. (CIBA medal 1966), Am. Soc. Biol. Chemists (hon.), Am. Acad. (hon. fgn. mem.), Australian Acad. Sci. (corr.), Netherlands Acad. Sci. (fgn.), NAS (fgn. assoc.). Home: Saxon Down, Cuilfail, Lewes BN7 2BE England Office: U Sussex Sch Molecular Scis, Falmer, Brighton BN1 9QJ, England

CORNMAN, LARRY BRUCE, physicist; b. L.A., Jan. 4, 1958; s. Noel and Marion (Tessel) C. BA in Math., BA in Physics, U. Calif., Santa Cruz, 1979; MS in Physics, U. Colo., 1985. Assoc. scientist Nat. Ctr. Atmospheric Rsch., Boulder, Colo., 1985—; cons. FAA Windshear User Group, Washington, 1989—; content advisor Nat. Ctr. Atmospheric Rsch./NSF Windshear Exhibit, Boulder, 1992—. Contbr. articles to profl. jours. Recipient Laurel award Aviation Week and Space Tech., 1990. Mem. AIAA. Achievements include 4 patents for improved low-level windshear alert system; development of methods for detecting windshear from doppler radars and surface anemometers; development of methods for detecting atmospheric turbulence from aircraft. Office: Nat Ctr Atmospheric Rsch PO Box 3000 Boulder CO 80307

CORNWALL, JOHN MICHAEL, physics educator, consultant, researcher; b. Denver, Aug. 19, 1934; s. Paul Bakewell and Dorothy (Zitkowski) C.; m. Ingrid Linderos, Oct. 16, 1965. AB, Harvard U., 1956; MS, U. Denver, 1959; PhD, U. Calif., 1962. NSF postdoctoral fellow Calif. Inst. Tech., Pasadena, 1962-63; mem. Inst. Advanced Study, Princeton, N.J., 1963-65; prof. physics UCLA, 1965—; vis. prof. Niels Bohr Inst., Copenhagen, 1968-69, Inst. de Physique Nucléaire, Paris, 1973-74, MIT, 1974, 87, Rockefeller U., N.Y.C., 1988; cons. Inst. Theoretical Physics, Santa Barbara, Calif., 1979-80, 82, bd. dirs., 1979-83; assoc. Ctr. Internat./Strategic Affairs, UCLA, 1987—; cons. MITRE Corp., Aerospace Corp., Los Alamos Nat. Labs.; mem. dir's. adv. com. Lawrence Livermore Labs., 1991—; mem. Def. Sci. Bd., 1992—. Author: (with others) Academic Press Ency. of Science and Technology, other encys. and books; contbr. numerous articles to profl. jours. With U.S. Army, 1956-58. Grantee NSF, NASA; NSF pre/postdoctoral fellow 1960-63, A.P. Sloan fellow, 1967-71. Mem. Am. Phys. Soc., Am. Geophys. Union, N.Y. Acad. Sci. Avocations: jogging, bicycling, golf, bridge. Office: UCLA Dept Physics Los Angeles CA 90024

CORONEOS, RULA MAVRAKIS, mathematician; b. Crete, Greece, Oct. 24, 1961; d. Theodoros and Artemis Mavrakis; m. Emmanouel Coroneos, Aug. 27, 1985. BS in Math., U. Utah, 1984; MS in Math., Cleve. State U., 1990. Math. tutor, computer programmer Cleve. State U., 1986-87, acct., 1988-90; mathematician NASA Lewis Rsch. Ctr., Cleve., 1990—. Recipient Achievement award NASA Lewis Rsch. Ctr., Cleve., 1992. Office: NASA Lewis Rsch Ctr 21000 Brookpark Rd Cleveland OH 44135

COROTIS, ROSS BARRY, civil engineering educator, academic administrator; b. Woodbury, N.J., Jan. 15, 1945; s. A. Charles and Hazel Laura (McCloskey) C.; m. Stephanie Michal Fuchs, Mar. 19, 1972; children—Benjamin Randall, Lindsay Sarah. S.B., MIT, Cambridge, 1967, S.M., 1968, Ph.D., 1971. Lic. profl. engr., Ill., Md., structural engr., Ill. Asst. prof. dept. civil engring. Northwestern U., Evanston, Ill., 1971-74, assoc. prof. dept. civil engring., 1975-79, prof. dept. civil engring., 1979-81; prof. dept. civil engring. Johns Hopkins U., Balt., 1981-82, Hackerman prof., 1982-83, Hackerman prof., chmn. dept. civil engring., 1983-90, Hackerman prof., assoc. dean engring., 1990—; mem. bldg. research bd. Nat. Research Council, Washington, 1985-88; lectr. profl. confs. Editor in chief Internat. Jour. Structural Safety, 1991—; contbr. articles to profl. jours. Mem. Mayor's task force City of Balt. Constrn. Mgmt., 1985. Rsch. grantee NSF, Nat. Bur. Standards, U.S. Dept. Energy, 1973-90; recipient Engring. Teaching award Northwestern U., 1977; named Md. Engr. of Yr., Balt. Engrs. Week Coun., 1989. Fellow ASCE (chmn. safety bldgs. com. 1985-89, chmn. tech. adminstrv. com. structural safety and reliability 1988-92, Walter L. Huber rsch. prize 1984, mem. dept. heads coun., Civil Engr. of Yr. award Md. chpt. 1987, v.p. Md. chpt. 1987-88, pres. 1988-89, Outstanding Educator award Md. chpt. 1992); mem. Am. Concrete Inst. (chmn. structural safety com. 1986-88), Am. Nat. Standards Inst. (chmn. live loads com. 1978-84). Office: Johns Hopkins U Engring Deans Office Baltimore MD 21218

CORRADO, DAVID JOSEPH, systems engineer; b. Cleve., Mar. 22, 1960; s. Peter Joseph and Diana Jane (Labuda) C. BA in Chemistry, Baldwin-Wallace, 1983; BSChemE, Cornell U., 1983. Chem. engr. IBM, Endicott, N.Y., 1983-86; systems engr. IBM, Cleve., 1986—; rep. presenting tech. advances in raw board circuit packaging, Stuttgart, Fed. Republic Germany. Achievements include design of thin core conveyance system used during development of advanced circuit boards. Office: IBM Corp 1300 E 9th St Cleveland OH 44114-1502

CORREALE, STEVEN THOMAS, materials scientist; b. Morristown, N.J., Aug. 12, 1953; s. Salvatore Jerry and Edith Jean (Tillard) C.; m. Kimberly Anne Carroll, Nov. 18, 1978; children: Brett Robert, Dustin Thomas. BS in Physics, Muhlenberg Coll., 1975; MS in Material Scis., Stevens Inst. Tech., 1986. Lab. tech. Johanson Mfg. Inc., Booton, N.J., 1975-77; sr. lab. tech. Allied-Signal Corp., Morristown, 1977-79, rsch. physicist, 1979-90; x-ray lab. supr. fibers div. AlliedSignal Corp., Petersburg, Va., 1990—. Contbr. articles to Jour. Materials Sci., Jour. Applied Polymer Sci., Advances in X-ray Analysis, Macromolecules, Jour. Chem. Physics; contbr. chpt. to A Guide to Materials Characterization and Chemical Analysis, 1988. Mem. Am. Phys. Soc., Am. Chem. Soc., Digital Equipment Users Soc. Achievements include research in fiber morphology amorphous phase in semicrystalline polymer fibers. Home: 4749 Pompton Ln Chester VA 23831 Office: AlliedSignal Corp Fibers PO Box 31 Petersburg VA 23804

CORRELL, DONALD LEE, JR., physicist; b. San Pedro, Calif., June 16, 1947; s. Donald Lee and Agnes Goden (Essayian) C.; divorced; 1 child, Shanan Dee. BS in Physics, Calif. State U., Long Beach, 1969; MA in Physics, U. Calif., Irvine, 1972, PhD in Plasma Physics, 1975. Engr. Northrop Corp., 1970; assoc. project engr. Rockwell Internat., 1971-72; project engr. Hughes Aircraft, 1972-73; researcher U. Calif., Irvine, 1973-75; exptl. plasma physicist magnetic fusion energy div. Lawrence Livermore (Calif.) Nat. Lab., 1976-87, asst. program leader inertial confinement fusion div., 1988-89, assoc. program leader, 1990-91, dep. program leader, 1992—; lectr. dept. applied sci. U. Calif., Davis, 1987-88; guest lectr. STanford U., U. Calif., Berkeley, Calif. State U., Long Beach, Sonoma State U., Calif.-Poly San Luis Obispo; mem. fusion energy adv. bd. Chgo. Sci. and Industry Mus.; numerous presentations to orgns., such as Commonwealth Club of San Francisco, the Rotary Club and other civic orgns.; plasma physics and fusion energy spokesman Sci. Edn. Ctr. of Lawrence-Livermore Nat. Lab. Contbr. articles to profl. jours. NSF fellow, Hughes doctoral fellow, Regents of U. of Calif. fellow. Mem. AAAS, Am. Phys. Soc., Am. Assn. Physics Tchrs. Achievements include research in beam plasma neutron sources based on beam-driven mirror, stable operation of an effectively axisymmetric neutral beam driven tandem mirror, high performance beam-plasma neutron sources for fusion materials devel. Home: 1251 Whispering Oaks Dr Danville CA 94506 Office: Lawrence Livermore Nat Lab PO Box 5508 L-481 Livermore CA 94550

CORRIDORI, ANTHONY JOSEPH, engineering executive; b. Worcester, Mass., Sept. 27, 1961; s. Frank Joseph and Anita Louise (Ferdella) C. AS in Engring., Worcester Jr. Coll., 1983; BS in Engring, Cen. N.E. Coll., 1984; diploma, Worcester Indsl. Tech. Inst., 1981. Rsch. and devel. tech. Acumeter Labs., Marlboro, Mass., 1984-85; project engr. Hypertronics Corp., Hudson, Mass., 1985-86; sr. project engr. Augat Inc., Attleboro, Mass., 1986-92; process engr. Leach & Garner, North Attleboro, Mass., 1993—; coll. coord. Am. Soc. of Mech. Engrs., Worcester, 1983-84. Inventor floatable surface mount terminal. Mem. Am. Soc. of Metals, NRA, Knights of Columbus, Worcester Vocat. Tech. AlumniAssn. Roman Catholic. Avocations: sking, boating, target shooting (handgun, archery), sky diving. Home: 727 Franklin St Worcester MA 01604-1758 Office: Leach & Garner 57 John L Dietsch Sq North Attleboro MA 02760

CORRIGAN, JAMES JOHN, JR., pediatrician; b. Pitts., Aug. 28, 1935; s. James John and Rita Mary (Grimes) C.; m. Carolyn Virginia Long, July 2, 1960; children: Jeffrey James, Nancy Carolyn. B.S., Juniata Coll., Huntingdon, Pa., 1957; M.D. with honors, U. Pitts., 1961. Diplomate: Am. Bd. Pediatrics (hematology-oncology). Intern, then resident in pediatrics U. Colo. Med. Center, 1961-64; trainee in pediatric hematology-oncology U. Ill. Med. Center, 1964-66; asso. in pediatrics Emory U. Med. Sch., 1966-67; asst. prof. Emory U. Med. Sch., Atlanta, 1967-71; mem. faculty U. Ariz. Coll. Medicine, Tucson, 1971-90; prof. pediatrics U. Ariz. Coll. Medicine, 1974-90; chief sect. pediatric hematology-ongology, also dir. Mountain States Regional Hemophilia Center, U. Ariz., Tucson, 1978-90; chief of staff U. Med. Ctr. U. Ariz., Tucson, 1984-86; prof. pediatrics, vice dean for acad. affairs Tulane U. Sch. Medicine, New Orleans, 1990-93, interim dean, 1993—. Assoc. editor Am. Jour. Diseases of Children, 1981-89, 90-93, interim editor, 1993; contbr. numerous papers to med. jours. Grantee NIH, Mountain States Regional Hemophilia Ctr., Ga. Heart Assn., GE, Am. Cancer Soc. Mem. Am. Acad. Pediatrics, Am. Soc. Hematology, Soc. Pediatric Research, Western Soc. Pediatric Research; Internat. Soc. Toxinology, Western Soc. Pediatric Research (Ross award Pediatric research 1975), Am. Heart Assn. (council thrombosis), Internat. Soc. Thrombosis and Haemostasis, Am. Pediatric Soc., World Fedn. Hemophilia, Pima County Med. Assn. (v.p. 1986—, pres. 1988—), Alpha Omega Alpha. Republican. Roman Catholic. Office: Tulane U Sch Medicine Office of Dean 1430 Tulane Ave New Orleans LA 70112-2699

CORRIGAN, VICTOR GERARD, automotive technology executive; b. Cleve., Mar. 23, 1957. BS in Biology, Oberlin Coll., 1980. Mgr. product devel. PPG Industries, Cleve., 1987-90, mgr. automotive electrocoat tech., 1990—. Mem. Am. Chem. Soc., Engring. Soc. for Advanced Mobility Land Sea Air and Space. Achievements include patents in field. Office: PPG Industries Inc Cleve 3800 W 143rd St Cleveland OH 44111-4901

CORRIPIO, ARMANDO BENITO, chemical engineering educator; b. Mantua, Cuba, Mar. 6, 1941; came to U.S., 1961; s. Bernardo Manuel and Maria Teresa (Pedraja) C.; m. Consuelo Lucia Careaga, June 9, 1962; children: Consuelo T., Bernardo M., Mary A., Michael G. BChemE, La. State U., 1963, MChemE, 1967, PhD, 1970. Systems engr. Dow Chem. Co., Plaquemine, La., 1963-68; instr. La. State U., Baton Rouge, 1968-70, asst. prof., 1970-74, Disting. Faculty fellow, 1974, assoc. prof., 1974-81, prof. dept. chem. engring., 1981—; pvt. cons., 1968—; vis. engr. MIT, Cambridge, 1978-79. Author: Tuning of Industrial Control Systems, 1990; co-author: Automatic Process Control, 1985; contbr. numerous articles to profl. jours. Chmn. St. George Bd. Edn., Baton Rouge, 1975-77; lector St. Aloysius Cath. Ch., Baton Rouge, 1989—. Recipient Excellence in Instrn. award Exxon Co., 1986, Excellence in Teaching award Dow Chem. Co., 1989. Fellow Am. Inst. Chem. Engrs. (instr. 1977-87, chmn. Baton Rouge sect. 1990, Charles E. Coates Meml. award with Am. Chem. Soc. 1990); mem. Instrument Soc. Am. (sr. mem.; instr. 1977—), Tau Beta Phi, Phi Lambda Upsilon, Phi Kappa Phi, Sigma Xi. Avocations: tennis, swimming, reading, computer programming. Home: 9344 Bermuda Ave Baton Rouge LA 70810-1121 Office: La State Univ Dept of Chem Engring Baton Rouge LA 70803

CORSI, PATRICK, computer scientist; b. Arles, France, Aug. 11, 1951; s. Ettore and Dinora Alessandra (Guidi) C.; m. Chantal Chopin, Dec. 6, 1986; children: Alexandra, Jean-François. M Math., U. Marseilles, 1974; cert. in computer engring, Inst. Nat. Poly. Grenoble, France, 1977; D Computer Scis., Inst. Nat. Poly. Grenoble, 1979. Vis. scientist, rsch. lab. IBM Corp., San Jose, Calif., 1979-81; devel. engr. IBM France, La Gaude, 1981-84; dept. head Cognitech, Paris, 1984-86; tech. adviser Cognitech, 1986-88, European programs rep., 1988-89; dept. head, advanced studies sect. Syseca Temps Réel, St. Cloud, France, 1989—; dir. gen. telecommunications, info. industries and innovation Commn. of European Communites, Brussels, Belgium, 1990-93, gen. directorate III industry, 1993—; assoc. prof., U. Grenoble, U. Nice, France, U. Neuchatel, Switzerland, 1979-83; cons., Commn. European Communities, Brussels, 1986-90. Contbr. author: Speech Recognition, 1982; contbr. numerous articles, studies and reports to profl. publs. Mem. Am. Assn. Artificial Intelligence, IEEE, Internat. Com. Neural Networks. Roman Catholic. Home: 38 Rue de Tourville, 365 Allée de Barbeau, F-77530 Bois-Le-Roi France Office: Commn European Communities, 200 Rue de la Loi, 1049 Brussels Belgium

CORSON, JOHN DUNCAN, vascular surgeon; b. Burnley, Lancashire, Eng., Aug. 26, 1946; s. Thomas Charles and Kathleen Mary (Gillard) C.; m. Sarah Josephine Lamb, Feb. 12, 1971; children: Robert, Thomas. M.B.Ch.B., Edinburgh U., Scotland, 1968. Diplomate Am. Bd. Surgery. Intern Edinburgh Royal Infirmary, Scotland, 1968-69; resident in surgery Boston Univ. Med. Ctr., 1977-80; vascular surgery fellow Mass. Gen. Hosp., Boston, 1981; chief vascular surgery VA Med. Ctr., Albany, N.Y., 1981-86; chief vascular surgery VA Med. Ctr., Iowa City, 1986-88; dir. vascular surgery sect. U. Iowa Hosps. and Clinics, Iowa City, 1986—; prof. surgery U. Iowa, 1988—. Contbr. articles to Jour. Biomed. Engring., Archives of Surgery, Jour. Vascular Surgery. Hugh Greenwood Travel fellow, 1974. Fellow ACS, Royal Coll. Surgeons Eng.; mem. Midwestern Vascular Surg. Soc., Soc. Univ. Surgeons, Soc. Vascular Surgery. Achievements include research in prognostic value of doppler ankle, comparative analysis of vein and prothetic, relationship between vasa vasorum, polytetrafluoethylene grafts for aneurysm, finite analysis of artery-graft anastomoses. Office: U Iowa Hosps and Clinics 200 Hawkins Dr Iowa City IA 52242-1086

CORSON, WALTER HARRIS, sociologist; b. Phila., June 16, 1932; s. Bolton Langdon and Carolyn Reeves (Davis) C.; m. Sarah Peabody Lord, Sept. 21, 1964 (div. 1979); children: Trevor C. Corson, Ashley P. Corson; m. Ann Stevens Dusel, Oct. 24, 1981. BSChemE, Princeton (N.J.) U., 1954; PhD in Sociology, Harvard U., 1971. Rsch. engr. G&W.H. Corson, Inc., Plymouth Meeting, Pa., 1956-63; rsch. assoc. U. Mich., Ann Arbor, 1969-71; vis. scholar John Hopkins U., Washington, 1971-73; rsch. assoc. World Population Soc., Washington, 1974-77; ptnr. Corson Investment Co., Plymouth Meeting, 1977—; sr. assoc. Global Tomorrow Coalition, Washington, 1984—; v.p. Janelia Found., Washington, 1991—; bd. dirs. Rachel Carson Coun., Bethesda, 1992—; trustee Internat. Ctr., Washington, 1982—. Author: Measuring Conflict and Cooperation Intensity, 1971, Global Issues and Resource Management, 1992; editor: Citizen's Guide to Sustainable Development, 1989, The Global Ecology Handbook, 1990; contbr. articles to profl. jours. Steering com. Campaign for a Sustainable Energy Future, Washington, 1991—; task force mem. Blueprint for the Environ., Washington, 1987-88; bd. dirs. New Directions, Washington, 1977-81. With U.S. Army, 1954-56. Rsch. grant NSF, 1971; recipient Sustainable Devel. award Global Tomorrow Coalition, 1991. Mem. AAAS, Am. Geophys. Union, Ecol. Soc. of Am., U.S. Assn. for the Club of Rome (bd. dirs. 1992—), World Population Soc. (bd. dirs. 1992—). Achievements include development of a ranking of priorities for alleviating major global problems and achieving sustainable econ. and social devel.: reduced population growth, sustainable agriculture, sustainable energy use, forest protection, poverty reduction, sustainable water use, reduced waste generation. Home: 1399 Orchard St Alexandria VA 22302

CORTNER, HANNA JOAN, science administrator, research scientist, educator; b. Tacoma, Wash., May 9, 1945; d. Val and E. Irene Otteson; m. Richard Carroll Cortner, Nov. 14, 1970. BA in Polit. Sci. magna cum laude with distinction, U. Wash., 1967; MA in Govt., U. Ariz., 1969, PhD in Govt., 1973. Grad. teaching and rsch. asst. dept. govt. U. Ariz., Tucson, 1967-70, rsch. assoc. Inst. Govt. Rsch., 1974-76, rsch. assoc. forest-watershed and landscape resources divsns. Sch. Renewable Natural Resources, 1975-82, adj. assoc. prof. Sch. Renewable Natural Resources, 1983-89, adj. assoc. prof. renewable natural resources, assoc. rsch. scientist Water Resources Rsch. Ctr., 1988-89, prof., rsch. scientist Water Resources Rsch. Ctr., 1989-90, prof., rsch. scientist, dir. Water Resources Rsch. Ctr., 1990—; program analyst USDA Forest Svc., Washington, 1979-80; vis. scholar Inst. Water Resources, Corps of Engrs., Ft. Belvoir, Va., 1986-87; com. arid lands AAAS, 1987—; com. natural disasters NAS/NRC, 1988—; rev. com. nat. forest planning Conservation Found., Washington, 1987-90; chair adv. com. renewable resources planning techs. for pub. lands Office of Tech. Assessment, U.S. Congress, 1989-91; mem. policy coun. Pinchot Inst. Conservation Studies, 1991—; co-chair working party on evaluation of forest policies Internat. Union Forestry Rsch. Orgns., 1990—; vice-chair Man and the Biosphere Program, Temperate Directorate, US Dept. of State, 1991—; cons. Greeley and Hansen, Consulting Engrs., U.S. Army Corps Engrs., Ft. Belvoir, U.S. Forest Svc., Washington, Portland, Oreg., Ogden, Utah. Author: (with others) New Dimensions to Energy Policy, 1979, Energy Policy and Public Administration, 1980, Energy and the Western United States: Politics and Development, 1982, Borderlands Sourcebook, 1983, Climate Change and U.S. Water Resources, 1990; assoc. editor Society and Natural Resources, 1992—; book reviewer Western Polit. Science Quar., Am. Polit. Quar., Perspectives, Natural Resources Jour., Climatic Change, Society and Natural Resources; pub. papers and monographs; contbr. articles to profl. jours. Bd. dirs. Planned Parenthood Southern Ariz., 1992—, planning com., 1992; bd. dirs. Northwest Homeowners Assn., 1982-83, v.p., 1983-84, pres., 1984; vice chmn., chmn. Pima County Bd. Adjustment Dist. 3, 1984; exec. asst. Pima County Bd. Suprs., 1985-86; active Pima Assn. of Govts. Avra Valley Task Force, 1988-90, Tucson Tomorrow, 1984-88, water mgmt. com. Joint SAWARA-Tucson Tomorrow, 1983-85; environ. planning adv. com. Pima Assn. Govts., 1989-90, chmn., 1984, water quality subcom., 1983-84; bd. dirs. Southern Ariz. Water Resources Assn., 1984-86, 87-93, sec., 1987-89, com. alignment & terminal storage, 1990—, CAP com., 1989—, chair, 1989—, basinwide mgmt. com., 1983-86, 93; active Ariz. Interagy. Task Force on Fire and the Urban/Wildland Interface, 1990-92, wastewater mgmt. adv. com. Pima County, 1988-92, subcom. on effluent reuse Joint CWAC-WWAC, 1989-92, citizens water adv. com. Water Resources Plan Update Subcom., 1990-91; bd. dirs. Ctrl. Ariz. Water Conservation Dist., 1985-90; active adv. coun., forum Colo. River Salinity Control, 1989-90, fin. com., 1987-88, spl. studies com., 1987-88, nominating com., 1987; chair adv. com. Tucson Long Range Master Water Plan, 1988-89; active water adv. com. City of Tucson, 1984. Travel grantee NSF/Soc. Am. Foresters; Rsch. grantee US Geol. Survey, US Army Corps of Engrs., USDA Forest Svc., Soil Conservation Svc., Utah State U., Four Corners Regional Commn., Office of Water Rsch. & Tech.; Sci. & Engring. fellow AAAS, 1986-87; recipient Copper Letter Appreciation cert. City of Tucson, 1985, 89, SAWARA award, 1989. Mem. Am. Water Resources Assn. (nat. award com. 1987—, statuses and bylaws com. 1989-90, tech. co-chair ann. meeting 1993), Am. Forestry Assn. (forest policy ctr. adv. coun. 1991—), Soc. Am. Foresters (task force on sustaining long-term forest health and productivity, 1991-92), Am. Polit. Sci. Assn., Western Psn. (com. constn. and by-laws 1976-80, chair 1977-79, exec. coun. 1980-83, com. profl. devel. 1984-85, com. on the status of women 1984-85), Nat. Fire Prevention Assn. (tech. com. on forest and rural fire protection 1990—), Phi Beta Kappa. Democrat. Achievements include research in political and socioeconomic aspects of natural resources policy, administration, and planning, water resources management, wildland fire policy and management. Home: 1425 Calle Tiburon Tucson AZ 85704 Office: University of Arizona Water Resources Research Ctr 350 N Campbell Tucson AZ 85721

CORWIN, WILLIAM, psychiatrist; b. Boston, Oct. 28, 1908; M.D., Tufts Coll., 1932; m. Frances M. Wetherell (dec.) m. Joyce S. Newman, 1965. Intern Wesson Meml. Hosp., Springfield, Mass., 1932-33; physician Met. State Hosp., Waltham, Mass., 1933-37, asst. supt., 1937-42; research fellow Harvard, 1937-46; practice medicine, specializing in psychiatry, Springfield, Mass., 1946-54, Miami, Fla., 1954-88, Ocala, Fla., 1988—; mem. staff Charter Springs Hosp., Ocala., Marion Community Hosp., Ocala; instr. psychiatry Boston U., 1937-46, Tufts Coll., 1941-46; clin. assoc. prof. psychiatry U. Miami, 1955-70, clin. prof., 1970-88. Past mem. State Fla. Adv. Com. on Mental Health; agy. ops. com. United Fund. Bd. dirs. Family and Childrens Svcs. Miami. Served to lt. col. M.C., USAAF, 1942-46. Diplomate Am. Bd. Psychiatry and Neurology, Am. Bd. Forensic Psychiatry. Fellow Am. Psychiat. Assn. (life), Am. Coll. Psychiatrists; mem. AMA, S. Fla. Psychiat. Soc. (councillor), Fla. Psychiat. Soc. Contbr. articles on physiology of schizophrenia to profl. publs. Office: 1111 NE 25th Ave Ocala FL 34470

CORY, LESTER WARREN, electrical engineering educator; b. Tiverton, R.I., July 25, 1939; s. Harold R. and Margaret (Grant) C.; m. Patricia L. Barrett, May 23, 1981; children from previous marriage: Stephen, Dyan, Michael, Ann. MSEE, Northeastern U., 1970; MEd, Bridgewater State Coll., 1974. Prof. elec. engring. U. Mass.-Dartmouth, North Dartmouth, 1963—, dir. Ctr. for Rehab. Engring., 1988—. Co-author: Electrical Measurements for Engineers, 1970. Col. (ret.) R.I. Air NG, 1957-92. Decorated Legion of Merit; recipient Pres. Vol. Action award Pres. Ronald Reagan, 1985, Meritorious Achievement award Johns Hopkins U., 1985, Others award Nat. Salvation Army, 1985, R.I. Star, Gov. of R.I., 1992, citation R.I. State Legis., 1992, Mass. State Legis., 1992. Mem. IEEE, Soc. for Shape Found., Inc., Human Advancement through Rehab. Engring. (pres. 1982—, founder). Home: 45 Summit Ave Tiverton RI 02878 Office: U Mass Dartmouth Dept Elec Engring North Dartmouth MA 02747

COSCAS, GABRIEL JOSUE, ophthalmologist, educator; b. Tunis, Mar. 1, 1931; s. Jules Joseph and Gilda (Guez) C.; M.D., U. Paris, 1963; m. Gisele Nataf, Mar. 23, 1957; children—Florence, Brigitte. Chief ophthalmology clinic Hotel Dieu, Paris, 1963-70; maitre conf. agrege Univ. Paris-Val de Marne, 1970-79, prof. ophthalmology, chmn. dept. Univ. Eye Clinic de Creteil, 1979—; founder Conf. Angiographic de Creteil, 1972—. Served as med. lt. French Army, 1958-60. Decorated chevalier de l'Ordre de Palmes Academiques, 1984, chevalier de l'Ordre Nat. de la Legion d'Honneur, 1985. Mem. Internat. Orgn. Against Trachoma (pres. 1977), French Soc. Photocoagulation (pres. 1988), Acad. Internat. Ophthalmology. Author books, articles in field. Home: 203 Vaugirard, 75015 Paris France Office: 40 Ave Verdun, 94010 Creteil France

COSENS, KENNETH WAYNE, sanitary engineer, educator; b. Fairgrove, Mich., June 22, 1915; s. Omer and Sarah Emma (Keyser) C.; m. Arlene M. Hilsinger, Sept. 5, 1936; children: Barry L., Karen D., Brian K., Denise A. BS, Mich. State U., 1938, MS, 1946, degree in civil engring., 1948. Registered profl. engr., Mich., Ohio. Instr., then asst. prof. engring. Mich. State U., East Lansing, 1941-47; asst. prof. hydraulics and sanitary engring. U. Tex., Austin, 1947-49; assoc. prof. engring., cons. engr. Ohio State U., Columbus, 1949-70; ptnr. A.E. Stilson-Consulting Engrs., Columbus, 1970-80; adminstr. supply and treatment Divsn. of Water, City of Columbus, 1980-84, ret., 1984—; cons. Battelle Meml. Inst., Columbus, 1963-68, Burgess & Niple Engrs., Columbus, 1950-61. Contbr. over 40 tech. articles to profl. jours. Recipient Disting. Svc. medal Divsn. of Fire, City of Columbus, 1988. Fellow ASCE; mem. NSPE (pres. local chpt.), Am. Water Works Assn. (George W. Fuller award 1980), Water Pollution Control Assn. Methodist. Home: 2620 Chester Rd Columbus OH 43221

COSGRIFF, STUART WORCESTER, internist, consultant; b. Pittsfield, Mass., May 8, 1917; s. Thomas F. and Frances Deford (Worcester) C.; m. Mary Shaw, Jan. 23, 1943; children: Mary, Thomas, Stuart, Richard, Robert. B.A. cum laude, Holy Cross Coll., 1938; M.D., Columbia U., 1942, D.Med. Sci., 1948. Diplomate: Am. Bd. Internal Medicine. Intern Presbyterian Hosp., N.Y.C., 1942-43; asst. resident in medicine, 1943, 46-47, chief resident, 1947-48; instr. in medicine Columbia U., N.Y.C., 1948-50, clin. asst. prof. medicine, 1951-63, clin. assoc. prof., 1963-73, clin. prof. medicine, 1973-83, clin. prof. emeritus, 1983—; attending physician Presbyn. Hosp., N.Y.C., 1948-83; cons. emeritus Presbyn. Hosp., 1984—; individual practice medicine, specializing in internal medicine and vascular diseases, 1948—; cons. in medicine to dir. Selective Svc., N.Y.C., 1957-73, N.Y. Giants Baseball Club, 1951-57, San Francisco Baseball Club, 1958-61; dir. thrombo-embolic clinic Vanderbilt Clinic, N.Y.C., 1948-83. Contbr. articles to med. jours. Served to capt. M.C., U.S. Army, 1943-45, ETO. Fellow ACP, Pan Am. Med. Assn.; mem. Am. Heart Assn., N.Y. Heart Assn., Alpha Omega Alpha. Roman Catholic. Club: Knickerbocker Country (Tenafly, N.J.). Home and Office: 11 Park St Tenafly NJ 07670-2217 Office: 161 Ft Washington Ave New York NY 10032-3713

COSGROVE, BENJAMIN A., retired aerospace company executive. Former sr. v.p. engring. and tech. Boeing Comml. Airplane Group, Seattle. Recipient Wright Brothers Lectureship in Aeronautics AIAA, 1991. Office: Boeing Comml Airplane Group N 8th & Park Ave N Box 3707 Seattle WA 98124*

COSGROVE, DANIEL JOSEPH, biology educator; b. Westover AFB, Mass., Sept. 15, 1952; s. Alfred Kevin and Irene B. (Roullier) C.; m. Leandra Throughton, Sept. 29, 1979; 1 child, Kevin. BA, U. Mass., 1974; PhD,

Stanford U., 1980. Vis. scientist Nuclear Rsch. Ctr., Julich, West Germany, 1980-81; postdoctoral assoc. botany dept. U. Wash., Seattle, 1981-82; asst. prof., assoc. prof. biology dept. Pa. State U., University Park, 1983-91, prof., 1991—. Contbr. articles to profl. jours. Named Presidential Young Investigator NSF, 1984, 89; recipient award McKnight Found., 1986, 89, Sr. Prof. award Fulbright/Hayes Commn., 1989-90; J.S. Guggenheim Found. fellow, 1989. Mem. AAAS, Am. Soc. Plant Physiologists (Charles Albert Shull award 1991), Am. Soc. Gravitational and Space Biology, Am. Soc. Photobiology (councillor 1987-90), Am. Inst. Biol. Scis. Office: Pa State U Biology Dept 208 Mueller Lab University Park PA 16802-5301

COSGROVE, RAYMOND FRANCIS, science company executive; b. Wallasey, Merseyside, Eng., Dec. 17, 1947; s. Francis and Patricia Marjorie (Dodwell) C.; m. Kathleen Susan Plant, Oct. 16, 1971; children: Simon Francis, Matthew James. PhD, London U., 1978. Chartered biologist. Microbiologist E.R. Squibb & Sons, Moreton, Eng., 1966-86; rsch. mgr. Shandon Sci. Ltd., Runcorn, Eng., 1986-89, bus. devel. dir., 1989-93; exec. dir. Microbiological Assocs. Internat. Ltd., Stirling, Scotland, 1993—. Treas. Wallasey Civic Soc., 1983-84. Recipient R&D 100 award R&D Mag., Ill., 1990, award Brit. Design Coun., London, 1991. Mem. Inst. Biology (com. mem. N.W. br. 1983). Roman Catholic. Achievements include patents for chemotherapy and clin. lab. automation; research on fundamental action of polyene antibiotics. Office: Microbiological Assocs Internat Ltd, Hillfoots Rd, Stirling FK9 4NF, Scotland

COSMI, ERMELANDO VINICIO, obstetrics/gynecology educator, consultant; b. Perugia, Umbria, Italy, June 10, 1937; s. Dante and Evarista (Galletti) C.; m. Viktoria Bastianon, Dec. 26, 1971; children: Kristian, Erich. MD summa cum laude, U. Perugia, 1962; degree in Ob-Gyn., U. Florence, Italy, 1968; degree in Anethesiology, U. Pisa, Italy, 1966; degree in Pediatrics, U. Perugia, 1972; LD of Anesthesiology (hon.), Ministry for Edn., Rome, 1968, LD of Ob-Gyn., 1972. Asst. surgery dept. U. Perugia and Pisa, 1962-64; asst. ob-gyn. U. Florence, Italy, 1964-65; NATO rsch. fellow ob-gyn. Columbia U., N.Y.C., 1965-66; rsch. fellow ob-gyn. Einstein Coll. Yeshiva U., N.Y.C., 1966-67; asst. prof. ob-gyn. U. Rome, 1967-72; assoc. prof. ob-gyn. Einstein Coll. Yeshiva U., N.Y.C., 1973-74; assoc. prof. ob-gyn. and anesthesiology U. Rome, 1976-79; chmn. ob-gyn. dept. U. Perugia, 1980; chmn. reproductive medicine dept. U. Rome, 1990—, chmn. Inst. Ob-Gyn., 1992; cons. Italian Rsch. Coun., Rome, 1976—; perinatal rsch. coord. U. Perugia, 1986—; mem. FIGO 1983—; mem. EEC Commn., 1980—. Author: Obstetric Anesthesia and Perinatology, 1981, (with Scarpelli) Pulmonary Surfactant System, 1983 (with Di Renzo and Hawkins) Progress in Perinatal Medicine, 1990. Mem. Italian Soc. Gynecology and Obstetrics, Internat. Assn. for Maternal and Neonatal Health, Italian Soc. for Interdisciplinary Medicine (founder, 1st chmn. 1985—), Soc. of Perinatal Obstetricians (USA), European Assn. Perinatal Medicine (chmn. 1986-88), Soc. for Gynecologic Investigation (USA), World Assn. Perinatal Medicine (pres. elect 1991—). Avocations: tennis, underwater fishing, skiing. Home: Via Monte Madonna No 23, Rome 00060, Italy Office: U Rome Dept Reproductive Medicine, V Le Regina Elena 314, Rome 00161, Italy

COSNER, CHRISTOPHER MARK, engineer; b. Balt., July 26, 1961; s. Donald Lester and Shirley Marie (Ryan) C.; m. Christine Tina Itano, July 5, 1992. BS with highest distinction, U. Va., 1983; MSME, U. Calif., Berkeley, 1990. Mem. tech. staff Calif. Inst. Tech. Jet Propulsion Lab., Pasadena, 1983-88; tech. staff Hughes Aircraft Co., L.A., 1990—. Contbr. articles to IEEE Robotics and Automation Jour., IEEE Aerospace Controls Systems Jour. Regents fellow U. Calif., Berkeley, 1988-89; named Outstanding Aerospace Student Sigma Gamma Tau, 1983; recipient Jefferson Scholar award U. Va. Alumni Assn., Charlottesville, 1979-83, Achievement award NASA, Washington, 1989-90, Hughes Tech. Achievement award 1993. Mem. AIAA (student chpt. pres. 1983), Tau Beta Pi. Office: Hughes Aircraft Co PO Box 92919 MS EO-E1-D110 Los Angeles CA 90009

ĆOSOVIĆ, M. BOŽENA, chemist, researcher; b. Zagreb, Croatia, June 8, 1940; d. Mijo and Anica (Tapšanji) Peh; m. Čedomir Ćosović. July 13, 1963; 1 child, Ksenija. BSc in Chem. Tech., U. Zagreb, Croatia, 1963, MSc in Chemistry, 1965, PhD in Chemistry, 1967. From asst. to sr. scientist Rudjer Bošković Inst., Zagreb, 1963-90—; head of lab. Radjer Bošković Inst., Zagreb, 1977—, dir. ctr. for Marine Rsch. at, 1991—; prof. in oceanology U. Zagreb, 1971—; cons. Ministry Water Mgmt., Zagreb, Croatia, 1983—; mem. Editorial Bd. Croatica Chemica Acta, 1991—; mem. European Rsch. Conf. on Natural Waters and Water Tech., vice chairperson, 1990, chairperson, 1992. Contbr. 70 sci. articles to profl. jours.; author (book chpts.) Chemical Processes in Lakes, 1985, Aquatic Chemical Kinetics, 1990. Mem. Internat. Assn. Water Rsch. Pollution Control, Am. Chem. Soc. (environ. div.), Internat. Com. for Sci. Exploration of Mediterranean, European Environ. Rsch. Orgn., Croatian Chem. Soc. Achievements include development of new analytical methods for rsch. and control of marine, freshwater and estuarine systems; pollution monitoring and legislation; internat. sci. collaboration. Home: Mazuranicev trg 5, 41000 Zagreb Croatia Office: Rudjer Bošković Inst, PO Box 1016 Bijenička 54, 41000 Zagreb Croatia

COSSINS, EDWIN ALBERT, biology educator, academic administrator; b. Havering, Eng., Feb. 28, 1937; came to Can., 1962; s. Albert Joseph and Elizabeth H. (Brown) C.; m. Lucille Jeannette Salt, Sept. 1, 1962; children: Diane Elizabeth, Carolyn Jane. B.Sc., U. London, 1958, Ph.D., 1961, D.Sc., 1981. Research assoc. Purdue U., Lafayette, Ind., 1961-62; from asst. prof. to prof. U. Alta., Edmonton, Can., 1962—, acting head dept. botany, 1965-66, assoc. dean of sci., 1983-88; mem. grant selection panel Natural Scis. and Engring. Research Council, Ottawa, Ont., Can., 1974-77, 78-81. Author: (with others) Plant Biochemistry; 1980, 1988, Folates and Pterins, 1984. Assoc. editor Can. Jour. Botany, 1969-78. Contbr. numerous articles to profl. jours. Recipient Centennial medal Govt. of Can., 1967. Fellow Royal Soc. Can. (life); mem. Can. Soc. Plant Physiologists (western dir. 1968-70, pres. 1976-77), Japanese Soc. Plant Physiologists, Am. Soc. Plant Physiologists. Clubs: Faculty (U. Alta.), Derrick Golf and Winter. Avocations: gardening; golf; curling; cross-country skiing. Home: 99 Fairway Dr, Edmonton, AB Canada T6J 2C2 Office: U Alta, Dept Botany, Edmonton, AB Canada T6G 2E9

COSTA, LUÍS CHAVES DA, engineering executive; b. Lisbon, July 7, 1952; s. Tomas Chaves da and Maria Rita da C.; m. Isabel Borges (div.); 1 child, Rita; m. Maria de Lurdes Sabino Chaves, Dec. 27, 1983; 1 child, André. Diploma in Civil Engring., Inst. Superior Técnico, Lisbon, 1976; hydrologist, Cen. de Estudos Hidrógraficos, Madrid, 1977, Lomonosov Univ., Moscow, 1978. Registered profl. engrs. Civil engr. Ministry of Pub. Works, Lisboa, Portugal, 1976-80; mng. dir. Vortice, Equipamentos Científicos, Lda., Lisboa, 1980—; fiscal coun. mem. Assn. Portuguesa Empresas Tech. Ambiental, Lisboa, 1991—. Author: Formula Simplificada para a determinacao dos caudais maximos de cheia nas regioes situadas a sul da bacia do tejo, 1975. Fellow Ordem dos Engenheiros, Assn. Portuguesa dos Recursos Hidricos. Avocations: lecturing, tennis. Office: Vortice Equipamentos Cientitificos Lda, Rua de Xabregas 20 Piso, 2 E 2 04 1900 Lisbon Portugal

COSTAKOS, DENNIS THEODORE, neonatologist, researcher; b. N.Y.C., Mar. 7, 1958; s. Nicholas and Pauline (Mary) C.; m. Anne Elizabeth Merideth, Nov. 12, 1989; 1 child, Chloe Elizabeth. AB in Biochemistry, Columbia U., 1980; MD, Dartmouth Coll., 1984. Diplomate Am. Bd. Med. Examiners; cert. regional neonatal resuscitation instr. Am. Heart Assn., Am. Acad. Pediatrics. Pediatrics resident Mount Sinai Hosp., N.Y.C., 1984-87; neonatal fellowship Cornell-N.Y. Hosp., 1987-89; dir. neonatal intensive care St. Francis Hosp., La Crosse; instr. St. Francis-Mayo Clinic Residency Program, La Crosse, 1989—; apptd. subcom. on metabolic diseases Newborn Screening Adv. Bd., Wis., 1990; apptd. The Perinatal Group-La Crosse County Health Soc.; presenter and lectr. in field. Contbr. articles to Psychiatry Research, Acta Pharmacol, et Toxicol, Biochemical Pharmacology, Critical Care Medicine, Jour. Rheumatology, Am. Jour. Perinatology, Am. Jour. Diseases Children. Hon. chairperson March of Dimes, La Crosse, 1992. Fellow Am. Acad. Pediatrics (bd. cert.); mem. AAAS, Wis. Assn. Perinatal Care, Hellenic Med. Soc. N.Y., Fedn. for Clin. Rsch. Greek Orthodox. Home: N 3589 Peters Rd La Crosse WI 54601 Office: St Francis-Mayo 700 West Ave S La Crosse WI 54601

COSTA NETO, ADELINA, chemistry educator, consultant; b. Rio de Janeiro, Jan. 29, 1935; arrived in U.S., 1956-57; d. Kleber Augusto and Nair Analia (Rocha) de Moraes; m. Claudio Costa Neto, Jan. 28, 1956 (div. 1987); children: Cristina Costa Neto, Marcelo Costa Neto. Degree chemistry, Fed. U., Rio de Janeiro, 1959, MS, 1965, DSc, 1974. Asst. rschr. Nat. Rsch. Coun., Rio de Janeiro, 1959-62; acad. staff mem. Fed. U., Rio de Janeiro, 1963-90; coord. grad. sch. organic chem. dept. Inst. of Chemistry, Fed. U., 1981-84; mem. of com. for approval of grad sch. of chemistry Coordenação de Aperfeiçoamento de Pessoal de Nivel Superior U. of São Carlos, CAPES, Brasilia, 1984; postdoctoral U. Bristol, U. Sussex, Eng., 1990-91; invited collaborative prof. grad. sch. Inst. of Chemistry, Rio de Janeiro, 1992-93; cons., rsch. coord. Salgema Industrias Quimicas S/A, Rio de Janeiro, 1984-89; cons. projects, reports Financiadora de Estudos e Projetos, Rio de Janeiro, 1984-89; cons. projects Nat. Rsch. Coun., Rio de Janeiro, 1984-89. Author: What is Chemistry/Utile Chemistry and Futile Chemistry, 1989, Interesting Problems in Organic Analysis, 1989; contbr. articles to profl. publs. Home: Rua Timoteo da Costa, 297/602, 22450-130 Rio de Janeiro Brazil

COSTILL, DAVID LEE, physiologist, educator; b. Feb. 7, 1936. BSEd, Ohio U., 1959; MA, Miami U., 1961; PhD in Physiology, Ohio State U., 1965. Instr. physiology Ohio State U., 1963-64; asst. prof. SUNY, Cortland, 1964-66; prof. phys. edn. and biology human performance lab. Ball State U., Muncie, Ind., 1966—, dir.; vis. faculty Desert Rsch. Inst., 1968; hon. lectr. U. Sulford, Eng., 1972; rsch. assoc. Gymnastik-och idrottshogskolan, Stockholm, 1972-73. Mem. Am. Physiol. Soc., Am. Soc. Zoologists. Achievements include research in alterations in skeletal muscle water and electrolytes following prolonged exercise and dehydration in man, glycolytic-oxidative enzyme and fiber composition in human skeletal muscle. Office: Ball State University Human Performance Lab Muncie IN 47306*

COSTIN, DANIEL PATRICK, mechanical engineer; b. Evergreen Park, Ill., June 18, 1962; s. Edward James and Mary Rita (Barrett) C.; m. Kimberly Sue Stump, Sept. 3, 1988; 1 child, Claire. BS, U. Ill., 1984, MS, 1987; PhD, U. Tex., Arlington, 1991. Registered profl. engr., Tex. Engr. GM, Warren, Mich., 1984-85; engr. Gen. Dynamics, Ft. Worth, 1986-89, sr. engr., 1989-92, engring. specialist, 1992-93; sr. scientist BorgWarner, 1993—. Mem. ASME.

COSTNER, CHARLES LYNN, civil engineer; b. Banner, Miss., Aug. 15, 1928; s. Charles Arthur and Clyde Margarite (Head) C.; m. Sara Lynn McGuire, May 26, 1951; 1 child, Jeffrey Lynn. BSCE, U. Miss., 1951, MSCE, 1955. Registered profl. engr., Ark., La., Miss., Tex.; registered land surveyor. Engr. E.I. Dupont, Wilmington, Del., 1951-53, Ross E. Cox, Baton Rouge, 1955-65, Brown and Butler Cons. Engrs., Baton Rouge, 1965-83; pres. Brown & Butler Inc., Baton Rouge, 1983—. Contbr. articles to mags. Airport Services Management, Ports 83, ASCE. With U.S. Army, 1946-48, Korea. Mem. Am. Cons. Engrs. Coun., Am. Concrete Inst., Assn. State Dam Safety Ofcls., Soc. Am. Mil. Engrs. Republican. Baptist. Home: 8476 S Parkland Dr Baton Rouge LA 70806 Office: Brown & Butler Inc Ste D 8335 Kelwood Ave Baton Rouge LA 70806

COTÉ, GERARD LAURENCE, biomedical engineering educator; b. North Tonawanda, N.Y., May 3, 1963; s. Lionel A. and Irma G. (Peola) C.; m. Iris E. Allen, June 7, 1986; children: Spencer Harrison, Mason Allen. BSEE, Rochester Inst. Tech., 1986; MS, U. Conn., 1987, PhD, 1990. Registered profl. engr., Tex. Engr. Delco Products div. G.M., Rochester, N.Y., 1983; quality control engr. IBM, Boca Raton, Fla., 1984; test and design engr. Fed. Systems div. IBM, Owego, N.Y., 1984; elec. engr. Bell Aerospace/Textron, Niagara Falls, N.Y., 1985-86; teaching and rsch. asst. U. Conn., Storrs, 1986-90, postdoctoral rschr., 1991; asst. prof. biomed. engring. Tex. A&M U., College Station, 1991—; cons. InoMet, Inc., Roseville, Minn., 1992. Contbr. articles to IEEE Trans. on Biomed. Engring., Biology and Medicine, others. Mem. IEEE (Engrng. in Medicine and Biology Soc.), Biomed. Optics Group, Biomed. Engring. Soc., Sigma Xi, Tau Beta Pi, Eta Kappa Nu, Phi Kappa Phi. Achievements include patent pending for noninvase glucose sensor. Home: 1204 Neal Pickett Dr College Station TX 77840 Office: Tex A&M U 233 Zachry Engring Ctr College Station TX 77843-3120

CÔTÉ, JEAN-CHARLES, research molecular biologist; b. Cap-Chat, Que., Can., Aug. 31, 1956; s. Adrien and Lucienne (St.-Laurent) C.; m. Pauline Frève, May 19, 1984; children: Philippe, Mélissa. BSc, U. Que., Rimouski, 1980; MSc, Sherbrooke (Que.) U., 1983; PhD, Cornell U., 1988. Rsch. scientist Agr. Can., St-Jean-sur-Richelieu, Que., 1989—; invited rsch. scientist U. Montreal, Que., 1992—; reviewer NSERC-Biotech., 1991—, CORPAP-Biotech., 1992—. Contbr. articles to sci. jours. Scholar Fonds pour la Formation de Chercheurs et L'Aide à la Recherche, 1981-83, Agr. Can., 1984-88. Mem. Coun. for Responsible Genetics (assoc.), Soc. for Invertebrate Pathology, Assn. Adv. Biotech., Planetary Soc. Roman Catholic. Achievements include research on molecular evolutionary genetics, Bacillus thuringiensis as microbial insecticides, pest resistance to pesticides. Home: 880 Lanoue, Saint Jean-sur-Richelieu, PQ Canada J3B 7M4 Office: Agr Can, 430 Gouin Blvd, Saint Jean-sur-Richelieu, PQ Canada J3B 3E6

COTÉ, MARC GEORGE, electrical engineer; b. Lewiston, Maine, Sept. 4, 1959; s. Romeo Maurice Coté and Therese Doris (Lavertu) Boardman. BS, U. Maine, Orono, 1982, MS, 1989. Rsch. scientist Rome Lab., Hanscom AFB, Mass., 1983—. Contbr. articles to IEEE Transactions on Antennas and Propagation and on Instrumentation and Measurement. Vol. AIDS Action Com., Boston, 1988—; vol. tutor Boston Ptnrs. in Edn., 1991—. Mem. IEEE, Antenna Measurement Tech. Assn. Office: Rome Lab RL/ERCT 31 Grenier St Hanscom AFB MA 01731-5000

COTNEY, CAROL ANN, research aerospace engineer; b. Huntsville, Ala., July 23, 1957; d. John Walter and Helen Maxine (Bechtold) C. BS in Family Resource Mgmt., Auburn U., 1979; BSME, U. Ala., Huntsville, 1986, student. Clk. Jack Eckerd Drug Co., Huntsville, 1979-83; engring. student trainee U.S. Army Missile Command RDEC, Redstone Arsenal, Ala., 1983-86; rsch. aerospace engr. U.S. Army Missile Command, RDEC, Redstone Arsenal, Ala., 1988-93; U.S Army rep. Autoignition Temperature Experimental Standard, 1991—; mem. structures and mech. behavior subcom. material and properties and characterization panel, 1987-88, mem. structural analysis panel, 1987-88, acting program chmn., 1987-88, mem. profl. devel. com. originator, 1988, mem. propulsion systems hazards subcom. cookoff hazard tech. panel, 1991-93; U.S. Army rep., 1991-93; mem. cookoff workshop originator and planning com., 1991-92, mem. propulsion industry cookoff tech. long range planning com., 1991. Contbr. articles on propulsion mechanics and insensitive munitions to profl. jours. Sr. high sch. advisor Covenant Presbyn. Ch., Huntsville, 1980-83; chmn. by-laws com. Christian Singles Fellowship, Huntsville, 1984, v.p., pres., 1985-86; singles coord. Hope Presbyn. Ch., Huntsville, 1984-87, trustee, 1987-88; treas. United Meth. Women, Circle 9, Lathan United Meth. Ch., Huntsville, 1991-92, mem. missions com., 1992—, sec., 1992; tutor Seminole Svc. Ctr., Huntsville, 1991; judge local, regional, state Ala. Sci. and Engring. Fair. Recipient award of commendation U.S. Army Missile Command, 1988, 92. Mem. AIAA (U.S. Army Missile Command facility rep. Ala.-Miss. sect. 1991-92, assoc. dir. edn. 1991-92, dir. telemetry newsletter 1992, region II co-dep. dir. precoll. outreach 1992—, nat. engrs. week planning com. region II rep. 1993), Nat. Space Soc. Achievements include 2 patents pending in field; avocations: contemporary Christian and classical instrumental music, travel, collecting Hallmark Christmas ornaments and spoons, bicycling. Home: 13909 Clovis Cir SW Huntsville AL 35803-2509 Office: Comdr USA Missile Command AMSMI-RD-PR-M Redstone Arsenal AL 35898

COTT, DONALD WING, aerospace engineer; b. San Francisco, July 17, 1941; s. William Warner and Eula Vesta (Sullivan) C.; m. Sandra Ann Bales, Jan. 25, 1963; children: Craig, Damon, Jamalea, Kamalea. MS, Okla. State U., 1966; PhD, U. Tenn., 1969. Rsch. assoc. Arnold Engring. Devel. Ctr., Tullahoma, Tenn., 1966-69; asst. prof. N.C. State U., Raleigh, 1969-72; assoc. prof. Clemson (S.C.) U., 1972-75; rsch. scientist Reynolds Metals Co., Sheffield, Ala., 1975-77; staff scientist MSE, Inc., Butte, Mont., 1977—; staff scientist, pres. Blue Sky Rsch., Inc., Idaho Falls, Idaho, 1992—; cons. R. S. Noonan, Inc., Greenville, S.C., 1975, Malcolm Jones and Assocs., Alexandria, Va., 1977, High Country Engring., Inc., Butte, 1981-91. Mem. AIAA, ASME (Outstanding Paper award 1991). Baha'i. Achievements include first new applications of magnetohydrodynamics to electromagentic pulse warheads, submarine and space propulsion; introduction of sacraficial ion injection to electrode walls of MHD generators for improved performance. Home: 630 Crestview St Idaho Falls ID 83402 Office: Blue Sky Rsch Inc Box 2501 Idaho Falls ID 83403-2501

COTTEN-HUSTON, ANNIE LAURA, psychologist, educator; b. Oxford, N.C., Nov. 18, 1923; d. Leonard F. and Laura Estelle (Spencer) Cotten; diploma Hardbarger Bus. Coll., 1944; AB, Duke U., 1945; MEd, U. Hartford, 1965; PhD, Union Grad. Sch., 1979; children: Hollis W., Rebecca Ann, Laura Cotten. Diplomate Am. Bd. Sexology. Asst. to pres. So. Meth. U., 1953; rsch. asst. Duke U., 1947-49; exec. sec. Ohio Wesleyan U., 1955-56, Conn. Coun. Chs., 1958-60; adj. prof. U. Hartford, 1976-78; clin. pastoral counselor Hartford Hosp., 1962-65; asst., then asso. dir. social svcs. Hartford Conf. Chs., 1965-67; teaching fellow U. N.C., 1970-71; adj. prof. U. Hartford, 1976-78; assoc. prof. Cen. Conn. State U., New Britain, 1967—; dir. elderhostel programs, Cen. Conn. State U., 1989—; organizer ctr. adult learners Cen. Conn. State U., 1991—; cons. Somers Correctional Ctr. (Conn.), 1980-81, instr./researcher, 1980-81; cons. Life Ins. Mktg. Rsch., 1981—; amb. to China, spring, 1986; presenter 3d Internat. Interdisciplinary Cong. on Women, 1987; vis. prof., scholar Duke U., 1989; vis. prof. Conn. Coll. New London, Conn., 1990; dir. Ctr. Adult Learners Cen. Conn. State U., 1991—. Elder hostel dir. Cen. Conn. State U., 1987—. Mem. AAUW, Am. Personnel and Guidance Assn., Am. Assn. Marriage and Family Therapists (cert.), Am. Psychol. Assn. (presenter conf. 1987), Nat. Coun. Family Rels., Am. Assn. Sex Educators, Counselors and Therapists (cert.), Conn. Psychol. Assn., Conn. Council Chs. (dir.), Hartford Women's Network. Contbr. articles to profl. jours. Home: 193 Westland Ave West Hartford CT 06107-3057 Office: Ctrl Conn State U Dept Psychology New Britain CT 06050

COTTER, JOHN BURLEY, ophthalmologist, corneal specialist; b. Zanesville, Ohio, Sept. 14, 1946; s. John Burley and Evelyn Virginia (Ross) C.; m. Perrine Abauzit, Aug. 17, 1977; children: Neils John, Jeremy Pierre. BA, U. Kans., 1968; med. degree, U. Kans., Kansas City, 1968-72. Ophthalmology resident U. Mo., Kansas City, 1976-79; family practice Ashland (Kans.) Hosp., 1973-74; emergency room physician Providence-St. Margaret Hosp., Kansas City, Kans., 1974-75; family orthopedic practice Mountain Med. Assocs., Vail, Colo., 1975-76; ophthalmologist, pvt. practice Duluth, Minn., 1979-82; surgeon-chief out-patient clinic King Khaled Eye Specialist Hosp., Riyadh, Saudi Arabia, 1983-90, mem. exec. com., 1985-90; asst. clin. prof. King Saud U., Riyadh, Saudi Arabia, 1985-90; corneal specialist Simel Eye Assocs., Greensboro, N.C., 1990—; seminar chmn. Status of Refractive Surgery, Riyadh, 1986; active Nat. Survey Eye Disease and Ea. Province Survey Coun., Saudi Arabia, 1984, 90. Author: (booklet) Radial Keratotomy, 1986; contbr. articles to profl. jours. Rsch. grantee Contact Lens Assn. of Ophthalmology, 1981, Lasers Steering Com. King Khalid Eye Hosp. at Hosp. Hotel Dieu, Paris, 1988; ORBIS fellow Baylor U., Houston, 1982. Fellow Am. Acad. Ophthalmology: mem. AMA, Internat. Assn. Occular Surgeons, Internat. Soc. Refractive Keratoplasty, Societe Francaise D'Ophtalmologie, Saudi Ophthalmologisl Soc. Independent. Avocations: wind surfing, scuba diving, running, math games. Office: Simel Eye Assocs 300 W Northwood St Greensboro NC 27401-1380

COTTINGHAM, MARION SCOTT, computer scientist, educator; b. Paisley, Scotland, Apr. 3, 1948; d. James Paterson and Janet Rankine (Scott) Reynolds; m. Andrew Cottingham, Sept. 25, 1975 (div. May 1979); children: Steven, Alan. BSc, U. Glasgow, Scotland, 1981, PhD, 1988. Tech. programmer Britoil PLC, Glasgow, 1981-83; lectr. computer sci. dept. Monash U., Melbourne, Australia, 1986-89, U. Western Australia, Perth, 1989—. Chmn. Ausgraph 88 Video Evening & Art Exhbn. sub-com., Melbourne, 1988; convener Computers in Soc., Anzaas Centenary Congress, Sydney, Australia, 1988. Mem. IEEE, Eurographics. Achievements include research in computer graphics, intelligent tutoring systems, virtual reality. Office: Univ Western Australia, Computer Sci Dept, Nedlands Western Australia 6009, Australia

COTTLE, EUGENE THOMAS, aerospace engineer; b. Quincy, Wash., Sept. 3, 1958; s. Wallace A. and Melva (Ravsten) C.; m. Margret James, Dec. 14, 1982; children: Andrew, Stephanie, Jennifer. BSME, Utah State U., 1984; MS in Aerospace Engring., Air Force Inst. Tech., 1990. Cons. design engr. Hale Motor Co., Logan, Utah, 1983-84; commd. 2d lt. USAF, 1984, advanced through grades to capt., 1987; mech. engr. Air Force Armament Lab. USAF, Eglin AFB, Fla., 1984-89; structural loads and dynamics engr. Aero. Systems Ctr. USAF, Wright Patterson AFB, Ohio, 1990-92; resigned USAF, 1992; project engr. wheels and brakes divsn. Allied Signal Aircraft Landing Systems, South Bend, Ind., 1993—. Author conf. procs. in field. Mem. AIAA. Republican. Mem. LDS Ch. Office: Allied Signal ALS 3520 Westmoor St South Bend IN 46628-1391

COTTLE, ROBERT DUQUEMIN, plastic surgeon, otolaryngologist; b. Montreal, Que., Can., May 10, 1935; s. Melvin Wheeler and Lilian Louise (Butt) C.; m. Mildred Isabel Cave, 1960 (div. 1968); children: Stephen, Michael, Sean, Scott; m. Suzanne Kern, 1969; 1 child, Melanie Catherine. BA magna cum laude, Loyola Coll., Montreal, 1956; MD, C.M., McGill U., Montreal, 1960. Diplomate Am. Bd. Otolaryngology. Intern, then resident in gen. surgery Henry Ford Hosp., Detroit, 1960-62; rsch. fellow in otology Columbia U., N.Y.C., 1964; asst., resident surgeon in otolaryngology/head and neck surgery Columbia-Presbyn. Med. Ctr., N.Y.C., 1965-67; pvt. practice medicine specializing in facial plastic surgery, head and neck surgery, otolaryngology Stamford, Conn., 1967—; pres., med. dir. Stamford Hearing and Speech Ctr., Stamford; pres. SKC AIR, Inc., White Plains, N.Y., 1972—; chmn. bd. dirs., chief exec. officer Maple Leaf Petroleum, Dallas, 1982-85; chief div. otolaryngology/head and neck surgery St. Joseph Hosp., Stamford, 1979-92; sr. attending staff Stamford Hosp. Bd. dirs. United Way Stamford, PSRO Conn., Fairfield Health Plan, Physicians Health Svcs. Conn. (bd. chmn. IPA 1986-87). Capt. M.C., USAF, 1962-64. Recipient Gov.-Gen. of Can. medal, 1956. Fellow ACS, Am. Acad. Cosmetic Surgery, Am. Acad. Otolaryngology/Head and Neck Surgery, Am. Acad. Facial Plastic Surgery; mem. AMA, Am. Coll. of Surgeons (sec. Conn. state coun. 1987-88, pres. elect 1989-90, pres 1990-92), N.Y. Acad. Scis., Fairfield County Med. Assn. (trustee 1986—, chmn. bd. trustees 1988-90), Conn. State Med. Soc. (del. 1979—, assoc. councilor 1990—), Stamford Med. Soc. (pres. 1978-79), Internat. Platform Assn., Landmark Club. Republican. Office: 22 Long Ridge Rd Stamford CT 06905-3803

COTTLER, LINDA BAUER, epidemiologist; b. St. Louis, Oct. 30, 1950; d. James F. and Dolores (Bultas) Bauer; m. Matthew R. Cottler, Mar. 5, 1977; children: Emma, Laura, Sara. MPH in Epidemiology, Boston U. Sch. of Pub. Health, 1980; PhD in Sociology, Epidemiology, Washington U., 1987. RN. Project coord. Drug Epidemiology Unit, Birth Defects Interview Study, Boston, 1974-80, Epidemiology Catchment Area Program, St. Louis, 1980-86; project coord. Washington U. Sch. of Medicine, St. Louis, 1986-87, instr. dept. psychiatry, 1987—, rsch. instr., 1988-89, asst. prof., 1990-93; assoc. prof. Washington U. Sch. of Medicine, 1993—; mem. DSM-IV Task Force on Substance Use, 1991—; profl. rev. com. Mo. Dept. Mental Health, 1991-95. Contbr. articles to profl. jours. Mem. Nat. Adv. Coun. on Drug Abuse. Recipient Nat. Inst. Drug Abuse Travel scholarship, 1987, Washington U. Grad. of Arts & Scis. Tuition scholarship, 1981-85. Mem. Am. Coll. Epidemiology, Mo. Pub. Health Assn., Am. Sociol. Soc., Am. Psychopathol. Assn. Home: 524 Par Ln Ct Kirkwood MO 63122 Office: Washington U Sch Medicine Dept of Psychiatry 4940 Children's Pl Saint Louis MO 63110-1093

COTTON, FRANK ALBERT, chemist, educator; b. Phila., Apr. 9, 1930; s. Albert and Helen (Taylor) C.; m. Diane Dornacher, June 13, 1959; children: Jennifer Helen, Jane Myrna. Student, Drexel Inst. Tech., 1947-49; A.B., Temple U., 1951, D.Sc. (hon.), 1963; Ph.D., Harvard U., 1955; Dr. rer. Nat. (hon.), Bielefeld U., 1979; D.Sc. (hon.), Columbia U., 1980, Northwestern U., 1981, U. Bordeaux, 1981, St. Joseph's U., 1982, U. Louis Pasteur, 1982, U. Valencia, 1983, Kenyon Coll., 1983, Technion-Israel Inst. Tech., 1983, U. Cambridge, 1986, Johann Wolfgang Goethe Universität, 1989, U. S.C., 1989, U. Rennes, 1992, Lomonosov U., 1992. Instr. chemistry M.I.T., 1955-57, asst. prof., 1957-60, assoc. prof., 1960-61, prof., 1961-71; Robert A. Welch Distinguished prof. chemistry Tex. A&M U., 1971—, dir. Lab. for Molecular Stucture and Bonding, 1983—; Cons. Am. Cyanamid, Stamford, Conn., 1958-67, Union Carbide, N.Y.C., 1964—; Todd prof. U. Cambridge, 1985-

86. Author: (with G. Wilkinson) Advanced Inorganic Chemistry, 5th edit, 1988, Basic Inorganic Chemistry, 1987, Chemical Applications of Group Theory, 3d edit, 1989, (with L. Lynch and C. Darlington) Chemistry, An Investigative Approach; editor: (with L. Lynch and C. Darlington) Progress in Inorganic Chemistry, Vols. 1-10, 1959-68, Inorganic Syntheses, Vol. 13, 1971, (with L.M. Jackman) Dynamic Nuclear Magnetic Resonance Spectroscopy, (with R.A. Walton) Multiple Bonds between Metal Atoms, 2d edit., 1993. Recipient Michelson-Morley award Case Western Res. U., 1980, Nat. Medal of Sci., 1982; (hon.) fellow Robinson Coll., U. Cambridge. Mem. NAS (chmn. phys. scis. 1985-88, coun. 1991—, gov. bd. NRC 1992—, Cosepup 1992—, chem. scis. award 1990), King Faisal Internat. prize 1990, Am. Soc. Biol. Chemists, Am. Chem. Soc. (awards 1962, 74, Baekeland medal N.J. sect. 1963, Nichols medal N.Y. sect. 1975, Pauling medal Oreg. and Puget Sound sect. 1976, Kirkwood medal N.Y. sect. 1978, Gibbs medal Chgo. sect. 1980, Richards medal N.E. sect. 1986), Am. Acad. Arts and Scis., N.Y. Acad. Scis. (hon. life), Göttingen Acad. Scis. (corr.), Royal Soc. Chemistry (hon.), Royal Danish Acad. Scis. and Letters (hon.), Societa Chimica Italiana (hon.), Indian Acad. Scis. (hon.), Indian Nat. Acad. Sci. (hon.), Royal Soc. Edinburgh (hon.), Am. Philos. Soc., Acad. Europe (hon.). Home: 4101 Sand Creek Rd Bryan TX 77803-9616 Office: Tex A&M Univ Dept of Chemistry College Station TX 77843

COTTON, JEAN-PIERRE AIMÉ, physicist; b. Lyon, France, June 10, 1941; s. Eugene and Marie-Louise Cotton; m. Claude-Marie Brigitte Cabotte, Oct. 10, 1964; children: Christine, Antoine. Lic. in physics, U. Marseille, France, 1964; grad., Ecole Superieure d'Electricite, Paris, 1966; DSc, U. Paris, 1973. Rsch. assoc. Commissariat à l'Energie Atomique, Saclay, France, 1968-72, physicist, 1972—. Editor Jour. Physics, 1979-83; contbr. articles to profl. jours. Lt. French Navy, 1966-68. Mem. Soc. Fransaise de Physique, European Phys. Soc., Am. Phys. Soc. Avocations: garden, bicycle, movies. Home: 5 Allee du Bois Comtesse, 91440 Bures Sur Yvette France Office: Lab Leon Brillouin, CEN Saclay, 91191 Gif sur Yvette France

COTTRELL, ALAN, materials scientist; b. Birmingham, Ala., July 17, 1919; s. Albert and Elizabeth C.; m. Jean Elizabeth Harber, 1944. Educator Moseley Grammar Sch., 1943-49, U. Birmingham, U. Cambridge; lectr. in metallurgy U. Birmingham, 1943-49, prof. physical metallurgy, 1949-55; dep. head metallurgy divsn. A.E.R.E., Harwell, 1955-58; Goldsmith's prof. metallurgy Cambridge U., 1958-65; dep. chief sci. adviser Ministry of Defense, 1965-67, cheif adviser, 1967, dep. chief sci. adviser to H.M. Govt., 1968-71, chief sci. adviser, 1971-74; master Jesus Coll, 1974-86; vice chancellor Cambridge U., 1977-79. Recipient J. Herbert Holloman award ACTA Metallurica, 1991. Fellow Am. Soc. Metals, Royal Soc., Royal Swedish Acad. Scis.; mem. Academia Europaea, Inst. Meatals (Rosenhain medal, 1961), and numerous others. Home: 40 Maids Causeway, Cambridge CB5 8DD, England*

COTTRELL, DANIEL EDWARD, systems engineer; b. Anchorage, May 19, 1951; s. Julian Stearns and Jean Evelyn (Connelly) C.; m. Ann Marie Folsom, Sept. 17, 1981; 1 child, Michelle Anne. BS, U. Wash., 1982, USAF Acad., Colorado Springs, Colo., 1973. Quality control engr. Graco, Inc., Mpls., 1980; sales rep. Western Union, Inc., Seattle, 1981; intelligence officer CIA, Washington, 1982-87; systems engr. The Analytic Scis. Corp., Rosslyn, Va., 1987—. Asst. treas. High Point Pool, Falls Church, Va., 1992-93. 1st lt. USAF, 1973-78. Achievements include research in fiber optics - generic indsl. process model, worldwide survey of electronic warfare capabilities. Home: 615 Lincoln Ave Falls Church VA 22046 Office: The Analytic Scis Corp 1101 Wilson Blvd Ste 1500 Arlington VA 22209

COUCH, JAMES VANCE, psychology educator; b. Kansas City, Kans., Mar. 16, 1946; s. Clifton Elwood and Betty (Vance) C.; m. Linda Sue Beavers, June 1, 1968; children: Christopher, Emily, Gregory. BA, Trinity U., San Antonio, 1968; MS, U. Mass., 1971, PhD, 1972. From asst. prof. to assoc. prof. psychology James Madison U., Harrisonburg, Va., 1972-81, asst. dept. head, 1978-80; prof. 1981—; head dept. James Madison U., Harrisonburg, Va., 1980-86, 88-92. Author: Fundamentals of Statistics for the Behavioral Sciences, 1987, Hardware, Software, and the Mental Health Professional, 1991, Computer Use in Psychology, 1992. Mem. Rockingham County (Va.) Planning Commn., 1987-92; supr. Rockingham County, 1991—. Mem. Va. Psychol. Assn. (pres. 1979-80), Ruritan (pres. Dayton, Va. 1989-90), Rotary (bd. dirs.), Elks. Republican. Methodist. Home: 775 Hillview Dr Dayton VA 22821

COUCH, ROBERT BARNARD, physician, educator; b. Guntersville, Ala., Sept. 25, 1930; s. Ezekiel Harvey and Frances Jane (Barnard) C.; m. Katherine Frances Klein, Apr. 23, 1955; children—Robert Steven, Leslie Ann, Colleen Frances, Elizabeth Lee. B.A., Vanderbilt U., 1952, M.D., 1956. Diplomate: Am. Bd. Internal Medicine. Intern Vanderbilt U. Hosp., Nashville, 1956-57, chief resident phys., 1960-61; resident in medicine Vanderbilt U. Hosp., 1959-60, chief resident in medicine, 1960-61; clin. asso. NIH, Washington, 1957-59; sr. investigator NIH, 1961-65, head clin. virology sect., 1965-66; asso. prof. Baylor Coll. Medicine, Houston, 1966-71; dir. influenza research center Baylor Coll. Medicine, 1974—, prof. microbiology and immunology and medicine, 1971—, head infectious diseases sect. medicine, 1987-92, chmn. dept. microbiology and immunology, 1989—; mem. rsch. rev. panels infectious diseases; cons. NIH, Dept. Def., FDA. Contbr. articles to med. jours. Served to sr. surgeon USPHS, 1957-66. Mem. ACP, AAAS, Soc. Exptl. Biology and Medicine, Am. Soc. Microbiology, Infectious Diseases Soc. Am., Am. Assn. Immunologists, Am. Fedn. Clin. Rsch., Am. Soc. Clin. Investigation, So. Soc. Clin. Investigation, Am. Assn. Physicians, Am. Soc. Epidemiology, Am. Soc. Virology. Office: Baylor Coll Medicine 1 Baylor Plz Houston TX 77030

COULDWELL, WILLIAM TUPPER, neurosurgeon; b. Vancouver, B.C., Can., Dec. 15, 1955; s. William John and Janet Mary (Tupper) C.; m. Marie Francoise Simard; children: Sandrine, Mitchell. MD, McGill U., 1984, PhD, 1991. Resident in neurosurgery U. So. Calif., L.A., 1984-89; fellow neuroimmunology Montreal Neurol. Inst., 1989-91, fellow epilepsy surgery, 1990; fellow neurosurgery CHUV, Lausanne, Switzerland, 1990-91; asst. prof. dept. neurol. surgery U. So. Calif., L.A., 1991—; ICU dirs. com. U. So. Calif. Med. Ctr., L.A., 1991—. Contbr. articles to profl. jours. Recipient Preuss award Am. Assn. Neurol. Surgeons, 1991; Med. Rsch. Coun. Can. Centennial fellow, 1990; McGill U. scholar, 1984, Wood Gold medal. Mem. Am. Assn. Neurol. Surgeons (joint sect. on tumors 1991—), Congress of Neurol. Surgeons. Achievements include demonstration of the dominant role of the protein kinase C system in growth regulation of malignant gliomas/ demonstration of protein Kinase C as a successful chemotherapeutic target; use of staurosporine and high-dose Tamoxifen as chemotherapeutic agent. Office: USC Univ Hosp 1510 San Pablo Los Angeles CA 90033

COULL, BRUCE CHARLES, marine ecology educator; b. N.Y.C., Sept. 16, 1942; s. Charles and Ida Louise (Lind) C.; m. Judith Mapletoft, June 3, 1967; children: Brent Andrew, Robin Lind. BS in Biology, Moravian Coll., 1964; MS, Lehigh U., 1966, PhD, 1968. Post doctoral fellow Duke U., Beaufort, N.C., 1968-70; asst. prof. Clark U., Worcester, Mass., 1970-73; assoc. prof. U. S.C., Columbia, 1973-78, prof., 1978-86, Carolina rsch. prof., 1986--; chmn. Marine Sci. Program, U.S.C., 1982-94, Internat. Assn. Meiobenthologists, 1974-77, Invertibrate Zoology Am. Soc. Zoologists, 1983-85; pres. Am. Microscopical Soc., 1987-89. Editor: Ecology or Marine Benithos, 1977; co-editor: Marine Bentic Dynamics, 1980; contbr. articles to profl. jours. Coordinator Habitat for Humanity, Columbia, S.C., 1990--. Recipient Outstanding Researcher award U. S.C., 1982, Outstanding Advisor award, 1991, Excellence in Teaching awward Mortar Bd. U.S.C., 1991, Mungo Teaching Excellence award U. S.C., 1993; Nat. Sci. Found. Rsch. grantee, 1972-92. Mem. Marine Rsch. Fedn., Am. Microscopical Soc. (pres. 1987-89), Ecological Soc. Am. Methodist. Office: Marine Sci U SC Columbia SC 29208

COULOURES, KEVIN GOTTLIEB, process development researcher; b. Novato, Calif., Oct. 17, 1964; s. Irvin Gottlieb Kettler and Susan Alice (Ford) Couloures. BS, U. Calif., Davis, 1987. Fisheries observer Nat. Marine Fisheries Svc., Seattle, 1987; sr. tech. operator Genentech, Inc., San Francisco, 1987-90; sr. rsch. assoc. Calif. Biotechnology, Inc., Mt. View, 1990—. Mem. AAAS, Am. Chem. Soc., No. Calif. Soc. Indsl. Microbiologists. Achievements include design of reconstructed skeletons of Dall's porpoise and beaked whale (on display at Bodega Marine Laboratory). Office: Calif Biotechnology Inc 2450 Bayshore Pky Mountain View CA 94043

COULSON, KINSELL LEROY, meteorologist; b. Hatfield, Mo., Oct. 7, 1916; s. Charles Samuel and Nora Madge (Swank) C.; m. Vera Vivien Vainer, Mar. 23, 1947. B.S., Northwest Mo. State Tchrs. Coll., 1942; M.A., UCLA, 1952, Ph.D., 1959. Jr. meteorologist U.S. Weather Bur., Chgo., 1942; meteorologist UN, Shanghai, China, 1946-47, Naval Civil Service, China Lake, Calif., 1950-51; assoc. research meteorologist UCLA, 1951-59; meteorologist Stanford Research Inst., Menlo Park, Calif., 1959-60; mgr. geophysics Gen. Electric Space Scis. Lab., Phila., 1960-65; prof. meteorology U. Calif., Davis, 1965-79, prof. emeritus, 1984—; dir. Mauna Loa Obs., Hilo, Hawaii, 1979-84; cons., lectr. Author: Polarization and Intensity of Light in the Atmosphere, 1988, Solar and Terrestrial Radiation: Methods and Measurements, 1975 (with J.V. Dave and Z. Sekera) Tables Related to Radiation Emerging from a Planetary Atmosphere with Rayleigh Scattering, 1960; contbr. articles to profl. jours.; patentee atmospheric density calculator, polarization perception device. Served with USN, 1943-46. Recipient numerous research grants. Fellow Am. Meteorol. Soc.; mem. Am. Geophys. Union, Am. Solar Energy Soc., AAAS, No. Calif. Energy Assn., Planetary Soc., Mauna Kea Astron. Soc., Sigma Xi. Home: 119 Bryce Way Vacaville CA 95687-3405

COULSTON, GEORGE WILLIAM, chemical engineer, researcher; b. New Haven, Apr. 21, 1963; s. George William and Florence (Cofrancesco) C.; m. Donna Lynn Esposito, June 17, 1989; 1 child, Christina Marie. BS in Engring., U. Conn., 1985; MS, MPhil, PhD, Yale U., 1990. Rschr. Du Pont Cen. Rsch., Wilmington, Del., 1991—; cons. Novametric Med. Instruments, Wallingford, Conn., 1988-89. Author: Catalysis: Science & Technology, vol. 9, 1991; contbr. articles to profl. jours. Tchr. ARC/Operation Desert Storm, Kaiserslautern, Germany, 1990. NATO postdoctoral fellow NSF, 1990; High Tech. fellow State of Conn., 1989; Olin fellow Yale U., 1985-90. Mem. AICE, Am. Chem. Soc., Phila. Catalysis Club, Sigma Xi, Tau Beta Pi, Omega Chi Epsilon, Phi Kappa Phi. Achievements include first observations of the dependence of reaction cross-section on the relative phase of vibrational motions; showed for the first time how to calculate eigenenergies for 4 quantized levels interacting with 2 lasers off-resonance. Office: Du Pont Exptl Sta Wilmington DE 19880-0262

COULTER, ELIZABETH JACKSON, biostatistician, educator; b. Balt., Nov. 2, 1919; d. Waddie Pennington and Bessie (Gills) Jackson; m. Norman Arthur Coulter Jr., June 23, 1951; 1 child, Robert Jackson. A.B., Swarthmore Coll., 1941; A.M., Radcliffe Coll., 1946, Ph.D., 1948. Asst. dir. health study Bur. Labor Stats., San Juan, P.R., 1946; research asst. Milbank Meml. Fund, N.Y.C., 1948-51; economist Office Def. Prodn., 1951-52; research analyst Children's Bur.-HEW, 1952-53; from statistician to chief statistician Ohio Dept. Health, 1954-65; lectr. econs., then clin. asst. prof. preventive medicine Ohio State U., 1954-65; asst. clin. prof. biostats. U. Pitts. Sch. Pub. Health, 1958-62; assoc. prof. biostats. U. N.C., Chapel Hill, 1965-72, assoc. prof. econs., 1965-78, prof., 1972-90; adj. assoc. prof., hosp. adminstr. Duke U., 1972-79; assoc. dean undergrad. pub. health studies U. N.C., Chapel Hill, 1979-86, prof. emerita, 1990—. Contbr. articles to profl. jours. Mem. AAAS, AAUP, APHA (governing coun. 1970-72), Am. Econ. Assn., Am. Statis. Assn., Am. Acad. Polit. and Social Sci., Biometric Soc., Am. Evaluation Assn., Assn. for Health Svcs. Rsch., Sigma Xi, Delta Omega. Methodist. Home: 1825 North Lake Shore Dr Chapel Hill NC 27514-6742 Office: U NC Sch Pub Health Chapel Hill NC 27599

COURANT, ERNEST DAVID, physicist; b. Goettingen, Germany, Mar. 26, 1920; came to U.S., 1934, naturalized, 1940; s. Richard and Nina (Runge) C.; m. Sara Paul, Dec. 9, 1944; children: Paul N., Carl R. B.A., Swarthmore Coll., 1940; M.S., U. Rochester, 1942, Ph.D., 1943; M.A. (hon.), Yale U., 1962; D.Sc. (hon.), Swarthmore Coll., 1988. Scientist Atomic Energy Project, Montreal, Que., Can., 1943-46; research asso. physics Cornell U., 1946-48; mem. staff Brookhaven Nat. Lab., 1947—, sr. physicist, 1960-89, disting. scientist emeritus, 1990—; Brookhaven prof. physics Yale U., 1962-67, vis. prof., 1961-62; prof. physics and engring. SUNY-Stony Brook, 1967-85; vis. asst. prof. Princeton, 1950-51; cons. Gen. Atomic div. Gen. Dynamics Corp., 1958-59; vis. physicist Nat. Accelerator Lab., 1968-69. Co-originator strong-focusing particle accelerators. Fulbright research fellow Cambridge (Eng.) U., 1956; recipient Fermi award U.S. Dept. of Energy, 1986. Fellow Am. Phys. Soc. (R.R. Wilson prize 1987), AAAS; mem. Nat. Acad. Scis., N.Y. Acad. Scis. (Boris Pregel prize 1979). Home: 109 Bay Ave Bayport NY 11705-2003

COURET, RAFAEL MANUEL, electrical engineering executive; b. Pinar del Rio, Cuba, June 17, 1941; came to U.S., 1958; s. Bienvenido and Regla (Zubizarreta) C.; m. Irmina E. Cabarrouy, Nov. 28, 1964; children: Ivonne Lisa, Rafael Gustavo, Ileana Rosa, Ivette Anne. BEE, U. Fla., 1962, B. Indsl. Engring., 1964. Registered profl. engr., Fla., N.C., S.C., Ala., Ga., Va. Engr. trainee Phillips Petroleum Co. Inc., Bartlesville, Okla., 1964-65, elec. engr., 1965-69; sr. elec. engr. Diaz, Seckinger & Assocs. Inc., Tampa, Fla., 1969-74; elec. mgr. Black, Crow & Einsness Inc., Gainesville, Fla., 1974-76; elec. regional mgr. CH2MHill SE Inc., Gainesville, 1976-81; v.p. DSA Group, Tampa, 1981—. Mem. Fla. Engring. Soc. (ethics com. 1990-91, design build com., Publicity award 1973). Republican. Roman Catholic. Home: 3901 Empedrado Tampa FL 33629-6832 Office: DSA Group Inc 2005 Pan Am Cir Tampa FL 33607-2361

COURSAGET, PIERRE LOUIS, virologist; b. Blois, France, May 19, 1947; s. Paul and Louise (Poy) C.; m. Dominique Gaumain, Aug. 14, 1981. Pharmacist degree, U. Tours (France), 1972, PhD, 1974, M. in Microbiology, 1977, Habilitation Diriger des Recherches, 1986. Attaché Tours Hosp., 1975-82; chargé de recherche INSERM, Tours, 1979-89, dir. de recherche, 1989—; pres. Inst. de Virologie, Tours, 1981—. Editor: Progress in Hepatitis B Immunization, 1990; patentee Hepatitis B vaccine, Non-A, non B hepatitis virus. Maj. French Air Force, 1975-76. Avocation: fly fishing. Home: 6 Place de Richemont, 37550 Saint Avertin France Office: Inst de Virologie, 2 Bis Bld Tonnellé, 37042 Tours France

COURT, ARNOLD, climatologist; b. Seattle, June 20, 1914; s. Nathan Altshiller and Sophie (Ravitch) C.; m. Corinne H. Feibelman, May 27, 1941 (dec. Feb. 1984); children: David, Lois, Ellen; m. Mildred Futor Berry, Apr. 6, 1988. BA, U. Okla., 1934; postgrad., U. Wash., 1938, MS, 1949; PhD, U. Calif., Berkeley, 1956. Reporter and city editor Duncan (Okla.) Banner, 1935-38; observer, meteorologist U.S. Weather Bur., Albuquerque, Washington, Little Am., Los Angeles, 1938-43; chief meteorologist U.S. Antarctic Service, 1939-41; climatologist office Q.M. Gen. U.S. Army, Washington, 1946-51; research meteorologist U. Calif., Berkeley, 1951-56; meteorologist U.S. Forest Service, Berkeley, 1956-60; chief applied climatology, Cambridge Research Labs. USAF, Bedford, Mass., 1960-62; sr. scientist Lockheed-Calif. Co., Burbank, 1962-65; prof. climatology San Fernando Valley State Coll. (now Calif. State U.), Northridge, 1962-85, chmn. dept. geography, 1970-72, prof. emeritus, 1985—; part-time prof. Calif. State U., Northridge, 1986-87, UCLA, 1987-90. Editor: Eclectic Climatology, 1968; assoc. editor Jour. Applied Meteorology, 1978-88; chmn. editorial bd. Jour. Weather Modification, 1978-86; contbr. articles and revs. to profl. jours. Served to 1st lt. USAAF, 1943-46. Recipient Spl. Congl. medal, 1944. Fellow AAAS, Am. Meteorol. Soc., Royal Meteorol. Soc.; mem. Am. Geophys. Union (life), Am. Statis. Assn., Assn. Am. Geographers, Assn. Pacific Coast Geographers (pres. 1978-79), Calif. Geog. Soc., Weather Modification Assn. (trustee 1973-76), Western Snow Conf., Sigma Xi, Phi Beta Kappa. Home: 17168 Septo St Northridge CA 91325-1672 Office: Calif State U Dept Geography Northridge CA 91330

COURT, JOHN HUGH, psychology educator; b. Hounslow, Middlesex, Eng., Aug. 10, 1934; came to U.S. 1989; s. Reginald Victor and Margaret Clare (Longland) C.; m. Pamela Clare Arthaud, Aug. 22, 1959; children: Penelope, Rowena, Peter, Philip. BA with honors, Reading U., Eng., 1956; PhD, Adelaide U., Australia, 1970. Chartered psychologist, U.K.; lic. psychologist, South Australia. Clin. psychologist Crichton Royal Hosp., Dumfries, Scotland, 1961-64; sr. lectr. Adelaide U., Australia, 1964-71; assoc. prof. psychology Flinders U., Bedford Park, South Australia, 1971-82; psychologist Spectrum Ctr., Adelaide, 1982-89; prof. psychology Fuller Sem., Pasadena, Calif., 1989—; mem. South Australia Psychol. Bd., Adelaide, 1974-77. Writer, presenter film: Man with a Problem, 1965; author: Law Light and Liberty, 1975, Pornography, 1980; co-author test manual: Raven's Progressive Matrices, 1991; editor: Kinsey, Sex and Fraud, 1991. Pres. Bible Coll. of South Australia, 1985-88, Inst. of Pvt. Practising Psychologists, South Australia, 1985-88. Recipient Harley Brindal prize Australian Soc. for Hypnosis, 1986, Clin. Hypnosis Diploma, 1990. Fellow Australian Psychol. Soc. (coun. mem. 1987-88, chair ethics com. 1985-88), Brit. Psychol. Soc.; mem. APA. Achievements include research in manic-depressive psychosis. Office: Grad Sch Psychology 180 N Oakland Ave Pasadena CA 91101

COURT, WILLIAM ARTHUR, chemist, researcher; b. Canmore, Alta., Can., Feb. 14, 1943; s. William Edgar and Violet (Loder) C.; m. Alison Sara Newman, Oct. 8, 1966; children: William Bruce, Tamara Dawn. BSc, Carleton U., Ottawa, Ont., 1965, MSc, 1968; PhD, U. New Brunswick, Fredericton, 1970. Asst. prof. U. Va., Charlottesville, 1970-72; postdoctoral fellow divsn. biol. scis. Nat. Rsch. Coun. Can., Ottawa, 1972-73; rsch. scientist Agriculture Can. Rsch. Sta., Delhi, Ont., 1973—, acting dir., 1990-92. Contbr. over 80 articles & abstracts to profl. publs. Achievements include patents for Diterpenoid Triepoxides, Their Isolation and Use; Antileukemic Ansa Macrolides; Ansa Macrolides; Novel Antileukemic Diterpenoid Triepoxides. Office: Agriculture Canada Research Stn, PO Box 186, Delhi, ON Canada N4B 2W9

COURTENAY, IRENE DORIS, nursing consultant; b. Regina, Sask., Can., July 1, 1920; d. Thomas Greer and May Elizabeth (York) C. BS in Nursing, U. Western Ont., 1956, MPH (IH), U. Mich., 1957. RN. Occupational health nurse Chrysler of Can., 1948-50, 52-55; cons. occupational health nursing N.C. Bd. Health, 1958-61; occupational health nursing specialist Nat. League for Nursing, 1961-64; cons. occupational health nursing Dept. Nat. Health and Welfare, Ottawa, Ont., 1966-69; asst. prof., dir. grad. program occupational health nursing U. N.C., Chapel Hill, 1971-75; assoc. prof., dir. grad. program occupational health profls. NYU, 1975-78; dir. grad. program occupational health profls.; pvt. practice cons. occupational health nursing, 1978—. Author several publs. Mem. Am. Nurses Assn., Nat. League Nursing, Am. Assn. Occupational Health Nurses, Am. Bd. Occupational Health Nurses (bd. dirs.), Am. Indsl. Hygiene Assn. (assoc.), Permanent Commn. and Internat. Assn. Occupational Health, AAUP, Am. Pub. Health Assn., Can. Council Occupational Health Nurses (bd. dirs., chmn. exam. com.). Mem. Anglican Ch. Home: # X14, 5110 Wyandotte St E, Windsor, ON Canada N8S 1L2

COURTNEY, JAMES MCNIVEN, chemist; b. Glasgow, Scotland, Mar. 25, 1940; s. George and Margaret (McNiven) C.; m. Ellen Miller Copeland, June 26, 1965; children: Margaret Ellen Louise, James George, David William. BSc in Applied Chemistry, U. Glasgow, 1962; PhD in Chemistry, U. Strathclyde, Glasgow, 1969; Dr.sc.nat., U. Rostock, Germany, 1984. Chartered chemist U.K. Rubber technologist MacLellan Rubber Ltd., Glasgow, 1962-65, Uniroyal Ltd., Dumfries, 1965-66; lectr. bioengring. U. Strathclyde, Glasgow, 1969-81, sr. lectr., 1981-86, reader, 1986-89, prof. bioengring., 1989—; dir. Bio-Flo Ltd., Glasgow. Editor: Artificial Organs, 1977, Biomaterials in Artificial Organs, 1984, Progress in Bioengineering, 1989; contbr. articles to profl. jours. Rudolf Virchow prize, Acad. Sci. of GDR, Berlin, 1986. Fellow Royal Soc. Chemistry, Inst. Materials; mem. Internat. Soc. for Artificial Organs, European Soc. for Artificial Organs, Nat. Conf. Univ. Profs. Avocation: football. Home: 8 Maple Dr Lenzie, Glasgow Scotland G66 4EA Office: U Strathclyde Bioengring, 106 Rottenrow, Glasgow Scotland G4 0NW

COURTOIS, YVES, biologist; b. Saulieu, France, Dec. 4, 1940; s. André Courtois; married, Oct. 11, 1966; children: Sophie, Annabelle. Ingénieur, Ecole Supevieure de Chimie Industrielle de Lyon, Lyon, France, 1964; PhD, U. Orsay, Paris, 1969. Rsch. fellow Com. Energy Atomique Gif/Yvette, Paris, 1964-69; rsch. fellow Med. Sch. Harvard U., Boston, 1969-71; chargé de recherche Assn. Claude-Bernard, Paris, 1971-74; maître de recherche INSERM, Paris, 1974-82, dir. recherche, 1982, dir. l'unité, 1983—, mem. sci. com., 1973—. Contbr. articles to sci. jours. Rsch. fellow ALCON Inst., 1984. Achievements include research on growth factors in the eye. Home: 59 rue Thiers, 92100 Boulogne France Office: INSERM U.118, 29 Rue Wilhem, 75016 Paris France

COURY, LOUIS ALBERT, JR., chemistry educator, researcher; b. Bluefield, W.Va., Aug. 31, 1960; s. Louis Albert and Wanda Lee (Slater) C. BS cum laude, Miami U., 1982; PhD in Chemistry, U. Cin., 1988. Analytical tech. Shepherd Chem. Co., Cin., 1984; grad. teaching asst. U. Cin., Ohio, 1984-88; post-doctoral assoc. U. N.C., Chapel Hill, 1988-90; asst. prof. Duke U., Durham, N.C., 1990—; rev. panel mem. Nat. Inst. Environ. Health and Safety/Nat. Toxicology Program, Research Triangle Park, N.C., 1991-92; presenter in field. Contbr. articles to profl. jours. Ohio State Bd. Regents scholar, Oxford, 1978-82; Lowenstein-Twitchell fellow U. Cin., 1987-88; N.C. Biotechnology Ctr. grantee, Research Triangle Park, 1991—. Mem. Am. Chem. Soc. (Petroleum Rsch. Fund grant 1991—), Am. Inst. Chemists, Internat. Union Pure and Applied Chemistry, The Electrochemical Soc. Achievements include rsch. in electrochemistry, electrocatalysis, sensor design, biol. electron transfer, sonochemistry, gamma irradiation, polymer chemistry, chemically-modified electrodes. Home: #53E 200 Westminster Dr Chapel Hill NC 27514 Office: Duke U Dept Chemistry Durham NC 27706

COUSTEAU, JACQUES-YVES, marine explorer, film producer, writer; b. St. Andre-de-Cubzac, France, June 11, 1910; s. Daniel P. and Elizabeth ((Duranthon) C.; m. Simone Melchoir, July 11, 1937 (dec. 1990); children: Jean-Michel, Diane Elizabeth, Pierre-Yves Daniel, Philippe (dec.); m. Francine Triplet, 1992. Bachelier, Stanislas Acad., Paris, 1927; midshipman, Brest Naval Acad., 1930; D.Sc., U. Calif., Berkeley, 1970, Brandeis U., 1970. Founder Groupe d'etudes et de recherches sous-marines, Toulon, France, 1946; founder, pres. Campagnes oceanographiques francaises, Marseille, 1950, Centre d'etudes marines avancees (formerly Office Francais de recherche sous marine), Marseille, 1952; leader Calypso Oceanographic Expdns.; dir. Oceanographic Mus., Monaco, 1957-88; ret., 1988, promoted Conshelf saturation dive program, 1962; gen. sec. I.C.S.E.M., 1966; leader sci. cruise around world, 1967, basis for TV series The Undersea World of Jacques-Yves Cousteau; leader expdn. to Antarctic and Chilean coast, 1972, expdn. to Amazon, 1982, Mississippi, 1983, Rediscovery of World (Haiti, Cuba, Marquesas Islands, New Zealand, Australia), 1985—; lectr., speaker at numerous worldwide confs. and assns. Recipient numerous awards, including: Motion Picture Acad. Arts and Scis. award (Oscar) for best documentary feature, The Silent World, also for The World Without Sun, 1965, for best short film The Golden Fish 1960, Grand Prix, Gold Palm, Festival Cannes for The Silent World 1956; author and producer documentary films which received awards at Paris, Cannes and Venice film festivals; producer over 100 films for TV; TV series include The World of Jacques-Yves Cousteau, 1966-68, The Undersea World of Jacques-Yves Cousteau, 1968-76, Oasis in Space, 1977, The Cousteau Odyssey Series, 1977-81, The Cousteau/ Amazon Series, 1984—; TV spls. include The Tragedy of the Red Salmon, The Desert Whales, Lagoon of Lost Ships, The Dragons of Galapagos, Secrets of the Southern Caves, The Unsinkable Sea Otter, A Sound of Sea Dolphins, South to Fire and Ice, The Flight of Penguins, Beneath the Frozen World, Blizzard of Hope Bay, Life at the End of the World, (film series) Cousteau's Rediscovery of the World, 1985—; author: Par 18 metres de fonds, 1946, La Plongee en scaphandre, 1950, The Silent World, 1952, (editor with James Dugan) Captain Cousteau's Underwater Treasury, 1959, (with James Dugan) The Living Sea, 1963, World Without Sun, 1965, (with Philippe Cousteau) The Shark: Splendid Savage of the Sea, 1970, (with Philippe Cousteau) Life and Death in a Coral Sea, 1971, Diving for Sunken Treasure, 1971, The Whale: Mighty Monarch of the Sea, 1972, Octopus and Squid, 1973, Three Adventures: Galapagos- Titicaca- the Blue Holes, 1973, The Ocean World of Jacques Cousteau, 1973, Diving Companions, 1974, Dolphins, 1975, Jacques Cousteau: The Ocean World, 1979, A Bill of Rights for Future Generations, 1980, The Cousteau Almanac of the Environment, 1981, Jacques Cousteau's Calypso, 1983, Jacques Cousteau's Amazon Journey, 1984, (with Yves Pacalet) Jacques Cousteau--Whales, 1985; contbr. articles to Nat. Geographic Mag.; co-inventor: (with Emile Gagnon) aqualung, (with Malavard and Charrier) turbosail system, 1985. Served as lt. de vaisseau French Navy, World War II. Decorated comdr. Legion of Honor, Croix de Guerre with palm, Merite Agricole, Merite Maritime, officier des Arts et des Lettres; recipient Potts medal Franklin Inst., 1970, Gold medal Grand Prix d'oceanographie Albert I, 1971, Presdl. medal of

Freedom, 1985, Founders award Internat. Council Nat. Acad. Arts and Scis., 1987, Centennial awd., Nat. Geog. Soc., 1988, Third Internat. Catalan Prize, 1991; inducted into TV Hall of Fame, 1987. Fgn. assoc. Nat. Acad. Scis. U.S.A.; mem. Academie Francaise. Office: Équipe Cousteau, 233 rue du Faubourg St Honoré, 75405 Paris France also: The Cousteau Soc Ste 402 870 Greenbrier Chesapeake VA 23320

COVAULT, LLOYD R., JR., hospital administrator, psychiatrist; b. Troy, Ohio, Feb. 3, 1928; s. Lloyd R. and Anne Marie (Grisez) C.; m. Janet Eileen Davidson, June 12, 1951; children: Sheryl Ann, Jane Helen, Michael Lee, Roger Ken. BA, Miami U., Oxford, Ohio, 1950; MD, Ohio State U., 1954. Extern Orient (Ohio) State Inst., 1953-54, staff physician, 1954-57, clin. dir., until 1966, asst. supt., 1968-70; psychiat. trainee Cen. Ohio Psychiat. Hosp., Columbus, 1966-68, psychiatrist, 1982-85; supt. Columbus State Inst., 1970-74; med. dir. North Cen. Community Mental Health Ctr., Columbus, 1974-79; assoc. prof. psychiatry Ohio State U. Med. Sch., 1975-76; cons. psychiatry North Cen. Community Mental Health Ctr., Columbus, 1985-90; dir. S.E. Mental Health Ctr., Columbus, 1979-82, asst. med. dir., 1985-86, cons. psychiatrist, 1986—; med. dir. Charles B. Milles Mental Health Ctr., Marysville, Ohio, 1989—, Madison County Mental Health Ctr., London, Ohio, 1984-85; cons. psychiatrist Madison County Mental Health Ctr., London, 1984-85; pvt. practice psychiatry Columbus, 1968-75; mem. Franklin County Mental Health and Retardardation Bd., 1970-74, Ohio Dept. Mental Health, ret. 1983; cons. psychiatrist Ohio Correction Complex, Orient, Ohio, 1988-89; 1st med. coord. Netcare Admission Unit Ctrl. Ohio Psychiatric Hosp., 1985—; founding father Physicians Assn. Ohio Dept. Mental Health, 1956-68, pres. 1957. Fellow Am. Assn. Mental Retardation (life, chmn. adminstrn. state chpt. 1974-75), Am. Psychiat. Assn.; mem. Ohio Psychiat. Assn. (mem. coun. 1975), Neuropsychiat. Soc. Cen. Ohio (pres. 1973-74), Mental Health Supts. Assn. (pres. Ohio dept. 1973-74). Home: 11096 Darby Creek Rd Orient OH 43146-9747 Office: SE Mental Health Ctr 1455 S 4th St Columbus OH 43207-1013 also: Charles B Mills Mental Health Ctr 715 S Plum St Marysville OH 43040

COVELLO, ALDO, physics educator; b. Naples, Italy, Jan. 3, 1935; s. Mario and Rita (Gabrieli) C.; m. Renata Moro, June 25, 1966; children: Rita, Paola, Sandra. Maturità Classica, Liceo Umberto I, Naples, Italy, 1953; Laurea in Phys. Chemistry, U. Naples, Italy, 1958, postgrad., 1959-60. Prof. incaricato U. Naples, Italy, 1962; researcher Istituto Nazionale di Fisica Nucleare, Naples, Italy, 1963-64; rsch. assoc. Rutgers U., Dept. of Physics, New Brunswick, N.J., 1965-67; prof. incaricato U. Naples, Italy, 1968-80, libero docente of theoretical physics, 1970; prof. nuclear physics U. Naples Federico II, Italy, 1981—; adviser on nuclear physics to Govt. of Uruguay Internat. Atomic Energy Agy., Montevideo, Uruguay, 1971. Mem. editorial bd. Internat. Rev. Nuclear Physics, 1984, Il Nuovo Cimento A, 1990; contbr. over 40 articles to profl. jours. Mem. Italian Phys. Soc., Am. Phys. Soc. Avocations: reading, music. Home: Via Posillipo 382, 80123 Naples Italy Office: Dept di Scienze Fisiche, Mostra D'Oltremare Pad 19, 80125 Naples Italy

COVERT, EUGENE EDZARDS, aerophysics educator; b. Rapid City, S.D., Feb. 6, 1926; s. Perry and Eda (Edzards) C.; m. Mary Solveig Rutford, Feb. 22, 1946; children: David H., Christine J., Pamela M., Steven P. BS, U. Minn., 1946, MS, 1948; ScD, MIT, 1958. Registered profl. engr., Mass.; chartered engr., U.K. Preliminary design group USNADC, Johnsville, Pa., 1948-52; mem. staff MIT Aerophysics Lab., 1952-63, assoc. dir., 1963-75, assoc. prof. aeronautics and astronautics, 1963-68, prof., 1968—, T. Wilson prof. aeronautics, 1993—, head dept. aeronautics and astronautics, 1985-90; cons. Bolt, Beranek & Newman, Inc., Govt. Israel, Pratt and Whitney Aircraft div. United Techs, Hercules, Inc., MIT Lincoln Lab., Sverdrup Tech., U.S. Army Research Office; chief scientist USAF, 1972-73; mem. panel Naval Aeroballistic Adv. Com., 1965-75; mem. NASA Aeronautical Adv. com., 1985-89, Aeronautics and Space Engring. Bd., 1986—; mem., chmn. USAF Sci. Adv. Bd., 1975-86, 90—; chmn. Power, Energetics and Propulsion panel Adv. Group for Aerospace Research and Devel. NATO, 1982-86; aero. policy com. Office Sci. and Tech. Policy, 1976—; dir. Rohr Industries Inc., Chula Vista, Calif., Allied Signal Inc., Morristown, N.J., Phys. Scis. Inc., North Andover, Mass.; mem. Pres.' commn. for investigation of space shuttle accident. Served with USNR, 1943-47. Recipient Exceptional Civilian Sci. award USAF, 1973, '86, Univ. Educator of Yr. award, Am. Soc. Aerospace Edn., 1980, Tech. Leadership award Univ. of Minn. Alumni Assocs., 1993, Pub. Svc. award NASA, 1991, von Karman medal Adv. Group for Aerospace R & D, 1980. Fellow AIAA (bd. dirs., Ground Testing award 1990, W.F. Durand lectureship for pub. svc. 1992), AAAS, Royal Aero. Soc.; mem. NAE, N.Y. Acad. Scis., Sigma Xi. Office: MIT Rm 33-215 77 Massachusetts Ave Cambridge MA 02139-4307

COVINGTON, STEPHANIE STEWART, psychotherapist, author; b. Whittier, Calif., Nov. 5, 1942; d. William and Bette (Robertson) C.; children: Richard, Kim. BA cum laude, U. So. Calif., 1963; MSW, Columbia U., 1970; PhD, Union Inst., 1982. Pvt. practice psychotherapy La Jolla, Calif., 1981—; instr. U. Calif., San Diego, 1981—, Calif. Sch. Profl. Psychology, San Diego, 1982-88, San Diego State U., 1982-84, Southwestern Sch. Behavioral Health Studies, 1982-84, Profl. Sch. Humanistic Psychology, San Diego, 1983-84, U.S. Internat. U., San Diego, 1983-84, UCLA, 1983-84, U. So. Calif., L.A., 1983-84, U. Utah, Salt Lake City, 1983-84; cons. L.A. County Sch. Dist., N.C. Dept. Mental Health, others; designer women's treatment, cons. Betty Ford Ctr.; presenter at profl. meetings; lectr. in field. Author: Leaving the Enchanted Forest: The Path from Relationship Addiction to Intimacy, 1988, Awakening Your Sexuality: A Guide for Recovering Women and Their Partners, 1991; contbr. articles to profl. jours. Mem. Am. Assn. Sex Educators, Counselors and Therapists, Am. Bd. Med. Psychotherapists (diplomate), Am. Bd. Sexology (diplomate), Am. Pub. Health Assn., Assn. Women in Psychology, Calif. Women's Commn. on Alcoholism (Achievement award), Ctr. for Study of the Person, San Diego Network Women Leaders, Friends of Jung, Internat. Coun. on Alcoholism and Addictions (past chair women's com.), Kettil Brun Soc. (Finland), San Diego Soc. Sex Therapy and Edn., Soc. for Study of Addiction (Eng.). Avocations: reading, theater, raising orchids. Office: 7946 Ivanhoe Ave # 201B La Jolla CA 92037

COWAN, GEORGE ARTHUR, chemist, bank executive, director; b. Worcester, Mass., Feb. 15, 1920; s. Louis Abraham and Anna (Listic) C.; m. Helen Dunham, Sept. 9, 1946. BS, Worcester Poly. Inst., 1941; DSc, Carnegie-Mellon U., 1950. Research asst. Princeton U., 1941-42, U. Chgo., 1942-45; mem. staff Columbia U., N.Y.C., 1945; mem. staff, dir. rsch. Los Alamos (N.Mex.) Sci. Lab., 1945-46, 49-88, sr. fellow emeritus, 1988—; teaching fellow Carnegie Mellon U., Pitts., 1946-49; chmn. bd. dirs. Los Alamos Nat. Bank, Trinity Capital Corp., Los Alamos; pres. Santa Fe Inst., 1984-91; mem. The White House Sci. Coun., Washington, 1982-85, cons., 1985-90, Air Force Tech. Applications Ctr., 1952-88; bd. dirs. Applied Tech. Assocs., Inc., Title Guaranty, Inc., Universal Properties, Inc. Contbr. sci. articles to profl. jours. Bd. dirs. Santa Fe Opera, 1964-79; treas. N.Mex. Opera Found., Santa Fe, 1970-79; regent N.Mex. Inst. Tech., Socorro, 1972-75; bd. dirs. N.Mex. Symphony Orch., N.Am. Inst., Santa Fe Inst. Recipient E.O. Lawrence award, 1965, Disting. Scientist award N.Mex. Acad. Sci., 1975, Robert H. Goddard award Worcester Poly. Inst., 1984, Enrico Fermi award, Presdl. Citation, Dept. Energy, 1990. Fellow AAAS, Am. Phys. Soc.; mem. Am. Chem. Soc., N.Mex. Acad. Sci., Cosmos Club(Washington), Sigma Xi. Avocation: skiing. Home: 721 42D St Los Alamos NM 87544 Office: Santa Fe Inst 1660 Old Pecos Trl # A Santa Fe NM 87501-4768

COWAN, HENRY JACOB, architectural engineer, educator; b. Glogow, Poland, Aug. 21, 1919; s. Arthur and Erna (Salisch) C.; B.S. with honors, U. Manchester, 1939, M.S., 1940; Ph.D., U. Sheffield (Eng.), 1952, D.Eng., 1963; M.Arch., U. Sydney, 1984; DArch (hon.), U. Sydney, 1987; m. Renate Proskauer, June 22, 1952; children: Judith Anne, Esther Katherine. Mem. faculty dept. archlt. sci. U. Sydney (Australia), 1953—, prof., 1953—, head dept., 1953-84, dean architecture, 1966-67, pro-dean, 1968-84; vis. prof. Cornell U., 1962, Kumasi (Ghana) U., 1973, Trabzon (Turkey) U., 1976; pres. Bldg. Sci. Forum of Australia, 1969-71. Served with Royal Engrs., 1941-45. Decorated officer Order of Australia, 1983. Fellow Royal Australian Inst. Architects (hon. 1979), Royal Soc. Arts, Inst. Structural Engrs. (Spl. Svc. award 1988), Instn. Engrs. Australia (recipient R.W. Chapman medal 1956), ASCE. Author over 23 books including: The Master Builders, Science and Building, Architectural Structures, Environmental Systems, The Science and Technology of Building Materials, Handbook of Architectural Technology; editor Archtl. Sci. Rev., 1958—, Vestes, 1966-78. Home: 6 Hale Rd Apt 57, Mosman New South Wales 2088, Australia Office: U Sydney, Sydney New South Wales 2006, Australia

COWAN, MARY ELIZABETH, medical technologist; b. Solon, Ohio, Mar. 25, 1907; d. Charles Frederick and Edna May (Van Valkenburg) C. BS, Denison U., 1929; degree in med. tech., Mt. Sinai Hosp., 1936. Med. tech. E. 100th St. Med. Bldg. Lab., Cleve., 1936-37, Cleve. Clinic, 1938-39; med. tech., supr. trace evidence dept. Cuyahoga County Coroner's Lab., Cleve., 1939—; lectr. Case Western Res. U. Law-Med. Ctr., Cleve., 1953-92, police and attys., Cleve., 1953—; mem. People to People Forensic Sci. Project to Soviet Union, 1988; expert witness Ct. of First Instance, Manila, 1968, Cts. of Common Pleas, various counties, Ohio. Editorial bd.: Jour. Forensic Scis., 1961-66; contbr. articles to Jour. Radioanalytical Chemistry, Jour. Forensic Scis., Am. Jour. Clin. Pathology, Postgrad. Medicine; contbr. chpt.: Forensic Science, 1986. Named Disting. Alumna Denison U., 1969; recipient Law Enforcement Commendation medal Sons Am. Revolution, 1988, Disting. Svc. award Cuyahoga County Ombudsmen, 1989; inducted in Hall of Fame Bedford High Sch., 1992. Fellow Am. Acad. Forensic Scis. (ret.); mem. Zonta Internat. (sec. Cuyahoga Valley club 1959-60, dir. 1960-61), Alpha Phi (Disting. Mem. 1972), Sigma Xi,. Baptist.

COWAN, MICHAEL JOHN, civil engineer; b. Louisville, Ky., Sept. 20, 1953; s. John William and Jane (Starkey) C.; m. Barbara Ann Gdaniec, Aug. 16, 1980; 1 child, Sophia Marie. BSCE, U. Louisville, 1981. Registered profl. engr. Field engr. Gust K. Newberg Constrn., Chgo., 1982-84; chief engr. U.S. Housing Components, Louisville, 1984-89; regional engr. Trus Joist Corp., Nashville, 1989-90; sr. engr. Project Mgmt. Engring., Birmingham, Ala., 1991—. With USN, 1973-76. Mem. ASCE, NSPE, Exptl. Aircraft Assn., Am. Concrete Inst., Southern Bldg. Code Congress, VFW. Presbyterian. Home: 5220 Kirkwall Ln Birmingham AL 35242 Office: Project Mgmt Engring 2509 Keith Dr Columbia TN 38401

COWAN, NELSON, cognitive psychologist, researcher; b. Washington, Mar. 7, 1951; s. Arthur and Shirley B. (Frink) C.; m. Priscilla Roth, 1982 (div. 1985); m. Jean Mona Ispa, Aug. 16, 1987; 1 child, Alexander; stepchildren: Simone, Zachary. BS, U. Mich., 1973; PhD, U. Wis., 1980. Fellow NYU, N.Y.C., 1981-82; asst. prof. U. Mass., Amherst, 1982-85; asst. prof. U. Mo., Columbia, 1985-89, assoc. prof., 1989—. Editorial bd. Psychonomic Bull. and Review, 1993—, Jour. exptl. Psychology: Learning, Memory and Cognition, 1993—. Democrat. Jewish. Achievements include observation of effects of the duration of speech output on verbal short-term memory; effects of attention on sensory memory. Office: U Mo Dept Psychology 210 McAlester Hall Columbia MO 65211

COWAN, PENELOPE SIMS, materials engineer; b. Chipley, Fla., Aug. 6, 1966; d. James Thompson Sims and Ann (Harris) Harrison; m. John Russell Cowan, Aug. 31, 1991. B in Materials Engring., Auburn U., 1989, MS in Materials Engring., 1991. Rsch. asst. Auburn (Ala.) U., 1989; engr. Marshall Space Flight Ctr. NASA, Huntsville, Ala., 1989—. Office: NASA Marshall Space Flight Ctr EP-62 Huntsville AL 35812

COWAN, WILLIAM MAXWELL, neurobiologist; b. Johannesburg, South Africa, Sept. 27, 1931; s. Adam and Jessie Sloan (Maxwell) C.; m. Margaret Sherlock, Mar. 31, 1956; children: Ruth Cowan Eadon, Stephen Maxwell, David Maxwell. B.Sc., Witwatersrand U., Johannesburg, 1951, B.Sc. (hon.), 1952; D.Phil., Oxford U., 1956, MB, B.Ch., 1958, M.A., 1959. From demonstrator to lectr. anatomy Oxford U., 1953-66; fellow Pembroke Coll., 1958-66; vis. prof. anatomy Washington U. Med. Sch., St. Louis, 1964-65; assoc. prof. U. Wis. Med. Sch., Madison, 1966-68; prof., chmn. dept. anatomy and neurobiology Washington U. Med. Sch., 1968-80; research prof., dir. Weingart Lab. Devel. Neurobiology, Salk Inst. Biol. Studies, La Jolla, Calif., 1980-86; v.p. Salk Inst. Biol. Studies, 1982-86; provost and exec. vice chancellor Washington U., St. Louis, 1986-87; v.p., chief sci. officer Howard Hughes Med. Inst., Bethesda, Md., 1988—; mem. Inst. Medicine, Nat. Acad. Scis., 1978; fgn. assoc. Nat. Acad. Scis., 1981; disting. adj. prof. neuroscience Johns Hopkins Sch. Medicine, 1988—. Editor-in-chief Jour. Neurosci., 1980-87; editor: Ann. Revs. Neurosci. Fellow Am. Acad. Arts and Scis., Royal Soc. (London), Royal Soc. S. Africa; mem. AAAS, Internat. Brain Research Orgn. (exec. council), Anat. Soc. Gt. Britain and Ireland, Royal Micros. Soc., Am. Assn. Anatomists, Soc. Neurosci. (pres. 1977-78), Norwegian Acad. Sci. (fgn.), Am. Philosophical Soc., Sigma Xi, Alpha Omega Alpha, Phi Beta Kappa. Home: 6337 Windermere Cir Rockville MD 20852-3550 Office: Howard Hughes Med Inst 4000 Jones Bridge Rd Chevy Chase MD 20815-6789

COWDEN, RONALD REED, biomedical educator, cell biologist; b. Memphis, July 9, 1931; s. Robert and Martel (Braswell) C.; m. Beverly Marie Louise Sherwood, Aug. 23, 1956. BS, La. State U., 1953; PhD, U. Vienna, Austria, 1956. USPHS postdoctoral fellow Oak Ridge (Tenn.) Nat. Lab., 1956-57; asst. prof. biology Johns Hopkins U., Balt., 1957-59; asst. mem. cell biology USPHS, N.Y.C., 1960-61; asst. prof. pathology U. Fla. J.H. Miller Health Ctr., Gainesville, 1962-66; assoc. prof. anatomy La. State U. Med. Ctr., New Orleans, 1966-68; prof., chmn. biol. Sci. U. Denver, 1968-72; prof., chmn. anatomy Albany (N.Y.) Med. Coll., 1972-75; assoc. dean basic sci. East Tenn. U. Coll. Medicine, Johnson City, 1975-80, chmn. dept. biophysics, 1980-86, retired, 1986; trustee Bermuda Biolog. Sta., St. Georges, Bermuda, 1968-78. Assoc. editor: Histochemistry Jour. (London), 1968-74, Jour. Exptl. Zoology, 1969-73, 78-82, Transactions the Am. Microscopical Soc., 1971—, Jour. Morphology, 1972—, Histochemistry (Berlin), 1978-86, Acta Embryologia et Morphologiae Experimentalis, Nova Seria (Palermo), 1978—, European Jour. Histochemistry, 1984—; co-editor: Aspects of Sponge Biology, 1976, Devel. Biology of Freshwater Invertebrates, 1982, Advances in Microscopy, 1984; contbr. over 120 articles to profl. jours. Recipient Career Devel. award USPHS, 1962-66. Fellow Royal Microscopical Soc., Am. Assn. Zoologists, Am. Soc. Cell Biology, Soc. for Invertebrate Pathology, Assn. Southeastern Biologists, Soc. Western Naturalists, Soc. for Devel. Biology, Histochem. Soc. N.Am. (program chmn. 1981-84). Republican. Roman Catholic. Achievements include rsch. in cytochemistry of marine invertebrate embryos, quantitative cytochemistry of nuclei, fluoresence cytochemistry. Home: 74272 Military Rd Covington LA 70433-6121

COWDRY, REX WILLIAM, physician, researcher; b. Des Moines, Feb. 12, 1947; s. Edmund William and Jane Kirkwood (Rex) C.; m. Donna Elizabeth Patterson, Aug. 29, 1981; 1 child, Ryan Elizabeth. BA, Yale Coll., 1968; MD, Harvard U., 1973, MPH, 1973. Diplomate Am. Bd. Psychiatry and Neurology. Resident Mass. Mental Health Ctr., Boston, 1973-76; staff psychiatrist NIMH, Bethesda, Md., 1976-83, clin. dir., 1983-86; acting dep. dir. NIMH, Rockville, Md., 1986-88; chief exec. officer NIMH Neuropsychiat. Research Hosp., Washington, 1988—; assoc. clin. prof. Georgetown U. Med. Sch., Washington, 1983—. Contbr. articles to profl. jours and coauthor books. Recipient Presdl. scholar U.S. Govt., 1964, Adminstrs. award Alcohol, Drug Abuse and Mental Health Adminstrn. 1986, 88; fellow U.S. Pub. Health Svc. 1970. Fellow Am. Psychiat. Assn. (Falk fellow 1975). Episcopalian. Office: NIMH Neuropsychiat Rsch Hosp 2700 Martin L King Ave SE Washington DC 20032

COWEN, BRUCE DAVID, environmental services company executive; b. Springfield, Mass., Jan. 19, 1953; s. Irving Abraham and Pearl (Glushien) C.; m. Judith Paterson, July 2, 1983. BS in Bus. Adminstrn., Am. Internat. Coll., 1974. CPA, Conn. Audit mgr. Price Waterhouse & Co., Hartford, Conn., 1974-79; controller TRC Environ. Cons., Inc., East Hartford, Conn., 1979-85; treas., sec. TRC Cos., Inc., Windsor, Conn., 1980-92; exec. v.p., CFO TRC Cos., Inc. and TRC Environ. Cons., Inc., 1992—; pres. TRC Environ. Cos., Inc., 1987—, TRC Cos., Inc., 1992—; dir. TRC Cos., Inc., Halcyon, Ltd., Hartford, Conn., 1984; dir. U.S. Chamber of Commerce; chmn. Sammy Davis, Jr. Greater Hartford Open PGA Golf Tournament, 1984, Conn. Gaming Policy Bd., 1987—; commr. Conn. Cen. Commn. for Statue of Liberty and Ellis Island, 1985-87; bd. dirs. Greater Hartford Chpt. of Am. Red Cross. Mem. Am. Inst. CPAs, Nat. Assn. Accts., Nat. Assn. Security Dealers (info. com. of bd. govs.), U.S. C. of C. (small bus. bd.), Hartford Jaycees. Republican. Jewish. Home: 117 Knob Hill Rd Glastonbury CT 06033-3709 Office: TRC Env Corp 5 Waterside Crossing Windsor CT 06095

COWEN, DONALD EUGENE, physician; b. Ft. Morgan, Colo., Oct. 8, 1918; adopted s. Franklin and Mary Edith (Dalton) C.; BA, U. Denver, 1940; MD, U. Colo., 1943; m. Hulda Marie Helling, Dec. 24, 1942; children: David L., Marilyn Marie Cowen Dean, Theresa Kathleen Cowen Cunningham Byrd, Margaret Ann Cowen Koenigs. Intern, U.S. Naval Hosp., Oakland, Calif., 1944; gen. practice medicine, Ft. Morgan, 1947-52; resident internal medicine U. Colo. Med. Ctr., Denver, 1952-54; practice medicine specializing in allergy, Denver, 1954-90, ret.; mem. staff Presbyn. Med. Ctr., Denver, Porter, Swedish hosps., Englewood, Colo.; clin. asst. prof. medicine U. Colo. Med. Center, 1964—; postgrad. faculty U. Tenn. Coll. Medicine, Memphis, 1962-82; cons. Queen of Thailand, 1973, 75, 77. Pres. Community Arts Symphony Found., 1980-82. Served to lt. M.C., USN, 1943-47. Fellow ACP, Am. Coll. Chest Physicians (vice chmn. com. on allergy 1968-72, 75-87, sec.-treas. Colo. chpt. 1971-77, pres. 1978-80), Am. Coll. Allergy and Immunology, Acad. Internat. Medicine, West Coast Allergy Soc., Southwest Allergy Forum, Am. Acad. Otolaryngic Allergy, Colo. socs. internal medicine, Colo. Allergy Soc. (past pres.), Ill. Soc. Opthalmology and Otolaryngology (hon.), Denver Med. Soc. (chmn. library and bldg. com. 1963-73), Arapahoe Med. Soc. (life emeritus mem.). Presbyterian (ruling elder 1956—). Club: Lions. Contbr. numerous articles to profl. jours. Home: Cherry Hills Village 1501 E Quincy Ave Englewood CO 80110

COWEN, JOSEPH EUGENE, mechanical engineer; b. Joplin, Mo., Nov. 5, 1946; s. James Wesley and Dorothy (Berin) C.; m. Sue Baker, June 6, 1970; children; Clinton Wesley, Christina Michelle. BS, U. Mo., 1969. Registered profl. engr., Mo., Kans. Okla., Tex. Jr. engr. ARCO Pipeline Co., Independence, Kans., 1969, Blaw-Knox Chem. Plants, Pitts., 1970-71; mech. engr. Bruce Williams Consulting Engrs., Joplin, Mo., 1971-72; engring. foreman B.F. Goodrich Tire Co., Miami, Okla., 1972-75; sr. chem. engr. Gulf Oil Chems. Co., Pittsburg, Kans., 1976—; cons. John S. Hill & Assocs., Joplin, 1974-86, Able Body Co., Inc., Joplin, 1974-86, Atlas Powder Co., Duenueg, Mo., 1983-86. Precinct committeeman Dem. Party, Joplin, 1976-86, county chmn., Jasper County, Mo., 1982-86; mem. Residential Design Rev. Com., The Woodlands, Tex., 1990—. Recipient Curator's award U. Mo., 1964. Mem. ASME, NSPE, Soc. Mfg. Engrs., Nat. Fire Protection Assn. Democrat. Achievements include approval from Assn. of Am. Railroads for specially designed intermodal container for shipping hazardous waste. Office: USPCI Inc 515 W Greens Rd Houston TX 77067

COWEN, STANTON JONATHAN, air pollution control engineer; b. Atlanta, June 17, 1954; s. Arnold Edwin and Audrey Corinne (Batkin) C.; m. Linda Jean Grifford, Nov. 27, 1987; children: Rachel, Cassandra. BS, Columbia U., 1976; MS, Calif. Tech. Inst., 1977. Rsch. scientist Meteorology Rsch. Inc., Altadena, Calif., 1977-81; environ. engr. Kaiser Steel, Fontana, Calif., 1981-83; air pollution engr. Ventura (Calif.) County Air Pollution Control Dist., 1983—. Contbr. articles to profl. jours. Mem. Air Waste Mgmt. Assn. Achievements include rsch. on measurement of light absorption of fly ash, the collection of submicron fly ash by fabric filtration, the size dependent penetration of trace chemicals through fabric filters. Office: Ventura County APCD 702 County Square Dr Ventura CA 93003

COWERN, NICHOLAS EDWARD BENEDICT, physical scientist; b. Brighton, Sussex, Eng., Apr. 26, 1953; s. Raymond Teague and Margaret Jean (Trotman) C.; m. Mary-Barr Le Messurier, Sept. 1, 1979; children: Oliver, Beatrice. Degree in physics with honors, Oxford (Eng.) U., 1975, PhD, 1980, MA, 1980. Sr. sci. officer exptl. atomic collisions physics U.K. Atomic Energy Authority, Harwell and JET Labs., Oxfordshire, Eng., 1979-85; rsch. scientist Hirst Rsch. Ctr. Gen. Electric Co., London, 1985-87; sr. scientist semiconductor process physics, modeling, cons. Philips Rsch. Labs., Eindhoven, The Netherlands, 1987—. Referee Nuclear Instr. and methods in Physics Rsch., Jour. Applied Physics and Applied Physics Letters; contbr. 50 articles to profl. jours. Mem. Am. Phys. Soc. Office: Philips Rsch Labs, PO Box 80.000, 5600 JA Eindhoven The Netherlands

COWGILL, URSULA MOSER, biologist, educator, environmental consultant; b. Bern, Switzerland, Nov. 9, 1927; came to U.S., 1943, naturalized, 1945; d. John W. and Mara (Siegrist) Moser. A.B., Hunter Coll., 1948; M.S., Kans. State U., 1952; Ph.D., Iowa State U., 1956. Staff MIT, Lincoln Lab., Lexington, Mass., 1957-58; field work Doherty Found., Guatemala, 1958-60; research assoc. dept. biology Yale U., New Haven, 1960-68; prof. biology and anthropology U. Pitts., 1968-81; environ. scientist Dow Chem. Co., Midland, Mich., 1981-84; assoc. environ. cons. Dow Chem. Co., 1984-91; environ. cons., 1991—; mem. environ. measurements adv. com. Sci. Adv. Bd. EPA, 1976-80; Internat. Joint Commn., 1984-89. Contbr. numerous articles on ecology, biology and minerology to sci. publs. Trustee Carnegie Mus., Pitts., 1971-75. Grantee NSF 1960-78, Wenner Gren Found., 1965-66, Penrose fund Am. Philos. Soc., 1978; Sigma Xi grant-in-aid, 1965-66. Mem. Soc. Environ. Geochemistry and Health, Mineral Soc. Gt. Britain, Am. Soc. Limnology and Oceanography, Internat. Soc. Theoretical and Applied Limnology. Home and Office: PO Box 1329 Carbondale CO 81623

COWIN, STEPHEN CORTEEN, biomedical engineering educator, consultant; b. Elmira, N.Y., Oct. 26, 1934; s. William Corteen and Bernice (Reidy) C.; m. Martha Agnes Eisel, Aug. 10, 1956; children: Jennifer Marie, Thomas Burrows. BCE (Md. State scholar, Ambrose Howard Carner scholar), Johns Hopkins U., 1956, MCE (Univ. fellow), 1958; Ph.D. in Engring. Mechanics, Pa. State U., 1962. Registered profl. engr., La. Prof. mech. engring. Tulane U., 1969-77, prof. mechanics dept. biomed. engring., 1977-85, adj. prof. orthopedics, 1978-88, prof.-in-charge Tulane-Newcomb Jr. Yr. Abroad program, 1974-75, chmn. applied math. program, 1975-79, prof. applied stats., 1979-88, Alden J. Laborde prof. engring., 1985-88; Disting. prof. CUNY, 1988—; Sci. Research Council Gt. Brit. sr. vis. fellow U. Strathclyde, 1974, 80; vis. research prof. Instituto de Matematica, Estatistica e Ciencia de Computanao, Universidade Estadual de Campinas, Brazil, 1978; participant U.S. Nat. Acad. Scis. interacad. exchange program with Bulgaria, 1983; fellow Japan Soc. for the Promotion Sci., 1987. Editor: (with M. Satake) Continuum Mechanical and Statistical Approaches in the Mechanics of Granular Materials, 1978, Mechanics Applied to the Transport of Granular Materials, 1979, (with M.M. Carroll) The Effects of Voids on Material Deformation, 1976, Bone Mechamics, 1988; assoc. editor: Jour. Applied Mechanics, 1974-82, Jour. Biomech. Engring., 1982-88; editorial adv. bd. Handbook of Materials, Structures and Mechanics, 1981—, Handbook of Bioengineering, 1981, Acta Biomechanica, 1986—; editorial bd. Annals Biomed. Engring., 1985—; editorial cons. Jour. Biomechanics, 1988—. Served to capt. U.S. Army, 1957-64. Research grantee, NSF, NIH, U.S. Army Research Office, Edward G. Schlieder Found. Fellow AAAS, ASME (Melville medal 1993), Am. Acad. Mechanics; mem. Orthopedic Rsch. Soc., Soc. Rheology, Soc. Natural Philosophy (treas. 1977-79), Soc. Engring. Sci., Math. Assn. Am., N.Y. Acad. Scis., Sigma Xi. Home: 107 W 86th St Apt 4F New York NY 10024-3409

COWING, THOMAS WILLIAM, architectural, civil engineer; b. Jamestown, N.Y., Apr. 6, 1951; s. John Henry and Eleanor Charlotte (Olsen) C.; divorced; children: Nathan T., Emily E. BSCE, SUNY, Buffalo, 1973; MSCE, U. Colo., 1984. Registered prof. engr. Colo.; cert. energy mgr. Comml. contractor pvt. practice, Sherman, N.Y., 1974-76; product devel. engr. Earthwise Shelters Inc., Lyndonville, Vt., 1977-78; constrn. mgr. Energywise Homes Inc., St. Johnsbury, Vt., 1978-79; engring. cons. pvt. practice, Chautauqua, N.Y., 1980-81; staff rsch. engr. Solar Energy Rsch. Inst., Golden, Colo., 1981-83; engring. cons. pvt. practice, Boulder, Colo., 1983; sr. prof. engr. U. Colo., Boulder, 1983—; speaker bldg. systems sem. Coll. Engring. U. Colo., Boulder, 1986—. Contbr. articles to profl. jours. Sunday sch. tchr. Unity of Boulder, 1990—; cub scout pack asst. Boy Scouts Am., Niwot, Colo., 1990—. Regents of State of N.Y. scholar, 1969-73. Mem. ASHRAE, Am. Waterworks Assn., Internat. Dist. Heating and Cooling Assn., Rocky Mt. Assn. Energy Engrs. (pres./bd. dirs. 1984-89, Energy Engr. of Yr. 1990, Corp. Energy award 1987), Assn. Energy Engrs. Achievements include co-development of manufactured system of pre-engineered energy-efficient structures; implementation of a national monitoring and assessment program for passive solar homes in U.S. Office: U Colo Dept Facilities Mgmt Box 53 Boulder CO 80309

COWLEY, JOHN MAXWELL, physics educator; b. Peterborough, South Australia, Feb. 18, 1923; came to U.S., 1970; s. Alfred Ernest and Doris (Milway) C.; m. Roberta Joan Beckett, Dec. 15, 1951; children—Deborah Suzanne, Jillian Patricia. B.Sc., U. Adelaide, Australia, 1942, M.Sc., 1945, D.Sc., 1957; Ph.D., Mass. Inst. Tech., 1949. Research officer Commonwealth Sci. and Indsl. Research Orgn., Melbourne, Australia, 1945-62; chief research officer, head crystallography sect. Commonwealth Sci. and Indsl. Research Orgn., 1960-62; prof. physics U. Melbourne, Australia, 1962-70; Galvin prof. physics Ariz. State U., Tempe, 1970—, Regents' prof., 1988—; mem. U.S. Nat. Com. for Crystallography, 1973-78, 84-86. Author: Diffraction Physics, 1975; editor: (with others) Acta Crystallographica, 1971-80; contbr. (with others) articles to profl. jours. Fellow Australian Acad. Sci., Inst. Physics (London), Australian Inst. Physics, Royal Soc. (London), Am. Phys. Soc.; mem. Internat. Union Crystallography (mem. exec. com. 1963-69, chair commn. on electron diffraction 1987-93, Ewald Prize 1987), Am. Inst. Physics, Am. Crystallographic Assn., Electron Microscope Soc. Am. (dir. 1971-75). Home: 2625 E Southern Ave #C 90 Tempe AZ 85282-7670 Office: Dept Physics and Astronomy Ariz State U Tempe AZ 85287-1504

COWLEY, SCOTT WEST, chemist, educator; b. Logan, Utah, Oct. 18, 1945; s. Joseph Enos and June (West) C.; m. Diane Winger, Aug. 20, 1971; children: Matthew Scott, Lisa Loraine, Katherine June, Preston Ira. BS in Chemistry, Utah State U., 1967, MS, 1971; PhD in Organic Chemistry, So. Ill. U., 1975. Postdoctoral fellow fuels engring. U. Utah, Salt Lake City, 1975-76, asst. prof., 1976-79; asst. prof. Colo. Sch. Mines, Golden, 1979-84, assoc. prof., 1984—; indsl. fellow Marathon Oil Co., Littleton, Colo., 1980. Contbr. articles to Indsl. Engring. Chemistry Fund, Jour. Catalysis, Progress in Tech. Series. Mem. North Am. Catalysis Soc., Am. Chem. Soc., Rocky Mountain Fuel Soc., Western States Catalysis Club. Republican. Ch. Latter Day Saints. Achievements include patent for enhanced performance of alcohol fueled engines during cold start conditions. Office: Colo Sch Mines Dept Chemistry Golden CO 80401

COWSIK, RAMANATH, physics educator; b. Nagpur, Madhya, India, Aug. 29, 1940; came to U.S., 1970; s. Ramakrishna K. and Saraswati C. (Ayyar) C.; m. Shyamala Balasubrahmanian, Aug. 20, 1979 (div. Feb. 1989); 1 child, Siddhartha. BS, Mysore U., Bangalore, India, 1958; PhD, Bombay U., 1968. Jr. rsch. assoc. Tata Inst. of Fundamental Rsch., Bombay, 1961—, reader, 1975—, assoc. prof., 1977—, prof., 1984—; asst. prof. U. Calif., Berkeley, 1970-73; vis. scientist Max-Planck Inst. Extension Physik, Munich, 1973-74; vis. prof. Washington U., St. Louis, 1987—. Contbr. articles to Jour. Physics Rev., Astrophys. Jour. Recipient Sarabhai award Hari om Soc./Phys. Rsch. Lab., 1981, Group Achievement award NASA, 1986. Fellow Indian Acad. Scis., Indian Nat. Sci. Acad. (Bhatnagar award 1984); mem. Am. Phys. Soc. (life), Internat. Astron. Union (life). Achievements include development of the theory that weakly interacting particle relicts from the big bang are the constituents of dark matter and set the upper bound on the sum of their masses, in particular of neutrinos; recognized the cosmological significance of the hard x-ray background; derived the leaky box and nested leaky box models for cosmic rays; research in high energy astrophysics of nonthermal emissions from quasars and supernova remnants and in astroparticle physics and experimental gravitation; measurement of the double beta decay life-time of the tellurium-128 nucleus as 7.7x10 24 years the longest, implying the Majorana mass of the neutrino to be less than 1 eV. Home: 6351 Waterman Ave Saint Louis MO 63130-4708 Office: Washington U Box 1105 1 Brookings Dr Saint Louis MO 63130-4862

COX, CHAD WILLIAM, medical technologist; b. Wauseon, Ohio, Feb. 4, 1949; s. Ross Marion and Claudene Ruth (Smith) C.; m. Claris Threet, May 22, 1982; children: Lisa, Kelli, Chad, Melissa. BA in Biology, Albion Coll., 1971. Cert. med. technologist. Lab. aid Albion (Mich.) Community Hosp., 1971-72; med. technologist Mercy Hosp., Toledo, 1972-73, Bixby Med. Ctr., Adrian, Mich., 1973-84; lab. supr. Bixby Med. Ctr., Adrian, 1984—. Dist. com. mem. Boy Scouts Am., Adrian, 1984—, troop coord., treas., 1984—. Mem. Clin. Lab. Mgmt. Assn. (treas. We. Lake Erie chpt. 1988-91). Methodist. Office: Bixby Med Ctr 818 Riverside Ave Adrian MI 49221

COX, CHARLES SHIPLEY, oceanography researcher, educator; b. Paia, Hawaii, Sept. 11, 1922; s. Joel Bean and Helen Clifford (Horton) C.; m. Maryruth Louise Melander, Dec. 23, 1951; children: Susan (dec.), Caroline, Valerie, Ginger, Joel. BS, Calif. Inst. Tech., 1944; PhD, U. Calif., San Diego, 1955. From asst. researcher to prof. U. Calif., San Diego, 1955—. Researcher in field. Fellow AAAS, Am. Geophys. Union (Maurice Ewing medal 1992), Royal Astron. Soc. Democrat. Office: U Calif San Diego Scripps Inst of Oceanography Dept Oceanograpay La Jolla CA 92093

COX, DAVID LEON, telecommunications company executive; b. Lima, Ohio, Sept. 9, 1952; s. Leon Hamiln and Mildred Marie (Johnson) C.; m. Carolle Marie Mallette, July 14, 1977; children: Paul David, Elizabeth Christine. BS in Chemistry, Mich. State U., 1975, BS in Computer Sci., 1976. Registered profl. engr., Va. Asst. v.p. engring. KollMorgan Corp., Newburgh, N.Y., 1975-76; staff mgr. AT&T, Bedminster, N.J., 1976-79; asst. v.p. Satellite Bus. Systems, McLean, Va., 1979-83; devel. mgr. MCI, Washington, 1983-84; chief engr. Harris Corp., Melbourne, Fla., 1984. Contbr. articles to profl. jours. Active Friends of the Palm Bay (Fla.) Libr., Space Coast Sci. Ctr., 1000 Friends of Fla., Tallahassee, Turkey Creek Homeowners Assn., Turkey Creek Santuary Bd., Palm Bay PTA; vice chmn. pub. rels. Boy Scouts of Am., 1991—, dist. com. mem., 1991—, unit commr., 1990—, troop com. mem., 1992—, park com. mem., 1987—; mem. Comprehensive Plan Com., Palm Bay, Fla., 1986-87. Mem. IEEE, Am. Chem. Soc., Am. Inst. Plant Engrs., Mensa, Assn. for Computing Machinery, N.Y. Acad. Sci., Nat. Fire Protection Assn., Building Industry Cons. Svc. Internat., Am. Radio Relay League, Nat. Eagle Scout Assn. (life), Nat. Coun. Boy Scouts of Am., Mich. State U. Alumni Assn., Lyman Briggs Coll. Alumni Assn., Mason (3d deg.), Orlando Scottish Rite (32nd deg., Master of Royal Secret), Alpha Phi Omega (Beta Beta chpt.). Republican. Presbyterian. Achievements include 4 patents in Integrated Services Digital Network technology. Home: 900 Mandarin Dr NE Palm Bay FL 32905-4707 Office: Harris Corp 2400 Palm Bay Rd NE Palm Bay FL 32905

COX, DENNIS JOSEPH, engineer; b. Longview, Wash., Jan. 13, 1951; s. Rhodes Joseph and Marian Ruth C.; m. Nancy Jo Neumayer, July 28, 1979; children: Kelly L., Daniel J. BS, Wash. State U., 1974; MBA, U. Wash., 1989. Engr. E.I. Dupont, Wilmington, Del., 1974-76; tech. rep. Drew Chem., Parsippany, N.J., 1976-77; tech. specialist Betz Labs., Trevose, Pa., 1977-79; engr. svc. mgr. Boise Cascade Paper, Wallula, Wash., 1979—. Mem. Assn. Energy Engrs., Tech. Assn. Pulp and Paper, Jaycees (sec. Burbank, Wash. chpt. 1987), Alpha Chi Sigma. Office: Boise Cascade Paper PO Box 500 Wallula WA 99363-0500

COX, DONALD CLYDE, electrical engineer; b. Lincoln, Nebr., Nov. 22, 1937; s. Elvin Clyde and C. Gertrude (Thomas) C.; m. Mary Dale Alexander, Aug. 27, 1961; children: Bruce Dale, Earl Clyde. BS, U. Nebr., 1959, MS, 1960, DSc (hon.), 1983; PhD, Stanford U., 1968. Registered profl. engineer Ohio, Nebr. With Bell Tel. Labs., Holmdel, N.J., 1968-84, head radio and satellite systems rsch. dept., 1983-84; mgr. radio and satellite systems rsch. divsn. Bell Comm. Rsch., Red Bank, N.J., 1984-91, exec. dir. radio rsch. dept., 1991—; mem. commns. U.S. nat. com. Internat. Union of Radio Sci.participant FCC enbanc hearing on PCS, 1991. Contbr. articles to profl. jours.; patentee in field. Johnson fellow, 1959-60; recipient Guglielmo Marconi prize in Electromagnetic Waves Propagation, Inst. Internat. Comm., 1983. Fellow IEEE (Morris E. Leeds award 1985, Alexander Graham Bell medal 1993), AAAS, Bellcore 1991, Radio Club Am.; mem. Comm. Soc. of IEEE (Leonard G. Abraham Prize Paper award 1992, Comms. Mag. Prize Paper award 1990), Vehicular Tech. Soc. of IEEE (Paper of Yr. award 1983), Antennas and Propagation Soc. of IEEE (elected mem. adminstrn. com. 1986-88), Sigma Xi. Achievements include research in low-power tetherless personal portable communication systems, cellular radio systems, radio propagation. Home: 101 Peter Coutts Circle Stanford CA 94305-2515 Office: Bell Comm Rsch Inc Bellcore Rm 3Z325 331 Newman Springs Rd Red Bank NJ 07701

COX, FRANK D. (BUDDY COX), oil company executive, exploration consultant; b. Shreveport, La., Dec. 20, 1932; s. Ohmer M. and Beulah O. (Scott) C.; m. Betty Jean Hand, June 19, 1956; children: Cynthia Dell, Carolyn Diane, Frank D. Jr. BS in Bus. Adminstrn., La. Tech. U., 1956; postgrad., Centenary Coll., 1958-59. Cert. profl. landman; lic. real estate, Fla. Various postition Exxon Corp., Houston, 1955-86, chief landman, v.p. coal resources, 1980-86; pvt. practice Houston, 1986-89; sr. v.p. Energy Exploration Mgmt. Co., Houston, 1989—. Capt. USAF, 1956-58. Named disting. mil. grad. La. Tech. U., Ruston, 1955. Mem. Am. Assn. Petroleum Landmen, Houston Assn. Petroleum Landmen, W. Houston Assn. Petroleum Landmen, W. Houston Exxon Annuitant Club, Omicron Delta Kappa. Republican. Baptist. Avocations: golf, tennis, amateur radio. Home: 14830 Carolcrest St Houston TX 77079-6312 Office: Energy Exploration Mgmt Co 2929 Briarpark Dr Ste 314 Houston TX 77042-3709

COX, HOLLACE LAWTON, electrical engineering educator; b. Oak Park, Ill., Nov. 17, 1935; s. Hollace Lawton Sr. and Frances Marion (Murray) C.; m. Alice Sue Burdon, June 25, 1983; 1 child, Alison Elizabeth. BA, U. Rochester, 1959; PhD, Ind. U., 1967. Mem. tech. staff Tex. Instruments Inc., Dallas, 1967-70; postdoctoral fellow Baylor U., Waco, Tex., 1970-73, M.D. Anderson Hosp. and Tumor Inst., Houston, 1973-76; instr. MaVinckrodt Inst. Radiology, St. Louis, 1976-77; asst. prof. U. Kans. Med. Ctr., Kansas City, 1977-80; assoc. prof. U. Louisville Med. Ctr., 1980-83; assoc. prof. elec. engring. dept. U. Louisville, 1983—; mem. summer faculty program U.S. Army Missile Command, Huntsville, Ala., 1986, 87, 88, 91, U.S. Army Strategic Def. Co., Huntsville, 1990. Contbr. articles to Jour. Chem. Physics, Bull. Am. Phys. Soc., Phys. Rev. A., Med. Physics, Jour. Light Wave Tech. Robert A. Welch Found. fellow, 1970-73, 73-76. Mem. IEEE Lasers and Electro-Optical Soc., SPIE Internat. Soc. Optical Engring., Am. Phys. Soc., Optical Soc. Am. Republican. Episcopalian. Office: U Louisville Dept Elec Engring Louisville KY 40292

COX, JAMES CARL, JR., chemist, researcher, consultant; b. Wolf Summit, W.Va., June 17, 1919; s. James Carl and Maggie Lillian (Merrells) C.; m. Alma Lee Tenney, Sept. 8, 1945; children: James Carl III, Joseph Merrells, Alma Lee, Elizabeth Susan Cox Unger, Albert John. BS summa cum laude, W.Va. Wesleyan Coll., 1940; MS in Organic Chemistry, U. Del., 1947, PhD in Phys. Organic Chemistry, 1949; postgrad. in law, Am. U., summer 1953, George Washington U., summer 1954; JD with honors, U. Md., 1955. Bar: Md. 1955; registered profl. sanitarian, Tex. Rsch. chemist E.I. duPont de Nemours Corp., Belle, W.Va., 1940-43; grad. instr. chemistry U. Del., Newark, 1946-49; prof. chemistry, head dept. chemistry Wesleyan Coll., Macon, Ga., 1949-51; prof. U.S. Naval Acad., Annapolis, 1951-55, Md., 1951-55; prof., rsch. dir. Lamar U., Beaumont, Tex., 1955-65; prof., head dept. chemistry, dir. div. sci. and math. Oral Roberts U., Tulsa, 1965-68; prof., head dept. chemistry Wayland Baptist U., Plainview, Tex., 1968-76; v.p., rsch. dir. Agrl. & Indsl. Devel., Inc., Plainview, 1976-79; environ. health expert Tex. Dept. Health, Plainview, 1979-84; mem. U. Tex indsl. planning commn., commdr. 19th dist. Dept. Tex.; field rep. Bureau of Census, 1990-92; cons., quality assurer Agri-Search Corp.; lectr. U. London, U. Dublin, Heidelberg U., summer 1976; cons. in field; vis. prof. organic chemistry Middle Tenn. State U., Murfreesboro, summer 1950, U. Baghdad, Iraq, 1956-57; quality assurer Agri-Search, Inc., 1992—. Author: Lives of Splendor, 1970; Patterson's German-English Chemical Dictionary, rev. edit., 1985; editor The Condenser, 1957-65; contbr. articles to profl. jours., also abstracts. Precinct chmn. Hale County Rep., Plainview, 1983-84; bd. dirs. Plainview chpt. ARC, 1969-73, United Way, Plainview, 1972-75. Served to cpl. Combat Engrs., U.S. Army, 1943-45, ETO; field rep. Bureau Labor Stats., 1992—. Named Outstanding Prof., Lamar U., 1963-64, Wayland Bapt. U., 1971-74; fellow DuPont Endowment Found., 1947-49, Carnegie Found., 1949-51, State of Tex., 1957-59. Fellow Tex. Acad. Sci.; mem. Am. Chem. Soc., AAAS, AAUP, Tex. Pub. Health Assn., Tex. Environ. Health Assn. (governing council 1982-84), Am. Legion (adj.) DAV (chmn. Americanism, judge adv.), VFW (vice commdr.), Rotary (pub. relations officer 1969-84), Confederate Air Force (col.). Methodist. Current work: Novel fuels for industry; agricultural chemicals. Subspecialties: Organic chemistry; Polymer chemistry.

COX, JAMES GRADY, chemist; b. LaGrange, Ga., Sept. 23, 1949; s. James Grady and Ruth (Brown) C.;m. Kimberley Kay Reagan, Nov. 29, 1985. AS, So. Union State Jr. Coll., 1979; BA, LaGrange Coll., 1981. Cert. drinking water analyst, Ohio. Environ. specialist State of Ga., Brunswick, Ga., 1982-83; contract chemist Continental Tech. Svcs., Augusta, Ga., 1985-88; chemist ETTC Labs., Findlay, Ohio, 1988-89, B.E.C. Labs., Toledo, 1989-90; method devel. chemist Aquatech Environ. Labs., Melmore, Ohio, 1990—. With U.S. Army, 1968-72, Vietnam. Mem. Am. Chem. Soc., Ducks Unltd. Mem. Ch. of Christ. Achievements include development of method for G.C. determination of nitriles, method for determination of ethylene oxide in water samples. Office: Aqua Tech Environ Labs 6873 S State Rte 100 Melmore OH 44845

COX, NORMAN ROY, engineer, educator, consultant; b. Ft. Worth, Jan. 30, 1949; s. Robert Leonard and Anna Louise (Anderson) C.; m. Georgia Law, June 15, 1986; children: Tyrone, Emily, Karen. BSEE, U. Tex., Arlington, 1972, MSEE, 1975, PhD, 1980. Registered profl. engr., Mo. Engr. trainee Del Norte Tech., Inc., Euless, Tex., 1972-73; grad. teaching asst. U. Tex., Arlington, 1974-77, grad. rsch. assoc., 1978-80; asst. prof. elec. engring. U. Mo., Rolla, 1981—; Author tech. papers. Mem. IEEE, Toastmasters Internat. Methodist. Achievements include research on high-speed digital video technology, developments applicable to electric vehicles. Office: Univ Mo Rolla 113 Electrical Engineering Rolla MO 65401

COX, ROBERT HAMES, chemist, scientific consultant; b. Toronto, Mar. 23, 1923; came to U.S., 1951; s. Giffard and Lavinia Sarah (Hames) C.; m. Dora Maria Forstrom, Sept. 5, 1953; children: William H., Frederick G., Irene M. PharmB, U. Toronto, 1946; BSP, U. Sask., Saskatoon, Can., 1948, MSc, 1950; PhD in Medicinal Chemistry, U. Mich., 1954. Lic. pharmacist, Ont. Head dept. pharm. chemistry U. B.C., Vancouver, Can., 1949-51; asst. to mgr. product devel. Mallinckrodt Chem. Works, St. Louis, 1954-56; tech. dir. Vick Internat. div. Richardson-Merrell, N.Y.C., 1956-60; assoc. dir. tech. svcs., 1960-64; v.p. rsch. and devel. Walker Labs. Richardson-Merrell, Mt. Vernon, N.Y., 1964-66; dir. new products Winthrop Labs. div. Sterling Drug, N.Y.C., 1966-75; co-founder, pres. New Eng. Pharms., Inc., Randolph, Mass., 1978-82; pres. Robert H. Cox & Co., Scarsdale, N.Y., 1975—, Cox & Fay, Inc., Scarsdale, 1991—; cons. Drug Enforcement Adminstrn., Washington, 1976-78, Nat. Cancer Inst., Bethesda, Md., 1980-81, Indonesian Govt., Jakarta, Java, 1991—. Co-editor-in-chief: Medicinal Chemistry, Vol. III, 1956, Vol. IV, 1959. Leader Jamaica Mission, UN Adv. Svcs., 1988; mem. U.S. Exchanges del. to China, 1990. Recipient Roberts medal Ont. Coll. Pharmacy, 1955, George E. Parke medal, 1957. Fellow Am. Inst. Chemists (pres. N.Y. 1986-87, leader sci. del. to China 1986, co-leader to USSR 1989); mem. Am. Chem. Soc. (treas. medicinal chemistry div. 1962-63), Parenteral Drug Assn., Cen. Atlantic States Assn. Food and Drug Ofcls., Chemists Club (trustee). Episcopalian. Achievements include patents for drugs (sympatholytics/cycloplegics) and medical devices including hemodialysis; conducted practical synthesis of suberone precursor of early antihypertensive, guanethidine; early evaluation (1940s) of oxidized cholesterols in etiology of experimental atherosclerosis. Office: 33 Ferncliff Rd Scarsdale NY 10583-5955

COX, ROBERT HAROLD, physiology educator; b. Phila., Sept. 10, 1937. BS, Drexel Inst. Tech., 1961, MS, 1962; PhD in Biochemical Engring., U. Pa., 1967. Assoc. physiologist U. Pa., Phila., 1967-69, asst. prof., 1969-72, assoc. prof. physiology, 1972—, prof., assoc. prof. biomechanics, 1973—, assoc. dir. Bockus rsch. inst., 1970—. Mem. IEEE, AAAS, Am. Physiol. Soc., Sigma Xi. Achievements include research in vascular smooth muscle mechanics, arterial wall physiology, hypertension, carotid sinus reflex. Office: Bockus Rsch Inst Grad Hosp 415 S 19th St Philadelphia PA 19146*

COX, TERRENCE GUY, manufacturing automation executive; b. Revere, Mass., Feb. 29, 1956; s. Thomas Ambrose and Jennie Costance (Meli) C.; m. Therese Marie Paone, Sept. 15, 1979. BS in Fin. cum laude, Babson Coll., 1976, MBA, 1977. Asst. to pres. Standard Bldg. Systems, Inc., Point of Rocks, Md., 1977-80; sr. fin. analyst Nortek, Inc., Cranston, R.I., 1980-82; mgr. new bus. devel. Compo Industries, Waltham, Mass., 1982-83; founder, v.p., chief fin. officer, treas. CAD/CAM Integration, Inc., Woburn, Mass., 1983—, also bd. dirs.; treas., bd. dirs. Encode, Inc., Nashua, N.H. Contbr. articles to profl. jours. Founder Revere track league, 1974, pres., 1974-77; coach Revere little league, 1969-77; bd. dirs. Revere Parks and Recreation Commn., 1972-74, Boy's Club of Revere, 1974-77. Roman Catholic. Avocations: reading, travel, diving. Office: CAD/CAM Integration Inc 76 Winn St Woburn MA 01801-2898

COX, THOMAS PATRICK, biological psychology educator; b. Billings, Mont., Oct. 18, 1946. BS, Eastern Mont. Coll., 1968; PhD, U. Manitoba, Winnipeg, Can., 1978. Rsch. asst. Oreg. Zool Rsch. Ctr., Portland, Oreg., 1972-73; instr. U. man., Winnipeg, 1976-77; rsch. asst. U. Oreg. Health Sci. Ctr., Portland, 1978-79; asst. prof. Coll. Great Falls, Lewistown, Mont., 1979-81; adj. prof. Eastern Mont. Coll., Billings, 1982-84; asst. prof. SUNY Helath Sci. Ctr., Bklyn., 1984-87; rsch. assoc. Hunter Coll., N.Y.C., 1987-88; assoc. prof. Black Hills State U., Spearfish, S.D., 1988--. Contbr. articles to profl. jours. Recipient Individual Nat. Rsch. Svc. award Nat. Inst. Neurological and Communicative Disorders. Mem. Am. Soc. Mammalogists (life), Animal Behavior Soc., S.D. Acad. Scis., Sigma Xi. Office: Black Hills State U Box 9066 1200 University Spearfish SD 57799

COX, VERNE CAPERTON, psychology educator; b. Newport, R.I., Aug. 31, 1938; s. Lorlys Verne and Nathalie Patricia (Sheehan) C.; m. Mary Hazel Wilkins, Aug. 20, 1960; children: Verne Jr., Kevin, Sean. PhD, U. Houston, 1964. Rsch. assoc. Fels Rsch. Inst., Yellow Springs, Ohio, 1964-70; prof. U. Tex., Arlington, 1970—, dept. chair, 1982-88. Contbr. articles to profl. jours. Bd. pres. Creative Arts Theatre & Sch., Arlington, 1988-90. Mem. Soc. for Neurosci., Am. Psychol. Soc. Office: U Tex Arlington S Cooper St Arlington TX 76019

COX, WILLIAM ANDREW, cardiovascular thoracic surgeon; b. Columbus, Ga., Aug. 3, 1925; s. Virgil Augustus and Dale Jackson C.; m. Nina Recelle Hobby, Jan. 1, 1948; children: Constance Lynn Cox Rogers, Patricia Ann Cox Brown, William Robert, Janet Elaine. Student Presbyn. Coll., 1942, Harvard U., 1944-45, Cornell U., 1945, BS, Emory U., 1950, MD, 1954, MS in Surgery, Baylor U., 1961. Active duty USN, 1943-46; lt. (sr.g.) USNR, 1946-54; commd. 1st lt. M.C., U.S. Army, 1954, advanced through grades to col., 1969; intern Brooke Army Med. Ctr., San Antonio, 1954-55, resident in gen. surgery, 1956-60; resident in cardiovascular thoracic surgery Walter Reed Army Med. Ctr., Washington, 1960-62, staff cardiothoracic surgeon, 1962; asst. chief cardiothoracic surgery Letterman Gen. Hosp., 1962-65; chief dept. surgery and cardiothoracic surgery 121 Evacuation Hosp., Seoul, Korea, cons. cardiothoracic surgery Korean Theatre, 1965-66; asst. chief cardiothoracic surgery Brooke Army Med. Ctr., 1966-69, chief, 1969-73, bd. dirs. thoracic surgery residency programs, 1966-73, ret., 1973; clin. prof. cardio-thoracic surgery U. Tex. Sch. Medicine, San Antonio, 1971—; practice medicine specializing in cardiovascular thoracic surgery, Corpus Christi, Tex., 1973—; cons. cardio-thoracic surgery Brooke Army Med. Ctr., San Antonio, 1977—; chief staff Meml. Med. Ctr., 1980; dir. disaster med. care region 3A Tex. State Dept. Health, 1973-88; mem. Coastal Bend council Gov.'s Emergency Med. Service Commn., 1979—; mem. adv. bd. on congenital heart disease Tex. Dept. Health, 1980-88; participant joint confs. on cardiovascular surgery and thoracic surgery Am. People Ambassador Program, Leningrad, Moscow, Bucharest, Romania, Belgrade, Yugoslavia, Prague, Czechoslovakia, 1987; del. Vanderbilt U. joint conf. vascular surgery Dublin, Ireland, Edinburgh, Scotland, London, 1986; participant joint confs. cardiovascular surgery and thoracic surgery Am. Ambassador People to People Program Singapore, Kuala Lumpur, Malayasia, Hanoi, Vietnam, DaNang, Vietnam, Hue, Vietnam, Saigon, Vietnam, Hong Kong, Feb.-Mar., 1993. Decorated Legion of Merit; recipient A Prefix award Surgeon Gen. U.S. Army; diplomate Am. Bd. Surgery, Am. Bd. Thoracic Surgery. Fellow Am. Coll. Chest Physicians; mem. AMA, Soc. Thoracic Surgeons, Denton A. Cooley Cardiovascular Surgery Soc., Tex. Med. Assn. (del. conf. infectious diseases Bankgkok, Hong Kong, Beijing, Shanghai, 1983), So. Thoracic Surgery Assn., Nueces County Med. Soc., Corpus Christi Surg. Soc., 38th Parallel Med. Soc., U.S. Power Squadron, People to People Internat., Internat. Platform., USN League (life), Retired Officers Assn. (life), Republican. Presbyterian. Clubs: Yacht (past commodore presidio San Francisco); T-Bar-M Racquet; Corpus Christi Country, Corpus Christi Athletic, Corpus Christi Town; Ft. Sam Houston Officers. Contbr. numerous articles in field to profl. jours. Home: 5214 Wooldridge Rd Corpus Christi TX 78413-3833 Office: Spohn Tower Ste 302 613 Elizabeth St Corpus Christi TX 78404-2220

COX, WILLIAM WALTER, dentist; b. Abingdon, Va., Jan. 17, 1947; s. Walter Roy and Beatrice Ellen (Woodward) C.; m. Neva Duncan Herzog, Apr. 23, 1976 (div. Sept. 1985); 1 child, Lesley Ellen. BS in Chemistry, U. Richmond, 1969; DDS, Med. Coll. Va., 1973. Dentist Taylor, Bisese, Kail & Cox, Inc., Portsmouth, Va., 1976—. Capt. U.S. Army, 1973-76. Recipient Disting. Svc. medal U.S. Army Dental Corps, 1976. Mem. Tidewater Dental Assn. (chmn. patient rels. com. 1984—), Tidewater Dental Study Club, Portsmouth-Suffolk Dental Study Club (sec.-treas. 1990-91, pres. 1991-92). Republican. Methodist. Avocations: golf, snow skiing. Home: 3209 Bruin Dr Chesapeake VA 23321-4601 Office: Taylor Bisese Kail & Cox 5717 Churchland Blvd Portsmouth VA 23703-3308

COY, WILLIAM RAYMOND, civil engineer; b. Omaha, Nov. 28, 1923; s. Vern Elmer and Edna Mae (Seymour) C.; m. Geraldine Petra Zaback, July 31, 1943; children: Carol Sue, William R. Jr., Russell B., Steven D., Marcus R. Student, Omaha Mcpl. U., 1944, 45. Registered profl. engr., N.D. Lab. technician U.S. Army CE, 1946-57; paving engr. CE, Albuquerque Dist., Roswell, N.Mex., 1958-61; chief materials testing CE Ballistic Missile Constrn., Minot and Grand Forks, N.D., 1961-65; engr.-in-charge CE Green River Dam, Campbellsville, Ky., 1965-68; chief concrete sect. CE Div. Lab., Omaha, 1968-78; chief materials engr. CE Missouri River Div., Omaha, 1978-87; assessor, constrn. technology Nat. Inst. Standards and Tech., Omaha, 1987—. With USN, 1945-46. Mem. ASTM, Am. Concrete Inst. Home: 3226 S 44th Ave Omaha NE 68105

COYKENDALL, ALAN LITTLEFIELD, dentist, educator; b. Hartford, Conn., Jan. 11, 1937; s. Linley Robert and Mary Helen (Peck) C.; m. Betty J. Drum, Sept. 6, 1959; 1 child, John. BS, Bates Coll., 1959; DMD, Tufts U., 1963; MS, George Washington U., 1970. Asst. dental officer USS Coral Sea CVA43, 1965-67; microbiologist Naval Med. Rsch. Inst., Bethesda, Md., 1967-71, Naval Dental Rsch. Inst., Great Lakes, Ill., 1971-72; rsch. assoc. VA Hosp., Newington, Conn., 1972-78; asst. prof. Sch. Dental Medicine U. Conn., Farmington, 1978-84, assoc. prof. Sch. Dental Medicine, 1984—. Contbr. articles to profl. publs. Trustee Farmington Libr., 1974-92; mem. Farmington Bd. Edn., 1979-87. NIH grantee, 1977-87. Mem. Am. Soc. for Microbiology (jour. editorial bd. 1990—). Achievements include introduction of DNA hybridization to the study of cariogenic streptococci and other streptococci; presenter theory of evolution of dental caries. Office: U Conn Health Ctr 263 Farmington Ave Farmington CT 06030-1605

COYNE, PATRICK IVAN, physiological ecologist; b. Wichita, Kans., Feb. 26, 1944; s. Ivan Lefranz and Ellen Lucille (Brown) C.; m. Mary Ann White, Aug. 22, 1964; children: Shane Barrett, Shannon Renee. BS, Kans. State U., 1966; PhD, Utah State U., 1970. R & D coord. U.S. Army Cold Regions Rsch. and Engring. Lab., Hanover, N.H., 1970-72; asst. prof. forestry U. Alaska, Fairbanks, 1973-74; plant physiologist, environ. scientist Lawrence Livermore (Calif.) Nat. Lab., 1975-79, cons., 1980—; rsch. plant physiologist USDA/ Agrl. Rsch. Svc., Woodward, Okla., 1979-85; prof., head Ft. Hays Experiment Sta. Kans. State U., Hays, 1985—; mem. adv. coun. Kans. Geol. Survey, Lawrence, 1986-91. Contbr. 29 articles to profl. jours. Capt., U.S. Army, 1970-72. Mem. AAAS, Am. Soc. Agronomy, Soil Sci. Soc. Am., Crop Sci. Soc. Am., Soc. Range Mgmt., Coun. Agriculture Sci. and Tech., Hays Area C. of C. (bd. dirs. 1988-90), Rotary, Phi Kappa Phi, Gamma Sigma Delta, Sigma Xi. Republican. Mennonite Brethren Ch. Office: Kans State U Ft Hays Experiment Sta 1232 240th Ave Hays KS 67601-9228

COZZARELLI, NICHOLAS R., molecular biologist, educator; b. Jersey City, Mar. 26, 1938; s. Nicholas and Catherine (Meluso) C.; m. Linda Angela Ambrosini, July 28, 1967; 1 child, Laura Amelia. AB, Princeton U., 1960; PhD, Harvard U., 1966. Postdoctoral fellow dept. biochemistry Stanford U. Med. Sch., Palo Alto, Calif., 1966-68; asst. prof. dept. biochemistry U. Chgo., 1968-70, asst. prof. depts. biochemistry, biophysics and theoretical biology, 1970-74, assoc. prof., 1974-77, prof., 1977-82; prof. dept. molecular biology U. Calif., Berkeley, 1982-89, prof., div. biochemistry and molecular biology, dept. molecular and cell biology, 1989—, chmn. dept., 1986-89, dir. virus lab., 1986—. Contbr. articles to profl. jours.; mem. editorial bd. Jour. Biol. Chemistry, 1988—, Cell, 1983-86; mem. editorial adv. bd. Biochemistry, 1982-86. Mem. NAS, AAAS, Am. Soc. for Microbiology, Am. Soc. for Biol. Chemistry. Democrat. Avocations: reading, theatre. Office: U Calif Divsn Biochem & Molecular Biology Dept Molecular & Cell Biology Berkeley CA 94720

CRABTREE, LEWIS FREDERICK, aeronautical engineer, educator; b. Keighley, Yorkshire, Eng., Nov. 16, 1924; s. Lewis and Susan (Wilson) C.; m. Averil Joan Escott, Mar. 16, 1955; children: Elizabeth Clare, Richard Gareth Lewis. BScME, Leeds U., Eng., 1945; diploma in aeronautics, Imperial Coll., 1947; PhD in Aero. Engring., Cornell U., 1952. Air engr. officer Royal Naval Vol. Res., Eng., 1945-46; grad. apprentice Saunders-Roe Ltd., East Cowes, Isle of Wight, Eng.; various engring. assignments Royal Aircraft Establishment, Farnborough, Eng., 1953-73; Sir George White prof. aero. engring. Bristol (Eng.) U., 1973-85, prof. emeritus, 1985—; Co-author: Elements of Hypersonic Aerodynamics, 1965; contbr. (books) Incompressible Aerodynamics, 1960, Laminar Boundary Layers, 1963, Engineering Structures, 1983. Fellow AIAA, Royal Aero. Soc. (pres. 1978-79). Home: Brecon, Dan-Y-Coity, LD3 7YN Talybont-on-Usk LD3 7YN, England

CRABTREE, ROBERT HOWARD, chemistry educator, consultant; b. London, Apr. 17, 1948; came to U.S., 1977, naturalized, 1985; s. Arthur and Marguerite (Vaniere) C. B.A., Oxford U., 1970; Ph.D., Sussex U., Eng., 1973, D.Sc. (hon.), 1985. Attache De Recherche Centre Nationale de la Recherche Scientifique, Paris, 1975-77; asst. prof. chemistry, Yale U., New Haven, Conn., 1977-83, assoc. prof., 1983-85, prof., 1985—; cons. Air Products, Allentown, Pa., 1984—. Contbr. articles to sci. jours. Fellow Royal Chemistry Soc. (Corday-Morgan medal 1984, Organometallic Chemistry medal 1991); mem. Am. Chem. Soc. (Organometallic Chemistry award 1993).

CRAFT, BENJAMIN COLE, III, accelerator physicist; b. Baton Rouge, Jan. 6, 1955; s. Benjamin Cole Craft Jr. and Carolyn (Daniels) Stottlemyer. BS in Physics and Math., Tex. A&M U., 1976; PhD in Physics, MIT, 1982. Postdoctoral fellow Bates Lab. MIT, Middleton, 1983; ring mgr. Nat. Synchrotron Light Source Brookhaven Nat. Lab., Upton, N.Y., 1985-88; assoc. dir. Ctr. for Advanced Microstructures and Devices La. State U., Baton Rouge, 1988—. Mem. IEEE, AAAS, Am. Phys. Soc., Soc. Photo-Optical Instrumentation Engrs. Office: La State U Ctr Advanced Microstructures & Devices 3990 W Lakeshore Dr Baton Rouge LA 70803

CRAFT, DAVID WALTON, clinical microbiologist, army officer; b. Roanoke, Va., June 25, 1957; s. Robert Lee and Charlotte Ann (Walton) C.; m. Janice Gayle Kinzie, June 20, 1981; children: Sarah Gayle, Nathan David. BS, U. Ala., Tuscaloosa, 1979, MS, 1981; PhD, U. Ga., 1989. Commd. 2d lt. U.S. Army, 1981, advanced through grades to maj., 1991; chief reference bacteriology Letterman Army Med. Ctr., San Francisco, 1981-83, chief virology, 1983-85; chief HIV-immunology 10th Med. Lab., Landstuhl, Germany, 1989-91, chief microbiology, 1991-92; chief microbiology and immunology Eisenhower Army Med. Ctr., Augusta, Ga., 1992—; microbiology cons. U.S. Army in Europe, Landstuhl, 1991-92; adj. asst. prof. Med. Coll. Ga., Augusta, 1992—. Contbr. articles to sci. jours. Coach mil., ch. and youth svcs., 1983—; deacon Bapt. chs., Germany, Augusta, 1991—. Decorated Meritorious Svc. medal. Mem. AAAS, Am. Soc. for Microbiology, Soc. Armed Forces Med. Lab. Scientists, Assn. U.S. Army, Sigma Xi. Republican. Achievements include first publication of A549 human cell line as a broad spectrum line in the clinical virology laboratory. Home: 890 Chase Rd Evans GA 30809 Office: Eisenhower Army Med Ctr Dept Clin Investigation Fort Gordon GA 30905

CRAFT, TIMOTHY GEORGE, utility company executive; b. Racine, Wis., Apr. 22, 1960; s. George Irving and Delores Lillian (Heidenreich) C. BA, Carthage Coll., 1982. Assoc. Mgmt. Aides Inc., Racine, 1982-83; asst. dir. River Bend Nature Ctr., Racine, 1983-86; spl. projects coord. Wis. So. Gas Co., Lake Geneva, 1986—. Editor: (newsletter) Contact, 1986—. Mem. Assn. Energy Engrs., Wis. Utilities Assn. (chmn. mkt. rsch. com. 1992—, vice chmn. 1991-92), Exptl. Aircraft Assn. (bd. dirs. Racine chpt. 1986—), Nat. Ski Patrol. Lutheran. Achievements include research in market manager energy analysis software. Office: Wis So Gas Co Inc 120 E Sheridan Springs Rd Lake Geneva WI 53147

CRAGG, GORDON MITCHELL, government chemist; b. Cape Town, Cape, South Africa, Sept. 4, 1936; came to U.S., 1979; s. Ernest Lynn and Doris Jessie (Mitchell) C.; m. Jacqueline Claire Tuers, Dec. 30, 1966. BSc with honors, Rhodes U., Grahamstown, South Africa, 1956; D.Phil., Oxford U., England, 1963. Sr. lectr. U. South Africa, Pretoria, 1966-72, U. Cape Town, 1973-79; sr. rsch. chemist Cancer Rsch. Inst. Ariz. State U., 1979-80, asst. to dir. Cancer Rsch. Inst., 1980-84; expert natural products br. Nat. Cancer Inst., Bethesda, Md., 1984-87, chemist, 1988-89; chief chemist Nat. Cancer Inst., Frederick, Md., 1989—. Author: Organoboranes in Organic Synthesis, 1973; co-author: Biosynthetic Products for Cancer Chemotherapy, 1985; contbr. over 60 articles to profl. jours. Recipient Merit award NIH, 1991. Mem. Am. Chem. Soc., Am. Soc. Pharmacognosy, South African Chem. Inst., Internat. Union Pure and Applied Chemistry, Soc. Econ. Botany. Independent. Achievements include patents for Prostratin, Michellamine B, Calanolides, Conocurvone; research in natural products isolation and structural elucidation with emphasis on novel antineoplastic and antiviral agents, steroid reactions, particularly the synthesis of vicinal diamines for platinum complex formation and the reaction of steroidal alkenes with dihaloketenes and organoborn reagents. Office: National Cancer Institute Natural Products Branch Fairview Ctr Ste 206 PO Box B Frederick MD 21702-1201

CRAGGS, ANTHONY, mechanical engineer; b. Wallsendupoh Tyne, U.K., Mar. 19, 1938; s. William Newrick and Mary Elizabeth (Youens) C.; m. Rosalind Patricia Goode, Aug. 15, 1964; children: Rachael Tamsin, John Newrick, Luke Anthony, Amy Megan. BS with hons., Mech. Engring. U., Durham, U.K., 1962; PhD, U. Southampton, 1969. Rsch. engr. Motor Industry Rsch. Assn., U.K., 1962-65; rsch. fellow U. Southampton, U.K., 1965-69; asst. prof. U. Alberta, Can., 1969-71, assoc. prof., 1971-78, prof., 1978—; cons. Gen. Motors Rsch. Labs, 1976-81. Author: Acoustics, Finite Elements Rotor Dynamics, 1970—; referee Jour. of SoundVibration. Mem. Acoustical Soc. Am., Canadian Machinery Vibration Assn. (pres. 1991—). Achievements include first application of finite element methods in acoustics treatment of absorption; design of mufflers and duct acoustics, sound transmission in vehicles. Home: 5603-143 St, Edmonton, AB Canada Office: U Alta, Dept Mech Engring, Edmonton, AB Canada

CRAGHEAD, JAMES DOUGLAS, civil engineer; b. Petersburg, Va., Nov. 27, 1950; s. William Douglas and Edith Marcia (Smith) C.; m. Vicki Lynn Taylor, June 5, 1970; 1 child, Jeffrey Taylor. BS, N.Mex. State U., Las Cruces, 1976. Design engr. Black & Veatch Cons. Engrs., Kansas City, Mo., 1976-77; engr. Frank Henri & Assocs., Las Cruces, N.Mex., 1977-78; sr. engr. Hughes Aircraft Co., Tucson, 1978-80, supr. engring., 1980-84, head environ. engr., 1985-90, sr. tech. specialist, 1990—. Inventor in field. Bd. dirs. Our Town Family Svcs., 1987-91. Sgt. U.S. Army, 1970-73. Scholar ROTC, 1969, N.Mex. Mil. Inst., 1969. Mem. ASCE, Am. Legion, Chi Epsilon. Republican. Episcopalian. Avocations: reading, genealogy, modeling. Home: 8221 E Kenyon Dr Tucson AZ 85710-4225 Office: Hughes Aircraft Co PO Box 11337 Tucson AZ 85734-1337

CRAIB, KENNETH BRYDEN, resource development executive, physicist, economist; b. Milford, Mass., Oct. 13, 1938; s. William Pirie and Virginia Louise (Bryden) C.; m. Gloria Faye Lisano, June 25, 1960; children—Kenneth Jr., Judith Diane, Lori Elaine, Melissa Suzanne. BS in Physics, U. Houston, 1967; MA in Econs., Calif. State U., 1982; postgrad. Harvard U., 1989. Aerospace technologist NASA, Houston, 1962-68; staff physicist Mark Systems, Inc., Cupertino, Calif., 1968-69; v.p. World Resources Corp., Cupertino, 1969-71; dir. resources devel. div. Aero Service Corp., Phila., 1971-72; dir. ops. Resources Devel. Assocs., Los Altos, Calif., 1972-80, pres., chief exec. officer, Diamond Springs, Calif., 1980-85; owner Sand Ridge Arabians, 1980—; chmn., dir. Resources Devel. Assocs., Inc., 1982—, Devel. Support Internat. Inc., Placerville, Calif., 1981—; pres., chmn., dir. RDA Internat., Inc., 1986—; dir. Sierra Gen. Investments, 1985—. Contbr. articles to profl. jours. Served with USAF, 1957-61. Recipient Sustained Superior Performance award NASA, 1966. NASA grantee, 1968. Mem. Am. Soc. Photogrammetry, Soc. Internat. Devel., Agrl. Research Inst., Calif. Select Com. Remote Sensing, Internat. Assn. Natural Resources Pilots, Remote Sensing Soc. (council), Am. Soc. Oceanography (charter), Aircraft Owners and Pilots Assn., Gulf and Caribbean Fisheries Inst., Placerville C. of C., Harvard Alumni Assn. Republican. Universalist. Home: 6431 Mary Ann Ln Placerville CA 95667-8167 Office: RDA Internat Inc 801 Morey Dr Placerville CA 95667-4411

CRAIG, DAVID PARKER, science academy executive, emeritus educator; b. Sydney, NSW, Australia, Dec. 23, 1919; s. Andrew Hunter and Mary Jane (Parker) C.; m. Veronica Bryden-Brown, Aug. 29, 1948; children: Andrew, David, Mary Louise, Douglas. BSc with honors, U. Sydney, 1940, MSc, 1941; PhD, U. Coll. London, 1949, DSc, 1956; DChem (hon.), U. Bologna, Italy, 1985; DSc (hon.), U. Sydney, 1989. Lectr. chemistry U. Sydney, 1945; Turner and Newell rsch. fellow and lectr. U. Coll. London, 1946-52; prof. phys. chemistry U. Sydney, 1952-56; prof. theoretical chemistry U. Coll. London, 1956-67; prof. theoretical and physical chemistry Australian Nat. U., Canberra, Australian Capital Ter., 1967-85; pres. Australian Acad. Sci., Canberra, 1990—; exec. mem. Commonwealth Sci. & Indsl. Rsch. Orgn., Canberra, 1980-85. Author: Excitons in Molecular Crystals, 1968, Molecular Quantum Electrodynamics, 1984; contbr. articles to profl. jours. Named officer Order of Australia, Govt. of Australia, 1985; fellow Royal Soc. London, 1968, Australian Acad. Sci., 1969. Home: 199 Dryandra St, O'Connor ACT 2601, Australia Office: Australian Nat U, Rsch Sch Chemistry, Canberra ACT 2600, Australia

CRAIG, GEORGE BROWNLEE, JR., entomologist; b. Chicago, Ill., July 8, 1930; s. George Brownlee and Alice Madeline (McManus) C.; m. Elizabeth Ann Pflum, Aug. 7, 1954; children: James Francis, Mary Catherine (dec.), Patricia Ann, Sarah Lynn. BA, Ind. U., 1951; MS, U. Ill., 1952, PhD, 1956. Rsch. asst. entomology U. Ill., 1951-53; entomologist Mosquito Abatement Dist., summers 1951-53; rsch. entomologist Chem. Corps Med. Labs., Army Chem. Ctr., Md., 1955-57; asst. prof. biology dept. U. Notre Dame, Ind., 1957-61, assoc. prof., 1961-64, prof., 1964-74, George and Winifred Clark disting. prof. biology, 1974—, dir. WHO Internat. Reference Ctr. for Aedes; vis. rsch. dir. Internat. Ctr. for Insect Physiology and Ecology, Nairobi, Kenya, 1968-76; mem. NIH study sect. on tropical medicine and parasitology, 1970-77; mem. med. rsch. adv. com. U.S. Army., 1980-88. Contbr. 450 articles on Aedes mosquitoes to profl. publs. Chmn. Ind. Vector Control Adv. Coun., 1975—. Served as 1st lt., Med. Svc. Corps U.S. Army, 1954-55. Rsch. grantee NIH, Ctr. Disease Control, AEC, WHO, NSF, Dept. Def. Fellow NAS (rsch. grantee) Am. Acad. Arts and Scis., Ind. Acad. Scis., Entomol. Soc. Am. (bd. govs. 1969-75, Disting. Teaching medal 1975, fellow 1986, Meml. lectr. 1988); mem. Am. Mosquito Control Assn. (Outstanding Rsch. award 1976, meml. lectr. 1984, pres. 1987), Am. Soc. Tropical Medicine and Hygiene (councilor 1978-83, councilor Am. com. on med. entomology 1985-90, Walter Reed medal 1993), Ind. Pub. Health Found. for Accomplishment Environ. Health (Hulman medal 1991), Sigma Xi. Home: 19645 Glendale Ave South Bend IN 46637-1817 Office: U Notre Dame Vector Biology Laboratory Notre Dame IN 46556

CRAIG, HURSHEL EUGENE, agronomist; b. Chrisman, Ill., May 18, 1932; s. Thomas Hurshel and Letha Mae (Short) C.; m. Zada Pauline Honnold, Dec. 29, 1954; children: Toni Jane, Tina Jean. Student, Ea. Ill. U., 1951, Ill. State U., 1956; BS, U. Ill., 1958, MS, 1970, postgrad., 1974. Mgr. Lime Svc. Co., Chrisman, Ill., 1959-61; br. mgr. Remole Soil Svc., Inc., Potomac, Ill., 1961-64, home office mgr., 1966-67; ptnr., agronomist Harris Fertilizer, Inc., Farmer City, Ill., 1967-69; farm cons. Gifford (Ill.) State Bank, 1964-66; instr. agr. Danville (Ill.) Area Community Coll., 1970-80; agronomy cons. and ptnr. C & S Pro-Farm Svcs., Ridge Farm, Ill., 1977-80; agronomy cons. Ag-Vantage, Westerville, Ohio, 1980-85; soils analysis sales CLC Labs Inc., West Lafayette, 1985-90, agronomy cons., 1987—; soil and plant tissue analysis sales Cal-Mar Soil Testing Lab., West Lafayette, 1990—. Co-author: Career Awareness Test for Agriculture Students and Prospective Spouses, 1974. Chmn. adminstrv. coun. Bismarck (Ill.) United Meth. Ch., 1989-91. With U.S. Army, 1952-54. Mem. Ill. Fertilizer and Chem. Assn., Profl. Crop Cons. Ill. (pres. 1992), Ill. Soil Testing Assn. Methodist. Avocations: reading, photography, fishing, shooting. Home and Office: RR 3 Box 256 Danville IL 61832-9531

CRAIG, JAMES CLIFFORD, JR., chemical engineer; b. Connellsville, Pa., Aug. 10, 1936; s. James Clifford and Virginia Marie (Nicholson) C.; m. Mary Elizabeth Hogan; children: Anne Barbara, Elizabeth Jean, James Clifford III. BSCE, Johns Hopkins U., 1958; postgrad., U. Pa., 1960-66. Chem. engr. Ea. Rsch. Ctr., USDA, Wyndmoor, Pa., 1959-79, rsch. unit head, 1979—; mem. tech. adv. com. Binational Agrl. Rsch. Devel. Fund, 1990-93; reviewer rsch. grants AID, NAS, 1991—. Contbr. over 50 articles to profl. jours. With U.S. Army, 1959. Recipient Rohland D. Isker award R&D Assocs., 1991. Mem. AICE, Am. Soc. Agrl. Engrs., Inst. Food Technologists (reviewer Jour. Food Sci. 1982—), Orgn. Profl. Employees of USDA, Sigma Xi. Achievements include 6 patents in food processing. Office: USDA 600 E Mermaid Ln Philadelphia PA 19118

CRAIG, JAMES LYNN, physician, consumer products company executive; b. Columbia, Tenn., Aug. 7, 1933; s. Clifford Paul and Maple (Harris) C.; m. Suzanne Anderson, July 20, 1957; children: James Lynn, Margaret; m. Roberta Anne, May 17, 1980. Ed., Mid. Tenn. State U., 1953; MD, U. Tenn., 1956; MPH, U. Pitts., 1963. Diplomate Am. Bd. Preventive Medicine, Am. Bd. Family Practice. Intern U. Tenn. Meml. Hosp., Knoxville, 1957; resident in occupational medicine U. Pitts., 1962-64; resident in occupational medicine TVA, Chattanooga, 1964-65, physician, 1966-69, chief med. officer, 1969-74; corp. med. dir. Gen. Mills Corp., Mpls., 1974-76, v.p. corp. med. dir., 1976-80, v.p., dir. health and human svcs., 1980—; clin. instr. U. Tenn., Memphis, 1970-74, Meharry Med. Sch., Nashville, 1972-74; clin. prof. U. Minn., Mpls., 1979—. Contbr. articles to profl. jours. Bd. dirs. Mpls. Blood Bank, 1976-88, Minn. Bible Coll., Rochester, 1978-83, Minn. Safety Coun., 1981-90, Minn. Heart Assn., Mpls., 1976-87, Children's Heart Fund, 1976-88, Meth. Hosp. Found., 1979-87, Park Nicolett Med. Found., 1987-93, Altcare, 1983—, Meth. Hosp. Health Assn., 1987-93, Minn. Wellness Coun., 1986-91, Health System Minn. Assocs., 1993—. Capt. USAF, 1958-61. Recipient Physician Recognition award AMA, 1975, 78, 81, 85, 89. Fellow Am. Occupational Medicine Assn. (bd. dirs. 1974-78), Am. Acad. Occupational Medicine (treas. 1982-83, sec. 1983-84, v.p. 1984-85, pres. 1986-87), Health Achievement in Occupational Medicine, Am. Acad. Family Practice; mem. AMA (alt. del. Ho. Dels. 1990-92, del. 1992—), Occupational Health Inst. (chmn. 1983-84), North Cen. Occupational Medicine Assn. (pres. 1977), Mpls. Acad. Medicine (sec. 1983-85, pres. 1985-86), Emergency Physicians Assn. (bd. dirs. 1984-92), U. Minn. Health Scis. (mem. adv. coun. 1992—). Home: 10008 S Shore Dr Minneapolis MN 55441-5011 Office: Gen Mills Corp General Mills Blvd Minneapolis MN 55426

CRAIG, ROBERT GEORGE, dental science educator; b. Charlevoix, Mich., Sept. 8, 1923; s. Harry Allen and Marion Ione (Swinton) C.; m. Luella Georgine Dean, Sept. 29, 1945; children: Susan Georgine, Barbara Dean, Katherine Ann. BS, U. Mich., 1944, MS, 1951, PhD (E.I. du Pont research fellow 1952-53), 1955; MD (hon.), U. Geneva, Switzerland, 1989. Rsch. chemist Linde Air Products Co., 1944-50, Texaco, Inc., Beacon, N.Y., 1954-55; rsch. assoc. U. Mich. Engring. Rsch. Inst., 1955-57; faculty dept. dental materials Sch. Dentistry, U. Mich., Ann Arbor, 1957-87, asst. prof., 1957-60, assoc. prof., 1960-64, prof., 1964-87, chmn. dept., 1969-87, prof. biologic and material sci., 1987-93, Marcus Ward prof. dentistry, 1990-93, prof. emeritus, 1993—; dir. Specialized Materials Sci. Ctr. Nat. Inst. Dental Rsch., Ann Arbor, 1989-93; prin. investigator Nat. Inst. Dental Rsch., 1989—; mem. exec. com. Sch. Dentistry, U. Mich., Ann Arbor, 1972-75; mem. budget priorities com. U. Mich., Ann Arbor, 1978-81; chmn. U. Mich. Budget Priorities Com., 1979-81; mem. sci. adv. com. Dental Research Inst., 1980-89, chmn., 1984-89; cons. Walter Reed Army Hosp., 1969-75; assessor for Nat. Health and Med. Rsch. Coun., Commonwealth Australia. Author: Restorative Dental Materials, 9th edit., 1993, (with K.A. Easlick, S.I. Seger and A.L. Russell) Communicating in Dentistry, 1973, (with W.J. O'Brien, J.M. Powers) Dental Materials-Properties and Manipulation, 5th edit, 1992, (with J.M. Powers) Workbook for Dental Materials, 1979; editor, contbr. Dental Materials Rev., 1977, Dental Materials-A Problem Oriented Approach, 1978; asst. editor: Jour. Biomed. Materials Research, 1983—; cons. editor: Jour. Dental Research, 1971-73, 77-80, Jour. Dental Edn, 1971-76, Jour. Oral Rehab, 1974—, Mich. State Dental Jour, 1973-77, Jour. Oral Implantology, 1988—; mem. adv. bd. Saudi Dental Jour., 1989—; contbr. articles to profl. jours. Prin. investigator specialized material Scis. Rsch. Ctr. (funded by NIDR 1989-94). Research grantee Nat. Inst. Dental Research, 1965-76, 84—; Nat. Scis. Res. Service Tng. grantee, 1976—. Mem. Am. Nat. Standards Inst. (chmn. spl. com. 1968-77), Internat. Assn. Dental Research (pres.-elect dental materials group 1972-73, pres. 1973-74, Wilmer Souder award 1975), Am. Assn. Dental Schs. (chmn. biomaterials sect. 1977-79), Am. Chem. Soc., ADA (cons. council on dental materials and devices 1983—), Soc. Biomaterials (Clemson award for basic research in biomaterials 1978, program chmn. 1983), Acad. Operative Dentistry (George Hollenbach Meml. prize 1991), Phi Kappa Phi, Phi Lambda Upsilon, Sigma Xi (sec. U. Mich. chpt. 1978-81), Omicron Kappa Upsilon. Home: 1503 Wells St Ann Arbor MI 48104-3914 Office: U Mich Sch Dentistry 1011 N Univ Ann Arbor MI 48109-1078

CRAIG, THOMAS FRANKLIN, electrical engineer; b. Indpls., Feb. 26, 1943; s. Robert Watson and Iris Evelyn (Evans) C.; BSEE (Charles M. Malott scholar), Purdue U., 1965, MSIA, 1970; m. Ester Annelle Cantrell, Aug. 3, 1968; children: Amy Delynne, Josie Leigh. Reliability engr. SCI Systems, Inc., Huntsville, 1968-71; components engr. Safeguard System Command, Huntsville, Ala., 1971-73; software engr. U.S. Army Guidance and Control Labs., Redstone Arsenal, Ala., 1973; systems engr. Pershing Project, 1974-77, communications equipment program mgr., 1977-81, chmn. Pershing II Communications Working Group; project engr. Stinger Project Office, 1981-82; prin. engr. Boeing Aerospace, Huntsville, Ala., 1982-89; program mgr. The EC Corp., Huntsville, Ala., 1989-91; prin. engr. space sta. freedom program Boeing Co., Huntsville, Ala., 1991—. Asst. treas., bd. dirs., mem. adv. bd. Huntsville Depot Mus., 1976-84; chmn. bd. elders 1st Christian Ch., 1979-81, 88, chmn. ofcl. bd., 1985-86; treas., trustee Helion Temple, 1978-84; treas. Huntsville Community Ballet Assn., 1986-88. Served to capt. ordnance, USAR, 1965-68. Engr.-in-tng., Ind., 1965. Mem. AAAS, IEEE, IEEE Computer Soc., Armed Forces Communications and Electronics Assn., Am. Def. Preparedness Assn., Ala. Acad. Sci., Assn. Unmanned Vehicle Systems, Assn. Old Crows, Mensa. Republican. Mem. Disciples of Christ. Ch. Clubs: Mountain Springs Swim, N. Ala. R.R. Mus., Sierra. Lodges: Shriners, Masons (past master, York Rite Coll. Gold Honor award 1981, Order of Purple Cross 1991). Editor The North Star, 1979—, Ala. Supplement to KT mag., 1985—. Ala. sec. Quatuor Coronati Corr. Circle, London, 1980—. Home: 1000 Lexington St SE Huntsville AL 35801-2533 Office: PO Box 240002 Huntsville AL 35824

CRAIGE, DANNY DWAINE, dentist; b. Okla., Mar. 25, 1946; s. William and Ruby G. (Sinor) C.; m. Mary Ann Thompson, Dec. 22, 1970. BS in Math., Southeastern Okla. State U., 1967; MS, Okla. U., 1968, DDS, 1980. Tchr. Yuba (Okla.) Pub. Schs., 1968-69, Sherman (Tex.) Pub. Schs., 1971-73; asst. mgr. Thompson Book and Supply Co., Durant, Okla., 1973-76; pvt. practice Durant, 1980—; cons. Med. Ctr. of S.E. Okla., Durant, 1980—, Bryan County (Okla.) Nursing Homes, 1980—. Bd. dirs. Texoma chpt. Boy Scouts Am., Durant and Denison, Tex., 1983-90, chmn. sustaining membership drive, Durant, 1989, 90; bd. dirs. Durant Western Days Talent Contest, 1989, 90, 91. Capt. USNR. Okla. U. fellow, 1968. Mem. ADA, Okla. Dental Assn., Naval Res. Assn. (life), Naval Order of U.S., Texoma Dental Study Club (chmn. 1980—). Home: 601 W Pine St Durant OK 74701-3735 Office: 203 N 16th Ave Durant OK 74701-3620

CRAIGHEAD, FRANK COOPER, JR., ecologist; b. Washington, Aug. 14, 1916; s. Frank Cooper and Carolyn (Johnson) C.; m. Esther Melvin Stevens, Nov. 9, 1943 (dec. 1980); children: Frank Lance, Charles Stevens, Jana Catherine; m. Shirley Ann Cocker, July, 1987. AB, Pa. State U., 1939; MS, U. Mich., 1940, PhD, 1950. Sr. rsch. assoc. Atmospheric Scis. Rsch. Ctr., N.Y., 1967-77; wildlife biologist, cons. U.S. Dept. Interior, Washington, 1959-66; wildlife biologist U.S. Forest Svc., Washington, 1957-59; mgr. desert game range U.S. Dept. Interior, Las Vegas, 1955-57; cons. survival tng. Dept. Def., Washington, 1950-55; pres. Craighead Environ. Research Inst., Moose, Wyo., 1955—; research assoc. U. Mont., Missoula, 1959—, Nat. Geographic Soc., Washington, 1959—; lectr. in field. Author: Track of the Grizzly, 1979, A Field Guide to Rocky Mountain Wildflowers, 1963, Hawks, Owls and Wildlife, 1956, How to Survive on Land and Sea, 1943, Hawks in the Hand, 1937. Mem. Pryor Mountain Wild Horse Adv. Com., Dept. Interior, 1968; mem. Horizons adv. group Am. Revolution Bicentennial Commn., 1972. Recipient citation Sec. of Navy, 1947; recipient letter of commendation U.S. Dept. Interior, 1963, Disting. Alumnus award Pa. State U., 1970; alumni fellow Pa. State U., 1973; recipient John Oliver LaGorce Gold medal Nat. Geog. Soc., 1979, U. Mich. Sch. Natural Resources Alumni Soc. award for Disting. Service, 1984, Centennial award Nat. Geog. Soc., 1988. Mem. AAAS, Wilderness Soc., Wildlife Soc., Explorers Club, Phi Beta Kappa, Sigma Xi, Phi Sigma, Phi Kappa Phi. Home: PO Box 156 Moose WY 83012-0156 Office: Craighead Environ Rsch Inst PO Box 156 Moose WY 83012-0156

CRAIGHEAD, HAROLD G., physics educator. BS, U. Md., 1974; PhD, Cornell U., 1980. Mem. tech. staff Bell Telephone Labs., Holmdel, N.J., 1979-84; rsch. mgr. Bell Comms. Rsch., Red Bank, N.J., 1984-89; prof. Cornell U., Ithaca, N.Y., 1989—; dir. Nat. Nanofabrication Facility, Ithaca, 1989—. Contbr. articles to profl. jours. Office: Cornell U Applied Physics Clark Hall Ithaca NY 14853

CRAIGHEAD, JOHN JOHNSON, wildlife biologist; b. Washington, Aug. 14, 1916; married; 3 children. BA, Pa. State U., 1939; MS, U. Mich., 1940, PhD, 1950. Biologist N.Y. Zool. Soc., 1947-49; dir. survival tng. for armed forces U.S. Dept. Def., 1950-52; wildlife biologist, leader Mont. coop. rsch. unit Bur. Sport Fisheries and Wildlife, 1952-77; prof. zoology and forestry U. Mont., 1952-77; bd. dirs., founder Craighead Wildlife-Wildlands Inst., Missoula, Mont., 1978—. Recipient Conservation award Am. Motors, 1978, John Oliver LaGorce Gold medal Nat. Geog. Soc., 1979, Centennial award, 1988, Lud Browman award Scientific and Tech. Writing, U. Mont., 1990; grantee NSF, AEC, NASA, U.S. Forest Svc., Mont. Fish and Game Dept. Mem. AAAS, Wildlife Soc., Wilderness Alliance. Office: Craighead Wildlife-Wildlands Inst 5200 Upper Miller Creek Rd Missoula MT 59803

CRAIGO, GORDON EARL, engineer; b. Glasgow, W.Va., June 17, 1951. BS in Nuclear Engring., U. Fla., Gainesville, 1977. Field engr. General Electric, Oak Brook, Ill., 1978-84; pres., owner Craigo Tech. Svc., Inc., Woodridge, Ill., 1984—. Office: Craigo Tech Svcs Inc Ste 12 G 65235 Steeple Run Dr Naperville IL 60540

CRAIN, BARBARA JEAN, pathologist, educator; b. Long Beach, Calif., Sept. 18, 1950; d. Gerald Clough and Reva Jean (Dahms) C.; m. Michael Joseph Borowitz, Dec. 29, 1978; children: Jeffrey Adam, David Douglas. BS, U. Calif., Irvine, 1972; PhD, Duke U., 1978, MD, 1979. Diplomate Am. Bd. Neuropathology. Asst. prof. pathology and neurobiology Duke U. Med. Ctr., Durham, N.C., 1983-92; asst. clin. prof. pathology, asst. rsch. prof. neurobiology, 1992-93; assoc. prof. pathology Johns Hopkins Sch. of Medicine, Balt., 1993-93; cons. John Unstead State Hosp., Butner, N.C., 1989-93; neurobiology merit rev. bd. VA, Washington, 1989-93; staff physician VA Hosp., Durham, 1983—. Contbr. articles to profl. jours. Pharmacology-morphology fellow Pharm. Mfrs. Assn., 1982; NIH Ctr. grantee, 1990. Fellow Coll. Am. Pathologists (neuropathology com.); mem. Am. Assn. Neuropathologists, U.S.-Can. Acad. Pathology, Soc. Neurosci. Office: Johns Hopkins Sch Medicine Dept Pathology 600 N Wolfe St Baltimore MD 21287

CRAIN, CULLEN MALONE, electrical engineer; b. Goodnight, Tex., Sept. 10, 1920; s. John Malone and Margaret Elizabeth (Gunn) C.; m. Virginia Raftery, Jan. 16, 1943; children—Michael Malone, Karen Elizabeth. B.S. in Elec. Engring, U. Tex., Austin, 1942, M.S., 1947, Ph.D., 1952. From instr. to asso. prof. elec. engring. U. Tex., 1943-57; group leader communications and electronics Rand Corp., Santa Monica, Calif., 1957-69; assoc. head and head engring. and applied scis. Rand Corp., 1969-88; cons. to govt., 1958—. Author numerous papers in field. Pres. Austin chpt. Nat. Exchange Club, 1954, Santa Monica chpt., 1975. Served with USNR, 1944-46. Recipient Disting. Grad. award U. Tex., 1987. Fellow IEEE (life); mem. Nat. Acad. Engring. Achievements include inventing microwave atmospheric refractometer. Home: 463 17th St Santa Monica CA 90402-2235 Office: 1700 Main St Santa Monica CA 90401-3297

CRAIN, DANNY B., microbiologist; b. Lynwood, Calif., Feb. 2, 1964; s. John Dennis and Doreen Aglaie (LaPierre) C.; m. Deborah Kay Burisch, July 17, 1990; 1 child, Tiffany Nicole Burisch-Crain. BA, U. Nev., Las Vegas, 1986; PhD, NYU, 1989. Lab. asst. Bunyan Rsch., Inc., Dover, Eng., 1980-81, U. Nev., Las Vegas, 1982-85; owner, biologist Crain Environ. Rsch., Las Vegas, 1987—; adj. cons. Space Biospheres, Tucson, 1989—; coord. Planetary Soc., Las Vegas, 1989-90; speaker Young Astronauts Nev., Las Vegas, 1990—, del. NASA Space Sta. Freedom Utilization Conf., 1993. Tchr. Jr. Girl Scouts Las Vegas, 1988-89. Recipient Hon. Young Astronaut award, cert. appreciation Young Astronauts Nev., Las Vegas, 1990; fellow Huntingdon (Eng.) Rsch. Inst., 1980; scholar Harry S Truman Scholar Found., Las Vegas, 1984. Mem. AAAS, Aris. Acad. Sci., Am. Inst. Biol. Scis., Nev. Acad. Sci., Am. Soc. Naturalists, N.Y. Acad. Scis. Achievements include research in elucidated roles of protein complexes on cell surface membranes.

CRAM, DONALD JAMES, chemistry educator; b. Chester, Vt., Apr. 22, 1919; s. William Moffet and Joanna (Shelley) C.; m. Jane Maxwell, Nov. 25, 1969. BS, Rollins Coll., 1941; MS, U. Nebr., 1942; PhD, Harvard U., 1947; PhD (hon.), U. Uppsala, 1977; DSc (hon.), U. So. Calif., 1983, Rollins Coll., 1988, U. Nebr., 1989, U. Western Ontario, 1990, U. Sheffield, 1991. Rsch. chemist Merck & Co., 1942-45; asst. prof. chemistry UCLA, 1947-50, assoc. prof., 1950-56, prof., 1956-90, S. Winstein prof., 1985—, univ. prof., 1988-90, univ. prof. emeritus, 1990—; chem. con. Upjohn Co., 1952-88, Union Carbide Co., 1960-81, Eastman Kodak Co., 1981-91, Technicon Co., 1984-92, Inst. Guido Donegani, Milan, 1988-91; State Dept. exch. fellow to Inst. de Quimica, Nat. U. Mex., 1956; guest prof. U. Heidelberg, Fed. Republic Germany, 1958; guest lectr. S. Africa, 1967; Centenary lectr. Chem. Soc. London, 1976. Author: From Design to Discovery, 1990, (with Pine, Hendrickson and Hammond) Organic Chemistry, 1960, 4th edit., 1980, Fundamentals of Carbanion Chemistry, 1965, (with Richards and Hammond) Elements of Organic Chemistry, 1967, (with Cram) Essence of Organic Chemistry, 1977, (with Cram) Molecular Container Chemistry, 1994; contbr. chpts. to textbooks, articles in field of host-guest complexation chemistry, carbanions, stereochemistry, mold metabolites, large ring chemistry. Named Young Man of Yr. Calif. Jr. C. of C., 1954, Calif. Scientist of Yr., 1974, Nobel Laureate in Chemistry, 1987, UCLA medal, 1993; recipient award for creative work in synthetic organic chemistry Am. Chem. Soc., 1965, Arthur C. Cope award, 1974, Richard Tolman medal, 1985, Willard Gibbs award, 1985, Roger Adams award, 1985, Herbert Newby McCoy award, 1965, 75, Glenn Seaborg award, 1989, Nat. Medal of Science, Nat. Sci. Found., 1993; award for creative rsch. organic chemistry Synthetic Organic Chem. Mfrs. Assn., 1965; Nat. Rsch. fellow Harvard U., 1947, Am. Chem. Soc. fellow, 1947-48, Guggenheim fellow, 1954-55. Fellow Royal Soc. (hon. 1989); mem. NAS (award in chem. scis. 1992), Am. Acad. Arts and Scis., Am. Chem. Soc., Royal Soc. Chemistry, Surfers Med. Assn., San Onofre Surfing Club, Sigma Xi, Lambda Chi Alpha. Office: UCLA Dept Chemistry Los Angeles CA 90024

CRAM, LAWRENCE EDWARD, astrophysicist; b. Nowra, N.S.W., Australia, Aug. 26, 1948; s. Laurence N. and Nellie (Wright) C.; m. Barbara J. Noble, July 15, 1973; childen: Edward D., Andrew T. BSc with honors, U. Sydney, 1970, BE with honors, 1972, PhD, 1975. CSIRO postdoctoral fellow Paris, 1975-76; astronomer Kiepenheuer Inst., Freiburg, Germany, 1976-78, Sacramento Peak Obs., Sunspot, N.Mex., 1978-81; prin. rsch. scientist CSIRO div. of Applied Physics, Lindfield, N.S.W., 1981-87; prof. astrophysics U. Sydney, 1987—, head Sch. Physics, 1991—; chmn. Steering Com. of Australia Telescope, Sydney, 1990—; bd. dirs. Anglo-Australian Obs., 1990—; dir. Sci. Found. for Physics, 1991—. Co-editor: The Physics of Sunspots, 1981, FGK and T Tauri Stars, 1990, Living with the Environment, 1991; co-author: Plasma Loops in the Solar Corona, 1991; contbr. over 100 articles to profl. jours. Fellow Australian Inst. Physics; mem. Am. Astron. Soc., Australian Astron. Soc., Internat. Astron. Union. Office: U Sydney, Sch Physics, Sydney NSW 2006, Australia

CRAMER, FRANK BROWN, engineering executive, combustion engineer, systems consultant; b. Long Beach, Calif., Aug. 29, 1921; s. Frank Brown and Clara Bell (Ritzenthaler) C.; m. Hendrika Van der Hulst, 1948 (div. 1962); children: Frieda Hendrika, Eric Gustav, Lisa Monica, Christina Elena; m. Paula Gil, Aug. 3, 1973; children: Alfred Alexander, Consuelo F., Peter M. BA, U. So. Calif., 1942, postgrad., 1942-43, 46-51. Rsch. fellow U. So. Calif., L.A., 1946-51; supr. engring. Rocketdyne, Canoga Park, Calif., 1953-63; pres. Multi-Tech, Inc., San Fernando, Calif., 1960-69; systems cons. Electro-Optical Systems, Pasadena, Calif., 1969-70, McDonnell-Douglas Astronautic, Huntington Beach, Calif., 1971-72; pres. Ergs Unltd. Inc., Mission Hills, Calif., 1973-89, Acquisition, Mission Hills, 1988—; instr. engring. stats. U. So. Calif., L.A., 1955-57, systems cons. dept. medicine, 1959-68; systems cons. Jet Propulsion Lab., Pasadena, 1964-68. Author: Statistics for Medical Students, 1951, Combustion Processes/Liquid Rocket Engineering, 1968; contbr. articles to profl. jours. Committeeman Libertarian Party, San Fernando Valley, Calif., 1966, Rep. Party, Mission Hills, 1967-68; pres. San Fernando Rep. Club, 1967-68. Achievements include patents for coal liquefication, for solid propellant processing, for pulse-meter/watch. Office: Acquisition 14800 Alexander St Mission Hills CA 91345

CRAMER, HOWARD ROSS, geologist, environmental consultant; b. Chgo., Sept. 17, 1925; s. Don William and Esther Natalia (Johnson) C.; m. Ardis V. Lahann, Dec. 15, 1950 (dec. 1980); m. Themis Poulos, Dec. 5, 1982. B.S. (with honors), U. Ill., 1949, M.S., 1950; Ph.D., Northwestern U., 1954. Registered geologist, Ga. Mem. faculty Franklin and Marshall Coll., 1953-58, asst. prof. geology Emory U., Atlanta, 1958-62, assoc. prof., 1962-76, prof., 1976-87, chmn. dept., 1981-87; cons. geology Ga. State U., Atlanta, 1988-91; chmn. Ga. Bd. Registration Geologists, 1977-79; mem. Ga. Natural Areas Council, 1968-72. Contbr. articles to sci. jours., chpts. to books on geology. Served with AUS, 1943-46, to lt. USAR, 1948-53. Decorated Bronze Star; recipient Holgate prize Northwestern U., 1953, Cert. Commendation, Am. Assn. State and Local History, 1974, Honor award Am. Fedn. Mineralogy and Lapidary Socs., 1986. Fellow Geol. Soc. Am.; mem. Am. Assn. Petroleum Geologists, Paleontol. Soc., Nat. Assn. Geology Tchrs. (pres. Southeastern sect. 1971-73), Ga. Acad. Sci. (pres. 1964-65), Lambda Chi Alpha. Greek Orthodox. Lodge: Ahepa. Home: 2047 Deborah Dr NE Atlanta GA 30345-3917

CRAMER, JACQUELINE MARIAN, environmental scientist, researcher; b. Amsterdam, The Netherlands, Apr. 10, 1951; d. Johan and Betty (Nebig) C.; m. Sabinus Luitzen Sander Kooistra; children: Tessa Lianne, Daniël. MS in Biology, U. Amsterdam, 1976, PhD in Sci. Dynamics, 1987. Staff mem. dept. biology and soc. U. Amsterdam, 1976-80, assoc. prof. dept. sci. dynamics, 1980-86; sr. researcher TNO Ctr. for Tech. and Policy Studies, Apeldoorn, The Netherlands, 1986—; prof. dept. environ. sci. U. Amsterdam, 1989—; vice chmn. Adv. Coun. for Rsch. on Nature and Environment, Rijswijk, 1991, mem. The Netherlands Coun. Traffic and Waterworks. Author: Ecology and Policy-Making, 1983, Mission-orientation in Ecology, 1987, The Green Wave, 1989; mem. adv. bd. Jour. Milieu, 1986—; mem. editorial bd. Jour. Clean Tech. and Environ. Scis., 1991—; contbr. over 100 articles to profl. jours. Co-founder, bd. dirs. Sci. Shop, U. Amsterdam, 1977-82; chmn. Environ. League, Amsterdam, 1985-87, Nat. Environ. Platform, Utrecht, The Netherlands, 1988-89. Recipient scholarship Dutch-Am. Inst., 1969-70; Eisenhower exch. fellow, 1992. Mem. Netherlands Soc. Environ. Scis. (bd. dirs. 1986-91), Alliance for Sustainable Devel. (adv. bd. 1990—), Univ. Edn. Environ. Scis. (adv. com. 1988—). Avocations: squash, hiking. Office: TNO Policy Rsch, Laan van Westenenk 501, 7300 Am Apeldoorn The Netherlands

CRAMER, JAMES PERRY, association executive, publisher, educator, architectural historian; b. Aberdeen, S.D., Aug. 7, 1947; s. Harry John and Carol B. (Bickel) C.; m. Corinne M. Aaker, Dec. 21, 1969; children: Ryan James, Austin Michael. BS, No. State U., Aberdeen, 1969; MA, St. Thomas U., St. Paul, 1974; planning cert., U. Minn., Mpls., 1976; bus. mgmt. cert., Wharton Sch. Bus., U. Pa., 1987. Dir., teaching faculty U. Minn., Mpls., 1974-76; dir. St. Louis Park Community Svcs., Minn., 1976-78; exec. v.p. Minn. Soc. Architects, Mpls., 1978-82; pres., chief exec. officer AIA Svc. Corp., Washington, 1982-86, also bd. regents; pres. Am. Archtl. Found. and Octagon Mus., Washington, 1986-89; chief exec. officer AIA, Washington, 1988—; group pub. Architecture Mag., 1982-89, pub. chmn., 1990—; with Archtl. Tech. Mag., 1983-89. Pres. Coun. Archtl. Components Washington, 1980-81; pres. Greenway Civic Assn., McLean, Va., 1986-88; bd. trustees Nat. Bldg. Mus., Washington, 1989—; chmn. United Way Assn., Washington div., 1992; White House liason, 1989—. Recipient Disting. Alumnus award No. State U., 1992. Mem. AIA (hon., chmn. 1981-82, chief exec. officer 1989—, spl. award 1982), Am. Soc. Assn. Execs. (cert. assn. exec.), Mag. Pubs. Am., Octagon Soc. (life hon.), Am. Archtl. Found. (life, pres. 1986-89, regent 1981-82, 86—), Am. Design Coun. (founder, bd. dirs. 1988—). Avocations: gardening, tennis, squash, design. Home: 8300 Riding Ridge Pl Mc Lean VA 22102-1316 Office: AIA 1735 New York Ave NW Washington DC 20006-5209*

CRAMER, JOHN GLEASON, JR., physics educator, experimental physicist; b. Houston, Oct. 24, 1934; s. John Gleason and Frances Ann (Sakwitz) C.; m. Pauline Ruth Bond, June 2, 1961; children: Kathryn Elizabeth, John Gleason III, Karen Melissa. B.A., Rice U., 1957, M.A., 1959, Ph.D. in Physics, 1961. Postdoctoral fellow Ind. U., Bloomington, 1961-63, asst. prof., 1963-64; asst. prof. physics U. Wash., Seattle, 1964-68, assoc. prof., 1968-74, prof., 1974—, dir. nuclear physics lab., 1983—; guest prof. W. Ger. Bundesministerium and U. Munich, 1971-72; mem. program adv. com. Los Alamos Meson Physics Facility, Los Alamos Nat. Lab., 1976-78, Nat. Superconducting Cyclotron Lab., 1983-87, TRIUMF (UBC), 1985-88; program adviser-cons. Lawrence Berkeley Lab., Calif., 1979-82; Gast prof. Hahn-Meitner Inst., West Berlin, 1982-83. Columnist Analog mag, 1983—; author: Twistor, 1989; contbr. articles to tech. and popular publs. Fellow AAAS, Am. Phys. Soc. (chmn. nuclear sci. tech. com. div. nuclear physics 1977-82, exec. com. div. nuclear physics 1981-83). Home: 7002 51st Ave NE Seattle WA 98115-6132 Office: Dept Physics FM-15 U Wash Seattle WA 98195

CRAMER, MERCADE ADONIS, JR., computer company executive; b. Charleston, S.C., Mar. 24, 1932; s. Mercade Adonis and Catharine (McTeer) C.; m. Nancy Caldwell; children: Gay Heyward, Catherine Ruth, Scott Tolman. B.S.E., U.S. Naval Acad., 1953; M.B.A., U.S. Air Force Inst. Tech., 1958. With Space Div, Gen. Electric Co., King of Prussia, Pa., 1969-84, with mil. programs dept., 1976-80, with mil. data systems, 1980-84; sr. v.p., gen. mgr. optical group Perkin-Elmer Corp., Danbury, Conn., 1984-87; pres., chief exec. officer Vitro Corp., Silver Spring, Md., 1987—; bd. dirs. Pensions for Profls., Media, Pa., 1978-80; mem. council of execs. Western Conn. State U., Danbury, 1985—. Commr. Nether Providence Twp., Pa., 1966-69; trustee U.S. Naval Acad. Found., Annapolis, Md., 1975—; v.p. bd. dirs. Riddle Meml. Hosp., Media, Pa., 1978-84. Served to capt. USAF, 1953-58. Recipient cert. significant achievement U.S. Air Force, 1977. Fellow AIAA (assoc.); mem. Am. Def. Preparedness Assn. (vice chmn. bd. 1987—), Security Affairs Support Assn. (chmn. bd. 1987—), Nat. Space Club, Am. Astron. Soc., Air Force Assn. Republican. Episcopalian. Avocations: antique collecting; refinishing; tennis; boating. Office: Vitro Corp 14000 Georgia Ave Silver Spring MD 20906-2960

CRANDALL, CHRISTIAN STUART, social psychology educator; b. Seattle, Aug. 18, 1959; s. Gordon Francis and Gail Louise (Schmitz) C.; m. Monica R. Biernat, Mar. 7, 1992. BS, U. Wash., 1980; PhD, U. Mich., 1987. Postdoctoral fellow Yale U., New Haven, 1987-88, lectr., 1988-89; asst. prof. U. Fla., Gainesville, 1989-92, U. Kans., Lawrence, 1992—. Contbr. articles to Jour. Exptl. Psychology, Am. Psychologist, Jour. Personality and Social Psychology, Personality and Social Psychology Bull., Psychol. Sci. Participant MacArthur Found. U. Mich., 1985-86, program for internat. peace and security rsch. Soviet-Am. Detente. Grantee Mich. Alumni Coun., 1985, Pediatric AIDS Found., 1991. Mem. APA, Am. Psychol. Soc., Soc. Exptl. Social Psychologists, Soc. for Psychol. Study of Social Issues (grantee 1987, 89), Soc. for Behavioral Medicine. Democrat. Presbyterian. Achievements include rsch. in delineating role of social influence in addictive behaviors. Office: U Kans Dept Psychology 426 Fraser Hall Lawrence KS 66045-0001

CRANDALL, DAVID HUGH, physicist; b. Chgo., July 9, 1942; s. Hugh John and Nellie Grace (Alborn) C.; m. Sidney Kathryn Horn, June 22, 1963 (div. Nov. 1975); children: Kathryn Ann, Christine Rene; m. Ellen Joy Dorros, Mar. 7, 1976; 1 child, Brian David. BS, Sioux Falls Coll., 1964; MS, U. Nebr., 1967, PhD, 1970. Vis. asst. prof. U. Mo., Rolla, 1970-71; rsch. assoc. Joint Inst. for Lab. Astrophysics, Boulder, Colo., 1971-74; rsch. scientist Oak Ridge (Tenn.) Nat. Lab., 1974-79, program mgr. physics, 1979-83; br. chief exptl. plasma rsch. Office of Fusion Energy Dept. of Energy, Washington, 1983-87, dir. applied plasma physics, 1987—; liason between Office of Fusion, Dept. Energy and profl. socs. including Nat. Acad. Scis. and Am. Phys. Soc., 1988—. Contbr. 100 articles to profl. jours. Coach Montgomery County Youth Basketball, Germantown, Md., 1988-91; official Mongomery County Swim League, Montgomery County, Md., 1989-91. Fellow Am. Phys. Soc.; mem. AAAS. Achievements include research on original measurements on capture of electrons by highly charged ions during collision with atoms and molecules demonstrating the importance initial ion charge in this basic process, excitation of electrons with ions by collision with free electrons is important in basic physics and within plasmas.; organized large scientific endeavors such as the current initiative to understand transport of energy and particles within takomak plasmas. Office: US Dept Energy ER54 GTN Washington DC 20545

CRANDALL, STEPHEN HARRY, engineering educator; b. Cebu, Philippines, Dec. 2, 1920; s. William Harry and Julia Josephine (Kuenemann) C.; m. Patricia Estelle Stickel, Jan. 21, 1949; children: Jane S., William S. M.E., Stevens Inst. Tech., 1942; Ph.D., MIT, 1946. Registered profl. engr. Mem. staff radiation lab MIT, Cambridge, 1942-43; instr. math MIT, 1944-46, asst. prof. mech. engring., 1947-51, assoc. prof., 1951-58, prof., 1958—, Ford prof. engring., 1975-91, prof. emeritus, 1991—, head div. applied mechanics, 1957-59, 61-67, head. div. mechanics and materials, 1968-71; vis. prof., Marseille, France, 1960, U. Nat. Autonoma Mex., Mexico City, 1967, Ecole Nationale Superieure de Mecanique, Nantes, France, 1978, Technion, Israel, 1987, Fla. Atlantic U., 1993. Author: Engineering Analysis, 1956, Random Vibration in Mechanical Systems, 1963, (with others) Dynamics of Mechanical and Electromechanical Systems, 1968; editor: Random Vibration vol. 1, 1958, Random Vibration vol. 2, 1963, (with others) Mechanics of Solids, 1959, author (with others), 3d edit., 1978; contbr. artcles to profl. jours. Fulbright fellow, exchange prof. Imperial Coll., London, 1949; NSF sci. faculty fellow, vis. scholar U. Calif.-Berkeley, 1964-65; hon. research assoc. Harvard U., 1971-72; Lady Davis vis. prof. Technion, Israel, 1987; recipient Von Karman medal ASCE, 1984, Alexander von Humboldt Sr. U.S. Scientist award, 1989; Schmidt Disting. Vis. Prof. Fla. Atlantic U., 1993. Fellow AAAS, ASME (Worcester Reed Warner medal 1971, v.p. 1978-80, hon. mem. 1988, Timoshenko medal 1990, Den Hartog award 1991), Am. Acad. Arts and Scis., Am. Acoustical Soc. (Trent-Crede medal 1978), Am. Acad. Mechanics; mem. NAS, NAE, NSPE, Soc. Indsl. and Applied Math., Am. Math. Soc., Am. Soc. for Engring. Edn., Internat. Union Theoretical and Applied Mechanics (chmn. U.S. del. 1974). Home: 25 Tabor Hill Rd Lincoln MA 01773-2906 Office: MIT/3-360 Dept Mech Engring Cambridge MA 02139

CRANE, HORACE RICHARD, educator, physicist; b. Turlock, Calif., Nov. 4, 1907; s. Horace Stephen and Mary Alice (Roselle) C.; m. Florence Rohmer LeBaron, Dec. 30, 1934; children—Carol Ann, Janet (dec.), George Richard. B.S., Calif. Inst. Tech., 1930, Ph.D., 1934. Research fellow Calif. Inst. Tech., 1934-35; mem. faculty U. Mich., Ann Arbor, 1935—; prof. physics U. Mich., 1946—, chmn. dept. physics, 1965-72, George P. Williams Univ. prof., 1972-78, emeritus, 1978—; Research asso. (radar) Mass. Inst. Tech., 1940-41; physicist Carnegie Inst. Washington, 1941; project dir., proximity fuze project U. Mich., 1941-43, atomic energy project, 1943-45; cons. NDRC, 1941-45; mem. standing com. on controlled thermonuclear research AEC, 1969-72; Vice pres. Midwestern Univs. Research Assn., 1956-57, pres., 1957-60; mem. policy bd. Argonne Nat. Lab., 1957-67; Bd. govs. Am. Inst. Physics, 1964-71, chmn., 1971-75; mem. Commn. on Human Resources, 1977-80, Council for Internat. Exchange of Scholars, 1977-80. Author: (monthly series) How Things Work in the Physics Teacher, 1983—, (books) How Things Work, 1992, Exhibits Guide, 1992; inventor, designer exhibits for hands-on type museums, 1981—; contbr. sci. articles to profl. mags. Recipient Davisson-Germer prize, 1967, Disting. Alumni medal Calif. Inst. Tech., 1968, Disting. Svc. award U. Mich., 1957, Nat. medal of sci., 1986; Henry Russel lectr., 1967. Fellow Am. Phys. Soc., AAAS, Am. Acad. Arts and Scis.; mem. Nat. Acad. Scis., Am. Assn. Physics Tchrs. (pres. 1965, Oersted medal 1977, Melba Newell Phillips award 1988), Sigma Xi. Clubs: Research Univ. of Mich. (pres. 1956-57); Science Research (U. Mich.) (v.p. 1946-47, pres. 1947-48). Inventor of Race Track, a modified form of synchrotron for nuclear studies, 1946; made early discoveries in field of artificially produced radioactive atoms, 1934-39; measurements of magnetic moment of free electron, 1950. Home: 830 Avon Rd Ann Arbor MI 48104-2738

CRANE, JACK WILBUR, agricultural and mechnical engineer; b. Flint, Mich., Sept. 21, 1932; s. Wilbur and Marvel (Gray) C.; m. Marilyn L. Hudson, Dec. 27, 1952; children: David L., Cathy C., Scott E. BS in Agrl. Engring., Mich. State U., 1956, MS, 1957. Registered profl. engr., Pa. Design engr. Sperry New Holland, Pa., 1957-61, project engr., 1961-69, engring. devel. mgr., 1969-73, product design mgr., 1973-77, dir. R&D engring., 1977-85; rsch. coord. Ford New Holland, Pa., 1985-91; cons. engr. Engring. Cons. Svcs., New Holland, 1991—. Contbr. articles to profl. jours. Fellow Am. Soc. Agrl. Engrs., NSPE (v.p. local chpt. 1978). Republican. Methodist. Achievements include 26 patents. Home and Office: 117 Greentree Dr New Holland PA 17557

CRANE, RICHARD TURNER, otolaryngologist; b. Ann Arbor, Mich., Sept. 3, 1951; s. Frederick Loring and Helen Marguerite (Eggerth) C.; m. Sharon Louise Gormley, May 6, 1978; children: Kevin Gormley, Ryan Anders, Chelsea Connell. BS, U. Mich., 1973; postgrad., Purdue U., 1973-74; MD, Ind. U., Indpls., 1978. Diplomate Am. Bd. Otolaryngology; lic. physician, Ind. Intern in surgery Naval Regional Med. Ctr., Portsmouth, Va., 1978-79; winter-over med. officer McMurdo Sta., Antarctica, 1979-80; resident in otolaryngology Naval Hosp., Portsmouth, Va., 1981-85, attending otolaryngologist, 1985-87; pvt. practice otolaryngology Ashland, Wis., 1987-88, Green Bay, Wis., 1988-89; fellow in neurotology House Ear Clinic, L.A., 1990; neurotologist Indpls., 1991-92; pvt. practice otolaryngology Chesapeake, Va., 1992—; asst. prof. clin. otolaryngology Ea. Va. Med. Sch., 1993—. Contbr. articles to profl. jours. Lt. comdr. USN, 1978-87. Fellow Am. Acad. Otolaryngology Head & Neck Surgery (mem. com. on history and archives, chmn. 1991-92); mem. AMA, Explorers Club of N.Y. Achievements include Antarctic search for meteorites; leader Back River arctic canoeing expedition. Home: 437 Broad Bend Cir Chesapeake VA 23320 Office: Chesapeake ENT Assocs 200 Medical Pkwy #303 Chesapeake VA 23320

CRANE, ROGER ALAN, engineering educator; b. Des Moines, Nov. 27, 1942; s. Elmer Lee and Treva Lee (Pope) C.; m. Susan Lorraine Jackson, Apr. 30, 1961 (div. Jan. 1988); children: Vincent Paul, Donna Jean Watt, Sarah Elizabeth; m. Marika Tali, Jan. 4, 1993. BS, U. Mo., Rolla, 1964, MS, 1966; PhD, Auburn U., 1973. Registered profl. engr., Fla. Engr. Newport News (Va.) Shipbuilding Co., 1964-65; sr. engr. Babcock & Wilcox, Lynchburg, Va., 1966-74; prof. U. South Fla., Tampa, 1974—. Grantee NASA, Lewis Rsch. Ctr., 1987-92, Kennedy Space Ctr., 1991—. Mem. AAAS, ASME, Am. Soc. for Engring. Edn. Democrat. Baptist. Home: 14003 Briardale Ln Tampa FL 33618 Office: U South Fla 4202 Fowler Ave ENG118 Tampa FL 33620

CRANEFIELD, PAUL FREDERIC, pharmacology educator, physician, scientist; b. Madison, Wis., Apr. 28, 1925; s. Paul Frederic and Edna (Rothnick) C. Ph.B., U. Wis., 1946, Ph.D., 1951; M.D., Albert Einstein Coll. Medicine, 1964. Fellow biophysics Johns Hopkins U., 1951-53; from instr. to assoc. prof. physiology State U. N.Y. Downstate Med. Center, N.Y.C., 1953-62; research fellow psychiatry Albert Einstein Coll. Medicine, 1960-64; exec. sec. com. publs. and med. information, editor bull. N.Y. Acad. Medicine, 1963-66; adj. assoc. prof. pharmacology Columbia Coll. Physicians and Surgeons, 1964-75, adj. prof., 1975—; assoc. prof. Rockefeller U., 1966-75, prof., 1975—. Author: (with Hoffman) The Electrophysiology of the Heart, 1960, Paired Pulse Stimulation of the Heart, 1968, (with C. McC. Brooks) The Historical Development of Physiological Thought, 1959, The Way In and the Way Out, 1974, The Conduction of the Cardiac Impulse, 1975, Claude Bernard's Revised Edition of his Introduction a L'Etude de la Médicine Expérimentale, 1976, (with Aronson) Cardiac Arrhythmias, The Role of Triggered Activity and Other Mechanisms, 1988, Science and Empire: East Coast Fever in Rhodesia and the Transvaal, 1991; also numerous articles; editor: Two Great Scientists of the Nineteenth Century, 1982, Jour. Gen. Physiology, 1966—; mem. editorial bd.: Circulation Research, Spl. Collections, Jour. of Electrocardiology; cons. editor: Internat. Microform Jour. Legal Medicine, 1969-77. Chmn. bd. dirs. LaMama Exptl. Theatre Club, 1965-69; chmn. bd. dirs. Circle Repertory Co., 1970-76, The Working Theatre; trustee Milton Helpern Library Legal Medicine. Recipient Einthoven medal U. Leiden, 1983. Fellow N.Y. Acad. Medicine (medal 1988), Internat. Acad. History of Medicine; mem. Am. Physiol. Soc., Biophys. Soc., Am. Assn. History Medicine, Bibliog. Soc., Episcopal Actors Guild (mem. coun. 1990—). Clubs: Century, Players, Nat. Arts, Grolier, Coffee House (N.Y.C.); Cosmos (Washington); Savile (London). Home: 310 E 9th St New York NY 10003-7901 Office: Rockefeller U 1230 York Ave New York NY 10021-6341

CRANKSHAW, JOHN HAMILTON, mechanical engineer; b. Canton, Ohio, Aug. 29, 1914; s. Fred. Weir and Mary (Lashels) C.; m. Wilma Chaffee Thurlow, June 5, 1940; children: Wilma Jean, John H., Geoffrey Thurlow. B.S. in Mech. Engring., MIT, M.S., 1940. Rotating engr. Gen. Electric Co., 1940-41, sect. engr., mech. design sect. Motor Engr. div. Locomotive Car Equipment, Erie, Pa., 1946-52; exec. engr. J.A. Zurn Mfg. Co., Am. Flexible Coupling Co., 1952-54; v.p. engring., 1954; exec. v.p., dir. Zurn Industries, Inc., mng. dir. Zurn Research and Devel. Div., until 1957; pres., dir. Dynetics, Inc., Erie, 1957—, Dynetic Systems, Inc., Erie, 1970—; expert witness numerous product liability cases. Mem. adv. council Gannon Coll.; chmn. Erie Sewer Authority. Served to maj. Ordnance Dept., AUS, 1941-46. Registered profl. engr., Pa. Mem. ASME, Soc. Automotive Engrs., Assn. Iron and Steel Engrs., Soc. Exptl. Stress Analysis, Soc. Naval Architects and Marine Engrs., Am. Soc. Metals, ASTM, Am. Soc. Lubricating Engrs. Erie Engring. Socs. Council (pres. 1955-57), Pa. Soc. Profl. Engrs., Sigma Xi. Clubs: MIT (N.Y.); Erie. Author several tech. papers. Achievements include 25 US and 5 foreign patents; invention and design of main propulsion couplings and clutches for nuclear powered submarines and Navy and Coast Guard surface ships. Home: 439 Shawnee Dr Erie PA 16505-2433 Office: Dynetics Inc 439 Shawnee Dr Erie PA 16505-2433

CRANNELL, HALL LEINSTER, physics educator; b. Berkeley, Calif., Feb. 23, 1936; s. Clarke Winslow and Alice Benjamin (Durant) C.; m. Carol Jo Argus, June 17, 1961; children: Annalisa, Francesca, Tasha. BA, Miami U., Oxford, Ohio, 1956, MA, 1958; PhD, Stanford U., 1964. Postdoctoral rsch. assoc. High Energy Physics Lab., Stanford (Calif.) U., 1964-67; assoc. prof. physics Cath. U. Am., Washington, 1967-72, prof., 1972—, chmn. dept., 1978-80, 86-89; vis. prof. physics Westfield Coll., U. London, 1973-74, MIT, 1974, Calif. Inst. Tech., Pasadena, 1990; vis. sr. rsch. scientist Rutherford Lab., London, 1973-74; guest sr. scientist Nat. Bur. Standards, 1980-81; sr. vis. scientist Nat. Inst. for Nuclear and High Energy Physics, Amsterdam, The Netherlands, 1981; program dir. for intermediate energy nuclear physics NSF, 1982-84. Contbr. over 60 articles to Phys. Rev. Letters, Phys. Rev. C, Nuclear Instrumentation and Methods. Vice pres. Springbrook High Sch. PTA, Silver Spring, Md., 1983-84; pres. Hillandale Citizens Assn., Silver Spring, 1991-93. Fellow Am. Phys. Soc.; mem. AAAS, AAUP, Am. Assn. Physics Tchrs., Bates Linear Accelerator Users Group (chmn. 1972-73, bd. dirs. 1982), Continuous Electron Beam Accelerator Facility Users Group (chmn. 1984-85, bd. dirs. 1985-86), Los Alamos Meson Physics Facility Users Group, Sigma Xi. Achievements include research on improved method for pressing thin targets from powders, design and testing of apparatus for 180 degree electron scattering at low momentum-transfer. Home: 10000 Branch View Ct Silver Spring MD 20903 Office: Cath U Am Dept Physics Washington DC 20064

CRANSTON, WILBER CHARLES, engineering company executive; b. Wenatchee, Wash., Nov. 4, 1933; s. Wilber James and Dorothy June (Thompson) C.; m. Martha Elisabeth Nyqvist, Dec. 26, 1975; children: John Albert, Cinthia Carrol, Barbara June; 1 stepchild: Petra Elisabeth. Student pub. schs. Mech. draftsman Cranston Machinery Co., Inc., Portland, Oreg., 1950-52, supt. shop, 1954-60; indsl. engr. Norris Thermador Corp., Riverbank, Calif., 1952-54; mgr. dept. engring. C. Tennant, Sons & Co. of N.Y., Warren, Ohio, 1960-70; cons. A. Ahlstrom Osakeyhtio, Karhula Engring. Works, Finland, 1971-77; founder, mng. dir. Cranston Ky, Kotka,

Finland, 1977—; Lectr. in field. Patentee automatic binding machines, semi-automatic hooks, one-trip load lifting straps; author: Forest ProductsCargo Handling Methods. Recipient Entrepreneur of Yr. award City of Kotka, 1981, Productive Idea award Finnish Jaycees, 1983, KAUPPALEHTI award Finnish Min. Industry and Commerce, 1983. Mem. Internat. Cargo Handling Coordination Assn., Internat. Forest Products Transport Assn. Office: Cranston Ky, PO Box 14, 48400 Kotka Finland

CRAVENS, GERALD MCADOO, telecommunications engineer; b. San Antonio, Apr. 16, 1918; s. Edward Oscar and Meta Louise (Laue) C.; m. Peggy Abbott, June 7, 1941 (div. 1953); 1 child, Gerald William; m. Mary Rita Bastow, May 15, 1963. BS, Tex. A&M U., 1939; MS, MIT, 1948. Instrument engr. Magnolia Petroleum Co. (now Mobil), Beaumont, Tex., 1939-41; commd. 2nd lt. U.S. Army, 1939, advanced through grades to col., 1964, ret., 1969; mem. tech. staff/sr. prin. engr. Computer Scis. Corp., Falls Church, Va., 1969-89, ret., 1989; systems analyst Combat Devel. Agy., Ft. Leavenworth, Kans., 1962-65; chief JCS Comm. Security & EW Div., Washington, 1967-69; cons. Computer Scis. Corp., Falls Church, 1983-88. Contbr. articles to profl. jours. Area coord. MIT Alumni Fund, Fairfax County, Va., 1977-79. Decorated Legion of Merit, Bronze Star (2), Joint Svc. Commendation medal, Army Commendation medal, French Croix de Guerre. Mem. IEEE (life), Armed Forces Comm. and Electronics Assn., Assn. U.S. Army, Sigma Xi (emeritus assoc.). Episcopalian. Achievements include conceptual development and engineering of complex military, other government emergency and civilian multi-media telecommunications systems and integration into synergistic cost-effective systems. Home: 4118 Whispering Ln Annandale VA 22003-2058

CRAWFORD, CLAUDE CECIL, III, biomedical consultant; b. Atlanta, Sept. 4, 1943; s. Claude C. Jr. and Lucille (Brown) C.; m. Marsha Ann Dietz, Apr. 2, 1972. BS, N.C. State U., 1966, MS, 1968, PhD, 1971. Zoologist HNTB Engrs., Alexandria, Md.; assoc. rsch. scientist Johns Hopkins U., Balt.; dir. lab. Dynatech Labs., Alexandria, 1976-86; mgr. R&D Ohmeda Infant Care Div., Columbia, Md., 1986-88; biomed. cons. Annapolis, Md., 1988-93. Capt. USAR, 1972. Mem. AAAS, Mensa, Sigma Xi. Home and Office: 709 Whitehall Plains Rd Annapolis MD 21401

CRAWFORD, E(DWIN) BEN, psychologist; b. Seattle, Nov. 6, 1944; s. Ben Crawford and Brideen (O'Brien) Milner; m. Virginia Roy Crawford, July 26, 1991. BA, U. San Francisco, 1971, MA, 1979; PhD, Loyola U. Chgo., 1985. Licensed psychologist, Alaska. Postdoctoral intern El Dorado County Mental Health, South Lake Tahoe, Calif., 1985-86; staff psychologist Calif. Mens Colony, San Luis Obispo, 1986-87; sr. psychologist Anchorage Community Mental Health, 1987-92; pvt. practice Showhegon, Maine, 1992; child and adolescent therapist Thad Woodard, M.D., P.C., Anchorage, 1992—; affiliate prof. U. Alaska, Anchorage, 1990-92. Author profl. publs. and presentations. Mental health coord. for disaster plan Municipality of Anchorage, 1990-92. With U.S. Army, 1965-67. Recipient Achievement award VA Hosp., San Francisco, 1976, Alaska Psychol. Assn.award, 1992. Mem. APA, Alaska Psychol. Assn. (treas. 1988-90, mem. at large 1993), Am. Group Psychotherapy Assn., Alpha Sigma Nu, Phi Delta Kappa. Achievements include research in mass disaster, assessment, post trauma stress. Office: 1200 Airport Heights Blvd Ste 140 Anchorage AK 99508

CRAWFORD, IAN DRUMMOND, business executive, electronic engineer; b. Dunfermline, Fife, Scotland, Jan. 21, 1945; came to U.S., 1978; s. John and Mary Ann (Drummond) C.; m. Hazel Ann Arkill, June 13, 1970; children: Lynne, Tracey, Douglas. BSc, Edinburgh U., 1966. Design engr. Ferranti Ltd., Edinburgh, 1966-74, chief engr., 1974-78; engring. mgr. Internat. Laser Systems, Orlando, Fla., 1978-79; pvt. cons. Longwood, Fla., 1979-82; pres. Analog Modules Inc., Longwood, 1980—. Contbr. articles to profl. jours. Mem. IEEE (sr.), Inst. Elec. Engrs. (sr.). Achievements include several patents in electro-optics and power supplies. Office: Analog Modules Inc 126 Baywood Ave Longwood FL 32750

CRAWFORD, ISIAAH, clinical psychologist; b. Washington, Nov. 29, 1960; s. Isiaah Sr. and Arthurine (Knox) C. MA, DePaul U., 1985, PhD, 1987. Lic. psychologist, Ill. Assoc. prof. Loyola U., Chgo., 1987—; pres. Crawford Cons. Svcs., Chgo., 1990—; bd. dirs. ROPA, St. Louis. Contbr. articles to profl. jours. Fellow APA, Soc. for the Scientific Study of Sex. Democrat. Roman Catholic. Home: 4324 N Dayton #C Chicago IL 60613 Office: Loyola U 6525 N Sheridan Ln Chicago IL 60626

CRAWFORD, KEVAN CHARLES, nuclear engineer, educator; b. Salt Lake City, Utah, Jan. 26, 1956; s. Paul Gibson and Norma Irene (Christiansen) C. MS, U. Utah, 1983, PhD, 1986. Lic. Sr. reactor oper. U.S. NRC. V.p. Computer Mktg. Corp., Salt Lake City, 1977-81; sr. reactor engr. U. Utah Nuclear Engring. Lab., Salt Lake City, 1981-86; mgr. reactor ops. Tex. A&M U. Nuclear Sci. Ctr., College Station, Tex., 1986-88; prof. U. Utah, Salt Lake City, 1988—, Idaho State U., Pocatello, 1991—; ex officio mem. reactor safety com. U. Utah, Salt Lake City, 1981-91, Idaho State U., Pocatello, 1992—; mem. radiation safety com. U. Utah, 1988-91, Idaho State U., 1992—; cons. Envirocare of Utah, Inc., Salt Lake City, 1989, Westinghouse Idaho Nuclear, 1992—. Cadet Air Force Acad., 1974-75. Recipient S.S. Kisler scholarship U. Utah, Salt Lake City, 1975-78. Mem. Am. Nuclear Soc., Am. Soc. Engring. Educators, Phi Kappa Phi, Alpha Nu Sigma. Mormon. Achievements include research contributions in nuclear reactor dynamics and control, and neutron diffraction for use in beam condensers and neutron microscopes. Office: Idaho State Univ Coll Engring Pocatello ID 83209 also: Univ Utah 3209 MEB Salt Lake City UT 84112

CRAWFORD, LESTER MILLS, JR., veterinarian; b. Demopolis, Ala., Mar. 13, 1938; s. Lester Mills and Susan Doris (Mitchell) C.; m. Catherine Walker, July 27, 1963; children: Catherine Leigh, Mary Stuart. D.V.M., Auburn U., 1963; Ph.D., U. Ga., 1969; M.D.V. (hon.), Budapest U., Hungary, 1987. Pvt. practice vet. medicine Meridian, Miss. and Birmingham, Ala., 1963-64; research and devel. staff Agrl. div. Am. Cyanamid Co., Princeton, N.J., 1964-66; also cons. Am. Cyanamid Co.; assoc. dean Coll. Vet. Medicine, U. Ga., 1970-75, head dept. physiology-pharmacology, 1980-82; dir. Bur. Vet. Medicine, FDA, HEW, Rockville, Md., 1978-80, 82-85; assoc. adminstr. food safety and inspection service USDA, Washington, 1986-87, adminstr., 1987-91; exec. v.p. sci. affairs Nat. Food Processors Assn., Washington, 1991-93; exec. dir. Assn. Am. Veterinary Medical Colls., Washington, 1993—; cons. pharm. industry, agribus FDA, WHO; mem. Health Professions Commn., Pew Meml. Trust, 1990—; bd. dirs. Embrex Inc. Contbr. sci. articles to profl. jours. Vice chmn. Codex Alimentarius Commn., 1991—; bd. dirs. Food and Drug Law Inst., 1988—, Toxicology Forum, 1991—, Agribus Rsch. Inst. Recipient A.M. Mills award, 1979, K.F. Meyer award, 1980, U.S. Presdl. Rank award of Meritorious Exec., 1988, Disting. Alumnus award Auburn U., 1989, Wooldridge Meml. medal Brit. Vet. Assn., 1991; Commrs. Spl. citation FDA, award of merit, 1983. Mem. AVMA (Aux. award), Nat. Acad. Practice, D.C. Vet Med. Assn., French Acad. Vet. (hon.), Cosmos Club (Washington), Sigma Xi, Phi Zeta, Phi Kappa Phi. Republican. Home: 5815 Highland Dr Chevy Chase MD 20815 Office: Assn Am Veterinary Med Colls 1401 New York Ave NW 1101 Vermont Ave Ste 710 Washington DC 20005

CRAWFORD, MARIA LUISA BUSE, geology educator; b. Beverly, Mass., July 18, 1939; d. William Theodore Buse and Barbara (Kidder) Aldana; m. William A. Crawford, Aug. 29, 1963. B.A., Bryn Mawr Coll., 1960; postgrad., U. Oslo, 1960-61; Ph.D., U. Calif., 1965. Asst. prof. Bryn Mawr (Pa.) Coll., 1965-73, assoc. prof., 1973-79, prof., 1979-92, prof. environ. studies and sci., 1992—, MacArthur fellow, 1993—, William R. Kenan Jr. prof., 1985-92, chmn. dept. geology, 1976-88, prof. in sci and environ. studies, 1992—; chmn. women geoscientists com. Am. Geol. Inst., 1976-77; mem. U.S. Nat. Com. Geochemistry, 1980-82; organizing com. 28th Internat. Geol. Cong., 1987-89. NASA grantee, 1973-76; NSF grantee, 1967—; MacArthur Found. fellow, 1993. Fellow Geol. Soc. Am. (councillor 1982-85), Mineral Soc. Am. (councillor 1989-92);mem. Mineral Assn. Can. (councilor 1985-87), Am. Geophys. Union, Norwegian Geol. Soc., Phila. Geol. Soc., Assn. Women in Sci. Office: Bryn Mawr Coll Dept Geology Bryn Mawr PA 19010

CRAWFORD, MARK HARWOOD, technology and energy journalist; b. Washington, Aug. 2, 1950; s. Sterling and Patricia (Moore) Lee; m. Jean Burke, Feb. 2, 1974; 1 child, Emily Lee. BA in Polit. Sci. and Comm., The Am. U., 1973. Reporter Suffolk Life, Westhampton, N.Y., 1974-75; assoc. editor Fairfax Jour., Springfield, Va., 1975-77; mng. editor Coal Week, Washington, 1978-81; assoc. editor Inside Energy, Washington, 1981-83; corr. McGraw Hill World News, Washington, 1983-85; sr. writer Science Mag., Washington, 1985-90; freelance reporter Washington, 1990; reporter New Tech. Week, Washington, 1991—. Co-editor, developer (book) Federal Coal Leases, 1981. Bd. dirs. Hillbrook-Tall Oaks Civic Assn., Annandale, Va., 1991—. Recipient Commendation letter Fairfax (Va.) Police Dept., 1976, 1st place for investigative reporting Va. Press Assn., 1977. Mem. Soc. Profl. Journalists (v.p. Washington chpt. 1989-90, bd. dirs. 1990—). Home: 4604 Monterey Dr Annandale VA 22003

CRAWFORD, RICHARD M., botanist. With Alfred Wegener Inst., Brenerhaven, Germany. Recipient Gerald W. Prescott award Phycological Soc. of Am., 1991. Office: Alfred Wegener Inst, PO 120161, Bremerhaven D 2850, Germany*

CRAWFORD, THOMAS MARK, laser/optics physicist, business owner, consultant; b. Balt., Apr. 19, 1954; s. Lynn Baker and Ruth Elizabeth (Anderson) C.; m. Virginia Houston Delany, Aug. 9, 1975; children: Daniel, Andrew, Ruth. BS in Physics, U. N.C., Charlotte, 1978; MS in Physics, Calif. State U., Northridge, 1984; postgrad. in Physics, U. N.Mex., 1986—. Rsch. asst. dept. physics U. Va., Charlottesville, 1979; sr. engr. Litton Guidance and Control, Woodland Hills, Calif., 1979-86; staff physicist Hughes Aircraft, Albuquerque, 1986-88; prin. scientist Lockheed Engring. & Scis. Corp., Las Cruces, N.Mex., 1988-90; engring. specialist EG&G-Idaho, Idaho Falls, 1990—; pres., chief scientist, owner Optical Physics Techs., Idaho Falls, 1988—. Author copyrighted software in optical instrumentation field. Precinct committeeman Bonneville County (Idaho) Rep. Party, 1992—, dist. treas., 1992—; mem. ednl. steering com. Skyline High Sch., Idaho Falls, 1992—. Recipient fellowship Hughes Aircraft, 1987, Advanced Class Amateur Radio Lic. Mem. Internat. Soc. Optical Engrs., Am. Soc. Precision Engring., Am. Radio Relay League, Sigma Pi Sigma, Pi Mu Epsilon. Achievements include a patent in the optical instrumentation field; 2 patents pending in the optical instrumentation field. Office: Optical Physics Techs PO Box 51483 Idaho Falls ID 83405

CREAGER, JOE SCOTT, geology and oceanography educator; b. Vernon, Tex., Aug. 30, 1929; s. Earl Litton and Irene Eugenia (Keller) C.; m. Barbara Clark, Aug. 30, 1951 (dec.); children: Kenneth Clark, Vanessa Irene; m. B. J. Wren, Sept. 5, 1987. B.S., Colo. Coll., 1951; postgrad., Columbia, 1952-53; M.S., Tex. A. and M. U., 1953, Ph.D., 1958. Asst. prof. dept. oceanography U. Wash., Seattle, 1958-61; assoc. prof. U. Wash., 1962-66, prof. oceanography, 1966-91, prof. geol. scis., 1981-91, prof. emeritus, 1991—, asst. chmn. dept. oceanography, 1964-65, assoc. dean arts and scis. for earth and planetary scis., 1966—, assoc. dean for rsch., 1966-91; program dir. for oceanography NSF, 1965-66; chief scientist numerous oceanographic expdns. to Arctic and Sub-arctic including Leg XIX of Deep Sea Drilling project, 1959-91; vis. geol. scientist Am. Geol. Inst., 1962, 63, 65; U.S. Nat. coord. Internat. Indian Ocean Expedition, 1965-66; vis. scientist program lectr. Am. Geophys. Union, 1965-72; Battelle cons., advanced waste mgmt., 1974; cons. to U.S. Army C.E., 1976, U.S. Depts. Interior and Commerce, 1975; exec. sec., exec. com., chmn. planning com. Joint Oceanographic Insts. Deep Earth Sampling, 1970-72, 76-78; mem. evaluation com. Northwest Assn. Schs. and Colls., 1989—. Mem. editorial bd. Internat. Jour. Marine Geology, 1964-91; assoc. editor Jour. Sedimentary Petrology, 1963-76; asst. editor Quaternary Research, 1970-79; contbr. articles to profl. jours. Skipper Sea Scout Ship, Boy Scouts Am., Bryan, Tex., 1957; coach Little League Baseball, Seattle, 1964-71, sec., 1971; cons. sci. curriculum Northshore Sch. Dist., 1970; mem. Seattle Citizens Shoreline Com., 1973-74, King County Shoreline Com., 1980. Served with U.S. Army, 1953-55. Colo. Coll. scholar, 1949-51; NSF grantee, 1962-82; ERDA grantee, 1962-64; U.S. Army C.E. grantee, 1975-82; Office of Naval Research grantee; U.S. Dept. Commerce grantee; U.S. Geol. Survey grantee. Fellow Geol. Soc. Am., AAAS; mem. Internat. Assn. Quaternary Research, Am. Geophys. Union, Internat. Assn. Sedimentology, Internat. Assn. Math. Geologists, Soc. Econ. Paleontologists and Mineralists, Marine Tech. Soc. (sec.-treas. 1972-75), Sigma Xi, Beta Theta Pi, Delta Epsilon. Club: Explorers. Home: 6320 NE 157th St Bothell WA 98011-4345 Office: U Wash Dept Oceanography WB-10 Seattle WA 98195

CREASIA, DONALD ANTHONY, toxicologist, researcher; b. Milford, Mass., Mar. 28, 1937; s. Dominic and Minnie (Bufalo) C.; m. Joan Labelle, June 29, 1963; children: Karen Joan, Tracey Dawn. BS in Biology, U. Vt., 1961; DSc, Harvard U., 1967; PhD, U. Tenn., 1981. Rsch. assoc. Sch. Pub. Health, Harvard U., Cambridge, Mass., 1963-69; toxicologist Oak Ridge (Tenn.) Nat. Lab., 1970-77; program dir. Frederick (Md.) Cancer Rsch. Ctr., 1977-83; rsch. chemist U.S. Army R&D, Frederick, 1983—; cons. toxicology, 1963—. Author: (chpts. in books with others) Internat. Symposium on the Biological Effects of Ozone and Related Photochemical Oxidents, 1983, Trycothecine Mycotoxicosis: Pathophysiological Efffects, 1989; contbr. over 120 articles to profl. jours. NSF scholar, 1965-67; NRC fellow, 1981-83. Mem. AAAS, Soc. Toxicology, Am. Coll. Toxicology, Soc. Govt. Toxicologists, Internat. Soc. Toxicology, Sigma Xi. Achievements include patents pending for use of castor bean protein as an immunological adjuvant, for nose-only and body plethismograph animal holder used in inhalation toxicology studies, and for discovery that insulin is equally effective in lowering blood glucose when inhaled into deep lung as when it is administered intramuscularly. Home: 6187 Viewsite Dr Frederick MD 21701-6750 Office: US Army R&D Ft Detrick Frederick MD 21702

CREASY, WILLIAM RUSSEL, chemist, writer; b. Catawissa, Pa., Apr. 27, 1958; s. Donald Eric and Geraldine Ruth (Billig) C. BA, Franklin & Marshall Coll., 1980; MS, U. Rochester, 1982, PhD, 1986. Rsch. assoc. Naval Rsch. Lab., Washington, 1986-87; staff chemist IBM Corp., Endicott, N.Y., 1987-92; freelance writer Endicott, 1992—; mem. adv. editorial bd. Fullerene Sci. and Tech., Budapest, Hungary, 1992—. Contbr. articles to profl. publs. Mem. Am. Phys. Soc., Am. Chem. Soc., Am. Soc. Mass Spectrometry. Achievements include studies of laser microprobe mass spectrometry, fullerene formation. Home and Office: 315 Garfield Ave Endicott NY 13760-5457

CREECH-EAKMAN, MICHELLE JEANNE, physicist, educator; b. Fargo, N.D., May 25, 1967; d. James F. and Claudia J. (Batterman) Creech; m. Tyson A. Eakman, Mar. 12, 1989. BS in Physics and Math. cum laude, U. N.D., 1990, MS in Physics, 1992. Grader, teaching asst. depts. math. and physics U. N.D., Grand Forks, 1988-90, grad. teaching asst. dept. physics, 1990-92; tchr. physics Rainy River C.C., International Falls, Minn., 1992-93; grad. rsch. asst. U. Denver, 1993—. Jay and Marie Bjerkaas scholar, U. N.D., 1987-88, Lawrence Welk Music scholar, 1985-87, High ACT Scores scholar, 1985-87. Mem. Soc. Physics Students, Am. Astron. Soc., Minn. C.C. Faculty Assn., Sigma Pi Sigma. Home: 5401 E Warren Ave Apt 205 Denver CO 80222

CREELMAN, MARJORIE BROER, psychologist; b. Toledo, Dec. 5, 1908; d. William F. and Ethel (Griffin) Broer; m. George Douglas Creelman, June 29, 1932 (div. Dec. 1958); children: Carleton Douglas, Stewart Elliott, Katherine George Skrobela. AB, Vassar Coll., 1931; MA, Columbia U., 1932; PhD, Western Res. U., 1954. Asst. psychologist N.Y. Psychiat. Inst., N.Y.C., 1932-33, Sunny Acres Sanitorium, Cleve., 1947-48; clin. asst. dept. psychology Case Western Res. U., Cleve., 1947-49, supr. field work, dir. practicum tng. program, 1949-54; asst. to dir. parent edn. program Children's Aid Soc., Cleve., 1952-53; ptnr., sr. assoc. Creelman Assocs., Cleve., 1954-58; pvt. practice psychology Cleve., 1954-64, Washington, 1964-69; research psychologist behavioral studies St. Elizabeth's Hosp., Washington, 1963-65, dir. psycho-physiology, clin. and behavioral studies, 1965-67; dir. psychol. services Alexandria Community Mental Health Ctr., 1967-69; mem. policy and planning com. Midwest Tng. Ctr. in Human Relations, 1954-60; prof. psychology Cleve. State U., 1969-76, prof. emeritus, 1976—; pvt. practice psychology, 1969—; mem. profl. staff Gestalt Inst. Cleve., 1969-81, hon. fellow, 1978—. Author: The Experimental Investigation of Meaning, 1965; editor: Ohio Psychologist, 1956-59; contbr. articles to profl. jours. Mem. Citizen's Adv. Bd. Case Western Res. U. Psychiat. Habilitation Ctr., 1979-81. Fellow Internat. Council Psychologists (sec. 1964-65), Am. Soc. Group Psychotherapy and Psychodrama, Ohio Psychol. Assn.; mem. Internat. Soc. Gen. Semantics (v.p. Cleve. chpt. 1950-61), Am. Acad. Psychotherapists (life, publs. com. directory editor 1963-67), Cleve. Acad. Cons. Psychologists (pres. 1957-58), Sigma Delta Epsilon, Psi Chi, Alpha Chi Omega. Club: Cleve. Skating.

CREMEENS, DAVID LYNN, soil scientist, consultant; b. St. Louis, Oct. 3, 1957; s. Virgil V. and Mildred D. (Brandon) C.; m. Cindy L. Wind, June 2, 1979; children: Melissa, Brian. AA in Life Sci., St. Louis Community Coll., 1977; BS in Agriculture, U. Mo., 1979; MS in Pedology, Mich. State U., 1983; PhD in Pedology, U. Ill., 1989. Cert. profl. soil scientist. Rsch. technologist dept. agronomy U. Mo., Columbia, 1977-79; sr. rsch. technologist Utah State U., Fillmore, 1979-80; grad. rsch. asst. Mich. State U., East Lansing, 1980-83; grad. rsch. asst. U. Ill., Urbana, 1983-88, rsch. specialist, 1988-89; staff soil scientist GAI Cons., Inc., Monroeville, Pa., 1989—. Contbr. articles to profl. jours. Mem. Soil Sci. Soc. Am., Geol. Soc. Am., Geochemical Soc. Home: 106 Brandywine Dr Irwin PA 15642 Office: GAI Cons Inc 570 Beatty Rd Monroeville PA 15146

CRENSHAW, MICHAEL DOUGLAS, chemist, researcher; b. Cin., Apr. 1, 1956; s. John Bent and Edith (Allen) C. BS, U. Cin., 1978, PhD, 1984. Rsch. assoc. Univ. Kans., Kans. City, 1984-86, TVA, Muscle Shoals, Ala., 1986-89; rsch. analyst Boeing Co., Huntsville, Ala., 1989-91; sr. scientist Kaman Scis. Corp., Alexandria, Va., 1991-92; prin. rsch. scientist Battelle Meml. Inst., Columbus, Ohio, 1992—; adj. asst. prof. U. Ala., Huntsville, 1991-92. Contbr. articles to Organic Preparations and Procedures, Jour. Organic Chemistry, Jour. Heterocyclic Chemistry. Mem. Am. Chem. Soc. (sect. sec. 1988), Am. Soc. for Mass Spectrometry, Sigma Xi. Republican. Achievements include patent in Thiopyridine-N-oxides, thiopyridines, and thiopyrimidines as urease enzyme inhibitors. Office: Battelle Meml Inst Rm 7247 505 King Ave Columbus OH 43201-2693

CREPET, WILLIAM LOUIS, botanist, educator; b. N.Y.C., Aug. 10, 1946; s. Louis Henry and Adaire Elaine (Richardson) C.; m. Laura Marie Stewart, July 29, 1972 (div. 1978); m. Ruth Chadab, July 27, 1980. BA, SUNY, Harpar Coll., 1969; MPh (Wadsworth fellow), Yale U., 1972, PhD (Cullman fellow), 1973. Cons. to Grad. Sch. U. Tex., Austin, 1972-73; lectr. Ind. U., 1973-75; asst. prof. U. Conn., 1975-78, assoc. prof., 1979-84, prof., 1985—, head dept., 1985-90; chmn., prof. Bailey Hortorium Cornell U., 1990—. N.Y. State Regents scholar SUNY, 1969. Fellow Explorers Club; mem. Bot. Soc. Am. (chmn. paleobotany sect. 1979-80, Paleobot. award 1972), Am. Inst. Biol. Scis., Beta Chi Sigma. Achievements include research in Mesozoic and Tertiary genera. Office: Cornell U LH Bailey Hortorium 467 Mann Libr Ithaca NY 14853

CRESSER, MALCOLM STEWART, soil scientist, educator; b. Ilford, Essex, England, Apr. 17, 1946; s. Edward Ernest and Doris Ann (Ferrari) C.; m. Louise Elizabeth Blackburn, Sept. 2, 1967; children: Frances Louise, Adam John, Laura Anne. BS, Imperial Coll., 1967, PhD, 1970. Lectr. in soil sci. Aberdeen (Scotland) U., 1970-82, sr. lectr., 1982-86, reader, 1986-88, reader in plant and soil sci., 1988-89, prof., 1989—, dept. head, 1989-90; vis. prof. Ga. Tech. U., Atlanta, 1979; mem. critical loads adv. group Dept. Environ., United Kingdom, 1988—; mem. plant and environ. funding panel Agriculture and Food Rsch. Coun., Swindon, United Kingdom, 1992—, mem. rsch. bd., 1991—. Contbr. over 170 articles to profl. jours. Fellow Royal Soc. Chemistry (Silver medal 1984); mem. Brit. Soc. Soil Sci. Home: 92 Bonnymuir Pl, Aberdeen AB2 4NP, United Kingdom Office: U Aberdeen, Dept Plant and Soil Sci, Aberdeen AB9 2UE, Scotland

CRESSON, DAVID HOMER, JR., pathologist; b. Danville, Va., May 23, 1955; s. David Homer and Virginia Mae (Swink) C.; m. Lisa Carol Meadors, June 23, 1979; 1 child, Carol Christine. BA in Chemistry magna cum laude, Duke U., 1977; MD, U. Tenn., 1981. Diplomate Am. Bd. Anatomic and Clin. Pathology, Dermatopathology and Cytopathology. Intern, then resident in pathology U. N.C., Chapel Hill, 1981-85, Am. Cancer Soc. surg. pathology fellow, 1985-86; fellow in dermatopathology Stanford (Calif.) U., 1986-87; pathologist, dermatopathologist Va. Bapt. Hosp.-Lynchburg Hosp., 1987—, chmn. dept. pathology, mem. exec. com. med. staff, 1991—; pathologist, dermatopathologist, cytopathologist Pathology Cons. of Cen. Va. Contbr. articles to profl. jours. Recipient Wiley Forbus Pathology Resident Rsch. award N.C. Pathology Soc., 1985. Fellow Coll. Am. Pathologists; mem. Internat. Acad. Pathology, Am. Soc. Clin. Pathologists, Sigma Xi, Alpha Omega Alpha. Achievements include research on synthetic amidino esteroprotease inhibitors, laryngeal carcinomas evaluation of DNA ploidy. Office: Pathology Cons Cen Va 1905 Atherholt Rd Lynchburg VA 24501-1198

CREVELING, CYRUS ROBBINS, chemist, neuroscientist; b. Washington, May 30, 1930; s. Cyrus Robbins and Edith Lois (Hill) C.; m. Cornelia Mills Rector, Sept. 3, 1954; children: Victoria Anne Mariano, Diana Rector Mears. BS, George Washington U., 1954, MS, 1955, PhD, 1962. Chemist Naval Ordinance Lab., Washington, 1953-54; med. tech. Sibley Meml. Hosp., Washington, 1955-57; chemist Nat. Heart Inst., Bethesda, Md., 1957-62; rsch. assoc. Harvard U., Boston, 1963-64; chemist NIH, Bethesda, 1964—; prof. of pharmacology Howard U. Sch. Medicine, Washington, 1967-86; coord. technology dept. NIH, Bethesda, 1989—; prof. pharmacology and toxicology Med. Coll. Va., Richmond, 1985—; mem. divsn. rsch. grants NIH, 1966-70; lectr. in field; mem. task force on environ. cancer and heart and lung disease, Project Group on standardization, measurements and tests, EPA, 1979-82; project adv. bd. screening food additives, FDA, 1979-83. Contbr. numerous articles to profl. jours., chpts. to books; editor: Transmethylation Series I, 1979, II, 1982, III, 1986; reviewer Analytical Biochemistry, Archives of Biochemistry and Biophysics, Biochem. Jour., Biochem. Pharmacology, Brain Rsch., Drug Metabolism and Disposition, Endocrinology, European Jour. Pharmacology, Life Scis., Jour.Neurosci., Jour. Neurochemistry, Jour. Molecular Pharmacology, Jour. Medicinal Chemistry, Jour. Biol. Chemistry, Jour. Am. Chem. Soc., others. Chmn. adv. bd. Montgomery Libr., Kensington, 1984-86; mem. Bethesda HELP, 1987—; mem. BB Telecasts, Channel 7, Washington, 1982-91. Fellow Am. Diabetes Found., 1955, Mass. Gen. Hosp., 1963-64; recipient Disting. Sci. award Soc. Exptl. Biology and Medicine, 1979, Pub. Health Svc. award NIH, 1991; rsch. grantee Eli Lilly, 1984, E.I. duPont de Nemours & Co., 1985. Mem. Am. Fedn. Scientists, Am. Chem. Soc., Am. Soc. Pharmacol. Exptl. Therapeutics, Soc. Exptl. Biology and Medicine (pres. 1977-78, councilor 1975—), Washington Acad. Scis. (fellow, bd. mgrs. 1979-80, chmn. biol. scis. panel 1980—, v.p. membership affairs 1991—), Fdn. Advanced Edn. in Scis. (chair scholarship com. 1989—), Catecholamine Club (pres. 1980-81), Soc. Neurosci., Gordon Conf. on Cyclic Nucleotides, Chem. Soc. Washington, Internat. Soc. Study Xenobiotics, Internat. Union Physiol. Scis., Internat. Union Pure and Applied Chemistry (affiliate), Internat. Narcotics Rsch. Conf. 1989—, Internat. Union Physiol. Sci., many others. Republican. Methodist. Achievements include discovery and synthesis of 6-hydroxydopamine; discovery and characterization of "synaptoneurosomes"; measurement of sodium channels with batrachotoxin, catechol-O-methyl thransferase in prevention of estrogen induced carcinogenesis. Home: 4516 Amherst Ln Bethesda MD 20814-4008 Office: NIH 8A/1A27 9500 Rockville Pike Bethesda MD 20892

CREWE, KATHERINE, engineer. Gen. mgr. CYEX Med. Tech. Inc., Mossissauga, Ont., Can. Recipient Young Engr. Achievement award Can. Coun. Profl. Engrs., 1992. Office: CYEX Med Tech Inc, 5080 Timberiea Blvd, Mississauga, ON Canada L4W 4M2*

CREWS, JOHN ERIC, rehabilitation administrator; b. Marion, Ind., Mar. 4, 1946; s. Odis Earl and Beatrice True (Wright) C.; m. Nancy J. Murphy, Aug. 9, 1975; 1 child, Katherine. BA in English, Franklin Coll., 1969; MA in English, Ind. U., 1971; MA in Blind Rehab. with honors, Western Mich. U., 1977, D in Pub. Adminstrn., 1990. Mem. English faculty Ball State U., Muncie, Ind., 1971-73, S.W. Mo. State U., Springfield, 1973-76, Western Mich. U., Kalamazoo, 1976-77; rehab. tchr. Mich. Commn. for the Blind, Saginaw, 1977-80; program mgr. Sr. Blind Program, Saginaw, Southeastern Mich. Ctr. for Ind. Living, Detroit, 1980-92, Ind. Living Rehab. Program, 1986-92; rsch. health scientist Rehab. R&D Ctr. Va. Med. Ctr., Decatur, Ga., 1992, chief behavioral sect. Rehab. R&D Ctr., 1992; acting dir. Rehab. Rsch. and Devel. Ctr. VA Med. Ctr., Decatur, Ga., 1993—; v.p. bd. Midland County Coun. on Aging, 1982-84; bd. dirs. Saginaw Valley Spl. Needs Vision Clinic, 1981-88; mem. adv. bd. rehab. continuing edn. program So. Ill. U., Carbondale, 1985-90, Lighthouse Nat. Ctr. on Vision and Aging, 1992—; sec. Statewide Ind. Living Coun., 1987-92; mem. editorial bd. Jour. Visual

Impairment and Blindness, 1984-90. Contbr. to books and profl. publs. Mem. exec. coun. Am. Found. for the Blind, 1990—; mem. bd. dirs. Midland Community Concert Soc., 1988-92. Recipient Grant award Ind. Living Svcs. for Older Blind Rehab. Svcs. Adminstrn., 1986, Community Svc. award Saginaw Valley Rehab. Ctr., 1992; grantee Ctr. Ind. Living U.S. Dept. Edn., 1980, 82, Ind. Living for Elderly Blind, 1986; All-Univ. grad. rsch. and creative scholar Western Mich. U., 1988. Mem. Nat. Coun. Aging, Assn. Retarded Citizens (pres. Midland 1981-87, Ann. Appreciation award 1981). Methodist. Home: 5287 Candleberry Dr Lilburn GA 30247 Office: VA Med Ctr Rehab R&D 1670 Clairmont Rd Decatur GA 30033

CREWS, PATRICIA COX, textile scientist, educator; b. Lexington, Va., Oct. 26, 1948; d. Harry Hamilton and Sarah Viola (Gregory) Cox; m. David William Crews, June 7, 1969; 1 child, David Seth. BS, Va. Poly. Inst. and State U., 1971; MS, Fla. State U., 1973; PhD, Kans. State U., 1984. Instr. Oreg. State U., Corvallis, 1973-74, Va. Western Community Coll., Roanoke, 1975, Bluefield (W.Va.) State Coll., 1976-77, Kans. State U., Manhattan, 1977-84; asst. prof. U. Nebr., Lincoln, 1984-89, assoc. prof. dept. textiles, clothing and design, 1989—. Contbr. articles to profl. publs. Recipient Manufactured Fibers Rsch. award Am. Fiber Mfrs. Assn., 1991. Mem. Am. Assn. Textile Chemists and Colorists, Am. Chem. Soc. (cellulose, paper and textile div.), Internat. Textile and Apparel Assn. (treas. 1987-89), Am. Inst. Conservation. Office: Dept Textiles Clothing and Design Univ Nebr Lincoln NE 68583-0802

CRICK, FRANCIS HARRY COMPTON, biologist, educator; b. June 8, 1916; s. Harry and Annie Elizabeth (Wilkins) C.; m. Ruth Doreen Dodd, 1940 (div. 1947); 1 son; m. Odile Speed, 1949; 2 daus. B.Sc., Univ. Coll., London; PhD, Cambridge U., Eng. Scientist Brit. Admiralty, 1940-47, Strangeways Lab., Cambridge, Eng., 1947-49; biologist Med. Rsch. Coun. Lab. of Molecular Biology, Cambridge, 1949-77; Kieckhefer Disting. prof. Salk Inst. Biol. Studies, San Diego, 1977—, non-resident fellow, 1962-73; adj. prof. psychology U. Calif., San Diego; vis. lectr. Rockefeller Inst., N.Y.C., 1959; vis. prof. chemistry dept. Harvard U., 1959, vis. prof. biophysics, 1962; fellow Churchill Coll., Cambridge, 1960-61; Korkes Meml. lectr. Duke U., 1960; Henry Sedgewick Meml. lectr. Cambridge U., 1963; Graham Young lectr., Glasgow, 1963; Robert Boyle lectr. Oxford U., 1963; Vanuxem lectr. Princeton U., 1964; William T. Sedgwick Meml. lectr. MIT, 1965; Cherwell-Simon Meml. lectr. Oxford U., 1966; Shell lectr. Stanford U., 1969; Paul Lund lectr. Northwestern U., 1977; Dupont lectr. Harvard U., 1979, numerous other invited, meml. lectrs. Author: Of Molecules and Men, 1966, Life Itself, 1981, What Mad Pursuit, 1988, The Astonishing Hypothesis: The Scientific Search for the Soul, 1994; contbr. papers and articles on molecular, cell biology and neurobiology to sci. jours. Recipient Prix Charles Leopold Mayer French Academies des Scis., 1961; (with J.D. Watson) Rsch. Corp. award, 1961, (with J.D. Watson & Maurice Wilkins) Nobel Prize for medicine, 1962, Gairdner Found. award, 1962, Royal Medal Royal Soc., 1972, Copley Medal, 1976, Michelson-Morley award, 1981, Benjamin P. Cheney medal, Spokane, Wash., 1986, Golden Plate award, Phoenix, 1987, Albert medal Royal Soc. of Arts, London, 1987, Wright Prize VIII Harvey Mudd Coll., Claremont, Calif., 1988, Joseph Priestly award Dickinson Coll., 1988. Fellow AAAS, Royal Soc.; mem. Acad. Arts and Scis. (fgn. hon.), Am. Soc. Biol. Chemistry (hon.), U.S. Nat. Acad. Scis. (fgn. assoc.), German Acad. Sci., Am. Philos. Soc. (fgn. mem.), French Acad. Scis. (assoc. fgn. mem.), Indian Acad. Scis. (hon. fellow), Order of Merit. Office: Salk Inst Biol Studies PO Box 85800 San Diego CA 92186-5800

CRIDER, HOYT, health care executive; b. Arley, Ala., June 5, 1924; s. Lindsey C. and Bessie P. C.; student Ga. Sch. Tech., 1942-43; B.S. in Naval Sci. and Tactics, U. S.C., 1946; M.A. in Polit. Sci., U. Ala., 1949; D.Pub. Adminstrn., U. So. Calif., 1954; m. Judie Watkins, Nov. 2, 1951; children—Kim, Marc. Vis. asst. prof., dir. research U. So. Calif. team, Iran, 1954-56; adminstrv. analyst Chief Adminstrv. Offices Los Angeles County, 1956-59; v.p. Watkins & Watkins Constrn. Co., Hanford and Morro Bay, Calif., 1959-64; co-owner, adminstr. Kings Convalescent Hosp., Hanford, Calif., 1964-66; adminstr. Villa Capistrano Convalescent Hosp., Capistrano Beach, Calif., 1966-68; partner Hunt and Crider, San Diego Convalescent Hosp., 1968-70; pres., chief exec. officer Health Care Enterprises, Inc., San Clemente, Calif., 1970—; mem. Regional Health Planning Commn. Kings County, 1963-64. Served with USNR, 1941-46. Fellow Am. Coll. Nursing Home Adminstrs. (pres. 1976-77), Am. Coll. Health Care Adminstrs.; mem. Calif. Assn. Health Facilities (past v.p. local chpt. 1964), Gerontol. Soc., AAAS. Club: San Clemente Kiwanis (Kiwanian of Yr. 1970, pres. 1970-71). Home: 2215 Avenida Oliva San Clemente CA 92673

CRILLY, EUGENE RICHARD, engineering consultant; b. Phila., Oct. 30, 1923; s. Eugene John and Mary Virginia (Harvey) C.; m. Alice Royal Roth, Feb. 16, 1952; ME, Stevens Inst. Tech., 1944, MS, 1949; MS, U. Pa., 1951; postgrad. UCLA, 1955-58. Sr. rsch. engr. N.Am. Aviation, L.A., 1954-57; sr. rsch. engr., Canoga Park and Downey, Calif., 1962-66; process engr. Northrop Aircraft Corp., Hawthorne, Calif., 1957-59; project engr., quality assurance mgr. HITCO, Gardena, Calif., 1959-62; sr. rsch. specialist Lockheed-Calif. Co., Burbank, 1966-74; engrng. specialist N.Am. aircraft ops. Rockwell Internat., El Segundo, Calif., 1974-89. Author tech. papers. Served with USNR, 1943-46; comdr. Res. ret. Mem. Soc. for Advancement Material and Process Engrng. (chmn. L.A. chpt. 1978-79, gen. chmn. 1981 symposium exhbn., nat. dir. 1979-86, treas. 1982-85, Award of Merit 1986), Soc. Mfg. Engrs. (sr.), Naval Inst., Am. Soc. for Composites, ASM Internat., Naval Res. Assn., VFW, Mil. Order World Wars (adj. San Fernando Valley chpt. 1985, 2d vice comdr. 1986, commdr. 1987-89, vice comdr. West, Dept Cen. Calif., 1988-89, comdr. Cajon Valley-San Diego chpt. 1990-92, adj./ROTC chmn. region XIV 1990-91, comdr. Dept. So. Calif. 1991-93, vice comdr. region XIV, 1992-93, dep. comdr. Gen. Staff Officer region XIV 1993—, Disting. Chpt. Cmdr. Region XIV 1990-91), Former Intelligence Officers Assn. (treas. San Diego chpt. one 1990—), Ret. Officers Assn. (treas. Silver Strand chpt. 1992—), Navy League U.S., Naval Order U.S., Naval Intelligence Profls. Assn., Brit. United Svc. Club L.A., Marines' Meml. Club (San Francisco), Sigma Xi, Sigma Nu. Republican. Roman Catholic. Home and Office: 276 J Ave Coronado CA 92118-1138

CRIMINALE, WILLIAM OLIVER, JR., applied mathematics educator; b. Mobile, Ala., Nov. 29, 1933; s. William Oliver and Vivian Gertrude (Sketoe) C.; m. Ulrike Irmgard Wegner, June 7, 1962; children: Martin Oliver, Lucca. B.S., U. Ala., 1955; Ph.D., Johns Hopkins U., 1960. Asst. prof. Princeton (N.J.) U., 1962-68; asso. prof. U. Wash., Seattle, 1968-73; prof. oceanography, geophysics, applied math. U. Wash., 1973—, chmn. dept. applied math., 1976-84; cons. Aerospace Corp., 1963-65, Boeing Corp., 1968-72, AGARD, 1967-68, Lenox Hill Hosp., 1967-68, NASA Langley, 1990—; guest prof., Can., 1965, France, 1967-68, Germany, 1973-74, Sweden, 1973-74, Scotland, 1985, 89, 91, Eng., 1990, 91, Stanford, 1990, Brazil, 1992; Nat. Acad. exch. scientist, USSR, 1969, 72. Author: Stability of Parallel Flows, 1967; Contbr. articles to profl. jours. Served with U.S. Army, 1961-62. Boris A. Bakmeteff Meml. fellow, 1957-58, NATO postdoctoral fellow, 1960-61, Alexander von Humboldt Sr. fellow, 1973-74, Royal Soc. fellow, 1990-91. Mem. AAAS, Am. Phys. Soc., Am. Geophys. Union, Fedn. Am. Scientists, Soc. Indsl. and Applied Math. Home: 1635 Peach Ct E Seattle WA 98112-3428 Office: U Wash Dept Applied Math FS 20 Seattle WA 98195

CRINO, MARJANNE HELEN, anesthesiologist; b. Rochester, N.Y., Aug. 18, 1933; d. Michael Jay and Helen Barbara (Kennedy) C.; m. Michael Anthony La Iuppa, Nov. 12, 1960; children: James Michael, Barbara Anne, John Christopher. BS, Coll. St. Teresa, 1955; MD, Med. Coll. Wis., 1959; MA in Theology, St. Bernard's Inst., 1991. Diplomate Nat. Bd. Med. Examiners. House staff Genesee Hosp., Rochester, 1959-61; perinatal mortality rsch., resident in anesthesiology Jackson Meml Hosp.-U. Miami, 1962-65; attending staff in anesthesiology Genesee Hosp., Rochester, N.Y., 1969—; mem. exec. com., med. staff sec., 1980, 82; acting chmn. dept. anesthesiology Genesee Hosp., Rochester, N.Y., 1989, 91, chmn. pain control com., 1989—; clin. instr. anesthesiology U. Rochester Sch. Medicine, 1983—; cons. anesthesiology Rochester Psychiat. Ctr., 1975-85; instr. anesthesiology U. Miami Sch. medicine, 1966, 67; attending staff anesthesiology Jackson Meml. Hosp., Miami, 1966, 67. Mem. com. Pittsford (N.Y.) Republican Party, 1970s-80s; vol. chaplain Genese Hosp. Mem. N.Y. State Soc. Anesthesiologists (bd. dirs., vice speaker 1983-86), Am. Soc. Anesthesiologists (del. 1979-86), AMA, N.Y. State Med.Soc., Med. Soc. County of Monroe, Rochester Acad. Medicine, Cath. Physicians Guild Rochester (pres. 1988-89), Margaret Roper Guild (pres. 1975-76). Roman Catholic. Avocations: reading, gardening, music. Office: Genesee Hosp Dept Anesthesiology 224 Alexander St Rochester NY 14607-4050

CRIPPEN, ROBERT LAUREL, naval officer, former astronaut; b. Beaumont, Tex., Sept. 11, 1937; s. Herbert W. and Ruth C. (Andress) C.; m. Pandora Lee Puckett, Nov. 7, 1987; children from previous marriage: Ellen Marie, Susan Lynn, Linda Ruth. BS in Aerospace Engrng., U. Tex., 1960; grad., USAF Aerospace Rsch. Pilot Sch., 1965. Commd. ensign USN, 1960, advanced through grades to capt., 1980; assigned to flight tng. USN, Whiting Field, Fla., 1961, Chase Field, Beeville, Tex., 1961; attack pilot (Fleet Squadron VA-72 aboard U.S.S. Independence), Chase Field, Beeville, 1962-64; instr. USAF Aerospace Rsch. Pilot Sch., Edwards AFB, Calif., 1965-66; rsch. pilot USAF Manned Orbiting Lab. Program, L.A., 1966-69; NASA astronaut Johnson Space Ctr., Houston, 1969—; crew mem. Skylab Med. Experiments Altitude Test, 1972; mem. astronaut support crew Skylab 2, 3 and 4 missions, Apollo-Suyez Test Project mission, 1973-74; pilot Space Shuttle Columbia STSI, 1981; comdr. Space Shuttle Challanger STS-7, STS-41C, STS-41G, 1983-84; dep. dir. Shuttle Ops. Nat. Space Transp. System Ops., NASA, Kennedy Space Ctr., Fla., 1987-89; dir. Space Shuttle, NASA, Washington, 1989-92; pres., dir. NASA John F. Kennedy Space Ctr., 1992—. Recipient Exceptional Svc. medal NASA, 1972, Disting. Svc. award Dept. Defense, 1981, Achievement award Am. Astronautical Soc. Flight, 1981, Gardiner Greene Hubbard medal Nat. Geographic Soc. 1981, Disting. Svc. award FAA, 1982, Goddard Meml. trophy, Harmon trophy, 1982, 4 NASA Space Flight medals; named to Aviation Hall of Fame. Mem. Soc. Exptl. Test Pilots. Office: NASA Kennedy Space Ctr Kennedy Space Center FL 32899*

CRISP, JOY ANNE, research scientist; b. Colorado Springs, Apr. 10, 1958; d. Harold Huber and Patricia Ann (Monson) Miller; m. David Crisp, June 13, 1981. BA, Carleton Coll., 1979; PhD, Princeton U., 1984. Postdoctoral fellow, adj. lectr. dept. earth & space scis. U. Calif., L.A., 1984-86; nat. rsch. coun. rsch. assoc. Jet Propulsion Lab., Pasadena, Calif., 1987-88, rsch. scientist, 1989—; mem. lunar exploration sci. working group NASA, 1992—. Contbr. articles to Jour. Volcanol. and Geothermal Rsch., Contbns. to Mineralogy and Petrology, Jour. Geophys. Rsch., Icarus. Rsch. grantee Sigma Xi, 1981, 83, NASA, 1989—, sci. grantee Explorers Club, 1981. Mem. Internat. Assn. Volcanology and Chemistry of Earth's Interior, Am. Geophys. Union, Geol. Soc. Am. (rsch. grantee 1981, 83), Mineral. Soc. Am. Achievements include rsch. in remote sensing of volcanic gases, emplacement dynamics of lava flows on Earth and Mars, and visible/near-infrared/mid-infrared spectroscopy of volcanic rocks. Office: Jet Propulsion Lab MS183-501 4800 Oak Grove Dr Pasadena CA 91109

CRISP, POLLY LENORE, psychologist; b. Atlanta, May 20, 1952; d. John Pershing and Dorotha Amelia (Hogan) C. BA, U. Tenn., 1976; MA, Mich. State U., 1981, PhD, 1984. Psychotherapist Arbours Ctr., London, 1983-85; clin. psychologist Kennebec Valley Mental Health Ctr., Augusta, Maine, 1987-90, Overlook Mental Health Ctr., Maryville, Tenn., 1990—. Contbr. articles to profl. publs. Mem. APA (membership com. div. clin. psychology 1990—), Brit. Psychol. Soc., Soc. Psychotherapy Rsch., N.Y. Acad. Scis., Phi Beta Kappa, Phi Kappa Phi, Alpha Lambda Delta. Avocations: woodworking, stained glass. Office: Overlook Mental Health 219 Court St Maryville TN 37801

CRIST, B. VINCENT, chemist; b. East Liverpool, Ohio, June 2, 1953; s. Paul Adam and Sarah Jane (Williams) C.; m. Fumiyo Saito, Feb. 8, 1979; 1 child, Rachel Sara. BS, Ohio State U., Reno, 1976; PhD, U. Nev., 1981. Postdoctoral assoc. Stanford (Calif.) U., 1981-83; sr. surface scientist Surface Sci. Labs., Mountain View, Calif., 1983-86; sr. tech. staff Hakuto Co., Ltd., Tokyo, 1986—; mem. steering com. Versailles Project for Advancement of Materials and Stds.-Surface Chem. Analysis, Tokyo, 1989—; mem. subcom. for standardization of XPS and AES data mgmt. and treatment Internat. Stds. Orgn. Author: Handbook of Monochromatic XPS Data, Vols. I-IV, 1991; inventor X-ray Photoelectron Spectroscopy-Laser System, 1991. Mem. ASTM (com. for surface analysis), Am. Chem. Soc., Surface Sci. Soc. Japan. Avocations: tennis, inventing, backpacking. Office: Hakuto Co Ltd, 1-13 Shinjuku 1-chome, Tokyo 160, Japan

CRIST, THOMAS OWEN, ecologist; b. Wenatchee, Wash., Aug. 2, 1960; s. Burton Wayne and Dorothy Elizabeth (Jones) C.; m. Candace Lucille Witmer, July 17, 1982. BA in Biology, McPherson Coll., 1982; M of Forest Ecology, Yale U., 1984; PhD in Biology Ecology, Utah State U., 1990. Teaching asst. Yale U., New Haven, Conn., 1983-84, rsch. asst., 1984-85; teaching asst. Utah State U., Logan, 1985-89; rsch. assoc. Colo. State U., Ft. Collins, 1990—. Contbr. articles to profl. jours. Rsch. grantee NSF, 1987; academic scholar Yale U., 1982-84. Mem. Ecol. Soc. Am., Entomol. Soc. Am., Am. Inst. Biol. Scis., Sigma Xi. Home: 912 Ponderosa Dr Fort Collins CO 80521 Office: Colo State Univ Dept Biology Fort Collins CO 80523

CRISTESCU, ROMULUS, mathematician, educator, science administrator; b. Ploiesti, Romania, Aug. 4, 1928; s. Ioan and Ecaterina (Georgescu) C.; m. Eufrosina Barbu, May 20, 1957. D of Math., U. Bucharest, Romania, 1955. Asst. prof. math. U. Bucharest, 1950-55, lectr. math., 1955-60, assoc. prof., 1960-66, prof., 1966—, dir. Inst. Math., 1973-75; now chmn. math. Academia Română, Bucharest. Author: Functional Analysis, 1965, 4th edit., 1983, Ordered Vector Spaces and Linear Operators, 1976, Topological Vector Spaces, 1977, others; contbr. articles to profl. jours. Mem. Am. Math. Soc., Romanian Math. Soc., Romanian Acad. (prize 1966). Home: Intrarea Dridu 2, 78416 Bucharest Romania Office: Academia Română, Calea Victoriei 125, 71102 Bucharest Romania also: U Bucharest Inst Math, Str Academiei 14, Bucharest 12, Romania

CRISTIANI, VINCENT ANTHONY, counseling psychology educator; b. Boston, Apr. 10, 1932; s. Michael and Angelina (Catino) C.; m. Jean T. DiStasio, Sept. 3, 1962; children: Angela, Vincent Jr., John. BS, Boston State Coll., 1953; EdM, Boston U., 1954, EdD, 1960. Nat. cert. sch. psychologist, hypnotherapist; lic. ednl. psychologist, allied mental health profl. Tchr. Boston Pub. Schs., 1953-55; instr., counselor IT G Quartermaster Sch., Petersburg, Va., 1955-57; elem. sch. tchr. Cohasset (Mass.) Pub. Schs., 1957-58; teaching fellow Boston U., 1958-60; vis. prof. U. N.H., Durham, 1959, 60; adminstr., faculty Boston U., 1960-66; prof. psychology Boston State Coll., 1966-82; prof. counseling psychology U. Mass., Boston, 1982—; chairperson counseling psychology dept. U. Mass. Grad. Coll. Edn., Boston, 1982—; grad. program dir. Sch. Psychology, 1982—; assoc. dir. Inst. for Learning and Teaching, 1984-88, chairperson counseling psychology dept., 1986-89; assoc. dean Coll. of Edn., 1988-89. Contbr. articles to profl. jours. Bd. dirs. Quincy Interfaith Sheltering Coalition, Fr. Bill's Shelter for Homeless, Quincy, Little Bros. of St. Francis, Parker St., Mission Hill, Roxbury, Mass. Recipient Doctoral Teaching fellowship Boston U., 1958-60, Citizen of Yr. award Quincy Civic Inst., 1971, Black Student Assn. award for Outstanding faculty, 1975, 76, Disting. svc. award for Excellence in Teaching, Boston State Coll., 1979, Knight of Yr., 1968, 72, 76, 82, 84, Outstanding Sch. Psychology Trainer of Yr. award, 1987, Pres.'s award Outstanding Faculty Mem., 1988, Cert. of Recognition award Mass. Sch. Psychologists, 1988, 90. Mem. AAAS, NEA, AAUP, APA, Nat. Assn. Sch. Psychologists, Mass. Sch. Psychologists Assn., Mass. Psychol. Assn., Am. Ednl. Rsch. Assn., Am. Assn. Sch. Adminstrs., Internat. Reading Assn., Nat. Inst. for Advanced Study in Teaching Disadvantaged Youth, Mass. Tchrs. Assn., Dante Alighieri Soc., Com. on Urban Edn., Nat. Commn. on Tchr. Edn. and Profl. Standards, Mass. Audubon Soc., Nat. Detroit Sci. Club, Psi Chi, Phi Delta Kappa. Democrat. Roman Catholic. Home: 39 Sturtevant Rd Quincy MA 02169-1818 Office: U of Massachusetts Grad Coll of Edn Wheatley 1-077K Wheatley 3-016 Boston MA 02125-3393

CRISTINO, JOSEPH ANTHONY, electrical engineer; b. Bridgeport, Conn., June 5, 1947; s. Joseph and Raffaela (Muccherino) C.; m. Lois Ann Buchanan, Oct. 21, 1988; children: Joseph Anthony Jr., Jason, Natalie, Nicole, Jody. AS in Electro/Mech. Engrng., Norwalk (Conn.) State Tech. U., 1967; BSEE, U. Bridgeport, 1982. Registered profl. engr., Conn., Mass., Maine. Test technician Fermont Dynamics, Bridgeport, 1966-69; engrng. estimator N.E. Utilities, Norwalk, 1969-72; test technician N.E. Utilities, Berlin, Conn., 1972-78; regional test supr. N.E. Utilities, Norwalk, 1978-82, Bethel, Conn., 1982-87; prin. engr. Cristino Assocs., Inc., Newtown, Conn., 1983—. Author/prodr. (video tape) Keeping the Heart Healthy, 1984. Vol. Norwalk Seaport Assn., 1985—; commr., lt. Redding (Conn.) Fire Co. #1, 1988—. Recipient Outstanding Vol. award Norwalk Seaport Assn., 1987, Cert. of Recognition, 1989. Mem. IEEE, NSPE, Internat. Elec. Testing Assn. (affiliate), Conn. Soc. Profl. Engrs., Moose (sgt. of arms 1987-89), Sons of Italy. Republican. Roman Catholic. Avocations: reading, music. Office: Cristino Assocs Inc 75 Glen Rd Ste 315 Sandy Hook CT 06482

CRISTOL, STANLEY JEROME, chemistry educator; b. Chgo., June 14, 1916; s. Myer J. and Lillian (Young) C.; m. Barbara Wright Swingle, June 1957; children: Marjorie Jo, Jeffrey Tod. BS, Northwestern U., 1937; MA, UCLA, 1939, PhD, 1943. Rsch. chemist Standard Oil Co., Calif., 1938-41; rsch. fellow U. Ill., 1943-44; rsch. chemist U.S. Dept. Agr., 1944-46; asst. prof., then assoc. prof. U. Colo., 1946-55, prof., 1955—, Joseph Sewall Disting. prof., 1979—, chmn. dept. chemistry, 1960-62, grad. dean, 1980-81; vis. prof. Stanford U., summer 1961, U. Geneva, 1975, U. Lausanne, Switzerland, 1981; with OSRD, 1944-46; adv. panels NSF, 1957-63, 69-73, NIH, 1969-72. Author: (with L.O. Smith, Jr.) Organic Chemistry, 1966; editorial bd., Chem. Revs., 1957-59, Jour. Organic Chemistry, 1964-68; contbr. rsch. articles to sci. jours. Guggenheim fellow, 1955-56, 81, 82; recipient James Flack Norris award in phys.-organic chemistry, 1972, Alumni Merit award Northwestern U., 1987. Fellow AAAS (councilor 1986-92), Chem. Soc. London; mem. NAS, AAUP, Am. Chem. Soc. (chmn. organic chemistry div. 1961-62, adv. bd. petroleum rsch. fund 1963-66, council policy com. 1968-73), Colo.-Wyo. Acad. Sci., Phi Beta Kappa, Sigma Xi, Phi Lambda Upsilon. Home: 2918 3d St Boulder CO 80304 Office: U Colo Dept Chemistry & Biochemistry CB 215 Boulder CO 80309

CRISWELL, CHARLES HARRISON, analytical chemist, evironmental and forensic consultant and executive; b. Springfield, Mo., Jan. 9, 1943; s. John Philip and Elba Anne (Denton) C.; m. Joyce LaVonne Louth, Apr. 26, 1968; 1 child, Christina Rachel. AB in Chemistry and Biology, Drury Coll., 1967; postgrad., U. Mo., 1967-68. Cert. hazardous materials and waste specialist, profl. environ. health specialist, cert. profl. chemist; registered hazardous substances profl. Dir. Water Pollution Control Labs City of Springfield, 1968-72, chief Water Pollution Sect., 1972-80; pres., chmn. bd. dirs. Consulting Analytical Svcs. Internat., Springfield, 1979—; assoc. Environ. Planning Assocs., Inc., 1985—; appointed by gov. mem. Mo. Hazardous Waste Mgmt. Commn., 1978; mem. Mo. Joint Commn. on Hazardous Waste Mgmt. Legis., statewide Ad-hoc Com. on Regulations; speaker in field nationwide. Contbr. numerous articles to profl. jours. Active Springfield Employees Activities Club, ARC, Friends of Zoo; vice chmn., 1990-93, chair 1993—, mem. numerous subcoms. Greene County Local Emergency Planning Com.; ruling elder 1st and Calvary Presbyn. Ch., elected for life, 1974, deacon, sr. high sch. youth advisor, active numerous coms.; mem. permanent jud. commn. John Calvin Presbytery, 1977-85, treas., 1975—; alumni bd. dirs. Greenwood Lab. Sch., 1992—, pres. 1993—; mem. spl. adminstrv. commns. Presbytery Synod Gen. Assembly Inter-judicatory Consultation on Long Range Ch. Fin., also clk. and other offices. Fellow Am. Inst. Chemists; mem. Am. Inst. Biol. Scis., Am. Chem. Soc. (charter Ozarks sect., com. environ. analytical methodology), Nat. Environ. Health Assn., Internat. Union Pure and Applied Chemistry (affiliate), Assoc. Industries Mo. (environ. com., hazardous waste task group), Mo. Acad. Sci., Mo. Water and Sewerage Conf. (sec. pres. 1975), Mo. Water Pollution Control Assn. (pres. 1979, exec. com. 1977-83, chmn. 1979-80, newsletter assoc. editor, chmn. numerous coms. and confs., Award Merit, 1991, 92, 93), Water Environment Fedn. (chmn. ann. nat. conf. 1982, 83, 90, 91, 92, 93, asst. chmn. 1980, 81, 84, 88, 89, active numerous other coms. 1976—, Arthur Sidney Bedell award); mem. Am. Mensa, Ltd. (life), Springfield Area C. of C. (environ. com., chair emergency preparedness and community right to know sub com.), Beta Beta Beta, Phi Mu Alpha, Gamma Alpha. Republican. Avocations: music, tennis, other sports. Office: Cons Analytical Svcs Internat 2804 E Battlefield Rd Springfield MO 65804-4014

CRISWELL, MARVIN EUGENE, civil engineering educator, consultant; b. Chappell, Nebr., Oct. 31, 1942; s. Wilbur Arthur and Evelyn Lucille (Jeffries) C.; m. Lela Louise Kennedy, Sept. 5, 1965; children: Karin Lee, Glenn Alan, Dianne Marie, Melanie Anne. BSCE, U. Nebr., 1965; MSCE, U. Ill., 1966, PhD in Civil Engrng., 1970. Registered profl. engr., Colo. Structural engr. Clark & Enerson, Olson, Burroughs & Thompson, Lincoln, Nebr., 1965; research structural engr. U.S. Army Corps Engrs. Waterways Experiment Station, Vicksburg, Miss., 1967-69; asst. prof. civil engring. Colo. State U., Ft. Collins, 1970-75, assoc. prof., 1975-84, prof., 1984—, acting assoc. dept. head, 1989-91, assoc. head dept. Acad. Affairs, 1991—; program visitor Accreditation Bd. Engrng. and Tech., 1981-89; bd.dirs., sec. Engrng. Data Mgmt., Inc., Ft. Collins, 1983-85. Co-author: Properties and Tests of Engineering Materials, 1978. Recipient Abel Faculty Teaching award Colo. State U., 1984, 1988. Mem. ASCE (com. on wood, safety of bldgs., design of engineered wood constrn. standards), ASTM, Am. Soc. Engrng. Edn. (chmn. civil engring. div. 1981-82, chmn. Rocky Mt. sect. 1987-88, bd. dirs. and zone IV chmn. 1990-92, Dow Outstanding Young Faculty award 1978, civil engrng. div. G Wadlin award), Am. Concrete Inst. (coms. on fiber concrete, lunar concrete, connections in monolithic concrete and shear and torsim). Methodist. Avocations: photography, hiking, travel. Home: 1536 Freedom Ln Fort Collins CO 80526-1707 Office: Colo State U Dept Civil Engrng Fort Collins CO 80523

CRITCHFIELD, HAROLD SAMUEL, retired civil engineer; b. Somerset, Pa., Jan. 30, 1922; s. John Earl and Marian Vera (Geary) C.; m. Hazel Rhea Etter, Oct. 22, 1942 (dec.); children: Richard, Peggy. BSCE, U. Pitts., 1947. Registered profl. engr., Pa., Del. Hwy. bridge constrn. mgr. E.F. Goetz Constrn. Co., Chambersburg, Pa., 1948-50; bridge constrn. project mgr. C. J. Langerfelder Constrn. Co., Balt., 1950-62, W.P. Dickerson & Son Constrn. Co., Youngwood, Pa., 1962-89; ret., 1989—; project engr. C.J. Langerfelder Inc., Balt., 1950-62, W.P. Dickerson & Son Inc., Youngwood, 1962-89. Vol. advisor Somerset Community Hosp., 1991—. Officer Coast Artillery, 1942-44. Mem. NSPE, Pa. Profl. Engr., Am. Soc. Hwy. Engrs. Home: RD #2 Box 257 Somerset PA 15501

CRITCHFIELD, JEFFREY MOORE, immunology researcher; b. Kendallville, Ind., Jan. 23, 1965; s. H.H. and Hilary (Moore) C.; m. Laura Elizabeth Dinwiddie, Oct. 11, 1992. BAS in Classics and Biology, Stanford U., 1987; postgrad., U. Calif., San Francisco, 1993—. Rsch. assoc. dept. structural and cell biology Stanford (Calif.) U., 1986-88; rsch. fellow Howard Hughes Med. Inst. at the NIH, Bethesda, Md., 1991—. Editorial cons.: (textbooks) Biochemistry, 3d edit., 1988, Review of Medical Physiology, 1990, Handbook of Essential Diagnosis, 1992. Mem. AAAS, Phi Beta Kappa. Achievements include patent pending for a project in T lymphocyte biology: interleukin-4 predisposed T cell receptor mediated apoptosis of lymphocytes as a strategy to treat Auto Immune Diseases and allergy.

CRITCHLOW, B. VAUGHN, research facility administrator, researcher; b. Hotchkiss, Colo., Mar. 5, 1927; s. Arthur Burtis and Nancy Gertrude (Lynch) C.; m. Janet Lee Howell, Mar. 1, 1987; children from previous marriage: Christopher, Eric, Jan, Carey. AA, Glendale Coll., 1946-49; BA, Occidental Coll., 1951; PhD, UCLA, 1957. Instr. to prof. anatomy, acting chmn. anatomy Baylor Coll. Medicine, Houston, 1957-72; prof., chmn. anatomy Oreg. Health Scis. U., Portland, 1972-82; mem. sci. adv. com. Oreg. Regional Primate Rsch. Ctr., Beaverton, 1973-82, mem. ad hoc adv. com., 1981-82, dir., 1982—; trustee Med. Rsch. Found. Oreg., 1982—; vis. investigator Nobel Inst. Neurophysiology, Karolinska Inst., Stockholm, 1961-62; invited speaker 2d Internat. Cong. Hormonal Steroids, Milan, Italy, 1966, 3d Internt. Cong. Endocrinology, Mexico City, 1968, others; mem. NIH reproductive biology study sects., 1969-73, 75-77. Contbr. numerous articles to profl. jours. Served with USN, 1945-46. NIH rsch. career devel. awardee, 1959-69; NIH rsch. grantee, 1958—. Mem. Am. Assn. Anatomists, Endocrine Soc., Soc. for Neurosci., Internat. Soc. Neuroendocrinology, Internat. Brain Rsch. Orgn. Office: Oreg Regional Primate Rsch Ctr 505 NW 185th Ave Beaverton OR 97006-3499

CRITOPH, EUGENE, retired physicist, nuclear research company executive; b. Vancouver, B.C., Can., Mar. 29, 1929; s. Dennis Basil and Lillian Sarah Critoph; m. Mary Elizabeth Ivens, Feb. 9, 1952; children: Christopher Michael, Stephen Bard, Eugene Mark, Boyd. B in Applied Sci., U. B.C., 1951, M in Applied Sci., 1957. Physicist Chalk River (Ont., Can.) Nuclear Labs., Atomic Energy of Can. Ltd., 1953-67, br. head, reactor physics, 1967-

75, dir. fuels and materials div., 1975-76, dir. advanced projects and reactor physics div., 1976-79, v.p., gen. mgr., 1979-86; v.p. strategic tech. mgmt. Atomic Energy of Can. Ltd. Research Co., Ottawa, Ont., 1986-92; mem., sec., chmn. European-Am. Com. on Reactor Physics, 1962-69. Mem. Can. Nuclear Soc. (W.B. Lewis medal 1986).

CRITTENDEN, CALVIN CLYDE, retired engineering executive; b. Bonham, Tex., Aug. 9, 1928; s. George W. and Lulu Lee (Mann) C.; m. Laurin Donaldson, Jan. 28, 1949; children: Rosanna, Lucinda. BS, Tex. A&M U., 1947. Various positions Gulf Oil Corp., Tex., Pa., 1947-80; mgr. project engring. Gulf Oil Corp., Houston, 1980-83; engring. mgr. Petrolite Corp., St. Louis, 1986-91. Pres. Port Arthur (Tex.) Sch. Bd., 1969-72. Mem. ASME (sect. chmn. 1952), Am. Petroleum Inst. Baptist. Home: 509 Wedgefield Rd Granbury TX 76048

CRITTENDEN, MARY LYNNE, science educator; b. Detroit, Oct. 27, 1951; d. William and Marie (Ryall) C. BS, Wayne State U., 1974; MS, U. Detroit, 1984. Tchr. sci. Detroit Bd. Edn., 1974-77, Highland Park (Mich.) C.C., 1980—; faculty researcher Air Force program Wright Patterson AFB, Dayton, Ohio, 1991; speaker Mich. Ednl. Occupational Assn., 1989, Liberal Arts Network Devel., Lansing, Mich., 1990. Author ednl. materials; contbr. to profl. publs. Mem. AAAS, Am. Chem. Soc., Civic Ctr. Optimist Club (bd. dirs. 1991—, coord. scis. 1990—), Mich. Community Coll. Biologists. Achievements include development of successful paradigm and teaching methods to make science palatable to urban community college students, modeling normal values in humans and some rodents applicable to physiologically-based pharmokinetics. Home: 15386 Alden St Detroit MI 48238 Office: Highland Park C C Glendale at 3d Ave Highland Park MI 48203

CROCKER, ALLEN CARROL, pediatrician; b. Boston, Dec. 25, 1925. Student, MIT, 1942-44; MD, Harvard U., 1948. Lab. house officer Children's Hosp., Boston, 1948-49, jr. asst. resident medicine, 1949-51, fellow pathology, 1953-56, from asst. to assoc. physician, 1956-62, rsch. assoc. pathology, 1956-68, assoc. medicine, 1962-66, sr. assoc., 1966—, dir. devel. evaluation ctr., 1967—; rsch. assoc. pathology Harvard Med. Sch., 1956-60, rsch. assoc. pediatrics, 1960-66, tutor med. sci., 1964-79, asst. prof., 1966-69, assoc. prof., 1969—;. Mem. Am. Assn. Mental Retardation (v.p. medicine 1980-82), Am. Assn. U. Affliates (pres. persons devel. disabilities 1982-83), Nat. Down Syndrome Congress (v.p. 1984-85), Soc. Behavioral Pediatrics (pres. 1987-88). Achievements include research in clinical investigation, pediatric metabolic diseases, biochemistry of the lipids, mental retardation. Office: Devel Evaluation Ctr Children's Hosp 300 Longwood Ave Boston MA 02115*

CROCKER, MALCOLM JOHN, mechanical engineer, noise control engineer, educator; b. Portsmouth, Eng., Sept. 10, 1938; came to U.S., 1963, naturalized, 1975; s. William Edwin and Alice Dorothy (Mintram) C.; m. Ruth Catherine, July 25, 1964; children: Anne Catherine, Elizabeth Claire. B.Sc. in Aeros. with honors, Southampton (Eng.) U., 1961, M.Sc. in Noise and Vibration, 1963; Ph.D. in Acoustics, Liverpool (Eng.) U., 1969. Co-op. apprentice, Vickers scholar Brit. Aerospace Co., Weybridge, Surrey, Eng., 1957-62; rsch. asst. Southampton U., 1962-63, vis. rsch. fellow, 1976; scientist Wyle Labs. Rsch., Huntsville, Ala., 1963-66; rsch. fellow U. Liverpool, 1967-69; assoc. prof. mech. engring. Purdue U., West Lafayette, Ind., 1969-73; prof. Purdue U., 1973-83; asst. dir. acoustics and noise control Herrick labs., 1977-83; prof., head dept. mech. engring. Auburn (Ala.) U., 1983-90, disting. univ. prof., 1990—; vis. prof. U. Sydney, 1976; cons. to industry; lectr. in field; gen. chmn. acoustics confs. including Inter-Noise 72, Washington, Noise-Con 79, Nat. Conf. Noise Control Engring., West Lafayette, Ind., 1979, Internat. 1st and 2d Congresses on Recent Devel. in Air and Structure-Borne Sound and Vibration, Auburn, 1990, 92; Internat. Conf. Noise and Vibration Control, St. Petersburg, Russia, 1993. Author: Noise and Noise Control, 2 vols, 1975, 82, Benchmark Papers in Acoustics: Noise Control, 1984; editor: Noise and Vibration Control Engineering, 1972, Reduction of Machinery Noise, 1974, rev. edit., 1975, others; editor-in-chief: Noise Control Engineering Jour., 1973—; mem. editorial bd.: Archives Acoustics, Warsaw, Poland, 1979—; contbr. numerous articles to profl. jours. Grantee NSF, 1972-74, 75-77, U.S. Dept. Transp., 1972-73, 79-81, EPA, 1976-80, NASA, 1980-83, 84—, Dept. Def., 1984-90, others; Acoustical Soc. India hon. fellow, 1985. Fellow Acoustical Soc. Am.; mem. Inst. Noise Control Engring./U.S.A. (dir., v.p. for communications, pres. 1981), Inst. Acoustics (London), Am. Soc. Engring. Edn. (chmn. engring acoustics & vibration 1986-88), ASME, Am. Nat. Standards Inst. (com. chmn.). Home: 454 Pinedale Dr Auburn AL 36830-7404 Office: Auburn U Dept Mech Engring Auburn AL 36849

CROFFORD, OSCAR BLEDSOE, JR., internist, medical educator; b. Chickahsa, Okla., Mar. 29, 1930; married, 1957; 3 children. AB, Vanderbilt U., 1952, MD, 1955. Intern medicine Vanderbilt U. Hosp., 1955-56, asst. resident, 1956-57; USPHS research fellow clin. physiology, 1959-62, resident in medicine, 1962-63; USPHS fellow clin. biochemistry U. Geneva, 1963-65; from asst. prof. to assoc. prof., 1965-74; prof. medicine Vanderbilt U. Sch. Medicine, 1974—, assoc. prof. physiology and biophysics, 1970—; investor Howard Hughes Med. Inst., 1965-71; mem. metabolism study sect. NIH, 1970-74, chmn. 1972-74; Addison B. Scoville jr. Chair Diabetes & Metabolism, Vanderbilt U., 1973—, div. head diabetes & metab., dept. medicine, 1973—, dir. Diabetes-Endocrin. Ctr., 1973-78; chmn. Nat. Commn. Diabetes, 1975-76; dir. Diabetes Research & Tng. Ctr., 1978—. Mem. Am. Diabetes Assn. (pres. 1981, Lilly award 1970, Charles H. Best award 1976, Banting medal 1982, Scoville award 1987, Outstanding Educator award 1989), Am. Physiol. Soc., Endocrine Soc., Am. Soc. Clin. Investigation, Assn. Am. Physicians. Research in hormone control of metabolism in adipocytes; mechanism of action of insulin; sugar transport; pathophysiology and treatment of Diabetes Mellitus; chairman diabetes control and complications trial. Office: Vanderbilt U-School of Medicine Dept Metabolism 21st Ave S & Garland Nashville TN 37232

CROFTS, ANTONY RICHARD, biophysics educator; b. Harrow, Eng., Jan. 26, 1940; came to U.S., 1978; s. Richard Basil Iliffe and Vera Rosetta (Bland) C.; m. Paula Anne Hinds-Johnson, June 7, 1969 (div. 1981); 1 child, Charlotte Victoria Patricia; 1 adopted child, Rupert Charles; m. Christine Thompson Yerkes, Dec. 23, 1982; children: Stephanie Boynton, Terence Spencer. BA, U. Cambridge, Eng., 1961, PhD, 1965. Asst. lectr. dept. biochemistry U. Bristol, Eng., 1964-65, lectr., 1966-72, reader, 1972-78; prof. biophysics U. Ill., Urbana-Champaign, 1978—, prof. microbiology, 1992—, chmn. biophysics div., 1978-91; Mem. organizing com. 4th Internat. Congress Photosynthesis, Reading, Eng., 1977, 7th Internat. Congress Photosynthesis, Providence, 1986, Table Ronde, Rousel-UCLA Forum, Paris, 1985; vis. prof. Coll. de France, 1983; Melandri lectr. European BioEnergetics Conf., Lyon, France, 1982. Contbr. over 160 articles, revs., etc., in area of biophysics, photosynthesis and bioenergetics; mem. editorial bd. Biochem. Jour., U.K., 1971-72, Biochimica Biophysica Acta, Holland, 1972-77, jour. Bacteriology, 1979-83, Archives Biochemistry and Biophysics, 1980-85. Major scholar nat. sci. U. Cambridge, 1967, U. Ill. scholar, 1989-92; grantee U.S. Dept. Energy, 1982, 88, 90, 92, Guggenheim Found., 1985, NSF, NIH, U.S. Dept. Agr., 1979-92. Fellow AAAS; mem. Biophys. Soc., Am. Soc. Biochemistry and Molecular Biology, Am. Soc. Plant Physiologists (Charles F. Kettering award 1992). Avocations: windsurfing, skiing, fishing, gardening. Office: U Ill Program in Biophysics 156 Davenport Hall 607 S Matthews Urbana IL 61801-3704

CRONBAUGH, KURT ALLEN, industrial electrician; b. Cedar Rapids, Iowa, July 8, 1967; s. Lowell David and Nelda Ruth (Ranfeld) C.; m. Beth Ann Stoelk (div.); children: Danielle Marie, Ryan Curtis; m. Janell Kay Saling; 1 child, Jennifer Lee Foster. AAS in Comm. Electronics., Kirkwood Community Coll., Cedar Rapids, Iowa, 1987. Cert. electronic tech. Telecomm. engring. tech. KTS-TV 13, Kirkwood Community Coll., Cedar Rapids, 1985-86; control booth asst. engr. KGAN-TV 2, Cedar Rapids, 1986; elec. supply staff Van Meter Elec. Supply, Inc., Cedar Rapids, 1986-87; indsl. electrician North Star Steel Iowa, Wilton, 1987—; ind. elec. contr. K's Elec. Inc., Wilton, 1988—. Mem. Radio Electronics Assn., Elec. Servicing and Mfg. Assn., Electronic Tech. Assn. Democrat. Ch. of God. Home: PO Box 836 Wilton IA 52778-0836 Office: K's Elec Inc 215 W Prairie Wilton IA 52778-0836

CRONBERG, STIG, infectious diseases educator; b. Malmö, Sweden, Jan. 20, 1935; s. Nils Ebbe and Gerd (Carlander) C.; m. Gertrud Hellsten, May 28,1960; children: Nils, Hans, Olof, Per, Truls, Barbro, Cecilia. MD, Lund (Sweden) U., 1960. Intern Dept. of Medicine, Malmö, 1960-68; scientist Hôpital Saint-Louis, Paris, 1969-70; cons. Dept. of Infectious Diseases, Malmö, 1970-76; asst. prof. dept. infectious diseases Malmö Gen. Hosp., 1976—. Author: Infektioner, 5 edits., 1976-91, Maladies Infectieuses, 1988; co-author: Platelets: Physiology and Pathology, 1977. Home: Tygelsjövägen 127, S 23042 Tygelsjö Sweden Office: Malmö Gen Hosp, Dept Infectious Diseases, S 21401 Malmö Sweden

CRONEMEYER, DONALD CHARLES, physicist; b. Chanute, Kans., Nov. 10, 1925; s. Theodore Harry and Annette (Zook) C.; m. Anita Grace Schulle, June 13, 1953; children: Paul David, Timothy James, Lois Ann, Mark Stephen, Mary Beth. BS in Engring. Physics, U. Kans., 1945; ScD in Physics, MIT, 1951. Physicist Electronics Lab. GE, Liverpool, N.Y., 1951-60; physicist Rsch. Lab. Bendix, Southfield, Mich., 1960-66; instr. physics Wayne State U., Detroit, 1962-66; assoc. prof. physics Wheaton (Ill.) Coll., 1966-67; physicist ITT Rsch. Inst., Chgo., 1966-67; physicist T.J. Watson Rsch. Ctr. IBM, Yorktown Heights, N.Y., 1967-91. Lt. (j.g.) USNR, 1943-48, PTO. Fellow Am. Phys. Soc.; mem. IEEE (sr.). Republican. Baptist. Home and Office: Antennas and Propation Lab Inc PO Box 4237 Greenville SC 29608-4237

CRONENWETT, WILLIAM TREADWELL, electrical engineering educator, consultant; b. Texarkana, Tex., Jan. 3, 1932; s. John E. and Frances P. (Treadwell) C.; m. Carolyn E. Somers, June 8, 1963 (div. Oct. 1976); children: Will J., Carrie. BS, Tex. A&I U., 1954; MS, U. Tex., 1960, PhD, 1966. Registered profl. engr., Tex., Okla. Rsch. engr. Electro Mechanics Co., Austin, Tex., 1959-62; Welch Found. fellow U. Tex., Austin, 1964-66; rsch. fellow U. Leicester, Eng., 1966-68; prof. U. Okla., Norman, 1968—; expert in field of electrical accidents, equipment malfunctions, investigations. Contbr. articles to profl. jours. Recipient Dist. 1st prize AIEE, Dist. 15, 1952, Wonders of Engring. award Ctrl. Okla. Soc. Profl. Engrs., Tulsa, 1972. Mem. IEEE (sr. mem.), Am. Welding Soc., Sigma Xi. Achievements include patent for electrosurgical apparatus. Office: Univ of Okla Elec Engring 202 W Boyd St Norman OK 73019

CRONHJORT, BJORN TORVALD, systems analyst; b. Åbo, Finland, Sept. 11, 1934; s. Harald and Edit (Salonen) C.; m. Ulla Margareta Rönnholm, Mar. 24, 1962; children: Marika, Mikael, Andreas, Marcus. MS, Helsinki (Finland) U. Tech., Finland, 1959; PhD, Helsinki (Finland) U. Tech., 1964. Cert. European Engr., 1993. Project engr. IBM Nordic Labs., Lidingö, Sweden, 1960-70; mgr. of sci. and edn. rels. IBM Finland, Helsinki, 1970-77; program mgr. Nat. Swedish Bd. for Tech. Devel., Stockholm, 1977-82; program engr. Ericsson Info. Systems, Stockholm, 1982-86; program mgr. Nat. Swedish Bd. for Spent Nuclear Fuel, Stockholm, 1986-92; assoc. prof. Dept. Automatic Control, Royal Inst. Tech., Stockholm, 1986—; prof. info. tech. U. Oulu, Finland, 1964-65, 70; staff sci. IBM rsch. divsn., San Jose, Calif., 1967-68; chmn. and mem. of numerous program coms. for postgrad. edn. in Finland, Finnish Engrs. Postgrad. Inst., Helsinki, 1970-77; lectr. Helsinki U. Tech., control engring., Helsinki, 1971-77; expert software tech. throughout Europe, Commn. European Communities; computer expert Purdue-Europe Workshop, Brussels, 1976-83; prof. U. Turku, Finland, 1986; hon. mem. adv. coun. IBC, Cambridge, England, rsch. bd. advs. ABI, Raleigh, N.C. Mem. editorial bd. Computers and Security, 1981-92, Engring. Applications Artificial Intelligence, 1987—; contbr. articles to profl. jours. Vice-chmn. govt. com. Sch. of Computing, No. Finland, Helsinki, 1970-71; rsch. and devel. com. Delegation for Sci. and Tech. Info., Stockholm, 1979-82; mem. Swedish Assn. of Mems. of Parliament and Researchers, Stockholm, 1983—, Swedish Engrs. for the Prevention of Nuclear War, Stockholm, 1987—. Fellow Swedish Acad. Engring. Scis. in Finland, Helsinki, 1987—. Mem. IEEE, Assn. for Computing Machinery, Spl. Interest Group on Software Engring., Internat. Fedn. Automatic Control (application com., working group on automation in mining, mineral and metal processing, computer com., social effects of automation com., systems engring. com., working group on supplemental ways for improving internat. stability), Soc. for Risk Analysis, Engring. Soc. in Finland. Avocations: arts and crafts of the Southwest Indians, windsurfing. Office: Royal Inst Tech, Dept Automatic Control, S-10044 Stockholm Sweden

CRONIN, JAMES WATSON, physicist, educator; b. Chicago, Ill., Sept. 29, 1931; s. James Farley and Dorothy (Watson) C.; m. Annette Martin, Sept. 11, 1954; children: Cathryn, Emily, Daniel Watson. A.B., So. Methodist U. (1951); Ph.D., U. Chgo. Asso. Brookhaven Nat. Lab., 1955-58; mem. faculty Princeton, 1958-71, prof. physics, 1965-71; prof. physics U. Chgo., 1971—; Loeb lectr. physics Harvard U., 1967; participant early devel. spark chambers; co-discoverer CP-violation, 1964. Recipient Research Corp. Am. award, 1967; John Price Wetherill medal Franklin Inst., 1976; E.O. Lawrence award ERDA, 1977; Nobel prize for physics, 1980; Sloan fellow, 1964-66; Guggenheim fellow, 1970-71, 82-83. Mem. Am. Acad. Arts and Scis., Nat. Acad. Sci. (council mem.). Home: 5825 S Dorchester Ave Chicago IL 60637-1764 Office: U Chgo Enrico Fermi Inst 5630 S Ellis Ave Chicago IL 60637-1433

CRONIN, VINCENT SEAN, geologist; b. L.A., Jan. 18, 1957; s. Gilbert Francis and Dorothy Mary (Fahey) C.; m. Cynthia Ellis, Apr. 16, 1988; 1 child, Kelly Elizabeth. BA in Geology, Pomona Coll., 1979; AM in Earth Scis., Dartmouth Coll., 1982; PhD in Tectonophysics, Tex. A&M U., 1988. Exploration geologist Phillips Uranium Corp., Casper, Wyo., 1980; engring. geologist Slosson and Assocs., L.A., 1982-84; computer geologist Exxon, Houston, 1986; asst. prof. geoscis. U. Wis., Milw., 1988-; faculty assoc. Argonne (Ill.) Nat. Lab., 1991-92. Contbr. articles to profl. jours.; reviewer numerous books and articles. U. Wis. Milw. Rsech grantee, 1990, Sigma Xi Rsch. grantee, 1987. Mem. Am. Geophysical Union, Geological Soc. Am., Am. Assn. Petroleum Geologists. Achievements include pioneering work in Himalayan geology and development of mathematical model to describe the relative motion of plates across the Earth's surface over extended time intervals (cycloid model). Office: U Wis Dept Geoscis PO Box 413 Milwaukee WI 53201

CRONKLETON, THOMAS EUGENE, physician; b. Donahue, Iowa, July 22, 1928; s. Harry L. and Ursula Alice (Halligan) C.; BA in Biology, St. Ambrose Coll., 1954; MD, Iowa Coll. Medicine, 1958; m. Wilma Agnes Potter, June 6, 1953; children: Thomas Eugene, Kevin P., Margaret A., Catherine A., Richard A., Robert A., Susan A., Phillip A. Diplomate Am. Bd. Family Practice. Rotating intern St. Benedict's Hosp., Ogden, Utah, 1958-59; Donahue, Iowa, 1959-61, practice family medicine, Davenport, Iowa, 1961-66, Laramie, Wyo., 1966—; asso. The Davenport Clinic, 1961-63, partner, 1963-66; active staff St. Luke's Hosp., Mercy Hosp., Davenport; staff physician U. Wyo. Student Health Service, 1966-69, 70-71, 74-75, 76—, acting dir., 1988-89; staff physician outpatient dept. VA Hosp., Iowa City, 1969-70: staff physician outpatient dept. VA Hosp., Cheyenne, Wyo., 1971-74, chief outpatient dept., 1973-74; dir. Student Health Service Utah State U., Logan, 1975-76; physician (part-time) dept. medicine VA Hosp., Cheyenne, 1976-81. Active Long's Peak council Boy Scouts Am., 1970—, scout chaplain Diocese of Cheyenne, 1980—, mem. Diocesan Pastoral Council, 1982-85. Served with USMC, World War II, Korea. Recipient Dist. Scouter award Boy Scouts Am., 1974, St. George Emblem, Nat. Cath. Scouter award, 1981. Recipient 5, 10, and 15-yr. service pins Boy Scouts Am. Fellow Am. Acad. Family Practice; mem. Wyo. State Med. Soc., Albany County (Wyo.) Med. Soc., Iowa Med. Soc., Johnson County (Iowa) Med. Soc. Democrat. Roman Catholic. Club: K.C. (4 deg.). Home: 2444 Overland Rd Laramie WY 82070-4808 Office: U Wyo Student Health Svc Laramie WY 82071

CROOK, TROY NORMAN, geophysicist, consultant; b. Wall, Tex., May 24, 1928; s. Otis Allen and Callie Viola (Aylor) C.; m. Ruby Mae Keel, June 5, 1949; children: David Preston, Larry Norman. BEE, Tex. A&M U., 1949; BS in Geology, U. Houston, 1961. Seismic operator Humble Oil & Refining Co., Miss., Ala., Fla., La., 1949-54; rsch. geophysicist Humble Oil & Refining Co., Houston, 1955-64, asst. div. geophysicist, 1968-69, div. geophysicist, 1969-71; supr. rsch. Esso Prodn. Rsch. Co., Houston, 1965-67, mgr. basic geophysics, 1967-68; mgr. exploration systems div. Exxon Prodn. Rsch. Co., Houston, 1971-75, mgr. div. stratigraphic exploration, 1975-84, mgr. long-range rsch., 1984-86; prin. T. Norman Crook Cons., Houston, 1986—. Contbr. numerous articles to profl. jours.; patentee in field. Mem. Bethel Ind. Presbyn. Ch. Mem. IEEE, Am. Assn. Petroleum Geologists, European Assn. Exploration Geophysicists, Soc. Exploration Geophysicists (pres. 1978-79), Geophys. Soc. Houston. Home: 5527 Sylmar Rd Houston TX 77081-7423

CROON, GREGORY STEVEN, nuclear engineer; b. Park Ridge, Ill., Nov. 11, 1960; s. Refert Dirks and Blanche Francis (Long) C.; m. Janet Elizabeth Flood, June 25, 1983; children: Allison Elizabeth, Lauren Eileen. BS in Nuclear Engring., U. Ill., 1983; MS in Nuclear Engring., Air Force Inst. Technology, 1985. Commd. 2d lt. USAF, 1983, advanced through grades to capt., 1991; particle beam rsch. engr. Air Force Weapons Lab., Albuquerque, 1985-87, sect. chief, 1987-88; tech. liaison USAF, Wiesbaden, Germany, 1988-91; reactor core engr. Commonwealth Edison Co., Chgo., 1991—. Author: (report) Particle Beam Vacuum Chamber Propagation, 1986, Particle Beam Induced Fluorescense, 1987, Particle Beam Effects on Electronics, 1988, Particle Beam Discrimination Handbook, 1989. Decorated Air Force Commendation medal; recipient AFWL Tech. Officer award, 1985, AFROTC Disting. Grad. award, 1985. Mem. Am. Nuclear Soc. (internat. rels. commn. 1990—, pub. edn. commn. 1991—), German Nuclear Soc., Masons. Republican. Lutheran. Achievements include development of theoretical description of neutral particle beam interactions with upper ionosphere and beam induced fluorescence, of computer code simulating production and detection of spontaneous/induced materials emissions, of thermal hydraulic transient analyses of CECo BWR NSSS; work with Dresden isolation condenser break evaluation with extended valve isolation valve closure times, spent fuel pool time to boil analysis, time to boil analysis for Quad-cities at Dresden and La Salle nuclear stations, computer modelling of reactor control systems, thermal hydraulic transient analysis of boiling water reactor systems. Home: 22638 Lake Shore Dr Richton Park IL 60471 Office: Commonwealth Edison Co 125 S Clark St Chicago IL 60690

CROPPER, ANDRÉ DOMINIC, electrical engineering educator; b. Port-of-Spain, Trinidad and Tobago, Aug. 4, 1961; came to U.S., 1978; s. Anthony and Vilma V. (Skinner) C. BSEE, Howard U., 1984, MSEE, 1987; post-grad., Va. Tech. Instr. Norfolk (Va.) U., 1989-91, asst. prof. elec. engring., 1991-92; rsch. assoc. Morgan State U., 1993—; cons. Advance Controls and Equipment Svcs., Freeport, Bahamas, summer 1990; dir. engring. enrichment program NASA/Morgan State U., Balt., summer 1990, 92; mem. CISE Infrastructure Planning Task Force, Norfolk, 1989-90; presenter in field. Contbr. articles to profl. publs. Speaker St. Paul's High Sch., Freeport, 1991; mem. worship com. Christ the King Ch., Norfolk, 1991-92. Mem. Tidewater Water Polo Club, Alpha Phi Alpha, Tau Beta Pi, Beta Kappa Chi. Roman Catholic. Home: 750 Tall Oaks Dr # 13100G Blacksburg VA 24060

CROSBIE, ALFRED LINDEN, mechanical engineering educator; b. Muskogee, Okla., Aug. 1, 1942; s. Alfred Henry and Jacquetta Hope (Stoneburner) C.; M. Ann Frances Cirou, July 18, 1963; children: Mark, Jacqueline. BS in Mech. Engring., U. Okla., d1964; MS in mech. engring., Purdue U., 1966, PhD in mech. engring., 1969. Asst. prof. U. Mo., Rolla, 1968-72, assoc. prof., 1972-75, prof., 1975-91, curators' prof., 1991—. Editor: Aerothermodynamics and Planetary Entry, 1981, Heat Transfer and Thermal Control, 1981; editor-in-chief Jour. Thermophysics and Heat Transfer, 1986—; assoc. editor Jour. Quantitative Spectroscopy and Radiative Transfer, 1979—; contbr. 70 articles on radiative heat transfer to profl. jours. Fellow AIAA (chmn. thermophysics com. 1984-86, tech. program chmn. 15th Thermophysics Conf. 1980, assoc. editor AIAA Jour. 1981-83, Thermophysics award 1987, Tech. Contbn. award, 1988), ASME (heat transfer com. on theory and fundamentals 1983—, Heat Transfer Meml. award 1990); mem. AAAS, Optical Soc. Am., Phi Eta Sigma, Sigma Pi Sigma, Tau Beta Pi, Pi Tau Sigma, Sigma Tau, Pi Mu Epsilon, Sigma Xi. Lutheran. Avocation: fishing. Home: 8 Mcfarland Dr Rolla MO 65401-3805 Office: U Mo Dept Mech Engring 233 Mechanical Engring Rolla MO 65401

CROSBY, MARSHALL ROBERT, botanist, educator; b. Jacksonville, Fla., June 3, 1943; s. Robert Gilbert and Anne (Respess) C.; m. Carol Anderson (div.); children: Matthew Turner, Sara Elizabeth. BS, Duke U., 1965, PhD, 1969. Curator of cryptograms Mo. Botan. Garden, St. Louis, 1969-74, chmn. dept. botany, 1974-79, dir. rsch., 1977-86, dir. botan. info. resources, 1986-88, asst. dir., 1988-92, acting chmn. dept. edn., 1989, sr. adviser to dir., sr. botanist, 1992—; rsch. assoc. dept. botany Washington U., St. Louis, 1968-69, adj. asst. prof., 1970-73, faculty assoc., 1974-79, adj. prof. biology, 1980-87; adj. assoc. prof. biology U. Mo., St. Louis, 1983—; hon. assoc. curator bryophyta Museo Nacional de Costa Rica, 1978—. Editor: Annals Mo. Botan. Garden, 1969-74, 90-91, Monographs in Systematic Botany from Mo. Botan. Garden, 1978-88, 91—, Novon, 1991—; co-editor: Herbarium News, 1981-88; assoc. editor: The Bryologist, 1974-78; mem. editorial com. Systeamtic Botany Monographs, 1978-82, Flora of North Am., 1991—, Bryoflora of China, 1990—; contbr. articles to numerous profl. jours. Mem. AAAS, Am. Byological and Lichenological Soc. (bus. mgr. 1972-79, sec.-treas. 1973-79), Am. Inst. Biol. Scis., Am. Soc. Plant Taxonomists (councilor 1982-84), Botan. Soc. Am., British Bryological Soc., Internat. Assn. Bryologists (coun. mem. 1979—, v.p. 1987—), Internat. Assn. for Plant Taxonomy, Nordic Bryological Soc., Sigma Xi. Office: Mo Bot Garden Po Box 299 Saint Louis MO 63166

CROSET, MICHEL ROGER, electronics engineer; b. Woking, Surrey, Eng., Aug. 16, 1945; s. Louis Felix and Eidith (Fiever) C.; m. Maureen Mullins, Mar. 28, 1967 (div. Jan. 1984); children: Gillian, Lisa, Pauline. Cert., Wandsworth Tech. Inst., London, 1963. Engr. Philips, London, 1965-71, De la Rue, Portsmouth, Eng., 1971-88; chief engr. Method Tech., Portsmouth, 1988—; cons. engr. Wessex Tech., Portsmouth, 1986—, Redshore Electronics, Portsmouth, 1988—. Patentee in field; inventor sheet presenting assembly, counting apparatus, sheet delivery assembly. Mem. Tory Party. Anglican. Avocations: sailing, flying. Home: 2 St Johns Ave, Portsmouth PO7 5PG, England

CROSS, GEORGE ALAN MARTIN, biochemistry educator, researcher; b. Cheadle, Cheshire, Eng., Sept. 27, 1942; s. George Bernard and Beatrice Mary (Horton) C.; m. Nadia Maria Nogueira, Feb. 26, 1986; 1 child, Julia Elizabeth. BA, Cambridge (Eng.) U., 1964, PhD, 1968. Scientist Med. Research Council, Cambridge, 1970-77; dept. head Wellcome Found. Research Labs., Kent, Eng., 1977-82; Andre and Bella Meyer prof. molecular parasitology Rockefeller U., N.Y.C., 1982—; cons. Wellcome Found., Eng., 1982-87, World Health Orgns., Geneva, 1983-87, New Eng. Biolabs., Beverly Mass., 1985—. Contbr. articles to profl. jours. Recipient Paul Ehrlich prize, 1984, Chalmers medal Royal Soc. of Tropical Medicine, 1983; named Fleming Lectr. Soc. for Gen. Microbiology, 1978. Fellow The Royal Soc. Office: The Rockefeller Univ 1230 York Ave New York NY 10021-6341

CROSS, GLENN LABAN, engineering company executive, development planner; b. Mt. Vernon, Ill., Dec. 28, 1941; s. Kenneth Edward and Mildred Irene (Glenn) C.; m. Kim Lien Duong, Aug. 30, 1968 (div. Oct. 1975); m. Tran Tu Thach, Dec. 26, 1975; children: Cindy Sue, Cristy Luu, Crystal Tu, Cassandra Caitlynn; BA, Calif. Western U., 1981, MBA, 1982. Hosp. adminstr. pub. health div. USAID, Dept. State, Washington, 1966-68; pers. mgr. Pacific Architects and Engrs., Inc., L.A., 1968-70, contract adminstr., 1970-73, mgr. mgmt. svcs., 1973-75; contracts adminstr. Internat. Svcs. div., AVCO, Cin., 1975-77; sr. contract adminstr. Bechtel Group, Inc., San Francisco, 1977-80, Arabian Bechtel Co. Ltd.; contract adminstrv. supr. Bechtel Civil, Inc., Jubail Industrial City, Saudi Arabia, 1980-85; cons. Bechtel Western Power Corp., Jakarta, Indonesia, Pacific Engrs. and Constructors, 1985-90, prin. contract adminstr. Ralph M. Parsons Co., Pasadena, Calif., 1990-93, contract adminstr. Parsons-Brinckerhoff, Costa Mesa, Calif., 1993—. Author: Living With a Matrix: A Conceptual Guide to Organizational Variation, 1983. Served as sgt. 1st spl. forces group, airborne, AUS, 1962-65; Okinawa, Vietnam. Decorated Combat Infantryman's Badge. Mem. Nat. Contract Mgmt. Assn., Construction Mgmt. Assn. Am., Internat. Pers. Mgmt. Assn., Assn. Human Resource Systems Profls., Human Resource Planning Soc., Assn. MBA Execs., Am. Mgmt. Assn., Am. Arbitration Assn., Internat. Records Mgmt. Coun., Adminstrv. Mgmt. Soc. Republican. Avocations: swimming, reading. Home: 25935 Faircourt Ln Laguna

Hills CA 92653-7517 Office: Parsons-Brinckerhoff 345 Clinton St Costa Mesa CA 92626-6011

CROSS, HAROLD ZANE, agronomist, educator; b. Portales, N.Mex., Dec. 25, 1941; s. Guy Edner and Hagabelle (Lawson) C.; m. Glenda Faye Wilhoit, Nov. 24, 1961; children: Carter Dale, Carson Lee, Curtis Don, Cathryn Faye. BS with honors, N.Mex. State U., 1965, MS, 1967; PhD, U. Mo., 1971. Rancher Elida, N.Mex., 1965-67; grad. rsch. asst. N.Mex. State U., Las Cruces, 1965-67; NDEA fellow U. Mo., Columbia, 1967-71; asst. prof. N.D. State U., Fargo, 1971-77, assoc. prof., 1977-82, prof., 1982—; cons. Agrl. Inst. Osijek, Yugoslavia, 1984, CIMMYT, Mexico City, 1984, Eli Lilly Co., Indpls., 1987. Contbr. numerous articles to profl. jours. Crops judge N.D. Winter show, Valley City, 1973-93. Santa Fe Rwy. scholar, 1961-62; NDEA fellow, 1967-71; recipient Outstanding Sr. Rsch. award N.D. State U. Coll. Agr., 1992. Mem. Crop Sci. Soc. Am. (editor for maize germplasm 1989-92), Am. Soc. Agronomy, Sigma Xi, Phi Kappa Phi, Gamma Sigma Delta, Alpha Zeta. Achievements include devel. and release of 33 inbred parental lines of maize and 26 synthetic varieties of maize; devel. of maize breeding procedures to genetically improve grain drying rates, procedues to improve leaf growth rates, kernel growth. Office: ND State Univ Crop and Weed Sci Dept Fargo ND 58105

CROSS, TREVOR ARTHUR, physicist; b. Chelmsford, Essex, Eng., May 14, 1960; s. Arthur and Joan Cross; m. Maureen Cross, Oct. 3, 1987. BSc in Solid State Physics, U. Bath, Eng., 1982; PhD in Semicondr. Physics, U. Lancaster, Eng., 1986. Devel. engr. EEV Ltd., Chelmsford, 1985-87, prin. engr., 1988-90, mgr. GaAs solar cells, 1990-92, bus. devel. mgr. GaAs solar cells, 1992—; contbr. Space Tech. Adv. Bd. to U.K. British Nat. Space Ctr., London, 1992; mem. tech. rev. com. European Space Power Conf., 1992. Author conf. publs. Recipient grants and contracts for solar cell R&D, 1989—. Mem. Brit. Assn. for Crystal Growth, Inst. of Physics. Achievements include patent in thin bifacial solar cell technology; production of first European gallium arsenide solar panel to be used in space; record for most efficient indium phosophide solar cell. Office: EEV Ltd, 106 Waterhouse Ln, Chelmsford CM1 2RH, England CM1 2QU

CROSSLEY, FRANCIS RENDEL ERSKINE, engineering educator; b. Derby, Eng., July 21, 1915; came to U.S., 1937; s. Erskine Alick and Edith Mary (Helme) C.; m. Mary Eleanor de Lacy Coyne, Aug. 23, 1941; children: Phyllis de Lacy Crossley Mervine, Michael Francis Erskine Crossley. BA, Cambridge (Eng.) U., 1937, MA, 1941; D of Engring., Yale U., 1949. Lab. asst. GM Rsch. Labs., Detroit, 1937-38; designer Ford Motor Co., Dearborn, Mich., 1939-41; instr. in drafting U. Detroit, 1942-44; asst. prof. Yale U., New Haven, 1944-55, assoc. prof., 1955-65; vis. fellow U. Manchester (Eng.)/Inst. of Sci. & Tech., 1965; prof. mech. engring. Ga. Inst. Tech., Atlanta, 1966-69; prof. mech. and civil engring. U. Mass., Amherst, 1970-80; adj. prof. U. Fla., Gainesville, 1988-91; Fulbright lectr. Tech. U., Munich, 1962-63; U.S. mem. delegation forestry energy div. Internat. Energy Agy., Washington, 1977-79. Author: (textbook) Dynamics in Machines, 1954; editor, founder Jour. of Mechanisms, 1966-71; editor-in-chief Jour. Mechanism and Machine Theory, 1971-73. Member Town of Branford (Conn.) Bd. of Edn., 1987-92, chmn. solid waste mgmt. com., 1986-87; staff scientist, legis. fellow Conn. State Legislature, Hartford, 1981-83; bd. dirs. U.S. nat. com. Internat. Assn. for Exch. Students for Tech. Experience, 1963-65, 68-70. Von Humboldt Found. Sr. Scientist award, 1975-76. Fellow ASME (life, chmn. mechanisms conf. Atlanta 1967-68, policy bd. gen. engring. 1979-80, Centennial medal 1980, Machine Design award 1991); mem. Internat. Fedn. for Theory of Machines and Mechanisms (hon., founding com. 1967, 1st v.p. 1969-75, constitution author 1969), Verein Deutscher Ingenieure (corr., hon.). Republican. Episcopalian. Avocations: gardening, day care.

CROSSMAN, STAFFORD MAC ARTHUR, agronomist, researcher; b. Basseterre, St. Kitts, W.I., Sept. 23, 1953; came to U.S., 1979; s. Arthur Crulilee Crossman and Ivy Malvina (Berkeley) Fergus; m. Lona Burnetta Austin, Apr. 27, 1985; children: Desiree Tamara, Devon Jermaine, Nicole Ashima. Diploma, Jamaica Sch. Agr., 1978; BS, Tuskegee (Ala.) U., 1981, MS, 1984. Tchr. Ministry of Edn., Basseterre, 1970-74; field asst. Nat. Agrl. Corp., Basseterre, 1975-79; field trainee Jamaica Coffee Bd., Kingston, 1978; rsch. technician Tuskegee U., 1983-84; agronomist St. Kitts Sugar Mfg. Corp., Basseterre, 1984-87; rsch. specialist U. V.I., St. Croix, 1988—; subject matter specialist U. W.I., Basseterre, 1984-87, Midwestern Univs. Consortium for Internat. Activities, 1984-87, Caribbean Agrl. Extension Project, Basseterre, 1984-87; mem. US/AID project adv. bd. Ministry of Agr., Basseterre, 1985-87; v.p. Caribbean Agrl. Tech., St. Croix, 1988—. Co-contr. articles to Jour. Hortsci., Can. Jour. Microbiology, Agronomy Jour., W.I. Sugar Technologists Procedings, Caribbean Food Crop Soc. Proceedings, Internat. Soc. Tropical Root Crops Proceedings. Class leader Bethel Meth. Ch., St. Croix, 1989-92, steward, 1991—, chapel steward, 1992—, sec. 1992—, treas., 1993—; treas. Claude O. Markoe Sch. PTA, St. Croix, 1989, 90; bd. dirs. Community United Meth. Sch., St. Croix, 1989—, v.p., 1992—. Carver Rsch. fellow Tuskegee U., 1982-84; named Student of the Yr., Jamaica Sch. Agr., 1978; recipient Cert. Food Agrl. Orgn./Internat. Atomic Energy Agy., 1986, microirrigation design and installation Internat. Irrigarion Supply, 1993. Mem. AAAS, Am. Soc. Horticultural Scis., Am. Soc. Agronomy, Crop Sci. Soc. Am., Soil Sci. Soc. Am., Internat. Soc. Tropical Root Crops, Caribbean Food Crop Soc. (v.p. V.I. chpt. 1988-92), Internat. Herb Growers and Marketers Assn. Achievements include isolation and characterization of nitrogen fixing bacteria from fibrous roots of sweet potato; conducted first field inoculation trial involving sweet potato and nitrogen fixing bacteria. Home: PO Box 2261 Kingshill VI 00851-2261 Office: U VI RR 2 Box 10000 Kingshill VI 00850

CROSSON, JOSEPH PATRICK, metallurgical engineer, consultant; b. N.Y.C., Mar. 17, 1950; s. Joseph T. and Mary E. (McClafferty) C.; m. Valerie Cario, July 7, 1973. BS, Poly. Inst. Bklyn., 1971, MS, 1975; MBA, NYU, 1986. Registered profl. engr., N.Y., N.J. Metall. engr. Lucius Pitkin Inc., N.Y.C., 1970-81, sr. metall. engr., 1981-92, v.p., 1992—. Contbg. author: Metals Handbook, vol. 17, 1989. Staff sgt. USAR, 1971-77. Mem. NSPE, Am. Soc. Metals, Am. Soc. for Nondestructive Testing, Am. Welding Soc. Republican. Roman Catholic. Office: Lucius Pitkin Inc 50 Hudson St New York NY 10013

CROSTON, ARTHUR MICHAEL, pollution control engineer; b. Salford, Lancashire, Eng., Sept. 18, 1942; came to U.S. 1988; s. Arthur and Lydia (Preston) C.; m. Andrea Marina Morris, Dec. 21, 1974 (div. 1989); children: Faye Lizabeth, Joanna Elise; m. Sherrin Louise Wismer, July 28, 1991. Student, U. Manchester, Eng., 1968-70; Mech. Engr., U. Salford, Eng., 1972. Works mgr. John Hamilton & Sons, Manchester, 1965-72; design engr. Ames Crosta Mills, Toronto, Ont., Can., 1972-73, U.S. Filters, Toronto, Ont., Can., 1973-74; owner/designer CMS Rotordisk Inc., Toronto, Ont., Can., 1973-88; owner/designer Amicros Inc., Toronto, Ont., Can., 1985—, East Aurora, N.Y., 1989—; cons. Region of Peel, Mississauga, Can., 1985-87. Achievements include patents on rotordisk W.W.T. pollution control device, recirc/attenuation devise, upflow filtration devise. Home: 19 Tolland Bore East Aurora NY 14052 Office: Amicros Inc 19 Tolland Bore East Aurora NY 14052

CROTHERS, DONALD MORRIS, biochemist, educator; b. Fatehgarh, India, Jan. 28, 1937; came to U.S., 1939; s. Morris K. and Eunice F.C.; m. Leena Kareoja, June 24, 1960; children: Nina H., Kirstina A. BS, Yale U., 1958; BA, Cambridge U., 1960; PhD, U. Calif., San Diego, 1963. Postdoctoral fellow Max Planck Inst., Gottingen, Germany, 1963-64; asst. prof. Yale U., New Haven, 1964-68, assoc. prof., 1968-71, prof. chemistry and molecular biophysics and biochemistry, 1971—, chmn. dept. chemistry, 1975-81; chmn. biophysics, biophys. chemistry B study sect. NIH, 1972-76; co-chmn. nucleic acids Gordon Conf., 1975. Author: Physical Chemistry of Nucleic Acids, 1974, Physical Chemistry with Application to the Life Sciences, 1979; mem. editorial bd. Jour. Molecular Biology, 1971-75, Nucleic Acids Research, 1973-82, Biochemistry, 1975-78, Biopolymers, 1977-90; contbr. articles to profl. jours. Recipient Sci. and Engring. award Yale U., 1977; Alexander von Humboldt Sr. Scientist award, 1981; Mellon fellow Clare Coll. Cambridge U., 1958-60; Guggenheim fellow, 1978. Fellow AAAS, Am. Acad. of Arts and Scis.; mem. Biophys. Soc. (council 1979-82), Nat. Acad. of Scis., Am. Soc. Biol. Chemists. Office: Yale Univ Dept of Chemistry New Haven CT 06520

CROUCH, STEVEN L., mining engineer; b. L.A., Apr. 25, 1943. BS, U. Minn., 1966, MS in Mineral Engring., 1967, PhD in Mineral Engring., 1970. Rsch. officer Mining Rsch. Lab. Chamber of Mines of South Africa, Johannesburg, 1968-70; from asst. to assoc. prof. civil and mineral engring. U. Minn., Mpls., 1970-81, prof., 1981—, acting head dept., 1987-88, head, 1988—; vis. lectr. dept. applied math. U. Witwatersrand, Johannesburg, South Africa, 1976-77, People's Republic of China, 1983; mem. U.S. NAS Com. on Feasibility of Returning Coal Mine Waste Underground, 1973; mem. NAS Task Force on Underground Engring. at Basalt Waste Isolation Project, 1987; active Sandia Nat. Labs. Yucca Mtn. Site Characterization Project Rock Mechanics Rev. Panel, 1989—; cons. in field. Author: (with A.M. Starfield) Boundary Element Methods in Solid Mechanics, 1983; contbr. articles to profl. jours. Recipient U.S. Nat. Com. for Rock Mechanics Applied Rsch. award, 1992. Mem. AIME (Rock Mechanics award 1991), ASCE, Internat. Soc. Roch Mechanics, Minn. Soc. Surveyors and Engrs., Engrs. Club Mpls. Office: U Minn Civil & Mineral Engring Dept Minneapolis MN 55455*

CROW, NEIL BYRNE, geologist; b. Chgo., July 13, 1927; s. Robert Neil and Helen (Byrne) C.; m. Fidel Schilling, Sept. 28, 1957; 1 child, Charles N. BS, Northwestern U., 1949; postgrad., Colo. Sch. Mines, 1949-50, U. Colo., 1951-52, 53-54. Registered geologist, Calif. Groundwater geologist U.S. Geol. Survey, Boise, Idaho, 1952-53; petroleum geologist Amoco Prodn. Co., Shreveport, La., 1954-61; supr. tech. library Aerospace Corp., El Segundo, Calif., 1961-66; supr. tech. library Lawrence Livermore (Calif.) Nat. Lab., 1966-74, environ. geologist, hydrogeologist, 1974—. Contbr. articles on environ. geology and hydrogeology to profl. jours. Mem. Am. Assn. Petroleum Geologists, Geol. Soc. Am., No. Calif. Geol. Soc. Republican. Roman Catholic. Avocations: travel, computers. Home: 2341 Willet Way Pleasanton CA 94566 Office: Lawrence Livermore Nat Lab PO Box 808 L-453 Livermore CA 94551-0808

CROWE, DEVON GEORGE, physicist, engineering consultant; b. Portland, Oreg., Mar. 11, 1948; s. Frank Irving and Jeannie Campbell (Scott) C.; m. Kelly Anne Novia, June 27, 1992. BS in Astronomy and Math., U. Ariz., 1971, MBA in Ops. Mgmt., 1977, MS in Optical Scis., 1980. Sr. rsch. asst. Kitt Peak Nat. Obs., 1975-76; chief systems devel. and ops. Bell Tech. Ops. Corp., Tucson, 1978-80; mgr. tech. analysis div., sr. scientist Sci. Applications, Inc., Tucson, 1980-87; pres. Desert Cat Software, Ltd., 1984-87; dir. Electromagnetics Lab. Ga. Inst. Tech., Atlanta, 1987-90; assoc. v.p. strategic planning and rsch. Ga. Inst. Tech., 1990—, chief scientist, 1991—, adj. prof. elec. engring; cons. in field. Lt. USAF, 1971-74. Fellow Brit. Interplanetary Soc. Fellow Optical Soc. Am. (past pres. Tucson sect.); mem. AAAS, IEEE (sr.), Photo-Optical Instrumentation Engrs., Am. Astron. Soc. Co-author: Optical Radiation Detectors, 1984; contbr. articles in field to profl. jours. Home: 825 Nile Dr Alpharetta GA 30202-5445 Office: Ga Inst Tech Electromagnetics Lab Atlanta GA 30332-0800

CROWE, HAL SCOTT, chiropractor; b. Atlanta, Apr. 19, 1953; s. Hugh Lee and Dorothy Elizabeth (Cooke) C.; m. PiHsiou Hsu, Mar. 29, 1980; children: Hal Scott Jr., Colleen Jao. Student, Johns Hopkins U., 1971-72, Ga. State U., 1973-76, 78-80; D of Chiropractic, Life Chiropractic Coll., 1983; post D. in Chiropractic Neurology, Logan Chiropractic Coll., 1992. Diplomate Nat. Bd. Chiropractic Examiners; cert. chiropractic orthospinologist and cert. chiropractic instr. Life Coll.; cert. chiropractic neurologist Logan Chiropractic Coll. Radiol. technician Crowe Chiropractic Offices, College Park, Ga., 1979-83; chiropractic practitioner Crowe Chiropractic Offices, College Park and Brunswick, Ga., 1983—; clin. rschr. Sweat Found., Atlanta, 1984—; resident in neurology Am. Coll. of Chiropractic Neurology, 1989—; mem. postgrad. faculty Life Chiropractic Coll.; preceptor faculty mem. Palmer Chiropractic Coll.; participant 4th through 9th Ann. Upper Cervical Confs., 1987=92, mem. rev. com. for 10th conf., 1993. Contbr. articles to profl. jours.; 2d chair trombonist Jekyll Island Big Dance Band, 1985—. Host Columbus Ship Replica exhibit, 1992; missionary United Meth. Ch., Petite Goave, Haiti, 1986; mem. coun. on ministries McKendree United Meth. Ch., Brunswick, 1986-87. Recipient Appreciation award Grostic Study Club, Life Chiropractic Coll., 1985. Mem. Internat. Chiropractors Assn., Soc. Chriopractic Orthospinologists (cert. doctor, instr.), Chiropractic Atlas Orthogonists, Nat. Upper Cervical Chiropractic Assn., Lions (bd. dirs. 1986-88, Lion Tamer, 1986-87, presdl. appreciation award 1986, pres. 1988-89, Tail Twister 1991). Avocations: golf, water sports, Caribbean history, archaeology. Home: 20 Ogden St Jekyll Island GA 31527-0638 Office: Crowe Chiropractic Offices 2321 Parkwood Dr Brunswick GA 31520-4720

CROWLEY, JOSEPH MICHAEL, electrical engineer, educator; b. Phila., Sept. 9, 1940; s. Joseph Edward and Mary Veronica (McCall) C.; m. Barbara Ann Sauerwald, June 22, 1963; children: Joseph W., Kevin, James, Michael, Daniel. B.S., MIT, 1962, M.S., 1963, Ph.D., 1965. Vis. scientist Max Planck Inst., Goettingen, W.Ger, 1965-66; asst. prof. elec. engring. U. Ill., Urbana, 1966-69, assoc. prof., 1969-78, prof., dir. applied electrostats. research lab., 1978-88; pres. JMC Inc., 1981-91, Electrostatic Applications, 1986—; Piercey Disting. prof. chem. engring. U. Minn., 1993; adj. prof. U. Ill., Urbana, 1988—; Piercey disting. prof. chem. eug. U. Minn., 1993; cons. to several corps. Contbr. articles to profl. jours.; patentee ink jet printers. Pres. Champaign-Urbana Bd. Cath. Edn., 1978-80. Recipient Gen. Motors scholarship, 1958-62; AEC fellow, 1962-65; NATO fellow, 1965-66. Mem. IEEE (sr.), Electrostats. Soc. Am., Am. Phys. Soc., Soc. Inf. Display, Mensa. Roman Catholic.

CROWLEY, MICHAEL JOSEPH, mechanical engineer; b. Wilmington, Del., Aug. 7, 1962; s. Robert Joseph and Rita Francis (Mulcahy) C. BS in Engring., Widener U., Chester, Pa., 1984. Maintenance engr. E.I. duPont, Aiken, S.C., 1984-86; power cons. E.I. duPont, Wilmington, 1986-91, power design, 1991—. Achievements include patent on recreational vehicle with 2 rigidly coupled sects. Home: 104 Dewalt Rd Newark DE 19711 Office: E I duPont 633 Paper Mill Rd Newark DE 19711

CROWLEY, MICHAEL SUMMERS, ceramic engineer; b. Chgo., Dec. 24, 1928; s. John Laurance and Roslyn (Summers) C.; m. Elaine Agnes Reynolds, June 17, 1950; children: Veronica Crowley Parish, Michael Reynolds. BS in Ceramic Engring., Iowa State U., 1953; PhD in Geochemistry, Pa. State U., 1959. Registered profl. engr., Ill., Tex., Calif. Rsch. engr. Standard Oil Ind., Whiting, 1953-55; API fellow Pa. State U., State College, 1955-59; sr. rsch. assoc. Amoco Corp., Naperville, Ill., 1959-89; cons. M.S. Crowley & Assocs., Crete, Ill., 1989—; chmn. refractories Metals Prop Coun., N.Y.C., 1972-73; cons. U.S. Dept. Energy, Washington, 1977-80, Argonne (Ill.) Nat. Lab., 1980; adj. prof. dept. ceramic engring. Clemson U., 1992—. Author: Guidelines for Refractories, 1991; contbr. articles to profl. jours. Fellow Am. Ceramic Soc. (chmn. St. Louis sect. T.J. Planje award 1992); mem. Ill. Soc. Profl. Engrs. (pres. 1990-91). Achievements include 7 patents in refractory technology. Home and Office: 1270 Deer Run Tr Crete IL 60417

CROWLEY, WILLIAM FRANCIS, JR., medical educator; b. Meriden, Conn., Dec. 28, 1943; s. William Francis and Kathryn (Kiernan) C.; m. Nancy Marie Colwell; children: William Francis III, Sean Timothy, Regan Elizabeth, Colin Colwell. BA (honors curriculum), Holy Cross Coll., Worcester, Mass., 1965; MD, Tufts U., 1969. Diplomate Am. Bd. Internal Medicine. Intern Mass. Gen. Hosp., Boston, 1969-70, asst. resident in medicine, 1970-71, sr. resident in medicine, 1973-74, clin. and rsch. fellow in endocrinology, 1974-76; instr. medicine Harvard Med. Sch., Mass. Gen. Hosp., 1976-80; asst. prof. medicine Harvard Med. Sch., Boston, 1980-84, assoc. prof., 1984-92, prof., 1992—; chief Reproductive Endocrine Unit, Mass. Gen. Hosp., 1984—, attending physician, 1988—, dir. Vincent Rsch. Labs., 1987-90; adv. bds. NIH, FDA, 1979—; lectr. in sci. writing, Harvard U., 1974-76; vis. prof. Yale U., 1982, Duke U., 1983, N.Y. Obstet. Soc., 1983, George Washington U., 1985, U. Miami, 1989; Goldfarb lectr., Vanderbilt U., 1989; Leathem lectr., Rutgers U., 1981; Israel Mackler lectr. Albert Einstein Coll. Medicine, 1982; invited speaker Laurentian Hormone Conf., 1984, 90; Winkler Meml. lectr. U. Buffalo, 1989; cons. Study Sect., Ctr. for Population, 1979-80, Contract Adv. Bd. Contraceptive Devel. Br., Nat. Inst. Child Health and Human Devel., NIH, 1979—; dir. Mass. Gen. Hosp. Reproductive Endocrine Scis. Ctr.-NIH Ctrs. of Reproductive Excellences, 1991, Nat. Ctr. Infertility Rsch., Mass. Gen Hosp., 1991, NIH Tng. Grant in Reproductive and Devel. Biology, 1991. Editor: (with J.G. Hofler) The Episodic Secretion of Hormones, 1987; contbr. articles to profl. jours.; mem. editorial bds. numerous profl. and sci. jours. including Jour. Clin. Endocrinology and Metabolism, 1983-87, Neuroendocrinology, 1987—, Acta Endocrinologica, 1983—, Internat. Jour. Fertility, 1985—, Annals Internal Medicine, 1986—, Endocrinology, 1988—, Molecular and Cellular Neuroscis., 1989—. Mem. Union of Concerned Scientists, Planned Parenthood, RESOLVE (physicians' bd.). Lt. USNR, 1971-73. Daland fellows Am. Philos. Soc., 1975-78; grantee NIH. Mem. ACP, Soc. for Study Reprodn., Am. Soc. Clin. Investigation, Assn. Am. Physicians, Endocrine Soc., Mass. Med. Soc., Am. Fertility Soc., Am. Fedn. Clin. Rsch., N.Y. Acad. Scis., Hyannisport Country Club (Mass.). Avocations: tennis, skiing, walking, reading. Office: Mass Gen Hosp Reproductive Endocrine Scis Ctr Bartlett Hall Ext # 5 Boston MA 02114

CROWTHER, RICHARD LAYTON, architect, consultant, researcher, author, lecturer; b. Newark, Dec. 16, 1910; s. William George and Grace (Layton) C.; m. Emma Jane Hubbard, 1935 (div. 1949); children: Bethe Crowther Allison, Warren Winfield, Vivian Crowther Tuggle; m. 2d Pearl Marie Tesch, Sept. 16, 1950. Student, Newark Sch. Fine and Indsl. Arts, 1928-31, San Diego State Coll., 1933, U. Colo., 1956. Registered architect, Colo. Prin. Crowther & Marshall, San Diego, 1946-50, Richard L. Crowther, Denver, 1951-66, Crowther, Kruse, Landin, Denver, 1966-70, Crowther, Kruse, McWilliams, Denver, 1970-75, Crowther Solar Group, Denver, 1975-82, Richard L. Crowther FAIA, Denver, 1982—; vis. critic, lectr. U. Nebr., 1981; holistic energy design process methodology energy cons. Holistic Health Ctr., 1982-83; adv. cons. interior and archtl. design class U. Colo., 1982-83, Cherry Creek, Denver redevel., 1984-88, Colo. smoking control legislation, 1985, interior solar concepts Colo. Inst. Art, 1986, Bio-Electro-Magnetics Inst., 1987-88; mentor U. Colo. Sch. Architecture, 1987-88. Author: Sun/Earth, 1975 (Progressive Architecture award 1975), rev. edit., 1983, Affordable Passive Solar Homes, 1983, Paradox of Smoking, 1983, Women/Nature/Destiny: Female/Male Equity for Global Survival, 1987, (monographs) Context in Art and Design, 1985, Existence, Design and Risk, 1986, Indoor Air: Risks and Remedies, 1986, Human Migration in Solar Homes for Seasonal Comfort and Energy Conservation, 1986, 88, Ecologic Architecture, 1992, Ecologic Digest, 1993, and others. NSF grantee, 1974-75. Fellow AIA (commr. research, edn. and environ. Colo. Central chpt. 1972-75, bd. dirs. chpt. 1973-74 AIA Research Corp. Solar Monitoring Program contract award). Achievements include research in solar, air, earth, water, and electromagnetic energies in reference to architecture and human biologic response.

CROZIER, DAVID WAYNE, mechanical engineer; b. Cleve., May 24, 1958; s. Howard Wayne and Carol Marie (Smith) C.; m. Helene Marie Kneier, July 7, 1984; children: Katelyn Julia, Matthew David, Allison Marie. BSME, U. Cin., 1981; MBA, John Carroll U., 1990. Registered profl. engr., Ohio. Design engr. Whitey Co., Highland Heights, Ohio, 1981-84, mfg. engring. supr., 1984-86; project engr. Kinetico Inc., Newbury, Ohio, 1986-87; prodn. engring. mgr. Kinetico Inc., Newbury, 1987-89; mfg. engring. mgr. Teledyne Hyson, Brecksville, Ohio, 1989-92, v.p. engring., 1992—. Trustee Huntington's Disease Soc., Cleve., 1984-88; bd. mem. Greater Cleve. Spina Bifida Soc., 1992—. Mem. NSPE. Achievements include research in the field of valves, resulting in patents for employers. Home: 825 Rose Blvd Highland Heights OH 44143 Office: Teledyne Hyson 10367 Brecksville Rd Brecksville OH 44141

CRUESS, RICHARD LEIGH, surgeon, university dean; b. London, Ont., Can., Dec. 17, 1929; s. Leigh S. and Martha A. (Peever) C.; m. Sylvia Crane Robinson, May 30, 1953; children: Leigh S., Andrew C. B.A., Princeton U., 1951; M.D., Columbia U., 1955. Diplomate Am. Bd. Orthopedic Surgery. Intern Royal Victoria Hosp., Montreal, Que., 1955-56; resident surgery Royal Victoria Hosp., 1956-57; resident surgery N.Y. Orthopedic Hosp., 1959-60, asst. resident orthopedic surgery, 1960-61, resident orthopedic surgery, 1961-62, Annie C. Kane fellow orthopedic surgery, 1961-62; research asso. depts. orthopedic surgery and biochemistry Columbia U., N.Y.C., 1962-63; John Armour Travelling fellow, 1962-63, Am.-Brit.-Can. Travelling fellow, 1967; practice medicine specializing in orthopedic surgery Montreal, 1963—; orthopedic surgeon Royal Victoria Hosp., orthopedic surgeon-in-charge, 1968-81, asst. surgeon-in-chief, 1970-81; chief surgeon Shriner's Hosp. for Crippled Children, Montreal, 1970-82; prof. surgery McGill U., Montreal, 1970—; chmn. div. orthopedic surgery McGill U., 1976-81, dean faculty medicine, 1981—; hon. cons. orthopedic surgery Queen Elizabeth Hosp., 1972—; mem. clin. grants com. Med. Research Council, 1972-75, mem. council, 1980-86, mem. exec., 1983-86. Contbr. articles on surgery to profl. jours.; editorial bd.: Jour. Internat. Orthopedics, 1976-85, Jour. Bone and Joint Surgery, 1977-83, Current Problems in Orthopedics, 1977—, Jour. Orthopaedic Rsch., 1986-88. Served to lt. M.C., USN, 1957-59. Fellow Royal Coll. Physicians and Surgeons Can. (chief examiner orthopedic surgery 1970-72), ACS, Am. Acad. Orthopedic Surgeons, Royal Soc. Can.; mem. Can. Orthopedic Assn. (sec. 1971-76, pres. 1977-78), Can. Orthopedic Research Soc. (pres. 1971-72), Am. Orthopedic Research Soc. (pres. 1975-76), Am. Orthopedic Assn., Assn. Orthopedic Surgeons Province Que. (treas. 1971-72), Société Française de Chirurgie Orthopedique (hon.), McGill Osler Reporting Soc., Assn. Can. Med. Colls. (pres. 1987-89). Home: 526 Mount Pleasant Ave, Montreal, PQ Canada H3Y 3H5 Office: McGill U, 3655 Drummond St, Montreal, PQ Canada H3G 1Y6

CRUM, ALBERT BYRD, psychiatrist, consultant; b. Omaha, Nov. 17, 1931; s. J. Rufus and Alberta (McCreary) C.; m. Rosa Maria Hennessy y Sinclair; children: Rosa Maria Crum O'Brien, Elsie Crum Coy, Alberta Crum Fousek. BS, U. Redlands, Calif., 1953; MD, Harvard U., 1957; MS, NYU, 1987; DS (hon.), U. Redlands, 1974. Med. intern Columbia U. div. Bellevue Med. Ctr., N.Y.C., 1957-58; rsch. fellow, psychiat. resident Creedmoor Inst. for Psychol. Studies, Queens Village, N.Y., 1958-59; chief, neuropsychiatric svcs., Continental Air Command USAF Hosp., 1959-61; psychiat. resident Columbia U. Psychiat. Inst. of Columbia-Presbyn. Hosp., N.Y.C., 1961-63; pvt. practice Brooklyn Heights, N.Y., 1963—; active attending staff Gracie Sq. Hosp., N.Y.C., 1963—; med. dirs. Psychiatric Svcs. Internat. P.C., Brooklyn Heights, 1980—; ednl. dir. med. and health seminars Internat. Inst. for Human Behavior, Inc., Brooklyn Heights, 1983—; advisor Office of Tibet, N.Y.C., 1984—; clin. prof. behavioral scis. NYU, N.Y.C., 1987—; pres., dir. behavioral scis. Way of Life/N.Y., Ltd., Brooklyn Heights, 1989—; pres. Y.F. One/N.Y., Ltd., Brooklyn Heights, 1991—, Y.F. Nationwide, Inc., Brooklyn Heights, 1991—; co-chmn. U.S. Coordinating Commn. for Nomination of His Holiness the Dalai Lama of Tibet for the Nobel Peace Prize, Brooklyn Heights, 1986—; adj. prof. anatomy and neuroanatomy, NYU, 1987—; ptnr. Burdick Assocs. Investment Firm, Brooklyn Heights, 1976—; pres. Burdick Assocs. Owners Corp., Brooklyn Heights, 1983—; chmn. Human Behavior Found., Brooklyn Heights, 1968—; chmn. selection com. Human Behavior Found.'s Albert Schweitzer Humanitarian Award, Brooklyn Heights, 1986—. Author (chpt.) The Triumphant Person, 1989. Bd. dirs. Albert Schweitzer Fellowship, N.Y.C., 1982—; chmn. William James Found., Brooklyn Heights, 1989—; bd. dirs. Burdick Internat. Ancestry Library, Sarasota, Fla., 1985—; mem., chn., bd. advisors NYU's Coll. of Dentistry, N.Y.C., 1988—; mem. Brooklyn Heights Assn., 1970—. Capt. USAF, 1959-61. Recipient Disting. Svc. award Bklyn. Jr. C. of C., 1966, Bicentennial award Nat. Jogging Assn., 1976. Fellow Royal Coll. Physicians and Surgeons in Psychiatry; mem. Pan Am. Med. Assn., Nat. Bd. Med. Examiners, Med. Coun. of Can., Am. Acad. Clin. Psychiatrists, Am. Orthopsychiatric Assn., Am. Psychiat. Assn., AMA, Med. Soc. State of N.Y., Kings County (N.Y.) Med. Soc., World Med. Assn., World Fedn. Mental Health, Am. Physicians Art Assn., Harvard Med. Soc., English Speaking Union, Harvard Club of N.Y., Bklyn. Club, Heights Casino and Racquet Club, MENSA (life, nat. coord. 1980-84), Phi Beta Kappa (councillor 1981-84). Avocations: jogging, studying world religions, history, leadership. Home and Office: Psychiat Svcs Internat PC 77 Remsen St Brooklyn NY 11201-3400

CRUM, JOHN KISTLER, chemical society director; b. Brownsville, Tex., July 28, 1936; s. John Mears and Mary Louise (Kistler) C. B.S., U. Tex., 1960, Ph.D., 1964; grad. Advanced Mgmt. Program, Harvard U., 1975. Research fellow Robert A. Welch Found., 1962-64; asst. editor Am. Chem. Soc., Washington, 1964-65; assoc. editor Am. Chem. Soc., 1966-68, mng. editor, 1969-70, group mgr. jours., 1970, dir. books and jours. div., 1971-75, treas., chief fin. officer, 1975-80, dep. exec. dir. and chief operating officer, 1981-82, exec. dir., 1983—; chmn. bd. Centcom Ltd., chmn. governing bd. Chemical Abstracts Svc., 1991—; mem. U.S. nat. com. Internat. Union Pure and Applied Chemistry; mem. Nat. Com. for Edn. in Space; regular mem. Con. Bd.; bd. dirs. Consumers Union of U.S. Mem. editorial adv. bd. Am. Men and Women of Sci.; contbr. articles to profl. jours. Fellow Washington Acad. Scis.; mem. Chem. Soc. (London), Am. Chem. Soc., Coun. Engring. and Sci., Soc. Execs., Assn. Sci. Soc. Editors, N.Y. Acad. Scis., Chem. Soc. Washington, Cosmos CLub, City Club, Univ. Club (Washington), Sigma Xi, Phi Theta Kappa. Republican. Home: 1701 N Kent St Arlington VA 22209-2112 Office: Am Chem Soc 1155 16th St NW Washington DC 20036

CRUMPTON, EVELYN, psychologist, educator; b. Ashland, Ala., Dec. 23, 1924; d. Alpheus Leland and Bernice (Fordham) Crumpton. AB, Birmingham So. Coll., 1944; MA, UCLA, 1953, PhD in Psychology, 1955. Lic. psychologist, Calif.; diplomate Am. Bd. Profl. Psychology. Rsch. psychologist VA Hosp., Brentwood, L.A., 1955-77; asst. chief psychology svc., dir. clin. tng. VA Adminstrn. Med. Ctr. West Los Angeles, 1977-88; clin. prof. dept. psychology UCLA, assoc. rsch. psychologist dept. psychiatry, UCLA Sch. Med., 1957—; cons. chief of staff Brentwood div., VA Adminstrn. Med. Ctr. Contbr. numerous articles to profl. jours. Recipient Profl. Svc. award, Assn. Chief Psychologists VA, 1979. Fellow Soc. Personality Assessment; mem. APA, Western Psychol. Assn., Sigma Xi.

CRUTZEN, YVES ROBERT, engineer, scientist, educator; b. Verviers, Liege, Belgium, May 16, 1950; arrived in Italy, 1973; s. Robert and Marguerite Crutzen; m. Giovanna Cammilleri, Apr. 9, 1980. MSCE, Poly. U. Brussels, 1973; M Structural Engring., Poly. U. Milan, Italy, 1976; PhD in Applied Scis., U. Brussels, 1979. Designer Studio Wagner, Milan, 1974-76; European boursier Commn. European Communities/Joint Rsch. Ctr., Brussels and Ispra, Italy, 1976-79; safety expert Bur. D'Etudes Pirnay & Schwachhoffer, Charleroi, Belgium, 1980; CAD/CAE software analyst, software cons. Control Data Italia, Milan and Turin, 1981-85; scientist NASA Rsch. Ctr., 1984; sci. officer Commn. European Communities/Joint Rsch. Ctr., Ispra, Italy, 1985-90, project leader, 1990—; tchr. U. Brussels, U. Milan, U. Bologna, U. Pisa, Italy, U. Okayama, Japan, USAF Acad., 1979—; internat. lectr. Europe, U.S.A., Japan. Author, editor: Computers in Information Science, Eurocourses Blue Book, Vol. I, 1990; contbr. articles to profl. jours. (NATO prize 1980). Civil svc. officer Belgian Nat. Govt., 1973-78; sci. com. observer, various profl. and trade unions and orgns., JRC, CEC, 1990-92. Mem. Associazione Italiana di Meccanica Teorica ed Applicata (mem. doctorate steering com.), Association des Ingenieurs de Bruxelles (Verdeyen prize 1973), Structural Mechanics in Reactor Tech. (conf. sect. chmn. 1979-89, Travel award in U.S. 1983), Testing Electromagnetic Analysis Methods (sec. workshop 1988-93 Europe and Japan, editor 1989-93), Internat. Thermonuclear Exptl. Reactor-Internat. Atomic Energy Agy. (Publ. award 1990). Roman Catholic. Avocations: lyrical music, art, skiing, tennis, racing Ferrari cars. Home: Via G Deledda 5, 21020 Ranco Italy Office: Commn European Communities, Joint Rsch Ctr TP210, 21020 Ispra Italy

CRUZ, JOSE BEJAR, JR., electrical engineering educator, dean; b. Bacolod City, Philippines, Sept. 17, 1932; came to U.S., 1954, naturalized, 1969; s. Jose P. and Felicidad (Bejar) C.; m. Stella E. Rubia; children by previous marriage: Fe E. Cruz Langdon, Ricardo A., Rene L., Sylvia C., Loretta C. Cruz Reynolds. B.S. in Elec. Engring. summa cum laude, U. Philippines, 1953; M.S., MIT, 1956; Ph.D., U. Ill., 1959. Registered profl. engr., Ill., Ohio. Instr. elec. engring. U. Philippines, Quezon City, 1953-54; research asst. MIT, 1954-56, vis. prof., 1973; instr. U. Ill., Urbana-Champaign, 1956-59; asst. prof. U. Ill., 1959-61, assoc. prof., 1961-65, prof. elec. engring., 1965-86, assoc. mem. Ctr. Advanced Study, 1967-68; research prof. Coordinated Sci. Lab., 1965-86; prof. dept. elec. and computer engring. U. Calif., Irvine, 1986-92, chmn. dept., 1986-90; prof. elec. engring., dean Coll. Engring. Ohio State U., Columbus, 1992—; vis. assoc. prof. U. Calif., Berkeley, 1964-65; vis. prof. Harvard, 1973; pres. Dynamic Systems; mem. theory com. Am. Automatic Control Council, 1967; gen. chmn. Conf. on Decision and Control, 1975; mem. profl. engring. exam. com. State of Ill., 1984-86; mem. Nat. Council Engring. Examiners, 1985-86. Author: (with M.E. Van Valkenburg) Introductory Signals and Circuits, 1967, (with W.R. Perkins) Engineering of Dynamic Systems, 1969, Feedback Systems, 1972, System Sensitivity Analysis, 1973, (with M.E. Van Valkenburg) Signals in Linear Circuits, 1974; Assoc. editor: Jour. Franklin Inst, 1976-82, Jour. Optimization Theory and Applications, 1980—; series editor Advances in Large Scale Systems Theory and Applications; contbr. articles on network theory, automatic control systems, system theory, sensitivity theory of dynamical systems, large scale systems, dynamic games and dynamic scheduling in mfg. systems to sci., tech. jours. Recipient Purple Tower award Beta Epsilon U., Philippines, 1969, Curtis W. McGraw Rsch. award Am. Soc. for Engring. Edn., 1972, Halliburton Engring. Edn. Leadership award, 1981, Most Outstanding Alumnus award U. of the Philippines Alumni Assn. Am., 1989, Most Outstanding Overseas Alumnus Coll. Engring., U. of the Philippines Alumni Assn., 1990. Fellow AAAS (sect. com. for sect. on engring. 1991—), IEEE (chmn. linear systems com., group on automatic control 1966-68, assoc. editor Trans. on Circuit Theory 1962-64); mem. IEEE Control Systems Soc. (adminstrv. com. 1966-75, 78-80, v.p. fin. and adminstrv. activities 1976-77, pres. 1979, chmn. awards com. 1973-75, ednl. activities bd. 1973-75, editor Trans. on Automatic Control 1971-73, mem. tech. activities bd. 1979-83, chmn. 1982-83, v.p. tech. activities 1982-83, edn. med. com. 1977-79, dir. 1980-85, vice chmn. publs. bd. 1981, chmn. 1984-85, chmn. panel of tech. editors 1981, chmn. TAB periodicals com. 1981, chmn. PUB Soc. publs. com. 1981, v.p. publ. activities 1984-85, exec. com. 1982-85, Richard M. Emberson award 1989), Philippine Engrs. and Scientists Orgn., Soc. Indsl. and Applied Math., AAUP, Am. Soc. for Engring. Edn., U.S. Nat. Acad. Engring., Philippine-Am. Acad. Sci. and Engring. (founding mem. 1980, pres. 1982), Internat. Fedn. Automatic Control (chmn. theory com. 1981-84, vice chmn. tech. bd. 1984-87, policy com. 1987-93, vice chmn. 1993, congress internat. program com.), Sigma Xi, Phi Kappa Phi, Eta Kappa Nu. Achievements include introduction of concept of comparison sensitivity in dynamical feedback systems, of leader-follower strategies in hierarchical engineering systems; development of synthesis methods for time-varying systems. Office: Ohio State U Coll Engring Columbus OH 43210-1275

CRUZE, ALVIN M., research institute executive; b. Maryville, Tenn., Apr. 18, 1939; s. Gifford G. and Kathryn B. (McNutt) C.; children: Sidney, Warren; m. Karen H. Winton, Aug. 8, 1991. B.S., U. Tenn., 1961; M.S., Rutgers U., 1962; Ph.D., N.C. State U., 1972. Economist Research Triangle Inst., Research Triangle Park, N.C., 1965-70, 71-75, ctr. dir., 1975-83, v.p., 1983-88, exec. v.p., 1989—; lctr. U. N.C., Chapel Hill, 1970-71. Served to lt. U.S. Army, 1963-65. Democrat. Methodist. Avocations: jogging; gardening. Home: 211 S Bank Dr Cary NC 27511-9760 Office: Rsch Triangle Inst PO Box 12194 Durham NC 27709-2194

CRUZEN, MATT EARL, research biochemist; b. Upland, Calif., Apr. 14, 1962; s. Archie Jerome and Natalie (Damon) C.; m. Kathleen Sullivan, Apr. 4, 1992. BS magna cum laude., Calif. Poly., 1985; PhD, Univ. Calif., 1993. Rsch. assoc. Oncotech Inc., Irvine, Calif., 1993—. Author: Gene Amphimion in Mamalia Cells, 1992; contbr. articles to profl. jours. Home: 530 S B St Tustin CA 92680

CRUZ-PINTO, JOSE JOAQUIM C., chemical and materials engineering educator, researcher; b. Lisboa, Portugal, June 6, 1948; s. Joao and Susana Maria (Costa) C. Chem. Engr., Inst. Sup. Tecnico, Lisboa, Portugal, 1971; Ing. Gen. Chim., U. Nancy, France, 1972; PhD, U. Manchester, U.K., 1979. Asst. lectr. U. Lourenco Marques, 1972-75; asst. lectr. U. Minho, Braga, Portugal, 1978-79, assoc. prof., 1980-92, prof. polymer sci., 1992—; pres. Centro de Quimica Pura e Aplicada, Braga, 1983-85; head Dept. Polymer Engring., Braga, 1985-87; mem. Nat. Rsch. Coord. Com., Lisboa, 1990—. Contbr. articles to profl. and sci. jours. Pres. CHCUM, Local Housing Coop., Braga. Mem. Sociedade Portuguesa de Quimica, Sociedade Portuguesa de Materials, Associazione Italiana di Scienza e Tecnologia delle Macromolecole, Internat. Confedn. for Thermal Analysis, Internat. Union Pure and Applied Chemistry (Commn. IV2 Polymer Characterization and Properties), N.Y. Acad. Scis. Roman Catholic. Avocations: music, computing, photography. Office: Univ do Minho, Largo do Paco, 4719 Braga Portugal

CRYDER, CATHY M., plant geneticist; b. Twin Falls, Idaho, June 29, 1953; d. Oris Donald and Ruth Carol (Strickland) C.; m. Daryl R. Wilson, July 18, 1992. BS in Biology, Boise State U., 1975; MS in Plant Pathology, U. Idaho, 1978; PhD in Plant Molecular Genetics, N.Mex. State U., 1988. Rsch. assoc. II U. Idaho, Moscow, 1979; rsch. asst. plant breeder Asgrow Seed Co., San Juan Bautista, Calif., 1979-80, plant breeder, 1980-85; grad. teaching asst. dept. agronomy and hort. N.Mex. State U., Las Cruces, 1985-88, grad. rsch. asst., 1986-88; plant breeder, sta. mgr. Nickerson-Zwaan Seed Co., Las Cruces, 1988-91; plant breeder, sta. mgr., onion seed prodn. mgr. Shamrock Seed Co., Las Cruces, 1991—; program com. chmn. Tomato Breeders Round Table, 1983-85. Contbr. article to Theor. Applied Genetics, Phytopathology News. Lay minister First Evang. Free Ch., Las Cruces, 1990-92. Mem. European Assn. for Rsch. on Plant Breeding, Am. Soc. Hort. Sci. (hall of fame com. 1989-90, scholarship com. 1990-91, devel. com. 1989-90), Phi Alpha Xi, Gamma Sigma Delta, Phi Sigma. Achievements include development of hybrid tomato varieties XPH896, XPH898, Allegro, Bridgade, Centurion and onion variety Caribou. Office: Shamrock Seed Co N Mex Rsch Sta Drawer F Mesilla NM 88046

CRYER, DENNIS ROBERT, pharmaceutical company executive, researcher; b. Dearborn, Mich., Mar. 30, 1944; s. Earl Wilton and Marguerite Gladys (Root) C.; m. Leilani Chen, 1974 (div.); 1 child, Jonathan Eric; m. Sharon Therese Kniezewski, July 26, 1986; children: Catherine Grace, Laura Rose. BA in Biology, Johns Hopkins U., 1968; MD, Albert Einstein Coll. Medicine, 1977. Intern Children's Hosp. Phila., 1977-78, resident, 1978-79, 80-81; fellow in pathology and molecular biology U. Pa. Sch. Medicine, Phila., 1979-80; fellow in human genetics Sch. Medicine U. Pa., Phila., 1981-84, clin. asst. prof. pediatrics Sch. Medicine, 1983-84, asst. prof. pediatrics Sch. Medicine, 1984-87; assoc. clin. rsch. dir. E.R. Squibb and Sons, Princeton, N.J., 1987-89; assoc. med. devel. dir. Squibb U.S. Pharm. Group, Princeton, 1989-90, med. ops. dir., 1990-91; med. dir. Bristol-Myers Squibb U.S. Pharm. Divsn., Princeton, 1991—; corp. rep., corp. affairs com. Am. Soc. Hypertension, 1991—. Author: with others Cold Spring Harbor Symposium on Quantitative Biology, 1974, Methods in Cell Biology, 1975; contbr. articles Jour. of Molecular Biology, Jour. Lipid Rsch., Jour. Clin. Investigation. Grantee Nat. Heart, Lung, and Blood Inst. , NIH, 1986, Am. Heart Assn., 1987; recipient Merck Faculty Devel. award Merck, Sharp, and Dohme, 1984. Fellow Am. Heart Assn. (arteriosclerosis coun.); mem. AAAS, Am. Fedn. Clin. Rsch., Am. Soc. Human Genetics, Am. Soc. Hypertension (corp. rep., corp. affairs com.), N.Y. Acad. Scis., Alpha Epsilon Delta. Achievements include pioneering development of evidence that eukaryotic chromosomes contain a single, double-stranded DNA molecule; demonstration of a gene dosage effect for mitochondrial DNA (using mating strains of yeast); development of methods using stable isotopes and gas chromatography-mass spectrometry to study human lipoprotein metabolism; demonstration of accurate measurement of hepatic lipoprotein synthesis using these methods; demonstration of a powerful autosomal dominant human gene which lowers cholesterol in a family with coexistent familial hypercholesterolemia. Home: 7 N Circle Ct Yardley PA 19067 Office: Bristol-Myers Squibb Co PO Box 4500 Princeton NJ 08543

CRYER, THEODORE HUDSON, ophthalmologist, educator; b. Chgo., May 8, 1946; s. Arthur William and Maxine Ritter C.; A.B. in Chemistry, Taylor U., 1968; M.D., U. Md., 1972; children: Timothy Hudson, Jordan Tinley, Megan Elizabeth, Rebecca Jeanne. Straight med. intern South Balt. Gen. Hosp., 1972-73; jr. asst. resident, 1973-74; asst. resident U. Md. Hosp., Balt., 1974-76, resident, 1976-77; practice medicine specializing in ophthalmology, Waynesboro, Pa., 1977—, Westminster, Md., 1977-85; instr. U. Md. Sch. Medicine, 1979-91, clin. asst. prof. medicine, 1991—; chmn. com. on ethics Waynesboro Hosp., 1984, chmn. com. quality assurance, 1987—, v.p. med. staff, 1987-89, pres. med. staff, 1989-91, trustee, 1991—. Clk. session Westminster Reformed Presbyn. Ch., 1980-83. Mem. AMA, Am. Acad. Ophthalmology, Pa. Med. Soc., Franklin County Med. Soc., Md. Eye Physicians and Surgeons, Pa. Acad. Otolaryngology and Ophthalmology, AAAS, Nat. Soc. to Prevent Blindness (charter mem.), Ophthal. Assn. Rsch. to Prevent Blindness, Ind., 1978. Republican. Methodist. Office: 45 Roadside Ave Waynesboro PA 17268

CRYMBLE, JOHN FREDERICK, chemical engineer, consultant; b. N.Y.C., Oct. 18, 1916; s. Hugh and Hannah (Knecht) C.; B.A., Columbia U., 1938; B.S., 1939, Chem. E., 1940; m. Mary Alenda Smith, June 24, 1944; 1 dau., Joanne Lee (Mrs. Donald L. Gilmore). Prodn. supr. E.I. duPont de Nemours and Co., Chambers Works, Deepwater, N.J., 1940-73, sr. prodn. engr., 1973-76; cons., 1977—. Past pres. Salem City Bd. Edn., also rep. N.J. Sch. Bds. Assn.; bd. dirs. Salem Free Libr. Mem. Am. Chem. Soc., Am. Inst. Chem. Engrs., John Jay Assos. Columbia, Thomas Egleston Assocs. Columbia Sch. Engring. and Applied Sci., Columbia U. Alumni Assn. (alumni medalist 1988), Sigma Xi, Phi Lambda Upsilon, Tau Beta Pi. Methodist (trustee, past lay leader). Clubs: DuPont Country (Wilmington, Del.); Columbia U. Alumni (past exec. com., v.p. Phila.). Home and Office: 65 W Broadway Salem NJ 08079-1329

CRYMES, RONALD JACK, structural steel detailer, draftsman; b. Atlanta, Jan. 13, 1935; s. Earl Arlington and Lillian May (Carlton) Crymes; m. Josephine Florence Boyd, June 1965; children: Jennifer Lynn, Jonathan Boyd, Joseph Earl. Student, Ga. Inst. Tech., 1957-61, U. Ga., Atlanta, 1961-62. Draftsman Jack Zwecker & Co., Engrs., Atlanta, 1961-75; draftsman checker Atlanta, 1975-84, L.N. Ross Engring., Atlanta, 1984-87; chief draftsman Addison Steel, Atlanta, 1987; pres. Accurate Detailing Assn., Atlanta, 1987—. USAF, 1954-57. Presbyterian. Avocations: electric trains, computer programming. Home: 5084 Faye Ct Norcross GA 30071-3252

CSANADY, GABRIEL TIBOR, oceanographer, meteorologist, environmental engineer; b. Budapest, Hungary, Dec. 10, 1925; s. Arpad Kalman and Elizabeth (Marosi) C.; m. Ada Luige, Sept. 3, 1954 (div. 1968); 1 son, Andrew John; m. Joyce Eva Stever, Jan. 19, 1969. Diploma Ing., Technische Hochschule, Munich, W.Ger., 1948; Ph.D., U. New South Wales, Sydney, Australia, 1958. Engr. State Electric Co., Victoria, Melbourne, 1952-54; sr. lectr. U. New South Wales, Sydney, 1954-61; assoc. prof. mech. engring. U. Windsor, Ont., Can., 1961-63; prof. mech. engring. U. Waterloo, Ont., Can., 1963-73; sr. scientist Woods Hole Oceanographic Instn., Mass., 1972-87; Slover prof. of Oceanography Old Dominion U., Norfolk, Va., 1987—. Author: Theory of Turbomachines, 1964, Turbulent Diffusion in the Environment, 1973, Circulation in the Coastal Ocean, 1982. Recipient Pres.'s prize Can. Meteorol. Soc., 1970, Chandler-Misener award Internat. Assn. Great Lakes Research, 1977, A.G. Huntsman award Bedford Inst. Oceanography, 1991; Sherman Fairchild disting. scholar Calif. Inst. Tech., 1982. Fellow Royal Meteorol. Soc. London; mem. Am. Meteorol. Soc. (Editor's award 1975), Am. Geophys. Union, ASME. Office: Old Dominion U Norfolk VA 23529

CSÁNGÓ, PÉTER ANDRÁS, microbiologist; b. Budapest, Hungary, May 14, 1942; s. Ferenc and Magda (Herczeg) C.; m. Gerda Patricia Dub, Oct. 25, 1969; children: Monica, Miriam, Michelle. MD, U. Medicine, Pécs, Hungary, 1967; postgrad., U. Birmingham, Eng., 1972-73. Registrar County Lung Hosp., Szombathely, Hungary, 1967-68, Semmelweis Hosp., Budapest, 1968-69, State Lung Hosp., Tromsø, Norway, 1969-70; registrar Nat. Inst. Pub. Health, Oslo, 1970-73, sr. registrar, 1973-78, virologist, 1978; dir. Dept. Clin. Microbiology Ctr. Hosp., Kristiansand, 1978—. Contbr. articles to profl. jours. Grantee Royal Ministry Social Affairs, Norway, 1972, Royal Ministry Fgn. Affairs, Norway, 1986, Mid. East Eye Rsch. Inst., Israel, 1987. Mem. Norwegian Med. Assn., Norwegian Soc. Infectious Diseases, Norwegian Soc. Med. Microbiology (chmn. 1989-91), Scandinavian Soc. Genitourinary Medicine (bd. dirs. 1986—), European Soc. Chlamydia Rsch. (bd. dirs. 1987—), Am. Soc. Microbiology, European Group Rapid Viral Diagnosis, Internat. AIDS Soc. Avocations: reading, traveling, languages. Home: PO Box 2126, Kristiansand N-4602, Norway Office: Vest-Agder Ctrl Hosp, Dept Clin Microbiology, N-4604 Kristiansand Norway

CSÁSZÁR, ÁKOS, mathematician; b. Budapest, Feb. 26, 1924; s. Károly Császár and Gizella Szücs; m. Klára Cseley. Student, Budapest U. Prof. Loránd Eötvös U., Budapest, 1957—, dir. Inst. Math., 1983-86; vis. prof. Technische Universität, Stuttgart, 1975, Università di Torino, 1979, Technische Universität Graz, 1983. Author: Foundations of General Topology, 1960, General Topology, 1974, Valós analízis (Real Analysis), 1983 (2 vols.); contbr. articles to internat. prof. jours.; past mem. editorial bd. Acta Mathematica Hungarica, Periodica Mathematica Hungarica, Studia Mathematica Hungarica; mem. editorial bd. Applied Categorical Structures; chief editor Annales U. Scientiarum Budapestinensis Sectio Mathematica. Recipient Kossuth prize 1963. Mem. János Bolyai Math. Soc. (gen. sec. 1966-80, pres. 1980-90, hon. pres. 1990—), Hungarian Acad. Scis. (corr. 1970, full 1979—). •

CSENDES, ERNEST, chemist, corporate and financial executive; b. Satu-Mare, Romania, Mar. 2, 1926; came to U.S., 1951, naturalized, 1955; s. Edward O. and Sidonia (Littman) C. m. Catharine Vera Tolnai, Feb. 7, 1953; children: Audrey Carol, Robert Alexander Edward. BA, Protestant Coll., Hungary, 1944; BS, U. Heidelberg (Ger.), 1948, MS, 1950, PhD, 1951. Rsch. asst. chemistry U. Heidelberg, 1950-51; rsch. assoc. biochemistry Tulane U., New Orleans, 1952; rsch. fellow chemistry Harvard U., 1952-53; rsch. chemist organic chems. dept. E. I. Du Pont de Nemours and Co., Wilmington, Del., 1953-56, elastomer chems. dept., 1956-61; dir. rsch. and devel. agrl. chems. div. Armour & Co., Atlanta, 1961-63; v.p. corp. devel. Occidental Petroleum Corp., L.A., 1963-64, exec. v.p. rsch., engring. and devel., mem. exec. com., 1964-68; chief operating officer, exec. v.p., dir. Occidental Rsch. and Engring. Corp., L.A., London, Moscow, 1964-68; mng. dir. Occidental Rsch. and Engring. (U.K.) Ltd., London, 1964-68; pres., chief exec. officer TRI Group, London, Amsterdam, Rome and Bermuda, 1968-84; chmn., chief exec. officer Micronic Techs., Inc., L.A., 1981-85; mng. ptnr. Inter-Consult Ltd., Pacific Palisades, Calif., 1984—; internat. con. on tech., econ. feasibility and mgmt., 1984—. Contbr. 250 articles to profl. and trade jours., studies and books; achievements include 29 patents; research in area of elastomers, rubber chemicals, dyes and intermediates, organometallics, organic and biochemistry, high polymers, phosphates, plant nutrients, pesticides, process engineering, design of fertilzer plants, sulfur, potash and phosphate ore mining and metallurgy, coal burning and acid rain, coal utilization, methods for aerodynamic grinding of solids, petrochemicals, biomed. engring., consumer products, also acquisitions, mergers, internat. fin. related to leasing investment and loans, trusts and ins.; regional devel. related to agr. and energy resources. Recipient Pro Mundi Beneficio gold medal Brazilian Acad. Humanities, 1975; Harvard U. fellow, 1953. Fellow AAAS, Am. Inst. Chemists, Royal Soc. of Chemistry (London); mem. AIAA, Am. Chem. Soc., German Chem. Soc., N.Y. Acad. Sci., Am. Inst. Chem. Engrs., Am. Concrete Inst., Acad. Polit. Sci., Global Action Econ. Inst., Am. Def. Preparedness Assn., Sigma Xi. Home: 514 N Marquette St Pacific Palisades CA 90272-3314

CSERMELY, THOMAS JOHN, computer engineer, physicist; b. Szombathely, Hungary, June 25, 1931; s. Janos and Maria (Szarvas) C.; diploma in engring. Tech. U. Budapest, 1953; Ph.D., Syracuse U., 1968; m. Tiiu Vaharu, June 17, 1962; 1 son, Erik Thomas. Instr., Inst. Theoretical Physics, Tech. U. Budapest (Hungary), 1953-56; nuclear engring. cons. Design Bur. Power Stas., Budapest, 1956; research engr. Carrier Corp., Syracuse, N.Y., 1957-67; research assoc. physics Syracuse U., 1967-68, assoc. prof. elec. and computer engring., 1976-88, assoc. prof. bioengring., 1988—; asst. prof. physiology SUNY Upstate Med. Center, Syracuse, 1968-76; asst. prof. physics LeMoyne Coll., Syracuse, 1976-77. Recipient Wolverine Diamond Key award ASHRAE, 1964. Mem. Am. Phys. Soc., IEEE, Biophys. Soc., N.Y. Acad. Scis., Am. Assn. Physics Tchrs., Soc. Computer Simulation, AAAS, Sigma Xi. Club: Tech. Syracuse. Contbr. articles to profl. publs. and orgns. on control, heat exchange dynamics, quantum biochemistry, brain functions and computer simulation neuronal network dynamics, computer applications in medicine. Home: 149 Humbert Ave Syracuse NY 13224-2251

CSERNAK, STEPHEN FRANCIS, structural engineer; b. Bethlehem, Pa., Sept. 9, 1952; s. John Alexander and Mary Clara (Schultz) C.; m. Carla Diann Treadway, June 29, 1974; children: Holly Elizabeth, Amy Katherine, Mary Frances. BSCE, Clemson U., 1974, MSCE, 1976. Registered profl. engr., S.C. Structural engr. Delon Hampton & Assocs., Washington, 1976-77; structural engr. to structural dept. head Enwright Assocs., Inc., Greenville, S.C., 1977-84; dir. structural engring. Stevens & Wilkinson, Inc., Columbia, S.C., 1984-91; structural group leader O'Neal Engring., Greenville, S.C., 1991—; vis. prof. Clemson U., 1984-85. Alt. del. S.C. Dem. Party State Conv., Columbia, 1984; team capt. United Way of the Midlands, Columbia, 1989-90; Sun. sch. tchr. Our Lady of the Hills Ch., Columbia, 1988-90. Mem. ASCE (sec. com. on profl. registration 1989-93, state pres. 1982-83), NSPE, S.C. Soc. of Engrs. (pres. 1989-91, S.C. Young Engr. of the Yr. 1983), KC. Roman Catholic. Office: O'Neal Engring 850 S Pleasantburg Dr Greenville SC 29607

CSIZA, CHARLES KAROLY, veterinarian, microbiologist; b. Pacsony, Hungary, Apr. 4, 1937; came to U.S., 1966; s. Istvan and Anna (Nemeth) C.; m. Colette Marcoux, Sept. 5, 1964; children: Andrew, Kathleen, Stephen. DVM, U. Toronto, Ont., Can., 1963; PhD, Cornell U., 1970. Rsch. asst. Connaught Med. Rsch. Lab. U. Toronto, 1963-66; teaching asst. dept. microbiology N.Y. State Vet. Coll./Cornell U., Ithaca, 1966-70; rsch. scientist Wadsworth Ctr. Labs. and Rsch. N.Y. State Dept. Health, Albany, 1970—. Author: Pathogenisis of Feline Panleukopenia Virus, 1971, Myelin Deficient Mutation in Rats, 1979, Lipid Class Analysis of Myelin Deficient Rats, 1982. Mem. Am. Coll. Vet. Microbiologists, Am. Vet. Med. Assn., N.Y. Acad. Scis. Roman Catholic. Office: Wadsworth Ctr Labs/Rsch NY State Dept Health ESP Albany NY 12201-0509

CUELLO, AUGUSTO CLAUDIO GUILLERMO, medical research scientist, author; b. Buenos Aires, Argentina, Apr. 7, 1939; came to Can., 1985; s. Juan Andres and Rita Maria (Sagarra) Cuello-Freyre; m. Martha Maria J. Kaes, Mar. 10, 1967; children: Paula Marcela, Karina Rosa. MD, U. Buenos Aires, 1965; MA (hon.), Oxford (Eng.) U., 1978, DSc, 1986; hon. degree, U. Fed. do Ceará, 1992. Asst. prof. Sch. Biochemistry, U. Buenos Aires, 1974-75; scientist MRC Neurochem. Pharmacology, Cambridge, Eng., 1975-78; lectr. depts. pharmacology and human anatomy U. Oxford, 1978-85; med. tutor, E.P. Abraham sr. research fellow Lincoln Coll., Oxford, 1978-85; chmn., prof. pharmacology and therapeutics McGill Univ., Montreal, Que., Can., 1985—; cons. Seralab Ltd., Sussex, Eng., 1981-85, Sandoz Ltd., Basel, Switzerland, 1982-84, Medicorp-Immunocorp, Montreal, 1985—, Synthelabo, Paris, 1977-78, Fidia Rsch. Labs., 1991-93, UN Inst. Biotech. and Genetic Engring., Italy, 1993—; internat. advisor Cajal Inst., Madrid, 1983-88. Editor: Co-Transmission 1, 1982, Immunohistochemistry, 1983, Brain Microdissection Techniques, 1983, Substance P and Neurokinins, 1987, Immunohistochemistry II, 1993, NeuroReport, Jour. of Chem. Neuroanatomy; mem. editorial bd. profl. jours. Recipient Estela A. de Goytia prize Argentinian Assn. Advancement Sci., 1968, Prof. A. Rosenblueth award Grass Found., 1979, Robert Feulgen prize Gesellschaft für Histochemie, 1981; NIH Postdoctoral fellow, 1970-72; named Hon. Prof. faculty of Pharmacy and Biochemistry, Buenos Aires U., 1992, Norman Bethune U. Med. Scis., China., Hon. Citizen of New Orleans, 1992. Mem. Am. Soc. for Neuroscience, , Brit. Pharm. Soc., Brain Research Assn., Brit. Physiol. Soc., Can. Coll. Neuropsychopharmacology, Can. Soc. for Clin. Pharmacology, European Neurosci. Assn., Internat. Soc. Neuroendocrinology, Physiol. Soc., Soc. for Neurosci., Assn. Med. Sch. Pharmacology, Can. Assn. Neurosci., Can. Pain Soc., Internat. Soc. Neurochemistry, Basal Ganglia Internat. Soc., Am. Soc. Pharmacology and Exptl. Therapeutics, Oxford Soc., Physiol. Soc. Gt. Britain, Pharm. Soc. of Can., Gesellschaft fur histochemie, Corr. Pharma. Soc. of Argentina, Oxford Soc., Oxford and Cambridge United Club, Univ. Club Montreal. Avocations: tennis, reading, theatre, history, Spanish and Latin American literature. Office: McGill U Dept Pharmacology, 3655 Drummond St, Montreal, PQ Canada H3G 1Y6

CUEVAS, DAVID, psychologist; b. Mt. Vernon, Ohio, Sept. 28, 1947; s. Robert Myron Ryan and Wanda Mae Carter; m. Belia Cuevas, Apr. 1, 1980; 1 child, Angela Marie. AA in Gen. Edn., Lassen Coll., 1975; BS in Psychology, U. Md., 1984; MA in Psychology, Webster U., 1986; PhD in Psychology, Columbia Pacific U., 1989. Drafted USN, 1964-67; enlisted U.S. Army, 1971, advanced through ranks to sgt., specialist behavioral sci., 1966-86, ret., 1986; faculty El Paso (Tex.) Community Coll., 1986—; cons. Cuevas Cons. Svcs., El Paso, 1986—; sch. dir. Computer Career Ctr., El Paso, 1987-89; speaker conf. Fed. Women's Assn., Ft. Bliss, Tex., 1990. Advisor, vol. Paralyzed Vets. Am., El Paso, 1991—. 2d jr. vice-comdr. DAV. Mem. APA, ASCD, VFW (life). Avocations: boating, camping, travel. Home: 11140 Tenaha Ave El Paso TX 79936-1108 Office: El Paso Community Coll PO Box 20500 El Paso TX 79998-0500

CUFFEY, ROGER J., paleontology educator; b. Indpls., May 2, 1939; s. James and Rita Letitia (Paraboschi) C.; m. 1964 (div. 1987); children: Clifford, Kurt. BA, Ind. U., 1961; postgrad., U. Mich., 1962-63; MA, Ind. U., 1965, PhD, 1966. Paleontologist-stratigrapher Kans. Geol. Survey, Lawrence, 1962; geographic pathologist Armed Forces Inst. Pathology, Washington, 1965-67; asst. prof. dept. geology/geosci. Pa. State U., University Park, 1967-73, assoc. prof., 1973-79, prof., 1979—; mem. adv. panel Treatise on Invertebrate Paleontology, U. Kans., 1980-90. Contbr. approximately 200 articles and monographs to profl. jours. Capt. U.S. Army, 1965-67. Recipient Bahamian Bryozoan Reef Study grant NSF, 1978-80. Fellow Geol. Soc. Am.; mem. Internat. Bryozoology Assn. (pres. 1986-89), Paleontol. Soc. (book rev. editor Jour. Paleontology 1970-75), Soc. Vertebrate Paleontology, Soc. for Sedimentary Geology. Achievements include recognition and delineation of the roles played by ancient and modern bryozoans in reefs and reef-building and bryozoan evolutionary lineages. Office: Pa State Univ Dept Geosci 412 Deike Bldg. University Park PA 16802

CUFFNEY, ROBERT HOWARD, electro-optical engineer; b. Syracuse, N.Y., May 14, 1959; s. William Charles and Ellen Ann (Holmes) C.; m. Judith Anne DiGuiseppe, Apr. 7, 1990; children: Rachel Marie, Joseph William. BS in Physics, SUNY, Cortland, 1980; BSEE, Clarkson U., 1982; MS in Optical Engring., U. Rochester, 1986. Sr. rsch. engr. Eastman Kodak, Rochester, N.Y., 1982—. Tchr. St. Paul of the Cross Ch., Honeoye Falls, N.Y., 1990—. Mem. Optical Soc. Am. (rep. 1990—). Roman Catholic. Achievements include patents in the field of laser electr-optic applications. Office: Eastman Kodak D 309 MS35305 7 York St Honeoye Falls NY 14472-1157

CUK, SLOBODAN, engineering educator. Assoc. prof. Calif. Inst. Tech., Pasadena. Recipient Edward Longstreth medal Franklin Inst., 1991. Office: Calif Inst Tech Div Engring & Applied Sci Pasadena CA 91125*

CULBERSON, WILLIAM LOUIS, botany educator; b. Indpls., Apr. 5, 1929; s. Louis Henry and Lucy Helene (Hellman) C.; m. Chicita Forman, Aug. 24, 1953. BS, U. Cin., 1951; Diplome d'Etudes Supérieures, U. de Paris, 1952; PhD, U. Wis., 1954. NSF postdoctoral fellow Harvard U., Cambridge, Mass., 1954-55; instr. Duke U., Durham, N.C., 1955-58, asst. prof., 1958-64, assoc. prof., 1964-70, prof., 1970-84, Hugo L. Blomquist prof., 1984—; vis. research prof. Mus. Nat. d'Histoire Naturelle, Paris, 1980. Author over 100 rsch. papers. Dir. Sarah P. Duke Gardens at Duke U., Durham, 1978—. Grantee NSF, 1957—. Mem. Am. Bryological and Lichenological Soc. (pres. 1987-89), Bot. Soc. Am. (pres. 1991-92), Am. Soc. Plant Taxonomists, Mycol. Soc. Am. Avocations: greenhouse gardening. Home: Box 297 King Rd Rt 7 Durham NC 27707 Office: Duke U Dept Botany Durham NC 27706

CULBERTSON, JAMES CLIFFORD, physicist; b. Portsmouth, Ohio, Mar. 14, 1953; s. Harvey Carl and Jean (Foos) C. BS in Engring. Physics, U. Calif., Berkeley, 1975, PhD, 1983. Physicist Naval Rsch. Lab., Washington, 1983-; Contbr. artilces to profl. jours. Mem. Am. Phys. Soc. Achievements include measurement of the electron hole correlation function for electron-hole-liquid in Germanium as function of density; discovery of powerful new probe of electron-hole-liquid free exciton system used to study nucleation theory; research on ballistic phonon scattering to measure symmetry of EL2 defect in GaAs, Kosterlitz-Thouless transition in high temperature superconductors; development of optical probes of superconducting microstrip lines that allow measurement of local surface resistance and the local current density of microwave frequency superconducting currents. Office: Naval Rsch Lab Code 6874 Washington DC 20375

CULICK, FRED ELLSWORTH CLOW, physics and engineering educator; b. Wolfeboro, N.H., Oct. 25, 1933; s. Joseph Frank and Mildred Beliss (Clow) C.; m. Frederica Mills, June 11, 1960; children—Liza Hall, Alexander Joseph, Mariette Huxham. Student, U. Glasgow, Scotland, 1957-58; S.B., MIT, 1957, Ph.D., 1961. Rsch. fellow Calif. Inst. Tech., Pasadena, 1961-63, asst. prof., 1963-66, assoc. prof., 1966-70, prof. mech. engring. and jet propulsion, 1970—; cons. to govt. agys. and indsl. orgns. Fellow AIAA; mem. Am. Phys. Soc., Internat. Fedn. Astronautics (corr.). Home: 1375 Hull Ln Altadena CA 91001-2620 Office: Calif Inst Tech 1201 E California Blvd Pasadena CA 91125

CULLARI, SALVATORE SANTINO, clinical psychologist, educator; b. Caroniti, Calabria, Italy, Apr. 1, 1952; came to U.S., 1955; s. Carmelo and Carmela (Cullari) C.; m. Kathryn Plesce, Apr. 26, 1985; children: Catherine, Dante. BA, Kean Coll., 1974; MA, Western Mich. U., 1976, PhD, 1981. Lic. psychologist, Pa., W.Va. Dir. psychology White Haven (Pa.) Ctr., 1982-83; psychologist Danville (Pa.) State Hosp., 1983-84; coord. of psychology Harrisburg (Pa.) State Hosp., 1984-86; assoc. prof. Lebanon Valley Coll., Annville, Pa., 1986—; cons. Harrisburg State Hosp., 1986—, Bur. Disability Determination, Harrisburg, 1987—. Author acad. questionnaire acad. social evaluation scales, 1990; contbr. numerous articles to profl. jours. Mem. APA, Assn. Advancement of Behavior Therapy, Assn. Behavior Analysts. Office: Lebanon Valley Coll Psychology Dept Annville PA 17003

CULLEN, ERNEST ANDRÉ, chemist, researcher; b. Saint-Angèle, Que., Can., Jan. 11, 1926; came to U.S., 1978; s. John P. and Alma M. (Cyr) C.; m. Clairette M. Langlois, Sept. 16, 1954; children: Peter, Johanne. BA, Laval U., Gaspé, Que., 1950, BSc, 1955, PhD, 1960. Postdoctoral fellow Worcester (Mass.) Found., 1960-62; rsch. chemist Can. Armament R&D Establishment, Que., 1962-63, Delmar Chem., LaSalle, Que., 1963-65; sr. prin. scientist Boehringer Ingelheim Pharms., Inc., Ridgefield, Conn., 1965—. Author: Anti-Viral; Anti-Inflammation, 1992, Anti-Viral, 1991; contbr. articles to profl. jours. Achievements include research in immunomodulators, anti-inflammation, anti-viral agents (AIDS). Home: Taunton Lake Rd Newtown CT 06470

CULLEN, JOHN KNOX, JR., hearing science educator; b. Denver, Jan. 2, 1936; s. John Knox and Mary Register (Prather) C.; m. Judith Anne Dumars, Sept. 7, 1957 (div. Apr. 1977); children: Janice Lynn, Raymond Charles; m. Deirdre Mary Rafferty, May 17, 1987. BSEE, U. Md., 1960, MS, 1969; PhD, La. State U., 1975. Engr. NIH, Bethesda, Md., 1960-64; instr. The Johns Hopkins Univ. Sch. Medicine, Balt., 1964-68; prof. La. State U. Sch. Medicine, New Orleans, 1968-86; instr. acoustics, physiology and psychoacoustics La. State U., Baton Rouge, 1986—; Contbr. chpts. to books and articles to profl. jours. Mem. Preservation Resource Ctr., New Orleans, 1987, Irish Channel Neighborhood Assn., New Orleans. Recipient Spl. fellowship NIH, New Orleans, 1975, Rsch. grants NIH, New Orleans, 1977-91, Rsch. grant La. Bd. Regents, Baton Rouge, 1990, Rsch. fellowship Philips Corp., Eindhoven, Netherlands, 1991. Mem. AAAS, Acoustical Soc. Am., Assn. for Rsch. in Otolaryngology, Sigma Xi (La. State U. Med. Ctr. club pres. 1983-84). Achievements include speech, complex acoustic signal perception by humans, evoked auditory potentials. Office: Music & Dramatic Arts Bldg Rm 163 La State Univ Baton Rouge LA 70803

CULLEN, MARION PERMILLA, nutritionist; b. Pitts., Dec. 3, 1931. BS, Pa. State U., 1954, MS, 1956; PhD, U. Calif., Berkeley, 1968. Biochemist Am. Meat Inst. Found., Chgo., 1956-60; predoctoral fellow Harvard U. Sch. Pub. Health, Boston, 1960-62; rsch. asst., rsch. assoc. U. Calif., Berkeley, 1962-68; fellow Cancer Rsch. Sch. Medicine Johns Hopkins U., Balt., 1969-74; sr. rsch. scientist N.Y. State Dept. Health Div. Labs. and Rsch., Albany, 1971-76; asst. prof. biochemistry Albany Med. Sch., 1971-76; exec. adminstr. cons. Subsitarary U.S.A. Corp., London, 1976-81; sr. scientist, project officer pediatric renal div. Guy's Hosp., London, 1981-85; health scientist cons. Washington, 1985-87; nutrition data liaision Tenn. State Dept. Health Maternal and Child Health, Nashville, 1988—; Prin. investigator, biochemist Albany Med. Sch., 1974-76, mem. coun. setting up nutrition program for med. students, 1974-76; cons. set up for renal study Harvard U., Boston, 1986. Contbr. articles to profl. jours. Elder Presbyn. Ch., Albany, 1971-76. Predoctoral fellow Harvard U., Boston, 1960-62; fellow Finney Howell Lab. Johns Hopkins U., Balt., 1969-71; travel grantee Fisson Pharm., Crewe, Eng., 1984, Marks & Spencers, London, 1984. Fellow N.Y. Acad. Sci.; mem. AAAS, Am. Chem. Soc., Royal Soc. Medicine, Harvard Alumni Assn. Achievements include research in lab. assay tech. as related to viral transmission, functional aspects of nutrients as applied to different age groups, role of foods in cancer prevention, proposed, designed and brought on stream a high pressure liquid chromatography system to assess the functional status of kidney pore size, in premature infants; developed and instituted a State wide prenatal surveillance system. Home: 865 Bellevue Rd C 18 Nashville TN 37221

CULLER, FLOYD LEROY, JR., chemical engineer; b. Washington, Jan. 5, 1923; s. Floyd Leroy Culler; m. Della Hopper, 1946; 1 son, Floyd Leroy III. B. Chem. Engring. cum laude, Johns Hopkins, 1943. With Eastman Kodak and Tenn. Eastman at Y-12, Oak Ridge, 1943-47; design engr. Oak Ridge Nat. Lab., 1947-53, dir. chem. tech. div., 1953-64, asst. lab. dir., 1965-70, dep. dir., 1970-77; pres. Electric Power Research Inst., Palo Alto, Calif., 1978-88, pres. emeritus, 1988—; research design chem. engring. applied to atomic energy program, chem. processing nuclear reactor plants, energy research. Mem. sci. adv. com. Internat. Atomic Energy Agy., 1974-87; mem. energy research adv. bd. Dept. Energy, 1981-86. Recipient Ernest Orlando Lawrence award, 1964; Atoms for Peace award, 1969, Robert E. Wilson award in nuclear chem. engring., 1972, Engring. Achievement award East Tenn. Engrs. Joint Council, 1974, Outstanding Scientist award State of Tenn., 1988. Fellow AAAS, Am. Nuclear Soc. (dir. 1973-80, spl. award 1977, Walter Zinn award 1988), Am. Inst. Chemists, Inst. Chem. Engrs.; mem. Am. Chem. Soc., Nat. Acad. Engring. Office: Electric Power Rsch Inst Inc 3412 Hillview Ave Box 10412 Palo Alto CA 94303

CULLUM, COLIN MUNRO, psychiatry and neurology educator; b. Freeport, Tex., Mar. 28, 1959; s. Colin George and Helen Mae Cullum. BA, Pacific Luth. U., 1981; PhD, U. Tex. at Austin, 1986. Postdoctoral fellow U. Calif., San Diego, 1986-88, asst. rsch. neuropsychologist, 1988-89; asst. prof. psychiatry and neurology U. Colo. Sch. of Medicine, Denver, 1989—; consulting neuropsychologist Fort Lyon VA Med. Ctr., Las Animas, Colo., 1989—. Fellow Nat. Acad. Neuropsychology (exec. sec. 1992—); mem. Internat. Neuropsychol. Soc., Am. Psychol. Assn., Am. Psychol. Soc., Assn. Med. Sch. Profs. Psychology. Achievements include research in clinical neuropsychology of memory disorders, neuropsychological functions in aging and dementia, neuropsychological and neuroimaging correlates in neurological and psychiatric disorders. Office: Univ Colo Sch Medicine Univ North Pavilion 4455 E 12th Ave Denver CO 80220

CULTER, JOHN DOUGHERTY, chemical engineer; b. Tulsa, May 1, 1937; s. Gail Curtis and Virginia Belle (Dougherty) C.; m. Patricia Ann Woodall, June 20, 1957 (div. 1968); children: Karen Gail, Shawna Lynn; m. Shirley Arlene Seiler, Nov. 27, 1968; children: Judith Ann Leavesley, Kirk Harris Leavesley. BS in Petroleum Engring., U. Tulsa, 1959, MS in Petroleum Engring., 1961; PhD in Chem. Engring., U. Mo., Rolla, 1976. Rsch. engr. Continental Oil Co., Ponca City, Okla., 1960-69; sr. rsch. engr. Am. Enka Co., Ashville, N.C., 1973-76; mgr. polymer and converting tech. St. Regis Corp., West Nyack, N.Y., 1976-84; prin. scientist Gen. Mills, Inc., Mpls., 1984—; pres. Advanced Material Engring. Inc., Edina, Minn., 1988—; cons. on packaging materials and polymer processing. Contbr. articles to profl. jours.; patentee in field. Sgt. USAF, 1960-61. Mem. TAPPI, Soc. Rheology, Am. Chem. Soc., Soc. Plastics Engrs. Avocations: sailing, electronics, computers, hunting, fishing. Home: 7116 Gleason Rd Edina MN 55439 Office: Gen Mills Inc 9000 Plymouth Ave N Minneapolis MN 55427-3899

CULTON, SARAH ALEXANDER, psychologist, writer; b. Burwell, Nebr., Nov. 12, 1927; d. James Claude and Frances Ann (Evans) Alexander;m. Verlen Ross Culton, June 19, 1949; children: James Verlen, Sarah Ann. BA in Edn., Ea. Wash. U., 1953, MA in Edn., 1956; EdD in Psychology, U. Idaho, 1966. Tchr. pub. schs. Kennewick, Northport, Wash., Potlatch, Idaho, 1946-56; prof. Lewis-Clark U. of Idaho, Lewiston, 1956-59, North Idaho Jr. Coll., Coeur d'Alene, 1961-66; sch. psychologist Sch. Dist. 81, Spokane, Wash., 1966-67; prof. psychology Spokane Falls Community Coll., 1967-88; author Colville, Wash., 1988—; sch. psychologist, sch. counselor vol. Northport Schs., 1989-92; presenter convs. in field. Author: Psychology of Stress and Nutrition, 1991. Doctoral fellow Wash. State U., 1959, U. Idaho, 1964; recipient Faculty Achievement award Burlington No. Found., 1988. Fellow Am. Inst. Stress; mem. NEA, APA, Intenat. Coun. Psychologists, Internat. Stress Mgmt. Assn. (editor newsletter), Nat. Stroke Assn., Western Psychol. Assn., Am. Counseling Assn. (writer invitation 1992), Alpha Delta Kappa. Avocations: travel, painting, photography, genealogy, writing. Home and Office: 717 Prouty Corner Loop Colville WA 99114

CULVER, DAVID ALAN, aquatic ecology educator; b. Oak Ridge, Tenn., Feb. 14, 1945; s. Joseph Simpson and Ella Elizabeth (Smart) C.; m. Virginia Ruth Nagel, Sept. 7, 1967; children: Timothy David, Cynthia Diane. BA, Cornell U., 1967; MS, U. Wash., 1969, PhD, 1973. Asst. prof. biology dept. Queen's U., Kingston, Ont., Can., 1973-75; asst. prof. zoology dept. Ohio State U., Columbus, 1975-81, assoc. prof., 1981—, co-dir. young scholars minority program in biology, 1988—; vis. scientist U. Adelaide, Australia, 1984-85; mem. exec. com. Ohio Sea Grant Coll. Program, Columbus, 1983—; dir. tchrs. tng. program Howard Hughes Med. Inst., Columbus, 1989—; mem. Rsch. Group on Zebra Mussels in Gt. Lakes Basin, Columbus, 1989—; cons. Va. Electric Power Co., Mt. Storm, W.Va., 1991—. Contbr. articles to sci. jours. With U.S. Army, 1969-71. Fellow Coop. Inst. for Limnological and Ecosystems Rsch., NOAA; numerous rsch. grants, including U.S. Dept. Interior, 1976-82, EPA, 1977-78, Dept. Commerce, 1980-83, 92—, Ohio Dept. Natural Resources and Ohio Biol. Survey, 1982-83, U.S. Fish and Wildlife Svc., 1983-84, 87-92, 93—, Ohio Sea Grant Program, 1983-92, North Cen. Regional Aquaculture Ctr., USDA, 1990-91; also tng. grants. Mem. Ecol. Soc. Am., Am. Soc. Limnology and Oceanography, Am. Fisheries Soc., Internat. Assn. for Gt. Lakes Rsch., Internat. Assn. for Theoretical and Applied Limnology, Australian Soc. for Fish Biology, Sigma Xi. Democrat. Avocations: boating, fishing, hiking. Office: Ohio State U Dept Zoology 1735 Neil Ave Columbus OH 43210

CULVERN, JULIAN BREWER, chemist, educator, writer-naturalist; b. July 23, 1919; m. Shirley Bowman, 1946; children: Janine Amelia, David Bowman, Linda Hazel. BS, N.C. State U., 1942; MSc, Ohio State U., 1948; postgrad., U. Tenn., 1970-72. Assay chemist Haile Gold Mine, 1940-41; asst. mgr. Chem. & Microscopical Lab., 1949-61; sr. process engr. Am. Enka Corp., Lowland, Tenn., 1961-69; instr. gen. chemistry, earth and space sci., environ. sci. Morristown (Tenn.) Coll., 1969-76, chmn. div. natural sci., 1969-73.; conducted libr. rsch. in field sci. and religion Templeton Found., 1970; chemist atomic bomb project Corps of Engrs., Manhattan Dist., Oak Ridge, Tenn. Columnist Daily Gazette-Mail, Morristown, 1960-74; contbr. articles to Sci. of Mind mag., others. Chmn. Cherokee dist. Boy Scouts Am.; ruling elder 1st Presbyn. Ch., Morristown, Tenn., Marshall, N.C. With U.S. Army, 1944-46. Mem. AAUP, Am. Chem. Soc., Tenn. Acad. Scis., Gamma Sigma Epsilon, Phi Lambda Upsilon. Home: Birdsong Hill 2832 Indian Trail Morristown TN 37814-5824

CUMINGS, EDWIN HARLAN, biology educator; b. Washington, Mar. 2, 1933; s. Glenn Arthur and Winifred (Wenkheimer) C. AB magna cum laude, DePauw U., Greencastle, Ind., 1954; AM in Zoology, Dartmouth Coll., 1956; AM in Biology, Harvard U., 1958, postgrad., 1978—. Teaching asst. in zoology DePauw U., 1953-54; teaching fellow in Zoology Dartmouth Coll., Hanover, N.H., 1954-56; teaching fellow in Biology Harvard U., Cambridge, Mass., 1956-57; tutor in zoology George Washington U., Washington, 1963-64, teaching assoc. in zoology, 1964; lectr. in Biology Boston U., 1965-66, 67, 68-73; instr. in Biology Northeastern U., Boston, 1966-68; asst. in Biology Harvard U., Cambridge, 1976-77. Rector scholar DePauw U., 1950-54; Jeffries Wyman scholar in Anatomy Harvard U., 1956-57; predoctoral research fellow NIH, 1957-61. Mem. AAAS, AAUP, Am. Mus. Natural History, N.Y. Smithsonian Inst., Mus. Fine Arts, Harvard U. Art Mus., N.Y. Acad. Scis., Soc. Study Amphibians and Reptiles, Nat. Geog. Soc., New England Hist. Geneal. Soc., Faculty Club, Harvard, Phi Sigma, Sigma Xi, Phi Eta Sigma, Phi Beta Kappa. Democrat. Presbyterian. Avocations: foreign languages. Home: 28 Wendell St Apt 4 Cambridge MA 02138-1825

CUMMING, JANICE DOROTHY, clinical psychologist; b. Berkeley, Calif., Nov. 20, 1953; d. Gordon Robertson and Helen (Stanford) Cumming; m. Philip J. Keddy, Aug. 2, 1985; 1 child, Shauna Cumming Keddy. BA, U. Calif., Davis, 1975; MA, Calif. State U., Sacramento, 1980; PhD, Calif. Sch. Profl. Psychology, Berkeley, 1985. Lic. psychologist, Calif. Counselor and instr. Serendipity Diagnostic/Treatment, Citrus Hts., Calif., 1978-79; reg. psychologist asst. John Gibbins, PhD, Castro Valley, Calif., 1984-87, Enrico Jones, PhD, Berkeley, Calif., 1985-86; asst. rsch. specialist U. Calif., Berkeley, 1985-90; clin. cons. Family Guidance, Children's Hosp., Oakland, Calif., 1987-90; clin. supr. psychiat. svcs. Children's Hosp., San Francisco, 1987-90; pvt. practice psychology Castro Valley, 1987-90, San Francisco, 1987-90, Oakland, Calif., 1990—; mem. researcher San Francisco Psychotherapy Rsch. Group, 1987—, instr., 1991, conf. chair, 1992-93; conv. chair Calif. State Psychol. Assn., Sacramento, 1985, 86; lectr. U. Calif., San Francisco, 1992—. Mem. APA, Calif. Psychol. Assn. (continuing edn. com. chair 1986, co-chair 1987), Psychologists for Social Responsibility (bd. dirs. 1984-87, chair 1987-89), Soc. for Psychotherapy Rsch., No. Calif. Soc. for Psychoanalytical Psychology, Alameda County Psychol. Assn., Phi Beta Kappa. Avocations: walking, gardening, reading, piano. Office: 5625 College Ave Ste 310 Oakland CA 94618-1585

CUMMINGS, BELINDA, construction engineer; b. Garden City, Kans., June 27, 1963; d. Everett Lee and Karin Kaye (Coerber) Glenn; m. Bryan Cummings, Aug. 14, 1993. BS in Constrn. Sci., Kans. State U., 1986. Surveyor Coleman Indsl. Constrn. Co., Wichita, Kans., 1986-87, foreman, 1987; field engr. Herzog Contracting Corp., L.A., 1987-89; project engr. Herzog Contracting Corp., Long Beach, Calif., 1989-90, Sacramento, 1990-92; asst. constrn. engr. Sacramento (Calif.) Regional Transit Dist., 1992—. Mem. Kans. State U. Alumni Assn. Lutheran. Avocations: swimming, softball. Home: PO Box 1038 Gridley CA 95948-1038 Office: Sacramento Regional Transit Dist 2811 O St Sacramento CA 95816

CUMMINGS, JOHN CHESTER, JR., research professional; b. San Diego, May 27, 1947; s. John Chester Sr. and Mary Lucy (Wino) C.; m. Ellen Mary Curtin, Aug. 17, 1968; children: Brian Lodi, Scott Michael. BS, Calif. Inst. Tech., 1969, PhD, 1973. Mem. tech. staff TRW Inc., Redondo Beach, Calif., 1973-75; Mem. tech. staff Sandia Nat. Labs., Albuquerque, 1975-83, div. supr., 1983-90, dept. mgr., 1990—; co-chmn. Div. Fluid Dynamics Local Meeting Com., Albuquerque, 1990-93. Contbr. articles to profl. jours. Youth tchr. Our Lady of Annunciation Ch., Albuquerque, 1976-92; coach Am. Youth Soccer Assn., Albuquerque, 1976-82; pres. Duke City Soccer League, Albuquerque, 1986-88. GM scholar, 1965-69; Nat. Defense Rsch. fellow, 1969-73; recipient D.S. Clark award, 1967. Mem. Am. Phys. Soc. (exec. com. 1979-80), Tau Beta Pi, Sigma Xi. Roman Catholic. Achievements include demonstration of very high mach number shocks in a cryogenic shock tube, demonstration of strong second sound shocks in liquid helium. Home: 3008 Vermont NE Albuquerque NM 87110

CUMMINGS, NICHOLAS ANDREW, psychologist; b. Salinas, Calif., July 25, 1924; s. Andrew and Urania (Sims) C.; m. Dorothy Mills, Feb. 5, 1948; children—Janet Lynn, Andrew Mark. AB, U. Calif., Berkeley, 1948; MA, Claremont Grad. Sch., 1954; PhD, Adelphi U., 1958. Chief psychologist Kaiser Permanente No. Calif., San Francisco, 1959-76; pres. Found Behavioral Health, San Francisco, 1976—; chmn., chief exec. officer Am. Biodyne, Inc., San Francisco, 1985-93; chmn., CEO Kendron Internat., Ltd., Reno, Nev., 1993—, Reno, 1992—; co-dir. South San Francisco Health Ctr., 1959-75; pres. Calif. Sch. Profl. Psychology, Los Angeles, San Francisco, San Diego, Fresno campuses, 1969-76; chmn. bd. Calif. Community Mental Health Ctrs., Inc., Los Angeles, San Diego, San Francisco, 1975-77; pres. Blue Psi, Inc., San Francisco, 1972-80, Inst. for Psychosocial Interaction, 1980-84; mem. mental health adv. bd. City and County San Francisco, 1968-75; bd. dirs. San Francisco Assn. Mental Health, 1965-75; pres., chmn. bd. Psycho-Social Inst., 1972-80; dir. Mental Rsch. Inst., Palo Alto, Calif., 1979-80; pres. Nat. Acads. of Practice, 1981-93. Served with U.S. Army, 1944-46. Fellow Am. Psychol. Assn. (dir. 1975-81, pres. 1979); mem. Calif. Psychol. Assn. (pres. 1968). Office: Kendron Internat Ltd 561 Keystone Ave Ste 212 Reno NV 89503

CUMMINS, NANCYELLEN HECKEROTH, electronics engineer; b. Long Beach, Calif., May 22, 1948; d. George and Ruth May (Anderson) Heckeroth; m. Weldon Jay, Sept. 15, 1987; stepchildren: Tracy Lynn, John Scott, Darren Elliott. Student avionics, USMC, Memphis, 1966-67. Tech. publ. engr. Lockheed Missile and Space Div., Sunnyvale, Calif., 1973-76, engring. instr., 1977; test engr. Gen. Dynamics, Pomona, Calif., 1980-83; quality assurance test engr. Interstate Electronics Co., Anaheim, Calif., 1983-84; quality engr., certification engr. Rockwell Internat., Anaheim, 1985-86; sr. quality assurance programmer Point 4 Data, Tustin, Calif., 1986-87; software quality assurance specialist Lawrence Livermore Nat. Lab., Yucca Mountain Project, Livermore, Calif., 1987-89, software quality mgr., 1989-90; sr. constrn. insp. EG&G Rocky Flats, Inc., Golden, Colo., 1990, sr. quality assurance engr., 1991, engr. IV software quality assurance, 1991-92, instr., developer environ. law and compliance, 1992—; customer engr. IBM Gen. Systems, Orange, Calif., 1979; electronics engr. LDS Ch. Exhibits Div., Salt Lake City, 1978; electronics repair specialist Weber State Coll., 1977-78. Author: Package Area Test Set, 6 vols., 1975, Software Quality Assurance Plan, 1989. Vol., instr. San Fernando (Calif.) Search and Rescue Team, 1967-70; instr. emergency preparedness and survival, Clairmont, Calif., 1982-84, Modesto, Calif., 1989; mem. Lawrence Livermore Nat. Lab. Employees Emergency Vols., 1987-90, EG&G Rocky Flats Bldg. Emergency Support Team, 1990—. Mem. NAFE, NRA, Nat. Muzzle Loading Rifle Assn., Am. Soc. Quality Control, Job's Daus. (majority mem.). Republican. Mem. LDS Ch. Avocations: living history, survival, weapons, camping, native Am. crafts. Home: PO Box 334 2282 Country Rd 87 Jamestown CO 80455-0334 Office: EG&G Rocky Flats Inc PO Box 464 Golden CO 80402-0464

CUNEFARE, KENNETH ARTHUR, mechanical engineer, educator; b. Decatur, Ill., Jan. 11, 1961; s. Jerry Lee and Patricia Louise (Nix) C.; m. Stephanie Ann Rosson, Aug. 21, 1982; 1 child, Taylor Jayne. BS in Mech. Engring., U. Ill., 1982; MS in Acoustical Engring., U. Houston, 1987; PhD, Pa. State U., 1990. Engr. in tng. Project engr. Exxon Gas System, Inc., Houston, 1982-84; sr. project engr. Exxon Co. U.S.A., Midland, Tex., 1984-86; sr. engr. Exxon Co. U.S.A., Houston, 1986-87; rsch. fellow U. Houston, 1987-88, Pa. State U., State College, 1988-90, asst. prof. Ga. Inst. Tech., Atlanta, 1990—; cons. Pa. Power & Light, 1989, Sheesly Cement Co., State Coll., 1990, Von Roll Inc., Savannah, Ga., 1992. Contbr. articles to profl. jours. NASA GSRP fellow, 1987-90. Mem. ASME, AIAA, Acoustical Soc. Am. (F.V. Hunt fellow 1990-91). Achievements include development of boundary element method for active noise control, exterior acoustic modal representation, finite element/boundary element method for acoustic radiation, boundary element method for design sensitivity. Office: Ga Inst Tech Mech Engring 219 SSTC Atlanta GA 30332-0405

CUNHA-VAZ, JOSE GUILHERME FERNANDES, ophthalmologist; b. Coimbra, Portugal, Nov. 5, 1938; s. Antonio Manso and Maria Izilda (Calado) Cunha-Vaz; m. Teresa Maria Dinis Coutinho, Dec. 6, 1962; chldren: Ricardo Jose, Maria Cecilia, Eduardo Henrique. MD, U. Coimbra, 1962, D Med Scis., 1967; PhD, U. London, 1966. Rsch. and clin. asst. Inst. Ophthalmology, Morfields Eye Hosp., London, 1963-66; asst. prof. ophthalmology U. Coimbra, 1966-74, assoc. prof., chmn. dept., 1974-79, dean faculty medicine, 1976; prin. investigator L.F.E.N. Biology, Lisbon, Portugal, 1968-69; prof. ophthalmology U. Ill., U. Ill. Hosp., Chgo., 1979-81, 84-86; prof. ophthalmology, chmn. dept. ophthalmology U. Coimbra and Hosp., 1981; dir. Ctr. for Ophthalmology Rsch., Coimbra, 1975—; pres. Assoc. Biomed. Inst. Rsch. in Light and Image, Coimbra, 1989—; adj. prof. U. Ill., Chgo., 1986—. Editor Experientia Ophthalmologica, Jour. Francais D'Ophthalmologie, Internat. Ophthalmology, others; editor 3 books; co-inventor Vitreous Fluorophotometry, 1975; contbr. articles to sci. jours., chpts. to books. Mem. Coordinating Commn. Health Rsch., Jnict, Portugal, 1990—, Nat Coun. Sci. and Tech., Portugal, 1990—; Portuguese del. to European Community Biomed. Rsch. Program, 1991—. Lt. Portugal mil., 1969-71. Recipient medicine prize U. Coimbra, 1962, prize Inst. Alta Cultura, 1965, prize Soc. Med. Scis. Lisbon, 1975; Internat. rsch. scholar Prevention of Blineness; NIH rsch. grantee, 1979, 81, 84, 89, Junta Nacional Investigacao Cientif, Portugal. Fellow Am. Acad. Ophthalmology, Coll. of Ophthamologists (U.K.); mem. Portuguese Soc. Ophthalmology (pres. 1981-83, 90-92), Internat. Soc. Ocular Fluorometry (pres. 1992—), European Soc. Engring. and Medicine (bd. dirs. 1991—), European Soc. Cataract Refractive Surgery (bd. dirs. 1991—), Easd-Diabetic Eye Complications (v.p. 1989), Internat. Ocular Microsurgery Study Group Club Jules Gonin. Roman Catholic. Home: R Penedo da Saudade 30, 3000 Coimbra Portugal Office: Hosp U Coimbra, Dept Ophthalmology, Coimbra Portugal

CUNKELMAN, BRIAN LEE, mechanical engineer; b. Indiana, Pa., Jan. 27, 1969; s. Paul Robert and Sally Ann (Clawson) C. ScBME, Brown U., 1991. Design/test engr. Westinghouse Air Brake Co., Wilmerding, Pa., 1990—. Mem. ASME, NSPE, AIAA, Pa. Soc. Profl. Engrs. (apprenticeship com. 1992—), Air Brake Assn., Pitts. Soc. Profl. Engrs. (rec. sec.). Democrat. Methodist. Home: 710 Air Brake Ave Wilmerding PA 15148 Office: Westinghouse Air Brake Co Staton St Wilmerding PA 15148

CUNNINGHAM, ATLEE MARION, JR., aeronautical engineer; b. Corpus Christi, Aug. 17, 1938; s. Atlee Marion and Carlos Dean (Shepherd) C.; m. Diana Wahl Bonelli, July 17, 1976; children by previous marriage: Christopher Atlee, Scott Patrick, Sean Michael. BS in Mech. Engring., U. Tex., 1961, MS in Mech. Engring., 1963, PhD, 1966. Research scientist Def. Research Lab., Austin, Tex., 1965; engring. staff specialist Gen. Dynamics Corp., Ft. Worth, 1965-93, Lockheed Corp., Ft. Worth, 1993—; vis. indsl. prof. So. Meth. U. Inst. Tech., Dallas, 1969-70; vis. assoc. prof. aero. engring. U. Tex., 1978—; lectr. in aeroelasticity Nat. Cheng Kung U., Taiwan, 1984, U. Tex., Arlington, 1990—; cons. NASA, USAF, USN, U. Tex. Vice pres. Tex. Fine Arts Assn., Fort Worth, 1972. Served with USN, 1962-64. Welding Rsch. Assn. fellow, 1961-62; NATO fellow, 1964-65; recipient NASA Cert. of Recognition for tech. publ., 1980, Extraordinary Achievement award Gen. Dynamics, 1980, 83, 89. Fellow AIAA (assoc.; tech. reviewer jours.); mem. Sigma Xi. Contbr. articles to profl. jours. and AGARD publs.; innovator in subsonic, transonic and supersonic steady and oscillatory aerodynamics method; developer new methods for predicting high angle of attack aerodynamics in subsonic and supersonic flows. Major contbr. to aeroelastic developments and improvements for Gen. Dynamics F-16 and F-111 aircrafts. Pioneer in new technology development for unsteady separated flows and buffeting on aircraft maneuvering at high angle of attack involving support of Air Force, Navy, NASA, National Aerospace Laboratory (Netherlands), General Dynamics and University of Texas at Austin. Developer of steady and unsteady force testing techniques for aerodynamic investigations using water tunnels. Home: 4932 Black Oak Ln Fort Worth TX 76114-2936

CUNNINGHAM, DOROTHY JANE, physiology educator; b. Jersey City, Nov. 7, 1927; d. John Henry and Alice Geraldine (McCabe) C. A.B., Caldwell Coll., 1949; M.S., Cath. U., 1951; Ph.D., Yale U., 1966. Asst. prof. dept. biol. scis. Montclair (N.J.) State Coll., 1958-62; postdoctoral research fellow Yale U. Sch. Medicine, New Haven, 1966-67; lectr. environ. physiology Yale U. Sch. Medicine, 1967-69, asst. prof. epidemiology (environ. physiology), 1969-70; assoc. prof. physiology Sch. Health Scis., Hunter Coll., CUNY, 1970-75, prof., 1975—; lectr. div. environ. medicine, dept. community medicine Mt. Sinai Sch. Medicine, 1971-82; adj. prof. environ. medicine Inst. Environ. Medicine, N.Y. U. Med. Center, 1981-92; research affiliate Yale U. Sch. Medicine, 1985—; chmn. com. on Grad. Sch., Yale U. Council, 1981-86; vis. fellow environ. health Yale U. Sch. Medicine and John B. Pierce Found. Lab., 1982-83. Editorial bd.: The Sciences, 1978-83. Trustee Caldwell Coll, 1971-77. Fellow N.Y. Acad. Scis. (v.p. 1977-81); mem. Am. Physiol. Soc., Harvey Soc., Am. Fedn. for Clin. Research, AAAS, AAUP, Yale Sci. and Engring. Assn. (exec. bd. 1975—, v.p. met. dist. 1982—), Assn. of Yale Alumni (gov. 1979-82, Yale medal 1985), Sigma Xi (pres. Hunter Coll. chpt. 1980-82, grants-in-aid of research com. 1979-84). Office: CUNY Hunter Coll Sch Health Scis 425 E 25th St New York NY 10010-2590

CUNNINGHAM, JOHN RANDOLPH, systems analyst; b. Alexandria, La., July 17, 1954; s. John Adolphus and Zelma Audrey (Cox) C.; m. Teresa Ellen Toms, Jan. 22, 1977. BS in Computer Sci., La. Tech. U., 1976. Customer support specialist South Ctrl. Bell Tel. Co., New Orleans, 1977-81; data communication designer Weyerhaeuser, Tacoma, 1981-87, acct. rep., 1987-89, planning mgr., 1989—; mem. adv. bd. U. Wash., Seattle, 1989—. Contbr. articles to profl. jours. Vol. Big Bros., Tacoma, 1989—. Mem. Computer and Automated Systems Assn. (treas. 1991), Indsl. Computing Soc., Instrument Soc. Am. Republican. Baptist. Home: 319 SW 328th St Federal Way WA 98023

CUNNINGHAM, KEITH ALLEN, II, computer services company executive; b. Belington, W.Va., Aug. 1, 1948; s. Keith A. and Jeanne Antionette (Viquesney) C.; m. Barbra Anne McCoy, 1991. Student Oakland U., 1972; BA, Mich. State U., 1973. Asst. mgr. Joseph Lucas N.Am., Inc., Detroit, 1973; distbn. ctr. mgr. Lucas Industries, Inc., San Francisco, 1974-78; export mgr. Primark, Inc., San Mateo, Calif. and Reno, Nev., 1978-79; operational planning mgr. United Nuclear Corp., Falls Church, Va., 1979-80; pres., chief exec. officer, dir. Unicore, Inc. subs. United Nuclear Corp., North Haven, Conn., 1980-82; dir. bus. devel. UNC Resources, 1982; pres., chief exec. officer, dir. UNC Teton Exploration Drilling Co., a UNC Resources co., 1982-84; prin., dir. Atlis Systems, Inc., Vienna, Va., v.p. fin. and adminstrn., chief fin. officer, 1985-87, pres., chief oper. officer, 1987—; dir. McFarland Graphics and Design, Dilsburg, Pa., 1989. Mem. Mich. State U. Alumni Assn. Home: 6 Coral Gables Ct N Potomac MD 20878 Office: Atlis Systems Inc 6011 Executive Blvd Rockville MD 20852-3804

CUNNINGHAM, LEMOINE JULIUS, physicist; b. Carthage, Mo., Feb. 8, 1934; s. LeMoine Edward Cunningham and Thelma Ann (Meredith) Rogers; m. Harriet Grace McNerney, Aug. 20, 1955; children: Michael Jay, Julia Elizabeth Cunningham Weir, Harriet McNerney Cunningham Coleman, Charles Martin. AB, U. Mo., 1956; postgrad., Kans. U., 1958-62. Rsch., teaching asst. radiation biophysics dept. Kans. U., Lawrence, 1958-62; health physicist Phillips Petroleum Co., Idaho Falls, Idaho, 1962-63; supr. Westinghouse, Idaho Falls, Idaho, 1963-73; br. chief U.S. Nuclear Regulatory Com., Rockville, Md., 1973—. With U.S. Army, 1956-58. Named Health Physicist of Yr., 1971. Mem. Health Physics Soc., Balt./Washington Chpt. Health Physics Soc., Eastern Idaho Chpt. Health Physics Soc. Home: 8 Cullinan Dr Gaithersburg MD 20878 Office: US Nuclear Regulatory Commn Washington DC 20555

CUNNINGHAM, R. WALTER, venture capitalist; b. Creston, Iowa, Mar. 16, 1932; s. Walter Wilfred and Gladys (Backen) C.; children from previous marriage: Brian Keith, Kimberly Ann. B.S. in Physics, UCLA, 1960, M.A., 1961; advanced mgmt. program, Harvard Grad. Sch. Bus., 1974. Research asst. Planning Research Corp., Westwood, Calif., 1959-60; physicist RAND Corp., Santa Monica, Calif., 1960-64; astronaut NASA, 1964-71; crew member of first manned Apollo spacecraft Apollo 7; sr. v.p. Century Devel., 1971-74; pres. Hydrotech Devel. Co., Houston, 1974-76; sr. v.p. 3D/International, Houston, 1976-79; founder The Capital Group, Houston, 1979-86; mng. ptnr. Genesis Fund, 1986—; bd. dirs. numerous tech. based cos. Author: The All American Boys, 1977. Judge Rolex awards for enterprise, 1984. Served with USNR, 1951-52; as fighter pilot USMCR, 1952-56; col. Res., ret. Recipient NASA Exceptional Service medal, also; Haley Astronautics award; Profl. Achievement award U. Calif. at Los Angeles Alumni, 1969; Spl. Trustee award Nat. Acad. Television Arts and Scis., 1969; medal of valor Am. Legion, 1975; Outstanding Am. award Am. Conservative Union, 1975; named to Internat. Space Hall of Fame, Houston Hall of Fame. Fellow Am. Astronautical Soc.; mem. Soc. Exptl. Test Pilots, Am. Inst. Aeros. and Astronautics, Assn. Space Explorers-U.S.A., Am. Geophys. Union, Sigma Pi Sigma. Office: Acorn Ventures Inc 520 Post Oak Blvd Ste 130 Houston TX 77027-9405

CUNNINGHAM, RAYMOND LEO, research chemist; b. Easton, Ill., Jan. 5, 1934; s. Raymond J. and Minnie G. (Vaughn) C. BA, St. Ambrose U., Davenport, Iowa, 1955. Phys. sci. aid in chemistry Nat. Ctr. Agrl. Utilization Rsch USDA Agrl. Rsch. Svc., Peoria, Ill., 1957-61, chemist Nat. Ctr. Agrl. Utilization Rsch., 1961-78, rsch. chemist Nat. Ctr. Agrl. Utilization Rsch., 1978—. Contbr. articles to profl. jours. With U.S. Army, 1958. Co-recipient R&D 100 award R&D mag., 1988. Fellow Am. Inst. Chemists; mem. AAAS, Am. Assn. Cereal Chemists, Am. Chem. Soc. Home: 1108 W Macqueen Ave Peoria IL 61604-3310 Office: USDA Nat Ctr Agrl Utilization Rsch 1815 N University St Peoria IL 61604-3902

CUNNINGHAM, THOMAS B., aerospace engineer; b. Washington, May 8, 1946. BS, U. Nebr., 1969; MS, Purdue U., 1972, PhD in Engring., 1973. Dir. rsch. engring. automatic control Honeywell, Inc., 1973—; adj. prof. U. Minn., 1978—. Mem. IEEE, Am. Inst. Aeronaut. & Astronaut., Sigma Xi. Achievements include applications of modern control and estimation theory to aerospace and industrial problems. Office: Honeywell Systems & Rsch Ctr 3660 Technology Dr Minneapolis MN 55418-1096*

CUNTZ, MANFRED ADOLF, astrophysicist, researcher; b. Landau, Rheinland-Pfalz, Fed. Republic of Germany, Apr. 21, 1958; came to U.S., 1988; s. Gerhard Hermann and Irene Emma (Messerschmitt) C.; m. Anne-Gret Vera Friedrich, Sept. 19, 1988; 1 child, Heiko Benjamin. Diplom in Physics, U. Heidelberg, Fed. Republic of Germany, 1985, PhD in Astronomy, 1988. Postdoctoral, rsch. assoc. Joint Inst. Lab. Astrophysics-U. Colo., Boulder, 1989-91; postdoctoral, rsch. assoc. High Altitude Obs. div. Nat. Ctr. Atmospheric Rsch., Boulder, 1992—. Contbr. articles to Astrophys. Jour., Astronom. Jour., Astronomy and Astrophysics. Grantee German Rsch. Found., NASA, NSF, Dutch Nat. Sci. Orgn. Mem. Internat. Astron. Union (com. 36), Am. Astron. Soc., Deutsche Astronomische Gesellschaft, Deutsche Physikalische Gesellschaft, Vereinigung der Sternfreunde. Achievements include research in theoretical astrophysics, stochastic radiation hydrodynamics in stellar atmospheres. Office: High Altitude Obs Bldg 2 3450 Mitchell Ln Boulder CO 80301

CUOMO, JEROME JOHN, materials scientist; b. N.Y.C., Sept. 30, 1936; s. Gennaro and Rose (Gentile) C.; m. Rita Cossa, June 20, 1959; children: Stephanie, Gennaro, Andrea. BS in Chemistry, Manhattan Coll., 1958; MS in Phys. Chemistry, St. Johns U., 1960; PhD in Physics, Odense U., Denmark, 1979. Chief chemist Secon Metal, 1960-63; staff mem. spl. techniques cen. sci. svcs. IBM, Yorktown Heights, N.Y., 1963-68, mgr. materials processing group cen. sci. svcs., 1968-75, sr. mgr. materials lab. cen. sci. svcs., 1975—; mem. adv. com. materials rsch. lab. Pa. State U., 1990; adj. prof. elec. engring. dept. Mich. State U., 1990, Cornell U., 1984—; past adj. prof. Colo. State U.; affiliate prof. dept. materials sci. and engring. Cornell U., 1983—; elected mem. Japanese and US Workshop on Diamond Tech.; mem. adv. bd. materials rsch. lab. on diamonds Case Western U., Ohio, 1990; material sci. advisor dept. applied physics Unidad Merida, Yucatan, Mex., 1989; mem. adv. com. material sci. dept. N.C. State U., 1988—. Co-editor: (with S. Rossnagel, H. Kaufman) Handbook of Ion Beam Processing Technology, 1989, (with S. Rossnagel, W. Westwood) Handbook of Plasma Processing Technology, 1989; contbr. more than 300 articles and papers to Jour. Applied Physics, Jour. Vacuum Sci. & Tech., Applied Physics Letters, Phys. Rev., Jour. Materials Physics, AIP Conf. Proceedings, Material Rsch. Soc. Symposium Proceedings, Brit. Jour. Vacuum TAIP, others. Organizer-chmn. sputtering topical symposium Am. Vacuum Soc., 1986, mem. program com. thin film div., 1983-84, program chmn. thin film div., 1977, 82, bd. dirs., 1981-82, mem. steering com., 1973-79; organizer-chmn. 1st tech. symposium on silicon nitride Electrochem. Soc., 1966. Recipient Indsl. Rsch. IR-100 award, 1974, 75, Outstanding Paper award Am. Soc. Metals, 1985, Morris N. Liebman Field award IEEE, 1992. Mem. Nat. Acad. Engring. Roman Catholic. Achievements include 77 parents in field. Home: PO Box 353 78 Lovell St Lincolndale NY 10540-0353 Office: IBM T J Watson Rsch Ctr Kitchawan Rd Yorktown Heights NY 10598

CUPP, JON MICHAEL, environmental scientist; b. Bluffton, Ind., Sept. 15, 1955; s. Willard Otto and Alice (Mosure) C.; m. Barbara Jo Keezer, Mar. 10, 1979; children: John Paul, Mary Elisabeth, Rachel Marie, Martha May. BBA, Ind. U., 1977; MPA, Ind. U., Ft. Wayne, 1992. Project planner Region III-A Coordinating Com., Albion, Ind., 1978-79; environ. sanitarian Steuben County Health Dept., Angola, Ind., 1979-84; adminstr., chief environ. scientist Kosciusko County Health Dept., Warsaw, Ind., 1984—; vice chmn. Local Emergency Planning Com., Warsaw, 1991—; tech. advisor Kosciusko County Solid Waste Bd., 1990—, USDA Upper Tippecanoe Watershed Project, 1991—. Author county solid waste disposal ordinance, county pvt. water well ordinance. Mem. Nat. Environ. Health Assn., Ind. Environ. Health Assn., Am. Soc. for Pub. Adminstrn., Am. Water Works Assn., Greater Warsaw C. of C. (solid waste task com. 1991—), Pi Alpha Alpha. Mem. Wesleyan Ch. Avocations: gardening, boating. Home: 106 2d St Winona Lake IN 46590 Office: Kosciusko Co Health Dept 3d Fl Rm 2 County Courthouse Warsaw IN 46580-2377

CUPPAGE, FRANCIS EDWARD, physician, educator; b. Cleve., Aug. 17, 1932; s. Frank Edward and Eunice Agnes (Bartels) C.; m. Virginia Lee Bartch, Aug. 18, 1956; children: Lisa Kay, Peter John, Sharon Elizabeth. BS, Case Western Res. U., 1954; MD, Ohio State U., 1959, MS in Pathology, 1959. Diplomate Am. Bd. Pathology. Intern U. Hosps. of Cleve., 1959-60, resident in pathology, 1960-64, instr. pathology, 1964-65; asst. prof. pathology Ohio State U. Sch. Medicine, Columbus, 1965-67; asst. prof. U. Kans. Med. Ctr., Kansas City, 1967—, Prof., 1973—; prof., chmn. pathology, 1984-85, 90-92; cons. in field. Contbr. articles to profl. jours. Mem. com. Civic Arts Commn. City of Shawnee, Kans., 1970-73; lay leader Luth. Ch. Orgns., 1967—; bd. dirs. Trinity Manor Nursing Home, Merriam, Kans., 1980-89, Bethany Coll., Lindsborg, Kans., 1983-91. Teaching award, U. Kans., 1972; NIH grantee, 1967-75; Fogarty Internat. fellow, 1979. Mem. Am. Assn. Pathologists, Kansas City Soc. Pathologists (pres. 1980-81), Am. Soc. Nephrology, Internat. Acad. Pathologists, AAUP, Group for Rsch. in Pathology Edn. Avocations: woodcarving, hiking, canoeing, photography, ship building. Home: 4740 Black Swan Dr Shawnee Mission KS 66216-1235 Office: U Kans Med Ctr Dept Pathology 3901 Rainbow Blvd Kansas City KS 66160-7410

CURETON, CLAUDETTE HAZEL CHAPMAN, biology educator; b. Greenville, S.C., May 3, 1932; d. John C. and Beatrice (Washington) Chapman; m. Stewart Cleveland, Dec. 27, 1954; children: Ruthye, Stewart II, S. Charles, Samuel. AB, Spelman Coll., 1951; MA, Fisk U., 1966. Tchr. North Warren High Sch., Wise, N.C., 1952-60; tchr. Sterling High Sch., Greenville, 1960-66, Wade Hampton High Sch., Greenville, 1967-73; instr. Greenville Tech. Coll., 1973—; bd. dirs. State Heritage Trust, 1978-91; commr. Basic Skills Adv. Program, Columbia, 1990—; mem. adv. bd. Am. Fed. Bank, NCNB Bank, Greenville, 1991—. Mem. Greenville Urban League, NAACP. Recipient Presdl. award Morris Coll., 1987, 91, Svc. award S.C. Wildlife and Marine Dept., 1986, Outstanding Jack & Jill of Am. citation, 1986, Excellence in Teaching award Nat. Inst. for Staff and Orgnl. Devel., U. Tex.-Austin, 1992-93. Mem. Nat. Assn. Biology Tchrs., Delta Sigma Theta (v.p. Greenville Alumnae chpt.). Democrat. Baptist. Office: Greenville Tech Coll PO Box 5616 Greenville SC 29606-5616

CURET-RAMOS, JOSÉ ANTONIO, internist; b. Caguas, P.R., Aug. 8, 1957; s. José Antonio Curet-Crespo and Enriqueta (Del Carmen) Ramos. Diploma, Sch. Medicine, Galicia, Spain; Lic. in Medicine, Santiago De Compostela, 1985. Rotating intern Humocao (P.R.) Regional Hosp., 1986-87; pub. health svc. physician Diagnostic and Treatment Ctr., Aibonito, P.R., 1987-88; pvt. practice, 1988-89; resident in internal medicine VA Med. Ctr., San Juan, P.R., 1989-92; pvt. practice Guabo, P.R., 1992—; staff physician Hosp. Regional Caguas, 1992—; mem. Family Therapy and Orientation Inst., Caguas. Home: Terralinda 21 Sevilla St Caguar PR 00725 Office: Ste 155 St Andres Arus Rivera Gurabo PR 00778

CURFMAN, DAVID RALPH, neurological surgeon, musician; b. Bucyrus, Ohio, Jan. 2, 1942; s. Ralph Oliver and Agnes Mozelle (Schreck) C.; m. Blanche Lee Anderson, June 6, 1970. Student, Capital U., 1960-62; AB, Columbia Union Coll., 1965; MS, George Washington U., 1967, MD, 1973. Diplomate Nat. Bd. Med. Examiners. Asst. organist, choirmaster Peace Luth. Ch., Galion, Ohio, 1956-62; bus. mgr. Mansfield/Galion Ambulance Svc., Galion, Ohio, 1962-66; with news div. Sta. WTOP-TV (CBS), Washington, 1965; choirmaster, assoc. organist Grace Luth. Ch., Washington, 1966-73, historian, curator, 1969—; teaching fellow in anatomy George Washington U., Washington, 1966-67, gen. surgery intern, 1973-74, resident in neurol. surgery, 1974-78; resident in neuropathology Armed Forces Inst. Pathology, Washington, 1975; resident in pediatric neurol. surgery Children's Hosp. Nat. Med. Ctr., Washington, 1976; teaching fellow in anatomy Georgetown U., Washington, 1967-69, clin. instr. neurol. surgery, 1978—, practice medicine specializing in neurol. surgery, 1978—; chief Div. Neurol. Surgery, Jefferson Hosp., Alexandria, Va., 1989-93, Wash. Hosp. Ctr. Soc., 1992—; vice chmn. bylaws com. Providence Hosp., 1987—; panelist for Am. Assn. Neurol. Surgery ann. meeting "Ethical Issues in Neurol. Surgery.". Chmn., chief author: Physician's Reference Guide for Medicolegal Matters, 1982. Elected mem. D.C. Rep. Com., 1988—; bd. dirs. The Christmas Pageant of Peace, Inc., Washington, Washington Columbus Celebration Assn. Mem. AMA (Phys. Recognition award 1983—), Assn. Am. Med. Colls. (nat. student chmn. rules and regulations com. 1971-73), Med. Soc. D.C. (chmn. medicine and religion com. 1981-83, chmn. medico-legal com. 1986-88), Pan Am. Med. Soc. (mem. exec. bd. 1993—), Congress Neurol. Surgeons, Am. Coll. Legal Medicine, Washington Acad. Neurosurgery, Assn. Mil. Surgeons U.S., Galion Hist. Soc. (charter), Children Am. Revolution (pres. Ohio 1963-64, hon. pres.), SR, U.S. Capitol Hist. Soc. (founding supporting mem.), Nat. Cathedral Assn., Cathedral Choral Soc. (v.p. bd. trustees 1981-83, pres. 1984-86, repertoire chmn. 1981-92), Am. Guild Organists (dean D.C. chpt. 1974-76, publicity chmn. nat. conv. 1982, state chmn. 1984-91), Internat. Congress Organists (Washington program chmn. 1977), Royal Sch. Ch. Music (Eng.), English-Speaking Union, Luth. Laymen's Fellowship, Pilgrim Soc. (Plymouth chpt.), Hymn Soc. Am., Sovereign Mil. Order Temple of Jerusalem, Mil. Order Loyal Legion U.S., Sons of Union Vets. Civil War, Crawford County Coin Club, Am. Polit. Items Collectors Assn., George Washington U. Club, Elks (Galion Lodge No. 1191, Sigma Xi (pres. chpt. 1981-82), Phi Delta Epsilon. Home: 4201 Massachusetts Ave NW Washington DC 20016-4701 Office: 3301 New Mexico Ave NW Ste 210 Washington DC 20016-3658

CURFMAN, FLOYD EDWIN, engineering educator; b. Gorin, Mo., Nov. 16, 1929; s. Charles Robert and Cleo Lucille (Sweeney) C.; m. Eleanor Elaine Fehl, Aug. 5, 1950; children: Gary Floyd, Karen Elaine. BSCE, U. Mo., 1958; BA in Edn., Mt. Mary Coll., 1988, BS in Edn., Math., 1989. Registered profl. engr., Wis., Mo.; cert. tchr., Wis. Forest engr. U.S. Forest Svc., Rolla and Harrisburg, Mo., Ill., 1958-70; engring. dir. U.S. Forest Svc., Milw., 1970-84; chief tech. engr. U.S. Forest Svc., Washington, 1984-86; teacher Wauwatosa (Wis.) High Sch., 1987-89; tchr. Our Lady of Rosary, Milw., 1989—. Author: (booklet) Forest Roads-R-9, 1973; co-author: (tng. manual) Transportation Roads, 1966. Co-leader Boy Scouts Am., Harrisburg, 1958-62; activities coord. Community Action Com., Brookfield, 1970-76; bike and hiking trails com. City of Brookfield (Wis.), 1982-83; program chair Math Counts, 1982. With U.S. Army, 1952-54. Mem. ASCE (program chair, Letter Nat. award 1970), NSPE (coms. 1970-86), Nat. Coun. Tchrs. Math., Wis. Soc. Profl. Engrs. (pres. Milw. chpt. 1982-83, State Recognition award 1983). Avocations: travel, auto trips, reading. Home: 1755 N 166th St Brookfield WI 53005-5114

CURIEN, HUBERT, mineralogy educator; b. Cornimont, France, Oct. 30, 1924; s. Robert and Berthe (Girot) C.; m. Anne-Perrine Dumézil, Dec. 19, 1949; children: Nicolas, Christophe, Pierre-Louis. D., U. Paris, 1951. Prof. U. Paris, 1954; dir. gen. Ctr. Nat. Recherche, 1969-73; del. gen. Del. Gal. Recherche, 1973-76; pres .Ctr. Etudes Spatiales, 1976-84; chmn. European Space Agy., 1981-84; minister for rsch. and tech. French Cabinet, Govt. of France, 1984-86, 88-93; chmn.-elect Organisation Europeene pour la Recherche Nucleaire. Named Grand Officer Légion d'honneur, France. Home: 24 Rue des Fosses Saint Jacques, F 75005 Paris France

CURL, RANE LOCKE, chemical engineering educator, consultant; b. N.Y.C., July 5, 1929; s. Herbert Clarence and Erna (Locke) C.; m. Katherine Ide, June 26, 1954 (div. 1961); children: Stefan Luther, Jocelyn Chandler; m. Shirley Richardson, Sept. 26, 1963 (div. 1976); m. Alice Rolfes, Feb. 27, 1982; 1 child, Vittoria Sarah. SB, MIT, 1951, ScD, 1955. Engr. Shell Devel. Co., Emeryville, Calif., 1955-61; hon. rsch. assoc. Univ. Coll. London, 1961-62; rsch. assoc. Technische Hogeschool, Eindhaven, The Netherlands, 1962-64; prof. chem. engring. U. Mich., Ann Arbor, 1964—; cons. in field. Contbr. more than 60 articles on chem engring. and karst geomorphology to profl. jours.; patentee in field. Pres. Mich. Karst Conservancy, Ann Arbor, 1983—. Fellow Explorers Club (sec. Gt. Lakes chpt. 1985-91); mem. AAAS, Am. Chem. Soc., Am. Inst. Chem. Engrs., Am. Soc. for Engring. Edn., Internat. Assn. for Math. Geology, Mich. Basin Geol. Soc., Cave Rsch. Found., Karst Waters Inst. (bd. dirs. 1991—, exec. sec. 1993—), Nat. Speleological Soc. (hon., bd. dirs. 1958-61, 67-70, 74-89, pres. 1970-74, treas. sect. geol.-geog. 1975-85), Sigma Xi (v.p. chpt. 1989-91), Tau Beta Pi, Alpha Chi Sigma. Avocations: skiing, sailing, music, amateur radio (N8REG), caving. Home: 2805 Gladstone Ave Ann Arbor MI 48104-6432 Office: U Mich Dept Chem Engring Dow Bldg Ann Arbor MI 48109-2136

CURL, ROBERT FLOYD, JR., chemistry educator; b. Alice, Tex., Aug. 23, 1933; s. Robert Floyd and Lessie (Merritt) C.; m. Jonel Whipple, Dec. 21, 1955; children—Michael, David. BA, Rice U., 1954; PhD, U. Cal. at Berkeley, 1957. Research fellow Harvard U., Cambridge, Mass., 1957-58; asst. prof. chemistry Rice U., Houston, 1958-63, assoc. prof., 1963-67, prof., 1967—, chmn. dept. chemistry; master Lovett Coll., 1968-72; vis. research officer NRC Can., 1972-73; vis. prof. Inst. for Molecular Sci., Okazaki, Japan, 1977, U. Bonn, 1985. Contbr. articles profl. jours. Fellow NSF, Alfred P. Sloan fellow, 1961-63; NATO postdoctoral fellow, 1964; recipient Clayton prize Instn. Mech. Engrs., London, 1958, Internat. New Materials prize Am. Phys. Soc., 1992; Alexander von Humboldt sr. U.S. scientist award, 1984. Mem. Am. Chem. Soc., Phi Beta Kappa, Sigma Xi. Methodist. Home: 1824 Bolsover St Houston TX 77005-1728 Office: Rice University PO Box 1892 6100 South Main Houston TX 77251

CURL, SAMUEL EVERETT, university dean, agricultural scientist; b. Ft. Worth, Dec. 26, 1937; s. Henry Clay and Mary Elva (Watson) C.; m. Betty Doris Savage, June 6, 1957 (div.); children: Jane Ellen, Julia Kathleen, Karen Elizabeth; m. Mary Behrends Reeves, Sept. 11, 1993; stepchildren: Ryan Andrew, Shelly Lynn. Student, Tarleton State Coll., 1955-57; BS, Sam Houston State U., 1959; MS, U. Mo., 1961; PhD, Tex. A&M U., 1963. Mem. faculty Tex. Tech U., Lubbock, 1961, 63-76, 79—, tchr., researcher animal physiology and genetics, 1963-76, asst., assoc. and interim dean Coll. Agrl. Sci., 1968-73, assoc. v.p. acad. affairs, 1973-76, dean Coll. Agrl. Scis., prof., 1979—; pres. Phillips U., Enid, Okla., 1976-79; agrl. cons., 1964—; bd. dir. Agrl. Workers Mut. Auto Ins. Co.; mem. Gov.'s Task Force on Agrl. Devel. in Tex., 1982-83, 88; mem. Tex. Crop and Livestock Adv. Com., 1985-91; mem. Tex. Agrl. Resources Protection Authority, 1989—; trustee Water, Inc.; del. Eisenhower Consortium for Western Environ. Forestry Rsch., 1979-84; mgmt. com S.W. Consortium on Plant Genetics and Water Resources, 1984—, chmn. 1989—; mem. USDA Nat. Planning Com. on Hispanic Minority Recruitment, 1988—; trustee Consortium for Internat. Devel., mem. exec. com., 1981-84, 86-87, 89-90; mem. High Plains Rsch. Coordinating Bd.; former mem. So. Regional Coun., U.S. Joint Coun. Food and Agrl. Scis.; chmn. agrl. and natural resources program rev. task force Sam Houston State U., 1982-83; mem. adv. com. Sch. Agriculture Angelo State U., 1989—. Author: (with others) Progress and Change in the Agricultural Industry, 1974, Food and Fiber for a Changing World, 1976, 2d edit., 1982; contbr. 95 articles to profl. jours. Pres. Lubbock Econ. Coun., 1982; bd. overseers Ranching Heritage Assn. 2d lt. U.S. Army, 1959; capt. USAR. Recipient Disting. Alumnus award, Faculty-Alumni Gold medal U. Mo., 1975, Disting. Svc. to Tex. Agriculture award Profl. Agrl. Workers Tex., 1985, Outstanding Agriculture Alumnus award Sam Houston State U., 1986, Tex. Citation for Outstanding Svc. award Tex. 4-H Found., 1987, Disting. Svc. award Vocational Agrl. Tchrs. Assn. Tex., 1987, Blue and Gold Meritorious Svc. award Tex. Future Farmers of Am., 1988, Tex. State Degree Future Farmers Am., 1988, Area Disting. Svc. award Vocat. Agr. Tchrs., 1987, Disting. Alumnus award Sam Houston State U., 1993, Tex. 4-H Alumni award, 1993; Danforth Assoc. fellow, 1964-76; Am. Coun. Edn. fellow, 1972-73. Mem. Am. Soc. Animal Sci. (program com. Biennial Symposium on Animal Reprodn. 1972-76, reviewer Jour. Animal Sci.), Am. Assn. Univ. Agrl. Adminstrs., Assn. U.S. Univ. Dirs. Internat. Agrl. Programs, So. Assn. Agrl. Scientists, Nat. Assn. of State Univs. and Land-Grant Colls., Coun. Adminstrv. Heads of Agr., Tex. Agrl. Leadership Coun., Profl. Agrl. Workers Tex., Tex. Tech. Ex-Students Assn. Century Club, West Tex. C. of C. (bd. dirs., chmn. agrl. and ranching com.), Lubbock C. of C. (bd. dirs., 1988-92, chmn. agriculture task force, chmn. rsch. com. 1981-86, mem. water com., legis. affairs com., agriculture com., gubernatorial appointments task force), Rotary (bd. dirs., 1st v.p. Lubbock club), Farmhouse Fraternity (assoc.), Omicron Delta Kappa, Sigma Xi, Phi Kappa Phi, Gamma Sigma Delta. Home: 5613 83d Ln Lubbock TX 79424 Office: Tex Tech U Office Dean Agrl Scis 108 Goddard Bldg Lubbock TX 79409

CURLE, ROBIN LEA, computer software industry executive; b. Denver, Feb. 23, 1950; d. Fred Warren and Claudia Jean (Harding) C.; m. Lucien Ray Reed, Feb. 23, 1981 (div. Oct. 1984). BS, U. Ky., 1972. Systems analyst 1st Nat. BAnk, Lexington, Ky., 1972-73, SW BancShares, Houston, 1973-77; sales rep. Software Internat., Houston, 1977-80; dist. mgr. Uccell, Dallas, 1980-82; v.p. Info. Sci., Atlanta, 1982-83; v.p. sales TesserAct, San Francisco, 1983-85, Foothill Rsch., San Francisco, 1986-87; v.p. sales and field ops. Natural Lang., Inc., Berkeley, Calif., 1987-89; pres., founder Curle Cons. Group, San Francisco, 1987-89; mgr. strategic mktg. MCC, Austin, Tex., 1989-91; founder, exec. v.p. Evolutionary Tech., Inc., Austin, 1991—. Mem. U. Ky. Alumni Assn., Delta Gamma (pres. 1969). Avocations: scuba diving, running, skiing, cooking. Home: 709 Hidatas Cove Austin TX 78748

CURLIS, DAVID ALAN, civil engineer; b. Tiffin, Ohio, Feb. 22, 1950; s. Meric Alan and Francis Jeanne (Ludwig) C.; m. Suzanne Marie Arbogast, Aug. 7, 1976; children: Trisha Lee, Adrienne Marie. BS in Civil Engring., Ohio No. U., 1973. Registered profl. engr., Ohio. Resident engr. Finkbeiner Bettis & Strout, Ltd., Toledo, Ohio, 1973-78, facilities engr., 1978-79; plant engr. Union Carbide Corp., Fostoria, Ohio, 1979-86; project engr. Nat. Elec. Carbon Corp., Fostoria, 1986—; cons. engr. SEC, Inc., Sycamore, Ohio, 1986-88. Twp. trustee Tymochtee Twp., Wyandot County, 1991—. Mem. ASCE, Masons (32 degree). Home: 6228 Twp Hwy 36 Sycamore OH 44882 Office: Nat Elec Carbon Corp 200 N Town St Fostoria OH 44830

CURRAN, DIAN BEARD, physicist, consultant; b. Woodland, Calif., Aug. 8, 1956; d. David Breed and Eileen Mona (Hersey) Beard; m. Terrence Antony Whelan, June 28, 1985 (div. June, 1990). BS in Physics, U. Kans., 1983; MS in Physics, U. Iowa, 1987. Cons. S.W. Rsch. Inst., San Antonio, 1988—; rsch. asst. astronomy dept. U. Tex., Austin, 1992—. Contbr. articles to Jour. Geophys. Rsch., Geophys. Rsch. Letters. Mem. San Antonio Coun. Native Ams., 1989-93, bd. dirs, 1991-93; Dem. del. to Iowa State Conv., 1984, 88; bd. dirs. Pet Helpers, San Antonio, 1992—. Travel fellow NASA, 1987, 91. Mem. Am. Geophys. Union, Am. Astron. Soc. (assoc.). Democrat. Home: 1901 E Anderson Ln # 215 Austin TX 78752 Office: U Tex Astronomy Dept Austin TX 78712

CURRAN, THOMAS, molecular biologist, educator; b. Broxburn, West Lothian, Scotland, Feb. 14, 1956; came to U.S., 1982; s. Thomas and Jane Holden (McGovern) C.; m. Frances Ko-Fang Yao, Dec. 27, 1979; 1 child, Sean Philip. BS, U. Edinburgh, Scotland, 1978; PhD, U. Coll. London, 1982. Postdoctoral fellow Salk Inst., San Diego, 1982-84; sr. scientist Hoffman-La Roche Inc., Nutley, N.J., 1984-85; asst. mem. Roche Inst. Molecular Biology, Nutley, 1985-86, assoc. mem., 1986-87, full mem., 1987-88, head dept., 1989—, assoc. dir., 1991—; adj. prof. Columbia U., N.Y.C., 1989—; mem. adv. bd. study sect. NIH, Washington, 1991—, Damon Runyan/Walter Winchell Cancer Rsch. Fund., N.Y.C., 1992—; Merton F. Utter Meml. lectr. Case We. Res. U., 1992. Editor: The Oncogene Handbbok, 1988, Origins of Human Cancer, 1991; contbr. over 100 articles to sci. jours. and books. Recipient Young Scientist award Passano Found., 1992, Rita Levi Montalcino Lecture award Fidia Rsch. Found., 1992, Glasgow U.-Tenovus-Scotland medal, 1992; Imperial Cancer Rsch. Fund grantee. Mem. AAAS, Am. Soc. for Microbiology, Am. Assn. for Cancer Rsch. (Rhoads award 1993), Am. Soc. for Cell Biology, Am. Soc. Biochemistry and Molecular Biology, Soc. for Neurosci., Harvey Soc. Roman Catholic. Achievements include discovery and characterization of fos oncogene which causes bone tumors in mice; demonstration that fos gene expression in increased rapidly in many cell types treated with agents associated with mitogenesis, differentiation and stimulation of neurons, fos encodes DNA binding protein that functions in transcriptional regulation in association with the product of the jun oncogene. Office: Roche Inst Molecular Biology 340 Kingsland St Nutley NJ 07110

CURRERI, JOHN ROBERT, mechanical engineer, consultant; b. N.Y.C., July 20, 1922; s. Girolomo and Genoveffa (Dasaro) C.; m. Margaret McHugh, June 2, 1946 (dec.); children: Eileen, Ellen, Joan; m. Ann Jurgenson, Oct. 1, 1976. BME cum laude, Bklyn. Poly. Inst., 1944; MME, Poly. U., Bklyn., 1948. Registered profl. engr., N.Y. Assoc. prof. Bklyn. Poly. Inst., 1948-55; head dynamics sect. ARMA Corp., Garden City, N.Y., 1955-57; prof. Poly. U., Bklyn., 1957-64, head mech. dept., 1964-74, prof., 1974-90; vis. prof. CUNY, 1990-91; prof. emeritus Poly. U., Bklyn., 1992—; cons. Brookhaven Nat. Lab., Upton, N.Y., 1974-90, Sperry, Great Neck, N.Y., 1966-70. Author: (textbook) Vibration of Structures, 1961; co-author: (textbook) Vibration Control, 1958; contbr. articles to more than 60 profl. jours. Recipient Sci. Faculty fellowship NSF, 1963. Mem. ASME, NSPE, Am. Soc. Engring. Edn., N.Y. Soc. Profl. Engrs., Sigma Xi, Tau Beta Pi, Pi Tau Sigma. Home and Office: John R Curreri 10 San Carlos Ct Toms River NJ 08757

CURRERI, PETER ANGELO, materials scientist; b. N.Y.C., June 22, 1952; s. Angelo Paul and Consiglia (Felicelia) C.; m. Linda Lou Atzel, July 12, 1975; 1 child, John. PhD, U. Fla., 1977, U. Fla., 1979. Postdoctoral fellow U. Fla., Gainesville, 1979-80, asst. prof., 1980-81; materials scientist NASA MSFC, Huntsville, Ala., 1981—; mission scientist U.S. Micro Gravity Payload space shuttle missions NASA. Editor (NASA book) Material Science on Parabolic Aircraft, 1992; author (handbook chpt.) Low Gravity Effects on Solidification, 1988; contbr. articles to profl. jours. V.p. Acad. Sci. Fgn. Lang. PTA, Huntsville, 1991. Recipient Japanese Govt. Rsch. award Nat. Inst. Metals, Tsukuba, Japan, 1990. Mem. AIAA, Soc. for Advancement of Materials and Process Engring. (chmn. materials in space subcom. 1990-92). Roman Catholic. Achievements include over 100 flight hours in aircraft low-gravity experiments. Office: NASA MSFC ES75 Huntsville AL 35812

CURRIE, BRUCE LAMONTE, pharmaceutical sciences educator, medicinal chemistry researcher; b. Pasadena, Calif., Mar. 1, 1945; s. Paul Quentin and Ellen Irene (Gifford) C.; m. Lynda Marc Thompson, July 2, 1965; children: Paul, Charles, Kathren. BS in Chemistry, Ariz. State U., 1966; PhD in Organic Chemistry, U. Utah, 1969. Postdoctoral rsch. assoc. Inst. for Biomed. Rsch., U. Tex., Austin, 1969-74; asst. prof. U. Ill., Chgo., 1974-81, assoc. prof., 1981-92; prof., chmn. dept. pharm. scis. Chgo. Coll. Pharmacy, Downers Grove, Ill., 1992—; curriculum cons. Midwest Univs. Coop. Internat. Activities, Indonesia, 1985-86. Contbr. over 65 articles to profl. jours. Recipient Rsch. grant NIH, 1980-92. Mem. Internat. Soc. Heterocyclic Chemistry, Am. Chem. Soc. (div. biol., organic and medicinal chemistry), Endocrine Soc., Am. Assn. Colls. of Pharmacy, Am. Assn. Pharm. Scientists. Baptist. Office: Chgo Coll of Pharmacy 555 31st St Downers Grove IL 60515

CURRIER, ROBERT DAVID, neurologist; b. Grand Rapids, Mich., Feb. 19, 1925; s. Frederick Plummer and Margaret (Hoedemaker) C.; m. Marilyn Jane Johnson, Sept. 1, 1951; children: Mary Margaret, Angela Maria. AB, U. Mich., 1948, MD, 1952, MS in Neurology, 1956; postgrad., Nat. Hosp., U. London, 1955; postgrad. Medico-Social Research Bd, Dublin, Ireland, 1972. Intern, then resident in neurology Univ. Hosp., Ann Arbor, 1952-56; from instr. to asso. prof. U. Mich. Med. Sch., 1956-61; mem. faculty U. Miss. Med. Ctr., Jackson, 1961—, prof. neurology, 1971—, chief div., 1961-77, chmn. dept., 1977-90, H.F. McCarty prof., 1987—; mem. adv. bd. Nat. Ataxia Found., rsch. dir., 1985-93; mem. clin. adv. coun. Amyotrophic Lateral Sclerosis Soc. Am., 1979—; mem. Ataxia com. World Fedn. Neurology, 1981—, sec., 1985-93. Co-editor: Yearbook of Neurology and Neurosurgery, 1981-88, editor, 1989-92; co-editor (jour.) Key Quar. Neurology and Neurosurgery, 1986-92; asst. editor for history Archives of Neurology, 1983—; assoc. editor Jour. Neuroscis., 1990—; contbr. articles to med. jours. Served with USAAF, 1943-45, ETO. Decorated Air medal with 2 oak leaf clusters; NIH grantee, 1961-74. Fellow Am. Acad. Neurology (chmn. history com. 1980-82, treas. 1991—); mem. Am. Neurol. Assn., Central Soc. Neurol. Research (pres. 1971), Sigma Xi, Alpha Omega Alpha. Home: 5529 Marblehead Dr Jackson MS 39211-4249 Office: 2500 N State St Jackson MS 39216-4505

CURRY, NORVAL HERBERT, retired agricultural engineer; b. St. Francis, Kans., Oct. 10, 1914; s. Charles Edward and Florence Eleanor (Ward) C.; m. Helen Maurine Smith, June 8, 1938; children: Sharon Gay Curry Morgret, Janice Kay Curry Dalal. Student, Ft. Hays State U., 1934-36; BS in Archtl. Engring., Iowa State U., 1940, MS in Agrl. Engring., 1946. Field engr. Structural Clay Products Inst., Ames, Iowa, 1940-44; rsch assoc. Iowa State U., Ames, 1944-46, instr., 1946-47, asst. prof., 1947-49, assoc. prof., 1949-54, prof. agrl. engring., 1954-59; pvt. practice Ames, 1959-79; founder, chief exec. officer Curry-Wille Consulting Engrs. P.C., Ames, 1979-80; sci. aide Rockefeller Found., Colombia, 1955-56, Chile, 1961. Named life mem. Iowa Engring. Soc., 1980. Fellow Am. Soc. Agrl. Engrs. (life; nat. pres. 1969-70, Cyrus Hall McCormick Gold medal 1980); mem. NAE, Iowa Engring. Soc. (life), Izaak Walton League (pres. 1953), Lions Club (dir. 1987-89). Republican. Avocations: woodworking, hunting, fishing. Home: 227 Campus Ave Ames IA 50014-7407

CURRY, ROBERT MICHAEL, broadcast engineer; b. Indpls., Jan. 15, 1947; s. William Archie Curry and Lois Maxine (Miller) Ward. Grad. with highest honors, U.S. Army Signal Sch., 1968. Lic. radiotelephone operator, FCC; amateur radio operator, KC3VO. Customer engr. I.B.M. Corp., Arlington, Va., 1966-74; license examiner FCC, Washington, 1974-75; service technician Sony Corp., Langley Park, Md., 1975-76; communications technician Montgomery County Govt., Rockville, Md., 1976-76; Maintenance Engr. Sta. WJLA-TV, Washington, 1978-80; chief operator Sta. WHMM-TV, Howard U., Washington, 1980—; cons. Tele-Trek Prodns., Washington, 1985—; technical asst. Potomac Valley Amateur Athletic Union, Washington area, 1974—. Served with U.S. Army, 1968-71. Mem. Armed Forces Communications and Electronics Assn. (honor award), Am. Amateur Radio Relay League, Nat. Assn. Broadcast Employees and Technicians. Democrat. Achievements include provided uplink transmitter for 1st live Ham color tv transmission from earth to the space shuttle mission STS37, traveling to former USSR and The Ukraine volunteering technical services to improve civilian amateur radio digital communications network, donating computers, and other equiptment. Office: Howard U Sta WHMM-TV 2222 4th St Washington DC 20059-0001

CURTIN, DAVID YARROW, chemist, educator; b. Phila., Aug. 22, 1920; s. Ellsworth Ferris and Margeretta (Cope) C.; m. Constance O'Hara, July 1, 1950; children—Susan McLean, David Ferris, Jane Yarrow. A.B., Swarthmore Coll., 1943; Ph.D., U. Ill., 1945. Pvt. asst. Harvard, 1945-46; instr., then asst. prof. chemistry Columbia U., 1946-51; mem. faculty U. Ill., Urbana, 1951—; prof. chemistry U. Ill., 1954-86, Fuson prof. emeritus, 1988—, head div. organic chemistry, 1963-65; vis. lectr. Inst. de Quimica, Mexico, summer 1955, U. Tex., 1959; Reilly lectr. U. Notre Dame, 1960. Mem. editorial bd.: Organic Reactions, 1954- 64; adv. bd., 1965—; mem. bd. editors: Jour. Organic Chemistry, 1962-66. Einstein fellow Israel, 1982. Mem. Am., Brit., Swiss chem. socs., Nat. Acad. Sci., Am. Crystallographic Assn. Achievements include special research organic reaction mechanisms, stereochemistry, exploratory organic chemistry, reactions in solid state. Home: 3 Montclair Rd Urbana IL 61801-5823 Office: Univ of Ill care Dept of Chemistry Urbana IL 61801

CURTIS, BILL, software engineering researcher; b. Meridian, Tex., Sept. 3, 1948; s. Willard H. Curtis Jr. and Virginia Mae (White) Stedman; m. Janell Johnston, Jan. 3, 1981; children: Crystal Eden, Catherine Anne. MA, U. Tex., 1974; PhD, Tex. Christian U., 1975. Rsch. asst. prof. U. Wash., Seattle, 1975-76; staff psychologist Weyerhaeuser, Federal Way, Wash., 1977; mgr. GE, Arlington, Va., 1978-80, ITT Corp., Stratford, Conn., 1980-83; dir. Microelectronics and Computer Tech. Corp., Austin, Tex., 1984-90, Software Engring. Inst., Pitts., 1991—. Editor: Human Factors in Software Development, 1985, (conf. proceedings) Human Factors in Computer Systems-II, 1985; contbr. articles to Communications of the ACM. Mem. IEEE (sr.), Assn. for Computing Machinery, APA, Human Factors Soc. Home: 3644 Ranch Creek Austin TX 78730 Office: Software Engring Inst Carnegie Mellon U Pittsburgh PA 15213-3890

CURTIS, CLARK BRITTEN, software engineer; b. Lake Forest, Ill., July 12, 1951; s. Edwin Martin and Mary Kathryn (Iversen) C.; m. Kimberly Ann Robinson, July 24, 1976; 1 child, April Brittany. BS in Computer Engring., U. Ill., 1974; MBA with honors, Lake Forest Coll., 1986. Programmer A.S.C. Tabulating Corp., Lake Bluff, Ill., 1972-73; elec. engr. Motorola Inc., Schaumburg, Ill., 1974-77; tech. analyst Adminstrn./Systems/Communication, Lake Bluff, 1977-78; sr. systems cons. Am. Hosp. Supply Corp., McGaw Park, Ill., 1978-81; software devel. mgr. Chgo. Laser Systems, Inc., Chgo., 1981-89; prin. Curtis Cons., Waukegan, Ill., 1989—; software devel. engr. Clear Communications Corp., Lincolnshire, Ill., 1989-90; pres., owner Systematic Communications Inc., 1990—; tech. cons. Am. Bus. Systems, Northbrook, Ill., 1989, Hewitt Assocs., Lincolnshire, Ill., 1989-90, Clear Communications Corp., Lincolnshire, 1989, TriLAN Systems Corp. (subs. Dukane Corp.), St. Charles, Ill., 1990-91, Abbott Labs., Abbott Park, Ill., 1991—. Ch. organist Ecclesia Fellowship, North Chgo., Ill., 1973-76, deacon, 1974-76. Mem. Internat. Platform Assn., Assn. Computing Machinery. Avocations: relativity physics, guitar, pilot, organ. Home and Office: 3503 Country Club Ave Waukegan IL 60087-4130

CURTIS, EDWARD JOSEPH, JR., gas industry executive, management consultant; b. Boston, May 26, 1942; s. Edward Joseph and Violet Ella (Upton) C.; m. Virginia Carolyn Fye, May 6, 1976; children: Jane Mercedes, Sherri Jean; 1 stepchild, Virginia Amy. BSChemE, Worcester Polytech., 1964, MSChemE, 1966. Engr. Cabot Corp., Boston, 1966-68; mgr. corp. devel. Distrigas Corp., Boston, 1968-72; pres. E.J. Curtis Assocs., Inc., York Harbor, Maine, 1972—; pres. Pine Hill Assocs., Inc., Hollis, N.H., 1976-80; ptnr. ABC Mgmt. Systems, Bellingham, Wash., 1977-82; mng. ptnr. Essex Cons. Svcs., Boston, 1981-82; bd. dirs. Essex County Gas Co., Maine Mfg. Co. Pres. York Harbor Neighborhood Assn., 1989—. Mem. Am. Inst. Chem. Engrs., Am. Gas Assn., New Eng. Gas Assn. (dir. 1988-91), Soc. Gas Lighting, Soc. Energy Engrs., Guild of Gas Mgrs., York Golf and Tennis Club, Twenty Assocs. Club (pres. 1985-86), Agamenticus Yacht Club, Theta Chi. Republican. Mem. Congl. Ch. Avocations: sailing, skiing, golf, computer sci., music. Office: E J Curtis Assocs Inc Box 1000 York Harbor ME 03911

CURTIS, LAWRENCE ANDREW, biologist, educator; b. Hartford, Conn., Apr. 14, 1942; s. George Walter and Phillis (Andrew) C. BA in Biology, Nasson Coll., 1965; MS in Zoology, U. N.H., 1967; PhD in Biol. Sci., U. Del., 1973. Lectr. biology Fairleigh Dickinson U., Madison, N.J., 1967-68; rsch. assoc. U. N.H., Durham, 1968; instr. U. Del., Newark, 1972-73, asst. prof., 1973-83, assoc. prof., 1983—. Contbr. articles to sci. publs. Mem. AAAS, Am. Soc. Zoologists, Ecol. Soc. Am. Achievements include discovery of a unique effect of a parasite on the behavior of its host. Office: U Del Cape Henlopen Lab Lewes DE 19958

CURTISS, JOSEPH AUGUST, electrical engineer, consultant; b. Bklyn., Aug. 29, 1938; m. Barbara Magdeline Perez, Apr. 17, 1960; children: Jacqueline Ann, Jonathan August, Jennifer Arlene. BSEE, Poly. Inst. Bklyn., 1963, MSEE, 1970. Registered profl. engr., N.Y. Devel. engr., Alternating Gradient Synchrotron Brookhaven Nat. Lab., Upton, N.Y., 1963-80; instrumentation engr. Burns & Roe, Inc., Cons. Engrs., Oradell, N.J., 1980-85; sr. project engr. dept. advanced tech. Brookhaven Nat. Lab., Upton, N.Y., 1985—; pvt. practice, cons. engr. Miller Place, N.Y., 1975—. Project mgmt., design supervision, tech. reports. Dir. Miller Place (N.Y.) Badminton Club, 1978-85. Mem. IEEE (sr.), Instrument Soc. Am. (sr.), Nat. Soc. Profl. Engrs. (sec., treas. 1975-78), Lambda Chi Alpha (pledge trainer 1960). Republican. Roman Catholic. Home: 78 Hempstead Ave Miller Place NY 11764-2610 Office: Brookhaven Nat Lab Dept Advanced Tech Upton NY 11973

CURTNER, MARY ELIZABETH, psychologist; b. Hampton, Va., Oct. 31, 1957; d. Myron Lester II and Mary Joan (Corley) C. BS, Baylor U., 1980; MS, U. North Tex., 1982; PhD, U. N.C., Greensboro, 1991. Tchr. Clear Creek Ind. Sch. Dist., League City, Tex., 1982-86; rsch. asst. U. N.C., Greensboro, 1986-90; asst. prof. U. Ala., Tuscaloosa, 1990—; presenter in field. Mem. Soc. for Rsch. in Child Devel., Soc. for Rsch. in Adolescence, Am. Psychol. Assn. Office: U Ala PO Box 870158 201 Child Devel Ctr Tuscaloosa AL 35487

CUSANO, CRISTINO, mechanical engineer, educator; b. Sepino, Italy, Mar. 22, 1941; s. Crescenzo and Carmela (D'Anello) C.; m. Isabella Pera, Aug. 7, 1974. B.S., Rochester Inst. Tech., 1965; M.S., Cornell U., 1967, Ph.D., 1970. Asst. prof. mech. engring. U. Ill., Urbana, 1970-74; assoc. prof. U. Ill., 1974-83, prof., 1983—; cons. Mattison Machine Works, Whirlpool Corp. Contbr. articles to profl. jours. NSF fellow, 1965-69, ASME fellow; recipient Capt. Alfred E. Hunt award, Al Sonntag award, Xerox award. Mem. Soc. Tribologists and Lubrication Engrs., Am. Soc. Engring. Edn., Sigma Xi, Phi Kappa Phi, Pi Tau Sigma. Roman Catholic. Home: 1303 Belmeade Dr Champaign IL 61821-5027 Office: Univ Ill Dept Mech Engring 1206 W Green St Urbana IL 61801-2906

CUSATIS, JOHN ANTHONY, chemist; b. Hazleton, Pa., July 22, 1951; s. Anthony John and Elizabeth Catherine (Jacobs) C.; m. Patricia Ockovic, June 24, 1972 (div. July 1974); m. Marsha Anne Petzold, Aug. 17, 1974; 1 child, Joshua Michael. AS in Chemistry, Lehigh County C.C., 1971; BS in Geology magna cum laude, Bloomsburg U., 1991. Sr. lead technician Air Products and Chems., Hometown, Pa., 1970—. Author inhouse procedure manuals. Advisor explorer troop Boy Scouts Am., Hometown, 1992; advisor Sci.-by-Mail, Hometown, 1991-92; advisor, leader class play stage crew Marian High Sch., Hometown, 1990—. Named to Order of the Arrow, Boy Scouts Am., 1991. Mem. AAAS, Geol. Soc. Am., Phi Kappa Phi. Achievements include work with supercritical CO2 developing purity levels. Office: Air Products & Chems Box 351 RD 1 Tamaqua PA 18252-9475

CUSHING, STEVEN, educator, researcher, consultant; b. Brookline, Mass., June 25, 1948; s. Alfred Edward and Evelyn (Kaufman) C. SB, MIT, 1970; MA, UCLA, 1972, PhD, 1976. Rsch. asst. MIT, 1967-70, UCLA, 1973-74; instr. U. Mass., Boston, 1974-75, Roxbury Community Coll., Boston, 1975-77; rsch. staff Higher Order Software Inc., Cambridge, Mass., 1976-82; rsch. assoc. Rockefeller U., N.Y.C., 1979; lectr. Northeastern U., Boston, 1983-86; master lectr. Boston U., 1986-89, assoc. prof., 1989—; rsch. fellow NASA-Ames Rsch. Ctr., Mountain View, Calif., 1987-88, Stanford U., Palo Alto, Calif., 1987-88, NASA-Langley Rsch. Ctr., Hampton, Va., 1989; asst. prof. St. Anselm Coll., Manchester, N.H., 1983-85, Stonehill Coll., North Easton, Mass., 1985-89; mem. bd. editorial commentators The Behavioral and Brain Scis., 1978—; chmn. software design Internat. Conf. System Scis., Honolulu, 1978; mem. 1st fgn. del. USSR Acad. of Scis., 1989; participant workshops. Author: Quantifier Meanings: A Study in the Dimensions of Semantic Competence, 1982; contbr. articles to profl. jours. Mem. nat. exec. coun. Nat. Ethical Youth Orgn., 1965-66; violist Brockton (Mass.) Symphony Orch., 1987—; vol. scientist sci.-by-mail program Mus. Sci., Boston, 1988—; fiddler Boston Scottish Fiddle Club, 1990—. Recipient New Eng. Regional award Future Scientists of Am., 1965; NSF grantee, 1965, 70-71, NIMH grantee, 1970-71, NDEA grantee, 1970-73; Woodrow Wilson Found. fellow, 1970-71, NASA Summer Faculty fellow, 1987-89; rsch. affiliate MIT, 1978-79, Boston U., 1986-88. Mem. N.Y. Acad. Scis., Linguistic Soc. Am., Assn. Symbolic Logic, Am. Math. Soc., Am. Assn. for Applied Linguistics, Internat. Coun. Psychologists (profl. affiliate), Soc. for Computers in Psychology, Assn. for Computers and Humanities, Cognitive Sci. Soc., Internat. Cognitive Linguistics Assn., Math. Assn. Am., Assn. Computational Linguistics, Internat. Pragmatics Assn. Home: 90 Bynner St Apt 4 Jamaica Plain MA 02130-1045 Office: Boston U 755 Commonwealth Ave Boston MA 02215-1400

CUSICK, JOSEPH DAVID, science administrator, retired; b. Chgo., Oct. 18, 1929; s. Joseph M. and Rose (Gerrity) C.; m. Kathryn Vermilya Moore, Feb. 2, 1952; children: Stephen, Anne, Eileen, Michael, Joseph R., Mary, James, John. BA, Stanford U., 1951; postgrad. in Law, U. San Francisco, 1956-58, U. Santa Clara, 1956-58; MBA, U. Santa Clara, 1963; postgrad. fellow in Bus., Stanford U., 1972-73; MS in Cybernetic Systems, San Jose State U., 1976; postgrad., Def. Systems Mgmt. Coll., Air Force Inst. Tech. Tech. writer Magna Power Tool Corp., Menlo Park, Calif., 1956, McGraw-Hill Publishing Co., San Francisco, 1956-57; adminstrv. asst. Lockheed Missiles and Space Co., Sunnyvale, Calif., 1958-61; supr. Satellite Test Ctr., Sunnyvale, 1962-68; civilian mgr./exec., chief dir., dep. dir. Air Force Consol. Space Test Ctr., Sunnyvale, 1968-91 (ret.); mgr. Air Force Consol. Space Ops. Ctr., Colorado Springs, Colo. Editor Libr. Assocs. Newsletter, Stanford Assocs. Report; contbr. articles to Def. Mgmt. Jour., The Lamp. Bd. dirs. Los Gatos (Calif.) Mus. Assn.; Lector St. Mary's Ch., Los Gatos; vol. fundraiser Stanford U.; active Stanford Hist. Soc. With USN, 1951-56, lt. commdr. Res. ret. Recipient Gold Spike award Stanford U., 1973, Block S pin, 1986. Mem. AIAA, Stanford Arms Control and Internat. Security Group, Stanford Alumni, Stanford and Santa Clara Bus. Sch. Alumni Assns., Stanford U. Libs. (bd. dirs. 1976-88, chmn. 1984-86), Stanford Music Guild, Commonwealth Club Calif., Saratoga Men's Club (bd. dirs.), Stanford Faculty Club, Stanford Block S Soc. and Buck Club, Nat. Assn. Ret. Fed. Employees, VFW, No. Calif. Golf Assn., Sigma Delta Chi (past chpt. pres.). Achievements include development of operational concepts, satellite support policies, technical procedures and test plans for the world-wide network supporting DoD satellites, systematic resource schedule planning for ground support equipment, statistical analysis of network resource utilization; research in science policy making. Home: 163 Eastridge Dr Los Gatos CA 95032

CUSUMANO, JAMES ANTHONY, chemical company executive, former recording artist; b. Elizabeth, N.J., Apr. 14, 1942; s. Charles Anthony and Carmella Madeline (Catalano) C.; m. Jane LaVerne Melvin, June 15, 1985; children: Doreen Ann, Polly Jean. BA, Rutgers U., 1964, PhD, 1967; grad. Exec. Mktg. Program, Stanford U., 1981, Harvard U., 1988. Mgr. catalyst rsch. Exxon Rsch. and Engring. Co., Linden, N.J., 1967-74; pres., chief exec. officer, founder Catalytica Inc., Mountain View, Calif., 1974-85, chmn., 1985—, also bd. dirs.; lectr. chem. engring. dept. U. Stanford U., 1978; advisor to Inst. Internat. Edn. Fulbright Scholar Program; lectr. Rutgers U., 1966-67; Charles D. Hurd lectr. Northwestern U., 1989-90; speaker to chem. and physics grads. U. Wis., 1992; mem. NSF Com. on Catalysts and Environment; exec. briefings with Pres. George Bush and Cabinet mems., 1990, 92; fellow Churchill Coll., Cambridge U., 1992—. Author: Catalysis in Coal Conversion, 1978, (with others) Critical Materials Problems in Energy Production, 1976, Advanced Materials in Catalysis, 1977, Liquid Fuels from Coal, 1977, Kirk-Othmer Encyclopedia of Chemical Technology, 1979, Chemistry for the 21st Century, Perspectives in Catalysis, 1992; contbr. articles to profl. jours., chpts. to books; founding editor Jour. of Applied Catalysis, 1980; rec. artist with Royal Teens and Dino Take Five for ABC Paramount, Capitol and Jubilee Records, 1957-67; single records include Short Shorts, Short Shorts Twist, My Way, Hey Jude, Rosemarie, Please Say You Want Me, Lovers Never Say Goodbye; albums include The Best of the Royal Teens, Newies But Oldies; appeared in PBS TV prodn. on molecular engring., Little by Little, 1989. Recipient Surface Chemistry award Continental Oil Co., 1964; Henry Rutgers scholar, 1963, Lever Bros. fellow, 1965, Churchill Coll. fellow Cambridge Univ., 1992. Mem. Am. Chem. Soc., Am. Inst. Chem. Engrs., Am. Phys. Soc., N.Y. Acad. Sci., Am. Mus. Natural History, Pres.'s Assn., Smithsonian Assocs., Sigma Psi, Phi Lambda Upsilon (hon.). Republican. Roman Catholic. Achievements include 20 patents in catalysis and surface science; avocations: skiing, hiking, sailing, swimming, travel. Home: 3111 Bandera Dr Palo Alto CA 94304-1341 Office: Catalytica Inc 430 Ferguson Dr Bldg 3 Mountain View CA 94043-5215

CUTHBERT, ROBERT LOWELL, product specialist; b. Bay City, Mich., June 28, 1939; s. Lowell Robert and Katherine Ann (Popp) C.; m. Carol Ann Barcia, Apr. 23, 1960; children: Steven Robert, Douglas Brian, Kristi Ann. Student, Bay City Jr. Coll., 1957-59, Delta Coll., 1966-67. Lab. tech. coatings Dow Corning Corp., Midland, Mich., 1964-70, silicone acrylic rsch., 1970-72, electronic tech., 1972-78, solar cell rsch., 1978-81, electrical prodn. tech. rep., 1981-88, masonry products tech. rep., 1988-90, product specialist, 1990—. Contbr. articles to profl. jours. With USAF, 1959-63. Mem. Am. Soc. Testing and MAterials, Am. Radio Relay League, Elks. Democrat. Methodist. Achievements include patents for masonry water repellent compositions and research in field. Office: Dow Corning Mail C02230 Midland MI 48686-0997

CUTHBERT, VERSIE, chemical engineer; b. Chgo., July 7, 1960; d. Luther Hunter and Arilla (Mays) Epting; m. Michael Cuthbert, Nov. 19, 1983. BSChemE, Ill. Inst. Tech., 1989. Rsch. asst. Argonne Nat. Lab., Chgo., 1981-83; environ. engr. EPA, Chgo., 1990-91; chem. engr. III, Texaco, Port Arthur, Tex., 1991—. Recipient Texaco Rsch. and Devel. Vol. Action award. Mem. AICE, Nat. Soc. Women Engrs., Nat. Soc. Black Engrs., Golden Triangle Tex. Alliance Minority Engrs. (dist. coord. 1991—).

CUTKOSKY, RICHARD EDWIN, physicist, educator; b. Mpls., July 29, 1928; s. Oscar F. and Edna M. (Nelson) C.; m. Patricia A. Klepfer, Aug. 28, 1952; children: Mark, Carol, Martha. B.S., Carnegie Inst. Tech., 1950, M.S., 1950, Ph.D., 1953. Asst. prof. physics Carnegie-Mellon U., Pitts., 1954-61, prof. physics, 1961—, Buhl prof., 1963—. Fellow Am. Phys. Soc., AAAS. Home: 1209 Wightman St Pittsburgh PA 15217-1220

CUTLER, CASSIUS CHAPIN, physicist, educator; b. Springfield, Mass., Dec. 16, 1914; s. Paul A. and Myra B. (Chapin) C.; m. Virginia Tyler, Sept.

27, 1941; children: (Cassius) Chapin, William (Urban) (dec.), Virginia Cutler Raymond. B.Sc., Worcester Poly. Inst., 1937, D.Eng. (hon.), 1975. With Bell Telephone Labs, 1937-78; asst. dir. electronics and radio research Bell Telephone Labs, Murray Hill, N.J., 1959-63; dir. electronic and computer systems research lab. Bell Telephone Labs, Holmdel, N.J., 1963-78; prof. applied physics Stanford U., 1979—. Contbr. articles to profl. jours. Mem. 1st Ch. of Christ Scientist, Keyport, N.J., 1966-78, Menlo Park, Calif., 1979—, reader, chmn. bd., Plainfield, N.J., 1946-66. Recipient Robert H. Goddard Disting. Alumni award Worcester Polytechnic Inst., 1982. Fellow IEEE (Edison medal 1981, Centennial medal 1984, Alexander Graham Bell medal 1991), AAAS; mem. Nat. Acad. Engring. Nat. Acad. Scis. Patentee numerous devices. Home: 106 Peter Coutts Cir Palo Alto CA 94305-2516 Office: Stanford U Ginzton Lab Stanford CA 94305

CUTLER, LEONARD SAMUEL, physicist; b. Los Angeles, Jan. 10, 1928; s. Morris and Ethel (Kalech) C.; m. Dorothy Alice Pett, Feb. 13, 1954; children: Jeffrey Alan, Gregory Michael, Steven Russell, Scott Darren. BS in Physics, Stanford U., 1958, MS, 1960, PhD, 1966. Chief engr. Gertsch Products Co., Los Angeles, 1948-56, v.p. research and devel., 1956-57; with Hewlett-Packard Co., Palo Alto, Calif., 1957—, dir. physics research lab., 1969-85, dir. instruments and photonics lab., 1985—, dir. superconductivity lab., 1987-89; disting. contbr., 1989—; mem. adv. panels Nat. Bur. Standards; cons. Kernco, Inc., Danvers, Mass., 1982—, others. Patentee in field. Served with USNR, 1945-46. Recipient Achievement award Indsl. Rsch. Inst., 1990. Fellow IEEE (Morris Leeds award 1984, Rabi award 1989); mem. AAAS, NAE, Am. Phys. Soc., Sigma Xi. Home: 26944 Almaden Ct Los Altos CA 94022-4316 Office: Hewlett-Packard Co PO Box 10350 Palo Alto CA 94303-0867

CUTLER, NEAL EVAN, gerontologist, educator. Cert. Polit. Instns., Cambridge U., Eng., 1964; BA in Polit. Sci. cum laude, U. So. Calif., 1965; MA in Polit. Sci., Northwestern U., 1966, PhD in Polit. Sci., 1968. Postdoctoral rsch. sci. dirs. divsn. Civil Defense Rsch. Project Oak Ridge Nat. Lab., 1968-69; asst. prof. polit. sci. U. Pa., 1969-72; from assoc. prof. to prof. polit. sci., grad. dir. dept. polit. sci., prof. gerontology U. So. Calif., 1973-89; prof. gerontology and social policy Am. Coll., Bryn Mawr, Pa., 1989-92; pres., dir. Boettner Inst. Fin. Gerontology U. Pa., 1992—; founding dir. Social Sci. Data Ctr. U. Pa., 1969-72, sr. rsch. assoc. Fgn. Policy Rsch. Inst., 1969-72; profl. staff mem. spl. com. aging U.S. Senate, Washington, 1979-81; sr. rsch. assoc. Gerontology Rsch. Inst. U. So. Calif., 1973-89, chair devel. com. new PhD program gerontology, lab. chief Social Policy Lab., 1973-81, co-dir. Andrus Gerontology Ctr. Inst. Advanced Study Gerontology and Geriatrics, 1986-89; dir. Boettner Rsch. Inst., Am. Coll., 1989-91; U.S. rep. Demography Resource Panel No. Am. Regional Tech. Conf. World Assembly Aging, 1981; invited seminar Kennedy Sch. Govt. Harvard U., 1986; prin. investigator in field; mem. older consumer adv. bd. Bell Atlantic, 1992—, bd. dirs. Sr. Mark Svcs., Sr.-Net; mem. adv. com. Phila. Corp. Aging, 1990—, PULSE Am. Soc. CLU-ChFC, 1990—; sr. advisor Am. Bd. Family Practice, 1990-91; mem. task force stats. health policy aging soc. NAS, 1984-87, other adv. bds., coms.; reviewer in field. Author: (with Davis W. Gregg and M. Powell Lawton) Age, Money, and Life Satisfaction: Aspects of Financial Gerontology, 1992, (with R. Doyle, K. Tacchino, T. Kurlowicz, J. Schnepper) Can You Afford to Retire?, 1992, (with Albert Tedesco and Robert Frank) The Differential Encoding of Political Images: A Content Analysis of Network Television News, 1972, The Alternative Effects of Generations and Aging Upon Political Behavior: A Cohort Analysis of American Attitudes Toward Foreign Policy, 1946-66; bi-monthly rsch. columnist J. Am. Soc. CLU & ChFC, 1990—; edit. bds. Am. Jour. Alzheimer's Care and Research, 1988—, Aging and Health, 1987—, Experimental Aging, 1979-87, The Gerontologist, 1983-85; presenter papers in field;contbr. book chpts., articles to profl. jours. Recipient Virginia Little Meml. lectr. U. Vt., 1991; Nat. Inst. Aging post-doctoral rsch. fellow, 1986-87, Fulbright Rsch. fellow U. Glasgow, Scotland, 1988, Helsinki U., 1972-73; Woodrow Wilson fellow, 1965-66, Woodrow Wilson Dissertation fellow 1967-68. Fellow Gerontological Soc. Am. (exec. coun. behavioral and social sci. sect. 1982-83, rep. to planning com. Internat. Assn. Gerontology mtg. 1991-93, econs. aging spl. interest group 1989—, pub. policy com. 1983-84, chair fellow selection com. behavioral and social sci. sect. 1982-83, program com. 1980); mem. Am. Soc. Aging (bd. dirs. 1990—, edit. bd. Aging Today, 1990—, chair rsch. com. 1990—, co-organizer annual rsch. symposium, 1989-90, edit. bd. Generations, 1982-87, pub. com. 1982-86), Population Assn. Am., Am. Risk and Ins. Assn. (program planning com. 1992), Internat. Ins. Soc. Office: U Pa Boettner Inst 3718 Locust Walk Philadelphia PA 19104

CUTLIP, RANDALL BROWER, retired psychologist, former college president; b. Clarksburg, W.Va., Oct. 1, 1916; s. M.N. and Mildred (Brower) C.; m. Virginia White, Apr. 21, 1951; children: Raymond Bennett, Catherine Baumgarten. AB, Bethany Coll., 1940; cert. indsl. personnel mgmt., So. Meth. U., 1944; M.A., East Tex. U., 1949; Ed.D., U. Houston, 1953; LL.D., Bethany Coll., 1965, Columbia Coll., 1980; L.H.D., Drury Coll., 1975; Sc.D., S.W. Bapt. U., 1978; Litt.D., William Woods Coll., 1981. Tchr., adminstr. Tex. pub. schs., 1947-50; dir. tchr. placement U. Houston, 1950-51, supr. counselling, 1951-53; dean students Atlantic Christian Coll., Wilson, N.C., 1953-56, dean, 1956-58; dean personnel, dir. grad. div. Chapman U., Orange, Calif., 1958-60; pres. William Woods Coll., Fulton, Mo., 1960-81, pres. emeritus, 1981—; trustee William Woods U., Fulton, Mo., 1981-85, 92—; chmn. bd. dirs. Mo. Colls. Fund, 1973-75; chmn. Mid-Mo. Assn. Colls., 1972-76; bd. dirs. Marina del Sol, bd. pres., 1985-90, 92—. Mem. bd. visitors Mo. Mil. Acad., 1966-90, chmn. bd., 1968-72; trustee Schreiner Coll., Kerrville, Tex., 1983-92, Amy McNutt Charitable Trust, 1983—; bd. dirs. U. of Ams., Mexico City, 1984—, exec. v.p., 1985—; elder 1st Christian Ch.; trustee Permanent Endowment Fund, Scholarship Found. and Res. Fund. Served with AUS, 1943-45. Recipient McCubbin award, 1968, Delta Beta Xi award, 1959. Mem. Am. Personnel and Guidance Assn., Alpha Sigma Phi, Phi Delta Kappa, Kappa Delta Pi, Alpha Chi. Address: 1400 Ocean Dr Corpus Christi TX 78404

CYNAR, SANDRA JEAN, electrical engineering educator; b. Chgo., Aug. 7, 1941; d. Lionel Thomas and Dorothy Adeline (Swain) Bowers; m. Raymond John Cynar, Mar. 6, 1965; 1 child, Mark Jon. BSEE, Long Beach (Calif.) State U., 1963; MSEE, Calif. State U., Long Beach, 1978; PhD in Elec. Engring., U. Calif., Irvine, 1986. Controls engr. Gen. Dynamics, Pomona, Calif., 1963-64; mgmt. trainee Pacific Telephone, Alhambra, Calif., 1964-65; sci. programmer N. Am. Rockwell, Downey, Calif., 1965-68, McDonnell Douglas, Long Beach, 1968-70; prof. engring./computer engring. Calif. State U., Long Beach, 1977—; faculty advisor Calif. State U. IEEE Computer Soc., 1988-93, Soc. Women Engrs., 1988-93; program chmn. Simulation and Engring. Edn., San Diego, 1988-89, conf. chmn., 1989-90, session chmn., 1991-92; faculty advisor CSULB Micro Mouse Team, 1987-93. Author: Numerical Methods for Engineers; contbr. articles to profl. jours.; creator animated films: Tuned Pendulum, 1988, Solution of Ode's, 1989. Mem. IEEE (sr. mem.), Soc. Computer Simulation, Nat. Computer Graphics Assn., Am. Soc. Engring. Edn., ACM, Soc. Women Engrs. Republican. Methodist. Avocations: gardening, reading, computer graphics, travel. Office: Calif State U 1250 N Bellflower Blvd Long Beach CA 90840-0001

CYPHERS, DANIEL CLARENCE, aerospace engineer; b. Skokie, Ill., July 1, 1964; s. Daniel Clay and Joan Marie (Arkenburg) C.; m. Jana Sue Henderson, Aug. 12, 1989. B in Mech. Engring., U. Dayton, 1986, MS in Aerospace Engring., 1991. Engring. intern Inland Div. GM, Dayton, 1985-86; engr. Booz, Allen & Hamilton, Inc., Dayton, 1986-88, sr. engr., 1988-89; program engr. II W.J. Schafer Assocs., Inc., Dayton, 1989-91, program engr. III, 1991—; session chmn. Damping '91 Conf., San Diego, 1991. Author: (govt. document) A Summary of Space Environmental Effects Facilities Vol. 1: Atomic Oxygen Test Facilities, 1991; co-author: (govt. documents) Vulnerability Reduction Design Guide for Military Aircraft, 1989., Component Vulberability Ballistic Resistance Data Base, 1989. Mem. AIAA, Am. Astronautical Soc., Sports Car Club Am. Republican. Roman Catholic. Achievements include designing single sta. brake hose fatigue test machine; assisting in design of diagnostic method for antenna and radar arrays using infrared imaging.; performing vulnerability assessment of A-10 aircraft in close-air support role. Home: 55 Green Valley Enon OH 45323

CZACHURA, KIMBERLY ANN NAPUA, electrical engineer; b. Dayton, Ohio, Oct. 31, 1963; d. William Nohokaiu and Eileen Adele (Konicki) Kama; m. Steven Jonathan Czachura, Apr. 29, 1989; 1 child, Zachary Ahiokalani Kama. BSEE, Ohio State U., 1985. Devel. engr. Loral Def. Systems, Akron, Ohio, 1986—. Head advisor Loral Computer Explorer Post Great Trails coun. Boy Scouts Am., Akron, 1990—; youth group advisor Nativity Cath. Ch. Religious Edn. Program, Akron, 1990—. Mem. IEEE, Ohio State U. Alumni Club. Roman Catholic. Office: Loral Def Systems 1210 Massillon Rd Akron OH 44315

CZAJKOWSKI, CARL JOSEPH, metallurgist, engineer; b. Bklyn., May 23, 1948; s. John Anthony and Helen Elizabeth (Petliski) C.; m. Donna Margery Walcott, June 5, 1971; children: Kimberly, Lori. BS in Metall. Engring., U. Mo., Rolla, 1971; MS in Metall. Engring., Poly. Inst. N.Y., 1982. Quality control engr. United Nuclear Inc., Uncasville, Conn., 1972-73; quality assurance engr. Ebasco Svcs. Inc., N.Y.C., 1973-75; chief welding supr. L.I. Lighting Co., Hicksville, N.Y., 1975-80; sr. rsch. engr. Brookhaven Nat. Lab., Upton, N.Y., 1980—; mem. Joint Coordinating Com. on Civilian Nuclear Reactor Safety Mission, Dept. State/USSR, 1989-90. Contbr. chpts. to books, articles to profl. jours. Mem. Am. Soc. Metals (chair L.I. chpt. 1992-93), Am. Nuclear Soc. Home: PO Box 129 South Jamesport NY 11970 Office: Brookhaven Nat Lab Bldg 830 Upton NY 11973

CZAJKOWSKI, EVA ANNA, aerospace engineer, educator; b. New Britain, Conn., Sept. 4, 1961; d. Jan Wiktor and Weronika Janina (Nadolny) C. Student, Yale U., 1978; BS cum laude in Aero. Engring., Rensselaer Poly. Inst., 1983, M in Aero. Engring., 1983; MS in Aeronautics and Astronautics, MIT, 1985; PhD in Aerospace Engring., Va. Poly. Inst. and State U., 1988. Registered profl. engr., N.Y. Student trainee U.S. Govt., Washington, 1981-82; intern N.Y. State Assembly, Albany, 1983; teaching asst. Rensselaer Poly. Inst., Troy, N.Y., 1983, rsch. asst. U.S. Army Rsch. Office Ctr. Excellence, 1982-83; engring. analyst Pratt & Whitney Aircraft, West Palm Beach, Fla., 1984; rsch. asst. Gas Turbine and Plasma Dynamics Lab., Cambridge, 1984-85; rsch. asst., teaching asst. Dept. of Aerospace and Ocean Engring. Va. Poly. Inst. and State U., Blacksburg, 1985—; participant several U.S. dels. to six European nations, 1991-92. Contbr. papers to confs., articles to profl. jours. and ency. Vol. New Britain Gen. Hosp., 1977-79. Assoc. mem. Nat. Air and Space Mus., Am. Mus. Natural History. Recipient Commemorative Medal of Honor, 1987; named Woman of Yr., 1990, Internat. Woman of Yr., 1991-92; Amelia Earhart fellow Zonta Internat., 1983-84, 84-85; scholar Am. Helicopter Soc. Vertical Flight Found., 1983, Unico Nat., 1979-80; fellow Prat Presdl. Engring. Program, 1985-88. Mem. AIAA, NAFE, N.Y. Acad. Scis., Am. Astronaut. Soc., Am. Helicopter Soc., The Planetary Soc., Internat. Platform Assn., World Found. Successful Women, Nat. Space Soc., Confederation Chivalry, Sigma Xi, Sigma Gamma Tau, Tau Beta Pi, Phi Kappa Phi, Gamma Beta Phi. Avocations: art, horseback riding, piano, flying pvt. plane, skiing. Home: 170 Carlton St New Britain CT 06053-3106

CZARNECKI, GREGORY JAMES, ecologist; b. Bristol, Conn., Sept. 27, 1959; m. Karen Lynn Herman, Aug. 27, 1983. BS in Biology, Gannon U., 1989; MS in Biology, Edinboro U., 1992. Chem. technologist Lord Corp., Erie, Pa., 1980-92; writer, photographer Czarnecki Environ. Communications, Erie, 1989—; environ. specialist GAI Cons., Pitts., 1992—; instr. Pa. State U., Erie, 1992—, Gannon U., Erie, 1991—; chmn. adv. com. Presque Isle State Park, Erie, 1991—. Chmn. Millcreek Solid Waste Com., Erie, 1992; mem. Coastal Zone Mgmt. Steering Com., Erie, 1991--. Mem. Pa. Acad. Sci., Am. Soc. Mammalogists, Sigma Xi. Achievements include a European patent for Stable Butadiene Polymer Latices. Home: 509 Strathmore Ave Erie PA 16505

CZERNILOFSKY, ARMIN PETER, biochemist; b. Puchberg, Noe, Austria, Mar. 28, 1945; came to U.S., 1976; s. Josef and Paula (Gut) C.; m. Barbara Baker, 1982 (div. 1989); 1 child, Daniel Josef. Univ. dozent, PhD, U. Vienna, 1974, 91. Biochemist U. Vienna, Austria, 1972-77; postdoctoral fellow, adj. asst. prof. U. Calif., San Francisco, 1976-77, 78-81; postdoctoral fellow U. Vienna, 1977-78; staff scientist Acad. Sci., Salzburg, Austria, 1981-82, Max Planck Soc., Köln, Federal Republic of Germany, 1982-87; cons. Cal-Bio and AGS, Calif., 1987-88; biochemist Boehringer Ingelheim/Bender, Vienna, 1989—; prof. biochemistry Rheinisch-Westf. Tech. U., Aachen, Fed. Republic of Germany, 1986-87, pvt. dozent molecular biology, 1987—, dozent biochemistry, U. Vienna, 1991. Contbr. articles to profl. jours. on biochemistry and molecular biology. Rsch. grantee Am. Rsch. Fund, 1977, 81, Austrian Sci. Fund, 1981, European Community, 1986; fellow European Molecular Biology Orgn., 1973, 74-75, 81; recipient Honor award for Sci. and Arts The Körner Stiftung, 1974. Mem. AAAS, N.Y. Acad. Scis., Soc. Neurosci., Am. Microbiology Soc., Austrian Biochemistry Soc., German Cancer Soc. Office: Bender & Co, Dr Boehringer-Gasse 5-11, Vienna Austria A-1121

CZIRBIK, RUDOLF JOSEPH, cell biologist; b. Budapest, Hungary, July 4, 1953; came to U.S., 1957; s. Rudolf and Frances (Bucs) C.; m. Joan Marie Buonafide, July 2, 1983; children: Richard Joseph, Michael James. BS, Fordham U., 1975, PhD, 1984. Sr. scientist Alfacell Corp., Bloomfield, N.J., 1983-87; sr. scientist instrumentation lab. Fisher Diagnostics, Orangeburg, N.J., 1987-89; mgr. hybridoma lab. Roche Diagnostic Systems, Inc., Belleville, N.J., 1989—. Brookdale scholar Gerontology Ctr. Fordham U., N.Y.C., 1981-83. Mem. AAAS, Am. Chem. Soc., Tissue Culture Assn., Sigma Xi. Roman Catholic. Office: Roche Diagnostic Systems Inc 11 Franklin Ave Belleville NJ 07109-3597

DAAY, BADIE PETER, electrical engineer; b. Baghdad, Apr. 18, 1940; came to the U.S., 1967; s. Butrus Elias and Almas (Oshana) D.; m. Youliah Paul, Nov. 21, 1981; children: Fiona, Ramson. MS, U. Mo., 1972; PhD, Iowa State U., 1985. Control and instrumentation supr. Black & Veatch Cons., Kansas City, Mo., 1972-76; control and instrumentation engr. Sargent & Lundy Engrs., Chgo., 1976-77; rsch. asst. Iowa State U., Ames, 1977-80; elec. engr. Ill. Power Co., Decatur, 1980-87; asst. prof. U. System Ga.-Columbus Coll., 1987-91; sr. engr., scientist Systems Maintenance & Tech., Washington, 1991—; cons. in field, Columbus, 1987-91. Editor: Introduction to Engring. Profession, 1990; author: Ferroresonance Destroys Transformers, 1991. Recipient Cert. of Recognition Jr. Engring. Tech. Soc., 1990. Mem. IEEE, Nat. Soc. Profl. Engrs. Home: 10333 Colony Park Dr Fairfax VA 22032 Office: Systems Maintenance Technol 500 E St SW Ste 940 Washington DC 20024

DABKOWSKI, JOHN, electrical engineer, consultant, researcher; b. Chgo., Feb. 15, 1933; s. John and Harriet (Sierakowski) D.; m. Mary A Walkosz, Aug. 15, 1959 (dec. Apr. 1973); 1 child, Colette A.; m. Cecilia Klonowski, June 26, 1976; 1 child, Katherine A. BSEE, Ill. Inst. Tech., 1955, MSEE, 1960, PhD in Elec. Engring., 1969. Sr. rsch. engr. Ill. Inst. Tech. Rsch. Inst., Chgo., 1957-79; ops. mgr. Sci. Applications Internat. Corp., Hoffman Estates, Ill., 1979-85, dir. EM effects rsch., 1985-87, div. mgr., 1987-88; pres. Electro Scis., Inc., 1988—; instr. Grad. Sch., Ill. Inst. Tech., Chgo., 1962-79. Author rsch. publs. in field. With U.S. Army, 1955-57. Mem. IEEE (sr.), Nat. Assn. Corrosion Engrs., Sigma Xi. Republican. Roman Catholic. Home: 7021 Foxfire Dr Crystal Lake IL 60012-1641

DABROWSKI, ADAM MIROSLAW, digital and analog signal processing scientist; b. Łódz, Poland, Apr. 30, 1953; s. Mirosław and Zofia Anna (Siemierkowska) D.; m. Agata Brodnicka, Oct. 31, 1981; 1 child, Marcin. MScEE, Tech. U. Poznan, Poland, 1976; PhD in Electronics, Tech. U. Poznań, Poland, 1982; Habil. in Telecommunications, Tech. U. Poznan, Poland, 1989; MSc in Math., Adam Mickiewicz U., Poznań, 1977. Asst. Inst. Electronics and Telecommunication Tech. U. Poznań, 1977-82, asst. prof., 1982-88, assoc. prof., 1988-91, prof., 1991—; designer electronic equipment Lab. Electronics U. Oslo, 1978; vis. scientist Tech. U. Vienna, Austria, 1981-82; Alexander von Humboldt fellow Ruhr-U. Bochum, Fed. Republic Germany, 1984-86; vis. prof. ETHZ, Inst. Signal and Info. Processing, Zurich, Switzerland, 1987-89; cons. Teleelectronics Works, Teletra, Poznan, 1989—; organizer Commn. of European Communities, Trans-European Mobility Scheme for Univ. Studies, 1990—. Author: Recovery of Effective Pseudopower in Multirate Signal Processing, 1988 (award 1990); contbr. articles to profl. publs.; patentee in field. Recipient Minister award Ministry of Sci. and Tech., Warsaw, 1981, Ministry of Nat. Edn., Warsaw, 1990. Fellow Soc. Polish Elec. Engrs. (trustee 1987), Polish Soc. Theoretical and Applied Elec. Engring.; mem. IEEE (seminar organizer Polish chpt. Circuits and Systems Soc. 1991, chmn. 1992—), N.Y. Acad. Scis., Lions Club (v.p. Poznan 1991—). Roman Catholic. Avocations: movie making, piano, gardening, tinkering, travel. Home: Wiedeńska 35, 60 688 Poznań Poland Office: Tech U Poznań, Piotrowo 3a, 60 965 Poznan Poland

DA COSTA, NEWTON CARNEIRO AFFONSO, mathematics educator; b. Sao Paulo, Brazil, Sept. 16, 1929; s. Dymas C.A. and Sylvia E.C. (Carneiro) da C.; m. Neusa C. Feitosa, Jan. 15, 1955; children: Newton C.A., Sylvia Lucia F.A., Marcelo F.A. B in Civil Engring., U. Parana, Curitiba, PR, 1952, B in Math., 1955, PhD in Math., 1960. Asst. prof. U. Parana, Curitiba, PR, 1956-60; prof. U. Parana, Curitiba, 1961-67, U. Sao Paulo, 1968—; vis. prof., lectr., researcher at many Univs. including U. Calif., Stanford, Paris, Turin, Milan, Warsaw, Sofia, Sydney and Buenos Aires. Author: Foundations of Mathematics, Introduction to Logic; contbr. over 200 articles to profl. jours. Home: Rua Joao Cachoeira 272/51, 04535 São Paulo SP, Brazil Office: U Sao Paulo, University City, São Paulo SP, Brazil

DAEHN, GLENN STEVEN, materials scientist; b. Chgo., July 4, 1961; s. Ralph Charles and Beverly S. (Shanske) D.; m. Margaret A. Burkhart, Oct. 25, 1987; children: Andrew Joseph, Katrin Ellen. BS, Northwestern U., 1983; MS, Stanford U., 1985, PhD, 1988. Rsch. asst. Stanford U., Palo Alto, Calif., 1983-87; asst. prof. dept. materials sci. and engring. Ohio State U., Columbus, 1987-92, assoc. prof. dept. materials sci. and engring., 1992—; pres. BFD, Inc., 1992—. Co-editor: Modeling the Definition of Crystalline Solids, 1991. Named Nat. Young Investigator, NSF, 1992; recipient Young Investigator award Army Rsch. Office, 1992, R.L. Hardy Gold medal TMS, 1992, Marcus Grossman award ASM Internat., 1990. Mem. ASM Internat., Am. Ceramic Soc., Materials Rsch. Soc., Minerals, Metals and Materials Soc. (Robert Lansing Hardy Gold medal 1993). Achievements include description and practical application of how temperature changes accelerate the deformation of composite materials; co-development of new class of ceramic-metal composites. Home: 2076 Fairfax Rd Upper Arlington OH 43221 Office: Ohio State U Materials Sci Dept 116 W 19th Ave Columbus OH 43210

DAFERMOS, CONSTANTINE MICHAEL, applied mathematics educator; b. Athens, Greece, May 26, 1941; came to U.S., 1964; s. Michael Constantine and Sophia (Raptarchis) D.; m. Stella Theodoracopoulos, Sept. 6, 1964; children: Thalia, Michael. Diploma, Athens Nat. Tech. U., 1964; PhD, Johns Hopkins U., 1967. Fellow Johns Hopkins U., 1967-68; asst. prof. Cornell U., 1968-71; assoc. prof. Brown U., 1971-76, prof. applied math., 1976—, Univ. prof., 1988—, dir. Lefschetz Ctr. for Dynamical Systems, 1988—. Mem. editorial bd. Archive for Rational Mechanics and Analysis, 1972—, Jour. of Thermal Stresses, 1978—, Quar. Applied Math., 1985—, Math. Modelling and Numerical Analysis, 1986—, Proc. Royal Soc. Edinburgh, 1987—, Advanced Math. Applied Sci., 1989—, Math. Models and Methods, 1990—; contbr. articles in field to profl. jours. NSF grantee, 1970—, Office Naval Rsch. grantee, 1972-80, USAF grantee, 1972-73, U.S. Army grantee, 1973—, NASA grantee, 1972-73. Mem. Soc. Natural Philosophy (treas. 1975-76, chmn. 1977-78), Am. Math. Soc. Office: Brown U Div Applied Math Providence RI 02912

DAFTARI, INDER KRISHEN, physicist, researcher; b. Srinagar, Kashmir, India, Dec. 20, 1947; came to U.S., 1981; s. Som Nath and Som Rani (Zutshi) D.; m. Pratibha Kaul, June 2, 1974; children: Manish, Naveen. MSc in Physics, Roorkee (India) U., 1969; PhD in Sci., Jadavpur U., Calcutta, India, 1974. Vis. scientist in physics Syracuse (N.Y.) U., 1981-83, rsch. assoc., 1983-86; postdoctoral fellow in radiation therapy Thomas Jefferson U., Phila., 1986-88; staff scientist II Lawrence Berkeley (Calif.) Lab., 1988-93; sr. physicist U. Calif. Cancer Ctr., Davis, 1993—; lectr. in field. Contbr. over 75 articles and abstracts to sci. jours., including Phys. Rev. Letters, Med. Physics. Fellow Univ. Grants Commn., New Delhi, 1970. Mem. Am. Assn. Physicists in Medicine (travel grantee 1986, 91), Radiation Rsch. Soc., Sigma Xi. Achievements include discovery of zeta, others; research on proton, anti-proton, heavy ion, electron and photon beams with nuclear emulsion, bubble chamber, wire chamber and water phantoms; suggested numerous novel techniques in high energy and medical physics. Office: U Calif Lawrence Berkeley Lab 55-121 1 Cyclotron Rd Berkeley CA 94720

DAGGETT, WESLEY JOHN, electrical engineer; b. Massena, N.Y., Mar. 22, 1963; s. Walter Raymond and Phyllis (Bell) D.; m. Margaret Cathrine Bachman, Oct. 18, 1986; 1 child, Adam Wesley. BSEE, Rochester Inst. Tech., 1986; MBA, Chapman U., 1991. Registered intern engr., N.Y. Elec. test technician Gulf and Western, Union Springs, N.Y., 1984-86; elec. engr.-test Wickes Mfg. Co., Union Springs, N.Y., 1986-90; test engring. supr. Wickes Mfg. Co., Auburn, N.Y., 1990; mgr. quality assurance Transp. Electronics divsn. TRW, Auburn, N.Y., 1990—. Mem. IEEE, N.Y. State Profl. Engrs., Am. Soc. for Quality Control (cert. quality engr.). Office: TRW Trans Electronics Divsn 2240 Cranebrook Dr Auburn NY 13021

D'AGOSTINO, PAUL ANTHONY, electrical engineer; b. Boston, Nov. 10, 1963; s. Ralph John and Janet (Bordeau) D'A. MS, Northeastern U., 1987; MS in Indsl. Engring., N.Mex. State U., 1990, MS in Elec. Engring., 1992; MBA, Western New England Coll., 1990. Registered profl. engr., N.Mex. BS U. Lowell, 1985; MSME Air Force Plant Rep. Office, Wilmington, Mass., 1985, project mgr., 1985-88; MSEE USAF Contract Mgmt. Div., Albuquerque, 1988-90; aircraft systems analyst USAF Operational Test and Evaluation Ctr., Albuquerque, 1990—. Capt. USAF, 1985—. Mem. N.Mex. State U. Soc. Engring., Operation Rsch. Soc., Soc. Logistics Engrs. Republican. Roman Catholic. Home: 24 Farnham St Brockton MA 02402

D'AGOSTINO, RALPH BENEDICT, mathematician, statistician, educator, consultant; b. Somerville, Mass., Aug. 16, 1940; s. Bennedetto and Carmela (Piemonte) D'A.; m. Lei Lanie Carta, Aug. 28, 1965; children: Ralph Benedict, Lei Lanie Maria. A.B., Boston U., 1962, M.A., 1964; Ph.D., Harvard U., 1968. Lectr. math. Boston U., 1964-68, asst. prof., 1968-71, assoc. prof., 1971-76, prof. math. and stats., 1976—, chmn. dept. math., 1986-91, dir. stats. cons. unit, 1986—, dir. stats. unit Framingham Heart Study, 1985—, dir. Biostats MA/PhD Program, 1988—, lectr. law, 1975-91, prof. pub. health, 1982—, assoc. dean Grad. Sch., 1976-78, prof. law, 1991—; vis. prof. biostats. clin. epidemiology unit Univ. Hosp., Geneva, Switzerland, 1993; vis. lectr. Am. Statis. Assn., 1975-86, 88—; vis. prof. biostats. clin. epidiology unit Univ. Hosp., Geneva, 1993; vis. scientist Nat. Heart, Lung and Blood Inst., 1993; spl. scientist Boston City Hosp., 1981—, New Eng. Med. Ctr., 1990—; mem. Health Inst. New Eng. Med. Ctr., 1991—; cons. stats. United Brands, 1968-76, Diabetes and Arthritis Control Unit, Boston, 1971-75, City of Somerville, Mass., 1972, ednl. div. Bolt, Beranek & Newman, 1971, Harvard U. Dental Sch., 1969, Lahey Clinic Found., 1973-85, Walden Rsch., 1974-79, FDA Biometrics Div. and Over-the-Counter Div., 1975—, Cardio and Renal Div. FDA, 1987—, Arnold & Porter, 1980, Bedford Rsch. Assn., 1976-81, Corneal Scis., 1976, Biotek, 1979-88, GCA, 1979-87, Lever Bros., 1982-87, Conrail, 1981, FBI, 1984, Ctr. Psychiat. Rehab., Boston U., 1985—, NIMH, 1985, Dade Clin. Assays, 1986-90, Millipore, 1983—, VLI Corp., 1985-90, New England Coll. Optometry, 1985—, Dupont corp., 1985, Bristol Myers, 1986, Cheseborough Ponds, 1987—, med. decision making div. and health svcs. rsch. unit Tufts New England Med. Ctr., 1986—, Am. Inst. Rsch. in Social Scis., 1983-88, New England Rsch. Insts., 1987—, Thompson Med., 1987—, Merck, Sharpe and Dohme, 1988—, Carter Ctr. Emery U., 1989—, Unilever, 1991—, Miles, 1991—, Ultra Fem., 1991—, Health Effects Inst., 1992—, Forsyth Dental Clinic, 1992—, Bard Vascular, 1990—, Block Med., 1993—, Bayer Pharmaceutical, 1993—, Astra Pharmaceutical, 1993—, other rsch. insts.; mem. various FDA coms. including fertility and maternal health drugs adv. com., 1978-81, life support subcom., 1979-81, drug abuse adv. com., 1987-90, gastrointestinal drugs adv. com., 1990—; mem. task force on design and analysis in dental and oral rsch., 1979—, health tech. com. Harvard U., 1986-90; mem. Honolulu Heart Study adv. com. NIH, 1989—, Balt. Longitudinal Study of Aging adv. com., 1990, NIH Consensus Panel on Liver Transplantation, 1983, consensus Panel on Fresh Frozen Plasma, 1984, Consensus Panel on Geriatric Assessment Methods for Clin. Decision Making, 1987; mem. task force Office Tech. Assessment, 1980; mem. consensus panel on intraoral techniques ADA, 1990; mem. study sect. Agy. for Health Care Policy and Rsch., 1990—; prin., co-prin. investigator or sr. statistician on grants Nat. Ctr. Health Svcs. Rsch., 1976-82, Nat. Heart Lung and Blood Inst., 1982—, USAF, 1980-85, Nat. Cancer Inst., 1985—, Nat. Inst. Criminal Justice, 1982-85, Nat. Ctr. Child Abuse and Neglect,

1982-85, Robert Wood Johnson Found., 1981-85, Social Security Adminstrn., 1982-86, 90—, Motor Vehicle Mem. Assn., 1987, NIOSH, 1985, Nat. Insts. Aging, 1986—, Agency for Health Care Policy and Rsch., 1989—; grant and contract reviewer NAS, 1979—, Nat. Ctr. Health Svcs. Rsch., 1976, 89, NIH, 1983, NSF, 1987—, AHCPR, 1990—. Author: (with E.E. Cureton) Factor Analysis, An Applied Approach, 1983, (with Shuman and Wolfe) Mathematical Modeling, Applications in Emergency Health Services, 1984, (with Stephens) Goodness of Fit Techniques, 1986; assoc. editor: Am. Statistician, 1972-76, Statistics in Medicine, 1981-91, 93—, Jour. Am. Statis. Assn., 1993—; editor: Emergency Health Service Rev., 1981-88; mem. editorial bd. Biostatistica, 1990—; book reviewer Houghton Mifflin, Ho; contbr. articles to profl. jours.; co-developer instrument for predicting acute ischemic heart disease and stoke health risk appraisal function. Recipient Spl. citation FDA Commr., 1981, Metcalf awrd for excellence in teaching Boston U., 1985; Am. Heart Assn. fellow, 1991; pre-doctoral fellow NIH, 1962-68. Fellow Am. Statis. Assn. (pres. Boston chpt. 1972, v.p. 1971, mem. nat. coun. 1973-75, vis. lectr. 1976-78, 80—, Statistician of the Yr. Boston chpt. 1993); mem. Am. Heart Assn. (cardiovascular epidemiology coun.), Inst. Math. Stats., Am. Soc. Quality Control, Biometrics Soc. (regional adv. com. 1989—), Am. Pub. Health Assn. (chmn. sect. emergency health svcs. 1982-83, governing coun. 1983-85), Phi Beta Kappa, Sigma Xi. Home: 5 Everett Ave Winchester MA 01890-3523 Office: Boston U Dept Math 111 Cummington St Boston MA 02215-2411

DAGUM, CAMILO, economist, educator; b. Argentina, Aug. 11, 1925; arrived in Can., 1972, naturalized, 1978; s. Alexander and Nazira (Hakim) D. PhD (Gold medal summa cum laude), Nat. U. Cordoba, 1949, (hon.) U. Bologna, 1988, Nat. U. Cordova, 1988, m. Estela Bee, Dec. 22, 1958; children: Alexander, Paul, Leonardo. Mem. faculty Nat. U. Cordoba, 1950-66, prof. econs., 1956-66, dean Faculty Econ. Scis., 1962-66; sr. rsch. economist Princeton U., 1966-68; prof. Nat. U. Mex., 1968-70; vis. prof. Inst. d'Etudes du Devel. Econ. and Social, U. Paris, 1967-69, U. Iowa, 1970-72; prof. econs. U. Ottawa (Ont., Can.), 1972-91, chmn. dept., 1973-75, mem. acad. senate, 1981-84, bd. govs., 1983-84, prof. emeritus U. Ottawa, 1992—; prof. stats. and econs. U. Milan, Italy, 1990—, chmn. Inst. Quantitative Methods, 1993—; pres. Cordoba Inst. Social Security, 1962-63; cons. to govt. and industry, 1956—; rsch. prof. U. Rome, 1956-57, London Sch. Econs., 1960-62, Inst. Sci. Econmique Appliquée, Coll. France, 1965; vis. fellow Birkbeck Coll., U. London, 1960-61, Australian Nat. U., 1985; guest scholar Brookings Instn., 1978-79; vis. prof. U. Siena, Italy, 1987, 88, U. Rome, 1989; speaker in field. Mem. Acad. Coun. Rsch. Ctr. on Income Distbn., U. Siena, 1986—, Sci. Com. of Econ. Rsch. and Analysis Program, U. Montreal, 1992—. Res. officer Argentine Army, 1948. Decorated Pro-Patria Gold medal, 1948; hon. prof. Inst. Advanced Studies, Salta, Argentina, 1972; extraordinary prof. Cath. U. Salta, 1981—. Elected mem. Accademia di Scienze e Lettere, Istituto Lombardo, 1992—. Mem. Internat. Inst. Sociology, Internat. Statis. Inst., Statis. Soc., Econ. Soc., Econ. History Soc. Argentina, U.S. Eastern Econ. Assn., Econometric Soc., Am. Statis. Assn., Am. Econ. Assn., Can. Econ. Assn., Can. Statis. Soc., Assn. Social Econs., Acad. of Scis. of the Inst. of Bologna, Academia Lombarda di Scienze e Lettere Italy. Roman Catholic. Author books on econ. theory; editor econ. and statis. jours.; contbr. articles to profl. jours. Home: 408 Buena Vista Rd, Rockcliffe Park, ON Canada K1M 0W3 Office: U Ottawa Faculty Social Scis, Dept Econs, Ottawa, ON Canada K1N 6N5

DAHIYA, JAI BHAGWAN, chemist; b. Badhkhalsa, Haryana, India, June 1, 1956; came to U.S., 1978; s. Vijai Singh Dahiya and Lachhmi (Antil) Devi; m. Sharda Grewal Dahiya, June 30, 1982; children: Anita, Vicki, Neil. BS, Kurukshetra U., India, 1977, East Tex. State U., 1982; MS, East Tex. State U., 1985. Rsch. asst. East Tex. State U., Commerce, 1979-80, rsch. scholar, 1980-81; lab. asst. Wesleyan Coll., Buckhannon, W.Va., 1983; mgr. Texmart Inc., Commerce, 1984-86; chemical analyst E-Systems Inc., Greenville, Tex., 1987-89; technologist in Chemistry Barnes Hosp. Washington U. Med. Ctr., St. Louis, 1990-91; chemist Internat. Tech. Corp., St. Louis, 1991—. Robert A. Welch Rsch. Scholar East Tex. State U., 1980-81. Mem. Am. Chem. Soc. Avocations: tennis, jogging, swimming, reading, photography. Home: 12332 Inletridge Dr # B Maryland Heights MO 63043

DAHIYA, JAI NARAIN, physics educator, researcher; b. Badh Khalsa, Haryana, India, Oct. 15, 1946; came to U.S., 1973; s. Vijay Singh Dahiya and Lachhmi (Antil) Devi; m. Leela Grewal, Jul. 4, 1967; children: Rajiv S., Meera, Vijay N., Anjuli. MSc, Meerut Univ., 1968; PhD, North Tex. State U., 1980; BSc, Panjab Univ., 1965. Rsch. assoc. North Tex. State Univ., Denton, 1974-80; fellow Tex. A&M Univ., College Station, 1981; asst. prof. Wesleyan Coll., Buckhannon, W.VA., 1981-84; asst. prof. Southeast Mo. State Univ., Cape Girardeau, Mo., 1984-89, assoc. prof., 1989-93; lectr. Hindu Coll., Sonepat, Haryana, 1968-72; vis. asst. prof. North Tex. State Univ., Denton, 1980; mem. faculty senate Southeast Mo. State U., 1989-92; mem. adv. bd. Internat. Students, Southeast Mo. State U., 1992—; Sci. Symposium com., Southeast Mo. State U., 1984-88, Nat. Sci. Steering com., 1988-91, mem. scholarship com., 1988-91, college coun., 1992-95. Author: Solution Manuals for Physics, 1993; contbr. articles to profl. jours. Recipient Faculty Achievement award Southeast Mo. State Univ., 1991-92, Faculty Senate award, 1992, Svc. Recognition award, North Tex. State Univ., 1980. Fellow Soc. Physics Student (pres. 1966-68, treas. 1978-79),Mo. Acad. Sci., Am. Physical Soc., Am. Soc. Engring. Edn., Pi Sigma Pi. Achievements include research interests include microwave spectroscopy, dielectric relaxation in polar molecules, liquid crystals, superconductors, and biological samples at microwave frequencies. Home: 3914 Carolewood Dr Cape Girardeau MO 63701 Office: Southeast Mo State Univ MS 6600 One University Plaza Cape Girardeau MO 63701

DAHIYA, RAJBIR SINGH, mathematics educator, researcher; b. Rattangarh, Haryana, India, Dec. 3, 1940; came to U.S., 1968; s. Ram S. and Kesar (Devi) D.; m. Krishna Tavathia, Dec. 11, 1966; children: Madhu, Ranjan. PhD, Birla Inst. Sci. and Tech., Pilani, India, 1967. Lectr. Birla Inst. Sci. and Tech., 1967 68; asst. prof. math. Iowa State U., Ames, 1968-72, assoc. prof., 1972-78, prof., 1978—; reviewer math. revs. Zentrallblat; referee applied math. jours. Contbr. over 100 articles on delay and advanced differential equations, transform theory and spl. functions to U.S., European and Australian profl. jours. Mem. Am. Math. Soc., Soc. Indsl. and Applied Math. Democrat. Hindu. Home: 3305 Taft Ct Ames IA 50010-4340 Office: Iowa State U Math Dept Ames IA 50011

DAHL, ANDREW WILBUR, health services executive; b. N.Y.C., Feb. 19, 1943; s. Wilbur A. and Margret L. Dahl; BS, Clark U., 1968; MPA, Cornell U., 1970; ScD, Johns Hopkins U., 1974; m. Janice White, Sept. 4, 1965; children: Kristina, Jennifer, Meredith. Staff asst. Md. Comprehensive Health Planning Agy., Balt., 1970-72; dir. planning St. John Hosp., Detroit, 1972-79; exec. v.p., chief oper. officer St. John Health Corp., Detroit, 1979-85; pres., chief exec. officer United Health System, Detroit, 1983-88; v.p. devel. Hosp. Corp. Am. Mgmt. Co., 1988-90; pres., chief exec. officer IVF America Inc., Greenwich, Conn., 1990—; instr. U. Mich. Bur. Hosp. Adminstrn., 1981-88. Bd. dirs. Detroit Sci. Ctr., 1984-91; mem. Nat. Com. for Quality Health Care, Washington, 1984-89; bd. dirs. Forum Health Care Planning. Served with USN, 1965-67. Recipient Disting. Service award Mich. Jaycees, 1977, Outstanding Contbns. to Profl. Mgmt. award, Cornell U., 1980. Mem. AAAS, Am. Coll. Hosp. Adminstrs., Am. Hosp. Assn., Am. Pub. Health Assn., Internat. Health Econs. and Mgmt. Inst., Am. Fertility Soc., Detroit Athletic Club, Cornell Club N.Y. Methodist. Office: 500 W Putnam Ave Greenwich CT 06830-6096

DAHL, GERALD LUVERN, psychotherapist, educator, consultant, writer; b. Osage, Iowa, Nov. 10, 1938; s. Lloyd F. and Leola J. (Painter) D.; m. Judith Lee Brown, June 24, 1960; children: Peter, Stephen, Leah. BA, Wheaton Coll., 1960; MSW, U. Nebr., 1962; PhD in psychotherapy (Hon.), Internat. U. Found., 1987. Juvenile probation officer Hennepin County Ct. Services, 1962-65; cons. Citizens Council on Delinquency and Crime, Mpls., 1965-67; dir. patient services Mt. Sinai Hosp., Mpls., 1967-69; clin. social worker Mpls. Clinic of Psychiatry, 1969-82, G.L. Dahl & Assocs., Inc., Mpls., 1983—; assoc. prof. social work Bethel Coll., St. Paul, 1964-83; spl. instr. sociology Golden Valley Luth. Coll., 1974-83; pres. Strategic Team-Makers, Inc., 1985—; adj. prof. U. Wis., River Falls, 1988—. Founder Family Counseling Service, Minn. Baptist Conf.; bd. dirs. Edgewater Baptist Ch., 1972-75, chmn., 1974-75. Mem. AAUP, Am. Assn. Behavioral Therapists, Pi Gamma Mu. Author: Why Christian Marriages Are Breaking Up, 1979; Everybody Needs Somebody Sometime, 1980, How Can We Keep Christian Marriages from Falling Apart, 1988. Office: Ste 140 4825 Hwy 55 Minneapolis MN 55422-5155

DAHL, HILBERT DOUGLAS, mining company executive; b. Wilkinsburg, Pa., July 1, 1942; s. Hilbert Greiner and Melissa (Jones) D.; m. Joanne Hallman, Apr. 14, 1965; children: James, Michael, Melynda, Gregg. B in Mining Engring., Pa. State U., 1965, M in Mining Engring., 1967, PhD in Mining Engring., 1969; postgrad., MIT, 1981. Rsch. scientist Conoco, Ponca City, Okla., 1969-78; dir. underground mine engring. Consolidation Coal Co., Pitts., 1978, v.p. dept. exploration and land, 1979, v.p. Moundsville (W.Va.) ops., 1980-82, exec. v.p. engring., exploration and environ. affairs and non-mining, 1982-90; pres., chief oper. officer Drummond Co., Inc., Birmingham, Ala., 1990—. Contbr. articles to profl. jours.; holder 11 patents. Recipient Disting. Alumni award Pa. State U., 1985. Mem. AIME (Rock Mechanics award 1990, Disting. Mem. award 1991), Nat. Coal Assn. (bd. dirs.), Nat. Coal Coun. (bd. dirs.), Am. Coke and Coal Chems. Inst. (bd. dirs. 1991), Soc. for Mining, Metallurgy and Exploration (Daniel C. Jackling award 1991), Ala. Coal Assn. (bd. dirs. 1991), Old Timers Club, King Coal Club, Shoal Creek Country Club, Musgrove Country Club. Republican. Office: Drummond Co Inc PO Box 10246 Birmingham AL 35202-0246

DAHL, LAWRENCE FREDERICK, chemistry educator, researcher; b. Evanston, Ill., June 2, 1929; s. Lawrence Gustave and Anne (Stuessy) D.; m. June Lomnes, Sept. 1, 1956; children: Larry, Eric, Christopher. BS in Chemistry, U. Louisville, 1951; PhD, Iowa State U., 1956; DSc (hon.), U. Louisville, 1991. Postdoctoral fellow Ames (Iowa) Lab. AEC, 1957; from instr. to assoc. prof. chemistry U. Wis., Madison, 1957-64, prof., 1964-78, R. E. Rundle chair, 1978—, Hilldale chair and prof., 1991—; Brotherton rsch. prof. U. Leeds, 1983. Recipient Inorganic Chemistry award Am. Chem. Soc., 1974, Disting. Alumnus award Coll. Letters and Sci., U. Louisville, 1983, Sr. U.S. Scientist Humboldt award Alexander von Humboldt-Stiftung, 1985, R.S. Nyholm medal Royal Soc. Chemistry, 1985, P. Chini medal. Societa Chimica Italiana, 1989, J.C. Bailar, Jr. medal U. Ill., 1990; named to Honorable Order of Ky. Cols., 1982; Alfred P. Sloan fellow, 1963-65, Guggenheim fellow, 1969-70; 1st Alumnus fellow of coll. letters and sci. U. Louisville, 1990. Fellow AAAS, N.Y. Acad. Sci., Am. Acad. Arts and Scis.; mem. NAS. Home: 4817 Woodburn Dr Madison WI 53711-1345 Office: Univ of Wis Dept of Chemistry 1101 University Ave Madison WI 53706-1396

DAHLIN, ROBERT STEVEN, chemical engineer; b. Charleston, W.Va., June 30, 1953; s. Robert J. and Grace A. (Cochran) D.; m. Deborah A. Craig, June 24, 1978; 1 child, Matthew. BS in Chem. Engring., W.Va. U., 1975; MS in Chem. Engring., U. Ky., 1976; PhD in Chem. Engring., U. Ala., 1985. Registered profl. engr., Ala. Process engr. FMC Corp., Charleston, W.Va., 1975; grad. rsch. asst. U. Ky., Lexington, 1975-79; rsch. chem. engr. So. Rsch. Inst., Birmingham, Ala., 1979-85, sect. head, 1986-90, program mgr., 1990—; cons., project mgr. EPA, Dept. Energy, Electric Power Rsch. Inst., pvt. utilities, industries. Contbr. over 40 articles to profl. publs.; presenter in field. Mem. Air and Waste Mgmt. Assn. (tech. coms. 1976—), Am. Inst. Chem. Engrs., Alpha Phi Omega, Tau Beta Pi, Omega Chi Epsilon. Republican. Achievements include one patent in field; research in fine particle control and measurement. Home: 3705 Keswick Circle Birmingham AL 35242 Office: So Rsch Inst 2000 9th Ave S Birmingham AL 35205

DAHLSTROM, NORMAN HERBERT, engineering executive; b. Chgo., Oct. 22, 1931; s. Herbert D. and Myrtle C. (Papenthein) D.; m. Salome B. Filipiak, Oct. 25, 1952; children: Kenneth F., Dennis J. Susan M. Diploma, Chgo. Vocat. Sch., 1950; student, Northwestern U., 1952-56. Registered profl. engr., cert. mfg. engr. Mech. draftsman Halicrafters, Zenith Corps., Chgo., 1950-52; designer Stewart Warner Corp., Chgo., 1952-56; sr. design engr. Cook Rsch. Lab., Chgo., 1956-59; chief prodn. engr. Gen. Am. Transp. Corp., Chgo., 1959-69; pres., owner KDK Corp., Chgo., 1969-70; v.p., gen. mgr. KDK Corp. div. Pam Am. Resources, N.Mex., 1970-71; indsl. engring. mgr. bus. machine div. The Singer Co., Albuquerque, 1971-76; bus. tech. cons. Albuquerque, 1976-80; mfg. engring. mgr. Boeing Mil. Airplanes, Wichita, Kans., 1980-88, ops. program mgr., 1988-90, mgr. Work Transfer Mfg. Engring., 1990-93; engring. mgr. comml. spares prodn., 1993—; cons. in field, 1976-80; advisor Wichita State U., 1987-88; chmn. Wichita Indsl. Trade Show, 1988. Inventor three patents. Capt. Rep. party, Morton Grove, Ill. 1968; lectr., tchr. St. Mary's Ch., Derby 1988. Mem. Soc. Mfg. Engrs. (vice chmn. region 10, past chpt. chmn.), KC, Elks. Roman Catholic. Avocations: golf, fishing, gardening.

DAHM, ARNOLD JAY, physicist, educator; b. Oskaloosa, Iowa, Sept. 12, 1932; s. Henry and Minnie Henrietta (Van Roekel) D.; m. Susan Margaret, June 29, 1968; children: Amy, Kristi. BA, Cen. Coll., 1958; PhD in Physics, U. Minn., 1965. Postdoctoral fellow U. Pa., Phila., 1966-68; asst. prof. Case Western Res. U., Cleve., 1968-73, assoc. prof., 1973-80, prof., 1980—; Fulbright Hays fellow U. Sussex, Brighton, Eng., 1976-77, U. Mainz, Germany, 1983-84. Active Coalition to Support Human Svcs., Cleve., 1970-76; pres. Your Schs., Cleveland Heights, Ohio, 1981-83; presbytery Peacemaking Task Force, 1990-91. Cpl. U.S. Army, 1953-55. Fellow Am. Phys. Soc.; mem. AAUP, Sigma Xi. Presbyterian. Achievements include study of Josephson junctions, charge motion in solid helium, two-dimensional electrons on the surface of liquid helium, two-dimensional melting directional solidification. Office: Case Western Res U Dept Physics Cleveland OH 44106

DAHOTRE, NARENDRA BAPURAO, materials scientist, researcher, educator; b. Poona, India, Dec. 2, 1956; came to U.S., 1981; s. Bapurao B. and Latika B. Dahotre; m. Anita Thangan, Dec. 6, 1984; 1 child, Shreyas. BS in Metall. Engring., U. Poona, 1980; MS in Metallurgy, Mich. State U., 1983, PhD in Materials Sci., 1987. Instr. metallurgy and materials sci. Mich. State U., East Lansing, 1985-86; postdoctoral fellow, instr. materials sci. U. Wis., Milw., 1987 88; rsch. metallurgist U. Tenn. Space Inst., Tullahoma, 1988-91, adj. asst. prof. engirng. sci. and mechanics, 1991—. Mem. internat. editorial bd. Indsl. Laser Handbook, 1992-94; reviewer Jour. Metall. Transactions, 1991—, Jour. Materials and Mfg. Processes, 1991—; contbr. articles to profl. jours. and conf. procs. Rsch. grantee Internat. Lead Zinc Rsch. Orgn., 1990-91, Energy Conversion Program, U. Tenn. Space Inst., 1992—, NASA Marshal Flight Ctr., 1990-91. Mem. Am. Soc. for Metals, The Metall. Soc. of AIME, Laser Inst. Am., Materials Rsch. Soc., Sigma Xi. Achievements include rsch. on laser processign of composites, ceramics and intermetallic compounds, phase transformations, characterization of materials using analytical techniques. Office: U Tenn Space Inst BH Goethert Pkwy Tullahoma TN 37388

DAI, CHARLES MUN-HONG, mechanical engineer, consultant; b. Shanghai, China, Feb. 19, 1953; came to U.S., 1972; s. Wing-Fong and So-Ching (Chung) D.; m. Yi-Ying Zhang, May 1, 1990. PhD, U. Md., 1988. Mech. engr. GE, Schenectady, 1976-77, Atlantic Rsch., Gainesville, Va., 1981-82, David Taylor Model Basin, Bethesda, Md., 1982—. Contbr. articles to profl. jours. Recipient Oustanding Svc. award USN, 1989. Mem. ASME, Soc. for Indsl. and Applied Math. Achievements include fundamental rsch. in turbulence modeling using vorticity representation, new propulsor concept for marine propulsion and cavitation modeling. Office: David Taylor Model Basin Bethesda MD 20084

DAI, HAI-LUNG, physical chemist, researcher; b. Taiwan, China, Feb. 25, 1954; came to U.S., 1976; s. Chuan-yen and Cheng-hua (Liu) Tai; m. Tien Tien, 1976 (div. 1984); m. Surrina Mi-Na Hu, 1992. BS, Nat. Taiwan U., 1974; PhD, U. Calif., Berkeley, 1981; MS (hon.), U. Pa., 1989. Asst. prof. U. Pa., Phila., 1984-89; assoc. prof., 1989-92; prof. 5, Phila., 1992—; vis. prof. Nat. Taiwan U., Taipei, 1991-92; adj. prof. Nat. Tsing-Hua U., Hsing-chu, Taiwan, 1991-92. Editor: Vibrational Dynamics and Spectroscopy by Stimulated Emission Pumping, 1993; contbr. articles to profl. jours. Conductor Chinese Musical Voices, Phila., 1988—; pres. Nat. Taiwan Alumni Assn., Greater Phila. area, 1989. Postdoctoral fellow MIT, Cambridge, Mass., 1981-84, Sloan Found. fellow, 1988-90, Disting. New Faculty fellow Camille and Henry Dreyfus Found., 1985, Tchrs. Scholar award, 1989, Coblentz award in spectroscopy, 1990. Fellow Am. Phys. Soc.; mem. Am. Chem. Soc., Chinese Am. Chem. Soc., Chinese Am. Phys. Soc. (exec. coun.). Achievements include developed several laser spectroscopic techniques for studying highly vibrationally exited radicals, the low frequency intermolecular vibrational levels of molecular clusters and species adsorbed on surfaces. Office: U Pa Dept Chemistry 231 S 34th St Philadelphia PA 19104

DAI, PETER KUANG-HSUN, government official, aerospace executive; b. Shanghai, Republic of China, Dec. 14, 1934; came to the U.S., 1959; s. Ying Shen and Esther Ya-Ying (Huang) D.; m. Janie Ko-Tsen Chen; children: Diane P., Frederick J. BS in Engring., Cheng Kung U., Tainan, Taiwan, 1957; MS in Engring., U. Ill., 1960, PhD in Engring., 1963. Rsch. assoc. U. Ill., Urbana, 1962-63; rsch. engr. Air Force Materials Lab., Dayton, Ohio, 1963-65; sect. head TRW Systems, Redondo Beach, Calif., 1965-68, 69-74; prin. engr. Ralph M. Parsons Co., L.A., 1968-69; dept. mgr. TRW Def. and Electronics, Redondo Beach, 1974-80; asst. program mgr. ballistic missile div. TRW Def. and Space, San Bernardino, Calif., 1980-87, dep. program mgr., 1987-91; dir. nat. space program Republic of China, 1992—; lectr. U. So. Calif., L.A., 1967-68, Calif. State U., Fullerton, 1968-69, Long Beach, 1971-72; tech. rev. panel mem. Underground Nuclear Test Def. Nuclear Agy., Washington, 1988-89. Mem. ASCE (chmn. subcom. 1978), AIAA, Chinese-Am. Engrs. and Scientists Assn. So. Calif. (pres. 1976, chmn. bd. 1977-79). Avocations: tennis, ballroom dancing. Office: Nat Space Program Office Rm 2510, 25F 333 Keelung Rd, Sec 1 Taipei 10548, Taiwan

DAIE, JALEH, science educator, researcher, administrator; b. Iran, July 17, 1948; came to U.S., 1973; d. M.A. Daie and D. Zahiroleslam Zadeh; m. Roger E. Wyse, Dec. 27, 1986. BS, U. of Ahwaz, Iran, 1970; MS, U. Calif., Davis, 1975; PhD, Utah State U., 1980. Postdoctoral fellow Agrl. Rsch. Svc. U.S. Dept. Agr., Logan, Utah, 1980-82; rsch. asst. prof. Utah State U., Logan, 1982-85; assoc. prof. Rutgers U., New Brunswick, N.J., 1985-89, prof. dept. crop sci., 1989-93, dir. interdisciplinary plant biology grad. program, 1989-92, chmn. crop sci. dept., 1989-92; dir. Ctr. for Interdisciplinary Studies in Turfgrass Scis. Rutgers U., New Brunswick, 1991-93; also chmn. George H. Cook Honors Rutgers U., New Brunswick, N.J., 1987-91; sr. sci. advisor to v.p. for acad. affairs U. Wis. System, Madison, 1993—; prof. botany U. Wis., Madison, 1993—; invited speaker. Editor, author in field. Grantee NSF, others; named Outstanding Young Woman Am., 1982; Henry Rutgers rsch. fellow, 1985-87. Mem. Sigma Xi (pres. 1989-90), Phi Kappa Phi. Avocations: interior design, opera, skiing. Office: U Wis Dept Botany Birge Hall Madison WI 53706

DAIGLE, RONALD ELVIN, medical imaging scientist, researcher; b. Lake Charles, La., Oct. 14, 1944; s. Elvin and Dorothy Helen (Mayo) D.; m. Ann Jean Lassman, Dec. 30, 1966; children: Janah Wryn, Jon Kim. BA in Physics, U. Calif., Santa Barbara, 1967; MSEE, Colo. State U., 1971, PhD in Physiology and Biophysics, 1974. Rsch. assoc. dept. physiology and biophysics Colo. State U., Ft. Collins, 1974; postdoctoral fellow cardiovascular tng. program Ctr. for Bioengring., U. Wash., Seattle, 1974-75; rsch. assoc. Ctr. for Bioengring., U. Wash., 1975-76; sr. engr. cen. rsch. Varian Assocs., Palo Alto, Calif., 1976-81; dir. advanced devel. Advanced Tech. Labs., Bothell, Wash., 1981-86, sr. scientist, 1986-88, chief sr. tech. staff, 1988-89, dir. advanced products, 1990—, lectr. Doppler physics, 1987—, tech. fellow in perpetuity, 1988—; cons. in cardiovascular instrumentation NASA, Houston, 1987—; mem. microsensor rev. bd. Wash. Tech. Ctr., Seattle, 1987—. Mem. editorial bd. Jour. Cardiovascular Ultrasonography, 1983—; contbr. articles to profl. jours., chpts. to books; patentee in field. Recipient IR-100 award Indsl. Rsch. mag., 1978; Nat. Inst. Heart and Lung grantee, 1976. Mem. Am. Inst. Ultrasound in Medicine, Amiga Users Group (pres. 1988-89). Democrat. Avocations: music, triathalons, llama ranching. Home: 22126 NE 62d Pl Redmond WA 98053 Office: Advanced Tech Labs PO Box 3003 Bothell WA 98041-3003

DAILEY, FRANKLYN EDWARD, JR., electronic image technology company executive, consultant; b. Rochester, N.Y., Feb. 5, 1921; s. Franklyn Edward and Isabel Louise (Lasher) D.; m. Marguerite Virginia Parker, Apr. 1, 1944; children: Franklyn III, Michael, Philip, Elizabeth, John, Paul, Thomas, Vincent. BS, U.S. Naval Acad., 1942; BSEE, U.S. Naval Postgrad. Sch., 1950; MS in Applied Physics, UCLA, 1951. Commd. ensign USN, 1942, advanced through ranks to capt.; mgr. planning and engring. ops. Stromberg-Carlson Co., Rochester, N.Y, San Diego, 1956-61; treas. Stati-Systems Inc., Springfield, Mass., 1962-65; dir. mfg. Tecnifax Corp., Holyoke, Mass., 1965-66; asst. dir. rsch. The Plastic Coating Corp., South Hadley, Mass., 1966-67; asst. v.p. mktg. Scott Graphics Inc., South Hadley, 1967-68, v.p. new bus. devel., 1968-70, v.p. rsch., 1970-76; cons. Image Tech. & Application, Wilbraham, Mass., 1977—; pres. Photon Chroma Inc., Westfield, Mass., 1982, 84; image cons. McGraw-Hill, N.Y.C., 1978, Isomet Corp., Springfield, Va., 1980; v.p. mfg. Coulter System Corp., Bedford, Mass., 1981; chmn. Electronic Imaging Conf., Boston, Anaheim, 1985-90; speaker in field. Author three current reports on electronic imaging. Pres. Pioneer Valley chpt. Am. Diabetes Assn., 1986-88. Roman Catholic. Avocations: tennis, biking. Home and Office: 19 Brookside Cir Wilbraham MA 01095-2101

DAILEY, VICTORIA ANN, economist, policy analyst; b. San Antonio, Aug. 30, 1945; d. John Thomas and Helen (Bass) D. BA, Swarthmore Coll., 1967; PhD, U. Va., 1973. Economist FTC, Washington, 1972-79, U.S. Dept. Transp., Washington, 1979—. Brookings Econ. Rsch. fellow Brookings Instn., 1971-72. Office: US Dept Transp 400-7th St SW Washington DC 20590

DAILY, AUGUSTUS DEE, JR., psychologist, retired; b. St. Louis, Aug. 13, 1920; s. Augustus D. and Delta Antoinette (Baker) D.; m. Natalie Maravic, June 3, 1955(div.); children: Marla Dee, Gordon Ames, Brian Ellis. BA, Washington U., St. Louis, 1942, PhD, 1951. Instr. Washington U., 1947-50; rsch. asst. Princeton (N.J.) U., 1950-51; asst. prof. Carnegie Inst. of Tech., Pitts., 1951-54; exec. scientist Am. Inst. for Rsch., Santa Barbara, Calif., 1959-61; sr. mem. tech. staff System Devel. Corp., Santa Monica, Calif., 1961-63; lectr., rsch. assoc. U. So. Calif., L.A., 1968—; cons. in human factors and indsl. psychology, 1968-86; cons. in field; human factors experience include mem. tech. staff Rockwell Space Sta. Div., 1984-86, Rockwell B-1 B Div., 1981-84, Autonetics Marine System Div., 1978-80, others; faculty U. Calif., Santa Barbara, Calif. State U., Fla. Inst. Tech. Contbr. articles to tech. publs. Pres. Hampton Twp. Civic Assn., Allison Park, Pa., 1956-58. Lt. USNR, 1942-46, PTO. Decorated Bronze Star. Mem. Am. Psychol. Assn., Sigma Xi. Unitarian. Home: 784Q Via Los Altos Laguna Hills CA 92653

DAILY, LOUIS, ophthalmologist; b. Houston, Apr. 23, 1919; s. Louis and Ray (Karchmer) D.; B.S., Harvard U., 1940; M.D., U. Tex. at Galveston, 1943; Ph.D., U. Minn., 1950; m. LaVerl Daily, Apr. 5, 1958; children: Evan Ray, Collin Derek (dec.). Intern, Jefferson Davis Hosp., Houston, 1943-44; resident in ophthalmology Jefferson Davis Hosp., 1944-45, Mayo Found., Rochester, Minn., 1947-50; individual practice medicine, specializing in ophthalmology, Houston, 1950—; clin. assoc. prof. ophthalmology U. Tex-Houston, 1972-86, Baylor Med. Sch., Houston, 1950—. Vice pres. bd. dirs. Mus. Med. Sci., 1973-85, pres., 1980-82. Served as lt. (j.g.) USNR, 1945-46. Diplomate Am. Bd. Ophthalmology. Fellow A.C.S., Internat. Coll. Surgeons; mem. Soc. Prevention of Blindness (med. chmn. Tex. 1968-70), Contact Lens Assn. Ophthalmologists (exec. bd. 1976-78), Tex. Ophthal. Assn. (pres. 1963-64), Houston Ophthal. Soc. (pres. 1970-71), numerous other med. socs., Sigma Xi, Alpha Omega Alpha. Jewish. Clubs: Doctors, Harvard (dir. 1965-66) (Houston). Editorial bd. Jour. Pediatric Ophthalmology, 1964-68; asso. editor Eye, Ear, Nose and Throat Monthly, 1962-65, Jour. Ophthalmic Surgery, 1970; contbr. numerous articles to profl. publs., also contbr. to books. Home: 2523 Maroneal St Houston TX 77030-3117 Office: 1517 Med Towers 1709 Dryden Rd Houston TX 77030

DAILY, WILLIAM ALLEN, retired microbiologist; b. Indpls., Nov. 10, 1912; s. Thomas Alvin Daily and Mary Bernice Swengel; m. Eva Fay Kenoyer, June 24, 1937. BS, Butler U., 1936; MS, Northwestern U., 1938. Asst. sr. microbiologist Eli Lilly Co., Indpls., 1941-77. Co-author: Coccoid Myxophceae, 1956, History of Indiana Academy of Science 1885-1984, 1984. Curator cryptogamic bot. herbarium biology dept. Butler U., Indpls. 1941—. Mem. Ind. Acad. Sci. (pres. 1958), Phycological Soc. Am. (pres. 1958), Bot. Soc. Am., Sigma Xi. Republican. Home: 5884 Compton St Indianapolis IN 46220-2653

DAJANI, ESAM ZAPHER, pharmacologist; b. Jaffa, Palestine, May 30, 1940; came to U.S., 1958; s. Zapher Rageb and Mamdouha (Dajani) D.; m.

Najwa Said Beidas, July 16, 1966; children: Mona, Zapher, Nora. BS in Pharmacy, U. Mo., 1963; MS in Pharmacology and Med. Chemistry, Auburn U., 1966; PhD in Pharmacology, Purdue U., 1968. Sr. pharmacologist Rohn and Hass Co., 1968-72; group leader G.D. Searle and Co., Chgo., 1974-80, chmn. G.I. diseases, 1974-80, sect. head, 1980, asst. dir., 1980-82, assoc. dir., 1982-85, dir. cytotec, 1985-87, dir. clin. rsch., 1987-93; pres. Internat. Drug Devel. Cons., Long Grove, Ill., 1993—; editorial adv. bd. Drug Devel. Rsch., Dallas, 1983—, Jour. Assn. Acad. Minority Physicians, Bklyn., 1992—, Jour. Physiology and Pharmacology, Krakow, Poland, 1993—; adj. prof. Chgo. Med. Sch., 1983-90; prof. medicine UCLA, 1984-91. Editor: Gastrointestinal Cytoprotection, 1987; author: (with others) Prostaglandins and GI Mucosa, 1987, Pharmacology of Misoprostol, 1989, Prostaglandins and Esophagus, 1991, Pharmaceutical Industry Perspective, 1991; contbr. 195 rsch. papers and presentations in field; patentee in field. Mem. Arab-Am. Anti-discrimination Com., Washington, 1972, Arab-Am. U. Grads., Washington, 1991. Recipient Edward M. Queeny award, Monsanto Corp., 1991; named Disting. Alumnus Purdue U., 1991. Fellow Am. Coll. Gastroenterology; mem. Am. Soc. Pharmacology and Exptl. Therapeutics, Am. Gastroent. Assn., Am. Pharm. Assn., European Soc. Gastroenterology and Endoscopy, Assn. Acad. Minority Physicians, N.Y. Acad. Sci., Rho Chi, Phi Kappa Phi. Achievements include co-discovery and development of Misoprostol, first commercial prostaglandin anti-ulcer drug; 8 patents; directed pre-clinical and clinical research at major multi-national pharmaceutical companies. Office: Mid Gulf USA Inc Divsn Internat Drug Devel Cons 1549 RFD Long Grove IL 60047-9532

DAKE, KARL MANNING, research psychologist; b. Redwood City, Calif., Sept. 16, 1954; s. Frank Joseph and Gene Maree (Newland) D. BS, U. Calif., Davis, 1977; PhD, U. Calif., Berkeley, 1990. Rsch. cons. Profl. Risk Mgmt. Calif., Oakland, 1988-89, Inst. Med. Risk Studies, Sausalito, Calif., 1988—; prin. investigator Unilever, London, 1990-92; asst. rsch. psychologist U. Calif., Berkeley, 1990—; adj. faculty Calif. Sch. Profl. Psychology, Alameda, 1991—; mem. adv. bd. Am. Acad. Arts and Scis., Cambridge, Mass., 1989-90. Co-editor: Cultural Theory Applied to the Global Environment, 1993. Bradley Found. fellow, 1992-93; grantee Scaife Family Found., Pitts., 1993—. Mem. APA, Am. Polit. Sci. Assn., Soc. Risk Analysis (founding bd. mem. No. Calif. chpt. 1986). Achievements include findings suggesting worldviews-patterns of deeply held beliefs and values-are predictive of wide variety of risk perceptions, from concern about war, to social deviance, to econ. troubles and to dangers associated with techs. Office: U Calif Survey Rsch Ctr 2538 Channing Way Berkeley CA 94720

DAKEN, RICHARD JOSEPH, JR., electrical engineer; b. Paterson, N.J., Nov. 23, 1947; s. Richard Joseph and Ruth Maria (Gorman) D.; m. Jeanine Theresa Blanchfield, Sept. 5, 1970; children: Peter Michael, Timothy Jacob, Alice Jeanine, Andrew Paul. BEE, N.J. Inst. Tech., 1975, MEE, 1982; MBA, William Paterson Coll., 1990. Registered profl. engr., N.J. Biomed. engr. Dover (N.J.) Gen. Hosp., 1976-80; dir. biomed. engring. Newark (N.J.) Beth Israel Med. Ctr., 1980-84; dir. clin. engring. Morristown (N.J.) Meml. Hosp., 1984-88, NYU Med. Ctr., N.Y.C., 1988—; asst. prof. County Coll. Morris, Randolph, N.J., 1975-77, adj. asst. prof., 1977—, tech. adv. bd., 1977—; exec. v.p., chief operating officer Applied Med. Engring., Morristown, 1986-88; tech. adv. bd. N.J. Inst. Tech., Newark, 1992. Contbr. articles to profl. jours. Coach Holy Spirit Basketball Team, Pequannock, N.J., 1991-92. With USAF, 1969-73. Mem. Nat. Soc. Profl. Engrs., Assn. for Advancement Med. Instrumentation, Internat. Cert. Commn., Clin. Engring. Assn. N.J. (pres. 1986-92, 82-84), Am. Soc. Hosp. Engrs. Home: 118 W Parkway Pompton Plains NJ 07444 Office: NYU Med Ctr 560 1st Ave New York NY 10016

DAKIN, ROBERT EDWIN, electrical engineer; b. Bennington, Vt., July 1, 1949; s. Robert James and Rose (DeLuca) D.; m. Maureen Pietryka, Aug. 14, 1971; children: Melanie Jean, Patrick Robert. BSEE, Norwich U., 1971. Registered profl. engr., Vt., N.H., Maine, N.Y., N.J., Mass., Ireland (EC). Distrn. engr. Burlington (Vt.) Electric Dept., 1971-75, system engr., 1975-79; v.p. engring. Jennison Engring., Burlington, 1979-89; group mgr. Thermo Cons. Engrs., Williston, Vt., 1989—; mem., chmn. elec. exam com. Nat. Coun. Examiners for Engring. and Surveying, Clemson, S.C., 1984-91; mem. Vt. Profl. Engr. Registration Bd., Vt. Sec. of State, Montpelier, 1974-91. Mem. Colchester (Vt.) Planning Commn., 1973-80; capt. Chittenden County United Way, Burlington, 1984-87. Named Vol. of Yr., Burlington YMCA, 1976. Mem. IEEE (sr. mem., sec., treas., v.p., pres., PACE Achievement award 1983), NSPE, Vt. Soc. Profl. Engrs. (Young Engr. of Yr. 1979), Vt. Soc. Engrs. (sec., treas.). Roman Catholic. Home: 6 Ponderosa Dr Colchester VT 05401 Office: Thermo Cons Engrs 7 Park Ave Williston VT 05446

DAL, ERIK, former editorial association administrator; b. Grenaa, Denmark, Dec. 20, 1922; s. Johannes and Karen (Andersen) D.; m. Estrid Bruun Jörgensen, July 19, 1949; children: Ea, Ebbe. Grad. Danish and Musicology, U. Copenhagen, 1949, D. in Philosophy, 1960. Libr. Royal Libr. Copenhagen, 1953-63, head libr., 1963-67; prof. Royal Danish Sch. Librarianship, Copenhagen, 1967-74; adminstr. The Danish Soc. Lang. and Lit., Copenhagen, 1977-91. Bd. dirs. Internat. Folk Music Coun., 1956-70. Named Comdr. of the Order of Dannebrog, 1992. Mem. The Royal Danish Acad. Sci. and Letters (editor 1975-88, pres. 1988—). Lutheran. Home: Forchhammersvej 1, Frederiksberg, DK 1920 Copenhagen Denmark Office: Kongelige Danske Videnskabernes Selskab, H C Andersens Blvd 35, DK-1553 Copenhagen V, Denmark

DALDERUP, LOUISE MARIA, medical and chemical toxicologist, nutritionist; b. Rotterdam, Netherlands, Oct. 22, 1925; d. Clemens Bernardus Franciscus Maria and Jacoba Diedericka Louise (Wever) D. PhD in Physiol. Chemistry, U. Amsterdam, 1949, MD, 1957. Cert. social medicine, pub. health, Amsterdam, 1966, occupational health, Leiden, 1974. From jr.rsch. asst. to mgr. lab. and animal dept. Netherlands Inst. Nutrition, Amsterdam, 1948-70; sr. med. inspector to factory inspectorate Ministry of Social Affairs and Employment, Amsterdam, 1971-90; ret.; mem. Netherlands Nutrition Coun. and its expert coms. on dental caries (fluoride), cardiovascular diseases, 1958-85; mem. expert com. pesticide first aid measures for gen. practitioner Netherlands Pub. Health Coun., 1980-90; mem. expert com. prevention of radioactive contamination of food and feed Ministry of Agr. and Fishery, 1966-90; cons. Netherlands Office Nutritional Edn., 1970-79, Netherlands Heart Assn., 1972-80; expert com. Treshold Limit Values for Chem. in working environ., 1976-91. Author: Nutrition and Dental Caries, 1959, Myocardial Infarction and Risk Factors and Prevention, 1974; contbr. over 150 articles to profl. jours. Mem. med. com. drafting safety cards, com. safety and hygiene in hosps. Decorated Officer of Order of Oranje-Nassau, 1990. Mem. Netherlands Med. Assn., Netherlands Chem. Assn.

DALE, CHARLENE BOOTHE, international health administrator; b. Washington, June 10, 1942; d. John Edward and Frances Elizabeth (Jett) Boothe; children: Cynthia Lee, Anthony John, Jennifer Elizabeth. AA with high honors, Howard Community Coll., 1977; BA magna cum laude, U. Md., 1979. Asst. dir. univ. rels., alumni dir. U. Md., Catonsville, 1977-81; assoc. dir. univ. rels. and devel. U. Md., College Park, 1982-83; sr. devel. officer Internat. Ctr. Diarrhoeal Disease Rsch., Dhaka, Bangladesh, 1984-86; exec. v.p. Internat. Child Health Found., Columbia, Md., 1985—; organizer internat. symposium on food-based oral rehydration therapy Aga Khan U., Pakistan, 1989; cons. to organize symposium oral rehydration therapy Nat. Coun. Internat. Health, Washington, 1987. Author: (tng. manual) Prevention and Treatment of Childhood Diarrhea with Oral Rehydration Therapy, Nutrition and Breastfeeding, 1992; editor: Proceedings, Oral Rehydration Therapy Symposia, 1987, 89; contbr. articles to profl. jours. Pub. affairs chmn. United Way, Washington Capital Area, Prince Georges County, 1981-83; pres. Windstream Assn., 198-89; mem. pub. relations com. Md., Del. Cable TV Assn., Balti., 1981-83. Mem. Nat. Council Internat. Health, Am. Pub. Health Assn., AAUW, U. Md. Balti.County Alumni Assn. (bd. dirs. 1979-83), Women's Internat. Pub. Health Network. Democrat. Club: Columbia Assn. Athletic (Md.) (capt. women's traveling racquetball team 1979-83). Avocations: racquetball, windsurfing, skiing, oil painting. Home: 10441 Waterfowl Ter Columbia MD 21044-2465 Office: Internat Child Health Found Am City Bldg PO Box 1205 Columbia MD 21044-0205

DALE, CHARLES JEFFREY, operations research analyst; b. Cleve., Oct. 12, 1944; s. Charles John and Emily (Tecl) D.; m. Rosemarie Workman, Sept. 20, 1980. BS, Kent State U., 1966; MS, U. Ga., 1970; MDS, Ga. State U., 1972, PhD, 1978. Physicist U.S. Naval Rsch. Lab., Washington, summer 1968; fin. economist U.S. Dept. Treasury, Washington, 1978-79; internat. economist U.S. Dept. Commerce, Washington, 1979-82; rsch. economist U.S. Dept. of the Army, Alexandria, Va., 1982-89; assoc. Booz, Allen & Hamilton, Bethesda, Md., 1989-90; ops. rsch. analyst U.S. Dept. Energy, Washington, 1990—; adj. prof. Am. U., Washington, 1986-91. Contbr. articles to profl. jours. Mem. Nat. Economists Club (v.p. 1986-87), Am. Phys. Soc., Ops. Rsch. Soc., Am. Econ. Assn., Soc. Govt. Econs. Achievements include extended theories of agrl. commodity futures pricing to energy future pricing. Office: US Dept Energy EI-432 1000 Independence Ave SW Washington DC 20585

DALE, MARTHA ERICSON, clinical psychologist, educator; b. Chgo., Apr. 29, 1914; d. Ivar George and Caroline (Nelson) Ericson; m. Edward C. Dale, June 25, 1949; children: Peter David and John Daniel (twins), Edward Charles. BS, Northwestern U., 1938; MS, Iowa State U., 1941; PhD, U. Chgo., 1945. Assoc. prof. Merrill Palmer Sch., Detroit, 1944-50; psychologist Grosse Point (Mich.) Sch. System, 1957-59; assoc. prof. Mich. State U., East Lansing, 1961-68, Lake Superior State Coll., Sault Ste. Marie, Mich., 1968-69; cons. psychologist Ea. U.P. Mental Health Clinic, Sault Ste. Marie, 1969-71; dir. psychol. svcs. Newberry (Mich.) State Hosp., 1971-73. Mem. Psi Chi, Lambda Theta. Home: 10521 Tropicana Circle Sun City AZ 85351

DALE, PHILIP SCOTT, psychologist educator; b. Indpls., Feb. 15, 1943; s. Harry and Evylen (Straus) D.; m. Beverly Ann Goodell, June 27, 1965 (div.); children: Jonathan Harry, Jessica Louise; m. Joan Louise Dacres, July 26, 1992. BS in Math., U. Chgo., 1963; PhD in Comm. Scis., U. Mich., 1968. Lectr. psycholinguistics U. Mich., Ann Arbor, 1967-68; asst. prof. psychology and linguistics U. Wash., Seattle, 1968-74, assoc. prof. psychology and linguistics, 1974—; mem. NIH study sect. on human devel. Author: Language Development: Structure and Function, 1972, 76; editor: Child Language: An International Perspective, 1981; editorial bd. mem. Child Development, 1977-84; contbr. articles to profl. jours. Mem. Internat. Assn. for Study Child Lang. (sec. 1990-93), Soc. for Rsch. Child Devel. (program chmn. 1989-91), Linguistic Soc. Am., Am. Speech-Lang.-Hearing Assn. Jewish. Achievements include co-developing "Macarthur Communicative Development Inventories"; research language development in normal, handicapped, and precocious children. Office: Psychology Dept U Washington Seattle WA 98195

DALESSIO, ANTHONY THOMAS, industrial and organization psychologist; b. Cleve., Dec. 22, 1952; s. Anthony Charles and Marilyn (Farinacci) D.; m. Lisa Kuller, Aug. 27, 1989; 1 child, Emily Kuller. BS, Denison U., 1974; MS, Ill. Inst. Tech., 1976; PhD, Bowling Green State U., 1983. Asst. prof. U. Mo., St. Louis, 1981-84, Old Dominion U., Norfolk, Va., 1984-86; dir. LIMRA Internat., Farmington, Conn., 1986-92; staff dir. NYNEX Corp., White Plains, N.Y., 1992—; cons. various orgns., 1981-86, NASA Langley Rsch. Ctr., Hampton, Va., 1986. Contbr. articles to Pers. Psychology, Jour. Applied Psychology, Acad. Mgmt. Jour., IEEE Trans. on Engring. Mgmt., Human Rels., Human Performance. Summer faculty fellow NASA-Am. Soc. Engring. Edn., 1984, 85. Mem. APA, Am. Psychol. Soc., Soc. for Indsl. and Orgnl. Psychology (profl. affairs com. 1991-92), Acad. Mgmt. Home: 46 Wilton Rd W Ridgefield CT 06877 Office: NYNEX Corp Rm 1250 1111 Westchester Ave White Plains NY 10604

DALEY, GEORGE QUENTIN, internist, biomedical research scientist; b. Catskill, N.Y., Nov. 13, 1960; s. Frank Leonard and Natalie Alcine (Evans) D. AB, Harvard U., 1982; PhD in Biology, MIT, 1989; MD summa cum laude, Harvard U., 1991. Diplomate Am. Bd. Internal Medicine. Postdoctoral rsch. scientist MIT, Cambridge, Mass., 1989-90; resident in internal medicine Mass. Gen. Hosp., Boston, 1991-93; chmn. pre-med. adv. com. Quincy House, Harvard U., Cambridge, 1987-92. Contbr. articles to sci. jours. Recipient rsch. award for Clin. Trainees NIH, 1992; nat. scholar Harvard U., 1978-91. Mem. AAAS. Achievements include demonstration of molecular basis of human leukemia, chronic myelogenous leukemia; creation of mouse model for chronic myelogenous leukemia.

DALEY, HENRY OWEN, JR., chemist, educator; b. Quincy, Mass., June 18, 1936; s. Henry Owen and Edyth M. (Osgood) D.; m. Rosemary Tansey, Aug. 12, 1961; children: Owen F., Suzanne M., Sean P. BS, Bridgewater State Coll., 1958; PhD, Boston Coll., 1964. Sr. rsch. chemist Am. Cyanamid, Bound Brook, N.J., 1963-64; prof. Bridgewater (Mass.) State Coll., 1964—. Author: Fundamentals of Microprocessors, 1983, Problems in Chemistry, 1988. NSF fellow, 1961-63. Mem. Am. Chem. Soc., Puritan Bridge Club (treas. 1980—), Sigma Xi. Roman Catholic. Home: 115 Robinswood Rd South Weymouth MA 02190 Office: Bridgewater State Coll Dept Chemistry Bridgewater MA 02325

DALEY, JOHN PATRICK, physicist; b. Bklyn., July 1, 1958; s. John Patrick and Dorothy (Vovchik) D.; m. Ellen Marie Vovchik, June 21, 1993; 1 child, Alex Ryan. Grad. high sch., Centereach, N.Y. Founder, pres. Emrad Corp., Centereach, N.Y., 1991-92; mem. particle beam group Am. Phys. Soc., 1988-90. Author: A Physlosopher's Garage Sale!, 1990, God, 1992. Achievements include research on theoretical cosmological and quantum physics, information theory, experimental electrochemically induced thermonuclear fusion, effects of electromagnetic radiation on biological systems. Home: 8 Lolly Ln Centereach NY 11720-3802

DALFERES, EDWARD ROOSEVELT, JR., biochemical researcher; b. New Orleans, Nov. 4, 1931; s. Edward R. and Ray M. (Jones) D.; m. Anita Yvonne Bush, Aug. 1, 1959; children: Edward René, Anthony Renard. BS in Biology, Xavier U., 1956; various postgrad. courses, 1957-87. Rsch. assoc. Dept. Medicine, La. State U. Med. Ctr., New Orleans, 1957-75, instr. of medicine, 1975-92, asst. prof. rsch. dept. medicine, 1992; instr., rschr. dept. applied health scis. Sch. Pub. Health and Tropical Medicine, Tulane U. Med. Ctr., New Orleans, 1992—. Conbr. articles to profl. jours. including Jour. Exptl. Medicine, Biochim and Biophys. Acta, Anal Biochem, Am. Jour. Cardiology, Atherosclerosis, Molecular Cell Biochem., Clin. Chim Acta, and many others. Pres. parents club, St. Augustine High Sch., New Orleans; sch. bd. mem. St. Leo the Great. With U.S. Army, 1950-52. Recipient numerous grants. Mem. AAAS, N.Y. Acad. Sci., So. Connective Tissue Soc. for Complex Carbohydrates, Sigma Xi. Democrat. Roman Catholic. Achievements include rsch. on biochemistry of C-V connective tissue, particularly glycosaminoglycan, proteoglycan and glycoproteins. Home: 2053 Treasure St New Orleans LA 70122 Office: Tulane U Med Ctr 1430 Tulane Ave New Orleans LA 70112

DALIGAND, DANIEL, engineer, association executive, expert witness; b. Lyon, France, June 14, 1942; s. Maurice and Pierina Maria (Grandelli) D.; m. Christiane Elisabeth H. Agobian, Dec. 17, 1966. M. in Chem. Enging., ICPI, 1965. Cert. engr. Engr., SNIP, Paris, 1967-72, gen. sec., 1972—; adminstrv. sec. Eurogypsum, Paris, 1971—; sec. 13 mem. jury best craftsmanship in plaster work soc., Paris, 1982, 86; counsellor for technol. edn., 1986—; chmn. French Com. for Standardization of Gypsum and Gypsum Products, 1990; gov. 3 task forces for European standardization of gypsum products. Author: (with J. Gibaru) Le Platre, 1981; (with others) Le Platre, Physico chimie, fabrication, 1982; Le Platre—Techniques de l'Ingenieur, 1986; editor: Platre Info. Jour. Office: SNIP, 15 Av du Recteur Poincaré, 75016 Paris France

DALRYMPLE, GARY BRENT, research geologist; b. Alhambra, Calif., May 9, 1937; s. Donald Inlow and Wynona Edith (Pierce) D.; m. Sharon Ann Tramel, June 28, 1959; children: Stacie Ann, Robynne Ann, Melinda Ann Dalrymple McGurer. AB in Geology, Occidental Coll., 1959; PhD in Geology, U. Calif., Berkeley, 1963; DSc (hon.), Occidental Coll., Los Angeles, 1993. Registered geologist, Calif. Rsch. geologist U.S. Geol. Survey, Menlo Park, Calif., 1963-81, 84—, asst. chief geologist we. region, 1981-84; vis. prof. sch. earth scis. Stanford U., 1969-72, cons. prof., 1983-85, 90—; disting. alumni centennial speaker Occidental Coll., 1986-87; expert witness ACLU/State of Calif.; presenter workshops Calif. State Dept. Edn., San Mateo County Bd. Edn., San Francisco Sch. Dist. Author: Potassium-Argon Dating, 1969, Age of Earth, 1991; contbr. over 150 papers, abstracts, book reviews and essays to profl. and refereed jours., chpts. to books. Fellow NSF, 1961-63; recipient Meritorius Svc. award U.S. Dept. Interior, 1984. Fellow Am. Geophys. Union (pres. elect 1988-90, pres. 1990-92), Geol. Soc. Am.; mem. AAAS, NAS, Am. Inst. Physics (bd. govs. 1991—), Am. Acad. Arts Scis. Achievements include discovery that the earth's magnetic field reverses polarity and determination of time scale of these reversals for the past 3.5 million years; development of ultra-fast high-sensitivity thermoluminescence analyzer for studying lunar surface processes; development and refinement of K-Ar and 40 Ar/39 Ar dating methods and instrumentation, Menlo Park continuous laser probe for determining ages of microgram-sized mineral samples; research on volcanoes in the Hawaiian-Emperor volcanic chain, chronology of lunar basin formation, development and improvement of isotopic dating techniques and instrumentation, geomagnetic field behavior, plate tectonics of the Pacific Ocean basin, evolution of volcanoes, impact history of early lunar surface, various aspects of Pleistocene history of the western U.S. Office: US Geol Survey Br Isotope Geology 345 Middlefield Rd Menlo Park CA 94025-3591

DALRYMPLE, GLENN VOGT, radiologist; b. Little Rock, Dec. 23, 1955; s. Clyde William and Sarah (Darnall) D.; m. Mary Jo Jung, Dec. 3, 1955; children: Anne Theresa, Mark Gregory. BS, U. Ark., 1956, MD, 1958. Diplomate Am. Bd. Radiology, Am. Bd. Nuclear Medicine. Rotating intern and resident in radiology U. Arks. Med. Ctr., Little Rock, 1959-60; resident in radiology U. Colo. Med. Ctr., Denver, 1960-61, chief resident in radiology and rsch. assoc., 1961-62; asst. prof. to prof. radiology U. Ark. for Med. Scis., Little Rock, 1965-73, chmn. dept. radiology, 1973-76, prof. radiology, 1987-90; cons. in radiology Little Rock, 1976-87; prof. radiology and internal medicine U. Nebr. Med. Ctr., Omaha, 1990—; med. radiation adv. com. FDA, Washington, 1978-80; attending staff dept. radiology U. of Nebr. Med. Ctr., Omaha, 1990; courtesy staff dept. radiology St. Joseph Hosp., Omaha, 1990—, VA Med. Ctr., Omaha, 1990—. Co-author: Medical Radiation Biology, 1972, Basic Science Principles of Nuclear Medicine, 1974, Radiology in Primary Care, 1975, Practical Radioimmunoassay, 1975; contbr. book chpts. and articles to profl. jours.; assoc. editor Radiation Research Soc., 1974-76. Prin. french horn U. Nebr. Orch., Omaha, 1990—; french horn/trombone Ark. Symphony Orch., Little Rock, 1965-89, bd. dirs., 1979-83. Capt. USAF, 1963-65. Grantee NASA, USAEC, NIH, Am. Cancer Soc., VA Med. Rsch. Fellow Am. Coll. Radiology (commn. pub. health radiol. units and standards 1984-90, Ark. chpt. pres. 1979); mem. AMA, Soc. of Nuclear Medicine, Radiation Rsch. Soc., Am. Assn. of Physics in Medicine, Am. Coll. Nuclear Physicians, Am. Roentgen Ray Soc., Assn. of Univ. Radiologists, Health Physics Soc., Internat. Wound Ballistics Assn., Nebr. Med. Assn., Radiation Rsch. Soc., Radiol. Soc. of N.Am., Am. Rifle Assn. (life), U.S. Practical Shooting Assn. (life), Sigma Xi, Alpha Omega Alpha. Office: Univ Nebr Med Ctr 600 S 42nd St Omaha NE 68198-1045

DALTO, MICHAEL, medical microbiologist; b. N.Y.C., Nov. 19, 1956; s. Cono and Emanuela (Asaro) D. BS, St. John's U., Jamaica, N.Y., 1978, MS, 1982, M in Philosophy, 1986, PhD in Biology, 1989. Cert. of qualification for clin. lab. dir., N.Y.C. Biology dept. technician St. John's U., 1981-82, instr. gifted children program, 1981-88; asst. attending microbiologist Queens Hosp. Ctr., Jamaica, 1989-93; adj. faculty mem. Elizabeth Seton Coll., Yonkers, N.Y., 1987-88; substitute tchr. St. John's U., 1988. Capt. United Way Fund, L.I. Jewish affiliation Queens Hosp. Ctr., 1990. Mem. AAAS, Am. Soc. Microbiology (N.Y.C. br.), Am. Soc. for Med. Tech., Am. Soc. for Clin. Pathologists, N.Y. Acad. Sci., Soc. for Indsl. Microbiology, Phi Theta Kappa. Avocations: computer programming, photography, tennis. Home: 1494 Sweetman Ave Floral Park NY 11003-3012

DALTON, OREN NAVARRO, mathematician; b. Leeds, Utah, Apr. 26, 1929; s. Orin and Valhalla (Angell) D.; m. Dixie Ann Brown, June 2, 1962 (dec. June 1963); m. Marilee Willson, Aug. 31, 1966; children: Marilee Ellen, James Willson. BA, U. Utah, 1955. Mathematician FL Simulation, White Sands Missile Range, N.Mex., 1955-60, Simulation & Devel. Space Tech. Labs., L.A., 1960-65; mathematical programmer and developer White Sands Missle Range, N.Mex., 1966-92; session co-host Ops. Rsch. Soc. Am., Las Vegas, Nev., 1983. Contbr. articles to profl. jours. Home: PO Box 4435 El Paso TX 79914-4435

DALTON, PETER JOHN, electronics executive; b. Balt., Feb. 5, 1944; s. Peter J. and Rita Dalton; m. Pat Hubbard, Sept. 29, 1991; children: Kelley, Kathy, Amy. B.S. in Acctg., LaSalle U., 1966. Cost acct. Johnson & Johnson, New Brunswick, N.J., 1966-68; div. adminstr. M/A Com., Sunnyvale, Calif., 1968-70; controller, treas. Wilsey Foods, L.A., 1970-75; pres. U.S. Textile Corp., Oakland, Calif., 1975-80; chmn., pres. KLM Electronics, Inc., Morgan Hill, Calif., 1980-85, Harlan & Dalton, Burlingame, Calif., 1986-91, EPRO Corp., San Jose, Calif., 1988-91, RJE Communications, Sunnyvale, Calif., 1990-92; pres., CEO Am. Quality Mfg., 1991-93; dir. First Fin. Savs. Bank, Santa Clara, Calif.; mng. ptnr. Dalton Ptnrs., 1991—. Contbr. articles to popular mags. Home and Office: 17880 Andrews St Monte Sereno CA 95030-4231

DALTON, ROBERT EDGAR, mathematician, computer scientist; b. Boston, May 2, 1938; s. Robert Evelyn and Mildred Louise (Zoellick) D.; m. Sally Turner, Sept. 12, 1961 (div. 1977); children: Stephen Howard, Alena Lynn; m. Judith Eyges, July 17, 1993. BS in Math., U. Chgo., 1959; MS in Applied Math., N.C. State U., 1961, PhD in Applied Math., 1964; MS in Computer Sci., Fla. State U., 1982. Systems analyst RCA Svc. Co., Cocoa Beach, Fla., 1964-65; mem. tech. staff TRW Systems Group, Cocoa Beach, 1965-71; ops. rsch. analyst Naval Underwater Systems Ctr., West Palm Beach, Fla., 1971-79; grad. teaching asst. Fla. State U., Tallahassee, 1980-81; asst. prof. Am. U., Washington, 1981-83; mem. tech. staff Mitre Corp., Greenbelt, Md., 1983-85; prin. investigator Vitro Corp., Silver Spring, Md., 1985—. Contbr. chpts. to books, articles to jours. Sec. U.S. Jaycees, Boynton Beach, Fla., 1974; chmn. U. Chgo. Alumni Fund, Palm Beach County, Fla., 1975-79. NSF coop. fellow, 1962-63. Mem. IEEE Computer Soc., Math. Assn. Am. (vice-chmn. Fla. Goldcoast sect. 1976-79, moderator computer sci. 1979), Am. Assn. Artificial Intelligence, Assn. Computing Machinery, North Am. Fuzzy Info. Processing Soc. Achievements include research in knowledge acquisition and learning, computer games, pattern recognition, knowledge-based system development, and decision support with fuzzy logic. Office: Vitro Corp 14000 Georgia Ave Silver Spring MD 20906-2972

DALTON, ROBERT LOWRY, JR., petroleum engineer; b. Okmulgee, Okla., June 6, 1931; s. Robert Lowry and Helen Ester (Calhoun) D.; m. Billie Lou Ricards, Sept. 10, 1952; children: Stephen Calhoun, Allen Keith. BS in Chem. Engring., Rice U., 1954. Rsch. engr. Exxon Prodn. Rsch. Co., Tulsa and Houston, 1954-69; rsch./engring. supr. Exxon Prodn. Rsch. Co., Houston, 1969-80, rsch. and engring. div. mgr., 1980-86, tng. div. mgr., 1986-91. Mem. Soc. Petroleum Engrs. (continuing edn. com. 1988-90, tech. editor 1987—). Achievements include co-development of use of chem. tracers to determine residual oil saturation in petroleum reservoirs; patent on use of tracers to measure flow rate of fluids in petroleum reservoirs; methods to interpret pilot water flood recovery behaviour in petroleum reservoirs; methods to use in applying oil reservoir simulators in field situations; measurement of three phase relative permeability in porous media.

DAM, A. SCOTT, nuclear/mechanical engineer; b. Palo Alto, Calif., Jan. 21, 1946; s. Francis Scott and Florawood (Smith) D.; m. Carol Diane Peao, June 8, 1968; children: Christie Lynne, Robin Michelle, Meredith Lee. BME, U. Louisville, 1968, M Engring., 1973; MBA, Rutgers U., 1982. Registered profl. engr., Va., Pa., N.J. Chief Windsor (Conn.) Field Office Naval Reactors US AEC, 1974-75; project engr., project mgr. Burns and Roe, Inc., Oradell, N.J., 1975-79; mgr. projects Burns and Roe, Inc., Paramus, N.J., 1979-83; mgr. waste tech. svc. nuclear power div. Babcock & Wilcox, Lynchburg, Va., 1983-88; assoc. program mgr., project dir. Roy F. Weston Inc., Washington, 1988-91; mgr. projects BNFL Inc., Fairfax, Va., 1991—; dir., treas., Waste Mgmt. Symposia Inc., Tucson, 1992—. Contbr. articles to profl. pubs. Chmn. Ridgewood (N.J.) Bicycle Bd., 1981-82. Lt. USN, 1968-73. Mem. ASME, Nuclear Waste Brokers and Processors Assn. (pres. 1987-88), Am. Nuclear Soc. (sec. treas. divsn. fuel cycle and waste mgmt. 1991-92, vice chair,1992-93, chair 1993—, chair spl. com. site cleanup and restoration standards, 1992—, Best Paper award 1984), Beta Gamma Sigma, Tau Beta Pi. Office: BNFL Inc 9302 Lee Hwy Ste 950 Fairfax VA 22031

D'AMATO, ANTHONY SALVATORE, food products company executive; b. Bklyn., 1930. BSChemE, Poly. Inst. Bklyn., 1952. With Borden, Inc., N.Y.C., 1959—, exec. v.p., from 1985; pres. Borden Packaging and Indl.

Products (formerly Borden Chem. Div.), Columbus, Ohio, from 1985; chmn., chief exec. officer Borden Packaging and Indl. Products (formerly Borden Chem. Div.), Columbus, until 1990; pres. Borden, Inc., N.Y.C., 1990—, CEO, 1991—, chmn. bd., 1992—. Office: Borden Inc 180 E Broad St Columbus OH 43215-3707 also: Borden Inc 277 Park Ave New York NY 10172

DAMATO, DAVID JOSEPH, electrical engineer; b. Chgo., Aug. 7, 1953; s. Joseph Charles and Evelyn (Darling) D.; m. Elizabeth Earle Foster, Aug. 14, 1976; children: Matthew David, Ashley Elizabeth. BSEE, DeVry Inst.of Tech., 1974; MSEE, U. Ala., 1981. Registered Engr., Ala., Ill. Field svc. tech. Taylor Instrument Co., Rochester, N.Y., 1974-75; control systems design engr. B.E.& K. Inc., Birmingham, Ala., 1974-75, mgr. CAD, 1979-81; sr. systems Engr. Scott Paper Co., Mobile, Ala., 1981-82; sr. process engr. S.D. Warren Div. Scott Paper Co., Mobile, Ala., 1982-86; staff engr. Rust Internat., Birmingham, 1986-91; cons. engr. OPTEC div. Rust, Internat., 1991—; paper machine winder and machine drive specialist Rust Internat. Author: The Coated Process, 1993. Leader Boy Scouts Am,. Chgo., Birmingham, N.Y.C. and Mobile, 1974—. Mem. IEEE, NSPE (officer 1978-88; young engr. of yr. 1979); sr. mem. Instrument Soc. of Am. (officer 1974-78; young engr. of yr. 1977), Tech. Assn. Pulp Paper Industry, Inverness Country Club. Republican. United Methodist. Avocations: amatur radio, cycling, photography. Office: Rust Internat 100 Meadowbrook Corp Pky Birmingham AL 35242

DAME, THOMAS MICHAEL, radio astronomer; b. Winthrop, Mass., Oct. 16, 1954; s. Chester Thomas and Claire J. (White) D.; m. Geraldine Ann Healey, Aug. 23, 1985. BA, Boston U., 1976; PhD, Columbia U., 1983. Rsch. assoc. Goddard Inst. for Space Studies NASA, N.Y.C., 1983-84; rsch. assoc. Columbia U., N.Y.C., 1985-86; radio astronomer Smithsonian Astrophys. Obs., Cambridge, Mass., 1986—; lectr. astronomy Harvard U., Cambridge, 1989—. Contbr. articles to Astrophys. Jour., Sky and Telescope Mag., NASA Publ. NAS fellow, 1983-84. Mem. Internat. Astronomical Union, Am. Astron. Soc. Achievements include publication of first and only complete map of molecular gas in Milky Way Galaxy; catalogued largest molecular clouds in Milky Way; calibrated conversion from CO intensity to molecular mass. Home: 187 Nichols St Everett MA 02149-5338 Office: Ctr for Astrophysics 60 Garden St Cambridge MA 02138-1596

D'AMICO, GIUSEPPE, nephrologist; b. Messina, Sicily, Italy, Sept. 6, 1929; s. Gaetano and Gaetana (Trifiletti) D'A.; m. Anna Maria Allegri, July 11, 1957; 1 child, Stefano. M.D., State U. Milan, 1952; L. Docenza in Internal Medicine, State U. Milan, 1964. Traineeship in endocrinology Chgo. Med. Sch., 1957-58; asst. prof. dept. internal medicine State U. Med. Sch., Milan, Italy, 1964-67, head div. nephrology San Carlo Hosp., Milan, 1967—. Contbr. articles to profl. jours. Recipient Ganassini Internat. award Ganassini Found., 1961; Fulbright travel grantee, 1957; NIH grantee, 1957. Mem. Italian Soc. Nephrology (pres. 1980-83), European Dialysis Transplant Assn. (exec. coun.), Internat. Soc. Nephrology (coun.), Internat. Soc. Nutrition in Renal Diseases (coun.), N.Y. Acad. Sci. Home: Viale Papiniano, 22/B, Milan Italy Office: San Carlo Hosp, Via Pio II 3, Milan Italy

D'AMICO, WILLIAM PETER, JR., mechanical engineer; b. San Jose, Calif., Oct. 20, 1944. BSME, Santa Clara U., 1966, MSME, 1968; PhD, U. Del., 1977. Supervisory mech. engr. U.S. Army Rsch. Lab., Weapons Tech. Directorate, Aberdeen Proving Ground, Md., 1968—, chief advanced munitions concepts br., 1992—. Coach youth sports Harford County (Md.) Recreation programs, 1980—. Recipient Meritorious Civilian Svc. award; named Engr. of Yr. Am. Soc. Phys. Engrs., 1987; BRL fellow. Mem. AIAA (assoc., contbr. to jours. and presentations). Achievements include 2 patents. Office: Dir US ARL AMSRL-WT-WB Aberdeen Proving Ground MD 21005-5066

DAMJI, KARIM SADRUDIN, environmental toxicologist, consultant; b. Dar Es Salaam, Tanzania, July 17, 1959; came to U.S., 1973; s. Sadrudin P. and Saker S. Damji. MS, U. So. Calif., 1987, PhD, 1989. Environ. toxicologist AeroVironment Inc., Monrovia, Calif., 1990-91, Warzyn Inc., Pasadena, Calif., 1991—; asst. clin. prof. pathology, U. So. Calif., L.A., 1990—. Contbr. articles to profl. jours. Mem. sci. adv. panel Santa Barbara (Calif.) County Air Pollution Control Dist., 1992. Recipient fellowship U. So. Calif., 1985, 86, 87, 88, fellowship Bowles Meml. Fellowship, 1989. Mem. AAAS, Soc. Risk Analysis, Soc. Toxicology, Soc. Immunotoxicology. Office: Warzyn Inc 320 N Halstead Ste 240 Pasadena CA 91107

DAMKEN, JOHN AUGUST, computer systems engineer; b. Hackensack, N.J., June 13, 1950; s. John August and Margaret Ann (Kearney) D.; m. Sheryl Elizabeth Lancaster, Sept. 8, 1979; children: Shanna Elizabeth, Stacey Michelle. BS in Elec. Engring., U. Mich., 1972, MS in Indsl. Engring., 1973; MBA in Fin., Ga. State U., 1983. Computer programmer E.I. duPont, Martinsville, Va., 1974-78; project mgr. Am. Express, Atlanta, 1978-90; dir. Glasrock Home Health Care, Atlanta, 1990-91; computer systems engr. GTE Mobile Comm., Atlanta, 1992—; cons., Atlanta, 1991-92. Capt. USAF, 1972-80. Mem. Digital Equipment Users Soc. Home: 235 Chiswick Close Alpharetta GA 30202 Office: GTE Mobile Comm 245 Perimeter Center Pkwy Atlanta GA 30346

DAMMANN, DAVID PATRICK, electronics engineer; b. Huron, S.D., June 14, 1965; s. Gerald Dale Dammann and Janice Marie (Lillie) Barry. BSEE, S.D. Sch. Mines and Tech., 1987. Registered profl. engr., Calif. Systems engr. Naval Warfare Assessment Ctr., Corona, Calif., 1987—. Mem. Nat. Soc. Profl. Engrs. Home: PO Box 70212 Riverside CA 92513-0212

DAMSBO, ANN MARIE, psychologist; b. Cortland, N.Y., July 7, 1931; d. Jorgen Einer and Agatha Irene (Schenck) D. B.S., San Diego State Coll., 1952; M.A., U.S. Internat. U., 1974, Ph.D., 1975. Diplomate Am. Acad. Pain Mgmt. Commd. 2d lt. U.S. Army, 1952, advanced through grades to capt., 1957; staff therapist Letterman Army Hosp., San Francisco, 1953-54, 56-58, 61-62, Ft. Devers, Mass., 1955-56, Walter Reed Army Hosp., Washington, 1958-59, Tripler Army Hosp., Hawaii, 1959-61, Ft. Benning, Ga., 1962-64; chief therapist U.S. Army Hosp., Ft. McPherson, Ga., 1964-67; ret. U.S. Army, 1967; med. missionary So. Presbyterian Ch., Taiwan, 1968-70; psychology intern So. Naval Hosp., San Diego, 1975; pre-doctoral intern Naval Regional Med. Ctr., San Diego, 1975-76, postdoctoral intern, 1975-76, chief, founder pain clinic, 1977-86; chief pain clinic, 1977-86; adj. tchr. U. Calif. Med. Sch., San Diego; lectr., U.S., Can., Eng., France, Australia; cons. forensic hypnosis to law enforcemnt agys. Contbr. articles to profl. publs., chpt. to book. Tchr. Sunday sch. United Meth. Ch., 1945—; Rep. Nat. Candidate Trust Pres. adv. com.; mem. Lonta Internat. Presdl. Adv. Com. Fellow Am. Soc. Clin. Hypnosis (psychology mem. at large, exec. bd. 1989-90); mem. San Diego Clin. Hypnosis (pres. 1980), Am. Phys. Therapy Assn., Calif. Soc. Clin. and Hypnosis (bd. govs.), Am. Soc. Clin. Hypnosis Edn. Rsch. Found. (trustee 1992-94), AAUW, Internat. Platform Assn., Am. Soc. Clin. Hypnosis (exec. bd.) Ret. Officers Am., Retired Officers Assn. (rep. presdl. task force, pres. adv. com.), Toastmasters (local pres.), Job's Daus. Republican. Home and Office: 1062 W 5th Ave Escondido CA 92025-3802

DAMUTH, JOHN ERWIN, marine geologist; b. Dayton, Ohio, Nov. 22, 1942; s. Jason Donald and Sarah Maxine (Simpson) D.; m. Patricia Jane Keenan, Oct. 8, 1971 (div. July 1990). BS in Geology, Ohio State U., 1965; MA in Geology, Columbia U., 1968, PhD in Geology, 1973. Grad. rsch. asst. Lamont-Doherty Geol. Observatory Columbia U., 1965-73, rsch. scientist, 1973-74, rsch. assoc., 1974-82, sr. rsch. assoc., 1982-83; rsch. geologist Dallas Rsch. Lab. Mobil R&D Corp., 1983-84, sr. rsch. geologist, 1984-92; sr. rsch. scientist Earth Rsch. and Environment Ctr. U. Tex. at Arlington, 1992—; adj. rsch. scientist Lamont-Doherty Geol. Observatory Columbia U., 1983-91; instr. ecology adult edn. N.J. High Sch., 1977-83; mem. Nat. Site Assessment Com Subseabed Disposal High-Level Nuclear Waste, 1978-83; cons. crustal evolution project Nat. Assn. Geology Tchrs., 1978; lectr. in field. Contbr. articles to profl. jours. Texaco scholar, 1964-65; Eugene Higgins fellow, 1965-66, Pan Am. Oil Co. fellow, 1967, Pres.'s fellow, 1968-69, Nat. Lord Britton fellow, 1967-68. Fellow Geol. Soc. Am.; mem. Am. Assn. Petroleum Geologists, Soc. Econ. Paleontologists and Mineralogists, Am. Geohphys. Union, Sigma Xi. Avocations: fishing, skiing, travel, exercise. Office: U Tex at Arlington Earth Resource & Environ Ct PO Box 19049 Arlington TX 76019

DANAHER, FRANK ERWIN, transportation technologist; b. Montclair, N.J., Mar. 5, 1936; s. Frank E. and Mildred (Acquino) D.; m. Joan Marie Donovan, Apr. 12, 1986; children: Maria, Frank, Heather (dec.). BA in Math., Rutgers U., 1961; MBA, Fairleigh Dickinson, 1982. Supr. programming ITT, Paramus, N.J., 1961-66; mgr. systems Lummus, Bloomfield, N.J., 1966-83; rsch. specialist Dun & Bradstreet, Basking Ridge, N.J., 1983-87; technologist Met. Transp. Auth., N.Y.C., 1987—; cons. in field. Contbr. articles to profl. publs. Area gov. Toastmasters, N.Y.C., 1984-85; speaker in field. With U.S. Army, 1959. Urban Mass Transit Authority grantee, 1988, 90. Mem. Assn. for Systems Mgmt. (pres. 1983), Geog. Info. System Users (chmn. 1992), Computer Aided Design and Drafting, User Group Met. Transp. Authority (chmn. 1990-92), Delta Mu Delta. Republican. Roman Catholic. Home: 454 Prospect Ave # 147 West Orange NJ 07052 Office: Met Transit Auth 460 W 34th St Rm 615 New York NY 10001

DANBURG, JEROME SAMUEL, oil company executive; b. Houston, Dec. 21, 1940; s. August and Rosalie (Bornstein) D.; m. Gudrun Ella Ernestine Scholz, Sept. 8, 1965; children: Aron Ralf, Andrea Leda, Sylvia Freia, Sonja Rebecca. BS in Physics, MIT, 1962; Diplom in Physics, Freie Universität Berlin, 1964; PhD in Physics, U. Calif., Berkeley, 1969. Assoc. physicist Brookhaven Nat. Lab., Upton, N.Y., 1969-72; sr. rsch. geophysicist Shell Devel. Co., Houston, 1973-81, rsch. mgr., 1981-86, rsch. dir., 1992—; mgr. Shell Oil Co., Houston, 1986-92; physics dept. vis. com. mem. U. Tex., Austin, 1990—. Contbr. articles to profl. jours. Fulbright scholar, Freie Universität Berlin. Mem. Am. Phys. Soc., Soc. Exploration Geophysicists, Fulbright Alumni Assn. Home: 7611 Burning Hills Dr Houston TX 77071-1413 Office: Shell Devel Co PO Box 481 Houston TX 77001-0481

DANCE, WILLIAM ELIJAH, industrial neutron radiologist, researcher; b. Montezuma, N.Mex., Sept. 30, 1930; s. James Claude and Maude Lee (Fullerton) D.; m. Elizabeth Ann Hennessee, Aug. 20, 1955; children: James Christopher, William Alton. BS, Carson Newman Coll., 1952; MS, La. State U., 1955, Phd, 1959. Phys. chemist Tenn. Eastman Co., Kingsport, 1952; grad. asst. physics dept. La. State U., Baton Rouge, 1953-59; sr. scientist LTV Rsch. Ctr., Dallas, 1963-67; sr. scientist Advanced Tech. Ctr., Inc., Dallas, 1967-78, mgr. space projects, 1972-74; mgr. neutron radiology LTV Aerospace and Def. Co., Dallas, 1979—. Contbr. articles to profl. jours. Bd. mem. Northwest Dallas Improvement League, 1992. Mem. IEEE, ASTM (subcom. vice chmn. 1983-92), Am. Soc. Nondestructive Testing. Achievements include seven patents in the realm of development of neutron radiology advanced devices and techniques. Home: 3306 Whitehall Dr Dallas TX 75229 Office: Loral Vought Systems Corp P O Box 650003 M S EM-16 Dallas TX 75265-0003

DANDASHI, FAYAD ALEXANDER, applied scientist; b. Damascus, Syria, July 20, 1959; s. A.K. and Ghada (Bahnasi) D.; m. Mami Tazaki, Apr. 12, 1989. BArch, U. Aleppo, 1982; BSCE, George Washington U., 1987, MSME in Aeronautics & Astronautics, 1989, degree of applied sci., 1992. Cert. advanced engring., mgmt. edn., plant engr. Designer/engr. engring. design div. George Washington U., Washington, 1985-89, head engring. design div., 1989-91, fellow, 1991—. Mem. AIAA (sr. mem.), Ops. Rsch. Soc. Am., Am. Inst. Plant Engrs., Wash. Soc. Engrs., Inst. Mgmt. Sci., Fedn. Am. Scientists, Omega Rho, Sigma Gamma Tau, Tau Beta Pi, Sigma Xi. Achievements include research in stochastic and mathematical modeling, mathematical programming and optimization, numerical analysis, nonlinear dynamics and complexity, design and analysis of aerospace, air-breathing single stage earth-to-orbit vehicles and rockets with APS/RAMJET/SCRAMJET/LACE propulsion systems using slush hydrogen and slush oxygen propellant, advanced aerospace technologies, science and technology policy and the mathematical and statistical theory of probability.

DANDO, NEAL RICHARD, chemist; b. Pottsville, Pa., Aug. 29, 1957; s. Lewis Ray and Joanne Patricia (Pellish) D.; m. Jonell M. Kerkhoff, Jan. 3, 1987; 1 child, Kerrick Robert. BS, Shippensburg U., 1979; MA, PhD, U. Del., 1984. NMR spectroscopist E.I. DuPont de Nemours & Co., Inc., Wilmington, Del., 1983-84; sr. rsch. chemist PPG Inc., Allison park, Pa., 1984-86; tech. specialist Aluminum Co. of Am., New Kensington, Pa., 1986—. Contbr. articles to profl. jours. U. Del. grad. fellow, 1983, NSF fellow, 1974. Mem. Am. Chem. Soc., Spectroscopy Soc. Pitts. (membership chair, Pitts. conf. com. chair). Achievements include development and implementation spectroscopic gauges for organic coating thickness measurements on aluminum sheet. Office: Alcoa Labs 100 Technical Dr New Kensington PA 15069-0001

DANDONA, PARESH, endocrinologist; b. Peshawar, India, May 25, 1943; came to U.S. 1991; s. Som Nath and Sham Kumari (Mehra) D.; m. Oct. 8, 1970; children: Sonny, Kabir. BSc, Allahabad U., India, 1960; M.B.B.S., All India Inst. Med. Scis., 1965; PhD, Oxford U., U.K., 1974. Rsch. fellow (Rhodes Scholar) MRC Neuroendocrinology Unit, U. Oxford, 1966-69; house officer Northwick Park Hosp. Brampton, London, 1971-73; sr. registrar Royal Free Hosp. Sch. Medicine, London, 1975-78, dir. metabolic unit, 1978-91; prof. medicine U. Buffalo, 1991—. Editorial bd. Platelets, Prostaglandins, Leucotrienes and Essential Fatty Acids; contbr. articles to profl. jours. Vol. screening for vascular disease risks Buffalo, 1992—. Wellcome fellow, 1969-71; recipient numerous grants. Fellow Royal Coll. Physicians U.K.; mem. Am. Diabetes Assn. (chpt. bd. dirs. 1992—), Endocrine Soc. Achievements include research in abnormal vascular reactivity in diabetes; abnormal exocrine function in diabetes; calcium deficiency rickets in Africa. Office: Millard Fillmore Hosp 3 Gates Cir Buffalo NY 14209

D'ANDREA, MARK, radiation oncologist; b. Palos Park, Ill., May 21, 1960; s. Anthony E. and Adriene (Boka) D'A. BA in Chemistry, Religion, and Biology, Luther Coll., 1981; MD, Ponce (P.R.) Sch. Medicine, 1985. Diplomate Am. Acad. Pain Mgmt. Resident in internal medicine Cabrini Med. Ctr., N.Y.C., 1985-86; resident in radiation oncology Meth. Hosp., Bklyn., 1986-89; radiation oncologist East Tex. Cancer Ctr., Tyler, 1989—, Mother Frances Hosp., Tyler, Med. Ctr. Hosp., Tyler, U. Tex. Health Ctr., Tyler, St. Josephs Hosp., Paris, Tex., McCuistion Hosp., Paris, Tex., Longview (Tex.) Radiation Oncology Ctr.; resident U. Tex. Med. Branch, Galveston, 1991-92; prof. radiation biology Tyler Jr. Coll., 1990—; chief residen in radiation oncology Meth. Hosp., Bklyn., 1988-89; cons. Longview Regional, Good Shepherd Hosp.; pres., chmn. bd. DANHUL Corp., 1992. Chmn. Com. Pub. health Kings and Bklyn. County, N.Y., 1988-89. Named One of Outstanding Young Men Am., 1987; recipient Outstanding award Ill. Jr. Acad. Sci., 1978. Fellow Am. and Internat. Coll. Angiology, InterAm. Coll. Physicians and Surgeons; mem. Am. Inst. Chemists (ethics com.), Am. Chem. Soc., Am. Soc. Clin. Oncology, Am. Soc. Therapeutic Radiology and Oncology, AMA, Radiol. Soc. N.Am., Med. Soc. N.Y. State, Kings County Med. Soc., Acad. Medicine Bklyn., Smith County Med. Soc., Tex. Med. Assn., Circilo de Radioterapeutas ibero Latino Americanos. Office: E Tex Cancer Ctr 721 Clinic Dr Tyler TX 75701-2000

DANDRIDGE, WILLIAM SHELTON, orthopedic surgeon; b. Atoka, Okla., May 21, 1914; s. Theodore Oscar and Estelle (Shelton) D.; m. Pearl Sessions, Feb. 3, 1941; children: Diana Dawn, James Rutledge. B.A., U. Okla., 1935; M.D., U. Ark., 1939; M.S., Baylor U., 1950. Intern, St. Paul's Hosp., Dallas, 1939-40; surg. residence Med. Arts Hosp., Dallas, 1940; commd. 1st lt. USAF, advanced through grades to lt. col., 1950; chief reconditioning svc. and reconstructive surgery Ashburn Gen. Hosp., McKinney, Tex., 1945-46; neurosurg. resident Brooke Army Med. Center, San Antonio, 1946-47; orthopedic surg. resident, 1947-50; chief orthopedic svc. and gen. surgery Francis E. Warren AFB, Cheyenne, Wyo., Travis AFB, Susan, Calif., 1950-51; chief orthopedic svc. and gen. surgery Shepherd AFB, 1951-52; comdg. officer, chief orthopedic svc., chief gen. surgery Craig AFB Hosp., Selma, Ala., 1952-53; pvt. practice medicine specializing in orthopedic surgery, Muskogee, Okla., 1954-69, 72—; courtesy staff Muskegon Gen. Hosp.; orthopedic cons. McAlester (Okla.) Gen. Hosp., VA Hosp., Muskogee. Exec. mem. Eastern Okla. council Boy Scouts Am. Fellow ACS, Internat. Coll. Surgeons; mem. Am. Fracture Assn., Nat. Found. (adviser 1958-61), N.Y. Acad. Scis., Okla. State, Pan-Am., So., Aerospace med. assns., AMA, So. Orthopaedic Assn., Eastern Okla. Counties med. socs., S.W. Surg. Congress, Am. Rheumatology Soc., Air Force Assn. (life). Republican. Methodist. Masons, K.T. Shriners, Jesters, Lions. Club: Muskogee Country. Contbr. articles to profl. jours.; research and evaluation of various uses of refrigerated homogenous bone. Home: 3504 University St Muskogee OK 74403-1843 Office: 1601 W Okmulgee St Muskogee OK 74401-6745

DANEHOLT, PER BERTIL EDVARD, molecular geneticist; b. Borås, Sweden, Nov. 25, 1940; s. Alva Frisk, Nov. 25, 1967. B Medicine, U. Gothenburg (Sweden), 1962; MD, Karolinska Inst., Stockholm, 1970. Asst. prof. Karolinska Inst., Stockholm, 1970-77, prof. dept. molecular genetics, 1981—, chmn. dept., 1987—; researcher Swedish Natural Sci. Rsch. Coun., Stockholm, 1978-80; mem. faculty bd. Karolinska Inst., Stockholm, 1984-90, mem. Nobel Com., 1990—; bd. mem. Swedish Natural Sci. Rsch. Coun., Stockholm, 1983-89; mem. Coun. of European Molecular Biology Lab., Heidelberg, 1983-89. Contbr. articles to profl. jours. Mem. Swedish Hybrid DNA Delegation, Stockholm, 1983—; chmn. ethical com. Royal Swedish Acad. Scis., Stockholm, 1990—; bd. mem. Göran Gustafsson Found., Stockholm, 1989—. Recipient Eric K. Fernström prize Karolinska Inst., Stockholm, 1984; E. Roosevelt Internat. Cancer fellow Internat. Union Against Cancer, Geneva, 1987-88. Mem. Royal Swedish Acad. Scis. (mem. bd.), Royal Physiographic Soc., Academia Europaea, European Molecular Biology Orgn. Avocations: bird watching, mountain walking, skiing. Home: Vallstigen 13, S-17246 Sundbyberg Sweden Office: Karolinska Inst, Box 60400, S-10401 Stockholm Sweden

DANĔK, VLADIMIR, chemistry scientist; b. Vienna, Austria, Apr. 6, 1940; s. Vladimír Daněk and Franziska (Halbmayer) Daňkova; m. Maria Ševčíková, July 5, 1969. Engr., T.U., Bratislava, 1962, PhD, 1973, DSc., 1991. Asst. prof. Inst. of Inorganic Chemistry Slovak Acad. of Scis., Bratislava, 1963-73; sr. scientist Inst. of Inorganic Chemisty Slovak Acad. of Scis., Bratislava, 1973-90, leading scientist, 1990-91, dir. of inst., 1991—, assoc. prof., 1993—; head of dept. Inst. Inorganic Chemist, Slovak Acad. Scis., 1990—, chmn. of sci. bd., 1990-91; mem. Sci. Coun. Slovak Acad. Scis. Contbr. over 80 articles to profl. jours. Lt. Fuel, 1962-63. Recipient Award Czechoslovak Acad. Sci., 1968, Slovak Acad. of Sci., 1986, 89, Silver medal D. Stur Slovak Acad. Sci., 1990. Roman Catholic. Achievements include 5 patents for thermodynamic model of silicate melts, dissociation model of molten salts mixtures. Office: Inst Inorganic Chem Slovak Acad Scis, Dubravska cesta 9, 842 36 Bratislava Slovak Republic

DANFORTH, DAVID NEWTON, JR., physician, scientist; b. N.Y.C., June 25, 1942; s. David Newton and Gladys Margaret (Blaine) D.; m. Anne Walker Nickson, Apr. 13, 1985. BA, Northwestern U., Evanston, Ill., 1965; MD, Northwestern U., Chgo., 1971; MS, U. N.Mex., Albuquerque, 1967. Diplomate Am. Bd. Surgery. Intern, then resident Cornell Med. Ctr., N.Y.C., 1971-74, 77-79; clin. assoc. NIH, Bethesda, Md., 1974-77; surg. fellow M.D. Anderson Hosp., Houston, 1979-80; sr. staff fellow NIH, Bethesda, 1980-82; sr. investigator Nat. Cancer Inst., NIH, Bethesda, 1982—. Editor: Diagnosis and Management of Breast Cancer, 1988; contbr. articles to profl. jours. Served to lt. comdr. USPHS, 1974-76. Fellow Am. Cancer Soc., 1979-80. Fellow ACS, Soc. Surg. Oncology, Am. Soc. Clin. Oncology, Am. Assn. Cancer Research, Endocrine Soc. Republican. Episcopalian. Avocations: travel, sports, reading. Home: 4701 Willard Ave Apt 536 Bethesda MD 20815-4611 Office: Nat Cancer Inst Surgery Br Bldg 10 Rm 2B38 Bethesda MD 20892

D'ANGELO, ANDREW WILLIAM, civil engineer; b. Bklyn., Jan. 23, 1924; s. William and Filomena (Soviero) D'A.; m. Filomena Margaret Loiero, June 26, 1949; children: Carol Lorraine Mauch, William Andrew. BSCE, Bklyn. Poly. Inst., 1952, MCE, 1956. Lic. engr., N.Y., N.J., Pa., Md., Conn., Mass., Fla. Project engr. D.B. Steinman, N.Y.C., 1952-56, Merritt Chapman & Scott Corp., N.Y.C., 1956-67; v.p. engring. Murphy Pacific Marine Salvage Co., N.Y.C., 1967-74; chief engr. Internat. Underwater Contractors, City Island, N.Y., 1974-76; cons. self-employed N.Y.C., 1976-77; pres. D'Angelo, Schoenewaldt Assoc. Inc., Floral Pk., N.Y., 1977-85; project mgr. North Star Contracting, New Rochelle, N.Y., 1985-88; project engr. Yonkers Contracting Co. Inc., Yonkers, N.Y., 1988—; cons. in field. Author: Salvage of Coastwise #1, 1975. Coun. mem. Incarnation Parish, Bellaire, N.Y., 1981-83, exec. bd. mem., 1970—; v.p. Incarnation Parish Holy Name Soc., 1990—. Sgt. U.S. Army, 1943-45. Recipient Bronze star U.S. Army, 1945, Commendation USN, 1965. Mem. ASCE, Nat. Soc. Profl. Engrs. Republican. Roman Catholic. Achievements include patent for Mooring Apparatus. Home: 19618 Keno Ave Jamaica NY 11423-1428 Office: Yonkers Contracting Co Inc 969 Midland Ave Yonkers NY 10704-1086

DANIEL, CHARLES TIMOTHY, transportation engineer, consultant; b. N.Y.C., Aug. 3, 1958; s. John Carl and Eleanor (Sauer) D. BA in Engring., Lafayette Coll., 1980; MS in Transp., MIT, 1982; MBA, NYU, 1991. Staff engr. George Beetle Co., Phila., 1983-84; project engr. Transamerica Leasing, White Plains, N.Y., 1984-87, mgr. tech. svcs., 1987-89, engring. cons., 1989—; mem. domestic freight container standards subcom. Internat. Standardization Orgn. Tech. Com. on Freight Containers, 1986-88. Mem. alumni planning coun. Rutgers Preparatory Sch., Somerset, N.J., 1985—; county committeeman Middlesex County (N.J.) Dem. Orgn., 1992—. Mem. ASCE, Sigma Xi, Beta Gamma Sigma. Lutheran. Achievements include development of code structure for electronic data interchange of freight container chassis repair data. Home: 34 North Dr East Brunswick NJ 08816 Office: Transamerica Leasing 100 Manhattanville Rd Purchase NY 10577

DANIEL, DONALD CLIFTON, aerospace engineer; b. Atlanta, May 21, 1942; s. Ben Melton and Jimmie Elizabeth (Dobbs) D.; m. Donna Maria Brown, June 13, 1976; 1 child, Jennifer Sim. BS in Aerospace Engring., U. Fla., 1964, MS in Aerospace Engring., 1965, PhD in Aerospace Engring., 1973. Chartered engr., U.K. Rsch. engr. The Boeing Co., Huntsville, Ala., 1965-68; rsch. assoc. U. Fla., Gainesville, 1971-72; aerospace engr. Air Force Armament Lab., Eglin AFB, Fla., 1973-88; chief scientist Arnold Engring. Devel. Ctr., Arnold AFB, Tenn., 1988—; bd. dirs. U. Tenn. Space Inst. Nat. Adv.Bd., Tullahoma, 1984—, Von Karman Inst., Brussels, Belgium, 1991—; mem. editorial adv. bd. AIAA Jour. of Spacecraft and Rockets, Washington, 1988— Contbr. articles to profl. jours. Mem. Tullahoma Rotary club, 1988—. Fellow AIAA (assoc.), Royal Aero. Soc. Office: Arnold Engring Devel Ctr AEDC/CN Arnold AFB TN 37389

DANIEL, EDDY WAYNE, mechanical engineer; b. Amarillo, Tex., Nov. 24, 1960; s. William Everett and Glenda Sue (Johnson) D.; Aug. 18, 1984; children: Courtney Michelle, Chase Wayne. BS in Mechincal Engring., Tex. A & M U., 1983. Registered profl. engr. Tex.; cert. operator Tex. Water Commn. Gen. ptnr. Susnshine Design Landscape Co., Farmersville, Tex., 1983-86; staff engr. Hopewell Water Supple Corp., Caddo Mills, Tex., 1986-87, gen. mgr., 1987-89; gen. mgr. Caddo Basin Spl. Utility Dist., Greenville, Tex., 1989--; state dir. Tex. Rural Water Assn., Austin, 1992--; bd. sec. Collin County Water Auth., McKinney, Tex., 1990--; mem. Water Utility Coun. Tech. Action Workgroup, Vorhees, N.J., 1991-92. Trustee Farmersville Ind. Sch. Dist., 1992--; pres. Farmersville C. of C., 1991-92; mem. Hunt County Rural Devel. Adv. Coiun., Farmersville Indsl. and Comml. Devel. Com. Mem. Nat. Soc. for Profl. Engrs., Am. Water Works Assn., Am. Soc. Civil Engrs., Am. Soc. Mechanical Engrs., Rotary (pres. 1990-91). Republican. Baptist. Home: Rt 1 Box 324-H Farmersville TX 75442 Office: Caddo Basin Spl Utility Rt 3 Box 300 Greenville TX 75401

DANIEL, JOHN MAHENDRA KUMAR, biomedical engineer, researcher; b. Nagercoil, India, Feb. 27, 1964; came to U.S., 1986; s. John Samson Nathanial and Rukmani Daniel. B of Engring., Anna U., Madras, India, 1985; MS, Iowa State U., 1988, PhD, 1991. Tutor Iowa State U., Ames, 1987-89, lab. asst. racing chemistry, 1990-91, grad. rsch. asst. biomed. engring., 1987-91; R&D project engr. Scimed Life Systems, Maple Grove, Minn., 1991—; senator grad. student senate Iowa State U., Ames, 1990-91; grad. student rep. grad. coun. Iowa State U., Ames, 1989-90; treas. biomedical engring. interest group Iowa State U., Ames, 1987-88. Mem. Soc. Biomaterials, Biomed. Engring. Soc., Sigma Xi, Phi Kappa Phi. Achievements include development of a new polymeric balloon material for coronary angioplasty balloon catheters; design and development of a multiple-lumen silicone rubber nerve cuff to bridge a 5mm gap in the adult rat sciatic nerve. Home: 6301 Quinwood Ln # 213 Maple Grove MN 55369 Office: Scimed Life Systems 6655 Wedgwood Rd Maple Grove MN 55311

DANIEL, MARK PAUL, fiber optics network technician; b. Fairmont, W.Va., Apr. 8, 1962; s. Thomas Sasseen and Martha Charley (Thomas) D. Grad., Marion County Vo-Tech., 1980. Systems installer Am.

Telecommunications Co., Orlando, Fla., 1984-85; with network staff Electra Communications, Austin, Tex., 1985—. Sgt. USAF, 1980-84. Democrat. Roman Catholic. Home: 1403 Church St Bastrop TX 78602-2913 Office: Electra Communications 621 N Pleasant Valley Rd Austin TX 78702-3944

DANIEL, RAMON, JR., psychologist, consultant, bilingual educator; b. Phoenix, Oct. 30, 1936; s. Ramon Sr. and Rosario (Lopez) D.; m. Lydia Cadriel, June 4, 1960; children: Lynda Ruth, Michael Ray, Patricia Lynn. BA in Edn., Ariz. State U., 1964, MA in Edn., 1966; PhD, U.S. Internat. U., 1990. Lic. psychologist, Calif.; cert. bilingual and math. tchr., Calif. Tchr. Phoenix Sch. Dist., 1964, Garden Grove (Calif.) Unified Sch. Dist., 1964-75, Cypress (Calif.) Coll., 1975-77; psychologist Santa Ana (Calif.) Unified Sch. Dist., 1978—; dropout prevention program specialist, student success teams coord. Santa Ana Unified Sch. Dist.; mem. adj. faculty Nat. U., Irvine, Calif., 1991—; with Mexican Consulate, Orange County (Calif.) Office Ednl. Affairs, 1991. Columnist (newspapers) La Conexion Humana, 1988. Mem. Least Restrictive Edn. Task Force, Santa Ana, 1990, Task Force on Linguistic and Cultural Differences, Santa Ana, 1990-91, Community Svc. Bd., Anaheim, Calif., 1990—. Mem. Calif. Assn. Sch. Psychologists, Coun. for Exceptional Children, Assn. Mex.-Am. Educators, Phi Delta Kappa (pres. Calif. chpt. 1988-89, Svc. Key 1989). Avocations: travel, music, sports, photography, writing. Office: Santa Ana Unified Sch Dist 1405 French St Santa Ana CA 92701-2499

DANIELE, JOAN O'DONNELL, clinical psychologist; b. Queens, N.Y., May 6, 1958; d. James and Joan (Cullen) O'Donnell; m. Richard James Daniele, May 19, 1991. BA in Dance, Hunter Coll., 1979, MS in Dance Therapy, 1983; MA in Sch. Psychology, Adelphi U., 1987, PhD in Clin. Psychology, 1989. Lic. psychologist, N.Y.; cert. psychoanalyst. Dance therapist Bellevue Hosp., N.Y.C., 1982-85; sch. psychologist N.Y. Bd. Edn., Bklyn., 1987-90; psychology intern Postgrad. Ctr. for Mental Health, N.Y.C., 1988-89, asst. coord. substance abuse unit, staff therapist, 1989-91, supr., instr., psychology intern, 1991—; pvt. practice psychology N.Y.C., 1991—; with Psychoanalytic Inst. Postgrad. Ctr. for Mental Health, N.Y.C., 1993—. Pres. Staten Island (N.Y.) Clearwater Friends, 1976-78. Hunter Coll. Dance Co. scholar, 1978. Mem. Postgrad. Psychoanalytic Soc. (assoc.). Democrat. Roman Catholic. Avocations: horseback riding, listening to music, ice skating, writing, reading. Office: 250 W 90th St 6I New York NY 10024

DANIELL, LAURA CHRISTINE, pharmacology and toxicology educator; b. Atlanta, Sept. 16, 1957; d. Sidney Shalar and Rosemary (Hughes) D. BA, NYU, 1978; PhD, U. Tex., 1985. Postdoctoral fellow U. Colo. Health Scis. Ctr., Denver, 1985-88, instr., 1988; asst. prof. pharm. and toxicology Med. Coll. Ga., Augusta, 1988—; lectr. in field. Ad hoc reviewer various sci. jours.; contbr. articles to pubs. Ad hoc reviewer various granting orgns., 1988—. Student fellow U. Ala., 1981, U. Colo., 1985-88. Mem. AAAS, Am. Translators Assn., Am. Assn. Advancement Slavic Studies, AM. Soc. Pharm. and Exptl. Therapeutics, N.Y. Acad. Sci., Rsch. Soc. on Alcoholism, Soc. for Neurosci., Southeastern Pharm. Soc., Internat. Anesthesia Rsch. Soc., Phi Kappa Phi. Achievements include research in alcohol. Home: 516 Cambridge Rd Augusta GA 30909 Office: Med Coll Ga Dept Pharm & Toxicology Augusta GA 30912-2300

DANIELS, CHARLES JOSEPH, III, electrical engineer; b. L.A., July 1, 1941; s. Charles Joseph and Madeline Anna (Baldassare) D.; m. Joan Reichhold, Oct. 10, 1965; 1 child, Charles Joseph. BSEE, U. Md., 1971, MS in Physics, 1976. Draftsman Hazen & Sawyer Engrs., N.Y.C., 1960-62; technician EMR Corp., Greenbelt, Md., 1966-68; rsch. engr. Naval Rsch. Lab., Washington, 1971-73; sanitary engr. FDA, Rockville, Md., 1973-77; head of prodn. testing Digital Communications Corp., Germantown, Md., 1977-78; mem. tech. staff Watkins-Johnson, Gaithersburg, Md., 1978-79; sr. engr. Digital Communications Corp., Germantown, 1979-80; sr. staff engr. Frederick (Md.) Electronics, 1980-81; disting. mem. tech. staff AT&T Bell Labs., Allentown, Pa., 1981—. Patentee in field; contbr. over 20 articles to profl. jours. Cpl. USMC, 1962-66. Mem. Am. Phys. Soc., Tau Beta Pi (pres. 1970-71), Sigma Xi, Phi Kappa Phi, Omicron Delta Kappa (sec. 1970-71), Sigma Pi Sigma. Office: AT&T Bell Labs 555 Union Blvd Allentown PA 18103-1229

DANIELS, CINDY LOU, space agency executive; b. Moline, Ill., Sept. 24, 1959; d. Ronald McCrae and Mary Lou (McLaughlin) Guthrie; m. Charles Burton Daniels, June 19, 1982. Student, Augustana Coll., Rock Island, Ill., 1977-78; BS cum laude, No. Mich. U., 1981. Field engr. Ford Aerospace, Houston, 1982-83; engr. flight ops. McDonnell Douglas Corp., Houston, 1983-85; electronics engr. Johnson Space Ctr. NASA, Houston, 1985-89; project mgr. multiple program control ctr. NASA, 1989-90; project mgr. NASA, Houston, 1989-91, mission control ctr. upgrade project mgr., 1990-91; mgr. program control office NASA, 1991—; dynamics contr. NASA Johnson Space Ctr., 1982-83; payload data engr. NASAa, 1983-84, earth radiation budget satellite joint ops. integration plan mgr., 1984; mem. payload assist module team NASA-McDonnell Douglas Corp., 1984-85. Avocations: skiing, sailing. Home: 3703 Pine Trail La Porte TX 77571-7266 Office: Johnson Space Ctr Engring Directorate ET14 Houston TX 77058

DANIELS, FREDERICK THOMAS, reactor engineer; b. Ontario, Oreg., Sept. 7, 1947; s. Frederick Aaron Daniels and Maxine Virginia (Harris) Marrs; m. Judy Rose Kajmowicz, Oct. 7, 1979; children: Tami Ann, Thomas Aaron, Tara Ashley. AA, Olympic Coll., 1975; BA, Western Ill. U., 1983; MS in Energy Econs., Websters U., 1986. Cert. naval nuclear engr., internat. safeguards insp. Enlisted USN, 1965, advanced through grades; ret. Gen. Physics/Engring. Diverse Svcs. Nuclear, 1977; cons. Gen. Physics/EDS Nuclear, 1977-78; sr. resident insp. U.S. Nuclear Regulatory Commn., Chgo., 1978-82; safeguards insp. IAEA, Vienna, Austria, 1982-89; sr. reactor engr., team leader U.S. Nuclear Regulatory Commn., Washington, 1989-91, br. chief Dept. of Energy, 1991—; mem. affirmative action adv. com. bd. U.S. Nuclear Regulatory Commn., Washington, 1990-91. Author: Safeguard Practices for Research Reactors, 1987, Safeguard Practices for Critical Assemblies, 1987. Vice-chmn. staff coun. Internat. Atomic Energy Agy., Vienna, 1985; asst. dist. commr. Boy Scouts Am., Balt., 1990—. Mem. Am. Nuclear Soc. (founder Austria local sect., pres. 1985-86, cert. of governance 1985, exec. bd. 1985-88), Am. Legion (activities chmn. 1981), Elks (outer guard 1974, 75, nat. patron 1989). Republican. Mem. LDS Ch. Achievements include co-development of cerenkov view device used to monitor spent nuclear fuel elements. Home: 15513 Bushy Tail Run Woodbine MD 21797-8025

DANIELS, STEPHEN BUSHNELL, geneticist; b. Middletown, Conn., July 28, 1950; s. Willard Herbert Sr. and Jessie (Bushnell) D. BA, U. Conn., 1975, MS, 1983. Rsch. biologist U. Ariz., Tucson, 1985-87, U. Conn., Storrs, 1980-85, 87—. Contbr. articles on evolution, genetics and molecular genetics to profl. pubis. Home: PO Box 797 Storrs CT 06268

DANIELS, WILLIAM BURTON, physicist, educator; b. Buffalo, Dec. 21, 1930; s. William C. and Sophia (Penner) D.; m. Adriana A. Braakman, Sept. 2, 1958; children: Charlotte, William Fredrik, Donald Christopher. BS in Physics, U. Buffalo, 1952; MS, Case Inst. Tech., 1955; PhD, Case Inst. Tech., 1957. Instr. to asst. prof. Case Inst. Tech., 1957-59; rsch. scientist Union Carbide Corp., 1959-61; mem. faculty Princeton U., 1961-72, prof. solid state scis., 1967-72; Unidel prof. physics U. Del., Newark, 1972—; rsch. collaborator Brookhaven nat. Lab.; cons. U.S. Army Rsch. Lb.; guest scientist rsch. facility, Denmark, 1976; invité Coll. France, 1977; exch. prof. U. Paris, 1977; guest scientist IBM Zurich Lab., 1977; guest scientist Max Planck Inst. for Festkörperforschung. Recipient Alexander von Humboldt Sr. Scientist award, 1981, 92; John S. Guggenheim Meml. fellow, 1976-77. Fellow Am. Phys. Soc.; mem. AAAS. Rsch. and publs. properties materials at high pressure, equation of state of solids, experimentation on solidified permanent gases, electronic structure of compressed solids, instrumentation high pressure rsch., non-linear optics. Office: U Del Physics Dept Newark DE 19716

DANIELSON, NEIL DAVID, chemistry educator; b. Ames, Iowa, July 25, 1950; s. Gordon Charles and Dorothy Elisabeth (Thompson) D.; m. Elizabeth Moore, Aug. 4, 1979 (dec. July 28, 1986); 1 child, Glenn James; m. Kami Lee Park, Oct. 7, 1990; 1 child, Kenneth Park. BS, Iowa State U., 1972; MS, Nebr. U., 1974; PhD, Ga. U., 1978. Asst. prof. Miami U., Oxford, Ohio, 1978-83, assoc. prof., 1983-91, prof., 1991—; vis. scientist E.I. DuPont Co., Wilmington, Del., 1985-86; cons. Interaction Chems., Inc., Mountain View, Calif., 1983-91; sec. Ohio Valley Chromatography Symposium, 1988—. Contbr. articles to Analytical Chemistry, Jour. Chromatography, HPLC in Food Sci., Ency. Sci. & Tech. Achievements include research in high performance liquid chromatography and capillary electrophoresis. Office: Miami U Dept Chemistry Oxford OH 45056

DANIELS-RACE, THEDA MARCELLE, electrical engineering educator; b. New Orleans; d. William Mack and Valda Marie (Wilder) D.; m. Paul A. Race, May 11, 1985. BS, Rice U., 1983; MS, Stanford U., 1985; PhD in Elec. Engring., Cornell U., 1989. Asst. prof. elec. engring. Duke U., Durham, N.C., 1990—. Nat. Consortium of Grad. Engring. Minorities fellow, 1983-85; AT&T Coop. Rsch. fellow, 1985-89. Mem. IEEE, Am. Phys. Soc., Materials Rsch. Soc., Soc. Black Engrs. (advisor). Office: Duke U Dept Elec Engring PO Box 90291 Durham NC 27708-0291

DANKO, EDWARD THOMAS, mechanical and industrial engineer; b. Uniontown, Pa., Feb. 2, 1952; s. Edward John Danko and Julia Lucille (Kelemen) McGrath; m. Rebecca Lynn Huey, June 17, 1972. BS in Occupational/Ednl., Wayland Bapt. Coll., 1991. Machinist Westinghouse Electric Corp., Large, Pa., 1977-84, renewal parts engr., 1984-90; renewal parts engr. Westinghouse Savannah River Co., Aiken, S.C., 1991—; instr. C.C. Allegheny County, West Mifflin, Pa., 1987-89. Author, editor: Corporal Edward J. Danko WWII Diary, 1992. Sgt. USAF, 1972-76. Mem. Sons of Italy, Am. Vets. Democrat. Roman Catholic. Office: Westinghouse Savannah River Co Bldg 245-6F Aiken SC 29803

DANKO, JOSEPH CHRISTOPHER, metals engineer, university official; b. Homestead, Pa., Jan. 12, 1927; s. John and Anna Danko; m. Laverne Elizabeth Uramey, June 20, 1951; children: Christopher, Kimberly, Mark. BS in Metals Engring., Carnegie-Mellon U., 1951; MS in Metals Engring., Lehigh U., 1954, PhD in Metals Engring., 1955. Engr. GE Co., Schenectady, 1951-52; mgr. GE Co., San Jose, Calif., 1964-78; instr. Lehigh U, Bethlehem, Pa., 1952-56; mgr. Westinghouse Electric Corp., Pitts., 1956-62; program mgr. Electric Power Rsch. Inst., Palo Alto, Calif., 1978-84, 1978-84; v.p. Am. Welding Inst., Knoxville, Tenn., 1984-86; dir. Ctr. for Materials Processing U. Tenn., Knoxville, 1986—; mem. corrosion adv. com. Electric Power Rsch. Inst., Palo Alto, Calif., 1974-78; mem. tech. adv. com. MPC, N.Y.C., 1974-91; cons. Dupont Savannah River Lab., Aiken, S.C., 1986-91; cons. in field. Contbr. over 100 tech. papers to publs. Pres. Knoxville Dismas Project, Knoxville, 1987-89. Fellow Am. Soc. Metals; mem. AAAS, Nat. Assn. Corrosion Engrs., Am. Welding Soc., Am. Nuclear Soc. Achievements include patents in material processing; first demonstration of solar thermionic power system; first operation of nuclear thermionic thermoelectric device. Office: U Tenn Coll Engring Knoxville TN 37996-2000

DANKS, ANTHONY CYRIL, detector scientist, research astronomer; b. Willenhall, Eng., June 28, 1945; came to U.S., 1985; s. Cyril Edward and Mary Lillian (Taylor) D.; m. Eva-Ann Marie Johansson (div.); 1 child, Jean-Paul Austin; m. Maria Cristina Bories, July 19, 1985; children: Nicole Alejandra, Julio-Antonio Enrique. BSc in Applied Physics, Hull U., 1967; MSc in Astrophysics, Sussex U., Brighton, Eng., 1968; PhD in Astronomy, Manchester (Eng.) U., 1970. Rsch. scientist Kapteyn Sterrewacht, Roden, The Netherlands, 1975-77; sr. staff astronomer ESO, La Silla, Chile, 1977-85; vis. assoc. prof. Mich. State U., East Lansing, 1985-86; asst. prof. Fla. Internat. U., Miami, 1986-87; sr. scientist Applied Rsch. Corp., Landover, Md., 1987-90; chief scientist Hughes STX, Lanham, Md., 1990—; editor newsletter, founder Ctr. Astronomy and Space Physics ST Systems Corp., Lanham, Md., 1990-91; founer Ctr. for Astronomy and Space Physics/Hughes STX; co-I and detector scientist Space Telescope Imaging Telescope, NASA Goddard Space Flight Ctr., Greenbelt, Md., 1989—; instr. Prince Georges C.C., Lago, Md., 1988-91. Contbr. over 60 articles to sci. jours. Royal Soc. fellow English Speaking Union, U. Liege, Belgium, 1970-72, Lindeman fellow U. Tex., 1973-75. Fellow Am. Astron. Soc.; mem. Internat. Astron. Union (commns. 15, 34, 28). Republican. Achievements include research on molecular spectroscopy of comets, chemistry of the interstellar medium, galactic structure, star formation in SO galaxies, detector science related to CCDs and MAMA detectors, diffraction gratings, high temperature superconductivity, and remote sensing using passive radiometry. Home: 1315 Peach Tree Ct Bowie MD 20721-3000 Office: Hughes STX/ Goddard Space Flight Ctr Code 683/O Bldg 21 Greenbelt MD 20771

DANNENBERG, KONRAD K., aeronautical engineer; b. Weissenfels, Germany, Aug. 5, 1912; came to U.S. 1945.; s. Hermann and Klara (Kittler) D.; m. Ingeborg M. Kamke, Apr. 8, 1944 (dec.); 1 child, Klaus Dieter; m. Jacquelyn E. Staiger, Mar. 31, 1990. MS Engring., Techn. U., Hannover, Ger., 1938. Asst. Tech. U., Hannover, 1938; engr. Tech. U., Frankfurt, Ger., 1939; researcher HAP-Peenemuende, Germany, 1940-45; mgr. U.S. Army Ordnance, Ft. Bliss, Tex., 1945-50, ABMA, Huntsville, Ala., 1950-60, NASA/MSFC, Huntsville, 1960-73; assoc. prof. UTSI-U. Tenn., Tullahoma, 1973-78; cons. The Space & Rocket Ctr., Huntsville, 1978—. Author: In Memory of H. Oberth, 1990, Vahrenwald to Dresden, 1990. Lt. German Army, 1939-40. Recipient Meritorious Svc. award, U.S. Army, 1960, Exceptional Svc. award, NASA, 1969. Assoc. Fellow AIAA (chpt. chmn. 1967, Durand lectr. pub. svc. 1990); mem. Hermann Oberth Soc. (hon. mem.), Nat. Space Soc. (charter mem.), Am. Rocket Soc. (chmn. 1962), Internat. Assn. Educators for World Peace (treas.). Lutheran. Achievements include patents in rocket engine design. Home: 64 Revere Way Huntsville AL 35801-2846 Office: Space & Rocket Ctr 1 Tranquility Base Huntsville AL 35805-3399

DANNEWITZ, STEPHEN RICHARD, emergency physician, consultant, toxicologist; b. Decatur, IL, Sept. 14, 1948; s. Richard Wayne and Irma Lorine (Gaffron) D.; m. Gretchen Hellerberg, Aug. 22, 1976 (div.); m. Julie Valbuena, May 25, 1978 (div.); children: Susan, Michael, Katherine. BS, Valparaiso U., 1970; MD, Case Western Res. U., 1976. Diplomate Am. Bd. Emergency Medicine, Am. Bd. Med. Toxicology. Intern Henning County Med. Ctr., 1976-77, resident, 1977-79; Pvt. practice Robbindale, Minn.; examiner, team leader Am. Bd. Emergency Medicine, 1981—. Contbr. articles to profl. jours. Fellow Am. Coll. Emergency Physicians. Lutheran. Home: 5400 Vernon Ave S # 203 Edina MN 55436 Office: North Meml Med Ctr 3300 Oakdale Ave N Robbinsdale MN 55422-2926

DANSE, ILENE HOMNICK RAISFELD, physician, educator, toxicologist; b. Bklyn., June 24, 1940; d. Jack and Henrietta (Poverstein) Homnick; m. James Atherton Danse, Aug. 10, 1982; children: Arthur Raisfeld, Robin Raisfeld. BS, CUNY, 1960; MD, NYU, 1964. Diplomate Nat. Bd. Med. Examiners, Am. Bd. Internal Medicine, Am. Bd. Toxicology. Assoc. prof. internal medicine SUNY, Stony Brook, 1975-83, assoc. prof. pharmacology, 1977-83, dir. clin. pharmacology and toxicology Sch. Medicine, 1978-83; acting chairperson clin. pharmacology Northport VA Hosp., L.I., N.Y., 1978-83; sr. advisor Chevron Environ. Health Ctr., San Pablo, Calif., 1982-84; prin. ENVIROMED Health Svcs., Inc., San Rafael, Calif., 1985—; ind. med. examiner toxicology and internal medicine Dept. Indsl. Rels., State of Calif., 1985—; assoc. clin. prof. dept. medicine div. occupational and environ. medicine U. Calif., San Francisco, 1991—; cons. in fields of environ., occupational and internal medicine, toxicology and pharmacology, 1984—. Author: Common Sense Toxics, 1991; contbr. articles to sci. publs. Fellow ACP, Am. Coll. Clin. Pharmacology; mem. AAAS, Am. Acad. Clin. Toxicology, Am. Chem. Soc. (environ. health and safety sect.), Am. Coll. Occupational Medicine, Am. Indsl. Hygiene Assn. (occupational medicine sect.), Am. Coll. Toxicology, Am. Soc. Pharmacology and Therapeutics, Soc. Toxicology, Western Occupational Med. Assn., San Rafael C. of C. Achievements include patent for epithelial cell growth-regulating composition containing polyamines, and method of its use. Office: ENVIROMED Health Svcs Inc 705 Mission Ave San Rafael CA 94901-2919

DANTUONO, LOUISE MILDRED, obstetrician/gynecologist; b. N.Y.C., July 29, 1916; d. Anthony and Margaret (Cogliano) D.; m. John David Angelides, May 28, 1955 (dec. 1970); children: Helen, Elaine. BA magna cum laude, Adelphi U., 1937; MD, Womens Med. Coll. Pa., 1942. Diplomate Am. Bd. Ob-Gyn. Dir. ob-gyn. Cabrini Hosp., N.Y.C., 1959-62; attending ob-gyn. Booth Meml. Hosp., N.Y.C., 1957-59, French Hosp., N.Y.C., 1946-68, U. Hosp., N.Y.C., 1949—, Bellevue Hosp., N.Y.C., 1946—; attending ob-gyn. cons. N.Y. Infirmary, N.Y.C., 1947—; attending ob-gyn. emeritus Doctors Hosp., N.Y.C., 1948—. Contbr. articles to profl. jours. Recipient Disting. Alumnae award Bellevue Ob-Gyn. Soc., 1978. Fellow ACS, Am. Coll. Ob-Gyn.; mem. AMA, N.Y. County Med. Soc., N.Y. State Med. Soc., N.Y. Acad. Scis., Delta Tau Alpha, Alpha Omega Alpha. Roman Catholic. Office: 35 E 35th St New York NY 10016

DANTZIG, GEORGE BERNARD, applied mathematics educator; b. Portland, Oreg., Nov. 8, 1914; s. Tobias and Anja (Ourisson) D.; m. Anne Shmuner, Aug. 23, 1936; children—David Franklin, Jessica Rose, Paul Michael. A.B. in Math., U. Md., 1936; M.A. in Math., U. Mich., 1937; Ph.D. in Math., U. Calif.-Berkeley, 1946; hon. degree, Technion, Israel, Linkoping U., Sweden, U. Md., Yale U., Louvain U., Belgium, Columbia U., U. Zurich, Switzerland, Carnegie-Mellon U. Chief combat analysis br. Statis. Control Hdqrs. USAAF, 1941-46, math. advisor, 1946-52; research mathematician Rand Corp., Santa Monica, Calif., 1952-60; prof., chmn. Ops. Research Ctr., U. Calif.-Berkeley, 1960-66; C.A. Criley prof. ops. research and computer sci. Stanford U., Calif., 1966—; chief methodology Internat. Inst. Applied System Analysis, 1973-74; cons. to industry. Author: Linear Programming and Extensions, 1963; co-author: Compact City, 1973; contbr. articles to profl. jours.; assoc. editor Math. Programming, Math. of Ops. Research, others. Recipient Exceptional Civilian Svc. medal War Dept., 1944, Nat. medal of Sci., 1975, Von Neumann theory prize in ops. rsch., 1975, award Nat. Acad. Scis., 1975, Harvey prize Technion, 1985, Silver Medal Operational Rsch. Soc., Gt. Britain, 1986, Coors Am. Ingenuity award, 1989. Fellow Am. Acad. Arts and Scis., Econometric Soc., Inst. Math. Stats.; mem. Nat. Acad. Scis., Nat. Acad. Engring., Ops. Research Soc. Am., Am. Math. Soc., Math. Programming Soc. (chmn. 1973-74), Inst. Mgmt. Sci. (pres. 1966), Phi Beta Kappa, Sigma Xi, Phi Kappa Phi, Pi Mu Epsilon. Home: 821 Tolman Dr Stanford CA 94305-1025 Office: Stanford Univ Ops Research Dept Stanford CA 94305

DANTZIG, JONATHAN A., mechanical engineer, educator; b. Balt., Aug. 14, 1951; s. Henry P. and Mildred B. (Krieger) D.; m. Anne H. Tifford, Oct. 31, 1976; 1 child, Erika H. BS in Engring., Johns Hopkins U., 1972, MS, 1975, PhD, 1977. Rsch. scientist Olin Metals Rsch. Lab., New Haven, Conn., 1977-79, sr. rsch. scientist, 1979-82; asst. prof. mech. engring. U. Ill., Urbana, 1982-87, assoc. prof., 1987—; asst. adj. prof. U. New Haven, 1980-82; with NASA Materials Sci., Washington, 1985—. Editor: Modeling of Casting and Welding Processes, 1984, Nature and Properties of Semi-Solid Materials, 1992; author: (monograph) Thermal Stress Development in Metal Casting Processes, 1989. Mem. Am. Inst. Mining, Metall. and Petroleum Engrs. (chair solidification com. 1987-88), ASM. Achievements include patents for degassing of molten metal, semi-solid materials processing, electromagnetics and materials processing. Office: U Ill Dept Mech Engring 1206 W Green St Urbana IL 61801

DANZIN, CHARLES MARIE, enzymologist; b. Paris, Sept. 8, 1944; s. Andre and Nicole (De Freminville) D.; m. Elisabeth Evrard, Aug. 25, 1970; 1 child, Claire. Degree in Engring., Ecole Speciale de Meca. Elec., Paris, 1966; PhD in Biochemistry, U. Paris, 1974. Predoctoral fellow dept. biophysics Commissariat A L'Energie Atomique, Saclay, France, 1969-74; biochemist Merrell Internat. Rsch. Ctr., Strasbourg, France, 1975-77; rsch. biochemist Merrell Dow Rsch. Inst., Strasbourg, 1977-87; group leader, head enzymology Marion Merrell Dow Rsch. Inst., Strasbourg, 1988-92, dir. dept. enzymology, 1993—; mem. overseas adv. bd. Internat. Symposia on Vitamin B6 and Carbonyl Catalysis, 1987—; speaker several internat. scientific meetings and scientific orgns. Contbr. 110 articles to sci. publs.; jour. reviewer Biochem. Pharmacology, Analytical Biochemistry, British Jour. Cancer, Jour. Am. Chem. Soc., Eur. Jour. Med. Chemistry. Mem. AAAS, Am. Chem. Soc., N.Y. Acad. Scis., Soc. Francaise de Chimie. Office: Marion Merrell Dow Rsch Ins, 16 Rue D'Ankara BP 067, F 67046 Strasbourg France

DANZL, DANIEL FRANK, emergency physician; b. Cin., Apr. 2, 1950; s. Frank Bernard and Mary Ellen (Doerger) D.; m. Joanna Colosimo Danzl, Nov. 25, 1978; children: Maggie, Julia. BS magna cum laude, U. Cin., 1972; MD, Ohio State U., 1976. Diplomate Am. Bd. Emergency Medicine. Intern St. Francis Med. Ctr., Peoria, Ill., 1976-77; resident in emergency medicine U. Louisville, 1977-79, asst. prof. emergency medicine, 1979-83, assoc. prof. emergency medicine, 1983-89, prof. emergency medicine, 1989-91, prof., chair, 1991—; bd. dirs., councilman-at-large Univ. Assn. for Emergency Medicine, 1988-89, indsl./govtl. relations com., 1984-85, nominating com., 1987-88; bd. dirs. Soc. for Academic Emergency Medicine, 1989, mem. annals of emergency medicine task force, 1989; bd. dirs. Am. Bd. Emergency Medicine, mem. ad hoc com., oral examiner , 1982—; mem. Com. to Advise the Nat. Am. Red Cross, 1984-85, 85-86, 86-87; reviewer for various med. jours. Author book chpts., monographs and textbooks including Airway Management in the Trauma Patient in the Clinical Practice of Emergency Medicine, 1991; editorial bd. Jour. Emergency Medicine, 1983—, Poisindex-Emergindex, 1982—, Jour. Wilderness Medicine, 1991—; contbr. more than 70 articles to Jour. Wilderness Medicine, Jpur. Emergency Medicine, Annals of Emergency Medicine, Am. Jour. Emergency Medicine, others. Mem. Water Safety Com. Nat. Safety Coun.-Pub. Safety Div., 1981-84; alternate med. dir. Jefferson Vocat. Edn.-Louisville EMS Paramedic Training Program, 1989-90, 90-91. Recipient Silver Tongue Orator award Soc. Tchrs. of Emergency Medicine, 1986, 88; grantee Office of Naval Resources, 1983-85, Key Pharmaceuticals, 1985, Hoffman-LaRoche, Inc., 1988, 89. Fellow Am. Coll. Emergency Physicians (nat. coun. mem. 1981-93, reference com. mem. 1981, 85, 89, rsch. com. mem. 1982-83, 83-84); mem. AMA (Physician's Recognition awards), NAS, Am. Soc. Circumpolar Health, Soc. for Academic Emergency Medicine (bd. dirs. 1989, task force 1989), Nat. Rsch. Coun., Undersea and Hyperbaric Oxygen Med. Soc., Ky. Chpt. Am. Coll. Emergency Physicians (councillor 1981-93, sec.-treas. 1983-84, pres.-elect 1984-85, pres. 1985-86), Wilderness Med. Soc., Phi Beta Kappa, Beta Theta Pi. Roman Catholic. Achievements include research on hypothermia. Home: 4804 Smith Rd Floyds Knobs IN 47119-9214 Office: Univ of Louisville Dept Emergency Medicine 530 S Jackson St Louisville KY 40292-0001

DAO, LARRY, computer software engineer; b. Saigon, Vietnam, Dec. 23, 1960; came to U.S., 1982; s. Xuong V. and Tu Nguyen D.; m. Cindy Dang, Mar. 12, 1988; 1 child, David T. BS in Computer Engring., Nat U., 1988. Programmer analyst Electro Plasma, Irvine, Calif., 1987-88, Volt Info. Scis., Anaheim, Calif., 1988-90; computer software engr. Fisons Instruments Inc., Valencia, Calif., 1990—.

DAOUD, MOHAMED, physicist; b. Tunis, Tunisia, Mar. 31, 1947; came to France, 1964; s. Hassen Daoud and Beya (Bahri) D.; m. Aïcha Mehiri; 1 child, Hassen. Maitrise, Ecole Normale Superieure, Saint Cloud, 1968-70, D.E.A., 1972, (hon.) Agregation, 1971; These, U. Paris VI, 1977. Research assoc. Commissariat a l'Energie Atomique, C.E.A. Saclay, 1974-78, physicist, 1980—; physicist Boston U., 1978-80. Editor: J. de Physique. Recipient Grand Prix, Groupement Français de Polymeres, 1978, prix Commissariat ā l'Energie Atomique, 1986. Mem. Societe Française de Physique, Am. Phys. Soc., N.Y. Acad. of Scis. Office: CEN Saclay, Lab Leon Brillouin, 91191 Gif sur Yvette France

DAPPLES, EDWARD CHARLES, geologist, educator; b. Chgo., Dec. 13, 1906; s. Edward C. and Victoria (Gazzolo) D.; m. Marion Virginia Sprague, Sept. 2, 1931; children—Marianne Helena, Charles Christian. B.S., Northwestern U., 1928, M.S., 1934; M.A., Harvard, 1935; Ph.D., U. Wis., 1938. Geologist Ziegler Coal Co., 1928; geologist Truax-Traer Coal Co., 1928-32, mine supt., 1932; instr. Northwestern U., 1936-41, asst. prof., 1941, asso. prof., 1942-50, prof. geol. scis., 1950-75, prof. emeritus, 1975—; geologist Ill. Geol. Survey, 1939, Sinclair Oil Co., 1945-50, Pure Oil Co., 1950; dir. Evanston Exploration Corp., 1954-84; sr. vis. scientist U. Lausanne, Switzerland, 1960-61; vis. prof. U. Geneva, Switzerland, 1970. Author: Basic Geology for Science and Engineering, 1959, Atlas of Lithofacies Maps, 1960. Fellow Geol. Soc. Am., Soc. Econ. Geologist; mem. Am. Inst. Mining Engrs. (Legion of Honor), Assn. Petroleum Geologists, Internat. Assn. Sedimentologists, Soc. Econ. Paleontologists and Mineralogists (pres. 1970, hon. mem. 1974), Am. Inst. Profl. Geologists (pres. Ill.-Ind. sect. 1979, pres. Ariz. 1982, hon. mem. 1986), Assn. Engring. Geologists. Home: 13035 98th Dr Sun City AZ 85351

DAR, MOHAMMAD SAEED, pharmacologist, educator; b. Lahore, Pakistan, Dec. 10, 1937; came to U.S., 1979; s. Mohammad Usaf and Amir Begum (Amir) D.; m. Parveen Saeed, Mar. 16, 1969; children: Mahammed

Mujtaba, Moahad Saeed. MS in Pharmacy, U. Med. Scis., Bangkok, 1966; PhD in Pharmacology, Med. Coll. Va., 1970. Rsch. assoc. Rockefeller Found., Bangkok, 1970-72; asst. prof., then assoc. prof. Shiraz (Iran) U. Med. Sch., 1972-79, acting chmn., 1977-79; asst. prof. dept. pharmacology East Carolina U. Sch. Medicine, Greenville, N.C., 1979-85, assoc. prof., 1985-90, prof. pharmacology, 1990—. Contbr. rsch. articles to sci. jours. Mem. Am. Soc. Pharmacology and Exptl. Therapeutics, Soc. Neurosci., Rsch. Soc. Alcoholism, N.Y. Acad. Scis., Internat. Soc. Biomed. Rsch. on Alcoholism. Home: 115 Heritage St Greenville NC 27858 Office: East Carolina U Sch Medicine Greenville NC 27858

DARDEN, WILLIAM HOWARD, JR., biology educator; b. Tuscaloosa, Ala., Apr. 25, 1937; s. William Howard and Jannie Belle (Herring) D.; m. Caroline Jackson, July 15, 1959; children: Leanne Carol, Michael Howard. B.S., U. Ala., Tuscaloosa, 1959, M.S., 1961; Ph.D., Ind. U., 1965. Asst. prof. biology U. Ala., Tuscaloosa, 1965-68, assoc. prof., 1969-73, prof., assoc. chmn. dept. biology, 1973-74, prof., chmn. dept. biology, 1974—. Contbr. articles to sci. jours. Bd. dirs. Springhill Lake Assn., 1980-85, So. Grass Tennis Club, 1979-81, Ala. Credit Union, 1987—. Predoctoral fellow NIH, 1963-65; grantee NSF, 1972, U. Ala., Tuscaloosa, 1965-71. Mem. Assn. Southeastern Biologists, Southeastern Assn. Edn. Tchrs. of Sci., Sigma Xi, Beta Beta Beta, Omicron Delta Kappa, Phi Kappa Phi. Am. Baptist. Home: 3628 Rainbow Dr Tuscaloosa AL 35405-5331 Office: U Ala PO Box 870344 Tuscaloosa AL 35487-0344

DARE, CHARLES ERNEST, civil engineer, educator; b. Peoria, Ill., June 3, 1938; s. Edwin Joyner and June Elmira (Loveless) D.; m. Jane Ann Hugen, Sept. 2, 1967; children: Robert E., Elizabeth J. BA, U. Iowa, 1961, BS, 1962, MS, 1963, PhD, 1968. Registered profl. engr., Mo. Environ. protection engr. Ill. EPA, Springfield, 1973-75; assoc. prof. U. Colo., Denver, 1975-77; prof. civil engring. U. Mo., Rolla, 1977—. Contbr. articles to profl. jours. Commr. Rolla Planning and Zoning Commn., 1988—. Mem. ASCE, Transp. Rsch. Bd., Am. Soc. Engring. Edn., Inst. Transp. Engrs. (past pres.'s award 1969). Office: U Mo Rolla Rm 223 Civil Engring UMR Rolla MO 65401

DARLING, CHERYL MACLEOD, health facility administrator, researcher; b. Detroit, Feb. 23, 1949; d. Norman Duncan and Elsie Ruth (Howland) MacLeod; m. Jeffery F. Hunter, Jan. 1, 1976 (div. 1982); m. Richard W. Darling, Dec. 31, 1986; 1 stepchild, Erin Marie. AA, U. S.C., 1979; BS, No. Ill. U., 1982, MS, 1984. Medic, clin. diagnostician U.S. Army, 1972-80; grad. rsch. asst. Sch. Allied Health Profls. No. Ill. U., DeKalb, 1982-84; planning aide Comprehensive Health Planning N.W. Ill., Rockford, 1982-83, health planner, cons., 1982-85; dir. community health edn. and safety svc. Dane County Chpt. ARC, Madison, Wis., 1985-88; coord. rsch., adminstrn. Ctr. for Clin. Ethics Luth. Gen. Hosp., Park Ridge, Ill., 1988—; rsch. interviewer Ctr. for Clin. Med. Ethics U. Chgo., 1990. Bd. dirs. Vet's House, Madison, 1986-88; profl. vol. bereavement team, counselor Rainbow Hospice, Park Ridge, 1989, mem. quality assurance com., 1990—, mem. profl. adv. bd., 1991—; mem. planning com. Chgo. Women's AIDS Project, 1989-90. Mem. Soc. for Health and Human Values, Applied Rsch. Ethics Nat. Assn., Kennedy Inst. Ethics, Profl. Responsibility in Medicine and Rsch. Mem. Unity Ch. Avocations: camping, metaphysics, gardening, crafts. Home: 1264 Cedar Ave Elgin IL 60120-2205 Office: Luth Gen Hosp Ctr for Clin Ethics 1775 Dempster St Park Ridge IL 60068-1173

DARLINGTON, JULIAN TRUEHEART, biology educator retired; b. Barboursville, W.Va., Mar. 18, 1918; s. Urban V.W. and Virginia Lee (Bourne) D.; m. Jeanne Matthews, Dec. 5, 1942; children: Virginia B., Patricia M. AB, Emory U., 1940, MS, 1941; PhD, U. Fla., 1952. Asst. prof. U. Ga., Atla., 1953-54; prof. biology Shorter Coll., Rome, Ga., 1954-58, Furman U., Greenville, S.C., 1958-63, Rhodes Coll., Memphis, 1964-85. Contbr. articles to profl. jours. Commdr. Mid South Ex-POWs, Memphis, 1992-93. Lt. USAAF, 1941-45, ETO. Decorated Disting. Flying Cross, 1943, Bronze medal, 1945. Mem. Exchange Club, Sigma Xi, Phi Beta Kappa. Episcopalian.

DARLINGTON, RICHARD BENJAMIN, psychology educator; b. Woodbury, N.J., Nov. 16, 1937; s. Charles Joseph and Eleanor (Collins) D.; m. Elizabeth Day, June 13, 1959; children: Jean Susan, Lois Heather. BA, Swarthmore Coll., 1959; PhD, U. Minn., 1963. Asst. prof. psychology Cornell U., Ithaca, N.Y., 1963-68, assoc. prof., 1968-80, prof., 1980—. Author: Radicals and Squares, 1975, (with others) Lasting Effects of Early Education, 1982, (with Patricia M. Carlson) Behavioral Statistics: Logic and Methods, 1987, Regression and Linear Models, 1990; contbr. articles to profl. jours.; contbr. chpts. to books. Project dir. Am. Friends Service Com., 1960, 61. Fellow NSF, 1959-60; fellow Woodrow Wilson Found., 1959-60; grantee HEW, 1977-81, Office of Edn., 1966-67, 70-71, Dept. of Labor, 1980-81. Fellow AAAS; mem. Phi Beta Kappa. Quaker. Home: 204 Fairmount Ave Ithaca NY 14850-4804 Office: Cornell Univ Dept Psychology Uris Hall Ithaca NY 14853

DARNELL, ALFRED J(EROME), chemical engineer, consultant; b. Denton, Tex., Aug. 20, 1924; s. Thomas Dixon and Jessie Mae (Harvey) D.; m. Eunice Irene Johnson, Dec. 24, 1947; 1 child, Thomas Edwin. BA, San Diego State U., 1950; PhD, UCLA, 1964. Chem. engr. Calif. Inst. Tech./Jet Propulsion Lab., Pasadena, 1950-55, Mobile Oil Corp., Santa Fe Springs, Calif., 1955; project engr. Rockwell Internat., Canoga Park, Calif., 1955-80; project scientist Energy Tech. Engring. Lab., Canoga Park, 1980-84; vis. prof. Calif. Poly. U., San Luis Obispo, 1990; cons. Hughes Rsch. Lab., Malibu, Calif., 1962, Energy Tech. Engring. co., Canoga Park, 1985—; presenter at regional and nat. profl. confs. Contbr. over 150 articles, reports to sci. publs. With USNR, 1942-46, PTO. Rsch. fellow N.Am. Aviation, L.A., 1962-64, UCLA, 1963-64. Fellow Am. Inst. Chemists; mem. Am. Chem. Soc., Am. Phys. Soc., Am. Legion, Sigma Xi. Achievements include 6 patents in field of vapor pressure measure and superconductor materials; synthesization of first artificial metal. Home: 23030 Burbank Blvd Woodland Hills CA 91367-4205

DARNELL, JAMES EDWIN, JR., molecular biologist, educator; b. Columbus, Miss., Sept. 9, 1930; s. James Edwin and Helen (Hopkins) D.; m. Jane Roller, 1957; children: Christopher, Robert, Jonathan. BS, U. Miss., 1951; MD, Washington U., 1955. Intern Barnes Hosp., 1955-56; asst. to sr. surgeon USPHS, Bethesda, Md., 1957-60; asst. and assoc. prof. MIT, Cambridge, 1961-64; prof. Albert Einstein Coll. Medicine, N.Y.C., 1967; prof. Columbia U., 1968-74, chmn. dept. biol. scis., 1971-74; Vincent Astor prof. Rockefeller U., N.Y.C., 1974—, v.p. acad. affairs, 1990-91. Co-author: (textbooks) General Virology, 1967, 77, Molecular Cell Biology, 1986, 90. Recipient Am. Acad. Arts and Scis. award, 1973, H.T. Rickets award U. Chgo., 1979, internat. award Gairdner Found., Toronto, 1986. Mem. NAS, Am. Acad. Arts and Scis. Office: Rockefeller U Molecular Cell Biology 1230 York Ave New York NY 10021-6341

DARNELL, LONNIE LEE, electronic engineer; b. Marlinton, W.Va., Nov. 11, 1928; s. George Washington and Carrie Elizabeth (Neighbors) D.; m. Ruby Pearl Moore, Oct. 10, 1929 (div. May 1986); children: Linda Sharon, Lonnie II, Brenda, Donna, Bridgette, Deborah; m. Constance Sue Smith-Ramsey, Aug. 20, 1989. BS in Electrical Tech., U. Md., 1963. Cert. engr. Supr. cent. office United Tel. Co. Ohio, Warren, 1969-76; amateur astronomer Gettysburg (Pa.) Coll., 1988--. Cons., advisor Adams County Homeless Shelter, Gettysburg, 1991—, Adams County AIDS Com., 1990—; chmn., bd. dirs. Adams County United Way, Gettysburg, 1988-90; bd. dirs. South Cent. Community Action Program, Gettysburg, 1992—; cmmdr., life mem. Disabled Am. Vets., Carlisle, Pa., 1978-79. With U.S. Army 1947-69, Korea. Decorated Bronze Star, 1953, 3 Purple Hearts, 1951. Mem. VFW (life, cmmdr. DAV 1969), Kiwanis (v.p. 1991-92). Democrat. Baptist. Achievements include slow release relay circuit for classified electronic encryrtion system, trickle charge circuit N cadmium power source for NIKE Hercules guided missile. Home: 51 Pettigrew Gettysburg PA 17325

DARON, HARLOW H., biochemist; b. Chgo., Oct. 25, 1930; s. Garman H. and Gulah V. (Hoover) D.; m. Carol Fields, Aug. 8, 1969; children: Barbara P., Charles E.; children: Ruth A., Leslie S. BS, U. Okla., 1956; PhD, U. Ill., 1961. NSF postdoctoral fellow Calif. Inst. Technology, Pasadena, 1961-63; asst. prof. Tex. A&M U., College Station, Tex., 1963-67; asst. prof. Auburn (Ala.) U., 1967-72, assoc. prof., 1972-80, prof., 1980—. Mem. Am. Soc. Biochemistry and Molecular Biology, AAAS, Am. Chem. Soc., Sigma Xi. Office: Dept Animal and Dairy Scis Auburn Univ Auburn AL 36849

DARRAH, JAMES GORE, physicist, financial executive, real estate developer; b. Milford, Mich., Nov. 28, 1928; s. Carl Williard and Marie (Rathburn) D.; m. Maud Gray, June 27, 1953; children: Kimberley D. Bohan, Sandra Gray Capalongan. BS in Metall. Engring., Rensselaer Poly. Inst., 1952, MS in Metall. Engring., 1953; PhD, Lehigh U., 1955. Project engr. Gen. Motors Corp., Warren, Mich., 1956-58; mgr. nuclear research and devel. United Aircraft Corp., East Hartford, Conn., 1958-64; div. mgr. Eimac-Varian, San Carlos, Calif., 1964-65; gen. mgr. Teledyne-Monolith, Mountain View, Calif., 1965-68; pres. Stratamet Corp., Sunnyvale, Calif., 1968-84; also chmn. bd. dirs. Stratamet Corp., Fremont, Calif.; pres. Darrah Capital Corp., Menlo Park, Calif., 1984—; chmn. bd. dirs. Ceramic Products Corp., Fremont, Calif., 1975-88, Gold Mind of N.Y., Buffalo, 1985-89, Darco Leasing Co., Menlo Park, 1984—; bd. dirs. Cambridge Laser labs., 1989—. Patentee in field. Served as cpl. U.S. Army, 1946-47. Mem. Sigma Xi, Tau Beta Pi. Republican. Presbyterian. Avocations: tennis, traveling, amateur playwriting, golf. Home and Office: 927 Continental Dr Menlo Park CA 94025-6622

DARROW, GEORGE F., natural resources company owner, consultant; b. Osage, Wyo., Aug. 13, 1924; s. George Washington and Marjorie (Ord) D.; m. Elna Tannehill, Oct. 23, 1976; children by previous marriage: Roy Stuart, Karen Josanne, Reed Crandall, John Robin. AB in Econs., U. Mich., 1945, BS in Geology, 1949. Geologist Amerada Petroleum Corp., Billings, Mont., 1949-50; v.p. Northwest Petroleum Co., 1951-58; resource cons. Resource Consultants, Billings, 1959-78; pres., CEO Crossbow Corp., Billings, 1962—; v.p. Kootenai Galleries, Bigfork, Mont., 1976—; sr. ptnr. Crossbow Assocs., resource mgrs.; chmn. Mont. Environ. Quality Coun., Helena, 1971-73; bd. dirs. Ord Ranch Corp., Lusk, Wyo., Mont. Pvt. Capital Network. Contbr. articles on resource mgmt. and econs. to various publs. Elected mem. Mont. Ho. of Reps., 1967-69, 71-73, Mont. Senate, 1973-75; bd. dirs. Bigfork Ctr. Performing Arts, 1980—. Lt. (j.g.) USNR, 1943-46, PTO. Mem. Am. Assn. Petroleum Geologists (past pres. Rocky Mountain sect.), Am. Inst. Profl. Geologists (charter), Mont. Geol. Soc. (founding mem., charter), Billings Petroleum Club. Home and Office: Crossbow Corp 2014 Beverly Hills Blvd Billings MT 59102-2014 also: Paladin Farms 924 Chapman Hill Dr Bigfork MT 59911

DARSEY, JEROME ANTHONY, chemistry educator, researcher; b. Houma, La., Aug. 26, 1946; s. Elmer Joseph and Arline (Houghton) D.; m. Patricia Ann Bukowski, June 10, 1989; children: Brittany Angele, Joseph Anthony. BS in Physics, La. State U., 1970, PhD in Chemistry, 1982. Asst. prof. chemistry and physics Gordon Coll. U. Ga. System, Barnsville, 1983-84; asst. prof. Tarleton State U./Tex. A&M U., Stephenville, Tex., 1984-88, assoc. prof., 1988-90; asst. prof. U. Ark., Little Rock, 1990-93, assoc. prof., 1993—; univ. scholar natural scis. Tarleton State U./Tex. A&M U., 1989-90; cons. Oak Ridge (Tenn.) Nat. Lab., 1990-93. Contbr. articles to profl. jours. Am. Chem. Soc. grantee, 1986, 90. Fellow AAAS; mem. Am. Chem. Soc. (chmn. Ark. sect. 1993), Am. Phys. Soc., Ark. Acad. Sci., S.W. Theoretical Chemistry Conf. (chmn. 1986-87), Tex. Acad. Sci. (vice chmn. chemistry divsn. 1986-87, chmn. 1987-88). Home: 1514 Alberta Dr Little Rock AR 72207-5803 Office: U Ark Dept Chemistry 2801 S University Ave Little Rock AR 72204-1000

DARTEZ, CHARLES BENNETT, environmental consultant; b. Lafayette, La., Apr. 15, 1951; s. Jerry Olivier and Rita (Howard) D.; m. Ann Myers, Aug. 3, 1974; children: Cynthia Evalyn, Benjamin Armstrong. BS, U. Southwestern La., 1974. Registered environ. profl.; registered environ. auditor; cert. hazardous materials mgr. Staff scientist Gulf South Rsch. Inst., New Iberia, La., 1974-77; lab. technician B.J. Hughs, Lafayette, 1977-78; environ. coord. TERA Corp., Baton Rouge, 1978-83; sr. staff scientist Woodward-Clyde Cons., Baton Rouge, 1983-87, sr. project scientist, 1987-89, assoc. scientist, 1989-92, sr. assoc. scientist, 1992—; recertification bd. Acad. Hazardous Material Mgrs., Indpls., 1992—. Mem. Nat. Water Well Assn., Am. Water Resource Assn., La. Cert. Hazardous Material Mrs. (pres.-elect 1992—), Water Pollution Control Fedn. Office: Woodward Clyde Cons 2822 ONeal Ln Baton Rouge LA 70816

DARVENNES, CORINNE MARCELLE, mechanical engineering educator, researcher; b. Algiers, Algeria, Jan. 19, 1961; came to U.S., 1985; d. René Joseph and Solange Marie (Rosso) D.; m. Glenn B. Focht, Jan. 24, 1989; 1 child, Angeline S. Focht. Engring. Degree, Université Technologie de Compiègne, France, 1984; MSc, Inst. Sound and Vibrations Rsch., Southampton, Eng., 1985; PhD, U. Tex., 1989. Rsch. asst. U. Tex., Austin, 1985-89, rsch. engr., 1989-90; asst. prof. mech. engring. Tenn. Technol. U., Cookeville, 1990—. Contbr. articles to Jour. of Acoustical Soc. Am., Frontiers of Nonlinear Acoustics. Mem. AIAA, ASME (sect. treas. 1992-93, sect. sec. 1993-94), Acoustical Soc. Am., Sigma Xi, Tau Beta Pi. Office: Tenn Tech U Box 5014 Cookeville TN 38505

DARVILL, ALAN G., biochemist, botanist, educator; b. Redditch, Worchester, U.K., Jan. 27, 1952; came to U.S., 1976; s. Bryan Richard and Pamela Mary Darvill; m. Janet Elizabeth Jones, July 12, 1975; 1 child, Sarah Jayne. BS in Plant Biology, Wolverhampton Poly., U.K., 1973; PhD in Plant Physiology, Univ. Coll. Wales, Aberystwyth, 1976. Postdoctoral assoc. U. Colo., Boulder, 1976-78, sr. rsch. assoc., 1978-83, asst. prof. dept. molecular, cellular & devel. biology, 1983-84, assoc. prof., 1984-85; assoc. prof. dept. biochemistry & botany U. Ga., Athens, 1985-87, prof., 1987—, assoc. dir. Complex Carbohydrate Rsch. Ctr., 1985-87, dir., 1987—, co-dir. Ctr. for Plant & Microbial Complex Carbohydrates, 1987—. Contbr. over 125 articles to profl. jours. Mem. AAAS, Am. Chem. Soc. (exec. com. divsn. carbohydrate chemistry 1993—), Soc. for Complex Carbohydrates. Office: University of Georgia Complex Carbohydrate Rsch Ctr 220 Riverbend Rd Athens GA 30602

DARWIN, CHRISTOPHER JOHN, experimental psychologist; b. Ripon, Yorkshire, Eng., Nov. 11, 1944; s. John Robert and Dorothy Aileen (Hick) D.; m. Catherine Margaret Axton, Aug. 17, 1968; children: Clare, Ruth Margaret Grace, Thomas Henry. BA, U. Cambridge, Eng., 1966; PhD, U. Cambridge, 1969. Harkness fellow Commonwealth Fund, New Haven, 1969-71; rsch. fellow SERC/NATO, Montreal, 1971; lectr. U. Sussex, Eng., 1971-78, reader, 1978-85, prof. psychology, 1985—; mem. animal scis. and psychology com. SERC, 1983-86, neuroscis. grants com. MRC, London, 1987-91, neuroscis. bd. 1993—. Editor: Processing of Complex Sounds by the Auditory System, 1992. Office: U Sussex, Brighton England BN1 9QG

DAS, DILIP KUMAR, chemical engineer; b. Khulna, Bangladesh, Aug. 23, 1941; came to U.S., 1969; s. Murari Mohan and Sudha Rani (Roy) D.; m. Mala Mazumder, June 20, 1972; children: Shamik, Alina. BSc with honors, Rajshahi U., Bangladesh, 1961; BChE with honors, Jadavpur U., Calcutta, 1966; MSChemE, U. Wash., 1971. Profl. engr., La. Chem. engr. Kuljian Corp., Calcutta, 1966-67, A.P.V. Engring. Co. Ltd., Calcutta, 1967-69; sr. chem. engr. C.F. Braun & Co., Alhambra, Calif., 1973-75; sr. process engr. Stauffer Chem. Co., Dobbs Ferry, N.Y., 1975-80; sr. project engr. Rhône-Poulenc, Inc., Princeton, N.J., 1980-84; prin. process engr. Ciba-Geigy Corp., St. Gabriel, La., 1984—. Co-author: Chemical Engineering for Professional Engineers Examination, 1984; contbr. articles to profl. jours. Mem. AIChE. Hindu. Achievements include patent for fail-safe diazotization dip tube. Home: 17523 W Muirfield Dr Baton Rouge LA 70810 Office: Ciba-Geigy Corp PO Box 11 Saint Gabriel LA 70776

DAS, ISHWAR, chemistry educator; b. Gorakhpur, U.P., India, June 26, 1952; s. Balram and Manorma Devi; m. Namita Agrawal, March 9, 1984; children: Abhishek Agrawal, Pranav Agrawal. BSc, Gorakhpur U., 1970, MSc, 1972, PhD, 1977. Lectr. U. Gorakhpur, 1979—, reader on deputation Acad. Staff Coll., 1989-91. Contbr. articles to Jour. Phys. Chemistry, Jour. Chem. Edn.; producer films on oscillatory chem. reactions and pattern formation. Promotion of Sci. Interest in Youths grantee, 1985. Mem. Internat. Union for Pure and Applied Chemistry U.K. (affiliate), Am. Chem. Soc. (affiliate), Nat. Environ. Assn. (pres.). Home: Arya Nagar North, Gorakhpur 273001, India Office: U Gorakhpur, Dept Chemistry, Gorakhpur 273009, India

DAS, KAMALENDU, chemist; b. Sylhet, Bangladesh, Feb. 2, 1944; came to U.S. 1971; s. Kailash and Prabhashini Das; m. Shyamali Chowdhury, Dec. 6, 1969; children: Mrinal Kanti, Sampa. BSc with honors, U. Dhaka, 1964, MSc, 1966; PhD, U. Houston, 1975. Asst. prof. Women's Coll. & Tolaram Coll., Bangladesh, 1966-71; postdoctoral instr. U. Houston, 1976-78; rsch. scientist, group leader Baker Sand Control, Houston, 1979-85; rsch. chemist, physical scientist, project mgr. U.S. Dept. Energy, Morgantown, W.Va., 1985—; Contbr. articles to profl. jours. Tennis com. chmn. Homeowners Assn., Houston, 1982-84. Govt. Bangladesh talent scholar, 1964-65; Welch Found. fellow, 1973-74. Fellow Am. Inst. Chemists; mem. Am. Chem. Soc., Soc. Petroleum Engrs., Internat. Union Pure and Applied Chemistry. Hindu. Home: 465 Lawnview Dr Morgantown WV 26505-2130 Office: US Dept of Energy-METC MS C05 PO Box 880 Morgantown WV 26507-0880

DAS, KUMUDESWAR, food and biochemical engineering educator; b. Calcutta, India, Dec. 1, 1932; came to U.S., 1985; s. Mahendra Kumar and Bhabatara (Mazumder) Das; m. Suchitra Mukherjee, Apr. 16, 1961; children: Sandip Kumar, Rupa. BChemE, Jandavpur U., Calcutta, 1955, MChemE, 1957, PhD in Engring., 1969. Rsch. fellow Food and Biochem. Engring. div. Jadavpur U., Calcutta, 1955-58; project mgr. Food Processing Factory, Malda, West Bengal, 1958-63; head, sr. lectr. Dept. of Biochem. Engring. and Food Tech., Kanpur U., India, 1963-68; lectr., asst. prof. Chem. Engring. dept. Indian Inst. of Tech., New Delhi, 1968-81; prof. chem. engring. dept. Indian Inst. of Tech., New Delhi, 1982-85, prof., 1986-87; prof. Rensselaer Poly. Inst., Troy, N.Y., 1985-86; MSM chair, prof. Sch. of Indsl. Tech., U. Sci. Malaysia, Penang, 1988—; vis. lectr. chem. and biochem. engring. dept., U. Coll., London, 1971-72; vis. expert, asst. prof. Food Tech. Dept., Basrah U., Iraq, 1974-76; vis. prof. UNESCO Internat. Ctr. of Biotech., Dept. of Fermentation, Osaka U., 1981. Author: (with others) Advances in Biochemical Engineering, 1971, Methods in Enzymology, 1980, Fermentation Research: Acetone-Butanol Fermentation; contbr. articles to profl. jours. Fellowship Coun. of Scientific and Indsl. Rsch., 1955-57. Fellow Indian Inst. of Chem. Engrs. (life); mem. AICE (food and bioengring. div.), Am. Chem. Soc. (BIOT div.), Assn. of Food Scientists and Technologists, Ramsay Soc. of Chem. and Biochem. Engrs. UCL. Achievements include rsch. on the area of enzymic and fermentation process kinetics and prodn. of high value chems. and biomass protein foods from cellulosic wastes, work on enzyme extraction and purification by high resolution techniques. Home: 214 Holmes St Syracuse NY 13210 Office: Sch of Indsl Tech, U Sci Malaysia, Penang 11800, Malaysia

DAS, PURNA CHANDRA, physics and mathematics educator; b. Kurujang, Orissa, India, Jan. 5, 1954; came to U.S., 1977; s. Banchhanidhi and Kamala (Mishra) D.; m. Anupama Tripathy, June 27, 1983; children: Anshuman, Smita. PhD, CUNY, N.Y.C., 1983. Postdoctoral fellow U. Calif., Santa Barbara, 1982-85; asst. prof. physics Pa. State U., Erie, 1985-91, Purdue U. North Ctrl., Westville, Ind., 1991—. Contbr. articles to sci. jours. Mem. Am. Phys. Soc., Am. Assn. Physics Tchrs. Achievements include contribution to development of the theory of surface enhanced spectroscopy. Office: Purdue U North Ctrl 1401 S US 421 Westville IN 46391

DAS, SAJAL KUMAR, computer science educator, researcher; b. Dainhat, India, Jan. 3, 1960; came to U.S., 1985; s. Baidya Nath and Bimala (Dhani) D.; m. Nandini Dutta, Dec. 15, 1989. BS in Computer Sci., Calcutta U., 1983; MS in Computer Sci., Indian Inst. Sci., 1984; PhD in Computer Sci., U. Ctrl. Fla., 1988. Teaching asst. computer sci. dept. Wash. State U., Pullman, 1985-86; rsch. asst. dept. computer sci. U. Ctrl. Fla., Orlando, 1986-88; asst. prof. dept. computer sci. U. North Tex., Denton, 1988-92, assoc. prof. dept. computer sci., 1993—, faculty Ctr. for Rsch. in Parallel & Distributed Computing, 1990—; faculty advisor U. North Tex. Badminton Club, Denton, 1989-90; founding advisor India Student Assn. of U. North Tex., Denton, 1990—; mem. steering com. Internat. Conf. on Computing and Info., Can., 1992; speaker and presenter in field. Mem. editorial bd.: Parallel Processing Letters, 1991—, Jour. of Parallel Algorithms & Applications, 1992—; contb. articles to profl. jours. Recipient Summer Rsch. fellowship U. North Tex., Denton, 1990, Honor Prof. award U. North Tex., Denton, 1991; Rsch. grantee Higher Edn. Coordinating Bd., Tex., 1991-93, Travel grantee Goethe U., Germany, 1992, grantee No. Bell Rsch., Dallas, 1993. Mem. IEEE Computer Soc., Assn. for Computing Machinery, N.Y. Acad. Scis., Sigma Xi. Office: Univ North Tex Dept Computer Sci PO Box 13886 Denton TX 76203-3886

DAS, SALIL KUMAR, biochemist; b. Rangoon, Burma, Dec. 21, 1940; came to U.S., 1962; s. Santi R. and Provabati D.; 1 child, Shouvik. MS, Calcutta U., Calcutta, India, 1961; ScD, MIT, 1966; DSc, Calcutta U., 1974. Rsch. asst. MIT, Cambridge, Mass., 1962-66; rsch. assoc. MIT, Food Sci. and Nutrition, Cambridge, Mass., 1966; rsch. assoc. dept. physics U. Ariz., Tucson, 1966-67; rsch. assoc. dept. chemistry U. Ark., Little Lock, 1967-68, Duke U., Durham, N.C., 1968-69; asst. prof. Meharry Med. Coll., Nashville, 1969-74, assoc. prof., 1974-81, prof., 1981—. Contbr. articles to profl. jours. Recipient Cressy Morrison award in Natural Sci., N.Y. Acad. Sci., 1967, Calcutta U. Medal for Top Position in MS Exam, 1961, Travel award, IPA award NIH, 1985; grantee in field. Office: Meharry Med Coll Dept Biochemistry Nashville TN 37208

DAS, SUJIT, policy analyst; b. Calcutta, India, May 11, 1958; came to U.S. 1980; s. Hari Mohan and Sandhya Rani Das; m. Suchita De, June 16, 1987; 1 child, Sreetham. BTech, Indian Inst. Tech., 1979; MS, U. Tenn., 1982, MBA, 1984. Rsch. staff Oak Ridge (Tenn.) Nat. Lab., 1984—; vis. fellow Tata Energy Rsch. Inst., New Delhi, India, 1992-93. Contbr. articles to profl. jours. Mem. Ops. Rsch. Soc. Am., Soc. for Internat. Devel. Achievements include research in plastics recycling, oil vulnerability, flood damage estimation, uranium assessment, assessment of advanced materials technologies. Office: Oak Ridge Nat Lab PO Box 2008 Bldg 4500N Oak Ridge TN 37831-6205

DAS, SUMAN KUMAR, plastic surgeon, researcher; b. Calcutta, India, May 6, 1944; came to U.S., 1980; s. Bisweswar and Devi Rani (Ghosh) D.; m. Carole Ellen Simmons, July 10, 1976 (div. Apr. 1984); children: Louise Angelique, Natalie Krishna; m. Rosyln Tanner, Mar. 22, 1991. B of Medicine and Surgery, Calcutta (India) U., 1967; MD, Ednl. Commn. Fgn. Med. Grad., 1981. Diplomate Am. Bd. Plastic Surgery. Intern R.G. Kar Med. Coll. and Hosp., Calcatta, 1966-67, resident in gen. surgery, house officer, 1967-68; sr. house officer in accident and emergency, orthopaedics Royal Infirmary, Bolton, Lancs, Eng., 1968-69, house surgeon in gen. surgery, 1969-70; sr. house officer in gen. surgery Royal United Hosp., St. Martins's Hosp., Bath, Eng., 1970-72; house officer in medicine Whiston Hosp., Prescot, Liverpool, Eng., 1970; registrar in gen. surgery Frenchay Hosp., Bristol, Eng., 1972-73, sr. house officer in plastic surgery, 1973-74; registrar in plastic surgery Frenchay Hosp., Bristol, Eng., 1974, Royal Victoria Infirmary, Fleming Meml. Children's Hosp., Newcastle-Upon-Tyne, Eng., 1974-77; fellow in plastic and reconstructive surgery Hosp. for Sick Children, Toronto, Ont., Can., 1978; fellow in micro and hand surgery St. Vincent's Hosp., Melbourne, Australia, 1979-80, asst. plastic surgeon, 1979-80; rsch. assoc. in plastic surgery UCLA Med. Ctr., 1980-82; co-dir. microsurgery tng. program Harbor/UCLA Med. Ctr., 1980-82; dir. plastic surgery rsch. VA Wadsworth Med. Ctr., L.A., 1980-82; resident in plastic surgery U. Miss. Med. Ctr., Jackson, 1982-83, sr. and chief resident in plastic surgery, 1983-84; pvt. practice Jackson, 1984-86; chief and asst. prof. div. plastic surgery U. Miss. Med. Ctr., Jackson, 1986-87, chief and assoc. prof. div. plastic surgery, 1987-90, prof. plastic surgery, chief div. plastic surgery, chief, 1990—; attending plastic surgeon VA Med. Ctr., Jackson; asst. prof. orthopaedic surgery U. Miss. Med. Ctr., Jackson; cons. plastic surgery Meth. Med. Ctr., Miss. Bapt. Med. Ctr., River Oaks Hosp., Womans Hosp., Doctors Hosp., Rankin Gen. Hosp., Jackson, Miss.; vis. prof. dept. surgery div. plastic surgery U. Calif., San Francisco, 1981, U. Ala., 1992; mem. patient care com. U. Miss., Jackson, 1990—; presenter and exhibitor in field at numerous profl. meetings. Author: (with others) Manual of Operative Plastic and Reconstructive Surgery, 1980, Textbook of Surgery, 2nd edit., 1988, Encyclopedia of Flaps, 1990; contbr. articles to Brit. Jour. Surgery, Brit. Jour. Plastic Surgery, Indian Jour. Dermatology, Hand, Plastic Surgery Forum, Jour. Singapore Acad. Sci., Jour. Oral Surgery, Plastic Reconstrn. Surgery, Acta Anatomica, Jour. Clin. Pathology, others. Donor Miss. Symphony Orch., Jackson, 1991, Indian Assn., 1991; archangel New Stage Theatre, 1991. Recipient prize North Eng. Surg. Soc., 1977, Plastic Surgery Ednl. Found. Rsch. grant 1983-84, other grants Eli Lilly 1989, Tyra, 1989,

Collagen Corp. 1989, 90-91, NIH, 1989, Am. Soc. Aesthetic Plastic Surgery, 1990, 91. Fellow ACS, Royal Coll. Surgeons London, Royal Coll. Surgeons Edinburgh (traveling scholarship 1976); mem. AMA, AAAS, Am. Fedn. for Clin. Rsch., Am. Assn. Hand Surgery (rsch. grant com. 1990-91, chmn. rsch. grant com. 1992), Am. Assn. Acad. Plastic Surgeons (fellowship com. 1990), Am. Soc. Plastic and Reconstructive Surgeons, Am. Assn. Plastic Surgeons, Internat. Soc. Burn Injuries, Internat. Soc. Reconstructive Microsurgery, Internat. Soc. Surgery, Internat. Soc. Emergency Medicine and Critical Care (charter), Brit. Assn. Plastic Surgeons (best prize and cert. 1967), Brit. Soc. Surgery of Hands (European traveling scholarship 1977), Soc. N.Am. Skull Base Surgery (founding), Miss. State Med. Assn., Plastic Surgery Rsch. Coun., N.Y. Acad. Sci., S.E. Soc. Plastic and Reconstructive Surgeons (program com. 1990—), Miss. Acad. Scis. (chmn. 1992), Southern Med. Assn. (chmn. elect 1991, chmn. 1992), Sigma Xi. Achievements include discovery that silicone does not elicit anyange in T cell population; that capsular contracture with silicone implant is not an immunological effect; research on best treatment for finger tip amputation in children, size and lengthening of human omentum, muscle transplantation by microvascular technique fatigue like normal muscle. Home: 155 Olympia Fields Jackson MS 39211-2510 Office: U Miss Med Ctr 2500 N State St Jackson MS 39216-4505

DAS, UTPAL, electrical engineer, researcher; b. Calcutta, India, Aug. 26, 1955; came to U.S., 1982; MS, Oreg. State U., 1983; MS, PhD, U. Mich., 1987. Rsch. assoc. U. Mich., Ann Arbor, 1987-88; asst. prof. elec. engring. U. Fla., Gainesville, 1988—; NSF reviewer, Washington, 1990. Contbr. articles to Applied Physics Letters, Optics Letters, Physics Rev. B, Jour. Vacuum Sci. Tech., others. Govt. of India scholar, 1976; Tau Beta Pi eminent rschr., 1990. Mem. IEEE, Am. Phys. Soc., Am. Vacuum Soc., U.S. Geog. Soc. Achievements include research on strained layer superlattice, optical waveguiding in strained layer superlattices in compound semiconductors. Office: U Fla Dept Elec Engring Bldg 722 Center Dr Gainesville FL 32611

D'ASARO, LUCIAN ARTHUR, physicist; b. Buffalo, N.Y., Jan. 20, 1927; s. Lucio and Viola (Wassman) D'A.; m. Barbara Sachs, Mar. 27, 1953; children: Eric, Andrea, Lisa, Joanna. BS, Northwestern U., 1949, MS, 1950; PhD, Cornell U., 1955; MBA, Fairleigh Dickinson U., 1981. Disting. mem. tech. staff AT&T Bell Labs., Murray Hill, N.J., 1955—. Mem. IEEE (sr. life mem.), Am. Phys. Soc., Sigma Xi. Achievements include publications and patents on semiconductor devices and processing: silicon diffusion, tunnel diodes, microwave schottky barrier mixers, avalanche multiplication photodiodes, dark line defects in junction lasers, via transistors and FET-SEED photonic switches. Home: 7 Woodcliff Dr Madison NJ 07940 Office: AT&T Bell Labs 600 Mountain Ave Murray Hill NJ 07974

DASGUPTA, PURNENDU KUMAR, chemist, educator; b. Calcutta, India, Dec. 5, 1949; came to U.S., 1973; s. Nirmal Kumar and Mina (Sen) D.; div.; 1 child, Michael Akash. MSc, U. Burdwan, India, 1970; PhD, La. State U., 1977. Grad. asst., then instr. La. State U., Baton Rouge, 1973-78; asst. rsch. chemist U. Calif., Davis, 1979-81; asst. prof., then assoc. prof. Tex. Tech. U., Lubbock, 1981-88, prof. chemistry, 1988-92, Paul Whitfield Horn prof., 1992—; chmn. sulfur subcom. Intersoc. on Analysis, Phila., 1986—; assoc. editor Atmospheric Environ., 1990—; editorial bd. Talanta, 1990—. Co-author: Ion Chromatography, 1987, Measurement Challenges in Atmospheric Chemistry, 1993; contbr. over 125 articles to peer-reviewed jours. Recipient Frank Blood award Soc. Toxicology and Pharmacology, 1981, Traylor Creativity award Dow Analytical Scis., 1989. Mem. Am. Chem. Soc., Sigma Xi. Achievements include patent on continuously rejuvenated ion exchangers, annular dual permselective device and method, apparatus and method for automated microbatch reaction, method and apparatus for generating high-purity chromatographic eluent. Office: Tex Tech U Dept Chemistry Lubbock TX 79409-1061

DASGUPTA, RANJIT KUMAR, virologist; came to U.S., 1972; s. Manoranjan and Labanya (Sengupta) D.; m. Purabi Sengupta, Feb. 6, 1976; 1 child, Ankur. MS, U. Calcutta, 1966, PhD, 1972. Asst. scientist Inst. for Molecular Virology, Madison, Wis., 1978-82, assoc. scientist, 1982-87, sr. scientist, 1987—; co-prin. investigator NIH, 1986—. Contbr. articles to profl. Nature, Cell, Pro. NAS USA, Jour. Mol. Biol. Mem. Am. Soc. Virology, India Student's Assn. Achievements include research on eukaryotic messenger RNA translation, capping at the ends of messenger RNAs, development of infectious clones from plant or insect viral RNAs, methods for sequencing nucleic acids, recombinant DNA techniques, biotechnology. Office: U Wis Inst Molecular Virology 1525 Linden Dr Madison WI 53706-1596

DASHEVSKY, SIDNEY GEORGE, clinical and industrial psychologist; b. N.Y.C., Aug. 18, 1934; s. Sam and Ann (Orenstein) D.; m. Catherine Searles, Feb. 5, 1965 (div. 1973); children: Anne S., Karen S.; m. Virginia Waters, Mar. 5, 1983; children: Adam Waters-Dashevsky, Hallie Waters-Dashevsky. BA magna cum laude, N.Mex. Highlands U., 1958; PhD, U. Rochester, 1968. Rsch. psychologist U.S. Army Proving Ground, Aberdeen, Md., 1962-63, The Psychol. Corp., N.Y.C., 1965; orgn. psychologist PPG Industries, Harmarville, Pa., 1967; assoc. prof. York Coll. of Pa., 1969; dir. Behavior Therapy Ctr., York, 1968-81; clin. dir. N.J. Ctr. for Counseling & Psychotherapy, North Plainfield, 1977-81; co-dir. Cranford (N.J.) Counseling Ctr., 1981—; consulting psychologist Taylor Manor Hosp., Ellicott City, Md., 1969-74, Meml. Hosp., York, 1974-76, York Police Dept., 1973-81, Union County Prosecutor's Office, Elizabeth, N.J., 1982-85, AT&T Internat. Corp., Basking Ridge, N.J., 1986, others. Contbr. articles to profl. jours. With USAF, 1952-56. Mem. Am. Psychol. Soc., N.J. Psychol. Assn., Children with Attention-Deficit Disorders. Office: Cranford Counseling Ctr 347 Lincoln Ave E Cranford NJ 07016

DASSIOS, GEORGE THEODORE, mathematician, educator, researcher; b. Patras, Greece, Jan. 22, 1946; s. Theodore and Helen (Kalliafas) D.; m. Eleni Dassios, Sept. 6, 1970; children: Constantine, Theodore. Diploma, U. Athens, Greece, 1970; MS, U. Ill., Chgo., 1972, PhD, 1975; Habilitation, Nat. Tech. U. Athens, 1980. Assoc. prof. Nat. Tech. U. Athens, 1977-81; asst. prof. U. Patras, 1975-77, prof., 1981—; vis. asst. prof. Brown U., Providence, 1978-79; vis. prof. U. Tenn., Knoxville, 1986-87. Rsch. grantee EEC, 1984—. Avocation: stamp collecting. Home: 186 Korinthou St, 262 21 Patras Greece Office: Univ Patras, 261 10 Patras Greece

DATILES, MANUEL BERNALDES, III, ophthalmologist; b. Manila, Feb. 26, 1951; came to U.S., 1979; s. Roberto Aguiling and Loretta (Bernaldes) D.; m. Jacqueline Romero, Mar. 13, 1976; children: Michelle, Joyce, Margaret, Jennifer, Manuel IV, Michael. BS cum laude, U. Santo Tomas, Manila, 1970; MD cum laude, U. Santo Tomas, 1974. Rsch. fellow Philippine Eye Rsch. Inst.-U. Philippines, Manila, 1975-76; resident in ophthalmology U. Philippines-Philippine Gen. Hosp., Manila, 1976-79; rsch. scholar, vis. scientist Lab. Vision Rsch. Nat. Eye Inst.-NIH, Bethesda, Md., 1979-82; clin. fellow corneal and cataract surgery Wilmer Eye Inst.-Johns Hopkins U. Hosp., Balt., 1982-83; sr. staff ophthalmologist Nat. Eye Inst.-NIH, Bethesda, 1983-88; acting chief cornea and cataract sect., clin. svc. br. Nat. Eye Inst.-NIH, 1989-92, chief cornea and cataract sect., clin. svcs. br., 1992—; vis. lectr. Wilmer Eye Inst.-Johns Hopkins U., Balt., 1984, Osaka (Japan) U., 1986, U. Munich, 1988. Editor cataract sect. Duane's Clinical Ophthalmology Textbook Series; contbr. articles to profl. jours., chpts. to ophthalmol. books. Mem. Assn. Rsch. in Vison and Ophthalmology, Am. Acad. Ophthalmology, Castroviejo Soc. Corneal Surgeons, Johns Hopkins Med. Surg. Assn., Internat. Assn. Ocular Surgeons, Washington Acad. Ophthalmology, Md. Soc. Eye Physicians and Surgeons. Roman Catholic. Avocations: drawing, soap carving, target shooting, guitar, chess. Office: NIH Nat Eye Inst Bldg 10 Rm 10N226 Bethesda MD 20892

DATIRI, BENJAMIN CHUMANG, soil scientist; b. Sho, Plateau, Nigeria, May 1, 1953; came to U.S., 1983; s. Chumang Dangyang and Antele (Pam) D.; m. Roseline Chundung Gwott, Apr. 5, 1980; children: Simidarwei, Teyeidarwei, Ninratdarwei, Yeipyeng. BSc, Ahmadu Bello U., Zaria, Nigeria, 1978, MSc, 1982; PhD, U. Wis., 1989. Mem. Nigerian Youth Svc. Program, Onicha-Olona, 1978-79; agrl. officer II Plateau State Ministry of Agr., Lafia, Nigeria, 1979; grad. asst. Ahmadu Bello U., 1979-80, asst. lectr., 1980-82; rsch. asst. U. Wis., Madison, 1983-84; rsch. assoc. Tuskegee (Ala.) U., 1990—. Contbr. articles to profl. publs. Grad. scholarship Plateau State Govt. of Nigeria, 1983-86; recipient Ch. Recreation League of Leadership award Macon County of Ala., 1992. Mem. Am. Soc. Agronomy, Soil Sci. Soc. Am., Sigma Xi (Tuskegee U. chpt. Outstanding leadership 1992, Dedicated Svc. award 1992). Achievements include discovery that chisel and no-till tillage systems do reduce surface runoff and increase infiltration, that wetting front migrations were found to be much deeper in soil profile than in conventional moldboard plowing; hence, although surface water pollution may be reduced by conservation tillage, it could present a potential for groundwater pollution. Office: Tuskegee U 308 Patterson Hall Tuskegee AL 36088

DATLOWE, DAYTON WOOD, space scientist, physicist; b. N.Y.C., Mar. 16, 1942; s. Samuel A. and Marghretta (Wood) D. m. Karen Janine Mc Caffrey, Aug. 3, 1974; children: Nicholas, Elizabeth, Peter. SB in Physics, MIT, 1964; PhD in Physics, U. Chgo., 1970. Scientist U. Calif., San Diego, 1970-76, Lockheed R & D, Palo Alto, Calif., 1976—. Contbr. articles to Jour. Geophys. Rsch., Astrophys. Jour., Solar Physics, Nuclear Instruments and Methods, Geophys. Rsch. Letters. Mem. IEEE, Am. Geophys. Union, Am. Astron. Soc. Achievements include research on X-rays and relativistic electrons from solar flares, electrons in the near-earth space environment, and X-rays from the earth's auroral zone. Office: Lockheed R&DD D91-20 B255 3251 Hanover St Palo Alto CA 94304-1121

DATTA, INDRANATH, naval architect, educator; b. Calcutta, India, Sept. 11, 1951; arrived in Can., 1982; s. Gobinda Lal and Snehalata (Kundu) D.; m. Mita Dawn, Aug. 26, 1983; children: Iman Anwar Michael, Lalit Christopher. B of Tech. with honors, Indian Inst. Tech., Kharagpur, India, 1973; PhD, U. Strathclyde, Glasgow, Scotland, 1982. Registered profl. engr., Nfld., Can. Asst. naval architect Garden Reach Shipbuilding & Engring. Ltd., Calcutta, 1973-76; rsch. asst. U. Strathclyde, 1979-82; with Nordco Ltd., St. John's, Nfld., 1982-84; from asst. to assoc. rsch. officer Inst. for Marine Dynamics Nat. Rsch. Coun., St. John's, 1984, sr. rsch. officer Inst. for Marine Dynamics, 1992; mem. panel rsch. study group on slamming and green seas loadings NATO RSG, Europe and North Am., 1984-90, mem. panel full scale wave measurement, 1984-90; adj. asst. prof. Meml. U. Newfoundland, St. John's, 1988—. Contbr. articles to profl. jours. Nat. Merit scholar West Bengal Govt., 1968-73, scholar U. Strathclyde, 1977-79, Brit. Coun., 1976. Mem. Soc. Naval Architects and Marine Engrs. (mem. panel HS-2 on impact loadings 1985—), Royal Inst. Naval Architects (London), Amnesty Internat. (activist 1978-79). Home: 452 Newfoundland Dr, Saint John's, NF Canada A1A 4E3 Office: Nat Rsch Coun, Inst for Marine Dynamics, PO Box 12093, Saint John's, NF Canada A1B 3T5

DATTA, PRASANTA, mechanical engineer; b. Calcutta, West Bengal, India, June 6, 1946; came to U.S., 1976; s. Dhirendra Lal and Snehalata (Datta) D.; m. Sikha Ghosh, Feb. 18, 1980; children: Monica, Sidhrtha, Monisha. MSc, U. Birmingham, England, 1971, PhD, 1974. Cert. engr. Asst. prof. Northeastern U., Boston, 1977-78; sr. engr. No. Rsch. and Engring. Corp., Cambridge, Mass., 1978-80; prin. engr. Stone and Weber Engring. Corp., Boston, 1980-83; sr. engr. United Techs., East Hartford, Conn., 1983-90; prin. engr. Dresser-Rand Co., Olean, N.Y., 1990—; Stone and Webster Engring. Corp.; spl. lectr. U. New Haven, Conn., 1990. Mem. AIAA (sr.), ASME (technical paper reviewer Internat, Gas Turbine Inst. 1991—), Nat. Mgmt. Assn. Republican. Hindu. Achievements include design and development of process for metal cutting by CO_2 lasers with gasjet assist. Office: Dresser-Rand Co Paul Clark Dr Olean NY 14760

DATTA, SYAMAL KUMAR, medical educator, researcher; b. Cuttack, Orissa, India, Sept. 21, 1943; came to U.S., 1967; s. Jitendra Nath and Kalyani (Hazra) D.; m. Tapati Chaudhury, 1976; 1 child, Ronjon. BS, U. Calcutta, India, 1960, MB, BS, 1966. Diplomate Am. Bd. Internal Medicine. Resident in medicine Cook County Hosp., Chgo., 1969-71, fellow in hematology, 1971-72; rsch. assoc. Tufts U./New Eng. Med. Ctr., Boston, 1972-74, instr. in medicine, 1974-76, asst. prof. medicine, 1976-79, assoc. prof. medicine, 1979-85, prof. medicine, 1985-93; Solovy Arthritis-Rsch. Soc. prof., prof. medicine Northwestern U. Med. Sch., Chgo., 1993—; sr. faculty mem. grad. program immunology Sackler Sch., Boston, 1975-93; mem. study sects. NIH, Bethesda, 1987—. Assoc. editor Jour. Immunology, 1984—. Leukemia Soc. Am. fellow, 1972-74; Am. Cancer Soc. scholar, 1975-78. Mem. AAAS, Am. Assn. Immunologists, N.Y. Acad. Scis., Am. Coll. Rheumatology. Achievements include development of mouse model for human systemic autoimmune disease; identification of retroviral genes and definition of their relationship to autoimmune disease; identification and isolation of pathogenic anti-DNA autoantibodies and definition of the sequences of their genes; identification of unique sequences shared by the T cell receptor genes expressed by the pathogenic T helper cells that allow specific therapy designs; devised a method to isolate T helper cell clones that induce pathogenic autoantibodies in lupus; identified new subset of T cellsin man that play an important role in autoimmune disease. Office: Northwestern Univ Med Sch Arthritis Div Ward 3-315 303 E Chicago Ave Chicago IL 60611-3008

DATTA, TIMIR, physicist, solid state/materials consultant; b. Calcutta, India, Sept. 14, 1947; came to U.S., 1970; s. Bhola N. Datta and Usha R. Neogy; m. Elizabeth Guillen, Jan. 21, 1982. PhD, Tulane U., 1979. Dir. Inst. for Superconductivity, U. S.C., Columbia, 1990—; prof. physics and astronomy U. S.C., 1991—; vis. rsch. prof. Physics at Sci. and Tech. Ctr. for Superconductivity, U. Ill., Urbana, 1991. Co-author: Copper Oxide Superconductors, 1988. Mem. Am. Phys. Soc. Achievements include being first to observe Meissner effect and confirm high-Tc superconductivity in the Tl and Hg-copper-oxide compounds. Office: U SC Inst for Superconductivity Physics & Astronomy Dept Columbia SC 29208

DATZ, ISRAEL MORTIMER, infosystems specialist; b. N.Y.C., Feb. 11, 1928; s. A. Mark and Lillian (Barkin) D.; BS, CCNY, 1950; postgrad. U. Bergen (Norway), 1951-55; m. Gerd Elin Alme-Torkildsen, Apr. 30, 1956. Chief programming group Internat. Inst. Meteorology, Stockholm, Sweden, 1958-59; head support svcs. sect. NASA Goddard Space Flight Ctr., Greenbelt, Md., 1959-61; mathematician Army Strategy and Tactics Analysis Group, Bethesda, Md., 1961-63; acting chief div. ops. analysis Dept. Commerce Maritime Adminstrn. Washington, 1963-64; head computer div. marine engring lab. Annapolis (Md.) div. Naval Ship R & D Ctr., 1964-68, rsch. coord. math., 1968-72, tech. adv. ops. rsch. 1972-79; ind. cons., 1979-84; chief, studies and analysis, U.S. Army Engr. Sch., Ft. Leonard Wood, Mo., 1984-92; ind. cons., 1992—. Recipient summer stipend Woods Hole Oceanographic Instn., 1949, rsch. stipend The Geophysics Inst., Bergen, Norway, 1953. Fellow AAAS; mem. N.Y. Acad. Sci., Ops. Rsch. Soc. Am., Assn. Computing Machinery, Am. Def. Preparedness Assn., Am. Soc. Naval Engrs., Marine Tech. Soc., Soc. Naval Architects and Marine Engrs., U.S. Naval Inst. Author: Power Transmission and Automation for Ships and Submersibles; Planning Tools For Ocean Transportation. Contbr. articles to profl. jours. in U.S., Eng., Norway, Sweden, Germany. Home and Office: 1343 California Dr Rolla MO 65401-4529

DAU, PETER CAINE, neurologist, immunologist, educator; b. Fresno, Calif., Feb. 25, 1939; s. Julius Jensen and Elva (Caine) D.; m. Barbara Joan Berry, Jan. 31, 1965; children: Birgitt, Kirstin, Rikke. BA, Stanford U., 1960, MD, 1964. Diplomate Am. Bd. Psychiatry and Neurology. Intern Los Angeles County Gen. Hosp., L.A., 1964-65; resident in neurology U. Wis., Madison, 1965-66; resident in neurology U. Chgo., 1966-68, USPHS fellow in immunology, 1968-69; rsch. asst. German Cancer Rsch. Ctr., Heidelberg, 1972-74; asst. clin. prof. medicine U. Calif., San Francisco, 1977-81; prof. neurology Northwestern U. Med. Sch., Chgo., 1989—; mem. staff Evanston (Ill.) Hosp.; cons. Cobe Labs., Lakewood, Colo., 1979-88, Alpha Therapeutics Corp., L.A., 1981-83, W.R. Grace & Co., Lexington, Mass., 1987—. Editor: Plasmapheresis and the Immunobiology of Myasthenia Gravis, 1978; contbr. articles to profl. jours. V.p. Muscular Dystrophy Assn., 1987—, grantee, 1976-87, 88—. Maj. USAF, 1969-71. Recipient Peter Bassoe Meml. award Chgo. Neurol. Soc., 1969, Nat. Rsch. Svc. award NIH, 1976-77. Mem. Am. Assn. Immunologists, Am. Acad. Neurology. Achievements include demonstration of lymphocyte blastogenesis with myelin basic protein in EAE and multiple sclerosis, non-histone chromatin proteins associated with lymphoblastoid transformation; pioneered therapeutic plasmapheresis in U.S. and introduced its use in treatment of myasthenia gravis, multiple sclerosis, polymyositis, dermatomyositis, Eaton-Lambert Syndrome, chronic inflammatory demyelinating poly-neuropathy and scleroderma. Office: Evanston Hosp 2650 Ridge Ave Evanston IL 60201-1718

DAUB, MARGARET E., plant pathologist, educator. Prof. plant pathology N.C. State U., Raleigh. Recipient CIBA-GEIGY award Am. Phytopathological Soc., 1992. Office: North Carolina State University Box 7616 Plant Pathology Raleigh NC 27695*

DAUBECHIES, INGRID, mathematics educator; b. Houthalen, Limburg, Belgium, Aug. 17, 1954; d. Mariel and Simmone (Duran) D.; m. A. Robert Calderbank, May 9, 1987; children: Michael, Carolyn. BS in Physics, Vrije U., 1975, PhD in Physics, 1980. Rsch. asst. Belgium NSF, 1975-84, rsch. assoc., 1984-87; mem. rsch. staff AT&T Bell Labs., Murray Hill, N.J., 1987-92; prof. Rutgers U., New Brunswick, N.J., 1992—; vis. lectr. Princeton U., N.J., 1981-83; vis. prof. NYU, 1986-87. Author: Ten Lectures on Wavelets, 1992; contbr. rsch. articles to profl. jours. McArthur Found. fellow, 1993; recipient Louis Enpain proze Belgian NSF, 1984. Mem. IEEE, Am. Math. Soc., Am. Women Mathematicians, Math. Assn. Am. Office: Rutgers U Math Dept Busch Campus New Brunswick NJ 08903

DAUBEN, WILLIAM GARFIELD, chemist, educator; b. Columbus, Ohio, Nov. 6, 1919; s. Hyp J. and Leilah (Stump) D.; m. Carol Hyatt, Aug. 8, 1947; children: Barbara, Ann. AB, Ohio State U., 1941; AM, Harvard U., 1942, PhD, 1944; PhD (hon.), U. Bordeaux, France, 1980. Edward Austin fellow Harvard U., 1941-42, teaching fellow, 1942-43, research asst., 1943-45; instr. U. Calif. at Berkeley, 1945-47, asst. prof. chemistry, 1947-52, assoc. prof., 1952-57, prof., 1957—; lectr. Am.-Swiss Found., 1962; mem. med. chem. study sect. USPHS, 1959-64; mem. chemistry panel NSF, 1964-67; mem. Am.-Sino Sci. Cooperation Com., 1973-76; NRC, 1977-80. Mem. bd. editors Jour. of Organic Chemistry, 1957-62; mem. bd. editors Organic Syntheses, 1959-67, bd. dirs., 1971—; editor in chief Organic Reactions, 1967-83, pres., 1967-84, bd. dirs. 1967—; mem. edit. bd. Steroids, 1989—; contbr. articles profl. jours. Recipient citation U. Calif., Berkeley, 1990; Guggenheim fellow, 1951, 66, sr. fellow NSF, 1957-58, Alexander von Humboldt Found. fellow, 1980. Fellow Royal Soc. Chemistry, Swiss Chem. Soc.; mem. NAS (chmn. chemistry sect. 1977-80), Am. Chem. Soc. (chmn. div. organic chemistry 1962-63, councilor organic div. 1964-70, mem. coun. publ. com. 1965-70, mem. adv. com. Petroleum Research Fund 1974-77, award Calif. sect. 1959, Ernest Guenther award 1973, Arthur C. Cope scholar 1990), Am. Acad. Arts and Scis., Pharm. Soc. Japan (hon.), Phi Beta Kappa, Sigma Xi, Phi Lambda Upsilon, Phi Eta Sigma, Sigma Chi. Club: Bohemian. Home: 20 Eagle Hl Kensington CA 94707-1408 Office: U Calif Berkeley Dept Chemistry Berkeley CA 94720

DAUCHEZ, PIERRE GUISLAIN, roboticist; b. Lille, France, Sept. 24, 1958; s. Francis Jean and Marie Alix (Thuillier) D.; m. Edith Emilie Dupont, Oct. 29, 1984; 1 child, Sarah. Agregation, ENS Cachan, France, 1981; PhD, USTL, Montpellier, France, 1983; These d'Etat, U. Montpellier, France, 1990. Engring. Eleve prof. ENS Cachan, Cachan, France, 1978-83; prof. agrege Min. Edn., France, 1983-84; sr. rscher. 1st level U. Montpellier II, Montpellier, France, 1984-89; vis. roboticist U. Calif., Santa Barbara, Calif., 1986-87; sr. rscher. 2nd level U. Montpellier II, Montpellier, France, 1989—. Contbr. articles to profl. jours; co-guest editor: International Journal of Robotics and Automation, 1991. Mem. Inst. Elec. and Electronics Engrs., Robotics Soc. of Japan. Avocations: jogging, swimming, photography. Home: 10 Plan Verdi, 34970 Lattes France Office: LIRMM U Montpellier II, 161 Rue ADA, 34392 Montpellier France

DAUGHENBAUGH, RANDALL JAY, chemical company executive; b. Rapid City, S.D., Feb. 10, 1948; s. Horace Allan and Helen Imogene (Reder) D.; m. Mary R. Wynja, Aug. 25, 1973; children: Jason Allan, Jill Christen. BS, S.D. Tech. U., 1970; PhD, U. Colo., 1975. Rsch. chemist Air Prod. and Chem., Allentown, Pa., 1975-80; rsch. dir. Chem. Exchange Industries, Boulder, Colo., 1980-83; pres. Hauser Chem. Rsch., Inc., Boulder, 1983-93, chief tech. officer, exec. v.p., 1993—. Contbr. articles to profl. jours.; patentee in field. Recipient R&D Mag. IR-100 award, 1993; named Inc. Mag. Entrepreneur of Yr., 1992. Mem. Am. Chem. Soc. Home: 11022 N 66th St Longmont CO 80503-9163 Office: Hauser Chem Rsch Inc 5555 Airport Blvd Boulder CO 80301-2339

DAUKSHUS, A. JOSEPH, systems engineer; b. Tamaqua, Pa., Oct. 17, 1948; s. Anna Daukshus. BS in Aerospace Engring., Pa. State U., 1975. Devel. engr. Carl Zeiss Inc., Thornwood, N.Y., 1984-88; cons. Panasonic, Secaucus, N.J., 1988, Pratt & Whitney, E. Hartford, Conn., 1989; cons. AT&T, Largo, Fla., 1990, Somerset, N.J., 1990-91; cons. Torrington (Conn.) Co., 1990, Trecom Bus. Systems, Edison, N.J., 1990-91; systems engr. Canberra Industries, Meriden, Conn., 1991—. Mem. N.Y. Acad. Sci. Home: PO Box 8916 New Fairfield CT 06812-1776

DAUSSET, JEAN, immunologist; b. Toulouse, France, Oct. 19, 1916; s. Henri and Elizabeth D.; m. Rose Mayoral, Mar. 17, 1962. AB, Lycee Michelet, 1939; MD, U. Paris, 1945. Intern, then resident in internal medicine and hematology Paris Mcpl. Hosps., 1946-50; dir. lab. Nat. Transfusion Ctr., 1950-63; prof. immunohematology U. Paris, 1963-77; prof. exptl. medicine Coll. de France, Paris, 1977-87; dir. research unit on immunogenetics Hopital Saint-Louis, Paris, 1969-84; dir. Human Polymorphism Study Ctr., 1984—; researcher in field of man's histocompatibility system anbd human genome. Served to capt., World War II. Recipient Nobel prize in physiology and medicine, 1980, Honda prize Honda Found. Japan, 1987. Mem. Academie des Sciences de l'Institut de France, Am. Acad. Arts and Sci., NAS (Washington). Home: 9 rue Villersexel, 75007 Paris France Office: 27 rue Juliette Dodu, 75010 Paris France

DAUSSMAN, GROVER FREDERICK, industrial engineer, consultant; b. Warrick County, Ind., May 6, 1919; s. Grover Cleveland and Madeline (Springer) D.; m. Elli Margrite Kilian, Dec. 27. 1941; children—Cynthia Louise Daussman Quinn, Judith Ann, Margaret Elizabeth Daussman Davidson Cooper. Student, U. Cin., 1936-38, Carnegie Inst. Tech., 1944-45, George Washington U., 1948-56; BSEE, U. Ala., 1963, postgrad., 1963-64, 77; postgrad., Indsl. Coll. Armed Forces, 1955, 63; PhD (hon.), Hamilton State U., 1973. Registered profl. engr. , Ala., Va., D.C.; cert. fallout shelter analyst. Coop. engr. Sunbeam Elec. Mfg. Co., Evansville, Ind., 1936-38; engr., draftsman Phila. Navy Yard, 1941-42; resident engr., supr. shipbldg. USN, Neville Island, Pa., 1942-45; engr. Pearl Harbor Navy Yard, 1945-48; sect. head Bur. Ships USN, Washington, 1948-56; head guidance and control tech. liaison Army Ballistic Missile Agy., Huntsville, Ala., 1956-58, chief program coordination Guidance and Control Lab.,, 1958-60; chief program coordination Astrionics Lab., Marshall Space Flight Ctr., Huntsville, 1960-63, dir's staff asst. for advanced rsch. and tech., 1963-70, engring. cons., 1970—; project dir. fallout shelter surveys Mil Dept. Tenn., 1971-73; head drafting dept. Alverson-Draughon Coll., Huntsville, 1974-77; instr. Ala. Christian Coll., 1977-79; engring. draftsman Reisz Engring., 1979; chief engr. Sheraton Motor Inn, 1979; sr. engr. Sperry Support Services, 1980; assoc. Techni-Core Profls., Huntsville, 1980-81; elec. engr. Reisz Engring., Huntsville, 1981-86; tutor in mathematics, scis. and engring. North Ala. Ctr. for Ednl. Excellence, Huntsville, 1986—. Chmn. community spl. gifts com. Madison County Heart Assn., 1965; mem. Population Action Coun. Recipient cert. of recognition, 1945, cert. of service USN, 1946; performance award cert. U.S. Army, 1960; certs. of appreciation AIEE, 1960, 61, 62; IEEE Centennial Medal, 1984, IEEE Honor Role of Outstanding Vols., 1986, IEEE Ednl. Activities Award, 1987, award for disting. services Huntsville sect. IEEE, 1964, Engr. of Yr. award, 1969; award for contbn. to successful launch of 1st Saturn V, George C. Marshall Space Flight Center, 1967, also award for contbn. to 1st manned lunar landing, 1969; Apollo achievement award NASA, 1969; cert. of Appreciation North Ala. Ednl. Opportunity Ctr., Inc., 1987, 88, 89, 90, 91. Fellow Explorers Club, Redstone Arsenal Officers Club; mem. AAAS, AIAA, AARP, IEEE (life, sr., sect. chmn. No. Ala. Sect. 1961-62, founder and chmn. engring. mgmt. chpt. 1964-65, mem. Region 3 exec. com. 1969-79, mem. inst. rsch. com. 1965-67, mem. adminstrv. com. engring. mgmt. soc. 1966-86, sec. soc. 1968-85, regional del.-dir. S.E. region, mem. inst. bd. dirs. 1972-73), Planetary Soc. (charter), Hellenic Profl. Assn. Am. (hon.), U. Ala. Alumni Assn., Ala. (state dir. 1962-65, 68-71, 85-91, chpt. pres. 1966-67, regional mathcounts coord. 1981-91, state mathcounts coord. 1988-89, Engr. of Yr. 1968, 82, 89), Nat. Socs. Profl. Engrs., Ala. Soc. Profl. Engrs. (Cert. Appreciation 1982, Pres.'s award 1989), Am. Inst. Urban and Regional Affairs, The Cousteau Soc.,

Am. Def. Preparedness Assn. (post dir. Tenn. Valley 1963-66), Internat. Platform Assn., Nat. Assn. Retarded Children, Huntsville Assn. Tech. Socs. (founder; sec. 1969-70; v.p. 1970-71), Am. Soc. Naval Engrs., U.S. Naval Inst., Assn. U.S. Army, Missile, Space and Range Pioneers (life), Jr. Engring. Tech. Soc. (organizer local high sch. chpts.), Nat. Assn. of Retired Fed. Employees, NASA Retirees Assn. (v.p. 1973-74, pres. 1974—). Democrat. Mem. United Ch. of Christ (treas. 1959-61, ch. council 1964-66; sec. ch. council, program com. chmn. ch. council 1965-66; vice moderator Ala.-Tenn. assn. 1965-68; bd. dirs. Southeast conf. 1965-66, mem. budget and finance com. 1965-66). Home: 1910 Colice Rd SE Huntsville AL 35801-1675 Office: 1525 Sparkman Dr NW # B Huntsville AL 35816-2607

DAVÉ, VIPUL BHUPENDRA, polymer engineer; b. Rajkot, Gujarat, India, June 3, 1962; came to U.S., 1986; s. Bhupendra Maneklal and Hansa Bhupendra (Bhatt) D.; m. Shruti Vipul Kusumgar, Dec. 18, 1990. BS, U. Baroda, India, 1986; MS, U. Mass., 1990; PhD, Va. Poly. Inst., 1992, postdoctoral. Rsch. asst. U. Mass., Lowell, 1986-87, Mich. State U., East Lansing, 1987; project asst. Va. Poly. Inst., Blacksburg, 1988-92; vis. scientist Stazione Sperimentale Per La Cellulosa, Milan, Italy, 1992-93; speaker in field. Contbr. articles to profl. jours. Sec. Grad. Students Assn., U. Mass., 1987. Coll. Engring. scholarship U. Baroda, 1983; Pratt Engring. fellowship Va. Tech. Inst., 1989. Mem. Am. Chem. Soc., Soc. Plastics Engrs., Soc. for Advancement of Material and Process Engring. Achievements include design of fabricated fiber spinning machine. Home: 5211 RFD Long Grove IL 60047

DAVENPORT, CLYDE MCCALL, development engineer; b. Sevier County, Tenn., Dec. 15, 1938; s. Clyde Macaulay and Maymie Estelle (Williams) D.; m. Iris Lynn McMillin, Sept. 11, 1976; children: Kathryn, Ryan. MS in Physics, U. Tenn., 1969, MS in Math., 1978. Devel. engr. Union Carbide Corp., Oak Ridge, Tenn., 1969-84; devel. staff Martin Marietta Energy Systems, Oak Ridge, Tenn., 1984—; Author: A Commutative Hypercomplex Calculus, with Applications to Special Relativity, 1991. With USAF, 1959-63. Recipient award for tech excellence Dept. Energy, 1989. Mem. AAAS, Math. assn. Am., Mensa, Tau Beta Pi, Sigma Pi Sigma, Phi Kappa Phi. Republican. Achievements include patents on interpolator for numerically controlled machine tools and improved beam/seam alignment control for electron beam welding. Home: 4124 Guinn Rd Knoxville TN 37931 Office: Martin Marietta Energy Sys MS-8084 PO Box 2009 Oak Ridge TN 37831

DAVENPORT, FRANCIS LEO, chemical engineer; b. Troy, N.Y., Jan. 11, 1951; s. Bernard Washington and Clara Lilian (Welch) D.; m. Eileen Marie Barry, Aug. 12, 1972; children: Eric, Sarah, Simon. BSChemE, Clarkson U., 1973. From design engr. to chief design mgr. Press Fabrics Div. Albany (N.Y.) Internat., 1974-86; dir. R&D Albany (N.Y.) Internat., East Greenbush, N.Y., 1986—. Mem. Tech. Assn. of Pulp and Paper Industry. Achievements include patents for Dynamic Filtration Simulator, and Papermaking Fabric with Pin Seam Composed of Braided Yarns. Office: Albany Internat Press Fabrics Div 253 Troy Rd East Greenbush NY 12144

DAVEY, CHARLES BINGHAM, soil science educator; b. Bklyn., Apr. 7, 1928; s. Francis Joseph and Mary Elizabeth (Bingham) D.; m. Elizabeth Anne Thompson, July 11, 1952; children: Douglas Alan, Barbara Lynn, Andrew Martin. B.S., Syracuse U., 1950; M.S., U. Wis., 1952, Ph.D., 1955. Soil scientist Research Service, Dept. Agr., Beltsville, Md., 1957-62; assoc. prof. N.C. State U., Raleigh, 1962-65; prof. N.C. State U., 1965—, head dept., 1970-78, Carl Alwin Schenck Disting. prof., 1978—, Alumni Disting. prof., 1989. Editor: Tree Growth and Forest Soils, 1970; asso. editor: Soil Sci. Soc. Am. proc, 1967-72; contbr. articles to profl. jours. Served with AUS, 1955-57. Fellow AAAS, Am. Soc. Agronomy, Soil Sci. Soc. Am. (pres. 1975-76); mem. Soc. Am. Foresters (Barrington Moore Research award), Internat. Soil Sci. Soc., Internat. Soc. Tropical Foresters, Sigma Xi (Research award), Phi Kappa Phi, Gamma Sigma Delta, Xi Sigma Pi. Achievements include patents in field. Home: 3704 Bryn Mawr Ct Raleigh NC 27606-2515

DAVEY, JAMES JOSEPH, process control engineer; b. N.Y.C., Mar. 9, 1945; s. James Francis Davey and Gertrude (LaRose) Miskel; m. Brenda Newman, Apr. 20, 1968 (div. Nov. 1975); children: Theresa, Mary Katherine; m. Janet Calonge, Nov. 2, 1979; children: Rachel, John. BS in Engring., U. Tenn., 1974. Operator City of Knoxville, Tenn., 1969-72; cons. engr. Chaney Engring., Memphis, Tenn., 1974-76; dir. of wastewater City of Gatlinburg, Tenn., 1976-81, City of Columbia, Tenn., 1981-86; dir. utilities City of Walterboro, S.C., 1986-87; mgr. plant ops. ANFLOW, Inc., Oak Ridge, Tenn., 1987-88; process control engr. Schreiber Corp., Birmingham, Ala., 1988—. Bd. dirs. Clean Community System, Columbia, 1982-86; negotiator Saturn Task Force, Columbia; rep. for city Upper Duck River Devel., D.C.; mem. Tenn. Mcpl. League Ins. Pool, Pub. Works Adv., Fed. Republic Germany. Recipient Outstanding Operated Plant, Columbia, Ky.-Tenn. Water Pollution Control Assn., 1984, Outstanding Operated Plant, Gatlinburg, 1981. Mem. Am. Legion, VFW (quartermaster 1982-85), K.C. Democrat. Roman Catholic. Home: 6012 Cayce Ln Columbia TN 38401-5009 Office: Schreiber Corp 100 Schreiber Dr Trussville AL

DAVID, GEORGE, psychiatrist, economic theory lecturer; b. N.Y.C., Feb. 19, 1940; s. Norman and Jennie (Danziger) D. BA, Yale U., 1961; MD, NYU, 1965. Intern Children's Hosp., San Francisco, 1965; resident in psychiatry Colo. Psychiat. Hosp., Denver, 1965-66; practice medicine specializing in psychiatry San Francisco; staff Mt. Zion Hosp., San Francisco, 1966-67, San Mateo County (Calif.) Mental Health Svcs., 1968-71; lectr. on application of econ. theory to personal decision making. Mem. San Francisco Clin. Hypnosis (v.p. 1973-74). Libertarian. Home: 2334 California St San Francisco CA 94115-2705 Office: 3527 Sacramento St San Francisco CA 94118-1846

DAVID, LOURDES TENMATAY, librarian; b. Manila, July 24, 1944; d. Augusto Lontoc and Juana Villaruz (Avelino) Tenmatay; children: Laura, Ariel Noel, Arnel Rauel, Rommel Aries, Clara Riza. BS in Food Tech., U. of The Philippines, Quezon City, 1965, MLS, 1985. Instr. dept. chemistry U. of The Philippines, Los Baños, Laguna, 1966-73; rsch. asst. Main Libr. U. of The Philippines, Diliman, Quezon City, 1973-80, libr., 1982-87, supr. rsch. and extension svcs. documentation ctr., 1985—, chief libr. Coll. of Sci., 1987—; asst. dir. info. Regional Ctr. for Edn. in Sci. and Math., Penang, Malaysia, 1980-82; cons. libr. Nat. Sci. Rsch. Inst. U. of The Philippines, Diliman, Quezon City, 1983-87; mem. nat. interagy. group energy info. Philippine Nat. Oil Co., Diliman, 1985—; info. coord. BITNET project Nat. Computer Ctr., Diliman, 1992; UNESCO cons. for ednl. and health librs., Myanmar, 1992; sr. lectr. inst. libr. sci. U Philippines, 1993. Editor Jour. of Sci. and Math. Edn. in S.E. Asia, 1980-82. Vol. Santa Elena Found., Novaliches, 1985. Libr. sci. fellow UNESCO, 1966, COLOMBO, 1977, Brit. Coun., 1989, grad. teaching fellow U. of The Philippines, 1965-67. Mem. Philippine Assn. Food Technologists (overall chmn. commodity desk 1991—), Assn. for Childhood Edn. Internat. (U. Philippines chpt. 1987-91), Philippine Acad. and Rsch. Librs., Assn. Reflexology Cons. Internat. Roman Catholic. Office: U of The Philippines, Coll Sci Libr, Quezon City 1010, The Philippines

DAVIDS, ROBERT NORMAN, petroleum exploration geologist; b. Elizabeth, N.J., Apr. 27, 1938; s. William Scheible and Anna Elizabeth (Backhaus) D.; AB in Geology, U. Va., 1960; MS, Rutgers U., 1963, PhD, 1966; m. Carol Ann Landauer, Apr. 20, 1957; 1 child, Robert Norman. With Exxon Co. USA, 1966—, micropaleontologist, New Orleans, 1965-71, uranium geologist, Denver, 1971-72, Albuquerque, 1972-78, supervisory geologist Tex. area exploration, Corpus Christi, 1978-80, N.W. area supr., 1981, dist. geologist so. dist., New Orleans, 1981-84, div. exploration tng. coord., spl. trades unit geologist, 1984-86, geol. tng. advisor, Houston, 1986-89; rsch. geologist Exxon Prodn. Rsch. Co., 1989-92; exploration geologist Exxon Exploration Co., 1992—. Formerly active local Little League Baseball, Jr. Achievement. NSF grad. fellow, 1964-65. Mem. Geol. Soc. Am., AIME, Soc. Econ. Paleontologists and Mineralogists (treas. Gulf Coast sect. 1971), Am. Assn. Petroleum Geologists, Explorers Club, Krewe of Endymion, Sigma Xi, Beta Theta Pi. Author papers. Home: 173 Golden Shadow Cir Spring TX 77381-4162 Office: PO Box 4279 Houston TX 77210-4279

DAVIDSON, ERNEST ROY, chemist, educator; b. Terre Haute, Ind., Oct. 12, 1936; s. Roy Emmette and Opal Ruth (Hugunin) D.; m. Reba Faye Minnich, Jan. 27, 1956; children: Michael Collins, John Philip, Mark Ernest, Martha Ruth. B.Sc. (Union Carbide fellow), Rose-Hulman Inst., 1958; Ph.D. (NSF fellow), Ind. U., 1961. NSF Postdoctoral fellow U. Wis.-Madison, 1961-62; asst. prof. chemistry U. Wash., 1962-65, assoc. prof., 1965-68, prof., 1968-84; prof. Ind. U., Bloomington, 1984-86, disting. prof., 1986—; disting. vis. prof. Ohio State U., 1974-75; vis. prof. IMS, Japan, 1984, Technion, Israel, 1985; cons. Lawrence Livermore Labs. Editor: Jour. Computational Physics, 1975—, Internat. Jour. Quantum Chemistry, 1975—, Jour. Chem. Physics, 1976-78, Chem. Physics Letters, 1977-84, Jour. Am. Chem. Soc, 1978-83, Jour. Phys. Chemistry, 1982—, Accounts of Chem. Research, 1984—, Theoretica Chimica Acta, 1985—, Chem. Revs., 1986—; contbr. numerous articles on density matrices and quantum theory of molecular structure to profl. jours. Battelle Meml. Inst. Sloan fellow, 1967-68; Guggenheim fellow, 1974-75; laureate l'Academie Internationale des Sciences Moleculaires Quantiques, 1971. Mem. Am. Chem. Soc. (Computers in Chemistry award 1992), Am. Phys. Soc., Nat. Acad. Scis., Internat. Acad. Quantum Molecular Scis., Sigma Xi, Phi Lambda Upsilon, Tau Beta Pi. Home: 1013 S Woodbine Ct Bloomington IN 47401-5445 Office: Ind U Chemistry Dept Bloomington IN 47405

DAVIDSON, FRANK PAUL, macro-engineer, lawyer; b. N.Y.C., May 20, 1918; s. Maurice Philip and Blanche (Reinheimer) D.; m. Izaline Marguerite Doll, May 19, 1951; children: Roger Conrad, Nicholas Henry, Charles Geoffrey. BS, Harvard U., 1939, JD, 1948, DHL (hon.), Hawthorne Coll., 1987. Bar: N.Y. 1953, U.S. Dist. Ct. (so. dist.) N.Y. 1953. Dir. mil. affairs Houston C. of C., 1948-50; contract analyst Am. Embassy, Paris, 1950-53; assoc. Carb, Luria, Glassner & Cook, N.Y.C., 1953-54; pvt. practice law, N.Y.C., 1955-70; rsch. assoc. MIT, Cambridge, 1970—, also chmn. system dynamics steering com. Sloan Sch. Mgmt., coord. macro-engring. Sch. Engring.; pres. Tech. Studies, Inc., N.Y.C., 1957—; vice chmn. Inst. for Ednl. Svcs., Bedford, Mass. 1980-84; spl. lectr. Société des Ingénieurs et Scientifiques de France, 1991; Nat. Acad. Scis. del. to Renewable Resources Workshop, Katmandu, Nepal, 1981; governing bd. Channel Tunnel Study Group, 1957—; co-founder Channel Tunnel Study Group, London and Paris, 1957; appointed to NASA Exploration Task Force, Washington, 1989. Author: Macro: A Clear Vision of How Science and Technology Will Shape our Future, 1983, Macro: Big Is Beautiful, 1986; editor series of AAAS books on macro-engring., Tunneling and Underground Transport, 1987; co-editor Macro-Engineering, Global Infrastructure Solutions, 1992, Solar Power Satellites, 1993; mem. editorial bd. Interdisciplinary Sci. Revs., 1985—; mem. adv. bd. Tech. in Soc., 1979—, Mountain R & D, 1981—, Project Appraisal, 1986—. Bd. dirs. Internat. Mountain Soc., Boulder, Colo., 1981—; trustee Norwich (Vt.) Ctr., 1980-83; mem. steering com. Am. Trails Network, 1986-88; bd. dirs. Am. Trails. Washington, 1988-90; apptd. NASA Exploration Task Force, Washington, 1989. Capt. RCAC, 1941-46, ETO. Decorated Bronze Star; recipient key to City of Osaka, Japan, 1987; Lewis Mumford fellow Rensselaerville Inst., 1982. Mem. ABA, Internat. Assn. Macro-Engring. Socs. (bd. dirs. 1987—), Am. Soc. Macro-Engring. (bd. dirs. 1982—, vice chancellor 1983—), Assn. Bar of City of N.Y. (internat. law com. 1959-62). Clubs: Knickerbocker (N.Y.C.); St. Botolph (Boston). Home: 140 Walden St Concord MA 01742-3613 Office: MIT E40-294 Cambridge MA 02139

DAVIDSON, JAMES MADISON, III, engineer, technical manager; b. San Antonio, Feb. 24, 1930; s. James Madison Jr. and Ella Louise (Wehmeyer) D.; m. Geneva Upchurch, Aug. 28, 1949; children: Robert John, William Allen, James Brian. BS, S.W. Tex. State U., 1951. Registered profl. engr., Wash. Engr., then sr. engr. GE Co., Richland, Wash., 1951-65; mgr. fast flux test facility, materials and tech. dept. Battelle-Pacific N.W. Lab., Richland, 1965-67, sr. adviser to lab. dir., 1967-72; mgr. office nat. security tech. Battelle-Pacific N.W. Lab., 1972-89; staff mem. Los Alamos (N.Mex.) Nat. Lab., 1989-90, acting group leader, 1990-91, group leader, 1992—; tech. adviser Coordinating Com. on Munitions, Paris, 1987—. Exec. bd. Boy Scouts Am., 1970-75. Recipient Silver Beaver award Boy Scouts Am., 1972. Home: 18 W Wildflower Dr Santa Fe NM 87501

DAVIDSON, JOAN GATHER, psychologist; b. Long Branch, N.J., Jan. 26, 1934; d. Ralph Paul and Hilde (Bresser) Gather; m. Harry Gene Davidson, Sept. 14, 1957; children; Guy, Marc, Kelly. BA, Shorter Coll., 1956; BA cum laude, U. South Fla., 1982; MS, Fla. Inst. Tech., 1986, PsyD, 1987. Lic. psychologist, Fla., RN, Ga. Clin. instr. Ga. Bapt. Sch. Nursing, Atlanta, 1956-59; dir. nurses Aidmore Hosp., Atlanta, 1959-60; dir. insvc. edn., asst. dir. nurses Bayfront Med. Ctr., St. Petersburg, Fla., 1960; instr. St. Petersburg Jr. Coll., 1971-76; pvt. practice St. Petersburg-Clearwater, 1987—. Mem. Am. Psychol. Assn., Fla. Psychol. Assn., Nat. Register Health Svc. Providers in Psychology, Assn. for Advancement Psychology, Am. Assn. Christian Counselors, Psi Chi, Phi Kappa Phi. Republican. Baptist. Home: 11600 87th Ave Largo FL 34642-3613 Office: 2329 Sunset Point Rd # 204 Clearwater FL 34625-1426

DAVIDSON, JOHN HUNTER, agriculturist; b. Wilmette, Ill., May 16, 1914; s. Joseph and Ruth Louise (Moody) D.; m. Elizabeth Marie Boynton, June 16, 1943; children—Joanne Davidson Hildebrand, Kathryn Davidson Bouwens, Patricia. B.S in Horticulture, Mich. State U., 1937, M.S. in Plant Biochemistry, 1940. Field researcher agrl. chems. Dow Chem. Co., Midland, Mich., 1936-42, with research and devel. dept. agrl. products, 1946-72, tech. adviser research and devel. agrl. products, 1972-80, tech. adviser govt. relations, 1980-84, cons., 1984—. Served to lt. USNR, 1945. Mem. Am. Chem. Soc., Am. Soc. Hort. Sci., Weed Sci. Soc., Am. Pathol. Soc., Phi Kappa Phi, Alpha Zeta. Republican. Presbyterian. Club: Exchange of Midland. Contbr. articles on plant pathology and weed control to profl. jours. Home: 4319 Andre St Midland MI 48642-3779

DAVIDSON, KAREN SUE, computer software designer; b. Chgo., July 24, 1950; d. Woodrow Wilson and Velma Louise (Dickinson) D. BS in Comm., U. Ill., 1972; MBA, De Paul U., 1977. News producer Sta. WIND, Westinghouse Broadcasting Co., Chgo., 1973-75; mktg. rep. div. data processing IBM, Chgo., 1977-80, process industry specialist, 1980; industry applications specialist IBM, White Plains, N.Y., 1981-83; sr. sales rep. Wang Labs., Chgo., 1983-84; ptnr. KDA-K Davidson & Assocs., Centralia, Ill., 1984—; pres. KDA Software Inc., Centralia, 1988-92; cons. desktop pub. Greater Centralia C. of C., 1987-88; instr. Belleville (Ill.) Area Coll., 1992; mem. rev. bd. State of Ill. Pvt. Enterprise. Author/designer software programs; contbr. articles to profl. pubs. State of Ill. Small Bus. Adv. Bd., Internat. Trade/Export Rep., 1990-93; WordPerfect cert. resource instr. WordPerfect Corp., 1991—. Named Outstanding Working Woman of Ill. Fedn. Bus. & Profl. Women's Clubs, 1990, Word Perfect Cert. Resource, Word Perfect Corp. Mem. Soc. Profl. Journalists, Ind. Computer Cons. Assn., Ill. Software Assn., Chgo. High Tech. Assn. Assn. St. Louis Info. Systems Trainers (v.p. 1988), Centralia Cultural Soc., Inventors' Assn. St. Louis, Greater Centralia C. of C. (bd. dirs. 1990-93, good will amb. 1990), Rotary, Zeta Tau Alpha. Methodist. Office: KDA Software Inc PO Box 1163 315 E 3d St Centralia IL 62801

DAVIDSON, KEITH DEWAYNE, corrosion engineer; b. Aurora, Colo., Dec. 30, 1955; s. Dennie Jethro and Vineta Rose (Randolph) D.; m. Gayle Lynn Savage, Aug. 21, 1976; children: Amber, Shayn, Megan. BS in Chemistry, Harding U., 1977; MS in Ocean Engring., Fla. Atlantic U., 1988. Rsch. chemist Reilly Chem., Indpls., 1977-79; engr. United Space Boosters, Inc., Kennedy Space Ctr., Fla., 1981-84, Pan Am. World Svcs., Cape Canaveral AFS, Fla., 1985-88; sr. engr. Computer Scis. Raytheon, Patrick AFB, Fla., 1988-93. Author: (manual) Protective Coatings Procedure for SRB's, 1982, Corrosion Control Program for Air Force Eastern Range, 1985. Tchr. children's gymnastics programs, Melbourne, Fla. Mem. AIAA (sr.), Nat. Assn. Corrosion Engrs. (cert., com. on aerospace corrosion 1990), Steel Structures Painting Coun. (com. on zinc paints 1988—), Espl. Aircraft Assn. Republican. Mem. Ch. of Christ. Achievements include design of cathodic protection of space shuttle solid rocket boosters; design of mus. displays; corrosion control programs. Home: 3145 Hilliard Ct Melbourne FL 32934

DAVIDSON, MICHAEL, psychiatrist, neuroscientist; b. Sept. 18, 1949. MD, U. Statale, Milan, 1976. Lic. N.Y.; diplomate Am. Bd. Psychiatry and Neurology. Intern, then resident in cardiology and internal medicine Ihilov Hosp., Tel Aviv, 1976-81; resident in psychiatry N.Y. Med. Coll., 1981-82, Mount Sinai Sch. Medicine, N.Y.C., 1982-83; chief resident Bronx (N.Y.) VA Med. Ctr., 1983-84, dir. psychiatric rsch. unit, 1985-88; dir. divsn. geriatric psychiatry Mount Sinai Sch. Medicine, N.Y.C., 1988-89, dir. divsn. clin. rsch. dept. psychiatry, 1987—, assoc. clin. dir. Pilgrim Psychiatric Ctr., 1990—, prof. psychiatry, 1993—. Author: The Psychiatric Clinics North America: Alzheimer's Disease, 1991; contbr. articles to Am. Jour. Psychiatry, Jour. Psychiatry Rsch. and others. Recipient Heyman Rsch. award, 1985, Maed-Johnson Rsch. award 1986, Schizophrenia Rsch. award Nat. Inst. Mental Health. Mem. Am. Coll. Neuropsychopharmacology, Soc. Biological Psychiatry (co-chair program com. 1991). Home: 61-63 W 106th St New York NY 10025 Office: Mount Sinai Schl of Medicine Psychiatry Service 116A 130 W Kingsbridge Rd Bronx NY 10468

DAVIDSON, RICHARD ALAN, data communications company executive; b. Chgo., June 25, 1946; s. Jacob Aaron and Belle Rina (Feldman) D.; m. Sharyn Gail Ellman, Aug. 19, 1973; children: Kevin Scott, Caryl Elise. BSEE, U. Mich., 1970; MBA, Northwestern U., 1975. Project engr. Motorola, Inc., Schaumburg, Ill., 1967-74; ptnr. Feature Film Svcs., Skokie, Ill., 1974-77; mgr. planning Motorola, Inc., Schaumburg, Ill., 1977-78, mgr. mktg., 1978-79; tech. dir. Voice & Data Systems, Chgo., 1979-82; engring. mgr. Infolink Corp., Northbrook, Ill., 1982-84; pres. Davidson Data Communications, Deerfield, Ill., 1984—; v.p. engring. Feature Film Svcs., Skokie, 1976—. Inventor pay TV system; contbr. articles to profl. jours. Unit commr. Boy Scouts Am., Lake County, Ill., 1989—; comms. officer USAF Aux. CAP, 1991—. Recipient Cert. of Appreciation Boy Scouts Am., 1990. MEm. IEEE, Assn. for Computing Machinery, Assn. for MBA Execs., North Shore Radio Club (tech. dir.), Tau Delta Phi. Republican. Jewish. Avocations: amateur radio, electronics, photography. Home and Office: 1900 S Millburne Rd Lake Forest IL 60045-4112

DAVIDSON, RONALD CROSBY, physicist, educator; b. Norwich, Ont., Can., July 3, 1941; s. William Crosby and Annie Beatrice (Caley) D.; m. Jean Farncombe, May 18, 1963; children: Cynthia Christine, Ronald Crosby Jr. B.Sc., McMaster U., 1963; Ph.D., Princeton U., 1966. Mem. faculty dept. physics U. Md., 1968-78; vis. scientist Los Alamos Sci. Lab., 1974-75; asst. dir. for applied plasma physics Office of Fusion Energy Dept. Energy, Washington, 1976-78; prof. physics, 1978-91; dir. Plasma Fusion Center MIT, Cambridge, Mass., 1978-88; chmn. magnetic fusion adv. com., 1982-86; prof. astrophys. scis. Princeton U., 1991—; dir. Princeton Plasma Physics Lab., Princeton, 1991—. Author: Methods in Nonlinear Plasma Theory, 1972, Theory of Nonneutral Plasmas, 1974, 89, Physics of Nonneutral Plasmas, 1991. Recipient Disting. Assoc. award Dept. Energy, 1986, Leadership award Fusion Power Assocs., 1986; Ford Found. fellow, 1963-64; Imperial Oil fellow, 1963-66; Sloan Research Found. fellow, 1970-72. Fellow AAAS, Am. Phys. Soc. (chmn. div. plasma physics, 1983-84). Office: Princeton Plasma Physics Lab PO Box 451 Princeton NJ 08543-0451

DAVIDSON, TERRY LEE, experimental psychology educator; b. Vassar, Mich., June 1, 1951; s. George Louis and Helen Marsha (Laskey) D.; m. Cheryl Joan Gohm, June 23, 1973; children: Tyler Louis, Dena Lynn. BA, Mich. State U., East Lansing, 1973; PhD, Purdue U., West Lafayette, Ind., 1981. Asst. prof. St. Olaf Coll., Northfield, Minn., 1981-83; rsch. scientist Inst. Neurol. Sci., Phila., 1983-86; lectr. U. Pa., Phila., 1984-86; asst. prof. Va. Mil. Inst., Lexington, 1986-90; assoc. prof. exptl. psychology Purdue U., West Lafayette, 1990—; cons. Cancer Rsch. Ctr., Children's Hosp., Phila., 1984-87. Editor Va. Mil. Inst. Undergrad. Rsch. Rev., Lexington, 1987-90. Grantee NIH, 1991, 92; recipient Acad. Rsch. Enhancement award, 1987, Nat. Rsch. Svc. award NIH, 1987. Mem. AAAS, Soc. for Study Ingestive Behavior, Psychonomic Soc., Va. Acad. Scis. (vice chmn./sec. 1988-90), NSF (affiliate mem. Neurobiology of Learning and Memory). Achievements include development of learning paradigm to study associative biological controls of feeding; provided evidence, based on learning mechanisms, that anxiolytic (tranquilizing) drugs can reduce tolerance to stressors in animal subjects and that anxiogenic drugs can have opposite effects; (with L. Jarrard) provided evidence that hippocampus is not required to solve conditional learning problems and that learning deficits involve extrahippocampal structures. Office: Purdue U Dept Psychol Scis Pierce Hall West Lafayette IN 47907

DAVIDSON, THOMAS FERGUSON, chemical engineer; b. N.Y.C., N.Y., Jan. 5, 1930; s. Lorimer Arthur and Elizabeth (Valentine) D.; m. Nancy Lee Selecman, Nov. 11, 1951; children: Thomas Ferguson, Richard Alan, Gwyn Ann. BS in Engring., U. Md., 1951. Sr. project engr. Wright Air Devel. Ctr., Dayton, Ohio, 1951-58; dep. dir. Solid Systems Div., Edwards, Calif., 1959-60; mgr. govt. ops. Thiokol Chem. Corp., Ogden, Utah, 1960-64; dir. aerospace mktg. Thiokol Chem. Corp., Bristol, Pa., 1965-67; dir. tech. mgmt. Thiokol Chem. Corp., Ogden, 1968-82; v.p. tech. Morton Thiokol Inc., Chgo., 1983-88, Thiokol Corp., Ogden, 1989-90; cons. Ogden, 1990—; mem. subcom. lubrications and wear NACA, Washington, 1955-57; chmn. Joint Army, Navy, NASA, Air Force exec. com., 1959-60. Editor: National Rocket Strategic Plan, 1990; contbr. articles to profl. jours. Chmn. bd. Wesley Acad., Ogden, 1990—; mem. Utah State Bd. Edn., 1993—; trustee Family Counselling Svc., Ogden, 1991—; dir. Habitat for Humanity Internat.; vice moderator Shared Ministry of Utah; mem. Rep. Presdl. Task Force, Washington, 1987-92, Am. Security Coun., Washington, 1975—. Lt. USAF, 1951-53. Fellow AIAA (assoc., sect. chmn. 1979-80, chmn. AIA rocket propulsion com. 1987-90, mem. AIA aerospace tech. coun. 1987-90, Wyld Propulsion award 1991); mem. Am. Newcomers Soc., Smithsonian Instn., Exchange Club, Ogden Golf and Country Club. Republican. Methodist. Home: 4755 Banbury Ln Ogden UT 84403-4484

DAVIDSON MOORE, KATHY LOUISE, psychologist; b. Kansas City, Kans., July 2, 1949; d. Eskridge Earl and Bonnie Lee (Rider) Davidson; m. Dieter Hans Eberl, Oct. 6, 1979 (div. July 1990); m. Edward Alan Moore, Aug. 18, 1990. BA, U. Kans., 1973; MA, U. Colo., 1978. Behavior modification cons., Divsn. Devel. Disabilities State of Colo., Denver, 1974-76; CETA grant coord. Ridge Regional Ctr., Wheatridge, Colo., 1977-78, unit psychologist Ridge Regional Ctr., Wheatridge, 1978-81; behavioral cons. Boulder, Colo., 1981-87; dir. psychology Country View Care Ctr., Longmont, Colo., 1987-91; dir. behavioral, psychol. svcs. Health Concept Corp., Greeley, Colo., 1990-91; lead community integration coord. Health Concept Corp., Greeley, 1991-92; dir. family resource ctr. Yavapai Regional Med. Ctr., Prescott, Ariz., 1992—. Fundraiser Spl. Olympics, Denver, 1975-87. Assoc. mem. APA., Nat. Assn. of the Dually Diagnosed (bd. dirs. Columbine chpt. 1989-92, Valuable Svc. award 1991). Office: Yavapai Regional Med Ctr 1003 Willow Creek Rd Prescott AZ 86303

DAVIES, KELVIN JAMES ANTHONY, research scientist, medical educator, consultant; b. London, Oct. 15, 1951; came to U.S., 1975; s. Alfred B. and Phyllis (Garcia) D.; m. Joanna Davies, Sept. 14, 1980; children: Sebastian, Alexander. BEd, Liverpool (Eng.) U., 1974; BS, MS, U. Wis., 1976, 77; C.Phil., U. Calif., Berkeley, 1979, PhD, 1981; DSc (hon.), U. Moscow, Russia, 1983. Instr. Beal Sch. for Boys, London, 1974-75; rsch. asst. U. Wis., Madison, 1975-77, U. Calif., Berkeley, 1979-80; rsch. assoc. U. So. Calif., L.A., 1980-81; instr. Harvard U., Boston, 1981-83; asst. prof. biochemistry U. So. Calif., L.A., 1983-86, assoc. prof. biochemistry, 1986-90, prof. biochemistry, 1990; prof. biochemistry and molecular biology Albany (N.Y.) Med. Coll., 1991—, chmn. dept. biochemistry and molecular biology, 1991—; hon. dist. prof. Russian State Med. U., Moscow, 1983. Editor in chief: (jour.) Free Radical Biology and Medicine, 1981—; contbr. articles to profl. jours. and books. Recipient Chancellors award for Rsch. U. Calif., Berkeley, 1981. Mem. Am. Physiolog. Soc. (Harwood S. Belding award 1982), Am. Soc. for Biochemistry & Molecular Biology, The Oxygen Soc. (fellow, sec. gen. 1987-90, pres.-elect 1990-92, pres. 1992—), Phi Beta Kappa. Avocations: opera, symphony. Office: Albany Med Coll Dept Biochemistry/Molecular New Scotland Ave Albany NY 12208

DAVIES, PETER JOHN, plant physiology educator, researcher; b. Sudbury, Middlesex, Eng., Mar. 7, 1940; came to U.S., 1966; s. William Bertram and Ivy Doreen (Parmentier) D.; m. Linda Kay DeNoyer, Aug. 2, 1976; children—Kenneth DeNoyer, Caryn Parmentier. B.Sc. with honors, U. Reading, Eng., 1962; M.S., U. Calif.-Davis, 1964; Ph.D., U. Reading, 1966. Instr. Yale U., New Haven, 1966-69; asst. prof. plant physiology Cornell U., Ithaca, N.Y., 1969-75, assoc. prof., 1975-83, prof., 1983—, chmn. sect. plant biology, 1992—. Author: (with others) The Life of the Green

Plant, 1980, Control Mechanisms in Plant Development, 1970; editor: Plant Hormones and Their Role in Plant Growth and Development, 1987; editor-in-chief Plant Growth Regulation, 1987-92. Mem. Am. Soc. Plant Physiology, Internat. Plant Growth Substance Assn. (coun. 1991—). Office: Cornell U Plant Biology Plant Sci Bldg Ithaca NY 14853

DAVIES, ROGER, geoscience educator; b. London, Aug. 29, 1948; came to U.S., 1972, naturalized, 1985; s. Trevor Rhys and Gracie Rhys (Beaton) D.; m. Corinne Marie Scofield, Oct. 29, 1977; children: Colin, Gavin. BS with honors, Victoria U., Wellington, N.Z., 1970; PhD, U. Wis., 1976. Meteorologist, New Zealand Meteorol. Service, Wellington, 1971-77; scientist U. Wis., Madison, 1977-80; asst. prof. atmospheric sci. Purdue U., West Lafayette, Ind., 1980-86, assoc. prof. 1986—; mem. Earth Radiation Budget Expt. Sci. Team, 1980—, First Internat. Satellite Cloud Climatology Project, Regional Exptl. Sci. Team, 1984—. Assoc. editor Jour. Geophys. Research, 1987—; contbr. articles and book revs. to profl. publs. Research grantee NASA. Mem. Am. Meteorol. Soc., Am. Geophys. Union, Optical Soc. Am. Avocation: sailing. Office: McGill Univ-Radar Weather Observ, MacDonald College PO Box 241, Sainte Anne de Bellevue, PQ Canada H9X 1C0

DAVIS, ALAN ROBERT, molecular biologist; b. Stillwater, Okla., Jan. 25, 1947; s. Robert R. Davis and Ernestine M. (Wood) Warneke; m. Sandra Jorn, June 15, 1968; children: Scott, Laura. AB in Bacteriology, U. Calif., L.A., 1968, PhD in Molecular Biology, 1973. Postdoctoral fellow U. Calif., L.A., 1974-78, from asst. to assoc. rsch. virologist, 1978-82; unit supr. Wyeth Labs., Radnor, Pa., 1983-85, mgr., 1985-87; assoc. dir. molecular biology Wyeth-Ayerst Rsch., Wayne, Pa., 1987-88, dir. molecular biology, 1988—. Contbr. articles to profl. jours. Recipient fellowship Nat. Cancer inst., 1976-78, Leukemia Soc. Am., 1974-76. Achievements include patents in oral vaccines; in microbial expression of influenza virus hemagglutinin. Home: 631 Timber Dr Wayne PA 19087 Office: Wyeth Ayerst Rsch PO Box 8299 Philadelphia PA 19101

DAVIS, ALVIN ROBERT, JR., structural engineer; b. Pensacola, Fla., July 7, 1954; s. Alvin Robert Davis and Jane (Hodgkinson) Eddy; m. Deborah Ann Bushell, Sept. 9, 1978; 1 child, Amy Elizabeth. BS in Archtl. Engring., Roger Williams Coll., 1977; MSCE, Worcester Poly. Inst., 1992. Assoc. engr. Philip S. Mancini, Jr., P.E., Providence, R.I., 1977-78, L.A. Garofalo Inc., Warwick, R.I., 1978-80; chief engr. NAS Inc., Canton, Mass., 1980-83; structural engr. Shawmut Metal Products Inc., Swansea, Mass., 1983-87; project engr. Denis R. Samson AIA, Warwick, 1987-90; resident engr. Commonwealth of Mass., Boston, 1990-91; chief engr. Isaacson Structural Steel Inc., Berlin, N.H., 1991—. Author: Structural Steel Fabrication: A KBES to Assist Constructability in Structural Steel Design and the Fabrication Industry, 1992. Mem. ASCE (com. on metals 1991—), Nat. Soc. Profl. Engrs. (legis. com. 1991—), Am. Inst. Steel Constrn., Am. Welding Inst. Episcopalian. Achievements include development of computer program to design heavy braced connections based on research provided by Dr. Thornton. Office: Isaacson Structural Steel Jericho Rd Berlin NH 03570

DAVIS, BARRY ROBERT, biometry educator, physician; b. N.Y.C., June 2, 1952; s. Hyman Israel Davis and Judith Estelle (Ahrend) Block; m. Wallis Ann Lowenthal, June 30, 1974; children: Hillel Asher, Rachel Shira. MD, U. Calif., San Diego, 1977; PhD, Brown U., 1982. Diplomate Am. Bd. Preventive Medicine. Asst. prof. applied math. & medicine Brown U., Providence, 1982-83; asst. prof. biometry U. Tex. Sch. Pub. Health, Houston, 1983-87, assoc. prof. biometry, 1987-92, prof. biometry, 1992—; cons. NIH, Bethesda, Md., 1985—; com. mem. Nat. Inst. Allergy & Infectious Disease, Bethesda, 1990—; mem. adv. bd. VA, New Haven, 1991—. Contbr. articles to profl. jours. Coach Odyssey of the Mind, Sugar Land, Tex., 1992. Regents scholar U. Calif., 1973-77; Univ. fellow Brown U., 1978-87. Fellow Am. Coll. Preventive Medicine, Am. Heart Assn. (coun. on epidemiology, chair com. on criteria & methods 1988-91); mem. Biometric Soc., Am. Statis. Assn. (rep. 1992—), Soc. Clin. Trials (bd. dirs. 1993—). Democrat. Jewish. Office: U Tex Sch Pub Health 1200 Herman Pressler St Houston TX 77030

DAVIS, BRIAN RICHARD, chemical engineer; b. Calcutta, Bengal, India, Oct. 31, 1934; came to U.S., 1969; s. William Richard and Eulalia (Rule) D.; m. Dilys Julia Morgan, Dec. 30, 1961; children: Martin Richard, Fiona June. Diploma ChemE, Loughborough (Eng.) Coll., 1956; PhD in ChemE, U. B.C., 1966. Devel. engr. DuPont of Can., Kingston, Ont., 1956-61; tech. mgr. Cyanamid of Can., Niagara Falls, 1965-71; dir. engring. NL Chems., Hightstown, N.J., 1971-83; v.p. mfg. Calabrian Chems., Houston, 1984-89; v.p. tech. XYTEL-Bechtel Inc., Houston, 1989—. Contbr. articles to sci. jours. Achievements include patents for chlorination of impure magnesium chloride, extracting titanium values from titaniferous ores, process for manufacturing titanium compounds, process for manufacturing titanium dioxide, process for manufacturing of sulfur oxide; research in rate of isomerization of cyclopropane in a flow reactor, measurement of the effective diffusivity of porous pellets. Office: XYTEL-Bechtel 1400 Brittmoore Rd Houston TX 77043

DAVIS, CHARLES ALEXANDER, electrical engineer, educator; b. Petersburg, Va., Aug. 20, 1936; s. Charles Alexander and Bernice (Goodwyn) D.; m. Clemetine Johnson, Sept. 17, 1961; children: Lisa Kathleen, Karen Andrea, Glen Anthony. MEE, U. Mich., 1963; PhD, Mich. State U., 1975. Registered profl. engr., Mich. Engring. asst. Sandia Labs., Albuquerque, 1958-59; rsch. assoc. U. Mich. Willow Run Labs, Ann Arbor, 1960-63; engr. Bendix Corp., Southfield, Mich., 1963-64; process engr. Ford Motor Co., Ypsilanti, Mich., 1964-67; prof. Western Mich. U., Kalamazoo, 1967—; mem. tech. staff AT&T Bell Labs., Holmdel, N.J., 1982-83; engr. Kellogg Corp., Battle Creek, Mich., 1978-79; commr. Mich. OSHA, Lansing, 1979-82. Author: Industrial Electronics: Design and Application, 1973, Handbook for New College Teachers, 1992; contbr. articles to profl. jours. Dir. Civic Black Theatre, Kalamazoo, 1991—; bd. pres. Douglass Community Assn., Kalamazoo, 1981. GM scholar, 1956-59. Mem. IEEE, Mich. Soc. Profl. Engrs. (pres.-elect S.W. chpt. 1993—), Phi Kappa Phi, Tau Beta Pi. Democrat. Baptist. Home: 816 Newgate Rd Kalamazoo MI 49006 Office: Western Mich U Dept Elec Engring Kalamazoo MI 49008

DAVIS, DEAN EARL, aerospace engineer; b. Cheyenne, Wyo., Sept. 13, 1952; s. Vernon Lewis and Myrtle Elizabeth (Cary) D.; m. Victoria Marie Anderson, June 28, 1986; 1 child, Celeste Marie. BA in Physics, math. and Geology, U. Colo., 1976, MS in Astrogeophysics, 1977, MS Aerospace Engring., 1977; BS in Computer Sci., Met. State Coll., 1982; MS in Systems Engring. Mgmt., U. Denver, 1990. Assoc. engr. to engr. Boeing Aerospace, Seattle, and Boeing Mil. Airplane Co., Wichita, Kans., 1978-79; sr. engr. Martin Marietta Strategic Sys. and Space Sys., Denver, 1979-82; sr. engring. specialist Gen. Dynamics Convair and Spce Systems, San Diego, Calif., 1982-84; advanced sr. systems specialist Lockhead Missiles and Space Co., Sunnyvale, Calif., 1984-86; sr. staff engr. Ultra Systems Defense and Space Systems, Sunnyvale, 1986-88, Martin Marietta Astronautics, Denver, 1988-90; pres. and chief cons. Star Tech. Internat. Corp., Littleton, Colo., 1990--; long range planning cons. Nat. Space Policy Bd. on Manned Space, Denver, 1989-90; systems archieture cons. Stratigic Defense Initiative Blue Ribbon Panel, San Diego, 1983-84. Author: Handbook Two for Aerospcae Education, 1991; author, editor: Spacecraft Design, 1990; contbr. articles to profl. jours. Mem. Am. Inst. Aeronautics and Astronautics (sect. pres. 1977-78, Outstanding Achievement award 1988), Am. Astronautical Soc. (pub. rels. chmn. 1988-89, Most Prolific Speaker award 1990), Nat. Space Soc. (guest lectr. 1988-90, Outstanding Speaker award 1989), Mil. Ops. Rsch. Soc. (conf. organizer 1983-84, Most. Creative Simulation award 1989), Air Force Assn. (tutor, lectr.), Sigma Gama Tau, Sigma Pi Sigma, Mu Alpha Theata. Republican. Home: 642 W Acacia Ave El Segundo CA 90245 Office: Star Tech Internat Corp 1601 N Sepulveda Blvd #113 Manhattan Beach CA 90266

DAVIS, DWIGHT, cardiologist, educator; b. Winston-Salem, N.C., Apr. 11, 1948; s. James C. Davis; m. Lorna Jean Enck, July 30, 1988. BS, N.C. A&T State U., 1970; MD, U. Rochester, 1975. Rsch. asst. U. Rochester, N.Y., 1970-71; intern in medicine Boston U. Hosp., 1975-76, resident in medicine, 1976-78; cardiology fellow Duke U. Med. Ctr., Durham, N.C., 1978-81; asst. prof. medicine, cardiology div. Pa. State U., Hershey, 1981-87, assoc. prof., 1987-92, Disting. lectr., 1986, prof. medicine, 1992—, cardiology dir. heart transplantation, artificial organs and preclin. teaching program, dir. cardiology preclinical tng. program, 1984—, dir., cardiology fellow tng. program, 1984-87, dir. cardiac catheterization lab., 1987—, med. dir. cardiac rehab. program, 1988—, dir. clin. cardiology program, 1991—; vice chmn. faculty affairs faculty senate Pa. State U., University Park, 1988—; mem. med. alumni coun. U. Rochester Sch. Medicine and Dentistry, 1992—; various disting. lectureships. Contbr. numerous articles to profl. jours.; editorial reviewer Annals Internal Medicine, 1983—; editorial adv. bd. Primary Cardiology, 1985—. Mem. Pa. Coun. on Aging, Harrisburg, 1989—. Recipient Outstanding Physician award Pa. State U. Sch. Medicine, 1984, Disting. Teaching awards, 1988-89, Tchr. of Yr. award, 1991, Disting. Prof. award for teaching, 1991, Outstanding Tchr. of Yr. award med. sch. class of 1995, 1993; Alumni Excellence award N.C. A&T State U., 1986, Disting. Alumni award Nat. Assn. Equal Opportunity in Higher Edn., 1987. Fellow Am. Coll. Cardiology, Am. Coll. Angiology; mem. AAAS, Am. Heart Assn. (fellow coun. on clin. cardiology, rsch. com. Pa. affiliate 1992—), Am. Fedn. Clin. Rsch., Am. Assn. Cardiovascular and Pulmonary Rehab. (expert panel cardiac rehab. guidelines project 1992—), N.Y. Acad. Scis., Alpha Omega Alpha. Mem. United Ch. of Christ. Achievements include discovery that abnormalities of the sympathetic nervous system in patients with heart failure is due to an increase in norepinephrine spillover and a decrease in norepinephrine clearance from the circulation. Avocations: chess, drama, reading. Office: Pa State U Coll Medicine Divsn Cardiology PO Box 850 Hershey PA 17033-0850

DAVIS, EDGAR GLENN, science and health policy executive; b. Indpls., May 12, 1931; s. Thomas Carroll and Florence Isabelle (Watson) D. m. Margaret Louise Alandt, June 20, 1953; children: Anne-Elizabeth Davis Polestra, Amy Alandt, Edgar Glenn Jr. AB, Kenyon Coll., 1953; MBA, Harvard U., 1955. With Eli Lilly & Co., Indpls., 1958-91, mgr. budgeting and profit planning, 1963-66, mgr. econ. studies, 1966-67, mgr. Atlanta sales dist., 1967-68, dir. market research and sales manpower planning, 1968-69, dir. mktg. plans, 1969-74, exec. dir. pharm. mktg. planning, 1974-75, exec. dir. corp. affairs, 1975-76, v.p. corp. affairs, 1976-90, v.p. health care policy, 1990; pres., chmn. bd. dirs. Centre for Health Sci. Info., Boston, 1990—; fellow Ctr. for Bus. and Govt. Kennedy Sch. of Govt. Harvard U., 1991—; pres. Eli Lilly and Co. Found., 1976-88; mem. Inst. Ednl. Mgmt., Harvard U. Grad. Sch. Edn., 1987; mem. Inst. Medicine NAS, 1981—; chmn. staff Bus. Roundtable Task Force on Health, 1981-85; U.S. rep. UN Indsl. Devel. Orgn. Conf., Lisbon, 1980; participant UNIDO meeting of experts on pharms., 1981; rep. to UN Commn. on Narcotic Drugs, Vienna, 1981, UN Econ. and Social Council, N.Y.C., 1981, UN Indsl. Devel. Orgn. conf. Casablanca, 1981, Budapest, 1983, Madrid, 1987; trustee Boston Biomed. Rsch. Inst., 1991—; fellow Ctr. for Bus. and Govt., Kennedy Sch. Govt., Harvard U.; co-chmn. Harvard Conf. on Govt. Role in Civilian Technology, 1992, Harvard Conf. Pharmaceutical Rsch., Innovation and Pub. Policy, 1993; vis. scholar and advisor Health and Welfare Unit, Inst. for Econ. Affairs, London; lectr. in field. Contbr. articles to profl. jours. Pres., chmn. bd. Indpls. Health Inst., 1988-91; trustee Kenyon Coll. Gambier, Ohio, Boston Biomed. Rsch. Inst.; bd. dirs. Martha's Vineyard Hist. Preservation Soc., Inc.; bd. advisors Christian Theol. Sem.; bd. dirs. Carnegie Council on Ethics and Internat. Affairs and accredited nongovtl. observer rep. to UN, Goodwill Industries Found. Cen. Ind., Inc., Sta. WFYI Pub. TV, Indpls., 1983-91, Indpls. Mus. of Art, Am. Symphony Orch. League, 1987-92, Nat. Health Council, 1984-92, Pub. Affairs Council, Washington, Nat. Fund for Med. Edn.; chmn. bd. dirs. Ind. Repertory Theatre, 1979-85; bd. visitors Bishops Sch., La Jolla, Calif.; vice chmn., mem. exec. com., bd. dirs. Indpls. Symphony Orch. and Ind. State Symphony Soc., 1977-91; chmn. task force on fine arts, Commn. for Future of Butler U.; chmn. exec. com. Pan Am.. Econ. Leadership Conf., 10th Pan Am. Games, Indpls; bd. visitors N.C. Sch. of Arts; mem. Chgo. Council on Fgn. Relations; trustee Eiteljorg Mus. Am. Indian and Western Art. Served to lt., USN, 1955-58. Fellow The Hudson Inst. (sr. adj. Indpls.); mem. NAM (bd. dirs., vice-chmn. health policy com. 1987-91), Met. Club (Washington), Overseas Press Club, Edgartown Yacht Club, Chgo. Yacht Club, Naples (Fla.) Yacht Club Woodstock Club, Contemporary Club, Lambs Club, Crooked Stick Golf Club, Weston Golf Club, Chappaquiddick Beach Club, N.Y. Yacht Club, Traders Point Hunt Club, Reform Club London, Country Club of Naples (Fla.). Office: Harvard U Ctr for Bus and Govt Kennedy Sch of Govt Weil Hall 79 JFK St Cambridge MA 02138

DAVIS, ELIZABETH EMILY LOUISE THORPE, vision psychophysicist, psychologist and computer scientist; b. Grosse Pointe Farms, Mich., Aug. 11, 1948; d. Jack and Mary Alvina (McCarron) Thorpe; student U. Calif.-Irvine, 1966-69; BS, U. Ala., 1972; MA, Columbia U., 1975, MPhil, 1976, PhD in Exptl. Psychology, 1979, MS in Computer Sci., 1987; m. Ronald Wilson Davis, May 16, 1969. Lectr. Am. Lit. and English composition Nei Ming Inst., Lamtin, Hong Kong, 1969-71; research fellow Columbia U., 1973-77; postdoctoral fellow N.Y.U., 1979-81, adj. asst. prof., 1981; asst. prof. exptl. psychology Oberlin (Ohio) Coll., 1981-82; rsch. asst. prof., mem. grad. faculty Inst. for Vision Rsch., SUNY Coll. Optometry, 1983-87; assoc. prof. dept. visual scis., 1987-90; assoc. prof. sch. psychology Ga. Inst. Tech., 1990—. Recipient Nat. Rsch. Svcs. award; fellow Hertz Found., 1983; NIH grantee, 1979-81, 84-90; grantee Sigma Xi, 1977, Oberlin Coll., 1981. Mem. AAAS, Am. Psychol. Assn., Assn. Rsch. Vision and Ophthalmology, Soc. Neuroscis., Optical Soc. Am. (co-feature editor jour. 1987, session chair 1984), N.Y. Acad. Scis., Psychonomic Soc. (session chair 1988), Human Factors Soc., Soc. Photo-Optical Instrumentation Engrs., Sigma Xi, Pi Mu Epsilon. Author papers in field. Office: Ga Inst Tech Sch Psychology Atlanta GA 30332-0170

DAVIS, GORDON DALE, II, biochemist; b. Honolulu, Apr. 21, 1956; s. Gordon Dale Sr. and Marilyn (Mitchell) D. BS in Chemistry, Okla. Christian Coll., 1978; PhD in Biochemistry, Okla. State U., 1993. Clin. toxicologist Med. Arts Lab., Oklahoma City, 1978-84; rsch. asst. dept. biochemistry Okla. State U., Stillwater, 1985—. Co-author: Biosynthesis of Secondary Metablize Products, 1992; contbr. articles to profl. jours. Predoctoral fellowship Phillips Petroleum, Battles of the Okla., 1987-90. Mem. Am. Chem. Soc., Phi Kappa Phi, Sigma Xi, Phi Lambda Upsilon (sec. local chpt. 1987-88). Achievements include rsch. in isolation and purification of enzyme activity induced during bacterial blight of cotton. Home: 1608 McGraw Ponca City OK 74601 Office: Dept of Biochemistry Okla State U Phys Scis II 565 Stillwater OK 74078

DAVIS, GREGORY JOHN, mathematician, educator; b. Green Bay, Wis., May 29, 1959; s. Robert Eugene and Theresa May (Schneider) D.; m. Jennifer Susan Hoff, May 25, 1984; children: Meagan Riane, Kyle Mackenzie. BS in Math./Physics, U. Wis., Green Bay, 1981; MA in Math., Northwestern U., 1985, PhD, 1987. Ad hoc instr. dept. math. Northwestern U., Evanston, Ill., 1984-87, Ill. Benedictine Coll., Lisle, 1986-87; instr. dept. math. LakeLand Coll., Lifelong Learning, Green Bay, 1988—; asst. prof. U. Wis., Green Bay, 1987-92, assoc. prof. dept. natural and applied scis., dept. math., 1992—; chair dept. math. U. Wis., 1993—. Contbr. articles to Theoretical Population Biology, Biol. Conservation, Am. Math. Soc. Transactions, Publicacions Matematiques. Mem. Am. Math. Soc., London Math. Soc., Math. Assn. Am., Wildlife Sanctuary Green Bay. Democrat. Achievements include research in population dynamics in metapopulation models; phenomena near homoclinic tangencies in low dimensional maps. Office: Univ Wis Green Bay 2420 Nicolet Dr Green Bay WI 54311

DAVIS, JAMES EVANS, general and thoracic surgeon, parliamentarian, author; b. Goldsboro, N.C., Mar. 2, 1918; s. Daniel Wilborn and Annie Maude (Evans) D.; m. Margaret Royall, June 14, 1943; children James Evans Jr. (dec.), Kenneth Royall, George Harrison. AB in Chemistry, U. N.C., 1940; MD, U. Pa., 1943; DSc (hon.), U. N.C., 1988. Diplomate Am. Bd. Surgery. Intern N.Y. Hosp. Cornell U. Med. Ctr., 1944-45, resident in surgery, 1946-50, chief surg. resident, 1950-51; instr. surgery Med. Sch., Cornell U., N.Y.C., 1945-51; prof. clin. surgery Sch. Medicine, U. N.C., Chapel Hill, 1954—; assoc. prof. clin. surgery Med. Sch., Duke U., Durham, N.C., 1954—; chmn. dept. surgery Watts and Durham Regional Hosp., 1954-80; chmn. bd. dirs. N.C. Inst. Medicine, Durham, 1983—; pres. Med. Mut. Ins. Co. N.C., Raleigh, 1975-88. Author: Major Ambulatory Surgery, 1987, Rules of Order, 1992; mem. editorial bd. Jour. Ambulatory Care Mgmt., 1980—; contbr. articles to profl. jours. Chmn. City of Medicine Program, Durham, 1981—; mem. Bipartisan Commn. on Comprehensive Health Care, 1989-90; trustee Durham Gen. Hosp., 1991—. Lt. (j.g.) USNR, 1945-46, lt. comdr. res. ret., Korea. Recipient Disting. Svc. award U. N.C. Sch. Med., Chapel Hill, 1960, Disting. Svc. award Gen. Alumni Assn. U. N.C., 1975, Man of Yr. civic award, Durham, 1988. Mem. AMA (pres. 1988-89, speaker ho. dels. 1983-87), ACS (gov. 1980-86, pres. N.C. chpt. 1978-79, Disting. Surgeon award 1991), N.C. Med. Soc. (pres. 1975-76), N.C. Surg. Soc. (pres. 1980-81), Am. Soc. Gen. Surgeons (founding pres. 1992—), Med. Speakers Assn., Greater Durham C. of C. (pres. 1983-84), Hope Valley Country Club, Country Club N.C., Coral Bay Beach Club. Democrat. Episcopalian. Home: 7 Beverly Dr Durham NC 27707-2223 Office: Drs Davis & Loehr PA 2609 N Duke St Ste 402 Durham NC 27704-3085

DAVIS, JOHN ADAMS, JR., electrical engineer, roboticist, executive; b. Winston-Salem, N.C., May 26, 1944; s. John Adams and Jean Elizabeth (Bowles) D.; m. Sharon Kay Hammons, Dec. 19, 1965; 1 child, Heather Noelle. BSEE with honors, N.C. State U., 1971; MS in Engring., Fla. Tech. U., 1976; MBA, Loyola Coll., Balt., 1980. Design engr. Martin Marietta Corp., Orlando, Fla., 1971-76; sr. program mgr. Martin Marietta Corp., Glen Burnie, Md., 1988-89; sr. engr., project mgr. Gould, Inc., Glen Burnie, Md., 1976-79, program mgr., 1984-88; mgr. data systems Bendix Corp., Columbia, Md., 1979-81; corp. ops. mgr. Vector Automation, Inc., Balt., 1981-82; dir., div. mgr. Ill. Inst. Tech. Rsch. Inst., Chgo., 1982-83; gen. mgr. Marine Systems div., dir. corp. bus. area Eastport Internat., Inc., Upper Marlboro, Md., 1989-93; pres. JADE Rsch. Corp., Severna Park, Md., 1993—; pres., cons. Bustech Co., Severna Park, 1982-84; speaker profl. confs. Contbr. articles to profl. jours. Bd. dirs. Cape Arthur Community Improvement Assn., Severna Park, 1981-82, 89-90; mem. Md. Gov.'s Com. to Elect Sch. Bd., 1982. Sgt. USAF, 1965-68, Vietnam. Recipient USAF Commendation medal. Mem. AIAA, IEEE (bd. dirs. 1978-79), Am. Def. Preparedness Assn., Assn. U.S. Army, Am. Legion, Hazardous Materials Controls Rsch. Inst., Am. Nuclear Soc., Marine Tech. Soc. (remotely operated vehicle com.), Hazardous Materials Controls Resources Inst., Assn. Unmanned Vehicle Systems (Mem. of Yr. 1991, tech. co-chmn. nat. conf. and exhibition, 1991, 92, 93, mem. bd, trustees), Tau Beta Pi, Eta Kappa Nu. Republican. Presbyterian. Avocations: snow and water skiing, boating, swimming, ch. choir, chess. Office: JADE Rsch Corp 5 Linda Ln Severna Park MD 21146

DAVIS, JOHN STAIGE, IV, physician; b. N.Y.C., Oct. 28, 1931; s. John Staige, III and Camilla Ruth (Cole) D.; m. Frederica Abbott, June 22, 1956; children—Susan, John, Stewart, Frederica, Rufus. B.A., Yale U., 1953; M.D., U. Pa., 1957. Diplomate Am. Bd. Internal Medicine. Intern Hosp. U. Pa., Phila., 1957-58; resident in medicine U. Va. Hosp., Charlottesville, 1958-60; fellow in rheumatology, 1960-62; mem. faculty U. Va. Med. Sch., 1961—, prof. medicine, 1972—, Trolinger prof. rheumatology, 1983—, chief div. rheumatology, 1967-92; attending physician U. Va. Hosp., 1961—; vis. prof., chmn. immunology dept. U. Milan (Italy) Faculty Medicine, 1966-67; vis. prof. WHO Immunology Research and Tng. Center, Geneva, 1978-79; cons. WHO, 1979. Contbr. articles to med. jours., chpts. to books. Markle scholar acad. medicine, 1964-69; sr. investigator Arthritis Found., 1964-66, 67-70; Fogarty sr. internat. fellow, 1978-79. Master ACP; mem. Am. Bd. Internal Medicine-Rheumatology, Am. Fedn. Clin. Research, So. Soc. Clin. Investigation, Am. Clin. and Climatological Assn., Am. Coll. Rheumatology (bd. dirs.), Am. Assn. Immunology, Albemarle County Med. Soc., Brit. Soc. Immunology, Med. Soc. Va., Soc. for Preservation and Encouragement of Barber Shop Quartet Singing in Am., U.S. Squash Racquet Assn., Noconomo Yacht Club, Boar's Head Sports. Republican. Episcopalian. Home: 325 Kent Rd Charlottesville VA 22903-2409 Office: U Va Med Sch Dept Internal Medicine Div Rheumatology Box 412 Charlottesville VA 22908

DAVIS, JOHN STEWART, medical educator, endocrinology researcher; b. Fargo, N.D., Nov. 11, 1952; s. Thomas D. and Elizabeth (Odegard) D.; m. Renda Jane Peterson, July 24, 1974; children: Nicholas, Benjamin, Tyler. MS in Physiology and Pharmacology, U. N.D., 1977, PhD in Physiology, 1979. Postdoctoral fellow dept. biochemistry Endocrine Lab., U. Miami, 1979-81; asst. prof. dept. ob-gyn. U. Miami, 1981-83; asst. prof. dept. internal medicine U. South Fla., Tampa, 1983-88, assoc. prof. internal medicine, 1988; assoc. prof. dept. ob-gyn. and internal medicine U. Kans. Sch. Medicine, Wichita, 1988—; assoc. career rsch. scientist VA Med. Ctr., Wichita, 1988—; sr. scientist Women's Rsch. Inst., Wichita, 1988—. Author 10 book chpts.; ast. editor Biology of Reproduction Jour., 1990—; contbr. over 70 articles to profl. jours. Capt. USAR, 1971—. Recipient NIH grant, 1983—; VA Merit Rev. grantee, 1983—. Mem. AAAS, Endocrine Soc., Soc. for Study of Reproduction. Achievements include research in mechanism of action of hormones and growth factors. Office: Womens Rsch Inst 2903 E Central Wichita KS 67214

DAVIS, JUNE LEAH, psychologist; b. Craigsville, W.Va., Nov. 10, 1922; d. Ernest Layton and Bessie May (Bostic) Taylor; m. Charles William Heasley, Jan. 16, 1943 (div. 1961); children: Denasse Ann Heasley Dugan, Wanda Lori Heasley Schwartz; m. Theodore R. Davis, Nov. 20, 1971. BA, Glenville (W.Va.) State Coll., 1962; MA, Ohio State U., 1967. Cert. psychologist, Ohio. Tchr. Nicholas County Bd. Edn., Summersville, W.Va., 1943-46, 1958-60; tchr. Columbus (Ohio) Bd. Edn., 1962-70, sch. psychologist, 1970—. Active First Baptist Ch., Columbus, 1975—. Mem. Sch. Psychologists Cen. Ohio (pres., Best Practice award 1987), Ohio Sch. Psychologists Assn. (mem. various coms.), Cen. Ohio Psychologists Assn. (Hueslman award for outstanding svc. 1989), Nat. Assn. Sch. Psychologists, Heart of Ohio Smocking Club. Democrat. Avocations: smocking, camping, walking. Home and Office: 432 S Weyant Ave Columbus OH 43213-2262

DAVIS, KEITH ROBERT, plant biology educator; b. Watervliet, Mich., June 11, 1957; s. Robert Wayne and Alora May (Howard) D.; m. Julie Nelson, June 20, 1981; children: Jonathon, Sarah, Joshua. BA in Biology cum laude, Albion Coll., 1979; PhD in Biology, U. Colo., 1985. Lab. teaching asst. chemistry and biochemistry Albion (Mich.) Coll., 1977-79; teaching asst. U. Colo., Boulder, 1980-81, grad. teaching asst. molecular, cellular and devel. biology, 1980-85, supr. undergrad. honors rsch., 1984-85; postdoctoral rsch. assoc. complex carbohydrate ctr. U. Ga., Athens, 1985-86; rsch. fellow dept. molecular biology Mass. Gen. Hosp., 1986-89; rsch. fellow dept. genetics Harvard Med. Sch., Boston, 1986-89; asst. prof. plant biology, plant pathology Biotech. Ctr. Ohio State U., Columbus, 1989—; tchr. cellular biochemistry lab. Emmanuel Coll., Boston, 1987; mem. rev. panel Tobacco and Health Rsch. Inst., U. Ky., 1991-92, also NSF, Dept. Edn., USDA, Mont. State U.; lectr., presenter in field. Author: Ohio Science Workbook: Biotechnology, 1991; contbr. articles to profl. publs., chpts. to books. Grantee Ohio State U., 1990-91, Pioneer Hi-Bred Internat. Rsch. Grants Program, 1990-92, Nat. Inst. Gen. Med. Scis., 1991—, Midwest Plant Biotech. Consortium, 1992—, NSF, 1991—, U.S. EPA, 1991—. Mem. Am. Phytopathol. Soc. (mem. biochemistry, physiology and molecular biology com.), Am. Soc. Plant Physiologists, Internat. Soc. Plant Molecular Biology, Internat. Soc. Plant-Microbe Interactions. Office: Ohio State Biotech Ctr 1060 Carmack Rd Columbus OH 43210

DAVIS, KENNETH EARL, SR., environmentalist; b. Dallas, July 30, 1937; s. William Earl Davis and Ruby June (Wilson) Kelley; m. Lucille Marie Currey, Jan. 10, 1959 (div. Dec. 1980); children: Kenneth Earl, Deborah Darleen, Laura Denise, Michael Lynn; m. Priscilla Jane Herring, May 9, 1984. BS in Chemistry, U. Houston, 1965. Rsch. technologist Baroid Div. Nat. Lead Co., Houston, 1958-67; tech. rep. Baroid Div. Nat. Lead Industries, Houston, 1967-69; v.p. Subsurface Disposal Corp., Houston, 1969-78, pres., 1976-78; pres. Ken E. Davis Assocs., Inc., Houston, 1978-89; pvt. practice environ. cons. Houston, 1989—. Contbr. numerous articles to profl. jours. Technical advisor EPA, Washington, 1974, instr. and lectr., Washington and Atlanta, 1983, Phila. and San Francisco, 1984, Seattle, 1985. Mem. Nat. Assn. Corrosion Engrs., Nat. Water Well Assn., Underground Injection Practices Coun. (chmn. rsch. com. 1985-87, chmn. div. I tech. com. 1987-91). Republican. Roman Catholic. Avocations: hunting, fishing, boating, classical cars, traveling.

DAVIS, KENNETH LEON, psychiatrist, pharmacologist, medical educator; b. N.Y.C., Sept. 10, 1949; married, 1972; 2 children. BA, Yale U., 1969; MD, Mt. Sinai Med. Sch., 1973. Diplomate Am. Bd. Psychiatry and Neurology. Resident assist. dept. pharmacology Mt. Sinai Sch. Medicine, 1971-73, assoc. prof. psychiatry and pharmacology, 1979—; intern Stanford U., 1973-74, resident, 1973-76; life sci. rsch. assoc. Sanford U., 1975-76; clin. psychiat. cons. Santa Clara Valley Med. Ctr., 1976-79; chief dept. psychiat. VA Med. Ctr., 1979—; cons. Nat. County Mental Health Bd., 1969-73; dir., founder Stanford Comprehensive Care Clinic, Stanford Hosp., 1974-79; asst. dir. Stanford Psychiat. Clin. Rsch. Ctr. VA Med. Ctr., 1975-79, rsch. assoc.

1974-79, dir. schizophrenia biol. rsch. ctr., 1981—. Recipient A. E. Bennett Clin. Sci. Rsch. award, 1977, Saul Horowitz Jr. Meml. award, 1977-78, Solomon Silner award, 1981. Mem. Am. Psychiat. Assn., Soc. Biol. Psychiatry, Acad. Psychosomatic Medicine, Am. Neuropsychopharmacology, N.Y. Acad. Medicine. Achievements include research in the biological basis of senile dementia of the Alzheimers' type, depression, and schizophrenia. Office: Mount Sinai School of Medicine Alzheimers Disease Rsch Ctr 1 Gustave Levy Pl Box 1230 New York NY 10029*

DAVIS, KEVIN JON, chemical engineer, consulting engineer; b. Bristol, Pa., Dec. 12, 1963; s. Bertram George and Dorothy Ann (Rusinko) D.; m. Angela Amadio, Jul. 1, 1989; 1 child, Alexandra Nicole. BSCE, Drexel Univ., 1986. Registered profl. engr., Pa, Mass. Environ. engr. Monsanto Chem. Co., Bridgeport, N.J., 1984-85; production supr. Stonhard, Maple Shade, N.J., 1986-87; assoc. project engr. Roy F. Weston Inc., West Chester, Pa., 1987-90; sr. project engr. The Earth Tech. Corp., Cherry Hill, N.J., 1990—. Co-author: Types of Remedial Alternatives within SARA Mandates, 1991. Mem. Nat. Soc. Profl. Engrs. Office: The Earth Tech Corp Ste 316 53 Haddonfield Rd Cherry Hill NJ 08002

DAVIS, LAWRENCE WILLIAM, radiation oncologist; b. N. Braddock, Pa., Sept. 5, 1935; s. William Paul Davis and Julia Helen Zukas; children: James G., Karen E. BS, Juniata Coll., Huntington, Pa., 1957; MA, U. Pa., 1969; MBA, Temple U., 1984; MD, Georgetown U., 1961. Diplomate Am. Bd. Radiology (trustee 1981—); lic. physician Pa., Md., N.Y., Ga. Asst. instr. radiology U. Pa., Phila., 1962-66, instr. radiology, 1966, 68-69, asst. prof. radiology, 1969-72, assoc. prof. radiology, 1972-75; prof. radiation therapy Thomas Jefferson Sch. Medicine, 1975-84; prof. and chmn. radiation oncology Albert Einstein Coll. Medicine, Bronx, 1984-91, Emory U., Atlanta, 1991—; cons. Armed Forces Radiobiology Rsch. Inst., Bethesda, 1968-70; exec. com. of med. staff Montefiore Med. Ctr., 1984-87, 1990-91, div. coun., 1988-89; prof. svc. com. Phila. div. Am. Cancer Soc., 1970-75. Contbr. numerous articles to profl. jours.; assoc. editor Internat. Jour. Radiation Oncology, 1986—; editorial bd. Neuro Oncology, 1989—, assoc. editor, 1991—; editorial bd Am. Jour. Clin. Oncology, 1991—. Capt. USAF, 1966-68. Fellow Am. Cancer Soc., Phila., 1963-64, NIH, 1964-66, Am. Cancer Soc. traineeship, 1968-71. Fellow Am. Coll. Radiology; mem. AMA, AAAS, Am. Assn. Cancer Rsch., Am. Coll. Radiology (commn. on radiation oncology 1981-90, bd. chancellors 1993—), Am. Soc. Therapeutic Radiology and Oncology (chmn. bd. 1988-89, pres. 1987-88), Am. Coll. Hosp. Adminstrs., Am. Mgmt. Assn., Am. Radium Soc. (pres. 1992-93), Am. Soc. Clin. Oncology, Med. Assn. Atlanta, John Morgan Soc. of U. Pa., N.Y. Acad. Scis., Ga. State Med. Soc., Ga. State Radiol. Soc., Radiation Rsch. Soc., Radiol. Soc. N.Am., Alpha Omega Alpha. Office: Emory Clinic 1365 Clifton Rd NE Atlanta GA 30322

DAVIS, LESLIE SHANNON, research chemist, chemistry educator; b. Statesboro, Ga., Dec. 6, 1963; d. Larry Eugene and Marcia Anne (McClurd) D. BS in Chemistry cum laude, Ga. So. Coll., 1984; PhD in Inorganic Chemistry, U. Fla., 1988. Lab. technician Braswell Food Co., Statesboro, 1982; sr. rsch. chemist Monsanto Chem. Co., Pensacola, Fla., 1988—; mentor biotech. Monsanto-Escambia County Sch. Dist., Pensacola, 1990—; adj. prof. chemistry Pensacola Jr. Coll., 1989—. Contbr. articles to profl. jours.; patentee in field. Program chair Expanding Your Horizons, Pensacola, 1990-91. Ty Cobb Found. scholar, 1982-84. Mem. AAUW (chair pub. policy com. 1991), Am. Chem. Soc. (chair-elect 1992, chair 1993), Pensacola Women's Alliance. Office: Monsanto Chem Group PO Box 12830 Pensacola FL 32575

DAVIS, MARK AVERY, ecologist, educator; b. Sturgeon Bay, Wis., June 25, 1950; s. Nathan Smith and Naomi (Nichols) D.; m. Jean Marie Hanske; children: Hazel Dean, Benjamin Maitland, Zachary Nathan. AB, Harvard Coll., 1972, EdM, 1974; PhD, Dartmouth Coll., 1981. Asst. prof. biology Macalester Coll., St. Paul, 1981-87; assoc. prof. Macalester Coll., 1987-92, prof., 1992—. Mem. AAAS, Ecol. Soc. Am., Am. Inst. Biol. Scientists, Soc. Conservation Biology. Office: Macalester Coll Dept Biology 1600 Grand Ave Saint Paul MN 55105

DAVIS, MARK M., microbiologist, educator. Prof. microbiology and immunology Stanford U.Med. Sch., Stanford, Calif.; investigator Howard Hughes Med. Inst. Recipient Passano Found. award, 1985, Gairdner Found. award, 1989, Nat. Acad. Scis., 1993. Office: Stanford U Sch Medicine Howard Hughes Med Inst Beckman Bldg Stanford CA 94305

DAVIS, PAMELA BOWES, pediatric pulmonologist; b. Jamaica, N.Y., July 20, 1949; d. Elmer George and Florence (Welsch) Bowes; m. Glenn C. Davis, June 28, 1970 (div. Mar. 1987); children: Jason, Galen. AB, Smith Coll., 1968; PhD, Duke U., 1973, MD, 1974. Internal medicine intern Duke Hosp., 1973-74, resident in internal medicine, 1974-75; sr. investigator NIAMD/NIH, Bethesda, Md., 1977-79; asst. prof. U. Tenn. Coll. Medicine, Memphis, 1979-81; asst. prof. Case Western Res. U. Sch. Medicine, Cleve., 1981-85, assoc. prof., 1985-89, prof., 1989—, chief, pediatric pulmonary div., 1985—; pres. Am. Fedn. for Clin. Rsch., Thorofare, N.J., 1989-90; trustee Rsch Am., Arlington, Va., 1989-90; mem. adv. coun. Nat. Inst. Diabetes, Digestive and Kidney Diseases, 1992—. Contbr. articles to profl. jours. Chmn., med. adv. coun. Cystic Fibrosis Found., Bethesda, 1988-90. Fellow Am. Coll. Physicians; mem. Am. Physiol. Soc., Am. Thoracic Soc., Soc. for Pediatric Rsch., Phi Beta Kappa, Sigma Xi. Office: Rainbow Babies/Child Hosp 2101 Adelbert Rd Cleveland OH 44106-2624

DAVIS, PATRICIA MAHONEY, software engineer; b. Pitts., Dec. 26, 1957; d. John Francis and Lillian Rosemary (Peck) Mahoney; m. Larry Allen Davis, Dec. 1, 1989; 1 child, Mark Benjamin Mahoney. BS, Towson State U., 1980; postgrad., U. Md., 1980-83, Johns Hopkins U., 1993—. Computer programmer FBI, Washington, 1984-88; software engr. Quality Systems, Inc., Tysons Corner, Va., 1988-89; sr. software engr. Martin Marietta Corp., Washington, 1989—. Vol. Big Bros. and Big Sisters, Balt., 1982. Recipient Md. State Senatorial scholarship, 1981. Mem. Assn. for Computing Machinery, Nat. Student Speech and Hearing Assn., Omicron Delta Kappa (Student Leader of Yr. 1981). Republican. Roman Catholic. Avocations: horseback riding, swimming. Home: 9610 Sparrow Ct Ellicott City MD 21042-1773

DAVIS, PHILLIP EUGENE, oil company executive, chemical engineer; b. Ft. Wayne, Ind., June 24, 1933; s. Ora Merle and Alice Louise (Fox) D.; m. Patsy Ann Smith, July 28, 1985; 1 child, Alex Steven. B.S. in Chem. Engring., U. Mich., 1955; postgrad. Ind. State U., 1965-67, Marylhurst Coll., Portland, Oreg., 1979—. Registered profl. engr., Tex. Chemist, tech. engr., prodn. engr. Shell Chem. Corp., Deer Park, Tex., 1955-62; project engr., prodn. engr., plant engr. Velsicol Chem. Co., Marshall, Ill., 1962-70; project engr. Rhone Poulenc, Inc., New Brunswick, N.J., 1970-76, successively plant engr., prodn. supt., tech. mgr., Portland, Oreg., 1976-80; chief engr. Houston plant Baker Performance chems. div. Baker Internat., 1980-82, div. engr. mfg. div., 1982-92, site engr. constrn. Baytown Polymers Ctr. Exxon Chems., 1991—, also cons. Served with AUS, 1955-57; maj. M.P. Res., 1957-76. Mem. Am. Inst. Chem. Engrs., Am. Inst. Plant Engrs., Res. Officers Assn. U.S., Alumni Assn. U. Mich. Republican. Mem. Christian and Missionary Alliance. Home: 1308 Town Cir Baytown TX 77520-3431 Office: 5200 Bayway Dr Baytown TX 77522

DAVIS, RANDY L., soil scientist; b. L.A., Nov. 23, 1950; s. Willie Vernon and Joyce Catherine (Manes) D. AA, Yuba Community Coll., 1972; BS in Soils and Plant Nutrition, U. Calif., Berkeley, 1976. Vol. soil scientist U.S. Peace Corps, Maseru, Lesotho, 1976-79; soil scientist Hiawatha Nat. Forest, Sault Saint Marie, Mich., 1979-86; forest soil scientist Bridger-Teton Nat. Forest, Jackson, Wyo., 1986—; detailed soil scientist Boise (Idaho) Nat. Forest, 1989, 92. Editor Soil Classifiers newsletter; contbr. articles to profl. jours. Pres. Sault Community Theater, Sault Saint Marie, 1984-86. Mem. Am. Chem. Soc., Soil Sci. Soc. Am., Soil and Water Conservation Soc. Am. (bd. dirs. 1991-92, chpt. pres. 1993), Am. Water Resources Assn., Internat. Soc. Soil Sci., Soc. for Range Mgmt. Methodist. Home: PO Box 7795 Jackson WY 83001-7795 Office: Bridger-Teton Nat Forest PO Box 1888 Jackson WY 83001-1888

DAVIS, RAYMOND, JR., chemist, researcher, educator; b. Washington, Oct. 14, 1914; s. Raymond and Ida Rogers (Younger) D.; m. Anna Marsh Torrey, Dec. 4, 1948; children: Andrew Morgan, Martha Safford, Nancy Elizabeth, Roger Warren, Alan Paul. BS, U. Md., 1937, MS, 1939; PhD, Yale U., 1942; DSc, U. Pa., 1990. Chemist Dow Chem. Co., Midland, Mich., 1938-39, Monsanto Chem. Co., Dayton, Ohio, 1946-48; sr. chemist Brookhaven Nat. Lab., Upton, N.Y., 1948-84; Brookhaven Nat. Lab.; research prof. dept. astronomy U. Pa., Phila., 1984—. Contbr. articles to profl. jours. Served with USAAF, 1942-46. Recipient Boris Prejel prize N.Y. Acad. Scis., 1955, award for nuclear applications in chemistry Am. Chem. Soc., 1979. Mem. AAAS, NAS (Comstock prize 1978), Am. Phys. Soc. (Tom W. Bonner prize 1988, W.K.H. Panofsky prize 1992), Am. Geophys. Union, Am. Astron. Soc. Office: U Pa Dept Astronomy Philadelphia PA 19104

DAVIS, ROBERT DRUMMOND, SR., chemist, researcher; b. Durham, N.C., Nov. 15, 1955; s. James Pleasant Jr. and Elaine Patricia (Rochelle) D.; m. Verma Ellen Hamilton, Aug. 25, 1976; children; Robert Jr., Karen, Evelyn. BS in Organic Chemistry, U. N.C., 1976, PhD in Organic Chemistry, 1982. Group leader Reilly Industries, Inc., Indpls., 1982—. Achievements include patent in high temperature process for selective production of 3-methylpyridine. Office: Reilly Industries Inc Indianapolis IN 46242

DAVIS, ROBERT GLENN, research scientist; b. Pitts., May 16, 1951; s. Glenn Ruthven and Roberta (McClurg) D. BS in Chemistry, Pa. State U., 1973; PhD in Organic Chemistry, Purdue U., 1979. Postdoctoral assoc. U. Vt., Burlington, 1979-81; asst. prof. chemistry Temple U., Phila., 1981-84; editor, indexer Inst. Sci. Info., Phila., 1984-85; sr. rsch. chemist Uniroyal Chem. Co., Inc., Middlebury, Conn., 1985—. Active The Nature Conservancy, Middletown. John and Elizabeth Holmes Teas scholar Pa. State U., University Park, 1973. Mem. AAAS, Am. Chem. Soc. (chmn. New Haven sec. 1990-91), Phi Lambda Upsilon, Phi Kappa Phi, Nat. Honor Soc., Sigma Xi. Achievements include 5 patents; discovery of herbicide Pantera; 2 patents pending for herbicide synthesis. Home: 151 Andrew Ave Apt 49 Naugatuck CT 06770-4327 Office: Uniroyal Chem Co Inc R-G-40 Benson Rd Middlebury CT 06749-0002

DAVIS, ROBIN L., pharmacy educator; b. Mar. 18, 1953. PharmD, U. Wash., 1984. Pharmacy resident U. Wash., Seattle, 1982-84, infectious disease fellow, 1985; asst. prof. Coll. Pharmacy U. N.Mex., Albuquerque, 1985-91, assoc. prof. Coll. of Pharmacy, 1991—; mem. infectious disease adv. panel Annals of Pharmacotherapy, 1989—; presdl. lectr. U. N.Mex., 1988-90. Contbr. articles to profl. jours. Recipient Superior Achievement in Clin. Pharmacy award Smith, Kline & French, 1982. Mem. Soc. Hosp. Pharmacists (infectious diseases specialty practice gorup), Soc. Infectious Disease Pharmacists, Am. Assn. Colls. of Pharmacy. Office: U NMex Coll Pharmacy 2502 Marble NE Albuquerque NM 87131

DAVIS, ROGER L., aeronautical engineer. BS in Aero. Engring., Ohio State U., 1974, MS in Aero. Engring., 1975; PhD in Mech. Engring., U. Conn., 1982. Sr. analytical engr. Pratt and Whitney, East Hartford, Conn., 1975-81; with United Techs. Rsch. Ctr., East Hartford, 1983—, sr. rsch. engr. theoretical and computational fluids dynamics. Contbr. over 20 articles to profl. jours. Mem. ASME (Gas Turbine award 1990), AIAA. Office: Theoretical & Computational United Tech Research Ctr East Hartford CT 06108*

DAVIS, RUTH C., pharmacy educator; b. Wilkes-Barre, Pa., Oct. 27, 1943; d. Morris David Davis and Helen Jane Gillis. BS, Phila. Coll. Pharmacy and Sci., 1967. Cert. pharmacist, Pa., Md. Mgr. pharmacist Fairview Pharmacy, Etters, Pa.; mgr., pharmacist Neighborcare Pharmacy, Balt.; dir. ambulatory svcs. Rombro Health Svcs., Balt.; tchr., pharmacist Boothwyn Pharmacy, Phila. Republican. Baptist. Avocations: training and raising American quarter horses, music, reading. Home and Office: 75 Lion Dr Hanover PA 17331-3847

DAVIS, STEPHAN ROWAN, aerospace engineer; b. Athens, Ohio, Aug. 21, 1960; s. Barbara Ann (Zeigler) Ramsay; m. Bonnie Ellen Bennett, Aug. 3, 1985; children: Brooke, Virginia Ann. BS in Math., Siena Coll., 1982; MS in Engring. Sci., U. Tenn., 1985; MBA, Ala. A&M, 1991. Sr. engr. Teledyne Brown Engring., Huntsville, Ala., 1985-89; project engr. Marshall Space Flight Ctr. NASA, Huntsville, 1989—. Named an Outstanding Young Huntsvillian, Jaycees 1988. Mem. AIAA, Huntsville Archaeol. Soc. Office: NASA EJ 23 Marshall Space Flight Ctr Huntsville AL 35812

DAVIS, TERRY LEE, communications systems engineer; b. Enid, Okla., Aug. 18, 1950; s. Walter Joseph and Bessie Lee (McDaniel) D.; m. Jennie Sue Petrik, Jan. 21, 1972; children: Mistie Rae, Brandon Scott. BSCE, Okla. State U., 1972. Registered profl. engr., Okla., Colo., Wash. Civil engr. U.S. Army Corps of Engrs., Webbers Falls, Okla., 1973, Omaha, 1974; water resources engr. U.S. Bur. Reclamation, Grand Junction, Colo., 1974-75, Amarillo, Tex., 1974-75, Montrose, Colo., 1976; water resources engr. U.S. Bur. Reclamation, AID, Dubai, United Arab Emirates, 1977-78; communication and control systems engr. Western Area Power Adminstrn., U.S. Dept. Energy, Loveland, Colo., 1979-84, Boeing, Seattle, 1984—. Mem. Issaquah (Wash.) Devel. Commn., 1987—, Issaquah Rivers and Streams Bd., 1985-86, Issaquah Basin Planning Team, 1990-92. Mem. Masons (jr. warden lodge 108 1993—), Order Eastern Star (chaplain century chpt. 66). Republican. Office: Boeing Computer Svcs PO Box 3707 Seattle WA 98124-2207

DAVIS, THOMAS PINKNEY, mathematics educator; b. Seminole, Okla., Oct. 10, 1956; s. George Pinkney and Flora Elizabeth (Bollinger) D.; m. Leslie Anne Workman, Jan. 26, 1990; children: Brianna Elizabeth, Mary Katherine; stepchildren: Christopher, Jennifer, Matthew, Jacob, Joshua Beene. BS with Honors, East Cen. U., Ada, Okla., 1979, BA with Honors, 1979. Dir. math. lab. East Cen. U., 1991-92; tchr., chair math. dept. Roosevelt (Okla.) High Sch., 1992-93; tchr. math. Keota (Okla.) High Sch., 1993—. Reviewer Sci. Books and Films, 1986—. Fellow Brit. Interplanetary Soc.; mem. AIAA, Am. Astronautical Soc., Asns. Lunar and Planetary Observers, Alpha Chi, Pi Gamma Mu. Republican. Episcopalian. Home: Box 4112 Rt 2 Stigler OK 74462-9633 Office: Keota High Sch PO Box 160 Keota OK 74941

DAVIS, THOMAS WILLIAM, college administrator, electrical engineering educator; b. Belvidere, Ill., Mar. 14, 1946; s. Thomas William and Charlotte Ann (Schildgen) D.; m. Lyndel Etta Schuettpelz, Apr. 3, 1971; 1 child, Bryan William. BSEE, Milw. Sch. Engring., 1968; MSEE, U. Wis., Milw., 1971. Registered profl. engr., Wis. From asst. prof. to assoc. prof. elec. engring. Milw. Sch. Engring., 1971-75, head computer engring. tech., 1975-77, prof. 1976—, chmn. dept. elec. engring., 1977-84, dean rsch., 1981-84, dean acads. and rsch., 1984-87, v.p. academics, dean faculty, 1987-89, sr. v.p. 1989—; lectr. U. Wis., Milw., 1973-76. Author: Problems in Measurements, 1968, (textbooks) Computer Aided Analysis, 1973, Introduction to Interactive Programs, 1978, Experimentation with Microprocessor Applications, 1981; patentee in field. Warning and communications officer Ozaukee County Emergency Govt., Wis., 1981-82; sgt. reserves Grafton Police Dept., Wis., 1976—; corp. mem. Curative Rehab. Ctr, Gov.'s Quality Improvement Task Force, Gov.'s Sci. and Tech. Coun.; bd. dirs. Jordan Controls, Inc., Jagemann Stamping, Michaels Machine Co., Mechanical Industries Inc.; past pres. Milw. Coun. Engring. and Sci. Socs. Mem. IEEE (sr., student activity dir. 1972-73), Engrs. and Scis. Milw. (pres.), Robotics Internat., Soc. Mfg. Engrs. (sr.), Assn. Computing Machinery, Am. Soc. Engring. Edn. (membership policies com.), Milw. Sch. Engring. Alumni Assn. (achievement award 1968), Phi Kappa Phi, Tau Alpha Pi, Eta Kappa Nu. Avocation: flying, amateur radio. Home: 5590 Gray Log Ct Grafton WI 53024-9622 Office: Milw Sch Engring 1025 N Broadway Milwaukee WI 53202-3109

DAVIS, WALTER BARRY, quality control professional; b. New London, Conn., July 10, 1942; s. Luna Alonzo and Mary Elizabeth (Shell) D.; 1 child, Alexandra Elizabeth; m. Carol Michael Fairchild, Nov. 24, 1984. Student, U. Utah, 1959-60; BS in Physics, U.S. Naval Acad., 1964; MA in Bus., Cen. Mich. U., 1978. Cert. nuclear plant operator, nuclear weapons mgmt. Commd. ensign USN, 1964, advanced through grades to comdr., 1978; exec. officer Nuclear Submarine Repair Ship, Kings Bay, Ga., 1981-83; ep. head missile br. Strategic Systems Program Office, Washington, 1983-85; ret., 1985; program mgr. missile chaff Tracor Aerospace, Austin, Tex., 1985-86, dir. mfg. stratetic Counter Measures, 1986-89, program mgr. indsl. modernization, 1989-92, mgr. employee programs, 1991-92; program mgr. supplier quality Intertek Svcs., Fairfax, Va., 1992—; presenter in field. Examiner Am. Radio Relay League, Austin, 1990-92. Mem. AIAA, Assn. for Mfg. Excellence, Soc. for Mfg. Engrs. Home: 401 Meadow Creek Dr Pflugerville TX 78660 Office: Intertek Svcs 9900 Main St Ste 500 Fairfax VA 22031

DAVIS, WILLIAM CHARLES, veterinary microbiology educator; b. Red Bluff, Calif., Feb. 12, 1933; s. Maurice Trow and Emily Ann (Newton) D.; m. June 18, 1956; children: Jennifer Elizabeth, Valerie Anne. BA, Chico State Coll., 1955; PhD, Stanford U., 1967. Fellow U. Calif./San Francisco Med. Ctr., 1966-68; from asst. prof. to prof. dept. vet. microbiology Coll. Vet. Medicine, Wash. State U., Pullman, 1968—; program mgr. USDA Competitive Grants, Washington, 1985-86. With U.S. Army, 1956-58. Recipient many grants. Achievements include research in monoclonal antibody technology and flow cytometry to characterize the immune system in domestic animals. Office: Wash State Univ Dept Vet Microbiology Pullman WA 99164-7040

DAVIS-BRUNO, KAREN L., pharmacologist, researcher; b. Cornwall, N.Y., June 1, 1961; d. Arthur Palmer and Audrey Ann (Butwell) Davis; m. Robert Stephen Bruno, May 24, 1991. BS in Biology, Fordham U., 1983; MS in Pharmacology, N.Y. Med. Coll., 1990, PhD in Pharmacology, 1991. Clin. lab. technologist Sloan-Kettering Cancer Ctr., N.Y., 1983-84; rsch. fellow N.Y. U. Med. Ctr., 1984-85; rsch. tech. Dept. Medicine N.Y. Med. Coll., Valhalla, N.Y., 1985-87, rsch. fellow, 1987-91; post doctoral fellow Med. U. S.C, Charleston, 1991-93; staff scientist Progenics Pharmaceuticals Inc., Tarrytown, N.Y., 1993—. Contbr. articles to Jour. Biological Chemistry, 1988, Experimental Eye Research, 1991, Investigative Ophthalmology and Experimental Eye Research, 1989-91. Vol. N.Y. State EMT, Town of Newburgh (N.Y.) Vol. Ambulance Corps, 1986-91. Mem. AAAS, N.Y. Acad. Sci. Office: Progenics Pharmaceuticals Inc 777 Old Saw Mill River Rd Tarrytown NY 10591

DAVISON, ARTHUR LEE, scientific instrument manufacturing company executive, engineer; b. Burlington, Iowa, May 8, 1936; s. John Earl and Helen Medera (Jones) D.; m. Dorothea Ellen Jones, June 14, 1958; children: Ken, Ron, Greg. BA, Monmouth Coll., 1958; MS, Purdue U., 1960. Registered profl. engr., Calif. R & D engring. physicist Baird Corp., Bedford, Mass., 1960-65; rsch. engr. Bethlehem (Pa.) Steel, 1965-69; project engring. mgr. Applied Rsch. Labs., Sunland, Calif., 1969-79; sr. v.p., gen. mgr. Applied Rsch. Labs., Valencia, Calif., 1985-91; dir. engring. Berkey Colortron, Burbank, Calif., 1979-80; pres., chief exec. officer Labtest Equipment Co., L.A., 1981-85; pres. Kevex Instruments, San Carlos, Calif., 1991-92; pres. X-ray div. Fisons Instruments, Santa Clarita, Calif., 1992—. Bd. dirs. Valencia Indsl. Assn., 1988—; vol. worker Boy Scouts Am., various youth groups. Mem. Am. Optical Soc., Am. Phys. Soc., Soc. Applied Spectroscopy. Democrat. Methodist. Office: Kevex Instruments 24911 Avenue Stanford Valencia CA 91355

DAVISON, GLENN ALAN, podiatric surgeon; b. Newark, Sept. 18, 1963; s. Phillip Davison and Arlene Joan (Schwartz) Dubrovsky; m. Anne Frances Zebrowski, Oct. 7, 1990. BA, Skidmore Coll., 1985; D Podiatric Medicine, N.Y. Coll. Podiatric Medicine, 1989. Lic. in podiatric medicine, N.J., N.Y. Resident in podiatric medicine and surgery Union (N.J.) Hosp., 1989-90; assoc. in podiatry Hillside, N.J., 1990-91; pvt. practice Elizabeth, N.J., 1991—; clin. dir. Foot Care Clinic; mem. med. staff Union Hosp., Elizabeth Gen. Med. Ctr., Runnells Hosp., St. Elizabeth Hosp.; mem. intern and resident tng. com., preceptor of podiatry out-patient tng., mem. resident teaching staff, podiatric continuing med. edn. credit adminstr., mem. podiatric core faculty Union Hosp.; guest lectr. various community hosps., nursing homes, civic orgns.; participant pedal screening programs and local health fairs. Cable TV appearances include Hello Cranford, 1990, Union Hosp. Health Scenes, 1991; contbr. article on foot surgery to profl. jour., numerous articles on diabetic foot care and common pedal ailments to newspapers. Active, com. mem. Shoes for the Homeless project, N.J., 1991. Recipient 1st prize OUM Group Podiatric Writing Contest, 1990, PICA Podiatric Writing Contest, 1990. Mem. Am. Podiatric Med. Assn., N.J. Podiatric Med. Soc. (cert. of achievement 1991, treas. exec. com., v.p. ea. div.), Am. Coll. Foot Surgeons, Am. Podiatric Med. Postgrad. Soc., Alumni Assn. of N.Y. Coll. Podiatric Medicine. Office: 1308 Morris Ave Union NJ 07083

DAVISSON, MURIEL TRASK, geneticist; b. Tremont, Maine, Apr. 19, 1941; d. Charles Orville and Esther (Moore) Trask; m. Farrell Robert Davisson, Apr. 20, 1966; 1 child, Sven. AB cum laude, Mount Holyoke Coll., 1963; PhD in Genetics, Pa. State U., 1969. Rsch. assoc. The Jackson Lab., Bar Harbor, Maine, 1971-80, assoc. staff scientist, 1980-85, staff scientist, 1985-92, sr. staff scientist, 1992—. Assoc. editor Genomics jour., Jour. of Heredity. Mem. AAAS, Genetics Soc. Am., Am. Genetics Assn., Am. Soc. Human Genetics. Office: The Jackson Lab 600 Main St Bar Harbor ME 04609-1500

DAVOODI, HAMID, mechanical engineering educator; b. July 31, 1959; s. Djavid Davoodi and Mansoureh Navidi-Kasmaei. MS, Worcester Polytech. Inst., 1983, PhD, 1989. Scientific programmer Morgan Constrn. Co., Worcester, Mass., 1989; asst. prof. Mech. Engring. U. P.R., Mayaguez, 1989—; co-organizer short course on random vibration Worcester Polytechnic Inst., 1987. Contbr. articles to profl. jours. Active Animal Protection Assn., Mayaguez. Mem. Am. Acad. Mechs., Sigma Xi, Pi Tau Sigma, Sigma Xi, Tau Beta Pi.

DAVY, MICHAEL FRANCIS, civil engineer, consultant; b. Springfield, Mo., Mar. 24, 1946; s. Philip Sheridan and Caecilia Magdelen (Thiemann) D.; m. Joyce Kay Young, Aug. 17, 1968; children: Mark Sheridan, Katherine Ann, Jennifer Mary. BS, U. Wis., 1969. Diplomate Am. Acad. Environ. Engrs. Project engr. Davy Engring. Co., La Crosse, Wis., 1969-74; v.p. Davy Engring. Co., La Crosse, 1975-88; mgr. Davy Labs., La Crosse, 1975—; pres. Davy Engring. Co., La Crosse, 1989—. Bd. dirs. Gateway Area Coun. Boy Scouts Am., La Crosse, 1973— (pres. exec. bd., 1989-91); mem. Gov.'s Clean Water Task Force, 1988-89. Disting. Eagle Scout, 1988, Silver Beaver Gateway Area Coun., 1987. Mem. NSPE (nat. bd. dirs. 1987-93), ASCE (Young Engr. Yr. 1980), Wis. Soc. Profl. Engrs. (Engr. Yr. 1987, pres. 1984-85, sec. 1980-82, Young Engr. Yr. 1976), Wis. Assn. Consulting Engrs. (bd. dirs. 1987-90), Profl. Engrs. in Pvt. Practice (vice chmn. 1981-83, Merit award 1990). Roman Catholic. Avocations: swimming, boating. Home: 615 N 23rd St La Crosse WI 54601 Office: Davy Engring Co 115 S 6th St La Crosse WI 54601

DAWDY, DAVID RUSSELL, hydrologist; b. San Antonio, July 1, 1926; s. Gladys Bird Johnston Dawdy; m. Doris Ostrander, Feb. 21, 1951. BA in History, Trinity U., San Antonio, 1948; MS in Statistics, Stanford U., 1962. Cert. hydrologist. Rsch. hydrologist U.S. Geol. Survey, various locations, 1951-76; chief hydrologist Dames & Moore, Bethesda, Md., 1976-80, Nortec, Irvine, Calif., 1980-82; pvt. hydrology cons. San Francisco, 1982—; adj. prof. civil engring. U. Miss., Oxford, 1978-87; cons. to AID, 1972, Govt. of India, 1980, World Bank, 1983, Nat. Inst. Hydrology, U. Roorkee, India, 1983, Govt. of Colombia, 1978, 80, World Meteorol. Orgn., Guatemala, 1978, Switzerland, 1969, Govt. of Venezuela, 1978, 82; mem. numerous ad hoc coms. Nat. Acad. Scis./NRC, other fed. agys.; lectr. in field. Contbr. numerous articles to profl. jours. With U.S. Army, 1943-46. Recipient Meritorious Svc. award U.S. Dept. Interior, 1969. Fellow Am. Geophys. Union; mem. ASCE, Am. Inst. Hydrology, Internat. Assn. Hydrol. Scis. (sec. commn. on water resources rels. 1972-80, chmn. U.S. nat. com. 1980-84), Internat. Union of Geodesy and Geophysics (U.S. nat. com. 1980-84). Democrat. Home and Office: 3055 23d Ave San Francisco CA 94132

DAWICKI, DOLORETTA DIANE, research biochemist, educator; b. Fall River, Mass., Sept. 13, 1956; d. Walter and Stella Ann (Olszewski) D. BS, S.E. Mass. U., 1978; PhD, Brown U., 1986. Rsch. assoc. Meml. Hosp. R.I., Pawtucket, 1986-92; asst. prof. Brown U., Providence, 1986—; asst. prof. med. rsch. VA Med. Ctr., Providence, 1992—. Contbr. articles to profl. jours. Mem. AAAS, Am. Soc. for Biochemistry and Molecular Biology, Sigma Xi. Achievements include research on in vivo antiplatelet mechanism of action of the clinical agent dipyridamole, the role of tyrosine phosphorylation in controlling platelet physiology. Home: 1201 S Main St Fall River MA 02724-2753 Office: VA Med Ctr Pulmonary Sect 83 Chalkstone Ave Providence RI 02908

DAWID, IGOR BERT, biologist; b. Czernowitz, Romania, Feb. 26, 1935; came to U.S., 1960, naturalized, 1977; s. Josef and Pepi (Druckmann) D.; m. Keiko Naito Ozato, Apr. 5, 1976. Ph.D., U. Vienna, 1960. Fellow dept. biology MIT, 1960-62; fellow dept. embryology Carnegie Instn. of Washington, Balt., 1962-66; mem. staff Carnegie Instn. of Washington, 1966-78; chief devel. biochemistry sect. Lab. Biochemistry, Nat. Cancer Inst., Bethesda, Md., 1978-82; chief lab. molecular genetics Nat. Inst. Child Health and Human Devel. (NIH), Bethesda, 1982—; vis. scientist Max Planck Inst. for Biology, 1964-67; asst. prof. to prof. dept. biology Johns Hopkins U., 1967-78. Editor: Devel. Biology, 1971-75, Cell, 1977-80; editor-in-chief: Devel. Biology, 1975-80, adv. editor, 1980-85. Mem. NAS, AAAS, Am. Soc. Biol. Chemists, Am. Soc. Cell Biology, Soc. Devel. Biology, Internat. Soc. Devel. Biologists. Office: NIH Bldg 31 9000 Rockville Pike Bethesda MD 20892-0001

DAWN, FREDERIC SAMUEL, chemical, textile engineer; b. Shanghai, Republic of China, Nov. 24, 1916; s. Keith Frederic and Paula (Yui) D.; m. Marie Dunn; children: Robert, William, Victoria. BS in Chemistry summa cum laude, China Inst. Tech., 1936; MS in Textile Engring., U. Lowell and N.C. State U., 1938, 39; PhD in Chemistry, Nat. Inst. Tech., 1948; DSc in Chemistry with honors, China Inst. Tech., 1967. Prof. China Inst. Tech., Shanghai, 1939-49; dir. rsch. China Textile Ind., Inc., Shanghai, 1945-49; postdoctoral rsch. fellow and assoc. U. Wis., Madison, 1950-55; dir. rsch. Decar Plastic Corp., Madison, 1955-60; supervisory rsch. engr. USAF, Aeronautical Systems Div., Wright Patterson AFB, 1960-62; chief matls. rsch. lab. NASA Manned Spacecraft Ctr., Houston, 1962-76; dir. adv. matls. rsch. and devel. NASA Johnson Space Ctr., Houston, 1976-89, chief engr., 1989—; adv. bd. Nat. Elec. Mfrs. Assn., N.Y.C., 1956-60. Tech. adv. com. Jour. Indsl. Fabrics, 1985-87; contbr. articles to profl. jours. Bd. dirs. Houston Sister City Soc., 1970—; mem. Presdl. Task Force, Washington, 1982—. Named to Space Tech. Hall of Fame, U.S. Space Found., 1989; recipient numerous awards various govt. agy., profl. and community orgns., 1961—, Exceptional Engring. Achievement medal NASA, 1984, Sci. medal, 1991, Tech. award, 1991. Fellow NSPE, Am. Inst. Chemists, AIAA; mem. Am. Chem. Soc. (sr. mem.), Phi Lambda. Achievements include research in high-temperature and flame resistant materials; development of nonflammable Beta fibers; development and design of high temperature, flame resistant and thermal/micrometeoroid protective polymeric materials and coatings for Apollo, Space Shuttle and Advanced extravehicular space suit, intravehicular flight suit, related flight equipment and spacecraft thermal protection system; patents in field. Home: 1615 Richvale Ln Houston TX 77062-5420

DAWOOD, MOHAMED YUSOFF, obstetrician/gynecologist; b. Singapore, Singapore, Sept. 13, 1943; came to U.S., 1974; s. Sheikh and Fatimah (Hussein) D.; m. Firyal Sultana Khan, July 14, 1978; children: Fatimah Sultana, Fauzia Sultana, Firdaus Sultana, Hassan Yusoff. MB, ChB, U. Sheffield, Yorkshire, Eng., 1968, MD; M of Medicine, U. Singapore, 1972. Diplomate Am. Bd. Obstetrics and Gynecology, Am. Bd. Reproductive Endocrinology. First asst. in ob-gyn. U. Melbourne, 1974; from instr. to assoc. prof. ob-gyn. Cornell U. Med. Coll., N.Y.C., 1974-79; prof. ob-gyn. U. Ill., Chgo., 1979-90, U. Tex. Med. Sch., Houston, 1990—; lectr. U. Singapore, 1973-74; cons., editorial cons., reviewer in field. Author: Green's Gynecology, 1990, Dysmenorrhea, 1981, Premenstrual Syndrome and Dysmenorrhea, 1985, Oxytocin, vol. 2, 1984, Prostaglandin Inhibition in Obstetrics and Gynecology, 1983; contbr. articles to profl. jours. Recipient Gold medal Jr. C. of C. Singapore, 1973. Fellow ACS, ACOG, Am. Gynecol. & Obstet. Soc., Royal Coll. Ob-Gyn. (Edgar Gentilli prize 1974, Gold medal 1973); mem. Endocrine Soc. Achievements include research in prostaglandins in the causation of menstrual cramps and relief by blocking prostaglandins; role of oxytocin in human partuition, bone-depleting effect of GnRH agonists during treatment of endometriosis; presence of neurohypophyseal peptides in primate and human ovaries. Office: Univ Texas Medical School 6431 Fannin Ste 3.204 Houston TX 77030

DAWSON, BRIAN ROBERT, chemist, environmentalist; b. Detroit, June 4, 1947; s. Alan Donald and Jean Helen (Oelschlegel) D.; m. Faye Ann Kindle, Dec. 18, 1973; children: Melissa Ayn, Marc Jarvis. Student, U. Toledo, 1964-69. From lab. technician to plant mgr. Inland Chem. Corp., Toledo, Orange, Calif., 1967-76; plant mgr. Inland Chem. Corp., Newark, 1976-78; area sales rep. Inland Chem. Corp., Toledo, 1978-80; v.p., gen. mgr. Inland Chem. Corp., Manati, P.R., 1980-81; gen. mgr. McKesson Envirosystems, Manati, 1981-83; mktg. mgr. McKesson Envirosystems, Ft. Wayne, Ind., 1983-84, resource mgr., 1984-86; gen. mgr. Chem. Waste Mgmt., West Carrollton, Ohio, 1986-88; bus. devel. mgr. Chem. Waste Mgmt., Princeton, N.J., 1988-89; v.p. ops. and new project devel. Cemtech, Westchester, Ill., 1989-91; v.p. sales and mktg. bus. devel., 1991-93; speaker in field. Contbr. articles to profl. jours. With U.S. Army, 1969. Grantee U.S. EPA, 1987, Goskompriroda, USSR, 1990. Mem. ASTM, Nat. Assn. Chem. Recyclers (v.p. 1990-92, bd. dirs. 1992-93, exec. com. 1990-92), Cement Kiln Recycling Coalition (bd. dirs. 1990-91, sec. 1991-92), Am. Mensa, Brown's Run Country Club, Alpha Sigma Phi. Home: 8486 Point O Woods Ct Springboro OH 45066-9600

DAWSON, CHANDLER R., opthalmologist, educator; b. Denver, Aug. 24, 1930; married; 3 children. AB, Princeton U., 1952; MD, Yale U., 1956. USPHS epidemiologist Communicable Disease Ctr., 1957-60; resident ophthalmologist Sch. Medicine U. Calif., San Francisco, 1960-63, asst. clin. prof., 1963-66, asst. rsch. prof., 1966-69, assoc. prof. in residence, 1969-75, prof. ophthalmology, 1975—; fellow Middlesex Hosp. Med. Sch., London, 1963-64; dir. WHO Colloborative Ctr. Prevention Blindness & Trachoma; assoc. dir. Francis I. Proctor Found., 1970—. Recipient Knapp award AMA, 1967, 69, Medaille Trachome, 1978. Mem. Am. Soc. Microbiology, Am. Acad. Ophthalmology & Otolaryngology, Assn. Rsch. Vision & Ophthalmology. Achievements include rsch. in epidemiology of infectious eye diseases and cataracts; prevention of blindness; pathogenesis of virus diseases of the eyes; electron microscopy of eye diseases. Office: U Calif Francis I Proctor Found Rsch Opthalmology San Francisco CA 94143-0412*

DAWSON, EARL BLISS, obstetrics and gynecology educator; b. Perry, Fla., Feb. 1, 1930; s. Bliss and Linnie (Calliham) D.; BA, U. Kans., 1955; student Bowman Gray Sch. Medicine, 1955-57; MA, U. Mo., 1960; PhD, Tex. A. & M. U., 1964; m. Winnie Ruth Isbell, Apr. 10, 1951; children: Barbara Gail, Patricia Ann, Robert Earl, Diana Lynn. Rsch. instr. dept. ob-gyn. U. Tex. Med. br., Galveston, 1963-65, rsch. asst. prof., 1965-68, rsch. assoc. prof., 1968-89, assoc. prof. dept. ob-gyn., 1989—; cons. Interdeptl. Com. on Nutrition for Nat. Def., 1965-68; cons. Nat. Nutrition Survey, 1968-69. Scoutmaster Boy Scouts Am., 1969—. With USNR, 1951-52. Nutrition Rsch. fellow, 1960-61; NSF scholar, 1961-62; NIH Rsch. fellow, 1962-63. Mem. Tex., N.Y. Acads. Scis., Am. Fert. Soc. Am. Inst. Nutrition, Am. Soc. Clin. Nutrition, Am. Coll. Nutrition, Am. Fertility Soc., Soc. Exptl. Biology and Medicine, Soc. Environ. Geochemistry and Health, Sigma Xi, Phi Rho Sigma. Baptist. Mason. Club: Mic-O-Say (Kansas City, Mo.). Author: Effect of Water Borne Nitrites on the Environment of Man; contbr. numerous articles to profl. jours., chpts. to books. Achievements include research on prenatal nutrition, male fertility, epidemiology of lithium in Texas. Home: 15 Chimney Corners Dr La Marque TX 77568-5274 Office: U Tex Med Br Dept Ob-Gyn Galveston TX 77550

DAWSON, GERALD LEE, engineering company executive; b. Santa Ana, Calif., July 6, 1933; s. Harold Guy and Violet Jean (Swanson) D.; m. Shirley Jean Webb, Dec. 28, 1966; children: Debbi Lynn, John Guy. Grad. high sch., Santa Ana. Technician Beckman Instruments, Costa Mesa, Calif., 1954-55, mgr. quality control, 1955-57; mgr. Nat. Theaters, Santa Ana, 1958-59; customer engr. IBM Corp., Santa Monica, Calif., 1959-63; engring. specialist IBM Corp., Lexington, Ky., 1963-65, engr., 1965-70, engring. mgr., 1970-75, prodn. engr., 1975-82, sr. engr., 1982-89; pres. MAS-HAMILTON Group/MAS-HAMILTON Security Internat., Lexington, 1990—. Patentee electronic keyboard, 5 electronic combination locks, access control systems. Bus. chmn. United Way, Calif., 1958; chmn. Rep. campaign, Calif., 1957. Mem. Robotics Internat., Soc. Mfg. Engrs., Elks, Moose (prelate 1969-70). Avocations: flying, golf, fishing, travel.

DAWSON, GERALDINE, psychologist, educator; b. Cobleskill, N.Y., Mar. 29, 1951; d. Frank Gates Dawson Jr. and Beta (Holmes) Dale; m. Charles Joseph Coates, July 21, 1985; 1 child, Christopher Staats. BS in Psychology, U. Wash., 1974, PhD in Psychology, 1979. Asst. prof. psychology U. N.C., Chapel Hill, 1980-85; assoc. prof. U. Wash., Seattle, 1985-87, prof. psychology, 1987—, dir. child clin. psychology program, 1985-91. Editor: Autism: Nature, Diagnosis and Treatment, 1989, Human Behavior and the Developing Brain, 1993; contbr. to profl. jours. Bd. dirs. Autism Soc. Wash., Seattle, 1991—. Grantee NIH. Mem. APA, Soc. Rsch. in Child Devel. Achievements include research in areas of autism and childhood depression. Office: U Wash Dept Psychology NI 25 Seattle WA 98195

DAWSON, JEFFREY ROBERT, immunology educator; b. Lakewood, Ohio, Oct. 5, 1941; s. Robert Eugene and Elva Rose (Lincks) D.; m. Linda Elizabeth Issler, June 14, 1964; children: Amy Elizabeth, Mary Catherine, Michael Jeffrey. BS in Biology, Rensselaer Poly. Inst., 1964; PhD in Biochemistry, Case Western Res. U., 1969. Instr. Duke U. Med. Ctr., Durham, N.C., 1971-72, assoc., 1972-74, asst. prof., 1974-78, assoc. prof., 1978-90, prof., 1990—; acting chief div. immunology Duke U. Med. Ctr., Durham, 1991—; mem. Duke U. Comprehensive Cancer Ctr., Durham, 1976—. Author: Key Facts in Immunology, 1985, Zinsser Microbiology, 20th edit., 1991; (monograph) Immunology-BSCS Series, 1985; contbr. articles to profl. jours. Recipient Golden Apple award Am. Med. Students Assn., 1987; NIH grantee, 1986—. Mem. Am. Assn. Immunologists, Am. Soc. for Histocompatibility and Immunogenetics. Democrat. Episcopalian. Achievements include rsch. on regulation of natural killer cells. Home: 902 Clarion Dr Durham NC 27705-1731 Office: Duke U Med Ctr Div Immunology Box 3010 Durham NC 27710*

DAWSON, JOHN MYRICK, plasma physics educator; b. Champaign, Ill., Sept. 30, 1930; s. Walker Myrick and Wilhelmina Emily (Stephan) D.; m. Nancy Louise Wildes, Dec. 28, 1957; children: Arthur Walker, Margaret Louise. B.S., U. Md., 1952, M.S., 1954, Ph.D., 1957. Fulbright scholar Inst. Plasma Physics, Nagoya, Japan, 1964-65; research physicist Plasma Physics Lab. Princeton U., 1956-73, head theoretical group, 1965-73; prof. plasma physics UCLA, 1973—, assoc. head Inst. for Plasma Physics & Fusion Engring., 1976-88; cons. in field; John Danz lectr. U. Wash., 1974; guest Russian Acad. Scis., 1971; invited lectr. Inst. Plasma Physics, Nagoya, Japan, 1972. Contbr. articles in field to profl. jours. Recipient Exceptional Sci. Achievement award TRW Systems, 1977; James Clerk Maxwell prize in Plasma Physics, 1977; named Calif. Scientist of the Year, 1978. Fellow AAAS, Am. Phys. Soc. (chmn. plasma div. 1970-71); mem. Nat. Acad. Scis., N.Y. Acad. Scis., N.J. Acad. Scis., Sigma Pi Sigma, Phi Kappa Phi, Sigma Xi. Unitarian. Patentee in field. Home: 359 Arno Way Pacific Palisades CA 90272-3348 Office: Univ Calif 405 Hilgard Ave Los Angeles CA 90024-1301

DAWSON, WALLACE DOUGLAS, JR., geneticist; b. Louisville, Mar. 15, 1931; s. Wallace Douglas and Ida Belle (Hieatt) D.; m. Victoria Hollowell; 3 children. B.S., Western Ky. U., 1954; M.S., U. Ky., 1959; Ph.D. (NSF Coop. fellow), Ohio State U., 1962. Asst. prof. biology U. S.C., 1962-66, asso. prof., 1966-71, prof., 1971—, chmn. dept. biology, 1974-77, George Bunch prof. biology, 1977-81; vis. scientist div. mammals Smithsonian Instn., 1979. Served to 1st lt. USAF, 1955-57. Recipient Disting. Teaching award S.C. Honors Program, 1977, 85; NIH grantee, 1964, 71; NSF grantee, 1985, 90. Mem. AAAS, Am. Genetic Assn., Am. Soc. Mammalogists, Assn. Southeastern Biologists, Genetics Soc. Am., Soc. Study Evolution, S.C. Acad. Sci., Sigma Xi. Rsch. and publs. in field. Office: Dept Biol Scis U SC Columbia SC 29208

DAWSON, WILLIAM RYAN, zoology educator; b. Los Angeles, Aug. 24, 1927; s. William Eldon and Mary (Ryan) D.; m. Virginia Louise Berwick, Sept. 9, 1950; children: Deborah, Denise, William. Student, Stanford, 1945-46; B.A., UCLA, 1949, M.A., 1950, Ph.D., 1953; D.Sc., U. Western Australia, 1971. Faculty zoology U. Mich., Ann Arbor, 1953—; prof. U. Mich., 1962—, D.E.S. Brown prof. biol. scis., 1981—, chmn. div. biol. scis., 1974-82, dir. mus. zoology, 1982-93; Lectr. Summer Inst. Desert Biology, Ariz. State U., 1960-71, Maytag prof., 1982; researcher Australian-Am. Edn. Found., U. Western Australia, 1969-70; mem. Speakers Bur., Am. Inst. Biol. Sci., 1960-62; Mem. adv. panel NSF environ. biology program, 1967-69; mem. adv. com. for research NSF, 1973-77; adv. panel NSF regulatory biology program, 1979-82; mem. R/V Alpha Helix New Guinea Expdn., 1969; chief scientist R/V Dolphin Gulf of Calif. Expdn., 1976; mem. R/V Alpha Helix Galapagos Expdn., 1978. Editorial bd.: Condor, 1960-63, Auk, 1964-68, Ecology, 1968-70, Ann. Rev. Physiology, 1973-79, Physiol. Zoology, 1976-86; co-editor: Springer-Verlag Zoophysiology and Ecology series, 1968-72; assoc. editor: Biology of the Reptilia, 1972. Served with USNR, 1945-46. USPHS Postdoctoral Research fellow, 1953; Guggenheim fellow, 1962-63; Recipient Russell award U. Mich., 1959, Distinguished Faculty Achievement award, 1976; Wheeler Lectr. U. N.D., 1986. Fellow AAAS (council del. 1984-86), Am. Ornithol. Union (Brewster medal 1979); mem. Am. Soc. Zoologists (pres. 1986), Am. Physiol. Soc., Ecol. Soc. Am., Cooper Ornithol. Soc. (hon., Painton award 1963), Phi Beta Kappa, Sigma Xi, Kappa Sigma. Home: 1376 Bird Rd Ann Arbor MI 48103-2351

DAX, SCOTT LOUIS, chemist, researcher; b. Allentown, Pa., July 12, 1959; s. Irving F. and Frances (Friedman) D.; m. Carrie Milza, Apr. 27, 1991; 1 child, Kristy Lynn. BA, Shippensburg State Coll., 1981; MS in Chemistry, U. Mich., 1983, PhD in Chemistry, 1986. Teaching asst., lectr. U. Mich., Ann Arbor, 1981-86; postdoctoral fellow U. Wis., Madison, 1986-88, NIH, Washington, 1987-88; assoc. rsch. investigator Hoffmann-LaRoche, Inc., Nutley, N.J., 1988-92; adj. prof. Chemistry Montclair (N.J.) State Coll., 1992—; sr. rsch. scientist DuPont Merck, Wilmington, Del., 1993—; NIH/Alchohol, Drug Abuse and Mental Health Assn. peer rev. cons., 1992—; reviewer Journal of the American Chemical Society, Journal of Organic Chemistry, Journal of Medicinal Chemistry, Tetrahedron Letters, 1986—. Author: (textbook) Antibacterial Chemotherapeutic Agents, 1994; contbr. articles to profl. jours. Recipient Mayor's citation City of Allentown, 1977, Dow-Britton fellowship U. Mich., 1984, 85. Mem. AAAS, Am. Chem. Soc. (Soc. award 1981, Analytical Chemistry award 1980), N.Y. Acad. Scis. Achievements include 3 U.S. patent applications. Home: 3 Quail Dr Landenberg PA 19350 Office: DuPont-Merck Pharm Co The Experimental Sta E353/342 Wilmington DE 19880-0353

DAY, AGNES ADELINE, microbiology educator; b. Plains, Ga., July 20, 1952; d. Robert David and Annie Lee (Harvey) Laster; m. John Henry Day Jr., Mar. 10, 1973; 1 child, Teresa Denise. BS, Bethune Cookman Coll., 1974; PhD, Howard U., 1984. Staff fellow NIH, Nat. Inst. Dental Rsch., Bethesda, Md., 1984-88; asst. prof. cell and molecular biology program Howard U. Grad. Sch., Washington, 1988-90; asst. prof. Howard U. Coll. of Medicine, Washington, 1991—; mem. biomed. rsch. and tech. study sect. NIH, Bethesda, 1991-95. Contbr. articles to profl. jours. Mentor AAAS Sci. Edn. Directorate, Washington, 1988—. Named Outstanding Rsch. scholar Howard U., 1983, MARC fellow NIH, 1982-84, Outstanding Rsch. advisor Howard U. Coll. of Medicine, 1991; grantee NSF, 1990-93, Orthopaedic Rsch. and Edn. Found., 1990-92; recipient E.E. Just Rsch. award Sigma Xi Soc. Mem. Am. Soc. for Microbiology, Am. Soc. for Bone and Mineral Rsch., Am. Soc. Biochem. and Molecular Biol., Sigma Xi. Democrat. Baptist. Achievements include being the first to clone the small proteoglycan II protein of bovine bone and connective tissue; demonstration of coordinated regulation and gene expression of several connective tissue proteins. Office: Howard U Coll of Medicine 520 W St NW Washington DC 20059

DAY, CECIL LEROY, agricultural engineering educator; b. Dexter, Mo., Oct. 4, 1922; s. Cecil Lawrence and Katherine (Kleffer) D.; m. Peggy Eunice Thrower, Aug. 29, 1948; children: Stanley K., Thomas L. BS in Agrl. Engring., U. Mo., 1945, MS, 1948; PhD, Iowa State U., 1957. Mem. faculty U. Mo. at Columbia, 1945-85, prof. agrl. engring., 1962-85; prof. emeritus, 1985—, chmn. dept., 1969-82; vis. prof. U. Thessaloniki, Greece, 1972; pres. Penreico, Inc., 1968-79. Author articles, bulls. Chmn. elec. appeals bd., Columbia, 1966-76. Fellow Am. Soc. Agrl. Engrs. (Outstanding Individual of Yr. Mo. sect. 1982); mem. Agrl. Engrs of Mo. Inc. (pres. 1987—), Lakeshore Villa Homes Assn. Inc. (treas. 1985—). Mem. Ch. of Christ. Home: 806D Bourn Ave Columbia MO 65203-1470

DAY, DELBERT EDWIN, ceramic engineering educator; b. Avon, Ill., Aug. 16, 1936; s. Edwin Raymond and Doris Jennings (Main) D.; m. Shirley Ann Foraker, June 2, 1956; children: Lynne Denise, Thomas Edwin. BS in Ceramic Engring., Mo. Sch. Mines and Metallurgy, 1958; MS in Ceramic Tech., Pa. State U., 1960, PhD in Ceramic Tech., 1961. With U. Mo., Rolla, 1961—, dir. Indsl. Rsch. Ctr., 1965-72, dir. Grad. Ctr. Materials Rsch., 1983-92, Curators' prof. ceramic engring., 1981—; vis. prof. chemistry Miss. Coll., 1963, Eindhoven Tech. U., The Netherlands, 1971; mem. tech. staff Sandia Nat. Labs., Albuquerque, 1981, 91; sr. vis. faculty scientist Battelle Pacific N.W. Labs., Richland, Wash., 1990; asst. dean grad. studies Mo. Sch. Mines and Metallurgy, 1979-81; chmn. acad. coun. U. Mo., Rolla, 1978-79, active numerous other coms.; cons. Los Alamos Nat. Labs., 1983—, NASA, 1974-88, numerous other glass and refractories cos., 1958—; vice-chmn. Gordon Rsch. Conf. on Glass, 1990-92, chmn., 1992—; tech. program dir. confs. on glass including Baden-Baden, Germany, 1973, Rolla, 1975, XII Internat. Glass Congress, Albuquerque, 1980, Internat. and 7th U. Conf. Glass Sci., Clausthal-Zellerfeld, Germany, 1983. Contbr. more than 185 articles to profl. jours. Bd. chmn. Wesley Found.; chmn. United Ministries Higher Edn. Bd. Dirs., 1969; adv. Explorer Scout Post 82, 1964-69; bd. dirs. Rolla Community United Fund, 1975-81, Mo. Incutech Found., 1984-87; mem. bd. adjustment City of Rolla, 1973-79; fin. chmn. United Meth. Ch., 1978-80; pres., bd. dirs. Rolla Community Devel. Corp., 1967-61, 82-90. 1st lt. CE US Army, 1958-64. Recipient Outstanding Young Man award Clinton (Miss.) Jaycees, 1963, Mo. Jaycees, 1968, Community Builder award Fraternal Order of Eagles, 1971. Fellow Am. Ceramic Soc. (Outstanding Educator award ednl. coun. 1991, v.p. rsch. 1990-91, trustee 1986-91, trustee glass divsn. 1986-89, chmn. glass divsn. 1982-83, fellows com. 1987-92, publs. com. 1980-82, 90—, v.p. Publications 1992-93, treas. 1993—, others); mem. ASTM, Am. Soc. Engring. Edn. (chmn. mineral engring. div. 1968-69, program chmn. mineral engring. div. 1967-68), Nat. Inst. Ceramic Engrs. (Profl. Achievement in Ceramic Engring. award 1971), Brit. Soc. Glass Tech., Materials Rsch. Soc., Mo. Acad. Sci. (corp. membership com. 1989-90), Keramos, Blue Key, Tau Beta Pi, Phi Kappa Phi, Sigma Gamma Epsilon, Sigma Xi (treas. U. Mo.-Rolla chpt. 1966-67, sec. 1967-68, v.p. 1968-69, pres. 1969-70). Achievements include 30 U.S. and foreign patents (with others) for Alumina Zircon Bond for Refractory Grains, Chemically Durable Nitrogen Containing Phosphate Glasses Useful for Sealing to Metals, Fabrication of Precision Glass Shells by Joining Glass Rods, Glass Microspheres, Fiber Filled Dental Porcelain, Composition and Method for Radiation Synovectomy of Arthritic Joints, Process for Glass Microspheres, Transparent Composite Material, Ammonia Treated Phosphates Useful for Sealing to Metals, Radioactive Biologically Compatible Glass Microspheres, Radioactive Glass Microspheres, others; invention of Therasphere used for treatment of liver cancer. Home: PO Box 357 Rolla MO 65401-0357 Office: U Mo-Rolla Grad Ctr Material Rsch 101 Straumanis Hall Rolla MO 65401

DAY, JOHN H., physicist; b. Savannah, Ga., June 5, 1952; s. John H. and Elsie M. (Gilliard) D.; m. Agnes A. Lasiter, Mar. 10, 1973; 1 child, Teresa D. BS in Physics, Bethune-Cookman Coll., Daytona Beach, Fla., 1973; MS in Physics, Howard U., Washington, 1976, PhD in Physics, 1982. Cert. total quality mgmt. and leadership edn. Engr. Martin Marietta Aerospace Corp., Orlando, Fla., 1973; physicist Nat. Bur. Stds., Gaithersburg, Md., 1974-78, U.S. Geol. Survey, Reston, Va., 1979-82; engr. energy conversion sect. NASA/Goddard Space Flight Ctr, Greenbelt, Md., 1982-88, sect. head, 1988-90, asst. br. head space power br., 1990-92, br. head, 1992—; mem. Interagy. Advanced Power Group, Washington, 1983—; mem. NASA Historically Black Colls. Working Group, Washington, 1991-92; mem. NASA/Goddard Space Flight Ctr. Recruitment Team, Greenbelt, 1990—. Mem. Pub. Schs. Math. Task Force, Prince George's County, Md., 1991-92. Grad. fellow Howard U., 1973, 74, 75, 79, NSF fellow, 1976, 77, 78; recipient Internat. Cometary Explorer Group award NASA, 1985, Internat. Sun-Earth Explorer Group award NASA, 1987, NASA Performance Mgmt. and Recognition System awards, 1989, 90, 91, 92, Cosmic Background Explorer Group Achievement award NASA, 1990, Roentgen Satellite Group Achievement award NASA, 1991, Gamma Ray Observatory Group award NASA, 1992, Upper Atmosphere Rsch. Satellite Team award NASA, 1992, Goddard Exceptional Achievement award, 1993. Mem. IEEE Power Engring. Soc., AAAS, Am. Phys. Soc. Forum on Physics and Soc., Nat. Soc. Black Physicists, Phi Beta Sigma. Achievements include design and development of solar power systems for numerous NASA sci. satellites. Home: 9711 Bald Hill Rd Mitchellville MD 20721 Office: NASA Goddard Space Flight Ctr Space Power Br/Code 734 Greenbelt MD 20771

DAY, LUCILLE ELIZABETH, laboratory administrator, educator, author; b. Oakland, Calif., Dec. 5, 1947; d. Richard Allen and Evelyn Marietta (Hazard) Lang; m. Frank Lawrence Day, Nov. 6, 1965; 1 child, Liana Sherrine; m. 2nd, Theodore Herman Fleischman, June 23, 1974; 1 child, Tamarind Channah. AB, U. Calif., Berkeley, 1971, MA, 1973, PhD, 1979. Teaching asst. U. Calif., Berkeley, 1971-72, 75-76, research asst., 1975, 77-78; tchr. sci. Magic Mountain Sch., Berkeley, 1977; specialist math. and sci. Novato (Calif.) Unified Sch. Dist., 1979-81; instr. sci. Project Bridge, Laney Coll., Oakland, Calif., 1984-86; sci. writer and mgr. precollege edn. programs, Lawrence Berkeley (Calif.) Lab., 1986-90, life scis. staff coord., 1990-92, mgr. Hall of Health, Berkeley, Calif., 1992—. Author numerous poems, articles and book reviews; author: (with Joan Skolnick and Carol Langbort) How to Encourage Girls in Math and Science: Strategies for Parents and Educators, 1982; Self-Portrait with Hand Microscope (poetry collection), 1982. NSF Grad. fellow, 1972-75; recipient Joseph Henry Jackson award in lit. San Francisco Found., 1982. Mem. AAAS, No. Calif. Sci. Writers Assn., Nat. Assn. Sci. Writers, Women in Communications, Phi Beta Kappa, Iota Sigma Pi. Home: 1057 Walker Ave Oakland CA 94610-1511 Office: Hall of Health 2230 Shattuck Ave Berkeley CA 94704

DAY, MARY JANE THOMAS, cartographer; b. Connors, New Brunswick, Can., Oct. 12, 1927; d. Angus and Delina (Michaud) Thomas; m. Howard M. Day, July 1, 1949; children: Laurie Anne Day Greene, Angus Howard. BS in Geography, U. Md., 1974, BS in Bus. & Mgmt., 1977. Meteorol. aide Hangar 8 Eastern Airlines, N.Y.C., 1946-47, U.S. Weather Bur., Washington, 1948-50; cartographic aide U.S. Navy Hydrographic Office, Suitland, Md., 1950-57, cartographer, 1957-62; cartographer U.S. Navy Oceanographic Office, Suitland, 1962-72, Def. Mapping Agy., Suitland/Brookmont, 1972-93; cartographer USNS Harkness, 1978, Indonesian Naval Personnel, Jakarta, Indonesia, 1981-82. Compiled, wrote and published: The Descendants of John Thomas of Connors, N.B., 1988. Mem. Nat. Aeronautic Assn., Am. Soc. Photogrammetry & Remote Sensing. Club: Andrews Officers (Md.). Avocations: ice skating, sky diving, traveling, genealogy, foreign languages. Home: 3532 28th Pky Temple Hills MD 20748-2922

DAY, MELVIN SHERMAN, information company executive; b. Lewiston, Maine, Jan. 22, 1923; s. Israel and Frances (Goldberg) D.; m. Louisa Walker; children: Cynthia Day Solganick, Wendy Day Johnson, Robert Marshall. BS, Bates Coll., 1943; postgrad., U. Tenn., 1953-54. Chemist Metal Hydrides Inc., Beverly, Mass., 1943-44, Tenn. Eastman Corp., Oak Ridge, 1944-46; sci. analyst AEC, Oak Ridge, 1946-48, asst. chief tech. info. svc. extension, 1950-56, chief, 1956-58; dir. tech. info. div. AEC, Washington, 1958-60; dep. dir. Tech. Info. and Ednl. Programs Office, NASA, Washington, 1960-61, dir. Sci. and Tech. Info. div., 1961-67, dep. asst. adminstr. tech. utilization, 1967-70; head Office Sci. Info. NSF, Washington, 1970-72; dep. dir. Nat. Libr. Medicine, HEW, Bethesda, Md., 1972-78; dir. Nat. Tech. Info. Svc. Dept. Commerce, 1978-82; v.p. Info. Tech. Group, 1982-84, Rsch. Publs., 1984-86; sr. v.p. Herner & Co., 1986-88; pres. M. Day Cons. Internat., Inc., Arlington, 1988—; exec. v.p. BIIS Corp., Herndon, 1991—; cons. Internat. Atomic Energy Agy., 1960; adviser OECD, 1970, 75; U.S. mem. info. policy group; U.S. mem. NATO Tech. Info. Panel, 1960-70, 79-82, chmn., 1970; chmn. com. on sci. and tech. info. Fed. Coun., 1970-72, chmn. com. on intergovtl. sci. rels., 1969-70, chmn. sci. info. exch. adv. bd., 1963-69, mem. chem. abstracts adv. bd., 1964-68; mem. Fed. Libr. Com., 1968-78, chmn. exec. bd., 1973-75; trustee Found. Ctr. 1972-78, trustee emeritus, 1991—; U.S. mem. adv. com. on librs., documentation and archives UNESCO; pres. abstracting bd. Internat. Coun. Sci. Unions, 1977-83; bd. dirs. Internat. Coun. for Sci. and Tech. Info., 1983—, Inst. for Internat. Info. Programs, 1985—; trustee Engring. Info., Inc., 1981-84, bd. dirs., 1993—; del. numerous panels; cons., adviser and lectr. in field; mem. adv. com. HHS

Health Svcs. Rsch. Dissemination and User Liaison, 1990-92, also mem. dissemination com. Mem. editorial bd. Health Comm. and Informatics, 1977-80, Infomediary, 1990—, Yearbook of the Database Info. Industry, 1990-91, Bull. of Am. Soc. Info. Sci., 1977-80. Bd. visitors U. Pitts. Grad. Sch. Info. Sci., 1977-83. With U.S. Army, 1944-46. Recipient Exceptional Svc. medal NASA, 1971, Superior Svc. award USPHS, 1976. Fellow Am. Soc. Advancement Sci.; mem. ALA, Am. Soc. Info. Sci. (chmn. internat. rels. com. 1972-75, pres. 1975-76, coun. 1975-77, editorial bd. bull.), Am. Chem. Soc., Spl. Libr. Assn., Am. Soc. Cybernetics (bd. dirs. 1975-79), Venezuelan Acad. Scis. (hon. corr.), Internat. Coun. Sci. and Tech. Info. (hon.), Cosmos Club. Office: 620 Herndon Pky Herndon VA 22070

DAY, PETER RODNEY, geneticist, educator; b. Chingford, Essex, Eng., Dec. 27, 1928; came to U.S., 1963; m. Lois Elizabeth Rhodes, May 26, 1951; children: Susan Catherine, Rupert Peter, William Rodney. BS in Botany, Birbeck Coll., Eng., 1950; PhD, U. London, 1954. Sr. scientific officer John Innes Inst., Hertford, Eng., 1957-63; assoc. prof. Ohio State U., Columbus, 1963-64; chief, genetics dept. Conn. Agrl. Expt. Sta., New Haven, 1964-69; dir. Plant Breeding Inst., Cambridge, Eng., 1979-87; prof. genetics, dir. Rutgers U., New Brunswick, N.J., 1987—; sec. Internat. Genetics Fedn., 1984-93; trustee Internat. Ctr. for Maize and Wheat Improvement, Mexico City, 1986-92; chmn. Mng. Global Genetic Resources Bd. on Agrl., NAS, Washington, 1986—. Author: Fungal Genetics, 1963, Genetics of Host-Parasite Interaction, 1974. Commonwealth Fund fellow U. Wis., 1954-56; Guggenheim Meml. fellow U. Queensland, 1972. Home: 394 Franklin Rd North Brunswick NJ 08902 Office: AgBiotech Ctr Rutgers U/Coll Farm Rd Cook Coll PO Box 231 New Brunswick NJ 08903-0231

DAY, ROBERT MICHAEL, oil company executive; b. Winnfield, La., Jan. 28, 1950; s. Robert Neal and Virginia Ruth (Franklin) D.; m. Noelie Barron, Dec. 20, 1975; children: Robert Michael Jr., Brionne. BS, La. State U., 1976; MBA, U. Houston-Clear Lake, 1989. Roustabout Global Marine Drilling Co., Houston, 1976-77; sales engr. NL Baroid Petroleum Svcs., Houston, 1977-78; drilling technician East Tex. div. Exxon Co., USA, Houston, 1978-79; sr. drilling technician Southeastern div. Exxon Co., USA, New Orleans, 1979-81, drilling supt., 1981-84; drilling supt. hdqrs. Exxon Co., USA, Houston, 1984-89; drilling supt. Offshore div. Exxon Co., USA, New Orleans, 1989-91; ops. supr. hdqrs. drilling Exxon Co., Internat., Houston, 1991—. Contbr. articles to profl. jours. Ruling elder Clear Lake Presbyn. Ch., Houston, 1987-88. With U.S. Army, 1969-73. Mem. Soc. Petroleum Engrs., Soc. of the 1st Div., Masons. Republican. Home: 20730 Chappell Knolls Dr Cypress TX 77429-5510

DAY, ROBERT WINSOR, research administrator; b. Framingham, Mass., Oct. 22, 1930; s. Raymond Albert and Mildred (Doty) D.; m. Jane Alice Boynton, Sept. 6, 1957 (div. Sept. 1977); m. Cynthia Taylor, Dec. 16, 1977; children: Christopher, Nathalia. Student, Harvard U., 1949-51; MD, U. Chgo., 1956; MPH, U. Calif., Berkeley, 1958, PhD, 1962. Intern USPHS, Balt., 1956-57; resident U. Calif., Berkeley, 1958-60; research specialist Calif. Dept. Mental Hygiene, 1960-64; asst. prof. sch. medicine UCLA, 1962-64; dep. dir. Calif. Dept. Pub. Health, Berkeley, 1965-67; prof., chmn. dept. health services Sch. Pub. Health and Community Medicine, U. Wash., Seattle, 1968-72, dean, 1972-82, prof., 1982—; dir. Fred Hutchinson Cancer Rsch. Ctr., Seattle, 1981-91, pres., 1991—; mem. Nat. Cancer Adv. Bd., 1992—; cons. in field. Pres. Seattle Planned Parenthood Ctr., 1970-72. Served with USPHS, 1956-57. Fellow Am. Pub. Health Assn., Am. Coll. Preventive Medicine; mem. Am. Soc. Clin. Oncology, Soc. Preventive Oncology, Assn. Schs. Pub. Health (pres. 1981-82), Am. Assn. Cancer Insts. (bd. dirs. 1983-88, v.p. 1984-85, pres. 1985-86, chmn. bd. dirs., 1986-87). Office: Fred Hutchinson Cancer Rsch Ctr LY-301 1124 Columbia St Seattle WA 98104

DAY, STACEY BISWAS, physician, educator; b. London, Dec. 31, 1927; came to U.S. 1955, naturalized 1977.; s. Satis B. and Emma L. (Camp) D.; m. Ivana Podvalova, Oct. 18, 1973; 2 children. M.D., Royal Coll. Surgeons, Dublin, Ireland, 1955; Ph.D., McGill U., 1964; D.Sc., Cin. U., 1971. Intern King's County Hosp., SUNY Downstate Ctr., 1955-56; resident fellow in surgery U. Minn. Hosp., 1956-60; hon. registrar St. George's Hosp., London, Eng., 1960-61; lectr. exptl. surgery McGill U., Montreal, Que., Can., 1964; asst. prof. exptl. surgery U. Cin. Med. Sch., 1968-70; assoc. dir. basic med. research Shriner's Burn Inst., Cin., 1969-71; from asst. to assoc. prof. pathology, head Bell Mus. Pathobiology U. Minn., Mpls., 1970-74; dir. biomed. communications and med. edn. Sloan-Kettering Inst., N.Y.C., 1974-80; mem. Sloan-Kettering Inst. for Cancer Research, 1974-80; mem. adminstrv. council, field coordinator, 1974-75; prof. biology Sloan Kettering div. Grad. Sch. Med. Sci. Cornell U., 1974-80, ret., 1980; clin. prof. medicine div. behavioral medicine N.Y. Med. Coll., 1980-92; prof. biopsychosocial medicine, chmn. dept. community health U. Calabar (Nigeria) Sch. Medicine, 1982-85; prof. internat. health, dir. Internat. Ctr. for Health Scis. Meharry Med. Coll., Nashville, 1985-89, dir. WHO Collaborating Ctr. ICHS, 1987-89; founding dir. WHO Collaborating Ctr., Nashville, 1987-89, emeritus dir., 1989; adj. prof. family and community medicine U. Ariz. Coll. Med. Scis., Tucson, 1985-89; univ. prof. internat. health U. Calabar, Nigeria, 1989—; permanent vis. prof. med. edn. Oita Med. Univ., Japan, 1992—; Arris and Gale lectr. Royal Coll. Surgeons, Eng., 1972, vis. lectr., Ireland, 1972; vis. prof. U. Bologna, 1977, Saga, Japan, 1992; vis. prof. health communications U. Santiago, Chile, 1979-80; vis. prof. Oncologic Research Inst., Tallinn, Estonia, 1976, All India Insts. Health, 1976, Univ. Maiduguri, 1982, Kyushu, Japan, 1990; vis. prof. internat. health U. Mauritius, 1991; vis. prof. Bratislava U., 1991; moderator med. cartography and computer health Harvard U., 1978, Acad. Scis., Czechoslovakia, 1987, Australia, 1988; Fulbright prof. Charles U., Czechoslovakia, 1989, vis. prof. U. Mauritius, 1991, Kyushu, Japan, 1990, Saga, Japan, 1992, Oita, Japan, 1992—, Hokkaido, Japan, 1992, Asahikawa, Japan, 1992; vis. academic, Oxford Univ., 1993; vis. prof. Kyoto, 1993, Beijing, China, 1993; cons. Pan Am. Health Assn., 1974-90, U.S.-USSR Agreement for Health Cooperation, 1976, WHO Collaborating Centre Meharry Med. Coll., Nashville, 1983, liaison officer NAFEO/AID, 1986-89; mem. expert com. for health, manpower devel., WHO, 1986-90, cons. div. strengthening health care resources WHO, Geneva, 1987-90; cons. to UN-FSSTD, 1987; AID/Joint Memorandum of Understanding cons. West Africa, Kenya, Sudan. Southern Africa, 1985-89; pres., chmn., pub. Cultural and Ednl. Prodns., Montreal, U.S.A., 1966-85; advisor to dean Med. Coll., Faculty Medicine and Health Scis., ABHA, Province of Asir, Saudi Arabia, 1981; cons. advisor to rector Universidad Autonoma Agraria Antonio Narro, Saltillo, Mexico, 1987-89; cons. U.S. Dept. Edn. Office Spl. Edn. Region X, San Francisco, 1986-89; cons. Dictionary of Sci. Biography, Ency. Britannica; cons., advisor to dir. High Tatras symposia Post Grad. Med. Inst., Bratislava, 1990—; bd. dirs. Internat. Health, African Health Consultancy Service, Nigeria; bd. dirs., v.p. Am. sci. activities Mario Negri Research Found., 1975-80; hon. founding chmn., bd. dirs. Lambo Found. U.S.; v.p., trustee Cancer Relief Found., Calabar; pres., exec. dir. Internat. Found. for Biosocial Devel. and Human Health, 1978-86, chmn., 1986—; cons. Inst. Health, Lyfford Cay, Bahamas, 1981, Govt. Cross River State, Nigeria, Itreto State and H.H. Obong of Calabar, Nat. Bd. Advs., Am. Biog. Inst., 1982—; cons. community health and health communications Navaho Nation, Sage Meml. Hosp., Ganado, Ariz., 1984; founder, cons. Primary Self-Health Clinics, Ekpo, Ikot Oku Okono, and Ik, Nigeria, 1982-84; cons. High Tatras Internat. Health Symposia, Slovakia, 1990—; appointed ambassador Gov. State of Tenn., 1986—; adj. clin. prof. medicine N.Y. Med. Coll.; researcher in field. Writer, 1965—; author: verse Collected Lines, 1966; play By the Waters of Babylon, 1966; verse American Lines, 1967; play The Music Box, 1967; Three Folk Songs Set to Music, 1967, Poems and Etudes, 1968; novel Rosalita, 1968; The Idle Thoughts of a Surgical Fellow, 1968, Edward Stevens-Gastric Physiologist, Physician and American Statesman, 1969; novella Bellechasse, 1970; A Leaf of the Chaatim, 1970, Ten Poems and a Letter from America for Mr. Sinha, 1971, Curling's Ulcer: An Experiment of Nature, 1972, Tuluak and Amaulik: Dialogues on Death and Mourning with the Innuit Eskimo of Point Barrow and Wainwright, Alaska, 1974, East of the Navel and Afterbirth: Reflections from Rapa Nui, 1976, Health Communications, 1979, The Biopsychosocial Imperative, 1981, What Is Survival: The Physician's Way and the Biologos, 1981; editor: Death and Attitudes Toward Death, 1972, Membranes, Viruses and Immune Mechanisms in Experimental and Clinical Disease, 1972, Ethics in Medicine in a Changing Society, 1973, Communication of Scientific Information, 1975, Trauma: Clinical and Biological Aspects, 1975, Molecular Pathology, 1975; (with Robert A. Good) series Comprehensive Immunology, 9 vols., 1976-80; Cancer Invasion and Metastasis-Biologic Mechanisms and Therapy, 1977, Some Systems of Biological Communication, 1977, Image of Science and Society, 1977, What Is a Scientist, 1978, Sloan Kettering Inst. Cancer Series, 1974-80, (with K. Inokouchi) Selections From the Chronicle of The Hagakure as Wisdom Literature: The Way of The Samurai of Saga Domain, 1993; editor-in-chief, mem. editorial bd.: Health Communications and Informatics, 1974-80; editor in chief: The American Biomedical Network: Health Care System in America Present and Past, 1978, A Companion to the Life Sciences, Vol. 1, 1979, A Companion to the Life Sciences, Vol. 2, Integrated Medicine, 1980, A Companion to the Life Sciences, Vol. 3: Life Stress, 1981, Advance to Biopsychosocial Health, 1984; editor in chief, mem. editorial bd. Health Communications and Biopsychosocial Health; editor: (with others) Cancer, Stress and Death, 1979, 2d edit., 1986, Computers for Medical Office and Patient Management, 1981, Readings in Oncology, 1980, Biopsychosocial Health, 1981; editor: Primary Health Care Guidelines: A Training Manual for Community Health, 2d edit., 1986, (with T.A. Lambo) Contemporary Issues in International Health, 1989; sr. editor (with Salat and others): Health and Quality of Life in Changing Europe in the Year 2000, 1992, Hagakure-Spirit of Bushido, (with H. Koga), 1993, (with K. Inokuchi) Selections from the Chronicles of the Hagakure as Wisdom Literature: The Way of the Samurai of Saga Domain, 1993; mem. editorial bd.: Annual Reviews on Stress; also co-editor various publs.; contbr. articles to profl. lit.; producer TV and radio health edn. programs, Nigeria, TV film River Blindness (Onchocerciasis) in Africa, 1988. Served with Brit. Army, 1946-49. Recipient Moynihan medal Assn. Surgeons Gt. Britain and Ireland, 1960, Reuben Harvey triennial prize Royal Coll. Physicians, Ireland, 1957, disting. scholar award Internat. Communication Assn., 1980, Sama Found. medal, 1982, disting. citation Hagakure Soc., 1992; named to Hon. Order Ky. Cols., 1968; named Chieftan Ntufam Ajan of Oban Ejagham People, Cross River State, Nigeria, 1983; recipient Chieftan Obong Nsong Idem Ibibio Nigeria, 1983, Mgbe (Ekpe) honor Nigeria, commendation WHO address Fed. Govt. Nigeria, Calabar, 1983, Leadership in Internat. Med. Health citation Pres. U.S., 1987, WHO medal, 1987, Agromedicine citation Commr. of Agr., State of Tenn., 1987, Assembly citation State of N.Y., 1987, Citation Congl. Record., 1987; Maestro Honorifo, U. Autonoma Agraria, Coahuila, Mex., 1987; presented Key to the City of Nashville, 1987; ipient Vice-Chancellor's Citation and Presentation for Primary Health Care Teaching in Nigeria, U. Calabar, 1988; Pamétni medal Postgrad. Med. Coll., Prague, 1991, Gold medal U. of Bratislava, 1991, Disting. Citation Hagakure Rsch. Soc., Japan, 1992; addresses presented by people of Ikot Imo, Nsit Anyang, Oban, 1982-84, Commendation from King of Calabar, 1984; Ciba fellow Can., 1965; Stacey Day Ward named in his honor by Fed. Min. and Gov. of Cross River State, Calabar Med. Ctr., Nigeria, 1986; charter mem. U.S. Normandy Com., 1988; 1st fgn. hon. mem. Hagakure Res. Soc. (Samurai), Kyushu, Japan, 1991;. Fellow Zool. Soc. London Royal Micros. Soc., Royal Soc. Health, World Acad. Arts and Scis., Japanese Found. for Biopsychosocial Health (internat. hon. fellow and most disting. mem.), African Acad. Sci., African Acad. Med. Scis. (founder); mem. AAS, AMA, Am. Burn Assn., Internat. Burn Assn., Can. Authors Assn., N.Y. Acad. Scis., Am. Assn. History Medicine, Am. Inst. Stress (bd. dirs.), Am. Anthrop. Assn., Am. Rural Health Assn. (v.p. internat. sci. affairs, bd. dirs.), Soc. Med. Geographers USSR. Home: 6 Lomond Ave Chestnut Ridge NY 10977-6901

DAYAL, SANDEEP, marketing professional; b. Jaipur, India, Dec. 13, 1960; s. Vireshwar Dayal and Damyanti M.; m. Sujata Tyagi, mar. 17, 1987; 1 child, Ashwin V. BSEE with hons., Birla Inst. Tech. and Sci., Pilani, India, 1983; M in Pub. and Pvt. Mgmt., Yale U., 1989. Project mgr. Kalsan Industries, PLC, Sirsi, India, 1983-85; sales mgr. Teknix Internat., Jaipur, 1985-87; cons. Chesebrough Pond's, Inc., Clinton, Conn., 1988-89; interal cons. Dexter Corp., Windsor Locks, Conn., 1989-91; mktg. mgr. Dexter Automotive Divsn., Charlotte, N.C., 1991-92; dir. mktg. Dexter Automotive Divsn., Waukegan, Ill., 1992—. Mem. IEEE. Achievements include design of systems selling indsl. engring. sales tng. program. Office: Dexter Automotive Divsn E Water St Waukegan IL 60085-5652

DAYAL, VINAY, aerospace engineer, educator; b. Meerut, India, Oct. 30, 1950; came to U.S., 1982; s. Vishnu D.S. and Savitri Bhatnagar; m. Amita Bhatnagar, Feb. 9, 1976; children: Tuhina, Tushar. B in Aeronautical Engring., I.T.T. Knp., India, 1972; MSME, U. Mo., 1983; PhD, Tex. A&M U., 1987. Scientist Aero. Devel. Establishment, Bangalore, India, 1972-82; rsch. asst. prof. N.C. A&T State U., Greensboro, 1987-88; asst. prof. Iowa State U., Ames, 1989—. Contbr. articles to profl. jours. and chpts. to books. V.p. India Cultural Assn., Ames, 1992-93. Mem. AIAA, ASTM, Soc. Exptl. Mechanics (Hetenyi award 1991). Achievements include researching strain measurements by optical fibers and smart structures; ultrasonic studies on thin composites. Office: Iowa State Univ 304 Town Engring Bldg Ames IA 50011

DAYANIM, FARANGIS, occupational and environmental medicine consultant. Student, Pahlavi U. Sch. Medicine, Shiraz, Iran, 1960-68; MS in Environ. and Occupational Health Sci., Hunter Coll., 1992. Diplomate Am. Bd. Preventive Medicine, Am. Bd. Occupational Medicine; lic. medicine, surgery, N.Y., Conn.; cert. Med. Rev. Officer Cert. Coun. Staff physician N.Y. Tel. Co., 1980-89; resident occupational medicine Mt. Sinai Sch. Medicine, N.Y.C., 1988-89; med. dir. Consol. Edison Co. N.Y., 1989-92; cons. occupational and environ. medicine N.Y.C., 1992—; expert witness. Fellow Am. Coll. Occupational Environ. Medicine, Am. Coll. Preventive Medicine, N.Y. Acad. Medicine; mem. AMA (Physician's Recognition award), Am. Pub. Heath Assn., N.Y. County Med. Soc. (apptd. grievance com. peer rev., subcom. worker's compensation, com. govt. affairs), N.Y. State Med. Soc., N.Y. Occupational Med. Assn., N.Y. Com. Occupational Safety Health, N.Y. Acad. Scis., N.Y.C. Health Hosps. Corp. Office: 208 E 51st St Ste 357 New York NY 10022

DAYNES, RAYMOND AUSTIN, immunology educator. PhD, Purdue U., 1972. Prof. immunology and head experimental pathology Sch. Medicine, U. Utah, Salt Lake City, 1975—. Achievements include rsch. in immunology of ultraviolet radiation carcinogenesis; lymphocyte recirculation dynamics; interleukin-1. Office: U Utah Radiobiology Div Bldg 351 Salt Lake City UT 84112 also: 1747 Orchard Dr Salt Lake City UT 84106*

DAYSON, RODNEY ANDREW, chemical engineer; b. Torrance, Calif., Feb. 25, 1964; s. Robert Andrew and Delores Yvonne (Ford) D.; m. Maria Del Carmen Aravena, Sept. 17, 1982; 1 child, Isabel. BSChemE, U. Calif., San Diego, 1991. Process engr. Kelco div. Merck, East San Diego, 1992—; bd. dirs. Mesa Minority Engring. Program Adv. Bd., La Jolla, Calif. Howard Hughes fellow, 1990. Mem. AICE. Office: Kelco Div Merck 2145 E Belt St San Diego CA 92113

DAYTON, DEANE KRAYBILL, computer company executive; b. Marion, Ind., May 24, 1949; s. Wilber Thomas and Donna Irene (Fisher) D.; m. Carol Mae Noggle, June 2, 1969; 1 child, Christopher Thomas. BA in Chemistry Edn., Ind. Wesleyan U., 1970; MS in Teaching, Randolph-Macon U., 1974; MS in Instrnl. Tech., 1976, PhD in Instrnl. Tech., 1976. Sci. tchr., chair sci. dept. Jessamine County Jr. High Sch., Nicholasville, Ky., 1970-73; asst. prof. instructional tech. sch. edn. U. Va., Charlottesville, 1976-77; grad. asst., teaching asst. Ind. U., Bloomington, 1973-76, asst. prof. instrnl. tech. Sch. Edn., 1977-83, dir. prodn. svcs. audio-visual ctr., 1979-83; dir. media div. CDC, Atlanta, 1981; v.p. cons. svcs. Ednl. Techs., Inc., Charlotte, N.C., 1983-85; exec. mgr. corp. publ. svcs. Intergraph Corp., Huntsville, Ala., 1985—; cons. trainer George Meany Ctr. Labor Studies, Silver Spring, 1978-82; cons., developer Discover Pl., Charlotte, 1984-86; tng. developer First Union Bank, Charlotte, 1982-85, United Carolina Bank, Monroe, N.C., 1984; cons. Anacomp, Sarasota, Fla., 1982-84. Co-author: Planning and Producing Instructional Media, 1985; producer (film) Computer Graphics for Communication, 1982; contbr. articles to profl. jours. Chairperson exhibits com. North Ala. Sci. Ctr., Inc., Huntsville, 1990-92. Recipient Young Scholar award AV Comm. Rev./Ednl. Resources Info. Ctr., 1977. Mem. ASTD, Nat. Soc. for Performance and Instrn., Soc. for Tech. Comm., Assn. for Ednl. Comm. and Tech. (pres. media design and prodn. divsn. 1982, James W. Brown award 1986). Avocation: developing computerized science museum exhibits. Home: 301 Lincoln St SE Huntsville AL 35801 Office: Intergraph Corp Huntsville AL 35894-0001

DAYTON, JOHN THOMAS, JR., computer researcher; b. Columbus, Ohio, Nov. 10, 1955; s. John Thomas Sr. and Patricia Ann (Findley) D. BA in Experimental Psychology, New Coll., 1981; PhD in Experimental Psychology, U. Okla., 1989. Postdoctoral fellow IBM T.J. Watson Rsch. Ctr., Hawthorne, N.Y., 1989-90; mem. tech. staff Bell Communications Rsch., Piscataway, N.J., 1990-. Co-author: Taking Software Design Seriously, 1991; contbr. articles to profl. jours. NSF Grad. fellow, 1984-87. Mem. Am. Psychology Soc., Assn. for Computing Machinery. Home: 41 C Franklin Greens S Somerset NJ 08873 Office: Bellcore RRC 1H-226 444 Hoes Ln Piscataway NJ 08854

D'CRUZ, JONATHAN, aeronautical engineer; b. Singapore, Sept. 15, 1963; arrived in Australia, 1981; s. Bastine Augustine and Topsy Marie Therese (Smith) D'C.; m. Lynne Maree Garner, Mar. 15, 1992; 1 child, Caitlin Alexandra. B Aero. Engr., Royal Melbourne Inst. Tech., Australia, 1986; PhD, Monash U., Australia, 1991. Aero. engr. Aero. Rsch. Lab. Def. Sci. and Tech. Orgn., Melbourne, Australia, 1985-86; rsch. scientist Def. Sci. and Tech. Orgn., Melbourne, 1991—; postgrad fellow Australia Dept. Def., 1986-91; vis. scientist Langley Rsch. Ctr. NASA, Hampton, Va., 1992-93. Contbr. articles to profl. jours. Mem. AIAA, IEEE, Royal Aero. Soc. Roman Catholic. Achievements include working in the area of active control with an emphasis on smart structures, flutter suppression and gust load alleviation; devised novel techniques for the solution of non-linear inverse problems in force identification; worked on holographic interferomerty and in-plane displacement measurement using Moire Fringe techniques. Office: DSTO Aeronautical Rsch Lab, 506 Lorimer St, Fishermen's Bend Victoria 3207, Australia

DE AGUIAR, RICARDO JORGE FRUTUOSO, research geophysicist; b. Caldas da Rainha, Portugal, Feb. 17, 1963; s. Asdrúbal João amd Maria Fernanda (Frutuoso) De Aguiar; m. Rute Mendes, Nov. 23, 1991. Licentiate in physics, U. Lisbon, Portugal, 1984. Asst. U. Aveiro, Portugal, 1985-86; jr. rsch. asst. Nat. Inst. Engring. and Tech. for Industry, Lisbon, 1986-89, rsch. asst., 1990—; cons. Portuguese Ministry Environ., 1991. Co-author: White Book on Portuguese Environment, 1991; also articles in Solar Energy. Recipient best article award Portuguese Physics Soc., 1990. Mem. Internat. Solar Energy Soc. (bd. dirs. Portuguese br. 1989-91, best article award Portuguese and Spanish sects. 1987). Home: Rua Andrade 57 Apt 3D, 1100 Lisbon Portugal Office: INETI-DER, Estrada do Paco do Lumiar, 1699 Lisbon Portugal

DEAK, CHARLES KAROL, chemist; b. Budapest, Hungary, Sept. 26, 1928; s. Karoly and Ida (Benes) D.; came to U.S., 1955, naturalized, 1961; B.S., Eotvos Coll., Budapest, 1948; student Sorbonne, Paris, 1949; postgrad. Wayne State U., 1957-61; m. Jenny Bocinski, Apr. 9, 1958; children—James, Christine. With Frankel Co., Inc., Detroit, 1957-73, quality control mgr., 1968-71, mgr. tech. services, 1971-73; pres. Analytical Assocs., Inc., Detroit, 1973-92; pres. C.K. Deak Tech. Svcs., Inc., 1992—. Cert. profl. chemist. Fellow Am. Inst. Chemists; mem. Am. Chem. Soc., ASTM, Am. Soc. Metals, Assn. Analytical Chemists, Photog. Soc. Am. Roman Catholic. Patentee in chem. firefighting agts. and dense metal separation. Club: Internat. Brotherhood Magicians. Home: 29844 Wagner Dr Warren MI 48093-8635

DEÁK, PETER, physicist, educator; b. Budapest, Hungary, Feb. 11, 1952; s. Peter and Eva (Gritzbach) D.; 1 child, Peter-András; m. Maria Tóth, Feb. 11, 1984. Postgrad., Hungarian Acad. Scis., Budapest, 1983; D in Physics, Lóránd Eötvös U. Scis., Budapest, 1984. Postdoctoral rsch. assoc. SUNY, Albany, 1985-87; asst. prof. physics Tech. U. Budapest, 1976-85, assoc. prof., 1987—, head surface physics lab., 1993—; vis. prof. U. Kaiserslautern, Germany, 1992. Author: Introduction to Semiconductor Physics, 1990; contbr. articles to sci. publs., 1978—. Named Outstanding Inventor, Govt. of Hungary, 1982; Humboldt fellow Max Planck Inst. Solid State Rsch., Stuttgart, Germany, 1991. Fellow Inst. for Study of Defects in Solids (sr.); mem. Am. Phys. Soc., German Physics Assn., Lóránd Eötvös Phys. Soc. (Gyulai prize Solid State Rsch.). Home: Gyakorlō utca 4/A-6.26, H-1106 Budapest Hungary Office: Tech U Budapest, Phys Inst, Budafoki ut 8, H-1111 Budapest Hungary

DEAK, TIBOR, microbiologist; b. Szeged, Hungary, Aug. 23, 1935; s. Ferencz and Margit (Schwartz) D.; m. Anna Petsy, Aug. 4, 1961; 1 child, Susanne. MS, U. Szeged, 1957; PhD, U. Eotvos, Budapest, 1970; DSc, Hungarian Acad. Sci., 1989. Microbiologist Duna Canning Factory, Budapest, 1957-63; researcher Rsch. Inst. for Canning, Budapest, 1963-65; asst. prof. Coll. for Food Industry, Budapest, 1965-70; prof. U. Horticulture, Budapest, 1970—, chmn., dean, 1986-91, rector, 1993—; postdoctoral rsch. assoc. U. Ga., Griffin, 1991-93. Author: Encyclopaedia of Food Science, 1992, Microbiology of Canning and Freezing, 1980; editor: Microbial Associations in Food, 1984; contbr. articles to profl. jours.; assoc. editor in chief: Internat. Jour. Food Microbiology, 1989-91. Fulbright fellow, 1986, 91. Mem. Am. Soc. Microbiology, Hungarian Sci. Soc. Food Industry (v.p. 1990—, Sigmund award 1981), Hungarian Acad. Sci. (chmn. food microbiology sect. 1987—), Am. Acad. Microbiology, World Fedn. Culture Collections (exec. mem. 1989—), Internt. Commn. Food Microbiology Hygiene (mem.-at-large 1989—), Internat. Commn. Yeasts (bd. dirs. 1989—), Hungarian Soc. Microbiology (Manninger award 1990), Inst. Food Tech., N.Y. Acad. Sci. Office: U Horhiculture Dept Microbiology, Somloi ut 14-16, H1118 Budapest Hungary also home: 66 Bimbo ut, 1022 Budapest Hungary also office: U Horticulture, Dept Microbiology, Somloi ut 14-16, 1118 Budapest Hungary

DEAL, JO ANNE MCCOY, quality control professional; b. Farmville, N.C., Nov. 4, 1953; d. James Richard Jr. and Helen Ruth (Holloman) McCoy; m. David Hilton Goins, June 23, 1973 (div. Sept. 1983); children: Michael Stacy Goins, Brent Justin Goins; m. Wesley Kelvin Deal, May 6, 1988. BS in Chemistry, East Carolina U., 1978, MBA, 1988. Control scientist I Burroughs Wellcome Co., Greenville, N.C., 1978-84, control scientist II, 1984-90, quality assurance validation specialist, 1990-91, quality assurance tng. adminstr., 1991—. Mem. Am. Soc. Quality Control (cert.), Southeastern GMP Tng. and Edn. Assn., East Carolina U. Chemistry Profl. Soc. (bd. dirs., treas. 1990—). Republican. Presbyterian. Home: 115 1st St Farmville NC 27828 Office: Burroughs Wellcome Co PO Box 1887 Greenville NC 27835-1887

DE ALBA-AVILA, ABRAHAM, plant ecologist; b. San Jose, Costa Rica, Mar. 26, 1955; s. Jorge and Consuelo (Esparza) de Alba; m. M.E. Lina Flores de Alba, Dec. 20, 1984; children: Cristina De Alba-Flores, Santiago de Alba-Flores. BS in Agr., Cornell U., 1978; MS in Range Mgmt., U. Ariz., 1983. Rsch. asst. Instituto de Ecologia, Mexico City, 1978-86; rsch. asst. and rsch. leader INIFAP, Jalpa, 1986-90; prin. researcher INIFAP, Aguascalientes, Mexico, 1990—. Contbr. articles to profl. jours. Mem. Ecol. Soc. Am., Brit. Ecol. Soc., Soc. for Range Mgmt., Sociedad Mexicana de Manejo de Pastizales A.C. (mem. and editor-referee). Office: CIFAP-AGS, Apartado Postal #20, Pabellon-Arteaga Mexico 20660

DE ALMEIDA, ANTONIO CASTRO MENDES, surgery educator; b. Fronteira, Alentejo, Portugal, June 15, 1934; s. Manuel Mendes de Almeida and Margarida Vitoria F.C. Mendes de Almeida; m. Maria Teresa Sá Lopo de Carvalho, Nov. 1, 1974 (dec. 1992). MD, U. Lisbon, Portugal, 1960. Intern Univ. Hosp. Santa Maria, Lisbon, 1963-64, intern in surgery, 1964-65, resident in surgery, 1965-66, cons. surgeon, 1973-81, chief cons. surgeon, 1981—; intern in surgery Kings County Hosp./Downstate Med. Ctr., Bklyn., 1966-67, resident in surgery, 1967-71, chief surg. resident, 1971-72; asst. prof. surgery U. Lisbon Med. Sch., 1973-91, assoc. prof. surgery, 1991—; vis. prof. Tulane U. Med. Ctr., New Orleans, 1989, U. Guadalajara, Mex., 1990, Cleve. Clin. Found., 1991; disting. guest Zapopan, Guadalajara, 1990. Contbr. articles to Coloproctology, Am. Jour. Gastroenterology, Am. Jour. Surgery, Mt. Sinai Jour. Medicine, Digestive Surgery. Capt. med. officer, Portuguese mil., 1958-63. Recipient Physician's Recognition award AMA, 1972. Fellow ACS; mem. Portuguese Assn. Hosp. Career Physicians (exec. 1991), Internat. Soc. Surgery, Internat. Biliary Assn., Internat. Hepato-Pancreatico-Biliary Assn. (sci. program com. 1995 European congress), World Assn. Hepato-Pancreatic-Biliary Surgeons, Am. Soc. Colon and Rectal Surgeons, Internat. Soc. Univ. Colon and Rectal Surgeons (sci. program com.), Soc. for Surgery Alimentary Tract, Cordoba Surg. Soc. (hon.). Roman Catholic. Avocations: tennis, bridge. Home: Praca Principe Real 23-3, 1200 Lisbon Portugal Office: U Hosp Santa Maria, Ave Egas Moniz, 1699 Lisbon Portugal

DEAN, CLEON EUGENE, physicist; b. Lubbock, Tex., Dec. 30, 1957; s. Clifford Lonnie and Paula Dianne (Fix) D.; m. Marjorie Carol Lockwood, July 25, 1992. BS summa cum laude, Tex. A&M U., 1980, MS in Physics, 1982; PhD in Physics, Wash. State U., 1989. Rsch. asst. dept. atmospheric scis. U. Ariz., Tucson, 1982-83, teaching asst. dept. physics, 1983-86; rsch. asst. dept. physics Wash. State U., Pullman, 1986-89, tech. asst. dept. physics, 1989-90; rsch. fellow Naval Oceanographic and Atmospheric Rsch. Lab. NOARL, Stennis Space Ctr., 1990-92; rsch. fellow Naval Rsch. Lab., Stennis Space Ctr., Stennis Space Ctr., 1992; asst. prof. physics Ga. So. U., Statesboro, 1992—; chair Novel Algorithm Session, Soc. for Photonics & Instrumentation in Engring. Automatic Object Recognition Conf., Orlando, Fla., 1991. Contbr. articles to profl. jours. Nat. Merit scholar, 1976; Office Naval Tech. fellow, 1990. Mem. Am. Assn. Physics Tchrs., Acoustical Soc. Am., Optical Soc. Am., Soc. for Photonics and Instrumentation in Engring. Unitarian Universalist. Achievements include analogies between acoustical, optical and elastodynamic scattering; side scattering resonances in interactive scattering; applications of differential geometry and catastrophe theory to scattering; used modified Sommerfeld-Watson transform to explain critical angle scattering from air bubbles in water; developed shell theories to explain fluid-loading in mid-frequency range for thin spherical shells. Office: Ga So Univ Dept Physics Landrum Box 8031 Statesboro GA 30460

DEAN, FRANK WARREN, JR., chemist, pet food company executive; b. Connellsville, Pa., July 6, 1954; s. Frank Warren and Jessie Louise (Weideman) D.; m. Sherry Sue Welling, Feb. 4, 1974; children: Nicole Diane, Rebekah Sue. BS in Biology, Houghton Coll., 1977; postgrad., California U. Pa., 1991—. Quality control technician Foseco Minesep, Mt. Braddock, Pa., 1977-78; chemist, shift supr. Wayne Feeds, Everson, Pa., 1978-83; prodn. and maintenance coord. Wayne Pet Foods, Continental Grain Co., Everson, 1983-85; analytical chemist Pa. Dept. Agr., Meadowlands, 1985-87; mgr. quality assurance, safety dir. Royal Canin-USA, Everson, 1987-90; dir. quality and regulatory affairs Best Feeds and Farm Supplies, Inc., Oakdale, Pa., 1990—. Mem. Am. Chem. Soc., Assn. Ofcl. Analytical Chemists, Am. Soc. Quality Control, Nat. Fire Protection Assn., Nat. Safety Coun., Rotary Club. Republican. Methodist. Office: Best Feeds & Farm Supplies 106 Seminary Ave Oakdale PA 15071

DEAN, GARY NEAL, architect; b. Alexandria, Va., Sept. 19, 1953; s. Louie Franklin D. BS in Architecture, U. Ill., 1977. Registered architect, Calif., Tex., Ill., N.Y., Fla., Vt., N.C., Wis., Iowa, Ark., Ga., Md.; registered interior designer, Ill., Tex. Designer, draftsman SRGF, Inc. Architects, Springfield, Ill., 1977-79; project mgr. Sarti-Huff Archtl. Group, Inc., Springfield, 1979-82; architect Henningson, Durham & Richardson, Inc., Dallas, 1982-84; prodn. mgr. Bogard, Guthrie & Ptnrs., Inc., Dallas, 1984-85; mgr., project architect Archtl. Designers, Inc., Dallas, 1985-87; pvt. practice architecture Dallas, 1987—; prin. Kaiser Gochnauer Ltd., 1987—; v.p., ptnr. Designhaus, Inc., 1989—. Prin. works include State of Ill. Capitol, LaCima Club, Las Colinas, Tex., Hackberry Creek Country Club, Las Colinas, Renaissance Club, Phoenix, Creve Coeur Club, Peoria, Ill., Hard Rock Cafe Retail Store, N.Y.C., Hard Rock Cafe, Singapore, Atlanta, Miami, Fla., Planet Hollywood, N.Y.C. Mem. Nat. Coun. Archtl. Registration Bds. Home: 6606 Shady Brook Ln # 3170 Dallas TX 75206-1116

DEAN, JOHN AURIE, chemist, author, chemistry educator emeritus; b. Sault Ste. Marie, Mich., May 9, 1921; s. Aurie Jerome and Gertrude (Saw) D.; m. Elizabeth Louise Cousins, June 20, 1943 (div. 1981); children: Nancy Elizabeth, Thomas Alfred, John Randolph, Laurie Alice, Clarissa Elaine; m. Peggy DeHart Beeler, Oct. 23, 1981; stepchildren: Diane Barbara, Lisa Lynn, James Edward, Jonathan Curtis. B.S. in Chemistry, U. Mich., 1942, M.S. in Chemistry, 1944, Ph.D., 1949. Teaching fellow in chemistry U. Mich., Ann Arbor, 1942-44, 45-46; lectr. in chemistry U. Mich., 1946-48; chemist X-100 Phase Manhattan Project Chrysler Corp., Detroit, 1944-45; assoc. prof. chemistry U. Ala., Tuscaloosa, 1948-50; asst. prof. chemistry U. Tenn., Knoxville, 1950-53; assoc. prof. U. Tenn., 1953-58, prof. chemistry, 1958-81, prof. emeritus, 1981—; cons. Union Car Nuclear Div., Oak Ridge, 1953-74, Stewart Labs., Knoxville, 1968-81; vis. lectr. Peoples Republic of China, 1985. Author: Instrumental Methods of Analysis, 1948, 7th edit., 1988, Flame Photometry, 1960, Chemical Separation Methods, 1969, Flame Emission and Atomic Absorption Spectrometry, vol. 1, 1969, vol. 2, 1971, vol. 3, 1975, Lange's Handbook of Chemistry, 14th edit., 1992, Handbook of Organic Chemistry, 1986, Solutions Manual for Instrumental Methods of Analysis, 7th edit., 1988, The Chemist's Ready Reference Handbook, 1990; contbr. articles to profl. jours. and chpts. to books. Mem. Am. Chem. Soc. (Charles H. Stone award Carolina-Piedmont sect. 1974), Soc. Applied Spectroscopy (chmn. S.E. sect. 1971-73, editor newsletter 1984—, Disting. Svc. award 1991), Archaeol. Inst. Am., East Tenn. Soc. (pres. 1980-81), U.S. Naval Inst. (life), Oriental Inst., U. Mus. Pa., Am. Guild Organists, Sigma Xi, Phi Kappa Phi. Presbyterian. Address: 201 Mayflower Dr Knoxville TN 37920-5871

DEAN, JOHN FRANCIS, astronomer, researcher; b. Liverpool, Eng., Jan. 25, 1946; arrived in South Africa, 1948; s. Geoffrey Joseph and Norah Mary (Devlin) D.; m. Ann Maureen Wamsteker, Apr. 17, 1978; children: Geoffrey Richard, Pamela Mary. BSc, Capetown (South Africa) U., 1966; BSc in Physics with honors, Rhodes U., Grahamstown, South Africa, 1968; postgrad. in radio astronomy, Manchester U., Eng., 1969, PhD, 1973. Astronomer Coun. Sci. Indsl. Rsch., Capetown, 1973-80; sr. officer main office Electricity Supply Commn./ESKOM, Johannesburg, South Africa, 1981-86, sr. scientist modelling and devel., prodn. optimisation, system ops., 1987—. Contbr. articles to profl. jours. Mem. South African Inst. Physics, Ops. Rsch. Soc. South Africa, South Africa Prodn. and Inventory Control Soc., South AfricanCoun. Natural Scientists, N.Y. Acad. Scis. Achievements include development of method of determining interstellar absorption of light from stars using cepheid variables. Home: 15 Craig Ave, Randburg Transvaal 2194, South Africa

DEAN, RICHARD ANTHONY, mechanical engineer; b. Bklyn., Dec. 22, 1935; s. Anthony David and Anne Mylod Dean; m. Sheila Elizabeth Grady, Oct. 5, 1957; children: Carolyn Anne, Julie Marie, Richard Drews. BSME, Ga. Inst. Tech., 1957; MSME, U. Pitts., 1963, PhDME, 1940. Registered profl. engr., Calif. From jr. engr. to mgr. thermal and hydraulic engring. Westinghouse Nuclear Energy Systems, 1959-70; v.p., tech. dir. water reactor fuels General Atomics, San Diego, 1970-74, v.p. uranium and light water reactor fuel, 1974-80, sr. v.p., 1980—; cons. U.S. Congress Office Tech. Assessment. 1st lt. U.S. Army, 1957-59. Mem. AAAS, ASME (former chmn. nuclear fuels tech. com.), Am. Nuclear Soc. (gen. chmn. annual meeting 1993), Global Found. (bd. advisors), Internat. Thermonuclear Experimental Reactor. Achievements include the develpoment of commercial nuclear power stations; advanced the understanding of boiling heat transfer phenomena; invention of advanced nuclear fuel assembly. Home: 6699 Via Estrada La Jolla CA 92037 Office: General Atomics PO Box 85608 San Diego CA 92186

DEAN, ROBERT BRUCE, architect; b. Brockton, Mass., Jan. 15, 1949; s. Robert George and Marjorie Gertrude (O'Donnell) D.; m. Mary Hood Hoskinson, June 18, 1977; children: Robert Maxwell, Anne, Claire. BA, U. Pa., 1971; MArch, Columbia U., 1976. Registered architect, N.Y., Conn. Staff architect Skidmore, Owings & Merrill, Architects, N.Y.C., 1976-77; job capt. Stephen Jacobs & Assn., N.Y.C., 1977-78; staff architect Johnson-Burgee Architects, N.Y.C., 1978-79; pvt. practice architecture N.Y.C. and Syracuse, 1979-85; project architect Robert A.M. Stern Architects, N.Y.C., 1985-86; pres. Dean Design, Inc., New Canaan, Conn., 1986—; adj. assoc. prof. Columbia U., N.Y.C., 1978-83; asst. prof. Syracuse U., 1980-84. Contbr. articles to profl. jours. Trustee North Stamford Congl. Ch., 1986-87; governing coun. Redding (Conn.) Congl. Ch., 1991—; mem. Planning Commn. Town of Redding, Conn. Grantee Syracuse U., 1982, grantee Nat. Endowment Arts, 1983-84; William Kinne Fellow, 1976. Mem. AIA, Conn. Soc. Architects. Democrat. Congregationalist. Avocations: American cultural and commercial history. Office: Dean Design Inc 111 Cherry St New Canaan CT 06840-5530

DEANE, GRANT BIDEN, physicist; b. Te Awamute, New Zealand, Oct. 26, 1961; came to U.S., 1990; s. John David and Fay Barbara (Sinclair) D.; m. Bonnie L. St. John, June 24, 1989. BS in Physics, U. Auckland, New Zealand, 1983; MSc, U. Auckland, 1985; PhD, U. Oxford, Eng., 1989. Rsch. fellow Math. Inst. U. Oxford, Eng., 1989-90; Mellow fellow Scripps Inst. Oceanography U. Calif., LaJolla, 1991-92, rsch. asst., project scientist, 1992—. Mem. Acoustical Soc. Am., Rotary. Achievements include researching new theoretical technique describing three dimensional sound fields in the ocean; application of newly developed theory of mass and energy transport in fusion plasmas. Office: Marine Phys Lab Scripps Inst Oceanography Mail Code 0238 La Jolla CA 92093-0238

DEANE, JOHN HERBERT, technologist; b. L.A., Feb. 3, 1952; s. Theodore E. and Marion Gail (Blake) D.; m. Linda Zampier, Nov. 22, 1973 (div. 1982); m. Anne N. Butler, June 13, 1987; children: Devon, Jennifer. BS, SUNY, Albany, 1976; MBA, SUNY, 1978. Sr. cons. Arthur Young & Co., N.Y.C., 1978-81; project mgr. Morgan Stanley & Co., N.Y.C., 1981-86; v.p. Asia Morgan Stanley & Co., Tokyo, Japan, 1986-87; v.p., dir. Salomon Bros. Asia Ltd., Tokyo, Japan, 1987-90; ptnr. Price Waterhouse, N.Y.C., 1990—. Sgt. USAF, 1970-74. Mem. Tokyo Am. Club, Glen Ridge Country Club, DAV (life), Nat. Eagle Scout Assn. (life), N.Y. Yacht Club. Methodist. Avocation: sailing. Home: 126 Forest Ave Glen Ridge NJ 07028-2414 Office: Price Waterhouse 153 E 53d St New York NY 10022

DEANGELIS, CATHERINE D., pediatrics educator; b. Scranton, Pa., Jan. 2, 1940; m. James C. Harris. BA, Wilkes Coll., 1965; MD, U. Pitts., 1969; MPH, Harvard U., 1973. RN, Pa., N.Y.; diplomate Nat. Bd. Med. Examiners, Am. Bd. Pediatrics. Intern in pediatrics Children's Hosp., Pitts., 1969-70; resident in pediatrics Johns Hopkins Hosp., Balt., 1970-72, teaching fellow pediatrics dept. internat. health Sch. Pub. Health, 1972; pediatrician Roxbury Comprehensive Health Clinic, Boston, 1972-73; asst. prof. pediatrics Coll. Physicians and Surgeons, asst. prof. health svc. adminstrn. Sch. Pub. Health Columbia U., 1973-75; mem. staff divsn. pediatric ambulatory care, dir. med. edn. Child Care Project Columbia Presbyn. Med. Ctr., 1973-75; asst. prof. pediatrics Sch. Medicine U. Wis., 1975-77, assoc. prof. pediatrics Sch. Medicine, 1977-78; dir. ambulatory pediatric svcs. U. Wis. Hosps., 1975-78; assoc. prof. pediatrics Johns Hopkins Sch. Medicine, 1978-85; dir. pediatric primary care and adolescent medicine Johns Hopkins Hosp., 1978-84, co-dir. adolescent pregnancy program, 1979-82; with dept. health svcs. adminstrn. and dept. internat. health Johns Hopkins Sch. Hygiene and Pub. Health, 1980—; dir. residency tng. dept. pediatrics Johns Hopkins Hosp., 1983-90, dir. divsn. gen. pediatrics and adolescent medicine, 1984-90; deputy chmn. dept. pediatrics Johns Hopkins Sch. Medicine, 1983-90, prof. pediatrics, 1986—, assoc. dean acad. affairs, 1990-93, sr. assoc. dean acad. affairs and faculty, 1993—; mem. Gov.'s Task Force to Evaluate Health Care in Wis. State Prisons, 1975-78; chmn. ambulatory care com. U. Wis. Hosp., 1976-78; mem. med. sch. admissions com. U. Wis. Sch. Medicine, 1976-78, chmn., 1977-78; mem. exec. coun. dept. pediatrics and Children's Ctr., Johns Hopkins U. Sch. Medicine 1982-90, chmn. fin. com. dept. pediatrics, 1984-85, chmn. assoc. prof.'s promotion com., 1985-88; chmn. com. developing Women's Health Ctr. at Johns Hopkins Med. Instns., 1993—; mem. Gov.'s Task Force on Women's Health, Md., 1993—; mem. search com. U. Wis., 1976, Johns Hopkins Sch. Medicine, 1984, 88, 92, 93; mem. nat. review com. for accreditation of nurse practitioners Am. Nurses' Assn., 1975-79, co-chmn., 1977; mem. peer review com. nurse practitioner programs divsn. nursing Health Resources Agy., Dept. Health, Edn. and Welfare, 1979-81; mem. Nat. Commn. on Nursing, 1985-86, Physician Consortium on Substance Abuse Edn., 1989—; mem. clin. scholar's adv. com. Robert Wood Johnson Found., 1992—; mem. Assn. Health Svcs. Rsch., 1993—; with immunization team, Nicaragua, 1969; subintern Harbel Hosp., Liberia, West Africa, 1969; organizer immunization program Peru, 1972, West Indies Sch. Nursing, 1977; mem. editorial bd. The Hosp. Med. Staff, 1982—, Pediatrician, 1984—, Jour. of Pediatrics, 1986—, Pediatric Annals, 1990—, Pediatrics in Review, 1990—, Archives of Pediatrics and Adolescent Medicine, 1993—; reviewer Acad. Medicine, Am. Jour. Diseases of Children, Am. Jour. Medicine, Clin. Pediatrics, Jour. Pediatrics, Med. Care, Pediatrics; writer weekly column Balt. Sun, 1987-90. Author: Basic Pediatrics for the Primary Health Care Provider, 1975, Pediatric Primary Care, 1984; editor: An Introduction to Clinical Research, 1990; editor: (with others) Principles and Practice of Pediatrics, 1990; assoc. editor Pediatric Annals, 1990—; editor Archives of Pediatrics and Adolescent Medicine, 1993—. Mem. steering com. Rural Health Planning, Wis.; cons. Robert Wood Johnson Found., 1973—; mem. adv. group on improving outcomes for children Pew Charitable Trusts, 1991-92; mem. adv. panel medicine Pew Health Profession's Commn.; mem. nat. adv. com. Robert Wood Johnson Clin. Scholars Program, 1992—. NIH fellow, 1973; recipient George Armstrong award Ambulatory Pediacatric Assn., Acad. Adminstrn. and Health Policy scholarship Assn. Acad. Health Ctrs., 1993. Fellow APHA, Am. Acad. Pediatrics (govt. affairs com. 1984-88, chpt. III youth com. N.Y. chpt. 1974-75, chmn. adolescent c-84); mem. Am. Pediatric Soc. (sec., treas. 1989—), Am. Bd. Pediatrics (examiner 1986—, long range planning com. 1990-91, chmn. long range planning com. 1992—, bd. dirs. 1990—, fin. com. 1991—, sec., treas. 1993—, search com. 1990), Soc. Adolescent Medicine, Alpha Omega Alpha. Office: Am Pediatric Soc PO Box 675 Elk Grove Village IL 60009-0675*

DEANGELIS, LISA MARIE, neurologist, educator; b. New Haven, Mar. 5, 1955; d. Daniel and Antoneta (Cocca) DeA.; m. Peter M. Okin, June 26, 1977; children: Daniel Andrew, Stephen David. BA, Wellesley Coll., 1977; MD, Columbia U., 1980. Diplomate Am. Bd. Psychiatry and Neurology. Intern in medicine Presbyn. Hosp., N.Y.C., 1980-81; resident in neurology Neurologic Inst. Presbyn. Hosp., N.Y.C., 1981-84, fellow clin. neuro-oncology, 1984-85; fellow neuro-oncology Meml. Sloan-Kettering Cancer Ctr., N.Y.C., 1985-86; clin. asst. Meml. Sloan-Kettering Cancer Ctr., 1986-89; asst. mem. Meml. Sloan-Kettering Cancer Ctr., N.Y.C., 1989-93, assoc. mem., 1993—; asst. prof. neurology Cornell U., N.Y.C., 1986-92, assoc. prof. neurology, 1992—. Contbr. articles to profl. jours. Recipient Clin. Oncology Career Devel. award Am. Cancer Soc., 1986-89, Boyer Young Investigator award Meml. Sloan-Kettering Cancer Ctr., 1992. Mem. Am. Acad. Neurology, Am. Neurol. Assn. Office: Meml Sloan Kettering 1275 York Ave New York NY 10021

DEANGELIS, THOMAS P., chemist; b. Apr. 2, 1951; married; two children. BS in Chemistry cum laude, Villanova U., 1973; MS in Analytical Chemistry, U. Cin., 1976, PhD in Chemistry, 1977. Tech. analyst Mobil Rsch. Corp., Paulsboro, N.J., 1973; grad. teaching asst. U. Cin., 1973-76; sr. scientist Corning (N.Y.) Inc., 1976-78, sr. rsch. scientist, 1979-83, rsch. assoc., supr. dept. ceramic rsch., 1984-87, mgr. advanced materials rsch., 1988-90; dir. ceramic tech. Carborundum Co., Niagara Falls, N.Y., 1990—; invited speaker 1st U.S.-Japanese workshop on combustion synthesis. Author (with others): Combustion and Plasma Synthesis of High Temperature Materials, 1990; contbr. articles to profl. jours. J. W. Richards fellow, 1975, Proctor and Gamble fellow, 1975-76; U. Cin. Rsch. Coun. grantee, 1975. Mem. Am. Chem. Soc. (Analytical Chemistry award 1973), Am. Ceramic Soc., Am. Def. Preparedness Assn., Navy League of U.S., Sigma Xi. Achievements include 7 patents on processes for making advanced ceramic materials, others; 2 patents pending.

DEASON, JONATHAN P., environmental engineer, federal agency administrator; b. Charleston, S.C., Feb. 8, 1948; married; 3 children. BS in Civil Engring., U.S. Mil. Acad., 1970; MBA in Mgmt., Golden Gate U., 1975; MS in Environ. Engring., Johns Hopkins U., 1978; PhD in Environ. Systems, U. Va., 1984. Registered profl. engr., Va. Commd. U.S. Army, 1970, advanced through grades to capt.; engr. officer U.S. Army Corps of Engrs., 1970-75, civil engr. North Atlantic Divsn., 1975-78; chief water resources program U.S. Bureau Indian Affairs, 1978-82; sr. policy advisor office of water policy U.S. Dept. Interior, 1982-83; spl. asst. Office Asst. Sec. of Army, 1983-86; mgr. Nat. Irrigation Water Quality Program U.S. Dept. of Interior, 1986-89, dir. Office Environ. Affairs, 1989—; adj. prof. environ. and energy mgmt. George Wash. U., 1984—; chmn. fed. liaison group Bd. Environ. Studies and Toxicology Nat. Rsch. Coun./NAS, 1990-91; mem. nat. panel of experts U.S. Com. Irrigation and Drainage, 1987; chmn. Pres.'s Task Force Indian Water Resources Devel., 1978-80. Author: (with others) Risk Based Decision Making in Water Resources, 1989; contbr. over 24 articles to profl. jours. Col. USAR. Recipient Engring. Achievement award Va. Engring. Found., 1993, Founder's medal and Fed. Engr. of Yr. award Nat. Soc. Profl. Engrs., 1992, Arthur S. Flemming award Jr. C. of C., 1984. Mem. Am. Soc. Civil Engrs. (bd. trustees scholarship trust 1992-93, pres. nat. capital sect. 1990-91, Meritorious Svc. award 1988), Am. Water Resources Assn. (dir. Chesapeake region 1989-91). Home: 7001 Petunia St Springfield VA 22152 Office: US Dept of the Interior Office of Environmental Affairs 1849 C St NW Washington DC 20240*

DEATON, JOHN EARL, aerospace experimental psychologist; b. San Diego, June 12, 1949; s. John Hoover and Ethel Lanore (Fry) D.; m. Mau Thi Vu, June 5, 1982; children: Lara Phuong Khanh, Sierra Phuong Thao. BA, San Diego State U., 1972, MA, 1975; PhD, Cath. U. Am., 1988. Commd. lt. (j.g.) U.S. Navy, 1980, advanced through the grades to comdr., 1993; project mgr. Office Naval Rsch., Arlington, Va., 1983-86; adj. prof. Embry Riddle Aero. U., Daytona Beach, Fla., 1981-92, Villanova (Pa.) U., 1988-91, U. Cen. Fla., Orlando, 1991—; engring. psychologist Naval Air Warfare Ctr., Warminster, Pa., 1988-91; aerospace psychologist Naval Tng. Systems Ctr., Orlando, 1991—; Navy rep. Human Factors Engring. Tech. Group, 1992—. Editor Human Factors Jour., 1993—; contbr. articles to profl. jours. Recipient Herman Salmon Publ. award Soc. Exptl. Test Pilots, 1992. Mem. Am. Psychol. Soc., Human Factors Soc., Sigma Xi. Home: 297 Saxony Ct Winter Springs FL 32708 Office: Naval Tng Systems Ctr 12350 Research Pky Orlando FL 32826

DEATON, LEWIS EDWARD, biology educator; b. Washington, July 5, 1949; s. James Edward and Bonnie Lovell (Condee) D.; m. Ellen Sue Humphreys, Oct. 17, 1981; children: James William, Thomas Edward. BS, William & Mary, 1970, MA, 1974; PhD, Fla. State U., 1979. Post-doctoral assoc. U. Tenn., Knoxville, 1979-81, SUNY, Stony Brook, 1981-83, U. Fla. Whitney Lab., St. Augustine, 1983-85; adj. asst. prof. Iowa State U., Ames, 1985-86; asst. prof. U. Southwestern La., Lafayette, 1986-91, assoc. prof., 1992—. Contbr. articles to profl. jours. 1st lt. U.S. Army, 1970-72, Viet Nam. Mem. Am. Soc. Zoology, Am. Soc. Cell Biology. Achievements include research on cellular volume regulation and comparative physiology. Office: Univ Southwestern La Biology Dept Lafayette LA 70504

DEATON, TIMOTHY LEE, computer systems integrator; b. Warsaw, Ind., Jan. 21, 1951; s. Wilfred Lowell and Patty Louise (Jones) D.; m. Marie Elaine Jones, De. 12, 1976. BS in Mechanical Engring., Purdue U., 1973, MS in Indsl. Adminstrn., 1974; diploma in Data Comm., MIT, 1986. Prodn. supr. RR Donnelley and Sons, Warsaw, Ind., 1972; project engr. Zimmer USA, Warsaw, 1973; gen. mgr. JI Case, Racine, Wis., 1974-76; dir. mktg. KCL Corp., Shelbyville, 1976-78; prin. Tim Deaton and Assocs., Shelbyville, 1976—; v.p. Cent. Computer Co., Inc., Shelbyville, 1977—. Author: How to Sell More, 1981, 2d edit., 1992. Office: Cent Computer Co Inc 1641 S Riley Hwy Shelbyville IN 46176-2855

DE BAETS, MARC HUBERT, immunologist, internist; b. Ghent, Belgium, Apr. 2, 1950; arrived in The Netherlands, 1981; s. Roland and Julia (Vertommen) De B.; m. Aldegonde Janssens, Sept. 24, 1977; children: Marleen, Patrick, Karolien. MD, U. Ghent, Belgium, 1974; PhD, U. Limburg, Maastricht, The Netherlands, 1984. Intern Univ. Hosp., Ghent, 1973-74, resident, 1974-79; rsch. fellow Scripps Clinic and Rsch. Found., La Jolla, Calif., 1979-81; asst. prof. U. Limburg, Maastricht, 1981-86, assoc. prof., 1986—; adj. head immunology dept. U. Limburg, Maastricht, 1986—. Author: Myasthenia Gravis, 1988, 2d edit., 1993; contbr. articles to profl. jours. Fogarthy fellow NIH, 1979, Fullbright fellow, 1985. Mem. Am. Assn. Immunologists, Dutch Assn. Immunologists, Dutch Assn. Endocrinologists, N.Y. Acad. Sci., 51 Internat. Home: Wijnantslaan 14, 3630 Maasmechelen Belgium Office: U Limburg Dept Immunology, PO Box 616, 6200 MD Maastricht Limburg, The Netherlands

DEBAKER, BRIAN GLENN, mechanical engineer; b. Green Bay, Wis., Nov. 13, 1961; s. Glenn Maurice and Jacqueline Ann (Frisbie) DeB.; m. Amy Teresa Heck, Sept. 9, 1989; 1 child, Daniel Andrew. BSME, U. Wis., 1986. HVAC project engr. G.M. DeBaker & Assocs., Inc., Manitowoc, Wis., 1980-88; regional supr. ANCO Cons. Group Inc., Milw., 1988-92; HVAC applications engr. Gustave A. Larson Co., New Berlin, Wis., 1992—. Mem. ASHRAE, Assn. of Energy Engrs., Triangle Fraternity (v.p. 1985-86). Home: 1729 Michigan Ave Waukesha WI 53188 Office: Gustave A Larson Co 2425 S 162nd Ave New Berlin WI 53151

DEBAKEY, MICHAEL ELLIS, cardiovascular surgeon, educator; b. Lake Charles, La., Sept. 7, 1908; s. Shaker Morris and Raheeja (Zorba) DeB.; m. Diana Cooper, Oct. 15, 1936; children: Michael Maurice, Ernest Ochsner, Barry Edward, Denis Alton, Olga Katerina; m. Katrin Fehlhaber, July 1975. B.S., Tulane U., 1930, M.D., 1932, M.S., 1935, LL.D., 1965; Docteur Honoris Causa, U. Lyon, France, 1961, U. Brussels, 1962, U. Ghent, Belgium, 1964, U. Athens, 1964; D.H.C., U. Turin, Italy, 1965, U. Belgrade, Yugoslavia, 1967; LL.D., Lafayette Coll., 1965; M.D. (hon.), Aristotelean U. of Thessaloniki, Greece, 1972; D.Sc., Hahnemann Med. Coll., 1973; Docteur honoris causa, U. Louis Pasteur, Paris, 1991. Diplomate Nat. Bd. Med. Examiners, Am. Bd. Surgery, Am. Bd. Thoracic Surgery. Intern Charity Hosp., New Orleans, 1932-33, asst. surgery, 1933-35; asst. surgery U. Strasbourg, France, 1935-36, U. Heidelberg, Fed. Republic of Germany, 1936; instr. surgery Tulane U., New Orleans, 1937-40, asst. prof., 1940-46, assoc. prof., 1946-48; prof., chmn. dept. surgery Baylor Coll. Medicine, 1948-93, v.p. med. affairs, 1968-69, chief exec. officer, 1968-69, pres., 1969-79, chancellor, 1979—; dir. Nat. Heart and Blood Vessel Research and Demonstration Ctr. Baylor (Tex.) Coll. Medicine, 1975-85, dir. DeBakey Heart Ctr., 1985—; surgeon-in-chief Ben Taub Gen. Hosp., 1963—; sr. attending surgeon Meth. Hosp.; clin. prof. surgery U. Tex. Dental Br., Houston; cons. surgery VA Hosp., St. Elizabeth's Hosp., M.D. Anderson Hosp., St. Luke's Hosp., Tex. Children's Hosp., Tex. Inst. Rehab. and Research Brooke Gen. Hosp., Brooke Army Med. Ctr., Ft. Sam Houston, Tex., Walter Reed Army Hosp., Washington.; mem. med. adv. com. Office Sec. Def., 1948-50, Ams. for Substance Abuse Prevention, 1984; Med. Adv. Bd., Internat. Brotherhood Teamsters, 1985—; chmn. com. surgery NRC, 1953, mem. exec. com., 1953; mem. com. med. services Hoover Commn.; Friends of Nat. Library of Medicine (founding bd. dirs.), 1985—; chmn. bd. regents Nat. Library Medicine, 1959; past mem. nat. adv. heart council NIH; mem. Nat. Adv. Health Council, 1961-65, Nat. Adv. Council Regional Med. Programs, 1965—, Nat. Adv. Gen. Med. Scis. Council, 1965, Program Planning Com., Com. Tng., Nat. Heart Inst., 1961—; mem. civilian health and med. adv. council Office Asst. Sec. Def.; chmn. Pres.'s Commn. Heart Disease, Cancer and Stroke, 1964; mem. adv. council Nat. Heart Lung and Blood Inst., 1982-87; mem. Tex. Sci. and Tech. Council, 1984-86; chmn. Found. Biomed. Rsch., 1988—, Physicians for Health in the Middle East, 1991—. Author: (with Robert A. Kilduffe) Blood Transfusion, 1942, (with Gilbert W. Beebe) Battle Casualties, 1952, (with Alton Ochsner) Textbook of Minor Surgery, 1955, (with T. Whayne) Cold Injury, Ground Type, 1958, A Surgeon's Visit to China, 1974, The Living Heart, 1977, The Living Heart Diet, 1985, The Living Heart Brand Name Shopper's Guide, 1992; editor: Yearbook of Surgery, 1958-70; chmn. adv. editorial bd.: Medical History of World War II; editor Jour. Vascular Surgery, 1984-88; contbr. over 1200 articles to med. jours. Mem. Tex. Constl. Revision Commn., 1973. Col. Office Surgeon Gen., AUS, 1942-46; now Col. Res.; cons. to Surgeon Gen., 1946—; disting. mem. US Army Med. Dept. Rgt., 1989. Decorated Legion of Merit, 1946, Independence of Jordan medal 1st class, Merit order of Republic 1st Class Egypt, comdr. Cross of Merit Pro Utiliate Hominum Sovereign Order Knights of Hosp. of St. John of Jerusalem in Denmark; recipient Rudolph Matas award, 1954, Internat. Soc. Surgery Disting. Service award, 1957, Modern Medicine award, 1957, Roswell Park medal, 1959, Leriche award Internat. Soc. Surgery, 1959, Great medallion U. Ghent, 1961, Grand Cross, Order Leopold Belgium, 1962, Albert Lasker award for clin. research, 1963, Order of Merit Chile, 1964, St. Vincent prize med. scis. U. Turin, 1965, Orden del Libertador Gen. San Martin Argentina, 1965, Centennial medal Albert Einstein Med. Ctr., 1966, Gold Scalpel awat. Cardiology Found., 1966, Disting. Faculty award, 1973, Eleanor Roosevelt Humanities award, 1969, Civilian Service medal Office Sec. Def., 1970, USSR Acad. Sci. 50th Anniversary Jubilee medal, 1973, Phi Delta Epsilon Disting. Service award, 1974, La Madonnina award, 1974, 30 Yr. Service award Harris County Hosp. Dist., 1978, Knights Humanity award honoris causa Internat. Register Chivalry, Milan, 1978, diploma de Merito Caja Costarricense de Seguro Social, San Jose, Costa Rica, 1979, Disting. Service plaque Tex. Bd. Edn., 1979, Britannica Achievement in Life award, 1979, Medal of Freedom with Distinction Presdl. award, 1969, Disting. Service award Internat. Soc. Atherosclerosis, 1979, Centennial award ASME, 1980, Marian Health Care award St. Mary's U., 1981, Clemson U. award, 1983, Humana Heart Inst. award, 1985, Theodore E. Cummings award, 1987, Nat. Med. of Sci. award 1987, Markowitz award Acad. Surg. Rsch., 1988, Assn. Am. Med. Colls. award 1988, Crille award Internat. Platform Assn., 1988, Thomas Alva Edison Found. award, 1988, first issue Michael DeBakey medal ASME, 1989, Inaugural award Scripps Clinic and Rsch. Found., 1989, DeBakey-Bard Chair in Surgery, Baylor Coll. of Medicine, 1990, Disting. Svc. award

Am. Legion, 1990, Lifetime Achievement award Found. for Biomed. Rsch., 1991, Jacobs award Am. Task Force for Lebanon, 1992, Maxwell Finland award Nat. Found. for Infectious Diseases, 1992, Lifetime Achievement award Acad. Med. Films, 1992, Order of Independence First Class medal United Arab Emirates, 1992, Academy of Athens award, 1992, Cmdrs. Cross Order of Merit (Fed. Germany), 1992, Gibbon award Am. Soc. Extracorporeal Tech., 1992, Michael E. DeBakey Libr. Svc. Outreach award Friends of the Nat. Libr., 1993, others; named Dr. of Yr., Med. World News, 1965, Med. Man of Yr., 1966, Disting. Service Prof., Baylor U., 1968, Humanitarian Father of Yr., 1974, Tulane U. Alumnus of Yr., 1974, Tex. Scientist of Yr., Tex. Acad. Sci., 1979; Michael E. DeBakey Heart Inst. Wis. named in his honor Kenosha Hosp. and Med. Ctr., 1992; Michael E. DeBakey, M.D. award for Excellence in Visual Edn. named in his honor, 1993. Fellow ACS (Ann. award Southwestern Pa. chpt. 1973), Inst. of Medicine Chgo. (hon.), Royal Coll. Physicians and Surgeons of U.S. (hon., disting. fellow 1992), Am. Inst. Med. and Biol. Engring. (founding fellow 1993); mem. AAAS, Am. Coll. Cardiology (hon. fellow), Royal Soc. Medicine, Halsted Soc., Am. Heart Assn., So. Soc. Clin. Research, Am. Coll. of Health Care Execs. (hon. fellow 1990), Southwestern Surg. Congress (pres. 1952), Soc. Vascular Surgery (pres. 1953), Soc. Vascular Surg. Lifeline Found. (pres. 1988), AMA (Disting. Service award 1959, Hektoen Gold medal), Am. Surg. Assn. (Disting. Service award 1981), So. Surg. Assn. (pres. 1989-90), Western Surg. Assn., Am. Assn. Thoracic Surgery (pres. 1959), Internat. Cardiovascular Soc. (pres. 1958, pres. N.Am chpt. 1964), Assn. Internat. Vascular Surgeons (pres. 1983), Mex. Acad. Surgery (hon.), Soc. Clin. Surg., Nat. Acads. Practice Medicine, Soc. Univ. Surgeons, Internat. Soc. Surgery, Soc. Exptl. Biology and Medicine, Hellenic Surg. Soc. (hon.), Bio-med. Engring. Soc. (bd. dirs. 1968), Houston Heart Assn. (mem. adv. council 1968-69), Soc. Nacional de Cirugia (Cuba), C. of C., Japanese Assn. Thoracic Surgery (first fgn. hon. mem. 1989), Assn. Francaise de Chirurgie (hon.), University Club (Washington), Acad. of Athens, Sigma Xi, Alpha Omega Alpha. Episcopalian. Achievements include development of roller pump universally used in the heart-lung machine, of Dacron artificial arteries and Dacron-velour arteries as surgical replacement of diseased arteries now used throughout the world, of first successful patch-graft angioplasty, of fundamental concept of therapy in arterial disease, of left ventricular bypass pump for cardiac assistance and first successful clinical application; first successful resection and graft replacement of fusiform aneurysm of descending thoracic aorta; first successful carotid endarterectomy for cerebrovascular insufficiency; first successful resection and graft replacement of aneurysm of distal aortic arch; first successful resection of dissecting aneurysm of thoracic aorta; first successful resection of aneurysm of thoracoabdominal aorta with replacement by graft including celiac, superior mesenteric and both renal arteries; first successful resection and graft replacement of aneurysm of ascending aorta; first successful resection with graft replacement of fusiform aneurysm of entire aortic arch; first successful aorto-coronary bypass with autogenous saphenous vein graft; many others. Office: Baylor Coll Medicine 1 Baylor Plz Houston TX 77030 also: Tex Med Ctr 6535 Fannin Houston TX 77030

DEBARI, VINCENT ANTHONY, medical researcher, educator; b. Jersey City, Feb. 1, 1946; s. Vincent and Josephine C. (Buzzanco) DeB.; m. Margaret A. Danning, Feb. 28, 1970; children: Michele, Christopher V., Jillanne. BS, Fordham U., Bronx, N.Y., 1967; MS, Newark Coll. Engring., 1970; PhD, Rutgers U., 1981. Rsch. and devel. chemist Witco Chem. Corp., Oakland, N.J., 1967-73; rsch. chemist St. Joseph's Hosp. & Med. Ctr., Paterson, N.J., 1973-81, dir. renal lab., 1981-89, dir. rheumatol lab., 1989—, dir. rsch., 1989—; assoc. prof. medicine Seton Hall U. Sch. Grad. Med. Edn., South Orange, N.J., 1988—, dir. rsch. internal medicine, 1989—; cons. Rutgers U., 1981, Biomed. Clin. Labs., Wayne, N.J., 1985, Micro-Membranes Inc., Newark, 1986-89, GenCare Biomed. Rsch. Corp., Mountainside, N.J., 1989—; med. staff affiliate St. Joseph's Hosp. & Med. Ctr., St. Michael's Med. Ctr., Newark. Contbr. over 50 articles to profl. jours. Bd. dirs. Lupus Erythematosus Found. N.J., Elmwood Park, 1983-91; trustee Paquannock Twp. (N.J.) Bd. Edn., 1986-91, Bay Head Shores Club, Point Pleasant, N.J., 1993. Grantee Lupus Found., Elmwood Park, N.J., 1978—, Lions Found., 1981-85; recipeint Boston Biomedica award Clin. Ligand Assay Soc., 1989. Fellow Nat. Acad. Clin. Biochemistry, Am. Inst. Chemists; mem. Am. Assn. Clin. Chemistry (chmn. N.J. 1990-92, chair clin. and diagnostic immunology div. 1991-93, Disting. Svc. award 1989, Clin. Chem. Recognition award 1984, 87, 90, 92, Bernard F. Gerulat award 1990), Nat. Coun. Univ. Rsch. Adminstrn., Am. Fedn. Clin. Rsch., Am. Coll. Rheumatology. Roman Catholic. Achievements include recognition of neutrophil defects in chronic hemodialysis patients, studies of autoantibodies in systemic autoimmune diseases and in AIDS, investigation of pathophysiologic effects of endotoxins, studies of relationship between surface electrochemistry and phagocytosis; rsch. on electrophoretic methods to study clonotype restrictions. Office: St Josephs Hosp & Med Ctr 703 Main St Paterson NJ 07503

DEBAS, HAILE T., gastrointestinal surgeon, physiologist, educator; b. Asmara, Eritrea, Ethiopia, Feb. 25, 1937; came to U.S., 1980; s. Tesfaye and Keddes (Gabre) D.; m. Ignacia Kim Assing, May 23, 1969. BS in Biology, U. Coll., Addis Ababa, Ethiopia, 1958; MD,CM, McGill U., Montreal, Que., Can., 1963. Intern Ottawa (Ont.) Civic Hosp., Can., 1963-64; resident in surgery U. B.C., Vancouver, Can., 1964-69, asst. prof. surgery, 1970-76, assoc. prof., 1976-79; fellow in gastrointestinal physiology UCLA, 1972-74, assoc. prof. surgery, 1979-81, prof., 1981-85; chief gastrointestinal surgery U. Wash., Seattle, 1985-87; prof., chmn. dept. surgery U. Calif., San Francisco, 1987—; key investigator Ctr. for Ulcer Rsch. and Edn., UCLA, 1980—; cons. NIH, Bethesda, Md., 1983-87, Bd. Med. Quality Assurance, Calif., 1983-85; bd. dirs. Am. Bd. Surgery. Mem. editorial bd. Am. Jour. Physiology, Am. Jour. Surgery, Jour. Surg. Rsch., Western Jour. Medicine, Gastroenterology; contbr. over 250 articles and abstracts to jours. and chpts. to books. Recipient Merit awards, Va., 1981-87; fellow Med. Rsch. Coun. of Can., 1972-74, rsch. grantee NIH, 1970—. Fellow ACS, Royal Coll. Physicians and Surgeons Can.; mem. AAAS, Am. surg. Assn., Am. Gastroent. Assn., Am. Assn. Endocrine Surgeous, Collequium Internat and Chirugiae Digestivae, Soc. Univ. Surgeons, Soc. Surgeons Alimentary Tract (trustee 1985-89), Inst. Medicine NAS, Inst. Medicine, Am. Acad. Arts and Scis., Internat. Hepato-Biliary Pancreatic Assn. (pres. 1991-92), Assn. Minority Acad. Physicians (pres. 1993—). Office: Univ Calif Dept Surgery 505 Parnassus Ave # 320S San Francisco CA 94143-0001

DEBENEDETTI, PABLO GASTON, chemical engineer; b. Buenos Aires, Argentina, Mar. 30, 1953; came to U.S., 1980; s. Sergio Isaias and Francine Fanny (Lehmann) D.; m. Silvia Irene Strauss, July 11, 1987; 1 child, Gabriel Alejandro. BS in Chem. Engring., Buenos Aires U., 1978; MS, MIT, 1981, PhD, 1984. Rsch. engr. O de Nora Impianti Elettrochimici, Milan, Italy, 1978-80; asst. prof. Dept. Chem. Engring. Princeton (N.J.) U., 1985-90, assoc. prof., 1990—; edit. adv. bd. Jour. Supercritical Fluids, 1988—. Contbr. articles to profl. jours. including Jour. Chem. Phys., Jour. Phys. Chem., others. European Econ. Community fellow, 1978, Camille and Henry Dreyfus Tchr. scholar, 1989, Guggenheim fellow, 1991; named NSF Presdl. Young Investigator, 1987. Mem. Am. Inst. Chemical Engrs., Am. Chemical Soc., Am. Physical Soc., AAAS, Sigma Xi. Achievements include development of the concept of attractive, repulsive and weakly attractive supercritical mixtures; formation of polymer microparticles for controlled drug release by rapid expansion of supercritical solutions, formation of protein powders using supercritical antisolvents; theory of thermodynamic stability of supercooled liquids; computer simulation of molecular interactions in dilute supercritical mixtures. Office: Princeton U Dept Chem Engring Princeton NJ 08544

DEBEYSSEY, MARK SAMMER, molecular and cellular biologist; b. Putnam, Conn., Mar. 24, 1966; s. Ghaleb and Widad Debeyssey. BS cum laude, U. Conn., 1988, MS in Moledular and Cell Biology, 1992. Sr. toxicologist Ciba-Geigy, Farmington, Conn., 1988-89; rsch. scientist U. Conn. Health Ctr., Farmington, 1989-90; molecular and cellular biologist VA Med. Ctr., West Haven, Conn., 1990—. Contbr. articles to profl. jours. Chmn. Am. Druze Soc., Conn., 1990-91. Recipient Superior Performance award Dept Vets. Affairs, 1991. Mem. AAAS, Am. Chem. Soc., Planetary Soc. (assoc.), Am. Mus. Natural History, Archaeol. Inst. Am. (assoc.), Smithsonian Nat. Assocs. Office: VA Med Ctr 950 Campbell Ave West Haven CT 06516-2700

DE BIASI, RONALDO SERGIO, materials science educator; b. Rio de Janeiro, Mar. 11, 1943; s. Renato and Ruth Freitas (Pinho) de B.; m. Marilia Villa-Forte Coutinho, June 15, 1967; children: Sérgio, Cláudio. BSEE, Pontifical Cath. U., Rio de Janeiro, 1965; MSEE, PUC/RJ, 1967; PhD, U. Wash., 1971. Asst. prof. materials sci. Inst. Mil. Engring., Rio de Janeiro, 1971-74, prof., 1974—. Author: The World of Electronics, 1966, Fast Guide to DOS 6.0, 1993, also 5 others; editor Isaac Asimov Mag.; regional editor Revista Latino-Am. de Metalurgia y Materiales; mem. editorial bd. Revista Militar de Ciência e Tecn.; contbr. numerous articles to profl. jours. Mem. Am. Phys. Soc., Brazilian Soc. Physics. Avocations: soccer, playing piano. Home: Rua Pinheiro Machado 99/901, 22231 Rio de Janeiro RJ, Brazil Office: Inst Mil Engring, Pr Gen Tiburcio 80, 22290-270 Rio de Janeiro Brazil

DE BLAS, ANGEL LUIS, biologist, educator; b. Madrid, Jan. 31, 1950; came to U.S., 1974; s. Clemente De Blas and Mercedes Ortega; m. Celia Pilar Miralles; 1 child, Celia Aurora. BS in Biology, U. Madrid, 1972, MS in Biochemistry, 1972; PhD in Biochemistry, Ind. U., 1978; postgrad., NIH, Bethesda, Md., 1978-81. Asst. prof. SUNY, Stony Brook, 1981-86, assoc. prof., 1986-89; prof. U. Mo., Kans. City, 1989—; grant reviewer NSF, Washington, 1984—, NIH, Washington, 1985—, Human Frontier Sci. Program, Strasbourg, France, 1988—; sci. advisor Internat. Found. for Sci., Stockholm, 1991—. Contbr. articles to profl. jours. Recipient John E. Fogarty fellowship, Washington, 1978-81, Esther A. and Joseph Klingenstein fellowship award in neuroscience, N.Y.C., 1985-88, faculty fellowship U. Kans. City Bd. Trustees, 1992. Mem. Internat. Soc. for Neurochemistry, Internat. Brain Rsch. Orgn., Society forNeuroscience, Am. Soc. for Biochemistry & Molecular Biology, Am. Soc. for Neurochemistry, N.Y. Acad. Sciences. Office: U Mo Div Molecular Biology 109 Biol Scis Bldg Kansas City MO 64110-2499

DEBLASI, ROBERT VINCENT, school system administrator; b. Lodi, N.J., Sept. 23, 1936; s. Patrick Archimedes and Nancy (Pecoraro) DeB.; m. Marie Sarah Cerini, June 30, 1963; children: Glenn Robert, Lori Marie, Keith Charles. BA in Edn., Paterson State Coll., 1963, MA in Edn., 1968. Cert. elem. tchr., sci. specialization, supr., prin., N.J. Tchr. Paramus (N.J.) Pub. Schs., 1963-68, sci. cons., 1968-69, acting prin., 1984, acting administrv. asst., 1985, sci. supr., 1989—; mem. staff Ctr. Elem. Sci., Madison, N.J., 1985—; adv. bd. Liberty Science Ctr., Jersey City, 1990—, Hackensack Water Co., 1990-92; mem. writing com. Elem. Sci. Curriculum Guide N.J. State Dept. Edn. Mem. editorial com. The Assembly Guide for Self-Study, Evaluation, and Accreditation of Elementary Middle Schools and School Systems, 2d edit., 1981; contbr. articles to profl. jours. wtih USN, 1955-59. Grantee NSF, 1979, N.J. State Dept. Edn., 1989-90; recipient Svc. award N.J. Sci. Conv., 1981, 86, 91, commendation for exemplary elem. edn. sci. programs, N.J. State Bd. Edn., 1985;. Mem. Nat. Sci. Suprs. Assn. (com. to elect Nation's Outstanding Sci. Supr., 1978-79, 84-85, chair planning com. Summer Leadership conf., 1978, issues and action com., 1984-85, futures com. 1989-90, at. large, 1981-84, steering com. Leadership Inst., 1987, cofounder Nat. Outstanding Supr. Award, recipient Excellence in Sci. Edn. 1988, bd. dirs. Region B 1992-93), Nat. Sci. Tchrs. Assn.(Excellence in Sci. Edn. award 1984), Nat. Assn. Atomic Vets., N.J. Sci. Suprs. Assn. (charter 1968—, sec., treas. 1971-72, v.p. 1972-73, chair sci. materials exhibit 1975, 76, pres. 1975-76, liaison to Nat. Sci. Suprs. Assn., 1977—, Disting. Svc. award 1981, Pres.'s award 1984, 93, historian 1984—, cofounder N.J. Sci. Conv.), N.J. Sci. Tchrs. Assn., Disabled Am. Vets., Am. Legion, Kappa Delta Pi. Home: 16 Kings Rd Rockaway NJ 07866

DEBLIEU, IVAN KNOWLTON, plastic pipe company executive, consultant; b. Opelousas, La., Aug. 21, 1919; s. Ivan Knowlton and Lucile (Wells) DeB.; m. Helen Louise Snider, Dec. 2, 1950; children: Kenneth A., Janice K., Douglas J. BSChemE, U. Fla., 1942. Tech. engr. E.I. duPont de Nemours, Childersburg, Ala., 1942-45, Parkersburg, W. Va., 1947-56; product engr. E.I. duPont de Nemours, Wilmington, Del., 1956-60; product specialist E.I. duPont de Nemours, Wilmington, 1961-76, product program mgr., 1976-84, sr. product program mgr., 1984-86; pres. DeBlieu & Assocs., Wilmington, 1986—. Contbr. articles to Modern Plastics Encyclopedia, Modern Plastic Mag., Am. Gas Assn. Distbn. Proceedings, Symposia on Plastic Piping for Gas Distbn., Plastic Pipe Inst. Plastic Piping Manual.; author (with others) Am. Gas Assn. Manual for Gas Service, 1985. Bd. mgrs. YMCA, Wilmington, 1960-66; scout master Boy Scouts Am., 1963-66. chmn. troop com., 1975-79. With U.S. Army, 1946-47. Recipient award of Merit, Am. Gas Assn., 1978, cert. Appreciation, 1986. Fellow ASTM (chmn. awards com. 1984-90, Internat. Standards Orgn. liaison 1988-90, Merit award 1988, Svc. award 1990, Marshall Hall award 1993); mem. Plastic Pipe Inst. (chmn. tech. bd. 1975-76, v.p. 1977, pres. 1978-81, bd. dirs. 1975-83, hydrostatic stress bd. 1971-90, Plaque Appreciation 1986, hon. life mem. 1990), Soc. Plastic Industries (bd. dirs. 1978-81), Internat. Standards Orgn. (chmn. U.S. del. to subcom. 1973-90, chmn. U.S. del. to full com. 1986-90). Republican. Presbyterian. Achievements include initiation or improvement of 23 ASTM specifications applicable to plastic piping which upgraded safety and quality of the plastic piping industry; incorporation into specification for polyethylene pipe critical requirement which validated safe long term performance in pressure applications; development of standard molding and extrusion industry acrylic plastic composition, acrylic plastic composition resistant to high level ultraviolet emmissions making possible application of acrylic refractors in outdoor use with mercury vapor lights, and highway surface light reflectors. Home and Office: DeBlieu and Assocs Inc 127 Murphy Rd Wilmington DE 19803-3046

DEBNATH, LOKENATH, mathematician, educator; b. Hamsadi, Bengal, India, Sept. 30, 1935; came to U.S., 1968; s. Jogesh Chandra and Surabala (Nath) D.; m. Sadhana Bhowmik, Aug. 1, 1969; 1 child, Jayanta. BS in Pure Math., U. Calcutta, India, 1954, MSc in Pure Math., 1956, PhD in Pure Math., 1965; DIC and PhD in Applied Math., Imperial Coll. Scis., London, 1967. Prof. math. East Carolina U., Greenville, N.C., 1968-82, prof. physics, 1972-82; chair, prof. math. U. Cen. Fla., Orlando, 1983—, prof. mech. and aero. engring., 1991—; presenter seminars, lectr. in field; presenter at profl. confs. Co-author: Introduction to Hilbert Spaces with Applications, 1990, Partial Differential Equations for Scientists and Engineers, 1987, Nonlinear Waves, 1983, Advances in Nonlinear Waves, 1984, Elements of the Theory of Elliptic and Associated Functions with Applications, 1965, Elements of General Topology, 1964, Nonlinear Dispersive Wave Systems, 1992; editorial bd. Internat. Jour. Math. and Math. Scis., 1976—, Internat. Jour. Math. Edn. in Sci. and Tech., 1983-89, Internat. Jour. Math. Stats. Sr. Fulbright fellow, USSR, 1978; sr. scientist U.S.-India Exch. Scientist program NSF, 1975; grantee NSF, USOAS, Internat. Ctr. Theoretical Physics, others. Mem. Am. Math. Soc., Am. Phys. Soc., Math. Assn. Am. , Soc. Indsl. and Applied Math., Calcutta math. Soc., Indian Sci. Congress Assn., Indian Assn. Theoretical and Applied Mechanics, Inst. Math. and its Applications (Eng.), Calcutta Math Soc. (pres. 1987-90), Sigma Xi, Phi Kappa Phi, Pi Mu Epsilon. Home: 3601 TCU Blvd Orlando FL 32817 Office: Univ Cen Fla PO Box 161364 Orlando FL 32816-1364

DE BOER, PIETER CORNELIS TOBIAS, mechanical and aerospace engineering educator; b. Leiden, Netherlands, May 21, 1930; s. Pieter and Willemina (Zuydam) deB.; m. Joan Lieshout, June 7, 1956; children: Maarten P., Claire E., Yvette E. MechE degree, Delft U. Tech., 1955; PhD in Physics, U. Md., 1962. Rsch. asst., assoc. Tech. U. Delft, 1954-55; rsch. assoc. U. Md., 1957-62, rsch. asst. prof., 1962-64; asst. prof. Cornell U., 1964-68, assoc. prof., 1968-74; prof. Sibley Sch. Mech. and Aerospace Engring., Cornell U., 1974—, assoc. dir., 1982-91; mem. tech. staff Aerospace Corp., summer 1963, 65, 67, 92-93; vis. prof. von Karman Inst. for Fluid Dynamics, Belgium, 1968, Cornell Aero. Lab., Buffalo, 1969, Tech. U. Delft, 1985-86; mem. tech. staff Ford Motor Co., 1971-73, gas turbine div. GE Co., 1978-79; cons. Conelec, Elmira, N.Y., Allied Chem., Inc., Buffalo Inst. for Def. Analyses, Arlington, Va., others. Assoc. editor N.Am. Applied Scientific Rsch., 1987—; contbr. articles to profl. jours. With Dutch Army, 1955-57. NATO fellow, 1968. Assoc. fellow AIAA; mem. ASME, Am. Phys. Soc., Am. Soc. Engring. Edn., Internat. Assn. Hydrogen Energy, Royal Inst. Ingenieurs Netherlands, Royal Netherlands Acad. Scis. (corr.), Golden Key Soc., Rsch. Club Cornell U., Finger Lakes Cycling Club, Finger Lakes Runners Club, Cayuga Nordic Ski Club (pres.), Sigma Xi, Pi Tau Sigma, Sigma Pi Sigma. Office: Cornell U Sibley Sch Mech Aerospace Upson Hall Ithaca NY 14853

DE BOKX, PIETER KLAAS, research chemist; b. Pematang-Siantar, Sumatra, Indonesia, Oct. 5, 1955; came to U.S., 1964; s. Jacob Anné and Saakje (Wassenaar) De B.; m. Carola Cornelia Stallen, Mar. 31, 1981; children: Thomas Geleyn, Carl Adriaan. MSc, U. Utrecht (the Netherlands), 1981, PhD, 1985. Rsch. scientist Philips Rsch. Labs., Eindhoven, the Netherlands, 1985—. Contbr. over 25 articles to Jour. Catalysis, Jour. Phys. Chemistry, Jour. Chromatography and others. Pres. Home-Owners Assn., Utrecht, 1982-84. Mem. Royal Dutch Chem. Assn. Achievements include research on kinetics of catalytic decomposition, on catalyst deactivation (growth of Filamentous Carbon), on compensating mixture theory for multicomponent mixtures; on theory of Chromatography. Office: Philips Rsch Labs, PO Box 80000, 5600 JA Eindhoven The Netherlands

DE BOLD, ADOLFO J., pathology and physiology educator, research scientist; b. Paraná, Argentina, Feb. 14, 1942; arrived in Can., 1968; s. Adolfo E.G. and Ana (Patriarca) deB.; m. Mercedes L. Kuroski; children: Adolfo A., Alejandro J., Cecilia I., Gustavo A., Pablo G. B.Sc. (hon.), Faculty Chem. Sci., Cordoba, Argentina, 1968; M.Sc. in Pathology, Queen's U., Kingston, Ont., 1971, PhD in Pathology, 1973. Cert. clin. chemist. Demonstrator in physics Nat. U. Cordoba, 1961-62, demonstrator normal and path. histology, 1964-67; resident, chief resident Nat. Hosp., Clinicas, Cordoba, 1966-68; asst. prof., lab. scientist Queen's U. and Hotel-Dieu Hosp., Kingston, 1974-82, assoc. prof., 1982-85, prof., 1985-86; prof. pathology and physiology U. Ottawa, Ont., Can., 1986—; bd. dirs. research U. Ottawa Heart Inst. at Ottawa Civic Hosp., 1986—. Discovered Atrial Natriuretic Hormone, 1981, patented, 1986; contbr. over 100 sci. articles and chpts. to books in field. Bd. dirs. Heart Inst., Ottawa, 1986—. Recipient Gairdner Internat. award Gairdner Found., Toronto, Ont., 1986, Manning Prin. award Manning Found., Alta., Can., 1986, Sci. Achievement award Am. Soc. Hypertension, N.Y., 1986, Rsch. Achievement award Can. Cardiovascular Soc., Ottawa, 1986, Disting. Rsch. Prof. award Ont. Heart and Stroke Found. Fellow Royal Soc. Can.(McLaughin medal of excellence in rsch. 1988), Royal Coll. Physicians and Surgeons (Can.); mem. Can. Hypertension Soc., Am. Soc. for Hypertension, Internat. Soc. Hypertension (Rsch. Achievement award), Internat. Soc. Heart Rsch., Am. Sect. Can. Fedn. Biol. Socs., Histochem. Soc., U.S. Acad. Pathology, Can. Acad. Pathology, AAAS, Am. Soc. Cell Biology, Can. Soc. Cell Biology, Internat. Acad. Pathology, Am. Assn. Pathology, Fedn. Am. Soc. Exptl. Biology, Microscopial Soc. Can., Soc. Exptl. Biology and Medicine, Can. Soc. Anatomy, N.Y. Acad. Sci., Order of Can. (officer). Roman Catholic. Avocation: classical guitar. Office: U Ottawa Heart Inst, 1053 Carling Ave, Ottawa, ON Canada K1Y 4E9

DEBONO, KENNETH GEORGE, psychology educator; b. N.Y.C., July 28, 1958; s. Samuel Charles and Joan Marie (Daley) DeB.; m. Shirley Katheryn Bourquin, Aug. 29, 1981; children: Craig, Jillian. BA, Grinnell Coll., 1980; PhD, U. Minn., 1985. Asst. prof. Mich. State U., East Lansing, Mich., 1985-86; assoc. prof. Union Coll., Schenectady, N.Y., 1986—. Contbr. articles to profl. jours. Mem. APA, Soc. for Exptl. Social Psychology, Am. Psychol. Soc., Ea. Psychol. Assn., Phi Beta Kappa (treas. Alpha of N.Y. 1989—). Office: Union Coll Dept Psychology Schenectady NY 12308

DEBOO, BEHRAM SAVAKSHAW, microbiologist; b. Navsari, Gujarat, India, Sept. 4, 1931; came to U.S., 1955; s. Savakshaw Faramji and Cooverbai B. (Awari) D.; m. Manijeh Gushtaspa Nikmanesh, May 23, 1965; children: Kuma, Cyrus. BSc, U. Gujarat, 1954; BS, U. Wash., 1972. Lab. technician St. Luke's-Presbyn. Hosp., Chgo., 1959; med. technologist Providence Hosp., Everett, Wash., 1959-62; chief microbiologist Everett Clinic, 1962—; mem. adv. bd. Prepared Media Lab., Tualatin, Oreg., 1978-79. Author booklet: Parsis and Inbreeding, 1990. Mem. Am. Soc. Microbiology. Zoroastrian. Achievements include first to isolate pseudomonas vesicularis and vancomycin sensitive strains of gonococci from cervical cultures. Home: 2934 Panaview Blvd Everett WA 98203-6915 Office: Everett Clinic 4004 Colby Ave # 101 Everett WA 98201-4929

DE BOOR, CARL, mathematician; b. Stolp, Germany, Dec. 3, 1937; m. Matilda C. Friedrich, Feb. 6, 1960 (div. Sept. 12, 1984); children—C. Thomas, Elisabeth, Peter, Adam; m. Helen L. Bee, Jan. 2, 1991. Student, Universitaet Hamburg, 1956-59, Harvard U., 1959-60; Ph.D., U. Mich., 1966. Rsch. mathematician Gen. Motors Research Labs., 1960-64; asst. prof. math., computer sci. Purdue U., 1966-68, assoc. prof., 1968-72; prof. math., computer sci. U. Wis.-Madison, 1972—; vis. staff mem. Los Alamos Sci. Labs., 1970—. Author: (with S. Conte) Elementary Numerical Analysis, 1972, 80, A Practical Guide to Splines, 1978, (with J.B. Rosser) Pocket Calculator Supplement for Calculus, 1979. Fellow Am. Acad. Arts and Scis.; mem. NAE, Soc. Indsl. and Applied Math., Phi Beta Kappa. Office: U Wis Depts Computer Scis & Math Madison WI 53706

DEBROECK, DENNIS ALAN, design engineer; b. Van Nuys, Calif., Jan. 29, 1957; s. Justin Bernard and Janice Arlene (Buzzell) DeB.; m. Kathi Lynn Cole, June 17, 1978; children: Katie Lynn, Daniel Alan. Grad. high sch. with honors, Milton-Freewater, Oreg., 1975. Cons. Hanford Nuclear Plant, Richland, Wash., 1983, Oreg. D.E.Q., Portland, 1984; design engr. 3-D Tank & Petroleum Equipment, Milton-Freewater, 1975-86, DTEK Corp., Milton-Freewater, 1986—. Deacon Ch. of Christ, 1989-90. Achievements include design of over 100 petroleum fueling sites, an infrared detector, a volcanic filter for consoles and programmable dispensers for card systems. Office: DTEK Corp 3475 Powerline Rd Walla Walla WA 99362

DE BROUWER, NATHALIE, librarian; b. Montreal, Nov. 27, 1962. MLS, Montreal U., 1987, cert., 1991. Asst. libr. McCarthy Tetrault, Montreal, 1987-91, Nordic Labs., Laval, Canada, 1990-91; info. specialist, libr. Noranda Tech. Ctr., Pointe-Claire, Canada, 1991—. Author: Index of Courrier du Sud Newspapers, 1987. Office: Noranda Tech Ctr, 240 Hymus Blvd, Pointe Claire, PQ Canada H9R 1G5

DEBUSK, GEORGE HENRY, JR., paleoecologist; b. Goose Bay AFB, Canada, Oct. 10, 1966; s. George Henry and Clea (Moraes) Vieira DeB. BS, Clemson U., 1987. Teaching asst. zoology dept. Duke U., Durham, N.C., 1990—. NSF fellow Duke U., 1987-90; grantee Sigma Xi, 1992. Mem. AAAS. Republican. Home: 1242 Hillside Dr Hanahan SC 29406 Office: Duke U Zoology Dept 243 Bio-Sci Box 90325 Durham NC 27708-0325

DE CEUSTER, LUC FRANS, avionics educator; b. Antwerp, Belgium, Mar. 28, 1959; s. Victor and Augusta (Huybrechts) De C. M in Engring. Sci., Royal Mil. Acad., Brussels, Belgium, 1982. Enlisted Royal Engrs. Belgium Army, 1977, advanced through grades to capt., 1983; platoon commdr. Royal Engrs. Belgium Army, Cologne, Germany, 1983-84, com. commd., 1984-85; architect Mil. Constr. Agy., Leopoldsburg, Belgium, 1985-89; co. adj. Royal Engr. Belgium Army, Arolsen, Germany, 1989-90; educator avionics power plants Belgian Aviation Sch., Haren, 1990—. Mem. AIAA, Koninklyke Vlaamse Ingenieursvereniging. Home: Hof Ter lLo 8/31, 2140 Borgerhout Belgium Office: Belgium Aviation Sch, Raketstraat 90, 1130 Haren Belgium

DECHER, RUDOLF, physicist; b. Wuerzburg, Ger., Aug. 22, 1927; came to U.S., 1960, naturalized, 1967; s. Hermann Alexander and Karola (Krenig) D.; m. Christa Anna Hort, Jan. 7, 1956; children—Peter H., Marianne C. M. in Physics, U. Wuerzburg, W. Ger., 1950, Ph.D. in Physics, 1954. Research scientist Dynamit AG, Troisdorf, W. Ger., 1955-60; with NASA Marshall Space Flight Ctr., Huntsville, Ala., 1960—; chief astrophysics div., space sci. lab. NASA Marshall Space Flight Ctr., 1970-86, asst. dir. space sci. lab., 1986-89, chief astrophysics div., space sci. lab., 1989—. Recipient Exceptional Service medal NASA, 1977. Mem. Am. Phys. Soc., AIAA. Roman Catholic. Home: 718 Owens Dr SE Huntsville AL 35801-2034 Office: ES61 Marshall Space Flight Ctr Huntsville AL 35812

DE CHERNEY, ALAN HERSH, obstetrics and gynecology educator; b. Phila., Feb. 13, 1942; s. William Aaron and Ruth (Hersh) DeC.; m. Deanna Faith Saver, June 26, 1966; children: Peter, Alexander, Nicholas. BS in Natural Scis., Muhlenberg Coll., 1963; MD, Temple U., 1967; MA (hon.), Yale U., 1985. Diplomate Am. Bd. Ob-Gyn (examiner 1984—), am. Bd. Reproductive Endocrinology (bd. dirs. 1988—), Nat. Bd. med. Examiners (examiner 1987-90). Intern in gen. medicine U. Pitts., 1967-68; resident in ob-gyn. U. Pa., Phila., 1968-72, instr. dept. ob-gyn, 1970-72; asst. prof. ob-gyn. Yale U. Sch. Medicine, New Haven, 1974-78, assoc. prof., 1979-84, prof., 1984-91, John Slade Ely prof. ob-gyn, 1987-92, dir. div. reproductive endocrinology, dept. ob-gyn, 1982-92, lectr. dept. biology, 1985-92; Louis E. Phaneur prof., chmn. dept. ob-gyn. Tufts U. Sch. Medicine, 1992—. Maj. U.S. Army, 1972-74. Recipient Disting. Alumni award Temple U., 1989. Fellow ACOG, Am. Fertility Soc., Soc. for Assisted Reproductive Tech. (pres. 1987-88), Am. Assn. History of Medicine, Soc. Reproductive Endocrinologists (pres. 1988), Soc. Reproductive Surgeons (charter, pres. 1991), Endocrine Soc., European Soc. Human Reproduction and Embryology, Soc. Gynecologic Surgeons, Soc. for Study of Reproduction. Democrat. Jewish. Office: Tufts U Dept Ob-gyn 750 Washington St Boston MA 02111

DECHERNEY, GEORGE STEPHEN, research scientist, research facility administrator; b. Wilmington, Del., June 16, 1952; s. Herman George and Grace Antoinette (Lewis) DeC.; m. Cleonice Anne DiSabatino, June 9, 1992; children: Elizabeth, Constance, Sarah, Elliot. BA, Columbia U., 1974; MD, Temple U., 1978. Instr. Vanderbilt U., Nashville, 1983-84; asst. prof. Uniformed Svcs. U., Bethesda, Md., 1985-89; dir. Diabetes and Metabolic Ctr., Wilmington, 1989—; clin. assoc. prof. Thomas Jefferson Univ., Phila., 1989—; chief clin. pharmacology, endocrinology Med. Ctr. Del., Newark, 1990—; dir. Med. Rsch. Inst. Del., Newark, 1990—; assoc. prof. physiology U. Del., 1991—; invited internat. speaker 50th anniversary Greenslopes Hosp., Brisbane, Australia. Maj. USAF, 1986-89. Mem. AMA, Am. Diabetes Assn. (pres. Del. affiliate 1990—), N.Y. Acad. Scis., Endocrine Soc. Achievements include research in endocrinology, in diabetes mellitus, in aerospace medicine. Office: Med Rsch Inst Del 4755 Ogletown-Stanton Rd Newark DE 19718

DE CHINO, KAREN LINNIA, engineering association administrator; b. Hartford, Conn., Dec. 31, 1955; d. George Arthur and Carol Ann (Nelson) Holmelund; m. Frank Louis De Chino, Mar. 22, 1979; 1 child, Brittanie Francis. BA in Psychology, Montclair State Coll., 1978; MS in Applied Psychology, Stevens Inst. Tech., 1992. Counselor Livingston (N.J.) Youth Svcs. Bur., 1974-77; adminstrv. asst. Bamberger's, Newark, N.J., 1978-80; nursing scheduler St. Barnabas Med. Ctr., Livingston, 1981-83; tng. adminstr. Singer-Kearfott, Little Falls, N.J., 1983-86; human resources rep. Kearfott Guidance & Navigation Corp., Little Falls, N.J., 1986-89; mgr. bus. & adminstrn. IEEE, Piscataway, N.J., 1989-92, dir. tech. programs, 1992—. Mem. Stas. Engring. Soc. Office: IEEE 445 Hoes Ln PO Box 1331 Piscataway NJ 08855-1331

DECHMANN, MANFRED, psychotherapist; b. Düsseldorf, Fed. Republic of Germany, Nov. 7, 1942; s. Johann Dechmann and Josefine (Buttermann) Inge; m. Heide Birgit Bofinger, Dec. 24, 1969; children: Dina Kea Noanoa, Caspar Daniel, Anna Nele. Degree in sociology, U. Zürich, Switzerland, 1971, PhD, 1976, lic. phil. psychology, 1978. Asst. Deutsche Bundespost, 1960-64; rsch. asst. Sociol. Inst., Zürich; lectr. U. Zürich, 1971-81, Kaderschule SRK, Zürich; lectr., supr. Sch. Social Work, Zürich; supr. Couple & Family Tng.; psychotherapist, cons. Zürich; cons. Zurich; pres. Couple & Family Therapists Schweizerische Gesellschaft fur Gruppenpsychologie und Gruppendynamik, Switzerland, 1981-88; CEO IMPACT Selbstverteidigung AG. Author: Teilnahme und Beobachtung; Sprache, Denken, Wissenschaft; Aktive Patienten; contbr. articles to profl. jours. Mem. Greenpeace, European Soc. Med. Sociology, Schweizerische Gesellschaft für Hypnotherapie. Avocations: dogs, sailing, aikido, cities, biking. Office: PO Box 29, 8042 Zurich Switzerland

DECKER, ARTHUR JOHN, optical physicist, researcher; b. Butte, Mont., Oct. 16, 1941; s. Lester Paul and Louise Constance (Kraft) D.; m. Marilyn Ann Goe, Nov. 7, 1970; 1 child, Bruce Lester. BS in Physics, U. Wash., 1963; MA in Physics, U. Rochester, 1966; PhD, Case We. Res. U., 1977. Rsch. scientist NASA-Lewis Rsch. Ctr., Cleve., 1966—. Contbr. articles to profl. jours. Mem. Optical Soc. Am. Achievements include methods for using pulsed laser holographic interferometry for measurement of the properties of aerospace flows and structures; incorporation of artificial neural network technology in optical measurement systems. Office: NASA-Lewis Rsch Ctr MS 77-1 21000 Brookpark Rd Cleveland OH 44135

DECKER, CHRISTIAN LUCIEN, research chemist; b. Guebwiller, Alsace, France, June 10, 1940; s. Lucien and Lucie (Kastler) D.; m. Danielle Freyss, Sept. 4, 1964; children: Luc, Sylvaine. M in Chemistry, Univ., Strasbourg, France, 1961, Dr es-Sci., 1967. Postdoctoral fellow Stanford Rsch. Inst., Menlo Park, Calif., 1971-72; rsch. asst. CNRS-Univ., Strasbourg, France, 1967-80; rsch. dir. CNRS-Univ., Mulhouse, 1981-92; head of polymer photochemistry lab. U. Mulhouse; internat. adv. bd. RadTech-Asia, Tokyo, 1990-93. Author: (book chpts.) Degradation and Stabilization of PVC, 1984, Handbook of Polymer Science, 1989, Polymers for Microelectronics, 1990, Lasers in Polymer Science, 1990, Laser-Assisted Processing, 1991, Radiation Curing-Science and Technology, 1992. Named Best Sci. paper RadTech Corp., 1985, 87, 92. Achievements include development of highly reactive monomers for radiation curing applications; study of laser-induced polymerization of acrylic resins; real time monitoring of ultrafast polymerizations; direct laser writing of microcircuits.; photostabilization of polymers. Office: CNRS Univ, 3 Rue Werner, 68200 Mulhouse France

DECOSTA, PETER F., chemical engineer; b. New Bedford, Mass.; s. Anthony and Deolinda (DeSouza) DeC. BS with distinction, U. Mass., Dartmouth, 1962; MA, U. Conn., 1966; cert., U. R.I., 1968; MS, MIT, 1970; cert., Boston U., 1979, Northeastern U., 1980; profl. engring. degree, U. Wis., 1981. Chemist Portsmouth Naval Shipyard, N.H., 1962-64; quality assurance engr. U.S. FDA, Boston, 1966-71; systems analyst U.S. EPA, Washington, 1971-75; ops. rsch. analyst U.S. Army Natick (Mass.) Rsch., Devel. and Engring. Ctr., 1976-83, phys. scientist, 1983-86, gen. engr., 1986—; part-time instr. Northeastern U., Boston, 1987—, Worcester (Mass.) Poly. Inst., 1984—, Mass. Bay Community Coll., Wellesley, 1983—. Contbr. articles to profl. jours. Planner, Planning Bd. City of New Bedford, Mass., 1975; mem. ch. activities, charities; mem. Leadership Cir. WGBH-TV (PBS), Boston, 1992; charter mem. U.S.S. Constn. Mus., 1992. Commonwealth scholar U. Mass., 1958-62, City of New Bedford scholar, 1958-62, Allied Chem. Co. scholar, 1960-62; U. Conn. teaching fellow, 1964-65, NSF fellow Boston Coll., 1965, Advanced Engring. fellow MIT, 1969-70, NSF/AAAS fellow MIT, 1984; recipient Natick Comdr.'s award U.S Army Natick Rsch., Devel. and Engring. Ctr., 1986. Fellow Am. Inst. Chemists; mem. AAUP, Am. Chem. Soc., Am. Inst. Chem. Engrs. (cert. 1990), Am. Assn. Univ. Profs., Mus. Sci., Children's Mus., MIT Faculty Club, MIT Club Southeastern Mass., Natick Officers Club, Sigma Xi, Alpha Chi Sigma, Phi Lambda Upsilon. Avocations: gemology, coin collecting. Office: US Army Natick Rsch Devel & Engring Ctr Kansas St Natick MA 01760-5015

DECOSTA, WILLIAM JOSEPH, manufacturing engineer; b. Honolulu, Aug. 2, 1960; s. James Anthony and Emalois (Potter) DeC.; m. Jacqueline Marie Cunningham, Apr. 5, 1980; 1 child, Dena Nicole. AS in Electronics, Allan Hancock Coll., 1984; BS in Mfg. Engring., Nat. U., Sacramento, 1991. Technician AB Systems, Rancho Cordova, Calif., 1985; prodn. mgr. Insta Cool, Rancho Cordova, 1985-87; quality assurance profl. Coherent Inc., Auburn, Calif., 1987-91; engr. Coherent Inc., Auburn, 1991—. Mem. Soc. Mfg. Engrs., Am. Soc. Quality Control. Office: Coherent Inc 2301 Lindburegh St Auburn CA 95603

DECOURSEY, WILLIAM LESLIE, engineer; b. Mpls., Nov. 9, 1931; s. Fred Perry and Edna May (Ferrel) D.; m. Mertle Joyce Bird, Nov. 24, 1950; children: Constance, William D., David F. BSBA, U. Minn., 1971, MBA, 1973. Photo tech. Nat. Sch. Studios, Inc., Mpls., 1947-56; prodn. mgr. Lifetouch/NSS, Inc., Mpls., 1956-89; dir. environ. affairs Lifetouch, Inc., Mpls., 1989—. Mem. Photo Mktg. Assn. (SPFE com. 1984-89), Soc. for Imaging Sci. and Tech., Soc. Photo Finishing Engrs. (adv. com.). Home: 4648 Fillmore NE Minneapolis MN 55421 Office: Lifetouch Inc 7831 Glenroy Rd # 400 Minneapolis MN 55439

DECOWSKI, PIOTR, physicist; b. Lwow, Poland, Jan. 17, 1940; came to U.S., 1990; s. Marian and Bronislawa (Goettel) D.; m. Hermine (Ineke) ter Meulen, Jan 4, 1972; children: Patrick, Olaf. MSc, Univ. Warsaw, Poland, 1961, PhD, 1967. Rsch. asst., assoc. prof. Univ. Warsaw, 1961-75, assoc. prof., 1977-81; asst. prof. Mich. State Univ., East Lansing, 1975-77; rschr. Cen. Nuclear Rsch., Juelich, Germany, 1981-84; assoc. prof. Univ. Warsaw, 1984-86; researcher, prof. Univ. Utrecht, The Netherlands, 1986-90; prof. Smith Coll., Northampton, Mass., 1990—. Author: The Atomic Nucleus, 1987, 88; co-author: Encyclopedia of Technics, 1972, Encyclopedia of Physics, 1971; co-author, co-editor: Encyclopedia of Contemporary Physics, 1982; contbr. articles to profl. jours. Recipient Second Class award Polish Atomic Energy Comn., 1965, Third Class award Polish Ministry of Edn., 1968, 74, Award and Cert. USSR State Com. Inventions, 1982, Nat. Edn. Com. medal, Polish Ministry of Edn., 1974, Golden Cross of Merit Pres. Polish Republic, 1985. Mem. Am. Physical Soc., N.Y. Acad. Scis., European Physical Soc.(coun. 1978-81), Polish Physical Soc. (sec. gen. 1974-75, 77-81), Sigma Xi. Achievements include discovery of the phenomenon of "deep inelastic nuclear transfer in heavy ion reactions". Office: Smith Coll Clark Sci Cen Northampton MA 01063

DECROSTA, EDWARD FRANCIS, JR., former paper products company executive, consultant; b. Hudson, N.Y., Sept. 20, 1926; s. Edward F. and Anna Ruth (Crisci) DeC.; m. Annette Mae Powell, Sept. 20, 1953; children: Donna Marie, Lisa Ann. BCE, Rensselaer Poly. Inst., 1950; MS in Phys. Chemistry, Siena Coll., 1960; MBA, Rensselaer Poly. Inst., 1978. Plant chemist (process engr.) Universal Match Corp., Hudson, 1951-64; chemist Albany Felt Co., N.Y., 1965-69, mgr. tech. devel., 1969-72, dir. rsch. and devel., 1972-76; dir. rsch. and tech. devel. Papermaking Products Group, Albany Internat. Corp., N.Y.C. 1977-82, sr. scientist, 1982-83. Author: (booklet) Chemical and Physical Properties of the Elements, 1956; contbr. articles on tech. of paper to profl. jours.; patentee in field. With USAAF, 1944-45, 1950-51. Fellow Tech. Assn. Pulp and Paper Industry (chmn. engring. div. 1982-84, E.H. Neese award engring. div. 1987); mem. Am. Chem. Soc., Internat. Assn. Sci. Papermakers, N.Y. Acad. Sci., Am. Legion, KC, Elks. Roman Catholic. Home and Office: 28 James St Hudson NY 12534-1310

DE CUSATIS, CASIMER MAURICE, fiber optics engineer; b. Hazleton, Pa., Dec. 23, 1964; s. Casimer Maurice and Helen (Paytas) De C.; m. Carolyn Jean Sher, Aug. 5, 1990. BS, Pa. State U., 1986; MS, Rensselaer Poly Inst., 1988, PhD, 1990. Cert. engr.-in-tng. Pa. Soc. Profl. Engrs. Rsch. asst. Rensselaer Poly. Inst., Troy, N.Y., 1986-90; staff engr. IBM, Kingston, N.Y., 1990-93, Poughkeepsie, N.Y., 1993—; rsch. assoc. Apollo Lightcraft Program NASA-Univ. Space Rsch. Assn., 1988-90; speaker Internat. Conf. on Acousto-Optics, Leningrad, USSR, 1990, invited speaker 5th Internat. Conf. on Acousto-Optics and Applications, Gdansk, Poland, 1992; session chmn. IBM Inter-divisional Tech. Liaison meeting, Raleigh, N.C., 1992. Co-author: Acousto-Optics: Fundamentals and Applications, 1991; contbr. articles to Applied Optics, Jour. Selected Areas in Communication, Ultrasonics Symposium, Optics News, Optical Engring. Fellow GE Found., 1988. Mem. IEEE, Optical Soc. Am. (book pub. com. 1993—), Sigma Xi, Tau Beta Pi, Eta Kappa Nu. Achievements include 4 fiber optic and electrical transmission patents pending on line interface to multichip modules. Office: IBM Dept 37GA MS 451 Dept F60N MS P343 522 South Rd Poughkeepsie NY 12601

DECYK, VIKTOR KONSTANTYN, research physicist, consultant; b. Ellwangen, Baden-Wurtemburg, Germany, Feb. 6, 1948; came to U.S., 1952.; s. Wolodymyr and Taissa (Osinska) D.; m. Betsy Bradford Newell, June 23, 1973; 1 child, Marika Norlander. BA in Physics, Amherst Coll., 1970; MS in Physics, UCLA, 1972, PhD in Physics, 1977. Asst. rsch. physicist UCLA, 1977-83, assoc. rsch. physicist, 1983-88, rsch. physicist, 1988—; cons. IBM, Milford, Conn., 1988—. Contbr. articles to profl. jours. Active bd. dirs. Calif. Assn. to Aid Ukraine, L.A., 1992—. Recipient Cert. Recognition, NASA, Pasadena, Calif., 1990, Jet Propulsion Lab., 1991; Edwin Gould Found. scholar Amherst (Mass.) Coll., 1966-70. Mem. Am. Phys. Soc. Democrat. Ukrainian Catholic. Achievements include development of algorithms for parallel processing of particle-in-cell codes for plasma modeling. Office: UCLA Dept Physics 405 Hilgard Ave Los Angeles CA 90024

DE DATTA, SURAJIT KUMAR, soil scientist, agronomist, educator; b. Shwebo, Upper Burma, Burma, Aug. 1, 1936; came to U.S., 1991; s. Dinanath and Birahini De Datta; m. Vijayalakshmi L., April 20, 1967; 1 child Raj Kumar De Datta. BS in Agrl., Banaras Hindu U., 1956; MS Soil Sci. and Agrl. Chem., Indian Agrl. Rsch. Inst., New Delhi, India, 1958; PhD Soil Sci., U. Hawaii, 1962. Postdoctoral agrl. experiment station Ohio State U., Columbus, 1962-63; prof. agronomy and soil sci. U. Philippines, Los Banos, Philippines, 1964—; assoc. agronomist Internat. Rice Rsch. Inst., Manila, Philippines, 1964-69, agronomist, 1969-85, radiological health and safety officer, 1967-78, acting head dept. soil chem., 1975-76, dept. head, agronomy, 1967-89, principle scientist, 1986-91, program leader, 1990-91; dir., office internat. rsch. and devel. Va. Tech., Blacksburg, Va., 1991—, assoc. dean internat. agrl., 1993—, prof. crop and soil environ. sciences, 1991—; bd. dirs. Southeast Consortium for Internat. Devel., Washington, D.C., SANREM-CRSP Univ. Ga., Toxicology Center Va. Tech; chmn. Internat. Agronomy Div. Am. Soc. Agronomy, Wis., 1982-83; vis. prof. Purdue U. 1971-72, Kasetsart Univ., Thailand, 1984—, Central Luzon State U., Nueva Ecija, Philippines, 1983-91; vis. scientist Univ. Calif. Davis, 1978-79. Author: Principles and Practices of Rice Production, 1981; consulting editor: Plant & Soil Jour. 1978—; contbr. numerous articles to profl. jours. Recipient Internat. Soil Sci. award Soil Sci. Soc. Am., 1986, Best Paper award Weed Sci. Pest Control Coun. Philippines, 1986, Eminence award Bureau of Plant Industry, Philippines, 1987, Best Paper award Asian-Pacific Weed Sci. Soc., Taiwan, 1987, Second Best Paper award Asian-Pacific Weed Soc., Korea, 1989, Agronomic Rsch. award Am. Soc. Agronomy, 1990, Norman Borlaug award Coramandal, New Delhi, India, 1992. Fellow Am. Soc. Agronomy, Soil Sci. Soc. Am., Indian Soc. Soil Sci., Internat. Svc. in Agronomy; mem. Crop Sci. Soc. Am., Internat. Soil Sci. Soc., Soil Sci. of Japan, Crop Sci. Soc. Philippines, Weed Sci. Soc. Philippines, Indian Soc. Weed Sci., Asian-Pacific Weed Sci. Soc., Internat. Weed Sci. Soc., Pakistan Weed Sci. Soc. Hindu. Home: 512 Floyd St Blacksburg VA 24060 Office: Va Poly Tech Inst Internat Rsch Devel Dept 1060 Litton Reaves Hall Blacksburg VA 24061-0334

DEDRICK, KENT GENTRY, retired physicist, researcher; b. Watsonville, Calif., Aug. 9, 1923; s. Frederick David and Matilda (Redman) D.; 1 child, Susan Marie. BS in Chemistry and Physics, San Jose (Calif.) State U., 1946; MS in Phys. Scis., Stanford U., 1949, PhD in Theoretical Physics, 1955. Rsch. assoc. U. Mich., Ann Arbor, 1954-55, Stanford U., 1955-62; math. physicist Stanford Rsch. Inst., Menlo Park, Calif., 1962-75; cons. scientist Atty. Gen.'s Office State of Calif., Sacramento, 1976-80; with marine tech. safety dept. State Lands Commn., Sacramento, 1980-81, rsch. specialist, 1981-92; cons. scientist phys. and environ. scis., 1992—. Contbr. articles to profl. jours.; composer instrumental and vocal works, 1978—. Pres. Com. for Green Foothills, Palo Alto, Calif., 1972-74; founding co-chmn. So. Crossing Action Team, San Francisco Bay area, 1970-72, chmn. Bayfront com. Sierra Club, Palo Alto, 1967-72. Mem. Am. Phys. Soc., Am. Geophys. Union, Soc. Wetland Scientists, Sigma Xi. Achievements include co-discovery of mathematical theorem on Lagrange and Taylor series. Home: 1360 Vallejo Way Sacramento CA 95818-3450

DEDRICK, ROBERT LYLE, toxicologist, biomedical engineer; b. Madison, Wis., Jan. 12, 1933; m. 1955; 3 children. BE, Yale U., 1956; MSE, U. Mich., 1957, PhD in Chem. Engring., 1965. Asst. prof. mech. engring. George Washington U., 1959-62, asst. prof. engring. and applied sci., 1962-63, assoc. prof., 1965-66, actg. chief, 1966-67; chief chem. engring. sect. Biomedical Engring. and Instrumentation Br., NIH, 1967—. Recipient Founders' award Chem. Industry Inst. Toxicology, 1992. Mem. AAAS, AICE, Am. Soc. Engring. Edn., Am. Chem. Soc., Am. Soc. Artificial Internal Organs. Achievements include research in pharmacokinetics, in cancer chemotherapy, in transport, thermodynamics and kinetics in living systems, in biomaterials, in artificial internal organs and in risk estimation. Home: 1633 Warmex Ave Mc Lean VA 22101*

DE DUVE, CHRISTIAN RENÉ, chemist, biologist, educator; b. Thames-Ditton, Eng., Oct. 2, 1917; s. Alphonse and Madeleine (Pungs) de D.; m. Janine Herman, Sept. 30, 1943; children: Thierry, Anne, Françoise, Alain. M.D., U. Louvain, Belgium, 1941, Ph.D., 1945, M.Sc., 1946; D honoris causa, U. Turin, 1969, U. Leiden, 1970, U. Sherbrooke, 1970, U. Lille, 1973, Cath. U. Santiago, Chile, 1974, U. René Descartes, Paris, 1974, State U. Liege, 1975, State U. Ghent, 1975, Gustavus Adolphus Coll., St. Peter, Minn., 1975, U. Rosario, Argentina, 1975, U. Aix-Marseille II, 1979, U. Keele, 1982, Katholieke U. Leuven, 1984, Karolinska Inst., Stockholm, 1986, U. Montreal, 1992. Lectr. physiol. chemistry faculty medicine Cath. U. Louvain, 1947-51, prof., head dept. physiol. chemistry, 1951-85, emeritus prof., 1985—; prof. biochem. cytology Rockefeller U., N.Y.C., 1962-74, Andrew W. Mellon prof., 1974-88, prof. emeritus, 1988—; vis. prof. Albert Einstein Coll. Medicine, Bronx, N.Y., 1961-62, Chaire Francqui State U. Ghent, 1962-63, Free U. Brussels, 1963-64, State U. Liège, 1972-73, Facultés Universitaires Notre-Dame de la Paix, Namur, 1990-91; Mayne guest prof. U. Queensland, Brisbane, Australia, 1972; pres. Internat. Inst. Cellular and Molecular Pathology, Brussels, 1974-91. Mem. editorial bd. Subcellular Biochemistry, 1971-87, Preparative Biochemistry, 1971-80, Molecular and Cellular Biochemistry, 1973-80. Mem. Conseil d'Adminstrn. du Fonds Nat. de la Recherche Scientifique, 1958-61; mem. Conseil de Gestion du Fonds de la Recherche Scientifique Medicale, 1959-61; mem. Commn. Scientifique du Fonds de la Recherche Scientifique Medicale, 1958-61; mem. Comite des Experts du Conseil Nat. de la Politique Scientifique, 1958-61; mem. adv. bd. Ciba Found., 1960-85; mem. adult devel. and aging research and tng. rev. com. Nat. Inst. Child Health and Devel., NIH, 1970-73; mem. adv. com. for med. research WHO, 1974-79; mem. sci. adv. com. Max Planck-Inst. für Immunbiologie, 1975-78, Ludwig Inst. Cancer Research, 1985-91 , Mary Imogene Bassett Research Inst., 1986-90, Clin. Research Inst. Montreal, 1986—; mem. biology adv. com. N.Y. Hall of Sci., 1986—; adv. sci. com. Basel Inst. for Immunology, 1989—. Recipient Prix des Alumni, 1949, Prix Pfizer, 1957, Prix Francqui, 1960, Prix Quinquennal Belge des Sciences Médicales, 1967 (Belgium); Gairdner Found. Internat. award merit (Can.), 1967; Dr. H.P. Heineken prize (The Netherlands), 1973; Nobel prize for physiology or medicine, 1974; Harden award Biochem. Soc. (Gt. Britain), 1978; Theobald Smith award Albany Med. Coll., 1981; Jimenez Diaz award, 1985. Fellow AAAS; mem. NAS, Royal Acad. Medicine, Royal Acad. Belgium, Am. Chem. Soc., Biochem. Soc., Am. Philos. Soc., Am. Soc. Biol. Chemists, Pontifical Acad. Sci., Am. Soc. Cell Biology (coun. 1966-69, E.B. Wilson award 1989), Am. Philosophical Soc., Soc. Chimie Biologique, Soc. Belge Biochim. (pres. 1962-64), Deutsche Akademie der Naturforscher Leopoldina, Koninklyke Akademie voor Geneeskunde (Belgium), European Assn. Study Diabetes, European Molecular Biology Orgn., European Cell Biology Orgn., Internat. Soc. Cell Biology, N.Y. Acad. Scis., Soc. Belge de Physiologie, Sigma Xi; fgn. assoc. Am. Acad. Arts and Scis., Royal Soc. London, Académie des Sciences de Paris, Académie des Sciences d'Athénes, Academia Europaea, Deutsche Gesellschaft für Zellbiologie; numerous hon. memberships. Office: Rockefeller U 1230 York Ave New York NY 10021-6341 also: ICP, 75 Avenue Hippocrate, B-1200 Brussels Belgium

DEEDS, WILLIAM EDWARD, physicist, educator; b. Lorain, Ohio, Feb. 23, 1920; s. Dean Dalton and Mary Frances (Updike) D.; m. Alice Marie Brandt, July 26, 1950 (dec. 1970); children: Dean Alan, Eric Edward, Amy Diana, Holly Anne. BA, Denison U., 1941; MS, Calif. Inst. Tech., 1943; PhD, Ohio State U., 1951. Grad. asst., teaching fellow Calif. Inst. Tech., Pasadena, 1941-43; jr. physicist NDRC, Pasadena, 1943-46; asst. prof. Denison U., Granville, Ohio, 1946-48; univ. scholar, Texas Company fellow Ohio State U., Columbus, 1948-51, postdoctoral fellow, 1951-52; prof. physics U. Tenn., Knoxville, 1952-89, prof. emeritus, 1989—; cons. Oak Ridge (Tenn.) Nat. Lab., 1962-89, Redstone Arsenal, Huntsville, Ala., 1952-55, Chemstrand Corp., Raleigh, N.C., 1956-59. Contbr. articles to Phys. Rev., Jour. Applied Physics, Jour. Phys. Chemistry, others. Mem. Am. Phys. Soc. Achievements include patents in field, N-waveform of acoustic shock waves, energy distribution of cosmic rays; research on analytic theory of eddy currents, rocket ballistics, vibrational analysis of chain molecules. Office: U Tenn Dept Physics Knoxville TN 37996

DEEGEN, UWE FREDERICK, marine biologist; b. Freising, Fed. Republic Germany, Mar. 27, 1948; came to U.S., 1953; s. Friedrich Rudolf and Maria Magdalena (Dyrda) D.; m. Barbara Lynn Cannon, Aug. 7, 1982; 1 child, Jennifer Marie. BS, U. So. Miss., 1970, MS, 1972; PhD, U. Tex., 1979. Cert. fed. grants adminstr. Dept. Commerce, shellfish sanitarian FDA. Grad. rsch. assoc. U. So. Miss., Hattiesburg, 1970-72; instr. biology. Biloxi (Miss.) Sr. High Sch., 1972-73; instr. biology Pensacola (Fla.) Jr. Coll., 1973-74; grad. fellow U. Tex., Austin, 1974-76; staff scientist Miss. Marine Resources Coun., Long Beach, 1976-80; chief saltwater fisheries Miss. Bur. Marine Resources, Long Beach, 1980-89; sr. systems analyst Miss. Dept. Wildlife, Fisheries, and Parks, Biloxi, 1989—; mem. Gulf Mex. Fishery Mgmt. Coun., Tampa, Fla., 1980-89; mem. tech. rev. com. U.S./Israel Binat. Agrl. R&D Fund, 1980-86; vice chmn. statis. subcom., chmn. rec. fisheries com. Gulf States Fisheries Commn., Ocean Springs, 1983-87; mem. freshwater inflow adv. com. EPA/Gulf Mex. Program, 1990—. Author: Mathematical Modeling of Oxygen Distribution in Streams, 1972; columnist Coastal Fishing, 1983—; contbr. articles to profl. jours. Tournament weightmaster Miss. Trout Invitational Tournament, Pass Christian, 1985—, Long Beach Jaycees Fishing Tournament, 1989—. Recipient King Neptune award Miss. Deep Sea Fishing Rodeo Bd. Dirs., Gulfport, Miss., 1991. Mem. Am. Fisheries Soc. (cert.), Estuarine Rsch. Fedn., Internat. Gamefish Assn., Kappa Mu Epsilon, Beta Beta Beta. Lutheran. Avocations: fishing, backpacking, raquet sports, landscape gardening, nature photography. Home: 121 E 4th St Long Beach MS 39560-6107 Office: Miss Dept Wildlife Fisheries & Parks 2620 Beach Blvd Biloxi MS 39531-4501

DEEIK, KHALIL GEORGE, economist, financing company executive; b. Bethlehem, Jordan, Nov. 12, 1937; s. George Said Diek and Wadiea (Jalil) Lama; m. Jalileh Mary Marzouka, Aug. 22, 1965; children: George, Ramzi, Nader. BA, Sacramento State U., 1961, MA, 1964; PhD, U. So. Calif., 1972. Prin., adminstr. Manzanita Sch., Hyampom, Calif., 1964-65; mgr. Gen. Trading Co., Alkhobar, Saudi Arabia, 1966-69; program dir., instr. Krebs Coll., North Hollywood, Calif., 1969-72; chief investment officer, mng. dir., v.p., sr. advisor, exec. asst. to chmn. Olayan Saudi Investment Co., Olayan Financing Co., Jeddah, Saudi Arabia, 1973—, The Olayan Group of Cos.; bd. dirs. Saudi Polyester Products Co., Jeddah, 1984—; exec. com. mem. Saudi Arabian Constrn. and Repair Services Co., Jeddah, 1984—; hon. lectr. King Abdulaziz U., Saudi Arabia, 1979; faculty mem., program coord. Century U., Calif., 1978—. Mem. Internat. Educators Assn. (v.p. 1970-72), Marquis Club, Phi Delta Kappa, Phi Delta Epsilon. Club. Office: Olayan Fin Co, PO Box 8772, Riyadh 11492, Saudi Arabia

DEEPAK, ADARSH, meteorologist, atmospheric optician; b. Sialkot, India, Nov. 13, 1936. BS, Delhi U., 1956, MS, 1959; PhD in Aerospace Engr., U. Pa., 1969. Lectr. physics DB & KM Cols., Delhi U., 1959-63; instr. phys. sci. U. Fla., 1965-68, rsch. assoc. physics, 1970-71; fellow Nat. Rsch. Coun. Marshall Space Flight Ctr., NASA, 1972-74; rsch. assoc. prof. physics & geophys. sci. Old Dominion U., 1974-77; pres. Inst. Atmospheric Optics & Remote Sensing, Hampton, Va., 1977-84, Sci. & Tech. Corp., Hampton, Va., 1979—; cons. engring. sci. Wayne State U., 1970-72; mem. panel remote sensing & data acquisition NASA/OAST Technol. Workshop, 1975; NSF travel grant to visit Indian insts., 1976; adj. prof. physics Coll. William & Mary, 1979-80; leader U.S. Del. Internat. Workshop Appln. Remote Sensing Rice Prod., India, 1981. Mem. AAAS, Optical Soc. Am., Am. Meteorol. Soc., Am. Chem. Soc., Am. Geophys. Union. Achievements include research in remote sensing of atmospheric particulate and gaseous pollutants and motions, using laser doppler, optical scattering and photographic techniques from space airborne and ground platforms, theory of radiative transfer in scattering atmospheres, fogs and clouds, inversion methods for remotely sensed data. Office: Sci & Tech Corp 101 Research Dr PO Box 7390 Hampton VA 32666*

DEER, JAMES WILLIAM, physicist; b. Carney, Okla., Mar. 22, 1922; s. Walter Reese and Mary Ellen (Betzer) D.; m. Bonita Evelyn Simmons, Sept. 2, 1948; children: Linda, John, Martha. BS in Physics, U. Ark., 1949. Field engr. Naval Rsch. Lab., Washington, 1942; sect. leader Sandia Lab., Albuquerque, N.Mex., 1949-54; physicist Electronic Specialty, Portland, Oreg., 1954-75; engr. Tektronix, Beaverton, Oreg., 1975-86; adminstr., lectr., bd. dirs. Oregonians Concerned About Addiction Problems, 1987-91; pres., bd. dirs. Internat. Affairs Coord. Coun., 1987-91; rsch. assoc. dept. physics Linfield Coll., McMinnville, Oreg., 1992—. With USN, 1942-46. Mem. Am. Phys. Soc. Democrat. Unitarian. Achievements include research in understanding of momentum transfer in resonance fluorescence. Home: 11905 SW Settler Way Beaverton OR 97005 Office: Linfield Coll Physics Dept Mcminnville OR 97128

DEES, TOM MOORE, internist; b. Dallas, Mar. 4, 1931; s. Tom Hawkins and Maida Elizabeth (Board) D.; m. Suzanne Settle, Feb. 20, 1971; children: Tom Moore III, David Walsh. BA, Johns Hopkins U., 1952; MD, Southwestern Med. Sch., 1956. Intern Bellevue Hosp., N.Y.C., 1957, resident, 1958-59; rsch. fellow in cardiology Southwestern Med. Sch., Dallas, 1961; internist, ptnr. pvt. practice med. office Dallas, 1962—; dir. Swiss Ave Med. Bldg., Dallas, 1984—; clin. asst. prof. medicine Southwestern Med. Sch., Dallas, 1962—; assoc. attending physician Baylor Med. Ctr., Dallas, 1962—. Mem. dist. commn. Boy Scouts Am., Dallas, 1963-72; mem. ofcl. bd. Highland Park Meth. Ch., Dallas, 1963-72. Capt. USAF, 1959-61. Mem. ACP (life), AMA, Am. Soc. Internal Medicine, Johns Hopkins U. Alumni Assn. (pres. North Tex. chpt 1964-68), Tex. Club of Internists (pres. 1992-93). Republican. Avocations: hunting, fishing, gardening. Home: 3649 Stratford Ave Dallas TX 75205-2810 Office: 3434 Swiss Ave # 304 Dallas TX 75204-6282

DEETS, DWAIN AARON, aeronautical research engineer; b. Bell, Calif., Apr. 16, 1939; s. Kenneth Robert and Mildred Evelyn (Bergman) D.; m. Catherine Elizabeth Meister, June 18, 1961; children: Dennis Allen, Danelle Alaine. AB, Occidental Coll., 1961; MS in Physics, San Diego State U., 1964; ME, UCLA, 1978. Rsch. engr. Dryden Flight Rsch. Ctr., NASA, Edwards, Calif., 62-78, 79-85; hdqrs. liaison engr. NASA, Washington, 1978-79; mgr. NASA, Edwards, 1979-85, div. chief Dryden Flight Rsch. Facility, 1990—; hdqrs. mgr. flight rsch. NASA, Washington, 1988-89. Contbr. articles to tech. publs. Recipient Exceptional Svc. medal NASA, 1988. Fellow AIAA (assoc., Wright Bros. lectr. aeros. 1987, tech. com. on society and aerospace tech.); mem. Soc. Automotive Engrs. (chmn. aerospace control and guidance systems com. 1988-90), World Future Soc., Masons. Republican. Mem. Christian Ch. (Disciples of Christ). Office: NASA Ames-Dryden PO Box 273 Edwards CA 93523

DEEVI, SEETHARAMA C., materials scientist; b. Guntur, India, Aug. 11, 1955; came to U.S., 1981; s. Ramacharyulu and Satyavathi (Agnihotram) D.; m. Ariprala D. Sarojini, Oct. 27, 1984; children: Suleka, Sathish C. BS, Andhra U., 1975; MS, Mysore U., 1977; PhD, Indian Inst. Sci., 1981. Project asst. Indian Inst. Sci., Bangalore, India, 1981; postdoctoral fellow Worcester (Mass.) Poly. Inst., 1982-85; rsch. assoc. in engring. Brown U., Providence, 1985-88; sr. rsch. engr. U. Calif., Davis, 1985-88; rsch. scientist Philip Morris U.S.A., Richmond, Va., 1988-89, project leader, 1989—. Contbr. articles to Jour. Materials Sci., Materials Sci. and Engring., Ceramics Internat., Combustion Sci. and Tech., Jour. Solid State Chemistry. Fellow Am. Inst. Chemists; mem. Internat. Soc. Hybrid Microelectronics (advt. chmn. Tarheel chpt. 1992-93), Am. Soc. Metals (mem. chmn. Richmond chpt. 1992-93), Am. Ceramic Soc. Achievements include patents (with others) for electrically powered linear heating element, composite heat source comprising metal carbide, metal nitride and metal, chemical heat source compromising metal nitride, metal oxide and carbon. Office: Philip Morris USA Rsch & Devel Richmond VA 23264

DE FABO, EDWARD CHARLES, photobiology and photoimmunology, research scientist, educator; b. Wilkes-Barre, Pa., June 10, 1937; s. Giovanni and Anna (Marconi) De F.; m. Athena Macris, Aug. 17, 1967 (dec. June 1985); m. Frances Patricia Noonan. BS, Kings Coll., 1958; PhD, George Washington U., 1974. Rsch. scientist USDA, Beltsville, Md., 1974-75, NCI-Frederick (Md.) Cancer Rsch. Ctr., 1978-81; scientist, adminstr. U.S. EPA, Washington, 1975-78; asst. rsch. prof. dept. dermatology George Washington U., Washington, 1981-86, assoc. rsch. prof. dept. dermatology, 1986-92, rsch. prof. dept. dermatology, 1992—; chmn. project Sci. Com. on Problems of Environ. SCOPE ozone depletion and UV radiation, Paris, 1989-92; cons. U.S. EPA, 1984-85. Editor, organizer publ. Sci. Com. on Problems of Environment, 1992; contbr. articles to rsch. jours.; author: Immunology Today, 1992. Dir. congl. sci. fellowship program Am. Soc. Photobiology, Bethesda, 1981-85. Grantee Internat. Union Against Cancer, 1983, Am. Cancer Soc., 1987-89, U.S. EPA, 1987—, NIH, 1989—; fellow Smithsonian Inst., 1970-74, NSF, 1963-64. Mem. AAAS, Am. Soc. Photobiology (councilor 1980-83). Achievements include discovery (with F.P. Noonan)of a sunlight-activated immune-regulating photoreceptor on skin-urocanic acid; designer of unique UV monochromator for in vivo action spectrum studies. Office: George Washington U Med Ctr Ross Hall Rm 101-B 2300 Eye St NW Washington DC 20037

DEFILIPPES, MARY WOLPERT, pharmacologist; b. Sioux City, Iowa, Dec. 13, 1939; d. Paul Louis and Katherine (Block) Wolpert; m. Frank Michael DeFilippes, June 29, 1973. BS in Pharmacy, Creighton U., 1963; PhD, U. Mich., 1969. Postdoctoral fellow Yale U., New Haven, 1969-71; staff fellow Nat. Cancer Inst., Bethesda, Md., 1971-76, pharmacologist, 1976—. Mem. Am. Assn. for Cancer Rsch. Roman Catholic. Home: 4507 Sleaford Rd Bethesda MD 20814 Office: Nat Cancer Inst DCT/EPN 832 9000 Rockville Pike Bethesda MD 20892

DEFILIPPIS, CARL WILLIAM, engineer, meteorologist; b. East Orange, N.J., Mar. 4, 1931; s. Carl Alfredo and Anna Maria (Petriccone) DeF.; m. Antoinette Christina Lauer, Aug. 23, 1958; children: Don C., John M. BSME, N.J. Inst. Tech., 1954; student in meteorology, NYU, 1954-55; MBA, Seton Hall U., 1965. Registered profl. engr., N.J. Test engr. Curtis Wright Aero. Corp., Woodridge, N.J., 1953-54; supv. gen. engr. U.S. Army Munitions Command, Dover, N.J., 1958-90; cons. engr., West Caldwell, N.J., 1973—. Author: How to Live with Inflation, 1990; tech. advisor film Maintainability of Nuclear Warhead Sections, 1971. Active St. Aloysius Ch. Choir, Caldwell, N.J., 1988—, Caldwell High Sch. Wrestling Club, 1985-88. Lt. USAF, 1954-58. Mem. Am. Meteorol. Soc., N.J. Soc. Profl. Engrs., Am. Legion, N.J. Jazz Soc., Train Collectors Assn., Nat. Railroad Hist. Soc., Friends of N.J. Railroad Mus. Roman Catholic. Home: 130 Hillside Ave West Caldwell NJ 07006

DEFLORIO, MARY LUCY, physician, psychiatrist; b. Chgo.; d. Anthony Ralph and Bernice (B. Bounell) D. m. Robert Y. Shapiro, Dec. 27, 1986. BA with distinction, U. Wis.; MD, MPH, U. Ill., Chgo., 1984; cert. writing program, Columbia U., 1988-91. Cert. emergency med. technician. Adjudicator Fed. Disability Program, Ill. and Mass.; vocat. counselor U. Ill., Chgo.; resident internal medicine Mercy Hosp., Chgo., 1984-85; med. examiner Dept. Pub. Aid State of Ill., Chgo., 1985-87; resident psychiatrist St. Vincent's Hosp., N.Y.C., 1987-90; fellow cons. liaison psychiatry Meml. Sloan Kettering/Cornell Med. Ctr., N.Y.C., 1991-93; chief fellow Meml. Sloan Kettering, N.Y.C., 1992-93; attending physician Div. Psychiatry/Dept. Neurology Meml. Sloan Kettering and Cornell Med. Coll., N.Y.C., 1993—. Recipient Med. Econs. Writing award, 1987; James scholar U. Ill., Gen. Assembly scholar. Mem. AMA (Nutritional scholar 1983-84), Am. Women's Assn., Mass. Assn. Examiners (membership chmn.), Nat. Rehab. Assn., Assn. Acad. Psychiat. (Mead-Johnson fellow 1990), Am. Psychiat. Assn. (Br. Rsch. award 1990), Am. Psychiat. Arts Assn. (black and white photography and poetry award 1993). Roman Catholic. Avocations: writing, photography. Office: Meml Sloan Kettering Rm 767C 1275 York Ave New York NY 10021

DE FOREST, SHERWOOD SEARLE, agricultural engineer, agribusiness services executive; b. Ames, Iowa, Sept. 20, 1921; s. Frank Ray and Clara Maud (Searle) De F.; m. Virginia Mary Flynn, June 20, 1947; children: David, Debra, Denise, Kimberly. Student, U. Cin., 1939-40; B.S., Iowa State U., 1943, M.S., 1947. Registered profl. engr., Iowa. Instr. agrl. engring. Iowa State U., 1946-47, extension agrl. engr., 1947-52; engring. editor Successful Farming mag., Des Moines, 1952-59; with USX, Pitts., 1959-77; mgr. agrl. equipment mktg. USX, 1964-70, indsl. rep., 1970-77; v.p. Walt Montgomery Assocs., Inc., Tallahassee, Fla., 1977—; owner De Forest Agri-Services, Tallahassee, Fla., 1977—; pres. Ginande Corp., 1986-91; tech. transfer project leader No. Agrl. Energy Center, Sci. and Edn. Adminstrn., U.S. Dept. Agr., Peoria, Ill., 1980-81; cons. Pakistan, 1984, Portugal, 1985, 86; mem. indsl. and profl. adv. com. Coll. Engring. Pa. State U., 1966-71; Mem. NE Regional Agrl. Research Planning Com., 1970-72. Contbg. author: Power to Produce, U.S. Dept. Agr. Yearbook, 1960, Steel in Agriculture, 1966; Pub. TravelHost of Pitts. mag., 1982-83; tech. editor Soc. Automotive Engrs., 1987-89; numerous articles to Successful Farming Mag. Served to 1st lt. USAAF, 1943-46. Recipient Am. Soc. Agrl. Engrs.-Metal Bldg. Mfrs. Assn. award for disting. work in advancing knowledge and sci. of farm bldgs., 1964. Fellow Am. Soc. Agrl. Engrs. (pres. 1975-76). Presbyterian (ruling elder). Achievements include patents in field. Home and Office: 4173 Covenant Ln Tallahassee FL 32308

DE FRIES, JOHN CLARENCE, behavioral genetics educator, institute administrator; b. Delrey, Ill., Nov. 26, 1934; s. Walter C. and Irene Mary (Lyon) De F.; m. Marjorie Jacobs, Aug. 18, 1956; children—Craig Brian, Catherine Ann. B.S., U. Ill., Urbana, 1956, M.S., 1958, Ph.D., 1961. Asst. prof. U. Ill., 1961-66, assoc. prof., 1966-67; research fellow U. Calif.-Berkeley, 1963-64; assoc. prof. behavioral genetics and psychology U. Colo., Boulder, 1967-70, prof., 1970—, dir., 1981—. Author: (with G.E. McClearn) Introduction to Behavioral Genetics, 1973, (with Plomin and McClearn) Behavioral Genetics: A Primer, 1980, 2d edit., 1990, (with R. Plomin) Origins of Individual Differences in Infancy, 1985, (with R. Plomin and D.W. Fulker) Nature and Nurture During Infancy and Early Childhood, 1988; co-founder Behavior Genetics jour., 1970; cons. editor: Jour. Learning Disabilities; mem. editorial adv. bd. Behavior Genetics. Served to 1st lt. U.S. Army, 1957-65. Grantee in field. Fellow Internat. Acad. for Rsch. in Learning Disabilities; mem. AAAS, Am. Soc. Human Genetics, Behavior Genetics Assn. (sec. 1974-77, pres. 1982-83, Th. Dobzhansky award for outstanding rsch. in field of behavior genetics 1992), Am. Psychol. Soc., Internat. Soc. for the Study of Individual Differences, Orton Dyslexia Soc., Rodin Remediation Acad., Soc. for Rsch. in Child Devel., Soc. for Study of Social Biology. Office: U Colo Inst Behavioral Genetics CB447 Boulder CO 80309-0447

DEGADY, MARC, chemical engineer; b. Cairo, Egypt, Jan. 12, 1957; came to U.S., 1981; s. Maamoun Hassan and Mireille (Cohen) D.; m. Judy Mayo, Sept. 4, 1992; 1 stepchild, Meghan. BSChE, Cairo U., 1979; MSChE, U. N.Mex., 1982. Rsch. engr. Becton-Dickinson Rsch. Ctr., Research Triangle Park, N.C., 1983-85; product/process engr. Procter & Gamble, Cin., 1985-87; rsch. assoc. Warner Lambert Co., Morris Plains, N.J., 1987—. Contbr. articles to profl. jours. Vol. Big Bros./Big Sisters, Cin., 1986. Mem. Am. Inst. Chem. Engrs., Controlled Release Soc. (co-chmn. local com. 1992), Soc. Plastic Engrs. Achievements include patents on automatic liquid (plasma/blood) component separator, continuous production of chewing gum using twin screw extruder, reducing sugar lumps by dual gum base injection in a corotaling twin screw extruder. Patent pending: method for bleaching sucrose polyesters, and a fire-safe method for high content flavor encapsulation. Home: 91 Patriots Rd Morris Plains NJ 07950 Office: Warner Lambert 182 Tabor Rd Morris Plains NJ 07950

DÉGEILH, ROBERT, retired chemist; b. Magnac sur Toure, France, Apr. 19, 1927; s. Jean and Amelia (Boursier) D.; m. Josette Joliff, Sept. 5, 1963; children: Francoise, Marianne, Jean-Marie. Ingenieur, ENSCP, Paris, 1952; PhD, Ind. U., 1955. Rsch. fellow Calif. Tech. Inst., Pasadena, 1955-57; rsch. chemist Saint Gobain, Antony, France, 1958-61, Pechiney-Saint Gobain, Antony, 1961-69; group leader Saint Gobain, Aubervilliers, France, 1969-87, ret., 1987; cons. ECTI, Paris, 1987—. Councillor City Coun., Port-Marly, France, 1983-89. Scholarship State of France, 1939-52. Mem. ACS, ACA, Sigma Xi. Achievements include research in molecular structure of chelates and proteins, patents for production of polyvinyl butyral, process for production of pure silicon and semiconductors, research on study of metal surfaces and catalysts-Leed, deposition of metals on glass surfaces, extrusion of polymers. Home: 29 Rte de Versailles, Port-Marly 78560, France

DE GENNES, PIERRE GILLES, physicist, educator; b. Paris, 1932; m. Anne-Marie Rouet, 1954. Ed., Ecole Normale Superieure; PhD Rsch. scientist Centre d'Etudes Nucleaires de Saclay, 1955-59; prof. solid state physics U. Paris, Orsay, 1961-71; prof. Coll. de France, Paris, 1971—, dir. Ecole de Physique et Chimie, Paris, 1976—; sci. dir. for chem. physics Rhone Poulenc, France, 1988—. Author: Superconductivity of Metals and Alloys, 1965, The Physics of Liquid Crystals, 1973, Scaling Concepts in Polymer Physics, 1979, Simple Views on Condensed Matter, 1992. Ensign French Navay, 1959-61, Recipient Hollweck prize, 1968, prix Cognac-Jay, 1970, prix Ampere, 1977, Gold medal CNRS, 1981, Matteuci medal Sci. Acad. Rome, 1988, Harvey prize Technion, Israel, 1989, Sci. and Art prize L.V. Mh Group, Paris, 1989, Wolf Found. prize in physics, 1990, Nobel prize in physics, 1991. Mem. AAAS, Académie des Sciences, Dutch Acad. Scis., Royal Soc., Nat. Acad. Scis. Avocations: skiing, kayaking, windsurfing. Office: Ecole de Physique et Chimie, 10 rue Vauquelin, 75005 Paris France

DE GIORGI, ENNIO, mathematics educator; b. Lecce, Italy, Feb. 8, 1928. Prof. math. analysis Scuola Normale Superiore, Pisa, Italy. Recipient Wolf Found. Math. prize, 1990. Mem. Acad. Nat. Lincei. Office: Collegio Timpano, Lungarno Pacinotti 51, I-56100 Pisa Italy

DEGNAN, JOHN JAMES, III, physicist; b. Phila., Dec. 10, 1945; s. John James Jr. and Ruth Dolores (Vece); m. Adele Susan Henry, June 27, 1969; children: Adam John, Andrew Paul. BS in Physics, Drexel U., Phila., 1968; MS in Physics, U. Md., 1970, PhD in Physics, 1979. Student trainee NASA Goddard Space Flight Ctr., Greenbelt, Md., 1964-67, physicist, 1968-72, sr. physicist, 1972-79, sect. head., 1979-89, dep. mgr. crystal dynamics project, 1989-93, head space geodesy and altimetry projects office, 1993—; instr. Drexel U., Phila., 1967-68; assoc. mem. Adv. Group on Electron Devices, 1980-85, dep. mem. 1985-89; adj. prof. physics Am. U., Washington, 1988—; chmn. CSTG SLR Subcommn., 1992—. Contbr. articles to profl. publs. Mem. Common Cause, Annapolis, Md., 1970—; mem., v.p., treas. Pasadena Theatre Co., Md., 1982—. Drexel Bd. Trustees scholar, 1963; recipient Marple-Newtown Sch. Dist. Hall of Fame award, Disting. Alumnus, 1989. Mem. Optical Soc. Am., Am. Phys. Soc., Am. Geophys. Union, Planetary Soc., Laser Comm. Soc. (charter), Sigma Pi Sigma, Sigma Pi. Roman Catholic. Home: 928 Barracuda Cove Ct Annapolis MD 21401 Office: NASA Goddard Space Flight Ctr Greenbelt MD 20771

DEGRAVE, ALEX G., computer engineer, consultant; b. Lokeren, Belgium, Aug. 3, 1957; came to U.S., 1988; s. Andre and Mia (Zaman) D.; m. Francine P. Leroy, Oct. 29, 1985; children: Aurelien, Colas, Jeremie. BA in Grad. Mgmt., Leuven U., Belgium, 1984, MBA in Fin. & Internat. Mgmt., 1986; BA in Ops. Rsch., Leuven U., 1988. Industrial engr. Electronics Telecommunications, 1979; systems engr. IBM Belgium, Brussels, 1980-87; mktg. supr. IBM L.Am., Gaithersburg, Md., 1988-92; cons. IBM Consulting Group, Brussels, 1993—. Avocations: sailing, trekking, music, history. Home: Groeningelaan 13, 1933 Sterrebeek Belgium Office: IBM Consulting Group, Plantsoen 1 Victoria Regina, B-1210 Brussels Belgium

DEGROAT, WILLIAM CHESNEY, pharmacology educator; b. Trenton, N.J., May 18, 1938; s. William Chesney and Margaret (Welch) deG.; m. Dorothy Marion Albertson, June 13, 1959; children: Allyson L., Cynthia L., Jennifer L. BSc, Phila. Coll. Pharmacy and Sci., 1960, MSc, 1962; Ph.D., U. Pa., 1965, postdoctoral, 1965-66; postdoctoral, Australian Nat. U., Canberra, 1966-67. Vis. research fellow John Curtin Sch. Med. Research, Canberra, 1967-68; asst. prof. U. Pitts. Med. Sch., 1968-72, assoc. prof., 1972-77, prof. pharmacology, 1977—, acting chmn. dept. pharmacology, 1978-80, adj. prof. pharmacy, 1978-88, prof. psychology, 1982-86, mem. ctr. of neurosci., 1984—, prof. in dept. of behavioral neurosci., 1986—; mem. neurobiology study sect. NIH, 1983-88; vis. scientist NIAAA-NIH, 1989-90. Mem. editorial bd. Jour. Pharmacology and Exptl. Therapeutics, 1975—, Jour. Autonomic Nervous System, 1979—, assoc. editor, 1985—, Neurology and Urodynamics, 1982—, Am. Jour. Physiology, 1983—, Life Scis., 1993—; editorial cons. profl. jours.; contr. articles to profl. jours., chpts. in books. NSF predoctoral fellow, 1962-63; pharmacology fellow Riker Pharm. Co., 1966-67; NSF postdoctoral fellow, 1966-67; recipient research Career Devel. award NIH, 1972-77. Mem. AAAS, N.Y. Acad. Scis., Am. Soc. Pharmacology and Exptl. Therapeutics, Soc. for Neurosci., Internat. Brain Rsch. Orgn., Am. Gastroent. Assn., Urodynamics Soc., Internat. Med. Soc. of Paraplegia, Soc. for Basic Urologic Rsch., Am. Motility Soc., Sigma Xi, Rho Chi. Republican. Methodist. Home: 6357 Burchfield Ave Pittsburgh PA 15217-2732 Office: U Pitts Med Sch W-1352 Biomed Sci Tower Terrace St Pittsburgh PA 15261-0001

DEGUIRE, MARK ROBERT, materials scientist, educator; b. Chgo., Apr. 4, 1958; s. Robert LeRoy and Eleanor Marie (Perrella) DeG.; m. Eileen Ann Joyce, May 23, 1981; children: Audrey, Jeannette, Adam, Ruth. BS, U. Ill., 1980, MS, 1982; PhD, MIT, 1987. Teaching asst. U. Ill., Urbana-Champaign, 1980-82; rsch. asst. MIT, Cambridge, 1982-87; affiliated faculty mem. Case Ctr. Electrochem. Scis., Cleve., 1987—; Nord asst. prof. Case Western Res. U., Cleve., 1987-90, asst. prof. dept. materials sci. and engring., 1990-93, assoc. prof., 1993—; session chmn. ann. meetings Am. Ceramic Soc., 1989, 91, 93; invited mem. peer rev. panel Conservation and Renewable Energy Rsch. program Dept. Energy, 1992; cons. Eveready Battery Co., Cleve. Fluid Systems, PCC Airfoils, BP Am. Corp., Reliance Electric Co., Erico Products, others; reviewer Jour. Am. Ceramic Soc., Ceramic Abstracts, Solid State Ionics, other profl. jours. Contbr. articles to refereed jours. Mem. Am. Ceramic Soc., Materials Rsch. Soc., Keramos. Roman Catholic. Achievements include patent for Solution Synthesis of Ceramics; co-discovery of phenomenon of normal-state magnetic alignability of oxide superconducting particles. Office: Case Western Res Univ Dept Materials Sci/Engring 10900 Euclid Ave Cleveland OH 44106-7204

DE GUZMAN, ROMAN DE LARA, chemical engineer, consultant; b. Manila, Oct. 6, 1941; arrived in Can., 1969; s. Fernando B. deGuzman and Isidra A. de Lara; m. Erlinda L. Alano, July 4, 1970; children: Romalynn M., Philip H. BS in Chem. Engring., Far Eastern U., 1962; postgrad., U. Philippines, 1966-67, U. Alta., Edmonton, Can., 1975-76. Registered engring. technologist, Can. Plant lube engr. Ducon Fibres, Inc., Manila, 1963-64; instr. chemistry Nueva Eeja High Sch., Cabanatuan City, Philippines, 1964-66; lab. technologist Can. Western Natural Gas Co., Calgary, Alta., 1969-75; supr. lab. svcs. Northwestern Utilities Ltd., Edmonton, Alta., 1976—; pres., cons. Noram Chems. Ltd., Edmonton, 1985—. Named Most Outstanding Filipino Scientist Philippine Cultural Soc., Edmonton, 1981. Mem. Am. Soc. Testing Materials, Can. Inst. Mining and Metallurgy, Alta. Hydrocarbon and Analysis Com. Roman Catholic. Home: 10851 33d Ave, Edmonton, AB Canada T6J 2Z3

DEHAVEN, KENNETH LE MOYNE, retired physician; b. The Dalles, Oreg., Mar. 28, 1913; s. Luther John and Dora (Beeks) DeH.; m. Ledith Mary Ewing, Jan. 11, 1937; children: Marya LeMoyne DeHaven Keith, Lisa Marguerite DeHaven Jordan, Camille Suzanne DeHaven. BS, North Pacific Coll. Oreg., 1935; MD, U. Mich., 1946. Intern USPHS Hosp., St. Louis, 1947; intern Franklin Hosp., San Francisco, 1947-48, resident, 1949; clinician Dept. Pub. Health, City San Francisco, Dept. V.D., 1949-51; practice gen. medicine, Sunnyvale, Calif., 1955-87; mem. staff El Camino Hosp., Mt. View, Calif., San Jose (Calif.) Hosp. Pres. Los Altos Hills Assn. Served to capt., USAF, 1952-55. Fellow Am. Acad. Family Practice; mem. Calif. Med. Assn., Santa Clara County Med. Soc., Astron. Soc. Pacific, Sunnyvale C. of C. (bd. dirs. 1955-56), Book Club (San Francisco), Masons, Alpha Kappa Kappa. Republican. Home: 9348 E Casitas Del Rio Dr Scottsdale AZ 85255-4313

DE HERTOGH, AUGUST ALBERT, horticulture educator, researcher; b. Chgo., Aug. 24, 1935; s. Frank Joseph and Marie Louise (Van Cauwenbergh) De H.; m. Edna Faye Kipp, June 5, 1957 (div. Mar. 1985); children: Christopher Mark, Michelle Louise, Jennifer Leigh; m. Mary Belle Moore Shurling, Aug. 23, 1986. BS, N.C. State U., 1957, MS, 1961; PhD, Oreg. State U., 1963. Asst. plant physiologist Boyce Thompson Inst., Yonkers, N.Y., 1964-65; asst. prof. Mich. State U., East Lansing, 1965-69, assoc. prof., 1969-72, prof., 1972-78; prof. N.C. State U., Raleigh, 1978—, dept. head, 1978-88; advisor Dutch Bulb Exporters Assn., Hillegom, The Netherlands, 1965—. Author: Holland Bulb Garden Guide, 1982, Spring Flowering Bulbs, 1986, Holland Bulb Forcers' Guide, 1989; author computer software. 1st lt. U.S. Army, 1957-59. Named to Floriculture Hall of Fame Soc. Am. Florists, 1988; Japan Soc. for the Promotion Sci. rsch. fellow, 1988; recipient Medal of Honor The Netherlands Ministry Agrl. and Fisheries, 1985, Nicholaas Dames medal, The Netherlands, 1990, Golden Pin, Dutch Bulb Exporters Assn., 1990. Fellow Am. Soc. for Hort. Sci. (v.p. rsch. 1986-87); mem. Internat. Soc. for Hort. Sci., Gamma Sigma Delta, Phi Kappa Phi, Sigma Xi. Achievements include research in the physiology of growth and development of flowering bulbs (geophytes). Home: 117 Rosewall Ln Cary NC 27511-6639 Office: NC State U Dept Hort Sci Raleigh NC 27695-7609

DEHMELT, HANS GEORG, experimental physicist; b. Germany, Sept. 9, 1922; came to U.S., 1952, naturalized, 1962; s. Georg Karl and Asta Ella (Klemmt) D.; 1 child from previous marriage, Gerd; m. Diana Elaine Dundore, Nov. 18, 1989. Grad., Graues Kloster, Berlin, Abitur, 1940; D Rerum Naturalium, U. Goettingen, 1950; D Rerum Naturalium (hon.), Ruprecht Karl-Universitat, Heidelberg, 1986; DSc (hon.), U. Chgo., 1987. Postdoctoral fellow U. Goettingen, Germany, 1950-52, Duke U., Durham, N.C., 1952-55; vis. asst. prof. U. Wash., Seattle, 1955; asst. prof. physics U. Wash., 1956, asso. prof., 1957-61, prof., 1961—; cons. Varian Assocs., Palo Alto, Calif., 1956-76. Contbr. articles to profl. jours. Recipient Humboldt prize, 1974, award in basic research Internat. Soc. Magnetic Resonance, 1980, Rumford prize Am. Acad. Arts and Scis., 1985, Nobel prize in Physics, 1989; NSF grantee, 1958—. Fellow Am. Phys. Soc. (Davisson-Germer prize 1970); mem. Am. Acad. Arts and Scis., Nat. Acad. Scis. Co-discoverer (with Hubert Krüger) nuclear quadrupole resonance, 1949; other achievements include leader groups of physicists who first isolated at rest in a vacuum an individual subatomic particle, 1973, a charged atom 1980, an antimatter particle in 1981, measured magnetism and size of electron and positron with precisions 1000 times higher than previously in 1987, proposed triton-proton model of subatomic particles in 1987. Home: 1600 43d Ave E Seattle WA 98112 Office: U Wash Physics Dept FM 15 Seattle WA 98195

DE HOFF, JOHN BURLING, physician, consultant; b. Balt., May 28, 1913; s. George William and Pearle Ann (Burling) De H.; m. Mabelle Audrey Dunn, July 9, 1938; children: Susan De Hoff Montgomery, John Howard. MD, Johns Hopkin's U., 1939, MPH, 1967. Diplomate Am. Bd. Preventive Medicine, Am. Bd. Pub. Health. Med. intern Mt. Sinai Hosp., N.Y.C., 1939-41; asst. resident in psychiatry N.Y. Hosp., 1946; pvt. practice internal medicine Balt., 1947-65; asst. commr. Balt. City Health Dept., 1965-69, resident in pub. health, 1966-68, dep. commr., 1969-75, commr., 1975-84; advisor to mayor City of Balt., 1984-87; med. cons. Bd. Phys. Quality Assurance, State of Md., Balt., 1987—; med. cons. Social Security Adminstrn., 1986-89. Contbr. articles to profl. jours. and community health publs. Col. USAR, 1935-65, ETO. Commonwealth fellow N.Y. Hosp., 1946; Jacobi medalist Mt. Sinai Med. Ctr., N.Y.C., 1983. Fellow APHA (mem. governing coun.); mem. Am. Assn. Pub. Health Physicians (pres. 1977), U.S. Conf. Local Health Officers (pres. 1979), Md. Med. Soc. (councillor, del.), Balt. City Med. Soc. (pres. 1974). Democrat. Presbyterian. Home: 13801 York Rd Unit N-7 Cockeysville MD 21030 Office: State Md Bd Physician Quality Assurance 4201 Patterson Ave Baltimore MD 21215-2299

DEHORATIUS, RAPHAEL JOSEPH, rheumatologist; b. Phila., Sept. 16, 1942; s. Pasquale P. and Edith R. DeH.; m. Kathleen M. Carson, Aug. 21, 1965; children: Nicole, Danielle. BS, St. Joseph's U., Phila., 1964; MD, Jefferson Med. Coll., 1968. Med. intern Jefferson Med. Coll., Phila., 1968-69, asst. prof. medicine, 1976-78, assoc. prof. medicine, 1978-82; med. resident U. N.Mex., Albuquerque, 1969-70, rheumatology fellow, 1972-74, asst. prof. medicine, 1974-76; prof. medicine Hahnemann U., Phila., 1982-92, Jefferson Med. Coll./Thomas Jefferson U., Phila., 1992—; chmn. profl. meetings Am. Coll. Rheumatology, Atlanta, 1988-91, edn. coun., 1988-91. Contbr. articles to profl. jours./publs. Maj. USAF, 1970-72. Recipient Lupus Rsch. grant Commonwealth of Pa., Arthritis Rsch. grant. Fellow Am. Coll. Physicians, Am. Coll. Rheumatology; mem. Assn. Am. Immunologists, Am. Fedn. Clin. Rsch. Home: 667 Sproul Rd Villanova PA 19085-1216 Office: Thomas Jefferson Univ 613 Curtis Bldg 1015 Walnut St Philadelphia PA 19107

DEHOUSSE, JEAN-MAURICE, federal official; b. Liège, Belgium, Oct. 11, 1936; 4 children. Faculty of law degree, U. Liège, 1960; cert. higher internat. studies, John Hoskins U., 1961; cert. higher fed. studies, U. Coll. Aoste, 1963, diploma higher fed. studies, 1964; diploma Am. Sec. Edn., LLD. Asst. faculty mem. law U. Liège, 1962; with Nat. Belgian Assn. Sci. Rsch., 1962-65; asst. Faculty of Law Inst. European Law, Liège, 1966-71; lectr. Higher Sch. Transaltors and Interpreters, Brussels, 1965-71; with Svc. for Polit. Sci. Policy, 1969-71; v.p. Signs and Letters Ltd. and Le Grand Liège Ltd.; from dep. leader to leader Cabinet for Minister Communal Rels., 1970, 71; Dep. Liège, 1971-81; pres., regional exec. Walloon, 1972-81, 82—; local cllr. Liège, 1977—; Minister of French Culture, 1977-79, Minister of Walloon Region, 1979-81; senator Liège, 1981—; Minister Walloon Economy, 1982—; ministerial pres. Walloon Region; mem. senate commns. Walloon Regional Coun.; mem. PS Commn., Liège and Ctrl. Commn., USC, Liège, Fed. Assembly adn Fed. Com. of PS; Brussels and PS exec. Grantee

U. Liège, 1961. Mem. Socialist Party. Home: 17 Rue Saint-Pierre, 4000 Liège Belgium Office: Ministry of Education, 68 rue du Commerce, 1040 Brussels Belgium*

DEIBER, JULIO ALCIDES, chemical engineering educator; b. Vera, Santa Fe, Argentina, July 19, 1945; s. Julio Heraclio and Lidia Agustina (Gonzalez) D.; m. Beatriz Susana Yakas, July 25, 1975; 1 child, Pablo Lucas. Chem. engr. degree, U. Nacional del Litoral, Santa Fe, Argentina, 1971; MA, Princeton U., 1977, PhD in Chem. Engring., 1979. Cert. chem. engr. Postdoctoral rsch. assoc. Princeton U., 1979; prof. chem. engring. Inst. Desarrollo Tech. Industria Química U. Nacional Litoral, Santa Fe, 1980—; dir. rsch. Consejo Nacional Investigaciones Científicas y Técnicas, Argentina, 1982—; cons. researcher FATE S.A.I.C.I., Buenos Aires, 1987—, PASA S.A., Rosario, Argentina, 1991—. Contbr. articles to profl. jours.; mem. editorial com., founder Latin Am. Applied Rsch., 1987—. Pres. Latin Am. Com. Heat and Mass Transfer, 1986-88. Mem. Am. Soc. Rheology, Am. Inst. Chem. Engrs., Argentinian Com. Heat and Mass Transfer (pres. 1983-88), Argentinian Soc. Chem. Engring. and Applied Chemistry, Argentinian Soc. Rheology (founder 1984). Home: Sargento Cabral No 1350, Dpto 2, 3000 Santa Fe Argentina Office: INTEC, Güemes No 3450, 3000 Santa Fe Argentina

DEISENHOFER, JOHANN, biochemistry educator, researcher; b. Zusamaltheim, Bavaria, Germany, Sept. 30, 1943; came to U.S., 1988; s. Johann and Thekla (Magg) D.; m. Kirsten Fischer-Lindahl, June 19, 1989. Diploma in Physics, Technische U., Munich, 1971, PhD, 1974, Doctor habilis, 1987. Postdoctoral fellow Max-Planck Inst. Biochemie, Martinsried, Fed. Republic of Germany, 1974-76, staff scientist, 1976-88; investigator Howard Hughes Med. Inst., Dallas, 1988—; prof. biochemistry U. Tex., Dallas, 1988—. Contbr. over 50 sci. papers to profl. publs. Recipient Nobel prize for chemistry, 1988; co-recipient Biol. Physics prize Am. Phys. Soc., 1986, Otto Bayer prize, 1988; decorated The Knight Commander's Cross (Badge and Star) Of the Order of Merit of Germany, 1990, Bavarian Order of Merit, 1992. Mem. AAAS, Am. Crystallographic Assn., German Biophys. Soc., The Protein Soc., Biophysical Soc., Academia Europaea.

DEITER, NEWTON ELLIOTT, clinical psychologist; b. N.Y.C., Dec. 12, 1931; s. Benjamin and Anna (Leibowitz) D. BS, UCLA, 1957; MS, Leland Stanford, 1960; PhD in Clin. Psychology, U. Chgo., 1965. Cert. in clin. psychology. Pvt. practice clin. psychology L.A., 1965-90; exec. dir. Nat. Family Planning Coun., L.A., 1965-76, Gay Media Task Force, L.A., 1976—; staff cons. Aaron Spelling Prodns., L.A., 1980-90, spl. cons. NBC, L.A., 1970-79, cons. broadcast standards dept. CBS, L.A., 1968-82, cons. City Coun., City of L.A., 1975-85. Columnist Bottomline Mag., 1992—. Mem. Dem. Cen. Com., L.A., 1972-76; bd. dirs. Gay Community Svcs. Ctr., L.A., 1970-75, Am. Cancer Soc., L.A., 1972-77, Palm Springs Gay Tourism Coun., 1993—; commr. L.A. Probation Commn., 1977-85; bd. advisors San Francisco Sheriffs Dept., 1969-79; pres. Internat. Gay Travel Assn., 1991-92. Lt. col. USAFR, 1950-75. Mem. Acad. TV and Scis., Press Club L.A., Internat. Gay Travel Assn. (bd. dirs. 1986-93, pres. 1991-92), Desert Bus. Assn. (v.p. 1993, bd. dirs. 1992), Air Force Assn., Am. Mensa, Masons. Avocations: photography, wine making, travel writing. Home: 71426 Estelita Dr Rancho Mirage CA 92270-4215 Office: Rancho Mirage Travel 71-428 Hwy 111 Rancho Mirage CA 92270

DE JESÚS, NYDIA ROSA, physician, anesthesiologist; b. Humacao, P.R., Sept. 8, 1930; d. Manuel Aurelio De Jesus and Luz Maria González. BS, U. P.R., 1949, MD, 1955; cert. med. tech., Sch. Tropical Medicine, San Juan, P.R., 1950; cert. anesthesiology, Columbia Presbyn. Med. Ctr., 1958. Diplomate Am. Bd. Anesthesiology. Dir. dept. anesthesiology U. Hosp. & Sch. Medicine, San Juan, 1960-65; dir. div. anesthesiology P.R. Med. Ctr., San Juan, 1965-76; vis. prof. anesthesiology Harvard Med. Sch., Boston, Mass., 1973-74; dean acad. affairs Med. Scis. Campus, U. P.R., San Juan, 1976-78; dir. cardiovascular surg. ctr. P.R. Med. Ctr., San Juan, 1980-85; prof. anesthesiology U. P.R. Sch. Medicine, San Juan, 1965-90, dean, 1986-90; ret., 1990; staff mem. anesthesiology Las Americas Amb. Surg. Ctr., San Juan, P.R., 1990—; dir. intensive care unit Univ. Hosp., San Juan, 1974-75; chief sect. anesthesiology VA Hosp., San Juan, 1963-76; mem. cardiovascular commn. Sec. of Health, Commonwealth of P.R., 1985; cons. div. medicine, health resources adminstrn. Bur Health Manpower, USPHS, 1977-79; pres. cons. bd. Pediatric U. Hosp., San Juan, mem. bd. dirs., 1986. Fellow Am. Coll. Anesthesiology; mem. ACP, N.Y. Acad. Scis., AAAS, Am. Soc. Anesthesiology. Avocations: music, gardening, reading. Home: No 8 Jardines De Vedruna San Juan PR 00927 Office: Las Americas Amb Surg Ctr PO Box 4236 Hato Rey San Juan PR 00919-4236

DE JONG, GARY JOEL, chemist; b. Bellingham, Wash., June 21, 1947; s. Gerald J. and Marjorie (Kok) De J.; married; children: Amity Lynn, Anna Darlyne. BS, U. Calif., Riverside, 1969; PhD, Oreg. State U., 1973. Chem. researcher Rohm and Haas Co., Phila., 1973—. Co-founder Human Growth Ctr., Holland, Pa., 1980, pres., bd. dirs. 1986-92. Office: Rohm and Haas Co 727 Norristown Rd Spring House PA 19477

DEJONGE, CHRISTOPHER JOHN, obstetrics/gynecology educator; b. Evanston, Ill., Sept. 16, 1958; s. John Edward DeJonge and Lois Jean (Kleinofen) Mengarelli. BS, So. Ill. U., 1976-80; MS, Roosevelt U., 1981-84; PhD, Rush U., Chgo., 1984-89; post doctorate, Rush Med. Coll., Chgo., 1989-90. Lab asst. Roosevelt U., Chgo., 1981-84; rsch. asst. Rush Med. Coll., Chgo., 1984-89, instr., 1989-90, dir., asst. prof., 1990-92; dir., asst. prof. U. Nebr. Med. Ctr., Omaha, 1992—. Numerous contbr. articles to profl. jours. Named Finalst Chgo. Assn. Reproductive Endocrinologists Young Investigator, 1988-89; recipient The Grad. Coll. award, Grad. Coll. Rush U., 1989. Mem. Am. Assn. for Advancement of Sci., Am. Fertility Soc., Am. Soc. Andrology, Am. SOc. Cell Biology, SOc. for Study of Reproduction, Reproductive Biology Spl. Interest Group, Sigma Xi. Office: Univ Nebr Medical Ctr 600 S 42nd St Omaha NE 68198-3255

DÉKÁNY, IMRE LAJOS, chemistry educator; b. Szeged, Hungary, Dec. 9, 1946; s. Imre and Julianna (Kiss) D.; m. Klara, July 4, 1970 (div. Sept. 1980); children: Gyorgy, Andrea; m. Henriette Edit, Aug. 12, 1983. BS in Chemistry, U. Szeged, 1965-70; PhD in Chemistry, Hungary Acad. Sci., 1979, DSc in Chemistry, 1989. Assoc. prof. chemistry U. Szeged, 1980-90, prof., 1990—. Recipient Buzágh proze Hungary Acad. Sci., Budapest, 1979. Mem. Am. Chem. Soc., Deutsche Kolloidgesellschaft, European Chem. at Interfaces Conf., Iupac Commn. Roman Catholic. Office: U Szeged, Aradi Vértanuk Tere 1, 6720 Szeged Hungary

DE KRASINSKI, JOSEPH STANISLAS, mechanical engineering educator; b. Mszana, Cracow, Poland, June 15, 1914; arrived in Can., 1968; s. Henryk Edward and Maria Gertrude (Leska) de K.; m. Patricia Rosamunda Stansfield, Sept. 3, 1947; children: Maria, Patricia, Elizabeth, Frances. BSc in Aero. Engring., London Engring. U., 1944, PhD in Engring., 1964. Rsch. scientist Royal Aircraft Establishments, Farnborough, Eng., 1944-47, Instituto Aerotechnico, Cordoba, Argentina; prof. U. Nacion de Cordoba, 1948-68; prof. mech. engring. U. Calgary, Alta., Can., 1968-78, prof. emeritus, 1978—. Contbr. articles to profl. publs. F/lt. RAF, 1940-42. Recipient DFC, RAF, 1943, Virtuti Militari, Polish Air Force, 1943. Mem. AIAA, Can. Soc. Mech. Engring. Roman Catholic. Achievements include patent for radial diffuser, research on shock wave attenuation, liquid foams, environmental aerodynamics. Home: 3319 24th St, Calgary, AB Canada T2M 3Z8 Office: U Calgary Dept Mech Engring, 2500 University Dr, Calgary, AB Canada T2N 1N4

DEKSTER, BORIS VENIAMIN, mathematician, educator; b. Leningrad, USSR, Oct. 8, 1938; arrived in Can., 1974; s. Veniamin Moisey Zeigerman and Faina Aron Dekster; m. Nadezhda Sergey Prokopets, Feb. 7, 1969 (div. May 1985); 1 child, Sonya; m. Monika Bargiel, Dec. 14, 1990. Master's degree Leningrad U., 1962; Ph.D., Steklov Inst., Leningrad, 1971. Research assoc. U. Toronto, Ont., Can., 1974-78, asst. prof., 1981-86; asst. prof. U. Notre Dame, Ind., 1979-81; assoc. prof. Mt. Allison U., New Brunswick, Can., 1986-90, prof., 1990—. Contbr. articles on differential geometry and convexity to profl. jours. NSF grantee, 1980-81; Can. Natural Scis. Research Council grantee, 1981—. Mem. Am. Math. Soc. Home: 7 Raworth Heights, Sackville, NB Canada E0A 3C0 Office: Mount Allison U, Dept Math, Sackville, NB Canada E0A 3C0

DE LA CADENA, RAUL ALVAREZ, physician, pathology and thrombosis educator; b. Mex., Mex., Dec. 31, 1959; came to U.S., 1985; s. Raul A. and Florencia (Garnica) De La C; m. Donna Lynn Schlam, Aug. 18, 1985. BS in Biology, Centro Universitario Mexico, 1979; MD, LaSalle U., Mexico City, Mex., 1984. Med. intern Nutrition Nat. Inst., Mexico City, 1983-84, rsch. assoc., 1984-85; rsch. tng. Temple U. Sch. Medicine, Phila., 1985-86, postdoctoral trainee, 1986-89, asst. prof., 1989—, minority mentor program, 1988—. Contbr. articles to Jour. Lab. Clin. Med., Blood, Am. Jour. Physiology., Jour. Clin. Invest., Am. Jour. Pathology; contbr. chpt. to book. Recipient Sol Sherry award Temple U. 1988, Young Scientists Travel award Internat. Soc. Hemostasis and Thrombosis, 1989, Clin. Investigator award NIH, Rsch. Svc. award NIH, 1986-89, 88-89, AHA grant in aid, 1991-92, 93—. Mem. ACP, N.Y. Acad. Scis., Am. Fedn. Clin. Rsch., Am. Soc. Hematology, Am. Soc. Investigative Pathology. Jewish. Office: Temple U Sch Medicine OMS-403 3400 N Broad St Philadelphia PA 19140-5196

DELACOUR, YVES JEAN CLAUDE MARIE, technology information executive; b. Pamiers, Ariege, France, June 24, 1943; s. Jacques and Marie-Louise (Rambaud) D.; m. Elisabeth Peu-Duvallon; children: Thibault, Gauthier. Engr. French Naval Acad., 1965; BS in Econs., Fin., Institut d'Etudes Politiques, 1972; M.B.A., Stanford U., 1975. Commd. ensign, 1965, advanced through grades to lt. de Vaisseau, 1971; naval officer Helicopter Carrier Jeanne d'Arc, French Navy, 1965-69; in charge of communications dept. Destroyer, Toulon, 1966-69; staff mem. Navy Dept., Paris, 1969-73; credit officer Banque de l'Indochine et de Suez, Paris, 1975-77; pres. Transasia Corp., Paris, 1978—, pres. Internat. Data Group, France, 1980-86, v.p. France Internat. Data Group, Inc., 1986-92; pres. Leonardo, 1989—. Mem. Stanford Bus. Club (Paris) (pres. 1981-86), American Club Paris. (hon. sec. 1986-88). Home: 22 rue de la Federation, 75015 Paris France Office: Transasia Corp, 12 Ave George V, 75008 Paris France

DELAHAY, PAUL, chemistry educator; b. Sas Van Gent, The Netherlands, Apr. 6, 1921; came to the U. S., 1946, naturalized, 1955; s. Jules and Helene (Flahou) D.; m. Yvonne Courroye, 1962. B.S. in Gen. Engring, U. Brussels, 1941, M.S. in Chemistry, 1945; M.S. in Elec. Engring, U. Liege, 1944; Ph.D. in Chemistry, U. Ore., 1948. Instr. chemistry U. Brussels, 1945-46; research assoc. U. Oreg., 1948-49; faculty La. State U., 1949-65, prof. chemistry, 1955-56, Boyd prof. chemistry, 1956-65; prof. chemistry NYU, N.Y.C., 1965-87; Frank J. Gould prof. sci. NYU, 1974-87. Author: New Instrumental Methods in Electrochemistry, 1954, Instrumental Analysis, 1957, Double Layer and Electrode Kinetics, 1965; Editor: Advances in Electrochemistry, 1961-74. Guggenheim fellow Cambridge (Eng.) U., 1955-56; Guggenheim fellow N.Y. U., 1971-72; Fulbright prof. Sorbonne, Paris, France, 1962-63; Recipient medal U. Brussels, 1963, Heyrovsky medal Czechoslovak Acad. Sci., 1965. Mem. Am. Chem. Soc. (award pure chem. 1955, Southwest award 1959), Electrochem. Soc. (Turner prize 1951, Palladium award 1967, chmn. theoretical div. 1957-59), AAAS, Am. Phys. Soc., Internat. Union Pure and Applied Chemistry (chmn. commn. electrochem. data 1959-63, titular mem. analytical sect. 1961-65), Sigma Xi. Home: 6 Rue Benjamin Godard, 75116 Paris France

DELALOYE, BERNARD, nuclear medicine physician; b. Martigny, VS, Switzerland, Dec. 20, 1928; s. Léon and Marie-Thérèse (Ducrey) D.; m. Angelika Bischof, June 27, 1970; children: Sibylle, Raphael. MD, U. Lausanne, Switzerland, 1957, thesis, 1958; specialization, U. Lausanne, 1972. Fellow internal medicine, clin. medicine Univ. Hosp., Lausanne, Switzerland, 1957; fellow expl. medicine Prof. B. Halpern et al, Paris, 1957-58; fellow internal medicine, clin. medicine Hosp. St. Antoine, Lausanne, 1958-59; fellow hepatology, gastroenterology Curie Found., Paris, 1959-60; fellow Hosp. Frederic Joliot, Orsay, France, 1960-61; head Radioisotopes Lab. Isotopes Lab. Internal Medicine Clinic, Lausanne, 1962-65; head Div. Nuclear Medicine, Lausanne, 1966-72, Autonomous Div. Nuclear Medicine, Lausanne, 1972—; privat-docent Nuclear Medicine, Lausanne, 1967—; prof. Div. Nuclear Medicine, Lausanne, 1979—; pres. Commn. Nuclear Medicine and Biology Swiss Acad. Med. Scis., 1973-80; mem. Interkantonale Kontroll Stelle fur Heilmittel for Radiopharmaceuticals, Bern, 1976-85; pres. 1993 Congress European Assn. Nuclear Medicine, Lausanne; editorial bd. Internat. Jour. Nuclear Med. and Biology, 1975, European Jour. Nuclear Medicine, 1976. Author: Introduction à la Scintigraphie Clinique, 1966. Recipient grands, Swiss League Against Cancer, 1987-89, Swiss Acad. Medicine, 1959-61, Tossizza Found., 1959-62, Sylvano Found., 1964, French Govt., 1957-58, Soc. Acad. Vaudoise, 1970. Mem. European Nuclear Medicine Soc. (pres. 1978-80), Am. Heart Assn., Am. Soc. Nuclear Medicine, Assn. Nuclear Medicine, Soc. Nuclear Medicine, European Assn. for Nuclear Medicine (pres. European Nuclear Medicine Congress 1993, Lausanne), Collegio Brasileira de Radiologia (hon.), Soc. Brasileira de Radiologia (hon.), Purkynje Soc. CSSR (hon.), Soc. Italiana de Biologia e Medicina Nuclear (hon.). Achievements include research in nuclear cardiology, measurements of blood flow, lymphoscintigraphy, and radioimmunoscintigraphy. Home: Rte de Converney 52, VD, VD 1093 La Conversion Switzerland Office: DAMN-CHUV, VD, 1011 Lausanne Switzerland

DE LANEROLLE, NIMAL GERARD, process engineer; b. Colombo, Sri Lanka, Nov. 26, 1945; came to U.S., 1980; s. Eustace Joseph and Pearl Norberta (Jayasundera) de L.; m. Suranganee Mary Amarasingha, Sept. 8, 1973. BSc in Engring., U. Sri Lanka, Peradeniya, 1970; Master of Tech., Brunel U., Uxbridge, Eng., 1977; PhD, SUNY, Stony Brook, 1987. Chartered engr., U.K. Instr. U. Moratuwa, Sri Lanka, 1970-71; engr. Sri Lanka Transp. Bd., Werahera, 1971-73; lectr. U. Moratuwa, 1976-79; teaching and rsch. asst. SUNY, Stony Brook, 1980-85; process engr. Standard Microsystems Corp., Hauppauge, N.Y., 1985—; cons. Samuel Sons Ltd., Colombo, Sri Lanka, 1978-79, Brookhaven Nat. Lab., Upton, N.Y., 1991; rsch. adv. materials sci. dept. SUNY, 1987-90; editorial advisor Jour. of Metals, 1988—; adj. prof. dept. mech. engring. SUNY, 1992—. Editor: Microstructural Science for Thin Film Metallizations in Electronic Applications; contbr. articles to profl. jours. Fulbright scholar, 1979; recipient rsch. award Nat. Sci. Coun., Sri Lanka, 1978, rsch. assistantship SUNY, 1983-85. Mem. Metall. Soc. (mem. com. electronic and photonic device materials 1979), Instn. Mech. Engrs. (U.K.), Sigma Xi. Achievements include patent for Method for Fabricating Reliable Semiconductor Devices; mechanism for the degradation of titanium silicide thin films. Home: 101 Nadia Ct Prt Jefferson NY 11777-1444 Office: Standard Microsystems Corp 35 Marcus Blvd Hauppauge NY 11788-3791

DELANO, A(RTHUR) BROOKINS, JR., civil engineer, consultant; b. Burlington, Vt., Mar. 15, 1921; s. Arthur Brookins and Mabelle (George) D.; m. Laura Holt Coleman, Aug. 23, 1953; children: Marion Elizabeth, James Brookins, Linda Louise, George Coleman. BSCE, U. Vt., 1943. Registered profl. engr., Vt. Farmer Shoreham, Vt., 1943-54; civil engr. Vt. State Hwy. Dept., Montpelier, Vt., 1954-82; pvt. practice Barre, Vt., 1982—. Mem. Masons, Shriners. Home: RD 3 Box 6598 Barre VT 05641

DELAP, BILL JAY, engineer, consultant; b. Paola, Kans., Nov. 3, 1931; s. Wilbur Jay and Wilma Pauline (Carpenter) D.; m. Angela Ellen Irwin, Feb. 21, 1957; children: Deven Kelly, Dawn Denise. BS in Petroleum Engring., Kans. U., 1954; LLB, LaSalle Extension U., 1966. Registered profl. engr. Engr. Panhandle Eastern Corp., Kansas City, Mo., 1954-56; product engr. Panhandle Eastern Corp., Liberal, Kans., 1956-57; div. engr. Panhandle Eastern Corp., Springfield, Ill., 1957-69; reg. engr. Panhandle Eastern Corp., Liberal, 1969-71, area supt., 1971-73; region mgr. Panhandle Eastern Corp., Brighton, Colo., 1973-86; prin. engr. 4-D Triangle Co., Hudson, Colo., 1986—; bd. mem. Cottonwood Svcs. Corp., Denver, 1989—. Co-author: Natural Gas Storage Fields, 1964. Mem. Greater Brighton Area COFC, 1978. Mem. Nat. Soc. Profl. Engrs. (pres. northern chpt. 1976), Soc. Petroleum Engrs., Grand Lodge AF&AM. Republican. Baptist.

DELBARRE, BERNARD, pharmacologist, consultant; b. Trith St Leger, France, July 19, 1932; s. Robert and Clemence (Vilain) D.; m. Gisele Reynaud, May 8, 1958; children: Sylvie Anne, Jean Bernard, Savine. MD, Faculte De Medecine, 1962; Sci. Dr., Faculte Des Scis., 1975. Head dept. pharmacology Pfizer, France, 1966-71; head dept. rsch. physiopath. Faculte De Medecine de Tours, France, 1971—. Editor: Psychopharmacologie, 1990, Hypertension, 1993. Medical officer Armed Services, 1959-61. Mem. ASPET, Soc. Neuroscis., N.Y. Acad. Scis., Lions Club (pres. 1969). Achievements include patent in field. Home: 68 rue des Carnaux, 37510 Ballan Mire France

DEL CARMEN, RENE JOVER, process and environmental engineer; b. Manila, June 5, 1934; came to U.S., 1960; s. Vicente F. and Aurea (Jover) Del C.; m. Imelda B. Aquino, June 23, 1962; children: Dana, Stephanie, Jay, Darlene. BSCE, Mapua Inst. Tech., Manila, 1955; MSChem, Marquette U., 1962. Registered chem. and environ. profl., U.S., The Philippines. Rsch./applications engr. Green Bay (Wis.) Packaging, 1962-65; sr. process engr. Esso Fertilizer, Manila, 1965-68; sr. chem. engr. C.F. Braun & Co., Alhambra, Calif., 1968-70; process/project engr. Valley Nitrogen Prodrs., Fresno, Calif., 1970-74; process/environ. engr. CAPCO Mgmt. Group, Chowchilla, Calif., 1987-89; process engr. Jacobs Engring. Group, Martinez, Calif., 1989-90; prin. process engr. Morrison Knudsen Corp., Boise, Idaho, 1990—; cons. in engring. Indsl. Process Consulting, Fresno, 1975-87; presenter tech. papers on chem. stabilization solidification, 1992. Mem. AICE, Am. Soc. Agrl. Engrs., Nat. Registry Environ. Profls., Philippine Inst. Chem. Engrs. Roman Catholic. Achievements include rsch. and design in hazardous waste remediation techs., stabilization/solidification, water treatment techs., fertilizer ops., catalyst systems. Home: 2800 Bogus Basin Rd # A108 Boise ID 83702 Office: Morrison Knudsen Corp 720 Park Blvd Boise ID 83712

DEL CASTILLO, JAIME, economic planning consultant, educator; b. Santander, Cantabria, Spain, Apr. 14, 1951; s. Jaime del Castillo and Gabriela Hermosa; m. Mercedes Ugedo, Sept. 12, 1980; children: Gabriela, Jaime. Cert. in econs., Facultad Economicas, Bilbao, Spain, 1975; MA, IREP, Grenoble, France, 1981, PhD in Econ. Sci., 1985. Asst. prof. Bus. Sch., Santander, Spain, 1974-75; asst. prof. Facultad Economicas, Bilbao, 1975-80, 80-85, prof., 1985—; with dept. studies Sociedad Promocion Indsl., Bilbao, 1984-85; chmn. Info. y Desarrollo, Bilbao, 1988—; cons. EEC, Brussels, 1987—, OECD, Paris, 1987-88; expert Social Com. EEC, Brussels, 1989-90. Author: Cambio Económico y Cambio Espacial: perspectivas desde el eje atlántico, 1990, (with others) Medidas ayuda a la inversion, 1986; editor: Regional Development Policy, 1989, Spatial Aspects of Technological Change, 1989, Aspectos Estratégicos de Gestión Empresarial, 1989; econs. columnist El Correo, 1986—, TVE, 1987-90; contbr. articles to profl. jours. Mem. European Assn. Devel. Rsch. and Tng. Insts. (convenor working group European devel.), Assn. Regional Sci. Basque Country (pres. 1988-90).

DEL CASTILLO, JULIO CESAR, neurosurgeon; b. Havana, Cuba, Jan. 21, 1930; came to U.S., 1961, naturalized, 1968; s. Julio Cesar and Violeta (Diaz de Villegas) Del C.; m. Rosario Freire, Sept. 18, 1955; children: Julio Cesar, Juan Claudio, Rosemarie. B.S., Columbus Sch., Havana, 1948; M.D., U. Havana, 1955; Diploma Am. Coll. of Surgeons, 1971. Intern, Michael Reese Hosp., Chgo., 1955-56; resident Cook County Hosp., Chgo., 1957, Lahey Clinic, Boston, 1957-58, U. Pa. Grad. Hosp., 1958-60; research asst. dept. gen. surgery Jackson Meml. Hosp., Miami, Fla., 1962-64; practice medicine, specializing in neurosurgery, Havana, 1960-61, Quincy, Ill., 1965—; mem. staff Blessing Hosp., Quincy, pres. staff, 1972-74; mem. staff Blessing at 14th, Quincy, trustee, 1987; owner Top Hat Hobbies, Inc., Quincy. Bd. dirs. Western Ill. Found. for Med. Care, 1970-73; trustee Blessing Hosp., 1972-74, St. Mary's Hosp., 1988-90. Mem. Am. Acad. Model Aeros., Congress Neurol. Surgeons, AMA, A.C.S., Adams County Med. Soc. (sec., treas. 1966-75, pres.), Ill. Med. Soc., Exptl. Aircraft Assn. Rotarian (dir. 1970-72, pres. 1976-77). Home: 14 Curved Creek Rd Quincy IL 62301-6526 Office: 126 N 5th Quincy IL 62301

DE LEENHEER, ANDREAS PRUDENT, medical biochemistry and toxicology educator; b. Zele, Belgium, May 16, 1941; s. Petrus Hermanus and Maria (de Brauwer) de L.; m. Magy Marie-Louise van Decraen, Nov. 30, 1968; children: Patrick, Els, Marianne, Marc. Degree in pharmacy, Rijksuniversiteit Gent, Belgium, 1963, D. in Pharm. Sci., 1968, aggregé de l'enseignement supérieur, 1971; postgrad., U. Mich., 1969-70. Asst. dept. pharm. chemistry Rijksuniversiteit Gent, 1962-63, asst. dept. toxicology, 1963-67, sr. asst. dept. toxicology, 1967-71, docent, head dept. med. biochemistry and clin. analysis, 1971-77, prof., head dept. med. biochemistry and clin. analysis, 1977—, dean faculty pharm. scis., 1984-92; dir. Van Hauwaert Lab. Inst. Moderne, Ghent, 1978—; chmn. com. clin. biology Ministry Health, Brussels, 1984—. Editor and reviewer for numerous publs.; contbr. numerous articles to profl. jours. Recipient Vichy prize, 1962, Medal of Merit City of Ghent, 1974. Mem. AAAS, Am. Acad. Pharm. Scis., Am. Chem. Soc., Am. Assn. Clin. Chemistry, Am. Coll. Toxicology, Forensic Sci. Soc., Belgian Soc. for Pharm. Scis. (Biannual prize 1973), Internat. Assn. Forensic Toxicologists, Belgian Soc. Legal Medicine, Belgian Assn. for Clin. Chemistry, Fedn. Internat. Pharmaceutique, Cercle des Alumni de la Fondation Universitaire, Societe de Chimie Biologique, Flemish Chem. Assn., Can. Soc. Clin. Chemists, Dutch Assn. for Clin. Chemistry, Assn. Clin. Biochemists, Belgisch Genootschap voor Farmaceutische Wetenschappen, Belgisch Genootschap voor Gerechtelijke Geneskunde, Belgische Vereniging voor Klinische Chemie, Fedn. Internat. Pharm., Kring der Alumni van de Universitaire Stichting, Soc. de Chimie Biologique, Belgische Vereniging voor Biochemie, Vlaamse Chemische Vereniging, Nederlandse Vereninging voor Klinische Chemie, Soc. Francaise de Biologie Clinique, Deutsche Gesellschaft Für Klinische Chemie, Am. Assn. Mass Spectrometry, Belgische Vereniging voor Medische Info., Belgisch Genootschap voor Kerngeneeskunde, Am. Soc. for Bone and Mineral Rsch., N.Y. Acad. Scis., Soc. Environ. Toxicology and Chemistry, Assn. Ofcl. Analytical Chemists, Acad. Clin. Lab. Scis., Internat. Assn. Radiopharmacology, Clin. Ligand Assay Soc., Soc. for the Study of Inborn Errors of Metabolism, Am. Assn. Pharm. Scis., Van Slyke Soc., Acad. Europaea. Avocations: classical music, piano. Home: Ryvisschepark 18, B9052 Gent Belgium Office: Labs Medische Biochemie en Toxicologie, Harelbekestraat 72, B9000 Ghent Belgium

DELEO, RICHARD, engineering executive; b. Lackawanna, N.Y., Dec. 6, 1960; s. Louis Frank and Lorraine (Burnhart) D.; m. Patricia Lynn Moorman, Feb. 12, 1983; children: Stephanie Marie, Richard Nicholas. BSEE, U. Buffalo, 1982; MBA, William and Mary Coll., 1989. Engr. Newport News (Va.) Shipbldg., 1982-88, engring. supr., mgr., 1988—. Author: (software program) Artifical Intelligence Expert System for use on P.C., numismatic grading system; co-author cable mgmt. system software for design engring. use. Mem. Beta Gamma Sigma (life). Achievements include design and development of advanced/automated weapons handling system for next generation attack submarine, managing prototype devel. of fault tolerate submarine automated ship control sta. Home: 470 Blount Pt Rd Newport News VA 23606

DE LEON, PABLO GABRIEL, aerospace engineer; b. Canuelas, Buenos Aires, Argentina, Nov. 12, 1964; s. Bernardo Alberto and Celia (Ponce) De L.; div. Assoc. Elec. Engr., St. Martin Coll., Buenos Aires, 1983; Aerospace Engr., Interamerican Inst. High Tech., Buenos Aires, 1989; BS in Aerospace Engring., Pacific Western U., 1989. pvt. pilot Don Torcuato Flight Sch., 1987. With photovoltaic dept. Argentine Lab. Solar Energy, Buenos Aires, 1980-82; designer Sincorp Computer, Buenos Aires, 1984-86; dir. Channel 5 TV Magdalena, Argentina, 1989-92. Author: The First Human Travel to Orbit, 1991; (publs.) Design of a Low Cost Sc Sattellite, 1990, Construction of Launch Facilities 1991, Technical Manual of 4S-A1 Space Suit, 1992; introducer TV program Testing and Using a Space Suit, 1992. Mem. AIAA, Argentine Assn. Space Tech. (pres. 1989—), The Planetary Soc. Roman Catholic. Achievements include design of real time TV satellite for resources evaluation, design and constrn. of 1st Argentine microcomputer, Sincorp SBX, constrn. family of solid propellant rockets, constrn. of 4S-A1 simulation space suit for underwater tng. Office: Argentine Assn Space Tech, CC 142-Suc 28 b, 1428 Buenos Aires Argentina

DE LERNO, MANUEL JOSEPH, electrical engineer; b. New Orleans, Jan. 8, 1922; s. Joseph Salvador and Elizabeth Mabry (Jordan) De L.; BE in Elec. Engring., Tulane U., 1941; MEE, Rensselaer Poly. Inst., 1943. Registered profl. engr., Ill., Mass; m. Margery Ellen Eaton, Nov. 30, 1946 (div. Oct. 1978); children—Diane, Douglas. Devel. engr. indsl. control dept. Gen. Electric Co., Schenectady, 1941-44; design engr. Lexington Electric Products Co., Newark, 1946-47; asst. prof. elec. engring. Newark Coll. Engring., 1948-49; test engr. Maschinenfabrik Oerlikon, Zurich, Switzerland, 1947-48; application engr. Henry J. Kaufman Co., Chgo., 1949-55; pres. Del Equipment Co., Chgo., 1955-60; v.p. Del-Ray Co., Chgo., 1960-67; pres. S-P-D Svcs. Inc., Forest Park, Ill., 1967-81, S-P-D Industries, Inc., Berwyn, Ill., 1981—; mem. standards making coms. Nat. Fire Protection Assn. Internat. Lt. (j.g.) USNR, 1944-45, to lt. comdr., 1950-52. Fellow Soc. Fire Protection Engrs.; mem. IEEE (sr., life), Ill. Soc. Prof. Engrs., Am. Water Works Assn. Home:

36 W 760 Stonebridge Ln Saint Charles IL 60175 Office: 3105 Ridgeland Ave Berwyn IL 60402-3568

DELFINO, JOSEPH JOHN, environmental engineering sciences educator; b. Port Chester, N.Y., Oct. 6, 1941; s. John J. and Frances C. (Santoro) D.; m. Dorothy Janelle Justin. BS in Chemistry, Holy Cross Coll., 1963; MS in Chemistry, U. Idaho, 1965; PhD in Water Chemistry, U. Wis., 1968. Sect. head, tech. mgr. IBT & Nalco Environ. Sci., Northbrook, Ill., 1972-74; sect. head environ. scis. Wis. State Lab. Hygiene, Madison, 1974-82; from asst. prof. to assoc. prof. U. Wis., Madison, 1974-80; assoc. dir. water resources ctr. U. Wis., 1977-78, prof. civil and environ. engring., 1980-82; prof. environ. engring. sci. U. Fla., Gainesville, 1982—, chmn. dept. environ. engring. sci., 1990—; mem. adj. faculty U. Colo., Colorado Springs, 1969-71, Ill. Inst. Tech., Chgo., 1973. Writer, co-originator documentary Fla. Water Story, Sta. WEDU-TV, Tampa, Fla.; contbr. articles on water chemistry and environ. scis. to profl. publs. Mem. Citizens Environ. Quality Coun., Northbrook, 1972-74; mem. Mercury Tech. Adv. Com., State of Fla., 1991-93. Capt. USAF, 1968-72. Recipient Pub. Svc. award Univs. Coun. on Water Resources, 1990. Fellow AAAS; mem. Am. Chem. Soc. (mem. exec. com. on environ. chem. divsn. 1973-76), Assn. Environ. Engring. Profs., Am. Soc. Engring. Edn. Office: U Fla Dept Environ Engring Sci 217 Black Hall Gainesville FL 32611

DEL FOSSE, CLAUDE MARIE, aerospace software executive; b. Paris, June 27, 1940; came to the U.S., 1963; s. Guy and Gabrielle (Bouyges) D.F.; m. Genevieve Juliette Des Devises, Dec. 23, 1971; children: Laurent Fabrice, Olivier Andre, Oriane Gabrielle. Diploma in Enging. Ecole Nat. Superieure, d'Arts et Metiers, Paris, 1963; MS, Calif. Inst. Tech., 1964; MBA, U. Paris, 1966. Software engr. Soc. d'Info. Appliquee, Paris, 1964-67, Control Data Corp., L.A., 1968-69; sr. tech. staff CACI, Inc., L.A., Washington, 1969-78; v.p., div. mgr. CACI, Inc., Washington, 1979-84; cons., chief scientist Bite, Inc., Washington, 1984-86; mgr. program devel. Software Productivity Consortium, Reston, Va., 1986-89; v.p. tech. transfer Software Productivity Consortium, Herndon, Va., 1989—; bd. dirs. Winter Simulation Conf., 1979-82, gen. chmn., 1981. Bd. dirs. Lincolnia Park Recreational Club, Alexandria, Va., 1981, 82, 88. NATO fellow, 1964, Fulbright fellow, 1964. Mem. AFCEA, AIAA, Tech. Transfer Soc. Avocations: tennis, skiing. Home: 5229 Chippewa Pl Alexandria VA 22312-2023

DELGADO-BARRIO, GERARDO, physics educator, researcher; b. Santiago, Galicia, Spain, Apr. 9, 1946; s. Fernando Delgado and Maria Barrio; m. Marina Tellez de Cepeda, July 31, 1972; children: Marina, Ana Maria, Laura, Maria Del Mar. MSc in Physics, Complutense U., Madrid, 1968, PhD in Physics, 1973; MSc in Computer Sci., Politecnica U., Madrid, 1973. Rsch. assoc. U. D'Orsay, Paris, 1973-75; assoc. prof. C.S.I.C., Madrid, 1979-81, '82-86, prof., 1986—; assoc. prof. U. Complutense, Madrid, 1981-82; adviser in physics UNESCO, Colombia, 1975; chmn. Atomic and Molecular Physics Group, Madrid, 1985—, 1st So. European Sch. of Physics, Spain, 1991; 1st chmn. Spanish Atomic and Molecular Physics Div., Spain, 1988-92. Editor: Dynamical Processes in Molecular Physics, 1992; editor spl. issue Atomic and Molecular Physics, Real Soc. Española de Física, 1993; contbr. over 100 sci. papers on molecular physics to profl. publs. Mem. Real Soc. Española de Física (Young Scientist 1977, Gold medal 1983), Am. Phys. Soc., European Phys. Soc. Avocations: piano, marathon. Home: Virgen del Pilar 11, 28230 Las Rozas Madrid, Spain Office: Inst Fisica Fundamental, CSIC, serrano 123, 28006 Madrid Spain

DELIGIANIS, ANTHONY, chemical engineer; b. Athens, Greece, Mar. 20, 1951; came to U.S., 1956; s. Nicholas and Antonia (Kerastaris) D.; m. Debra Ruth Samuels, June 24, 1978; 1 child, Adam. BS in Chem. Engring., Northeastern U., Boston, 1975. Plant engr. Millipore Corp., Bedford, Mass., 1975-78; prodn. supt. Millipore Corp., Cidra, P.R., 1978-80; project engr. Herzog-Hart Corp., Boston, 1980-89, mgr. chem. engring., 1989-91, dir. chem. and mech. engring., 1991—. Mem. AICE (tech. mem. process data exch. inst. com.). Office: Herzog-Hart Corp 200 Berkeley St Boston MA 02116

DELIGNÉ, PIERRE R., mathematician; b. Brussels, Oct. 3, 1944; s. Albert and Renée (Bodart) Deligne; m. Elena Vladimirovna Alexeeva, Sept. 9, 1980; children: Natalia, Alexis. Licence en mathématiques, ULB (Université Libre de Bruxelles), Brussel, 1966, PhD in Mathematics, 1968. Jr. scientist Fond National de la Recherche Scientifique Belgium, Brussel, 1967-68; vis. mem. Institut des Hautes Etudes Scientifiques, Bures Sur Yvette, France, 1968-70; permanent mem. Institut Des Hautes Etudes Scientifiques, Bures sur Yvette, France, 1970-84; prof. Inst. for Advanced Study, Princeton, N.J., 1984. Editor Pub. Math. Institut des Hautes Etudes Scientifiques, 1970, Annals of Math, 1990; contbr. articles to profl. jours. Recipient Fields medal Internat. Math. Union, 1978; Crafoord prize, 1988. Mem. Associé Etranger Academie des Sciences, foreign honorary mem. Am. Acad. Art and Sci. Office: Inst for Advanced Study Sch Mathematics Olden Ln Princeton NJ 08540

DELISTRATY, DAMON ANDREW, toxicologist; b. Ft. Knox, Ky., July 13, 1952; s. John and Mary Delistraty; m. Jema Gail Allen, June 24, 1989; 1 child, Cody. BA in Biology, U. Calif. San Diego, La Jolla, 1974; MS in Biology, San Diego State U., 1976; PhD in Marine Biology, Coll. of William and Mary, 1982. Rsch. assoc. Scripps Instn. of Oceanography, La Jolla, 1977-78; rsch. asst. Va. Inst. of Marine Sci., Gloucester Point, Va., 1978-82; postdoctoral fellow Western Rsch. Inst., Laramie, Wyo., 1982-84; cardiac rehab. specialist U. Wyo., Laramie, 1984-86; rsch. physiologist Ea. Wash. U., Spokane, 1987-90; toxicologist Wash. State Dept. of Ecology, Spokane, 1990—. Contbr. articles to profl. jours. including Bull. Environ. Contamination and Toxicology, Aquatic Botany, Jour. Applied Biochemistry, Med. Sci. Sports Exercise, Environ. Conservation. Mem. Soc. of Environ. Toxicology and Chemistry, Phi Kappa Phi, Sigma Xi. Office: Wash State Dept Ecology Ste 100 N 4601 Monroe Spokane WA 99205

DELIYANNIS, CONSTANTINE CHRISTOS, economist, mathematician, educator; b. Kallithea, Mesolongion, Greece, July 7, 1938; came to U.S., 1964; s. Christos Constantine and Theodora Constantine (Merantzis) D. BA in Econs. summa cum laude, Athens Grad. Sch. Econs. and Bus. Scis., Greece, 1960; MA in Econs., U. Notre Dame, 1965; PhD in Econs., Cath. U. Am., 1982. Lic. life ins. agt. Prodn. supr. Piraiki-Patraiki, Cotton Mfg. Co., Inc., Athens, Greece, 1962-64; prof. econs. So. Ill. U., Edwardsville, 1965-66; cons. in stats. Bur. Social Rsch., Cath. U. Am., Washington, 1967; economist Internat. Bank for Reconstrn. and Devel., Washington, 1967-69; prof. econs. George Mason U., Fairfax, Va., 1970-71; economist The Urban Inst., Washington, 1971-72; prof. math.. U. D.C., 1972-74; cons. health econs. Transcentury Corp., Washington, 1975; cons. on bus. investment , stabilization instruments, urban pub. fin., poverty and inequality, and fin. planning Washington, 1976-92; econ. advisor Ministry of Agriculture, Athens, 1992—; prof. econs. No. Va. Community Coll., Annandale, 1986-88, Hellenic Mgmt. Assn., Athens, 1992—, Orgn. for Conflict Resolution, Athens, 1992—, European U., Athens, 1993—; ins. exec. Am. Fidelity Life Ins. Co., Washington, 1986-90; guest scholar Carlson Sch. of Mgmt., U. Minn., Mpls., 1992; resident guest scholar and tech. advisor Ctrl. Bank of Indonesia, Jakarta, 1992. Author: (book) Capital and Growth: Theory and A Case Study of Greece, 1962; contbr. numerous articles on efficient exchs., employment, taxation, internat. competition, factor shares, and minimum wages to profl. publs. 2d lt. Greek Army, 1960-62. Fellow Cath. U. Am., U. Notre Dame, Soc. Greek Studies. Mem. Am. Econ. Assn., Econometric Soc. Greek Orthodox. Avocations; tennis, swimming. Home: Papadiamantopoulou 72, 157 71 Athens Greece Office: Ministry of Agriculture, Acharnon 2, 101 76 Athens Greece

DELL, RALPH BISHOP, pediatrician, researcher; b. Mt. Village, Alaska, July 31, 1935; s. Elwin B. and Elizabeth B. (Bishop) D.; m. Kathryn M. Bownass, June 17, 1957 (div. Dec. 1982); children: Laura, Kenneth; m. Karen K. Hein, Aug. 28, 1983; stepchildren: Ethan Hein, Molly Hein. BA, Pomona Coll., 1957; MD, U. Pa., 1961. Diplomate Am. Bd. Pediatrics. Intern and resident Children's Hosp. Med. Ctr., Boston, 1961-63; NIH postdoctoral fellow Coll. Physicians and Surgeons, Columbia U., N.Y.C., 1963-66, assoc., 1966-67, asst. prof. pediatrics, 1967-72, assoc. prof., 1972-78, prof., 1978—. Author 3 books, 100 research papers; co-inventor amino acid solution. Recipient Research Career Devel award NIH, 1966-71, Career Scientist award Health Research Council N.Y., 1972-75; Fogarty Sr. Internat. fellow NIH, 1975-76. Mem. Am. Pediatric Soc., Am. Physiologic Soc., Am. Soc. Clin. Investigation, Soc. for Pediatric Research, Assn. for Computing Machinery. Democrat. Home: 116 Pinehurst Ave New York NY 10033-1755 Office: Columbia U Coll Physicians & Surgeons 630 W 168th St New York NY 10032-3702

DELLAVECCHIA, MICHAEL ANTHONY, ophthalmologist. BA in Physics, LaSalle Coll., Phila., 1970; MS in Biomedical Sci. and Engring., Drexel U., 1972, PhD in Biomedical Sci. and Engring., 1984; MD, Temple U., 1976. Diplomate Am. Bd. Med. Examiners; lic. physician, Pa., N.J. Resident in anatomical and clin. pathology Temple U. Hosp., Phila., 1977-80, chief resident, 1979-80, fellow in surg. pathology, 1980-81, resident in ophthalmology, 1981-84; fellow in ophthalmology Project Orbis, Inc., N.Y.C., 1985; v.p., med. dir., co-founder Mega Med. Electronics, Hatfield, Pa., 1984-86; assoc. John Reichel MD, Ltd., Bryn Mawr, Pa., 1984—; assoc. staff, clin. instr. Temple U. Hosp., Phila., 1986—; instr. Wills Eye Hosp., Phila., 1986—, Scheie Eye Inst., Phila., 1986—; prof. dept. biomed. engring. Drexel U., Phila., 1991—; med. dir. Interstate Blood Bank Inc., 1977-80, Information Mgmt. Corp., 1984-87, Sonic Technologies Inc., 1984-86; med. dir., adv. bd. Lehigh Ultrasonics Group, 1985-87; pres., founder Dell Med. Inc., 1985—; pres., treas., co-founder Med. Design Assoc., 1985-86; co-founder, med. dir. Omega Nutrients Inc., 1987-89; tech. adv. Project Orbis Inc., 1986—; clin. instr. ophthalmology svc. Wills Eye Hosp., 1986—, U. Pa., 1986—, Temple U., 1984—; dir. labs. Am. Clin. Laboratories, 1985—; dir. labs. Phila. Union Health Ctr., 1988—; radiolocial officer Emergency Mgmt. Assn., State of Pa., 1993—; cons. in field. Contbr. articles to profl. jours. including Clin. Rsch., Jour. of the AMA, Procs. of SPIE. Rsch. fellow Drexel U., 1976-77, Surg. Pathology fellow Temple U. Hosp., 1980-81; Pa. State Senatorial scholar; recipient numerous grants for rsch. Mem. IEEE, AMA (Physician Recognition award 1990-93, 93—), Internat. Bioelectrochem. Soc., Internat. Soc. Photoinstrumentation Engrs., Internat. Biomedical Optics Soc. (inaugural), Laser and Electr-Optics Soc., Engring. in Medicine and Biology, Am. Soc. Clin. Pathology, Am. Acad. Ophthalmology, N.Y. Acad. Scis., Del. Valley Ophthal. Soc., Intercounty Ophthal. Soc., Phila. County Med. Soc., Pa. Med. Soc., Montgomery County Med. Soc., Newtonian Soc., Chymian Soc., Sigma Xi, Kappa Mu Epsilon, Alpha Epsilon. Achievements include numerous patents in engineering and medical devices, including ophthalmic shield with removable compression device, ultrasonic dosing device, medicament delivery systems. Home: 6131 Grays Ave Philadelphia PA 19142-3207

DELLI COLLI, HUMBERT THOMAS, chemist, product development specialist; b. Utica, N.Y., July 8, 1944; s. Cyril Thomas and Carol Dolores (Fragola) D.; m. Judith Eleanor Maloney, June 24, 1967; 1 child, Kristin Anne. BS in Chemistry, Clarkson Coll., 1966, PhD, 1971. Physical chemist Edgewood Arsenal, Md., 1971-73; research scientist Westvaco Corp., Charleston, S.C., 1973-75, devel. mgr. agrichemicals, polychemicals dept., 1975-84, devel. mgr. new technology, polychemicals dept., 1984-91; mgr. agrl. chem. devel. Westavco Polychems., Charleston, S.C., 1991—; commd. lt. col., 1991; deputy div. chief studies and analysis div. Operation Desert Storm, 1991; cons. U.S. Army Chem. Systems Lab., Aberdeen, Md., 1971—. Contbr. articles to profl. jours.; patentee in field. Pres. Mcpl. Bd. Health, Goose Creek, S.C., 1974-75; mem. Berkeley County (S.C.) Water and Sewer Authority, 1978-79; mem. vestry bd. St. Thomas Ch., North Charleston, S.C., 1986—; with Operation Desert Storm, 1991. With U.S. Army, 1970-73; capt. USAR, 19773-92. Grantee NDEA, 1966, Clarkson Coll., 1962, 66; Uniroyal Undergrad. scholar, 1962, N.Y. State Regents scholar, 1962. Mem. AAAS, N.Y. Acad. Sci., Am. Chem. Soc., So. Weed Sci. Soc., N. Cen. Weed Control Conf., S.C. Sci. Council (adv. bd. mem.), Weed Sci. Soc. Am. Republican. Roman Catholic. Lodges: KC, Ducks Unltd. Avocations: black powder firearms, hunting, fishing, woodworking. Home: 7 Campanella Ct Charleston SC 29406-8606 Office: Westvaco Polychems PO Box 70848 Charleston SC 29415-0848

DELLUC, GILLES, physician, researcher; b. Périgueux, Dordogne, France, Aug. 22, 1934; s. Paul and Geneviève D.; m. Brigitte Antoine, Sept. 15, 1962; 1 child, Sophie. MD, Sch. of Medicine, Paris, 1967; D in Quaternary Geology, Anthropology, Prehistory, Paris VI Univ., 1985. Intern, asst. dr. various hosps., Paris, 1958-70; clinic chief Medical Sch., Paris, 1968-70; chief dept. of medicine Périgueux (France) Hosp., 1970—; researcher, lab. of prehistory Mus. of l'Homme, Paris, 1985—; dir. of clinic teaching Sch. of Medicine of Bordeaux, 1980—; v.p. Hist. Fedn. of the S.W. of France, 1988—, European Ctr. Prehistoric Researches, Vezere Valley. Author: Lascaux inconnu, 1979, Lascaux un nouveau regard, 1986, Les Chasseurs de la Préhistoire, 1979, Connaitre Lascaux, 1990, L'Art Pariétal archaique en Aquitaine, 1991, Connaitre la Préhistoire en Périgord, 1991, Découvrir le Périgord, 1993. Mem. Commn. des Sites (Dordogne), Périgueux, 1981—, Environmental Commn., Périgueux, 1981—. Lt. Navy, 1961-63. Mem. Spéléoclub de Périgueux, Groupe de recherche pédagogique en diabétologie, Soc. Prehistorique francaise, Soc. History and Archeology Périgord, Fedn. francaise d'Archéologie, Historic and Archeologic Soc. of Perigord (pres. 1981—). Roman Catholic. Avocations: speleology, research of caves with prehistoric paintings or engravings, photography. Home: 31 Boulevard de Vésone, 24000 Périgueux France Office: Centre Hosp, 24000 Périgueux France

DELONG, ALLYN FRANK, biochemist; b. Reading, Pa., Apr. 12, 1942; s. Warren F. and Florence (Calkins) D.; m. Gloria Renninger, Dec. 26, 1965; children: Audrey L., Wayne A. BS in Chemistry, Ursinus Coll., 1964; PhD in Biochemistry, Loyola U., Chgo., 1968. Biochemist W.H. Rorer Inc., Fort Washington, Pa., 1970-82; clin. biochemist Eli Lilly & Co., Indpls., 1982—; judge Ind. State Sci. Fair, Indpls., 1986 . Contbr. articles to profl. jours. Grantee, NIH, 1968-70. Mem. Am. Soc. Pharmacology and Exptl. Therapeutics, Am. Chem. Soc., Am. Assn. Pharm. Scientists. Office: Eli Lilly & Co 1001 W 10th St Indianapolis IN 46202

DE LONG, DALE RAY, chemicals executive; b. Oelwein, Iowa, Dec. 11, 1959; s. Jack Rollis De Long and Shirley Jean (Follett) Miller; m. Joyce Lynn Bazan, Aug. 15, 1981; children: Nicolas, Kymberly, Sabrina. Office mgr. 3D Inc., St. Joseph, Mich., 1978-82; v.p. 3D Inc., Benton Harbor, Mich., 1982—; pres. Darci Corp., Benton Harbor, 1987—. Republican. Club: Exchange. Avocation: golf. Office: Darci Corp 2053 Plaza Dr Benton Harbor MI 49022-2211

DE LOOF, JEF EMIEL ELODIE, general physician; b. Hofstade, Belgium, June 14, 1927; s. Laurent and Paula (Van Sande) De L.; m. Marie-Lousie De Meersman, July 5, 1950 (dec. May 1982); children: Annemie, Pieter, Geert, Hilde; m. Chris Van der Steen, Dec. 6, 1985; children: Bart, Tom. MD, Kath. Univ., Leuven, Belgium, 1951. Gen. practitioner Aalst, Belgium, 1951—; pres. Syndikaat der Belgische Huisartsen, Belgium, 1966-86, Vlaams Huisartsen Inst., Belgium, 1972-90, Soc. for Rsch. on Environment and Health, 1991—; editor Hisarts Nu, Belgium, 1990-93. Author: En Niemand hoort je huilen, 1983, Artsen tegen Oorlog, 1986; contbr. numerous articles to profl. jours. Pres. Medische Vereniging ter Preventie van een Atoomoorlog, 1981—, Vredeshuis Aalst, 1987—; v.p. 'Vlaamse Artsen voor het Milieu, 1992—. Recipient SIMG award Societas Internationalis Medicinae Generalis, Austria, 1986. Mem. Wetenschappelijke Vereniging der Vlaamse Huisartsen (com. mem. 1966—, Prijs Vlaamse Huisarts award 1984), Algemeen Syndikaat der Geneesheren van Belgie (com. mem. 1960—), European Gen. Practitioners Rsch. Workshop. Home and Office: KVd Woestijnestraat 18, B-9300 Aalst Belgium

DELP, EDWARD JOHN, III, electrical engineer, educator; b. Cin., Jan. 1, 1949; s. Edward John Jr. and Rosemary Helen (Bonaventura) D.; m. Marian I. Howe, Sept. 5, 1992; 1 child, Edward J. IV. MS, U. Cin., 1975; PhD, Purdue U., 1979. Registered profl. engr., Ohio. Asst. prof. elec. and computer engring. U. Mich., Ann Arbor, 1980-84; prof. elec. engring. Purdue U., West Lafayette, Ind., 1984—; expert witness in patent infringement litigation. Assoc. editor IEEE Transactions on Pattern Analysis and Machine Intelligence, 1991-93; assoc. editorial bd. Pattern Recognition Soc., 1992—, Internat. Jour. Cardiac Imaging, 1984-91; contbr. articles to tech. publs. Fulbright fellow, 1990; recipient D.D. Ewing and Honeywell teaching awards. Mem. IEEE (sr.), Optical Soc. Am., Pattern Recognition Soc., Soc. Photo-Optic and Instrumentive Engrs., Soc. Imaging Sci. and Tech. Roman Catholic. Office: Purdue Univ Sch Elec Engring 1285 Elec Engring Bldg West Lafayette IN 47907-1285

DEL REGATO, JUAN ANGEL, radio-therapeutist and oncologist, educator; b. Camaguey, Cuba, Mar. 1, 1909; came to U.S., 1937, naturalized, 1941; s. Juan and Damiana (Manzano) del R.; m. Inez Johnson, May 1, 1939; children: Ann Cynthia del Regato Jaeger, Juanita Inez del Regato Peters, John Carl. Student, U. Havana, Cuba, 1930; M.D., U. Paris, France, 1937, Laureat, 1937; D.Sc. (hon.), Colo. Coll., 1969, Hahnemann Med. Coll., 1977, Med. Coll. Wis., 1981. Diplomate Am. Bd. Radiology. Asst. radiotherapist Radium Inst., U. Paris, 1934-37; assoc. radiotherapist Chgo. Tumor Inst., 1938; radiotherapeutist Warwick Cancer Clinic, Washington, 1939-40; researcher Nat. Cancer Inst., Balt., 1941-43; chief dept. radiotherapy Ellis Fischel State Cancer Hosp., Columbia, Mo., 1943-48; dir. Penrose Cancer Hosp., Colorado Springs, Colo., 1949-73; prof. clin. radiology U. Colo. Med. Sch., 1950-74; prof. radiology U. South Fla., Tampa, 1974-83, prof. emeritus, 1983—; mem. Nat. Adv. Cancer Coun., Bethesda, Md., 1967-71; mem. med. adv. com. Milheim Found., Denver; David Gould lectr. Johns Hopkins U., 1983. Author: (with L.V. Ackerman, M.D.) Cancer: Diagnosis Treatment and Prognosis, 1947, 54, 62, 70, (with H.J. Spjut and J.D. Cox), 1985, Radiological Physicists, 1985; editor: Cancer Seminar, 1950-82; adv. editor Internat. Jour. Radiation Oncology, 1975—; contbr. articles to profl. jours. Active Colo. Springs Round Table, 1950-74. Decorated Order of Carlos Finlay of Cuba; Order Francisco de Miranda Republic of Venezuela; Béclère medal àtitre exceptionnel, 1980; recipient Gold medal Radiol. Soc. North Am., 1967; Gold medal Inter-Am. Coll. Radiology, 1967; Gold medal Am. Coll. Radiology, 1968; Gold plaque, 1975; Grubbe gold medal Ill. Radiol. Soc., 1973; Prix Bruninghaus French Acad. Medicine, 1979; Disting. Scientist award U. South Fla. Coll. Medicine, 1980; Disting. Service award Am. Cancer Soc., 1983; Ellis Fischel Disting. Achievemnt Oncology award Cancer Commn. State of Mo. and Fischel State Cancer Ctr., 1990; named Disting. Physician VA, 1974. Fellow Am. Coll. Radiology (bd. chancellors, chmn. commn. radiation therapy, com. awards and honors), AMA (coord. radiology jour. 1979—, Sci. Achievement award 1992); mem. Nat. Acad. Medicine of France (Laureat 1948), Radiol. Soc. N.Am. (v.p. 1959-60, Arthur Erskine lectr. 1978), Am. Roentgen Ray Soc., Am. Radium Soc. (v.p. 1963-64, treas. 1966-68, pres. 1968-69, chmn. exec. com. 1971-72, historian 1969—, Janeway gold medal 1973), Assn. Am. Med. Colls., Internat. Club Radiotherapists (pres. 1962-65), Inter-Am. Coll. Radiology (pres. 1967-71, U.S. counselor 1971-79), Am. Soc. Therapeutic Radiology (founder, sec. 1956-68, historian 1968—, pres. 1974-75, chmn. bd. dirs. 1975-76, gold medal 1977), Fedn. Clin. Oncologic Socs. (pres. bd. dirs. 1976-77); hon. mem. Rocky Mountain, Pacific N.W., Tex., Oreg., Minn. radiol. socs., radiol. socs. Cuba, Mex., Panama, Ecuador, Peru, Paraguay, Can., Argentina, Buenos Aires (Argentina), Am. Inst. Radiology (historian 1978-91, hon. mem. 17th Internat. Congress Radiology, Paris 1989), Arthur Purdy Stout Soc. Pathologists, Am. Soc. Physicists in Medicine. Achievements include discovery of dental lesions following irradiation of pharynx; invention of light localizer and columator for x-ray units adopted by cobalt and super voltage units; demonstration of curability of inoperable prostate cancer. Home: 3101 Cocos Rd Carrollwood Tampa FL 33618 Office: U South Fla Coll Medicine Dept Radiology Tampa FL 33618 also: VA Med Ctr 13000 Bruce Downs Blvd Tampa FL 33612

DEL TITO, BENJAMIN JOHN, JR., pharmaceuticals researcher; b. Darby, Pa., Apr. 23, 1955; s. Benjamin John Sr. and Elma Lillian (Tererri) Del T.; m. Cahterine Marie Golden, Aug. 4, 1978; 1 child, Gina Michele. BA, Millersville U., 1977; MS, Western Ky. U., 1980. Clin. tocixologist Bio Sci. Labs., Inc., Cleve., 1980-82; rsch. asst. Case Western Res. U., Cleve., 1982-84; product devel. assoc. Centocor, Inc., Malvern, Pa., 1984-85; sr. scientist Smith Kline Beecham Pharms., King of Prussia, Pa., 1985-89, analyt. svcs. and quality control, 1989—; mem. adv. bd. Northampton C.C., Allentown, Pa., 1985-88; chmn. Del. Valley Automation Group, Phila., 1986-89. Contbr. articles to profl. jours. Mem. Com. Environ. Concerns, Kimberton, Pa., 1989—. Full rsch. assistantship Leinch Corp., 1978-80. Mem. AAAS, Am. Chem. Soc., Am. Assn. Official Analyt. Chemists. Democrat. Achievements include research in integrated automation for an analytical services organization in R&D environment, consisting of sample requesting via bar coding into a VAX-based database system, robotics for sample preparations and manipulations, VAX-interfaced analytical instrument analysis for data collection, automatic VAX data analysis and reporting to same database for electronic distribution of results to requesters. Office: Smith Kline Beecham Pharms PO Box 1539 MSUE 3837 King Of Prussia PA 19406-0939

DELUCA, JOHN, neuropsychologist; b. Boonton, N.J., Mar. 22, 1956; s. Pasquale and Guiliana (D'Annuntis) DeL.; m. Deborah Kay Dillard, Oct. 23, 1983; children: Jessica, Danielle, Robert. BA in Psychology with honors, William Paterson Coll. N.J., 1979; MA in Psychology, SUNY, Binghamton, 1983, PhD in Psychobiology, 1988. Lic. psychologist, N.J. Human factors engr. Sperry Corp., Great Neck, N.Y., 1982-85; postdoctoral fellow JFK-Johnson Rehab. Inst., Edison, N.J., 1987-88, neuropsychologist, 1988-89; neuropsychologist Kessler Inst. for Rehab., West Orange, N.J., 1989—; asst. prof. U. Medicine and Dentistry of N.J., Newark, 1990—; vis. lectr. Rutgers U., New Brunswick, N.J., 1988; adj. prof. depts. biology and psychology William Paterson Coll., 1989—; instr. in psychology SUNY-Binghamton, 1985, 86; presenter in field. Cons. editor Neurorehab., 1992—; contbr. articles to profl. publs. Grantee Nat. Inst. Allergy and Immunological Diseases, 1991, Henry H. Kessler Found., 1990, NSF, 1978, Dept. Edn., 1990; recipient cert. of rsch. recognition Psi Chi, 1978. Mem. APA, Soc. Neurosci., Internat. Neuropsychol. Soc. Achievements include documenting the nature of memory impairments in a variety of disorders of the brain, study of human cerebral asymmetry and its functional implications, genetic studies of brain growth and development, other studies in field.

DELUCA, MARY, telecommunications engineer; b. Bronx, N.Y., Dec. 11, 1960; d. Dante and Julia (Ruotolo) DeL. BEE, Manhattan Coll., 1983. Traffic engr. MCI Internat., Rye Brook, N.Y., 1983-84; ops. supr. MCI Internat., Stamford, Conn., 1984; pvt. line engr. MCI Internat., Rye Brook, 1984-85; switch engr. MCI Telecommunications, Washington, 1985-86; test engr. MCI Telecommunications, Reston, Va., 1986-88; devel. compliance engr. MCI Telecommunications, McLean, Va., 1989; sr. engr. network systems planning MCI Telecommunications, Richardson, Tex., 1989-91; sr. engr. FCC, Washington, 1991—. Mem. IEEE, Communications Soc. IEEE. Office: FCC Common Carrier Bur Domestic Facilities Divsn Domestic Svcs Br Rm 6008 Washington DC 20554

DELUCA, PATRICK PHILLIP, pharmaceutical scientist, educator, administrator; b. Scranton, Pa., Sept. 7, 1935; m. Judy Beitzel, June 16, 1956; children—Paul, Thomas, Patrick, Donald, Michelle, Michael. B.S. in Pharmacy, Temple U., 1957, M.S. in Pharmacy, 1960, Ph.D. in Pharmacy (SKF W.G. Karr fellow), 1963. Analytical chemist SKF Co., 1957-59; instr., research asso. Temple U., 1959-62; sr. research pharmacist CIBA Co., Summit, N.J., 1963-66; plant mgr. CIBA Co., 1966-69, dir., 1969-70; dir. Cormedics Corp., Somerville, N.J.; mem. faculty U. Ky., 1970—; prof., asso. dean U. Ky. (Coll. Pharmacy), 1972-87, dir. ctr. for pharmaceutical sci. and tech., 1987-88; cons. to pharm. industry and FDA. Contbr. over 135 articles to profl. jours. Recipient Leo G. Penn award Temple U., 1957, Lunsford-Richardson Pharmacy Rsch. award Richardson Merrell Co., 1960, 62, Best Paper Toward Advancement Indsl. Pharmacy award N.J. Pharmacy Discussion Group, 1965, Disting. Alumni award Temple U., 1989, Outstanding Educator award in U.S., 1974; also numerous grants. Fellow Am. Assn. Pharm. Scientists (bd. dirs. 1986-88, Rsch. Achievement award 1988), Acad. Pharm. Sci. (pres. 1979-80), Inst. for Advanced Biotech. (sr.); mem. Am. Pharm. Assn., Parenteral Drug Assn. (Rsch. Achievement award 1975), Am. Soc. Hosp. Pharmacists (Rsch. award 1975), N.Y. Acad. Sci., Am. Soc. Enteral and Parental Nutrition, Sigma Xi, Rho Chi. Research, publs. in pharm. tech. Home: 3292 Nantucket Rd Lexington KY 40502-3269 Office: U Ky Coll Pharmacy Rose St Lexington KY 40536-0082

DE LUCIA-WEINBERG, DIANE MARIE, systems analyst; b. Huntington, N.Y., Apr. 29, 1964; d. Salvatore Joseph and Margaret Ann (Baker) De L.; m. Bruce Weinberg. BA in Econs. and Spanish, Clemson U., 1986. Cons. O.I.T., Inc., Huntington, 1986-87; systems analyst Blackbaud Micro Systems, Huntington, 1987-88, Datability Software Systems, N.Y.C., 1988-89, Aetna Life & Casualty Co., Lake Success, N.Y., 1989-90; pres. Lazer Tech., 1990—. Mem. Dem. Exec. Com. Town of Huntington. Mem. NAFE, AAUW, Am. Mgmt. Assn., Nat. Assn. Women Bus. Owners, L.I. Networking Entrepreneurs, L.I. Assn. Avocations: basketball, triathlons, computing, soccer. Office: Lazer Tech 11 W 23d St South Huntington NY 11746

DELUREY, MICHAEL WILLIAM, mechanical engineer; b. Bay Shore, N.Y., Oct. 4, 1962; s. Albert Joseph and Anne Marie (McGlynn) D.; m. Mary Frances Briening, Aug. 16, 1986; children: Elizabeth Anne, Madeleine Frances. B in Engring., SUNY, Stony Brook, 1986; MS in Mech. Engring., George Washington U., 1990, D of Engring. Mgmt., 1992. Cert. energy mgr.; engr. in tng. Mech. engr. Naval Surface Warfare Ctr., Dahlgren, Va., 1987-90, br. head, 1990, energy systems coord., 1991-92, energy systems analyst warfare systems dept., 1992—. Contbr. articles to Proceedings of 1991 and 1993 World Energy Engring. Congresses. Chairperson Pub. Works EEOC, Dahlgren, 1990; tutor Colonial Beach (Va.) High Sch., 1989-91; judge sci. fairs, King George, Va., 1990-91. Mem. Nat. Fire Protection Assn., Assn. Energy Engrs. Home: 1341 Wilson Rd Waldorf MD 20602 Office: Naval Surface Warfare Ctr Code J 42 Dahlgren VA 22448

DEMAIN, ARNOLD LESTER, microbiologist, educator; b. N.Y.C., Apr. 26, 1927; s. Henry and Gussie (Katz) D.; m. Joanna Kaye, Aug. 2, 1952; children: Pamela Robin (Demain) McCloskey, Jeffrey Brian. BS, Mich. State U., 1949, MS, 1950; PhD, U. Calif., Berkeley, 1954. Rsch. asst. U. Calif., Davis, 1952-54; rsch. microbiologist Merck & Co., Inc., Danville, Pa., 1954-56, Rahway, N.J., 1956-65; founder, head of dept. ferm. microbiology Merck & Co., Inc., Rahway, 1965-69; prof. of ind. microbiology MIT, Cambridge, 1969—; bd. dirs. Internat. Biotech. Labs., Cambridge, Mass. Editor: 7 books; contbr. 350 articles to profl. jours., 1954—. With U.S. Navy, 1945-47. Recipient Hotpack award Can. Soc. Microbiology, 1978, Rubro award Australian Soc. Microbiology, 1978, Indsl. Microbiology award Italian Pharm. Assn., 1989, Hans Knoll meml. award, 1990. Mem. Soc. Indsl. Microbiology (pres. 1990, Waksman award N.J. br. 1975, Cetus Biotech. award 1990), French Soc. Microbiology (hon.). Achievements include 16 patents; elucidation of biosynthetic pathway to penicillins and cephalosprins; recognition of phenomenon of biochemical regulation of secondary metabolism; discovery of role of lysine and amino adipic acid in penicillin biosynthesis. Office: MIT Biology Dept Rm 56-123 77 Massachusetts Ave Cambridge MA 02139-4307

DEMARCHI, ERNEST NICHOLAS, aerospace engineering administrator; b. Lafferty, Ohio, May 31, 1939; s. Ernest Costante and Lena Marie (Cireddu) D.; B.M.E., Ohio State U., 1962; M.S. in Engring., UCLA, 1969; m. Carolyn Marie Tracz, Sept. 17, 1960; children—Daniel Ernest, John David, Deborah Marie. Registered profl. cert. mgr. With Space div. Rockwell Internat., Downey, Calif., 1962—, mem. Apollo, Skylab and Apollo-Soyuz missions design team in electronic and elec. systems, mem. mission support team for all Apollo and Skylab manned missions, 1962-74, mem. Space Shuttle design team charge elec. systems equipment, 1974-77, in charge Orbiter Data Processing System, 1977-81, in charge Orbiter Ku Band Communication and Radar System, 1981-85, in charge orbitor elec. power distbr., displays, controls, data processing, 1984-87, in charge space based interceptor flt. exper., 1987-88, kinetic energy systems, 1988-90, ground based interceptor program, 1990—. Recipient Apollo Achievement award NASA, 1969, Apollo 13 Sustained Excellent Performance award, 1970, Astronaut Personal Achievement Snoopy award, 1971; Exceptional Service award Rockwell Internat., 1972, Outstanding Contbn. award, 1976; NASA ALT award, 1979; Shuttle Astronaut Snoopy award, 1982; Pub. Service Group Achievement award NASA, 1982; Rockwell Pres.'s award, 1983, 87; registered profl. engr., Ohio. Mem. AIAA, ASME, Nat. Mgmt. Assn., Varsity O Alumni Assn. Home: 25311 Maximus St Mission Viejo CA 92691 Office: 12214 Lakewood Blvd Downey CA 90241

DEMARCO, PETER VINCENT, standards assurance executive; b. Jersey City, Oct. 10, 1955; s. Vincent William and Yolanda (Stanziani) DeM.; (widowed Oct. 1990); children: Jessica E., Christopher V.; m. Susan K. Kinney, June 13, 1992. Technician Am. Standard, Inc., Piscataway, N.J., 1975-79, sr. technician, 1979-83; sr. devel. tech. Am. Standard, Inc., Edison, N.J., 1983-85; engring. asst. Am. Standard, Inc., Trenton, N.J., 1985-88, tech. svc. rep., 1988-90, mgr. codes and standards, 1990—. Mem. ASME (vice chmn. A112 panel 19 plumbing fixtures and working groups 1 cast iron and 4 enamel steel), Am. Soc. Sanitary Engrs., Internat. Assn. Plumbing and Mech. Ofcls., Nat. Fire Protection Assn., Bldg. Ofcls. and Code Adminstrn. Internat. Roman Catholic. Office: Am Standard Inc PO Box 8305 Trenton NJ 08650-0305

DEMAREE, ROBERT GLENN, psychologist, educator; b. Rockford, Ill., Sept. 20, 1920; s. Glenn and Ethel Mae (Champion) D.; BS, U. Ill., 1941, MA, 1948, PhD (univ. fellow 1949-50), 1950; m. Alyce Anisia Jones, Sept. 4, 1948; children— Dee Anne, Marta, James, David. Chief performance br. Pers. and Tng. Rsch. Ctr., Lowry AFB, Denver, 1951-57; dir. human factors Martin Space Flight div. Bell Aircraft Corp., Balt., 1958-60; dir. programs Matrix Corp., Arlington, Va., 1960-61; dir. office instructional rsch. U. Ill., 1961-63; projects dir. Life Scis., Inc., Hurst, Tex., 1963-66; mem. faculty Tex. Christian U., Ft. Worth, 1966-85, prof. psychology, prof. Inst. Behavioral Rsch., 1970-85, emeritus, 1985—; cons., 1985—. Served with AUS, 1941-46. Mem. APA, Am. Statis. Assn., Psychometric Soc., Soc. Multivariate Exptl. Psychologists, Sigma Xi. Home: 4813 Eldorado Dr Fort Worth TX 76180-7227

DEMARS, BRUCE, naval administrator. Dep. asst. sec. Naval Reactors Dept. of Energy, Arlington, Va. Office: Dept Energy Naval Reactors 2521 Jefferson Davis Washington DC 20362-0001 also: Navy Dept Naval Nuclear Propulsion Washington DC 20362

DE MAURET, KEVIN JOHN, geologist; b. Bayonne, N.J., Feb. 21, 1957; s. Ferdinand and Mary (Kryprius) de M. BA in Geoscis., Jersey City State Coll., 1978; MS in Geology, Rutgers U., 1984. Cert. profl. geologist, Del., Ind., Tenn., Wyo.; registered environ. profl. Rsch. asst. Lamont-Doherty Geol. Observatory, Palisades, N.Y., 1982-84, Ocean Drilling Program, College Station, Tex., 1984-86; project supr. IT Corp., Edison, N.J., 1986-88; project mgr. EFP Assocs., Inc., Matawan, N.J., 1988-90; pres., chief exec. officer Innovative Environ. Cons., Inc., Morristown, N.J., 1990—. Mem. Am. Assn. Petroleum Geologists, Am. Geophys. Union, Am. Inst. Profl. Geologists, Am. Water Resources Assn., N.J. Water Pollution Control Assn., N.Y. Acad. Scis., Soc. Econ. Paleontologists and Mineralogists, Assn. Ground Water Scientists and Engrs., Water Pollution Control Fedn., Assn. Environ. Health of Soils. Office: Innovative Environ Cons Inc PO Box 9210 Morristown NJ 07963-9210

DEMBEK, ZYGMUNT FRANCIS, epidemiologist; b. Hartford, Conn., June 15, 1950; s. Zygmunt and Bertha Ann (Petrauckas) D.; m. Carol Ann Wynglarz, Nov. 21, 1981. BA in Biology, Eastern Conn. State U., 1973; MS in Biomed. Sci., Hood Coll., 1982; postgrad., U. Conn., 1987—. Registered sanitarian, Conn.; water pollution control operator, Conn. Operator, chemist Windsor Locks (Conn.) Pollution Control, 1983-86; environ. sanitarian Conn. Dept. Health Svcs., Hartford, Conn., 1986-87, sr. environ. sanitarian, 1987-90, epidemiologist III, 1990—; lectr. Asnuntuck Community Coll., Enfield, Conn., 1985-86, Manchester (Conn.) Community Coll., 1983; group leader, divemaster Bermuda Biol. Sta. for Rsch., Ferry Reach, Bermuda, 1979-80; acad. investigator U.S. Army Rsch. Inst. of Infectious Disease, Fort Detrick, Md., 1981. Contbr. chpt. to book. Mem. Sara Title III Local Emergency Planning Com., Suffield, 1988—; rep., trustee North Ctrl. Health Dist., Enfield, 1989—. Capt. USAR, 1976—. Recipient Innovative Project grant EPA Region I, 1991. Mem. Am. Chem. Soc., Am. Inst. Chemists, N.Y. Acad. Scis., Res. Officer Assn. of U.S., Soc. of Armed Forces Med. Lab. Scientists, Knights of King Jan III Sobieski. Democrat. Roman Catholic. Office: Conn Dept of Health Svcs 150 Washington St Hartford CT 06106-4474

DEMENT, FRANKLIN LEROY, JR., aerospace engineer; b. Memphis, Tenn., Nov. 4, 1966; s. Franklin Leroy and Corinne Yvonne (Jones) D. BS in Aero. Engring., U. Ala., 1989. Structural analyst turbine engines Aerospace Propulsion and Power Lab., Wright Patterson AFB, Dayton, Ohio, 1989; commd. capt. USAF, 1989; projects officer LAAFB USAF, El Segundo, Calif., 1989-90; launch vehicle structural analyst The Aerospace Corp., El Segundo, 1990-91; small launch vehicles systems engr. space test program LAAFB USAF, El Segundo, 1991—. Mem. AIAA. Democrat. Baptist. Home: 4528 W 159th St Lawndale CA 90260-2513 Office: SMC/CULL LAAFB LAAFB PO Box 92960 El Segundo CA 90009-2960

DE MENT, JAMES ALDERSON, soil scientist; b. Haughton, La., Dec. 22, 1920; s. Ben Alderson and Myrtie Inez (Rounsavall) DeM.; m. Ruby Mae Weaver, June 2, 1941; children: James Alderson Jr., David W. BS, La. State U., 1941; PhD, Cornell U., 1962. Cert. profl. soil scientist. Soil scientist Soil Conservation Svc., Okla., 1941-48, La., 1948-56; grad. asst. Cornell U., Ithaca, N.Y., 1956-62; rsch. soil scientist Soil Conservation Svc., Lincoln, Nebr., 1962-65; soil scientist Soil Conservation Svc., Ft. Worth, Tex., 1965-77; ret.; colns. Lockheed Electronics, Clear Lake, Tex, 1977-78, Espey-Huston, Dallas, 1978-84, DeMent & Assocs., Haughton, La., 1984—. Lt. col. USAF, 1942-45; 1951-52. Recipient 3 Certs. of Merit, Soil Conservation Svc. Mem. Soil and Water Conservation Soc., Am. Soc. Agronomy, Sigma Xi, Alpha Zeta (chancellor 1940-41). Achievements include development of rate of loess deposition in Alaska through radio carbon dating, of system for mapping minesoils in Texas; first to recognize selected minesoils as prime farmland. Home: RR 1 Box 651 Haughton LA 71037-9731

DEMENTIS, KATHARINE HOPKINS, interior designer; b. Indpls., Dec. 20, 1922; d. Stephen Francis and Margaret Bell (Yeager) Hopkins; m. Gilbert X. Dementis, Feb. 1, 1953; children: Mary Margaret Dementis O'Dwyer, Stephen Ezra Hall. Student, John Herron Art Sch., 1941-44; BS, U. Wis., 1971. Interior designer L.S. Ayres and Co., Indpls., 1945-51; pres. Ariz. Questers, 1991—. Mem. DAR, Lakes Club, Union Hills Country Club, Passport Club. Republican. Presbyterian. Avocations: artist, decorator, antique collecting. Home: 12830 Castlebar Dr Sun City West AZ 85375-3270 Office: Questers 210 S Quince St Philadelphia PA 19107-5534

DEMERJIAN, KENNETH L., atmospheric science educator, research center director; b. Sept. 10, 1945; married; 2 children. BA in Chemistry, Northeastern U., 1968; MS in Phys. Chemistry, Ohio State U., 1970, PhD in Phys. Chemistry, 1973. Grad. teaching asst. Ohio State U., 1968-71, rsch. asst., 1972-73; rsch. assoc. Calspan Corp., 1973-74; phys. scientist, acting chief Model Devel. Br. EPA, 1974-75, supervisory phys. scientist, chief Atmospheric Modeling and Assessment Br., 1976-81, dir. meteorology divsn. ARL, 1981-85; prof. dept. environ. health and toxicology, environ. chemistry Sch. Pub. Health SUNY, N.Y. State Dept. Health, Albany, 1988—; assoc. divsn. applied scis. Harvard U., 1986—; dir. Atmospheric Scis. Rsch. Ctr. SUNY, Albany, 1986—, prof. dept. atmospheric sci., 1988—, chmn. dept. atmospheric sci., 1988-90; vis. scientist environ. sci. and physiology Sch. Pub. Health, rsch. fellow Energy and Environ. Policy Ctr. J.F. Kennedy Sch. Govt. Harvard U., 1984-85; adj. prof. chemistry SUNY, Albany, 1986—; lectr. in field. Contbr. articles to profl. jours. Mem. AAAS, Am. Chem. Soc., Am. Meteorol. Soc., Am. Geophys. Union, Air and Waste Mgmt. Assn. Achievements include research in chemical kinetics and mechanistic pathways of elementary atmospheric reactions and the development of reaction mechanisms of polluted and clean atmospheres, instrumentation development and measurement of atmospheric trace gas constituents, diagnostic analysis of atmospheric processes and air quality simulation modeling, experimental and theoretical studies of actinic solar flux and photolytic rate constants of atmospheric species, sources and evaluation of uncertainty in theoretical models of atmospheric processes, air quality and pollutant exposures, the articulation and effective use of scientific uncertainty in the decision making process. Office: SUNY Atmospheric Scis Rsch Ctr 100 Fuller Rd Albany NY 12205-5796

DEMERS, LAURENCE MAURICE, science educator; b. Lawrence, Mass., May 9, 1938; s. Laurence Onezime and Doris Corrine (Goulet) D.; m. Susan Ruth Bernard, Sept. 29, 1962; children: Laurence H., Michele L., Marc B., Christopher J., Andrew J. AB, Merrimack Coll., 1960; PhD, SUNY Upstate Med. Ctr., Syracuse, 1970. Postdoctoral fellow Med. Sch., Harvard U., Boston, 1970-72, instr., 1972-73; asst. prof. M.S. Hershey Med. Ctr., Pa. State U., Hershey, Pa., 1973-76, assoc. prof., 1976-80, prof., 1980—; cons. Robert Wood Johnson Pharm Rsch. Inst., Raritan, N.J., 1978—; bd. dirs. dBi Labs. Inc., Harrisburg, Pa.; mem. adv. bd. Abbott Labs., Chgo., 1984—; vis. prof. U. Oxford, Eng., 1981-82. Editor: Liver Function Testing, 1978, Premenstrual Syndrome, 1985, Premenstrual Syndrome and Menopausal Mood Disorders, 1989; editorial editor Clin. Chemistry Jour., 1990—. Eucharistic minister St. Joan of Arc Catholic Ch., Hershey, 1981—, mem. Knights of Malta, 1990—. Served to capt. Med. Svc. Corps, U.S. Army, 1961-65. Recipient Lalor award Lalor Found., 1973, Fogarty Internat. award Fogarty Ctr., NIH, 1981, Pharm. Mfrs. Assn. award, 1974. Fellow Nat. Acad. Clin. Biochemistry (pres. 1984-85 Dubin award 1991); mem. Endocrine Soc., Am. Assn. Clin. Chemistry (Ames award 1986), Am. Soc. Clin. Pathology, N.Y. Acad. Sci., Assn. Clin. Scientists, Hershey Country Club. Avocations: golf; tennis. Home: 1175 Stonegate Rd Hummelstown PA 17036-9776 Office: Pa State U MS Hershey Med Ctr University Dr Hershey PA 17033

DEMETRESCU, MIHAI CONSTANTIN, computer company executive, scientist; b. Bucharest, Romania, May 23, 1929; s. Dan and Alina (Dragosescu) D.; M.E.E., Poly. Inst. of U. Bucharest, 1954; Ph.D., Romanian Acad. Sci., 1957; m. Agnes Halas, May 25, 1969; 1 child, Stefan. Came to U.S., 1966. Prin. investigator Research Inst. Endocrinology Romanian Acad. Sci., Bucharest, 1958-66; research fellow dept. anatomy UCLA, 1966-67; faculty U. Calif.-Irvine, 1967-83, asst. prof. dept. physiology, 1971-78, assoc. researcher, 1978-79, assoc. clin. prof., 1979-83; v.p. Resonance Motors, Inc., Monrovia, Calif., 1972-85; pres. Neurometrics, Inc., Irvine, Calif., 1978-82; pres. Lasergraphics Inc., Irvine, 1982-84, chmn., chief exec. officer, 1984—. Mem. com. on hon. degrees U. Calif.-Irvine, 1970-72. Postdoctoral fellow UCLA, 1966. Mem. Internat. Platform Assn., Am. Physiol. Soc., IEEE (sr.). Republican. Contbr. articles to profl. jours. Patentee in field. Home: 20 Palmento Way Irvine CA 92715-2109 Office: 20 Ada Irvine CA 92718-2303

DEMETRIOU, IOANNES CONSTANTINE, mathematics educator, researcher; b. Ioannina, Epirus, Greece, Feb. 29, 1956; s. Constantine I. and Catherine (Papaioannou) D.; m. Emilia C. Rokomou, May 17, 1987; 1 child, Katerina. Diploma in math., U. Ioannina, 1979; MPhil in Computer Sci., U. Tech., Eng., 1981; postgrad., Pembroke Coll., Eng., 1981-85; PhD in Applied Math. and Theoretical Physics, Cambridge U., Eng., 1985. Math. asst. U. Ioannina, 1981; specialist in math. U. Ioannina, 1986-87; specialist in computer sci. Athens U. of Econs., 1990-93; asst. prof. U. Athens, 1991—; mng. dir. Ctr. of Computer Edn., Greece, 1986—; vis. prof. Hellenic Air Force Acad., Athens, 1989-90; invited asst. prof. Athens U. Econs., 1993—. Mem. editorial bd. Mathematical Review; contbr. articles on numerical approximations and other topics to profl. jours. Lt. Army, 1985-87. Grantee State Scholarship Found., 1974-78, 81-84, Pembroke Coll., 1985; recipient J.T. Knight prize Cambridge U., 1983. Mem. Am. Math. Soc., Inst. Math. and Its Applications (Eng.), Nat. Statistical Inst. (Greece), Greek Math. Soc., Ops. Rsch. Soc. (Greece). Orthodox Christian. Avocations: jogging, hiking. Home: Meg Alexander 53, 45333 Ioannina Greece Office: U Athens, Dept Econs Pesmazoglou 8, Athens Greece

DE MEYERE, ROBERT EMMET, civil engineer; b. N.Y.C., June 9, 1951; s. George Gaston and Mary Dolores (Carpenter) De M.; m. Marguerite Mary Milazzo, Jan. 11, 1981; 1 child, Meghan Maria. BS in Biology, SUNY, Stony Brook, 1976; BS in Civil Engring., Northeastern U., Boston, 1990. Draftsman Stone & Webster, Boston, 1981-83; designer Wormser Engring., Woburn, Mass., 1983-88; project engr. Alpine Am. Corp., Natick, Mass., 1988-89; project mgr. Meridian Power Corp., Boston, 1989-91; sr. advisor Gas Ventures Advisers, Boston, 1991—; mem. U.S. Trade Mission to Malaysia, Kuala Lumpur, 1991, U.S. Trade Mission to Indonesia, Jakarta, 1991; chmn. Conf. on Cogeneration and Combined Cycle Systems, Jakarta, 1992. Mem. ASCE, Chi Epsilon. Roman Catholic. Achievements include analysis of electric tariff structures in S.E. Asia; cons. on the regulation the privatization of electric power industry in S.E. Asia; identification and conceptualization of pvt. power facilities and opportunities in East Asia. Home: 137 Stock Farm Rd Sudbury MA 01776 Office: Gas Ventures Advisers 73 Tremont St Boston MA 02108

DEMING, WENDY ANNE, mental health professional; b. Boise, Idaho, Jan. 20, 1968; d. Ronald L. and Kathleen A. (Devlin) D. BS in Psychology, U. So. Colo., 1990. Psychiat. aide Colo. State Hosp., Pueblo, 1987—; mental health worker II Spanish Peeks Mental Health Ctr., Pueblo, 1990; clin. therapist children-adolescent unit Parkview Episcopal Med. Ctr., Pueblo, 1991—. Mem. Psi Chi. Office: Parkview Episcopal Med Ctr 400 W 16th St Pueblo CO 81003*

DEMING, W(ILLIAM) EDWARDS, statistics educator, consultant. BS, U. Wyo., 1921, LLD (hon.), 1958; MS, U. Colo., 1924, LLD (hon.), 1987; PhD, Yale U., 1928, hon. degree, 1991; ScD (hon.), Rivier Coll., 1981, Ohio State U., 1982, Md. U., 1983, Clarkson Inst. Tech., 1983; D in Engring. (hon.), U. Miami, 1985; LLD (hon.), George Washington U., 1986, D in Engring. (hon.), 1987; DSc (hon.), U. Colo., 1987, U. Ala., 1988, Fordham U., 1988, U. Oreg., 1989, U. S.C., 1991, Am. U., 1991, Howard U., 1993, Boston U., 1993, Harvard U., 1993. Instr. engring. U. Wyo., 1921-22; asst. prof. physics Colo. Sch. Mines, 1922-24, U. Colo., 1924-25; instr. physics Yale U., 1925-27; math. physicist USDA, 1927-39; adviser in sampling Bur. of Census, 1939-45; prof. stats Stern Sch. Bus., NYU, N.Y.C., from 1946; Disting. prof. in mgmt. Columbia U., 1986—; cons. research, industry, 1946—; statistician Allied Mission to Observe Greek Elections, 1946; cons. sampling Govt. India, 1947, 51, 71; adviser in sampling techniques Supreme Command Allied Powers, Tokyo, 1947-50, High Commn. for Germany, 1952, 53; mem. UN Sub-Commn. on Statis. Sampling, 1947-52; lectr. various univs., Germany, Austria, 1953, London Sch. Econs., 1964, Institut de Statistique de U. Paris, 1964; cons. Census Mex., 1954, 55; cons. Statistisches Bundesamt, Wiesbaden, Fed. Republic Germany, 1953, Central Statis. Office Turkey, 1959—, China Productivity Ctr., Taiwan, 1970, 71; Inter Am. Statis. Inst. lectr., Brazil, Argentina. Author: Statistical Adjustment of Data, 1943, Some Theory of Sampling, 1950, Statistical Design in Business Research, 1960, Quality, Productivity, and Competitive Position, 1982, Out of the Crisis, 1986, The New Economics, 1993; contbr. numerous articles to profl. publs. Decorated 2d Order medal of the Sacred Treasure (Japan), 1960; elected Most Disting. Grad., U. Wyo., 1972; recipient Taylor Key award Am. Mgmt. Assn., 1983, Nat. Medal Tech., Pres. of U.S., 1987, Edison award. 1989; enshrined in the Engring. and Sci. Hall of Fame, 1986. Fellow Am. Statis. Assn., Royal Statis. Soc. (hon.), Inst. Math. Stats.; mem. NAE, ASTM (hon.), Am. Soc. Quality Control (hon. life, Shewhart medal 1955), Internat. Statis. Inst. (hon. life), Philos. Soc. Washington, World Assn. Pub. Opinion Rsch., Market Rsch. Coun., Biometric Soc. (hon. life), Union Japanese Scientists and Engrs. (hon. life) (tchr. and cons. to Japanese industry 1950-52, 55, 60, 65—, honored in establishment of Deming prizes in Japan), Japanese Statis. Assn. (hon. life), Deutsche Statistische Gesellschaft (hon. life), Ops. Rsch. Soc. Am., Dayton Hall of Fame. Home and Office: 4924 Butterworth Pl Washington DC 20016

DEMIRBILEK, ZEKI, research hydraulic engineer; b. Istanbul, Turkey, Apr. 10, 1949; came to U.S., 1974; s. Yusuf and Hatice (Gulden) D.; m. Kay Kloiber, Nov. 17, 1984; 1 child, David Z. BS in Naval Architecture, Naval Acad., Istanbul, 1970; BS, MS, Naval Postgrad. Sch., Monterey, Calif., 1975, 76, DEng in ME, 1977; PhD in Civil Engring./Ocean Engring., Tex. A&M U., 1982. Registered profl. engr., Tex. Naval officer, diplomat Turkish Navy and NATO, Ankara, Turkey, 1970-74; prin. rschr. Hydromechanic Rsch. Inc., Pebble Beach, Calif., 1974-79; lectr., cons. Tex. A&M U., College Station, 1979-82; sr. ocean engr. Conoco Inc., Ponca City, Okla., 1982-88; pvt. cons. Houston, 1988-89; rsch. hydraulic engr. Coastal Engring. Rsch. Ctr., Vicksburg, Miss., 1989—; cons. Noble Denton Assocs., Houston, 1988-89; adj. prof. Tex. A&M U., 1989—, Miss. State U., Starkville, 1989—. Editor, contbr.: Tension-Leg-Platform, 1989, Handbook of Hydrology, 1993; contbr. numerous articles to profl. jours. Adv. mem. U.S. Coast Guard, Paramus, N.J., 1985-89. Recipient Pres.'s award Turkish Govt., 1967, 69, 70, Prime Minister's medal, 1967, 69, 70, Chief of Staff medal, 1967, 69, 70; NATO postgrad. scholar, 1974-77. Mem. ASCE (chmn. and/or mem. 8 coms. 1982—), Internat. Soc. Offshore and Polar Engring., Soc. Naval Architects and Marine Engrs. Republican. Muslim. Achievements include patents for a nonlinear theory of liquid sloshing; an added-mass stabilizer system for offshore platforms; the heave compensated tension leg platform concept for oil exploration in ultradeep waters. Office: Coastal Engring Rsch Ctr 3909 Halls Ferry Rd Vicksburg MS 39180-6199

DEMIRGIAN, JACK CHARLES, analytical chemist; b. N.Y.C., Dec. 5, 1947; s. Arakel and Mildred (Donerian) D.; m. Rose, Sept. 6, 1971; children: Anna, Abigail. AB in Chemistry, Hunter Coll., 1968; PhD in Chemistry, SUNY, Buffalo, 1976. Assoc. prof. Christopher Newport Coll., Newport News, Va., 1976-81; chemist Argonne (Ill.) Nat. Lab., 1981—; vis. prof. SUNY, Plattsburgh, 1975-76. Contbr. to books and articles to profl. jours. Mem. Am. Chem. Soc., Air & Waste Mgmt. Assn., Tech. Assn. of Pulp & Paper Ind. Achievements include environmental research to extend limits of detection using Fourier transform infrared spectroscopy; developed field testing methods to detect toxic organics in soil, developed instrumentation to detect ollegal drugs/laboratories. Office: Argonne Nat Lab 9700 S Cass Ave Argonne IL 60516

DEMOKAN, MUHTESEM SÜLEYMAN, electrical engineering educator; b. Tire, Izmir, Turkey, May 27, 1948; arrived in Hong Kong, 1988; s. Enver and Rezzan Belma (Sözdener) D.; m. Gülsen Tanak, Mar. 19, 1974; children: Emre, Simin. BS, Mid. East Tech. U., Ankara, Turkey, 1970; MS, King's Coll., U. London, 1972, PhD, 1976. Asst. prof. Mid. East Tech. U., Gaziantep, Turkey, 1977-81, assoc. prof., 1981, head dept., 1982-83; dean faculty of engring. Mid. East Tech. U., Gaziantep, Turkey, 1979-82; vis. sr. rsch. fellow Imperial Coll., King's Coll. U. London, 1983-84; head dept., chief engr. Hirst Rsch. Ctr. Hirst Rsch. Ctr. GEC, Wembley, Middlesex, Eng., 1984-88; head dept., chmn. bd. postgrad. studies, prof. elec. engring Hong Kong Polytechnic, 1988—. Author: Mode-Locking in Solid-State and Semiconductor Lasers, 1982; contbr. articles to profl. jours. in the areas of optoelectronics and fiber optics. Fulbright rsch. award, 1983; recipient Brit. Coun. Acad. Links Scheme award, 1977-80, postdoctoral rsch. award, 1976, Turkish Sci. Rsch. Coun., 1979. Fellow Hong Kong Inst. Engrs., Inst. Elec. Engrs. (U.K.); mem. IEEE (sr.), Hong Kong Inst. Sci. Office: Hong Kong Polytechnic, Dept Elec Engring, Hung Hom, Kowloon Hong Kong

DEMONSABERT, WINSTON RUSSEL, chemist, consultant; b. New Orleans, June 12, 1915; s. Joseph Francis and Davida Elizabeth (Gullett) deM.; m. Eleanor Ray Ranson, Aug. 8, 1955; 1 child, Winston Russel. BS in Chemistry, Loyola U., New Orleans, 1937; MA in Edn., Tulane U., 1945, PhD in Chemistry, 1952. Asst. prof. Loyola U., New Orleans, 1948-49, assoc. prof., 1949-55, prof., 1955-66; chief chemist Nat. Center for Disease Control, Dept. Health and Human Services, Atlanta, 1966-69, chief contract liaison br. Nat. Center for Health Services Research, 1969-73, chief extramural programs Bur. Drugs, FDA, Rockville, Md., 1973-79, scientist adminstr. office of interagy sci. coordination, office of commr. FDA, after 1979; now cons., govt. liaison environ. chemistry and toxicology; assoc. prof. Tulane U., 1957-58; research chemist Am. Cyanamid Co., 1957-58; vice-chmn. Interagy. Testing Com., 1982. Committeeman Boy Scouts Am., New Orleans and Atlanta; mem. curriculum coms. New Orleans Pub. Sch. Bd., 1965. Fellow AAAS, Am. Inst. Chemists (chmn. La. chpt. 1958-60, chmn. Ga. chpt. 1968-69, pres. D.C. chpt. 1982-83); mem. Am. Chem. Soc. (past chmn. La. sect.). Roman Catholic. Contbr. to Ency. Americana, Ency. Chemistry, also profl. jours. Achievements include research in environmental effects (detection, prevention and treatment) of toxic wastes, pesticides and air pollution, and zirconium chemistry. Home and Office: 4317 Lake Trail Dr Kenner LA 70065-1541

DEMPSEY, RAYMOND LEO, JR., radio and television producer, moderator, writer; b. Providence, June 18, 1949; s. Raymond Leo Sr. and Louise Veronica (Gambuto) D.; m. Patricia Batchelder (div. 1984); children: Joab, Jahdeam, Deezsha, Nathaniel, Talitha. BA in Liberal Arts, R.I. Coll., 1973; Cert. in Bus., U. R.I., 1979; cert., Blake Computer Programming Inst., 1977, Billy Graham Sch. Evangelism, 1989. Lic. real estate agt., R.I.; lic. radio sta. operator FCC; cert. secondary tchr., videographer, contractor, R.I. Writer local and nat. publs., 1980—; producer, moderator Chapter & Verse TV, RICA-TV, Providence, 1983—; tchr. R.I. Pub. High Schs., Providence and Cranston, 1988—; producer, moderator radio programs Ch. Focus and People, WRIB AM Radio, East Providence, R.I., 1989—; bd. dirs. Blessing, Inc., Providence; spl. corr. Songtime, U.S.A. Radio Network, 1988—, spl. reporter, spl. contbr., 1991; host Straight Talk, Sta. WKRI, 1989, dir. World Exch., 1991-93; co-host The Lighthouse, Sta. KDFL, Tacoma; prodr., co-host The Bible Answer Program, Sta. WARV, 1986; judge The Ace Awards, 1992, Cable Ace Awards, 1992; interviewer Gallup Poll, 1987; trainee N.E. Law Enforcement Officers Assn., 1991; elector Radio Hall Fame, 1993,

Stellar awards, 1993. Bd. dirs. R.I. Right to Life, Cranston, 1973—; witness R.I. Gen. Assembly, 1973—, R.I. Bd. Health, 1973—; vol. ARC, R.I. Hosp.; registrar voters, State of R.I., 1980, 91; del. Rep. Nat. Conv. 1980; sponsor World Vision, Pasadena, Calif., 1981—, Compassion Internat., Colorado Springs, Colo., 1989—; chief boys instr. Mattson Acad. Karate, Providence, 1969-71; del. Gov.'s Conf. on Libr. and Info. Svcs., 1991; elector White House Conf. on Libr. and Info. Svcs.; Justice of the Peace, 1991; R.I. state voter registrar, 1991, 92; regional rep. Students Against Vietnam War, 1971, Taxpayers Action Network, 1991; ptnr. Food for the Hungry, 1984—; del. Ellen McCormack for Pres., 1976. Named One of Top 4 Local Cable TV Prodrs. in Nation, Nat. Assn. Local Cable Programming, 1987, ofcl. Jerusalem Pilgrim, State of Israel, 1990; recipient 2 Internat. Angel awards for excellence in cable TV presentations, 1991, cert. U.S. Small Bus. Adminstrn., 1990, Diamond award, 1992, 1st prize for excellence in pub. affairs in R.I. and Mass., 1992, Achievement award Dale Carnegie Orgn., 1992, 1st place award Mastermedia; The Spotlight award, 1993. Mem. AAAS, ASCD, NRA, Am. Math Soc., Nat. Assn. High Sch. Tchrs. English, Nat. Assn. Edn. of Young Children, Am. Soc. Oriental Rsch., Archaeol. Inst. Am., R.I. Assn. for Edn. Young Children, R.I. Assn. for Supervision and Curriculum Devel., Mental Health Assn. R.I., N.Y. Acad. Scis., Internat. Press Assn. (founding mem.), Nat. Geog. Soc., Nat. Assn. Broadcasters, Nat. Assn. Radio Talk Show Hosts, Nat. Acad. Cable Programming, Near East Archaeol. Soc., Internat. Platform Assn., Jewish TV Inst. (charter), Smithsonian Air and Space Mus., Smithsonian Inst. (assoc.), Royal Inst Public Health and Hygiene London (assoc.), Royal Soc. Health London (affiliate), Bread for the World, Evangs. for Social Action, Mus. Heritage Soc., Mensa, USCG Aux., Abraham Lincoln Soc., Rel. Heritage Am., Providence Athenaeum, Toastmasters Internat., Phi Theta Kappa. Avocations: scuba diving, marksmanship, bibl. archeology. Home and Office: PO Box 41000 Providence RI 02940-1000

DEMUTH, JOSEPH E., physicist, research administrator; b. Ridgewood, N.J., Apr. 21, 1946; s. Joseph and Carolyn M. Demuth; m. Barbar M. Noble, Dec. 21, 1968; children: Bradley, Kimberly, Michael. BS, Rensselaer Poly. Inst., 1968; PhD, Cornell U., 1972. Postdoctoral fellow IBM Rsch. Divsn., Yorktown Heights, N.Y., 1972-74, rsch. staff mem., 1974-78, asst. to dir. rsch., 1979-80, sr. mgr., 1987—; vis. scientist KFA-Juezich, West Germany, 1978-79; fin. officer Noble Devel. Corp., Chappaqua, N.Y., 1988—. Recipient N.J. State Award in Physics N.J. Tchrs. Assn., 1964. Fellow Am. Phys. Soc. (Davisson-Germer prize 1992). Home and Office: 17 Alta Ln Chappaqua NY 10514

DENABURG, CHARLES ROBERT, metallurgical engineer, retired government official; b. Birmingham, Ala., Apr. 23, 1935; s. Simon and Mary Edith (Roseblum) D.; m. Sara Rose Lepp, Aug. 12, 1956; children: Elisa Jan, Cheryl Lyn, Danial A. BS in Metall. Engring., U. Ala., 1959. Registered profl. engr., Ala. Extraction metallurgist U.S. Bur. Mines, U. Ala., Tuscaloosa, 1959; metall. engr. Lazarov Surplus, Memphis, 1959-60; aerospace technologist for materials Marshall Space Flight Ctr.-NASA, Huntsville, Ala., 1960-61, 62-67; aerospace technologist for materials Kennedy Space Ctr. (Fla.)-NASA, 1967-83, chief malfunction analysis br., 1983-90, ret., 1990; chief exec. officer C.R. Denaburg & Assocs., Inc., Indian Harbour Beach, Fla., 1978—; mem. materials and processes working group NASA, Washington, 1987—, mem. space shuttle tech. steering group, 1969-81, mem. wind tunnel evaluation team, 1986. Contbr. articles to profl. publs. Mem. Indian Harbour Beach Planning and Zoning Bd., 1973-74, Windward Cove (Fla.) Archtl. Rev. Bd., 1989—. With U.S. Army, 1961-62. Recipient cert. of commendation Kennedy Space Ctr., 1973, NASA Exceptional Svc. medal, 1986; Snoopy award NASA Astronaut Office, 1970. Mem. Am. Soc. for Metals (editing team on corrosion failures Metals Handbook 1972-73, exec. com. Cen. Fla. chpt. 1974-76), Am. Ceramic Soc., Nat. Mgmt. Assn. Avocations: scuba diving, swimming, tennis, woodworking, amateur radio, camping. Home: 140 Windward Way Indian Harbour Beach FL 32937

DEN BREEJEN, JAN-DIRK, computer integrated manufacturing educator; b. Hardinxveld, The Netherlands, Feb. 18, 1963; s. Teunis and Johanna (de Smoker) den B.; m. Tanja Klysen. M in Internat. Rels., Ryks U., Utrecht, The Netherlands, 1987; M in Econs. and Stats. Edn., Ryks U., Utrecht, the Netherlands, 1987; M. in Bus. Edn., Erasmus U., Rotterdam, The Netherlands, 1990. Cert. in prodn. and inventory mgmt. Am. Prodn. & Inventory Control Soc., cert. quality engr. Am. Soc. Quality Control; register Informaticus; nederlands vereniging van register informatici. Lectr. bus. and econs. Gemeentelyke Scholen-Gemeenschap Noordendyk, Dordrecht, the Netherlands, 1987-89; lectr. bus. and logistics Hogeschool Rotterdam en Omstreken Polytechnische Faculteit, 1989-90, head computer integrated mfg. dept., 1991—; conf. mgr. HBO-Raad, Den Haag, The Netherlands, 1989-91; edn. mgr. ISW/Opleidingen, 1992—; freelance instr. logistics and quality mgmt. Wolters-Kluwer (I.S.W.) & Elsevier Opleidingen, (S.B.C.), 1990—. Bd. dirs. local history soc., Hardinxveld, The Netherlands, 1982-88. Mem. Dutch Soc. for Bus. Logistics, Dutch Soc. for Quality Control. Conservative Party. Avocations: scuba diving, archaeology. Home: Willem Buytewechstraat 160 b, 3024 VG Rotterdam The Netherlands Office: ISW/Opleidingen Omstreken GJ de Johghweg 4-6, Schuttersveldg, 2316XG Leiden The Netherlands

DENCE, MICHAEL ROBERT, research director; b. Sydney, Australia, June 17, 1931; s. Robert Cecil and Barbara Sidney (Laurence) D.; m. Carole E. Paintin, Sept. 24, 1967; children: Alexandra C., Victoria C. B.Sc., Sydney U., 1953. Geologist Falconbridge Nickel Mines Ltd., 1953-54; research asst. U. Toronto, Ont., Can., 1955-58; tech. officer Geol. Survey Can., 1959-61; sci. officer Dominion Obs., Ottawa, 1962-65; research scientist earth physics br. Can. Dept. Energy, Mines and Resources, Ottawa, 1966-81, dir. gravity and geodynamics div., earth physics br., 1981-82, dir. gravity, geothermics and geodynamics div., earth physics br., 1982-86; exec. sec. designate Royal Soc. Can., 1987, exec. dir., 1987-93, assoc. to pres., 1993—; prin. investigator NASA Lunar Sample Program. Recipient Public Service award Can. Public Service, 1973. Fellow AAAS, Royal Soc. Can., Meteoritical Soc. (Barringer medal 1988), Geol. Soc. Can.; mem. Am. Geophys. Union, Can. Geophys. Union. Home: 824 Nesbitt Pl, Ottawa, ON Canada K2C 0K1 Office: Royal Soc Can, PO Box 9734, Ottawa, ON Canada K1G 5J4

DENCOFF, JOHN EDGAR, research technologist; b. Albuquerque, Apr. 17, 1968; s. William Martin and Gloria Dean (Myers) D. BS summa cum laude, U. N.Mex., 1991. Student supr. fine arts libr. U. N.Mex., Albuquerque, 1988-90, rsch. technician, 1990-92, rsch. technologist, 1992—. Psychobiology summer fellow U. Calif., 1990. Mem. Internat. Brain Rsch. Orgn., Soc. for Neurosci., Golden Key Nat. Honor Soc., Sigma Xi. Home: 2909 Carolina St NE Albuquerque NM 87110 Office: U NM Med Sch Dept Neurology Albuquerque NM 87131

DENDY, ROGER PAUL, communications engineer; b. Phoenix, Sept. 21, 1964; s. Joe Ben and Sarah Jane (Tussey) D.; m. Teresa Elaine Arnaud, Aug. 18, 1990. BSEE, Ariz. State U., 1986. Electronics engr. Naval Warfare Assessment Ctr., Corona, Calif., 1987-91; communications engr. Analex Corp., Cleve., 1991—. Mem. IEEE, AIAA, Tau Beta Pi, Eta Kappa Nu. Presbyterian. Home: 19636 Scottsdale Blvd Shaker Heights OH 44122 Office: Analex Corp Mail Stop 54-6 21000 Brook Park Rd Cleveland OH 44135

DENEBERG, JEFFREY N., engineering executive; b. Chgo., Mar. 21, 1943; s. Joseph B. and Beatrice P. Denenberg; married June 25, 1967; children: Stacey, Scott. BSEE, Norhtwestern U., 1966; MSEE, Ill. Inst. Tech., 1968, PhD in Elec. Engring., 1971. Design engr. Motorola Inc., Chgo., 1966; sr. design engr. Warwick Electronics Inc., Niles, Ill., 1969, 70; tech. staff Bell Telephone Labs., Naperville, Ill., 1971-76; mgr. system design ITT Advanced Tech. Ctr., Shelton, Ct., 1977-81, dir. learning systems, 1982-85, dir. tech. planning, 1985-86, dir. systems architecture, 1986-87; mgr. systems tech. Prodigy Svcs. Co., White Plains, N.Y., 1978-89; v.p. rsch. devel., chief tech. officer Noise Cancellation Techs., Stamford, Conn., 1991—. Contbr. articles to profl. jours. Mem. IEEE (sr. mem. 1986, Outstanding Paper award 1970). Achievements include eleven patents and six patents pending in field. Home: 345 Putting Green Rd Trumbull CT 06611 Office: Noise Cancellation Tech Inc 800 Summer St Ste 500 Stamford CT 06901

DE NEVERS, ROY OLAF, retired aerospace company executive; b. Strasburg, Sask., Can., Dec. 30, 1922; s. Edouard Albrecht V.V. and Christy Helen (Hunt) de N.; divorced; children Gregory Frank (dec.), Sara Dianne. BS in Econs., U. London, 1963; BA in Econ. History, U. Winnipeg (Can.), 1971. Served to lt. comdr. Royal Can. Navy, 1946-67; chief contract adminstr., aircraft repair and overhaul Bristol Aerospace Ltd., Winnipeg, MB, Can., 1968-83; originator co. operating procedure Bristol Aerospace Ltd., Winnipeg, Man., Can., 1983-86. Editor aero. mag., 1956-60. Founder, mem. Commonwealth Air Tng. Plan Mus.; mem. We. Can. Aviation Mus. Served to flight lt. Royal Can. Air Force, 1941-45. Decorated DFC, 1945, Aircrew Europe Star, 1945, France and Germany Clasp, 1945, Def. medal CD, 1945. Mem. Naval Officers Assn. Can. (pres. Winnipeg br.), Canadian Naval Air Group Assn., Three Score Plus Circle, Can. Aeros. and Space Inst. (assoc. fellow), Masons. Mem. Adventist Ch. Clubs: Royal Air Force (London), Fleet Air Arm Officers (London). Address: Group 2 Box 9 Rte 1, Anola, MB Canada R0E 0A0

DENGLER, MADISON LUTHER, psychologist, educator; b. Berks County, Pa., Mar. 18, 1935; s. Madison Luther and Mae Esther (Bergman) D.; m. Beverly Jane, June 26, 1960; children: James, Susan, Leslie. BA, So. Meth. U., 1956; MA, U. Denver, 1968, PhD, 1968. Instr. U. Denver, 1967-68; asst. prof. psychology Muhlenberg Coll., Allentown, Pa., 1968-70; prof. psychology Luther Coll., Decorah, Iowa, 1970—. Office: Luther Coll Dept Psychology Decorah IA 52101

DENHARDT, DAVID TILTON, molecular and cell biology educator; b. Sacramento, Feb. 25, 1939; s. David Burton and Edith (Tilton) D.; m. Georgetta Louise Harrar, July 1, 1961; children—Laura Jean, Kristin Ann, David Harrar. B.A. in Chemistry with high honors, Swarthmore Coll., 1960; Ph.D. in Biophysics, Calif. Inst. Tech., 1965. Instr. biol. labs Harvard U., 1964-66, asst. prof., 1966-70; assoc. prof. biochemistry McGill U., Montreal, Que., Can., 1970-77; prof. McGill U., 1977-80; prof. biochemistry, microbiology and immunology, dir. Cancer Research Lab., U. Western Ont., London, 1980-88; prof., chmn. dept. biolog. scis. Rutgers U., New Brunswick, N.J., 1988—, dir. Bur. Biol. Rsch., 1988—; mem. sci. adv. bd. Ctr. for Advanced Biotech. and Medicine, Piscataway, N.J., 1988-91, 1988-91. Editor: Jour. Virology, 1977-87, Gene, 1985-93; mem. editorial bd. Jour. Cancer Rsch. and Clin. Oncology, In Vivo Internat. Jour. Fellow AAAS, Royal Soc. Can.; em. Am. Cancer Soc., Am. Soc. Biol. Chemists, Am. Microbiol. soc., N.Y. Acad. Scis., Am. Soc. Cell Biology, Phi Beta Kappa. Office: Rutgers U Nelson Labs PO Box 1059 Piscataway NJ 08855-1059

DE NIL, LUC FRANS, speech-language pathologist; b. Aalst, Belgium, Sept. 20, 1957; s. Bernard and Elizabeth (Hofmans) De N.; m. Lieve Maria Meirhaeghe, Apr. 4, 1983; children: Lien, Louis. Licentiate, U. Leuven, 1980; PhD, So. Ill. U., 1988. Orthopedagoog Speech & Hearing Rehab. Ctr., Gent, Belgium, 1980-84; rsch. asst. So. Ill. U., Carbondale, 1984-88; postdoctoral fellow U. Wis., Madison, 1988-90; asst. prof. speech/lang. pathology U. Toronto, Ont., 1990—. Contbr. articles to profl. jours. Grantee NSERC, Can., 1991—, U. Toronto Dean's Fund, 1991-92; fellow Ont. Ministry Health, 1992—. Mem. Am. Speech-Lang.-Hearing Assn., Internat. Fluency Assn. (treas. 1991—), Acoustical Soc. Am., Ont. Assn. Speech-Lang. Pathologists and Audiologists. Achievements include research of sensory deficiency (oral) in people with stuttering problems which may help clarify the pathophysiology of stuttering. Office: Univ of Toronto, 88 College St, Toronto, ON Canada M5R 1L4

DENIS, PAUL-YVES, geography educator; b. Montreal, Quebec, Canada, Aug. 19, 1932; s. Paul R. and Marie-Ange (Viger) D.; m. Madeleine Lépine, 1956; children—Marie Jose, Gontran. B.A., Coll. Ste.-Marie, 1952; M.A., U. Montreal, 1955; Ph.D., U. Cuyo, 1967. Town planner City of Montreal, Can., 1955-60; dir. B.A. for adults U. Montreal, Can., 1960-65; researcher, tchr. U. Cuyo, Mendoza and Nordeste, Resistencia, Argentina, 1965-68; asst., then full prof. U. Laval, Ste.-Foy, Que., Can., 1973—; cons. Contbr. articles to profl. jours. Served to capt. Ordnance, Can. Army Res., 1955-67. Recipient Cert. of Honor Conf. Latin Am. Geographers, 1984, Cert. of Honor La Societe de Geographie, Quebec, 1982. Mem. Royal Can. Soc. (assoc.), Can. Assn. Geographers (pres. 1982-85, Scholarly Distinction in Geography award 1990). Roman Catholic. Home: 4005 Chemin St Louis, Cap-Rouge, PQ Canada G1Y 1V7 Office: U Laval, Dept Geography, Sainte-Foy, PQ Canada G1K 7P4

DENKER, JOHN STEWART, physicist; b. Tucson, Sept. 15, 1954; s. Russell Ernest and Jean Stewart (Wasserman) D. BS, Calif. Inst. Tech., 1975; MS, Cornell U., 1981, PhD, 1986. Cert. comml. pilot, flight instr. (with airplane and instrument ratings), ground instr. (with advanced and instrument ratings), FAA accident prevention counselor. Propr. APh Technol. Consulting, Pasadena, Calif., 1974-82; teaching asst., rsch. asst. Cornell U., Ithaca, N.Y., 1977-85; mem. tech. staff AT&T Bell Labs., Holmdel, N.J., 1984-91; vis. prof. Inst. for Theoretical Physics, U. Calif., Santa Barbara, 1986-87; disting. mem. tech. staff AT&T Bell Labs., Holmdel and Murray Hill, N.J., 1991—; mem. organizing com. several internat. sci. confs., gen. rsch. colloquium com. AT&T, grad. rsch. program for women fellowship com. AT&T. Editor: Neural Networks for Computing, 1986; contbr. over 50 articles to profl. jours. NSF fellow, 1979-81. Mem. Am. Phys. Soc., Aircraft Owners and Pilots Assn., Exptl. Aircraft Assn., Monmouth Area Flying Club, Inc. (v.p., bd. trustees). Achievements include 7 patents in field; neural networks research in learning from examples (fundamental theory and practical applications); in foundations of quantum mechanics; in unconventional computing architectures. Office: 4g330 AT&T Bell Labs Holmdel NJ 07733

DENKO, JOANNE D., psychiatrist, writer; b. Kalamazoo, Mich., Mar. 29, 1927; d. John S. and Marian Mildred (Boers) Decker; m. Charles Wasil Denko, June 17, 1950; children: Christopher Charles, Nicholas Charles, Timothey Charles. BA summa cum laude, Hope Coll., 1947; MD, Johns Hopkins U., 1951; MS in Psychiatry, U. Mich., 1963. Lic. psychiatrist Md., Ill, Mich., Ohio. Pvt. practice Columbus, Ohio, 1961-68; staff psychiatrist Fairview Gen. Hosp., Cleve., 1968—; pvt. practice Rocky River, Ohio, 1968—; cons. Juvenile Diagnostic Ctr., Columbus, 1967-68, VA Hosp., Cleve., 1968-72, Community Mental Health Ctrs., Greater Cleve., 1974-80; clin. instr. Case Western Res. U., Cleve., 1981-83. Author: Through the Keyhole at Gifted Men and Women, 1977, (monograph) The Psychiatric Aspects of Hypoparathyroidism, 1962; contbr. articles to profl. jours.; author poetry, 1960—. Mem. AAAS (reviewer children's books), Cleve. Astron. Soc. (bd. dirs. 1984-86), Mensa Soc. (Cleve. area br. pres. 1986-87), Great Books Discussion Group (Rocky River, chmn. 1985—). Russian Orthodox. Achievements include naming sexual deviance klismaphilia; research in special problems of adults of high intelligence, educating gifted girls, teenage alcoholism, mental illness in pre-literate peoples, psychiatric aspects of lupus. Home and Office: 21160 Avalon Dr Cleveland OH 44116-1120

DENN, MORTON MACE, chemical engineering educator; b. Passaic, N.J., July 7, 1939; s. Herbert Paul and Esther (Taub) D; m. Vivienne Roumani; children: Matthew Philip, Susannah Rachel, Rebekah Leah. BS in Engring., Princeton U., 1961; PhD, U. Minn., 1964. Postdoctoral fellow U. Del., Newark, 1964-65, from asst. prof. to prof. Chem. Engring., 1965-77, Allan P. Colburn prof., 1977-81; prof. U. Calif., Berkeley, 1981—, chmn. dept. chem. engring., 1991—; Harry Pierce prof. chem. engring. Technion, Israel Inst. Tech., Haifa, Sept. 1979-Jan. 1980; Chevron Energy prof. chem. engring. Calif. Inst. Tech., Feb.-July, 1980; vis. prof. chem. engring. U. Melbourne, Australia, Jan.-June, 1985; program leader for polymers, Ctr. for Advanced Materials Lawrence Berkeley Lab., 1983—. Author: Optimization by Variational Methods, 1969, (co-author) Introduction to Chemical Engineering Analysis, 1972, Stability of Reaction and Transport Processes, 1975, Process Fluid Mechanics, 1980, Process Modeling, 1986; co-editor Chemical Process Control, 1976; contbr. numerous articles to profl. jours., author book chpts. Guggenheim fellow, 1971-72; William M. Lacey lectr. Calif. Inst. Tech., 1979, Fulbright lectr., 1979-80; Peter C. Reilly lectr. Notre Dame U., 1980; Bicentennial Commemoration lectr. La. State U., 1984; Arthur Kelly lectr. Purdue U., 1987; Stanley Katz lectr. CCNY, 1990. Fellow AICE (editor jour. 1985-91, Profl. Progress award 1977, William H. Walker award 1984); mem. NAE, Am. Soc. Engring. Edn. (chem. engring. divsn. lectureship 1993), Soc. Rheology (Bingham medal 1986), Brit. Soc. Rheology, Polymer Processing Soc., Sigma Xi. Office: U Calif Dept Chem Engring Berkeley CA 94720

DENNARD, ROBERT HEATH, engineering executive, scientist; b. Terrell, Tex., Sept. 5, 1932; s. Buford Leon and Loma (Heath) D.; children—Robert, Amy, Holly. BSEE, So. Methodist U., 1954, MSEE, 1956; PhD, Carnegie Inst. Tech., 1958. Staff engr. IBM, Yorktown Heights, N.Y., 1958-63; research staff mem. IBM Research Ctr., Yorktown Heights, N.Y., 1963-71, group mgr., 1971-79, fellow, 1979—. Contbr. articles to profl. jours.; patentee in field including basic dynamic RAM memory cell. Recipient Nat. Medal of Tech. Pres. of U.S., 1988, Indsl. Rsch. Inst. Achievement award, 1989, Harvey prize Technion-Israel Inst. Tech., 1990. Fellow IEEE; mem. NAE. Avocation: Scottish country dancing. Office: IBM Rsch Ctr PO Box 218 Yorktown Heights NY 10598-0218

DENNE-HINNOV, GERD BOËL, physicist; b. Landskrona, Sweden, Mar. 21, 1954; arrived in Eng., 1984; d. Nore Einar Vilhelm and Gerd Rigmor (Anderson) Denne; m. Einar Hinnov, July 21, 1990. BSc, Lund (Sweden) U., 1976, PhD, 1981. Teaching asst. physics dept. Lund U., 1981; vis. scientist Princeton (N.J.) Plasma Physics Lab., 1982-84; exptl. physicist Jet Joint Undertaking, Abingdon, Eng., 1984—; sr. lectr. (docent) Lund U., 1985—; mem. staff reps. com. Jet Joint Undertaking, Abingdon, 1988-89. Contbr. articles to profl. jours. Mem. AAAS, N.Y. Acad. Scis., Swedish Phys. Soc., Am. Phys. Soc., Sigma Xi. Lutheran. Office: Jet Joint Undertaking, Abingdon OX14 3EA, England

DENNIN, ROBERT ALOYSIUS, JR., pharmaceutical research scientist; b. Newark, Mar. 5, 1951; s. Robert Aloysius Sr. and Elizabeth Jane (Cooney) D. B in Biology, Montclair State Coll., 1975, M in Biology, 1976. From asst. scientist II to clin. project coord. Hoffmann La Roche, Inc., Nutley, N.J., 1977-88; sr. clin. rsch. coord. Hoffmann La Roche, Inc., Nutley, 1988—; rschr. in AIDS and cancer. Contbr. chpt. in book and articles to profl. jours. Mem. AAAS, Am. Soc. for Microbiology, N.Y. Acad. Scis. Avocations: piano, guitar. Office: Hoffmann-LaRoche Inc 340 Kingsland St Nutley NJ 07110

DENNING, MICHAEL MARION, computer company executive; b. Durant, Okla., Dec. 22, 1943; s. Samuel M. and Lula Mae (Waitman) D.; m. Suzette Karin Wallance, Aug. 10, 1968 (div. 1979); children: Lila Monique, Tanya Kerstin, Charlton Derek; m. Donna Jean Hamel, Sept. 28, 1985; children: Caitlin Shannon, Meghan O'Donnell. Student, USAF Acad., 1963; BS, U. Tex., 1966, Fairleigh Dickinson U., 1971; MS, Columbia U., 1973. Mgr. systems IBM, White Plains, N.Y., 1978-79; mgr. svc. and mktg. IBM, San Jose, Calif., 1979-81; nat. market support mgr. Memorex Corp., Santa Clara, Calif., 1979-81, v.p. mktg., 1981-82; v.p. mktg. and sales Icot Corp., Mountain View, Calif., 1982-83; exec. v.p. Phase Info. Machines Corp., Scottsdale, Ariz., 1983-84, Tricom Automotive Dealer Systems Inc., Hayward, Calif., 1985-87; pres. ADS Computer Svcs., Inc., Toronto, Ont., Can., 1985-87, Denning Investments, Inc., Palo Alto, Calif., 1987—, Pers. Solutions Group, Inc., Menlo Park, Calif., 1990—. With USAF, 1962-66; Vietnam. Mem. Rotary, English Speaking Union, Phi Beta Kappa, Lambda Chi Alpha (pres. 1965-66). Republican. Methodist. Home: H-2140 Camino de los Robles Menlo Park CA 94025 Office: Denning Investments Inc 2730 Watson Ct San Francisco CA 94104

DENNIS, JOHN EMORY, JR., mathematics educator; b. Coral Gables, Fla., Sept. 24, 1939; s. John Emory and Hazel Violet (Penny) D.; m. Ann Watson, Mar. 1, 1960; 1 child, John Emory III. BS in Engring., U. Miami, 1962, MS in Math., 1964; PhD in Math., U. Utah, 1966. Asst. prof. math. U. Utah, Salt Lake City, 1966-68; lectr. computer sci. U. Essex, Colchester, Eng., 1968-69; vis. asst. prof. computer sci. Cornell U., Ithaca, N.Y., 1969-70, assoc. prof. computer sci., 1970-76, prof. computer sci., 1976-79, dir. ctr. for applied math., 1978-79; prof. math. scis. Rice U., Houston, 1979-84, Noah Harding prof. math. scis., 1984—, chmn. math. sci., 1989-92, chmn. computer sci., 1992—; instr. math. Stillman Coll., Tuscalusa, Ala., 1967; Fulbright lectr., prof. U. Buenos Aires, 1986; rsch. assoc. Nat. Bur. Econ. Rsch. and U.K. Atomic Energy Rsch. Establishment, Harwell, Oxfordshire, Eng., 1975-76. Editorial bd. Math. Programming, 1974—, co-editor, 1986—; founding mng. editor SIAM Jour. for Optimization, 1989—; contbr. articles to sci. jours. NSF grantee, 1970—, U.S. Army Rsch. Office grantee, 1976—, USAF Office Sponsored Rsch. grantee, 1985—, DOE grantee, 1986—. Mem. Spl. Interest Group for Numerical Math. (bd. dirs. 1975-77, vice chmn. 1979-81), Am. Math. Soc. (joint applied math. com. with Soc. for Indsl. and Applied Math. 1980-82), Soc. for Indsl. and Applied Math. (editorial bd. Jour. for Numerical Analysis 1976-84, bd. dirs. 2d summer rsch. conf. in numerical analysis, mem. Siam coun. 1985—, exec. com. 1985-88, chair optimization activity group), Math. Programming Soc. (mem. internat. program com. 1987—). Avocations: reading, baseball. Home: 3107 Jarrard St Houston TX 77005-3013 Office: Rice U Dept Applied Math PO Box 1892 Houston TX 77251

DENNIS, RONALD MARVIN, chemical engineering educator, consultant; b. West Chester, Pa., May 23, 1950; s. Raymond Miller and Mary Jane (Wootten) D.; m. India Elizabeth Culpepper, Mar. 26, 1977; children: David Miller, Robert Forrest. BSChemE, Ga. Inst. Tech., 1973, MS, 1975; PhD in Chem. Engring., Vanderbilt U., 1984. Profl. engr., Tenn; cert. quality engr. Environ. engr. II Ga. Dept. Natural Resources, Atlanta, 1975-77; engr. E.I. du Pont de Nemours & Co., Inc., Memphis, 1977-80; rsch. engr. E.I. du Pont de Nemours & Co., Inc., Brevard, N.C., 1984-89; sr. project engr. Roy F. Weston, Inc., West Chester, 1989-91; asst. prof. Lafayette Coll., Easton, Pa., 1991—. Contbr. articles to profl. jours. Mem. AICE, NSPE, Am. Chem. Soc., Hazardous Materials Control Resources Inst., Am. Soc. Quality Control, Sigma Xi, Tau Beta Pi, Lambda Chi Alpha. Republican. Presbyterian. Achievements include research in hydrogen peroxide decomposition with Fenton's reagent and soil remediation with soil washing. Home: 417 Windsong Ln Exton PA 19341-3045 Office: Dept Chem Engring Lafayette Coll Easton PA 18042-1775

DENNISON, DANIEL B., chemist; b. Gainesville, Fla., June 4, 1947; s. Raymond A. and Mary Louise (Grumbein) D.; m. Janet Gannanay. BS, Furman U., 1969, MS, 1974; PhD, Mich. State U., 1978. Rsch. scientist Coca-Cola Co., Atlanta, 1978-81; mgr. product devel. Coca-Cola USA, Atlanta, 1981-85, v.p. rsch. and quality assurance, 1985-88; dir. R & D Coca-Cola Co., Atlanta, 1988—. 1st lt. U.S. Army, 1969-72. Mem. AAAS, Am. Chem. Soc., Inst. Food Technologists, Soc. Soft Drink Technologists. Office: Coca Cola Co Tech Div Coca Cola Plaza NW Atlanta GA 30303

DENNISON, ROBERT ABEL, III, civil engineer; b. Herkimer, N.Y., Sept. 2, 1951; s. Robert A. and Ruth (Friesen) D.; m. Marilyn Smith, June 21, 1981; children: Andrew, Christopher. AAS, Westchester Community Coll., 1974; BCE, Manhattan Coll., 1978. Registered profl. engr., N.Y., N.H. Dep. commr. Putnam County Dept. Hwy., Carmel, N.Y., 1983-85; mgmt. engr. region 8 N.Y. State Dept. Transp., Poughkeepsie, 1985-92; bur. dir. N.Y. State Dept. Transp., Albany, 1992—; town engr. Town of Kent, N.Y., 1983-90. Contbr. articles to Pub. Works Mag. Warden St. John's Episc. Ch., Kingston, N.Y., 1989—; arbitrator Am. Arbitration Assn., White Plains, N.Y., 1986—. Mem. N.Y. State Transp. Engrs. Home: 122 Wilson Ave Kingston NY 12401

DENNY, ROBERT WILLIAM, JR., electrical and electronics engineer; b. Darby, Pa., Aug. 2, 1953; s. Robert William and Doris Ann (Hecker) D.; m. Nancy Carol Kujda, July 12, 1975 (div. Apr. 1984); m. Darlene Mary Tremblay, Jan. 30, 1986; children: Michelle Darlene, Brian Robert. BSEE, Drexel U., Phila., 1976. Registered profl. engr., D.C., Md., N.C., Miss. Design engr. Harris Corp, RF Communications, Rochester, N.Y., 1976-78; chief engr. WVOR-FM, Rochester, 1978-80, WBT/WBCY FM, Charlotte, N.C., 1980-86; plans reviewer Charlotte (N.C.) Bldg. Standards, 1986-87; sr. engr. McCracken & Lopez, P.A., Charlotte, 1987; dir. Jules Cohen & Assocs. P.C., Washington, 1987-91, pres., 1991—; instr. Cen. Piedmont Community Coll., Charlotte, 1986-87. Mem. IEEE, Internat. Assn. Elec. Inspectors, Assn. Fed. Communications Cons. Engrs., Soc. Broadcast Engrs. Roman Catholic. Home: Rt 1 Box 287 Cardinal Ln White Plains MD 20695 Office: Jules Cohen & Assocs PC PO Box 18415 1725 Desales St NW Ste 600 Washington DC 20036-8415

DENNY, THOMAS ALBERT, product development specialist, researcher; b. Bklyn., Oct. 18, 1933; s. Bruno Jacob and Margaret Jeanette (White) D.; m. Janet Adrienne Weeden, June 2, 1962; 1 child, Adrienne Wilma. BS, Bklyn. Coll., 1955. Technician pollution unit Bur. of Fish N.Y. Conservation Dept., Rochester, 1955; technician Fedders Quigan Co., Maspeth, N.Y., 1956; tech. assoc. Arabol Adhesives, Bklyn., 1961-63; scientist Johnson & Johnson, North Brunswick, N.J., 1963-77, sr. scientist, 1977-84; prin. scientist Johnson & Johnson Patient Care, North Brunswick, 1984-89; sr. tech. svcs. assoc. Johnson & Johnson Med. Inc., Arlington, Tex., 1990—; sr. rsch. assoc. Johnson & Johnson Med., Inc., Arlington, Tex., 1991—. With USN, 1956-60. Mem. Sigma Xi. Presbyterian. Achievements include patents in sterile surgical packaging, disposable diaper with puff bonded facing layer, knitted surgical swabs, surgical laparotomy pads. Home: 3905 Murphy Ct Arlington TX 76016-3857 Office: Johnson & Johnson Med Inc 2500 E Arbrook Blvd Arlington TX 76014-3899

DEN OTTER, CORNELIS JOHANNES, animal physiologist, educator; b. Leiden, The Netherlands, May 25, 1935; s. Cornelis Johannes and Maartje Adriana (Maas) Den O.; m. Augusta Bolhuis, Dec. 22, 1960; children: Marjanke Johanna, Peter Albert Cornelis. BS, U. Leiden, 1957, MS cum laude, 1960; PhD, U. Groningen, The Netherlands, 1971. Asst. prof., head sensory physiology group U. Leiden, 1961-66; asst. prof., head sensory physiology group U. Groningen, 1966-71, assoc. prof., head sensory physiology group, 1971—; head sensory physiology rsch. unit Internat. Ctr. Insect Physiology and Ecology, Nairobi, Kenya, 1981-84; vis. prof. Max Planck-Inst., Seewiesen, Germany, 1974, U. Ouagadougou, Burkina Faso, 1985—, U. Campobasso, Italy, 1993—. Contbr. articles to profl. jours. Fellow Royal Entomol. Soc.; mem. European Soc. Comp. Physiology and Biochemistry, Royal Netherlands Phys. Soc., Netherlands Entomol. Soc., European Chemoreception Rsch. Orgn., Netherlands Found. for Biophysics, Netherlands Zoolog. Soc. Mem. Dutch Labour Parth. Avocations: reading, traveling, swimming. Home: Rijksstraatweg 377, Haren 9752 CH, The Netherlands Office: U Gronningen Animal Physiol, Kerklaan 30, Haren 9751 NN, The Netherlands

DENT, ERNEST DUBOSE, JR., pathologist; b. Columbia, S.C., May 3, 1927; s. E. Dubose and Grace (Lee) D.; m. Dorothy McCalman, June 16, 1949; children: Christopher, Pamela; m. 2d, Karin Frehse, Sept. 6, 1970. Student, Presbyn. Coll., 1944-45; M.D., Med. Coll. S.C., 1949. Diplomate clin. pathology and pathology anatomy Am. Bd. Pathology. Intern U.S. Naval Hosp., Phila., 1949-50; resident pathology USPHS Hosp., Balt., 1950-54; chief pathology USPHS Hosp., Norfolk, Va., 1954-56; assoc. pathology Columbia (S.C.) Hosp., 1956-59; pathologist, dir. labs. Columbia Hosp., S.C. Baptist Hosp., 1958-69; with Straus Clin. Labs., L.A., 1969-72; staff pathologist Hollywood (Calif.) Community Hosp, St. Joseph Hosp., Burbank, Calif., 1969-72; dir. labs. Glendale Meml. Hosp. and Health Ctr., 1972—; bd. dirs. Glendale Meml. Hosp. and Health Ctr. Author papers nat. med. jours. Mem. Am. Cancer Soc., AMA, L.A. County Med. Assn. (pres. Glendale dist. 1980-81), Calif. Med. Assn. (councillor 1984-90), Am. Soc. Clin. Pathology, Coll. Am. Pathologists (assemblyman S.C. 1965-67; mem. publs. com. bull. 1968-70), L.A. Soc. Pathologists (trustee 1984-87), L.A. Acad. Medicine, S.C. Soc. Pathologists (pres. 1967-69). Lutheran. Home: 1526 Blue Jay Way Los Angeles CA 90069-1215 Office: S Central and Los Feliz Aves Glendale CA 91225-7036

DENTEL, STEVEN KEITH, environmental engineer, educator; b. Washington, Nov. 4, 1951; s. Keith E. and Marcene A. (Chudomelka) D.; m. Carol Ann Post, July 2, 1983; children: Colin R., Aaron S. ScB in Mech. Engring., Brown U., 1974; MS in Environ. Engring., Cornell U., 1979, PhD in Environ. Engring., 1983. Registered profl. engr., Del. Student asst. Max Planck Inst. Stromungsforschung, Göttingen, Germany, 1974; mech. engr. Petro-Tex, Inc., Houston, 1975, Versar, Inc., Springfield, Va., 1976; rsch. asst., teaching asst. dept. civil, environ. engring. Cornell U., Ithaca, 1976-83; asst. prof. dept. civil engring. U. Del., Newark, 1983-89, assoc. prof., 1989—; vis. rsch. prof. Inst. Nat. Poly. de Lorraine, Nancy, France, 1989; mem. coagulation and flocculation chem. task group Nat. Sanitation Found., 1987-88, chair sludge joint task group, standard methods for examination of water and wastewater, 1988—; tchr. courses on water, wastewater, sludge treatment and environ. colloid chemistry. Author: Procedures Manual for Polymer Selection, 1989; editor newsletter specialist group on sludge mgmt., Internat. Assn. Water Quality, 1992—; contbr. articles to profl. jours. including Jour. Water Pollution Control Fedn., Jour. Am. Water Works Assn. Grantee NSF, 1985-88, Am. Water Works Assn. Rsch. Found., 1985-86, Del. Water Resources Ctr., 1991-93, U.S. Geol. Survey, 1991-93, Watern Environ. Rsch. Found., 1992-93. Mem. ASCE, Internat. Assn. Water Quality, Am. Chem. Soc. (div. colloid chemistry), Am. Water Works Assn., Water Environ. Fedn., Sigma Xi, Chi Epsilon. Achievements include origination of quantitative model of coagulation, leading study on control systems for coagulant and sludge conditioner optimization, leading research on electrokinetic analyses, especially streaming current. Office: Univ Del Dept Civil Engring Newark DE 19716

DENTINE, MARGARET RAAB, animal geneticist, educator; b. Great Falls, Mont., Feb. 25, 1947; d. James S. and Florence (Renz) Raab. BA, U. Calif., Santa Cruz, 1969; MS, N.C. State U., 1981, PhD, 1985. Asst. prof. U. Wis., Madison, 1985-91, assoc. prof., 1991—; mem. com. Nat. Animal Genome Mapping Com., USDA, 1992—, mem. organizing chair regional project on marker-assisted selection on dairy cattle, 1992. Mem. editorial bd. J. Dairy Sci., Am. Dairy Sci. Assn.; contbr. articles to profl. jours. Faculty advisor Assn. of Women in Agr., Madison, 1988—; univ. rep. genetic advancement com. Holstein-Friesian Assn., Brattleboro, Vt., 1990-92. Recipient Pound Rsch. award Coll. Agrl. and Life Scis., U. Wis., 1992; grantee Apple Computers, 1986. Mem. AAAS, Am. Dairy Sci. Assn. (mem. dairy cattle improvement com. 1990—), Sigma Xi, Gamma Sigma Delta, Phi Kappa Phi. Achievements include patent for genetic marker for superior milk production in dairy cattle; patent pending prediction of milk production in dairy animals; development of bovine gene map, marker assisted selection of livestock; other interests include molecular and biochemical markers linked to quantitative traits. Office: U Wis Dairy Sci Dept 1675 Observatory Dr Madison WI 53706

DENTON, DAVID LEE, laboratory manager, chemical engineer; b. Abingdon, Va., Aug. 6, 1952; s. Kenneth Oren and Patricia Ann (Kilgore) D.; m. Martha Lynn Jobe, Nov. 26, 1976; children: Jonathan David, Jennifer Lynn, Matthew Joseph. BSChemE, Va. Poly. Inst. and State U., 1974. Registered profl. engr., Tenn. Chem. engr. Eastman Chem. Co. Rsch., Kingsport, Tenn., 1974-80, sr. rsch. chem. engr., 1980-84, asst. to dir. engring. rsch., 1984-86, lab. coord., 1986-88, mgr. engring. svcs., 1988-89, asst. to v.p. rsch., 1989-90, lab. head, 1990—. Royal Amb. dir., deacon Colonial Heights Bapt. Ch., Kingsport, 1980—. Mem. Am. Inst. Chem. Engrs. (treas. 1988-89, co-chmn. crystallization session 1982), Blue Ridge Morgan Horse Assn. (pres. 1992-93). Republican. Achievements include patent in field. Office: Eastman Chem Co PO Box 1972 Kingsport TN 37662

DENTON, DOROTHEA MARY, electronics company manager; b. Bridgeport, Conn., Aug. 12, 1938; d. George Ernest and Alice Mary (DeLibro) Squibb; m. Fillmore Edgar Denton, June 8, 1963; children: Alan Scott, Brian Wayne. BA, St. Mary's Coll., 1985; cert., De Anza Coll., 1985, UCLA, 1990. Purchasing facilities mgr. Calif. Eastern Labs., Santa Clara, 1981-89; materials mgr. San Jose (Calif.) Med. Ctr., 1989-92; distribution mgr. Radionics, Salinas, Calif., 1992—; mem. Employee Relations Com., San Jose, 1986-88, Mgmt. Devel. Team, SAn Jose, 1986-88; chair Supr.'s/mag.'s Group, San Jose, 1986-88. Chair dist. tng., coun. commn. Boy Scouts Am., Santa Clara County, Calif., 1972. Mem. Am. Production & Inventory Control Soc., Nat. Purchasing Assn., Silicon Valley Purchasing Assn. Home: 1073 El Solyo Heights Dr Felton CA 95018

DENTON, M. BONNER, research chemistry educator; b. Beaumont, Tex., June 15, 1944; s. Harold and Julia (Bonner) D. BS in Chemistry, BA in Psychology, Lamar State Coll. Tech., 1967; PhD in Chemistry, U. Ill., 1972. Asst. prof. chemistry U. Ariz., Tucson, 1971-76, assoc. prof., 1976-80, prof., 1980—; cons. Lawrence Livermore (Calif.) Lab., Baird Corp., Bedford, Mass., DOW Chem. Corp., Midland, Mich.; bd. dirs. Root Corp., Houston, Denco Research, Tucson. Editor: Jour. Automatic Chemistry, Chemometrics and Intelligent Lab. Systems, 1986; contbr. articles to profl. jours.; inventor in field. Named one of Outstanding Young Men of Am., 1978; Alfred P. Sloan Research fellow, 1976-80. Mem. Am. Chem. Soc., Soc. Applied Spectroscopy (chmn. Ariz. chpt. 1977-78, nat. program com. 1978-79, Lester W. Strock award 1991), Instrument Soc. Am., Sigma Xi, Phi Lambda Upsilon, Alpha Chi Sigma. Avocations: scuba diving, skiing, spleunking, racing autos.

DENUZZO, RINALDO VINCENT, pharmacy educator; b. Cleve., Oct. 21, 1922; s. Luigi and Domenica Mary (Razzano) DiNuzzo; m. Lucy Bernadine Sneed, June 29, 1946; 1 child, Lisa Ann. BS, Albany Coll. Pharmacy, 1952; MS in Edn., SUNY-Albany, 1956. Prof. pharmacy N.Y. Coll. Pharmacy, Albany, 1952—, adminstrv. asst., 1963-80; cons. pharmacist N.Y. State Dept. Health, 1968—, chmn. tech. pharmacy adv. com.; lectr. drug product substitution and generic drugs. Author: Ann. Albany Coll. Pharmacy Prescription Survey, 1956-84, Substitution, the New York State Experience, 1980, RX Services, XIII Winter Games, 1980, Annual DeNuzzo Prescription Survey, 1985—, Impact of One-Line Presciption Form on Generic Drug Use, 1987, Using the Right Tools to Achieve Personal Success, 1990, Personal Selling, 1991, Cipro, Vasotec, Voltaren Post Biggest Gains, 1987, Annual Survey Tracks Drug Prescribing Trends, 1990, Consumer Prescription Prices Increase, 1991, How to Reduce Prescription Medical Costs, 1992, Changes in Dental Prescribing, 1991, Are Dental Prescriptions a Viable Target for RPhs?, 1992; editor Albany Coll. Pharmacy Alumni News, 1961-86; mem. editorial bd. MMM, 1977-80. Instr. first aid, responding to emergencies, CPR, ARC; mem. East Greenbush Cen. Sch. Dist. Bd. Edn., 1974-92, v.p., 1975-76, pres., 1976-78, 91-92, (svc. citation 1992), East Greenbush Edn. Found. (plaque); mem. adv. bd. Merrell-Dow Hosp., 1987; sec.-treas. Union U. Pharmacy Coll. Coun., 1970-80; cons. pharmacist, coord. pharm. svcs. XIII Winter Olympic Games, Lake Placid, N.Y., 1980; chmn. Albany Coll. Pharmacy Faculty, 1987-89, com. on coms., 1984-87, promotions com., 1989-92, faculty affairs, chmn., and rev., 1990, faculty ombudsman; mem. N.Y. State Sch. Bd. Assn., Albany Vis. Nurses Assn. (profl. adv. com.); mem. rev. panel on prescription payment rev. commn. of Office of Tech. Assessment, U.S. Congress, 1988; mem. ethics panel Siena Coll., 1992. With U.S. Army, 1941-46, USAF, 1946-47, capt. M.C. USAFR, 1948-63; ret. 1982. Recipient Francis J. O'Brien Pharmacay Man of Yr. award, 1979, 25-Yr. Svc. citation ARC, 30-Yr. Svc. citation ARC, Svc. plaque East Greenbush Ctrl. Sch. Dist., Svc. plaque East Greenbush Edn. Found. Fellow Am. Coll. Apothecaries; mem. Am. Assn. Colls. Pharmacy (sec.-treas. council of faculties 1979-80, council, chmn. elect 1982-83, chmn. 1984-87, dir. 1984-89), Am. Pharm. Assn., N.Y. State Pharm. Soc., N.Y. Sch. Bd. Assn., AARP, N.Y. State Pub. Employees Fedn., Albany Coll. Pharmacy Alumni Assn. (exec. dir. 1965-86, disting. service medal 1975), AAUP (pres. 1978—), 2 Disting. Svc. plaques, 1988, Kappa Psi (Cert. of Commendation 1989, Beta Delta ann. Rinaldo V. DeNuzzo luncheon 1988—), 46th and 72d Recon Assn., Albany Coll. Pharmacy Pres.'s Club (chmn. bd. 1962-87), Army Varsity Club, Kappa Psi (dep. grand coun. Beta Delta chpt., Albany grad., sec.-treas.). Republican. Roman Catholic. Clubs: Albany Coll. Pharmacy Pres.'s (chmn. bd.1962-87); Officers (West Point, N.Y.). Home: 19 Alva St East Greenbush NY 12061-2027 Office: 106 New Scotland Ave Albany NY 12208-3492

DEORIO, ANTHONY JOSEPH, surgeon; b. Chgo., June 27, 1945; s. Joseph John and Catherine Marie Deorio; m. Janet Ann Balskus, Jan. 10, 1970; children: Joseph, Catherine. BS, Loyola U., Chgo., 1967; MD, Loyola U., Maywood, Ill., 1971. Diplomate Am. Bd. Surgery. Intern St. Joseph Hosp., Chgo., 1971-72; resident in surgery Loyola Med. Ctr., Maywood, 1972-76, clin. instr. surgery, 1976-77, asst. prof. surgery, 1977—; pvt. practice Resurrection Hosp., Chgo., 1977—, dir. surg. edn., 1977—, chmn. dept. surgery, 1984-88, sec. med. staff, 1986-88; assoc. examiner Am. Bd. Surgery, 1993. Contbr. articles to profl. jours. Fellow ACS; mem. AMA, Ill. Med. Soc., Chgo. Med. Soc., Chgo. Surg. Soc., Ill. Surg. Soc., Alumni Assn. Stritch Sch. Medicine (bd. govs.), Columbian Club, KC, Blue Key, Alpha Omega Alpha. Roman Catholic. Avocations: model railroads, sports, fishing. Office: 7447 W Talcott Ave Chicago IL 60631-3745

DEOUL, NEAL, electronics company executive; b. N.Y.C., Feb. 27, 1931; s. George and Pearl (Hirschfield) D.; B.S. in Physics, Coll. City N.Y., 1952; postgrad. Rutgers U., 1954-55; JD, Bklyn. Law Sch., 1959; m. Bernice Kradel, Dec. 25, 1955 (div.); children: Cara Jan, Stefani Neva, Evan Craig; m., Kathleen B. Davis, June 20, 1982; 1 child, Shannon Rae. Engr., Signal Corps, U.S. Army, Evans Signal Lab., Belmar, N.J., 1952-55; engr. Airborne Instruments Lab., Deer Park, N.Y., 1955-56; sales mgr. FXR, Inc., Woodside, L.I., 1956-60; admitted to N.Y. State bar, 1960; pres. Microwave Dynamics Corp., Plainview, L.I., 1960-61, Paradynamics, Inc., Huntington Station, N.Y., 1961-64; mgr. Servo Corp. Am., Hicksville, N.Y., 1964-66; v.p. Trio Labs., Inc., Plainview, N.Y., 1966-69; exec. v.p. Microlab/FXR, Livingston, N.J., 1969-74; pres. Neal Deoul Assocs., Balt., Md., 1974—. Mem. IEEE (sr.), N.Y. State Bar Assn., Md. Bar Assn., Young Pres.'s Orgn., Profl. Group Engring. Mgmt., Am. Arbitration Assn. Home and Office: 2 Bellchase Ct Baltimore MD 21208

DEPALMA, RALPH GEORGE, surgeon, educator; b. N.Y.C., Oct. 29, 1931; s. Frank and Maria (Sibilio) DeP.; m. Maleva Tankard, Sept. 23, 1955; children: Ralph L., Edward F., Maleva B., Malinda G. A.B., Columbia U, 1953; M.D., NYU, 1956. Diplomate Am. Bd. Surgery, Am. Bd. Vascular Surgery. Resident in surgery Univ. Hosps., Cleve., 1962-64; instr. to assoc. prof. surgery Case Western Res. U., Cleve., 1964-80, prof. surgery, 1973-80; prof., chmn. surgery U. Nev., Reno, 1980-82, George Washington U. Sch. Medicine, Washington, 1982-92; Lewis B. Saltz prof. of surgery George Washington U. Med. Ctr., Washington, 1992—, Lewis B. Saltz prof. surgery, 1992—. Editor: (with J.M. Giordano) Reoperative Vascular Surgery, 1987, Basic Science of Vascular Surgery, 1988; assoc. editor: Haimovici Vascular Surgery: Principles and Techniques, 1989; mem. editorial bd. Internat. Vascular Surgery, Internat. Jour. Impotence Rsch.; contbr. articles to profl. jours. Stroke liaison nat. chpt. Am. Heart Assn. Served to capt. USAF, 1958-61. Grantee USPHS, 1974-82. Fellow ACS; mem. Cleve. Vascular Soc. (pres. 1977-78), Rocky Mt. Vascular Soc. (pres. 1981-82), Am. Surg. Assn, Soc. Vascular Surgery, Washington Acad. Surgery (sec. 1991-92, v.p. 1992-93, pres. 1993-94), Am. Venous Forum (sec. 1991—). Clubs: West River Sailing Club (Galesville, Md.); Cosmos (Washington) (membership com.).

DEPAOLO, DONALD JAMES, earth science educator; b. Buffalo, Apr. 12, 1951; s. Dominic James and Lorraine Marie (Nassiff) DeP.; m. Geraldine Sue Adler, Apr. 14, 1973 (div. Oct. 1984); 1 child, Tara Michelle; m. Bonnye Lynn Ingram, Aug. 24, 1985; 1 child, Daniel James. BS in Geology, SUNY, Binghamton, 1973; PhD in Geology, Calif. Inst. Tech., 1978. Asst. prof. geology UCLA, 1978-81, assoc. prof., 1981-83, prof., 1983-88; prof. geology, dir. Berkeley Ctr. for Isotope Geochemistry U. Calif., Berkeley, 1988—, chmn. dept. geology and geophysics, 1990-93; mem. various research coms. NRC, 1983—; mem. sci. adv. com. DOSECC, 1985-87. Contbr. articles to profl. jours. Fellow Am. Geophys. Union (chmn. award coms., Macelwane award 1983), Mineral. Soc. Am. (award 1987); mem. AAAS, Geol. Soc. Am. (Clarke medal 1978), Nat. Acad. Scis., Geochem. Soc. (chmn. awards coms.). Avocations: guitar, skiing, tennis. Office: U Calif Dept Geology & Geophysics Berkeley CA 94720

DEPASQUALE, FRANCESCO, aerospace engineer; b. Milazzo, Italy, Apr. 24, 1954; s. Giuseppe and Maria (La-Malfa) D.; m. Liana Viacava, Sept. 25, 1982. Engring. degree, U. Milan, 1978. Cert. aeronautic and aerospace engr. Project engr. Locatelli SPA, Bergamo, Italy, 1978-80, project mgr., 1981-83; project mgr. BPD Difesa e Spazio, Rome, 1983-85, head design group, 1986-89, head engring. dept., 1990—. Recipient Gold medal Rotary, 1972, award NASA Mission Space Shuttle, 1992. Mem. AIAA, Ordine Degli Ingegneri, Assn. Italiana di Astronautica. Achievements include design, development and manufacture of small hovercraft and European patent for device to glue posters. Home: Vincenzo Monti 2, I-00034 Colleferro Rome Italy Office: BPD Difesa e Spazio, Corso Garibaldi 22, 00034 Colleferro Rome Italy

DEPETRILLO, PAOLO BARTOLOMEO, medical educator; b. Rome, Italy, Nov. 24, 1956; m. Sandra B. McDougall, June 15, 1985; children: James Benjamin, Jordan Birks, Seth Henry. ScB in Biology, Brown U., 1978, MD, 1981. Diplomate Am. Bd. Internal Medicine, Am. Bd. Clin. Pharmacology. Resident medicine Roger Williams Hosp., Brown U., Providence, 1981-84, fellow clin. pharmacology, 1984-85; sr. asst. surgeon USPHS, Lewisburg, Pa., 1985-88; instr. Brown U., Providence, 1988-90, asst. prof. medicine, 1990—. Contbr. articles to Jour. of Chromatography, Alcoholism: Clin. and Exptl. Rsch. Lt. USPHS, 1985-88. Pfizer fellow Roger Williams Hosp., 1985; decorated Commendation medal USPHS, 1988; recipient Faculty Devel. award Pharm. Mfrs. Assn., Roger Williams Hosp., 1990. Mem. AAAS, ACP, Am. Med. Informatics Assn. Office: Brown U Sch Medicine Box G Providence RI 02912

DEPIANTE, EDUARDO VICTOR, nuclear engineer; b. Córdoba, Argentina, Feb. 20, 1959; s. Marco Osvaldo Antonio and Corina (Berón) D. BS in Nuclear Engring., Nat. U. Cuyo, Argentina, 1981; PhD, MIT, 1988. Rsch. and devel. engr. Nat. AEC, S.C. de Bariloche, Argentina, 1981-84; teaching asst. Nat. U Cuyo, S.C. de Bariloche, Argentina, 1981-84; rsch. asst. MIT, Cambridge, Mass., 1985-88; researcher Oak Ridge (Tenn.) Nat. Lab., 1988; postdoctoral assoc. MIT, 1988-89; researcher Argonne (Ill.) Nat. Lab., 1990—. Contbr. articles to profl. jours. Mem. Am. Nuclear Soc., Soc. for Computer Simulation, Sigma Xi, Alpha Nu Sigma. Achievements include development of parity simulation methodology for single and two-phase nuclear thermal hydraulic systems, of a procedure for automatic generation of state equations of dynamic systems in symbolic form, stability analysis of passively safe reactor designs. Office: Argonne Nat Lab 9700 S Cass Ave RA/208-G224 Argonne IL 60439-4842

DEPRISTO, ANDREW E., chemist, educator; b. Newburgh, N.Y., Nov. 3, 1951; s. Salvatore and Loretta DeP. BS in Physics, SUNY, Oneonta, 1972; PhD in Chem. Physics, U. Md., 1976. NSF postdoctoral assoc. Princeton (N.J.) U., 1977-79; prof. theoretical chemistry U. N.C., Chapel Hill, 1979-82, Iowa State U., Ames, 1982—; program dir. Ames Lab., 1989—. Contbr. articles to profl. jours. NSF postdoctoral fellow, 1977, Alfred P. Sloan fellow, 1984-88, John Guggenheim Found. fellow, 1987-88; Camille and Henry Dreyfus scholar, 1983-88. Fellow Am. Phys. Soc.; mem. Am. Vacuum Soc., Material Rsch. Soc., Am. Chem. Soc. (sec.-treas. div. phys. chemistry 1991—). Office: Ames Lab/Iowa State Univ 303 Wilhelm Hall Ames IA 50011

DEQUASIE, ANDREW EUGENE, chemical engineer; b. Edgeworth, Pa., Dec. 31, 1929; s. Arthur Ellis and Helen (McManus) D.; m. Clara Phyllis Ciraolo, Nov. 28, 1959; children: Linda, David, Diane, Arthur. BSChemE, U. Pitts., 1950; postgrad., Williams Coll., 1962-69. Rsch. asst. Freedom (Pa.) Valvoline Oil Co., 1950-51; specification writer Army Chem. Corps, Balt., 1951-53; pilot plant engr. Olin Mathieson Chem. Corp., Malta, N.Y., 1953-55; pilot plant supr. Olin Mathieson Chem. Corp., Niagara Falls, N.Y., 1955-60; devel. engr. Olin Mathieson Chem. Corp., New Haven, 1960-61; capacitor engr. Sprague Electric Co., Barre, Vt., North Adams, Mass., 1962-87; cons. Pownal, Vt., 1987—. Author: The Dragonslayers, 1973, Thirsty, 1983, The Green Flame, 1991. Recipient Medicine Pipe award Western Writers Am., 1984. Fellow Am. Chem. Soc. Achievements includes 2 patents and 4 co-patents on capacitor design. Home and Office: Box 211 Oak Dr Pownal VT 05261

DE QUEIROZ, KEVIN, zoologist; b. L.A., June 4, 1956; s. Richard and Kristine Toshiye (Kawaguchi) de Q. BA in Biology, UCLA, 1978; MS in Zoology, San Diego State U., 1985; PhD in Zoology, U. Calif., Berkeley, 1989. Postdoctoral fellow Calif. Acad. of Scis., San Francisco, 1989-91; zoologist, curator Nat. Mus. of Natural History Smithsonian Instn., Washington, 1991—; cons. Calif. Acad. Scis., San Francisco, 1989-91, rsch. assoc., 1991—; mem. editorial bd. Soc. Systematic Biologists, 1992—. Contbr. articles to profl. jours. Recipient Dissertation Yr. fellowship U. Calif., 1988, Annie Alexander scholarship Mus. Vertebrate Zoology, 1987; grantee Soc. Sigma Xi, 1987, Am. Mus. Natural History, 1987; named Outstanding Grad. Student instr. U. Calif., 1988. Mem. Am. Soc. Ichthyologists and Herpetologists (gov. 1990-92), Am. Soc. Zoologists, Soc. for Study of Evolution, Soc. Systematic Biologists (councillor 1991-93), Sigma Xi. Achievements include working out various details of the role of the principle of evolution in systematic biology. Office: Mus of Natural History Smithsonian Instn Div of Amphibians/Reptiles Washington DC 20560

DE RANTER, CAMIEL JOSEPH, chemistry educator; b. Hoboken, Antwerp, Belgium, Aug. 27, 1937; s. Antoon August and Clementine Leonie (Santens) DeR.; m. Monique Anna Blaton, Apr. 9, 1965; children: Carl, Johan. Lic. Chem. Scis., cert. in nuclear scis., Katholieke U. Leuven, Belgium, 1960; DSc, Katholieke U. Leuven, 1964. Postdoctoral fellow So. Ill. U., Carbondale, 1969; docent in analytical chemistry Katholieke Univ. Leuven, Belgium, 1970-74, prof., 1974-76, ordinary prof., 1976—; dean Inst. Pharm. Scis., Katholieke Univ. Leuven, Belgium, 1983-92, vice dean, 1992—. Editor, author: X-Ray Crystallography and Drug Action, 1984. Mem. Belgian Pharm. Soc., Belgian Biophys. Soc., Am. Crystallographic Assn., Koninklyke Nederlandse Chemische Vereniging, Koninklyke Vlaamse Chemische Vereniging, Quantitative Structure Activity Relationships Soc. (intern), Molecular Graphics Soc. Office: Lab Analytical Chemistry, Van Evenstraat 4, 3000 Leuven Belgium

DERBY, CHRISTOPHER WILLIAM, civil engineer; b. Massena, N.Y., July 1, 1963; s. Joseph Patrick and Joanna (Kulback) D.; m. Maire Nanette Graziano, Nov. 18, 1989. AAS, Canton (N.Y.) Coll. Tech., 1984; B Tech., Rochester Inst. Tech., 1987; MS, U. Buffalo, 1989. Civil engr. Lotz Designer Engrs. & Constructors, Horsham, Pa., 1989—. Assoc. mem. ASCE. Home: 607 Essex Circle King of Prussia PA 19406 Office: Lotz Designers Engrs & Constructors 601 Dresher Rd Horsham PA 19406

DERE, WILLARD HONGLEN, internist, educator; b. Sacramento, Jan. 8, 1954; s. William Janson and Bessie Lon (Joe) D.; m. Julia Mei Lum, June 18, 1978; children: Melissa Ellen, Kathryn Elizabeth. AB, U. Calif., Davis, 1975, MD, 1980. Diplomate Nat. Bd. Med. Examiners. Intern Health Sci. Ctr., U. Utah, Salt Lake City, 1980-81; resident Health Scis. Ctr., U. Utah, Salt Lake City, 1981-83; instr. internal medicine, geriatrics U. Utah, Salt Lake City, 1985-87, asst. prof., 1987-89; rsch. fellow U. Calif., San Francisco, 1983-85; asst. prof. Ind. U., Indpls., 1989—; clin. rsch. physician Lilly Rsch. Labs., Indpls., 1989-91, dir. European regulatory affairs, 1991; dir. emergency rm. VA Med. Ctr., Salt Lake City, 1985-86; cons. U. Utah Student Health Svc., Salt Lake City, 1985-89, acting dir., 1987-88. Editor: (book) Practical Care of the Ambulatory Patient, 1989; Contbr. articles to profl. jours. Vol. account exec. United Way, Ind., 1990. Hon. assoc. investigator VA, San Francisco, 1984. Mem. ACP, AAAS, Am. Geriatrics Soc., Sigma Xi. Presbyn. Achievements include research in TSH and cAMP enhanced expression of the MYC-oncogene, TSH and cAMP regulation of RAS-oncogene, adrenocortical function in AIDS, promoter region of the RAS-oncogene, oncogene regulation in thyroid cells; multi-center antibiotic trials. Home: 4720 Hazelwood Cir Carmel IN 46033-4633 Office: Lilly Rsch Labs Lilly Corp Ctr Indianapolis IN 46285

DERECHIN, MOISES, biochemistry educator; b. Basavilbaso, E.R., Argentina, Dec. 19, 1924; came to U.S., 1963; divorced; 1 child, Vivianna Maia. Physician degree, Buenos Aires U., 1952; PhD, London U., 1960. Diplomate Am. Bd. Clin. Pathology. Rsch. fellow U. Chile, Santiago, 1955-57; rsch. fellow Broodbank Cambridge (Eng.) U., 1960-63; asst. prof. SUNY, Buffalo, 1964-70, assoc. prof., 1970-93; vis. scientist NIH, Bethesda, Md., 1963-64. Scholar Britis Coun., 1957-60. Office: SUNY Buffalo Dept Biochemistry 3435 Main St 102 Cary Hall Buffalo NY 14214

DERELANKO, MICHAEL JOSEPH, toxicologist; b. Elizabeth, N.J., Apr. 21, 1951; s. Frank and Anne (Laskodi) D.; m. Patricia Rahoche, Sept. 6, 1975; children: Michael David, Robert Ross. BS, St. Peter's Coll., 1973; MS, NYU, 1976, PhD, 1978. Diplomate Am. Bd. Toxicology. Predoctoral fellow NIH, 1973-75; postdoctoral fellow Schering-Plough Corp., Bloomfield, N.J., 1978-80; rsch. toxicologist AlliedSignal Inc., Morristown, N.J., 1980-83, sr. toxicologist, 1983-88, mgr. toxicology, 1988-91, sr. mgr. toxicology, 1991—; chmn. Meko test rule sci. adv. bd. Pitts., 1989—; mem. content adv. com. Liberty Sci. Ctr., N.J., 1991—; reviewer OECD test guidelines chem. Mfrs. Assn., Washington, 1992; speaker in field. Contbr. articles to Jour. Occupational Medicine, Fundamental and Applied Toxicology, Jour. Pharm. Exptl. Therapeutics, Digestive Diseases and Scis., Exptl. Hematology. Mem. organizing com. North Jersey Regional Sci. Fair, 1986—. Mem. AAAS, Soc. Toxicology, Soc. Exptl. Biology and Medicine, Sigma Xi. Achievements include research in effects of chemical agents on the blood, prostaglandin-mediated gastrointestinal cytoprotection, erythropoiesis in experimental leukemias. Office: AlliedSignal Inc Dept Toxicology 101 Columbia Rd Morristown NJ 07962-1139

DE REMER, EDGAR DALE, aviation educator; b. L.A., May 1, 1935; s. Edgar Merton and Mary Margaret (Shuck) De R.; children: Beth M., Jolene, E. Lyle, Jeffry A. BS, Calif. State Poly. U., San Luis Obispo, 1957; MS, Utah State U., 1959, PhD, 1961. Dir. dept. agribus. Northeastern Jr. Coll., Sterling, Colo., 1963-66; pres. Amorydale Farms, Inc., Sterling, 1966, Agronomics, Inc., Phoenix, 1970-80; v.p. for land devel. Ceres Land Co., Sterling, 1966-69; asst. prof. Embry Riddle Aero. U., Prescott, Ariz., 1981-83; assoc. prof. aviation Ctr. for Aerospace Scis., U. N.D., Grand Forks, 1983-90, prof., 1990—. Author: Water Flying Concepts, 1989, Aircraft Systems for Pilots, 1991, Global Naviation for Pilots, 1993; also numerous articles, including AIAA Jour. Advisor Wilderness Pilots Assn., Grand Forks, 1987—; pres. Forx ARC, Grand Forks, 1989. Recipient outstanding faculty advisor award U. N.D., 1990. Mem. Seaplane Pilots Assn. (field dir. 1988—), Sigma Xi. Achievements include research on use of delta ratio in determining takeoff performance. Home: 1323 Noble Cove Grand Forks ND 58201-8416 Office: U ND PO Box 9007 Grand Forks ND 58202

DERESIEWICZ, HERBERT, mechanical engineering educator; b. Brno, Czechoslovakia, Nov. 5, 1925; s. William and Lotte (Rappaport) D.; m. Evelyn Altman, Mar. 12, 1955; children: Ellen, Robert, William. BME, CCNY, 1946; MS, Columbia U., 1948, PhD, 1952. Sr. staff engr. Applied Physics Lab., Johns Hopkins U., 1950-51; mem. faculty Columbia U., N.Y.C., 1951—, prof. mech. engring., 1962—, chmn. dept. mech. engring., 1981-87, 90—; cons. stress analysis, vibrations, elastic contact, wave propagation, mechanics granular and porous media., Fulbright sr. research scholar, Italy, 1960-61, Fulbright lectr., Israel, 1966-67; vis. prof., Israel, 1973-74. Editor Columbia Engring. Rsch., 1975-92; contbr. articles to profl. jours. Served with AUS, 1946-47. Univ. fellow Columbia U., 1949-50. Mem. AAAS, Sigma Xi. Home: 336 Broad Ave Englewood NJ 07631-4304 Office: Columbia U 220 SW Mudd Bldg New York NY 10027

DERESIEWICZ, ROBERT LESLIE, molecular biologist, physician, educator; b. N.Y.C., May 23, 1958; s. Herbert and Evelyn (Altman) D.; m. Kathryn Sue Kirshenbaum, July 16, 1989; 1 child, Hannah Yael. AB in biochemistry summa cum laude, Columbia U., 1979; MD, Mt. Sinai Med. Sch., 1983. Diplomate Am. Bd. Internal Medicine subspecialty in infectious diseases. Intern, then resident in medicine Columbia Presbyn. Med. Ctr., N.Y.C., 1983-86; vis. rsch. fellow Hebrew U. Hadassah Med. Sch., Jerusalem, 1986-87; clin. fellow in medicine, infectious diseases Harvard Med. Sch., Boston, 1987-88, rsch. fellow in medicine, microbiology, molecular genetics, 1988-90, instr. medicine, 1990—; assoc. in medicine Beth Israel Hosp., Boston, 1990-91; assoc. physician Brigham & Women's Hosp., Boston, 1990—. Contbr. articles to sci. jours. Recipient Am. Inst. Chemists' medal, 1979, Bowen Brooks fellowship N.Y. Acad. Medicine, 1986, Squibb clin. fellowship Nat. Found. Infectious Diseases, 1988. Mem. ACP, Infectious Disease Soc. Am., Mass. Med. Soc., Mass. Infectious Disease Soc., Am. Soc. Microbiology, Phi Beta Kappa, Alpha Omega Alpha, Phi Lambda Upsilon. Achievements include research in structure, function and regulation of bacterial exotoxins. Office: Channing Lab 180 Longwood Ave Boston MA 02115

DEREUS, HARRY BRUCE, mechanical engineer, consultant; b. Warner Robbins, Ga., Oct. 13, 1964; s. Robert William and Genevieve Grace (Waddell) DeR. BSME, U. South Fla., 1991. Mech. systems engr. Engring. Matrix, St. Petersburg, Fla., 1991-92; plant engr. Verlite Co., 1993—. Asst. scoutmaster Boy Scouts of Am., Tampa, Fla., 1982—. Mem. AIAA, ASME. Home: 405 Broxburn Ave Temple Terrace FL 33617

DE ROE DEVON, THE MARCHIONESS See DEVONTINE, JULIE E(LIZABETH) J(ACQUELINE)

DE ROSE, SANDRA MICHELE, psychotherapist, educator, supervisor, administrator; b. Beacon, N.Y.; d. Michael Joseph Borrell and Mabel Adelaide Edic Sloane; m. James Joseph De Rose, June 28, 1964 (div. 1977); 1 child, Stacey Marie. Diploma in nursing, St. Luke's Hosp., 1964; BA in Child and Community Psychology, Albertus Magnus Coll., 1983; MS in Counseling Psychology with honors, Century U., 1986, PhD in Counseling Psychology with honors, 1987. Gen. duty float nurse St. Luke's Hosp., Newburgh, N.Y., 1964-65; supr. nurses Craig House Hosp., Beacon, N.Y., 1965-70; staff devel., team dir., divsn. outpatient treatment svc. Conn. Mental Health Ctr., New Haven, 1970—; clin. instr. Sch. Nursing Yale U., New Haven, 1979-84, clin. instr. dept. psychiatry, 1989—; pvt. practice, 1976—. Mem. ANA (cert.), Conn. Nurses Assn., Conn. Soc. Psychoanalytic Psychologists, Conn. Soc. Nurse Psychotherapists, Assn. for Advancement Philosophy and Psychiatry, Sigma Theta Tau, Delta Mu, Alpha Sigma Lambda. Avocations: travel, interior design, antiquing, movies, plays, music. Office: Conn Mental Health Ctr 34 Park St New Haven CT 06519-1187 also: 210 Prospect St New Haven CT 06511

DE ROUX, TOMAS E., electrical engineer; b. Panama, Jan. 4, 1961; s. Juan Ramon and Licia Isabel (Guardia) DeR.; m. Eyda Estela Alvarado, Jan. 4, 1986; 1 child, Cristina. BSEE, U. Santa Maria, Panama, 1986; MSEE, Northwestern U., 1989. Electronics mechanic 193rd Infantry Brigade U.S. Army, Cocosolo, Panama, 1979-85; electronics engr. Devices Unlimited Corp., Panama, Panama, 1986-88; electronics instr. Panama Canal Commn., Balboa, Panama, 1985—; electronics engr. SESISA, Panama, 1990—; cons. TRIDEX S.A., Panama, 1987-88; bd. dirs. Joint Engring. and Architecture Com., Panama, 1992-93, SESISA, Panama, 1990—. Author: (tng. manual) Yagi Antennas & Related Arrays, 1990, Electronic Shop Practices, 1991; contbr. articles to profl. jours. Recipient Fulbright scholarship U.S. Govt., 1989. Mem. IEEE, Panamenian Soc. Engrs. Home: Box 871192, Panama 7, Panama Office: Panama Canal Commn Unit 2300 Indsl Tng Br APO AA 34011

DERRICKSON, JAMES HARRISON, astrophysicist; b. Bywood, Pa., June 15, 1944; s. Robert Murray and Mary Virginia (Coddington) D. BS in Physics, Drexel U., 1967; MS in Physics, U. Ariz., 1978; PhD in Physics, U. Ala., Huntsville, 1983. Astrophysicist Cosmic Ray Br. Space Sci. Lab., Huntsville, Ala., 1967--. Author: Advances in Space Research, 1989, Lecture Notes in Physics, 1985. Recipient Antarctica Svc. medal The Nat. Sci. Found, 1991, Antarctica Svc. medal Dept.. Navy, 1991. Mem. Am. Phys. Soc., Sigma Pi Sigma. Achievements include contribution to the analysis of the space radiation environment as it affects various spacecraft; direct measurement of the cosmic ray spectra at ultrarelativistic energies. Office: NASA/MSFC Space Sci Lab ES64 Huntsville AL 35812

DERRICKSON, WILLIAM BORDEN, business executive; b. Milford, Del., May 30, 1940; m. Patricia Jean Hayes, Feb. 1, 1964; children—Stephen Russel, Michael Scot. BSEE, U. Del., 1964; diploma, Harvard Bus. Sch., 1979. Registered profl. engr. Supr. elec. maintenance Delmarva Power, Salisbury, Md., 1964-68; instrumentation engr. Hercules, Inc., Wilmington, Del., 1968-69, Sun Shipbldg., Chester, Pa., 1969-70; dir. project Fla. Power & Light Co., Juno Beach, Fla., 1970-84; sr. v.p. Pub. Service Co. N.H., Manchester, 1984-85; pres. New Hampshire Yankee Electric Co., Seabrook, 1985-87; pres., chief ops. officer WPD Assocs., Inc., 1986-88; pres., chief oper. officer Quadrex Corp., Campbell, Calif., 1988-89, chmn. bd., chief exec. officer, 1989—; also chmn. bd. dirs.; nuclear advisor Tenn. Valley Authority Bd. Dirs., 1987. Contbr. articles to profl. publs. Named Constrn. Man of Yr. ENR/McGraw-Hill Publs., 1984. Mem. NSPE, Am. Nuclear Soc., Project Mgmt. Inst., N.H. Soc. Profl. Engrs., Internat. Platform Assn., Rep. Senatorial Inner Circle. Republican. Avocations: golf, travel, numismatics, piano. Home: 1864 SW Saint Andrews Dr Palm City FL 34990 Office: Quadrex Corp 1940 NW 67th Pl Gainesville FL 32606-1692

DERSTADT, RONALD THEODORE, health care administrator; b. Detroit, June 9, 1950; s. Theodore Edward and Dorothy J. (Semko) D.; m. J. Gail Adamson, June 9, 1990. BA, U. Detroit, 1971; M of Hosp. Healthcare Adminstn., Xavier U., 1975. Mgr. shared svcs. Bethesda Hosp. North, Cin., 1975-76; asst. adminstr. McCullough-Hyde Meml. Hosp., Oxford, Ohio, 1977-79; pres. Hospice of Cin., Inc., 1979-82; dir. strategic planning St. Francis-St. George Hosp., Cin., 1982-84; v.p. Mgmt. Dynamics, Inc., Cin., 1984-85; sr. v.p. St. Francis-St. George Mgmt. Co., Cin., 1986-88; v.p. Franciscan Health System of Cin., 1988-91; dir. hosp. affairs ChoiceCare, Cin., 1991—; vice-chmn., bd. dirs. Franciscan Health Network, Cin., Franciscan Health Ventures, Cin. Treas., bd. dirs. Ohio Easter Seals Soc., Columbus, 1987—; bd. dirs. S.W. Ohio Easter eal Soc., Cin., 1986-92; adv. bd. Dater Jr. High Sch., Cin., 1984-88. Fellow Am. Coll. Healthcare Execs.; mem. Healthcare Fin. Mgmt. Assn., Am. Hosp. Assn., Ohio Hosp. Assn. Avocations: boating, golf, radio control model building. Home: 7137 Pickway Dr Cincinnati OH 45233-4243 Office: 655 Eden Park Dr Cincinnati OH 45202

DERSTINE, MARK STEPHEN, mechanical engineer; b. Chambersburg, Pa., July 24, 1961; s. William Abram and Dolores Virgie (Lamm) D.; m. Ronda Kay Hahn, Sept. 6, 1980 (div. Nov. 1989); m. Barbara Marie Bert, May 22, 1993. BS in Mech. Engring., Wichita State U., 1983; MS in Engring. Mechanics, Va. Poly. Inst. & State U., 1988. Engr.-in-tng. cert., Kans. Design engr. Cessna Aircraft Co., Wichita, 1983-86; rsch. asst. Va. Poly. Inst. and State U., Blacksburg, 1986-88; project engr. Atlantic Rsch. Corp., Gainesville, Va., 1988—. Mem. AIAA, ASME, Soc. for Exptl. Mechanics. Home: 5918 Wild Brook Ct Centreville VA 22020 Office: Atlantic Rsch Corp 5945 Wellington Rd Gainesville VA 22065

DERVAN, PETER BRENDAN, chemistry educator; b. Boston, July 28, 1945; s. Peter Brendan and Ellen (Comer) D.; m. Jackqueline K. Barton; children: Andrew, Elizabeth. BS, Boston Coll., 1967; PhD, Yale U., 1972. Asst. prof. Calif. Inst. Tech., Pasadena, 1973-79, assoc. prof., 1979-82, prof. chemistry, 1982-88, Bren prof. chemistry, 1988—; adv. bd. ACS Monographs, Washington, 1979-81. Mem. adv. bd. Jour. Organic Chemistry, Washington, 1981—; mem. editorial bd. Bioorganic Chemistry, 1983—, Chem. Rev. Jour., 1984—, Nucleic Acids Res., 1986—, Jour. Am. Chem. Soc., 1986—, Acct. Chem. Res., 1988—, Bioorg. Chem. Rev., 1988—, Bioconjugate Chemistry, 1989—, Jour. Med. Chemistry, 1991—, Tetrahedron, 1992—, Biorganic and Med. Chemistry, 1993—, Chemical and Engineering News, 1992—; contbr. articles to profl. jours. A.P. Sloan Rsch. fellow, 1977; Camille and Henry Dreyfus scholar, 1978; Guggenheim fellow, 1983; Arthur C. Cope Scholar award 1986. Fellow Am. Acad. Arts and Scis.; mem. NAS, Am. Chem. Soc. (Nobel Laureate Signature award 1985, Harrison Howe award 1988, Arthur C. Cope Scholar award 1993, Willard Gibbs medal, 1993, Rolf Sammet prize 1993, Nichols medal 1994). Office: Calif Inst Tech 1201 E California Blvd Pasadena CA 91125-0001

DESAI, MANUBHAI HARIBHAI, surgeon; b. Kosamba, Bulsar, India, Aug. 21, 1933; came to U.S., 1976; m. Sudha. Nathubhai; 3 children. MBBS, U. Bombay, 1957, M in Surgery, 1962. Lic. surgeon, Tex., Mass., N.Y., N.J.; cert. Edn. Coun. for Fgn. Med. Students, Fed. Licensing Exam. Bd. Intern Grant Med. Coll., Allied J.J. Group Teaching Hosps., Bombay, 1958; resident house surgeon in gen. surgery Allied J.J. Group Teaching Hosps., Bombay, 1959, resident house surgeon in ENT, 1959-60, resident surg. registrar in gen. surgery, 1960-62; locum thoracic registrar in exptl. thoracic surgery J.J. Group Teaching Hosps., Bombay, 1962; resident surg. registrar Bombay Hosp., 1962-63; govt. med. officer Ministry of Health, Zambia, 1963-64; surg. specialist Kitwe Cen. Hosp., Zambia, 1965-72; chief surgeon Comml. and Indsl. Med. Aid Hosps., Kitwe, Ndola, Zambia, 1972-75; clin. fellow Cornell Med. Ctr., 1977-78; assoc. dir. critical care svcs. Baystate Med. Ctr., Springfield, Mass., 1978-83; assoc. prof. surgery divsn. gen. surgery U. Tex. Med. Br., Galveston, 1983-91, prof., 1992—, dir. Blocker Burn Ctr., 1992—; asst. chief of staff Shriners Burns Inst., Galveston, 1983—; cons. surgeon to Nchanga Consolidated Group of Copper Mind Hosps., Rokana divsn., Zambia, 1972-75; guest speaker in field; prin. investigator rsch. U. Tex. Med. Br., 1989—, Baystate Med. Ctr., Springfield, Mass., 1982-84. Author: (with others) Art and Science of Burn Care, 1987; contbr. articles to Med. Jour. Zambia, Maharashtra Med. Jour., Current Med Practice (Bombay), Arch. Surg., Nutritional Support Svcs., Cancer Treatment Reports, Am. Surgeon, Jour. Burn Care Rehab., Critical Care Quar., Jour. Trauma, Pediatric Clinics in N.Am., Emergency Care Quar., Surg. Forum, Annals of Surgery, Metabolism, Emergency Medicine, Jour. Pediatric Surgery, Postgrad. Medicine, Burns, Surg. Gynecology and Obstetrics. Recipient Appreciation placque Critical Care Founders Circle, 1989; grantee Shriners Hosps. for Crippled Children, 1986-89; HHS grantee Baystate Med. Ctr., 1982-84. Fellow ACS, Am. Coll. Critical Care Medicine, Internat. Coll. Surgeons, East African Assn. Surgeons, Royal Coll. Surgeons of Edinburgh; mem. AMA (Physician Recognition award 1980-82, 84-90), Surg. Infection Soc., Galveston County Med. Soc., Tex. Med. Assn., Singleton Surg. Soc., Pan Am. Med. Assn., Am. Trauma Soc., Soc. Critical Care Medicine, Mass. Med. Soc., Internat. Soc. for Burn Injuries, Hampden Dist. Med. Soc., Assn. for Acad. Surgery, Am. Burn Assn., Zambia Med. Assn., East African Assn. Office: Shriners Burns Inst 815 Market St Galveston TX 77550-2725

DESAI, MUKUND RAMANLAL, research and development chemist; b. Surat, India, June 4, 1946; came to U.S., 1984; s. Ramanlal Dolatrai and Kusumben R.; m. Bharti M. Desai, July 15, 1972; children: Hari, Om, Falguni. Pre-med. degree, Vidyasagar Coll., Calcutta, India, 1963; BS, Bombay U., 1967. Prodn. chemist En Gandhi & Co., Bombay, 1967-72; owner, chief exec. officer Amber Chemietron, Bombay, 1972-84; dir. Kafko Internat., Chgo., 1984-85; R & D chemist Worldwide Chems., Indpls., 1985—; ptnr., cons. Sterling Chem. Industries, Bombay, 1976-84. Patentee in field. Mem. Am. Chem. Soc. Hindu. Avocations: reading, travel. Office: Worldwide Chems 1910 S State Ave Indianapolis IN 46203-4146

DE SALVA, SALVATORE JOSEPH, pharmacologist, toxicologist; b. N.Y.C., Jan. 14, 1924; s. Nicola Carlo and Frances Agnes (Caldarella) De S.; m. Elaine Mae Radloff, June 14, 1948; children: Salaine Claire De Salva Bonanne, Christopher Joseph, Stephanie De Salva Farrelly, Steven William, Gregory Vincent, Peter Nicholas, Philip Anthony, Deidre De Salva Berry. BS, Marquette U., 1947, MS, 1949; postgrad., U. Ill., Chgo., 1951-53; PhD, Stritch Sch. Medicine, Loyola U., Chgo., 1958. Research and teaching asst. Marquette U., Milw., 1947-49; research biochemist Milw. County Gen. Hosp., 1954; instr. U. Ill., Chgo., 1951-52; asst. prof. Chgo. Coll. Optometry, 1951-53; pharmacologist Armour Pharm. Lab., Chgo, 1953-59; sect. head Colgate Palmolive Co., Piscataway, N.J., 1959-66, sr. research assoc., 1966-72, mgr., 1972-76, assoc. dir. research for pharmacology and toxicology, 1976-83, dir. research pharmacology and toxicology, 1983-88, worldwide ops. dir., 1988-90, corp. dir. human and environ. safety worldwide, 1990-92; pres. Salva Cons. Svcs., Somerset, N.J., 1992—; lectr. Loyola U., 1957-59; mem. technician tng. N.J. Council for Research and Devel., Rutgers U., 1969-72. Editor: Symposium for Biomedical Electronic Instrumentation, 1965; contbr. articles to profl. jours.; patentee in field; current work in pharmaco-toxicology of flourides, sequestering agts. and surfactants, nitrosamine risk assessment, alternative safety testing method devel. Mem. Park Forest (Ill.) Mosquito Abatement Program, 1952-55, Franklin Twp. (N.J.) Sch. Bd., 1969-70, Somerset (N.J.) Bd. Health, 1965-67, Cath. Youth Orgn., Somerset; v.p. Cedar Hill Swim Club, Somerset; active Boy Scouts Am., Somerset, 1965-67; trustee Franklin Twp. Day Care Ctr., 1969. Served with USN, 1942-46. Mem. AAAS, Soc. Exptl. Biology and Medicine, Am. Soc. Pharmacology and Exptl. Therapeutics, Soc. Toxicology, Internat. Union Pharmacology (toxicology sect.), N.Y. Acad. Scis., Internat. Soc. Regulatory Pharmacology and Toxicology, Internat. Soc. Study of Xenobiotics. Roman Catholic.

DESANTIAGO, MICHAEL FRANCIS, mechanical engineer; b. N.Y.C., Feb. 20, 1956; s. Michael and Carmen DeS.; m. Carmen Devivies, June 10, 1989; 1 child, Sabrina. BSME, U. Ill., Chgo., 1979. Registered profl. engr., Ill. Project engr. Sargent & Lundy, Chgo., 1979-87; co-founder, prin. Primera Engrs., Ltd., Chgo., 1987—. Mem. ASME, NSPE, ASHRAE, Northshore Toastmasters (pres. 1986-87), Latin Am. C. of C. (v.p. 1987-89). Achievements include the co-founding of Primera Engrs. Ltd., a 100% Hispanic-owned company. Office: Primera Engrs Ltd 25 E Washington #510 Chicago IL 60602

DE SANTIS, JAMES JOSEPH, clinical psychologist; b. L.A., Dec. 16, 1958; s. Pete John and Elsie Grace (Kerr) DeS. BA, U. So. Calif., L.A., 1981; MA, Calif. Sch. Profl. Psychology, L.A., 1983, PhD, 1985. Lic. psychologist, Calif. Registered psychol. asst. Royale Therapeutic Residential Ctr., Santa Ana, Calif., 1986-88, psychology dept. head, 1988-89; clin. dir. LaCasa Mental Health Ctr., Norwalk, Calif., 1989-90; staff psychologist Child Guidance Ctr. of Orange County, Huntington Beach, Calif., 1990, Calif. Sch. of Profl. Psychology, Alhambra, 1990—; pvt. practice Glendale, Calif., 1991—; allied health staff Glendale Adventist Med. Ctr., 1992—. Mem. Am. Psychol. Assn., Calif. Psychol. Assn., L.A. County Psychol. Assn., Nat. Register of Health Svc. Providers in Psychology, Phi Beta Kappa, Phi Kappa Phi. Achievements include rsch. on effects of the intensive Zen Buddhist mediation retreat on Rogerian congruence as real-self/ideal-self disparity on the Calif. Q-sort. Home: 234 N Kenwood St Apt 209 Glendale CA 91206 Office: Ste 300 138 N Brand Blvd Glendale CA 91203

DESAUTELS, EDOUARD JOSEPH, computer science educator; b. Winnipeg, Man., Can., Jan. 18, 1938; came to U.S. 1961; s. Ovide Joseph and Yvette (Aubin) D.; m. Jeannine Marie Blanchette, Aug. 12, 1961; children: Francine, Nicole, Philip. BSc with honors, U. Man., 1960; PhD, Purdue U., 1969. Rschr. MIT, Cambridge, 1963-64; instr. Purdue U., West Lafayette, 1964-69; prof. computer sci. U. Wis., Madison, 1969—; founder, dir. Computer Systems Lab., U. Wis.; cons. Entry Systems div. IBM, Boca Raton, Fla., 1984. Cons. editor W.C. Brown Pubrs., Dubuque, Iowa, 1980-87; author: Understanding and Using Computers, 1989, Introduction to Computer Architecture, 1989, others. With RCAF, 1955-60. Mem. ACM, IEEE. Home: 415 Virginia Ter Madison WI 53705 Office: Univ of Wis Computer Sci Dept 1210 W Dayton Madison WI 53706

DESBRANDES, ROBERT, petroleum engineering educator, consultant; b. Villefranche, Allier, France, Sept. 25, 1924; came to U.S., 1952; s. Francois and Suzanne (Berthelat) D.; m. Marie Joseph Jardiller, Oct. 21, 1955; children: Marc, Franck, Frederique. BS in Mech. Engring., Arts & Metiers Sch., Cluny, France, 1944; MS in Physics, Lyons U., France, 1962; DSc in Physics, Lyons U., 1965. Profl. engr., France. Field engr. Schlumberger, Venezuela, 1947-52; research engr. Schlumberger, Houston, 1952-58; sr. research engr., prof. French Petrolem Inst. and French Petroleum Sch., Paris, 1958-82; vis. prof. U. Houston, 1982-84; LSU Found. Disting. prof. La. State U., Baton Rouge, 1984—; cons. Kaman Corp., Colorado Springs, Colo., 1984; prof. Rike Ednl. Services, New Orleans, 1985-91; mem. Joint Oceanographic Instns. for Deep Earth Sampling, 1991—. Author: Theorie et interpretation des diagraphies, 1968, Ency. of Well Logging, 1985; contbr. over 75 tech. and sci. papers to profl. jours.; holder 37 patents. Assoc. mem. Smithsonian Instn., Washington, 1985—; judge La. State Fair, Baton Rouge, 1985. Lt. French Army, 1944-46, ETO. Grantee Amoco Found., 1986, Gas Rsch. Inst., 1989, Conoco, 1990-92, La. Bd. Regents, 1989-90. Mem. French Soc. Log Analysts (pres. 1970-71), Soc. Profl. Log Analysts (disting. speaker 1988-89), Soc. Petroleum Engrs. (co-recipient Ferguson award 1992, rev. editor for SPE Jours. 1992—), Joint Oceanographic Instns. for Deep Earth Sampling, Pi Epsilon Tau (faculty advisor 1988—). Avocations: tennis, skiing, equestrian sports, yoga, bee keeping. Office: La State U Coll of Engring Baton Rouge LA 70803

DESCHENES, JEAN-MARIE, agriculturist, researcher; b. St. Jean-Port-Joli, Que., Can., Aug. 9, 1941; s. Charles H. and Pauline (Bourgault) D.; m. Ghislaine Douville, Aug. 29, 1964; children: Jean-Francois, Daniel-Andre. BSc in Agr., Laval U., Can., 1963; PhD, Rutgers U., 1968. Rsch. scientist Rsch. Sta. Agr. Can., La Pocatiere, Que., 1967-70, Ste-Foy, Que., 1970-83; spl. advisor Rsch. Sta. Agr. Can., Ottawa, Ont., 1983-84, dep. dir. Rsch. Sta., 1984-89; dir. Rsch. Sta. Agr. Can., Lennoxville, Que., 1989—; cons. Can. Internat. Devel. Agy. Haiti, 1979, Rwanda, 1983. Recipient Best Paper award Agrl. Inst. Can., 1978. Roman Catholic. Office: Agr Can Rsch Sta, 2000 Rte 108 Est, Lennoxville, PQ Canada J1M 1Z3

DESCHUYTNER, EDWARD ALPHONSE, biochemist, educator; b. Chelsea, Mass., Sept. 3, 1944; s. Alphonso and Josephine Elizabeth (Kiewlicz) D.; m. Carolyn Ann McGraw, Aug. 1, 1971; children: Brian Charles, Matthew Edward. BA, Northeastern U., 1967; PhD, Boston Coll., 1972. Asst. in floriculture Waltham Exptl. Field Sta. U. Mass., 1963-64; lab. technician Mass. Soldiers Home, Chelsea, 1964-65; rsch. asst. New Eng. Med. Ctr. Hosps., Boston, 1965-67; asst. Cancer Rsch. Inst., Boston Coll., 1967-71; mem. faculty No. Essex Community Coll., Haverhill, Mass., 1971-85, prof. biology, 1985—; grant rev. panelist, NSF, 1976-80; chmn. Dept. Natural Scis. No. Essex Community Coll., 1988—; program coord., Eisenhower Title II Math and Sci. grant, 1989—. Author: (software) Biology in Action series, 1983, (with others) Principles of Biology, 2nd. ed., 1986, A Study and Laboratory Guide for Anatomy and Physiology, 2nd. ed., 1990. Eisenhower Title II Math. and Sci. grantee, 1989-90, 91-92, 92-93, 93-94, Nat. Edn. Act fellow Boston Coll., 1968-71; recipient citation for Outstanding Performance, Commonwealth of Mass., 1991. Mem. AAAS, Am. Soc. for Microbiology, N.Y. Acad. Scis., Mass. Assn. Sci. Tchrs., Mass. Assn. of Sci. Suprs. Office: No Essex Community Coll 100 Elliott St Haverhill MA 01830-2397

DE SÉGUIN DES HONS, LUC DONALD, physician, medical biologist; b. Paris, Nov. 4, 1919; s. Gabriel and Florence Louisa (Payne) de S. des H.; m. Macha Plaoutine, Dec. 12, 1952; children: Michel, Andre, Cyril. MD, U. Algiers (Algeria), 1943; biologist, Ordre Nat. Medecins, Paris, 1952. Master rsch. C.S.F. and Pasteur Inst., Paris, 1946-53; pres. Labs. Seguin, Drancy, France, 1948—; researcher biol. properties of microwaves French Acad. Scis., 1945-52. Patentee automatic analyzers. Served with Free French Forces, 1943-45. Mem. Union Syndicats Medicaux Region Parisienne (founder, pres. 1966-70), Union 93 (founder, pres. 1967—). Avocation: political philosophy.

DESER, STANLEY, educator, physicist; b. Rovno, Poland, Mar. 19, 1931. BS summa cum laude, Bklyn. Coll., 1949; MA, Harvard U., 1950, PhD, 1953; DPhil (hon.), Stockholm U., 1978. Mem. Inst. Advanced Study, Princeton, 1953-55, 93-94, Parker fellow, 1953-54; Jewett fellow Inst. for Advanced Study, Princeton, 1954-55; NSF postdoctoral fellow, mem. Inst. Theoretical Physics, Copenhagen, 1955-57; lectr. Harvard U., 1957-58; mem. faculty Brandeis U., Waltham, Mass., 1958—; prof. physics Brandeis U., 1965—, chmn. dept., 1969-71, 76-77, Ancell prof. physics, 1979—; vis. scientist European Center Nuclear Research, Geneva, 1962-63, 76, 80-81; mem. physics adv. com. NSF, 1982-86; Fulbright and Guggenheim fellow, vis. prof. Sorbonne, Paris, 1966-67, 71-72; Loeb lectr. Harvard U., 1975; S.R.C. sr. fellow Imperial Coll., 1976; vis. prof. College de France, Paris, 1976, 84; vis. fellow All Souls' Coll., Oxford (Eng.) U., 1977; investigator titular ad honorem CIDA (Venezuela), 1983. Mem. editorial bd. Jour. Geometry and Physics, Phys. Rev., Annals of Physics, Classical and Quantum Gravity; mem. sci. bd. I.H.E.S., France; chair sci. bd. Inst. Theoretical Physics, Santa Barbara. Fellow Am. Phys. Soc., Am. Acad. Arts and Scis. Spl. research on theoretical physics, field theory, gravitation. Office: Brandeis U Physics Dept Waltham MA 02254

DESHAZER, JAMES ARTHUR, agricultural engineer, educator, administrator; b. Washington, July 18, 1938; s. Grant Arthur and Velma (Morton) DeS.; m. Alice Marie Burton, Apr. 5, 1969; children: Jean Marie, David James. BS in Agr., U. Md., 1960, BSME, 1961; MS, Rutgers U., 1963; PhD, N.C. State U., 1967. Profl. engr., Idaho. Assoc. prof. U. Nebr., Lincoln, 1967-75, prof., 1975-91, asst. dean, 1988-89; head agrl. engring. dept. U. Idaho, Moscow, 1991—; chair animal care & use com. U. Nebr., 1989-90; program coord. North Cen. Sustainable Agrl., Washington, 1988-89; nat. chair Modeling Responses of Swine-CSRS, Washington, 1989-90, Systems Approach to Poultry Prodn.-CSRS, Washington, 1990-91. Contbr. chpt. to book; editor procs. Optics in Agr., 1990, Optics in Agr. & Forestry, 1992. Recipient Livestock Svc. award Walnut Grove, Iowa, 1988. Fellow Am. Soc. Agr. Engrs. (chair 1984-93, nat. medal 1979); mem. Am. Soc. for Engr. Edn. (chair 1993—), Nat. Soc. Profl. Engrs. (chpt. chair 1986-87, 93—, Young Engr. award 1974), Internat. Soc. Biometeorology, Lions. Home: 819 Nylarol St Moscow ID 83843 Office: Univ of Idaho Agr Eng Dept Moscow ID 83843

DESHPANDE, PRADEEP BAPUSAHEB, chemical engineer, educator, consultant; b. Hyderabad, A.P., India, Dec. 12, 1942; came to U.S., 1962; s. Bapusaheb Kishanrao and Subhadrabai (Deshpande) D.; m. Meena Pradeep Dhadphale, Dec. 6, 1967; children: Abhay Pradeep, Aneet Pradeep. BSc in Chemistry, Karnatak U., Dharwar, India, 1962; BS in Chem. Engring., U. Ala., 1965, MS in Chem. Engring., 1967; PhD in Chem. Engring., U. Ark., 1969. Registered profl. engr., Calif. Engr. Northrop Space Labs., Huntsville, Ala., 1965-67; asst. prof. U. Ark., Fayetteville, 1969-70, Howard U., Washington, 1970-72, Indian Inst. Tech., Kampur, India, 1972-73, Drexl U., Phila., 1973-74; control systems engr. Bechtel, Inc., San Francisco, 1974-75; prof. chem. engring. dept. U. Louisville, 1975—, dept. chmn., 1985-90; cons. E.I. du Pont de Nemours & Co., Wilmington, Del., Louisville, 1977-78, Exxon Chem. Co., Baton Rouge, La., 1987—; vis. scientist Nat. Chem. Lab., Pune, India, 1983-84; Dr. G.P. Kane vis. prof. U. Bombay dept. chem. tech., India; co-organizer U.S. India Conf. on Chem. Engring. Edn. for the future,

Bangalore, India, Jan. 1988. Author: (textbooks) Computer Process Control with Advanced Control Applications, 1983, rev. 2nd edit., 1988, Distillation Dynamics and Control, 1985, Multivariable Process Control, 1989; contbr. over 50 articles to profl. jours. including Chem. Engring. Edn., CACHE News, Indsl. Engring. Chem. Processing, Hydrocarbon Processing, Chem. Engring. Communications and others. Recipient Diamond Shamrock Corp. fellowship, 1968, Donald P. Eckman award in Process Control Edn., Instrument Soc. Am., 1990; finalist Kentuckiana Metroversity's Grawmeyer award in Instructional Devel., 1988. Mem. Am. Inst. Chem. Engrs. (bd. dirs. Louisville area 1976-77), Instrument Soc. Am. (sr. mem., assoc. dir. Automatic Control System Div. 1978-79, rep. summer computer simulation conf. 1976), Am. Chem. Soc., Sigma Xi. Achievements include development of new courses in process control, 4 new digital control algorithms. Office: U Louisville S Third St Louisville KY 40292

DE SIMONE, LIVIO DIEGO, diversified manufacturing company executive; b. Montreal, Que., Can., July 16, 1936; s. Joseph D. and Maria E. (Bergamin) De S.; m. Lise Marguerite Wong, 1957, children: Daniel J., Livia D., Mark A., Cynthia A. B.Chem. Engring., McGill U., Montreal, 957. With Minn. Mining & Mfg. Co., St. Paul; now exec. v.p. Minn. Mining & Mfg. Co. Office: Minn Mining & Mfg Co 3M Center Dr Saint Paul MN 55144

DESJARDINS, RAOUL, medical association administrator, financial consultant; b. Montreal, Quebec, Can., Oct. 8, 1933; came to U.S., 1962; s. Elso and Blanche (Lemieux) D.; m. Regina Turgeon, Oct. 10, 1961; 1 child, Bryan-Claude. BA, U. Montreal, 1953, MD, 1958; MS, Baylor U., 1964, PhD, 1966; MBA, Rutgers U., 1990. Diplomate Am. Bd. Medicine. Chief intern, resident St. Joan of Arc Hosp., Montreal, 1958-59; med. dir. Candiac (Can.) Med. Clinic, 1953-62, Ortho Research Found., Raritan, N.J., 1966-72; pres. Raoul Desjardins Assocs. Inc., Mendham, N.J., 1972-83, Research Cons. Inc., Mendham, 1983—, APG Internat., Inc., 1991—; med. dirs. Iroquois Class Co., Candiac, 1959-62; asst. prof. Hahnemann Hosp. and U., Phila., 1976-80; bd. govs. Internat. Medicines Exch. and Devel., Georgetown, Ga., 1986—; chmn. bd. advisors Fed. Inst. Health, 1991—; chmn. bd. govs. Grand Masters Found., 1989—. Recipient physician's recognition award AMA, 1969. Fellow Am. Coll. Angiology, The Royal Soc. Health, Am. Coll. Clin. Pharmacology, N.Y. Acad. Medicine; mem. Doctors Club, Met. Club (membership com. 1991—), Med. Execs. Club, Sigma Xi, Beta Gamma Omega. Roman Catholic. Avocations: safaris, medieval history, economic theory, tennis. Home and Office: Rsch Cons Inc 135 Talmage Rd Mendham NJ 07945-1508

DESOER, CHARLES AUGUSTE, electrical engineer; b. Ixelles, Belgium, Jan. 11, 1926; came to U.S., 1949, naturalized, 1958; s. Jean Charles and Yvonne Louise (Peltzer) D.; m. Jacqueline K. Johnson, July 21, 1966; children: Marc J., Michele M., Craig M. Ingenieur Radio-Electricien, U. Liege, Belgium, 1949, DSc (hon.), 1976; ScD in Elec. Engring, MIT, 1953. Rsch. asst. M.I.T., 1951-53; mem. tech. staff Bell Telephone Labs., Murray Hill, N.J., 1953-58; assoc. prof. elec. engring. and computer scis. U. Calif., Berkeley, 1958-62; prof. U. Calif., 1962-91, prof. emeritus, 1991—, Miller research prof., 1970-71. Author: (with L. A. Zadeh) Linear System Theory, 1963, (with E. S. Kuh) Basic Circuit Theory, 1969, (with M. Vidyasagar) Feedback Systems: Input Output Properties, 1973, Notes for a Second Course on Linear Systems, 1970, (with F. M. Callier) Multivariable Feedback Systems, 1982, (with L.O. Chua and E.S. Kuh) Linear and Nonlinear Circuits, 1987, (with A.N. Gündes) Algebraic Theory of Linear Feedback Systems with Full and Decentralized Compensation, 1990, (with F.M. Callier) Linear System Theory, 1991; contbr. numerous articles on systems and circuits to profl. jours. Served with Belgian Arty., 1944-45. Decorated Vol.'s medal; recipient Best Paper prize 2d Joint Automatic Control Conf., 1962, Univ. medal U. Liège, 1976, Disting. Teaching award U. Calif., Berkeley, 1971, Prix Montefiore Inst. Montefiore, 1975; award for outstanding paper IEEE, 1979, Field award in control sci. and engring., 1986, Am. Automatic Control Council Edn. award, 1983, Berkeley Citation, 1992; Guggenheim fellow, 1970-71. Fellow IEEE (Edn. medal 1975), AAAS; mem. Nat. Acad. Engring., Am. Math. Soc., Math. Assn. Am., Soc. Indsl. and Applied Math. Office: U Calif Dept Elec Engring and Computer Sci Berkeley CA 94720

DE SOFI, OLIVER JULIUS, data processing executive; b. Havana, Cuba, Dec. 26, 1929; came to U.S., 1956; naturalized, 1961; s. Julius A. and Edith H. (Zsuffa) DeS. B.S. in Math. and Physics, Ernst Lehman Coll., 1950; postgrad. in agronomy U. Havana, 1952, B.S. in Aero. Engring., 1956; m. Phyllis H. Dumich, Feb. 14, 1971; children: Richard D., Stephen R., Kerri L. Dir. EDP tech. svcs. and planning Am. Airlines, N.Y.C., 1968-70; dir. Sabre II, Tulsa, 1970-72; v.p. data processing and communications Nat. Bank of N. Am., Huntington Station, N.Y., 1972-76, sr. v.p. data processing and communications, 1976-78, sr. v.p. systems and ops., 1978-79, sr. v.p. adminstrn., N.Y.C., 1979-80, exec. v.p. adminstrn. group, 1980-83; exec. v.p. data processing methodologies and architecture Anacomp, Inc., Ft. Lee, N.J. and Sarasota, Fla., 1983-84; v.p. corp. devel. Computer Horizons Corp., N.Y.C., 1984-86; pres., chief exec. officer Coast to Coast Computers Inc., Sarasota, 1986-91; chief data processing cons. Arab Nat. Bank, Riyadh, Kingdom of Saudi Arabia; bd. dirs. The Bentley Group, San Francisco; lectr. program for women Adelphi Coll. Mem. Data Processing Mgmt. Assn., Computer Exec. Round Table, Am. Mgmt. Assn., Sales Execs. Club, Bank Adminstrn. Inst., AAAS, Internat. Platform Assn., Nat. Rifle Assn. Republican. Club: Masons (Havana).

DE SOTO, SIMON, mechanical engineer; b. N.Y.C., Jan. 8, 1925; s. Albert and Esther (Eskenazi) Soto; 1 dau., Linda Jane. B.M.E., CCNY, 1945; M.M.E., Syracuse U., 1950; Ph.D., UCLA, 1965. Lic. profl. engr., Calif., N.Y. Engr. Johns-Manville Corp., N.Y.C., 1946-48; instr. in engring. Syracuse U., 1948-50; research engr. Stratos-Fairchild Corp., Farmingdale, N.Y., 1950-54; research specialist Lockheed Missile Systems div. Lockheed Corp., Van Nuys, Calif., 1954-56; sr. tech. specialist Rocketdyne Rockwell Internat., Canoga Park, Calif., 1956-69; asso. prof. mech. engring. Calif. State U., Long Beach, 1969-72; prof. Calif. State U., 1972—; lectr. UCLA, 1954-70; cons. engr.; dir., sec.-treas. Am. Engring. Devel. Co.; mem. tech. planning com. Public Policy Conf.: The Energy Crisis, Its Effect on Local Govts., 1973; founding mem. Calif. State U. and Colls.; Statewide Energy Consortium and cons. tech. assistance program. Author: Thermostatics and Thermodynamics: An Instructor's Manual, 1963; author: research publs. in field. Served with U.S. Mcht. Marine, 1945-46. Recipient Outstanding Faculty award UCLA Engring. Student Body, 1962, Outstanding Faculty award Calif. State U., Long Beach, 1971, 73, 76, 89, 90. Mem. AAAS, SAG, Am. Soc. Engring. Edn. (prof. of recipients of Outstanding Design award 1990), Tau Beta Pi, Pi Tau Sigma. Avocation: acting on stage and film. Office: Calif State U Dept Mech Engring Long Beach CA 90840

DE SOUZA-MACHADO, SERGIO GUILHERME, physics research assistant; b. Nairobi, Kenya, Jan. 1, 1966; came to U.S., 1985; s. William Machado and Marie Antoinette De-Souza. BA, Coll. of Wooster, 1988; MS, U. Md., 1990. Rsch. intern dept. physics Ohio U., Athens, 1987; rsch. intern Materials Sci. Inst. U. Oreg., Eugene, 1988; rsch. grad. asst. Goddard Space Flight Ctr. NASA, Greenbelt, Md., 1991; grad. rsch. asst. lab. for plasma rsch. U. Md., College Park, 1990—. Contbr. articles to Phys. Rev. B, Am. Jour. Physics; co-programmer Computers in Physics simulation software, 1987. Treas. Newman Cath. Students Assn. Coll. Wooster, 1987-88, co-chair Soc. Physics Students, 1987-88. Mem. Am. Physics Soc. Roman Catholic. Home: 3429 Stanford St Hyattsville MD 20783 Office: U Md Dept Physics College Park MD 20742

DESROCHERS, GERARD CAMILLE, surgeon; b. Marlboro, Mass., June 8, 1922; s. Emery Hector and Eliane (Lemire) DesR.; m. Ellen Franklin, Sept. 27, 1959; children: Gerard, Emery, Lewis, Anthony. AB, Coll. of Holy Cross, 1944; MD, Tufts Coll., 1947. Diplomate Nat. Bd. Med. Examiners. Gen. rotating intern St. Mary's Hosp., Waterbury, Conn., 1947-48; teaching fellow in pathology Tufts Med. Sch., 1948-49; straight surg. intern Boston City Hosp., 1949-50, asst. resident surgeon, 1950-51; resident in surgery New Eng. Med. Center, Boston, 1955-57; practice medicine specializing in surgery, Manchester, N.H.; gen. surgeon staff Cath. Med. Center, Manchester; med. dir. Sea Supply Corp., Bangkok, Thailand, 1953-54; asst. chief surgery VA Hosp., Manchester, 1971-78. Contbr. articles to profl. jours. Incorporator Cath. Med. Ctr., Thomas More Found., Merrimack, N.H.; adv. bd. Lincoln Inst.; mem. N.H. Right to Life Com.; mem. bd. of policy Liberty Lobby; mem. New Eng. Med. Ethics Forum. Served as 1st lt. M.C., U.S. Army, 1970. Named Disting. Physician Am., 1989. Mem. AAAS, Manchester Med. Soc., Hillsboro County Med. Soc., Disting. Physicians Am., Internat. Coll. Physicians and Surgeons. Home: 402 Sagamore St Manchester NH 03104-3937 Office: 648 Belmont St Manchester NH 03104

DESSLER, ALEXANDER JACK, space physics and astronomy educator, scientist; b. San Francisco, Oct. 21, 1928; s. David Alexander and Julia (Shapiro) D.; m. Lorraine Hudek, Apr. 18, 1952; children: Pauline Karen, David Alexander, Valerie Jan, Andrew Emory. B.S., Calif. Inst. Tech., 1952; Ph.D., Duke, 1956. Sect. head Lockheed Missiles & Space Co., 1956-62; prof. Grad. Research Center, Dallas, 1962-63, prof. space physics and astronomy, 1963-82, 86—; chmn. dept. Rice U., Houston, 1963-69, 79-82, 87-92, campus bus. mgr., 1974-76; dir. space sci. lab. MSFC NASA, Huntsville, Ala., 1982-86; sci. adviser Nat Aeros. and Space Council, 1969-70; pres. Univs. Space Research Assn., 1975-81. Editor Jour. Geophys. Research, 1965-69, Revs. of Geophysics and Space Physics, 1969-74, The John Wiley Space Science Text Series, 1968-76, Geophys. Research Letters, 1986-89, Atmospheric and Space Science Series; adv. bd.: Planetary and Space Sci.; assoc. editor Space Solar Power Rev., 1980-85. Served with USN, 1946-48. Recipient Outstanding Young Scientist award Tex. wing Air Force Assn., 1963, medal for contbns. to internat. geophysics Soviet Geophys. Com., 1984, Stellar award for acad. devel., Rotary Nat., 1988. Fellow Am. Geophys. Union (Macelwane award 1963, John Adam Fleming medal); mem. Am. Astron. Soc., Internat. Assn. Geomagnetism and Aeronomy (v.p. 1979-83), Sigma Xi. Home: 5126 Loch Lomond Dr Houston TX 77096-2617 Office: Rice University PO Box 1892 6100 South Main Houston TX 77251

DE STASIO, ELIZABETH ANN, biology educator; b. Milw., Mar. 19, 1961; d. John Joseph and Meredith June (Russell) Dugan; m. Bart Thomas De Stasio Jr., July 30, 1983; 1 child, Matthew. BA, Lawrence U., 1983; PhD, Brown U., 1988. Asst. prof. dept. biology Lawrence Univ., Appleton, Wis., 1988-89, 92—; postdoctoral fellow dept. genetics U. Wis., Madison, 1989-92. Contbr. chpts. to books and articles to EMBO Jour., Jour. Molecular Biology, Biochimica Biophysica Acta, Biochimie. Recipient cash award rsch. competition Am. Soc. Microbiology, Boston, 1988; postdoctoral fellow NIH, 1990, 93, Am. Cancer Soc., 1990, Robert G. Sampson, 1990, NSF, 1993; NIH grantee 1993. Mem. Sigma Xi, Phi Beta Kappa, Phi Sigma. Unitarian Universalist. Office: Lawrence U Dept Biology Appleton WI 54911

DESTEFANO, PAUL RICHARD, optical engineer; b. Chgo., Nov. 7, 1967; s. James and Sara (Galfano) DeS. BS in Physics, Rose-Hulman, 1990, MS in Applied Optics, 1992. Asst. physics technician Rose-Hulman, Terre Haute, Ind., 1986-89, lab. technician, 1989-90, adj. prof., 1991-92; student engr. Hughes Optical Products, Inc., Des Plaines, Ill., 1990; mem. tech. staff GCA Tropel, Fairport, N.Y., 1992—. Mem. Soc. Photo Instrument Engrs., Am. Mensa Ltd. Republican. Roman Catholic. Achievements include research in design review and characterization of an infrared phase-shifting interferometer. Home: 1587-4 Elmwood Ave Rochester NY 14620 Office: GCA Tropel 60 O'Connor Rd Fairport NY 14450

DETELS, ROGER, epidemiologist, physician, former university dean; b. Bklyn., Oct. 14, 1936; s. Martin P. and Mary J. (Crooker) D.; m. Mary M. Doud, Sept. 14, 1963; children: Martin, Edward. BA, Harvard U., 1958; MD, NYU, 1962; MS in Preventive Medicine, U. Wash., 1966. Diplomate Am. Bd. Preventive Medicine. Intern U. Calif. Gen. Hosp., San Francisco, 1962-63; resident U. Wash., Seattle, 1963-66; med. officer, epidemiologist Nat. Inst. Neurol. Diseases, Bethesda, Md., 1969-71; assoc. prof. epidemiology Sch. Pub. Health UCLA, 1971-73, prof. Sch. Pub. Health, 1973—, dean, 1980-85, head div. epidemiology Sch. Pub. Health, 1972-80; guest lectr. various univs., profl. confs. and med. orgns., 1969—; mem. sci. adv. com. Am. Found. AIDS Rsch.; dir. UCLA/Fogarty Internat. Tng. Program in HIV/AIDS, 1988—; cons. Ministries of Health, Thailand, Myanmar, The Philippines, 1989—, St. Thomas' Med. Sch., London, 1993; mem. Nat. Adv. Environ. Health Scis. Coun., 1990—. Editor: Oxford Textbook of Public Health, 1985, 2d edit. 1991; contbr. articles to profl. jours. Lt. comdr. M.C. USN, 1966-69. Grantee in field. Fellow Am. Coll. Preventive Medicine, Am. Coll. Epidemiology (coun. 1987-89), (hon.) Faculty Pub. Health Medicine Royal Coll. Physicians of United Kingdom, 1992; mem. AAAS, Am. Epidemiol. Soc., Soc. Epidemiologic Resch. (pres. 1977-78), Assn. Tchrs. Preventive Medicine, Calif. Acad. Preventive Medicine (chmn. essay com. 1971), Am. Pub. Health Assn., Am. Assn. Cancer Edn. (membership com. 1979-80), Internat. Epidemiol. Assn. (treas. 1984-90, pres. 1990—), Assn. Schs. Pub. Health (sec.-treas. 1984-85), Sigma Xi, Delta Omega. Office: UCLA Sch Public Health Los Angeles CA 90024

DE THIBAULT DE BOESINGHE, LÉOPOLD BARON, physician; b. Gent, Belgium, Sept. 4, 1943; married; children: Isabel, Fernando. MD. Dir. Dept. of Occupational Health U. Gent; prof. radiotherapy and nuclear medicine Univ. Clinic of Gent; chief dept. Aalsters Stedelijk Ziekenhuis ASZ. Contbr. articles to profl. jours. Mem. Assoc. Europ. de Thermologie (past pres.), Provincial Chamber of the Bd. of Physicians of Oost-Vlaanderen (past pres.), Internat. Coll. of Thermology (past pres.), Oost Vlaanderen Belgisch Werk Tegen Kanker, Belgian Assn. of Radiationprotection (past pres.), Gand Club, Sté des redoutés Club, Univ Found. Club, Flemisch Ligue against Cancer O.VL. BWK. (pres.). Avocation: photography. Home: Sint-Martensstraat 10, B9000 Ghent Belgium Office: Univ Ziekenhuis, De Pintelaan 185, B9000 Ghent Belgium

DETLEFSEN, WILLIAM DAVID, JR., chemist, administrator; b. Scottsbluff, Nebr., Nov. 14, 1946; s. William David Sr. and Janette Fern (Tuttle) D.; m. Melba Kay Cunningham, Nov. 12, 1982; children: Michael David, Erika Lee, Whitney Anne. BS in Forestry, U. Idaho, 1970; PhD in Chemistry, U. Oreg., 1993. Chemist, applied technologist Borden, Adhesives and Resins, Springfield, Oreg., 1972-76, coord. tech. svc., 1976-78, supr. phenolic resins divsn., 1983-87, mgr. rsch. and devel., 1987—; sr. divsn. chemist Ga.-Pacific Resins, Crossett, Ark., 1978-83. Contbr. articles to sci. jours. 1st. lt. U.S. Army, 1970-72, Germany. Mem. AAAS, Am. Chem. Soc., Forest Products Rsch. Soc. Republican. Achievements include 3 U.S. patents; co-discoverer first commercially feasible resins for gluing high moisture veneers into phenolic-bonded plywood. Office: Borden Adhesives & Resins 610 S 2nd St Springfield OR 97477-5398

DETOFSKY, LOUIS BENNETT, secondary education educator; b. Phila., Nov. 1, 1944; s. Milton and Fae Minerva (Familant) D. BA, Rutgers U., 1968; MA, Glassboro (N.J.) State Coll., 1976. Cert. secondary edn. tchr., N.J. Instr. geosci., biology and anthropology Washington Twp. High Sch., Sewell, N.J., 1968—; club advisor Washington Twp. High Sch. Geology Club, Sewell, 1968-93; advisor high adventure geology Southern N.J. Coun. Author: Traversing New Jersey's Geology, 1976, Hawaiian Genesis, 1984, Our Appalachian Heritage, 1985, Jewels of the West Part I: Rocky Mountains, 1987, Jewels of the West Part II: Cascades, 1988. Advisor Explorer Post 8, Boy Scouts Am., 1986-93; sec. B'nai B'rith, Gloucester County, N.J., 1987-93. Mem. AAAS, Nat. Assn. Geology Tchrs., Nat. Earth Sci. Tchrs. Assn., Geol. Assn. of N.J., N.J. Earth Sci. Tchrs. Assn., Nat. Sci. Tchrs. Assn., Delaware Valley Paleontol. Soc., Delaware Valley Earth Sci. Soc. Jewish. Home: 12 Bedford Ter Turnersville NJ 08012-2102 Office: Washington Twp High Sch Hurffville-Cross Keys Rd Sewell NJ 08080

DEUBLE, JOHN L., JR., environmental science and engineering services consultant; b. N.Y.C., Oct. 2, 1932; s. John Lewis and Lucille (Klotzbach) D.; m. Thelma C. Honeychurch, Aug. 28, 1955; children: Deborah, Steven. AA, AS in Phys. Sci., Stockton Coll., 1957; BA, BS in Chemistry, U. Pacific, 1959. Cert. profl. chemist, profl. engr., environ. inspector; registered environ. profl., registered environ. assessor. Sr. chemist Aero-Gen Corp., Sacramento, Calif., 1959-67; asst. dir. rsch. Lockheed Propulsion Co., Redlands, Calif., 1968-73; asst. div. mgr. Systems, Sci. and Software, La Jolla, Calif., 1974-79; gen. mgr. Wright Energy Nev. Corp., Reno, Nev., 1980-81; v.p. Energy Resources Co., La Jolla, 1982-83; dir. hazardous waste Aerovironment Inc., Monrovia, Calif., 1984-85; environ. cons. Encinitas, Calif., 1986-88; sr. program mgr. Ogden Environ. and Energy Svcs., San Diego, 1989—. Contbr. articles profl. jours. With USAF, 1951-54. Recipient Tech. award Am. Ordnance Assn., 1969, Cert. of Achievement Am. Men and Women of Sci., 1986, Envrion. Registry, 1992. Fellow Am. Inst. Chemists; mem. ASTM, Am. Chem. Soc., AM. Inst. Chem. Engrs., Am. Meteorol. Soc., Am. Nuclear Soc., Am. Def. Preparedness Assn., Air and Waste Mgmt. Assn., Calif. Inst. Chemists, Hazardous Materials Control Rsch. Inst., N.Y. Acad. Scis., Environ. Assessors Assn. Republican. Lutheran. Achievements include development and pioneering use of chemical (non-radioactive) tracers--gaseous, aqueous, and particulate in environmental and energy applications. Home: 369 Cerro St Encinitas CA 92024-4805 Office: Ogden Environ & Energy Svcs 5510 Morehouse Dr San Diego CA 92121-3720

DEUTCH, JOHN MARK, federal official, chemist, academic administrator; b. Brussels, Belgium, July 27, 1938; came to U.S., 1940, naturalized, 1946; s. Michael Joseph and Rachel Felicia (Fisher) D.; children—Philip, Paul, Zachary. B.A., Amherst Coll., 1961, D.Sc. and Humane Letters (hon.), 1978; B. Chem. Engring, M.I.T., 1961, Ph.D. in Phys. Chemistry, 1965; D.Litt. (hon.), U. Lowell, 1986. System analyst Office Sec. Def., 1961-65; fellow Nat. Acad. Scis./NRC, Nat. Bur. Standards, 1966-67; asst. prof. Princeton U., 1967-70; mem. faculty MIT, 1970—, prof. chemistry, 1971—, chmn. dept., 1976, dean sci., from 1982, provost, 1982-90, inst. prof., 1990—; under sec. for acquisition and technology Dept. of Defense, Washington, DC, 1993—; chmn. adv. panel on chemistry NSF, 1974; mem. Def. Sci. Bd., 1977—, Pres.'s Nuclear Safety Oversight Com.; dir. Office Energy Rsch., U.S. Dept. Energy, Washington, 1977-79, acting asst. sec. for energy tech., 1979, under sec., 1979-80; mem. Army Sci. Adv. Panel, 1975-78, Pres.'s Commn. on Strategic Forces, 1983, The White House Sci. Coun., 1985-89; mem. Pres.'s Fgn. Intelligence Adv. Bd., 1990—. Author research articles. Sloan fellow, 1969-71; Guggenheim fellow, 1974. Mem. Am. Phys. Soc., Am. Chem. Soc., Council Fgn. Relations, Am. Acad. Arts and Scis. Avocations: tennis, squash, reading. Office: Under Sec for Acquisition Dept of Defense The Pentagon Washington DC 20301

DEUTSCH, CLAUDE DAVID, physicist, educator; b. Paris, July 20, 1936; s. David and Caroline (Petrover) D.; m. Nimet Elabed, July 9, 1962; children: Alain, Eric. Degree in chem. engring., Ecole Nat. Supérieure, Paris, 1959; M in Theoretical Physics, U. Paris XI, Orsay, 1961; DSc, U. Paris VI, 1969. Mem. staff Nuclear Physics Lab., Centre Nat. de la Recherche Sci., Orsay, 1959-60, Inst. Henri Poincaré, Paris, 1960-63; engr. CEA-EURATOM Nuclear Energy Ctr., Fontenay-aux-Roses, France, 1965-71; chief rsch. Plasma Physics Lab., Centre Nat. de la Recherche Sci., Orsay, 1973-80, dir., 1980-85, head rsch., 1983—; vis. scientist U. Montreal, Can., 1973, U. Gainesville, Fla., 1974, MIT, Cambridge, Mass., 1976-78, Stanford U., Palo Alto, Calif., 1980, 83; vis. prof. Okayama (Japan) U., 1978-86, 89, ICTP, Trieste, Italy, 1981, Weizmann Inst., Rehovot, Israel, 1983, GSI, Darmstadt, Germany, 1987; dir. Paris-Sud Info., Orsay, 1985-93; dir. Ion-Plasma Interaction Rsch. Gathering, CNRS, Orléans, Orsay, 1989—; mem. Nat. Com. on Plasma Sci., 1982; mem. editorial bd. Jour. Physique, 1985-88. Editor proc. for internat. confs. in field; contbr. articles to profl. publs. Fellow Japan Soc. Promotion of Sci., 1978, 86, 89, NATO, 1980. Recipient bronze and silver C.N.R.S. medals. Fellow French Phys. Soc. (prizes com. 1979-83, mem. coun. 1979-83); mem. Am. Phys. Soc., European Com. Heavy Ion Fusion, 1993. Avocations: jogging, piano. Home: 6 Ave Marie-Thérèse, 91400 Orsay France Office: Plasma Physics Lab, Bât 212, 91405 Orsay France

DEUTSCH, HANS-PETER WALTER, physicist; b. Bad Kreuznach, Fed. Republic Germany, Aug. 11, 1962; s. Johann Rudolf and Irmgard (Lukas) D.; m. Michaela Dorothea Brauburger, Apr. 3, 1992. Diploma, U. Mainz, 1989, doctorate summa cum laude, 1992. Software devel. Siemens Ag, Munich, 1986, Cardiology of Klinikum, Mainz, 1987-89; rsch. asst. U. Mainz, 1989-91; rschr./physics U. Forschungsanlage KFA, Juelich, 1991-92; cons. Andersen Cons., Arthur Andersen & Co., S.C., Frankfurt, 1992—; rschr. HLRZ-Grossforschungs project, 1992. Exhibits in group photography shows, Mainz, 1988—; contbr. aticles to profl. jours. including Jour. of Chem. Physics, Jour. of Non Crystalin Solids, Jour. of Statis. Physics, Macromolecules, Jour. de Physique, and Europhysics Letters. Mem. FIUTS, Seattle, 1986-87. Scholar Deutscher Akademischer Auslandsdienst, Bonn, 1986. Mem. Deutsch Physikalische Gesellschaft. Home: Dietrich Bonhoeffer Str 7, D55131 Mainz Germany Office: Andersen Consulting, Otto Volger Str 15, D-65843 Sulzbach Frankfurt, Germany

DEUTSCH, THOMAS FREDERICK, physicist; b. Vienna, Austria, Apr. 24, 1932; came to U.S., 1939; s. George and Sabina (Edel) D.; m. Judy Foreman, May 5, 1990. B. Engring Physics, Cornell U., 1955; AM, Harvard U., 1956, PhD, 1961. Prin. rsch. scientist Raytheon Co., Lexington, Mass., 1960-74; staff mem. Mass. Inst. Tech., Lexington, 1974-84; physicist Mass. Gen. Hosp., Boston, 1984—; assoc. prof. Harvard Med. Sch., Boston, 1987—. Contbr. articles to profl. jours.; patentee in field. Fellow Am. Phys. Soc., Optical Soc. Am. (Wood prize 1991), Am. Soc. for Lasers in Medicine; mem. IEEE (sr.). Office: Wellman Labs 50 Blossom St 2d Fl Boston MA 02114-2600

DEVADOSS, CHELLADURAI, physical chemist; b. Nazareth, India, Dec. 9, 1950; came to U.S., 1985; s. Samadhanam and Thnaam (Devairakkam) C.; m. Mallimalar Solomon, Sept. 4, 1978; children: Evangeline, Duraisingh. MSc, U. Madras, India, 1972; PhD, U. Notre Dame, 1990. Demonstrator in chemistry Govt. Arts Coll., Vriddhachalam, India, 1973-75; asst. prof. chemistry Pachiyappa's Coll. for Men, Kanchipuram, 1975-81, Pachaiyappa's Coll., Madras, 1981-85; grad. rsch. asst. U. Notre Dame, Ind., 1985-90; postdoctoral rsch. assoc. Ariz. State U., Tempe, 1990-91, U. Ill., Urbana-Champaign, 1991—. Contbr. articles to profl. jours. including Jour. Phys. Chemistry, Chem. Phys. Letters, Jour. of am. Chem. Soc., and Inorganic Chemistry. Mem. Am. Chem. soc., Inter-Am. Photochem. Soc. Home: 1906B Orchard St Urbana IL 61801 Office: U Ill Dept Chemistry Mail 32-5 Urbana IL 61801

DEVAULT, WILLIAM LEONARD, orthopedic surgeon; b. Elizabethton, Tenn., Dec. 4, 1950; s. Robert Moulton and Betty (Cloyd) DeV.; m. Anna Lee Homewood, Apr. 12, 1986; children: Jonathan, Elizabeth. BS in Chemistry, Birmingham-So. Coll., 1979; MD, Wake Forest U., 1983. Diplomate Nat. Bd. Med. Examiners. Resident in orthopaedic surgery U. S.C., Columbia, 1983-88; pvt. practice, Beaufort, S.C., 1988-89, Fredericksburg, Va., 1989-91, Clemson, S.C., 1991—; fellow in sports medicine and reconstructive surgey knee-hip Jewett Orthopaedic Clinic, Orlando, Fla., 1991—. Mem. Am. Acad. Orthopaedic Surgeons (candidate), Am. Running and Fitness Assn. (cons. 1989—). Baptist. Avocation: restoring automobiles. Office: Blue Ridge Orthopaedic Asso 1101 Tiger Blvd Clemson SC 29631

DEVEAUGH-GEISS, JOSEPH, psychiatrist; b. Rochester, N.Y., Apr. 7, 1946; s. Joseph Paul and Gilda (Sposato) Geiss; m. Joanne DeVeaugh, Aug. 17, 1974; children: Diana, Angela, Joseph, Victoria. BA, Syracuse U., 1968; MD, SUNY, Syracuse, 1972. Diplomate Nat. Bd. Med. Examiners, Am. Bd. Psychiatry and Neurology. From asst. to assoc. prof. Health Sci. Ctr. SUNY, Syracuse, 1975-85, rsch. assoc. prof. psychiatry Health Sci. Ctr., 1986—; dir. Ctrl. Nervous System clin. rsch. Ciba-Geigy Pharms., Summit, N.J., 1985-90, Glaxo Inc., Research Triangle Park, N.C., 1990—; assoc. cons. prof. psychiatry Med. Ctr. Duke U., Durham, N.C., 1991—; jr. asst. attending psychiatrist Upstate Med. Ctr., State U. Hosp., 1975-77, asst. attending psychiatrist, 1977-83, attending psychiatrist, 1983-85; staff physician Syracuse VA Med. Ctr., 1975-85, dir. Tardive Dyskinesia Clinic, 1978-85, acting chief psychiatry svc., 1984-85; cons. Rome (N.Y.) Devel. Ctr., 1978-82; chmn. Free Communications Sect. in Psychopharmacology, VI World Congress of Psychiatry, 1977; invited expert witness N.Y. State Assembly Com. on Mental Health, 1978. Jour. reviewer, referee Psychosomatics, 1984, Jour. Nervous and Mental Disease, 1984, Jour. Clin. Psychiatry, 1989, Life Scis., 1990, Psychopharmacology, 1991; cons., advisor on antipsychotic drug side effects N.Y. Times, CBS Evening News; author presentations, abstracts, jour. articles, audiovisual prodns. in field. Asst. mgr. Westfield (N.J.) Baseball League, 1989-90; scoutmaster pack #673 Cub Scouts, Westfield, 1989-90. Fellow Am. Psychiatric Assn. (Meet The Experts panelist Ann. Meeting 1983, chmn. paper session Adverse Effects of Neuroleptic Treatment 1985); mem. AMA, AAAS, Internat. Psychogeriatric Assn., N.Y. Acad. Scis., Soc. Biol. Psychiatry, Drug Info. Assn., Alpha Epsilon Delta. Achievements include development of first drug approved in

U.S. for treatment of obsessive compulsive disorder. Office: Glaxo Inc 5 Moore Dr Research Triangle Park NC 27709

DEVEREAUX, CHRISTIAN WINDSOR, III, computer scientist; b. Washington, Dec. 1, 1950; s. Phillip Ranier and Victoria Ann (Honfleur) D.; m. Katherine Louise Pitts, Sept. 22, 1990. MBA, Harvard U., 1974; PhD in Computer Sci., Stanford U., 1977. V.p. engring. svcs. Computer Pointe Corp., San Carlos, Calif., 1979-82; pres., gen. mgr. Advanced Computer Tech., Chgo., 1983-87; sr. v.p. advanced techs. C3I Techs. Inc., Key West, Fla., 1987—. Co-author: The Logical Computer and the Cognitive Mind, 1987; contbr. articles to profl. jours. State chair Cath. Homes for Children, Fla., 1990. Recipient Computer Security award Gibbling/Babbage Found., 1989, Advanced Computing in AI award UN Sci. Found., 1985, Applied Bus. Computing award IBM Found. for Applied Sci., 1982, Applied Computing in Physics award Max Planck Ctr. for Applied Physics, 1975. Fellow Nat Sci. Found., Nat. Acad. Scis. (Advances in AI award 1989), U.S. Naval Inst. (adv. bd. 1990-92), Mensa Soc. Am. Republican. Roman Catholic. Achievements include more than 150 patents, most dealing with massively parallel processors and distributed computing. Office: C3I Corp Drawer 4986 Key West FL 33041-4968

DEVEREAUX, WILLIAM A., engineer; b. Georgetown, Ont., Can., 1915. BASc in Mech. Engring., U. Toronto, Can., 1937. Engr., mgr. to dist. mgr. Bailey Meter Co. Ltd., Can., 1937-60; gen. mgr. Yarway Can. Ltd., Guelph, Ont., Can., 1961-62; mgr. Can. bus. devel. Montreal Engring. Co. Ltd., Can., 1962-72; mgr. Atlantic region Montreal Engring. Co. Ltd., Halifax, Can., 1972-75; mgr. profl. affairs divsn. Montreal Engring. Co. Ltd., Montreal, Can., 1975-82. Recipient John B. Sterling medal Engring. Inst. Can., 1992. Mem. Order of Engrs. Quebec, Assn. Profl. Engrs Ont., Am. Soc. Mech. Engrs., Engring. Inst. Can. (hon. treas., fin. com., chmn. life mem. group). Office: care Engring Inst of Canada, 202 280 Albert St, Ottawa, ON Canada K1P 5G8*

DEVEREUX, OWEN FRANCIS, metallurgy educator; b. Lexington, Mass., Aug. 23, 1937; s. George Francis and Mildred Anna (Gleeson) D.; m. Sally Williamson, June 15, 1957 (div. June 1969); children: Owen M., Amy L., Jonathan W., Nancy J.; m. Olivia Elaine Marin, June 13, 1969. BS, MIT, 1959, MS, 1960, PhD, 1962. Research chemist Chevron Research Co., La Habra, Calif., 1962-64, Corning (N.Y.) Glass Works, 1964-66, Chevron Oil Field Research Co., La Habra, 1966-68; assoc. prof. U. Conn., Storrs, 1968-76, prof., 1976—, head metallurgy dept., 1983—. Author: Topics in Metallurgical Thermodynamics, 1983; contbr. articles to profl. jours. Rsch. grantee NSF, 1970-76, U.S. Dept. Energy, 1976-86, NSF Industry/Univ. Corp. Rsch. Ctr. for Grinding Rsch. and Devel., 1990—. Mem. AIME, AAUP, Electrochem. Soc. (div. editor 1987-90), Nat. Assn. Corrosion Engrs. Avocations: raising quarter horses, carriage driving and restoration, saddle making. Home: PO Box 391 Storrs Mansfield CT 06268-0391 Office: Univ of Conn Box U-136 97 N Eagleville Rd Storrs CT 06268

DEVGAN, ONKAR DAVE N., technologist, consultant; b. Lahore, Panjab, India, Oct. 11, 1941; came to U.S., 1967; s. Thakar Dass Devgan and Sohag Wati Sharma; m. Veena Devgan, July 20, 1969; children: Sanjay, Pooja. BS, Panjab U., 1960; MS, Vikham U., 1963, PhD, 1966; MBA, Temple U., 1975. Instr., rsch. assoc. U. Pa., Phila., 1970-73; scientist C.E. Glass, Pennsauken, N.J., 1973-76; cons., vis. prof. U. Tex., Dallas, 1976-78; mgr. material devel., sr. engr. Tex. Inst., Dallas, 1978-83; engring. mgr. Fairchild Semiconductor, Palo Alto, Calif., 1983-84, 88; program mgr. Varian Assocs., Palo Alto, 1984-86; dir. microelectronics Northrup Corp., L.A., 1986-88; dir. tech. and ops. Polylithics Inc., Santa Clara, Calif., 1989-90; tech. and mgmt. cons. Devgan Assocs., Sunnyvale, Calif., 1991—; co-founder, pres. Paragon System Tech.; co-chmn. Semi GaAs Com., Mt. View, Calif., 1984-85; mem. Semi Automation Com., Mt. View, 1984-86; advisor Semi Equipment Uptime Com., Mt. View, 1985-86; presenter high tech. presentations to various corps., 1975-91; chmn. process monitoring and characterization session Internat. Semiconductor Mfg. Sci. Symposium, San Francisco, 1993. Contbr. articles to tech. and bus. jours. PhD fellow Govt. of India, 1963-66, Coun. of Sci. and Indsl. Rsch. sr. fellow, 1966-67; NIH postdoctoral fellow, 1967-70. Mem. IEEE, Am. Chem. Soc. Achievements include inventions in key proprietary tech. in the areas of infra-red transmitting materials and semiconductor processing. Home and Office: 161 Butano Ave Sunnyvale CA 94086-7025

DEVILBISS, JONATHAN FREDERICK, product marketing administrator; b. Saiburi, Pattani, Thailand, July 23, 1961; s. Frederick Henry and Iva Marie (Weidner) D. BS in Aero. Engring., Purdue U., 1984; BA in Liberal Arts, Wheaton (Ill.) Coll., 1984. Sales engr. Brit. Aerospace Inc., Herndon, Va., 1985-88, tech. sales engr., 1988-89, sr. tech. sales engr., 1989-91, sr. product engr., 1991-92; mgr. product mktg. Jetstream Aircraft subs. Brit. Aerospace, Washington, 1993—. Mem. AIAA, SAE (assoc.). Republican. Evangelical Christian. Home: 4906 Erie St Annandale VA 22003-5410 Office: Jetstream Aircraft Inc PO Box 16029 Washington DC 20041-6029

DEVINE, KATHERINE, publisher, environmental consultant; b. Denver, Oct. 15, 1951. BS, Rutgers U., 1973, MS, 1980; postgrad., U. Md., 1981-82. Lab. technician Princeton (N.J.) U., 1974-76; econ. and regulatory affairs analyst, program mgr. U.S. EPA, Washington, 1979-81, 82-89, cons., 1989—; exec. dir. Applied BioTreatment Assn., Washington, 1990-91; pres. DEVO Enterprises, Inc., Washington, 1990—; chair adv. bd. Applied Bioremediation Conf., 1993. Author: N.J. Agricultural Experiment Station of Rutgers University, 1980, Bioremediation Case Studies: An Analysis of Vendor Supplied Data, 1992, Bioremediation Case Studies: Abstracts, 1992; co-author: Biomediation and Field Experiences; founder, pub., editor (newspaper) Biotreatment News, 1990—; contbr. articles to profl. jours., chpts. to books. Mem. Women's Coun on Energy and the Environment. Recipient numerous fed. govt. and non- govt. awards. Mem. NAFE, Am. Chem. Soc., Met. Washington Environ. Profls., Futures for Children, Alpha Zeta. Office: DEVO Enterprises Inc 704 9th St SE Washington DC 20003-2804

DEVINEY, MARVIN LEE, JR., research institute scientist, program manager; b. Kingsville, Tex., Dec. 5, 1929; s. Marvin Lee and Esther Lee (Gambrell) D.; m. Marie Carole Massey, June 7, 1975; children: Marvin Lee III, John H., Ann-Marie K. BS in Chemistry and Math., S.W. Tex. State U., San Marcos, 1949; MA in Phys. Chemistry, U. Tex., Austin, 1952, PhD in Phys. Chemistry, 1956. Cert. profl. chemist. Devel. chemist Celanese Chem. Co., Bishop, Tex., 1956-58; rsch. chemist Shell Chem. Co., Deer Park, Tex., 1958-66; sr. scientist, head group phys. and radio-chemistry Ashland Chem. Co., Houston, 1966-68, mgr. sect. phys. and analytical chemistry, 1968-71; mgr. sect. phys. chemistry div. rsch. and devel. Ashland Chem. Co., Columbus, Ohio, 1971-78; rsch. assoc., supr. applied surface chemistry Ashland Ventures Rsch. and Devel., Columbus, 1978-84, supr. electron microscopy, advanced aerospace composites, govt. contracts, 1984-90; inst. scientist, mem. internal R & D com. SW Rsch. Inst., San Antonio, Tex., 1990—; adj. prof. U. Tex., San Antonio, 1973-75, Ohio State U., 1990-91; mem. sci. adv. bd. Am. Petroleum Inst. Rsch. Project 60, 1968-74. Contbr. numerous articles to profl. jours.; patentee in field. Mem. ednl. adv. com. Columbus Tech. Inst., 1974-84, Cen. Ohio Tech. Coll., 1975-82, Hocking Tech. Coll., 1989-91. Lt. col., USAR, retired. Humble Oil Rsch. fellow, 1954. Fellow Am. Inst. Chemists (pres. Ohio Inst. 1978-82); mem. Ohio, Tex. acads. scis., Am. Def. Preparedness Assn., Electron Microscopy Soc. Am., Materials Rsch. Soc., SAMPE Composite Soc., N.Am. Catalysis Soc., Am. Soc. Composites, Soc. Plastics Engrs., Soc. Automotive Engrs., Am. Chem. Soc. (chmn. chpt. exec. bd. 1969, bus. mgr. nat. div. Petroleum Chemistry, 1986-90, Best Paper award rubber div. 1967, 70, Hon. Mention awards 1968, 69, 73, symposia co-chmn., co-editor books on catalysis-surface chemistry 1985, carbon-graphite chemistry 1975), Engr.'s Coun. Houston (sr. councilor 1970-71), Sigma Xi, Phi Lambda Upsilon, Alpha Chi, Sigma Pi Sigma. Home: 15934 Alsace San Antonio TX 78232-2790 Office: SW Rsch Inst PO Box 28510 San Antonio TX 78228-0510

DEVITA, VINCENT THEODORE, JR., oncologist; b. Bronx, N.Y., Mar. 7, 1935; s. Vincent Theodore and Isabel DeV.; m. Mary Kay Bush, Aug. 3, 1957; children: Teddy (dec.), Elizabeth. BS, Coll. William and Mary, 1957; MD, George Washington U., 1961; DSc (hon.), N.Y. Med. Coll., 1987, Georgetown U., 1989. Diplomate: Nat. Bd. Med. Examiners, Am. Bd. Internal Medicine (subspecialty hematology, med. oncology). Intern U. Mich. Med. Center, Ann Arbor, 1961-62; resident in medicine George Washington U. Med. Service D.C. Gen. Hosp., 1962-63; clin. assoc. Lab. Chem. Pharmacology, Nat. Cancer Inst. NIH, Bethesda, Md., 1963-65; sr. resident in medicine Yale New Haven Med. Center, 1965-66; sr. investigator solid tumor service, medicine br. Nat. Cancer Inst. NIH, 1966-68, head solid tumor service, medicine br., 1968-71, chief med. br., 1971-74, dir. div. cancer treatment, 1974-80, clin. dir. inst., 1975-80; dir. Nat. Cancer Inst., Nat. Cancer Program, NIH, 1980-88; physician-in-chief Meml. Sloan-Kettering Cancer Ctr., N.Y.C., 1988-91, attending physician, mem., 1988-93, Benno C. Schmidt chair clin. oncology, 1988-93; prof. medicine Cornell U. Med. Coll., 1989-93; dir. Yale Comprehensive Cancer Ctr., New Haven, 1993—; prof. medicine Yale U. Sch. Medicine, New Haven, 1993—; assoc. prof. medicine George Washington U. Med. Sch., 1971-75, prof. medicine, 1975-89; vis. physician Rockefeller U. Hosp., 1989-93; mem. expert advisory panel WHO, 1976—; mem. Lasker Award Jury, 1974—; chmn. Com. French-Am. Agreement on Cancer Treatment Research, 1976—; vis. prof. Stanford U. Med. Sch., 1972; 1st ann. Clowes lectr. Roswell Park Meml. Inst. Buffalo, 1973; mem. sci. com. 4th Internta. Congress on Anti-Cancer Chemotherapy, 1991—; mem. nat. adv. com. Tobacco-Related Disease Rsch. Program State of Calif., 1991—; mem. adv. bd. Stop Cancer, 1991—; mem. sci. com. Italian-Am. Found. For Cancer Rsch., 1991—; bd. dirs. Imclone Systems Inc. Mem. editorial bd. Cancer Research, 1981-91, Gynecologic Oncology, 1981-91, Hematological Oncology, 1981—, Physicians' Drug Alert, 1982—, Cancer Investigation, 1982—, Jour. Clin. Oncology, 1983—; assoc. editor Online Jour. Current Clin. Trials, 1991—, Cancer Investigation, 1983-87, Am. Jour. Medicine, 1983—; mem. extramural bd. assoc. editors Physicians Desk Query (PDQ), Nat. Cancer Inst., 1989—; mem. editorial bd. or adv. editor numerous other med. jours.; contbr. numerous articles to med. jours. Mem. awards assembly Gen. Motors Cancer Research Found., 1981-85, adv. council, 1984—; mem. Armand Hammer Cancer Award Com., 1983—. Served with USMCR, 1955-61. Tobacco Research Industry fellow, 1959; decorated Oren del Sol en el Grando de Official, Govt. of Peru, 1970; recipient Albert and Mary Lasker Med. Research award, 1972; Superior Service award HEW, 1975; Esther Langer Found. award, 1976; Alumni medallion Coll. William and Mary, 1976; Jeffrey Gottlieb award, 1976; Bronze medal Am. Soc. Therapeutic Radiology, 1978, Karnofsky prize and lecture, 1979, Griffuel prize Assn. for Devel. Research on Cancer, 1980, James Ewing award Soc. Surg. Oncology, 1982; Meml. Sloan-Kettering Cancer Ctr. award, 1972; Disting. Service medal USPHS, 1983; Meyer and Anna Prentiss award, 1984; Second Emmanuel Cancer Found. award, 1984; Pierluigi Nervi award, Rome, 1985; Medal of Honor, Am. Cancer Soc., 1985; Barbara Bohen Pfeifer award Am.-Italian Found. Cancer Research, 1985; Stratton lectr. Am. Soc. Hematology, 1985, Leukemia Research Fund lecture, London, 1985; Tenth Richard and Hinda Rosenthal Found. award, Am. Assn. Cancer Research Inc., 1986; Stanley G. Kay Meml. award, D.C. Am. Cancer Soc., 1986, Sci. award Brady Cancer Res. Inst., 1987, Prix Cino del Duca, Paris, 1988, Pezcoller award Eur. Sch. Oncology, Trento, Italy, 1988, Surgeon Gen.'s Exemplary Svc. medal, 1988, Armand Hammer Cancer prize, 1990. Fellow ACP, N.Y. Acad. Medicine; mem. AMA, Am. Soc. Clin. Oncology (chmn. program com. 1972, dir. 1973-76, pres. 1977-78), Am. Cancer Soc., Am. Soc. Hematology, A Rsch. (dir. 1976-79), Am. Fedn. Clin. Rsch., Am. Soc. Clin. Investigation, Assn. Am. Physicians, Soc. Surg. Oncology, Smith-Reed-Russel Med. Soc., Internat. Coun. for Coordinating Cancer Rsch. (pres. Am. bd. 1989—), Alpha Omega Alpha. Office: Yale Comprehensv Cancer Ctr PO Box 3333 333 Cedar St Ste 205WWW New Haven CT 06510-8028

DE VITRY D'AVAUCOURT, ARNAUD, engineer; b. Versailles, France, July 27, 1926; s. Raoul and Nicole (LeBret) de V.; m. Henriette Doll, July 8, 1953; children: Anne, Catherine. Grad., Ecole Polytech., Paris; MS, MIT; MBA, Harvard U. Asst. to pres. Electro-Mech. Rsch. Inc., U.S.A., 1953-54; engr. Socony Mobil Oil Inc., U.S.A., 1954-58, chief engr., 1958-60; exec. dir. Mobil Chem. Co., U.S.A., 1960-62; dir. Dunlop S.Am., Paris, 1965, v.p., 1968, chmn., 1969-82; pres. Tech. Studies Inc., Luxembourg, 1957-82; pres., mng. dir. Eureka Sicav, Paris, 1980-90; bd. dirs. Digital Equipment Co., Ionics Inc., Schlumberger Ltd. Mayor Méré (Yvelines, France) Town Hall, 1986—; bd. dirs. French Library, Boston. Mem. Soc. Engrs. Civil France, French Am. Found., Fondation Franco-Americaine, Automobile Club France, Jockey Club, Brook Club, Somerset Club. Avocation: rare books. Home: 41 rue de l"Université, 75007 Paris France

DEVITT, JOHN LAWRENCE, consulting engineer; b. Denver, Sept. 27, 1925; s. Oliver Hinkley and Ellen Elizabeth (McPherson) D.; children: Jane, David, Ellen. BSEE, U. Colo., 1945, MS, 1949. Registered profl. engr., Colo. Engr. U.S. Bureau of Reclamation, Denver, 1947-50; plant mgr. AMF Corp., Colorado Springs, Colo., 1951-55; v.p., gen. mgr. Whittaker Corp. Power Sources div., Denver, 1955-61; chief engr. Metron Instrument Co., Denver, 1962-65; mgr. of electrochemistry Gates Corp., Denver, 1965-71; pvt. practice as a consulting engr. Denver, 1971—; profl. jazz musician (saxophone), Denver, 1946—. Co-inventor sealed lead-acid and lead-chloride batteries. Lt. USNR, 1943-52, PTO. Recipient Battery Research award, The Electrochem. Soc., 1986. Mem. The Electrochem. Soc., Am. Chem. Soc., Inst. Elect. and Electronic Engrs., Colo. Mountain Club (pres. 1975), Am. Alpine Club, New York. Avocation: mountaineer. Office: Consulting Engr 985 S Jersey St Denver CO 80224-1418

DEVITT, JOHN WILLIAM, physicist; b. Bayshore, N.Y., Oct. 25, 1959; s. John Gerald and Faith Mary (Fitzgerald) D. BS in Physics, SUNY, Stonybrook, 1981; MS in Physics, Ohio State U., 1984. Lab. asst. Physics Dept., SUNY, Stonybrook, 1980-81; grad. rsch. asst. Physics Dept. Ohio State U., Columbus, 1981-84; infrared evaluation specialist GE Aircraft Engines, Cin., 1984-86, infrared inspection specialist, 1986-88, lead scientist infrared measurements, 1988—. Contbr. articles to profl. jours.; numerous patents in field. Recipient Sanford Moss Meml. Engring. award, 1993. Mem. Soc. Photo Optical Engrs., ASTM. Avocations: running, volleyball, guitar. Home: 220 Cannonade Dr Loveland OH 45140-7104 Office: GE Aircraft Engines 1 Neumann Way J185 Cincinnati OH 45215-1988

DEVOE, LAWRENCE DANIEL, obstetrician/gynecologist educator; b. Chgo., Nov. 5, 1944; m. Anne Hester Devoe; 1 child, Laura. BA, Harvard U., 1966; MD, U. Chgo., 1970. Intern U. Chgo., 1971, resident, 1976, fellowship, 1979, asst. prof. ob/gyn., 1976-83; assoc. prof. Med. Coll. Ga., Augusta, 1983-89, prof. ob/gyn., 1989—, dir. maternal-fetal medicine, 1984—; cons. in obstetrics Dwight David Eisenhower Med. Ctr., Ft. Gordon, Ga., 1983—, Coliseum Park Hosp., Macon, Ga., 1984—, Meml. Med. Ctr., Savannah, Ga., 1984—. Author: Computer Applications for Perinatology, 1993; author chpts. in books. Bd. dirs. Augusta Opera Assn., 1987—, 2nd v.p., 1990—. Capt. M.C., U.S. Army, 1971-73. Recipient Purdue Frederick award Am. Coll. Ob/Gyn., 1981, 91. Fellow Am. Coll. Obstetricians and Gynecologists; mem. Ga. State Ob/Gyn. Soc., So. perinatal Assn. (pres. 1989), Ga. Perinatal Assn. (pres. 1992). Office: Med Coll of Ga 1120 15th St Augusta GA 30912

DEVONTINE, JULIE E(LIZABETH) J(ACQUELINE) (THE MARCHIONESS DE ROE DEVON), systems analyst, consultant; b. Edmund, Wis., Jan. 7, 1934; d. Clyde Elroy and Matilda Evangeline Knapp; m. Roe (Don Davis) Devon Gerringer-Busenbark, Sept. 30, 1968 (dec. Dec. 1972); student Madison Bus. Coll., 1952, San Francisco State Coll., 1953-54, Vivian Rich Sch. Fashion Design, 1955, Dale Carnegie Sch., 1956, Arthur Murray Dance Studio, 1956, Biscayne Acad. Music, 1957, L.A. City Coll., 1960-62, Santa Monica (Calif.) Jr. Coll., 1963; attended Hastings Coll. of Law, 1973, Wharton Sch., U. Pa., 1977, London Art Coll., 1979; Ph.D., 1979; attended Goethe Inst., 1985. Bar: Calif., 1965. Actress, Actors Workshop San Francisco, 1959, 65, Theatre of Arts Beverly Hills (Calif.), 1963, also radio; cons, and systems analyst for banks and pub. accounting agys.; artist, poet, singer, songwriter, playwright, dress designer. Pres., tchr. Environ Improvement, Originals by Elizabeth; atty. Dometrik's, JIT-MAP, San Francisco, 1973—; steering com. explorations in worship, ordained min. 1978. Author: The Cardinal, 1947, Explorations in Worship, 1965, The Magic of Scents, 1967, New Highways, 1967, The Grace of Romance, 1968, Happening - Impact-Mald, 1971, Seven Day Rainbow, 1972, Zachary's Adversaries, 1974, Fifteen from Wisconsin, 1977, Bart's White Elephant, 1978, Skid Row Minister, 1978, Points in Time, 1979, Special Appointment, A Clown in Town, 1979, Happenings, 1980, Candles, 1980, Votes from the Closet, 1984, Wait for Me, 1984, The Stairway, 1984, The River is a Rock, 1985, Happenings Revisited, 1986, Comparative Religion in the United States, 1986, Lumber in the Skies, 1986, The Fifth Season, 1987, Summer Thoughts, 1987, Crimes of the Heart, 1987, Toast Thoughts, 1988, The Contrast of Russian Literature Through the Eyes of Russian Authors, 1988, A Thousand Points of Light, 1989, The Face in the Mirror, 1989, Voices on the Hill, 1991, It's Tough to Get a Matched Set, 1991, Equality, 1991, Miss Geranium Speaks, 1991, Forest Voices, 1991, Golden Threads, 1991, Castles in the Air, 1991, The Cave, 1991, Angels, 1991, Real, 1991, An Appeal to Reason, 1992, We Knew, 1992, Like It Is, 1992, Politicians Anonymous, 1993. Mem. Assn. of Trial Lawyers of Am. Address: 1500 El Camino Ave # 382 Sacramento CA 95833

DEVOR, RICHARD EARL, mechanical and industrial engineering educator; b. Milw., Apr. 18, 1944; s. Robert Gerald DeV. and Betty (Hale) Roth; m. Jearnice Anna Luedtke, Apr. 5, 1968. BS, U. Wisc., 1967, MS, 1968, PhD, 1971. Asst. prof. U. Ill., Urbana, 1971-74, assoc. prof., 1974-84, prof., 1984—, assoc. head dept. mech. and indsl. engring., 1987-91, dir. mfg. rsch. ctr., 1989-90, exec. dir. inst. competitive mfg., mech. & indsl. engring., 1989—. Co-author: Statistical Quality Design and control: contemporary Concepts and Methods, 1991; contbr. articles to profl. jours. Mem. ASME (Blackall Machine Tool and Gage award 1983), Am. Soc. Quality Control, Soc. Mfg. Engrs. (North Am. Mfg. Rsch. Inst. sci. com. chmn. 1990-92, pres. 1993-94, Edn. award 1993). Achievements include being author or co-author of more than 70 publications in field of manufacturing and engineering statistics. Office: U Ill Dept Mech and Ind Eng 1206 W Green St Urbana IL 61801

DE VOS, ALOIS J., civil engineer; b. Moline, Ill., Mar. 3, 1947; s. John Robert and Josephine (Roberts) De Vos; children: Jonathan. BS in Civil Engring., U. Ill., 1975, MS in Civil Engring., 1976. Registered profl. engr., Iowa, Ill. Mgr. Adler Letter Co., L.A., 1971-72; insp. Terracon, Davenport, Iowa, 1976-77; supr. W.J. Reese & Assoc., Rock Island, Ill., 1977-84, U.S. Army Corps Engrs., Rock Island, 1984—. Co-author (rsch. project) Design of Emulsified Asphalt Concrete Pavement, 1976. Sgt. U.S. Army, 1969-71. Mem. ASCE. Home: 2636 Fair Davenport IA 52803 Office: US Army Corps Engrs Clock Tower Bldg Rock Island IL 61201

DEVRIES, FREDERICK WILLIAM, chemical engineer; b. N.Y.C., Feb. 5, 1930; s. Frederick and Maxine Celia (Sondheimer) DeV.; m. Mary Patricia Salva, Aug. 29, 1959; children: Margaret D. Poretz, Carol M. Donovan, F. Joseph. AB, Columbia U., 1949, BS, 1950, MS, 1951. Cert. profl. engr. Del. Prodn. supr. E.I. duPont Co., Wilimington, Del., 1951-59, supr. rsch. and devel., 1959-65, engring. supr., 1965-71, sr. tech. rep., 1971-77, tech. svc. cons., 1977-85, sr. cons., 1985-90; pres. Chem. Mining Consulting, Ltd., Chadds Ford, Pa., 1990—; trustee N.W. Mining Assn., Spokane, 1987-90. Editor: Advances in Gold and Silver Tech., 1992; contbr. articles to profl. jours. Pres., v.p., treas. Pennsbury Townwatch, Inc., 1983—. Mem. Soc. for Mining Metallurgy and Exploration, Am. Inst. Chem. Engrs., Am. Chemistry Soc., Nat. Assn. Environ. Profl. Republican. Jewish. Achievements include patents in field. Home: 25 Hillendale Rd Chadds Ford PA 19317 Office: Chem Mining Cons Ltd 25 Hillendale Rd Chadds Ford PA 19317

DEVRIES, GEERT JAN, science educator; b. Columbus, Ohio, Sept. 11, 1954; s. Geert and Maartje Hendrika (Kouwenhoven) De.; m. Lucie Henriette TerBorg, Oct. 11, 1974; children: Thijs, Paul David, Charlotte. Kandidaats, Free U., Amsterdam, The Netherlands, 1976, doktoraal, 1980; PhD, U. Amsterdam, 1985. Scientific staff mem. Netherlands Inst. for Brain Rsch., Amsterdam, 1980-84; adj. lectr. U. Calif., Irvine, 1985-86, lectr., 1987; asst. prof. U. Mass., Amherst, 1987-91, assoc. prof., 1991—. Editor: Progress in Brain Research Vol. 61: Sex Differences in the Brain. Relation Between Structure and Function, 1984; contbr. articles to Brain Research, Jour. of Neuroendocrinology, Jour. of Comparitive Neurology. Recipient Sandoz Travel Stipend award, Oss, The Netherlands, 1984; Twinning co-grantee European Tng. Program in Brain and Behavioral Rsch., 1984. Mem. NSF (brain and behavioral scis. chpt., adv. panel 1990, rsch. grantee 1988-91), NIMH (Nat. Rsch. Svc. award 1986-88, rsch. grantee 1991-94). Achievements include first to anatomically demonstrate sex differences in a specific neurotransmitter system in the mammalian brain, to anatomically demonstrate dramatic long term changes in specific neurotransmitter systems under influence of gonadal steroids; research on neural basis of paternal behavior. Office: U Mass Dept Psychology Amherst MA 01003

DE VRIES, KENNETH LAWRENCE, mechanical engineer, educator; b. Ogden, Utah, Oct. 27, 1933; s. Sam and Fern (Slater) DeV.; m. Kay M. DeVries, Mar. 1, 1959; children—Kenneth, Susan. AS in Civil Engring., Weber State Coll., 1953; BSME, U. Utah, 1959, PhD in Physics, Mech. Engring., 1962. Registered profl. engr., Utah. Rsch. engr. hydraulic group Convair Aircraft Corp., Fort Worth, 1957-58; prof. dept. mech. engring. U. Utah, Salt Lake City, 1962—, mem. faculty, 1969-76, prof. dept. mech. and indsl. engring., 1976-81, Disting. prof., 1991—, chmn. dept., 1970-81, assoc. dean rsch. Coll. Engring., 1983—; program dir. div. materials rsch. NSF, Washington, 1975-76; materials cons. Browning, Morgan, Utah, 1972—; cons. 3M Co., Mpls., 1985—; tech. adv. bd. Emerson Electric, St. Louis, 1978—; mem. Utah Coun. Sci. and Tech., 1973-77; trustee Gordon Rsch. Conf., 1989—, chair, 1992-93. Co-author: Analysis and Testing of Adhesive Bonds, 1978; contbr. chpts. to numerous books, articles and abstracts to profl. publs. Fellow ASME, Am. Phys. Soc.; mem. Am. Chem. Soc. (polymer div.), Soc. Engring. Scis. (nat. officer), Adhesion Soc. (nat. officer). Mem. LDS Ch. Office: U Utah 2220 Merrill Engring Bldg Salt Lake City UT 84112

DEVRIES, MARVIN FRANK, mechanical engineering educator; b. Grand Rapids, Mich., Oct. 31, 1937; s. Ralph B. and Grace (Buurma) DeV.; m. Martha Lou Kannegieter, Aug. 28, 1959; children: Mark Alan, Michael John, Matthew Dale. BS, Calvin Coll., 1960; BSME, U. Mich., 1960, MSME, 1962; PhD, U. Wis., 1966. Registered profl. engr., Wis. Teaching fellow mech. engring. U. Mich., Ann Arbor, 1960-62; instr. U. Wis., Madison, 1962-66, asst. prof., 1966-70, assoc. prof., 1970-77, prof. mech. engring., 1977—, chmn. dept. mech. engring., 1991—; vis. Fulbright prof. Cranfield (Eng.) Inst. Tech., 1979-80; dir Mfg. Systems Engring. Program, Madison, 1983-91; cons in field. Contbr. over 90 articles to profl. jours. Sr. program dir. NSF, Washington, 1987-90. Recipient Ralph Teetor award Soc. Auto. Engrs., 1967, Space Shuttle Tech. award NASA, 1984, Disting. Achievement award L.A. Coun. Engrs. and Scientists, 1985, Internat. Tech. Communications award Calif. Engring. Found., 1985. Fellow ASME, Soc. Mfg. Engrs. (pres. 1985-86, Olin Simpson award 1986), Instn. Prodn. Engrs. (U.K., life); mem. Internat. Instn. Prod. Engring. Rsch. (coun. mem. 1989-92). Avocations: travel, sports. Home: 901 Tompkins Dr Madison WI 53716-3267 Office: U Wis 1513 University Ave Madison WI 53706-1572

DE WANDELEER, PATRICK JULES, electronics executive; b. Wilrijk, Antwerp, Belgium, Sept. 29, 1949; s. Yvonne (Muylle) De Wandeleer; m. Sonja Spruyt; children: Laurent, Geraldine. Grad., St. Jan Berchmans Coll., Merksem/Antwerp, 1967; degree in Tech. Engring., Katholieke Industriele Hogesch. Antwerp, 1971; MBA, Fontainebleau, France, 1990. Project engr. Constructions and Entreprises Industrielles, Brussels, 1973-78, Stefens Electro, Antwerp, 1978-79; from gen. mgr. to mng. dir. GTI Belgium N.V., Antwerp, 1980-92; mng. dir. Tech. Contracting & Maintenance, Antwerp, Belgium, 1993—; dir. Energie Svc. Contract, Brussels, 1988-92; mng. dir. Ten. Tech. Installations Electro Thys, Antwerp, 1980-92, Genk, 1980-92, Chauffage Laurent Block, 1980-92, ESC, 1980-92. Lt. Belgium Mil., 1971-72. Mem. Flemish Econ. Assn., Nat. Confederation Bldg. Industry, Lions. Avocations: tennis, hockey. Home: Dopheidelaan 3, 2930 Brasschaat, 2000 Antwerp Belgium Office: Samberstraat 48-50, 2060 Antwerp Belgium

DEWANJEE, MRINAL KANTI, radiopharmaceutical chemist; b. Sitakund, Bangladesh, Mar. 19, 1941; s. Dinesh C. and Bidyut P. (Choudhury) D.; m. Urmila Dewanjee, July 15, 1970; 1 child, Sumit. BS, Comilla Victoria Coll., Bangladesh, 1960; MS, Dhaka U., Bangladesh, 1962; PhD, McGill U., Montreal, Que., Can., 1967. Cert. Am. Bd. Sci. in Nuclear Medicine. Instr. McGill U., 1963-67; asst. prof. Sir A.T. Coll., Bangladesh, 1962-63; rsch. assoc. Amherst Coll., Mass., 1968-69, Rensselaer Poly. Inst., Troy, N.Y., 1969-71; radipharmacist Peter Bent Brigham Hosp., Boston, 1972-76; dir. radiopharm. lab. Mayo Med. Sch., Rochester, Minn., 1976-87; dir. radiopharmacist lab. U. Miami Sch. Medicine, Fla., 1987—; vis. rsch. scientist SUNY Nuclear Accelerator Lab., Albany, 1969-71. Author: Procedure

Manual Radiopharmacy, 1988, Radioiodination, 1992, Cell Labeling, Experience and Clinical Applications, 1992. Recipient Shannon award NHLBI/NIH, 1992; nat. Rsch. Coun. scholar. Democrat. Achievements include patents for calcifying tissue value and radiolabeled oligonucliotide probe. Office: U Miami Sch Medicine PO Box 016960 Miami FL 33101

DEWAR, JAMES MCEWEN, agricultural executive, consultant; b. Williamsport, Pa., Aug. 4, 1943; s. James Livingston and Margaret Ann (McEwen) D.; B.S. in Internat. Affairs, Trinity U., 1965, postgrad. internat. law, 1966, postgrad. African studies, 1965-66; m. Margaret Cawley, Feb. 27, 1982; children: Alec, Porter, Leah. Mgr., Dash brand Procter & Gamble Corp., Cin., 1970-72; pres. DeLair & Dewar, Inc., Tucson, 1972-83; chmn. bd. Cabot South Asia, Inc., subs. Cabot Corp., 1982-87; pres., dir.-gen. ASI, Inc. subs. Boeing Co., 1987—; dir. Metz Constrn. Co., Marine Environ. Research Corp., Computational Analysis Corp. Bd. dirs. Casa de Los Ninos, Tucson, 1974—, Safari Club Internat., Tucson, 1974—, Internat. Marine Fisheries Corp., ; founding mem. Atty's. Victim/Witness Adv. Program; mem. White House Talent Pool, 1975-76, White House Nat. Cambodia Crisis Com., 1979-80, U.S. Aerospace Indsl. Reps. in Europe; adj. Mil. Order World Wars, Tucson, 1977-80, perpetual mem.; chmn. internat. bd. advs. Ariz.-Sonora Desert Mus.; bd. advs. guardian ad litem program Superior Ct. Ariz. Served to capt. USAF, 1966-70; Vietnam. Recipient Key to City Seoul (Korea), 1973, citation Pres. Korea, 1973, award for work with Mother Teresa The Cabot Found, 1982-87. Mem. Am. Soc. Agrl. Cons., Dirs. Guild Am. Republican, Assn. Old Crows. Roman Catholic. Clubs: Mountain Oyster; Australian/Asian Order Old Bastards (Sydney, Australia), L'Automobile Club de France, Maxim's Bus. Club, St. James, Chambers (New Delhi). Contbr. numerous articles to profl. publs. Home: 227 Blvd Saint Germain, 75007 Paris France Office: ASI, 5 Rue Du Faubourg Saint Honore, 75008 Paris France

DEWAR, MICHAEL JAMES STEUART, chemistry educator; b. Ahmednagar, India, Sept. 24, 1918; came to U.S., 1959, naturalized, 1980; s. Francis and Nan (Keith) D.; m. Mary Williamson, June 3, 1944; children: Robert Berriedale Keith, Charles Edward Steuart. B.A., Oxford (Eng.) U., 1940, D.Phil., 1942, M.A., 1943. Imperial Chem. Industries fellow Oxford U., 1945; phys. chemist Courtaulds Ltd., 1945-51; prof. chemistry, head dept. Queen Mary Coll., U. London, Eng., 1951-59; prof. chemistry U. Chgo., 1959-63; Robert A. Welch prof. chemistry U. Tex., 1963-90; grad. rsch. prof. U. Fla., Gainesville, 1990-93, emeritus prof., 1993—; Reilly lectr. U. Notre Dame, 1951; Tilden lectr. Chem. Soc. London, 1954; vis. prof. Yale U., 1957; Falk-Plaut lectr. Columbia U., 1963; Daines Meml. lectr. Western Res. U., 1963; Glidden C. lectr. U. Kans., 1964; William Pyle Phillips visitor Haverford Coll., 1964, 70; Arthur D. Little vis. prof. MIT, 1966; Marchon vis. lectr. U. Newcastle (Eng.), 1966; Glidden Co. lectr. Kent State U., 1967; Gnehm lectr. Eldg. Technische Hochschule, Zurich, Switzerland, 1968; Barton lectr. U. Okla., 1969; Disting. vis. lectr. Yeshiva U., 1970; Kahlbaum lectr. U. Basel, Switzerland, 1970; Benjamin Rush lectr. U. Pa., 1971; Kharasch vis. prof. U. Chgo., 1971; Venable lectr. U. N.C., 1971; Phi Lambda Upsilon lectr. Johns Hopkins U., 1972; Firth vis. prof. U. Sheffield, 1972; Foster lectr. SUNY-Buffalo, 1973; Five Colls. lectr., Mass., 1973; Sprague lectr. U. Wis., 1974; spl. lectr. U. London, 1974; lectr. chem. edn. Fla. State U., 1975; disting. Bicentennial prof. U. Utah, 1976; Bircher lectr. Vanderbilt U., 1976; Pahlavi lectr., Iran, 1977; Michael Faraday lectr. U. No. Ill., 1977; Priestley lectr. Pa. State U., 1981; research scholar lectr. Drew U., 1984; J. Clarence Karcher lectr. U. Okla., 1984, Coulson lectr. U. Ga., 1988; cons. to industry. Author: Electronic Theory of Organic Chemistry, 1949, Hyperconjugation, 1962, Introduction to Modern Organic Chemistry, 1965, Computer Compilation of Molecular Weights and Percentage Compositions of Organic Compounds, 1969, The Molecular Orbital Theory of Organic Chemistry, 1969, The PMO Theory of Organic Chemistry, 1975, A Semiempirical Life, 1991; also articles. Recipient Harrison Howe award Am. Chem. Soc., 1961, Southwest regional award, 1978, Robert Robinson Lecture, Chem. Soc., 1974, G.W. Wheland Meml. medal U. Chgo., 1976, Evans award Ohio State U., 1977, James Flack Norris award, 1984, Nichols medal, 1986, Auburn-G.M. Kosolapoff award Am. Chem. Soc., 1988, Tetrahedron prize, 1989, World Assn. Theoretical Organic Chemistry medal, 1989, Chemist Pioneer award Am. Inst. Chemists, 1990; hon. fellow Balliol Coll., 1974, Queen Mary Coll., 1993. Fellow Am. Acad. Art and Scis., Royal Soc. (Davy medal 1982), Chem. Soc. London; mem. NAS, Am. Chem. Soc. Home: 10520 NW 36th Ln Gainesville FL 32606-5075

DEWEY, ALAN H., electronic engineer; b. Freeport, N.Y., Apr. 4, 1957; s. Campbell G. and Elizabeth H. D.; 1 child, Ryan D. AAS in Communications, Parkland Coll., 1978; BS in Engring. Tech., So. Ill. U., 1981, MS in Mfg. Systems, 1991. Electronics journeyman Cen. Ill. Pub. Svc. Co., Marion, 1982-89; lead electronics engr. Laser Imaging Systems, Inc., Punta Gorda, Fla., 1990—. Mem. Soc. Mfg. Engrs., Am. Radio Relay League. Achievements include co-invention of photo-acoustic leak detection system and method. Office: Laser Imaging Systems Inc 204A E McKenzie St Punta Gorda FL 33950

DEWEY, CAMERON BOSS, civil engineer; b. Waynesville, N.C., July 19, 1961; d. Frederick Earl and Nancy Jane (Gilson) Boss; m. David Allen Dewey, June 25, 1988. BS in Engring., U. Central Fla., 1984, MS in Engring., 1986. Registered profl. engr. Grad. asst. U. Central Fla., Orlando, 1984-87; engr. St. Johns River Water Mgmt. Dist., Orlando, 1987—. Mem. ASCE, Am. Water Resources Assn. Home: 9268 Baton Rouge Dr Orlando FL 32818 Office: St Johns River Water Mgmt Dist 618 E South St Orlando FL 32801

DEWEY, CRAIG DOUGLAS, engineering executive; b. Milw., Apr. 8, 1950; s. RalphEarl and Suzanne Dewey; m. Madeline A. Reedy, Sept. 28, 1989. BSME, U. Wis., 1974. Registered profl. engr., Wis. Mgr. engring. Harnischfeger Corp., Milw., 1974-83, dir. engring., 1986-90; chief engr. Fairmont Railway Motors, Fairmont, Minn., 1983-86; v.p. engring. Pemco, Inc., Sheboygan, Wis., 1990-91; dir. engring. Quad/Tech, Sussex, Wis., 1991-92, dir. ops., 1992—. Patentee in field. Mem. Mensa. Avocations: personal computers, gardening. Home: 7155 N Green Tree Ct River Hills WI 53217-3708 Office: Quad/Tech N64 W23110 Main St Sussex WI 53089

DEWEY, JOHN F., geologist, educator; b. London, England, May 22, 1937; married; 2 children. BSc, London U., 1958, PhD, 1960, DIC, 1960; MA, Cambridge, 1965; ScD, Canterbury, 1987; MA, Oxford U., 1986, DSc, 1988. Lectr. U. Manchester, Eng., 1960-64; univ. lectr. Cambridge U., Eng., 1964-70; prof. SUNY, Albany, 1970-80, disting. prof., 1980-82; prof. geology U. Durham, Eng., 1982-86; prof. geology Dept. Phys. Scis. U. Oxford, Eng., 1986—. Recipient Lyell medal Geol. Soc. London, 1983, Academia Europaea, 1990, Arthur Holmes medal, 1993. Fellow Geol. Soc. Am. (Penrose medal 1992), Am. Geophys. Union, Royal Soc. Office: U Oxford Dept Phys Scis, University College, Oxford OX1 4BH, England

DEWHURST, CHARLES KURT, museum director, curator, folklorist, English educator; b. Passaic, N.J., Dec. 21, 1948; s. Charles Allaire and Minn Jule (Hanzl) D.;; m. Marsha MacDowell, Dec. 15, 1972; 1 dau., Marit Charlene. B.A., Mich. State U., 1970, M.A., 1973, Ph.D., 1983. Editorial asst. Carlton Press, N.Y.C., 1967; computer operator IBM, N.Y.C., 1968; project dir. Mich. State U. Mus., 1975, curator, 1976-83, dir., 1982—; guest curator Mus. Am. Folk Art, N.Y.C., 1978-83, Artrain, Detroit, 1980-83; visual arts panelist Mich. Council Arts, Detroit, 1981-83; cons. City of Cleve., 1983; dir. Festival of Mich. Folklife, 1987-91. Author: Reflections of Faith, 1983, Artists in Aprons, 1979, Rainbows in the Sky, 1978, Michigan Folk Art, 1976 (Am. Assn. State and Local History award 1977), Art at Work: Folk Pottery of Grand Ledge, Michigan, 1986, Michigan Quilts, 1987, Michigan Folklife Reader, 1988. Mus. profl. grantee Smithsonian Instn., Scandinavia, 1978, Fulbright, 1992; project grantee Mich. Council Humanities, 1990-92, NEH, 1991, Nat. Endowment Arts, 1982, 84-91. Mem. Am. Folklore Soc., Mich. Folklore Soc., Midwest Soc. Lit., Popular Culture Assn., Mich. Hist. Soc., Mich. Mus. Assn., Am. Assn. Museums (pres.), Internat. Council Museums. Democrat. Roman Catholic. Home: 212 N Harrison Rd East Lansing MI 48823-4141 Office: Mich State U Mus W Circle Dr East Lansing MI 48824

DEWHURST, PETER, industrial engineering educator; b. Great Harwood, Eng., Mar. 7, 1944. BSc, U. Manchester, Eng., 1970, MSc, 1971, PhD in Mech. Engring., 1973. With faculty mech. engring. U. Salford, 1973-80; vis. prof. mfg. group U. Mass., 1980-81, prof., 1980-85; prof. indsl. and mfg. engring. U. R.I., Kingston, 1985—, dir. grad. studies mfg. engring., 1985—. Referee Internat. Jour. Applied Mechanics, Internat. Jour. Prodn. Rsch., Am. Soc. Mech. Engrs., assoc. editor Mfg. Rev.; assoc. editor Jour. Design and Mfg., Internat. Jour. Systems Automation and Applications. Recipient Nat. medal Tech., U.S. Dept. Commerce Tech. Adminstrn., 1991. Mem. Soc. Mfg. Engrs. (sr.), Collegiate Internat. Pour Rsch. Prodn. (F. W. Taylor medal 1980). Achievements include research in mechanics of metal forming, in metal cutting, in computer aided numerical control of machine tools; design for manufacture and the application of robots in assembly procedures. Office: U RI Dept Indsl and Mfg Engring 103 Gilbreth Hall Kingston RI 02881*

DEWHURST, WILLIAM GEORGE, physician, psychiatrist, educator, researcher; b. Frosterley, Durham, Eng., Nov. 21, 1926; came to Can., 1969; s. William and Elspeth Leslie (Begg) D.; m. Margaret Dransfield, Sept. 17, 1960; children—Timothy Andrew, Susan Jane. B.A., Oxford U., Eng., 1947, B.M., B.Ch., 1950; MA, Oxford U., 1961; D.P.M. with distinction, London U., 1961. House physician, surgeon London Hosp., 1950-52, jr. registrar, registrar, 1954-58; registrar, sr. registrar Maudsley Hosp., London, 1958-62, cons. physician, 1965-69; lectr. Inst. Psychiatry, London, 1962-64, sr. lectr., 1965-69; assoc. prof. psychiatry U. Alta., Edmonton, Can., 1969-72, prof., 1972-92, prof. emeritus, 1992—, Hon. prof. pharmacy and pharm. scis., 1979—, chmn. dept. psychiatry, 1975-90, dir. emeritus neurochem. rsch. unit, 1990—, hon. prof. oncology, 1983—, chmn. med. staff adv. bd., 1988-90; mem. Atty. Gen. Alta. Bd. Rev., 1991, N.W.T. Bd. Rev., 1992; pres.'s coun. U. Alta. Hosps., 1988-90, quality improvement coun., 1988-90, ethics consultative com., 1984-88, planning com. Vision 2000, 1985-87, hosps.' planning com. and joint conf. com., 1971, 80, 87-90; cons. psychiatrist Royal Alexandra Hosp., Edmonton, Edmonton Gen. Hosp., Alberta Hosp., Ponoka, Ponoka Gen. Hosp.; chmn. med. coun. Can. Test Com., 1977-79, Royal Coll. Text Com. in Psychiatry, 1971-80, examiner, 1975-83. Co-editor: Neurobiology of Trace Amines, 1984, Pharmacotherapy of Affective Disorders, 1985; also conf. procs. Referee Nature, Can. Psychiat. Assn. Jour., Brit. Jour. Psychiatry; mem. editorial bd. Neuropsychobiology, Psychiat. Jour. U. Ottawa. Contbr. articles to profl. jours. Chmn. Edmonton Psychiat. Svcs. Steering Com., 1977-80; chmn. Edmonton Psychiat. Svcs. Planning Com., 1985-90; mem. Provincial Mental Health Adv. Coun., 1973-79, Mental Health Rsch. Com., 1973, Edmonton Bd. Health, 1974-76; Can. Psychiat. Rsch. Found., 1985— (also bd. dirs.); bd. dirs. Friends of Schizophrenics, 1980—, Alta., 1988; grant referee Health & Welfare Can., Med. Rsch. Coun. Can., Ont. Mental Health Found., Man. Health Rsch. Coun., B.C. Health Rsch. Found. Capt. Royal Army M.C., 1952-54. Fellow Can. Coll. Neuropsychopharmacology (pres. 1982-84, Coll. medal 1993), Am. Psychopathol. Assn., Am. Coll. Psychiatrists, Am. Psychiat. Assn., Royal Coll. Psychiatrist; mem. AAAS, Alta. Psychiat. Assn. (pres. 1973-74), Can. Psychiat. Assn. (pres. 1983-84), Alta Coll. Physicians and Surgeons, Alta. Med. Assn., Child and Adolescent Assn. (bd. dirs., v.p. 1992), Assn. for Acad. Psychiatry, Brit. Med. Assn., Faculty Club. Anglican. Avocations: music, hockey, football, chess, athletics. Office: U Alta Dept Psychiatry, 1E1 01 Mackenzie Ctr, Edmonton, AB Canada T6G 2B7

DEWITT, JAMES HOWARD, water treatment technician; b. Denver, Apr. 21, 1952; s. Fred Albert and Grace May (Strecker) DeW.; m. Linda R. Hite, Apr. 23, 1973 (div. Aug. 1990; 1 child, Frances Allie. AA, Community Coll. Red Rocks, Denver, 1978. Lic. water treatment plant operator. Lead shift operator water treatment plant City of Aurora, Colo., 1974-82; utilities plants supr. Town of Rangely, Colo., 1982--; pres. Colo. Water Waste Water Assn., 1975-78; mem. dialogue group Environ. Protection Agy., Washington, 1992--. Sgt. U.S. Army, 1970-73, Vietnam. Mem. VFW, Colo. Rural Water Assn. Office: Town of Rangely 209 E Main St Rangely CO 81648

DEWITT, SHEILA HOBBS, research chemist; b. Medina, N.Y., May 23, 1960; d. Arnold James and Marcia June (Cady) Hobbs; m. Joseph Arthur DeWitt, Sept. 5, 1992. BA in Chemistry, Cornell U., 1981; PhD in Chemistry, Duke U., 1986. Lab. rsch. asst. FMC Agrl. Chem. Divsn., Middleport, N.Y., 1979-81; rsch. chemist FMC Agrl. Chem. Divsn., Princeton, N.J., 1982, process chemist, 1986-88; prep lab. supr. Duke U., Durham, N.C., 1983, teaching asst., 1982-86, rsch. asst., 1983-86; scientist Parke-Davis Pharm. Rsch., Ann Arbor, Mich., 1988-89, sr. scientist, 1989-91, rsch. assoc., 1991-93, sr. rsch. assoc., chmn. molecular diversity project team, 1993—; chmn. seminar com., chemistry dept. Parke-Davis, 1992-93; grad. asst. Dale Carneigie Course, Ann Arbor, 1991; organic chemistry tutor, Office of Minority Affiars, Duke U., 1985-86; chmn. nat. symposia. Reviewer Bioorganic and Medicinal Chemistry Letters, Jour. Organic Chemistry; contbr. articles to sci. jours. Adv. com. Dexter (Mich.) Community Schs. Curriculum, 1989-93; mem. Mellon Project on Acad. Careers, Durham, 1985-86, Christian Edn. Com., Faith Alliance Ch., Durham, 1985-86, dir. Children's Ch., 1984-86. Recipient scholarship AAUW, 1978, N.Y. Bd. Regents, 1978-81. Mem. AAAS, N.Y. Acad. Sci., Am. Chem. Soc., Ann Arbor Art Assn., Ann Arbor Womens Painters, Phi Lambda Upsilon (v.p. 1985-86), Sigma Xi. Republican. Mem. Christian and Missionary Alliance. Achievements include 7 patents and 2 patents pending in field. Office: Parke-Davis 2800 Plymouth Rd Ann Arbor MI 48105

DE WOLFF, FREDERIK ALBERT, toxicology educator; b. Schiedam, The Netherlands, Sept. 19, 1944; s. Frederik A. and Andrea J. (Vogely) de W.; m. Deena Rouendaal, June 26, 1971; children: Jacob Frederik, Abraham Ruben. BSc in Biochemistry, U. Leiden, The Netherlands, 1966, MSc in Biochemistry, 1968, BA in Semitic Langs., 1971, PhD in Pharmacology, 1973, MA in Semitic Langs., 1984. Diplomate in clin. chemistry and toxicology. Resident U. Leiden, 1969-73, lectr. dept. Hebrew studies, 1981-87; resident in clin. biochemistry Univ. Hosp., Leiden, 1973-76, assoc. prof., head dept. human toxicology, 1976-91; vis. prof. U. Amsterdam (The Netherlands) Med. Ctr., 1979-91, prof., 1991—; Lady Davis vis. prof. Hebrew U., Jerusalem, 1993—; cons. various pharm. and other cos.; pres. Eurotox 91 Congress, Maastricht, The Netherlands, 1991. Editor: Therapeutic Relevance of Drug Assays, 1979, Medical Toxicology, 1992, Intoxications of the Nervous System I, 1993; contbr. over 250 sci. articles to profl. publs. Mem. Netherlands Soc. Clin. Chemistry (bd. mem. 1991—), Netherlands Soc. Toxicology (pres. 1988-90), Am. Assn. for Clin. Chemistry, Brit. Toxicology Soc., N.Y. Acad. Scis., Internat. Assn. Forensic Toxicologists. Jewish. Avocations: classical music, travel, Jewish languages and history.

DE WYS, EGBERT CHRISTIAAN, geochemist; b. Soerabaja, Netherlands East Indies, Apr. 9, 1924; came to U.S., 1925; s. Gerard H. L. and Agnita (Versteegh) de W.; m. Sheila Naulty, Dec. 12, 1973; children: Wendy, Tanya, Mark, Matthew. BA, Miami U., 1950, MA, 1951; PhD, Ohio State U., 1955. Scientist Owens-Corning Fiberglass Co., Newark, Ohio; sr. physicist, mgr. process control IBM, San Jose, Calif.; sr. scientist Jet Propulsion Lab., Pasadena, Calif.; prof. Tex. Tech. U., Lubbock; geologist Oasis Oil Co., Tripoli, Libya, Amoco Prodn. Co., Denver; v.p., dir. gen. Coastal Congo Corp., Point Noire, Congo; dep. dir. ESRI, Columbia, S.C.; pres. Exploration Geochemists, Santa Rosa, Calif., 1983—; adj. prof. San Jose U.; prof. U. Tripoli. Contbr. articles to profl. jours.; patentee in field. Sgt. maj. Royal Netherland East Indies Army; lt. Armee Secrete Belgium; paratrooper U.S. Army. Bownacker Scholar, Ohio State U. Fellow Am. Mineral. Soc., Jet Propulsion Lab. (sr. fellow); mem. Sigma Xi, Sigma Gamma Epsilon, Phi Sigma, Pi Delta Phi, Delta Phi Alpha. Avocations: computer programming, fencing, chess, gardening. Office: Exploration Geochemists 3600 Aaron Dr Santa Rosa CA 95404-1505

DEXTER, FRANKLIN, anesthesiologist; b. Huntington, N.Y., Aug. 4, 1964; s. Franklin Jr. and Rebecca (Shuman) D.; m. Elisabeth Uy, July 2, 1988. PhD in Biomed. Engring., Case Western Res. U., 1989, MD, 1990. Rsch. asst. dept. pathology R.I. Hosp., Providence, 1983; teaching asst. dept. biology and medicine Brown U., Providence, 1983, computer cons. Computer Ctr., 1983-84; rsch. assoc. dept. biomed. engring. Case Western Res. U., Cleve., 1989-90; resident anesthesiologist, researcher U. Iowa Hosps., Iowa City, 1990—. Contbr. articles to profl. jours. Mem. AMA, Am. Soc. Anesthesiologists, Am. Soc. Anesthesiologists, Sigma Xi, Alpha Omega Alpha. Democrat. Unitarian Universalist. Achievements include application of biomathematics to problems in physiology and anesthesiology. Home: 1937 Plaen View Dr Iowa City IA 52246 Office: U Iowa Hosps Dept Anesthesia Iowa City IA 52242

DEYONG, GREGORY DONALD, chemist; b. Rock Rapids, Iowa, Apr. 28, 1966; s. Donald Benny and Charlene Kay (Coy) DeY.; m. Penelope Alice Collins, Aug. 17, 1991. Student, Northwestern Coll., 1984-85; BS in Chemistry, Iowa State U., 1988. Chemist Hach Co., Ames, Iowa, 1988—. Author several conf. presentations. Math. and sci. scholar State of Iowa, 1984, Dow-Goetz scholar Iowa State U. and Dow Chems., 1986-88. Mem. Am. Chem. Soc. (edn. adv. group 1991-92). Republican. Lutheran. Achievements include invention of environmental analytical methods for chromium and zinc; co-invention of environmental analytical methods for lead and cadmium. Office: Hach Co 100 Dayton Ave Ames IA 50010-6400

DE YOUNG, DAVID SPENCER, astrophysicist; b. Colorado Springs, Colo., Nov. 29, 1940; s. Henry C. and Zona L. (Church) DeY.; m. Mary Ellen Haney. BA, U. Colo., 1962; PhD, Cornell U., 1967. Rsch. physicist Los Alamos Nat. Labs., Los Alamos, N. Mex., 1967-69; astronomer Nat. Radio Astronomy Obs., Charlottesville, Va., 1969-80; astronomer Kitt Peak Nat. Obs., Tucson, 1980-83, assoc. dir., 1983-88, dir., 1988—; mem. adv. bd. Aspen Ctr. Physics, Aspen, Colo., 1977—; exec. mem. steering com. San Diego Supercomputer Ctr., 1985—; bd. dirs. WIYN Telescope Consortium, Tucson, 1990—. Contbr. articles to profl. jours. NASA grantee. Mem. Am. Phys. Soc., Am. Astron. Soc., Internat. Astron. Union, Internat. Union Radio Sci. Office: Kitt Peak Nat Obs 950 N Cherry Ave Tucson AZ 85719-4933

DEYOUNG, RAYMOND, conservation behavior educator; b. Paterson, N.J., Oct. 24, 1952; s. Raymond Jospeh and Wanda Ann (Turbak) DeY.; m. Noreen Fran Horowitz, June 13, 1976; children: Jessica, Joshua. BS in Engring., Stevens Inst. Tech., 1974, MS in Ocean Engring., 1976; PhD, U. Mich., 1984. Engr. ITT, Midland Park, N.J., 1976-77, Camp, Dresser and McKee, Detroit, 1977-80; rsch. asst. U. Mich., Ann Arbor, 1980-82, rsch. assoc., 1982-84, rsch. fellow, 1984-90, asst. rsch. scientist, 1990-91, prof., 1991—; commr. Transp. Planning Com., Ann Arbor, 1992-93, Energy Commn., Ann Arbor, 1984-86. Contbr. articles to profl. jours. Mem. Nat. Recycling Coalition, Washington, bd. mem., 1988-90. Horace H. Rackham Predoct. fellow U. Mich., 1983-84. Mem. Am. Psychol. Soc., Environ. Design Rsch. Assn. Achievements include research in study of informational and motivational aspects of individual conservation behavior, the creation of psychologically-relevant environmental policy and clarity-based decision making. Office: U Mich Sch Nat Res and Environ 430 E University Ave Ann Arbor MI 48109-1115

DEYRUP-OLSEN, INGRITH JOHNSON, physiologist, educator; b. Englewood, N.J., Dec. 22, 1919; d. Alvin Saunders and Edith Henry (Henry) Johnson; m. Sigurd M. Olsen, Dec. 28, 1962 (dec.). BA, Barnard Coll., 1940; PhD, Columbia U., 1944. Instr. physiology Columbia Coll. Physicians and Surgeons, N.Y.C., 1942-47; asst. prof. to prof. zoology Barnard Coll., N.Y.C., 1947-64; rsch. prof. to prof. zoology U. Wash., Seattle, 1964-90, prof. emeritus, 1990—; exec. com. Biol. Scis. Curriculum Study, Boulder, Colo., 1960-70. Author: Metabolism, 1974; contbr. articles to profl. jours., chpts. to books. Recipient Medal of Distinction Barnard Coll.,1 992. Fellow AAAS; mem. Am. Physiol. Soc. (edn. com.), Am. Soc. Zoologists (edn. com.), Soc. Gen. Physiologists (edn. com.), Nat. Assn. Biology Tchrs. (hon.), Sigma Xi, Phi Beta Kappa. Office: Univ of Wash Dept Zoology NJ15 Seattle WA 98195

DHADWAL, HARBANS SINGH, electrical engineer, consultant; b. India, June 28, 1955; came to U.S., 1982; s. Inderjit Singh and Swaran Kaur (Najran) D.; m. Raj Rani Dub, June 9, 1979; children: Neetu, Anish, Rajiv. BSc with 1st class honors, U. London, 1976, PhD, 1980. Higher sci. officer Royal Aircraft Establishment, Farnborough, Eng., 1980-82; postdoctoral fellow dept. chemistry SUNY, Stony Brook, 1982-84, assoc. prof. elec. engring., 1984—; acad. visitor IBM Thomas Watson Rsch. Ctr., Yorktown Heights, N.Y., 1991; NASA summer faculty, 1992; cons. Andrew Corp., Chgo., Brookhaven Instruments Corp., L.I. Reviewer for Optical Engring., West Ednl. Pub., Holt, Winston and Rinehart, others; contbr. to sci. publs. Mem. IEEE (sr.), Internat. Soc. Optical Engrs., Optical Soc. Am. Achievements include patent for method and apparatus for determining viscosity, for laser light scattering and spectroscopic detector, for method and apparatus for determining physical properties of materials with dynamic light scattering. Home: 6 Ben Pl Setauket NY 11733 Office: SUNY Dept Elec Engring Stony Brook NY 11794-2350

DHANABALAN, PARTHIBAN, electrical engineer; b. Madurai, India, Sept. 10, 1959; s. Dhanabalan and Raja Malliga (Rathnasamy) S.M. Karruppanna Nadar; m. Muzhumathi Rajendran, May 26, 1988; children: Vikraman, Parthiban. BSEE, Madras U., 1981; MBA, Anna U., Madras, India, 1983; MSEE, U. Houston, 1989. Regional mgr. Aurelec Data Processing, Madras, India, 1984-86; rsch. asst. U. Houston, 1987-89; program analyst Shell Rsch., Houston, 1989-90; engr. analyst Amocams/Modular Inc., Houston, 1990—. Hindu. Office: Amocams/Modular Inc 2525 Bay Area Blvd III Houston TX 77058

DHARANI, LOKESWARAPPA RUDRAPPA, engineering educator; b. J. N. Kote, India, Oct. 25, 1947; came to U.S., 1979; s. Rudrappa and Badramma Dharani; m. Lalitha Hosangadi, Mar. 4, 1976; children: Vastsala, Shelia B. M Tech. Aero. Engring., Indian Inst. Tech., Kanpur, 1972; MSc in Aircraft Design, Cranfield Inst. Tech., 1975; PhD in Engring. Mechanics, Clemson U., 1982. Aero. engr. Hindustan Aeronautics, Ltd., Bangalore, India, 1971-76, dep. design engr., 1976-79; asst. prof. mech. engring. U. Mo., Rolla, 1982-88, assoc. prof. mech. engring., 1988-92, prof., 1992—; mem. damage tolerance subcom. Indian Dept. Def., New Delhi, 1978-79; vis. scientist materhals lab. Wright Patterson AFB, Dayton, Ohio, 1986. Contbr. articles to profl. publs., chpts. to book and ency. Mem. Soc. Automotive Engrs., Am. Soc. Engring. Edn., Soc. for Advancement of Materials and Process Engring. Achievements include development of micromechanics for the analsis cracks and interface debonding in high temperature composites, non-asbestos brake pads for automobiles, unified finite element code for processing modeling of ceramic matrix composites. Home: 1103 Sycamore Dr Rolla MO 65401 Office: U Mo Mech Engring Bldg Rolla MO 65401

DHIR, VIJAY K., mechanical engineering educator; b. Giddarbaha, Panjab, India, Apr. 14, 1943; came to U.S., 1969; s. Harnand Lal and Parsinni Devi (Sofat) D.; m. Komal Lata Khanna, Aug. 31, 1973; children: Vinita, Vashita. BScME, Punjab Engring. Coll., India, 1965; MTechME, Indian Inst. Tech., 1969; PhD in Mech. Engring., U. Ky., 1972. Asst. devel. engr. Jyoti Pumps, Ltd., Baroda, India, 1968-69; postgrad. engr. Engring. Rsch. Ctr. Tata Engring. & Locomotive Co., Poona, India, 1969; rsch. asst. U. Ky., Lexington, 1969-72, rsch. assoc., 1972-74; asst. prof. chem., nuclear & thermal engring. dept. UCLA, 1974-78, assoc. prof., 1978-82, prof. mech., aerospace & nuclear engring. dept., 1982—, vice chmn. mech., aerospace & nuclear engring. dept., 1988-91; cons. Nuclear Regulatory Commn., Seabulk Corp., Ft. Lauderdale, Fla., Argonne (Ill.) Nat. Lab., Pickard, Lowe & Garrick, Inc., Irvine, Calif., Rockwell Internat., Canoga Park, Calif., GE Corp., San Jose, Calif., Battelle N.W. Lab., Richland, Wash., Phys. Rsch., Inc., Torrance, Calif., Nat. Bur. Stds., Gaithersburg, Md., Los Alamos (N.Mex.) Nat. Lab., Sci. Applications Inc., El Segundo, Calif., Brookhaven Nat. Lab., Upton, N.Y.; chmn. numerous conf. sessions. Contbr. over 75 articles to profl. jours., over 65 papers to procs./conf. & symposia records; assoc. editor Applied Mechs. Rev., 1985-88, Jour. Heat Transfer, Transactions ASME, 1993—, ASME Symposium Vol., 1978; referee numerous jours. Fellow ASME (Heat Transfer Meml. award sci. category 1992). Office: Sch of Engring & Applied Sc Univ of Calif 38-137 W Engineering IV Los Angeles CA 90024

DHOM, ROBERT CHARLES, chemical engineer; b. Aurora, Ill., Dec. 30, 1964; s. Charles Theodore and Dorothy Ann (Cerven) D. BS in Chem. Engring., Iowa State U., 1987. Engr. Westinghouse, Aiken, S.C., 1988—. Mem. Am. Inst. Chem. Engrs. Lutheran. Home: 103 Groves Blvd North Augusta SC 29841

DIACONIS, PERSI W., mathematical statistician, educator; b. N.Y.C., N.Y., Jan. 31, 1945; s. Andrew and Syma (Meyerowitz) D. BS, CCNY, 1971; MS, Harvard U., 1972, PhD, 1974. Asst. prof. Stanford (Calif.) U., 1974-78, prof. stats., 1981-87; mem. tech. staff Bell Labs., 1979-80; prof. math. Harvard U., 1987—. Author: Use of Group Representations in Probability and Statistics, 1988. Recipient Rollo Davidson prize Cambridge

U. (Eng.), 1982, MacArthur Found. prize, 1982. Fellow Am. Acad. Arts and Scis., Inst. Math. Stats. Avocation: magic tricks. Office: Harvard U Dept Math, Science Ctr Cambridge MA 02138

DIAISO, ROBERT JOSEPH, civil engineer; b. Jersey City, N.J., Jan. 3, 1940; s. Dominick A. and Marie M. (Sarno) DiA.; m. Elaine Ricca, June 8, 1963; 1 child, Michael. BS, U.S. Naval Acad., 1962; M in Civil Engring., NYU, 1964; M in Urban and Regional Planning, U. Pitts., 1971; PhD, 1971, 1972. Engr. Clarke, Hartman & Dunn, 1955-57, 69; project dir. Inst. Urban Policy Analysis, 1970-71; assoc. partner Dewberry, Nealon & Davis, Annapolis, Md., 1971-81; sr. assoc. Dewberry & Davis, Annapolis, 1981-82; prin. Dewsberry & Davis, 1983-84; pres. Property Improvement Collaborative, Inc., 1984-90, LandScope, 1985—; dir. LandTech Corp., 1985-88, mng. dir., 1988—; chief exec. officer FM Tech Corp., 1991—; pres. Tech Group, 1993—; organizer, dir. Bay Nat. Bank; Land Tech. Corp., 1986—; bd. dirs. Scotts Seaboard Corp.; pres. Peacock Mgmt. Systems. Pres., Crofton Civic Assn., 1973; chmn. bd. trustees by 3 govs. Anne Arundel C.C., 1974-96; mem. County Zoning Adv. Task Force, 1983-84; mem. county coun. adv. com. on adequate facilities, 1977-78; bd. dirs. Anne Arundel Trade Coun.; chmn. Public Works Rev. Bd.; mem. Sewer Allocation Task Force; chmn. County Exec. Transition Task Force, 1982; mem. County Exec. Transition Team, 1991; mem. Gov's. com. on Affordable Housing, 1976-78; mem. adv. bd. Patuxent Water Reclamation Plant; mem. adv. com. Crofton on Municipal Incorp.; bldg. com. St. Elizabeth Ann Seton Ch. Served with USAF, 1962-69. Named Bus. Leader of Yr., Anne Arundel Trade Coun., 1982; HEW fellow, 1970-72. Mem. ASCE, Am. Planning Assn., Am. Inst. Certified Planners, Nat. Soc. Profl. Engrs., Assn. County Engrs. Roman Catholic. Office: 147 Old Solomons Island Rd Annapolis MD 21401-0904

DIAMOND, EDWARD, gynecologist, infertility specialist, clinician; b. Newark, N.J., Nov. 25, 1928; s. William D. and Ruth (Stier) Brief; m. Joan Marilyn Chase, June 16, 1951 (div. Apr. 1969); children: Gary Warren, Steven Michael, Stuart Eric; m. Linda Ruth Johnson, Sept. 4, 1969; 1 child, Mira Frieda. BS, Franklin and Marshall Coll., 1949; MB, MD, Chgo. Med. Sch., 1953. Diplomate Am. Bd. Obstetrics and Gynecology. Intern Beth Israel Med. Ctr., Newark, 1953-54, ob.-gyn. resident, 1954-56, attenting physician, 1956—; attenting physician Clara Maass Med. Ctr., Belleville, N.J., 1959—; idr. Diamond Inst. for Infertility, Irvington, N.J., 1968—; microsurgery cons. Am. Coll. Ob.-Gyn., Washington, 1979-82; design cons.Codman Med. Instruments, Johnson & Johnson, Waltham, Mass., 1973-91, Edward Weck Co. Med. Div., Chapel Hill, N.C., 1973-85; preceptorship in infertility Fertility Inst. of N.Y., N.Y.C., 1967. Author: Microsurgery in Gynecology I and II, 1977, 81, Endoscopy in Gynecology, 1978, Operative Perinatology, 1984; contbr. articles to profl. jours.; speaker in field. Trustee, assoc. Franklin and Marchall Coll., Lancaster, Pa., 1982—; assoc. Nat. Coun. on Arts and Scis., Washington, 1982—, Aspen (Colo.) Ctr. for Environ. Studies, 1985—. Capt. U.S. Army, 1956-58. Recipient Leading Pioneer in Tubal Microsurgery citation Prince George Med. Ctr., Cheverly, Md., 1976, Contbn. to Arts citation Rutgers U., Newark, 1984. Fellow Am. Coll. Ob.-Gyn.; mem. Am. Fertility Soc., Internat. Soc. Advancement of Humanistic Studies in Gynecology, N.Y. Soc. Reproductive Medicine. Hebrew. Achievements include participating in first seminar on microsurgery in infertility; recognition as one of first 3 gynecologists worldwide to conduct research and perform microsurgery on reproductive human organs; redesign of existing operating microscopes for use in depth of abdominal cavity for gynecological surgery; design of first set of microsurgical instruments for universal use in microsurgical specialties; establishment of first course and workshop in gynecological microsurgery; first gynecologist in world to perform gamete intra fallopian tubal transfer by culdoscopy, to achieve successful pregnancy by culdoscopy gamete intra fallopian tubal transfer. Home: P O Box 6353 Snowmass Village CO 81615

DIAMOND, IRVING T., physiology educator. James B. Duke Prof. physiology Duke U., Durham, N.C. Office: Duke U Dept of Psychology Durham NC 27710

DIAMOND, RICHARD MARTIN, nuclear chemist; b. L.A., Jan. 7, 1924; divorced; 4 children. BS, UCLA, 1947; PhD in Nuclear Chemistry, U. Calif. Berkeley, 1951. Instr. chemistry Harvard U., 1951-54; asst. prof. Cornell U., 1954-58; mem. sr. staff Lawrence Berkeley Lab., U. Calif., 1958—; Guggenheim fellow Denmark, 1966-67; mem. U.S. Physics del. to Russia, 1966, rev. com. physics divsn. Oak Ridge Lab., 1972-74, program adv. com. Ind. Cyclotron Facility, 1980-83; chmn. Gordon Conf. on Nuclear Chemistry, 1965, Conf. on Ion Exchange, 1969, rev. com. Univ. isotope seperator at Oak Ridge, Oak Ridge Nat. Lab., 1974-75, subcom. high spin and nuclei far from stability Dept. Energy-NSF, 1983; vis. fellow Japan Soc. for Promotion of Sci., 1981; co-organizer Int. Conf. Nuclear Physics, 1980, workshop Nat. Gamma-Ray Facility, 1987. Mem. Fellow Am. Phys. Soc. (Tom W. Bonner award), AAAS; mem. Am. Chem. Soc. (ACS award in Nuclear Chemistry 1993). Achievements include research in nuclear spectroscopy, coulomb excitation, high-spin nuclear structure. Office: Lawrence Berkeley Lab Nuclear Science Divsn Berkeley CA 94720*

DIAMOND, SEYMOUR, physician; b. Chgo., Apr. 15, 1925; s. Nathan Avruum and Rose (Roth) D.; m. Elaine June Flamm, June 20, 1948; children: Judi, Merle, Amy. Student, Loyola U., 1943-45; MB, Chgo. Med. Sch., 1948, MD, 1949. Intern White Cross Hosp., Columbus, Ohio, 1949-50; gen. practice medicine Chgo., 1950—; dir. Diamond Headache Clinic, Ltd., Chgo., 1970—; dir. inpatient headache unit Weiss Meml. Hosp., Chgo.; prof. neurology Chgo. Med. Sch., 1970-82, adj. prof. pharmacology and molecular biology, 1985—; clin. assoc. dept. medicine U. Chgo., 1989—; lectr. dept. community and family medicine Loyola U. Stritch Sch. Medicine, 1972-78; cons. mem. FDA Orphan Products Devel. Initial Rev. Group; lectr. Falconbridge lecture series Laurentian U., Sudbury, Ont., Can., 1987. Co-author: (with Donald J. Dalessio) The Practicing Physician's Approach to Headache, 5th edit., 1992, More Than Two Aspirin. Help for Your Headache Problem, 1976, Advice from the Diamond Headache Clinic, 1982, Coping with Your Headaches, 1982, 2d edit. (with Mary Franklin Epstein), 1987, Headache in Contemporary Patient Mgmt. series, 1983, Hope for Your Headache Problem, rev. edit., 1988, (with Diane Francis and Amy Diamond Vye) Headache and Diet, 1990, Sexual Aspects of Headaches (with Michael Maliszewski, 1992); editor: Migraine Headache Prevention and Management; editor in chief Headache Quar., 1990—; editorial cons. BIOSIS, 1986-90; contbr. numerous articles on headache and related fields to profl. jours., chpts. to books. Pres. Skokie (Ill.) Bd. Health, 1965-68. Recipient Disting. Alumni award Chgo. Med. Sch., 1977; Nat. Migraine Found. lectureship award, 1982; 1st recipient Migraine Trust lectureship, 1988; British Migraine Trust 7th Internat. Migraine Syposium, London. Fellow Royal Soc. Medicine; mem. AMA (Physicians Recognition awards 1970-73, 74, 77, 79, 82, 87, del. sect. clin. pharmacology and therapeutics 1987-89, mem. health policy agenda for Am. People, mem. Cost Effectiveness Conf., del. reference com. "C" on edn., reference com. C, 1988), mem. Bd. of Scientific and Policy Advs. for The Am. Council on Sci. and Health, So. Med. Assn., Am. Assn. Study of Headache (exec. dir. 1971-85, pres. 1972-74, regent 1984), Nat. Headache Found. (pres. 1971-77, exec. dir. 1977—), World Fedn. Neurology (exec. officer 1980—, research group on migraine and headache), Ill. Acad. Gen. Practice (chmn. mental health com. 1966-70), Ill. Med. Soc., Chgo. Med. Soc., Biofeedback Soc. Am., Internat. Assn. Study of Pain, Am. Soc. Clin. Pharmacology and Therapeutics (chmn. headache sect. 1982-89, mem. com. coordination sci. sects. 1983-89), Postgrad. Med. Assn. (pres. 1981). Office: 1040 N Lake Shore Dr Chicago IL 60625-2448

DIANA, GUY DOMINIC, chemist; b. New Haven, July 21, 1935; s. Guy A. and Margaret A. (Cinicola) D.; m. Joanna Elizabeth Rucker, June 5, 1960; children: Stephen, Laura, John, David. BS, Yale U., 1957; PhD, Rice U., 1961. Sr. rsch. chemist Sterling Winthrop Pharms. Rsch. Div., Rensselaer, N.Y., 1961-66, group leader medicinal chemistry, 1967-89, fellow medicinal chemistry, 1990-91, sr. fellow medicinal chemistry, 1992—. Mem. editorial bd. Antiviral Chemistry & Chemotherapy, 1989—; contbr. articles to profl. jours. Town supr. Stephentown (N.Y.) Town Bd., 1991-92; sch. bd. New Lebanon (N.Y.) Ctrl. Schs., 1974-82; bd. dirs. Columbia-Greene Bd. Coop. Edn., East Greenbush, N.Y., 1972-74. Mem. AAAS, Am. Chem. Soc., N.Y. Acad. Sci. Achievements include 61 patents in field; discovery of compounds which inhibit viral replication of common cold virus. Home: 1566 Glenmar Dr Pottstown PA 19464

DIANA, JOHN NICHOLAS, physiologist; b. Lake Placid, N.Y., Dec. 19, 1930; s. Alphonse Walton and Dolores (Mirto) D.; m. Anita Louise Harris, May 8, 1966; children: Gina Sue, Lisa Ann, John Nicholas. B.A., Norwich U., 1952; Ph.D., U. Louisville, 1965. Asst. prof. physiology Mich. State U. Med. Sch., 1966-68; assoc. prof., then prof. U. Iowa Med. Sch., 1969-78; prof. physiology, chmn. dept. La. State U. Med. Center, Shreveport, 1978-85; dir. cardiovascular research ctr. U. Ky., 1985-87, assoc. dean research and basic sci., 1987-88; dir. T&H Rsch. Inst., 1988—; cons. Nat. Inst. Neurol. Diseases and Stroke, 1973-75, Nat. Heart, Lung and Blood Inst., 1974—, mem. cardiovascular and renal study sect., 1980-85, mem. clin. scis. study sect., 1986-91, chmn. 1989-91; rsch. com. Iowa Heart Assn., 1974-77, bd. dirs., 1977-79; mem. cardiovascular study sect. Am. Heart Assn., 1981-84. Author papers, abstracts in field. Served with AUS, 1952-54; Served with USAR, 1961-62. NIH postdoctoral fellow, 1965-67. Mem. Am. Fedn. Clin. Research, Am. Physiol. Soc. (editorial bd. jour. 1974-78), Microcirculation Soc. (pres. 1977-78, editorial bd. jour. 1979-85), Am. Heart Assn. (fellow council circulation), N.Y. Acad. Scis., La. Heart Assn. (dir. 1979-81, research com. 1978-82), Sigma Xi. Democrat. Achievements include patent for coronary vasodilator. Home: 3656 Eleuthera Ct Lexington KY 40509-9525 Office: U Ky Coll Medicine Dean's Office 900 Rose St Rm MN140D Lexington KY 40546

DIAZ, A. MICHEL, computer and control science researcher, administrator; b. Carcassonne, France, July 27, 1945; s. Melchor and Pilar (Villa) D.; m. Monique Philipot, Apr. 10, 1969; children: Nicolas, Sandrine. Licences Physique I and II, Toulouse U., 1966; Dr.Sci., U. Paul Sabatier, 1974. Mem. Lab. de Automatique et d'Analyse des Systmes, du Centre National de la Recherche Scientifique (CNRS), 1968—; asst. prof. Inst. Tech., Toulouse U., 1969-70; researcher at CNRS, French Nat. Center for Sci. Rsch., Toulouse, France, 1970—, head team computer sci., 1975—, dir. rsch. CNRS; head European Esprit-Software Engring. Distributed Systems project, 1984-88; lectr. in field; mem. internat. sci. coms.; mem. & chmn. program coms. on specification, verification, and design of distributed systems. Author: Introd al Diseno Assistido por Ordenador de Circuitos Electronicos, 1977; editor: Protocol Specification, Testing and Verification V, North Holland, 1986; coeditor The Formal Description Technique Estelle, North Holland, 1989, The Formal Description Technique Lotos, North Holland, 1990; tech. editor Annals Telecoms., Network Rev. and Systems Distbn.; contbr. articles to profl. jours. and congresses. Sr. mem. IEEE (tech. editor Comms. mag.); mem. Assn. Francaise pour la Cybernetique Economique et Technique, Internat. Fedn. Info. Processing (co-editor Conf. Formal Description Technique-Forte V, North Holland, 1993, Silver Core distinction), Internat. Standardization Orgn., Assn. Francaise pour la Normalisation. Achievements include research in design of communication and cooperation distributed systems, with emphasis on formal specification and implementation of high speed, multimedia and cooperative systems. Office: LAAS du CNRS, 7 Av Cl Roche, 31077 Toulouse France

DIAZ, ALBERTO, biotechnologist; b. Buenos Aires, May 5, 1942; s. Antonio and Herminia Maria (Del'Bo) D.; m. Diana Lia Epstein, Mar. 21, 1970; children: Luciana, Ana Laura. B., Nat. Coll. Buenos Aires, 1960; lic. in chemistry, U. Buenos Aires, 1967. Fellow Nat. Rsch. Coun., Buenos Aires, 1968-74; prof. biochemistry U. Buenos Aires Sch. of Medicine, 1970-80; rsch. dir. Immunoquemia S.A., Buenos Aires, 1974-80; gen. dir. Biosidus, Buenos Aires, 1980-90, tech. mgmt. dir., 1990-92; dir. ALSZ Bioprojects, 1992—. Active Argentine Commn. Bolivar Program. Mem. Argentine Chemistry Assn., Internat. Soc. for Interferon Rsch. Home: Araoz 2305 2o A, 1425 Buenos Aires Argentina

DIAZ, FERNANDO GUSTAVO, neurosurgeon; b. Mexico City, Dec. 29, 1946; came to U.S., 1971; s. Fernando Diaz Calderon and Susana (Barriga) D.; m. Vicki M. Harrell, May 1, 1989; children: Fernando Austin, David Frederick, Jaime Marisa. BS, Centro Universitario Mex., 1963; MD, Univ. de Mex., 1969; MA, U. Kans., Kansas City, 1973; PhD, U. Minn., 1979; MA in Bus., Cen. Mich. U., Mt. Pleasant, 1987. Diplomate Am. Bd. Neurological Surgery; lic. physician and surgeon Mex., Can., Ill., Mich., Fla. Asst. gen. surgery Sanatario Santa Maria, Mexico City, 1964-68; intern Regina Gen. Hosp., Sask., Can., 1969-70, resident in anethesia, 1971; resident in gen. surgery U. Kans., Kansas City, 1971-73; resident in neurosurgery U. Minn. Hosps., Mpls., 1973-78; staff neurosurgeon Henry Ford Hosp., Detroit, 1978-87; chmn. Neurosci. Inst. Santa Fe, Gainesville, Fla., 1987-89; prof., chmn. dept. neurol. surgery Wayne State U., Detroit, 1990—; neurosurg. nat. cons. to U.S. Surgeon Gen., USAF, 1991; coord. neurosurgery resident edn. Henry Ford Hosp., 1979—; clin. assoc. prof. surgery U. Mich., 1986—. Editor: Neurosurgery Jour.; contbr. articles to profl. jours. Lt. col. USAFR. Recipient awards Lily Pharms., Merck, Sharp & Dome Pharms., Organon Labs. Fellow ACS, Interam. Coll. Physicians, Internat. Coll. Surgeons (vice regent U.S. sect. 1985); mem. AMA, Neurosurg. Soc. Am., Soc. Neurol. Surgeons, Mich. Med. Soc., Wayne County Med. Soc., Am. Assn. Neurol. Surgeons (cerebrovascular sect.), Congress of Neurol. Surgeons, Mich. Assn. Neurol. Surgeons (sec.-treas. 1984-86, v.p. 1986), Detroit Neurosurg. Acad. (v.p. 1986—), Soc. Critical Care Medicine, Mich. Heart Assn. (chmn. stroke com. 1984-86, community site ad-hoc com. 1984, community programs and edn. com. 1986), U. Minn. Alumni Assn. Republican. Roman Catholic. Office: Wayne State U Neurol Surg 4201 St Antoine Blvd Detroit MI 48201

DIAZ, NILS JUAN, nuclear engineering educator; b. Moron, Cuba, Apr. 7, 1938; came to U.S., 1961; s. Rafael Octavio Diaz and Rosa Dalia (Rojas) Chao; m. Zenaida G. Gonzalez, Oct. 9, 1960; children: NIls, Ariadne, Allene. BSME, U. Villanova, Havana, 1960; MS in Nuclear Engring. Sci., U. Fla., 1964, PhD in Nuclear Engring. Sci., 1969. Rsch. assoc. nuclear engring. sci. U. Fla., Gainesville, 1965-69, asst. prof., reactor supr., 1969-74, assoc. prof., dir. nuclear facilities, 1974-79, prof., dir. nuclear facilities, 1979-84; assoc. dean for rsch. Sch. of Engring. Calif. State U., Long Beach, 1984-86; prof. nuclear engring. scis. U. Fla., Gainesville, 1986—; dir. Innovative Nuclear Space Power and Propulsion Inst., Calif. and Fla., 1985—; sr. cons. Exxon Nuclear, Fla. Power and Light-Fla. Power Corp., Bellevue, Wash. and Fla., 1974-79; pres., chief engr. Fla. Nuclear Assocs., Inc., Gainesville, 1976—; prin. advisor Nuclear Safety Coun., Madrid, 1981-83. Contbr. articles to profl. jours. Chmn. Minority Engr. Program Adv. Bd., Long Beach, 1984-86. Recipient Disting. Svc. award Math. Engring. Sci. Achievements and Minority Engring. program State of Calif., Long Beach, 1983; named Hispanic Engr. of Yr. for Outstanding Tech. Contbns., Hispanic Engr. Nat. Achievement Com., Houston, 1990, Engr. of Yr., Cuban-Am. Engring. Soc., 1993. Mem. AAAS, ASME, Am. Soc. for Engring. Edn., Am. Nuclear Soc. Republican. Roman Catholic. Achievements include patents for heterogeneous gas core reactors, gamma ray flaw detection system; invention of vapor core propulsion system. Office: U Fla 202 Nuclear Scis Ctr Gainesville FL 32611

DIAZ VELA, LUIS H(UMBERTO), computer company executive; b. Acaponeta, Nayarit, Mex., Aug. 28, 1953; s. Oscar Angel Diaz Jimenez and Garciela (Vela) Lopez; m. Carmen Alicia Guerrero Diaz, June 16, 1990; children: Bertha Graciela, Diana Gisell. Student, Inst. Superior Computacion, Guadalajara, Jalisco, 1970-72, Inst. Tech. Estudios Superiore, Mexicali, Mex., 1986. Pres., proprietor Copymex (Copy Ctr.) #1, Mexicali, 1977—, Copymex (Copy Ctr.) #2, Mexicali, 1979—, Copy Print S.A. de C.V. (Copy Ctr.), Mexicali, 1991; pres. Grupo Proyeccion Internat., S.A. de C.V., Mexicali, 1979—; dir., proprietor Centro de Estudios Technologicos, Administrativos, Mexicali, 1983—, Tijuana, Mex., 1983—, Ensenada, Mex., 1983-86; dir. Copiadoras Univ. de Mexico, S.A., Mexicali, 1985-91; pres. Cetac, Escuela de Computacion A.C., 1991—; tchr. computers Inst. Fronterizo de Computacion Electronica, Mexicali, 1974-75. With Mexican mil., 1972. Recipient Aguila C Y P, Calidad y Prestigio A.C., Guadalajara, Jalisco, 1983, 84, 87, 1992, Estrella de Diamante Internacional a la Cuidad, Instituto Nacional de Mercadotecnia, 1992. Mem. Villafontana Country Club, Rotary Club. Roman Catholic. Office: Azueta No 281-303, 21100 Mexicali Mexico also: 233 Paulin Ste 7562 Calexico CA 92231-2615

DIBB, DAVID WALTER, research association administrator; b. Draper, Utah, July 4, 1943; s. Walter and Mary (Lisinsky) D.; m. Vivian Berrett, Dec. 15, 1966; children: Stephanie, Gregory, Steven, Rebecca. BS, Brigham Young U., 1970; PhD, U. Ill., 1974. Cert. profl. agronomist, cert. profl. soil scientist. Rsch. asst. U. Ill., Urbana, 1970-74, teaching asst., 1971-74; vis. asst. prof. N.C. State U., Raleigh, 1974-75; rsch. dir. Potash and Phosphate Inst., Atlanta, 1982-89; regional dir. Potash and Phosphate Inst., Columbia, Mo., 1975-82; coord. Latin Am. Potash and Phosphate Inst., Atlanta, 1982-85; v.p. North Am. Potash and Phosphate Inst., West Lafayette, Ind., 1985-86, sr. v.p., 1987-88; pres. Potash and Phosphate Inst., Atlanta, 1989—; pres. Agronomic Sci. Found., Madison, Wis., 1983-85; mem. fertilizer industry adv. com. Food and Agrl. Orgn. of UN, Rome, 1988—; exec. industry rev. group TVA, Muscle Shoals, Ala., 1989—; adj. prof. Purdue U., West Lafayette, 1985-88. Contbr. author: Potassium in Agriculture, 1985; editor: Fertilizer Research, 1989-90; contbr. articles to profl. jours. Instnl. rep. Boy Scouts Am., West Lafayette, 1982-85, asst. scoutmaster, Norcross, Ga., 1989-90; youth coach for basketball, baseball, and soccer, Mo., Ind., Ga., 1980-89; active PTA, Mo., Ind., Ga., 1978—. Fellow Am. Soc. Agronomy (chmn. budget and fin. com. 1988); mem. AAAS, Soil Sci. Soc. Am., Coun. for Agrl. Sci. and Tech., Internat. Soil Sci. Soc., Gamma Sigma Delta, Alpha Zeta. Office: Potash and Phosphate Inst 655 Engring Dr Ste 110 Norcross GA 30092-2821

DIBENEDETTO, ANTHONY THOMAS, engineering educator; b. N.Y.C., Oct. 27, 1933; s. Thomas and Mathilda DiB.; m. Rose Marie Lima, Feb. 12, 1955; children: Diane, Laura, Thomas, David, Stephen. B.Ch.E., CCNY, 1955; M.S., U. Wis., 1956, Ph.D., 1960. Chem. engr. Union Carbide Corp., 1954-55; prof. chem. engring. U. Wis., 1960-67; prof., dir. materials research lab. Washington U., 1967-71; head dept. chem. engring. U. Conn., 1971-77, v.p. grad. edn. and research, 1979-81, v.p. acad. affairs, 1981-86; Univ. prof. chem. engring. U. Conn., Storrs, 1986—, dir. Inst. Materials Sci., 1991—; cons. in field. Author: The Structure and Properties of Materials, 1967. Recipient Ednl. Service award Plastics Inst. Am., 1973; NSF profl. devel. award, 1977-79; Disting. Service award U. Wis., 1981, Outstanding Leadership award U. Conn., 1992. Mem. Soc. Plastics Engrs., Am. Inst. Chem. Engring., Am. Chem. Soc., Sigma Xi, Tau Beta Pi. Home: 1 Brookside Ln Mansfield Center CT 06250-1109 Office: U Conn Inst Materials Sci U-136 Storrs Mansfield CT 06268

DIBERARDINIS, LOUIS JOSEPH, industrial hygiene engineer, consultant, educator; b. Lawrence, Mass., July 2, 1947; s. Salvatore and Jane Marie (Lombari) DiB. BSChemE, Northeastern U., 1970; MS in Indsl. Hygiene, Harvard U., 1975. Diplomate Am. Acad. Indsl. Hygiene, Bd. Cert. Safety Profls. Rsch. asst. dept. environ. health scis. Harvard Sch. Pub. Health, Boston, 1966-68, staff indsl. hygienist, 1975-76, vis. lectr., 1986—, cons. dept. continuing edn., 1978—; asst. chemist div. occupational hygiene Mass. Dept. Labor and Industries, Boston, 1968-69; indsl. hygienist dept. environ. health and safety Harvard U. Health Svcs., Cambridge, Mass., 1976-86; indsl. hygiene engr. dept. health, safety-environ. affairs Polaroid Corp., Waltham, Mass., 1986-89; assoc. indsl. hygiene officer environ. med. svc. MIT, Cambridge, 1989-92, indsl. hygiene officer, 1992—; pres. DiBerardinis Assocs., Inc., indsl. hygiene cons., Wellesley, Mass., 1980—; mem. com. on safety codes for exhaust systems Am. Nat. Standards Inst., 1984—, also chmn. subcom. on lab. ventilation. Author: (with others) Guidelines for Laboratory Design: Health and Safety Considerations, 2d edit., 1992; contbr. articles to profl. jours., chpt. to book. Mem. sci. adv. com. City of Cambridge, 1984-86. Mem. Am. Conf. Govtl. Indsl. Hygienists, Brit. Occupational Hygiene Soc., Air Pollution Control Assn., Am. Soc. Safety Engrs., Am. Indsl. Hygiene Assn. (pres. New Eng. sect. 1985-86), Am. Acad. Indsl. Hygiene (sec.-treas. 1983-88, pres. 1993), Harvard Club of Boston. Avocations: winemaking, basketball, racquetball. Home: 4 Martin Rd Wellesley MA 02181-2440 Office: MIT Rm 20C-204 77 Massachusetts Ave Cambridge MA 02139-4307

DIBNER, MARK DOUGLAS, research executive, industry analyst; b. N.Y.C., Nov. 7, 1951; s. David Robert and Dorothy Joyce (Siegel) D.; m. M. Elaine Dibner, Jan. 1, 1983; 1 child, Ned Isaac. BA, U. Pa., 1973; PhD, Cornell U., 1977; MBA, Widener U., 1985. Rsch. fellow U. Colo. Med. Ctr., Denver, 1977-79; rsch. assoc. U. Calif., San Diego, 1979-80; prin. scientist E.I. DuPont, Wilmington, Del., 1980-86; dir. Inst. for Biotech. Info., Research Triangle Park, N.C., 1986—; v.p. for biotech. info. N.C. Biotech. Ctr., Research Triangle Park, 1986—; adj. assoc. prof. Fuqua Sch. of Bus., Duke U., Durham, 1990-91, N.C. State U., Raleigh, 1990—; bd. dirs. Assn. of Biotech Cos., Washington, 1989—, EDITEK, Inc.; chmn. Coun. of Biotech. Ctrs., Washington, 1988-90. Author: (with L.G. Davis and J.F. Battey) Basic Methods in Molecular Biology, 1986; Biotechnology Guide USA: Companies, Data and Analysis, 1988, 2d edit., 1991; (with R.S. White) Biotechnology Japan, 1989; (with R.T. Yuan) Japanese Biotechnology: A Comprehensive Study of Biotechnology in Japan, 1990; editor: (with T.J. Mabry and S.C. Price) Commercializing Biotechnology in the Global Economy, 1991; contbr. articles to profl. jours. Recipient Disting. Svc. award Assn. of Biotech. Cos., 1990, 93. Mem. Fedn. Am. Socs. Exptl. Biology, Am. Mgmt. Assn. Office: Inst for Biotech Info NC Biotech Ctr PO Box 13547 Research Triangle Park NC 27709

DIBONA, CHARLES JOSEPH, association executive; b. Quincy, Mass., Feb. 26, 1932; s. Guido Ralph and Helen Elizabeth (Pangraze) DiB.; m. Evelyn Rauch, July 2, 1959; children—Caroline Anne, Charles J. B.S., U.S. Naval Acad., 1956; M.A. (Rhodes scholar), Oxford U., Eng., 1962. Pres., chief exec. officer Center for Naval Analyses, 1967-73; spl. cons. to Pres. U.S., dep. dir.; White House Energy Policy Office, 1973-74; exec. v.p., chief oper. officer Am. Petroleum Inst., Washington, 1974-78; pres., chief exec. officer Am. Petroleum Inst., 1979—; vice chmn. U.S. nat. com. World Petroleum Congress; mem. Fed. City Council; bd. dirs. Logistics Mgmt. Inst. Served to lt. comdr. U.S. Navy, 1956-67. Mem. UN Assn. of U.S.A. (bd. dirs.), Am. Coun. for Capital Formation, World Econ. Forum (bd. govs.). Roman Catholic. Clubs: Cosmos, F Street, City Tavern, Met., Chevy Chase Country Club. Home: 9306 Georgetown Pike Great Falls VA 22066-2725 Office: Am Petroleum Inst 1220 L St NW Washington DC 20005-4018

DICAMILLO, PETER JOHN, software engineer; b. Needham, Mass., Mar. 18, 1953; s. Daniel Alfred and Eleanor Lucretia (Giurleo) DiC. BS, Brown U., 1975, MS, 1977. Lead systems programmer Brown U. Computing and Info. Svcs., Providence, 1979—. Author computer program. Mem. IEEE, Assn. for Computing Machinery, Tau Beta Pi. Home: 23 De Francesco Cir Needham MA 02192 Office: Brown U Computer Info Svcs PO Box 1885 Providence RI 02912

DICE, BRUCE BURTON, exploration company executive; b. Grand Rapids, Mich., Dec. 24, 1926; s. William and Wilma (Rose) D.; children: Karen, Kevin, Kirk. BS in Geology, U. Mich., 1950; MS in Geology, Mich. State U., 1956. With El Paso (Tex.) Natural Gas, 1956-62, Drilling and Exploration Co., 1962-63; chief geologist Ocean Drilling and Exploration, New Orleans, 1963-75; pres. Transco Exploration Co., Houston, 1975-82, Dice Exploration Co., Inc., Houston, 1982—; bd. dirs. Maxus Energy Corp., Dallas; cons. in field. Elder Northwoods Presbyn. Ch., Houston; mem. Republican Com., Houston. Mem. Am. Assn. Petroleum Geologists, Houston Geol. Soc., Forum Club (Houston). Home: 1907 Grand Valley Dr Houston TX 77090-1052 Office: Dice Exploration Co Inc 14405 Walters Rd Houston TX 77014-1309 also: PO Box 73507 Houston TX 77273

DICK, ELLEN A., computer engineer. MS in Computer Sci. and Math., U. Lowell, 1989. Software prin. engr. Digital Equipment Corp., Nashua, N.H., 1987—. Office: Digital Equipment Corp 110 Spit Brook Rd Nashua NH 03062

DICK, GARY LOWELL, agricultural consultant, educator; b. St. Paul, June 8, 1954; s. Donald M. and Betty J. (Perry) D.; m. Rhonda Sue Fuller, Jan. 8, 1977; children: Shannon, Jenette, Michelle. Lic. Ariz. pest control adv.; cert. Kans. Rsch./Demo Pesticide applicator. Agrl. cons. Sci. Crop Svc. Divsn. of Collingwood Grain, Greensburg, Kans., 1976-81, Agrl. Tech. Co., Inc., McCook, Nebr., 1981-82; agronomist Mehl Farming Hill Divsn., North Platte, Nebr., 1982; rsch. asst. Dept. Entomology Kans. State U., Garden City, Manhattan, 1983-87, asst. prof. Dept. Entomology, 1989-92; extension IPM specialist Dept. Entomology U. Ariz., Casa Grande, Tucson, 1987-89; pres., agrl. cons. GLD Agrl. Cons. Inc., Garden City, Kans., 1992—; chair Pesticide Bldg. Com. Southwest Rsch. Ext. Ctr., Garden City, 1990-92, mem. awards com., 1991-92; cons. Haas Farms, Wilcox, Bonita, Ariz., 1991-92. Contbr. articles to Mycologia, profl. jour., 1992. Deacon, Ch. clerk, First Baptist Ch., Greensburg, Kans., 1979-81, deacon, Garden City, 1990-91. Grantee various Agri-chemical cos., Garden City, 1989-92; Russian Wheat Aphid grantee Ariz. Commn. Agriculture and Horticulture, Casa

Grande, 1988-89; Corn Pest Mgmt. grantee Kans. Corn Commn., Garden City, 1992. Mem. Entomol. Soc. Am., Kans. Entomol. Soc., Phi Kappa Phi. Republican. Am. Baptist. Achievements include conducting first intensive study of fungal pathogen, Neozygites adjarica, in Banks grass mites Oligonychus pratensis; dicovered that a psocid Ectopsocopis cryptomeriae, previously thought to be of little consequence in field corn, may be an important spider mite predator; intensive study of alternate winter spider mite host plants in southwest Kans.; conducted initial rsch. on distbn. and impact of Russian Wheat Aphid in Ariz. Home: 1716 Pinecrest Garden City KS 67846

DICKE, ROBERT HENRY, educator, physicist; b. St. Louis, May 6, 1916; s. Oscar H. and Flora (Peterson) D.; m. Annie Henderson Currie, June 6, 1942; children: Nancy Jean Dicke Rapoport, John Robert, James Howard. A.B., Princeton U., 1939; Ph.D., U. Rochester, 1941, D.Sc. (hon.), 1981; D.Sc. (hon.), U. Edinburgh, 1972, Ohio No. U., 1981; DSc. (hon.), Princeton U., 1989. Microwave radar devel. Radiation Lab., MIT, 1941-46; physics faculty Princeton U., 1946—, Cyrus Fogg Brackett prof. physics, 1957-75, Albert Einstein prof. sci., 1975-84, Albert Einstein prof. sci. emeritus, 1984—, chmn. physics dept., 1967-70; mem. adv. panel for physics NSF, 1959-61; chmn. adv. com. atomic physics Nat. Bur. Standards, 1961-63; mem. com. on physics NASA, 1963-70, chmn., 1963-66; chmn. physics adv. panel Com. on Internat. Exchange of Persons (Fulbright-Hays Act), 1964-66; chmn. adv. com. on radio astronomy telescopes NSF, 1967, 69; mem. Nat. Sci. Bd., 1970-76; vis. com. Nat. Bur. Standards, 1975-79, chmn., 1979; vis. prof. Harvard U., 1954-55, Inst. Advanced Study, 1970-71; Sherman Fairchild Disting. scholar Calif. Inst. Tech., 1975; Walker Ames prof. U. Wash., Seattle, 1979; Jaynes lectr. Am. Philos. Soc., 1969; Scott lectr. Cambridge U. (Eng.), 1977. Author: (with Montgomery, Purcell) Principles of Micro-wave Circuits, 1948, (with J.P. Wittke) An Introduction to Quantum Mechanics, 1960, The Theoretical Significance of Experimental Relativity, 1964, Gravitation and the Universe, 1970. Trustee Assoc. Univs. Inc., 1980-88. Recipient Nat. Medal Sci., 1970, NASA medal for exceptional sci. achievement, 1973, Cresson medal Franklin Inst., 1974, Michelson Morley award Case Western Res. U., 1987, Pioneer award IEEE Microwave Theory and Techniques Soc., 1991. Mem. NAS (Comstock prize 1973), Am. Philos. Soc., Am. Geophys. Union, Am. Phys. Soc., Am. Astron. Soc. (Beatrice M. Tinsley prize 1992), Am. Acad. Arts and Scis. (Rumford medal 1967), Royal Astron. Soc. (assoc.).

DICKENS, DORIS LEE, psychiatrist; b. Roxboro, N.C., Oct. 12; d. Lee Edward and Delma Ernestine (Hester) D.; B.S. magna cum laude, Va. Union U., 1960; M.D., Howard U., 1966; m. Austin LeCount Fickling, Oct. 15, 1975. Diplomate Nat. Bd. Med. Examiners. Intern, St. Elizabeth's Hosp., Washington, 1966-67, resident, 1967-70; staff psychiatrist, dir. Mental Health Program for Deaf, St. Elizabeth's Hosp., Washington, 1970-87; clin. prof. Howard U. Coll. Medicine, 1982—. Co-founder Nat. Health Care Found. for Deaf (named now Deaf Reach); med. officer Region 4 Community Mental Health Ctr., Washington, Commn. on Mental Health, 1987—. Recipient Dorothea Lynde Dix award, 1980. Mem. Am. Psychiat. Assn. (achievement awards bd. 1988-89), Washington Psychiat. Soc., Alpha Kappa Mu, Beta Kappa Chi. Author: How and When Psychiatry Can Help You, 1972; You and Your Doctor; contbg. author: Hearing and Hearing Impairment, 1979, Counseling Deaf People, Research and Practice. Home: 12308 Surrey Circle Dr Tantallon MD 20744

DICKENS, THOMAS ALLEN, physicist; b. Radford, Va., May 15, 1959; s. George Thomas and Frances Nadene (Blair) D.; m. Gina Lynn King, May 5, 1984. BS in Physics, U. Va., 1981; PhD in Physics, Princeton U., 1987. Staff mem. MIT Lincoln Lab., Lexington, Mass., 1987-90; rsch. specialist Exxon Prodn. Rsch., Houston, 1990—; Mem. Circle K Club, Charlottesville, Va., 1978-79. Recipient Joseph Henry prize Princeton, 1981, Sr. Physics award U. Va., 1981. Mem. Am. Phys. Soc., Soc. Exploration Geophysicists. Home: 5454 Newcastle #1903 Houston TX 77081

DICKERSON, MICHAEL JOE, telecommunications engineer; b. Seattle, Aug. 11, 1967; s. Joe Steel and Heideltraut Ida (Schimanski) D. BSEE, Va. Poly. Inst. and State U., 1990. Systems engr. Lightwave Spectrum, Inc., McLean, Va., 1990—. Mem. IEEE. Republican. Office: Lightwave Spectrum Inc 8200 Greensboro Dr Ste 303 Mc Lean VA 22102

DICKEY, DAVID G., electrical engineer; b. Red Bank, N.J., Mar. 11, 1969; s. D. Gary and Suzanne (Smith) D.; m. Jennifer Kelli Langton, Aug. 17, 1991. BSEE, Worcester Poly. Inst., 1991. Sr. assoc. elec. engr. Ebasco Svcs., Inc., N.Y.C., 1990—. Coach Pop Warner Football, Aberdeen, N.J., 1987-90. Mem. IEEE, Am. Nuclear Soc. (assoc.). Home: 24 Weller Pl Holmdel NJ 07733 Office: Ebasco Svcs Inc 2 World Trade Ctr New York NY 10048

DICKEY, JOSEPH WALDO, physicist; b. Quantico, Va., Feb. 26, 1939; s. Joseph Lyle and Lillian Belle (Morlan) D.; m. Nancy E. Herr (div. 1975); children: Ellen, William. BS in Physics, Drexel U., 1963; MS in Physics, U. N.H., 1965; PhD in Physics, Catholic U. of Am., Washington, 1976. Sr. rsch. scientist David Taylor Rsch. Cntr., Annapolis, Md., 1965—; Congl. sci. fellow U.S. Congress, Washington, 1983-84; vis. assoc. prof. U.S. Naval Acad. Author: (with others) Technical Advances in Engineering and their Impact on Detection, Diagnostics and Prognostic Methods, 1983, numerous papers; contbr. articles to profl. jours. Pres., bd. dirs. Md. Fed. of Art, 1992—. Recipient Gov. Com. on Employment of the Handicapped award, 1985. Fellow Acoustical Soc. Am.; mem. Am. Inst. Physics., Am. Assn. Woodturners, Guild of Bookworkers, Sigma Xi. Achievements include 3 U.S. patents; development of sensing systems and performing experimental research in the areas of physical acoustics, optics and theoretical work in elastic scattering. Office: David Taylor Rsch Cntr Propulsion & Auxiliary Systems Dept Annapolis MD 21402

DICKINSON, KATHERINE DIANA, microbiologist; b. Albany, N.Y., May 31, 1954; d. Raymond Edward and Mary Jessica (Demott) Ball; m. David Paul Dickinson, Oct. 6, 1984; 1 child, Caitlin Emily. Cert. in med. tech., R.I. Hosp., Providence, 1980; BS, U. R.I., 1980. Microbiologist R.I. Hosp., 1979-81, Robert Wood Johnson Found. Hosp., New Brunswick, N.J., 1981-84; researcher Brown U., Providence, 1984-86; dir. tech. svcs. Analytical Biosystems Corp., Warwick, R.I., 1986-89; clin. rsch. assoc. CytoTherapeutics, Inc., Providence, 1989—; chair pub. issues com. R.I. div. Am. Cancer Soc., mem. exec. bd., 1991—. Contbr. to profl. publs. Vol. Am. Cancer Soc., Pawtucket, R.I., 1989—; tchr. Myron T. Francis Elem. Sch., East Providence, R.I., 1990-92. Mem. Am. Soc. Clin. Pathologists (cert. med. technician). Home: 218 Fair St Warwick RI 02888-2861

DICKINSON, WILLIAM RICHARD, geologist, educator; b. Nashville, Oct. 26, 1931; s. Jacob McGavock and Margaret Adams (Smith) D.; m. Margaret Anne Palmer, 1953 (div. 1968); children: Ben William, Edward Ross; m. Jacqueline Jane Klein, Feb. 20, 1970. BS in Petroleum Engring., Stanford U., 1952, MS in Geology, 1956, PhD in Geology, 1958. Prof. geology Stanford U., Palo Alto, Calif., 1958-79; prof. geoscis. U. Ariz., Tucson, 1979-91; retired, 1991. Contbr. and editor articles to profl. jours. Lt. USAF, 1952-54. Guggenheim Meml. fellow, 1965. Fellow Geol. Soc. Am. (Penrose medal 1991); mem. Am. Geophys. Union, Am. Assn. Petroleum Geologists, Nat. Acad. Sci., Nat. Assn. Geology Tchrs., Soc. for Sedimentary Geology.

DICKLER, HOWARD BYRON, biomedical administrator, research physician; b. Chgo., Jan. 2, 1942; s. Jerome Alvin and Josephine Rae (Sweet) D.; m. Leah Kayser, June 26, 1966 (div. Apr. 1986); 1 child, Joanna; m. Ana Isabel Martinez, Sept. 20, 1986; 1 child, Carl. BA, Johns Hopkins U., 1964; MD with honors, George Washington U., 1968. Diplomate Am. Bd. Internal Medicine, Nat. Bd. Med. Examiners. Lt. comdr. USPHS, 1972, advanced through grades to capt., 1985; intern in medicine N.Y. Hosp.-Cornell U. Med. Ctr., N.Y.C., 1968-69, resident, 1969-71; rsch. assoc. Rockefeller U., N.Y.C., 1971-72; clin. assoc. Nat. Cancer Inst., Bethesda, Md., 1972-74, sr. investigator, 1974-89, mem. instnl. rev. bd., 1982-84; acting dep. div. dir. Nat. Inst. Allergy and Infectious Disease, Bethesda, 1990-91, chief clin. immunology br., 1989—; com. vice chmn. WHO, Geneva, 1981-85; sci. lectr. at over 30 univs.; speaker, symposium chmn. at over 25 nat. and internat. sci. meetings. Assoc. editor Jour. Immunology, 1976-79; contbr. articles to Jour. Exptl. Medicine, Advances in Immunology. Recipient Commendation medal USPHS, 1985, Outstanding Svc. medal, 1991. Mem. Am. Assn. Immunologists, Am. Fedn. Clin. Rsch., Am. Soc. Clin. Investigation, Clin. Immunology Soc. (councilor 1991—), Alpha Omega Alpha. Democrat. Jewish. Achievements include discovery of receptors for antibody on human cells, interactions between various immune cell receptors, and regulatory mechanisms which control antibody production; pioneering research on classification of human immune cell populations. Home: 11009 Fawsett Rd Potomac MD 20854-1721 Office: NIH/NIAID Solar Bldg Rm # 4A-19 Bethesda MD 20892

DICKMAN, DEAN ANTHONY, mechanical engineer; b. Alton, Ill., June 9, 1961; s. John Quentin and Ann (Maley) D.; m. Rebecca Lyn Papesh, Oct. 6, 1984; children: Lindsay Rhianna, Justin Alexander. BSME, Ohio State U., 1985; MSME, Case Western Res. U., 1992. Engr. Loral Systems Group, Akron, Ohio, 1985-89; sr. engr. to engring. specialist Gen. Dynamics Corp., Ft. Worth, 1989—. Home: 8521 Charleston Ave Fort Worth TX 76123 Office: Gen Dynamics PO Box 748 Fort Worth TX 76101

DICKMAN, ROBERT LAURENCE, physicist, researcher; b. N.Y.C., May 16, 1947; s. Sidney and Eva (Goldberg) D.; m. Albertina Catharina Otter, Sept. 18, 1975; children: Joshua, Ilana. AB, Columbia U., 1969, PhD, 1976. Postdoctoral rsch. assoc. Rensselaer Poly. Inst., Troy, N.Y., 1975-77; mem. tech. staff The Aerospace Corp., L.A., 1977-80; faculty rsch. assoc. U. Mass., Amherst, 1980-85, assoc. prof., staff assoc., 1985-92; program mgr. Nat. Sci. Found., Washington, 1992—. Editor: Molecular Clouds in the Milky Way and External Galaxies, 1988; contbr. 80 articles to profl. jours. Recipient Ernest Fullam award Dudley Obs., 1986. Mem. Am. Phys. Soc., Am. Astron. Soc., Internat. Astron. Union. Office: NSF Div Astronomical Scis 1800 G St NW Washington DC 20550

DICKMAN, STEVEN GARY, science writer; b. Suffern, N.Y., Mar. 6, 1962; s. Barry and Carol (Hiller) D. BA in Biochemistry cum laude, Princeton U., 1984. Sci. writing intern Bus. Week mag., N.Y.C., 1986; ctrl. Europe corr. Nature Mag., Munich, 1987-91; freelance sci. writer Munich, 1991-92; freelance writer Science, The Economist, Cambridge, Mass., 1993—; guest lectr. U. Constance, Germany, U. Zurich, Switzerland, 1991—. Knight Sci. Journalism fellow MIT, 1992-93, German Acad. Exch. Svc. fellow Heidelberg, Germany, 1984-86. Mem. Internat. Sci. Writers Assn., Nat. Assn. Sci. Writers. Home: 8 Saint Paul St Apt 3 Cambridge MA 02139

DICKSON, BRIAN, physician, researcher, educator; b. Dearne, Yorkshire, U.K., Dec. 30, 1950; came to U.S., 1983; s. James Edward and Evelyn Maud (Windle) D.; m. Isobel Coleman, Mar. 23, 1974; children: Mark Christopher, Emily Clare, Antony Charles. B in Medicine and Surgery, Adelaide (South Australia) U., 1974. Intern Queen Elizabeth Hosp., Adelaide, South Australia, 1975-76; resident No. Regional Cardiothoracic Centre Shotley Bridge Hosp., Consett, Eng., 1976-77; physician Royal Adelaide Hosp., 1974-76; registrar Northern Cardiothoracic Ctr., Consett, U.K., 1976-78; group dir. Smith Kline Pharms., Phila., 1978-86; v.p. Searle Pharms., Chgo., 1986-88; sr. v.p. Warner Lambert-Parke Davis Rsch., Ann Arbor, Mich., 1988-90; chmn. Dickson Gabbay Corp., Berwyn, Pa., 1990—; mem. nutritional adv. bd. WHO, Geneva, Switzerland, 1980-82; adj. prof. medicine So. Ill. U., Springfield, 1987—. Editor-in-chief Jour. Pharm. Medicine, 1990—; contbr. over 66 articles to profl. jours. Fellow Royal Soc. Medicine; mem. Australian Med. Assn., Brit. Med. Assn., Faculty of Pharm. Medicine. Achievements include U.S. and European patent for Albendazole--Medical Use in Hydatid Disease; research in designing and conduct of appropriate development programs for pharmaceutical new chemical entities. Office: Dickson Gabbay Corp 1205 Westlakes Dr Ste 150 Berwyn PA 19312-2405

DICKSON, JAMES EDWIN, II, obstetrician/gynecologist; b. Pontiac, Mich., Feb. 18, 1943; s. James Edwin and Virginia (Farrar) D.; m. Joan Gayle Coonley, July 21, 1967; children: Alison, Andrew. BS, U. Mich., 1965; MD, Wayne State U., 1969. Diplomate Am. Bd. Ob-Gyn. Intern Harborview Med. Ctr., Seattle, 1969-70; resident U. Mich. Med. Ctr., 1972-75; pvt. practice Geneva, N.Y., 1975—; pres. Geneva Med. Assocs.; chmn. dept. ob-gyn. Geneva Gen. Hosp. Capt. M.C., USAF, 1970-72. Fellow Am. Coll. Obstetricians and Gynecologists; mem. Soc. Am. Laparoscopists, Miller Ob-Gyn Soc., N.Y. State Med. Soc., Buffalo Ob-Gyn Soc., Rotary (pres. Geneva 1988). Avocation: astronomy. Home: 16 Maplewood Dr Geneva NY 14456 Office: Geneva Med Assocs 324 W North St Geneva NY 14456-1596

DICKSON, PAUL WESLEY, JR., physicist; b. Sharon, Pa., Sept. 14, 1931; s. Paul Wesley and Elizabeth Ella (Trevethan) D.; m. Eleanor Ann Dunning, Nov. 17, 1952; children: Gretchen Ann, Heather Elizabeth, Paul Wesley. BS in Metall. Engring., U. Ariz., 1954, MS, 1954; PhD in Physics, N.C. State U., 1962. With Westinghouse Electric Corp., Large, Pa., 1963-84, mgr. weapon systems, 1965-68, mgr. advanced projects, 1969-72; mgr. reactor analysis and core design Westinghouse Electric Corp., Madison, Pa., 1972-79; tech. dir. Westinghouse Electric Corp., Oak Ridge, 1979-84; with EG & G Idaho, Idaho Falls, 1984-89, mgr. new tech. devel., 1984-87, mgr. reactor projects and programs, 1987-88; dir. Ctr. for Nuclear Engring. and Tech., 1988-89; tech. dir. reactor restart div. Westinghouse Savannah River Co., 1989-92; chief engr. nuclear materials processing div. Westinghouse, 1992—; mem. adv. com. on advanced propulsion systems NASA, Washington, 1970-72; mem. adv. com. reactor physics AEC/Dept. Energy, 1974-79; mem. rev. com. applied physics Argonne (Ill.) Nat. Lab., 1978-83, chmn., 1980; mem. rev. com. engring. physics Oak Ridge Nat. Lab., 1982-86, chmn., 1986; mem. fellow selection com. Dept. Energy, 1981-82; mem. rev. com. EBR II Argonne Nat. Lab., 1984, sci. and tech. adv. com., 1985-91. Contbr. numerous sci. articles to profl. publs. Capt. USAF, 1955-63. Fellow Am. Nuclear Soc.; mem. Am. Phys. Soc., N.Y. Acad. Scis., AIME, AAAS, Scabbord and Blade, Sigma Xi, Phi Kappa Phi, Tau Beta Pi, Phi Lambda Upsilon, Sigma Pi Sigma. Republican.

DIDLAKE, RALPH HUNTER, JR., surgeon; b. Albuquerque, Sept. 9, 1953; s. Ralph Hunter Sr. and Lorraine (McLaurin) D.; m. Millie Faith McDonald, Nov. 12, 1983; children: James Daniel, Jennifer Claire, Sarah Hunter. BS in Zoology with honors, U. Miss., 1975; MD, U. Miss., Jackson, 1979. Diplomate Am. Bd. of Surgery; lic. Miss. Gen. surgery resident U. Miss. Med. Ctr., Jackson, 1979-81, 83-84, resident in transplantation rsch. Exptl. Surgery Lab, 1981-83, chief resident in gen. surgery, 1984-85; transplantation fellow, instr. surgery U. Tex. Health Sci. Ctr., Houston, 1985-87; asst. prof. surgery U. Miss. Sch. Medicine, Jackson, 1987-90, assoc. prof., 1990—; mem. staff Hermann Hosp., Houston, 1985-87, Univ. Hosp., Jackson, 1987—, (also cons.) Med. Ctr., Jackson, 1987—, Miss. Meth. Rehab. Ctr., Jackson, 1987—; bd. dirs. ESRD Network 8, Inc., 1988—, med. rev. bd. mem., 1988-90; mem. teaching faculty, Geriatric Edn. Ctr.; mem. adv. com. Hinds Community Coll., Raymond, Miss.; invited lectr. Dept. Surgery, Ottawa (Ont., Can.) Civic Hosp., 1988; vis. prof., Dept. Surgery, Abington (Pa.), Meml. Hosp., 1990; presenter in field, internat. and U.S. Co-contbr. over 35 articles and 25 abstracts to sci. publs. Co-grantee NIH, 1984-89, Nat. Lung Inst., 1986-91; grantee Biomed. Rsch. Support, 1987-88, 89-90, Merck, Sharp and Dome, 1990-91. Fellow ACS (applicants com. 1990); mem. AMA, Acad. Surg. Rsch. (bd. dirs. 1990-91), Am. Soc. Transplant Surgeons, Soc. Am. Gastrointestinal Endoscopic Surgeons, Assn. Acad. Surgery, Am. Heart Assn. (Basic Sci. Coun.), Am. Soc. Gastrointestinal Endoscopy, Soc. for Leukocyte Biology, The Microcirculatory Soc., Miss. Med. Assn. (House Dels. 1990), Miss. Acad. Sci. (chmn. Health Sci. div. 1984), Southeastern Surg. Congress, Cen. Med. Soc., Transplantation Soc., Miss. Gerontol. Soc., Sigma Xi. Achievements include research in treatment of septic shock with fructose 1,6-diphosphate, in evaluation of hepatic ischemic injury by in vitro perfusion, in treatment of ARDS with fructose diphosphate, in hypothermic storage of hepatocytes, in neointimized, norfloxacin bonded PTFE. Office: U Miss Med Sch 2500 N State St Jackson MS 32916

DIDOMENICO, PAUL B., military officer; b. Washington, Feb. 28, 1968; s. Peter and Jean Marie (Halverson) DiD. BS in Astronautical Engring., USAF Acad., 1989; MS in Aeronautics & Astronautics, MIT, 1991. Commd. 2 lt. USAF, 1989—; project officer Global Positioning System Program Office, L.A. AFB, 1991—. C.S. Draper Lab. Rsch. fellow, 1989-91; recipient Bernard Kriegsman award, 1991. Mem. AIAA, MIT Club Southern Calif., Sigma Xi. Home: 7047 Alvern St #227 Los Angeles CA 90045 Office: SMC/CZS PO Box 92960 Los Angeles AFB CA 90009

DIDSBURY, HOWARD FRANCIS, futurist educator, lecturer, consultant; b. July 15, 1924; s. Howard Francis Didsbury and Blanche Mascot. BA, Yale U., 1947; MA, Harvard U., 1951; PhD, The Am. U., 1960. Rsch. analyst Am. Coun. Learned Socs., Washington, 1953-54; edn. officer Embassy of Pakistan, Washington, 1953-56; master of history Longfellow Sch. for Boys, Bethesda, Md., 1956-59; asst. prin. Maret Sch., Washington, 1959-60; from assoc. to full prof. Kean Coll. N.J. (formerly Newark State Coll.), Union, 1960-92; spl. studies dir. World Future Soc., Bethesda, 1977—; mem. N.J. commn. to study adolescent edn., Trenton, 1976-77. Editor: The Future: Opportunity, Not Destiny, 1988, Challenges and Opportunities: From Now to 2001, 1986, The Global Economy, 1985, Creating a Global Agenda, 1984; editor pre-conf. volumes World Future Soc., 1980—, coord. Prep 21 project, 1990—; writer, prodr. (26 part film series) Visions, Nightmares and Forecasts, 1991-92. Recipient Commendation Cert., U.S. Coun. for World Communications, 1983. Mem. Union of Concerned Scientists, World Futures Studies Fedn., U.S. Assn. for Club of Rome, Nat. Press Club. Democrat. Home: 2862 28th St NW Washington DC 20008 Office: World Future Soc 7910 Woodmont Ave Ste 450 Bethesda MD 20814

DIEBOLD, CHARLES HARBOU, seed company executive; b. Macedon, N.Y., Nov. 7, 1909; s. Frank Arnold and Emma Beneta (Harbou) D.; m. Elizabeth Strong, Dec. 21, 1935; children: Robert, James, David. BS in Forestry, Cornell U., 1930, PhD in Silviculture and Soils, 1936. Soil surveyor Cornell U., Ithaca, N.Y., 1930-33; soil scientist U.S. Soil Erosion Svc., Ithaca, 1934; soil scientist U.S. Soil Conservation Svc., Ellicottville, N.Y., 1942-44, Albuquerque, 1944-66; forester U.S. Forest Svc., Washington, Ft. Collins, Colo., 1936-42; dir. Appropriate Rural Tech. Assn. Inc., Peralta, N.Mex., 1984—; founder C.H. Diebold Missions, 1985—. Author: Estimating Soil Moisture by Feel. Precinct chrm. Peralta Dem. Com., 1982—; vol. missionary Presbyn. Ch. N.Y., Pakistan, Thailand, Mex., 1967-84. Recipient Outstanding Conservationist award Valencia Soil and Water Dist., 1992. Mem. Am. Soc. Agronomy, Soc. Range Mgmt., Soil Conservation Soc. Home and Office: 268 La Ladera Peralta NM 87042

DIECK, WILLIAM WALLACE SANDFORD, chemical engineer; b. Newark, May 23, 1924; s. George Ernst and Gladys Dorothy (Sandford) D.; m. Jane Elizabeth Humphreys, Sept. 30, 1950; children: Christopher, Gretchen, David, Louisa. BChemE, Rensselaer Polytech. Inst., 1948; MS in Engring., Princeton U., 1952. Registered profl. engr., N.Y. From engr. to asst. supt. film sensitizing Eastman Kodak Co., Rochester, N.Y., 1949-84; cons. Rochester, 1984—. Patentee in field. V.p. William Warfield Scholarship Eastman Sch. Music, 1986—. 1st lt. USAF, 1943-46. Mem. Am. Inst. Chem. Engrs. (sect. dir. 1984—), Am. Chem. Soc., Soc. for Imaging Sci. and Tech., N.Y. Acad. Scis., Princeton Club N.Y., Phi Lambda Upsilon, Sigma Xi. Presbyterian. Avocations: music, playing trombone, swimming, railroad history buff. Home and Office: 225 Idlewood Rd Rochester NY 14618

DIEHL, MYRON HERBERT, JR., engineer; b. York, Pa., Apr. 4, 1951; s. Myron H. and Elda (Hermann) D. BS in Indsl. Tech., Catonsville Coll., 1978; BS in Polit. Sci., U. Balt., 1982. Inspector Hartford (Conn.) Ins. Co., 1975-79, nuclear inspector, 1980-84, regional supr., 1984-88; chief boiler inspector State of Md., Balt., 1988—; adviser Md. Bd. Boiler Rules, 1988—, instr., 1988—. Author: Thoughts, 1982. Pres. Brookshire Condominium Assn., Reisterstown, Md., 1979. With USN, 1969-75. Mem. ASME (trustee nat. bd. boiler inspectors), VFW, Am. Nuclear Soc., Nat. Bd. Boiler and Pressure Vessel Inspectors, Engring. Soc. Balt., Am. Legion. Achievements include establishment of relations with safety inspection departments of several countries and Md. Dept. Licensing and Regulation. Office: State of Md 501 St Paul Pl Baltimore MD 21202-2208

DIEHL, RANDY LEE, psychology educator, researcher; b. Freeport, Ill., Apr. 2, 1949; s. Loren Emerson and Mildred Saloa (Clayton) D.; m. Mary Louise Wiley, June 13, 1970; children: Matthew Benjamin, Sarah Elizabeth. BS in Psychology, U. Ill., 1971; PhD in Psychology, U. Minn., 1975. Predoctoral assoc., ctr. for rsch. in human learning U. Minn., Mpls., 1971-75; asst. prof., dept. psychology U. Tex., Austin, 1975-81, assoc. prof., dept. psychology, 1981-88, prof., 1988—, fellow, ctr. for cognitive sci., 1989—; panel mem., dissertation and postdoctoral minority fellowships Ford Found., Washington, 1990, Nat. Rsch. Coun., Washington, 1991; vis. lectr. Instituto de techológico y de Estudios Superíores, Monterrey, Mex., summer 1991, 92. Consulting editor Jour. Exptl. Psychology: Human Perception and Performance; editor for North Am. Phonetica; contbr. articles to profl. jours., chpts. to books. Fellow NSF, 1971-74; Rsch. grantee NIH, 1977-80, 80-83, 84-87, 87-94. Mem. Internat. Soc. Ecol. Psychology, Am. Assn. Phonetic Scis., Linguistic Soc. Am., Acoustical Soc. Am. Democrat. Achievements include co-demonstration of human speech perceptual performance being explained in terms of general auditory mechanisms that apply equally to perception of speech and nonspeech signal mechanisms moreover that are not unique to humans; research in relationship between phonetics and phonology, auditory bases of speech perception and production, categorization of speech sounds, experimental psycholinguistics. Home: 9504 Topridge Dr Austin TX 78750 Office: U Tex Dept Psychology 330 Mezes Austin TX 78712

DIERCKS, FREDERICK OTTO, government official; b. Rainy River, Ont., Can., Sept. 8, 1912; s. Otto Herman and Lucy (Plunkett) D.; m. Kathryn Frances Transue, Sept. 1, 1937; children: Frederick William, Lucy Helena. B.S., U.S. Mil. Acad., 1937; M.S. in Civil Engring., MIT, 1939; M.S. in Photogrammetry, Syracuse U., 1950. Registered profl. engr., D.C. Commd. 2d lt. U.S. Army Corps Engrs., 1937; advanced through grades to col. U.S. Army, 1952; comdg. officer U.S. Army Map Service, Washington, 1957-61, Def. Intelligence Agy., 1961-63; dep. engr. 8th U.S. Army, Korea, 1963-64; dir. U.S. Army Coastal Engring. Research Ctr., 1964-67; ret., 1967; assoc. dir. U.S. Coast and Geodetic Survey, Rockville, Md., 1967-74; U.S. mem. commn. on cartography Pan Am. Inst. Geography and History, OAS, 1961-67, alt. U.S. mem. directing council, 1970-74, exec. sec. U.S. sect., 1975-87. Decorated Legion of Merit (U.S.), Grand Cross Order of King George II (Greece), Comdr. Most Exalted Order of White Elephant (Thailand), Bronze medal U.S. Dept. Commerce. Fellow ASCE, Soc. Am. Mil. Engrs. (Colbert medal); mem. Am. Soc. Photogrammetry (hon. mem., pres. 1970-71, Luis Struck award), Am. Congress on Surveying and Mapping, Sigma Xi. Republican. Presbyterian. Clubs: Army-Navy, Cosmos (Washington). Lodge: Hancock Lodge 311, A.F. and A.M. (Fort Leavenworth, Kans.). Home: 9120 Belvoir Woods Pkwy #216 Fort Belvoir VA 22060

DIERSING, ROBERT JOSEPH, computer science educator; b. Alice, Tex., Jan. 3, 1949; s. Phillip Theodore and Hulda Imelda (Reissig) D.; m. Julie Gillett, July 28, 1978; children: Julia Diane, Emily Catherine. BBA, Tex. A&I U., 1972, MS, 1974; MBA, Corpus Christi (Tex.) State U., 1983; PhD, Tex. A&M U., 1991. Programmer, analyst Tex. A&I U., Kingsville, Tex., 1972-75; asst. prof. computer sci. Tex. A&I U., Kingsville, 1990-92, assoc. prof. computer info. systems, 1992—; dir. computer svcs. Corpus Christi State U., 1975-83, asst. prof. computer sci., 1980-85, assoc. prof. computer sci., 1985-88, rsch. assoc., 1988-90. Contbr. articles to profl. jours. Recipient svc. awards Radio Amateur Satellite Corp., 1986, 87. Mem. IEEE, AIAA, Assn. for Computing Machinery, Phi Delta Kappa. Avocations: gardening, photography, amateur radio. Office: Tex A&I Univ Campus Box 184 Kingsville TX 78363

DIESEM, JOHN LAWRENCE, information systems specialist; b. Albuquerque, July 16, 1941; s. Walter Franklin and Glen Ethel (Helpbringer) D.; m. Barbara Jane Willmarth, Feb. 25, 1967 (div. Oct. 10, 1976); m. Kathleen Terese Walsh, Feb. 2, 1979. BA, George Washington U., 1964, MA, 1965; cert. in fin. mgmt., NYU, 1974; advanced profl. cert. in acctg., Pace U., 1992. Cert. in prodn. and inventory mgmt.; cert. mgmt. cons. Group mgr. Electronic Data Systems, N.Y.C., 1970-74; dep. commr. N.Y. State, 1974-75; sr. mgr. Arthur Andersen, N.Y.C., 1975-80; v.p. bus. systems devel. McGraw-Hill, N.Y.C., 1982-84; sr. mgr. Touche Ross & Co., N.Y.C., 1984-86; dir. strategic systems planning Peat, Marwick Main & Co., 1986-89; sr. v.p. systems tech. Am. Stock Exch., N.Y.C., 1989-92; group v.p. systems and tech. Simon & Schuster, 1992-93; COO Beta Systems/Kemper Securities Inc., Brookfield, Wis., 1993—. Dir. George Washington U. dean's alumni adv. bd., 1981-83. Served to capt. USAF, 1965-69, Fed. Republic Germany, Vietnam; to lt. col. USAFR; grad. Nat. Def. U., Air War Coll., Air Command and Staff Coll., Indsl. Coll. Armed Forces. Decorated Bronze Star, Air Force Commendation medal, Vietnamese Cross of Gallantry with Palm;

recipient Bus. Sch. Alumnus of the Yr. award George Washington U., 1991. Mem. N.Am. Coun. Info. Mgmt. Execs. (vice chmn.), Conf. Bd., N.Y. Athletic Club (N.Y.C.), Old Chatham (N.Y.) Hunt Club, Army and Navy Club (Washington), Columbia Golf and Country Club, Masons, Cercle Sportif (Saigon, Vietnam), Omicron Delta Kappa, Sigma Chi (life), Phi Eta Sigma, Alpha Kappa Psi. Democrat. Episcopalian. Home: 1150 Greenway Ter Brookfield WI 53005 also: Miller Rd Churchtown NY 12513 Office: Beta Systems Inc 350 N Sunnyslope Rd Brookfield WI 53005

DIESSNER, DANIEL JOSEPH, fiber optics engineer; b. Denver, July 19, 1963; s. Clement Paul and Ruth Anne (Nolan) D. BSEE, U. Colo., 1986, BS in Physics, 1986; MS in Physics, U. Wash., 1990. Software engr. Gates Corp., 1985-86; from engr. to lead engr. Boeing Comml. Fiber Optics Rsch., 1986-91; lead engr. Fiber Optics Group Boeing 777, Seattle, 1992—. Mem. Aeronaut. Radio Inc. (chmn. Fiber Optic Data Bus Working Group). Achievements include implementing the first fiber optic system on a commercial airplane. Office: Boeing 777 PO Box 3707 M/S 02-MP Seattle WA 98124

DIETER, GEORGE E., JR., university dean; b. Phila., Dec. 5, 1928; s. G. Ellwood and Emily (Muench) D.; m. Nancy Joan Russell, June 21, 1952; children: Carol Joan, Barbara June. B.S. in Metall. Engring, Drexel Inst. Tech., 1950; Sc.D., Carnegie Inst. Tech., 1953. Research engr. E.I. duPont Engring Research Lab., Wilmington, Del., 1955-59; research supr. E.I. duPont Engring Research Lab., 1959-62; prof., head dept. metall. engring. Drexel Inst. Tech., 1962-69; dean Coll. Engring. Drexel U., 1969-73; dir. Processing Research Inst., Carnegie-Mellon U., 1973-77; dean Coll. Engring., U. Md., 1977—; cons. in field. Author: Mechanical Metallurgy, 1961, 3d edit., 1986, Engineering Design, 1983, 2d edit., 1991. Mem. 1953-55, AUS. Fellow AAAS, Am. Soc. for Metals (A.E. White award 1986, Sauveur award 1992), Am. Soc. Engring. Edn. (pres. 1993), Minerals, Metal and Materials Soc.; mem. Nat. Acad. Engring., Am. Inst. Metal Engring., Soc. Mfg. Engrs. (Educator award 1987), Fedn. Materials Socs. (pres. 1990-92), Tau Beta Pi, Sigma Xi. Home: 1 Locksley Ct Silver Spring MD 20904-6321 Office: U Md College Park MD 20742

DIETERICH, DOUGLAS THOMAS, gastroenterologist, researcher; b. Queens, N.Y., Mar. 1, 1951; s. Albert Frederick and Florence Anna (Kilroy) D. BS, Yale U., 1973; M in Health Adminstrn., C.W. Post, 1974; MD, NYU, 1978. Diplomate Am. Bd. Internal Medicine and Gastroenterology. Intern, then resident Bellevue Hosp., N.Y.C., 1978-81, fellow gastroenterology, 1981-83; attending physician NYU Hosp., 1983—; cons. Rockefeller Ctr. Mgmt. Corp., N.Y.C., 1983-87, ITT World Hdqs., N.Y.C., 1983-86; asst. dir. medicine Gouverneur Hosp., N.Y.C., 1983-85; acting med. dir. ITT World Communications, Secaucus, N.J., 1985-86; attending physician Gouverneur Hosp. Walk In Clinic, 1981-83; teaching asst. NYU, 1979-83, clin. instr. medicine, 1983-88, clin. asst. prof., 1988-93, clin. assoc. prof., 1993—; mem. AIDS Clin. Trials Group NIH, 1986-94. Contbr. articles to profl. jours. Mem. Am. Liver Found., N.J. Fellow ACP, Am. Coll. Gastroenterology; mem. AMA, Am. Gastroent. Assn., Am. Soc. Gastrointestinal Endoscopy, Am. Soc. Internal Medicine, N.Y. County Med. Soc., N.Y. State Med. Soc., N.Y. Acad. Gastroenterology, Yale Club, Cherry Valley Club. Republican. Lutheran. Home: 62 St James St S Garden City NY 11530-6344 Office: 345 E 37th St New York NY 10016-3217

DIETERLE, ROBERT, chemist. Recipient Clare E. Bacon Person of Yr. award Soc. Plastics Industry, 1993. Home: 1216 Edward Rd Naperville IL 60540*

DIETZ, ALBERT GEORGE HENRY, engineering educator; b. Lorain, Ohio, Mar. 7, 1908; s. Peter and Adele (Grevsmuhl) D.; m. Ruth Avery, Sept. 9, 1936; children: Margaret, Henry Avery. A.B., Miami U., 1930; Sc.D., MIT, 1941. With dept. bldg. engring. and constrn. MIT, Cambridge, 1934-62; asst., instr. MIT, 1936-41, asst. prof., 1941-46, assoc. prof., 1946-50, prof., 1950-62, prof. bldg. engring. depts. civil engring. and architecture, 1962-73, emeritus, 1973—, dir. plastic rsch. lab., 1946-62; on leave of absence to Forest Products Labs. as sr. cons. engr., 1942; field svc. cons. Office Field Service OSRD, 1944-45; cons. constrn. and materials, 1940—; chmn. bldg. rsch. adv. bd. NAS-NRC. Author: Dwelling House Construction, 1946, 5th edit., 1991, Materials of Construction: Wood, Plastics, Fabrics, 1949, (with Marcia Koth, Julio Silva) Housing in Latin America, 1965, Plastics for Architects and Builders, 1970; editor: Engineering Laminates, 1949, Composite Engineering Laminates, 1970, (with Laurence Cutler) Industrialized Systems for Housing, 1971; contbr. articles to profl. jours. Mem. Engring. Edn. Mission to Japan, 1952. Sr. fellow East-West Ctr., 1973-74; recipient John Derham Internat. award Plastics Inst. Australia, 1962; New Eng. award Engring. Socs. New Eng., 1968; named Constrn. Man of Quarter Century Bldg. Rsch. Adv. Bd., Nat. Acad. Scis.-Nat. Acad. Engrs., 1977. Fellow AAAS, Am. Acad. Arts and Scis., ASCE (hon.), Royal Soc. Arts; mem. ASTM (hon., Richard L. Templin award 1948, Walter Voss award 1974, award merit 1957), Soc. Plastics Industry, ASME, Soc. Engring. Edn., Forest Products Research Soc., Soc. Plastics Engrs. (past nat. dir., Internat. Gold medal 1971), Boston Soc. Civil Engrs. (Desmond Fitzgerald award 1945, 56), Bldg. Research Inst. (past dir.), Phi Beta Kappa Assos., Phi Beta Kappa, Sigma Xi, Tau Beta Pi, Chi Epsilon (hon.). Home: 299 Cambridge St Unit 424 Winchester MA 01890-2389 Office: Mass Inst Tech Cambridge MA 02139

DIETZ, ALMA, microbiologist; b. Holyoke, Mass., Nov. 29, 1922; d. Louis William and Matilda Antoinette (Gotthardt) D. BA, Am. Internat. Coll., Mass., 1944; postgrad. in Botany, U. Mich., 1946-48. Registered Microbiologist. Lab. instr., biology Am. Internat. Coll., Springfield, Mass., 1944-46; lab. asst. Marine Biology Lab., Woods Hole, Mass., 1946; microbial taxonomist, culture curator, chem./biological screening unit Infectious Diseases Rsch, The Upjohn Co., Kalamazoo, Mich., 1948—. Contbr. to profl. jours. Recipient J. Roger Porter award Am. Soc. Microbiology, 1993. Mem. Am. Soc. Microbiology, U.S. Fed. for Culture Collections, Soc. Industrial Microbiology, Am. Assn. Advancement Sci., Medical Mycological Soc., Mycological Soc. Am., Nat. Registry Microbiologist, Phycological Soc. Am., World Fed. Culture Collections. Episcopal. Achievements include 5 Patents, and over 70 Taxonomy for U.S. Patents. Home: 2929 Memory Lane Kalamazoo MI 49007 Office: The Upjohn Co Chem Biological Screening Unit 301 Henrietta St Kalamazoo MI 49007*

DIETZ, THOMAS GORDON, chemist; b. Jersey City, Aug. 25, 1955. BS in Chemistry, U. Del., 1977; PhD, Rice U., 1982. Rsch. chemist Exxon Rsch. and Engring., Linden, N.J., 1981-82, sr. rsch. chemist, 1982-86, staff chemist, 1986-88, group leader, 1988-89, sect. head indsl. lubricants and specialties, 1989-91, sect. head gasoline quality, 1991-93; tech. svcs. coord., 1993—. Mem. Soc. Tribologists and Lubrication Engrs. (chmn. hydraulics and machine tool com. 1989-91), Soc. Automotive Engrs. Achievements include development of many lube and fuel products, primarily in areas of industrial lubricants and gasoline. Office: Exxon Rsch & Engring Co PO Box 51 Linden NJ 07036

DIETZ, THOMAS MICHAEL, human ecology educator and researcher; b. Kent, Ohio, Aug. 17, 1949; s. Glen Vernon and Nora Elizabeth (O'Brien) D.; m. Linda Elizabeth Henry, July 1, 1991; children: Alexandra, Adam. B Gen. Studies, Kent State U., 1972; PhD, U. Calif., Davis, 1979. Rsch. assoc. U. Calif., 1976-80; mem. faculty George Washington U., Washington, 1980-83; asst. prof. George Mason U., Fairfax, Va., 1983-85, assoc. prof., 1985-92, prof., 1992—, dir. No. Va. Survey Rsch. Lab., 1988-92; vis. fellow Swedish Collegium for Advanced Study in Social Sci., Uppsala, Sweden, 1988-91; mem. com. on human dimensions of global change NRC, Washington, 1989—; advisor to sec. HEW, Washington, 1972-76. Author: The Risk Professionals, 1987; editor: Handbook for Environmental Planning, 1977, Human Ecology: Crossing Boundaries. Del. Vt. Dem. Conv., 1992. Fellow Danforth Found., 1972, Inst. for Human Ecology, 1988. Mem. N.Y. Acad. Scis., Washington Statis. Soc. (bd. dirs. 1990-91), Sigma Xi. Roman Catholic. Home: 300 Rt 2 Grand Isle VT 05458 Office: Dept Sociology-Anthropology George Mason U Fairfax VA 22030

DIETZE, GERALD ROGER, chemist; b. Portland, Oreg., Aug. 13, 1959; s. Eckhardt Ludwig and Gertrud Herta (Hoopman) D.; m. Susan Elizabeth Morris, May 26, 1984. BS, Portland State U., 1982. Chemist Crystal Specialties, Portland, 1983-85; sr. epi process engr. SEH Am., Vancouver, Wash., 1985—; tech. advisor Clark County Fire Dist. 5 Hazmat Team, Vancouver, 1987—. Recipient Pub. Svc. award Clark County Washington, 1989, letters of commendation Clark County Fire Dist. 5. Mem. Am. Inst. Chemists, Electrochem. Soc., Am. Vacuum Soc. Office: SEH Am 4111 NE 112th Ave Vancouver WA 98682-6799

DÍEZ, JOSÉ ALBERTO, business educator; b. Málaga, Spain, Dec. 30, 1956; s. Tomás Díez and Esperanza De Castro; m. Carmen Redondo, Sept. 7, 1989. Grad., Facultad Econs., Málaga, 1978; D of Bus., Facultad Econs., Santiago, Spain, 1989. Assoc. prof. U. Santiago, 1984-90, titular prof., 1990—. Author: Modern Decision in Public Administration, 1991. Mem. Bus. European Assn., Decision Aid European Group, Esigma. Roman Catholic. Home: República Argentina 37-5o D, 15706 Santiago Spain Office: Facultad Econs, Avda Juan XXIII S/N, 15704 Santiago Spain

DIFINO, SANTO MICHAEL, hematologist; b. Newark, Nov. 24, 1948. BS, Fordham U., 1970; MD, N.J. Med. Sch., Newark, 1974. Diplomate Nat. Bd. Med. Examiners, Am. Coll. Internal Medicine, Am. Bd. Oncology, Am. Bd. Hematology. Intern, resident, fellow SUNY-Upstate Med. Ctr.; attending physician St. Joseph Crouse Irvins Meml. Community Gen. & SUNY Hosps., Syracuse, 1980—; clin. assoc. prof. SUNY Upstate Med. Ctr., Syracuse, 1980—; prin. investigator Community Clin. Outreach Program, Syracuse, 1985—; chief internal medicine St. Joseph's Hosp., Syracuse, 1984—; mem. exec. com. St. Joseph's Hosp., Syracuse, 1985—; trustee Leukemia Soc., Syracuse, 1988—. Contbr. articles to profl. jours. Mem. Cancer Acute Leukemia Group B, Nat. Surg. Adj. Breast and Bowel Project, Mid. Atlantic Oncology Program, Phi Beta Kappa. Office: 7209 Buckley Rd Liverpool NY 13088

DIFIORE, JULIANN MARIE, biomedical engineer; b. Cleve., May 27, 1962; d. Frank Walter and Nancy Marie (Fuchs) Drop; m. Jeff Goffredo DiFiore, Aug. 3, 1985; children: Jonathan Luca, Jason Matthew. BS, U. Toledo, 1984. Rsch. engr. Case Western Res. U., Cleve., 1984—; lectr. Pediatric Fellow's Rsch. Course, Cleve., 1990—. Mem. Perinatal Jour. Club. Achievements include research in founds. of respiratory care and continuous transcutaneous monitoring. Office: Rainbow Babies & Childrens Hosp 2101 Adelbert Rd Cleveland OH 44106

DIGHTON, ROBERT DUANE, military operations analyst; b. Delaware County, Iowa, May 23, 1934; s. Duane Roy and Rosanna Irene (Kirkpatrick) D.; m. Mabel Ann Hamblin, June 12, 1955; children: Mark Duane, Denise Ann Saunders. BS in Aero. Engring., Iowa State U., 1956; MS in Engring. Mechs., St. Louis U., 1968. Mgr. ops. analysis dept. McDonnell Aircraft Co., St. Louis, 1956, 60-89; dir. future systems requirements McDonnell Douglas Corp., Arlington, Va., 1989-90; mem. rsch. staff Inst. Def. Analyses, Alexandria, Va., 1990—. 1st lt. USAF, 1957-59. Recipient Profl. Achievement Citation in Engring. Iowa State U., Ames, 1991. Assoc. fellow AIAA (vice chmn. St. Louis sect. 1988-89); mem. Mil. Ops. Rsch. Soc., Air Force Assn., U.S. Naval Inst., Internat. Test and Evaluation Assn. Methodist. Achievements include research on cost-effectiveness and requirements analyses for Navy and Air Force tactical fighter and attack weapon systems over 30 year period. Home: 9332 Sibelius Dr Vienna VA 22182 Office: Inst Def Analyses 1801 N Beauregard St Alexandria VA 22311

DIGIACOMO, RUTH ANN, research scientist; b. Bellefonte, Pa., May 26, 1966; d. David John and Annie Fulton (Kelton) Krotchko; m. Thomas Allen DiGiacomo, Oct. 6, 1990; children: Jessica Elizabeth and Rebecca Kelton (twins). BS, Albright Coll., 1988; MS, East Stroudsburg U., 1992. Instr. microbiology Northampton C.C., Bethlehem, Pa., 1991; asst. scientist II Schering Plough Rsch. Inst., Kenilworth, N.J., 1991—. Mem. Albright Coll. Biology Club, Albright Coll. Pre-Med. Honor Soc. (historian 1987-88), Sigma Xi. Presbyterian. Office: Schering Plough Rsch Inst Rm D229C 2015 Galloping Hill Rd Kenilworth NJ 07033

DIHLE, ALBRECHT GOTTFRIED FERDINAND, classics educator, professional society administrator; b. Kassel, Germany, Mar. 28, 1923; s. Hermann and Frieda (von Reden) D.; m. Marlene S.L. Meier, Oct. 1, 1949; children: Franziska, Stefanie, Andreas, Barbara, Katharina. PhD, U. Göttingen, Germany, 1946; ThD (hon.), U. Bern, Switzerland, 1982; DPhil (hon.), U. Athens, Greece, 1988; DLitt, Macquarie U., Australia, 1993. Lectr. U. Göttingen, 1950-58; prof. classics U. Cologne, Germany, 1958-74, U. Heidelberg, Germany, 1974-89; pres. Heidelberger Akademie der Wissenschaften, Heidelberg, Germany; vis. prof. Harvard U., Cambridge, Mass., 1965-66, 89-90, Stanford (Calif.) U., 1968, U. Calif., Berkeley, 1973-74, Princeton U., 1983, Durban U., 1984, Perugia U., 1988. With German Army, 1940-42. Fellow Brit. Acad.; mem. Am. Acad. Arts and Scis., Düsseldorf Acad. Sci., Heidelberg Acad. Sci., Acad. Europara, Acad. Inscriptions (corr.), Rotary (pres. 1980-91). Office: U Heidelberg, Marstallhof 4, D-6900 Heidelberg Germany

DI JESO, FERNANDO, biochemistry educator; b. Cosenza, Italy, Oct. 29, 1931; s. Pasquale and Adelaide (Scopa) di J.; m. Andrée Marie Christine Roche, Oct. 13, 1962; 1 child, Vittorio. BA, Vittorio Emanuele II, Naples, Italy, 1949; MD, Univ. Med. Sch., Naples, 1955; PhD, U. Perugia (Italy), 1959; DSc, U. Paris, 1965. Researcher Biochem. Inst., U. Med. Sch., Naples, 1952-54; researcher biochem. dept. Cancer Inst., Naples, 1954-56; assoc. prof. phys. chemistry dept. U. Padua (Italy), 1956-57; researcher biochem. dept. Cancer Inst., Naples, 1957-60; prof. libero docente Biochemistry Inst., U. Med. Sch., Naples, 1961; assoc. prof. biochemistry dept. Coll. de France, Paris, 1961-63; assoc. prof. Biochemistry Inst., U. Med. Sch., Pavia, Italy, 1963-67; assoc. prof. biochemistry and molecular biology dept. Cornell U., Ithaca, N.Y., 1967-69; prof. biochemistry Biochemistry Inst., U. Med. Sch., Pavia, 1969—; bd. dirs. Postdoctoral Med. Sch. Neurophysiopathology, U. Pavia, 1990—; dir. of Cardiovascular Biochemistry Ctr. Editor, author: Membrane-Bound Enzymes, 1971; editor, pub. Medicina Democratica, 1976—, Glénans, 1985—, Aggiornament GISCRIS. Pres. Cooperativa Editoriale Pavese, Pavia, 1969-75, Laureati Cattolici, Pavia, 1970, CISL-Universita, Pavia, 1975-79, Medicina Democratica, Italy, 1976-90, Lega Navale Italiana, Pavia, 1990—. Abroad Rsch. grantee NATO, Tr. An Italian Com. of Nat. Rsch. Coun., 1961-63, Italian Nat. Rsch. Coun., 1967-69; recipient Concorso Nazionale di Poesia Amisani, Mede Municipality and Amisani Assn., Mede, Italy, 1988, Premio Letterario Casentino, Casentino Prize Com., Stia, Italy, 1981, 1988-93, Internat. D'Arte Moderna, Rome, 1992, Ungaretti, Sorrento, 1992, Luci di Poesi, Milan, 1993. Mem. Italian Soc. Biochemistry, Am. Chem. Soc., French Soc. Biology, Internat. Brain Rsch. Orgn., Italian Soc. Cardioneurology, Italian Soc. Pharmacology, Italian Soc. Neurosci., Assn. Fulbright Fellows (award 1967), Senator Micenei Internat. Acad. Office: Prima Cattedra di Chimica, Biologica 3 Viale Taramelli, 27100 Pavia Italy

DIJKSTRA, EDSGER WYBE, computer science educator, mathematician; b. Rotterdam, The Netherlands, May 11, 1930; came to U.S., 1984; s. Douwe Wijbe and Brechtje Cornelia (Kluyver) D.; m. Maria Cornelia Debets, Apr. 23, 1957; children: Marcus Joost, Femke Elisabeth, Rutger Michael. Candidaats degree, U. Leyden, The Netherlands, 1951; doctoral degree, U. Leyden, 1956; PhD, U. Amsterdam, 1959; DSc (hon.), Queen's U. Belfast, No. Ireland, 1976. Staff mem. Math. Centre, Amsterdam, The Netherlands, 1952-62; prof. math. Tech. U., Eindhoven, The Netherlands, 1962-73; rsch. fellow Burroughs Corp., Nuenen, The Netherlands, 1973-84; prof., Schlumberger Centennial chair in computer sci. U. Tex., Austin, 1984—. Editor Acta Informatica. Disting. fellow Brit. Computer Soc.; mem. Royal Netherlands Acad. Arts and Scis., Am. Acad. Arts and Scis. (hon. fgn.), Assn. for Computing Machinery (Turing award 1972). Home: 6602 Robbie Creek Cv Austin TX 78750-8138 Office: U Tex Dept Computer Scis Austin TX 78712-1188

DIKRANJAN, DIKRAN NISHAN, mathematics educator; b. Rousse, Bulgaria, Feb. 5, 1950; s. Nishan Kevorkian and Elis Onik (Shoulian) D.; m. Nevena Stefanova Georghieva, Nov. 8, 1980. Master degree, U. Sofia, Bulgaria, 1973, PhD, 1977; diploma di Perfezionamento, Scuola Normale Superiore, Pisa, Italy, 1983. Researcher Inst. Math., Bulgarian Acad. Sci., Sofia, 1978-80, 84-85, sr. researcher, 1986-91; lectr. U. l'Aquila, Italy, 1991-92; prof. Udine (Italy) U., 1992—; vis. prof. Padua (Italy) U., 1982-83, 91, 92, l'Aquila U., 1989, York U., Toronto, Can., 1989, 91, 93. Author: Topological Groups, 1990, Closure Operators, 1993; contbr. over 70 articles to profl. jours. Sgt. Bulgarian Infantry, 1974-75. Mem. Union Bulgarian Mathematicians. Avocation: swimming. Home: via Natisone 10, 33100 Udine Italy Office: Univ Udine, Dept Math, via Zanon 6, 33100 Udine Italy

DILALLA, LISABETH ANNE, developmental psychology researcher, educator; b. Bayshore, N.Y., July 6, 1959; d. David Elemelich and Leila Lois (Katz) Fisher; m. David Louis DiLalla, Aug. 12, 1984; children: Matthew Scott, Shaina Emily. BA, Brandeis U., 1981; PhD, U. Va., 1987. Nat. Inst. Child Health and Human Devel. Inst. Behavioral Genetics, U. Colo., Boulder, 1987-90; rsch. assoc. So. Ill. U., Carbondale, 1990-91, rsch. asst. prof., 1991-92, asst. prof. dept. behavioral and social scis., 1992—; cons. Joint Legislative Audit and Rev. Com., Richmond, Va., 1984; rsch. cons. Govt. of Bermuda, 1985; statis. cons. U. Va., Charlottesville, 1986-87; cons. Inst. Social Rsch., Boulder, Colo., 1988; workshop leader Women in Sci., Carbondale, 1992; tutor for sci. fair Giant City Sch., Carbondale, 1992. Grantee U. Colo. Health Scis., Ctr. Devel. Psychobiology Endowment Fund, 1988, NIH, 1992—; recipient award for manuscript Mensa Edn. and Rsch. Found., 1991. Mem. Soc. for Rsch. in Child Devel., Behavior Genetics Assn., Am. Psychol. Soc., Sigma Xi. Achievements include research on predictors of intelligence from infancy, genetic and environ. influences on inhibition, predictors of delinquency. Office: So Ill U Sch Medicine Dept Behavioral & Social Scis Carbondale IL 62901

DILEEPAN, KOTTARAPPAT NARAYANAN, biochemist, researcher, educator; b. Guruvayur, India, Oct. 17, 1947; came to U.S., 1975; s. Kottarappat and Kamaladevi (Manayil) Narayanan; m. Kanakam Dileepan, Dec. 25, 1978; children: Kavitha, Sangeetha. BS in Chemistry, U. Kerala, India, 1967; MS in Biochemistry, U. Baroda, India, 1970; PhD in Biochemistry, U. Delhi, India, 1974. Postdoctoral fellow Children Hosp. Med. Ctr., Oakland, Calif., 1975-77; rsch. assoc. Ind. U. Med. Ctr., Indpls., 1977-81; rsch. biochemist VA Med. Ctr., Kansas City, Mo., 1981-84; assoc. prof. U. Kans. Med. Ctr., Kansas City, 1985—. Contbr. papers to Nutrition Reports Internat. Jour. Nutrition, Life Sci., Jour. Cell Physiology, Biochem. Jour., Jour. Lab. Clin. Medicine, Jour. Molecular Cell Immunology, others; contbr. abstracts to numerous profl. jours. and confs. Pres. India Assn. of Kansas City, 1990, trustee, 1991-93. Rsch. fellow Indian Coun. Med. Rsch., 1971-74; rsch. grantee Am. Cancer Soc., Am. Heart Assn. Mem. Am. Soc. for Biochemistry and Molecular Biology, Am. Assn. Immunologists, Biochem. Soc. (U.K.), Soc. Biol. Chemists (India) Soc. for Exptl. Biology and Medicine, Soc. for Leukocyte Biology, Sigma Xi. Achievements include rsch. on mode of action of hormones and vitamins, enzyme structure and kinetics, immunomodulatory mechanisms, oxygen radical metabolism in phagocytic cells. Home: 10802 W 125th Pl Overland Park KS 66213 Office: U Kans Med Ctr 3901 Rainbow Blvd Kansas City MO 66160

DILIBERTO, PAMELA ALLEN, pharmacologist, researcher; b. Valparaiso, Ind., Mar. 13, 1956; d. Melvin Lloyd and Joyce Rosemary (Vietzke) Allen; m. Emanuel Joseph Diliberto Jr., Dec. 14, 1985. BA in Chemistry and Zoology, DePauw U., 1977; PhD in Pharmacology, Duke U., 1985. Rsch. asst. Wellcome Rsch. Labs., Research Triangle Park, N.C., 1979-80; predoctoral trainee dept. pharmacology Duke U., Durham, N.C., 1980-85; postdoctoral rsch. assoc. U. N.C., Chapel Hill, 1985-87, postdoctoral fellow, 1988-91, rsch. asst. prof. dept. cell biology, 1991—; teaching fellow U. N.C. Sch. Medicine, Chapel Hill, 1988-90. Contbr. articles to Jour. Biol. Chemistry, Jour. Cellular Biochemistry, Jour. Pharmacology and Exptl. Therapeutics. Mem. Am. Assn. for Cancer Rsch., Am. Soc. for Cell Biology, Internat. Assn. Women Bioscientists, N.C. Soc. for Neurosci., Sigma Xi, Phi Beta Kappa. Achievements include discovery of mechanisms of platelet-derived growth factor signal transduction, mechanisms of catecholamine and dopamine transport, identification of semidehydroascorbate as product of dopamine-B-hydroxylation. Office: U NC Dept Cell Biology CB # 7090 202 Taylor Hall Chapel Hill NC 27599

DILL, ELLIS HAROLD, university dean; b. Pittsburg County, Okla., Dec. 31, 1932; s. Harold and Mayme Doris (Ellis) D.; m. Cleone June Granrud, Sept. 12, 1953; children—Michael Harold, Susan Marie. A.A., Grant Tech. Jr. Coll., 1951; B.S. in Civil Engring, U. Calif. at Berkeley, 1954, M.S. in Civil Engring, 1955, Ph.D., 1957. Asst. prof. to prof. aeros. and astronautics U. Wash., 1956-77, chmn. dept. aeros. and astronautics, 1976-77; dean engring. Rutgers U., New Brunswick, N.J., 1977—. Mem. Soc. Natural Philosophy. Research, numerous publs. on mechanics of solids. Home: 436 Brentwood Dr Piscataway NJ 08854-3608 Office: Rutgers U Coll Engring New Brunswick NJ 08903

DILL, KENNETH AUSTIN, pharmaceutical chemistry educator; b. Oklahoma City, Dec. 11, 1947; s. Austin Glenn and Margaret (Blocker) D. S.B., Mass. Inst. Tech., 1971, S.M., 1971; Ph.D., U. Calif.-San Diego, 1978. Fellow Damon Runyon-Walter Winchell Stanford (Calif.) U., 1978-81; asst. prof. chemistry U. Fla., Gainesville, 1981-82; asst. prof. pharm. chemistry and pharmacy U. Calif., San Francisco, 1982-85, assoc. prof., 1985-89, prof., 1989—; adj. prof. pharmaceutics U. Utah, 1989—. PEW Found. scholar. Contbr. numerous sci. articles to profl. publs.; patentee in field. Fellow Am. Phys. Soc.; mem. AAAS, Am. Chem. Soc., Biophys. Soc. Office: Univ Calif Pharm Chemistry Dept San Francisco CA 94143

DILLE, JOHN ROBERT, physician; b. Waynesburg, Pa., Sept. 2, 1931; s. Charles Emanuel and Ruth Emma (South) D.; m. Joan Marie Sirtosky, Dec. 17, 1955; children: Paul Andrew, John Alan. BS, Waynesburg Coll., 1952; MD, U. Pitts., 1956; M in Indsl. Health, Harvard U., 1960. Diplomate Am. Bd. Preventive Medicine. Intern Akron City Hosp., 1956-57; resident in aerospace medicine USAF Sch. Aerospace Medicine, San Antonio, 1960-62; program adv. officer FAA Civil Aeromed. Rsch. Inst., Oklahoma City, 1961-61; regional flight surgeon FAA Civil Aeromed. Research Inst., Oklahoma City, 1965; chief FAA Civil Aeromed. Inst., U.S. Dept. Transp., Oklahoma City, 1966-87, ret., 1987; med. dir. Okla. Dept. Corrections, Oklahoma City, 1990-93; assoc. prof. U. Okla., 1961—, dir. tng. residency in aerospace medicine, 1967-72; state surgeon Okla. Army N.G., 1990-91. Assoc. editor: Ag Pilot Internat. mag., 1980—, Conservation Aeronautics mag., 1989-92, Above All mag., 1992—; mem. editorial bd. Aviation, Space and Environ. Medicine; contbr. articles to profl. jours. With USAF, 1957-59; col. M.C., U.S. Army N.G., 1976-91. Recipient Meritorious award William A. Jump Found., 1968; named Army N.G. Flight Surgeon of Yr. 1987, Master Flight Surgeon, 1987. Fellow Aerospace Med. Assn. (mem. exec. coun. 1978-81, chmn. history and archives com. 1982-90, chmn. sci. program com. 1985, 1st v.p., 1990-91, pres. 1992-93, Theodore C. Lyster award 1978, Harry G. Moseley award 1987), Am. Coll. Preventive Medicine (regent 1974-77); mem. Internat. Acad. Aviation and Space Medicine, Soc. U.S. Army Flight Surgeons (mem. bd. govs. 1990-92), Am. Air Mail Soc. (bd. dirs. 1990-92), Res. Officers Assn., Order Aeromed. Merit, Sigma Xi, Nu Sigma Nu. Presbyterian. Home and Office: 335 Merkle Dr Norman OK 73069-6429

DILLEY, DAVID ROSS, plant physiologist, researcher; b. South Haven, Mich., Mar. 10, 1934; s. Varnum M. and Marion (Dalquist) D.; m. Marion Olive Dilley, July 21, 1956; children: Catherine, Jane, Brenda. BS, Mich. State U., 1955, MS, 1957; PhD, N.C. State U., 1960. From asst. to assoc. prof. Mich. State U., East Lansing, 1960-67, prof., 1967—; cons. Grumman Allied Industries, Inc., Bethpage, N.Y., 1974-80, Frito-Lay, Dallas, 1982-84, Neogen Corp., Lansing, Mich., 1984—, Monsanto Corp., St. Louis, 1986—. Contbr. articles to profl. jours. Recipient Excellence in Rsch. award The Grower, 1986. Fellow AAAS, Am. Soc. Hort. Sci. (Dow Chem. award 1970, Disting. Grad. Teaching award 1981); mem. Am. Soc. Plant Physiologists, Sigma Xi. Republican. Achievements include 2 patents in field. Office: Mich State U Dept Horticulture East Lansing MI 48824

DILLINGHAM, CATHERINE KNIGHT, environmental consultant; b. Bklyn., Jan. 17, 1930; d. Donald Branch and Frances (Perry) Knight; m. Bruce E. Dillingham, June 30, 1951; children: Ruth, Laura Hargadon, Sallie Bowling. BA in Biology, Skidmore Coll., 1951; MA in Edn., Fairfield U., 1966; M of Natural Sci., Worcester Poly. U., 1984. With Roger Ludlowe High Sch., Fairfield, Conn., 1967-85, Environ. Defense Fund, N.Y.C., 1985—; adj. prof. environmental studies Fairfield U., 1989—; curriculum cons. Natural Resources Defense Coun., Washington, 1985-86; presenter in field. Contbr. articles to The Am. Biology Tchr., Environ. Action. Democrat. Home and Office: 247 Barlow Rd Fairfield CT 06430

DILLON, HOWARD BURTON, civil engineer; b. Hardyville, Ky., Aug. 12, 1935; s. Charlie Edison and Mary Opal (Bell) D.; m. Bonny Jean Garard,

May 19, 1962; 1 child, Robert Edward. BCE, U. Louisville, 1958, MCE, 1960; postgrad., Okla. State U., 1962, Mich. State U., 1962-65. Registered profl. engr., Ind. Instr. U. Louisville, Ky., 1958-60; from assoc. prof. to prof. Ind. Inst. Tech., Ft. Wayne, 1960-62; NSF fellow Okla. State U., Stillwater, 1962; NSF grantee, instr. Mich. State U., East Lansing, 1962-67; head civil engring. dept. MW Inc. Cons. Engrs., Indpls., 1967-83; project mgr. civil div. SEG Engrs. & Cons., Indpls., 1983-91; pvt. civil engring. cons. Howard B. Dillon, Cons. Engr., Indpls., 1991—; asst. dir. to local pub. road needs study for Ind., 1970; mem. design com. for dams in Ind., 1974—; spl. cons. to Ind. Dept. Nat. Resources on dams, 1980—; mem. infrastructure com. for State of Ind., 1984—. Contbr. articles to profl. jours. Committeeman Wayne 52 precinct, Indpls., 1972-86; vice-ward chmn. Wayne South Twp., Indpls., 1986-87. Hazelett and Erdal scholar, 1957-58, W.B. Wendt scholar U. Louisville; recipient Order of Engr. award Purdue U., 1993. Mem. ASCE (Outstanding Civil Engring. Grad. award 1958), NSPE, Am. Soc. Engring. Edn., ASTM, Internat. Soc. Found. Engrs., Mil. Engrs., Internat. Acad. Sci., Ind. Water Resources Assn., Nat. Audubon Soc., Optimists (pres. Suburban West chpt. 1972-74, bd. dirs. 1974-78, sec. 1992-93, lt. gov. ind. dist. 1972-74), Chi Epsilon. Democrat. Baptist. Avocations: fishing, traveling, photography, lecturing, coin collecting. Home and Office: 6548 Westdrum Rd Indianapolis IN 46241-1843

DILLON, JOSEPH FRANCIS, JR., physicist, consultant; b. N.Y.C., May 25, 1924; s. Joseph Francis and Ann Elizabeth (McElroy) D.; m. E. Lee Valentine, Sept. 21, 1946; children: Joseph F. III, Andrew Paul. BA, U. Va., 1944, MA, 1947, PhD in Physics, 1949. Physicist U.S. Dept. Agriculture, Pirbright, Surrey, U.K., 1949-52; mem. tech. staff dept. physics rsch. AT&T Bell Labs., Murray Hill, N.J., 1952-91; adj. prof. dept. applied physics Yale U., New Haven, Conn., 1991—; guest scientist Inst. Solid State Physics, U. Tokyo, 1966-67, vis. prof., 1976-77; adj. prof. physics dept. Toho U., Funabashi, Chiba, Japan, 1991-92; disting. lectr. Magnetic Soc., IEEE, 1981. Contbr. over 120 articles to profl. jours. Lt. (j.g.) USNR, 1943-46, PTO. Recipient Guggenheim fellow J. Simon Guggenheim Found., 1966-67. Mem. Am. Phys. Soc., Internat. Conf. Magnetism (sec. 1985), Conf. Magnetism and Magnetic Materials (gen. chmn. 1965). Achievements include 20 patents in field of magnetic behavior of materials; discovery of transparency and magnetooptical (MO) properties of rare earth iron garnets; of huge MO properties of the ferromagnet chromium tribromide; invention of optical modulator, circulator, compensation point memory, magnetic tunable laser and other devices; first to observe magnetostatic modes; research in and identification of losses and anisotropies of rare earth ions. Home: 45 Skyline Dr Morristown NJ 07960-5146

DILLON, KATHLEEN GEREAUX, dentist; b. Kankakee, Ill., Aug. 10, 1957; d. Marvin Jackand Marjorie Elizabeth (Bloch) G. BS cum laude, Ill. Benedictine Coll., 1979; DDS, U. Ill., 1983. Pvt. practice Manteno, Ill., 1983-84; assoc. Cortez Dental Clinic, Bradenton, Fla., 1984-86; pvt. practice Holmes Beach, Fla., 1987—; tchr. Santa Fe Community Coll. Hygiene Sch., Gainesville, Fla., 1984-85. State Rep. Christiansen scholar, 1980-81. Mem. Manatee Dental Soc., Anna Maria C. of C., Manatee C. of C., Acad. Gen. Dentistry. Democrat. Roman Catholic. Avocations: running, biking, pets. Office: 10414 Cavalcade St Great Falls VA 22066

DILLON, MICHAEL EARL, engineering executive; b. Lynwood, Calif., Mar. 4, 1946; s. Earl Edward and Sally Ann (Wallace) D.; m. Bernardine Jeanette Staples, June 10, 1967; children: Bryan Douglas, Nicole Marie, Brendon McMichael. BA in Math., Calif. State U., Long Beach, 1978, postgrad. Journeyman plumber Roy E. Dillon & Sons, Long Beach, 1967-69, ptnr., 1969-73; field supr. Dennis Mech., Long Beach, 1973-74; chief mech. official City of Long Beach, 1974-79; mgr. engr. Southland Industries, Long Beach, 1979-83; v.p. Syska & Hennessy, L.A. and N.Y., 1983-87; prin. Robert M. Young & Assoc., Pasadena, Calif., 1987-89; pres. Dillon Cons. Engrs., Long Beach, 1989—; mech. cons. in field; instr. U. Calif., Irvine & L.A., U. So. Calif., Calif. State U. at Long Beach. Contbr. over 160 poems to publs., 25 tech. articles to profl. jours; lectr. in field. Bd. examiners Appeals and Condemnations, Long Beach; mem. State Fire Marshals Adv. Bd., Sacramento, Calif.; adv. bd. City of L.A. Fellow ASHRAE (bd. dirs.), Inst. of Refrigeration Heating, Air Conditioning Engrs. of New Zealand, Inst. Advancement Engring.; mem. ASCE, NSPE, ASTM, Nat. Acad. Of Forensic Engrs., Nat. Inst. for Engring Ethics, Internat. Con. Bldg. Official, So. Bldg. Code Congress Internat. Inc., Nat. Fire Protection Assn., Calif. Cons. Engrs. Coun., Am. Cons. Engrs. Coun., Am. Soc. of Plumbing Engrs., Soc. of Fire Protection Engrs., Tau Beta Pi, Pi Tau Sigma, Chi Epsilon, others. Avocation: poetry. Home: 1107 E 46th St Long Beach CA 90807 Office: Dillon Cons Engrs 1165 E San Antonio Dr #D Long Beach CA 90807

DILLON, RAY WILLIAM, engineering technician; b. China Lake, Calif., Aug. 8, 1954; s. Duane L. and Audrey J. (Amende) D.; m. Kathy M. Shrum , Sept. 3, 1980; 1 child, Stephanie. Student, U. Okla., 1976-78, Oklahoma City Comm. Coll., 1986; BA, So. Nazarene U., Bethany, Okla., 1987. Cert. level IV Nat. Inst. for Certification in Engring. Techs. Technician B&B Fire Protection, Oklahoma City, 1977-78; lead technician Grinell Fire Protection Systems, Oklahoma City, 1978-80; gen. mgr. A.L. Fire Protection, Inc., Oklahoma City, 1980-87, exec. v.p., 1989-90; store mgr. Master Systems Ltd., Oklahoma City, 1987-89; region mgr. Casteel Automatic Fire Protection, Oklahoma City, 1990-92; gen. mgr. Allied Rubber & Gasket Co., Dillon, Colo., 1992—; instr. Okla. State U., 1990-91. Mem. Nat. Inst. Cert. Engring. Technicians (cert.), Oklahoma City IBM-PC Users Group, Okla. Fire Protection Contractors Assn. (sec. 1986-87, chmn. 1987, sec. 1990-91), Nat. Fire Protection Assn., Soc. of Fire Protection Engrs. Republican. Baptist. Home: PO Box 399 Silverthorne CO 80498

DILLY, RONALD LEE, civil engineer, construction technology educator; b. Sioux Falls, S.D., Feb. 3, 1956; s. Robert Lee and Marian Jennette (Linder) D.; m. Deborah Lin Dilly; 1 child, Amie Marie. BSCE, S.D. Sch. Mines, 1979; MSCE, Tex. A&M U., 1981. Materials engr. in tng. Law Engring. and Testing Co., Houston, 1981-82; asst. prof. civil tech. Coll. Tech. U. Houston, 1982-93, assoc. prof. civil tech. Coll. Tech., 1993—; concrete materials testing cons. McBride Ratcliff and Assocs., Houston, 1986, 90, 91, 93. Contbr. articles to profl. publs. Mem. ASCE, ASTM, Am. Concrete Inst. (sec. com. 1990—). Office: U Houston Coll Tech 4800 Calhoun Houston TX 77204

DILSAVER, STEVEN CHARLES, psychiatry educator; b. L.A., Mar. 5, 1953; s. Charles Joseph and Margaret Dominga (Sausedo) D. BS, Stanford U., 1975; postgrad., Trinity Evangel. Div. Sch., 1977-78; MD, U. Calif., San Diego, 1981. Diplomate Nat. Bd. Med. Examiners. Resident dept. psychiatry U. Mich., 1980-85, asst. prof. psychiatry, 1985-87; dir. psychopharmacology program, assoc. prof. psychiatry Ohio State U., 1987-90; prof. psychiatry and behavioral scis. U. Tex. Health Sci. Ctr., Houston, 1990—; dir. clin. rsch. unit Harris County Psychiat. Ctr., Houston, 1990—; dir., coord. rsch.; vis. prof. dept. psychiatry U. Pisa, Italy, 1989, U. N.C., Chapel Hill, 1990; invited lectr. Washtenaw County Child and Family Svcs., 1985, Ohio State U., 1987, Brown U., 1988, Eli Lilly Pharmaceuticals, Florence, Italy, 1988, and others; cons. Ontario Mental Health Adminstrn., 1986, 88, VA Merit Rev. Com., 1985, NIMH, 1985, N.C. Alcohol Rsch. Coun., 1990—. Editor: Jour. Clin. Psychopharmacology, 1983—, Psychosomatics, 1985—, Biol. Psychiatry, 1985—, Pharmacology Biochemistry and Behavior, 1987—, Hosp. and Community Psychiatry, 1987—, Jour. Clin. Psychiatry, 1988—, Schizophrenia Rsch., 1988—; and many others; contbr. over 100 articles to profl. jours. Recipient 1st prize Mich. Psychiat. Assn. Rsch. Paper Competition, 1982, Travel award Am. Coll. Neuropsychopharmacology, 1984, 1st prize Ohio Psychiat. Assn. Rsch. Paper Competition, 1989; Calif. State scholar, 1971-75, Orrin Dunn Honors scholar Stanford U., 1973-74; Summer Rsch. fellow in Neurosci. City of Hope, Duarte, Calif., 1974, Calif. State Grad. fellow 1975-79. Mem. Soc. for Advancement of Sci., Soc. for Neurosci., Soc. for Biol. Psychiatry, Internat. Behavioral Neurosci. Democrat. Achievements include treatment of severe disorders of mood (such as unipolar depression and bipolar disorder); basic and clin. studies relevant to psychobiology of disorders of mood; rsch. in treatments for depressed and manic phases of manic-depressive illness. Home: 2628 S Glen Haven Blvd Houston TX 77025-2130 Office: U Tex Harris County Psychiat Ctr 2800 S Macgregor Houston TX 77225

DILTS, DAVID ALAN, mechanical design engineer; b. N.Y.C., June 11, 1947; s. Walter Boden and Margret (Daleida) D.; m. Janet Ruth Bevers, July 11, 1970; children: James Jason, Jeremy Andrew. BSME, Clarkson U., 1970. Mech. engr. M/S div. Raytheon Co., Bedford, Mass., 1970-72; design engr., prin. Timeless Toys Co., Kingston, N.H., 1972-77; design engr., mgr. Exxon Solar Power Corp., Woburn, Mass., 1977-84; mgr. product devel. Standard Oil Corp., Cleve., 1984-87; mgr. renewable energy programs Mass. Div. Energy Resources, Boston, 1987—; mem. Robotics Internat., Boston, 1980-82, Solar/Electric Vehicle Conf. Com., Greenfield, Mass., 1989—; chmn. Electric Vehicle Steering Com., Boston, 1992—. Scoutmaster Gates Mills (Ohio) area Boy Scouts Am., 1985-87. Recipient Advanced Machine Design award Small Co., Inc., N.Y.C., 1982; grantee Fed. Hwy. Adminstrn., 1992. Mem. Am. Solar Energy Soc., Mass. Orgn. Scientists and Engrs. Achievements include patents for methods and apparatus for making photovoltaic modules, semirigid photovoltaic module assembly and structural support, flexible interconnected array of amorphous semiconductor photovoltaic cells. Home: 41 Brentwood Rd Exeter NH 03833 Office: Mass Div Energy Resources Rm 1500 100 Cambridge St Boston MA 02202

DILWORTH, STEPHEN JAMES, mathematics educator; b. Stockton, Eng., Mar. 27, 1959; came to U.S., 1984; BA, Trinity Coll., Cambridge, Eng., 1980, MA, 1984, PhD, 1985. Vis. asst. prof. U. Mo., Columbia, 1984-85; instr. math. U. Tex., Austin, 1985-87; asst. prof. math. U. S.C., Columbia, 1987-92, assoc. prof., 1992—. Contbr. articles to profl. jours. Grantee NSF 1986-91. Mem. Am. Math. Soc., Math. Assn. Am., London Math. Soc. Office: U SC Dept Math Columbia SC 29208

DI MAIO, VINCENT JOSEPH MARTIN, forensic pathologist; b. Bklyn., Mar. 22, 1941; s. Dominick J. and Violet (de Caprariis) Di M.; m. Theresa G. Richberg, Mar. 29, 1969; children: Dominick, Samantha. MD, SUNY, Bklyn., 1965. Diplomate Am. Bd. Pathology. Intern pathology Duke Hosp., Durham, N.C., 1965-66; resident pathology King's County Downstate Med. Ctr., Bklyn., 1966-69; fellow forensic pathology Office of Chief Med. Examiner, Balt., 1969-70; med. examiner Southwestern Inst. Forensic Scis., Dallas, 1972-81; chief med. examiner, dir. criminal investigation lab. Bexar County Forensic Sci. Ctr., San Antonio, 1981—; prof. dept. pathology UTHSC-SA. Author: Gunshot Wounds, 1985, Forensic Pathology, 1989; editor Am. Jour. Forensic Medicine & Pathology, 1992. Maj. U.S. Army, 1970-72. Recipient Commn. Continuing Edn. medal Am. Soc. Clin. Pthologists, Jean R. Oliver, MD Master Tchr. award SUNY Alumni Assn.-Downstate Med. ctr., 1990. Fellow Am. Acad. Forensic Scis.; mem. Nat. Assn. Med. Examiners, Internat. Wound Ballistics Assn. Office: Bexar County Forensic Sci Ctr 7337 Louis Pasteur San Antonio TX 78229-4565

DI MARIA, CHARLES WALTER, mechanical and automation engineer, consultant; b. Phila., May 3, 1927; s. Giuseppi and Antoinette Di Maria; m. Gloria Josephine Sarcone, June 14, 1958; children: Karen Marie, Lori Ann. BA in Physics, Temple U., 1968. Mech. designer Globe Holst Co., Wyndmoor, Pa., 1956-65; mfg. engr. Elco Corp., Willow Grove, Pa., 1965-72; project mgr. The Budd Co., Phila., 1972-81; adv. mfg. unit mgr. RCA Corp./GE Aerospace, Moorestown, N.J., 1981-93; mem. coop. edn. program Rensselaer Poly. Inst., Troy, N.Y., 1987-90. Contbr. articles to profl. jours. Mem. fund raising com. United Way, 1991. Mem. Soc. Mfg. Engrs. (cert.), Ind. Order Odd Fellows. Roman Catholic. Achievements include patent in Coax to Waveguide Transition. Home and Office: 2008 Grace Ln Flourtown PA 19031-1708

DIMITRIC, IVKO MILAN, mathematician, educator; b. Loznica, Serbia, Yugoslavia, Jan. 11, 1957; came to U.S. 1984; s. Milan Dobrivoje and Nadezda (Blagojevic) D. BS in Math., U. Belgrade, 1980, IMS in Math., 1984; PhD in Math., Mich. State U., 1989. Instr. Mich. State U., East Lansing, 1989-90; asst. prof. math. Pa. State U., Uniontown, 1990—. Contbr. articles to profl. jours. Rsch. grantee Pa. State U., 1990—. Mem. Am. Math. Soc., Math. Assn. Am., N.Y. Acad. Scis., London Math. Soc. Serbian Orthodox. Home: 50 W Main St N Uniontown PA 15401 Office: Pa State Univ 1 University Dr Rt 119N Uniontown PA 15401

DIMITRIOS, DON FEDON, civil engineer, land surveyor; b. Alexandria, Egypt, Sept. 6, 1928; came to U.S., 1947; s. Euripides and Alexandra (Kyrialidou) Dimitriades; m. Noreen Hall, Dec. 27, 1965 (div. 1978); children: Paul Fedon, Zoe Ann. BS in Bacteriology, U. Md., 1956, BS in Civil Engring., 1957; MS in Civil Engring., Tulane U., 1960. Registered profl. engr., land surveyor., La. Pub. health engr. III La. State Bd. of Health, New Orleans, 1957-66; design engr. DeFraites Assocs. Inc., New Orleans, 1966-67; sr. civil engr. New Orleans Sewerage and Water Bd., 1967—. Radio operator amateur radio Sta. WA5LQE, New Orleans, since 1965; instr. water safety ARC, New Orleans, since 1949. Recipient Cert. of Appreciation, Office of the Mayor New Orleans, 1981. Mem. ASCE, La. Conf. on Water, Wastewater and Indsl. Waste. Greek Orthodox. Office: Sewerage and Water Bd 8801 Spruce St New Orleans LA 70118

DIMITRIOU, MICHAEL ANTHONY, biochemist, sales and marketing manager; b. Ioannina, Epris, Greece, Apr. 29, 1951; came to U.S., 1956.; s. Anthony Fotos and Annastasia (Pappas) D.; m. Alexandra Jane Burger, June 13, 1972; children: Theodore, Katherine, Alexander. Ba, Knox Coll., 1974; MBA, U. Richmond, 1990. Lab. mgr. FS Svcs., Springfield, Ill., 1974-76, quality control rep., 1976-77; estimator Keene Corp., Aurora, 1977-78; sales engr. Carter Co., Atlanta, 1978-80; applications engr. Infilco Degrement, Richmond, Va., 1980-86, product. mgr., 1986-90; sales and mktg. mgr. Ozonia Nort Am., Richmond, Va., 1990—; mem. tech. adv. com. USEPA Disinfection By-Products, 1992. Editor: (book) Design Guidance Manual/ Ozone, 1990; contbr. articles to profl. jours. Dir. St. Constantine's and St. Helen's Chs., Richmond, 1989; adv. Chesterfield County, 1991. Fellow AICE; mem. Am. Water Works Assn. and Rsch. Found. (Tech. Achievement award 1991), Internat. Ozone Assn. (bd. dirs. 1986—, tech. dir. 1987-89, Tech. Achievement award 1989, sec. 1993). Greek Orthodox. Achievements include patent in sludge handling and dewatering; development of ozone decolouing of waters in U.S., European water treatment technology in the U.S.; research in advanced ozone technology, ozone system design, designing for ozone applications, ozone and water treatment. Home: 11700 Blakeston Ct Richmond VA 23236 Office: Ozonia N Am 2924 Emerywood Pkwy Richmond VA 23255

DIMMICK, KRIS DOUGLAS, civil engineer; b. North Tonawanda, N.Y., Mar. 7, 1963; s. Calvin A. and Judith G. (Baker) D. AAS, Paul Smith's (N.Y.) Coll., 1983; BS in Forest Engring., SUNY, Syracuse, 1985, MS in Environ. Engring., 1991. Registered profl. engr., N.Y. Forest technician Superior Forestry Svc., Leslie, Ark., 1983; constrn. inspector N.Y. State Dept. Transp., Binghamton, 1984-85; intern engr. Clough Harbour & Assocs., Albany, N.Y., 1985-86; project engr. John S. MacNeill, Jr., PC, Homer, N.Y., 1986-91; assoc. engr. Bernier, Carr & Assocs. PC, Watertown, N.Y., 1991—. Mem. ASCE (Key Alert mem. 1989-91, chmn. membership com. 1990-91), N.Y. State Soc. Profl. Engrs., Am. Motorcyclist Assn., Am. Pub. Wks. Assn. (exec. com. 1990-91). Office: Bernier Carr & Assocs PC 172 Clinton St Watertown NY 13601

DIMMIG, BRUCE DAVID, architect; b. Buffalo, N.Y., Apr. 19, 1956; s. George Clayton and June (Legler) D.; m. Robin S. Freese, Feb. 14, 1993. AAS with honors, Dutchess Community Coll., 1976; BArch cum laude, Kans. State U., 1980. Lic. architect Kans., N.J. Staff architect The Shaver Partnership, Salina, Kans., 1980-84; staff/project architect Fullerton, Carey & Oman, Kansas City, Mo., 1984-87; project architect The Ramos Group, Kansas City, 1987-88; archtl. project mgr. Toys "R" Us, Inc., Paramus, N.J., 1988-89; project mgr. Schenker, Schenker & Rabinowitz, Paterson, N.J., 1989-91; assoc. L. Michael Schenker, Architect, Paterson, 1991—; participant Govs. Task Force on Aging, New Brunswick, 1990, Monmouth Housing Alliance Competition, Marlboro, N.J., 1991, ADA video Kean Coll., 1992. Ch. treas. St. Mark's Evang. Luth. Ch., Salina, 1982; mem. ch. bd. edn. Gethsemane Evang. Luth. Ch., Kansas City, 1987, Sunday sch. tchr., 1987. Recipient Regents scholarship N.Y. State, 1974, Academic scholarship Phi Kappa Phi, Kans. State U., 1979, Neighborhood award Tenn. Town, Topeka, 1979. Mem. Nat. Coun. Archtl. Registration Bds., Bldg. Ofcls. and Code Adminstrs. Internat. Avocations: chess, reading, photography, world war II, sports. Home: 172 N Main St Apt 1 Pearl River NY 10965

DIMOND, ROBERTA RALSTON, psychology and sociology educator; b. Bakersfield, Calif., Mar. 25, 1940; d. Robert Leroy Vickers and Gail Anderson (Tritch) Ralston; m. James Davis, June 18, 1963 (div. 1970); 1 child, Jamie Amundsen Davis; m. Frederick Henry Dimond, Oct. 20, 1970; children: Frederick Ralston, Robert Vickers (div. 1991). BA in History and English, Stanford U., 1962, MAT in Edn., 1963; MS, U. Pa., 1970, EdD, 1973. Cert. secondary educator, ednl. specialist, counselor, coll. personnel adminstr. Thcr. Kamehameha Sch., Honolulu, 1965-67; asst. to dean of women U. Pa., Phila., 1969-70; asst. prof. Temple U., Ambler, Pa., 1970-87, Montgomery County Coll., Blue Bell, Pa., 1975-80; prof. psychology, speech, sociology Del. Valley Coll., Doylestown, Pa., 1987—; cons. ETS, Princeton, N.J., 1989—; speaker in field; lectr. on sexual responsibilities in the 90s and assertive affirmative action topics; researcher on athletics and aging females syngerism. Author: Gender & RAcial Bias by Vocational Counselors, 1973. Bd. dirs. Concerned Citizens of Upper Dublin, Maple Glen, Pa., 1980-91, Arrowhead Assn., Ambler, Pa., 1990-91. Recipient Fellowship, Newhouse Found., 1960-63, APA grant, 1969-70. Mem. AAUP, APA, U.S. Tennis Assn., Middle States Tennis Assn., MADD, Phila. Tennis Patrons, Phila. Tennis Assn. (v.p.). Democrat. Episcopalian. Avocations: tennis (ranked # 6 in U.S. in women's 35 and over tennis, # 1 in over 45, # 1 MSTA over 50), duplicate bridge. Home: 236 Amherst Dr Doylestown PA 18901 Office: Delaware Valley Coll Rte 202 Doylestown PA 18901

DINBERG, MICHAEL DAVID, industrial engineer; b. Ogdensburg, N.Y., June 28, 1944; s. Israel and Pauline (Karch) D. AS, Broome Community Coll., 1972; BS magn cum laude, Syracuse U., 1973; postgrad., Ga. Inst. Tech., 1973-74. Jr. field technician Singer-Link Div., Binghamton, N.Y., 1967-69, jr. field engr., 1969-70; health systems specialist Ga. Inst. Tech., Atlanta, Ga., 1973-74; sr. indsl. engr. Singer Link Div., Binghamton, 1974-75; commd. officer, capt. U.S. Pub. Health Svc., 1975—; mgmt. analyst NIH, Bethesda, Md., 1975-77, systems analyst, 1977-80; regulatory officer FDA, Rockville, Md., 1980-85, acting chief, evaluation br., 1985-86, assoc. dir. for evaluation, 1986-87, chief spl. studies & analysis staff, 1987; indsl. engring. officer Gillis W. Long Hansen's Disease Ctr., Carville, La., 1987-89; indsl. engr. health svcs. divsn., fed. bur. prisons Dept. Justice, Washington, 1989-90, mgmt. engr., 1990-91, sr. program mgmt. officer cons., 1991—; manpower subcom. mem. USPHS Engr. Career Devel. Com., Rockville, 1977-87; paramedic USPHS Disaster Med. Assistance Team, Rockville, 1984-87; mem. response team DHHS/FDA RAdiological Emergency Cadre, Rockville, 1986-87; USPHS rep. Fed. Interagency Com. on Emergency Med. Svcs., Washington, 1986-87; mem. response team Chernoble Nuclear Reactor Accident, Rockville, 1986; ops./tng. officer USPHS Disaster Med. Assistance Team, Rockville, 1989-92; rep. USPHS Engr. Profl. Adv. com., Rockville, 1990--. Vol. N.Y.-Pa. Health Planning Coun., Binghamton, N.Y., 1974-75; del. to Russia, Poland People to People Inst. Indsl. Engrs., 1992. With USNR, 1965-67. Mem. Am. Radio Relay League, Assn. Mil. Surgeons, Commd. Officers Assn. of the USPHS, Inst. Indsl. Engrs., N.Y. Acad. Sci., Res. Officers Assn., Soc. Am. Mil. Engrs., Soc. Health Systems, Tau Beta Pi. Jewish. Home: 1901 Stanley Ave Rockville MD 20851 Office: DOJ/FBOP/ Health Svcs Div 320 First St NW Ste 1000 Washington DC 20534

DING, AIHAO, medical educator, researcher; b. Chengdu, Sichuan, China, Nov. 20, 1945; came to U.S., 1980; d. Tepang and Susan (Zeng) Ting; m. Timothy Y. Chen, Jan. 1, 1976; 1 child, William. BS, Beijing U., 1969; PhD, SUNY, Bklyn., 1984. Tchr. high sch. Nanjing, People's Republic of China, 1970-80; teaching asst. SUNY Health Ctr., Bklyn., 1980-84; rsch. assoc. Rockefeller U., N.Y.C., 1984-86; asst. prof. Med. Coll. Cornell U., N.Y.C., 1986-91, assoc. prof. Med. Coll., 1991—. Contbr. articles to Jour. Exptl. Medicine, Jour. Biol. Chemistry, Jour. Sci. Grantee NIH. Mem. AAAS, Am. Assn. Immunologists, N.Y. Acad. Scis. Office: Cornell U Med Coll 1300 York Ave # 57 New York NY 10021-4896

DING, HAI, mathematician; b. Shanghai, People's Republic of China, Dec. 22, 1969; s. Ding Young Sheng and Du Chuan Xing; m. Hong Zhang. BS, Shanghai Jiao Tong U., 1991, B. Mgmt. (hon.), 1991. Diplomate computer software designing. Designer Shanghai Jiao Tong U., 1991—. Contbr. articles to profl. jours.; inventor optimize computer software, share market computer software. Scholar Hong Kong HuaXin Found., 1990. Mem. Math. Assn. Shanghai (bd. dirs. 1989-91), Math. Assn. Am. (Spl. Mem. award 1990). Avocations: bridge, basketball, computer games, classical music, singing. Home: 88 Anfu Rd, Shanghai 200031, People's Republic of China Office: Shanghai Jiao Tong U, Hua Shan Rd, Shanghai 200030, China

DING, MINGZHOU, physicist; b. Jinan, Shandong, China, Dec. 16, 1960; s. Lide and Tian Ling (Li) D.; m. Li Ding, Apr. 15, 1986; 1 child, Annie. BS, Bejing U., 1982; PhD, U. Md., 1990. Asst. prof. Fla. Atlantic U., 1990--. Contbr. articles to profl. jours. U. Md. Dissertation fellow, 1989; recipient First prize for Natural Scis. Chinese Acad. Scis., 1992. Mem. Am. Phys. Soc. Office: Ctr for Complex Systems 500 NW 20th St Boca Raton FL 33431

DING, NI, chemist; b. Hangzhou, Zheijiang, China, May 12, 1957; came to U.S., 1985; d. Zishang and Jiwen (Li) D.; m. Jiandong Huang, Aug. 11, 1990. MS, Dalian Inst. Chem. Physics, China, 1985; PhD, U. So. Calif., 1990. Rsch. asst. Dalian (China) Inst. Chem. Phys. Chinese Acad. Sci., 1982-85; rsch. scientist Dept. Chemistry, Zhejiang U., Hangzhou, China, 1985; rsch. teaching asst. Dept. Chemistry, U. So. Calif., L.A., 1985-90; sr. scientist Agrl. Dept., Rohm and Haas Co., Spring House, Pa., 1990-93, Schneider Inc., 1993—. Contbr. articles to profl. jours. Recipient Chem. Dept. Outstanding Teaching Asst. award U. So. Calif., 1987. Mem. Am. Chem. Soc., Am. Phys. Soc. Home: 2500 N Nathan Ln Apt 306 Plymouth MN 55441 Office: Schneider (USA) Inc Pfizer Hosp Products Group 5905 Nathan Ln Plymouth MN 55442

DINGEMAN, THOMAS EDWARD, wastewater treatment plant administrator; b. Detroit, July 12, 1950; s. Harry J. and Jane Bettina (Russell) D.; m. Ann Elizabeth Smits, July 22, 1983. AS, Northwestern Mich. Coll., 1970; BS, U. Mich., 1972; MA, No. Mich. U., 1975. Registered environ. profl., Colo. Instr. biology No. Mich. U., Marquette, 1975-76, tutor sci., 1976-77; maintenance technician Willow Apt. Complex, Greeley, Colo., 1977-78; wastewater treatment plant operator City of Greeley Water Pollution Control Facility, 1978-80, lab. technician, indsl. pre-treatment coord., 1980-81, lab. coord., 1981-89, supt. wastewater treatment plant, 1990—; mem. Colo. Biomonitoring Task Force, Denver, 1991—; task rev. com. Rocky Mountain Water Quality Analysts' Cert. Coun., Denver, 1989. Chair pinochle club St. Mary's Ch., Greeley, 1989, 92, sec., 1990. Recipient Ops. and Maintenance Excellence award U.S. EPA, 1990, 93. Mem. Colo. Wastewater Utility Coun., Rocky Mountain Water Quality Analysts' Assn., Rocky Mountain Water Pollution Control Assn. (pub. edn. com. 1992), Water Environ. Fedn., Colo. Biology Tchrs. Assn. Roman Catholic. Office: City of Greeley 300 E 8th St Greeley CO 80631

DINGLER, MAURICE EUGENE, civil engineer; b. Salina, Kans., Mar. 29, 1952; s. Herman Ludwig and Helen Josephine (Craig) D.; m. Shirley Jean Barnthson, Aug. 24, 1974; children: Eugene Edward, April Nicole. BS, Kans. State U., 1974. Registered profl. engr., Kans., Okla. Cons. engr. Reiss and Goodness Engrs., Wichita, Kans., 1974-83; sr. engr. Kans. Gas and Electric, Wichita, 1983-84, tech. staff engr., 1984-85; lead engr. Wolf Creek Nuclear Oper. Corp., Wichita, 1985-86; mgr. facilities engring. and analysis Wolf Creek Nuclear Oper. Corp., Burlington, Kans., 1986-90, mgr. nuclear plant engring. systems, 1990-92; mgr. nuclear plant engring. support, 1992—. Mem. ASCE (v.p. Wichita br. 1979-80, pres. 1980-81), Chi Epsilon. Office: Wolf Creek Nuclear Op Corp PO Box 411 Burlington KS 66839

DINGMAN, DOUGLAS WAYNE, microbiologist; b. Leon, Iowa, Sept. 12, 1953; s. Vernon Ralph and Elsie Mae (Broyles) D.; m. Sharon Douglas, Nov. 4, 1989. BS, U. Iowa, 1975, PhD, 1983. Teaching asst. U. Iowa, Iowa City, 1979-83; rsch. assoc. Tufts U., Boston, 1983-87; asst. rsch. scientist Conn. Agrl. Explt. Sta., New Haven, 1987—; participating scientist Sci.-by-Mail Mus. of Sci., Boston, 1991—. Contbr. articles to profl. jours. Mem. AAAS, Am. Soc. Microbiology, Boston Computer Soc. Office: Conn Agrl Exptl Sta 123 Huntington St New Haven CT 06511

DINGMAN, NORMAN RAY, engineering executive; b. Lincoln, Nebr., June 15, 1936; s. Waldo Emerson and Blossom (Lavone) D.; m. Sherry Carlene Schoneman, Apr. 12, 1957; children: David, Steven. Student, U.

Nebr., 1954-57. Elec. contractor Indsl. Elec. Svc., Lincoln, Nebr., 1959-69; constrn. supr. Cooper Nuclear Sta. and Nebr. City Power Sta., various locations, 1984-86; engring. technician Nebr. Pub. Power, Cooper Nuclear Sta., Brownville, 1984-86, motor operated valve engring. specialist, 1986-90, motor operated valve program supr., 1990—; vice chmn. Motor Operated Valve Users Group, 1988-90, chmn., 1990—. Bd. dirs. Brownville Hist. Soc., 1985-88. Home: Box 111 Peru NE 68421 Office: Nebr Pub Power Dist Box 98 Brownville NE 68321

DINKEVICH, SOLOMON, structural engineer; b. Leningrad, USSR, July 29, 1934; came to the U.S., 1978; s. Zusman and Sophiya (Yanovsky) D.; m. Irina Shubina, July 30, 1960; children: Mark, Eugene. MS in Petroleum Engring., Mining Inst., Leningrad, 1957; MS in Stress Analysis, Civil Engring. Inst., Leningrad, 1961; PhD in Structural Mechanics, Cen. Rsch. Inst., Moscow, 1967. Engr., supr. Cen. Rsch. Inst. Steel Constrn., Leningrad, 1957-63, sr. scientist, 1967-78; scientist Cen. Rsch. Inst. Structural Engring., Moscow, 1964-66; sr. prin. engr., supr. Ebasco Svcs. Inc., N.Y.C., 1979—. Author: Analysis of Cyclicly Symmetric Structures: The Spectral Method, 1978; contbr. articles to profl. jours. Mem. N.Y. Acad. Scis. Republican. Jewish. Achievements include research in structural dynamics and stability, application of group theory to analysis of symmetric systems, thermo-structural analysis of fusion machine's vacuum vessels, 3D plasma transport. Home: 6 Donner Ct Monmouth Junction NJ 08852-9603 Office: Princeton Plasma Physics PO Box 451 Princeton NJ 08543-0451

DINSTBER, GEORGE CHARLES, construction design engineer, consultant; b. N.Y.C., Jan. 31, 1966; s. George C. and Madelen Anne (Martin) D. Grad. high sch., Brentwood, N.Y. Driller Soil Mechs. Drilling Corp., Seaford, N.Y., 1984-86, Ind. Testing Labs. Inc., College Point, N.Y., 1987-88, Warren George Inc., Jersey City, 1988-89; chief exec. officer, owner Shawnee Falcon Drilling Corp., East Islip, N.Y., 1988—; constrn. insp., engr. Constrn. Techs. Inc., Garden City Park, N.Y., 1989—. Mem. Am. Concrete Industry, Nat. Waterwell Assocs., ASTM (engring. standards com), East Islip C. of C., L.I. Small Bus. Coun. Home and Office: 10 Jefferson St East Islip NY 11730-1808

DIONIGI, CHRISTOPHER PAUL, plant physiologist; b. Boulder, Colo., July 15, 1957. MS, U. S.W. La., 1982; PhD, Iowa State U., 1989. Rsch. asst. Iowa State U., Ames, 1983-89; postdoctoral food flavor quality rsch. U.S. Dept. Agr., Agrl. Rsch. Svc., New Orleans, 1989-91, plant physiologist, 1991—. Mem. Weed Sci. Soc. Am. Achievements include rsch. on the effects of farnesol on Streptomyces tendae and spore and geosmin formation; effects of clomazone on geosmin biosynthesis.

DIONNE, GERALD FRANCIS, research physicist; b. Montreal, Que., Can., Feb. 5, 1935; came to U.S., 1964, naturalized, 1980; s. Louis Philip and Clare Isabel (Flood) D.; m. Claudette Leblanc, June 29, 1963; 1 child, Stephen. BS summa cum laude, Concordia (Can.) U., 1956; B of Engring. magna cum laude, McGill U., 1958, PhD in Physics, 1964; MS, Carnegie-Mellon U., 1959. Jr. engr. IBM Corp., Poughkeepsie, N.Y., 1959-60; sr. engr. Sylvania Electric Products, Woburn, Mass., 1960-61; research asst., lectr. McGill U., Montreal, Que., Can., 1964; sr. research assoc. Pratt & Whitney Aircraft, North Haven, Conn., 1964-66; research staff mem. Lincoln Lab., MIT, Lexington, 1966—; guest lectr., grad student rsch. adv., 1961-63. Contbr. articles to sci. jours.; research in magnetism, magnetoelastic and magneto-optic phenomena, superconductivity theory, microwave, submillimeter-wave, optical and surface physics. NRC of Can. fellow. Mem. Am. Phys. Soc., IEEE (sr.), Corp. Profl. Engrs. Que., Sigma Xi. Home: 182 High St Winchester MA 01890-3366 Office: 244 Wood St Lexington MA 02173

DIONNE, OVILA JOSEPH, physicist; b. Brunswick, Maine, Dec. 8, 1947; s. Ovila Joseph Sr. and Eva Marie (Dube) D.; m. Elaine Francis Sabatinelli, Oct. 7, 1972; children: Michael T., Matthew K., Andrew J. BS, Ariz. State U., 1973; Fellow, Boston U., 1975. Cons. physicist Radiation Physics, Inc., Boston, 1975-82; mktg. mgr. Picker, Internat., Cleve., 1982-89; v.p. mktg., sales DDD Med. Imaging, St. David, Pa., 1989-91; pres. SDC Mktg. Cons., West Chester, Pa., 1991—. Co-author: Ultrasound in Cancer Research, 1983. Mem. fin. commn. Bellingham, Mass., 1979. Maj. U.S. Army Res., 1982—. Recipient multiple mil. and mktg. awards. Mem. Am. Assn. Physicists in Medicine, Am. Assn. Radiologic, Jaycees (pres. Bellingham chpt. 1977), Res. Officer Assn. Achievements include rsch. of ultrasound in determination of prostate cancer, mgmt. and orgn. studies; devel. employee first radiation therapy planning system, computerized tomography and other technologies in field. Home: 1654 Eldridge Dr West Chester PA 19380 Office: SDC Consulting 1654 Eldridge Dr West Chester PA 19380

DI PAOLO, JOSEPH AMADEO, geneticist; b. Bridgeport, Conn., June 13, 1924; s. John Anthony and Nancy (Montagano) Di P.; m. Arleta Mae Schrieb, June 14, 1952; children: Nancy, John. BA, Wesleyan U., 1948; MS, Western Res. U., 1949; PhD, Northwestern U., 1951; MD (hon.), U. Cagliari, Italy, 1991. Instr. genetics bacteriology dept. biology Loyola U., Chgo., Il., 1951-53; instr. clin. and exptl. pathology Northwestern U. Med. Sch., Chgo., 1953-55; sr. cancer research scientist Roswell Park Meml. Inst., Buffalo, 1955-63; research pharmacologist, cell biologist biology br., div. chem. and phys. carcinogenesis program Nat. Cancer Inst., Bethesda, Md., 1963-76, chief lab. biology, div. cancer etiology, 1976—; assoc. prof. lectr. anatomy George Washington U., Washington, 1973—; chmn. U.S.-Germany Cancer Program Area for Environ. Carcinogenesis, 1979-86, U.S.-USSR Mammalian Sometic Cell Genetics Related to Neoplesic Program, 1973-76. Editor, co-author: Chemical Carcinogenesis, 1971; assoc. editor: Jour. of Nat Cancer Inst., 1968-71, Cancer Rsch., 1970-78, Teratogenesis, Carcinogenesis, Mutagenesis, 1982-92; editorial acad. Internat. Jour. Oncology, 1992—. Served with USN, 1943-46. Fellow N.Y. Acad. Sci., AAAS; mem. Am. Assn. Cancer Rsch. (bd. dirs. 1983-86), Am. Soc. Human Genetics, Am. Soc. Exptl. Pathology, Genetics Soc. Am., Teratology Soc., Hamster Soc., Tissue Culture Assn., Am. Assn. Pathology, European Assn. for Cancer Rsch., Sigma Xi. Home: 6605 Melody Ln Bethesda MD 20817-3154 Office: Nat Cancer Inst Bldg 37-2A-19 Rockville Pike Bethesda MD 20892

DIPRIZIO, ROSARIO PETER, mathematics educator; b. Oak Park, Ill., Aug. 17, 1940; s. Peter and Lena (Allesandro) D.; m. Kathleen Doyle, July 4, 1978; children: Mark, Jeni, Mike, Carol, Dan. Student, Wright Jr. Coll., 1959-60; BS, Ill. Benedictine Coll., 1963; MS, Northeastern Ill. U., 1968. Tchr. math. Weber High Sch., Chgo., 1963-67, Glenbrook North High Sch., Northbrook, Ill., 1967-72; prof. math. Oakton Community Coll., Des Plaines, Ill., 1972—; manuscript and book reviewer Wadsworth, D.C. Heath, Houghton Mifflin, Wiley, Adison-Wesley, West, Brooks/Cole. Avocations: fishing, boating. Home: 532 Circle Dr Fox Lake IL 60020-1903

DIRECTOR, STEPHEN WILLIAM, electrical engineering educator, researcher; b. Bklyn., June 28, 1943; s. Murray and Lillian (Brody) D.; m. Lorraine Schwartz, June 20, 1965; children: Joshua, Kimberly, Cynthia, Deborah. BS, SUNY, Stony Brook, 1965; MS, U. Calif., Berkeley, 1967, PhD, 1968. Prof. elec. engring. U. Fla., Gainesville, 1968-77; vis. scientist IBM Rsch. Labs., Yorktown Heights, N.Y., 1974-75; prof. elec. and computer engring. Carnegie-Mellon U., Pitts., 1977—, U.A. and Helen Whitaker Univ. prof. electronics and elec. engring., 1980—, prof. computer sci., 1982—, head dept. elec. and computer engring., 1982-91; univ. prof., 1992—; dean Carnegie Inst. Tech. Carnegie-Mellon U., Pitts., 1991—; cons. Intel Corp., Santa Clara, Calif., 1977-84, Digital Equip. Corp., Hudson, Mass., 1982-88, Calma Corp., 1985-86, Mentor Graphics Corp., 1988-91; sci. adv. bd. Nextwave, Inc., 1990—; bd. dirs. OcCAD, Inc., 1991—, CAD Framework Initiative, Inc. 1991—, Aspect Devel. Corp., 1991-92; sr. cons. editor McGraw-Hill Book Co., N.Y.C., 1976—; dir. Rsch. Ctr. Computer-Aided Design, Pitts., 1982-89. Author: Introduction to System Theory, 1972, Circuit Theory, 1975, VLSI Design for Manufacturing: Yield Enhancement, 1989, Principles of VLSI System Planning: A Framework for Conceptual Design, 1990; editor: Computer-Aided Design, 1974. Recipient Frederick Emmons Terman award Am. Soc. Engring. Edn., 1976; named Outstanding Alumnus, SUNY, Stony Brook, 1984. Fellow IEEE (W.R.G. prize 1979, Centennial medal 1984); mem. IEEE Cirs. and Systems Soc. (pres. 1981, assoc. editor jour. 1973-75, Best Paper award 1970, 85, 92, Soc. award 1992), IEEE Computer Soc. Office: Carnegie-Mellon U CIT Dean's Office Schenley Pk Pittsburgh PA 15213-3830

DIRHEIMER, GUY, biochemist, educator; b. Basel, Switzerland, July 14, 1931; s. Charles and Ines (Marin) D.; m. Marguerite Mangin, Dec. 20, 1958; children: Florent, Bertrand, Pascale. Degree, Lycee Fustel de Coulanges, Strasbourg, France, 1950; PhD in Pharmacy, U. Strasbourg, 1961, PhD in Biochemistry, 1964. Attaché de recherche Nat. Ctr. Sci. Rsch. Strasbourg, 1955-61, chargé de recherche, 1961-64, asst. prof., 1964-69, prof. toxicology and molecular biology, 1969—, dean faculty pharmacy, 1969-70; dir. Inst. Molecular and Cellular Biology CNRS, Strasbourg, 1992—. Contbr. over 280 articles to profl. publs. Recipient Bonneau prize French Acad. Scis., 1973; named Officier des Palmes Académique, 1987. Mem. ACE, French Toxicology Soc., Brit. Toxicology Soc., French Soc. Genetic Toxicology, French Biochem. Soc., German biochem. Soc., French Nat. Acad. Medicine (laureat 1985), European Molecular Biology Orgn. Achievements include isolating and sequencing numerous transfer ribonucleic acids and determining the structure of their minor nucleotides, isolating and cloning aminoacyl-tRNA synthetases and several toxic glycoproteins from plants and mushrooms. Home: 34 sentier de l'Aubépine, 67000 Strasbourg France Office: Inst Biol Molec Cell, 15 rue Descartes, 67084 Strasbourg France

DIRLAM, DAVID KIRK, psychology educator; b. Corning, N.Y., Jan. 13, 1942; s. Arthur Clinton and Edith Lor (Kirk) D.; m. Annette Isaacs, Dec. 31, 1981; children: David, Djuna, Lydia, Gareth. BA, Northwestern U., 1964; MA, McMaster U., Can., 1967, PhD, 1970. Asst. prof. St. Norbert Coll., DePere, Wis., 1969-74; dir. Ednl. R&D Ctr. State Univ. Coll., Plattsburgh, N.Y., 1974-81; pres. Dirlam Data Systems, San Marcos, Calif., 1981-88; prof., chmn. Psychology Dept. King Coll., Bristol, Tenn., 1988—. Author: Standardized Developmental Ratings, 1978; contbr. (book) Toward a Theory of Psychological Development, 1980, (series of books) The Second "R": K-12 Writing Curriculum, 1980-81. Pres. Mt. Rogers Appalachian Trail Club, Abingdon, Va., 1990—; mem. bd. managers Appalachian Trl. Conf., Harpers Ferry, W.Va., 1993—. Mem. Am. Psychol. Soc. Episcopalian. Achievements include finding that the age of appearance of general drawing and discourse skills can be modeled by a sequence of exponential probability distributions with means equal to several years, yet individual children use skills from widely diverse parts of the sequence from one week to the next. Home: 84 Kingsbridge Bristol TN 24201 Office: King Coll 1350 King College Rd Bristol TN 37620

DI RUSSO, ERASMO VICTOR, aeronautical engineer, educator, consultant; b. Buenos Aires, May 29, 1955; s. Roque Di Russo and Victoria (Di Russo) Dangvilavityte; m. Sonia Miriam Papotto, Apr. 2, 1982; children: Natali Victoria, Michelle. Student, Nat. Tech. U., Buenos Aires, 1974-80; postgrad., U. de Belgrano, 1992—. Cert. aircraft engr. Aircraft technician Escuela Nacional de Educacion Tecnica No. 1 Haedo Nat. Tech. U., Buenos Aires, 1974-80; navigation instruments technician Argentine Airlines, Buenos Aires, 1976-79, Boeing 707 flight engr. instr., 1979-86; fire bombing equipment project engr. CATA SA, 1980; asst. prof. aerodynamics Nat. Tech. U., 1981-86, prof. aerodynamic design, 1986, dir. area fluid dynamics, 1989-92; Boeing 747 flight engr. instr., instrn. coord., fuel mgr. Argentine Airlines, 1990—; aero. project cons. Argentine Air Force, Buenos Aires, 1980—. With Argentine Army, 1973-76. Mem. AIAA, Planetary Soc. Home: La Cautiva 323, 1408 Buenos Aires Argentina Office: Paseo Colon 221 1er Piso, Buenos Aires Argentina

DISAIA, PHILIP JOHN, gynecologist, obstetrician, radiology educator; b. Providence, Aug. 14, 1937; s. George and Antoinette (Vastano) DiS.; divorced; children: John P., Steven D.; m. Patricia June; children: Dominic J., Vincent J. BS cum laude, Brown U., 1959; MD cum laude, Tufts U., 1963. Diplomate Am. Bd. Ob-Gyn. (examiner 1975—), Am. Bd. Gynecologic Oncology (bd. dirs. 1987—). Intern Yale U. Sch. Medicine, New Haven Hosp., 1963-64, resident in ob-gyn., 1964-67, instr. ob-gyn., 1966-67; fellow in gynecologic oncology U. Tex. M.D. Anderson Hosp. and Tumor Inst., Houston, 1969-70, NIH sr. fellow, 1969-70, instr. ob-gyn., 1969-71; asst. prof. ob-gyn. and radiology U. So. Calif. Sch. Medicine, L.A., 1971-74, assoc. prof., 1974-77; prof., chmn. dept. ob-gyn. U. Calif., Irvine Med. Ctr. Calif. Coll. Medicine, 1977-88, prof., 1977—, prof. radiology, radiation therapy div., 1978—, assoc. vice chancellor for health scis. Irvine Coll. Medicine, 1987-89, Dorothy Marsh chair of reproductive biology, 1989—; clin. prof. dept. ob-gyn. U. Nev. Sch. Medicine, Reno, 1985—; chmn. site visit team for surgery br. Nat. Cancer Inst. NIH, 1983, subcom. surg. oncology rsch. devel., 1982-83, mem. sci. counselors div. cancer treatment, 1979-83; mem. gov.'s adv. coun. on cancer State of Calif., 1980-85; vis. prof., lectr., speaker various sci. meetings, confs., courses, in Eng., Italy, Greece, Japan, others. Author: (with E.J. Quilligan) Ovarian Tumors, Current Diagnosis, 1974, (with others) Synopsis of Gynecologic Oncology, 1975, (with W.T. Creasman) Clinical Gynecologic Oncology, 1980, 4th Edit. 1993; contbr. numerous articles to profl. jours., book chpts.; assoc. editor Gynecologic Oncology, Endocuritherapy/Hyperthermia Oncology, Danforth's Textbook of Obstetrics & Gynecology; mem. editorial adv. bd. Am. Jour. Reproductive Immunology, Cancer Clinical Trials, The Female Patient, New Trends in Gynecology and Obstetrics (Italian publ.); reviewer Am. Jour. Ob-Gyn., Med. and Pediatric Oncology, New Eng. Jour. Medicine, Ob-Gyn. jour., Cancer; physician cons. Patient Care Standards jour.; sci. adv. bd. The Clin. Cancer Letter. Recipient Disting. Alumnus award M.D. Anderson Hosp. and Tumor Inst. U. Tex., 1980, Silver Apple award U. Calif. Med. Students, 1983, Lauds and Laurels Profl. Achievement award U. Calif. Alumni Assn., 1983, Hubert Haussel's award Long Beach Meml. Hosp., 1983, also various rsch. awards. Fellow Am. Coll. Obstetricians and Gynecologists (com. on human rsch. for cancer 1979—, chmn. 1984—, chmn. subcom. on gynecologic oncology 1984-85, prolog editorial and adv. com. 1986—, various others), ACS, Commn. on Cancer Liaison, Western Assn. Gynecologic Oncologists (founder 1971, pres. 1978-79), Am. Gynecol. and Obstet. Soc. (exec. coun. 1986—), Am. Gynecologic Soc., Pacific Coast Ob/Gyn Soc., South Atlantic Assn. Obstetricians and Gynecologists (hon.); mem. AMA, Am. Cancer Soc. (bd. dirs. L.A. County unit 1975-77, Orange County 1979, Calif. div. 1985—), Nat. Am. Cancer Soc. (dir.-at-large, bd. dirs. 1985—, chmn. program com. for nat. conf. 1986, active in others), Am. Coll. Radiology (commn. on cancer 1984-85), Am. Soc. Clin. Oncologists, Soc. Gynecologic Oncologists (exec. coun. 1975-80, pres. 1982-83), Internat. Gynecologic Oncology Cancer Soc., Italian Soc. Ob-Gyn., Calif. Med. Assn., other profl. orgns., Alpha Omega Alpha. Home: 12132 Skyline Dr Santa Ana CA 92705-3150 Office: U Calif Irvine Med Ctr 101 City Blvd W Bldg 26 Orange CA 92668-2901

DI SALVO, FRANCIS JOSEPH, chemistry educator; b. Montreal, Que., Canada, July 20, 1944; s. Francis J. and Rita (Doherty) Di S.; m. Barbara Anton, June 7, 1966; children: Pamela Marie, Katherine Marie. BS, MIT, 1966; PhD, Stanford U., 1971. Mem. tech. staff AT&T Bell Labs., Murray Hill, N.J., 1971-78, dept. head, 1978-86; prof. chemistry Cornell U., Ithaca, N.Y., 1986—. Contbr. over 200 articles to profl. jours. Fellow Am. Phys. Soc. (Internat. New Materials prize 1991); mem. NAS, Am. Chem. Soc., Am. Acad. Arts and Scis. Achievements include patents on rechargeable lithium batteries. Office: Dept Chemistry Cornell U Ithaca NY 14853

DISCHINGER, HUGH CHARLES, civil engineer; b. Chgo., Mar. 4, 1924; s. Irvin Ernest and Evelyn Martha (Bender) D.; m. Thelma Ann Brown, Dec. 30, 1950; children: Hugh Charles Jr., Martha Boynton, Joseph Brown, Amy Robins. BS, Va. Mil. Inst., 1950. Registered profl. engr.; cert. land surveyor. Field engr. Stone & Webster Engring. Corp., Richmond, Va., 1950-52; engr. Franklin R. Murray, Newport News, Va., 1952-54; ptnr. Murray & Dischinger, Engrs., Newport News, Va., 1954-71; owner Hugh C. Discinger, Engrs. & Surveyors, Hampton, Grafton, Va., 1971-84; pres. Hugh C. Discinger & Assocs., P.C., Grafton, 1984—; pres. Va. Sect. ASCE, 1972-73, Peninsula Chpt. Va. Soc. Profl. Engrs., Williamsburg, 1966-67. Pres. Peninsula Coun. Boy Scouts of Am., Hampton, Newport News, Williamsburg, Gloucester, Mathews, Va., 1976-79. Named Boss of Yr., Nat. Secs. Assn., 1976. Fellow ASCE; mem. ASCE, Am. Congress Surveying and Mapping, Va. Assn. Surveyors. Episcopalian. Home: PO Box 472 Gloucester VA 23061 Office: Hugh C Dischinger & Assocs 110-A Dare Rd Grafton VA 23692

DISHMON, SAMUEL QUINTON, JR., computer scientist; b. Dallas, Aug. 19, 1956; s. Samuel Q. and Mildred Artie (Phillips) D. AS in Computer Sci., Murray State Coll., 1989; BS, Southeastern Okla. State U., 1991. With quality and reliability assurance dept. Westinghouse Command and Control, Balt., 1978-86; contract expediter Bendix Environ. System, Towson, Md., 1986; with quality and reliability assurance dept. Spring Lake Rsch.-Dadelion, Frederick, Md., 1986; cons. Madill, Okla., 1986—; computer lab. asst. Murray State Coll., Tishomingo, Okla., 1987-89; computer asst. Southeastern Okla. State U., Durant, 1990, dorm mgr., 1990—; chief computer cons. Dishmon's, 1991—; grad. teaching asst. in physics U. Cen. Okla., 1992-93; security cons. Southeastern Okla. State U., 1990—. Contbr. articles to profl. jours. Mem. Soc. Physics Students, Southeastern Okla. State U. Computer Club.

DISTELBRINK, JAN HENDRIK, physicist, researcher; b. Terschelling, Friesland, The Netherlands, May 12, 1946; s. Johannes Theodorus and Anna Johanna (Houtkooper) D.; m. Katharina Josopandojo, Sept. 21, 1973; children: Nick, Martin. BSE, Tech. U. Delft, The Netherlands, 1969, MS in Physics, 1973; PhD in Nuclear Physics, Rensselaer Poly. Inst., 1991. Physicist Nat. Inst. for Nuclear and High Energy Physics, Amsterdam, The Netherlands, 1974-80; mem. rsch. staff Linear Accelerator Ctr. MIT, Cambridge, Mass., 1980-92; sr. rsch. scientist III U. N.H., Durham, 1992—; pres. Neologic Inc., Beverly, Mass., 1992—; chmn. scintillator working group Continuous Electron Beam Accelerator Facility, Newport News, Va., 1992—. Contbr. articles to Nuclear Physics Jour., Nuclear Instruments and Methods, Physics Letters , Phys. Rev. Achievements include development of large scale detector systems for use in nuclear and high energy physics, high speed digital delay line, high speed time to digital converters; research in charge movements in wire chambers and electron beam position monitors. Home: 7 Pinewood Rd Peabody MA 01960 Office: U NH Physics Dept Demeritt Hall Durham NH 03824

DITTBERNER, GERALD JOHN, engineer, meteorologist, space scientist; b. St. Paul, Minn., Oct. 24, 1941; s. Norbert R. and Emily B. (Tarr) D.; m. Mary K. Doerning, Sept. 11, 1965; children: Colleen M., Matthew J., Brigitte C. BEE, U. Minn., 1964; MS in Meteorology, Space Sci. & Engring., U. Wis., 1969, PhD in Meteorology, 1977. Bd. cert. cons. meteorologist. Commd. 2d lt. USAF, 1964, advanced through grades to lt. col.; weather forecaster 12th squadron USAF, Thule AB, Greenland, 1969-70; chief satellite data processing air force global weather cen. USAF, Offutt AFB, Nebr., 1970-74; vice comdr. environ. tech. applications ctr. USAF, Scott AFB, Nebr., 1977-81; chief aerospace scis. div. 2d weather wing USAF, Ramstein AB, Fed. Republic of Germany, 1981-84; program mgr. office sci. research USAF, Bolling AFB, D.C., 1984-85; sr. prin. engr. Harris Corp., Alexandria, Va., 1985-88; sr. scientist Kaman Scis. Corp., Alexandria, Va., 1988-93; scientist Mentor Techs., Inc., Lanham, Md., 1993—; scientist, field experimentor World Meteorol. Orgn., Barbados, West Indies, 1969; adj. prof. St. Louis U., 1976. Contbr. articles to profl. jours. Chmn. advancements Boy Scouts Am., Ramstein, 1981-84, outdoor chmn., Springfield, Va., 1984-89. Fellow AIAA (assoc., chmn. space ops. and support tech. com. 1993—), Royal Meteorol. Soc.; mem. IEEE, Am. Meteorol. Soc., Nat. Weather Assn. (v.p. 1979-80), Sigma Xi, Theta Tau. Roman Catholic. Avocations: camping, skiing, porsche racing, travel. Office: Mentor Techs Inc 10000 Aerospace Rd Ste N Lanham MD 20706-2254

DITTEMORE, DAVID H., aerospace engineer, engineering executive; b. 1953; 2 children. BS in Aero. and Astro. Engring., U. Wash., 1975. Project engr. TFE731 comml. engine project Garrett Engine Divsn., Phoenix, Ariz., 1975-81, supr. TFE731 projects, 1981-84, devel. and qualification test mgr. F109 engine program, 1984-87, component design and performance mgr. LHTEC T800 engine program, 1987-89, engring. mgr. propulsion controls and lubrication systems project, 1989-90; v.p. engring., chief engr. comml. wide body APU projects Garrett Aux. Power Divsn., Phoenix, 1990-93; v.p. engring. AlliedSignal Engines, Phoenix, 1993—. Office: AlliedSignal Engines 2739 E Washington St PO Box 5217 Phoenix AZ 85010

DITTRICH, HERBERT, mineralogist, researcher; b. Waiblingen, Germany, Aug. 29, 1955; s. Erwin and Emma (Poslushny) D.; m. Barbara Christine Bloss, Sept. 4, 1981; children: Eva Dorothea, Hannah Barbara. Abitur, Max-Planck-Gymnasium, Schorndorf, Germany, 1975, Diplom Mineralogie, Dr.rer.nat. Rschr. 4th Phys. Inst., U. Stuttgart, Germany, 1983-84, Inst. Phys. Electronics, U. Stuttgart, 1984-91; project coord. Ctr. for Solar Energy and Hydrogen Rsch., Stuttgart, 1991—. Contbr. articles to profl. jours. V.p. EURO 80, Schorndorf, 1983-86; pres. Folk Club Kith and Kin, Winterbach, 1992—. Mem. Deutsche Gesellschaft fuer Kristallzuchtung, Deutsche Mineralogische Gesellschaft. Achievements include basic research on selenization reactions for chalkopyrite thin film solar cells, development of XRD texture analysis on polycrystalline thin films, laser ablation as a deposition method for polycrystalline semicondr. thin films. Office: Solar and Hydrogen Rsch Ctr, Hessbruehlstrasse 21, 70565 Stuttgart Germany

DITTY, MARILYN LOUISE, gerontologist, educator. MS in Psychology, U. San Diego, 1976; postgrad., U. So. Calif., 1977-82; DPA, U. LaVerne, 1990. Lic. health care adminstr., Calif. Exec. dir. San Clemente (Calif.) Srs., Inc., South County Sr. Svcs., 1978—; assoc. prof. dept. continuing edn., coord. emeritus inst. Saddleback Coll., Mission Vievo, Calif., 1979—; dir. San Clemente Sr. Housing, Inc., 1980—; adj. prof. sociology, gerontology Orange Coast Coll., Costa Mesa, Calif., 1987—; cons. to developers of sr. housing; gov.'s appointee lontg-term care adv. com. Calif. Dept. Aging; past trustee Calif. Human Ins. Trust, bd. dirs. Sr. Housing Coun. Calif. Bldg. Indsutry; accreditation com. Calif. Higher Edn.; invitee White House Conf. on Aging, 1981, Internat. Congress on Gerontology, Eng., 1987. Presenter papers at profl. confs. Grantee Calif. Health Facilities Financing Authority, Calif. Sr. Bond Funds, Calif. Housing Community Devel. Funds, Calif. Dept. Edn., Calif. Dept. Health, Calif. Dept. Aging; recipient Award of Merit Orange County Community Svcs., 1979, Outstanding Svc. award Saddleback Coll., 1980, spl. cert. of merit Orange County Bd. Suprs., 1982, gerontology svc. cert. Nat. Assn. Adult Day Care, 1984, Long-Term Care Svcs. award Area Agy. on Aging. Mem. Calif. Assn. Adult Day Health Care Svcs. (past pres.), Calif. Assn. Nutrition Dirs. (past v.p.), Calif. Assn. Non-Profits. Office: San Clemente Srs Inc 2021 Calle Frontera San Clemente CA 92672

DIVAKARAN, SUBRAMANIAM, biochemist; b. Madras, India, Jan. 10, 1938; s. T.P. and Radha Subramaniam; m. Meena Divakar, Jan. 16, 1962; children: Amal Divakaran, Vasu. MVSc, U. Madras, 1965, PhD, 1977; diploma (hon.), City og Guilds, London, 1966; FIC, Inst Chemist, Calcutta, India, 1980. Cert. vet. Vet. surgeon Pasteur Inst. So. India, Nilgiris, 1959-63; scientist Cent. Leather Res. Inst., Madras, 1965-81; tech. dir. Cutfast Polymers Pvt. Ltd., Madras, 1982-84; rsch. scientist The Oceanic Inst., Waimanalo, Hawaii, 1984--; cons. Food and Agr. Orgn. U.N., Rome, 1978--, Vol. in Tech. Assistance VITA. Author: Animal Blood Its Use in Food Feed Fertilizer Industry and Medicine, 1979, Utilization of Dead Animals and Condemned Animal Offals, 1981, Animal By-Products Their Utilization, 1981, Animal Blood Processing and Utilization, 1982, Hand Book of Glue and Gelatine manufacture, 1984. Recipient Two Gold medals U. Madras, 1959, Invention award Gov. India, 1969. Mem. Am. Chem. Soc., Madras Astronomical Assn. (pres. 1978-79). Republican. Hindu. Achievements include manufacture of surgical sutukes, cosmetic collagen, glue from leather wastes. Office: Oceanic Inst Makapuu Point Waimanalo HI 96795

DIX, FRED ANDREW, JR., professional society executive; b. Connellsville, Pa., Aug. 27, 1931; s. Fred Andrew and Freda Pearl (Horton) D.; m. Jean Carol Bacon, July 18, 1953; children: Cynthia Carol, Jennifer Jean, Stephen Bacon. B.S., Rutgers U., 1953, M.S., 1957. Petroleum geologist Standard Oil Co. of Calif., Salt Lake City, 1957-62, Denver, 1962-64; Petroleum geologist Mobil Oil Corp., Jackson, Miss., 1965-66, Corpus Christi, Tex., 1966-72, Houston, 1972; div. exploration data processing coordinator Mobil Oil Corp., 1967-72; exec. dir. Am. Assn. Petroleum Geologists, Tulsa, 1973—. Contbr. articles to profl. jours. Served with USAF, 1954-55. Mem. Rutgers Geology Club (pres.), Geol. Soc. Am., Tulsa, Houston, Oklahoma City geol. socs., Council Engring. and Sci. Soc. Execs., Am. Assn. Petroleum Geologists (chmn. com. on statistics of drilling 1969-70, treas. 1972, Disting. Service award 1983, hon. mem.), Petroleum Club Tulsa, Tulsa C. of C., Kappa Sigma. Republican. Presbyn. Home: 6815 E 52d St Tulsa OK 74145 Office: Am Assn Petroleum Geologists PO Box 979 1444 S Boulder St Tulsa OK 74101

DIXON, BRIAN GILBERT, chemist; b. Berkeley, Calif., July 6, 1951; s. Joseph Ardiff and Marjorie Jane (Watts) D.; m. Karen Marie Marcavage, June 16, 1973; children: Matthew, Shannon. BS, Pa. State U., 1973; PhD, U. Ill., 1981. Chemist ICI U.S. Inc., Wilmington, Del., 1973-76; sr. rsch.

scientist Dow Chem., Inc., Wayland, Mass., 1980-85; v.p. Cape Cod Rsch., Inc., East Falmouth, Mass., 1985—. Contbr. articles to profl. jours. Mem. Bd. Health, Sandwich, Mass., 1989—. Grantee NSF, 1987, 88, 89, 1990, 92, Dept. Energy, 1990, 92, Dept. Health and Human Svcs., 1992. Mem. AAAS, Am. Chem. Soc. Achievements include patents for electroluminescent sensors, battery electrolyte, process & composition for inhibiting iron & steel corrosion. Home: 4 Nicholas Ln Sandwich MA 02563 Office: Cape Cod Rsch Inc 19 Research Rd East Falmouth MA 02536

DIXON, GORDON HENRY, biochemist; b. Durban, South Africa, Mar. 25, 1930; s. Walter James and Ruth (Nightingale) D.; m. Sylvia W. Gillen, Nov. 20, 1954; children: Frances Anne, Walter Timothy, Christopher James, Robin Jonathan. M.A. with honors, U. Cambridge, Eng., 1951; Ph.D., U. Toronto, 1956. Research asso. U. Wash., 1956-58; research asso. U. Oxford, Eng., 1958-59; asst. prof. biochemistry U. Toronto, 1959-61, asso. prof., 1961-63; prof. U. B.C., 1963-72; prof., chmn. dept. biochemistry U. Sussex, Eng., 1972-74; prof. med. biochemistry U. Calgary, Alta., Can., 1974—, chmn., 1983-88. Contbr. articles to profl. jours. Recipient Steacie prize Steacie Found., 1966, Killam Meml. prize Can. Coun., 1991; named Officer of the Order of Canada, 1993. Fellow Royal Soc. London, Royal Soc. Can. (Flavelle medal 1980); mem. Am. Soc. Biol. Chemists, Am. Soc. Cell Biology, Can. Biochemistry Soc. (pres. 1982-83, Ayerst award 1966), Pan-Am. Assn. Biochem. Socs. (v.p. 1984-87, pres. 1987-90, past pres. 1990—), Internat. Union Biochemistry (exec. coun. 1988—). Office: U Calgary Dept Med Biochemistry, 3330 Hospital Dr NW, Calgary, AB Canada T2N 4N1

DIXON, PATRICIA SUE, medical biotechnologist, researcher; b. Raleigh, N.C., Sept. 1, 1960; d. Richard Vaughn and Edna Ruth (Andersen) D. BS, U. Tex., San Antonio, 1982, MS, 1989. Rsch. asst. U. Tex. Health Sci. Ctr., San Antonio, 1983-89; rsch. biotechnologist USAF Wilford Hall Med. Ctr., Lackland AFB, Tex., 1989—. Author: (with others) Growth, Cancer and the Cell Cycle, 1984, Colon Cancer Cells, 1990; contbr. articles to profl. jours. Named Civilian of Yr. USAF, 1990. Mem. Am. Assn. Cancer Rsch., Tissue Culture Assn., Sigma Xi (assoc.). Achievements include research in the CD4 receptor on human lung fibroblasts that were primary cell cultures and growing skin grafts for treatment of Epidermulosa Bullosa. Office: Wilford Hall Med Ctr RD Clin Investigations Bldg 4430 1255 Wilford Hall Loop Lackland AFB TX 78236-5319

DIXON, ROBERT GENE, educator, mechanical company executive; b. Clatskanie, Oreg., Feb. 15, 1934; s. Hobart Jay and Doris Marie D.; m. Janice Lee Taylor, Sept. 19, 1954; children: Linda Dixon Johnson, Jeffrey, David. AS in Indsl. Tech., Chemeketa C.C., 1978, extended spl. courses, 1978-80. Journeyman machinist, cert. welder, Oreg., cert. master trainer, mfg. engr., trainer Devel. Dimension Internat. interaction mgmt., master trainer techniques for empowered workforce. Machinist apprentice to asst. mgr. AB McLauchlan Co., Inc., 1956-69; supt. design, rsch., devel. engrng. and prodn. Stevens Equipment Co., 1969-70; co-owner, operator Pioneer Machinery, 1970-72; supt. constrn. and repair Stayton Canning Co., 1972-73; mgr. Machinery div. Power Transmission, 1973-75; owner, operator Dixon Engrng., Salem, Oreg., 1975—; instr., program chair mfg. engrng. tech. Chemeketa C.C., 1975-92; apptd. tech. project coord. Oreg. Advanced Tech. Consortium. Tech. reviewer for major pubs. With U.S. Navy, 1952-56. Named Tchr. of Yr., Chemeketa Deaf Program, 1978, Outstanding Instr. of Yr., Am. Tech. Edn. Assn., 1983. Mem. ASTD, Am. Prodn. and Inventory Control Soc., Am. Vocat. Assn. (Outstanding Tchr. award 1981), Oreg. Vocat. Assn. (Instr. of Yr. 1980; pres. 1984), Oreg. Vocat. Trade Tech. Assn. (Instr. of Yr. 1979; pres. 1981; Pres.'s Plaque 1982), Soc. Mfg. Engrs. (cert., sr., chmn. Oreg. sect. 1988—, internat. dir. nominating com. 1992, Oustanding Internat. Faculty adv., 1989, 91), Am. Welding Soc., Am. Soc. Metals, Chemeketa Edn. Assn. (pres. 1979), Am Soc. Quality Control, Computer Automated Systems Assn., Phi Theta Kappa. Author: Benchwork, 1980, Procedure Manual for Team Approach for Vocational Education Special Needs Students, 1980, Smart Cam CNC/CAM Curriculum for Point Control Company; tech. reviewer textbook pubs., 1978—; designer, patentee fruit and berry stem remover. Home: 4242 Indigo St NE Salem OR 97305-2134 Office: PO Box 14007 Salem OR 97309

DIXON, ROBERT KEITH, plant physiologist, researcher; b. Kansas City, Mo., Apr. 16, 1955; s. Keith C. and Evelyn L. (Griffith) D. BS in Forest Mgmt., U. Mo., 1977, MS in Ecology and Soils, 1979, PhD in Plant Physiology, 1982. Asst. and assoc. prof. Univ. Minn., St. Paul, 1983-86; chief-of-party U.S. Agy. for Internat. Devel., Bangkok, Thailand, 1986-87; prof. Auburn (Ala.) Univ., 1987-89; program leader U.S. EPA, Corvallis, Oreg., 1989—; cons. World Bank, Washington, 1985-90, Office of Sci. and Tech. Policy of Exec. Office of U.S. Pres., Washington, 1986, Gov. Perpich Task Force on Biotech., Minn., 1985-86; adviser Internat. Found. Sci., Stockholm, 1986—; vis. prof. Karetsart U., Bangkok, 1986-87, Humboldt U., Berlin, 1985, Delhi (India) U., 1985; referee peer-rev. panels USDA, NSF, Dept. Energy; researcher intergovtl. panel on climate change Internat. Geosphere-Biosphere, NATO. Author: Process Modeling of Forest Response to Stress, 1989; editorial bd. Jour. New Forests; contbr. numerous articles to sci. jours. Smithsonian Inst. fellow, Delhi, 1985; Exxon scholar, 1983. Mem. AAAS, Soc. Am. Forests, Internat. Soc. Tropical Foresters, Sigma Xi, Phi Beta Kappa, Xi Sigma Pi, Gamma Sigma Delta. Achievements include patents in fertilizer technology; research in global climate change, natural resource management, environmental risk assessment. Office: U S EPA 200 SW 35th St Corvallis OR 97330

DIXON, SANDRA WISE, aerospace engineer; b. Savannah, Ga., May 30, 1964; d. Floyd Wallace Jr. and Mary Louise (Pierce) W.; m. Charles David Dixon, Jan. 7, 1989. BS in Aerospace Engrng., Miss. State U., 1988. Aero/performance engr. Gen. Dynamics, Ft. Worth, 1988-90; with Gulfstream Aerospace, Savannah, 1986-87, structural analysis engr., 1991—. Pres.'s scholar Miss. State U., 1985. Mem. AIAA (chmn. honors and awards 1991-92, chmn. tellor com.), Gulfstream Mgmt. Assn. (nominating com. 1991). Home: 204 Redan Dr Savannah GA 31410 Office: Gulfstream Aerospace Corp PO Box 2206 Savannah GA 31402

DIXON, VICTOR LEE, wire company executive; b. South Charleston, W.Va., Feb. 10, 1966; s. David L. and Barbara K. (Rule) D.; m. Melissa Ann Michael, July 28, 1992. BS, W.Va. U., 1989. In elec. mktg. McJunkin Corp., Charleston, W.Va., 1989-90; applications specialist Service Wire Co., Culloden, W.Va., 1990—. Mem. IEEE, ASTM, ASM Internat., Soc. Plastics Engrs., The Wire Assn. Internat., Internat. Mcpl. Signal Assn. Republican. Baptist. Avocations: soccer, tennis, volleyball. Home: 24 Warren Pl Charleston WV 25302 Office: Service Wire Co 310 Davis Rd Culloden WV 25510

DIZON, JOSE SOLOMON, planning engineer; b. Bethesda, Md., June 9, 1960; m. Shirley T. Watanabe, May 23, 1987; children: Jesselyn, Taylor. BS, U. Calif., Berkeley, 1982. Engr. Sacramento Mcpl. Utilities Dist., Calif., 1982; commd. ensign USN, 1982, advanced through grades to lt., 1987, resigned, 1991; sr. engr. Hawaiian Electric Co., Inc., Honolulu, 1991—. Supporting mem. Honolulu Zoo, 1988-91; mem. Waikiki Aquarium, Honolulu, 1988-91; bishop mus., Honolulu, 1991—. Lt. comdr. USNR, 1991—. Recipient award U.S. Naval Inst., 1982; Calif. Alumni scholar U. Calif., Berkeley. Mem. IEEE, Am. Nuclear Soc., Westloch Assn. (pres.), Toastmasters Internat. Home: 91-1007 Okupe St Ewa Beach HI 96706-3553

D'JAVID, ISMAIL FARIDOON, surgeon; b. Rasht, Iran, Apr. 10, 1908; arrived in Germany, 1926; came to U.S. 1946; s. Youssef and Khadidja D'Javid. Grad. high sch., Iran; MD, Friedrich Wilhelm U., Berlin, 1937, PhD, 1941. Diplomate Bd. Surgery. Intern and resident Urban-Krankenhous, Robert-Koch Krankenhous and Charite, Berlin, 1938-41; physician Imperial Iranian Embassy, Berlin, 1938-41; appointed by Shah as surgeon 500-bed hosp. Mash-had, Iran, 1941; chief surgeon Army Hdqrs., Tehran, 1941-43; surgeon-in-chief Iranian 4th Army, Kordestan, Iran; pvt. practice surgery Tehran, 1941-46; intern, resident various hosps. in U.S., 1947-49, chief resident in thoracic and abdominal surgery, 1950-63; major, surgeon U.S. Army M.C., 1953-55; pvt. practice specializing in surgery N.Y.C., 1957-78. Contbr. numerous articles to profl. jours.; patentee in field. Recipient Physician's Recognition award, AMA, 1977-80, 83. Sr. life fellow Am. Coll. Gastroenterology, Deutsche Gesellschraft fur Chrirurgie (Surg. Soc. Greater Germany); life fellow Am. Coll. Surgery, Acad. Sci., Assn. Mil. Surgeons of U.S., Am. Soc. Abdominal Surgeons, Am. Physician's Art Assn., Assn. Am. Physicians and Surgeons; mem. AMA, Coll. Surgeons, German Plastic Surg. Soc., Pan-Am. Med. Assn., Med. Soc. D.C., Calif. Med. Soc., World Med. Assn., N.Y. County Med. Soc., N.Y. State Med. Soc., Rudolf-Wirchow Med. Soc. Avocations: oil painting, portraits, cartons, calligraphy, develops surgico-technical and exploratory scientific novations, speaks 8 languages. Home: VOC 230 Fairway Oaks Dr Sedona AZ 86351

DJERASSI, CARL, chemist, educator, writer; b. Vienna, Austria, Oct. 29, 1923; s. Samuel and Alice (Friedmann) D.; m. Norma Lundholm (div. 1976); children: Dale, Pamela (dec.); m. Diane W. Middlebrook, 1985. A.B. summa cum laude, Kenyon Coll., 1942, D.Sc. (hon.), 1958; Ph.D., U. Wis., 1945; D.Sc. (hon.), Nat. U. Mex., 1953, Fed. U., Rio de Janeiro, 1969, Worcester Poly. Inst., 1972, Wayne State U., 1974, Columbia, 1975, Uppsala U., 1977, Coe Coll., 1978, U. Geneva, 1978, U. Ghent, 1985, U. Man., 1985, Adelphi U., 1993. Research chemist Ciba Pharm. Products, Inc., Summit, N.J., 1942-43, 45-49; asso. dir. research Syntex, Mexico City, 1949-52; research v.p. Syntex, 1957-60; v.p. Syntex Labs., Palo Alto, Calif., 1960-62; v.p. Syntex Research, 1962-68, pres., 1968-72; pres. of Zoecon Corp., 1968-83, chmn. bd., 1986-86; prof. chemistry Wayne State U., 1952-59, Stanford (Calif.) U., 1959—; D.Sc. (hon.); bd. dirs. Quidel, Inc., Affymax, N.V., Cortech, Inc.; pres. Djerassi Found. Resident Artists Program. Author: The Futurist and Other Stories, (novel) Cantor's Dilemma, The Bourbani Gambit, (poetry) The Clock Runs Backward, (autobiography) The Pill, Pygmy Chimps, and Degas' Horse, also 9 others; mem. editorial bd. jour. Organic Chemistry, 1955-59, Tetrahedron, 1958-92, Steroids, 1963—, Proc. of NAS, 1964-70, Jour. Am. Chem. Soc., 1966-75, Organic Mass Spectrometry, 1968-91; contbr. numerous articles to profl. jours., poems, memoirs and short stories to lit. publs. Recipient Intrasci. Rsch. Found. award, 1969, Freedman Patent award Am. Inst. Chemists, 1970, Chem. Pioneer award, 1973, Nat. Medal Sci. for first synthesis of oral contraceptive, 1973, Perkin medal, 1975, Wolf prize in chemistry, 1978, John and Samuel Bard award in sci. and medicine, 1983, Roussel prize, Paris, 1988, Discovers award Pharm. Mfg. Assn., 1988, Esselen award ACS, 1989, Nat. Medal Tech. for new approaches to insect control, 1991, Nev. medal, 1992; named to Nat. Inventors Hall of Fame. Mem. NAS (Indsl. Application of Sci. award 1990), NAS Inst. Medicine, Am. Chem. Soc. (award pure chemistry 1958, Baekeland medal 1959, Fritzsche award 1960, award for creative invention 1973, award in chemistry of contemporary tech. problems 1983, Priestley medal 1992), Royal Soc. Chemistry (hon. fellow, Centenary lectr. 1964), Am. Acad. Arts and Scis., German Acad. (Leopoldina), Royal Swedish Acad. Scis. (fgn.), Royal Swedish Acad. Engring. Scis. (fgn.), Am. Acad. Pharm. Scis. (hon.), Brazilian Acad. Scis. (fgn.), Mexican Acad. Sci. Investigation, Bulgarian Acad. Scis. (fgn.), Phi Beta Kappa, Sigma Xi, Phi Lambda Upsilon (hon.). Office: Stanford U Dept Chemistry Stanford CA 94305-5080

DJORDJEVIC, BORISLAV BORO, materials scientist, researcher; b. Brezice, Slovenia, Yugoslavia, Aug. 25, 1951; came to U.S., 1968; s. Branislav Branko and Cirila (Antolovic) D.; m. Nancy Grant, June 29, 1974; children: Julie Owen, Christine Antolovic. BS in Physics, Coll. William and Mary, 1973; MSE in Materials Sci., Johns Hopkins U., 1978, PhD in Materials Sci., 1979. Rsch. asst. Coll. William and Mary, Williamsburg, Va., 1971-72; grad. rsch. asst. Johns Hopkins U., Balt., 1973-79; tech. cons. Nat. Bureau of Standards, Gaithersburg, Md., 1979; postdoctural fellow Johns Hopkins U., Balt., 1979, prin. rsch. scientist Ctr. Nondestructive Evaluation, 1993—; scientist Martin Marietta Labs., Balt., 1980-84, sr. scientist, group leader, 1984-88, mgr. rsch., engring., 1988-93; dir. of Univ. Johns Hopkins U., 1980-93. Contbr. articles to profl. jours. Mem. ASTM, AAAS, IEEE, Am. Soc. for Nondestructive Testing (com. chmn. 1981-86), Am. Soc. Metals, Am. Phys. Soc. Achievements include Ultrasonic Liquid Jet Probe patent; development of high power ultrasonics for thick multilayer structures; development of through body ultrasonic test methods for solid rocket motors and robotic fully automated ultrasonic test system; research in area of optical probing of stress waves; engineering development of subsystem for launch vehicles, embedding of sensors for health monitoring of advanced composite structures, laser-optic sonar. Home: 1110 Bellevista Ct Severna Park MD 21146-4846 Office: Johns Hopkins U Ctr Nondestructive Evaluation 3400 N Charles St Baltimore MD 21218-2644

DLAB, VLASTIMIL, mathematics educator, researcher; b. Bzi, Czech Republic, Aug. 5, 1932; came to Can., 1968; s. Vlastimil Dlab and Anna (Stuchlikova) Dlabova; m. Zdenka Dvorakova, Apr. 27, 1959 (div.); children—Dagmar, Daniel Jan; m. Helena Briestenska, Dec. 18, 1985; children: Philip Adam, David Michael. R.N.Dr., Charles U., Prague, Czechoslovakia, 1956, C.Sc., 1959, Habilitation, 1962, D.Sc., 1966; Ph.D., U. Khartoum, Sudan, 1962. Research fellow Czechoslovak Acad. Scis., Prague, 1956-57; lectr., sr. lectr. Charles U., Prague, 1957-59, reader, 1964-65; lectr., sr. lectr. U. Khartoum, Sudan, 1959-64; research fellow, sr. research fellow Inst. Advanced Studies, Australian Nat. U., Canberra, 1965-68; prof. math. Carleton U., Ottawa, Ont., Can., 1968—, chmn. dept., 1971-74; vis. prof. U. Paris VI, Brandeis U., U. Bonn, U. Tsukuba, U. Sao Paulo, U. Stuttgart, U. Poitiers, Nat. U. Mex., U. Essen, U. Bielefeld, Hungarian Acad. Scis., Budapest, U. Warsaw, U. Normal Beijing, U. Vienna, UCLA, U. Va., Czechoslovak Acad. Scis., U. Trondheim, U. Paderborn. Author: Representations of Valued Graphs, 1980, An Introduction to Diagrammatical Methods, 1981; editor procs. internat. confs., 1974, 79, 84, 87, 90, 92, 93; contbr. numerous articles to profl. jours. Recipient Diploma of Honour Union Czechoslovak Mathematicians, 1962; Can. Council fellow, 1974; Japan Soc. Promotion of Sci. sr. rsch. fellow, 1981; sci. exchange grantee Nat. Scis. and Engring. Rsch. Coun. Can., 1978, 81, 83, 85, 88, 91. Fellow Royal Soc. Can. (convenor 1977-78, 80-81, coun. mem. 1980-81); mem. Am. Math. Soc., Math. Assn. Am., Can. Math. Soc. (coun., chmn. rsch. com. 1973-77, editor Can. Jour. Math.), European Math. Soc., London Math. Soc. Roman Catholic. Avocations: sports, music. Home: 277 Sherwood Dr, Ottawa, ON Canada K1Y 3W3 Office: Carleton U, Ottawa, ON Canada K1S 5B6

DLABACH, GREGORY WAYNE, mathematics educator; b. Tulsa, Oct. 7, 1964; s. Oscar James and Beverley Sue (Lofton) D.; m. Kathryn Anette Vacat, Aug. 8, 1992. BS in Math., Okla. State U., 1988; MA in Math., Southwest Mo. State U., 1992. Instr. math. and physics Spartan Sch. Aero., Tulsa, 1988-90; instr. math. Tulsa Jr. Coll., 1989; grad. asst. S.W. Mo. State U., Springfield, 1990-92; assoc. prof. math. South Ark. C.C., Eldorado, Ark., 1992—; adj. math. faculty Ozarks Tech. C.C., 1991-92. Active Springfield Spl. Olympics, 1990-92; vol. El Dorado Coun. Campfire, 1992—; participant Ark. Math Crusade, 1993; Phi Theta Kappa advisor South Ark. C.C. Recipient Danforth award Ralston-Purina, 1979, Woodman History award Woodman Ins. Co., 1980, 82, Blue Ribbon award for Outstanding Vol. Svc. to Campfire; Okla. Meml. Trust Found. ednl. grantee, 1982. Mem. Am. Math. Soc., Tulsa Assn. Physics Tchrs., Ark. Devel. Educators Assn. Avocations: golf, exercising. Office: S Arkansas Community Coll 300 SW Ave El Dorado AR 71730

DMYTRUK, MAKSYM, JR., mechanical engineer; b. New Haven, June 23, 1958; s. Maksym and Halia (Fysun) D.; m. Margaret Catherine Smith, May 13, 1989; children: Maksym Oleksyj III, Christopher Aleksander. Student, St. Basil Coll.; grad. machine drafting, Eli Whitney Tech. Sch., 1976; BSME, U. New Haven, 1981. Mech. engring. technician, sr. draftsman Hunt-Pierce Corp., Milford, Conn., 1976-79; v.p., co-founder D.F.M. Enterprises, Milford, 1979-83; chief engr., chief insp. Aerial Lift Inc., Milford, 1983—; pres., owner Maksym Corp.; mem. subcom. ANSI A92.2. Mem. machine drafting craft com. State of Conn. Vocat. Tech. Sch. Mem. ASM, SAE, ASME, Am. Welding Soc., Soc. of Mfg. Engrs., Fluid Power Soc., Am. Nat. Standards Inst., Soc. Advancing Mobility Land Sea Air and Space, Assn. Systems Mgmt. Orthodox. Achievements include patent pending for knuckle design; creation of standards and new designs for aerial man lift industry. Home: 220 Clifton St New Haven CT 06513-3320 Office: Aerial Lift Inc 571 Plains Rd Milford CT 06460-1796

DO, TAI HUU, mechanical engineer; b. Quang Binh, Vietnam, May 31, 1942; came to U.S., 1975; s. Mau Do and Thi Hai Nguyen; 1 child, Frederick Quan. BSME, U. Paris, 1970, MS, 1971. Rsch. engr. Soc. Automobile Engrs., Paris, 1970-71; test engr. Yanmar Diesel Co., Ltd., Osaka, Japan, 1971-72; prodn. mgr. Vietnam Products Co., Ltd., Saigon, Vietnam, 1972-75; chief engr. European Parts Exchange, Irvine, Calif., 1975-77; project mgr. Fairchild Fastener Group, Santa Ana, Calif., 1977—. Co-author: Literary Dissident Movement in Vietnam; editor: Khai Phong Mag.; patentee in field; contbr. articles to profl. jours. Mem. Soc. Automotive Engrs., Soc. Mfg. Engrs. Buddhist. Office: Fairchild Fastener Group 3130 W Harvard St Santa Ana CA 92704-3999

DOAN, HERBERT DOW, technical business consultant; b. Midland, Mich., Sept. 5, 1922; s. Leland Ira and Ruth Alden (Dow) D.; m. Donalda Lockwood, 1946 (div.); children: Jeffrey W., Christine Mary, Ruth Alden, Michael Alden; m. Anna Junia Cassell, July 16, 1979; 1 child, Alexandra Anne Alden. B Chem. Engring., Cornell U., 1949. Founder, owner Doan Assocs., Midland, 1971-85; chmn., dir. Neogen Corp., Lansing, Mich., 1983-90; pres. Mich. High Tech. Task Force, Lansing, 1981-90; nat. adv. com. dept. engrng. U. Mich., Ann Arbor, 1984—; chmn. Midland Molecular Inst., 1971—; dir. Mich. Materials and Processing Inst., Ann Arbor, 1984-92; trustee, sec. Herbert H. and Grace A. Dow Found., Midland, 1951—; researcher Dow Chem. Co., Midland, 1949-60, exec. v.p., 1960-62, pres., 1962-71; dir. Applied Intelligent Systems, Inc., Ann Arbor, Chem. Bank & Trust Co., Arch Devel. Corp., Chgo.; mem. engrng. coun. Cornell U., Ithaca, N.Y., 1964-85, emeritus, 1985—; mem. Nat. Sci. Bd., Washington, 1976-82, vice chmn., 1981-82; mem. Commn. on Phys. Scis., Maths. and Applications, NRC of NAS, Washington, 1987-91; bd. govs. Argonne Nat. Lab., U. Chgo., 1984-90; tech. assessment adv. coun. Office Tech. Assessment, Washington, 1992—. Staff sgt. USAF, 1942-45; PTO. Mem. Am. Inst. Chem. Engrs., Am. Chem. Soc., Sigma Xi. Home: 3801 Valley Dr Midland MI 48640 Office: PO Box 1431 Midland MI 48641

DOANE, THOMAS ROY, environmental toxicologist; b. Warsaw, N.Y., Feb. 8, 1947; s. Howard John and Janice Leanora (Whipple) D.; m. Donna Margarite Felix, July 20, 1968. BS, Oneonta State U., 1969; MS, Cornell U., 1971; MA, U. Tex., 1979; PhD, Va. Poly. Inst., 1984. Cert. sr. ecologist. Commd. 1st lt. USAF, 1974, advanced through grades to lt. col., 1989; environ. technician USAF Environ. Health Lab., McClellan AFB, Calif., 1971-74; environ. cons. USAF Environ. Health Lab., Kelly AFB, Tex., 1974-77, AF Occupation & Environ. Health Lab., Brooks AFB, Tex., 1977-81; chief ecology functions AF Occupation & Environ. Health Lab., Brooks AFB, 1984-87; dir. environ. protection Hdqrs. Human Systems divsn., Brooks AFB, 1974; assoc. dir. ops. Battelle Meml. Inst., San Antonio, 1991—; cons. Toxeco Consultants Inc., San Antonio, 1987-91. Author (book chpt.): Fresh Water Biological Monitoring, 1984; contbr. articles to N.Y. Fish and Game Jour., Procs. of 11th Conf. on Toxicology, Internat. Jour. Environ. Analytical Chemistry; editor: Land Based Environmental Monitoring-Herbicide Orange, 1980. Recipient Regent scholarship N.Y. State, 1965-69; grantee U.S. EPA, 1969. Mem. ASTM (mem. com. E-47), Soc. Environ. Chemistry and Toxicology, Ecol. Soc. Am., Air Force Assn., Phi Theta Kappa, Phi Kappa Phi, Sigma Xi. Home: RR 1 Box 62H Cibolo TX 78108-9601 Office: Battelle Meml Inst 4414 Centerview Ste 260 San Antonio TX 78228-1462

DOANE, WOOLSON WHITNEY, internist; b. Worcester, Mass., Mar. 21, 1939; s. Whitney Randall and Mary Helen (Woolson) D.; m. Patricia Louise Morse, June 21, 1962; children: Melinda L., Morse W., Seth J. BA, U. Vt., 1962, MD, 1965. Diplomate Am. Bd. Internal Medicine. Lt. USNR, 1966-68; gen. med. officer Force Troops Fleet, Camp Lejeune, N.C., 1966-68; pvt. practice Franklin Meml. Hosp., Greenfield, Mass., 1971-82; assoc. med. dir. respiratory therapy Maine Med. Ctr., Waterville, 1982-84; med. dir. Knolls Atomic Power Lab., Schenectady, N.Y., 1984-87; area med. dir. GE Plastics-GE Aerospace, Pittsfield, Mass., 1987-90; med. dir. GE Plastics, Pittsfield, 1991-93; corp. med. dir. Reynolds Metal Co., Richmond, Va., 1993—; bd. dirs. Ctr. and Mass. Am. Lung Assn., Worcester, 1973-82; trustee, bd. dirs. Franklin Meml. Hosp., Greenfield, 1975-79. Bd. dirs. Franklin County Community Action Coun., Greenfield, 1978-81; corporator Berkshire Med. Ctr., Pittsfield, 1987-93; mem. GE Elfan Soc., Pittsfield, Schenectady, 1984-93. Recipient Woodbury Alumni prize U. Vt., 1965. Mem. Am. Coll. Occupational and Environ. Medicine, Am. Coll. Chest Medicine, Am. Pub. Health Assn. Republican. Episcopalian. Achievements include communication program for lay persons on risks of polychlorinated biophenols; program devel. in respiratory therapy, cardiopulmonary lab., cardiac-rehab.

DOBAY, DONALD G., chemical engineer; b. Cleve., Sept. 12, 1924; s. Arnold W. and Hildegarde (Fekete) D.; m. Barbara Isham, 1966 (div. Dec. 1980); children: Carolyn Dobay Broadaway, Barbara Dobay Monroe, Deborah Herpel-Dobay, Amy, David. AB magna cum laude, Oberlin Coll., 1944; MS, U. Mich., 1945, PhD, 1948. Registered profl. engr., Conn. Sr. rsch. chemist Linde div. Union Carbide, Tonawanda, N.Y., 1948-51, B.F. Goodrich, Brecksville, Ohio, 1951-60; dir. latex tech. Goodrich Sponge Products, Shelton, Conn., 1960-74; sr. process engr. Crawford & Russell, Stamford, Conn., 1974-85; mgr. TRC, East Hartford, Conn., 1985; staff cons. Maquire Group, Inc., New Britain, Conn., 1986-90; pres. Conn. Environ. Engring. Svcs., Inc., Bloomfield, Conn., 1990—. Mem. Conn. Valley Mycological Soc. (pres. 1987-90), Conn. Rubber Group (pres. 1976), Masons (sr. warden 1993), Phi Beta Kappa, Sigma Xi, Phi Kappa Phi, Phi Lambda Upsilon. Home: 9 Kent Ln Bloomfield CT 06002 Office: Conn Environ Engring Svcs Inc 107 Old Windsor Rd Bloomfield CT 06002

DOBBEL, RODGER FRANCIS, interior designer; b. Hayward, Calif., Mar. 11, 1934; s. John Leo and Edna Frances (Young) D.; m. Joyce Elaine Schnoor, Aug. 1, 1959; 1½ child, Carrie Lynn. Student, San Jose State U., 1952-55, Chouinard Art Inst., L.A., 1955-57. Asst. designer Monroe Interiors, Oakland, Calif., 1957-66; owner, designer Rodger Dobbel Interiors, Piedmont, Calif., 1966—. Pub. in Showcase of Interior Design, Pacific edit., 1992; contbr. articles to mags. and newspapers. Decorations chmn. Trans. Pacific Ctr. Bldg Opening, benefit Oakland Ballet, and various other benefits and openings, 1982—; chmn. Symphonic Magic, Lake Marritt Plaza., Opening of Oakland Symphony Orch. Season and various others, 1985—; cons. An Evening of Magic, Oakland Hilton Hotel, benefit Providence Hosp. Foudn., bd. dirs., 1991. Recipient Cert. of Svc., Nat. Soc. Interior Designers, 1972, 74; recipient Outstanding Contbn. award, Oakland Symphony, 1986, Nat. Philanthropy Day Disting. Vol. award, 1991. Mem. Nat. Soc. Interior Designers (profl. mem. 1960-75, v.p. Calif. chpt. 1965, edn. found. mem. 1966—, nat. conf. chmn. 1966), Am. Soc. Interior Designers , Claremont Country, Diabetic Youth Found. Democrat. Roman Catholic. Avocations: travel, gardening.

DOBBIN, RONALD DENNY, federal agency administrator, occupational hygienist, researcher; b. Caldwell, Idaho, Aug. 18, 1944; s. Ronald Weir and Ileen Loraine (Stephen) D. BSEE, U. Idaho, 1967; MSc in Occupational Hygiene, Sch. Tropical Medicine-Hygiene, London, 1974. Cert. indsl. hygienist Am. Bd. Indsl. Hygienists. Rsch. occupational hygienist BOSH USPHS, Cin., 1967-71; liaison occupational hygiene NIOSH USPHS, Washington, 1971-72; chief environ. investigations NIOSH USPHS, Cin., 1974-78; chief policy analysis USPHS, Washington, 1978-82; analyst office tech. assessment U.S. Congress, Washington, 1983-85; liaison occupational hygiene office toxic substances U.S. EPA, Washington, 1985-88; adminstr. tng. grants program NIH, Research Triangle Park, N.C., 1988—; chair computer com. Am. Conf. Govt. Indsl. Hygiene, Cin., 1987—. Editorial bd. Applied Occupational and Environmental Health, 1991—; contbr. articles to profl. publs. Capt. USPHS, 1967—. Fellow Colleguim Ramazzini (Italy); mem. AAAS, APHA (program chair 1992), Soc. Occupational and Environ. Health (sec.-treas. 1988-93, chair-elect 1993—), Nat. Assn. Pub. Health Policy, Am. Acad. Indsl. Hygiene., Assn. Occupational and Environ. Clinics. Office: Nat Inst Environ Health Sci PO Box 12233 111 Alexander Dr Research Triangle Park NC 27709-2233

DOBBINS, JAMES TALMAGE, JR., analytical chemist, researcher; b. Chapel Hill, N.C., June 13, 1926; s. James Talmage and Lila (Shore) D.; m. Jacqueleene Bowen, Dec. 22, 1951; children: James Talmage III, Steven Earl. BS in Chemistry, U. N.C., 1947, PhD in Analytical Chemistry, 1958. Chief indsl. hygiene sect. Med. Gen. Lab., Tokyo, 1953-55, head dept. chemistry, 1955-6; rsch. chemist II R.J. Reynolds Tobacco Co., Winston-Salem, N.C., 1958-65; rsch. sect. head II R.J. Reynolds Tobacco Co., Winston-Salem, 1965-72; mgr. analytical rsch. div. R.J. Reynolds Industries, Winston-Salem, 1972-75; master scientist RJR Nabisco, Winston-Salem, 1975-83; master chemist Bowman Gray Tech. Ctr., Winston-Salem, 1983-89; retired, 1989. Contbr. articles to Jour. Assn. Official Agrl. Chemists, Jour. Assn. Official Analytical Chemists, Spectroscopy, Encyclopedia Ind. Chem. Analysis. Fellow Am. Inst. Chemists; mem. AAAS, Soc. for Applied Spectroscopy, N.Y. Acad. Sci., N.C. Acad. Sci., Am. Chem. Soc. (chmn. cen. N.C. sect. 1964-66), Sigma Xi (sec. Wake Forest chpt. 1986-90). Democrat.

Baptist. Achievements include conception of column-elutive sample prep for plant matter analysis; design of clean room facilities for ICP spectrometry of trace inorganics, and of first-of-its-kind computer intelligent auto-dilution by flow injection/ICP analysis. Home: 2838 Bartram Rd Winston Salem NC 27106-5105

DOBBS, CHARLES LUTHER, analytical chemist; b. Augusta, Ga., Aug. 25, 1952; s. Luther Daniel and Bruna Veronica (Yuzovich) D.; m. Michele Elaine Dobbs, July 12, 1986; children: Michael, Andrea. BS, Augusta Coll., 1973; PhD, Ga. Inst. Tech., 1979. Scientist Union Camp Corp., Princeton, N.J., 1979-80; scientist Alcoa Tech. Ctr., Alcoa Center, Pa., 1980-82, sr. scientist, 1982-83, staff scientist, 1983-86, group leader, 1986-90, sect. head, 1990—; mem. indsl. adv. bd. N.C. State U., Raleigh, 1990—, U. Ark., Little Rock, 1992—. Contbr. articles to profl. jours. Chmn. United Way, Alcoa Center, 1991. Mem. Am. Nuclear Soc., Am. Chem. Soc., Spectroscopy Soc. Pitts., Sigma Xi. Achievements include patent for Process Control and Monitoring of Hall Cells. Home: 150 N Washington Rd Apollo PA 15613-9607 Office: Alcoa Tech Ctr 100 Technical Dr Alcoa Center PA 15069

DÖBEREINER, JOHANNA, soil biology scientist; b. Aussig, Tchechoslovaquia, Nov. 28, 1924; came to Brazil, 1950; d. Paul and Margarethe (Schönhöfer) Kubelka; m. Jürgen Döbereiner, Mar. 26, 1950; children: Maria Luisa, Christian Erhard, Lorenz Döbereiner. MSc, U. Wis., 1963; DSc, U. Fla., 1975. Scientist Svc. Nac. Pesq. Agropec., Rio de Janeiro, 1952-73; head of soil biol. inst. EMBRAPA, Rio de Janeiro, 1974-89, scientist, 1989—. Author: N2-fixing Bacteria Nonleg. Crop Plants, 1987, (with Fabio O. Pedrosa) Springer Verl; contbr. over 300 articles to profl. jours. Recipient First Class Order of Merit German Govt., 1990, Bernardo Houssay Prize Orgn. Am. States, 1979, Sci. Prize UNESCO, 1989. Mem. Pontifical Acad. Sci., Third World Acad. Sci. (founding mem. 1981), Brazilian Acad. Sci. (1st sec. 1991). Achievements include research revealing that biological dinitrogen fixation is not restricted to the legume symbiosis but can be extended to cereals, forage grasses and especially sugar cane, the most promising alternative for bio-fuels where the replacement of nitrogenous fertilizers is the key to highly positive energy balances; description of seven new diazotrophic bacteria, all of which associate with gramineous plants and their role as endophytic diazotrophs in N2-fixation. Office: CNPAB/EMBRAPA, Seropédica, 23851-970 Rio de Janeiro Brazil

DOBES, WILLIAM LAMAR, JR., dermatologist; b. Atlanta, Apr. 16, 1943; s. William Lamar and Sara (Wilson) D.; B.A., Emory U., 1965, M.D., 1969; m. Martha Husmann, June 16, 1966; children—Margaret Alison, William Shane. Intern Grady Meml. Hosp., Atlanta, 1969-70; fellow dermatology Mayo Clinic, 1970-71; fellow U. Miami, 1971-73; clin. instr. Emory U. Sch. Medicine, Atlanta, 1973-77, asst. prof. dermatology, 1977-83, assoc. prof., 1983—, dir. immunofluorescense lab., 1978-85; mem. staff Crawford Long, West Paces Ferry, Grady Meml., Ga. Bapt., Piedmont hosps. (all Atlanta); dir. Skin Cancer Project, Emory Univ., 1981-89; chmn. profl. edn. unit Atlanta chpt. Am. Cancer Soc., 1980-86, also bd. dirs., pres., 1986-87, chmn. bd. dirs., 1987-88; pres. Carter's Atlanta, project chmn. Physicians Com., 1992—. Grantee Dermatology Found. Rsch. award, 1979; chmn. Ga. med. bd. Lupus Found., 1988. Diplomate Am. Bd. Dermatology. Mem. AMA, ACP, Soc. Investigative Dermatology, Am. Acad. Dermatology (chmn. com. quality assurance 1982-84; adv. council 1985—, ad coun. exec. com. 1991—, com. on standards of care 1987-91); So. Med. Assn. (vice chmn. 1983), Pan Am. Med. Assn., Am. Soc. Dermatologic Surgery, Ga. Dermatol. Assn. (pres. 1986-87), Atlanta Dermatol. Assn. (pres. 1979), N.Am. Clin. Dermatologic Soc., Soc. Tropical Dermatology, Med. Assn. Atlanta (bd. dirs. 1985-92, chmn. communications com. 1985-90, sec. 1988-89, pres.-elect, 1989-90, pres. 1990-91), Med. Assn. of Ga. (Intersplty. Council 1984—, com. on cancer, 1988—, pub. rels. com. 1988—, del. to Ga. Med. Assn. 1985—, Outstsnding Svc. award 1993), Atlanta Clin. Soc., Atlanta Olympic Medal Com., Emory U. Med. Alumni Assn. (pres. 1980, exec. com. 1992—), Phi Delta Theta (past pres.), Phi Chi (past pres.). Club: Cherokee Town & Country (Atlanta). Contbr. articles to profl. jours. and texts. Home: 2898 Rivermeade Dr NW Atlanta GA 30327-2010 Office: 478 Peachtree St NE Atlanta GA 30308-3103 also: Emory U Sch Medicine Dept Dermatology Atlanta GA 30308

DOBKIN, IRVING BERN, entomologist, sculptor; b. Chgo., Aug. 9, 1918; m. Frances Berlin, July 1, 1941; children: Jane, Joan, David, Jill. B.S. cum laude, U. Ill., 1940; postgrad., Ill. Inst. Tech., 1941-42. Chmn., pres. Dobkin Pest Control Co., Chgo., 1946-79; Pres., dir. Sculptors Guild Ill., 1964-86; lectr. schs. and assn.; life mem. Art Inst. Chgo., Chgo. Natural History Museum (life mem.), North Shore Art League, Evanston Art Center, Mus. Contemporary Art, Chgo. Exhibits include, Art Inst. Chgo., McCormick Pl., Chgo., Old Orchard. Assoc. mem. Smithsonian Instn.; assoc. mem. Peabody Mus. Natural History, Yale U.; pres. Suburban Fine Art Ctr., 1983-84; assoc. mem. Adler Planetarium Assn. Served to lt. USN, 1943-46, PTO. Mem. AAAS, AIA, Entomol. Soc. Am., Am. Registry Profl. Entomologists, Fedn. Am. Scientists, Am. Inst. Biol. Scis., Soc. Environ. Toxicology, UN Assn., Oceanographic Soc., Malacology Soc. Am., Sierra Club (assoc.), Chgo. Acad. Sci., Archeol. Inst. Am., Am. Defn. Mineral and Fossil Soc., Calif. Acad. Sci., Am. Schs. Oriental Rsch., Geog. Soc. Am., Nat. Audubon Soc., Ill. Audubon Soc., Am. Indian Affairs Fedn., S.W. Indian Fedn., Nat. Wildlife Soc., Internat. Wildlife Fedn., Save the Redwoods Soc., Wilderness Soc., Nat. Pks. and Conservation Assn., Am. Harp Soc., Chgo. Harp Soc., Archeol. Conservancy, Mid-Am. Paleontology Soc., Nature Conservancy, Chgo. Coun. on Fgn. Rels., UN Fencing Assn., Gen. Secretariat Orgn. U.S., Classical Art Soc., Primative Arts Soc., Internat. Flamenco Soc. (bd. dirs. ensemble Espanol), Ill. Arts Alliance, Explorers Club, Primitive Arts Soc., Lepidoperists Soc. Home: 306 Maple Ave Highland Park IL 60035-2057

DOBRIN, RAYMOND ALLEN, psychometrician; b. Bklyn., Aug. 10, 1942; s. Gilbert and Frances Gertrude (Rosen) D.; m. Marilyn Ruth Geller, Nov. 25, 1971; 1 child, Ellen. BA, Hunter Coll., 1964; MA, Columbia U., 1965. Personnel specialist N.Y.C. Dept. Personnel, 1966-67; testing cons. N.Y. State Dept. Labor, Bklyn., 1967—. Mem. Am. Psychol. Assn. Jewish. Home: 1770 E 27 St Brooklyn NY 11229 Office: NY State Dept Labor 1 Main St Brooklyn NY 11201

DOBROWOLSKI, FRANCIS JOSEPH, environmental engineer, consultant; b. N.Y.C., July 9, 1936; s. Francis J. and Cecylia (Biniewski) D.; m. Justine Marie Mazepa, June 29, 1963; children: James Francis, Laura Marie. BCE, Cooper Union, 1957. Registered profl. engr., N.J., N.Y. Engr. Buck, Seifert & Jost, N.Y.C., 1957-61; asst. town engr. Bridgewater (N.J.) Twp., 1961-62; project engr. Lee T. Purcell Assocs., Paterson, N.J., 1962-66; prin. assoc. Clinton Bogert Assocs., Englewood Cliffs, N.J., 1966—. Contbr. articles to profl. jours. Coach Little League Baseball, Ridgewood, N.J., 1973-76, 82. With USMC, 1959-67. Mem. ASCE, NSPE, Am. Acad. Environ. Engrs. (diplomate), Water Environ. Fedn., Am. Water Works Assn., N.J. Water Pollution Control Assn. (chmn. innovation awards com. 1975-77, interceptor design com. 1976). Achievements include development of innovative two-level secondary treatment with RBC's. Home: 240 Brookside Ave Ridgewood NJ 07450 Office: Clinton Bogert Assocs 270 Sylvan Ave Englewood Cliffs NJ 07632

DOBROWOLSKI, KATHLEEN, data processing executive; b. Pitts., July 5, 1954; d. Benedict and Alfreda (Miaczynski) Dobrowolski; m. James R. Monk, June 10, 1974 (div. 1978); 1 child, James R.; m. Dennis E. Beck, Sept. 1, 1990 (div. 1992); m. Reid O. Densmore, Feb. 14, 1993. BA in Sociology, LaRoche Coll., Pitts., 1987; MA, U. Pitts., 1987. Contract/cons. computers various orgns., Pa., Tex., 1975-80; programmer Union Nat. Bank, Pitts., 1980-81; sr. programming analyst Beecham Products U.S.A., Pitts., 1981-84; computer contract/cons. Pitts., 1984-85; sr. systems analyst Blue Cross of Western Pa., Pitts., 1985-86; computer contract/cons. Pitts. Bus. Cons., 1986-89; project leader Thrift Drug Co., Pitts., 1989-90; contract cons., instr. KCS Computer Svcs., Inc., Monroeville, Pa., 1990-93; owner Bus. Coll. Funds Rsch. Ctr., Pitts., 1992—; contract cons. SEI, Tampa, Fla., 1993—; adj. instr. Community Coll. Allegheny County, Pitts., 1989—, LaRoche Coll., Pitts., 1989—, Coll. of St. Francis, 1991; owner Coll. Funds Rsch. Ctr., Pitts. With U.S. Army, 1973-75. Recipient Outstanding Achievement award La Roche Coll. Alumni Assn., 1991. Mem. NEFE, Am. Anthropol. Assn. Soc. for Anthropology of Europe, E. European Soc., Nat. Geog. Soc., Nat. Student Anthropologists Soc., Nat. Assn. Practice of Anthropology, Assn. Political & Legal Anthropology, Smithsonian Instn., Nat. Audubon Soc., Nat. Trust Assn., Carnegie Mus., Allegheny County Police Assn., La Roche Coll. Alumni Assn. (Outstanding Achievement award), ABI Inner Circle of Achievement (Woman of the Yr. award 1992). Democrat. Roman Catholic. Home: 7507 Okeechobee Ct Temple Terrace FL 33637

DOBSON, F. STEPHEN, ecologist; b. Oakland, Calif., Aug. 21, 1949; s. Stuart Cromar and Beverly (Richardson) D.; m. Julia Dee Kjelgaard, Feb. 14, 1982. AB in Biology, U. Calif., Berkeley, 1975; MA in Biology, U. Calif., Sant Barbara, 1978; PhD in Biology, U. Mich., 1984. NATO postdoctoral fellow U. Alberta, Edmonton, Can., 1984-85, rsch. assoc., 1985; rsch. assoc. U. Lethbridge, Alta., 1986; vis. curator U. Mich., Ann Arbor, 1987-88; asst. prof. Auburn U., Ala., 1988--. Contbr. articles to profl. jours. NSF Rsch. grantee, 1990-92, Nat. Geographic Soc. rsch. grantee, 1992, The Ctr. for Field Rsch. rsch. grantee, 1990. Mem. Am. Soc. Naturalists, Animal Behavior Soc., Ecological Soc. Am., Soc. for the Study Evolution. Home: 550 Dumas Dr Auburn AL 36830 Office: Auburn U Dept Zool and Wildlife Auburn AL 36349

DOCHERTY, ROBERT KELLIEHAN, III, quality assurance engineer, computer instructor; b. Sterling, Kans., May 9, 1959; s. Robert Kelliehan II and Eileen Joyce (Rockefeller) D.; children: Robert K. IV, Elizabeth Ann. BS in Plastics Engring. Tech., Pittsburg (Kans.) State U., 1981. Computer specialist Allied Signal, Petersburg, Va., 1987-88; quality assurance engr. Rubbermaid-Winfield (Kans.) Inc., 1989—. Area coord. Boy Scouts Am., Winfield, 1990, 91, 92; pres. Winfield United Way, 1991-92, 92-93; treas., dir. Winfield Community Theatre, 1992. Capt. U.S. Army, 1981-87. Mem. Soc. Plastics Engrs., Am. Soc. Quality Control, Kiwanis Club, Am. Legion. Presbyterian. Home: 0114 Iowa St Winfield KS 67156 Office: Rubbermaid Specialty Products Inc 1616 Wheat Rd Winfield KS 67156

DOCKHORN, ROBERT JOHN, physician; b. Goodland, Kans., Oct. 9, 1934; s. Charles George and Dorotha Mae (Horton) D.; m. Beverly Ann Wilke, June 15, 1957; children: David, Douglas, Deborah. A.B., U. Kans., 1956, M.D., 1960. Diplomate: Am. Bd. Pediatrics. Intern Naval Hosp., San Diego, 1960-61; resident in pediatrics Naval Hosp., Oakland, Calif., 1963-65; resident in pediatric allergy and immunology U. Kans. Med. Center, 1967-69, asst. adj. prof. pediatrics, 1969—; resident in pediatric allergy and immunology Children's Mercy Hosp., Kansas City, Mo., 1967-69, chief div., 1969-83; practice medicine specializing in allergy and immunology Prairie Village, Kans., 1969—; clin. prof. pediatrics and medicine U. Mo. Med. Sch., Kansas City, 1972—; founder, chief exec. officer Internat. Med. Tech. Cons., Inc., Prairie Village, Kans., subs. Immuno-Allergy Tech. Cons., Inc., Clin. Research Cons., Inc. Contbr. articles to med. jours.; co-editor: Allergy and Immunology in Children, 1973. Fellow Am. Acad. Pediatrics, Am. Coll. Allergists (bd. regents 1976—, v.p. 1978-79, pres. 1981-82), Am. Assn. Cert. Allegists (pres. 1991—), Am. Acad. Allergy; mem. AMA, Kans. Med. Soc., Johnson County Med. Soc., Kans. Allergy Soc. (pres. 1976-77), Mo. Allergy Soc. (sec. 1975-76), Joint Coun. Socio-Econs. of Allergy (bd. dirs. 1976—, pres. 1978-79). Home: 8510 Delmar Ln Shawnee Mission KS 66207-1926

DODD, JOHN NEWTON, retired physics educator; b. Hastings, Hawkes Bay, New Zealand, Apr. 19, 1922; s. John Henry Dodd and Eva Elsie Weeks; m. Jean Patricia Oldfield, May 4, 1950; children: John Christopher, Nicholas George, Timothy Lewis, Catherine Jane. MS in Physics, Otago U., Dunedin, New Zealand, 1946, MS in Math., 1947; PhD in Physics, Birmingham U., 1952. Temporary lectr. in physics Otago U., Dunedin, 1943-44, lectr. in physics, 1952-56, sr. lectr. in physics, 1957-63, reader, assoc. prof., 1964-65, prof. physics, 1965-88, Beverly Prof., 1968-88, prof. emeritus, 1988—. Author: Atoms and Light, Interact, 1991; co-author: Einstein, 1981; contbr. papers on atomic and optical physics to profl. jours. Recipient New Zealand medal New Zealand Govt., Wellington, 1990. Fellow Inst. Physics, U.K., New Zealand Inst. of Physics; mem. Royal Soc. New Zealand (pres. 1989-93, Hector medal 1976). Achievements include research in light beats due to optical coherence, quantum beats in physical phenomena. Home: Royal Society of New Zealand, PO Box 598, 13 Malvern St, Dunedin 9001, New Zealand

DODD, STEVEN LOUIS, systems engineer; b. Gainesville, Ga., Aug. 19, 1953; s. Oscar Louis and Vivian Irene (King) D.; m. Laureen Matthias Tyler, Apr. 8, 1989. BS in Math., Davidson (N.C.) Coll., 1975; MS in Applied Math., N.C. State U., 1977, PHD in Ops. Rsch., 1982. Cons. EPA, Research Triangle Park, N.C., 1976-77; systems engr. AT&T Bell Labs., Holmdel, N.J., 1982-86, supr., systems engr. and developer, 1986-90; dist. mgr., tech. mktg. AT&T Bus. Communications Svcs., Bridgewater, N.J., 1990-91; dir. strategic planning Cin. Bell Info. Systems, Fairfax, Va., 1991-92, dir. platform mgmt., 1992; dir. gas and electric group Cin. Bell Info. Systems, 1993—. Contbr. articles to profl. jours. Mem. Phi Beta Kappa, Omicron Delta Kappa, Omega Ro, Upsilon Pi Epsilon, Pi Mu Epsilon. Achievements include research in performance evaluation review, IEEE computer graphics and applications and numerische mathematik. Office: Cin Bell Info Systems 600 Vine St Cincinnati OH 45202

DODDS, DALE IRVIN, chemicals executive; b. Los Angeles, May 3, 1915; s. Nathan Thomas and Mary Amanda (Latham) D.; m. Phyllis Doreen Kirchmayer, Dec. 20, 1941; children: Nathan E., Allan I., Dale I. Jr., Charles A. AB in Chemistry, Stanford U., 1937. Chem. engr. trainee The Texas Co., Long Beach, Calif., 1937-39; chemist Standard Oil of Calif., Richmond, 1939-41; chief chemist Scriver and Quinn Interchem., L.A., 1941-46; salesman E.B. Taylor and Co. Mfg. Rep., L.A., 1947-53, Burbank (Calif.) Chem. Co., 1953-57, Chem. Mfg. Co./ICI, L.A., 1957-68; pres., gen. mgr. J.J. Mauget Co., L.A., 1969—. Inventor: Systemic Fungicide, 1976; patentee in field; contributed to devel. Microinjection for Trees. Fellow Am. Inst. Chemists; mem. Am. Chem. Soc., L.A. Athletic Club, Sigma Alpha Epsilon Alumni (pres. Pasadena, Calif. chpt. 1973, 90). Republican. Christian Scientist. Office: JJ Mauget Co 2810 N Figueroa St Los Angeles CA 90065 1500

DODSON, GEORGE W., computer company executive, consultant; b. Danville, Ill., Jan. 21, 1937; s. Maurice Keith and Marjorie Ruth (Ingalsbe) D.; m. Evandra May Mendenhall, Aug. 4, 1957; children: Michael, Curtis, Janet. BS in Math., U. Ill., 1966; MS in Ops. Rsch., Union Coll., 1970. Statis. mgr. U. Ill., Urbana, 1960-66; sr. performance anayst IBM Corp., Poughkeepsie, N.Y., 1966-70, performance mgr., 1970-79; lab. performance mgr. IBM Corp., Tucson, 1979-85; program mgr. IBM Corp., Roanoke, Tex., 1987-91, prin. info. systems mgmt. cons., 1991—; dir. tech. svcs. Morino Assocs., Vienna, Va., 1985-86; dir. performance products UCCEL Corp., Dallas, 1986-87. Mem. Computer Measurement Group (chmn. 1983, 89, pres. 1983-85, bd. dirs. 1985-89, treas. 1990—). Avocations: softball, music, photography. Office: IBM Corp 40-E3-03 1 E Kirkwood Blvd Roanoke TX 76299

DODSON, JAMES NOLAND, geologist; b. Marianna, Fla., July 21, 1963; s. Loyce Joe and Lola (Noland) D. AA, Chipola Jr. Coll., Marianna, 1984; BS in Geology, Fla. State U., 1987. Geologist Pieco Miami, Inc., Miami, Fla., 1987-88; hydrogeologist Woodward-Clyde Cons., Tallahassee, 1988-93; profl. geologist I Fla. Dept. Environ. Protection, Tallahassee, 1993—. Mem. Am. Groundwater Assn., ASTM, Exptl. Aircraft Assn. (pilot), Tech. Assn. of Pulp and Paper Industry, Sports Car Club of Am. (driver). Democrat. Baptist.

DODSON, RONALD FRANKLIN, electron microscopist, administrator; b. Tyler, Tex., Feb. 14, 1942; s. Benjamin Franklin and Vera Inez (Eubank) D.; m. Sandra Jim Roberson, Nov. 13, 1965; children: Diana Lynn, Debra Kay. Degree in biology and chemistry, East Tex. State Coll., 1964, degree in biology and biochemistry, 1965; degree in biol. electron microscopy, Tex. A & M U., 1969. Instr. in neurology and pathology Baylor Coll. of Medicine, Houston, 1970-71, asst. prof. neurology and pathology, 1971-77, adj. asst. prof., 1977-79; chief exptl. pathology and environ. sci. U. Tex. Health Ctr. at Tyler, 1977-78, chmn., prof. cell. biology and environ. scis., 1978—, assoc. dir. for rsch., 1983—. Author: (with others) Mechanisms in Fibre Carcinogenisis, 1992; contbr. articles to profl. pubs. Mem. Heart Assn. fellow. Fellow Am. Coll. Chest Physicians. Office: U Tex Health Ctr at Tyler PO Box 2003 Tyler TX 75710

DOE, RICHARD PHILIP, physician, educator; b. Mpls., July 21, 1926; s. Richard Harding and Ruth Elizabeth (Schoen) D.; m. Shirley Joan Cedarleaf, Sept. 15, 1950; children—Nancy Jean, Charles Jeffrey, Robert Bruce. B.S., U. Minn., 1949, M.B., 1951, M.D., 1952, Ph.D., 1966. Intern Oakland (Cal.) Hosp., 1951-52; resident internal medicine Mpls. VA Hosp., 1952-54, fellow in endocrinology, 1954-55, chief chemistry sect., 1956-60, chief metabolic endocrine sect., 1960-69, endocrine staff, 1976-88; head metabolic endocrine sect. U. Minn. Hosp., 1969-76; faculty U. Minn. Med. Sch., Mpls., 1955-88; prof. medicine U. Minn. Med. Sch., 1969-88, prof. emeritus, 1988—; pres. MEMC Corp., Carmel, Calif., 1988—. Served with USNR, 1944-46. UPSHS grantee, 1958-88. Mem. Am. Soc. Clin. Investigation, Minn. Soc. Internal Medicine, Central Soc. Clin. Research, Endocrine Soc., Am. Fedn. Clin. Research. Achievements include isolation of hormone binding proteins CBG (transcortin), TBG and TBPA in highly purified form as proven by serial immunoelectrophoresis; development of methodology for measurement of non-protein bound cortisol and correlation with clinical status., of methodology for measurement of CBG activity in serum and showed concentration in different endocrine diseases, others. Home and Office: MEMC Corp Box 86 Corona Rd Carmel CA 93923-9610

DOEPKENS, FREDERICK HENRY, agriscience educator; b. Annapolis, Md., Dec. 2, 1958; s. William Phillip and Marjorie (Desmarais) D. BS in Agrl. Edn., U. Md., 1980. Cert. advanced profl. teaching agr. grades 7-12. Agr. tchr. Hereford High Sch., Parkton, Md., 1980-84; agrisci. tchr. Hereford Mid. Sch., Monkton, Md., 1984-86, agrisci. tchr.-in-charge, 1986-87, agrisci. dept. chmn., 1987—; workshop instr., presenter Md. Mid. Sch. Assn., Balt., 1988; workshop coord. hydrophonics pilot Baltimore County Pub. Schs., Towson, Md., 1989-91; co-coord./cons. Md. Agr. in the Classroom Conf., Annapolis, Md., 1990-91; bd. dirs. Md. Agr. Edn. Found., Inc., Balt., 1989—. Author: Hydroponics, 1989, Agriscience in the Middle School, 1990; inventor ednl. materials Portable Hydroponics Unit, 1990. Recipient Achievement in Curriculum award Md. State Dept. Edn., Balt., 1989; named Md. Agrisci. Tchr. of Yr., Nat. FFA Orgn., 1991; hydroponics grantee Tech. Edn. Assn. Md., Balt., 1989; agrl. engring. grantee Md. Agrl. Edn. Found., Inc., Balt., 1991. Mem. NEA, Nat. Vocat. Agrl. Tchrs. Assn. (outstanding young mem. 1984), Am. Vocat. Assn., Md. Agr. Tchrs. Assn. Md. Vocat. Assn., Md. Mid. Sch. Assn. Roman Catholic. Avocations: travel, photography. Home: 3702A Mt Carmel Rd Upperco MD 21155-9569 Office: Hereford Mid Sch 712 Corbett Rd Monkton MD 21111-1500

DOEPPNER, THOMAS WALTER, electrical engineer, educator, consultant; b. Berlin, May 22, 1920; came to U.S., 1939; s. August Friedrich and Ella Judith (Fraustädter) D.; m. Marjorie Ann Sloan, Sept. 16, 1944; children: Thomas Walter Jr., Ronald Sloan. Student, McPherson Coll., 1939-41; BSEE, Kans. State U., 1944; MSEE, U. Calif., Berkeley, 1959. Enlisted U.S. Army, 1944, advanced through grades to col., 1973, radio officer, 1945-65; program mgr. Def. Dept. Adv. Rsch. Proj. Agy., Washington, 1965-69; dir. electronics Army Gen. Staff U.S. Army, Washington, 1969-73; communications rsch. dir. Gen. Rsch. Corp., McLean, Va., 1973-76, dir. logistics engring. ops., 1976-84; prof. Def. Systems Mgmt. Coll., Ft. Belvoir, Va., 1984-86; cons. Alexandria, Va., 1986—; v.p. Washington Acad. Scis., 1991—; asst. prof. mil. sci. U. Calif., Berkeley, 1955-59; instr. math. U. Md., College Park, 1967-69. Contbr. numerous articles to profl. publs. Decorated Legion of Merit (2), Bronze Star; Medal of Honor (Republic of Vietnam). Fellow IEEE (Patron award 1982, Centennial medal 1984), Washington Acad. Scis. (v.p. 1991—). Home and Office: 8323 Orange Ct Alexandria VA 22309-2166

DOERING, WILLIAM VON EGGERS, organic chemist, educator; b. Ft. Worth, June 22, 1917; s. Carl Rupp and Antoinette (von Eggers) D.; m. Ruth Haines, 1947 (div. 1954); children: Christian, Peter, Margaretta; m. Sarah Cowles Bullitt, 1969 (div. 1981). B.S., Harvard U., 1938, Ph.D., 1943; D.Sc. (hon.), Tex. Christian U., 1974; D in Natural Sci. (hon.), Karlsruche U., Fed. Republic Germany, 1987. Faculty, Columbia U., 1943-52; prof. Yale U., 1952-67, dir. div. sci., 1962-65; prof. Harvard U., 1967-86, prof. emeritus, 1986; hon. prof. Fudan U., Shanghai, China, 1980; research chemist Nat. Def. Research Council, Harvard U., 1941-42, Polaroid Corp., 1943, Office Prodn. Research and Devel., 1944-45; dir. Hickrill Chem. Research Found., Katonah, N.Y., 1947-59. Contbr. articles to profl. jours.; hon. regional editor: Tetrahedron, 1958-60. Chmn. Council for Livable World, Washington, 1962-72, pres., 1973-78. Recipient John Scott award City of Phila., 1945; Pure Chemistry award Am. Chem. Soc., 1953; Synthetic Organic Chem. Mfrs. Assn.; medal for creative work in synthetic organic chemistry, 1966; Hofmann medal German Chem. Soc., 1962; William C. DeVane medal Yale Phi Beta Kappa, 1967; Theodore William Richards medal, 1970 and James Flack Norris award in phys. organic chemistry, 1989 both N.E. sect. Am. Chem. Soc.; Welch Found. award, 1990. Mem. Nat. Acad. Sci., Am. Acad. Arts and Scis. Office: Harvard U Dept Chemistry 12 Oxford St Cambridge MA 02138-2900 Home: 53 Francis Ave Cambridge MA 02138*

DOERR, STEPHEN EUGENE, research engineer; b. Myrtle Beach, S.C., July 30, 1959; s. Eugene Joseph and Eva Mary (Highlander) D.; m. Lisa Lynn Glazener, July 24, 1982; children: Kelsey Ann, Emily Christine. MS, U. Ill., 1984; PhD, U. Tex., 1990. Tech. staff Sandia Nat. Labs., Albuquerque, 1984-87; rsch. engr. U. Tex., Austin, 1987-90, Ktech Corp., Albuquerque, 1990-93; sr. engr. Sci. Applications Internat. Corp., Albuquerque, 1993—. Contbr. articles to profl. jours. Fellow U. Tex., 1987. Mem. AIAA (sr., sect. officer aerospace measurement tech. com.), Soc. Photo-Optical Engrs. Achievements include patent pending for dual plate micropositioner. Office: Sci Applications Internat Corp 2109 Air Park Rd SE Albuquerque NM 87106

DOERRY, NORBERT HENRY, naval engineer; b. Chicago Heights, Ill., Jan. 16, 1962; s. Wulf T. and Edith G. Doerry; m. Elizabeth Van Kirk, Oct. 24, 1992. BSEE, U.S. Naval Acad., 1983; naval engr., MSEE and Computer Sci., MIT, 1989, PhD, 1991. Cert. surface warfare officer. Commd. ensign USN, 1979, advanced through grades to lt. comdr., 1993; div. officer USN, Charleston, S.C., 1983-86; project engr. Naval Sea Systems Command, Annapolis, Md., 1991-92; teaching asst. MIT Profl. Summer, Cambridge, 1989-91. Recipient Steinmetz prize USNA, 1983, Capt. Boyd Alexander Honor award AFCEA, 1983, Navy Achievement medal, Naval Constrn. and Engring. award NAVSEA, 1989. Mem. IEEE (EASCON 82 award 1982), Naval Inst., Am. Soc. Naval Engrs., Soc. Naval Arch. and Marine Engring. Home: Apt 2B 2634 Fort Farnsworth Rd Alexandria VA 22303-2634

DOETSCH, PAUL WILLIAM, biochemist, educator; b. Cheverly, Md., Feb. 14, 1954; s. Raymond Nicholas and Janet Gray (Huddle) D.; m. Donna Lee Artz, Jul. 27, 1985. BS, Univ. Md., 1976; MS, Purdue Univ., 1978; PhD, Temple Univ., 1982. Rsch. fellow Harvard Medical Sch., Boston, 1982-85; asst. prof. Emory Univ. Sch. Medicine, Atlanta, 1985-91, assoc. prof., 1991—. Assoc. editor Radiation Research, 1990-94; editorial bd. Free Radical Biology and Medicine, 1992-96. Recipient Rsch. Career Devel. award NIH, 1989-94. Mem. Am. Chemical Soc., Am. Soc. Photobiology, Radiation Rsch. Soc., Am. Assn. Cancer Rsch. Achievements include coinventor patents in field. Office: Emory Univ Sch Medicine Rollins Rsch Ctr Atlanta GA 30322

DOETSCHMAN, DAVID CHARLES, chemistry educator; b. Aurora, Ill., Nov. 24, 1942; s. Charles F. and Mary (Thomas) D.; m. Evelyn Louise Siegel; children: Steven David, Christopher Randall. BS, No. Ill. U., 1963; PhD, U. Chgo., 1969. Rsch. fellow Australian Nat. U., Camberra, Australia, 1969-74, U. of Leiden, The Netherlands, 1974-75; asst. prof. SUNY, Binghamton, N.Y., 1975-82; assoc. prof. SUNY, Binghamton, 1982-93; prof., 1993—; scientist in residence Argonne (Ill.) Nat. Lab., 1982-83; cons. Union Carbide, Parina, Ohio, 1983-85; instrumentation ctr. coord. U. Coll. London, 1989-90. Contbr. over 35 articles to profl. publs. Named Outstanding Young Man of Am. Mem. Am. Chem. Soc., Am. Phys. Soc., Sigma Xi Soc. Office: SUNY at Binghamton Dept of Chemistry Vestal Pky East Binghamton NY 13902-6000

DOGANATA, YURDAER NEZIHI, research scientist; b. Izmir, Turkey, June 27, 1959; came to U.S. 1983; s. Kemal and Nezahat (Ozge) D.; m. Zinnur Oguz, Aug. 26, 1983; children: Zeynep Sebnem, Mehmet Erdener. BS, Middle East Tech. U., 1981, MS, 1983; MS, Calif. Tech. U., 1984, PhD, 1987. Postdoctoral rsch. fellow Calif. Tech. U., Pasadena, 1987-89; rsch. staff mem. IBM, Yorktown Heights, N.Y., 1989—. Contbr. articles to profl. jours. Mem. IEEE, Sigma Xi. Office: IBM H2-C20 Yorktown Heights NY 10598

DOHERTY, ROBERT FRANCIS, JR., aerospace industry professional; b. Quincy, Mass., Aug. 7, 1954; s. Robert Francis and Rose Virginia (Wheeler) D. BS in Mgmt., So. Mass. U., 1977. Sales mgr. Jordan Marsh Co., Boston, 1977-78; ops. mgr. Cramer Electronics, Newton, Mass., 1978-79; contracts mgr. Data Gen. Corp., Westboro, Mass., 1981-84, ops. analyst, 1979-81; sales ops. mgr. Printronix, Inc., Malden, Mass., 1984-87; contract adminstrn. mgr. M/A-Com, Inc., Burlington, Mass., 1987-89; mktg. mgr. M/A-Com, Inc., Chelmsford, Mass., 1989—; bd. dirs. M/A-Com Fed. Credit Union, Burlington, also newspaper correspondent, chair various restructuring coms. Mem. Nat. Contract Mgmt. Assn., Assn. of Old Crows, M/A-Com Mgmt. Club. Roman Catholic. Avocations: jogging, swimming, skiing, antiques, decorating. Home: 27 Dwight St #1 Boston MA 02118

DÖHLER, KLAUS DIETER, pharmaceutical and development executive; b. Weilburg, Germany, Dec. 29, 1943; s. Richard and Emma (Sauter) D.; m. Ursula Han Hua Fang, Feb. 26, 1972. B.A., Calif. State U.-Fresno, 1971, M.A., 1972; Ph.D., U. Göttingen (W.Ger.), 1974; Habilitation, Med. Sch. Hannover, W. Germany, 1978. Research asst. Med. Sch. Hannover, 1974-78, assoc. prof., 1981-84; asst. prof. Vet. Sch., Hannover, 1978-80; rsch. assoc. UCLA Sch. Med., Los Angeles, 1980-81; sci. dir. Bissendorf Peptide GmbH, Wedemark, Fed. Republic Germany, 1984-90; dir. devel. and prodn. Pharma Bissendorf Peptide GmbH, Hannover, Germany, 1990—; Recipient Scholler-Junkmann prize Soc. for Endocrinology, W.Ger., 1977; Heisenberg fellow German Research Soc., 1979. Mem. Endocrine Soc., N.Y. Acad. Scis., Soc. Exptl. Biology and Medicine, Sigma Xi. Lodges: Round Table, Rotary (pres. 1992-93). Contbr. articles to profl. jours. Home: Im Kamp 24, D-30657 Hannover Germany

DOHR, DONALD R., metallurgical engineer, researcher; b. Rio de Janeiro, Niteroi, Apr. 12, 1924; came to U.S., 1944; s. Nicholas and Candida (Caramuru) D.; m. Virginia Marion O'Donnell, Mar. 30, 1960 (dec. Feb. 1987). ME, Stevens Inst. Tech., 1952, MS in Metallurgy, 1968. Jr. metallurgist Crucible Steel Co., Harrison, N.J., 1952-54; metallurgist Engelhard Industries, Newark, 1954-56, Foster Wheeler Corp., Carterei, N.J., 1956-60, Weston Instruments, Inc., Newark, 1960-66; sr. metallurgist Singer Co., Denville, N.J., 1966-71; unit headmaterials and processes lab. Kearfott Guidance & Navigation Corp. (formerly Singer Co.), Little Falls, N.J., 1971—. Author: Liquid Phases Sintering Mechanisms, Magnetic Properties of Metals & Alloys. Staff sgt. U.S. Army, 1944-46, PTO. Mem. Am. Soc. Metals-Internat., Nat. Soc. Profl. Engrs., Soc. Mfg. Engrs. Republican. Roman Catholic. Achievements include patents in Magnetic Force Field Application and Thread Tensioner; patent pending for fixtures to measure magnetic properties of radialy oriented magnets. Home: 410A Troy Towers Bloomfield NJ 07003-3370 Office: Kearfott Guidance & Navigation 1225 Mcbride Ave Little Falls NJ 07424-2540

DOI, SHINOBU, research electrical engineer, consultant; b. Nara, Japan, July 10, 1964; came to U.S., 1984; s. Koichi and Masako Doi; m. Veronica Rivera, Dec. 23, 1989; 1 child, Danielle-Sakura. BS in Computer Engring., Fla. Inst. Tech., 1987, MS in Computer Engring., 1989, postgrad., 1992—. Lab. asst. Fla. Inst. Tech., Melbourne, 1988, teaching asst., 1988-90, R & D asst., 1992—; cons. F-Corp., Inc., Tokyo, 1988-92, N.B.F., Inc., Tokyo, 1991-92, Marks, Inc., Tokyo, 1991-92. Contbr. article to IEEE Proc., Internat. Soc. Optical Engring. Mem. IEEE, Internat. Soc. for Optical Engring., Eta Kappa Nu. Achievements include patent pending for optical neural network. Home: PO Box 1658 Melbourne FL 32902 Office: Fla Inst Tech PO Box 6074 Melbourne FL 32901

DOJKA, EDWIN SIGMUND, civil engineer; b. Niagara Falls N.Y., Dec. 20, 1924; s. Zygmunt Joseph and Felixa (Pasek) D.; BCE, Rensselaer Poly. Inst., 1951; m. Jean L. Keller, July 9, 1949; children: Paul, Gail Dojka Rutkowski, Jay. Structures engr. Bell Aircraft Corp., Wheatfield, N.Y., 1951-52; design engr. Hooker Electro Chem. Corp., Niagara Falls, N.Y., 1952-55; civil engr. City of Niagara Falls (N.Y.), 1955-58, asst. city engr., 1958-60, dep. city engr., 1960-63, city engr., 1963-79; city engr. City of North Tonawanda, 1979-85; mem. sewer commn., plumbing bd., 1963-85, mem. planning bd., 1963-66, bd. equalization rev., 1963-71; mem. Niagara County Planning Bd., 1978-91, Traffic Safety Commn., 1979-85. Mem. United Fund Community Budget Com., 1962-68; mem. Community Ambassador Gen. Com., 1938, 39; Fleet Safety adv. commr., Niagara Falls, 1960-68; bd. assocs. Mt. St. Mary's Hosp., 1969-70. With inf. AUS, World War II; ETO. Decorated Bronze Star, Purple Heart, Combat Infantryman's badge. Registered profl. engr., land surveyor, N.Y. Fellow ASCE; mem. Soc. Am. Mil. Engrs., Am. Pub. Works Assn., Am. Water Works Assn., Nat. Soc. Profl. Engrs., Water Pollution Control Fedn., Am. Planning Assn., Inst. for Engring., Am. Arbitration Assn. (comml. panelist 1978—), DAV, Am. Legion, 102d Inf. Div. Assn., AMVETS, Meml. Day Assn., Boys Club Alumni Assn., VFW, Pulaski Civic League, Polish Legion Am. Vets., Royal Canadian Legion (hon.), Mil. Order Purple Heart, 40 and 8, 2d Armored Div. Assn., 25th Bomb Group Assn., Hon. Order Ky. Cols., Kosciuszko Found., Dom Polski Club, First Friday Club, Echo Club, Sertoma Club, K.C. (hon.), Elks, Sigma Xi, Chi Epsilon, Tau Beta Pi. Roman Catholic. Home and Office: 509 80th St Niagara Falls NY 14304-2301

DOKIC, PETAR, chemist, educator; b. Novi Sad, Yugoslavia, Oct. 18, 1941; s. Pavle and Milica (Bijuklic) D.; m. Kler Miroslava, Oct. 2, 1965; children: Ljubica, Milica. Diploma in Engring., Faculty Tech., Novi Sad, 1964, MSc, 1971, PhD, 1974. Asst. Faculty Tech., Novi Sad, 1965-75, asst. prof., 1975-80, assoc. prof., 1980-85, prof. colloid chemistry, 1985, vice dean, 1975-77, pres. council Inst. Applied Chemistry, 1980-84, pro rector univ., 1983-87; head dept. applied chemistry Inst. Applied Chemistry; postdoctoral rsch. Queen Elizabeth Coll., London, 1977-78; pres. Commn. for Sci. and Edn. Union Univs. Yugoslavia, 1985-87; pres. Rsch. Resources Devel. Bd. Vojvodina, 1991. Contbr. over 100 articles in colloid chemistry to profl. jours. Pres. assembly for high and secondary edn. Province of Vojvodina, 1977-83; mem. exec. council, pres. com. for scis. and informatics Assembly Socialistic Autonomous Province of Vojvodina, 1987-90. Fellow Eisenhower Exchange, Phila., 1987; recipient Univ. Novi Sad award 1985, Silver Wreath decoration of labour Pres. Yugoslavia, 1986. Mem. Serbian Chem. Soc., Chem. Soc. Province Vojvodina, Fed. Com. for Sci. and Tech., Council Cen. for Comparative Studies on Technol. and Social Progress, Matica Srpska-Novi Sad, Internat. Assn. Colloid & Interface Scientists. Home: Sonje Marinkovic 21, 21000 Novi Sad Yugoslavia Office: 2 V Vlahovica, 21000 Novi Sad Yugoslavia

DOKLA, CARL PHILLIP JOHN, psychobiologist, educator; b. Bridgeport, Conn., Mar. 16, 1949; s. John Michael and Aldea Evelyn (Lavallee) D.; m. Janice Christine Passabet, June 14, 1986. BA, Fairfield U., 1971; MA, Fordham U., 1974, PhD, 1978. Neurology trainee Mt. Sinai Sch. Medicine, N.Y.C., 1973-74; lab. instr. Manhattan Coll., Bronx, N.Y., 1974-75; teaching fellow Fordham U., Bronx, 1975-76; adj. instr. So. Conn. State U., New Haven, 1978-79, Post Coll., Waterbury, Conn., 1981-84; rsch. assoc. Fairfield (Conn.) U., 1985-87; asst. prof. psychology St. Anselm Coll., Manchester, N.H., 1989—; rsch. cons. in neurology Albert Einstein Coll. Medicine, Bronx, 1984-85. Contbr. articles to Behavioral Neurosci., Brain Rsch., Psychopharmacology, others. Rep.-at-large Catchment Area Coun. 6/Community Mental Health Ctr., Ansonia, Conn., 1981-83. Westchester ADRDA rsch. grantee, 1984-86, Medicine Edn. Rsch. Fund grantee, 1985; Faculty Rsch. grantee St. Anselm Coll., 1990. Mem. AAAS, Soc. for Neurosci., Am. Psychol. Soc., New Eng. Psychol. Assn., Internat. Brain Rsch. Orgn., Sigma Xi. Achievements include research in the role of the basal (brain) nuclei in memory processes and as an animal model for Alzheimer's disease; research in efficacy of novel anticholinesterase series, RA compounds, in the treatment of Alzheimer's disease. Office: St Anselm Coll 87 St Anselm Dr Manchester NH 03102-1310

DOKULIL, MARIN, limnologist; b. Vienna, Austria, Feb. 26, 1943; came to U.S., 1974; s. Otto and Eleonore (Glück) D.; m. Maria-Elisabeth Milek, Apr. 4, 1970; 1 child, Simone Katharina. PhD, U. Vienna, 1970. Lectr. U. Vienna, 1984, prof., 1988; researcher Austrian Acad. Scis., Mondsee, Austria, 1970—, vice dir. inst. limnology, 1974—; sci. coord. Austrian Nat. Com. for Danube Rsch., Vienna, 1991—. Co-editor Limnologica, 1992—; editor Arch-Hydrobiol. Suppl. Large Rivers, 1991—; editor: Shallow Lakes, 1980; co-author: Neusiedlersee, 1979. Recipient Körner-Preis, Govt. of Austria, 1974. Mem. Freshwater Biol. Assn., Ecol. Soc. Am., Internat. Limnological Assn. (nat. rep.). Office: Inst für Limnologie, ÖAW Gaisberg 116, A 5310 Mondsee Austria

DOLAN, JOHN E., consultant, retired utility executive; b. N.Y.C., May 9, 1923; s. John A. and Marie C. (Comiskey) D.; m. Anne Dolan, Feb. 16, 1952; children—John E., Bryan, Vincent, Robert, Raymond, Philip, Lawrence, Paul. Student, Rensselaer Poly. Inst., 1946-47; B.S.M.E., Columbia U., 1950. With Am. Electric Power Service Corp., Columbus, Ohio, 1950-88, chief mech. engr., 1966, chief engr., 1967, sr. exec. v.p. engring., 1975-79, vice chmn. engring. and constrn., 1979-88; ret.; bd. dir., v.p. subs. cos. and Am. Electric Power Service Corp.; cons., 1988—. Served to 1st lt. USAAF, 1942-46. Decorated Air medal (4). Fellow ASME (James N. Landis medal 1990); mem. NAE, Tau Beta Pi. Roman Catholic. Home: 14448 Mark Dr Largo FL 34644-5102

DOLAN, JOHN PATRICK, psychiatrist; b. Bklyn., Jan. 23, 1935; s. James Francis and Agnes Barrett (Lane) D.; m. Mary McLaughlin, Dec. 21, 1964 (div. June 1972); children: Deborah Jean, John McLaughlin; m. Margaret Abel, Sept. 6, 1974. AB in History, Bklyn. Coll., 1958; MD, N.J. Coll. Medicine, 1962. Cert. Am. Bd. Psychiatry and Neurology. Resident Washington U., 1966-69, St. Louis U., 1970-72; intern medicine Kings County Hosp., Bklyn., 1963; resident neurology Washington U., St. Louis, 1969; resident psychiatry St. Louis U. Sch. Medicine, 1972, asst. prof. dept. psychiatry, 1972-78; chmn. dept. psychiatry Conemaugh Valley Meml. Hosp., Johnstown, Pa., 1979-83; clin. asst. prof. dept. psychiatry U. Pitts. Sch. Medicine, 1979-83; chmn. dept. psychiatry St. Vincent's Med. Ctr., Bridgeport, Conn., 1983—; clin. assoc. prof. psychiatry N.Y. Med. Coll., Valhalla, N.Y., 1987—; psychiat. cons. Pru-Care N.Y. and Conn., 1988—; past pres. Fairfield/Litchfield chpt. Conn. Psychiat. Soc., 1988-89; v.p. Alliance for Health IPA, Bridgeport, 1987—; counsellor Conn. Psychiat. Soc., 1985-89. Mem. Catchment Area Coun., Bridgeport, 1983-87, Southwest Regional Mental Health Bd., Fairfield County, Conn., 1983-87, legis. com. (Conn.) State Bd. Mental Health, 1985-86. Mem. Bridgeport Psychiat. Soc. (pres. 1987—), Am. Psychiat. Assn., Am. Acad. Med. Dirs., Am. Acad. Psychiatrists in Alcoholism and Addiction, N.Y. Acad. Scis., South Shore Music Club. Avocations: rock climbing, chess. Office: St Vincents Med Ctr 2800 Main St Bridgeport CT 06606-4292

DOLAN, THOMAS F., JR., pediatrician, educator; b. Cambridge, Mass., Mar. 2, 1928; s. Thomas and Agnes (Masterson) D.; m. Margaret Dolan, Jan. 13, 1953; children: Karen, Kevin, Maureen, Evelyn. BA, Harvard U., 1949, MD, 1953; MA, Yale U., 1973. Diplomate Am. Bd. Pediatrics. Intern Children's Med. Ctr., Boston, 1953-59; resident Boston City Hosp., 1956-58; sr. asst. surgeon NIH, NAID, Bethesda, Md, 1958-60; staff resident Ellsworth, Maine, 1958-62; attending physician L & M Hosp., New London, 1962-69; instr. Children's Med. Ctr., Boston, 1962-64; from asst. prof. to assoc. prof. dept. medicine Yale U., New Haven, 1969-75, prof. dept. medicine, 1975—. Sr. asst. surgeon USPHS, 1956-58. Mem. Conn. Acad. Pediatrics (pres. 1976-82). Home: 26 Skyton Dr Madison CT 06443 Office: Yale U Sch Med 333 Cedar St PO Box 3333 New Haven CT 06510

DOLAN-BALDWIN, COLLEEN ANNE, global technology executive; b. N.Y.C., June 1, 1960; d. Thomas C. and Julia (Joyce) Dolan; m. Thomas J. Baldwin, Sept. 12, 1987. BS, Fordham U., 1981; MBA, Pace U., 1985. Systems mgr. AT&T Communications, White Plains, N.Y. and, Morristown, N.J., 1981-85; mgr. tech. planning Mobil Oil Corp., N.Y.C., 1985-87; v.p. global tech. group J.P. Morgan & Co., N.Y.C., 1987—. Contbr. articles to profl. jours. Bd. dirs. ARC (N.Y.C.). Republican. Roman Catholic. Avocations: downhill skiing, tennis, travel. Home: 174 Belmont Rd Hawthorne NY 10532-2102 Office: JP Morgan & Co 60 Wall St New York NY 10260-0060

DOLBERG, DAVID SPENCER, business executive, lawyer, scientist; b. L.A., Nov. 28, 1945; s. Samuel and Kitty (Snyder) D.; m. Katherine Blumberg, Feb. 22, 1974 (div. 1979); 1 child, Max; m. Sarah Carnochan, May 23, 1992. BA in Biology with honors, U. Calif., Berkeley, 1974; PhD in Molecular Biology, U. Calif., San Diego, 1980; JD, U. Calif., Berkeley, 1989. Bar: Calif. 1989, U.S. Dist. Ct. (no. dist.) Calif. 1989, U.S. Patent and Trademark Office, 1990. Staff biologist, postdoctoral fellow Lawrence Berkeley Lab. U. Calif., 1980-85; assoc. Irell & Manella, Menlo Park, Calif., 1989-91; v.p. EROX Corp., Menlo Park, Calif., 1991-92; v.p. sci. and patents Pherin Corp., Menlo Park, Calif., 1992—; speaker in field. Contbr. articles to Jour. Gen. Virology, Jour. Virology, Nature, Science. Home: 360 Summit Dr Redwood City CA 94062-3330 Office: 535 Middlefield Rd Ste 240 Menlo Park CA 94025-3444

DOLBY, RAY MILTON, engineering company executive, electrical engineer; b. Portland, Oreg., Jan. 18, 1933; s. Earl Milton and Esther Eufemia (Strand) D.; m. Dagmar Baumert, Aug. 19, 1966; children—Thomas Eric, David Earl. Student, San Jose State Coll., 1951-52, 55, Washington U., St. Louis, 1953-54; B.S. in Elec. Engring., Stanford U., 1957; Ph.D. in Physics (Marshall scholar 1957-60, Draper's studentship 1959-61, NSF fellow 1960-61), Cambridge U., Eng., 1961. Electronic technician/jr. engr. Ampex Corp., Redwood City, Calif., 1949-53; engr. Ampex Corp., 1955-57, sr. engr., 1957; PhD research student in physics Cavendish Lab., Cambridge U., 1957-61, research in long wavelength x-rays, 1957-63; fellow Pembroke Coll., 1961-63; cons. U.K. Atomic Energy Authority, 1962-63; UNESCO adviser Central Sci. Instruments Orgn., Chandigarh, Punjab, India, 1963-65; owner, chmn. Dolby Labs. Inc., San Francisco and London, 1965—. Trustee Univ. High Sch., San Francisco, 1978-84; bd. dirs. San Francisco Opera; bd. govs. San Francisco Symphony; mem. Marshall Scholarship selection com., 1979-85. Served with U.S. Army, 1953-54. Recipient Beech-Thompson award Stanford U., 1956; Emmy award, 1957, 89; Trendsetter award Billboard, 1971; Top 200 Execs. Bi-Centennial award, 1976; Lyre award Inst. High Fidelity, 1972; Emile Berliner Maker of the Microphone award Emile Berliner Assn., 1972; Sci. and Engring. award Acad. Motion Picture Arts and Scis., 1979; Pioneer award Internat. Teleprodn. Soc., 1988, Edward Rhein Ring award Edward Rhein Found., 1988; Oscar award Acad. Motion Picture Arts and Scis., 1989; Life Achievement award Cinema Audio Soc., 1989; named Officer of the Most Excellent Order of the British Empire (O.B.E.), 1986, Man of Yr. Internat. Tape Assn., 1987; hon. fellow Pembroke Coll., Cambridge U., 1983. Fellow Audio Engring. Soc. (bd. govs. 1972-74, 79-84 Silver Medal award 1971, Gold medal award 1992, pres. 1980-81), Brit. Kinematograph, Sound, TV Soc., Soc. Motion Picture and TV Engrs. (Samuel L. Warner award 1979, Alexander M. Poniatoff Gold Medal 1982, Progress award 1983, hon. mem. 1992), Inst. Broadcast Sound; mem. IEEE, private pilot Inst. rating, Tau Beta Pi. Club: St. Francis Yacht. Inventions, research, publs. in video tape recording, x-ray microanalysis, noise reduction and quality improvements in audio and video systems; holder 50 U.S. patents. Office: Dolby Labs 100 Potrero Ave San Francisco CA 94103-4813

DOLCE, KATHLEEN ANN, health physicist, inspector; b. Willingboro, N.J., Mar. 10, 1964; d. Dominic Rudolph and Gene Regina (Sitek) D. BS, Rutgers U., 1986; MPH, U. N.C., 1988. Health physicist health and safety office Rutgers U., Piscataway, N.J., 1985, lab. technician chemistry dept., 1985-86, rsch. asst. radiation sci. dept., 1986; rsch. asst. U. N.C., Chapel Hill, 1986-87, researcher, 1987-88; health physicist NIH, Bethesda, Md., 1988-91; radiation safety officer Boehringer Ingelheim Pharm., Inc., Ridgefield, Conn., 1991-92; health physicist U.S. Nuclear Regulatory Commn., King of Prussia, Pa., 1992—; ex-officio mem. animal care and use com. Nat. Heart, Lung & Blood Inst., Bethesda, 1989-90; mem. U. N.C. Sch. Pub. Health Internal Rev. Bd., Chapel Hill, 1986-88; mem. radioactive drug rsch. com. NIH, Bethesda, 1990-91, lectr. radiation safety tng. program, 1988-91, project officer radon testing program, 1990-91. Author tech. memoranda. Honor guard, sr. officer escort USPHS, Washington, 1989-90, assoc. recruiter, 1989-91. Mem. Health Physics Soc. (N.J. chpt.), Appalachian Compact Users of Radioactive Isotopes. Roman Catholic. Home: 2 Orchard Ln Burlington NJ 08016-4214 Office: US NRC 475 Allendale Rd King Of Prussia PA 19406

DOLE, VINCENT PAUL, medical research executive, educator; b. Chgo., May 8, 1913; s. Vincent Paul and Anne (Dowling) D.; m. Elizabeth Ann Strange, May 23, 1942 (div. 1965); children—Vincent Paul, Susan, Bruce; m. Marie Nyswander, 1965. A.B., Stanford U., 1934; M.D., Harvard U., 1939. Intern Mass. Gen. Hosp., Boston, 1940-41; mem. staff Rockefeller U., N.Y.C., 1941—, mem. staff, prof., 1951—. Developer methadone maintenance treatment program for heroin addiction. Office: Rockefeller U care Dept of Medicine 1230 York Ave New York NY 10021-6341

DOLEJS, PETR, aquatic scientist, consultant; b. Tabor, Bohemia, Czechoslovakia, Feb. 19, 1952; s. Josef and Miluse (Krehlova) D.; m. Natasa Kalouskova, Feb. 19, 1988. MS, Prague Inst. Chem. Tech., 1975, PhD, 1980. Divsn. chief Czechoslovak Acad. Sci., Inst. of Ecology, Ceske Budejovice, 1980-86, vice dir., 1986, rsch. scientist, 1986-91; pres. Water & Environ. Tech. Team, Ceske Budejovice, 1991—; mem. of Senate Prague Inst. Chem. Tech., 1986-90; mem. of PhD Examining Com. Prague Inst. Chem. Tech., Charles Univ., Czechoslovak Acad. of Sci., Prague, 1986-90; mem. of adv. panel Fed. Com. of the Environment, Prague, 1990-92. Contbr. articles to profl. jours. Mem. Local Com. for the Environment, Ceske Budejovice, 1991—. Recipient rsch. fellowship French Ministry of Sci. and Tech., 1992-93, rsch. fellowship Govt. of Norway, 1987, Teplice Water award Czechoslovak Sci. Tech. Soc.,1985. Mem. Internat. Water Quality Assn. (spl. group co-chmn. 1986—), Internat. Water Supply Assn. (co-chmn. 1993), Internat. Humic Substances Soc., Joint Specialist Group on Reservoirs Protection, Mgmt. and Water Treatment (co-chmn. 1991—), Czechoslovak Waterworks Assn. (pres. 1991—), Am. Waterworks Assn. Achievements include method and apparatus for ammetric determination of oxidizing gases in air (2 patents); influence of different operational parameters of treatment of humic waters; development of new and simple method for optimization of treatment of humic waters by coagulation; research in partially neutralized aluminum sulphate coagulant for drinking water treatment, influence of temperature on coagulation, coagulation optimization in waterworks by centrifugation, new test method for optimum coagulant dosing, drinking water supplies from reservoirs, influence of algal exudates on coagulation. Office: Water & Environ Tech Team, Box 27, 37011 Ceske Budejovice Czech Republic

DOLLING, GERALD, physicist, research executive; b. Dunstable, Bedfordshire, Eng., Nov. 21, 1935; arrived in Can. 1961; m. Sheila Ann Peters, Aug. 29, 1959; children: Christine Susan, Jo-Anna, Jacqueline Sheila. BA, Cambridge U., Eng., 1957, PhD, 1961. Rsch. officer AECL Rsch., Chalk River, Ont., Can., 1961-78, br. mgr., 1979-85, divsn. mgr., 1985-89, v.p., 1989—. Author: (with others) Physics of Condensed Matter; editor: Physics in Canada; contbr. 115 articles to profl. jours. Fellow Am. Physical Soc. Office: AECL Rsch, Chalk River Labs, Chalk River, ON Canada K0J 1J0

DOLNEY, TABATHA ANN, physics educator; b. Park Ridge, Ill., Aug. 22, 1970; s. Leonard Leo and Pamella Diane (Newman) D. BS, Sam Houston State U., 1992. Lab. instr. Sam. Houston State U., Huntsville, Tex., 1989—; rsch. asst. Sam. Houston State U., Huntsville, 1992—, Oak Ridge (Tenn.) Nat. Lab., 1991, U. Tex., Austin, 1992. Active Girl Scouts U.S.A., 1976—. Mem. Soc. Physics Students (pres. 1991-92), Sigma Pi Sigma. Mem. Ch. of Christ. Home: Rt 3 Box 360-A Cleveland TX 77327

DOLNIKOWSKI, GREGORY GORDON, scientist; b. Huntingdon, Pa., May 19, 1958; s. George T. and Joanne (Phillips) D.; m. Edith Wilks, Dec. 30, 1980. BA, Coll. of Wooster, 1980; PhD, Mich. State U., 1987. Postdoctoral fellow dept. chemistry U. Warwick, Coventry, Eng., 1987-89, U. Nebr., Lincoln, 1989-90; scientist II USDA Human Nutrition Ctr. Tufts U., Boston, 1990—. Contbr. chpt. to book Mass Analyzers in Mass Spectrometry Methods in Enzyology, 1990, articles to Jour. Am. Chem. Soc., Anal. Chem., Int. Jour. Mass Spectrom. Ion Processes. Lector, co-leader altar guild St. Luke and St. Margaret's Ch., Boston, 1991—. Mem. Am. Chem. Soc., Am. Soc. Mass Spectrometry. Democrat. Episcopalian. Achievements include rsch. in ion trapping in quadrupoles, ion/molecule reactions, MS/MS of biomolecules, GC/MS of vitamins, isotope ratio MS. Office: Tufts U USDA-Human Nutrition Ctr 711 Washington St Boston MA 02111

DOMAE, TAKASHI, cereal chemist; b. Kurashiki, Okayama, Japan, July 30, 1947; s. Takeshi and Fusako (Uwakubo) D.; m. Fukumi Kuroda, Nov. 23, 1972; children: Rinako, Kayoko. Student, Kyoto Tech. Inst., Japan, 1972-75. Researcher Nagata Sangyo Co. Ltd., Hyogo, Japan, 1975—; tech. adviser FAO/WHO Codex Alimentarius Commn., Rome, 1987; speaker in field. Author: Daizu Geppo, 1986, High Polymers, Japan, 1990, Kou Bunshi Kakou, 1991. Mem. Japan Vegetable Protein Foods Assn. (tech. advisers 1980—), Japan Flour Separation Application Assn. (tech. adviser 1983—). Achievements include patents for Mfg. Process of Gluten Paste and Vital Gluten; for Method of Recovery on B-Amylase from Wheat Starch Waste Water; for Biodegradable Plastic Composed of Wheat Gluten; Characteristics of Wheat Gluten as Semiconductor. Home: 244 Yamasaki-Cho Kasho, Shiso-Gun 671-25, Japan Office: Laboratory, 215 Senbonya-Yamasaki-Cho, Shiso-Gun 671-25, Japan

DOMAN, ELVIRA, science administrator; b. N.Y.C.; d. Andrew and Lillian (McClary) Hand; m. John H. Holder (div.); children: Paula Holder Simpkins, Rodney M. BA in Chemistry, CUNY, 1955; MA in Biochemistry, Columbia U., 1959; MS in Molecular Biology, NYU, 1960; PhD in Physiology and Biochemistry, Rutgers U., 1965. Jr. tech. U. Hosp. N.Y.U. Bellevue Med. Ctr., 1955; postdoctoral fellow Sloan-Kettering Inst. Cancer Rsch., N.Y.C., 1965; rsch. asst. Coll. Physicians and Surgeons, N.Y.C., 1959-60; rsch. asst. Sloan-Kettering Inst. Cancer Rsch., N.Y.C., 1959-60, postdoctoral assoc., 1965; rsch. assoc. Rockefeller U., N.Y.C., 1965-68; lectr. Douglass Coll. Rutgers U., New Brunswick, N.J., 1970-73; asst. prof. Seton Hall U., South Orange, N.J., 1973-77; assoc. program dir. NSF, Washington, 1978-92, program dir., 1992—; sci. fair judge pub., pvt. schs., colls., Washington, 1975; vis. scientist Rutgers U., 1989. Bd. dirs. Math. Sci., Computer Learning Ctr. of Shiloh Bapt. Ch., Washington, 1989—. Recipient Achievement award NSF, 1986, 92; grantee Seton Hall U., 1975. Fellow Am. Inst. Chemists; mem. AAAS, Am. Chem. Soc., Assn. Women Sci., MinorityWomen Sci., Orgn. Black Sci. (pres. 1990-93). Office: NSF 1800 G St NW Rm 321 Washington DC 20550-0002

DOMAN, JANET JOY, association executive; b. Phila., Dec. 16, 1948; d. Glenn J. and Hazel Katie (Massingham) D. Student, U. Hull, England, 1969-70; BA, U. Pa., 1971. Cert. tchr. Clinician Inst. Achievement Human Potential, Phila., 1971-74; dir. English Early Devel. Assn., Tokyo, 1974-75; dir. Evan Thomas Inst. Early Devel., Phila., 1975-77, Inst. Achievement of Intellectual Excellence, 1977-80; vice dir. Inst. Achievement Human Potential, 1980-82, dir., 1982—; internat. lectr. treatment of brain injured children and superiority. Chair Child Brain Devel., United Steelworkers Am., 1987. Recipient Gold medal Centro de Reabilitacion Nosa Senhora da Gloria, Rio de Janeiro, 1974, Brit. Star Brit. Inst. Achievement Human Potential, 1976, Sakura Korosho medal Japanese Inst. Achievement Human Potential, 1977, statuette with pedestal Internat. Forum Human Potential, 1980. Office: Inst Achievement Human Potential 8801 Stenton Ave Philadelphia PA 19118-2397

DOMARADZKI, JULIAN ANDRZEJ, physics educator; b. Szczecin, Poland, June 7, 1951; came to U.S., 1981; s. Julian Domaradzki and Zofia Wukowicz; m. Anna Teresa Kulesza, Feb. 11, 1979; children: Mateusz Barnaba, Julia Jagna. MS, U. Warsaw, Poland, 1974; PhD, U. Warsaw, 1978. Asst. prof. U. Warsaw, 1978-80; von Humboldt fellow Essen (Germany) U., 1980-81; rsch. staff Princeton (N.J.) U., 1981-83; rsch. assoc. MIT, Cambridge, Mass., 1983-84; rsch. scientist Flow Industries, Inc., Kent, Wash., 1984-87; asst. prof. U. So. Calif., L.A., 1987-91, assoc. prof. aerospace engring., 1991—. Recipient Sr. Rsch. award Alexander von Humboldt Found., Bonn, Germany, 1992. Mem. AIAA, Am. Phys. Soc., Soc. Indsl. and Applied Math. Achievements include pioneer use of numerically simulated flows in investigations of nature of nonlinear interactions in turbulence, conduction of numerical and theoretical investigations of transition to turbulence and of fully developed turbulence. Office: U So Calif Aerospace Engring U Park Los Angeles CA 90089-1191

DOMBKOWSKI, JOSEPH JOHN, water treatment specialist; b. Detroit, Nov. 6, 1961; s. Stanley Anthony and Lucille Ann (Switalski) D.; m. Nancy Ann Black, Aug. 25, 1990; 1 stepchild, Kurt Cooper. BS with distinction in Environ. Sci., U. Mich., Dearborn, 1984. Analytical chemist Perolin Inc., Chattanooga, 1984-87, Burmah Tech. Svcs., Pontiac, Mich., 1987-89; tech. specialist Aquatec Chem. Internat., Pontiac, 1989-91; tech. dir. Chemco Products, Inc., Howell, Mich., 1992—. Mem. Natural Resources Def.

Coun., Am. Chem. Soc., Am. Soc. Sugar Beet Technologists. Roman Catholic. Avocations: fishing, upland hunting. Office: Chemco Products Inc 1349 Grand Oaks Dr Howell MI 48843

DOMBROSKI, LEE ANNE ZARGER, medical physicist; b. Pottstown, Pa., Mar. 17, 1954; d. Robert Samuel and Evelyn JoAnne (Applegate) Zarger; m. Dale Louis, Aug. 18, 1984; children: Alyssa, Tara. BS in Psychobiology, Albright Coll., 1976; MS in Radiation Sci., Rutgers U., 1985. Cert. in therapeutic radiol. physics Am. Bd. Radiology. Jr. physicist Meml. Sloan Kettering, N.Y.C., 1984-85; asst. physicist Valley Hosp., Ridgewood, N.J., 1985-88; med. physicist George Zacharopoulos Med. Physics Cons., Dover, N.J., 1988—; cons. physicist Radiology Nuclear Ultrasound Cons., P.A., Freehold, N.J., 1992. Mem. Am. Assn. Physicists in Medicine, Am. Coll. Radiology, N.J. Med. Physics Soc., N.J. Health Physics Soc., DAR. Democrat. Lutheran. Home: 154 Emmans Rd Flanders NJ 07836 Office: Dover Gen Hosp Radiol Oncology Jardine St Dover NJ 07801

DOMBROWSKI, ANNE WESSELING, microbiologist, researcher; b. Cin., Jan. 26, 1948; d. Robert John and Margaret Mary (Bell) Wesseling; m. Allan Wayne Dombrowski, Apr. 17, 1982; children: Amy, Alicia. BA summa cum laude, Xavier U., 1970; MS, U. Cin., 1972, PhD, 1974. Fellow Scripps Clinic & Rsch. Found., La Jolla, Calif., 1974-76; sr. rsch. microbiologist Merck & Co., Inc., Rahway, N.J., 1976-87, rsch. fellow, 1987—. Contbt. articles to profl. jours. Mem. AAAS, Soc. Indsl. Microbiology (sec. 1982-85), Am. Soc. Microbiology, Soc. Gen. Microbiology, Mycol. Soc. Avocations: reading, gardening.

DOMBROWSKI, FRANK PAUL, JR., pharmacist; b. Nashua, N.H., May 10, 1943; s. Frank Paul and Yvonne Joan (Paris) D.; B.S., Mass. Coll. Pharmacy, 1965, M.S., 1967; m. Eleanor Cassady, June 15, 1968; children—Michael, Peter, Laura, Cheryl, Douglas. Pharmacist, Androscoggin Valley Hosp., Berlin, N.H., 1974-75, Eastern Maine Med. Center, 1975-77; dir. pharm. services and central supply Concord Hosp., N.H., 1977-82; founder, pres. Hosp. Home Health Care of N.H., 1982-92, Hosp. Home Health Care of Maine, 1986-92; founder, pres. Weare Family Pharmacy, 1992—; commr. N.H. Bd. of Registration of Pharmacy; cons. nurse anesthetist sch. Concord Hosp. Served with U.S. Army, 1968-74. Decorated Combat Inf. badge, Bronze Star medal, Army Commendation medal. Fellow Am. Acad. Med. Adminstrs., Am. Coll. Apothecaries; mem. Am. Pharm. Assn., Am. Soc. Hosp. Pharmacists, N.H. Pharm. Assn., N.H. Soc. Hosp. Pharmacists, Nat. Assoc. Retail Druggists. Club: Lions (chpt. pres.). Home: 770 Broadcove Rd Contoocook NH 03229 Office: 425-8 S Stark Hwy Weare NH 03281

DOMBROWSKI, JOHN MICHEAL, architectural consulting engineer; b. Erie, Pa., Sept. 17, 1959; s. Edward Frances and Anne (Walters) D.; m. Patricia Jayne Nagrant, May 19, 11984; children: Craig James, Mark Edward. B. Archtl. Engring., Pa. State U., 1982. Registered profl. engr., Pa. Project engr., assoc. H.F. Lenz Co., Johnstown, Pa., 1982—. Bd. dirs. Sr. Activities Ctr., Johnstown, 1986-89; coach Spl. Olympics Bowling, Johnstown, 1987-89. Mem. ASHRAE, ASHE, NSPE, ISPE, Pa. Soc. Profl. Engrs. (Johnstown chpt., sec. 1991-92, pres.-elect 1992-93, pres. 1993-94), Am. Soc. Hosp. Engr. Office: H F Lenz Co 1407 Scalp Ave Johnstown PA 15904

DOMINGUE, RAYMOND PIERRE, chemist, consultant, educator; b. Berlin, N.H., Mar. 21, 1959; s. Robert F. and Doris R. (Lizzie) D.; m. Antoinette P. Ostop, June 14, 1986. BS, U. Vt., 1981; PhD, Stanford U., 1986. Sr. scientist Spectral Scis. Inc., Burlington, Mass., 1986-90; adj. prof. Daniel Webster Coll., Nashua, N.H., 1990-91; rsch. chemist Mobil R&D Corp., Paulsboro, N.J., 1991—. Contbr. articles to profl. jours. Recipient Cook Sci. award U.Vt., 1980, Am. Inst. Chemists award, U. Vt., 1981. Mem. AAAS, Am. Chem. Soc., Phi Beta Kappa, Sigma Xi. Achievements include research in Mercury analysis in natural gas. Home: 105 N Foxford Ln Mullin Hill NJ 08062 Office: Mobil R&D Corp Paulsboro NJ 08066

DOMINGUEZ ORTEGA, LUIS, medical educator, health facility administrator; b. Barcelona, Spain, Oct. 4, 1941; s. Jose Dominguez and Dolores Ortega (Araujo) Dominguez; m. Mercedes Sanchez Tamayo, Jan. 2, 1969; children: Elena, Jose Luis. Cert., Ramiro Maeztu Inst., Madrid, Spain, 1962; MD, Complutense U., Madrid, 1969, diploma in internal medicine, 1975. Postgrad. Clinico Hosp., Madrid, 1969-73; asst. physician emergency svc. Dept. of Internal Medicine, 12 de Octubre Hosp., Madrid, 1974-77, asst. physician, 1977—, asst. prof., 1977-86, assoc. prof., 1986—; dir., coord. sleep disorders unit 12 de Octobre Hosp., Madrid, 1990—; dir., founder sleep unit Ruber Clinic, Madrid, 1988-89, 91—; mem. faculty bd. 12 de Octubre Hosp., 1984-88; mem. hosp. bd. Med. Coll., Madrid, 1984-88, candidate to pres. hosp. bd., 1986; organizer, chmn. internat. meeting Advances in Sleep Disorders, Madrid, 1992. Member Club Liberal, Madrid, 1980-89; founder, v.p. Asociacion Nacional Medicos Empresarios, Spain, 1991—. FISS grantee, 1980-90, 92. Mem. Am. Sleep Disorders Assn., Nat. Assn. Internal Medicine, N.Y. Acad. Scis., Internat. Assn. Internal Medicine, European Sleep Rsch. Soc., European Assn. Internal Medicine. Roman Catholic. Avocations: music, literature, tennis, hunting. Office: Clinica Ruber, Juan Bravo no 49, 28006 Madrid Spain

DOMINICK, PAUL SCOTT, chemist, researcher; b. Kansas City, Mo., Apr. 13, 1962; s. Michael and Della B. (King) D.; m. Diane Maria Leardi, July 27, 1985; children: Anthony Michael, Paul Salvatore, Nicole Marie. BS in Chemistry, U. Mo., Kansas City, 1986. Rsch. asst. Sch. Engring., U. Mo.-Kansas City, Independence, 1985-86; asst. scientist Chemistry Sci. Lab., Lenexa, Kans., 1985-87; analytical chemist Marion Merrill, Kansas City, Mo., 1987-91; sales rep. LDC Analytical, Inc., Riviera Beach, Fla., 1991-92, Thermo Separation Products (formerly LDC Analytical, Inc.), Riviera Beach, Fla., 1992—. Ofcl. Sci. Olympiad, Kansas City, 1988—; active Friends of the Zoo, Kansas City, 1990—. Recipient Cert. of Achievement Bio Rad, Cambridge, 1988, Cert. of Apprecation, Marion Labs., 1988. Mem. Am. Chem. Soc., Soc. for Applied Spectroscopy (sec. 1992, treas. 1990), Nat. Geog. Soc. Republican. Home: 8212 Hardy Raytown MO 64138 Office: Thermo Separation Products 3661 Interstate Park Rd N Riviera Beach FL 33404

DONAHEY, REX CRAIG, structural engineer; b. Logan, Kans., Mar. 11, 1955; s. Lawrence I. and Irene (Brady) D.; m. Cynthia S. Maddy, May 28, 1977; children: Kelly, James. MSCE, U. Kans., 1983, PhD, 1986. Registered profl. engr., Okla. Project engr. Marley Cooling Tower Co., Mission, Kans., 1977-80; rsch. asst. Ctr. for Rsch., Inc., Lawrence, Kans., 1980-86; asst. prof. Okla. State U., Stillwater, 1986-89, U. Ill., Champaign, 1989-90; project engr. Ellerbe Becket, Inc., Kansas City, Mo., 1990—; reviewer ASCE, N.Y.C., 1986—; item writer, scorer Nat. Coun. Examiners Engring. and Surveying, Clemson, S.C., 1988-89; bd. dirs. Okla. chpt. Am. Concrete Inst., Oklahoma City, 1988-89; instr. U. Kans., Lawrence, 1992. Contbr. articles to Jour. Structural Engring., Jour. Am. Concrete Inst. Mem. ASCE (sec. task com. on design criter for composite structures in steel and concrete 1988-92, sec. task com. on design guide for composite semi-rigid connections 1993—). Achievements include development of design procedures for composite beams with web openings. Office: Ellerbe Becket Inc 605 W 47th St Kansas City MO 64112

DONAHOO, MELVIN LAWRENCE, aerospace management consultant, industrial engineer; b. Balt., Dec. 28, 1930; s. Lawrence E. and Margaret (Hartman) D.; m. Charlene B. Donahoo; children from previous marriage: Patricia Ann, Joseph, Teresa, Melvin Lawrence Jr. BS, U. Balt., 1954, MBA, 1974; postgrad., Am. Univ., 1964-67, George Washington U., 1969-71. Cons., v.p. L.E. Donahoo & Assoc., Phila., 1954-58; chief indsl. engring. Martin Marietta Corp., Balt. and Orlando, Fla., 1958-63; program mgr. NASA Goddard Space Flight Ctr., Greenbelt, Md., 1963-90; dir. ops. Idea, Inc., 1990-92; pvt. aerospace cons., 1992—; instr. indsl. mgmt. Author: Aircraft Learning Curves, 1959, Project Planning Handbook, 1970; also research papers. With USN. Mem. Soc. Mfg. Engrs. (sr., life, mem. coms.), Inst. Indsl. Engrs. (sr., life, mem. coms.), Am. Legion (post comdr. Md. 1983, county comdr. 1984-86), Kent Island Yacht Club, KC, Elks, Moose. Home: 2417 Pelham Ave Baltimore MD 21213-1036

DONAHUE, MARY ROSENBERG, psychologist; b. N.Y.C., Dec. 20, 1932; d. Lester and Ethel (Hyman) Rosenberg; children: Laurie, Rachel. BA, Adelphi U., 1954; MA, N.Y. U., 1958; PhD, St. John U., 1968. Tchr. Elmont, N.Y., 1954-57, sch. psychologist, 1957-63; cons. psychologist NIMH, 1964-65; sch. psychologist Mamaroneck, N.Y., 1966-67; pvt. practice psychology Bethesda, Md., 1971—; pres. Automated Psychol. Svcs.; bd. dirs. SPIFE, comprehensive testing svc.; expert witness local jurisdictions regarding domestic issues, womens issues, abuse, 1974—; speaker on custody evaluations and expert witness considerations. Co-author: On Your Own, 1993. NIMH grantee, 1962-63, 64-65. Mem. Am. Psychol. Assn., Md. Psychol. Assn., D.C. Psychol. Assn., Am. Orthopsychiat. Assn., Assn. Pvt. Practitioners, Nat. Assn. Women Bus. Owners. Home: 12017 Edgepark Ct Rockville MD 20854-2138 Office: 5902 Hubbard Dr Rockville MD 20852-4823

DONAHUE, ROBERT EDWARD, veterinarian, researcher; b. N.Y.C., Nov. 23, 1954; s. Robert Edward and Elizabeth (Tyrell) D.; m. Karen Elizabeth Wolski, Oct. 1, 1983; children: Kathryn Elizabeth Grace, Mary Elizabeth Brinn. BA in Anthropology and Cellular Biology, U. Pa., 1977, VMD, 1981; MS in Cancer Biology, Harvard U., 1985. Staff scientist Genetics Inst. Inc., Cambridge, Mass., 1985-87, prin. scientist, 1987-90; vet. med. officer Nat. Heart, Lung, and Blood Inst., Bethesda, Md., 1990—; mem. Commn. of the European Communities-Efficiency and Safety of Cytokines and Human Therapy, Brussels, Belgium, 1992. Contbr. articles to Nature, Sci., J. Exptl. Medicine. Mem. Am. Soc. Clin. Oncology, Am. Soc. Hematology, Am. Assn. Immunologists, Internat. Soc. Analytical Cytology, N.Y. Acad. Sci. Achievements include patents in treatment of AIDS-type disease and method of inducing leukocytes with a combination of IL-3 and GM-CSF. Office: NIH Bldg Rm 7C103 Clin Hematology Br Bethesda MD 20892

DONAHUE, THOMAS MICHAEL, physics educator; b. Healdton, Okla., May 23, 1921; s. Robert Emmett and Mary (Lyndon) D.; m. Esther Marie McPherson, Jan. 1, 1950; children: Brian M., Kevin E., Neil M. A.B., Rockhurst Coll., 1942, D.Sc. (hon.), 1981; Ph.D., Johns Hopkins U., 1947. Rsch. assoc., asst. prof. Johns Hopkins U., 1947-51; asst. prof. U. Pitts., 1951-53, assoc. prof., 1953-57, prof., 1957-74, dir. Lab. Atmospheric and Space Sci., 1966-74, dir. Space Rsch. Coordination Ctr., 1966-74; chmn. dept. atmospheric and oceanic sci. and Space Physics Rsch. Lab., U. Mich., Ann Arbor, 1974-81, prof., 1981-87, Edward H. White II disting. univ. prof. planetary sci. dept. atmospheric oceanic and space scis., dept. physics, 1987—; dir. ctr. for integrated study global change U. Mich., 1990-93; mem. phys. scis. com. NASA, 1972-77, adv. coun., 1982-88, solar system exploration com., 1981-82; mem. Arecibo adv. bd. Cornell U., 1971-76, 86-89, chmn. 1989; mem. Space Telescope Sci. Inst. Adv. Com., 1986-89, chmn., 1987-89; chmn. solar terrestrial rels. com. NAS, mem. atmospheric scis. com., mem. geophysics rsch. bd., mem. climate bd., mem. nominating com., 1982-88; chmn. space sci. in the twenty-first century study NAS, 1984-87, com. for U.S.-USSR workshop on planetary scis., 1988-91, com. on planetary and lunar exploration, 1992—; chmn. sci. steering groups Pioneer Venus multi-probe and orbital missions to Venus, 1974-93, pub. affairs com. Am. Geog. Union; trustee-at-large Upper Atmosphere Rsch. Corp., 1975-87; vice-chmn. exec. com., trustee Univ. Corp. for Atmospheric Rsch., 1978-85; chmn. bd. trustees Univs. Space Rsch. Assn., 1978-82; mem. vis. com. Max Planck Gesellschaft fur Aeronomie, 1989—; mem. nat. tech. adv. com. Nat. Inst. for Global Environ. Change, 1992; Marcel Nicolet lectr. Am. Geophys. Union, 1993. Editor: Space Research X, 1969; assoc. editor numerous publs., particularly specializing in atomic physics and properties of planetary atmospheres; editor: Venus, 1983; assoc. editor: Planetary and Space Sci. Served with AUS, 1944-46. Recipient Public Svc. award NASA, 1977, 88, 7 achievement awards, Disting. Public Svc. medal, 1980, Wellock Disting. Rsch. Accomplishments award U. Mich., 1980, Arctowski medal Nat. Acad. Sci., 1981, Fleming medal Am. Geophys. Union, 1981, Rsch. Excellence award Coll. Engring., 1981; Henry Russel lectr. U. Mich., 1987; Space Sci. award AIAA, 1988; 1st Space Sci. medalist Nat. Space Club, 1989; Marcel Nicolet lectr. Am. Geophys. Union, 1993; Gugghenheim fellow U. Paris, 1960. Fellow AAAS, Am. Phys. Soc., Am. Geophys. Union (pres. solar-planetary rels. 1972-75, v.p. 1969-72, chmn. pub. policies com. 1990-93, Marcel Nicolet lectr. 1993), Mich. Soc. Fellows; mem. NAS, Internat. Acad. Astronautics. Achievements include participation in Voyager mission to outer planets, Galileo mission to Jupiter, Cassini Mission to Saturn, Spacelab 1, Apollo 17, Apollo-Soyuz, chmn. sci. steering group Pioneer Venus multiprobe/orbiter missions. Home: 1781 Arlington Blvd Ann Arbor MI 48104-4105

DONALDSON, JAMES OSWELL, III, neurology educator; b. Butler, Pa., July 19, 1942; s. James Oswell Jr. and Estelle Mathilda (Unverzagt) D.; m. Mary Hoopingarner, Aug. 23, 1969 (div. Dec. 1983); 1 child, Andrew Robert; m. Susan McKernin, Nov. 3, 1984. BS, Haverford Coll., 1964; MD, U. Pa., 1968. Diplomate Am. Bd. Psychiatry and Neurology, Am. Bd. Internal Medicine. Intern in medicine Hosp. of U. Pa., Phila., 1968-69, resident, 1969-70, resident in neurology, 1974-76; hon. house physician Nat. Hosp. for Nervous Diseases, London, 1973-74, sr. vis. fellow, 1991; asst. prof. neurology U. Conn. Sch. Medicine, Farmington, 1977-82, assoc. prof., 1982-88, prof., 1988—. Author: Neurology of Pregnancy, 1978, 2d edit., 1989. Maj. M.C., U.S. Army, 1970-73. Fellow ACP, Am. Acad. Neurology; mem. Am. Neurol. Assn. Office: U Conn Health Ctr 263 Farmington Ave Farmington CT 06030-0001

DONALDSON, ROBERT LOUIS, computer systems professional; b. Boston, Dec. 11, 1961; s. Gilbert Young and Marilyn Mary (Knowles) D. BSBA, Suffolk U., 1981. With Mass. Dept. Revenue, Boston, 1984 , Lan mgr., 1990—; adv. bd. Networld, Boston, 1989-90. Democrat. Roman Catholic. Office: Mass Dept Revenue 100 Cambridge St Rm 704 Boston MA 02204

DONALDSON, STEVEN LEE, materials research engineer; b. Dayton, Ohio, Feb. 7, 1959; s. Charles Morgan and Charlotte (Davis) D.; m. Elizabeth Schwarzkopf, Dec. 27, 1980; 1 child, Chloe Youngsoon. BS, Purdue U., 1981; MS, U. Dayton, 1987; PhD, Stanford U., 1993. Engr. NASA Lewis Rsch. Ctr., Cleve., 1978-79; materials rsch. engr. USAF Materials Directorate, Wright-Patterson AFB, Ohio, 1982-91; rsch. group leader USAF Materials Direstorate, Wright-Patterson AFB, Ohio, 1991—; lectr. USAF, 1982—; vis. researcher Linkoping, Sweden, 1984. Contbr. articles to profl. jours. Mem. Am. Soc. Composites, Am. Inst. Aeronautics and Astronautics, Am. Soc. Testing and Materials, Soc. Advancement Materials. Methodist. Achievements include composites delamination failure criterion, fracture surface characterization, damage tolerance of composite structures. Office: USAF Materials Directorate Wright-Patterson AFB Dayton OH 45433

DONALDSON, WILLIS LYLE, research institute administrator; b. Cleburne, Tex., May 1, 1915; s. Charles Lyle and Anna (Bell) D.; m. Frances Virginia Donnell, Aug. 20, 1938; children: Sarah Donaldson Seaberg, Susan Donaldson Pollock, Sylvia Donaldson Nelson, Anthony Lyle. B.S., Tex. Tech. U., 1938. Registered profl. engr., Pa., Tex. Distbn. engr. Tex. Electric Service Co., 1938-42, supervisory engr., 1945-46; asst. prof. elec. engring. Lehigh U., 1946-51, assoc. prof., 1953-54; with S.W. Research Inst., San Antonio, 1954—; v.p. S.W. Research Inst., 1964-72, v.p. planning and program devel., 1972-74, sr. v.p. planning and program devel., 1974-85, sr. cons., 1985—. Bd. dirs. San Antonio Chamber Music Soc., pres., 1962-72, 87—, mem. 1954—. Served to capt. USNR, 1942-45, 51-53. Named Disting. Engr. Tex. Tech. U., 1969. Fellow IEEE, Am. Soc. Nondestructive Testing; mem. Armed Forces Communications and Electronics Assn. (disting. life), Sigma Xi, Tau Beta Pi, Eta Kappa Nu, Alpha Chi. Home: PO Box 160218 San Antonio TX 78280-2418 Office: 6220 Culebra Rd San Antonio TX 78284-0001

DONATI, GIANNI, chemical engineer, administrator; b. Cuggiono, Mi, Italy, Jan. 1, 1943; s. Pietro and Margherita (Re') D.; m. Paola Mariani, May 19, 1971; children: Christina, Raffaella. D in Chem. Engring., Politechico Milan, Italy, 1967. Asst. prof. Politechico di Milan, Italy, 1968-76; project leader AGIP Spa, Milan, 1976; head chem. engring. Guido Donegani SpA, Novara, Italy, 1976-82; mktg. mgr. Guido Donegani SpA, Novara, 1982-84; internat. mktg. mgr. Montedison SpA, Milan, 1984-85; bus. mgr. Montefluos Spa, Milan, 1985-86; R&D div. mgr. Emichem Spa, San Donato, Italy, 1986—; project leader ARS SpA, Milan, 1968-74; cons. ENI, Nuovopignone, SNPE, Total, 1968-74; Italian rep. EFChE-CRE, 1978-84. Contbr. 50 articles to profl. jours. including Chem. Engring. and Biotech. Bioengring. Mem. Assn. Chem. Engrs., Am. Chem. Soc., Italian Assn. Chem. Engring., Italian Chem. Soc. Chistian Dem. Achievements include 10 patents and improvement of many industrial processes; membership on task force of new himont polypropilene process development. Home: Via Meda 30, 20017 Rho MI, Italy Office: Emichem Spa, Via Maritano 26, 20097 San Donato MI, Italy

DONCHIN, EMANUEL, psychologist, educator; b. Tel Aviv, Apr. 3, 1935; came to U.S., 1961; s. Michael and Guta D.; m. Rina Greenfarb, June 3, 1955; children: Gill, Opher, Ayala. B.A., Hebrew U., 1961, M.A., 1963; Ph.D., UCLA, 1965. Teaching and research asst. dept. psychology Hebrew U., 1958-61; research asst. dept. psychology UCLA, 1961-63, research psychologist, 1964-65; research asso. div. neurology Stanford U. Med. Sch., 1965-66, asst. prof. in residence, 1966-68; research asso. neurobiology br. NASA, Ames Research Center, Moffett Field, Calif., 1966-68; asso. prof. dept. psychology U. Ill., Urbana-Champaign, 1968-72; prof. psychology and physiology U. Ill., 1972—, head dept. psychology, 1980—. Author: (with Donald B. Lindsley) Averaged Evoked Potentials, 1969; editor: Cognitive Psychophysiology, 1984, (with M.G.H. Coles and S.W. Porges) Handbook of Psychophysiology, 1986; contbr. articles to profl. jours. Served with Israeli Army, 1952-55. Fellow AAAS, Am. Psychol. Assn.; mem. Soc. Psychophysiol. Research (pres. 1980), Fedn. Behavioral, Cognitive and Psychol. Socs. (v.p. 1981-85), Am. EEG Soc., Psychonomic Soc., Soc. Neuroscis. Office: U Ill Dept Psychology 603 E Daniel St Champaign IL 61820-6267

DONES, MARIA MARGARITA, anatomist, educator; b. Chgo., Dec. 9, 1956; d. Nestor and Rafaela (Rivera) D.; m. Louis P. Dell; 1 child, Brent Eugene Nestor Smith. BS, U. P.R., Rio Piedras, 1978; PhD, U. Okla., 1986. Postdoctoral fellow U. Fla., Gainesville, 1987-89; prof. biology, mentorship coord. Okla. Sch. Sci. and Math., Oklahoma City, 1989—; bd. dirs. Okla. Lit. Coun., Oklahoma City, 1990—, v.p., 1992—; bd. dirs. Okla. Lit. Initiatives Commn., Oklahoma City, 1992—. Grad. Profl. Opportunity Program fellow scholar U. Okla., 1983-86. Mem. Soc. for Study of Reproduction, Gordon Rsch. Conf. (Jr. Rschr. award 1988), Rotary Internat. (amb. 1984-85), Sigma Xi. Roman Catholic. Office: Okla Sch Sci and Math 1141 N Lincoln Blvd Oklahoma City OK 73104

DONG, CHENG, bioengineering educator. BS in Engring. Mechanics, Shanghai Jiao-Tong U., China, 1982; MS in Civil Engring. and Engring. Mechanics, Columbia U., 1984, PhD in Bioengineering, 1988. Asst. research bioengineer U. California San Diego, La Jolla, Calif., 1988-91; prof. biology Pennsylvania State U., University Park, Penn., 1992—. Scholar Govt. of China, 1982-84. Mem. ASME (assoc., Best Paper award bioengineering divsn. 1989, Melville medal 1990), Biomedical Engring. Soc., Sigma Xi. Office: Pennsylvania State U Dept Biology 229 Hallowell Bldg University Park PA 16802*

DONG, DENNIS LONG-YU, biochemist; b. Zhou-Shan, Zhe-jiang, China, May 26, 1966; came to U.S., 1986; s. You-Xiang and Zhi-Qing (Chen) D.; m. Helen Hui-Hsin Sun, Aug. 22, 1992. Student, Zhe-Jiang Marine Inst., 1983-86; BS, SUNY, Stony Brook, 1989; postgrad., Johns Hopkins U., 1989—. Rsch. asst. SUNY, Stony Brook, 1988-89; biochemistry researcher Johns Hopkins U. Sch. Medicine, Balt., 1989—. Fellow Merck Found., 1991-92, NIH, 1989-91. Mem. Am. Soc. Cell Biology, Soc. Chinese Bioscientists in Am. Office: Johns Hopkins U Sch Med Dept Biol Chemistry 725 N Wolfe St Baltimore MD 21205

DONG, LINDA YANLING, optical engineer; b. Jilin, Peoples Republic China, Oct. 5, 1962; came to U.S., 1990; d. Lijuan Dong and Guizhi Zhong; m. Don Dongxiao Yu, Nov. 30,1 990; 1 child, Don Xiaodong Yu. BS, Tsinghua U., PhD in Engring. Sr. engr. Chinese Acad. Sci., 1989-90; sr. optical engr. Surface Optics Corp., San Diego, 1990-91; v.p. Catoctin Rsch. Corp., Germantown, Md., 1991—. Mem. Photonics Soc. Chinese-Ams.

DONGUY, PAUL JOSEPH, aerospace engineer; b. St. Brieuc, Bretagne, France, June 27, 1940; s. Paul Louis and Marie-Louise Nelly (Duvinage) D.; m. Marie-Annick Madeleine Laffitte, Aug. 19, 1964; children: Arnaud, Nicolas, Agnes. Engr., Ecole Nationale Supérieure de Mécanique et d'Aerotechnique, Poitiers, France, 1964; BS in Sci., U. Poitiers, 1964. Designer solid rocket nozzle European Soc. Propulsion, Bordeaux, France, 1966-76, group leader advanced solid rocket and ramjet design, 1976-84, head Solid Rocket Program Office,, 1984-86, head advanced propulsion program mktg. dept., 1986-89; liquid rocket stratetic planner European Soc. Propulsion, Vernon, France, 1989-90; dir. advanced propulsion, dep. gen. mgr. Hyperspace, European Soc. Propulsion, Suresnes, France, 1990—. Contbr. articles to Jour. Brit. Interplanetary Soc., Jour. Aircraft, Jour. Spacecraft and Rockets, Aeros. and Astronautics. Lt. French Army, 1965-66. Mem. AIAA (solid rocket tech. com. 1982-84, 86-90, best solid rocket paper award 1981), Aeros. and Astronautic Assn. France. Roman Catholic. Achievements include patents on flexible bearing joint, unfoldable divergent nozzle for a rocket engine, mechanical and insulating connection between a nozzle and the filament wound casing of the combustion chamber. Office: European Propulsion Soc BP 203, 24 Rue Salomon Rothschild, F-92156 Suresnes Cedex, France

DONNAY, ALBERT HAMBURGER, environmental health engineer; b. Paris, Sept. 28, 1958; came to U.S., 1960; s. Joseph D.H. and Gabrielle (Hamburger) D.; m. Yvonne Lynn Ottaviano, June 7, 1986; 1 child, Gabriel Francis. BA, McGill U., Montreal, Que., Can., 1980; M of Health Scis., Johns Hopkins U., 1982. Program analyst Radiation Safety Office, U. Md., Balt., 1981-82; exec. dir. Nuclear Free Am., Balt., 1982-90; mktg. dir. EcoWorks, Inc., Balt., 1990—; ind. mgmt. cons., Balt., 1990—; staff scientist Inst. for Energy and Environ. Rsch., Takama Park, Md., 1992—. Author: How to Save the World, 1970; editor: The Investor's Guide to the Military Industry, Fy 1987, 1988; editor (jour./newsletter) The New Abolitionist, 1982-90. Mem. APHA, Delta Omega. Avocation: sailing. Office: EcoWorks Inc 2326 Pickwick Rd Baltimore MD 21207

DONNELLY, BARBARA SCHETTLER, medical technologist; b. Sweetwater, Tenn., Dec. 2, 1933; d. Clarence G. and Irene Elizabeth (Brown) Schettler; A.A., Tenn. Wesleyan Coll., 1952; B.S., U. Tenn., 1954; cert. med. tech., Erlanger Hosp. Sch. Med. Tech., 1954; postgrad. So. Meth. U., 1980-81; children—Linda Ann, Richard Michael. Med. technologist Erlanger Hosp., Chattanooga, 1953-57, St. Luke's Episcopal Hosp., Tex. Med. Ctr., Houston, 1957-58, 1962; engring. R &D SCI Systems Inc., Huntsville, Ala., 1974-76; cons. hematology systems Abbott Labs., Dallas, 1976-77, hematology specialist, Dallas, Irving, Tex., 1977-81, tech. specialist microbiology systems, Irving, 1981-83, coord. tech. svc. clin. chemistry systems, 1983-84, coord. customer tng. clin. chemistry systems, 1984-87, supr. clin. chemistry tech. svcs., 1987-88, supr. clin. chemistry customer support ctr., 1988—. Mem. Am. Soc. Clin. Pathologists (cert. med. technologist), Am. Soc. Microbiology, Nat. Assn. Female Execs., U. Tenn. Alumni Assn., Chi Omega. Contbr. articles on cytology to profl. jours. Republican. Methodist. Home: 204 Greenbriar Ln Bedford TX 76021-2006 Office: 1921 Hurd St Irving TX 75061

DONNELLY, JOHN JAMES, III, immunologist, blood banker; b. Phila., June 26, 1954; s. John James Jr. and Erma Marie (Cocci) D.; m. Betsy Ann Burkhardt, Dec. 30, 1976; children: Ann Marie, James Arthur. BA, U. Pa., Phila., 1975, PhD, 1979. Postdoctoral fellow U. Cambridge, U.K., 1979-81, Johns Hopkins U., Balt., 1982-83; asst. prof. U. Pa., Phila., 1983-88; rsch. fellow Merck & Co., Inc., West Point, Pa., 1988—; adj. asst. prof. U. Pa., Phila., 1988—; cons. WHO, Geneva, 1983—, U.S. Agy. for Internat. Devel., Washington DC, 1988—. Author: (book chpt.) Molecular and Cellular Mechanisms of Hypersensitivity, 1989, Vaccines, '91, 1991; contbr. articles to profl. jours. Dir. Blood Donor Program, 79th U.S. Army Res. Command, Pa., 1987-90. Maj. USAR, 1984—. Decorated Bronze Star; NIH predoctoral fellow, 1975, Fight for Sight, Inc. postdoctoral fellow, 1980, NIH postdoctoral fellow, 1982. Fellow Royal Soc. Tropical Medicine and Hygiene; mem. Am. Assn. Immunologists, Assn. for Rsch. in Vision and Ophthalmology, British Soc. for Immunology. Achievements include patent for novel carrier protein for use in vaccines; research in antigen processing, regulation of transplantation antigen expression, transplantation and tumor immunity, and immunoregulation. Office: Merck Sharp Dohme Rsch Labs Sumneytown Pike West Point PA 19486

DONNELLY, KIM FRANCES, computer scientist; b. Phila., Feb. 4, 1960; d. Francis Edward and Gertrude Louise (Wooten) D.; m. Kurt Allen Gluck, June 14, 1987; children: Jonathan, Joshua. BS, East Stroudsburg U., 1979; MS, Rensselaer Poly. Inst., 1981. Mem. tech. staff Bell Labs, Bell Communication Rsch., Piscataway, N.J., 1981-86; dir. Bell Communication Rsch., Piscataway, 1986--. V.p. Friends of the Rutgers Ecological Preserve, Piscataway, 1991-92. Nat. Merit scholar, 1976. Mem. IEEE (computer soc.), AAAS. Achievements include patent for a method and apparatus for selectively post processing paginated output. Home: 72 N Ross Hall Blvd Piscataway NJ 08854 Office: Bell Communication Rsch PY4J305 33 Knightsbridge Rd Piscataway NJ 08855

DONOHO, LAUREL ROBERTA, industry analyst; b. San Jose, Calif., Oct. 8, 1952; s. Donald Frank Donoho and Flora (Stephen) Donoho Homan. BA in History, Pt. Loma Coll., 1974; MBA, San Jose State U., 1992. Cons. mktg. Southland Corp., San Jose, 1984-90; sr. industry rsch. analyst Frost & Sullivan, Mountain View, Calif., 1990—. Contbr. articles to profl. jours. Mem. Santa Clara County Human Rels. Task Force, San Jose, 1992—. Mem. Am. Mktg. Assn., Instruments Soc. Am., Bay Area Career Women (pres 1990—). Achievements include research in market intelligence, sensors and industrial instrumentation. Office: Frost & Sullivan Market Intelligence 2525 Charleston Rd Mountain View CA 94043

DONOHOE, ROBERT JAMES, spectroscopist; b. Phoenix, Dec. 3, 1956; s. Thomas Aquinas and Lillian Julia (Doerr) D.; m. Anne Reynolds, Apr. 14, 1985; children: Sean, Patrick. PhD, N.C. State U., 1985. Fellow Carnegie-Mellon U., Pitts., 1985-88; fellow Los Alamos (N.Mex.) Nat. Lab., 1988-91, staff mem., 1991—. Contbr. articles to profl. jours. Fellow NIH, 1987. Mem. Phi Lambda Upsilon. Achievements include discovering indirect communication of localized chromophores via coulombic and backbonding effects; first accurate description of vibrational modes in chlorophyll and bacteriochlorophyll; researched vibrational characteristics of defect states and defect mobilization in low-dimensional materials, vibrational characteristics of groundstates in low-dimensional materials under pressure. Office: INC-14 MS-C345 Los Alamos Nat Lab Los Alamos NM 87545

DONOHUE, MARC DAVID, chemical engineering educator; b. Watertown, N.Y., Sept. 10, 1951; s. Paul Francis and Beverly Gertrude (Hodge) D.; m. Mary Ann Chamberlain, July 20, 1974; children: Paul, Megan, Ian. BS, Clarkson Coll. Tech., 1973; PhD, U. Calif., Berkeley, 1977. Asst. prof. chem. engring. Clarkson Coll. Tech., Potsdam, N.Y., 1977-79; asst. prof. Johns Hopkins U., Balt., 1979-83, assoc. prof., 1983-87, prof., 1987—, chmn. dept., 1984—. Recipient Adminstr.'s Pollution Prevention award for Region III, U.S. EPA, 1992, Md. sect. Outstanding Engring. Achievement award NSPE, 1989. Mem. Am. Inst. Chem. Engrs., Am. Chem. Soc., Am. Soc. Engring. Edn. (Outstanding Young Engr. award 1984), Tau Beta Pi. Office: Johns Hopkins U Dept of Chem Engring 3400 North Charles St Baltimore MD 21218

DONOHUGH, DONALD LEE, physician; b. Los Angeles, Apr. 12, 1924; s. William Noble and Florence Virginia (Shelton) D.; m. Virginia Eskew McGregor, Sept. 12, 1950 (div. 1971); children: Ruth, Laurel, Marilee, Carol, Greg; m. Beatrice Ivany Redick, Dec. 3, 1976; stepchildren: Leslie Ann, Andrea Jean. BS, U.S. Naval Acad., 1946; MD, U. Calif., San Francisco, 1956; MPH and Tropical Medicine, Tulane U., 1961. Diplomate AM. Bd. Internal Medicine. Intern U. Hosp., San Diego, 1956-57; resident Monterey County Hosp., 1957-58; dir. of med. svcs. U.S. Depart. Interior, Am. Samoa, 1958-60; instr. Tulane U. Med. Sch., New Orleans, 1960-63; resident Tulane Svcs. V.A. and Charity Hosp., New Orleans, 1961-63; cons. Internat. Ctr. for Rsch and Tng., Costa Rica, 1961-63; asst. prof. medicine & preventive medicine La. State U. Sch. Medicine, 1962-63; assoc. prof., 1963-65; vis. prof. U. Costa Rica, 1963-65; faculty advisor, head of Agy. Internat. Devel. program U. Costa Rica Med. Sch., 1965-67; dir. med. svcs. Med. Ctr. U. Calif. (formerly Orange County Hosp.), Irvine, 1967-69; assoc. clin. prof. U. Calif., Irvine, 1967-79, clin. prof., 1980-85; pvt. practice Tustin, Calif., 1970-80; with Joint Commn. on Accreditation of Hosps., 1981; cons. Kauai, Hawaii, 1981—. Author: The Middle Years, 1981, Practice Management, 1986, Kauai, 1988, 3d edit., 1990; co-translator; Rashomon (Ryonosuke Akutagawa), 1950; also numerous articles. Lt. USN, 1946-52, capt. USNR, 1966-84. Fellow Am. Coll. Physicians (life); mem. Delta Omega. Republican. Episcopalian. Home: 4890 Lawai Rd Koloa HI 96756

DONOVAN, BRIAN JOSEPH, maritime industry executive; b. Paterson, N.J., Feb. 12, 1953; s. John Harold and Helen (Cheevers) D.; m. Rachael Cecile Couvillon, Jan. 16, 1982; children: Meaghan Marie, Michael John. Student, Villanova U., 1970-71; BS in Marine and Nuclear Engring., U.S. Merchant Marine Acad., 1975. Lic. chief engr. USCG. Marine engr. J. Ray McDermott, U.K., 1975-77; sr. project mgr. Offshore Logistics, Inc., various fgn. cities, 1977-82; prin. B. Donovan and Assocs., Inc., Lafayette, La., and Riyadh, Saudi Arabia, 1982—; chmn., chief exec. officer Internat. Drilling and Exploration, Inc., Lafayette, Montevideo, Uruguay, 1987—; del. U.S.-China Joint Session on Industry, Trade, and Econ. Devel., China, 1988. Author: Vessel Preservation, 1986; patentee oil and gas well blowout suppression system. Mem. Am. Soc. Naval Architects and Marine Engrs., U.S. Mcht. Marine Acad. Alumni Assn. Roman Catholic. Avocations: golf, skiing, karate, world travel. Home: 104 Canada Dr Lafayette LA 70506-6752 Office: INDEX Inc PO Box 30286 Lafayette LA 70593-0286

DONOVAN, JOHN EDWARD, psychologist; b. Englewood, N.J., Sept. 25, 1949; s. Francis James Jr. and Genevieve Frances (Keller) D.; m. Edith Kavanaugh Lightner, Aug. 18, 1972; children: Carrie Elizabeth, Sean Patrick, Colin James, Keegan Michael. BA magna cum laude, U. Colo., 1971, MA, 1974, PhD, 1977. Rsch. assoc. Inst. Behavioral Sci. U. Colo., Boulder, 1977-92; assoc. prof. psychiatry Sch. Medicine U. Pitts., 1992—; mem. alcohol psychosocial rsch. rev. com. Nat. Inst. Alcohol Abuse and Alchoholism, Rockville, Md., 1985-89. Co-author: Beyond Adolescence: Problem Behavior and Young Adult Development, 1991. Grantee Nat. Inst. Drug Abuse, 1984, Nat. Inst. Alcohol Abuse and Alcoholism, 1989. Mem. Rsch. Soc. Alcoholism. Achievements include research in psychosocial and behavioral correlates and antecedents of problem behavior in adolescence. Office: U Pitts Sch Medicine Psychiatry Dept 3811 O'Hara St Pittsburgh PA 15213

DONOVAN, LAWRENCE, physicist; b. Pitts., Apr. 8, 1952; s. William Ross and Nellie Joanne (Dobrowolski) D.; m. Patricia Louise Urban, Jan. 11, 1980. MA in Health Svcs. Mgmt., Webster U., 1982; MS in Physics, Southwest Tex. State U., 1989. Asst. health physicist Radiation Safety Office U. Pitts., 1975-78; radiol. physicist Mideast Ctr. Radiol. Physics, Pitts., 1978-80; commd. 2d lt. USAF, 1980, advanced through grades to maj., 1987; staff health physicist radioisotope com. USAF, Brooks AFB, Tex., 1985-90, chief licensing actions, 1990-92, chief radiation dosimetry Armstrong Lab., 1992—; presenter at profl. confs. Lector St. John Neumann Cath. Ch., San Antonio, 1986—, eucharistic minister, 1987—, organist, 1991—. Mem. Am. Assn. Physicists in Medicine, Health Physics Soc., Am. Nat. Standards Inst. (working group on performance requirements for pocket sized alarm dosimeters and alarm ratemeters). Democrat. Home: 8750 Ridgemoon Dr San Antonio TX 78239 Office: Armstrong Lab OEBD Brooks AFB TX 78235

DONOVAN, STEPHEN JAMES, mechanical engineer; b. Dorcester, Mass., Mar. 15, 1951; s. Timothy S. and Helen (Wapleton) D.; m. Jeanne M. Donovan, July 23, 1983; children: Kelly, Richard. BSME, Worcester Poly. Inst., 1972; MSME, Navy Postgrad. Sch., Monterey, Calif., 1973. Registered profl. engr., Mass. Maintenance engr. Tex. Instruments, Attleboro, Mass., 1976-79; mech. test engr. Stone and Webster Engring. Corp., Boston, 1979-90; prin. engr. Cleary Sta., Taunton (Mass.) Mcpl. Lighting Plant, 1990—. Chmn. zone 4 Boston Area Navy Recruiting Dist. Assistance Coun., 1991-92. With USN, 1973-76, comdr. USNR, 1976—. Mem. Am. Inst. Plant Engrs. (tour guide), Am. Soc. Naval Engrs., Barbershop Quartet, Barbershop Chorus. Home: 1526 West St Attleboro MA 02703 Office: Taunton Mcpl Light Plant PO Box 870 Taunton MA 02780-0870

DONSKOY, DIMITRI MICHAILOVITCH, physicist, researcher; b. Gorky, Russia, July 12, 1955; came to U.S., 1990; s. Michael V. and Anna I. Donskoy; m. Tatyana E. Selezneva, Feb. 8, 1986; 1 child, Nina D. BS/MS, Gorky State U., 1977; PhD, Inst. Applied Physics, USSR Acad Scis., 1984. Researcher Inst. Applied Physics/USSR Acad. Scis., Gorky, 1977-88, sr. rsch. scientist, 1988-90; sr. scientist Stevens Inst. Tech., Hoboken, N.J., 1991—; lectr. Gorky State Univ., 1985-90, Stevens Inst. Tech., 1992—. Contbr. articles to Soviet Physics-Acoustics, Soviet Physics-Doklady, Acoustic Letters U.K., Jour. Acoustical Soc. of Am. Recipient nat. prize for rsch. in nonlinear acoustics USSR Acad. Sci., Moskow, 1987. Mem. Acoustical Soc. Am. Achievements include six U.S. and USSR patents in field. Office: Stevens Inst Tech Castle Point Sta Hoboken NJ 07030

DOO, YI-CHUNG, aerospace engineer; b. Taipei, Taiwan, China, July 17, 1954; came to U.S., 1971; s. Yung Zung and Lein Y. (Wang) D.; m. Pei Syan Liu, Oct. 1, 1982; 1 child, Alex Deluen. BS, U. Md., 1977; MS, PhD, Mass. Inst. Tech., 1983. Product engr. Western Electric Co., North Andover, Mass., 1977-78; sr. engr. AVCO/Systems Div., Wilmington, Mass., 1982-89; mem. tech. staff Aerospace Corp. Fluid Mechanics Dept., El Segundo, Calif., 1984-89; mgr. The Aerospace Corp., El Segundo, Calif., 1989—. Contbr. articles to profl. jours. Grad. fellow Thomas Electric, 1982. Mem. AIAA (sr. mem.), ASME, Tau Beta Pi, Pi Tau Sigma, Sigma Xi. Achievements include research interests include computational fluid dynamics, turbulent flows, rarfied gas dynamics, fluid mechanics. Office: The Aerospace Corp PO Box 92957 M4/964 Los Angeles CA 90009

DOODY, DANIEL PATRICK, pediatric surgeon; b. Evergreen Park, Ill., July 19, 1952; s. Francis Xavier and Mary Therese (Neylon) D.; m. Scarlet Beverly Artruc, Nov. 28, 1981; children: Colin James, Shaylyn Claire, Evan Patrick. BS, U. Ill., Urbana, 1973; MD, U. Ill., Chgo., 1977. Intern surgery U. Ill., Chgo., 1977-78, resident surgery, 1978-79; rsch. fellow Mass. Gen. Hosp., Boston, 1979-81; resident surgery U. Ill./Cook County Hosps., Chgo., 1981-83, chief resident surgery, 1983-84; rsch. fellow Mass. Gen. Hosp., Boston, 1984-85, pediatric surgeon, 1987—; resident pediatric surgery Montreal (Que., Can.) Children's Hosp., 1985-87; instr. advanced trauma life support, Boston, 1988—, pediatric ALS, Boston, 1990—; asst. prof. surgery Harvard Med. Sch., 1990. Contbr. articles on basic sci., pediatric and pediatric surgery to profl. jours., chpts. to books. Recipient Golden Apple award U. Ill. Sch. Medicine, Chgo., 1984. Fellow ACS, Am. Acad. Pediatrics; mem. Am. Pediatric Surg. Assn., New Eng. Pediatric Surg. Soc., Warren Cole Soc., Karl Meyer Soc., Pediatric Oncology Group, Soc. Critical Care Medicine. Avocations: pencil and ink sketching, photography, oenology, phys. fitness. Home: 2 Fletcher Rd Lynnfield MA 01940 Office: Mass Gen Hosp Fruit St Boston MA 02114

DOORISH, JOHN FRANCIS, physicist, mathematician; b. Bklyn., Jan. 13, 1957; s. Thomas Joseph Anthony and Annunciata Ann (Longobardi) D. BS in Physics, St. John's U., 1980; MS in Applied Physics, Columbia U., 1985, EdD in Math. and Astrophysics, 1988. Rsch. physicist N.Y.C. Bur. Noise Abatement Dept. Environ. Protection, 1981; adj. asst. prof. Boro. Manhattan Community Coll., N.Y.C., 1991—; assoc. rsch. scientist Princeton U., N.J., 1992-93. Contbr. articles to profl. jours. and internat. sci. confs. Mem. ASCPA, N.Y.C., Internat. Fund Animal Welfare, Boston. St. John's scholar. Mem. AAAS, Am. Astronomical Soc., N.Y. Acad. Sci., Planetary Soc. Republican. Roman Catholic. Office: Manhattan CC/CUNY Dept Scis 199 Chambers St New York NY 10007

DORAN, ROBERT STUART, mathematics educator; b. Winthrop, Iowa, Dec. 21, 1937; s. Carl Arthur D. and Imogene (Ownby) Doran Nodurft; m. Shirley Ann Lange, June 27, 1959; children: Bruce Robert, Brad Christopher. BA with hons., U. Iowa, 1962, MA, 1964; MS, U. Washington, 1967, PhD, 1968. Instr. U. Wash., 1968; asst. prof. U. No. Iowa, Cedar Falls, 1968-69; asst. to prof. math. Tex. Christian U., Ft. Worth, 1969—, chmn. dept. math., 1990—; vis. prof. U. Tex., Austin, 1979; cons. in field. Author: Approximate Identities and Factorization in Banach Modules, 1979, Characterizations of C*-Algebras: The Gelfand-Naimark Theorems, 1986, Representations of Locally Compact Groups and Banach *-Algebraic Bundles, 1988; editor: Selfadjoint and Nonselfadjoint Operator Algebras and Operator Theory, 1991; contbr. articles to profl. jours. Editor: Cambridge U. Press, 1987—. Chmn. bd. deacons Birchman Bapt. Ch., Ft. Worth, 1987; vol. Van Cliburn Internat. Piano Competition, 1984—, Am. Cancer Soc., 1987—. Recipient Burlington No. award for disting. teaching, 1988, Top Ten Prof. award Ho. of Reps., 1986, 87, 91, Mortar Bd. Preferred Prof. award 1983, 87, 91, Gold medal for Prof. of Yr., Coun. for Advancement and Support of Edn., 1989, TCU Honors Prof. of Yr. award, 19993; vis. scholar MIT, 1981, Oxford U., 1988; Minnie Stevens Piper prof., 1989. Mem. Inst. Advanced Study (chmn. we. U.S. 1984—), Assn. Mems. Inst. for Advanced Study (pres. bd. trustees 1990—), Am. Math. Soc., Math. Assn. Am. (vis. lectr. 1990—, Beckenbach Book award prize com. 1990—), Phi Beta Kappa, Sigma Xi, Pi Mu Epsilon. Republican. Avocations: chess, running, swimming. Home: 4204 Ridglea Country Club Dr Fort Worth TX 76126-2224 Office: Tex Christian U Dept Math Fort Worth TX 76129

DORÉ, ROLAND, dean, science association director; b. Montreal, Que., Can., Feb. 16, 1938. Grad. in engring., Ecole Poly., Que., 1960; MSME, Stanford U., Que., 1965; PhD, Stanford U., 1969; hon. Can. Royal Military Coll., St.-Jean and McGill U., Can. Prof. mech. engring. Ecole Poly., Montreal, Que., Can., 1959-80, dir. dept., 1975-80, dean rsch., 1980-85, dean, dir., 1985-89, pres., 1989—; also chmn. bd. Ecole Poly.; pres. Can. Space Agency, St.-Hubert, Que., 1992—; cons. MLW-Worthington, Can. Vickers, Babcock and Wilcox, Dominion Bridge; v.p. Natural Scis. and Engring. Rsch. Coun. Can., lectr. internat. confs. Contbr. articles to profl. jour. Recipient Centré medal Centre Jacques Cartier, Grand prix de l'Excellence award Ordre des Ingénieurs du Québec, 1993. Mem. Nat. Rsch. Coun. Can. (adv. com.indusl. materials rsch. inst.), Engring. and Applied Scis. Can. (com. deans), Nat. Scis. and Engring. Rsch. Coun. Can. (v.p. 1988-92), Can. Soc. Mechanical Engring., Engring. Inst. Can. (Julian C. Smith medal 1992), Can. Acad. Engring., Ordre des Engénieurs du Québec, Order Can. (officer, 125th Birthday Commemorative medal 1992), Sigma Xi. Achievements include 30 major projects in applied engineering research and design. Office: Can Space Agency, 6767 route de l'Aéroport, Saint-Hubert, PQ Canada J3Y 8Y9

DOREMUS, ROBERT HEWARD, glass and ceramics processing educator; b. Denver, Sept. 16, 1928; s. Francis Heward and Elsie Marion (Segelke) D.; m. Germaine Briancon, Mar. 19, 1956; children—Marc Francis, Elaine, Carol, Natalie. B.S., U. Colo., 1950; M.S., U. Ill., 1951, Ph.D., 1953; Ph.D. (Fulbright fellow), U. Cambridge, Eng., 1956. Phys. chemist Gen. Electric Research and Devel. Ctr., Schenectady, 1956-71; N.Y. State prof. glass and ceramics Rensselaer Poly. Inst., Troy, N.Y., 1971—; cons. in field. Author: Glass Science, 1973, Rates of Phase Transformations, 1985. Co-editor: Growth and Perfection of Crystals, 1958; Contbr. articles to profl. jours. Bd. dirs. Phila. Luth. Sem., 1967-76. Fellow Am. Ceramic Soc.; mem. AAAS, Sigma Xi, Sigma Tau, Tau Beta Pi. Lutheran. Home: 1544 Keyes Ave Niskayuna NY 12309-5116 Office: Materials Dept Rensselaer Poly Inst Troy NY 12181

DORFF, GERALD J., physician; b. Milw., May 30, 1938; s. Joseph Louis and Mary Olive (La Perriere) D.; m. Sandra Jeanne Geyser, June 13, 1959; children: Elizabeth, Carol, Gregory, Anna, Gary, Joseph, Donald. BS, Marquette U., 1960, DMS, 1964. Diplomate Am. Bd. Internal Medicine, Am. Bd. Infectious Diseases. Intern Luth. Hosp. of Milw., 1964-65; svc. unit dir. U.S. Pub. Health Svc., Bellcourt, N.D., 1965-67; resident in internal medicine Milw. County Gen. Hosp., 1967-69, fellow in infectious diseases, 1969-70; with Harwood Med. Assocs., S.C., Wauwatosa, Wis., 1971-90; chief sect. infectious disease and rheumatology Deaconess Hosp., Milw., 1971-85, acting dir. family practice residency, 1973-74, clin. cons. microbiology, 1977-83; hosp. epidemiologist St. Joseph's Hosp., Milw., 1979—, chief sect. infectious diseases, hosp. epidemiologist, 1982-90, chmn. dept. medicine, 1987—; staff physician outpatient Milw. County Tb Clinic, 1974-78; asst. clin. prof. Med. Coll. Wis., Milw., 1974-80, assoc. clin. prof., 1981-87, clin. prof., 1988—; med. cons. Milw. County Zoo, 1981—. Fellow ACP; mem. Am. Fedn. Clin. Rsch., Am. Soc. Microbiology, Wis. Soc. Internal Medicine, Infectious Disease Soc. Milw. (pres. 1978-79, treas. 1978—), Infectious Disease Soc. Am., Milw. Internist Club. Achievements include research in causes of pneumonia in hospitals, immunological methods for detection of infectious diseases, and management of hospital infections. Home: 16035 Burleigh Pl Brookfield WI 53005 Office: Infectious Diseases Specialists SC 3070 N 51st St Ste # 208 Milwaukee WI 53210

DORGAY, CHARLES KENNETH, chemical engineer; b. Wilmington, Del., Feb. 6, 1956; s. Charles and Vivian Elvira (Freeman) D. BS in Chem. Engring., U. S.C., 1984, postgrad. Engr. S.C. Dept. Health and Environ. Control, Columbia, 1989—. 2d lt. U.S. Army N.G., 1987-90. EPA fellow, 1992. Mem. Phi Beta Kappa. Home: PO Box 12594 Columbia SC 29211 Office: SC Dept Health Environ Control 2600 Bull St Columbia SC 29201

DORIS, PETER A., biomedical scientist; b. Durham, Eng., July 1, 1956; came to U.S., 1976; m. Kinga Elzbieta Nurowska. BA, U. Calif., Riverside, 1979, PhD, 1981. Rsch. fellow M.R.C. Dunn clin. nutrition unit U. Cambridge, Eng., 1982-83; rsch. fellow U. Reading, Eng., 1982-83; asst. prof. Tex. Tech. U. Sch. Medicine, Lubbock, 1983-89, assoc. prof., 1989—. Recipient grant NIH, Am. Heart Assn. Mem. Am. Physiol. Soc. Office: Tex Tech Univ Sch of Medicine 3601 4th St Lubbock TX 79430

DORKO, ERNEST ALEXANDER, chemist, researcher; b. Detroit, Sept. 16, 1936; s. John and Julia Anne (Pala) D.; m. Betty Jane Kurtz, June 18, 1971; 1 child, Thomas. BSChE, U. Detroit, 1959; MS, U. Chgo., 1961, PhD, 1964. Rsch. chemist U.S. Army Missile Command, Huntsville, Ala., 1964-67; prof. chemistry Air Force Inst. Tech., Dayton, Ohio, 1967-86; rsch. chemist Phillips Lab., Albuquerque, 1986—; adj. prof. Air Force Inst. Tech., 1986—. Contbr. articles to Jour. Chem. Physics, Jour. Phys. Chemistry, Chem. Physics Letters, Tetrahedron, Jour. Am. Chem. Soc. Mem. Am. Chem. Soc., AAAS, K.C. (4th degree), Tau Beta Pi, Sigma Xi. Roman Catholic. Achievements include patent for deuteration of alcohols; research in spectroscopy, kinetics, lasers, synthesis of novel compounds, flow tube reactors. Office: Phillips Lab PL/LIDB 3550 Aberdeen Ave SE Albuquerque NM 87117-5776

DORMAN, CRAIG EMERY, oceanographer, academic administrator; b. Cambridge, Mass., Aug. 27, 1940; s. Carlton Earl and Sarah Elizabeth (Emery) D.; m. Cynthia Eileen Larson, Aug. 25, 1962; children: Clifford Ellery, Clark Evans, Curt Emerson. BA, Dartmouth Coll., 1962; MS, Navy Postgrad. Sch., 1969; PhD, MIT/WHOI Joint Prog. Oceanog., 1972. Commd. ensign USN, 1962, advanced through grades to rear admiral, 1987, ret., 1989; CEO Woods Hole (Mass.) Oceanographic Instn., 1989—; dir. Maritrans., Phila.; gov. Joint Oceanographic Insts., Washington, 1989—; advisor Dir. Naval Intelligence, Washington, 1989—. Trustee Mass. Maritime Acad., Buzzards Bay, 1989—; dir. Mass. Ctrs. Excellence Corp., Boston, 1989—; mem. MITRE Tech. Adv. Panel. Decorated Legion of Merit (2). Mem. N.Y. Yacht Club, St. Botolph Club, Cosmos Club. Episcopalian. Home: PO Box 164 Woods Hole MA 02543-0164 Office: Woods Hole Oceanographic Instn Woods Hole MA 02543

DORNBURG, RALPH CHRISTOPH, biology educator; b. Nuernberg, Fed. Republic Germany, Aug. 12, 1952; came to U.S., 1986; s. Robert and Freia (Puchtler) D.; m. Ute Pietrass, Aug. 12, 1980; children: Alex, Rebecca. Diploma, Ludwig-Maximilians U., Munich, 1982; postgrad., Max-Planck Inst. Biochemistry, Martinsried, Fed. Republic Germany, 1982-86; PhD, U. Munich, 1986. Deutsche Forschungs-Gemeinschaft postdoctoral fellow McArdle Lab. for Cancer Rsch., Madison, Wis., 1986-89; asst. prof. U. Medicine and Dentistry of N.J., Piscataway, 1989—. Author: Encyclopedia of Human Biology, 1991; contbr. articles to Molecular and Cellular Biology, Jour. Virology, others. Achievements include patents for targeted gene transfer and self-inactivating retroviral vectors. Office: U Medicine & Denistry NJ RWJMS 675 Hoes Ln Piscataway NJ 08854-5635

DORNFELD, DAVID A., engineering educator; b. Horicon, Wis., Aug. 3, 1949; s. Harlan Edgar and Cleopatra D.; Barbara Ruth Dornfeld, Sept. 18, 1976. BS in Mech. Engring. with Honors, U. Wis., 1972, MS in Mech. Engring., 1973, PhD in Mech. Engring., 1976. Asst. prof. dept. Systems-Design U. Wis., Milw., 1976-77; asst. prof. Mfg. Engring. U. Calif., Berkeley, 1977-83, assoc. prof. Mfg. Engring., 1983-89, vice-chmn. instrn. dept. Mech. Engring., 1987-88, dir. Engring. Systems Rsch. Ctr., 1989—, prof. Mfg. Engring., 1989—; assoc. dir. rsch. Ecole Nationale Superieure des Mines de Paris, 1983-84; invited prof. Ecole Nationale Superieure D'Arts et Metiers, Paris, 1992-93. Contbr. articles to profl. jours., chpts. in books; presenter numerous seminars, confs.; patentee in field. Fellow ASME (past editor, mem. editorial bd. Manufacturing Review Jour., chair honors com. prodn. engring. divsn., press advisory com., Blackall Machine Tool and Gage award 1986), Soc. Mfg. Engrs. (editorial bd. Jour. Mfg. Systems, Outstanding Young Engr. award 1982); mem. IEEE, Am. Welding Soc., Nat. Rsch. Coun. (unit mfg. process rsch. com.), Acoustic Emission Working Group, N.Am. Mfg. Rsch. Inst. (past pres., scientific com.), Japan Soc. Precision Engring., Coll. Internat. pour l'Etude Scientifique des Techniques de Production Mechanique (co-chair tool condition monitoring working group). Avocations: hiking, travelling, reading. Office: U Calif Dept Mech Engring Berkeley CA 94720

DORNING, JOHN JOSEPH, nuclear engineering, engineering physics and applied mathematics educator; b. Bronx, N.Y., Apr. 17, 1938; s. John Joseph and Sarrah Cathrine (McCormack) D.; m. Helen Marie Driscoll, July 27, 1963; children: Michael, James, Denise. B.S. in Marine Engring., U.S. Mcht. Marine Acad., 1959; M.S. (AEC fellow), Columbia U., 1963, Ph.D. (AEC fellow), 1967. Marine engr. U.S. Mcht. Marine, 1960-62; asst. physicist Brookhaven Nat. Lab., Upton, N.Y., 1967-69, assoc. physicist, group leader, 1969-70; assoc. prof. nuclear engring. U. Ill., Urbana, 1970-75, prof., 1975-84; Whitney Stone prof. nuclear engring., engring. physics and applied math. U. Va., Charlottesville, 1984—; NRC vis. prof. math. physics U. Bologna, Italy, 1975-76, 81, 85, 87; internat. prof. nuclear engring. Italian Ministry of Edn., 1983, 84, 86; physicist plasma theory group, div. magnetic fusion energy Lawrence Livermore (Calif.) Nat. Lab., 1977-78; cons. to U.S. nat. labs. and indsl. research labs., 1970—. Contbr. articles to various publs. Served as ensign USN, 1959-60. Recipient Ernest O. Lawrence award U.S. Dept. Energy, 1990. Fellow AAAS, Am. Phys. Soc., Am. Nuclear Soc. (Mark Mills award 1967); mem. Am. Soc. for Engring. Edn., (Glenn Murphy award 1988), Soc. Indsl. and Applied Math., N.Y. Acad. Scis., Sigma Xi. Office: U Va Reactor Facility Thornton Hall Charlottesville VA 22903-2442

DORRELL, VERNON ANDREW, engineering executive; b. Brownsville, Tex., June 13, 1932; s. Vernon A. Dorrell and Loma Alice (Evans) Recktenwald; m. Joan Elaine Lahar (div.); children: Donald Kent Dorrell, Lissa Sturdy Matthes; m. Lani Dean Tomberlin (div.); children: Amy Loma Dorrell, Alana Dean Dorrell, Allison Kathleen Dorrell-Diego; m. Margaret Anna Jones, July 5, 1974. BSEE, UCLA, 1956; MSEE, San Jose State U., 1980. Draftsman Thompson, Ramo, Woolridge, L.A., until 1956; engr. Telemetering Corp. of Am., Sepulveda, Calif., 1956-62; sub-contracts mgmt. North Am. Aviation, Downey, Calif., 1962-65; dir. mktg. Collins Radio Co., Newport Beach, Calif., 1965-70; dir. new bus. devel. Harris Radiation, Inc., Melbourne, Fla., 1970-73; engr., program mgr. Magnavox Rsch. Labs., Torrance, Calif., 1973-74; engr. Lockheed Missiles & Space Co., Sunnyvale, Calif., 1975-80; engr., program mgr. Martin Marietta, Denver, 1980-92; pres. engring./rsch. Sedona (Ariz.) Scientific, 1991—; mem. Ariz. Innovation Network. Mem. Am. Soc. Photogrammetry and Remote Sensing, Armed Forces Comms. and Electronics Assn. Office: Sedona Scientific Ste 3D 1785 W Hwy 89A Sedona AZ 86336-5559

DORROS, IRWIN, retired telecommunications executive; b. Bklyn., Oct. 3, 1929; s. Harry and Irene (Shapiro) D.; m. Janet Eve Levine, Oct. 1, 1930; children: Robert, Mark, Gail, Gerald. BS, MS, MIT, 1956; D in Engring. Sci., Columbia U., 1962. Exec. dir. Bell Telephone Labs., Holmdel, N.J., 1956-78; asst. v.p. AT&T, Basking Ridge, N.J., 1978-83; exec. v.p. tech. svcs. Bell Communications Rsch. (Bellcore), Livingston, N.J., 1984-93; bd. dirs. Microelectronics & Computer Tech. Corp., Austin, Tex., Vertex Industries, Clifton, N.J.; mem. com. on electronic mail NRC, 1979-81, mem. telecommunication bd. computer applications, 1980-82. Contbr. numerous articles to profl. jours.; holder 5 patents in telecommunication circuits and systems. Bd. dirs. Fair Haven (N.J.) Sch. Bd., 1972-75; trustee Congregation B'nai Israel, Rumson, N.J., 1979-82; mem. Jewish Community Housing Bd., 1991—. With U.S. Army, 1950-51. Recipient N.J. Sci. & Tech. medal N.J. R&D Coun., 1992. Fellow IEEE (Leadership recognition 1990, Founders medal 1990); mem. NAE, IEEE Comm. Soc. (policy bd. 1976-82), Fairmont

Country Club. Republican. Jewish. Avocations: golf, photography, gardening, philately, oenophile.

DORSETT, CHARLES IRVIN, mathematics educator; b. Lufkin, Tex., Sept. 25, 1945; s. C.B. and Dorothy Alice (Smith) D. BS, Stephen F. Austin State U., Nacogdoches, Tex., 1967, MS, 1968; PhD, N. Tex. State U., 1976. Cert. secondary sch. tchr., Tex. Teaching fellow Stephen F. Austin State U., 1967-68, instr., 1968-71; teaching fellow North Tex. State U., Denton, 1971-76, lectr., 1976-77, 78-79; asst. prof. La. Tech U., Ruston, 1977-78, assoc. prof., 1982-90, prof., 1990—; lectr. Tex. A&M U., College Station, 1979-82; reviewer Zentralblatt Für Mathematik und Ihre Grenzgebrite, 1983—; referee Indian Jour. Pure and Applied Math., 1986—, Indian Jour. Math., 1986—, Bull. of the Faculty of Sci., Assiut Univ.; Physics and Math., 1986—, Glasnik Matematicki, 1988—, Bull. of the Malaysian Math. Soc., 1989—; mem. bd. editors Pure Mathematics Manuscript, 1987—. Contbr. articles to profl. publs. Recipient Cert. for Excellence in Rsch., La. Tech U., 1984-85, La. Tech. Sigma Xi Rsch. award, 1987. Mem. Am. Math. Soc., Internat. Platform Assn., Indian Acad. Math., Bharata Ganita Parisad, Sigma Xi. Baptist. Avocations: mathematical rsch., farming. Home: 402 W Arizona Ave Lot 20 Ruston LA 71270-4362 Office: La Tech U Dept Math and Stats Ruston LA 71272

DORSEY, JAMES BAKER, surgeon, lawyer; b. Saratoga Springs, N.Y., Aug. 29, 1927; s. Francis Edward and Katherine (Baker) D.; m. Patricia Ann Walsh, June 10, 1950; children: Katherine, Mary Lee, Pamela, Suzanne, James B., Jr., Alison. BA, Brown U., 1949; LLB, Union U. Sch., 1952, JD, 1991; MD, N.Y. Med. Coll., 1957. Bar: N.Y. 1953, Mass. 1988, U.S. Supreme Ct. 1982; lic. physician, N.Y., Mass., Calif. Intern Greenwich Hosp., Conn., 1957-58; resident White Plains Hosp., N.Y., 1958-59, Lenox Hill Hosp., N.Y.C., 1961-64; chmn. dept. surgery Saratoga Hosp., Saratoga Springs, 1976-79, 85-87; cons. surgeon Wesley Nursing Homes, Saratoga Springs, 1964—. Bd. dirs. Saratoga YMCA, Saratoga Springs, 1971-72; pres. Saratoga Springs Hist. Soc., 1972-74. Diplomate Am. Bd. Surgery. Fellow ACS, Am. Coll. Legal Medicine; mem. Saratoga County Bar Assn., Saratoga County Med. Soc. (pres. 1982-85), Med. Soc. N.Y., Mass. Med. Soc., N.Y. State Bar Assn., Mass. Bar Assn., AMA. Republican. Roman Catholic. Lodge: Elks, K.C. Office: 112 S Broadway Adirondack Trust Bldg St 1 Saratoga Springs NY 12866

DORSEY, WILLIAM WALTER, aerospace engineer, engineering executive; b. Long Branch, N.J., Dec. 23, 1934; s. Walter Gorman and Esther (Smith) D.; m. Lorraine Shirley Sanders, June 26, 1962; children: William W., Suzanne E. BSME, George Washington U., 1958, MS in Engring., 1965. Aerospace engr. Nat. Bur. Standards, Washington, 1960-65; sr. engr. Fairchild Hiller Corp., Germantown, Md., 1965-69; spacecraft mgr. European Space Agy., Noordwijk ann Zee, Holland, 1970-76; prin. engr. Fairchild Industries, Germantown, 1977-79; mem. tech. staff INTELSAT, Washington, 1979-85; dir. engring. Fairchild Space & Def. Co., Germantown, 1985—. Contbr. articles to sci. jours. Capt. USAF, 1958-60. Mem. AIAA, ASME. Achievements include development of unique design for the deployment control of the GEOS spacecraft 20 meter cable boom, of an analytical approach to the station keeping problem of colocating communication satellites in the same orbital location; design of numerous spacecraft thermal control subsystems including ATS-F, SERT-II, NIMBUS-D, and IMP-I using large computer programs; management of design, development and manufacture of instrument module for TOPEX Scientific Spacecraft. Home: 11832 Goya Dr Rockville MD 20854-3307 Office: Fairchild Space & Defense Co 20301 Century Blvd Germantown MD 20874-1181

DORST, HOWARD EARL, entomologist; b. Pomeroy, Ohio, Sept. 19, 1904; s. Otto Henry and Clara Barbara (Kautz) D.; m. Martha Hauserman, Aug. 1, 1931; 1 child, Ronald Valison. AB, U. Kans., 1929, MA, 1930; DHL (hon.), Westminster Coll., 1984. Jr. entomologist USDA, Richfield, Utah, 1929; asst. entomologist USDA, Salt Lake City, 1930-36; commd. 2d lt. USAR, 1932; assoc. entomologist USDA, Logan, Utah, 1936-41,; 1932; advanced through grades to col. USAR, 1954; commd. capt. SN Corps, 1941; retired USAR, 1964; ret. lt. col. MSC, 1946; sr. entomologist USDA, Logan, Utah, 1946-65, ret., 1965; emeritus prof. zoology Utah State U., Logan, 1966. Contbr. articles to Jour. Econ. Entomology, Proceedings Am. Soc. Sugar Beet Tech. Chmn. Am. Cancer Soc. Cache County, Logan, 1967-71, ARC Cache County, Logan, 1985-86; moderator Presbytery Utah, Logan, 1980. Served to col. Med. Svc. Corps. USAR, 1964. Recipient Presdl. Citation award Utah State U., 1988; named to Old Main Soc., Utah State U., 1990. Republican. Presbyterian. Achievements include inventions of forecast methods for migrations of insects. Home: 1679 E 1030 N Logan UT 87321

DORST, NEAL MARTIN, meteorologist, computer programmer; b. Wayne, Nebr., July 17, 1955; s. Claire Vanderhoof and Mary Fairchild (Crowe) D. BS, Fla. State U., 1977; BFA, Fla. Atlantic U., 1981. Meteorologist NOAA/Hurricane Rsch. Divsn., Miami, Fla., 1979—. Sci. advisor Mus. of Discovery & Sci., Ft. Lauderdale, Fla., 1990-92; display designer Aldrin Planetarium & Mus., West Palm Beach, Fla., 1971. Mem. AAAS, Am. Meteorol. Soc. Republican. Congregationalist. Home: 618 NW High St Boca Raton FL 33432 Office: AOML/Hurricane Rsch Divsn 4301 Rickenbacker Cswy Miami FL 33149

DORWART, BRIAN CURTIS, geotechnical engineer, consultant; b. Corning, N.Y., June 9, 1949; s. Robert Moris and Nancy (Anderson) D.;m. Dana Joy Wallace, Dec. 8, 1979; children: Kelsey, Casey, Keeley, MacKenzie. BA in Geology, U. Rochester, 1972; MSCE, U. Mass., 1979. Registered profl. engr., Mass., N.H., Maine, Conn. Field technician, driller Rochester (N.Y.) Drilling Co., 1972-75; engr., sr. project mgr. GZA GeoEnviron. Inc., Newton, Mass., 1979-87; sr. project mgr. GZA GeoEnviron Inc., Manchester, N.H., 1987-91; assoc. Shannon & Wilson, Inc., Seattle, 1991—. Mem. ASTM, ASCE. Republican. Home: PO Box 92 Kingston WA 98346 Office: Shannon & Wilson Inc Box 300303 400 N 34th St Seattle WA 98106

DOSS, EZZAT DANIAL, mechanical engineer, researcher; b. Cairo, Egypt, Apr. 2, 1945; s. Danial Doss and Evon Nicola; m. Nadia M. Tanagho, July 18, 1968; children: Erini, Chistine, Angela. BSME, Cairo Univ., Egypt, 1965; MSME, Cairo U., Egypt, 1968; PhD in Engring., U. Calif., Davis, 1972. Registered profl. engr., Ill. Teaching asst. Cairo Univ., 1965-68; teaching and rsch. asst. Univ. Calif., Davis, 1968-71; postdoctoral researcher Univ. Calif., 1971-72; rsch. scientist and engr. STD Rsch. Corp., Arcadia, Calif., 1972-76; engr., group leader Argonne (Ill.) Nat. Lab., 1976-86; engr., sect. mgr. Argonne Nat. Lab., 1986—; cons. Nat. Inst. Stds. and Tech., U.S. Dept. Commerce, Gaithersburg, Md., 1988-89. Contbr. more than 80 articles to profl. jours. Mem. sch. bd., sec. and chmn. facilities com. Hinsdale (Ill.) High Sch. Dist., 1985—; treas., bd. dirs. St. Mary Orthodox Ch., Northbrook, Ill., 1985-88; Sun. sch. tchr., religious edn. dir. Orthodox Ch., Calif., Ill., 1974—. Mem. AIAA (plasma dynamics tech. com. 1979-83), ASME (Best Tech. Presentation 1990), Sigma Xi. Achievements include research in Magnetohydrodynamic advanced power generation and seawater propulsion systems, plasma and fluid dynamics, heat transfer, energy systems, and computational fluid dynamics. Home: 817 Comstock Ln Darien IL 60561 Office: Argonne Nat Lab 9700 S Cass Ave Argonne IL 60439

DOTSON, GERALD RICHARD, biology educator; b. Brownsville, Tex., Sept. 8, 1937; s. Jasper William and Mary Agnes (Courtney) D.; m. Rose Dolores Gonzales; children: Roberta Ana, Deborah Irene, Matthew Charles. BS, Coll. Santa Fe (N.Mex.), 1960; MS, U. Miss., 1966; PhD, U. Colo., 1974; postgrad., U. Tex., Loyola U. New Orleans, El Paso, 1960-61, Loyola U., New Orleans, 1962-63. Sci. tchr. Cathedral High Sch., El Paso, Tex., 1959-61; sci./math./music tchr. St. Paul's High Sch., Covington, La., 1961-62; sci./math./Spanish tchr. Christian Bros. Sch., New Orleans, 1962-63; sci. tchr., chmn. Hanson High Sch., Franklin, La., 1963-67; biology instr. Coll. Santa Fe (N.Mex.), 1967-69, U. Colo., Boulder, 1969-70, Community Coll. Denver, 1970-77; prof. biology and chmn. sci. Front Range Community Coll., Westminster, Colo., 1977—; mem. com. for teaching excellence FRCC in Westminster, 1988—, mem. curriculum devel. com., 1980—, mem. acad. standards com., 1980—. Reviewer biology text books, 1970—; contbr. articles to profl. jours. Mem. recreation dept. City of Westminster, 1971—. Named Master Tchr. of the Yr., Front Range Community Coll., 1985-86. Mem. Am. Microscopical Soc., Am. Soc. Limnology and Oceanography, Nat. Assn. Biology Tchrs., Nat. Sci. Tchrs. Assn. (regional sec. 1985), Human Anatomy and Physiology Soc., Eagles, KC (3rd and 4th deg.), Elks, Sigma Xi, Phi Sigma. Roman Catholic. Avocations: fishing, hunting, camping, golf, bowling, walking. Home: 8469 Otis Dr Arvada CO 80003-1241 Office: Front Range Community Coll 3645 W 112th Ave Westminster CO 80030-2199

DOTT, ROBERT HENRY, JR., geologist, educator; b. Tulsa, June 2, 1929; s. Robert Henry and Esther Edgerton (Reed) D.; m. Nancy Maud Robertson, Feb. 1, 1951; children—James, Karen, Eric, Cynthia, Brian. Student, U. Okla., 1946-48; B.S., U. Mich., 1950, M.S., 1951; Ph.D. (AEC fellow), Columbia U., 1956. Exploration geologist Humble Oil & Refining Co., Ariz., Oreg., Wash., 1954-56, So. Calif., 1958; mem. faculty U. Wis.-Madison, 1958—, prof. geology, 1966-84, Stanley A. Tyler Disting. prof., 1984—, chmn. dept. geology and geophysics, 1974-77; vis. prof. U. Calif., Berkeley, 1969; Cabot Disting. vis. prof. U. Houston, 1986-87; NSF sci. faculty fellow Stanford U. and U.S. Geol. Survey, 1978, U. Colo., 1979; acad. visitor Oxford U., Imperial Coll., London, 1985-86, Adelaide U., Australia, 1992; cons. Roan Selection Trust, Ltd., Zambia, 1967, Atlantic-Richfield Co., 1983-85, Hubbard Map Co., 1984—; lectr. Bur. Petroleum and Marine Geology, People's Republic of China, 1986; Erskine fellow and vis. prof. Canterbury U., N.Z., 1987. Co-author: Evolution of the Earth, 5th edit. 1994; contbr. articles on sedimentology, tectonics, geology of So. Andes and history of geology to profl. jours. Served to 1st lt. USAF, 1956-57. Recipient Outstanding Tchr. award Wis. Student Assn., 1969, Ben H. Parker award Am. Inst. Profl. Geologists, 1992, William H. Twenhofel medal Soc. Sedimentary Geology, 1993. Fellow Geol. Soc. Am. (chmn. history of geology div. 1990, councilor 1992—); mem AAAS, Am. Assn. Petroleum Geologists (Pres.'s award 1956, Disting. Svc. award 1984, Disting. Lectr. 1985), Soc. Econ. Paleontologists and Mineralogists (sec.-treas. 1968-70, v.p. 1972-73, pres. 1981-82, hon. mem. 1987, Twenhofel medal 1993, del. S.E.P.M.), Internat. Assn. Sedimentologists, History of Earth Sci. Soc. (pres. 1990), Sigma Xi (Disting. lectr. 1988-89). Unitarian. Office: U Wis Dept Geology & Geophysics 1215 Dayton Madison WI 53706

DOTY, RICHARD LEROY, medical researcher; b. Boulder, Colo., Oct. 14, 1944; s. George David and Frances Amelia (Bradley) D. BS, Colo. State U., 1966; MA, Calif. State U., 1968; PhD, Mich. State U., 1971; postgrad., U. Calif., Berkeley, 1973. Instr. dept. psychology Calif. State U., San Francisco, 1971-72, U. San Francisco, 1971-72; asst. mem. Monell Chem. Senses Ctr., Phila., 1974-76, assoc. mem., head human olfaction sect., 1976-78; dir. smell and taste ctr. Hosp. U. Pa., Phila., 1979—; sci. dir. clin. smell and taste rsch. ctr. Sch. Medicine, U. Pa., Phila., 1980—, asst. prof. dept. otorhinolaryngology & human communication, 1983-89, assoc. prof. dept. otorhinolaryngology & human communication, 1989—; cons. in field; lectr. in field; editorial cons. for numerous profl. jours.; external adv. bd. taste and smell ctr. U. Conn./Yale U., 1982—, Rocky Mountain Taste and Smell Ctr., U. Colo. Sch. Medicine, 1985, Mayo Found. Project, 1989; internat. adv. bd. 1st Internat. Congress on Food and Health, Salsomaggiore Terme, Italy, 1985. Author: The Smell Identification Test (TM) Administration Manual, 1983, 2nd edit., 1989; editor: Mammalian Olfaction, Reproductive Processes and Behavior, 1976; co-editor: (with T.V. Getchell, E.P. Koster) Chemical Senses, spl. edit., 1981, (with D.G. Laing, W. Breipohl) Human Olfaction, 1990, (with L.M. Bartoshuk, T.V. Getchell and J.B. Snow) Smell and Taste in Health and Disease, 1991, (with D. Müller-Schwartze) Chemical Signals in Vertebrates VI, 1992. NIH postdoctoral rsch. fellow, 1973-75; grantee Nat. Inst. on Aging, 1989-91, Nat. Inst. Neurol. and Communicative Disorders and Stroke, 1980-93. Mem. European Chemoreception Rsch. Orgn. (organizational com. 1981), Assn. for Chemoreception Scis. (program com. 1985, 87, elections com. 1987), AAAS, N.Y. Acad. Scis., Assn. for Rsch. in Otolaryngology, Am. Acad. Otolaryngology (head and neck surgery), Am. Psychol. Assn., Internat. Soc. for Chem. Ecology, Phila. Coll. Physicians (adv. com., sect. on geriatrics and gerontology). Home: 125 White Horse Pike Haddon Heights NJ 08035-1933 Office: U Pa Smell & Taste Ctr 5 Ravdin Bldg 3400 Spruce St Philadelphia PA 19104-4220

DOUBLEDEE, DEANNA GAIL, software engineer, consultant; b. Akron, Ohio, July 29, 1958; d. John Wesley and Elizabeth (Nellis) Doubledee; m. Philip Henry Simons, Jan. 1, 1986. BSc in Computer Sci., Ohio State U., 1981; MSc in Software Engring., Nat. U., Inglewood, Calif., 1988. Cons. Ohio State U., Columbus, 1980-81; engr. Ocean Systems div. Gould, Inc., Cleve., 1981-82, Aircraft div. Northrop Corp., Hawthorne, Calif., 1982-83; tech. staff SEDD, TRW, Inc., Redondo Beach, Calif., 1983-85; staff engr. MEAD, TRW, Inc., Redondo Beach, Calif., 1985-88, subproject mgr., 1988-89; project engr. SDD, TRW, Inc., Redondo Beach, Calif., 1989-91; CEO, pres. Innovatice Concepts Continuum, Redondo Beach, Calif., 1993—; dir. software engring. TWI Engring., Inglewood, Calif., 1991-92; cons. Microcosm, Inc., Torrance, Calif., 1990-91; sr. computer scientist IIT Rsch. Inst., 1993—; judge state sci. fair Ohio Acad. Sci., Columbus, 1988; active Orange County Venture forum, 1992, MIT Enterprise forum, Chgo., 1992-93. Chmn. bd. dirs. Fedn. of Presch. and Community Edn. Svcs. (Headstart), Carson, Calif., 1988-90, bd. dirs., 1990-91. Recipient award for outstanding vol. svc. Fedn. of Presch. and Community Edn. Ctrs., 1987; Exemplar Ohio Acad. Sci., 1987, 89, 90. Mem. IEEE, ACM, IEEE Computer Soc., Soc. Women Engrs. (awards chair 1987), Am. Astron. Soc. Avocations: scuba diving, flying, tennis, photography, travel.

DOUCETTE, PAUL STANISLAUS, environmental scientist; b. Everett, Mass., June 2, 1966; s. Alfred Joseph and Joan Katherine (Aylward) D. BS in Civil Engring., Merrimack Coll., North Andover, Mass., 1989. Mem. survey crew County of Essex Engring. Dept., Salem, Mass., 1985; jr. engr. Town of Burlington (Mass.) Engring. Dept., 1986; engring. aide Barnes & Jarnis, Inc., Boston, 1986; coop. engr. Camp, Drsser & McKee, Inc., Boston, 1987-89; environ. engr. Camp, Drsser & McKee, Inc., Cambridge, Mass., 1989-91; environ. scientist Briggs Assocs., Inc., Rockland, Mass., 1992—. Mem. ASCE. Roman Catholic. Office: Briggs Assocs Inc 400 Hingham St Rockland MA 02370

DOUDRICK, ROBERT LAWRENCE, research plant pathologist; b. Kansas City, Mo., May 29, 1950; s. Robert W. and Joanne (McLane) D.; m. Nancy S. Parsons, June 26, 1971; children: Scott R., Karey D. BS, U. Mo., 1972, MS, 1984; PhD, U. Minn., 1988. Instr. U. Minn., St. Paul, 1988, postdoctoral appointee, 1988-89; postdoctoral appointee So. Forest Experiment Sta. USDA Forest Svc., Gulfport, Miss., 1989-90; rsch. plant pathologist USDA Forest Svc. SFES, Gulfport, Miss., 1990—; cons. SymPol, Inc., Mpls., 1988—. Contbr. articles to profl. jours. Bd. dirs. Miss. Coast Fly Fishers, Gulfport, 1992; scoutmaster Troop 3, Boy Scouts Am., Columbia, Mo., 1983-84. Fellowship U. Minn., 1984, Potlatch Found., 1986. Mem. Am. Phytopathol. Soc. (membership forest pathology com. 1993-97), Genetics Soc. of Am., Mycol. Soc. Am., Sigma Xi. Achievements include rsch. in identifying mating types genes in mycorrhizal fungus genus, Laccaria, preparing genetic linkage map of the fusiform rust fungus. Office: USDA Forestry Scis Lab 1925 34th St PO Box 2008 GMF Gulfport MS 39505

DOUGAN, DEBORAH RAE, neuropsychology professional; b. Urbana, Ill., Jan. 22, 1952; d. Francis William and Barbara Belle (Ash) D. BA in Psychology, U. Ill., 1973; MA in Counseling, Gov.'s State U., 1978; PhD in Neuropsychology, Oreg. State U., 1982. Cert. psychol. assoc., Tex. Staff therapist Ozark Community Mental Health, Joplin, Mo., 1982-85; neuropsychol. cons. Tex. Commn. for the Blind, Austin, 1985-87; psychol. assoc. Warm Springs Rehab. Hosp., Gonzales, Tex., 1987-88, Rehab. Hosp. South Tex., Corpus Christi, 1988-89; psychosocial dir. New Medico Rehab. Ctr., Lindale, Tex., 1989-90; clin. coord. Rainbow Rehab. Ctrs., Ft. Worth, 1991-93; neuropsychologist Cypress Creek Rehab. Ctr., Houston, 1993—; predoctoral intern Ea. State Hosp., Vinita, Okla., 1981-82; Thia survivors group leader Corpus Christi Head Injury Chpt., 1988-89, Tyler (Tex.) Head Injury Chpt., 1989-90; survivors group leader Ft. Worth Head Injury Chpt., 1991-93. Mem. APA, Tex. Head Injury Assn., Toastmasters Internat. Avocations: jazzercise, computer, house plants. Home: 36 N Circlewood Glen The Woodlands TX 77381 Office: Cypress Creek Rehab Ctr 17750 Cali Dr Houston TX 77090

DOUGHERTY, ELMER LLOYD, JR., chemical engineering educator, consultant; b. Dorrance, Kans., Feb. 7, 1930; s. Elmer Lloyd and Nettie Linda (Anspaugh) D.; m. Joan Victoria Benton, Nov. 25, 1952 (div. June 1963); children—Sharon, Victoria, Timothy, Michael (dec.); m. Ann Marie Da Silva. Student, Ft. Hays State Coll., 1946-48; B.S. in Chem. Engring., U. Kans., 1950; M.S. in Chem. Engring., U. Ill., 1952, Ph.D. in Chem. Engring., 1955. Chem. engr. Esso Standard Oil Co., Baton Rouge, 1951-52; chem. engr. Dow Chem. Co., Freeport, Tex., 1955-58; research engr. Standard Oil of Calif., San Francisco, 1958-65; mgr. mgmt. sci. Union Carbide Corp., N.Y.C., 1965-68; cons. chem. engring. Stamford, Conn. and Denver, 1968-71; prof. chem. engring. U. So. Calif., Los Angeles, 1971—; cons. OPEC, Vienna Austria, 1978-82, SANTOS, Ltd., Adelaide, Australia, 1980—. Contbr. numerous articles to profl. jours. Mem. Soc. Petroleum Engrs. (Disting. mem., chmn. Los Angeles Basin sect. 1984-85, Ferguson medal 1964, J.J. Arps award 1989), Am. Inst. Chem. Engrs., Internat. Assn. Energy Economists, Inst. Mgmt. Sci. Republican. Clubs: El Niguel Country (bd. dirs. 1976-78) (Laguna Niguel, Calif.). Avocations: golf; poetry recitals. Home: 33531 Marlinspike Dr Monarch Beach CA 92629 Office: U So Calif Univ Park HED-306 Los Angeles CA 90089-1211

DOUGHERTY, HARRY MELVILLE, III, reliability, maintainability engineer; b. San Antonio, Aug. 7, 1959; s. Harry Melville Jr. and Wanda Lee (McLin) D. BS in Engring. Physics, U. Tulsa, 1982. Registered profl. engr., Tex., Okla. Reliability engr. design support engring. Tex. Instruments, Inc., Lewisville, 1982-93; reliability, maintainability engr., engring. support Allied-Signal Tech. Svcs. Corp., Colorado Springs, Colo., 1993—. Mem. NSPE, Am. Soc. Quality Control (cert. reliability engr. and quality engr.), Nat. Eagle Scout Assn., Lewisville Texins Assn. Radio Club (pres. 1990, v.p. 1991, 92), Sigma Pi Sigma. Home: 3525 Knoll Lane Apt 152 Colorado Springs CO 80917 Office: Allied-Signal Tech Svcs 1925 Aerotech Dr Ste 200 Colorado Springs CO 80916

DOUGHERTY, JAMES, orthopedic surgeon, educator; b. Lawrence, Mass., July 31, 1926; s. James A. and Maude D. (Dillard) D.; m. Rita Buchman; children: James (dec.), Charles, Janice, Jonathan, Christopher. BS, Trinity Coll., Hartford, Conn., 1950; MD, Albany Med. Coll., N.Y., 1951. Diplomate Am. Bd. Orthopaedic Surgery. Intern U. Chgo. Clinics, 1951-52, resident, 1951-56, instr., 1955-56; chmn. div. orthopaedic surgery SUNY, Syracuse, 1958-59; prof. clin. surgery Albany Med. Coll., 1960—; trustee, chmn. med. staff Albany Med. Ctr., 1987-89; trustee Albany Med. Ctr., 1993—; cons. Subacute Care Alternative Project, Washington. Contbr. articles to profl. jours. Mem. Bd. Edn. Ravena-Coeymans-Selkirk Central Schs., Ravena, N.Y., 1960-75; med. dir. N.Y. Sr. Games, 1986-89; trustee Schaeffer Meml. Libr., 1990-92, Albany Med. Ctr., 1993—; bd. dirs. Inst. for Study Aging, 1990—. With U.S. Army, 1944-46. Fellow Am. Acad. Orthopaedic Surgeons; mem. Crawford Cambell Soc. (founder, pres. 1978-84), Northeastern Regional Assn. Sports Medicine (chmn. 1984-89), Albany Med. Coll. Alumni Assn. (trustee 1990—, pres.-elect 1992—), Internat. Platform Assn., Alpha Omega Alpha, Sigma Psi. Baptist. Home: Onesquethaw Rd Feura Bush NY 12067 Office: 1 Executive Park Dr Albany NY 12203-3717

DOUGHERTY, PERCY H., geographer, educator; b. Kennett Square, Pa., Feb. 20, 1943; s. Percy H. Sr. and Anna (Cloud) D.; m. Anne Barbara Zinn, July 9, 1966; children: Thomas P., Robert J. BS in Geography, Biology, West Chester U., 1967, MEd in Phys. Geography, 1968; PhD in Phys. Geography, Geology, Boston U., 1980. Tchr. geography and earth sci. Plymouth (Pa.) Jr. High Sch., 1967-68; asst. prof. West Chester (Pa.) U., 1968-70, Trenton (N.J.) State Coll., 1972-77, CUNY, 1977-78, U. Cin., 1978-83; vis. prof. Ohio U., Athens, 1983-84; cons., vis. asst. prof. U. Ky., Lexington, 1984-85; assoc. prof. Kutztown (Pa.) U., 1985-90, prof. geography, 1990—. Contbr. articles to profl. publs. Chmn. Lower Macungie Twp. Planning Commn., 1991-92; bd. dirs. Sloans Valley Conservation Taks Force, Lexington, Ky.; chmn. comprehensive planning com. Joint Planning Commn. Lehigh and Northampton Counties, Pa., 1990—. NSF fellow, 1971, 80, 92, NASA fellow, 1981, NOAA fellow, 1982. Fellow Nat. Speleol. Soc. (life); mem. Assn. Am. Geographers (life, past officer), Mid. States div. Assn. Am. Geographers (pres., bd. dirs.), Delaware Valley Geog. Assn. (pres., bd. dirs.), Nat. Coun. Geographic Edn. (life), Am. Water Resources Assn., Am. Soc. Photogrammetry and Remote Sensing, Conf. Latin Am. Geography, Pa. Geog. Soc. (past v.p., bd. dirs.). Republican. Achievements include research on remote sensing, air photo, geomorphology, karst, climatic geomorphology, groundwater diffusion, water resources, geographic education, planning. Office: Kutztown Univ Dept Geography 115 Grim Hall Kutztown PA 19530

DOUGHERTY, RICHARD MARTIN, library and information science educator; b. East Chicago, Ind., Jan. 17, 1935; s. Floyd C. and Harriet E. (Martin) D.; m. Ann Prescott, Mar. 24, 1974; children—Kathryn E., Emily E.; children by previous marriage—Jill Ann, Jacquelyn A., Douglas M. B.S., Purdue U., 1959, LHD honoris causa, 1991; M.L.S., Rutgers U., 1961, Ph.D., 1963. Head acquisitions dept. Univ. Library, U. N.C., Chapel Hill, 1963-66; assoc. dir. libraries U. Colo., Boulder, 1966-70; prof. library sci. Syracuse U., N.Y., 1970-72; univ. librarian U. Calif-Berkeley, 1972-78; dir. univ. library U. Mich., Ann Arbor, 1978-88, acting dean. Sch. Library Sci., 1984-85, prof. info. libr. studies, 1978—; cons. libr., higher edn., computer info. systems; pres. Dougherty & Assocs. Inc., mgmt. cons.; founder, pres. Mountainside Pub. Corp., 1974—. Author: Scientific Management of Library Organizations, 2d edit., 1982; co-author: Preferred Futures for Libraries I, 1991, Preferred Futures for Libraries II, 1993; editor Coll. and Research Libraries jour., 1969-74, Jour. Acad. Librarianship, 1975—, Library Issues, 1981—. Recipient Esther Piercy award, 1968, Disting. Alumnus award Rutgers U., 1980, Acad. Librarian Yr., Assn. Coll. and Research Libraries, 1983, ALA Hugh C. Atkinson Meml. award, 1988, Blackwell Scholarship award, 1992; fellow Council on Library Resources. Mem. ALA (coun. 1969-76, 89—, exec. bd. 1972-76, 89—, endowment trustee 1986-89, pres. 1990-91), Assn. Rsch. Librs. (bd. dirs. 1977-80), Rsch. Librs. Group, Inc. (exec. com. 1984-88, chmn. bd. govs. 1986-87), Soc. Scholarly Pub. (bd. dirs. 1990—, exec. com. 1991-92), Internat. Fedn. Libr. Assns. (round table of editors of library jours. 1985-87, standing com. univ. libr. sect. 1981-87). Home: 6 Northwick Ct Ann Arbor MI 48105-1408 Office: U Mich Sch Info & Libr Studies Ann Arbor MI 48109

DOUGLAS, MICHAEL GILBERT, cell biologist, educator; b. Perth, Australia, June 9, 1945; came to U.S., 1948; s. Claude Earl and Margaret Mary (Anderton) D.; m. Joanne Beth Arnall, Nov. 24, 1973; children: Hannah Kathleen, Peter Mahan, Sarah Michael. BS, Southwestern U., 1967; PhD, St. Louis U., 1974. Postdoctoral fellow U. Tex. Health Sci. Ctr., Dallas, 1974-76, asst. prof. biochemistry, 1977-82, assoc. prof. biochemistry, 1982-85, prof. biochemistry, 1988-89; Swiss state fellow Biozentrum U. of Basel, 1976-77; prof., chair biochemistry and biophysics dept. Med. Sch., U. N.C., Chapel Hill, 1989—; prof. genetics U. N.C., Chapel Hill, 1990—; cons. Sandoz Pharm., Basel, 1984-85, U.S. Army R & D Ctr., Aberdeen, Md., 1983-85, Sigma Aldrich Fluka Chem. Co., St. Louis, 1992—. Contbr. articles to profl. publs., chpts. to books. Pres. parish coun. Prince of Peace Cath. Ch., San Antonio, 1979-85, chair bldg. com., 1980-83; mem. sch. bd. St. Rita Cath. Ch., Dallas, 1988-89. With U.S. Army, 1969-70, Vietnam. NIH grantee, 1988—, 89—. Mem. AAAS, Am. Soc. Cell Biology, Am. Soc. Biol. Chemists, Am. Soc. Microbiology, Genetics Soc. Am., Am. Biophys. Soc., Assn. Med. Sch. Depts. Biochemistry. Achievements include genetic mapping of the first structural genes on mitochondrial DNA, gene fusion studies, first demonstration of molecular chaperone regulators in the Eukaryote. Home: 209 Wood Cir Chapel Hill NC 27514 Office: U NC Med Sch Biochemistry & Biophysics CB # 7260 FLoB Chapel Hill NC 27599-7260

DOUGLAS, PATRICIA JEANNE, systems designer; b. Coats, Kans., Sept. 27, 1939; d. Curtis Claire and Pearl L. (Haney) Coe; divorced; children: Tricia Jeanne Douglas Nash, Robert Charles Jr. Student, Willamette U., 1958-59; BA, U. Ariz., 1961, MEd, 1973; PhD, Colo. State U., 1988; postgrad., Columbia U. Cert. tchr., Ariz., Colo. Tchr. Amphitheater Sch. Dist., Tucson, 1962-83; corp. trainer IBM, Boulder, Colo., 1983-86, systems analyst Internat. Purchasing and Distbn. Ctr., 1987-88, rsch. statistician, 1988-89, instnl. system designer, 1989—; asst. expense acct. analyst Colo. State U., Ft. Collins, 1986-87, rsch. assoc., 1988-89. Mem. tchr. workshops LDS Ch., Boulder, 1988; mem. Substance Prevention Project, Boulder, 1988; election judge Boulder County, 1988; vol. in schs., Wappingers Falls, N.Y., 1991. IBM grantee, 1988. Mem. Inter-Am. Orgn. for Higher Edn., Internat.

Coun. for Distance Edn., Consortium-Distance Edn. Network Orgn. U.S., Phi Delta Kappa, Omicron Tau Theta. Avocations: stained glass windows, jewelry making, walking, swimming, music. Office: 500 Columbus Ave Thornwood NY 10594-1900

DOUGLAS, ROBERT ANDREW, civil engineer, educator; b. Windsor, Ont., Can., Mar. 22, 1954; s. Frederick John and Margaret Helen (Gerry) D.; m. Elizabeth Edith Tidridge, Dec. 28, 1978; children: Neil, Alison, Gwen. BSCE, U. Windsor, 1976; PhD, Southampton (Eng.) U., 1980. Registered profl. engr., Ont., N.B. Engr.-in-tng. R.M. Hardy & Assocs., Edmonton, Alta., Can., 1976, 77; cons. engr. Golder Assocs., Windsor, 1979-82; rsch. assoc. Royal Mil. Coll., Kingston, Ont., 1982-83; asst. prof., then assoc. prof. civil engring. U. N.B., Fredericton, 1983—. Contbr. articles to refereed jours. Mem. Am. Soc. Agrl. Engring. (com. chmn. 1982), Can. Geotech. Soc., Internat. Soc. Geotextiles. Office: Forest Engring Univ NB, Box 4400, Fredericton, NB Canada E3B 5A3

DOUILLARD, PAUL ARTHUR, engineering and financial executive, consultant; b. Pittsfield, Mass., Nov. 5, 1927; s. Arthur and Anna (Champagne) D.; m. Melania Josephine Kubica, Nov. 28, 1953; children: Geoffrey Paul, Marie Suzanne, Ellen Michelle. BS in Physics/Math., Mass. State Coll., 1964; MS in Mgmt. Sci., Calif. Western U., 1970. Registered profl. engr.; cert. cost analyst. Project engr. Gen. Electric Co., Pittsfield, Mass., 1962-66, systems analyst, 1966-67; sr. rsch. engr. Gen. Dynamics Corp., San Diego, 1967-69; cons. San Diego, 1969-71; supervisory ops. rsch. USN, Dept. Def., China Lake, Calif., 1971-76, 81-85; tech. project officer U.S. Dept. Energy Strategic Petr. Reserve, Washington, 1976-81; supr. ops. rsch. for financial mgmt. Office of Asst. Sec. of Navy, Washington, 1985-89; dep. chief cost analysis EER Systems, Inc., Huntsville, Ala., 1989-91; cons. ops. rsch., CEO The Delphi Co., INc., Huntsville, 1991—. Contbr. articles to profl. jours. Chmn. collections March of Dimes, Stockbridge, Mass., 1966; neighborhood com. Boy Scouts Am., San Diego, 1967-68, asst. dist. com., 1969-71. With U.S. Army, 1946-47, USNR, 1951-52. Mem. IEEE (program chmn. 1964-67), AIAA, Soc. of Cost Estimating/Analysis, Internat. Soc. for Parametric Analysis, Redstone Arsenal Officers Club. Achievements include patent for hollow glass body for vacuum circuit breakers; introduced ultra-sonic meters to strategic petroleum res. Home: 1875 Shellbrook Dr Huntsville AL 35806 Office: The Delphi Co Inc PO Box 1202 West Sta Huntsville AL 35807

DOUMAS, BASIL THOMAS, chemist, researcher, educator; b. Argos Orestikon, Macedonia, Greece, July 16, 1930; came to U.S., 1957; s. Thomas and Iphigenia (Hatzopoulos) D.; m. Maria Kosma, June 2, 1957; children: Vasiliki Alafouzos, Thomas, Iphigenia Smith. BS, U. Thessaloniki, Greece, 1952; MS, U. Tenn., 1960, PhD, 1962. Clin. chemist Bapt. Meml. Hosp., Memphis, 1962-64; assoc. prof. clin. pathology U. Ala., Birmingham, 1964-70; prof. pathology Med. Coll. Wis., Milw., 1970—; dir. clin. chemistry labs. Milw. County Med. Complex. Bd. editors Jour. Clinica Chimica Acta, 1980—, Jour. Clin. Chemistry, 1981-90; contbr. numerous articles to profl. jours. Bd. dirs. Nat. Com. for Clin. Lab. Standards, Villanova, Pa., 1982-88. Mem. Am. Assn. for Clin. Chemistry (pres. 1992, Sam Natelson award Chgo. chpt. 1979, Roche award 1983), Nat. Acad. Clin. Biochemistry (pres. 1985-86), Acad. Clin. Lab. Physicians and Scientists, Hellenic Philatelic Soc. Am., Sigma Xi. Greek Orthodox. Avocation: stamp collector. Office: Med Coll Wis 8700 W Wisconsin Ave Milwaukee WI 53226-3595

DOUPNIK, CRAIG ALLEN, physiologist; b. Lincoln, Nebr., Feb. 7, 1962; s. Darrell Dean and Marilyn Yvonne (Laue) D.; m. Anita Marie Stahl, Sept. 12, 1987. BS, Ohio State U., 1984; MS, U. Cin., 1989, PhD, 1993. Rsch. asst. U. Cin., 1985-89, grad. asst., 1989-93; postdoctoral rsch. fellow Calif. Inst. Tech., Pasadena, 1993—. Predoctoral fellow NIH, 1989-92. Mem. AAAS, Am. Physiol. Soc., Biophys. Soc., Soc. Neurosci. Achievements include research in physiology and biophysics. Office: Calif Inst Tech Div Biology 156-29 Pasadena CA 91125

DOUSKEY, THERESA KATHRYN, health facility administrator; b. New Haven, Conn., Nov. 30, 1938; d. Stanley Anthony and Wadia (Mekdeci) D. RN, Grace New Haven Sch. Nursing, 1959; BS in Nursing, So. Conn. State U., 1962; MPA in Health Care, U. New Haven, 1979. Various positions Yale New Haven Hosp., 1959-80; asst. dir. nursing Meriden (Conn.) Wallingford Hosp., 1980-81; nurse Regional Visiting Nurse Agy., North Haven, Conn., 1983-87; case mgr., nurse Community Care, Inc., New Haven, 1988-90; home care coord. Milford (Conn.) Hosp., 1990 . Mem. Am. Nurses Assn., Conn. Nurses Assn. (nominating com. 1072-74), Conn. Assn. Continuity of Care, Sigma Theta Tau. Republican. Avocations: needle crafts, gardening, working with animal humane groups. Home: 412 Narrow Ln Orange CT 06477

DOUT, ANNE JACQUELINE, manufacturing company executive; b. Detroit, Mar. 13, 1955; d. George Edwin and Virginia Irene (Fisher) Boesinger; m. James Edward Dout, July 16, 1977; 1 child, Brian Ross. Student, Macomb Community Coll., 1972-74; BBA, Western Mich. U., 1976; MBA, Duquesue U., 1982. Cert. cash mgr. Internal auditor Koppers Co. Inc., Pitts., 1976-78, cash analyst, 1978-79, supr. cash ops., 1979-80, mgr. cash ops., 1980-81, mgr. cash ops., asst. treas., 1981-87, dir. treasury svcs., asst. treas., 1987-88; corp. staff v.p., asst. treas. IMCERA Group Inc., Northbrook, Ill., 1988-91; acting treas. IMCERA Group, Inc., Northbrook, Ill., 1990-91, v.p., treas., 1991—. Mem. allocating com. United Way, Pitts., 1979-83; bd. dirs. N.E. Lake County coun. Boy Scouts Am., v.p. adminstrn., 1989-92; bd. dirs. Barat Coll., Lake Forest, Ill., 1992. Mem. Pitts. Cash Mgmt. Assn. (co-founder 1980, v.p. 1980-81, pres. 1981-82), Treas. Mgmt. Assn. (exec. com. 1988-90, govt. rels. com. 1984-86, bd. dirs. 1986-89, strategic plan com. 1987-90, chmn. 1989-91), Fin. Exec. Inst., Mid Am. Com., Econ. Club, Exec. Club. Protestant. Office: IMCERA Group Inc 2315 Sanders Rd Northbrook IL 60062-6108

DOUTY, RICHARD THOMAS, structural engineer; b. Williamsport, Pa., June 12, 1930; s. Richard Otis and Helen Anna (Sauter) D.; Patricia Marie Hopkins, June 27, 1959; children: Richard K., Eric Thomas, Ellen M., Christopher. BSCE, Lehigh U., 1956; MS, Ga. Inst. of Tech., 1957; PhD, Cornell U., 1964. Registered profl. engr., Pa., Mo. Engr. Bethlehem Steel Corp., Rankin, Pa., 1956-57; prof. civil engring. U. Mo., Columbia, 1962-92; structural engr. MDX, Inc., Columbia, 1992—; mem. selection panel Nat. Rsch. Coun. Fellowships, Washington, 1972-88. Editorial bd. Engring. Optimization Jour., Eng., 1972-76; contbr. articles to profl. jours. With USN, 1950-54. Rsch. fellow U. Pa., 1970. Mem. ASCE. Home: 1412 Ridgemont Ct Columbia MO 65203-1955 Office: MDX Inc. 1412 Ridgemont Ct Columbia MO 65203

DOUVAN-KULESHA, IRINA, chemist; b. Krakow, Poland, Feb. 26, 1938; came to U.S., 1948; d. Boris Victor and Nadine (Sokolov) Douvan; m. Alexander Kulesha, Feb. 4, 1968; children: Eugene, Paul. BS in Chemistry, Wagner Coll., 1960; MS in Organic Chemistry, Rutgers U., 1963. Rsch. scientist Hoffmann La Roche, Nutley, N.J., 1963—. Co-author: Peptides, 1992; contbr. articles to profl. jours. Mem. Am. Chem. Soc., N.Y. Acad. Scis., Soc. of Russian Am. Scholars. Republican. Russian Orthodox. Office: Hoffmann La Roche Inc 340 Kingsland St Nutley NJ 07110

DOVALE, FERN LOUISE, civil engineer; b. Ft. Leavenworth, Kans., May 11, 1956; d. Riel Stanton and Beatrice Marie (Mayer) Crandall; m. Antonio Joseph DoVale Jr., Oct. 17, 1981. BSCE, MIT, 1978; MSCE, Columbia U., 1982. Registered profl. engr., N.J. Assoc. engr. M.W. Kellogg Co., Hackensack, N.J., 1978-80; engr. Nuclear Power Svcs., Inc., Secaucus, N.J., 1980-83, sr. engr., 1983-85, lead engr., 1985-86, project engr., 1986-88; project mgr. NPS Technologies Group, Inc., Secaucus, 1988-89; engring. mgr. NPS Technologies Group, Inc., Elmwood Park, N.J., 1989-92, Integrated Engring. Software, Inc., Englewood Cliffs, N.J., 1992—. Author, editor computer manuals. Mem. ASCE, ASME, NSPE, Am. Nuclear Soc. (exec. com. No. N.J. sect. 1986-89), MIT Alumni Club No. N.J. (bd. dirs. 1979-81, membership v.p. 1982-84, 89-92, program v.p. 1984-85, 92—, pres. 1985-86), MIT Ednl. Coun. Home: PO Box 865 Oakland NJ 07436-0865 Office: Integrated Engring Software Inc 560 Sylvan Ave Englewood Cliffs NJ 07632

DOVIDIO, JOHN FRANCIS, psychology educator; b. Medford, Mass., Oct. 3, 1951; s. John Guido and Florence (Langone) D.; m. Linda Ann Ravenelle, Aug. 12, 1977; children: Alison Hope, Michael John. AB, Dartmouth Coll., 1973; MA, U. Del., 1976, PhD, 1977. From asst. to assoc. to full prof. Colgate U., Hamilton, N.Y., 1977—; adv. bd. mem. N.Y. State Senate, N.Y., 1988-90; cons. USAF Human Rels. Educator, Washington, 1990—. Co-author: Emergency Intervention, 1981; co-editor: Power, Dominance, and Nonverbal Behavior, 1985, Prejudice, Discrimination, and Racism, 1986; contbr. articles to profl. jours. Recipient Gordon Allport Intergroup Rels. prize Am. Psychol. Assn., 1985. Fellow Am. Psychol. Assn., Am. Psychol. Soc., Soc. for Personality and Social Psychol.; mem. Soc. for Experimental Social Psychol. Achievements include development of techniques for understanding conscious and unconscious processes in prejudice. Office: Dept Psychology Colgate Univ Oak Drive Hamilton NY 13346

DOVRING, FOLKE, land economics educator, consultant; b. Rystad, Sweden, Dec. 6, 1916; came to U.S., 1960, naturalized, 1968; parents Karl Gustav and Naemi (Arnman) Ossiannilsson; m. Karin Dovring, May 30, 1943. PhD, Lund (Sweden) U., 1947. Assoc. prof. Lund U., 1947-53; statistician, economist FAO, UN, Rome, 1953-60; prof. land econs. U. Ill., Urbana, 1960-87, prof. emeritus, 1987—; cons. UN Econ. Commn. for Europe, Geneva, 1953, OECD, Paris, 1963-64, World Bank, AID, USDA, Dept. Energy, Washington, 1967-79. Author: (14 books including) Riches to Rags, 1984, Productivity and Value, 1987, Farming for Fuel, 1988, Inequality, 1991; contbr. over 200 other publs. to profl. jours. Mem. publicity campaign on fuel farming for Voice of Am., Washington, and other nationwide mass media, 1990-91. Rockefeller Found. fellow, 1953-54. Mem. Cosmos Club (Washington). Democrat. Avocations: gardening, entomology. Home: 613 W Vermont Ave Urbana IL 61801-4824

DOW, WILLIAM GOULD, electrical engineer, educator; b. Faribault, Minn., Sept. 30, 1895; s. James Jabez and Myra Amelia (Brown) D.; m. Edna Lois Sontag., Oct. 24, 1924 (dec. Feb. 1963); children—Daniel Gould, David Sontag; m. Katherine Bird Keene, Apr. 2, 1968; stepchildren—John S. Keene, Margaret Keene Hannan, Karen Keene Day. B.S., U. Minn., 1916, E.E., 1917; M.S.E., U. Mich., 1929; D.Sc. (hon.), U. Colo., 1980. Registered profl. engr., Mich. Diversified engring. and bus. experience, 1917-26; faculty, dept. elec. engring. U. Mich., Ann Arbor, 1926-65; prof. elec. engring. U. Mich., 1945-65, chmn. dept. elec. engring., 1958-64, prof. emeritus, 1966—; sr. research geophysicist Space Physics Research Lab., 1966-71; electronics cons. Nat. Bur. Standards, 1945-55; research staff Radio Research Lab., Harvard, 1943-45, assignment, U.K., winter 1944-45; sci. adv. com. Harry Diamond Labs., 1953-64; bus. mgr. Lang. Studies Abroad, Spain, summers, 1965-74; Mem. vacuum tube devel. com. NDRC, World War II; (European vacuum tube research survey), 1953; mem. rocket and satellite research panel, 1946-60; U.S. tech. panel on rocketry IGY, 1956-59; made world tour for space research and engring. edn. survey, 1969-70; Charter mem. bd. trustees Environmental Research Inst. Mich., 1972-90, trustee emeritus, 1990—. Author: Fundamentals of Engineering Electronics, 1937, rev. 1952, Very High Frequency Techniques (co-author), 2 vols, 1947; Contbr. tech. articles in field; patentee trochoidal nuclear fusion system. Served as lt. C.E. U.S. Army, World War I. Recipient medal, award in elec. engring. edn. IEEE, 1963. Fellow IEEE (bd. editors 1941-54), Engring. Soc. Detroit, AAAS; mem. AAUP, Am. Phys. Soc., Am. Inst. Aeros. and Astronautics, Am. Geophys. Union, Nat., Mich. socs. profl. engrs., Am. Astronautical Soc., N.Y. Acad. Scis., Am. Soc. Engring. Edn., Am. Welding Soc., Nat. Electronics Conf. (dir. 1949-52, chmn. bd. 1951), Sigma Xi, Tau Beta Pi, Eta Kappa Nu. Episcopalian. Clubs: Mason. (Minn.), Cosmos (Washington). Home: 915 Heatherway St Ann Arbor MI 48104-2833

DOWBEN, ROBERT MORRIS, physician, scientist; b. Phila., Apr. 6, 1927; s. Morris and Zena (Brown) D.; m. Carla Lurie, June 20, 1950; children—Peter Arnold, Jonathan Stuart, Susan Laurie. A.B., Haverford Coll., 1946; M.S., U. Chgo., 1947, M.D., 1949. Intern U. Chgo. Clinics, 1949-50; research fellow U. Oslo, 1950-51; fellow Johns Hopkins Hosp., 1951-52; resident in medicine U. Pa. Hosp., 1952-53; instr. medicine U. Pa. and dir. radioisotope unit VA Hosp., Phila., 1953-55; asst. prof. medicine Northwestern U. Med. Sch., 1957-62; asso. prof. biology M.I.T., 1962-68; lectr. medicine Harvard U. Med. Sch., 1962-68; prof. med. sci. Brown U., 1968-72; prof. biochemistry U. Bergen, Norway, 1972; prof. physiology and neurology, dir. grad. program in biophysics U. Tex. Health Sci. Center, Dallas, 1972-88; prof. neurology U. Tex. Health Sci. Ctr., Dallas, 1988-93; dir. Med. Cell Biology Lab. Baylor Rsch. Inst., Dallas, 1987-93; prof. physiology Brown U., Providence, R.I., 1993—; cons. neurologist Children's, Scottish Rite, Presbyn., Baylor hosps., 1972-93; Mem. corp. Haverford (Pa.) Coll., Marine Biol. Lab., Woods Hole, Mass.; bd. dirs. Greenhill Sch., Dallas, 1974-77. Author: Biol. Membranes, 1969, General Physiology, 1971, Cell Biology, 1972, also numerous articles; editor: Cell and Muscle Motility. Served to capt. M.C. USAF, 1955-57. Lalor fellow; recipient Disting. Service award Assn. Neuromuscular Diseases, 1964, Disting. Service award Alumni Assn. U. Chgo., 1980. Mem. Am. Physiol. Soc., Am. Soc. Biol. Chemists, Am. Chem. Soc., Soc. Exptl. Biology and Medicine, Biophys. Soc., Soc. Clin. Investigation, Central Soc. Clin. Research, Mass. Med. Soc., So. Med. Soc., Dallas County Med. Soc., Tex. Med. Assn., Biochem. Soc. London, Faraday Soc. (London), Phi Beta Kappa, Sigma Xi. Quaker. Office: Brown U Physiology Dept Box G-B3 Providence RI 02912

DOWDLE, WALTER REID, microbiologist, medical center administrator; b. Irvington, Ala., Dec. 11, 1930; s. Ruble C. and Rebecca (Powell) D.; m. Mable Irene Graham, Apr. 2, 1953; children—Greta Denise Dowdle Rackley, Robert Reid, Jennifer Leigh. B.S., U. Ala., 1955, M.S., 1957; Ph.D., U. Md., 1960. Chmn. respiratory virology Ctr. for Disease Control, Atlanta, 1964-73, dir. virology div., 1973-79, asst. dir. for sci., 1979-81, dir., Ctr. for Infectious Diseases, 1981-86, coordinator PHS AIDS, 1986, dep. dir., 1987—; chmn. US PHS Interagy. Reye Syndrome Task Force, 1982—. Co-author: Informed Consent, 1983; mem. editorial bd. Jour. Clin. Microbiology, 1974—; contbr. articles to profl. jours. Served with USAF, 1948-52. Recipient Presdl. Rank award HHS, Washington, 1984, 89, Disting. Svc. award, 1982, Congl. Excalibur award, 1985; named Fed. Exec. of Yr., Sr. Exec. Assn., Washington, 1983; John Curtain Sch. Medicine rsch. hon. fellow, Canberra, Australia, 1972-73. Mem. AAAS, Am. Acad. Microbiology (bd. govs. 1984-85), Am. Soc. for Microbiology (pres. 1989), Infectious Disease Soc. Am., Am. Soc. Virology. Home: 1708 Mason Mill Rd NE Atlanta GA 30329-4129 Office: Ctrs for Disease Control Bldg 1 1600 Clifton Rd NE Atlanta GA 30333

DOWELL, EARL HUGH, university dean, aerospace and mechanical engineering educator; b. Macomb, Ill., Nov. 16, 1937; s. Earl S. and Edna Bernice (Dean) D.; m. Lynn M. Cary, July 31, 1981; children: Marla Lorraine, Janice Lynelle, Michael Hugh. B.S., U. Ill., 1959; S.M., Mass. Inst. Tech., 1961, Sc.D., 1964. Rsch. engr. Boeing Co., 1962-63; rsch. asst. MIT, 1963-64, rsch. engr., 1964, asst. prof., 1964-65; asst. prof. aerospace and mech. engring. Princeton U., 1964-68, assoc. prof., 1968-72, prof., 1972-83, assoc. chmn., 1975-77, acting chmn., 1979; dean Sch. Engring. Duke U., Durham, N.C., 1983—; cons. to industry and govt.; mem. scientific adv. bd. USAF. Author: Aeroelasticity of Plates and Shells, 1974, A Modern Course in Aeroelasticity, 1978, 2d edit., 1989, Nonlinear Studies in Aeroelasticity, 1988; assoc. editor: AIAA Jour., 1969-72, Jour. Sound and Vibration, 1988—, Jour. Fluids and Structures, 1987—, Jour. Nonlinear Dynamics, 1990—; contbr. articles to profl. jours. Chmn. N.J. Noise Control Council, 1972-76. Named outstanding young alumnus U. Ill. Sch. Aero. and Astronautical Engring., 1973, disting. alumnus, 1975; recipient Alumni Honor award Coll. Engring. U. Ill. Fellow AIAA (Structures, Structural Dynamics and Material award 1980, v.p. publs. 1981-83), Am. Acad. Mechs., ASME; mem. NAE, Acoustical Soc. Am., Am. Helicopter Soc. Home: 2207 Chase St Durham NC 27707-2228 Office: Duke U Sch Engring Durham NC 27706

DOWLING, ANN PATRICIA, mechanical engineering educator, researcher; b. Taunton, Somerset, Eng., July 15, 1952; d. Mortimer Joseph and Joyce (Barnes) D.; m. Thomas Paul Hynes, Aug. 31, 1974. BA in Math., Cambridge (Eng.) U., 1973, MA in Math., 1977, PhD in Engring., 1978. Asst. lectr. engring. dept. Cambridge U., 1979-82, lectr., 1982-86, reader, 1986-93, dep. head dept., 1990—, prof., 1993—; cons. Rolls-Royce, Bristol, Eng., 1978—, Topexpress Ltd. and Admiralty, 1980—. Author: (with J.E. Ffowcs Williams) Sound and Sources of Sound, 1983, (with D.G. Crighton and others) Modern Methods in Analytical Acoustics, 1992; also numerous articles. Fellow Sidney Sussex Coll., Cambridge U., 1977—. Fellow Inst. Acoustics (A.B. Wood medal 1990), Instn. Mech. Engrs.; mem. AIAA. Achievements include advances to understanding and control of flow-induced noise and vibration; development of theory to predict occurrence and form of acoustically-coupled oscillations in afterburners of aeroengines and demonstration that active control can eliminate oscillations. Office: Cambridge U Engring Dept, Trumpington St, Cambridge CB2 1PZ, England

DOWNARD, DANIEL PATRICK, electrical engineer, consultant; b. Louisville, Aug. 15, 1946; s. Julian Patrick and Helen L. (Fensterer) D.; m. Helen M. Downard, Sept. 28, 1968; children: Cynthia, Patrick; m. Judy A. Daniel, Dec. 23, 1989. BEE, U. Louisville, 1972. Profl. engr., Ky. Sr. elec. engr. C&I Engring., Inc., Louisville, 1986—. Contbr. articles to profl. jours. Mem. Jefferson County Elec. Bd. Control, Louisville, 1991—. Office: C&I Engring 11003 Bluegrass Pky Ste 610 Louisville KY 40299

DOWNER, MICHAEL C., physicist; b. Rockville Centre, N.Y., July 19, 1954; s. William John and Margaret Elizabeth (Walter) D. BA in Physics, U. Rochester, 1976; MA in Physics and Philosophy, Oxford (Eng.) U., 1978; PhD in Applied Physics, Harvard U., 1983. Mem. tech. staff AT&T Bell Labs., 1983-85; asst. prof. of physics U. Tex., Austin, 1985-91, assoc. prof. physics, 1991—; vis. prof. engring. T.U. of Aachen, Fed. Republic Germany, 1990. Author: (with others) Laser Spectroscopy of Solids II, 1989; contbr. articles to Jour. Optical Soc. of Am., Phys. Rev. B., Phys. Rev. Letters and numerous others. Recipient Faculty Devel. award IBM Corp., 1987, Young Investigator award Office of Naval Rsch., 1988, Presdl. Young Investigator NSF, 1988. Mem. Am. Phys. Soc., Optical Soc. Am. Achievements include development of technique of spectral blueshifting for measuring femtosecond time scale ionization and plasma dynamics, first slow-motion movie on femtosecond time scale, showing melting and evaporation of photoexcited silcon surface; measurement of extensive two-photon absorption spectra of rare earth ions in crystals and solutions, dielectric properties of liquid phase of carbon. Office: U Tex at Austin Physics Dept Austin TX 78712

DOWNER, ROGER GEORGE H., biologist; b. Belfast, Northern Ireland, Dec. 21, 1942; arrived in Can., 1967; s. Leslie George and Edith Patricia (Hamill) D.; m. Jean Elizabeth Taylor, Apr. 6, 1966; children: Kevin George, Kathleen Patricia, Tara Siobhean. BSc, Queens U., Belfast, 1964, MSc, 1967; PhD, U. Western Ont., 1970; DSc, Queens U., Belfast, 1984. Asst. prof. U. Waterloo, Ont., 1970-76, assoc. prof., 1976-81, prof., 1981—, chair dept. biology, 1986-89, acting dean of sci., 1989, v.p., 1989—; cons. Am. Cyanamid, Princeton, N.J., 1984-88; dir. Insect Biotech. Can., Ont., 1989-90. Editor 4 books; contbr. 145 articles to profl. jours. Fellow Royal Soc. Can.; mem. Can. Soc. Zoologists (Fry medal 1991), Entomol. Soc. Can. (Gold medal 1991). Home: 15 Wildwood Pl, Waterloo, ON Canada N2L 4B2 Office: Univ of Waterloo, 200 University Ave West, Waterloo, ON Canada N2L 3G1

DOWNES, GREGORY, architectural organization executive; b. Cambridge, Mass., Mar. 17, 1939; s. Thomas M. and Jean (Gregory) D.; m. Sandra Motley Snow, June 9, 1962; children: Katharine Appleton, Elizabeth Amory. BA cum laude, Harvard U., 1961, MArch, 1965. Registered architect, Mass., Ala., Conn., D.C., Maine, N.J., Nev., R.I. With The Architects Collaborative, Inc., Cambridge, 1965—, v.p., 1980—, prin., 1982—, dir., 1987—; instr. Boston Archtl. Ctr., 1972-73, Harvard Grad. Sch. Design, 1973-76. Contbr. articles to profl. jours. Chmn. bd. trustees Capt. Robert Bennet Forbes Mus., 1988—; bd. dirs. Shirley Eustis Historic House, Roxbury, 1986—; seminarian Boston Mus. Fine Arts, 1988-90; corporator Mt. Auburn Hosp.; mem. vestry Ch. of Redeemer, Brookline, 1982-85, St. Michaels Ch., Milton, 1988-91. Recipient numerous awards of honor. Mem. AIA, Boston Soc. Architects, Am. Planning Assn., Nat. Assn. Indsl. and Office Parks, Urban Land Inst., Harvard Club of Boston, The Country Club, Harvard Faculty Club. Republican. Episcopalian. Avocations: tennis, hockey, squash, skeet, gardening. Home: 203 Adams St Milton MA 02186-4297 Office: The Architects Collaborative Inc 46 Brattle St Cambridge MA 02138-3700

DOWNEY, RICHARD MORGAN, lawyer; b. S.I., N.Y., Nov. 13, 1946; s. William Sexton and Marion (Herbert) D.; m. Judith Yestrumskas, May 31, 1980. BA, Fairfield U., 1968; JD, Georgetown U., 1971. Bar: D.C. Regional dir. Common Cause, 1975-77; atty. Pub. Citizen, 1973-74; with Dem. Nat. Com., 1972, Ams. for Indian Opportunity, 1970-72; dir. govt. and legal affairs dept. Am. Speech-Lang.-Hearing Assn., Rockville, Md., 1978-89; ptnr. Hoffheimer & Downey, 1989—; exec. dir. Internat. Neural Network Soc., 1990—; assoc. dir. Nat. Found. for Brain Rsch., 1989-91. Mem. Nat. Health Lawyers Assn., ABA , Am. Soc. Assn. Execs. (mem. legal sect. coun.). Democrat. Roman Catholic. Author: Health Insurance Manual for Speech-Language Pathologists and Audiologists, 1984, Legal Issues in Brain Science Advances, 1991. Home: 4411 42d St NW Washington DC 20016

DOWNING, MICHAEL WILLIAM, pharmaceutical company executive; b. Rock Island, Ill., Sept. 30, 1947; s. William Richard and Rita Louise (Shaughnessy) D.; m. Pamela Ludeman, Aug. 23, 1969; children: Katherine June, John Michael. BS, Iowa State U., 1970, MS, 1971. Adv. product devel. engr. Med. Products div. 3M, St. Paul, 1971-76; product devel. supr. Diagnostic Products dept. 3M, St. Paul, 1976-79; tech. mgr. Vision Care/3M, St. Paul, 1979-83, program mgr., 1983-88, lab. mgr., 1988-89; dir. product devel. 3M Pharms., St. Paul, 1989-91, dir. bus. devel., 1991-93; bus. mgr. drug deliver systems 3M Pharms, St. Paul, 1993—; bd. dirs. Alliance for Responsible CFC Policy, Arlington, Va., 1991—. Bd. dirs. YMCA, St. Paul, 1979-82; project bus. cons. Jr. Achievement, St. Paul, 1973, 78, 83. Mem. Knights of St. Patrick, Phi Kappa Phi, Phi Gamma Delta. Achievements include 2 patents. Home: 73 Deer Hills Ct North Oaks MN 55127 Office: 3M Pharms Bldg 275-3W-01 3M Ctr Saint Paul MN 55133-3275

DOWNS, FLOYD L., mathematics educator; b. Winchester, Mass., Jan. 21, 1931; s. Floyd L. and Emma M. (Noyes) D.; m. Elizabeth Lenci, Dec. 29, 1955; children: Karla C., John N. AB, Harvard U., 1952; MA, Columbia U., 1955. Lic. math. tchr. Math. tchr. East High Sch., Denver, 1955-60, Kent (Conn.) Sch., 1960-62, Newton High Sch., Newtonville, Mass., 1962-63; math. tchr., dept. chair Hillsdale High Sch., San Mateo, Calif., 1964-89; dir. undergrad. math. Ariz. State U., Tempe, 1988—; math. scis. adv. com. The Coll. Bd., N.Y., 1979-85; mem. U.S. nat. com. 2d Internat. Math. Study, 1979-86; Golden state math. com. Calif. State Dept. Edn., Sacramento, 1985-91; exec. dir. Ariz. Math. Coalition, 1991—. Co-author: Geometry, 1964, 91. With U.S. Army, 1952-54, Korea. Mem. Nat. Coun. Tchrs. Math., Nat. Coun. Suprs. Math., Math. Assn. Am., Calif. Math. Coun., Ariz. Assn. Tchrs. Math., Phi Delta Kappa. Home: 7753 E Bisbee Rd Scottsdale AZ 85258-3421 Office: Ariz State U Math Dept Tempe AZ 85287-1804

DOWNS, HARTLEY H., III, chemist; b. Ridgewood, N.J., Oct. 21, 1949; s. Hartley Harrison and Jennie Mae (Smith) D.; m. Cindy Marie Millen, June 19, 1976; children: Kathryn Marie, Jennifer Anne, Susanna Jayne. BS, Grove City Coll., 1971; MS, Indiana U. of Pa., 1973; PhD, W. Va. U., 1978; postgrad., U. Colo., 1976-77. Postdoctoral rsch. assoc. chemistry dept. U. So. Calif., L.A., 1977-78; staff chemist corp. rsch. labs. Exxon Rsch. and Engring. Co., Linden, N.J., 1978-81, Houston, 1981-83, Annandale, N.J., 1983-86; rsch. scientist, surface chemistry and corrosion sci. group supr. Baker Performance Chems., Houston, 1986-91, rsch. mgr., 1991-92, tech. dir., 1992—. Contbr. articles to profl. jours., chpt. to book; patentee in field. Recipient Union Carbide award W.Va. U., 1975, Stan Gillman award U. Colo., 1977, Tech. Merit award Baker-Hughes, 1989, 91, 93. Mem. Am. Chem. Soc., Soc. Petroleum Engrs., Offshore Operators Com. (task force on environ. sci.), Nat. Assn. Corrosion Engrs., Sigma Xi (award for grad. rsch. 1973), Phi Lambda Upsilon. Presbyterian. Office: Baker Performance Chems Inc PO Box 27714 Houston TX 77227-7714

DOYLE, MICHAEL PATRICK, food microbiologist, educator, researcher; b. Madison, Wis., Oct. 3, 1949; s. Donald Vincent and Evelyn (Bauer) D.; m. Annette Marie Ripple, Dec. 27, 1971; children: Michael Patrick, Patrick Matthew, Kristen Anne. BS in Bacteriology, U. Wis., 1973, MS in Food Microbiology, 1975, PhD in Food Microbiology, 1977. Sr. project leader Ralston Purina Co., St. Louis, 1977-80; asst. prof. U. Wis., Madison, 1980-84, assoc. prof., 1984-88, prof., 1988-91; prof., dir. U. Ga., Griffin, 1991—; mem. NAS Inst. Medicine, Food and Nutrition Bd., 1991-94, nat. adv. com. on microbiol. criteria for foods USDA, Washington, 1988-90; trustee Internat. Life Scis. Inst.-Nutrition Found., Washington, 1992-95, (sci. adv. 1987—); mem. Internat. Commn. on Microbiol. Specifications for Foods,

1989—; Fred W. Tanner lectr. Inst. Food Technologists, Chgo., 1986; Wis. Disting. prof. Bd. Regents U. Wis., Madison, 1989-91; James M. Craig Meml. lectr. Oreg. State U., Corvallis, 1990, Am. Soc. Microbiol. Found., 1991-93, Inst. Food Technologists, 1987-90. Editor: Foodborne Bacterial Pathogens, 1989; contbr. articles to Applied and Environ. Microbiology, Jour. Food Protection, Internat. Jour. Food Microbiology, Jour. Clin. Microbiology. Fellow Am. Acad. Microbiology, 1987. Mem. Internat. Assn. Milk, Food & Environ. Sanitarians (pres. 1992-93), Am. Soc. for Microbiology (chmn. food microbiology div. 1987-89), Inst. Food Technologists (Samuel Cate Prescott award for rsch. 1987), Phi Kappa Phi, Gamma Sigma Delta. Roman Catholic. Achievements include patent for monoclonal antibody to enterohemorrhagic E. coli; development of methods to control and detect foodborne pathogens. Office: U Ga Enhancement Lab Ga Expt Sta Griffin GA 30223

DOYLE, MICHAEL PHILLIP, civil engineer; b. Denver, May 7, 1955; s. Virgil Lee and Lela Virginia (Mercer) D.; m. Kathleen Mary Vice, Jan. 6, 1979; children: Kristina Michelle, Leland Colin. BSCE, U. Nev., 1985. Registered profl. engr., Calif. Staff engr. Omni Means Ltd., Reno, 1985, Sacramento, 1985-87; project engr. Carl Rodolf Engr., Sacramento, 1987; tech. dir. Sahuaro Petroleum and Asphalt Co., Phoenix, 1987-91; ptnr. Vinzoyl Petroleum Co., Phoenix, 1991—. Mem. ASCE, ASTM. Republican. Methodist. Achievements include development of a technology transfer business; research in modification of asphalt and asphalt mix to extend performance of pavements; in emulsions for cold paving without solvents; in use of recycled materials. Office: Neste Vinzoyl 1935 W McDowell Rd Phoenix AZ 85005

DOYLE, PATRICK FRANCIS, utility company executive; b. Arlington, Mass., July 17, 1948; s. Patrick Francis and Kathleen Therese (Thomas) D.; m. Paula A. Roy, Dec. 22, 1971; children: Michael P., Erin A. BSEE, Tufts U., 1970; MBA, Northeastern U., 1976, MSEE, 1980. Various positions Boston Edison Co., 1970-88, mgr. energy mgmt. dept., 1988—; mem. various coms. Elec. Coun. of N.E., Bedford, Mass., 1985—, Elec. Power Rsch. Inst., Palo Alto, Calif., 1985—; lectr. to various industry forums, seminars; participant various panels. Mem. Assn. Energy Engrs., Mass. Bldg. Congress, Assn. of Demand Side Mgmt. Profls. Home: 23 Hampshire Rd Peabody MA 01960 Office: Boston Edison Co 800 Boylston St Boston MA 02199

DRACHMAN, DANIEL BRUCE, neurologist; b. N.Y.C., July 18, 1932; s. Julian Moses and Emily (Deitchman) D.; m. Jephta Piatigorsky, Aug. 28, 1960; children: Jonathan Gregor, Evan Bernard, Eric Edouard. A.B. summa cum laude (N.Y. State scholar), Columbia Coll., 1952; M.D. (N.Y. State Med. scholar), NYU, 1956. Intern in internal medicine Beth Israel Hosp., Boston, 1956-57; asst. resident in neurology Harvard neurol. unit Boston City Hosp., 1957-58, resident in neurology, 1958-59; resident in neuropathology Harvard neurol. unit. and Mallory Inst. Pathology, 1959-60; teaching fellow in neurology Harvard U., 1957-60; clin. assoc. Nat. Inst. Neurol. Diseases and Blindness, NIH, Bethesda, Md., 1960-62, research asso. lab. neuroanat. scis., 1962-63; clin. instr. Georgetown U., 1961-63; asst. prof. neurology Tufts U., 1963-69; assoc. prof. Johns Hopkins U., 1969-73, prof., 1974—; attending neurologist Johns Hopkins Hosp.; mem. adv. bd. Multiple Sclerosis Soc., 1981-85; pres. med. adv. bd. Myasthenia Gravis Found.; mem. adv. bd. Familial Dysautonomia Found.; mem. bd. scientific councillors Nat. Inst. Neurol. and Communicative Disorders and Stroke, NIH, 1985—. Clarinetist.; author publs. on myasthenia gravis, muscular atrophy, muscular dystrophy, clubfoot, devel. disorders, neurology; mem. editorial bd. Muscle and Nerve jour., Exptl. Neurology, Autoimmunity. Served with USPHS, 1960-63. Recipient Founders' Day award NYU, 1956, Jacob Javits award, 1986; NIH grantee, 1963—, Muscular Dystrophy Assn. grantee, 1969—. Fellow Am. Acad. Neurology, N.Y. Acad. Scis.; mem. AAAS, Internat. Soc. Devel. Biology, Balt. Neurol. Soc., Phi Beta Kappa, Alpha Omega Alpha. Research on neurol. and neuromuscular diseases. Office: Johns Hopkins U Sch Medicine Dept Neurology 600 N Wolfe St Baltimore MD 21205-2104

DRĂGĂNESCU, MIHAI, electronic engineering educator; b. Făget, Romania, Oct. 6, 1929; s. Dumitru and Cornelia (Petrescu) D.; m. Nora Rebreanu, June 19, 1957. Engr., Poly. Inst., Bucharest, 1952, PhD, 1957, Dr Docent, 1974. Asst. prof. electronic engring. Poly. Inst., 1951-56, assoc. prof., 1956-65, prof., 1965—, dep. dean, dean dept. electronics, 1962-66, head dept., 1985-90; pres. commn. Nat. Coun. for Sci. and Tech., 1965-67, v.p. 1972-76; permanent sec. Commn. for Computers of Govt., 1967-71; dir. Rsch. Ctr. Electronic Components, 1969-70; gen. dir. Ctrl. Inst. for Informatics, 1976-85. Author: Electrons in Action, 1961, Transistor Circuits, 1961, Electronic Processes in Semiconductor Devices, 1961 (State prize 1964), Solid State Electronics, 1972, Work and Economy, 1974, Systems and Civilization, 1976, The Depths of the Material World, 1979, The Second Industrial Revolution, 1980, Science and Civilization, 1984, Orthophysics, 1985, Informatics and Society, 1987, Spirituality, Information, Matter, 1989, Information of Matter, 1990, Functional Electronics, 1991; editor: Global Problems of Mankind, 1979-85, books on Romanian sci. and tech. history; contbr articles on electronics to profl. jours., essays in philosophy. Dep. prime min. Romanian Govt., 1989-90; dep. mem. ctrl. com. Romanian Communist Party, 1969-74. Recipient Order Sci. Honor, Order Star of Republic, Romania, commandeur Legion of Honor, France, prize for sci. activity Romanian Ministry Edn., 1963. Mem. IEEE (sr.), Romanian Acad. (pres. 1990—), Ecuadorian Inst. for Natural Scis. (corr.). Avocation: philosophy. Office: Romanian Acad, Cales Victoriei 125, Bucharest Romania

DRAGO, ROBERT JOHN, psychologist; b. Queens, N.Y., Dec. 7, 1958; s. Michael Anthony and Adeline Barbara (Alberico) D. BA, St. John's U., 1980, MS, 1983, PhD, 1985; postdoctoral diploma, Adelphi U., 1991, addiction specialist cert., 1991; child abuse and domestic violence cert., Inst. Advanced Psychol. Studies, 1993. Lic. psychologist; cert. sch. psychologist, N.Y. Psychologist Blueberry Day Treatment Ctr., Bklyn., 1986-88; cons. Birch Early Childhood Ctr., Springfield Gardens, N.Y., 1990; sch. psychologist West Hempstead (N.Y.) Union Free Sch. Dist., 1988-92; pvt. practice Williston Park, N.Y., 1988—; therapist, supr. psychotherapy Advanced Ctr. Psychotherapy, Jamaica Estates, N.Y., 1990-93; assoc. psychologist Queens Children's Psychiat. Ctr., Bellerose, N.Y., 1992-93; sch. psychologist Mineola (N.Y.) Union Free Sch. Dist., 1993—; faculty Advanced Inst. Analytic Psychotherapy, 1993—. Office: 225 Hillside Ave Williston Park NY 11596

DRĂGOI, DĂNUT, physicist; b. Sărulesti, Buzău, Romania, May 8, 1952; came to U.S., 1990; s. Serban Radu and Ileana (Nichitelea) D.; m. Stefania-Cornelia Tudor, June 21, 1985. Degree in physics, U. Bucharest, Romania, 1972, postgrad., 1988. Physicist Enterprise for Ferrites, Urziceni, Romania, 1977-80; physicist Enterprise for Rsch. and Prodn. of Semicondr. Materials, Bucharest, 1980-84, prin. physicist, 1984-85, prin. scientist, 1989-90; vis. scientist U. Denver, 1990—; researcher Aluterv-FKI Budapest, Vishegrad, Hungary, 1986; diffractionist x-ray lab. U. Denver, 1990—. Author: Advances in X-Ray Analysis, 1992; contbr. articles to Jour. Applied Crystallography, Acta Crystallographica, Powder Diffraction. Achievements include rsch. in nonlinear equations for high accuracy of x-ray crystal orientation, nonlinear equations of linear crack patterns in silicon single crystles, x-ray crystal monochromator, peak broadening resulting from chi and psi tilts. Home: 2150 S Josephine St Denver CO 80210 Office: U Denver Dept Engring Denver CO 80208

DRAGT, ALEXANDER JAMES, physicist; b. Lafayette, Ind., Apr. 7, 1936; s. Gerrit and Beulah (Westra) D.; m. Lavonne Ann Wolters, Nov. 28, 1957; children: Alison Ann, Alexander James, William David. A.B., Calvin Coll., 1958; Ph.D. in Physics (NSF fellow), U. Calif., Berkeley, 1964. Sr. scientist Lockheed Missiles & Space Corp., Palo Alto, Calif., 1961-62; staff scientist Aerospace Corp., Los Angeles, 1963; mem. Inst. Advanced Study, Princeton, N.J., 1963-65; asst. prof. physics U. Md., 1965-68, assoc. prof., 1968-74, prof., 1974—, chmn. dept. physics and astronomy, 1975-78; mem. vis. staff Los Alamos Sci. Lab., 1978-79, cons., 1979—vis. prof. Tex. A&M U., 1984; mem. vis. staff Tex Accelerator Ctr., 1984; guest scientist Lawrence Berkeley Lab., 1985, cons., 1985—. Fellow Am. Phys. Soc.; Mem. Am. Geophys. Union, AAAS, Am. Math. Soc. Mem. Christian Reformed Ch. Research in theoretical physics, applied math. Office: U Md Dept Physics and Astronomy College Park MD 20742

DRAGUN, JAMES, soil chemist; b. Detroit, July 29, 1949; s. Henry George and Stella (Kubilus) D.; married, June 16, 1973; children: Nathan, Heather. BS, Wayne State U., 1971; MS, Pa. State U., 1975, PhD, 1977. Soil chemist U.S. EPA, Washington, 1978-82, Kennedy/Jenks, San Francisco, 1982-84, E. C. Jordan, Southfield, Mich., 1984-87, Stalwart Environ., Auburn Hills, Mich., 1987-88; soil chemist, pres. Dragun Corp., Farmington Hills, Mich., 1988—. Author: The Soil Chemistry of Hazardous Materials, 1988; editor-in-chief Sci. Jour.; contbr. over 70 articles to profl. publs. Recipient Disting. Svc. award Liquid Indsl. Control and Waste Mgmt. Assn., 1990. Mem. Sigma Xi, Phi Kappa Phi. Office: Dragun Corp Ste 260 30445 Northwestern Hwy Farmington Hills MI 48334

DRAIN, CHARLES MICHAEL, biophysical chemistry educator; b. St. Louis, July 30, 1959; s. Charles Michael and Marie (Kilgen) D. BS, U. Mo., 1980, BFA, 1981; PhD, Tufts U. Lab dir. Midco Products Co., St. Louis, 1980-84; postdoctoral fellow The Rockefeller U., N.Y.C., 1988-91; guest rschr. U. Louis Pasteur, Strasbourg, France, 1992—; Contbr. articles to jours. including Inorganic Chemistry, Biophys. Jour. and Procs. Nat. Acad. Sci. Recipient DuPont award DuPont, 1988, Chateaubriand French Scientific Mission, 1991; scholarship Internat. Sch. of Biophysics, 1989. Mem. AAAS, Am. Chem. Soc., Bioelectrochem. Soc. (Galvan Prize 1990). Office: U Louis Pasteur, 4 Rue Blaise Pascal, 67000 Strasbourg France

DRAKE, DAVID LEE, engineer; b. Campton, Ky., Mar. 15, 1960; s. Dudley and Sarah Ellen (Combs) D.; m. Bitha Mae Turner, June 10, 1983; 1 child, Thomas Shelton. AAS, Morehead State U., 1981, BS, 1983. Electronics lab. technician Morehead (Ky.) State U., 1979-81; quality control technician Computer Peripherals, Campton, 1981; robotics rsch. engr. Morehead State U., 1981-83; personal computer test technician Campton Electronics, 1984-86; chief engr. Automation Svcs., Lexington, Ky., 1986—. Contbr. articles to profl. jours. Mem. Sigma Tau Epsilon (parliamentarian 1982-83). Democrat. Home: PO Box 533 Campton KY 41301 Office: Automation Svcs Inc 2549 Richmond Rd Ste 400 Lexington KY 40509

DRAKE, FRANK DONALD, astronomy educator; b. Chgo., May 28, 1930; s. Richard Carvel and Winifred (Thompson) D.; m. Elizabeth Bell, Mar. 7, 1953 (div. 1977); children: Stephen, Richard, Paul; m. Amahl Zekin Shakhashiri, Mar. 4, 1978; children: Nadia, Leila. B in Engring. Physics, Cornell U., 1952; MA in Astronomy, Harvard U., 1956, PhD in Astronomy, 1958. Astronomer Nat. Radio Astron. Obs., Green Bank, W.Va., 1958-63; sect. chief Jet Propulsion Lab., Pasadena, Calif., 1963-64; prof. Cornell U., Ithaca, N.Y., 1964-84; dir. Nat. Astron. and Ionospace Ctr., Ithaca, 1971-81; dean natural sci. dept. U. Calif., Santa Cruz, 1984-88, prof., 1984—. Author: Intelligent Life in Space, 1962, Murmurs of Earth, 1978, Is Anyone Out There, The Scientific Search for Extraterrestrial Intelligence, 1992. Lt. USN, 1947-55. Fellow AAAS, Am. Acad. Arts and Scis.; mem. Nat. Acad. Scis., Internat. Astron. Union (chair U.S. nat. com.), Astron. Soc. Pacific (pres. 1988-90), Seti Inst. (pres. 1984—). Avocation: jewelry making. Office: U Calif Obs Santa Cruz CA 95064

DRAKE, JOHN WALTER, geneticist; b. Detroit, Feb. 10, 1932; s. John Alfred and Eleanor Bryan (Smith) D.; m. Pamela Elizabeth Grunau, Dec. 3, 1960; children: Juliet Anne, Jonathan Andrew Nicholas. BS magna cum laude, Yale U., 1954; PhD, Calif. Inst. Tech., 1958. Rsch. assoc., instr. microbiology U. Ill., Urbana, 1958-59, asst. prof., 1959-64, assoc. prof., 1964-69, prof., 1969—, chmn. genetics program, 1969-75; chief lab. environ. mutagenesis Nat. Inst. Environ. Health Scis., Research Triangle Park, N.C., 1977-78; chief lab. molecular genetics Nat. Inst. Environ. Health Scis., Research Triangle Park, 1979-82, head mutagenesis sect., 1983—, chief lab. molecular genetics, 1991—; adj. prof. biology U. N.C.; adj. prof. genetics Duke U.; vis. prof. Harvard U.; cons. in field. Author: Molecular Mechanisms of Mutation, 1970; editorial bd. Genetics, 1975—, editor-in-chief, 1982—; rsch. and publs. in embryology, virology, genetics. Fulbright fellow Weizmann Inst., Israel, 1957-58, Guggenheim fellow Lab. Molecular Biology, Cambridge, Eng., 1964-65, USPHS spl. fellow U. Edinburgh, Scotland, 1971-72. Mem. Genetics Soc. Am. (bd. dirs.), Environ. Mutagen Soc. (pres. 1976-77). Office: NIEHS Research Triangle Park NC 27709

DRAKE, RICHARD PAUL, physicist, educator; b. Washington, Oct. 25, 1954; s. Hugh Hess and Florence Jean (Steele) D.; m. Joyce Elaine Penner, Aug. 30, 1980; children: Katherine Anne, David Alexander. BA in Philosophy and Physics magna cum laude, Vanderbilt U., 1975; PhD in Physics, Johns Hopkins U., 1979. Physicist Lawrence Livermore (Calif.) Lab., 1979-89; assoc. prof. dept. applied sci. U. Calif., Davis, 1990-91, prof., 1991—; dir. Plasma Physics Rsch. Inst. Lawrence Livermore Nat. Lab.-U. Calif.-Davis, 1990—; ski instr. Squaw Valley (Calif.) USA, 1985—; chair Anomalous Absorption Conf., Tahoe City, Calif., 1987; referee NSF, Nature, Phys. Rev. Letters, other jours. Contbr. 80 articles to sci. publs. Fellow Am. Phys. Soc.; mem. AAAS, Phi Beta Kappa. Achievements include discovery of importance of time-dependence in parametric instabilities, original observation of stimulated Compton scattering, identification of absolute stimulated Raman scattering. Home: 2463 Covey Way Livermore CA 94550 Office: Plasma Physics Rsch Inst L 418 PO Box 808 Livermore CA 94551

DRAKE, ROGER ALLAN, psychology educator; b. Bremerton, Wash., July 25, 1943; s. Theodore Francis and Quinica Venetta (Cram) D.; m. Loretta Chen, July 14, 1979. Student, Calif. Inst. Tech.; BA, Western Washington U., 1966; MA, U. Iowa, 1969; PhD, U. Tenn., 1981. Prof. Western State Coll., Gunnison, Colo., 1969—; Fulbright lectr. Sheffield Polytech, Yorkshire, England, 1981-82; vis. prof. U. Colo., Boulder, 1984-85, sch. medicine Johns Hopkins U., Balt., 1988-89. Author: Cognitive Simplicity and Social Perceptual Errors, 1982; contbr. articles to profl. jours. Sci. rep. faculty senate Western State Coll., 1990-93; bd. trustees University Press Colo., 1992—. Sabbatical Rsch. grantee Charles A. Dana Found., 1988-89, Internat. Rsch. grantee NATO, Berlin, 1987-91; recipient Rsch. Opportunity grantee NSF, 1984-85. Mem. Soc. Advancement Social Psychology (pres. 1989-93), Psychonomic Soc., Soc. Neurosci. Achievements include research on effects of selective lateral brain activation on optimism, aesthetic judgments, risk, perceived control and causal attributions, understanding of persuasion and processing of arguments, including selective memory. Office: Western State Coll 103 Crawford Hall Gunnison CO 81231

DRAKE, SIMON ROBERT, inorganic chemistry; b. London, May 11, 1961; s. Arnold and Rhoda (Smith) D. BSc summa cum laude, U. Coll., Swansea, South Wales, 1980-84; PhD in Chemistry, Queens Coll., Cambridge, Eng., 1987. Rsch. fellow St. John's Coll., Cambridge, U.K., 1987-90; NSF rsch. fellow Dept. Chemistry, Ind. U., Bloomington, 1989-90; univ. lectr. Dept. of Chemistry, Imperial Coll., London, 1990—; cons. Halco Engring., Ltd., Didcot, U.K., 1992, Jica Electronic Systems, South Moreton, Oxon, 1992, Saffron Scientific, 1993—. Contbr. over 70 articles to profl. jours. Mem. Royal Soc. of Chemistry, Am. Soc. of Chemistry. Achievements include 2 patents in the area of low temperature routes to main group metal-non metal complexes and the application of the ammonium salt route to inorganic synthesis. Home: 164 Florence Rd, Wimbledon London SW19 8TN, England Office: Dept Chemistry, Imperial Coll of Sci, London SW 7 2AY, England

DRAKE, STEPHEN DOUGLAS, clinical psychologist, health facility administrator; b. Iola, Kans., Sept. 8, 1947; s. Harry Francis and Emojean (Price) D.; m. Rebecca Gonzalez. June 1, 1968; 1 child, Michael Paul. BA, U. Tex., 1970; PhD, U. North Tex., 1987. Lic. psychologist. Mental health worker Austin (Tex.) State Hosp., 1970-73; claims rep. Social Security Adminstrn., Galveston, Tex., 1974-77; ops. supr. Social Security Adminstrn., Dallas, 1977-79, staff asst., 1979-80; clin. psychologist Terrell (Tex.) State Hosp., 1987-89, Austin (Tex.) State Hosp., 1989-90; program dir. Austin State Hosp., 1990—. Contbr. articles to profl. jours. Vice-Chmn. bd. dirs. Galveston (Tex.) Island Mental Health/Mental Retardation Ctr., 1977; v.p. Grad. Assn. Students in Psychology U. North Tex., 1984, grad. rep. exec. com., 1984. Mem. APA, Assn. Advancement Behavior Therapy, Mensa, Phi Kappa Phi. Avocations: Tae Kwon Do, weightlifting, Eastern philosophy, foreign languages, travel. Office: Austin State Hosp 4110 Guadalupe St Austin TX 78751-4223

DRANCE, STEPHEN MICHAEL, ophthalmologist, educator; b. Bielsko, Poland, May 22, 1925; Can. citizen; MB,ChB, U. Edinburgh, Scotland, 1948, MD, 1949; Diploma in Ophthalmology, Royal Coll. Sci., London, 1953. Intern Western Gen. Hosp., Edinburgh, 1948-49; resident County Hosp., York, Eng., 1952-53, Edinburgh Royal Infirmary, 1953-55, Oxford Eye Hosp., Eng., 1955-57, Oxford U., 1955-57; asst. prof. and assoc. prof. medicine U. Sask., Saskatoon, Can., 1957-63; assoc. prof. ophthalmology U. B.C., Vancouver, Can., 1963-66, prof., 1966-90, dir. ophthalmologic research, 1967-73, head dept. ophthalmology, 1973-90; cons., lectr. medicine; vis. prof., lectr. numerous univs. Author: (with H. Reed) The Essentials of Perimetry, 2d edit., 1971, (with A. Neufeld) Applied Pharmacology of Glaucoma, 1984, (with D.R. Anderson) Automatic Perimetry in Glaucoma, 1985, (with A. Neufeld, M. van Buskirk) Applied Pharmacology of Glaucoma, 1991; assoc. editor Am. Archives Ophthalmology, 1961-74; mem. editorial bd. Can. Jour. Ophthalmology, 1966; mng. editor Albrecht von Graefe's Archive for Clin. and Exptl. Ophthalmology, 1979-90; contbr. articles to profl. jours., chpts. to books. With RAF, 1949-51. Decorated officer Order of Can., 1987; recipient numerous awards and grants for excellence in medicine. Mem. Can. Assn. Clin. Research, Assn. Ophthalmologic Research (U.K.), Assn. for Research in Vision and Ophthalmology, Can. Ophthalmol. Soc. (pres. 1974-75), B.C. Oto-Ophthalmol. Soc., Ophthal. Soc. U.K., Oxford Ophthalmol. Congress, Am. Acad. Ophthalmology, Can. Med. Assn., Brit. Med. Assn., B.C. Med. Assn., Internat. Perimetric Soc. (pres. 1982-88), Internat. Glaucoma Club, Pan-Am. Ophthalmol. Congress, Pan-Am. Glaucoma Soc., Pan-Am. Assn. Ophthalmology, Assn. N.Am. Glaucomatologists, N.Z. Ophthalmol. Soc. (hon.), N.Am. Glaucoma Club, Academia Ophthalmologica Internationalis, Internat. Congress Ophthalmology (pres. Glaucoma Soc. 1983-90), Concillium Ophthalmologica Universale (visual function com.), Royal Coll. Physicians and Surgeons Can. (sec. 1979), Royal Australian Coll. Ophthalmologists U.K. (hon. fellow). Office: U BC Dept Opthalmology, 2550 Willow St, Vancouver, BC Canada V5Z 3N9

DREA, EDWARD JOSEPH, pharmacist; b. Springfield, Ill., Jan. 24, 1954; s. Edward Francis and Doris Mae (Reynolds) D.; m. Lori Ann Urban, Dec. 6, 1985; children: Brandon Christopher, Bradley Joseph. BS in Pharmacy, U. Iowa, 1977, D in Pharmacy, 1986. Pharmacist The Pharmacy, Taylorville, Ill., 1977-84; rsch. assoc. U. Iowa Coll. Pharmacy and Medicine, Iowa City, 1985-86; dir. pharmacy St. Vincent Meml. Hosp., Taylorville, 1986-88; coord. clin. svcs. Meml. Med. Ctr., Springfield, Ill., 1988-92; asst. clin. prof. U. Ill. Coll. Pharmacy, Chgo., 1990-92; dir. clin. pharmacy svcs. CIGNA Healthplan Ariz., Phoenix, 1992—, HPI Health Care Svcs., Inc., 1992—; cons., Phoenix, 1990—. Assoc. editor book, 1990. Rsch. grantee The Uphjohn Co., 1988, Schering/Key Pharms, 1990, Xoma Corp., 1990. Mem. Am. Coll. Clin. Pharmacy, Am. Soc. Hosp. Pharmacists, Am. Pharm. Assn., Ill. Pharmacists Assn., Ariz. Pharmacy Assn., Acad. Managed Care Pharmacy. Avocations: family outings, cooking, gardening. Home: 15613 N 7th Dr Phoenix AZ 85023 Office: CIGNA Healthplan Ariz 8826 N 23d Ave Phoenix AZ 85021

DREBUS, RICHARD WILLIAM, pharmaceutical company executive; b. Oshkosh, Wis., Mar. 30, 1924; s. William and Frieda (Schmidt) D.; m. Hazel Redford, June 7, 1947; children—William R., John R., Kathryn L. BS, U. Wis., 1947, MS, 1949, PhD, 1952. Bus. trainee Marathon Paper Corp., Menasha, Wis., 1951-52; tng. mgr. Ansul Corp., Marinette, Wis., 1952-55; asst. to v.p. Ansul Corp., 1955-58, marketing mgr., 1958-60; dir. personnel devel. Mead Johnson & Co., Evansville, Ind., 1960-65; v.p. corporate planning Mead Johnson & Co., 1965-66, internat. pres., 1966-68; v.p. internat. div. Bristol-Myers Co. (merger Mead Johnson & Co. with Bristol-Myers Co.), N.Y.C., 1968-77, sr. v.p., 1977-78, v.p. parent co., 1978-85, sr. v.p. pharm. research and devel. div., 1985-89, ret., 1989. Past bd. dirs. Wallingford C. of C., Meriden C. of C., Wallingford Symphony, Jr. Achievement S.E. Conn., Meriden-Wallingford Mfrs. Assn., Meriden Silver Mus.; past bd. dirs. Meriden-Wallingford United Way, chmn. fund raising drive, 1988-89; trustee emeritus Quinnipiac Coll.; bd. dirs. Oshkosh Boys and Girls Club; mem. bd. visitors U. Wis., Madison. With inf. AUS, 1943-45. Decorated Combat Inf. Badge, Purple Heart, Bronze Star. Mem. APA, N.Y. Acad. Scis., U. Wis. Bascom Hill Soc., Oshkosh Country Club, Oskhosh Power Boat Club, North Shore Country Club, Phi Delta Kappa.

DREHER, LAWRENCE JOHN, mechanical engineer; b. Bklyn., Dec. 28, 1955; s. John Lawrence and Alice Martha (Baden) D.; m. Cynthia Sue Eichacker, May 27, 1978; children: Jeffrey S., Alison A., Stephen J. BS in Materials Engring., Rensselaer Poly. Inst., 1977; MS in Mech. Engring. and Ocean Engring., MIT, 1984. Registered profl. engr., Maine, Va. Commd. ens. USN, 1977, advanced through grades to lt. comdr., 1986, officer USS Arthur W. Radford, 1978-81; lt. comdr. USN supt. Norfolk Naval Shipyard USN, Portsmouth, Va., 1984-87; type desk-surface Norfolk Naval Shipyard USN, Portsmouth, Va., 1987-88; type desk-surface ships Comnavsurflant USN, Norfolk, Va., 1988-89; resigned USN, 1989; prin. engr. Bath (Maine) Iron Works, 1990-91, mgr. mech. engring., 1991—. Soccer coach Virginia Beach (Va.) Soccer Club, 1985-86; mem. traffic com. Town of Topsham, Maine, 1990—. Comdr. USNR, 1993. Mem. ASME, Am. Soc. Naval Engrs. (Brand award 1984), Sigma Xi, Tau Beta Pi. Lutheran. Home: 13 Sokokis Cir Topsham ME 04086 Office: Bath Iron Works 700 Washington St Bath ME 04530

DREIZEN, SAMUEL, oncologist; b. N.Y.C., Sept. 12, 1918; s. Charles and Rose (Schneider) D.; m. Ellie Jo Gilley, Aug. 3, 1956; 1 child, Pamela L. BA, Bklyn. Coll., 1941; DDS, Western Reserve U., 1945; MD, Northwestern U., 1958. Lic. MD Tex., Ill.; lic. DDS Ala., N.Y. Rsch. assoc. U. Cin. Med. Sch., Birmingham, Ala., 1945*47; instr. Northwestern Med. Sch., Chgo., 1947-49, asst. prof., 1949-58, assoc. prof., 1958-65; prof. U. Tex. Health & Sci. Ctr., Houston, 1965-89, prof. emeritus, 1989—; asst. dir. Nutrition Clinic Hillman Hosp., Brimingham, 1950-65. Contbr. articles to profl. jours. Capt. U.S. Army, 1953-60. Fellow Am. Assn. Advanced Sci.; mem. Am. Assn. Physical Anthropology, Internat. Assn. Dental Rsch., Soc. Rsch. Child Devel. Home: 5218 Dumfries Dr Houston TX 77096 Office: U Tex Health & Sci Ctr PO Box 20068 Houston TX 77225

DRELL, SIDNEY DAVID, physicist, educator; b. Atlantic City, N.J., Sept. 13, 1926; s. Tulla and Rose (White) D.; m. Harriet Stainback, Mar. 22, 1952; children: Daniel White, Persis Sydney, Joanna Harriet. AB, Princeton U., 1946; MA, U. Ill., 1947, PhD, 1949, DSc (hon.), 1981. Rsch. assoc. U. Ill., 1949-50; instr. physics Stanford U., 1950-52, assoc. prof., 1956-60, prof., 1960-63, Lewis M. Terman prof. and fellow, 1979-84; co-dir. Stanford U. Ctr. for Internat. Security and Arms Control, 1983-89; prof. Stanford Linear Accelerator Ctr., 1963—, dep. dir., 1969—, exec. head theoretical physics, 1969-86; research assoc. MIT, 1952-53, asst. prof., 1953-56, adv. bd. Lincoln Lab., 1985-90; vis. scientist Guggenheim fellow CERN Lab., Switzerland, 1961, U. Rome, 1972; vis. prof., Loeb lectr. Harvard U., 1962, 70; vis. Schrodinger prof. theoretical physics U. Vienna, 1975; cons. Office Sci. and Tech., 1960-73, Office Sci. and Tech. Policy, 1977-82, ACDA, 1969-81, Office Tech. Assessment U.S. Congress, 1975-91, House Armed Svcs. Com., 1990—, Senate Select Com. on Intelligence, 1990—, NSC, 1973-81; mem. high energy physics adv. panel Dept. Energy, 1973-86, chmn., 1974-82, mem. energy rsch. adv. bd., 1978-80; mem. Jason, 1960—; Richtmyer lectr. to Am. Assn. Physics Tchrs., San Francisco, 1978; vis. fellow All Souls Coll., Oxford, 1979; Danz lectr. U. Wash., 1983, Hans Bethe lectr. Cornell U., 1988; I.I. Rabi vis. prof. Columbia U., 1984; mem. Carnegie Commn. on Sci., Tech. and Govt., 1988—; chmn. U.C. pres. coun. on nat. labs., 1992—; chmn. internat. adv. bd. Inst. Global Conflict and Cooperation, U. Calif., 1990—; adj. prof. engring., pub. policy Carnegie Mellon U., 1989—. Author 6 books; contbr. articles to profl. jours. Trustee Inst. Advanced Study, Princeton, 1974-83; bd. govs. Weizmann Inst. Sci., Rehovoth, Israel, 1970—; bd. dirs. Ann. Revs., Inc., 1976—; mem. Pres. Sci. Adv. Com., 1966-70. Recipient Ernest Orlando Lawrence Meml. award and medal for research in theoretical physics AEC, 1972, Alumni award for distinguished service in engring. U. Ill., 1973, Alumni Achievement award, 1988; MacArthur fellow, 1984-89. Fellow Am. Phys. Soc. (pres. 1986, Leo Szilard award for physics in the public interest 1980); mem. Nat. Acad. Scis., AAAS (Hilliard Roderick Prize in Sci., Arms Control, Internat. Security, 1993), Am. Philo. Soc., Arms Control Assn. (bd. dirs. 1978—), Council on Fgn. Relations, Aspen Strategy Group (emeritus 1991). Home: 570 Alvarado Row Palo Alto CA 94305-8501 Office: Stanford Linear Accelerator Ctr PO Box 4349 Palo Alto CA 94309-4349

DRENTH, PIE:ER JOHAN DIEDERK, psychology educator, consultant; b. Appelscha, Friesland, The Netherlands, Mar. 8, 1935; s. Gerrit and

Froukje (Wouda) D.; m. Maria Annetta E. De Boer, 1959; children: Gerard D., Johannes Ch., Martin P. Candidate in psychology, Free U., Amsterdam, The Netherlands, 1955, doctoral in psychology, 1958, PhD in Psychology, 1960; D (hon.), State U. Ghent, 1981. Selection dept. Royal Dutch Navy, 1955-60; rsch. fellow Standard Oil Co. N.J., N.Y.C., 1960-61; sr. lectr. Free U., 1962-67, prof. psychology, 1967—, head dept. work and orgnl. psychology, 1967, vice chancellor, 1983-87; pres. Royal Netherlands Acad. Atrs Scis., 1990—; vis. prof. Washington U., St. Louis, 1966, U. Wash., Seattle, 1977; cons. Unilever, Rabo-Bank, Mandev, the Netherlands, 1975—; pres. 1st European Conf. Psychology, Amsterdam, 1989; mem. sci. com. adv. panel NATO, Brussels, 1969-83, chmn., 1980-83; mem. supervisory bd. Shell Nederland B.V., 1991—. Author 3 books, co-author 9 books; co-editor 9 books; contbr. over 100 articles to profl. jours., also tests and manuals. Bd. dirs. Netherlands-Am. Com. for Ednl. Exchange, 1986—, Fondation Praemium Erasmianum, Amsterdam, 1989—. 1st lt. Royal Dutch Navy, 1958-60. Mem. Royal Netherlands Acad. Arts and Sci. (gen. sec. 1987-90, pres. 1990—), Netherlands Inst. Psychologists (Heymans award 1986), European network Profs. in Indsl.-Orgnl. Psychology, Am. Psychol. Assn. (fgn. affiliate), Netherlands Orgn. for Advancement Pure Rsch. (coun. 1975-85), Internat. Assn. Applied Psychology (pres. div. orgnl. psychology 1982-86), Rotary. Home: Pekkendam 6, 1081 HR Amsterdam The Netherlands*

DRESCHER, EDWIN ANTHONY, chemical engineer; b. Ellwood, Pa., May 22, 1956; s. Victor Harold and Mildred (Hartung) D.; m. Patricia Elder, July 27, 1977 (div. 1980); 1 child, Anthony Leroy; m. Melissa Marie Ruiz, Jan. 2, 1982; children: Victor Myles, Briana Leigh. BS in Chemistry, Slippery Rock State U., 1978; MS in Chem. Engr., U. Dayton, 1981. Chemist Metcoa, Pulaski, Pa., 1977-78, Roessing Bronze, Evans City, Pa., 1978-79; rsch. asst. U.D. Rsch. Inst., DAyton, Ohio, 1980-81; field engr. Gearhart Industries, Lafayette, Ohio, 1981-84; process engr. Armant Metal Chlorides, Vacherie, La., 1984-88; chief chemist Ormet Corp., Burnside, La., 1988--. Mem. Am. Chem. Soc., Am. Soc. for Quality Control. Achievements include development of method of purifying bayer liquor,effeient granular aluninum chloride condenser, positive pressure solids feeder, increased oil yields from shale. Office: Ormet Corp PO Box 15 Burnside LA 70738

DRESCHLER, WOUTER ALBERT, audiologist, researcher; b. Amsterdam, The Netherlands, Sept. 19, 1953; s. Willem and Anna Maria (Pot) D.; m. Anne Marie Louise Overeem, June 7, 1974; children: Annemieke, Mark, Peter. EE, Tech. U., Delft, The Netherlands, 1977; PhD, Free U., Amsterdam, 1983. Researcher Free U., 1977-80; head audiol. ctr. U. Amsterdam, 1980—. Author: Relations Between Psychophysical Data and Speech Perception, 1983, (chpt.) Auditory Psychophysics: Spectro-temporal Representation of Signals, 1987. Fellow Internat. Collegium Rehab. Audiology, Acoustical Soc. Am., Dutch Soc. Audiology, Dutch Acoustical Soc. Dutch ENT Soc. Avocation: organ. Home: Beemdgras 4, 1441 WB Purmerend The Netherlands Office: Acad Med Ctr, Meibergdreef 9, 1105 AZ Amsterdam The Netherlands

DRESSELHAUS, MILDRED SPIEWAK, physics and engineering educator; b. Bklyn., Nov. 11, 1930; d. Meyer and Ethel (Teichteil) Spiewak; m. Gene F. Dresselhaus, May 25, 1958; children: Marianne Dresselhaus Cooper, Carl Eric, Paul David, Eliot Michael. BA, Hunter Coll., 1951, DSc (hon.), 1982; Fulbright fellow, Cambridge (Eng.) U., 1951-52; MA, Radcliffe Coll., 1953; PhD in Physics, U. Chgo., 1958; D Engring. (hon.), Worcester Poly. Inst., 1976; DSc (hon.), Smith Coll., 1980, N.J. Inst. Tech., 1984; Doctorat Honoris Causa, U. Catholique de Louvain, 1988; DSc (hon.), Rutgers U., 1989, U. Conn., 1992, U. Mass., Boston, 1992, Princeton U., 1992; DEngring, Colo. Sch. of Mines, 1993. NSF postdoctoral fellow Cornell U., 1958-60; mem. staff Lincoln Lab., MIT, Lexington, 1960-67; prof. elec. engring. MIT, Cambridge, 1967—, assoc. dept. head elec. engring., 1972-74, prof. physics, 1983—, Inst. prof., 1985—, Abby Rockefeller Mauze chair, 1973-85, dir. Ctr. for Materials Sci. and Engring., 1977-83; vis. prof. dept. physics U. Campinas, Brazil, summer 1971, Technion, Israel Inst. Tech., Haifa, 1972, 90, Nihon and Aoyama Gakuin Univs., Tokyo, 1973, IVIC, Caracas, Venezuela, 1977; vis. prof. dept. elec. engring. U. Calif., Berkeley, 1985; Graffin lectr. Am. Carbon Soc., 1982; chmn. steering com. on evaluation panels Nat. Bur. Standards, 1978-83; mem. Energy Rsch. Adv. Bd., 1984-90; bd. dirs. Rogers Corp., Quantum Chem. Corp. Contbr. articles to profl. jours. Bd. govs. Argonne Nat. Lab., 1986-89; mem. governing bd. NRC, 1984-87, 89-90, 92—. Recipient Alumnae medal Radcliffe Coll., 1973, Killian Faculty Achievement award, 1986-87, Nat. medal of sci. NSF, 1990; named Hunter Coll. Hall of Fame, 1972; Fulbright fellow Cambridge (Eng.) U., 1951-52. Fellow IEEE, AAAS (bd. dirs. 1985-89), Am. Phys. Soc. (pres. 1984), Am. Acad. Arts. and Scis.; mem. Nat. Acad. Engring. (coun. 1981-87), Soc. Women Engrs. (Achievement award 1977), Nat. Acad. Scis. (coun. 1987-90, chmn. engring. sect. 1987-90, chmn. class III 1990-93, treas. 1992—), Brazilian Acad. Sci. (corr.). Office: MIT Rm 13-3005 Cambridge MA 02139

DRESSICK, WALTER J., chemist; b. Windber, Pa., Mar. 8, 1955; s. Walter S. and Susan (Wargo) D. BS in Chemistry, U. Pitts., 1977; PhD in Inorganic Chemistry, U. N.C., 1981. Postdoctoral ast. U. Va., Charlottesville, 1981-83; rsch. scientist Colgate Palmolive Co., Piscataway, N.J., 1983-86; sr. chemist Igen, Inc., Rockville, Md., 1986-88; rsch. doctoral asst. U. Fribourg, Switzerland, 1988-89; rsch. chemist Geo-Centers, Inc., Ft. Washington, Md., 1989-91, Naval Rsch. Lab., Washington, 1991—. Contbr. articles to profl. jours. NSF fellow, 1977-80; recipient Ira R. Messer Chemistry award U. Pitts., 1977, Phillps medal 1977, Outstanding Performance award Dale Carnegie Inst., 1991. Mem. Am. Chem. Soc., SPIE. Roman Catholic. Achievements include patents on electrochemiluminescent assays and kits using ruthenium and osmium bipyridyl complexes as labels; electroless deposition employing a tin-free palladium catalyst. Office: Naval Rsch Lab Code 6900 4555 Overlook Ave Washington DC 20375-5348

DREVINSKY, DAVID MATTHEW, civil engineer; b. Worcester, Mass., Apr. 20, 1957; s. Peter Joseph and Dorothy Julia (Wayner) D. BSCE, Worcester Polytech, 1980; MSCE, Northeastern U., 1985. Lab. tech. Du Pont, New England Nuclear, Billerica, Mass., 1978; intern Dept. Environ. Protection, Lakeville, Mass., 1979; environ. engr. EPA, Boston, 1981; rsch. engr. Met. Area Planning Coun., Boston, 1985-86; assoc. engr. Bayside Engring., Somerville, 1986-87, Charles T. Main, Boston, 1987-89; project mgr. GSA, Boston, 1990—. Contbr. articles to profl. jours. Mem. Met. Chpt. Profl. Engrs., Air Waste Mgmt. Engrs. (met. chpt.), Boston Chpt. Civil Engrs., U.S. Chess Assn., Worcester Polytech Chess Club, Postal Chess. Roman Catholic. Achievements include research on improving efficiency of secondary clarifiers. Home: 219 Old Connecticut Path Wayland MA 01778-3121

DREW, RICHARD ALLEN, electrical and instrument engineer; b. Milw., Jan. 10, 1941; s. Frank Emmons and Irene Louise (Wollaeger) D. BSEE, Milw. Sch. Engring., 1970. Registered profl. engr., Wis. Instrument engr. Nekoosa Papers Inc., Port Edwards, Wis., 1970-74; sr. instrument engr. Nekoosa Papers Inc., Port Edwards, 1974-85, Specialty Systems Inc., Mosinee, Wis., 1985-87; chief elec. and instrument engr. Zimpro Environ. Inc., Rothschild, Wis., 1988—. With USAF, 1963-67. Recipient Outstanding Svc. award Pulp and Paper Industry Div., Instrument Soc. Am., 1983, Outstanding Alumnus award Milw. Sch. of Engring., 1985. Mem. Instrument Soc. Am. (sr. mem., chpt. pres. 1974-75), Am. Radio Relay League (life mem.), Milw. Sch. of Engring. Alumni Orgn. (chpt. pres. 1991—). Achievements include research in pulp and paper indsl. control systems and waste treatment control systems. Office: Zimpro Environ Inc 301 W Military Rd Rothschild WI 54474

DREWEK, GERARD ALAN, mechanical engineer; b. Milw., Sept. 3, 1963; s. Kenneth F. Drewek and Beverly M. (Brumer) Schellhaass; m. Sheri Ann Block, June 29, 1985; children: Matthew Gerard, Amanda Christine. B of Aerospace Engring. and Mechanics, U. Minn., 1986. Mech. engr. Control Data Corp., Mpls., 1986-92; sr. mech. engr. Computing Devices Internat., Bloomington, Minn., 1992—. Contbr. articles to Electronic Packaging and Prodn. Mem. IEEE, AIAA (sec. Digital Avionics Tech. com. 1990-93, mem. various coms.). Home: 4333 Garden Tr Eagan MN 55123 Office: Computing Devices Internat 8800 Queen Ave S Bloomington MN 55431

DREWS, JÜRGEN, pharmaceutical researcher; b. Berlin, Aug. 16, 1933; came to U.S., 1991; s. Walter and Charlotte (Schneider) D.; m. Helga Eberlein, July 24, 1963; children: Ulrike, Karoline, Bettina. MD, Free U. Berlin, 1959; Professorship, U. Heidelberg, Fed. Republic of Germany, 1973. Head chemotherapy Sandoz Rsch. Inst., Vienna, Austria, 1976-79, head of inst., 1979-82; head internat. pharm. rsch. and devel. Sandoz, Ltd., Basel, Switzerland, 1982-85; dir. pharm. rsch. F. Hoffmann-La Roche Ltd., Basel, 1985-86, chmn. rsch. bd., mem. exec. com., 1986-90; pres. internat. rsch. and devel., mem. exec. com. Hoffmann-La Roche Inc., Nutley, N.J., 1991—; prof. medicine, U. Heidelberg, 1973—; mem. sci. adv. bd. (jour.) Infection, München, Fed. Republic of Germany, 1973—, Drug News & Perspectives, Barcelona, Spain, 1988—, Klinische Pharmakologie, München, 1989—; bd. dirs. Genentech, Inc., South San Francisco; bd. dirs. and internat. bd. advisors Basel Inst. Immunology. Author: Immunpharmakologie, Grundlagen und Perspektiven, 1986, Immunopharmacology, Principles and Perspectives, 1990; editor: (with others) Topics in Infectious Diseases, vol. 1, 1975, vol. 2, 1977; also over 200 articles. Office: Hoffmann-La Roche Inc 340 Kingsland St Nutley NJ 07110-1199

DREYFUSS, PATRICIA, chemist, researcher; b. Reading, Pa., Apr. 28, 1932; d. Edmund T. and Anna J. (Oberc) Gajewski; m. M. Peter Dreyfuss, Jan. 30, 1954; children: David Daniel, Simeon Karl. BS Chemistry, U. Rochester, 1954; PhD, U. Akron, 1964. Postdoctoral fellow U. Liverpool (Eng.), 1963-65; rsch. chemist B.F. Goodrich, Brecksville, Ohio, 1965-71; rsch. assoc. Case Western Res. U., Cleve., 1971-73, sr. rsch. assoc., 1973-74; rsch. assoc. Inst. Polymers Sci., U. Akron, Ohio, 1974-84; sr. rsch. scientist, rsch. prof. Mich. Molecular Inst., Midland, 1984-90; vis. rsch. fellow U. Bristol, 1972; cons. in field, 1974—; vis. prof. Polish Acad. Scis., Poland, 1974; adj. prof. Cen. Mich. U., Mt. Pleasant, Mich. Tech U., Houghton, 1986-92, Mich. Molecular Inst., Midland, 1990-92. Author: Poly (Tetrahydrofuran), 1982; contbr. over 85 articles to profl. jours.; co-author books. Flutist West Suburban Philharmonic Orch., Lakewood, Ohio, 1969-75, Explorer advisor Explorer post 2069 Boy Scouts Am., Akron, 1975-81; sec., bd. dirs. Adhesion Soc., 1976-88; treas. LWV, 1959-60; mem. ensemble Blessed Sacrament Ch., Midland, Mich., occasional flute soloist. Centennial scholar U. Rochester, 1950-54; Sohio fellow U. Akron, 1960, NSF Coop. Grad. fellow, 1961-63, Internat. fellow AAUW, 1964-65, NIH Spl. fellow, 1972-73. Mem. Am. Chem. Soc. (cen. region mtg. chmn. 1984-90, loc. sec. chmn., vice chmn., sec. and bd. dirs. Akron chpt. 1974-84, bd. dirs. Midland chpt. 1985-89, Outstanding Leadership Performance award 1981, Disting. Svc. award Akron chpt. 1985), AAUW (bd. dirs. Akron chpt.). Achievements include 4 patents in field. Home: 3980 Old Pine Trl Midland MI 48642-8891

DREZNER, STEPHEN M., policy analyst, engineer; b. N.Y.C., Sept. 13, 1937; s. M. Harris and Freda (Rome) D.; m. Naomi Mutterperl, Aug. 25, 1957; children: Jeffrey A., David S., Melissa A., Caren B., Jonathan A. Student, Queens Coll., Flushing, N.Y., 1955-57; BS, Columbia U., 1959; postgrad., Rutgers U., 1959-61; student, UCLA, 1962-63; MS in Quantitative Sci., U. So. Calif., 1966, D in Bus. Administrn., Quantitative Sci., 1969. Registered profl. engr., Calif. Mathematician Standard Oil N.J., Linden, 1959-61; tech. staff mem. Hughes Aircraft, Culver City, Calif., 1961-63, Indsl. Dynamics div. Hughes Tool Co., Los Angeles, 1963-64; gen. mgr. Info. Transfer Corp., Santa Monica, Calif., 1969-72; analyst Rand Corp., Santa Monica, 1964-69, program dir., 1972-80, v.p. rsch., 1980—; cons. Family Service Assn., N.Y.C., 1969-80, Dept. Def., Washington, 1977-79; assoc. mem. Def. Sci. Bd., Washington, 1981-84; mem. UCLA Ctr. Internat. & Strategic Affairs; mem. Inter-Agy. Sem., Washington. Authro: A Planning Guide, 1979; author book chpt., numerous jour. articles and Rand Corp. pubs. Mem. Alpha Pi Mu, Tau Beta Pi. Avocations: backpacking, youth athletics. Home: 3934 Bon Homme Rd Calabasas CA 91302-5706*

DRICKAMER, HARRY GEORGE, retired chemistry educator; b. Cleve., Nov. 19, 1918; s. George Henry and Louise (Strempel) D.; m. Mae Elizabeth McFillen, Oct. 28, 1942; children: Lee Charles, Lynn Louise, Lowell Kurt, Margaret Ann, Priscilla. B.S., U. Mich., 1941, M.S., 1942, Ph.D., 1946. Chem. engr. Pan Am. Refining Corp., 1942-46; asst. prof. U. Ill. at Urbana, 1946-49, assoc. prof., 1949-53, prof. chemistry, chem. engring. and physics, 1953-90, prof. emeritus, 1990—. Recipient Bendix award, 1968, P.W. Bridgman award Internat. Assn. High Pressure Sci. and Tech., 1977; Michelson-Morley award Case Western Res. U., 1978, John Scott award City of Phila., 1984, Alexander von Humboldt award W.Ger., 1986, Robert A. Welch prize Welch Found., 1987, Disting. Profl. Achievement award U. Mich. Alumni Assn., Elliot Cresson medal Franklin Inst., 1988, Nat. Medal of Sci., 1989; Guggenheim fellow, 1952. Fellow Am. Acad. Arts and Scis., Am. Phys. Soc. (Buckley Solid State Physics award 1967), Am. Geophys. Union; mem. NAS, NAE, Am. Chem. Soc. (Ipatieff prize 1956, Langmuir award in chem. physics 1974, Debye award in physical chemistry 1987), Am. Inst. Chemists (Chem. Pioneers award 1983), Am. Inst. Chem. Engrs. (Colburn award 1947, Alpha Chi Sigma award 1967, Walker award 1972, W.K. Lewis award 1986), Faraday div. of Royal Soc. Chemistry (London),), Am. Philos. Soc., Center for Advanced Studies. Home: 304 E Pennsylvania Ave Urbana IL 61801-5129

DRIEDGER, PAUL EDWIN, pharmaceutical researcher; b. Oak Park, Ill., May 7, 1948; s. Edwin Wilfred and Elinor (Kester) D. BS, MS in Chemistry, Tufts U., 1970; PhD in Pharmacology, Harvard Med. Sch., 1979. Chemist Kodak Rsch. Labs., Rochester, N.Y., 1970-73; v.p., dir. rsch. Lifesystems Co., Inc., Newton, Mass., 1979-80; pres. LC Svcs. Corp., Woburn, Mass., 1980-92, Alder Rsch. Ctr. Corp., Woburn, 1983-92; pres., CEO and chief sci. officer Procyon Pharms., Inc., Woburn, Mass., 1992—. Mem. AAAS, Am. Soc. Microbiology, Am. Chem. Soc. Achievements include co-discovery of phorbol ester receptor on protein kinase C; patents issued for discovery of broad range of protein kinase C modulators with human therapeutic potential. Office: Procyon Pharms Inc 165 New Boston St Woburn MA 01801-6228

DRISCOL, JEFFREY WILLIAM, chemist, researcher; b. Davenport, Iowa, Aug. 18, 1961; married. BS in Chemistry, U. Wis., 1983; MS in Chemsitry, U. N.C., 1990. Rsch. asst. U. N.C., Chapel Hill, 1988—. Contbr. articles to Jour. Molecular Structure, Analytical chemsitry, Jour. Am. Chem. Soc. Treas. Brighton Sq. HOA, Carrboro, N.C., 1992. Lt. USNR, 1983-88. Achievements include research in conformational analysis of biological molecules, molecular modeling and laser ionization mass spectrometry. Office: U NC Venable Hall Chapel Hill NC 27599-3290

DRISCOLL, CHARLES F., research physicist; b. Tucson, Feb. 28, 1950; s. John Raymond Gozzi and Barbara Jean (Hamilton) Driscoll; m. Susan C. Bain, Dec. 30, 1972; children: Thomas A., Robert A. BA in Physics summa cum laude, Cornell U., 1969; MS, U. Calif. San Diego, La Jolla, 1972, PhD, 1976. Staff scientist Gen. Atomics, San Diego, 1969; rsch. asst. U. Calif. San Diego, La Jolla, 1971-76, rsch. physicist, sr. lectr., 1976—; staff physicist, cons. Molecular Biosystems, Inc., San Diego, 1981-82; assoc. dir. Inst. for Pure and Applied Scis., La Jolla, 1991; cons. Sci. Applications, Inc., 1980-81. Editor: Non-Neutral Plasma Physics, 1988; contbr. 30 articles to sci. jours. Fellow NSF, 1969-71. Fellow Am. Phys. Soc. (Excellence in Plasma Physics Rsch. award 1991); mem. AAAS, Math. Assn. Am., Phi Beta Kappa. Achievements include development of quantitative analysis of magnetic targeting of microspheres in capillaries, experiments and theory on magnetized electron plasmas, new electron plasma apparatus, laser-diagnosed ion plasma apparatus for in-situ particle transport measurements; establishment of magnetic containment characteristics of electron plasmas; measurement of collisional transport to thermal equilibrium; observation of new 2D fluid instability and quantified vortex formation, merger, and dynamical evolution. Office: U Calif San Diego Dept Physics 0319 9500 Gilman Dr La Jolla CA 92093-0319

DRISCOLL, HENRY KEANE, endocrinologist, researcher; b. Boston, Dec. 24, 1953; s. John Joseph and Marie Elizabeth (Keane) D. SB in Life Scis., Mass. Inst. Tech., 1975, SM, 1976; MD, U. Mass., 1981. Diplomate Am. Bd. Internal Medicine, Bd. on Endocrinology and Metabolism. Asst. prof. medicine Sch. Medicine, Marshall Univ., Huntington, W.V., 1987-91; assoc. prof. medicine Sch. Medicine, Marshall Univ., 1991—. Fellow ACP; mem. Am. Fedn. Clin. Rsch., Am. Diabetes Assn. (bd. dirs. W.Va. affiliate), Endocrine Soc., Juv. Diabetes Found. Achievements include: rsch. in immunology of insulin-dependent diabetes mellitus; hormone secretions from pancreatic islets; actions of retinoids. Office: Marshall U Sch Medicine Dept Medicine Huntington WV 25755-9410

DRISCOLL, TERENCE PATRICK, environmental engineer; b. Buffalo, Sept. 22, 1948; s. David Girard and Dorothy (Van Aernam) D.; m. Sandra Lynn Aponowich, May 14, 1983; children: Peter, Emily, Julia. BSME, Mich. State U., 1970, MS in Environ. Engring., 1975; MBA, Boston Coll., 1979; cert. advanced studies in internat. bus., Northeastern U., 1984. Registered profl. engr., Maine, Mass., N.C., Ga., Fla. Project mgr. Metcalf & Eddy, Inc., Boston, 1975-85; v.p. Metcalf & Eddy, Inc., Atlanta, 1985-88, Waste Purification Systems, Atlanta, 1988-90; gen. mgr. processing systems Bird Environ., Atlanta, 1990-92; prin. Terence P. Driscoll & Assocs., Atlanta, 1993—; mem. indsl. wastes com. Water Environ. Fedn., Alexandria, Va., 1991—. Author: Manual of Practice-Industrial Pretreatment, 1992; contbr. articles to profl. publs., chpt. to books. Big brother Big Bros. of Lansing, Mich., 1973-75; vol. Headstart, Lansing, 1970. Lt. USN, 1970-73. Home: 3091 Dale Dr Atlanta GA 30305

DRISKELL, CLAUDE EVANS, dentist; b. Chgo., Jan. 13, 1926; s. James Ernest and Helen Elizabeth (Perry) D., Sr.; B.S., Roosevelt U., 1950; B.S. in Dentistry, U. Ill., 1952, D.D.S., 1954; m. Naomi Roberts, Sept. 30, 1953; 1 dau., Yvette Michele; stepchildren—Isaiah, Ruth, Reginald, Elaine. Practice dentistry, Chgo., 1954—. Adj. prof. Chgo. State U., 1971—; dean's aide, adviser black students Coll. Dentistry U. Ill., 1972—; dental cons., supervising dentist, dental hygienists supportive health services Bd. Edn., Chgo., 1974. Vice pres. bd. dirs. Jackson Park Highlands Assn., 1971-73. Served with AUS, 1944-46; ETO. Fellow Internat. Biog. Assn., Royal Soc. Health (Gt. Britain), Acad. Gen. Dentistry; mem. Lincoln Dental Soc. (editor), Chgo. Dental Soc., ADA, Nat. Dental Assn., (editor pres.'s newsletter; dir. pub. relations, publicity; recipient pres.'s spl. achievement award 1969) dental assns., Am. Assn. Dental Editors, Acad. Gen. Dentistry, Soc. Med. Writers, Soc. Advancement Anesthesia in Dentistry, Omega Psi Phi. Author: The Influence of the Halogen Elements upon the Hydrocarbon, and their Effect on General Anesthesia, 1962; History of Chicago's Black Dental Professionals, 1850-1983. Asst. editor Nat. Dental Assn. Quar. Jour., 1977—. Contbr. articles to profl. jours. Home: 6727 S Bennett Ave Chicago IL 60649-1031 Office: 11139 S Halsted St Chicago IL 60628

DRITSAS, GEORGE VASSILIOS, electronics engineer; b. Markopoulo, Attica, Greece, Aug. 31, 1940; s. Vassilios George and Maria Spyridon (Stavrou) D.; m. Athena Aristides Paschou, Mar. 6, 1981; children: Aristides, Vassilios. BSEE, U. Ky., 1968; MSEE, Stevens Inst Tech., Hoboken, N.J., 1972. Design engr. RCA, Somerville, N.J., 1969-71; project engr. Fabri-Kal Co., Somerville, 1971-73; tech. cons. Univ. Computing Co., East Brunswick, N.J., 1973-75; sr. engr. Instrumentation Engring., Inc., Franklin Lakes, N.J., 1975-76; chief engr. Chatlos Systems, Inc., Whippany, N.J., 1976-78; sr. engr. Singer Co., Elizabeth, N.J., 1978-80; tech. cons. Anko S.A., Athens, Greece, 1980-82, Ministry of Defence CISD, Athens, 1982—. Contbr. technical articles to profl. jours. Mem. Am. Hellenic Ednl. Progressive Assn., N.J., 1971, Ednl. Progressive Assn., Markopoulo, Greece, 1980. 2nd lt. Greek Army, 1961-64. Recipient scholarship Rotary Internat., Lexington, Ky., 1966, U. Ky., 1968. Mem. IEEE, Tech. Chamber of Greece, Armed Forces Comm. and Electronics Assn., Hellenic Assn. Elec. Engrs., U. Ky. Electrical Engring. Honor Soc., U. Ky. Classical Studies Honor Soc. Mem. Greek Orthodox Ch. Avocations: painting, tennis, soccer, swimming. Home: Davaki 11, 19400 Koropi Greece Office: Ministry of Defence, Geetha/Dep, Athens Greece

DRIVER, RODNEY DAVID, mathematics educator, state legislator; b. London, July 1, 1932; came to U.S., 1945; s. William T. and Marjorie E. (Carter) D.; m. Carole J. Frandsen, Sept. 4, 1955; children: David M., Karen L., Bruce K. BSEE, U. Minn., 1953, PhD in Math., 1960. Postdoctoral fellow Rsch. Inst. for Advanced Studies, Balt., 1960-61; staff mem. Math. Rsch. Ctr., Madison, Wis., 1961-62, Sandia Labs., Albuquerque, 1962-69; assoc. prof. math. U. R.I., Kingston, 1969-74, prof., 1974—; mem. R.I. Ho. of Reps., 1987—. Author: Ordinary and Delay Differential Equations, 1977, Introduction to Ordinary Differential Equations, 1978, Why Math?, 1984; contbr. articles on functional differential equations to math. jours. Del. R.I. Constl. Conv., 1986 (state rep. 1987—). Mem. Am. Math. Soc., Amnesty Internat., Greenpeace. Democrat. Unitarian. Home: 37 Hoxsie Rd West Kingston RI 02892-1017 Office: URI Kingston RI 02881

DRNEVICH, VINCENT PAUL, civil engineering educator; b. Wilkinsburg, Pa., Aug. 6, 1940; s. Louis B. and Mary (Kutcel) D.; m. Roxanne M. Hosier, Aug. 20, 1966; children: Paul, Julie, Jenny, Marisa. BSCE, U. Notre Dame, 1962, MSCE, 1964; PhD, U. Mich., 1967. Registered profl. engr., Ky., Ind. Asst. prof. civil engring. U. Ky., Lexington, 1967-73, assoc. prof., 1973-78, prof., 1978-91; chmn. civil engring., 1980-84; acting dean engring. U. Ky., Lexington, 1989-90; prof., head sch. civil engring., dir. Purdue U., West Lafayette, Ind., 1991—; pres. Soil Dynamics, Instruments, Inc., West Lafayette, 1974—. Inventor in field. Fellow ASCE (Norman medal 1973, Huber rsch. prize 1980), ASTM (exec. com., tech. editor Geotech. Testing Jour. 1985-89, C.A. Hogentogler award 1979, Merit award 1993); mem. NSPE, Am. Soc. for Engring. Edn., Transp. Rsch. Bd., Earthquake Energy Rsch. Inst., Chi Epsilon (Harold T. Larson award 1985, James M. Robbins award 1989). Roman Catholic. Avocations: golf, fishing. Office: Purdue U 1284 Civil Engring Bldg West Lafayette IN 47907

DROESSLER, EARL G., geophysicist educator. Graduate, Loras College, Dubuque, Iowa; MS, meteorology, U.S. Naval Postgraduate Sch, Annapolis, Md., 1944. Various posts National Science Foundation, 1958-1966; prof., meteorology State U. New York, Albany, 1966-1971, North Carolina State U., 1971—; prof. emeritus geosciences; Former head, NSF, Atmospheric Sciences Division. Recipient Waldo E. Smith medal, Am. Geophysical Union, 1993. Office: North Carolina Univ Dept Geological Sciences Chapel Hill NC 27599*

DROLL, RAYMOND JOHN, engineer; b. Bklyn., Oct. 13, 1956; s. Raymond John and Rosemary (Donnelly) D.; m. Suzanne Joy Carlough, Feb. 26, 1983; children: Kyle Jeffrey, Connor Steven. B of Aero. Engring., Ga. Inst. Tech., 1978. Engr. Sperry Surveillance and Fire Control Systems, Great Neck, N.Y., 1983-85; sr. engr. integrated logistics support systems Grumman Aircraft Systems, Bethpage, N.Y., 1985-91; group leader integrated logistics support systems concepts Grumman Melbourne (Fla.) Systems, 1991—. Contbr. articles to profl. jours. Capt. USMC, 1978-82. Mem. Soc. of Logistics Engrs. (chmn. L.I. chpt. 1989-91, vice chmn. 1986-89), Sigma Gamma Tau. Home: 3570 Manassas Ave Melbourne FL 32934 Office: Grumman Melbourne Systems HO2-220 PO Box 9650 Melbourne FL 32902-9650

DROSSMAN, JAY LEWIS, aerospace executive; b. Bklyn., Dec. 30, 1932; s. Murray L. and Ruth (Cohen) D.; m. Sylvia F. Solomon, Dec. 26, 1954 (dec. July 1972); children: Bruce, Mark; m. Phyllis Cynthia Cross, July 3, 1974; stepchildren: Meri Wieder Sirkin, Drew Wieder. BA in Psychology, NYU, 1954; BSME, CCNY, 1960; MSME, CUNY, 1963. Dir. Trident Programs Kearfott Guidance & Navigation Corp., Wayne, N.J., 1978-91, dir. strategic programs, 1992—; mem. Fleet Ballistic Missile Programs, Washington, 1978—. Inventor optical means of measuring rotation. Bd. dirs. Temple Emeth, Teaneck, N.J., 1976-79. 1st lt. inf. USN, 1954-56, ETO. Mem. AIAA, IEEE, Jewish War Vets. Republican. Home: 57 Old Orchard Dr Hawthorne NJ 07506 Office: Kearfott Guidance Navigation Corp 150 Totowa Rd Wayne NJ 07474

DROZD, JOSEPH DUANE, aerospace engineer; b. Buffalo, June 15, 1957; s. Arthur Joseph and Phyllis Angeline (Jozwiak) D.; m. Tamera Sue Vaughan, July 4, 1991. BS in Aero. Engring., U. Ariz., 1983; MS in Systems Mgmt., U. So. Calif., 1989. Commd. 2d lt. USAF, Omaha, 1978; advanced through grades to capt. USAF, 1989; computer ops. engr. USAF, Omaha, 1978-81; ops. dir. satellite control network USAF, Sunnyvale, Calif., 1984-86, chief software engring. br., 1986-89; orbital ops. mgr. space systems div. USAF, L.A., 1989-90, chief launch readiness br., 1990-92, DSCS III program mgr., 1993—. Decorated Commendation and Air Force Achievement medals. Mem. AIAA. Republican. Roman Catholic.

DROZDZIEL, MARION JOHN, aeronautical engineer; b. Dunkirk, N.Y., Dec. 21, 1924; s. Steven and Veronica (Wilk) D.; m. Rita L. Korwek, Aug. 30, 1952; 1 child, Eric A. BS in Aero. Engring., Tri State U., 1947, BS in Mech. Engring., 1948; postgrad., Ohio State U., 1948, Niagara U., 1949-51, U. Buffalo, 1951-52. Stress analyst Curtiss Wright Corp., Columbus, Ohio, 1948; project engr. weight analysis Bell Aerospace Textron, Buffalo, 1949-52, stress analyst, 1952-60, asst. supr. stress analysis, 1960-64, chief stress analysis propulsion, 1964-79, chief engr. stress and weights, 1979-84, staff scientist, 1984-85, cons. structures and fractures mechanics, 1985—; mem. Am. Aerospace Materials Del. to USSR, 1989, Am. Aerospace Industries Del. to People's Republic China, 1991, Am. Aerospace Materials Del. to Czechoslovakia and Commonwealth Ind. States, 1992. Del. Internat. Citizens Ambassador Prog.; active Buffalo Fin Arts Acad. With U.S. Army, 1944-47. Recipient cert. of achievement NASA-Apollo, 1972; cert. commendation U.K. NATO program, 1982. Mem. AAAS, AIAA (Membership Chmn's award 1988, 89, 90, 92, 93), Soc. Reliability Engrs., U.S. Naval Inst., Am. Space Found., Nat. Conservancy, Nat. Audubon Soc., Am. Acad. Polit. and Social Sci., Acad. Polit. Sci., Union Concerned Scientists, Air Force Assn., Nat. Space Soc., Soc. Allied Weight Engrs., Planetary Soc., Am. Mgmt. Assn., Bibl. Archeology Soc., Archeol. Inst. Am., Cousteau Soc., Smithsonian Assocs., Buffalo Audubon Soc., Bell Mgmt. Club, Natural History Mus., Internat. Hypersonic Rsch. Republican. Roman Catholic. Achievements include development of criteria and methods of structural analysis extending analyses into the plastic and creep ranges for titanium and columbium rocket nozzle extensions; of criteria and methods of structural analysis for extendable rocket nozzle extensions, including rapid nozzle deployment involving plasticity; of methods of structural analysis for low strength, high ductility steels, aluminums, and teflons as positive expulsion devices for zero gravity application in propellant tanks including bellows, reversing heads, rolling diaphragms devices and collapsing or folding concepts; structural analysis on "X" series of aircraft, on Mercury, Gemini, and Apollo spacecraft reaction control and propulsion systems; structural and weight analysis of programs involving rocket engines, propulsion systems, aircraft, air cushion vehicles, surface-effect ships, laser systems avionics, airborne and ground antennae, Army tanks and fighting vehicles. Home and Office: 152 Linwood Ave Tonawanda NY 14150-4020

DRUCKER, DANIEL CHARLES, engineer, educator; b. N.Y.C., June 3, 1918; s. Moses Abraham and Henrietta (Weinstein) D.; m. Ann Bodin, Aug. 19, 1939; children: R. David, Mady Drucker Upham. BS, Columbia U., 1937, MCE, 1938, PhD, 1940; D Engring. (hon.), Lehigh U., 1976; DSc in Tech. (hon.), Technion, Israel Inst. Tech., 1983; DSc (hon.), Brown U., 1984, Northwestern U., 1985, U. Ill., 1992. Instr. Cornell U., 1940-43; supr. Armour Rsch. Found., Chgo., 1943-45; asst. prof. Ill. Inst. Tech., 1946-47; assoc. prof. Brown U., Providence, 1947-50, prof., 1950-64, L. Herbert Ballou Univ. prof., 1964-68, chmn. div. engring., 1953-59, chmn. phys. scis. coun., 1960-63; dean Coll. Engring. U. Ill., Urbana, 1968-84; grad. rsch. prof. engring. scis. U. Fla., Gainesville, 1984—; Marburg lectr. ASTM, 1966; mem., past chmn. U.S. Nat. Com. on Theoretical and Applied Mechanics; past chmn. adv. com. for engring. NSF; mem. National Sci. Bd., 1988—; hon. chmn. 3d SESA Internat. Congress on Exptl. Mechanics; rsch. in stress-strain rels., finite plasticity, stability, fracture and flow on macroscale and microscale. Author: Introduction to Mechanics of Deformable Solids, 1967; contbr. chpts. to tech. books, papers to mech. and sci. jours. Recipient Gustave Trasenster medal U. Liège, Belgium, 1979, Thomas Egleston medal Columbia U. Sch. Engring and Applied Sci., 1978, John Fritz medal Founder Engring. Socs., 1985, Nat. Medal of Sci., 1988, ASME medal, 1992; Guggenheim fellow, 1960-61; NATO Sr. Sci. fellow, 1968; Fulbright Travel grantee, 1968. Fellow AAAS (past chmn. sect. engring., past mem. coun.), AIAA (assoc.), ASME (hon. mem., chmn. applied mechanics div. 1963-64, v.p. policy bd. communications 1969-71, pres. 1973-74, Timoshenko medal 1983, Thurston lectr. 1986, Disting. lectr. 1987-89), ASCE (von Karman medal 1966, past pres. New Eng. coun., past pres. Providence sect., past chmn. exec. com. engring mechanics div.), Am. Acad. Mechanics (past pres.), Am. Acad. Arts and Scis. (past mem. membership com.); mem. NSPE, Ill. Soc. Profl. Engrs. (hon.), Soc. Exptl. Stress Analysis (hon.; past pres., W.M. Murray lectr. 1967, M.M. Frocht award 1971), Am. Technion Soc. (past pres. So. N.E. chpt.), Soc. Rheology, Am. Soc. Engring. Edn. (charter fellow mem., past 1st v.p., past chmn. engring. coll. coun., dir., pres. 1981-82, Lamme award 1967, Disting. Educator, Mechanics div. 1985, named to Hall oFame 1993), NAE (mem. com. on pub. engring. policy 1972-75, chmn. membership policy com. 1982-85), Soc. Engring. Sci. (William Prager medal 1982), Polish Acad. Scis. (fgn. mem.), Internat. Union Theoretical and Applied Mechanics (treas. 1972-80, pres. 1980-84, v.p. 1984-88, personal mem. Gen. Assembly 1988—), Internat. Coun. Sci. Unions (past mem. gen. com.), Sigma Xi (past pres. Brown U. chpt.), Phi Kappa Phi (past pres. U. Fla. chpt.), Tau Beta Pi, Pi Tau Sigma, Chi Epsilon, Sigma Tau. Office: U Fla 231 Aerospace Engring Bldg Gainesville FL 32611

DRUCKER, HARVEY, biologist; b. Chgo., Jan. 1, 1941; s. Adolph and Faye (Beezy) D.; m. Elizabeth Ann Baldridge, Aug. 21, 1965; children: Debi, Sheri. BS, U. Ill., 1963, PhD, 1967. Mgr. molecular biology and biophysics sect. Battelle Pacific N.W. Lab., Richland, Wash., 1972-79; mgr. biology dept. Battelle Pacific N.W. Lab., Richland, 1979-83; assoc. lab. dir. for biomed. and environ. rsch. Argonne (Ill.) Nat. Lab., 1983-86, assoc. lab. dir. for energy, environ. and biol. rsch., 1986—. Mem. indsl. adv. bd. Coll. Engring., Chgo., 1985—; bd. dirs. West Corp. Corridor Assn., Naperville, Ill., 1989—, Indsl. Adv. Bd. of DuPage, Oak Brook, Ill., 1990—. Mem. Am. Chem. Soc., Am. Soc. Microbiology, Am. Soc. Biologic Chemists, Sigma Xi. Office: Argonne Nat Lab EEBR-202 9700 S Cass Ave Argonne IL 60439

DRUCKMAN, WILLIAM FRANK, mechanical engineer; b. Boston, Apr. 19, 1939; s. Louis and Edith Ida (Boltin) D.; m. Marion Nancy Turin, July 11, 1961 (div. 1985); children: Scott J., Lawrence J.; m. Marlene Stepman, Dec. 22, 1985; stepchildren: Sherryl Gale Greenberg, Jay Allan Greenberg. AME, Northeastern U., Boston, 1964, BS in Indsl. Tech., 1968; MBA, Babson Coll., Wellesley, Mass., 1975. Registered profl. engr., Md., N.J. With Hermes Electronics div. Itek Corp., Cambridge, Mass., 1959-62, Gen. Communications Co., Boston, 1961-62, Adage, Inc., Boston, 1962-66, Computer Products Inc., Waltham, Mass., 1966-67, Atkins & Merrill, Boston, 1966-67, Raytheon Co., Wayland and Bedford, Mass., 1968-75, Vitro Labs., Silver Spring, Md., 1976-82; mech. engr. Martin Marietta Corp GES (formerly GE/RCA), Moorestown, N.J., 1982—; chmn. ins. com. Assn. Scientists and Profl. Engring. Pers. Advisor Boy Scouts Am., Moorestown, N.J., 1989—; chmn. promotional com. Aerospace and Def. Polit. Action Com. Mem. ASME, Am. Soc. Naval Engrs., Assn. of MBA Execs., Soc. Naval Architects and Marine Engrs., Sigma Epsilon Rho. Home: 1625 Plymouth Rock Dr Cherry Hill NJ 08003-2747 Office: Martin Marietta Corp 138-325 Corporate Ctr Moorestown NJ 08057

DRUEHL, LOUIS DIX, biology educator; b. San Francisco, Oct. 9, 1936; naturalized Can. citizen, 1974; s. Louis Dix and Charlotte (Primrose) D.; m. Jo Ann Reeve, Aug. 17, 1967 (div. 1974); m. Rae Kristanne Randolph, Aug. 11, 1983. BSc, Wash. State U., 1958; MSc, U. Wash., 1962; PhD, U. B.C., Vancouver, B.C., 1966. Rsch. advisor Brazil Navy, Cabo Frio, 1975-77; cons. biomass program GE, Catalina, Calif., 1981-83; from asst. prof. to assoc. prof. Simon Fraser U., Burnaby, B.C., 1966-88, prof. biology, 1988—, dir. Inst. Aquaculture Rsch., 1988-90; assoc. dir. Bamfield (B.C.) Marine Sta., 1992—; pres. Can. Kelp Resources Ltd., Bamfield, 1982—. Mem. editorial bd. European Jour. Phycology, 1993—; contbr. over 50 articles to profl. jours. Recipient Provasoli best paper award Jour. Phycology, 1988, Luigi Provasoli award Phycological Soc. Am., 1988. Mem. Western Soc. Naturalists (pres. 1988). Avocation: writing poetry. Home: 4 Port Desire, Bamfield, BC Canada V0R 1B0 Office: Bamfield Marine Sta, Bamfield, BC Canada V0R 1B0

DRUMHELLER, JERRY PAUL, laser scientist, development engineer; b. Bellfont, Pa., Oct. 3, 1948; s. Carl E. and Emma Elizabeth (Pielemeier) D.; m. Jayne Keyel, May 1, 1976; children: Daniel, Timothy, Julianna. BA in Physics, Wittenberg U., 1970. Process engr. resistor products div. Analog Devices, Rochester, N.Y., 1970-76; tech. assoc. Lab. for Laser Energetics U. Rochester, 1976-84; devel. engr. Hampshire Instruments, Rochester, 1984—. Contbr. numerous articles to profl. jours. and mags. Mem. Rochester Aeromodeling Soc. Lutheran. Achievements include patents on laser target geometry for an x-ray source for x-ray lithography. Home: 1245 Imperial Dr Webster NY 14580

DRUMMOND, ROGER OTTO, livestock entomologist; b. Peoria, Ill., Aug. 11, 1931; s. Jay Elmer and Edna Louise (Leben) D.; m. Ellen Peare, Sept. 6, 1953; children: Diane, Douglas. AB, Wabash Coll., Crawfordsville, Ind., 1953; PhD, U. Md., 1956. Registered profl. entomologist. Med. entomologist U.S. Dept. Agr., Agrl. Rsch. Svc., Kerrville, Tex., 1956-86; pvt. cons., 1986—. Author: Control of Arthropod Pests of Livestock, 1988, Ticks and What You Can Do About Them, 1990; contbr. articles to profl. jours. Recipient Coopers Achievement award in Livestock Entomology, 1988. Fellow AAAS, Entomol. Soc. Am. (br. pres. 1982-83), U.S. Animal Health Assn., Sigma Xi. Presbyterian. Achievements include research in livestock entomology. Home: 525 Drummond Dr Kerrville TX 78028-9140

DRUMMOND, WILLIAM ECKEL, physics educator; b. Portland, Oreg., Sept. 18, 1927; s. James Edgar and Blanche (Black) D.; m. Stephanie Jones, Nov. 28, 1953; children: Ellen, William E. Jr., Robin. BS in Physics, Stanford U., 1950, PhD in Physics, 1958. Physicist UCRL radiation lab, Livermore, Calif., 1954, Stanford (Calif.) Rsch. Inst., 1955-58; prin. scientist Avco Mfg. Co., Everett, Mass., 1958-59; physicist Gen. Atomic, San Diego, 1959-65; dir. fusion rsch. ctr. U. Tex., Austin, 1966-93, prof. physics, 1965—; lectr. Stanford U., 1955; dir. plasma turbulance lab. Gen. Atomic, 1961-65; chmn. Austin Rsch. Assocs., 1967-89. With USN, 1945-46. Fellow Am. Phys. Soc. Democrat. Avocation: tennis. Office: U Tex Fusion Rsch Ctr Robert L Moore Hall 11 222 Austin TX 78712

DRURY, KENNETH CLAYTON, biological scientist; b. Madera, Calif., Mar. 27, 1945; s. Carma and Alice (Zolinger) D.; m. Sandra Rosemary Hanlon, Apr. 28, 1972; children: Allison Hanlon, Vanessa Laura. BA, Westmont Coll., 1967; PhD, U. Geneva, Switzerland, 1979. NIH fellow U. Calif., Berkeley, 1979-82; rsch. scientist Codon Corp., South San Francisco, Calif., 1982-84; sr. scientist Microgenics Corp., Concord, Calif., 1984-86; dir. U. Louisville, 1986-92, In Vitro Fertilization and Gamete Physiology Labs. U. Fla., 1992—. Contbr. articles to profl. jours. 1st Lt. U.S. Army, 1969-72, Vietnam. Mem. Am. Fertility Soc., AAAS. Achievements include first investigator to directly implicate phosphorylation in mechanism of action of maturation promoting factor; one of first investigators to obtain a human live birth after ultra rapid freezing of embryos. Office: U Fla Dept Ob-Gyn Div Reprod Endocrinology PO Box 100294 Gainesville FL 32610

DRUSCHITZ, ALAN PETER, research engineer; b. Chgo., Dec. 7, 1955; s. Alexander and Marie Ann (Johnson) D.; m. Lorraine Martha Bonk, June 19, 1982; children: Edward Alan, Laurel Ann. BSME, Ill. Inst. Tech., 1978, PhDME, 1982. Sr. rsch. engr. GM Rsch. Labs., Warren, Mich., 1982-86, staff engr., 1986—. Contbr. articles to profl. jours. including Advanced Ceramic Materials, Jour. Am. Ceramic Soc., Ceramic Transactions. Mem. Zoning Bd. Appeals, Rochester Hills, 1990-92; precinct del. Rep. Party, Rochester Hills, 1990—. Mem. Am. Foundrymen's Soc. (exec. bd. Detroit sect. 1992—), Am. Soc. for Metals, Soc. of Automotives Engrs., Sigma Xi. Achievements include patent for melt containment apparatus with protective oxide melt contact surface. Office: GM Rsch Labs 30500 Mound Rd Warren MI 48090-9055

DRUSS, DAVID LLOYD, geotechnical engineer; b. Manhasset, N.Y., Nov. 14, 1953; s. Abe R. and Mildred (Schpeiser) D.; m. Joann Blanda, June 28, 1983; children: Judith, Nathan. BS in Civil Engring., SUNY, Buffalo, 1975; MS in Civil Engring., U. Pitts., 1984. Registered profl. engr., Pa., Mass. Structural engr. I. Thompson & Assocs., San Francisco, 1976-77; civil engr. U.S. Forest Svc., St. Anthony, Idaho, 1977-79; profl. assoc. Parsons, Brinckerhoff, Quade and Douglas, N.Y.C., 1980—; chief geotech. engr. Ctrl. Artery/Tunnel project, Boston. Contbr. papers to conf. proceedings., jours. Mem. ASCE. Home: 56 Edgehill Rd Providence RI 02906 Office: Bechtel Corp Parsons Brinckerhoff et al 1 South Station Boston MA 02110

DRUSS, RICHARD GEORGE, psychiatrist, educator; b. N.Y.C., Aug. 14, 1933; s. Joseph George and Ruth Druss; m. Margery Ellen Kramer, Aug. 28, 1960; children: Benjamin, Elizabeth. AB, Yale Coll., 1955; MD, Columbia U., 1959. Diplomate Am. Bd. Psychiatry and Neurology. Intern Mass. Meml. Hosps., 1959-60; resident Presbyn. Hosp. and N.Y. State Psychiat. Inst., 1960-63; tng. analyst Columbia Psychoanalytic Ctr., N.Y.C., 1975—, assoc. dir., 1981-91; clin. prof. dept. psychiatry Columbia U., N.Y.C., 1983—. Assoc. editor Internat. Jour. Psychoanalytic Psychotherapy, 1974-80, Bull. Assn. Psychoanalytic Medicine, 1974-83; contbr. articles and book revs. to profl. jours. Capt. U.S. Army, 1963-65. Recipient George E. Daniels award Assn. Psychoanalytic Medicine, 1990. Fellow Am. Psychiat. Assn., N.Y. Acad. Medicine; mem. AMA, Am. Psychoanalytic Assn. (cert.). Achievements include research in changes in body image following major surgical procedures, healthy and pathologic denial of illness in chronic medical patients. Office: 180 East End Ave New York NY 10128

DRYDEN, RICHIE SLOAN, physician; b. Robstown, Tex., Apr. 26, 1938; s. Edward Marseille and Lela Frances (Sloan) D.; m. Gladys Coleman, June 13, 1964; children: Warren Edward, Catherine Jeane, Donna Marsielle. BS Zoology, Tex. A&M U., 1960; MD, Baylor U., 1964; MPH, Harvard U., 1968. Commd. USAF, 1965, advanced through grades to col., 1980; intern Ben Taub Hosp., Jefferson Davis Hosp., Houston, 1964-65; resident Harvard Sch. Pub. Health, 1967-68, USAF Sch. Aerospace Medicine, Brooks AFB, Tex., 1968-70; chief aerospace medicine USAF Hosp., Cam Rahn Bay, Vietnam, 1970-71; chief comdr. Howard & Albrook AFB, Panama Canal Zone, 1971-74; chief flight medicine USAF Cons. Svc., Brooks AFB, Tex., 1975-81; chief aerospace medicine Tactical Air Command, Langley AFB, Va., 1981-87; project mgr. systems acquisition Air Force Office Med. Support, 1987-92. Com. chmn. Cub Scouts pack, Howard AFB, Panama, 1974, Brooks AFB, 1975. Decorated Bronze Star, Air medal. Assoc. fellow Aerospace Medicine Soc.; mem. AAAS, Assn. Mil. Surgeons U.S., Soc. USAF Flight Surgeons. Lutheran.

DRZEWIECKI, DAVID SAMUEL, mechanical engineer; b. South Bend, Ind., Dec. 22, 1953; s. Henry Joseph and Corinne Edith (Marek) D.; m. Cathy Jean Arndt, June 19, 1976; children: Yolanda Marie, Jennifer Lynn. BME, Rose-Hulman Inst. Tech., 1976; MBA, Ind. U., 1982. Mech. foreman Bethlehem Steel, Chesterton, Ind., 1978-81; project engr. Pfizer Inc., Valparaiso, Ind., 1981-83; project engr. Miles Inc., Elkhart, Ind., 1983-86, site utilities engr., 1986-88, mgr. utilities ops., 1988—; bd. dirs., officer Ind. Indsl. Energy Consumers, Indpls., 1986—. Contbr. articles to profl. jours. 1st lt. U.S. Army, 1976-82. Mem. NSPE, ASME, Am. Inst. Plant Engrs., Omicron Delta Epsilon (hon.). Achievements include development of natural gas transportation program; organized several highly successful public utility rate case interventions; introduced process engineering to utilities operation. Home: 51745 E Gatehouse Dr South Bend IN 46637 Office: Miles Inc PO Box 40 Elkhart IN 46515

DRZEWINSKI, MICHAEL ANTHONY, materials scientist; b. McKeesport, Pa., May 8, 1958; s. Stanley J. and Elvira (. (Vecchio) D.; m. Lori E. Dyer, Apr. 23, 1988; children: Alison M., Timothy J. BS, Rensselaer Poly. Inst., 1980; MS, U. Mass., 1982; ScD, MIT, 1986. Rsch. chemist Exxon, Linden, N.J., 1978-80; rsch. scientist Kimberly-Clark, Roswell, Ga., 1986-87; prin. scientist, sr. scientist, project leader EniChem Am., Princeton, N.J., 1987—; tech. advisor U. Conn. Inst. for Materials Sci., Storrs, 1989—; tech. expert/cons. Drexel Polymer Notes, Phila., 1988—; mem. indsl. liaison program MIT, Cambridge, Mass., 1987—. Contbr. over 15 articles on novel polymeric materials to profl. jours.; author: (with others) Polymer Blends & Mixtures, 1984. Founder Rensselaer Rsch. Squad, Troy, N.Y., 1978; mem. Rahway (N.J.) First Aid Squad, 1974-82. Rensselaer Acad. scholar, 1976-80. Mem. ACS, Am. Phys. Soc., AAAS, N.Y. Acad. Scis., Phi Kappa Tau (bd. dirs.), Sigma Xi, Phi Lamba Upsilon. Achievements include patents for Block Copolymers of Polybutadiene and Polypropylene, for Elastic, Melt-Blown, Nonwoven Fabrics and Their Use, for Dyeable, Nonwoven Polypropylene Fabric, for Polyolefin Blends Using Lactone Based Polymer Compatibilizers, for Tricyano Containing NLO Polymers, for Compatibilized Polycarbonate/Nylon Blends, for Improved Core-Shell Impact Modifiers for Polycarbonate, for Compatibilized Polystyrene/Polyethylene Terephthalate Blends, for Compatibilized Polystyrene/Nylon Blends, for Compatibilized Polystyrene/Polyethylene Blends, for Improved Polyphenylene Ether/Polyester Blends, for Polyethylene Blends Containing Polyphenylene Ethers, for Polyolefin Blends Containing Polyphenylen Ethers, for Solvent Resistant Styrenic Based Thermoplastic Elastomers, for Miscible Blends Based on Aromatic Methacrylate Polymers, for Modified Polycarbonates for Improved Polymer Blends, for Modified Methacrylates for Improved Polymer Blends. Office: EniChem Americas Inc 2000 Cornwall Rd Monmouth Junction NJ 08852-2409

D'SOUZA, MAXIMIAN FELIX, medical physicist; b. Bombay, India, Feb. 3, 1965; came to U.S., 1988; s. Ligori and Eliza D'S. BSc in Physics, St. Xavier's Coll., 1986; MS in Physics, Cen. State U., 1990; MS in Radiol. Sci., U. Okla., 1992. Computer programmer Bombay, 1987-88; teaching asst. Cen. State U., Edmond, Okla., 1989-90; rsch. asst. U. Okla., Oklahoma City, 1990—. Contbr. abstracts to profl. publs. Named to Dean's Honor roll U. Ctrl. Okla., 1988, Pres.'s Honor roll U. Ctrl. Okla., 1989. Home: 1830 1/2 NW 23 St Oklahoma City OK 73106

DU, DING-ZHU, mathematician, educator; b. Qigihaer, China, May 21, 1948; s. Jin-Gao and Ai-Hua (Xu) D.;m. Shu-Mei Li, Jan. 20, 1977; 1 child, Hong-Wei. MS, Chinese Acad. Scis., Beijing, 1982; PhD, U. Calif., Santa Barbara, 1985. Asst. prof. Inst. of Applied Math., Beijing, 1981-82; postdoctoral Math. Scis. Rsch. Inst., Berkeley, Calif., 1985-86; asst. prof. MIT, Cambridge, 1986-87; prof. Inst. of Applied Math., Beijing, 1987-90; rsch. assoc. Princeton (N.J.) U., 1990-91; assoc. prof. U. Minn., Mpls., 1991—. Author: Convergence Theory of Feasible Direction Methods; editor: Gradient Projection Methods in Linear and Nonlinear Programming, 1988, Combinatorics, Computing and Complexity, 1989; contbr. over 77 articles to profl. jours. Mem. Am. Math. Soc. Achievements include proof of Derman, Leiberman and Ross' conjecture on optimal consecutive-2-out-of-n system in 1982, proving Gilbert and Pollak's conjecture on Steiner ratio; solution to open problem on Rosen's method in nonlinear optimization; research on one-way function in complexity theory. Home: 1047 5th St SE Minneapolis MN 55414 Office: U Minn Computer Sci Dept Minneapolis MN 55455

DU, GONGHUAN, acoustics educator; b. Shanghai, People's Republic of China, Feb. 16, 1934; s. Shengyuan Du and Lianying Chen; m. Yiman Hua, May 18, 1965; children: Zheng, Rong. BA, Nanjing U., Jiangsu, People's Republic of China, 1957. From instr. to assoc. prof. acoustics Nanjing U., 1957-88, prof., 1989—; vis. researcher U. Tenn., Knoxville, 1984-85; vis. prof. U. Vt., Burlington, 1989-90; vis. scientist Nat. Ctr. for Phys. Acoustics, Oxford, Miss., 1990. Author: Fundamentals of Acoustics, 1981; contbr. articles to profl. jours.; inventor acoustical devices field. Mem. Acoustical Soc. China, Acoustical Soc. Am. Office: Nanjing U Acoustics Inst, 22 Hankou Rd, Nanjing Jiangsu 210008, China

DUARTE, CRISTOBAL G., nephrologist, educator; b. Concepción, Paraguay, July 17, 1929; s. Cristobal Duarte and Emilia Miltos; m. Norma Aquino, 1984. BS, Colegio de San José Asuncion, 1947; MD, Nat. U. Asunción, 1953. Intern De Goesbriand Meml. Hosp., Burlington, Vt., 1956; resident in medicine Carney Hosp. and St. Elizabeth's Hosp., Boston, 1956-58; fellow in medicine Lahey Clinc, Boston, 1959; fellow hypertension and renal medicine Hahnemann Hosp., Phila., 1960; assoc. in medicine U. Vt. Coll. Medicine, 1962-65; clin. investigator VA, 1966-68, staff physician, 1968-73; dir. Renal Function Lab., Mayo Clinic and Found., Rochester, Minn., 1973-77; asst. prof. lab. medicine Mayo Med. Sch., 1973-77; commd. lt. col. U.S. Army, 1977-84; assoc. prof. medicine and physiology Uniformed Svcs. U. Health Scis., Bethesda, Md., 1977-84, attending in medicine Walter Reed Army Med. Ctr., Washington, 1977-84; chief nephrology svc. Bay Pines VA Med. Ctr., 1984-87; assoc. prof. medicine U. South Fla., Tampa, 1984-87; med. officer cardio-renal drug products FDA, Rockville, Md., 1987—. Editor: Renal Function Tests, 1980; contbr. articles to profl. jours., chpts. to books. Recipient cert. of accomplishment VA, 1969; physician's recognition award AMA, 1993—; Cordell Hull Found. fellow, 1958-59. Fellow Am. Coll. Nutrition; mem. Nat. Kidney Found., Am. Fedn. Clin. Rsch., Am. Physiol. Soc., Am. Soc. Pharmacology and Exptl. Therapeutics, Midwest Salt and Water Club, Am. Soc. for Clin. Rsch., Inter-Am. Soc. Hypertension, Internat. Soc. Nephrology, Central Soc. for Clin. Rsch., Am. Soc. Nephrology, Sigma Xi. Roman Catholic. Current Work: Radiocontrast-induced renal failure. Subspecialty: Nephrology.

DUARTE, RAMON GONZALEZ, nurse, educator, researcher; b. San Fernando, Calif., Jan. 5, 1948; s. Salvador Revelez and Juanita (Gonzalez) D.; m. Sophia Constant Garabedian, Apr. 17, 1983; children: David Ramon, John Robert. AA in Nursing, Los Angeles Valley Coll., 1972; student, Calif. State U., Los Angeles, 1972-76. RN; Cert. Bd. Nephrology Examiners. Staff nurse hemodialysis unit U. So. Calif. Med. Ctr., L.A., 1971-75; charge nurse self care hemodialysis unit Kaiser Found. Hosp., L.A., 1976, Culver City (Calif.) Dialysis Svcs., Inc., 1981-82; adminstrv. head nurse hemodialysis unit Valley Presbyn. Hosp., Van Nuys, Calif., 1976-78; adminstrv. head nurse Kidney Dialysis Care Units, Lynwood, Calif., 1980-81; ind. nursing contractor Nursing Svcs. in Nephrology, Van Nuys and Santa Barbara, 1982—; clin. instr., rschr. Nursing Svcs. in Nephrology, Santa Barbara, 1980—, dir. rsch., 1988—; coord. clin. rsch. Valley Dialysis Assocs., Inc., Van Nuys, 1978-80; mem. rsch. com. Valley Presbyn. Hosp. Rsch.; founder, pres. Dialysis Mus. Coun., chmn. So. Calif. Dialysis Earthquake Preparedness Commn.; mem. coun. nephrology nurses and technicians, mem. allied profl. adv. com., chmn. allied health profl. rsch. grant com. Nat. Kidney Found., Inc.; mem. sci. adv. coun. Nat. Kidney Found. So. Calif.; coord. Airlift Armenia Dialysis Emergency Svcs.; team coord. Armenian Relief Soc. Editorial bd. Dialysis and Transplantation mag.; pubr., editor: Dialysis and the Earthquake Connection; contbr. articles to med. publs.; patentee biologicals. Founder Mus. Hope, Van Nuys; coord. Airlift Armenia, Armenia Relief Soc., 1988. Recipient Dedicated Svc. award Hemodialysis Found., 1976; named Allied Health Profl. of Yr., Nat. Kidney Found. So. Calif., 1986; scholar Am. G.I. Forum, 1966, 40 and 8, L.A. Valley Coll. Assoc. Students; grantee Santa Barbara Cottage Hosp., 1993. Mem. AACCN, Am. Assn. Artificial Internal Organs, Am. Assn. Nephrology Nurses and Technician, Am. Soc. Nephrology, N.Y. Acad. Scis. (cert.), Kidney Found. So. Calif., Ind. Nurses Assn., Nat. Assn. Patients on Hemodialysis and Transplantation Inc. Democrat. Roman Catholic. Achievements include development of the equations of projected fluid removal, of the first in vivo ultrafiltration rate monitor, of apheresis plasma charcoal hemoperfusion as a new treatment modality for the treatment of toxic substance overdoses, of the Heparin Wedge technique in anti-clotting mechanisms; solution of the problem of concentration polarization of protein molecules in interactions with hemodialyzers; introduction of the use of ascorbic acid to neutralize sodium hypochlorite in disinfection procedures. Home and Office: 3770 Torino Dr Santa Barbara CA 93105-4433

DUAX, WILLIAM LEO, biological researcher; b. Chgo., Apr. 18, 1939; s. William Joseph and Alice B. (Joyce) D.; m. Caroline Townsend Dowell, May 6, 1966; children: Julia, Sarah, William, Stephen. BA, St. Ambrose Coll., 1961; PhD, U. Iowa, 1967. Postdoctoral research fellow Ohio U., Athens, 1967-68; research assoc. Med. Found. Buffalo, 1968-69, head crystallography dept., 1969-70, head molecular biophysics dept., 1970-88, assoc. dir. research, 1983-88, research dir., 1988-93, exec v.p. rsch., 1993—, also bd. dirs.; adj. assoc. prof. Medicinal Chemistry Dept. SUNY, Buffalo, 1973—, assoc. research prof. Dept. Biochemistry, 1981—; dir. distbn. Cambridge Database in U.S., Buffalo, 1983—; lectr. various internat. confs. Editor: Atlas of Steroid Structure Vol. I, 1975, Vol. II, 1984, Molecular Structure and Biological Activity, 1982, Molecular Structure and Biological Activity of Steroids, 1992, Internat. Union of Crystallography Newsletter, 1993—. Mem. Am. Field Service, Amherst, N.Y. Served with USAR, 1961-67. Fulbright scholar Council for Internat. Exchange, 1987; grantee NIH, 1971—; recipient Spl. Merit award Inst. Arthritis and Metabolic Diseases NIH, 1987—, Disting. Alumni award, St. Ambrose Coll., 1983. Mem. AAAS, Am. Crystallographic Assn. (v.p. 1985, pres. 1986, exec. office 1988—), Am. Chem. Soc., Am. Cancer Soc., Biophys. Soc., Internat. Union Crystallography (charter mem., sec. com. on small molecules 1984—), Am. Inst. Physics (bd. govs. 1987—, exec. com. 1992), Coun. Sci. Soc. Pres. (govt. and pub. affairs com. 1987). Democrat. Roman Catholic. Club: Saturn (Buffalo). Office: Med Found of Buffalo Inc 73 High St Buffalo NY 14203-1196

DUBÉ, GHYSLAIN, earth scientist. Recipient H.T. Alrey award Can. Inst. Mining and Metallurgy, 1992. Office: care Xerox Tower Ste 1210, 3400 de Maisonneuve Blvd W, Montreal, PQ Canada H3Z 3B8*

DUBÉ, RICHARD LAWRENCE, landscape specialist, consultant; b. Portland, Maine, Mar. 25, 1950; s. Clarence Everett and Nancy Ann (Rowles) D.; m. Mary Louise Roberts, Sept. 7, 1974. AS in Forestry, Hocking Tech.

Coll., 1973, AS in Recreation and Wildlife, 1975. Interpretive naturalist Ohio Dept. Natural Resources, 1975-78; asst. mgr. Treeland, Portland, 1979-83; landscape designer Lucas Tree Experts, Portland, 1983-86, mgr. landscaping, 1986-88, mgr. landscape and tree depts., 1988—; speaker Celandine Info. Svc., Buxton, Maine, 1987—. Host for TV show Backyard Maine, 1992. Vice pres. Japan-Am. Soc. of Maine, Portland, 1989-90; bd. dirs China-Am. Friendship Assn. of Maine, 1991; founder, exec. dir. A Yr. of Peace, 1990. Mem. Associated Landscape Contractor of Am., Profl. Grounds Mgmt. Soc., Am. Assn. Nurserymen, Maine Nurseryman's Assn. Avocations: international market and political analysis, writing music. Office: Lucas Tree Experts 636 Riverside St # 958 Portland ME 04103-5901

DUBERG, JOHN EDWARD, aeronautical engineer, educator; b. N.Y.C., Nov. 30, 1917; s. Charles Augustus and Mary (Blake) D.; m. Mary Louise Andrews, June 11, 1943; children—Mary Jane, John Andrews. B.S. in Engring, Manhattan Coll., 1938; M.S., Va. Poly. Inst., 1940; Ph.D., U. Ill., 1948; grad., Fed. Exec. Inst., 1971. Engr. Cauldwell-Wingate Builders, N.Y.C., 1938-39; rsch. asst. U. Ill., 1940-43; rsch. engr. NASA, 1943-46; chief structures Langley Labs. NASA, Hampton, Va., 1948-56; mem. staff Langley Rsch. Ctr. NASA, 1959-79, assoc. dir. Langley Rsch. Ctr., 1968-79; rsch. engr. Standard Oil Co. Ind., 1946-48; with Ford Aeros., Glendale, Calif., 1956-57; mem. faculty U. Ill., 1957-59; rsch. prof. aeros. George Washington U., 1979-87; dir. Joint Inst. Advanced Flight Scis., 1971-79; mem. materials adv. bd. Nat. Acad. Sci., 1950; mem. subcom. profl. and sci. manpower Dept. Labor, 1971; mem. indsl. adv. com. U. Va., 1978-80; pres.'s adv. coun. Christopher Newport Coll., 1973-76, vice chmn., 1976; dir. Newport News Savs. Bank, 1966-92. Contbr. articles to profl. jours., chpts. to books. Trustee United Way Va., Peninsula, 1963-82; chmn. Hampton Roads Chpt. ARC, 1984-86. Fellow AIAA (DeFlorez award 1977), AAAS; mem. Va. Acad. Scis., N.Y. Acad. Scis., Am. Soc. Engring. Edn. (dir.), Engrs. Club Va. Peninsula (pres. 1955), Soc. Engring. Scis. (dir.), James River Country Club, Rotary (pres. Newport News chpt. 1967-68). Episcopalian. Home: 4 Museum Dr Newport News VA 23601-3621 Office: GWU/JIAFS NASA Langley Rsch Ctr M/S 269 Hampton VA 23665

DUBEY, RAM JANAM, inorganic and environmental chemist; b. Rampur, India, July 4, 1941; m. Manjula, Mar. 8, 1975; children: Gaytri, Girjesh, Arati, Anand. BSc, Agra (India) U., 1961, MSc, 1963; DSc, Laval U., 1973. Researcher Royal Inst. Great Britian, London, 1967-69; postdoct. rsch., 1973-78; dir. Canadian Internat. Devel. Agency-Instut Nat. de Recherche d'Agronomique du Niger, Niamay, Niger, 1978-80; asst. prof. U. Maiduguri (Nigeria), 1981-85; rsch. scientist emergency mines and resources Fed. Govt., Ottawa, Can., 1987-91, adv. dept. fisheries and oceans, 1992—. Contbr. articles to profl. jours. Mem. Royal Soc. Chem. (chartered), Inst. Physics (chartered), Can. Inst. Chem. Achievements include first Br N bond in literature; SO2 research work referred 41 times. Home: 2551 Needham Cres, Ottawa, ON Canada K1V 6K1 Office: Phys and Chem Scis DFO, 200 Kent St 1200, Ottawa, ON Canada K1A 0E6

DUBIEL, THOMAS WIESLAW, cardiothoracic surgeon, educator; b. Czestochowa, Poland, Feb. 14, 1929; arrived in Sweden, 1958; s. Leon and Sophia (Sawina) D.; m. Sonja Marianne Ellung, Dec. 8, 1963; children: Michael, Magdalena. MD, Med. Acad., Warsaw, Poland, 1955; PhD (hon.), Uppsala U., Sweden, 1974. Intern Inst. Tb, Warsaw, 1952-55, resident in cardiothoracic surgery, 1955-58; resident in cardiothoracic surgery Karolinska Hosp., Stockholm, Sweden, 1959-66, rsch. fellow, 1958; rsch. fellow Med. Coll. Georgia, Augusta, 1964; resident to chief cardiothoracic surgeon, assoc. prof., dept. vice chmn. Univ. Hosp., Uppsala, 1967—; sr. cons. in chest surgery Swedish Nat. Com. Against Tb, Stockholm, 1959-63, 70-73. Author: Aortic Valve Replacement with Frame-Supported Autologous Fascia Lata, 1974; contbr. articles to profl. jours. Fellow Scandinavian Assn. for Thoracic and Cardiovascular Surgery, Swedish Soc. Thoracic Surgeons; mem. Internat. Assn. Cardiac Biological Implants (founding), Swedish Med. Assn., Swedish Soc. Medicine, Assn. 2d World War Polish Underground Army Vets. Roman Catholic. Home: Urbergsvägen 5, S-752 41 Uppsala Sweden Office: Univ Hosp, Dept Thoracic & Cardiovascular Surgery, S-751 85 Uppsala Sweden

DUBOIS, ARTHUR BROOKS, physiologist, educator; b. N.Y.C., Nov. 21, 1923; s. Eugene Floyd and Rebeckah (Rutter) DuB.; m. Roberdeau Callery, June 21, 1950; children: Anne R., Brooks, James E.F. Student, Harvard U., 1941-43; M.D., Cornell U., 1946. Intern in medicine N.Y. Hosp., 1946-47; med. research fellow U. Rochester, 1949-51; asst. resident Peter Bent Brigham Hosp., Boston, 1951-52; asst. prof. to prof. physiology and medicine U. Pa., 1952-74; prof. epidemiology and physiology Yale U., 1974—; dir. John B. Pierce Found. Lab., 1974-88. Author: The Lung, 3d ed. 1986, Body Plethysmography, 1969; contbr. articles to profl. jours. Served with USNR, 1947-49. Recipient Rsch. Career award NIH, 1963-74; Edward Livingston Trudeau medal Am. Lung Assn., 1989. Mem. Am. Physiol. Soc., Am. Soc. Clin. Investigation, Assn. Am. Physicians, Undersea Med. Soc. Democrat. Clubs: Harvard, Cosmos. Home: 370 Livingston St New Haven CT 06511-1310 Office: 290 Congress Ave New Haven CT 06519-1403

DUBOIS, JANICE ANN, primatologist, educator; b. Woonsocket, R.I., Feb. 5, 1961; d. Raymond Rene and Helen (Recorvits) D. BA in Biology, R.I. Coll., 1983, MS in Biology, 1990. Med. asst. Dubois and Dubois, North Smithfield, R.I., 1982—; instr. C.C. R.I., Warwick, 1990—. Contbr. articles to profl. jours. Mem. Sigma Xi. Home: 372 E Beach Rd Charlestown RI 02813 Office: Community Coll Rhode Island 400 East Ave Warwick RI 02886

DUBOIS, JEAN GABRIEL, pharmaceutical executive, pharmacist; b. Bethisy, Oise, France, Dec. 27, 1926; s. Rene Edmond and Isabelle Francoise (Chauvel) DuB.; m. Francoise Pradille, Sept. 7, 1945; children: Anne Frederique, Jean Christophe, Isabelle, Caroline. PhD in Pharmacy, Faculte de Pharmacy, Paris, 1951; postgrad., U. N.Mex., Albuquerque, 1951-52. Gen. mgr. Comptoir Pharmaceutique de Cambodge, Phnom Penh, Cambodia, 1955-61; quality control and mfg. mgr. Laboratoire Nativelle, Longjumeau, France, 1962-63; plant mgr. Laboratoire ANA, Paris, 1963-64; tech. dir. Pfizer Egypt, Cairo, 1966-68; tech. dir. Pfizer France, Orsay, 1969-88, gen. and quality assurance dir., 1988-91; pharm. mfg. cons. Dubois Conseil, Paris, 1992. Capt. French M.C., 1953-54, Vietnam. Office: Dubois Conseil, 220 bd Raspail, 75016 Paris France

DUBOIS, NORMAND RENE, microbiologist, researcher; b. Lewiston, Maine, June 8, 1938; s. Rosario G. and Yvonne L. (Marcotte) D.; m. Eileen Catherine Barry, Oct. 17, 1964; children: Marc, David, Deborah. BA in Biology, Providence Coll., 1960; PhD in Plant and Soils, U. Mass., 1977. Biol. technician USDA Forset Svc., Hamden, Conn., 1961-69, rsch. microbiologist, 1970—; mem. sci. adv. bd. Entotech, Inc., Davis, Calif., 1990—. Reviewer 4 sci. jours.; contbr. 50 articles to profl. jours. Mem. Am. Soc. Microbiology, Soc. Invertebrate Pathology, Entomological Soc. Am. Democrat. Roman Catholic. Achievements include discovery of a new bacillus thuringesis bacteria. Office: USDA Forest Svc 51 Mill Pond Rd Hamden CT 06514

DU BOIS, PHILIP HUNTER, psychologist, educator; b. Newburgh, N.Y., July 8, 1903; s. Henry Reynolds and Hattie Aletha (Clough) DuB.; m. Margaret Eloise Barclay, Dec. 27, 1936; 1 child, Margaret (Mrs. Richard W. Watson). AB, Union Coll., 1925; MA, Columbia U., 1929, PhD, 1932. Diplomate: in counseling and guidance Am. Bd. Examiners in Profl. Psychology. Instr. English Am. U., Beirut, 1925-28; instr. psychology Columbia, 1930-33; intern psychology N.Y. State Psychiat. Inst. and Hosp., 1932-33; asst. prof. psychology Idaho State U., Pocatello, 1933-35; asst. prof., later asso. prof. and dir. bur. tests and records U. N.Mex., 1935-46; prof. Washington U., St. Louis, 1946-72, prof. emeritus, 1972—; dir. Psychol. Assos., 1958-84; rsch. psychologist Psychol. Assocs., 1972-89; test technician, later supr. N.Mex. Merit System Coun., 1940-42; psychol. cons. Dept. Police St. Louis, 1947-77; cons. human resources rsch. labs. Air R & DCommand, 1951-54, U.S. VA, 1947-70; mem. panel on pers., com. on human resources R & D Bd., Dept. Def., 1952-54; mem. panel on pers. and tng. Office Asst. Sec. Def. (R & D), 1954-57; cons. Cross-Cultural Rsch. Project in Social Psychology, Cairo, summer 1954, USAF Sch. Aviation Medicine, 1955-58, U.S. Office Edn., 1963-67, University City (Mo.) Schs., 1963-66, Ctr. for Nuclear Studies, Memphis State U., 1974-83; participant conf. on pers. rsch. NATO, Brussels, 1965; chmn. ETS Invitational Conf. on Testing Problems, 1969; Aviation psychologist USAAF, 1942-46; asst. dir. USAAF (Psychol. Rsch. Unit 2), 1943; advisor on psychol. screening of flying pers. Free French Air Force, North Africa, 1944; chief psychol. sect. USAAF (Med. and Psychol. Examining Unit 7), 1943-44; chief publs. unit, psychol. sect. USAAF (Hdqrs. A.A.F. Tng. Command), 1944-46. Author: Multivariate Correlational Analysis, 1957, An Introduction to Psychological Statistics, 1965, A History of Psychological Testing, 1970, (with G. Douglas Mayo) A Complete Book of Training: Theory, Principles and Techniques, 1987, A Catskills Boyhood, 1992; editor: A.A.F. Aviation Psychology Report 2, The Classification Program, 1947; co-editor: (monograph) Research Strategies for Evaluating Training; contbr. to profl. jours. With USAAF, 1942-46; Lt. col. USAFR, ret. Mem. AAAS, AAUP, Am. Psychol. Assn. (pres. mil. div. 1954-55, pres. div. evaluation and measurement 1968-69), Mo. Psychol. Assn. (pres. 1954-55), Psychometric Soc. (past sec., past pres.), Soc. Multivariate Exptl. Psychology., Nat. Coun. Measurement in Edn., Phi Beta Kappa, Sigma Xi, Phi Kappa Phi, Psi Upsilon. Home: 94 Aberdeen Pl Saint Louis MO 63105-2273

DUBOURDIEU, DANIEL JOHN, biotechnologist, researcher; b. Lake Forest, Ill., Dec. 15, 1956; s. Richard James and Virginia Rachel (Morris) DuB.; m. Rachel Huse Desley, June 5, 1984; children: Shauna, James. BA, Macalester Coll., 1979; PhD, U. Minn., 1987. Rsch. asst. U. Minn., Mpls., 1979-87; rsch. assoc. Hoffmann LaRoche Inc., Nutley, N.J., 1987-88; dir. mgr. DiaMed Inc., South Windham, Maine, 1988—; cons. Vets Plus Inc., Menomonie, Wis., 1992—; lectr. at colls. and univs., 1983—. Contbr. articles to Toxicology and Applied Pharmacology, Am. Jour. Pathology, Biochemica Biophysica Acta, Toxicon, Prostaglandins. T.H. Rowell grad. fellow Rowell Pharms., 1985. Mem. N.Y. Acad. Scis., Sigma Xi, Rho Chi. Achievements include research in cellular signal transduction, phospholipases, calcium in cell death, heart cell culture. Office: DiaMed Inc RR 1 Inland Farm Dr South Windham ME 04082

DUBOURG, OLIVIER JEAN, cardiology educator, researcher; b. Orleans, Loiret, France, Mar. 26, 1952; s. Jean Fortune and Arlette (Masson) D.; m. Sylvie Rolet, Sept. 21, 1979; 1 child, Benjamin. MD, U. Paris, 1981, Laureat, 1981. Cert. French Bd. Cardiology. Intern Salpetriere Hosp., Paris, 1975-76; resident in cardiology Pitie Hosp., Paris, 1977-81; fellow Ambroise Paré Hosp., Boulogne, France, 1981-89, prof. cardiology, 1989—; mem. steering com., tech. com., pres. human rels. Ambroise Paré Hosp. Contbr. articles to med. jours. Fellow Am. Coll. Chest Physicians, French Soc. Cardiology; mem. Am. Soc. Echocardiography, Am. Fedn. for Clin. Rsch., Am. Inst. Ultrasound, French Soc. Pharmacology. Avocation: golf. Home: 33 ave de Saxe, 75007 Paris France Office: Ambroise Paré Hosp, 9 ave Charles de Gaulle, 92100 Boulogne 1, France

DUBRIDGE, LEE ALVIN, physicist; b. Terre Haute, Ind., Sept. 21, 1901; s. Frederick Alvin and Elizebeth Rebecca (Browne) DuB.; m. Doris May Koht, Sept. 1, 1925 (dec. Nov. 1973); children—Barbara (Mrs. David MacLeod), Richard Alvin; m. Arrola Bush Cole, Nov. 30, 1974. A.B., Cornell Coll., Iowa, 1922, Sc.D., 1940; A.M., U. Wis., 1924, Ph.D., 1926; Sc.D., Wesleyan U., 1946, Poly. Inst. Bklyn., 1946, Washington U., 1948, U. B.C., 1947, Occidental Coll., 1952, U. Md., 1955, Columbia, 1957, Ind. U., 1957, U. Wis., 1957, Pa. Mil. Coll., De Pauw U., 1962, Pomona Coll., Rockefeller Inst., Carnegie Inst. Tech., 1965, Syracuse U., 1969, Rensselaer Poly. Inst., 1970; LL.D., U. Calif., 1948, U. Rochester, 1953, U. So. Calif., 1957, Northwestern U., 1958, Loyola U. of Los Angeles, 1963, U. Notre Dame, 1967, Ill. Inst. Tech., 1967; L.H.D., Redlands U., 1958, U. Judaism, 1958; D.C.L., Union Coll., 1961. Asst. in physics U. Wis., 1922-25, instr., 1925-26; NRC fellow Calif. Inst. Tech., 1926-28; asst. prof. physics Washington U., St. Louis, 1928-33; asso. prof. Washington U., 1933-34; prof. physics, chmn. dept. physics U. Rochester, 1934-46, dean faculty arts scis., 1938-41; investigator Nat. Def. Research Com.; dir. radiation lab. MIT, 1940-45; pres. Calif. Inst. Tech., 1946-69; pres. emeritus, 1970—; sci. adviser to Pres. U.S., 1969-70; Trustee Rand Corp., Santa Monica, Calif., 1948-61; mem. sci. adv. com. Gen. Motors, 1971-75; Mem. gen. adv. com. A.E.C., 1946-52; Naval Research Adv. Com., 1945-51, Air Force Sci. Adv. Bd., 1945-49; sci. advisor Weingart Found., 1979—; mem. Pres.'s Communications Policy Bd., 1950-51, Nat. Sci. Bd., 1950- 54, 58-64; chmn. sci. adv. com. Office Defense Moblzn., 1952-56. Author: (with A.L. Hughes) Photoelectric Phenomena, 1932, New Theories of Photoelectric Effect, 1935, Introduction to Space, 1960; Contbr. numerous sci. and ednl. articles to mags. Mem. Nat. Manpower Council, 1951-64; mem. Nat. Adv. Health Council, 1960-61; mem. distinguished civilan service awards bd. U.S. Civil Service Commn., 1963-65; chmn. Greater Los Angeles Urban Coalition, 1968-69; mem. Pres.'s Air Quality Adv. Bd., 1968-69, Pres.'s Sci. Adv. Com., 1970-72; bd. dirs. Nat. Merit Scholarship Corp., 1963-69, Nat. Ednl. TV, N.Y., 1962-69; Trustee Mellon Inst., 1958-67, Rockefeller Found., 1956-67, Nutrition Found., 1952-63, Carnegie Endowment Internat. Peace, 1951-57, Community TV So. Calif., Los Angeles, 1962-69, Henry E. Huntington Library and Art Gallery, 1962-69, Thomas Alva Edison Found., 1960-69, Pasadena Hall Sci., 1977-78. Recipient Research Corp. award, 1947; Medal for Merit U.S., 1948; King's Medal for Service Gt. Britain, 1946; Recipient Vannevar Bush award NSF, 1982; Benjamin Franklin fellow Royal Soc. Arts. Fellow Am. Phys. Soc. (pres. 1947); mem. AAAS, NAS, Am. Philos. Soc., Phi Beta Kappa, Sigma Pi Sigma, Eta Kappa Nu, Sigma Xi, Tau Kappa Alpha, Tau Beta Pi. Presbyterian. Home: 1763 Royal Oaks Dr Duarte CA 91010

DUBROFF, JEROME M., cardiologist; b. Bklyn., July 12, 1948. AB, Bklyn. Coll., 1970; MD, SUNY, 1974. Diplomate Am. Bd. Internal Medicine, Am. Bd. Cardiovascular Diseases. Intern Georgetown Med. Svc./D.C. Gen. Hosp., Washington, 1974-75, resident, 1975-76; resient Vets. Hosp., Washington, 1976-77; cardiology fellow U. Pitts., 1977-79; rsch. fellow Columbia U., N.Y.C., 1980-82; cardiologist Brookdale Hosp., Bklyn., 1982-89; staff cardiologist Queens Hosp. Ctr., Jamacia, N.Y., 1989-92, chief cardiology, 1992—; instr. medicine Albert Einstein Coll. Medicine, Bronx, N.Y., 1989-92; asst. prof. medicine Mt. Sinai Sch. Medicine, 1993—. Mem. AAAS, Am. Heart Assn., N.Y. Acad. Scis. Achievements include research on effects of heart valve replacement on LU function, use of 20 echo cardiography in heart transplantation. Office: Queens Hosp Ctr Divsn Cardiology 82-68 164th St Jamaica NY 11432

DUBUQUE, GREGORY LEE, medical physicist; b. Louisville, Feb. 6, 1948; s. Lionel Lee Dubuque and Mary Catherine (Bladen) Busija; m. Gloria Sue Horn, June 28, 1969 (div. May 1979); 1 child, Alison Lynn. BS in Physics, Ill. Inst. Tech., 1970; MS, U. Cin., 1972, PhD, 1974. Diplomate Am. Bd. Radiology in Therapeutic Radiology, Am. Bd. Radiology in Med. Nuclear Physics. NASA trainee U. Cin., 1970; instr. radiology U. Colo. Med. Ctr., Denver, 1974-80; chief physicist Nebr. Meth. Hosp., Omaha, 1980-86, Cape Cod Hosp., Hyannis, Mass., 1986—; clin. instr. U. Nebr. Med. Ctr., Omaha, 1982-86; presenter, chmn. numerous sci. symposia and meetings; reviewer Am. Jour. Roentgenology, Med. Physics. Contbr. articles to sci. and med. jours. Rsch. fellow NSF, 1969-70; fellow Bur. Radiol. Health, 1971-74. Mem. Am. Assn. Physicists in Medicine (pres. Rocky Mountain chpt. 1978-79), Am. Radium Soc., Am. Coll. Radiology. Achievements include co-development of SUAR ultrasound phantom test object. Office: Cape Cod Hosp 27 Park St Hyannis MA 02601-3276

DUCKWORTH, DONALD REID, oil company executive; b. Camp Shelby, Miss., Dec. 22, 1945; s. Joseph R. and Vivian M. (Chain) D.; m. Dannette D. Daniel, June 21, 1968; children: Robert, David. BBA, U. Ga., 1967; TEP, U. Va., 1984; MS, Command and Gen. Staff Coll./, U.S. Army, 1987. Commd. U.S. Army, 1967, advanced through grades to maj., retired, 1975; dir., corp. security Norton Co., Worcester, Mass., 1975-79; dir., corp. security Standard Oil, Cleve., 1979-85, dir. human resources, 1985-87; v.p. human resources and adminstrn. BP Exploration, Cleve., 1988-90; v.p. change and human resources BP America, Cleve., 1991—; bd. dirs. OMBAC, Inc., Cleve., Figmo, Inc., Cleve., Corp. Adv. Coun./Grad. Sch. Bus./U. Pitts. Co-author: The Art of Corporate Governance, 1985. Trustee Urban League of Cleve., 1989—, Greater Cleve. Literacy Coaliton, 1988—, Med. Edn. Found. Neoucom, Ruotstown, Ohio, 1988—; vice-chmn. Ohio Com. for Employer Support of the Guard and Res., Columbus, 1987—. Named Com. Chmn. of the Yr., Am. Soc. for Indsl. Security, Washington, 1984, Chmn.'s award, 1983. Mem. Beaver Creek Club, Hill 'n' Dale Club, Pine Lake Trout Club, Boston Big Game Fishing Club, Chagrin Valley Racquet Club. Avocations: fishing, hunting, golf, tennis, fencing. Office: BP Am 200 Public Sq Cleveland OH 44114-2301

DUCKWORTH, JERRELL JAMES, electrical engineer; b. Ft. Payne, Ala., July 22, 1940; s. James K. and Maggie Lee (Hartline) D.; m. Yvonne Cheryl Jones, Nov. 2, 1974; one child, Shelby Elizabeth. AAS in Elec. Engring., DeVry Inst. Tech., 1963. Gen. engr. McDonnell Aircraft Corp., St. Louis, 1963-66; sr. assoc. engr. IBM Corp. Space Systems Ctr., Huntsville, Ala., 1966-72; chief engr. Electric Systems Inc., Chattanooga, 1972-80; dir. elec. engring. Chattanooga Corp., 1980-91; dir. engring. Chattanooga Group Inc., 1991—. Vol. Golden for Congress campagn, Chattanooga, 1986. With U.S. Army, 1958-61. Recipient Apollo 8 medallion NASA, 1968, Apollo 11 medallion, 1969, Apollo Achievement award, 1970. Mem. IEEE, Engring. in Medicine and Biology Soc., Assn. for the Advancement of Med. Instrumentation, Instrument Soc. Am., U.S. Space Found. Mem. Ch. of God. Achievements include development of the Recent History Storage Unit used to locate both hardware and software faults, AC and DC drive systems and a therapeutic ultrasound generator with a multiple-frequency transducer. Home: 7916 Shallowmeade Ln Chattanooga TN 37421-1930 Office: Chattanooga Group Inc 4717 Adams Rd Hixson TN 37343

DUCKWORTH, WALTER DONALD, museum executive, entomologist; b. Athens, Tenn., July 19, 1935; s. James Clifford and Vesta Katherine (Walker) D.; m. Sandra Lee Smith, June 17, 1955; children: Clifford Monroe, Laura Lee, Brent Cullen. Student, U. Tenn., 1953-55; BS, Middle Tenn. State U., 1955-57; MS, N.C. State U., 1957-60, PhD, 1962. Entomology intern Nat. Mus. Nat. History, Washington, 1960-62, asst. curator, 1962-64, assoc. curator, 1964-75, entomology curator, 1975-78, spl.asst. to dir., 1975-78; spl. asst. to asst. sec. Smithsonian Inst., Washington, 1978-84; dir. Bishop Mus., Honolulu, 1984-86, pres., dir., 1986—; trustee Sci. Mus. Va., Richmond, 1982-86, bd. dirs. 1982-84; bd. dirs. Hawaii Maritime Mus., Honolulu, 1984—; mem. Sci. Manpower Commn., Washington, 1982-84. Co-editor: Amazonian Ecosystems, 1973; Am. editor: Dictionary of Butterflies and Moths, 1976; author, co-author numerous monographs and jour. articles in systematic biology. Pres. Social Ctr. for Psychosocial Rehab., Fairfax, Va., 1975. N.C. State U. research fellow, 1957-62; recipient numerous grants NSF, Am. Philos. Soc., Smithsonian Research Found. Assn., Exceptional Service awards Smithsonian Inst., 1973, 77, 80, 82, 84, Disting. Alumnus award Middle Tenn. State U., 1984. Mem. Am. Inst. Biol. Scis (pres. 1985-86, sec.-treas. 1978-84), Entomol. Soc. Am. (pres. 1982-83, governing bd. 1976-85, Disting. Svc. award 1981), Assn. Tropical Biology (exec. dir. 1971-84, sec.-treas. 1976-81), Hawaii Acad. Sci. (coun. 1985—), Arts Coun. Hawaii (legis. com. 1986-87), Assn. Sci. Mus. Dirs., Social Sci. Assn., Assn. Systematic Collections (v.p. 1988-89, pres. 1990-91, Disting. Svc. award 1992), Pacific Sci. Assn. (pres. 1987-91, pres. Pacific Sci. Congress, Honolulu 1991). Democrat. Presbyterian. Lodges: Rotary, Masons, Order Eastern Star. Office: Bishop Mus PO Box 190000A Honolulu HI 96817-9291

DUCOFF, HOWARD S., radiation biologist; b. N.Y.C., May 5, 1923; s. Dave and Tillie (Machinist) D.; m. Rose Hirsch, Aug. 25, 1946; children: Sandra Ducoff Garber, Barbara Ducoff Bruzas, Paul, Laura. BS in Biology, CCNY, 1942; PhD in Physiology, U. Chgo., 1953. Assoc. scientist Argonne (Ill.) Nat. Lab., 1946-57; prof. radiation biology U. Ill., Urbana, 1957-92, prof. emeritus, 1992—; vis. investigator zoology dept. U. Cambridge, Eng., 1964-65; staff cons. Argonne Univs. Assn., 1966-78; vis. investigator Lawrence Berkeley (Calif.) Lab., 1975-76, U. Sussex, Falmer, Eng., 1983-84. Staff sgt. U.S. Army, 1943-46. USPHS spl. fellow, 1964-65, travel grantee, 1961, 70, 79; NIH rsch. grantee. Mem. Radiation Rsch. Soc. (assoc. editor 1989-93), Am. Soc. Cell Biology, N.Am. Hypertherma Soc., Am. Soc. Zoologists, Oxygen Soc. Achievements include research in mitogenesis, causes of death in irradiated adult insects and radiation and longevity enhancement in tribolium. Office: U Ill Physiology Dept 407 S Goodwin Ave Urbana IL 61801

DUDERSTADT, JAMES JOHNSON, university president; b. Ft. Madison, Iowa, Dec. 5, 1942; s. Mack Henry and Katharine Sydney (Johnson) D.; m. Anne Marie, June 24, 1964; children: Susan Kay, Katharine Anne. B in Engring. with highest honors, Yale U., 1964; MS in Engring. Sci, Calif. Inst. Tech., 1965, PhD in Engring. Sci. and Physics, 1967. Asst. prof. nuclear engring. U. Mich., 1969-72, assoc. prof., 1972-76, prof., 1976-81; dean U. Mich. (Coll. Engring.), 1981-86; provost, v.p. acad. affairs U. Mich., 1986-88, pres. univ., 1988—; cons. in field. Author textbooks and; contbr. articles to profl. jours. Recipient E. O. Lawrence award U.S. Dept. Energy, 1986, Nat. Medal of Tech. award Pres. of U.S., 1991, NSPE award, 1991; AEC fellow, 1964-68; named Nat. Engr. of Yr., NSPE, 1991. Fellow Am. Nuclear Soc. (Mark Mills award 1968, Arthur Holly Compton award 1985), AAAS; mem. NAE, Am. Phys. Soc., Nat. Sci. Bd. (chair 1991—), Sigma Xi, Tau Beta Pi, Phi Beta Kappa. Office: U Mich Office of Pres 2074 Fleming Ann Arbor MI 48109

DUDIK, ROLLIE M., healthcare executive; b. Hartford, Conn., Sept. 29, 1935; s. Martin and Iola Maxine (Hamilton) D.; m. Nancy A. Slicner. PhD, Am. U., 1981. Gen. mgr. Freedman Artcraft Engring., Charlevoix, Mich., 1970-73; exec. v.p. Harrison Community Hosp., Cadiz, Ohio, 1973-77; dir. instl. rev. Dade Monroe PSRO, Miami, Fla., 1977-78; exec. dir. Fla. Keys Meml. Hosp., Key West, 1978-82; exec. v.p. Eisenhower Hosp., Colorado Springs, 1982-83; v.p. South Fla. Med. Mgmt., Inc., Miami, 1982—, Doctor's Health Systems Mgmt. Corp., Coral Gables, Fla., 1985—; pres. R.M. Dudik & Assocs., Inc., Miami, 1983—, Sydney, Australia, 1988—; med. projects cons. Orah Wall Med. Enterprises, Inc., San Antonio, 1985—; pres. Caribbean Diagnostic Ctrs., Inc., Coral Gables, 1986—; chief fin. officer Forenpsych Assocs., Inc., Coral Gables, 1990—. Served with U.S. Army, 1956-62. Mem. Am. Inst. Indsl. Engrs., Am. Acad. Med. Adminstrs., Am. Coll. Healthcare Executives, Hosp. Fin. Mgmt. Assn., Hosp. Mgmt. Systems Soc., Am. Soc. Law and Medicine, Nat. Assn. Flight Instrs., Nat. Assn. Accts., Nat. Counterintelligence Corps Assn., Assn. Former Intelligence Officers, Australian Intelligence Assn., Aerospace and Environ. Medicine Assn., Rotary, Kiwanis. Address: 26900 SW 192 Ave Homestead FL 33031-3752

DUDLEY, GEORGE WILLIAM, behavioral scientist, writer; b. Camden, N.J., July 18, 1943; s. Lester Allen and Dorothy Vernon (Hoopes) Boyd; m. Carol Ann Lorenzen, Sept. 14, 1968; 1 child, Suzanne Christine. BS, Baylor U., 1969; MS, N. Tex. State U., 1974. Dir. field testing, rsch. Southwestern Life Ins., Dallas, 1969-80; pres. Behavioral Scis. Rsch. Press, Dallas, 1980-92, chmn. bd. dirs., 1992—; discussant Southwestern Psychol. Assn., 1984. Editor: Handbook for Agent Selection in the Life Insurance Industry, 1980, Jour. Agt. and Mgmt. Selection and Devel., Dallas, 1984-87; co-author: The Psychology of Call Reluctance, 1986, Earning What You're Worth?, 1992; contbr. articles to mgmt. publs. Cpl. USMC, 1961-65. Mem. Am. Psychol. Soc. Achievements include pioneering research into inhibited contact initiation; discovery of 12 diagnostic categories and subclassifications; development of formal diagnostic/assessment procedures. Office: Behavioral Scis Rsch Press 2695 Villa Creek Dr Ste 100 Dallas TX 75234

DUDLEY, LONNIE LEROY, information scientist; b. Belding, Mich., Mar. 3, 1948; s. Edmond LeRoy and LuLa Madeline (Sloan) D. Student, Mich. State U., 1966-67, LaSalle Extension U., Chgo., 1968-69. Encoder III, inventory buyer, acctg. clk. III Brotman Med. Ctr., Culver City, Calif., 1984-88; materiels mgmt. system coord., dir. data processing Chino (Calif.) Community Hosp., 1988-91, dir. materials mgmt., 1988-91, dir. info. systems/materials mgmt./data processing, 1972-75. Scout master Boy Scouts Am., Dowagiac, Mich., 1972-75; adv. Order of the Arrow, Dowagiac, 1972-75, Med. Explorer Post, Dowagiac, 1972-75, Four-H, Dowagiac, 1975-79, Iven C. Kincheloe Jr. Chpt. Order of Demolay, Dowagiac, 1975-79. With USN, 1967-71. Mem. F&AM, Masons, Kinchelow Jr. Chpt. Order of Demolay (scribe and fin. adv., Lmp of Knowledge 1973).

DUDRICK, STANLEY JOHN, surgeon, educator; b. Nanticoke, Pa., Apr. 9, 1935; s. Stanley Francis and Stephania Mary (Jachimczak) D.; m. Theresa M. Keen, June 14, 1958; children: Susan Marie, Paul Stanley, Carolyn Mary, Stanley Jonathan, Holly Anne, Anne Theresa. B.S. cum laude, Franklin and Marshall Coll., 1957; M.D., U. Pa., 1961. Intern Hosp. of U. Pa., Phila., 1961-62, resident, 1962-67; pvt. practice specializing in surgery Phila., 1967-72, 88-90, Houston, 1972-88, 90—; chief surg. svcs. Hermann Hosp.,

Houston, 1972-80, surgeon in chief, dir. Ctr. Cardiovascular Disease, dir. nutritional support svcs., dir. Nutritional Sci. Ctr., 1990—; prof. surgery U. Tex. Med. Sch., Houston, 1972-82, clin. prof. surgery, 1982—, chmn. dept. surgery, 1972-80; cons. in surgery M.D. Anderson Hosp. and Tumor Inst., 1973-88, clin. prof. surgery, cons. to pres., 1982-88; sr. cons. surgery and medicine Tex. Inst. for Rehab. and Research, 1974-88; mem. Anatomical Bd., State of Tex., 1973-78; examiner Am. Bd. Surgery, 1974-78, bd. dirs., 1978-84, sr. mem., 1984—, also mem. and chmn. various coms.; chmn. sci. adv. com. Tex. Med. Ctr. Library, 1974; mem. food and nutrition bd. NRC-Nat. Acad. Scis., 1973-75; mem. sci. adv. com. Nat. Found. for Ileitis and Colitis; mem. surgery, anesthesia and trauma study sect. NIH, 1982-86; chmn. dept. surgery Pa. Hosp., Phila., 1988-90, surgeon in chief, 1988-91, hon. surgery staff, 1991—; clin. prof. surgery U. Pa., 1988—. Editor: Manual of Surgical Nutrition, 1975, Manual of Preoperative and Postoperative Care, 1983; assoc. editor Nutrition in Medicine, 1975—; editorial cons. Jour. of Trauma, 1972-76; mem. editorial bd. Annals of Surgery, 1975—, Infusion, 1978—, Nutrition and Cancer, 1980—, Nutrition Support Services, 1980-86, Jour. Clin. Surgery, 1980-83, Nutrition Research, 1981—, Intermed. Communications Nursing Services, 1981—, Postgraduate General Surgery, 1992—; others.; contbr. chpts. to books, articles to profl. jours.; inventor of new technique of intravenous feeding and anti-cholesterol therapy. Bd. dirs. Found. for Children, Houston, Harris County unit Am. Cancer Soc., Phila. chpt., 1988-90; trustee Franklin and Marshall Coll., 1985—, mem. student life and trusteeship coms., 1986—, mem. overseers bd., 1986—, exec. com. 1986—, alumni programs and devel. com., 1991—, pres. regional adv. coun., 1992—. Decorated knight Order St. John of Jerusalem Knights Hopitaller; recipient VA citation for significant contbn. to med. care, 1970; Mead Johnson award for research in hosp. pharmacy, 1972; Seale Harris medal So. Med. Assn., 1973; AMA-Brookdale award in medicine, 1975; Great Texan award Nat. Found. Ileitis and Colitis, 1975; Modern Medicine award, 1977; Disting. Alumnus citation Franklin and Marshall Coll., 1980; WHO, Houston, 1980; Stinchfield award Am. Acad. Orthopedic Surgery, 1981; Bernstein award Med. Soc. of State of N.Y., 1986 numerous others. Fellow ACS (vice chmn. pre and post operative com. 1975, gov. 1979-85, com. on med. motion pictures 1981-90, SESAP com. 1990—), Philippine Coll. Surgeons (hon.), Coll. Medicine and Surgery of Costa Rica (hon.), Am. Coll. Nutrition (Grace A. Goldsmith award 1982); mem. AMA (council on food and nutrition 1971-76, exec. com. 1975-76, council on sci. affairs 1976-81, Goldberger award in clin. nutrition 1970), Am. Surg. Assn., Am. Acad. Pediatrics (hon., Ladd medal 1988), Am. Pediatric Surg. Assn. (hon.), Am. Soc. Nutritional Support Services (bd. dirs. 1982-87, pres. 1984, Outstanding Humanitarian award 1984) Soc. Univ. Surgeons (exec. council 1974-78), Assn. for Acad. Surgery (founders group), Internat. Soc. Surg., Internat. Fedn. Surg. Colls., Internat. Soc. Parenteral Nutrition (exec. council 1975—, pres. 1978-81), Internat. Fedn. Surgery Soc., So. Med. Assn. (chmn. surgery sect. 1984-85), Houston Gastroent. Soc., Houston Surg. Soc., Tex. Surg. Soc., Tex. Med. Assn. (com. nutrition and food resources), Harris County Med. Soc., Am. Radium Soc., Am. Soc. Parenterald Enteral Nutrition (pres. 1977, bd. advs. 1978—, chmn. bd. advisers 1978, Vars award 1982, Rhoads lectr. 1985, Dudrick Rsch. Scholar award named in his honor), Am. Gastroent. Assn., Soc. Surg. Oncology, James Ewing Soc., Ravdin-Rhoads Surg. Assn., Excelsior Surg. Soc. (Edward D. Churchill lectr. 1981), Soc. Laparoendoscopic Surgery, Soc. Surg. Chairmen, So. Surg. Assn., Southwestern Surg. Congress, Southeastern Surg. Congress, Surg. Biology Club II, Surg. Infection Soc. (chmn. membership com. 1987-90), Western Surg. Soc., Halsted Soc., Allen O. Whipple Surg. Soc., Am. Inst. Nutrition, Soc. Clin. Surgery, Am. Soc. Clin. Investigation, Soc. for Surgery of Alimentary Tract, Am. Trauma Soc. (founders group), Am. Assn. for Surgery of Trauma, Soc. Clin. Surgery, Am. Soc. Clin. Nutrition, Fedn. Am. Soc. Exp. Biology, Am. Burn Assn., AAAS, AAUP, Coll. Physicians Phila., Phila. Acad. Surgeons, George Hermann Soc., Union League Phila., Med. Club Phila., Franklin Club Phila., Houston Drs. Club (gov. 1973-76), Cosmos Club, Athenaeum, Phi Beta Kappa, Phi Beta Kappa Assocs., Sigma Xi, Alpha Omega Alpha. (sec.-treas. Houston chpt. 1982-83). Home: 3050 Locke Lane Houston TX 77019 Office: Hermann Hosp 6411 Fannin St Houston TX 77030-1501

DUDZIAK, DONALD JOHN, nuclear engineer, educator; b. Alden, N.Y., Jan. 6, 1935; s. Joseph and Josephine Mary (Ratajczak) D.; m. Judith Ann Staib, Aug. 22, 1959; children: Alan Joseph, Matthew John, Karin Marie. BS in Marine Engring., U.S. Mcht. Marine Acad., 1956; MS in Radiol. Physics, U. Rochester, 1957; PhD in Applied Math., U. Pitts., 1963. Registered profl. engr., Calif. Commd. ensign USN, 1956, advanced through grades to capt.; sr. engr. Bettis Atomic Power Lab., Pitts., 1957-65; group leader U. Calif.-Los Alamos (N.Mex.) Nat. Lab., 1965-68, 69-81, group leader, sect. leader, 1982-88, lab. fellow, 1988—; ret. USN, 1987; prof., head dept. nuclear engring. N.C. State U., Raleigh, 1990—; vis. prof. U. Va., Charlottesville, 1968-69; guest scientist Swiss Fed. Inst. Reactor Rsch., Wuerenlingen, 1982; mem. lab. microfusion facility steering com. U.S. Dept. Energy, 1986-90, inertial confinement fusion adv. com., 1992—; chmn. fusion tech. working group, Neutronics, Brookhaven, N.Y., 1975; cons. nuclear power schs. USN, 1962-65. Editor: Reactor Principles, 1964, Radiation Shielding, 1964; contbr. editor Fusion Tech., 1987—; editor Progress in Nuclear Energy, 1992—; contbr. articles to profl. publs. Vice-chair Los Alamos County Planning and Zoning Commn., 1969-74. Fellow Am. Nuclear Soc. (divsn. chair 1972-73, 77-78, 92); mem. Am. Soc. Engring. Educators, U.S. Naval Inst., Los Alamos Sunrise Kiwanis (treas. 1981-90), Sigma Xi. Office: NC State Univ Dept Nuclear Engring BOX 7909 Raleigh NC 27695

DUDZIAK, WALTER FRANCIS, physicist; b. Adams, Nass, Berkshire, Jan. 7, 1923; s. Michael Casimer and Mary (Piekielniak) D.; m. Barbara Ann Campbell, June 25, 1954; children: Diane, Mary, Daniel, Suzanne. BS, Rensselaer Poly. Inst., 1946, MS in Physics, 1947; PhD in Physics, U. Calif., Berkeley, 1954. Instr. physics Rensselaer Poly. Inst., Troy, 1946-48; aero scientist NASA, Cleve., 1948-49; rsch. assoc. U. Calif., Berkeley, 1949-54; lectr. physics U. Calif., Santa Barbara, 1958-60; mgr. computer sci. Gen. Electric Co., Santa Barbara, 1958-64; exec. v.p. Pan Fax Corp., Santa Barbara, 1964-68; pres. Info. Sci., Inc., Santa Barbara, 1968—; rsch. analsyt Manhattan Dist. Project, Oak Ridge, Tenn., 1942-44, Carbon Carbide Corp., Oak Ridge, 1944-46. With U.S. Army, 1942-44. Recipient Adams Scholarhsip, City of Adams. Mem. KC (grand knight 1974—), Elks. Roman Catholic. Home: 1390 Camino Manadero Santa Barbara CA 93111-1048 Office: 123 W Padre St Santa Barbara CA 93105-3960

DUECK, JOHN, agricultural researcher, plant pathologist; b. Altona, Man., Can., Aug. 11, 1941; s. Gerhard H. and Elizabeth (Funk) D.; m. Mary Enns, Sept. 8, 1962; children: Harvey C., Cheryl E., Kenneth J. BSA, U. Man., Winnipeg, 1964; MSc, U. Minn., 1966, PhD, 1971. Agronomist Man. Dept. Agr., Winnipeg, 1966-68; plant pathologist Agr. Can., Harrow, Ont., 1971-73; plant pathologist, head tech. sect. Agr. Can., Ottawa, Ont., 1973-74; plant pathologist oil seeds Agr. Can., Saskatoon, Sask., 1974-81; dir. rsch. sta. Agr. Can., Regina, Sask., 1981-89, Summerland, B.C., 1989—; project coord. BARD Project, Pakistan, 1985-87, 89-91; pres.-elect Sask. Inst. Agrologists, Regina, 1988-89. Contbr. over 60 articles to publs. Recipient Disting. Alumni award Dept. Plant Pathology U. Minn., 1991. Mem. Am. Phytopath. Soc., Can. Phytopath. Soc., Agrl. Inst. Can. Office: Agr Can Rsch Sta, PO Box 5000, Summerland, BC Canada V0H 1Z0

DUER, ELLEN ANN DAGON, anesthesiologist; b. Balt., Feb. 3, 1936; d. Emmett Paul and Annie (Sollers) Dagon; m. Lyle Jordan Millan IV, Dec. 21, 1963; children: Lyle Jordan V, Elizabeth Lyle, Ann Sheridan Worthington.; m. T. Marshall Duer, Jr., Aug. 23, 1985. A.B., George Washington U., 1959; M.D., U. Md., 1964; postgrad., Johns Hopkins U., 1965-68. Intern Union Meml. Hosp., Balt., 1964-65; resident anesthesiology Johns Hopkins Hosp., Balt., 1965-68, fellow in surgery, 1965-68; practice medicine specializing in anesthesiology Balt., 1968—; faculty Church Home and Hosp., Balt., 1969—; attending staff Union Meml. Hosp., Church Home and Hosp., Frankling Sq. Hosp., Children's Hosp., James Lawrence Kernan Hosp., Balt., 1982—; co-chief anesthesiology James Kernan Hosp., 1983—, med. dir. out-patient surgery dept., 1987—; mem. med. exec. com. Kernan Hosp., 1988—; affiliate cons. emergency room Church Home and Hosp., Balt., 1969—, mem. med. audit and utilizaions com., 1970-72, mem. emergency and ambulatory care com., 1973-74, chief emergency dept., 1973-74; cons. anesthesiologist Md. State Penitentiary, 1971; fellow in critical care medicine Md. Inst. Emergency Medicine, 1975-76; mem. infection control com. U. Md. Hosp., 1975—; instr. anesthesiology U. Md. Sch. Medicine, 1975—; staff anesthesiologist Mercy Hosp., 1978—, audit com., 1979-80, 82; asst. prof. anesthegiology U. Md. Med. Sch., 1989—; mem. med. exec. com. Kernan Hosp., 1990—, v.p. 1990, chief of staff, 1992—. Mem. AMA, Am. Coll. Emergency Physicians, Met. Emergency Dept. Heads, Am., Md. Socs. Anaesthesiologists, Balt. County Med. Soc., Med. and Chior Faculty Md., Chiurgical Soc., Internat. Congress Anaesthesiologists, Internat. Anaesthesia Research Soc., Am., L'Hirondelle Club, Annapolis Yacht Club, Chesapeake Bay Yacht Racing Assn. Episcopalian. Address: 1011 Wagner Rd Ruxton MD 21204

DUFF, RONALD G., research scientist; b. Billings, Mont., Dec. 8, 1936; s. Ross I. and Alda M. (Markholt) D.; m. N. Darlene Duff, Aug. 23, 1962 (dec. June, 1987); children: Kelle Amber, Ross Alan; m. Dawn Newman, Aug. 2, 1988. MusB, U. Mont., 1959; PhD, U. Colo., 1967. Dir. music Deer Park (Wash.) Pub. Schs., 1959-62; postdoctoral fellow Baylor Coll. Medicine, Houston, 1967-68; assoc. prof. M.S. Hershey (Pa.) Med. Ctr., 1968-74; research fellow, head lab. Abbott Labs., North Chicago, Ill., 1974-83; prof. Chgo. Med. Sch., North Chicago, 1976-85; sr. v.p. research and devel. Damon Biotech, Inc., Needham Heights, Mass., 1983-87; sr. v.p. rsch. and devel. Curative Techs., Inc., Setauket, N.Y., 1987—; mem. sci. adv. bd. Biotheraphy Systems, Mountain View, Calif., 1984-86, Vivotech, Needham Heights, 1984-86. Contbr. over 100 articles to profl. jours. Mem. Am. Soc. Microbiology, Am. Assn. Cancer Research, Tissue Culture Soc., Sigma Xi, Phi Sigma. Avocations: music, golf, fishing. Office: Curative Techs 14 Research Way East Setauket NY 11733-3469

DUFF, WILLIAM GRIERSON, electrical engineer; b. Alexandria, Va., Dec. 16, 1936; s. Johnnie Douglas and Annetta Osceola (Rind) D.; BEE, George Washington U., 1959, postgrad., 1959-72; MS, Syracuse U., 1969; DSc in Elec. Engring., Clayton U., 1977; m. Sandra K. Via, June 25, 1983; children: Warren David, Valerie Lynn, Dawn Elizabeth, Deborah Arleen, Kelly Juanita. Chief engr. def. systems div. Atlantic Research Corp., Springfield, Va., 1959—; asst. prof. Capitol Inst. Tech., Greenbelt, Md., 1972—; instr. Interference Control Technologies, Don White Cons., Inc., Gainesville, Va. Counselor, Meth. Sr. High Youth Group, 1965-73. Recipient Good Citizenship award DAR, 1955; Math. award George Washington High Sch., Alexandria, 1955. Fellow IEEE (pres. EMC Soc., assoc. editor group newsletter 1970—); mem. AIEE (Best Paper award 1961), George Washington U. Engring. Alumni Assn. (pres. 1963-64, Engring. Alumni Svc. award 1980), Springfield Golf and Country Club, Occoquan Water Ski Club (pres. 1976), Sigma Tau, Theta Tau. Author: EMI Handbook, vol. 5, EMI Prediction and Analysis Techniques, 1972; Mobile Communications, 1976, Fundamentals of EMC, 1988, EMC in Telecommunications, 1988; contbr. articles to profl. jours. Home: 7601 S Valley Dr Fairfax VA 22039-2965 Office: Atlantic Rsch Corp 5501 Backlick Rd Ste 300 Springfield VA 22151-3938

DUFFELL, JAMES MICHAEL, computer systems analyst; b. Atlanta, Dec. 2, 1966; s. Gordon Michael and Carol (Jackson) D. AA, Oxford Coll., 1987; BS, Emory U., 1989. Programmer/analyst Emory U. Clinic, Atlanta, 1989-90, systems analyst, 1990—; advisor Emory/Digital Support Group, Atlanta, 1990—. Mem. Digital Equipment Corp. Users Soc. Achievements include design/implementation of pulmonary function lab.'s proprietary information management network. Home: 5412 M Vernon Way Dunwoody GA 30338-2814 Office: The Emory Clinic 1365 CLifton Rd NE Atlanta GA 30322-1104

DUFFEY, GEORGE HENRY, physics educator; b. Manchester, Iowa, Dec. 24, 1920; s. Henry Alfred and Marion Ella (Barr) D.; m. Helen Susie Hooper, Sept. 17, 1945; children: Ann Elizabeth, James Roy, Mary Kay. BA, Cornell Coll., 1942; PhD, Princeton (N.J.) U., 1945. Asst. prof. chemistry S.D. State Coll., Brookings, 1945-49, assoc. prof. chemistry, 1949-55, prof. chemistry, 1955-58; prof. chemistry physics U. Miss., Oxford, Miss., 1958-59; prof. physics S.D. State U., Brookings, 1959—; vis. prof. physics U. Western Australia, Perth, 1977. Author: Physical Chemistry, 1962, Theoretical Physics, 1980, A Development of Quantum Mechanics, 1984, Quantum States and Processes, 1992, Applied Group Theory, 1992; editor: Poems from the 1830s by a Poor Son of Ireland, 1977; contbr. articles to profl. jours. including Jour. Chem. Physics, Phys. Rev., Jour. Phys. Chemistry, Founds. of Physics and Jour. Chem. Edn. Recipient Excellence in Teaching award Western Elec. Fund, 1971-72. Mem. AAAS, Am. Phys. Soc., Am. Chem. Soc., Societa Italiana di Fisica, Philosophy of Sci. Assn. Baptist. Home: 628 11th Ave Brookings SD 57006-1526 Office: SD State U Dept Physics Brookings SD 57007-0395

DUFFIE, JOHN ATWATER, chemical engineer, educator; b. White Plains, N.Y., Mar. 31, 1925; s. Archibald Duncan and Lulie Adele (Atwater) D.; m. Patricia Ellerton, Nov. 22, 1947; children—Neil A., Judith A. Duffie Schwarzmeier, Susan L. Duffie Buse. B.Ch.E., Rensselaer Poly. Inst., 1945, M.Ch.E., 1948; Ph.D., U. Wis., 1951. Registered profl. engr., Wis. Instr. chem. engring. Rensselaer Poly. Inst., 1946-49; research asst. U. Wis., 1949-51; research engr. DuPont Co., 1951; sci. liaison officer Office Naval Research, 1952-53; mem. faculty dept. chem. engring. U. Wis.-Madison, 1954—, prof., 1957-88, prof. emeritus, 1988—, dir. solar energy lab., 1956-88; Fulbright scholar U. Queensland, Australia, 1964; sr. Fulbright-Hays scholar Commonwealth Sci. and Indsl. Research Orgn., Australia, 1977; hon. sr. research fellow U. Birmingham, 1984. Author: (with W.A. Beckman) Solar Energy Thermal Processes, 1974, Solar Engineering of Thermal Processes, 1980, 2d edit., 1991; (with W.A. Beckman, S.A. Klein) Solar Heating Design, 1977. Served with USN, 1943-46, 52-53. Fellow Am. Inst. Chem. Engrs.; mem. Internat. Solar Energy Soc. (past pres., Charles G. Abbot award Am. sect. 1976, editor Solar Energy jour. 1985-93, Farrington Daniels award 1987). Home: 5710 Dorsett Dr Madison WI 53711-3404 Office: Univ Wis 1500 Johnson Dr Madison WI 53706-1687

DUFFIE, NEIL ARTHUR, mechanical engineering educator, researcher; b. Madison, Wis., June 16, 1950; s. John Atwater and Patricia Lorain (Ellerton) D.; m. Colleen Marie Russell, July 9, 1972; children: Meghan, Patrick, Laura. BS in Computer Sci. with distinction, U. Wis. Madison, 1972, MSE, 1974, PhD in Mech. Engring., 1980. Cert. mfg. engr. Engring. specialist data acquisition and simulation lab. U. Wis. Madison, 1972-73, rsch. asst. engring. experiment sta., 1973-74, 76-80, engring. specialist solar energy lab., 1974-75, asst. prof. mech. engring. dept., 1980-86, assoc. prof., 1986-93, prof., 1993—; rsch. officer civil engring. dept. James Cook U., Townsville, Queenland, Australia, 1975-76; vis. lectr. dept. design of machine systems Cranfield (Eng.) Inst. Tech., 1982; assoc. dir. mfg. systems engring. program U. Wis. Madison, 1987-93; dir. robotic tech. devel. Wis. Ctr. for Space Automation and Robotics, U. Wis. Madison, 1987-92, Wis. Ctr. for Space Automation and Robotics, 1992—; mem. numerous profl. coms.; rschr. in field. Author: (with J. Bollinger) Computer Control of Machines and Processes, 1988; co-author: Control, Programming and Integration of Manufacturing, 1992; mem. editorial bd. Jour. Mfg. Systems, 1987—; contbr. numerous articles to profl. jours.; patent in self-guided welding machine. Co-pres. Van Hise PTO, Madison, 1991-92. Recipient TRW Post-Doctoral award Mfg. Engring., 1981. Mem. AIAA, ASME (assoc.), CIRP (corresponding), N. Am. Mfg. Rsch. Inst. (sr., scientific com. 1987—), Internat. Instn. Prodn. and Engring. Rsch. (sec. STC-P 1988-91), Soc. Mech. Engrs. (sr., Outstanding Young Mfg. Engr. award 1984). Avocations: ice hockey, running, folk music, gardening. Office: U Wis Mech Engring Mfg Systems Lab 1513 Univ Ave Madison WI 53706-1572

DUFFIELD, ALBERT J., mechanical engineer; b. Kalamazoo, May 8, 1924; s. Albert P. and Blanche Pearl (Kelly) D.; m. Lois Grace Gilson, Nov. 27, 1947; children: John Gilson, James Albert, Grace Ann. BSME, Mich. Tech. U., 1948. Project engr. Olin-Matheson Chem. Corp., East Alton, Ill., 1948-55; metals lab. supr. Olin Corp., New Haven, Conn., 1955-73; mgr. mfg. engr. shakeproof div. Ill. Tool Works, Elgin, Ill., 1973-74; operating mgr. extrusion Dresser Industries, Inc., Johnson City, Tenn., 1974-83; engring. mgr. Automatic Fastener Corp., Branford, Conn., 1983-84; mgr., mfg. engr. O.F. Mossberg and Sons, Inc., North Haven, Conn., 1984-85; engring. cons. A. J. Duffield, Guilford, Conn., 1985--; lectr. Hartford State Tech. Coll. Mem. Rescue 10 Team Marine Patrol, Guilford, 1985--, pres.; mem. Guilford Lakes Improvement Assn., 1985--, chmn., project oversight com.; bd. dirs. Conn. Bapt. Homes, Meriden, Conn., 1991--. With U.S. Army, 1943-46, PTO. Recipient Silver Beaver award Boy Scouts Am., 1963. Mem. Soc. Mfg. Engrs. (life). Republican. Baptist. Achievements include patents for Blanking Sheet Material, Ammunition, Stripper Mechanism, Explosive Welding, Metal Forming, Metal Cartridge Mfg., Shapping Tubular Shells, Pierce and Tap Drive Pin for Power Actuated Tool, Corrosion Resistant Fastening System & Method. Home and Office: 32 White Birch Dr Guilford CT 06437

DUFFIS, ALLEN JACOBUS, polymer chemistry extrusion specialist; b. N.Y.C., Oct. 13, 1939; s. Jacobo Alcides and Catherine Monica (Joseph) D.; m. Kathleen Mary Fountain, June 24, 1983; 1 child, Jacque Fountain. Student, CCNY, L.I. U., Pratt Inst. Technician Pepsi-Cola Corp., L.I., 1960-63; tech. specialist Allied Chem. Corp., Morristown, N.J., 1963-73; process engr. Berol Corp., Danbury, Conn., 1973-87; cons., owner AJD Enterprises Inc., New Preston, Conn., 1987—; tech. dir. Solmex Ag, Zug, Switzerland, 1988—, Makune Solmex (India) Ltd., Bombay, 1991—. Speaker Sci. Horizons for Youth, Danbury, 1985-87. Republican. Achievements include 3 patents for specialized high speed plastics co-extrusion processes and related systems (European). Home: 4 Hearthstone Terr New Milford CT 06776 Office: AJD Enterprises PO Box 2215 New Preston CT 06777

DUFFY, JAMES JOSEPH, engineer; b. Pawhuska, Okla., Aug. 28, 1917; s. James Leo and Margretta Marsden (Wittlinger) D.; m. Edna Jean Laramie, Aug. 15, 1953; children: Paul Edward, Donald Lawrence. BSME, Rice U., 1941; MS in Auto Engring., Chrysler Inst., 1948. Registered profl. engr., Mich. Jr. petroleum engr. Humble Oil and Refining Co., Houston, 1941-42; engine devel. engr. Chrysler Corp., Detroit, 1948-52, resident engr., 1952-54; auto transmission engr. Ford Motor Co., Dearborn, Mich., 1954-59, steering design engr., 1962-65, adv. steering design engr., 1965-90, tech. specialist advanced steering, 1990—; valve design engr. AiRsch., Phoenix, 1959-62. Patentee in field. 1st lt. USAF, 1943-45, SW Pacific. Recipient Disting. Inventor award Intellectual Property Owners, 1988. Mem. Soc. Automobile Engrs., Tau Beta Pi, Sigma Xi. Republican. Lutheran. Avocation: golf. Home: 35594 Orangelawn St Livonia MI 48150-2539 Office: Ford Motor Co PO Box 2053 Dearborn MI 48121-2053

DUFFY, LAWRENCE KEVIN, biochemist, educator; b. Bklyn., Feb. 1, 1948; s. Michael and Anne (Browne) D.; m. Geraldine Antoinette Sheridan, Nov. 10, 1972; children: Anne Marie, Kevin Michael, Ryan Sheridan. BS, Fordham U., 1969; MS, U. Alaska, 1972, PhD, 1977. Teaching asst. dept. chemistry U. Alaska, 1969-71, research asst. inst. arctic biology, 1974-77; postdoctoral fellow Boston U., 1977-78, Roche Inst. Molecular Biology, 1978-80; research asst. prof. med. br. U. Tex., 1980-82; asst. prof. neurology (biol. chemistry) Med. Sch. Harvard U., Boston, Mass., 1982-87, adv. biochemistry instr. Med. Sch., 1983-87; instr. gen. and organic chemistry Roxbury Community Coll., Boston, 1984-87; prof. chemistry and biochem. U. Alaska, Fairbanks, 1992—, coord. program biochem and molecular biology, summer undergrad. res. in chemistry and biochem., 1987—. Bd. dirs. Alzheimer Disease Assn. of Alaska, 1988—; mem. instnl. rev. bd. Fairbanks Meml. Hosp., 1990. Lt. USNR, 1971-73. NSF trainee, 1971; J.W. McLaughlin fellow, 1981; W.F. Milton scholar, 1983; recipient Alzheimers Disease and Related Disorders Assoc. Faculty Scholar award, 1987. Mem. Am. Soc. Neurochemists, Am. Soc. Biol. Chemists, N.Y. Acad. Sci., Am. Chem. Soc. (Analytical Chemistry award 1969), Internat. Soc. Toxicologists, Sigma Xi (pres. 1991 Alaska club), Phi Lambda Upsilon. Roman Catholic. Office: U Alaska Inst Arctic Biology Fairbanks AK 99775

DUFFY, MARTIN EDWARD, management consultant, economist; b. Fall River, Mass., May 24, 1940; s. Arthur Louis and Edna Marie (Cunneen) D.; m. Irene Patricia Daley, Aug. 24, 1968 (div. Jan. 1980); 1 child, Kathryn; m. Priscilla Claire Stieff, May 14, 1988; 1 child, Brianna. BS in History, BSEE, Tufts U., 1963; MBA, U. Pa., 1967, PhD, 1973. Asst. dean U. Pa., 1967-71; asst. dir. Fels Ctr., U. Pa., 1971-73; exec. asst. to fin. v.p. Harvard U., Cambridge, Mass., 1973-75; v.p. Data Resources, Lexington, Mass., 1975-84; v.p., gen. mgr. MRCA Info. Svcs., Cambridge, 1984-86; pres. The Perseus Group/RCG, Boston, 1986—; planning com. White House Conf. on Aging, Washington, 1981; cons. La. in 2001, Baton Rouge, 1982; lectr. in field. Author: The Elderly in Future Economy, 1981. Lt. USN, 1963-65. Mem. Am. Econs. Assn., Nat. Assn. Forensic Economists, Cambridge Sports Union, Tufts U. Alumni Assn. (pres. 1985, exec. com. 1982-87, mem. coun. 1978—), Nat. Bus. Travelers Assn. (bd. dirs. ednl. com. 1988—). Roman Catholic. Avocations: running marathons, mountain climbing, biking.

DUFFY, SALLY M., psychologist; b. Charleston, S.C., Mar. 16, 1953; d. Edward Baker and Mary Jane (Hutchins) D. BA, U. S.C., 1976; MS, Ea. Ky. U., 1979; PhD, U. Ky., 1989. Diplomate Am. Acad. Pain Mgmt. Psychologist, dir. Partial Hospitalization Program, Stanton, Ky., 1979-80; staff psychologist Frazier Rehab. Ctr., Louisville, 1981-84; psychologist Comprehensive Med. Rehab. Ctr., Lexington, Ky., 1987-89; postdoctoral fellow Med. U. S.C., Charleston, 1989; psychology svcs. coord. Carolinas Spine & Rehab. Ctr./HealthSouth, Charlotte, N.C., 1990-92; health psychologist The Rehab. Ctr., Charlotte, 1992—; adj. prof. psychology Georgetown Coll., 1988-89; lectr. in field. Contbr. articles to profl. jours. NIMH traineeship, 1985-86, 86-87; Counseling Psychology Departmental merit fellow U. Ky., 1984-85. Mem. APA (divsn. 17 counseling psychology, divsn. 38 health psychology, divsn. 35 psychology of women, divsn. 40 neuropsychology), Am. Bd. Med. Psychotherapists, Ky. Psychol. Assn., N.C. Psychol. Assn., Am. Pain Soc. Democrat. Avocations: camping, horseback riding, swimming, music. Office: The Rehab Ctr 2610 E 7th St Charlotte NC 28204

DUFFY, STEPHEN FRANCIS, civil engineer, educator; b. Detroit, Feb. 4, 1955; s. Frank Thomas and Kathleen Mildred (Flaherty) D.; m. Karen Sue Martin, Aug. 16, 1980; children: Megan, Erin. MSCE, U. Akron, 1980, PhD, 1987. Registered profl. engr., Ohio. Field engr. Messmore Testing and Supervision, Akron, Ohio, 1979-81; vis. instr. U. Akron, 1981-85; prof. civil engring. Cleve. State U., 1985—; rsch. assoc. NASA-Lewis Rsch. Ctr., Cleve. Co-editor: Life Prediction Methodologies and Data for Ceramic Materials, 1993; contbr. to profl. publs. Mem. ASCE, ASTM, Am. Ceramic Soc., Am. Soc. Materials. Roman Catholic. Office: Cleve State U Dept Civil Engring Euclid Ave at E 24th St Cleveland OH 44115

DUFRESNE, ARMAND FREDERICK, management and engineering consultant; b. Manila, Aug. 10, 1917; s. Ernest Faustine and Maude (McClellan) DuF.; m. Theo Rutledge Schaefer, Aug. 24, 1940 (dec. Oct. 1986); children: Lorna DuFresne Turnier, Peter, m. Lois Burrell Klosterman, Feb. 21, 1987. BS, Calif. Inst. Tech., 1938. Dir. quality control, chief product engr. Consol. Electrodynamics Corp., Pasadena, Calif., 1945-61; pres., dir. DUPACO, Inc., Arcadia, Calif., 1961-68; v.p., dir. ORMCO Corp., Glendora, Calif., 1966-68; mgmt., engring. cons., Duarte and Cambria, Calif., 1968—; dir., v.p., sec. Tavis Corp., Mariposa, Calif., 1968-79; dir. Denram Corp., Monrovia, Calif., 1968-70, interim pres., 1970; dir., chmn. bd. RCV Corp., El Monte, Calif., 1968-70; owner DUFCO, Cambria, 1971-82; pres. DUFCO Electronics, Inc., Cambria, Calif., 1982-86, chmn. bd. 1982-92; pres. Freedom Designs, Inc., Simi Valley, Calif., 1982-86, chmn. bd. dirs., 1982—; owner DuFresne Consulting, 1992—; chmn. bd., pres. DUMEDCO,Inc., 1993—. Patentee in field. Bd. dirs. Arcadia Bus. Assn., 1965-69; bd. dirs. Cambria Community Services Dist., 1976, pres., 1977-80; mem., chmn. San Luis Obispo County Airport Land Use Commn., 1972-75. Served to capt. Signal Corps, AUS, 1942-45. Decorated Bronze Star. Mem. Instrument Soc. Am. (life), Arcadia (dir. 1965-69), Cambria (dir. 1974-75) C. of C., Tau Beta Pi. Home: 901 Iva Ct Cambria CA 93428-2913

DUGAN, JOHN PATRICK, optical engineer; b. Indpls., Sept. 9, 1958; s. Thomas Patrick and Barbara Jean (Vogel) D. BS in Physics, Xavier U., 1980; MS in Optics, U. Rochester, 1982. Sr. optical engr. Itek Optical Systems, Lexington, Mass., 1982-88, Spectra Physics Laserplane, Dayton, Ohio, 1988—. Counselor disfunctional child Mass. Mentor, Boston, 1987-88; bd. mem., vol. Interfiath Food Pantry, Cin., 1990—. Mem. Internat. Soc. Optical Engrs. Roman Catholic. Achievements include patent for projection of two orthoginal reference light planes; patents pending for focus compensating lens for use in extreme temperatures, adjustable focus technique using a moveable weak lens. Home: 401 Wellesley Ave Cincinnati OH 45224 Office: Spectra Physics Laserplane 5475 Kellenburger Rd Dayton OH 45424

DUGAN, JOHN VINCENT, JR., research and development manager, scientist; b. Lost Creek, Pa., Oct. 22, 1936; s. John Vincent and May Ann

(Curley) D.; m. Joan Elaine Thomas, Dec. 26, 1964; children: John Edward, Paul Michael, Michael Thomas, Erin Elaine. BA in Chemistry, Lasalle Coll., 1957; PhD in Phys. Chem., U. Notre Dame, 1964. Scientist, adminstr. NASA Lewis Rsch. Ctr., Cleve., 1961-75; staff dir., tech. staff mem. on energy, space and aviation Com. on Sci., Space and Tech., U.S. Ho. of Reps., Washington, 1975-88; dir. Washington ops. Cortana Corp., Falls Church, Va., 1988-92; sr. phys. scientist ozone depletion, atmospheric chemistry Tech. Assessment Systems, Inc., Washington, 1992; v.p. congl. affairs Gen. Dynamics Space Systems Divsn., 1992—. Author: (with others) Dynamics of Ion Molecule Collisions, 1971; contbr. over 50 tech. papers and articles to profl. jours. Head physics judge Fairfax County Sci. Fairs, Va., 1985—; mem. adv. bd. Fairfax County Com. on Exceptional Children, 1977-79. Named Man of the Yr., Shenandoah (Pa.) C. of C., 1982. Mem. E.F. Sorin Soc. (U. Notre Dame), Founders Club (Lasalle Coll.), Sigma Xi. Roman Catholic. Achievements include first anlaytical and computer studies of directional forces in dynamics of ion-molecule collisions and predicted formation of long-lived collision complexes; development of computer movies of ion-molecule collisions involving orientation dependant forces; advanced research in concept integration, hydrodynamics, advanced materials, and automation. Home: 8301 Miss Anne Ln Annandale VA 22003-4619

DUGAN, PATRICK RAYMOND, microbiologist, university dean; b. Syracuse, N.Y., Dec. 14, 1931; s. Francis Patrick and Joan Irma (Clause) D.; m. Patricia Ann Murray, Sept. 22, 1956; children: Susan Eileen, Craig Patrick, Wendy Shawn, Carolyn Paige. B.S., Syracuse U., 1956, M.S., 1959, Ph.D., 1964. Asso. research scientist Syracuse U. Research Corp., 1956-63; mem. faculty Ohio State U., Columbus, 1964—, asso. prof., 1968-70, prof., chmn. dept. microbiology, 1970-73; acting dean Ohio State U. (Coll. Biol. Scis.), 1978-79, dean, 1979-85; prin. scientist EG&G Idaho Nat. Engring. Lab., Idaho Falls, 1987-91, sci. and engring. fellow, 1991—, dir. Ctr. for Bioprocessing Tech., 1987—. Author: Biochemical Ecology of Water Pollution, 1972. Trustee, Columbus Zool. Assn. and Zoo, 1982-87. Fellow Am. Acad. Microbiology; mem. Am. Soc. Microbiology (Ohio pres. 1968-70), AAAS, Soc. Indsl. Microbiology, Idaho Acad. Sci., Am. Chem. Soc., Soc. Mining Engrs.

DUGGAL, ARUN SANJAY, ocean engineer; b. London, June 22, 1962; came to the U.S., 1985; s. Manohar Lal and Lalita (Hora) D.; m. Anne Pimmel, Mar. 21, 1992. B of Tech., Indian Inst. Tech., 1985; MS, U. Miami, 1987; PhD, Tex. A&M U., 1992. Engr. in tng., Tex. Rsch. asst. U. Miami, Fla., 1986-87; teaching asst. dept. civil engring. Tex. A&M U., College Station, 1988-89; rsch. asst. Tex. A&I U., College Station, 1989-92; engring. rsch. assoc. Offshore Tech. Rsch. Ctr., College Station, 1992—. Contbr. articles to Jour. Waterways, others. Rosenstiel fellow U. Miami, 1985-86. Mem. Tau Beta Pi. Achievements include development of techniques and instrumentation to measure displacements of long flexible cyclinders in waves; research in wave-riser interaction; experimental research in second order wave diffraction around large floating bodies. Home: 3301 Providence Ave # 902 Bryan TX 77803 Office: Offshore Tech Rsch Ctr 1200 Mariner Dr College Station TX 77845

DÜHMKE, ECKHART, radiation oncology educator; b. Berlin, July 22, 1942; s. Martin Walter and Christa Anna Luise (Horrer) D.; m. Eva Leopoldine Herta Bagge, Dec. 31, 1970; children: Anna Katharina, Elisa Maria, Rudolf Martin, Victoria Christina. MD, U. Kiel, Germany, 1969, Habilitation in Med. Radiology, 1980. Resident dept. radiology U. Kiel, 1970-74, specialist in radiology, 1975-80, specialist in radiotherapy, lectr., 1980-85; prof. radiation oncology, chmn. dept. radiotherapy U. Göttingen, Germany, 1985-93, provisional chmn. dept. diagnostic radiology, 1985-87, provisional chmn. dept. radiation physics and biology, 1987-91, dir. div. radiology, 1989-93; chmn. dept. radiotherapy and radiation oncology U. Munich, 1993—; vis. prof. dept. radiation oncology and nuclear medicine Hahnemann U., Phila., 1984. Co-author: Handbook of Medical Radiology, 1974; author: Medicinal Radiography with Fast Neutrons, 1980; co-editor, editor: Function Preserving Therapy of Laryngeal Carcinoma Proc., 1990-91; also numerous articles on diagnostic and therapeutic radiology. Mem. German Roengten Soc., German Cancer Soc., Am. Soc. Therapeutic Radiology and Oncology, European Soc. Therapeutic Radiology and Oncology, Am. Soc. Clin. Oncology. Lutheran. Avocations: skiing, sailing. Office: Munich Dept Radiotherapy and Radiation Oncology, Marchioninistr 15, D-81377 Munich Germany

DUKE, CHARLES BRYAN, research and development manufacturing executive, physics educator; b. Richmond, Va., Mar. 13, 1938; s. Charles Joseph Jr. and Virginia (Welton) D.; m. Ann Evans, Jan. 1, 1961; children: Amy Dickerson, Emily Elizabeth. BS in Math., Duke U., 1959; PhD in Physics, Princeton U., 1963. Staff corp. rsch. GE, Schenectady, N.Y., 1963-69, cons., 1969-72; prof. of physics U. Ill., Urbana, 1969-72; mgr., sr. fellow Xerox Corp., Webster, N.Y., 1972-88; dep. dir., chief scientist Battelle Pacific Northwest Div., Richland, Wash., 1988-89; sr. rsch. fellow Xerox Corp., Webster, N.Y., 1989—; bd. govs. Am. Inst. Physics, N.Y.C., 1976-82, 84-87; adj. prof. physics, U. Rochester (N.Y.), 1972-88; affiliate prof. physics, U. Wash., Seattle, 1988-89. Author: Tunneling in Solids, 1969; editor in chief Jour. Materials Rsch., Pitts., 1985-86, Surface Sci., 1992—; contbr. over 300 articles to profl. jours. Named one of 1000 Most Cited Scientists, Inst. for Sci. Info., 1981. Fellow Am. Physical Soc., IEEE; mem. NAE, Am. Vacuum Soc. (hon., bd. dirs. 1983-76, 78-80, pres. 1979, M.W. Welch award in vacuum sci. and tech. 1977), Materials Rsch. Soc. (councillor 1988-90, treas. 1991-92), Am. Chem. Soc. Office: Xerox Webster Rsch Ctr 800 Phillips Rd # 114-38D Webster NY 14580-9791

DUKE, MICHAEL B., aerospace scientist; b. L.A., Dec. 1, 1935; s. Leon and Eva (Siegel) D.; m. Julia Elizabeth Bartram, 1958 (div. 1966); children: Lisa, Stuart; m. Mary Carolyn Creamer, July 17, 1967; children: Kenneth, Donna. BS Geology, Calif. Inst. Tech., 1957, MS Geochemistry, 1961, PhD Geochemistry, 1963. Rsch. scientist U.S. Geol. Survey, Washington, 1963-70; lunar sample curator NASA, Houston, 1970-77, chief solar system exploration div., 1977-90, dep. for sci., moon/Mars exploration, 1990—; bd. dirs. Spaceweek Houston, 1988—; adv. manned space exploration. Organizer: Lunar Bases Symposia, 1984-88; contbr. articles to profl. jours. Adult leader Boy Scouts Am. Houston, 1980-90. Recipient Nininger Meteorite award, 1963, Presdl. Meritorious award, 1988. Fellow AIAA (assoc., Space Sci. award 1991), AAAS, Meteoritical Soc.; mem. Internat. Acad. Astronautics, Geol. Soc. Washington (D.C.). Office: NASA Johnson Space Ctr Houston TX 77058

DUKE, STEPHEN OSCAR, physiologist, researcher; b. Battle Creek, Mich., Oct. 9, 1944; s. Oscar and Azalee Rosa (Tallant) D.; m. Barbara Alice Rowe, June 2, 1967; children: Gregory Ivan, Robin Anne. BS, Henderson State U., 1966; MS, U. Ark., 1969; PhD, Duke U., 1975. Plant physiologist So. Weed Sci. Lab. USDA, Stoneville, Miss., 1975-84; research leader USDA, Stoneville, 1984-87, lab. dir., 1987—; adj. prof. Miss. State U., Starkville, 1978—. Co-author: Physiology of Herbicide Action, 1993; editor: Weed Physiology, 2 vols., 1985, Pest Control with Enhanced Environmental Safety, 1993; contbr. numerous articles to profl. publs. Head referee Greenville Youth Soccer Assn. (Miss.), 1982—; soccer coach Washington Sch., Greenville, 1986-88. Served to 1st lt. U.S. Army, 1968-70, Vietnam. Decorated Bronze Star; recipient Edminster award USDA, 1986, Disting. Alumnus award Henderson State U., 1989, CIBA-GEIGY/Weed Sci. Soc. Am. award CIBA-GEIGY Corp., 1990. Fellow Weed Sci. Soc. Am. (assoc. editor 1978-83, Outstanding Young Scientist award 1984, Outstanding Article award 1987, 88, Rsch. award 1990); mem. Am. Soc. Plant Physiology (chmn. so. sect. 1985-86), Coun. for Agrl. Sci. and Tech. (bd. dirs. 1993—), Am. Chem. Soc., Scandinavian Soc. Plant Physiology, So. Weed Sci. Soc. (v.p. 1993). Avocations: soccer, writing. Home: 1741 W Azalea Dr Greenville MS 38701-7508 Office: USDA Agrl Research Service So Weed Sci Lab PO Box 350 Stoneville MS 38776

DUKERSCHEIN, JEANNE THERESE, aquatic biologist, educator, researcher; b. Menomonie, Wis., Aug. 17, 1951; d. Robert E. and Emma Jeanne (Kysilko) A.; m. Russell O. Dukerschein, Oct. 6, 1973; children: Erica, Jon. BS in Med. Tech., U. Wis., Eau Claire, 1973; MEPD, U. Wis., LaCrosse, 1988, MS in Biology, 1990. Cert. in med. tech., secondary edn. Bacteriologist U. Minn. Hosps., Mpls., 1973-74; med. technologist Missoula, Mont., 1974-77; free-lance writer Houston County News, others, La Crescent, Minn., 1982-85; substitute tchr. various locations, 1988, 91; rschr. River Studies Ctr. U. Wis., LaCrosse, 1988-90; field sta. team leader Wis. Dept. Natural Resources, Onalaska, 1992—; mem. water quality tech. com. Upper Mississippi River Conservation Commn., La. Crosse, 1991-93. Contbr. articles to jours. and newspapers. U. Wis.-La Crosse fellow, 1988-90; NSF scholar, 1968. Mem. Upper Mississippi River Consortium, Sigma Xi, Phi Delta Kappa. Achievements include discovery that male and female mayflies bio-accumulate different amounts of heavy metals and that the concentrations of cadmium and mercury accumulated decrease downstream from sites of point-source pollution. Office: Wis Dept Natural Resources 575 Lester Ave Onalaska WI 54650

DUKLER, ABRAHAM EMANUEL, chemical engineer; b. Newark, Jan. 5, 1925; s. Louis and Netty (Charles) D.; children—Martin Alan, Ellen Leah, Malcolm Stephen. B.S., Yale U., 1945; M.S., U. Del., 1950, Ph.D., 1951. Devel. engr. Rohm & Haas Co., Phila., 1945-48; research engr. Shell Oil Co., Houston, 1950-52; mem. faculty dept. chem. engring. U. Houston, 1952—, prof., 1963—, chmn. dept., 1967-73, dean engring., 1976-83; dir. State of Tex. Energy Council, 1973-75; cons. Schlumberger-Doll Research Co., Brookhaven Nat. Lab., Shell Devel. Co., Exxon, others. Contbr. chpts. to books, articles to profl. jours. Recipient Research award Alpha Chi Sigma, 1974. Fellow Am. Inst. Chem. Engrs. (Alpha Chi Sigma rsch. award 1974, D.Q. Kern rsch. award 1989), Nat. Acad. Engring., Am. Soc. Engring. Edn. (research lectureship award 1976); mem. Am. Inst. Chem. Engrs., ASME, AAAS, Am. Chem. Soc., AAUP, Sigma Xi, Tau Beta Pi. Office: Univ Houston Coll Engring Houston TX 77004

DULBECCO, RENATO, biologist, educator; b. Catanzaro, Italy, Feb. 22, 1914; came to U.S., 1947, naturalized, 1953; s. Leonardo and Maria (Virdia) D.; m. Gulseppina Salvo, June 1, 1940 (div. 1963); children: Peter Leonard (dec.), Maria Vittoria; m. Maureen Muir; 1 dau., Fiona Linsey. M.D., U. Torino, Italy, 1936; D.Sc. (hon.), Yale U., 1968, Vrije Universiteit, Brussels, 1978; LL.D., U. Glasgow, Scotland, 1970. Asst. U. Torino, 1940-47; research asso. Ind. U., 1947-49; sr. research fellow Calif. Inst. Tech., 1949-52, asso. prof., then prof. biology, 1952-63; sr. fellow Salk Inst. Biol. Studies, San Diego, 1963-71; asst. dir. research Imperial Cancer Research Fund, London, 1971-74; dep. dir. research Imperial Cancer Research Fund, 1974-77; disting. research prof. Salk Inst., La Jolla, Calif., 1977—, pres., 1989-92; pres. emeritus Salk Inst., La Jolla, 1993—; prof. pathology and medicine U. Calif. at San Diego Med. Sch., La Jolla, 1977-81, mem. Cancer Ctr.,; vis. prof. Royal Soc. G.B., 1963-64, Leeuwenhoek lectr., 1974; Clowes Meml. lectr. Atlantic City, 1961; Harvey lectr. Harvey Soc., 1967; Dunham lectr. Harvard U., 1972; 11th Marjory Stephenson Meml. lectr., London, 1973, Harden lectr., Wye, Eng., 1973, Am. Soc. for Microbiology lectr., L.A., 1979; mem. Calif. Cancr Adv. Coun., 1963-67; mem. vis. com. Case Western Res. Sch. Medicine; adv. bd. Roche Inst., 1968-71, Inst. Immunology, Basel, Switzerland, others. Trustee LaJolla Country Day Sch. Recipient John Scott award City Phila., 1958; Kimball award Conf. Pub. Health Lab. Dirs., 1959; Albert and Mary Lasker Basic Med. Research award, 1964; Howard Taylor Ricketts award, 1965; Paul Ehrlich-Ludwig Darmstaedter prize, 1967; Horwitz prize Columbia U., 1973; (with David Baltimore and Howard Martin Temin) Nobel prize in medicine, 1975; Targa d'oro Villa San Giovanni, 1978; Mandel Gold medal Czechoslovak Acad. Scis., 1982, Via de Condotti prize, 1990; Cavaliere di Gran Croce Italian Rep., 1991; named Man of Yr. London, 1975; Italian Am. of Yr. San Diego County, Calif., 1978; hon. citizen City of Imperia (Italy), 1983; Guggenheim and Fulbright fellow, 1957-58; decorated grand ufficiale Italian Republic, 1981; hon. founder Hebrew U., 1981. Mem. NAS (Selman A. Waksman award 1974, com. on human rights), Am. Assn. Cancer Rsch., Internat. Physicians for Prevention Nuclear War, Am. Philos. Soc., Accademia Nazionale dei Lincei (fgn.), Accademia Ligure di Scienze e Lettre (hon.), Royal Soc. (fgn. mem.). Home: 7525 Hillside Dr La Jolla CA 92037-3941

DULMES, STEVEN LEE, computer science educator; b. Sheboygan Falls, Wis., Apr. 29, 1957; s. Warren Lynn and LaVerne (Wensink) D.; m. Renee Kay VandeVrede, Dec. 2, 1989. BS in Mech. Tech., Purdue U., 1981; postgrad., DePaul, 1989—. Design engr. Harnischfeger Corp., Cedar Rapids, Iowa, 1981-84; mfg. specialist Ohmeda div. BOC, Madison, Wis., 1984-85; mfg. engr. EDS div. GM, Oak Creek, Wis., 1985-87; instr. computer aided design Coll. Lake County, Grayslake, Ill., 1987—; cons. Baxter Healthcare Corp., Round Lake, Ill., 1989—. Pres. Cedar Valley Running Assn., Cedar Rapids, 1981-83; adviser Jr. Achievement, Cedar Rapids, 1983; tchr. Wilwood Presbyn. Ch., 1989; active Calvary Meml. Ch., Racine, Wis., 1990. Republican. Mem. Ref. Ch. in Am. Avocations: running, skiing, swimming, computer programming. Home: 3531 Sherwood St Racine WI 53406 Office: Coll Lake County 19351 W Washington St Grayslake IL 60030-1198

DULSKI, THOMAS R., chemist, writer; b. Pitts., Nov. 17, 1942; s. Frank J. and Stephanie (Rakoczy) D.; m. Grace J. Verosky, Oct. 19, 1985. BS, U. Pitts., 1963. Rsch. chemist Crucible Materials Corp., Pitts., 1963-68, J&L Steel Corp., Pitts., 1968-72; chemistry supr. Numec Corp., Pitts., 1972; cons. Dravo Corp., Pitts., 1973; specialist in analytical chemistry Carpenter Tech. Corp., Reading, Pa., 1973—; mem. tech. adv. coun. Pa. State U. Berks Campus, Reading, 1983-87. Author: (book chpt.) Classical Wet Analytical Chemistry, Metals Handbook, 1986;author: (ency. entry) Metallurgical Analysis, Colliers, 1992; author of 13 pub. tech. articles and 6 pub. short stories and novelettes. Recipient Pharmacia prize Analytica Chimica Acta, 1983. Fellow ASTM (Cert. Appreciation 1979, Lundell-Bright award 1986, award of merit 1988, chmn. S-17 com. 1989—); mem. Am. Chem. Soc. Office: Carpenter Tech Corp R & D Ctr PO Box 14662 Reading PA 19612-4662

DUMAROT, DAN PETER, electrical engineer; b. N.Y.C., Mar. 7, 1956; s. Peter and Annita (Panzeri) D.; m. Catherine Ann Eberts, Jan. 18, 1975; children: James, Evan. BEE, Poly. Inst. N.Y., 1983. Mem. tech. staff Hughes Aircraft Co., Fullerton, Calif., 1983-86; sr. design engr. Diagnostic Retrieval Systems, Oakland, N.J., 1986-89; adv. engr. IBM T.J. Watson Rsch. Ctr., Hawthorne, N.Y., 1989-92, 1992—; cons. engr., N.Y.C., 1987-88. Achievements include patent for a high performance multi-bank global memory card for multi-processor systems.

DUMAS, JEFFREY MACK, lawyer; b. Corpus Christi, Tex., Sept. 29, 1945; s. Glenn Irven and Virginia (Jones) D.; m. Penny Mary Walter, June 5, 1971; children: Todd Glenn, Rebecca Hope. BS, U.S. Naval Acad., 1968, MSEE, 1969; JD, Harvard U., 1978; registered profl. engr., Colo., Wash., Mont., Calif. Bar: Wash. 1979, U.S. Ct. Appeals (9th cir.) 1979, Mont. 1980, Calif. 1983, U.S. Supreme Ct., 1984, U.S. Patents and Trademark Office, 1990. Fellow, UN, N.Y.C., 1978; corp. counsel Boeing Co., Seattle, 1978-80; atty. McClelland Law Office, Missoula, Mont., 1980-82; gen. counsel Briton-Lee, Inc., Los Gatos, Calif., 1982-83; sr. counsel Nat. Semicondr. Corp., Singapore, 1983-87; gen. counsel, sec. Cypress Semiconductor Corp., Santa Clara, Calif., 1987-91; assoc. gen. counsel Silicon Graphics, Inc. Santa Clara, Calif., 1991—. Author: (with Richard Gowan) Signals and Systems, 1975; contbr. articles to profl. jours. Chmn. Common Cause, Mont., 1982, Vietnam Vets. of Mont., 1982; chmn. Pikes Peak chpt. Sierra Club, 1975, treas. Cascade chpt., Seattle, 1980. Served with USN, 1968-75, aviator Southeast Asia, 1970-73. Decorated Dist. Flying Cross, Air medals, Bronze Star; recipient George Washington medal Freedoms Found., 1975. Mem. IEEE, AIAA, Council on Fgn. Relations, U.S. Naval Inst., Sigma Xi. Address: 1643 Hyde Dr Los Gatos CA 95030

DUMBACHER, JOHN PHILIP, evolutionary biologist; b. Cin., July 26, 1965; s. John Philip Dumbacher and Mary Joan Dieckhaus. BA in Gen. Biology, Vanderbilt U., 1987; student, Clemson U., 1988-90; MS in Ecology and Evolution, U. Chgo., 1993. Contbr. numerous publs. to sci. jours. Hinds grantee U. Chgo., 1992, grantee NIH, 1993, Nat. Geog., 1993; Centenial fellow U. Chgo., 1991. Mem. Am. Soc. Naturalists, Am. Ornithologists' Union, Soc. Study Evolution, Cooper Ornithological Soc., Papua New Guinea Bird Club. Achievements include co-discovery (with Bruce M. Beehler) of the Pitohui, a poisonous genus of bird that uses a powerful homobatrachotoxin as a chemical defense. Office: U Chgo Dept Ecology & Evolution 1101 E 57th St Chicago IL 60637

DUMITRESCU, LUCIEN Z., aerospace researcher; b. Bucharest, Romania, July 28, 1931; arrived in France, 1991; s. Zaharia D. and Natalia V. (Grigoriu) D.; m. Lucia A. Droc, June 21, 1952 (div. 1964); children: Michel-Paul. Dipl.engr., Poly. Inst. Fac. Aeron., Bucharest, 1954; D of Engring., Acad. Scis., Bucharest, 1969. Scientist Inst. Fluid Mechanics, Romanian Acad. Sci., 1952-70; sr. scientist, lab head Inst. of Aeronautics, Romania Ministry of Industry, Bucharest, 1970-80, sr. sci. counselor, 1980-90; prof. U. De Provence, Marseille, France, 1991—; v.p. quality assurance commn. Inst. of Aeronautics, Romanian Ministry of Industry, Bucharest, 1980-90. Author: Research in Shock Tubes, 1969; contbr. over 50 articles to profl. publs. Mem. AIAA (sr.), Internat. Adv. Com. for Shock Tube Symposia. Achievements include devel. of large aerodynamic test facilities; implementation of a quality assurance system for the design and devel. in the Romanian aeron. industry; rsch. on shock waves in gases and aircraft aerodynamics. Home: 29 Av Campagne Barielle, 13013 Marseille France Office: U Provence MHEQ, Centre St Jerome Case 321, 13397 Marseilles France

DUMONT, MARK ELIOT, biochemist, educator; b. N.Y.C., July 20, 1950; s. Allan Eliot and Joan (Auerbach) D.; m. Lynn Beth Mehlman, May 20, 1984; children: Anna Aline, Nora Matea. BA, Harvard U., 1972; MS, Western Wash. U., 1975; PhD, Johns Hopkins U., 1980. Postdoctoral fellow Yale U., New Haven, 1980-85; postdoctoral fellow U. Rochester, N.Y., 1985-89, asst. prof., 1989—. Achievements include research in intracellular sorting of proteins, yeast genetics and molecular biology, membrane proteins, import of cytochrome c into mitochondria. Office: U Rochester Sch Medicine Dept Biochemistry PO Box 607 Rochester NY 14642

DUMONT, MICHAEL GERARD, electro-optical engineer; b. Dover, N.H., Aug. 9, 1961; s. Gerard R. and Linda L. (Ramsdell) D.; m. Lisa Sherman, June 8, 1986. B of Engring. Tech., U. N.H., 1987; postgrad., Tufts U., 1991—. Devel. engr. USCI div. C.R. Bard Inc., Billerica, Mass., 1987-89, project engr., 1989-91; project engr. Bard Critical Care, Tewksbury, Mass., 1991—; cons. Rare Earth Med. Inc., Dennis, Mass., 1990—. Contbr. articles to profl. jours. Mem. Optical Soc. Am., Soc. Photographic and Instrumentation Engrs. Home: 29 Pinewood Dr Stratham NH 03885 Office: Bard Critical Care 200 Ames Pond Dr Tewksbury MA 01871

DUMOULIN, CHARLES LUCIAN, physicist; b. Japan, June 28, 1956; s. Charles Lucian and Jeanne (Lacoste) D.; m. Shelley Wynne Eckert, Mar. 17, 1979; children: Christine, Michelle, Peter, Paul. BS, Fla. State U., 1977, PhD, 1981. Ops. dir. NIH Resource Syracuse (N.Y.) U., 1981-84; chemist GE, Schenectady, N.Y., 1984—; adj. asst. prof. Syracuse U., 1982-84; assoc. prof. radiology Albany (N.Y.) Med. Coll., 1987—. Contbr. chpts. to books and articles to profl. jours. Named Disting. Inventor Intellectual Property Owners Inc. 1988, Inventor of Yr. 1991. Mem. Am. Chem. Soc., Soc. Magnetic Resonance Imaging, Soc. Magnetic Resonance in Medicine (editorial bd. mem. 1990—), Phi Beta Kappa, Phi Kappa Phi. Roman Catholic. Achievements include developing magnetic resonance angiography methods; application of high resolution NMR methods to magnetic resonance Imaging; patentee in field. Office: Gen Elec Co PO Box 8 Schenectady NY 12301

DUMSHA, DAVID ALLEN, chemical company executive; b. Phila., Feb. 11, 1957; s. K. Stanley and Stella M. (Kowalski) D. BS in Chem. Engring., Drexel U., 1983. Rsch. chemist Remond Co., Malvern, Pa., 1984-88; lab. dir. W.A. Reynolds Corp., Bristol, Pa., 1988-90, Coyne Chem. Co., Croydon, Pa., 1990—. Mem. NSPE, Am. Electroplaters Soc., Am. Chem. Soc., Am. Inst. Chem. Engrs., Home Bldg. Soc. (fin. com.). Achievements include development of process for continuous conversion of white phosphorous to red phosphorous, process for polymerization of polystyrene using novel reactor scheme, post-mortem analysis of galvanic oxygen sensors. Home: 5018 Ditman St Philadelphia PA 19124

DUNAU, ANDREW T., mathematics, science and technology consultant; b. Bethesda, Md., Nov. 29, 1959; s. Bernard and Anastasia (Thannhauser) D.; m. Robin Ann Marks, June 17, 1989; children: Peter, Casey. BA, Grinnell Coll., 1981; MPA, NYU, 1985. Rsch. assoc. NYU, N.Y.C., 1984-86; pres. Dunau Assocs., L.A., 1986-88, Spokane, Wash., 1988—; cons. GTE, L.A. United Sch. Dist., L.A. Ednl. Partnership, Bonneville Power, Washington Water Power, others. Co-author, mng. editor: Technology in Learning, 1992. Pres. bd. Inland Empire Discovery Ctr.; bd. dirs. Cannon Hill Children's Svcs. Office: Dunau Assocs N 1404 Thor Ct Ste 112 Spokane WA 99202

DUNBAR, BONNIE J., engineer, astronaut; b. Sunnyside, Wash., Mar. 3, 1949; d. Robert Dunbar; m. Ronald M. Sega. BS in Ceramic Engring., U. Wash., 1971, MS in Ceramic Engring. cum laude, 1975; PhD in Biomed. Engring., U. Houston, 1983. With Boeing Computer Svcs., 1971-73; sr. rsch. engr. space div. Rockwell Internat., Downey, Calif.; with NASA, 1978—, astronaut, 1981—, mission specialist flight STS 61-8, 1985, mission specialist flight STS-32, 1990, payload commander Shuttle Columbia Flight, 1992; spl. asst. to dep. assoc. adminstr. NASA, Washington, 1993; vis. scientist Harwell Labs., Oxford, Eng., 1975; adj. asst. prof. mech. engring. U. Houston, mem. bioengring. adv. group; participant space missions, 1985, 90. Recipient Nat. Engring. award Am. Assn. Engring. Socs., 1992. Mem. AAAS, Am. Ceramic Soc. (life, Greaves-Walker award 1985, Schwalt Zwalder PACE award 1990), Soc. Biomed. Engring., Materials Rsch. Soc., Nat. Inst. Ceramic Engrs., Arnold Air Soc. and Angel Flight (bd. dirs.), Keramos, Tau Beta Pi. Address: NASA 300 E St SW Houston TX 77058*

DUNBAR, GARY LEO, psychology educator; b. Cadillac, Mich., Feb. 5, 1949; s. Leo Arthur and Betty Jean (Norden) D.; m. Deborah Sue Prevost, Dec. 25, 1976; children: Darbi Sue, Gary Leo Jr. BA, Eckerd Coll., 1971, BS, 1975; MA, MS, Ctrl. Mich. U., 1976, 77; PhD, Clark U., 1988. Grad. asst. Ctrl. Mich. U., Mt. Pleasant, 1975-77, instr. psychology, 1977-83, 87-88, asst. prof., 1988-91, assoc. prof., 1991—; Clark U. scholar Worcester, Mass., 1983-85; rsch. fellow Clark U., Worcester, 1985-87. Author: Psychology and Human Behavior, 1978, book chpts.; contbr. articles to profl. jours. Mem. Am. Psychol. Soc., Soc. Neurosci. Achievements include rsch. in pharmacological treatment of behavioral deficits caused by damage to the brain or neurodegenerative diseases. Home: 906 W Hopkins Mount Pleasant MI 48858 Office: Ctrl Mich U Psychology Dept Mount Pleasant MI 48859

DUNBAR, MAXWELL JOHN, oceanographer, educator; b. Edinburgh, Scotland, Sept. 19, 1914; s. William and Elizabeth (Robertson) D.; m. Joan Jackson, Aug. 1, 1945; children: Douglas, William; m. Nancy Wosstroff, Dec. 14, 1960; children: Elisabeth, Andrew, Christine, Robyn. B.A., Oxford (Eng.) U., 1937, M.A., 1939; Ph.D., McGill U., 1941; DSc (hon.), Meml. U., Nfld., 1979, U. Copenhagen, 1991. Mem. faculty McGill U., Montreal, 1946—, prof., 1959—; also chmn. dept. marine sci., dir. Marine Sci. Ctr.; climate rsch. group dept. atmospheric and oceanic sis. McGill U, 1987—; dir. Eastern Arctic Investigations, Can., 1947-55. Author: Eastern Arctic Waters, 1951, Ecological Development in Polar Regions, 1968, Environment and Good Sense, 1971; contbr. articles profl. jours. Decorated officer Order of Can.; Guggenheim fellow Denmark, 1952-53; recipient Bruce medal Royal Soc. Edinburgh, 1950, Fry medal Can. Soc. Zoologists, 1979, Arctic Sci. prize North Slope Borough, 1986, No. Sci. award (Can.), 1987, J.P. Tully medal Can. Meteorol. and Oceanography Soc., 1988. Fellow Royal Soc. Can., Linnaean Soc. London, Arctic Inst. N.Am. (gov., past chmn., recipient Fellows award 1973). Home: 488 Strathcona Ave, Westmount, PQ Canada H3Y 2X1

DUNBAR, ROBERT COPELAND, chemist, educator; b. Boston, June 26, 1943; s. William Harrison and Carolyn (Roorbach) D.; m. Mary Asmundson, June 21, 1969; children: Geoffrey, William. AB, Harvard U., 1965; PhD, Stanford U., 1970. Rsch. assoc. Stanford (Calif.) U., 1970; from asst. prof. to prof. chemistry Case Western Res. U., Cleve., 1970—. Contbr. over 150 articles on chemistry of gas-phase ions to sci. publs. Sloan fellow, 1973-75, Guggenheim fellow, 1978-79. Mem. Am. Soc. Mass Spectrometry, Inter-Am. Photochem. Soc., Am. Chem. Soc., Sigma Xi. Home: 2880 Fairfax Rd Cleveland Heights OH 44118 Office: Dept Chemistry Case Western Res Univ Cleveland OH 44106

DUNBAVIN, PHILIP RICHARD, acoustical consultant; b. Liverpool, Eng., Feb. 1, 1953; s. Peter and Amelia Patricia (Wiseman) D.; m. Jane Ann Bowden, July 1, 1976; children: Rachel Helen, Peter James. MS in Indsl. Acoustics, Salford Univ., Eng., 1980. Scientific officer U.K. Atomic Energy Authority, Warrington, Eng., 1973-77; sales engr. Sension Ltd., Northwich,

Eng., 1977-78; sales mgr. Sension Scientific Ltd., Northwich, Eng., 1978-81; acoustic cons. Atmospheric Control Engrs., Ltd., Manchester, Eng., 1981-84, Sound Rsch. Labs. Ltd., Manchester, 1984-88; mng. dir. P.D.A. Ltd., Warrington, 1988—. Co-author: Noise Control in Building Services, 1988; contbr. articles to profl. jours. Gov. Thelwall Jr. Sch., Warrington, 1988-91. Mem. Inst. Acoustics, Soc. Environ. Engrs., Inst. Sound and Communication Engrs., Inst. Occupational Safety and Health, Inst. Sales and Mktg. Mgmt. Conservative. Avocations: 1st Dan Shukokai Karate, Sumo, music, theater. Office: PDA Ltd Vincent House, 212 Manchester Rd, Cheshire, WA1 3BD Warrington England

DUNCAN, CHARLES LEE, food products company executive; b. Waynesboro, Tenn., Oct. 10, 1939; s. Grady E. and D. Pearl (Dotson) D.; m. Barbara C. Woodburne, June 21, 1967; children: Stuart L., Andrew R., Amy L. BS, U. Tenn., Martin, 1961; MS, La. State U., 1963; PhD, U. Wis., 1967. Postdoctoral fellow dept. nutrition sci. U. Wis., Madison, 1966-68, asst. prof. Food Rsch. Inst. and dept. bacteriology, 1968-72, assoc. prof., 1972-76, prof., 1976-77; dir. food safety and nutrition Campbell Soup Co., Camden, N.J., 1977-79, v.p. food sci. and tech., 1979-81; v.p. R & D Hershey (Pa.) Foods Corp., 1981—; postdoctoral fellow dept. microbiology U. Wash., Seattle, 1975. Contbr. numerous articles to sci. jours. Recipient Rsch. Career Devel. award NIH, 1974-79. Mem. Am. Soc. Microbiology, Internat. Life Scis. Inst. (treas.), Inst. Food Technologists (mem. rsch. com. 1986-89), AAAS. Republican. Mem. Ch. of Christ. Avocation: orchid growing. Office: Hershey Foods Corp 1025 Reese Ave Hershey PA 17033-2298

DUNCAN, CONSTANCE CATHARINE, psychologist; b. Watertown, Wis., Nov. 2, 1948; d. Howard Burton and Mary Elizabeth (Fagan) Duncan; m. Allan Franklin Mirsky, July 4, 1986. BA, Northwestern U., 1970; AM, U. Ill., 1973, PhD, 1978. Sr. rsch. analyst Adolf Meyer Mental Health Ctr., Decatur, Ill., 1971-73; rsch. and teaching asst. Dept. Psychology, U. Ill., Champaign, 1974-78; postdoctoral fellow in neuroscis. Dept. Psychiat. and Behavioral Scis., Stanford U. Sch. Medicine, Palo Alto, 1978-81; rsch. psychologist VA Med. Ctr., Palo Alto, 1978-81; sr. staff fellow Lab. of Psychology & Psychopathology, NIMH, Bethesda, Md., 1981-88; chief unit on psychophysiology NIMH, Bethesda, Md., 1982-89, rsch. specialist, 1989—; pvt. practice psychology Bethesda, Md., 1981—; adj. assoc. prof. Johns Hopkins Sch. Hygiene and Pub. Health, Balt., 1987—. Assoc. editor Psychophysiology, 1987-91; cons. editor 15 sci. jours.; contbr. articles to profl. jours., chpts. to books. Found. assoc. Nat. Women's Econ. Alliance. Recipient Nat. Rsch. Svc. award, NIMH, 1978-81, Golden Anniversary Scholarship award, AAUW, 1974; USPHS fellow, 1970-74. Mem. APA (fellow 1992—), Soc. for Psychophysiol. Rsch. (dir. 1982-85, Disting. Sci. award for early career contbn. 1980, chmn. awards com. 1981-84, chmn. conv. com. 1983-87, chmn. program com. 1987, mem. Blue Ribbon panel on State of the Soc. in the Yr. 2000, 1990-93, chmn. enhancement com., 1992—), Soc. for Rsch. in Psychopathology (dir. 1986-88, membership com. 1987-88), Soc. for Neurosci., Internat. Neuropsychol. Soc., Am. Psychopathol. Assn., Mortar Bd., Shi-Ai, Sigma Xi, Phi Kappa Phi, Alpha Lambda Delta, Pi Mu Epsilon, Phi Beta Kappa. Achievements include research on event-related brain potential indices of neuropsychiatric disorders. Home: 6204 Perthshire Ct Bethesda MD 20817-3348 Office: Lab Psychology & Psychopathology NIMH NIH Bldg 10 Rm 4C110 Bethesda MD 20892

DUNCAN, DORIS GOTTSCHALK, information systems educator; b. Seattle, Nov. 19, 1944; d. Raymond Robert and Marian (Onstad) D.; m. Robert George Gottschalk, Sept. 12, 1971 (div. Dec. 1983). B.A., U. Wash., Seattle, 1967, M.B.A., 1968; Ph.D., Golden Gate U., 1978. Cert. data processor, systems profl., data educator. Communications cons. Pacific NW Bell Telephone Co., Seattle, 1968-71; mktg. supr. AT&T, San Francisco, 1971-73; sr. cons., project leader Quantum Sci. Corp., Palo Alto, Calif., 1973-75; dir. co. analysis program Input Inc., Palo Alto, 1975-76; dir. info. sci. dept. Golden Gate U., San Francisco, 1982-83, mem. info. systems adv. bd., 1983-85; lectr. acctg. and info. systems Calif. State U., Hayward, 1976-78, assoc. prof., 1978-85, prof., 1985—; cons. pvt. cos., 1975—; speaker profl. groups and confs. Author: Computers and Remote Computing Services, 1983; contbr. articles to profl. jours. Loaned exec. United Good Neighbors, Seattle, 1969; nat. committeewoman, bd. dirs. Young Reps., Wash., 1970-71; adv. Jr. Achievement, San Francisco, 1971-72; mem. nat. bd. Inst. for Certification of Computer Profls. Edn. Found., 1990—; mem. Editorial Rev. bd. Journal Info. Systems Edn., 1992—; bd. dirs. Computer Repair Svcs., 1992—. Mem. Data Processing Mgmt. Assn. (1982, Meritorious Service award, Bronze award 1984, Silver award 1986, Gold Award 1988, Emerald award 1992, Nat. grantee, 1984. dir., edn. chmn. San Francisco chpt. 1984-85, sec. and v.p. 1985, pres. 1986, assn. dir. 1987, bylaws chmn. 1987; nat. bd. dirs. spl. interest group in edn. 1985-87), Am. Inst. Decision Scis., 1982-83, Western Assn. Schs. and Colls. (accreditation evaluation team, 1984-85), Assn. Computing Machinery, 1984—. Club: Junior (Seattle). Subspecialties: Information systems (information science). Current work: curriculum development, professionalism in data processing field, professional certification, industry standards, computer literacy and user education, design of data bases and data banks. Office: Calif State U Sch of Bus and Econs Hayward CA 94542

DUNCAN, JOHN WILEY, mathematics and computer educator, retired air force officer; b. San Francisco, Aug. 8, 1947; s. Vernon Alexander and Nellie May (Shaw) D.; m. Trudy Rae Hirsch, Feb. 25, 1967; children: Amber Rose, John Anthony. BS in Math. and Physics, N.W. Mo. State U., 1969; MBA, So. Ill. U., 1973; MS in Computer Sci., U. Tex., San Antonio, 1982. Tchr. Savannah (Mo.) High Sch., 1969; enlisted USAF, 1969, advanced through grades to maj.; aeromed. officer 9AES USAF, Clark Air Base, The Philippines, 1978-80, student UTSA, San Antonio, 1981-82, systems implementation team leader Sch. of Health Care Scics., Sheppard AFB, Tex., 1982-83; asst. chief med. systems Hdqrs. Air Tng. Command, Randolph AFB, Tex., 1983-86; chief med. systems Hdqrs. Pacific AF, Hickham AFB, Hawaii, 1986-89, 15 Med. Group, Hickham AFB, Hawaii, 1989; tchr. Kapiolani Community Coll., Honolulu, 1989—; computer cons., 1983—; instr. Tex. Luth. Coll., Seguin, 1984-86, Hawaii Pacific Coll., Honolulu, 1987-89, Leeward Community Coll., 1989-91. Cons. Ronald McDonald House, San Antonio, 1986. Presbyterian. Avocations: computing, tennis, reading, travel. Home: 4486 Luapele Pl Honolulu HI 96818-1983

DUNCAN, LEWIS MANNAN, III, physicist, education administrator; b. Charleston, W.Va., July 11, 1951; s. Lewis Mannan Jr. and Anna Ruth (Painter) D.; m. Natasha Valentina Krylova, Mar. 10, 1987; children: Katya Anastasia, Lara Polina. BA, Rice U., 1973, MS, 1976, PhD, 1977. NSF postdoctoral fellow Rice U., Houston, 1977; from scientist to sect. head. Los Alamos (N.Mex.) Nat. Lab., 1977-87; Carnegie sci. fellow Stanford (Calif.) U., 1987-88; assoc. dean, prof. physics Clemson (S.C.) U., 1988-92; dean of engring. and applied scis., prof. physics U. Tulsa, 1992—; bd. visitors Office Naval Rsch., Washington, 1991—; dir. S.C. Space Grant Consortium, 1990-92. Co-author: The Upper Atmosphere, 1993; contbr. articles to profl. jours. Active Com. on Fgn. Rels., Tulsa, 1992—. Fellow Thurmond Inst. on Gov. & Pub. Affairs, 1987—; recipient Alan Berman award Naval Rsch. Lab., 1990. Mem. AAAS, Am. Geophys. Union, Am. Phys. Soc., Internat. Radio Sci. Union (U.S. nat. com. 1989—). Achievements include research in high-power radio wave propagation, internat. security and technology, coop. U.S.-Russian sci. Office: Univ of Tulsa 600 S College Ave Tulsa OK 74104-3189

DUNCAN, LISA SANDRA, engineer; b. Saigon, Vietnam, July 4, 1963; came to U.S. 1969; d. George Hughey and Kha Thi (Thai) Duncan. BS in Chemistry, U. Ala., Huntsville, 1985, MS in Mech. engring., 1993. Prdn. engr. Morton Thiokol, Inc., Huntsville, 1985-88; systems engr. Tech. Analysis, Inc., Huntsville, 1988-89; sr. systems engr. Wyle Labs., Huntsville, 1989-90; lead analyst, engr. The Analytic Scis., Corp., Huntsville, 1990-92; sr. systems engr. W.J. Schafer Assoc., Huntsville, 1992--. Mem. Am. Chem. Soc., Huntsville Assn. Tech. Socs., Am. Inst. Aeronautics and Astronaustics, Am. Soc. Mechanical Engrs. Office: WJ Schafer Assoc 1500 Perimeter Pky Ste 470 Huntsville AL 35806

DUNCAN, PHILLIP CHARLES, research scientist; b. Phoenix, Dec. 6, 1956; s. Louis Clifford and Gerlene Olive (Ogzewalla) D.; divorced; 1 child, Alexander Bryon. BS in Psychology, Brigham Young U., 1980, MS in Exptl. Psychology, 1982, PhD in Cognitive Psychology, 1987. Rsch. asst. Stanford (Calif.) U., 1981; programmer Microteacher, Inc., San Diego, 1981-82; instr. computer sci. and psychology Brigham Young U., Provo, Utah, 1982-87; knowledge engr. WICAT Edn. Inst., Orem, Utah, 1984-87; sr. staff scientist Search Tech., Norcross, Ga., 1987—; presented workshops Assn. for the Devel. of Computer-based Instrl. Systems, St. Louis, 1991, Computer Aided Multimedia and Presentations Show, L.A., 1990, Nat. Soc. for Performance and Instrn. Conf., Atlanta, 1990. Contbr. articles to Interactive Learning Environments, Jour. Artificial Intelligence in Edn., among others. Scout Leader LDS Ch., Norcross, 1989-92. Mem. AAAS, ASTD, APA, Am. Ednl. Researchers Assn., Sigma Xi, Phi Kappa Phi. Office: Search Tech # 200 4898 S Old Peachtree Rd Norcross GA 30071-4707

DUNCOMBE, RAYNOR LOCKWOOD, astronomer; b. Bronxville, N.Y., Mar. 3, 1917; s. Frederic Howe and Mabel Louise (Taylor) D.; m. Julena Theodora Steinheider, Jan. 29, 1948; 1 son, Raynor B. B.A., Wesleyan U., Middletown, Conn., 1940; M.A., State U. Iowa, 1941; Ph.D., Yale U., 1956. Astronomer U.S. Naval Obs., Washington, 1942-62; dir. Nautical Almanac Office, 1963-75; prof. aerospace sci. U. Tex., Austin, 1976—; research assoc. Yale U. Obs., 1948-49; lectr. dynamical astronomy U. Md., 1963, Yale Summer Inst., 1959-70, Office Naval Research Summer Inst. in Orbital Mechanics, 1971, NATO Advanced Study Inst., 1972; cons. orbital mechanics Projects Vanguard, Mercury, Gemini, Apollo, USN Space Surveillance System; mem. NASA space scis. steering com., NASA research adv. panel in applied math., 1967; adviser Internat. Com. on Weights and Measures, Internat. Radio Consultative Com., Internat. Telecommunications Union; mem. NAS-NRC astronomy survey com., 1970-72, Hubble Space Telescope Astrometry Team, 1976—. Author: Motion of Venus, 1958, Coordinates of Ceres, Pallas, Juno and Vesta, 1969; editor: (with V.G. Szebehely) Methods in Celestial Mechanics, 1966, Dynamics of the Solar System, 1979; (with D. Dvorak and P.J. Message) The Stability of Planetary Systems, 1984; assoc. editor: Fundamentals of Cosmic Physics, 1971; exec. editor: Celestial Mechanics, 1977-85; contbr. articles to profl. jours. Recipient Superior Achievement award Inst. Navigation, 1967. Fellow Royal Astron. Soc., AAAS (sect. chmn.); asso. fellow AIAA; mem. Internat. Astron. Union (pres. com. on ephemerides), Am. Astron. Soc. (chmn. div. dynamical astronomy 1970), Inst. Navigation (councillor 1960-64, v.p. 1964-66, pres. 1966-67, Hays award 1975), ASME (sponsor applied mechanics div. 1968-70), Internat. Assn. Insts. Nav. (v.p.), Assn. Computing Machinery, Sigma Xi. Home: 1804 Vance Cir Austin TX 78701-1035 Office: U Tex Dept Aerospace Engring Austin TX 78712

DUNFORD, ROBERT WALTER, physicist; b. N.Y.C., July 9, 1946; s. James Marshall and Virginia Louise (MacEachern) D.; m. Karen Sue Hawk, Aug. 20, 1977; 1 child, Michael. BS in Engring., U. Mich., 1969, PhD in Physics, 1978. Instr. Princeton (N.J.) U., 1978-80, asst. prof., 1980-86; physicist Argonne (Ill.) Nat. Lab., 1986—. Editor: Workshop on Polarized 3He Beams and Targets, 1985, Atomic Spectroscopy and Highly Ionized Atoms, 1987; contbr. articles to profl. jours. Lt. USN, 1969-73. Grantee Nat. Bur. Standards, 1983. Mem. AAAS, Am. Phys. Soc. (exec. com. topical group precision measurement and fundamental constants). Office: Argonne Nat Lab PHY-203 9700 S Cass Ave Argonne IL 60439

DUNHAM, GREGORY MARK, obstetrician/gynecologist; b. Trenton, N.J., Feb. 11, 1958; s. Robert Latham and Nancy Ann (Duncan) D.; m. Sarita Fawn Pennington, July 10, 1982; children: Taylor Erin, Seth Ian, Kambri Leigh. BS, Abilene Christian U., 1978; MD, U. Tex. Southwestern, 1983. Diplomate Am. Bd. Ob-Gyn. Intern St. Joseph Hosp., Denver, 1983-84, resident, 1984-86, chief resident, 1986-87; ob-gyn. Angelo Clinic Assn., San Angelo, Tex., 1987—; cons. Concho Valley Planned Parenthood, San Angelo, 1987—; advisor Home Econs. Adv. Coun., San Angelo, 1987—; med. dir. Child Advocacy Ctr., San Angelo, 1991—, Sexual Assault Nurse Examiner Program, San Angelo, 1992—; chmn. dept. ob-gyn. Shannon Med. Ctr., San Angelo, 1993—, mem. med. bd., chmn. laser safety com., 1992—. Bd. dirs. Angelo Clin. Assn., 1992—; deacon Johnson St. Ch. of Christ, San Angelo, 1989—. Recipient Physician's Recognition award AMA, 1986, 92. Fellow ACOG; mem. Am. Fertility Soc., Tex. Assn. Ob-Gyn, N.Am. Soc. for Pediatric and Adolescent Gynecology. Republican. Avocations: woodworking, hunting, fishing, landscape architecture, photography. Office: Angelo Clinic Assn 120 E Beauregard Ave San Angelo TX 76903-5919

DUNHAM, JEFFREY SOLON, physicist, educator; b. Caro, Mich., Oct. 29, 1953; s. Deane Alan and Etta Loyce (Cole) D. BS in Physics, U. Wash., 1975; MS in Physics, Stanford U., 1979, PhD in Physics, 1981. Asst., prof. physics Stanford (Calif.) U., 1981-83; assoc. prof. physics, chmn. dept. Middlebury (Vt.) Coll., 1983-93; prof., 1993—; tchr. Peace Corps, Nkwatia, Kwahu, Ghana, 1976-78. Mem. IEEE, Internat. Soc. for Optical Engring., Optical Soc. Am., Am. Phys. Soc. Office: Middlebury Coll Dept Physics Middlebury VT 05753

DUNIGAN, DAVID DEEDS, biochemist, virologist, educator; b. Evansville, Ind., July 4, 1951; s. Gerald Eugene and Margaret Cromwell (Deeds) D.; m. Carol Ann Quinley, Aug. 24, 1974; children: Matthew Daniels, Kate Reynolds, Anne Hempstead. BS, Ind. State U., 1977; PhD, U. Conn., 1985. Grad. asst., rsch. asst. U. Conn., Storrs, 1977-85; postdoctoral assoc. Cornell U., Ithaca, N.Y., 1985-88, rsch. assoc., 1988-89; asst. prof. dept. biology U. South Fla., Tampa, 1989—. Contbr. articles to sci. jours. Coach Odyssey of the Mind, Tampa, 1990-92. Mem. AAAS, Am. Soc. Microbiology (policy com. Southeastern br. 1991-92), Am. Soc. Virology (internat. travel award 1990), Sigma Xi. Achievements include research on replication mechanisms of RNA viruses. Office: Univ South Fla Dept Biology LIF 136 Tampa FL 33620-5150

DUNITZ, JACK DAVID, retired chemistry educator, researcher; b. Glasgow, Scotland, Mar. 29, 1923; s. William and Mildred (Gossman) D.; m. Barbara Steuer, Aug. 11, 1953; children: Marguerite, Julia. BS, Glasgow U., 1944, PhD, 1947; DSc (hon.), Technion Haifa, 1990; PhD (hon.), Weizmann Inst., 1992. Research fellow Oxford U., 1946-48, 51-53, Calif. Inst. Tech., Pasadena, 1948-51, 53-54; vis. scientist NIH, Bethesda, Md., 1954-55; sr. research fellow Royal Instn., London, 1956-57; prof. chem. crystallography Swiss Fed. Inst. Tech., Zurich, 1957-90, ret. 1990. Author: X-Ray Analysis and the Structure of Organic Molecules, 1979, (with others) Reflections on Symmetry...in Chemistry and Elsewhere, 1993; contbr. articles to profl. jours. Recipient Havinga medal U. Leiden, Netherlands, 1980, Paracelsus prize Swiss Chem. Soc., 1986, Aminoff prize Swedish Royal Acad., 1990; Bijvoet medal U. Utrecht, 1989; Churchill Coll. Overseas fellow, 1968. Fellow Royal Soc. London, Academia Europaea, AAAS; mem. NAS (fgn. assoc.), Deutsche Akademie Leopoldina, Swiss Chem. Soc., Am. Chem. Soc., Brit. Chem. Soc., Am. Crystallographic Assn. (Buerger award 1991), Brit. Crystallographic Assn., Swiss Crystallographic Soc. (hon.), Royal Netherlands Acad. Sci. (fgn.), European Acad. Scis and Arts. Home: 77 Obere Heslibach Str, Küsnacht Switzerland Office: ETH-Zentrum, 16 Universitatstrasse, Zurich Switzerland

DUNKLE, LISA MARIE, pediatrics educator; b. Ann Arbor, Mich., Oct. 31, 1946; d. Robert Henry and Dorothy Rose (Heagstedt) D.; m. Richard James Scheffler, Dec. 28, 1972; children: Richard James Scheffler III, Margaret Dorothy Scheffler. AB, Wellesley Coll., 1968; MD, Johns Hopkins U., 1972. Cert. Nat. Bd. Med. Examiners, Am. Bd. Pediatrics, Pediatric Infectious Disease Soc. Intern pediatrics Washington U., St. Louis, 1972-73, resident pediatrics, 1973-74, fellow infectious diseases, 1974-76; asst. prof. pediatrics St. Louis U., 1976-79, assoc. prof. pediatrics, 1979-85, assoc. prof. microbiology, 1979-85, prof. pediatrics, 1985-89; dir. antiviral clin. rsch. Bristol-Myers Squibb, Wallingford, Conn., 1989—; dir. pediatric infectious diseases St. Louis U., 1976-89, dir. infectious diseases lab. Cardinal Glennon Hosp., St. Louis, 1976-89, dir. infectious control program, 1976-89. Mem. editorial bd. Pediatric Infectious Disease Jour., 1989-92; contbr. articles to profl. jours. Fundraiser Wellesley (Mass.) Coll., 1975-88, The Forsyth Sch., St. Louis, 1988-89. Scholar Johns Hopkins U., 1972; rsch. grantee Cystic Fibrosis Found., 1984. Fellow Infectious Disease Soc. Am.; mem. Soc. for Pediatric Rsch., Midwest Soc. for Pediatric Rsch. (sec. 1983-88, pres. 1989-90), Interscience Conf. on Antimicrobial Agts. and Chemotherapy (program com. 1992—). Republican. Episcopalian. Achievements include identification of plasmid-mediated gene responsible for surface adhesion of S. aureus. Office: Bristol Myers Squibb 5 Research Pky Wallingford CT 06492-1996

DUNLAP, DALE RICHARD, civil engineer; b. Denver, Feb. 16, 1960; s. Daniel Richard and Dorothy (Nielsen) D.; m. Julie Ann Best, Sept. 7, 1985. BS, U. Colo., 1984. Registered profl. engr., Calif., Colo. V.p. Dunlap Architects/Engrs., Inc., Englewood, Colo., 1984-87; project mgr. Tait & Assocs.-Cons. Engrs., Orange, Calif., 1987-91; civil engr. Howard Needles Tammen & Bergendoff-Architects-Engrs.-Planner, Irvine, Calif., 1991—. Bd. dirs. Calvary Temple, Denver., 1986. Mem. ASCE. Republican. Office: Howard Needles Tammen Bergendoff 4 Executive Cir Ste 190 Irvine CA 92714

DUNLAP, LAWRENCE HALLOWELL, museum curator; b. Madison, Wis., Oct. 23, 1910; s. Frederick and Florence (Hallowell) D.; m. Elizabeth M. Suter, June 2, 1941; children: Mary Frances, Hallowell. AB, U. Mo., 1931, BS, 1933, MA, 1935; PhD, U. Ill., 1939. Chemist, Armstrong Cork Co., Lancaster, Pa., 1939-46, head sect., 1946-52, gen. mgr. chemistry div., 1952-62, sr. rsch. assoc., 1962-75; assoc. curator geology and mineralogy N. Mus. Franklin and Marshall Coll., Lancaster, 1975—. Patentee in fields of drying oils, polymers, floor coverings, fire retardance and rsch. mgmt.; contbr. articles to profl. jours. Bd. dirs. Lancaster County Council Chs. Served to lt. USNR, 1942-46. Fellow AAAS (chmn. Lancaster br. 1966-68), Am. Inst. Chemists (accreditation com.); mem. Am. Chem. Soc. (emeritus, chmn. Southeast Pa. sect., bd. dirs. polymer div. 1968-78), Lancaster Country Club, Cliosophic Soc., Dopagal Soc., Eagles Mere Country Club, Rotary, Sigma Xi, Alpha Chi Sigma. Republican. Episcopalian. Avocations: music, gardening. Home: 1315 Quarry Ln Lancaster PA 17603-2423 Office: N Mus Franklin and Marshall Coll Lancaster PA 17604

DUNLAP, RILEY E., sociologist; b. Wynne, Ark., Oct. 25, 1943; s. Riley W. Dunlap Jr. and F. Eugenia (Jones) Anderson; m. Lonnie Jean Brown, Aug. 25, 1966; children: Sara Jean, Christopher Eugene. MS, U. Oreg., 1969, PhD, 1973. Asst. prof. to prof. Wash. State Univ., Pullman, 1972—; mem. socioeconomic peer review panel Office of Exploratory Rsch., U.S. EPA, 1991; mem. panel on aesthetic attributes in water resources planning NRC/Nat. Acad. Scis., 1982; Gallup fellow in environment George H. Gallup Internat. Inst., 1992—. Editor, author: (jour. symposium) Am. Behavioral Scientist, 1980, Sociol. Perspectives, 1990; author book: American Environmentalism: The U.S. Environmental Movement, 1970-1990. Mem. AAAS (rural sociol. soc. rep. to sect. K 1986-89), Internat. Sociol. Assn. (bd. dirs. rsch. com. on environment and soc. 1992—), Am. Sociol. Assn. (chmn. sect. on environ. sociology 1981-83, disting. contbn. award 1986), Rural Sociol. Soc. (chmn. natural resources rsch. group 1978-79, award of merit 1985), Soc. for Study of Social Problems (chmn. environ. problems div. 1973-75). Achievements include being credited as co-founder of field of environmental sociology. Office: Washington State Univ Dept of Sociology Pullman WA 99164

DUNN, ANDREA LEE, biomedical researcher; b. Ft. Collins, Colo., Jan. 25, 1951; d. Aulden Neal and Ramona Alberta (Petersen) D.; m. Dennis Freeman, Dec. 28, 1971 (div. Aug. 1972). BA magna cum laude, Metro State Coll., 1982; MS, Va. Poly. Inst. and State U., 1986; PhD, U. Ga., 1990. With various nonprofit agys. and State of Wyo. Cheyenne, 1972-80; teaching asst. Va. Poly. Inst. and State U., Blacksburg, 1982-86; rsch. asst. U. Ga., Athens, 1986-89; NIMH postdoctoral fellow U. Colo. Health Sci. Ctr., Denver, 1990-92, instr., fellow, 1992—; part-time faculty mem. Metro State U., Denver, 1989—; cons. Nat. Jewish Hosp., Denver, 1992—. Presdl. scholar Metro State Coll., 1980, 81; grantee U.S. Olympic Com., 1987, Devel. Psychobiology Rsch. Group, 1992. Mem. AAAS, APA, Am. Coll. Sports Medicine, Soc. for Neurosci., Chalice Lighters. Democrat. Unitarian Universalist. Achievements include research in how cytokines such as interferon may lead to fatigue. Home: 762 Dexter St Denver CO 80220

DUNN, BRUCE SIDNEY, materials science educator; b. Chgo., Apr. 22, 1948; s. George Bernard and Goldye Rosalyn (Opper) D.; m. Wendy Joan Rader, June 7, 1970; 1 child, Julianne. BS in Ceramic Engring., Rutgers U., 1970; MS in Materials Sci., UCLA, 1972, PhD in Materials Sci., 1974. Staff scientist GE, Schenectady, N.Y., 1976-80; assoc. prof. materials sci. UCLA, 1981-85, prof., 1985—; cons. to numerous corps.; invited prof. U. Paris, 1986, 91, 92, 93. Contbr. articles to profl. jours. Fulbright fellow, 1985-86. Mem. Am. Ceramic Soc., Electrochem. Soc., Materials Rsch. Soc. Achievements include patents in field. Office: UCLA Dept Materials Scis & Engring 5731 Boelter Hall Los Angeles CA 90024

DUNN, GORDON HAROLD, physicist; b. Montpelier, Idaho, Oct. 11, 1932; s. Jesse Harold and Winifred Roma (Williams) D.; m. Donetta Dayton, Sept. 25, 1952; children: Jesse Lamont, Randall Dayton, Michael Scott, Brian Eugene, David Edward, Susan, Harold Paul, Richard Elzo. BS in Physics, U. Wash., 1954, PhD in Physics, 1961. NRC postdoctoral rsch. assoc. Nat. Bur. Standards, Washington, 1961-62; physicist Nat. Bur. Standards/Join Inst. for Lab. Astrophysics, Boulder, Colo., 1962-77, chief quantum physics div., 1977-85; sr. scientist Nat. Inst. Standards/Joint Inst. for Lab. Astrophysics, Boulder, 1985—; lectr. dept. physics U. Colo., Boulder, 1964-74, adj. prof., 1974—; commerce sci. fellow Commn. on Sci. & Tech., U.S. Ho. of Reps., Washington, 1975-76; chmn. com. on atomic molecular & optical sci. NRC, Washington, 1990—, vice-chair, 1989-90, mem. com., 1983-86; mem. NRC Panel on Instruments and Facilities, Washington, 1983-84, NRC Panel on Ion Storage Rings for Atomic Physics, 1985-88; mem. gen. com. Internat. Conf. on Physics of Electron & Atomic Collisions, 1969-73, mem. program com., 1974-75; chmn. Gaseous Electronics Conf., 1968-69, com. mem., 1966-70, sec., 1967-68; chmn. Atomic Processes in High Temperature Plasmas, 1980-81, com. mem., 1977-81, co-sec., 1978-79; co-organizer NATO Advanced Study Inst., 1984-85, U.S.-Japan Workshop, 1985-86; bd. editors Jour. Phys. & Chem. Reference Data, 1990—. Editor, author: Electron-Impact Ionization, 1985; contbr. over 85 articles to profl. jours. Scoutmaster Boy Scouts Am., Boulder, 1967-72, 87-88; coach, league pres. Little League, Boulder, 1966-71, 76-77, pres. Parent-Tchr. Orgns., Boulder, 1973, 75; bishop LDS Ch., Boulder, 1977-82. Recipient Gold medal U.S. Dept. Commerce, 1970. Fellow Am. Phys. Soc. (chmn. div. atomic & molecular physics 1989-90, vice-chair 1988-89, mem. exec. com. 1969, 85-88, 91, Davisson-Germer prize 1984), Joint Inst. for Lab. Astrophysics. Avocations: jogging, skiing, dancing, hunting, camping. Office: U Colo Joint Inst Lab Astrophysics Chmn Campus Box 440 Boulder CO 80309-0440*

DUNN, HORTON, JR., organic chemist; b. Coleman, Tex., Sept. 3, 1929; s. Horton and Lora Dean (Bryant) D. BA summa cum laude, Hardin-Simmons U., 1951; MS, Case Western Res. U., 1975, PhD, 1979. Instr. chemistry Hardin-Simmons U., 1951; ONR fellow Ohio State U., Columbus, 1951-52; teaching fellow in chemistry Purdue U., Lafayette, Ind., 1952-53; rsch. chemist Lubrizol Corp., Cleve., 1953-70, dir. tech. info. ctr., 1970-79, supr. rsch. div., 1980—; chmn. bd., bus. mgr. Isotopics, Cleve., 1964-67, editor, 1961-63. Contbr. articles to profl. jours.; patentee in field. Fellow Am. Inst. Chemists; mem. AAAS, SAR (life), Am. Chem. Soc. (treas. Cleve. chpt. 1968-70, chmn. 1987, bd. dirs. 1990—), Am. Soc. for Info. Sci. (chpt. pres. 1973-74), Royal Soc. Chemistry, Nat. Coun. Met. Opera, Royal Oak Soc. (life), Cleve. Tech. Soc. Coun. (treas. 1987), University Club, Cleve. Club, Cleve. Play House Club, Cleve. Cir. Decorative Arts Trust (treas. 1990-91, 93, v.p. 1992-93). Home: 530 Sycamore Dr Cleveland OH 44132-2150 Office: 29400 Lakeland Blvd Wickliffe OH 44092-2298

DUNN, KARL LINDEMANN, electrical engineer; b. Corning, N.Y., Apr. 29, 1942; s. Karl Lindemann and Elizabeth (Zilhaver) D.; m. Anne Clifford Roberts, June 24, 1967. BS in Physics, Rensselaer Poly. Inst., 1964, MS in Physics, 1972. Computer scientist Computer Scis. Corp., Huntsville, Ala., 1969-78; prin. engr. SCI Systems Inc., Huntsville, 1978-87; sr. design engr. Chrysler Mil/Pub/Pentastar, Huntsville, 1987-89; staff engr. Phoenix Microsystems Inc., Huntsville, 1989-91; sr. design engr. VME Microsystems Internat. Corp., Huntsville, 1991—. Mem. IEEE. Office: VME Microsystems Internat 12090 Memorial Pkwy SW Huntsville AL 35803

DUNN, MARVIN IRVIN, physician; b. Topeka, Dec. 21, 1927; s. Louis and Ida (Leibtag) D.; m. Maureen Cohen, Mar. 10, 1956; children—Jonathan Louis, Marilyn Paulette. B.A., U. Kans., 1950, M.D., 1954. Intern USPHS, San Francisco, 1954-55; resident U. Kans., 1955-58, fellow, 1958-59, instr. medicine, 1958-60, asso. in medicine, 1960-62, asst. prof. medicine, 1962-65, assoc. prof., 1965-70, prof., 1970—, Franklin E. Murphy Disting. prof., 1978—, dir. Cardiovascular Lab., head sect. Cardiovascular Disease Med. Center, 1963-92, dean Sch. of Medicine, 1980-84; cons. USAF, 1971—, FAA, 1990— . Author: Home Study Course: Difficult EKG Diagnosis, 1969, Translator Deductive and Polyparametric Electrocardiography, 1970, (with others) Clinical Vectorcardiography and Electrocardiography, 2d edit, 1977, Clinical Electrocardiography, 8th edit., 1989; editor in chief Cardiovascular Perspectives, 1985-89; mem. editorial bd. Am. Jour. Cardiology, 1970-75, Catheterization and Cardiovascular Diagnosis, 1980-87, AMA Archives Internal Medicine, 1984—, Jour. Am. Coll. Cardiology, 1983-89, Biomedicine and Pharmacotherapy, 1985—, Am. Jour. Noninvasive Cardiology, 1985-89, Chest, 1984-89, Practical Cardiology, 1980—, Heart and Lung, 1986-88, Bd-Advance in Therapy, 1992; cons., reviewer: Griffith Resource Library, 1980—, Am. Heart Jour., New Eng. Jour. Medicine, Jour. Acoustical Soc. Am. Bd. dirs. Hebrew Acad. Jewish Geriatric and Convalescent Center, Beth Shalom Synagogue. Served with AUS, 1946-47. Recipient Alumnus of Yr. award U. Kansas Sch. Medicine, 1987, silver medal U. Socrates, Thessaloniki, Greece, 1992. Fellow ACP, Am. Coll. Chest Physicians (mem. bd. regents; pres. 1988-89; gov. State of Kans.), Am. Coll. Cardiology (trustee), Am. Heart Assn., Royal Acad. Medicine (Ireland), Royal Coll. Physicians (Valencia, Spain); mem. Am. Physicians Fellowship (dir.), Univ. Cardiologists, Alpha Omega Alpha, Phi Chi. Home: 3205 Tomahawk Rd Mission Hills KS 66208 Office: U Kans Hosp 3901 Rainbow Blvd Kansas City KS 66160-7378

DUNN, ROBERT LELAND, energy engineer; b. Eugene, Oreg., May 21, 1946; s. Harold Marion and Avis Lillian (McLaughlin) D.; m. Sandra Lee Sanderfer, Aug. 12, 1976 (dec. Dec. 1982); m. Holly Elizabeth Rice, Apr. 12, 1987; children: Leland J., Jason T., Robert V. BA, Calif. State U., San Jose, 1977. Cert. energy mgr. Energy engr. U. Calif., Santa Cruz. Sgt. USAF, 1965-68. Mem. Assn. of Energy Engrs. (assoc.). Budhist. Home: 374 Vick Dr Santa Cruz CA 95060 Office: Univ Calif 1156 High St Santa Cruz CA 95064

DUNN, RONALD HOLLAND, civil engineer, management executive, railway consultant, forensic engineer; b. Balt., Sept. 15, 1937; s. Delmas Joseph and Edna Grace (Holland) D.; m. Verona Lucille Lambert, Aug. 17, 1958; children: Ronald H., Jr. (dec.), David R., Brian W. Student U. S.C., 1956-58; BS in Engring., Johns Hopkins U., 1969. Registered profl. engr., Va., D.C. Diplomate forensic engr. Field engr. Balt. & Ohio R.R., Balt., 1958-66; chief engr. yards, shops, trackwork DeLeuw, Cather & Co., Washington, 1966-73; mgr. engring. support Parsons-Brinckerhoff-Tudor-Bechtel, Atlanta, 1973-76; dir. railroad engring. Morrison-Knudsen Co., Inc., Boise, Idaho, 1976-78; v.p. Parsons Brinckerhoff-Centec, Inc., McLean, Va., 1978-83; v.p., area mgr., tech. dir. ry. engring., profl. assoc. Parsons Brinckerhoff Quade & Douglas, Inc., McLean and Pitts., 1983-84; dir. transp. engring. R.L. Banks & Assocs., Inc., Washington, 1984; pres. R.H. Dunn & Assocs., Inc., Fairfax, Va., 1984-91, Williamsburg, Va., 1991—; insp. ry. and rail transit facilities, Europe, 1980, 82, 84, China and Hong Kong, 1985; involved in engring. of 17 railroads and 17 rail transit systems throughout N. Am.; guest Japan Railway Civil Engring. Assn., 1972, French Nat. Railroads and Paris Transport Authority, 1988; mem. adv. com. track engrs. U.S. Dept. Transp., 1968-71. Chmn. Cub Scout Pack, Boy Scouts Am., 1972-73, committeeman, 1973-75, troop committeeman, 1979-85. Fellow ASCE, Inst. Transp. Engrs.; mem. Am. Arbitration Assn., Am. Mgmt. Assn., Am. Ry. Engring. Assn., Am. Public Transit Assn., Soc. Am. Mil. Engrs., Roadmasters and Maintenance of Way Assn. of Am., Am. Ry. Bridge and Bldg. Assn., Constrn. Specifications Inst., Nat. Soc. Profl. Engrs., Transp. Research Bd., Nat. Acad. Forensic Engrs., Nat. Assn. R.R. Safety Cons. and Investigators, Can. Soc. Civil Engring., Va. Soc. Profl. Engrs., Can. Urban Transit Assn., Ry. Tie Assn., Inst. of Rapid Transit, Phi Kappa Sigma. Methodist. Office: 123 Yorkshire Dr Williamsburg VA 23185-3984

DUNN, STANLEY MARTIN, biomedical engineering educator; b. Syracuse, N.Y., Aug. 11, 1956; s. Herbert and Rosanne (Balaban) D.; m. Margaret Breslin, Oct. 20, 1984; children: Laura Katherine, Jeffrey Randall. MS, U. Md., 1983, PhD, 1985. Rsch. assoc. Drexel U., Phila., 1979; grad. asst. Purdue U., West Lafayette, Ind., 1979-80; teaching asst., grad. asst., then lectr. U. Md., College Park, 1980-86; asst. prof. biomed. engring. Rutgers U., Piscataway, N.J., 1986-92, assoc. prof., 1992—; cons. Stuart Med., Inc., Columbia, Md., 1978-85, Maxillofacial Radiology Ctr., N.Y.C., 1992—, Block Drug Co., Jersey City, 1991—. Contbr. papers to profl. publs. Trustee N.J. Head Injury Assn., Edison, 1992—. Grantee NIH, 1992, U.S. Army, 1991, various corps. Mem. IEEE (sr.), Am. Acad. Oral and Maxillofacial Radiology (assoc.). Achievements include patent for skin imaging system, arrhythmia detection method, radiology robot; development of method for digital subtraction of intraoral x-rays for observing bone metabolism. Office: Rutgers Univ PO Box 909 Piscataway NJ 08855-0909

DUNN, STEVEN ALLEN, chemist; b. Laurens, Iowa, May 1, 1948; s. Lloyd and Avis (Nelson) D.; m. Diana R. Epply, Jan. 11, 1986; children: Mark William, Jeffrey Allen. BS in Phys. Sci., Peru (Nebr.) State Coll., 1980. Electronics technician Gen. Communications Co., Omaha, 1971-76; math. tutor, maintenance worker Peru State Coll., 1976-80; chemist power generation sect. Colo.-Ute Electric Assn., Hayden, 1980-92; sr. chemist power generation sec. Pub. Svc. Co. Colo., Hayden, 1992—. With USCG, 1967-71. Mem. Lions (sec. Hayden club 1986-87, pres. 1988). Democrat. Avocations: amateur radio, snow skiing. Office: Pub Svc Co of Colo PO Box C Hayden CO 81639

DUNPHY, EDWARD JAMES, crop science extension specialist; b. Frederick, Md., Nov. 14, 1940; s. Edward John and Marie W. (Barlow) D.; m. Judith Kay Mitchell, Aug. 18, 1962; children: Kevin James, Brian Patrick, Cory Edward. MS, U. Ill., 1964; PhD, Iowa State U., 1972. Rsch. asst. U. Ill., Urbana, 1962-64; agronomist Dunphy's Feed & Fertilizer, Sullivan, Ill., 1964-66; rsch. asst. Iowa State U., Ames, 1969-72; crop prodn. specialist Iowa State U., Des Moines, 1972-75; extension specialist soybeans N.C. State U., Raleigh, 1975—, prof. crop sci., 1986—; instr. soybean prodn. N.C., 1975—; mem. N.C. Land Use Value Adv. Bd., Raleigh, 1987—. Author 4 computer programs; contbr. numerous articles to profl. jours. Cubmaster Cub Scout Pack 398, Raleigh, 1976-81, troop com. chair, 1979-93; officer Athens Dr. Band Boosters, Raleigh, 1983-90. Sgt. U.S. Army, 1966-69. Recipient Meritorious Svc. award N.C. Soybean Producers. Mem. Am. Soc. Agronomy (com. chair, fellow), Crop and Soil Sci. Socs. Am., Am. Soybean Assn. (mem. S.Am. soybean mission), Coun. for Agrl. Sci. and Tech., Alpha Zeta, Epsilon Sigma Phi, Gamma Sigma Delta, Phi Eta Sigma, Phi Kappa Phi, Sigma Xi. Achievements include rsch. on soybean varieties, production, management and economics. Home: 1329 Swallow Dr Raleigh NC 27606-2414 Office: NC State U Box 7620 Raleigh NC 27695-7620

DUNSFORD, HAROLD ATKINSON, pathologist, researcher; b. Boston, Oct. 23, 1941; s. Reuben and Louise (Atkinson) D.; m. Helen Louise Lucia, Oct. 16, 1971; children: Harold Atkinson Jr., William R. BS, Dickinson Coll., 1963; MD, U. Md., Balt., 1969. Diplomate Am. Bd. Pathology. Intern U. Md. Hosp., 1969-70; resident anatomy and clin. pathology Peter Bent Brigham Hosp., Boston, 1970-74; staff pathologist U. Pitts. Med. Sch., 1974-77, Griffin Hosp., Derby, Conn., 1977-82; dir. autopsy svc. U. Tex. Med. Sch., Houston, 1983-88; dir. clin. chemistry U. Tex. Med. Br., Galveston, 1988-92; dir. anatomic pathology U. Miss. Med. Ctr., Jackson, 1992—; reviewer Am. Jour. Pathology, 1988—. Author: (monograph) Cellular Events during Hepatocarcinogenesis, 1987; mem. editorial bd. Human Pathology, 1990—; also articles. Recipient Chmn.'s Cup dept. pahology U. Tex., Houston, 1986; grantee NIH, 1985-87, Electric Power Rsch. Inst., 1989; fellowship Harvard Med. Sch., 1970-74. Fellow Coll. Am. Pathologists, Am. Soc. Clin. Pathologists; mem. AMA, Am. Assn. Pathologists, U.S. and Can. Acad. Pathology, Miss. Med. Assn., Jackson Soc. Model Engrs. (pres. 1993—), Sigma Xi. Congregationalist. Achievements include contribution to understanding that primitive stem cells exist in the liver and may be important precursors of liver cell carcinoma. Home: 103 Sunrise Ct Brandon MS 39042 Office: U Miss Med Ctr Dept Pathology 2500 N State St Jackson MS 39216

DUNSKI, JONATHAN FRANK, biologist, educator; b. Allentown, Pa., Dec. 10, 1969; s. Joseph and Gloria Ann (Groff) D. BS in Biology, Pa. State U., 1991, MS in Biology, 1992. Tchr. biology Török Ignác Gimnázium, Gödöllö, Hungary, 1992-93.

DUNSON, WILLIAM ALBERT, biology educator; b. Cedartown, Ga., Dec. 17, 1941; s. James Blake and Eleanor (Adams) D.; m. Margaret E. Kvashay, Aug. 19, 1963; children: Mary Elizabeth, William Albert, David Brian. B.S. in Zoology with honors, Yale U., 1962; MS, U. Mich., 1964, Ph.D., 1965. Teaching fellow U. Mich., Ann Arbor, 1962-63; mem. faculty Pa. State U., University Park, 1965—, prof. biology, 1974—; adj. prof. biology U. Miami; Old Dominion U.; chief scientist various internat. oceanographic expdns.; collaborator Everglades Nat. Park. Author: The Biology of Sea Snakes, 1975, 125 rsch. papers. Queens marine sci. fellow, 1972, hon. Fulbright fellow, 1972; grantee NSF, U.S. Dept. Interior, U.S. Geol. Survey, U.S. EPA. Mem. Am. Soc. for Environ. Toxicology and Chemistry, Am. Inst. Biol. Scis., Soc. for Study Amphibians and Reptiles (jour. editorial bd.), Am. Soc. Ichthyologists and Herpetologists, Herpetologists League, Ecol. Soc. Am., Atlantic Estuarine Rsch. Soc., Wetlands Soc., Soc. Conservation Biology, Fla. Acad. Sci. Achievements include study of ecotoxicology, physiological ecology and wetlands ecology. Home: 575 Brittany Dr State College PA 16803-1426 Office: Pa State U 208 Mueller Bldg University Park PA 16802

DUNWIDDIE, PETER WILLIAM, plant ecologist; b. Neenah, Wis., May 30, 1953; s. William Edgerton and Mary Jane (Vroman) D.; m. Elizabeth Alice Bell, Sept. 29, 1990. BA with honors, U. Wis., 1974, MS, 1976; PhD, U. Wash., 1983. Rsch. assoc. Lab. of Tree-Ring Rsch., U. Ariz., Tucson, 1976-79; plant ecologist Mass. Audubon Soc., Lincoln, 1983—; regional fire ecologist The Nature Conservancy, Boston, 1986-89; chmn. Nantucket (Mass.) Conservation Commn., 1987-89. Author: Seeing Green: A Layman's Guide to Field Ecology, 1982, Changing Landscapes, 1992. Mem. Nantucket Conservation Commn., 1984-92. Danforth fellow Danforth Found., 1980, grad. rsch. fellow U. Wash., 1979; dissertation grantee NSF, 1981. Mem. AAAS, Am. Quaternary Assn., Ecol. Soc. Am., Sigma Xi. Office: Mass Audubon Soc S Great Rd Lincoln MA 01773

DUONG, MINH TRUC, project engineer; b. HaNam, Vietnam, Dec. 9, 1938; came to U.S., 1975; s. Chung Quoc and Nhuan Thi (Pham) D.; m. Hoang Kim Thi Nguyen, June 12, 1962; children: Khanh Nguyen, Quyen Thanh, Thao Thanh. BA in Edn., Saigon U., Vietnam, 1975; AAS in Drafting & Design, J.E. Reynolds Community Coll., 1981; BS in Mech. Technology, Va. State U., 1990. Cert. in refrigeration and air conditioning. Sch. prin. ChoGom Elem. Sch., Quinhon, Vietnam, 1960-64; platoon leader, 1st lt. Viet Nam Army, Xuan Loc, 1964-69; high sch. tchr. Tu Duc High Sch., Saigon, Vietnam, 1969-75; air conditioning leadman Med. Coll. of Va., Richmond, 1975-81; sr. designer Philip Morris, U.S.A., Richmond, 1981-89, project engr., 1989—. Mem. Bldg. Code Bd. Appeals, City of Richmond, Va., 1993—. Home: 6105 Hermitage Rd Richmond VA 23228-5101

DUONG, TAIHUNG, anatomist; b. Dalat, Vietnam, Oct. 15, 1956; came to U.S. 1974; s. Trungtin and Ngocanh (Vothi) D.; m. Lisa Ann Giorgianni, Oct. 8, 1988; children: Peter Anthony, Patrick Michael. BA, Whittier (Calif.) Coll., 1977; PhD, UCLA, 1989. Anatomy instr. Santa Monica (Calif.) City Coll., 1984-91; staff rsch. assoc. Brain Rsch. Inst., UCLA, 1979-89; postdoctoral fellow Mental Retardation Rsch., UCLA, 1989-91; asst. prof. anatomy Ind. U. Sch. Medicine, Terre Haute, 1991—, Ind. State U., Terre Haute, 1991—; dir. Wabash Valley Neurol. Bank, Terre Haute, 1991—. Contbr. articles to profl. jours. Turken scholar, 1990. Mem. Am. Assn. Anatomists, N.Y. Acad. Scis., Soc. for Neurosci., Phi Alpha Theta. Achievements include first detection of amyloid P component in Alzheimer's disease intracellular lesions: the neurofibrillary tangles. Office: Terre Haute Ctr for Med Edn Ind State Univ Holmstedt Hall 135 Terre Haute IN 47809

DUONG, VICTOR (VIET) HONG, nuclear engineer; b. Vietnam, June 20, 1956; s. Huong Ngoc and Duy-So (Tran) D. BSc, U. Ottawa, 1979, MS cum laude, 1982. Registered profl. engr., Ontario, Canada. Nuclear engr. Combustion Engring., Ottawa, Ontario, 1979-84; lead engr. CAE Electronics, Montreal, Quebec, 1984-89; sr. engr. A Westinghouse Electric Corp., Pitts., 1989—. Contbr. articles to profl. jours. Recipient award Assn. Profl. Engrs. Mem. Profl. Engrs. Ontario, Am. Nuclear Soc. Achievements include development of mathematical models to simulate reactor coolant and primary heat transport systems for nuclear operation and transients accident analysis. Home: 637 College St Pittsburgh PA 15232 Office: Westinghouse Electric Corp PO Box 598 Pittsburgh PA 15230-0598

DUPREE, THOMAS ANDREW, forester, state official; b. Cambridge, Mass., Jan. 18, 1950; s. Glenn Stewart and Elvira (Pacifici) D.; m. Sandra Ann Becker, Aug. 31, 1975; 1 child, Steven. BS in Forestry, U. Mass., 1972. Svc. forester R.I. Div. Forest Environ., Hope Valley, 1974-76, sr. forester, 1976-78, prin. forester, 1978-86; chief R.I. Div. Forest Environ., Scituate, 1986—; past chmn. R.I. Tree Farm Com. Bd. mem. USS Mass. Meml. Com., Fall River, 1987—; vice chmn. Northea. Forest Fire Protection Commn., 1988-89, chmn., 1990-92; pres. So. New England Forest Consortium, Inc., 1990—; mem. forest productivity working group N.E. Govs.' and Ea. Can. Premiers, 1986-91. Mem. Soc. Am. Foresters (chmn. Yankee div. 1988), Am. Forestry Assn., Nat. Assn. State Foresters, New Eng. Soc. Am. Foresters (exec. com. 1982-90), Northeast Area Assn. State Foresters (sec.-treas. 1990, v.p. 1991, pres. 1992), R.I. Fire Chiefs Assn. Avocations: golf, hiking, hunting, fishing. Home: 7 Elmonte Dr Coventry RI 02816-7713 Office: RI Div Forest Environ 1037 Hartford Pike North Scituate RI 02857-1847

DURAN, EMILIO, molecular biologist; b. Oviedo, Asturias, Spain, Dec. 7, 1963; came to U.S., 1986; s. Jaime and Luisa Fernanda Duran. BS magna cum laude, U. Toledo, 1988, MS, 1990. Teaching asst. U. Toledo, 1988-90, rsch. asst., 1990—; course asst. MBL-Biology of Parasitism, Woods Hole, Mass., summer 1991, 92, 93. Contbr. articles to profl. jours. Coord. Am. Field Svc., Oviedo, 1984-86. Mem. AAAS, Am. Soc. Parasitologists, Spanish Profls. Am., Inc., Golden Key Nat. Honor Soc., Sigma Xi. Roman Catholic. Office: U Toledo Dept Biology Toledo OH 43606

DURAND, JAMES HOWARD, mechanical engineer; b. Chgo., Sept. 12, 1951; s. Thomas and Frances Muriel (Welles) D.; m. Elaine Linda Magurno, July 9, 1977; children: Michael, Matthew. BS in Energy Engring., U. Ill., Chgo., 1974; MS in Tech. and Policy, MIT, 1978, MSME, 1978; PhD in Mech. Engring., U. Mo., 1985. Registered profl. engr., Mo. Rsch. asst. Dept. Mech. Engring., Mass. Inst. Tech., Cambridge, 1974-76, Ctr. for Policy Alternatives, Mass. Inst. Tech., Cambridge, 1976-78; energy engr. div. energy State of Mo., Jefferson City, 1978-80; lectr. Univ. Mo., Columbia, 1980-85, vis. prof., 1985-87; staff engr. Mo. Pub. Svc. Commn., Jefferson City, 1987-92; sr. engr. Trax Corp., Lynchburg, Va., 1992—. Co-author: Alternative Energy in Agriculture, 1987; contbr. articles to profl. jours. Mem. Christian Fellow Ch., Columbia, 1983—; leader Christian Fellowship Boys Group, Columbia, 1989—. Mem. ASME, NSPE, Mo. Soc. Profl. Engrs. (dir. 1992—). Home: 101 Warwick Pl Forest VA 24551 Office: Trax Corp 242 Brook Park Pl Box 385-7250 Forest VA 24451

DURANT, FREDERICK CLARK, III, aerospace history and space art consultant; b. Ardmore, Pa., Dec. 31, 1916; s. Frederick Clark, Jr. and Cornelia Allen (Howel) D.; m. Carolyn Griscom Jones, Oct. 4, 1947; children: Derek C. (dec.), Carolyn M., William C., Stephen H. B.S. in Chem. Engring, Lehigh U., 1939; postgrad., Phila. Mus. Sch. Indsl. Arts, 1946-47. Registered profl. engr., D.C., Mass. Engr. E.I. duPont de Nemours & Co., Inc., 1939-41; rocket engr. Bell Aircraft Corp., 1947-48; dir. engring. Naval Air Rocket Test Sta., 1948-51; cons. Washington, 1952-53; mem. sr. staff Arthur D. Little, Inc., 1954-57; dir. Maynard Ordnance Test Sta., 1954-55; exec. asst. to dir. Avco-Everett Research Lab., 1957-59; dir. pub. and govt. relations, research and advanced devel. div. Avco Corp., Wilmington, Mass., 1959-61; sr. rep. Bell Aerosystems Co., Washington, 1961-64; asst. dir. and head astronautics dept. Nat. Air and Space Mus., Smithsonian Instn., Washington, 1964-80; cons., 1980—; dir. Nat. Space Soc., Washington, 1982-88; conservator Bonestell Space Art and Space Art Internat.; dir. Arthur C. Clarke Found. U.S. Inc.; cons. space mus. Nippon Steel Corp., Japan, 1989-92; participant ann. congresses Internat. Astronautical Fedn., 1951—, pres., 1953-56; mem. organizing com. Project Orbiter, 1954; cons. Astro Assocs. Author: First Steps toward Space, 1975, Worlds Beyond: The Art of Chesley Bonestell, 1983; Contbg. editor: Missiles and Rockets, 1956-58; contbr. to: Ency. Brit., 1958—, Funk & Wagnalls Year Book; contbr.: space terms Am. Heritage Dictionary. Served to comdr. as naval aviator USNR, 1941-46,48-52. Recipient spl. medal L'Assn. Pour l'Encouragement de l'Aeronautique et de l'Astronautique, 1963, Charles A. Lindbergh award Smithsonian Instn., 1976, hon. 6 Dan Karate-Do, Japan, 1978, Rathbone Alumni Achievement award Lehigh U., 1989. Fellow Am. Astronautical Soc., AIAA, Am. Rocket Soc. (pres. 1953); mem. Internat. Acad. Astronautics (co-chmn. history com. 1981-89), Nat. Space Club (gov. 1961), Nat. Space Club (Disting. Service award 1982), hon. fellow or mem. numerous fgn. rocket and space flight socs., Cosmos Club. Home and Office: 109 Grafton St Bethesda MD 20815

DURANT, GERALD WAYNE, materials engineer; b. Keokuk, Iowa, July 24, 1936; s. Martin John and Helen Eddith (Garrison) D.; m. Delores Jean Steele, Aug. 16, 1959 (div.); children: Douglas, Diana (dec.), Deborah, Daniel. BS, Northeast Mo. State U., 1962; MBA, Drake U., 1982. Instr. chemistry, physics Centerville (Iowa) Community High Sch., 1962-65; rsch. chemist Maytag Co., Newton, Iowa, 1965-70, sr. rsch. chemist, 1970-86, chief rsch. chemist, 1986-89, mgr. materials devel., 1989—. Mem. Newton Community Schs. Bd. Edn., 1975-78. Mem. ASTM, Soc. Plastics Engrs. Home: 3910 74th St Urbandale IA 50322 Office: Maytag Co 1 Dependability Sq Newton IA 50208

DURANT, JOHN RIDGEWAY, physician; b. Ann Arbor, Mich., July 29, 1930; s. Thomas Morton and Jean Margaret (deVries) D.; m. Mary Sue Avery Dillon, Jan. 13, 1990; children by previous marriage: Christine Joy, Thomas Arthur, Michele Grace, Jennifer Margaret. B.A., Swarthmore (Pa.) Coll., 1952; M.D., Temple U., Phila., 1956; hon. degree, UAB, 1993. Diplomate: Am. Bd. Internal Medicine. Intern, then jr. asst. resident in medicine Hartford (Conn.) Hosp., 1956-58; resident in medicine Temple U. Med. Center, 1960-62; spl. fellow med. neoplasia Meml. Hosp. for Cancer and Allied Diseases, N.Y.C., 1962-63; Am. Cancer Soc. advanced clin. fellow Temple U. Health Scis. Center, 1964-67, instr., then asst. prof. medicine, 1963-67; clin. assoc. chemotherapy Moss Rehab. Hosp., Phila., 1964-67; research assoc. Fels Research Inst., Phila., 1965-67; mem. faculty U. Ala. Med. Center, Birmingham, 1968-82; prof. medicine, dir. comprehensive cancer center U. Ala. Med. Center, 1970-82, prof. radiation oncology, 1978-82, chmn. Southeastern coop. cancer study group at univ., 1975-82, Disting. faculty lectr., 1980; pres. Fox Chase Cancer Ctr., Phila., 1982-88; sr. v.p. health affairs and dir. med. ctr. U. Ala., Birmingham, 1988—; chmn. coop. group exec. com. Nat. Cancer Inst., NIH, 1977-82, chmn. coop. group chairmen, 1979-82; cons. VA Hosp., Tuskegee, Ala., 1970-82; exec. com. Birmingham chpt. ARC, 1972-77; mem. Nat. Cancer Adv. Bd., 1986-92. Mem. editorial bd. Cancer Clin. Trials, 1979-82, assoc. editor, 1982—; editorial bd. Med. and Pediatric Oncology News, 1975-90; assoc. editor Cancer, 1984-92; contbr. numerous articles to med. jours. Served as officer M.C. USNR, 1958-60. Named Temple U. Med. Sch. Alumnus Yr., 1982. Fellow ACP, Coll. Physicians Phila.; mem. Am. Cancer Soc. (vice chmn. advanced clin. fellowship com. 1974-76, 85-87, mem. instl. rsch. grant com. 1979-82, pres. Ala. div. 1973-75, 77-79), Am. Assn. Cancer Rsch., Am. Radium Soc. (pres. 1984), Am. Bd. Med. Oncology (subcom. 1979-85, chmn. 1983-85), Assn. Am. Cancer Insts. (dir. 1978—, pres. 1982-83), Assn. Community Cancer Ctrs. (dir. 1979-81), Am. Soc. Clin. Oncology (chmn. pub. rels. com. 1976-79, bd. dirs. 1979-82, 84-87, pres. 85-86), more. Methodist. Office: Univ Ala University Sta Birmingham AL 35294

DUREK, THOMAS ANDREW, computer company executive; b. Sharpsville, Pa., July 1, 1929; s. Joseph Adam and Helen Barbara (Ondish) D.; m. Phyllis H. Norris, Aug. 1, 1987. BA, Pa. State U., University Park, 1953; MA, Baylor U., 1957; MS, Stanford U., 1959. Mgmt. scientist USAF, Pentagon, Washington, 1959-65; project engr. North Am. Rockwell Corp., Washington, 1965-68; systems engr. TRW, Inc., Washington, 1968-81, facility mgr., Patuxant, Md., 1981-82, project mgr., Washington, 1982-86; project mgr., prin. mem. tech. staff Software Productivity Consortium, Herndon, Va., 1986-89; sr. tech. staff, software technologist Systems Integration Group TRW Inc., Fairfax, Va., 1989-92; founder, prin. TAD Assocs., Bethesda, Md., 1992—; professorial lectr. George Washington U., 1960-66, George Mason U., 1991; chair software reusability conf. Nat. Inst. for Software Quality and Productivity, 1989-91, mem. adv. bd. , 1991—, chair info. systems engring. for downsizing conf., 1992; speaker in field. Contbr. articles to profl. jours. Mem. parish coun. Church of St. Stephen Martyr, Washington, 1970-78, pres., 1975-78, liturgical minister, 1973-87, chair Pastoral Coun., Continuing Edn. Com., Shrine Most Blessed Sacrament, Washington, 1993—. With USAF, 1953-65; to col. USAFR, ret., 1984. Decorated Meritorious Svc. medal, 1984. Roman Catholic. Home and Office: 7915 Quarry Ridge Way Riverhill Bethesda MD 20817

DURHAM, LAWRENCE BRADLEY, nuclear engineer, consultant, mediator; b. Decatur, Ala., Mar. 30, 1941; s. Robert Oron and Margaret Tipton (Jacks) D.; m. Anne Pauline Stimson, June 19, 1965; children: Margaret Frances, Bradley Franklin. BS in Math., Birmingham-Southern Coll., 1963; MA in Math., U. Ala., 1965; PhD in Behavioral Scis., U. Del., 1972; postgrad., Princeton U., 1975, Harvard U., 1977, MIT, 1993. Assoc. dir. Ednl. Devel. Ctr. U. Del., Newark, 1970-72; dir. devel. Birmingham (Ala.)-Southern Coll., 1972-74; dir. planning U. Ala., Tuscaloosa, 1974-76, dean, assoc. prof., 1976-84; accreditation team mgr. Inst. Nuclear Power Ops., Atlanta, 1984-89; nuclear tng. mgr. TVA, Chattanooga, 1989—; cons. to utility industry, bus., govt. and ednl. instns., 1970—; mem. operator issues working group NUMARC, Washington, 1992. Author: Log and Exponent Functions on Slide Rule, 1968; editor accreditation team reports, 1986-89; contbr. articles to profl. jours. Scoutmaster, committeeman Boy Scouts Am., 1965-89; div. chmn. United Way, Tuscaloosa, 1983; bd. dirs. Soc. Coll. and Univ. Planning, Ann Arbor, Mich., 1979-83; mem. So. Regional Coun., The Coll. Bd., Atlanta, 1982-84. Lt. USNR, 1966-70. Named Unidel fellow, 1970, Rotarian of the Quarter, East Cobb Rotary Club, Marietta, Ga. Mem. APA, Am. Nuclear Soc. (sect. chmn. 1992-93, edn. and tng. div. exec. com. 1993—), Phi Beta Kappa (chpt. sec. 1973-74), Omicron Delta Kappa, Sigma Alpha Epsilon (Merit award 1980). Methodist. Achievements include research in multiple strategies to successfully teach mathematical problem-solving, distance learning technology in U.S. nuclear utility industry. Home: 9114 Tennga Ln Chattanooga TN 37421-4563 Office: TVA 1101 Market St Chattanooga TN 37402

DURIGON, MICHEL LOUIS, pathologist, forensic medicine educator; b. Saint Cloud, France, Oct. 18, 1942; s. Ernest and Simonne (Etasse) D.; m. Geneviève Pascale Le Pont, May 13, 1971; children: Camille, Alice, Pauline. Physique biologie, Faculté Sciences, Paris, 1961; MD, U. Paris, 1969. Tng. in forensic medicine, Paris, 1969-72, tng. in pathology,, 1972-73; asst. Faculty of Médicine, Paris, 1974-77, asst. prof., 1977-84; prof., dept. head Hosp. R. Poincaré and Faculty of Medicine, Garches and Paris, France, 1985—; nat. expert Supreme Ct., Paris, 1986—; cons. ICPO, Lyon, France, 1982-91; coord. nat. edn. of forensic medicine, Paris. Author: Medecine Legale à Usage Judiciaire, 1979, Forensic Pathology, 1988. Mem. Forensic Sci. Soc., Acad. Internat. de Medecine Légale, Soc. d'Anthropology, Soc. de Medecine Légale (sec. 1978-89, v.p. 1989—), Sevilla Working Party. Roman Catholic. Achievements include research in sudden death syndrome, death timing. Home: 39 cours du 14 Juillet, 78300 Poissy France Office: Hospital R Poincaré, 92380 Garches France

DURKEE, JOE WORTHINGTON, JR., nuclear engineer; b. Albuquerque, Mar. 10, 1956; s. Joe Worthington Sr. and Hallie Mae (Payne) D. BS, Tex. A&M U., 1978, ME, 1981, PhD, 1983. Staff mem. Los Alamos (N.Mex.) Nat. Lab., 1983—; rsch. proposal reviewer LANL, 1986-87, Dept. Energy/ER Nuclear Engr. Proposal Rev. panel, 1988—. invited rsch. paper reviewer Jour. Nuclear Tech., 1987, Jour. Biomech. Engr., 1991; contbr. articles to Jour. Physics in Medicine and Biology, Progress in Nuclear Energy, Annuals Nuclear Energy, Jour. Nuclear Tech. Mem. Am. Nuclear Soc. (admissions com. 1986—, chair 1990—), Tex. A&M Former Student Assn., Nat. Space Soc. Achievements include development of Sn and Monte Carlo reactor physics design calculations for a number of thermal and fast nuclear reactor designs, of mathematical models depicting heat transport in the human body; notation of bifurcating behavior of multiregion bioheat and neutron diffusion equations and development of techniques to solve and computationally evaluate these expressions; research in space-time neutron diffusion and fission-product convective diffusion, reactor physics calculations in support of the LANL Omega West Reactor reconfiguration to produce radioisotopes for medical applications. Office: Los Alamos Nat Lab Group N-12 M/S K551 PO Box 1663 Los Alamos NM 87545-0001

DURKIN, ANTHONY JOSEPH, biomedical engineer, researcher; b. South Bend, Ind., May 18, 1963; s. Joseph Anthony and Ann Veronica (Gill) D. BS in Physics, Lamar U., 1985; MS in Physics, U. North Tex., 1988;

postgrad., U. Tex., Austin, 1990—. Teaching fellow, teaching asst. in physics U. North Tex., Denton, 1987-90, rsch. asst., 1988-90; rsch. asst. in biomed. engring. U. Tex., Austin, 1991—. Contbr. to profl. publs. Grantee Optical Soc. Am., 1992. Mem. Am. Phys. Soc., Optical Soc. Am., S.P.I.E., Sigma Xi, Sigma Pi Sigma (chpt.v.p. 1984-85). Office: Univ Tex Rm 610 Engring Sci Bldg Austin TX 78712

DURKIN, JOHN CHARLES, agriculturist; b. Garnett, Kans., May 10, 1951; s. James Ellsworth and Dorothy (Mundell) D.; m. Kathleen Ann Merskin, June 6, 1987. BS, Kans. State U., 1973; BBA, So. Meth. U., 1974; MS, Okla. State U., 1981. Acct. pvt. practice, 1974-75; tax auditor, 1976-77; internal auditor Kans. Dept. Revenue, 1978-79, Am. Agronomics Corp., Tampa, Fla., 1981-82; contr. Cal-Fresh Pak, L.A., 1982; treas. Citrus Growers, Inc., Weslaco, Tex., 1982-83; plant mgr. Myakka Processors, Inc., Arcadia, Fla., 1983-86; project mgr. Orange-co, Inc., Lake Hamilton, Fla., 1986-90; dir. project mgmt. Orange-co, Inc., Arcadia, 1990—; cons. PAX Rsch., Arcadia, 1974-90; instr. Bus. Tng. Internat., Palm Beach, Fla., 1986-88; presenter seminars. Mem. Fla. State Horticulture Soc., Selby Botan. Gardens, N.Y. Acad. Scis., Kiwanis (treas. Arcadia chpt. 1986). Methodist. Achievements include design of low-volume irrigation systems, hydraulic irrigation controller. Office: PAX Rsch 120 W Oak St Arcadia FL 33821

DURLAND, SVEN O., research and development engineer; b. Jonkoping, Sweden, Aug. 28, 1944; came to U.S., 1983; m. Maud I. Larson, Dec. 29, 1970; children: Carolina, Josephine. Engr. degree, Coll. Engring., Jonkoping, Sweden, 1964. From design engr. to project mgr. SAAB-Scania, Jonkoping, 1964-82; project mgr. SAAB Systems, Seattle, 1983-85; project mgr. COE Mfg., Portland, Oreg., 1986-87, r&d mgr., 1987—. Achievements include Patent Optical Scanning Method and Apparatus; developed machine vision system for grade scanning of lumber. Home: 14072 Taylors Crest Ln Lake Oswego OR 97035 Office: COE Mfg 7930 SW Hunziker Rd Portland OR 97223

DUROCHER, CORT LOUIS, aerospace engineer, association executive; b. Houghton, Mich., Aug. 30, 1946; s. Marshall Vincent and Mary (Hornick) D.; m. Beth Ann Walker, Dec. 29, 1968. BS, USAF Acad., 1968; MS, MIT, 1977. Commd. 2d lt. USAF, 1968, advanced through grades to lt. col., 1984, mil. pilot, 1968-76; astronautical engr. space div. USAF, L.A., 1977-81; ops. officer USAF, Tucson, 1981-84; dep. program mgr. advanced launch system USAF, L.A., 1984-88; ret. USAF, 1988; program mgr. Hughes Aircraft Co., L.A., 1988—; exec. dir. AIAA, Washington, 1989; comml. pilot FAA, 1968-88; study dir. Nat. Space Transp. Study, 1985-87. Contbr. articles to profl. jours. Athletic liaison USAF Acad., L.A., 1977-81. Decorated D.F.C., 7 Air medals. Fellow AIAA (tech. com. 1984-88); mem. Air Force Assn., MIT Club, Order of Daedalions. Home: PO Box 183 Middleburg VA 22117 Office: AIAA 370 L'Enfant Promenade SW Washington DC 20024*

DURRETT, ANDREW MANNING, industrial designer; b. Clarksville, Tenn., Jan. 7, 1924; s. Andrew Manning and Betty Ann (Empson) D.; m. Jean E. Lanning, May 14, 1949; children—Cheryl A. Durrett Yurs, Griffith Lynn, Susan Elizabeth Durrett Pantle, Robert Tracy, Vicki Jean Durrett Robinson. Student U. Tenn., 1946-48; B.F.A., Art Inst. Chgo., 1951. Mech. engr. Werthan Bag Corp., Nashville, 1951-52; designer C.F. Block, Chgo., 1952-53, Advt. Metal Display, Chgo., 1953-55; artist, poster designer Gen. Outdoor Advt. Co., Chgo., 1956-58; with sales, tng. films and scripts. Ross Wetzel Studios, Chgo., 1958-60; sales mgr. indsl. design Palma-Knapp Design, River Forest, Ill., 1961-65; founder, pres. Manning Durrett Design Assocs., R&D med. equipment, Spring Grove, Ill., 1965— . Author: (with Don Gilbert) Industrial Insectology, 1980; producer films on insect control; patentee in field, cancer and hepatitus diagnostics platelet collection, protein analyzing, life systems monitoring, patients wound site temperature control, non-ambulatory patient transfer, introvenius drug delivery; exhibited med. and indsl. products Mus. Sci. and Industry, Chgo., 1970. Bd. dirs., sec. Brookfield Citizens Mgmt. Assn. (Ill.), 1953-54; committeeman Oaks Assn., Libertyville, Ill., 1970-76, chmn. service com.; co-author flood plane ordinance, Libertyville, 1975. Served with USN, 1943-46. Recipient Excellence in Design awards Indsl. Design Rev., 1969, 69. Mem. Sch. Art Inst. Chgo. Alumni Assn., Chgo. Assn. Commerce and Industry, Internat. Platform Assn., Phi Kappa Phi. Republican. Methodist. Clubs: Oak Bus. Men's (Ill.); Cambridge Country (Libertyville). Office: Manning Durrett Design Assocs 38625 N Forest Ave Spring Grove IL 60081-9213

DUSCHL, WOLFGANG JOSEF, astrophysicist; b. Munich, May 17, 1958; s. Josef and Mathilde (Birkl) D. Diploma in Physics, Ludwig-Maximilians-U., Munich, 1982; D in Physics, Ludwig-Maximilians-U., 1985; Habilitation in Astronomy, Ruprecht-Karls-U., Heidelberg, Fed. Republic Germany, 1991. Researcher (Wissenschaftl Mitarbeiter) Max-Planck-Inst. for Astrophysics, Garching, Fed. Republic Germany, 1982-90; external fellow European Space Agy. Inst. Astronomy, U. Cambridge (Eng.), 1985-86; Wissenschaftlicher asst. Inst. for Theoretical Astrophysics, Ruprecht-Karls-U., 1988—, privatdozent, 1991—; rschr. Max-Planck-Inst. Radioastronomy, Bonn, Germany, 1993. Author: ... und über uns die Sterne, 1990. Fellow Royal Astron. Soc.; mem. Astronomische Gesellschaft, Am. Astron. Soc. (full), Vereinigung der Sternfreunde. Home: Leipziger Str 1, D-69181 Leimen Germany Office: Inst Theoret Astrophysik, Im Neuenheimer Feld 561, D-69120 Heidelberg Germany

DUSOLD, LAURENCE RICHARD, chemist, computer specialist; b. Chgo., Nov. 15, 1944; s. Henry E. and Colette M. Dusold; m. Karen A. Marsh, Aug. 29, 1970; children: Amy, Lauren, Patricia, Amanda. BS in Chemistry, Purdue U., 1966; MS, U. N.C., 1969; postgrad., Wayne State U., 1969-71. Rsch. chemist, residue analysis and methods investigation br. Bur. Foods FDA, Washington, 1971-75, chemist, computer specialist, div. chemistry and physics, 1975-81, sr. chemist, computer specialist, div. of chemistry and physics, 1981-86, chief telecommunications and scientific computer support, 1986—, mem. faculty, evening div. U. Md., 1973—; mem. fed. engring. planning group Dept. Health and Human Svcs., 1990-92. Mem. editorial bd. Scientific Computing & Automation, 1990-92; contbr. articles to profl. jours. and book chpt. Mem. AAUP, Computer Soc. of IEEE, Assn. Computing Machinery (chmn. SIGAPL, D.C. chpt. 1978-91, vice chmn. Potomac chpt. 1993), Greater Washington Fed. Agy. APL Users Group (co-chmn. 1977-87), Alpha Chi Sigma, Phi Lambda Upsilon. Republican. Roman Catholic. Office: FDA 200 C St SW Washington DC 20204-0002

DUTCHER, JANICE JEAN PHILLIPS, medical oncologist; b. Bend, Oreg., Nov. 10, 1950; d. Charles Glen and MayBelle (Fluit) Phillips; m. John Dutcher, Sept. 8, 1971 (div. 1980). BA with honors, U. Utah, 1971; MD, U. Calif., Davis, 1975. Diplomate Am. Bd. Internal Medicine, Am. Bd. Med. Oncology. Intern Rush-Presbyn. St. Luke's Hosp., Chgo., 1975-76, resident, 1976-78; clin. assoc. Balt. Cancer Rsch., Nat. Cancer Inst., 1978-81, sr. investigator, 1981-82; asst. prof. U. Md., Balt., 1982; asst. prof. Albert Einstein Coll. Medicine, N.Y.C., 1983-86, assoc. prof., 1986-92, prof., 1992—; course co-dir. Advances in Cancer Treatment Rsch. Albert Einstein Coll. Medicine, Manhattan, 1984—; chmn. biol. response mod. com. ECOG, Madison, Wis., 1989—; mem. data safety com. Nat. Heart Lung Blood Inst., Bethesda, Md., 1990—; mem. biologic response modifier study sect. Nat. Cancer Inst., Bethesda, 1988, 90. Editor: Handbook of Hematology/Oncology Emergencies, 1987, Modern Transfusion Therapy, 1990; mem. editorial bd. Jour. Immunotherapy; contbr. articles to Blood, Leukemia. Recipient Beecham award in Hematology So. Blood Club, 1983, Henry C. Moses Clin. Rsch. award Montefiore Med. Ctr., 1989, Outstanding Alumnus award U. Calif., Davis, 1989; recipient numerous grants. Fellow ACP; mem. Am. Soc. Clin. Oncology (mem. program com. 1988), Am. Assn. Cancer Rsch., Am. Soc. Hematology, Soc. for Biol. Therapy, Phi Beta Kappa (Presdl. scholar 1968), Alpha Lambda Delta, Phi Kappa Phi, Alpha Omega Alpha. Achievements include findings related to the management of alloimmunization to platelet transfusions, the intensive maintenance of patients with acute leukemia, and studies of new biologic response modifiers as antitumor drugs. Office: Albert Einstein Coll Med 1825 Eastchester Rd Bronx NY 10461-2301

DUTSON, THAYNE R., college dean; b. Idaho Falls, Oct. 3, 1942; s. Rollo and Thelma (Fugal) D.; m. Joyce Cook, Dec. 19, 1962 (div. 1980); 1 child, Bradley; m. Margaret McCallum, June 30, 1989; children: Taylor, Alexandra. BS, Utah State U., 1966; MS, Mich. State U., 1969, PhD, 1971. Postdoctoral fellow U. Nottingham, Sutton Bonnington, Eng., 1971-72; prof. Tex. A&M U., College Station, 1972-83; dept. head Mich. State U., East Lansing, Mich., 1983-87; dean and dir. Oreg. State U., Corvallis, 1987—. Editor: Advances in Meat Research (8 vols.) 1985-91; contbr. articles to profl. jours. Scoutmaster Boy Scouts Am., Mich., 1966-71. Mem. Inst. Food Technologists, Am. Meat Sci. Assn. (bd. dirs. 1979-81, Disting. Rsch. award 1985), Am. Soc. Animal Sci. (Meat Rsch. award 1981), Coun. for Agriculture Sci. and Tech. (pres. 1988), Phi Kappa Phi, Sigma Xi. Avocations: skiing, running, exercise, racquetball, rafting.

DUTT, DAVID ALAN, physicist; b. Bethlehem, Pa., Feb. 1, 1962; s. Darvin Lee and Myrtle Virginia D.; m. Cynthia Anderson, June 1, 1985; 1 child, Cassandra Leigh. BS, Moravian Coll., Bethlehem, 1984; MS, Lehigh U., 1986, PhD, 1989. Rsch. assoc. NRC-Naval Rsch. Lab., Washington, 1989-91; physicist Kopp Glass, Inc., Pitts., 1991—. Contbr. articles to profl. jours. Mem. Am. Phys. Soc., Am. Ceramic Soc. Achievements include patent pending for preparation of permanent photowritten optical diffraction gratings in irradiated glasses. Office: Kopp Glass Co 2108 Palmer St Pittsburgh PA 15218

DUTT, RAY HORN, reproduction physiologist, educator; b. Bangor, Pa., Aug. 26, 1913; s. Elmer James and Viola Belle (Horn) D.; m. Louise Elizabeth Gettys, June 22, 1946; children: Philip, Kathleen. BS, Pa. State U., 1941; MS, U. Wis., 1942, PhD, 1948. Asst. prof. U. Ky., Lexington, 1948-51, assoc. prof., 1951-58, prof. reprodn. physiology, 1958-81; ret., 1981. Editorial bd. Jour. Animal Sci., 1960-61, assoc. editor, 1961-63, editor, 1964-66; contbr. rsch. and rev. articles on physiology of reprodn. in farm animals to profl. jours. Maj. USMCR, 1942-46. Fulbright Found grantee, 1957. Fellow AAAS, Am. Soc. Animal Sci. (hon., v.p. 1967, pres. 1968), Sigma Xi, Phi Sigma, Phi Kappa Phi, Gamma Sigma Delta. Lutheran. Achievements include pioneering research on estrus synchronization in mammals. Home: 437 Bristol Rd Lexington KY 40502

DUTTA, ARUNAVA, chemical engineer; b. Calcutta, India, Mar. 10, 1958; came to U.S. 1980; s. Amiyamoy and Arati (Biswas) D. m. Malini Bose, Jan. 4, 1989. BTech with hons., IIT, Kharagpur, India, 1980; ScD, MIT, 1985. Advanced rsch. and devel. engr. GTE Products Corp., Danvers, Mass., 1985-90, engring specialist, 1990-92; mgr. Osram Sylvania Inc. (formerly GTE Products Corp.), Danvers, Mass., 1992--. Contbr. articles to profl. jours. Mem. Am. Inst. Chem. Engrs., Am. Chem. Soc., Electrochemical Soc., Sigma Xi. Achievements include patents for process for coating small solids, method for increasing cohesiveness of powders in fluid beds, method of reducing phosphor degradation, method of reclaiming lamp phosphor, method of detecting a degraded phosphor, method for fluidized bed discharge, apparatus for coating small solids, method for measuring the fluidity of fluidized beds. Home: 35 Westgate #6 Chestnut Hill MA 02167 Office: Osram Sylvania Inc 100 Endicott St Danvers MA 01923

DUTTA, PULAK, physicist, educator; b. Calcutta, India, Oct. 1, 1951. BSc, U. Calcutta, 1971; MSc, U. Delhi, India, 1973; PhD, U. Chgo., 1979. Postdoctoral assoc. Argonne (Ill.) Nat. Lab., 1979-81; asst. prof. physics Northwestern U., Evanston, Ill., 1981-87, assoc. prof. physics, 1987-92, prof. physics, 1992—. Contbr. articles to profl. publs., including Physical Review Letters, Jour. Chem. Physics, Jour. Applied Physics, Langmuir, others. Fellow Am. Phys. Soc.; mem. Am. Chem. Soc. Achievements include studies of resistance fluctuations in solids, neutron scattering studies of monolayers, synchrotron diffraction studies of organic monolayers. Office: Northwestern U Dept Physics & Astronomy 2145 Sheridan Rd Evanston IL 60208-3112

DUTTA, SUBIJOY, environmental engineer; b. Shillong, Meghlaya, India, Dec. 19, 1950; s. Subinoy and Santwana (Dasgupta) D.; m. Urmi Dutta, Jan. 15, 1986; 1 child, Sumit Dutta. BS in Mech. Engring., U. Gauhati, 1972; MS in Mech. Engring., U. Okla., 1981, MS in Geological Engring., 1984; postgrad., U. Kans., 1990. Registered profl. engr. Md., Okla., Ariz. Staff engr. Kay Kay & Assocs., Noble, Okla., 1981-84; mem. faculty, rsch. assoc. U. Okla., Norman, 1985-88; tech. dir. U.S. EPS Corp., Oklahoma City, Okla., 1988-89; environ. engr. Tinker Air Force Base, Oklahoma City, 1989-91, U.S. EPA, Washington, 1992—; edt. bd. Interant. Environ. Engring. jour, Norway, 1992--. Contbr. articles to profl. jours. Vol. Community Cleanup, Bowie, Md., 1992. Recipient award U.S Air Force, 1992. Mem. AWMA, Hazardous Material Control Rsch. Inst. Achievements include patents in field. Home: 1612 Portland Ln Bowie MD 20716-1865 Office: US EPA 5303 W 401M St SW Washington DC 20460

DUTTON, J. CRAIG, mechanical engineering educator; b. Williston, N.D., Mar. 1, 1951; s. John Lawrence and Charlotte Ruby (Larson) D.; m. Linda Marie Shelton, June 16, 1973; 1 child, Andrew Lawrence. BSME, U. Wash., 1973; MS, Oreg. State U., 1975; PhD, U. Ill., 1979. Cert. profl. engr., Tex. Mech. engr. Lawrence Livermore (Calif.) Nat. Lab., 1979-80; asst. prof. mech. engring. Tex. A&M U., College Station, 1980-84, assoc. prof., 1984-85; assoc. prof. U. Ill., Urbana, 1985-91, prof., 1991—; cons. Wright-Patterson AFB, Dayton, Ohio, 1983—, Rocketdyne Div. Rockwell Internat., Canoga Park, Calif., 1989—. Author: (with others) Encyclopedia of Fluid Mechanics, 1989; contbr. articles to profl. jours. Recipient Ralph R. Teetor Edn. award Soc. Automotive Engrs., 1986; Summer Faculty Rsch. fellow USAF Office Scientific Rsch., 1983. Fellow AIAA (assoc.), mem. ASME (assoc. editor 1987-91), Am. Soc. Engring. Edn. (AT&T Found. award 1989). Achievements include advanced research in field of compressible fluid dynamics. Office: U Ill Dept Mech and Ind Engring 1206 W Green St Urbana IL 61801

DUUS, PETER, neurology educator; b. Guderup, Denmark, Sept. 29, 1908; s. Christian D. and Ingeline (Dohsen) D.; m. Erika Müller (dec. 1988); children: Peter Christian, Barbara. Student, U. Kiel, Berlin, 1927-32; MD, J. W. Goethe U., Frankfurt, Germany, 1937; habil., J. W. Goethe U., Franfurt, Fed. Republic of Germany, 1943. Lectr., researcher U. Nervenklinik, Frankfurt, 1933-45; prof. neurology and psychiatry J. W. Goethe U. of Frankfurt, 1950—; founder neurol. dept., med. supt. St. Markus Hosp., Frankfurt, 1958-63; founder neurol. dept., dir. Academic Hosp., Frankfurt, 1963-74. Author: Neurologisch-Topishe Diagnostik, 1976 (transl. 10 langs.), 5th rev. edit., 1990; contbr. articles to profl. jours. Mem. German Assn. Neurology, German Assn. Neurosurgery, German Assn. Neuroradiology. Home: Thorwaldsenstr 33, D-6000 Frankfurt am Main, Germany

DUVDEVANI, ILAN, chemical engineer, researcher; b. Israel, May 10, 1938; came to U.S., 1963; m. Phyllis Hurwitz; children: Yael, Tamar. BS in Chem. Engring., Technion, Haifa, Israel, 1962; MS in Chem. Engring., Stevens Inst. Tech., 1965, PhD in Chem. Engring., 1969. Polymers processing rsch. Western Electric Co., Princeton, N.J., 1965-67; polymers cons., vis. scientist Stevens Inst. Tech., Linden, N.J., 1969-74; polymer tchr. Plastics Inst. Am., Hoboken, N.J., 1969-74; polymers researcher Exxon Chem. Co., Linden, 1974—; chmn. processability task force Rubber Mfrs. Assn., 1989-91. Contbr. articles to profl. jours. Recipient Ednl. Svc. award Plastics Inst. Am., 1973. Fellow Soc. Plastics Engrs. (engring. properties and structure div. chmn. tech. programs 1979-81); mem. Soc. Rheology, Am. Chem. Soc. Achievements include 44 patents in field; research on polymer extrusion, polymer rheology, thermoplastic elastomers, solfunated ionomers, polymeric fluids with unusual properties, butyl rubber. Office: Exxon Chem Co 1900 E Linden Ave Linden NJ 07036

DUVIVIER, JEAN FERNAND, management consultant; b. Niteroi, Brazil, Dec. 17, 1926; came to U.S., 1954; s. Herman Felix and Eugenie A. (Dits) D.; m. Barbara Johanne Doucet, June 9, 1956; children: Christine, Michele, John, Elizabeth, Marc. BSc, Boston U., 1955; SM, MIT, 1958, Engr. in Aeronautics & Astronautics degree, 1966. Project leader MIT Aeroelastic Lab., Cambridge, 1955-61; mem. sr. staff Ctr. Naval Analyses, Cambridge, 1961-66; cons. Rsch. Analysis Corp., McLean, Va., 1966; sr. engr. Electric Boat div. Gen. Dynamics, Quincy, Mass., 1966-68; mgr. strategy planning, R & D, dir. mktg. for Latin Am. Boeing Vertol Co., Phila., 1968-82; v.p. internat. mktg. Fairchild Republic, Farmingdale, N.Y., 1982-85; v.p. aerospace systems Lear Siegler Internat., Stamford, Conn., 1985-89; dir. systems mktg. Smiths Industries, Stamford, 1988-90; gen. mgr. Duvivier Assocs., Georgetown, Conn., 1990—. With Brazilian Air Force, 1944-45. Mem. AIAA (assoc. fellow, chmn. Conn. sect. 1991-92), Am. Helicopter Soc., Sigma Xi. Republican. Roman Catholic. Home: 244 Umpawaug Rd West Redding CT 06896-2213 Office: PO Box 755 Georgetown CT 06829-0755

DUVOISIN, ROGER CLAIR, physician, medical educator; b. Towaco, N.J., July 27, 1927; s. Roger Antoine and Louise (Fatio) D.; m. Winifred Theresa Murray, Feb. 21, 1948; children: Anne, Marc, Jacques, Jeanne. BA, Columbia U., 1950; MD, N.Y. Med. Coll., 1954. Diplomate Am. Bd. Neurology. Intern, rotating Lenox Hill Hosp., N.Y.C., 1954-55, resident in neurology, 1955-56; resident in neurology Presbyn. Hosp., N.Y.C., 1956-58; rsch. assoc. in neurology Coll. Physicians and Surgeons Columbia U., N.Y.C., 1962-64, asst. prof. neurology Coll. Physicians and Surgeons, 1964-69, assoc. prof. neurology Coll. Physicians and Surgeons, 1969-72, prof. neurology Coll. Physicians and Surgeons, 1972-73; prof. neurology Mt. Sinai Sch. Medicine, N.Y.C., 1973-79; chmn. dept. neurology Robert Wood Johnson Med. Sch. U. Medicine and Dentistry of N.J., New Brunswick, 1979—, William Dow Lovett prof. neurology, 1990—; cons. Fed. Air Surgeon, U.S. Dept. Transp., Washington, 1965-73; mem. sci. adv. bd. Parkinson Disease Found., N.Y.C., 1965-91, Ctr. for Advanced Biotech. and Medicine, Piscataway, N.J., 1987-91, Dystonia Med. Rsch. Found., L.A., 1989—, Am. Parkinson Disease Assn., 1990—; bd. sci. councilors Nat. Inst. Neurol. Diseases and Stroke, 1993-98. Author: Parkinson's Disease: A Guide for Patient and Family, 1978, 2nd edit., 1984, 3rd edit., 1990; editor: The Olivopontocerebellar Atrophies, 1989; contbr. numerous sci. articles and revs. to med. jours. With USNR, 1945-46, ETO; maj. M.C., USAF, 1955-62. NIH grantee, 1985—; recipient Disting. Alumni award N.Y. Med. Coll., 1992, Springer award Am. Parkinson Disease Assn., 1992. Fellow Am. Acad. Neurology, Am. Coll. Physicians; mem. Am. Neurol. Assn. (sec.-treas.), Soc. for Neuroscience. Office: UMDNJ-Robert Wood Johnson Medical Sch One Robert Wood Johnson Pl New Brunswick NJ 08903

DÜZGÜNES, NEJAT A., biophysicist; b. N.Y.C., Feb. 28, 1950; s. Orhan and Zeliha (Uygurer) D. BS, Mid. East Tech. U., Ankara, Turkey, 1972; PhD, SUNY, Buffalo, 1978. Postdoctoral fellow U. Calif., San Francisco, 1978-81, asst. rsch. biochemist, 1981-87, asst. adj. prof., 1985-87, assoc. rsch. biochemist, 1987—, assoc. adj. prof., 1987—; assoc. prof., chmn. Dept. Microbiology U. Pacific, San Francisco, 1990—; vis. prof. Kyoto (Japan) U., 1988. Editor: Membrane Fusion in Fertilization Cellular Transport and Viral Infection, 1988, Mechanisms and Specificity of HIV Entry into Host Cells, 1991, Membrane Fusion Techniques, Methods in Enzymology, Vols. 220 & 221, 1993. Vol. AFS Internat. Intercultural Programs, N.Y.C., 1969-86. Co-recipient Orgn. award U.S.-Japan Binational Seminar on Membrane Fusion, NSF, 1992; Japan Soc. Promotion of Sci. fellow, 1988; grantee Am. Heart Assn., 1983-87, Calif. Univ.-wide AIDS Rsch. Program, 1986-90, 92-93, NIAID/NIH, 1988—. Mem. Am. Soc. Cell Biology, Am. Soc. Microbiology, Internat. Soc. Antiviral Rsch., Internat. AIDS Soc., Am. Assn. Dental Schs., Am. Assn. Dental Rsch., Biophysics Soc. Office: U of Pacific Dept Microbiology 2155 Webster St San Francisco CA 94115-2399

DVORAK, CLARENCE ALLEN, microbiologist; b. Cedar Rapids, Iowa, July 6, 1942; s. Clarence Louis and Lily Ann (Duda) D. BS, Iowa State U., 1969. Microbiologist Penford Products Co., Cedar Rapids, 1969—, analytical chemist, 1970-81, sci. photographer. Mem. AAAS, Am. Soc. for Microbiology, Soc. for Indsl. Microbiology, Am. Phytopath. Soc. Home: 1231 Sierra Dr NE Apt 10 Cedar Rapids IA 52402-6541

DVORAK, GEORGE J., materials engineering educator; came to U.S., 1964; Degree in Civil Engring., Czech Technol. U., Prague, 1956; C.Sc., Czechoslovak Acad. Sci., Prague, 1964; PhD, Brown U., 1968. Rsch. assoc. divsn. engring. Brown U., 1964-67; with civil engring. and biomedical engring. dept. Duke U., Durham, N.C., 1967-79; prof., chmn. civil engring., prof. materials sci. U. Utah, Salt Lake City, 1979-84; prof. dept. civil and environ. engring. Rensselaer Poly. Inst., Troy, N.Y., 1984—, prof. mech. engring., aero. engring. and mechanics, chmn. civil and environ. engring.; sr. vis. fellow Brit. sci. rsch. coun. Cambridge U., Eng.; vis. fellow Clare Hall, Cambridge, 1975-76; vis. prof. Politecnico di Milano, Milan, Italy; with inst. ctr. composite materials and structures Rensselaer Poly. Inst., dir. univ. rsch. initiative Dept. Def. Assoc. editor Jour. Applied Mechanics, 1989—, Applied Mechanics Revs., 1989—. Recipient Citations for Accomplishment of Spl. Merit, Army Rsch. Office, 1977, 79. Fellow ASME (founding chmn. com. composite materials applied mechanics divsn., Arpard L. Nadai award 1992), ASCE, Am. Acad. Mechanics, Soc. Engring. Sci. Achievements include research in mechanics, physics of solids, micromechanics of heterogeneous media, mechanical behavior of composite materials. Office: Dept of Civil & Env Engring JEC 4049 Rensselaer Polytech Inst Troy NY 12180*

DWENGER, THOMAS ANDREW, engineer; b. Dayton, June 17, 1945; s. Ferdinand Bernard and Velma Theresa (Woeste) D.; m. Brenda Lee Langsdon, May 28, 1966; children: Kelly, Kevin. BSME, Ind. Inst. Tech., Ft. Wayne, 1969. Staff tire designer The Goodyear Tire and Rubber Co., Akron, Ohio, 1968-72; sr. tire designer The Goodyear Tire and Rubber Co., Luxembourg, 1972-77; sect. head aircraft tires sect. The Goodyear Tire and Rubber Co., Akron, Ohio, 1977-84, chief engr., 1984—. Inventor asymmetric tread aircraft tire, 1988. Mem. Stark County Conservation Dept., Canton, Ohio, 1989—, Evang. Luth. Ch. Am. Strategy for Akron Area, 1987-91; treas. Luth. Coun. Greater Akron, 1988-92; elder, fin. sec., mem. coun. St. John's Evang. Luth. Ch., pres. ch. coun., 1991-92. Mem. Soc. Automotive Engrs. (chmn. A-5C subcom. on aircraft tires 1988—), Tire and Rim Assn. (chmn. aircraft tire subcom. 1985-87, 89-90, 92, del. to Internat. Standards Orgn., 1986—), Rubber Mfrs. Assn., Aircraft Tire Engring. Com., Goodyear Antique Auto Club (treas. 1990-93, sec. 1993—), Farm Bur. Avocations: antique/unique autos, farming, softball. Home: 3611 Moonglo St NW Uniontown OH 44685-8027 Office: Goodyear Tire & Rubber Co D # 461B PO Box 3531 Akron OH 44309-3531

DWIGHT, HERBERT M., JR., optical engineer, manufacturing executive. Chmn., ceo., pres. Optical Coating Laboratory, Inc., Santa Rosa, Calif. Recipient Arthur L. Schawlow awd., Laser Inst. Am., 1990. Office: Optical Coating Lab Inc 2789 Northpoint Pky Santa Rosa CA 95407-7397*

DWORKIN, LARRY UDELL, electrical engineer, research director, consultant; b. Bklyn., Nov. 23, 1936; s. Joseph Henry and Mollie Susan (Hodas) D.; m. Gail Ann Cotton, Apr. 2, 1960; children: Stuart, Sharie Lee. BSEE with high distinction, Worcester Poly. Inst., 1958; MSEE, Rensselaer Poly. Inst., 1960; PhD in Elec. Engring., Bklyn. Poly. Inst., 1969. Teaching assoc. Rensselaer Poly. Inst., Troy, N.Y., 1958-60; electronic engineer IBM, Poughkeepsie, N.Y., 1960-61; assoc. lab. dir. Communication Electronics Command, Ft. Monmouth, N.J., 1962-88; pres. Larry Dworkin Cons., Holmdel, N.J., 1988-92; dir. ctr. of telematics Monmouth Coll., West Long Branch, N.J., 1992—; cons. Contel Corp., Fairfax, Va., 1988-90, Telos Corp., Shrewsbury, N.J., 1990-92, Mitre Corp., Eatontown, N.J., 1992—. Co-author: Atmospheric Channel Characterization, 1974; contrb. 43 articles to profl. jours. 1st lt. U.S. Army, 1961-62. Fellow IEEE (chmn. comm. chpt. 1991-93, spl. region award 1988); mem. Optical Soc. Am. (paper rev. com. 1980-82), Armed Forces Communication Electronics Assn., Sigma Xi, Tau Beta Pi, Eta Kappa Nu. Achievements include 3 patents in areas of optical communications, microwave radio, and communication system testing; developed fiber optic cables for the military. Home: 12 Stilwell Dr Holmdel NJ 07733 Office: Monmouth Coll West Long Branch NJ 07764

DWORKIN, MARTIN, microbiologist, educator; b. N.Y.C., Dec. 3, 1927; s. Hyman Bernard and Pauline (Herstein) D.; m. Nomi Rees Buda, Feb. 2, 1957; children—Jessica Sarah, Hanna Beth. B.A., Ind. U., 1951; Ph.D. (NSF predoctoral fellow), U. Tex., Austin, 1955. NIH research fellow U. Calif., Berkeley, 1955-57; vis. prof. U. Calif., summers 1958-60; asst. prof. microbiology Ind. U. Med. Sch., 1957-61, assoc. prof., 1961-62; assoc. prof. U. Minn., 1962-69, prof., 1969—; vis. prof. U. Wash., summer 1965, Stanford U., 1978-79; vis. scholar Oxford (Eng.) U., 1970-71; Found. for Microbiology lectr., 1973-74, 76-77, 81-82. Author: Developmental Biology of the Bacteria, 1985, Microbial Cell-Cell Interactions, 1991; contbr. numerous articles, revs. to profl. publs.; mem. editorial bd. Jour. Bacteriology, 1967-74, 86-88, Ann. Revs. Microbiology, 1975-79, The Prokaryotes, 2d edit. Alt. del. Democratic Nat. Conv., 1968; mem. Minn. Dem. Farm Labor Central Com., 1969-70. Served with U.S. Army, 1946-48. Recipient Career Devel. award NIH, 1963-68, 68-73; John Simon Guggenheim fellow, 1978-79. Mem. Am. Soc. Microbiology (vice chmn. div. gen. microbiology 1977-78, chmn. 1978-79, div. councillor 1980-82), Soc. Gen. Microbiology (Eng.). Home: 2123 Hoyt Ave W Saint Paul MN 55108-1314 Office: U Minn Microbiology Dept Minneapolis MN 55455

DWORZANSKI, JACEK PAWEL, analytical biochemist, researcher; b. Wloclawek, Poland, Jan. 5, 1952; came to U.S., 1987; s. Augustyn Franciszek and Cecylia (Piasecka) D.; m. Maria Teresa Siedlecka, June 12, 1987. MS, Silesian Med. Acad., Katowice, Poland, 1976; PhD, Jagiellonian U., Cracow, Poland, 1981. Instr., teaching fellow Silesian Med. Acad., Katowice, 1976-82, asst. prof., 1982-87; postdoctoral fellow U. Utah, Salt Lake City, 1987, Johns Hopkins U., Balt., 1987-88; rsch. assoc. U. Utah, Salt Lake City, 1989—. Co-author: (with H.L.C. Meuzelaar) Modern Techniques for Rapid Microbiological Analysis, 1991; contbr. articles to profl. jours. Recipient Internat. Fogarty fellowship NIH, 1987. Mem. AAAS, Am. Soc. Mass Spectrometry, Am. Chem. Soc., Polish Inst. Arts and Scis. Am., N.Y. Acad. Scis. Roman Catholic. Achievements include development of analytical methods based on pyrolytic derivatization for chromatographic and/or mass spectrometric analysis of complex organic and bioorganic materials, including whole bacterial cells, suitable for characterization and identification of polymers (melanins, sporopollenins), fossil resins, lipids as well as rapid detection and identification of microorganisms. Office: Ctr Micro Analysis Reaction Chem EMRL Bldg 61 Rm 214 Salt Lake City UT 84112

DWYER, DENNIS D., information technology executive; b. Oak Park, Ill., July 19, 1943; s. John J. and Jessie M. Dwyer; m. Carolyn R. Schultz, Apr. 29, 1967; children: David, Julianne. Various positions Harris Bank, Chgo., 1967-83, mgr. info. tech. planning, 1983-86, v.p. tech. mgmt., 1986—; resolutions chmn. Cooperating Users of Burroughs Equipment, Detroit, 1978-82; cons. Unisys mainframe computers. Pres. Hunting Ridge Homeowners Assn., 1983-85; mem. Palatine Plan Commn., 1984—, chmn., 1989—. Recipient Tom Grier award for Excellence Unisys Users Group, 1988. Home: 1032 Raven Ln Palatine IL 60067-6649 Office: Harris Bank PO Box 755 Chicago IL 60690-0755

DWYER, FRANCIS GERARD, chemical engineer, researcher; b. Phila., June 13, 1931; s. Francis George and Elizabeth Agnes (Foley) D.; m. Miriam Helen Hutelmyer, Jan. 28, 1961; children: Sharon, Timothy, Sean, Sheila, Colleen. B Chem. Engring., Villanova U., 1953; MSChemE, U. Pa., 1963, PhDChemE, 1966. Jr. technologist Mobil R & D Corp., Paulsboro, N.J., 1953-54; sr. rsch. engr. Mobile R & D Corp, Paulsboro, N.J., 1956-63, engring. assoc., 1969-78, sr. rsch. assoc., 1978-81, mgr. catalytic R & D group, 1981-82, sr. scientist, 1982—; mgr. catalytic R & D sect. Mobil R & D Corp., Paulsboro, 1985-93. Author: Shape Selection Catalysis, 1989; editor: Intrazeolite Chemistry, 1983; also articles; numerous patents in field. With U.S. Army, 1954-56. Recipient Personal Achievement in Chem. Engring. award Chem. Engring. mag., 1990; fellow Mobil Oil Corp., 1963. Mem. Am. Inst. Chem. Engrs. (Achievement award in chem. engring. practice 1990), Am. Chem. Soc., Internat. Zeolite Assn. (coun. 1986-92, treas. 1989-92), Catalysis Soc. N.Am. Roman Catholic. Avocations: gardening, music appreciation. Home: 1128 Talleyrand Rd West Chester PA 19382-7462 Office: Mobil R & D Corp PO Box 480 Paulsboro NJ 08066-0480

DWYER, GERALD PAUL, JR., economics educator, consultant; b. Pittsfield, Mass., July 9, 1947; s. Gerald Paul and Mary Frances (Weir) D.; m. Katherine Marie Lepiane, Jan. 15, 1966; children: Tamara K., Gerald P. III, Angela M., Michael J.L., Terence F. BBA, U. Wash., 1969; MA in Econs., U. Tenn., 1973; PhD in Econs., U. Chgo., 1979. Economist Fed. Res. Bank, St. Louis, 1972-74, vis. scholar, 1987-89; economist Fed. Res. Bank, Chgo., 1976-77; asst. prof. Tex. A&M U., College Station, 1977-81, Emory U., Atlanta, 1981-84; assoc. prof. U. Houston, 1984-89; prof. Clemson (S.C.) U., 1989—; acting head Dept. Economics, Clemson U., 1992—; sr. rsch. assoc. Law and Econ. Ctr. Emory U., Atlanta, 1982-84; vis. scholar Fed. Res. Bank, Atlanta, 1982-84; cons. FTC, Washington, 1983-84, Arthur Bros., Corpus Christi, Tex., 1980-81, Amerigas, Houston, 1985, Western Container Corp., 1987, Metrica, Inc., Bryan, Tex., 1989—; vis. fin. economist Commodity FUtures Trading Commn., Washington, 1990. Contbr. articles to profl. jours. NSF trainee U. Tenn., 1970-72; Weaver fellow Intercollegiate Studies Inst., 1974-75, Earhart Found. fellow 1975-77; rsch. grantee NSF, Earhart Found. Mem. Am. Econ. Assn., Am. Stats. Assn., Econometric Soc., Econ. History Assn., We. Econ. Assn., So. Econ. Assn., Beta Gamma Sigma, Phi Kappa Phi. Avocation: computers.

DWYER, JOHN JAMES, mechanical engineer; b. Jersey City, Mar. 1, 1928; s. John J. and Margaret (Casey) D.; m. Joan Catherine Hyde, June 26, 1954 (div. Jan. 1984); children: William J., Kathleen M., Barbara A.; m. JoAnna Mary Kuta, Feb. 4, 1989. BS, N.J. Inst. Tech., 1957; MBA, Lehigh U., 1972. Registered profl. engr., Pa., Tex. Machinery engr. Air Products and Chems., Inc., Allentown, Pa., 1957-63, mgr. machinery engring., 1963-83; cons. Houston, 1983—. Sgt. U.S. Army, 1950-52, Korea. Mem. ASME (mem. performance test code for centrifugal compressors com. 1975—), NSPE, Tex. Soc. Profl. Engrs. Roman Catholic. Home and Office: 8346 Silvan Wind Ln Houston TX 77040-1412

DYBVIG, DOUGLAS HOWARD, manufacturing executive, researcher; b. Bemidji, Minn., Feb. 14, 1935; s. Alfred Otto and Eva Ardis (Coleman) D.; m. Helen Corinne Dybvig, Aug. 24, 1957; children: Robert, Edith, Margaret, Kathryn, Andrew. BS, St. Olaf Coll., 1957; PhD, U. Ill., 1961. Sr. rsch. chemist 3M, St. Paul, 1960-64, lab. mgr., 1965-72, dir. rsch., 1972-84, 89—; assoc. prof. St. Olaf Coll., Northfield, Minn., 1964-65; mng. dir. Minn. 3M Ltd., Harlow, Eng., 1984-89. Recipient Bausch & Lomb award, 1953. Mem. Am. Chem. Soc. (contbr. jour. 1962-66), Soc. Photo Sci. and Engring. Lutheran. Achievements include co-invention of world's first color copier.

DYCK, GEORGE, psychiatry educator; b. Hague, Sask., Can., July 25, 1937; came to U.S., 1965; s. John and Mary (Janzen) D.; m. Edna Margaret Krueger, June 27, 1959; children: Brian Edward, Janine Louise, Stanley George, Jonathan Jay. Student, U. Sask., 1955-56; B. Christian Edn., Can. Mennonite Bible Coll., 1959; M.D., U. Man., 1964; postgrad., Menninger Sch. Psychiatry, 1965-68. Diplomate Am. Bd. Psychiatry and Neurology (added qualifications in geriatric psychiatry). Fellow community psychiatry Prairie View Mental Health Center, Newton, Kans., 1968-70; clin. dir. tri-county services Prairie View Mental Health Center, 1970-73; prof. U. Kans., Wichita, 1973—; chmn. dept. psychiatry U. Kans., 1973-80; med. dir. Prairie View, Inc., 1980-89; cons. Shenyang Psychiat. Hosp., People's Republic of China, 1990, Palestinian Mental Health Program, West Bank, 1990. Bd. dirs. Mennonite Mut. Aid, Goshen, Ind., 1973-85, Chmn., 1982-85; bd. dirs. Mid-Kans. Community Action Program, 1970-73, Wichita Council Drug Abuse, 1974-76. Fellow Am. Psychiat. Assn. (pres. Kans. chpt. 1982-84, dep. rep. 1984-86, rep. 1986—, cert. in adminstrv. psychiatry 1984); mem. AMA, Royal Soc. Physicians and Surgeons Can. (diplomate, cert.), Kans. Med. Soc., Kans.-Paraguay Ptnrs. (treas. 1986-89). Mennonite. Home: 1505 Hillcrest Rd Newton KS 67114-1340 Office: U Kans Sch Medicine-Wichita Dept Psychiatry 1010 N Kansas Wichita KS 67214-9999

DYE, JAMES LOUIS, chemistry educator; b. Soudan, Minn., July 18, 1927; s. Ray Ashley and Hildur Ameda (Limstrom) D.; m. Angeline Rosalie Medure, June 10, 1948; children: Roberta Rae, Thomas Anthony, Brenda Lee. AA, Virginia (Minn.) Jr. Coll., 1948; BA, Gustavus Adolphus Coll., 1949; PhD, Iowa State U., 1953; DSc (hon.), No. Mich. U., 1992. Rsch. assoc. Iowa State U., Ames, 1953; asst. prof. chemistry Mich. State U., East Lansing, 1953-60, assoc. prof., 1960-63, prof., 1963—, chmn. dept. chemistry, 1986-90; mem. U.S. Nat. Acad. Scis., 1989; vis. scientist Ohio State U., Columbus, 1968-69; cons. AT&T Bell Labs., Murray Hill, N.J., 1982-83. Author: Thermodynamics & Equilibrium, 1978; contbr. over 175 articles to profl. jours. Served with U.S. Army, 1945-46. NSF fellow, 1961-62, Guggenheim fellow, 1975-76, 90-91, Fulbright scholar, 1975-76; recipient Disting. Alumni award Gustavus Adolphus Coll., 1969. Fellow AAAS; mem. NAS, Am. Acad. Arts and Scis., Am. Chem. Soc., Am. Inst. Chemists (Chem. Pioneer award 1990), Am. Phys. Soc., Materials Rsch. Soc., Phi Kappa Phi, Sigma Xi (rsch. awards 1968, 87), Golden Key (teaching award 1986). Lutheran. Avocations: fishing, golf. Home: 2698 Roseland Ave East Lansing MI 48823-3847 Office: Mich State Univ Dept of Chemistry East Lansing MI 48824

DYE, ROBERT FULTON, chemical engineer; b. Gloster, Miss., Oct. 18, 1920; s. Curtis Marion and Ethel Mae (Bomar) D.; m. Rebecca Jane Gaston, June 17, 1947; children: Rebecca Jane, Margaret Gladys, Robin Elaine. BS in Chem. Engring., Miss. State U., 1943; MS in Chem. Engring., Ga. Inst. Tech., 1951, PhD in Chem. Engring., 1953. Registered profl. engr., La., Miss., Okla., Tex. Rsch. engr. Monsanto Co., Anniston, Ala., 1946-49; process engr. Phillips Petroleum Co., Bartlesville, Okla., 1953-62; dir. Miss. Indsl. and Tech. Rsch. Com., Jackson, Miss., 1962-65; sr. project engr. Shell Devel. Co., Emeryville, Calif., 1965-74; staff engr. Shell Oil Co., Houston, 1974-91; pres. Dye Engring. & Tech. Co., Sugar Land, Tex., 1991—; mem. So. Interstate Nuclear Bd., Atlanta, 1962-65; bd. mem. San Francisco Bay Area Engrs. Coun., 1970-72; cons. Dye Engring. & Tech. Co., Sugar Land, 1991—. Maj. U.S. Army, 1943-64. Fellow AAAS, Am. Inst. Chem. Engrs.; mem. Am. Chem. Soc., Internat. Union Pure and Applied Chemistry, Kiwanis (bd. mem. 1965—, pres. 1970-71, 75-76). Achievements include patents in field; research on ethylene oxide technology. Home: 3011 Fairway Dr Sugar Land TX 77478

DYKES, FRED WILLIAM, retired nuclear scientist; b. Pocatello, Idaho, Jan. 20, 1928; s. Fred Elmer and Lina Estelle (Dutton) D.; m. Peggy True, Oct. 8, 1950; children: Mark William, James Fred. BS in Chemistry, Idaho State U., 1952. Jr. chemist Am. Cyanamid Co., Idaho Falls, Idaho, 1952-53; chemist Phillips Petroleum Co., Idaho Falls, 1953-66, Idaho Nuclear Corp., Idaho Falls, 1966-71; engr. specialist Allied Chem. Corp., Idaho Falls, 1971-79; sr. scientist Exxon Nuclear Idaho Co., Idaho Falls, 1979-84; fellow scientist Westinghouse Idaho Nuclear Co., Idaho Falls, 1984-93; ret., 1993. Co-author: Progress in Nuclear Energy, Volume 10, 1972; contbr. U.S. Govt. reports and articles to profl. jours. Leader Boy Scouts Am., Pocatello, 1955-78. Sgt. U.S. Army, 1946-48, Korea. Mem. Idaho Hist. Soc., Oreg.-Calif. Trail Assn. (Idaho bd. dirs. 1990—). Achievements include development of layout and operating philosophy for a hot cell facility vital to U.S. Department of Energy's chemical analysis requirements at the Idaho Chemical Processing Plant, Idaho National Engineering Laboratory. Home: 964 Wayne Ave Pocatello ID 83201-3612

DYKMAN, ROSCOE ARNOLD, psychologist, educator; b. Pocatello, Idaho, Mar. 20, 1920; s. Henry Arnold and Mable (Balderston) D.; m. Virginia June Johnston, Sept. 17, 1941 (div. June 1975); children: Richard Arnold, Thomas Ross, Susan Lane, Laura Jane; m. Kathryn Donita Bowman, May 10, 1980. BS, George Williams Coll., 1946; PhD, U. Chgo., 1949. Instr. Ill. Inst. Tech., Chgo., 1948-50; postdoctoral fellow Johns Hopkins Med. Sch., Balt., 1950-52, instr., 1952-53; asst. dir. studies Assn. Am. Med. Colls., Chgo., 1953-55; assoc. prof. U. Ark. Med for Med. Scis., Little Rock, 1955-58, prof., dir. psychiat. rsch. lab., 1958-75, prof., head div. behavior scis., 1975-90; prof., head psychophysiology lab. Ark. Children's Hosp., Little Rock, 1990—; grant reveiwer NIMH, Nat. Inst. for Child Health and Devel., 1990—. Contbr. articles to profl. jours., chpts. to books; cons. editor Jour. Learning Disabilities, 1975—. Elder St. Andrews Ch., Little Rock, 1967-70; mem. Ark. State Planning Commn. for Mental Health, Little Rock, 1980-83. Recipient Rsch. award NIMH, 1961-72, Disting. Scientist award Pavlovian Soc. Am., 1970, Pioneer award Learning Disabilities Soc., 1991. Fellow APA; mem. Am. Psychol. Soc., Psychophysiology Soc. (bd. dirs., cons. editor). Democrat. Presbyterian. Achievements include research in acetonomic conditioning (heart rate and blood pressure), attentional deficit disorders and learning disabilities. Home: 3360 Woodard Rd Benton AR 72015 Office: Ark Children's Hosp Dept Pediatrics 800 Marshall St Little Rock AR 72202-3591

DYMICKY, MICHAEL, retired chemist; b. Synewidsko Wyzhne, Urkraine, Oct. 1, 1920; came to U.S., 1949; s. Mykola and Eva (Andrushkiw) D.; m. Olga Zhmurko, Jan. 22, 1943; children: Lida Dymicky Pakula, Oksana Dymicky Matla. Degree in chem. tech., Chem. Tech. Polytechnic, Lwiw, 1943; BS, U. Innsbruck, Austria, 1947, Doctorandum, 1949; PhD, Temple U., 1960. Chemist Am. Sugar Refining Co., Phila., 1949-52; rsch. chemist U. Pa. Med. Sch., Phila., 1952-53, Wyeth Inst. Med. Rsch., Radnor, Pa., 1953-56, 59-62; rsch. chemist Agr. Rsch. Svc. USDA, Phila., 1956-59, 66-89; assoc. prof. Kutztown (Pa.) U., 1962-65; gen. sec. Internat. Student Svcs., Innsbruck, 1947-49. Contbr. articles to profl. jours.; patentee in amino acid derivatives and anticlostridial agts. Recipient Citation of Merit DAV, 1970, Chem. Abstract Svc., 1971, USDA, 1989; Investor's award U.S. Dept. Commerce, 1987. Mem. Am. Chem. Soc. (student adviser 1963-65), Shevchenko Sci. Soc. (coun. 1968—). Avocations: swimming, skiing, tennis, volleyball, making perfumes. Home: 9653 Dungan Rd Philadelphia PA 19115-3221

DYNKIN, EUGENE B., mathematics educator; b. Leningrad, USSR, May 11, 1924; came to U.S., 1977, naturalized, 1983; s. Boris and Rebecca (Sheindlin) D.; m. Irene Pakshver, June 2, 1959; 1 child, Olga. B.A., Moscow U., 1945, Ph.D., 1948, D.Sc., 1951. Asst. prof. Moscow U., 1948-49, assoc. prof., 1949-54, prof., 1954-68; sr. research scholar Central Inst. Math. Econ. Acad. of Sci., Moscow, 1968-76; prof. math. Cornell U., Ithaca, N.Y., 1977—. Author: Theory of Markov Processes, 1960, Mathematical Conversations, 1963, Markov Processes, 1965, Mathematical Problems, 1969, Markov Processes-Theorems and Problems, 1969, Controlled Markov Processes, 1979, Markov Processes and Related Problems of Analysis, 1982; contbr. articles to profl. jours. Fellow AAAS, Inst. Math. Stats.; mem. Nat. Acad. Scis., Am. Math. Soc., Bernoulli Soc. Math. Stats. and Probability. Home: 107 Lake St Ithaca NY 14850-3855 Office: Cornell U Dept Math White Hall Ithaca NY 14853

DYREGROV, MICHAEL See BAKER, JOHN STEVENSON

DYSART, BENJAMIN CLAY, III, environmental management consultant, conservationist, engineer; b. Columbia, Tenn., Feb. 12, 1940; s. Benjamin Clay and Kathryne Virginia (Thompson) D.; m. Nancy Elizabeth McDonald, Dec. 28, 1991. BCE, Vanderbilt U., 1961, MS in San. Engring., 1964; PhD in Civil Engring., Ga. Inst. Tech., 1969. Staff engr. Union Carbide Corp., 1961-62, 64-65; from asst. prof. to prof. Clemson U., 1968-90, McQueen Quattlebaum prof. engring., 1982-83, dir. S.C. Water Resources Rsch. Inst., 1968-75, dir. water resources engring. grad. program, 1972-75, adj. prof., 1990-93; facility devel. mgr. Chem. Waste Mgmt., Inc., Marietta, Ga., 1990-91; regional facility devel. mgr. Chem. Waste Mgmt., Inc., Memphis, 1991; dir. project planning and integration Waste Mgmt. Inc., Washington, 1991-92; pres. Benjamin C. Dysart Environ. Issues Mgmt., Atlanta, 1992—; sci. advisor Office Sec. of Army, Washington, 1975-76; mem. EPA Sci. Adv. Bd., from 1983; sr. fellow The Conservation Found., 1985—; mem. adv. coun. Electric Power Rsch. Inst., 1989—; mem., chief of engrs. environ. adv. bd. U.S. Army Corps Engrs., 1988-92; mem. Glacier Nat. Park Sci. Coun., Nat. Park Svc., 1988-91; mem. S.C. Gov.'s Wetlands Forum, 1989-90; sec. appointee Outer Continental Shelf Adv. Bd. and OCS Sci. Com. Dept. Interior, 1979-82; mem. S.C. Environ. Quality Control Adv. Com., 1980-90, chmn., 1980-81; mem. Sci. Panel to Rev. Interagy. Rsch. on Impact of Oil Pollution NOAA, Dept. Commerce, 1980; mem. Nuclear Energy Ctr. Environ. Task Force Dept. Energy-So. States Energy Bd., 1978-81; mem. Nonpoint Source Pollutant Task Force EPA, 1979-80; mem. civil works adv. com. Office Sec. Army-Young Pres.'s Orgn., 1975-76; mem. S.C. Heritage Adv. Bd. S.C. Wildlife and Marine Resources Dept., 1974-76; cons. on environ. protection, pub. participation, corp. environmental leadership programs, water resources, facility siting, energy prodn. and public involvement matters to industry and govt. agys. Editor: (with Marion Clawson) Managing Public Lands in the Public Interest, 1988, Public Interest in the Use of Private Lands, 1989; contbr. articles on math. modeling in water quality and environ. mgmt. and pub. involvement to profl. jours.; author numerous profl. papers, reports. Trustee Rene Dubos Ctr. for Human Environs., 1985—, vice chmn., mem. exec. com., 1988—; bd. visitors Kanuga Episcopal Conf. Ctr., 1988—; trustee S.C. Conservation Edn. Found., 1987—, chmn. bd. trustees, 1989—. Recipient Tribute of Appreciation for Disting. Svc. EPA, 1981, 86, McQueen Quattlebaum Engring. Faculty Achievement award Clemson U., 1982, Order of Palmetto Gov. S.C., 1984; named Hon. Ky. Col., 1976. Mem. ASCE, Trout Unltd. (mem. bd. trustees 1990—), Nat. Wildlife Fedn. (bd. dirs. 1974-90, v.p. 1978-83, pres., chmn. bd. dirs. 1983-85), Am. Geophys. Union, Assn. Environ. Engring. Profs. (bd. dirs. 1978-83, pres., chmn. bd. dirs. 1981-82), Water Environ. Fedn. (bd. dirs. Rsch. Found. 1989-91), S.C. Wildlife Fedn. (bd. dirs. 1969—, pres., chmn. bd. dirs. 1973-74, S.C. Wildlife Conservationist of Yr.), Cosmos Club (Washington), Crescent Club (Memphis), Sigma Xi, Tau Beta Pi, Phi Kappa Phi,. Episcopalian. Office: Environ Issues Mgmt 224 Broadland Ct NW Atlanta GA 30342

DYSON, FREEMAN JOHN, physicist; b. Crowthorne, Eng., Dec. 15, 1923; s. George and Mildred Lucy (Atkey) D.; m. Verena Haefeli-Huber, Aug. 11, 1950 (div. 1958); children—Esther, George; m. Imme Jung, Nov. 21, 1958; children—Dorothy, Emily, Mia, Rebecca. B.A., Cambridge U., 1945. Operations research RAF Bomber Command, 1943-45; fellow Trinity Coll., Cambridge U., Eng., 1946-49; Commonwealth fellow Cornell U., Princeton, 1947-49; prof. physics Cornell U., 1951-53; prof. Inst. Advanced Study, Princeton, 1953—. Author: Disturbing the Universe, 1979, Weapons and Hope, 1984, Origins of Life, 1986, Infinite in all Directions, 1988, From Eros to Gaia, 1992. Fellow Royal Soc. London; mem. Am. Phys. Soc., Nat. Acad. Scis. Home: 105 Battle Road Cir Princeton NJ 08540-4904

DZIEWANOWSKA, ZOFIA ELIZABETH, neuropsychiatrist, researcher, physician; b. Warsaw, Poland, Nov. 17, 1939; came to U.S., 1972; d. Stanislaw Kazimierz Dziewanowski and Zofia Danuta (Mieczkowska) Rudowska; m. Krzysztof A. Kunert, Sept. 1, 1961 (div. 1971); 1 child, Martin. MD, U. Warsaw, 1963; PhD, Polish Acad. Sci., 1970. Asst. prof. of psychiatry U. Warsaw Med. Sch., 1969-71; sr. house officer St. George's Hosp., London, 1971-72; assoc. dir. Merck Sharp & Dohme, Rahway, N.J., 1972-76; vis. assoc. physician Rockefeller U. Hosp., N.Y.C., 1975-76; adj. asst. prof. of psychiatry Cornell U. Med. Ctr., N.Y.C., 1978—; v.p., dir. internat. therapeutic rsch. Hoffmann-La Roche, Inc., Nutley, N.J., 1976—. Contbr. articles to profl. publs. Bd. dirs. Royal Soc. Medicine Found. Recipient TWIN Honoree award for Outstanding Women in Mgmt., Ridgewood (N.J.) YWCA, 1984. Mem. AMA, Pharm. Mfrs. Assn. (sec. steering com. med. sect. 1984—), Am. Soc. Clin. Pharmacology and Therapeutics, Alumni Coun. Cornell Med. Ctr. Roman Catholic. Achievements include research on the role of the nervous system in the regulation of respiratory functions, on therapeutic uses of many new drugs including interferon efficacy in cancer and AIDS and drugs useful in cardiovascular, neuropsychiatric, infectious diseases and others, impact of different cultures on medical practices.

EADS, BILLY GENE, electrical engineer; b. Hilham, Tenn., June 20, 1940; s. Walter Herbert and Ruby Clyde (Cary) E.; m. Betty Gail Jenkins, Aug. 3, 1962; children: Wayne Alan, Scott Patrick, Christopher Brian. BSEE, Tenn. Tech. U., 1962; MSEE, U. Va., 1971. Elec. engr. Nat. Aerocautics & Space Adminstrn., Hampton, Va., 1964-67; devel. engr. Oak Ridge (Tenn.) Nat. Lab., 1967-77, program mgr., 1977-80, group leader, 1980-84, dir. instrumentation & controls div., 1984—; sci. and tech. advisor Tenn. Dept. Econ. and Community Devel., Nashville, 1993—; vice chmn. Indsl. Cluster for the Tenn. State U. Coll. Engring., Nashville, 1991-92; mem. Ctr. for Elec. Power adv. com. Tenn. Tech. U., Cookeville, 1988-93. With USAR, 1962-64. Mem. IEEE, Instrument Soc. Am., Tau Beta Pi. Office: Dept Econ and Community 320 Sixth Ave 6th Fl Nashville TN 37243-0405

EAGAR, THOMAS WADDY, metallurgist, educator; b. Chattanooga, Jan. 9, 1950; s. Harry Douglas Sr. and Emily Clarkson (Thompson) E.; m. Pamela Dozier Garrett, Apr. 17, 1973; children: Matthew, Rebekah, Linda, Karen, James, Anna, Thomas. BS in Metallurgy, MIT, 1972, ScD in Metallurgy, 1975. Registered profl. engr., Mass. Rsch. engr. Bethlehem (Pa.) Steel Corp., 1974-76; asst. prof., then assoc. prof. MIT, Cambridge, 1976-87, R.P. Simmons prof. metallurgy, 1987—; liaison scientist U.S. Office Naval Rsch., Tokyo, 1984-85; dir. Materials Processing Ctr., Cambridge, 1991—; adv. bd. Edison Welding Inst., Columbus, Ohio, 1989—; com. mem. Nat. Rsch. Coun., Washington, 1990—. Contbr. to tech. publs. Houdremont lectr. Internmat. Inst. Welding, Paris, 1990. Fellow Am. Soc. Metals (Howe medal 1992); mem. AIME (Mathewson Gold medal 1987), Am. Welding Soc. (Sparagen award 1991, Adams lectr. 1992, Jennings medal 1983, 91), Materials Rsch. Soc. Achievements include 10 patents. Office: MIT Rm 4-136 77 Massachusetts Ave Cambridge MA 02139

EAGER, ROBERT DONALD, computer science educator; b. Brighton, Sussex, Eng., Dec. 11, 1950; s. Donald George and Genevieve Elizabeth (Barbour) E.; m. Christina Jane Middleton, Nov. 24, 1979. BSc in Electronics, U. Kent, Canterbury, Eng., 1973; MSc in Computer Sci., U. Essex, 1974. Chartered European engr. Programmer U. Kent, Canterbury, 1976-78, lectr. computer sci., 1978-87, sr. lectr. computer sci., 1987—; Master Darwin Coll., 1992—. Author: Introduction to PC-DOS, 1985; co-author: Fundamentals of Operating Systems, 5th edit., 1993. Mem. Assn. for Computing Machinery, Brit. Computer Soc. Anglican. Avocations: theatre, music, reading. Office: University of Kent, Canterbury CT2 7NF, England

EAGLES, DAVID M., physicist; b. Edgware, Middlesex, England, May 27, 1935; came to U.S., 1992; BA, Cambridge U., 1956, MA, 1960; PhD, London U., 1965. Various rsch. positions, 1957-67; rsch. assoc. NASA Elec. Rsch. Ctr., Cambridge, Mass., 1967-69; prin. rsch. scientist Commonwealth scientific and Indsl. Rsch. Orgn., New South Wales, Australia, 1970-90; sr. researcher Found. Rsch. and Tech., Heraklion, Greece, 1990-91; rsch. scientist Czechoslovak Acad. Scis., Prague, 1991-92; sr. rsch. assoc. NASA Marshall Space Flight Ctr., Huntsville, Ala., 1992-93. Contbr. articles to profl. jours. Sr. Rsch. fellow Post Office Rsch. Sta., Dollis Hill, London, 1964-67; State scholar, England, 1952, St. John's Coll. scholar, 1955. Mem. Am. Phys. Soc. Achievements include research on evidence for superconductivity resembling that of a charged Bose gas and for transitions between polaron types in Zr-doped SrTi03. Office: NASA Marshall Space Flight Ctr Code Es 74 Huntsville AL 35812

EAGLETON, ROBERT DON, physics educator; b. Ladonia, Tex., Aug. 19, 1937; s. Winslow Frank and Bertha Mae (Hidler) E.; m. Barbara Francis Eagleton, Aug. 17, 1963; children: David, Jonathan, Jennifer, Elizabeth. B.S., Abilene Christian Coll., 1959; M.S., Okla. State U., 1962, Ph.D., 1968. Mem. faculty dept. physics Calif. State Poly. U., Pomona, 1968—; prof. physics. Calif. State Poly. U., 1978—. Served with USNR, 1962-65. NDEA fellow, 1959-62. Mem. AAAS, Am. Phys. Soc., Am. Assn. Physics Tchrs., Sigma Xi. Republican. Home: 1572 N Mountain Ave Claremont CA 91711-3439 Office: Calif State Poly U 3801 W Temple Ave Pomona CA 91768-2557

EAGLETON, ROBERT LEE, civil engineer; b. Salem, Ohio, Aug. 27, 1945; s. Jerome and Bertha H. (Pemberton) E.; m. Stephenie A. Hrinko, Nov. 27, 1969; children: Jeffrey M., James M. B of Engring., Youngstown U. Registered profl. engr. Engr. Mahoning Valley Sanitary Dist., Youngstown, 1969-73, asst. engr., 1973-85, water plant mgr. Cleve. Div. Water, 1985-89, chief civil engr., 1989-92, cons. engr., 1992–. Mem. ASCE (treas. Youngstown br., 1977-78, sec. 1978-79, v.p. 1979-80, pres. 1980-81), Am. Water Works Assn. (sec. N.E. dist. Ohio sect., 1974-81), Constrn. Specification Inst. (profl.). Home: 13501 Cormere Ave Cleveland OH 44120 Office: Cleve Div Water 1201 Lakeside Ave Cleveland OH 44114

EARLEY, CHARLES WILLARD, biologist; b. Oil City, Pa., Jan. 5, 1933; s. Willard Merle and Elaine Dorothy (Simon) E.; m. Patsy June Robinson, Jan. 10, 1958 (div. 1984); children: Charles, Ayla, Samuel, Daniel, Eric; m. Christine Arlene Valdovinos, Aug. 23, 1986. BS in Electrochemistry, Eltanin U., Israel, 1950; MS in Sci. Edn., Eltanin U., Ft. Meade, Md., 1955; PhD, Eltanin U., Murfreesboro, Ark., 1959; MSEE, Eltanin U., Port Arthur, Tex., 1963. Asst. city engr. City of Rouseville (Pa.), 1952-57; chem., bacteriol., radiol. warfare officer U.S. Army, 1953-55; chief electrochemist, electronic engr. U.S. Natural Resources, Port Arthur, Tex., 1957-66; chief engr. KPAS Radio, Banning Broadcasting Co., Banning, Calif., 1966-68; sci. editor Radio-TV News Svc. of Beamont, Calif., 1966-73; field archaeologist U.S. Natural Resources, Palo Verde, Calif., 1972-74; dir. rsch. U.S. Natural Resources, Ogden, Utah, 1974-81; technologist Simon Oil & Machine Co., Oil City, Pa., 1946-50; broadcast cons. Arcsine Zero Project, Banning, 1968—; lectr in field; Simon Post doctoral rsch. fellow Eltanin U., Banning, 1969-73. Author: De Magnete, 1977. Mem. IEEE, AAAS, Simon Sci. Fellowship Fund, Gold Prospectors Assn. Am., Lost Dutchmans Mining Assn., Beta Beta Beta, Alpha Gamma Sigma, Sigma Rho Theta. Republican. Ch. of Jesus Christ of Latter Day Saints. Achievements include development of first practical antigravity analog and diamagnetizer; discovery of Earley's Bicuñer Petroglyphs; application of heat conductance of diamonds in diamond detectors and semiconductor heat sinks; reported Cnidaria (jellyfish) in Tionesta Dam backwaters. Home: Earleys Wash Palo Verde CA 92266 Office: US Natural Resources PO Box 9873 Ogden UT 84409-0873

EARLOUGHER, ROBERT CHARLES, SR., petroleum engineer; b. Kans., May 6, 1914; s. Harry Walter and Annetta (Partridge) E.; m. Jeanne D. Storer, Oct. 6, 1937; children: Robert Charles, Jr., Janet Earlougher Craven, Anne Earlougher O'Connell. Grad., Colo. Sch. Mines, 1936. Registered profl. engr., Calif., Okla., Tex., Kans. Supr. core lab. The Sloan and Zook Co., Bradford, Pa., 1936-38; co-owner, cons. Geologic Standards Co., Tulsa, 1938-45; owner, cons. Earlougher Engring., Tulsa, 1945-73; chmn., cons. Godsey-Earlougher, Inc., Tulsa, 1973-76, Petroleum Cons. div. Williams Bros. Engring. Co., Tulsa, 1976-88, Reactivated Earlougher Engring., Inc., Tulsa, 1988. Patentee in field. Mem. AIME (hon., Anthony F. Lucas Gold medal 1980), Am. Petroleum Inst. (chmn. mid-continent dist. 1961-62, citation for service 1964), Ind. Petroleum Assn. Am. (bd. dirs. 9 yrs.), Interstate Oil Compact Commn. (oil recovery com. 1947—), Soc. Petroleum Engrs. (disting. svc. award 1973, disting. mem. award 1983, hon. mem. 1985, John Franklin Carll award 1990, enhanced oil recovery pioneer 1992), Soc. Petroleum Evaluation Engrs. (hon. life award 1993), Tulsa Club, Southern Hills Country Club (Tulsa), Masons, Tau Beta Pi. Republican. Episcopalian. Home: 2135 E 48th Pl Tulsa OK 74105-8764 Office: 2424 E 21st St Ste 440 Tulsa OK 74114-1741

EARLY, MARVIN MILFORD, JR., computer scientist; b. Kansas City, Kans., June 11, 1945; s. Marvin Milford and June Margaret (Brown) E.; m. Cheri Lee Hager, July 3, 1991; stepchildren: Jeffrey Laube, Janelle Laube; children by previous marriage: Debora, Darryl, Diana. BS with honors in Computer Sci., U. Kans., 1981, MS in Computer Sci., 1985. Avionics technician Trans World Airlines, Kansas City, Mo., 1965-81; software engr. Gen. Dynamics, Ft. Worth, 1981-82; staff mem. BDM Corp., Leavenworth, Kans., 1982-85; software engr. King Radio, Olathe, Kans., 1985; software engring. specialist Ford Aerospace, Leavenworth, Kans., 1985-86; sr. software engr. Bendix/King Radio, Olathe, 1986-88; software engring. specialist Wilcox Electric, Inc., Kansas City, 1988-93; cons., 1993—. Mem. ACM, Upsilon Pi Epsilon. Democrat. Roman Catholic. Achievements include devel. of real-time software design technique, real-time embedded scheduler designs for high order langs., network performance evaluation methods; derived complete equation set for first order low pass digital filtering software. Home and Office: 1817 E 152 Circle Olathe KS 66062

EARNHARDT, DANIEL EDWIN, automotive engineer; b. Charlotte, N.C., Nov. 25, 1950; s. Edwin Lee and Dolores (Roman) E.; m. Rita Faye Manning, Dec. 16, 1973; children: Matthew Blake, Holly Elizabeth. BS in Physics, East Carolina U., 1974. Sci. tchr. Rosewood High Sch., Goldsboro, N.C., 1974-77; devel. engr. A.C. Rochester (N.Y.) div. GMC, 1977-82; staff engr. Siemens Automotive, Newport News, Va., 1982—. Grantee Allied-Signal Corp., 1986, Siemens Automotive, 1991. Republican. Baptist. Achievements include patents for automatic metal-working machine part counter, production air-test of solenoid valves, others pending. Office: Siemens Automotive 615 Bland Blvd Newport News VA 23602

EASLEY, MICHAEL WAYNE, public health professional; b. Bryan, Ohio, Aug. 25, 1947; s. Warren Harding and Jeanne Ruth (Sargeant) E.; m. Carol Ann McCabe (div.); m. Lana Carol Curtis, May 4, 1990; 1 child by previous marriage, Alec Michael. DDS, Ohio State U., 1974; MPH, U. Mich., 1979; BS, U. of State of N.Y., Albany, 1980; DDS, USN, 1976. Dentist U.S. Navy, Pensacola, Fla., 1974-76, USPHS, Washington, 1976-78; dir. div. dental health Ohio Dept. Health, Columbus, 1980-86, Md. Dept. Health and Mental Hygiene, Balt., 1986-89; assoc. prof. Coll. Dentistry U. Detroit, 1989-90; rsch. coord. oral care div. Procter and Gamble Co., Cin., 1990-91, assoc. dir. profl. rels. health care div., 1991-92; commr. of health and environment Middletown (Ohio) City Dept. Health and Environment, 1992—; strategic planning com. Oral Health 2000, Am. Fund Dental Health, Chgo., 1992—. Author: (with R. Lichtenstein) Dental Health Programs for Correctional Institutions, 1979, (with others) Fluoridation: Litigation and Changing Public Policy, 1984, (with others) Abuse of the Scientific Literature, 1985; contbr. articles to profl. publs. Trustee Pub. Dental Svcs. Soc./United Way, Cin., 1990—; mem. Butler County Children's Svcs. Bd., Hamilton, Ohio, 1992—; mem. profl. adv. com. Middletown Regional Hosp., 1992—. Lt. USN, 1974-76, lt. commdr. USPHS, 1976-78. Fellow Am. Coll. Dentists; mem. Ohio Pub. Health Assn. (chair profl. affairs com. 1992—, Pub. Svc. award 1991), Assn. Ohio Health Commrs. (profl. affairs com. 1992—), Am. Assn. Pub. Health Dentistry (pres. 1987-88), Am. Oral Health Inst. (pres. 1985-90), Am. Pub. Health Assn. (sect. chair 1986-87). Democrat. Home: PO Box 62436 Cincinnati OH 45262 Office: City of Middletown Dept Health & Environment 1 City Centre Plz Middletown OH 45042

EAST, DONALD ROBERT, civil engineer; b. Kimberley, South Africa, June 2, 1944; came to U.S., 1985; s. Robert George and Gladys Enid (Macintyre) E.; m. Diana Patricia Ruske, Dec. 21, 1968 (div. Mar. 1993); children: Lisa Ann, Sharon Margaret. BSCE, U. Cape Town, 1969; MSc in Found. Engring., U. Birmingham, England, 1971. Jr. engr. Ninham Shand & Ptnrs., Cape Town, South Africa, 1968-71; mgr. Civilab Ltd., Johannesburg, South Africa, 1972-74; ptnr. Watermeyer, Legge, Piesold & Uhlmann, Johannesburg, 1975-85, Knight Piesold & Co., Denver, 1985—; Contbr. articles to profl. jours. Fellow South Africa Instn. Civil Engrs. (chmn. 1979-85); mem. ASCE, Soc. Mining Engrs. (com. mem. 1988-89). Home: 7902 E Iowa Ave Denver CO 80231 Office: Knight Piesold & Co 1600 Stout St Ste 800 Denver CO 80202-3107

EASTBURG, STEVEN ROGER, naval officer, aeronautical engineer; b. Oakland, Calif., Apr. 30, 1959; s. Paul Herbert and Barbara Jean (Rogers) E.; m. Catherine Rose D'Alessandro, Mar. 14, 1987; 1 child, Gregory David. BSME, U.S. Naval Acad., 1981; MS in Systems Mgmt., U. So. Calif., L.A., 1988; MS in Aero. Engring., Naval Postgrad. Sch., 1990, AeE in Aero. Engring., 1991; diploma, U.S. Naval Test Pilot Sch., 1991. Commd. USN, 1981—, advanced through grades to lt. comdr.; quality assurance officer VS-38, San Diego, 1983-86, operational test dir. Operational Test & Evaluation Force, San Diego, 1986-89; test pilot U.S. Naval Test Pilot Sch., Patuxent River, Md., 1990-91; S-3A/B project officer Force Warfare Aircraft Test Dir., Patuxent River, Md., 1991-92; UAV dep. program mgr. Strike Aircraft Test Directorate, Patuxent River, Md., 1992—. Deacon Presbyn. Ch. Am., Atlanta, 1988. Decorated Navy Commendation medal, White House Fellowship finalist, 1993. Mem. AIAA, Armed Forces Communications and Electronics Assn. (pres. South Md. chpt. 1993, chmn. edn. com. 1992, Meritorious Svc. award 1992, Disting. Young AFCEA of Yr. 1987), Assn. of Unmanned Vehicle Systems, Tau Beta Pi, Pi Tau Sigma. Republican. Home: 41 White Elm Ct California MD 20619 Office: Strike Aircraft Test Dir (SA-UAV) NAWC-AD Patuxent River MD 20670

EASTER, CHARLES HENRY, technical writer; b. Kingsport, Tenn., Aug. 17, 1958. BFA, Bowling Green State U., 1981. Tech. editor Bell Telephone Labs., Naperville, Ill., 1981-84; tech. writer, cons. AT&T End User Orgn., Bedminster, N.J., 1985-88; freelance tech./proposal writer Flemington, N.J., 1988—. Home and Office: 7 Park Ave Flemington NJ 08822

EASTER, STEPHEN SHERMAN, JR., biology educator; b. New Orleans, Feb. 12, 1938; s. Stephen Sherman and Myrtle Olivia (Bekkedahl) E.; m. Janine Eliane Piot, June 4, 1963; children—Michele, Kim. B.S., Yale U., 1960; postgrad., Harvard U., 1961; Ph.D., Johns Hopkins U., 1966. Postdoctoral fellow Cambridge U., Eng., 1967; postdoctoral U. Calif., Berkeley, 1968-69; asst. prof. biology U. Mich., Ann Arbor, 1970-74, assoc. prof., 1974-78, prof., 1978—, assoc. chmn., 1992-93, mem. LSA exec. com., 1993—, dir. neurosci. program, 1984-88. Editor Vision Rsch., 1978-85, Jour. Neurosci., 1989—, Visual Neurosci., 1990-92, Investigative Ophthalmology and Visual Sci., 1992—. Mem. Soc. Neurosci., Assn. Rsch. in Vision and Ophthalmology, Internat. Brain Rsch. Orgn., Soc. for Devel. Biology. Office: U Mich Dept Biology 3113 Natural Sci Bldg Ann Arbor MI 48109-1048

EASTERDAY, BERNARD CARLYLE, veterinary medicine educator; b. Hillsdale, Mich., Sept. 16, 1929; s. Harley B. and Alberta M. Easterday. D.V.M., Mich. State U., 1952; M.S., U. Wis., 1958, Ph.D., 1961. Diplomate Am. Coll. Veterinary Microbiologists. Pvt. practice veterinary medicine Hillsdale, Mich., 1952; veterinarian U.S. Dept. Def., Frederick, Md., 1955-61; assoc. prof., then prof. veterinary sci. U. Wis., Madison, 1961—, dean Sch. Vet. Medicine, 1979—; mem., chmn. com. animal health Nat. Acad. Sci.-NRC, Washington, 1980-83, mem. com. on sci. basis meat and poultry inspection program, 1984-85; mem. tech. adv. com. Binat. Agrl. Research and Devel., Bet-Degan, Israel, 1982-84; mem. expert adv. panel on zoonoses WHO, Geneva, 1978-84; mem. tech. adv. com. on avian influenza USDA, 1983-85; mem. sec. USDA adv. com. on fgn. animal and poultry diseases, 1991-92. Served to 1st lt. V.C. U.S. Army, 1952-54. Recipient Disting. Alumnus award Coll. Vet. Medicine, Mich. State U., 1975; named Wis. Veterinarian of Yr., Wis. Vet. Med. Assn., 1979. Mem. AVMA, Am. Assn. Vet. Med. Colls., Am. Assn. Avian Pathologists. Office: Sch Vet Medicine 2015 Linden Dr W Madison WI 53706-1102

EASTMAN, CAROLYN ANN, microbiology company executive; b. Potsdam, N.Y., Sept. 8, 1946; d. Frank Orvis and Irene (Rheaume) Eastman. BS in Biology, Nazareth Coll., 1968; AAS in Photography, Rochester Inst. Tech., 1976. Technician U. Rochester, N.Y., 1968-69; chemist Castle/Sybron, Rochester, 1969-79; owner, v.p. Sterilization Tech. Svcs., Rush, N.Y., 1979—; owner Fairfield Cosmetics, Rush, 1986—; ptnr. EFC Properties, 1983—; owner Microdispersions, Inc., 1988—, Medisperse L.P., 1988—; owner STS Ouotek Inc., 1991—, STS Particles Inc., 1991—, STS Biopolymers Inc., 1991. Contbr. articles to profl. jours.; patentee in field. Recipient various awards for photography, sculpture and painting. Mem. NOW, Assn. for Advancement of Med. Instrumentation, Sierra Club, Henrietta Art Club. Democrat. Roman Catholic. Avocations: stained glass, painting, sculpture, restoring antiques and old houses. Home: 6 Genesee St Scottsville NY 14546-1310 Office: 7500 W Henrietta Rd Rush NY 14543-9749

EASTMAN, G. YALE, aeronautical engineer; b. Summit, N.J., Sept. 9, 1928; s. Gardner Pettee and Marjorie Parks (Bell) E.; m. Jane Elizabeth Nickum, Jan. 9, 1954; children: Peter Yale, Roger Hawley. BA in Math., Amherst Coll., 1950. Engr. RCA, Harrison, Lancaster, N.J., Pa., 1950-65; lead engr., engring. mgr. RCA, Lancaster, 1965-75; pres., founder Thermacore, Inc., Lancaster, 1970-90; pres. DTX Corp., Lancaster, 1990—, also bd. dirs.; bd. dirs. Dynatherm Corp., Cockeysville, Md., Thermacore, Inc.; bd. dirs., vice chmn. Cen. Pa. Tech. Coun., Harrisburg. Contbr. articles to profl. jours. Bd. dirs., 1st v.p. Community Concert Assn., Lancaster, 1983-90. Fellow (assoc.) AIAA; mem. ASHRAE, Hamilton Club. Achievements include 15 patents in heat transfer field; developer heat pipes, thermionic energy converters, gas lasers, power-producing electrolytic cells. Office: DTX Corp 780 Eden Rd Lancaster PA 17601

EASTMAN, ROBERT EUGENE, electronic engineer, consultant; b. Lincoln, Neb., Sept. 14, 1929; s. Arthur Colgan and Lorraine Mary (Zimmer) E.; m. Dorothy G. Mitchell, Sept. 10, 1949 (div. May 1982); children: John, Ann Marie, Kathleen, Douglas, Caroline, David. AB in Physics and Math., U. Neb., 1956. Assoc. engr. Sperry Gyroscope Co., Great Neck, N.Y., 1956-57; rsch. engr. Schlumberger Well Surveying Corp., Ridgefield, Conn., 1957-58; engr. Electro-Mech. Rsch., Inc., Sarasota, Fla., 1958-61; sr. engr. Beckman Instruments, Inc., Fullerton, Calif., 1961-64; chief engr. Epsco, Inc., Westwood, Mass., 1964-70; v.p., dir. engring., co-founder Interface Engring. Inc., Stoughton, Mass., 1970-82; pres. Eastman Industries Inc., Imperial Beach, Calif., 1982-83; sr. mem. tech. staff GTE, Needham Hts., Mass., 1983-92; prin. engr. Dositec, Inc., Framingham, Mass., 1992—. Sgt. USAF, 1948-52. Physics scholar U. Neb., 1953-54, 54-55, 55-56. Mem. Lions Club, Am. Legion (svc. officer 1980-82). Republican. Roman Catholic. Achievements include patents in transition detector; in one-shot latch; in AC sample and hold. Home: 457 Center St Raynham MA 02767 Office: Dositec Inc 139 Newbury St Framingham MA 01701-4535

EASTRIDGE, MICHAEL DWAYNE, clinical psychologist; b. Martinsville, Va., June 14, 1956; s. James Edward Eastridge and Ann Marie (Stone) Beechum; m. Joyce Gayle Helms, Sept. 9, 1978; children: Philip Michael, Abigail Joyce. BA, Averett Coll., 1978; MS, Va. Tech., 1981, PhD, 1983. Diplomate Am. Acad. Pain Mgmt., Am. Bd. Profl. Disability Cons., Am. Bd. Vocat. Neuropsychology. Staff psychologist New River Valley Mental Health Svcs., Pulaski, Va., 1981-82; asst. prof., cons. Fla. Mental Health Inst., U. South Fla., Tampa, 1983-86; staff psychologist Humana Regional Pain Clinic, St. Petersburg, Fla., 1987-88; cons. Ctr. for Anxiety and Depressive Disorders, Largo, Fla., 1989-90, Recovery Bridge, Largo, 1990-91; dir. Fla. Neurobehavioral Inst., St. Petersburg, 1986-89, Biobehavioral Svcs., St. Petersburg, 1987—; pvt. practice psychology St. Petersburg, 1983—; dir. Fla. Neurobehavioral Inst., St. Petersburg, 1986-91; guest lectr. Pineallas County Profl. Guardian Assn., 1989-91, Am. Lung Assn., Pinellas County, 1990-91; vol. free svcs. for poor children 1989-91, free support groups for parents and handicapped children, St. Petersburg, 1990-91. Editor Ember, 1977; author various workshops stress, assertiveness, work efficiency, 1983-91. Mem. APA, Soc. Behavior Medicine, Fla. Psychol. Assn., Biofeedback Cert. Inst. Am., Alpha Chi, Omicron Delta Kappa. Avocations: scuba diving, tennis. Home: 214 S Trask St Tampa FL 33609-2537 Office: 9455 Koger Blvd N Ste 104 Saint Petersburg FL 33702-2431

EASTWOOD, SUSAN, medical scientific editor; b. Glens Falls, N.Y., Jan. 2, 1943; d. John J. and Della Eastwood; m. Raymond A. Berry. BA, U. Colo., 1964. Diplomate Bd. Editors in Life Scis. Rsch. adminstrn. assoc. Depts. Psychol., Psychiat., Stanford (Calif.) U., 1965-68; prin., tchr. Colegio Capitan Correa, Arecibo, P.R., 1968-70; sr. editor dept. lab. medicine U. Calif., San Francisco, 1971-77, prin. analyst sci. publs. dept. neurol. surgery and brain tumor rsch. ctr., 1977—; cons. Medtronic Inc., Mpls., 1987—, March of Dimes Calif. Birth Defects Monitoring Program, Emeryville, Calif., 1988—. Collaborating editor: Current Neurosurgical Practice, 1984-91, Brain tumor biology and therapy, 1984; editor: Brain tumors: A Guide, 1992; author: Guidelines on Research Data and Manuscripts, 1989. Recipient Pres. award Am. Med. Writers Assn., Bethesda, Md., 1989, Chancellors Outstanding Achievement award U. Calif., San Francisco, 1989, Cert. of award Nat. Brain Tumor Found., 1992, Am. Soc. Journalists and Authors, 1992. Mem. European Assn. Sci. Editors, Internat. Fedn. Sci. Editors, Am. Med. Writers Assn., N.Y. Acad. Scis., Coun. of Biology Editors. Office: Neurosurgery Editorial Office 1360 9th Ave Ste 210 San Francisco CA 94122

EATON, ALVIN RALPH, research and development administrator, aeronautical and systems engineer; b. Toledo, Ohio, Mar. 13, 1920; s. Alvin Ralph and Katherine (Hasel) E., A.B. in Physics (Miller scholar), Oberlin Coll., 1941; M.S. in Aeronautical Engring. Calif. Inst. Tech., 1943; m. Kathleen Steiner, Aug. 15, 1942 (div.); children: Eric Lloyd, Alan Ralph; m. 2d Ellen Griffiths Phillips, Oct. 3, 1970. Engr. So. Calif. Co-op. Wind Tunnel, Pasadena, 1944-45; with The Johns Hopkins U. Applied Physics Lab., Silver Spring, Md., 1945-75, Laurel, Md., 1975—, mem. prin. profl. staff, 1950—, supr. aerodynamics, dynamics and guidance analysis groups, 1949-54, program supr. supersonic missile and weapon system programs, 1954-64, supr. missile systems div., 1964-73, faculty evening coll. grad. sch. 1973-75, supr. fleet systems dept. 1973-83, asst. dir. for tactical systems Applied Physics Lab. 1973-79, asst. dir., 1979-86, assoc. dir., 1986-89, sr. fellow, dir. spl. programs, 1989—, mem. Johns Hopkins U. adv. bd. for Applied Physics Lab, 1963, 69-70, 73-89. Chmn. Def. Sci. Bd. Task Force, 1977-78, mem. task forces 1979-83; cons. to under sec. def. for rsch. and engring., 1977-83, chmn. and mem. special NATO and U.S. task forces 1977-92, mem. under sec. def. high energy laser rev. group, 1981-83, mem. under sec. def. durability of electronic countermeasures rev. group, 1983-86; mem. Navy planning and steering adv. Group for Surface Ship Security, 1979-82, chmn. and mem. subgroups 1979-82; cons. to Asst. Sec. of Army for research, devel. and acquisition, 1969-74, 80-86, chmn., Asst. Sec. of Army ind. rev. panel for major Army air def. system, 1980-86; mem. Army Sci. Bd., 1980-86, 89—; chmn. panel on adv. syst. test, 1980-81; dep. chmn. summer studies on sci. and engring. pers. and future devel. goals, 1982-83 , mem. subgroup on ballistic missile def., 1984-86, 89; chmn. atmospheric scis. lab. effectiveness rev., 1985, chmn. panel on electromagnetic/electrothermal gun tech. devel., 1989-92; chmn. subgroup on Army tactical space systems, 1991-92; mem. rsch. and new initiatives issue group, 1991—; mem. ad hoc study group on space systems and airland ops., 1992; mem. summer study on future army missile programs, 1993; chmn., asst. sec. army rsch., devel. and acquisition ind. rev. panel for anti-tactical missile program, 1986—; chmn. high attitude theater missile def. sensor panel Army Strategic Def. Command, 1992-93; dep. chmn., exec. bd. Air Armaments Systems Div. of the Am. Def. Preparedness Assn., 1984-90 (life mem.). Trustee Howard County (Md.) Gen. Hosp., 1977-85, chmn. fin. com., treas. 1979-81, vice-chmn., 1981-83, chmn., 1983-85, chmn. Community Rels. Coun., 1988—; mem. editorial bd. Jour. Def. Rsch., 1988-92. Recipient Meritorious Pub. Svc. award USN, 1957, Disting. Pub. Svc. award USN, 1975. Fellow The Hudson Inst.; mem. Balt. Coun. on Fgn. Affairs, N.Y. Acad. Scis., Phi Beta Kappa, Sigma Xi. Methodist. Clubs: Cosmos (Washington); Rolling Road Golf (Balt.); Country Club of Hilton Head, Long Cove, Oyster Reef Golf (Hilton Head Island, S.C.). Lodge: Rotary. Inventor in field; contbr. articles to profl. jours. Home: 6701 Surrey Ln Clarksville MD 21029 Office: Johns Hopkins Rd Laurel MD 20723-6099

EATON, GARETH RICHARD, chemistry educator, university dean; b. Lockport, N.Y., Nov. 3, 1940; s. Mark Dutcher and Ruth Emma (Ruston) E.; m. Sandra Shaw, Mar. 29, 1969. BA, Harvard U., 1962; PhD, MIT, 1972. Asst. prof. chemistry U. Denver, 1972-76, assoc. prof., 1976-80, prof., 1980—, dean natural scis., 1984-88, vice provost for rsch., 1988-89; organizer annual Internat. Electron-Paramagnetic Resonance Symposium. Author, editor 2 books; mem. editorial bd. 5 jours.; contbr. articles to profl. jours. Served to lt. USN, 1962-67. Mem. AAAS, Am. Chem. Soc., Royal Soc. Chemistry (London), Internat. Soc. Magnetic Resonance, Soc. Applied Spectroscopy, Am. Phys. Soc., Internat. Electron Paramagnetic Resonance Soc. Office: U Denver Denver CO 80208

EATON, GORDON PRYOR, geologist, research director; b. Dayton, Ohio, Mar. 9, 1929; s. Colman and Dorothy (Pryor) E.; m. Virginia Anne Gregory, June 12, 1951; children: Gretchen Maria, Gregory Mathieu. BA, Wesleyan U., 1951; MS (Standard Oil fellow), Calif. Inst. Tech., 1953, PhD, 1957. Instr. geology Wesleyan U., Middletown, Conn., 1955-57, asst. prof., 1957-59; asst. prof. U. Calif.-Riverside, 1959-63, assoc. prof., 1963-67, chmn. dept. geol. sci., 1965-67; with U.S. Geol. Survey, 1963-65, 67-81; dep. chief Office Geochemistry and Geophysics, Washington, 1972-74; project chief geothermal geophysics Office Geochemistry Geophysics, Denver, 1974-76; scientist-in-charge Hawaiian Volcano Obs., 1976-78; assoc. chief geologist Reston, Va., 1978-81; dean Coll. Geoscis. Tex. A&M U., 1981-83, provost, v.p. acad. affairs, 1983-86; pres. Iowa State U., Ames, 1986-90; dir. Lamont-Doherty Earth Obs. Columbia U., Palisades, N.Y., 1990—; mem. Commn. on Internat. Edn., Am. Coun. Edn.; mem. coun. advisors World Food Prize; mem. bd. earth scis. and resources; ocean studies bd., and com. on formation of nat. biol. survey, NRC, also mem. geophysics study com.; bd. dirs. Midwest Resources, Inc., Bankers Trust; mem., chair adv. com. U.S. Army Command and Gen. Staff Coll. Mem. editorial bd.: Jour. Volcanology and Geothermal Research, 1976-78; contbr. articles to profl. jours. Pres., bd. dirs. Iowa 4-H Found., 1986-90. NSF grantee, 1955-59. Fellow Geol. Soc. Am., AAAS; mem. NSF (mem., chair earth scis. adv. com., mem. Alan T. Waterman award com.), Am. Geophys. Union, Hawaii Natural History Assn. (bd. dir. 1976-78). Mailing address: Lamont-Doherty Obs Grounds PO Box 63 Palisades NY 10964 Office: Lamont-Doherty Earth Obs Office Dir Palisades NY 10964

EATON, HARVILL CARLTON, university administrator; b. Nashville, May 16, 1948; s. Robert Caldwell and Margaret Elizabeth (Stewart) E.; m. Lois Jean Acuff, June 28, 1969; children: Christopher Carlton, Mary Elizabeth. BS, Tenn. Tech. U., 1970, MS, 1972; PhD, Vanderbilt U., 1976. Asst. prof. La. State U., Baton Rouge, 1976-78, 1980-81, assoc. prof., 1981-87, assoc. dean. engring., 1986-88, prof., 1987--, assoc. vice chancellor, 1988-90, vice chancellor for rsch., 1991--; asst. prof. Tenn. Tech. U., Cookeville, 1978-80; bd. dirs. Baton Rouge Bank, La. Rsch. Pk. Corp., 1992--; tech. cons., 1976--. Contbr. articles to profl. jours. Bd. dirs. Boys and Girls Club, Baton Rouge, 1991—, La. Arts and Sci. Ctr., Baton Rouge, 1991—, Baton Rouge Urban League, 1992—. Numerous rsch. grants 1976-92. Mem. Am. Soc. for Mechanical Engrs., Am. Ceramic Soc., Sigma Xi, Theta Tau (Hall of Fame 1992). Home: 624 Hillgate Pl Baton Rouge LA 70808 Office: La State U Office Rsch and Econ Devel Baton Rouge LA 70803

EATON, MALACHY MICHAEL, computer systems educator; b. Galway, Ireland, July 10, 1960; s. Alexander Augustin and Kathleen Frances (Brennan) E.; m. Patricia Fitzgerald, Aug. 30, 1991. B in Engring., Nat. U. of Ireland, Galway, 1981; diploma in Applied Computing, U. Limerick, Ireland, 1982, M in Engring. Computer Systems, 1988. Computer specialist European Computer Aided Learning Ltd., Dublin, Ireland, 1982-83; assoc. CACI (Consol. Analysts and Cons. Inc.), Dublin, 1983-84; design engr. Tellabs, Shannon, Ireland, 1984-86; lectr. U. Limerick, 1986—; cons. pvt. practice, Limerick, 1986—. Inventor: (programming lang.) Protran, 1986; contbr. articles to profl. jours. Recipient 1st prize math. category Young Scientists Competition, Aer Lingus, Dublin, 1976. Mem. IEEE, Assn. for Computing Machinery, Internat. Neural Network, Inst. Engrs. Ireland. Avocations: philosophy, golf. Office: U Limerick, Limerick Ireland

EATON, RICHARD GILLETTE, surgeon, educator; b. Forty Fort, Pa., Dec. 3, 1929; s. Walter L. and Ruth (Shaw) E.; B.A., Franklin and Marshall Coll., 1951; M.D., U. Pa., 1955; m. Du Ree Hunter, June 13, 1954; children— Holly, Hillary. Intern, U. Pa. Grad. Hosp., 1956; gen. surg. resident Peter Bent Brigham Hosp., Boston, 1957; orthopedic resident Children's Hosp. Med. Center, Mass. Gen. Hosp. and Peter Bent Brigham Hosp., Boston, 1959-62; hand surgery fellow J.W. Littler, Roosevelt Hosp., N.Y.C., 1962, now attending orthopedic surgery and reconstrn., chief hand surgery service; prof. clin. surgery Columbia Coll. Physicians and Surgeons, N.Y.C. Ruling elder Huguenot Presbyn. Ch., Pelham, N.Y. Served to capt., M.C., U.S. Army, 1957-59. NIH fellow, 1963-64. Diplomate Am. Bd. Orthopedic Surgeons. Mem. Am. Acad. Orthopedic Surgery, Am. Orthopaedic Assn., Am. Soc. Surgery of Hand, A.C.S., Interurban Orthopedic Club, N.Y. Acad. Medicine, J.W. Littler Soc., N.Y. Soc. Surgery of Hand. Author: Joint Injuries of the Hand, 1971; also articles. Home: 640 Ely Ave Pelham NY 10803-2402 Office: Roosevelt Hosp 428 W 59th St New York NY 10019-1105

EATOUGH, CRAIG NORMAN, mechanical engineer; b. Provo, Utah, Sept. 11, 1958; s. Norman LeRoy and Marilyn (Buckner) E.; m. Marie Hayes Price, Jan. 2, 1985; children: Taylor, Sharla. BS, Calif. Poly. State U., 1983; PhD, Brigham Young U., 1991. Rsch. asst. Brigham Young U., Provo, 1987-91, mem. faculty, 1991—; cons. Geneva Steel, Ray, Quinney and Nebeker, Pacific Generation Co. Author: Fundamentals of Coal Combustion, 1992; Encyclopedia of Energy and Environment, 1993; contbr. articles to profl. jours. Cub master Boy Scouts of Am., Provo, 1990. Mem. ASME, Sigma Xi. Achievements include designing advanced controlled profile reactor for purpose of studying fossil fuel combustion. Office: Brigham Young U 270 CB Provo UT 84602

EAVES, ALLEN CHARLES EDWARD, hematologist, medical agency administrator; b. Ottawa, Ont., Can., Feb. 19, 1941; s. Charles Albert and Margaret Vernon (Smith) E.; m. Connie Jean Halperin, July 1, 1975; children—Neil, Rene, David, Sara. B.Sc., Acadia U., Wolfville, N.S., Can., 1962; M.Sc., Dalhousie U., Halifax, N.S., 1964, M.D., 1969; Ph.D., U. Toronto, Ont., Can., 1974. Intern Dalhousie U., Halifax, N.S., Can., 1968-69; resident in internal medicine Sunnybrook Hosp., Toronto, 1974-75, Vancouver Gen. Hosp., 1975-79; dir. Terry Fox Lab., Cancer Control Agy. B.C., Vancouver, Can., 1981—; asst. prof. medicine U. B.C., 1979-83, assoc. prof., 1983-88, head div. hematology, 1985—, prof., 1988—; bd. dirs. B.C. Cancer Found., Vancouver, Can., 1984—. Fellow Royal Coll. Physicians (Can.), ACP. Home: 2705 W 31st Ave, Vancouver, BC Canada V6L 1Z9 Office: Terry Fox Lab Cancer Rsch, 601 W 10th Ave, Vancouver, BC Canada V5Z 1L3

EBARA, RYUICHIRO, metallurgist, researcher; b. Nagasaki, Japan, Apr. 11, 1942; s. Chuichiro Hisamoto and Masako Ebara; m. Sumiko Kusumoto, Mar. 26, 1972; children: Kentaro, Kumiko. B in Engring., Kyusyu Inst. Tech., Kitakyushu, Japan, 1966; M in Engring., Nagoya (Japan) U., 1968, D in Engring., 1972. Lectr. Kyusyu Inst. Tech., 1971-72, asst. prof., 1972-74; postdoctoral fellow U. Conn., Storrs, 1972-73; rsch. engr. Hiroshima (Japan) R&D Ctr., 1974-79; sr. rsch. engr. Mitsubishi Heavy Industries, Ltd., Hiroshima, 1979-85, rsch. mgr., 1985-92, dep. chief rsch. engr., 1992—; part-time instr. Tokyo U., 1993, Shizuoka U., Hamamatsu, Japan, 1988, Ehime U., Matsuyama, Japan, 1990, Japan Internat. Coop. Agy., Nagoya, 1992, 93. Editor: Fractography; contbr. articles to jours. and books; patentee in field. Mem. ASTM, Japan Inst. Metals (trustee 1989—), Iron and Steel Inst of Japan (organizer internat. conf. 1987-90, Yamaoka prize 1990), Soc. Materials Sci. of Japan (Tech. prize 1992, Engring. award 1993), Japan Soc. Mech. Engrs., Japan Soc. Corrosion Engring., Japan Welding Soc., Japanese Soc. for Strength and Fracture of Metals (trustee 1988—), Soc. Naval Architects Japan. Home: 3-13-14 Ajinadai, Hatsukaichi Hiroshima 738, Japan Office: Mitsubishi Heavy Industries, Ltd Hiroshima R&D Ctr, 4-6-22, Kan-On-Shin-Machi, Hiroshima 733, Japan

EBELING-KONING, DEREK BRAM, nuclear engineer; b. Aruba, Netherlands, Antillies, Sept. 26, 1955; came to U.S., 1956; s. Jacob Johannes Ebeling-Koning and Nada Rebecca (Davies) Barry; m. Juliann Marie Mainzer, Sept. 29, 1990; 1 child, Natalie. BS in Nuclear Engring., Rensselaer Poly Inst., 1977; MS in Nuclear Engring., MIT, 1979, PhD in Nuclear Engring., 1984. Engr. Westinghouse Nuclear Power, Monroeville, Pa., 1983-85, sr. engr., 1985-91; mgr. licensing and safety analysis ABB Atom Inc., Windsor, Conn., 1991-92, ABB Combustion Engring. BWR Fuel, Windsor, Conn., 1992—. Contbr. articles to profl. jours. Sherman R. Knapp fellow, 1979-80. Mem. ASME, Am. Nuclear Soc. (text book reviewer 1988—), Orton Dyslexia Soc., Sigma Xi.

EBER, LORENZ, civil engineer; b. Bad Oldesloh, Germany, Jan. 30, 1963; came to U.S., 1980; s. Gerhard Clemens and Ursula (Eberhart) E.; m. Paula Susette Holmes, June 9, 1985; children: Anya C., Yvonne R. Student, Columbia U., 1981-83; BSCE, Northwestern U., 1986; postgrad., U. Tunis, Tunisia, 1986-87. Registered profl. engr., Wash. Coop. engr. Harza Engring. Co., Chgo., 1985-86; constrn. engr. Ill. Constructors Inc., St. Charles, 1987; civil engr. Howard Needles Tammen & Bergendoff, Chgo., 1988-90; project design engr. Andersen Bjornstad Kane Jacobs Inc., Seattle, 1990-93; owner, pres. Inventexx Co.; engring. vol. Navajo Indian Tribe, Window Rock, Ariz., 1984; product devel. cons. Ingenieur Büro Eber, Steinburg, Germany, 1983. Pres. Winslow Park Condominiums, Bainbridge Island, Wash., 1990-92; v.p. Northwestern Outing Club, Evanston, Ill., 1984-86. Mem. ASCE (treas. 1984-86), Inst. Transp. Engrs. Achievements include patent for surveying field book cover; invention of mech. cable drum lifter, air cushion hwy. cleaning machine, others in progress. Avocation: flying. Home: 927 1st Ave Grafton WI 53024

EBER, MICHEL, information technology company executive; b. Budapest, Hungary, Aug. 22, 1943; arrived in Belgium, 1965; s. Istvan and Ina (Sultson) E.; m. Anne-Veronique Jonas, Aug. 28, 1968; children: Pascal, Raphael. BS, Univ. of Scis., Budapest, 1965; MBA, U. Brussels, 1972. Cons. Advanced Decision Making, Ugrée, 1972-74; mktg. mgr. Monsanto Europe, Brussels, 1974-78, Pan European Pub., Brussels, 1979; country mgr. Chase Econometrics/Interactive Data, Brussels, 1979-88; regional mgr. Datastream Internat., Brussels, 1988-90; regional mgr. securities markets S.W.I.F.T., Brussels, 1991-92; mgr. risk mgmt. svcs. Chase Manhattan Bank, Brussels, 1991—; cons. Eurodeal-Systems, Brussels, 1987—, Softint, Luxembourg, 1988—, Screen Consulting, Hertogenbosch, 1990—. Author: Micromation, 1975 (Royale Belge award); editor Paninformatic-Ecopress, 1976—, Info. Tech., 1976—. Mem. Am. C. of C., Exec. Club. Avocations: classical music, fine arts, economics, finance, sports. Home: Rue d'Angoussart 125, 1301 Bierges Belgium Office: SWIFT, Adèle 1, 1310 La Hulpe Belgium

EBERHARD, MARC OLIVIER, civil engineering educator; b. Haies-les-Roses, France, Aug. 24, 1962; came to U.S., 1964; s. Philippe Henry and Jacqueline (Dentan) E. BS in Civil Engring./Material Sci., U. Calif., Berkeley, 1984; MS, U. Ill., 1987, PhD, 1989. Jr. civil engr. Calif. Dept. Transp., Sacramento, 1985; rsch. asst. U. Ill., Urbana-Champaign, 1985-89; asst. prof. civil engring. U. Wash., Seattle, 1989—. Contbr. articles to Transp. Rsch. Bd., Jour. Structural Engring., Mil. Engr., others. Asst. coach Spl. Olympics, Seattle, 1992. Kaiser scholar, 1983; U. Ill. fellow, 1985-89; Presdl. Young Investigator awardee NSF, 1991—. Mem. ASCE, Am. Concrete Inst., Earthquake Engring. Rsch. Inst., Tau Beta Pi. Home: 2525 Minor Ave E # 306 Seattle WA 98102 Office: U Wash 233 More Hall FX-10 Seattle WA 98195

EBERHARDT, ALLEN CRAIG, biomedical engineer, mechanical engineer; b. Cin., Aug. 30, 1950; s. Alfred John and Elfriede (Vollmer) E.; m. Mary Drake, June 9, 1973; children: William, Laura. PhD, N.C. State U., 1977. Prof. N.C. State U., Raleigh, 1977-89; pres. Structural Acoustics, Inc., Raleigh, 1981—. Contbr. articles to profl. jours. Grantee USDOT, 1984, Western Electric, 1980, IBM, 1982, Sci. and Tech., 1978. Mem. ASME, Acoustical Soc. Am., Soc. Automotive Engrs. (Teetor award 1980), Sigma Xi. Republican. Episcopalian. Achievements include patent for Pressure controlled left ventricle and Accelerated fatigue apparatus. Home: 624 Marlowe Road Raleigh NC 27609

EBERHART, STEVE A., federal agency administrator, research geneticist; b. S.D., Nov. 11, 1931; m. Laurel Lee Hammond, July 19, 1953; children: Lyndl Schuster, Paul Eberhart, Sally Cooley, Sue May. BS, U. Nebr., 1952, MS, 1958, DSc (hon.), 1988; PhD, N.C. State U., 1961. Rsch. geneticist USDA, Agrl. Rsch. Svc. Iowa State U., Ames, 1961-64, 69-75, U.S. Agy. for Internat. Devel./USDA, Agrl. Rsch. Svc., Kitale, Kenya, 1964-68; assoc. dir. rsch. Funk Seeds Internat., Bloomington, Ill., 1975-78, v.p. rsch., 1978-83, v.p. internat. tech., 1984-87; dir. Nat. Seed Storage Lab. USDA Agrl. Rsch. Svc., Ft. Collins, Colo., 1987—. Contbr. articles to profl. jours. 1st lt. USAF, 1952-56. Recipient Arthur S. Fleming award D.C. Jaycees, 1970. Fellow Am. Soc. Agronomy, Crop Sci. Soc. Am. (pres. 1990), Nat. Coun. Comml. Plant Breeders (1st v.p. 1986); mem. Sigma Xi. Avocations: skiing, hiking. Office: USDA Agrl Rsch Svc Nat Seed Storage Lab 1111 S Mason St Fort Collins CO 80521-4500

EBERL, JAMES JOSEPH, physical chemist, consultant; b. Dunkirk, N.Y., Oct. 7, 1916; s. George M. and Florece S. (Stedler) E.; m. Margaret Elizabeth Schill, June 3, 1941. BA, U. Buffalo, 1938, PhD, 1941; AMP, Harvard U., 1955. Asst. prof. chemistry U. Del., Newark, 1941-42; mgr. rsch. Paper Chem. Div. Hercules Inc., Wilmington, Del., 1942-43; sr. fellow Mellon Inst. Indsl. Rsch., Pitts., 1943-44; dir. spl. prodn. rsch. Johnson and Johnson, New Brunswick, N.J., 1944-48; asst. corp. v.p. Scott Paper Co., Chester, Pa., 1948-70; pres., CEO Newbold Inc., Phila., 1970-72; cons. Moylan, Pa., 1972—. Contbr. articles to profl. jours. Trustee The Franklin Inst., Phila., 1960-72; chmn. bd. dirs. rsch. fund Phila. Gen. Hosp., 1963-76; mem. adv. coun. Pa. Tech. Assistance Program Pa. State U., 1965-71. Mem. Am. Chem. Soc., Am. Inst. Chem. Engrs., N.Y. Acad. Scis., Empire State Paper Rsch. Assn. (pres. 1965-71), Sigma Xi. Achievements include 50 patents for dusting powder for surgical rubber gloves that does not produce abdominal adhesions; for single crystal whisker fibers; for new process for making hard coated plaster of Paris bandages; for polystyrene foam sheet process; for making soybean protein; for bleaching process for groundwood pulp; for chemical sterilization of microbes with epoxides; for hemostatic agents, for synthetic paper pulp fiberous extenders. Home: 7 Rose Hill Rd Moylan PA 19063-4024

EBERLY, RAINA ELAINE, psychologist, educator; b. Chambersburg, Pa., Sept. 17, 1952; d. Charles Alton and Betty Jane (Friese) E.; m. Brian Edward Engdahl, July 9, 1977; 1 child, Rebecca Raina. BS in Psychology, U. Pitts., 1973; PhD in Psychology, U. Minn., 1980. Lic. psychologist. Clin. asst. prof. psychology dept. U. Minn., Mpls., 1981-89, clin. assoc. prof., 1990—; psychologist, tng. dir. VA Med. Ctr., Mpls., 1980—. Contbr. articles to profl. jours. VA grantee, 1989-90, 90-93. Mem. APA, AAAS, Am. Psychol. Soc. Achievements include contbn. to understanding of chronic posttraumatic stress disorder and co-morbidity. Office: VA Med Ctr Psychology Svc 1 Veterans Dr Minneapolis MN 55417

EBERT, JAMES DAVID, research biologist, educator; b. Bentleyville, Pa., Dec. 11, 1921; s. Alva Charles and Anna Frances (Brundege) E.; m. Alma Christine Goodwin, Apr. 19, 1946; children—Frances Diane, David Brian, Rebecca Susan. AB, Washington and Jefferson Coll., 1942, ScD, 1969; PhD, Johns Hopkins U., 1950; ScD (hon.), Yale, 1973, Ind. U., 1975, Duke U., 1992; LL.D. (hon.), Moravian Coll., 1979. Jr. instr. biology Johns Hopkins U., 1946-49, Adam T. Bruce fellow biology, 1949-50, hon. prof. biology, 1956-86, hon. prof. embryology Sch. Medicine, 1956-86; instr. biology Mass. Inst. Tech., 1950-51; asst. prof. zoology Ind. U., 1951-54, assoc. prof., 1954-56, Patten vis. prof., 1963; dir. dept. embryology Carnegie Instn. of Washington, 1956-76, pres., 1978-87, trustee, 1987; prof. biology John Hopkins U., 1987-93, dir. Chesapake Bay Inst., 1987-92; vis. scientist med. dept. Brookhaven Nat. Lab., 1953-54; Philips vis. prof. Haverford Coll., 1961; instr. in charge embryology tng. program Marine Biol. Lab., summers 1962-66, trustee, 1964—, pres., 1970-78, 91—, dir., 1970-75, 77-78; mem. Commn. on Undergrad. Edn. in Biol. Scis., 1963-66; mem. vis. com. for biol. and phys. scis. Western Res. U., 1964-68; Mem. panels on morphogenesis and biology of neoplasia of com. on growth NRC, 1954-56; mem. adv. panel on genetic and developmental biology NSF, 1955-56, mem. divisional com. for biology and medicine, 1962-66, mem. univ. sci. devel. panel, 1965-70, adv. com. for instl. devel., 1970-72; mem. panel basic biol. research in aging Am. Inst. Biol. Sci., 1957-60; mem. panel on cell biology NIH, USPHS, 1958-62, mem. child health and human devel. tng. com., 1963-66; mem. bd. sci. counselors Nat. Cancer Inst., 1967-71, Nat. Inst. Child Health, 1973-77; mem. Com. on Scholarly Communication with People's Republic of China, 1978-81, chmn., 1989—; chmn. Nat. Com. on Sci. Edn. Stds. & Assessment, 1992—; mem. vis. com. to dept. biology Mass. Inst. Tech., 1959-68; mem. vis. com. biology Harvard, 1969-75, Princeton, 1970-76; chmn. bd. sci. overseers Jackson Lab., 1976-80; mem. Inst. Medicine; bd. dirs. Baxter Internat., Free Radical Sciences, Inc., 1993—. Author: (with others) The Chick Embryo in Biological Research, 1952, Molecular Events in Differentiation Related to Specificity of Cell Type, 1955, Aspects of Synthesis and Order in Growth, 1955, Interacting Systems in Development, 2d edit, 1970, Biology, 1973, Mechanisms of Cell Change, 1979; Mem. editorial bd.: (with others) Abstracts of Human Developmental Biology; editor: (with others) Oceanus; Contbr. (with others) articles to profl. jours. Trustee Worcester Found. Lt. USNR, 1942-46. Decorated Purple Heart. Fellow AAAS (v.p. med. scis. 1964), Am. Acad. Arts and Scis., Internat. Soc. Devel. Biology; mem. NAS (chmn. assembly life scis. 1973-77, v.p. 1981-93, chmn. Govt., Univ. and Industry Rsch. Roundtable 1987-92), Am. Philos. Soc., Am. Inst. Biol. Scis. (pres. 1964, President's medal 1972), Am. Soc. Naturalists, Am. Soc. Zoologists (pres. 1970), Soc. Study Growth and Devel. (pres. 1957-58), Phi Beta Kappa, Sigma Xi, Phi Sigma. Home: 4100 N Charles St Baltimore MD 21218-1065 Office: Marine Biol Lab Pres Office Woods Hole MA 02543

EBLING, GLENN RUSSELL, energy conservation executive; b. Reading, Pa., June 26, 1956; s. Kermit R. Ebling and Marilyn F. (Miller) Morrison; m. Ursula B. Wendt, Sept. 1, 1979; children: Stephanie, David, Ahren, Megan, Glenn, Hannah, Caleb. BS in Biochemistry, Pa. State U., 1978. Process chemist Dow Chem. Co., Midland, Mich., 1978-80; pres. Energy Engring. Inc., Lancaster, Pa., 1980—; engring. cons. U.S. Army, USN, USMC, Commonwealth of Pa. and others. Mem. The Rutherford Inst., Washington, 1990—. Govt. energy grants, 1982—. Mem. Assn. of Energy Engrs. Achievements include development, implementation and funding procurement of institutional energy mgmt. programs. Office: EEI-Energy Engring Inc 90 Miller Rd Willow Street PA 17584

EBNETER, STEWART DWIGHT, engineer; b. Ledgewood, N.J., Oct. 10, 1933; s. William and Emily Ann (Burd) E.; m. Evadna Grace Custer, Dec. 28, 1957; children: Stewart D. Jr., Steven D., Scott D. BSEE, Tri-State U., 1959; MBA, Athens State Coll., 1971. Registered profl. engr., Calif. System engr. Boeing Co., Seattle, 1959-61; reliability dept. head Spaco, Inc., Huntsville, Ala., 1961-70, v.p. engring., 1971-73; div. dir. br. chief U.S. Nuclear Regulatory Commn., Atlanta, King of Prussia, Pa., 1973-87; dir. office spl. projects U.S. Nuclear Regulatory Commn., Washington, 1987-88; dir. div. radiation safety U.S. Nuclear Regulatory Commn.; regional adminstr. U.S. Nuclear Regulatory Commn., Atlanta, 1989—. Allocation com. United Way, Huntsville, 1970-73; scout leader Boy Scouts Am., Huntsville, 1970-73. Sgt. USAF, 1953-57. Mem. Am. Soc. for Quality Control (sr.). Home: 107 Whitfield Run Peachtree City GA 30269 Office: US Nuclear Regulatory Commn 101 Marietta St NW Ste 3000 Atlanta GA 30323

EBY, DAVID W., research psychologist; b. Canoga Park, Calif., June 6, 1962; s. Hal H. and Patricia Sue (Parrott) E. MA, U. Calif., Santa Barbara, 1988, PhD, 1991. Grad. rsch. asst. U. Calif., Santa Barbara, 1984-85, teaching asst., 1986-89; rsch. assoc. Anacapa Scis., Inc., Santa Barbara, 1989-90; lectr. Western Wash. U., Bellingham, 1990-91; postdoctoral rsch. assoc. U. Calif., Irvine, 1991-92; lectr. Calif. State U., San Bernardino, 1992-93; asst. rsch. scientist U. Mich. Transp. Rsch. Inst., Ann Arbor, 1993—. Contbr. articles to profl. jours. Social Sci. Humanities Rsch. grantee, U. Calif., 1990. Mem. Assn. for Rsch. in Vision and Ophthalmology, Human Factors Soc., Am. Psychol. Soc., Psi Chi. Achievements include discovery that 3D scenes are perceptually foreshortened when enclosed by a frame; discovered many factors involved in seeing 3D structure in presence of 2D motion information. Home: 25752 Demeter Way Mission Viejo CA 92691 Office: The Univ of Michigan Transportation Rsch Inst 2901 Baxter Rd Ann Arbor MI 48109-2150

EBY, FRANK SHILLING, research scientist; b. Kansas City, Mo., Apr. 6, 1924; s. Frank Shilling and Irene (Trissler) E.; m. Nancy Rea Vinsonhaler, Sept. 2, 1958; children: Elizabeth, Susan, Carl. BS, U. Ill., 1948, MS, 1949, PhD, 1954. Group leader fusion research Lawrence Livermore (Calif.) Nat. Lab., 1954-58, group leader, 1958-66, div. head, 1967-72, sr. scientist, 1973—. Inventor classified mil. weaponry. Served to 1st lt. USAAF, 1942-46, PTO. Recipient Intelligence Community Seal medallion, 1992. Mem. AAAS. Avocations: hiking, camping, gardening, music. Home: 27 Castlewood Dr Pleasanton CA 94566 Office: Lawrence Livermore Nat Lab Dept Spl Projects Livermore CA 94550

ECCLES, SIR JOHN CAREW, physiologist; b. Melbourne, Australia, Jan. 27, 1903; s. William James and Mary (Carew) E.; m. Irene Miller, 1928; 9 children; m. Helena Táboríková, 1968. M.B., B.S., Melbourne U., 1925; M.A., Oxford U., 1929, D.Phil., 1929; LL.D., Melbourne U., 1965; D.Sc. (hon.), U. B.C., 1966, Cambridge U., 1960, U. Tasmania, 1964, Gustavus Adolphus, 1967, Marquette U., 1967, Loyola U., 1969, Yeshiva U., 1969, Charles U., Prague, 1969, Oxford U., 1974, U. Fribourg, 1981, U. Torino, 1983, Georgetown U., 1984, U. Tsukuba, Japan, 1986, U. Basel, 1990, U. Madrid, 1992. Research fellow Exeter Coll., Oxford U., 1927-34; tutorial fellow Magdalen Coll., 1934-37; dir. Kanematsu Meml. Inst. Pathology, Sydney (Australia) Hosp., 1937-43; prof. physiology Otago U., Dunedin, New Zealand, 1944-51, Australian Nat. U., Canberra, 1951-66; mem. AMA/E.R.F. Inst. Biomed. Research, Chgo., 1966-68; disting. prof. SUNY, Buffalo, 1968-75, emeritus, 1975—. Author: (with others) Reflex Activity of the Spinal Cord, 1932; The Neurophysiological Basis of Mind: The Principles of Neurophysiology, 1953; The Physiology of Nerve Cells, 1957; The Physiology of Synapses, 1964; (with Ito, Szentagothai) The Cerebellum as a Neuronal Machine, 1967; The Inhibitory Pathways of the Central Nervous System, 1968; Facing Reality, 1970; The Understanding of the Brain, 1973; (with Karl Popper) The Self and Its Brain, 1977; (with others) Molecular Neurobiology of the Mammalian Brain, 1978, 2d edit., 1987; (with W. Gibson) Sherrington, His Life and Thought, 1979; The Human Mystery, 1979; The Human Psyche, 1980; (with D.N. Robinson) The Wonder of Being Human: Our Brain, Our Mind, 1984; Evolution of the Brain: Creation of the Self, 1989, How the Self Controls its Brain, 1993. Decorated knight bachelor, 1958, Gold and Silver Stars Order of the Rising Sun, 1986, companion Order of Australia, 1990; recipient Royal medal Royal Soc., 1962, Cothenius medal Deutche Akademie der Naturforscher Leopoldina, Nobel prize in physiology and medicine (with A. L. Hodgkin and A. F. Huxley), 1963, Gold medal Charles U., Prague, 1993. Fellow Royal Soc., 1941, Australia Acad. Sci. (pres. 1957-61); mem. Pontifical Acad. Scis., Am. Philos. Soc. (hon.), Accademia Nazionale del Lincei (fgn. hon.), NAS (fgn. assoc.), Am. Physiol. Soc. (fgn. hon.), ACP (hon.), Am. Acad. Arts and Scis. (fgn. hon.), Max Planck Soc. (hon.). Research, numerous pubs. on the physiology of synapses of the nervous system and chemical transmitters, brain-mind problem. Home: Ticino, CH-6646 Contra Switzerland

ECHENIQUE, PEDRO MIGUEL, physicist, educator; b. Isaba, Navarra, Spain; s. Pedro Echenique and Felisa Landiribar. BS, U. Navarre (Spain), 1972; PhD, U. Cambridge (Eng.), 1976, U. Autonoma Barcelona (Spain), 1977. Prof. physics U. Barcelona, 1978-80; min. edn. Basque Govt., Spain, 1980-83, min. edn. and culture, govt. spokesman, 1983-84; prof. physics U. Basque Country, San Sebastian, Spain, 1986—; vis. prof. U. Cambridge, 1984-86. Co-author: Solid State Physics Series, vol. 43, 1990; contbr. articles to profl. jours. Overseas fellow Churchill Coll., Cambridge, 1985. Fellow Am. Phys. Soc.; mem. Spanish Acad. Scis., Royal Acad. Sic. and Arts Barcelona. Avocations: skiing, squash, literature. Home: Monasterio Aberin, 3-50 A, Pamplona Navarra, Spain Office: Dept Fisica, U Del Pais Vasco, 20080 San Sebastian Spain

ECHTENKAMP, STEPHEN FREDERICK, biomedical researcher; b. Fremont, Nebr., Dec. 13, 1951; s. Gilbert H. and Margaret Mary (Rice) E.; m. Mary Catherine Getzfrid, Oct. 14, 1979. BS, Nebr. Wesleyan U., 1974; PhD, U. Nebr., 1980. Grad. teaching asst. dept. physiology U. Nebr. Med. Ctr., Omaha, 1974-79; postdoctoral rsch. assoc. U. Mo., Columbia, 1979-81; asst. prof. dept. med. physiology Tex. A&M U., College Station, 1981-85; asst. prof. dept. physiology Ind. U. Sch. Medicine, Gary, 1985—. Contbr. articles to Am. Jour. Physiology, Circulation Rsch., Jour. Comparative Neurology, Jour. Autonomic Nervous System. Grantee Am. Heart Assn., 1982—, NIH, 1984—. Mem. Am. Physiol. Soc., Am. Heart Assn., Soc. for Neurosci., Internat. Brain Rsch. Orgn. Achievements include research in characterization of neurohormonal control of circulation in the primate. Office: Ind U Sch Medicine 3400 Broadway Gary IN 46408

ECK, KENNETH FRANK, pharmacist; b. Alma, Kans., Feb. 4, 1917; s. Clarence Joseph and Rosa Barbara (Noller) E.; m. Ouida Susie Landon, July 2, 1938 (dec. Sept. 1986); children: Alan Grantland, Mark Warren, Dana Landon; m. Lorraine B. Wooster Rubottom, Apr. 14, 1989. BS in Pharmacy summa cum laude, Southwestern Okla. State U., 1950. Ptnr., mgr. Taylor Drug Store, Healdton, Okla., 1950-51, Taylor-Eck Drug Store, Healdton, 1951-59, Johnson-Eck Drug Store, Healdton, 1959-72; pres. Eck Drug Co., Inc., Healdton, 1972-87, cons., relief pharmacist, 1979—; cons., relief pharmacist Eck Drug and Gift, Waurika, Okla., 1979—; affiliate instr. pharmacy Southwestern Okla. State U., Weatherford, 1970-87, mem. dean's adv. com. Sch. of Pharmacy, early 1980's; bd. dirs. med. adv. bd. Dept. Human Svcs. Okla., Oklahoma City, 1990—. Comdr. VFW Post 6374, Healdton, 1975-78, Am. Legion Post No. 203, Healdton, 1985-91; pres. Healdton C. of C., 1984-85; past mem. governing bd. Healdton Mcpl. Hosp.; mem. Okla. Profl. Responsiblity Tribunal of Okla. State Bar Assn., 1983-88, vice chief master, 1988; past bd. dirs. Carter County Red Cross, Ardmore, Okla., 1960's; bd. dirs. Okla. State Bd. Pharmacy, 1965-72. With USN, 1942-45, PTO. Recipient Achievement award Merck Sharp & Dohme, 1991, Bowl of Hygeia, 1985, outstanding svc. award Okla. Profl. Responsibility Tribunal of Okla. State Bar Assn. Mem. Okla. Pharm. Assn. (pres. 1990-91, plaque 1991), Healdton C. of C. (bd. dirs. 1975—), Southern Okla. Devel. Assn. (coun. mem. area agy on aging adv. bd. 1987—), Nat. Assn. Retail Druggists (profl. affairs com. 1990—). Democrat. Mem. Ch. of Christ (deacon). Avocations: photography, travel, fishing, boating, reading. Home: 1033 E Texas St Healdton OK 73438-3017

ECK, RONALD WARREN, civil engineer, educator; b. Allentown, Pa., May 11, 1949; s. Warren Edgar and Viola (Ruth) E.; m. Deborah Lynn Gregory, Oct. 14, 1989. BSCE, Clemson (S.C.) U., 1971, PhD, 1975. Registered profl. engr., W.Va. Asst. prof. civil engring. W.Va. U., Morgantown, 1975-80; assoc. prof. civil engring. W.Va. U., 1980-84, prof. civil engring., 1984—; cons. in field. Contbr. articles to profl. jours. Chmn. City Traffic Commn., Morgantown, 1989—; mem. Region 3, US DOT, Nat. Def. Exec. Res., 1982—. Recipient Dow Outstanding Young Faculty award Am. Soc. Engring. Edn., 1980, W.Va. U. Found. Outstanding Tchr. award, 1988, others. Mem. Am. Soc. Engring. Edn. (v.p., profl. interest coun. 1987-88), ASCE (pres. W.Va. sect. 1980), Inst. Transp. Engrs. (chmn. dept. 2 1987-90), Transp. Rsch. Bd. (chmn. com. on low volume rds. 1990—), Am. Soc. Safety Engrs., Am. Soc. Photogrammetry and Remote Sensing. Avocations: tennis, backpacking. Home: 609 Valley View Dr Morgantown WV 26505-2412 Office: West Virginia U PO Box 6101 Morgantown WV 26506-6101

ECKE, ROBERT EVERETT, physicist; b. L.A., Feb. 2, 1953; s. Harold E.; m. Cheryl Lynn Smith, Mar. 21, 1976; children: Laurel, Kevin. BS with distinction in Physics, U. Wash., 1975, PhD in Physics, 1982. Postdoctoral fellow U. Wash., Seattle, 1982-83; postdoctoral fellow Los Alamos (N.Mex.) Nat. Lab., 1983-86, staff mem., 1986-92; dep. dir. Ctr. for Nonlinear Studies, Los Alamos, 1990-92; com. mem. Calif. Coordinating Com. on Nonlinear Sci., U. Calif., 1991—. Author: Nonlinear Science: The Next Decade, 1992; contbr. articles to profl. jours. Recipient Fellows prize Los Alamos Nat. Lab., 1991. Mem. Am. Phys. Soc. Achievements include experiments on Chaos and Dynamical Systems-Universal Scaling at Transition to Chaos; Hydrodynamic Experiments on Rotating Thermal Convection; discovery of new states and instabilities. Office: Los Alamos Nat Lab P-10 MS-K764 Los Alamos NM 87545

ECKERS, CHRISTINE, mass spectrometry scientist; b. Newport, Gwent, U.K., Mar. 6, 1957; d. Frank and Grace (Heames) E. B of Pharmacy, U. Wales, Cardiff, 1978, PhD in Mass Spectrometry, 1981. Rsch. chemist Stauffer Chems., Dobbs Ferry, N.Y., 1982-83; applications engr. Hewlett Packard, Wokingham, Eng., 1983-87; sr. scientist Smith Kline Beecham, Welwyn, Eng., 1987—; Mem. organizing com. 10th Internat. Mass. Spectrometry Conf., Swansea, U.K., 1985. Author: (with others) Theraputic Drug Monitoring, 1985; contbr. articles to jours. including Biomedical Mass Spectrometry, Jour. Chromatography, Rapid Comm. in Mass. Spectrometry. Mem. Am. Chem. Soc., Am. Soc. for Mass Spectrometry, British Mass Spectrometry Soc., Soc. for Chem. Industry (com. mem. 1991—). Office: Smith Kline Beecham Pharms, The Frythe, AL6 9AR Welwyn England

ECKERT, JOHN ANDREW, chemist, technical consultant; b. Rochester, N.Y., Apr. 12, 1941; s. Elmer Edward and Jacomina Jeanette (Bommelje) E. BS with honors, Rochester Inst. Tech., 1964; PhD, MIT, 1970. Scientist Kodak, Rochester, N.Y., 1960-64; solar mgr. Exxon, Linden, N.J., 1970-79; sr. assoc. Exxon, Clinton, N.J., 1982-86; rsch. mgr. Enlighten, Stewartsville, N.J., 1986—; edn. dir. Exxon, Florham Park, N.J., 1979-82. Mem. editorial bd. Solar Energy Materials jour., 1979-80; editor Enlighten newsletter, 1991—. Recipient Indsl. Chemistry Affiliate award Am. Chem. Soc., 1960, 61. Mem. Water Gap Gliding Club (sec. 1987—). Achievements include patents for anti-mist jet fuels, semi-conductor photogalvanic effects in ceramic bulletin. Office: Enlighten PO Box 313 Stewartsville NJ 08886

ECKHART, MYRON, JR., marine engineer; b. South Bend, Ind., Mar. 29, 1923; s Myron Lester and Neva (Whitmer) E.; m. Joan Elizabeth Daniels, June 29, 1946; children: Joan Theresa, Michael Thomas, Jeri Anne. BS, U.S. Naval Acad., 1945; BSEE, MIT, 1949; MSEE, George Washington U., 1967. Commd. ensign USN, 1945; advanced through grades to capt. U.S. Navy, 1966; stationed at Norfolk (Va.) Naval Shipyard, 1950-55; project officer Regulus Missile (Underwater Sound Lab.), 1955-60; chmn. elec. sci. U.S. Naval Acad., 1962-65; dir. ship design div. Hdqrs. U.S. Navy, 1967-70; ret., 1970; mgr. advanced engring., marine systems div. Rockwell Internat., Anaheim, Calif., 1970-75; chief scientist Rockwell Internat., 1975-84, cons., 1985—. Contbr. articles to profl. jours. Mem. Soc. Naval Architects and Marine Engrs., Am. Soc. Naval Engrs., Am. Def. Preparedness Assn., U.S. Naval Inst. Achievements includes patent of fourier synthesis of complex waveforms; shipsinclude designs of Nimitz aircraft carriers, Trident strategic submarines, Los Angeles class submarines; prin. devel. roles include radar to shipboard displays, radar-based landing control of aircraft, in-helmet radio communications link, guidance system. Home: 1211 Belle Vista Dr Alexandria VA 22307-2016

ECKHART, WALTER, molecular biologist, educator; b. Yonkers, N.Y., May 22, 1938; s. Walter and Jean (Fairnington) E. B.S., Yale U., 1960; postgrad., Cambridge U., Eng., 1960-61; Ph.D., U. Calif.-Berkeley, 1965. Postdoctoral fellow Salk Inst., San Diego, 1965-69, mem., 1970-73, assoc. prof. molecular biology, 1973-79, prof., 1979—, dir. Armand Hammer Ctr. for Cancer Biology, 1976—; adj. prof. U. Calif.-San Diego, 1973—. Contbr. articles on molecular biology and virology to profl. jours. NIH research grantee, 1967—. Mem. AAAS, Am. Assn. Cancer Rsch., Am. Soc. Microbiology, Am. Soc. Virology. Home: 951 Skylark Dr La Jolla CA 92037-7731 Office: Armand Hammer Ctr Cancer Biology Salk Inst for Biol Studies PO Box 85800 San Diego CA 92186-5800

ECKMILLER, ROLF EBERHARD, neuroscientist, educator; b. Berlin, June 19, 1942; s. Hans Sigmund and Clara Bertha (Reese) E.; m. Marion Sangster, Nov. 3, 1978; children: Eva, Simon, Helen. MEE, Tech. U. Berlin, 1967, D Engring. summa cum laude, 1971. Sci. asst. in neurophysiology Free U. Berlin, 1967-72, asst. prof. in neurophysiology, 1971-78; prof., head div. biocybernetics U. Düsseldorf, Germany, 1978-92; prof., chair div. neuroinformatics U. Bonn, Germany, 1992—; vis. rsch. physiologist U. Calif., Berkeley, 1972-73, vis. prof., 1981; assoc. scientist Smith-Kettlewell Inst., San Francisco, 1977-78; cons. Rsch. Ministry, Düsseldorf, 1986—, Bonn, 1989—; chmn. Neural Networks Rsch. Consortium, 1991—. Editor: Neural Computers, 1988, Parallel Processes in Neural Systems and Computers, 1990, Advanced Neural Computers, 1990; contbr. articles to neurosci. and neural computing to sci. publs. Mem. IEEE Computer Soc., Soc. Neurosci., European Neurosci. Assn., Assn. Rsch. in Vision and Ophthalmology, Internat. Neural Networks Soc., Gesellschaft Informatik.

Avocations: history, human motivation, travel. Office: U Bonn Neuroinformatik, Römerstrasse 164, D 53117 Bonn Germany

ECKROAD, STEVEN WALLACE, electrical engineer; b. Xenia, Ohio, May 27, 1942; s. Wallace Bruce and Emily Elizabeth (Brodt) E.; m. Barbara Ann Brown, Dec. 12, 1969; children: Megan Elizabeth, Daniel Christopher. BA in Physics, Antioch Coll., 1965. Registered profl. engr., Calif. Jr. physicist Arthur D. Little, Cambridge, Mass., 1966-67; rsch. asst. Lawrence Berkeley (Calif.) Lab., 1975-79; engr. Bechtel Corp., San Francisco, 1979-85, project mgr., 1985-90; pres., prin. cons. EnerTech, Rolla, Mo., 1990-92; staff engr. U. Mo., Rolla, 1991-92; project mgr. Elec. Power Rsch. Inst., Palo Alto, Calif., 1992—. Contbr. articles to profl. jours. McMullin Engring. scholar Cornell U., 1960. Mem. IEEE. Achievements include design of world's largest battery energy storage plant, Chino, Calif. Office: Elec Power Rsch Inst 3412 Hillview Ave Palo Alto CA 94019

ECKSTEIN, JOHN WILLIAM, physician, educator; b. Central City, Iowa, Nov. 23, 1923; s. John William and Alice (Ellsworth) E.; m. Imogene O'Brien, June 16, 1947; children—John Alan, Charles William, Margaret Ann, Thomas Cody, Steven Gregory. B.S., Loras Coll., 1946; M.D., U. Iowa, 1950. Asst. prof. internal medicine U. Iowa, Iowa City, 1956-60; assoc. prof. U. Iowa, 1960-65, prof., 1965-92, prof. emeritus, 1993—; assoc. dean VA Hosp. affairs, 1969-70, dean coll. medicine, 1970-91, dean emeritus, 1993; chmn. cardiovascular study sect. NIH, 1970-72, Nat. Heart, Lung and Blood Adv. Council, 1974-78; gen. research support rev. com. NIH, 1980-84; mem. VA Manpower Study Group, 1988-92; mem. adv. com. to dir. NIH, 1990-93. Author papers and abstracts. Served with USAAF, 1943-45. Rockefeller Found. postdoctoral fellow, 1953-54; Am. Heart Assn. Research fellow, 1954-55; Nat. Heart Inst. spl. research fellow, 1955-56; Am. Heart Assn. established investigator, 1958-63; recipient USPHS Research Career award, 1963-70. Mem. Am. Heart Assn. (mem. publs. com. subcom. on heart 1986—, v.p. 1969, chmn. council on circulation 1969-71, pres. 1978-79), AMA (mem. health policy agenda panel 1982-86, governing council sect. on med. schs. 1985-92, mem. study sect. faculty and rsch. 1985-86, alt. del. Ho. of Dels. 1986-90, del. 1990-92), Am. Fedn. Clin. Rsch. (chmn. Midwestern sect. 1965), Central Soc. Clin. Research (sec.-treas. 1965-70, pres. 1973-74), Am. Soc. Clin. Investigation, Am. Clin. and Climatol. Assn., Assn. Am. Physicians, Assn. Am. Med. Colls. (exec. council 1981-82, adminstrv. bd. 1980-82, 85-86), Inst. of Medicine of Nat. Acad. Scis., Assn. Acad. Health Ctrs. (mem. sci. policy study group 1988—). Home: 1415 William White Blvd Iowa City IA 52245-4443 Office: U Iowa Coll Medicine 230 MAB Iowa City IA 52242-1101

ECONOMOU, ELEFTHERIOS NICKOLAS, physics educator, researcher; b. Athens, Greece, Feb. 7, 1940; s. Nickolas Vasilios and Sophia (Kordoni) E.; m. Athanasia E. Paganou, Feb. 17, 1966; 1 child, Sophia. Diploma, Tech. U. Athens, 1963; MS, U. Chgo., 1967, PhD, 1969. Rsch. assoc. U. Chgo., 1969-70; asst. prof. U. Va., Charlottesville, 1970-73, assoc. prof., 1973-79, prof., 1979-81; prof. U. Athens, 1978-81, U. Crete (Greece), 1981—; vis. prof. U. Chgo., U. Lausanne (Switzerland), Princeton (N.J.) U., Iowa State U.; cons. Naval Rsch. Lab., Washington, 1972-77, Exxon Rsch. Corp., Annandale, N.J., 1980-89; lectr. various internat. univs., rsch. ctrs. and sci. confs.; dir. gen., chmn. governing bd. Found. Rsch. and Tech., Heraklio, Crete, 1983—; mem. Com. European Devel. of Sci. and Tech., 1983-87; mem. Com. Evaluation of the European Framework Program for Sci and Tech., 1990-91; various administrv. positions at U. Crete. Author: Green's Functions in Quantum Physics, 1979, Nuclear Issues, 1985, Physics Today, 1986; contbr. numerous papers to sci. jours. Gen. sec. for rsch. and tech. Ministry of Industry, Energy and Tech., Athens, 1987-88; vice chmn. Nat. Adv. Bd. for R&D, Athens, 1988-90. Grantee NSF, Ministry Rsch. and Tech., NATO, Commn. European Communities. Mem. Am. Phys. Soc., Greek Soc. Engrs. Office: FORTH, PO Box 1527, 71110 Heraklio Crete Greece

EDDO, JAMES EKUNDAYO, industrial chemist, chemicals executive; b. London, Dec. 8, 1936; s. Johnson and Sarah Eddo; m. Elizabeth Thomas, May 5, 1963; children: James A., Margaret O. BSc, Thames Poly., London, 1965; diploma, Strathclyde U., Scotland, 1968; Phd, Pacific Western U., 1988; PhD, West London U., 1988. Prodn. mgr. Dresser MMM Ltd., London, 1969-74; rsch. chemist Makeridge Chems. Ltd., London, 1978-86, prodn. dir., 1989—. Inventor self bodying emulsifying agent. Mem. Royal Soc. of Chemistry (Great Britain chpt.), Soc. Cosmetic Scientists. Anglican. Avocations: golf, squash, swimming. Office: Makeridge Chems Ltd, Arena Indsl Estate Green Lanes Box 284, London N4 1UU, England

EDDY, CHARLES ALAN, chiropractor; b. Kansas City, Mo., Feb. 20, 1948; s. Sam Albert and Ella Louise (Gani) E.; m. Donna Darlene Perry, Oct. 23, 1971. Student, U. Mo., Kansas City, 1967; D in Chiropractic, Cleveland Chiropractic, Kansas City, 1970. Diplomate Nat. Bd. Chiropractic Examiners. Pvt. practice chiropractic Kansas City, 1970—; mem. peer rev. bd. Blue Cross and Blue Shield, Kansas City, 1972; pres. hon. bd. govs. Bapt. Hosp., Kansas City, 1987; cons. Quality Corp., Overland Park, Kans., 1988. Leader, profl. musician Chuck Eddy Band, Kansas City, 1966—; res. officer Kansas City Police Dept., 1970-77; sgt. 1977-82, capt. 1982—. Mem. Am. Chiropractic Assn., Mo. State Chiropractic Assn., Mo. Dist. II Chiropractic Assn., Cleve. Chiropractic Coll. (trustee 1990, vice chmn. 1992, 93), Cleve. Chiropractic Alumni Assn. (officer, bd. dirs. 1990—, ambassador's soc. 1983—, chmn. 1990-92), Optimist Club of Landing (pres. 1980, lt. gov. Mo. dist. 1982), Am. Lebanon Syrian Men's Club (pres. 1988-91, chmn. bd. 1992), St. Andrews Soc. (drummer in pipe band), DeMolay Legion Hon. (sec. 1988, treas. 1990, vice dean 1991, dean 1992), Pipes and Drums of Ararat (treas. 1977-90, pres. 1985, dir. 1989, 90), Elks, Shriners (divan of Ararat shrine temple, publicity chmn. 1991-92), Royal Order Jesters, Order Quetzalcoatl. Episcopalian. Avocations: photography, guns, stereo and video entertainment. Home: 406 W 109th St Kansas City MO 64114-4910 Office: 8301 State Line Rd # 108 Kansas City MO 64114-2019

EDELHAUSER, HENRY F., physiologist, ophthalmologist, medical educator; b. Dover, N.J., Sept. 9, 1937; married, 1961; 2 children. BA, Paterson State Coll., 1961; MS, Mich. State U., 1964, PhD in Physiology, 1966. Lab. technician Warner Lambert Pharm. Rsch. Inst., 1962; asst. physiologist Mich. State U., 1962-65; from instr. to prof. physiology and ophthalmology Med. Coll. Wis., 1966-89; prof. ophthalmology, dir. rsch. Emory U., Atlanta, 1989—, dir. grants dept. ophthalmology, 1990-93; bd. dirs. Am. Fight-for-Sight, Inc.; rsch. assoc. ophthalmology Marquette U., 1967-68; dir. tng. Nat. Eye Inst., 1975-90; prin. investigator Mt. Desert Island Biol. Lab., 1977; sci. cons. Alcon Labs, S.C. Johnson & Son, Am. Cyanamid; mem. rsch. com. Tissue Banks Internat. Fellow Marquette U., 1966-67; grantee Nat. Eye. Inst., 1969-92, Wis. Dept. Nat. Rsch., 1969-71; Olga K. Weiss Rsch. Scholar; named Marjorie and Joseph Heil prof. opthalmology, 1988, Ferst prof. ophthalmology, 1989. Mem. Am. Soc. Biol. Scis., Am. Soc. Zoology, Assn. Rsch. Vision and Ophthalmology (pres. 1990-91), Am. Physiol. Soc., Am. Acad. Ophthalmology (Honor award). Achievements include research in membrane physiology, pathophysiology of eye, fish physiology and eye disease, ocular toxicology, physiological effects of virectomy, cellular toxicology and ophthalmic drugs. Office: Emory University Lab for Ophthalmic Rsch 1327 Clifton Rd NE Rm 3704S Atlanta GA 30322*

EDELHERTZ, HELAINE WOLFSON, mathematics educator; b. Queens, N.Y., June 22, 1953; d. David and Sylvia Guttman Wolfson; m. Melvyn Paul, June 6, 1976; children: Allyson Leigh, Dustin Scott. BS, SUNY, Oneonta, 1977; MS, SUNY, New Paltz, 1985. Cert. tchr., N-6, N.Y. Substitute tchr. Roscoe (N.Y.) Cen. Sch., 1978-82; tchr. Yeshiva Sch., South Fallsburg, N.Y., 1982-83; substitute tchr. Middletown (N.Y.) City Sch. Dist., 1984-86, home tchr., 1984-87, math. specialist, 1987—; with Math Turnkey, Meml. Elem. Sch., Middletown, 1989—, Excellence and Accountability Program, 1989-92; math. com. Middletown Schs., 1989-92; bldg. com. Study Math. Portfolios, 1992-93. Bd. dirs. Wallkill Farms Homeowner's Assn., 1986-88, budget com., 1992—; asst. summer coord. Roscoe Free Libr., 1983. Mem. Am. Math. Tchrs. of N.Y. State, Nat. Coun. Tchrs. of Math., Middletown Tchrs. Assn. (sr. bldg. rep. 1992—). Jewish. Avocations: gardening, sewing, knitting, cooking, dance. Home: 118 Rolling Meadows Rd Middletown NY 10940-2611 Office: Meml Elem Sch 83 Linden Ave Middletown NY 10940-3794

EDELMAN, GERALD MAURICE, biochemist, educator; b. N.Y.C., N.Y., July 1, 1929; s. Edward and Anna (Freedman) E.; m. Maxine Morrison, June 11, 1950; children: Eric, David, Judith. B.S., Ursinus Coll., 1950, Sc.D., 1974; M.D., U. Pa., 1954, D.Sc., 1973; Ph.D., Rockefeller U., 1960; M.D. (hon.), U. Siena, Italy, 1974; DSc (hon.), Gustavus Adolphus Coll., 1975, Williams Coll., 1976; DSc Honoris Causa, U. Paris, 1989; LSc Honoris Causa, U. Cagliari, 1989; DSc Honoris Causa, U. degli Studi di Napoli, 1990, Tulane U., 1991. Med. house officer Mass. Gen. Hosp., 1954-55; asst. physician hosp. of Rockefeller U., 1957-60, mem. faculty, 1960-92, assoc. dean grad. studies, 1963-66, prof., 1966-74, Vincent Astor disting. prof., 1974-92; mem. faculty and chmn. dept. neurobiology Scripps Rsch. Inst., La Jolla, Calif., 1992—; mem. biophysics and biophys. chemistry study sect. NIH, 1964-67; mem. Sci. Council, Ctr. for Theoretical Studies, 1970-72; assoc., sci. chmn. Neurosciences Research Program, 1980—, dir. Neuroscis. Inst., 1981—; mem. adv. bd. Basel Inst. Immunology, 1970-77, chmn., 1975-77; non-resident fellow, trustee Salk Inst., 1973-85; bd. overseers Faculty Arts and Scis., U. Pa., 1976-83; trustee, mem. adv. com. Carnegie Inst., Washington, 1980-87; bd. govs. Weizman Inst. Sci., 1971-87, mem. emeritus; researcher structure of antibodies, molecular and devel. biology. Author: Neural Darwinism, 1987, Topobiology, 1988, The Remembered Present, 1989, Bright Air, Brilliant Fire, 1992. Trustee Rockefeller Bros. Fund., 1972-82. Served to capt. M.C. AUS, 1955-57. Recipient Spencer Morris award U. Pa., 1954, Ann. Alumni award Ursinus Coll., 1969, Nobel prize for physiology or medicine, 1972, Albert Einstein Commemorative award Yeshiva U., 1974, Buchman Meml. award Calif. Inst. Tech., 1975, Rabbi Shai Shacknai meml. prize Hebrew U.-Hadassah Med. Sch., Jerusalem, 1977, Regents medal Excellence, N.Y. State, 1984, Hans Neurath prize, U. Washington, 1986, Sesquicentennial Commemorative award Nat. Libr. Medicine, 1986, Cécile and Oskar Vogt award U. Dusseldorf, 1988, Disting. Grad. award U. Pa., 1990, Personnalité de l'année, Paris, 1990, Warren Triennial Prize award Mass. Gen. Hosp., 1992. Fellow AAAS, N.Y. Acad. Scis., N.Y. Acad. Medicine; mem. Am. Philos. Soc., Am. Soc. Biol. Chemists, Am. Assn. Immunologists, Genetics Soc. Am., Harvey Soc. (pres. 1975-76, Am. Chem. Soc., Eli Lilly award biol. chemistry 1965), Am. Acad. Arts and Scis., Nat. Acad. Sci., Am. Soc. Cell Biology, Acad. Scis. of Inst. France (fgn.), Japanese Biochem. Soc. (hon.), Pharm. Soc. Japan (hon.), Soc. Developmental Biology, Council Fgn. Relations, Sigma Xi, Alpha Omega Alpha. Office: Scripps Rsch Inst Dept Neurobiol SBR-14 10666 N Torrey Pines Rd La Jolla CA 92037

EDELSBRUNNER, HERBERT, computer scientist, mathematician; b. Graz, Styria, Austria, Mar. 14, 1958; s. Herbert and Berta Edelsbrunner; m. Ping Fu, Nov. 14, 1991. MS in Tech. Math., Tech. U. Graz, 1980, PhD in Tech. Math., 1982. Vertragsassistent Inst. Informationsverarbeitung Tech. U. Graz., 1981-84, Universitätsassistent Inst. Informationsverarbeitung, 1984-85; visitor IBM T.J. Watson Rsch. Ctr., Yorktown Heights, N.Y., 1987; asst. prof. dept. computer sci. U. Ill., Urbana-Champaign, 1985-87, assoc. prof. dept. computer sci., 1987-90, prof. dept. computer sci., 1990—; lectr. in field. Author: Algorithms in Combinatorial Geometry, 1987. Recipient Alan T. Waterman award NSF, 1991, Univ. Scholar award U. Ill. Found., 1990, Sr. Xerox award, 1989; grantee NSF, 1988-90, 90-92, 92—, Amoco Found. 1985-88. Achievements include research in data structures and algorithms, computational geometry, geometric visualization, discrete and combinatorial geometry, experimental geometric topology, a new approach to rectangle intersections, the shape of a set of points in the plane, optimal point location in a monotone subdivision, topologically sweeping an arrangement, simulation of simplicity: a technique to cope with degenerate cases in geometric algorithms, an acyclicity theorem for cell complexes in d dimensions, combinatorial complexity bounds for arrangements of curves and spheres, an optimal algorithm for intersecting line segments in the plane, an O(n2 log n) time algorithm for the minmax angle triangulation. Office: Univ of Ill Dept of Computer Sci 1304 W Springfield Ave Urbana IL 61801

EDELSON, JONATHAN VICTOR, entomologist; b. Memphis, June 5, 1952; s. Joe and JoAnn E.; m. Karen Orttman Maxey, Dec. 10, 1982; 1 child, Erica Ann. MS, Auburn U., 1978, PhD, 1982. Asst. prof. entomology Tex. A&M U., College Station, 1982-87, assoc. prof., 1987-89; assoc. prof. entomology and dir. Okla. State U., Stillwater, 1989—. Editor: Insecticide and Acaracide Tests, Entomol. Soc. Am., 1986-91; contbr. articles to profl. jours. Grantee Binational Agrl. Rsch. Devel., USA/Israel, 1988, U.S. Dept. Agr./SRIPM, 1989. Mem. AAAS, Entomol. Soc. Am. Achievements include development of pest management systems for vegetable crops that reduced damage to crops and reduced pesticide use on crops. Development a technique for field monitoring insecticide resistance in insect populations. Office: Okla State Univ PO Box 128 Lane OK 74555

EDELSTEIN, ALAN SHANE, physicist; b. St. Louis, June 27, 1936; s. Abe and Lilian (Ginesburg) E.; m. Maravene Gram, Aug. 12, 1964; children: Rachel, Rebecca. BS in Engring. Physics, Washington U., St. Louis, 1958; PhD in Physics, Stanford U., 1964. Mem. tech. staff IBM Rsch. Ctr., Yorktown Heights, N.Y., 1965-68; assoc. prof. U. Ill., Chgo., 1968-79; scientist Energy Conversion Devices, Detroit, 1979-80; rsch. physicist Naval Rsch. Lab., Washington, 1980-88, supervisory rsch. physicist, 1988—; cons. Argonne (Ill.) Nat. Lab., 1969-77; rev. panel Dept. of Energy, Washington, 1992. Contbr. 98 articles to profl. jours., chpts. in 2 books. Woodrow Wilson fellow, 1958, NSF postdoctoral fellow, 1964. Mem. Am. Phys. Soc., Materials Rsch. Soc. Achievements include discovery of giant gapless superconductivity due to Kondo impurities, Kondo effect in concentrated systems, and self-arrangement of Molybdenum particles into cubes. Office: Naval Rsch Lab Code 6371 Washington DC 20375

EDEMEKA, UDO EDEMEKA, surgeon; b. Ndon Eyo, Akwa Ibom, Nigeria, Sept. 11, 1944; came to U.S., 1973; s. Buddie Udo and Dinah Buddie (Ekwere) E.; m. Iboro Udo David Akpan, May 18, 1973; children: Ubong, Dinah, Idara, David, Dennis, Donald. MB and BS, U. Ibadan, Nigeria, 1970; diploma in anesthesia, U. Lagos, 1972. Diplomate Am. Bd. Surgery, Am. Bd. Emergency Physicians. Instr. surgery Downstate Med. Ctr., Bklyn., 1974-80; attending physician Kings County Hosp. Ctr., Bklyn., 1980-91, Meth. Hosp., Bklyn., 1988—. Leverhulme Exchange scholar Lever Bros. U. Coll. Hosp., London, 1969. Fellow N.Y. Acad. Scis., Internat. Coll. Surgeons, Am. Coll. Emergency Physicians. Office: Meth Hosp 506 6th St Brooklyn NY 11215-3645

EDEN, JAMES GARY, electrical engineering and physics educator, researcher; b. Washington, Oct. 11, 1950; s. Robert Otis and Joyce (West) E.; m. Carolyn Sue Thomas, June 10, 1972; children: Robert Douglas, Laura Ann, Katherine Joy. BS, U. Md., 1972; MS, U. Ill., 1973, PhD, 1975. Teaching asst. elec. engring. dept. U. Ill., Urbana, Jan.-June 1972, rsch. asst., 1972-75, asst. prof. elec. engring. dept., 1979-81, assoc. prof., 1981-83, prof. elec. engring. dept. and rsch. prof. Coordinated Sci. Lab, 1983—, mem. physics grad. rsch. faculty, asst. dean Coll. Engring., 1992-93; postdoctoral rsch. assoc. NRC, Washington, 1975-76; rsch. physicist Naval Rsch. Lab., Washington, 1976-79; assoc. mem. Ctr. for Advanced Study, U. Ill., 1987-88; program com. Conf. on Lasers and Electro-Optics, 1982, 83, 88, 89; chmn. Engring. Found. Conf. Ultraviolet Lasers, 1987; program chair ann. meeting IEEE Lasers and Electro-Optics Soc., 1990, conf. chair 1992; program vice chmn. Interdisciplinary Laser Conf. V, 1989, program chair ILS VI, 1990; conf. chair ILS VIII, 1992. Author: Photochemical Vapor Deposition, 1992; assoc. editor Photonics Technology Letters, 1988—; contbr. over 100 articles to profl. jours.; patentee for 12 inventions. Recipient Rsch. Publ. award Naval Rsch. Lab, 1978, Beckman Rsch. award U. Ill., 1988-89. Fellow IEEE, Optical Soc. Am., Am. Phys. Soc.; mem. Tau Beta Pi, Eta Kappa Nu, Sigma Xi, Phi Kappa Phi. Republican. Avocation: amateur radio. Home: 513 Taylor Dr Mahomet IL 61853-9246 Office: U Ill Everitt Lab 1406 W Green St Urbana IL 61801-2991

EDGAR, THOMAS FLYNN, chemical engineering educator; b. Bartlesville, Okla., Apr. 17, 1945; s. Maurice Russell and Natalie (Flynn) E.; m. Donna Jean Proffitt, July 15, 1967; children: Rebecca, Jeffrey. B.S. in Chem. Engring., U. Kans., 1967; Ph.D. in Chem. Engring., Princeton U., 1971. Registered profl. engr., Tex. Process engr. Conoco, Balt., 1968-69; prof. chem. engring. U. Tex., Austin, 1971—, chmn. dept., 1985-93, Abell chair, 1991—; prof. chem. engring. U. Calif., Berkeley, 1978; pres. CACHE Corp., Austin, Tex., 1981-84; pres. Am. Automatic Control Coun., Chgo., 1990-91; chair Coun. for Chem. Rsch., Washington, 1992-93. Author: Coal Processing and Pollution Control, 1983; co-author: Real Time Computing, 1982, Optimization of Chemical Processes, 1988, Process Dynamics and Control, 1989; editor: Chemical Process Control, 1981, In Situ (Marcel Dekker), 1977-89; also jours. Recipient Edn. award Am. Automatic Control Coun., 1992. Mem. Am. Inst. Chem. Engrs. (Outstanding Counselor award 1974 Colburn award 1980, editorial bd. jour. 1983-85, chmn. cast div. 1986, bd. dirs. 1989-92), Am. Soc. Engring. Edn. (Westinghouse award 1988, Meriam-Wiley Disting. Author 1990), Instrument Soc. Am., Am. Chem. Soc., Tau Beta Pi, Phi Lambda Upsilon, Omicron Delta Kappa, Phi Kappa Phi (Joe King award U. Tex. 1989, U. Kans. Disting. Engring. Svc. award 1990). Democrat. Methodist.

EDGE, JAMES EDWARD, health care administrator; b. Anacortes, Wash., Apr. 29, 1948; s. Edward and Carol Marie (Lian) E.; m. Nellie Ruth Horton, Mar. 21, 1970; children: Elissa Marie, Gina Dawn. BS in Pharmacy, U. Wash., 1971; MPH, U. Hawaii, 1979. Registered pharmacist. Commd. USPHS, 1969—, advanced through grades to capt.; staff pharmacist USPHS Indian Hosp., Albuquerque, 1971-73; chief pharmacy, lab/x-ray S.W. Indian Poly. Inst., Albuquerque, 1972-73, Neah Bay Indian Health Ctr., Wash., 1973-75; svc. unit dir. Neah Bay Svc. Unit, Indian Health Svc., 1975-78, Western Oreg. Service Unit, Indian Health Svc., Salem, 1980—; cons. in field. Active Combined Fed. Campaign, Salem, 1985—. John Quick Pharmacy scholar, U. Wash., 1967, Health Professions scholar, 1969. Mem. Am. Coll. Healthcare Adminstrs., Am. Acad. Med. Adminstrs., Assn. Mil. Surgeons of U.S., Res. Officers Assn., Commn. Offices of the USPHS, Am. Pub. Health Assn., Wash. Pharm. Assn., Nat. Coun. of Svc. Unit Dirs. (chmn. 1986-88). Avocations: running, scuba diving, sheep raising, computers. Office: PHS Indian Health Ctr 3750 Chemawa Rd NE Salem OR 97305-1111

EDGE, RONALD DOVASTON, physics educator; b. Bolton, Eng., Feb. 3, 1929; came to U.S., 1958, naturalized, 1968; s. James and Mildred (Davies) E.; m. Margaret Skulina, Aug. 14, 1956 (div. 1989); children: Christopher James, Michael Dovaston; m. Gertrude Hansen, Dec. 31, 1992. B.A., Cambridge U., 1950, M.A., 1952, Ph.D., 1956. Research fellow Australian Nat. U., Canberra, 1954-58; asst. then assoc. prof. physics U. S.C., Columbia, 1958-63; prof. U. S.C., 1964—; research assoc. Yale U., New Haven, 1963-64; vis. prof. Stanford U., Calif. Tech. Inst., U. Munich, U. Sussex, U. Witwatersrand, U. Aarhus, Oak Ridge Nat. Lab., Los Alamos Nat. Lab.; leader 1st Am. team Internat. Physics Olympiad, 1986. Author: Physics in the Arts, 1973, String and Sticky Tape Experiments, 1978; contbr. articles to profl. jours. Recipient Russell award U. S.C. Fellow Am. Phys. Soc. (James B. Pegram award 1979), Am. Assn. Physics Tchrs. (apparatus award 1973). Unitarian (past pres. Columbia fellowship). Home: 220 Jadetree Dr Hopkins SC 29061-9347 Office: U SC Physics Dept Columbia SC 29208

EDGERTON, ROBERT FRANK, optical scientist; b. Cambridge, Mass., May 10, 1935; s. Harold Eugene and Esther May (Garrett) E.; m. Elizabeth rose Lowe, Aug. 15, 1959; children: Eric Franz, Nina Adele, Sylvia Lowe. PhD, U. Rochester, 1962. Asst. prof. physics Carleton Coll., Northfield, Minn., 1963-66; assoc. prof. physics U. Maine, Orono, 1968-71; postdoctoral assoc. Cornell U., Ithaca, N.Y., 1966-68; tchr. physics Roeper City and Country Sch., Bloomfield Hills, Mich., 1971-74; assoc. prof. Lawrence Inst. Tech., Southfield, Mich., 1974-77; sr. scientist Energy Conversion Devices, Troy, Mich., 1980-87; cons. Ovonic Imaging Systems, Troy, 1987-90; staff scientist Airco Coating Tech., Concord, Calif., 1990—. Contbr. articles to profl. jours. Mem. Optical Soc. Am. (pres., v.p., program chmn. 1980-87). Mem. Soc. of Friends. Office: Airco Coating Tech Inc 4020 Pike Ln Concord CA 94524

EDIGER, MARK D., chemistry educator; b. Newton, Kans., July 26, 1957. BA in Chemistry and Math., Bethel Coll., 1979; PhD in Phys. Chemistry, Stanford U., 1984. Asst. prof. dept. chemistry U. Wis., Madison, 1984-90, assoc. prof., 1990—. Polymers Program grantee NSF, 1992-95, Petroleum Rsch. Fund. Type AC grantee, Am. Chemical Soc., 1992-94. Mem. Am. Chem. Soc. (mem. program com., divsn. polymer chemistry 1990—), Am. Phys. Soc. (Dillon medal, 1992). Office: U Wisconsin Dept Chemistry 1101 University Ave Madison WI 53706

EDLICH, RICHARD FRENCH, biomedical engineering educator; b. N.Y.C., Jan. 19, 1939; married, 1961; 3 children. MD, NYU, 1962; PhD, U. Minn., 1973. From instr. to assoc. prof. U. Va. Sch. Medicine, Charlottesville, 1971-76, prof. plastic surgery and biomed. engring., 1976-82, now disting. prof. plastic and maxillofacial surgery and biomed. engring.; dir. Emergency Med. Svc. and Burn Ctr., 1974-85; physician tech. adviser Bur. Emergency Svc., HEW, 1974—; cons. Div. Health Manpower and Nat. Ctr. Health Svc. Rsch., 1977—. Recipient U. Va. Pres.'s Report award, 1992. Mem. ACS, Soc. Univ. Surgeons, Am. Assn. Surg. Trauma, Am. Burn Assn., Am. Spinal Cord Injury Assn., Univ. Assn. Emergency Medicine, Am. Soc. Plastic and Reconstructive Surgeons, Soc. Surg. Infection Am., Coll. Emergency Physicians, Am. Surg. Assn. Research in biology of wound repair and infection. Office: U Va Dept Biomed Engring Charlottesville VA 22908

EDLUND, CARL E., physicist; b. El Campo, Tex., Mar. 14, 1936; s. Milton Carl and Sara Elizabeth (May) E.; m. Lynda Rahe (div.); children: Eric, Dana, Dean; m. Marjorie K. Reynolds, Aug. 5, 1983; children: Teresa, Angela, Joel. BS in Physics, St. Mary's U., San Antonio, 1963. Sr. electronics tech. S.W. Rsch. Inst., San Antonio, 1960-62, rsch. asst., 1962-65, rsch. physicist, 1965-68, sr. rsch. physicist, 1968-87, prin. scientist, 1987—. Contbr. articles to Jour. Applied Physics. With USN, 1955-59. Mem. AAAS. Republican. Lutheran. Achievements include patents for reciprocating engine-compressor indicator, analogs of reciprocating and centrifugal compressors, instrumentation for instable flow analysis; development of large sample magnetic resonance spectrometers, flow measurement computers, laser doppler velocity measurement of underwater explosive driven particles, wide area seismic surveillance system. Home: Rt 1 Box 23 Castroville TX 78009 Office: SW Rsch Inst 6220 Culebra Rd San Antonio TX 78238-5166

EDMONDS, ANDREW NICOLA, software engineer; b. London, Nov. 6, 1955; s. George Albert and Patricia Ethel (Battenti) E.; m. Anneke Himmele, Aug. 27, 1988; children: Alexander, Anna. Cert. electronic mus. instrument tech., London Coll. of Furniture, 1978. Elec. design engr. Hawker Siddeley Dynamics Engring., Hertfordshire, Eng., 1978-80; tech. dir. Guyvale Ltd., Hertfordshire, 1980-89, Neural Computer Scis., Olney, Eng., 1989-93; mng. dir. Prophecy Sys. Ltd., Olney, Eng., 1993—. Author: (computer software) Neurun, 1989, Neurun Light, 1990, Neural Desk, 1991, Neuforcast, 1992, Gaserver, 1992, Prophecy, 1993, Cerberus, 1993. Avocations: Romano-Celtic history, guitar, speculative software design, child rearing. Office: Prophecy Sys Ltd, 34b Market Pl, Olney Bucks MK46 4AJ, England

EDMONDS, RICHARD LEE, air force officer; b. Okmulgee, Okla., Nov. 23, 1953; s. R.V. John and Mary Lou (White) E.; m. Suzanne Elizabeth Merrick, Sept. 15, 1960. BSEE, Auburn U., 1981; MS in Space Ops., Air Force Inst. Tech., 1987. Commd. 2d lt. USAF, 1981, advanced through grades to maj., 1992, radar technician F4E, 1972-78; satellite ops. officer USAF, Sunnyvale, Calif., 1981-86; project engr. USAF-Def. Support Program, L.A., 1987-90, satellite system test mgr., 1990-92; Milstar flight comdr. Air Force Space Command, 4th Satellite Ops. Squadron, Colorado Springs, Colo., 1992; spacecraft flight dir. NASA Shuttle Mission STS-44. Named NASA Manned Flight Awareness Honoree, NASA-STS 44, 1991. Mem. Eta Kappa Nu. Baptist. Home: 2882 Shrider Rd Colorado Springs CO 80920

EDMONDSON, W(ALLACE) THOMAS, limnologist, educator; b. Milw., Apr. 24, 1916; s. Clarence Edward and Marie (Kelley) E.; m. Yvette Hardman, Sept. 26, 1941. BS, Yale U., 1938, PhD, 1942; postgrad., U. Wis., 1938-39; DSc (hon.), U. Wis., Milw., 1987. Research assoc. Am. Mus. Natural History, 1942-43, Woods Hole Oceanographic Instn., 1943-46; lectr. biology Harvard U., Cambridge, Mass., 1946-49; mem. faculty U. Wash., Seattle, 1949—, prof., 1957-86, prof. emeritus, 1986—, Jessie and John Danz lectr., 1987; R.E. Coker Meml. lectr. U. N.C., 1977; Brode lectr. Whitman Coll., 1988. Editor: Freshwater Biology (Ward and Whipple), 2d edit, 1959; contbr. articles to profl. jours. Recipient Einar Naumann-August Thienemann medal Internat. Assn. Theoretical and Applied Limnology, 1980, Outstanding Pub. Svc. award U. Wash., Seattle, 1987, commendation

State of Wash., 1987; NSF sr. postdoctoral fellow Italy, Eng. and Sweden, 1959-60, Wilbur Lucius Cross medal Yale U. Grad. Sch. Alumni Assn., 1993. Mem. NAS (Cottrell award 1973), AAAS, Am. Soc. Limnology and Oceanography (G. Evelyn Hutchinson medal 1990), Internat. Assn. Limnology, Ecol. soc. Am. (eminent Ecologist award 1983), Yale Grad. Sch. Alumni Assn. (Wilbur Lucius Cross medal 1993). Office: U Wash Dept Zoology NJ-15 Seattle WA 98195

EDMONSTON, WILLIAM EDWARD, JR., publisher; b. Balt., Nov. 20, 1931; s. William Edward and Helen (Mallonee) E.; m. Nellie Jane Kerley, Aug. 3, 1957; children—Kathryn Nell, Rebecca Jane, Owen William. B.A., Johns Hopkins U., 1952; M.A., U. Ala., 1956; Ph.D., U. Ky., 1960. Diplomate: Am. Bd. Psychol. Hypnosis. Instr., asst. prof. Washington U., St. Louis, 1960-64; mem. faculty Colgate U., Hamilton, N.Y., 1964-93, dir. neurosci. program, 1972-93, prof. psychology, 1973-93, prof. emeritus, 1993—; chmn. dept. psychology Colgate U., 1971-81; Gast prof. U. Erlanger, Nürnberg, Fed. Republic Germany, 1982; pub. Edmonston Pub., Inc., Hamilton. Author: Hypnosis and Relaxation: Modern Verification of an Old Equation, 1981, The Induction of Hypnosis, 1986, Unfurl the Flags: Remembrances of the American Civil War, 1989; editor: Am. Jour. Clin. Hypnosis, 1968-76; contbr. articles to profl. jours. Served with U.S. Army, 1952-54. Recipient Bernard E. Gorton award, 1961; Sloan Found. fellow, 1967, 69; sr. fellow U. Wash., 1971; USPHS grantee, 1964-65; Fulbright fellow, 1982; CASE N.Y. State Prof. of Yr., 1988. Fellow AAAS, APA, Internat. Soc. Clin. and Exptl. Hypnosis, Am. Psychopathol. Assn., mem. N.Y. Acad. Scis., Eastern Psychol. Assn., Soc. Neurosci., Sigma Xi. Home: 30 Maple Ave Hamilton NY 13346-1219

EDMUNDS, ROBERT THOMAS, retired surgeon; b. Toledo, Sept. 14, 1924; s. Marion Kenneth and Frances Ethel (McCauley) E.; widowed, 1983; children: Nancy, Priscilla, Elizabeth, Cynthia, Robert. BA, Harvard U., 1947; MD, Columbia U., 1951. Diplomate Am. Bd. Surgery. Intern St. Luke's Hosp., N.Y.C., 1951-52, asst. resident surgery, 1952-55, resident surgery, 1955-56, attending surgeon, 1956-78; clin. prof. surgery Columbia Coll. Physicians and Surgeons, N.Y.C., 1966-78; mini-residency in occupational medicine Inst. Environ. Health U. Cin. Coll Medicine, 1983; med. dir. U.S. Steel Corp., Pitts., 1978-89; ret., 1989; prin. investigator Cen. Oncology Group, Madison, Wis., 1956-70. Contbr. articles to profl. jours. Lt. (j.g.) USNR, 1942-46. Fellow ACS; mem. Union Club. Republican. Congregationalist. Achievements include enhancement of vision in albino children by use of contact lenses with opaque sclerae. Home: 400 E 71st St New York NY 10021

EDSON, HERBERT ROBBINS, hospital executive; b. Upper Darby, Pa., Dec. 26, 1931; s. Merritt Austin and Ethel Winifred (Robbins) E.; m. Constance Anne Lowell, May 20, 1961 (div. Nov. 8, 1967); m. Rose Anne McGowan, July 25, 1970; children: Patricia Anne, David William, Merritt Austin III, Herbert Robbins Jr. BA, Tufts U., 1955; MBA, U. Pa., 1972. Commd. 2d lt. USMC, 1955, advanced through grades to major, 1967, adminstr., mgr., supr. various orgns., 1955-72; controller III Marine Amphibious Force and 3d Marine Div. USMC, Camp Butler, Japan, 1972-73; dir. acctg. Marine Corps Supply Activity USMC, Phila., 1973-75; ret. USMC, 1975; cons. acctg. Ardmore, Pa., 1975-77; CFO Mercy Meml. Hosp. Corp., Monroe, Mich., 1977-92, Mercy Meml. Hosp. Found., Monroe, 1986-92, Monroe Health Ventures Inc., 1986-92, Monroe Community Health Svcs., 1989-92, Byerly Hosp., Hartsville, S.C., 1992—; assoc. Quorum Health Resources, Inc., Nashville, 1992—. Co-pres. Custer Elem. Sch. Parent Tchr. Orgn., Monroe, 1985-87; v.p., trustee Christ Evang. Luth. Ch., Monroe, 1981-86; treas., chmn. Taylor Endowment Fund com. St. Paul's Evang. Luth. Ch., Ardmore, Pa., 1974-76, trustee, chmn. property com., 1976. Decorated Purple Heart, Navy Commendation medal, Combat Action ribbon. Mem. Am. Hosp. Assn., Healthcare Fin. Mgmt. Assn., Inst. Mgmt. Accts., Monroe County C. of C. (bd. dirs. 1982-84), NRA (life), U.S. Naval Inst. (life), Marine Corps Assn. (life), 1st Marine Div. Assn. (life), Edson's Raiders Assn. (hon. life 1st Marine Raider Bn.), Ret. Officers Assn. (life), Am. Assn. Ret. Persons, Nat. Geog. Soc., Marine's Meml. Club, Hartsville Country Club, Army and Navy Club. Republican. Lutheran. Home: 1121 Pinelake Dr Hartsville SC 29550 Office: Byerly Hosp 413 E Carolina Ave Hartsville SC 29550

EDWARDS, CHARLES RICHARD, entomology and pest management educator; b. Lubbock, Tex., Jan. 22, 1945; s. Troy B. and Jeanette E. (McDermett) E.; m. Claudia Frances Henderson, Dec. 21, 1966; children: Cecily Elizabeth, Celeste Elaine. BS, Tex. Tech. U., 1968; MS, Iowa State U., 1970, PhD, 1972. Bd. cert. entomoloist. Prof. Entomology Purdue U., West Lafayette, Ind., 1972—; cons. Agri-Growth Rsch., Hollandale, Minn., 1984-89, Consortium for Internat. Crop Protection, Geneva, N.Y., 1985—. Contbr. articles to profl. jours. Mem. Entomol. Soc. Am. (Extension Achievement Award 1984, Award of Merit 1985), Royal Entomol. Soc. London, Kans. Entomol. Soc., Am. Soc. Agronomy, Sigma Xi, Alpha Zeta, Gamma Sigma Delta. Avocations: running, woodworking. Office: Purdue U 1158 Entomology Hall West Lafayette IN 47907-1158

EDWARDS, DARREL, psychologist; b. San Francisco, July 9, 1943; s. Darrus and Rose Pearl (Sannar) E.; children: Alexander Hugh, Peter David, James Royce. BS in Psychology and Philosophy, Brigham Young U., 1965, MS in Psychology and Philosophy, 1967, PhD in Clin. Psychology and Philosophy, 1968. Diplomate Am. Bd. Profl. Psychology. Postdoctoral fellow in psycholinguistics Pa. State U., 1969; commd. lt. (j.g.) USN, 1970, advanced through grades to lt. comdr., 1978; dir. psychologist Tri Community Svc. Systems, San Diego, 1973-78; prof. Calif. Sch. Profl. Psychology, San Diego, 1971-78; dir. Grid Rsch., San Diego, 1978—; pres. The Edwards Assoc., San Diego, 1983—; co-founder Summus Cons. Strategies, Darien, Conn., 1989—, Strategic Solutions, Washington and London; pres. Strategic Vision, 1973—; cons. strategist for govt. and pvt. sector, U.S. and Eng., 1978—. Co-inventor in field; contbr. articles to profl. jours. Mem. Am. Psychol. Assn. Achievements include creation of Values Centered research procedures. Office: The Edwards Assocs PO Box 420429 San Diego CA 92142-0429

EDWARDS, DONALD KENNETH, mechanical engineer, educator; b. Richmond, Calif., Oct. 11, 1932; s. Samuel Harrison and Georgette Marie (Bas) E.; m. Nathalie Beatrice Snow, Oct. 11, 1955; children: Victoria Ann, Richard Earl. B.S. with highest honors in Mech. Engring., U. Calif., Berkeley, 1954, M.S. in Mech. Engring., 1956, Ph.D. in Mech. Engring., 1959. Thermodynamics engr. missile systems div. Lockheed Aircraft Co., Palo Alto, Calif., 1958-59; asst. prof. engring. UCLA, 1959-63, assoc. prof., 1963-68, prof., 1968-81, chmn. dept. chem., nuclear and thermal engring., 1975-78; prof. U. Calif., Irvine, 1981—, chmn. dept. mech. engring., 1982-86, assoc. dean, 1986-89; dir., chmn. bd. Gier Dunkle Instruments, Inc., 1963-66. Author: (with others) Transfer Processes, 1973, 2d edit., 1979; assoc. editor: ASME Jour. Heat Transfer, 1975-80, Solar Energy, 1982-85; contbr. articles to profl. jours. Fellow AIAA (first Thermophysics award 1976), ASME (Heat Transfer Meml. award 1973); mem. Optical Soc. Am., Internat. Solar Energy Soc., Phi Beta Kappa, Sigma Xi, Pi Tau Sigma, Tau Beta Pi. Office: U Calif/Irvine Mech Engring Dept Irvine CA 92717

EDWARDS, GEORGE HENRY, technical writer; b. Hammond, Ind., Feb. 19, 1932; s. Samuel Finley and Eula Gertrude (Gruber) E.; m. Marian Joan Weiss, May 24, 1939; children: Susan, Judith, Sandra. BA in Math., Am. U., 1959. Sr. engr. Vitro Labs., Silver Spring, Md., 1973-76; mem. tech. staff Rockwell Internat., Anaheim, Calif., 1976-81; dep. program mgr. Lockheed Aircraft Svcs., Ontario, Calif., 1981-83; program mgr. Flight Systems Inc., Newport Beach, Calif., 1983-87; engring. mgr. Litton Applied Tech., San Jose, Calif., 1987-89; broker Edwards Real Estate, San Jose, 1989-91; tech. writer Comtech Svcs. Inc., San Jose, 1991-92, SK Writers, Santa Clara, Calif., 1993—. Pres. Presidents' Assn. Prince George's County, Md., 1971-72; bd. dirs. varisou civic assns., 1970-73; v.p. Toastmasters, Olney, Md., 1974-78; pres. Mission Club, Assn. Old Crows, Orange County, Calif., 1982-83, bd. dirs., 1983-84. With USN, 1951-54. Mem. Mensa. Achievements include co-invention of electronic counter-counter measure system. Home: 2921 Glen Darby Ct San Jose CA 95148 Office: SK Writers 3140 Dela Cruz Blvd Ste 200 Santa Clara CA 95054

EDWARDS, HAROLD HUGH, JR., aerospace engineer, management consultant; b. El Paso, Tex., Jan. 22, 1926; s. Harold Hugh Sr. and Alma Evelyn (McKnight) E.; m. LaRuth Buie, Aug. 28, 1948 (dec.); children: Richard, Margaret, Marianne; m. Donna Jane O'Steen, Oct. 14, 1966. BS, Tex. A&M Coll., 1948. Jr. engr. Chance Vought Airc., Stratford, Conn., 1948-49; design engr. Chance Vought Airc., Dallas, 1949-58; design specialist Vought Astronautics, Dallas, 1958-62; project engr. LTV Aerospace, Dallas, 1962-80, mgr. system effectiveness, 1980-88; pvt. practice cons. Dallas, 1988—; pres. Plano Conservatory for Young Artists, Dallas, 1991—. Vice pres. Preston Homeowners Assn., Dallas, 1991—; instr. profl. devel. engring. staff Mountain View Community Coll./LTV Aerospace, Dallas, 1958-80. With USAF, 1944-46, ETO. Mem. AIAA, Soc. Automotive Engrs. (chmn. Dallas chpt. 1961-62). Achievements include patent for air crew safety seat; pioneer in crew accomodations and rescue for spacecraft.

EDWARDS, JACK ELMER, personnel researcher; b. Charleston, W.Va., Mar. 3, 1955; s. Harold Kenneth and Mary Lee (Loomis) E. MS, Ohio U., 1979, PhD, 1981. Cons. Fed. Res. Bank Chgo., 1982-84; contractor Ryerson, an. Inland Steel Co., Chgo., 1984-85, Ameritech Svcs., Schaumburg, Ill., 1986-87; faculty researcher USN, San Diego, 1986-87, sabbatical faculty, 1988, contractor, 1986-89; contractor Murphy and Murphy Mktg. Communications, Chgo., 1983, 89; asst. prof. psychology Ill. Inst. Tech., Chgo., 1981-86, assoc. prof. psychology, 1986-89; pers. rsch. psychologist Navy Pers. Rsch. and Devel. Ctr., San Diego, 1989—; sci. advisor to Chief of Naval Personnel USN Navy Annex, Arlington, Va., 1992-93; tchr. Calif. Sch. Profl. Psychology, San Diego, 1990-92, San Diego State U., 1991, Calif. State U., San Marcos, 1990, Ohio U., Athens, 1978-81. Contbr. articles to profl. jours. Coord. United Way, 1981, 82, 84, 85. Recipient NPRDC Tech. Dir.'s Spl. award, 1990; Office Naval Tech. fellow, 1988. Mem. Soc. for Indsl. and Organizational Psychology, Inc., Am. Psychology Soc., Mil. Testing Assn., Pers. Testing Coun. San Diego, Beta Kappa Chi, Alpha Kappa Mu. Office: Navy Pers R&D Ctr 53335 Ryne Rd San Diego CA 92152-7250

EDWARDS, JIMMIE GARVIN, chemistry educator, consultant; b. Boswell, Okla., July 27, 1934; s. Lester Lee and Gladys Marie (Wright) E.; m. Carolyn Elaine Hatton, Oct. 29, 1956; children: Jon Timothy, Mary Susan, Elizabeth Anne. BS in Chemistry, U. Cen. Okla., 1956; PhD in Chemistry, Okla. State U., 1964. Instr. U. Nev., Reno, 1960-61; asst. prof. U. Mo., Rolla, 1965-66; from asst. prof. to prof. U. Toledo, 1967—; Contbr. articles to profl. jours. Democrat. Office: Univ Toledo Dept Chemistry Toledo OH 43606

EDWARDS, JOHN RALPH, chemist, educator; b. Streator, Ill., Feb. 27, 1937; s. Ralph E. and Ruth M. (Wilson) E.; m. Margaret E. Smith, July 15, 1961; children: Peter J., Sharon E., Susan D. BS, Ill. Wesleyan U., 1959; PhD, U. Ill., 1964. NIH postdoctoral fellow Tufts U., Boston, 1964-66; asst. prof. chemistry Villanova (Pa.) U., 1966-73, assoc. prof., 1973-80, prof., 1980—, chmn. dept. chemistry, 1980-90. Contbr. articles to profl. jours. Active Boy Scouts Am. NIH grantee, 1970-76. Mem. Am. Soc. Biochemistry and Molecular Biology, Am. Chem. Soc. (bd. dirs. Phila. sec.), U.S. Orienteering Fedn., NY Acad. Sci., Sigma Xi, Phi Kappa Phi. Office: Villanova U Dept Chemistry Villanova PA 19085

EDWARDS, JOHN WESLEY, JR., urologist; b. Ferndale, Mich., Apr. 9, 1933; s. John W. and Josephine (Wood) E.; m. Ella Marie Law, Dec. 25, 1954; children: Joella, John III. Student, Alma Coll., 1949-50; BS, U. Mich., 1954; postgrad., Wayne State U., 1954-56; MD, Howard U., 1960. Internship Walter Reed Gen. Hosp., 1960-61, surg. resident, 1962-63, urol. resident, 1963-66; asst. chief urology Tripler Army Med. Ctr., 1966-69; comdr. 4th Med. Battalion, 4th Infantry Div., Vietnam, 1969; chief profl. svcs., urology 91st Evacuation Hosp., Vietnam, 1969-70; urologist Straub Clinic, Inc., 1970-74; pvt. practice, 1974—; v.p. med. staff. svcs. Queen's Med. Ctr., Honolulu, 1993—; chief Dept. Surgery, Straub Clinic and Hosp., 1973; asst. chief Dept. Surgery Queen's Med. Ctr., 1977-79, chief, 1989-93; cons. in urology; chief Dept. Clin. Svcs., Kapiolani Women's and Children's Med. Ctr., 1981-83; clin. assoc. prof. U. Hawaii Sch. of Medicine. Contbr. articles to profl. jours. Bd. dirs. Am. Cancer Soc., Honolulu unit, 1977-79, Hawaii Med. Svc. Assn., 1979-85, Hawaii Heart Assn., 1977-79, Hawaii Assn. for Physicians Indemnification, 1980-86; commr. City and County of Honolulu Liquor Commn., 1986-89; mem. reorgn. commn. City and County of Honolulu, 1990-91. Recipient Howard O. Gray award for Professionalism, 1988, Leaders of Hawaii award, 1983. Fellow ACS (sec.-treas. Hawaii chpt. 1980-81, gov.-at-large 1986-92); mem. AMA, NAACP, Am. Urol. Assn. (alt. del. Western sect. 1991-92, gen. chmn. Western sect. 56th ann. meeting 1980, exec. com. 1983-84, del. dist. 1 1985-86, gen. chmn. 63d ann. meeting 1987, pres. 1989-90), Am. Coll. Physician Execs., Nat. Med. Assn., Hawaii Urol. Assn., Hawaii Med. Assn., Surgicare of Hawaii (v.p. 1983-86), Alpha Phi Alpha, Chi Delta Mu, Alpha Omega Alpha. Office: The Queen's Med Ctr 1301 Punch Bowl St Honolulu HI 96813

EDWARDS, JOHN WILLIAM, physicist; b. L.A., May 1, 1955; s. William Myron Edwards and Ann (Hopson) Houser. BA, Rice U., 1977; MPhil, Yale U., 1985, PhD, 1990. Asst. prof. physics U. Mo., Rolla, 1986-89, U. Nev., Las Vegas, 1989-90; rsch. scientist Nichols Rsch. Corp., Vienna, Va., 1992—. Vol. Arlington (Va.) Schs., 1992. Mem. Am. Phys. Soc. Office: Nichols Rsch Corp 1604 Spring Hill Vienna VA 22182

EDWARDS, KENNETH NEIL, chemist, consultant; b. Hollywood, Calif., June 8, 1932; s. Arthur Carl and Ann Vera (Gomez) E.; children: Neil James, Peter Graham, John Evan. BA in Chemistry, Occidental Coll., 1954; MS in Chem. and Metall. Engring., U. Mich., 1955. Prin. chemist Battelle Meml. Inst., Columbus, Ohio, 1955-58; dir. new products rsch. and devel. Dunn-Edwards Corp., L.A., 1958-72; sr. lectr. organic coatings and pigments dept. chem. engring. U. So. Calif., L.A., 1976-80; bd. dirs. Dunn-Edwards Corp., L.A.; cons. Coatings & Plastics Tech., L.A., 1972—. Contbr. articles to sci. jours. Mem. Am. Chem. Soc. (chmn. divisional activities 1988-89, exec. com. div. polymeric materials sci. and engring. 1963—), Alpha Chi Sigma (chmn. LA.A profl. chpt., 1962, pacific dist. counselor 1967-70, grand profl. alchemist nat. v.p. 1970-76, grand master alchemist nat. pres. 1976-78, nat. adv. com. 1978—). Achievements include patents for air-dried polyester coatings and application, for process and apparatus for dispensing liquid colorants into a paint can, and for mechanical mixers. Home: 2926 Graceland Way Glendale CA 91206-1331 Office: Dunn Edwards Corp 4885 E 52d Pl Los Angeles CA 90040

EDWARDS, KENNETH WARD, chemistry educator; b. Ann Arbor, Mich., Jan. 18, 1933; s. Daniel Ward and Florence Elizabeth (Bell) E.; m. Evelyn Ruth Gwinn Edwards, June 7, 1956; children: Jeffrey Thomas, Nancy Kay. BS in Chemistry, U. Mich., 1954; MA, Dartmouth U., 1956; PhD, U. Colo., 1963. Instr. Colo. Sch. Mines, Golden, 1957-60, asst. prof., 1966-71, assoc. prof., 1971-92, prof., 1992—; rsch. chemist U.S. Geol. Survey, Denver, 1960-66; pres. Natural Resources Lab., Golden, 1970-86; owner Quality Assurance Cons., Lakewood, Colo., 1991—. Mem. Am. Chem. Soc. Office: Colo Sch Mines Dept Chemistry & Geochem Golden CO 80401

EDWARDS, MARGO H., marine geophysicist, researcher; b. Omaha, Apr. 25, 1961; d. Stephen Jerome Gregorek and Rita Joy (Escover) Edwards; m. Roger Bowne Davis, Sept. 22, 1990. BS, Washington U., 1983, MA, 1986; PhD, Columbia U., 1992. Rsch. asst. Columbia U., N.Y.C., 1985-91; asst. researcher U. Hawaii, Honolulu, 1991—. Author (map series) Relief of the Surface of the Earth, 1985; contbr. articles to profl. jours. Vol. N.Y. AIDS Walk, N.Y.C., 1989, 90. Recipient Wheeler fellowship Washington U., 1984, Shell Oil fellowship, 1986. Mem. Am. Geophys. Union. Office: Univ Hawaii 2525 Correa Rd Honolulu HI 96822

EDWARDS, NED CARMACK, JR., agronomist, university program director; b. Wiggins, Miss., Aug. 28, 1942; s. Ned C. Sr. and Rosa Mae E.; m. Elvia E. Chambers, Aug. 21, 1965; children: April, Ned III. BS, Miss. State U., 1964, MS, 1966; PhD, U. Tenn., 1970. Asst. agronomist Miss. State U., Raymond, 1970-76; assoc. agronomist Miss. State U., Raymonn, 1976-83, agronomist, 1983-89; supt. Miss. State U., Poplarville, 1989—. Contbr. articles to profl. jours. Mem. Am. Soc. Agronomy, Miss. Forage & Grass Land Coun., Gama Sigma Delta, Sigma Xi, Alpha Zeta. Baptist. Office: Miss State Univ S Miss Expt Sta PO Box 193 Poplarville MS 39470

EDWARDS, PAUL BEVERLY, retired science and engineering educator; b. Ridge Spring, S.C., Nov. 12, 1915; s. Paul Bee and Chloe Agnes (Watson) E.; m. Sarah Dee Barnes, Apr. 10, 1943; 1 child, Susan Dee Edwards Von Suskil. BS, U. Tampa, 1937; EdM, Harvard U., 1958; EdD, George Washington U., 1972. Owner, operator Edwards' Hobbies, Tampa, Fla., 1938-54; tchr. math. Hillsborough High Sch., Tampa, 1955-60; head dept. math. King High Sch., Tampa, 1960-63; coord. Grad. Ctr., supr. edn. and tng. Johns Hopkins U. and Applied Physics Lab., Balt. and Laurel, Md., 1963-75; dir. Grad. Ctr., supr. edn. and tng. Johns Hopkins U. and Applied Physics Lab., Balt. and Laurel, 1975-81. Contbr. articles to profl. jours. Mem. Sun City Ctr. Voters League, 1989—, Community Assn., Sun City Ctr., 1987—; mem. Greenbriar Property Owners Assn., Sun City Ctr., 1987—. Lt. comdr. USNR, 1942-46. Named Meritorious Tchr., State of Fla., 1962; recipient various fellowships. Mem. Ret. Officers Assn., Naval Res. Assn., Sun City Ctr. Golf and Racquet Club. Avocations: swimming, computing, photography. Home: 1843 Wolf Laurel Dr Sun City Center FL 33573-6422

EDWARDS, RICHARD CHARLES, oral and maxillofacial surgeon; b. Oelwein, Iowa, Dec. 16, 1949; s. Charles Osborne Edwards and Eleanor Irene (Arness) Nardi; m. Celeste Mariel Frawley, Feb. 14, 1976; children: Travis Damien, Trina Demaris, Trent Dustin. BA, Adams State Coll., 1974; DDS, U. Colo., 1978. Diplomate Am. Bd. Oral and Maxillofacial Surgery. Intern, then resident in oral and maxillofacial surgery Wilmington (Dela.) Med. Ctr., 1978-81; oral and maxillofacial surgeon USAF Regional Med. Ctr., Clark AFB, Philippines, 1981-84; pvt. practice Rochester, Minn., 1984-85; chief oral and maxillofacial surgeon Air Univ. Regional Hosp., Montgomery, Ala., 1985-88; attending staff oral and maxillofacial surgery residency program David Grant Med. Ctr., Fairfield, Calif., 1988—, Highland Hosp., Oakland, Calif., 1989—; regional cons. for USAF, David Grant Med. Ctr., Fairfield, 1988—; lectr. Napa-Solano Dental Soc., Napa, Calif., 1989, Yosemite Dental Soc., Stockton, Calif., 1989, Panhandle Dental Soc., Panama City, Fla., 1987, Freeborne County Dental Soc., Albert Lea, Minn., 1984. Contbr. articles to profl. jours. Advanced cardiac life support instr. Am. Heart Assn., numerous states, 1979—. Lt. col. USAF, 1987—. Named Jr. Dental Officer of Yr., USAF, 1986; recipient David Grant Med. Ctr. Outstanding Educator award in oral and maxillofacial surgery, 1990-93. Fellow Am. Coll. Oral and Maxillofacial Surgeons; mem. Assn. Mil. Surgeons, USAF Clin. Surgeons Soc., Internat. Soc. Plastic, Aesthetic and Reconstructive Surgery, Lions (Rochester, Minn.). Republican. Avocations: fishing, skiing, scuba diving, golfing, sky diving. Home: 300 Bel Air Dr # 108 Vacaville CA 95687 Office: David Grant Med Ctr Travis AFB Fairfield CA 94535

EDWARDS, ROBERT MITCHELL, nuclear engineering educator; b. Dubois, Pa., Jan. 15, 1950; s. David Henry and Ethel Claire (Freedline) E.; m. Jacqueline Ferne Mentzer, Sept. 25, 1980; children: Eric Allen, Jonathan Robert. MS in Nuclear Engring., U. Wis., 1972; PhD in Nuclear Engring., Pa. State U., 1991. Registered profl. engr., Calif. Engr. Gen. Atomic, San Diego, Calif., 1972-76; sr. nuclear engr. Combustion Engring., Windsor, Conn., 1976-77; dir. software LeMont Scientific, State College, Pa., 1977-87; sr. project assoc. Pa. State U., University Park, Pa., 1987-89, rsch. asst., 1989-91, rsch. assoc., 1991, asst. prof. nuclear engring., 1991—; cons. Advanced Rsch. Instruments, Inc., Boulder, Colo., 1989—. Contbr. articles to profl. jours. 1st Lt. USAR, 1972-75. Mem. IEEE (com. mem.), Am. Nuclear Soc., Soc. Computer Simulation, Am. Soc. Engr. Edn. Home: 812 Wintergreen Circle State College PA 16801 Office: Pa State Univ 231 Sackett University Park PA 16802

EDWARDS, STEPHEN GLENN, air force officer, astronautical engineer; b. Berkeley, Calif., Sept. 17, 1964; s. Justin Sargent and Helen Louise (Creesy) E. BS in Astronautical Engring., USAF Acad., Colorado Springs, Colo., 1986; MS in Astronautical Engring., Air Force Inst. Tech., Dayton, Ohio, 1990. Commd. 2d lt. USAF, 1986, advanced through grades to capt., 1990; missile systems test engr. 6596th Test and Evaluation Group, USAF, Vandenberg AFB, Calif., 1986-88; sr. weapons systems analyst 6596th Test and Evaluation Group, USAF, Vandenburg AFB, Calif., 1988-89; project officer Ballistic Missile Orgn., Norton AFB, Calif., 1990-93, Pentagon, Va., 1993—. Mem. AIAA (treas. Vandenberg AFB chpt. 1987-89), Space Studies Inst., Nat. SpaceSoc. Achievements include design of new algorithm for maximizing grasp capability and minimizing slip for robotic grasps using a dextrous robotic hand. Office: USAF SAF/SP Sterling VA 20164

EDWARDS, VICTOR HENRY, chemical engineer; b. Galveston, Tex., Oct. 17, 1940; s. Philip Lacy and Margaret Ruth (Hopkins) E.; m. Mary Margaret Litzmann, June 10, 1963; children: Henry L., Mary E. BA, Rice U., 1962; PhD in Chem. Engring., U. Calif., Berkeley, 1967. Registered profl. engr., Tex. Asst. prof. chem. engring. Cornell U., Ithaca, N.Y., 1967-73; mgr. adv. tech. U.S. Nat. Sci. Found., Washington, 1971-72; research fellow Merck, Sharp, Dohme Research, Rahway, N.J., 1973-76; supr. research engring. United Energy Resources, Houston, 1976-79; vis. prof. environ. engring. Rice U., Houston, 1979-80; sr. process engr. Fluor Engrs. and Constructors, Houston, 1980-82; southwest editor Plant Services mag., Chgo., 1982-85; project engr. Allstates/BE&K, Inc., Houston, 1984-90, lead process engr., 1990-93, process engring. mgr., 1993—; cons., lectr. in field, 1968-92. Contbr. articles to profl. jours. Mem. organizing com. Woodlands (Tex.) Harvest Festival, 1979-86, Rice U. Alumni Assn., 1982, 87; chmn. industry adv. coun., dept. chem. engring. Prairie View A&M U., 1991—. Recipient Disting. Svc. award, dept. chem. engring. Prairie View A&M U., 1992. Mem. AIChE (South Tex. sect. Churchwell award 1981, chmn. Process Plant Safety Symposium-92, Disting. Svc. award 1991, program chmn. Process Plant Safety Symposium-94, Exec. Position One 1993), Am. Chem. Soc. (chmn. Ithaca sect. 1969, councilor 1970-77), Engrs. Coun. Houston (councilor 1987-92), N.Y. Acad. Scis. Methodist. Avocations: reading, tennis, sailing, golf. Office: Allstates Engring Co/BE&K Inc 140 Cypress Sta Dr Houston TX 77090

EDWARDS, WARD DENNIS, psychology and industrial engineering educator; b. Morristown, N.J., Apr. 5, 1927; s. Corwin D. and Janet W. (Ferriss) E.; m. Silvia Callegari, Dec. 12, 1970; children: Tara, Page. B.A., Swarthmore Coll., 1947; M.A., Harvard U., 1950, Ph.D., 1952. Instr. Johns Hopkins U., 1951-54; with Personnel and Tng. Research Center, USAF, Denver, 1954-56, San Antonio, 1956-58; research psychologist U. Mich., 1958-63, prof. psychology, 1963-73, head Engring. Psychology Lab., 1963-73; assoc. dir. Hwy. Safety Research Inst., 1970-73; prof. psychology and indsl. engring., dir. Social Sci. Research Inst., U. So. Calif., 1973-93; cons. in field. Author: (with J. Robert Newman) Multiattribute Evaluation, 1982, (with D.V. Winterfeldt) Decision Analysis and Behavioral Research, 1986; editor: (with A. Tversky) Decision Making: Selected Readings, 1967; editor: Utility Theories: Measurements and Applications, 1992; contbr. to Ency. Social Scis., 1968. Served with USNR, 1945-46. Recipient Franklin V. Taylor award Soc. Engring. Psychologists, 1978. Fellow Am. Psychol. Assn., Decision Scis. Inst.; mem. Western Psychol. Assn., Psychonomic Soc., Soc. Med. Decision-Making, Ops. Research Soc. Am. (Frank P. Ramsey medal 1988, pres.-elect Special Interest Group on Decision Analysis, 1992-94), Inst. Mgmt. Scis. (pres. Coll. Managerial Problem Solving 1987-88). Office: U So Calif Social Sci Rsch Inst Los Angeles CA 90089-1111

EDWARDS-HOLLAWAY, SHERI ANN, civil engineer; b. Iowa City, Iowa, Sept. 13, 1963; d. Glenn LaVerne and Joy Lou (Scott) Edwards; m. Joseph William Hollaway, Dec. 30, 1989; children: Lindsay Ann, Kristen Jane, Shelby Nicole. BS, Tex. A&M U., 1985, M in Engring., 1990. Registered profl. engr., Tex. Estimator II H.B. Zachry Co., San Antonio, 1985-87, engr., 1988; teaching asst. Tex. A&M U., College Station, 1987-88; engr. Becon Constrn. Co., Inc., Houston, 1988-90; asst. engr. Harris County Dept. Engring., Houston, 1990-91, engr., 1991-93; project engr. JNS Consulting Engrs., Inc., Houston, 1993—; assoc. prof. U. Houston, Downtown, 1991—. Adv. bd. San Jacinto Community Coll., Houston, 1992—. Mem. Tex. Soc. Profl. Engrs., NSPE. Home: 4019 Mission Valley Missouri City TX 77459

EDWARDSON, JOHN RICHARD, agronomist; b. Kansas City, Mo., Apr. 17, 1923; s. George Edward and Louise Marie (Sundstrom) E.; m. Betty Jo Cook, Aug. 24, 1948 (dec.); children: George, Elizabeth, Sarah; m. Mickie Newbill, Dec. 26, 1969. BS in Agr., Tex. A&M U., 1948, MS in Agronomy, 1949; PhD in Biology, Harvard U., 1954. Asst. agronomist U. Fla., Gainesville, 1953-60, assoc. agronomist, 1960-66, agronomist, 1966—. Author: Some Properties of the Potato Virus Y Group, 1974; co-author: Viruses Infecting Legumes, CRC Handbook, 1991; contbr. articles to Am. Jour. Botany. Staff sgt. U.S. Army, 1943-45, ETO. Fellow AAAS; mem.

Am. Genetic Assn., Am. Phytopathol. Soc. (Ruth Allen award 1993), Mediterranean Phytopathol. Union. Democrat. Achievements include research in describing the structure of potyvirus-induced cylindrical inclusions, using cylindrical inclusions for classification of potyviruses, using cylindrical inclusions in diagnosing infections induced by potyviruses. Home: 2721 SW 3d Pl Gainesville FL 32607 Office: U Fla Agronomy Dept PVL-Bldg #164 Gainesville FL 32611

EDWARDS-VIDAL, DIMAS FRANCISCO, mechanical engineer, consultant; b. Barahona, Dominican Republic, Jan. 5, 1965; came to U.S., 1974; s. Earl Wilbert Edwards and Zaida Maria Vidal. BSME, U. Puerto Rico, Mayaguez, 1987. HVAC designer Fluor Daniel Carbide Div., San Juan, P.R., 1987-88; resident engr. Basora & Lopez, San Juan, 1988; project engr. McNeil Pharm., San Juan, 1988-91, Pedro Panzardi and Assocs., San Juan, 1991—. Winner regional design contest ASME, U. P.R., Mayaguez Campus, 1985, 87, West Point, N.Y., 1987. Mem. Assn. Energy Engrs. Achievements include development and implementation of managerial concept of integrated environment, energy and maintenance management for Puerto Rico's industries. Home: Villa del Señorial low rise 50 apt 1E Rio Piedras PR 00926 Office: Pedro Panzardi & Assocs PO Box 2291 Hato Rey PR 00919-2291

EEROLA, OSMO TAPIO, research electrical engineer; b. Längelmäki, Finland, Sept. 19, 1956; s. Aimo Otto and Kerttu Lahja (Aho) E.; m. Lea Maria Heinonen, Sept. 5, 1981; children: Risto, Lauri, Jyrki. MSc in Elec. Engring., Tampere (Finland) U. Tech., 1980, Licentiate in Tech., 1993. Engr. LKB-Wallac, Turku, Finland, 1979-82; product specialist Internat. Mktg./LKB Wallac, Turku, 1983-84, LKB Instruments Ltd., Croydon, Eng., 1983; mng. dir. Biorec Oy, Kaarina, Finland, 1984-88, Integra Oy, Kaarina, Finland, 1986—; project mgr. Nokia Mobile Phones, Salo, Finland, 1988-90; rsch. engr. Turku U., 1990—; cons. Biorec Inc., Chgo., 1984-86; Nokia's rep. European Telecommunication Standards Inst./Group Speciale Mobile/Working Party 5 Intellectual property Rights, Sophia Antipolis, France, 1989-90. Contbr. sci. publs. in speech perception rsch. Hon. bd. dirs. Students' Assn. Dept. Elec. Engring., Tampere U. Tech., 1983. Sub lt., S.C., Finnish mil., 1982-83. Mem. Engring. Soc. Finland Assoc. bd. mem. 1991-93), Kaarina Jr. C. of C. (v.p. 1989). Lutheran. Avocations: piano, Lapland hiking, sci-fi, photography, caravan camping. Home: Metsätie 1 AS 8, SF-21620 Kuusisto Finland Office: Turku U, Cognitive Neuroscience, SF-20520 Turku Finland

EFFINGER, CHARLES EDWARD, JR., environmental/energy engineer; b. Louisville, Sept. 25, 1954; s. Charles Edward and Julia Catherine (Blanford) E.; m. Susan Adelaide Kaiser, July 9, 1976; children: Charles Edward III, Stacey Irene. ASME, Louisville Tech. Inst., 1979; BSBA in Fin., U. Louisville, 1993. Lic. energy mgr., water treatment plant operator, Ky., Ind.; backflow prevention device technician Ky., Ind. Design draftsman Am. Air Filter, Louisville, 1978, Chemetron Corp., Louisville, 1978-80; maint. planner ICI Americas, Inc., Charlestown, Ind., 1980-85, material coord., 1985-87, preventive maint. coord., 1987-88, energy engr., 1988-89, utilities engr., 1989-92; utilities engr. CC Joyce, Inc., Clarksville, 1992-93; project mgr. Reliable Mech., Inc., Louisville, 1993—; Author: Standard Operating Procedure/Maintenance Coatings Applications, 1986, Energy Conservation, 1989, Portable Water Distribution, 1992. Recipient Energy Systems Tech. award U.S. Army C.E., 1991, U.S. Army Installation Energy award U.S. Army Material Command, 1988, 89. Mem. Assn. Energy Engrs., Bluegrass Cross-Connection Prevention Assn., Am. Backflow Prevention Assn., Ky. Cols. Democrat. Home: 2251 Bradford Dr Louisville KY 40218

EFFROS, RICHARD MATTHEW, medical educator, physician; b. N.Y.C., Dec. 10, 1935. BA, Columbia U., 1957; MD, NYU, 1961. Intern, then resident NYU Sch. Medicine, 1961-64, fellow in nephrology, 1964-66, fellow in cardiopulmonary, 1966-68; instr. medicine Goldwater Meml. Hosp., N.Y.C., 1967-68; asst. prof. N.J. Coll. Medicine, 1968-71, assoc. prof., 1971-74; assoc. prof. Harbor-UCLA Med. Ctr., Torrance, Calif., 1974-80, prof., 1980-89; prof., chief pulmonary & critical care Med. Coll. Wis., Milw., 1989—; mem. adv. com. Nat. Heart, Blood, Lung Inst., 1975-93, Am. Heart Assn., 1977-89, Am. Lung Assn., 1986-90, VA Merit Rev., 1989-92. Contbr. articles to Sci., Jour. Applied Physiology, Jour. Clin. Investigations, Circulation Rsch. Recipient Career Devel. award NIH, 1973-78; grantee pulmonary edema NIH, 1974—. Mem. ACP, Am. Physiol. Soc., am. Thoracic Soc., Am. Coll. Chest Physicians. Achievements include research in the documentation of carbonic anhydrase activity on endothelium of lungs and other organs, in vivo measurements of intracellular pH in organs, active Na+ transport in lungs, detection of lung injury with scanning procedure, urea transporters in liver. Home: 9360 N Broadmoor Rd Bayside WI 53217 Office: Med Coll Wis 9200 W Wisconsin Ave Milwaukee WI 53226

EFRON, BRADLEY, mathematics educator; b. St. Paul, May 24, 1938; s. Miles Jack and Esther (Kaufman) E.; m. Gael Guerin, July 1969 (div.); 1 son, Miles James; m. Nancy Troup, June 1986 (div.). B.S. in Math., Calif. Inst. Tech., 1960; Ph.D., Stanford U., 1964. Asst. and assoc. prof. stats. Stanford (Calif.) U., 1965-72, chmn. dept. stats., 1976-79, 1991—, chmn. math. scis., 1981—, prof. stats., 1974—, assoc. dean humanities and scis., 1987-90, endowed chair Max H. Stein prof. humanities and scis., 1987—, chmn. dept. stats., 1991—; statis. cons. Alza Corp., 1971—, Rand Corp., 1962—, Aprex Corp., 1986. Author: Bootstrap Methods, 1979, Biostatistics Casebook, 1980. MacArthur Found. fellow, 1983; named Outstanding Statistician of Yr. Chgo. Statis. Assn., 1981; Wald and Rietz Lectr. Inst. Math. Stats., 1977, 81. Fellow Inst. Math. Stats. (pres. 1987), Am. Statis. Assn. (Wilks medal 1990); mem. Internat. Statis. Assn., Nat. Acad. Scis. Democrat. Office: Stanford U Dept Stats Sequoia Hall Stanford CA 94305

EFRON, ROBERT, neurology educator, research institute administrator; b. N.Y.C., Dec. 22, 1927; s. Alexander and Rose (Kunitz) E.; m. Mary Louise Snyder, June 6, 1948 (div. 1966); children: Carol, Paul, Sonni; m. Barbara Klein, Dec. 30, 1967. BA, Columbia U., 1948; MD cum laude, Harvard U., 1952. Med. house officer Peter Bent Brigham Hosp., Boston, 1952-53; Moseley traveling fellow Harvard U., Boston, 1953-54; rsch. assoc. Nat. Hosp. Queen Sq., London, 1956-60; asst. prof. Boston U. Sch. Medicine, 1960-70; assoc. chief staff R & D VA Med. Ctr., Martinez, Calif., 1970—; prof. neurology U. Calif. Sch. Medicine, Davis, 1974—; pres. East Bay Inst. for Rsch. and Edn., Martinez, 1989—; MacEachran lectr. U. Alta., Can., 1989. Author: Decline and Fall of Hemisphere Assymmetry, 1990; contbr. articles to profl. jours. Lt. USNR, 1954-56. Fellow Acoustical Soc. Am.; mem. Phi Beta Kappa, Alpha Omega Alpha. Home: 8 Honey Hill Rd Orinda CA 94563-1512 Office: VA Med Ctr Rsch Adminstrn Office 150 Muir Rd Martinez CA 94553-4695

EGAN, BRUCE A., engineering consultant. AB in Engring., Harvard Coll.; SM in Mech. Engring., Harvard U., ScD in Environ. Health Scis. With engring. and applied physics labs. Harvard U.; with Ednl. Devel. Ctr.; chief scientist ENSR Consulting and Engring.; vis. lectr. sch. pub. health Harvard U.; mem. Commonwealth Mass. Pesticide Bd.; cons. Office Tech. Assessment, Washington, 1975; panel leader workshop develop recommendations atmospheric dispersion models complex terrain U.S. EPA, Raleigh, N.C., 1979, panel leader workshop role atmospheric models regulatory decision making, Airlie House, Va., 1981, mem. peer rev. com. office rsch. and devel., 1985, mem. rsch. grant rev. panel, Durham, N.C., 1991; mem. Nat. Commn. Air Quality Atmospheric Dispersion Modeling Panel, Washington, 1979; mem. Argonne U. Assn. rev. com. energy and environ. systems divsn. Argonne Nat. Lab., 1980-82, mem. U. Chgo. rev. com. energy and environ. systems divsn., 1983-85; participant geochemical and hydrological processes and their protection expert panel NSF/Coun. Environ. Quality, 1984; mem. rev. com. Ill. state water survey program atmospheric scis. DOE, 1985, mem. rev. com. atmospheric scis. complex terrain program, 1986; chmn. task group III peer rev. panel completed work Nat. Acid Precipitation Assessment Program, 1988; co-chmn., speaker air toxics regulation conf. Exec. Enterprises, Inc., Washington, 1990; presenter in field. Mem. Am. Meterol. Soc. (cert., mem. com. meteorol. aspects air pollution, 1975-79, chmn. 1976, 77, program chmn. AMS/APCA joint conf. applications air pollution meteorology 1977, mem. steering group AMS/EPA coop. agreement 1979-82, subgroup chmn. AMS/EPA workshops 1980, 82, mem. awards com., chmn. AMS/EPA workshop dispersion complex terrain, 1983), Air and Waste Mgmt. Assn. (mem. AB-3 com., editorial rev. bd. jour. 90, 92). Achievements include research in air quality modeling and application. Office: ENSR Consulting & Engring 35 Nagog Park Acton MA 01720-3423*

EGAN, JOHN THOMAS, computer scientist; b. Troy, N.Y., Mar. 20, 1937. BS, St. Louis U., 1965; MS, SUNY, Buffalo, PhD, 1976. Programmer/analyst Bell Aerospace, Niagara Falls, N.Y., 1967-70; asst. prof. SUNY, Buffalo, 1970-77; computer systems engr. NASA-Ames Rsch. Ctr., Moffet Field, Calif., 1977-80; computer scientist U.S. Naval Rsch. Lab., Washington, 1980-87, staff computer scientist, 1987—; cons. Spawar/Navelex, Washington, 1980—, Opnav-94, Washington, 1990—; adj. prof. George Mason U., Fairfax, Va., 1982—. Contbr. articles to profl. publs. Grantee Am. Soc. Elec. Engrs., 1977, 80; postdoctoral fellow NASA-Ames, 1978-79. Mem. AAAS, Assn. Computer Machinery, N.Y. Acad. Scis. Achievements include work on naval sensor system, algorithms to support sensor integration, process for evaluating large-scale naval warfare architectures. Office: US Naval Rsch Lab 4555 Overlook Ave Washington DC 20375-5000

EGAR, WILLIAM THOMAS, information systems consultant; b. Evergreen Park, Ill., July 16, 1955; s. William Patrick and Mary Ann (Byrne) E.; m. Sarah Ann Robinson, Dec. 1, 1984; children: Gemma, Neil. BS in Computer Sci., Utah State U., 1978. Programmer Ore-Ida Foods, Boise, Idaho, 1978-80; software specialist Digital Equipment Corp., Bloomington, Minn., 1980-89; cons. Integrated Systems Engring., Inc., Excelsior, Minn., 1989-92; v.p. info. systems Clin. Pharmacy Systems, Edina, Minn., 1993—. Mem. Assn. Computing Machinery, Digital Equipment Corp. User Soc., IEEE Computer Soc., Nat. Coun. Prescription Drug Programs.

EGBERT, ROBERT IMAN, electrical engineering educator, academic administrator; b. May 25, 1950. BSEE, U. Mo., Rolla, 1972, MSEE, 1973, PhD, 1976. Registered profl. engr., Mo., Kans. Grad. teaching asst. U. Mo., Rolla, 1972-75, grad. instr., 1975-76; systems engr. power div. Black & Veatch Cons. Engrs., Kansas City, Mo., 1976-80; asst. prof. elec. engring. Wichita (Kans.) State U., 1980-86, assoc. prof., 1986—, dir. Ctr. for Energy Studies, 1987—. Contbr. articles to profl. jours. Mem. IEEE (sr.), NSPE, Am. Soc. Engring. Edn. (Dow Outstanding Young Faculty award 1982-83), Eta Kappa Nu, Phi Kappa Phi, Tau Beta Pi, Sigma Xi. Office: Wichita State U Ctr Energy Studies Campus Box 44 Wichita KS 67208

EGBOGAH, EMMANUEL ONU, petroleum engineer, geologist; b. Nigeria, Sept. 14, 1942; m. Chirota Okoli; children: Emy, Liza, Shirley. MSc in Applied Petroleum Geology, Friendship U., Moscow, 1969; MSc in Petroleum Engring., U. Alta., Edmonton, Can., 1975; PhD in Petroleum Engring., Imperial Coll., London, 1979. Cons. petroleum engr. oil and gas div. Dept. Indian and No. Affairs, Info. and Liaison, ACND, Ottawa, Ont., Can., 1973-77; cons. Porta-Test Mfg. Ltd., Edmonton, Alta., Can., 1975-76; assoc. cons. reservoir engr. Applied Geosci. and Tech. Cons. Ltd., Calgary, Alta., 1979; tech. svcs. mgr. United Petro Labs. Ltd., Calgary, 1980-81, mgr. reservoir engring., 1981-82; v.p. engring., dir. rsch. Applied Geosci. and Tech. Cons. Ltd., Calgary, 1981-83; pres. EMEG Engring. Ltd., Calgary, 1981—; petroleum engring. mgr., enhanced oil recovery specialist Amerigo Tech. Ltd., Calgary, Houston, 1983-85; enhance oil recovery advisor Nat. Oil Corp., Tripoli, 1985-90; tech. advisor petroleum engring. dept. Petronas Carigali SDN BHD subs. Petroliam Nasional Berhad, Kuala Lumpur, Malaysia, 1991—; teaching and rsch. asst. Sedimentary Geology Labs. dept. geol. scis. McGill U., Montreal, Que., Can., 1970-71, dept. mineral engring. U. Alta., Edmonton, 1973-76; lectr. in petroleum tech. U. Ibadan, Nigeria, 1971-72, Pioneer Nigerian lectr. in petroleum engring. Inst. Applied Sci. and Tech., 1974-75, sr. lectr. in petroleum and natural gas engring., 1979-80; teaching asst. Petroleum Engring. Lab., Royal Sch. Mines, Imperial Coll. Sci. and Tech., London, 1978-79; presenter in field. Contbr. articles, reports to profl. publs. Judge Calgary Youth Sci. Fair, 1981-85; Grantee U. Ibadan, 1977-79, Govt. of Can., 1981-83, Nat. Scis. and Engring. Rsch. Coun. Can., 1984, 85. Mem. AAAS, Soc. Petroleum Engrs., Petroleum Soc. of Can. Inst. mining and Metallurgy, Am Soc. Engring. Edn., Can. Rock Mechanics Assn., Internat. Rock Mechanics Assn., Assn. Profl. Engrs., Geologists and Geophysicists of Alta. Achievements include research on effects of heterogeneous wettability on fluid flow and fluids distribution in porous media, mechanics of fluid flow in porous media, reservoir sensitivity and completion procedures in sandstone reservoirs and integrated reservoir management. Office: Petronas Carigali Sdn Bhd, Petroleum Engring Dept, PO Box 12407, 50776 Kuala Lumpur Malaysia

EGBUONU, ZEPHYRINUS CHIEDU, civil engineer, electrical engineer; b. Onitsha, Nigeria, Mar. 4, 1965; came to U.S., 1984; s. Boniface and Cathrine Egbuonu. BS in Civil Engring., Prairie View A&M U., 1991, BSEE, 1991. Asst. resident engr. Caltran's, San Luis Obispo, Calif., 1991; asst. project engr. Collection Systems Engring. div. City of L.A., 1991-92, project engr. Bureau of Engring., 1993—. Mem. IEEE, ASCE (assoc.). Roman Catholic. Home: PO Box 862431 Los Angeles CA 90086-2431

EGDAHL, RICHARD HARRISON, surgeon, medical educator, health science administrator; b. Eau Claire, Wis., Dec. 13, 1926; s. Harry I. and Rebecca (Ball) E.; children: Scott, David, Bruce, Julie; m. 2d, Cynthia Taft, Apr. 1983. M.D., Harvard U., 1950; Ph.D., U. Minn., 1957. Intern U. Minn. Hosp., 1950-51, resident, 1956-57; prof. surgery Med. Coll. Va., 1957-64; prof., chmn. surgery Boston U. Med. Ctr., 1964-73, dir., 1973—; also acad. v.p. for health affairs, dir. Health Policy Inst., Boston U.; bd. dirs. Essex Investment Mgmt. Co. Inc., Peer Rev. Analysis Inc., Health Payment Rev. Inc., Mediplex, Pioneer Family of Mut. Funds. Editor: Comprehensive Manuals of Surgical Specialties; mem. editorial bd. Am. Jour. Surgery, World Jour. Surgery. Lt. USNR, 1952-55. Mem. ACS, Soc. Univ. Surgeons (pres. 1970-71), Am. Physiol. Soc., Am. Soc. Clin. Investigation, Am. Soc. Exptl. Pathology, Am. Surg. Assn. (1st v.p. 1980), Boston Surg. soc. (pres. 1977), Soc. Med. Adminstrs., Endocrine Soc. (CIBA award 1961), Inst. Medicine Nat. Acad. Scis., Internat. Assn. Endocrine Surgeons (pres. 1981-83), Comml. Club, Brookline Country Club, Algonquin Club, Badminton and Tennis Club, The Registry Resort, Phi Beta Kappa, Alpha Omega Alpha. Home: 333 Commonwealth Ave Apt 23 Boston MA 02115-1931 Office: Boston U Health Policy Inst 53 Bay State Rd Boston MA 02215-2197

EGEL, CHRISTOPH, computer scientist; b. Koblenz, Germany, Mar. 21, 1962; s. Felix and Hanni (Grünschlag) Hachenburg; m. Sylvia Schneider, Aug. 8, 1988; children: Sarah-Ruth, Ben-Marvin. Diploma in Computer Sci., Darmstadt (Germany) Tech. Hochschule, 1987. Programmer Linde AG, Mainz, Germany, 1981-83; system mgr. Linde AG, Mainz, 1984-85, mgr. tech. data processing, 1987-90, mgr. tech. orgn., 1991—; prototype developer IBM Wissenschaftliches Zentrum, Heidelberg, Germany, 1986. Author: Mengenorientierte I/O's, 1987. Named Studienstiftung des Deutschen Volkes, Bad Godesberg, 1980. Mem. Gesellschaft für Informatik, Assn. for Computing Machinery. Avocations: biking, billiards, reading, cooking. Home: Turmstrasse 10, 55120 Mainz Germany Office: Linde AG, Koslheimer Landstr 21, 55246 Mainz Kostheim, Germany

EGGEN, OLIN JEUCK, astrophysicist, administrator; b. Orfordville, Wis., July 9, 1919; s. Olin J. and Bertha Clara (Jeuck) E. B.A., U. Wis., 1940, Ph.D. in Astrophysics, 1948. Astronomer Lick Obs., Calif., 1948-56; chief asst. astronomer Royal Greenwich Obs., Eng., 1956-61; prof. Calif. Inst. Tech., Pasadena, 1961-63; chief sci. officer Royal Greenwich Obs., Eng., 1964; astronomer Mt. Wilson Obs., Calif., 1965; dir. Mt. Stromlo and Siding Spring Obs., 1966-77; prof. Australian Nat. U., Canberra, 1966-77; sr. astronomer, acting dir. Cerro Tololo Inter-Am. Obs., La Serena, Chile, 1977—. Contbr. numerous articles to profl. jours. Named Pawsey Meml. lectr. Australian Inst. Physics, 1967, Henry Norris Russell lectr. Am. Astron. Soc., 1985. Mem. Royal Astron. Soc. (v.p. 1961-62), Australian Soc. Astronomers (pres. 1971-72). Home and Office: Observatoria Interam de Cerro Telolo, Casilla 603, La Serena Chile

EGGERT, ROBERT JOHN, SR., economist; b. Little Rock, Dec. 11, 1913; s. John and Eleanora (Fritz) Lapp; m. Elizabeth Bauer, Nov. 28, 1935; children: Robert John, Richard F., James E. BS, U. Ill., 1935, MS, 1936; candidate in philosophy, U. Minn., 1938; LHD (hon.), Ariz. State U., 1988. Research analyst Bur. Agrl. Econs., U.S. Dept. Agr., Urbana, Ill., 1935; prin. marketing specialist War Meat Bd., Chgo., 1943; research analyst U. Ill., 1935-36; rsch. analyst U. Minn., 1936-38; asst. prof. econs. Kans. State Coll., 1938-41; asst. dir. marketing Am. Meat Inst., Chgo., 1941-43; economist, assoc. dir. Am. Meat Inst., 1943-50; mgr. dept. marketing research Ford div. Ford Motor Co., Dearborn, Mich., 1951-53; mgr. program planning Ford div. Ford Motor Co., 1953-54, mgr. bus. research, 1954-57, mgr. marketing research marketing staff, 1957-61; mgr. marketing research Ford div. Ford Motor Co. (Ford div.), 1961-64, mgr. internat. marketing research marketing staff, 1964-65, mgr. overseas marketing research planning, 1965-66; mgr. marketing research Ford div. Ford Motor Co. (Lincoln-Mercury div.), 1966-67; dir. agribus. programs Mich. State U., 1967-68; staff v.p. econ. and marketing research RCA Corp., N.Y.C., 1968-76; pres., chief economist Eggert Econ. Enterprises, Inc., Sedona, Ariz., 1976—; lectr. mktg. U. Chgo., 1947-49; adj. prof. bus. forecasting No. Ariz. U., 1976—; mem. econ. adv. bd. U.S. Dept. Commerce, 1969-71, mem. census adv. com., 1975-78; mem. panel econ. advisers Congl. Budget Office, 1975-76; interim dir. Econ. Outlook Ctr. Coll. Bus. Adminstrn. Ariz. State U., Tempe, 1985-86, cons., 1985—; mem. Econ. Estimates Commn. Ariz., 1979—; apptd. Ariz. Gov.'s Commn. Econ. Devel., 1991—, Investment Adv. Coun. Ariz. State Retirement System, 1993—; bd. trustees Marcus J. Lawrence Med. Ctr. Found., 1992—. Contbr. articles to profl. lit.; editor: monthly Blue Chip Econ. Indicators, 1976—; exec. editor Ariz. Blue Chip, 1984—, Western Blue Chip Econ. Forecast, 1986—, Blue Chip Job Growth Update, 1990—, Mexico Consensus Econ. Forecast, 1993—. Elder Ch. of Red Rocks. Recipient Econ. Forecast award Chgo. chpt. Am. Statis. Assn., 1950, 60, 68; Seer of Yr. award Harvard Bus. Sch. Indsl. Econs., 1973. Mem. Coun. Internat. Mktg. Rsch. and Planning Dirs. (chmn. 1965-66), Am. Mktg. Assn. (bd. dirs. 1949-50, pres. Chgo. chpt. 1947-48, v.p. mktg. mgmt. div. 1972-73, nat. pres. 1974-75), Am. Statis. Assn. (chmn. bus. and econ. stats. sect. 1957—, pres. Chgo. chpt. 1948-49), Fed. Stats. Users Conf. (chmn. trustees 1960-61), Conf. Bus. Economists (chmn. 1973-74), Am. Quarter Horse Assn. (dir. 1966-73), Nat. Assn. Bus. Economists (coun. 1969-72), Alpha Zeta, Ariz. Econ. Roundtable, Am. Econs. Assn., Phoenix Econ. Club (hon.), Ariz. C. of C. (bd. dirs.). Republican. Club: Poco Diablo Country. Office: Eggert Econ Enterprises Inc PO Box 2243 Sedona AZ 86339-2243

EGGLESTON, DRAKE STEPHEN, pharmaceutical researcher; b. Charleston, W.Va., June 18, 1954; s. Joseph William and Georgetta (Uhl) E.; m. Emily Turner, Dec. 29, 1979; children: Adam, Matthew, Michael. BA, Wake Forest U., 1976, MS, 1979; PhD, U. N.C., 1983. Assoc. sr. investigator Smith Kline Beecham, Phila., 1985-87, sr. investigator, 1987-90, assoc. fellow, 1990—; U.S. Nat. com. for cyrstallography Nat. Acad. Sci. Adv., 1992—, sec. treas., 1993—. Contbr. 112 articles to profl. jours. Named Outstanding Grad. Student fellowship U. N.C., 1982. Mem. AAAS, Am. Inst. Physics (devel. com 1988—), Am. Crystallographic Assn. (devel. officer 1988—, small molecule sig chair 1992), Am. Peptide Soc., Sigma Xi, Omicron Delta Kappa. Home: 939 Dogwood Hill Rd West Chester PA 19380 Office: Smith Kline Beecham 709 Swedeland Rd King Of Prussia PA 19406

EGHBAL, MORAD, geologist, lawyer; b. Tehran, June 7, 1952; s. Mohammad Ali and Fari Eghbal; m. Niloofar Sadjadi, July 17, 1983; 1 child, Elaheh. BA, George Washington U., 1975, MA, 1977, JD, Howard U., 1989, LLM, U. of the Pacific, 1991. Asst. George Washington U., Washington, 1972; asst. to dir. Smithsonian Instn., Washington, 1972-75; spl. advisor to dir. Georgetown U., 1975; cons. Leo A Daly, Washington, 1975, Kodak, Rochester, N.Y., 1976; chief exec. officer MERE Enterprises, Washington, 1976-87; fgn. assoc. Pestalozzi, Gmuer & Heiz, Zurich, Switzerland, 1989; law clk. Hon. William B. Bryant, U.S. Dist. Ct. D.C., 1990-91; trustee, chief fin. officer Riess Found., Washington, 1983—; dir., pres. The Grail Corp.; dir., v.p. exploration Gasco, Inc.; judge oral arguments and memls. internat. semi-finals Jessup Competition Internat. Law Students Assn., 1990-93. Researcher: The Divining Hand (E.P. Dutton), 1973-79. Keynote speaker symposium Dickinson Sch. Law, Carlisle, Pa., 1991. Recipient Cert. of Achievement, Pacific Energy & Mineral Resources Conf., 1978, Ga. U., 1980, 2d Place Nat. Roscoe Hogan Environ. Law Essay Contest award Assn. Trial Lawyers Am., 1988, finalist, 1989, Outstanding Student Advocate award, Met. Trial Lawyers Assn. Mem. ABA, Am. Assn. Petroleum Geologists (founding mem. energy minerals div.), Geol. Soc. Am., Soc. Econ. Paleontologists and Mineralogists, Potomac Appalachian Trails Club, Nat. Capital Area Paralegal Assn., Nat. Bar Assn., Internat. Law Soc., Nat. Lawyers Club, U.S. Japan Trade Coun., Am. Inns Ct. (Preftyman/Leventhal chpt.), Phi Delta Phi. Address: Riess Found PO Box 9555 Washington DC 20016

EGLE, DAVIS MAX, mechanical engineering educator; b. New Orleans, Jan. 31, 1939; s. Merlin Joseph and Leona (Pujol) E.; m. Judith Johanna Reynolds, June 1, 1963; children: Robert, William. BSME, La.State U., 1960; MS, Tulane U., 1962, PhD, 1965. Registered profl. engr., Okla. Asst. prof., then. assoc. prof. U. Okla., Norman, 1965-73, prof. mech. engring., 1973—, dir., 1981-90; mech. engr. NASA Langley Rsch. Ctr., Hampton, Va., 1966-67; rsch. engr. Lawrence Livermore (Calif.) Nat. Lab., 1979; biomed. engr. Sports Medicine Specialists, Oklahoma City, 1990-91. Contbr. articles to profl. jours. Fellow Acoustic Emission Working Group (chair 1982-84); mem. ASME, Am. Soc. Nondestructive Testing (chair tech. com. 1979-82, Achievement award 1980). Office: U Okla Sch AME 865 Asp Ave Norman OK 73019-0001

EGLER, FRANK EDWIN, ecologist, administrator; b. N.Y.C., Apr. 26, 1911; s. Charles John and Florence Edna (Wilshusen) E.; m. Happy Hitchel, June 6, 1968 (dec.). BS, U. Chgo., 1932; MS, U. Minn., 1934; PhD, Yale U., 1936. Asst. prof. N.Y. State Coll. Forestry, Syracuse, 1937-44; dir. Chicle Devel. Co. Exptl. Sta., Belize, 1941-44; pres. Aton Forest Inc., Norfolk, Conn., 1945—. Author 6 books; contbr. over 400 articles, reports, and revs. to profl. jours. With USAAF, 1941-45. Guggenheim fellow, 1956-58. Fellow AAAS, Am. Geog. Soc., Am. Mus. Natural History; mem. Ecol. Soc. Am., 40 internat., nat. and state orgns. Achievements include development of natural areas sensy stricto; founder of right-of-way vegetation management, esthetic landscape vegetation management. Home and Office: Aton Forest N Colebrook Rd Norfolk CT 06058

EGO-AGUIRRE, ERNESTO, surgeon; b. Lima, Peru, Dec. 16, 1928; came to U.S., 1956; s. Ernesto and Benjamina (Palma) E. B in Medicine, San Marcos U., 1947, MD, 1955; postgrad. in biometrics, Cornell U., 1965-66. Diplomate Am. Bd. Surgery. Intern in surgery Barnes Hosp. Washington U., St. Louis, 1956-57, resident asst. in surgery, 1957-60; asst. and sr. resident in surgery Ellis Fischel Cancer Hosp., Columbia, Mo., 1960-62; spl. fellow plastic and reconstructive surgery Meml. Sloan Kettering Cancer Ctr., N.Y.C., 1963-64, sr. resident, fellow in surgery, 1965-69; fellow in rsch. Sloan Kettering Inst., N.Y.C., 1964-69; attending in surgery N.Y. Infirmary, 1970-78, Doctors Hosp. (now Beth Israel Med. Ctr.-North), N.Y.C., 1978—. Contbr. articles to profl. jours. Fellowship grant Nat. Cancer Inst., 1965-68, Am. Cancer Soc., 1968-69; recipient 8 Continuous Med. Edn. award AMA, 1969—. Mem. N.Y. County Med. Soc., N.Y. Med. Soc., N.Y. Acad. Scis., Vet. Corps. of Artillery (N.Y.), Mil. Order of the Loyal Legion, Sovereign Orthodox Order Knights Hospitallier of St. John, Seventh Regiment Rifle Club (N.Y.), Down Town Assn. Republican. Roman Catholic. Achievements include research in heterotopic whole liver transplantation in dogs, liver regeneration after major hepatectomy in normal, and in azathioprine treated, dogs, compartmental analysis using I131 tagged Rose Bengal; research on human lung function subsequent to storage and transplantation into a xenograft system, on erlich ascitis tumor cell growth in the presence of gastric mucin. Avocations: opera, historic military societies, target shooting. Home: 209 E 56th St New York NY 10022-3705 Office: 135 William St Fl 10 New York NY 10038-3805

EGOLF, KENNETH LEE, chemistry educator; b. Carlisle, Pa., Jan. 12, 1938; s. Raymond Cyrus and Olive Louise (Mayhugh) E.; m. June Louise Enck, July 19, 1959; children: Debra Sue, Leanne Michelle, David Andrew. BS, Dickinson Coll., 1959; MS, Purdue U., 1966; PhD, U. Md., 1978. Cert. tchr., Pa. Rsch. chemist Pennsalt Chem. Corp., Chestnut Hill, Pa., 1960-61; tchr. chemistry Carlisle Area Sch. Dist., 1962—; asst. prof. Dickinson Coll., Carlisle, 1985—; pres. Carlisle Area Sci. Adv. Com., 1985-86, com. chair. Co-editor: REACTS 1973. Pres. coun. 1st Luth. Ch., Carlisle, 1969, 72-74. 2d lt. U.S. Army, 1959-60. Mem. Nat. Sci. Tchrs. Assn. (reviewer 1973—). Home: 809 Fairview Rd Carlisle PA 17013

EHLER, HERBERT, computer scientist; b. Passau, Bavaria, Germany, Sept. 8, 1958; s. Josef and Marianne (Sattler) E.; m. Sonja Katharina Hackl, Oct. 4, 1985; 1 child, Manuel Lorenz. Diplom. Vorprüfung, Technische U. München, Munich, Germany, 1981; MS, Ohio State U., 1983; Dr. Rer. Nat., Technische U. München, Munich, Germany, 1989. Teaching employee Ohio State U., Columbus, 1983-84; rsch. employee Technische U. Münich, 1984-89; rsch. asst. Technische U. Münich, München, 1989—; organizing com. Ferienakademie, Tech. U. München, 1988—; observer Internat. fedn. for Info. Processing, 1989—. Co-author: The Munich Project CIP, Vol. II, 1987; contbr. to books and articles in field. Recipient Franz-und-Maria Stockbauer grant, 1980, 82, Oskar-Karl-Forster grant Dist. of Niederbayern, 1976, grant State of Bavaria, 1979-83, Fulbright grant German Fulbright Commn., 1982-83. Mem. Gesellschaft Für Informatik, European Assn. for Theoretical Computer Sci., Bund der Freunde der Technischen Universität Müchen. Avocations: travel, water sports. Home: Rudlfinger Str 14a, D-85417 Marzling Bavaria, Germany Office: Inst für Informatik, Arcisstr 21, Arcisstr 21, D-80290 Bavaria Bavaria, Germany

EHMANN, CARL WILLIAM, consumer products executive, researcher; b. Buffalo, Aug. 30, 1942; s. Christian John and Grayce E. (Packer) E.; m. Elaine Ann O'Gorek, June 22, 1968; children: Elayne Grayce, Karen Beth. BA, SUNY, Buffalo, 1963, MD, 1967. Diplomate Am. Bd. Dermatology. Intern Buffalo Deaconess Hosp., 1967-68; resident U. Wash. Sch. Medicine, Seattle, 1970-72; asst. prof. dermatology SUNY, 1973-75, clin. asst. prof., 1983-87; pvt. practice, Virginia, Minn., 1975-79; dir. clin. rsch. in dermatology Hoffman-LaRoche, Nutley, N.J., 1979-83; v.p. rsch. in dermatology Bristol-Myers & Westwood, Buffalo, 1983-87; v.p. R & D, Johnson & Johnson Baby Products Co., Skillman, N.J., 1987-89; exec. v.p. R & D, Johnson & Johnson Consumer Products Worldwide, Skillman, 1989-92; exec. v.p. R & D R. J. Reynolds Co., Winston-Salem, N.C., 1992—; clin. assoc. prof. dermatology U. Medicine and Dentistry N.J., Piscataway, 1987—, U. Pa., Phila., 1988-92, Wake Forest U. Bowman Gray Sch. Medicine, Winston-Salem, N.C., 1993—; clin. asst. prof. skin and cancer NYU, N.Y.C., 1986-89; mem. adv. bd. Touch Rsch. Inst., U. Miami, Nat. Ichthyosis Found. Surgeon USPHS, 1968-70. Fellow ACP, Am. Acad. Dermatology, Am. Acad. Med. Dirs., Am. Soc. Clin. Pharmacology and Therapy, N.Y. Acad. Medicine. Lutheran. Office: R J Reynolds Co Bowman Gray Tech Ctr Winston Salem NC 27102

EHMKE, DALE WILLIAM, agriculturist; b. Pekin, Ill., July 31, 1944; s. Paul Wilhem and Thelma Francis (Drockelman) E.; m. Pamella Jean Smith, Sept. 18, 1979. Analyst Archer Daniels Midland Co., Mapleton, Ill., 1963-67; sr. analyst Ashland Chem. Co., Mapleton, 1967-76; phys. sci. technician USDA Agrl. Rsch. Svc., Nat. Ctr. for Agrl. Utilization Rsch., Peoria, Ill., 1976—. Served in U.S. Army, 1967-68. Vietnam. Office: USDA ARS NCAUR 1815 N University Peoria IL 61604

EHRENFELD, DAVID WILLIAM, biology educator, author; b. N.Y.C., Jan. 15, 1938; s. Irving and Anne (Shapiro) E.; m. Joan Gardner, June 28, 1970; children: Kate, Jane, Jonathan, Samuel. BA, Harvard Coll., 1959; MD, Harvard Med. Sch., 1963; PhD, U. Fla., 1968. From asst. prof. biology to assoc. prof. biology Barnard Coll. Columbia U., N.Y.C., 1967-74; prof. biology Cook Coll. Rutgers U., New Brunswick, N.J., 1974—. Author: Biological Conservation, 1970, The Arrogance of Humanism, 1978, 2d. edit., 1981, Beginning Again: People and Nature in the New Millennium, 1993; founder and editor jour. Conservation Biology, 1987—; columnist mag. Orion, 1989—; contbr. articles to jours. Tech. Rev., Hudson Rev., Animal Behav., Sci., others. Trustee E. F. Schumacher Soc., Great Barrington, Mass., 1979—, Caribbean Conservation Corp., Gainesville, Fla., 1980—, Ednl. Found. Am., Westport, Conn., 1987—. Fellow AAAS; mem. Ecol. Soc. Am., Internat. Union for the Conservation of Nature, Marine Turtle Specialist Group, Specialist Group on Sustainable Use of Wild Species. Jewish. Home: 44 N 7th Ave Highland Park NJ 08904-2931 Office: Rutgers U Cook Coll Box 231 New Brunswick NJ 08903

EHRENKRANTZ, DAVID, medical researcher; b. New Haven, Aug. 26, 1952; s. Harold Louis and Katherine (Russo) E. BA magna cum laude, U. Hartford, 1979; MSW, Adelphi U., 1982; MPH, N.Y. Med. Coll., 1987; ScD, U. Pitts., 1991. Cert. social worker, N.Y. Rsch. asst. U. Conn. Sch. Medicine, 1985; med. rsch. affiliate Genentech Corp., San Francisco. Contbr. articles to sci. jours. Mem. Alpha Chi. Office: NY Med Coll Div Pediatric Endocrinology Munger Pavillion 1st Fl Valhalla NY 10595

EHRENSTEIN, GERALD, biophysicist; b. N.Y.C., Sept. 27, 1931; s. Irving and Adele (Holzer) E.; m. Deborah Ploscowe, Dec. 17, 1960; children—Ruth, David, Steven. B.E.E., Cooper Union, 1952; M.A., Columbia U., 1958, Ph.D., 1962. Engr., Arma Corp., N.Y.C., 1952; chief sect. on molecular biophysics NIH, Bethesda, 1975—. Corp. mem. Marine Biol. labs., Woods Hole, Mass., 1970—. Mem. editorial bd. Biophys. Jour., 1980-83; editor Methods of Exptl. Physics-Biophysics, 1982. Served to lt. (jg) USCG, 1952-54. Mem. Biophys. Soc. (program com. 1981-84), Am. Phys. Soc., Sigma Xi. Democrat. Jewish. Avocation: birdwatching. Home: 7502 Nevis Rd Bethesda MD 20817-4742 Office: NIH 9000 Rockville Pike Bethesda MD 20892

EHRHARDT, ANTON F., medical microbiology educator; b. San Bernadino, Calif., Aug. 11, 1960; s. Horst A. and Selma (Golber) E.; m. Nancy L. Backenstoe. BA, Calif. State U., 1983, MS, 1988; PhD, Ariz. State U., 1990. Post-doctoral fellow med. microbiology dept. Creighton U. Sch. Medicine, Omaha, Nebr., 1991-92, asst. prof. med. microbiology dept., 1993—, asst. dir. antibiotics rsch. lab., 1993—; del. Collegium '91, La Jolla, Calif., 1991; faculty Infectious Disease Challenges in Primary Care, Chgo., 1992. Author: Recombinant DNA Methodology, 1990; contbr. articles to profl. jours. Miles Incorporated scholar Miles Pharmaceuticals, 1991, 92. Mem. AAAS, Am. Fedn. Clin. Rsch., Am. Soc. Microbiology, N.Y. Acad. Scis. Office: Creighton Univ Sch Medicine Med Microbiology Dept 2500 California Pla Omaha NE 68178

EHRICH, FREDRIC F., aeronautical engineer; b. N.Y.C., Dec. 17, 1928; s. William and Yetta (Benjamin) E.; m. Joan Collier, Sept. 5, 1955; children: Diane, Elliott, Noami. BS, MIT, 1947, ME, 1949, ScD, 1951; postgrad., Tech. Inst. at Delft, 1947-48. Engr., sr. engr., engrng. supr. Aircraft Gas Turbine div. Westinghouse Electric Co., Phila., 1951-55; resident rep. at Rolls Royce Westinghouse Electric Co., Derby, Eng., 1955-56; adv. engr. Westinghouse Electric Co., Kansas City, Mo., 1956-57; aerodynamics engr. aircraft engines GE, Lynn, Mass., 1957-58, mgr. preliminary design, 1958-59, mgr. T64 engine design, 1959-63, mgr. design tech. ops., 1963-69, mgr. tech. and advanced product plans, 1969-82, cons. engr., staff engr., 1982—; instr. Drexel Inst. Tech., Phila., 1952-53, U. Kans., Kansas City, 1957. Editor: Handbook fo Rotor Dynamics, 1991; assoc. editor Internat. Jour. Turbo and Jet Engines, 1988—; contbr. articles to profl. jours. and encys.; patentee aircraft engine tech. field. Mem. edn. coun. MIT, Cambridge, 1983—. Fellow ASME (chmn. design engrng. div. 1972-73, editor Jour. Mech. Design and Vibration 1979-84), AIAA (program chmn. 1988); mem. NAE, Am. Helicopter Soc., Sigma Xi, Tau Beta Pi, Pi Tau Sigma. Jewish. Office: GE Aircraft Engines 1000 Western Ave Lynn MA 01910

EHRIG, HARTMUT, computer science educator, mathematician; b. Angermünde, Germany, Dec. 6, 1944; s. Kurt and Gerda (Haufschild) E.; m. Gertraud Summerer; children: Karsten, Timo, Rita. PhD in Math., Tech. U. Berlin, 1971, Habilitation in Computer Sci., 1974. Asst. dept. math. Tech. U. Berlin, 1970-72, asst. prof. computer sci., 1972-76, assoc. prof., 1976-85, prof., 1985—, dir., vice dir. Inst. Software and Theoretical Computer Sci., 1978-93; vis. prof. IBM, Yorktown Heights, U. So. Calif., UCLA, U. Spain, U. Italy; lectr. in field. Author: Fundamentals of Algebraic Specifications, Vol. 1, 1985, Vol. 2, 1990; contbr. over 200 articles to proceedings, profl. jours. Grantee Deutsche Forschungsgemeinschaft, 1981—, ESPRIT 1 + 2, 1985-92, ESPRIT Basic Rsch., 1989—, BMFT Project Korso, 1992—. Mem. Soc. for Informatics, Assn. for Computing Machinery, European Assn. for Theoretical Computer Sci. Avocation: rowing. Home: Ambossweg 9, W-1000 Berlin 26, Germany Office: Tech U, Franklinstrasse 28/29, W-1000 Berlin 10, Germany

EHRINGER, WILLIAM DENNIS, membrane biophysicist; b. Jeffersonville, Ind., Dec. 16, 1964; s. Dennis Paul Ehringer and Mary Ellen (Schaefer) Williams; m. Sharon Jeanine Hensley, May 18, 1985; children: Daniel Scott, Kristen Elise. MS in Biology, Purdue U., 1990; PhD in Med. Biophysics, Ind. U., Indpls., 1993. Teaching asst. Ind. U.-Purdue U., Indpls., 1988-90; lab. dir. Ind. U., Indpls., 1989—, rsch. assoc., 1991—. Contbr. articles to profl. publs. Sci. judge Carmel (Ind.)-Clay High Sch., 1990. Mem. Am. Physics Soc., Biophys. Soc., Ind. Acad. Scis., Sigma Xi. Republican. Roman Catholic. Achievements include discovery of the role of omega-3 polyunsaturated fatty acids in promoting membrane fusion and permeability in artificial bilayer membranes. Home: 2214 Pine Hill Ct Jeffersonville IN 47130 Office: Ind U 723 W Michigan St Indianapolis IN 46202-5132

EHRLICH, DANIEL JACOB, optical engineer, optical scientist; b. Washington, June 9, 1951; s. Paul and Celia (Lesley) E. BS in Physics, U. Rochester, 1973, PhD in Optical Engring., 1977. Staff engr., then group leader Lincoln Lab., MIT, Lexington, 1977—; chmn. Microphysics Surfaces conf., Santa Fe, 1985, 87, NATO workshops, 1987, 92, MRS Symposia Laser Processing, Boston, 1981, 83, Internat. Symposium on Electron, Ion and Photon Beams, Seattle, 1991. Editor 5 books; contbr. numerous articles to profl. jours.; holder 11 patents. Fellow Optical Soc. Am. (R.W. Wood prize 1991); mem. Am. Vacuum Soc., Materials Rsch. Soc. Office: MIT Lincoln Lab 244 Wood St Lexington MA 02173-6499

EHRLICH, FREDERICK, surgery consultant, orthopedist, rehabilitation specialist; b. Czernowitz, Bukowina, USSR, Mar. 23, 1932; came to Australia, 1947; s. Alexander and Klara (Schneider) E.; m. Shirley Rose Eastbourne, Sept. 26, 1959; children: Paul, Rachel, Simon, Adam, Miriam, Mark. M.B., B.S. (hons.), Med. Faculty, U. Sydney, 1955; B.A., Macquarie U., Sydney, 1970, Ph.D., 1979; Dip.Phys. and Rehab. Medicine, Australian Postgrad. Fedn. in Medicine, Canberra, 1974. Intern, Royal Newcastle Hosp., N.S.W., 1955, rotating resident, 1955-57; resident surg. officer Charing Cross and Fulham Hosp., Hammersmith, London, 1958-59; surg. registrar Royal Newcastle Hosp., 1959-63; dir. surg. svc. State Psychiat. Svc., Sydney, 1962-75; prin. advisor State Geriatric and Rehab. Svcs., N.S.W., 1975-77; pvt. practice cons. surgeon, Sydney, Australia, 1979—; hon. cons. Sydney Hosp., 1977—; cons. geriatrics and rehab. Hornsby Hosp., Sydney, 1977—, St. George Hosp., Sydney, 1990—, orthopaedic surgery Spastic Ctr. New South Wales; vis. gen. and orthpaedic surgeon Marrickville Dist. Hosp., Sydney, 1977—; prof. rehabilitation, aged and extended care U. New South Wales, also chmn. med. bd., Chatswood Community Hosp., Sydney, 1978—, Concord Hosp, 1987—. Cons. Subnormal Children's Welfare Assn., Multiple Sclerosis Soc., 1968-75. Author: Chronic Illness in New South Wales, 1977 and more than 40 monographs, papers; editor: The Demography of Disability, 1969; New Thinking on Housing for the Aging, 1973; Aging in a Metropolis, 1974; Rehabilitation and Geriatric Services; Report of a Task Force, 1978. Fellow Royal Coll. Surgeons Eng., Royal Coll. Surgeons Edinburgh, Australian Coll. Rehab. Medicine (found. mem.), Royal Coll. Psychiatrists, Australasian Faculty of Rehab. Medicine, Total Care Found. (hon. life), Adv. Council on Visually Handicapped (hon. life), Australia Assn. Gerontology (pres. New South Wales div.), New South Wales Council on Aging (chmn.). Jewish. Home: Box E11, St James, Sydney 2000, Australia

EHRLICH, MARGARET ISABELLA GORLEY, systems engineer, mathematics educator, consultant; b. Eatonton, Ga., Nov. 12, 1950; d. Frank Griffith and Edith Roy (Beall) Gorley; m. Jonathan Steven Ehrlich. BS in Math., U. Ga., 1972; MEd, Ga. State U., 1977, EdS, 1982, PhD, 1987; postgrad. Woodrow Wilson Coll. of Law, 1977-78. Cert. secondary tchr., Ga. Tchr. DeKalb County Bd. Edn., Decatur, Ga., 1972-83; chmn. dept. math. Columbia High Sch., Decatur, 1978-83; with product devel. Chalkboard Co., Atlanta, 1983-84; math instr. Ga. State U., Atlanta, 1983—; pres. Testing and Tech. Svcs., Atlanta, 1983—; course specialist Ga. Pacific Co., Atlanta, 1984-86; systems engr. Lotus Devel. Corp., 1986-89; instr. math. Ga. State U., Atlanta, 1989-92; rsch. assoc. SUNY-Stony Brook, 1976; modeling instr. Barbizon Modeling Sch., Atlanta, 1991; instr. Ga. State Coll. for Kids, 1984-85; test-taking cons., hon. mem. Communication Workers of Am. Local 3204, Atlanta, 1985—; tng. cons. Lotus Devel. Corp. Author: (software user manual) Micro Maestro, 1983, Music Math, 1984, (test manual) The Telephone Company Test, 1991, AMI Pro Advanced Courseware, 1992; mem. editorial bd. CPA Computer Report, Atlanta, 1984-85. Active Atlanta Preservation Soc., 1985, Planned Parenthood; tchr. St. Phillips Ch. Sch., Atlanta, 1981-88; vol. Joel Chandler Harris Assn., Atlanta, 1984-87; St. Phillips drug and alcohol counseling, welcome com., HOPE counselor, 1988—. Named STAR Tchr. DeKalb County Bd. Edn., 1979, 80, 81, Most Outstanding Tchr., Barbizon Schs. of Modeling, 1980, Colo. Outward Bound, 1985, Disting. Educator, Ga. State U., 1987. Mem. LWV, Math. Assn. Am., Nat. Council Tchrs. Math., Ga. Council Tchrs. Math., Math. Assn. Am., Assn. Women in Math. (del. to China Sci. and Tech. Exch., 1989-90), Am. Soc. Tng. and Devel. Greater Atlanta, DeKalb Personal Computer Instr. Assn. (pres. 1984), Aux. Med. Assn. Ga., Daus. of Confederacy, Atlanta Track Club, N.Y.C. Track Club. Democrat. Episcopalian. Avocations: piano, jogging, fashion modeling, skiing, bonsai. Home: 240 Cliff Overlook Atlanta GA 30350-2601 Office: PO Box 500173 Atlanta GA 31150-0173

EHRLICH, PAUL RALPH, biology educator; b. Phila., May 29, 1932; s. William and Ruth (Rosenberg) E.; m. Anne Fitzhugh Howland, Dec. 18, 1954; 1 dau., Lisa Marie. AB, U. Pa., 1953; AM, U. Kans., 1955, PhD, 1957. Research assoc. U. Kans., Lawrence, 1958-59; asst. prof. biol. scis. Stanford U., 1959-62, assoc. prof., 1962-66, prof., 1966—, Bing prof. population studies, 1976—, dir. grad. study dept. biol. scis., 1966-69, 1974-76, pres. Ctr. for Conservation Biology, 1988—; cons. Behavioral Research Labs., 1963-67; corr. NBC News, 1989—. Author: How to Know the Butterflies, 1961, Process of Evolution, 1963, Principles of Modern Biology, 1968, Population Bomb, 1968, 2d edit., 1971, Population, Resources, Environment: Issues In Human Ecology, 1970, 2d edit., 1972, How to Be a Survivor, 1971, Global Ecology: Readings Toward a Rational Strategy for Man, 1971, Man and the Ecosphere, 1971, Introductory Biology, 1973, Human Ecology: Problems and Solutions, 1973, Ark II: Social Response to Environmental Imperatives, 1974, The End of Affluence: A Blueprint for the Future, 1974, Biology and Society, 1976, Race Bomb, 1977, Ecoscience: Population, Resources, Environment, 1977, Insect Biology, 1978, The Golden Door: International Migration, Mexico, and the U.S., 1979, Extinction: The Causes and Consequences of the Disappearance of Species, 1981, The Machinery of Nature, 1986, Earth, 1987, The Science of Ecology, 1987, The Birder's Handbook, 1988, New World/New Mind, 1989, The Population Explosion, 1990, Healing the Planet, 1991, Birds in Jeopardy, 1992; contbr. articles to profl. jours. Recipient World Wildlife Fedn. medal, 1987; co-recipient Crafoord prize in Population Biology and Conservation of Biol. Diversity, 1990; MacArthur Prize fellow, 1990—. Fellow Calif. Acad. Scis., Am. Acad. Arts and Scis., AAAS, Am. Philos. Soc., Entomology Soc. Am.; mem. Nat. Acad. Scis., Entomological Soc. Am., Soc. for Study Evolution, Soc. Systematic Zoology, Am. Soc. Naturalists, Lepidopterists Soc., Am. Mus. Natural History (hon. life mem.). Office: Stanford U Dept Biol Scis Stanford CA 94305

EHRMANN, ROBERT LINCOLN, pathologist; b. Boston, Sept. 2, 1922; s. Herbert and Sara (Rosenfeld) E.; m. Janice Lee Panella, Sept. 7, 1958; children: Lisa, Martha. AB, Swarthmore Coll., 1944; MD, NYU, 1946. Diplomate Am. Bd. Pathology, cert. in Anatomic Pathology. Resident in pathology Peter Bent Brigham Hosp. and Boston Lying In Hosp., Boston, 1956-59; asst. pathologist Free Hosp. for Women, Parkway Div. Boston Hosp. for Women, Brookline, Mass., 1959-70; pathologist Parkway Div. Boston Hosp. for Women, Brookline, 1970-80, Brigham and Women's Hosp., Boston, 1980—; teaching fellow, instr., clin. assoc. in pathology, Harvard Med. Sch., Boston, 1958-71; asst. prof. pathology Harvard Med Sch., Boston, 1971-89, assoc. prof. pathology, Harvard Med. Sch., Boston, 1981—. Author: Malignant Progression in the Cervix, 1993; contbr. articles to profl. jours. Exec. bd. mem. Boston chpt. Am. Jewish Com., 1975. Capt. U.S. Army, 1947-49. Grantee: Nat. Cancer Inst., 1964-69. Fellow Coll. Am. Pathologists, Internat. Acad. Cytology; mem. Internat. Soc. Gynecol. Pathologists, Sigma Xi (assoc.). Democrat. Jewish. Achievements include devel. of technique for coating tissue culture vessels with rat-tail collagen which promotes better growth; discovered hemoglobin stimulates growth of cell colonies on glass in roller tube tissue culture; discovered that choriocarcinoma in the hamster cheek pouch stimulates blood vessel dilation and proliferation; endometrium in organ culture spontaneously secretes glycogen within 48 hours; basement membranes of squamous cell carcinoma of the cervix and vulva are defective, as shown by laminin and collagen IV stains. Home: 315 Woodward St Waban MA 02168-2120 Office: Brigham and Womens Hosp Pathology Dept 75 Francis St Boston MA 02115

EHSANI, MEHRDAD, electrical engineering educator, consultant; b. Tehran, Iran, Oct. 9, 1950; came to U.S., 1968; s. Heshmat and Didar (Ahmadi) E.; m. Mildred Ann Holder, July 2, 1977; children: Evan Mancil, Nathaniel William. MS, U. Tex., 1974; PhD, U. Wis., 1981. Registered profl. engr., Tex. Rsch. engr. Fusion Rsch. Ctr. U. Tex., Austin, 1974-77; rsch. engr. Argonne (Ill.) Nat. Lab., 1977-81; prof. elec. engring. Tex. A&M U., College Station, 1981—. Author: Converter Circuits for Superconductive Magnetic Energy Storage, 1988; co-author: ANSI/IEEE Standards 936, 1987; contbr. over 75 articles to profl. jours. Named Outstanding Young Engr., Tex. Soc. Profl. Engrs., 1984. Mem. IEEE (sr.), Power Electronics Soc. of IEEE (adminstrv. coun. 1990—), Industry Applications Soc. of IEEE (exec. coun. 1989—), Sigma Xi. Bahai. Achievements include 1 patent in field; 2 patents pending; invention of frequency modulated sensorless variable reluctance motor drive, of two generations of high power dc-dc voltage converters. Office: Tex A&M U Dept Elec Engring College Station TX 77843

EICHBERGER, LEROY CARL, stress analyst, mechanical engineering consultant; b. Chgo., Oct. 26, 1927; s. Roy George and Phyllis Zena (Goss) E.; m. Mary Ann Teresa Bronars, Sept. 10, 1955; children: Charles David, David Paul, Scott Thomas. BSME, U. Ill., 1951, MS, 1955, PhD, 1959. Registered profl. engr., Tex. Assoc. prof. U. Houston, 1959-77; mgr. engring. Weatherford Lamb USA, Houston, 1977-80; mgr. R&D Atlas Bradford, Houston, 1980-89; ind. cons. Houston, 1989—; tech. cons. Reed Roller Bit Co., Houston, 1959-61, Exxon Co. USA, Houston, 1968-77; staff cons. H.O. Mohr Rsch. and Engring., Houston, 1989—. Author monographs on methods for dynamic calibration of pressure transducers. With USCG, 1946-47. Recipient Arthur Lubinski award of excellence Offshore Tech. Conf., Houston, 1984. Mem. ASME, Soc. Exptl. Stress Analysis. Christian Scientist. Home and Office: 5310 Dumfries Dr Houston TX 77096

EICHEN, MARC ALAN, computer support; b. N.Y.C., Aug. 15, 1949; s. Moe and Phyllis (Edelson) E.; m. Elise Marie Long, June 17, 1979. BA, Clark U., 1971, PhD, 1976. Lectr. Middlebury (Vt.) Coll., 1978-80; asst. to pres. Queens Coll., Flushing, N.Y., 1980-85, mgr. acad. computer ctr., 1985-91; dir. acad. computer svcs. Hunter Coll., N.Y.C., 1991—; bd. dirs. Nat. Ctr. for Tech. in Edn., N.Y., 1989—, Pioneering Ptnrs., Indpls., 1992—. Recipient Sci. for Citizens fellowship NSF, 1980, Fulbright fellowship Arab Republic of Egypt, 1985-86. Home: 131 Park Pl Brooklyn NY 11217 Office: Hunter Coll Acad Computing Svcs 695 Park Ave New York NY 10021

EICHHOLZ, GEOFFREY G(UNTHER), physics educator; b. Hamburg, Germany, June 29, 1920; s. Max and Adele Daisy (Elias) E. BS in Physics, U. Leeds, Eng., 1942, PhD in Physics, 1948, DSc, 1979. Asst. prof. physics U. B.C., Vancouver, Can., 1947-51; head physics and radiotracer subdiv. Can. Bur. Mines, Ottawa, Ont., 1951-63; Regents prof. nuclear engring. Ga. Inst. Tech., Atlanta, 1963-89; regional advisor Internat. Atomic Energy Agy, Vienna, 1968, 78; vis. prof. Universidad Nacional Autonoma de Mexico, 1972, 74. Author: Environmental Aspects of Nuclear Power, 1976, Nuclear Radiation Detection, 1979; author, editor: Radioisotope Engineering, 1972. Served with British Admiralty, 1942-46. Recipient Outstanding Tchr. award Ga. Inst. Tech., 1973. Fellow Am. Nuclear Soc. (chair isotope and radiation div. 1967-68), Health Physics Soc. (pres. Atlanta chpt. 1978-79); mem. Am. Phys. Soc., Can. Assn. Physicists, Sigma Xi (pres. Atlanta chpt. 1967). Avocations: history, genealogy, tennis. Home: 1784 Noble Dr NE Atlanta GA 30306-3142 Office: Ga Inst Tech Dept Nuclear Engring Atlanta GA 30332-0225

EICHNER, GREGORY THOMAS, computer engineer, consultant; b. Huntington, N.Y., Aug. 6, 1961; s. George Thomas and Ann (Shannon) E.; m. Mary Elizabeth Driscoll. BSE, U. Ctrl. Fla., 1983; MSE, U. CVen. Fla., 1986. Computer engr. Naval Tng. Systems Ctr., Orlando, Fla., 1983-90; system engr. Daedalean, Inc., Orlando, 1990-91; mgr. software dept. Daedalian, Inc., Orlando, 1991-92; founder, pres. Eichner Bradley Controls Corp., Orlando, 1992—. Author: A One-Step Collision Detection Method for Computer Graphics Programs, 1983. Mem. Sigma Xi. Roman Catholic. Home: 7726 Diamondstar Ct Orlando FL 32822 Office: Eichner Bradley Controls PO Box 678048 Orlando FL 32867-8048

EICHSTÄDT, HERMANN WERNER, cardiology educator; b. Alten-Buseck/Giessen, W. Ger., Feb. 15, 1948; s. Karl Heinz and Elisabeth Magdalena (Froese) E.; med. student U. Mainz, 1968; cand. med. U. Düsseldorf, 1971, M.D., 1974; children: Björn, Kerstin, Bastienne, Bernadette. Intern, Univ. Hosp., Düsseldorf, 1974, Augusta Hosp., Düsseldorf, 1974-75; researcher, Düsseldorf, 1972-74; resident cardiol. clinic, Bad Krozingen, W. Ger., 1975-77; resident U. Tübingen (W. Ger.), 1977-79; lectr. dept. cardiology, cons. internal medicine Free U. Berlin, 1979—; mem. directory internal medicine and radiology Universitätsklinikum Charlottenburg, Berlin, 1981—, prof. cardiology, 1985—. Fellow Am. Coll. Chest Physicians, Internat. Coll. Angiology; mem. N.Y. Acad. Scis., German Soc. Internal Medicine, Profl. Assn. Internists, German Soc. Cardiology, European Soc. Cardiology (working group on isotopes), German Soc. Cardiovascular Surgery, German Heart Found. Roman Catholic. Contbr. numerous articles to English, French and German jours. Editor books and monographs. Home: 61 Konstanzerstrasse, 10707 Berlin Germany Office: 130 Spandauer Damm, 14050 Berlin 19, Germany

EICKELBERG, W. WARREN BARBOUR, academic administrator; b. N.Y.C., Jan. 19, 1925; s. Graham Alexander and Lillian (Hayes) E.; student Harvard U., 1942-43; BA, Hope Coll., Holland, Mich., 1949; MA (Dennison fellow), Wesleyan U. Conn., 1951; children—William, Margaret, Robert, Janet. Prof., Adelphi U., Garden City, L.I., N.Y., 1952—, dir. devel., v.p., 1958-66, dir. premed. curriculum, 1967-89; cons. devel. planning Nat. Ctr. for Disability Svcs., Albertson, N.Y., Joseph Bulova Sch., Woodside, N.Y. 1958—; mem. biomechanics cons. group Pres.'s Com. on Disabled; mem. ad hoc com. N.Y. State Joint Legis. Com. on Transp. Chmn. Nassau County Museum Council, 1966-67; mem. founding com. Adelphi Suffolk Coll.; nat. cons. Nat. Council Cath. Men, 1969. Served to 1st lt. USAAF, 1943-46. Recipient Flambeau award Adelphi U., 1956, Disting. Service award, 1981, Disting. Teaching award, 1985; L.I. Gov.'s award, 1966, Alpha Epsilon Delta award, 1989, Joseph Serio award, 1990-91, 91-92, 92-93; certificate of distinction Dictionary Internat. Biography, 1968; Wisdom award, 1970. Mem. N.Y. Acad. Scis., Internat. Soc. Biomechanics (charter), Nat. Soc. Fund Raisers, Public Relations Soc. Am. (accredited, pres. L.I. chpt. 1980), L.I. Pub. Relations Assn. (past pres., dir.), L.I. Sci. Tchrs. Assn. (past pres., dir.). Sigma Xi, Kappa Eta Nu. Clubs: Lions; Wesleyan (N.Y.C.); Unqua Yacht (Amityville). Home: 38 Unqua Pl Amityville NY 11701-4231 Office: Adelphi U Garden City NY 11530

EIDELS, LEON, biochemistry educator; b. Jersey City, May 25, 1942; s. Benjamin and Mary (Lourie) E.; m. Eleanor Royster, Apr. 8, 1973; children: Heather, Shelley. BS with highest honors, U. Calif., Davis, 1964, MS, 1966, PhD, 1969. Postdoctoral fellow U. Conn., Farmington, 1970-74; asst. prof. microbiology dept. U. Tex. Southwestern Med. Ctr., Dallas, 1974-80, assoc. prof., 1980-90, prof., 1990—; chmn. grad. program in molecular microbiology U. Tex. Southwestern Med. Ctr., Dallas, 1986—, found. lectr. Am. Soc. Microbiology, 1985-86. Postdoctoral fellow Arthritis Found., Farmington, 1970-73, NIH grantee U. Tex. Southwestern Med. Ctr., Dallas, 1980—. Mem. AAAS, Am. Soc. Microbiology, Am. Soc. Biochemistry and Molecular Biology, Sigma Xi. Achievements include discovery of the nature of the diphtheria toxin receptor. Office: U Tex Southwestern 5323 Harry Hines Blvd Dallas TX 75235-9048

EIDEMILLER, DONALD ROY, mechanical engineer; b. Dayton, Ohio, July 11, 1943; s. Daniel Loren and Glenna Maria (Waddle) E.; m. Sandra Kay Dodd, Apr. 8, 1990; children from previous and current marriages: James, Jeff, Neal, Bob, Susan, Sherri Huffman, Kim Toplikar, Dawn Holder. BS in Mech. Engring., Ohio State U., 1967. Mech. and control engr. Black & Veatch, Kansas City, Mo., 1969-77, project engr., 1977-81, project mgr., 1982-89, gen. ptnr., 1989-92, sr. ptnr., 1993—. 1st lt. U.S. Army, 1967-69, Korea. Mem. ASME, Nat. Soc. Profl. Engrs., Toastmasters Internat. (pres. 1986-87). Office: Black & Veatch PO Box 8405 Kansas City MO 64114

EIDEN, MICHAEL JOSEF, aerospace engineer; b. Trier, Germany, Sept. 14, 1949. Dipl. engr., U. Stuttgart, Germany, 1978. Consulting engr. I Koss, Stuttgart, Germany, 1979-82; sr. structural engr. European Space Agy./ESTEC, Noordwijk, The Netherlands, 1983-89; head mechanisms sect. European Space Agy./ESTEC, Noordwijk, 1990—. Mem. AIAA.

EIFLER, CARL FREDERICK, retired psychologist; b. Los Angeles, June 27, 1906; s. Carl Frederick and Pauline (Engelbert) E.; m. Margaret Christine Aaberg, June 30, 1963; 1 son, Carl Henry; 1 adopted son, Byron Hisey. BD, Jackson Coll., 1956; Ph.D., Ill. Inst. Tech., 1962. Insp. U.S. Bur. Customs, 1928-35, chief insp., 1936-37, dep. collector, 1937-56; bus. mgr. Jackson Coll., Honolulu, 1954-56, instr., 1955-56; grad. asst. instr., research asst. Ill. Inst. Tech., Chgo., 1959-62; psychologist Monterey County Mental Health Services, Salinas, Calif., 1964-73; ret., 1973. Contbg. author Psychon. Sci., vol. 20, 1970; co-author: The Deadliest Colonel; author, pub.: Jesus Said. Served with U.S. Army, 1922-23, 40-47; col. ret. Decorated Combat Infantryman's Badge, Legion of Merit with 2 oak leaf clusters, Bronze Star medal, Air medal, Purple Heart; named to Military Intelligence Corps Hall of Fame, 1988. Mem. AAUP, Am. Psychol. Assn., Western States Psychol. Assn., Calif. Psychol. Assn., Res. Officers Assn. (Hawaii pres. 1947), Assn. Former Intelligence Officers (bd. govs., Western coord.), Pearl Harbor Survivors, 101 Assn., Assn. U.S. Army Vets. of OSS (past bd. govs., Western coord., v.p.), Ret. Officers Assn., Masons, KT, Shriners, Elks, Nat. Sojourners, Psi Chi. Home: 22700 Picador Dr Salinas CA 93908-1116

EIGEN, HOWARD, pediatrician, educator; b. N.Y.C., Sept. 8, 1942; s. Jay and Libbie (Kantrowitz) E.; m. Linda Hazzard; children—Sarah Elizabeth, Lauren Michelle. B.S., Queens Coll., 1964; M.D., Upstate N.Y. Med. Ctr., Syracuse, 1968. Diplomate Am. Bd. Pediatrics, Am. Bd. Pediatric Pulmonology, Am. Bd. Critical Care Medicine, Nat. Bd. Med. Examiners (mem. pediatric test com. 1986-90). Resident in pediatrics Upstate Med. Ctr., Syracuse, 1968-71; fellow in pediatric pulmonology Tulane U., New Orleans, 1973-76; asst. prof. pediatrics Ind. U., Indpls., 1976-84, prof., 1984—; assoc. chmn. of Pediatrics for Clin. Affairs, dir. pediatric intensive care, pulmonology sect. Riley Hosp. for Children, med. dir. ambulatory care, 1989—. Assoc. editor Pediatric Pulmonology, 1984-91. Contbr. articles to profl. jours. Served to maj. U.S. Army, 1971-73. Fellow Am. Acad. Pediatrics (pres. chest sect. 1983-85, pulmonology 1986—), Am. Thoracic Soc., Am. Bd. Pediatrics, Am. Lung Assn. (pres. Ind. 1984-85). Avocation: tennis. Office: Ind U Dept Pediatrics 702 Barnhill Dr Indianapolis IN 46223-5225

EIGEN, MANFRED, physicist; b. Bochum, Germany, May 9, 1927; s. Ernst E. and Hedwig (Feld) E.; m. Elfriede Müller; 2 children. Studies in physics and chemistry, U. Gottingen, Germany; hon. degrees, U. Göttingen, U. Chgo., Washington U., St. Louis, Nottingham U., Bristol U., U.K., Hebrew U., Jerusalem, Cambridge U., U.K., Debrecen U., Techn. U., Munich, Bielefeld U., Utah State U. Sci. asst. Inst. Phys. Chemistry U. Göttingen, 1951-53; mem. staff, then chmn. Max Planck Inst. Phys. Chemistry, Göttingen, 1953—; Utah State U.; vis. lectr. Cornell. Author tech. papers. Co-recipient Nobel prize in chemistry, 1967. Mem. Bunsen Soc. Phys. Chemistry (Bodenstein Preis), Faraday Soc., NAS. Achievements include studying evolution of biol. macromolecules; research on control of enzymes. Office: Max Planck Inst, Max Planck Inst, 37077 Göttingen Germany

EIGLER, DONALD MARK, physicist; b. L.A., Mar. 23, 1953; s. Irving Baer and Evelin Muriel (Baker) E.; m. Roslyn Winifred Rubesin, Nov. 2, 1986. BA, U. Calif., San Diego, 1975, PhD in Physics, 1984. Rsch. assoc. U. Köln (Fed. Republic Germany), 1975-76; rsch. assoc. U. Calif., San Diego, 1977-84, postdoctoral rsch. assoc., 1984, assoc. rsch. physicist dept. physics, 1986; postdoctoral mem. tech. staff AT&T Bell Labs., Murray Hill, N.J., 1984-86; mem. rsch. staff IBM, San Jose, Calif., 1986-93, IBM fellow, 1993—. Office: IBM Almaden Rsch Ctr 650 Harry Rd San Jose CA 95120-6099

EILTS, MICHAEL DEAN, research meteorologist, manager; b. La Chapelle, France, Aug. 22, 1959; (parents Am. citizens); s. Leonard Gene and Arlys Mamie (Ziegler) E. BS in Meteorology, U. Okla., 1981, MS in Meteorology, 1983, MBA in Fin. and Human Resource Mgmt., 1991. Rsch. asst., rsch. meteorologist Coop. Inst. for Mesoscale Meteorol. Studies, Norman, Okla., 1981-84; rsch. meteorologist Nat. Severe Storms Lab., Norman, 1984-87, mgr. weather hazards to aviation project, 1987-91, chief forecast applications rsch. group, 1991-93; asst. dir. for Stormscale Rsch. and Applications Stormscale Rsch. and Applications, 1993—; mem. exptl. forecast facility mgmt. group, 1991—; mem. Cleveland County YMCA Program Com.; spl. mem. adj. faculty U. Okla. Sch. Meteorology, Norman, 1989—; mem. Okla. Mesonet Steering Com. Contbr. articles to profl. jours. Recipient sustained superior performance award Nat. Severe Storms Lab., 1987, 89, 90, 91, 92, 93; grantee FAA, 1987-93, NASA, 1990-91, Nat. Weather Svc., 1991-93, Dist. Authorship award Environ. Rsch. Labs., 1991. Mem. Am. Meteorol. Soc. Lutheran. Avocations: golf, volleyball, softball. Home: 2405 Bonnybrook St Norman OK 73071-4329 Office: Nat. Severe Storms Lab 1313 Halley Cir Norman OK 73069-8493

EILTS, SUSANNE ELIZABETH, physician; b. Council Bluffs, Iowa, Oct. 12, 1955; d. Ervin Edwin and Mary Margaret (Leonard) E. BS, Nebr. Wesleyan U., 1976; MD, U. Iowa, 1980. Diplomate Am. Bd. Internal Medicine. Intern, resident U. Nebr. Med. Ctr., Omaha, 1980-83; pvt. practice, Omaha, 1983—; clin. instr. internal medicine U. Nebr., Omaha, 1983—; med. dir. Ambassador Nursing Home, Omaha, 1990-92; quality assurance reviewer Sunderbruch Corp. Nebr., Lincoln, 1990—. Med. columnist Omaha World Herald, 1986-88. Mem. ACP, Am. Geriatric Soc., Nebr. Med. Assn. (alt. del. 1990-91, del. 1992, young physician com. 1989-91), Am. Women's Med. Assn. Nebr. (sec. 1987), Dundee-Meml. Park Neighborhood Assn., Beta Beta Beta, Phi Kappa Phi, Phi Lambda Upsilon. Avocations: bicycling, playing the pennywhistle. Office: Internal Medicine Assocs PC 4242 Farnam St Ste 650N Omaha NE 68131

EIN, DANIEL, allergist; b. Liege, Belgium, Nov. 26, 1938; came to U.S., 1941; s. Max Motel and Sabine (Toeman) E.; m. Marion Hess, June 25, 1961 (div. 1978); children: Mark David, Jon Spencer; m. Marina Wallach, Apr. 10, 1988. AB, Columbia U., 1959; MD, Albert Einstein Coll. Medicine, 1964. Diplomate Am. Bd. Internal Medicine, Am. Bd. Allergy and Immunology. Intern Bronx Mcpl. Hosp., N.Y.C., 1964-65; staff assoc. Nat. Cancer Inst., Washington, 1965-67, clin. assoc., 1967-68; asst. resident Mass. Gen. Hosp., Boston, 1968-69; sr. investigator Nat. Cancer Inst., Washington, 1969-71; pvt. practice Washington, 1971—; clin. prof. medicine George Washington U., 1982—. Contbr. articles to profl. jours. and newspapers. Fellow ACP, Am. Acad. Allergy; mem. Med. Soc. of D.C. (pres. 1991), Greater Washington Allergy Soc. (pres. 1979), Cosmos Club. Jewish. Achievements include discovery of OZ factors on human immunoglobulin light chains.

EINAGA, HISAHIKO, chemistry educator; b. Gose, Nara, Japan, May 2, 1936; s. Kyujiro and Shigeko (Tsujimoto) E.; m. Yoko Kishigami, Oct. 22, 1965; children: Kuniko, Hisahiro. B in Engring., Nagoya (Japan) Inst. Tech., 1960; DSc, Tokyo Inst. Tech., 1969. Rsch. chemist Onoda Cement Co. Ltd., Tokyo, 1960-67, Nat. Inst. Rsch. in Inorganic Materials, Tokyo and Tsukuba, Japan, 1967-77; from assoc. to full prof. U. Tsukuba, 1977-89; prof. Nagoya Inst. Tech., 1989—; vis. rsch. chemist Nat. Inst. Rsch. in Inorganic Materials, Tsukuba, 1977—. Inventor purification of beryllium compounds. Mem. Am. Chem. Soc., Swiss Chem. Soc. Avocation: baseball. Home: 1155, Toge, Gose 639-22, Japan Office: Nagoya Inst Tech, Gokiso-Cho, Showa 466, Japan

EINAGA, YOSHIYUKI, chemist; b. Himeji, Hyogo, Japan, Jan. 28, 1945; s. Tetsuo and Tsuyako (Yamanaka) E.; m. Naoko Yagi, May 13, 1973; children: Hiroyuki, Yukiko, Michiko. BS, Kyoto U., 1967, MS, 1969, PhD, 1972. Instr. Osaka U., Toyonaka, Japan, 1973-88; assoc. prof. chemistry Kyoto U., 1988—; sr. rsch. chemist Carnegie-Mellon U., Pitts., 1979-81. Contbr. articles to profl. jours. Mem. Am. Chem. Soc., Soc. Polymer Sci. Japan, Soc. Rheology Japan. Avocation: gardening. Home: Nishikyoku, Ooenishishinbayashi 6-9-18, Kyoto 610-11, Japan Office: Kyoto U, Sakyoku, Yoshidahonmachi, Kyoto 606, Japan

EINAV, SHMUEL, biomedical engineering educator; b. Tel Aviv, Oct. 30, 1942; came to U.S., 1991; s. Meir and Bella (Shkolnik) E.; m. Chasia Korn; children: Gali, Noya, Rona. MSc, Technion, Haifa, Israel, 1968; PhD, SUNY, Stony Brook, 1972. Mng. dir. Ramot U. Authority, Tel Aviv, 1982-85; prof. engring. MIT, Cambridge, 1985-86; Miller prof. U. Calif., Berkeley, 1991-92; dir. biomed. engring. Tel Aviv U., 1988—, prof. engring., 1974—; rsch. prof. UCLA, 1991—, Cedars Sinai Med. Ctr., L.A., 1991—; chmn. bd. Cameran, Tel Aviv, 1990-91; cons. Foster Miller, Waltham, Mass., 1985-86. Contbr. articles to Jour. Biomech. Engring. and Med. Physics. Mem. Jewish Fedn., L.A., 1991-92. Capt. AF, 1964-67. Recipient Valve award Binational Sci. Found., 1988. Jewish. Achievements include patents for a removable heart valve prosthesis, ultrasound recanalization, MRI of blood flow. Office: Cedars Sinai Med Ctr 8700 Beverly Blvd D-6065 Los Angeles CA 90048

EINSEL, DAVID WILLIAM, JR., consultant, retired army officer; b. Tiffin, Ohio, Nov. 4, 1928; s. David William and Naomi Dorothy (Williams) E.; m. Elva yates Aylor, June 16, 1956; children: Susan Vagnier, Mary Kost. BA, MA in Chemistry, Ohio State U., 1950; MSc, U. Va., 1956. Commd. 2d lt. U.S. Army, 1950, advanced through grades to maj. gen., 1980; staff officer Orgn. of the Joint Chiefs of Staff, Washington, 1968-70; comdr. Harry Diamond Labs., Adelphi, Md., 1970-75; chief nuclear-chem. officer hdqrs. Dept. of the Army, Washington, 1975-76; dep. commanding gen. U.S. Army Armament R&D Command Picatinny (N.J.) Arsenal, 1976-80; asst. to the sec. of def. Office of the Sec. of Def., Washington, 1980-85; nat. intelligence officer Nat. Intelligence Coun., Washington, 1985-89; ret. U.S. Army, 1989; cons. Tiffin, Ohio, 1989—. Author many classified articles, two sects. to International Military Encyclopedia, 1991; contbr. article to Jour. Analytical Chemistry. Decorated Silver Star, Bronze Star, Purple Heart. Mem. AAAS, Assn. of the U.S. Army, Am. Def. Preparedness Assn., Kiwanis, Masons, Phi Beta Kappa, Sigma Xi. Republican. Methodist. Achievements include patent in automatic electrolytic apparatus for determining acid prodn. rates. Home and Office: 594 S Washington St Tiffin OH 44883-3320

EINSTEIN, FREDERICK WILLIAM BOLDT, chemistry educator; b. Auckland, New Zealand, Nov. 7, 1940; arrived in Can., 1965; s. Otto Simon and Franziska A.A. (Boldt) E.; m. Judith May Tripney, Apr. 19, 1965; children: David Michael, Susan Frances. MSc with honors, U. New Zealand, 1963; PhD, U. Canterbury, 1965. Postdoctoral fellow U. B.C., Vancouver, 1965-67; asst. prof. Simon Fraser U., Burnaby, 1967-71, assoc. prof., 1971-75, prof., 1975—. Contbr. articles to profl. jours. Fellow Can. Inst. Chemistry; mem. Am. Crysatllpgraphic Assn. Office: Dept Chemistry, Simon Fraser U, Burnaby, BC Canada

EINZIG, STANLEY, pediatric cardiologist, researcher; b. Bklyn., July 25, 1942; s. Louis and Sally (Weiser) E.; m. Gloria Einzig (div.); children: Deborah, Dana, David. MD, UCLA, 1967; PhD, U. Minn., 1977. Diplomate Am. Bd. Pediatrics, sub.-bd. Pediatric Cardiology. Intern, then resident dept. pediatrics U. Minn., Mpls., 1967-70, from instr. to assoc. prof., 1977-90; prof. physiology and pediatrics, chief pediatric cardiology W.Va. U., Morgantown, 1990—. Contbr. numerous articles to profl. jours. Lt. comdr. USN, 1971-73. NIH fellow, 1974, 75. Fellow Am. Coll. Cardiology; mem. Am. Phys. Soc., Soc. for Pediatric Rsch., Alpha Omega Alpha. Achievements include discovery of blood flow and antioxidant effects of anisodamine. Office: WVa U HSCN Dept Pediatric Cardiology Morgantown WV 26506

EISCH, JOHN JOSEPH, chemist, educator; b. Milw., Nov. 5, 1930; s. Frank Joseph and Gladys (Riordan) E.; m. Joan Terese Scheuerell, Sept. 5, 1953; children: Margaret (dec.), Karla, Paula, Joseph, Amelia. B.S. summa cum laude, Marquette U., 1952; Ph.D. (Procter and Gamble fellow 1955, Union Carbide fellow 1956), Iowa State U., 1956. Postdoctoral fellow Max Planck Inst. für Kohlenforschung, Mülheim, Germany, 1956-57; research assoc. European Research Assocs., Brussels, 1957; faculty St. Louis U., 1957-59, U. Mich., 1959-63, Catholic U. Am., Washington, 1963-72; chmn. dept. chemistry State U. N.Y., Binghamton, 1972-78; prof. State U. N.Y., 1972—, disting. prof., 1983—; cons. in field, 1957—. Author: The Chemistry of Organometallic Compounds, 1967, (with R.B. King) Organometallic Syntheses, Vol. I, 1965, Vol. II, 1981, Vol. III, 1986, Vol. IV, 1988; contbr. over 290 articles to profl. publs. Mem. Am. Chem. Soc., Am. Inst. Chemists, Sigma Xi, Phi Lambda Upsilon, Phi Kappa Phi. Research and publs. on the synthesis and properties of organometallic compounds (those with carbon-metal bonds) and heterocycles, with emphasis on the kinetics and stereochemistry of carbon-metal bond and hydrogen-metal bond additions to olefins, acetylenes; radical-anion, halogenation, nonbenzenoid aromatic studies, photochemistry of organometallics; catalytic processes of polymerization, heteroatom removal and isomerizations and materials science. Home: 212 Sheedy Rd Vestal NY 13850 Office: SUNY Binghamton Dept Chemistry Binghamton NY 13902-6000

EISELE, CAROLYN, mathematician; b. N.Y.C., June 13, 1902; d. Rudolph and Caroline (Wuest) E.; m. June 24, 1943 (dec. 1963). BA, Hunter Coll., CUNY, 1923; MA, Columbia U., 1925; HHD (hon.), Tex. Tech. U., 1980; DSc (hon.), Lehigh U., 1982. Instr. to prof. Hunter Coll., CUNY, 1923-72; advisor Peirce Edit. Project, Ind. U., Indpls., 1982—. Author, editor: Studies in the Scientific and Mathematical Philosophy of C.S. Peirce, 1976, The New Elements of Mathematics by Charles S. Peirce, 5 vols., 1976, Historical Perspectives on Peirce's Logic of Science (A History of Science), 2 vols., 1985; contbr. articles to profl. jours. Mem. Phi Beta Kappa. Home: 215 E 68th St Apt 27E New York NY 10021-5729

EISENBERG, ADI, chemist; b. Breslau, Germany, Feb. 18, 1935; emigrated to U.S., 1951; s. Oscar and Helene E.; m. Sandra M. Kloner, June 9, 1957 (div. 1985); 1 son, Elliot. B.Sc., Worcester Poly. Inst., 1957; M.A., Princeton U., 1959, Ph.D., 1960. Postdoctoral fellow U. Basel, Switzerland, 1961-62; asst. prof. chemistry UCLA, 1962-67; assoc. prof. chemistry McGill U., Montreal, Que., Can., 1967-74, prof. McGill U., 1975—, dir. Polymer McGill, 1991—, Otto Maass Prof. Chemistry, 1993—; cons. in field. Author 6 books in field; contbr. articles to profl. jours. NATO fellow, 1961-62; Killam Research fellow, 1987-88. Fellow Am. Phys. Soc. (chmn. div. high polymer physics 1975-76), Chem. Inst. Can. (Dunlop award 1988); mem. Am. Chem. Soc., Sigma Xi. Achievements include patents in field. Office: McGill University, 3480 University St, Montreal, PQ Canada H3A 2A7

EISENBERG, HOWARD EDWARD, physician, psychotherapist, educator, consultant; b. Montreal, Que., Can., Aug. 5, 1946; s. Harold and Elsie (Goldbloom) E.; m. Nancy Roberta Jeffries, Jan. 10, 1976; children: Taryn Noelle, Jory Michael, Meredith Kate, Tessa Chloe. B.Sc. with honors in Psychology, McGill U., 1967, M.Sc., 1971, M.D.C.M., 1972. Intern Sunnybrook Med. Ctr., U. Toronto, 1973; research asst. psychology dept. McGill U., 1966-69, research asst. gerontology unit Alan Meml. Inst. Psychiatry, McGill U., 1968, clin. fellow Clarke Inst. Psychiatry, U. Toronto, 1973; lectr. Centre for Continuing Edn., York U., 1973-78, Sheridan Coll., Oakville, 1974-76; supr. individual directed study Faculty Environ. Studies, York U., 1975; lectr. dept. interdisciplinary studies U. Toronto, 1975; instr. ind. studies program, Innis Coll., U. Toronto, 1975-78, lectr. 1976-81, spl. conf. coordinator, 1977-79, 88-89, lectr. Sch. Continuing Studies, 1977-89; lectr. continuing edn. U. Vt., 1990-92; assoc. dir. edn. and growth opportunities program York U., 1975-76, dir. E.G.O. program, 1976-78; lectr. Sch. Adult Edn., McMaster U., 1980-89; instr. profl. and mgmt. devel. Humber Coll., 1982-85; pvt. practice psychotherapy, Toronto, Ont., 1973-91, Stowe, Vt., 1991—; assoc. prof. dept. family practice, Coll. Med. Univ. Vt., 1993; pres. Synectia Cons., Inc., Toronto, 1980-84, Syntrek, Inc., Stowe, Vt., 1989—, Synectia Prodns., Inc., Toronto, 1974—. Author: Inner Spaces, 1977, The Tranquility Experience, 1987, Stress Mastery for the Real World, 1991; contbr. articles to profl. jours. McGill scholar, 1966-67; Quebec scholar, 1967-68; Earle C. Anthony fellow, 1967-68; Ont. Arts Council grantee, 1977. Mem. Ont. Med. Assn. (former chmn. sect. ind. physicians), Orgnl. Devel. Network, Am. Soc. for Tng. and Develop., Vt. State Med. Soc. Address: Syntrek Inc PO Box 1393 Stowe VT 05672

EISENHAUER, WILLIAM JOSEPH, JR., aerospace engineer; b. Dayton, Ohio, Feb. 6, 1964; s. William Joseph and Dorothy Ann (Michael) E.; m. Janice Elaine McCloskey, Apr. 20, 1991; 1 child, William Charles. BS in Aerospace Engring., U. Cin., 1987; MS in Aerospace Engring., U. Dayton, 1992. Commd. USAF, 1987, advanced through grades to capt., 1991; rsch. engr. USAF Aero. Propulsion & Power Lab., Wright-Patterson AFB, Ohio, 1988-90; flight test engr. 4950 Test Wing, Wright-Patterson AFB, Ohio, 1990—. Mem. AIAA. Office: Adv Range Instrumentation Aircraft Tech Oper Br 4950 Test Wright Patterson AFB OH 45433

EISENSTARK, ABRAHAM, research director, microbiologist; b. Warsaw, Poland, Sept. 5, 1919; came to U.S., 1922; s. Isadore and Sarah (Becker) E.; m. Roma Gould, Jan. 18, 1948 (dec. July 1984); children: Romalyn, David Allen, Douglas Darwin; m. Joan Weatherly, Apr. 6, 1991. BA, U. Ill., 1941, MA, 1942, PhD, 1948. Program dir., acting sect. head Molecular Biology Sect. NSF, Washington, 1969-70; assoc. prof. Okla. State U., Stillwater, 1948-51; prof. Kans. State U., Manhattan, 1951-71; prof., dir. divsn. biol. scis. U. Mo., Columbia, 1971-80, prof., 1980-90, prof. emeritus, 1990—; dir. Cancer Rsch. Ctr., Columbia, 1990—; sr. scientist Lab. and Environ. Tech., Inc., Columbia, 1990—. Contbr. over 100 articles to profl. jours. With U.S. Army, 1942-46. Fellow John Simon Guggenheim Found., 1958-59, USPHS, 1959; sr. postdoctoral fellow NSF, 1966-67; recipient Sigma Xi Rsch. award Kans. State U., 1954, Thomas Jefferson Faculty Excellence award U. Mo., 1986, Most Disting. Scientist award Mo. Acad. Sci., 1989; Byler Disting. Prof., 1990. Mem. AAAS, Am. Soc. Microbiology. Office: Cancer Research Center 3501 Berrywood Dr Columbia MO 65201

EISENSTEIN, BRUCE ALLAN, electrical engineering educator; b. Phila., Sept. 10, 1941; s. Ira S. and Rose (Futerman) E.; m. Toby Karet, Sept. 8, 1963; children: Eric, Andrew, Ilana. BEE, MIT, 1963; MEE, Drexel U., 1965; PhD, U. Pa., 1970. Registered profl. engr., Pa. Vis. asst. prof. Princeton (N.J.) U., 1970-71; asst. prof. Drexel U., Phila., 1971-77, assoc. dean grad. sch., 1976-78, prof., head dept. elect. engring., 1980—; cons. forensic engring. numerous lawyers; cons. electronics and signal processing projects various indsl. and legal firms, book pubs.; chmn. adv. com. for elect., communications and systems engring. div. NSF; mem. adv. com. for engring. directorate NSF; adv. coun. for engring. C.C. Phila., 1992—. Bd. mgrs. Cen. High Sch. Alumni Assn., Phila, 1972-79, chmn. com. on sch. standards, 1974-76. Ford Found. fellow, 1965, NASA-Am. Soc. Engring. Edn. summer faculty fellow Stanford U., 1969. Fellow IEEE (bd. dirs. 1993—, exec. com. Phila. sect., counselor student br. Drexel U. and student br. IEEE Computer Soc.), mem. Edn. Soc. IEEE (pres. 1984-85, bd. dirs. 1993—, mem. AdCom, Achievement award 1987), Am. Soc. Engring. Edn., Eta Kappa Nu (C. Holmes McDonald award 1976), Sigma Xi, Tau Beta Pi. Republican. Jewish. Avocations: squash, tennis, skiing, classical music-piano. Home: 7804 Pine Rd Philadelphia PA 19118-2527 Office: Drexel U Dept Elec & Computer Engring 32nd Chestnut Philadelphia PA 19118-3744

EISNER, HOWARD, engineering educator, engineering executive; b. N.Y.C., Aug. 8, 1935; s. Samuel Eisner and Mary (Isser) Wegodner; m. Joan Arlene Knopfer, Feb. 9, 1957; children: Seth Eric, Susan Rachel, Oren David. BEE, CCNY, 1957; MS, Columbia U., 1958; DSc, George Washington U., 1966. Teaching asst. Columbia U., 1957; lectr. dept. physics Bklyn. Coll., 1957-59; lectr., asst. professorial lectr. George Washington U., 1961-67; prof. U. Maryland, 1987-89; various engring. positions ORI, Inc., Rockville, Md., 1959-68, v.p., 1968-71, exec. v.p., 1971-84, corp. exec. v.p., 1984-85, also dir.; pres. Intercon Systems Corp. subs. ORI, Group, Inc., Rockville, 1985-89, Atlantic Research Services Corp., Alexandria, Va., 1987-89; Disting. rsch. prof. George Washington U., Washington, 1989—. Author: Advanced Algebra, 1960, Computer-Aided Systems Engineering, 1988; contbr. articles in field. Fellow IEEE, N.Y. Acad. Scis.; mem. AIAA, Ops. Rsch. Soc. Am., Inst. Mgmt. Sci., Sigma Xi, Tau Beta Pi, Eta Kappa Nu, Omega Rho. Avocations: personal computers, tennis, choral singing, writing. Office: George Washington U Gelman Libr Washington DC 20052

EISNER, THOMAS, biologist, educator; b. Berlin, June 25, 1929; s. Hans Edouard and Margarete (Heil) E.; m. Maria Lobell, June 10, 1952; children: Yvonne, Vivian, Christina. B.A., Harvard U., 1951, Ph.D., 1955; D.Sc. hon., U. Wurzburg, Fed. Republic Germany, 1982, U. Zurich, Switzerland, 1983, U. Göteborg, 1989, Drexel U., 1992. Postdoctoral fellow Harvard U., 1955-57; asst. prof. biology Cornell U., Ithaca, N.Y., 1957-62; assoc. prof. Cornell U., 1962-66, prof., 1966-76, Jacob Gould Schurman prof. biology, 1976—; vis. scientist dept. entomology Sch. Agr., Wageningen, Netherlands, 1964-65; vis. scientist Smithsonian Tropical Research Lab., Barro Colorado Island, C.Z., 1968; sr. vis. scientist Max Planck Inst. für Verhaltensphysiologie, Seewiesen, Fed. Republic Germany, 1971, Div. Entomology, CSIRO, Canberra, Australia, 1972-73; Rand fellow Marine Biol. Labs., Woods Hole, Mass., 1974; vis. research prof. U. Fla., Gainesville, 1977-78; research assoc. Archbold Biol. Sta., 1973—; vis. prof. Stanford U., 1979-80, U. Zurich, 1980-81. Co-author: Animal Adaptation, 1964, Life on Earth, 1973, and 3 other books; mem. editorial bd. Sci., 1970-71, Am. Naturalist, 1970-71, Jour. Comparative Physiology, 1974-80, Chem. Ecology, 1974—, Behavioral Ecology and Sociobiology, 1976—, Sci. Yr. World Books, 1979-82, Human Ecology Forum, 1981-85, Living Bird Quar., 1982-88, Experientia, 1982—, Quar. Review Biology, 1983-87; co-editor: Explorations in Chemical Ecology Series, 1987—; contbr. articles to profl. jours. Mem. adv. bd. Econ. Potential Rare and Threatened Plants, Ctr. Plant Conservation, 1992—. Recipient Founder's Meml. award Entomol. Soc. Am., 1969, Archie F. Carr medal, 1983, Procter prize Sigma Xi, 1986, Karl Ritter von Frisch medal, 1988, Centennial medal Harvard U., 1989, Tyler Environ. Achievement prize U. So. Calif., 1990, Esselen award, 1991, Silver medal Internat. Soc. Chem. Ecology, 1991; Guggenheim fellow, 1964-65, 72-73. Fellow Am. Acad. Arts and Scis., Royal Soc. Arts, Animal Behavioral Soc., Entomol. Soc. Am.; mem. NAS (film com. 1986—, com. on human rights 1987-90), Am. Philos. Soc., Am. Inst. Biol. Scis. (task force for 1990s 1990—), Explorers Club, Deutsche Akad. Naturforscher Leopoldina, Club of Earth, Zero Population Growth (bd. dirs. 1969-70), Nature Conservancy (nat. scientific adv. coun. 1969-74), Nat. Audubon Soc. (bd. dirs. 1970-75), Fedn. Am. Scientists (mem. coun. 1977-81), AAAS (chmn. biology sect. 1980-81, mem. com. on sci. freedom and responsibility 1980-87, chmn. subcom. sci. and human rights 1981-87, Newcomb Cleveland prize 1967), Ctr. on Consequences Nuclear War (steering com. 1983-90), World Wildlife Fund (mem. sci. adv. coun. 1983-91), Am. Soc. Naturalists (pres. com. 1989-90), Monell Chem. Senses Inst. (mem. adv. coun. 1988—), World Resources Inst. (mem. adv. coun. 1988—), Com. Concerned Scientists (nat. sponsor 1988—), Union Concerned Scientists (bd. dirs. 1993—), Xerces Soc. (sci. adv. com. 1990—, pres. 1992—). Office: Cornell U Dept Neurobiology and Behavior W347 Seeley Mudd Hall Ithaca NY 14853

EKELAND, ARNE ERLING, surgeon, educator; b. Voss, Norway, Oct. 14, 1942; s. Magne and Birgit (Birkeland) E.; m. Synnove Godvik; 1 child, Magnus. MD, U. Oslo, Norway, 1967; PhD, U. Oslo, 1981. Lic. gen. surgeon 1984,, orthopedic surgeon 1988. Resident Aker Hosp., Oslo, 1971-74, lectr., 1974-76; rsch. fellow Rikhosp., Oslo, 1976-81; resident Ullevaal Hosp., Oslo, 1981-86, sr. orthopedic surgeon, 1987-90; prof. U. Oslo, 1990—; chief surgeon Ullevaal Hosp., Oslo, 1990—; resident Sophies Minde Orthopedic Hosp., Oslo, 1986-87. Editorial bd. Scandinavian Jour. for Medicine and Sci. in Sports, 1990—, Knee Surgery, Sports Traumatology and Arthroscopy, 1992—. With Norwegian Army. Mem. Surg. Assn. Oslo (sec. 1984-86), Internat. Soc. for Fracture Repair (founding mem.), European Soc. for Calcified Tissue, Norwegian Soc. Sport Rsch. (bd. dirs. 1986-88), Internat. Soc. Skiing Safety (bd. dirs. 1981-89, v.p. 1989—). Avocations: skiing, jogging, reading. Office: Ullevaal Hosp, Surgical Clinic, N-0407 Oslo Norway

EKELÖF, TORD JOHAN CARL, elementary particle physicist; b. Uppsala, Sweden, Sept. 12, 1945; s. Per Olof and Marianne (Hesser) E.; m. Monique Francoise Gapany, June 23, 1979; 1 child, Nils. MSc, U. Uppsala, 1966, M. Engring., 1968, PhD, 1972. Rsch. fellow CERN, Geneva, 1972-75, rsch. assoc., 1977-79, staff physicist, 1979-83; researcher Swedish Sci. Rsch. Coun., Geneva, 1983-88; rsch. assoc. CERN, Geneva, 1988-90; researcher, reader Swedish Sci. Rsch. Coun., Geneva, 1990-91, Uppsala U., 1992—. Mem. Swedish Phys. Soc. (bd. dirs.), European Com. for Future Accelerators (sec.), CERN Com. of the Swedish Sci. Rsch. Coun., Instrumentation Panel of Internat. Com. for Future Accelerators (chmn.), High Energy Physics Bd. of European Phys. Soc., CERN Experimental Adv. Com., Super Proton Synchrotron Com., Large Electron Positron Collider Com. Home: Nybrogatan 40 B, S-11440 Stockholm Sweden Office: U Uppsala Dept Radiation Sci, Box 535, S-75121 Uppsala Sweden

EKIS, IMANTS, materials engineer, consultant; b. Riga, Latvia, May 4, 1943; came to U.S., 1949.; s. Nikolajs and Tekla (Skurulis) E.; m. Irene Biruta Zile, Sep. 17, 1966; children: Paul Andrew, Sandra Agate, Robert Michael. BS, Marquette U., 1977, postgrad., 1978-84. Products and materials engr. Milw. Electric Tool Co., Brookfield, Wis., 1974-81; assoc. project materials engr. J.I. Case Co., Hinsdale, Ill., 1981-88; mgr. gear metallurgy Mercury Marine Co., Oshkosh, Wis., 1988-90; sr. materials engr. Ford New Holland (Pa.) Co., 1990—. Editor: (book) Gear Design and Their Performance, 1984, contbr. articles to profl. jours. Mem. Internat. Metallographic Soc., Am. Soc. Agrl. Engrs., Soc. Tribologists and Lubrication Engrs., Soc. Automotive Engrs. (dir. group C-Iron, mem. steel tech. com. 1985-89), The Minerals Metals and Materials Soc., Gear Rsch. Inst. Lutheran. Avocations: classical music, reading, fishing. Home: 35 Aspen Dr Leola PA 17540-9616 Office: Ford New Holland Co 500 Diller Ave New Holland PA 17557-9301

EKIZIAN, HARRY, civil engineering consultant; b. N.Y.C., July 31, 1921; s. Harry and Nectar (Tatsian) E.; m. Justine Markarian, July 14, 1946 (dec.); 1 child, Michelle Lynne. BSCE, CCNY, 1942. Registered profl. engr., N.Y. Field engr. Johnson-Perini Corp., Chambersburg, Pa., 1942; rsch. engr. Nat. Com. for Aeronautics, Langley Field, Va., 1945-46; civil engr. Dept. Pub. Works, Westchester County, N.Y., 1946; assoc. TAMS Cons., Inc., N.Y.C., 1947-90; cons. engr. Mamaroneck, N.Y., 1991—. Contbr. articles to profl. jours. Pres. Prince Willow Civic Assn., Mamaroneck; co-chmn. tech. com. for constrn. St. Vartan Cathedral, N.Y.C., 1962-64; active various civic and ch. orgns. With U.S. Army, 1943-45. Fellow ASCE; mem. Tau Beta Pi. Achievements include design of ports, transport systems for industerial facilities. Home and Office: 8 Prince Willow Ln Mamaroneck NY 10543

EL-AGRAA, ALI M., economics educator; b. Wad Medani, Sudan, Jan. 1, 1941; s. Mohammed A. El-Agraa and Fardous (Raoda) Yaghoup; m. Diana Latham Moult, Oct. 20, 1979; children: Mark Stephen, Frances Hannah. BSc in Econs. with honors, U. Khartoum, 1964; MA in Econs. with distinction, U. Leeds, 1967, PhD, 1971. Tutor in Econs. Khartoum (Sudan) U., 1964-67, lectr., 1967-72; lectr. Leeds (Eng.) U., 1971-81, sr. lectr., 1981—; vis. prof. internat. econ. Internat. U. Japan, Yamato-Machi, 1984-86; vis. disting. scholar Fudan U., Shanghai, China, 1985; prof. internat. econs. and European/Am. econs. Fukuoka (Japan) U., 1988—; lifetime vis. prof. Wuhan U., China; chmn. bd. advisors Kyushu Conf. for Govt./Industry/Univs., 1989—; chmn. Fukuoka Internat. Forum, 1990—; cons. Kyushu Econ. Forum, Fukuoka, 1989—, Anglo-Japanese Econ. Inst., London, 1988—. Author, editor: The Economics of the European Community, 1981, 4th edit., 1993, International Economic Integration, 1982, 2d edit., 1988, Britain Within the European Community, 1983, Protection, Cooperation, Integration and Development, 1987, Public and International Economics, 1993; author: Theory of International Trade, 1983, Trade Theory and Policy, 1984, Japan's Trade Frictions, 1988, International Trade, 1989, Theory and Measurement of International Economic Integration, 1989; co-author: Theory of Customs Unions, 1981. Mem. Royal Econ. Soc., Am. Econ. Assn., Internat. Econs. Study Group (com. 1982-88). Avocations: jogging, tennis, classical music, bridge. Home: 1-31-17 Higashi Irube, Sawara-ku, Fukuoka 811-11, Japan Office: Fukuoka U Faculty of Commerce, Nanakuma, Jonan-ku, Fukuoka 814-01, Japan

ELBARBARY, IBRAHIM ABDEL TAWAB, chemist; b. Alexandria, Arab Republic of Egypt, Jan. 17, 1933; came to U.S., 1970; s. Abdel Tawab Ali Elbarbary and Tafida Ibrahim Basuni; m. Fatma Oweiss, Sept. 17, 1972; children: Meeral, Mohamed, Raney. MS in Chemistry, Cairo U., Cairo, Arab Republic of Egypt, 1969; PhD in Chemistry, SUNY, Binghamton, 1977. Labs. dir. chemistry dept. Ministry of Industry, Cairo, 1957-70; chief chemist Belmont Smelting and Refining Works, Bklyn., 1970-76; teaching fellow SUNY, Binghamton, 1976-85; assoc. prof. Ga. Inst. Tech., Atlanta, 1981-83; head environ. studies King Abdul Aziz U., Jeddah, Saudi Arabia, 1985—; project mgr. Bur. Engraving & Printing, U.S. Treasury, Washington, 1986—; cons. water programs U.S. AID, Washington, 1979—, UN Devel. Program, N.Y.C., 1982-83; chmn. dept. environ. studies King Abdul Aziz. U., Jeddah, 1981-83. Author 27 papers and reports. Recipient Postgrad. Study award Egyptian Govt., Max Planck Inst. for Iron Rsch., Dusseldorf, Fed. Republic of Germany, 1961-63. Fellow Am. Inst. Chemists; mem. Am. Soc. Metals, Am. Chem. Soc., Assn. Egyptian Am. Scholars. Achievements include research in analytical chemistry method development; in environment analysis and control; in waste water treatment. Home: 7707 Elgar St Springfield VA 22151-2515

ELBAUM, CHARLES, physicist, educator, researcher; b. May 15, 1926; married; 3 children. M.A.Sc., U. Toronto, Ont., Can., 1952; Ph.D. in Applied Sci., U. Toronto, 1954; M.A. (hon.), Brown U., 1961. Research fellow in metal physics U. Toronto, Ont., Can., 1954-57; research fellow in metal physics Harvard U., 1957-59; asst. prof. applied physics Brown U., Providence, 1959-61; assoc. prof. physics Brown U., 1961-63, prof. physics, 1963—, chmn. dept. physics, 1980-86, also Hazard prof. physics; cons. to industry. Fellow Am. Phys. Soc.; mem. AIME, Soc. Neurosci., AAAS. Office: Brown U Dept Physics Box 1843 Providence RI 02912

ELBAUM, MAREK, electro-optical sciences executive, researcher; b. Kovel, U.S.S.R., May 8, 1941; came to U.S., 1969; s. Isaak Elbaum and Maria Rajbenbach; m. Lia Krusin, Jan. 2, 1969; 1 child, Martin Krusin-Elbaum. MSc, Warsaw (Poland) Tech. U., 1966; PhD, Columbia U., 1977. Rsch. assoc. Polish Acad. Sci., Warsaw, 1966-68; mem. rsch. staff Riverside Rsch. Inst., N.Y.C., 1969-79, mgr. electro-optics div., 1979-82, rsch. dir., 1982-90; pres. Electro-Optical Scis., Irvington, N.Y., 1990—. Contbr. more than 60 articles to profl. jours. Achievements include patent for novel holographic technique, novel techniques for tracking space object in the visible and infrared; development of theory for direct detection laser radars; first demonstration of frequency diversity technique for laser speckle reduction and ultrasound applications. Office: Electro Optical Scis 1 Bridge St Irvington NY 10533-1543

EL-BAYYA, MAJED MOHAMMED, civil engineer; b. Gaza-Strip, July 8, 1963; came to U.S. 1989; s. Mohammed Ali and Rashida Mohammed (El-Ladah) El-B.; m. Faten Fathi Baroud, Aug. 2, 1991; 1 child, Lina. BSCE, BirZeit U., 1987; MSCE, U. Mo., 1991. Designer civil engr. Municipality of Gaza, Gaza-Strip, 1987-89; rsch. engr. Capsule Pipeline Rsch. Ctr., Columbia, Mo., 1991—. Contbr. articles to profl. jours. Mem. acad. coun. Engring. Coll., Birzeit, WestBank, 1985. Grantee AMIDEAST, 1989, UNRWA, Gaza, 1982. Mem. Assn. Engrs. (Gaza), Sigma Xi (assoc. mem.). Muslim. Achievements include predicting mathematical models for water-hammer in hydraulic capsule pipeline. Home: 208 Old 63 N Apt 22 Columbia MO 65201 Office: Capsule Pipeline Rsch Ctr E2421 Engring Bldg E Columbia MO 65211

EL-BAZ, FAROUK, program director, educator; b. Zagazig, Egypt, Jan. 1, 1938; came to U.S., 1960, naturalized, 1970; s. El-Sayed Mohammed and Zahia Abul-Ata (Hammouda) El-B.; m. Catherine Patricia O'Leary, 1963; children—Monira, Soraya, Karima, Fairouz. B.Sc., Ain Shams U., 1958; M.S., U. Mo., 1961; Ph.D., U. Mo. and Mass. Inst. Tech., 1964; DSc (hon.), New England Coll., 1989. Demonstrator geology dept. Assiut U., Egypt, 1958-60; lectr. Mineralogy-Petrography Inst., U. Heidelberg, Ger., 1964-65; geologist exploration dept. Pan Am.-UAR Oil Co., Egypt, 1966; supr. lunar exploration Bellcomm and Bell Telephone Labs., Washington, 1967-72; research dir. Center for Earth and Planetary Studies, Nat. Air and Space Mus., Smithsonian Instn., Washington, 1973-82; v.p. sci. and tech. Itek Optical Systems, Litton Industries, Lexington, Mass., 1982-86; cons. geology; prof. geology and geophysics U. Utah, 1975-77; prof. geology Ain Shams U., Egypt, 1976-81; sci. adviser Pres. Anwar Sadat of Egypt, 1978-81; dir. Ctr. for Remote Sensing Boston U., 1986—. Author or co-author: Say It in Arabic, 1968, Coprolites: An Annotated Bibliography, 1968, Glossory of Mining Geology, 1970, The Moon as Viewed by Lunar Orbiter, 1970, Astronaut Observations from the Apollo-Soyuz Mission, 1977, Apollo Over the Moon: A View from Orbit, 1978, Egypt As Seen by Landsat, 1979, Apollo-Soyuz Test Project Summary Science Report: Earth Observations and Photography, 1979, Desert Landforms of Southwest Egypt: A Basis for Comparison with Mars, 1982, Deserts and Arid Lands, 1984, The Geology of Egypt: An Annotated Bibliography, 1984, Physics of Desertification, 1986, Remote Sensing and Resource Exploration, 1989, Sand Transport and Desertification in Arid Lands, 1990; also author or co-author articles. Decorated Order of Merit 1st class Egypt; recipient certificate merit U.S. Bur. Mines, 1961, Exceptional Sci. Achievement medal NASA, 1971, Alumni Achievement award U. Mo., 1972, Honor citation Assn. Arab-Am. U. Grads., 1973. Fellow Royal Astron. Soc., Geol. Soc. Am. (certificate commendation 1973); mem. AAAS, Am. Assn. Adv. Sci. (Pub. Understanding of Sci. and Tech. award, 1992), World Aerospace Edn. Orgn. (cert. of merit, 1973), Internat. Inst. of Boston (Golden Door award 1992), Sigma Xi. Clubs: Explorers, University. Office: Boston U Ctr Remote Sensing 725 Commonwealth Ave Boston MA 02215-1401

ELBEN, ULRICH, chemist; b. Erlangen, Bavaria, Germany, Oct. 26, 1950; s. Robert George and Sophie (Heyl) E.; children: Sandra, Anreas Dean. PhD, U. Bonn, 1979. Rsch. chemist Hoechst AG Pharm. Devel., Wiesbaden, Germany, 1980-86; group leader Hoechst AG Pharm. Rsch., Wiesbaden, 1986-88; rsch. team leader Hoechst AG Pharm. Devel., Wiesbaden, 1989-91, project mgr., 1992—; lectr. chemistry Inst. for Pharmacy, U. Frankfurt, Germany, 1992—; presenter profl. symposiums. Contbr. articles to profl. jours., including Drug Rsch., Kontakte, Jour. Chem. Rsch. Mem. Internat. Soc. Heterocyclic Chemistry, European Soc. Neurochemistry. Home: Stettiner Str 15/4, D-65239 Hochheim Germany Office: Hoechst AG, Werk Kalle-Albert, Rheingaustr, D-65174 Wiesbaden Germany

ELCHUK, STEVE, chemist. Mem. Chem. Inst. Can. (Norman and Marion Bright Meml. award 1992). Home: 16 Sandy Dr, Petawawa, ON Canada K8H 3E6*

ELDER, BESSIE RUTH, pharmacist; b. Ovalo, Tex., June 28, 1935; d. William Kinsalow and Ima Ruth (Carter) Griffing; m. George Davis Elder, Sept. 15, 1950 (dec. Nov. 1991); children: Michael Davis, Linda Sue Elder Claborn. BS in Pharmacy, Southwestern Okla. U., 1989. Staff pharmacist Coleman Pharmacy, Dimmitt, Tex., 1989-90, United Pharmacy, Lubbock, Tex., 1990—, St. Mary's Hosp. Pharmacy, Lubbock, 1992—. Home: 8602-A Memphis St Lubbock TX 79423

ELDER, MARK LEE, university research administrator, writer; b. Littlefield, Tex., May 3, 1935; s. Mark Gray and Ethel Ruby (Hill) E.; m. Elizabeth Ellen Lovejoy, Dec. 19, 1992; 1 child, Staci Lee. BA in Journalism, U. Okla., 1965, MA in Comm., 1973. Tech. writer/editor rsch. inst. U. Okla., Norman, 1964-66; asst. dir., then dir. info. svcs. dept., rsch. inst. U. Okla., 1968-73, assoc. dir., sponsored programs adminstr./dep. dir., dir. office rsch. adminstrn., 1973-84, security supr., 1977-83; tech. writer/editor Los Alamos (N.Mex.) Sci. Lab., 1964-66; presentations writer/editor Martin-Marietta Corp., Orlando, Fla., 1966-67, Collins Radio Co., Richardson, Tex., 1967-68; dir. office rsch. devel. and adminstrn. Ariz. State U., Tempe, 1984-85; univ. patent/copyright officer, dir. sponsored projects U. North Tex., Denton, 1986—, asst. v.p. rsch., 1987—; U. North Tex. primary rep. Tex. consortium participants, chair preaward issues subgroup, Tex. state issues task force, mem. nat. state issues com. Fed. Demonstration Project, 1988—; dep. dir., corp. sec. North Tex. Rsch. Inst., 1989—; cons. to pres. U. Tex., San Antonio, 1991; mem. various coms. U. Okla., Ariz. State U., U. North Tex.; speaker in field. Author: Jedcrow, 1974; (with R. Martin) Handbook for Effective Writing, 1975; Wolf Hunt, 1976, Swedish edit., 1977, Italian edit., 1978, The Prometheus Operation, 1980, Brit. edit., 1981; (with others) Reasearch Administration and Technology Transfer, 1988; tech. editor: Energy Alternatives: A Comparative Analysis (D. Kash., et al.), 1975, Energy From the West: A Progress Report of a Technology Assessment of Western Energy Resource Development (I. White, et al.), 1977. With U.S. Army, 1954-56. Mem. Nat. Coun. Univ. Rsch. Adminstrs. (chair nat. publs. com., 1980-81, v.p. 1981-82, pres. 1982-83, exec. com. 1986-88, co-chair regional congress 1988, mem. various coms., editor newsletter, 1986-87, Past Pres.'s award 1983, Spl. Citation region V 1984, Spl. Citation newsletter editor 1987), Assn. Univ. Tech. Mgrs., Tex. Tech. Transfer Assn. (charter, co-founder 1988—, interim sec.-treas. 1988, v.p., pres.-elect 1988-89, pres. 1989-90, bd. dirs. 1990-91, Bob G. Davis award 1991). Office: Univ North Texas UNT Box 5396 Denton TX 76203

ELDERING, HERMAN GEORGE, physicist; b. Newark, N.J., Apr. 8, 1930; s. George Jacob and Marie Eldering; m. Barbara Jane Eldering, Sept. 16, 1954; children: Eric Charles, Joyce Ann. BS in Physics, CCNY, 1953; student, U. Denver, 1955-56. Project engr. Radio Corp. Am., Burlington, Mass., 1956-61; staff scientist U. Mich., Maui, Hawaii, 1967-69; chief engr. Thermo Electron, Waltham, Mass., 1986-87; prin. scientist Baird Corp.-Imo Industries, Bedford, Mass., 1961-92; chief scientist Specialty Instruments, Bedford, Mass., 1992—; chmn. Electro Optical Info Process Iris, Boston, 1965-67. Co-Leader Explorer Post, Maui, 1967-69, leader, Chelmsford, Mass., 1975. With U.S. Army, 1953-55. Mem. Optical Soc. Am., Internat. Soc. Optical Engring. Unitarian. Acheivements include patents for Solar Spectro Radiometer, Disturbed Gun Mount. Home: PO Box 273 Nutting Lake MA 01865-0273 Office: Specialty Instruments 19A Crosby Dr Bedford MA 01730-1419

ELDRED, KENNETH MCKECHNIE, acoustical consultant; b. Springfield, Mass., Nov. 25, 1929; s. Robert Moseley and Jean McKechnie (Ashton) E.; m. Helene Barbara Koerting Fischer, May 31, 1957; 1 dau., Heidi Jean. B.S., MIT, 1950, postgrad., 1951-53; postgrad., UCLA, 1960-63. Engr. in charge vibration and sound lab. Boston Naval Shipyard, 1951-54; supervisory physicist, chief phys. acoustics sect. U.S. Air Force, Wright Field, Ohio, 1956-57; v.p., cons. acoustics Western Electro-Acoustics Labs., Los Angeles, 1957-63; v.p., tech. dir. sci. services and systems group Wyle Labs., El Segundo, Calif., 1963-73; v.p., dir. div. environ. and noise control tech. Bolt Beranek and Newman Inc., Cambridge, Mass., 1973-77; prin. cons. Bolt Beranek and Newman Inc., 1977-81; dir. Ken Eldred Engring.; mem. exec. standards coun. Am. Nat. Standards Inst., 1979-89, vice-chmn., 1981-83, chmn., 1985-87, bd. dirs., 1983-87; mem., past chmn. Acoustical Standards Bd.; mem. com. hearing, bioacoustics and biomechanics NRC, 1963—. Served with USAF, 1954-56. Fellow Acoustical Soc. Am. (standards dir. 1987—, past chmn. coordinating com. environ. acoustics); mem. NAE, Inst. Noise Control Engring. (pres. 1976, bd. dirs. 1987-91), Soc. Automotive Engrs., Soc. Naval Architects and Marine Engrs., U.S. Yacht Racing Union. Clubs: Down East Yacht, Blue Water Sailing. Home: Meadow Cove East Boothbay ME 04544 Office: Box 501 East Boothbay ME 04544

ELDRED, NELSON RICHARDS, chemist, consultant; b. Oberlin, Ohio, Mar. 6, 1921; s. Arthur Newell and Zell Mittilene (Richards) E.; m. Alice Byer, June 15, 1945 (div. 1967); children: Janice, Jean Weaver, David, Elaine Weller, Lois; m. Penny Gibson, Mar. 30, 1968. BA, Oberlin Coll., 1943; MS, Wayne State U., 1947; PhD, Pa. State U., 1951. Lab. asst. Parke Davis & Co., Detroit, 1943-46; group leader Union Carbide Corp., South Charleston, W.Va., 1951-69; asst. mgr. devel. Buckman Labs., Memphis, Tenn., 1969-70; mgr. Graphic Arts Tech. Found., Pitts., 1970-89; cons. New Port Richey, Fla., 1989—. Author: Solving Offset Ink Problems, 1981, Chemistry for the Graphic Arts, 1992, (with others) What the Printer Should Know About Ink, 1990, Package Printing, 1993. Mem. Am. Chem. Soc. (chmn. Kanawha Valley sect. 1960), Tech. Assn. Pulp and Paper Industry (chmn. printing com. 1976-77), Tech. Assn. Graphic Arts. Home and Office: 4901 Bellemede Blvd New Port Richey FL 34655

ELDREDGE-THOMPSON, LINDA GAILE, psychologist; b. Lubbock, Tex., Apr. 3, 1959; d. Jerry Greever and Madge (Harshbarger) Eldredge; m. Paul Edward Thompson, May 26, 1979. BS, Howard Payne U., 1980; MA, Tex. Woman's U., 1981; EdD, Baylor U., 1989. Lic. psychologist, chem. dependency counselor, Tex.; cert. chem. dependency specialist; tchr. hearing impaired; sch. counselor, spl. edn. sch. counselor, Tex. Tchr. hearing impaired Waco (Tex.) Ind. Sch. Dist., 1982-85, spl. edn. sch. counselor, cons. hearing impaired, 1986-87; doctoral teaching fellow Baylor U., Waco, 1985-87; dir. regional alcohol and drug abuse svcs. Heart of Tex. Coun. Govts., Waco, 1987; psychotherapist Clin. Psychology Assocs., Houston-Webster, Tex., 1989-91; psychologist Clin. Psychology Assocs., Houston-Webster, 1991-93, Tex. Sch. for the Deaf, 1993—, Austin Psychotherapy Assocs., 1993—. Mem. APA, World Found. of Successful Women, Nat. Assn. Alcoholism and Drug Abuse Counselors, Am. Deafness and Rehab. Assn., Gentle Art of Verbal Self-Defense Trainers Network, Tex. Assn. Alcoholism and Drug Abuse Counselors (sec.-treas. Waco chpt. 1988-89). Avocations: collecting gems, minerals and seashells, jazz fusion, reading, art, water sports. Office: Austin Psychotherapy Assocs 7719 Wood Hollow Ste 152 Austin TX 78731 Also: Tex Sch for the Deaf 1102 S Congress Ave Austin TX 78764

ELDRIDGE, PETER JOHN, fishery biologist; b. New Bedford, Mass., Feb. 6, 1937; s. Carlton Calvin and Catherine Frances (O'Connell) E.; m. Joan Tyler, Dec. 20, 1968 (div. 1986); children: Kelly Michelle, Elizabeth Christine. BS in Zoology, U. Mass., 1959; MA in Marine Sci., Coll. William and Mary, 1962; PhD, U. Wash., 1975. Biologist S.C. Wildlife and Marine Resources Dept., Charleston, 1972-78, Nat. Marine Fish Svc., Charleston, 1978-90; fishery mgr. Nat. Marine Fish Svc., St. Petersburg, Fla., 1990—. Author tech. reports; contbr. sci. articles to profl. publs. Capt. U.S. Army, 1962-65. Fellow Am. Inst. Fishery Rsch. Biologists (Carolina dist. dir. 1975-77), Am. Fisheries Soc., South Atlantic Fisheries Mgmt. Coun. (chmn. scientific stats. com. 1984-87), Atlantic States Marine Fisheries Commn. (chmn. statistics com. 1974-78), Sigma Xi (pres. Charleston chpt. 1987-88). Roman Catholic. Home: 8921 118th Way N Seminole FL 34642 Office: Nat Marine Fishery Svc Southeast Regional Office 9450 Koger Blvd Saint Petersburg FL 33702

EL-DUWEINI, AADEL KHALAF, clinical pharmacologist, information scientist; b. Cairo, Dec. 26, 1945; Arrived in Can., 1988.; s. Atallah Khalaf El-Duweini and Flora Philobos Sedrah; m. Salwa Michael Rizk, Nov. 19, 1970; children: Salam, Hadi. BSc in Pharmacy, Cairo U., 1967, M Pharm. Sci., 1976; postgrad. in info. mtkg., Cath. U. Am., 1981; MBA, Am. U., 1981. Bibliographer Nat. Info. and Document Ctr., Cairo, 1967-73, abstractor, indexer, 1974-76, staff coord., 1977-81; mgr. mktg., communications and sci. rels. Acad. Sci. Rsch. and Tech., Cairo, 1982-87; rsch. scientist Nat. Rsch. Ctr., Cairo, 1968-88; asst. to pharmacist Cumberland Drug, Dorval, Can., 1988—; mktg. cons. Cow and Gate baby foods, Cairo, 1976-80; mgmt. cons. Bureau Joseph, Patents & Trade Marks Atty., Cairo, 1980-88; info. cons. Nat. Cancer Inst., Cairo U., 1983-88; instr. info. sci. Am. U. in Cairo, 1984-88. Co-author: The Infrastructure of an Information Society, 1984; contbr. articles to profl. jours. Mem. Ecumenical Youth Com., Cairo, 1970-88, PTA German Schs., Cairo, 1982-88; Sunday sch. coord. St. Mark Ch., Montreal, Que., Can., 1990—. Scholar Goethe Inst., Fed. Republic Germany, 1973, Internat. Agy. for Rsch. on Cancer, 1976, AID, 1980, 81, Brit. Coun., 1980, 84. Mem. Egyptian Assn. Sci. and Tech. Librs. and Info. Ctrs. (founding sec. gen. 1985—), Egyptian Sci. Soc. for Group Tng. (coun. 1982—), Am. Soc. for Info. Sci., Libr. Assn. Gt. Britain, Egyptian Soc. for Pharmacology and Exptl. Therapeutics, Can. Pharm. Assn., Can. Cancer Soc., Egyptian Pharm. Assn. Coptic Orthodox. Coptic Orthodox. Avocations: photography, violin, table tennis, stamp and coin collecting. Home: 8 Oriole Dr, PQ, Kirkland, PQ Canada H9H 3X3 Office: Cumberland Drugs, Dorval, PQ Canada H9S 1B3

ELEJALDE, CESAR CARLOS, food engineer; b. Lima, Peru, June 19, 1960; came to U.S. 1988; s. Cesar and Adriana (Vignolo) E. MS, Rutgers U., 1991, PhM, 1992. Tech. mgr. ASA Alimentos S.A., Lima, 1984-87; prodn. mgr. Promisac S.A., Lima, 1987-88; rsch. scientist M & M Mars, Hackettstown, N.J., 1992—; tech. cons. ASA Alimentos S.A., 1987-88. Recipient Endel Karmas Excellence in Teaching award Food Sci. Dept., Rutgers U., 1990. Mem. Inst. Food Technologists, Soc. Rheology. Home: 52D The Village Green Budd Lake NJ 07828 Office: M & M Mars High St Hackettstown NJ 07840

ELEQUIN, CLETO, JR., retired physician; b. Antique, Philippines, Oct. 18, 1933; s. Cleto and Enriqueta (Tengonciang) E.; m. Nancy Johnson, May 14, 1958; children: Tracy, Thomas Kyle, Stuart Scott. M.D., Far Eastern U., Philippines, 1957. Rotating intern Good Samaritan Hosp., Lexington, Ky., 1957-58; gen. practice resident Central Bapt. Hosp., Lexington, 1958-59; psychiat. resident State Hosp., Danville, Pa., 1959-60, 61-62; psychiat. resident with child psychiatry State Hosp., New Castle, Del., 1962-63; staff physician Eastern State Hosp., Lexington, 1960-61, dir. Fayette County Project, dir. intensive treatment service, 1964-67, supt., 1969-71; dep. commr. Dept. Mental Health, State Ky., 1967-69; practice medicine, specializing in family practice Pecos, Tex., 1971-72, Austin, Tex., 1974-89; ret.; vis. lectr. in medicine and psychiatry Am. U. of the Caribbean, Plymouth, Montserrat; asst. dep. commr. Tex. Dept. Mental Health and Mental Retardation, Austin, 1973-74, dep. commr. mental health, 1974; attending psychiatrist U. Ky. Med. Ctr., 1964-71, Good Samaritan Hosp., 1969-71, Central Bapt. Hosp., 1966-71; cons. psychiatrist U. Ky. Student Health Service, 1965-71, Peace Corps, 1966-68, Bur. Rehab., State Ky., 1965-71, Blue Grass Community Care Ctr., 1967-71, Covington (Ky.) Community Care Ctr., 1969-71, Hazard Community Care Ctr., 1969-71, Danville (Ky.) Community Care Ctr., 1969-71, Maysville (Ky.) Community Care Ctr., 1969-71; clin. instr., asst. clin. prof. dept. psychiatry U. Ky. Med. Ctr., 1964-69, assoc. clin. prof., 1969-71. Mem. Profl. Adv. Coun. Community Mental Health-Retardation Ctr., Lexington, 1967-71; mem. Lexington Hosp. Coun., 1969-71. Mem. AMA, Am. Psychiat. Assn., Tex. Med. Assn., Travis County Med. Soc., Austin Psychiat. Soc., Assn. Med. Supts. Mental Hosps., Am. Acad. Family Physicians. Home: 10101 Jupiter Hills Dr Austin TX 78747-1312

ELESH, DAVID BERT, sociology educator; b. Chgo., Oct. 20, 1940; s. Joseph and Beatrice (Lichtenstein) E.; m. Estella McEwan Munson, May 27, 1967; 1 child, Benjamin Frost. BA, Reed Coll., 1962; PhD, Columbia U., 1968. Instr. U. Wis., Madison, 1967-68, asst. prof., 1968-74; assoc. prof. Temple U., Phila., 1974—, dir. Social Scis. Data Lab., 1979—; CEO David Elesh Cons., Merion, Pa., 1989—; mem. adv. com. Roper Pub. Opinion Rsch. Ctr., Hartford, Conn., 1967; mem. Mng. Dirs. Com. on Geocoded Databases, Phila, 1984; mem. review com. NIH/SBIR, Washington, 1990-92. Co-author: Philadelphia Neighborhoods: Division and Conflict in a Postindustrial City, 1991; contbr. chpt. to book. Grantee Selo Fund, 1982, Nat. Archives, 1984, NIH, 1989, Legis. Office Rsch. Liason, 1990; fellow NSF, 1962-64. Mem. Am. Sociol. Assn., Am. Statis. Assn., Ea. Sociol. Assn. Office: Temple U Dept Sociology Philadelphia PA 19122

ELFSTROM, GARY MACDONALD, aerospace engineer, consultant; b. Vancouver, B.C., Can., Aug. 14, 1944; s. Roy Harold and Vera Marie (Macdonald) E.; m. Carol Ann Skelton, Sept. 25, 1969; children: David Roy, Julie Ann. BS, U. B.C., 1968; PhD, U. London, 1971. Registered profl. engr., Ont., Can. Asst. prof. U. Tenn., Tullahoma, 1971-73; assoc. rsch. officer NRC, Ottawa, Ont., 1973-81; chief aerodyns. DSMA Internat. Inc., Toronto, Can., 1981—; mgr. bus. devel. DSMA-Babcock Inc., Toronto, 1992—; chmn. adv. com. aerodyn. NRC, Ottawa, 1978-79; mem. adv. com. Ryerson Poly. Aerospace, Toronto, 1989—. Contbr. articles to profl. jours. Recipient Athlone fellowship U.K. Bd. Trade, 1968-71. Fellow Can. Aero. Space Inst. (chmn. aero. edn. 1991—), AIAA (chmn. ground test tech. com. awards 1987-91); mem. Royal Aero. Soc. (assoc. mem.). Achievements include development of physical model of hypersonic turbulent flow field interactions, test technique for deducing performance of airfoils at transonic speeds, viz wavedgrag. Office: DSMA Babcock Inc, 6285 Northam Dr Ste 200, Mississauga, ON Canada L4V 1X5

EL-GAMMAL, ABDEL-AZIZ MOHAMED, aero-mechanical engineer; b. Kafr El-Shiek, Egypt, Feb. 10, 1946; s. Mohamed Mostafa and Asmaa El-Bateaa (El-Gammal) E.; m. Somaya Osman El-Deeb, Jan. 30, 1978; children: Mohamed, Zeyad. BS in Mech. Engring./Aeronautics, Mil. Tech. Coll., Cairo, 1967, diploma in aero. engring. air frames, 1968, MS in Mech. Engring., 1983; PhD in Mech. Engring., Belgrade U., Yugoslavia, 1991. Cert. Aero-Mech. Engr. Mech. engr. Air Force, Egypt, 1967—, advanced through ranks to maj. gen., aircraft maintenance engr., 1967-68, workshop comdr., 1968-72, dep. chief engr., 1972-73, chief engr., 1973-75; chief fighter bomber dept. Engring. Authority, Air Force, Eqypt, 1975-80; chief rschr. Armed Forces Rsch. Ctr., Egypt, 1980-85, dep. dir. R & D dept., 1985-87; dir. Integrated Logistics Monitoring & Analysis Ctr., Egypt, 1991—; cons. Armed Forces Rsch. Ctr., Cairo, 1985-87; vis. lectr. Higher Tech. Inst., Ramadan Tenth City, 1991—. Moslem. Achievements include rsch. in and devel. of modeling criteria for model performance measure, systematic modeling and real-time simulation of gas turbines; condition monitoring and diagnostics; co-development of weapons integration and upgrading of air crafts; co-establishment of Air Force Research Center. Address: 41 Dr Mohamed Sakr Kafaga St, Madinet Nasr Cairo Egypt also: care Dr Atef Faghry, PO Box 1180, Jeddah 21431, Saudi Arabia

ELGAVISH, ADA, biochemist; b. Cluj, Romania, Jan. 23, 1946; came to U.S. 1979; d. David and Malca (Neuman) Simchas; m. Gabriel A. Elgavish,

Dec. 28, 1968; children: Rotem, Eynav. BSc, Tel-Aviv U., 1969, MSc, 1972; PhD, Weizmann Inst. Sci., Rehovot, Israel, 1978. Postdoctoral vis. fellow NIH, Balt., 1979-81; instr. U. Ala. Sch. Medicine, Birmingham, 1981-82, rsch. assoc., 1982-84, rsch. asst. prof., 1984-89, asst. prof. comparative medicine, 1989-92, assoc. prof., 1992—; assoc. scientist Cystic Fibrosis Ctr., Birmingham, 1984—. Grantee Cystic Fibrosis Found., 1986-90, Am. Lung Assn., 1987-92, NIH, 1989—. Mem. Am. Physiol. Soc., AAAS, N.Y. Acad. Sci., Ala. Acad. Sci., Southeastern Pharmacology Soc., Am. Thoracic Soc., Sigma Xi. Home: 1737 Valpar Dr Birmingham AL 35226-2343 Office: Univ of Ala Sch Medicine Dept Comparative Medicine Birmingham AL 35294

ELGAVISH, GABRIEL ANDREAS, physical biochemistry educator; b. Budapest, Hungary, July 29, 1942; arrived in Israel, 1957, came to U.S., 1979; s. László and Katalin Barbara (Szentmiklóssy) Schwarcz; m. Ada Stephanie Simcas, Dec. 28, 1967; children: Rotem László Abraham, Eynav Elgavish. BSc, Hebrew U., Jerusalem, 1967; MSc, Tel-Aviv U., 1972; PhD, Weizmann Inst. of Sci., 1978. Vis. fellow NIH, Balt., 1979-81; asst. prof. U. Ala., Birmingham, 1981-87, assoc. prof., 1987—. 1st lt. Israeli Army, 1961-64. Mem. Am. Chem. Soc., Am. Soc. for Biochemistry and Molecular Biology, Am. Heart Assn./Basic Sci., Soc. Magnetic Resonance in Medicine. Jewish. Achievements include 2 patents on Contrast Agents for Nuclear Magnetic Resonance Imaging; research in nuclear magnetic resonance spectroscopy. Office: U Ala 1900 University Blvd THT 336 Birmingham AL 35294-0006

EL-GEWELY, M. RAAFAT, biology educator; b. Damanhour, Behira, Egypt, June 2, 1942; s. Ahmed Abdelsalam and Aziza (Makey) El-G.; m. Sara Ann Harma, Nov. 16, 1981; children: Noor, Tarek. BSc, Alexandria (Egypt) U., 1963; PhD, U. Alberta, Edmonton, Can., 1971. Postdoctoral fellow McGill U., Montreal, 1971-73; asst. prof. genetics Higher Inst. Cotton Affairs, Alexandria, 1973-74; asst. prof. molecular and biochem. genetics Cairo U., Giza, Egypt, 1974-77; assoc. rsch. scientist U. Mich., Ann Arbor, 1983-88; prof. biotech. U. Tromsø, Norway, 1988—; vis. scholar U. Mich., Ann Arbor, 1977-83, dir. Tromso Biotech. Ctr., 1990—; vis. scientist Parke-Davis Pharm., Ann Arbor, 1992-93. Editor: Site-directed Mutagenesis and Protein Engineering, 1991. UNESCO fellow, 1977; grantee NAVF, Oslo, 1988—, Aakre-Fund, Tromsø, 1989, DNK, Oslo, 1989—, NFFR, Trondheim, 1989—. Mem. AAAS, Genetics Soc. Can., Genetics Soc. Am., Am. Soc. Microbiology, Egyptian Soc. Genetics, Am. Soc. Biochemistry and Molecular Biology, Norwegian Soc. Biochemistry, N.Y. Acad. Scis. Home: Uranusveien 38, Tromsø Norway Office: U Tromsø IMB, U Tromsø MB, Tromsø Norway

ELGHAMMER, RICHARD WILLIAM, psychologist; b. Sitka, Alaska, Nov. 18, 1951; s. W. Robert and Doris (Postelwaite) E.; m. Mona Kay Collins, June 1, 1985. BS cum laude, Knox Coll., 1975; MS, Ea. Ill. U., 1983; PhD, Utah State U., 1989. Staff researcher U. Stockholm, 1977-78; psychology intern Ind U. Sch. Medicine, Indpls., 1988-89; child psychologist Wabash Valley Hosp., Crawfordsville, Ind., 1988-89, Elghammer Family Ctr., Danville, Ill., 1989—; cons. VA Med. Ctr., Danville, 1990—, Family Crisis Shelter, Crawfordsville, 1991—, Crosspoint Mental Health Ctr., Danville, 1991—. Contbr. articles to profl. jours. Recipient Rsch. award U. Stockholm, 1977; Utah State U. fellow, 1984-85, scholar, 1988. Mem. APA, Ill. Psychol. Assn., Ind. Psychol. Assn. Office: Elghammer Family Ctr 723 N Logan Danville IL 61832

ELGIN, CHARLES ROBERT, chief technology officer; b. Columbus, Ohio, Jan. 23, 1956; s. Charles Joseph and Beatrice Helen (Hagman) E.; m. Sharon Ann, June 28, 1980. BS, Ohio State U., 1978. CPA, Ohio; registered profl. engr., Ohio; cert. data processor, systems profl. Store mgr. Minelli's Restaurant, Columbus, Ohio, 1972-78; staff cons. Anderson Consulting, Columbus, Ohio, 1978-81, sr. cons., 1981-84, mgr., 1984-89, sr. mgr., 1989-91; chief tech. officer Chem. Mortgage Co., Worthington, Ohio, 1991—. Bd. dirs. St. Matthew's Parish Coun., Gahanna, Ohio, 1992. Mem. Athletic Club of Columbus. Democrat. Roman Catholic. Home: 667 Parkedge Dr Gahanna OH 43230 Office: Chem Mortgage Co 200 Old Wilson Bridge Rd Worthington OH 43085

ELGIN, GITA, psychologist; b. Santiago, Chile; came to U.S., 1968, naturalized 1987; d. Serafin and Regina (Urízar) Elguin; BS in biology summa cum laude, U. Chile, Santiago, DPs, 1964; PhD in Counseling Psychology, U. Calif., Berkeley, 1976; m. Bart Bödy, Oct. 23, 1971; children: Dio Christopher Károly, Alma Ilona Raia Julia. Clin. psychologist Barros Luco-Trudeau Gen. Hosp., Santiago, 1964-65; co-founder, co-dir. Lab. for Parapsychol. Rsch., Psychiat. Clinic, U. Chile, Santiago, 1965-68; rsch. fellow Found. Rsch. on Nature of Man, Durham, N.C., 1968; researcher psychol. correlates of EEG-Alpha waves U. Calif., Berkeley, 1972-76; originator holistic method of psychotherapy Psychotherapy for a Crowd of One, 1978; co-founder, clin. dir. Holistic Health Assos., Oakland, Calif., 1979—; lectr. holistic health Piedmont (Calif.) Adult Sch., 1979-80; hostess Holistic Perspective, Sta. KALW-FM, Nat. Public Radio, 1980. Author: (video documentary) Taking the Risk: Sharing the Trauma of Sexual & Ritualistic Abuse in Group Therapy, 1992. Lic. psychologist, Chile, Calif. Chancellor's Patent Fund grantee U. Calif., 1976, NIMH fellow, 1976. Mem. Am. Psychol. Assn., Am. Holistic Psychol. Assn. (founder 1985—), Alameda County Psychol. Assn., Calif. State Psychol. Assn., Montclair Health Profls. Assn. (co-founder, pres. 1983-85), Sierra Club, U. Calif. Alumni Assn. Contbr. articles in clin. psychology and holistic health to profl. jours. and local periodicals. Presenter Whole Life Expo, 1986. Office: Montclair Profl Bldg 2080 Mountain Blvd Ste 203 Piedmont CA 94611-2817

ELGORT, ANDREW CHARLES, school psychologist; b. N.Y.C., Mar. 27, 1953; s. George and Janet T. E.; m. Virginia B. Elgort, June 24, 1979; children: Mayan G., Ari N. BS, West Chester (Pa.) State Coll., 1975; MEd, U. Va., 1977; EdS, James Madison U., 1985; EdD, Coll. of William and Mary, 1992; postgrad., 1992. Nat. cert. sch. psychologist. Tchr. emotionally disturbed Greenbank Sch., Glenmore, Pa., 1975-76; asst. dir., project dir. TREES project Augusta County Pub. Schs., Fisherville, Va., 1977-78; tchr. emotionally disturbed City of Manassas (Va.) Pub. Schs., 1978-79, ednl. diagnostician, 1979-80; tchr. emotionally disturbed Albemarle County Pub. Schs., Charlottesville, Va., 1980-83; staff psychologist Child Devel. Ctr. James Madison U., Harrisburg, Va., 1984-85; sch. psychologist Henrico County Pub. Schs., Richmond, Va., 1985-93, Howard County Pub. Schs., Ellicott City, Md., 1993—. Mem. edn. com. Rudlin Torch Acad., Richmond, 1989-91, chmn. edn. com., 1989-90, v.p. adminstrn., 1991-92, sec., 1992-93; bd. dirs. Keneseth Beth Israel Synagogue, Richmond, 1987-93; bd. dirs. Sir Moses Mantefiore Cemetery Corp., 1991-93, sec., 1992-93. Postgrad. scholar, 1992. Mem. Nat. Assn. Sch. Psychologists, Va. Psychol. Assn., Coun. Exceptional Children, Va. Acad. Sch. Psychologists, Kappa Delta Phi. Office: Howard County Pub Schs 10910 Rt 108 Ellicott City MD 21042-6198

EL-HAMALAWAY, MOHAMED-YOUNIS ABD-EL-SAMIE, computer engineering educator; b. Cairo, Egypt, Aug. 21, 1947; s. Abd-El-Samie Mohamed and Fatma Younis (Al-Batriq) El-H.; m. Safaa Abd-El-Hamid Osman, Aug. 7, 1982; children: Mahmoud, Fatma. BSEE, Ain Shams U., Cairo, 1968; B.e.i. in Elec. Engring., U. Ghent, Belgium, 1973, PhD in Elec. Engring., 1976. Registered profl. engr. Rsch. engr. Tech. Tng. and Rsch. Inst., Cairo, 1968-72; rsch. asst. U. Ghent (Belgium), 1973-76; microcontroller systems engring. cons. Nat. Semiconductor GmbH, Furstenfeldbruck, Federal Republic of Germany, 1977; lectr. in computer engring. Al Azhar U., Cairo, 1977-84, assoc. prof. computer engring., 1984-89, prof., 1989—; cons. Wang Europe S.A., Ottergem, Belgium 1974-75, Misr Consulting Office, Cairo, 1977—, Egyptian Mil. com. for Mfg. Computers, Cairo, 1984-86, Misr Fatramo Computers and Electronic Equipment Co., Arab Republic of Egypt, 1985—; lectr. Ain Shams U., Cairo, 1977-79; coord. Nat. Com. for Setting an Egyptian Standard for Using Arabic Characters in Computers, Cairo, 1984-90; external assoc. prof. Mil. Tech. Coll., Cairo, 1984-86; mng. bd. dirs. El-Nasr Television and Electronics Co., Cairo, 1992—. Author: Relaxation Oscillators, 1970, Transistors-A Programmed Text, 1971, Electronic Engineering, 1971, Logic Laboratory Notebook, 1991; patentee in field; contbr. articles to profl. jours.; mem. editorial bd.: Microcomputer Applications Jour., 1985-88. Co-sec. Internat. Students Assn. of Ghent, Belgium, 1974; treas. Egyptian Students Union in Belgium, 1974, sec., 1975. Recipient Golden medal in Tng., Cen. Tng. Orgn., Arab Republic of Egypt, 1971, scholarship, Brit. Coun., Eng., 1972, U. Ghent, Belgium, 1973-76. Mem. IEEE, Nat. Soc. for Technol. and Econ. Devel, Egyptian Syndicate Engrs., Egyptian Soc. Engrs., Egyptian Inventors and Innovators Soc. (treas. 1989-90, vice chmn. 1990—), Egyptian Computer Soc., Chamber of Egyptian Engring. Industries (bd. dirs. electronics com.), Egyptian Specialist Group in Computational Linguistics and Arabization (founding mem.), Egyptian Assn. Ednl. Tech. Moslem. Avocations: reading, calligraphy, photography. Home: 5 Siliman Mohamed Abazah St, 2d Dist Heliopolis, Cairo 11351, Egypt Office: Al Azhar U, Faculty Engring Systems & Computers, Nasr City Egypt

EL-HUSBAN, TAYSEER KHALAF, internist, consultant; b. Hamamah, Mafvag, Mafraq, May 25, 1955; s. Khalaf Eyadeh and Kamayel (Hussein) El-H.; m. Angeleki Panayotis Dimakakou, May 20, 1987; children: Faris, Samir. BM, Athens (Greece) U., 1980. Cert. specialist in internal medicine. House officer St. Georg Hosp., Pireaus, Greece, 1980-81, St. Helen Hosp., Athens, 1981-85; med. specialist King Khalid Hosp., Riyadh, Saudi Arabia, 1985-87, Nat. Ins. Found., Athens, 1987-90; cons. State Hosp. Pireas, Greece, 1990—. Home: Aer Papanastasiou 23, M Asias S1, 115-27 Athens Greece Office: Elefsis Med Ctr, Kimonos 12, 19200 Elefsing Attiki, Greece

ELIA, MICHELE, mathematics educator; b. Berzano, Asti-Piemonte, Italy, Jan. 2, 1945; s. Luigi and Cristina (Fogliatti) E. Dr. engr., Politecnico di Torino, 1970. Researcher FIAT, Torino, Italy, 1970-71, Politecnico di Torino, 1971-77, assoc. prof. math., 1977-90, prof., 1990—. Author: (with others) The Information Theory Approach to Communications, 1977. Contbr. articles to profl. jours. Mem. Unione Matematica Italiana, Am. Math. Soc., Math. Assn. Am., Soc. Indsl. and Applied Math., IEEE (sr.). Roman Catholic. Home: Via G Marconi 3, Castiglione Torinese, 10090 Torino Italy Office: Politecnico di Torino, Dipartimento di Elettronica, Corso Duca degli Abruzzi 24, 10129 Torino Italy

ELIAS, ANTONIO L., aeronautical engineer, aerospace executive. BS, MIT, MIT; EAA, MIT, PhD in Aeronautics and Astronautics. Rsch. asst. and staff mem. Space Guidance and Navigation Divsn., CS Draper Lab., 1972-80; asst. prof. aeronautics and astronautics MIT, 1980-86; sr. v.p. engring. Orbital Sci. Corp., Fairfax, Va., 1986—; technical dir. Pegasus Air-Launched Orbital Booster Program, Orbital Sci. Corp. Contbr. articles to profl. jours. Recipient Nat. medal of Tech., U.S. Dept. Commerce Tech. Adminstrn., 1991. Mem. AIAA (Aircraft Design award 1991, Engr. of Yr. award 1992), Inst. Navigation. Achievements include patents in field. Office: Orbital Sciences Corp Chief Engineer 12500 Fair Lakes Circle Fairfax VA 22033*

ELIAS, DONALD FRANCIS, environmental consultant; b. Cleve., Aug. 8, 1949; s. Richard Joseph and Marie Terese (Sievers) E. BS in Chemistry, U. S.C., 1971; cert. in meteorology, St. Louis U., 1972; MS in Environ. Engring., Wash. State U., 1977. Chemist S.C. Dept. of Health and Environ. Control, Columbia, 1974-75; rsch. asst. Wash. State U., Pullman, 1975-77; sr. assoc. scientist I.I.T. Rsch. Inst., Chgo., 1977-78; mgr. Northrop Svcs., Research Triangle Park, N.C., 1978-80; prin. Dames & Moore, Houston and Bethesda, Md., 1980-83; mgr. Camp, Dresser & McKee, Denver and Edison, N.J., 1982-86; founding prin. Research Triangle Park Environ. Assocs., Inc., Green Brook, N.J., 1978-86, pres., prin., 1986—. Contbr. articles to profl. jours. Lt. Martinsville (N.J.) Rescue Squad, 1984-90, pres., 1991—; Eucharistic min., lectr. Blessed Sacrament, Martinsville, 1986—; mem. Green Brook Rescue Squad, 1988-93. Lt. USAF, 1971-74. Mem. Am. Chem. Soc., Natural Resources Def. Coun., Assn. Energy Engrs. (sr.), Air and Waste Mgmt. Assn. (vice chmn. waste source group 1989—), Environ. Def. Fund, Amnesty Internat., Sierra Club (life). Avocations: playing golf, playing tennis, reading. Office: RTP Environmental Assoc Inc 239 US Route 22E Green Brook NJ 08812

ELIAS, JACK ANGEL, physician, educator; b. Fayetteville, Ark., Apr. 10, 1951; s. Gabriel and Alma (Kowalsky) E.; m. Sandra Gross, Jan. 3, 1981; 1 child, Lauren Rachel. BA, U. Pa., 1973, MD, 1976. Intern internal medicine Tufts- New England Med. Ctr., Boston, 1976-77, resident internal medicine, 1977-78; sr. resident internal medicine Hosp. of U. Pa., Phila., 1975-79, joint fellow allergy and immunology, pulmonary and critical care medicine, 1979-82; prof., chief pulmonary and critical care medicine Yale U. Sch. Medicine, New Haven, 1990—; asst. prof. U. Pa. Sch. Medicine, Phila., 1979-82, 82-88, assoc. prof., 1988-90; mem. coun. Am. Lung Assn., 1992. Contbg. editor Jour. Applied Physiology, 1988-91; assoc. editor Am. Jour. Physiology Lung, 1991—, Jour. Lab. Clin. Medicine, 1991—; contbr. articles to Jour. Immunology, Jour. Clin. Investigation, others. Grantee Nat. Heart-Lung-Blood Inst./NIH, 1988—, 1991. Fellow Am. Coll. Chest Physicians; mem. Am. Soc. Clin. Investigation, Am. Thoracic Soc. (bd. dirs. 1992, pres. sci. assembly on allergy, immunology and inflammation 1990-91). Achievements include discovery of Interleuken-I-tumor necrosis factor synergy, Interleukin-I-Interleukin 6 synergy, lung fibroblast hyaluronidase, Interleukin-6. Office: Yale U Sch Medicine 105 LCI PO Box 3333 333 Cedar St New Haven CT 06510

ELIAS, SAMY E. G., engineering executive; b. Cairo, June 28, 1930; came to U.S., 1956, naturalized, 1964; s. Elias Girgis and Tahia N. (Kassabgy) E.; m. Janice Lee Craig, Aug. 21, 1960; children: Mona Lee, Tresa Jean, Cecilia Ruth. BS in Aero. Engring., Cairo U., 1955; MS in Aero. Engring., Tex. A&M U., 1958; PhD in Indsl. Engring. and Mgmt., Okla. State U., 1960. Grad. asst. Tex. A&M U., College Station, 1957-58; grad. asst. Okla. State U., Stillwater, 1958-60; asst. prof., indsl. engring. Kans. State U., Manhattan, 1960-61; exec. asst. to chmn. bd. Orgn. of Mil. Factories, Egypt, 1961-62; asso. prof. indsl. engring. Kans. State U., 1962-65; assoc. prof. indsl. engring. W.Va. U., Morgantown, 1965-67; prof. W.Va. U., 1967-79, chmn. dept. indsl. engring., 1969-76, spl. asst. to univ. pres. for personal rapid transit, 1970-77, Claude Worthington Benedum prof. transp., 1976-82; dir. Harley O. Staggers Nat. Transp. Ctr., 1980-82; dir. transit engring. and safety Washington Met. Area Transit Authority, 1982-84; v.p. Transp. and Distbn. Assocs., Inc. subs. Day & Zimmermann, Phila., 1984-87; prin. FAI Assocs., Inc., McLean, Va., 1987—; assoc. dean engring. rsch. U. Nebr., Lincoln, 1988—; cons. Kansas City Transit, N.Y. Transit Authority, N.Y. Transit Authority Police Dept., Omaha Transit Co., Cin. Transit Co., W.C. Gilman & Co., Inc., Brown Engring., Transp. and Distbn. Assocs., PRC Harris, Arab Petroleum Cons., Urban Transp. Devel. Corp., World Bank, also others. Contbr. over numerous publs. to profl. jours. Recipient Americanism medal DAR, 1977. Fellow Chartered Inst. Transp., Inst. Indsl. Engrs. (Transp. and Distbn. award 1979); mem. Soc. Am. Value Engrs., Am. Soc. Engring. Edn. (chmn. indsl. engring. divsn. 1972-73), Soc. for Computer Simulation, Nat. Soc. Propl. Engrs., Accreditation Bd. Engring. Tech. (engring. accreditation com. 1987-92, W.Va. Soc. Profl. Engrs. Coptic Orthodox. Home: 8111 Dorset Dr Lincoln NE 68510-5209 Office: U Nebr Coll Engring & Tech W 150 Nebraska Hall Lincoln NE 68588-0502

ELIAS, THOMAS ITTAN, mechanical engineer; b. Piramadom, Kerala, India, Mar. 16, 1947; came to U.S.A., 1975; s. Varkey and Sara (Paul) I.; m. Gracy Thomas, Sept. 3, 1979; children: Joseph, John. BS, U. Kerala, Trivandrum, 1970, MS, 1972; MS, U. Cin., 1977; PhD, U. Minn., 1981. Registered profl. engr. Devel. engr. Vikram Sarabhai Space Ctr., Trivandrum, 1972-75; rsch. asst. U. Cin., 1976-77; teaching assoc., rsch. asst. U. Minn., 1977-80; asst. prof. Ind. Inst. Tech., Ft. Wayne, 1980-82; asst. prof. Youngstown (Ohio) State U., 1982-87, assoc. prof., 1987-88; lead engr. McDonnell Douglas Corp., St. Louis, 1988-91; project mgr. U.S. Dept. Energy, Idaho Nat. Engring. Lab., Idaho Falls, 1991—; tech. cons. Westinghouse Electric Corp. Contbr. articles to profl. jours. Mem. India Assn. of Greater Youngstown, 1982-88. Mem. ASME, AIAA, Full Gospel Bus. Men's Fellowship (v.p. 1988), Sigma Gamma Tau, Phi Kappa Phi. Home: 1335 Herring St Idaho Falls ID 83404 Office: Dept Energy System Analysis Div 785 Doe Pl Idaho Falls ID 83402

ELIAS, THOMAS SAM, botanist, author; b. Cairo, Ill., Dec. 30, 1942; s. George Sam and Anna (Clanton) E.; m. Barbara Ana Boyd (dec.); children: Stephen, Brian. BA in Botany, So. Ill. U., 1964, MA in Botany, 1966; PhD in Biology, St. Louis U., 1969. Asst. curator Arnold Arboretum of Harvard U., Cambridge, Mass., 1969-71; adminstr., dendrologist Cary Arboretum, N.Y. Botanical Garden, Millbrook, 1971-73, asst. dir., 1973-84; dir., chief exec. officer Rancho Santa Ana Botanic Garden, Claremont, Calif., 1984—; chmn., prof. dept. botany Claremont Grad. Sch., 1984—; lectr. in extension Harvard U., 1971; adj. prof. Coll. Environ. Science and Forestry, Syracuse, N.Y., 1977-80; coord. U.S.A/U.S.S.R. Botanical Exch., Program for U.S. Dept. of Interior, Washington, 1976—, U.S.A./China Botanical Exch., Program for U.S. Dept. of Interior, 1988—. Editor: Extinction is Forever, 1977 (one of 100 Best Books in Sci. and Tech. ALA 1977), Conservation and Management of Rare and Endangered Plants, 1987; author: Complete Trees of North America, 1980 (one of 100 Best Books in Sci. and Tech. ALA 1980), Field Guide to Edible Wild Plants of North America (one of 100 Best Books in Sci. and Tech. ALA 1983). Recipient Cooley award Am. Soc. Plant Taxonomist, 1970, Disting. Alumni award So. Ill. U., 1989. Home: 2447 San Mateo Ct Claremont CA 91711-1652 Office: Rancho Santa Ana Botanic Garden 1500 N College Ave Claremont CA 91711-3157

ELIASSON, KERSTIN ELISABETH, science and education policy advisor; b. Stockholm, Sweden, Jan. 30, 1945; d. Sixten Elov and Greta Irene Solveig (Marker) Englesson; m. Jan Kenneth Eliasson, Nov. 18, 1967; children: Anna, Emilie, Johan. BA, U. Stockholm, 1967; MA, Am. U., Washington, 1973. Project dir. Swedish Inst. for Opinion Rsch., Vallingby, 1974-75; sec. Govt. Commn. on Tng., Stockholm, 1975-77; advisor Swedish Ministry of Edn., Stockholm, 1977-79, head of sect., 1979-88; advisor Prime Minister's Office, Stockholm/N.Y.C., 1988-91, Swedish Ministry of Edn. and Sci., Stockholm/N.Y.C., 1992—; bd. dirs. Scandinavian Inst. for African Studies, Uppsala, rsch. grants com. Swedish Inst., Stockholm; chmn. Com. for Sci. and Tech. Policy, OECD, Paris, 1987-91. Contbr. articles to profl. jours. Mem. AAAS, N.Y. Acad. Scis. Office: Consulate Gen of Sweden 885 Second Ave 45th Fl New York NY 10017

ELIEL, ERNEST LUDWIG, chemist, educator; b. Cologne, Germany, Dec. 28, 1921; came to U.S., 1946, naturalized, 1951; s. Oskar and Luise (Tietz) E.; m. Eva Schwarz, Dec. 23, 1949; children: Ruth Louise, Carol Susan. Student, U. Edinburgh, Scotland, 1939-40; degree in phys.-chem. sci., U. Havana, Cuba, 1946; Ph.D., U. Ill., 1948; D.Sc. (hon.), Duke U., 1983, U. Notre Dame, 1990. Mem. faculty U. Notre Dame, South Bend, Ind., 1948-72, prof. chemistry, 1960-72, head dept., 1964-66; W.R. Kenan Jr. prof. chemistry U. N.C., Chapel Hill, 1972-93; Sir C.V. Raman vis. prof. U. Madras, India, 1981; Le Bel Centennial lectr., Paris, 1974; Sir C.V. Raman vis. prof. U. Madras, India, 1981; Geoffrey Coates lectr. U. Wyo., 1989; Smith, Kline & French lectr. U. Ill., 1990. Author: Stereochemistry of Carbon Compounds, 1962, Conformational Analysis, 1965, Elements of Stereochemistry, 1969, From Cologne to Chapel Hill, 1990; co-editor: Topics in Stereochemistry, vols. I-XXI, 1967-94. Pres. Internat. Relations Council, St. Joseph Valley, 1961-63. Recipient Coll. Chem. Tchrs. award Mfg. Chemists Assn., 1965, Laurent Lavoisier medal French Chem. Soc., 1968, N.C. award in Sci., 1986; NSF sr. rsch. fellow Harvard U., 1958, Calif. Inst. Tech., 1958-59; E.T.H. Zurich, Switzerland, 1967-68, Guggenheim fellow Stanford U., Princeton U., 1975-76, Duke U., 1983-84. Fellow AAAS (chmn. chemistry sect. 1991-92), Royal Soc. Chems.; mem. NAS, AAUP (chpt. pres. 1971-72, 78-79), Am. Acad. Arts and Scis., Am. Chem. Soc. (chmn. St. Joseph Valley sect. 1960, councillor 1965-73, 75—, chmn. com. publs. 1972, 76-78, dir. 1985-93, chmn. bd. dirs. 1987-89, pres. 1992, Morley medal Cleve. sect. 1965, Harry and Carol Mosher award Santa Clara Valley sect. 1982, Herty medal Ga. sect. 1991, So. Chemist award Memphis sect. 1991, Madison Marshall award North Ala. sect., 1993), Royal Spanish Chem. Soc. (hon.), Argentine Chem. Assn. (corr.), Peruvian Chem. Soc. (corr.), Mex. Chem. Soc. (hon.), Mex. Acad. Scis. (corr.), Sigma Xi (chpt. pres. 1968-69), Phi Lambda Upsilon, Phi Kappa Phi. Home: 725 Kenmore Rd Chapel Hill NC 27514-2019

ELIEN, MONA MARIE, air transportation professional; b. Atwood, Kans., June 13, 1932; d. Lawrence Wallace Berry and Adele Rosina (Gulzow) Wright; m. R.J. Wright, Jan. 1952 (div. 1957); m. J.P. Kobus, Nov. 1968 (div. 1991); m. Robert Louis Tour, Oct. 3, 1992. BS, U. Ariz., 1961; grad., Swiss Mountain Climbing Inst., Rosenalui, 1963; postgrad., No. Ariz. U., 1966-67, Ariz. State U., 1967-69, 86-87; MPA, Ariz. State U., 1981. Customer rels. rep. Ariz. Pub. Svc. Co., Casa Grande, Flagstaff, Ariz., 1961-67; owner/operator Mona's Clipping Svc., Phoenix, 1969-74; various positions City of Phoenix, 1974—; contract mgr. Phoenix CETA/PSE/PNP, 1978-81; planning and devel. asst. Phoenix Sky Harbor Internat. Airport, 1986—; staff asst. 1988 Citizens Bond Com. for Aviation, Phoenix, 1987-88. Compiler, editor: Aviation Acronyms and Abbreviations, 1987, 2d rev. edit., 1992; editor, writer (newsletter) Rapsheet, 1972-75; author profl. columns, 1961-67. Pres. state home econs. occupations adv. bd. Ariz. State U., 1983-84; mem. City Mgr.'s Women's Issues com., Phoenix, 1989-91, Zonta Internat., 1962-65; pres.-elect, Tri-City (Ariz.), 1964-65; vol. speaker's bur. Phoenix Community Alliance and Prep. Acad. Partnership, 1992—; mem. exec. com. Svc. Fund Dr., Phoenix, 1984-86; precinct com. Yuma County, Ariz., 1958; employee-of-yr. com. mem. City of Phoenix Aviation Dept., 1993, 94. Recipient Recognition Pub. Svc. award Ariz. Dept. Econs. Security, 1975, Heart and Soul award Barry M. Goldwater Terminal 4, 1990, PHXcellence award, 1993; named one of Outstanding Young Women of Am., 1966. Mem. ASPA (life; Phoenix chpt. awards banquet com. 1991, 92, nat. com. 1990-91), Am. Home Econs. Assn. (life), Ariz. Home Econs. Assn. (pres. no. region 1965-67), Sinagua Soc. Museum No. Ariz., Satisfied Frog Gold Mountain Club, Flagstaff C. of C. (chmn. Indian princesses, retail mchts. sect. 1965-67), So. Ariz. Hiking Club, Desert Bot. Gardens, Swinging Stars Square Dance Club, Delta Delta Delta. Republican. Lutheran. Home: 2201 E Palmaire Ave Phoenix AZ 85020-5633 Office: Phoenix Sky Harbor Internat Airport 3400 E Sky Harbor Blvd Phoenix AZ 85034-4403

ELIEZER, ISAAC, chemistry educator; b. Sofia, Bulgaria, Jan. 19, 1934; m. Naomi Eliezer. MSc, Hebrew U., Jerusalem, 1956, PhD, 1960. Prof. chemistry, assoc. dean Oakland U., Rochester, Mich., 1979—. Contbr. over 100 articles to profl. jours. Fellow Royal Soc. Chemistry, Royal Inst. Chemistry; mem. AAAS, Internat. Union Pure and Applied Chemistry, Am. Assn. Colls. for Tchr. Edn., European Acad. Arts, Humanities and Scis., Internat. Studies Assn., Am. Chem. Soc., N.Y. Acad. Scis., Bd. Govs. of William Beaumont Hosp. Rsch. Inst., Sigma Xi. Avocations: books, music. Office: Oakland U Rochester MI 48309

ELIGON, ANN MARIE PAULA, physicist; b. Port-of-Spain, Trinidad, West Indies, Apr. 2, 1957; came to U.S., 1989; d. Joseph William and Pearl Umilta (Pierre) Samaroo; m. Ronnie Evrol Eligon, Aug. 19, 1976; children: Michael Adrian, John Warren, Alex David. BS in Physics & Math., U. West Indies, 1979, PhD in Physics, 1988. Teaching asst. U. West Indies, St. Augustine, Trinidad, 1979-85, asst. lectr., 1985-88, lectr., 1988-89; faculty EMBY-Riddle Aeronautical U., Daytona Beach, Fla., 1990, Valencia Community Coll., Orlando, Fla., 1990-92; posdoctoral fellow Ctr. Rsch. Electro-Optics & Lasers, Orlando, Fla., 1992—; adj. instr. U. Ctrl. Fla., ORlando, 1990. Contbr. articles to profl. jours. NSF Career Advancement award, 1992. Mem. Am. Phys. Soc. Roman Catholic. Office: Ctr Rsch Electro-Optics & Lasers 12424 Research Pky Orlando FL

ELIN, RONALD JOHN, pathologist; b. Mpls., Apr. 14, 1939; s. John Matthew and Helen Sophia (Lind) E.; m. Susan May Krogh, June 14, 1969; children: Derek, Justin. BA, U. Minn., 1960, BS, 1962, MD, 1966, PhD, 1969. Diplomate Am. Bd. Pathology, Am. Bd. Clin. Chemistry. Intern U. Hosp. Calif., San Diego, 1969-70; commd. med. officer USPHS, 1970, advanced through grades to med. dir., 1975; staff assoc. Nat. Inst. Allergy and Infectious Diseases NIH, Bethesda, Md., 1970-73, resident clin. pathology dept., 1973-74, chief clin. pathology dept., 1975—, chief chemistry svc., 1977—; clin. prof. Uniformed Svcs. Univ. of Health Scis., Bethesda, 1978—; intiator, first chmn. Gordon Rsch. Conf. on Magnesium in Biomed. Processes in Medicine, 1978. Contbr. over 140 articles to profl. jours. Decorated Commendation medal USPHS, 1980, Meritorious Svc. medal USPHS, 1984. Fellow Am. Coll. Nutrition, Coll. Am. Pathologists, Am. Soc. Clin. Pathologists; mem. Am. Assn. Pathologists, Am. Assn. Clin. Chemisty, Acad. Clin. Lab. Physicians and Scientists (sec./treas. 1985-87, pres. 1990-91). Lutheran. Achievements include research on magnesium metabolism, properties of endotoxin. Home: 11401 Marcliff Rd Rockville MD 20852-3635 Office: NIH Clin Pathology Dept 9000 Rockville Pike Rm 2c306 Bethesda MD 20892-0001

ELION, GERTRUDE BELLE, research scientist, pharmacology educator; b. N.Y.C., N.Y., Jan. 23, 1918; d. Robert and Bertha (Cohen) E. AB, Hunter Coll., 1937, DSc (hon.), 1989; DSc (hon.), NYU, 1989; DMS (hon.), Brown U., 1969; DSc (hon.), U. Mich., 1983, N.C. State U., 1989, Ohio State U., 1989, Poly. U., 1989, U. N.C., 1990, Russell Sage Coll., 1990, Duke U.,

1991, MacMaster U., 1992, SUNY, Stony Brook, 1992, George Washington U., 1969, Columbia U., 1992, Washington Coll., 1993, U. South Fla., 1993, U. Wis., 1993. Lab. asst. biochemistry N.Y. Hosp. Sch. Nursing, 1937; rsch. asst. in organic chemistry Denver Chem. Mfg. Co., N.Y.C., 1938-39; tchr. chemistry and physics N.Y.C. secondary schs., 1940-42; food analyst Quaker Maid Co., Bklyn., 1942-43; rsch. asst. in organic synthesis Johnson & Johnson, New Brunswick, N.J., 1943-44; biochemist Wellcome Rsch. Labs., Tuckahoe, N.Y., 1944-50; sr. rsch. chemist Wellcome Rsch. Labs., 1950—, asst. to assoc. rsch. dir., 1955-62, asst. to the rsch. dir., 1963-66, head exptl. therapy, 1966-83, sci. emeritus, 1983—; adj. prof. pharmacology and exptl. medicine Duke U., 1970, rsch. prof. pharmacology, 1983—; adj. prof. pharmacology U. N.C., Chapel Hill, 1973; chmn. Gordon Conf. on Coenzymes and Metabolic Pathways, 1966; mem. bd. sci. counselors Nat. Cancer Inst., 1980-84; mem. coun. Am. Cancer Soc., 1983-86; mem. Nat. Cancer Adv. Bd., 1984-91. Contbr. articles to profl. jours.; patentee in field. Recipient Garvan medal, 1968, Pres.'s medal Hunter Coll., 1970, Medal of Honor Am. Cancer Soc., 1990; Disting. Chemist award N.C. Inst. Chemists, 1981, Judd award Meml. Sloan-Kettering Cancer Ctr., 1983, Bertner award M.D. Anderson Hosp., 1989, Third Century award Fedn. for Creative Am., 1990, Discoverers award Pharm. Mfg. Assn., 1990, City of Medicine award Durham, N.C., 1990; co-recipient Nobel prize in medicine, 1988, Nat. Medal of Sci. NSF, 1991; inductee Hunter Coll. Hall of Fame, 1973, Nat. Inventors Hall of Fame, 1991, Nat. Women's Hall of Fame, 1991, Engring. and Sci. Hall of Fame, 1992; named Dame, Order of St. John of Jerusalem Ecumenical Found. (Knights of Malta) 1992. Fellow N.Y. Acad. Scis.; mem. AAAS, Am. Chem. Soc., Nat. Acad. Scis., Am. Acad. Arts and Scis., Inst. of Medicine, Chem. Soc. London, Am. Soc. Biol. Chemists, Am. Assn. Cancer Rsch. (bd. dirs. 1981, 83, pres. 1983-84, Cain award 1984), Am. Soc. Hematology, Transplantation Soc., Am. Soc. Pharmacology and Exptl. Therapeutics. Home: 1 Banbury Ln Chapel Hill NC 27514-2504 Office: Burroughs Wellcome Co 3030 W Cornwallis Rd Research Triangle Park NC 27709

ELION, HERBERT A., optoelectronics and bioengineering executive, physicist; b. N.Y.C., Oct. 16, 1923; s. Robert and Bertha (Cohen) E.; m. Sheila Thall, June 16, 1945; children: Gary Douglas, Glenn Richard, Jonathon Lee, Maxine Yael Gold. BSME, CCNY, 1945; MS, Bklyn. Poly. Inst., 1949, grad. in physics, 1954; grad. cert. X-ray Microanalysis, MIT, 1960; PhD (hon), Hamilton State U., 1973; cert., Cambridge U., Eng., U. Bordeaux, France, Pa. State U., Rutgers U., M.I.T., Northeastern U.; MA, U. Calif., Santa Barbara, San Francisco and Davis; postgrad., Coll. of Marin, Calif., Revere Acad. Jewelry Arts, San Francisco; cert., MacDowell Labs., Aldie, Va. Registered profl. engr., Mass., Pa., N.Y.; MacDowell Labs. VA Cert. Group leader RCA, Camden, N.J., 1957-59; pres. Elion Instruments, Inc., Burlington, N.J., 1959-64; assoc. dir. space sci. GCA Corp., Bedford, Mass., 1965-67; mng. dir. electro-optics Arthur D. Little Inc., Cambridge, Mass., 1967-79; pres., chief exec. officer Internat. Communications and Energy, Inc., Framingham, Mass., 1979—; pres. Aetna Telecommunications Cons., Centerville, Mass., 1981-85; also ptnr. Aetna Telecommunications Cons., Hartford, Conn., 1981-85; pres., chief exec. officer Internat. Optical Telecommunications, Mill Valley, Calif., 1981—; co-founder Kristallchemie M & Elion GmbH, Meudt, Fed. Fepublic Germany, 1961-64; lectr. on communications to Japanese, French, Can., Korean and Brazilian govts., 1970—; lectr. on optical communication to govt. depts. in Japan, France, Can., Korea, Brazil, 1970—; cons. on data communications Exec. Office of Pres., Washington, 1978-79; cons. Ministry Internat. Trade and Industry, Tokyo, 1975-88; chmn. internat. conf. European Electro-optics Conf., Heeze, The Netherlands, 1972-78; Mont. amb. Clean Coal Energy com., 1990—; internat. lectr. in field. Author, editor 27 books including 11 on lightwave info. networks; co-editor: Progress in Nuclear Energy in Analytical Chemistry Series, 1964-75; mem. adv. bd. Photonics Mag.; several Japanese and internat. world records in geothermal energy devel. activities in clean energy by GeoGas (R) process and clean air by elimination of methane and carbon dioxide gases and econ. prodn. of methyl alcohol from geothermal gases and econ. prodn. of methyl alcohol from geothermal gases; devel. supervision in 100% high strength organically biodegradable or smokeless burning plastic; contbr. articles to profl. jours. Pres. Elion Found., Princeton, N.J., 1960-67; founder Rainbow's End Camp, Ashby, Mass., 1960; elder Unitarian Ch., Princeton, 1963-64, Wellesley Soc. of Friends, 1970. With USN, 1944-46. Decorated Chevalier du Tastevin (France); recipient Presdl. awards Arthur D. Little Inc. Fellow Am. Phys. Soc.; mem. AAAS, IEEE (life mem., sr.), Optical Soc. Am., Soc. Photo Instrumentation Engrs., Am. Vacuum Soc., Nat. Security Industrial Assn., Soc. for Nondestructive testing, Am. Chem. Soc., Geothermal Resources Coun., Sigma Xi, Epsilon Nu Gamma, United Fedn. of Doll Clubs. Office: Internat Comms and Energy PO Box 2890 Sausalito CA 94966-2890 also: Box 789 Chatham MA 02633

ELIZARDO, KELLY PATRICIA, chemical engineer; b. Fayetteville, N.C., June 26, 1963; d. Paul Howard and Brenda Aline (Southerland) Westbrook; m. Homer Elizardo. B in Chem. Engr., Ga. Inst. Tech., 1986; MBA, U. Houston, 1991. Devel. mgr. Solvay Interox, Houston, 1986-. Contbr. articles to profl. jours. Chmn. Ga. Tech. Scholarship Com., Houston, 1987-91; v.p., sec., bd. dirs. Ga. Tech. Houston Alumni Club, 1987-93. Mem. Am. Inst. Chem. Engrs. Office: Solvay Interox 3333 Richmond Houston TX 77098

EL-KHATIB, SHUKRI MUHAMMED, biochemist; b. Ein Yabroud, Ramallah, Palestine, Aug. 20, 1931; s. Muhammed Abdel Latif and Salmeh Sulaiman (Subih) El-K.; m. Joan Ann Sullivan,Apr. 10, 1960 (div.); children: Issam, Runda, Hala, Mona; m. Naheel Jaser Salameh, Aug. 8, 1991; 1 child, Muhammed Shukri. PhD, Tex. A&M, 1968. Instr. Baylor Coll. of Medicine, Houston, 1968-70; asst. prof. U. Tenn. Med. Sch., Memphis, 1970-73; assoc. prof. U. P.R. Med. Sch., San Juan, 1974-76; v.p. U. Cen. Del Caribe, Cayey, 1984-88, chmn., 1976-91; prof. U. Cen. Del Caribe, Bayamon, 1991—; cons. Lastra Ednl. Experts, San Juan, 1980. Author: Polytoxin and Tumor, 1976, Nutrition and Colon Cancer, 1983, 4th edit., 1986. Pres. Islamic Ctr. of P.R., Rio Piedras, 1985-90. Mem. N.Y. Acad. Sci., Sigma Xi. Republican. Muslim. Home: C-16 Calle 3 Versalles Urb Versalles Urb Bayamon PR 00959 Office: Univ Cen Del Caribe Med Sch Bayamon PR 00960

ELKIES, NOAM D., mathematics educator. Prof. dept. math. Harvard U., Cambridge, Mass. Recipient NAS award for Initiatives in Rsch. Nat. Acad. Sci., 1991. Office: Harvard Univ Dept of Math Cambridge MA 02138*

ELKIND, JEROME ISAAC, computer scientist, company executive; b. N.Y.C., Aug. 30, 1929; s. Samuel and Rose (Klion) E.; m. Linda Valenstein, Jan. 18, 1959; children: James, Sarah, Kenneth. SB in Elec. Engring., MIT, 1951, SM in Elec. Engring., 1952, ScD in Elec. Engring., 1956. Staff mem. Lincoln Lab., MIT, Lexington, 1952-56; group leader RCA, Waltham, Mass., 1956-58; sr. scientist Bolt Beranel & Newman, Cambridge, Mass., 1958-62, v.p., then sr. v.p., 1962-69; vis. prof. MIT, Cambridge, 1969-70; lab. mgr. Palo Alto (Calif.) Rsch. Ctr., Xerox Corp., 1971-78, v.p., 1978-88; chmn. Lexia Inst., Palo Alto, 1989—; mem. com. on human factor NAS, Washington, 1985-91. Editor, author: Human Performance Models for Computer Aided Engineering, 1989; also articles. Pres. Family Svc. Mid-Peninsula, Palo Alto, 1991-92. Fellow Human Factors Soc.; mem. IEEE (editor Trans. on Human Factors in Electronics 1961-66), Assn. for Computing Machinery. Achievements include research on characteristics of simple manual control systems. Home: 2040 Tasso St Palo Alto CA 94301 Office: The Lexia Inst 766 Raymundo Los Altos CA 94022

ELKIND, MORTIMER MURRAY, biophysicist, educator; b. Bklyn., Oct. 25, 1922; s. Samuel and Yetta (Lubarsky) E.; m. Karla Annikki Holst, Jan. 27, 1960; children—Sean Thomas, Samuel Scott, Jonathan Harald. B.M.E., Cooper Union, 1943; M.M.E., Poly. Inst. Bklyn., 1949; M.S. in Elec. Engring, Mass. Inst. Tech., 1951, Ph.D. in Physics, 1953. Asst. project engr. Wyssmont Co., N.Y.C., 1943; project engr. Safe Flight Instrument Corp., White Plains, N.Y., 1946-47; head instrumentation sect. Sloan Kettering Inst. Cancer Research, 1947-49; physicist Nat. Cancer Inst. on assignment to Mass. Inst. Tech., 1949-53; on assignment to Donner Lab., U. Calif. at Berkeley, 1953-54; physicist Lab. Physiology, Nat. Cancer Inst., Bethesda, Md., 1954-67; sr. research physicist Lab. Physiology, Nat. Cancer Inst., 1967-69; sr. biophysicist biology dept. Brookhaven Nat. Lab., Upton, L.I., N.Y., 1969-73; guest scientist MRC exptl. radiopathology unit Hammersmith Hosp., London, 1971-73; sr. biophysicist, div. biol. and med. research Argonne (Ill.) Nat. Lab., 1973—, asst. dir., 1976-78, head mammalian cell biology group, 1978-81; prof. radiology U. Chgo., 1973-81; prof., chmn. dept. radiology and radiation biology Colo. State U., 1981-89, U. Disting. prof. dept. radiological health scis., 1986—; mem. radiation study sect. NIH, 1962-66, molecular biology study sect., 1970-71; mem. developmental therapeutics com. Nat. Cancer Inst., 1975-77. Author monograph. Served with USNR, 1944-46. Recipient E.O. Lawrence award AEC, 1967, Superior Service award HEW, 1969, L.H. Gray medal Internat. Com. Radiation Units and Measurements, 1977, E.W. Bertner award M.D. Anderson Hosp. and Tumor Inst., 1979, A.W. Erskine award Radiol. Soc. N. Am., 1980, Albert Soiland Meml. award Albert Soiland Cancer Found., 1984, 1st Henry S. Kaplan Disting. Scientist award Internat. Assn. Radiation Research, 1987, Charles F. Kettering prize Gen. Motors Cancer Rsch. Found., 1989, Honor Scientist award Colo. State U. chpt. Sigma Xi, 1991; Outstanding Investigator grantee Nat. Cancer Inst., 1988; Nat. Cancer Inst. Spl. fellow, 1972-74. Fellow Am. Coll. Radiology (hon.); mem. AAAS, Biophys. Soc., Radiation Research Soc. (council 1965-66, assoc. editor jour. 1965-68, pres.-elect and pres. 1980-81, G. Failla Meml. award 1984), Am. Assn. Cancer Research (assoc. editor jour. 1980-81), Am. Soc. Therapeutic Radiology and Oncology (gold medalist 1983), Tau Beta Pi. Office: Colo State U Dept Radiol Health Scis Fort Collins CO 80523

ELKINS, ROBERT N., association executive; b. N.Y.C., June 5, 1943; s. Jacob B. and Lee (Marcus) E.; m. Mary Beth Ackerley (div. 1991). BA, U. Pa., 1965; MD, SUNY, N.Y.C., 1975. Gen. ptnr. Hampstead (N.H.) Hosp. Physician Group, 1976-79; pres. Cen. Md. Health Systems, Inc., 1978-80; gen. ptnr., co-founder Continental Care Group, Md., 1980-86; chmn., chief exec. officer Integrated Health Svcs., Inc., Hunt Valley, Md., 1986—. Active Associated Jewish Charities, Balt. Recipient Entrepreneur of Yr. award Ernst & Young Inc. Mag., Merrill Lynch, 1991. Mem. Am. Entrepreneurs for Econ. Growth (co-chair), Caves Valley Club. Office: Integrated Health Svcs Inc 10065 Red Run Blvd Owings Mills MD 21117

EL KODSI, BAROUKH, gastroenterologist, educator; b. Cairo, Aug. 24, 1923; s. Moussa and Zohra (Aslan Cohen) El K.; came to U.S., 1957, naturalized, 1963; M.D., Cairo U., 1945; m. Marie Menasha, Mar. 26, 1960; children—Sylvia, Robert, Karen. Intern, Univ. Hosp. Cairo Sch. Medicine, 1946; resident in gen. medicine Jewish Hosp., Cairo, 1947-50, attending physician, 1950-57; intern, Miriam Hosp., Providence, 1958; resident in internal medicine, Boston City Hosp., 1959-61, chief resident, 1961-62, fellow in gastroenterology, 1962-64; asst. dir. medicine Union Hosp., Framingham, Mass., 1964-65; asso. dir. medicine Maimonides Med. Center, Bklyn., 1965-67, dir. gastroenterology, 1968—; chief gastroenterology Coney Island Hosp., N.Y.C., 1967-68; instr. Boston City Hosp., 1962-65; instr. Downstate Med. Center, SUNY, Bklyn., 1965-69, asst. prof. medicine, 1969-76, asso. prof., 1976—. Chmn. Bklyn. physicians com. United Jewish Appeal. Fellow Am. Coll. Gastroenterology, ACP; mem. Am. Fedn. Clin. Research, Am. Gastroent. Assn., Am. Soc. Gastrointestinal Endoscopy, Am. Soc. Study of Liver Disease, AMA, N.Y. Gastroenterologic Assn. (pres. 1985-86), Ostomy Club (mem. exec. council). Contbr. articles to profl. jours. Home: 118 Girard St Brooklyn NY 11235-3010 Office: 925 48th St Brooklyn NY 11219-2919

ELKOMOSS, SABRY GOBRAN, physicist; b. Elkoussia, Egypt, Apr. 2, 1925; immigrated to France, 1957, naturalized, 1959; s. Gobran Bishay and Rifka Morcos Elkomoss; B.Sc. in Math. with distinction, Alexandria U., 1949; M.Sc. in Physics, 1953; D.Scis. Physiques (French Govt. scholar 1951-52), U. Strasbourg (France), 1955; m. Arlette Meyer, Dec. 11, 1957; children—Anita, Alexander. Asst., Alexandria U., 1949-56; mem. staff Nat. Center Sci. Research, 1952-62; sr. research scientist, exec. adv. space and missile div. Douglas Corp., Santa Monica, Calif., 1963-64; sr. staff mem. space and missile div. Litton Industries, Beverly Hills, Calif., 1964-66; research scientist plasma div. McDonnell Corp., St. Louis, 1966-67; maitre recherches Nat. Center Sci. Research, Strasbourg, 1967—; lectr. U. Ein Shams, Cairo, also prof. physics Lycee Francais, Alexandria, 1956-57. Fulbright advanced scholar, 1959-61; postdoctoral asso. research U. Notre Dame, 1959-63. Mem. Am. Phys. Soc., New York Acad. of Sci. (sec. European study group solid state spectroscopy), Sigma Xi. Mem. Coptic Orthodox. Ch. Contbr. articles to profl. jours. Home: 4 rue de Stockholm, 67000 Strasbourg France

ELKOURIE, PAUL, telecommunications company manager; b. Birmingham, Ala., May 13, 1947; s. Fred Joseph and Virgie Teresa (Joseph) E.; m. Wanda Joy Skinner, Nov. 22, 1972; 1 child, Jessica Ellen. BSME, U. Ala., 1970, MBA, 1972. Sr. engr. Stockham Valves & Fittings Co., Birmingham, Ala., 1979-82; mgr. Apex Woodworks, Birmingham, 1982-84; bldg. engr. South Cen. Bell Telephone Co., Birmingham, 1984-87; staff mgr. Bell South Telecomm., Birmingham, 1987—; seminar instr. U. ala., Tuscaloosa, 1991—, Deep South Ctr. for Occupational Health, Birmingham, 1988—; ptnr. Apex Woodworks, Birmingham, 1991—. Author: Temperature and Humidity Requirements (nat. std.), 1989, Fire Resistance Criteria (nat. std.), 1990, (manual) Asbestos Operation and Maintenance, 1992. Comdr. 926th Engr. Bat., Birmingham, 1991—. Capt. U.S. Army, 1972-79. Mem. ASHRAE (chmn. 1990-91, historian 1991-93, bd. dirs. 1993—, Gold Ribbon 1992, 93), Assn. of Energy Engrs. (historian 1991—), Nat. Inst. Bldg. Scis., Capstone Engring. Soc., Ala. Zool. Soc., Phi Tau Sigma, Mu Alpha Theta. Roman Catholic. Achievements include rsch. in probablistic decision model for evaluating earthquake protection costs and risks. Home: 305 22nd Ave S Birmingham AL 35205 Office: Bell South Telecomm 3700 Colonnade Pkwy Birmingham AL 35243

ELKOWITZ, ALLAN BARRY, information systems manager; b. San Antonio, Mar. 30, 1948; s. Israel and Rina (Cohen) E.; m. Peggy Wingard. BS, Calif. Tech. Inst., 1970; PhD, U. Tex., 1975. Dir. info. tech. Children's Hosp., Boston, 1987—. Mem. IEEE, Am. Phys. Soc. Office: Children's Hosp 300 Longwood Ave Boston MA 02115

ELKOWITZ, LLOYD KENT, dental anesthesiologist, dentist, pharmacist; b. Bklyn., Jan. 26, 1936; s. Paul and Lillian (Applebaum) E.; m. Deanna A. Weinger; children: Sheryl, Andrew, Marc. BS in Pharmacy, Columbia U., 1956; DDS, Case Western Res. U., 1960, postgrad., 1961. Resident in anesthesiology U. Ctr. Hosp, Pitts., 1961, fellow in anesthesiology, 1966; anesthesiologist Walson Army Hosp., Fort Dix, N.J., 1962-64; pvt. practice Queens, N.Y., 1964—; dir. div. dental anesthesiology dept. dentistry Nassau County Med. Ctr., East Meadow, L.I., 1975—; pres. dental adv. coun. Adelphi U., Tufts U., Garden City, N.Y., 1986—; adj. prof. dept. biology Adelphi U., 1982—; chmn. dept. dental anesthesiology Flushing (N.Y.) Hosp. Med. Ctr., 1989—. Trustee Kings Point (N.Y.) Civic Assn., 1978—. Capt. U.S. Army, 1962-64. Recipient Callahan Meml. award Ohio State Dental Assn., 1960. Fellow Am. Dental Soc. Anesthesiology, Acad. Gen. Dentistry, Am. Soc. for Advancement Anesthesia in Dentistry; mem. ADA, Am. Pharm. Assn., N.Y. State Dental Assn., Queens Dental Assn., Internat. Anesthesia Rsch. Soc., Am. Soc. Dentistry for Children, Queens Inst. for Continuing Dental Edn. (charter) Alpha Zeta Omega, Alpha Omega, Alpha Epsilon Delta. Avocations: piano, snow skiing, sailing, boating, tennis. Office: 42-60 Main St 107 Northern Blvd Great Neck NY 11021-4309

ELLER, THOMAS JULIAN, aerospace company executive, astronautical engineer, computer scientist; b. Pelham, Ga., Oct. 19, 1937; s. Eugene Robert and Frances Elizabeth (Greer) E.; m. Beverly Anne Lafitte, June 7, 1963; children: Julie Anne Eller Schake, Elizabeth Jean, Robert Lafitte. Student, Furman U., 1955-57; BS in Engring. Scis., USAF Acad., 1961; MS in Aero. and Astronautics, Purdue U., 1969; PhD in Aerospace Engring., U. Tex., 1974. Commd. 2nd lt. USAF, 1961, advanced through grades to col., 1981, pilot, 1961-69; prof., asst. dean faculty USAF Acad., Colorado Springs, Colo., 1969-78, comdr. 2nd group, 1978-79, prof., head dept. astronautics and computer sci., 1979-81; ret. USAF, 1981; program mgr. Internat. Tng. & Edn. Co., Boston, 1981-82; chief contracts requirements Martin Marietta Aerospace, Denver, 1982-85; bus. devel. mgr. Kaman Scis. Corp., Colorado Springs, 1985-86, space applications and astrodynamics mgr., 1986—; chmn. Mil. Space Doctrine Symposium USAF Acad., 1981; mem. space adv. com. Colo. 5th U.S. Congl. Dist., 1987—, chmn. 1992—. Contbr. articles on astronautics to profl. jours. Trustee Colorado Springs Fine Arts Ctr., 1986-92; bd. dirs. The Falcon Found., Colorado Springs, 1983—, Acad. Rsch. and Devel. Inst., Colorado Springs, 1986—; founder, pres. Colo. Vietnam Vets. Leadership Program, Denver, 1983-84. Decorated D.F.C. (2), Airman's medal, Air medal (6), Vietnamese Cross of Gallantry with palm. Fellow AIAA (assoc., chmn.-elect astrodynamics tech. com. 1991-92, chmn. 1992—, astrodynamics standards com. 1991—), Accreditation Bd. Engring. and Tech. (engring. visitor 1981—), Assn. Grads. of USAF Acad. (v.p. 1975-79, bd. dirs. 1969-75, pres., chmn. 1979-83). Republican. Presbyterian. Achievements include co-development of magnetic momentum dumping system for GPS NAVSTAR satellite, of first SPECTRE model of C-130E gunship. Avocation: gardening. Office: Kaman Scis Corp PO Box 7463 Colorado Springs CO 80933-7463

ELLERBUSCH, FRED, environmental engineer; b. Germany, Mar. 5, 1951. BSCE in Environ. Engring., N.J. Inst. Tech., 1973, MS in Environ. Engring., 1980. Registered profl. engr., N.J. Staff engr. Elson T. Killam Assocs., Inc., Millburn, N.J., 1973-74; staff environ. engr. Indsl. Environ. Rsch. Lab. U.S. EPA, Edison, N.J., 1974-77; environ. systems engr. METREK div. MITRE Corp., McLean, Va., 1977-78; regulatory conformance coord. Bristol-Myers Products div. Bristol-Myers Squibb Co., Bridgewater, N.J., 1978-83, mgr. regulatory compliance and govt. affairs, 1983-85, dir. safety, security and environ. affairs, 1985-89; dir. corp. environ. affairs Rhone-Poulenc Inc. affiliate Rhone-Poulenc SA, Monmouth Junction, N.J., 1989—; adj. prof. environ. engring. grad. div. N.J. Inst. Tech., Newark, 1980-83; seminar leader div. continuing edn. N.J. Inst. Tech., Newark, 1977-90. Co-author: Electrotechnology Applications in Manufacturing, Vol. 2, 1978, Industrial/Hazardous Waste Impoundment, 1979, Biomass Applications and Technology, 1980; co-editor: Carbon Adsorption Handbook, 1978, Guide for Industrial Noise Control, 1982; contbr. articles to profl. publs. Mem. nat. panel consumer arbitrators Better Bus. Bur. Mem. Acad. Hazard Control Mgmt., Acad. Hazardous Materials Mgmt. (bd. examiners 1984-85), Am. Indsl. Health Coun. (govt. affairs com. 1987, sci. policy com. 1988—, vice chmn. 1989,), Am. Sci. Affiliation, Chem. Mfrs. Assn. (mem. various coms.), N.Y. Acad. Scis., Soc. for Risk Analysis, N.J. Water Control Assn., N.J. Inst. Tech. (indsl. adv. bd. 1986—), Nat. Environ. Tng. Assn., N.J. Acad. Scis., Pharm. Mfrs. Assn. (environ. control resource com. 1985-89), others. Home: 73 Ferguson Rd Warren NJ 07059 Office: Rhone-Poulenc Inc 125 Black Horse Lane Monmouth Junction NJ 08852

ELLIGOTT, LINDA A., environmental scientist; b. Troy, N.Y., Jan. 11, 1955; d. Patrick J. and Elinor A. (DuFour) E. BS, SUNY, Syracuse, 1980, MS, 1982. Hydrologist U.S. Geol. Survey, Jackson, Miss., 1978-80, Ithaca, N.Y., 1982-84; rsch. assoc. dept. natural resources Cornell U., Ithaca, 1984-87; asst. editor Wingtips Jour. Ornithology, Ithaca, 1987-89; sr. scientist, cons. biologist Blasland and Bouck Engrs., Syracuse, N.Y., 1989—; presenter at profl. confs. Mem. AAAS, Soc. Environ. Toxicology and Chemistry. Office: Blasland and Bouck Engrs PC 6723 Towpath Rd Syracuse NY 13214

ELLINGTON, JAMES WILLARD, mechanical design engineer; b. Richmond, Ind., May 26, 1927; s. Oscar Willard and Leola Lenora (Sanderson) E.; m. Sondra Elaine Darnell, Dec. 6, 1952; children: Ronald, Roxanna. BSME summa cum laude, West Coast U., L.A., 1978. Designer NATCO, Richmond, Ind., 1954-67; design engr. Burgmaster, Gardena, Calif., 1967-69; sr. mfg. engr. Xerox Co., El Segundo, Calif., 1969-84; cons. mem. engring. staff Xerox Co., Monrovia, 1984-87; staff engr. Photonic Automation, Santa Ana, Calif., 1987-88; sr. mech. engr. Optical Radiation Co., Azusa, Calif., 1988; sr. staff engr. Omnichrome, Chino, Calif., 1988—. With USN, 1945-52. Mem. Soc. Mfg. Engrs. (sec. 1984), West Coast U. Alumni Assn. (bd. dirs. 1988—, v.p. budget and fin.). Republican. Baptist. Avocation: gardening. Office: Omnichrome 13580 5th St Chino CA 91710-5113

ELLIOT, DOUGLAS GENE, chemical engineer, engineering company executive, consultant; b. Medford, Oreg., June 3, 1941; s. Don Joseph and Eleanor Joan (Sheets) E.; m. Noma Warnken, July 16, 1966 (div. 1979); 1 child, Jennifer M.; m. Patricia Jean Nichols, Mar. 15, 1980; children: Steven V. Bates, Michael A. Castillo. BSChemE, Oreg. State U., 1964; MS, U. Houston, 1968, PhD, 1971. Reservoir/prodn. engr. Humble Oil & Refining Co., Beaumont, Tex., 1964-66; co-founder, v.p. and bd. dirs. S.W. Wire Rope, Inc., 1967-70; cons. Gas Processors Assn., Houston, 1971; process/project engr. Hudson Engring. Co., Houston, 1971-78; mgr. process engring. Davy-McKee Corp., Houston, 1978-83, v.p. oil and gas, 1983-85; pres. D. G. Elliot & Assocs. Inc., Houston, 1985-86; sr. v.p., gen. mgr. Internat. Process Svcs. Inc., Houston, 1986—, also bd. dirs.; adj. prof. Rice U., Houston, 1976-77; mem. indsl. adv. com. Okla. State U., Stillwater, 1979-83; mgmt. cons. Norsk Hydro Oil & Gas Div., Oslo, 1984—. Contbr. articles to profl. jours.; mem. editorial rev. bd. Energy Process mag., 1981—; patentee in field. Mem. Tex. Energy Adv. Com., Austin, 1978; founding mem., bd. dirs. Tex. Solar Energy Soc., Austin, 1978; mem. Ctr. of Excellence R&D rev. panel Okla Ctr. for Advancement of Sci. & Tech., Oklahoma City. Recipient Outstanding Achievement award Tex. Soc. Profl. Engrs., 1978, citation for merit Bechtel Corp., 1991. Mem. Am. Inst. Chem. Engrs. (sec./treas. South Tex. sect. 1979, chmn. elect 1980, chmn. 1981, bd. dirs. fuels and petochem. div. 1982-85), Soc. Petroleum Engrs., Gas Processors Assn. (sec./treas. 1983), Sigma Zi. Avocations: hunting, fishing. Home: 12114 Green Glade Dr Houston TX 77099

ELLIOTT, JARRELL RICHARD, JR., chemical engineer, educator; b. Newport News, Va., Oct. 25, 1958; s. Jarrell Richard Sr. and Miriam (Huggins) E.; m. Tevhide Guliz Arf, Nov. 29, 1985; 1 child, Serra. MS in Chem. Engring., Va. Tech. Inst., 1982; PhD in Chem. Engring., Pa. State U., 1985. Teaching asst. Va. Tech. Inst., Blacksburg, 1980-82; rsch. asst. Pa. State U., University Park, 1982-85; asst. prof. U. Akron, Ohio, 1986-91; assoc. prof. U. Akron, 1991—; cons. Berty Reaction Engrs. Ltd., Akron, 1987-89, Cryotherm, Akron, 1988—. Contbr. articles to profl. jours. Chmn. Midwest Thermodynamics Symposium, Cleve., 1990. Mem. Am. Inst. Chem. Engrs., Am. Chem. Soc., Am. Soc. Engring. Educators, Phi Lambda Upsilon. Achievements include research in a practical engineering equation of state for multiphase equilibria of complex mixtures including oil, water, alcohols and ketones; in a novel process for methanol synthesis based on three-phase heterogeneous catalysis; in a novel approach to manipulating microstructure at the nanometer scale; observed key differences between chain molecules and spheres. Home: 7250 Honeydale Dr Northfield OH 44067-2610 Office: U Akron Chem Engr Dept Akron OH 44325-3906

ELLIOTT, JOHN EARL, water plant administrator; b. Oakland, Iowa, Oct. 1, 1946; s. Boyd Martin and Priscilla Maxine (Thompson) E.; m. Dixie Sue Hall, July 4, 1969 (div. 1984); children: Susan Mae, Boyd Kenneth; m. Ann Marie DeBasio, Oct. 13, 1984. Student, Iowa State U., 1977-78, Kirkwood C. C., 1989. Machinist Easton Corp., Shenandoah, Iowa, 1972-76; dir. pub. works City of Emerson, Iowa, 1976-78, City of Sidney, Iowa, 1978-82, City of Villisca, Iowa, 1982-89; supt. water plant City of Shenandoah, 1989—. Sgt. USAF, 1964-68, Vietnam. Mem. Am. Water Works Assn. (regional chair 1992). Office: City of Shenandoah Box 338 500 W Clarinda St Shenandoah IA 51601

ELLIOTT, MARC ELDON, civil engineer; b. Three Rivers, Mich., June 10, 1955; s. Harvey J. Elliott and Mary Jane Zerbe Ahlgrim; m. Laurie Anne Russell, Oct. 15, 1989. BA in Math., Kalamazoo Coll., 1977; BSCE magna cum laude, U. Mich., 1983, MSE in Pub. Works Adminstrn., 1984. Registered profl. engr., Tex.; cert. pub. works profl. Quality assurance specialist U.S. Army Material and Readiness Command, Lexington, Ky., 1978-82; rsch. asst. dept. civil engring. U. Mich., Ann Arbor, 1983-84; environ. planner North Cen. Tex. Coun. Govts., Arlington, 1984-88; adminstrv. sec. pub. works adv. com. North Cen. Tex. Coun. Govts., 1985-89; sr. pub. works planner North Cen. Tex. Coun. Govts., Arlington, 1988-89; exec. dir. Cass County Planning & Econ. Devel. Commn., Cassopolis, Mich., 1989-91; dir. strengthening instns. program Glen Oaks C.C., Centerville, Mich., 1992; project engr. United Environ. Techs., Inc., Kalamazoo, Mich., 1992—; co-chmn. procedures com. Urban Mgmt. Assts. of North Tex., Arlington, 1986-87; sec. Indsl. Devel. Corp. of Cass County, Cassopolis, 1989-91. Editor: Standard Specificaitons for Public Works Construction, 1985-89; editor, co-author: Solid Waste Plan for Cass County, 1990. EPA grantee, 1987; Mich. Dept. Natural Resources grantee, 1990. Mem. ASCE, NSPE, Am. Pub. Works Assn. (congress del. 1985-87), North Tex. Hazardous Materials Assn. (founding mem. and exec. com. 1985), Community Growth Alliance of Cass County (sec. 1989-91), Mich. Soc. Planning Ofcls., Chi Epsilon, Tau Beta Pi. Home: 4396 E Tu Ave Vicksburg MI 49097 Office: United Environ Tech Inc 5066 Sprinkle Rd Kalamazoo MI 49002

ELLIOTT, ROBERT BETZEL, physician; b. Ada, Ohio, Dec. 8, 1926; s. Floyd Milton and Rose Marguerite (Betzel) E.; m. Margaret Mary Robichaux, Aug. 26, 1954; children: Howard A., Michael D., Robert Bruce, Douglas J., John C., Joan O. BA, Ohio No. U., 1949; MD, U. Cin., 1953. Diplomate Am. Bd. Family Pracitice. Intern Charity Hosp., New Orleans, 1953-54; resident in pathology Bapt. Meml. Hosp., Memphis, 1958-59; practice medicine specializing in family practice Ada, 1959—; mem. staff Ohio No. U. Health Service, Ada, 1960-70; coroner Hardin County, 1973-93. Mem. Ada Exempted Village Sch. Bd., 1960—, pres., 1966-69, 72—, v.p. 1971—. Named Ohio Family Physician of Yr., 1985. Mem. AMA, Ohio State Med. Assn., Hardin County Med. Soc. (pres. 1964), Am. Acad. Family Physicians, Ohio Acad. Family Physicians, Lima Acad. Family Physicians, Am. Coll. Health Assn. Democrat. Presbyterian. Lodges: Masons, Elks. Home: 4429 State Route 235 Ada OH 45810-9509 Office: 302 N Main St Ada OH 45810-1198

ELLIOTT, W(ILLIAM) CRAWFORD, geology researcher; b. Abington, Pa., July 13, 1955; s. William H. Jr. and Patricia (Seitz) E. BA in Geology, Franklin & Marshall Coll., 1978; MA in Geology, Temple U., 1981; PhD in Geology, Case Western Res. U., 1988. Technologist Republic Steel Corp., Cleve., 1981-82; rsch. assoc. Case Western Res. U., Cleve., 1987-89, sr. rsch. assoc., 1990—; asst. prof. Mich. State U., East Lansing, 1989-90; cons. Exxon Prodn. Rsch. Co., Houston, 1990. Referee proposal and jour. Clay & Clay Minerals, NSF, Petroleum Rsch. Found., 1990—; contbr. articles to Bull. Am. Ceramic Soc., Geochimica, AAPG Bull., GSA Bull., Geology, Econ. Geology. Mem. fin. com. Hallinan Ctr., Cleve., 1991—. NSF grantee, 1990-92. Mem. AAAS, No. Ohio Geol. Soc., Clay Minerals Soc. Achievements include research in formation of illite in Appalachian basin relatable to Alleghanian Orogeny; research on the origin of the clay minerals and discovery of pyroclastic labradorite at Cretaceous/Tertiary boundary at Stevns Klint, Denmark; determined kinetics of smectite-to-illite trans in the Denver Basin; determined effects of several cations and hydrogen-containing gases on rate of fayalite reduction. Office: Case Western Res U Dept Geol Sci Cleveland OH 44106-7216

ELLIOTT-WATSON, DORIS JEAN, psychiatric, mental health and gerontological nurse educator; b. Caney, Kans., Dec. 6, 1932; d. Alva Orr and Mary Amelia (Boyns) Elliott; children Marsha Jean Watson, Sherwood Elliott Watson. BE, U. Miami, Fla., 1952, MEd, 1954; EdD, Pacific Western U., 1982; BSN, U. Kans., 1985; AS in Psychology, Kansas City (Kans.) C.C., 1989. RN, Kans., Mo.; cert. clin. specialist gerontology nurse, gerontology nurse generalist, psychiat.-mental health nurse, med.-surg. nurse, ANCC; cert. elem. to jr. coll. tchr., Kans., Mo.; lic. adult care home adminstr., Kans. Tchr. learning disabled, gifted, emotionally disturbed Shawnee Mission, Kans., 1961-76; instr. hospitalized psychiat. and med.-surg. children U. Kans. Med. Ctr., Kansas City, 1979-82; pvt. practice, gerontol. nursing educator Bonner Springs, Kans., 1985—; libr. U. Miami, 1952, Kans. U., 1978; nurse ARC, Kansas City, 1985—; nurse educator Am. Heart Assn., Kansas City, 1985—; program designer mainstreaming spl. needs children into regular classrooms, 1969; sex needs fulfillment geriatric patients in nursing homes, 1986. Editor Park Stylus, Parkville, Mo., 1952; author, speaker Kansas City area, 1950—. Tutor/organizer Tutoring Vol. Orgn. for Inner City Children, 1965-68; life mem., patron Wyandotte County Hist. Mus./Soc. Kansas City, 1975; sustaining mem. Rep. Nat. Com., Washington, 1978—, Rep. Congl. Com., 1978—, Rep. Senatorial Com., 1978—; pres. Young Reps., Kansas City, 1960; patron, charter mem. Kaw Valley Community Choir, 1990-92, mem. Kansas City Community Choir, 1992—; vol. working with alcohol, drugs, food abuse teenagers, 1976—. Inducted Rep. Nat. Hall of Honor, Rep. Nat. Conv., 1992. Mem. ANA (coun. on gerontol. nurses), NEA (life, del. state conv. 1980, nat. conv. 1973), Kans. Nurses Assn., U. Kans. Alumni Assn., Bus. and Profl. Women, Order Ea. Star (electra 1982), Order Rainbow for Girls (worthy advisor 1950), Am. Volkssport Assn. (Tri-Athlete 1993, 1500 Km Walking award 1993, Sunflower State Games Athlete 1993), Tiblow Trailblazers Walking Club (pres. 1993), Kappa Delta Pi, Pi Delta Epsilon, Phi Theta Kappa, Alpha Kappa Delta, Phi Alpha Theta. Avocations: holistic healing, gardening, camping, community involvement. Home and Office: 231 Sheidley Ave Bonner Springs KS 66012-1410

ELLIS, ARTHUR BARON, chemist, educator; b. Lakewood, Ohio, Apr. 4, 1951; s. Nathan and Carolyn Joan (Agulnick) E.; m. Susan Harriet Trebach, Nov. 9, 1975; children: Joshua, Margot. BS, Calif. Inst. Tech., 1973; PhD, MIT, 1977. Asst. prof. chemistry U. Wis., Madison, 1977-82, assoc. prof., 1982-84, prof., 1984-86, Meloche-Bascom prof., 1986—. Editor: Chemistry and Structure at Interfaces, 1986; patentee in field; contbr. articles to profl. jours. Fellow A.P. Sloan Found., 1981, H.I. Romnes fellow U. Wis., 1985, Guggenheim fellow, 1989. Mem. Am. Chem. Soc. (Exxon fellow 1980), Electrochem. Soc. Democrat. Jewish. Achievements include creating 1-2-3 levitation kit based on high-temperature superconductors. Office: Univ of Wis Materials Science Research Ctr 1500 Johnson Dr Madison WI 53706-1396

ELLIS, BRENDA LEE, mathematician, computer scientist, consultant, educator; b. Norfolk, Va., Jan. 4, 1965; d. Lester and Annie Mae (Leak) E. BS cum laude, Norfolk State U., 1987; postgrad., Old Dominion U., 1988-89, Cleve. State U., 1991. Computer clk. Naval Electronics Systems Engring. Ctr., Portsmouth, Va., 1986-88; rsch. asst. Old Dominion U., Norfolk, 1988-89; computer analyst Dept. Def., Ft. Meade, Md., 1989; mathematician NASA Lewis Rsch. Ctr., Cleve., 1989—; lab. dir., instr. Norfolk State U., 1993—; panelist Sonia Kovalevsky Math. Day, Cleve. State U., 1990-92; guest speaker Math. Counts Workshop, 1991. Usher, Mt. Calvary Bapt. Ch., Cleve., 1991; vol. Combined Fed. Campaign, Cleve., 1991, Norfolk Community Hosp., 1982; tutor, mentor, sci. fair coord. East Tech. High Sch., Cleve., 1991-92; meet dir. Lake Erie Indoor Track Field Championship, 1992. Recipient scholarships and fellowship, first place trophy North Coast Relays, 1991, 1992. Mem. IEEE, NAFE, Am. Bus. Women's Assn., Assn. Computing Machinery (acting sec. 1986-87), Over the Hill Track Club (v.p. 1991-92), Beta Kappa Chi. Baptist. Avocations: reading, track and field, volleyball, softball. Office: NASA Lewis Rsch Ctr 21000 Brookpark Rd Cleveland OH 44135

ELLIS, BRIAN NORMAN, engineering executive; b. Bristol, U.K., July 9, 1932; s. Jack Jefferson and Lilian Margaret (Long) E.; m. Margaret Kilgour Hepburn, Oct. 17, 1959; 1 dau., Catrina Irwin. S.L.C., George Watson's Coll., 1948; F.T.C., Heriot-Watt U., 1951. Chartered engr. Research engr. Welwyn Electric Ltd., Bedlington, U.K., 1955-56; comml. engr. Fielden Electric Ltd., Stockton, U.K., 1956-63; research engr. Paillard SA, Yverdon, Switzerland, 1963-64; tech. dir. Kudelski SA, Paudex, Switzerland, 1964-74; gen. dir. Protonique SA, Romanel, Switzerland, 1974—; cons. GESO PC Standards Working Group, chmn., 1975-81, delegate for Swiss Fed. Office of the Environ. (UN Environ. Programme Solvents Tech Com. 1989—, chmn. Electronics Chapt. 1991);. Author: Handbook of Contamination, 1981; Cleaning and Contamination of Electronic Components and Assemblies, 1986, Replacement of CFC-113 in Industry, 1990; (with others) Circuits Imprimés, 1983; contbr. articles to profl. jours.; patentee in field. Fellow Inst. Circuit Tech.; mem. IEEE. Home: Route d'Echallens 3, CH-1032 Romanel Switzerland Office: Protonique SA Case Postale 78, Romanel CH-1032, Switzerland

ELLIS, BRUCE W., electrical engineering educator; b. Jamestown, N.D., June 15, 1939; s. Harold A. and Mary M. (Cole) E.; m. Kathleen A. Monroe, June 12, 1965; children: Heather, Stuart. AB, Jamestown (N.D.) Coll., 1961; PhD, U. Minn., 1973. Instr. U. N.D., Grand Forks, 1962-64; instr., asst. prof., assoc. prof., prof. engring. St. Cloud (Minn.) State U., 1965—. Author: An Introduction to Spice, 1988. Mem. Health and Housing Adv. Bd., St. Cloud, 1988—. Mem. IEEE, Am. Soc. Engring. Edn. Home: 424 Birch Dr Saint Cloud MN 56304

ELLIS, CLIFFORD AUBREY, civil engineer; b. New Brunswick, N.J., Oct. 14, 1935; s. Sedric E. and Helen (Mitchell) E.; m. Carol L. Bailey, June 27, 1959; children: Kim Ellis Shidlowski, Kathy Ellis Eskander. BCE, Rutgers U., 1958, cert. pub. mgr., 1986; grad., Hwy Mgmt. Inst. at Miss., 1970. Licensed profl. engr., land surveyor, N.J. From asst. engr. to prin. engr. N.J. State Hwy. Dept., Metuchen, 1959-67; project engr. N.J. Dept. Transportation, Trenton, 1967-69, supervising engr. II, 1969-72, supervising engr. I, 1972-81, regional design engr., 1981-88, chief engr. regional design, 1988-92. Co-author (with Robins, Lutin, Hart and Kirkyla): Integrating LRT into Devel. Projects on Hudson River Waterfront, 1988. Chmn. Neighborhood Watch, Ewing Twp., N.J., 1985; jr. warden St. Luke's Episcopal Ch., Trenton, 1975; vestryman St. James Episcopal Ch., Trenton, 1980, All Saints Episcopal Ch., Highland Park, N.J., 1960. Capt. USAR, 1958-67. Recipient Edward Fuller Meml. prize in civil engring. Rutgers U., 1958, N.J. Transp. Constrn. Community Planning Right of Way and Design award 1981, Am. Assn. State Hwy. and Transp. Offcls., cert. Appreciation for Meritorious Svc., 1985. Mem. ASCE (Robert Ridgeway award Met. sect. 1958), NSPE, Cert. Pub. Mgrs., Am. Soc. Hwy. Engring., Tau Kappa Epsilon. Home: 58 Bakun Way Trenton NJ 08638-1537

ELLIS, DAVID H., biologist, research behaviorist; b. Hayward, Calif., Apr. 7, 1945; s. Harry S. Ellis and Margaret Lowe; m. Catherine Hunt, Dec. 20, 1968; children: Merlin, Jared, Karina. BS in Zoology magna cum laude, Brigham Young U., 1969; PhD in Behavioral Ecology, U. Mont., 1973. Biologist Mont. Game and Fish Dept., Helena, 1974; founder, rsch. dir. Inst. for Raptor Studies, Oracle, Ariz., 1977-85; instr. Ctrl. Ariz. Coll., Aravapa, 1981-82; biologist-in-charge Ariz. Field Sta. Patuxent Wildlife Rsch. Ctr., Tucson, 1974-77, coord. whooping crane and Miss. sand crane rsch. projects, 1985—, rsch. behaviorist, 1985—, acting leader propagation and lab. investigations sect., 1987; leader raptor survey of the Noatak River, Alaska, 1988; participant seabird censuses in Pacific Ocean and quantification of fauna and flora of Caroline Atoll as part of first Soviet/U.S. Oceanographic Expedition to Pacific, 1988, prin. investigator of satellite telemetry study of migration routes of cranes summering in Siberia, 1990-92; chmn. inter-agy. team on inspection tours of Calif. Condor propagation facilities at San Diego Wild Animal Park and L.A. Zoo, 1985; cons. endangered species recovery teams U.S. Fish and Wildlife Svc., 1975—; orinthological cons. enforcement raids U.S. Fish and Wildlife Svc./Ariz. Game and Fish Dept.; mem. World Working Bird of Prey Study Group, Internat. Coun. Bird Preservation; advisor Gruidae Adv. Group, Am. Assn. Zool. Parks and Aquariums, 1988; presenter rsch. reports to various sci. groups. Contbr. over 90 articles on migratory and mating habits of birds to Auk, Condor, BioScience, Jour. of Raptor Rsch., Environmental Pollution, others; English editor for publs. Japanese Soc. for Rsch. of Golden Eagle; publ. over 200 photographs to sci. and conservation jours. including Wildlife Monographs, Nat. Geographic Mag., Audubon. Troop com. chmn., com. mem. Boy Scouts Am., Md., 1989-93. Recipient Spl. Achievement award U.S. Fish and Wildlife Svc., 1979, 90; scholar Brigham Young U., 1963-69; NSF grad. rsch. fellow, 1969, 69-72, Internat. Found. for Telemetering grad. fellow, 1972-73, Mont. Coop. Wildlife Rsch. Unit fellow, 1972-73; rsch. grantee Nat. Geographic Soc., 1973, 81, Peregrine Falcon rsch. grantee U.S. Forest Svc., Maricopa Audubon Soc., Tucson Audubon Soc., 1977, jaguar rsch. grantee Safari Club Internat., 1979; also numerous contracts for studies. Mem. AAAS, Am. Ornithologist's Union (peer rev. com., grantee for travel to Internat. Ornithol. Congress 1990), Assn. for the Study of Animal Behavior, Nat. Audubon Soc., Nat. Geographic Soc., N.Am. Crane Working Group, Behavioral Ecology Soc., Cooper Ornithol. Soc. (peer rev. com.), The Nature Conservancy, Raptor Rsch. Found., The Raptor Soc. (peer rev. com.), Whooping Crane Conservation Assn., The Wildlife Soc., Phi Kappa Phi, Sigma Xi, Beta Beta Beta. Achievements include leadership of satellite telemetry project which revealed home of cranes from northwestern Siberia; development of behavior classification scheme for world's 15 species of cranes which provides organizational framework for behavior of all cranes and to a degree all birds; the most extensive and intensive study available of raptor responses to military jet aircraft; resolution of the taxonomic status of the elusive pallid falcon through fieldwork in Argentina and Chile; the most extensive raptor surveys available for Latin America; the first statewide habitat preferences inventory for the species Peregrine Falcon through fieldwork in Arizona; monograph on Golden Eagles behavior credited as the most complete descriptive developmental study for any vertebrate and which served as a model for subsequent studies; research of and reintroduction efforts with the endangered Masked Bobwhite Quail leading to eventual establishment of the Buenos Aires National Wildlife Refuge. Home: U S Fish & Wildlife Service Rt 2 Box 263 Laurel MD 20708 Office: US Fish and Wildlife Svc Patuxent Wildlife Rsch Ctr Laurel MD 20708

ELLIS, DAVID R., aeronautical researcher; b. Rapid City, S.D., Aug. 7, 1935; s. Ivan F. and Grace G. Ellis; m. Eunice M. Shideler, Mar. 26, 1960; children: David Jr., Catherine C. BS in Aero. Engring., U. Colo., 1957; MS in Aero. Engring., Princeton U., 1962. Mgr. flight dynamic rsch. Princeton (N.J.) U., 1967-77; mgr. adv. design and system rsch. Cessna Aircraft Co., Wichita, Kans., 1977-87; assoc. prof. U. Kans., Lawrence, 1988-89; v.p. engring. Comdr. Aircraft Co., Bethany, Okla., 1988-91; dir. R & D Nat. Inst. for Aviation Rsch. Wichita State U., 1991—; cons. Aeronautical Rsch. Assn. Princeton, 1967-77; adj. assoc. prof Wichita State U., 1982-85; pres. Omega-D Inc., Wichita, 1981—; mem. Congl. Adv. Com. on Aeronautics, Washington, 1984-85, NASA Aerospace Rsch. and Tech. Subcom., Washington, 1986-87, NASA Ad Hoc Com. on Gen. Aviation, Washington, 1987-87; participant FAA, Gen. Aviation Mfg. Assn., FAR/JAR, others, 1991-93. Contbr. more than 50 tech. publs. to profl. jours. Recipient NASA Group Achievement award, 1986; NSF Sci. faculty fellow, Princeton U., 1964. Fellow AIAA (assoc., Space Shuttle Banner award 1984, Gen. Aviation award 1985); mem. Soc. Automotive Engrs., Tau Beta Pi. Achievements include patent (with other) for integrated lift/drag controller; leader of preliminary design teams for Cessna Crusader and Cessna Caravan. Home: Wichita State Univ Natl Inst for Aviation Research 14505 Willowbend Cir Wichita KS 67230 Office: Wichita State U Nat Inst for Aviation Rsch Wichita KS 67620-0093

ELLIS, GEORGE EDWIN, JR., chemical engineer; b. Beaumont, Tex., Apr. 14, 1921; s. George Edwin and Julia (Ryan) E.; B.S. in Chem. Engring., U. Tex., 1940; M.S., U. So. Calif., 1950, M.B.A., 1965, M.S. in Mech. Engring., 1968, M.S. in Mgmt. Sci., 1971, Engr. in Indsl. and Systems Engring., 1979. Research chem. engr. Tex. Co., Port Arthur, Tex., 1948-51, Long Beach, Calif., Houston, 1952-53, Space and Information div. N.Am. Aviation Co., Downey, Calif., 1959-61, Magna Corp., Anaheim, Calif., 1961-62; chem. process engr. AiResearch Mfg. Co., Los Angeles, 1953-57, 57-59; chem. engr. Petroleum Combustion & Engring. Co., Santa Monica, Calif., 1957, Jacobs Engring. Co., Pasadena, Calif., 1957, Sesler & Assos., Los Angeles, 1959; research specialist Marquardt Corp., Van Nuys, Calif., 1962-67; sr. project engr. Conductron Corp., Northridge, 1967-68; information systems asst. Los Angeles Dept. Water and Power, 1969-92. Instr. thermodynamics U. So. Calif., Los Angeles, 1957. Served with USAAF, 1943-45. Mem. ASTM, ASME, Nat. Assn. Purchasing Mgmt., Nat. Contract Mgmt. Assn., Am. Inst. Profl. Bookkeepers, Am. Soc. Safety Engrs., Am. Chem. Soc., Am. Soc. Materials, Am. Electrophaters and Surface Finishers Soc., Am. Inst. Chem. Engrs., Am. Electroplaters Soc., Inst. Indsl. Engrs., Am. Prodn. and Inventory Control Soc., Am. Soc. Quality Control, Am. Inst. Plant Engrs., Am. Soc. Engring. Mgmt., Inst. Mgmt. Accts., Soc. Mfg. Engrs., Assn. Proposal Mgmt. Profls., L.A. Soc. Coating Tech., Assn. Finishing Processes, Pi Tau Sigma, Phi Lambda Upsilon, Alpha Pi Mu. Home: 1344 W 20th St San Pedro CA 90732-4408

ELLIS, GEORGE FRANCIS RAYNER, astronomy educator; b. Johannesburg, South Africa, Aug. 11, 1939; s. George Rayner and Gwendoline (MacRobert) E.; m. Sue Parkes (div.); children: Margaret, Andrew; m. Mary Roberts MacDonald. BSc, U. Cape Town, South Africa, 1961, BA, 1982; PhD, Cambridge (eng.) U., 1964. Lectr. Cambridge U., 1968-73; prof. U. Cape Town, 1973-87, SISSA, Trieste, Italy, 1988-93, Cape Town U., 1990—; vis. prof. Hamburg, Fed. Republic of Germany, Chgo., Boston, Tex. and Alta. Author: (with S. Hawking) Large Scale Structure of Space Time, 1973, (with O. Dewar) Low Income Housing Policy, 1980, Before the Beginning, 1993. Chmn. Friends of the Ciskei People, Cape Town, 1978-83; clk. S.A. Yearly Meeting of Quakers, Cape Town, 1982-86; chmn. Quaker Svc., Cape Town, 1978-86, South African Inst. Race Rels., Cape Town, 1986-88. Peterhouse fellow U. Cambridge, 1965-67, U. Cape Town fellow, 1978. Fellow Royal Soc. South Africa (v.p. 1990-92, pres. 1992—, Herschel medal 1978), Royal Astron. Soc.; mem. Internat. Soc. Gen. Relativity and Gravitation (pres. 1987-91), Internat. Astron. Union. Home: 3 Marlow Rd Kenilworth, Cape Town 7700, South Africa Office: U Cape Town, Rondebosch, Cape Town 7700, South Africa

ELLIS, JOHN TAYLOR, pathologist, educator; b. Lufkin, Tex., Dec. 27, 1920; s. John Taylor and Rowena (McCurdy) E.; m. Marian A. Caldwell, Dec. 26, 1942; children: Evelyn Floy, George Caldwell, John Taylor. BA, U. Tex., 1942; MD, Northwestern U., 1946. Diplomate Nat. Bd. Med. Examiners, Am. Bd. Pathology. Rotating intern St. Luke's Hosp., Chgo., 1945-46, asst. resident in pathology, 1946; rsch. asst. William Buchanan Blood Ctr., Baylor Hosp., Dallas, 1984; resident in pathology N.Y. Hosp., N.Y.C., 1948-49; asst. in pathology Cornell U. Med. Coll., N.Y.C., 1948-49, instr. in pathology, 1949-50, asst. prof., 1950-56, assoc. prof., 1956-62; prof., chmn. dept. pathology Med. Coll., Emory U., N.Y.C., 1962-67, Med. Coll., Cornell U., N.Y.C., 1968—; attending pathologist, pathologist in chief N.Y. Hosp., 1968—; attending pathologist Meml. Sloan-Kettering Cancer Ctr., N.Y.C., 1973—; chief pathology dept. pathology N.Y. Downtown Hosp., N.Y.C., 1991—; mem. adv. bd. Office of Chief Med. Examiner, N.Y.C., 1988. Capt. USMC, 1946-48. Recipient Milton Helpern Meml. award. Milton Helpern Libr. Legal Medicine, 1989. Mem. AMA, Am. Soc. Investigative Pathology, Coll. Am. Pathology, Assn. Pathology Chmn., Internat. Acad. Pathology, Arthur Purdy Stout Soc., Harvey Soc., N.Y. Path. Soc. Democrat. Avocation: bird watching. Home: 180 E End Ave New York NY 10128-7763 Office: NY Hosp-Cornell Med Ctr 525 E 68th St New York NY 10021-4873

ELLIS, MICHAEL DAVID, aerospace engineer; b. Sacramento, July 13, 1952; s. John David and Priscilla Agnes (Tupper) E.; m. Virginia Katherine Hanlon, Mar. 27, 1976; children: Gwendolyn Dawn, January Marie, Jennifer Noel. BS in Space Sci., Fla. Inst. Tech., 1975. With satellite ops., orbit analyst Western Union, Sussex, N.J., 1976-77; with satellite ops., 3 axis RCA Americom, Sussex, N.J., 1977-78; with satellite ops., Land Sat ATS-6 Goddard Space Flight Ctr., Greenbelt, Md., 1978-79; with Voyager System Lead Jet Propulsion Lab., Pasadena, Calif., 1979-82; STS ground ops. analyst Applied Rsch. Inc., El Segundo, Calif., 1982-83; mission ops. Aerospace Corp., El Segundo, Calif., 1983-88; space sta. mission ops. Johnson Spaceflight Ctr., Houston, 1988—. Mem. troop asst. San Jacinto Girl Scouts, Houston, 1988—, Confraternity Christian Doctrine tchr. St. Bernadette, Clear Lake, 1990. Mem. Soc. Automotive Engrs. (chmn. spacecraft com. 1986-87), Am. Inst. Aeronautics and Astronautics (chmn. 1972-75).

ELLIS, ROBERT JEFFRY, health facility executive; b. Augusta, Ga., Aug. 15, 1935; s. Herbert Monroe and Dorothy Louise (Doney) E.; m. Ann Marie Jarvis, July 19, 1969; 1 child, Tonya Dawn. BA in Mktg./Creative Writing, Columbia Pacific U., San Rafael, Calif., 1980, MA in Mktg./Creative Writing, 1981. Asst. mgr. Publix Employees Fed. Credit Union, Lakeland, Fla., 1970-75; mktg. rep. L.B. Sowell Corp., Tampa, Fla., 1977-82, Cooper Distbrs., Inc., Orlando, Fla., 1983-86; exec. v.p. Nat. Orthotic Labs., Inc., Winter Haven, Fla., 1987-93; co-founder, v.p. mktg. Data Trends, Inc., Auburndale, Fla., 1993—; film producer Focus Prodns., Hollywood, Calif., 1991; corp. sec. The CAM Group, Inc., St. Petersburg, Fla., 1988—. Composer film soundtrack recording, 1973. Mem. Am. Radio League (pub. svc. awards 1970, 76). Avocation: amateur radio. Home: PO Box 7349 Winter Haven FL 33883-7349 Office: Data Trends Inc PO Box 888 Auburndale FL

ELLIS, STEPHEN ROGER, research scientist; b. Berlin, Germany, May 1, 1947; came to U.S., 1947; s. Milton Robert and Ruth (Burg) E.; m. Aglaia Panos, Dec. 30, 1972; children: Matthew, Peter, Mariela. AB, U. Calif., Berkeley, 1969; MA, McGill U., 1971, PhD, 1974. Postdoctoral fellow Brown U., Providence, 1974-75; lectr. Calif. State U., Turlock, 1975-76; rsch. assoc. NAS, Moffett Field, Calif., 1978-80; rsch. scientist NASA Ames Rsch. Ctr., Moffett Field, 1980—; rsch. asst. U. Calif., Berkeley, 1976-78, asst. clin. prof., 1981-92. Editor, author: Pictorial Communication in Virtual and Real Environments, 1991, 2d edit., 1993; mem. editorial bd. Human Factors, 1986—, Presence, 1991—; contbr. to NASA CP 10032. Recipient Univ. medal Kyushu Sanyo U., 1992, Ind. Group award Meckler. Mem. AAAS, Assn. Computing Machinery, Psychonomic Soc., Human Factors Soc. Achievements include development of perspective displays for aerospace applications virtual environmental displays. Office: NASA Ames Rsch Ctr MS 262-2 Moffett Field CA 94035-1000

ELLIS, WAYNE ENOCH, nurse, anesthetist, air force officer; b. Reno, Jan. 24, 1945; s. Willard Edward and Thelma Miriam (Patterson) E.; m. robin Marie Mumme, Dec. 25, 1987; children: Wayne II, Sharon Peisel, Terri Lynn, Michael W., Melissa D., Rebekah J. Stube, Christopher H. Stube, Marina Noél Ellis, Peter Enoch Ellis. Diploma, L.S. Kaufmann Sch. Nursing, 1965; BS, Chapman Coll., 1982, MS, 1984; PhD in Edn., Tex. A&M U., 1990. Commd. 2d lt. U.S. Army, 1966; trans. U.S. Air Force, 1980, advanced through grades col., 1992; asst. operating room supr., instr. regional ctr. U.S. Army, Ft. Bragg, N.C., 1987; dir. inhalation therapy regional ctr. U.S. Army, Ft. Benning, Ga., 1970-72; chief anesthesia U.S. Air Force Hosp. U.S. Air Force, Edwards AFB, Calif., 1980-83; asst. edn. dir. sch. anesthesia Wilford Hall Med. Ctr. U.S. Air Force, Lackland AFB, Tex., 1983-86; liaison officer Inst. Tech. U.S. Air Force, Wright Patterson AFB, Ohio, 1986-89; instr., coord. paramedical programs Cochise Coll., Sierra Vista, Ariz., 1975-79; pvt. practice Ellis Enterprises, Sierra Vista, 1972-79; chief nurse anesthetist Hobart (Ind.) Anesthesia Assocs., 1979-80; clin. dir. anesthesia program sch. nursing U. Tex. Health Sci. Ctr., San Antonio, 1989-93, asst. prof. sch. nursing, 1993—; dir. nurse anesthesia clin. tng. USAF David Grant Med. Ctr., Travis AFB, Calif., 1990-93; program dir., facilitator UTHSCSASN/USAF Nurse Anesthesia Major, 1993—; cons., program coord., Northwest Anesthesia Seminars, Pasco, Wash., 1983—. Author tng. manual for respiratory therapy, emergency medicine; contbg. author Nurse Anesthesia Practice; contbr. articles to numerous publs. Decorated Cross of Gallantry with Palms (Republic Vietnam). Mem. Am. Assn. Nurse Anesthetists, Am. Assn. Adult and Continuing Edn., Calif. Assn. Nurse Anesthetists (bd. dirs. 1990-93), Air Force Assn. of Nurse Anesthetists. Republican. Roman Catholic. Avocations: computers, photography, mountain climbing.

ELLISON, CAROL RINKLEIB, psychologist, educator; b. Santa Barbara, Calif., Sept. 7, 1938; d. Edwin Henry and Margaret Bovard (Skinner) E.; m. William Robert Rinkleib (div. 1974); children: Randy Rinkleib, Teresa Rinkleib, Karen Marie Rinkleib, Monica Anne Rinkleib; life ptnr. Alexander Weiss. AB in Human Devel., U. Chgo., 1960; MS in Child Devel. and Family Studies, U. Calif., Davis, 1970, PhD in Med. Psychology, 1975. Lic. psychologist, marriage, family and child counselor, Calif. Rep. human relations program. coop. extension, U. Calif., Davis, 1971-75; instr. psychology Sacramento City Coll., 1973-78, Calif. State U., Sacramento, 1975-76, Calif. Sch. Profl. Psychology, Berkeley, 1979-80, Alameda, Calif., 1991, U. Calif. Ext., Berkeley, 1983—, John F. Kennedy U., Orinda, Calif. 1986-88; intern. human sexuality program U. Calif., San Francisco, 1976-77, assoc. mem staff, 1977-81, coord. didactic tng., 1978-79, asst. clin. prof. dept. psychiatry, 1989—; pvt. practice psychotherapy, Oakland and San Francisco, 1976—; vis. prof. dept. psychology U. Calif., Berkeley, 1992. Author: (with others) Understanding Sexual Interaction, 1977, 2d edit., 1981; Understanding Human Sexuality, 1980, contbr. chpt. to Principles and Practice of Sex Therapy, 1980, Questions & Answers in the Practice of Hypnosis, 1986, Sexology, 1988; co-host Love Matters, Sta. KALW, San Francisco, 1989; advice columnist Dear Penelope, 1984-89. Founder, 1st pres. Atascadero (Calif.) Pre-sch. Assn., 1962-68; vol. counselor Planned Parenthood, Sacramento, 1972-73; coordinator Family Life Edn. Council, Sacramento, 1973-75; chmn. counseling sect. Western Council Family Relations, 1976-77; mem. program evaluation com. Big Sisters Orgn., Oakland, 1983-85. Fellow Soc. Sci. Study of Sex (treas. San Francisco Bay area 1983-84, nat. admissions com. 1983-84, nat. profl. sex therapy edn. com., 1985-86, treas. We. region 1986, chair nat. student grants program com. 1987-89); mem. APA, Soc. for Menstrual Cycle Rsch., Living Tao Found. Office: 1957 Oak View Dr Oakland CA 94602-1945

ELLISON, LUTHER FREDERICK, oil company executive; b. Monroe, La., Jan. 2, 1925; s. Luther and Gertrude (Hudson) E.; m. Frances Z. Williams, July 17, 1948; children: Constance Elizabeth, Carolyn Williams. Student, Emory U., 1943-44; B.S. in Petroleum Engring., Tex. A&M U., 1949, B.S. in Geol. Engring., 1950. Registered profl. engr., Tex., La. Jr. petroleum engr. Sun Prodn. Co., Kilgore and McAllen, Tex., 1950-52; area petroleum engr. Sun Prodn. Co., Garcia Field, Tex., 1952-54; Delhi (La.) unit engr. Sun Prodn. Co., 1954-60; asst. region supt. Sun Prodn. Co., Dallas, 1960-62; dist. drilling engr. Sun Prodn. Co., Corpus Christi, 1962-63; dist. engr. Sun Prodn. Co., McAllen, 1963-65; supr. engring. Sun Prodn. Co., Dallas, 1965-66, div. chief petroleum engr., 1966-70, regional mgr. engring., 1970-75, region mgr., 1975-78, dir. devel., 1978-80, v.p. devel., 1980-84; div. v.p., dir. Sun Exploration and Prodn. Co., 1984-86, pres., bd. dirs., 1986—; pres., chief exec. officer Oil & Gas Experts, Inc., Dallas, 1986—, Am. Energy

Enterprises Inc., Dallas, 1988—; pres., dir., mem. exec. com. Nabors-Sun Drilling Co.; dir., mem. exec. com. East Tex. Salt & Water Disposal Co.; pres. Oil & Gas Experts Inc., 1986; speaker in field. Vice pres. Northwood Jr. High Sch. PTA, Dallas, 1967-68, pres., 1968-69. Served with USNR, 1943-46. Mem. Tex.-Mid-Continent Oil and Gas Assn. (Outstanding Achievement award 1964, chmn. area 1964-65, mgr. north region, operating com., Outstanding Performance award 1985—), Am. Petroleum Inst., Soc. Petroleum Engrs., Dallas Engrs. Club, Petroleum Engrs. Club, Dallas Petroleum Club, Dallas Energy Club, Northwood Club (Dallas), Lions Club, Premier Club (Dallas), Parents League, Sigma Alpha Epsilon (pres. 1944-45). Presbyterian (elder). Home: 3 Castlecreek Ct Dallas TX 75225-1808 Office: Ste 613 6440 N Central Expy Dallas TX 75206

ELLISON, THORLEIF, consulting engineer; b. Lyngdal, Norway, May 13, 1902; s. Andreas Emanuel and Gemalie (Svensen) E.; CE, Christiania Coll. Tech., 1924; postgrad. George Washington U., U. Va.; m. Reidun Ingeborg Skonhoft, Jan. 1, 1932; children: Earl Otto, Thorleif Glenn, Sonja Karen. Came to U.S., 1928, naturalized, 1933. Supervising engr. GSA, Washington, 1948-57; supervising airport and airways svc. engr. FAA, 1957-61; chief airways engring. AID, Iran, West Pakistan, Turkey, 1961-67; cons. engr., Washington and Va., 1971-82; supervising structural engr. for reconstrn. of The White House, 1949-52; mission dir. Bethlehem, Israel, Holy Land Christian Mission, Kansas City, 1968-71. Recipient U.S. Navy commendation, 1945. Active Christian Bus. Men's Com., Washington, Boy Scouts Am. Registered profl. engr. Mem. Nat. Soc. Profl. Engrs. (dir.), Sons of Norway (pres. Washington chpt.), Norwegian Soc. (treas.). Presbyterian (ruling elder). Home: Svennevik Rosfjord, Lyngdal 4580, Norway also: 6324 Telegraph Rd Alexandria VA 22310

ELLNER, PAUL D., clinical microbiologist; b. N.Y.C., May 2, 1925; s. George and Cele (Weis) E.; m. Estelle Ziswasser, 1948 (div. 1960); 1 child, Diane; m. Cornelia Johns, Jan. 15, 1965; children—David, Jonathan. B.S., L.I. U., 1948; M.S., U. So. Calif., 1952; Ph.D., U. Md. Coll. Medicine, 1956. Diplomate Am. Bd. Med. Microbiology; cert. clin. lab. dir. N.Y.C. Dept. Health. Clin. bacteriologist Los Angeles hosps., 1948-52; research asst. Mt. Sinai Hosp., N.Y.C., 1952-53; instr. microbiology U. Fla. Coll. Medicine, 1956-60; asst. prof. U. Vt. Coll. Medicine, 1960-63; asst. prof. Columbia U. Coll. Physicians and Surgeons, N.Y.C., 1963-66, assoc. prof., 1966-70, prof. microbiology, 1971-78, prof. microbiology and pathology, 1978-89, prof. emeritus, 1989, dir. clin. microbiology service, 1971-89; assoc. microbiologist Presbyn. Hosp., N.Y.C., 1966-70, attending staff, 1971-89; cons. in field; vis. prof. N.Y. Med. Coll., Valhalla, 1979; ASM Latin Am. vis. prof., Medellín, Colombia, 1982; Am. Bur. Med. Advancement in China vis. prof., Taiwan, 1982; regional coordinator Nat. Disaster Med. System; v.p. Am. BioSci. Cons. Author: Current Procedures in Clinical Bacteriology, 1978, Understanding Infectious Disease, 1992; editor: Infectious Diarrheal Diseases: Current Concepts and Laboratory Procedures, 1984; mem. editorial bd. Sexually Transmitted Diseases, 1982-84, European Jour. Clin. Microbiology, 1985-89; contbr. chpts. to books, numerous articles to sci. jours. Served with AC, USN, 1943-44; to capt. USPHS Res., 1956—; health project officer USCG, 1982-91. U.S. Navy research fellow, 1954-56. Fellow Am. Acad. Microbiology, Assn. Clin Scientists, N.Y. Acad. Medicine (assoc.), Infectious Diseases Soc. Am.; mem. AMA (spl. affiliate), Am. Soc. Microbiology (chmn. clin. divsn. 1980-81, Sonnenwirth Meml. award 1992), Acad. Clin. Lab. Physicians and Scientists, Am. Venereal Disease Assn., Sigma Xi. Republican. Jewish. Avocations: flying, fishing, gardening, photography.

ELLSAESSER, HUGH WALTER, retired atmospheric scientist; b. Chillicothe, Mo., June 1, 1920; s. Charles Theobald and Louise Minerva (Bancroft) E.; m. Lois Merle McCaw, June 21, 1946; children: Corbin Donald, Adrienne Sue. AA, Bakersfield (Calif.) Jr. Coll., 1941; SB, U. Chgo., 1943, PhD, 1964; MA, UCLA, 1947. Commd. 2d lt. USAF, 1943, advanced through grades to lt. col., 1960; weather officer USAF, Washington, Fla., Eng., 1942-63; ret., 1963; physicist Lawrence Livermore (Calif.) Nat. Lab., 1963-86, guest scientist, 1986—. Editor: Global 2000 Revisited, 1992; contbr. numerous articles to profl. jours. Mem. Am. Meteorol. Soc., Am. Geophysics Union. Republican. Presbyterian. Avocation: languages. Home: 4293 Stanford Way Livermore CA 94550-3463 Office: Lawrence Livermore Nat Lab PO Box 808 Livermore CA 94551-0808

ELLSWORTH, PETER CAMPBELL, entomologist; b. Springfield, Mass., Nov. 15, 1960; s. Gerald Merle and Jane (Campbell) E.; m. Donna Marie DiFrancesco, July 10, 1982. BS in Entomology, U. N.H., 1981; MS in Entomology, U. Mo., 1985; PhD in Entomology, N.C. State U., 1990. Co. entomologist W.B. McCloud & Co., St. Louis, 1982; rsch. asst. U. Mo., Columbia, 1982-85; rsch. asst. N.C. State U., Raleigh, 1985-89, insectary mgr., 1990-91; integrated pest mgmt. specialist U. Ariz., Maricopa (Ariz.) Agrl. Ctr., 1991—; edn. subcom. chair Nat. Cotton Coun., Pink Bollworm Action Com., Memphis, 1992—; state IPM coord. U. Ariz. Coop. Ext., Tucson, 1992—; mem. Ariz. Cotton Growers, Univ. Rsch. Coun., Phoenix, 1991—. Contbr. articles to profl. jours. Recipient Ext. Communicator award U. Ariz. Coop. Ext., 1992, Outstanding Doctoral Dissertation award N.C. State U., 1990, Outstanding Grad. Student Teaching award N.C. State U., 1989, Outstanding Grad. Student award U. Mo., 1985. Mem. Entomol. Soc. Am. (Pres.'s Prize for Outstanding Presented Poster 1988), Ga. Entomol. Soc., Ctrl. States Entomol. Soc., Ariz. County Agts. Assn., Sigma Xi. Achievements include devel. of Integrated Pest Mgmt. Programs for arid land prodn. of cotton; devel. mgmt. guidelines for Sweet Potato Whitefly; evaluation of novel means of insect control (genetically engineered cottons) for lepidopteran pests. Office: U of Arizona Maricopa Agrl Ctr 37860 W Smith-Enke Rd Maricopa AZ 85239

ELLWOOD, BROOKS BERESFORD, geophysicist, educator; b. Chgo., July 18, 1942; s. John F.F. and Doris (Hammill) E.; m. Suzanne Higgins, Feb. 25, 1965; children: Amber B., Robin E., John Richard. BS, Fla. State U., 1970; PhD, U. R.I., 1976. Rsch. asst. U. R.I., Narragansett, 1970-76; rsch. assoc. Ohio State U., Columbus, 1976-77; asst. prof. U. Ga., Athens, 1977-81, assoc. prof., 1981-83; assoc. prof. U. Tex., Arlington, 1983-88, prof., 1988—, acting chair, 1989-92; dir. Ctr. for Geoenviron. and Geoarcheol. Studies, Arlington, 1987—. Contbr. articles to Jour. Geophys. Rsch., Eart and Planetary Sci. Letters, Geology; co-producer: (video) Applied Geoarcheology, 1988. With U.S. Army, 1964-66. Fellow Geol. Soc. Am., Geol. Soc. Can.; mem. Am. Geophys. Union, Soc. Exploration Geophysicists, Sigma Xi. Office: U Tex Arlington Dept Geology PO Box 19049 Arlington TX 76019

ELMAN, HOWARD LAWRENCE, aeronautical engineer; b. N.Y.C., Dec. 18, 1938; s. Dave and Pauline (Reffe) E.; m. Joan Carter, Dec. 28, 1974 (div. 1985); children: David Lawrence, Elizabeth Nadine. SB, M.I.T., 1960; M in Aerospace Engring., U. Okla., 1962; postgrad. Rensselaer Poly. Inst., 1963-65. Registered profl. engr., Conn., Ohio, Tex., Calif. Rsch. engr. United Aircraft Rsch. Labs., East Hartford, Conn., 1963-68; sr. rotor dynamics engr. Sikorsky Aircraft, Stratford, Conn., 1968-70; sr. analytical engr. Pratt & Whitney Aircraft, East Hartford, 1970-71; rsch. engr. Kaman Aerospace, Bloomfield, Conn., 1972-76; sr. mission equipment coord. Hughes Helicopters, Culver City, Calif., 1977-80; sr. engr., sect. leader in ops. analysis Fairchild-Republic, Farmingdale, N.Y., 1980-85; sr. engr. Gould, Inc., 1985-86; prin. engr. in ops. analysis Grumman Corp., 1986-92; cons. mus. and hist. socs. Contbg. author: The Changing World of the American Military, 1978; contbr. articles to various jours. Served to 1st lt. USAF, 1960-63, to col. Res., 1964-90. Mem. Conn. Aero. Hist. Assn. (life; dir. 1968-76, v.p. rsch. 1969-73, exec. v.p. 1973-76), Inst. Aero. Scis., AIAA, Am. Helicopter Soc., Nat. Soc. Profl. Engrs., Am. Aviation Hist. Soc., Soc. World War I Aero. Historians, Air Force Assn., U.S. Naval Inst. Home: 31 Old Post Rd E Port Jefferson NY 11777

ELMORE, STANLEY MCDOWELL, orthopaedic surgeon; b. Raleigh, N.C., Dec. 17, 1933; s. Kelly Lee and Isabel (McDowell) E.; m. Emily Ruth Rustin, June 20, 1958; children: Kelly Rustin, Stephen Mark, Andrew Glenn, Peter Sean. BA, Vanderbilt U., 1955, MD, 1958. Diplomate Am. Bd. Orthopaedic Surgeons. Intern Vanderbilt U., Nashville, 1958-59, asst. resident, 1959-60, resident in orthopaedics, 1962-65, acad. tng. fellow NIH, 1965-66; sr. asst. surgeon NIH, Bethesda, Md., 1960-62; assoc. prof., chmn. dept. orthopaedic surgery Md. Coll. Va., Richmond, 1966-73; chmn. dept. orthopaedic surgery Chippenham Hosp., Richmond, 1988-90. Contbr. chpts. in books, several articles to jours. Recipient Borden award, 1958. Fellow Am. Acad. Orthopaedic Surgeons (fellow 1971), Am. Bd. Orthopaedic Surgeons; mem. Va. Orthopaedic Soc., Richmond Acad. Medicine, Richmond Orthopaedic Club (pres. 1987-88). Avocations: sailing, photography. Home: 315 Clubview Ct Richmond VA 23229-7609 Office: West End Orthopaedic Clinic 7135 Jahnke Rd Richmond VA 23225-4017

EL-MOURSI, HOUSSAM HAFEZ, civil engineer; b. Cairo, Egypt, Oct. 28, 1944; s. Hafez El-Moursi and Anwar (Mohamed) Wahdan; m. Samia Mohamed El-Sayed, Apr. 13, 1977; children: Suzanne, Dalia. BS, Cairo U., 1967; Postgrad. Diploma, Internat. Inst. Earth Sci., Delft, Holland, 1971; PhD, Northwestern U., 1975. Registered profl. engr., Ill., Ind., Wis., Ohio, Okla., Tex., Va., Ky. Instr. Cairo U., Egypt, 1967-70; rsch. asst. Northwestern U., Evanston, Ill., 1971-75; project engr. STS Consultants, Inc., Cedar Rapids, Iowa, 1975-77; asst. prof. U. Petroleum and Minerals, Dhahran, Saudi Arabia, 1977-81; gen. mgr. Al-Thara Consultants, Jeddah, Saudi Arabia, 1981-86; prin. Terracon Consultants, Inc., Oklahoma City, 1986-91, Naperville, Ill., 1991-93; mgr. geotech. engring. Engrs. Internat., Inc., Naperville, 1993—. Contbr. profl. publs. to profl. jours. Fellow ASCE; Egyptian Inst. of Engring. Profls. Republican. Moslem. Home: 68 Sterling Circle #208 Wheaton IL 60187

EL NAHASS, MOHAMMED REFAT AHMED, physiology educator, researcher; b. Damietta, Egypt, Dec. 12, 1938; s. Ahmed Mohammed and Zeniab Yosef (Ali) El N.; m. Fatma Ahmed Kamal, Nov. 7, 1968; children: Ahmed, Mohammed. M in Biochemistry, Ein Shams U., Cairo, Egypt, 1963; MSc in Physiology, Cairo U., Egypt, 1966, MD in Physiology, 1969. Demonstrator faculty of medicine Monusra (Egypt) U., 1963, lectr. in physiology, 1969-74, asst. prof. physiology, 1974-78, prof. physiology, 1978—, head of physiology dept., 1982; vice dean Faculty of Medicine, Monsura U., 1988-91, dean, 1991—; mem. permanent sci. com. for physiology profs. and asst. profs., supreme coun. of univ., 1982—. Author: Central Nervous System, 1985, Biophysics for Medical Students, 1989, Autonomic Nervous System, 1990, Special Senses, 1991. Pres. Monsura Univ. Club, 1987—. Moslem. Home: El Bedely St, Monsura DeKahlia, Egypt Office: Monsura Univ Faculty of Medicine, El Gomhoria St, Monsura DeKahlia, Egypt

ELNOMROSSY, MOKHTAR MALEK, aeronautical engineer; b. Cairo, Dec. 4, 1946; s. Malek Mahmoud and Aisha Mohamed (Tewfik) E. BSc, Mil. Tech. Coll., Cairo, 1969, MSc, 1973; PhD, VAAZ, Tchekoslovakia, 1977. Tchr. Mil. Tech. Coll., 1970-73, prof., 1977-83; engring. mgr. Teledyne Ryan Aeronautical, U.S.A., 1984-88; researcher Armament Rsch. Inst., Cairo, 1988-91; dir. Airforce Rsch. Inst., Cairo, 1991—; cons. Aircraft Factiry, Helwan, Egypt, 1988-90; mem. adv. bd. Arab Brit. Helicopter Factory, Cairo, 1989-92. Author: Effect of Aerodynamic Heating of Aircraft Structures, 1982; contbr. articles to profl. publs. Gen. Egyptian Air Force, 1969-70. Mem. AIAA (sr.), Nat. Com. Theoretical and Applied Mechanics, Nat. Geog. Soc., Internat. Tech. Inst. Muslim. Office: Airforce Rsch Inst, Khalifa Elmamoun, Heliopolis Cairo Egypt

ELORANTA, EDWIN WALTER, meteorologist, researcher; b. Owen, Wis., Nov. 26, 1943; s. Edwin Jonas and Elma H. Ruonavarra; m. Jean Burris, Nov. 25, 1986. MS, U. Wis., 1967, PhD, 1972. Asst. scientist U. Wis., Madison, 1972-78, assoc. scientist, 1978-83, sr. scientist, 1983—. Contbr. articles to Jour. Applied Meteorology, Applied Optics, other profl. jours. Fellow Am. Meteorol. Soc.; mem. Optical Soc. Am. Achievements include patent in weather facsimile and rapid teletype decoder. Home: 2520 Lunde Ln Mount Horeb WI 53572 Office: Univ Wis 1225 W Dayton St Madison WI 53706

ELROD, DAVID WAYNE, computational chemist, information scientist; b. Niles, Mich., Sept. 25, 1952; s. James W. Jr. and Doris I. (Biederman) E.; m. Nikki Jean Charles, Aug. 17, 1974; 1 child, Timothy D. BA, Kalamazoo Coll., 1974; PhD in Chemistry, We. Mich. U., 1992. Cancer rsch. chemist Upjohn Co., Kalamazoo, 1974-85, computational chemist, 1985-90, info. scientist in computational chemistry, 1990—. Contbr. articles to Jour. Organic Chemistry, Jour. Medicinal Chemistry, Jour. Chem. Info. Computer Sci., others. Mem. Am. Chem. Soc., Molecular Design Ltd. Software Users Group, Sun Users Group. Methodist. Achievements include several patents, work on neural nets in chemistry. Office: The Upjohn Co Upjohn Labs 301 Henrietta St Kalamazoo MI 49007-4940

EL-SAIEDI, ALI FAHMY, science foundation administrator; b. Egypt, Jan. 17, 1936. BSc in Mech. Engring., Cairo U., 1957; MSc in Reactor Physics and Tech., Birmingham U., U.K., 1961; PhD in Nuclear Engring., U. Ill., 1968; Fellow, Nat. Def. Coll., Nasser Supreme Mil. Acad. Mech. engr. dept. physics and reactors Atomic Energy Authority Egypt, 1957-59, sr. mech. engr. reactors dept., 1961-63, sr. rsch. officer, 1970-73, assoc. prof. nuclear engring., dep. head reactors dept., 1973-77; asst. prof. nuclear engring. Kans. State U., 1968-70; head divsn. studies, projects and nuclear affairs Nuclear Power Plants Authority Egypt, 1977-81, exec. vice chmn. studies and nuclear affairs, 1981-85, exec. chmn., 1985-93; dir. divsn. tech. co-operation implementation, dept. tech. co-operation Internat. Atomic Energy Agy., Vienna, Austria, 1993—; mem. tech. secretariat Supreme Coun. Energy, 1980-93; mem. nat. negotiating team experts Bilateral Agreements Cooperation in Peaceful Uses Nuclear Energy between Egypt and USA, France, Can., UK, others, 1981-88; head, mem., rapporteur coordinating com. Internat. Cooperation Peaceful Uses Nuclear Energy, 1983-87; co-chmn. various coms. Nuclear Cooperation between Egypt and Other Countries, 1985-93; sec. Supreme Coun. Peaceful Uses Nuclear Energy, 1986-93; head nuclear energy br. Egyptian Acad. Rsch. and Tech., 1987-93, mem. coordinating office, 1987-93, mem. new and renewable energy br., 1987-93, mem. energy planning br., 1987-93; mem., rapporteur interministerial coordination and follow-up Com. Peaceful Uses Nuclear Energy, 1987-93; head steering com. Devel. and Localization Indsl. Capability in Egypt, 1988-93; mem. nat. br. World Energy Coun., 1989-93; mem. nat. team experts Nuclear Free Zone in Mid. East, 1993; tchr. in field; expert, adviser on nuclear power. Editor-in-chief Jour. Electricity and Energy, bd. dirs. Mem. Egyptian Soc. Mech. Engrs., Egyptian Soc. Nuclear Scis. and their Applications, Egyptian Soc. Health and Environ. Legis., Nat. Soc. Technol. and Econs. Devel., OKO Hon. Soc. Achievements include research in reactor physics, reactor engineering, radioisotope and radiation applications, nuclear power generation, planning, nuclear power and the environment.

EL-SAYED, AHMED FAYEZ, aeronautical engineering educator; b. Zagazig, Sharkia, Egypt, Jan. 16, 1948; s. Abdel Azim El-Sayed and Fatma (Mohamed) Ahmed; m. Amany Abdel Azim, June 30, 1981; children: Mohamed, Abdallah, Khalid. BSc in Aero. Engring., Cairo U., 1970, MSc in Aero. Engring., 1976, PhD in Aero. Engring., 1980. Rsch. engr. Egypt Air Co., Cairo Airport, 1975-81, sr. rsch. engr., 1982-83; postdoctoral rsch. fellow Carnegie-Mellon U., Pitts., 1981-82; asst. prof. Zagazig U., 1983-85, assoc. prof. engring., 1987-92, prof., 1993—; vis. prof. Royal Mil. Coll. Sci., Swindon, Wiltshire, Eng., 1985, Vrije Universtiet Brussels, 1986; cons. Sci. R & D Acad. Cairo, 1983-85, Egyptian Air Acad., Cairo, 1988. Author: (text and workbook) Engineering Drawing, 1986, Machine Design with Practical Applications, 1987; contbr. articles to profl. publs. Capt. Egyptian Engring. Corps, 1970-75. Mem. ASME, AIAA, Internat. Tech. Inst., Internat. Assn. Hydrogen Energy, Internat. Assn. Sci. and Tech. for Devel., Egyptian Soc. Aeronautical Engrs., Sigma Xi. Achievements include investigation of performance deterioration of both axial and centrifugal compressors of aero-engines, supersonic/hypersonic flow around blunt noses of rockets, aerodynamics of horizontal axis wind turbines, erosion of turbomachines used in coal fired power plants, pneumatic transport, aeroelastic behavior of vertical axis wind turbines, ways of reducing aerodynamic drag and fuel consumption of road vehicles. Home: 3 El-Lais St, Cairo 11321, Egypt Office: Zagazig U, Mech Power Engring, Sharkia Egypt

EL-SAYED, KARIMAT MAHMOUD, physics and crystallography educator; b. Meet El-Kholy Mokmen, Dakahlia, Egypt, Dec. 10, 1933; d. Mahmoud El-Sayed and Nazla Ali (Keiwan) Soliman; divorced; 1 child, Mohamed El Shinnawi; m. Salah Ahmed Tahoun, Aug. 28, 1969; children: Hussein, Sarah. BSC in Physics with honors, Ain Shams U., Cairo, Egypt, 1957; PhD in Crystallography, London U., 1965. Demonstrator Cairo U., 1957-59; lectr. Ain Shams U., Cairo, 1965-72, asst. prof., 1972-78, prof., 1978—, chmn. dept. physics, 1992—. Referee Egyptian Jour. Solid, Cairo, 1980—; editor (procs.) The Third International School of Crystallography (Cairo), 1990; contbr. 35 articles on crystallography and material sci. to profl. jours.; author several local notes for teaching physics. Developer courses for teaching physics to Ain Shams U. undergrad. students UNESCO Regional Office for Sci. and Tech. Recipient 1st Medal Award of Sci. and Art, Pres. of Egypt, 1971. Mem. Nat. Com. Crystallography, Nat. Com. Pure and Applied Physics, Nat. Com. for Promoting Profs. (chmn. 1985-89). Home: PO Box 8014 Masaken Nassr, Nassr City, Cairo 11371, Egypt Office: Faculty of Sci, Ain Shams U Physics Dept, Cairo Egypt

EL-SAYED, MOSTAFA AMR, chemistry educator; b. Zifta, Egypt, May 8, 1933; s. Amr and Zakia (Ahmed) El-Sayed; m. Janice Jones, Mar. 15, 1957; children—Lyla, Tarric, James, Dorea Jehan, Ivan Homer. B.Sc., Ein Shams U., Cairo, 1953; Ph.D., Fla. State U., 1959. Research fellow Yale U., 1957; research fellow Harvard U., 1959-60, Calif. Inst. Tech., 1960, 61; asst. prof. chemistry UCLA, 1961-64, assoc. prof. chemistry, 1964-67, prof. chemistry, 1967—; vis. prof. Am. U. Beirut, 1967-68; fgn. prof. U. So. Paris, Orsay, 1976; Sherman Fairchild disting. scholar Calif. Inst. Tech., 1980; cons. Space Tech. Lab, 1962-63, Electro-Optical System, 1963-66, N.Am. Aviation, 1964-65, Navy Electronics Labs., 1969-73, Ford Research Labs., 1970, Northrop Corp., 1979-81; mem. adv. bd. Alexandria Research Ctr., 1979-83; trustee Associated Univs., 1988; mem. steering com. Internat. Ctr. Pure and Applied Chemistry, Trieste, Italy, 1988. Mem. adv. bd. Chem. Physics, Chem. Physics Letters and Accounts of Chem. Research; contbr. numerous articles to profl. jours., chpts. to books. Mem. chemistry grant selection com. NRC of Can.; mem. chemistry research evaluation panel for directorate of chem. scis. Air Force Office of Sci. Research; mem. rev. com. San Francisco Laser Ctr., radiation lab Notre Dame U., dept. energy and environment Lawrence Berkeley Lab.; mem. NRC com. to survey opportunities in chemistry; mem. vis. com. Brookhaven Nat. Lab., 1986—. Recipient Disting. Teaching award UCLA, 1964; Fresenius nat. award in pure and applied chemistry, 1967; McCoy Research award, chemistry dept. UCLA, 1969; Alexander von Humboldt Sr. U.S. Scientist award Fed. Republic Germany, 1982; King Faisal Internat. Prize in Sci. (Chemistry), 1990. Mem. Am. Chem. Soc. (Gold Medal award Calif. sect. 1971, editor in chief Jour. Phys. Chemistry 1980—and editor Internat. Revs. Phys. Chemistry 1984-90, Tolman award 1990), NAS (elected), Am. Acad. Arts and Scis. (elected), AAUP, AAAS, Assn. for Harvard Chemists, Western Spectroscopy Assn., N.Y. Acad. Scis., Third World Acad. Scis. (elected), Phys. Chemistry Div. Internat. Union Pure and Applied Chemistry (elected, vice chmn. U.S. NRC com. 1987, chmn. 1992). Home: 3325 Colbert Ave Los Angeles CA 90066-1213 Office: UCLA Dept Chemistry Los Angeles CA 90024

ELSHERBENI, ATEF ZAKARIA, electrical engineering educator; b. Cairo, Egypt, Jan. 8, 1954; came to U.S., 1987; s. Zakaria Mostafa Elsherbeni; m. Magda Mohammed Elshemy, Sept. 15, 1980; children: Dalia, Donia, Tamer. BS in Elec. Engring., Cairo (Egypt) U., 1976, MS in Engring. Physics, 1983; PhD in Elec. Engring., U. Manitoba, Winnipeg, Can., 1987. Part time systems engr. United Group Co., Cairo, 1977-79; software and systems design engr. Automated Data System Ctr., Cairo, 1979-82; rsch. asst. applied physics dept. Cairo U., 1978-82, elec. engring. dept U. Manitoba, Winnipeg, Can., 1983-87; postdoctoral fellow elec. engring. dept U. Manitoba, Winnipeg, 1987; asst. prof. elec. engring. dept. U. Miss., University, 1987-91, assoc. prof., 1991—. Contbr. 32 articles to profl. and scholarly jours, 3 book chpts.; 50 publs. in procs. of sci. confs. or seminars; 10 tech. reports; 6 invited presentations. Recipient Grad. fellowship U. Manitoba, Winnipeg, 1984-86, postdoctoral fellow, 1987; grantee U. Miss., U.S. Army Rsch., USAF Found., NSF. Mem. IEEE (sr.), Electromagnetics Acad. (Invited), Sigma Xi. Achievements include development of three software packages for undergraduate electromagnetic education WGVMAP, ARRAYS, and APV. Office: U Miss Elec Engring Dept University MS 38677

ELSNER, FREDERICK HARVEY, chemical scientist; b. Phoenix, Oct. 4, 1963; s. Harvey Julian and Marilyn Carmel (Mentemeyer) E.; m. Darlene Carol Hand, July 19, 1992. BS, Harvey Mudd Coll., 1985; MS, U. Calif., San Diego, 1986. Sr. scientist Gen. Atomics, San Diego, 1987—. Contbr. articles to profl. jours. including Jour. of Am. Chem. Soc., Jour. of Organometallic Chemistry, Organometallics, Applied Phys. Letters. Achievements include preparation of the first stable formilsilane R3SiCHO, 12 patents in high temperature superconductivity. Home: PO Box 372 Cardiff by the Sea CA 92007

ELSON, HANNAH FRIEDMAN, research biologist; b. Lublin, Poland, July 10, 1943; came to U.S., 1949; m. Edward C. Elson; 2 children. BA, Vassar Coll., 1964; PhD, MIT, 1970. Arthritis Found. postdoctoral fellow Med. Rsch. Coun. Lab Molecular Biology, Cambridge, Eng., 1970-72; asst. prof., then asst. rsch. biologist U. Calif.-San Diego, La Jolla, 1972-79; rsch. pathologist VA Med. Ctr., La Jolla; rsch. scientist MEDSAT Rsch. Co., Bethesda, Md., 1986-90; sr. resident rsch. assoc. Nat. Rsch. Coun. Walter Reed Army Inst. Rsch., Washington, 1988-90; sr. staff fellow Nat. Heart, Lung, Blood Inst., NIH, Bethesda, 1990-92; expert Nat. Cancer Inst., NIH, 1992—. Contbr. articles to sci. jours. Mem. AAAS, Am. Chem. Soc. Am. Soc. Cell Biology, N.Y. Acad. Scis., Sigma Xi. Achievements include research on protein synthesis, membrane changes during development of skeletal muscle, membrane fusion by HIV, and gene therapy. Office: Nat Cancer Inst Sect Membrane Struc/Func 9000 Rockville Pike Bethesda MD 20892-0001

ELTON, RICHARD KENNETH, polymer chemist; b. Amsterdam, N.Y., Mar. 19, 1952; s. George Truman and Dorothy Adelaide (Moeller) E.; m. Cheryl Ann Avy, July 9, 1983; children: Daniel Christopher, Matthew Thomas. BS, St. Lawrence U., 1974; MS, U. Vt., 1977. Product devel. chemist GE, Schenectady, N.Y., 1977-86; sr. polymer chemist C.R. Bard Inc., Glens Falls, N.Y., 1986—. Mem. Am. Chem. Soc., Soc. Plastics Engrs. Achievements include 12 patents in field of biomaterials and high voltage dielectric materials. Office: CR Bard PO Box 787 Glens Falls NY 12801-0787

ELVIUS, AINA MARGARETA, retired astronomer; b. Stockholm, June 26, 1917; d. Axel Markus and Sigrid Augusta (Wiman) Eriksson; m. Tord Elvius, Jan. 3, 1940; children: Torbjörn, Margareta, Ragnhild. Fil. mag., Stockholm U., 1945, Fil. lic., 1951, Fil. Doktor, 1956. Asst. prof. Stockholm U., 1956-58, Uppsala (Sweden) U., 1958-63; researcher Swedish Natural Sci. Rsch. Coun., Uppsala, 1963-68; assoc. prof. Royal Swedish Acad. Scis., Saltsjöbaden, Sweden, 1968-73; prof. Stockholm U., 1973-81; dir. Stockholm Obs., Saltsjöbaden, 1977-81; Swedish rep. ESO Instrumentation com., Europe, 1971-74; priority com. mem. Swedish Natural Sci. Rsch. Coun., Stockholm, 1977-80. Editor: From Plasma to Planet, 1972; contbr. articles to profl. jours. Sec. Swedish Union Asst. Profs., Stockholm, 1959-60; chmn. Swedish Astron. Assn., Stockholm, 1986-90, Swedish Nat. Com. for Astronomy, 1977-82. Recipient Zorn award Sweden-Am. Found., 1961. Mem. Internat. Astron. Union, Am. Astron. Soc., Kungl. Fysiografiska Sällskapet, Royal Swedish Acad. Scis., Zonta Internat. Home: Norrlandsgatan 34 F, S 752 29 Uppsala Sweden Office: Stockholm Obs, S 133 36 Saltsjöbaden Sweden

ELY, WAYNE HARRISON, broadcast engineer; b. Alliance, Ohio, Aug. 31, 1933; s. Dwight Harrison and Mable Evellen (Jones) E.; m. Roslyn Rose Ambrose, June 14, 1964 (div. Nov. 1981): children: Eric (dec.), Kevin, Gayle, Mitchell; m. Linda Kay Grubb, July 22, 1989. Student Mount Union Coll., 1955-56, Ohio U., 1956-62. Transmitter engr. Sta. WOUB-FM-TV, Ohio U., Athens, 1958-62; studio field engr. ABC, N.Y.C., 1962-66, 67-72; studio engr. CBS, N.Y.C., 1966-67; transmitter supr. Sta. WOUC-FM-TV, Ohio U., Quaker City, 1972-91; retired, 1991—; technical adv./vol. Muskingum Perry Career Ctr. Radio-TV Dept., 1991—; tchr. radio tech. Ohio U., Zanesville. Served with C.E., U.S. Army, 1952-54. Mem. Soc. Broadcast Engrs. (profl. broadcast engrs. cert.). Home: 140 Riley Rd Norwich OH 43767-9722

EMANUEL, KERRY ANDREW, earth sciences educator; b. Cin., Apr. 21, 1955; s. Albert II and Mary Catherine (Schonegevel) E.; m. Susan Boyd-Bowman, Dec. 29, 1990. SB in Earth and Planetary Scis., MIT, 1976, PhD in Meteorology, 1978. From adj. asst. prof. to asst. prof. dept. atmospheric scis. UCLA, 1978-81; postdoctoral fellow Coop. Inst. Mesoscale Meteorological Studies, U. Okla., 1979; asst. prof. dept. meteorology and physical oceanography MIT, 1981-83, from asst. prof. to assoc. prof. ctr. meteorology and physical oceanography, dept. earth, atmospheric & planetary scis., 1983-

87, prof., 1987—, dir., 1989—. Contbr. articles to profl. jours., textbooks, monographs. Mem. Am. Meteorol. Soc. (Meisinger award 1986, Banner I. Miller award with Richard Rotunno, 1992), Sigma Xi, Phi Beta Kappa. Avocations: sailing, classical music. Office: MIT Dept of Earth Sciences 77 Massachusetts Ave Cambridge MA 02139

EMBLETON, TOM WILLIAM, horticultural science educator; b. Guthrie, Okla., Jan. 3, 1918; s. Harry and Katherine (Smith) E.; m. Lorraine Marie Davidson, Jan. 22, 1943; children: Harry Raymond (dec.), Gary Thomas, Wayne Allen, Terry Scott, Paul Henry. BS, U. Ariz., 1941; PhD, Cornell U., 1949; Diploma de Honor al Ingeniero Agronomo, Coll. Engring. Agronomy, Santiago, Chile, 1991. Jr. sci. aide Bureau Plant Industry USDA, Indio, Calif., 1942, horticulturist Bureau Plant Industry, 1942, 1946; asst. horticulturist Wash. State Coll., Prosser, 1949-50; asst. horticulturist to prof. hort. sci. U. Calif., Riverside, 1950-86, prof. hort. sci. emeritus, 1987—; cons. in field, 1973—. Contbr. over 200 articles to profl. jours. Leader Riverside Boy Scouts of Am., 1952-74. Recipient Citrograph Rsch. award, Citrograph Mag., 1965, Award of Honor, Lemon Men's Club, 1987, Calif. Avocado Soc., 1987, Chancellor's Founders' award U. Calif, 1990, Honor award Am. Soc. Agronomy, 1993. Fellow AAAS, Am. Soc. Hort. Sci. (Wilson Popenoe award 1985, chmn. western region 1958-59); mem. U. Calif. Riverside Faculty Club (pres. 1958), Sigma Xi (pres. Riverside chpt. 1981-82), others. Achievements include research on use of leaf analysis as guide for citrus and avocado fertilization; on providing a means of substantially reducing nitrate pollution of ground-waters from citrus fertilization. Home: 796 Spruce St Riverside CA 92507-2501 Office: U Calif Dept Botany & Plant Scis Riverside CA 92521-0124

EMBODY, DANIEL ROBERT, biometrician; b. Ithaca, N.Y., July 10, 1914; s. George Charles and Mary Madeline (Riceman) E.; m. Margaret Constance Gran, Mar. 21, 1946 (dec. Mar. 1961); children: James Michael, Daniel Robert, David Richard. BS, Cornell U., 1938, M.S., 1939, postgrad., 1939-42; postgrad., N.C. State Coll., summer 1940. Instr. limnology Cornell U., Ithaca, N.Y., 1940-42; sr. math. analyst Arnold Bernard & Co., N.Y.C., 1947-48; statistician Wash. Water Power Co., Spokane, 1949-53; head statistics sect. E.R. Squibb & Sons-Olin, New Brunswick, N.J., 1953-57, mgr. electronic data processing svc. ctr., 1958-63, coord. sci. computations, 1964-65; math. statistician Bur. Ships, Navy Dept., Washington, 1965-67; biometrician Dept. Agr., Beltsville, Md., 1967-72; staff biometrician animal and plant health inspection svc. Dept. Agr., Hyattsville, Md., 1972-87; sr. ptnr. EIC Assocs., Hyattsville, 1981—; cons. Idaho Fish and Game Dept., 1950-60, U.S. Geol. Survey, 1953-58, N.J. Dept. Fish and Game, 1953-60. Contbr. articles to profl. jours. Lt. comdr. USNR, 1942-46, ETO. Mem. NRA, Am. Statis. Assn., Biometric Soc., Entomol. Soc. Am. (cert., emeritus), N.Y. Acad. Scis., Assn. Computing Machinery, Am. Legion, Am. Fisheries Soc., Sigma Xi, Gamma Alpha. Home: 5025 Edgewood Rd College Park MD 20740-4603

EMBREE, NORRIS DEAN, chemist, consultant; b. Kemmerer, Wyo., Nov. 29, 1911; s. Royal Howard and Mary Wallace (Scott) E.; m. Jane Irma Meyers, May 20, 1937 (dec. 1978); children: William Norris, Suzanne Scott, Thomas Dean; m. Barbara Dixon Ely, June 1, 1980. BA, U. Wyo., 1931; PhD, Yale U., 1934. Chemist DPI div. Eastman Kodak Co., Rochester, N.Y., 1934-47, dir. rsch., 1948-58, v.p. rsch. and mfg., 1959-70; dir. rsch. div. health and nutrition Tenn. Eastman Co., Kingsport, 1971-74; ind. chem. cons. Kingsport, 1975—; com. chmn. Nat. Rsch. Coun., Washington, 1960-75; head U.S. del. Internat. Commn. on Fats and Oils, London, 1962-75. Contbr. articles on vacuum technique, fats and vitamins to sci. publs. Bd. dirs. local symphony orch. and other civic orgns., Kingsport, 1970—; chmn. Svc. Corps Ret. Execs., Greeneville and Kingsport, Tenn., 1975—. Mem. AICE, Am. Soc. Biochemistry and Molecular Biology, Am. Oil Chemists Soc. (pres. 1959), Am. Chem. Soc. (sect. chair 1945), Phi Beta Kappa, Sigma Xi. Home and Office: 89 Crown Cir Kingsport TN 37660

EMEAGWALI, DALE BROWN, molecular biologist; b. Balt., Dec. 24, 1954; d. Leon Robert and Johnnie Doris (Baird) Brown; m. Philip Emeagwali, Aug. 15, 1981; 1 child, Ijeoma. BA in Biology, Coppin State Coll., 1976; PhD in Microbiology, Georgetown U., 1981. Teaching asst. sch. medicine Georgetown U., Washington, 1977-80; postdoctoral fellow Nat. Inst. Allergy and Infectious Diseases, Bethesda, Md., 1981-84, Uniformed Svc., U. Health Sci., Bethesda, 1985-86; rsch. assoc. U. Wyo., Laramie, 1986-87; sr. rsch. fellow U. Mich., Ann Arbor, 1987-88, asst. rsch. scientist, 1989-92; rsch. assoc. U. Minn., St. Paul, 1992—. Contbr. articles to sci. jours., chpts. to books. Vol. Sci. Mus. Minn., St. Paul. 1993. Grantee NSF, 1990, Am. Cancer Soc., 1990. Mem. AAAS, Sigma Xi. Achievements include discovery of isoforms for the enzyme kynurenine foramidase in the bacteria Streptomyces parvulus; demonstration of sequence homology between the structural proteins of the Kilham rat virus, and that ras oncogene expression could be inhibited by antisense analogues. Office: U Minn 250 Biosci Ctr 1445 Grotner Ave Saint Paul MN 55108-1095

EMERICK, JOSEPHINE L., engineer. Recipient Young Engineer of the Yr. award Nat. Soc. Profl. Engrs., 1992. Home: 14051 Calcutta Dr Chesterfield MO 63017 Office: Booker Assocs Inc 1139 Olive St Saint Louis MO 63101*

EMERSON, SUSAN, oil company executive; b. Bryan, Tex., Nov. 2, 1947; d. Joseph Nathanial and Lorraine Parks; m. John S. Emerson, June 5, 1970 (div. 1984); children: John H., Christopher P.; m. Gerald W. Parker, Apr. 4, 1985. Owner Emerson Ins. Agy., San Antonio, 1970-84, Emerson Oil Co., San Antonio, 1970—; bd. dirs. Washington Hosp. Ctr. Mem. Washintgon Hosp. Ctr. Women's Aux., 1988—; mem. D.C. Rep. Com., 1991—, alt. del. Rep. Nat. Conv., Washington, 1992, 4th ward committeewoman, 1992; commr. Adv. Neighborhood Commn., Washington, 1990—; 2n v.p. 4D commn., Washington, 1990; founder Boarder Baby Project, 1991—; Rep. candidate for D.C. del. to Congress, 1992. Recipient Sr. Adv. Silver Fox award Wash. Hosp. Women's Aux., 1989, Vol. award, 1990. Mem. LWV, D.C. Hosp. Assn. (trustee 1989), Am. Hosp. Assn. (D.C. del. 1990-92), Vis. Nurses Assn. (bioethic com. 1991—), League Rep. Women, Tex. Breakfast Club. Lutheran. Avocations: travel, gourmet cooking, gardening, needlepoint. Home: Quarters # 6 USSAH Washington DC 20317 Office: Emerson Oil Co Box 497 Washington DC 20317

EMERSON, WILLIAM KARY, engineering company executive; b. Enid, Okla., July 15, 1941; s. Kary Cadmus and Mary Rebecca (Williams) E.; m. Marcie Louise Stogner, Mar. 13, 1965; children: Rebecca A., Phillip W. BS, Okla. State U., 1965, MS, 1974; diploma, Command and Gen. Staff Coll., 1979, Defense Systems Mgmt. Coll., 1980. Commd. 2d lt. U.S. Army, 1965; advanced through grades to lt. col., 1985; prin. program mgr. Honeywell, Inc., Minnetonka, Minn., 1985-90; sr. program mgr. Alliant Techsystems, Inc., Minnetonka, 1990-92; dep. dir. engring. Teledyne Brown Engring. Co., Huntsville, Ala., 1992—. Author: Chevrons, 1983, Encyclopedia of Insignia, 1993; contbr. articles to profl. jours. and ency. Apptd. mem. dist. 281 Sch. Bd. Adv. Coun., Minn., 86-88, Summer Sch. Concept Com., 1988-89. Decorated with Legion of Merit, Bronze Star, 3 Purple Hearts. Fellow Co. Mil. Historians (bd. dirs. 1983-86, editor 1986-92, Miller award 1977); mem. Am. Soc. Mil. Insignia Collectors (Best Nat. Display award 1984, editor jour. 1993—), Am. Def. Preparedness Assn., Assn. U.S. Army, Heritage Club. United Methodist. Avocations: running, fishing, racquet ball, gardening. Home: 124 Kensington Dr Madison AL 35758 Office: Teledyne Brown Engring 300 Sparkman Dr Huntsville AL 35807

EMERSON, WILLIAM STEVENSON, retired chemist, consultant, writer; b. Boston, Mar. 25, 1913; s. Natt Waldo and Marion (Stevenson) E.; m. Flora Millicent Carter, Dec. 12, 1958. A.B., Dartmouth Coll., 1934; Ph.D., MIT, 1937. DuPont fellow U. Ill., Urbana, 1937-38, instr. chemistry, 1938-41; research chemist Monsanto Co., Dayton, Ohio, 1941-44, research group leader, 1944-51, asst. dir. cen. research dept., 1951-54, asst. dir. gen. devel. dept. St. Louis, 1954-56; mgr. cen. research dept. Am. Potash & Chem. Corp., Whittier, Calif., 1956-60; sr. staff assoc. Arthur D. Little, Inc., Cambridge, Mass., 1960-72, ret., 1972. Author: Guide to the Chemical Industry, 1983. Contbr. numerous articles to profl. jours. Patentee in field. Mem. Am. Chem. Soc. (chmn. Dayton sect. 1952), Am. Ornithologists Union, Am. Birding Assn., Phi Beta Kappa, Sigma Xi, Phi Lambda Upsilon, Delta Kappa Epsilon, Alpha Chi Sigma. Republican. Club: Chemists (N.Y.C.). Avocations: fly fishing; birding; golf; squash; reading; philately. Home: Box 30 HC 61 Damariscotta ME 04543-8904

EMERT, GEORGE HENRY, biochemist, academic administrator; b. Tenn., Dec. 15, 1938; s. Victor K. Emert and Hazel G. (Shultz) Ridley; m. Billie M. Bush, June 10, 1967; children: Debra Lea Lipp, Ann Lanie Taylor, Laurie Elizabeth, Jamie Marie. BA, U. Colo., 1962; MA, Colo. State U., 1970; PhD, Va. Tech. U., 1973. Registered profl. chem. engr. Microbiologist Colo. Dept. Pub. Health, Denver, 1967-70; post doctoral fellow U. Colo., Boulder, 1973-74; dir. biochem. tech. Gulf Oil Corp., Merriam, Kans., 1974-79; prof. biochemistry, dir. biomass rsch. ctr. U. Ark., Fayetteville, 1979-84; exec. v.p. Auburn (Ala.) U., 1984-92; pres. Utah State U., Logan, 1992—; adj. prof. microbiology U. Kans., Lawrence, 1975-79. Editor, author: Fuels from Biomass and Wastes, 1981; author book chpt.; contbr. articles to profl. jours. Mem. So. Tech. Coun., Raleigh, N.C., 1985-92; dir. Ala. Supercomputer Authority, Montgomery, 1987-92. Capt. U.S. Army, 1963-66, Vietnam. Named to Educators Hall of Fame, Lincoln Meml. U., 1988. Fellow Am. Inst. Chemists; mem. Auburn Arts Assn., Rotary (Paul Harris fellow, pres., v.p. 1989-90). Republican. Achievements include patent for method for enzyme reutilization. Office: Utah State Univ Old Main Logan UT 84322

EMERY, ALAN ROY, museum director; b. Trinidad, West Indies, Feb. 21, 1939; s. Roy W. and Ruth I. (Jackson); m. Frances D. Ruttan, June 23, 1962; children: Katherine, Timothy. BSc with honors, U. Toronto, Ont., Can., 1962; MSc, McGill U., Montreal, Que., Can., 1964; PhD, U. Miami, 1968. Rsch., teaching asst. various positions Toronto and Montreal, 1959-65; rsch. asst. Inst. of Marine Scis., Miami, Fla., 1965-68; rsch. scientist Ont. Ministry of Natural Resources, Maple, 1968-72; rsch. assoc. Royal Ont. Mus., Toronto, 1969-73, scis. coordinator, 1976-78, assoc. curator, 1973-80, curator, Ichthyology and Herpetology, 1980-83; assoc. prof. U. Toronto, 1976-83; dir. Can. Mus. Nature, Ottawa, 1983—. Author: The Coral Reef, 1981; contbr. articles and chapters to profl. jours. Recipient Citation Sports Fishing Inst., Washington, Marine Environ. award Found. for Ocean Rsch., Toronto, 1986. Mem. Assn. Systematics Collections (pres. 1987-89), Royal Can. Inst. (pres. 1983), Am. Soc. Ichthyologists and Herpetologists (editor, bd. govs. 1976-86). Avocations: photography, writing, music. Office: Canadian Museum of Nature, Box 3443 STN D, Ottawa, ON Canada K1P 6P4

EMERY, MARK LEWIS, civil engineer; b. El Paso, Tex., Dec. 18, 1957; s. Arthur L. and Lavanne (Gardner) E.; m. Shauna L. Bolling; 1 child, Jessica L. BSCE, Tex. Tech U., 1982. Registered profl. engr., Tex. Logging engr.; design engr. Tex. Dept. Transp., Tyler, 1985—. Mem. ASCE.

EMERY, PAUL EMILE, psychiatrist; b. Montreal, May 2, 1922; came to U.S., 1951; s. Esdras Fernand and Julia (Benoit) E.; m. Virginia Olga B. Kennick, July 27, 1979. BA, U. Montreal, 1942, MD, 1948. Diplomate in gen. psychiatry and forensic psychiatry. Staff psychiatrist Austen Riggs Ctr., Stockbridge, Mass., 1958-60; chief mental hygiene VA, Bridgeport, Conn., 1960-62; staff psychiatrist, chief of psychiatry VA, Manchester, N.H., 1988—; pvt. practice Concord, N.H., 1962-85; clin. dir. Ctr. for Stress Recovery, Brecksville, Ohio, 1985-87, dir., 1987-88; med. dir. forensic unit N.H. Hosp., Concord, 1980-82; cons. VA med. Ctr., Manchester, 1962-64, 82-85, pub. health State of N.H., Concord, 1962-71, St. Paul's Sch., Concord, 1971-78; mem. faculty Dartmouth Coll. Med. Sch., 1971—, Western Res. Sch. Medicine, 1985—. Contbr. articles to profl. jours.; author: Trauma Psychology Model of the Mind, 1993. Sec. adv. commn. health and welfare State of N.H., Concord. Capt. U.S. Army, 1953-55. Recipient Salutation plaque N.H. Program on Alcoholism, 1971. Fellow Am. Psychiat. Assn. (life); mem. N.H. Med. Soc. (cert. commendation 1972). Office: VA Med Ctr 718 Smyth Rd Manchester NH 03104-4098

EMERY, VIRGINIA OLGA BEATTIE, psychologist, researcher; b. Cleve., Apr. 9, 1938; d. Joseph P. and Antoinette Pauline (Misja) Kennick; m. Paul E. Emery; children: Tamsan Beattie Tharin, Paul Beattie. BA, U. Chgo., 1962, PhD, 1982; MA, Ind. U., 1973. Lic. psychologist, N.H., Ohio. Adj., clin. asst. prof. psychiatry Dartmouth Med. Sch., Lebanon, N.H., 1983-85; asst. prof. psychology Case Western Res. U., Cleve., 1986-89, asst. clin. prof. psychiatry, 1986-89; clin. assoc. prof. psychiatry Dartmouth Med. Sch., Lebanon, N.H., 1989—; dir. Ctr. on Aging, Health and Soc., Concord and Hanover, N.H., 1989—; mem. com. human devel. Nat. Inst. Mental Health, Adult Devel. & Aging Traineeship, U. Chgo., 1974-76; sub-project dir. Case Western Reserve U. Sch. Medicine, 1986-90; sec. women's faculty assn. Case Western Reserve U., 1987-89. Author: Language and Aging, 1985, Pseudodementia, 1988; contbr. articles to profl. jours. Bd. dirs., pres. Frontiers of Knowledge Trust, Concord, N.H., 1990—. Recipient Rsch. prize Am. Aging Assn., 1983, Havighurst Prize for Aging Rsch., U. Chgo., 1984, Rsch. grant Western Reserve Coll., 1986-87, NIMH Mental Health Clin. Rsch. Ctr. grant. Fellow N.H. Psychol. Orgn. (bd. dirs. 1991-93, chair com. acad. rsch. interests 1992—, Riggs Disting. Contbn. award 1991); mem. Internat. Psychiat. Rsch. Soc., Internat. Psychogeriatric Assn., Am. Psychol. Assn. (student rsch. award 1984), Gerontol. Soc. Am. (disting. creative contbn. award behavioral and social sci. sect. 1989), Boston Soc. Gerontol. Psychiatry, Acad. Psychosomatic Medicine, 1993—. Home: 15 Buckingham Dr Bow NH 03304-5207 Office: Dartmouth Med Sch Dept Psychiatry Lebanon NH 03756

EMGE, THOMAS MICHAEL, electrical engineer; b. Milw., Apr. 7, 1957; s. Thomas Ray and Bernice Margaret (Weiss) E. BSEE, U. Evansville, 1981. Elec. design engr. Pan Am. World Svcs., Patrick AFB, Fla., 1981-84, Tex. Instruments, Inc., McKinney, 1984—. Vol. St. Judes Children's Hosp., 1992—. Recipient Spl. Merit award U. Evansville, 1981. Mem. AAAS, N.Y. Acad. Scis. Home: 1106 Leland St McKinney TX 75069 Office: Tex Instruments PO Box 801 MS8042 McKinney TX 75069

EMLING, WILLIAM HAROLD, metallurgical engineer; b. Erie, Pa., Feb. 28, 1957; s. William John and Joan Marie (Mahoney) E.; m. Jean Marie Herzog, July 4, 1981; children: Brian William, Patrick George, Jennifer Marie. BS in Metall. Engring., Case Western Res. U., 1979. Quality control metallurgist Nat. Steel Corp., Granite City, Ill., 1979-81; rsch. engr. LTV Steel Co., Pitts., 1981-83; sr. rsch. engr. LTV Steel Co., East Chgo., Ind., 1983-86, Cleve., 1986-89; mgr. continuous casting LTV Steel Co., Independence, Ohio, 1989—; adv. bd. Dept. Material Sci., Case Inst. Tech., 1992—. Contbr. articles to profl. jours. Chmn. Pack 3322 Boy Scouts Am., Hudson, Ohio, 1990—; Little League coach, Hudson, 1988-89, 93. Mem. AIME (Charles H. Herty award 1989, 91, Robert W. Hunt award, 1990), Am. Iron and Steel Inst., Case Inst. Tech. Alumni Assn. (class agt. 1987—). Office: LTV Steel Co Tech Ctr 6801 Brecksville Rd Independence OH 44131

EMMANUEL, JORGE AGUSTIN, chemical engineer, environmental consultant; b. Manila, Aug. 28, 1954; came to U.S., 1970; s. Benjamin Elmido and Lourdes (Orozco) E.; 1 child, Andres Layanglawin. BS in Chemistry, N.C. State U., 1976, MSChemE, 1978; PhD in Chem. Engring., U. Mich., 1988. Registered profl. engr., Calif., environ. profl.; cert. hazardous materials mgr. Process engr. Perry Electronics, Raleigh, N.C., 1973-74; rsch. asst. N.C. State U., Raleigh, 1977-78; rsch. chem. engr. GE Corp. R & D Ctr., Schenectady, N.Y., 1978-81; Amoco rsch. fellow U. Mich., Ann Arbor, 1981-84; sr. environ. analyst TEM Assocs., Inc., Emeryville, Calif., 1988-91; pres. Environ. & Engring. Rsch. Group, Hercules, Calif., 1991—; environ. cons. to the Philippines, UN Devel. Program, 1992; rsch. assoc. U. Calif., Berkeley, 1988-90. Contbr. articles to profl. jours. Mem. Assn. for Asian Studies, Ann Arbor, 1982-88; sec. Alliance for Philippine Concerns, L.A., 1983-91; assoc. Philippine Resource Ctr., Berkeley, 1988-92; bd. dirs. Arms Control Rsch. Ctr., San Francisco, 1990—. N.C. State U. grantee, 1976, Phoenix grantee U. Mich., 1982. Mem. NSPE, AAAS, Air and Waste Mgmt. Assn., Calif. Acad. Scis., N.Y. Acad. Sci., Filipino-Am. Soc. Architects and Engrs. (exec. sec. 1989-90, Svc. award 1990). Avocations: classical guitar, ethnomusicology, Asian studies. Office: The Environ & Engring Rsch Group PO Box 5544 Hercules CA 94547

EMMEL, BRUCE HENRY, mathematics educator; b. St. Cloud, Minn., Jan. 8, 1942; s. Henry Joseph and Mary Ann Emily (Kangas) E.; m. Phyllis Wanda Campbell, Aug. 29, 1982; children: Debra Lynn Huber, Kathi Marie, Brent Boyd, Daniel Henry Huber, Brandi Rose. BS, St. Cloud State U., 1967; MA in Edn., Ball State U., 1973. Cert. vocat. tchr., Minn. Tchr. Lincoln Jr. High Sch., Hibbing, Minn., 1967-70, West Concord (Minn.) High Sch., 1970-72; vocat. tchr. Moorhead (Minn.) Tech. Coll., 1972-90; tchr. Moorhead Pub. Schs., 1984—; mem. Dist. Math. Com., Moorhead, 1978—. Comdr. Fargo (N.D.) CAP, 1984-86, 89; dir. pub. affairs N.D. CAP, Mandan, 1986-89; precinct chmn. Dem. Com., Moorhead, 1968-90. Mem. Nat. Coun. Tchrs. Math., Kappa Delta Pi. Congregationalist. Avocations: flying, aircraft builder, hunting, walking, traveling. Home: 1121 3d St S Moorhead MN 56560-4015 Office: Moorhead Sr High Sch 2300 4th Ave S Moorhead MN 56560-3298

EMMERT, RICHARD EUGENE, professional association executive; b. Iowa City, Iowa, Feb. 23, 1929; s. Frank Thomas and Okie Leona (Seydel) E.; m. Marilyn Ruth Marner, June 19, 1949; children: Debra Sue Emmert Warrington, Andrea Gale Emmert Mazzuca, Lisa Alison Emmert Grant. B.S., U. Iowa, 1951; M.S., U. Del., 1952, Ph.D., 1954; DSc (hon.), Manhattan Coll., 1992. Supt. mfg. textile fibers dept. E.I. du Pont de Nemours & Co., Martinsville, 1966-67; mfg. mgr. textile fibers dept. E.I. du Pont de Nemours & Co., Wilmington, Del., 1967-69; mgr. engring. tech. and materials rsch. E.I. du Pont de Nemours & Co., Wilmington, 1969-73, dir. rsch. and devel. pigments dept., 1973-75, dir. instrument products, dept. photo products, 1975-77, dir. electronic products, photo products dept., 1977-79, gen. mgr. textile fibers dept., 1979-80, v.p. corp. plans, 1980-83, v.p. electronics dept., 1984-87; exec. dir. AICE, 1988—; Trustee U. Del. Rsch. Found., Newark, 1987—. Author: Gas Absorption and Solvent Extraction, 1963; contbr. articles to profl. jours. Vice chmn. Stanton Sch. Bd., Del., 1961-64; chmn. adv. bd. Coll. Engring., U. Iowa, Iowa City, 1974-80; chmn. adv. bd. dept. chem. engring. U. Calif., Berkeley, 1978-87, chmn., 1982-83; co-chmn. adv. bd. dept. chem. engring. U. Del., Newark, 1984-88; bd. trustees Med. Ctr. Del., Wilmington, 1983—; dir. Del. Found. for Phys. Edn., Wilmington, 1984—. With U.S. Army, 1954-56. Recipient 1st Disting. Engring. Alumni award U. Del., 1984, Medal of Distinction, 1993, Disting. Alumni award U. Iowa, 1988. Mem. NAE, Am. Inst. Chem. Engrs., Am. Chem. Soc., Del. Tennis Assn. (pres. 1982-83), Tau Beta Pi, Sigma Xi, Phi Eta Sigma. Republican. Presbyterian. Avocation: tennis. Home: 162 75th St Avalon NJ 08202-1007 Office: AICE 345 E 47th St New York NY 10017-2330

EMMONS, WILLIAM DAVID, chemist; b. Mpls., Nov. 18, 1924; married; 3 children. BS, U. Minn., 1947; PhD in Chemistry, U. Ill., 1951. Sr. chemist Rohm & Haas Co., Spring House, Pa., 1951-52, group leader organic chemistry, 1952-57, lab. head, 1957-61, rsch. supr., 1961-72, dir. pioneering rsch., 1973—. Editor Organic Syntheses, 1961-69. Mem. Am. Chem. Soc. (Earle B. Barnes Leadership in Chem. Research Mgt. award 1993). Achievements include research in peracids, small ring heterocycles, organophosphorus chemistry, polymers and surface coatings. Office: Rohm Haas Com Research Lab Spring House PA 19477*

EMONT, GEORGE DANIEL, healthcare executive; b. Columbus, Ohio, July 23, 1958; s. Milton David and Marietta (Gruenbaum) E.; m. Jill Ellen Preminger, July 6, 1986; children: Margo, Jacob. AB, Oberlin Coll., 1980; MBA, U. Chgo., 1985. Cons. Data Resources Inc., N.Y.C., 1980-82, mng. cons., 1982-83; bus. assoc. Baxter Healthcare Corp., Deerfield, Ill., 1985-86, mktg. mgr., 1986-87, sr. mktg. mgr., 1987-88; assoc. IAI Venture Capital Group, Mpls., 1988-90; v.p. ops. OncoTherapeutics Inc., Mpls., 1990-91, acting pres., 1991-92, exec. v.p. bus. devel., 1992—. Home: 179 Lake St Metuchen NJ 08840 Office: OncoTherapeutics Inc 1002 Eastpark Blvd Cranbury NJ 08512

EMORI, RICHARD ICHIRO, engineering executive; b. Tokyo, May 12, 1924; s. Yasuhei and Yae Emori; m. Hiroko Hirade, Mar. 6, 1966; 1 child, John Y. BSME, U. Tokyo, 1949, DrEngrME, 1966; MSME, U. Mich., 1952, MS in Engring. Mechanics, 1958. Registered profl. engr., Calif. Stress analyst Clark Equipment Co., Jackson, Mich., 1953-55; sr. project engr. GM Corp., Detroit and Santa Barbara, Calif., 1955-67; staff engr. IBM, San Jose, Calif., 1967-68; asst. prof. UCLA, 1968-73; prof. Seikei U., Tokyo, 1973-90, prof. emeritus, 1990—; pres. Kalok Japan, Tokyo, 1990—; auto accident reconstructionist for traffic cts., Japan, 1973—; safety cons. Honda Motors Co., Tokyo, 1975—. Author: Scale Models in Engineering, 1974, Automobile Accidentology, 1975, Applied Accidentology, 1983. Recipient Ministry of Transp. award Sumitomo Found., 1987. Mem. ASME, Soc. Automotive Engineering, Japan Soc. Mech. Engring., Soc. Automotive Engring. Japan. Avocations: surfing, motorcycling, skating, jogging. Office: 1-1-5 Kami-Renjaku, #212 Mitaka, Tokyo 181, Japan

EMPLIT, RAYMOND HENRY, electrical engineer; b. Darby, Pa., May 2, 1948; s. Henry Raymond and Caroline Winifred (Parker) E.; m. Patricia Jean Jezl, Aug. 7, 1976; children: Eric, Susan. BS summa cum laude in Engring., U. Pa., 1978, MS in Engring., 1979. Engr. Custom Controls Co., Broomall, Pa., 1972-75, tech. dir., 1975-78, v.p., 1979-82; chief engr. Robertshaw Controls, Havertown, Pa., 1982-87; pres. Electronic Devel. Corp., Edgemont, Pa., 1987-89; project engr Gt. Lake Instruments, Edgemont, Pa., 1989-91; pres. Interlink Techs., Broomall, Pa., 1991. Patentee indsl. level instrumentation in U.S. and Can. With U.S. Army, 1968-71. Recipient Hugo Otto Wolf Meml. prize U. Pa., 1978. Mem. IEEE, U.S. Power Squadron, Instrument Soc. Am., Eta Kappa Nu, Tau Beta Pi. Republican. Avocations: reading, wine, investing, boating. Home: 71 Sweetwater Rd Glen Mills PA 19342-1710 Office: Interlink Techs 1005 Sussex Blvd Broomall PA 19008

EMPSON, CHERYL DIANE, validation engineer; b. Wichita, Kans., Jan. 13, 1962; d. Charles Lee Empson and Luella Lorajean (Peterson) Eshelman. BS in Chem. Engring., U. Mo., 1984, MDA, Baker U., 1992. Mem. tech. svcs. staff Genentech, South San Francisco, Calif., 1984-86; rsch. asst. Sanoti Animal Health, Lenexa, Kans., 1989-90; process devel. engr. Pfizer Animal Health, Lee's Summit, Mo., 1990-93; sr. qualification validation specialist Triad Techs., Inc., New Castle, Del., 1993—. Ct. appointed children's advocate, Jackson County, Kansas City, Mo., 1992. Monsanto fellow, Columbia, 1986; Mag. fellow U. Mo., Kansas City, 1988. Mem. Internat. Soc. Pharm. Engrs. Avocations: golf, tennis. Office: Triad Techs Inc 101 Centerpoint Blvd New Castle DE 19720

EMRICK, DONALD DAY, chemist, consultant; b. Waynesfield, Ohio, Apr. 3, 1929; s. Ernest Harold and Nellie (Day) E.; B.S. cum laude, Miami U., Oxford, Ohio, 1951; M.S., Purdue U., 1954, Ph.D., 1956 Grad. teaching asst. Purdue U., Lafayette, Ind., 1951-55; with chem. and phys. research div. Standard Oil Co. Ohio, 1955-64, research asso., 1961-64; cons., sr. research chemist research dept. Nat. Cash Register Co., Dayton, Ohio, 1965-72, chem. cons., 1972—. Mem. AAAS, Am. Chem. Soc., Phi Beta Kappa, Sigma Xi. Patentee in Field. Contbr. articles to profl. jours. Home: 4240 Lesher Dr Dayton OH 45429-3042

EMSLIE, WILLIAM ARTHUR, electrical engineer; b. Denver, Oct. 30, 1947; s. William Albert and Hazel Esther (Niles) E.; m. Tracey Jane Palmer, Feb 22, 1975; children: David Barrett, Andrew Niles, Charles William, Alexis Claire. BSEE, U.S. Naval Acad., 1971; MSEE, Mich. State U., 1972. Registered profl. engr., Colo. Commd. ensign USN, 1971, advanced through grades to lt., 1975; with USNR, 1978—, advanced through grades to cmdr., 1992; energy conversion engr. Pub. Svc. Co. of N.Mex., Albuquerque, 1978-79; mgr. engr. Horizon Tech., Ft. Collins, Colo., 1979-80; staff engr. Platte River Power Authority, Ft. Collins, 1980-85, planning supr., 1985-89, mgr. quality improvement, 1989-92, exec. engr., 1992—; mem. renewable task force Electric Power Rsch. Inst., Palo Alto, Calif., 1982-85, mgt. com. Western Energy Supply and Transmission Assocs., Albuquerque, 1991—; chmn. Am. Pub. Power Assn. Demonstration of Energy Efficient Devels. Bd., Washington, 1992—. Chmn. campaign Ft. Collins Area United Way, 1986, pres. bd. dirs., 1988; chmn. Sch. Mill Levy Tax Com., Ft. Collins, 1988. Grantee State of Colo., Am. Public Power Assn./Demonstration of Energy Efficient Devels., Western Energy Supply and Transmission Assocs., U.S. Dept. Energy, City of Colorado Springs, 1986-90. Mem. IEEE, Foothills Rotary of Ft. Collins. Achievements include research in photovaltics which provided a solar insolation assessment that is more accurate than the typical meterological year and a comprehensive evaluation of the effectiveness of 4 types of photovaltic systems. Home: 825 E Pitkin St Fort Collins CO 80524 Office: Platte River Power Authority 2000 E Horsetooth Rd Fort Collins CO 80525

ENDAHL, LOWELL JEROME, retired electrical cooperative executive; b. Jerauld County, S.D., July 2, 1922; s. John Martin and Olga A. (Bunde) E.; m. Vronna Belle Lee, Oct. 16, 1948; children: John Raymond and Jay Jerome (twins), Mark Arnold. B.S. in Agrl. Engring., S.D. State U. Power use adviser Tri-County Electric Assn., Plankinton, S.D., 1948-51; mgr. power use dept. Sioux Valley Empire Electric, Colman, S.D., 1951-54; mgr., mem. services Nat. Rural Electric Coop. Assn., Washington, 1954-75; mgr. energy research and devel. Nat. Rural Electric Coop. Assn., 1975-89; ret. Nat. Rural Electric Coop. Assn., Washington, 1989; cons., vol. Nat. Rural Electric Coop. Assn., 1990—; U.S. rep. UN Working Party on Rural Electrification, Belgium and Netherlands, 1968; cons. Internat. Program div. NRECA, Ecuador and Colombia, 1969, cons. Internat. program div., Vietnam, 1970. Columnist: rural Electrification, 1975-89. Pres. Luther Pl. Meml. Ch., Washington, 1976-78. Capt. USMC, 1943-46, PTO. Fellow Am. Soc. Agrl. Engrs. (chmn. EPP div. 1873-74 George W. Kable Electrification award, vice-chmn. editorial bd. Agrl. Engr. 1981-82, past pres. Md.-D.C. chpt.). Office: Nat Rural Electric Coop Assn 1800 Massachusetts Ave NW Washington DC 20036-1806

ENDERS, ALLEN COFFIN, anatomy educator; b. Wooster, Ohio, Aug. 5, 1928; s. Robert Kendal and Abbie Gertrude (Crandell) E.; m. Alice Hay, June 15, 1950 (div. Dec. 1975); children: Robert H., George C., Richard S., Gregory H.; m. Sandra Jean Schlafke, Aug. 5, 1976. AB, Swarthmore Coll., 1950; AM, Harvard U., 1952, PhD, 1955. From asst. prof. to assoc. prof Rice Inst., Houston, 1954-63; from assoc. prof. to prof. Washington U., St. Louis, 1963-75; prof., chmn. dept. human anatomy U. Calif., Davis, 1976-86; cons. NIH, Bethesda, Md., 1964-68, 70-73, 76-80, 83-93. Author: (with others) Bailey's Microscopic Anatomy, 1984; editor: Delayed Implantation, 1964; contbr. numerous articles on anatomy and reproduction to profl. jours. Nat. pres. Perinatal Research Soc., 1981. Grantee NIH, 1959—. Fellow AAAS; mem. Am. Assn. Anatomists (v.p. 1980-82, pres. 1983-84), Soc. Study Reprodn., Am. Soc. Cell Biology. Home: 39707 Barry Rd Davis CA 95616-9415 Office: U Calif Sch of Medicine Cell Biology and Human Anatomy Davis CA 95616

ENDO, BURTON YOSHIAKI, research plant pathologist; b. Castroville, Calif., Feb. 5, 1926; s. Wakichi and Dan (Kato) E.; m. Joyce Stephens, Apr. 29, 1933 (dec. Dec. 1985); children: Martha Hess, Carol Bowen; m. Helen McGovney, Mar. 3, 1990. BS, Iowa State U., 1951; MS, N.C. State U., 1955, PhD, 1958. Nematologist USDA, ARS, Jackson, Tenn., 1958-63; rsch. plant pathologist, Crops Protection Rsch. Br. USDA, ARS, Beltsville, Md., 1963-74; chief nematology lab. Plant Protection Inst., USDA, Beltsville, 1964-75, chmn., 1975-88; rsch. plant pathologist Nematology Lab. Plant Scis. Rsch. Inst., Beltsville, 1988—. Author: (with others) Plant Parasitic Nematode, 1971, Cyst Nematode, 1986, Introduction to Crop Protection, 1979, Electron Microscopy of Plant Pathogens, 1990, The Soybean Cyst Nematode, 1991. With U.S. Army, 1951-53. Fellow Soc. Nematologists (sec. 1968-71, v.p. 1971-72, pres. 1972-73), Washington Acad. Sci., Rotary Internat. Home: 1010 Jigger Ct Annapolis MD 21401 Office: Nematology Lab PSI Rm 153 B011A BARC-W Beltsville MD 20705

ENDO, HAJIME, research chemist; b. Tateyama, Chiba, Japan, May 21, 1950; s. Mamoru and Teiko (Akiyama) E.; m. Ann Carol Slone, Jan. 2, 1981; children: Akira, Makoto, Takashi. BA, Utsunomiya U., Tochigi, Japan, 1973; MA in Chemistry, Rice U., Houston, 1977, PhD in Chemistry, 1977. Grad. fellow Rotary Internat. Found., 1973, R. A. Welch Found., 1974; rsch. fellow chemistry Rice U., 1977; rsch. fellow Inst. Phys. Chemistry Göttingen (Fed. Republic of Germany) U., 1977-78; staff chemist Procter & Gamble Co., Cin., 1978-81; tech. brand mgr. Procter & Gamble Far East, Inc., Osaka, Japan, 1981-85; sect. chief Industry Unit Hoechst Japan Ltd., Tokyo, 1985, dep. mgr. Industry Unit, 1988-91; assoc. lab dir. Advanced Tech. Unit Hoechst Japan Ltd., Kawagoe, Saitama, Japan, 1991-92, lab. dir., 1993—; sect. chief Hoechst AG/GB-E, Frankfurt, Fed. Republic of Germany, 1986-87. Author procs. 16th Symposium on Combustion, 1976; contbr. articles to profl. jours. Fellow Alexander von Humboldt Found.; mem. Am. Chem. Soc., Japan Oil Chemists Soc. Avocations: photography, driving, cycling, traveling. Home: 5-7-30 Kosenba-cho, Kawagoe-shi Saitama-ken 350, Japan Office: Hoechst Japan Ltd/ATL, 1-3-2 Minamidai, Kawagoe-Shi Saitama-ken 350, Japan

ENDOH, RYOHEI, cardiologist; b. Omachi, Nagano, Japan, Apr. 1, 1954; s. Shinzi Endoh and Miharu Momose. M of Medicine, Shinshu U., Matsumoto, Japan, 1981, MD, 1990. Intern Shinshu U. Hosp., Matsumoto, 1981-82, resident, 1983-84, cardiologist, 1985-87; intern Matsumoto Nat. Hosp., 1982-83; cardiologist Nagano Red Cross Hosp., 1984-85; head physician internal medicine Suwa Red Cross Hosp., 1987; head physician cardiology Showa Inan Gen. Hosp., Komagane, 1987—. Contbr. articles to Japan Circulation Jour., Artery, Japan Jour. Applied Physiology, Tex. Heart Inst. Jour. Fellow Japanese Soc. Internal Medicine, Japanese Circulation Soc.; mem. Japanese Assn. Acute Medicine, Japanese Coll. Angiology. Office: Showa Inan Gen Hosp, 3230 Akaho, Komagane-shi Nagano, Japan 399 41

ENEGESS, DAVID NORMAN, chemical engineer; b. Winchester, Mass., Aug. 25, 1946; s. Norman Leonard and Shirley Mildred (Lewis) E.; m. Jane Deborah Enegess, June 20, 1970; children: Deborah Marie, Christine Kerry. BSChemE, Tufts U., 1968, MSChemE, 1971. Registered profl. engr., Conn., Mich. Nuclear systems engr. Combustion Engring. Inc., Windsor, Conn., 1972-75, supr. radwaste systems devel. and design, 1975-77; project engr. Hazardous Waste Systems Group, WP Corp., Ramsey, N.J., 1977-79, mgr. projects, 1979-80, gen. mgr., 1980-82; co-founder, v.p. Waste Chem. Corp., Paramus, N.J., 1982-88, Envirogen, Inc., Princeton, N.J., 1988—; mem. EPA Bioremediation Action Adv. Com., Washington, 1990—, Waste Mgmt. Symposia Tech. Program Com., Tucson, 1980-88, Oak Ridge (Tenn.) Nat. Lab. Waste Form Adv. Com., 1988-90; adviser Tufts U. Career Adv., Medford, Mass., 1990—; bd. dirs. Vapex Corp. Contbr. articles to profl. jours. Mem. Wyckoff (N.J.) Environ. Commn., 1986-88, Com. on Extended Learning, Wyckoff, 1984-85; mem., co-chair High Sch. Music Program Fund Raising, Wyckoff, 1987-90; coach Jr. Soccer League, Wyckoff, 1981-88. 1st lt. U.S. Army, 1969-72. NSF fellow, 1968-69. Mem. ASME (mem. radwaste systems com. 1978-85), Am. Inst. Chem. Engrs., Atomic Indsl. Forum (rep. 1976-77), Am. Nuclear Soc., Hazardous Materials Control Rsch. Inst., Applied Biotreatment Assn. (dir. 1989-90). Achievements include patents for a static device for separations of gas mixtures into individual components, system using membrane device for radwaste processing, and process involving thin-film evaporation for radioactive wastewaters. Office: Envirogen Inc 4100 Quakerbridge Rd Lawrenceville NJ 08648-4702

ENFLO, ANITA MARGARITA, physicist, chemist; b. Helsinki, Finland, Aug. 2, 1943; arrived in Sweden, 1972; d. Allan and Edit (Björk) Henriksson; m. Bengt Enflo, July 29, 1972; children: Björn, Kerstin, Helena. PhD in Chemistry, U. Helsinki, 1970; PhD in Physics, U. Stockholm, 1986. Asst. prof. U. Helsinki, 1966-71; rsch. assoc. Inst. Theoretical Physics U. Stockholm, 1972—; prin. physicist Swedish Radiation Protection Inst., Stockholm, 1986—. Office: U Stockholm, Inst Theoretical Physics, S-11346 Stockholm Sweden

ENG, RICHARD SHEN, electrical engineer; b. Canton, Peoples Republic of China, Dec. 11, 1930; came to the U.S., 1947; s. Gooy Yu and Tuck Yin (Chin) Ng; m. Annette Kathleen Chang, Dec. 28, 1958; children: Victor John, Douglas Walter, Patricia Mona Lan. BEE cum laude, CCNY, 1954; MEE, MIT, 1955; PhD, Poly. Inst. N.Y., 1971. Sr. engr. Sperry Gyroscope Corp., Great Neck, N.Y., 1955-65; rsch. engr. Grumman Aerospace Corp., Bethpage, N.Y., 1965-69; mem. staff MIT Lincoln Lab., Lexington, 1971-76, mem. tech. staff, 1986—; prin. scientist Laser Analytics Inc., Bedford, Mass., 1976-81, Raytheon Co., Sudbury, Mass., 1981-86. Contbr. articles to profl. jours. Active election presdl. com. Dem. Cen. Com., Jackson Heights, N.Y.C., 1960. NSF fellow, 1969-71. Mem. Optical Soc. Am. (organizer, jour. article reviewer 1971—), Photonics Soc. Chinese Ams. (treas. 1992—), Hanscom Tennis Club (sec. 1991—). Achievements include patents for birefringent lenses, long life CO2 laser cathode, spin flip laser, differential isotopic CO2 absorption spectrometer for metabolic rate detection. Office: MIT Lincoln Lab 244 Wood St Lexington MA 02173

ENGDAHL, BRIAN EDWARD, psychologist; b. Owatonna, Minn., July 3, 1952; s. Gilbert Donald and Marion Eloise (Scofield) E.; m. Raina Elaine Eberly, July 9, 1977; 1 child, Rebecca Raina. PhD, U. Minn., 1980. Psychologist, coord. VA Med. Ctr., Mpls., 1980—; clin. asst., assoc. prof. Dept. Psychology U. Minn., Mpls., 1980—. Contbr. chpts. to books and articles to profl. jours. Grantee VA 1989-91, 91-94. Mem. AAAS, APA, Am. Psychol. Soc. Home: 1376 Summit Ave Saint Paul MN 55105 Office: VA Med Ctr Psychology One Veterans Dr Minneapolis MN 55417

ENGEL, ANDREW GEORGE, neurologist; b. Budapest, Hungary, July 12, 1930; s. Alexander and Alice Julia (Gluck) E.; m. Nancy Jean Brombacher, Aug. 15, 1958; children: Lloyd William, Andrew George. B.Sc., McGill U., 1953, M.D., 1955. Diplomate: Am. Bd. Internal Medicine, Am. Bd. Psychiatry and Neurology. Intern Phila. Gen. Hosp., 1955-56; sr. asst. surgeon, clin. asso. USPHS, NIH, Bethesda, Md., 1958-59; fellow in neuropathology Columbia U., N.Y.C., 1962-64; with Mayo Clinic, 1956-57, 60-62; cons., 1965—; prof. neurology Mayo Med. Sch., Rochester, Minn., 1973—; William L. McKnight-3M prof. neurosci. Mayo Med. Sch., 1984—; mem. sci. adv. com. Muscular Dystrophy Assn., 1973—; mem. rev. com. NIH, 1977-81. Mem. editorial bd.: Neurology, 1973-77, Annals of Neurology, 1978-84, 1990—, Muscle and Nerve, 1978—, Jour. Neuropathology, 1981-83; Contbr. articles to profl. jours. Served with USPHS, 1957-59. Mem. Am. Acad. Neurology, Am. Neurol. Assn., Am. Assn. Neuropathologists, Am. Soc. Cell Biology, Soc. Neurosci., AAAS. Home: 2027 Lenwood Dr SW Rochester MN 55902-1051 Office: Mayo Clinic 100 1st Ave SW Rochester MN 55905-0001

ENGEL, CHARLES ROBERT, chemist, educator; b. Vienna, Austria, Jan. 28, 1922; s. Jean and Lucie (Fuchs) E.; m. Edith H. Braillard, Aug. 6, 1951; children: Lucie Tatiana Engel Berthoud, Christiane Simonne, Francis Pierre, Marc Robert. BA, U. Grenoble, 1941; MSc, Swiss Fed. Inst. Tech., Zurich, 1947, DSc, 1951; State-DSc, U. Paris, 1970. Research fellow, asst. Swiss Fed. Inst. Tech., Zurich, 1948-51; asst. prof. med. research Collip Med. Research Lab. U. Western Ont., London, 1951-55, asso. prof. med. research, 1955-58; hon. spl. lectr. chemistry, dept. chemistry U. Western Ont., 1951-58; prof. chemistry Laval U., Quebec, Que., 1958-90; vis. prof. Inst. de Chimie des Substances Naturelles CNRS, Gif-sur-Yvette, France, 1966-67. Mem. editorial bd. Steroids, 1964-91; hon. editorial bd. Current Abstracts of Chemistry, 1971-72, Index Chemicus, 1971-72; mem. editorial adv. bd. Can. Jour. Chemistry, 1974; editor Can. Jour. Chemistry, 1986-91.; mem. French govtl. commn. chem. terminology, 1992—. Lt. for Can.-Que., Equestrian Order of Holy Sepulchre of Jerusalem, 1970-89, mem. Grand Magisterium, Vatican, 1989-93; bd. dirs. Cath. Culture Ctr., London, Ont. Decorated comdr. Equestrian Order of Holy Sepulchre of Jerusalem, 1964, comdr. with star, 1970, knight grand cross, 1973; knight Legion of Honour, France, 1985; medal Austrian Ministry of Edn. Fellow Chem. Inst. Can. (chmn. organic div. 1965-66, exec. med. div. 1968-79), Royal Soc. Chemistry (London); mem. Am., Swiss, French chem. socs., Canadian Biochem. Soc., French-Can. Assn. for the Advancement of Scis., N.Y. Acad. Scis., Order Chemists Que., Association Canadienne-française pour l'Avancement des Sciences, Sigma Xi. Office: Laval U, Dept Chemistry, Quebec, PQ Canada G1K 7P4

ENGEL, JAMES HARRY, computer company executive; b. Rahway, N.J., June 11, 1946; s. August Joseph and Laura Ellen (Rigright) E. AAS in Computer Sci., Union County Tech. Inst., 1966. Systems analyst State of N.Mex., Santa Fe, 1977; programmer analyst Los Alamos (N.Mex.) Nat. Lab., 1977-80; database analyst EG&G, Morgantown, W.Va., 1980-82; systems programmer U.S. Army AMCCOM, Dover, N.J., 1982-83, Cray Rsch. Inc., Mendota Heights, Minn., 1983-88; mgr. ECF support Grumman Data Systems, Houston, 1988—. Active Clean Water Action, Houston, 1990—, Assn. for Community TV, Houston, 1990—, Whale Adoption Project, Houston, 1990—, Tex. State Trooper Assn., Houston, 1990—. With U.S. Army, 1966-68, Vietnam. Mem. Nat. Mgmt. Assn. Avocation: boating. Home: 16811 Dale Oak Way Houston TX 77058 Office: Grumman Data Systems 12000 Aerospace Ave Houston TX 77034-5567

ENGEL, JEROME, JR., neurologist, neuroscientist, educator; b. Albany, N.Y., May 11, 1938; s. Jerome and Pauline (Feder) E.; m. Catherine Margaret Lambourne, Feb. 26, 1967; children: Sean, Jesse, Anasuya. BA, Cornell U., 1960; MD, Stanford U., 1965, PhD in Physiology, 1966. Diplomate Nat. Bd. Med. Examiners, Am. Bd. Qualification in EEG, Nat. Bd. Psychiatry and Neurology. Intern in medicine Ind. U., Indpls., 1966-67; resident in neurology Albert Einstein Coll. Medicine, Bronx, N.Y., 1967-68, 70-72; resident in EEG Nat. Hosp. Nervous and Mental Disease Queen Sq., London, 1971, Maudsley Hosp., London, 1972; attending neurologist, dir. electroencephalography labs. Bronx Mcpl. Hosp. Ctr., Hosp. Albert Einstein Coll. Medicine, 1972-76; attending neurologist, chief of epilepsy, clin. neurophysiology UCLA Hosp. and Clinics, 1976—; assoc. investigator lab. nuclear medicine of Lab. Biomed. and Environ. Scis. UCLA Med. Ctr., 1981—; staff assoc. NINDS NIH Lab. of Perinatal Physiology, San Juan, P.R., vis. asst. prof. dept. physiology and biophysics Sch. Medicine U. P.R., 1968-69, Lab. of Neural Control, Bethesda, Md., 1969-70; asst. prof. neurology Albert Einstein Coll. Medicine, Bronx, 1972-76, neuroscience 1974-76; assoc. prof. neurology Sch. Medicine UCLA, 1976-80, anatomy, 1977-80, prof. neurology, anatomy and cell biology, 1980—; assoc. investigator Lab. Nuclear Medicine Lab. Biomedical and Environ. Scis., 1981—; chmn. Internat. and Coop. Projects Study Sect. NIH, 1989-90, mem. Biomed. Scis. Study Sect., 1985-89, chmn. 1988-89; vis. prof. dept. anatomy Sydney U., 1984. Author: Epilepsy and Positron CT, Clinical Relevance for Diagnosis of Epilepsy, 1985, Surgical Treatment of the Epilepsies, 1987, Seizures and Epilepsy, 1989, Surgical Treatment of Epilepsies, 1993, (with others) Neurotransmitters, Seizures and Epilepsy II, 1984, Neurotransmitters, Seizures and Epilepsy III, 1986, The Epileptic Focus, 1987, Fundamental Mechanisms of Human Brain Function, 1987, Clinical Use of Emission Tomography in Focal Epilepsy, Current Problems in Epilepsy, Vol. 7, 1990, Neurotransmitters in Epilepsy, 1992, Molecular Neurobiology and Epilepsy, 1992; contbr. over 77 chpts. to books including Functional Brain Imaging, 1988, Anatomy of Epileptogenesis, 1989, EEG Handbook, rev. series vol. 4, 1990, Comprehensive Epileptology, 1990, Generalized Epilepsy, 1990, Neurotransmitters in Epilepsy, Epilepsy Research (Supplement), 1992, Molecular Neurobiology and Epilepsy; contbr. over 440 articles to profl. jours. including, New Issues in Neurosci., Neurology, Jour. Neurosurg., Jour. Epilepsy, Epilepsia, Can. Jour. Neurol. Sci., Radiology, Jour. Cerebral Blood Flow Metabolism, Acta Neurochirugica, Jour. Clin. Psychiatry; chief editor Advances in Neurobiology of Epilepsy, 1989-91; assoc. editor Jour. Clin. Neurphysiology, 1983—, Epilepsy Rsch., 1985—, Epilepsy Advances, 1985-87, Brain Topography, 1990—. Active profl. adv. bd. Epilepsy Found. Am., 1979-87; chmn. organizing com. Workshop on Neurobiology of Epilepsy Internat. League Against Epilepsy, 1988-91. Lt. comdr. USPHS, 1968-70. Recipient N.Y. State Regents scholarship, 1956-60, NIH traineeship, summer 1962, predoctoral fellowship, 1964, postdoctoral fellowship, 1965-66, career devel. award 1972-76, Epilepsy Found. Am. award, 1963, Stiftung Michael prize, 1982; named Fulbright scholar, 1971-72, fellow in neurology Sch. Medicine Stanford U., 1965-66, Lab. Applied Neuophysiology, C.N.R.S., Marseilles, France, 1966, Dagan Lectr. Winter Conf. on Brain Rsch., 1981, John Guggenheim fellow, 1983-84, Hanna lectr. Case-Western Reserve, 1983, First Aird lectr. U. Calif. San Francisco, 1985, First Cox lectr. Albert Einstein Coll. Medicine, 1985, First Vaajasalo lectr. and award, Kuopio, Finland, 1987, Aring lectr. U. Cin. Med. Ctr., 1987, First Hans Berger lectr. Internat. Congress of EEG and Clin. Neurophysiology, 1990. Fellow Am. Acad. Neurology (self assessment epilepsy task force chair 1990—); mem. AAAS, Am. EEG Soc. (or 1984-87, chmn. rsch. fellowship com. 1988-91, pres. elect 1991-92, pres. 1992-93), Am. Epilepsy Soc. (sec. 1979-82, 2nd v.p. 1982-83, 1st v.p. 1983-84, pres. 1984-85, councillor 1985-86, v.p. to Internat. League Against Epilepsy 1990—, William G. Lennox lectr. 1990), Am. Neurol. Assn. (mem. program com. 1987-90), Am. Physiol. Soc., Internat. Brain Rsch. Orgn., Internat. Fedn. EEG and Clin. Neurophysiology Socs. (program com. 1988-90, chmn. com. on guidelines for long-term monitoring for epilepsy 1989—), Internat. League Against Epilepsy (program com. 1986-88, commn. on epilepsy surgery 1988—, chmn. commn. on neurobiology of epilepsy 1989—, amb. for epilepsy award 1991), Internat. Soc. Cerebral Blood Flow and Metabolism, Ea. Assn. Electroencephalographers, Nat. Assn. Epilepsy Ctrs. (bd. dirs. 1988—, treas. 1990—), Soc. for Neurosci. (neurobiology of disease workshop organizing com 1989—), Australian Assn. Neurologists (hon.), Western Electroencephalography Soc. Achievements include research on basic mechanisms of epilepsy and epilepsy related behavior, particularly involving surgical treatment of partial seizures and use of new technology such as positron emission tomography and advanced EEG telemetry. Home: 791 Radcliffe Ave Pacific Palisades CA 90272-4334 Office: UCLA Sch Medicine Reed Neurol Rsch Ctr # 1250 710 Westwood Plz Los Angeles CA 90024-1769

ENGEL, JOANNE NETTER, internist, educator; b. Rockville, Md, Oct. 30, 1955; d. Herman Wolfgang and Doris Renate (Netter) E. BS, Yale U., 1976; MD, Stanford (Calif.) U., 1983, PhD, 1983. Diplomate Am. Bd. Internal Medicine and Infectious Disease. Asst. prof. U. Calif., San Francisco, 1990—. Markey Biomed. scholar Lucille Markey Charitable Trust, 1988. Mem. AAAS, Am. Soc. Microbiology, Am. Fedn. for Clin. Rsch., Infectious Disease Soc. Am.

ENGEL, JUERGEN KURT, chemist, researcher; b. Gerbitz, Saxonia, Germany, Aug. 31, 1945; s. Kurt and Irmgard (Hastaedt) E.; m. Rita Busset, Jan. 7, 1972; children: Joerg Bernhard, Kirsten Rita. Diploma in engring., Naturwissenschaf-Technische Acad., Isny-Allgau, Fed. Republic Germany, 1969; diploma in chemistry, Tech. U. Braunschweig, Fed. Republic Germany, 1972, D Natural Scis., 1975; Habilitation in Pharmacy, U. Regensburg, Fed. Republic Germany, 1985. Lab. leader pharm. div. Degussa, Frankfurt, Fed. Republic Germany, 1976-80, head rsch. coordination, 1980-87, head medicinal chemistry synthesis, 1982-87, head chem. rsch., 1984-87; head chem. and pharm. R & D ASTA Medica AG subs., Frankfurt, 1987-93, head rsch. and devel., v.p., 1993—; prof. sch. pharmacy U. Regensburg, 1990-93; prof. Tech. U., Dresden, Germany, 1993—; mem. Bundesverband der Pharmazeutischen Industrie Drug Synthesis Com., 1982—, BHA Drug Com. and Phytochemistry, Bonn. Author: Pharmazeutische Wirkstoffe, 1982, 2d edit., 1987; editor: Arzeinmittel, 1987; contbr. over 100 articles to profl. jours., chpts. to books; patentee in field. Recipient Galileo Galilei silver medal 5th Internat. Symposium on Platinum and Other Substances, Padua, Italy. Mem. German Chem. Soc., German Pharm. Soc., Internat. Soc. Heterocyclic Chemistry, N.Y. Acad. Sci. Lutheran. Avocations: literature, history. Office: ASTA Medica AG, Weismuellerstrasse 45, D-6000 Frankfurt Germany

ENGEL, THOMAS GREGORY, electrical engineer, educator; b. Houston, Nov. 16, 1959; s. Oscar Joseph and Lizzie Bernice (Dunlap) E.; m. Mary Christine Flores, Dec. 14, 1985; children: Victoria Alexe, Gregory Aleksander. BSEE, Tex. Tech. U., 1985, MSEE, 1987, PhD in Elec. Engring., 1990. Machinist Lubbock (Tex.) Machine Tool, 1980-85; rsch. engr. Enfitek, Inc., Lubbock, 1988-90, v.p. R & D, 1990-91; rsch. asst. prof. Tex. Tech. U., Lubbock, 1992—; cons. Lubbock Machine Tool, 1986—, Technicare Dental Lab, Lubbock, 1991—, FOA-Sweden Def. Rsch. Establishment, Stockholm, 1991—. Author: SPEAR II: High Power Space Insulation; contbr. articles to 6 jours.; 25 conf. papers, and publs. Mem. IEEE Plasma Sci. Soc., IEEE Dielectrics Soc., Tau Beta Pi, Eta Kappa Nu, Sigma Xi. Achievements include research in channel expansion in hydrogen arcs with oscillatory current pulses; HYTAG-III Disclosure Document No 244620. Home: 1715 27th St Lubbock TX 79411 Office: Tex Tech U PO Box 43102 Lubbock TX 79411-3102

ENGELBRECHT, RICHARD STEVENS, environmental engineering educator; b. Ft. Wayne, Ind., Mar. 11, 1926; s. William C. and Mary Elizabeth (Stevens) E.; m. Mary Condrey, Aug. 21, 1948; children: William, Timothy. A.B., Ind. U., 1948; M.S., M.I.T., 1952, Sc.D., 1954. Teaching asst. Ind. U. Sch. Medicine, Indpls., 1949-50; research asst. M.I.T., Cambridge, 1950-52, instr., 1952-54; asst. prof. U. Ill., Urbana-Champaign, 1954-57, assoc. prof., 1957-59, prof. environ. engring., 1959—; dir. Advanced Environ. Control Tech. Research Center, 1979-91; Ivan Racheff prof., 1987; cons. EPA, WHO; mem. Ohio River Valley Water Sanitation Commn., 1976—, chmn., 1980-82. Named Ernest Victor Balsom Commemoration Lectr., 1978; recipient Eric H. Vick award Inst. Public Health Engrs., U.K., 1979; George J. Schroepfer award Central States Water Pollution Control Assn., 1985; Benjamin Garver Lamme award Am. Soc. Engring. Edn., 1985. Mem. AAAS, NAE, Internat. Assn. Water Quality (hon., pres. 1980-86), Am. Water Works Assn. (George W. Fuller award 1974, Publ. award 1975), Water Environment Fedn. (Eddy medal 1966, Arthur Sidney Bedell award 1973, pres. 1978, hon. mem. 1986, Fair medal 1987), Am. Soc. Microbiology, Abwasser Technische Vereinigung (hon., Germany). Home: 2012 Silver Ct W Urbana IL 61801-6331 Office: U Ill 3230 Newmark Civil Engring Lab 205 N Mathews Ave Urbana IL 61801-2350

ENGELMAN, MELVIN ALKON, retired dentist, business executive, scientist; b. Waterbury, Conn., July 27, 1921; s. Herman B. and Marion (Halpern) E.; m. Muriel Phillips, Aug. 27, 1949; children: Curtis Land, Suzanne Ruth. AB, Ohio U., 1942; DDS, Western Res. U., 1944. Diplomate: Am. Bd. Oral Electrosurgery. Pvt. practice dentistry Wappingers Falls, N.Y., 1949-89; chmn. oral diagnosis and oral pathology sect., dir. oral diagnostic ctr. St. Francis Hosp., Poughkeepsie, N.Y., 1963-77, attending dentist, 1963-89, dir. dept. dentistry, 1967, 71-74, 78, hon. staff, 1989—; pres. Di-Equi Dental Products Inc., 1980—, Dentifax Internat. Inc., 1982—; observer Meml. Hosp. Cancer and Allied Diseases, N.Y.C., 1962-66; mem. adv. bd. Dutchess Community Coll., 1963-69, lectr. dental assts. program, 1960-63; dir. 1st regional sci. fair, Dutchess County, N.Y., 1960-61; project dir. USPHS community cancer demonstration project, St. Francis Hosp., 1963-66; asst. chief med. officer Dutchess County N.Y. CD, 1963-68; cons. Nat. Cancer Inst., mem. clin. cancer tng. com., 1968-71, Profl. edn. com. for cancer control, 1972-73; attending dentist Central Dutchess Nursing Home, 1970-85; cons. VA Hosp., Castle Point, N.Y., 1976-77, Lactona Corp., div. Warner Lambert, 1976-80; internat. lectr. on fixed prosthodontics, premedication, oral cancer, metallurgy. Co-author: Oral Cancer Examination Procedure, 16 edits., 1967-83; contbr. articles to profl. jours.; patentee for feeder bar, sprue pin, and spruing assembly. Chmn. Wappinger Red Cross Fund Drive, 1956; committeeman Troop 6, Boy Scouts Am., Chelsea, N.Y., 1963-67; pres. Dutchess County unit Am. Cancer Soc., 1969-71. From ensign to lt. Dental Corps, USNR, FMF PAC, 1942-46; lt. comdr. ret. Fellow AAAS (life), Royal Soc. Health (Eng.), Am. Pub. Health Assn., Acad. Gen. Dentistry; mem. ADA (life), Internat. Assn. Dental Research, Assn. Mil. Surgeons (life mem.), 9th Dist. Dental Soc. (life mem.), Dutchess County Dental Soc. (pres. 1965), Am. Acad. Dental Electrosurgery (pres. 1983), Wappinger Conservation Assn. (v.p. 1970-71), Wappingers Falls C. of C. (pres. 1952-54), Alpha Omega. Clubs: Masons (32 deg.), Shriners, B'nai B'rith (pres. So. Duchess lodge 1963-64). Address: Nutmeg Hill 76 Old State Rd RD7 Wappingers Falls NY 12590

ENGELMANN, PAUL VICTOR, plastics engineering educator; b. Ann Arbor, Mich., Jan. 15, 1958; s. Manfred David and Patricia (Park) E.; m. Martha Ann Heystek, Aug. 14, 1983. AS in Geology, Lansing Community Coll., 1980; BS in Indsl. Edn., Western Mich. U., 1982, MA in Vocat. Edn., 1984, EdD in Ednl. Leadership, 1988. Owner H.L. & S. Auto Restoration & Fabrication, Lansing, Mich., 1977-82; teaching asst. dept. Engring. Tech. Western Mich. U., Kalamazoo, 1982-83, part time instr., 1983-87, instr., 1987-89, asst. prof. plastics, 1989-93, assoc. prof., 1993—; prin. investigator Rsch. and Tech. Inst., Grand Rapids, Mich., 1988—; researcher Robert Morgan & Co., Battle Creek, Mich., 1990—; cons. plastics Parker Hannifin Corp., Otsego, Mich., 1990—. Author (book) Manufacturing Technology, 1989; contbr. articles to profl. jours.; patentee in field. Pres. Plainwell (Mich.) Hist. Preservation Soc., 1990-91; bd. dirs. Pipp Found., 1992—, sec. 1992—. Recipient Protective Package of the Yr. award Children's Hosp. of Birmingham, 1990, Teaching Excellence award Western Mich. U., 1990. Mem. Soc. Plastics Engrs. (past pres. 1992-93, pres. 1991-92, pres.-elect 1990-91, v.p. Western Mich. sec. 1989-90, sec. 1988-89, edn. chmn. 1985-88, Sectional award 1986, 87, 88, Best Paper award 1992). Methodist. Avocations: antique auto restoration, old house preservation, environmental preservation. Home: 311 E Chart St Plainwell MI 49080-1703 Office: Western Mich U Dept Engring Tech Kalamazoo MI 49008-5064

ENGELMANN, RUDOLPH HERMAN, electronics consultant; b. Hewitt, Minn., Mar. 5, 1929; s. Herman Emil Robert and Minna Louise (Kniep) E. BA, U. Minn., 1953. Electronic designer Lawrence Livermore (Calif.) Lab., 1959-61; cons. Atlantic Rsch. Corp., Manchester, N.H., 1961-64, Gen. Radio Co., West Concord, Mass., 1963-69, Possis Engring., Mpls., 1970—, 3M Co., St. Paul, 1977-78, Pako Photo, Mpls., 1977-78, Litton Microwave, Mpls., 1977-79; Presenter papers at confs., 1988-89, 89-90. Contbr. articles to profl. jours. 1st lt. USAF, 1946-53. Achievements include patents in stealth penetrating radar, high-efficiency shape memory alloy modulation, linear circuitry, digital pico second counting and logic, indicia prodn., digital power transformers, digital tracking transformers, digitally accurate switching power supplies, digitally accurate battery chargers, high efficiency power amplifiers, power factor maintenance system, internal combustion engine with complex two-stroke cycle, lighting management for electrostatic copying machines, proportional temperature control for color photo processing-high speed. Office: PO Box 117 Atlantic Beach FL 32233

ENGELSTAD, STEPHEN PHILLIP, mechanical engineer; b. Ames, Iowa, May 20, 1957; s. Orvis Phillip and Marlys Marie (Sargent) E. BS in Indsl. Engring., Auburn U., 1979, MSME, 1983; PhD in Engring. Mechanics, Va. Polytech. Inst. & State U., 1990. Process engr./team mgr. Buckeye Cellulose, Huntsville, Ala., 1979-80; grad. teaching asst. Auburn (Ala.) U., 1981-83; sr. analytical engr. Pratt & Whitney, West Palm Beach, Fla., 1983-86; grad. teaching asst. Va. Polytech. Inst., Blacksburg, 1987-88, grad. rsch. asst., 1989-90, asst. prof. Engring., 1991; specialist engr. Lockheed Aero. Systems Co., Atlanta, 1991—. Contbr. articles to profl. jours. Boeing summer fellow, Blacksburg, 1988. Mem. AIAA (sr.), ASME, Am. Soc. Engring. Educators. Office: Lockheed Aero Systems Co zone 0685 86 S Cobb Dr dept 73-47 Marietta GA 30063

ENGHETA, NADER, electrical engineering educator, researcher; b. Tehran, Iran, Oct. 8, 1955; came to U.S. 1978; s. Abdollah and Meymanat (Mesghali) E.; m. Susanne Hoshyar, Oct. 15, 1983; children: Alex Cameron, Sarah Katherine. BSEE, U. Tehran, 1978; MSEE, Calif. Inst. Tech., 1979, PhD in Elec. Engring., 1982. Grad. rsch. asst. Calif. Inst. Tech., Pasadena, 1979-82, postdoctoral rsch. fellow, 1982-83; sr. rsch. scientist Kamam Scis. Corp., Santa Monica, Calif., 1983-87; asst. prof. elec. engring. U. Pa., Phila., 1987-90, assoc. prof. elec. engring., 1990—; gen. chmn. Benjamin Franklin Symposium, Phila., 1990-91; vis. lectr. UCLA, 1986; condr. seminars in field; lectr. in field. Guest editor spl. issue of Jour. of Electromagnetic Waves and Applications on wave interaction with chiral and complex media, Vol. 6, No. 5/6, 1992; contbr. over 40 articles to profl. jours., chpts. to books. NSF Presdl. Young Investigator, 1989; AT&T Spl. Purpose grantee, 1988; U. Pa. Rsch. Found. grantee, 1988, 90. Mem. AAAS, IEEE (chmn. antennas and propagation/microwave theory and technique Phila. chpt. 1990-91), Optical Soc. Am., Am. Phys. Soc., Sigma Xi. Achievements include two patents (with others) for method of measuring chiral parameters of chiral materials and novel electromagnetic shielding reflection and scattering control using chiral materials; patents pending (with others) for electromagnetically non-reflective material, novel waveguides using chiral materials, novel printed-circuit antennas using chiral materials, novel antenna arrays using chiral materials, novel radomes using chiral materials, novel lenses using chiral materials; research on applied electromagnetics, optics, complex and coupeling materials, microwave, polarization-difference imaging and vision, role of continuous dimensionality and non-integral (fractional) calculus in electrodynamics. Office: Univ of Pa 200 S 33d St Philadelphia PA 19104

ENGLAND, ANTHONY WAYNE, electrical engineering and computer science educator, astronaut, geophysicist; b. Indpls., May 15, 1942; s. Herman U. and Betty (Steel) E.; m. Kathleen Ann Kreutz, Aug. 31, 1962. S.B. S.M., MIT, 1965, Ph.D., 1970. With Texaco Co., 1962; field geologist Ind. U., 1963; scientist-astronaut NASA, 1967-72, 79-88; prof. elec. engring. and computer sci. U. Mich., Ann Arbor, 1988—; with U.S. Geol. Survey, 1972-79; crewmember on Spacelab 2, July, 1985; adj. prof. Rice U., 1987-88. Asso. editor: Jour. Geophys. Research. Recipient Sci. Achievement medal, Antarctic medal, Spaceflight medal NASA, Spaceflight award Am. Astron. Soc., Outstanding Scientific Achievement medal NASA; NASA grantee. Mem. Am. Geophys. Union, IEEE. Home: 7949 Ridgeway Ct Dexter MI 48130-9700 Office: U Mich Dept EECS Ann Arbor MI 48109-2122

ENGLAND, GARY ALAN, television meteorologist; b. Seiling, Okla., Oct. 3, 1939; s. Leslie Elwood and Hazel Wanda (Stong) E.; m. Mary Helen Carlisle, Aug. 27, 1961; 1 child, Molly Michelle. BS in Math. and Meteorology, U. Okla., 1965. Cert. profl. meteorologist. V.p. mktg. Southwestern Weather Svc., Oklahoma City, 1965-67; cons. meteorologist A.H. Glenn & Assocs., New Orleans, 1967-71; v.p. mktg. England & May, Oklahoma City, 1971-74; v.p. meteorology U.S. Weather Corp., Oklahoma City, 1974-78; pres. The Gary Co., Oklahoma City, 1978—; dir. meteorology Griffin TV, Oklahoma City, 1971—; cons. meteorologist Techrad, Okla., 1977-78; forensic meteorologist legal field, Okla., 1971—. Author: Oklahoma Weather, 1975, United States Weather, 1976, Those Terrible Twisters, 1987. Fundraiser The Christmas Connection, Oklahoma City, 1982-93, Harvest Fund Drive, Oklahoma City, 1982-93; mem. adv. bd. Make-A-Wish Found., Oklahoma City, 1993. With USN, 1957-61. Named to Western Okla. Hall of Fame Western Okla. Hist. Soc., 1983, Dewey County Hall of Fame Dewey County Hist. Soc., 1989, Best Meteorologist Okla. Gazette Newspaper, 1987-93, Outstanding Young Men of Am., 1976; recipient Ptnrs. for Excellence award Okla. Sch. Pub. Rels. Assn., 1989. Mem. AAAS, Am. Meteorol. Soc., Am. Bus. Club, N.Y. Acad. Sci. Mem. Christian Ch. Achievements include initiation of development with Enterprise Electronics Corp. of the world's first comml. Doppler radar; first person to use Doppler radar for direct warnings to the public; developed an automated weather warning system called First Warming; developed an automated severe weather tracking and projection computer system called Storm Tracker, now in use nation wide. Office: Sta KWTV PO Box 14159 Oklahoma City OK 73113

ENGLAND, MARTIN NICHOLAS, astro/geophysicist; b. Pretoria, South Africa, Dec. 17, 1954; came to U.S., 1979; s. Michael George and Maureen (McCabe) E.; m. Sheila Mary Murphy, Feb. 20, 1982; children: Kathryn Maureen, Maureen Bridget. BSc, Victoria U., Wellington, New Zealand, 1976; PhD, U. Fla., 1986. Scientist, geophysicist Interferometrics Inc./NASA-Goddard Space Flight Ctr., Vienna, Va., 1986-91; astronomer Computer Sci Corp/Iue Obs, NASA Goddard Space Flight Ctr., Greenbelt, Md., 1991—. Contbr. articles to profl. jours. Mem. Am. Astron. Soc., Am. Geophys. Soc., Royal Astron. Soc. New Zealand, Phi Beta Kappa. Office: Computer Scis Corp Iue Obs Code 648.9 NASA/Goddard Space Flt Ctr Greenbelt MD 20771

ENGLEBIENNE, PATRICK P., biochemist, consultant; b. Charleroi, Hainaut, Belgium, Apr. 3, 1949; s. André and Isabelle C. (Pierard) E.; m. Anne Van Hoonacker, Sept. 29, 1976; children: Gwenn, Maykin, Kerridwen, Joan. B in Pharm., Paul Pastur U., Charleroi, 1969; PhD in Biochemistry, Pacific Western U., 1985, BS in Chemistry (hon.), 1985. Chemistry lectr. Mureke (Burundi) Coll., 1969-70; scientist Free U., Brussels, 1973-76, Queen Fabiola Hosp., Charleroi, 1976-78; sci. dir. CTR Rsch. Diagnostic Endocrinology, Kain, Belgium, 1978-86, Sopar Diagnostics, Brussels, 1987-91; with APE Assocs., Ghislenghien, Belgium, 1991—; cons. Wellcome, Aalst, Belgium, 1986-87, Private Clinical Labs., Belgium, 1986-87, Carbochimica Italiana, Fidenza, Italy, 1987-90. Author: The Serum Steroid Transport Proteins, 1984; contbr. articles to Jour. Steroid Biochemistry, Jour. Immunological Methods, Clin. Chemistry, Clin. Biochemistry, Anticancer Rsch. Civil servant Belgian Gov., Burundi, 1969-70. Mem. AAAS, Am. Chem. Soc. (biol. chemistry div. medicinal chemistry div., polymers div.), Biochem. Soc., Royal Soc. Chemistry (assoc.), N.Y. Acad. Scis. Achievements include identification and characterization of new protein in cancer patients; original designs for proteins and steroids; research on application of colloids in biological and pharmaceutical systems, development of new imaging agents; patent on use of soluble conducting polymers in immunoassay technology, 1993. Home: Strijpstraat 21, B9750 Zingem Belgium Office: APE Assocs, Industrial Estate, B-7822 Ghislenghien Belgium

ENGLEHART, EDWIN THOMAS, metals company executive; b. Johnstown, Pa., Aug. 7, 1921; s. Edwin T. and Genevieve (Conley) E.; m. Genevieve Whittaker, May 26, 1951; 1 child, Daria. BS, Pa. State U., 1943; postgrad., U. Pitts., 1946-49. Corrosion engr. Alcoa Rsch. Labs., New Kensington, Pa., 1943-59; asst. chief chem. metals div. Alcoa Rsch. Labs., New Kensington, 1959-77; sect. head corrosion alloy tech. div. Alcoa Tech. Ctr., Alcoa Ctr., Pa., 1977-83; advisor, com. mem. ASTM, Am. Welding Soc., Nat. Assn. of Corrosion Engrs. Author: Aluminum Updated, 1983, Vol. III, 1967; contbr. over 20 articles to profl. jours. Mem. K.C., Lions, Sigma Xi. Republican. Home: 450 Dakota Dr Lower Burrell PA 15068

ENGLEMAN, DENNIS EUGENE, electrical engineer; b. Falls City, Nebr., July 24, 1948; s. Eugene Adolf and Mary Alice (Franklin) E.; m. Deborah Faye Paulson, May 4, 1985; children: John Nicholas, Lily Eugenia, Mary Victoria. Student, U. Nebr., 1966-70; B in Engring. Tech., Wichita State U., 1982. Ordained to ministry Holy Order of Mans, 1977; ordained acolyte/reader Eastern Orthodox Ch., 1988. Minister parishes various locations, 1977-81; elec. technician Kans. Gas & Elec. Co., Wichita, 1981-82; elec. engr. Boeing Mil. Airplane Co., Wichita, 1982; design engr. Nat. Data Corp., Atlanta, 1983, Raymond Carousel Corp., Atlanta, 1984; elec. designer Cons. & Designers, Inc., Atlanta, 1984; project engr. Nordson Corp., Norcross, Ga., 1984-85, sr. engr., 1985-90, engring. supr., 1990-92, sr. engring. supr., 1992—; pres. Holy Mountain Imports, Atlanta, 1987—, Engleman Family Photography, Atlanta, 1989—; appeared on various radio broadcasts, Cin.; mem. dept. of ministry and evangelism Greek Orthodox Archdiocese of Vasiloupolis, 1992. Producer of programming People TV, Pub. Access TV, 1990—; singer, songwriter, arranger Condor Classix Records., 1992, Pristine Records, 1993; contbr. articles ot profl. publs.; co-editor Tree of Life, 1992; producer various slide shows; writer, producer folk opera. Pres. Christian Community Alanta, 1985-87; ordained acolyte/reader Blessed John the Wonderworker Ea. Orthodox Ch., 1988; dir. St. Cyril's Village Orch., 1987—; mem. Atlanta Balalaika Soc. Orch., 1984-87. Republican. Orthodox Christian. Avocations: folk music, photography. Office: Nordson Corp 11475 Lakefield Dr Tech Pk/Johns Creek Duluth GA 30136

ENGLEMAN, EPHRAIM PHILIP, physician; b. San Jose, Calif., Mar. 24, 1911; s. Maurice and Tillie (Rosenberg) E.; m. Jean Sinton, Mar. 2, 1941; children—Ephraim Philip, Edgar George, Jill. B.A., Stanford U., 1933; M.D., Columbia U., 1937. Intern Mt. Zion Hosp., San Francisco; resident U. Calif., San Francisco, Jos. Pratt Diagnostic Hosp., Boston; research fellow Mass. Gen. Hosp., Boston, 1937-42; practice medicine specializing in rheumatology San Francisco, 1948—; mem. faculty U. Calif. Med. Center, San Francisco, 1949—; clin. prof. medicine U. Calif. Med. Center, 1965—; dir. Rosalind Russell Arthritis Center, 1979—; prin. investigator Nat. Inst. Dental Rsch., 1989—; mem. staff U. Calif., Mills Meml., Peninsula hosps.; Chmn. Nat. Commn. Arthritis and Related Diseases, 1975-76. Author: The Book on Arthritis: A Guide for Patients and Their Families, 1979; also articles, chpts. in books. Served to maj. M.C. USMCR, 1942-47. Nat. Inst. Arthritis grantee; recipient citation Arthritis Found., 1973; Ephraim P. Engleman Disting. Professorship in Rheumatology established in his honor U. Calif., San Francisco, 1991. Fellow ACP; Mem. Internat. League Against Rheumatism (pres. 1981-85), Am. Coll. Rheumatology (founding fellow, master, pres. 1962-63), Nat. Soc. Clin. Rheumatologists, AMA, Am. Fedn. Clin. Research; hon. mem. Japanese Rheumatism Soc., Spanish Rheumatism Soc., Uruguay Rheumatism Soc., Australian Rheumatism Assn., Chinese Med. Assn., French Soc. Rheumatology, Internat. League against Rheumatism, Gold-Headed Cane Soc. (U. Calif., San Francisco). Republican. Jewish. Club: Family (San Francisco). Office: U Calif Rosalind Russell Med Rsch Ctr Arthritis 350 Parnassus Ave #600 San Francisco CA 94117

ENGLISH, BRUCE VAUGHAN, museum director and executive, environmental consultant; b. Richmond, Va., Aug. 6, 1921; s. Pollard and Lucy Kelly (Rice) E.; m. Virginia Tejas McCall Shaw, Feb. 6, 1949. BS in Physics and Math., Randolph-Macon Coll., 1942; MS in Physics and Math., Ind. U., 1943; PhD in Physics, U. Va., 1958. Grad. asst. physics Ind. U., Bloomington; rsch. asst., instr. army specialized tng. program Manhattan Dist. Engrs. Project, Ashland, Va.; asst. prof. physics Randolph-Macon Coll., Ashland, Va., 1943-44, assoc. prof., acting chmn. dept. physics, 1948-58, prof., chmn. dept., 1958-64; physicist, head high pressure lab. U.S. Navy Underwater Sound Reference Lab., Orlando, Fla., 1946-48; physicist, cons. historic preservation, pollution control and environment Ashland, 1964—; dir. Poe Found., Inc., Richmond, 1969—, pres., 1973-92; pres., dir. Edgar Allan Poe Mus., Richmond, 1973-92; pres. Pollution Control Assocs., Richmond, 1967-70. Co-pub.: Poe's Richmond, 1978; columnist Herald-Progress, 1971—; contbr. numerous articles to Poe Messenger mag. Founding mem. Richmond Symphony, 1956; mem. Patrick Henry Scotchtown Com., Hanover County, Va., 1958—; pres. Hist. Richmond Found., 1967-70; bd. dirs. Church Hill Model Neighborhood Bd., Richmond, 1968-73; chmn. Bicentennial Com. for Hanover County, 1974-92, Drainage Com., Ashland, 1980s, Courthouse Com. for Hanover County, 1985—; lay reader, mem. vestry St. John's Ch., Church Hill, Richmond, Va., 1969-70. With USN, 1944-45. Named Hon. Citizen State of Md., 1990; Ford Faculty fellow, 1951-52, Danforth fellow, 1956-57, du Pont fellow, 1957-58. Mem. AAAS, Am. Phys. Soc., Va. Acad. Sci., Va. Hist. Soc., Soc. Archtl. Historians, Nat. Trust for Hist. Preservation, Irish Georgian Soc., Cousteau Soc. (founding), Air and Waste Mgmt. Assn., Nat. Soc. for Clean Air Gt. Britain, Soc. Descendants of Peter Francisco (founder, advisor), City Tavern Club, Commonwealth Club, Farmington Country Club, Downtown Club, Phi Beta Kappa, Sigma Xi, Omicron Delta Kappa, Chi Beta Phi, Pi Delta Epsilon. Episcopalian. Achievements include research for project developing atomic bomb; increasing awareness of hazards of pollution since 1955, of Edgar Allan Poe's cosmology, cryptography, and other scientific writings.

ENGLISH, FLOYD LEROY, telecommunications company executive; b. Nicholas, Calif., June 10, 1934; s. Elvan L. and Louise (Corliss) E.; m. Wanda Parton, Sept. 8, 1955 (div. 1980); children: Roxane, Darryl; m. Elaine Ewell, July 3, 1981; 1 child, Christine. A.B. in Physics, Calif. State U.-Chico, 1959; M.S. in Physics, Ariz. State U., 1962, Ph.D. in Physics, 1965. Div. supr. Sandia Labs., Albuquerque, 1965-73; gen. mgr. Rockwell Internat.-Collins, Newport Beach, Calif., 1973-75; pres. Darcom, Albuquerque, 1975-79; cons in energy mgmt. and acquisitions Albuquerque, 1979-80; v.p. U.S. ops. Andrew Corp., Orland Park, Ill., 1980-82; pres., chief operating officer Andrew Corp., 1982-83, pres., chief exec. officer, 1983—, dir., 1982—. Contbr. articles to profl. jours. Served to 1st lt. U.S. Army, 1954-57; served to capt. USAR, until 1969. Mem. IEEE, Execs. Club of Chgo. (bd. dirs.), Exec. Adv. Coun. Nat. Comms. Forum. Republican. Presbyterian. Office: Andrew Corp 10500 W 153D St Orland Park IL 60462

ENGLISH, FRANCIS PETER, ophthalmologist, educator; b. Cairns, Australia, May 31, 1932; s. Peter Bede and Mona (Elliott) E.; m. Leonie Therese Jones, May 31, 1975; children: Lawrence, James. M.B., B.S., U. Queensland, Australia, 1957. Glaucoma fellow Howe Lab., Harvard U., Cambridge, Mass., 1962-63; clin. asst. Moorfields Hosp., London, 1963-66; vis. prof. ophthalmology U. Okla., 1969; fellow retina svc. U. Tex., Houston, 1967-68; fellow in oculoplastic surgery Manhattan Eye and Ear Hosp., N.Y.C., 1969-70; fellow cornea svc. Retina Found., Boston, 1970-71; ophthalmic surgeon Repatriation Dept., Brisbane, Australia, 1972-88; instr. ophthalmology U. Queensland, 1972-88; cons. Australian Govt., 1979—; rsch. assoc. Queensland Inst. of Med. Rsch., 1989—. Author: Reconstructive and Plastic Surgery of the Eyelids, 1975; contbr. to Current Ocular Therapy (Fraunfelder), 1985, 2nd edit., 1988, 3rd. edit., 1989, Techniques in Ophthalmic Plastic Surgery (Wesley), 1986; contbr. articles to profl. jours. Recipient ophthalmic study tour award Grenfell Found., Can., 1968. Fellow Internat. Coll. Surgeons, Royal Coll. Surgeons, Royal Australian Coll. Ophthalmologists; mem. Internat. Oculoplastic Soc. (bd. dirs. 1982—). Clubs: Tattersalls, United Svcs. Home: 41 Charlton St Ascot, Brisbane 4007, Australia Office: 113 Wickham Terr, Brisbane 4000, Australia

ENGLISH, GARY EMERY, military officer; b. Williamsport, Pa., Jan. 10, 1962; s. Curtis Riegel and Janet Laura (Emery) E.; m. Rebecca Cole Stevenson, Nov. 1, 1986. BS in Applied Sci., U.S. Naval Acad., 1984; MS in Applied Sci., Naval Postgrad. Sch., 1992. Commd. ensign USN, 1984—. Active Monterey (Calif.) Bay Aquarium, 1991-92. Mem. U.S. Naval Inst., Acoustical Soc. Am., Surface Navy Assn., Nat. Eagle Scout Assn. Office: USS John Rodgers (DD-983) FPO AA 34092

ENGLISH, ROBERT EUGENE, manufacturing engineering educator; b. Paris, Ill., July 9, 1953; s. James P. and Exia Mae (Reagin) E.; m. Kathy L. Stout, Mar. 19, 1976; children: Christopher M., Jeremy D. MS, Ind. State U., 1981; EdD, Ind. U., 1992. Mfg. engr. Zenith Radio Corp., Paris, Ill., 1976-77, mgr. mfg. engring., 1977-82; asst. prof. Ind. State U., Terre Haute, 1982-87, assoc. prof. engring., 1987—; tech. cons. Hane Indsl. Tng., Terre Haute, 1986-88, MicroLink, Inc., Indpls., 1985-86; rsch. cons. Beeco, Inc., Indpls., 1985-86. Mem. Instrument Soc. Am. (sect. edn. chair 1992—), Nat. Assn. Indsl. Tech. (chair student membership 1986-87). Home: 13826 State Rte 159 Lewis IN 47858 Office: Ind State Univ N 6th & Cherry Terre Haute IN 47809

ENGLOT, JOSEPH MICHAEL, structural engineer; b. N.Y.C., May 5, 1950; s. Joseph and Mary (Yanoschak) E.; m. Jane Frances Mines, Oct. 9, 1982; children: Brendan Joseph, Michael Edmund. BSCE, Poly. Inst. Bklyn., 1972, MSCE, 1972. Registered profl. engr., N.Y. Asst. engr. Port Authority N.Y. and N.J., N.Y.C., 1972-75, assoc. engr., 1975-80, engr., 1980-84, sr. engr., 1984-88, prin. engr., 1988-89, asst. chief structural engr., 1989-92, chief structural engr., 1992—; lectr. at civil engring. seminars and constrn. trade assns. Contbr. articles to profl. publs. Mem. ASCE. Achievements include discovery of rehabilitation and repair techniques for bridges, tunnels and transportation terminal buildings. Office: Port Authority NY and NJ 1 World Trade Ctr Rm 72S New York NY 10048-0202

ENGLUND, JOHN ARTHUR, research company executive; b. Omaha, June 4, 1926; s. Arthur D. and Marquerite E. (Welsh) E.; m. Marilyn Ann Miller, Aug. 9, 1952; children: John A. Jr., Ann E., George A., James M., Edward M. BS, Creighton U., 1949; MS, Mass. Inst. Tech., 1951. Teaching fellow MIT, Cambridge, 1949-51; asst. prof. Creighton U., Omaha, 1951-56; ops. analyst Strategic Air Command, Omaha, 1956-62; military systems analyst U.S. Arms Control and Disarmament Agy., Washington, 1962-63; mathematician, branch chief, div. mgr. Analytic Svcs. Inc., Falls Church, Va., 1963-76; exec. v.p. Analytic Svcs. Inc., Arlington, Va., 1976-81, pres., 1981-91; trustee Analytic Svcs., Inc., 1991—.

ENGVILD, KJELD CHRISTENSEN, plant physiologist; b. Ars, Jutland, Denmark, Mar. 25, 1940; s. Hans and Ragnhild Egekvist (Sorensen) Christensen; m. Ruth Wohlert, July 16, 1984; 1 child, Sara. Degree, U. Copenhagen, Denmark, 1965. Soil microbiologist Govt. Lab. Soil and Crop Rsch., Lyngby, Denmark, 1967-69; plant physiologist Riso Nat. Lab., Roskilde, Denmark, 1969—. Mem. jour. com. Scandinavian Soc. Plant Physiology, 1985—; contbr. articles to profl. jours. including Plant Tissue Culture, Chloro Indole Auxins, Symbiotic N Fixation. Achievements include rsch. on chlorine-containing plant hormones in vicia and lathyrus. Office: Riso Nat Lab, Environ Sci and Tech Dept, DK-4000 Roskilde Denmark

ENHORNING, GORAN, obstetrician/gynecologist, educator; b. Birkdale, Eng., Mar. 18, 1924; came to U.S. 1986; s. Emil Augustin and Maria Rosina (von Haartman) E.; m. Louise Christina Carlberg, Apr. 16, 1955; children: Ulf, Dag and Peder (twins), Marianne. MD, Karolinska Inst., Stockholm, 1952, PhD in Physiology, 1961. Asst. prof. ob/gyn. Karolinska Inst., Stockholm, 1952-61; Fulbright scholar U. Utah, Salt Lake City, 1961-63, UCLA, 1963-64; assoc. prof. ob/gyn. Karolinska Inst., 1964-71; assoc. prof. ob/gyn. U. Toronto, Ont., Can., 1971-75, prof. ob/gyn., 1975-86; prof. ob/gyn. SUNY, Buffalo, 1986—. Contbr. articles to profl. jours. Achievements include contribution to understanding of urinary bladder's closure mechanism; methods for evaluating surface properties of pulmonary surfactant; initiation of concept that neonatal respiratory distress syndrome can be prevented/treated by instillation of pulmonary surfactant into upper airways, and concept that symptoms of asthma may be caused by a surfactant dysfunction. Home: 21 Oakland Pl Buffalo NY 14222

ENNIS, JOEL BRIAN, physicist; b. Miami, Fla., Mar. 25, 1959; s. Harry Esley and Mary Ethel (Smith) E.; m. Kristal Lee Welch, May 12, 1990. BS, Calif. Inst. Tech., 1981. Mem. tech. staff Hughes Aircraft Co., El Segundo, Calif., 1981-84; staff engr. Maxwell Labs., Inc., San Diego, 1984-86, sr. staff engr., 1986-89, chief engr., 1990-91, asst. div. mgr., 1991-92, div. mgr., 1992—. Contbr. articles to profl. jours. Mem. IEEE, Am. Phys. Soc., Elec. Insulation Soc. Achievements include development of high energy density film capacitors. Office: Maxwell Labs Inc 4949 Greencraig Ln San Diego CA 92123

ENNS, KEVIN SCOTT, architect, artist; b. Hutchinson, Kans., Nov. 9, 1959; s. Victor Dale Enns and Francis Geraldean (McPherson) Jones. Student, Pitts. State U., 1977-78; BArch, Kans. State U., 1985; M of Community Planning and Urban Design, U. Cin., 1993. Draftsperson Engring. Con., PA, Hutchinson, Kans., 1979-81; student intern Ebert-Keating and Phinney, Bartlesville, Okla., 1983-84; architect's asst. TDE/Sweet and Assoc., AIA, Mahattan, Tex., 1984-85; jr. designer James R. Pratt Architecture & Urban Design, Dallas, 1985-86; archtl. designer Jeff Krehbiel Assoc., AIA, Wichita, Kans., 1986-88; project designer Charles F. McAfee FAIA Noma PA, Wichita, 1988—; rsch. asst. Ctr. for Urban Design, U. Cin., Cin., 1991-93; mem. group study exch. team Rotary Internat., South Korea, 1988. Drawing and art works exhibited at Kans. State Capital Bldg., 1988, Wichita Arts Coun., 1988, Kans. State Amateur Exhbn., 1988-91, built work Kans. State U. Vietnam Vets. Meml. (1st Prize 1987, dedicated 1989). Med. asst. Feed the Children Med. Missions Team, Honduras, C. Am., 1990, 91, Guatemala C. Am., 1989; chmn. missions and Outreach, Mt. Hope, Kans., 1990-91. Univ. scholar U. Cin.; named Outstanding Grad. Student in Coll. of Design, Art, Architecture and Planning, U. Cin., 1993. Mem. AIA, Am. Planning Assn., Nat. Hist. Preservation Soc., Kans. State U. Alumni Assn. Democrat. Protestant.

ENOS, PAUL, geologist, educator; b. Topeka, July 25, 1934; s. Allen Mason and Marjorie V. (Newell) E.; m. Carol Rae Curt, July 5, 1958; children—Curt Alan, Mischa Lisette, Kevin Christopher, Heather Lynne. B.S., U. Kans., 1956; postgrad., U. Tubingen, W.Ger., 1956-57; M.S., Stanford U., 1961; Ph.D., Yale U., 1965. Geologist Shell Devel. Co., Coral Gables, Fla., 1964-68; research geologist Shell Devel. Co., Houston, 1968-70; from assoc. prof. to prof. geology SUNY, Binghamton, 1970-82; Haas Disting. prof. geology U. Kans., Lawrence, 1982—; cons. to industry; sedimentologist Ocean Drilling, 1975, 92; rsch. vis. Oxford U., 1989; fgn. scientist Ministry Geology, People's Republic China, 1988; co-convener Working Group 4, 1992—; with Global Sedimentary Geology Project, 1988—. Co-author: Quaternary Sedimentation of South Florida, 1977, Mid-Cretaceous, Mexico, 1983; editor: Field Trips: South-Central New York, 1981, Deep-Water Carbonates, 1977; contbr. articles to sci. jours. Served to 1st lt. C.E., U.S. Army, 1957-59. U. Liverpool fellow, 1976-77; NSF fellow, 1959-62; Fulbright fellow, 1956-57; Summerfield scholar, 1954-56. Mem. Soc. Econ. Paleontologists and Mineralogists (assoc. editor 1976-80, 83-87), Internat. Assn. Sedimentologists (assoc. editor 1983-87), Am. Assn. Petroleum Geologists, AAAS, Sigma Xi, Omicron Delta Kappa. Avocations: photography, diving, cycling, history. Home: 2032 Quail Creek Dr Lawrence KS 66047-2139 Office: U Kans Dept Geology 120 Lindley Hall Lawrence KS 66045

ENRICO, DAVID RUSSELL, mechanical engineer; b. Seoul, Feb. 20, 1959; (parents Am. citizens); s. Dominic John and Dorothy Irene (Ervin) E.; m. Dana Christine Flowers, Aug. 26, 1989. BSME, Rose-Hulman Inst. Tech., 1981. Registered profl. engr., Ill. Staff project engr. A.E. Staley Mfg. Co., Decatur, Ill., 1981-83, assoc. project engr., 1984-85, project engr., 1986-88; sr. project engr. Henkel Corp., Decatur, 1988-89, Ambler, Pa., 1989-90; plant project engr. Henkel Corp., Crosby, Tex., 1990-92; maintenance/project engr. Henkel Corp., Cin., 1993—. Roman Catholic. Home: 8167 Autumn Ln West Chester OH 45069 Office: Henkel Corp 4900 Este Ave Cincinnati OH 45232

ENRIGHT, JOHN CARL, occupational health engineer; b. Terre Haute, Ind., June 15, 1948; s. John F. and C. Rosemary (Lundstrom) E. BS in Engring., Purdue U., 1970; MBA in Bus. Adminstrn., U. Dayton, 1975. Cert. indsl. hygienist; cert. safety profl.; diplomate Am. Acad. Indsl. Hygiene. Adminstr. hazardous materials control GM, Warren, Mich., 1969-88; sr. cons. Occusafe, Inc., Wheeling, Ill., 1988-91; loss control cons. Richard Oliver Risk Mgrs., Arlington Heights, Ill., 1991—; tech. content cons. United Auto Workers-GM Hazard Comm. Program. Recipient 6 internat. film awards Chgo. Internat. Film Festival, 1986. Mem. Mich. Indsl. Hygiene Soc. (bd. mem. 1985-87, pres. elect 1987-88, pres. 1988-89), Am. Indsl. Hygiene Soc., Soc. Toxicology, N.Y. Acad. Scis. Republican. Roman Catholic. Achievements include establishment of basis for evaluating passenger risk to contaminants released from deployment of inflatable restraints (air bags); development of video and laser-disc interactive tng. formats. Office: Richard Oliver Risk Mgrs 1590 N Arlington Heights Rd Arlington Heights IL 60004

ENRIGHT, MICHAEL JOSEPH, radiologist; b. Richmond, Va., Mar. 27, 1955; s. Wlliam Joseph and Margaret (O'Connell) E.l (div.); children: Kelly Ann, Margeaux Elizabeth; m. Susan Ross Lemon, June 29, 1991; 1 child, Darby Michelle. BS in Pharmacy, Ohio State U., 1978; MD, Ea. Va. Med. Sch., 1981. Diplomate Nat. Bd. Med. Examiners, Am. Bd. Radiology, Va. Bd. Pharmacy. Resident in radiology Ea. Va. Grad. Sch. Medicine, Norfolk,

1981-85; radiologist U.S. Navy, Charleston, S.C., 1985-88; body imaging fellow U. Va. Med. Ctr., Charlottesville, 1988-89; radiologist Radiology Assocs. of Roanoke, Va., 1989—; treas. Low Country Imaging Soc., Charleston, 1986-87; sect. head Body and Musculoskeletal Magnetic Resonance Imaging, Radiology Assn. Roanoke, 1993—. Author (exhibit) Scrotal Ultrasonography at Am. Roentgen Ray Soc. Meeting, 1989. Lt. commdr. U.S. Navy, 1985-88. Mem. AMA, Am. Coll. Radiology, Radiol. Soc. N. Am., Roanoke Soc. Medicine. Republican. Avocations: running, tennis, racquetball, golf, boating. Home: PO Box 18006 Roanoke VA 24014-5756 Office: Radiology Assocs of Roanoke Ste 113 2037 Crystal Spring Ave SW Roanoke VA 24014-2494

ENRIQUEZ, FRANCISCO JAVIER, immunologist; b. Mexico City, June 24, 1955; came to U.S., 1982; s. Camilo Javier and Maria de Lourdes (Serralde) E.; m. Lisa Barricklow, May 30, 1987; 1 child, Alesi Brie. Diploma, London Sch. Hygiene & Tropical, 1984; MD, La Salle U., Mexico City, 1979; PhD, Cornell U., 1986. Clinician Santa Fe Hosp., Mexico City, 1979-81; asst. prof. Sch. of Medicine La Salle U., 1980-82; med. researcher Ministry of Health, Mexico City, 1980-82; postdoctoral fellow U. Ariz., Tucson, 1987-88, dir. hybridoma tech., 1988—; cons. NIH, Bethesda, Md., 1989—, Health Corp., 1992—, NCR, Washington, 1990—; sci. cons. Aegeria Inc., 1992—, Igx Inc., 1992—, Bristol Meyers, Squibb, Mead Johnson, 1992—. Author: (novel) Las Primas Segundas, 1991, (short stories) La evolucion del Pensamiento, 1992; editorial reviewer sci. jours., 1991—; contbr. numerous articles to profl. jours. Avocations: tennis, Tae Kwon Do, classical guitar, writing. Home: 339 E University Blvd Tucson AZ 85705-7848 Office: U Ariz Ariz Health Scis Ctr Tucson AZ 85724

ENROUGHTY, CHRISTOPHER JAMES, nuclear chemistry technician; b. Richmond, Va., Mar. 2, 1961; s. James Warren and Johannah Margaret (Southward) E.; m. Silvia Ann Chopelas, May 25,1 985; children: Matthew Christopher, Michael James. BA in Managerial Econs., Hampden-Sydney Coll., 1983; cert., Naval Nuclear Power Sch., 1985; grad. with honors, Svc. Sch. Command, 1985. Enlisted USN, 1984; engring. lab. technician USN, Charleston, S.C., 1984-90; resigned USN, 1990; asst. broker, radiol. and health physics Chem-Nuclear Systems, Columbia, S.C., 1990-91; nuclear chemistry technician Va. Power Surry Power Sta., 1991—. Office: Va Power Surry Power Sta Surry VA 23883

ENSMINGER, DALE, mechanical engineer, electrical engineer; b. Mt. Perry, Ohio, Sept. 26, 1923; s. Charles Henry and Mary Elpha (Koehler) E.; m. Lois Elizabeth Hamilton, Mar. 25, 1948; children: Martha Jean, Laura Lee, Charles Robert, Jonathan Dale, Mary Ann, Daniel Joseph. BSME, BSEE, Ohio State U., 1950, postgrad., 1950-53. Registered profl. engr., Ohio. Researcher Battelle Meml. Inst., Columbus, Ohio, 1950; prin. researcher Battelle Meml. Inst.; sr. researcher Battelle Columbus Labs., mgr. ultrasonics, sr. rsch. scientist, 1988—; cons. in field. Author: Ultrasonics, 1973, 2d edit. 1988; contbr. articles to profl. jours., chpts. to books; patentee in field; contbr., reviewer Am. Soc. Non-Destructive Testing Handbook, 1989—. Sec. Columbus Prison Assn., Columbus, 1950—; dean, dir. Columbus Bible Inst., 1952—. With U.S. Army, 1943-46. Recipient Cert. of Recognition, NASA, 1975. Mem. Acoustical Soc. Am., Soc. for Non-Destructive Testing, Ultrasonic Industry Assn. Home: 198 E Longview Ave Columbus OH 43202-1236 Office: Battelle 505 King Ave Columbus OH 43201-2681

ENSMINGER, MARION EUGENE, animal science educator, author; b. Stover, Mo., May 28, 1908; s. Jacob and Ella (Belt) E.; m. Audrey Helen Watts, June 11, 1941; children: John Jacob, Janet Aileen (dec.). B.S., U. Mo., 1931, M.S., 1932; Ph.D., U. Minn., 1941. Field agt. U. Mo., summers 1929-30; instr. Mo. State U., Marysville, summers 1931-32; asst. to supt. U.S. Soil Erosion Sta., Bethany, Mo., 1933; soil erosion specialist U.S. Dept. Interior, U.S. Dept. Agr., Ill., 1934; mgr. Dixon Springs (Ill.) project U.S. Dept. Agr., 1934-37; asst. prof. U. Mass., 1937-40; teaching asst. U. Minn., 1940-41; prof., chmn. dept. animal sci. Wash. State U., 1941-62; owner, pres. Consultants-Agriservices, Clovis, Calif., 1962—; Distinguished prof. U. Wis.-River Falls, 1963—; collaborator U.S. Dept. Agr., 1965—; adj. prof. Calif. State U., Fresno, 1973—, U. Ariz., Tucson, 1977; cons. nucleonics dept. Gen. Electric Co., AEC, 1947-66; Mem. nat. bd. field advisers SBA, 1959-60; hon. prof. Huazhong Agrl. U., China, 1984. Author: books including The Stockman's Handbook, Animal Science, Beef Cattle Science, Sheep & Goat Science, Swine Science, Dairy Cattle Science, Poultry Sci., Horses and Horsemanship, Horses & Tack, The Complete Encyclopedia of Horses, Horses! Horses! Horses!, Tack! Tack! Tack!, Feeds & Nutrition, Food and Animals: A Global Perspective, The Complete Book of Dogs, (with others) China-The Impossible Dream, Foods & Nutriton Encyclopedia, Food for Health; syndicated columnist: The Stockman's Guide, 1956—, Horses, Horses, Horses, 1962—; works transl. into fgn. langs., also books on record for blind. Mem. adv. bd. People-to-People Found.; pres. Agriservices Found., Pegus Co., Inc. Named to Dept. Agrl. Scis. Hall of Fame, Wash. State U., 1958, Hon. Mem., Indian Council for Farmers, New Delhi., Hon. State Farmer, Future Farmers Am., Hon. Prof., Huazhong Agrl. Coll., Wuhan, Peoples Republic of China; recipient Humanitarian of Yr. award, Acad. Dentistry Internat., 1987, Faculty-Alumni Gold medal, citation of merit U. Mo., 1975, Outstanding Achievement award U. Minn., 1991; Ensminger Beef Cattle Ctr., Wash. State U. named in his honor, 1984; honoree, commd. portrait Saddle and Sirloin Portrait Gallery, Louisville, 1985; recipient Disting. Svc. award AMA, 1993. Fellow AAAS, Am. Soc. Animal Sci. (sec.-treas., v.p., pres. Western sect., Disting. Tchr. award); mem. Am. Genetic Assn. (plaque), Soil Conservation Soc. Am., Am. Soc. Range Mgmt., Am. Dairy Sci. Assn., Am. Soc. Agrl. Cons. (1st pres.), Agrl. Inst. Can., CATEC France (hon. v.p.), Assn. Spanish Purebred Horse Breeders Guatemala (hon.), Sigma Xi, Alpha Zeta, Lambda Gamma Delta. Clubs: Boots and Spurs (hon. life) (Calif. State U. San Luis Obispo) 1973); Dairy (Calif. State U. Fresno) (hon. 1973). Address: 648 W Sierra Ave Clovis CA 93612

ENTHOVEN, DIRK, clinical pharmacologist, researcher; b. Hillegom, The Netherlands, Apr. 26, 1924; came to U.S., 1953; s. Jan and Geertruida (Marseille) E.; m. Maria Anna Zuidema, Sept. 20, 1952; children: Jan Theodore, Pauline Gertrude, Anne Marie, Dirk Henry. MD, U. Leiden, Holland, 1952. Staff physician Sanatorium for Chest Diseases, Catawba, Va., 1955-56; pvt. practice medicine Buena Vista, Va., 1958-65; staff physician Hoffmann-La Roche Inc., Nutley, N.J., 1965-72, asst. med. dir., 1972-76, dir. clin. rsch., 1976-81, dir. specialty team, 1981-91, dir. clin. rsch. mgmt., 1991—, mem. human investigations protocol rev. com., 1980—, head, 1985—. Author 2 book chpts.; contbr. articles to profl. jours. Clk. of session Presbyn. Ch., Wayne, N.J., 1972-78, 84-89; pres. Lions, Buena Vista, Va., 1964. Capt. U.S. Army, 1956-58. Fellow Am. Coll. Gastroenterology; mem. Am. Gastroent. Assn. (coun. 1983-93), Pharm. Mfrs. Assn., Am. Acad. Pharm. Physicians (trustee). Achievements include new drug approvals for Valium as muscle relaxant anti-convulsant, Bactrim in complicated infections, Carprofen as NSAID in rheumatoid arthritis. Office: Hoffmann La Roche Inc 340 Kingsland St Bldg 115/5 Nutley NJ 07110-1199

ENTIAN, KARL-DIETER, microbiology educator; b. Mainz, Fed. Republic Germany, Oct. 4, 1952; s. Karl and Anneliese Mueller; m. Heike Grueter-Entian. Diploma in biology, TH Darmstadt, Fed. Republic Germany, 1977; postgrad., U. East Anglia, Norwich, Eng., 1978, U. Tuebingen, Fed. Republic Germany, 1985. Asst. U. Tuebingen, 1978-87, pvt. docent, Heisenberg scholar, 1987-88; prof. microbiology U. Frankfurt, Fed. Republic Germany, 1988—. Contbr. articles to profl. publs. Deutsche Forschungsgemeinschaft grantee, 1987. Mem. Soc. Biol. Chemistry Germany, Soc. Gen. and Applied Microbiology Germany. Avocations: music, sports, literature. Office: U Frankfurt Inst Microbiol, Niederurseler Hang, D-6000 Frankfurt Germany

EOGA, MICHAEL GERARD, network systems programmer; b. Bronx, N.Y., Oct. 11, 1966; s. Anthony B.J. and Rita (Perri) E. BSEE, Lehigh U., 1984-88; MSEE, Polytechnic U., 1989-91. Mvs systems programmer IBM, Somers, N.Y., 1988-91; newtork systems programmer ISSC/IBM, Sterling Forest, N.Y., 1991-93, distributed computing specialist, 1993—. Mem. IEEE, Eta Kappa Nu. Republican. Roman Catholic. Home: 20 Jinella Ct Boonton NJ 07005-2329 Office: ISSC IBM P O Box 700 Suffern NY 10901

EOM, KIE-BUM, computer engineering educator; b. Seoul, Korea, June 15, 1954; came to U.S., 1981; s. In-Bok and Yoon-Won (Chang) E.; m. Ock-Ja Park, Aug. 1, 1981; children: Robert Joowon, Elizabeth Jooyun. MS, Korea Advanced Inst. Sci., Seoul, 1978; PhD, Purdue U., 1986. Sr. engr. Taihan Elec. Co., Seoul, 1977-81; asst. prof. Syracuse (N.Y.) U., 1986-89; asst. prof. The George Washington U., Washington, 1989-92, assoc. prof., 1992—; cons. Enerlog Systems, Syracuse, 1988-89, Niagara Mohawk Power Corp., Syracuse, 1988-92; session chmn. Midwest Symposium on Circuits and Systems, 1987, Internat. Conf. on Pattern Recognition, 1990; rev. panel mem. NSF, 1991. Author: (with others) Advances in Electronics and Electron Physics, 1988; contbr. articles to profl. jours. Rsch. grantee NSF, 1988. Mem. IEEE (sr. mem., Best Paper award Syracuse br. 1989), Sigma Xi, Tau Beta Pi, Eta Kappa Nu. Roman Catholic. Home: 13823 S Springs Dr Clifton VA 22024 Office: The George Washington Univ Dept Elec Engring/Comp Sci Washington DC 20052

EPEL, LIDIA MARMUREK, dentist; b. Buenos Aires, Argentina, Sept. 30, 1941; came to U.S., 1966; d. Israel and Ita Rosa (Sonabend) Marmurek; children: Diana, Bryan. BS, Buenos Aires U., 1959, DDS, 1964. Lic. dentist, N.Y. Gen. practice dentistry Argentina, 1965-66, Long Beach, N.Y., 1967-70, Lynbrook, N.Y., 1970-73, Rockville Centre, N.Y., 1973—. Bd. dirs. Rosa Lee Young Childhood Ctr., Rockville Centre, 1982—, Rockville Ctr. Edn. Found., 1990—. Mem. ADA, Am. Assn. Gen. Dentistry, Fedn. Dentaire Internat., Nassau County Dental Soc. (bd. dirs., chair com. on pub. and profl. rels. 1990—, chairperson com. on health., treas. exec. com., 1993, chair membership com.), Overseas Dentists Assn. (pres. N.Y. chpt. 1968-72), Dental Soc. of State of N.Y. (coun. for pub. and profl. rels. 1990—, chair children's dental health month campaign 1991), Hadassah Club (bd. dirs. Rockville Ctr. 1983-84, 92-93). Democrat. Jewish. Avocations: painting, traveling. Office: 165 N Village Ave Rockville Centre NY 11570-3701

EPLER, KATHERINE SUSAN, chemist; b. Hershey, Pa., Nov. 25, 1962; d. Jay Elmer and Wilma Ann (Smiley) Epler; m. Perry Neil Sharpless, Dec. 29, 1991. BA, Hood Coll., 1984; PhD, U. Md., 1989. Teaching asst. U. Md., College Park, 1984-86; rsch. asst. U. Md./Nat. Inst. Stds. and Tech., College Park, 1986-89; rsch. chemist Nat. Inst. Stds. and Tech., Gaithersburg, Md., 1989—. Am. Chem. Soc., Soc. for Applied Spectroscopy (editor New Product News 1986—). Home: 4998 Linganore View Dr Monrovia MD 21770 Office: Nat Inst Stds & Tech Chemistry B156 Gaithersburg MD 20899

EPLEY, MARION JAY, oil company executive; b. Hattiesburg, Miss., June 17, 1907; s. Marion Jay and Eva (Quin) E.; m. Dorris Glenn Ervin, Feb. 12, 1934; children: Marion Jay III, Sara Perry (Mrs. Richard H. Davis). LL.B., Tulane U., 1930. Bar: La. 1930. Practiced in New Orleans, 1930-42, 45-47; gen. atty. Texaco, Inc., New Orleans, N.Y.C., 1948-58; v.p., asst. to chmn. bd. Texaco, Inc., N.Y.C., 1958-60; sr. v.p. Texaco, Inc., 1960-61, exec. v.p., 1961-64, pres., 1964-70, chmn. bd., 1970-71; also dir., chmn. bd. Dormar Ltd., 1986-88. Served as lt. USNR, 1942-45. Decorated officer Ordre de la Couronne, Belgium. Mem. ABA, La. Bar Assn., Boston Club, Everglades Club, Bath and Tennis Club, Gov.'s Club. Address: 340 S Ocean Blvd Palm Beach FL 33480

EPPERSON, VAUGHN ELMO, civil engineer; b. Provo, Utah, July 20, 1917; s. Lawrence Theophilus and Mary Loretta (Pritchett) E.; m. Margaret Ann Stewart Hewlett, Mar. 4, 1946; children: Margaret Ann Epperson Hill, Vaughn Hewlett, David Hewlett, Katherine (Mrs. Franz S. Amussen), Lawrence Stewart. BS, U. Utah, 1953. With Pritchett Bros. Constrn. Co., Provo, 1949-50; road design engr. Utah State Road Commn., Salt Lake City, 1951-53, bridge design engr., 1953-54; design engr. Kennecott Copper Corp., Salt Lake City, 1954-60, office engr., 1960-62, sr. engr., 1962, assigned concentrator plant engr., 1969-73, assigned concentrator project engr., 1973-78; cons. engr. Vaughn Epperson Engring. Service, Salt Lake City, 1978-87; project engr. Newbery-State Inc., Salt Lake City, 1980, geneal. computerized research programs, 1983-88, ancestral file programs family history dept. Ch. Jesus Christ of Latter-Day Saints, 1989—. Scoutmaster Troop 190, Salt Lake City, 1949-51. Served to capt. AUS, 1941-45; maj. N.G., 1951; col. Utah State Guard, 1952-70. Decorated Army Commendation medal; recipient Service award Boy Scouts Am., 1949, Community Service award United Fund, 1961, Service award VA Hosp., Salt Lake City, 1977. Mem. ASCE, Am. Soc. Mil. Engrs., Sons of Utah Pioneers. Republican. Mormon. Home: 1537 Laird Ave Salt Lake City UT 84105-1729

EPPINK, ANDREAS, psychologist; b. Gendringen, Gelderland, The Netherlands, Aug. 9, 1946; s. Jan and Henriette (Jansen Spittmann) E. BA in Sociology, U. Amsterdam, The Netherlands, 1968; BA in Psychology, U. Amsterdam, 1969, MA in Psychology, 1971, PhD, 1977, Psychoterapist, 1981. Social researcher U. Koln, Germany, 1972; tchr. Social Acad., Rotterdam, The Netherlands, 1972-77; dir. Averroes Stichting, Amsterdam, 1978-83, Psychologisch Buro, Hilversum, The Netherlands, 1983—; cons. Amsterdam, 1972-78; dir. Omroepadvies, Hilversum, 1990—, Dr. Eppink Adviesgroep, Hilversum, 1991—; expert Intergovernmental Com. European Migration, Geneva, 1978-79; forensic expert Dist. Ct., Amsterdam, 1979—. Author: Cultuurverschillen en Communicatie, 1981, Het Masker van de Pijn, 1990; editor: Kind Zijn in Twee Culturen, 1981, Cultuurcontact en Cultuurconflict, 1988. Fellow Assn. Francaise de Psychiatrie et de Psychopathologie Sociales; mem. Nederlands Inst. van Psychologen, Internat. Coun. Psychologists. Avocations: Andalucian and Arabic culture, my Spanish home. Office: Dr Eppink Adviesgroep, Burg Lambooylaan 11, 1217 LB Hilversum The Netherlands

EPRIGHT, CHARLES JOHN, engineer; b. Bklyn., Jan. 11, 1932; s. Charles and Margaret Mary (Tripoli) E.; m. Mary Lucy Bono, May 29, 1954; children: Daniel John, Michael James, Marisa R. Becker, Victoria Epright Carmona, Maria Carmela. BS in Math., U. Nev., 1965; MS in Engring. Mgmt., Northeastern U., 1971. Sr. engr. Raytheon, Andover, Mass., 1970-78, Delmo-Victor, Belmont, Calif., 1978-79; advanced systems engring. specialist Lockheed Missile & Space Co., Austin, Tex., Sunnyvale, Calif., 1979-87; engring. scientist Tracor Aerospace, Austin, Tex., 1987-89, staff engr. Lockheed Engring. and Sci. Co., Houston, 1989—; cons. Colo. Electronic Tng. Ctr., Colorado Springs., 1968-69. Civic adv. Salem-in-Action, N.H., 1977-79; dir. Reachout, Salem, 1976-79; community action com. mem. N.H. Com. for Adopted and Foster Children, Manchester, 1978-79, Runaway Hotline, Austin, 1984-88, Middle Earth Spectrum Shelter, 1987-89; mem. pub. responsibility com. Mental Health/Mental Retardation, Austin, 1988-89; bd. dirs., treas. Assn. Retarded Citizens, 1989—; mem. outreach Covenant House Tex., Houston, 1990—. With USAF, 1950-70. Decorated Legion of Merit, 1969; recipient Family of Yr. award Sons of Italy, 1968, 69. Mem. Internat. Assn. Elec. and Electronic Engrs., Air Force Assn. (life), Assn. Old Crows, Am. Inst. Am. Scientists, DAV. Roman Catholic. Lodge: KC (grand knight 1968-69). Avocations: stamp collecting; photography; collecting old books. Home: 2012 Fairfield Ct N League City TX 77573-3504 Office: Lockheed Engring and Sci Co 1150 Gemini St # 23A Houston TX 77058-2742

EPSTEIN, DAVID AARON, biochemist; b. Phila., Dec. 30, 1942; s. Benjamin and Molly (Hoffman) E.; m. Eugenia Gurfinkiel, Aug. 18, 1968; children: Miriam, Adva. BA, U. Calif., Berkeley, 1965; MA, Brandeis U., 1970; DSc, The Technion Sch. Medicine, 1976. Rsch. asst. U. Hosp. Dept. Pathology, Boston, 1968-69; rsch. assoc. St. Vincent's Hosp. Dept. Medicine, Worcester, Mass., 1969-71; instr. Biochemistry Dept. Med. Sch. Technion, Haifa, Israel, 1972-76; project leader Internat. Biotechnologies Ltd., Jerusalem, Israel, 1983-84, Biol. Inds. Ltd., Beth Haemek, Israel, 1984-85; sr. scientist Life Techs. Inc., Grand Island, N.Y., 1985-90, prin. scientist, 1990—. Contbr. articles to profl. jours. NIH Staff fellow, 1976-80, Sr. Staff fellow, 1980-82; Brandeis U. scholar, 1965-68; recipient David Coffin award, 1988, 90, 92. Mem. AAAS, Internat. Soc. Hort. Sci., N.Y. Acad. Sci., Soc. Cell Transplantation. Home: 7 Shady Grove Dr East Amherst NY 14051 Office: Life Techs Inc 2086 Grand Island Blvd Grand Island NY 14072

EPSTEIN, GERALD LEWIS, technology policy analyst; b. Washington, Dec. 13, 1956; s. Joseph Bernard and Rosalie E.; m. Ellen Mika, June 30, 1985; children: Alanna, Nathan. SB, MIT, 1978; MA, U. Calif., Berkeley, 1980, PhD in Physics, 1984. Analyst Office Tech. Assessment, Washington, 1983-87, sr. analyst, 1987-89, 91—; project dir. Kennedy Sch. Govt., Harvard U., Cambridge, Mass., 1989-91. Co-author: Beyond Spinoff: Military and Commercial Technologies in a Changing World, 1992; project dir.: Starpower: The U.S. and the International Quest for Fusion Energy, 1987. Fannie and John Hertz Found. fellow, 1978; Congl. fellow Office Tech. Assessment, 1983. Mem. AAAS, Am. Phys. Soc., Phi Beta Kappa, Sigma Xi, Tau Beta Pi. Home: 6008 Anniston Rd Bethesda MD 20817 Office: Office Tech Assessment US Congress Washington DC 20510

EPSTEIN, JONATHAN STONE, engineering executive; b. White Plains, N.Y., May 11, 1957; s. Gerald Samual Epstein and Mary Holt (Griffen) Wilson; m. Susan Ann Chavez, July 3, 1985. BS in ME, Colo. State U., Ft. Collins, 1980; PhD, Va. Poly. Inst., Blacksburg, 1983. Postdoctoral fellow Oxford (Eng.) U., 1984; engring. specialist EG&G Idaho, Inc., Idaho Falls, 1984-88, sr. engring. specialist, 1990—; asst. prof. Ga. Inst. Tech., Atlanta, 1988-90; cons. Rockwell Missile Div., Atlanta, 1988-90, EG&G Idaho, Inc., 1988-90; reviewer U.S. Dept. Energy, Washington, 1984-90. Editor: Experimental Technique in Fracture, 1990, Optics and Lasers in Engineering, 1990; mem. editorial bd. Experimental Mechanics, 1990; contbr. articles to profl. jours. Mem. ASME, Soc. Engring. Sci., Soc. Exptl. Mechanics, Sigma Chi. Achievements include patents in Dynamic Moire Interferometry, High Temperature Moire, Protective Space Shielding, and Armor/Anti-Armor Devices. Office: EG&G Idaho Inc PO Box 1625 Idaho Falls ID 83415-0001

EPSTEIN, JOSEPH ALLEN, neurosurgeon; b. Bklyn., Oct. 3, 1917; s. Alexander and Sarah (Schwartz) E.; m. Natalie Rhoda, Nov. 15, 1951; children: Janet, Nancy. BS, CCNY, 1938; MD, L.I. Coll. of Medicine, 1942. Diplomate Am. Bd. Neurosurgery. Intern Jewish Hosp. of Bklyn., May Land, N.Y., 1943; resident Mt. Sinai Hosp., N.Y.C., N.Y., 1947-49, Beth Israel Hosp., N.Y.C., N.Y., 1949-51; attending neurosurgeon North Shore Cornell Hosp., Manhasset, N.Y., 1960—, L.I. Jewish Hosp., New Hyde Park, N.Y., 1960—; clin. prof. Albert Einstein Coll. Medicine, Bronx, N.Y., 1986—; clin. assoc. prof. Cornell Med. Sch./North Shore Hosp. Divsn., 1970—. Contbr. articles to profl. jours. Capt. U.S. Med. Corps, 1942-46. Fellow Cervical Spine Rsch. Soc. (past pres.), Internat. Soc. for Study of Lumbar Spine. Jewish. Home: 4 Kings Dr Old Westbury NY 11568 Office: LI Neurosurgical Assn 410 Lakeville Rd New Hyde Park NY 11042

EPSTEIN, SAMUEL SETH, computer scientist, researcher; b. Morristown, N.J., Sept. 12, 1948; s. Benjamin and Edith Florence (Lobel) E.; m. Michele Miller, Sept. 5, 1974 (div. 1993); children: Aaron Daniel, Rachel Elisabeth. SB in Math., MIT, 1970; MS in Math., Stanford U., 1973; PhD in Linguistics, U. Calif., San Diego, 1976. Postdoctoral fellow IBM Thomas J. Watson Rsch. Ctr., Yorktown Heights, N.Y., 1976-77, U. Paris, 1977-78, Ecole des Hautes Etudes, Paris, 1978-79; mem. tech. staff Bell Labs., various locations, N.J., 1979-83; mem. tech. staff, prin. investigator Bell Communications Rsch., Morristown, N.J., 1984—; referee various confs., jours. and granting instns.; invited lectr., discussant various profl. orgns. Contbr. articles on computer lang. design and implementation, mechanized lang. processing natural lang. syntax and semantics to profl. jours., books, conf. procs. Vol. lectr., cons. Mendham (N.J.) Borough Schs., 1989—; counselor MIT Ednl. Coun., Cambridge, Mass., 1991—. NATO postdoctoral fellow NSF, 1978; exch. fellow Ctr. Nat. Rsch. Sci., Paris, 1977. Mem. Assn. Computing Machinery, Assn. Computational Linguistics, Am. Assn. Artificial Intelligence, Linguistic Soc. Am. Achievements include research in computer languages with potential for user fluency, extensions to expressive power of computer languages, application of government-binding theory to language processing technology, syntax, semantics of natural langs. Office: Bell Communications Rsch 445 South St Morristown NJ 07960

EPSTEIN, SANDRA GAIL, psychologist; b. Boston, July 19, 1939; d. Mischa and Frances (Greenfield) Schneiderman; 1 child, Suanne Charyl. AB, Boston U., 1962; MA, U. Conn., 1969, diploma, 1978, PhD, 1979. Sch. psychol. examiner various pub. schs., Conn., 1970-73, sch. psychologist, 1973—; staff psychologist Day Kimball Hosp. Mental Health Clinic, Putnam, Conn., 1971-74; psychologist Thompson (Conn.) Med. Ctr., 1973-80; pvt. practice Woodstock, Farmington, Conn., 1982-85, Putnam, Farmington, 1985—; instr. Annhurst Coll., Woodstock, Conn., 1971-76; cons. N.E. Area Regional Ednl. Svc., Wauregan, Conn., 1978-79, Capitol Region Ednl. Coun., West Hartford, Conn., 1980—, Ctr. for Interpersonal Rels., Putnam, 1985-88, Hebrew Acad. Greater Hartford, Bloomfield, Conn., 1983-89. Mem. Am. Psychol. Assn., Internat. Soc. Hypnosis, Internat. Psychosomatics Inst., N.Y. Acad. Sci., Conn. Psychol. Assn., Am. Soc. Clin. Hypnosis, Am. Acad. Pain Mgmt., N.Y. Acad. Sci. Office: 365 Woodstock Ave Putnam CT 06260-1015

EPSTEIN, SCOTT MITCHELL, engineering executive; b. Boston, Feb. 2, 1958; s. Martin Epstein and Irene Silver. BSME, Wentworth Inst., Boston, 1978. Thermedics engr. Woburn, Mass., 1988; engr. Glens Falls, N.Y., 1989; owner, engr. Med. Device Labs., Holliston, Mass., 1992, Bodyguard Polymer Protection Products, Holliston, 1992, SME Design, Inc., Holliston, 1984—; cons. SME Design, Holliston, 1989—. Mem. ASME, Soc. Plastic Engrs. Achievements include patents for ctrl. veins cathets-reduce distal tip thrombosis due to stasis, long term percutaneous access site tech., radiation resistant polymer and methods of mfg., organ transplant cortainer, polyurethene condoms, gloves and methods of manufacturing. Office: SME Design 118 Washington St Holliston MA 01746

EPSTEIN, SETH PAUL, immunologist, researcher; b. N.Y.C., Sept. 11, 1958; s. Donald and Eileen (Schulman) E.; m. Elizabeth Koppekin, June 15, 1991. BA in Chemistry with high honors, Brandeis U., 1980; MD, Autonomous U. Guadalajara, Mex., 1984. Med. extern Pontiac (Mich.) Gen. Hosp., 1984; postdoctoral rsch. fellow Mich. Cancer Found., Detroit, 1985-86; postdoctoral rsch. fellow NYU Med. Ctr., N.Y.C., 1987-91, asst. rsch. scientist, 1991; rsch. asst. Mt. Sinai Med. Ctr., N.Y.C., 1991—. Contbr. articles to profl. jours. Tng. fellow NIH, 1987; grantee Dermatology Found., Inc., 1990. Mem. Assn. Rsch. in Vision and Ophthalmology, Phi Beta Kappa. Achievements include research on Cyclosporine A relating to cytokine-induced upregulation of Langerhans cells, cell chemotaxis into the cornea and skin and treatment of herpetic keratitis; sunscreen prevention of Ultraviolet-activated herpes simplex. Office: Mount Sinai Med Ctr Dept Ophthalmology 1 Gustave L Levy Pl New York NY 10029

EPSTEIN, WILLIAM LOUIS, dermatologist, educator; b. Cleve., Sept. 6, 1925; s. Norman N. and Gertrude (Hirsch) E.; m. Joan Goldman, Jan. 29, 1954; children—Wendy, Steven. A.B., U. Calif., Berkeley, 1949, M.D., 1952. Mem. faculty U. Calif., San Francisco, 1957—; assoc. prof. div. dermatology U. Calif., 1963-69, prof. div. dermatology, 1969—, dir. dermatol. rsch., 1957-70, acting chmn. div. dermatology, 1966-69, chmn. dept. dermatology, 1970-85; cons. dermatology Outpatient Dept.; cons. various hosps. Calif. Dept. Public Health; cons. Food and Drug Adminstrn., Washington, 1972—, Dept. Agriculture, 1979; dir. div. research Nat. Program Dermatology, 1970-73; Dohi lectr., Tokyo, 1982; Beecham lectr., 1988-89; Nippon Boehringer Ingelheim lectr. 18th Hakone Symposium on Respiration, Japan, 1990. Mem. AAAS, AMA, Am. Soc. Cell Biology, Am. Acad. Dermatology and Syphlology (nominating com. 1984), Pacific Dermatologic Assn., Am. Fedn. Clin. Rsch., Soc. Investigative Dermatology (bd. dirs., pres. 1985), Am. Dermatol. Assn., Assn. Profs. Dermatology (sr. mem.), Dermatology Found. (pres. 1986-87), Phi Beta Kappa, Sigma Xi. Home: 267 Goldenhind Passage Corte Madera CA 94925

ERB, DORETTA LOUISE BARKER, polymer applications scientist; b. Upper Darby, Pa., June 21, 1932; d. Ralph Merton and Pauline Kaufman (Isenberg) B.; m. Robert Allan Erb, June 27, 1953; children: Sylvia Ann, Susan Doretta, Carolyn Joy. BS in Pharmacy, Phila. Coll. Pharmacy and Sci., 1954. Registered pharmacist, Pa. Pharmacist Borland's Pharmacy, Upland, Pa., 1954-65; assoc. scientist Franklin Rsch. Ctr., ATC div. Calspan Corp., Norristown, Pa., 1974-93; sole propr., pres. Siliclone Studio, Valley Forge, Pa., 1982—. Mem. Am. Anaplastology Assn., Sigma Xi (chpt. sec. 1990-93). Presbyterian. Achievements include co-invention of intrinsic coloration techniques for highly realistic external prostheses. Home and Office: PO Box 86 Jug Hollow Rd Valley Forge PA 19481-0086

ERB, KARL ALBERT, physicist, government official; b. Chgo., June 30, 1942; s. Edgar Gillette and Dorothy (Carsten) E.; m. Betty G. Hesse, June 22, 1963; children: Janet, Margaret. BA, NYU, 1965; MS, U. Mich., 1966, PhD, 1970. Instr. U. Pitts., 1970-72; instr., asst. prof., assoc. prof. Yale U.,

New Haven, 1972-80; staff scientist Oak Ridge (Tenn.) Nat. Lab., 1980-86; program dir. NSF, Washington, 1986-89, dep. dir. physics div., 1991; asst. dir. White House Office Sci. and Tech. Policy, Washington, 1989-91; acting assoc. dir. for phys. scis. & engring. White House Office of Sci. and Tech. Policy, Washington, 1991-92, assoc. dir. for phys. scis. engring., 1992-93; sr. sci. advisor NSF, 1993—; mem. U.S. Nuclear Sci. Adv. Com., Washington, 1983-86. Contbr. over 50 articles to physics jours. and encys., chpts. to books. Mem. Am. Phys. Soc. Achievements include exptl. research on molecular resonances nuclear reactions of heavy nuclei and nuclear structure and accelerator technology. Office: NSF Office of Dir Washington DC 20550

ERBEL, RAIMUND, physician, educator; b. Baesweiler, Fed. Republic of Germany, Mar. 9, 1948; s. Peter and Maria (Loogen) E.; m. Hldegard Schürmann, Aug. 3, 1978; children: Susanne, Christian, Sebastian, Matthias. MD, U. Düsseldorf, 1974. Asst. St. Josef Krankenhaus, Leverkusen, 1973-74, U. Düsseldorf, 1974-75; stabsarzt Bundeswehrzentralkrankenhaus, Koblenz, 1975-77; asst. RWTH, Aachen, 1977-82; lectr. U. Mainz, 1982—, prof. medicine, 1983-93; dir. dept. cardiology U. Essen, 1993—. Co-editor: Bildgebende Vertahren in der Diagnostik von Herzerkrankungen, 1991, Transeiophageal Ecto Cardiography, 1989. Recipient Paul Beiersdorf award, 1983. Fellow Am. Coll. Cardiology, European Soc. Cardiology; mem. Am. Soc. Echocardiography, German Soc. Biomedicine. Office: U Essen Divsn Internal Med, Dept Cardiology Hufelandstr 55, D-45122 Essen Germany

ERBIL, AHMET, physics educator; b. Zile, Tokat, Turkey, Dec. 8, 1955; came to U.S., 1977; s. Ismail and Ismahan E.; m. Amy Church, Oct. 8, 1978; children: William Kaya, Suna Anne. BS, Ankara U., 1976; PhD, MIT, 1983. Grad. teaching asst. MIT, 1980; assoc. mem. tech. staff Bell Telephone Labs., Murray Hill, 1980; grad. rsch. asst. MIT, 1980-83; cons. IBM-Watson Rsch. Ctr., 1985, Tex. Instruments, 1985—, Ionic Atlanta, 1985-87; asst. prof. Ga. tech., 1985-91, assoc. prof., 1991—; vis. scientist IBM-Watson Rsch. Ctr., 1983-85, Tex. Instruments, 1985; vis. prof. Ecole Polytech Fed. de Lausanne, Switzerland, 1991. Contbr. articles to profl. jours. speaker and presenter in field. NATO fellow, 1976-80, Scientific and Tech. Rsch. Coun. Turkey fellow, 1972-76, Alfred P. Sloan Rsch. fellow, 1985-87; recipient Materials Rsch. Soc. Outstanding Grad. Student Rsch. award, 1982. Mem. Am. Ceramic Soc., Am. Phys. Soc., Internat. Soc. Optical Engring., Materials Rsch. Soc. Achievements include 5 patents in chemical vapor deposition. Office: Ga Tech Sch Physics Atlanta GA 30332-0430

ERCE, IGNACIO, III, aerospace engineer; b. Zaragoza, Spain, June 25, 1964; s. Ignacio Erce II and Carmela Garcia; m. Itziar Urrutia, Mar. 2, 1991. B. in Aeronautics and Astronautics, U. Wash., 1987, M. in Aeronautics and Astronautics, 1989; MBA, U. Deusto, Spain, 1992. Computing engr. wind tunnel testing Boeing Comml. Airplane Co., U. Wash., Seattle, 1985-87; aerospace engr., project mgr. Sener Tecnica Indsl. y Naval S.A., Bilbao, Spain, 1989—. Mem. AIAA, Planetary Soc., Real Aero Club de Espana, Tau Beta Pi. Home: Ormetxe 19, 48990 Getxo Spain Office: Sener, Avd Zugazarte 56, 48930 Las Arenas Spain

ERDBERG, MINDEL RUTH, psychiatrist; b. N.Y.C., Sept. 20, 1916; d. Samson Erdberg and Sarah (Shapiro) Erdberg; m. J. Levitz (div. 1975); 1 child, Beverly Lee Levitz Vosko. MD, Lausanne U., Switzerland, 1942. Diplomate Am. Bd. Psychiatry and Neurology. Intern Bellevue Hosp., N.Y.C., 1942-43, asst. resident in medicine, 1943-44; resident in medicine Mt. Sinai Hosp., N.Y.C., 1944-45, rsch. fellow in medicine, 1945-46; resident in psychiatry Cen. Islip (N.Y.) State Hosp., 1959-62; sr. psychiatrist Cen. Islip (N.Y.) State Hosp., Woodmere, N.Y., 1962-63; staff psychiatrist Ctrl. Island Community Mental Health Ctr., Uniondale, L.I., 1963-67; supervising psychiatrist, asst. dir. Suffolk Psychiatric Hosp., Ctrl. Islip, L.I., 1967-72; sr. psychiatrist II Creedmoor Psychiat. Ctr., 1972-83; pvt. practice Woodmere, N.Y., 1983—; vol. attending L.I. Jewish Med. Ctr.; attending psychiatrist Franklin Hosp. Med. Ctr., Valley Stream, N.Y., South Nassau Communities Hosp., Oceanside, N.Y., St. John's Episcopal Hosp.; cons. in psychiatry Peninsula Hosp. Ctr., Far Rockaway, N.Y. Contbr. articles to profl. jours. Mem. Am. Psychiatric Assn., Nassau Psychiatric Soc., League Women Voters. Home: 366 Longacre Ave Woodmere NY 11598-2417

ERDEMIR, ALI, materials scientist; b. Kadirli, Adana, Turkey, July 2, 1954; came to U.S., 1979; s. Haci Ali and Hasibe (Tamer) E.; m. Ruth Marie Haegele, Nov. 25, 1981; children: Altan, Kenan. MS in Metallurgy, Ga. Inst. Tech., 1982, PhD in Materials Engring., 1986. Metall. engr. Iskenderun Iron and Steel Works, Iskenderun, Hatay, Turkey, 1977-79; rsch. asst. Ga. Inst. Tech., Atlanta, 1983-86, postdoctoral fellow, 1986-87; asst. rsch. scientist Argonne (Ill.) Nat. Lab., 1987-90, assoc. rsch. scientist, 1990—. Recipient R&D 100 award R&D Mag., Chgo., 1991. Mem. Am. Soc. Metals Internat., Soc. Tribologist and Lubrication Engrs. (Al Sonntag award 1992), Am. Ceramic Soc., The Metall. Soc., Sigma Xi. Achievements include patent pending for discovery of solid lubrication mechanism of boric acid; devel. of solid and liquid lubricants for high-temperature applications, Synergistic lubricants for ceramics. Home: 439 Westglen Dr Naperville IL 60565 Office: Argonne Nat Lab 9700 S Cass Ave Argonne IL 60439

ERDMANN, JOHN BAIRD, environmental engineer; b. Mpls., Feb. 24, 1950; s. Robert Keith and Mary Ann (Baird) E.; m. Ann Marie Freed, June 12, 1969 (div. 1976); 1 child, Rachel; m. Diane Lee Johnson, Apr. 24, 1982; children: Saleha, Daniel. AB in Engring. and Applied Physics, Harvard U., 1972; MS in Civil Engring., U. Minn., 1990. Registered profl. engr., Minn. Rsch. assoc. New Eng. Aquarium Rsch. Dept., Boston, 1972-73; asst. engr. Mass. Div. Water Pollution Control, Westborough, 1973-77; environ. engr. Eugene A. Hickok & Assocs., Inc., Wayzata, Minn., 1977-85; prin. environ. engr. Wenck Assocs., Inc., Maple Plain, Minn., 1985—. Contbr. articles to profl. jours. Dir., water quality chair Charles River Watershed Assn., Newton, Mass., 1974-76. Mem. ASCE, Assn. Ground Water Scientists and Engrs., Water Environ. Fedn., Minn. Lake Mgmt. Fedn. (sec. 1989-90). Democrat. Episcopalian. Achievements include new methods of river water quality data analysis including visualization of dissolved oxygen concentrations in space and time, and calculation of photosynthesis and respiration rates; new understanding of groundwater pumpout capture zone dimensions, important in groundwater contamination cleanup. Office: Wenck Assocs Inc 1800 Pioneer Creek Ctr Maple Plain MN 55359

ERDREICH, JOHN, acoustician, consultant; b. N.Y.C., Mar. 13, 1943; s. Harry and Esther E.; m. Linda Schuman, Aug. 7, 1966; children: David, Lauren. BSEE, N.J. Inst. Tech., 1965; PhD, U. Mich., 1977. Engr. Nat. Security Agy., Ft. Meade, Md., 1965-66, Vitro Labs., Silver Spring, Md., 1966-68; teaching asst. U. Mich., Ann Arbor, 1968-73; asst. prof. otolaryngology U. Okla., Oklahoma City, 1973-80; rsch. engr. Nat. Inst. Occupational Safety & Health, Cin., 1980-86; prin. Ostergaard Acoustical Assocs., West Orange, N.J., 1986—; cons. editor Am. Inst. Physics-Modern Acoustics & Signal Processing, N.Y.C., 1992—; bd. dirs. Nat. Coun. Acoustical Cons., Springfield, N.J., 1991—; cons. numerous ANSI Acoustical Stds. groups, N.Y.C., 1981—. Contbr. articles to profl. jours., chpt. to book. Fellow Acoustical Soc. Am. (chmn. tech. com. 1984-87); mem. IEEE (sr.), Assn. for Rsch. in Otolaryngology. Achievements include research on classification of occupational noise exposures and the relationship between exposure and hearing loss. Office: Ostergaard Acoustical Assoc 100 Executive Dr West Orange NJ 07052

ERFANI, SHERVIN, electrical engineer; b. Tehran, Iran, Mar. 28, 1948; came to U.S. 1982; s. Ibrahim and Rashedeh (Naghari) Erfani; m. Janet E. Kovar, Dec. 30, 1982. MSEE, U. Tehran, Iran, 1971; MS, So. Meth. U., 1974, PhD in EE, 1976. Asst. prof. Nat. U. Iran, Eveen, 1978-82; research assoc. So. Meth. U., Dallas, 1982-83; asst. prof. U. Mich., Dearborn, 1983-85; mem. tech. staff AT&T Bell Labs., Holmdel, N.J., 1985—; vis. prof. U. Puerto Rico. Translator: Elec. Engring. textbook, Circuit Design & Synthesis, 1985; assoc. editor Computers and Elec. Engring.: An Internat. Jour., Jour. of Network and Systems Mgmt.; contbr. articles to profl. jours. 2nd lt. Signal Corps Iran Army, 1972-73. Mem. IEEE (sr. mem., v.p. S.E. Mich. chpt. 1985), N.Y. Acad. Scis., Tau Beta Pi, Eta Kappa Nu. Islam. Avocations: flying, numismatics, antiques, philately. Home: 82 Statesir Pl Red Bank NJ 07701-6128 Office: AT&T Bell Labs 200 Laurel Ave Middletown NJ 07748

ERGAS, ENRIQUE, orthopedic surgeon; b. Santiago, Chile, Oct. 8, 1938; came to U.S., 1964; s. Jaime and Rebecca E.; m. Joscelyn Krauss, June 20, 1955; children: Eileen, Jamie, Arielle. MD, U. Chile, 1964. Diplomate Am. Bd. Orthopaedic Surgery. Intern Methodist Hosp., Bklyn., 1964-65; resident in gen. surgery Mt. Sinai Hosp., N.Y.C., 1965-67; resident in orthopaedic surgery Albert Einstein Coll. Medicine, N.Y.C., 1967-71; practice medicine specializing in orthopaedic surgery N.Y.C., 1971—; asst. clin. prof. orthopaedic surgery Albert Einstein Coll. Medicine, N.Y.C., 1973-87, NYU, 1987—; dir. Latin Am. programs, Latin Am. fellowships Hosps. for Joint Diseases Orthopaedic Inst.; lectr., moderator various hosps. and colls. Sci. exhibit Am. Acad. Orthopaedic Surgery, New Orleans, 1982; contbr. articles to profl. jours. Bd. dirs., sec. med. bd. Trafalgar Hosp., 1976-77; mem. panel determination mal practice Supreme Ct., N.Y., 1987. Recipient Masada award Hadassah Hosp., 1974, Order of Merit cum laude, Orthopaedic Research Soc., 1985. Fellow Am. Acad. Orthopaedic Surgeons, ACS; mem. Arthroscopy Assn. N.Am., Internat. Arthroscopy Assn., Gericare (pres. 1987-89), AMA, N.Y. State Soc. Orthopaedic Surgeons, Med. Soc. County New York, N.Y. Acad. Medicine, Soc. Latim Am. Orthopedia Traumatologia, Spanish Am. Med. Soc., Chilean Soc. Orthopaedic and Traumatology (corr.), Am. Assn. French Speaking Health Profls., Montefiore Orthopaedic Alumni Assn. (regional bd. dirs.). Office: 1056 5th Ave New York NY 10028-0112

ERICKSEN, JERALD LAVERNE, educator, engineering scientist; b. Portland, Oreg., Dec. 20, 1924; s. Adolph and Ethel Rebecca (Correy) E.; m. Marion Ella Pook, Feb. 24, 1946; children: Lynn Christine, Randolph Peder. B.S., U. Wash., 1947; M.A., Oreg. State Coll., 1949; Ph.D., Ind. U., 1951; D.Sc. (hon.), Nat. U. Ireland, 1984, Heriot-Watt U., 1988. Mathematician, solid state physicist U.S. Naval Research Lab., 1951-57; faculty Johns Hopkins U., 1957-83, prof. theoretical mechanics, 1960-83; prof. mechanics and math. U. Minn., Mpls., 1983-90; cons. Florence, Oreg., 1990—. Served with USNR, 1943-46. Recipient Bingham medal, 1968, Timoshenko medal, 1979, Engring. Sci. medal, 1987. Mem. Nat. Acad. Engring., Soc. Rheology, Soc. Natural Philosophy, Soc. Interaction Mechanics and Math., Soc. Engring. Sci. Home and Office: 5378 Buckskin Bob Dr Florence OR 97439-8320

ERICKSON, BRICE CARL, chemist; b. Mora, Minn., Dec. 12, 1957; s. Reynold Milton and Marian (Markgren) E.; m. Shirley Jane Howe, Oct. 22, 1983; children: Jared Carl, Grant Christian, Collin Brice. BA magna cum laude, Concordia Coll., Moorhead, Minn., 1979; PhD, U. Wash., 1988. Rsch. tech. N.D. State U., Fargo, 1980; chemist Henkel Corp., Mpls., 1980, N.D. Dept. Health, Bismarck, 1981-84; sr. chemist Chemolite Ctr., 3M Co., St. Paul, 1990—; vis. asst. prof. Macalester Coll., St. Paul, 1988-90. Contbr. articles to profl. jours. Mem. Am. Chem. Soc., Minn. Chromatography Forum. Achievements include research in chemometrics, infrared emission spectroscopy, flow injection analysis. Office: 3M Co 3M Ctr MS 236-2B-11 Saint Paul MN 55144-1000

ERICKSON, HOMER THEODORE, horticulture educator; b. Pulaski, Wis., Mar. 8, 1925; s. Elmer and Luella (Thorson) E.; m. Carolyn J. Cochran, Sept. 10, 1955; children—Ann, Jean, Charles, Neal. B.S., U. Wis., 1951, M.S., 1953, Prof. Honoris Causa, 1954; Prof. Honorus Causa, Fed. U., Vicosa, Minas Gerais, Brazil, 1963. Asst. prof. U. Maine, 1954-56; mem. faculty Purdue U., 1956, 1956—; prof. horticulture Purdue U., 1964-92; prof. emeritus Purdue, U., 1993—; head dept. Purdue U., 1967-75; research center coordinator with Spanish govt., Zaragoza, 1975-76; cons. in horticulture, Ghana, 1972, Brazil, 1982, Liberia, 1984, Zimbabwe, Mozambique and Jamaica, 1985, Zaire, 1986. Served with AUS, 946-47. Mem. Am. Soc. Hort. Sci., Nat. Audubon Soc., Delta Theta Sigma; founder, hon. mem. Brazilian Hort. Soc. Lutheran. Club: Optimist Internat. Home: 1409 N Salisbury St West Lafayette IN 47906-2419

ERICKSON, JOHN RONALD, research administrator; b. Sidnaw, Mich., Jan. 8, 1934; s. John August and Aileen (Hendrickson) E.; m. Marion Stefani Erickson, Feb. 28, 1935; children: Kim Maki, John, Beth, James. BSME, Mich. Tech. U., 1956, MSME, 1968. Engr. Ladish Co., Cadahy, Wis., 1956-60, Kenosha, Wis., 1960-62; rsch. Forest Svc. USDA, Houghton, Mich., 1962-75; rsch. adminstr. Forest Svc. USDA, Washington; rsch. adminstr. forest products lab. USDA, Madison, Wis., 1983-85, 1985--. Mem. Soc. Am. Foresters, Am. Forestry Assn., Hardwood Rsch. Council, Forest Producs Rsch. Soc., Internat. Union Forestry Orgns, Fellow Internat. Acad. Wood Sci. Office: Forest Products Lab Forest 1 Gifford Pinchot Dr Madison WI 53705-2398

ERICKSON, ROBERT ANDERS, optical engineer, physicist; b. Benson, Minn., Aug. 6, 1962; s. Wilton Robert and Irene Dorothy (Fenstra) E. BS in Physics, S.D. Sch. Mines and Tech., 1985; MS in Physics, U. Mo., St. Louis, 1989. Instr. physics S.D. Sch. Mines and Tech., Rapid City, 1984-85; optical engr., physicist McDonnell Douglas Corp., St. Louis, 1985—. Twin Cities scholar, 1981, Frank & Portia Vanlueve scholar, 1983. Mem. IEEE, Optical Soc. Am., Am. Inst. Physics, Sigma Pi Sigma (chpt. pres. 1984-85). Republican. Lutheran. Home: 5426 Forest Creek Dr Apt K Hazelwood MO 63042-3745 Office: McDonnell Douglas Corp Box 516 Dept 257 MC 1022172 Saint Louis MO 63166

ERIKSEN, CHARLES WALTER, psychologist, educator; b. Omaha, Feb. 4, 1923; s. Charles Hans and Luella (Carlson) E.; m. Garnita Tharp, July 22, 1945 (div. Jan. 1971); children—Michael John, Kathy Ann; m. Barbara Becker, Apr. 1971. BA, U. Omaha, 1943; PhD, Stanford, 1950. Asst. prof. Johns Hopkins, 1949-53, research scientist, 1954-55; lectr. Harvard, 1953-54; faculty U. Ill., Urbana, 1956—; prof. U. Ill., 1959—; Prop. farm; research cons. VA, 1960-80; mem. psycho-biology panel NSF, 1963; mem. expt. psychology study sect. NIH, 1958-62, 66-70. Author: Behavior and Awareness, 1962; Editor: Am. Jour. Psychology, 1968; prin. editor: Perception and Psycho Physics, 1971—; cons. editor: Jour. Exptl. Psychology, 1965-71, Jour. Gerontology, 1980—; Contbr. articles to profl. jours. Recipient Stratton award Am. Psychopath. Assn., 1964, NIMH Research Career award, 1964. Mem. Am. Psychol. Soc., Psychonomic Soc., Soc. Exptl. Psychologists, Midwestern Psychol. Assn., AAAS, Sigma Xi. Home: RR 1 Box 67 Oakland IL 61943-9713 Office: U Ill Psychol Bldg 603 E Daniel St Champaign IL 61820-6267

ERIKSEN, CLYDE HEDMAN, ecology educator; b. Santa Barbara, Calif., May 1, 1933; s. Hedman Carl and Virginia Alice (LaSource) E.; divorced; children: Helene Elizabeth, Maria Catherine; m. Lillian Marie Pfannenstiel, Sept. 8, 1990. BA, U. Calif., Santa Barbara, 1955; MS, U. Ill., 1957; PhD, U. Mich., 1961. Asst. prof. zoology Calif. State U., L.A., 1960-63; assoc. prof. zoology U. Toronto, 1963-67; prof. biology The Claremont (Calif.) Colls., 1967—, chmn. joint sci. dept., 1968-72, dir. Bernard biol. sta., 1977—; ecological & adminstrv. specialist U.S. Forest Svc., Dillon, Mont., 1973; vis. rsch. entomologist U. Calif., Berkeley, Calif., 1981; cons. in land mgmt. U.S. Forest Svc., 1971-79, 84-91; accreditation teams mem. Western Assn. Schs. & Colls., Oakland, Calif., 1970-91; curriculum devel. evaluator Linfield Coll., McMinnville, Oreg., 1978, 81. Contbr. chpts. to books and articles to profl. jours. Spokesman Environ. Resources Task Force City of Claremont, 1970-71; cons. endangered species County of Riverside, Calif., 1988-91. Rsch. grantee, 1956—. Mem. N.Am. Benthological Soc., Ecological Soc. Am., Internat. Assn. Theoretical & Applied Limnology, Conservation Assn., Crustacean Soc. Office: The Claremont Colls Keck Sci Ctr Claremont CA 91711

ERIKSON, GEORGE EMIL (ERIK ERIKSON), anatomist, archivist, historian, educator, information specialist; b. Palmer, Mass., May 3, 1920; s. Emil and Sofia (Gustafson) E.; m. Suzanne J. Henderson, Apr. 23, 1950; children: Ann, David, John, Thomas. BS, Mass. State Coll. (now U. Mass.), 1941; MA, Harvard U., 1946, PhD, 1948. Reader in history of sci. and learning Harvard U., 1943-45, asst. prof. gen. edn. in biology, 1949-52; instr. anatomy Harvard Med. Sch., 1947-49, rsch. fellow anatomy, 1949-52; res. fellow in anatomy, 1949-52; assoc. in anatomy Harvard Med. Sch., 1952-55, asst. prof. anatomy, 1955-65, assoc. curator Warren Anat. Mus., 1961-65; prof. med. sci. Brown U., Providence, 1965-90, prof. emeritus, 1990—, chmn. sect. morphology, 1968-85, co-chmn. sect. population biology, morphology & genetics and chmn. for anatomy, 1985-90; visiting lectr. in surgery med. sch. Harvard U., 1990—; anatomist dept. surgery Mass. Gen. Hosp., Boston, 1990—; pres. Erikson Biographical Institute, Inc., Providence, 1990—; anatomist various Boston hosps., 1952-82, Mass. Gen. Hosp. Sch. Med. Illus., 1947-60, Mass. Gen. Hosp., 1990—, Lahey Clinic, Boston, 1947-60; anatomist depts. surgery, orthopedics & rehab., and neurosurgery R.I. Hosp.; vis. prof. dept. anatomy and cellular biology, Harvard U. Med. Sch., 1989-90,; cons. anatomist Surg. Techniques Illus., 1976-80; cons. Dorlands Illus. Med. Dictionary; Rockefeller Found. cons. med. and pub. health, S. Am., 1959; specialist State Dept., Brazil, 1962, (Fulbright Fellow); adj. mem. faculty R.I. Sch. Design, 1970—; Kate Hurt Mead lecturer Coll. Physicians Phila., 1977; Raymond C. Truex lecturer Hahnemann U. Sch. Med., 1985. Sheldon traveling fellow, Cent. Am., 1946; Guggenheim fellow, S. Am., 1949. Mem. AAAS, Am. Assn. Phys. Anthropologists (archivist and co-historian 1981—, mem. com. on history and honors, 1978—), Am. Soc. Mammalogists (emeritus), HistorySci. Soc. (life mem.), Am. Soc. Zoologists, Am. Assn. Anatomists (historian and archivist 1972-86, 1990—, archivist 1986-90), Am. Assn. History Medicine (council 1972-74), Am. Assn. Clin. Anatomists, Anat. Soc. Gt. Britain and Ireland, Oral Hist. Assn., Anatomische Gesellschaft, Assn. of Anatomy Chairmen (emeritus), Sigma Xi. Achievements include special research in new world primates and gen. intellectual history, especially biology and medicine, developing database on over 250,000 careers with extensive statis. and graphical analyses. Home: 153 Bay Rd Norton MA 02766-3029 Office: Brown U Div Biology and Medicine Providence RI 02912

ERIKSON, RAYMOND LEO, biology educator; b. Eagle, Wis., Jan. 24, 1936. BS, U. Wis., 1958, MS, 1961, PhD in Molecular Biology, 1963. Asst. prof. to assoc. prof. U. Colo., Denver, 1965-72, prof. pathology, 1972-82; Am. Cancer Soc. prof. cellular and devel. biology Harvard U., Cambridge, Mass., 1982—. USPHS fellow, 1963-65; recipient Papaicolau award, 1980, Albert Lasker award, 1982, Robert Koch prize, 1982, Alfred P. Sloan Jr. prize GM Cancer Rsch. Found., 1983. Mem. NAS, Am. Academia of Arts and Scis., Am. Soc. Biol. Chemists, Am. Soc. Microbiology. Office: Harvard U Dept Cellular & Dev Biology 16 Divinity Ave Cambridge MA 02138-2020

ERIKSSON, KARL-ERIK LENNART, biochemist, educator; b. Bohus-Malmön, Sweden, May 27, 1932; came to U.S., 1988; s. Erik Mårten and Lilly Kristina (Lund) E.; m. (Aina) Gunilla Strand, Dec. 31, 1958; children: Mats Erik Rudolf, Pia Gunilla, Aina Lisa Kristina. BS, U. Uppsala, Sweden, 1958, PhD (fil. lic.), 1963; Dr. Sci., U. Stockholm, 1967. Rsch. asst. Swedish Forest Products Rsch. Lab., Stockholm, 1958-64, head biol. rsch., 1964-88; postdoctoral fellow Calif. Inst. Tech., Pasadena, 1968-69; prof. biochemistry, eminent scholar biotechnology U. Ga., Athens, 1988—; adj. prof. biochemistry Inst. Paper Sci. and Tech., Atlanta, 1990—; adj. prof. Biol. Ag. Engring. U. Ga., 1992—; cons. UN FAO, UNIDO for India, 1977-84; cons. to pulp and paper industry worldwide, 1960—; organizer sci. coop. between Sweden and developing countries, 1984—. Co-author: Microbial and Enzymatic Degradation of Wood and Wood Components, 1990; contbr. articles to profl. publs. Fulbright fellow Calif. Inst. Tech., 1968-69; Am.-Scandinavian Found. fellow Calif. Inst. Tech., 1968-69; named hon. Gadolin lectr. Chem. Soc., Turku, Finland, 1982; recipient Internat. M. Wallenberg prize STORA, Falun, Sweden, 1985; named Cameron-Gifford lectr. U. Newcastle, Eng., 1986. Fellow Royal Swedish Acad. Engring. Sci. (bd. dirs. 1982-85, chmn. forestry and forest industry scis. sect. 1982-85), Internat. Acad. Wood Sci., World Acad. Arts and Scis. Lutheran. Achievements include patent on method for producing cellulose pulp, 5 Swedish patents in field, discovery of new enzymes involved in cellulose and lignin degradation, discovery of cellobiose quinone oxidoreductase, cellobiose oxidase. Home: 145 Great Oak Dr Athens GA 30605 Office: Univ Ga B 304 Life Sci Bldg Athens GA 30602-7229

ERNEST, J. TERRY, ocular physiologist, educator; b. Sycamore, Ill., June 26, 1935; married, 1965; 2 children. BA, Northwestern U., 1957; MD, U. Chgo., 1961, PhD in Visual Sci., 1967. Prof. ophthalmology U. Wis., 1977-79; prof., chmn. ophthalmology Ind. U., 1980-81; prof. ophthalmology U. Ill., 1981-85; prof., chmn. ophthalmology U. Chgo., 1985—; mem visual sci. A study sect., NIH, 1975-78, chmn. 1978-79, chmn. visual disorders study sect., 1979-80; rsch. prof. Rsch. to Prevent Blindness, Ind., 1981-84; mem. Vision Rsch. Program Com., 1982-84. Editor Investigative Ophthalmology and Visual Sci., 1988—. Recipient Rsch. Career Devel. award NIH, 1972. Mem. AAAS, Am. Ophthalmol. Soc., Am. Acad. Ophthalmology (Honor award 1982), Assn. Rsch. Vision and Ophthalmology. Achievements include research in ocular circulation with special emphasis on glaucoma and diabetic retinopathy using various methods of in vivo blood flow measurements. Office: University of Chicago Visual Sciences Ctr 939 E 57th St Chicago IL 60637*

ERNSBERGER, FRED MARTIN, former materials scientist; b. Ada, Ohio, Sept. 20, 1919; 4 children. AB, Ohio Northern U., 1941; PhD in Phys. Chemistry, Ohio State U., 1946. Rsch. chemist U.S. Naval Ordnance Test Sta., 1947-54, S.W. Rsch. Inst., 1954-56, Mellon Inst. of Indsl. Rsch., 1957; rsch. chemist Glass Rsch. Ctr. PPG Industries, Inc., 1958-82; ret., 1982; adj. prof. dept. material sci. and engring. U. Fla., Gainesville, 1982—. Recipient IR-100 award, 1981, Scholes award, 1989. Fellow Am. Ceramic Soc. (Frank Forrest award 1964, Toledo Glass and Ceramic award 1970, G.W. Morey award 1974, Albert Victor Bleininger award 1993); mem. Am. Chem. Soc., Soc. Glass Tech. Achievements include research in surface chemistry, surface structure, mechanical properties of glass and glass-ceramics, chemistry of float glass. Office: 1325 NW 10th Ave Gainesville FL 32605*

ERNSBERGER, PAUL ROOS, research biologist, neuropharmacologist; b. Yonkers, N.Y., Apr. 30, 1956; s. David Jaques Ernsberger and Deborah Scott; m. Debra Lynne Bieber, June 24, 1984; 1 child, Timothy Scott. BA in Psychology summa cum laude, Macalester Coll., 1978; PhD in Neuroscience, Northwestern U., 1984. N.Y. Heart Assn. postdoctoral fellow Cornell U. Med. Coll., N.Y.C., 1984-86, instr. neurobiology in neurology, 1987, asst. prof. neurobiology in neurology, 1988-89; lectr. in field; chmn. adv. bd. Nat. Assn. to Advance Fat Acceptance, Sacramento, 1985—; cons. hypertension sect. Abbott Labs., N. Chgo., 1986, Laboratoires Therapeutique Moderne, Paris, 1989—, Solvay Pharma, Hannover, Germany, 1989—, cardiovascular divsn. UpJohn Pharms., Kalamazoo, 1990, cardiovascular pharmacology dvsn. Neurogenetic Corp., Paramus, N.J., 1991; workshop leader N.Y. Assn. Nutritionists, Albany, 1989; vis. prof. coll. pharmacy Ohio State U., Columbus, 1991; Merck, Sharpe & Dhome vis. prof. depts. pharmacology and medicine U. Western Ontario, Canada, 1991; lab. instr. neuroanatomy series Brain and Mind Com., 1991, small group co-leader, 1991. Co-author: Rethinking Obesity: An Alternative View of its Health Implications, 1988. Alliss scholar, Macalester Coll., 1977; predoctoral fellow Nat. Sci. Found., 1981; Dissertation Year grantee Northwestern U., 1984; M. Robert Gallop fellow N.Y.Heart Assn., 1984; recipient Young Investigator award Ea. Hypertension Soc., 1987, First Independent Research Support and Transition award Nat. Heart, Lung and Blood Inst., 1990. Mem. Internat. Brain Rsch. Orgn., Am. Soc. Hypertension (charter mem. 1986), Am. Soc. Pharmacology and Exptl. Therapeutics (neuropharmacology dvsn 1988—), Am. Fedn. Clin. Rsch., Am. Heart Assn. (high blood pressure coun.), N.Y. Acad. Sci., Soc. Neuroscience, Catecholamine Club. Jewish. Achievements include discovery of a new form of hypertension produced by refeeding after fasting, due to cycles of weight loss and regain (caused by activation of sympathetic nervous system), a new receptor protein in the brainstem, kidney and adrena medulla important in the regulation of blood pressure (lead to treatment of hypertension through new drugs in use in France and Germany); research in numerous areas including: open field behavior in two models of genetic hypertension and behavioral aspects of salt excess; muscarinic acetylcholine receptors of the brainstem and the developing trachea; nervous system control of the kidney; receptor subtypes in brain and kidney. Office: Case Western Res U Sch Medicine Dvsn Hypertension W147A Cleveland OH 44106

ERNST, EDWARD WILLIS, electrical engineering educator; b. Great Falls, Mont., Aug. 28, 1924; s. Paul Wilson and Grace Vio (Woodmore) E.; m. Helen Kitty Todd, Jan. 29, 1950 (dec. Mar. 1975); children: Deborah Kitty, Thomas Edward (dec.); m. Margaret Frances Patton, Sept. 13, 1975; children: Alan Harmon, Ruth Margaret, Betty Carol. BS, U. Ill., 1949, MS, 1950, PhD, 1955. Rsch. engr. GE, Syracuse, N.Y., 1955, Stewart-Warner, Chgo., 1955-58; assoc. prof. U. Ill., Urbana, 1958-68, prof., 1968-89, assoc. head elec. engring., 1970-85, assoc. dean engring., 1985-89; Allied-Signal prof. engring. U. S.C., Columbia, 1990—; bd. dirs. Nat. Engring. Consortium; program dir. NSF, Washington, 1987-90; chmn. engring. accreditation commn. Accreditation Bd. for Engring. Tech., N.Y., 1989-90. Pres. Mckinley Found., Champaign, Ill. 1968-72. Fellow IEEE (v.p. 1981-82, Centenial medal 1984, EAB Meritorious Achievement award in Accreditation Activities, 1985, Edn. Soc. Achievement award 1989, ASEE (editor Jour. Engring. Edn. 1992—), AAAS, ABET (Linton Grinter award 1992). Presbyterian. Avocations: photography, hiking, reading. Office: U SC Swearingen Engring Ctr Columbia SC 29208

ERNST, GREGORY ALAN, energy consultant; b. Owatonna, Minn., Mar. 2, 1960; s. William Bernard and Darlene (Scherer) E. BSChemE, U. Minn., 1982. Engr.-in-tng. Gen. contractor Ernst Constrn. Co., Alexandria, Minn., 1974-83; gen. mgr. Northwest Restaurants, Mpls., 1983-84; process engr. North Am. Chem. Co., Austin, Minn., 1984-86; energy cons. G.A. Ernst & Assocs., Inc., Eagan, Minn., 1986—. Office: GA Ernst & Assocs Inc 4247 Sunrise Rd Eagan MN 55122-2246

ERNST, RICHARD ROBERT, chemist, educator; b. Winterthur, Zurich, Switzerland, Aug. 14, 1933; s. Robert and Irma (Brunner) E.; m. Magdalena Kielholz, Oct. 9, 1963; children: Anna Magdalena, Katharina Elisabeth, Hans-Martin Walter. Diploma Chemistry, ETH-Zurich, 1956, DSc in Tech., 1962; PhD (hon.), ETH-Lausanne, Switzerland, 1986, Technische Hochschule, Munich, 1989. Scientist ETH-Zurich, 1962-63, privatdozent, 1968-70, asst. prof., 1970-72, assoc. prof., 1973-76, prof., 1976—; scientist Varian Assocs., Palo Alto, Calif., 1963-68; cons. Spectrospin AG, Fällanden, Switzerland, 1978—, v.p. bd. dirs. Numerous inventions, patents in field. 1st lt. ACS-Dienst, 1953-88, Swiss mil. Recipient Silver medal ETH-Zurich, 1962, Ruzicka prize, 1968, Gold medal Soc. Magnetic Resonance in Medicine, San Francisco, 1983, Benoist prize Swiss Fed. Confedn., Berne, 1986, Kirkwood award Yale U., 1989, Ampere prize, 1990, Wolf prize in chemistry, 1991, Louisa Gross Horowitz prize Columbia U., 1991, Nobel prize in chemistry, 1991, award for Achievements in Magnetic Resonance EAS, 1992. Mem. NAS (India), Deutsche Akademie Leopoldina, Acad. Europaea, Schweizerische Chemische Gesellschaft, Royal Soc. London, Österreichische Gesellschaft für Analytische Chemie, Am. Phys. Soc., U.S. Nat. Acad. Sci., Schweizerische Akademie d. Tech. Wiss. Avocations: Tibetan art, music. Office: Lab F Phys Chem ETH-Zentrum, 8092 Zurich Switzerland

ERNST, WALLACE GARY, geology educator, dean; b. St. Louis, Mo., Dec. 14, 1931; s. Fredrick A. and Helen Grace (Mahaffey) E.; m. Charlotte Elsa Pfau, Sept. 7, 1956; children: Susan, Warren, Alan, Kevin. B.A., Carleton Coll., 1953; M.S., U. Minn., 1955; Ph.D., Johns Hopkins U., 1959. Geologist U.S. Geol. Survey, Washington, 1955-56; fellow (Geophys. Lab.), Washington, 1956-59; mem. faculty UCLA, 1960-89, prof. geology and geophysics, 1968-89, chmn. geology dept. (now earth and space scis. dept.), 1970-74, 78-82, dir. Inst. Geophysics and Planetary Physics, 1987-89; prof. geology and geophysics Stanford (Calif.) U., 1989—, dean Sch. Earth Scis. Author: Amphiboles, 1968, Earth Materials, 1969, Metamorphism and Plate Tectonic Regimes, 1975, Subduction Zone Metamorphism, 1975, Petrologic Phase Equilibria, 1976, The Geotectonic Development of California, 1981, The Environment of the Deep Sea, 1982, Energy for Ourselves and Our Posterity, 1985, Cenozoic Basin Development of Coastal California, 1987, Metamorphic and Crustal Evolution of the Western Cordillera, 1988, The Dynamic Planet, 1990. Trustee Carnegie Instn. of Washington, 1990—. Mem. NAS (chmn. geology sect. 1979-82), AAAS, Am. Geophys. Union, Am. Geol. Inst., Geol. Soc. Am. (pres. 1985-86), Am. Acad. Arts and Sci., Geochem. Soc., Mineral. Soc. Am. (recipient award 1969, pres. 1979-80), Mineral. Soc. London. Office: Stanford U Sch Earth Scis Stanford CA 94305-2210

ERRAMPALLI, DEENA, molecular plant pathologist; b. Machilipatnam, India, Mar. 7, 1958; came to U.S. 1984; d. Stephen Devadatham and Mary Bharathi (Kondaveti) E. BS, Andhra U., India, 1976; MS, Banaras Hindu U., India, 1979; PhD, Okla. State U., 1990. Rsch. assoc. Interant. Crops Rsch. Inst. for the Semi Arid Tropics, Andhra Pradesh, India, 1980-84; rsch. asst. Okla. State U., Stillwater, 1985-89, teaching asst., 1985, postdoctoral rsch. assoc., 1989-92; postdoctoral rsch. fellow U. Toronto, Ont., Can., 1992--. Contbr. articles to profl jours. Mem. Social Svc. League, Vijayawada, India, 1973-76. Electron Microscopy grantee Okla. State U., 1986-89. Mem. Am. Phytopathological Soc. (co-moderator 1986), Internat. Orgn. for Mycoplasomology, Okla. Acad. Sci., Sigma Xi. Home: 907-505 Locust St, Burlington, ON Canada L7S 1X6 Office: U Toronto, Erindale Campus Botany Dept, Mississauga, ON Canada L5L 1C6

ERSKINE, DAVID JOHN, physicist; b. South Bend, Ind., May 19, 1957; s. John Robert and Diane Patricia (Bergquist) E. BS, U. Ill., 1979; PhD, Cornell U., 1984. Postdoctoral researcher dept. Physics U. Calif., Berkeley, 1984-87; physicist Lawrence Livermore (Calif.) Nat. Lab., 1987—. Contbr. articles to profl. jours. Mem. Am. Phys. Soc. Avocation: concert pianist. Office: Lawrence Livermore Nat Lab 5000 East Ave Livermore CA 94550

ERSLEV, ERIC ALLAN, geologist, educator; b. Harvard, Mass., Jan. 30, 1954; s. Allan Jacob and Betsy (Lewis) E.; m. Kathryn Tweedie, June 19, 1976; children: Peter Tweedie, Brett Covey. BS, Wesleyan U., 1976; AM, Harvard U., 1978, PhD, 1981. Asst. prof. Lafayette Coll., Easton, Pa., 1981-83; asst. prof. Colo. State U., Fort Collins, 1983-89, assoc. prof., 1989—. Editor: Laramide Basement Deformation, 1993; contbr. articles to profl. jours. Mem. Am. Assn. Petroleum Geologists, Geol. Soc. Am., Colo. Sci. Soc. (councilor 1992), Rocky Mountain Assn. Geologists (structure series editor 1988—). Unitarian. Achievements include normalized Fry method, macroprobe analysis of cleavage, basement balancing of Laromide structures, triangular shear zone folding. Office: Colo State Univ Dept Earth Resources Fort Collins CO 80523

ERTEM, ÖZCAN, aeronautical engineer; b. Ankara, Turkey, Dec. 10, 1962; s. Orhan and Sevim (Çora) E.; m. Çiğdem Oran, Mar. 19, 1990. BS, Middle East Tech. U., Ankara, 1984, MS, 1987. Maintenance mgr. Turkish Air League, Ankara, 1987-90; sr. mfg. engr. Turkish Aerospace Industries, Ankara, 1990--; tech. team mem. Turkish Air Force Primary Trainer Aircraft, Ankara, 1988-89; team mem., test pilot TAI Unmanned Aerial Vehicle prototype, UAV-X1. Mem. Am. Inst. Aeronautics and Astronautics. Office: Turkish Aerospace Inds, Mürted, 06372 Ankara Turkey

ERTEZA, IREENA AHMED, electrical engineer; b. Albuquerque, June 13, 1965; d. Ahmed and Sohela Erteza; m. Brian K. Bray, Sept. 8, 1990. BSEE, U. N.Mex., 1986; MSEE, Stanford U., 1988, PhD, 1993. Rsch. asst. Thin Films Lab., U. N.Mex., Albuquerque, 1985-86; sr. tech. asst. AT&T Bell Labs., Murray Hill, N.J., 1986; rsch. engr. Applied Tech. Assocs., Albuquerque, 1987; rsch. asst. Elec. Engring. Dept., Stanford (Calif.), 1987-88, 90—; rsch. engr. Almaden Rsch. Ctr., IBM, San Jose, 1989; sr. mem. tech. staff Sandia Nat. Labs, Albuquerque, 1993—. Contbr. articles to profl. jours. Army Rsch. Office fellow, 1992, IBM Doctoral fellow, 1988-92, Stanford Engring. Grad. fellow, 1986-87; AT&T Grad. Rsch. Program for Women grantee, 1986-92. Mem. Optical Soc. Am. (Stanford chpt. founding mem. 1991, treas. 1991-92), Golden Key, Tau Beta Pi, Eta Kappa Nu. Achievements include research in a variational analysis of rectangular dielectric waveguides using gaussian modal approximations.

ERTL, PETER, chemist, researcher; b. Bratislava, Czechoslovakia, June 28, 1959; s. Milan Ertl and Eva (Lichardova) Ertlova; m. Eva Fajnerova Ertlova; children: Monika, Pavol. Grad., Comenius U., Bratislava, 1983, PhD, 1989. Researcher Comenius U. Inst. Chemistry, Bratislava, 1983—; exec. bd. Molecular Computing, Ltd., Bratislava, 1989—. Contbr. articles to profl. jours. Mem. WATOC, Molecular Graphics Soc. Roman Catholic. Achievements include development of molecular modeling software used in academic sites. Office: Comenius Univ Inst Chem, Mlynska Dolina Ch-2, 84215 Bratislava Slovak Republic

ERTL, RONALD FRANK, research coordinator; b. Oak Park, Ill., Apr. 16, 1946; s. Alexander Felix and Shirley Delores (Kral) E. BS, U. Wis., Whitewater, 1970; cert. in mgmt., U. Nebr., 1992. Cert. tchr., Calif. Tchr. sci. Arlington Heights (Ill.) Sch. Dist., 1970-73, Lakeside (Calif.) Sch. Dist., 1976-82; staff rsch. assoc. U. Calif., San Diego, 1982-84; rsch. coord. U. Nebr. Med. Ctr., Omaha, 1984—, cons., 1987—. Author: The Airway Epitheliuam, 1991; contbr. articles to Am. Jour. Respiratory Cell and Molecular Biology. Mem. Soc. Cell Biologists. Achievements include research in role of airway epithelium in cellular migration. Office: U Nebr Med Ctr 600 S 42d St Omaha NE 68198-5130

ERYUREK, EVREN, nuclear engineer, researcher; b. Eskisehir, Turkey, Aug. 19, 1963; came to U.S., 1987; s. Mete and Senay (Koksal) E. BS, Hacettepe U., Ankara, Turkey, 1986; MS, U. Tenn., 1991, postgrad., 1992—. Cons. engr. EKA div. Westinghouse, Istanbul, Turkey, 1986-87; asst. dir. Kurt & Kurt div. Hitachi, Istanbul, 1987; computer system mgr. nuclear engring. U. Tenn., Knoxville, 1990—, rsch. nuclear engring., 1988—; cons. MODICO, Knoxville, 1991—; vis. scientist ECN, Petten, The Netherlands, 1991. Contbr. articles to Nuclear Tech., Nuclear Sici., Nuclear Europe Worldscan, numerous others. Grad. rep. faculty senate engring. rep. student govt. U. Tenn., 1991—; pres. Turkish Student Assn., U. Tenn., 1992—. Mem. IEEE (control soc. 1992—), Am. Nuclear Soc. (grad. student rep. 1992—, Outstanding Paper award 1990, 93, 1st Pl. Reactor Design Competition 1989), Sigma Xi (2d Pl. award 1990). Achievements include copyright of code for neural network applications. Home: 1064 Forest Heights Dr Knoxville TN 37919 Office: U Tenn Nuclear Engring Dept 305 Pasqua Engring Bldg Knoxville TN 37996-2300

ESAKI, LEO, physicist; b. Osaka, Japan, Mar. 12, 1925; came to U.S., 1960; s. Soichiro and Niyoko (Ito) E.; m. Masako Kondo, May, 31, 1986; children from previous marriage: Nina Yvonne, Anna Eileen, Eugene Leo. B.S., U. Tokyo, 1947, Ph.D., 1959. With Sony Corp., Japan, 1956-60; with Thomas J. Watson Research Center, IBM, Yorktown Heights, N.Y., 1960-92; IBM fellow Thomas J. Watson Research Center, IBM, 1967-92, mgr. device research, 1965-92; dir. IBM-Japan, 1975-92; pres. U. Tsukuba, Ibaraki, Japan, 1992—. Recipient Stuart Ballantine medal Franklin Inst., 1961, Japan Acad. award, 1965, Nobel prize in physics, 1973; decorated Order of Culture Govt. of Japan, 1974. Fellow Am. Phys. Soc. (Internat. prize for new materials 1985, councillor-at-large 1971-74), IEEE (Morris N. Liebman Meml. prize 1961, Medal of Honor 1991), Japan Phys. Soc., Am. Vacuum Soc. (dir. 1973-74); mem. Am. Acad. Arts and Scis., NAS (fgn. assoc.), NAE (fgn. assoc.), Academia Nacional de Ingenieria Mex. (corr.), Japan Acad. Achievements include invention of Esaki tunnel diode, 1957. Home: Takezono 3-772, Tsukuba Ibaraki 305, Japan Office: U Tsukuba, Tsukuba Ibaraki 305, Japan

ESAKI, TOSHIYUKI, pharmaceutical chemist; b. Nagoya, Aichi, Japan, Nov. 10, 1947; s. Hidekata and Saku (Hayashi) E. BS in Pharmacy, Kyoto (Japan) U., 1970, PhD, 1975. Faculty engring., dept. applied chemistry Nagoya U., 1976-85; dir. Esaki Rubber Co., Nagoya, 1985—; abstractor The Japan Inf. Ctr. of Sci. and Tech., Tokyo, 1982—; lectr. in computer sci. Shotoku Gakuen Women's Jr. Coll., Gifu, Japan, 1990-92, Chukyo U., Nagoya, 1992—. Mem. Pharm. Soc. of Japan, Chem. Soc. of Japan, Am. Chem. Soc. Achievements include rsch. in computer system for theoretical prediction of new drug molecules, quantitative drug design, quantum pharmacology, drug-receptor interaction. Office: Medicinal La Esaki Rubber Co, 1-7 Omiya-cho Nakamura-ku, Nagoya 453, Japan

ESCH, GERALD WISLER, biology educator; b. Wichita, Kans., June 22, 1936; s. Frank Day and Ruby Fern (Wisler) E.; m. Ann Speir, Dec. 22, 1958; children: Craig, Elisabeth, Charles. BS, Colo. Coll., 1958; MS, PhD, U. Okla., 1963. Postdoctoral fellow U. N.C., 1963-65; from asst. prof. to assoc. prof. biology Wake Forest U., Winston-Salem, N.C., 1965-75, prof. biology, 1975-92, chmn. dept. biology, 1975-84, 90-92, dean grad. sch., 1984-90. Author: A Functional Biology of Parasitism: Ecological & Evolutionary Implications, 1993; editor: Thermal Ecology II, 1972, Regulation of Parasite Populations, 1977, Parasite Communities: Patterns & Processes, 1991. Active N.C. Bd. Sci. & Tech., Raleigh, 1986-90, 90—. Rsch. fellow U. London, 1971-72, U. Ga., 1974-75; recipient Louis T. Benezet award Colo. Coll., 1992. Mem. Am. Soc. Parasitologists (coun. mem. 1984-87, v.p. 1993, editor jour. 1994—). Office: Wake Forest Univ Dept Biology Winston Salem NC 27109

ESCOBAR, LUIS FERNANDO, pediatrician, geneticist; b. Guatemala, Guatemala, Apr. 5, 1957; came to U.S. 1984; s. Augusto Guillermo and Maria Teresa (Vargas) E.; m. Michelle L. Wagner, June 1, 1991. MS, Ind. U., 1987; MD, San Carlos U., Guatemala, 1984. Postdoctoral fellow oral facial genetics Ind. U., Indpls., 1985-88, asst. rsch. scientist, 1987-88; house staff officer pediatrics Ind. U., 5, 1990--; vis. scholar biology U. Notre Dame, South Bend, 1989-90; edit. reviewer Am. Jour. Med. Genetics, Helena, Mont., 1990--, Obstetrics and Gynecology, Indpls., 1990--. Contbr. articles to profl. jours. Chmn. orientation Am. Field Svc., Guatemala, 1978; physician for vols. Peace Corps, Guatemala, 1983. Recipient Nat. Rsch. Svc. award NIH, 1985, Work Toward World Peace award Am. Field Svc., 1976; Oral Facial Genetics grantee NIH, 1988. Mem. Am. Soc. Human Genetics, Am. Soc. Craniofacial Genetics, Am. Acad. Pediatrics, Sigma Xi. Roman Catholic. Achievements include development of technique of fetal cephalometry antropometry in utero by ultrasound. Office: Ind U Indianapolis IN 46224

ESCOTT, SHOOLAH HOPE, microbiologist; b. Stamford, Conn., May 20, 1952; d. Robert R. and Fanny (Levy) E.; m. Joseph J. Sulmar, Sept. 6, 1992. Cert. med. tech., St. Vincent's Hosp., Bridgeport, Conn., 1974; BS, U. Conn., 1974; MS, Northeastern U., Boston, 1985. Cert. med. technologist., 1976; Med. technologist St. Elizabeth's Hosp., Boston, 1976, Harvard U. Health Svcs., Cambridge, Mass., 1976-79; med. technologist microbiology lab. New England Deaconess Hosp., Boston, 1979-84; supr. microbiology Norwood (Mass.) Hosp., 1984-87; supr. microbiology lab. Meml. Microbiology Med. Ctr. Cen. Mass., Worcester, 1987-91; adminstrv. supr. microbiology labs. Meml. and Hahnemann Microbiology Med. Ctr. Cen. Mass., Worcester, 1991—. Named Nat. Merit Scholar, 1970; grantee, 1970. Mem. Am. Soc. Clin. Pathologists, Am. Soc. for Microbiology and Infectious Disease (bd. dirs. Mass. chpt. 1989-91, treas. 1991-93, pres.-elect 1993—). Avocation: travelled extensively in U.S., Can., Europe, 1974-76. Office: Med Ctr Cen Mass Microbiology Lab 119 Belmont St Worcester MA 01605-2903

ESFANDIARI-FARD, OMID DAVID, microbiologist; b. Tehran, Iran, Mar. 28, 1961; came to U.S., 1977; s. Mustafa and Mehan (Maghzi) E. BS in Biology, Calif. State U., Northridge, 1984; MS in Microbiology, 1987; postgrad. in Bus. Adminstrn., Pepperdine U., 1990—. Lab. mgr. Home Medix, Inc., Long Beach, Calif., 1988-89; med. coord. Unicare, Inc., L.A., 1986-89; pres. Euro-Sports-Medicine, Mar Vista, Calif., 1990—. Co-author: (book) Petroleum Microbiology in the Persian Gulf, 1989-90; editor Health Industry Today, 1990; contbr. articles to Microbiology Revs., Sporting Goods Bus., Applied Microbiology, Chiropractic Health. Mem. Assn. for Advancement Medicine, Assn. Microbiology, Instrumentation, Assn. of Med. Design and Mfg., Nat. Assn. Med. Equipment and Suppliers, Am. Assn. Sports Medicine. Achievements include patents for Pro Inversion, a motorized back therapy and injury preventing device. Home: PO Box 251528 Los Angeles CA 90025-1528 Office: Unicare Inc 4355 W Pico Blvd Los Angeles CA 90019-3136

ESHAM, RICHARD HENRY, internal medicine and geriatrics educator; b. Maysville, Ky., Oct. 6, 1942; s. Elwood and Ruth (Opfer) E.; children: Ashley Ruth, Richard Henry II, Clay Hamlet. MD, U. Louisville, 1967. Resident in internal medicine U. Ala. Hosps. and Clinics, Birmingham, 1968-71, straight med. intern., 1967-68; pvt. practice Mobile, Ala., 1974-90; chief, prof. div. gen. internal medicine and geriatrics U. South Ala., Mobile, 1990—; chmn. Bd. Med. Examiners, Montgomery, Ala., Ala. Bd. Health, Montgomery. Contbr. med. articles to profl. jours. Bd. dirs. Arthritis Found., Mobile, 1974-81. Maj. U.S. Army, 1972-74. Fellow ACP; mem. Med. Assn. State of Ala. (officer, counselor, chmn. bd. censors), Mobile Area C. of C. (bd. dirs. 1991-92, med. svcs. com. 1991—). Avocations: fishing, hunting. Office: U South Ala Health Svc Bldg 201H HSB USA Campus Mobile AL 36688-0002 also: PO Drawer 8569 Mobile AL 36688

ESHBAUGH, W(ILLIAM) HARDY, botanist, educator; b. Glen Ridge, N.J., May 1, 1936; s. William Hardy Eshbaugh Jr. and Elizabeth (Wakeman) Henderson; m. Barbara Keller, Sept. 6, 1958; children: David Charles, Stephen Hardy, Elizabeth Wendy, Jeffrey Raymond. BA, Cornell U., 1959; MA, Ind. U., 1961, PhD, 1964. Lectr. in botany Ind. U., Bloomington, 1962; spl. asst. to chief ecology and epidemiology br. Dugway (Utah) Proving Ground, 1964-65; asst. prof., curator botany So. Ill. U., Carbondale, 1965-67; asst. prof., curator botany Miami U., Oxford, Ohio, 1967-71, assoc. prof., 1971-77; prof. botany, curator Willard Sherman Turell Herbarium, Oxford, Ohio, 1977—; chmn. dept. botany Miami U., Oxford, Ohio, 1983-88; assoc. program dir. NSF, Washington, 1982-83. Co-author: The Vascular Flora of Andros Island, Bahamas, 1988; contbr. papers, book revs. to profl. publs. Troop com. Oxford area Boy Scouts Am., 1986-90. Capt. U.S. Army, 1964-65. Fellow AAAS, Ohio Acad. Sci.; mem. Am. Soc. Plant Taxonomists (pres. 1991-92), Soc. Econ. Botany (v.p. 1982-83, pres. 1983-84), Botan. Soc. Am. (pres. 1988-89, BSA Merit award 1992), Nature Conservancy (vice-chmn. Ohio chpt. 1970-75, trustee 1970-77, 89—), Assn. Systemics Collections (bd. dirs. 1981-84, rep.-at-large), Internat. Orgn. Plant Biosystematists (coun. 1987-89, ad hoc com. 1989-92, N.Am. treas. 1992—), Internat. Field Studies (trustee 1989). Methodist. Avocations: camping, sailing, skiing, fly-fishing, photography. Home: 209 Mckee Ave Oxford OH 45056-9025 Office: Miami University Willard Sherman Turrell Herbarium Dept of Botany Oxford OH 45056

ESHLEMAN, VON RUSSEL, electrical engineering educator; b. Darke County, Ohio, Sept. 17, 1924; married; 4 children. BEE, George Washington U., 1949; MS, Stanford U., 1950, PhD in Elec. Engring., 1952. Rsch. assoc. Radio Propagation Lab. Stanford (Calif.) U., 1952-56, from instr. to prof. elec. engring., 1956-61, prof. elec. engring., co-dir. Ctr. Radar Astronomy, 1961—, dir. Radioscience Lab., 1974-83; cons. NAS, Nat. Bur. Standards, SRI Internat., Jet Propulsion Lab.; mem. Internat. Astronaut Congress, Internat. Astron. Union, Internat. Sci. Radion Union; dir. Watkins-Johnson Co.; mem. radio sci. team Galileo Mission to Jupiter, 1979-80. Fellow AAAS, IEEE, Am. Geophys. Union, Royal Astronomy Soc.; mem. NAE. Achievements include rsch. in radar astronomy, planetary exploration, ionospheric and plasma physics, radio wave propagation, astronautics. Office: Stanford U Ctr for Radar Astronomy Durand Bldg 221 Stanford CA 94305*

ESIN, JOSEPH OKON, computer information systems educator; b. Ofi Uda-Mbo, Oron, Akwa Ibom, Nigeria, May 1, 1953; s. Maurice Okon and Felicia (Nkoyo) E. BS, St. Louis U., 1983, MA, 1985; EdD, U.S. Internat. Univ., 1989. Computer educator Acad. For Computers, San Diego, Calif., 1986-88; prof. computer info. systems Nat. Univ., Stockton/Sacramento, Calif., 1989-91; cons., software analyst Jenco Computer Software Tech., Dallas, 1989-93; vice chancellor for instnl. rsch., strategic planning and mgmt. info. systems Mgmt. Info. Systems/Paul Quinn Coll., Dallas, 1991—; prof., spl. asst. dir. instnl. rsch. and planning, mgmt. info. systems Paul Quinn Coll., Dallas, 1991—; adj. prof. computer sci. Chapman Coll., Stockton, 1989; adj. prof. computer info. system Trinity Valley C.C., Terrell, Tex., 1991—. Mem. KC (advocate). Office: PO Box 41792 Dallas TX 75241-0792

ESLAMI, MOHAMMAD REZA, mechanical engineering educator; b. Tehran, Iran, Mar. 8, 1945; came to U.S., 1969; s. Mohammad Sadegh and Khadijeh (Shahrestani) E.; m. Minoo N. Nezami, July 21, 1970; children: Saam, Golnaz. BS, Tehran Poly., 1968; PhD, La. State U., 1973. Teaching asst. La. State U., Baton Rouge, 1969-70; rsch. asst. La. State U.-NASA Space Shuttle Team, Baton Rouge, 1970-73; asst. prof. Amirkabir U. Technology, Tehran, Iran, 1973-77; assoc. prof. La. State U., Baton Rouge, 1977; assoc. prof. Amirkabir U. Technology, Tehran, 1978-89, prof., 1989—; minister advisor Ministry of Mine and Industry, Tehran, 1978; rsch. adv. bd. Ministry of Def., Tehran, 1983—. Contbr. articles to Jour. Thermal Stress, AIAA Jour.; author tech. papers. Candidate Nat. Sci. award Office of Pres., 1991; named Disting. Grad. Ct. of Iran, 1968, Recognized scholar Office of Pres., Islamic Republic of Iran, 1990. Mem. AIAA, ASME, Acad. Sci. (head mech. engring. branch), Iranian Soc. Mech. Engring. (honor). Office: Amirkabir U Technology, Hafez Ave, Tehran 15914, Iran

ESPARZA, EDWARD DURAN, mechanical engineer; b. Aguascalientes, Mex., Oct. 19, 1942; came to U.S., 1955; s. Jesus Reyes and Teresa (Duran) E.; m. Sonia Elizabeth Chavez, Oct. 8, 1966; children: Elizabeth, Carlos, Edward. BSME, Tex. A&M U., 1966; MSME, U. N.Mex., 1971. Registered profl. engr., Tex., N.Mex. Aerospace technologist NASA-Johnson Space Ctr., Houston, 1966; instrument engr. Celanese Chem. Co., Kingsville, Tex., 1966-67; with Southwest Rsch. Inst., San Antonio, 1972—, sr. rsch. engr., 1979-92, prin. engr., 1992—. Contbr. to profl. publs. Youth sports coach Cath. Youth Orgn., San Antonio, 1981-91; coach Math Counts, St. Paul Sch., San Antonio, 1989-91. Capt. USAF, 1967-72. Mem. ASME, Instrument Soc. Am., Soc. Exptl. Mechanics. Achievements include development of equations and methods for estimating buried pipeline stresses from nearby blasting, measurement of blast parameters at very small scaled distances. Office: Southwest Rsch Inst 6220 Culebra Rd San Antonio TX 78238

ESPINOLA, AÏDA, chemical engineer, educator; b. Rio de Janerio, Brazil, Apr. 18, 1920; d. Alvaro Bustamante and Margarida (Costa Neves) de Oliveira; m. Cesar Godinho Espinola, Oct. 2, 1943 (div. 1954). BS in indsl. chemistry, Sch. Chemistry, Rio de Janeiro, 1941, BAChemE, 1954; PhD in Chemistry, Pa. State U., 1974. Chemist, technologist Lab. Mineral Prodn., Rio de Janeiro, 1942-71; assoc. prof. Fed. U. Rio de Janeiro, 1955-81, full prof., 1981-90; researcher Tech. Aerospace Ctr., Sän José Dos Campos, 1974-75; full prof. Coord. Grad. Programs in Engring., Rio de Janeiro, 1975—; vis. prof. Fla. Atlantic U., Boca Raton, 1966, Fed. U. Pernambuco, Recife, Brazil, 1973, Fed. U. Bahia, Salvador, Brazil, 1973. Author: Analytical Separations and Pre-Concentration, 1989; author: (with others) Voltammetry of Water in Fused Alkali Nitrates, 1978; editor: Proceedings of the V Brazilian Energy Congress, 1990; translator: Chemistry (Quagliano and Vallarino), 1979, Inorganic Analytical Chemistry (Bassett), 1981; contbr. articles to profl. jours. Fellow Pa. State U., 1969. Mem. Am. Chem. Soc., Brazilian Chem. Soc. (v.p. 1956, pres. 1957), Brazilian Soc. for Progress of Sci., Sindicate Chemists of Rio de Janeiro (Gold Retort award 1979). Achievements include development of fuel cell prototype. Home: Copacabana, Rua Sousa Lima 289 Apt 1002, 22081-010 Rio de Janeiro Brazil Office: Coppe Fed U Rio de Janeiro, PO Box 68505, 21945-970 Rio de Janeiro Brazil

ESPY, JAMES WILLIAM, chemicals executive; b. Denver, July 27, 1948; arrived in Saudi Arabia, 1980; s. James Bruce and Marian (Honan) E.; m. Marjorie Claire Wismer, Aug. 9, 1975; 1 child, James Paul. BS in Chemistry, Beloit (Wis.) Coll., 1970; MS in Chemistry, U. Wis., 1973; MS in Engring. Mgmt., U. Mo., Rolla, 1985. Chemist Dow Chem. USA, Midland, Mich., 1973-78; from advanced rsch. chemist to sr. rsch. chemist 3M Co., St. Paul, Minn., 1978-80; from indsl. chemist to lab. sci. I ARAMCO, Dhahran, Saudi Arabia, 1980-85; pres. Espy, Inc., Boulder, Colo., 1985-86; from lab. sci. I to supr. lab support svcs. Saudi Aramco, Dhahran, 1986-90, supr. chem. quality assurance, 1990-92, sr. supr. lab. support svcs. sect., 1993, supt. lab support svcs. divsn., 1993—. Editor: (jour.) The Midland Chemist, 1976-78. Mem. Am. Inst. Chemists, Am. Chem. Soc. (chmn. Saudi Arabian chpt. 1988-89), Kiwanis (sec. Midland chpt. 1977-78), Lions (dir. Boulder chpt. 1985-86). Roman Catholic. Avocations: swimming, skiing. Home: Saudi Aramco Box 2410, Dhahran 31311, Saudi Arabia Office: Saudi Aramco Labs Dept, Box 62, Dhahran 31311, Saudi Arabia

ESPY-WILSON, CAROL YVONNE, electrical engineer, educator; b. Atlanta, Apr. 23, 1957; d. Matthew and Mattie Pearl (Cooper) Espy; m. John Silvanus Wilson, June 15, 1985; children: Ayana, Ashia. BS, Stanford U., 1979; MS, MIT, 1981, PhD, 1987. Postdoctoral fellow MIT, Cambridge, 1987-88, rsch. scientist, 1988-90; mem. tech. staff Bell Labs., Murray Hill, N.J., 1987-88; asst. prof. Boston U., 1990—. Contbr. articles to Jour. Acoustical Soc. Am., IEEE Transactions on Acoustics, Speech and Signal Processing. Active Coalition 100 Black Women, Boston, 1991. Named Clare Boothe Luce prof. Henry Luce Found., 1990; grantee NSF, 1990. Mem. Acoustical Soc. Am., IEEE, AAAS, Sigma Xi. Democrat. Methodist. Achievements include development of speech recognition system for sounds w y r l in American English demonstrating viability of feature-based approach. Office: Boston U ECS Dept 44 Cummington St Boston MA 02215

ESQUIBEL, EDWARD VALDEZ, psychiatrist, clinical medical program developer; b. Denver, May 28, 1928; s. Delfino C. and Beatrice (Solis) E.; m. Elaine F. Telk (div. 1961); children: Roxanne, Cyndi, Allen, James; m. Lillian D. Robb, 1961; children: Amanda, Ramona. MD, U. Colo., 1958. Diplomate Am. Bd. Psychiatry and Neurology. Assoc. chief svc. Ill. State Psychiat. Inst., Chgo., 1964-66; dir. undergrad. program psychiatry, asst.

prof. psychiatry Chgo Med. Sch., 1966-68; cons. and supr. group therapy Lake County Mental Health Clinic, Gary, Ind., 1968-72; pvt. practice Daytona Beach, Jacksonville, Fla., 1972-82; chief forensic svcs., dir. div. maximum security and inst. rsch. Colo. State Hosp., Pueblo, 1981; assoc. clin. prof. psychiatry Quillen-Dishner Coll. Medicine, Johnson City, Tenn., 1982-84; clin. psychiatrist VA Outpatient Clinic, Riviera Beach, Fla., 1984-86; mental health coord., supr. VA, Pensacola, Fla., 1986-88; assoc. chief staff, ambulatory care VA Med. Ctr., Ft. Lyon, Colo., 1988-90, VA Carl Vinson Med. Ctr., Dublin, Ga., 1990-91; staff physician VA Med. Ctr., Sheridan, Wyo., 1993—. Contbr. articles to profl. jours. Sgt. U.S. Army, 1952-58. Recipient Plaque Recognition award Southeastern Psychiat. Inst., 1964, Internat. Pers. Creative award, 1972, Key to City Daytona Beach, 1975, Hosp. Dirs. commendation VA, 1991. Mem. Am. Soc. Psychoanalytic Physicians. Avocations: gardening, arts and crafts, reading. Home: 801 Gospel Island Rd Inverness FL 34450-3582

ESQUINAZI, PABLO DAVID, physicist; b. Tucuman, Argentina, May 25, 1956; s. Samuel and Perla Eva (Salomon) E.; m. Gudula Maria Borgers, Aug. 2, 1985; children: Marco David, Clara Elisabeth. BS, Juan Carlos Davalos, Salta, Argentina, 1973; Licenciado, Inst. Balseiro, Bariloche, Argentina, 1979; PhD, Instituto Balseiro, Bariloche, Rio Negro, 1983; habilitation, Bayreuth (Germany) U., 1991. Fellow Nat. Commn. Atomic Energy, Bariloche, 1976-79, fellow A1, 1980-83, rsch. scientist A11, 1986-87; rsch. scientist Heidelberg (Germany) U., 1983-85; acad. cons. Bayreuth U., 1988—; cons. in field. Contbr. articles to profl. jours. Sponsor UNICEF. Mem. N.Y. Acad. Scis. Jewish. Avocations: music, windsurfing, tennis. Home: Elbering 3, 95445 Bayreuth Germany Office: Exp Physik V, Bayreuth U, 95440 Bayreuth Germany

ESSER, ARISTIDE HENRI, psychiatrist; b. Padalarang, Java, Indonesia, May 11, 1930; came to U.S., 1961; s. Samuel Jonathan and Anganita (Tawalujan) E.; m. Ada Reif; children: Jonathan Hendrik, Jessica. MD, U. Amsterdam, The Netherlands, 1955. Diplomate Am. Bd. Psychiatry and Neurology. Med. dir. N.S. Kline Inst., Orangeburg, N.Y., 1962-69; dir. research Letchworth Village, Thiells, N.Y., 1969-71; dir. Cen. Bergen Community Mental Health Ctr., Paramus, N.J., 1971-77; med. dir. Mission for Immaculate Virgin, S.I., N.Y., 1977-80; dir. quality assurance Bronx (N.Y.) Psychiat. Ctr., 1980-85; unit chief for supportive rehab. Rockland Psychiat. Ctr., Orangeburg, 1985-88, chief geriatrics div., 1988-90; psychiatrist St. Dominic's Home, Orangeburg, 1990—; attending psychiatrist Rye (N.Y.) Hosp. Ctr., 1990—, Good Samaritan Hosp., Suffern, N.Y., 1990—; rsch. prof. NYU Med. Ctr., N.Y.C., 1985—. Co-author: Mental Illness: A Homecare Guide, 1989, Chi Gong: The Ancient Chinese Way to Health, 1990; co-editor: Behavior and Environment, 1971, Design for Communality and Privacy, 1978; editor Jour. Man-Environment Systems, 1969— (Internat. Design award 1973). Recipient travel grant City of Leyden, The Netherlands, 1960; Lederle Labs. fellow Yale U., 1961. Fellow AAAS, Am. Psychiat. Assn.; mem. Soc. for Biol. Psychiatry, Soc. for Gen. Systems Research, Am. Acad. Acupuncture (founding), Assn. for Study Man-Environment Relations (founding). Home: 435 S Mountain Rd New City NY 10956-5731 Office: 21 N Broadway Nyack NY 10960-2621

ESSEX, DOUGLAS MICHAEL, optical engineer; b. Milw., June 26, 1961; s. Joe Michael and Mary (Davidson) E. BS in Physics, U. Ill., 1983. Instr. Triton Coll., River Grove, Ill., 1983-84; thin film engr. Laser Power Optics, San Diego, 1984-89, Hughes Optical Products, Des Plaines, Ill., 1989-91; coating lab. mgr. Virgo Optics, Port Richey, Fla., 1991—. Contbr. article to conf. procs. Dir. Optical Soc. San Diego, 1988. Mem. Optical Soc. Am. Office: Virgo Optics 6736 Commerce Ave Port Richey FL 34668

ESTEBAN, ERNESTO PEDRO, physicist; b. Huancayo, Junin, Peru, Feb. 13, 1951; came to U.S. 1977; s. Ernesto Flavio and Felicita Berta (Avila) E.; m. Ana Maria Bustamante, Feb. 18, 1977. MS, U. Notre Dame, 1979, PhD, 1983. Auxiliar prof. U. P.R., Humacao, 1983-86; assoc. prof. physics Univ. P.R.-, Humacao, 1986-91; prof. physics U. P.R., Humacao, 1991—; vis. rsch. scientist U. Ill., Urbana, 1989-90. Contbr. articles to profl. jours. NSF grantee, 1987, 90, 92, Resource Ctr. for Sci. grantee, 1985, 91. Mem. Am. Phys. Soc. Home: PO Box 10100 Humacao PR 00792 Office: Physics Dept Univ PR Humacao PR 00791

ESTERMAN, BENJAMIN, ophthalmologist; b. Vilna, Lithuania, May 6, 1906; came to U.S., 1908; s. Marcus and Bella (Shirling) E.; m. Sophie Milgram, Sept. 5, 1935 (dec. Oct. 1968); children: Mark, Daniel, Laura; m. Cinnabelle Morris, Dec. 3, 1972; 1 stepchild, Errol Morris. AB, Columbia U., 1927; MD, Cornell U., 1931. Diplomate Am. Bd. Ophthalmology. Resident in ophthalmology N.Y. Post-Grad. Hosp., N.Y.C., 1931-32; resident in ophthalmologic surgery Knapp Meml. Eye Hosp., N.Y.C., 1933-35, ophthalmic surgeon, 1935-40; ophthalmic surgeon Manhattan Eye & Ear Hosp., N.Y.C., 1940-80, cons. opthalmology surgeon, 1980—; dir. ophthalmology Peninsula Hosp., L.I., 1940-80, St. John's Episcopal Hosp., L.I., 1940-80, L.I. Jewish Hosp., 1940-80; pres. med. bd. Peninsula Hosp., 1961, St. John's Episc. Hosp., 1961. Author: The Eye Book, 1977; contbr. articles to profl. jours. Trustee Temple Israel, Lawrence, N.Y., 1945—; founding mem. Aircraft-Noise Abatement Nassau County, L.I., 1961. Fellow ACS, Am. Acad. Ophthalmology, Internat. Coll. Surgeons, N.Y. Acad. Medicine; mem. N.Y. Acad. Sci., N.Y. Soc. for Clin. Ophthalmology (pres. 1955), Manhattan Ophthal. Soc. (pres. 1976), Alpha Omega Alpha. Achievements include patent for Esterman Scale for functional evaluation of peripheral vision.

ESTES, JACOB THOMAS, JR., pharmacist, consultant; b. Dallas, Sept. 2, 1944; s. Jacob Thomas and Durgenia Mae (Kelly) E.; m. Susan Joan Rador, Mar. 7, 1980; 1 child, Amy Dianne. B.S. in Pharmacy, U. Tex., 1967. Hosp. pharmacist St. Joseph Hosp., Ft. Worth; pharmacist Park Row Pharmacy, Arlington, Tex., Plaza Pharmacy, Ft. Worth; pharmacist in-charge, mgr. Whitten Pharmacy; pharmacist in-charge K-Mart Pharmacy, Ft. Worth, Pla. Pharmacy, Granbury, Tex.; owner Queen of Angels Cath. Books and Gifts, Ft. Worth; owner Ctrl. and North Ctrl. Tex. Relief Pharmacists, 1990—; cons. nursing homes, hospice; dist. intervenor Pharmacy Helpline for Impaired; dir. Tex. Bd. Pharmacy, Ft. Worth; ins. agt. N.Y. Life; pharm. cons. to surgery ctrs. in the Dallas and Ft. Worth areas. Bd. dirs. Mother and Unborn Baby, Ft. Worth, 1983—, Unborn Child Clinic Care, 1984—. Served to capt. USAFR, 1968-74. Mem. Am. Soc. Cons. Pharmacists, Tex. Pharm. Assn., Tarrant County Pharm. Assn., U. Tex. Ex Students Assn., Kappa Psi. Roman Catholic. Club: Holy Family Chor., Sierra Internat., Serra Club. Lodge: K.C. Avocation: golfing. Achievements include research in formulation and compounding for chemical peels with fruit acid, and T.C.A. products; compounds formulated and compounds for local anesthetic gel to be used in conjunction with T.C.A. solutions; pharmaceutical compounding of new local anesthetics to be used in treatment of ears, of retinoic acid for reduction of wrinkles, of progesterone products for PMS syndrome in female patients, of special eye solutions, of ointment combinations for treatment of herpes simplex virus, of acne compound solutions, of sympathomimet compounds for urinary retention, of special ophthalmologic allergy solutions; rsch. in formulation and compounding antibiotic topical preparations for infection. Home: 2541 Coldstream Dr Fort Worth TX 76123-1241 Office: K-Mart 4346 Pharmacy 1701 S Cherry Ln Fort Worth TX 76108-3699 also: Whitten Pharmacy 3700 E Rosedale Fort Worth TX 76105 also: Plz Pharmacy 800 8th Ave Forth Worth TX 76104

ESTES, L(OLA) CAROLINE, aquarium store owner, operator; b. Harlingen, Tex., May 17, 1959; d. Fred W. Etter and Margaret (Smith) Caldwell; m. Stephen P. Estes, July 2, 1982 (div. 1985). BA in Anthropology, U. Tex., 1982. Mgr. Gingham Dog Kennel, Austin, Tex., 1977-82, Docktor Pet Ctr., Dallas, 1982-84; maintenance technician Living Interiors Aquarium Maintanence, Dallas, 1984-85; mgr. Kingfish Aquarium, Austin, 1985-92; owner Amazonia Aquariums, Austin, 1992—; involved in an endangered species maintenance program for N. Am. fish. Contbr. articles to profl. jours. Vol. coord., tchr. 4-H Aquatic Maestro, Austin, 1990—; activist Earth First, Austin, 1988—. Recipient Altruism award, Fedn. Tex. Aquarium Socs., 1991; Bicentennial Youth Debate scholar, 1976. Mem. Capitol Aquarium Soc. (pres., founder 1986—, Best in Show Fish award 1988, organizer ann. aquarium conv. 1988—), Tex. Aquaculture Assn., Tex. Cichlid Assn. (bd. dirs. 1992, 93), Am. Cichlid Assn., Aquatic Conservation Network, Fedn. Tex. Aquarium Socs. (del.), Tex. Fish Judges Registry, Alpha Lambda Delta. Avocations: aquariums, aquaculture, backpacking, hiking, fish shows, travel. Home: 1100 Payne Ave Austin TX 78757-3024 Office: Amazonia Aquariums 4631 Airport Blvd # 118 Austin TX 78751

ESVELT, LARRY ALLEN, environmental engineer; b. Spokane, Wash., Oct. 19, 1938; s. Howard A. and Clara C. (Beck) E.; m. Sherry Kay Maize, Aug. 12, 1962; children: Kathy Lynn, Mark Howard, Erik Frederick. BS in Civil Engring., Wash. State U., 1961; MS in Civil Engring., U. Calif., Berkeley, 1964, PhD in Engring., 1971. Lic. profl. engr., Wash., Idaho, Calif., Oreg., Mont., N.Mex., Alaska; bd. cert. specialist Am. Acad. Environ. Engrs. Engr. J.P. Esvelt, Cons. Engr., Spokane, Wash., 1961; rsch. asst. San. Engring. Lab. U. Calif., Berkeley, 1964, rsch. assoc., 1970-71; san. engr. Gray & Osborne, Cons. Engrs., Yakima, Wash., 1964-69, Esvelt Engrs./STR Engrs., Spokane, 1972-73; sr. assoc. Bovay Engrs., Spokane, 1973-76; prin., owner Esvelt Environ. Engring., Spokane, 1976—; mem., chmn. civil engring. adv. bd. Wash. State U., Pullman, 1980—; evaluator Accreditation Bd. for Engring. and Tech., N.Y.C., 1985—; mem. Water Reuse Adv. Bd., Wash. Dept. Health, 1992—. Contbr. articles to profl. jours. Lt. USPHS, 1961-63. Recipient Cert. of Merit for aquifer protection Spokane County, Wash., 1991, Cert. of Appreciation Wash. State U., Pullman, 1992. Mem. ASCE (sect. pres. 1979-80, engr. of merit Inland Empire sect. 1981), Water Environ. Fedn., Am. Water Works Assn., Internat. Assn. Water Quality, Rotary Club of Spokane Valley. Achievements include research in toxicity removal from muncipal wastewaters, aerobic treatment of fruit processing wastes, reuse of treated wastewater, food processing waste activated sludge as a cattle feed. Office: Esvelt Environ Engring 7605 E Hadin Dr Spokane WA 99212-1710

ETGES, WILLIAM JAMES, biologist; b. Cin., Jan. 6, 1955; s. Frank Joseph Etges; m. Elizabeth C. Siddall, June 18, 1977; children: Lauren, Suzanne, Peter, Katherine. BS, N.C. State U., 1976; MS, U. Ga., 1979; PhD, U. Rochester, 1984. Vis. asst. prof. U. Ariz., Tucson, 1984-85, NSF postdoctoral fellow, 1985-87; asst. prof. U. Ark., Fayetteville, 1987—. Author: (with others) Ecological and Evolutionary Genetics of Drosophila, others; contbr. articles to profl. jours. including Am. Naturalist, Evolution, Physiol. Zoology, Heredity. Recipient BRSG award NIH, U. Ark., 1988-92; rsch. grants Ark. Game and Fish Commn., U. Ark., 1992—, 93—. Fellow Soc. of Am. Naturalists; mem. AAAS, Soc. for the Study of Evolution (life), Genetics Soc. of Am. Achievements include rsch. on evolutionary mechanisms. Office: U Ark Dept Biol Scis SCEN 632 Fayetteville AR 72701

ETHRIDGE, LOYDE TIMOTHY, hydraulic engineer, consultant; b. Anderson, S.C., Aug. 17, 1953; s. James Loyde and Doris Irene (Dunn) Ethridge; m. Elizabeth Ann Waring, June 11, 1981; children: Angela Renee, Amy Elizabeth. BS, U. Miss., 1976, MS, 1979. Registered profl. engr., Miss.; cert. profl. geologist, Ind. Seismologist U. Miss. Seismological Obs., Oxford, 1975-76; rsch. geologist Miss. Mineral Resources Inst., Oxford, 1976-78; rsch. geologist Sedimentation Lab. USDA, Oxford, 1978-80; hydraulic engr. lower Miss. div. U.S. Army C.E., Vicksburg, 1980—; topographic mapping coord. Miss. River Commn., Vicksburg, 1986—; cons. Waterways Experiment Sta., Vicksburg, 1989-90. Co-author: A Study of Vegetation on Revetments, Sacramento River, (chpt.) History of Bank Protection Through the Use of Revetments, 1985. Mem. ASCE, Am. Assn. Petroleum Geologists, U. Miss. Geol. Soc. Achievements include design of computer program used as design criteria for designing and constructing dikes for channel improvement. Home: 3024 Rose Ln Vicksburg MS 39180-4753

ETHRIDGE, MAX MICHAEL, civil engineer; b. Aurora, Mo., Aug. 7, 1949; s. James and Muriel (Boswell) E.; m. Martha Inez Thomason, Jan. 2, 1970; children: Marcia Inez, Marguerite Muriel. BSCE, U. Mo., 1970; MSCE, Purdue U., 1975, PhD, 1977; MPA, Golden Gate U., 1983. Registered profl. engr., Ind.; cert. photogrammetrist. Chief photo party 65 Nat. Ocean Svc., Norfolk, Va., 1972-74; exec. officer NOAA Ship Whiting, Norfolk, Va., 1977-79; chief coastal mapping Nat. Ocean Svc., Norfolk, Va., 1981-82; policy analyst Nat. Ocean Svc., Washington, 1982-84, liaison officer to def. mapping agy., 1984-85; deputy chief nat. geodetic survey Nat. Ocean Svc., Rockville, Md., 1985-87, chief nat. geodetic survey, 1987-89; sr. sci. advisor U.S. Geol. Survey, Reston, Va., 1989-90, chief office of strategic analysis, 1990—; adj. assoc. prof. Old Dominion U., Norfolk, 1979-82. Contbr. articles to profl. jours. deacon Presbyn. Ch., West Lafayette, Ind., 1975-77; asst. U.S. Girl Scouts, Virginia Beach, Va., 1980-82. Recipient Unit Citation, Nat. Oceanic and Atmospheric Adminstrn., 1971, letters of commendation, 1972-85, dirs. ribbons, 1987, Commendation medal, 1990, U.S. Dept. Interior Superior Svc. award 1993. Fellow Am. Congress on Surveying and Mapping (tech. tour coord. 1990-91, fed. govt. rep. on presdl. task force 1991-92); mem. ASCE, Am. Geophysical Union, Am. Soc. Photogrammetry and Remote Sensing, Am. Cartographic Assn. (bd. dirs. 1993—), Nat. Assn. Hispanic Fed. Execs. Avocations: home improvements, football, basketball, dancing, investments. Office: US Geol Survey 514 National Ctr Reston VA 22092

ETO, MORIFUSA, chemistry educator; b. Isahaya, Nagasaki, Japan, Feb. 20, 1930; s. Soroku and Hatsu (Kikuchi) E.; m. Tadako Ishida, Dec. 28, 1958; children: Nozomu, Megumu. MS, Kyushu U., Fukuoka, Japan, 1952, PhD, 1962. Instr. Kyushu U., Fukuoka, 1957-63, assoc. prof., 1963-77, prof. agrl. chemistry, 1977-93, prof. emeritus, 1993—; pres. Miyakonojo (Japan) Nat. Coll. Tech., 1993—; vis. assoc. prof. U. Calif., Berkeley, 1973-74. Author: Organophosphorus Pesticides, 1974; author, editor Bioorganic Chemistry of Pesticides, 1985, A New Turn in Pesticide Sciences, 1987. Mem. task group WHO, IPCS Environ. Health Criterir, Geneva, 1990; chmn. Com. for Pesticide Use Guidelines, Kumamoto, Japan, 1990; mem. Com. for Environ. Hazards, Fukuoka, 1992. Recipient Japan Agrl. Sci. award Fedn. Japanese Agrl. Sci. Socs., Tokyo, 1981. Mem. Agrl. Chem. Soc. Japan (chmn. West Japan br. 1981-83, agrl. chemistry award 1963, Agrochemicals Rsch. award 1993), Pesticide Sci. Soc. Japan (pres. 1989-91), Soc. Synthetic Organic Chemistry. Achievements include invention of insecticide salithion. Home: 5-3 Hanaguri-cho, Miyakonojo 885, Japan Office: Miyakonojo Nat Coll Tech, 473-1 Yoshio-cho, Miyakonojo 885, Japan

ETTRICK, MARCO ANTONIO, theoretical physicist; b. Panama City, Panama, July 17, 1945; came to U.S., 1963; s. Clemente Adolfo and Olga Rosa (Birmingham) E.; m. Adys Marie Hippolyte, Oct. 22, 1966 (div. Mar. 1977); children: Rudolphe Antoine, Marc Edouard. BS in Math., Poly. Tech. U., Bklyn., 1968; MS in Math., Poly. Tech. U., 1986, doctoral study. Programmer analyst Citibank, N.Y.C., 1969-71; lic. bacteriologist Lincoln Hosp., N.Y.C., 1975-76; lectr. in math. Queens (N.Y.) Coll., 1980-81, L.I. U., Bklyn., 1981-82, N.Y. Tech. Coll., Bklyn., 1982-84, Medgar Evers Coll., Bklyn., 1986-87; mem. staff Poly. U., 1990—; mem. NASA Langley Ctr. and Washington Hdqrs. Contbr. articles to sci. jours. Mem. AAAS, Am. Fedn. Scientists, Am. Phys. Soc., N.Y. Acad. Scis., Math. Assn. Am., Pi Mu Epsilon. Roman Catholic. Avocations: swimming, boxing. Home: 79 Sterling St Brooklyn NY 11225-3318

ETZ, LOIS KAPELSOHN, architectural company executive; b. Newark, Feb. 7, 1944; d. Sol D. and Matilda (Zlotnick) Kapelsohn; m. Leonard Etz, Dec. 4, 1967 (dec. May 1976); children: Rachel Jennie, Rebecca Sarah. BA, Mount Holyoke Coll., 1966; MA, Seton Hall U., 1968. Counselor N.J. Rehab. Commn., Trenton, 1966-68; pvt. antique dealer Princeton, N.J., 1968-78; pres. Nat. Code Cons., Princeton, 1971-78; dir. purchasing, aux. svcs. Mercer County Community Coll., Trenton, 1978-81; v.p. Hillier Group Architects, Princeton, 1981—. Bd. dirs. Vols. in Probation, Princeton, 1981, N.J. Printmaking Coun., Princeton Arts Coun., Mercer County Spl. Svc. Com., Hadassah; v.p. McCarter Theatre Assocs., Princeton, 1986-89; bd. dirs. McCarter Theatre Trustees, Princeton, 1989-91; past v.p., bd. dirs. Jewish Ctr. Commendation Chief Justice N.J. Supreme Ct., 1982. Commendation Chief Justice N.J. Supreme Ct., 1982. Mem. Mt. Holyoke Alumnae Assn. (past pres. Princeton chpt.), Record Mgmt. Assn. (founding officer), Princeton Pers. Assn. Democrat. Jewish. Home: 1038 Princeton-Kingston Rd Princeton NJ 08540 Office: The Hillier Group CN23 500 Alexander Pk Princeton NJ 08540

EUL, WILFRIED LUDWIG, chemist; b. Koblenz, Germany, Feb. 19, 1955; came to U.S., 1989; s. Albert M. and Gertrud K. (Engel) E.; m. Ursula M. Mocha, Sept. 6, 1980; children: Matthias J., Andreas B. MS in Chemistry, Johannes Gutenberg U., Mainz, Germany, 1980, PhD in Chemistry, 1985. Rsch. chemist Degussa AG, Hanau, Germany, 1985-86, mgr. pulp and paper, 1986-89; mgr. active oxygen chems. Degussa Corp., Allendale, N.J., 1989-90, dir. active oxygen chems., 1990-92, dir. indsl. chems., 1992-93; dir. bus. devel. paper chems. & chem. industry Degussa Corp. Peroxygen Chem. Div., Ridgefield Park, N.J., 1993—; mem. divsn. bus. devel. paper chems. and chem. industry Degussa Corp., Ridgewood Park, N.J., 1993—; bd. dirs. Panam. com. Internat. Ozone Assn., Norwalk, Conn., 1990—; mem. indsl. adv. bd. Hazardous Substance Mgmt. Rsch. Ctr., Newark, 1991—. Contbr. articles to profl. jours. Mem. TAPPI, Can. Pulp and Paper Assn., Am. Water Works Assn., Water Environ. Fedn., Air and Water Mgmt. Assn., Am. Mgmt. Assn., Internat. Ozone Assn., Internat. Chem. Assn. Roman Catholic. Achievements include patents for process for stabilization of viscosity in chem. pulp, multistage process for bleaching of mech. pulp, color-stripping from mass-dyed papers and carbonless copy papers. Home: 4 Crescent Hollow Ramsey NJ 07446 Office: Degussa Corp 65 Challanges Rd Ridgefield Park NJ 07660

EVANCHO, JOSEPH WILLIAM, engineering executive, metallurgical engineer; b. Pitts., Feb. 20, 1947; m. Linda M. Kapples, Nov. 23, 1968; children: Michael, Michelle, William, Jeffrey, Matthew, Joseph, James, Brian. BS in Metall. and Materials Engring., U. Pitts., 1969. Engr. div. alloy tech. Alcoa, Alcoa Center, Pa., 1969-80; mgr. dept. tech. Alcoa, Pitts., 1980-82; v.p. Alcoa Steamship Co., Pitts., 1982-86; pres. Alcoa Steamship Co., N.Y.C., 1986-87; gen. mgr. Alcoa ARALL Laminates Co., New Kensington, Pa., 1987-91; pres. Structural Laminates Co., New Kensington, 1991—. Contbr. articles to profl. jours. Achievements include patents for aluminum structural members for vehicles, for aluminum brazing sheet. Office: Structural Laminates Co 510 Constitution Blvd New Kensington PA 15068

EVANS, ALAN GEORGE, electrical engineer; b. Upland, Pa., June 8, 1942; s. Thomas Leslie and Jennie (Lewis) E.; m. Barbara Lee Kilhefner, June 26, 1965; children: Christopher Alan, Jennifer Lee. BSEE, Widener U., 1964; MSEE, Drexel U., 1967, PhD, 1972. Asst. engr. Phila. Electric Co., 1964-70; computation analyst Material Scis. Corp., Blue Bell, Pa., 1970-72; teaching asst. Drexel U., Phila., 1965-72; assoc. engr. Calspan Corp., Cheektawaga, N.Y., 1972-74; asst. prof. U.S. Naval Acad., Annapolis, Md., 1983-84; electronic engr. Naval Surface Warfare Ctr., Dahlgren, Va., 1974—; mem. symposium tech. com. Inst. Navigation, Alexandria, Va., 1989-93, U.S. Def. Mapping Agy., Arlington, Va., 1986-92, U.S. Nat. Genetic Survey, Rockville, Md., 1985. Contbr. articles to profl. jours. Sec. Sch. Adv. Coun., LaPlata, Md., 1984-91; asst. leader 4-H, LaPlata, 1982-89. Recipient R&D award U.S. Mapping Agy., 1988, Disting. Alumni award Chichester High Sch., Pa., 1986, fellowship and teaching assistantship Drexel U., 1965-71. Fellow Internat. Assn. Geodesy (spl. study group 1986—); mem. IEEE, Inst. Navigation, Sigma Xi. Republican. Achievements include patents in field; research in the application of global positioning system satellite in areas of relative positioning; in signal multipath, signal processing, receiver development and geodetic measurements. Home: 5036 Woodhave Dr LaPlata MD 20646 Office: Naval Surface Warfare Ctr Code K13 Dahlgren VA 22448

EVANS, CHARLIE ANDERSON, chemist; b. Columbus, Ga., Dec. 29, 1945; s. James William and Mollie Ree (Carter) E.; m. Phyllis Angela Roberts, Dec. 16, 1967 (div. 1992); children: Timothy Anderson, Laurin Stephen, Paul Thomas. BS, Ga. Inst. Tech., 1968; PhD, U. Ga., 1974. Postdoctoral fellow Centre d'Etudes Nucleaire, Grenoble, France, 1973-74, U. Western Ont., London, 1974-76; applications chemist Varian Assocs., Florham Park, N.J., 1976-80; applications chemist JEOL, Cranford, N.J., 1980-81, mgr. applications lab., 1981-84; scientist Berlex Labs., Cedar Knolls, N.J., 1984-87; sr. prin. scientist Schering-Plough Corp., Bloomfield, N.J., 1987-90, devel. fellow, 1990—; part-time insr. Ga. Inst. Tech., Atlanta, 1967-68; adj. asst. prof. Drew U., Madison, N.J., 1978; adj. prof. Fairleigh Dickinson U., 1988—. Contbr. articles to profl. jours. With U.S. Army, 1969-71. Muscogee Found. scholar, 1964-68; NSF summer fellow, 1967; NDEA Title IV fellow, 1971-73; Fulbright-Hays fellow, 1973-74. Mem. AAAS, Am. Chem. Soc. (chmn. NMR discussion group 1988, 94), N.Y. Acad. Sci., Internat. Soc. Magnetic Resonance. Democrat. Presbyterian. Office: Schering Plough Rsch Inst #0450 2015 Galloping Hill Rd Kenilworth NJ 07033

EVANS, DAVID A., plant geneticist; b. Apr. 9, 1952; m. Kitty Ann Renninger, Dec. 19, 1978. BSc in Plant Genetics, Ohio State U., 1973, MSc in Plant Genetics, 1975, PhD, 1977; MA in Mgmt. Devel., Harvard U., 1987. Rockefeller Found. postdoctoral fellow NRC/Prairie Regional Lab., Sask., Can., 1977-78; asst. prof. dept. biol. scis. SUNY, Binghamton, 1978-80; mgr. cellular genetics Campbell Inst. for Rsch. and Tech., Cinnaminson, N.J., 1981; v.p. tech. and prodn. devel. DNA Plant Tech. Corp., Cinnaminson, 1981-89, v.p. bus. devel., 1989—; adj. prof. divsn. biology grad. program plant scis. Rutgers U., New Brunswick, N.J., 1980—, bd. mgrs., 1992—; mem. bd. advisors coll. agriculture Ohoi State U., 1988—; mem. departmental vis. com. dept. botany U. Tex., 1989—. Co-author: Handbook of Plant Cell Culture, Vols. 1-6, 1983-89, Biotechnology of Plants and Microorganisms, 1986; contbr. articles to sci. jours. Office: DNA Plant Tech Corp 2611 Branch Pike Cinnaminson NJ 08077

EVANS, DAVID A(LBERT), chemistry educator; b. Washington, D.C., Jan. 11, 1941; s. Albert Edward and Iris (Hill) Evans Yohe; m. Selena Anne Welliver, Dec. 27, 1962; 1 child, Bethan Hill. AB, Oberlin Coll., 1963; PhD, Calif. Inst. Tech., Pasadena, 1967; MA (hon.), Harvard U., Cambridge, 1983. Asst. prof. chemistry UCLA, 1967-72, assoc. prof., 1972-73, prof., 1974; prof. chemistry Calif. Inst. Tech., Pasadena, 1974-83, Harvard U., Cambridge, 1983—; mem. com. on chem. scis. NRC; cons. to pharm. industry; lectr. in field. Contbr. more than 125 articles of profl. jours.; hon. editor Tetrahedron and Tetrahedron Letters, 1981—; mem. editorial adv. bd. Jour. Am. Chem. Soc., 1983—. Recipient Camille and Henry Dreyfus Tchr.-Scholar award Dreyfus Found., 1971-76, Alfred P. Sloan Found. fellow, 1972-74; Disting. Teaching award UCLA Alumni Assn., 1973. Mem. Am. Chem. Soc. (award for creative work in synthetic organic chemistry 1982), Nat. Acad. Scis. (award 1984). Home: 39 Pine Hill Ln Concord MA 01742-4414 Office: Harvard U Dept of Chemistry 12 Oxford St Cambridge MA 02138-2900

EVANS, DAVID MYRDDIN, aerospace engineer; b. Kansas City, Kans., Aug. 9, 1946; s. Myrddin and Chessie Grace (Riley) E.; m. Margaret A. Stack (div. June 1980); children: Heather C., Jeremy D. BS, Kans. U., 1968. Aerodyns. staff McDonnell Douglas, St. Louis, 1968-87, dep. program mgr. proprietary programs, 1987-91, group mgr. F-15 aerodyns., 1991—; best practice chmn. McDonnell Douglas, St. Louis, 1991—. Mem. AIAA (sr.). Office: McDonnell Douglas Box 516 Saint Louis MO 63166

EVANS, DENIS JAMES, research scientist; b. Sydney, NSW, Australia, Apr. 19, 1951; s. Maxwell Leslie and Lila Mary (McHattan) E.; m. Valda Ozols, Dec. 23, 1972; 1 child, Kathryn Nicole. BS with honors, Sydney U. (Australia), 1972; PhD, Australian Nat. U., Canberra, 1975; postgrad., Oxford U., England, 1976. Rsch. assoc. Cornell U., N.Y., 1977; prof. Australian Nat. U., 1989—, acad. dir. supercomputer facility, 1989-92. Author: Statistical Mechanics of Nonequilibrium Liquids, 1990; contbr. articles to profl. jours. Rsch. fellow Australian Nat. U., 1978-81, fellow, 1982-84, Sr. fellow, 1985-88, Fulbright fellow Nat. Bur. Stds., 1980; recipient Young Disting. Chemist Fedn. Asian Chem. Socs., 1989, Frederick White prize Australian Acad. Scis., 1990. Mem. Australian Rheological Soc., Am. Phys. Soc., Royal Australian Chem. Inst. (Rennie medal 1982). Avocations: surfing, snorkelling, reading, ancient history. Office: Rsch Sch Chemistry, GPO Box 4, Canberra 2601, Australia

EVANS, DENNIS HYDE, chemist, educator; b. Grinnell, Iowa, Mar. 28, 1939; s. Leonard Hyde and Clara Ethel (Parmley) E.; m. Ruth Elizabeth Turnbull, June 28, 1958 (div. July 1986); children: Susan Katherine, John Hyde, Andrew Turnbull; m. Mary Jean Wirth, Aug.2, 1986. B.S., Ottawa U., 1960; A.M., Harvard U., 1961, Ph.D., 1964. Instr. chemistry Harvard U., Cambridge, 1964-66; asst. prof. chemistry U. Wis., Madison, 1966-70, asso. prof., 1970-75, prof., 1975-84, Meloche-Bascom prof. chemistry, 1984-86, chmn. dept., 1977-80, assoc. dean Coll. of Letters and Sci., 1983-86; prof. chemistry U. Del., Newark, 1986—. Contbr. articles to profl. jours. Named Danforth fellow, 1960-64, NIH fellow, 1961-64; recipient C.N. Reilley award Soc. for Electroanalytical Chemistry, 1993. Mem. Am. Chem. Soc., Internat. Soc. Electrochemistry, Electrochem. Soc. for Electroanalytical Chemistry (pres. 1993—). Baptist. Home: 26 E Parkway Pky Elkton MD 21921-2042 Office: U Del Dept Chemistry Newark DE 19716

EVANS, DONALD FOSTER, electronics technician, computer consultant; b. Atlanta, Sept. 27, 1949; s. Marlin Donald Evans and Winnefred Irene (Pelton) Yow; divorced; 1 child, Randi Leigh. Grad., USAF Sch. Electronics, Biloxi, Miss., 1969; student, Evergreen Community Coll., San Jose, Calif., 1980, San Jose City Coll., 1986. Enlisted USAF, 1968, advanced to staff sgt., 1973, resigned, 1979; tng. instr. Sylvania Tech. Systems, Sunnyvale, Calif., 1979-80, Ampex Inc., Redwood City, Calif., 1980-82; master technician Convergent Techs., San Jose, 1983-84; sr. electronics technician Atari, Sunnyvale, 1982-83, Forté Data Systems, San Jose, 1984-86, Alp's Electric Inc., San Jose, 1986—. Mem. Silicon Valley Gay Men's Chorus, San Jose, 1986—; bd. dirs. Necisities & More, 1989, Project Disseminate Info. on AIDS by Phone Lines, San Francisco, 1990. Mem. Assn. Computing Machinery, Math. Assn. Am. Avocations: reading, music, computers, fishing. Home: 945 S 3d St San Jose CA 95112 Office: Alps Electric Inc 3553 N 1st St San Jose CA 95134-1898

EVANS, DWIGHT LANDIS, psychiatrist, educator; b. Lancaster, Pa., Mar. 27, 1947; m. Janet Elaine Strickler, Sept. 2, 1970; children: Elizabeth Anne, Meredith Lee, Benjamin Dwight, Christopher William. MS, Bucknell U., 1972; MD, Temple U., 1976; postgrad., U. N.C., 1976-79, psychoanalytic tng., 1976-83. Diplomate Am. Bd. Psychiatry and Neurology; cert. tchr., Fla. From instr. psychiatry to prof. psychiatry and medicine U. N.C., Chapel Hill, 1979-89; prof. psychiatry, medicine and neurosci., chmn. psychiatry dept. U. Fla., Gainesville, 1992—; chmn. study sect. NIMH. Contbr. articles to profl. jours. Grantee NIH 1983—. Mem. Ctr. for Neurobiol. Sci. (assoc.), Psychoneuroimmunology Adv. Coun., Ctr. for Alcohol Rsch., Am. Psychiat. Assn., Am. Coll. Psychiatry, Am. Coll. Neuropsychopharmacology. Office: Univ Fla Dept Psychiatry PO Box 100256 Gainesville FL 32610

EVANS, EDWARD SPENCER, JR., entomologist; b. Woodbury, N.J., Aug. 7, 1943; s. Edward Spencer and Hazel Louise (Flagg) E.; m. Marilyn Dale Kernohan, Aug. 13, 1966 (div. 1981); children: Tracey Lynn, Edward Spencer III; m. Sandra Ruth Ehrhardt, June 9, 1984. BS, Rutgers U., 1965, MS, 1967, PhD, 1975. Cert. entomologist. Asst. wildlife biologist N.J. Div. Fish and Game, Tuckahoe, 1964; grad. asst. dept. entomology Rutgers U., New Brunswick, N.J., 1965-67, 69-73; entomologist U.S. Army Environ. Hygiene Agy., Aberdeen Proving Ground, Md., 1973-76; pesticide coord. U.S. Army Environ. Hygiene Agy., Md., 1976-83, supervisory entomologist, 1983—; chmn. Armed Forces Pest Mgmt. Bd., Washington, 1988-92; adj. asst. prof. Uniformed Svcs. U. Health Scis., Bethesda, Md., 1986—. Co-author: Pesticides, 1991; contbr. articles and tech. reports to profl. jours. Chmn. long range planning com. Bel Air (Md.) United Meth. Ch., 1988, mem. bldg. com., 1990—; coach youth baseball Recreation Coun., Joppatowne, Md., 1980-85. Capt. U.S. Army, 1967-69, Korea. Mem. ASTM (chmn. com. E-35, 1983-89), Am. Mosquito Control Assn., Entomol. Soc. Am., Sigma Xi, Alpha Zeta. Home: 1309 Beckett Ct Bel Air MD 21014-2736 Office: US Army Environ Hygiene Agy Entomol Sci Div Aberdeen Proving Ground MD 21010-5422

EVANS, EDWIN CURTIS, internist, educator, geriatrician; b. Milledgeville, Ga., June 30, 1917; s. Watt Collier and Bertha Chambers E.; m. Marjorie Claire Wood, Nov. 27, 1945; children: Nancy, Edwin, Marjorie, Jane and Jill (twins), Carol. B.S., U. Ga., 1936; M.D., Johns Hopkins U., 1940. Diplomate: Am. Bd. Internal Medicine, also recertified. Intern Hartford (Conn.) Hosp., 1940-42; resident in medicine Balt. City Hosp., 1946-47; fellow in pathology Hosp. of U. Pa., Phila., 1947-48; practice medicine specializing in internal medicine Atlanta, 1948-87; dir. geriatrics Ga. Bapt. Med. Ctr., 1987-91; clin. assoc. prof. medicine Emory U. Sch. Medicine, 1972-87, clin. prof. emeritus medicine, 1987—; clin. prof. medicine Sch. Pharmacy, Mercer U., 1980-90; chief of staff Ga. Baptist Med. Center, Atlanta, 1973-79; pres. Atlanta Blue Shield, 1968-70. Contbr. articles to profl. jours. Served to maj. M.C. AUS, 1942-46. Fellow ACP (gov. Ga. 1972-76), Am. Coll. Chest Physicians; mem. Am. Geriatrics Soc., Diabetes Assn. Atlanta (pres. 1958), Ga. Diabetes Assn. (pres. 1965), Med. Assn. Atlanta (pres. 1973, many awards), Ga. Soc. Internal Medicine (pres. 1963), Am. Soc. Internal Med. (pres. 1972-73), So. Med. Assn. (pres. 1981-82, cert. appreciation 1979), Med. Assn. Ga. (cert. meritorious/disting. svc. 1990), AMA (physicians recognition award 1978, 81, 84, 86), Am. Heart Assn., Inst. Medicine Nat. Acad. Scis., Phi Chi (pres. Kappa Delta chpt. 1939). Republican. Methodist. Clubs: Cherokee Town and Country, Vinings. Home: 500 Westover Dr NW Atlanta GA 30305-3538

EVANS, ESSI H., research scientist; b. Bad-Schwalbach, W.Ger., Jan. 12, 1950; came to U.S., 1951, naturalized, 1957; d. John H. (b. Horst H. Jahn) and Jean E. (von Schwerin); m. Everett M. Turner, Jr., Aug. 16, 1974. BS in Agr. (James Harris scholar), U. Md., 1972; MS in Animal Sci., U. Guelph, 1974, PhD in Animal Sci., 1976. Polymer chemist Monarch Rubber Co., Balt., 1972; rsch. asst., teaching asst. U. Guelph (Ont., Can.), 1972-76; project dir. animal nutrition Can. Packers Inc., Toronto, Ont., 1976-85; tech. mgr. animal nutrition and animal health Can. Packers, Inc., Toronto, 1986-89, rsch. mgr., 1989-90, gen. rsch. and nutrition mgr. shur-gain div., 1990—; farm cons.; guest lectr. Hubbard Farms fellow, 1975-76; NRC Indsl. postdoctoral fellow, 1976-79; recipient Hamilton Milk Producers award, 1973, 74; Ont. Ministry of Agr. and Foods Provincial Lottery grantee, 1980-83. Mem. Am. Soc. Animal Sci., Am. Dairy Sci. Assn., Ont. Comml. Rabbit Growers Assn., Am. Assn. Vet. Nutritionists, Coun. for Agrl. Sci. and Tech., Nat. Feed Industry Assn., Can. Feed Industry Assn. Republican. Contbr. to sci. publs. and confs. Home: 64 Scugog St, Bowmanville, ON Canada L1C 3J1 Office: Shur-Gain Div Maple Leaf Foods Inc, 2700 Matheson Blvd E Ste 600 E, Mississauga, ON Canada L4W 4V9

EVANS, GARY WILLIAM, human ecology educator; b. Summit, N.J., Nov. 22, 1948. AB in Psychology with high honors, Colgate U., 1971; MS, U. Mass., 1973, PhD, 1975. Rsch. asst. in psychology Colgate U., 1969-71; rsch. assoc. Inst. for Environ. Studies, U. Mass., 1971-73; instr. psychology U. Mass., Amherst, 1973-75; assoc. dir. for undergrad. studies, social ecology U. Calif., Irvine, 1978-81, asst. prof. social ecology, 1975-80, assoc. prof., 1980-85, prof., 1985-92; prof. design and environ. analysis Cornell U., Ithaca, N.Y., 1992—. Contbr. articles to Jour. Environ. Psychol., Psychol. Bull., Jour. Applied Psychology, Jour. Expt. Psychol., Environ. & Behavior, Jour. Personal and Soc. Psychology, others; author/co-author 4 books. Recipient numerous grants in field. Fellow APA, Am. Psychol. Soc., Soc. Psychol. Study of Social Issues; mem. Environ. Design Rsch. Assn., Internat. Assn. Applied Psychology, Internat. Assn. of People and Their Surroundings, Internat. Soc. Complex Environ. Studies, Soc. Exptl. Social Psychology, Soc. for Human Ecology. Office: Cornell U Coll Human Ecology Design and Environ Analysis Ithaca NY 14853-4401

EVANS, GEORGE FREDERICK, consulting engineer; b. Erie, Pa., Sept. 17, 1922; s. Frederick Brandon and Mary (MacCleod) E.; m. Doris Anne Stickle, Feb. 19, 1944; children: Margaret, George. BS, U. Pitts., 1943. Registerd profl. engr., Ohio and 20 other states. Staff engr. Harrison Radiator Div. GM, Lockport, N.Y., 1943-46; constrn. supt. Stickle, Kelly and Stickle, Cleve., 1946-47; prof. engring. Fenn Coll. Engring., Cleve., 1947-57; pres. Evans and Assocs., Inc., Cleve., 1950—; chmn. adv. bd. Cleve. State U. Sch. Engring., 1970-74, Ohio State Bd. Registration for Profl. Engrs., Columbus, 1976-86; trustee Engrs. Found. Ohio, Columbus, 1977—; pres. Ohio Soc. Profl. Engrs., Columbus, 1964-65, Consulting Engrs. Ohio, Columbus, 1962; liaison rep. Nat. Coun. Engring. Examiners, Clemson, S.C., 1988—; bd. dirs. Cleve. Water Conditioning Corp.; arbitrator Am. Arbitration Assn., Cleve. and Pitts., 1980—. Author: (text book) Water Supply and Conditioning, 1980; contbr. articles to profl. jours. Vice chmn. City of Independence (Ohio) Charter Commn., 1978. Recipient Award of Merit, Nat. Coun. for Profl. Devel., 1976, Meritorious Svc. award Ohio Ho., 1982, 83, Cert. of Appreciation, Nat. Coun. Engring. Examiners, 1977, 79; named Engr. of Yr., Cleve. Soc. Profl. Engrs., 1968, Ohio Soc. Profl. Engrs., 1973. Fellow ASRACE; mem. Kiwanis (chmn. 1974—, bd. dirs. 1978-82, Legion of Honor award 1984), Univ. Club Cleve. (bd. dirs. 1983-85), Phi Tau Sigma, Sigma Tau, Tau Omega. Presbyterian. Achievements include design of over 2,500 building projects throughout the U.S., Can., S.Am. and Africa. Home: 7448 Oval Dr Independence OH 44131 Office: Evans and Assocs Inc # 315 6100 Rockside Woods Blvd Independence OH 44131

EVANS, GEORGE LEONARD, microbiologist; b. Wilkes-Barre, Pa., Aug. 3, 1931; s. George Leonard and Anna M. (Check) E.; m. Joan Marie Snyder, Feb. 8, 1958; children: Paula Jean, Gregory Allen, Christopher Thomas. BS, King's Coll., 1954; MS, Fordham U., 1957; PhD, Temple U., 1962. Specialist microbiologist Am. Acad. Microbiology. Sr. scientist Warner-Lambert Pharm. Co., Morris Plains, N.J., 1961-64; rsch. virologist Univ. Labs., Inc., Highland Park, N.J., 1964-65; group chief Hoffmann-La Roche, Inc., Nutley, N.J., 1965-70; dir. diagnostic rsch. Schering Corp., Bloomfield, N.J., 1970-75; rsch. fellow Becton Dickinson & Co., Hunt Valley, Md., 1975—; chmn., advisor subcom. Nat. Com. Clin. Lab. Standards, Villanova, Pa., 1980—. Contbr. articles to Jour. Immunology, Jour. Reticuloendothelial Soc., Jour. Bacteriology, Am. Jour. Med. Tech., Jour. Clin. Microbiology. Mem. Am. Soc. Microbiology, Am. Assn. Clin. Chemistry. Achievements include 10 patents for Process for Preparation of a Soluble Bacterial Extract, Diagnostic Preparation and Process for Detection of Acetylmethylcarbinol, Paper Strip Test for Citrate Utilization, Colorimetric Method for Determining Iron in Blood, Growth Inhibition of Selected Mycoplasmas, Polyanionic Compounds in Culture Media, Novel Diagnostic System for Differentiation of Enterobacteriaceae, Diagnostic Test for Determination of Sickling Hemoglobinopathies, Serologic Test for Systemic Candidiasis, Selective Medium for Growth of Neisseria. Office: Becton Dickinson 250 Schilling Cir Cockeysville Hunt Valley MD 21031-1103

EVANS, JOHN ROBERT, former university president, physician; b. Toronto, Ont., Can., Oct. 1, 1929; s. William Watson and Mary Evelyn Lucille (Thompson) E.; m. Jean Gay Glassco, 1954; children: Derek, Mark and Michael (twins), Gillian, Timothy, Willa. MD, U. Toronto, 1952; DPhil (Rhodes scholar), Oxford U., 1955; LLD (hon.), Dalhousie U., McMaster U., McGill U., 1972, Queen's U., 1974, Wilfred Laurier U., 1975, York U., 1977, U. Toronto, 1980, U. Western Ont., 1982, Yale U., 1978; DSc (hon.), Meml. U., 1973, U. Montreal, 1977, Royal Mil. Coll., 1989; DHL (hon.), Johns Hopkins U., 1978; D Univ. (hon.), U. Ottawa, 1978, U. Limbourg, The Netherlands, 1980. Intern Toronto Gen. Hosp., 1952-53, chief resident physician, 1958-59; practice medicine specializing in cardiology Toronto, 1961—; assoc. dept. medicine U. Toronto Med. Sch., 1961-65, prof., 1972—, pres. univ., 1972-78; dir. population, health and nutrition dept. World Bank, Washington, 1979-83; chmn. Allelix Inc., Mississauga, Ont., 1983—; physician Toronto Gen. Hosp., 1961-65; dean Faculty Medicine McMaster U., Hamilton, Ont., 1965-72, v.p. health scis., 1967-72; bd. dirs. Dofasco, Inc., Hamilton, Ont., Torstar Ltd., Toronto, Royal Bank of Can., Alcan Aluminum Ltd., Montreal, Trimark Fin. Corp., Toronto, MDS Health Group, Toronto, Connaught Labs. Inc., Toronto, Pasteur Mérieux Serums and Vaccines, Lyon, France; hon. fellow London Sch. Hygiene and Tropical Medicine, Univ. Coll., Oxford, Eng. Trustee Rockefeller Found., N.Y.C., 1982—, chmn., 1988—; chmn. African Med. Rsch. Found., Can., 1986-90. Decorated companion Order of Can.; Order of Ontario, 1991; Markle scholar, 1960-65; recipient Gairdner Foundation Wightman Award, Gairdner Foundation, 1992. Master ACP; fellow Royal Soc. Can., Royal Coll. Physicians and Surgeons Can., Royal Coll. Physicians (London). Home: 58 Highland Ave, Toronto, ON Canada M4W 2A3 Office: Allelix Inc, 6850 Goreway Dr, Mississauga, ON Canada L4V 1P1

EVANS, JOHN VAUGHAN, satellite laboratory executive, physicist; b. Manchester, Eng., July 5, 1933; came to U.S., 1960; s. Cyril John and Gertrude Veronica (Bayliss) E.; m. Maureen Vervain Patrick, Oct. 19, 1958; children: Carol, David, Lesley. BS in Physics, Manchester U., 1954, PhD, 1957. Leverhulme research fellow Jodrell Bank Exptl. Sta., U.K., 1957-60; staff mem. Lincoln Lab., MIT, Lexington, 1960-66, 67-70, assoc. group leader surveillance techniques, 1970-72, group leader, 1972-74, assoc. div. head Aerospace div., 1974-77, asst. dir., 1977-83; dir. Haystack Obs., prof. meteorology MIT, Cambridge, 1980-83; v.p. research and devel. COMSAT Labs., Clarksburg, Md., 1983, v.p., dir., 1983—; pres., 1992—; G.A. Miller vis. prof. U. Ill., Urbana, 1966-67; trustee Univ. Corp. for Atmospheric Research, Boulder, Colo., 1980-87. Editor: (with T. Hagfors) Radar Astronomy, 1968; contbr. numerous articles to profl. jours. Served with Brit. Territorial Army, 1951-57. Recipient Appleton prize Royal Soc. London, 1954. Fellow IEEE; mem. Nat. Acad. Engring., Am. Geophysical Union, Internat. Astron. Union, AIAA. Unitarian. Club: Cosmos (Washington). Office: COMSAT Labs 22300 Comsat Dr Clarksburg MD 20871-9470

EVANS, JUDY ANNE, health center administrator; b. Elmira, N.Y., Mar. 29, 1940; d. Hugh Kenneth and Mary (Faul) Leach; m. Nolly Seymour Evans, Feb. 18, 1965; children: Samantha, Meredydd, Clelia, Nolly III. BS, Cornell U., 1962; MBA, Syracuse U., 1990-92. Fin. analyst Morgan Guaranty Trust Co., N.Y.C., 1962-66; bus. adminstr. SUNY Health Sci. Ctr., Syracuse, 1983-89, adminstr. dept. pediatrics, 1990—. Mem. allocations com. Children Miracle Network, Syracuse, 1990—; children's hosp. steering com. Crouse Irving/Univ. Hosp., Syracuse, 1990—; bd. dirs. Syracuse Friends of Chamber Music, 1983-89, Syracuse Camerata, 1982-88. Mem. Assn. Adminstrs. of Acad. Pediatrics. Avocations: sailing, cooking. Home: 26 Lyndon Rd Fayetteville NY 13066-1016 Office: SUNY Health Sci Ctr 750 E Adams St Syracuse NY 13210-2306

EVANS, LOUISE, psychologist; b. San Antonio; d. Henry Daniel and Adela (Pariser) E.; m. Thomas Ross Gambrell, Feb. 23, 1960. BS, Northwestern U., 1949; MS in Clin. Psychology, Purdue U., 1952, PhD in Clin. Psychology, 1955. Lic. Marriage, Family and Child Counselor Calif., Nat. Register of Health Svc. Providers in Psychology; lic. psychologist N.Y. (inactive), Calif.; diplomate Clin. Psychology, Am. Bd. Profl. Psychology. Intern clin. psychology Menninger Found.-Topeka (Kans.) State Hosp., 1952-53, USPHS-Menninger Found. postdoctoral fellow clin. child psychology, 1955-56; staff psychologist Kankakee (Ill.) State Hosp., 1954; head staff psychologist child guidance clinic Kings County Hosp., Bklyn., 1957-58; dir. psychology clinic Barnes-Renard Hosp., instr. med. psychology Washington U. Sch. Medicine, 1959; clin. rsch. cons. Episc. City Diocese, St. Louis, 1959; pvt. practice clin. psychology, 1960-92; approved fellow Internat. Coun. Sex Edn. and Parenthood, 1984; psychol. cons. Fullerton (Calif.) Community Hosp., 1961-81; staff cons. clin. psychology Martin Luther Hosp., Anaheim, Calif., 1963-70; nat., internat. lectr. clin. psychology schs. and profl. groups, 1950—; chairperson, participant psychol. symposiums, 1956—; guest speaker clin. psychology civic and community orgns., 1950—. Elected to Hall of Fame, Central High Sch., Evansville, Ind., 1966; recipient Service award Yuma County Head Start Program, 1972, Statue of Victory Personality of the Yr. award Centro Studi E. Ricerche Delle Nazioni, Italy, 1985; named Miss Heritage, Heritage Publs., 1965. Fellow APA (clin. divsn., psychology of women divsn., cons. divsn., dir.exec. bd. 1976-79), Am. Assn. Applied and Preventative Psychology (charter), Royal Soc. Health England (emeritus), Internat. Council of Psychologists (dir. 1977-79, sec. 1962-64, 73-76), AAAS (emeritus), Am. Orthopsychiat. Assn. (life), World Wide Acad. of Scholars of N.Z. (life), Am. Psychol. Soc. (charter); mem. AAUP (emeritus), Nat. Register Health Svc. Providers in Psychology, Los Angeles Soc. Clin. Psychologists (exec. bd. 1966-67), Calif. State Psychol. Assn. (life, ins. com. 1961-65), Los Angeles County Psychol. Assn. (emeritus), Orange County Psychol. Assn. (charter founding mem., exec. bd. 1961-62), Orange County Soc. Clin. Psychologists (founder, exec. bd. 1963-65, pres. 1964-65), Am. Public Health Assn. (emeritus), Internat. Platform Assn., N.Y. Acad. Scis. (emeritus), Purdue U. Alumni Assn. (life, Citizenship award 1975, Disting. Alumni award 1993, Old Master 1993), Center for Study of Presidency, Soc. Jewelry Historians USA, Alumni Assn. Menninger Sch. Psychiatry, Soc. Sigma Xi Nat. Rsch. Hon. (emeritus), Pi Sigma Pi (pres. 1947-48, sec. 1946-47). Contbr. articles on clin. psychology to profl. publs. Achievements include development of innovative theories and techniques of practice; acknowledged pioneer in devel. psychology as sci. and profession both nat. and internat., and pioneer in marital and family therapy. Office: PO Box 6067 Beverly Hills CA 90212-1067

EVANS, MICHAEL LEIGH, physiologist; b. Detroit, July 26, 1941; s. Kenneth Lynn and Martha Agnes (Paquette) E.; m. Linda Lee Sims, Sept. 1, 1962; children: Todd, Amber, Matthew. MS, U. Mich., 1965; PhD, U. Calif., Santa Cruz, 1967. Asst. prof. Kalamazoo (Mich.) Coll., 1967-70; NATO fellow U. Freiburg, Germany, 1970-71; asst. prof. Ohio State U., Columbus, 1971-73, assoc. prof., 1973-78, prof., 1978—. Co-author: Plants: Their Biology and Importance, 1989; contbr. articles to Planta, Plant Cell Physiology, Plant Physiology, Protoplasma. Grantee NASA, NSF. Mem. Am. Soc. Plant Physiologists, Japanese Soc. Plant Physiologists, Am. Soc. Gravitational and Space Biology. Office: Ohio State U Dept Plant 1735 Neil Ave Columbus OH 43210

EVANS, PAUL, osteopath; b. Nutley, N.J., May 23, 1950; m. Roxanne Romack. BS cum laude in Biology, U. Miami, 1972; DO, Phila. Coll. Osteopathic Med., 1979. Diplomate Am. Bd. Family Practice, Nat. Bd. Osteo. Examiners, Am. Acad. Pain Mgmt.; cert. Am. Osteo. Bd. Gen. Practice. Commd. 2d lt. U.S. Army, 1972, advanced through grades to lt. col., 1989; asst. chief mil. pers. U.S. Army Med. Svc. Corps, Frankfurt, Fed. Republic Germany, 1972-75; intern Letterman Army Med. Ctr., San Francisco, 1979-80; resident in family practice Womack Army Community Hosp., Ft. Bragg, N.C., 1980-82; dir. family practice quality assurance Tripler Army Med. Ctr., Hawaii, 1984-86, dir. residency tng. dept. family practice, 1984-86; asst. prof. family practice, physician Uniformed Svcs. U. Health Scis., F. Edward Hebert Sch. Med., Bethesda, Md., 1986-92, clerkship dir., 1986-88, dir. continuing med. edn., 1987-91, asst. prof. mil. and emergency medicine, 1991-92; chief family practice Reynolds Army Community Hosp., Ft. Sill, Okla., 1992—, chmn. rsch. com., dir. hosp. continuing med. edn., 1992—, dir. physicians asst. tng. program, dir. quality improvement, 1992—; presenter, lectr., cons. in field; part-time clin. faculty mem., family practice residency DeWitt Army Hosp., Ft. Belvoir, Va., 1986-89, 91-92, Malcolm Grow USAF Med. Ctr., Andrews AFB, Md., 1989-91; reviewer Patient Care jour., 1988—. Contbr. articles to profl. publs. Asst. med. dir. Old Dominion 100 Mile Run, Front Royal, Va., 1990, med. dir., 1991; asst. med. dir. Am. Diabetes Assn. Youth, Honolulu, 1984, med. dir., 1985. USUHS grantee. Fellow Am. Acad. Family Physicians; mem. Uniformed Svcs. Acad. Family Physicians (chmn. edn. com. 1993—), Soc. Tchrs. Family Medicine (mem. genogram rsch. com. 1989—), Am. Acad. Pain Mgmt., Assn. Mil. Surgeons, Am. Osteo. Acad. Sports Medicine, Phila. Coll. Osteo. Medicine Alumni Assn. (life), Omicron Delta Kappa, Alpha Epsilon Delta. Avocations: triathlons, downhill skiing, scuba diving, collecting salt water fish, finch breeding. Home: 2725 NW Denver Ave Lawton OK 73505 Office: Reynolds Army Community Hosp Fort Sill OK 73503

EVANS, PAULINE D., physicist, educator; b. Bklyn., Mar. 24, 1922; d. John A. and Hannah (Brandt) Davidson; m. Melbourne Griffith Evans, Sept. 6, 1950; children: Lynn Janet Evans Hannemann, Brian Griffith. BA, Hofstra Coll., 1942; postgrad., NYU, 1943, 46-47, Cornell U., 1946, Syracuse U., 1947-50. Jr. physicist Signal Corps Ground Signal Svc., Eatontown, N.J., 1942-43; physicist Kellex Corp. (Manhattan Project), N.Y.C., 1944; faculty dept. physics Queens Coll., N.Y.C., 1944-47; teaching asst. Syracuse U., 1947-50; instr. Wheaton Coll., Norton, Mass., 1952; physicist Nat. Bur. Standards, Washington, 1954-55; instr. physics U. Ala., 1955, U. N.Mex., 1955, 57-58; staff mem. Sandia Corp., Albuquerque, 1956-57; physicist Naval Nuclear Ordnance Evaluation Unit, Kirtland AFB, N.Mex., 1958-60; programmer Teaching Machines, Inc., Albuquerque, 1961; mem. faculty dept. physics Coll. St. Joseph on the Rio Grande (name changed to U. Albuquerque 1966), 1961—, assoc. prof., 1965—, chmn. dept., 1961—. Mem. AAUP, Am. Phys. Soc., Am. Assn. Physics Tchrs., Fedn. Am. Scientists, Sigma Pi Sigma, Sigma Delta Epsilon. Achievements include patents on mechanical method of conical scanning (radar), fluorine trap and primary standard for humidity measurement Home: 730 Loma Alta Ct NW Albuquerque NM 87105-1220 Office: U Albuquerque Dept Physics Albuquerque NM 87140

EVANS, PETER YOSHIO, ophthalmologist, educator; b. Tokyo, Dec. 19, 1925; came to the U.S., 1957; s. Paul Yuzuru Kawai and Vicki (Wichgraf) Evans; m. Helga Kemp, Sept. 19, 1953; children: Johannes, Marina, Michael, Andre, Thomas, Ursula, Christiane. MD, Innsbruck U., 1951. Resident Innsbruck & Frankfurt (Germany) Univs., 1951-55; intern Sisters Charity Hosp., Buffalo, N.Y., 1957-58; chief dept. ophthalmology D.C. Gen. Hosp., 1958-63; fellow Georgetown U., Washington, 1958-59, program dir. div. ophthalmology, 1963-69, chmn., 1969-83, prof., 1973-92, prof. emeritus, 1992—; cons. D.C. Columbia Lighthouse for the Blind, 1959-63; sr. cons. D.C. Child and Maternal Welfare Dept., 1961-74; exec. v.p. Joint Commn. Allied Health Personnel in Ophthalmology, St. Paul, 1983—. Author, producer scientific films; contbr. articles to profl. jours.; editor numerous jours. Fellow Am. Acad. Ophthalmology (Disting. Svc. award 1982), Austrian Ophthalm. Soc. (Ernst Fuchs Meml. Lectr. 1975), German Ophthalm. Soc., Am.-Austrian Soc. (pres. 1989-91), Cosmos Club D.C. Lutheran. Avocations: skiing, violin, philately, photography, bridge. Home and Office: 3113 Lewis Pl Falls Church VA 22042

EVANS, RICHARD ALEXANDER, research chemist; b. Brisbane, Queensland, Australia, May 15, 1965; s. George William and Charlotte Clair (Mercer) E. BSc with honors, U. Queensland, 1987, PhD, 1992. Postdoctoral fellow Commonwealth Scientific and Indsl. Rsch. Orgn., Victoria, Australia, 1992—. Contbr. articles to profl. jours. Mem. Inst. Modern Art, Brisbane, 1984-93, Australian Ctr. for Contemporary Art, Melbourne, 1992—, Nat. Gallery Victoria Soc., 1992—. Recipient Commonwealth Postgrad. Rsch. award Commonwealth of Australia, 1988-89, Australian Postgrad. Rsch. award U. Queensland, 1990-92. Mem. Royal Australian Chem. Inst., Am. Chem. Soc. Achievements include research in the field of organic reactive intermediates and ring opening free radical polymerisation. Office: CSIRO Div Chem & Polymers, Bayview Ave, Clayton Victoria 3168, Australia

EVANS, RICHARD JAMES, mechanical engineer; b. Wabash, Ind., Nov. 26, 1960; s. Tommy Lewis and Joyce Anne (Leckrone) E.; m. Marcia Lee Winters, July 3, 1985; children: Matthew Thomas, Kari Lynn, Jenna Marie. BSME, Rose-Hulman Inst. Tech., 1983; MBA, Ind. U., 1993. Registered profl. engr., Ind.; cert. lighting efficiency profl. Sales engr. Johnson Controls, Inc., Indpls., 1983-90, sales team leader in healthcare mktg., 1990—. Active Cicero (Ind.) United Meth. Ch., 1988—, Sons of Am. Legion, Wabash, 1989—. Mem. ASHRAE (ctrl. Ind. chpt. pres. 1991-92, bd. mem. 1992-93, presdl. award excellence 1992), NSPE, Assn. Energy Engrs. (cert. energy mgr.), Am. Soc. Hosp. Engrs., Ind. Soc. Profl. Engrs., Ind. Soc. Hosp. Engrs. Home: 579 Shore Ln Cicero IN 46034 Office: Johnson Controls Inc 1255 N Senate Ave Indianapolis IN 46202

EVANS, ROBERT VINCENT, engineering executive; b. Mobile, Ala., Sept. 21, 1958; s. William Alexander Evans and Katherine Barbara (Doerr) Davidson; m. Debra Marie Winters, July 27, 1984; children: James Vernon, Chelsea Marie. BS in Computer Info. Systems, Regis U., Denver, 1987, BS in Tech. Mgmt., 1987. Electrician Climax (Colo.) Molybdenum Co., 1978-82; applications engr. Honeywell, Inc., Englewood, Colo., 1982-83, sales engr., 1983-87; systems engr. Apple Computer, Inc., Seattle, 1987-88; regional systems engring. mgr. Apple Computer, Inc., Portland, Oreg., 1988—. Author: Anthology of American Poets, 1981. Dir. Operation Lookout, Seattle, 1989; mem. Rep. Nat. Com. Recipient USMC Blues award, Marine Corps Assn. Leatheanger award, 1977, Denver Post Outstanding Svc. award, 1983, N.Y. Zool. Soc. Hon. medal. Mem. Am. Mgmt. Assn., Mensa. Republican. Mem. Vineyard Ch. Avocations: reading, church ministry, family activities. Office: Apple Computer Inc 10210 NE Points Dr Ste 310 Kirkland WA 98033

EVANS, ROBLEY DUNGLISON, physicist; b. University Place, Nebr., May 18, 1907; s. Manley Jefferson and Alice (Turner) E.; m. Gwendolyn Elizabeth Aldrich, Mar. 10, 1928; children: Richard Owen, Nadia Ann, Ronald Aldrich; m. Mary Margaret Shanahan, Feb. 24, 1990. B.S., Calif. Inst. Tech., 1924-28, M.S., 1929, Ph.D., 1932. Cert. health physicist, Am. Bd. Health Physics, 1961. With rsch. lab. C.F. Braun & Co., Alhambra, Calif., 1929-31; instr. Poly. Sch., Pasadena, Calif., 1931-32; nat. rsch. fellow U. Calif., Berkeley, 1932-34; asst. prof. MIT, Cambridge, 1934-38, assoc. prof., 1938-45, prof., 1945-72, prof. emeritus, 1972—; dir. Radioactivity Center, MIT, 1935-72, cons., 1972-81; advisor U. N.Mex Uranium Epidemiology Study, 1978-89; vis. prof. Ariz. State U., 1966-67; staff cons. Peter Bent Brigham Hosp., Boston, 1945-72; cons. surgeon gen. Dept. Army, 1962-69, USN Radiol. Def. Lab., 1952-69; cons. div. biology and medicine AEC, 1950-75; mem. Internat. Union of Pure and Applied Physics, 1947, Joint Commn. Standards, Units, and Constants of Radioactivity, Internat. Council of Sci. Unions, 1948-55, Mixed Commn. on Radiobiology, 1961; spl. project asso. Mayo Clinic, 1973-81; cons. div. biol. and environ. research ERDA, Dept. Energy, 1975-81; cons. physics Mass. Gen. Hosp., 1948-73, USPHS, 1961-71, Fed. Radiation Council, 1965-69, Roger Williams Hosp., Providence, 1965-72; Chmn. Internat. Conf. Applied Nuclear Physics, Cambridge, 1940; vice chmn. com. on nuclear sci. Nat. Research Council, 1946-72; mem. Nat. Acad. Scis.-Nat. Research Council panel 231 adv. to Nat. Bur. Stds. on radiation physics, 1963-66, chmn., 1964; chmn. standing com.

for radiation biology aspects of supersonic transport FAA, 1967; mem. com. on radioactive waste mgmt. Nat. Acad. Scis., 1968-70; adviser U. Chgo., 1964-68; sci. adv. bd. New Eng. Deaconess Hosp., 1963-69; sr. U.S. del. Internat. Assn. Radiation Research, Cortina, 1966; mem. organizing com. U.S. Nat. Com. Med. Physics, 1966-69; vis. com. med. dept. Brookhaven Nat. Lab., 1965-68; cons. Blood Research Inst., 1967-74; vice chmn. adv. com. to U.S. Transuranium and Uranium Registries, 1968-86; mem. tech. adv. com. Ariz. Atomic Energy Commn., 1971-72. Author: The Atomic Nucleus, 1955; Editorial bd.: Internat. Jour. Applied Radiation and Isotopes, 1955-69; hon. mem. editorial bd., 1976—; editorial bd.: Mt. Washington Obs. Bull, 1962-70, Health Physics, 1962-70, Physics in Medicine and Biology, 1963-66; editor physics: Radiation Research, 1959-62; Contbr. sci. research papers to various publs. Vice pres. Found. for Study and Aid of Emotionally Unstable, 1948—. Recipient Theobald Smith medal in Med. Scis., AAAS, 1937; U.S. Presdl. Cert. of Merit, 1948; Hull award and Gold medal AMA, 1963; Silvanus Thompson medal Brit. Inst. Radiology, 1966; William D. Coolidge award Am. Assn. Physicists in Medicine, 1984, Enrico Fermi award U.S. Dept. Energy, 1990. Fellow AAAS, Am. Phys. Soc., Am. Acad. Arts and Scis., N.Y. Acad. Scis., Am. Assn. Physicists in Medicine; mem. Radiation Research Soc. (v.p., pres. 1965-67), Am. Roentgen Ray Soc. (asso.), Am. Indsl. Hygiene Assn., Am. Assn. Physics Tchrs., Am. Nuclear Soc., Health Physics Soc. (pres. 1972-73, Disting. Achievement award 1981, fellow 1984), Nat. Council Radiation Protection and Measurements (council 1965-71, hon. mem. 1975—), Soc. Nuclear Medicine (hon. mem.), Royal Sci. and Lit. Soc. (hon.), Kungliga Vetenshapoch Vitterhets-Samhallet (Goteborg, Sweden), Sigma Xi, Kappa Gamma, Tau Beta Pi, Pi Kappa Delta. Republican. Home: 4621 E Crystal Ln Paradise Valley AZ 85253-2939 Office: MIT Physics Dept Cambridge MA 02139

EVANS, ROGER LYNWOOD, scientist, patent liaison; b. Ipswich, Suffolk, Eng., June 25, 1928; came to U.S., 1953; s. Evelyn Jesse and Ethel Jane (Woods) E.; m. Jane Adelaide Baird, Nov. 24, 1954 (div. 1976); children: Robert Malcolm Baird, Roderick Lawrence Woods, Alison Clare; m. Wendy Dorothy Grove, Apr. 11, 1977. BA in Natural Sci., Oxford (Eng.) U., 1953, MA, 1955, DPhil in Natural Sci., 1958; MS in Inorganic Chemistry, U. Minn., 1955. With chem. and radiopharm. R & D dept. 3M Co., St. Paul., 1958-77, patent liaison, 1977-91; developer intellectual property initiative, tech. devel. dept., 1992—. Founder, editor Newsletter of the Tech. Forum, 1971—; inventor, writer, producer series of videos on intellectual property topics. Mem., chmn. Mendota Heights Planning Commn., 1962-68, Sunfish Lake Planning Commn., 1968-84, Dakota County Planning Commn., Minn., 1965-72. 2d lt. Brit. Army, 1946-49, Eng. Anglican. Avocations: photography, amateur opera singer, travel, writing. Home: 9965 Rich Valley Blvd Inver Grove Heights MN 55077-4529 Office: 3M Co 3M Ctr 255-2N-1 Saint Paul MN 55144

EVANS, ROGER MICHAEL, acoustical engineer; b. New London, Conn., July 24, 1960; s. Roger James and Barbara Florence (Pfaffenschalger) E.; m. Victoria Lynn Thomas, May 2, 1992. BSME, N.C. State U., 1983; M Mech. Engring., Cath. U., 1988. Engr. David Taylor Rsch. Ctr., Bethesda, Md., 1984-88, Newport News Shipbuilding, Arlington, Va., 1988—; mem. vibrations panel SNAME, HS-7, Arlington, 1991—. Recipient commendation Office of Naval Rsch., Arlington, 1988. Mem. Sigma Xi, Pi Tau Sigma. Democrat. Unitarian. Home: 5163 N Washington Blvd Arlington VA 22205 Office: Newport News Shipbuilding 2711 Jefferson Davis Hwy Arlington VA 22202

EVANS, WALLACE ROCKWELL, JR., mechanical engineer; b. El Paso, Tex., Aug. 23, 1914; s. Wallace Rockwell and Margaret (Strickland) E.; m. Evelyn Lucille Osborne, Feb. 12, 1944; children: Lynda Lucille, Wallace Rockwell III. BS in Mech. Engring., Va. Poly. Inst. and State U., 1941. Registered profl. engr., N.C. Constrn. engr. Weigel Engring. Co., Chattanooga, 1941-42; air conditioning engr. Page Air Conditioning Co., Charlotte, N.C., 1946-48, Celanese Fibers Co., Rock Hill, S.C., 1948-54; utilities mgr. Celanese Fibers Co., Charlotte, 1954-79; pres. Evans Engring. Co., Rock Hill, 1979—. Contbr. articles on refrigeration, air conditioning and radiant heat to profl. jours. Pres. PTA, Rock Hill, 1952-53. Lt. USN, 1942-47. Mem. ASME (life, power generation div. 1942), ASHRA, Am. Soc. Profl. Engrs. (pres. Piedmont chpt. 1969, Engr. of Yr. 1976). Repubican. Methodist. Achievements include patents in refrigeration and absorption solvent recovery systems; invention of dry-coil air conditioning system; design of the air management solvent recovery system for solvent spun textile and solvent cast photographic film plants. Home and Office: 411 Lakeside Dr Rock Hill SC 29730-6105

EVANS, WAYNE EDWARD, environmental microbiologist, researcher; b. Indpls., Nov. 17, 1962; s. Warren J. II and Susan Carol (Winans) E.; m. Kathleen Francis Herbst, Dec. 30, 1989. BS in Biology, BA in Chemistry, Purdue U., 1986; MS in Microbiology, Ball State U., 1992. Supr. lab. microbiology and chemistry Moseley Labs./Vivolac Cultures, Indpls., 1986-89; environ. chemist Ind. Dept. Environ. Mgmt., Indpls., 1989-90; chemist technician Hoosier Microbiol. Labs., Muncie, Ind., 1991-92; micro lab supr. Ball State U., Muncie, 1990-92; environ. microbiologists ASCI Corp. at Waterways Exptl. Sta., Vicksburg, Miss., 1992—. Fellow AAAS, Am. Soc. for Microbiology, Am. Chem. Soc., Soc. for Indsl. Microbiology. Office: 6811 Old Canton Rd # 1905 Ridgeland MS 39157

EVENSON, KENNETH M., physicist; b. Waukesha, Wis., June 5, 1932; m. 1955; 4 children. BS, Mont. State Coll., 1955; MS, Oreg. State U., 1960, PhD in Physics, 1964. Physicist Nat. Inst. Standards and Tech., Boulder, Colo., 1963—. Mem. Am. Phys. Soc. (Earle K. Plyzer prize 1991). Achievements include research in quantum electronics, atomic and molecular structure, chemical kinetics, electron paramagnetic resonance, microwave and optical spectroscopy. Office: Natl Inst of Technology & Stnds Gaithersburg MD 20899*

EVENSON, MICHAEL DONALD, software/hardware engineer; b. Yokosuka, Japan, July 6, 1961; s. David Donald and Teresita (Perez) E. BS, Marquette U., 1984. Engr. Grumman Tech. Svcs., Kennedy Space Ctr., Fla., 1984—, sr. engr., 1991—. Pres. St. Teresa Young Adult Club, Titusville, Fla., 1990-91, Cath. Young Adults, Brevard County, Fla., 1992—, diocese coord., 1993—; tutor Grumman Adopt-a-Sch. program. Recipient awards Grumman, NASA, 1987, 88, 91. Republican. Roman Catholic. Avocations: computer games, utilities creation, cycling, swimming, camping. Office: Grumman Tech Svcs GTS-653 Kennedy Space Center FL 32899

EVERETT, JESS WALTER, environmental engineering educator, researcher; b. Dover, Del., Dec. 21, 1962; s. Walter Earle and Nancy Charlotte (Schlichting) E.; m. Denise Coats, Aug. 9, 1986; children: Brady Summer, Caleb Walter. BS in Engring., Duke U., 1984, MS, 1986, PhD, 1991. Asst. prof. environ. engring. U. Okla., Norman, 1991—. Contbr. articles to Resources Conservation and Recycling, Jour. Environ. Engring., ASCE, Jour. Resource Mgmt. and Tech., Jour. Urban Planning Devel., Waste Mgmt. and Rsch. Mem. ASCE (assoc.), Nat. Recycling Assn. Coalition, Chi Epsilon, Sigma Chi. Office: U Okla 202 W Boyd Rm 334 Norman OK 73019

EVERETT, ROBERT WILLIAM, environmental engineer; b. L.A., Mar. 19, 1947; s. Wilhelm Sydow and Geneva Pauline (Brown) E.; m. Anita Lynn Highland, Apr. 25, 1992; 1 child, Joshua Marcus. BA, U.S. Internat. Univ., 1970. Engring. technician Pulsation Controls Corp., Santa Paula, Calif., 1965-71; prodn. and application engr. Am. Air Filter, Louisville, 1971-78; pres. Environ. Tech. Cons., Louisville, 1978-89; v.p. Gale/Jordan Assocs., Redondo Beach, Calif., 1989—; bd. dirs. Arboleda Corp., Santa Paula; co-chmn. Biosafety Conf., Lexington, Ky., 1980-81. Mem. ASHRAE, Am. Indsl. Hygiene Assn., Assn. Energy Engrs. Achievements include patents for ventilation system, for plenum air flow. Office: Gale/Jordan Assocs 1650 S Pacific Coast Hwy Ste 201 Redondo Beach CA 90277

EVERETT, ROYICE BERT, obstetrician/gynecologist; b. Erick, Okla., Sept. 12, 1946; s. George Bert Everett and Mildred Inez (Gilchriest) Isham; m. Dian Brandt, Jan. 15, 1977; children: Heather, Joshua. BS, Okla. Bapt. U., 1968; MD, U. Okla., 1972, postgrad., 1972-76. Diplomate Am. Bd. Ob-Gyn. Intern U. Okla., Oklahoma City, 1972-73, resident, 1973-76; fellow in reproductive endocrinology Health Sci. Ctr. U. Tex., Dallas, 1976-78; Pvt. practice Oklahoma City, 1978—; med. dir. Bapt. Laser Inst., 1987—, Bennett Fertility Inst., 1985—. Contbr. numerous articles to profl. jours.; patentee method for use of high intensity light and heat in female sterilization; inventor lateral firing laser fiber. Mem. First Bapt. Ch., Oklahoma City. Maj. USAR, 1972-80. Recipient Disting. Alumni Award, Okla. Bapt. U., 1987. Fellow ACOG; mem. AMA, Okla. Med. Assn., Oklahoma County Med. Assn., Oklahoma City Ob-Gyn. Soc. (past pres.), Cen. Assn. Ob-Gyn., Gynecologic Laser Soc. (dist. chmn.), Am. Fertility Soc., Soc. Reproductive Surgeons, Soc. for Assisted Reproductive Tech., Internat. Soc. Gynecologic Endoscopy, Am. Soc. for Lasers in Medicine and Surgery, Am. Soc. Gynecologic Laparascopists, Okla. C. of C., Kerr McGee Swim Club (pres. 1990). Republican. Baptist. Avocations: hunting, fishing, golf. Office: Royice B Everett MD Inc 3433 NW 56th St Ste 870 Oklahoma City OK 73112-4959

EVERHART, FRANCIS GROVER, JR., manufacturing company executive; b. Winston-Salem, Sept. 2, 1947; s. Francis G. and Grace (Merritt) E.; m. Kathy Cline, Aug. 27, 1979; children: Timmothy Wade, Frankie. Machinist, Forsyth Tech. Coll., Winston-Salem, 1976. Machinist Briggs-Schaffner Co., Winston-Salem, 1965-66; mechanic Piedmont Airlines, Winston-Salem, 1969-71; with Westinghouse Turbine Components, Winston-Salem, 1971-74; machinist Bahnson Co., Winston-Salem, 1974-75, Hayes-Albion Co., Winston-Salem, 1975-80; sr. metrology tech. AMP Inc., Winston-Salem, 1980—. With USMC, 1966-69. Moravian Ch. Office: AMP Inc Mail Stop 109-04 3441 Myer-Lee Rd Winston Salem NC 27102

EVERHART, THOMAS EUGENE, university president, engineering educator; b. Kansas City, Mo., Feb. 15, 1932; s. William Elliott and Elizabeth Ann (West) E.; m. Doris Arleen Wentz, June 21, 1953; children—Janet Sue, Nancy Jean, David William, John Thomas. A.B. in Physics magna cum laude, Harvard, 1953; M.Sc., UCLA, 1955; Ph.D. in Engring., Cambridge U., Eng., 1958. Mem. tech. staff Hughes Research Labs., Culver City, Calif., 1953-55; mem. faculty U. Calif., Berkeley, 1958-78, prof. elec. engring. and computer scis., 1967-78, Miller research prof., 1969-70, chmn. dept., 1972-77; prof. elec. engring., Joseph Silbert dean engring. Cornell U., Ithaca, N.Y., 1979-84; prof. elec. and computer engring., chancellor U. Ill., Urbana-Champaign, 1984-87; prof. elec. engring. and applied physics, pres. Calif. Inst. Tech., Pasadena, 1987—; fellow scientist Westinghouse Rsch. Labs., Pitts., 1962-63; guest prof. Inst. Applied Physics, U. Tuebingen, Germany, 1966-67, Waseda U., Tokyo, Osaka U., 1974; vis. fellow Clare Hall, Cambridge, U., 1975; chmn. Electron, Ion and Photon Beam Symposium, 1977; cons. in field; mem. sci. and ednl. adv. com. Lawrence Berkeley Lab., 1978-85, chmn., 1980-85; mem. sci. adv. com. GM, 1980-89, chmn., 1984-89, bd. dirs., 1989—; bd. dirs. Hewlett Packard Corp., 1991—; tech. adv. com. R.R. Donnelly & Sons, 1981-89. NSF sr. fellow, 1966-67, Guggenheim fellow, 1974-75. Fellow IEEE, AAAS, Royal Acad. Engring.; mem. NAE (ednl. adv. bd. 1984-88, mem. com. 1984-89, chmn. 1988—, coun. 1988—), Microbeam Analysis Soc. Am., Electron Microscopy Soc. Am. (coun. 1970-72, pres. 1977), Coun. on Competitiveness (vice-chmn. 1990—), Assn. Marshall Scholars and Alumni (pres. 1965-68), Athenaeum Club, Sigma Xi, Eta Kappa Nu. Home: 415 S Hill Ave Pasadena CA 91106-3407 Office: Calif Inst Tech Office of Pres 1201 E California Blvd Pasadena CA 91125-0001

EVERITT, HENRY OLIN, III, physicist; b. Huntsville, Ala., May 6, 1963; s. Henry O. and Janice E. (Shores) E. BS, Duke U., 1985, PhD, 1990. NASA fellow Duke U., Durham, N.C., 1987-90; physicist Sparta, Inc., Huntsville, Ala., 1990-91; adj. asst. prof. Duke U., Durham, 1991—; program mgr., physicist U.S. Army Rsch. Office, Research Triangle Park, N.C., 1991—. Editor: Development and Application of Photonic Band of Structures, 1993; author: Applications of Photonic Band Gap Structures, 1992; co-author: Dynamics and Tunability of Optically Pumped FIR Lasers, 1986. Mem. Am. Phys. Soc., Optical Soc. of Am., Sigma Xi. Achievements include unprecedented operation of optically pumped far infrared lasers; discovery of fundamental collisional progress in methyl fluoride. Office: US Army Rsch Office PO Box 12211 Research Triangle Park NC 27709

EVERS, MARTIN LOUIS, internist; b. Newark, Dec. 17, 1957; s. George and Eva (Auslander) E. BA in Biochemistry, Rutgers U., 1979; MD, U. Medicine and Dentistry N.J., 1985. Diplomate Am. Bd. Internal Medicine. Emergency medicine resident Hershey (Pa.) Med. Ctr., 1985-86; internal medicine resident Raritan Bay Med. Ctr., Perth Amboy, N.J., 1986-89; critical care fellow Presbyn. Hosp., Pitts., 1989; emergency room physician Shadyside Hosp., Pitts., 1989-90; assoc. program dir. St. Francis Med. Ctr., Trenton, N.J., 1990—; ACLS instr. Ctr. for Emergency Medicine, Pitts., 1990, St. Francis Med. Ctr., Trenton, 1991—. Fellow Acad. Medicine N.J., ACP; mem. AMA, Soc. Critical Care Medicine. Jewish. Office: St Francis Med Ctr 601 Hamilton Ave Trenton NJ 08629-1986

EVERSTINE, GORDON CARL, mathematician, educator; b. Balt., Mar. 30, 1943; s. Carl Nicholas and Barbara (Schilling) E.; m. Virginia Elizabeth Flad, Aug. 24, 1968; children: Karen, Eric. BS in Engring. Mechs., Lehigh U., 1964; MS in Engring., Purdue U., 1966; PhD, Brown U., 1971. Mem. tech. staff Bell Telephone Labs., Indpls., 1964-66; mathematician David Taylor Model Basin, Bethesda, Md., 1969—; professorial lectr. George Washington U., Washington, 1979—. Contbr. articles to profl. jours. Recipient Douglas Michel award NASA, 1989. Fellow Wash. Acad. Scis. (Scientific Achievement award 1984). Achievements include development of computational structural acoustics techniques. Office: David Taylor Model Basin Bethesda MD 20084

EVERT, RAY FRANKLIN, botany educator; b. Mt. Carmel, Pa., Feb. 20, 1931; s. Milner Ray and Elsie (Hoffa) E.; m. Mary Margaret Maloney, Jan 2, 1960; children: Patricia Ann, Paul Franklin. B.S., Pa. State U., 1952, M.S., 1954; Ph.D., U. Calif. at Davis, 1958. Mem. faculty Mont. State U., 1958-60; mem. faculty U. Wis.-Madison, 1960—, prof. botany, 1966-77, prof. botany and plant pathology, 1977-88, Katherine Esau prof. botany and plant pathology, 1988—, chmn. dept. botany, 1973-74, 77-79; vis. prof. U. Natal, Pietermaritzburg, S. Africa, winter, spring 1971, U. Göttingen, W.Ger., summer 1971, 74-75, summer 1988; mem. gen. biology and genetics fellowship rev. panel NIH, 1964-68, NSF Adv. Com. for Biol. Research Ctrs. Program, 1987-88; forensic plant anatomy cons. Co-author: Biology of Plants; sci. editor Physiol. Plantarum, 1983—; mem. editorial bd. Trees, 1991—, Internat. Jour. Plant Scis., 1991—; contbr. articles on food conducting tissue in higher plants and leaf structure-function relationships. Recipient Alexander von Humboldt award, 1974-75, Emil H. Steiger award for excellence in teaching U. Wis., 1981, Bessey Lectr. award Iowa State U., Ames, 1984, Benjamin Minge Duggar lectureship award Auburn U., 1985, Disting. Service citation Wis. Acad. Scis., Arts and Letters, 1985; Guggenheim fellow, 1965-66. Fellow Am. Acad. Arts and Scis., AAAS; mem. Bot. Soc. Am. (pres. 1986-87, Merit award 1982), Am. Inst. Biol. Scis., Wis. Acad. Scis., Arts and Letters, Am. Soc. Plant Physiol., Internat. Assn. Wood Anatomists, Deutschen Botanischen Gesellschaft, Golden Key Nat. Honor Soc., Sigma Xi, Phi Kappa Phi, Phi Sigma, Phi Epsilon Phi., Pi Alpha Xi. Home: 810 Woodward Dr Madison WI 53704-2238

EWANKOWICH, STEPHEN FRANK, JR., mechanical engineer; b. Clearwater, Fla., Oct. 14, 1966; s. Stephen Frank and Marlene Ann (Snyder) E.;m. Linda Beth Nossaman, Apr. 11, 1992. BSME, Va. Poly. Inst. and State U., 1988. Product specialist Square D Co., Raleigh, N.C., 1988-91, regional mktg. specialist, 1991-92, mgr. regional mktg., 1992—; team mem. Vision Mission Team, Raleigh, 1990-92. Editor brochure: Lighting Control Equipment Application Guide, 1989; author product brochure: Lighting Control Equipment, 1989. Bd. dirs., pres. Villages East Homeowners Assn., 1992. Mem. ASME. Republican. Methodist.

EWEL, KATHERINE CARTER, ecologist, educator; b. Glens Falls, N.Y., Sept. 30, 1944; d. Robert C. Jr. and Carolyn H. (Ferres) Carter; m. John J. Ewel, Aug. 25, 1969. AB, Cornell U., 1966; PhD, U. Fla., 1970. Instr. Duke U., Durham, N.C., 1969-71; postdoctoral rsch. assoc. U. Fla., Gainesville, 1971-73, interim asst. prof., 1973-75, asst. rsch. scientist, 1975-77, asst. prof., 1977-81, assoc. prof., 1981-85, prof. ecology, 1985—; vis. scientist Cornell U., Ithaca, N.Y., 1982; vis. scholar U. Cambridge, England, 1985; NRC fellow NASA/Ames Rsch. Ctr., Moffett Field, Calif., 1986. Co-author: A Model Menagerie, 1972, Population and Energy, 1975; sr. editor: Cypress Swamps, 1984; author: (software) Simulation of Ecological Models, 1989; contbr. articles to profl. jours. Mem. Internat. Soc. Ecol. Modeling, Am. Inst. Biol. Scis., Ecol. Soc. Am. (program chair 1990-92), Soc. Wetland Scientists. Achievements include research in terrestrial forests and wetland management. Office: Univ Fla Dept Forestry Gainesville FL 32611-0420

EWELL, ALLEN ELMER, JR., naval officer; b. Washington, Sept. 26, 1960; s. Allen Elmer Sr. and Rose Marie (Gibbs) E.; m. Allesa Jean Bird, July 25, 1992. BSEE, Va. Mil. Inst., 1982; MS, U. R.I., 1992. Prodn. officer Pub. Works Dept., Newport, R.I., 1978-90; staff civil engr. Nat. Security Agy., 1993—. Lt. USN, 1982-87, Operation Dessert Storm 1990-91. Mem. ASCE, Soc. Am. Mil. Engrs. (pres. 1989, 90), Nat. Ground Water Assn. Republican. Methodist. Home: 11805 N Marlton Ave Upper Marlboro MD 20772 Office: Nat Security Agy LS2 Fort Meade MD 20755

EWELL, WALLACE EDMUND, transportation engineer; b. Fort Worth, Sept. 3, 1942; s. Wallace Mortimer and Claire Eleanor (Kiker) E.; m. Judy Lee King, Sept. 30, 1972; children: George T. Finkle, James Roy, Melissa Anne. BSCE, Tex. A&M U., 1976; MS in Systems Mgmt., U. So. Calif., 1980; MSCE, U. Tex., 1984. Registered profl. engr., Tex. Electronics systems maintenance supr. USMC, 1960-76; supr. constrn. subjects USMC Engr. Sch., N.C., 1976-80; constrn. inspector Tex. Dept. Hwys., Fort Worth, 1980-84, bridge inspection supr., 1984-85, roadway design engr., 1985-86; dist. traffic engr. Tex. Dept. Transp., Ft. Worth, 1986-92, dir. transp. ops., 1992—; mem. ops. sub-com. on engring. edn., Austin, 1988—, joint tech. transp. sub-com. North Cen. Tex. Coun. Govts., 1986—. Contbr. articles to Transportation Quarterly, 1989, 92 and other jours. on traffic engring. and mgmt. systems. Mem. NSPE, Tex. Soc. Profl. Engrs. Inst. Transp. Engrs. (Tex. sect.). Achievements include supervision of design, construction, operation and maintenance of unique traffic management system in Ft. Worth, an element of the new intelligent-vehicle highway systems. Office: Tex Dept of Transportation PO Box 6868 Fort Worth TX 76115

EWEN, H.I., physicist; b. Chicopee, Mass., Mar. 5, 1922; s. Arthur and Ruth Frances (Fay) E.; m. Mary Ann Whitney, Feb. 10, 1956; children: Donald, Jim, Bruce, Mark, David, Deborah, Daniel, Rebecca. BA, Amherst Coll., 1943; MA, Harvard U., 1948, PhD, 1951. Mem. faculty Amherst Coll., 1943; co-dir. Harvard Radio Astronomy Program, 1952-58, rsch. assoc. astronomy dept., 1958-65, assoc., 1965-80; gen. mgr. Militech Corp., South Deerfield, Mass., 1993—; pres. Ewen Knight Corp., Weston, Mass., 1952-88, Ewen Dae Corp., 1958-88, E.K. Assocs., 1993—; sci. advisor to Cin. Electronics Corp. for USAF Air Weather Svc.; mem. Global Solar Radio Telescope Network, 1977-86; exec. v.p. Millitech Corp., South Deerfield, Mass., 1989-93. Contbg. author: Advances in Microwaves, vol. 5, 1970, Electromagnetic Sensing of the Earth from Satellites, 1967, Geoscience Instrumentation, 1974, also articles; co-discoverer 21 cm interstellar hydrogen line, 1951; remote sensing of atmospheric ozone distribution (resonant line at 102 GHz), 1966. Served to lt. USNR, 1943-46. NRC fellow, 1946-49; recipient svc. award Harvard Coll., 1977. Fellow AAAS, IEEE (Morris E. Leeds award 1970), Am. Acad. Arts and Scis.; mem. Am. Astron. Soc. (Tinsley prize 1988), Phi Beta Kappa, Sigma Xi.

EWIG, CARL STEPHEN, chemist; b. Elmira, N.Y., May 28, 1945; s. John and Wilma (Lamb) W.; m. Patricia Laichu, Nov. 10, 1977; children: Kevin, Celeste, Melvin. BS, U. Rochester, 1967; PhD, U. Calif., Santa Barbara, 1973. Rsch. asst. prof. Vanderbilt U., Nashville, 1976-82, assoc. prof., 1982-91; sr. scientist Biosym Technologies, Inc., San Diego, 1991—. Contbr. over 70 sci. publs. to profl. jours. Mem. AAAS, Am. Chem. Soc., Am. Phys. Soc., Sigma Xi. Achievements include rsch. interests in several novel applications of quantum chemistry and computational chemistry to molecular electronic structure, energetics and biomolecular simulation. Home: 5961 Stresemann St San Diego CA 92122 Office: Biosym Technologies Inc 9685 Scranton Rd San Diego CA 92121

EWING, CRAIG MICHAEL, aerospace engineer; b. Glasgow, Mont., June 12, 1964; s. Donald Allen and Joanne Frances (Harper) E.; m. Darsi McDowell, May 22, 1993. BS in Aero./Astron. Engring., Ohio State U., 1986; MS in Aerospace Engring., U. Fla., 1992. Aerospace engr. Simulation and Assessment Br., Eglin AFB, Fla., 1987—; pres. Civilian Employees Adv. Coun., 1992—. Pres. Stage Crafters Community Theatre, Ft. Walton Beach, Fla., 1992. Mem. AIAA (guest lectr. Conf. on Estimation). Achievements include named national expert on SDIO algoritm panel. Office: WL/MNSH Eglin A F B FL 32542

EWING, JAMES FRANCIS, biochemist, researcher; b. Syracuse, N.Y., June 7, 1962; s. Jack Arthur and Judith Elaine (Sansone) E.; m. Julie Shore Palmer, June 27, 1987; children: Jack Andrus, Chelsea Shore, Taylor Jane. BS, Union Coll., 1984; MS, Albany Med. Coll., 1987, PhD, 1990. Instr. Achilles Figure Skating Club, Schenectady, N.Y., 1980-87; lab. supr. dept. urology U. Rochester, N.Y., 1987-90, NIH postdoctoral fellow dept. biophysics, environ. medicine, 1990-93; scientist dept. biophysics U. Rochester, 1993—. Contbr. articles to Jour. Neurochemistry, Jour. Clin. Endocrinology and Metabolism, Proceedings NAS, others. Ofcl. Empire State Games, N.Y., 1983-87. Double Gold medalist U.S. Figure Skating Assn., 1980, Gold medal Can. Figure Skating Assn., 1982; recipient rsch. svc. award Nat. Inst. Environ. Health Scis., 1992. Mem. AAAS, Soc. Neurosci., N.Y. Acad. Scis., Soc. of Toxicology, Bucks Harbor Yacht Club. Achievements include mapping enzymes of heme degradation throughout brain. Office: Univ Rochester Dept Biophysics 601 Elmwood Ave Rochester NY 14642

EWING, RODNEY CHARLES, mineralogist, geology educator, materials scientist; b. Abilene, Tex., Sept. 20, 1946; s. Charles Thomas and Maria Luisa (Cobos) E.; m. Jerrilyn A. Harris, June 17, 1973 (div. June 1988); m. Helga G. Rosenthal, Nov. 20, 1992; children: Travis Russell, Allison Christine. BS, Tex. Christian U., 1968; MS, Stanford U., 1972, PhD, 1974. From asst. to assoc. prof. U. N.Mex., Albuquerque, 1974-84, chmn. dept. geology, 1979-84, prof., 1984—; mem. waste isolation pilot plant rev. panel NRC-NAS, Washington, 1984—. Contbg. author, editor: Radioactive Waste Forms for the Future, 1988; guest editor Jour. Nuclear Materials, Vol. 190, 1992. Mem. Amnesty Internat., Albuquerque, 1985-91. Sgt. U.S. Army, 1969-70, Vietnam. Recipient Major Equipment award NSF, 1992. Mem. Internat. Union Materials Rsch. Socs. (sec. 1990—), Materials Rsch.Soc. (councillor 1982-89), Rotary. Democrat. Achievements include the authoring of articles which twice appeared on the cover of Science magazine. Office: U New Mex Dept Earth and Planetary Sci Albuquerque NM 87131

EXE, DAVID ALLEN, electrical engineer; b. Brookings, S.D., Jan. 29, 1942; s. Oscar Melvin and Irene Marie (Mattis) E.; m. Lynn Rae Roberts; children: Doreen Lea, Reana Lynn. BSEE, S.D. State U., 1968; MBA, U. S.D., 1980; postgrad. Iowa State U., 1969-70, U. Idaho, 1978-80. Registered profl. engr., Idaho, Oreg., Minn., S.D., Wash., Wyo., Utah, N.Y., Ind., Wis. Applications engr. Collins Radio, Cedar Rapids, Iowa, 1969-70; dist. engr. Bonneville Power Adminstrn., Idaho Falls, Idaho, 1970-77; instr. math U. S.D., Vermillion, 1977-78; chief exec. officer EXE Assocs., Idaho Falls, Idaho, 1978-83, Bloomington, Minn., 1985—; safety mgr. CPT Corp., Eden Prairie, Minn., 1983-85; owner, chief exec. officer Exe Inc., Eden Prairie, 1983—; chmn. bd. Applied Techs. Idaho, Idaho Falls, 1979—; chmn., chief exec. officer Azimuth Cons., Idaho Falls, 1979-81; v.p. D & B Constrn. Co., Idaho Falls, 1980-83; bd. dirs., v.p., chief ops. officer Nat. Multi-Housing Corp., 1989. Technical advisor Nat. Earth Day, 1991; apptd. Minn. State Bd. Profl. Engrs., 1991. With USN, 1960-64. Mem. IEEE, Am. Cons. Engrs., Nat. Soc. Profl. Engrs., Nat. Contrcts Mgrs. Assn., IEEE Computer Soc., Mensa, Am. Legion, VFW, Masons, Elks. Office: Exe Assocs Inc 10740 Lyndale Ave S Minneapolis MN 55420-5615

EXLEY, SHECK, geologist. Recipient William J. Stephenson Outstanding Svc. award Nat. Speleological Soc.. 1992. Office: Cathedral Cayon Rte 8 Box 374 Live Oak FL 32060*

EYKHOFF, PIETER, electrical engineering educator; b. The Hague, The Netherlands, Apr. 9, 1929; s. Hendrik and Henderika (Strating) E.; m. Johanna N.F. Pabon, Dec. 1, 1955; children: André H., Gerard P. BSc, Delft (The Netherlands) U. Tech., 1955, MSc, 1956; PhD, U. Calif., Berkeley, 1961; doctorate degree (hon.), Free U., Brussels, 1990. Sci. officer Delft U. Tech., 1956-64; prof. elec. engring. Eindhoven (The Netherlands) Tech. U., 1964—, dean dept. elec. engring., 1977-80; vis. prof. various countries including U.S., Can., USSR, Japan, People's Republic of China, Chile,

Brazil; vis. rsch. fellow U.S. Nat. Acad. Scis., 1958-60; hon. prof. Xi'an Jiaotong U., People's Republic of China. Author: System Identification, Parameter and State Estimation, 1974, 2d edit., 1979 (Russian, Chinese, Polish, Romanian transls.); editor: Trends and Progresses in System Identification, 1980 (Russian transl.); contbr. articles to profl. jours. Knight Order of the Lion of The Netherlands, 1991. Fellow IEEE; mem. Royal Netherlands Acad. Arts and Scis., Royal Instn. Engrs. in The Netherlands, Internat. Fedn. Automatic Control (exec. coun. 1975-81, publs. mng. bd. 1976-90, hon. editor 1972-75, editor-in-chief workshop procs. 1988—, chmn. publs. com. 1981-87, Outstanding Svc. award 1990), Friendship Soc. Netherlands-China (chmn. 1984-91, hon. chmn. 1991—), Sigma Xi. Home: Vermeerstraat 11, NL-5691 ED Son The Netherlands Office: Eindhoven U Tech, PO Box 513, NL-5600 MB Eindhoven The Netherlands

EYLER, EDWARD EUGENE, physicist, educator; b. Akron, Ohio, Mar. 8, 1955; s. Eugene Bartholomew and Mary Phyllis (Lang) E.; m. Karen Greer, June 12, 1982. SB, MIT, 1977; PhD, Harvard U., 1982. Postdoctoral fellow Harvard U., Cambridge, Mass., 1982-83; asst. prof. physics Yale U., New Haven, 1983-88, assoc. prof. physics, 1988-89; assoc. prof. physics U. Del., Newark, 1989—; mem. exec. com. Topical Group on Fundamental Constants and Precise Tests of Phys. Law, 1991—. Contbr. articles to profl. publs. Grantee NSF, 1984—, Nat. Inst. Standards of Tech., 1990-92. Mem. Am. Phys. Soc., Optical Soc. Am. Office: U Del Dept Physics of Astronomy Newark DE 19716

EYRING, EDWARD MARCUS, chemical educator; b. Oakland, Calif., Jan. 7, 1931; s. Henry and Mildred (Bennion) E.; m. Marilyn Murphy, Dec. 28, 1954; children—Steven C., Valerie, David W., Sharon K. B.A., U. Utah, 1955, M.S., 1956, Ph.D., 1960. Asst. prof. chemistry U. Utah, 1961-65, asso. prof., 1965-68, prof., 1968—, chmn. dept., 1973-76, 84-85. Author: (with H. Eyring) Modern Chemical Kinetics, 1963, (with others) Statistical Mechanics and Dynamics, 1964; Contbr. (with others) numerous articles to sci. jours. Served to lt. USAF, 1955-57. NSF Postdoctoral fellow to U. Goettingen, 1960-61; NATO sr. sci. fellow, 1976; J.S. Guggenheim Found. fellow, 1982-83. Mem. Am. Chem. Soc., AAAS, Phi Beta Kappa, Sigma Xi, Phi Kappa Phi. Mem. Ch. Jesus Christ of Latter-day Saints (Bishop, 1966-69). Home: 4570 Sycamore Dr Salt Lake City UT 84117-4351

EZIN, JEAN-PIERRE ONVÊHOUN, mathematician; b. Guezin, Mono, Benin, Dec. 7, 1944; s. Irené Ehouéton (Fiogbe) E.; m. Victoire Mocheboratan Akele, Oct. 7, 1972; children: Maxellende, Wallerend, Franz-Olivier, Majoric. B of Math., U. Dakar (Sénégal), 1963, Maîtrise Math., 1968; D of 3e Cyc, U. Lille (France) I, 1972, D of d'Etat, 1981. Cert. d'Aptitude à l'Adminstrn. Emterprises. Asst. prof. U. Nat. Bénin, Cotonou, 1973-77; assoc. prof. U. Lille I, 1977-81; prof. U. Nat. Bénin, 1982-86; researcher Internat. Ctr. Theolretical Physics, Trieste, Italy, 1986-88; prof. U. Nat. Bénin, 1988—; dir. Inst. Math. Scis., Physics, Proto-Novo, Bénin, 1989—; head math. dept. U. Nat. Dénin, 1982-86, recteur, 1990-92; rep. Internat. Ctr. Theoretical Physics, 1988. Contbr. articles to profl. jours. Founder, mem. Groupe d'Etude Recherche pour Dem. Devel. Econ. Social, Cotonou, 1990, Notre Cause Commune, Porto-Novo, 1990, also bd. dirs. Named Officer Palmes Acad., France Embassy, Cotonou, 1991. Mem. Am. Math. Soc., Lions. Roman Catholic. Avocation: tennis. Home: Carré 3 N, Cotonou Benin Office: U Nat Bénin, PO Box 526, Cotonou Benin

FABER, SANDRA MOORE, astronomer, educator; b. Boston, Dec. 28, 1944; d. Donald Edwin and Elizabeth Mackenzie (Borwick) Moore; m. Andrew L. Faber, June 9, 1967; children: Robin, Holly. B.A., Swarthmore Coll., 1966, D.Sc. (hon.), 1986; Ph.D., Harvard U., 1972. Asst. prof., astronomer Lick Obs., U. Calif., Santa Crux, 1972-77, assoc. prof., astronomer, 1977-79, prof., astronomer, 1979—; mem. NSF astronomy adv. panel, 1975-77; vis. prof. Princeton U., 1978, U. Hawaii, 1983, Ariz. State U., 1985; Phillips visitor Haverford Coll., 1982; Feshbach lectr. MIT, 1990; Darwin lectr. Royal Astron. Soc., 1991; Marker lectr. Pa. State U., 1992; Bunyar lectr. Stanford U., 1992; Tomkins lectr. U. Calif., San Francisco, 1992; mem. Nat. Acad. Astronomy Survey Panel, 1979-81; chmn. vis. com. Space Telescope Sci. Inst., 1983-84; co-chmn. sci. steering com. Keck Observatory, 1987—; mem. wide field camera team Hubble Space Telescope, 1985—, user's com., 1990-92; mem. Calif. Coun. on Sci. and Tech., 1989—. Assoc. editor: Astrophys. Jour. Letters, 1982-87; editorial bd.: Ann. Revs. Astronomy and Astrophysics, 1982-87; contbr. articles to profl. jours. Trustee Carnegie Instn., Washington, 1985—; bd. dirs. Ann. Revs., Inc., 1989—. Recipient Bart J. Bok prize Harvard U., 1978, Director's Distinguished Lectr. award Livermore Nat. Lab., 1986, Carnegie Lectr. Carnegie Inst. Washington, 1988; NSF fellow, 1966-71; Woodrow Wilson fellow, 1966-71; Alfred P. Sloan fellow, 1977-81; listed among 100 best Am. scientists under 40, Sci. Digest, 1984; Tetelman fellow, Yale U., 1987. Mem. NAS, Am. Astron. Soc. (councilor 1982-84, Dannie Heineman prize 1986), Internat. Astron. Union, Nat. Acad. Arts and Scis., Phi Beta Kappa, Sigma Xi. Office: U Calif Lick Obs Santa Cruz CA 95060

FABIAN, JOHN M., former astronaut, air force officer; b. Goosecreek, Tex.; m. Donna Kay Buboltz; 2 children. B.S., Wash. State U., 1962; M.S., Air Force Inst. Tech., 1964; Ph.D., U. Wash., 1974. Commd. officer U.S. Air Force, advanced through grades to col.; astronaut NASA, Houston, 1978-86, mission specialist Challenger flight 2, 1983; mission specialist Discovery flight, 1985; ret. USAF, 1987; exec. v.p. ANSER Corp. (Analytical Services Inc.), 1987-91, pres., CEO, 1991—; served as pilot USAF, Vietnam. Decorated Def. Superior Service medal, Legion of Merit, NASA Space Flight medal, French Legion of Honor, Saudi Arabian King Abdul-Aziz medal, Vietnam Cross of Gallantry. Fellow Am. Inst. Aeronautics and Astronautics (assoc.); mem. Assn. Space Explorers (pres. 1990-93), Internat. Acad. Astronautics (corr.). Home: 3303 Circle Hill Rd Alexandria VA 22305-1709

FABRE, RAOUL FRANÇOIS, electronics company executive; b. Arles, France, Oct. 24, 1925; s. Louis Noël and Eugenie (Chagnolleau) F.; m. Simone Rosine Allex, Apr. 26, 1948; children: Frederic, Sylvie, Didier. Lic. Scis., U. Toulouse (France), 1946; degree in engring., Tech. Inst. Toulouse, 1946. Tech. engr. Alsthom Co., Paris, 1947-65; chief engr. CGEE Alsthom Co., Paris, 1965-80, mgr. in power electronics, 1980—. Mem. Echanges et Consultations Tech Internat., active Algeria Group. Mem. Internat. Electotech. Com. Roman Catholic. Home: 52 Ave Jean Jaures, 92290 Chatenay-Malabry France

FABRICANT, JILL DIANE, medical technology company executive; b. L.A.; d. I. Robert and Lillian (Solid) F.; m. Johan K. Trautmann, Mar. 18, 1989. BA, Mills Coll., 1971; MA, Occidental Coll., 1971; PhD, McGill U., 1976. Postdoctoral fellow Pasteur Inst., Paris, 1976-78; scientist NASA-Johnson Space Ctr., Houston, 1978-79; asst. prof. U. Tex. Med. Br., Galveston, 1979-82; pres. Biosyne Corp., Houston, 1982-88; v.p. bioscis. KVM Techs., Inc., Houston, 1989-90; pres. OvTex Corp., Houston, 1991—; dir. Tech. Commercialization Ctr. NASA, Johnson Space Ctr., Houston, 1993—. Asst. fund raiser Am. Heart Assn., Houston, 1992. Mem. Tech. Transfer, Environ. Mutation Soc., Am. Soc. Cell Biologists, Sigma Xi. Achievements include patents in field of sperm sexing and early-embryo sexing, development of ovulation detection device. Home: 18230 Lakeside Ln Houston TX 77058

FABRIKANT, CRAIG STEVEN, psychologist; b. Buffalo, Jan. 8, 1952; s. Benjamin and Laurine Miriam (Zucker) F.; B.A., Fairleigh Dickinson U., 1974, M.A., 1977; Ph.D., Fla. Inst. Tech., 1983; m. Carol Diane Golub, Nov. 6, 1977; children—Chad Adam, Carly. Intern in psychology N.J. Dept. Human Services, Trenton, 1977-78; clin. psychologist North Jersey Devel. Ctr., Totowa, 1978-85, Cedar Grove Residential Ctr.; chief psychologist Hackensack Med. Ctr., N.J., 1985—; adj. instr. Montclair State Coll., 1980-82; part-time instr. Fairleigh Dickinson U.; cons. psychology N.J. Dept. Labor and Industry, Newark, 1980—. Mem. Assn. Advancement Behavior Therapy, Am. Psychol. Assn., N.J. Psychol. Assn. Author profl. papers. Home: 750 Martin Ave Oradell NJ 07649-2300 Office: 106 Old Hook Rd Westwood NJ 07675-2421

FABRYCKY, WOLTER JOSEPH, engineering educator, author, industrial and systems engineer; b. Queens County, N.Y., Dec. 6, 1932; s. Louis Ludwig and Stephanie (Wadis) F.; m. Luba Swerbilow, Sept. 4, 1954; children: David Jon, Kathryn Marie. BS, Wichita State U., 1957; MS, U. Ark., 1958; PhD, Okla. State U., 1962. Instr. indsl. engring. U. Ark., 1957-60; from asst. to assoc. prof. indsl. engring. and mgmt. Okla. State U., 1962-65; prof. indsl. and systems engring. Va. Poly. Inst. and State U., Blacksburg, 1965-88, John L. Lawrence prof., 1988—, chmn. systems engring., 1968-75, assoc. dean engring., 1970-76, dean rsch. div., 1976-81; mem. Engring. Edn. Del. to People's Republic of China, 1978. Author: (with G.J. Thuesen) Economic Decision Analysis, 1974, 2d edit., 1980, (with B.S. Blanchard) Systems Engineering and Analysis, 1981, 2d edit., 1990, (with G.J. Thuesen) Engineering Economy, 1950, 8th edit., 1993, (with P.M. Ghare and P.E. Torgersen) Applied Operations Research and Management Science, 1984, (with J. Banks) Procurement and Inventory Systems Analysis, 1987, (with B.S. Blanchard) Life-Cycle Cost and Economic Analysis, 1991; editor: (with J.H. Mize) Prentice-Hall International Series in Industrial and Systems Engineering, 1972—. Recipient Lohmann medal Okla. State U., 1992; Ethyl Corp. doctoral fellow Okla. State U., 1962. Fellow AAAS, Inst. Indsl. Engrs. (exec. v.p. 1982-84, Book of Yr. award 1973, Outstanding Educator award 1990); mem. Am. Soc. Engring. Edn. (v.p. 1977-78), Nat. Coun. on Systems Engring., Ops. Rsch. Soc. Am., Sigma Xi, Alpha Pi Mu, Sigma Tau. Home: 1200 Lakewood Dr Blacksburg VA 24060 Office: Va Poly Inst and State U 302 Whittemore Hall Blacksburg VA 24061

FABUNMI, JAMES AYINDE, aeronautical engineer; b. Ile-Ife, Nigeria, Sept. 1, 1950; came to U.S. 1974; s. Michael Ajayi and Wuraola Abeni (Adeniji) F.; children: Michael, Okunlola, Odunayo, Monisade. MSc in Aero. Engring., Kiev Inst., Ukraine, 1974; PhD, MIT, 1978. Registered profl. engr., Conn., 1980. Rsch. engr. Kaman Aerospace Corp., Bloomfield, Conn., 1978-80; proprietor Fabunmi & Assocs., Windsor, Conn., 1979-81; asst. prof. aero. engring. U. Md., College Park, 1981-87; pres. Aedar Corp., Landover, Md., 1985—. Contbr. articles to profl. jours. Mem. AIAA, ASME, Am. Helicopter Soc., Internat. Soc. of African Scientists. Home: 8302 Stanwood St New Carrollton MD 20784 Office: Aedar Corp 8401 Corporate Dr #460 Landover MD 20785

FACCINI, ERNEST CARLO, mechanical engineer; b. Livo, Trento, Italy, May 28, 1949; parents Am. citizens; s. Carlo and Elena Agnes (Pancheri) F. AA, Western Wyo. Community Coll., 1969; BS, U. Wyo., 1972, MS, 1976. Registered profl. engr. Wyo., Md., N.Mex. Engring. technician Laramie (Wyo.) Energy Rsch. Ctr., 1968-71; field engr. Mountain Fuel Supply Co., Rock Springs, Wyo., 1972; research engr. Aberdeen (Md.) Proving Grounds, 1972-73; rsch. asst. mech. engring. U. Wyo., Laramie, 1973-76; engring. asst. Bridger Coal Co., Rock Springs, Wyo., 1973; mech. engr. Naval Explosive Ordnance Disposal Facility, Indian Head, Md., 1976-85; sr. scientist TERA/NMIMT, Socorro, N.Mex., 1986-89; prin. sientist Textron Def. Systems, Wilmington, Mass., 1989—. Contbr. articles to profl. jours. Mem. ASME (chmn. student sect. 1971-72), Am. Phys. Soc., Am. Soc. Metals. Roman Catholic. Achievements include research in rapid explosive excavation techniques, underwater non-explosive excavation, surface/subsurface ordnance clearance vehicle design, remote fuse disassembly, multi-fuel combuster design, internal ballistics, blast effects, design of shaped charges and of grenades for special applications, counter-mine methods, use and fabrication of Ta metal for warhead liners, application of orbital forging to warhead liners, use of powdered metallurgical techniques to obtain starting material for forging liners, use of end-game analysis; vulnerability lethality analysis codes in the design of warheads, design of space shielding concepts (multiple plate armor design), use of reactive/energetic materials and insensitive explosives applications to warheads; patents for specialty shaped charge, nonmagnetic/nonmetallic excavator with integral shaft liner, four-bar manifold for burn bar, combuster; patent pending for warhead liner material. Home: 9 Spring Rd Londonderry NH 03053-2912 Office: Textron Def Systems 201 Lowell St Wilmington MA 01887-2969

FACTOR, RONDA ELLEN, research chemist; b. Boston, June 16, 1953; d. Milton and Roslyn (Gordon) Factor; m. Eric R. Schmittou, May 5, 1985. BS in Chemistry, U. Mass., 1975; PhD in Chemistry, U. Wis., 1980. Rsch. scientist Eastman Kodak Co., Rochester, N.Y., 1980-83, sr. rsch. scientist, 1984-86, rsch. lab. mgr., 1987—. Mem. Am. Chem. Soc. Achievements include 12 patents in areas of dye chemistry, colloid chemistry, polymer chemistry. Office: Eastman Kodak Co Bldg 59 RL 1669 Lake Ave Rm 116 Rochester NY 14650-1708

FADDEEV, LUDWIG D., theoretical mathematician. Recipient Paul Adrian Maurice Dirac medal Internat. Ctr. Theoretical Physics, Italy, 1990. Office: Steklov Math Inst, Naberezhnaya Fontanki 27, 191011 Saint Petersburg D 11, Russia*

FADDEN, DELMAR MCLEAN, electrical engineer; b. Seattle, Nov. 10, 1941; s. Gene Scott and Alice Elizabeth (McLean) F.; m. Sandra Myrene Callahan, June 22, 1963; children: Donna McLean, Lawrence Gene. BSEE, U. Wash., 1963, MSEE, 1975. Lic. comml. pilot, Wash. With Boeing Comml. Airplane Co., Seattle, 1969—, chief engr. 737/757 avionics/flight systems, 1990—. Contbr. articles to IEEE Proceedings. Capt. USAF, 1963-69. Mem. AIAA, IEEE, Human Factors Soc., Soc. Automotive Engrs. (vice chmn. G-10 com. 1981-91, chmn. systems integration task group 1990—), Mountaineers (pres. 1984-85). Achievements include 2 patents in field. Home: 5011 298th Ave SE Preston WA 98050 Office: Boeing Comml Airplane Co PO Box 3707 M/S 6X-MF Seattle WA 98124-2207

FADDEYEV, LUDVIG DMITRIYEVICH, mathematician, educator; b. Leningrad, Russia, Mar. 23, 1934. Student, Leningrad U. Sr. rsch. fellow Inst. of Math., Leningrad, 1965—, dep. dir., 1976—; mem. staff Leningrad State U., 1967—, prof. math.-mech. faculty, 1969—; pres. Internat. Math. Union, 1986—. Recipient D. Heinemann prize Am. Phys. Soc., 1975, State prize USSR, 1971. Mem. Am. Acad. Arts and Scis., Russian Acad. Scis. Office: St Petersburg Br of Steklov Inst, Naberezhnaya Fontanki 27, 191011 Saint Petersburg Russia*

FADEL, GEORGES MICHEL, mechanical engineering educator; b. Beirut, Lebanon, Jan. 12, 1954; came to U.S., 1977; s. Michel Georges and Huguette (Kerba) F.; m. Lettie E. (Johnnie) Coleman, Sept. 8, 1979; children: Monica Ann, Michael. Diploma mech. engring., Swiss Fed. Inst. Tech., Zurich, 1976; MS in Computer Sci., Ga. Inst. Tech., 1978, PhD in Mech. Engring., 1988. Rsch. assoc. Kuwait Inst. Sci. Rsch., 1980-83; instr. Am. U. Beirut, Lebanon, 1983-85; computer operation supr. Ga. Inst. Tech., Atlanta, 1985-87, rsch. engr., 1987-88, asst. prof., 1989-92; asst. prof. mech. engring. Clemson (S.C.) U., 1992—; exec. v.p. Engring. Measurements & Modeling, Atlanta, 1990-92. Mem. ASME, AIAA, Assn. Computing Machinery, Sigma Xi. Achievements include research in energy, structural optimization, applied math. modeling, concurrent engring., design. Office: Clemson U Dept Mech Engring Clemson SC 29634

FADEM, BARBARA H., psychobiologist, psychiatry educator; b. N.Y.C., May 18, 1943; children: Jennifer, Daniel, Jonathan. MS, Rutgers U., 1977, PhD, 1979. Assoc. prof. psychiatry U. Medicine and Dentistry of N.J., Newark, 1980—. Rsch. grantee NSF, 1987, 90, March of Dimes, 1990. Office: U Medicine and Dentistry NJ NJ Med Sch 185 S Orange Ave Newark NJ 07103

FAETH, GERARD MICHAEL, aerospace engineering educator, researcher; b. N.Y.C., July 5, 1936; s. Joseph and Helen (Wagner) F.; m. Mary Ann Kordich, Dec. 27, 1959; children: Christine Louise, Lorraine Vera, Elinor Jean. BME, Union Coll., 1958; MS, Pa. State U., 1961, PhD, 1964. Instr. mech. engring. Pa. State U., University Park, 1958-59, research asst., 1959-64, asst. prof., 1964-68, assoc. prof., 1968-74, prof., 1974-85, prof. emeritus, 1985—; Modine prof., head gas dynamics labs. U. Mich., Ann Arbor, 1985—; vis. prof. Air Force Office Sci. Rsch., Washington, 1983-84; cons. GM, Warren, Mich., 1977—, Applied Rsch. Lab., Pa. State U., 1964-85; prof.-in-residence GM Inst., Detroit, 1983. Mem. editorial bd. Combustion Sci. and Tech., 1979—, Ann. Rev. Numerical Fluid Mechanics and Heat Transfer, 1985—, Atomization and Sprays, 1989—; contbr. numberous articles to profl. jours. Rep. Precinct Chmn. Centre County, Pa., 1977-84; bd. dirs. Eagles Mere (Pa.) Assn., 1982-88, Eagles Mere Park Assn., 1978-85. Recipient rsch. award Pa. State U., 1979, Outstanding Engring. Alumnus award, 1990; rsch. award U. Mich., 1988, Stephen S. Atwood award, 1993. Fellow ASME (tech. editor 1981-85, sr. tech. editor, 1985-90, Meml. award heat transfer div. 1988), AIAA, AAAS; mem. Combustion Inst. (dep. editor 1984-90, tech. editor 1990-96, bd. dirs. 1988-94), Nat. Acad. Engring., Am. Phys. Soc., Sigma Xi, Pi Tau Sigma, Phi Kappa Phi. Episcopalian. Home: 2665 Overridge Dr Ann Arbor MI 48104-4039 Office: U Mich 218 Aerospace Engring Bldg Ann Arbor MI 48109-2140

FAETH, LISA ELLEN, chemical engineer; b. Hackensack, N.J., June 13, 1955; d. Robert Francis and Dolores (Sexton) F.; m. Ghulam Ali, Aug. 8, 1984. BA in Chemistry/Chem. Engring., Barnard Coll., 1978; MSChemE, MIT, 1980, ChE in Chem. Engring., 1984. Polymer rschr. Westvaco Corp., Charleston, S.C., 1979, E.I. du Pont de Nemours & Co., Inc., Wilmington, Del., 1980; environ. compliance monitor U.S. Dept. Def., Washington, 1984-86; regulation writer on hazardous waste U.S. EPA, Washington, 1986-89, internat. activities specialist, 1988—. Contbr. articles to profl. jours. Mem. Sigma Xi (assoc.). Office: US EPA 401 M St SW TS-799 Washington DC 20460

FAGERHOL, MAGNE KRISTOFFER, immunologist; b. Aalesund, Norway, Mar. 10, 1935; s. Reidar and Margot (Husby) F.; m. Liv Anne Myklebust, June 25, 1960; children: Pia, Cathrine, Magne. MD, U. Bergen (Norway), 1960; PhD, U. Oslo, 1970. Med. diplomate. Resident dept. immunology Nat. Inst. Pub. Health, Oslo, 1963-69; dir. blood bank and dept. immunology Ullevaal Univ. Hosp., Oslo, 1970—; prof. transfusion medicine U. Oslo, 1992—; med. cons. Norwegian Def. Med. Corps, 1990—; Norwegian del. Coun. of Europe, 1990—; grant application reviewer NIH, Bethesda, Md., 1971. Author chpts. in sci. books; contbr. articles to profl. publs. Mem. Norwegian Blood Coun. (sec. 1970-82), Soc. Immunohaematology (sec. 1964-68), Soc. Immunology, N.Y. Acad. Scis., Vestheim Rotary Club. Home: Konsul Schjelderupsvei 7, 0286 Oslo Norway Office: Ullevaal Hosp Blood Bank, Dept Immunology, 0407 Oslo Norway

FAGG, WILLIAM HARRISON, retired infosystems specialist; b. Indpls., Mar. 28, 1924; s. Lloyd Ralph and Jeannette (Marker) F.; B.S., Purdue U., 1951; postgrad. Butler U., 1948-49, SUNY, Albany, 1970-71; m. Marie Aurora Schecton, Aug. 24, 1952; children—William H., Anthony Scott, Michael Lloyd. Engr., Gen. Electric Co., 1951-53, value analyst, 1953-56; sr. systems analyst Sperry Rand Univac, Indpls., 1956-63; application engr. Gen. Electric, Schenectady, 1963-65; supr. data processing N.Y. State Dept. Mental Hygiene, 1968-69; asst. dir. R/T systems N.Y. State Dept. Motor Vehicles, Albany, 1968-69; asso. EDP cons. N.Y. State Div. Budget, 1969-70; dir. adminstrv. and organizational analysis N.Y. State Dept. Motor Vehicles, 1970-74, dir. R/T systems, 1974-77, dir. systems coordination and control, 1977-81, dir. div. systems planning, 1981-85, dir. electronic data processing and adminstrv. analysis, 1985-86, dir. records mgmt. ops., 1986-89, dir. office records mgmt., 1989-90; cons., svc. MIS systems Masonic Orgns., 1990—. Active Boy Scouts Am. Served with USMC, 1943-46. Mem. Assn. Records Mgrs. and Adminstrn. Internat., Soc. Advancement Mgmt., IEEE, Adminstrv. Mgmt. Soc., Am. Ordnance Soc. Methodist. Clubs: Masons, Scottish Rite, York Rite, Shriners, Royal Order of Jesters, Elks. Home: 34 Crestwood Dr Schenectady NY 12306-3325

FAGHRI, ARDESHIR, civil engineering educator; b. Esfahan, Iran, May 19, 1959; came to U.S., 1975; s. Baderdean and Parvin (Safavi) F. BSCE, U. Wash., 1981, MSCE, 1983; MS, U. Va., 1985, PhD, 1987. Rsch. scientist fellow Va. Transp. Rsch. Coun., Charlottesville, 1984-88; sr. transp. engr. Va. Dept. Transp., Richmond, 1987-88; sr. systems analyst KLD Assoc., Inc., L.I., N.Y., 1988-90; asst. prof. U. Del., Newark, 1990-; chmn. neural networks subcom. of the artificial intelligence Tranps. Rsch. Bd., Washington, 1991--. Recipient Paper award Inst. Transp. Engrs., 1985, 86, 87. Office: U Del Dept Civil Engrs Newark DE 19716

FAHEY, JOHN LESLIE, immunologist; b. Cleve., Sept. 8, 1924. MS, Wayne State U., 1949; MD, Harvard U., 1951. Intern medicine Presbyn. Hosp., N.Y., 1951-52, asst. resident, 1952-53; clin. assoc. Nat. Cancer Inst., NIH, 1953-54, sr. investigator metabolism, 1954-63, chief immunology br., 1964-71; prof. microbiology & immunology, chmn. dept. Sch. Med. UCLA, 1971-81, dir. Ctr. Interdisciplinary Rsch. Immunological Diseases, 1978—. Mem. Am. Physiology Soc., Soc. Exptl. Biology & Medicine, Am. Assn. Cancer Rsch., Am. Fedn. Clin. Rsch., Am. Soc. Clin. Investigators. Achievements include rsch. in immunology, oncology. Office: UCLA Sch Medicine Ctr Interdisciplinary Rsch 12-262 Factor Bldg Los Angeles CA 90024*

FAHEY, PAUL FARRELL, college administrator; b. Lock Haven, Pa., July 2, 1942; s. Paul F. and Margaret (Hennessey) F.; m. Rosemarie Corallo, Sept. 4, 1965; children: Paul III, Salvatore, Augustine, Bridget. BS, U. Scranton, Pa., 1964; PhD, U. Va., 1968. Asst. prof. U. Scranton, 1968-73, assoc. prof., 1973-78, prof., 1978—, dean Coll. Arts and Scis., 1989—; vis. assoc. prof. Cornell U., Ithaca, N.Y., 1975-76; resident visitor AT&T Bell Labs., Murray Hill, N.J., 1982. Contbr. articles to profl. jours. Sci. faculty fellow NSF, 1975-76, 82. Mem. AAAS, IEEE, Acoustical Soc. Am., Coun. Undergrad. Rsch. Achievements include research in biophysics of model cell membrane and biophysics of hearing. Office: U Scranton Scranton PA 18510

FAHIM, MOSTAFA SAFWAT, reproductive biologist, consultant; b. Cairo, Egypt, Oct. 7, 1931; came to U.S., 1966; s. Mohamed and Amna (Hussin) F.; m. Zuhal Fahim, Feb. 23, 1959; 1 child, Ayshe. B.S. in Agrl. Chemistry, U. Cairo, 1953; M.S., U. Mo., 1958, PhD in Reproductive Biology, 1961. Research assoc. Sch. Medicine, U. Mo., Columbia, 1966-68, asst. prof., 1968-71, assoc. prof., 1971-75, prof., 1975—, chief reproductive biol. rsch., 1971-87, dir. Ctr. Reproductive Sci. and Tech., 1987—; prof. environ. trace substances rsch. U. Mo., Columbia, 1981—; cons. in field. Contbr. articles to profl. jours.; patentee in field. Mem. Am. Pub. Health Assn., Mo. Pub. Health Assn., Nutrition Today Soc., Internat. Andrology Soc., Internat. Toxicology Soc., Am. Coll. Clin. Pharmacology, Am. Soc. Pharmacology and Exptl. Therapeutics, Internat. Fertility Soc., Am. Fertility Soc., Fedn. Am. Socs. Exptl. Biology, N.Y. Acad. Scis., Soc. Environ. Geochemistry and Health, Soc. Study Reprodn., AAAS, Sigma Xi, Gamma Alpha. Achievements include patents for Minerals in Bioavailable Form, for Composition and Process for Promoting Epithelial Regeneration, for Intraprostatic Injection of Zinc Ions for Treatment of Inflammatory Conditions and Benign and Malignant Tumors of the Prostate, for Method of Inhibiting Generation, Maturation, Motility and Viability of Sperm With Minerals in Bioavailable Form. Office: U Mo Sch Medicine Ctr Repro Sci and Tech 111 Allton Bldg Columbia MO 65212

FAHNING, MELVYN LUVERNE, veterinary educator; b. St. Peter, Minn., Apr. 28, 1936; s. Otto Ernest and Esther Gladys (Reeck) F.; m. Marlene Vanda Emrud, Dec. 22, 1956; children: Mark L., Mitchell L., Matthew L., Marcy L. BS, U. Minn., 1958, MS, 1960, DVM, PhD, 1964. Asst. prof. U. Minn., St. Paul, 1966-72; v.p. Internat. Cryogenics Biol. Svcs., St. Paul, 1972-76; pres. OvaTech, River Falls, Wis., 1976-84; prof. U. Minn. Vet. Coll., St. Paul, 1980—, divsn. head theriogenology, 1991—; pres. CryovaTech Internat., Inc., Hudson, Wis., 1984—. Named one of 10 Outstanding Young Men in Minn., Outstanding Jaycee, Outstanding Regional Jaycee, U.S. Jaycees, 1969. Mem. Internat. Embryo Transfer Soc. (charter), Am. Embryo Transfer Soc. (charter), Minn. Vet. Med. Assn., Soc. of Theriogenology. Lutheran. Achievements include research in the current status of cryogenic preservation of mammalian embryos and the status of embryo transfer in bovines. Office: CryovaTech Internat Inc 592 Hwy 35 S Hudson WI 54016 also: U Minn Coll Vet Medicine 385 An Sci/Vet Med Bldg Saint Paul MN 55108

FAHRENTHOLD, ERIC PAUL, engineering educator; b. Phila., Sept. 13, 1952; s. Gerald Herman and Grace Lynell (Blankenship) F. BS, U.S. Mil. Acad., 1974; MS, Rice U., 1981, PhD, 1984. Profl. engr., Tex. Asst. prof. U. Tex., Austin, 1984-90, assoc. prof., 1990—. Contbr. articles to profl. jours. Capt. U.S. Army, 1974-79. Recipient Terry de la Mesa Allen award U.S. Mil. Acad., 1974; Temple Found. fellow U. Tex., 1991; Ednl. Asst. grantee Mobil Oil, 1983. Mem. ASME (meritorious svc. award Cen. Tex. sect. 1989), AIAA, IEEE, Soc. Petroleum Engrs. Office: Dept Mech Engring U Tex Austin TX 78712

FAIG, WOLFGANG, survey engineer, engineering educator; b. Crailsheim, Germany, Apr. 27, 1939; married; 3 children. Diploma Ing, Technion U. Stuttgart, 1962; Dr Ing, U. Stuttgart, 1969; MScE, U. N.B., 1965. Rsch. assoc. photogrammetry dept. civil engring. U. N.B., 1965, Inst. Applied

Geodesy, Stuttgart, Germany, 1966-69; asst. prof. civil engring. U. Ill., Champaign-Urbana, 1970-71; from asst. prof. to assoc. prof. survey and photogrammetry U. N.B., 1971-78, prof. survey engring., 1978—, assoc. dean engring., 1981-90, dean engring., 1990—; chmn. Working Group V-2 Internat. Soc. Photogrammetry and Remote Sensing, 1972-76, nat. reporter, 1980—; vis. prof. sch. survey U. NSW, Sidney, Australia, 1984-85, Faculty Engring. Survey Wuhan Tech. U. Survey and Mapping, 1986; active Internat. Rels., Nat. Sci. and Engring. Rsch. Coun. Can., 1988—. Mem. Am. Soc. Photogrammetry and Remote Sensing, Can. Inst. Survey and Mapping. Achievements include rsch. in self-calibration of amateur cameras and their use for precision photogrammetry; modeling of systematic errors, combined adjustment of vastly different observables; four-dimentional photogrammetry in deformation studies. Office: U New Brunswick, PO Box 4400, Fredericton, NB Canada E3B 5A3*

FAIN, JOHN NICHOLAS, biochemistry educator; b. Jefferson City, Tenn., Aug. 18, 1934; s. Samuel Clark and Virginia Manson (Hunt) F.; m. Ann Duff, June 7, 1958; children: Margaret Ann, John Nicholas Jr., James Clark. BS magna cum laude, Carson-Newman Coll., 1956; PhD in Biochemistry, Emory U., 1960. Research assoc. Emory U., Atlanta, 1960-61; NSF fellow NIH, Bethesda, Md., 1961-62, postdoctoral fellow USPHS, 1962-63; biochemist NIH and Nat. Inst. Arthritis and Metabolic Diseases, Bethesda, 1963-65; asst. prof. Brown U., Providence, 1965-68, assoc. prof., 1968-71, prof., 1971-85, chmn. biochemistry, 1975-85; Van Vleet prof., chmn. U. Tenn., Memphis, 1985—. Contbr. numerous articles to sci. jours. Del. gen. assembly United Presbyn. Ch., Providence, 1972. Recipient Disting. Alumnus award Carson-Newman Coll., 1986; fellow Cambridge U., 1977-78; NIH Fogarty fellow, 1984-85; Macy Faculty scholar, 1977-78. Mem. Am. Soc. Biol. Chemists, Biochem. Soc. Democrat. Office: U Tenn Coll Medicine Dept Biochemistry 800 Madison Ave Memphis TN 38163-0002

FAINBERG, ANTHONY, physicist; b. London, Jan. 14, 1944; came to U.S., 1947; s. Benjamin and Elizabeth (Martelli) F.; m. Louise Vasvari (div. 1986); m. Diane August, Sept. 7, 1986. AB, NYU, 1964; PhD, U. Calif., Berkeley, 1969. Physicist INFN U. of Turin, Italy, 1970-72; rsch. prof. Syracuse (N.Y.) U., 1973-78; physicist Brookhaven Nat. Lab., Upton, N.Y., 1978-83; legis. aide Office of Senator Bingaman, Washington, 1983-84; sr. assoc. Office of Tech. Assessment, Washington, 1985—; fellow Ctr. for Internat. Security & Arms Control, Stanford, 1991-92. Editor: (book) The Energy Source Book, 1991. Mem. AAAS, Am. Phys. Soc. (mem. panel on pub. affairs 1990-92, congl. fellow 1983-84). Office: Office Tech Assessment 600 Pennsylvania Ave SE Washington DC 20003-4316

FAINTICH, STEPHEN ROBERT, chemical engineer; b. L.A., Mar. 28, 1963; s. Jerome Arnold Faintich and Lynn Margo (Federbush) Bishop; m. Myra B. Chunn, Mar. 4, 1989; 1 child, Courtney Anne. BS, Northwestern U., 1985; postgrad., Fla. State U., 1988—. Process engr. Cincinnati Milacron, Maineville, Ohio, 1985-88; R & D engr. Olin Corp., St. Marks, Fla., 1988—. Recipient Olin Corp. chem. engr. rsch. grant, 1984. Mem. AIAA. Achievements include patented ball powder propellants, process to deter ball powder using a polyester deterent; design of analytical method to quantitavely measure gun muzzle signature in small caliber applications. Home: 3150 Whirlaway Trail Tallahassee FL 32308 Office: Olin Corp PO Box 222 Saint Marks FL 32355

FAIR, CHARLES MAITLAND, neuroscientist, author; b. N.Y.C., Sept. 18, 1916; s. Charles Maitland Fair and Gertrude Modora (Bryan) Knapp; m. Mary Katherine Ruddy, Feb. 2, 1952 (div. 1980); children: Ellen, Katherine, Charles (dec.); m. Louise Sadler Kiessling, May 5, 1980. Guggenheim fellow Brain Rsch. Inst., UCLA, 1963-64; resident scientist MIT Neuroscis. Rsch. Program, 1964; lab. scientist Mass. Gen. Hosp., Boston, 1966-67, MIT, 1967; officer Synax, Somerville, Mass., 1970-72. Author: The Physical Foundations of the Psyche, 1963, The Dying Self, 1969, From the Jaws of Victory, 1971, The New Nonsense, 1974, Memory and Central Nervous Organization, 1988, Cortical Memory Functions, 1992; contbr. articles and revs. to profl. jours. With USNR, 1938. Am. Acad. Arts. and Scis. grantee, 1961. Mem. AAAS, N.Y. Acad. Sci. Democrat. Avocations: jazz, piano, sailing. Home: Jerry Brown Farm 110 Fire Lane 1 Wakefield RI 02879-5460

FAIR, JAMES RUTHERFORD, JR., chemical engineering educator, consultant; b. Charleston, Mo., Oct. 14, 1920; s. James Rutherford and Georgia Irene (Case) F.; m. Merle Innis, Jan. 14, 1950; children: James Rutherford III, Elizabeth, Richard Innis. Student, The Citadel, 1938-40; B.S., Ga. Inst. Tech., 1942; M.S., U. Mich., 1949; Ph.D., U. Tex., 1955; D.Sc. (hon.), Wash. U., 1977; HHD (hon.), Clemson U., 1987. Research engr. Shell Devel. Co., Emeryville, Calif., 1954-56; with Monsanto Co., 1942-52, 56-79; engring. dir. corp. engring. dept. Monsanto Co. (World hdqrs.), St. Louis, 1969-79; McKetta chair chem. engring. U. Tex., Austin, 1979—; dir., v.p. Fractionation Research, Inc., Bartlesville, Okla., 1969-79; pres. James R. Fair Inc., 1981—. Author: North Arkansas Line, 1969, Distillation, 1971; Contbr. numerous articles to profl. publs. Recipient profl. achievement award Chemical Engineering mag., 1968, King award U. Tex., 1987. Fellow AICE (bd. dirs. 1965-67, Walker award 1973, Practice award 1975, Founders award 1977, Inst. lectr. 1979); mem. NSPE, Am. Chem. Soc. (Separation Sci. and Tech. award 1993), NAE, Am. Soc. Engring. Edn., Faculty Club U. Tex., Headliners Club (Austin), Sigma Nu. Republican. Presbyterian. Home: 2804 Northwood Rd Austin TX 78703-1603 Office: U Tex Dept Chem Engring Separations Rsch Program Austin TX 78712

FAIRALL, RICHARD SNOWDEN, mechanical engineer; b. Chgo., Nov. 25, 1927; s. Arlo Herman and Marian Josephine (Schaffer) F.; m. Mary J. Bjorling Fairall, Jan. 2, 1955; children: Victoria Lynn, Saundra Jo., Richard Michael, Timothy Jon. BS, Calif. Inst. Tech., 1950; MS, Stanford U., 1951. Registered profl. engr. Calif. Devel. engr. Garrett Mfg. Co., L.A., 1951-58; asst. chief engr. Precision Equipment Co., Torrance, Calif., 1958-60; sr. engr. Aerojet Gen. Corp., Sacramento, Calif., 1960-73; project leader U.S. Bur. Reclamation, Sacramento, 1974-92; instr. U. Calif., L.A., 1955-60; cons. Sacramento County Suprs., 1973-75; acoustic cons. Presbyterian Ch., Rancho Cordova, CAlif., 1985--. Contbr. articles to profl jours. Scoutmaster Boy Scouts Am., Rancho Cordova, 1965; deacon PResbyterian Ch., RAncho Cordova, 1981; coach football Rancho Cordova Pk. Youth, 1973-76. Republican. Office: US Bur Reclamation 2800 Cottage Way Sacramento CA 95825

FAIRBANKS, MARY KATHLEEN, data analyst, researcher; b. Manhattan, Kans., June 4, 1948; d. Everitt Edsel and Mary Catherine (Moran) F. BS, St. Norbert Coll., 1970; postgrad., Calif. Family Study Ctr., 1981-82. Neuropsychology researcher U.S. VA Hosp., Sepulveda, Calif., 1970-76; mgr. print shop Charisma In Missions, City of Industry, Calif., 1976-77; neuropsychology researcher L.A. County Women's Hosp., 1977-79; mem. tech. staff Computer Scis. Corp., Ridgecrest, Calif., 1979-81; systems programmer Calif. State U., Northridge, 1982-84; bus. systems analyst World Vision, Monrovia, Calif., 1984-86; configuration analyst Teledyne System Co., Northridge, 1986-87; applications system analyst Internat. Telephone and Telegraph/Fed. Electric Corp., Altadena, Calif., 1987-88; supr. data analysts OAO Corp., Altadena, 1988—. Co-author, contbr.: Serotonin and Behavior, 1973, Advances in Sleep Research, vol. 1, 1974. Mem. OAO Mgmt. Assn., So. Calif. Application System Users Group, Digital Equipment Computer Users Soc. Roman Catholic. Avocations: photography, reading, music, hiking, camping. Home: 37607 Lasker Ave Palmdale CA 93550-7721 Office: OAO Corp 787 W Woodbury Rd Ste 2 Altadena CA 91001-5368

FAIROBENT, DOUGLAS KEVIN, computer programmer; b. Detroit, Jan. 10, 1951; s. Jack Edward and Doris Kathleen (Kennedy) F.; m. Paulette Marie Gillig, June 13, 1981. BS in Physics, U. Mich., 1972, MS in Physics, 1975, PhD in Physics, 1978. Engr. Ford Motor Co., Allen Park, Mich., 1978-80; lectr. physics Ohio State U., Columbus, 1980-82; sr. systems programmer Rockwell Internat., Columbus, 1982-85, Cin. Milacron, 1985-90, Quantum Chem. Co., Cin., 1990—. Contbr. articles to Phys. Rev., other publs. Mem. Am. Phys. Soc., Planetary Soc., Nat. Taxpayers Union, Citizens Against Govt. Waste. Libertarian. Office: Quantum Chem Co 11500 Northlake Dr Cincinnati OH 45249

FAJARDO, JULIUS ESCALANTE, plant pathologist; b. Los Banos, Laguna, Philippines, July 16, 1959; came to U.S. 1987; s. Pelagio Sevilla and Basilia Daria (Escalante) F. BS in Agr., U. Philippines, 1980, MS in Plant Pathology, 1985; PhD in Plant Pathology, Tex. A&M U., 1992. Cert. plant pathologist. Rsch. asst. Nat. Crop Protection Ctr., Los Banos, 1980-82; rsch. assoc. Nat. Plant Genetic Resources Lab., Los Banos, 1982-87; grad. rsch. asst. Tex. A&M U., College Station, 1988—; postdoctoral internship U.S. Agy. for Internat. Devel./Peanut Collaborative Rsch. Support Program, College Station, 1992—. Natural Scis. and Engring. Rsch. Coun. of Can. vis. fellow, 1993; Rotary Found. internat. scholar, 1987. Mem. N.Y. Acad. Scis., Am. Phytopathol. Soc., Am. Peanut Rsch. and Edn. Soc., Assn. Ofcl. Analytical Chemists (Mid-Can. Region), Sigma Xi, Phi Beta Delta. Roman Catholic.

FALCI, KENNETH J., food and nutrition scientist. BA in Organic Chemistry, Marist Coll., 1968; PhD, Fordham U., 1976. Sr. rsch. chemist Olin Corp., New Haven, Conn., 1976-77; consumer safety officer, food and color additives FDA, Washington, 1977-78, dep. program mgr., 1978-80, supr. petitions control branch, 1980-83, supr. generally recognized as safe branch, 1983-85, supr. consumer safty officer, indirect additive branch, 1985-91, chief regulatory affairs staff, 1991—; guest speaker various nat. and internat. orgns. Contbr. articles to profl. jours., chpts. in books. Active Sci. and Tech. Commn., Rockville, Md., 1990—, chmn. 1990-92. Mem. AAAS, Am. Chemical Soc., Inst. Food Technologists, Sigma Xi. Office: FDA-Center for Food Safety & Applied Nutrition 200 C St SW Washington DC 20204

FALCON, JOSEPH A., mechanical engineering consultant; married; two children. BSME, Poly. U., 1943; MSME, Stevens Inst. Tech., 1947; postgrad., Columbia U., UCLA. Registered profl engr., Calif., N.Y. Former mem. teaching staff, dir. power generation programs UCLA; mgr. project engring. Bechtel Power Corp., 1970-87; sr. ptnr. J. A. Falcon & Assocs., Huntington Beach, Calif., 1987—. Bd. dirs. Newport Found. Fellow ASME (bd. govs. 1986-90, past bd. on minorities and women, v.p. energy conservation group, chmn. Mex. sect., rep. to Pan. Am. Engring. Assn., various positions power divsn., pres. 1992-93, Ch. medal 1991), Inst. for Advancement Engring.; mem. L.A. Power Producers Assn. (bd. dirs.), Geothermal Resource Coun. (bd. dirs.). Office: J A Falcon & Assocs 17155 Roundhill Dr Huntington Beach CA 92649*

FALES, HENRY MARSHALL, chemist; b. N.Y.C., Feb. 12, 1927; s. Henry Marshall and Cecile Marie (Vatet) F.; m. Caroline Eleanor McCullagh, Dec. 20, 1947; children: Marsha Kent Fales Mazz, Suzanne Kent Fales Palmer, Henry Richard. BSc in Chemistry, Rutgers U., 1948, PhD in Organic Chemistry, 1953. Instr. Rutgers U., New Brunswick, N.J., 1953; rsch. chemist, lab. chief Nat. Heart, Lung and Blood Inst., NIH, Bethesda, Md., 1953—. With USN, 1944-46. Recipient Superior Svc. award U.S. Govt., 1973, Cert. of Merit, U.S. Govt., 1978, Meritorious Svc. award U.S. Govt., 1987, Profl. Svc. award Wash. Chpt. of Alpha Chi Sigma. Mem. Am. Chem. Soc., Am. Soc. Mass Spectrometry (mem.-at-large, sec., v.p. programs, pres.-elect). Avocation: fishing. Home: 63 Orchard Way N Rockville MD 20854-6127 Office: NIH Rm 7N318 Bethesda MD 20892

FALEY, ROBERT LAWRENCE, instruments company executive; b. Bklyn., Oct. 13, 1927; s. Eric Lawrence and Anna (Makahon) F.; B.S. cum laude in Chemistry, St. Mary's U., San Antonio, 1956; postgrad. U. Del., 1958-59; m. Mary Virginia Mumme, May 12, 1950; children: Robert Wayne, Nancy Diane. Chemist, E.I. Dupont de Nemours & Co., Inc., Wilmington, Del., 1956-60; sales mgr. F&M Sci., Houston, 1960-62; pres. Faley Assos., Houston, 1962-65; sales mgr. Tech. Inc., Dayton, Ohio, 1965-70; biomed. mkt. mgr. Perkin-Elmer Co., Norwalk, Conn., 1967-69; mktg. dir. Cahn Instruments, Los Angeles, 1970-72; pres. Faley Internat., El Toro, Calif., 1972-93; pres. Status Internat., Las Vegas, Nev., 1993—. Internat. speaker in field; dir. Whatman Lab. Products Inc., 1981-82, Status Instrument Corp., 1985-87; tech. mktg. cons. Whatman Ltd., Abbott Labs., OCG Tech., Inc., Pacific Biochem., Baker Commodities, Bausch & Lomb Co., Motorola Inc., Whatman Inc., Filtration Scis. Corp., PMC Industries. Mem. adv. com. on Sci., tech., energy and water U.S. 43d Congl. Dist., 1985-87. With USMS, 1944-47, 1st lt. USAF, 1948-53. Charter mem. Aviation Hall Fame. Fellow Am. Inst. Chemists, AAAS; mem. ASTM, Am. Chem. Soc. (sr.), Instrument Soc. Am. (sr.), Inst. Environ. Scis. (sr.), Aircraft Owners and Pilots Assn., U.S. Power Squadrons, Delta Epsilon Sigma. Club: Masons. Contbr. articles on technique of gas chromatography to profl. jours. Home: 27850 Espinoza San Juan Capistrano CA 92692-2156 Office: PO Box 43267 Las Vegas NV 89116-1267

FALGIANO, VICTOR JOSEPH, electrical engineer, consultant; b. San Francisco, Nov. 25, 1957; s. Victor Anthony and Frances Mary Falgiano; m. Linda Maxine Owens, July 24, 1982; children; Gregory Joseph, Nicholas Rexford. BS in Elec. Engring. Tech. magna cum laude, Cogswell Coll., 1989, BS in Computer Engring. magna cum laude, 1989. Sr. design engr. Amdahl Corp., Sunnyvale, Calif., 1978-93; staff system design engr. Nat. Semiconductor Corp., Santa Clara, Calif., 1993—; mem. steering com. System Design and Integration Conf., Santa Clara, Calif.; mem. acad. adv. com. Cogswell Coll., Cupertino, Calif., 1991. Advisor to high sch. students Jr. Achievement. Mem. IEEE (sr.), Assn. Computing Machinery. Achievements include development of computer program pre-reading children, automobile digital instrumentation, speech recognition user interface for automotive applications, data aquisition circuitry used in mainframe computer power systems, high performance connector system for mainframe computers. Office: Nat Semiconductor Corp PO Box 58090 M/S 10-225 2900 Semiconductor Dr Santa Clara CA 95052

FALK, DEAN, anthropology educator; b. Seattle, June 25, 1944; children: Sarah Falk Schofield, Adrienne Jane. Student, Antioch Coll., 1962-63, U. Wash., 1964-65; BA with honors, U. Ill., 1970, MA, 1972. Asst. prof. Rollins Coll., Winter Park, Fla., 1976-77, So. Ill. U., Carbondale, 1977-79; asst. prof. health scis. Boston U., 1979-80; investigator Caribbean Primate Rsch. Ctr., 1980-84, curator Cayo Santiago primate skeletal collection, 1982-86, assoc. researcher, 1984-86; asst. prof. anatomy Sch. of Medicine U. P.R., 1980-82, assoc. prof. anatomy Sch. of Medicine, 1982-86; assoc. prof. Purdue U., 1986-88; prof. SUNY, Albany, 1988—. Author: External Neuroanatomy of Old World Monkeys, Primate Brain Evolution: Methods and Concepts, 1982, Braindance: What New Findings Reveal About Human Origins and Brain Evolution, 1992; mem. panel referees Human Evolution; mem. editorial bd. Man, Advances in Human Evolution; contbr. articles to Jour. Neuroscis., Jour. Neurosci. Methods, Am. Jour. Phys. Anthropology, Internat. Jour. Primatology. Antioch Alumni Assn. scholar, 1962, Ctr. for Continuing Edn. of Women scholar, 1975, Alice C. Lloyd scholar, 1975; grantee NSF, 1970-71, 71-72, 78, 82, 84-86, 90-92, U. Mich., 1976, So. Ill. U., 1979, NIH, 1986-90. Fellow Am. Anthropol. Assn. (elected, sec., treas. biol. anthropology sect. 1989-91); mem. AAAS, Am. Assn. Phys. Anthropologists, N.Y. Acad. Scis., Lang. Origins Soc., Sigma Xi. Home: 173 S Main Ave Albany NY 12208-2411 Office: SUNY Dept of Anthropology Albany NY 12222

FALK, JOEL, electrical engineering educator; b. N.Y.C., Feb. 12, 1945; s. Benjamin and Mina (Goldberg) F.; m. Victoria Lee McKesson, July 31, 1983; children: Leah, Joshua. BEE, CCNY, 1965; MS, Stanford U., 1967, PhD, 1971. Engring. specialist GTE Sylvania, Mountain View, Calif., 1970-75; assoc. prof. elec. engring. U. Pitts., 1975-83, prof. elec. engring., 1983—, coord. engring. physics program, 1984—; physicist Lawrence Livermore (Calif.) Nat. Lab., 1981-82; program chmn. SPIE Symposium on High Power Lasers, L.A., 1992. Contbr. over 50 articles ot profl. publs. Mem. IEEE (sr.), Optical Soc. Am., Am. Phys. Soc. Office: U Pitts Dept Elec Engring 348 Benedum Pittsburgh PA 15261

FALK, ROBERT BARCLAY, JR., anesthesiologist, educator; b. Lancaster, Pa., July 1, 1945; s. Robert Barclay and Miriam (Neff) F.; BA, Franklin and Marshall Coll., 1967; MD, Jefferson Med. Coll., 1971; m. Carol Anne Gundel, May 30, 1970; 1 child, Juliana Gundel. Intern, Conemaugh Valley Meml. Hosp., Johnstown, Pa., 1971-72; resident in anesthesiology M.H. Hershey Med. Sch. Hosp., 1974-77; partner Anesthesia Assocs., Lancaster, 1977—, sr. v.p. 1993—; staff anesthesiologist Lancaster Gen. Hosp., 1977—; clin. asst. prof. dept. anesthesiology Hershey (Pa.) Med. Sch., 1977—, vice chmn. dept. anesthesiology, 1984-85, chmn., 1985-92. Participant alumni phonathon Franklin and Marshall Coll., 1978-81, vice chmn., 1981, chmn., 1983, mem. alumni admissions com., 1977-79, chmn., 1980-87, chmn. 20th reunion gift com.; mem. Lancaster Regional Alumni Council, 1987-91, trustee athletic com., 1988—; mem. Lancaster Area Arts Coun., 1989-91; Sunday sch. tchr. Trinity Lutheran Ch., Lancaster, 1977-80; bd. dirs. Lancaster Summer Arts Festival, 1981—, v.p., 1982-84, pres., 1985-90; bd. dirs. Pa. Acad. Music, 1991—, vice-chmn., 1991-92, chmn., 1993—. Lt. M.C., USNR, 1972-74. Diplomate Am. Bd. Anesthesiology. Mem. AMA, Am. Soc. Anesthesiologists, Pa. Soc. Anesthesiologists, Internat. Anesthesia Rsch. Soc., Pa. Med. Soc., Lancaster Country Club, Hamilton Club, Masons, Shriners. Republican. Contbr. articles in field to profl. jours. Home: 1025 Marietta Ave Lancaster PA 17603-3106 Office: Anesthesia Assocs 133 E Frederick St Lancaster PA 17602-2294

FALK, STEVEN MITCHELL, biomedical engineer, consultant; b. Lausanne, Switzerland, May 29, 1961; (parents Am. citizens); s. Maurice and Carolyn (Rubin) F. BS in Engring., U. Pa., 1983; MS in Engring., Cath. U. Am., 1986. Registered profl. engr., Md. Rsch. asst. Johnson Rsch. Found., Phila., 1981-82; rsch. asst. neurosurgery svc. Grad. Hosp., Phila., 1982-83; patent examiner U.S. Patent Office, Washington, 1983-85; engring. mgr. GMS Engring. Corp., Columbia, Md., 1985-91; sr. engr. LT Industries, Rockville, Md., 1991-92, Ohmeda, Columbia, 1992—. Contbr. articles to Current Eye Rsch., Proc. Med. Chemistry Biosci. Rev., Proc. Am. Conf. for Engrs. in Medicine and Biology. Programmer Steven Silberfarb for State Senate, Bethesda, Md., 1990. Mem. ASME, Assn. for Advancement of Med. Instrumentation, Broadcast Music Inc., N.Y. Acad. Sci. Achievements include patent on measurement of blood pressure, also patents pending. Home: 137 Kinsman View Cir Silver Spring MD 20901-1654 Office: Ohmeda 9065 Guilford Rd Columbia MD 21046

FALKINGHAM, DONALD HERBERT, oil company executive; b. Lexington, Ill., Dec. 13, 1918; s. William Bishop and Violet (Ashabran) F.; m. Mary Margaret Chalmers, Aug. 23, 1947; children: Deanna Beth Falkingham Worst, Janis Kay Falkingham Fenwick. BS, Mo. Sch. Mines, 1941; Profl. Engr., U. Mo., Rolla, 1973. Registered profl. engr., land surveyor, Wyo. Field engr. Amoco, Rangely, Colo., 1951-53; dist. engr. Amoco Producing Co., Cody, Wyo., 1953-59; div. engr. Amoco Producing Co., Casper, Wyo., 1959-61; dist. supt. Amoco Producing Co., New Orleans, 1961-68; pres. Amoco UK Exploration Co., London, 1968-70, Amoco Iran Oil Co., Tehran, 1970-71; co-chmn. bd. dirs. Pan Am. Iran Oil Co., Teheran, 1970-71; gen. mgr. producing dept. Amoco Internat. Oil Co., Chgo., 1972-77; pres. Amoco Drilling Svcs., Chgo., 1975-77, Oceanwide Constrn. Co., St. Helier, Isle of Jersey, 1977-78; chmn. bd. World Maritime, Bermuda, 1977-78; ptnr., co-owner Falcar Energy Co., Houston, 1978—; dist. chmn. Am. Petroleum Inst., 1961; com. mem. Am Bur. Shipping, Bldg. and Classing Offshore Drilling Units, N.Y.C., 1966-68; chmn. exploration and production forum, Oil Industry Internat., London, 1974-77. Pres. bd. trustees, Presbyterian Ch., Cody. Pilot U.S. Army, 1942-45, ETO, maj. ret. Decorated D.F.C., Air medal with oak leaf cluster. Mem. Soc. Petroleum Engrs. (dist. chmn. 1967), Petroleum Club (pres. Casper chpt.), Cody Country Club (pres.), Masons, Shriners. Republican. Avocation: travel. Home: 5918 S Atlanta Pl Tulsa OK 74105-7506 Office: Falcar Energy Co PO Box 1323 Montgomery TX 77356

FALLET, GEORGE, civil engineer; b. Berlin, Pa., May 18, 1920; s. John and Anna (Hrobak) F.; m. Sybil Lorene DeLoach, Apr. 30, 1949; children: George Michael, Carol Ann, Mary Jane. BCE, Poly. Inst. Bklyn., 1957, MSCE, 1963. Registered profl. engr., N.H., N.Y.; registered land surveyor, N.Y. Asst. engr. Balt. & Ohio R.R., S.I. Rapid Transit Rwy., Staten Island, N.Y., 1946-53; structural designer H.K. Ferguson Co., N.Y.C., 1953-58; civil engr. U.S. Corps Engrs., N.Y.C., 1958-60; structural engr. U.S. Naval Facilities Command, N.Y.C., 1960-68, Fed. GSA, N.Y.C., 1968-85; dep. dir. bldg. dept. City of Nashua, N.H., 1986-89; cons. engr. Nashua, 1989—; mem. subcom. U.S. Com. on Seismic Safety, Washington, 1978-85. Active Community Planning Bd., Staten Island, 1975-81. Recognized for Outstanding Citizenship Borough of Staten Island, 1980. Fellow ASCE (Robert Ridgway award 1957), NSPE (GSA Nat. Engr. of the Yr. 1981), Soc. Am. Mil. Engrs., Chi Epsilon, Tau Beta Pi. Republican. Home: 32 Watersedge Dr Nashua NH 03063-1120 Office: PO Box 3233 Nashua NH 03061-3233

FALLETTA, JOHN MATTHEW, pediatrician; b. Arma, Kans., Sept. 3, 1940; s. Matthew John and Norma (Luke) F.; m. Carolyn Ontjes, June 22, 1963; children: Elizabeth, Matthew. A.B., U. Kans., 1962, M.D., 1966. Diplomate: Am. Bd. Pediatrics, with subsply. hematology-oncology. Intern in mixed medicine Kans. U. Med. Ctr., Kansas City, 1966-67; surgeon Epidemic Intelligence Svc., Tex. Children's Hosp. USPHS, Houston, 1967-69; asst. instr. pediatrics Baylor Coll. Medicine, Houston, 1967-69, resident in pediatrics, 1969-71, chief resident Tex. Children's Hosp., 1971, postdoctoral fellow hematology-oncology, 1971-73, asst. prof. pediatrics, 1973-76; assoc. prof. pediatrics Duke U., Durham, N.C., 1976-83, prof. pediatrics, 1984—, chief div. hematology-oncology, dir. Clin. Pediatric Lab., 1976—; Chmn. transfusion com. Duke U. Med. Ctr., 1978—, mem. exec. com. med. staff, 1978—, instl. rev. bd. human rsch., 1979—; assoc. dir. clin. cancer edn. program Duke Comprehensive Cancer Ctr., 1978—; mem. instl. rev. bd. human rsch. Baylor Coll. Medicine, 1974-76; mem. acad. coun. Duke U., 1982-86, 87—, exec. com., 1988, faculty compensation com., 1988, faculty com. on univ. governance, 1988, trustee-faculty com. to rev. pres., 1989; cons. pediatric hematologist-oncologist Charlotte (N.C.) Meml. Hosp., 1978—. Contbr. more than 105 articles to Nature, Am. Jour. Ophthalmology, Pediatrics, New Eng. Jour. Medicine, Clin. Pediatric Oncology, others. Cons. pediatric hematologist-oncologist Project Hope, Pediatric Inst., Krakow, Poland, 1979—; prin. investigator Pediatric Oncology Group, 1981—, chmn. epidemiology com., mem. prin. investigator's exec. com., new agts. and pharmacology com.; chmn. prophylactic penicillin study I Nat. Heart, Lung and Blood Inst., NIH, 1982-86, chmn. study II, 1987—; active cancer ctr. support rev. com. Nat. Cancer Inst., NIH, 1986-90, NIH Reviewers Res., 1990—; trustee Ronald McDonald Children's Charities, 1986-90;. Mem. Am. Assn. Cancer Rsch., Am. Acad. Pediatrics, Am. Pediatric Soc., Am. Soc. Clin. Oncology, So. Soc. Pediatric Rsch (pres. 1981-82), Soc. Pediatric Rsch., N.C. Pediatric Soc., N.C. Med. Soc., Phi Beta Kappa, Alpha Omega Alpha. Office: Duke U Med Ctr PO Box 2916 Durham NC 27710-0001

FALLON, DAVID MICHAEL, mechanical engineer; b. Passaic, N.J., Oct. 9, 1946; s. L. Joseph and Ellen Catherine (Power) F.; m. Joan Elizabeth Gerber, Sept. 20, 1970; children: David Michael Jr., Daniel James. BSME, N.J. Inst. Tech., 1977, MS in Engring. Mgmt., 1992. Registered profl. engr., N.J. Asst. project engr. Automatic Switch Co., Florham Park, N.J., 1977-82, product engr., 1982-90, sr. product engr., 1990—. Author: (publ.) Pressure/Temperature Switch Qualification for Nuclear Power Generating Stations, 1983. Pres. Rutherford (N.J.) Youth Soccer, 1989-92. Decorated Bronze Star, Purple Heart. Mem. NSPE, Bergen County Soc. Profl. Engrs. (sec. 1983-84, v.p. 1984-85), Pi Tau Sigma. Home: 30 Hawthorne St Rutherford NJ 07070

FALLON, L(OUIS) FLEMING, JR., public health educator, researcher; b. Jersey City, Mar. 21, 1950; s. Louis Fleming and Patricia (Nelson) F. AB in Math. and Psychology, Colby Coll., 1972; MS in Microbiology, Wagner Coll., 1977; MBA in Econs., U. New Haven, 1979; MD, St. George's (Grenada) U., 1984; MPH, Columbia U., 1986. Cert. health officer, N.J., Pa. Bus. rsch. analyst So. New Eng. Telephone Co., New Haven, 1978-80; assoc. A.T. Kearney, Inc., N.Y.C., 1980-81; v.p. Medcon of Am., Chester, N.J., 1984-87; resident in environ. and occupational medicine Columbia U., N.Y.C., 1985-88, lectr. Sch. Pub. Health, 1986-88, asst. prof. clin. pub. health, 1988—; assoc. prof. pub. health Slippery Rock (Pa.) U., 1990—; cons. psychologist Psychol. Rsch. Svcs., Cleve., 1973-77; lectr. grad. program St. Joseph's Coll., Windham, Maine, summer 1990, adj. prof., 1990—; cons. in pub. health adminstrn. and occupational medicine, 1988—; presenter in field. Author: The Penultimate Muffin; editor, author: The Management Perspective, 1989; contbr. articles to Brit. Jour. Indsl. Medicine, Jour. Soc. Occupational Medicine, Jour. Perinatal Medicine, New Eng. Jour. Medicine, Indian Jour. Indsl. Medicine, Med. Problems of Performing Artists, HMO Mgmt., also others. Mem. exec. coun. Boy Scouts Am., N.J., Maine, Pa., 1986—; pres., mem. Chester (N.J.) Bd. Health, 1987-90; chmn. bd. trustees United Meth. Ch., Gladstone, N.J., 1988-90; mem. Environ. Protection Commn. for Chester Twp., N.J., 1987-90; mem. environ. health adv. com. Maine Bur. Health, Augusta, 1990; exec. bd. Women's Shelter/Rape Crisis Ctr. Lawrence County.; institutional review bd. protection of human subjects Slippery Rock U. Fellow Am. Cancer Inst., 1985-88. Fellow Royal Soc.

Medicine (U.K.), Soc. Occupational Medicine (U.K.); mem. APHA, Am. Coll. Occupational Medicine (com. on ethical practice, 1989—, adv. bd. com. for internat. edn. 1991—), Am. Coll. Preventive Medicine, Nat. Eagle Scout Assn., Psi Chi. Achievements include establishment of a linkage between noise induced hearing loss and hypertension; creation of personal computer based system linking physicians, hospitals and a state department of health. Office: Slippery Rock U Allied Health Dept Slippery Rock PA 16057

FALLS, ELSA QUEEN, biologist, educator; b. Charlottesville, Va., Jan. 25, 1942; d. Vernon Oswald and Wilma Emilie (Wagner) Queen; m. Donald Parker Falls, July 6, 1963; children: Melissa Arnold, Mark Parker. BA, Westhampton Coll., 1964; MA, U. Richmond, 1972. Biology tchr. Douglas Freeman High Sch., Richmond, Va., 1984-85; lab. instr. U. Va. Extension, Roanoke, 1965-66; instr. in biology U. Richmond, 1972-76, J. Sargeant Reynolds Community Coll., Richmond, 1976-78; from instr. to assoc. prof. biology Randolph-Macon Coll., Ashland, Va., 1978—. Contbr. articles to profl. jours. Trustee U. Richmond, 1980-84; elder River Road Presbyn. Ch., Richmond, 1986—. Mellon Found. fellow, 1982, 85, 89. Mem. AAAS, Va.Acad. Sci. (treas. 1989-90, sec. 1990-91, pres.-elect. 1993-94), Nat. Sci. Tchrs. Assn., Westhampton Coll. Alumnae Assn. (pres. 1972), Sigma Xi, Phi Beta Kappa, Omicron Delta Kappa, Beta Beta Beta. Home: 1515 Helmsdale Dr Richmond VA 23233 Office: Randolph-Macon Coll Ashland VA 23005

FAN, CHANGXIN, electrical engineering educator; b. Beijing, Sept. 12, 1931; s. Jiqing and Lizhen (Chen) F.; m. Xinru Lu, Jan. 19, 1957; 1 child, Hongmin. Grad., Beijing U., 1952. Teaching asst. Xidian U., Xi'an, People's Republic China, 1952-62, lectr. elec. engring., 1962-78, assoc. prof., 1978-82, prof., 1982—, dir. Info. Sci. Inst., 1990—. Author: Principles of Communications, 1984 (nat. award 1988), Engineering Matrix Methods, 1988, Introduction to Digital Communications, 1977; contbr. articles to profl. jours. Bd. dirs. Shaanxi br. China Internat. Culture Exch., Xi'an, 1985. Recipient Excellent Prof. award Xidian U., 1985. Fellow Chinese Inst. Electronics; mem. IEEE (sr.), China Inst. Communications (bd. dirs. 1980—). Avocation: travel. Office: Xidian U Dept Info Engring, 2 Taibai Nan Rd, Xi'an Shaanxi 710071, China

FAN, J.D. (JIANGDI), physics educator; b. Sichuan, China, Aug. 19, 1941; came to U.S., 1983; s. Quanpu and Zhong Yu (Han) F.; m. Kaiyuan Yu, Oct. 1, 1970; 1 child, Wei. MS in Physics, U. Houston, 1986, PhD in Physics, 1989. Rsch. assoc. 21st Inst., Beijing, China, 1963-70; engr. Post & Telecomm. Factory, Chongqing, 1970-78; instr. applied physics Chongqing U., 1978-83; asst. prof. physics Brown U., Providence, 1983-84; rsch. asst. U. Houston, 1984-89; asst. prof. So. U., Baton Rouge, 1989—. Contbr. articles to profl. jours. Vice-chmn. Friendship Assn. of Chinese Students and Scholars U. Houston, 1985-87; pres. Chinese Culture Soc., Houston, 1988-89. Faculty scholar U. Houston, 1984; grantee USAF Office Sci. Rsch., 1991—, Ctr. for Energy & Environ. Studies, 1992. Mem. Am. Phys. Soc., Acad. La. Sci. Achievements include research in molecular dyanmics simulation study of liquid Rb in graphite: dynamical structure factor S(q,w); MD simulation of 2D Rb liquid and solid phases in graphite; structure and modeling 2D alkali liquids in graphite; pair interaction potential in a layered two-dimensional system; superconductivity and phase transition in a layered two dimensional correlated Fermi-liquid; insulator, metal and superconductor: phase transitions and mechanism; electron-phoron interaction and quasiparticle damping in a layered two-dimensional metal. Home: 2223 Oakdale Dr Baton Rouge LA 70810 Office: So U Dept Physics PO Box 10554 Baton Rouge LA 70813

FAN, JIAXIANG, economics educator, researcher; b. Hefei, Anhui, Republic China, May 6, 1924; s. Siyu and Kunduo (Hu) F.; m. Mengwan Dong, Apr. 4, 1953; children: Yuzhao, Yuxin, Yuhui. Degree in econs., Peking U., Beijing, 1950. Asst. dept. econs. Peking U., Beijing, 1950-54, lectr. dept. econs., 1954-79, assoc. prof. dept. econs., 1979-85, prof. dept. econs., 1985—; vis. scholar U. Cambridge, Eng., 1987-88. Author: Economic Growth, 1982, The Theory of International Trade, 1985 (award 1987), The Contemporary Western Economic Doctrines, Vols. I and II, 1989 (award 1991), The Western Economics, Vols. I, II and III, 1990; exec. editor Handbook of Chinese Banking Practices, 1991, (translation into Chinese) The New Palgrave: A Dictionary of Economies Vols. I, II, III, and IV, 1993; editor-in-chief An English-Chinese Dictionary of International Finance, 1993; vice editor-in-chief Encyclopedia of the Chinese Economy, 1991; contbr. articles to profl. jours.; exec. editor Econ. Sci., Bimonthly, 1984—. Hon. fellow U. Wis.-Madison, 1991. Mem. Fgn. Econ. Doctrine Soc. China (dir. 1978—), Fgn. Econ. Doctrine Soc. Beijing (dep. chmn. 1979—). Home: Peking U, 35 Yen Dong Yuan, Beijing 100080, China Office: Peking U Coll Econs, Haidian, Beijing 100871, China

FAN, TIAN-YOU, applied mathematician, educator; b. Nanjing, Jiang Su, China, July 1, 1939; s. Chun-Seng and Yue-Hua (Ma) F.; m. Zi-Tong Li, feb. 22, 1973; children: Jing, Le. Grad. diploma, Peking U., Beijing, Peoples Republic China, 1963. Asst. lectr. Beijing Inst. Tech., 1963-79, assoc. prof., 1980-85, prof., 1986—; rsch. fellow Alexander von Humboldt Found. U. Kaiserslautern, Fed. Republic Germany, 1981-83, 1986.; vis. prof. U. Waterloo, Can., 1987, U. Tokyo, 1990-91; reviewer Math. Reviews Am. Math. Soc., 1989—. Author: Foundation of Fracture Mechanics, 1978, Introduction to the Theory and Applications of Fracture Dynamics, 1990 (recipient 2d class prize China Book 1991, outstanding monograph prize China Edn. Commn. 1992), Foundation of Applied Fracture Dynamics, 1992; contbr. articles to profl. jours.; translator: Fracture Mechanics, 1985, Contact Problems in the Classical Theory of Elasticity, 1991. Recipient Second Grade Prize of China book prize. Mem. Am. Math. Soc. Home: 7 Baishiqiao Rd, Beijing 100081, China Office: Inst Mechs, Univ Kaiserslautern, D-67663 Kaiserslautern Germany

FAN, XIYUN, polymer engineer; b. Beijing; came to U.S., 1986; m. Zaigong Wang; 1 child, Jiuzhu Wang. MS, Beijing Inst. Chem. Tech., 1981; PhD, U. Akron, 1990. Engr. Beijing Plastic Product Inc., 1976-77; rsch. engr. Beijing Inst. Plastics, 1977-78; lectr. Beijing Inst. Chem. Tech., 1981-86; rsch. engr. W.R. Grace Washington Rsch. Ctr., Columbia, Md., 1990—. Contbr. articles to Jour. Rheology, Jour. Material Sci. Mem. Soc. Plastic Engring., Soc. Rheology, Soc. Internat. Polymer Processing. Achievements include patents for disclosures on formulations of honeycomb substrate for emmision control; research in particle filled polymer melt rheology and processing, constitutive equation of particle filled polymer melt. Office: WR Grace Washington Rsch 7379 Rt 32 Columbia MD 21044

FANG, CHUNCHANG, physical chemist, chemical engineer; b. Taiwan, Republic of China, July 9, 1955; came to U.S., 1980; s. Chuan Jyue and Chuen Huei (Lin) F.; m. Inghwa Suen, Jan. 24, 1986; children: Phyllis, Miranda. BS in Chemistry, Tung Hai U., Taichung, Taiwan, 1978; MS in Petroleum Engring., U. Houston, 1982; MS in Chem. Engring., Ohio State U., 1985, PhD in Phys. Chemistry, 1989. Rsch. asst. environ. engring. program U. Houston, 1982-83; rsch. assoc. dpet. chem. engring. Ohio State U., Columbus, 1984-85, rsch. assoc. dept. chemistry, 1986-89; process rsch. chemist QO Chems., Inc., Belle Glade, Fla., 1989-91; process devel. engr. QO Chems., Inc., Belle Glade, also Memphis, 1991—. 2d lt. Taiwan Army, 1978-80. Mem. Am. Chem. Soc., Am. Inst. Chem. Engrs. Achievements include modification and optimization of furfural process, yield increased by 25%; discovery, isolation and identification of reaction intermediates of xylose to furfural; removal of sulfur compounds in furfural, saving millions of dollars on furan catalyst cost annually. Home: 20609 Drexel Dr Walnut CA 91789 Office: QO Chems Inc 3324 Chelsea Ave Memphis TN 38108

FANG, JOSEPH PE YONG, chemistry educator; b. Lie-yong, Jiangsu, China, June 30, 1911; came to U.S., 1941; s. Foo-tze Fang and Ling-tseng Huong; m. Yu Hou Liu, Mar. 9, 1940; children: Helen C., Elizabeth-Linda, Josephine-Ann, Catherine, Mou-yi. PhD, Poly. of Milan, 1941. Registered profl. engr. Rsch. engr. Bell Tel. Labs., Summit, N.J., 1942-47; prof. Nantung (China) Coll., 1947-50; prof., dept. head China Textile U., Shanghai, 1951-87; prof., sr. advisor Super Material Rsch. Inst., Rockville, Md., 1987—; founder faculty environ. sci. and tech. China Textile U.; cons. editor Chinese Etymolog. Dictionary, Comprehensive Ency. Author: (Italian) Chemistry of Rayon, 1941, (Chinese) Organic Nomenclature Handbook, 1954, Man-made Fibers Technology, 1959, Isotactic Polypropylene Fibers, 1960; contbr. articles to profl. jours. Recipient Letter of Appreciation and Insignia for Disting. Svc. from Pres. Roosevelt, Bronze medal Gen. Alumni Assn. Poly. Milan, 1991, cert. of honor for disting. svc. China Textile U. Mem. AAAS, Am. Chem. Soc. (sr. mem.), Am. Alumni Assn. China Textile U. (pres. 1992—), Shanghai Environ. Sci. Soc. (co-founder, past editor, spl. adviser jour.). Achievements include patents for high-yield synthesis of tungsten carbonyl, efficient processes for treating kevlar wastewater and treating wastewater containing high PVA; research in knitting factory wastewater, printing and dyeing wastewater.

FANG, PEN JENG, engineering executive and consultant; b. Tainan, Taiwan, July 13, 1931; came to U.S., 1958; s. Den Chuang and Wu Tien (Su) F.; m. Elizabeth Meiling Yang, Aug. 4, 1962; children: Kenneth, Terry, Shona. BS, Nat. Taiwan U., 1955; MS, Okla. State U., 1960; PhD, Cornell U., 1966. Registered profl. engr., R.I.; registered structural engr., Ill. Sr. engr. Inar D. Hillman & Assocs., Chgo., 1960-63; sr. rsch. engr. Applied Rsch. Lab., U.S. Steel Corp., Monroeville, Pa., 1965-68; asst. prof. engring. Concordia U. (formerly Sir George Williams U.), Montreal, Que., Can., 1968-70; asst. to assoc. prof. U. R.I., Kingston, 1970-81; mgr. engring. analysis ITT-Grinnell Corp., Providence, 1981-84; pres. EngiTek, Inc. and CadMetrix, Inc., Cranston, R.I., 1985—; head div. engring. and tech. Roger Williams Coll., Bristol, R.I., 1987-90; engr. Town of West Warwick, R.I., 1989-92; exec. cons. Promon Engring. Co., Rio de Janeiro, 1975-78, Natron Engring. Co., Rio de Janeiro, 1978-81. Contbr. over 20 tech. articles to profl. jours. Vice chmn. bd. dirs. R.I. Assn. Chinese-Ams., Providence, 1978-86; commr. State Fire Safety Bd. Appeal and Rev., Providence, 1988—; mem. R.I. Heritage Commn., Providence, 1988—; mem. Zoning Bd. Rev., North Kingstown, R.I., 1991—. Mem. ASCE (structural div., Collingwood prize 1967), Assn. Energy Engrs., Nat. Fire Protection Assn., Am. Concrete Inst., Am. Water Works Assn., Rotary. Avocations: bridge, reading, swimming. Home: 95 Sedgefield Rd North Kingstown RI 02852-3838 Office: EngiTek Inc/CadMetrix Inc 1370 Plainfield St Cranston RI 02920-2549

FANG, SHU-CHERNG, industrial engineering and operations research educator; b. Nantou, Republic of China, June 14, 1952; came to U.S., 1976; s. Shao-Han and Lei Fang; m. Chi-Hsin Chao. BS in Math., Nat. Tsing-Hua U., Republic of China; MS in Math., Johns Hopkins U.; PhD in Indsl. Engring., Northwestern U. Asst. prof. U. Md., Balt., 1979-80; sr. staff mem. AT&T Engring. Rsch. Ctr., Princeton, N.J., 1980-85; supr. AT&T Bell Labs., Holmdel, N.J., 1985-88; dept. chief AT&T Corp. Hdqrs., Berkeley Heights, N.J., 1986-87; prof. N.C. State U., Raleigh, 1988—, dir. ops. rsch., 1990—; cons. AT&T Advanced Decision Support Systems, Berkeley Heights, 1988-90, Rsch. Triangle Inst., Raleigh, 1988-90. Author: Introduction to Fiber Optical Communications, 1986; Linear Optimization: Theory and Algorithms, 1993; contbr. more than 70 articles to profl. jours. Cray Rsch. Inc. grantee, 1990, 91, 92; Murphy fellow Northwestern U., 1977, Hopkins fellow Johns Hopkins U., 1976; recipient AT&T Tech. Achievement award, 1984. Fellow N.C. Supercomputing Ctr.; mem. Ops. Rsch. Soc. Am., Inst. Indsl. Engrs. (sr. mem.), Assn. Chairpersons Ops. Rsch. Depts. (sec.-treas. 1990-91, v.p. 1991-92, pres. 1992—), Honor Soc. Ops. Rsch. (faculty mem.). Office: NC State U Box 7913 OR Program Raleigh NC 27695-7913

FANG, YUE, statistician; b. Beijing, China, Dec. 16, 1960; came to U.S., 1989; m. Lan Zhou, Jan. 1, 1988; 1 child, Demetrius Z. Fang. BS, Qinghua U., Beijing, 1984, MS, 1987. Asst. prof. Chinese People's U., Beijing, 1987-89; teaching asst. Syracuse (N.Y.) U., 1989-91, U. Md., College Park, 1991-92; rsch. asst. MIT, Cambridge, 1992—. Contbr. articles to profl. jours. Syracuse U. fellow, 1989. Mem. Am. Statis. Assn., Soc. Indsl. and Applied Math. Office: MIT Ops Rsch Ctr Rm E40-149 Cambridge MA 02139

FANGER, BRADFORD OTTO, biochemist; b. Lafayette, Ind., May 14, 1956; s. Gene Otto and Carolynn (Tate) F. BA in Chemistry, U. Colo., 1979; PhD in Biochemistry, U. Vt., 1984. Lab. aid dept. molecular, cellular and devel. biology U. Colo., 1976-78; teaching asst. dept. biochemistry Coll. Medicine U. Vt., 1979-80, technician Given Analytical Facility, 1980-82; rsch. asst. dept. physiology U. N.C., 1982-84; guest researcher lab. chemoprevention Nat. Cancer Inst., NIH, 1984-86; rsch. fellow dept. biochemistry Coll. Medicine Vanderbilt U., 1986-87; assoc. scientist-biochemist dept. cancer biology Marion Merrell Dow Rsch. Inst., Cin., 1987—; vol. asst. prof. dept. anatomy and cell biology U. Cin. Med. Ctr., 1988—; reviewer Analytical Biochemistry, Clin. Chemistry, Regulatory Peptides; coord. sci. program Growth Factors in Cancer and Cardiovascular Disease, Jan. 1993. Author: (with others) Receptor Phosphorylation, 1989, Methods in Enzymology, 1989; contbr. articles to Biochemistry, Anal. Biochemistry, Biochemistry, FASEB, Regulatory Peptides. Mem. AAAS, Am. Chem. Soc., Am. Soc. for Biochemistry and Molecular Biology. Mem. Unitarian Ch. Achievements include patents in peptides, bombesin analogs having modified peptide bonds, and phenylalanine analogs of peptides; showed that receptors for epidermal growth factor exist in high and low affinity state and are regulated by glucocorticoids, that epidermal growth factor induces receptor dimerization; identification of the size and structure of betaglycan; characterization of the receptor for bombesin; development of receptor antagonists which are being tested for treating small cell lung cancer. Office: Marion Merrell Dow Rsch Ins 2110 E Galbraith Rd Cincinnati OH 45215-6300

FANJUL, RAFAEL JAMES, JR., engineer, mathematician; b. West Palm Beach, Fla., May 28, 1963; s. Rafael James and Blanca Estela (Linares) F. BSME, Ga. Inst. Tech., 1985, MSEE, 1986; MS in Math. Sci., U. Cen. Fla., 1992; postgrad., U. Fla., 1992—. Sr. guidance and controls engr. Martin Marietta Electronics, Info. and Missile Group, Orlando, Fla., 1986-92. NSF fellow, U. Fla., 1992. Mem. IEEE, AIAA, Hispanic engring. Soc. Roman Catholic.

FANNING, RONALD HEATH, architect, engineer; b. Evanston, Ill., Oct. 5, 1935; s. Ralph Richard and Leone Agatha (Heath) F.; m. Jenine Vivian Schnelle, Jan. 9, 1960; children: Anthony Lee, Traycee Anne. BArch, Miami U., Oxford, 1959. Registered architect in 21 states, Nat. Coun. Archtl. Registration Bd.; registered engr. in 8 states Nat. Coun. Engring. Examiners. Pres. Fanning, Howey Assoc., Inc., Celina, Ohio, 1959--; mng. ptnr. Manning Partnership, Celina 1978—, FFH Ltd. Partnership, 1986-91. Chmn. Mercer County Young Reps., Celina 1962-65. Mem. Am. Inst. Architects, Soc. of Architects, Nat. Soc. Profl. Engrs., Ohio Soc. Profl. Engrs., Ohio Soc. Architects, Soc. Mktg. Profl. Services, Elks Club. Methodist. Avocations: tennis, bowling, golf. Home: 422 Magnolia St Celina OH 45822-1254 Office: Fanning Howey Assoc Inc PO Box 71 Celina OH 45822-0071

FANNING, WILLIAM HENRY, JR., computer scientist; b. N.Y.C., Feb. 12, 1917; s. William Henry and Therese Genevieve (Moloney) F.; m. Mary Major Winter, Sept. 5, 1940; children: Hugh M. (dec.), Helen A. Smith, Mary M., Gerard, William Henry III. BS in Polit. Sci., Fordham U., 1940; postgrad., Cath. U., 1940-41, Jersey City State Coll., 1977, Pace U., 1989-91. Exch. editor N.Y. Times, 1938-40; Greek and German instr. Gonzaga High Sch., Washington, 1940-41; news editor Nat. Cath. News Svc., Washington, 1941-55; dir. Rome News Bur., Radio Free Europe, 1955-57; dir. news and info. svcs. Radio Free Europe, Munich, 1957-59; dir. Paris News Bur. Radio Free Europe, 1959-60; editor The Cath. News, N.Y.C., 1960-66; freelance writer CBS-TV, N.Y.C., 1966-68; writer Harcourt Brace Jovanovich, N.Y.C., 1967-72; v.p. promotion and advt. Diamond Prodns., Ltd., N.Y.C., 1967-69; analyst CGA Computer Assocs., Holmdel, N.J., 1969-73; sr. systems specialist Equitable Life Assurance Soc. U.S., N.Y.C., 1973-87; computer and network mgr. Mayor's Office of Midtown Enforcement, N.Y.C., 1988—; cons. Bill Fanning Productivity Systems, N.Y.C., 1966—; lectr. journalism Good Counsel Coll., White Plains, N.Y., 1967-69; head U.S. Cath. Bishops Press Rels. Office, Rome-2d Vatican; mem. pres.'s com. Employment of the Handicapped, 1947-66. Bd. dirs. Westchester Cath. Edn. Coun., N.Y.C., 1963-69; mem. Archdiocese Edn. Commn., N.Y.C., 1961-66, 45th St. Block Assn., N.Y.C., 1988—. Lt. comdr. USN, 1942-45, ATO, PTO, ETO. Mem. N.Y. Acad. Scis., Writers Guild Am., Phi Kappa Theta (hon.). Democrat. Roman Catholic. Office: Mayor's Office of Midtown Enforcement 330 W 42nd St New York NY 10036-6902

FANUELE, MICHAEL ANTHONY, electronics engineer, research engineer; b. Bronx, N.Y., Feb. 24, 1938; s. Joseph A. and R. Fanny (Rubino) F.; m. Joyce L. Cassidy, May 23, 1964; children: Gina M., Peter A. BEE, NYU, 1959; MSEE, Rutgers U., 1968. Electronics engr. U.S. Army Combat Surveillance & Target Acquisition Lab., Fort Monmouth, N.J., 1960-72, sr. electronics engr., 1972-80, project officer, 1980-81, dir. ISTA systems div., 1981-85; chief systems and signals analysis div. U.S. Army Electronic Warfare, Reconnaissance Surveillance and Target Acquisition Ctr., Fort Monmouth, N.J., 1985-88; sr. rsch. engr. Ga. Tech. Rsch. Inst., Ga. Inst. Tech., 1988—; cons. in field; chmn. dept. electromagnetic engring. U.S. Army Internal Tng. Program, Ft. Monmouth, 1968-78, advisor, 1978-88; Army chmn. Tri-Service Radar Symposium Steering Group, Ft. Monmouth, 1973-88; Army mem. Internat. Tech. Group, 1977-81, Internat. Radar panel, 1984-88. Patentee in field; contbr. articles to profl. jours. Served to 2d lt. U.S. Army, 1959-60. Mem. IEEE (sr.), Assn. Old Crows. Roman Catholic. Lodge: KC (treas. Brickton, N.J. 1968-70). Avocations: boating, photography, woodworking, collecting records. Home: 440 Colleen Ct Toms River NJ 08755-7376

FAPOHUNDA, BABATUNDE OLUSEGUN, energy specialist; b. Effon-Alaiye, Ondo, Nigeria, Aug. 19, 1952; came to U.S., 1980; s. Festus Ojo and Dorcas 'Wunmi (Osunrayi) F.; m. Abimbola Omolola Mabayoje, May 12, 1984; children: Ayobami Babasanmi, Omowunmi Oyebo. Profl. diploma elec. engring., The Poly., 1978; MSEE, Okla. State U., 1982; M in Energy Resources, U. Pitts., 1988. Registered profl. engr., U.K.; cert. energy mgr. Sr. engr. Nigerian TV Authority, Ibadan, Nigeria, 1982-84; energy specialist South Western Pa. Energy Ctr., Indiana, 1989—; planning com. mem. Affordable Comfort Conf. VI, Pitts., 1991-92; judge Pa. Coll. Energy Debates, Greenburg, Pa., 1992, 93. Mem. IEE (U.K., assoc.), Assn. Energy Engrs., Environ. Energy Mgrs. Inst., Indiana-Midday Rotary Club. Office: South West Pa Energy Ctr 650 S 13th St Indiana PA 15705-1087

FARACCA, MICHAEL PATRICK, engineer; b. Buffalo, June 4, 1958; s. Pasquale Angelo and Lucille (Clancy) F.; m. Mary Doreen DiVita, June 12, 1982; children: Lauren Margaret, Michael Robert. BS, SUNY, Buffalo, 1984. System engr. Comptek Rsch., Buffalo, 1984-86; sr. engr. Moore Rsch. Ctr., Grand Island, N.Y., 1986—; pres. Smartware Applications, Amherst, N.Y., 1990—. With USN, 1976-80. Mem. Western N.Y. Cons. Assn. Achievements include creation of business form on personal computer in Microsoft Windows environment. Office: Moore Rsch Ctr 300 Lang Blvd Grand Island NY 14072

FARACH-CARSON, MARY CYNTHIA, biochemist, educator; b. Galveston, Tex., Jan. 4, 1958. BS in Biology, U. S.C., 1978; PhD in Biochemistry, Med. Coll. Va., 1982. Postdoctoral fellow Johns Hopkins U., Balt., 1983, M.D. Anderson Cancer Ctr., Houston, 1983-86; instr. Baylor Coll. Medicine, Houston, 1986-87; asst. prof. U. Tex. Dental Br., Houston, 1988—. Reviewer various sci. jours.; contbr. sci. articles to major profl. jours. Active Vols. in Pub. Schs., Houston Ind. Sch. Dist., 1986—. Grantee NIH, 1988—, Muscular Dystrophy Assn., 1988-92. Mem. Am. Soc. Cell Biology, Assn. for Women in Sci., Endocrine Soc. Achievements include research in regulation of bone cell function and structure, vitamin D3 and bone matrix. Office: Dept Biol Chemistry Univ Tex Dental Br Houston TX 77030

FARADAY, BRUCE JOHN, scientific research company executive, physicist; b. N.Y.C., Dec. 9, 1919; s. Timothy John and Amelia (Vydra) F.; m. Beverly Anne Hartzell, Feb. 4, 1950; children: Diana, Deborah, Richard, Christopher, Thomas. AB magna cum laude, Fordham Coll., 1940, MS, 1947; PhD, Cath. U., 1963. Rsch. physicist Naval Rsch. Lab., Washington, 1948-63, head, semiconductor sect., 1963-70, cons. radiation effects, 1972-73, head radiation effects br., 1974-78, program mgr., low observables, 1980-86; vis. scientist Office of Naval Rsch., Arlington, Va., 1970-71; program adminstr., electronics Naval Material Command, Arlington, 1973; pres. Faraday Assocs., Inc., Annandale, Va., 1986—, also chmn. bd. dirs.; lectr. math. Prince Georges C.C., Largo, Md., 1960-79; lectr. physics U. Md., College Park, 1967-70; adj. prof. math. No. Va. C.C., Annandale, 1980-87; postdoctoral advisor NRC-NAS, 1967-83. Inventor battery holder satellites, solar cell mounting tech; contbr. articles to profl. jours. Staff sgt. U.S. Army, 1943-46. Recipient Superior Civilian Svc. award Dept. Navy, 1981, Improved Gov.'s Ops. award Office Pers. Mgmt., 1985; apptd. hon. mem. Internat. Biog. Ctr. Adv. Coun., Cambridge, U.K., 1992. Fellow Am. Phys. Soc.; mem. N.Y. Acad. Scis., Assn. Old Crows, Fordham U. Club, Am. Legion, Internat. Soc. for Hybrid Microelectronics, Sigma Xi (chpt. pres. 1974-76). Roman Catholic. Avocations: reading history, hiking. Office: Faraday Assocs Inc 8607 Sinon St Annandale VA 22003-4237

FARAZDEL, ABBAS, chemist; b. Tehran, Iran, Apr. 18, 1943; came to U.S. 1968; s. Abdolhossein and Moloud (Kanipour) F.; children: Farshad, Farnaz; m. Susan Jane Raymond, Aug. 2, 1991. BSc in Chemistry, U. Tehran, 1966; MSc in Physics, U. Mass., 1975, PhD in Chemistry, 1975. Prof. chemistry Nat. U. Iran, Tehran, 1979-83; acad. guest ETH, Zurich, Switzerland, 1983-84; prof. chemistry Queens U., Kingston, Ont., 1984-85, Amherst (Mass.) Coll., 1985-89; rsch. cons. IBM, Edison, N.J., 1989—; cons. in field; lectr. in field. Referee Phys. Rev., Jour. Am. Chem. Soc.; contbr. articles to profl. jours. Mem. AAAS, IEEE Computer Soc., Am. Chem. Soc., Am. Phys. Soc. Islam. Achievements include pioneer work in quantum mechanical treatment of positronic and other exotic systems; developed Farazdel-Epstein simulation procedure instrumental in observations of mobility edge in motion of charged particles in random media. Home: 45 Magellan Way Franklin Park NJ 08823 Office: IBM 399 Thornall St Edison NJ 08818

FARBER, EUGENE MARK, psoriasis research institute administrator; b. Buffalo, July 24, 1917; s. Simon and Mathilda Farber; m. Ruth Seiffert, Mar. 4, 1944; children: Nancy, Charlotte, Donald. BA, Oberlin (Ohio) Coll., 1939; MD, U. Buffalo, 1943; MS, U. Minn., 1946; DSc, Calif. Coll. Podiatric Medicine, 1973. Clin. asst., prof. dermatology Stanford (Calif.) U., 1949-50, asst. prof. pathology, 1949-50, clin. prof., dir. div. dermatology, 1950-598, prof., chmn. emeritus, 1959-86; pres. Psoriasis Rsch. Inst., Palo Alto, Calif., 1973—; cons. Pacific Med. Ctr., San Francisco, 1982-84; nat. cons. to surgeon gen. USAF, Washington, 1957-64; cons. in dermatology Calif. State Dept. of Pub. Health, Sacramento, 1963-66. Contbr. chpts. to book and articles to profl. jours. Recipient Physician's Recognition award AMA, 1982-85, 83-85, 91-94, Jose Marie Vargas award Cen. U. of Caracas, 1972, Mr. and Mrs. J.B. Taub Internat. Meml. award for Psoriasis Rsch., 1974, Disting. Svc. meda. Bd. Regents Uniformed Svcs. U., 1984, Order of Andres Bello, Banda de Honor, 1984, City of Paris medal, 1991. Mem. Am. Acad. Dermatology (bd. dirs. 1957-60, others), Am. Dermatol. Assn. (bd. dirs. 1974, hon. membership com. 1983-87), Assn. of Profs. of Dermatology (exec. bd., sec. 1977-80, pres. 1977-80, chairperson fin. com. 1980), Pacific Dermatology Assn. (bd. dirs. 1965-68, pres.-elect 1979-80, pres. 1980-81), Soc. for Investigative Dermatology (bd. dirs. 1957-62, pres. 1966-67, com. on hon. membership 1979—), Nat. Program for Dermatology, Space Dermatology Found. (v.p. 1986, pres. 1989-90), others. Home: 157 Ramoso Rd Portola Valley CA 94028 Office: Psoriasis Rsch Inst 600 Town and Country Palo Alto CA 94301-2326*

FARBER, NEAL MARK, biotechnologist, molecular biologist; b. N.Y.C., Oct. 23, 1950; s. Sol Z. and Nettie (Handelman) F.; m. Varda E. Farber, Aug. 19, 1973; children: Dani, Arielle. BSc with honors, Hebrew U., Jerusalem, 1973; MA, Columbia U., 1975, PhD, 1979; postgrad., Harvard U., 1979. Rsch. fellow Harvard U., Cambridge, Mass., 1980-82; rsch. scientist Biogen, Inc., Cambridge, 1982, project leader, 1983-85, coord. new projects, 1986-87, mgr. bus. devel., 1988, product mgr., 1989-93; dir. bus. devel. T Cell Scis., Inc., Cambridge, 1993—. Contbr. articles to profl. jours. Pres. SSDS Day Sch., Boston, 1989-91. Recipient Rsch. Svc. award NIH, 1975-78, Hammett award Columbia U., 1979; Helen Hay Whitney Found. fellow, Boston, 1980-82; Wexner Heritage Found. fellow, 1991-93. Mem. AAAS, N.Y. Acad. Scis., Sigma Xi. Office: T Cell Scis Inc 38 Sidney St Cambridge MA 02139

FARCUS, JOSEPH JAY, architect, interior designer; b. McKeesport, Pa., June 17, 1944; s. Howard E. and Fannie (Meyers) F.; m. Jeanne Cohen, Dec. 31, 1983. BArch, U. Fla., 1967. Registered architect, Fla.; cert. Nat. Coun. Archtl. Registration Bds. Designer Morris Lapidus Assocs., Miami Beach, Fla., 1967-77; prin. Joseph Farcus Architect, Miami, Fla., 1977—. Published in newspapers and mags. including Hotel and Restaurant Design, Fabrics & Architecture, Travel Weekly; patentee ship funnel design. Bd. dirs. Am. Jewish Com., Miami, 1991—. Mem. Constrn. Specifications Inst. Home and Office: 5285 Pine Tree Dr Miami FL 33140

FARGE, YVES MARIE, physicist; b. Pontoise, France, May 15, 1939; s. Francois and Jeanne (Callies) F.; m. Arlette Eliet, Feb. 15, 1965; children:

Emmanuel, Benjamin. Maitrise Physique, U. Paris, 1960, PhD in Physics, 1967. Researcher Centre Nat. de la Recherche Scientifique, U. Paris-Sud, Orsay, France, 1963-80; civil servant Ministere de la Recherche de le Technologie, Paris, 1980-84; R&D v.p. Pechiney, Paris, 1984—; rsch. nat. coun. Ministry Rsch. and Tech., Paris, 1984-86; chmn. IRDAC, EEC, Brussels, 1987—; founder, dir. French Synchotron Radiation Lab., LURE, 1971-80; chmn. European panel European Synchotron Radiation Facility, 1978-86. Author: Electronic and Vibrational Properties of Points Defect in Ionic Crystals. Chmn. Nat. Rsch. and Tech. Panel X Plan, Paris, 1989. Recipient Chevallier Legion D'Honneur, Paris, 1988, Officer Ordre Du Merite, Paris, 1990. Mem. Am. Soc. Materials, Am. Chem. Soc., Internat. Union Pure and Applied Physics (v.p.), Soc. Francaise de Physique. Home: 151 Rue L M Nordmann, 75013 Paris France Office: Pechiney, 10 Pl Des Vosges, 92048 Paris La Defense France

FARGNOLI, GREGORY E., safety engineer; b. Hartford, Conn., Sept. 2, 1966; s. Enrico and Ruth (Bail) F. BS in Environ. Engring., Norwich U., 1989. Cons. CON-TEST, Inc., East Longmeadow, Mass., 1989-92; rsch. indsl. hygiene & safety engr. United Technologies Rsch. Ctr., East Hartford, Conn., 1992—. Mem. NSPE, Am. Soc. Safety Engrs., Air & Waste Mgmt. Assn., Greater Hartford Jr.C. of C. (Cornerstone award 1992).

FARHO, JAMES HENRY, JR., mechanical engineer, consultant; b. Omaha, June 28, 1924; s. James Henry and Mary (Mena) F.; m. Dummer Ree Mitchem, Nov. 12, 1946; children: Sandra, Joann, Wayne. BSME, U. Nebr., 1965. Enlisted USN, 1942, advanced through grades to sr. aviation chief machinist, 1942-62, ret., 1962; engr. Exxon Rsch. & Engring. Co., Florham Park, N.J., 1965-66, project engr., 1966-68, sr. project engr., 1968-70, engring. group head, 1970-71, engring. sect. head, 1971-78; sr. staff advisor Exxon Rsch. & Engring. Co., Clinton, N.J., 1978-85; cons. engr. Lighthouse Point, Fla., 1985—; cons. Exxon Prodn. Rsch., Houston, 1985, Swiki Anderson & Assocs., Bryan, Tex., 1986-87, Glaxo Pharms., Research Triangle, N.C., 1988-91. Mem. VFW, Fleet Res. Assn., Am. Legion, Elks, Sigma Xi. Republican. Roman Catholic. Home and Office: 2401 NE 33d St Lighthouse Point FL 33064

FARIAS, FRED, III, optometrist; b. Alexandria, La., Feb. 9, 1957; s. Fred F. and Mamie (Avila) F. BS in Speech Communication, U. Tex., 1980; BS in Biology, So. Coll. Optometry, Memphis, 1985, OD, 1987. Cert. optometrist, Tex., La. Tech. mgr. profl. svcs. Bausch & Lomb, Inc., Rochester, N.Y., 1988; pres., owner 20/20 Vision Care, Inc., McAllen, Tex., 1988—; internat. lectr. and profl. cons. to major contact lense mfrs. Contbr. articles to profl. jours. Bd. dirs. RGV chpt. Am. Cancer Soc., McAllen, 1991—, Arthritis Found., McAllen, 1991—; active Leadership McAllen Class IX, mem. steering bd. dirs., 1993. Recipient Ptnrs. in Excellence award McAllen Ind. Sch. Dist., 1989-93. Mem. Am. Optometric Assn. (Nat. Optometric Recognition award 1991, contact lens sect., sports vision sect., low vision sect.), Tex. Optometric Assn. (specialties com. 1990-91, bd. dirs. 1993), Rio Grande Valley Optometric Soc. (v.p. 1990-91). Roman Catholic. Office: 20/20 Vision Care 1305 S 10th St Mcallen TX 78501

FARKAS, DANIEL FREDERICK, food science and technology educator; b. Boston, June 20, 1933; m. Alice Bridgetta Brady, Jan. 25, 1959; children: Brian Emerson, Douglas Frederick. BS, MIT, 1954, MS, 1955, PhD, 1960. Lic. chem. engr., Calif. Commd. U.S. Army., 1954, advanced through grades to major, ret., 1974; staff scientist Arthur D. Little, Cambridge, Mass., 1960-62; asst. prof. Cornell U. Agrl Expt. Sta., Geneva, N.Y., 1962-66; rsch. leader we. regional rsch. ctr. USDA, Albany, Calif., 1967-80; prin. Daniel F. Farkas Assocs., Corvallis, Oreg., 1976—; prof., chair dept. food sci. U. Del., Newark, 1980-87; v.p. process rsch. and devel. Campbell Soup Co., Camden, N.J., 1987-90; prof., head dept. food sci. and tech. Oreg. State U., Corvalis, 1990—. Contbr. more than 50 articles to peer-reviewed sci. and tech. jours. Fellow Inst. Food Technologists (councelor); mem. AICE, Am. Chem. Soc. (profl.), Sigma Xi. Achievements include 5 U.S. patents for centrifugal fluidized bed food drying system, application of ultra-high hydrostatic pressure to food preservation. Office: Oregon State Univ Dept Food Sci &Tech Corvallis OR 97331-6602

FARKAS, EDWARD BARRISTER, airport administrator, electrical engineer; b. Bklyn., Jan. 17, 1954; s. Willhelm and Esther (Davidovic) F.; m. Marianna Safarian, Feb. 16, 1989. AA, LaGuardia Coll., 1973; AS, CUNY, 1985; BS, Inst. Tech., 1990; PhD. (hon.), U. Calif., Modesto, 1991. Dir. campus activities CUNY, Queens, 1970-72; asst. dir. fin. aid CUNY, L.I., 1972-73; prodn. mgr. LAGZ Ltd., Cali, Colombia, 1974-85; dir. engring. North Techtronics, Queens, 1980-85; dir. project mgmt. Port Authority of N.Y.-N.J., Queens, 1985—; utility liaison Port Authority of N.Y.-N.J., N.Y.C., 1985-86; test engr. Port Authority of N.Y.-N.J., Jersey City, 1986-87; airport adminstr. Port Authority of N.Y.-N.J., Queens, 1987—; bd. dirs. Briarwood (N.Y.) Engring. Group. Editor NewsNet Jour., 1987-91; author, researcher: Power Engineering Review, Ground Faults, 1984; inventor various patents electrochem. oxidation inhibitor, thermal electrolyte, etc.; editor (newsletter) Monitor, 1980-87. Fellow Elec. Engring. Tech. Assn., Electrotech. Engring. Soc. (Engring. Rsch. award 1987); mem. IEEE (chmn. gov. act. com. 1987-90, Engring. Professionalism award 1985, Profl. leadership award 1990), AIAA, N.Y. Acad. Sci., Am. Assn. Airport Execs., Nat. Elec. Testing Assn., Am. Inst. Corrosion Control (bd. dirs. N.Y.C. chpt. 1989—). Avocations: reading, golf. Office: Port Authority of NY-NJ LaGuardia Airport Sta # H7C Queens NY 11371

FARKAS, GYÖRGY-MIKLÓS, chemist, researcher; b. Sf. Gheorghe, Covasna, Romania, June 18, 1941; s. György and Anna P.; m. Maria M. Imreh, June 1, 1970; children Balázs, György. B in Chem., Babes Bolyai Univ., Cluj, Romania, 1962; PhD in Macromol. Chem., Politech. Inst., Iasi, Romania, 1978. Chemist Petrochem. Co., Onesti, Romania, 1962-65; researcher Petrochem. Co., Onesti, 1965-74, Rsch. Inst. for Non-metallics, Cluj, 1974—. Contbr. articles to profl. jours. Mem. Am. Chem. Soc., Romanian Polymer Soc. Achievements include patents pending for ABS graft copolymer; for surface treated nonmetallic fillers for plastics and paints; for fluidifier for ceramic slips, for organoclays. Home: Iezer No 1 Ap 43, 3400 Cluj-Napoca Romania Office: Inst.de Cercetări, Miniere, T Vladimirescu 15-17, 3400 Cluj-Napoca Romania

FARKAS, JULIUS, chemist; b. Brownsville, Pa., Apr. 9, 1958; s. Julius and Marcella (Stanko) F. BA, Washington and Jefferson Coll., Washington, Pa., 1980; PhD, Pa. State U., University Park, 1985. Teaching/rsch. asst. Washington and Jefferson Coll., 1979-80; lab. technician Stauffer Chem. Co., Washington, 1979-80; teaching asst. Pa. State U., University Park, 1980-82, grad. rsch. assoc., 1981-85; advanced R & D chemist B.F. Goodrich Co., Brecksville, Ohio, 1985-86; R & D assoc. B.F. Goodrich Co., Avon Lake, Ohio, 1987—. Mem. Am. Chem. Soc., Am. Inst. Chemists, Soc. Plastics Engrs. Presbyterian. Office: BF Goodrich Co Tech Ctr Moore & Walker Rds Avon Lake OH 44012

FARKAS, PAUL STEPHEN, gastroenterologist; b. N.Y.C., 1952; s. Benjamin J. and Ellen (Tanner) F.; m. Esta Miriam Cantor, June 24, 1973; children: Melanie Sharon, Joshua David. AB magna cum laude with distinction in psychology, Brandeis U., 1972; MD, Tufts U., 1976. Diplomate Am. Bd. Internal Medicine, Am. Bd. Gastroenterology. Intern Baystate Med. Ctr., Sprinfield, Mass., 1976-77, resident in internal medicine, 1977-79; fellow in gastroenterology Albert Einstein Coll. Medicine, Bronx, N.Y., 1979-81; asst. clin. prof. medicine Tufts U., Boston, 1985—; med. advisor Med. Assist Program Springfield Tech. C.C., 1989—; adj. asst. prof. clin. pharmacology Mass. Coll. Pharmacy, Boston, 1982—. Author: Diagnostic Diagrams Gastroenterology, 1985; contbr. book chpts., articles and revs. in field. Co-dir. med. edn. Mercy Hosp., Springfield, 1990—, dir. libr., 1988—; med. adv. bd. VNA, Springfield, 1984-88; bd. dirs. B'nai Jacob Synogogue, Springfield, 1987-88, Com. for Longmeadow, Mass., 1989. Fellow ACP; mem. AMA, Am. Coll. Gastroenterology, Am. Gastroent. Assn., Am. Soc. Gastrointestinal Endoscopy, New England Soc. Gastrointestinal Endoscopy. Office: 299 Carew St Springfield MA 01104

FARKAS, THOMAS, secondary education educator; b. South Bend, Ind., May 6, 1937; s. Frank Joseph and Velma (Bodnar) F.; m. Janice Anne Prough, Nov. 28, 1959; children: Alan Paul, Karen Anne. BS, Purdue U., 1960; MA, We. Mich. U., 1964; postgrad., Notre Dame U., 1974-75; student and postgrad., Ind. U., 1955-56, 81-82. Cert. tchr. Tchr. South Bend (Ind.) Community Sch. Corp., 1960—; math. dept. head Edison Mid. Sch., South Bend, 1988-91, team leader, 1985-88; leader Jr. High Math. League, Notre Dame, South Bend, 1970-80. South Bend Community Sch. Corp. Minigrantee, 1982. Mem. Nat. Coun. Tchrs. Math., Mich. Area Tchrs. Math., Ind. State Tchrs. Assn., South Bend Ednl. Assn., Rotary (pres., sec., treas. Mishawaka, Ind. 1960—, Paul Harris fellow 1990, Svc. Above Self 1984). Avocations: computers, playing cards, making cotton candy. Home: 1246 Catherwood Dr South Bend IN 46614-2752

FARLEY, MARTIN BIRTELL, geologist; b. Boston, Dec. 29, 1958; s. Belmont G. and Elizabeth (Billhime) F. BS in Geoscis., Pa. State U., 1980; MA in Geology, Ind. U., 1982; PhD in Geology, Pa. State U., 1987. Postdoctoral fellow Smithsonian Instn., Washington, 1988-90; instr. Mary Washington Coll., Fredricksburg, Va., 1989; from rsch. geologist to sr. rsch. geologist Exxon Prodn. Rsch. Co., Houston, 1990—. Contbr. articles on palynology, paleoecology and sedimentology to peer-reviewed jours. Grad. fellow NSF, 1981-84. Mem. AAAS, Am. Assn. Stratigraphic Palynologists (short course com. chmn. 1991—, dir.-at-large 1992-94), Geol. Soc. Am. Office: Exxon Prodn Rsch Co PO Box 2189 Houston TX 77252

FARLEY, ROSEMARY CARROLL, mathematics and computer science educator; b. N.Y.C., Apr. 7, 1952; d. Joseph William and Nancy (Flaherty) C.; m. Dennis Michael Farley, Oct. 10, 1976; children: Christopher, Mary Ann, Brian, Nancy. BS, Coll. Mt. St. Vincent, 1974; MS, NYU, 1976, PhD, 1991. Instr. Fordham U., Bronx, N.Y., 1976-79; instr. Manhattan Coll., Bronx, N.Y., 1979-82, assoc. prof., 1989—; asst. prof. Coll. Mt. St. Vincent, Bronx, N.Y., 1982-88. Mem. Math. Assn. Am., Am. Math. Soc., Am. Statis. Assn. Democrat. Roman Catholic. Home: 120 Bennett Ave Yonkers NY 10701-6310 Office: Manhattan Coll Manhattan Coll Pky Bronx NY 10471-3913

FARLEY, WAYNE CURTIS, mechanical engineer; b. Beckley, W.Va., Mar. 4, 1960; s. Ivan and Rosetta Ann (Lowery) F. BSME, W.Va. Inst. Tech., 1982. Mech. engr. U.S. Army Rsch. and Devel. Ctr., Dover, N.J., 1984-85; tech. mgr. Hercules Aerospace Co., Rocket Center, W.Va., 1985-87; regional rep. Hercules Aerospace Co., Huntsville, Ala., 1987; resident mgr. Hercules Aerospace Co., Mt. Arlington, N.J., 1988-90; sr. regional rep. Hercules Aerospace Co., Huntsville, 1990—; chmn. AUSA-Armtment Systems, Dover, 1990. Mem. Assn. U.S. Army (chmn. armtment systems 1990), Am. Def. Preparedness Assn., Assn. Unmanned Vehicle Systems, Sigma Phi Epsilon. Republican. Achievements include management of the tech. demonstration of an advanced rocket assisted kinetic energy antitank round of ammunition. Office: Hercules Inc Ste 303 600 Blvd South Huntsville AL 35802

FARMAR, ROBERT MELVILLE, medical scientist, educator; b. Bonthe Sherbro, Sierra Leone, Nov. 18, 1942. BA in Biology and Chemistry, Ind. Cen. U., 1961; MSc in Medicine, Temple U., 1973. Rsch. assoc. Temple U., Phila., 1973-78, 83-85; biochem. rsch. cons. Pa. State Govt., Harrisburg, 1978-79; rsch. assoc. Immunopathology Wills Eye Hosp., Phila., 1982-85; scientist Rorer Cen. Rsch., King of Prussia, Pa., 1987-89; drug info. specialist Rhone-Poulenc Rorer Pharm., Collegeville, Pa., 1989-93; adj. prof. Biol. Scis. Ursinus Coll., Collegeville, 1992—. Contbr. articles to profl. jours. Fellow USPHS 1962-63, Coll. Am. Pathologists 1966. Mem. Am. Med. Writers Assn., Drug Info. Assn., N.Y. Acad. Scis., Sigma Zeta. Methodist. Achievements include research in molecular bases of disease: biochemical pathology; epidemiology of disease as determinants of health policy formulation in developing countries. Home: 237 Riverview Rd King of Prussia PA 19406 Office: Rhone-Poulenc Rorer Pharm 500 Arcola Rd Collegeville PA 19406

FARMER, CROFTON BERNARD, atmospheric physicist; b. Cardiff, Wales, May 30, 1931; came to U.S., 1967; s. Francis Herbert and Cicely (Arnott) F.; m. Roberta Josephine Stewart, June 20, 1956; (div); children: Louise Josephine, Joanna Cicely, Philippa Bernice, Christopher Llewellyn; m. Christine Louise Conaway, Feb. 29, 1992. B.S., U. London, 1952, Ph.D., 1968. Research physicist EMI Electronics, Ltd., Eng., 1952-60; head infrared research dept. EMI Electronics, Ltd., 1960-62; led sci. expdns. to Bolivian Andes, 1962, 64; sr. research scientist Jet Propulsion Lab., Calif. Inst. Tech., Pasadena, 1967-72; mgr. planetary atmospheres Jet Propulsion Lab., Calif. Inst. Tech., 1972-75; prin. investigator NASA Viking Mars, 1975-77, Shuttle Spacelab, from 1977; vis. prof. Calif. Inst. Tech. divsn. Geol. and Planetary Sci., 1978-81; v.p. San Juan Capistrano Rsch. Inst.; mem. subcoms. on planetary atmospheres and stratospheric research NASA; cons., lectr. remote sensing of atmospheres. Contbr. articles on solar-terrestrial spectroscopy and composition of planets' atmospheres to sci. jours. Recipient Exceptional Sci. Achievement medal NASA, 1975, 77, 87, Antartica Svc. medal, 1987; named Disting. Vis. Sci. Jet Prop. Lab, 1989—. Home: 2525 Hollister Ter Glendale CA 91206-3039

FARMER, JIM L(EE), civil engineer, consultant; b. Edinburg, Tex., Jan. 9, 1935; s. James Nathan and Johnny Lois (Smith) F.; m. Zeta Lu Jones, June 11, 1954; children: David Leigh, Susan Lynet. BSCE, U. Tex., 1965; A in Mid-mgmt., Eastfield Coll., 1977. Registered profl. engr., Tex.; registered profl. surveyor, Tex. Foreman family farm Rio Grande Valley, Tex., 1954-56; rt. salesman Rainbo Baking Co., Harlingen, Tex., 1956-57; hwy. draftsman Tex. Hwy. Dept., Pharr, 1957-59; jr. engr. J.G. Threadgill and Assocs., Dallas, 1965; staff engr. Zetterlund Boynton, 1965-67; co-owner Bledsoe Farmer & Roderick, 1967-70; supervising engr. Army and Air Force Exch. Svc., Dallas, 1970—. With U.S. Army N.G., 1954-62. Mem. ASCE, ASTM. Nazarene. Home: 618 Athenia Way Duncanville TX 75137 Office: Army and Air Force Exchange Svc Attn CF E/S 2727 LBJ Fwy Dallas TX 75234

FARMER, JOE SAM, petroleum company executive; b. Hot Springs, Ark., Mar. 2, 1931; s. Walter L. and T. Naomi F.; m. Elizabeth Jean Keener, Dec. 27, 1952; children: J. Christopher, David E., Kathryn L. Student, Ohio State U., 1950-51; B.Sc., Tex. A & M U., 1955. Cert. petroleum geologist. Geologist Lion Oil Co., Shreveport, La., 1955; geologist, then asst. chief geologist Placid Oil Co., New Orleans, Shreveport and Dallas, 1958-68; exploration mgr. N.Am. div. Union Carbide Petroleum Corp., Houston, 1968-71; v.p. domestic exploration and prodn. Ashland Exploration Inc., Houston, 1971-73, exec. v.p., 1973-77; adminstrv. v.p. Ashland Oil Inc., Ashland, Ky., 1977-79; v.p. Mesa Petroleum Co., Houston, 1979-80; pres. chief operating officer Union Tex. Petroleum Corp., Houston, 1980-83; pres. JSF Interests, Inc., Houston, 1983—; bd. dirs. De Kalb Energy Co. Served with USAF, 1955-57. Mem. Am. Assn. Petroleum Geologists, Assn. Profl. Geol. Scientists, Am. Petroleum Inst., Ind. Petroleum Assn. Am. (bd. dirs.), Mid-Continent Oil and Gas Assn., Houston Geol. Soc. Clubs: April Sound Country (Conroe, Tex.); Petroleum. Office: JSF Interests Inc 1201 Louisiana Ste 3320 Suite 1400 Houston TX 77002

FARNSWORTH, MICHAEL EDWARD, mechanical engineer; b. Springfield, Mass., Dec. 8, 1963; s. Edward Ellis and Dorothy May (Hamel) F. BSME, U. Maine, 1986; MSME, Western New England Coll., 1992. Applications engr. Terry Steam Turbine, Windsor, Conn., 1986-87; product engr. Conval Inc., Somers, Conn., 1987-88, product engring. mgr., 1988-89, engring. mgr., 1989-92; engring. mgr. Neles-Jamesbury, Inc., Glens Falls, N.Y., 1992—. Mem. ASME, Am. Soc. Testing and Materials, Am. Materials Soc. Republican. Roman Catholic.

FAROKHI, SAEED, aerospace engineering educator, consultant; b. Tehran, Feb. 26, 1953; came to U.S., 1971; s. Nasrollah and Shamselmolouk (Yamani) F.; m. Mariam A. K. Esfahani, Feb. 15, 1971; children: Kamelia, Parisa, Farima. BS, U. Ill., 1975; MS, MIT, 1976, PhD, 1981. Asst. prof. U. of Sci. & Tech., Tehran, 1979-80; design and devel. engr. Brown, Boveri & Co., Baden, Switzerland, 1981-84; asst. prof. U. Kans., Lawrence, 1984-87, assoc. prof., 1987-92, dir. Flight Rsch. Lab., 1990—, acting chmn. aerospace engring. dept., 1991, prof. aerospace engring. dept., 1992—; cons. and proprietor Aerotech Engring. and Rsch., Lawrence, 1988—, GE-Aircraft Engines, Cin., 1989, Marion Merell-Dow, Kansas City, 1991. Contbr. articles to AIAA Jour. of Propulsion and Power, AIAA Jour., AIAA Jour. of Aircraft, Internat. Jour. of Turbo and Jet Engines, ASME Jour. of Fluids Engring. Recipient Innovative Rsch. award USAF, USN and Def. Advanced Rsch. Project Agy.; prin. investigator on rsch. grants NASA-Lewis Rsch. Ctr., NASA-Langley Rsch. Ctr., GE-Aircraft Engines, Kans. Tech. Enterprise Corp. Mem. AIAA (faculty project adv. 1988, 89, 90, 91, 92), ASME (tech. com. 1991, 92), Soc. Automotive Engrs., Am. Acad. Mechanics, Am. Soc. Engring. Edn. Office: Flight Rsch Lab Kansas Univ 2291 Irving Hill Dr Lawrence KS 66045

FARR, LEE EDWARD, physician; b. Albuquerque, Oct. 13, 1907; s. Edward and Mabel (Heyn) F.; m. Anne Ritter, Dec. 28, 1936 (dec.); children: Charles E., Susan A., Frances A.; m. Miriam Kirk, Jan. 22, 1985. BS, Yale U., 1929, MD, 1933. Asst. pediatrics Sch. Medicine, Yale U., 1933-34; asst. medicine Hosp. of Rockefeller Inst. Med. Research, 1934-37, assoc. medicine, 1937-40; dir. research Alfred I. duPont Inst. of Nemours Found., Wilmington, Del., 1940-49; vis. assoc. prof. pediatrics Sch. Medicine, U. Pa., 1940-49; med. dir. Brookhaven Nat. Lab., 1948-62; prof. nuclear medicine U. Tex. Postgrad. Med. Sch., 1962-64, prof. nuclear and environ. medicine Grad. Sch. Bio-Med. Scis., U. Tex. at Houston, 1965-68; chief sect. nuclear medicine U. Tex.-M.D. Anderson Hosp. and Tumor Inst., 1962-67, prof. environ. health U. Tex. Sch. Pub. Health, Houston, 1967-68; head disaster health services Calif. Dept. Health, 1968, chief emergency health services unit, 1968-70, 1st chief bur. emergency med. services, 1970-73; Lippitt lectr. Marquette U., 1941; Sommers Meml. lectr. U. Oreg. Sch. Med., Portland, 1960; Gordon Wilson lectr. Am. Clin. and Climatol. Assn., 1956; Sigma Xi nat. lectr., 1952-53; guest scientist Institut fur Medizinder Kernforschungsanlage, Julich, Germany, 1966; Brookhaven Nat. Lab. lectr., 1990. Mem. NRC adv. com. Naval Med. Res., 1953-68; chmn. NRC adv. com. Atomic Bomb Casualty Commn., 1953-68; mem. adv. com. Naval Res. to Sec. of Navy and CNO, 1969-71; NRC adv. com. on medicine and surgery, 1965-68, exec. com., 1962-65; Naval Research Mission to Formosa, 1953; tech. adviser U.S. delegation to Geneva Internat. Conf. for Peaceful Uses Atomic Energy, 1955; mem. N.Y. Adv. Com. Atomic Energy, 1956-59; mem. AMA Com. Nuclear Medicine, 1963-66; mem. com. med. isotopes NASA Manned Spacecraft Ctr., 1966-68; mem. expert adv. panel radiation WHO, 1957-79; mem. Calif. Gov.'s Ad Hoc Com. Emergency Health Service, 1968-69; mem. sci. adv. bd. Gorgas Meml. Inst., 1967-72; numerous other sci. adv. bds., panels; cons. TRW Systems, Inc., 1966-70, Consol. Petroleum Co., Beverly Hills, Calif., 1946-70. Mem. alumni bd. Yale, 1962-65, mem. alumni fund, 1966-76. With USNR, 1942-46; capt. (M.C.) USNR, ret. Recipient Mead Johnson award for pediatric research, 1940, Gold Cross Order of Phoenix, Greece, 1960, Verdienstkreuz 1st class Fed. Republic Germany, 1963, guest scientist Institut für Medizin der Kerforschungsanlage, Julich, Germany, 1966; named Community Leader in Am., 1969, Disting. Alumni Yale U. Med. Sch., 1989. Diplomate Nat. Bd. Med. Examiners, Am. Bd. Pediatrics. Fellow AAAS, Royal Soc. Arts, Am. Acad. Pediatrics, N.Y. Acad. Scis., Royal Soc. Health, Am. Coll. Nuclear Medicine (disting. fellow); mem. Soc. Pediatric Research, Soc. Exptl. Biology, Harvey Soc., Am. Pediatric Soc., Soc. Exptl. Pathology, Am. Soc. Clin. Investigation, Radiation Research Soc., AMA (mem. council on sci. assembly 1960-70, chmn. 1968-70), Med. Soc. Athens (hon. mem.), Alameda County Med. Assn., Sigma Xi, Alpha Omega Alpha, Phi Sigma Kappa, Nu Sigma Nu, Alpha Chi Sigma. Club: Commonwealth (San Francisco). Author articles on nuclear medicine, protein metabolism, emergency med. services, radioactive and chem. environ. contaminants, environ. noise. Home: 2502 Saklan Indian Dr Apt 2 Walnut Creek CA 94595-3001

FARRAR, RICHARD BARTLETT, JR., secondary education educator, wildlife biology consultant; b. Penn Yan, N.Y., Apr. 25, 1939; s. Richard B. and Margaret M. (Stevenson) F.; m. Gayle G. Green, Aug. 23, 1963; children: Michelle, Marc. BS, Houghton Coll., 1960; MEd, Frostburg (Md.) State U., 1990. Cert. wildlife biologist. Sci. tchr. Hinckley (Maine) Sch., 1960-61, Concord (Mass.) High Sch., 1962-64; program dir. Mass. Audubon Soc., Lincoln, 1964-65; instr. U. Ill., Chgo., 1966-68; chair sci. dept. Woodstock County Sch., 1968-73; exec. dir. Vt. Inst. Natural Sci., Woodstock, 1971-73, N.J. Audubon Soc., Franklin Lakes, 1974-78; field exec. Nat. Wildlife Fedn., Washington, 1979-81; wildlife biology cons. Washington, 1982-86; lead sci. tchr. Garrett County Bd. Edn., Oakland, Md., 1987—; rsch. advisor Coastal Facilities Rev. Act, State of N.J., Trenton, 1977-78; mem. State of N.J. Natural Resources Coun., 1978-79; advisor Savage River State Forest Coun., 1991-92. Author: Birds of East-Central Vermont, 1971, The Hungry Snowbird, 1975, The Birds' Woodland, 1976; editor Vt. Natural History mag., 1970-73, N.J. Audubon mag., 1974-78; contbr. articles to popular and sci. publs. Treas. League for Conservation Legis., N.J., 1978; dir. Mid.-Atlantic Naturalist Soc., Md., 1981-82. Recipient Outstanding Biology Tchr. award Nat. Assn. Biology Tchrs., 1971, Conservation award Connecticut River Watershed Coun., 1971, Children's Sci. Book award Children's Libr. Coun., 1975, NSTA, 1976. Mem. Rotary (treas. Friendsville, Md. 1988). Home: RR 1 Box 219 Friendsville MD 21531-9772 Office: Garrett County Bd Edn Oakland MD 21550

FARRAR, THOMAS C., chemist, educator; b. Independence, Kans., Jan. 14, 1933; s. Otis C. and Agnes K. F.; m. Friedemarie L. Farrar, June 22, 1963; children: Michael, Christian, Gisela. BS in Math., Chemistry, Wichita State U., 1954; PhD in Chemistry, U. Ill., 1959. NSF fellow Cambridge U., Eng., 1959-61; prof. chemistry U. Oregon, Eugene, 1961-63; chief, magnetism sect. Nat. Bur. Standards, Washington, 1963-71; dir. R & D Japan Electron Optics Lab., Cranford, N.J., 1971-75; dir. instr. NSF, Washington, 1975-79; prof. chemistry U. Wis., Madison, 1979—; chmn. adv. com. MIT Nat. Magnetics Lab., Cambridge, Mass., 1979-84; mem. adv. com. NIH High Field Nuclear Magnetic Resonance Lab., Pitts., 1985—. Author: Introduction to Pulse NMR Spectros, 1989, Density Matrix Theory, 1991; contbr. over 100 articles to profl. jours. Recipient Silver medal Dept. Commerce, Washington, 1971, Silver medal Nat. Science Found., Washington, 1979. Fellow Wash. Acad. Science; mem. Am. Chem. Soc. (sec.-treas. Wis. sect. 1986-89), Am. Physical Soc. Office: Univ Wis Dept Chemistry 1101 University Ave Madison WI 53706 1396

FARRELL, EDWARD JOSEPH, mathematics educator; b. San Francisco, Mar. 28, 1917; s. Christopher Patrick and Ethel Ann (Chesterman) F.; m. Pearl Philomena Rongone, Aug. 21, 1954; children: Paul, Paula. B.Sc., U. San Francisco, 1939; M.A., Stanford U., 1942. Mem. faculty U. San Francisco, 1941—, prof. math., 1968-82, prof. emeritus, lectr., 1982—; Guest lectr. regional and nat. meetings Nat. Council Tchrs. Math., 1966, 67, 69; cons. math. text pubs. Mem. adv. panels NSF, 1966—, dir. summer and in-service insts., 1960-75, dir. confs. geometry, 1967, 68, 70-75; mem. rev. panel Sci. Books. Author math. reports; editor studies teaching contemporary geometry. Served with AUS, 1944-46. NSF faculty fellow, 1956-57. Mem. AAAS, Am. Assn. Physics Tchrs., Nat. Council Tchrs. Math., Sch. Sci. and Math. Assn. Republican. Roman Catholic. Home: 2526 Gough St San Francisco CA 94123-5013

FARRELL, GREGORY ALAN, biomedical engineer; b. Bklyn., May 12, 1942; s. Edmond William and Edna Florence (Williams) F.; BS in Mech. Engring., Cooper Union, 1964; MS in Biomed. Enggring., Columbia U., 1972, postgrad., 1972—; m. Mary Louise Lupiani, Sept. 3, 1966; children: Juliana Eden, Cristina Elizabeth. Mech. engr. Gen. Dynamics, San Diego, 1964-65, Rochester, N.Y., 1965-67; research asst. Columbia U. Med. Sch., N.Y.C., 1968-69; instr. pathology N.Y. Med. Coll., 1969-72; research engr. Technicon Instruments Corp., Tarrytown, N.Y., 1972-82; mgr. mech. engring. Baker Instruments Corp., Allentown, Pa., 1982-84, prin. mech. engr., 1984-86; prin. engr. Nat. Patent Devel. Corp., N.Y.C., 1986-87; project engr. Miles Diagnostics (formerly Technicon Instruments), Tarrytown, N.Y., 1987-90, project mgr. 1990—. Patentee in field; achievements include product devel. of several automated clin. hematology and other instruments; contbr. articles to profl. jours. Democrat. Roman Catholic. Home: 447 Hillcrest Rd Ridgewood NJ 07450-1520 Office: Miles Diagnostics 511 Benedict Ave Tarrytown NY 10591-5005

FARRELL, MARK DAVID, environmental engineer; b. Buffalo, N.Y., Mar. 5, 1955; s. John Arthur and Laura Joan (Ricci) F.; m. Susan Elizabeth Voltz, July 7, 1979; 1 child, Kaitlyn Elizabeth. BS in Civil Engring., W.Va. U., 1977, MS in Environ. Engring., 1978; MBA, U. Pitts., 1983. Registered profl. engr. Sr. environ. engr. Republic Corp., Pitts., 1978-85; mgr. engring. S.W. Fla. Water Mgmt. Dist., Brooksville, 1985-86, dir. resource mgmt., 1986-89, asst. exec. dir., 1989-; chmn. Sarasota Bay Nat. Estuary Program, Sarasota, Fla., 1989-90; prin. cons. U.S. Egypt Water Conservation Project, Tampa, Fla., 1992; chmn. Electric Power Rsch. Inst. Desalination com., San Diego, 1992. Contbr. articles to profl. jours. Pres. Northdale Taxing Dist., Tampa, 1992; grad. Leadership Tampa Bay, 1992, Harvard Program Senior

Execs. Recipient Profl. Devel. award Fla. Soc. Engring., 1989; Nat. Sci. Found. scholar, 1977. Mem. Am. Soc. Civil Engrs., Nat. Soc. Profl. Engrs., Am. Water Resources Assn., Internat. Desalination Assn. Home: 16603 E Course Dr Tampa FL 33624 Office: SW Fla Water Mgmt Dist 2379 Broad St Brooksville FL 34609

FARRELL, PATRICIA ANN, psychologist, educator; b. N.Y.C., Mar. 11, 1945; d. Joseph Alexander and Pauline (Loth) F.; BA, Queens Coll., 1976; MA, N.Y. U., 1978, PhD, 1990. Lic. psychologist, N.J. Assoc. editor Pubs. Weekly Mag., N.Y.C., 1968-72; editor Bestsellers Mag., N.Y.C., 1972-73; assoc. editor King Features Syndicate, N.Y.C., 1973-78; staff psychologist, intake coord. Mid-Bergen Community Mental Health Ctr., Paramus, N.J., 1978-84; instr. Bergen Community Coll., Paramus, 1978—; adj. prof. The Union Inst., Cin.; resident clin. psychology Am. Inst. for Counseling, N.J., 1990-91; cons. Family Counseling Svc. of Ridgewood, N.J., 1984; clin. psychology intern Marlboro Psychiat. Hosp., N.J., 1984-85, staff psychologist, 1985-87; rsch. analyst Mt. Sinai Sch. of Medicine, 1987-88; account exec., sr. sci. writer Manning, Selvage and Lee, N.Y.C., 1988-90; sr. clin. psychologist, mem. med. staff Greystone Park (N.J.) Psychiat. Hosp., 1990—; pvt. practice psychology, Englewood Cliffs, N.J.; cons. Intensive Weight Loss Program, Cath. Med. Ctr. Bklyn. and Queens; cons. pharm. clin. protocols; guest radio shows Sta. WWDJ, Hackensack, N.J., Maury Povich TV Show. Contbr. articles to Writer's Digest, Real World, Postgrad. Medicine, Psychotherapy, New England Jour. Medicine, and newspapers, author manual Alzheimer's Disease Assessment Scale test. Mem. Bergen County Task Force on Crimes Against Children, Bergen County Task Force on Alcoholism and Drunken Driving, 1984. McDonald's rsch. grantee; recipient Good Citizen award DAR, Sci. award Rotary Club. Mem. AAAS, APA, Assn. for Advancement Psychology, Assn. for Indep. Video and Filmmakers, Soc. Behavioral Medicine, Eastern Psychol. Assn., Biofeedback and Self-regulation Soc. N.J., N.J. Psychol. Assn., Nat. Alliance for Patient Welfare (exec. v.p.), Nat. Register Health Providers in Psychology. Avocations: fitness, racquetball, kite-flying. Office: PO Box 1283 Englewood Cliffs NJ 07632

FARRELL, WILLIAM JAMES, JR., mathematician; b. Balt., Mar. 23, 1952; s. William James and Marie Irene (Schley) F.; m. Juliana Bertoldi, Aug. 30, 1971; children: William James III, Heather Elizabeth. BS in Math. and Physics magna cum laude, Towson (Md.) State Coll., 1975; MSEE, Johns Hopkins U., 1983. Math. analyst Bendix Field Engring. Corp., Columbia, Md., 1977-80; assoc., sr. mathematician Johns Hopkins U. Applied Physics Lab., Laurel, Md., 1980—; adj. math. instr. Howard C.C., Columbia, 1992—. Mem. AIAA, Sigma Pi Sigma. Roman Catholic. Home: 5182 Stone House Village Ct Sykesville MD 21784 Office: Johns Hopkins U Applied Physics Lab Johns Hopkins Rd Laurel MD 20707

FARRER-MESCHAN, RACHEL (MRS. ISADORE MESCHAN), obstetrics/gynecology educator; b. Sydney, Australia, May 21, 1915; came to U.S., 1946, naturalized, 1950; d. John H. and Gertrude (Powell) Farrer; m. Isadore Meschan, Sept. 3, 1943; children—David Farrer, Jane Meschan Foy, Rosalind Meschan Weir, Joyce Meschan Lawrence. MB, BS, U. Melbourne (Australia), 1940; MD, Wake Forest U., 1957. Intern Royal Melbourne Hosp., 1942; resident Women's Hosp., Melbourne, 1942-43, Bowman-Gray Sch. Medicine, Wake Forest U., Winston-Salem, N.C., 1957-73, asst. clin. prof. dept. ob-gyn, 1973—; also marriage counselor. Co-author (with I. Meschan): Atlas of Radiographic Anatomy, 1951, rev., 1959; Roentgen Signs in Clinical Diagnosis, 1956; Synopsis of Roentgen Signs, 1962; Roentgen Signs in Clinical Practice, 1966; Radiographic Positioning and Related Anatomy, 1968; Analysis of Roentgen Signs in General Radiology, 1973; Roentgen Signs in Diagnostic Imaging, Vol. III, 1986, Vol. IV, 1987. Home: 305 Weatherfield Ln Kernersville NC 27284-8337

FARRIES, JOHN KEITH, petroleum engineering company executive; b. Cardston, Alta., Can., July 9, 1930; s. John Mathew and Gladys Helen (Adams); B.S. in Petroleum Engring., U. Okla., 1955; postgrad. Banff Sch. of Advanced Mgmt., 1963; m. Donna Margaret Lloyd, Dec. 30, 1960; children—Gregory, Bradley, Kent. Engr., dist. engr., joint interest supt. Pan Am. Petroleum Corp., Calgary, Edmonton, Tulsa, Drayton Valley, 1955-65; pres. Tamarack Petroleums Ltd., Calgary, Alta., 1965-70, Canadian Well Services & Tank Co. Ltd., Calgary, 1968-70, Farries Engring. Ltd., 1970—, Wave Internat. Engring., Inc., Israel, 1975-88, Westridge Petroleum Corp, 1985—, Muskeg Oilfield Services, 1987—; v.p. Bobby Burns Petroleum Ltd., 1983—; dir. Westgrowth Petroleums, 1983-88; dir. Pension Fund Energy Resources, 1990—, Trisol Inc., 1990—. Mem. AIME, Canadian Inst. Mining and Metallurgy (dir. petroleum soc. 1966-68), Assn. Profl. Engrs. of Alta., B.C. and Sask., Canadian Assn. of Drilling Engrs. (pres. 1977-78). Clubs: Calgary Petroleum, Willow Park Golf and Country. Past pub. chmn. Jour. Canadian Petroleum Tech. Home: 10819 Willowglen Pl, Calgary, AB Canada T2J 1R8

FARRINGTON, JOSEPH KIRBY, microbiologist; b. Jacksonville, Fla., Oct. 5, 1948; s. Joseph Allison Jackson and Eleanor Francis (High) F.; m. Mary Ellen Harris, June 8, 1976. BS, LaGrange Coll., 1971; MS, Clemson U., 1973; PhD, Auburn U., 1976. Microbiologist Kellogg & Co., Atlanta, 1975-77; mgr. Plough Inc., Memphis, 1977-84; assoc. dir. quality control U.S. consumer products Schering-Plough Co., Memphis, 1984-91, dir. quality control health care products, 1991—; v.p. Plough Labs. Schering-Plough HCP, Memphis, 1984—; adj. prof. Sch. Pharmacy U. Tenn., Memphis, 1984—; prof. Memphis State U., 1982—; mem. microbiology exec. com. Cosmetic, Toiletries, Fragrance Assn., Washington,1 978—. Mem. Am. Soc. Microbiology, Soc. Cosmetic Chemists, Inst. Food Technologists. Republican. Methodist. Office: Schering Plough Health Care PO Box 377 Memphis TN 38151

FARRIOR, GILBERT MITCHELL, metallurgical engineer; b. Raleigh, N.C., Dec. 21, 1923; s. John Alexander and Dora Thorn (Mitchell) F.; m. Patricia Fowler, May 10, 1952; children: Barbara Linda Farrior Kovalsky, Patricia Ann, Marian Louise. BChE, N.C. State U., 1948, MEngr Physics, 1949; MS, U. Tenn., 1962, PhD in Metall. Engring., 1965; grad., Army War Coll., 1973. Registered profl. engr.; cert. logistician. Engr. smelting div. ALCOA, Alcoa, Tenn., 1953-55; rsch. metallurgist U.S. Bur. of Mines, Norris, Tenn., 1959-65, Battelle Meml. Inst., Columbus, Ohio, 1965-68; sr. rsch. metallurgist Kawecki-Berylco, Boyertown, Pa., 1968-81; prin. metall. engr. ALUMAX Engineered Metal Processes, St. Louis, 1981—; cons. Milward Alloys, Lockport, N.Y., 1981-86. Co-author sect. on semisolid metal forming ASM Handbook, Vol. 15, 1988; contbr. articles to profl. jours. Cpl. USMC, 1944-46; sgt. N.C. N.G., 1946-48; sgt. ORC, 1948-51; 1st lt. U.S. Army, 1951-53; capt. Tenn. N.G., 1955-64; col. USAR, 1964-78. Mem. Am. Foundry Soc., Am. Soc. for Metals (sr.), Metall. Soc. (sr.), Sigma Xi. Achievements include research and development of transition metal diborides, thermoelectrics, aluminum grain refiners, and semi-solid metal forming. Home: 513 Meadow Creek Ln Saint Louis MO 63122-1656 Office: ALUMAX Engineered Metal Processes 1277 N Warson Rd Saint Louis MO 63132-1895

FARRIS, THOMAS N., engineering educator, researcher; b. Daisetta, Tex., Sept. 29, 1959; s. Robert Quentin and Kathleen Ruth (Kelling) F.; m. Bernadette Paulson, May 9, 1982; children: Joanna K., John T., Steven Q., Andrew B. BSME, Rice U., 1982; MS in Theoretical And Applied Mechanics, Northwestern U., 1984, PhD, 1986. Asst. prof. engring. Purdue U., West Lafayette, Ind., 1986-91, assoc. prof., 1991—; reviewer in field. Contbr. articles to profl. jours. Roy scholar Rice U., 1981; Cabell fellow Northwestern U., 1982; NSF Presdl. Young Investigator, 1990; Japan Soc. for Promotion of Sci. fellow, 1991, ASME Newkirk award, 1992. Mem. ASME (rsch. com. on tribology, Burt L. Newkirk award 1992), AIAA (advisor 1988-93, assoc. editor Jour. Aircraft 1992—), Soc. Tribologists and Lubrication (sec. lubrication fund com. 1990—). Avocations: running, tennis, reading, family. Home: 2208 Miami Trail West Lafayette IN 47906-1924 Office: Purdue U Sch Aeronautics & Astronautics West Lafayette IN 47907

FARRUKH, USAMAH OMAR, electrical engineering educator, researcher; b. Beirut, Lebanon, Aug. 24, 1944; came to U.S., 1969; s. Omar Abdullah and Amenah (Helmi) F.; m. Samar M. Hussami, 1980; children: Muna, Omar, Marwa. BSEE, Am. U. Beirut, 1967; PhD in Elec. Engring., U. So. Calif., 1974. Lead analyst Wolf R&D Group, Riverdale, Md., 1975-76; sci. specialist Phoenix Corp., McLean, Va., 1976-77; sci. cons. Applied Sci. and Tech. Inc., Rosslyn, Va., 1978; staff scientist Inst. Atmospheric Optics and Remote Sensing, Hampton, Va., 1978-85; assoc. prof. dept. elec. engring. Hampton U., 1985-93, prof., 1993—; cons. U.S. Army Rsch. Office, 1990-91. Contbr. articles to profl. jours. Bd. dirs. Hypohidrotic Ectodermal Dysplasia and Related Disorders Found., Hampton, 1987—. Recipient Prin. Investigation grant NASA, 1986—. Mem. IEEE, Optical Soc. Am. Office: Hampton Univ Olin Engring Bldg Hampton VA 23601

FARST, DON DAVID, zoo director, veterinarian; b. Wadsworth, Ohio, Feb. 25, 1941; s. Walter K. and Ada (Stetler) F.; m. Jan Rae Harber, June 17, 1980; children: Julie K., Jenny Lynn, John David. D.V.M., Ohio State U., 1965. Veterinarian, mammals curator Columbus Zoo, Ohio, 1969-70; assoc. dir. Gladys Porter Zoo, Brownsville, Tex., 1970-74, dir., 1974—. Editor: Jour. Zoo Animal Medicine, 1973-77. Mem. Am. Assn. Zool. Parks and Aquariums (pres. 1979-80, chmn. ethics bd. 1990-91, Edward H. Bean Meml. award 1992), Am. Assn. Zoo Veterinarians, Internat. Union Dirs. Zool. Gardens. Home: 640 Edgewater Isle San Benito TX 78586-9209 Office: Gladys Porter Zoo 500 Ringgold St Brownsville TX 78520-7918

FARTHING, G. WILLIAM, psychology educator; b. New Orleans, July 18, 1943; s. Gene William and Nancy (Luster) F.; m. Karon R. Jack, 1963 (div. 1971); children: William, Katherine; m. Carol Hershkowitz, June 1, 1974; children: David, Michael. BA, Grinnell Coll., 1965; PhD, U. Mo., 1969. Asst. prof. U. Maine, Orono, 1969-74, assoc. prof., 1974-82, prof., 1982-93; ret. Author: The Psychology of Consciousness, 1992. Mem. Am. Psychol. Soc. Office: U Maine Dept Psychology 301 Little Hall Orono ME 04469

FASELLA, PAOLO MARIA, general science researcher, development facility director; b. Rome, Dec. 16, 1930; s. Felice and M. (Parazzoli) F.; m. Sheila Dionisi; children: Maria, Caterina, Margherita, Elisabetta. MD, U. Rome; PhD in Biol. Chemistry, Ministry Edn., Rome, PhD in Applied Biochemistry; Laurea Honoris Causa, Nat. U., Dublin, Ireland, 1990. Asst. prof., then assoc. prof. in biol. chemistry U. Rome, 1959-65, prof. biol. chemistry, 1971-81; assoc. prof., then prof. biochemistry U. Parma, 1965-71; rsch. assoc. MIT, Cambridge, Mass., 1961-62, vis. scientist, 1963-64; vis. prof. dept. chemistry Cornell U., Ithaca, N.Y., 1966; dir.-gen. of directorate, gen. for sci., rsch. and devel. Joint Rsch. Centre, Commn. of European Communities, Brussels, 1981—; chmn. Joint European Torus Coun., Culham, Eng., 1990—; rep. C.E.C. in Eureka Group of High Ofcls., 1985—. Contbr. articles, papers to profl. publs. Fellow Inst. Biology Cambridge (hon.), Belgian Royal Acad. Medicine (hon.); mem. Acad. Nazionale delle Scienze. Office: Commn European Communities, 200 rue de la Loi, 1049 Brussels Belgium

FASH, WILLIAM LEONARD, architecture educator, retired college dean; b. Pueblo, Colo., Feb. 9, 1931; s. James Leonard and Jewel Dean (Rickman) F.; m. Maria Elena Shaw, June 5, 1982; children—Cameron Shaw, Lauren Victoria; children by previous marriage—Victoria Ruth, William Leonard. B.Arch., Okla. State U., 1958, M.Arch., 1960; postgrad., Royal Acad. Fine Arts, Copenhagen, 1960-61. Asst. prof. U. Ill., Urbana, 1961-64, assoc. prof., 1967-70, prof., 1970-76; assoc. prof. Okla. State U., Stillwater, 1964-66, U. Oreg., Eugene, 1966-67; prof., dean coll. architecture Ga. Inst. Tech., Atlanta, 1976-92, prof., 1992-93; vis. prof. Chulalongkorn U., Bangkok, Thailand, 1973-74; bd. dirs. Nat. Archtl. Accreditation Bd., 1981-85 (commendation award 1985); mem. edn. com. adv. bd. Nat. Coun. Archtl. Registration Bds., 1982-85; profl. cons. U.S. Navy Trident Submarine Base, Kings Bay, Ga., 1980-89; mem. adv. bd. Atlanta Urban Design Commn.; cons. Atlanta High Mus. Art Bldg. Com., 1982-84. Author, editor monographs. Chmn. bd. SE Energy Tech. Group; chmn. tech. adv. com. Gov.'s Commn. State Growth Policy, 1982-83; mem. Mayor's transition team for housing, Atlanta, 1989, Atlanta Symphony New Hall Com., 1991. Recipient awards for design, recognition citations for teaching excellence, 1975, 76; Spl. Recognition cert. Atlanta Coll. Architecture, 1986, Recognition award Indsl. Designers Soc. Am., 1990; Fulbright-Hays fellow, Copenhagen, 1960-61. Mem. Assn. Collegiate Schs. Architecture (recognition award 1985), AIA (award juries 1968, 75, 76, 79), Phi Kappa Phi, Sigma Tau, Pi Mu Epsilon, Alpha Rho Chi, Omicron Delta Kappa. Club: Bent Tree Country (Jasper, Ga.). Home: 2854 Ridgemore Rd NW Atlanta GA 30318-1448 Office: Ga Inst Tech Coll Architecture Atlanta GA 30332

FASMAN, GERALD DAVID, biochemistry educator; b. Drumheller, Alta., Can., May 28, 1925; came to U.S., 1955, naturalized, 1964; s. Morris and Sarah (Stauffer) F.; m. Jean Schalit, Dec. 27, 1953; children—Michael, Daniel, Jonathan. B.S., U. Alta., 1948; Ph.D., Calif. Inst. Tech., 1952; postgrad., Cambridge (Eng.) U., 1951-53, Eidg. Technische Hochschule, Zurich, Switzerland, 1953-54, Weizmann Inst. Sci., Rehovoth, Israel, 1954-55. Research asst. Children's Cancer Research Found., Children's Med. Center, Boston, 1955-56; research asso. pathology Children's Med. Center and Children's Cancer Research Found., Boston, 1957-61; asst. in pathology Harvard U. Med. Sch., 1957-58, research asso. pathology, 1958-60, research asso. biol. chemistry, 1960-61; lectr. protein chemistry Boston U., 1958-59; asst. head biophys. chemistry lab. Children's Cancer Research Found., Boston, 1959-61; tutor in biochem. sci. Harvard, 1960-62; established investigator Am. Heart Assn., 1961-66; asst. prof. biochemistry Brandeis U., Waltham, Mass., 1961-63, assoc. prof., 1963-67, prof., 1967—, Rosenfield prof. biochemistry, 1971—; cons. African Primary Sci. Program, Ednl. Services, Inc., Dar es Salam, Tanzania, 1966, mem. program steering com., Accra, Ghana, 1967, mem. adv. group, 1968-69; mem. sci. adv. com. Am. Cancer Soc., 1979-83; mem. molecular biology adv. panel NSF, 1980-82. Editor: CRC Critical Revs. in Biochemistry, 1972—, Chemtracts, Biochemistry and Molecular Biology, 1990—; adv. bd.: Biopolymers, 1975—; editorial bd.: Internat. Jour. Peptide and Protein Research, 1976-82, Biophys. Jour., 1976-79, Cell Biophysics, 1982—, Jour. Protein Chemistry, 1982—, CRC Critical Revs. in Eukaryotic Gene Expression, 1987—. NSF sr. postdoctoral fellow Protein Inst., Osaka (Japan) U. and Weizmann Inst. Sci., 1967-68; Guggenheim fellow, 1974-75, 88-89; research fellow Japan Soc. for Promotion of Sci., 1979. Fellow AAAS, Am. Inst. Chemists; mem. Am. Chem. Soc., Biophys. Soc., Am. Soc. Biochemistry and Molecular Biology, Chemists, Royal Soc. Chemistry (London), N.Y. Acad. Sci., Sigma Xi. Home: 69 Kingswood Rd Newton MA 02166-1013 Office: Brandeis U Waltham MA 02254

FASS, ROBERT J., epidemiologist, academic administrator; b. N.Y.C., Feb. 23, 1939. BS in Chemistry, Biology, Tufts U., 1960, MD, 1964; MS in Med. Microbiology, Ohio State U., 1971. Diplomate Am. Bd. Internal Medicine; licensed physician, Ohio, N.Y., Pa. Intern in mixed medicine Montefiore Hosp., N.Y.C., 1964-65, resident in medicine, 1965-66; resident in medicine Ohio State U., Columbus, 1968-69, fellow in infectious diseases, rsch. asst. in med. microbiology, 1969-71, clin. instr. medicine, 1970-71, asst. prof. medicine, 1971-75, asst. prof. med. microbiology, 1971-76, assoc. prof. medicine, 1975-80, assoc. prof. med. microbiology, 1976-80, prof. internal medicine, medical microbiology and immunology, 1980—, Samuel Saslaw prof. infectious diseases, 1991—, dir. divsn. infectious diseases, 1987—; dir. infectious diseases fellowship tng. program Ohio State U., 1987-93, mem. task force on program evaluation Coll Medicine, 1973-75, search com. chmn. dept. med. microbiology, 1973-76, clin. cirriculum devel. project, 1875-77, profl. adv. com. to med. illustations, 1975-85, vice chmn. practice plan com., 1979-85, trustee Med. Rsch. and Devel. Found., 1979-86, treas., 1979-85, assocs. medicine exec. com. dept. medicine, 1973-76, chmn., 1974-75, chmn. libr. com., 1973-78, bd. dirs. Dept. Medicine Found., 1987—, finance com., 1979-86, 91—, chmn., 1981-86, phase III module com. Med. Microbiology and Immunology divsn., 1973, 75, 76, 81, dir., 1976, 81, mem. infection control com. Univ. Hosp., 1972-77, exec. com., 1976-77, pharmacy and therapeutics com., 1981—, chmn. pharmacy and therapeutics antimicrobial com., 1984—, exec. com., 1984—; mem. infectious disease conf. Riverside Meth. Hosp., 1971-76; prin. investigator for Ohio State U. NIH AIDS Clin. Trials Group, 1987—, opportunistic info. com., 1987—, instl. evaluation com., 1989—; mem. internat. adv. bd. Bayer AG Auinolone Bd., 1985-93, Miles Inc. External Adv. Bd., 1988-93. Mem. editorial bd. Antimicrobial Agents and Chemotherapy, 1982-91, Quinolene Bulletin, 1989—; reviewer Am. Jour. Medicine, Am. Soc. Hosp. Pharmacists Drug Info., Antimicrobial Agents and Chemotherapy, Annals of Internal Medicine, Archives of Internal Medicine, Chest, Clin. Microbiology Revs., Jour. Infectious Diseases, Infectious Diseases in Clin. Practice, N.Y. State Jour. Medicine, Revs. Infectious Diseases, New England Jour. Medicine. Med. officer USAF, 1966-68. Fellow ACP, Infectious Diseases Soc. Am.; mem. AMA (reviewer AMA Drug Evaluations, Jour. AMA), Am. Fedn. Clin. Rsch., Am. Soc. for Microbiology, Am. Thoracic Soc., Ctrl. Soc. for Clin. Rsch., Inter-Am. Soc. for Chemotherapy, Brit. Soc. Antimicrobial Chemotherapy, Columbus Soc. Internal Medicine (sec.-treas. 1979, pres. 1980). Achievements include research in laboratory predictors of antimicrobial efficacy, pathogenesis of anaerobic and mixed bacterial infections, antibiotic susceptibility testing and resistance, antimicrobial agents in clinical, pharmacological and laboratory studies, biology and significance of cell-wall defective bacteria, infective endocarditis, AIDS. Office: Ohio State U Med Ctr Divsn Infectious Diseases 410 W 10th Ave N-1148 Doan Hall Columbus OH 43210-1228

FASTL, HUGO MICHAEL, acoustical engineer, educator; b. Muenchen, Bavaria, Germany, Apr. 18, 1944; s. Michael and Mathilde (Nester) F. Diploma in engring., Tech. U. Muenchen, 1970, D in Engring., 1974. Rsch. assoc. Tech. U. Muenchen, 1970-74, asst. prof., 1974-78, assoc. prof., 1978-87, acad. dir., 1987—, prof., 1991—; guest prof. Osaka (Japan) U., 1987; mng. dir. SFB 204 Hearing, Muenchen, 1983-91; head electroacoustics ITG, Frankfurt, Germany, 1983—; head hearing DEGA, Bonn, Germany, 1991—. Author: Dynamic Hearing Sensations, 1982, (with others) Psychoacoustics, 1990; contbr. over 100 articles to profl. jours. Fellow Acoustical Soc. Am.; mem. Soc. for Info. Tech. Bd., Acoustical Soc. Japan, Acoustical Soc. Germany. Office: Tech U, Arcisstr 21, 80333 Munich Germany

FATEMI, SEYYED HOSSEIN, cell biologist, physician; b. Tehran, Iran, May 1, 1952; came to U.S., 1970; s. Seyyed Mehdi and Fatemeh (Parsa-Moghaddam) F.; m. Maryam Jalali-Mousavi, May 22, 1991; 1 child, Neelufaar. BS in Biology, Baylor U., 1974; MS in Anatomy, U. Nebr., 1978, PhD in Anatomy, 1979; MD, Case Western Res. U., 1991. Lic. physician, Ohio, Iowa. Postdoctoral fellow dept. pharmacology U. Tex., Houston, 1980-81, postdoctoral fellow Cyclic Nucleotide Labs., 1981-82; vis. fellow NIDR, NIH, Bethesda, Md., 1983; postdoctoral fellow dept. anatomy McGill U., Montreal, Que., Can., 1982-83, asst. prof., 1983-84; sr. instr. Inst. Pathology Case Western Res. U., Cleve., 1984-86, asst. prof., 1986-87; resident radiology U. Cin. Hosps., 1991-92; resident psychiatry U. Hosps. Cleve., 1992—. Co-author: Lipid Anchored Proteins, 1990; contbr. articles to profl. jours. Regents tuition fellow U. Nebr. Med. Ctr., Omaha, 1979-80; spl. fellow Leukemia Soc. Am., N.Y., 1986-87; Biomed. Rsch. Support grantee NIH, Case Western Res. U., 1986; recipient Citation for Teaching Contbn. in Anat. Scis., Case Western Res. U. Sch. Medicine, 1988, Paul Curtis Muskuloskeletal Essay award Case Western Res. U., 1989. Mem. AMA, AAAS, Am. Psychiat. Assn. Islam. Achievements include first description of mutant lymphoma cells defective in biosynthesis of Thy-1 glycophospholipid anchor; first publication of a mammalian cell protein (Thy-1) with such a lesion; research in sulfation of basement membrane proteoglycans, role of secretory granules in collagen transport, intestinal cell proliferation and camp synthesis and degradation. Home: 1414 Som Center Rd Apt 802 Mayfield Heights OH 44124

FATH, GEORGE R., electrical engineer, communications executive; b. Cin., July 9, 1938; s. Charles and Rebecca (Gilreath) F.; m. Nancy Morris, June 1959; 1 child, George R. Jr. BSEE, U. Miami, Coral Gables, Fla., 1960; MEE, Syracuse U., 1966, PhD, 1969. Mgr. advanced tech. GE, Arlington, Va., 1975-77; mgr. avionics devel. GE, Utica, N.Y., 1977-83; mgr. digital engring. GE, Lynchburg, Va., 1983-85, mgr. pub. svc. engring., 1985-90; mgr. product and process engring. Ericsson GE Mobile Communications, Lynchburg, 1990, mgr. engring., 1990-91, v.p., gen. mgr., 1991—; presenter at profl. confs.; panel chmn. Space Div. Info. Processing Systems, 1975, 76. Adviser New Hartford High Sch. Jr. Achievement, 1979-80; bd. dirs. spl. activities Utica (N.Y.) chpt. Elfun Soc., 1979, 80. Mem. IEEE, AIAA, Air Force Communication Electronics Assn., Nat. Assn. Bus. and Ednl. Radio, Elks, Piedmont Club, Radio Club Am., Assn. Old Crows, Sigma Psi. Home: 102 Millview Terr Forest VA 24551 Office: Ericsson GE Mobile Communications Mountain View Rd Lynchburg VA 24502

FATHAUER, THEODORE FREDERICK, meteorologist; b. Oak Park, Ill., June 5, 1946; s. Arthur Theodore and Helen Ann (Mashek) F.; m. Mary Ann Neesan, Aug. 8, 1981. BA, U. Chgo., 1968. Cert. cons. meteorologist. Rsch. aide USDA No. Dev. Labs., Peoria, Ill., 1966, Cloud Physics Lab., Chgo., 1967; meteorologist Sta. WLW Radio/TV, Cin., 1967-68, Nat. Meteorol. Ctr., Washington, 1968-70, Nat. Weather Svc., Anchorage, 1970-80; meteorologist-in-charge Nat. Weather Svc., Fairbanks, Alaska, 1980—; instr. U. Alaska, Fairbanks, 1975-76, USCG aux., Fairbanks and Anchorage, 1974—. Contbr. articles to weather mags. Bd. dirs. Fairbanks Concert Assn., 1988—, Friends U. Alaska Mus., 1993—, exec. com.; mem. coll. of fellows U. Alaska, 1993—. Recipient Outstanding Performance award Nat. Weather Service, 1972, 76, 83, 85, 86, 89, Fed. Employee of Yr. award, Fed. Exec. Assn., Anchorage, 1978. Fellow Am. Meteorol. Soc. (TV and radio seals of approval), Royal Meteorol. Soc.; mem. Am. Geophys. Union, AAAS, Western Snow Conf., Arctic Inst. N.Am., Oceanography Soc. Republican. Lutheran. Avocations: reading, music, skiing, canoeing. Home: 1738 Chena Ridge Rd Fairbanks AK 99709-2612 Office: Nat Weather Svc Forecast Office 101 12th Ave Box 21 Fairbanks AK 99701-6266

FATLAND, CHARLOTTE LEE, chemist; b. Madelia, Minn., June 3, 1944; d. Arlo Arther and Elaine Irene (Pfeil) Becker; m. Robert Glen Fatland, Sept. 28, 1968. BS, Mankato State U., 1966. Tchr. math. and chemistry Dodge Center (Minn.) Pub. Sch., 1966-67; chemist Bioscis. Rsch. Lab., ARS, USDA, Fargo, N.D., 1967—. Contbr. articles to Jour. Chem. Engring., Insect. Biochemistry, Jour. Chem. Ecology, others. Mem. Am. Soc. for Mass Spectrometry, Minn. Chromatography Forum, Minn-Mass. Office: USDA/ARS Bioscis Rsch Lab 1605 Albrecht Blvd Fargo ND 58102

FATTINGER, CHRISTOF PETER, physicist; b. Brazil, Oct. 3, 1954; arrived in Switzerland, 1969; s. Volker Hermann and Doris (Stanke) F.; m. Regula Veronika Hegetschweiler, June 18, 1982; children: Felix Simon, Sara Corinna, Nicolas Benjamin, Stefan Alexander. Diploma in physics, Swiss Fed. Inst. Tech., Zurich, 1980, PhD in Physics, 1987. Rsch. asst. optics lab. Swiss Fed. Inst. Tech., 1981-87; rsch. scientist Watson Rsch. Ctr. IBM, Yorktown Heights, N.Y., 1987-89; mem. rsch. staff pharm. rsch. new techs. F. Hoffmann-La Roche Ltd., Basel, Switzerland, 1989—. Inventions in biochem. sensor tech. and optical surface sensing, including the bidiffractive grating coupler, phase mapping of guided optical surface waves; pioneering contbns. to terahertz optoelectronics; contbr. articles to profl. jours. Mem. AAAS, Optical Soc. Am., Swiss Phys. Soc., Swiss Optical Soc. Home: Emmengasse 7, CH-4249 Blauen Switzerland Office: F Hoffmann-La Roche Ltd, Grenzacherstr. 124, CH-4002 Basel Switzerland

FAUBEL, GERALD LEE, agronomist, golf course superintendent; b. Normal, Ill., Feb. 14, 1941; s. Elmer Joseph and Agnes (Alexander) F.; m. Sally Sue Shook, Feb. 22, 1973; 1 child, Sarah. AAS, Iowa State U., 1963, MS, 1969. Golf course supt. Lawsonia Golf Course, Green Lake, Wis., 1962-63, South Hills Club, Fond du Lac, Wis., 1963-68, Saginaw (Mich.) Country Club, 1969—; pres. Exec. Golf Search, Inc.; dir. Mich. Turfgrass Found., Lansing, 1978-81, sec.-treas., v.p., 1982, pres., 1984. Parks commn. Saginaw (Mich.) Twp., 1972-78. Mem. Golf Course Supt. Assn. Am. (bd. dirs. 1985-92, sec.-treas. 1988, v.p. 1989, pres. 1990), U.S. golf Assn., Mich. Assn. Agr. (exec. sec. 1991—), Nat. Inst. Golf Mgmt. (bd. regents), Nat. Golf Found. Republican. Mennonite. Home: 699 Westchester Rd Saginaw MI 48603-6232

FAUBEL, MANFRED, physicist; b. Alzey, Germany, Mar. 27, 1944; s. August and Elisabeth (Buerckel) F.; m. Christine Pilz, Apr. 9, 1976; children: Friedrich, Gerhard, Regina, Leonhard. Diploma in physics, Gutenberg U., Mainz, Fed. Republic Germany, 1969; DSc, U. Göttingen, Fed. Republic Germany, 1976. Scientist Max-Planck Inst. für Strömungsforschung, Göttingen, 1972—. Contbr. articles on inelastic ion scattering, crossed beam studies of molecular rotation and free evaporation of volatile liquids to sci. jours. Mem. German Phys. Soc. (Stern-Gerlach prize 1989), Bunsen Soc., Royal Soc. Chemistry (molecular beams group). Home: Steinbreite 32, W-3405 Sieboldshausen Germany Office: Max-Planck-Inst Stroemungs, Bunsenstrasse 10, W-3400 Göttingen Germany

FAUCI, ANTHONY STEPHEN, health facility administrator, physician; b. Brooklyn, N.Y., Dec. 24, 1940; s. Stephen A. and Eugenia A. Fauci. AB, Coll. of Holy Cross, 1962; MD, Cornell U., 1966; DSc (hon.), Coll. Holy Cross, 1987, Georgetown U., 1990, Hahnemann U., 1990, Mt. Sinai Sch. Medicine, 1990, Universita di Roma, 1990, St. John's U., 1991, Long Island U., 1992. Diplomate Am. Bd. Internal Medicine, Am. Bd. Allergy and Immunology (bd. dirs. 1984—), Am. Bd. Infectious Diseases. Intern N.Y. Hosp.-Cornell Med. Ctr., 1966-67, asst. resident in medicine, 1967-68, chief resident dept. medicine, 1971-72; clin. assoc. Nat. Inst. Allergy and Infectious Diseases-NIH, Bethesda, Md., 1968-70, sr. staff fellow, 1970-71, sr. investigator, 1972-74, head, clin. physiology sect., 1974-80, dep. clin. dir., 1977-80, chief Lab. Immunoregulation, 1980—, dir. Nat. Inst. Allergy and Infectious Diseases, 1984—; dir. Office of AIDS Rsch., NIH,. and assoc. dir. NIH for AIDS Rsch., 1988—; cons. Naval Med. Ctr., Bethesda, 1972—. Contbr. numerous articles to med. jours. Served with USPHS, 1968—. Recipient USPHS meritorious service award, 1979, Arthur S. Fleming award, 1983, Squibb award Infectious Diseases Soc., 1983, Commrs. spl. citation FDA, 1984, Clemons von Pirquet award Georgetown U. Med. Ctr., 1986, Disting. Clin. Educator award NIH Clin. Ctr., 1988, Leadership award Columbus Citizens Found., Inc., 1988, spl. award for rsch. in AIDS Nat. Hemophilia Fedn., 1989, Lee P. Brown Nat. Pub. Svc. award Nat. Acad. Pub. Adminstrn. and Nat. Soc. for Pub. Adminstrn., 1989, numerous awards Duke U., AMA, Children's Hosp. Nat. Med. Ctr., Surgeon Gen., Am. Assn. Physicians for Human Rights, Nat. Health Coun., Nat. Found. Infectious Diseases, Helen Hayes award for med. rsch., 1989, Excellence in Pub. Svc. award Com. for Support of Pub. Svc., 1990, Lifetime Sci. award Inst. Advanced Studies in Immunology and Aging, 1990, Internat. Chiron prize, 1990, Pres. award N.Y. Acad. Sci., 1990, Thomas H. Ham-Louis R. Wasserman award Am. Soc. Hematology, 1992, Dr. Nathan Davis award AMA, 1992. Fellow AAAS, Am. Acad. Allergy, ACP, N.Y. Acad. Med. (hon. 1991), Am. Acad. Arts and Scis.; mem. Am. Fedn. for Clin. Research, AAAS (Westinghouse award 1988), Am. Assn. Immunologists (program chmn. 1982-85, Kober lectr. 1988), Am. Fedn. Clin. Rsch (pres. 1980-81), Commd. Officers Assn. USPHS (Pub. Health Leader of Yr. award), Infectious Diseases Soc. Am., Am. Soc. for Clin. Investigation, Assn. of Am. Physicians (recorder 1988—), Collegium Internationale Allergologicum, Inst. Medicine, NAS USA, Royal Danish Acad. Sci. and Letters (fgn.). Roman Catholic. Avocations: running; tennis. Home: 3012 43d St NW Washington DC 20016 Office: Nat Inst Allergy and Infectious Diseases Bldg 31 Rm 7A03 Bethesda MD 20892*

FAUDREE, EDWARD FRANKLIN, JR., military officer; b. Jacksonville, Fla., Oct. 15, 1955; s. Edward Franklin Sr. and Janet Curry (Wade) F.; m. Victoria Gray Wandres, Mar. 8, 1980; children: Rachel Christine, Andrew Bradford, Joel Prescott, Daniel Hawthorne. BA, U. Va., 1978; MS, Air Force Inst. Tech., 1985. Commd. 2d lt. USAF, 1978, advanced through grades to maj., 1989; orbital analyst Aerospace Def. Command, Colorado Springs, 1979-81; chief ops. tng. 3422 Sch. Squadron, ATC, Denver, 1981-84, Second Space Wing (AFSPACECOM), Colorado Springs, 1986-88; chief integration div. Hdqr. Air Force Space Command, Colorado Springs, 1988-91; polit. analyst Space and Missile Systems Ctr. (AFMC), L.A., 1991-92, dep. dir. customer svc. Office Space Test & Experimentation, 1992—. Mem. Air Force Assn. Home: 28706 Mt Langley Ct Rancho Palos Verdes CA 90732-1820 Office: Space and Missiles Sytems Ctr 160 Skynet St Ste 1536A Los Angeles AFB CA 90245-4683

FAUGHNAN, WILLIAM ANTHONY, JR., electrical/electronics engineer; b. Augusta, Ga., Dec. 8, 1944; s. William Anthony and Dorothy Ann (Boyle) F.; m. Carol Ann Palladino, Nov. 23, 1976; 1 child, William Anthony III. B in Indsl. Engring., Ga. Inst. Tech., 1968, BEE, 1979; MBA, Ga. State U., 1969. Engr. Ga. Power Co., Atlanta, 1962-78; commd. 2nd lt. USAR, 1968, advanced through grades to lt. col., 1987; nuclear engr. USN-Portsmouth (N.H.) Naval Shipyard, 1979—. Office: USN Portsmouth Naval Shipyard Code 2330 NEPD Portsmouth NH 03801

FAUL, GARY LYLE, electrical engineering supervisor; b. Clarksburg, W.Va., Nov. 5, 1939; s. Lyle Joseph and Irene (Hadden) F.; m. Edith Uvenzia Kelly, Nov. 26, 1966. BSEE, AS in Instrumentation and Control, Cleve. State U., 1966; AS in Mktg. Mgmt., ICS, 1969. Lic. FCC 1st class. Engr., project mgr. Bailey Controls, Wickliffe, Ohio, 1962-77; engr. Davey McKee, Independence, Ohio, 1977-78; engr., supr. Ariz. Pub. Svc., Phoenix, 1978—; prin., cons. engr. Advanced Systems Design, Phoenix, 1982—. Contbr. tech. papers to jours. Vol. Child Ctr., Phoenix, 1986, Kids Voting, Phoenix, 1992; parade marshall Fiesta Bowl, Phoenix, 1991. With U.S. Army, 1959-65. Mem. Instruments Soc. Am. Republican. Office: Ariz Pub Svc Co PO Box 53999 411 N Central Ave Phoenix AZ 85072

FAULK, WARD PAGE, immunologist; b. Ruston, La., Nov. 14, 1937; s. Clarence E. and Louise B. (Page) F.; children: Robin, Saskia, Josie, Holly. BS, U. South, 1959; MD, Tulane U., 1964; MRC, Royal Coll. Pathology, 1973, FRC, 1983. Fellow Netherlands Red Cross, Amsterdam, 1965-66, Mayo Clinic, Rochester, Minn., 1966-67, U. Calif. Med. Ctr., San Francisco, 1967-69; mem. staff British Med. Rsch. Coun., London, 1973-76; prof. Med. U. S.C., Charleston, 1976-79, Royal Coll. Surgeons, East Crinstead, England, 1979-81; prof., dir. Purdue U./Meth. Hosp., Indpls., 1986—; vis. prof. faculty medicine U. Nice, France, 1982-84, dept. obstetrics So. Ill. U., Springfield, 1984-85; cons. hematology and oncology Pembury (England) Hosp., 1982-92. Founder, co-editor: Placenta, 1979-89; co-editor: Immunological Obstetrics, 1992; contbr. articles to more than 300 publs.; co-editor for 11 jours. Med. officer WHO, Geneva, 1970-73; mem. NATO Advanced Study Group, Cordoba, Spain, 1988, NIH Study Group, Bethesda, Md., 1990. Recipient award Calif. Trudeau Soc., 1965, Metchnikoff medal Bulgarian Acad. Scis., 1971. Fellow Royal Coll. Pathology; mem. Internat. Soc. for Immunology Reproduction, Am. Soc. Immunology Reproduction (Munksgaard award 1992), Internat. Soc. Heart and Lung Transplantation. Republican. Episcopalian. Achievements include patents for use of amnion in wound healing, tumor localization and drug targeting, development of study of effects of nutrition on immune response in Africa and Middle East, immunocolloidal gold technique for electron microscope. Office: Methodist Hosp 1701 N Senate Indianapolis IN 46202

FAULKNER, FRANK DAVID, mathematics educator; b. Humansville, Mo., Apr. 6, 1915; s. Marion Alexander and Bertha Ellanora (Pfandler) F.; m. Theresa Alice Hellmer, Jan. 5, 1941 (dec. 1978); children: Frank David Jr., Harold George, Mary Alice Faulkner Kirk, William Marion, Robert Gordon, Andrew Wayne. BS in Edn., Kans. State Tchrs. Coll., 1940; MS, Kans. State U., 1942; PhD, U. Mich., 1969. Engr. U.S. Rubber Co., Detroit, 1942-43; instr. dept. math. U. Mich., Ann Arbor, 1943-44; engr. applied physics lab. Johns Hopkins U., Silver Spring, Md., 1944-46; rsch. mathematician Engring. Rsch. Inst. U. Mich., Ann Arbor, 1946-50; from asst. prof. to disting. prof. math. USN Postgrad. Sch., Monterey, Calif., 1950-80; ret., 1980; cons., engr. Boeing Airplane Co., Seattle, 1956-64. Contbr. papers to profl. publs. Mem. Am. Math. Soc., Math. Assn. Am., Moose, Sierra Club, Sigma Xi. Democrat. Home: PO Box 3835 Carmel CA 93921-3835 Office: US Naval Postgrad Sch Math Dept Monterey CA 93943

FAULKNER, LLOYD C., veterinary medicine educator; b. Longmont, Colo., Oct. 24, 1926; s. Earl Dickerson and Verna Virginia (Sonner) F.; m. Elaine Mae Wagner, June 11, 1954; children: Chad B., Kurt L., Vickie L., Earl H., Ron E. DVM, Colo. State U., 1952; PhD, Cornell U., 1963. Diplomate Am. Coll. Theriogenologists. Staff vet. Animal Hosp., Longmont, 1952-55; successively asst. prof., assoc. prof., prof. and head physiology and biophysics Colo. State U., Ft. Collins, 1955-78; assoc. dean U. Mo., Columbia, 1979-81; assoc. dean vet. rsch. and grad. edn., asst. dir. agrl. experiment sta. Okla. State U., Stillwater, 1981-93; interim dir. Okla. Animal Disease Diagnostic Lab., 1991-92; bd. dirs. Western Vet. Conf., pres., 1990-91; bd. trustees Kerr Ctr. Sustainable Agriculture, Poteau, Okla., 1986—, chmn. 1991—; bd. dirs. Internat. Alliance for Sustainable Agriculture; chmn. Kerr Conf. Brucellosis Eradication, 1982-83; past cons. Rockefeller Found., Quaker Oats Co., Ft. Dodge Labs, Boehringer Ingelheim. Author: Veterinary Endocrinology and Reproduction, 1975, Current Therapy in Theriogenology, 1980; editor: Abortion Diseases in Livestock, 1968; contbr. articles to profl. jours. Sponsor Christian Vet. Fellowship, 1985-93. Served with USN, 1944-46. NIH fellow, 1960-62, Congl. Sci. fellow with Senator Charles Mathias, 1976-77. Mem. AVMA (exec. bd. 1976-78, rsch. coun. 1971-74), Intermountain Vet. Med. Assn. (bd. dirs. 1984-91, pres. 1990-91), Am. Assn. Accreditation Lab. Animal Care (exec. com. 1978-81), Phi Zeta, Phi Kappa Phi, Gamma Sigma Delta (Disting. Svc. award Okla. chpt. 1993). Republican. Presbyterian. Avocations: reading, gardening, gospel music. Office: Okla State U 308 Vet Med Bldg Stillwater OK 74078

FAUNCE, MARK DAVID, product design engineer; b. Delanco, N.J., Feb. 4, 1967; s. Edward Steetle and Anna Veronica (Belli) F. BSME, Tri-State U., 1990. Cert. engr.-in-tng. Sr. product design engr. Nishikawa Std. Co., Dearborn, Mich., 1990—. Office: Nishikawa Std Co 2401 S Gulley Rd Dearborn MI 48124

FAUNCE, WILLIAM DALE, clinical psychologist, researcher; b. Lansing, Mich., Dec. 4, 1947; s. Lucius Dale and Wilhelmina (Hall) F. BA, Mich. State U., 1972; MA, Calif. State U., L.A., 1978; PhD in Clin. Psychology, U. So. Calif., 1983. Lic. psychologist, Alaska, N.C. Psychology intern Brentwood (Calif.) VA, 1981-82; clin. psychologist UCLA Neuropsychiat. Inst., Westwood, Calif., 1983, Coldwater Canyon Hosp., North Hollywood, Calif., 1983-84, So. Peninsula Community Mental Health Ctr., Homer, Alaska, 1984-86; pvt. practice Homer, Alaska, 1986-87; cons. Santa Cruz, Calif., 1987-90; clin. psychologist, program dir. Broughton State Hosp., Morganton, N.C., 1990—; mem. faculty Appalachian State U., Boone, N.C., 1992—. Co-author: (chpt.) Imagery, 1984; contbr. articles to profl. jours. Fellow NIMH; mem. APA, Union Concerned Scientists. Avocations: creative writing, guitar, travel, foreign languages. Home: PO Box 241 Jonas Ridge NC 28641-0241 Office: Broughton State Hosp Morganton NC 28655 also: Appalachian State U Boone NC 28608

FAURE, FRANÇOIS MICHEL, metallurgical engineer; b. Paris, July 16, 1947; s. Michel François and Alice Marie-Louise (Loiseau) F. Ingenieur Civil, Ecole des Mines, Nancy, France, 1969; PhD, Stanford U., 1972. Mktg. engr. B.P. France, Paris, 1972-73; head welding rsch. IRSID, St. Germain en Laye, France, 1974-78; various positions, now head metallurgy Framatome, Paris, 1978—. Contbr. articles to profl. jours. and confs. Mil. engr. French Army, 1973-74. Mem. Inst. de Soudure (sci. coun.), Welding Inst. (rsch. bd.). Home: 4 Ave Florent Schmitt, F-92210 Saint Cloud France Office: Framatome, Tour Fiat Cedex 16, F92084 Paris La Defense France

FAUSETT, ROBERT JULIAN, engineering geologist, consultant; b. Winfield, Kans., Apr. 9, 1923; s. Carl Alva and Bessie Opal (Jones) F.; m. Patricia Mae Probst, Feb. 15, 1946; children: Rory, Teresa, Carl. Student, U. Pa., Utah State U., Westminster Coll.; PhD in Mineral Engring. and Geology, U. Wis., 1982. Cons. depts. mining and mineral engring. and geology U. Wis., Madison, 1946; devel. engr. Pickands-Mather (Bethlehem Ores), Hibbing, Minn., 1951-52; sr. engr. A.C.O.E., Baraboo, Wis., 1952-53; cons. Madison, 1953—; arbitrator Am. Arbitration Assn., Chgo., Mpls., St. Paul, 1975—; critic, engr., tech. publs. Prentice-Hall, Wiley, U. Wis.; lectr. U. Wis. Acad., Madison, 1980—. Author: Interdisciplinary Approach to Resource Inventory, 1981, Stability of Soils and Rock, 1979, Specifications: Their Derivation of Use, 1978. Active C. of C. Mesabi Range, Hibbing, 1952; commr. Boy Scouts Am., Madison, 1971. Master sgt. USMC, 1941-46, PTO. Recipient Nat. Hon. award Boy Scouts Am., 1972, dist. awards S.C.S. USDA, 1969, 71, 78. Mem. ASTM, Geol. Soc. Am., Soc. Mining, Metallurgy and Exploration, Am. Geol. Inst., Soc. Rsch. Scientists, Sigma Xi. Roman Catholic. Home: S7398A Stone's Pocket Rd North Freedom WI 53951

FAUST, JOHN WILLIAM, JR., electrical engineer, educator; b. Pitts., July 25, 1922; s. John William and Helen (Crowther) F.; m. Mary Claire Barton, June 7, 1947; children: Mary Faust Baumert, Elizabeth Wickham Kemp, John William III, Charles Barton, Ann Louise Faust Spires, Susan Bosley Rossell, Helen Crowther, Thomas McCullough. BSChemE, Purdue U., 1943; MA, U. Mo., 1949, PhD, 1951. Research scientist Westinghouse Research Labs., Pitts., 1951-63; mgr. materials characterization lab. Westinghouse Research Labs., 1963-65, project mgr. crystal growth, 1965-67; prof. materials sci. Pa. State U., State College, 1967-69; prof. elec. & computer engring. U. S.C., Columbia, 1969—, Disting. prof., 1992—, assoc. chmn. elect. and computer engring dept., 1990—, grad. dir. elect. and computer engring. dept., 1989-92, disting. prof., 1992—, faculty senator, 1989-91; disting. lectr. Naval Rsch. Labs., 1979, rsch. physicist, 1980-81; cons. Wright-Patterson Air Force Rsch. Ctr., 1957, Corning Glass Rsch. Labs., 1968-70, semicondr. div. Dow Corning, 1967-69, Gen. Tel. & Tel. Labs., 1968, materials div. Sylvania, 1968-70, Langley Air Force Rsch. Labs., 1970, Air Force Materials Lab., Wright-Patterson AFB, 1977, Borg-Warner Corp., 1982-84, Silaq Corp., 1979-80, Morgan Semicondr. Corp., 1982-88; co-chmn. Internat. Com. on Silicon Carbide, 1969-75; chmn. tech. adv. panel on solar energy S.C. Ho. of Reps., 1979-90. Editor: The Surface Chemistry of Metals and Semiconductors, 1960, Marcel Dekker, Inc., 1967-71, Silicon Carbide-1973, 1974; contbr. articles to profl. jours.; patentee in field. Vol. Food Program for Needy. Served with USNR, 1943-46. Recipient Outstanding Prof. of Yr. award IEEE. Fellow Am. Inst. Chemists; mem. AAAS (coucillor 1967-69), The Minerals, Metals and Materials Soc., Electrochem. Soc. (editorial com. 1971-91, divsn. editor jour. 1971-91), Am. Phys. Soc., Am. Soc. Metals, Internat. Com. on Crystal Growth, Am. Assn. for Crystal Growth, Internat. Soc. Hybrid Microelectronics, Am. Chem. Soc., Materials Rsch. Soc., Robert Burns Soc., Sigma Xi, Eta Kappa Nu, Tau Beta Pi. Home: 2455 Robin Crest Dr West Columbia SC 29169-5450

FAVARO, MARY KAYE ASPERHEIM (MRS. BIAGINO PHILIP FAVARO), pediatrician; b. Edgerton, Wis., Sept. 30, 1934; d. Harold Wilbur and Genevieve Catherine (Hyland) Asperheim; B.S., U. Wis., 1956; M.S., St. Louis Coll. Pharmacy, 1965; M.D., U. Wis., 1969; m. Biagino Philip Favaro, May 31, 1969; children—Justin Peter, Gina Sue. Instr. pharmacology St. Louis U. and St. Mary's Hosp. Sch. Practical Nurses, 1959-64; staff pharmacist U. Hosps., Madison, Wis., 1964-65; intern Albany (N.Y.) Med. Center, 1969-70, resident, 1970-71; resident in pediatrics U. S.C., Charleston, 1971-72, asst. prof. pediatrics, 1973-75; pvt. practice pediatrics, 1974—. Mem. A.M.A., Am. Med. Women's Assn. Roman Catholic. Author: Pharmacology, an Introductory Text, 1992; The Pharmacologic Basis of Patient Care, 1985. Home: 1866 Capri Dr Charleston SC 29407-7600 Office: 5390 Dorchester Rd North Charleston SC 29418

FAVERO, KENNETH EDWARD, medical administrator; b. Bklyn., Aug. 7, 1946; s. Edward and Mary Anna (Lang) F.; m. Violet Mae Brown, Oct. 24, 1992; children: Anna Marie, Angela Lisa, Jason Edward, Craig George, Matthew Demetrios. BS, SUNY, Old Wesbury, 1977; MS, L.I. U., 1978; MPH, St. John's U., 1984, PhD, 1985. Cert. clin. lab. dir.; med. microbiologist. Assoc. lab. dir. Lindenhurst (N.Y.) Med. Lab., 1968-86; asst. prof. health scis. L.I. U.-C.W. Post Campus, Greenvale, N.Y., 1979-86; dir. diagnostic svcs. Melbourne (Fla.) Internal Medicine Assocs., 1988—; cons. in field. Bd. dirs. Big Bros./Big Sisters, Melbourne, 1989-91; mentor-scis. fair Brevard County (Fla.) Sch. Bd., 1986—. Lt. comdr. MSC, USNR. Grantee U.S. Dept. Health, 1977. Mem. AAAS, Am. Soc. for Microbiology, Am. Soc. for Med. Tech., Am. Assn. Biolanalysts. Office: Melbourne Internal Medicine 200 E Sheridan Rd Melbourne FL 32901

FAVRE, ALEXANDRE JEAN, physics educator; b. Toulon, France, Feb. 23, 1911; s. Auguste Edouard and Mélanie Jeanne (Mercure) F.; m. Luce Jeanne Palombe; children: Christian, Elyette, Nadine. Degree in Engring., Engrs.' Sch., Marseilles, France, 1931; BS, Faculty of Scis., Marseilles, 1932; DSc, Faculty of Scis., Paris, 1938. Asst. Faculty of Scis., Marseilles, 1932-38, chief of labor, 1938-45, lectr. fluid mechanics, 1945-51, prof. fluid mechanics, 1951-80; founder, dir. Turbulence Inst. of Statis. Mechanics, Marseilles, 1960-80, hon. dir., 1981—; mem. French Acad. Scis., Paris, 1977—; vis. prof. Johns Hopkins U., Balt., 1962-63; sci. cons. Air Ministry, Paris, 1932-60; Nat. Sci. Aerospatial Rsch., Paris, 1947-71, Commr.'s Office of Atomic Energy, Paris, 1958-71, Adv. Com. Univs., 1975-77; cons., rschr. USAF Office Sci. Rsch., 1963-70. Inventor space-time correlations, supersonic centrifugal compressor, time-correlation apparatus; co-author: La Turbulence en Mécanique des Fluides, 1976, De la Causalité à la Finalité a propos de la Turbulence, Maloine, 1988, Chaos and Determinism, 1993; contbr. 85 articles to sci. jours. Decorated chevalier, officer and comdr. Acad. Palmes; chevalier and officer Legion of Honor; officer Nat. Order of Merit. Mem. Nat. Com. Sci. Rsch. (cons. 1963-70), Acad. Fedn. Mechanics (pres. 1970-71), Nat. Com. Mechanics (v.p. 1980-88), Rotary (hon.). Roman Catholic. Home: 122 rue du Commandant Rolland, Le Chambord #1, 13008 Marseilles France Office: Inst Mechanics Stats Turbulence, 12 Ave General 1ECLERC, 13003 Marseilles France

FAW, RICHARD EARL, nuclear engineering educator; b. Ohio, June 22, 1936; s. Robert Harvey and Mary Elizabeth (Baird) F.; m. Beverly A. Giltner, Mar. 25, 1961; children: Jennifer, Andrew. BSChemE, U. Cinn., 1959; PhD in Chem. Engring., U. Minn., 1962. Cert. chem. engr., Ohio, nuclear engr., Kans. Prof. nuclear engring. Kans. State U., Manhattan, 1968—. Author: Radiological Assessment, 1992; co-author: Principles of Radiation Shielding, 1984; contbr. articles to profl. jours. Capt. U.S. Army, 1962-64. Mem. Am. Nuclear Soc. (profl. excellence award 1986), Health Physics Soc., Soc. Nuclear Medicine. Methodist. Office: Kans State Univ Nuclear Engring Dept Manhattan KS 66506-2503

FAWCETT, HOWARD HOY, chemical health and safety consultant; b. McKeesport, Pa., May 31, 1916; s. Harry Garfield and Ada (Deetz) F.; m. Ruth Allen Bogan, Apr. 7, 1942 (dec. Oct. 1986); children: Ralph Willard, Harry Allen. BS in Indsl. Chemistry, U. Md., 1940; postgrad. U. Del., 1945-47. Registered profl. engr., Calif. Rsch. chemist Manhattan project E.I. DuPont de Nemours & Co., Inc., Chgo., Hanford, Wash., 1944-45, rsch. and devel. chemist organic chemistry div., Deepwater, N.J., 1945-48; cons. engr. GE, Schenectady, N.Y., 1948-64; tech. sec. com. on hazardous materials Nat. Acad. Scis.-NRC, Washington, 1964-75; staff scientist, project mgr. Tracor Jitco, Inc., Rockville, Md., 1975-78; sr. chem. engr. Equitable Environ. Health, 1978-81; pres., sr. engr. Fawcett Consultations, Inc., 1981—; mem. adv. com. study on socio-behavioral preparations for, responses to and recovery from chem. disasters NSF, 1977-82; adj. prof. Fed. Emergency Mgmt. Agency Acad., 1983—; cons. to industry and govt. agys. Author Am.-Can. supplement Hazards in Chemical Lab., 1983, Hazardous and Toxic Materials, Safe Handling and Disposal, 1984, 2d edit., 1988; co-editor: Safety and Accident Prevention in Chemical Operations, 1965, 2d edit., 1982, (with others) Hazards in the Chemical Laboratory, 4th and 5th edits.; mem. editorial adv. bd. Jour. Safety Rsch., 1968—, Transp. Planning and Tech., 1972—; N.Am. regional editor Jour. Hazardous Materials, 1975—; also book chpt. Chief radiol. sect. Schenectady County CD, 1953-63; bd. dirs. Safety sect. Schenectady C. of C., 1957-64; tech. advisor Hazmat Emergency Response Team, Montgomery County, Md., 1988—. Deacon Warner Meml. Presbyn. Ch., Kensington, Md., 1990—. Recipient Disting. Svc. to Safety citation Nat. Safety Coun., 1966, Cameron award, 1962, 69, Profl. Svc. award, 1992. Fellow Am. Inst. Chemists; mem. Am. Chem. Soc. (sec. com. chem. safety, chmn. council com. on chem. safety 1974-77, chmn. div. chem. health and safety 1977-79, 91, vice-chair, 1990-91, chair, 1991—, councilor 1980-82, archivist, 1984—, author audio course on hazards of materials 1977, CHAS award 1993), ASTM (membership sec. 1972—, sub-chmn. D-34 com.), Am. Inst. Chem. Engrs. (com. on occupational health and safety 1977—, editor newsletter 1988-89), Internat. Platform Assn., Am. Indsl. Hygiene Assn. (dir. Balt.-Washington chpt. 1975-77), Alpha Chi Sigma (contbr. video tapes on chemical hazards). Home and Office: PO Box 9444 12920 Matey Rd Wheaton MD 20916

FAY, ROBERT CLINTON, chemist, educator; b. Kenosha, Wis., Mar. 14, 1936; s. Clinton Edward and Selma (Lenz) F.; m. Carol Lee Baker, Aug. 25, 1960. A.B., Oberlin Coll., 1957; postgrad., Wheaton Coll., 1957-58; M.S., U. Ill., 1960, Ph.D., 1962. Teaching fellow Wheaton (Ill.) Coll., 1957-58; teaching asst. U. Ill., Urbana, 1958-59; inorganic chemist Nat. Bur. Standards, Washington, summers 1957-60; asst. prof. chemistry Cornell U., Ithaca, N.Y., 1962-68; assoc. prof. Cornell U., 1968-75, prof., 1975—; vis. prof. chemistry Harvard U., Cambridge, Mass., 1990-91, contract prof. U. Bologna, Italy, 1992. Contbr. articles to profl. jours. NSF fellow, 1960-62; NSF faculty fellow U. East Anglia, U. Sussex, Eng., 1969-70; Sci. and Engring. Research Council vis. fellow and NATO/Heineman sr. fellow Oxford (Eng.) U., 1982-83; NSF grantee, 1964-80; recipient Clark Disting. Teaching award Cornell U., 1980. Mem. Am. Chem. Soc., Royal Soc. Chem. (London), Am. Crystallographic Assn., Sigma Xi, Phi Kappa Phi, Phi Beta Kappa, Phi Lambda Upsilon, Pi Mu Epsilon. Home: 318 Eastwood Ave Ithaca NY 14850-6202 Office: Cornell Univ Dept Chemistry Ithaca NY 14853

FAYAD, NABIL MOHAMED, chemist, researcher; b. Jaffa, Palestine, July 25, 1947; s. Mohamed Mohamed Fayad and Sauad Ahmed (Dahaweer) F.; m. Kawkab Mohamed Abu Gobara, June 22, 1974; children: Belal, Nedal, Ahmed, Amal. BSc, Alexandria U., 1969; PhD, U. Tech., Loughborough, Eng., 1979. Tchr. sci. Ministry of Edn., Makkah, Saudi Arabia, 1971-76; rsch. scientist Rsch. Inst. King Fahd U. Petroleum & Minerals, Dhahran, Saudi Arabia, 1980—. Contbr. articles to Internat. Jour. Environ. Analyt. Chemistry, Bull. Environ. Contamination and Toxicology. Mem. Am. Chem. Soc., Am. Water Work Assn., Water Polution Control Fedn. Achievements include research in understanding and assessment of environmental pollution problems in Saudi Arabia in the fields of drinking water contamination; oil pollution. Home and Office: King Fahd U, PO Box 1479, Dhahran 31261, Saudi Arabia

FEARN, JEFFREY CHARLES, biochemist; b. Ogdensburg, N.Y., July 11, 1960; s. William John Rogers and Brigitte (Esser) F. BS, Cornell U., 1982; PhD, U. Ill., Chgo., 1987. Teaching asst., rsch. asst. Cornell U., Ithaca, N.Y., 1981-82; grad. fellow U. Ill., Chgo., 1982-87; postdoctoral fellow Boyce Thompson Inst. Plant Rsch./Cornell U., Ithaca, 1987-91; rsch. assoc. N.C. State U., Raleigh, 1992-93; group leader, cell biology Upstate Biotech., Inc., Lake Placid, N.Y., 1993—. Contbr. articles to sci. jours. Nat. Agrl. Biotech. Coun. fellow, 1990-91. Mem. Am. Soc. Plant Physiologists, Sigma Xi. Republican. Episcopalian. Achievements include demonstration of protein kinase C modulated epidermal growth factor receptor binding via phosphorylation by using an in vitro assay system; characterization of nitrogen-fixing pea mutant, confirmation that mutant was ethylene-sensitive. Home: 59 1/2 Sentinel Rd # 4 Lake Placid NY 12946 Office: Upstate Biotech Inc 199 Saranac Ave Lake Placid NY 12946

FEARON, LEE CHARLES, chemist; b. Tulsa, Nov. 22, 1938; s. Robert Earl and Ruth Belle (Strothers) F.; m. Wanda Sue Williams, Nov. 30, 1971. Student, Rensselaer Polytech. Inst., 1957-59; BS in Physics, Okla. State U., Stillwater, 1961, BA in Chemistry, 1962, MS in Analytical Chemistry, 1969. Rsch. chemist Houston process lab. Shell Oil Co., Deer Park, Tex., 1968-70; chief chemist Pollution Engring. Internat., Inc., Houston, 1970-76; rsch. chemist M-I Drilling Fluids Co., Houston, 1976-83; cons. chemist Profl. Engr. Assocs., Inc., Tulsa, 1983-84; chemist Anacon, Inc., Houston, 1984-85; scientist III Bionetics Corp., Rockville, Md., 1985-86; sr. chemist L.A. County Sanitation Dist., Whittier, Calif., 1986; chemist Enseco-CRL, West Sacramento, Calif., 1986-87; consulting chemist Branham Industries, Inc., Conroe, Tex., 1987-89; adv. laboratorian EILS, QA sect. Wash. State Dept. Ecology, Manchester, 1989—; cons. chemist Terra-Kleen, Okmulgee, Okla., 1988—, Excel Pacific, Inc. , Camarillo, Calif., 1993—. With U.S. Army, 1962-65. Fellow Am. Inst. Chemists; mem. AAAS, Am. Chem. Soc., Am. Inst. Chemists. Avocations: photography, travel. Home: PO Box 514 Manchester WA 98353-0514 Office: PO Box 488 Manchester WA 98353-0488

FEAST, WILLIAM JAMES, polymer chemist; b. Birmingham, Eng., June 25, 1938; s. William Edward and Lucy Mary (Willis) F.; m. Jenneke Elizabeth C. van der Kuijl, Aug. 8, 1967; children: Saskia, Marieke. BSc in Chemistry with honors, Sheffield U., 1960; PhD in Chemistry, Birmingham U., 1963. Postdoctoral fellow Birmingham U., 1963-65; lectr. Durham U., Eng., 1965—, sr. lectr., 1975—; prof. chemistry Durham U., 1975—; vis. prof. Max Planck Inst. for Polymer Forschung, Mainz, 1985-88; Gillett Internat. Rsch. fellow Leuven, Belgium, 1968-69. Contbr. 150 articles to profl. publs. Fellow Royal Soc. Chemistry; mem. MacroGroup (Eng., chmn. 1989-92). Achievements include research on polymer synthesis, conjugated polymers, fluorinated polymers and functionalized polymers via ring opening matathesis polymerization. Office: Durham U, South Rd, Durham DH1 3LE, England

FEDAK, MITCHEL GEORGE, chemistry educator; b. McKeesport, Pa., Jan. 2, 1952; s. George and Emma Mary (Kadar) F. BA in Chemistry, Ind. U. Pa., 1973; postgrad., Carnegie-Mellon U., 1974-76. Chemist R & D PPG Industries Inc., Springdale, Pa., 1974-79; prodn. supr. U.S.X. Corp., Clairton, Pa., 1979-83; analytical chemist Alkav Labs., Pitts., 1988-89; prof. chemistry, physics and math. C.C. Allegheny County, Monroeville, Pa.,

1983—. Reviewer and editor Tesxtbook Reviewer, 1987—. Vol. Am. Legion, White Oak, Pa., 1978—, AmVets, McKeesport, Pa., 1988—. Sgt. USMC, 1977-78. Mem. ASCD, Am. Assn. Physics Tchrs., Am. Chem. Soc. Roman Catholic. Home: 1345 Walnut Rd North Versailles PA 15137 Office: C C Allegheny County 595 Beatty Rd Monroeville PA 15146

FEDOR, GEORGE MATTHEW, III, industrial engineer; b. Bridgeport, Conn., Nov. 17, 1967; s. George Matthew and Joan Patricia (Hammond) F. AS in Mech. Engring., Waterbury State Tech. Coll., 1987; BS in Indsl. Engring., Western New Eng. Coll., 1991. Engr. Avco Lycoming Textron, Stratford, Conn., 1987; advanced devel. engr. Black & Decker (U.S.) Inc., Shelton, Conn., 1988—. Mem. ASME, Inst. Indsl. Engrs., Am. Soc. Quality Control, Am. Prodn. and Inventory Control Soc., Tau Beta Pi, Sigma Beta Tau. Avocations: photography, camping, travel, skiing, bicycling. Home: 51 Applewood Dr Shelton CT 06484

FEDOROWICZ, JANE, information systems educator; b. Derby, Conn., Mar. 5, 1955; m. Michael Golibersuch, Sept. 3, 1984; 1 child, Andrew. BS in Health Systems, U. Conn., 1976; MS, Carnegie-Mellon U., 1978, PhD in Systems Sci., 1981. Instr., lectr. Carnegie-Mellon U., Pitts., 1978-80; asst. prof. Northwestern U., Evanston, Ill., 1980-85; assoc. prof. Boston U., 1985—; gen. chair DSS-92, Chgo., 1992; mem. editorial bd. Info. Systems Rsch., Providence, 1987-93, Mgmt. Info. Systems Quarterly, 1989-94; chair coll. info. system Inst. Mgmt. Scis., Providence, 1986-88. Contbr. articles to profl. jours. Named J.L. Kellogg Rsch. Prof., Northwestern U., 1984-85. Mem. Assn. Computing Machinery, Inst. Mgmt. Scis. (coll. chair 1986-88). Roman Catholic. Achievements include rsch. on the impact of info. tech. on individuals and orgns. Office: Boston U 25 Salem End Ln Framingham MA 01701

FEE, GERALDINE JULIA, psychophysiologist; b. Perth Amboy, N.J., Apr. 26, 1937; d. John Clarence and Julia (Elek) Krisak; m. John James Fee, July 4, 1959 (dec. July 1980); children: John P., Brian D., Timothy M., Kevin J., Sean T. BS, Kean Coll., 1958; MA, Seton Hall U., 1962; Phd, Calif. Coast U., 1992. Cert. tchr., N.J.; group leader cert. N.J. State Dept. Health; nat. biofeedback cert. Dir. West Orange (N.J.)/Princeton Biofeedback Clinic, 1984-87; biofeedback therapist Neurology Group of Bergen County, Ridgewood, N.J., 1987—; psychology educator Middlesex County Coll., Edison, N.J., 1987—; assoc., intern Behavioral Medicine & Biofeedback Clinic, Princeton, 1982-84. Mem. Am. Assn. Behavioral Therapist, Assn. Applied Psychophysiology & Biofeedback, N.Y. Neuropsychol. Group, N.Y. Acad. Scis., N.Y. Biofeedback Soc., N.J. Biofeedback Soc. Home: 799 Ridgedale Ave Woodbridge NJ 07095-3638

FEEMAN, JAMES FREDERIC, chemist, consultant; b. Lebanon, Pa., June 1, 1922; s. Edwin L. and Florence A. (Wenrich) F.; m. June Permilla Zartman, Apr. 12, 1947; children—James Frederic, Jane Elizabeth, Joan Ann, John Harry. B.S., Muhlenberg Coll., 1945; M.S., Lehigh U., 1947, Ph.D. in Chemistry, 1949. Rsch. assoc. Ohio State U. Rsch. Found., Columbus, 1949-50, Althouse Chem. Co., Reading, Pa., 1950-54; rsch. assoc. Crompton & Knowles Corp., Reading, 1954-68, asst. dir. rsch., 1968-72, assoc. dir. rsch., 1972-74, dir. rsch., 1974-80, v.p. R&D, 1980-86, sr. scientist, 1986-88; cons. Lawrence Livermore Nat. Lab., 1989—. Patentee in field (30); author: (with others) The Chemistry of Synthetic Dyes, Vol. VIII, 1978; contbr. articles to sci. jours.; researcher on textile, paper, leather dyes; inventor first neutral dyeing series of dyes for nylon having outstanding properties. Instr., CD, 1955-65; dir. Reading-Berks Sci. Fair, 1955-70. Served with Signal Corps, U.S. Army, 1942-46; ETO, PTO. Muhlenberg Coll. scholar, 1940-43; Lehigh U. fellow, 1947-49. Fellow AAAS, Am. Inst. Chemists; mem. Am. Chem. Soc. (sect. chair 1967-68, nat. councillor 1973-75), Am. Assn. Textile Chemists and Colorists, N.Y. Acad. Scis., Reading Chemists Club, Sigma Xi. Republican. Lutheran. Club: Torch (pres. 1977-78) (Reading). Lodge: Rotary (West Reading and Wyomissing). Avocations: art; woodworking; photography. Home and Office: 6 Oriole Dr Reading PA 19610-2841

FEENEY, CRAIG MICHAEL, chemist; b. Rochester, N.Y., Apr. 24, 1956; s. William Patrick and Phyllis Dorsey (Hromowyk) F.; m. Paula May Riley, May 20, 1977; 1 child, Daniel Patrick. BS, U. Ariz., 1979. Chemist Mountain States Rsch. and Devel., Tucson, 1980-84; mem. tech. staff Hughes Aircraft Co., Tucson, 1984-91, process engr., 1991—. Achievements include research in atomic spectroscopy. Office: Hughes Missile Systems Co Old Nogales Hwy Tucson AZ 85734

FEENEY, ROBERT EARL, research biochemist; b. Oak Park, Ill., Aug. 30, 1913; s. Bernard Cyril and Loreda (McKee) F.; m. Mary Alice Waller, Dec. 3, 1954; children: Jane, Elizabeth. Student, Rochester (Minn.) Jr. Coll., 1932-33; BS in Chemistry, Northwestern U., 1938; MS in Biochemistry, U. Wis., 1939, PhD in Biochemistry, 1942. Diplomate Am. Bd. Nutrition. Rsch. assoc. Harvard U. Med. Sch., Boston, 1942-43; rsch. biochemist USDA Lab., Albany, Calif., 1946-53; prof. chemistry U. Nebr., Lincoln, 1953-60; prof. dept. food sci. and tech. U. Calif., Davis, 1960-84, prof. emeritus, rsch. biochemist, 1984—, interim dir. protein structure lab., 1990-91; bd. dirs. Creative Chemistry Cons., Davis. Author: (with Richard Allison) Evolutionary Biochemistry of Proteins, 1969, (with Gary Means) Chemical Modification of Proteins, 1971, Professor On the Ice, 1974; editor: (with John Whitaker) Protein Tailoring for Food and Medical Uses, 1986; editor jour. Comments on Agr. and Food Chemistry, 1985—. Capt. wound rsch. team M.C., U.S. Army, 1943-46. Recipient Superior Svc. award USDA, 1953, ; Feeney Peak, Antarctica named in his honor U.S. Bd. on Geog. Names, 1968. Mem. Am. Chem. Soc. (chmn. div. agrl. and food chemistry, 1978-79, award for disting. svc. in agrl. and food chemistry, 1978), Am. Soc. for Biochemistry and Molecular Biology, Inst. of Food Technologists, Explorers Club. Democrat. Avocations: polar sci., polar exploration lit. Home: 780 Elmwood Dr Davis CA 95616-3517 Office: U Calif Dept Food Sci and Tech Cruess Hall Davis CA 95616

FEFFERMAN, CHARLES LOUIS, mathematics educator; b. Washington, Apr. 18, 1949; s. Arthur Stanley and Liselott Ruth (Stern) F.; m. Julie Anne Albert, Feb. 1975; children: Nina Heidi, Elaine Marie. BS, U. Md., 1966, hon. doctorate, 1979; PhD, Princeton U., 1969; hon. doctorate, Knox Coll., 1981, Bar-Ilan U., Israel, 1985. Instr. math. Princeton (N.J.) U., 1969-70, prof. math., 1974—; mem. faculty U. Chgo., 1970-74, prof. math., 1971-74. Author research papers. Recipient Salem prize for oustanding work in fourier analysis by young mathematician, 1978, Alan T. Waterman award, 1978, Fields medal Internat. Cong. Mathematicians, 1978, 84. Mem. Nat. Acad. Scis., Am. Math. Soc., Am. Acad. Arts and Scis. Home: 234 Clover Ln Princeton NJ 08540-4051 Office: Fine Hall Princeton U Princeton NJ 08540

FEHER, GEORGE, physics and biophysics scientist, educator; b. Czechoslovakia, May 29, 1924; s. Ferdinand and Sylvia (Schwartz) F.; m. Elsa Rosenvasser, June 18, 1961; children—Laurie, Shoshanah, Paoli. B.S. in Engring. Physics, U. Calif.-Berkeley, 1950, M.S. in Elec. Engring., 1951, Ph.D. in Physics, 1954. Research physicist Bell Telephone Labs., Murray Hill, N.J., 1954-60; vis. assoc. prof. Columbia U., N.Y.C., 1959-60; prof. physics U. Calif.-San Diego, 1960—; vis. prof. biology MIT, Cambridge, 1967-68; William Draper Hawkins lectr. U Chgo., May 1986; Raymond and Beverly Sackler disting. lectr. U. Tel-Aviv, June 1986; vis. prof. Hebrew U. of Jerusalem, Israel, spring 1989, 93; bd. govs. Weizmann Inst. Sci., Rehovot, Israel, 1988; Bruker lectr. Oxford U., Eng., 1992. Author: Electron Paramagnetic Resonance with Applications to Selected Problems in Biology, 1970; contbr. numerous articles to profl. jous., chpts. to books. Bd. govs. Technion-Israel Inst. Tech., 1968. Recipient Oliver E. Buckley Solid State Physics prize, 1976; NSF fellow, 1967-68, Inaugural Annual award Internat. Electron Spin Resonance Soc., 1991, Rumford Medal Am. Acad. Arts and Scis., 1992. Fellow AAAS; mem. Am. Phys. Soc. (prize 1960, biophysics prize, 1982), Biophys. Soc. (nat. lectr. 1983), Nat. Acad. Scis., Am. Acad. Arts and Scis., Sigma Xi. Office: U Calif Dept Physics 0319 9500 Gilman Dr La Jolla CA 92093-0001

FEHL, BARRY DEAN, civil engineer; b. Abilene, Tex., Mar. 23, 1957; s. Georg H. and Josephine D. (Hensiek) F.; m. Colleen S. Murphy, Aug. 15, 1980; children: Leah, Blake, Kristen. BSCE, U. Mo.-Rolla, 1980; MS in Engring., U. Tex., 1987. Registered profl. engr., Mo. Structural engr. St. Louis Dist. Corps of Engrs., 1980-91; civil engr. Waterways Experiment Sta.-Corps of Engrs., Vicksburg, Miss., 1991—. Contbr. articles to profl. jours. Recipient Gustave Willems award Permanent Internat. Assn. Navigation Congress, 1989. Mem. ASCE. Lutheran. Office: Waterways Experiment Sta 3909 Halls Ferry Rd Vicksburg MS 39180

FEHLNER, THOMAS PATRICK, chemistry educator; b. Dolgeville, N.Y., May 28, 1937; s. Herman Joseph and Mary (Considine) F.; m. Nancy Lou Clement, July 28, 1962; children: Thomas P., Anne Marie. BS, Siena Coll., Loudonville, N.Y., 1959; PhD, Johns Hopkins U., 1963. Rsch. assoc. Johns Hopkins U., Balt., 1963-64; asst. prof. U. Notre Dame, Ind., 1964-67, assoc. prof., 1967-75, prof., 1975—, chmn. dept. chemistry, 1982-88, Grace Rupley chair chemistry, 1988. Author: (with others) Inorganic Chemistry; contbr. articles to profl. jours. Guggenheim fellow, 1988-89. Fellow AAAS; mem. Am. Chem. Soc., Materials Rsch. Soc., Internat. Union Pure & Applied Chemistry. Democrat. Roman Catholic. Office: U Notre Dame Dept Chemistry Notre Dame IN 46556

FEIGENBAUM, ABRAHAM SAMUEL, nutritional biochemist; b. N.Y.C., Mar. 11, 1929; s. Benjamin and Pearl Feigenbaum; m. Hannah Devries, Aug. 17, 1952; children: Benjamin, Josef, Miriam. BS, Rutgers U., 1951, MS, 1959, PhD, 1962. Chemist E.R. Squibb, New Brunswick, N.J., 1954-57; rsch. asst. Rutgers U., New Brunswick, N.J., 1957-61; rsch. scientist, chief neuroendocrinology N.J. Bur. Rsch. in Neurology and Psychiatry, Skillman, 1961-73; dir. clin. nutrition, dir. clin. coordination Warren-Teed Pharm./Adria Labs., Inc., Columbus, Ohio, 1973-81; clin. project dir., dir. rsch. Pharm. Rsch. Inst., sub. Akzo, Columbus, Ohio, 1981—; dir. clin. devel. Organon, Inc., West Orange, N.J., 1981—; guest lectr. in nutrition Hahnemann Med. Coll., Phila., 1977-80. Mem. Bd. Edn., Highland Park, N.J., 1969-73, pres., 1970-72, v.p., 1972-73. NSF fellow, 1961. Mem. Am. Inst. Nutrition, Am. Chem. Soc., Soc. Exptl. Biology and Medicine. Home: 304 N 4th Ave New Brunswick NJ 08904 Office: Organon Inc 375 Mt Pleasant Ave West Orange NJ 07052-2798

FEIGENBAUM, EDWARD ALBERT, computer science educator; b. Weehawken, N.J., Jan. 20, 1936; s. Fred J. and Sara Rachman; m. H. Penny Nii, 1975; children: Janet Denise, Carol Leonora, Sheri Bryant, Karin Bryant. BEE, Carnegie Inst. Tech., 1956, Ph.D. in Indsl. Adminstrn., 1960. Asst., then assoc. prof. bus. adminstrn. U. Calif. at Berkeley, 1960-64; assoc. prof. computer sci., then prof. Stanford U., 1965—; prin. investigator heuristic programming project, 1965—; dir. Computation Ctr. Stanford U., 1965-68, chmn. dept. computer sci., 1976-81; pres. Intelli Genetics Inc., 1980-81; chmn., dir. Teknowledge, Inc., 1981-82; mem. tech. adv. bd. Intelli Genetics Inc., 1983-86; dir. IntelliCorp, 1984—; cons. to industry, 1957—; mem. computer and biomath. scis. study sect. NIH, 1968-72, mem. adv. com. on artificial intelligence in medicine, 1974—; mem. Math. Social Sci. Bd., 1975-78; computer sci. adv. com. NSF, 1977-80; mem. Internat. Joint Coun. on Artificial Intelligence, 1973-83; bd. dirs. Design Power Ind. Author: (with others) Information Processing Language V Manual, 1961, (with P. McCorduck) The Fifth Generation; author: (with R. Lindsay, B. Buchanan, J. Lederberg) Applications of Artificial Intelligence to Organic Chemistry: the Dendral Program; Editor: (with J. Feldman) Computers and Thought, 1963, (with A. Barr and P. Cohen) Handbook of Artificial Intelligence, 1981, 82, 89, (with Pamela McCorduck and H. Penny Nii) The Rise of the Expert Company: How Visionary Companies are using Artificial Intelligence to Achieve Higher Productivity and Profits; mem. editorial bd.: Jour. Artificial Intelligence, 1970—. Feigenbaum medal established in his honor World Congress on Expert Systems; Fulbright scholar Gt. Britain, 1959-60. Fellow AAAI, AAAS, Am. Coll. Med. Informatics; mem. Nat. Acad. Engring., Assn. Computing Machinery (nat. coun. 1966-68, chmn. spl. interest group on biol. applications 1973-76), Am. Assn. Artificial Intelligence (pres. 1980-81), Cognitive Sci. Soc. (coun. 1979-82), Sigma Xi, Tau Beta Pi, Eta Kappa Nu, Pi Delta Epsilon. Home: 1017 Cathcart Way Palo Alto CA 94305-1048 Office: Stanford U Knowledge Systems Lab 701 Welch Rd # C Palo Alto CA 94304-1709

FEIGENBAUM, MITCHELL JAY, physics educator; b. Phila., Dec. 19, 1944; s. Abraham Joseph and Mildred (Sugar) F. B.E.E., CCNY, 1964; Ph.D., MIT, 1970. Research assoc., instr. physics dept. Cornell U., Ithaca, N.Y., 1970-72, prof. physics, 1982-87; research assoc. physics dept. Va. Poly. Inst., Blacksburg, 1972-74; staff mem. theory div. Los Alamos Nat. Lab., 1974-81, lab. fellow, 1981-82; Toyota prof. physics Rockefeller U., N.Y.C., 1987—; vis. mem. Inst. Advanced Study, Princeton, N.J., 1978, Institute des Hautes Recherches Scientifiques, Bueres-sur-Yvette, France, 1980—. Mem. editorial bd. Advances in Applied Math. Recipient Disting. Performance award Los Alamos Nat. Lab., 1980, Ernest O. Lawrence award U.S. Dept. Energy, 1983, MacArthur Found. award, 1984, Wolf Prize for Physics, 1986. Fellow Am. Phys. Soc.; mem. NAS, Am. Acad. Arts and Scis., N.Y. Acad. Scis., Sigma Xi. Discovered theory period doubling route to turbulence, 1976-79. Home: 450 E 63d St 10L New York NY 10021 Office: Rockefeller U Physics Dept York Ave and 66th St New York NY 10021

FEIN, WILLIAM, ophthalmologist; b. N.Y.C., Nov. 27, 1933; s. Samuel and Beatrice (Lipschitz) F.; m. Bonnie Fern Aaronson, Dec. 15, 1963; children: Stephanie Paula, Adam Irving, Gregory Andrew. BS, CCNY, 1954; MD, U. Calif., Irvine, 1962. Diplomate Am. Bd. Ophthalmology. Intern L.A. County Gen. Hosp., 1962-63, resident in ophthalmology, 1963-66; instr. U. Calif. Med. Sch., Irvine, 1966-69; mem. faculty U. So. Calif. Med. Sch., 1969—, assoc. clin. prof. ophthalmology, 1979—; attending physician Cedars-Sinai Med. Ctr., L.A., 1966—, chief ophthalmology clinic svc., 1979-81, chmn. div. ophthalmology, 1981-85; attending physician Los Angeles County-U. So. Calif. Med. Ctr., 1969—; chmn. dept. ophthalmology Midway Hosp., 1975-78; dir. Ellis Eye Ctr., L.A., 1984—. Contbr. articles to med. publs. Chmn. ophthalmology adv. com. Jewish Home for Aging of Greater L.A., 1993—. Fellow Internat. Coll. Surgeons, Am. Coll. Surgeons; mem. Am. Acad. Ophthalmology, Am. Soc. Ophthalmic Plastic and Reconstructive Surgery, Royal Soc. Medicine, AMA, Calif. Med. Assn., L.A. Med. Assn. Home: 718 N Camden Dr Beverly Hills CA 90210-3205 Office: 415 N Crescent Dr Beverly Hills CA 90210-4860

FEINENDEGEN, LUDWIG EMIL, retired hospital and research institute director; b. Garzweiler, Germany, Jan. 1, 1927; s. Ludwig and Rosa (Klauth) F.; m. Jeannine Gemuseus; children: Dominik, Christophe. Student med. edn., U. Cologne, Germany, 1946-52; postgrad. in internal medicine, U. Hosp., Cologne. Med. tng. St. Peter's Gen. Hosp., New Brunswick, N.Y.; St. Cornelius Hosp., Viersen, Germany; med. tng. St. Vincents Hosp., N.Y.C., 1952-58; asst. physician, scientist med. dept. Brookhaven Nat. Lab., Upton, N.Y., 1958-63; sci. officer European Atomic Energy Commn., Brussels, Belgium, 1962-67; sci. officer lab. Pasteur de l'Inst. du Radium, Paris, France, 1964-67; dir. inst. medicine rsch. ctr. Jülich GmbH., Fed. Republic Germany, 1967-93; prof. nuclear medicine U. Hosp. Düsseldorf, Düsseldorf, Germany, 1967-93, prof. emeritus, 1993—; appt. rsch. collaborator med. dept. Brookhaven Nat. lab., Upton, N.Y., 1963—; mem. Nat. Coun. on Radiation Protection and Measurements, Com. 24, U.S., 1969-79; mem. adv. coun. Fed. Ministry Health, Bonn, Germany, 1972-93; mem. adv. coun. Fed. Ministry Def., Bonn., Germany, 1983—; mem. Internat. Commn. on Radiol. Protection, Com. 2, Eng., 1973-85; mem. Interna. Comm. on Radiation Units an dMeasurements, 1982—; mem. sci. coun. dept. for radiation hygiene Fed. Bd. Health, Berlin, 1973—; mem. coun. sci. & lit. Goethe Inst., Munich, 1978-87. Mem. civil def. commn. Fed. Ministry of Interior, Bonn, Germany, 1974—; mem. commn. experts "Rsch. Baden-Württemberg," Ministry Sci. and Arts, Stuttgart, Germany, 1988-89. Recipient Order of Merit, 1982; Dr. Robert K. Match disting. scholar L.I. Jewish Med. Ctr., N.Y., 1989. Mem. Rhine-Westfalian Acad. Scis. (class for natural scis., engring. and econs. 1971—), Fed. German Med. Assn. (sci. coun. 1982—), Kuratorium (ann. Lidau meetings Nobel Laureates 1978—, G.V. Hevesy Lecture medal 1990, C.W. Röntgen medal 1991, state prize Northrhine-Westfalia, 1991). Home: Wolfshover Str 197, D 5170 Jülich Stetternich, Germany

FEINGOLD, DANIEL LEON, physician, consultant; b. Boston, May 19, 1958; s. Macey Gerson and Hélène Sultana (Benlolo) F. BS with distinction, U. Ill., Chgo., 1980; MD, U. Health Scis., Chgo. Med. Sch., 1984. Intern Weiss Meml. Hosp., Chgo., 1984-85; resident in anesthesiology U. Ill. Hosps. and Clinics, Chgo., 1986-89; anesthesiologist Hosp. Anesthesia Group, Chgo., 1989—. Contbr. articles to profl. publs. Mem. AMA, AAAS, Am. Soc. Anesthesiologists, Am. Soc. Regional Anesthesia, Ill. State Med. Soc. Home: 4180 N Marine Dr Chicago IL 60613-2219 Office: PO Box 25678 Chicago IL 60625-0678

FEINGOLD, MARK LAWRENCE, electronic warfare officer; b. Paterson, N.J., Mar. 9, 1963; s. Martin Herbert and Carole Fern (Stenchever) F.; m. Carol Ann Vician, July 17, 1988. BS in Aero./Astro. Engring., U. Ill., 1985; MS in Engring. Mgmt., U. Alaska, 1991. Commd. 2d lt. USAF, 1985, advanced through grades to capt., 1989; stationed at Mather AFB, 1986-87; standardization evaluation instr. electronic warfare officer 24th Reconnaissance Squadron, Eielson AFB, 1987-91; reconnaissance crew commdr. instr., electronic warfare officer 343 Reconnaissance Squadron, Offutt AFB, Nebr., 1991—. Vol. Meals on Wheels, Omaha, Nebr., 1992. Decorated Air medal with one oak leaf cluster, Aerial Achievement medal with three oak leaf clusters. Mem. AIAA, Assn. of Old Crows, Air Force Assn. Jewish. Home: 16315 Decatur Cir Omaha NE 68118 Office: 343 Reconnaissance Squadron Offutt AFB NE 68113

FEINSTEIN, ALVAN RICHARD, physician; b. Phila., Dec. 4, 1925; s. Joel B. and Bella (Ukasz) F. B.S., U. Chgo., 1947, M.S. in Math, 1948, M.D., 1952; M.A. (hon.), Yale U., 1969. Intern, then resident Yale-New Haven Hosp., 1952-54; research fellow Rockefeller Inst., 1954-55; resident Columbia-Presbyn. Med. Center, N.Y.C., 1955-56; clin. dir. Irvington House, N.Y.C., 1956-62; instr., then asst. prof. N.Y. U. Sch. Medicine, 1956-62; chief clin. pharmacology VA Hosp., West Haven, Conn., 1962-64; chief clin. biostatistics VA Hosp., 1964-74; mem. faculty Sch. Medicine, Yale U., 1962—, prof. medicine and epidemiology, 1969—, dir. clin. scholar program, 1974—, Sterling prof., 1991—; chief Eastern Research Support Ctr. VA, 1967-74; pres. New Haven area chpt. Assn. Computing Machinery, 1968-69. Author: Clinical Judgment, 1967, Clinical Biostatistics, 1977, Clinical Epidemiology, 1985, Clinimetrics, 1987; editor Jour. Clinical Epidemiology; also articles. Served with AUS, 1944-46. Recipient Francis G. Blake award for outstanding teaching Yale Med. Sch., J. Allyn Taylor Internat. prize, awards Soc. for Gen. Internal Medicine, U. Chgo., Ludwig Heilmyer Soc. (Europe), Gairdner Found. Internat. award 1993 (Can.). Mem. ACP (award), AMA, Assn. Am. Physicians, Am. Soc. Clin. Investigation, Am. Epidemiol. Soc., Inst. Medicine, Am. Bd. Internal Medicine, Am. Fedn. Clin. Research, Am. Soc. Clin. Pharmacology Therapeutics, Am. Statis. Assn., Assn. Computing Machinery, Biometric Soc., Am. Assn. History Medicine, Alpha Omega Alpha. Home: 164 Linden St New Haven CT 06511-2400 Office: Yale U Sch Medicine 333 Cedar St New Haven CT 06510-0825

FELAK, RICHARD PETER, electric power industry consultant; b. Newark, Aug. 1, 1945; s. Peter and Helen (Kazuba) F.; m. Joan Marie Carpentier, July 15, 1972. BSEE, Rensselaer Poly. Inst., 1966, MSEE, 1967. Registered profl. engr., N.Y. Various mgmt. positions GE Co., Schenectady, N.Y., 1967-90; pres. Richard P. Felak-Svcs. for the Power Industry, Schenectady, 1990—. Author: (with others) Cogeneration, 1986; contbr. articles to profl. publs. Mem. IEEE (sr.), Task Force on Transmission Access and Non-Utility Generation, Phi Sigma Kappa, Eta Kappa Nu. Achievements include rsch. on the necessity of including financial simultaion in long range generation planning, cutting fuel costs with optimal power flows, analysis of wheeling costs shows impact on system, ferc testimony, cogeneration project development, least cost system studies. Home: 27 Norwood Way Schenectady NY 12309-4856 Office: Svcs for the Power Industry 27 Norwood Way Schenectady NY 12309

FELCMAN, JUDITH, chemistry educator; b. Rio de Janeiro, Brazil, May 3, 1941; d. Jacob and Dora Ostrower; m. Elias Felcman, Mar. 10, 1962; children: Gisele, Rosane. BS, U. Brazil, Rio de Janeiro, 1962; MS, Cath. U. of Rio, Rio de Janeiro, 1980, ScD, 1983. Pharmacist Jewish Hosp. of Rio, Rio de Janeiro, 1973-77; auxiliar prof. Cath. U., Rio de Janeiro, 1978-84, asst. prof., 1984-88, assoc. prof. chemistry, 1988—; research in bioinorganic chemistry. Author: Cromo, 1988; contbr. sci. articles to profl. publs., chpts. to books. Mem. Wizo-Women Internat. Zionist Orgn., Israel, 1963—. Mem. Am. Chem. Soc., Royal Soc. Chemistry, Brazilian Soc. Chemistry. Office: PUC-RJ Dept Chemistry, Rua Marques de S Vincente 225, 22453900 Rio de Janeiro Brazil

FELDMAN, BRUCE ALAN, psychiatrist; b. St. Louis, Apr. 21, 1959; s. Jerome Stanley and Arlene (Greenberg) F.; m. Kathryn Matilda Estill, May 25, 1990. BA in Biology, U. Mo., Kansas City, 1982, MD, 1985. Diplomate Nat. Bd. Med. Examiners, Am. Bd. Psychiatry and Neurology. Resident in psychiatry So. Ill. U., Springfield, 1986-89, adminstrv. chief resident, 1988, geriatric psychiat. chief resident, 1989; pvt. practice Psychiat. Assocs., Springfield, 1990—; resident cons. psychiatrist Alzheimers Ctr. for Decatur, Ill., 1989, Alzheimers Ctr. and Memory Disorders Ctr., Springfield, 1989; cons. psychiatrist Taylorville (Ill.) Mental Health Ctr., 1990—, Jacksonville (Ill.) Mental Health Ctr., 1990—, Country View Living Ctr., Decatur, 1990—, Jacksonville Terr. Nursing Home, 1990—, Walnut Ridge N.H., 1992—, Springfield Mental Health Ctr., 1992—; hosp. affiliate Doctors Hosp., 1990—, Meml. Med. Ctr., 1990—, St. Johns Hosp., 1990—, St. Vincent Meml. Hosp., 1990—, Passavant Hosp., 1990—; mem. utility rev. com. Drs. Hosp., 1990-91, pharm. & therapeutical com. St. John's Hosp., 1992—; ltd. ptnr., owner Drs. Hosp., Springfield, 1990—, Drs. Hosp., Wentzville, Mo., 1991—; contbr. THA clin. studies SIV Alzheimers Clinic, 1989, Prosom Clin. Study, 1992. Mem. Jewish Fedn., Springfield, Ill., Temple Israel Synagogue, Springfield; exec. prodr., dir. Miss K.C. Pageant, 1981; mem. Nat. Rep. Congress com., 1988—, Rep. Nat. Candidate Trust, 1992—, Rep. Inner Circle, 1993—. Mem. AMA, Am. Psychiat. Assn., N.Y. Acad. Scis. (life), So. Med. Assn., Am. Assn. for Geriatric Psychiatry, Am. Geriatrics Soc., U.S. Senatorial Club, Bnai Brith (exec. bd. mem.), Delta Chi (past pres. Kansas City chpt. 1979, nat. v.p., chpt. advisor 1981-82, Alumnus of Yr. 1982). Republican. Avocations: reading, politics, sports, stamps, travel. Office: Psychiat Assocs 1124 S 6th St Springfield IL 62703-2406

FELDMAN, BRUCE ALLEN, otolaryngologist; b. Washington, Mar. 22, 1941; s. Irvin and Miriam Thelma (Rothstein) F.; m. Sharon Lee Pearlman, Dec. 25, 1966; children: Kathryn Ellen, Michael Aaron. AB, Dartmouth Coll., 1962, B Med. Sci., 1963; MD, Harvard U., 1965. Diplomate Am. Bd. Otolaryngology. Intern Hosp. of U. Pa., Phila., 1965-66, resident in surgery, 1966-67; resident in otolaryngology Mass. Eye and Ear Infirmary-Harvard U., Boston, 1967-70; pvt. practice Washington, 1972—; clin. prof. Surgery, Otolaryngology and Health Care Scis George Washington U., Washington, 1990—. Contbr. articles to med. jours., chpt. to book. Lt. comdr. M.C., USNR, 1970-72. Mosby scholar, 1963; recipient Physician's Recognition award Children's Hosp. Washington, 1991. Fellow ACS, Am. Laryngol., Rhinol. and Otol. Soc. (Mosher award 1981); mem. AMA, Med. Soc. D.C., Jacobi Med. Soc. (pres. 1986-87), Washington Met. Ear, Nose and Throat Soc. (pres. 1978-79), Woodmont Country Club (Rockville, Md.), Phi Beta Kappa, Alpha Omega Alpha, Phi Delta Epsilon (pres. grad. club 1979-80). Jewish. Office: 1145 19th St Washington DC 20036

FELDMAN, CHARLES, physicist; b. Balt., Mar. 20, 1924; s. Maurice and Trixie Lucile (Rubenstein) F.; m. Estelle Harriet Rosenberg, Aug. 26, 1946; children: Kay Emil, John Isaac. AB, Johns Hopkins U., 1944, MA, 1949; PhD cum laude, Sorbonne, U. Paris, 1952. Sr. physics master Wellingborough Grammer Sch., Eng., 1950; fellow Ctr. Nat. Rsch. Scientfic, Paris, 1951-52; section head U.S. Naval Rsch. Lab., Washington, 1953-59; lab. head Melpar, Inc., Falls Church, Va., 1959-67; group supr. Applied Physics Lab./Johns Hopkins U., Balt., 1967-85; cons. Washington, 1985—; physics cons. to industries, 1956-59; adj. prof. George Washington U., Washington, 1969-74; vis. prof. Nat. U. Singapore, 1983; com. mem. Army Basic Rsch., Nat. Rsch. Coun., Washington, 1969-75; adv. group election devices, U.S. Govt., Washington, 1967-80. Contbr. physics articles, referee to profl. jours.; contbg. author book chpts. in field. Com. mem. Dem. Party, Ward 3, Washington, 1991-93. With U.S. Army, 1943-46, ETO. Decorated Bronze Star; recipient Naval Rsch. Applied Sci. award Rsch. Soc. of Am., Washington, 1958, NASA Invention award, 1969. Mem. Am. Phys. Soc., Optical Soc. Am., Am. Vacuum Soc., Sigma Xi. Achievements include pioneering work in thin solid films; 13 U.S. and fgn. patents; covering thin film elec. devices leading to present day integrated cirs.; amorphous semiconductors, thin film solar cells, early work on cathode ray tube screens. Home and Office: 2855 Davenport St NW Washington DC 20008

FELDMAN, ELAINE BOSSAK, medical nutritionist, educator; b. N.Y.C., Dec. 9, 1926; d. Solomon and Frances Helen (Fania) Nevler Bossak; m.

Herman Black, Dec. 23, 1951 (div. 1957); 1 child, Mitchell Evan; m. Daniel S. Feldman, July 19, 1957; children: Susan, Daniel S. Jr. AB magna cum laude, NYU, 1945, MS, 1948, MD, 1951. Diplomate Am. Bd. Internal Medicine, Am. Bd. Med. Examiners; cert. in Clin. Nutrition. Rotating intern Mt. Sinai Hosp., N.Y.C., 1951-52, resident in pathology, 1952, asst. resident, 1953, fellow in medicine, resident in metabolism, 1954-55, rsch. asst. in medicine, 1955-58, clin. asst. physician Diabetes Clinic, 1957; asst. vis. physician Kings County Hosp., Bklyn., 1958-66, assoc. vis. physician, 1966-72; asst. attending physician Maimonides Hosp., Bklyn., 1960-68; spl. fellow USPHS Dept. of Physiol. Chemistry U. of Lund, Sweden, 1964-65; attending physician Eugene Talmadge Meml. Hosp., Augusta, Ga., 1972—; attending physician Univ. Hosp., Augusta, 1972—, cons., 1973; prof. Med. Coll. Ga., Augusta, 1972-92, prof. emeritus, 1992—, chief sect. of nutrition, 1977-92, chief emeritus, 1992—, acting chief sect. of metabolic/endocrine disease, 1980-81, prof. physiology and endocrinology, 1988-92, prof. emeritus physiology and endocrinology, 1992—; instr. medicine SUNY Downstate Med. Ctr., 1957-59, asst. prof. medicine, 1959-68, assoc. prof. medicine, 1968-72; teaching fellow dept. zoology, U. Wis. Grad. Sch., 1945-46, dept. biology NYU Grad. Sch., 1946-47; cons. N.Y.-N.J. Regional Ctr. for Clin. Nutrition Edn., 1983-92; vis. prof. and Harvey Lectr. Northeastern Ohio Sch. of Med., Youngstown, 1985; cons., vis. prof. U. Nev. Sch. of Medicine, Reno, 1985, 86; cons. U. Nevada Sch. Medicine (NCI Grant) 1989-94; mem. Nat. Adv. Com. Nutrition fellowship program, Nat. Med. Fellowship Inc., 1988—; dir. Ga. Inst. Human Nutrition, 1978-92, dir. emeritus, 1992—, Clin. Nutrition Rsch. Unit, 1980-86. Author: Essentials of Clinical Nutrition, 1988; (with others) Conference on Biological Activities of Steroids in Relation to Cancer, 1969, Nicotinic Acid, 1964, The Menopausal Syndrome, 1974, Hyperlipidemia, Medcom Special Studies, 1974, Medcom Famous Teaching in Modern Medicine, 1979, Harrison's Principles of Internal Medicine, 1980, Health Promotion: Principles and Clinical Applications, 1982, The Encyclopedic Handbook of Alcoholism, 1982, The Climacteric in Perspective, 1986, Selenium in Biology and Medicine, Part A., 1987, Medicine for the Practicing Physician, 1988, Clinical Chemistry of Laboratory Animals, 1989, Ency. Human Biology, 1991; editor: Nutrition and Cardiovascular Disease, 1976, Nutrition in the Middle and Later Years, 1983 (paperback edit. 1986), Nutrition and Heart Disease, 1983; mem. editoraial adv. bd. Contemporary Issues in Clin. Nutrition, 1980-92; mem. editorial bd. Am. Jour. Clin. Nutrition, 1983-91, 92—, Jour. Clin. Endocrinology and Metabolism, 1984-88, MidPoint: Counseling Women through Menopause, 1984-85, Jour. Nutrition, 1985-89; cons. editor Jour. Am. Coll. Nutrition, 1982—; reviewer Jour. Lipid Rsch., Biochm. Pharmacology, Sci., The Physiologist, Jour. Am. Acad. Dermatology, Israel Jour. Med. Scis., N.Y. State Jour. Medicine, Jour. of Nutrition Edn., Jour. Am. Dietetic Assn., Am. Jour. Medicine, Am. Jour. Med. Sci., So. Med. Jour.; author 141 published articles in field, numerous abstracts and presentations. Mem. tech. adv. com. for sci. and edn. Rsch. Grants Program, Human Nutrition Grants Peer Panel, USDA, 1982, mem. bd. sci. counselors human nutrition; Community Svc. Block Grant Discretionary Program Panel; vice chmn. Urban and Rural Econ. Devel. Panel, Dept. HHS, 1982, grant reviewer, 1983; mem ad hoc and spl. rev. coms. and groups NIH, 1979-93, mem. nutrition study sect., 1976-80; mem. Rev. Panel Nat. Nutrition Objectives, Life Scis. Rev. Office, Fed. Am. Socs. Exptl. Biology, 1985-86; mem. subcom. Women's Health Trial Nat. Cancer Inst., 1987, mem. bd. sci. counselors cancer prevention and control program, 19 adv. com. Clin. Nutrition Rsch. Unit, U. Ala., 1986—, Ga. Nutrition Steering Com., 1974-75, Ctrl. Savannah River Area Nutrition Project Coun. 1974-75, ednl. adv. com. Health Central, 1980; mem. geriatrics and gerontology rev. com. Nat. Inst. on Aging, 1986-90. N.Y. Heart Assn. rsch. fellow, 1955-57. Fellow Am. Heart Assn. Coun. on Atherosclerosis (nominating com. 1978, chmn. nominating com., mem. exec. com. 1979-80); mem. Am. Coll. Nutrition (chmn. com. pub. affairs), Am. Soc. for Clin. Nutrition (com. on nutrition edn. 1982, chmn. subcom. on nutrition edn. in med. schs. 1983-84, chmn. com. on med./dental residency edn., 1985-87, com. on subsplty. tng. 1988-92, nominating com. 1982, 90, com. on clin. practice issues in health and disease 1989-92, Nat. Dairy Coun. award 1991, rep. coun. acad. socs., 1990—), Am. Inst. Nutrition (grad. nutrition edn. com. 1980-83, 89-93), Fedn. Am. Socs. Exptl. Biology. Am. Oil Chemists Soc., Am. Physiol. Soc., Endocrine Soc., Soc. Exptl. Biology and Medicine, So. Soc. Clin. Investigation, Am. Diabetes Assn., Am. Fedn. Clin. Rsch., Am. Gastroent. Assn.. AMA (Joseph B. Goldberger award 1990), Am. Med. Women's Assn. (profl. resources com. 1975-76, med. edn. and rsch. fund com. 1976-79, chmn. 1978-90, chmn. student liaison subcom. of membership com. 1981-84, pres. Br. 51, Augusta 1977-80, treas. 1980—, Calcium Nutrition Edn. award 1991), Am. Soc. Parenteral and Enteral Nutrition, Am. Heart Assn. (Ga. affiliate, nutrition com., chmn. sci. session for nutritionists, 1978, chmn. nutrition com. 1979-90, mem. long range planning com. 1980-81, rsch. com. 1980-83, bd. dirs. 1987-90, profl. edn. task force, 1988-89, found. on aging com.), Richmond Country Med. Assn., Augusta Opera Assn. (bd. dirs. 1973—, recording sec. 1973-74, pres. 1974-75, coord. audience devel. 1975-77), Augusta Sailing Club (women's com. 1973), Greater Augusta Arts Coun. (Arts Festival Collage 1982 chmn. promotion and publicity com., Festival coms. 1983-86, 89, 90, mem. bd. dirs. 1984—), Gertrude Herbert Inst. Art (bd. dirs. 1987-92), Authors Club Augusta, Ctrl. Savannah River Area Found. Aging (v.p.), Philomathic Club, Phi Beta Kappa, Sigma Xi (chpt. sec. 1982-83, pres. elect 1983-84, pres. 1984-85), Alpha Omega Alpha. Avocations: opera, wine tasting, travel. Home: 2123 Cumming Rd Augusta GA 30904 Office: Med Coll Ga Sch Medicine 1120 15th St Rm 3207A Augusta GA 30912-1003

FELDMAN, FELIX, pediatric hematologist, oncologist; b. Bklyn., June 16, 1919; s. Saul and Mollie (Klein) F.; m. Judith Milberg, Dec. 19, 1943 (dec. 1986); children: Lenore Feldman Rosenberg, Susan; m. Eleanore Perlmutter, Mar. 6, 1988. BA, NYU, 1939; MD, SUNY, 1943. Diplomate Am. Bd. Pediatrics, subbd. Pediatric Hematology-Oncology. Rotating intern Jewish Hosp. of Bklyn., 1943; fellow pediatrics N.Y. Med. Coll., 1946-47; pediatric resident King's County Hosp., Bklyn., 1947-48; pvt. practice pediatrics Bklyn., 1948-64; chief pediatrics Coney Island Hosp., Bklyn., 1964-74; dir. pediatrics Maimonides Med. Ctr., Bklyn., 1971-91, dir. pediatric hematology/oncology, 1960—, pres. med. bd. Coney Island Hosp., 1969-74; vis. prof. Rockefeller U., 1984; prof. clin. pediatrics SUNY Health Sci. Ctr., Bklyn. Contbr. articles to profl. jours., chpts. to books. Capt. U.S. Army, 1944-46; ETO. Grantee USPHS, 1960, Rsch. Coun. of City of N.Y., 1965; recipient Richard L. Day Master Tchr. award in pediatrics SUNY, 1989. Fellow Am. Acad. Pediatrics, Internat. Soc. Hematology; mem. Am. Soc. Hematologists, Am. Soc. Pediatric Hematology and Oncology, Am. Soc. ChuGenetics. Democrat. Jewish. Office: Maimonides Med Ctr 4802 10th Ave Brooklyn NY 11219-2999

FELDMAN, GARY MARC, nutritionist, consultant; b. Bklyn., Dec. 3, 1953; m. Debra Lynn Bieler, Sept. 21, 1984. Diploma in Sci. of Nutritional Cons., Am. Nutrition Cons. Assn., 1986. Pres. Steps In Health, Ltd., Douglaston, N.Y., 1986-88, Margate, Fla., 1988-90, Nesconset, N.Y., 1990—; educator for children in sci. of food and nutritional supplementation; ind. tester household products for lead content. Developer: Steps in Health Ltd.'s Catalogue of Name-Brand Nutritional Supplements and Health Products. Vol. listen to children program Mental Health Assn. and Vol. Program Broward County (Fla.) Pub. Schs., 1989; mem. Ctr. for Sci. in the Pub. Interest, Washington. Mem. Am. Nutrition Cons. Assn., Life Extension Found., Pub. Citizen Health Rsch. Group, People for Ethical Treatment of Animals, Doris Day Animal League, Humane Soc. Broward County, Ctr. for Sci. in the Pub. Interest, Internat. Platform Assn., N.Y. State Sheriff's Assn., Inc., Tri-City Jaycees, Better Bus. Bur. South Fla. (arbitration participant 1989-90), L.I. Assn., Inc. Avocations: reading and data collection in health field, bodybuilding. Office: PO Box 83 Nesconset NY 11767-0083

FELDMAN, JACOB ALEX, electrical engineer; b. Kishinev, Russia, Sept. 17, 1954; came to the U.S., 1984; s. Isaak and Raisa (Aizenberg) F.; m. Irina V. Reznik, Dec. 23, 1984; 1 child, Avital L. BSEE, Poly. Inst., USSR, 1976. Sr. engr. VNH Industries, Inc., Rochester, N.Y., 1992—; mfg. engr. Eastman Kodak Co., Rochester, 1991-92. Mem. Mfg. Engring. Soc., Surface Mount Tech. Assn.

FELDMAN, KATHLEEN ANN, microbiologist, researcher; m. Daniel S. Feldman; children: Aaron C., Amanda M., Sarah C. BS, U. R.I., 1977; MS, U. Conn., 1980; PhD, Purdue U., 1985. Postdoctoral fellow Purdue U., West Lafayette, Ind., 1985-86, rsch. microbiologist, 1986-87; postdoctoral fellow U. Hartford, West Hartford, Conn., 1987-90; dir. R&D Earthgro, Inc., Lebanon, Conn., 1990—. Contbr. articles to Applied Environ. Microbiology, Biotech. Letters, Enzyme and Microbial Tech., Estuaries. Small Bus. Innovative Rsch. grantee Dept. Agr., 1992. Mem. Am. Soc. for Microbiology, Am. Soc. for Quality Control, Sigma Xi. Office: Earthgro Inc Rte 207 PO Box 143 Lebanon CT 06249

FELDMANN, EDWARD GEORGE, pharmaceutical chemist; b. Chgo., Oct. 13, 1930; s. Edward Louis and Vera (Arnesen) F.; m. Mary J. Evans, Aug. 30, 1952; children: Ann Marie, Edward William, Robert George, Karen Lynn. B.S. in Chemistry, Loyola U., Chgo., 1952; M.S. in Pharmacy (research fellow Am. Found. Pharm. Edn. 1953-55), U. Wis., 1954, Ph.D. in Pharm. Chemistry-Biochemistry, 1955; postgrad., Northwestern U., 1956, U. Chgo., 1958. Teaching asst. Loyola U., Chgo., 1951-52; research asst. U. Wis., 1952-53; sr. chemist Am. Dental Assn., 1955-58, dir. div. chemistry, 1958-59; assoc. dir. sci. div. Am. Pharm. Assn., 1959-60, dir., 1960-85, assoc. editor sci. edit. assn. jour., 1959-60, editor, 1960, assoc. exec. dir. for sci. affairs, 1970-83, v.p. sci. affairs, 1983-85, project dir. Handbook of Non-Prescription Drugs, 1985-86, mng. editor, 1989-90; project cons. Handbook of Non-Prescription Drugs, 1991-93; exec. sec. Acad. Pharm. Scis., 1983-85; pvt. pharm. cons., 1985—; assoc. dir. revision Nat. Fomulary, 1959-60, dir. revision, 1960-70; mem. adv. panel dental drugs Nat. Formulary, 1955-60; reviewer Internat. Pharmacopeia, WHO, 1958; spl. lectr. drug standards George Washington U., 1960-64; del. conf. on fellowships Nat. Health Council, 1960; mem. coordinating com. Nat. Conf. Antimicrobial Agts., Soc. Indsl. Microbiology, 1960-63; mem. adv. panel pharm. nomenclature A.M.A.-Am. Pharm. Assn.-U.S. Pharmacopeia, 1961-66, mem. nomenclature com., 1962-66; sec. U.S. Com. Internat. Drug Standards, 1964-65; adv. panel food chems. codex Nat. Acad. Scis.-NRC, 1961-71, liaison rep. to drug research bd., 1968-76; spl. liaison rep. to Commn. of Life Scis., NAS-NRC, 1973-85; mem. lab. com. Am. Pharm. Assn. Found., 1961-75; mem. com. Ebert prize, 1961-75; judge Lunsford-Richardson Pharmacy Awards, 1962-69; cons. Council on Drugs, A.M.A., 1962; vis. scientist Am. Assn. Colls. of Pharmacy, NSF, 1963-66; mem. expert adv. panel on internat. pharmacopeia and pharm. preparation World Health Orgn., 1963-75; drug cons. Office Sec., U.S. Dept. Health, Edn. and Welfare, 1967-70; nomenclature cons. to Commr., U.S. Food and Drug Adminstrn., 1968-71; mem. expert working group Indsl. Devel. Orgn., UN, 1969; mem. organizing com. 31st Internat. Congress Pharm. Scis., 1970-71; mem. NRC, 1971-85; del. U.S. Pharmacopeia, 1970-85, 90—; mem. Nat. Council on Drugs, 1976-83; mem. scientific adv. bd. Biodecision Labs., Inc., 1987-90; scientific cons. Am. Assn. Pharmaceutical Scientists, 1986-93; pharm. scis. cons. ERGO Inc., 1992—; mem. steering com. Japan-U.S. Pharmaceutical Scis. Congress, 1987; expert witness congressional drug legis. hearings and civil litigation cases; lectr. in field. Assoc. editor Drug Standards, 1959-60, editor, 1960; chmn. (1960-70) Nat. Formulary Bd.; editor Jour. Pharm. Scis., 1961-75, cons. editor, 1975-85, 87-89, interim editor, 1991, editor in chief, 1991—; editor APS Acad. Reporter, 1983-85; author more than 420 articles in field, editor or co-editor 24 ref. books; mem. editorial adv. bd. Index Chemicus, 1968-71; med. contbr. World Book Ency., 1986-88. Mem. membership com. Ravenwood Park Citizens Assn., Falls Church, Va., 1962, mem. nominating com., 1971-72; mem. Lake Barcroft Community Assn., 1975—. Recipient Man of Yr. award Nat. Assn. Pharm. Mfrs., 1970; Distinguished citation U. Wis., 1971; G.A. Bergy Lectr. award U. W.Va., 1975, Commr.'s citation FDA, 1975. Acad. Fellow Acad. Pharm. Scis.; life mem. Am. Pharm. Assn.; mem. Am. Chem. Soc., Am. Assn. Pharm. Scis. (charter mem. fellow, fellows selection com. 1989), N.Y. Acad. Scis., Nat. Soc. Med. Research (council 1961-69), Am. Testing Materials, Council Biology Editors, A.M.A. (affiliate), Fedn. Internat. Pharm., U.S. Tennis Assn., Sigma Xi, Rho Chi, Lambda Chi Sigma. Roman Catholic. Clubs: Sleepy Hollow Bath and Racquet (Falls Church, Va.); Arlington Tennis and Squash; 4-Seasons Tennis; Fairfax Golden Racquets. Lodge: K.C. Home and Office: 6306 Crosswoods Cir Falls Church VA 22044-1302

FELDMANN, HERMAN FRED, chemical engineer; b. Chgo., July 17, 1935; s. Herman F. and Marie (Oparka) F.; m. Marilyn Louise Yusup, Aug. 28, 1965; children: Lisa, Karl, Kurt. BSChemE, Ill. Inst. Tech., 1957, MSChemE, 1962. Project engr. Underwriters Lab. Inc., Chgo., 1957-58; rsch. engr. Hotpoint Co., Chgo., 1958-60; gasification rsch. mgr. Inst. Gas Tech., Chgo., 1960-65; project engr. Dept. Energy, Pitts., 1965-73; program mgr. Battelle Meml. Inst., Columbus, Ohio, 1973-90; sr. scientist Ill. Clean Coal Inst., Carterville, 1990—; mem. steering com. Internat. Med. Rsch. Com. AIME, 1970-72; cons. U.S. AIME Adv. Groups for Renewable Rsch.; mem. steering com. coal data book for Dept. Energy, 1986. Contbr. over 30 articles to profl. jours. Recipient Tech. Achievement award Dept. Energy, 1984. Mem. AICE. Achievements include 15 patents in energy conversion technology; development of proprietary coal gasification and biomass conversion processes. Home: RR2 Box 333-P Carbondale IL 62801 Office: Ill Clean Coal Inst PO Box 8 Coal Devel Park Carterville IL 62918

FELDMAR, GABRIEL GABOR, psychologist, educator; b. Budapest, Hungary, Apr. 14, 1947; came to U.S. 1962; s. Tibor and Clara (Adania) F.; m. Suzanne Maros, Oct. 26, 1969; 1 child, Monique. MA, CUNY, 1974; PhD, Hofstra U., 1983. Lic. psychologist, N.Y.; cert. in treatment of addictions Am. Assn. Psychologists Treating Addiction. Psychometrician/rschr. L.I. Jewish-Hillside Med. Ctr., Glen Oaks, N.Y., 1974-77; psychologist, dir. rsch. Creedmoor Psychiat. Ctr., Queens Village, N.Y., 1977-86; dir. psychology The Holliswood (N.Y.) Hosp., 1986-90; clin. supr./dir. rsch. Cath. Med. Ctr. of Bklyn. and Queens, Jamaica, N.Y., 1991—; adj. asst. prof. psychology St. John's U., Flushing, N.Y., 1980—; cons. Biol. Psychiat. Rsch. unit Columbia U.-Creedmoor div., 1986-90, Elmcor (Drug Alcohol Rehab. Ctr.), Corona, N.Y., 1987—; established psychology and neuropsychology dept. at Holliswood Hosp., neuropsychology svcs. at L.I. Jewish-Hillside Med. Ctr. Founding editor newsletter Symmetry, 1981-83; founding co-editor newsletter Progress Note, 1986-88; founding and exec. editor Jour. Urban Psychiatry, 1979-83; contbr. articles to profl. jours. Pres. Soc. for Obsessive-Compulsive Disorders, Great Neck, N.Y., 1981-83. Mem. APA, World Fedn. for Mental Health, Internat. Neuropsychol. soc. Achievements include establishment of biological psychiatry research unit at Creedmoor Psychiatric Center; discovered (collaboratively) a new neuropsychiatric symptom (prosopo-affective agnosia). Office: St Marys Hosp 170 Buffalo Ave Brooklyn NY 11213

FELDSIEN, LAWRENCE FRANK, civil engineer; b. New Ulm, Minn., Nov. 8, 1939; s. Frank and Viola (Buboltz) F.; m. Norene B. Johnson, June 26, 1964; children: John, Craig. BSCE, U. Minn., 1962, MSCE, 1964. Registered profl. engr., Ohio, Minn., N.D. Engr. Battelle Meml. Rsch., Columbus, Ohio, 1966-68, Pillsbury Co., Mpls., 1968-69, Bonestroo Rosene Anderlik, St. Paul, 1969-79; v.p. Geotech. Engring., St. Paul, 1979-89; sr. engr. Braun Intertec, St. Paul, 1989-91; field engr. Sinclair Mktg. Co., Mpls., 1992—. Mem. Shoreview (Minn.) Planning Com., 1980—. 1st lt. U.S. Army, 1964-66, Vietnam. Mem. ASCE (mem. com. 1976-78), VFW, Soc. Am. Military Engrs. (mem. com. 1986-88), Soc. Profl. Engrs. (state pres. 1991-92). Office: Sinclair Mktg Co 6602 Portland Ave S Richfield MN 55423

FELDSTEIN, STANLEY, psychologist; b. Bklyn., Dec. 13, 1930; s. Mark and Dora (Kruger) F.; m. Joyce Lister, Sept. 27, 1964; children: Heather M. Quay, Judd Thomas Markham. BA, CUNY, 1953; MA, Columbia U., 1954, PhD in Clin. Psychology, 1961. Cert. psychologist, N.Y. Teaching fellow English dept. Bklyn. Coll. CUNY, 1954; rsch. asst. Columbia U., N.Y.C., 1957-60, rsch. assoc. in psychiatry, 1964-68; clin. intern Franklin D. Roosevelt VA NP Hosp., Montrose, N.Y., 1957-58, East Orange (N.J.) VA GM&S Hosp., 1958-59, Newark VA Mental Health Clinic, 1959-60; rsch. psychologist rsch. dept. William Alanson White Inst., N.Y.C., 1961-67, mem. faculty, 1964-69; assoc. prof. psychiatry div. biol. psychiatry N.Y. Med. Coll., 1964-72; prof. psychology U. Md., Baltimore County, 1971—, assoc. chmn. dept. psychology, 1971-75, 75-78, acting chmn., 1975; rsch. and clin. cons. drug addiction svcs. Beth Israel Med. Ctr., 1961-68; rsch. cons. bur. rsch. in neurology and psychiatry N.J. Neuropsychiat. Inst., Princeton, 1965-68; vis. prof. Clarke Inst. Psychiatry, U. Toronto, Ont., Can., 1978-79; lectr. coll. physicians and surgeons Columbia U., 1984—; vis. scholar dept. psychology Brigham Young U., Provo, Utah, 1986; cons., presenter in field; vis. scholar German Acad. Exch. Svc. at U. Giessen, Fed. Republic Germany, 1980. Co-author: Rhythms of Dialogue, 1970, computer Aided Interactive Psychiatric Diagnosis Programs, 1971, Nonverbal Behavior and Communication, 1978, others, also books chpts.; sect. editor Clin. Psychologists, 1967-71; cons. editor Profl. Psychology, 1969-76; mem. editorial bd. Jour. Psycholinguistic Rsch., 1976—, Jour. Lang. and Social Psychology, 1982—, Jour. Comm. Disorders, 1984-91, Jour. Asian Pacific Comm., 1988—; reviewer, contbr. to publs. in field/. Chmn. bd. trustees, acting pres. Blue Bird Sch., Ruxton, Md. 1977-78, chmn. bd. dirs., 1979—. Grantee NIMH, Laidlow Found., Nat. Heart, Lung and Blood Inst. of NIH, March of Dimes Birth Defects Found., Md. Psychiat. Rsch. Ctr., DRIF Fund, others. Mem. AAAS, APA (charter), Soc. Personality and Social Psychology, Internat. Soc. for Infant Studies, N.Y. Acad. Scis., Acoustical Soc. Am., Eastern Psychol. Assn., Sigma Xi, Kappa Delta Pi. Home: 244 Blenheim Rd Baltimore MD 21212 Office: U Md Baltimore County Dept Psychology Baltimore MD 21228

FELKER, PETER, chemistry educator. Prof. dept. chemistry UCLA. Recipient Colbenz award Colbenz Soc., 1993. Office: UCLA Dept of Chemistry 405 Hilgard Ave Los Angeles CA 90024*

FELL, BARRY (HOWARD BARRACLOUGH FELL), marine biologist, educator; b. Lewes, Sussex, Eng., June 6, 1917; came to U.S., 1964; s. Howard Towne and Elsie Martha (Johnston) F.; m. Renee Clarkson, Oct. 10, 1942; children: Roger Barraclough, Francis Julian, Veronica Irene. B.Sc. U. Zealand, 1938, M.Sc., 1939; Ph.D., Edinburgh U.-Scotland, 1941, D.Sc., 1955; A.M. (hon.), Harvard U., 1965. Sr. lectr. zoology Victoria U., Wellington, N.Z., 1946-56, assoc. prof., 1956-64; curator Mus. Comp. Zoology, Harvard U., Cambridge, Mass., 1964-77; prof. biology Harvard U., Cambridge, 1965-77; prof. emeritus, 1977—; vis. prof. U. Tripoli, Libya, 1978; cons., author Geol. Soc. Am., Lawrence, Kans., 1963-70; cons. AAAS Com. on Panama Seaway, Washington, 1968; cons., author Woods Hole Oceanographic Inst., 1967. Author: America B.C., 1976, 2nd edit., 1989, also Japanese, Spanish and British edits.; Saga America, 1980, Arabic edit., 1988; Bronze Age America, 1982. Contbr. articles to profl. jours. Served to maj., Brit. Army, 1941-46. Fellow Royal Soc. N.Z. (Hector medal 1959, Hutton medal 1962), Am. Acad. Arts and Scis. (fellow emeritus); Mem. N.Z. Assn. Scientists (pres. 1948-49), Epigraphic Soc. (pres. emeritus), Sociedad Portuguesa de Antropologia e Etnologia (hon. fellow), Societe d'Etude des anciens peuples (membre honoris causae), Inst. for the Study of Am. Cultures (gold medalist 1990). Home: 6625 Bamburgh Dr San Diego CA 92117-5105 Office: Museum of Comparative Zoology Harvard U Cambridge MA 02138

FELL, JAMES CARLTON, scientific and technical affairs executive, consultant; b. Buffalo, Nov. 1, 1943; s. Carlton Joseph and Marion Rose (Bernhard) F.; m. Kimberly Ann DiBernardo, Sept. 26, 1981; children: Todd James, Brandon Paul, Donde Stephen. BS, SUNY, Buffalo, 1966, MS, 1967. Asst. systems engr. Calspan, Inc., Buffalo, 1967-69; phys. scientist Nat. Hwy. Traffic Safety Adminstrn., Washington, 1969-83, program mgr., 1983-90, sci. advisor, 1990—; U.S. rep. Orgn. for Econ. Cooperation and Devel., 1973-75. Contbr. articles to profl. jours. Recipient Outstanding Performance award U.S. Dept. Transp., 1972, 77, 81, 85, 89. Mem. Human Factors Soc. (A.R. Lauer award for contbn. to traffic safety 1992), Assn. Advancement Automotive Medicine (pres. 1987-88, Best Sci. Paper award 1979, 83, Svc. award 1985). Achievements include development of motor vehicle accident causal system; publication of alcohol-involvement rates per unit of exposure in fatal crashes, background paper for U.S. Surgeon General's workshop on drunk driving. Home: 4313 Guinea Rd Annandale VA 22003 Office: US DOT NHTSA NTS 20 400 Seventh St SW Washington DC 20590

FELLER, DENNIS RUDOLPH, pharmacology educator; b. Monroe, Wis., Oct. 27, 1941; s. Rudolph and Helen Clara (Regez) F.; m. Grace Leone Zimmerman, Aug. 17, 1963; children: Renée Marie Feller Hallam, Jay Dennis. BS, U. Wis., 1963, MS, 1966, PhD, 1968. Postdoctoral researcher NIH, Bethesda, Md., 1967-69; asst. prof. Ohio State U., Columbus, 1969-74, assoc. prof., 1974-80, prof., 1980—, chair divsn. pharmacology, 1987—; vis. prof. U. Cen. Venezuela, Caracas, 1982, 88, 91, U. Zagazig (Egypt), 1983, 87; profl. leave NIH, Bethesda, 1984. Co-editor: (monograph) Clofibrate and Related Analogs, 1977, Antilipidemic Drugs, 1991; author: (with others) Hepatic Peroxisome Proliferation Induced by Hypolipidemic Drugs, 1991, Chemical Carcinogenicity of Peroxisome Proliferating Agents, 1986. Mem. Am. Soc. for Pharmacology and Exptl. Therapeutics, Soc. for Exptl. Biology and Medicine, Am. Heart Assn., Internat. Soc. for Thrombosis and Haemastasis. Achievements include research characterizing the stereochemical requirements of molecules interacting with subtypes of alpha- and beta-adrenoceptors and thromboxane alpha-2 receptor systems; investigated the biogransformation of drugs, in vivo and in vitro; also research into drugs wich exhibit antilipidemic, antiasthmatic and antiplatelet activities. Home: 2846 Wellesley Dr Columbus OH 43221 Office: Ohio State U Divsn Pharmacology 500 W 12th Ave Columbus OH 43210-1291

FELLER, WILLIAM FRANK, surgery educator; b. St. Paul, Nov. 2, 1925; s. William and Eva Caroline (Nordstrom) F.; children: William Frank III, Elizabeth Susan. BA magna cum laude, U. Minn., 1948, BS, 1952, MD, 1954, PhD, 1962. Diplomate Am. Bd. Surgery. Intern U. Minn., Mpls., 1954-55; asst. prof. Georgetown U., Washington, 1964-69, assoc. prof., 1969—. Contbr. articles to profl. jours. Warden St. John's Episc. Ch., Chevy Chase, Md., 1975-76. Mem. AAAS, ACS, Am. Assn. Cancer Rsch., Am. Scandinavian Found. (chpt. pres. 1969-71), Med. Soc. D.C., Am. Cancer Soc. (D.C. div. pres. 1984-85), Washington Acad. Medicine, Southeastern Surg. Congress, N.Y. Acad. Sci. Achievements include 2 patents for Cancer Detection Methods. Office: Georgetown U 3800 Reservoir Rd NW Washington DC 20007-2196

FELLER, WINTHROP BRUCE, physicist; b. Cleve., Nov. 1, 1950; s. Robert William and Virginia Adele (Winther) F.; m. Lydia M. Conca, Aug. 14, 1988; 1 child, Daniel James. SB, MIT, 1974; postgrad., Yale U., 1974-75. Lectr. in physics and astronomy Northwestern U., Evanston, Ill., 1977-83; scientist Galileo Electro-Optics Corp., Sturbridge, Mass., 1984-85, sr. scientist, 1985-90, assoc., 1990—; v.p., chief scientist Nova Sci., Inc., 1993—. Contbr. articles to profl. jours. Recipient R&D 100 award R&D mag., 1989; grantee NASA, Smithsonian Instn., ARPA. Mem. AAAS, Am. Phys. Soc., Am. Philos. Assn., Optical Soc. Am., Soc. Photo-Optical Instrumentation Engrs. (session co-chmn. detector conf.), Fedn. Am. Scientists, Union Concerned Scientists. Achievements include patents in microchannel plate field; development of low noise, conductively cooled, neutron-, hard x-, and gamma ray sensitive microchannel plate detectors, of digital readout systems, x-ray and neutron focusing microchannel lens, others; research on detectors for EUV, x-ray and gamma ray astronomy, high resolution PET, med. radiography, optical signal processing, time-of-flight mass analysis of biomolecules, and the philosophy of science. Home: 81 Fiske Hill Rd Sturbridge MA 01566-1231

FELLNER-FELDEGG, HUGO ROBERT, scientific consultant; b. Vienna, Austria, Feb. 27, 1923; s. Heinrich and Lilli (von Fabini) Fellner von Feldegg; m. Ruth Elisabeth Debelli, Mar. 28, 1953. MS in Chemistry, U. Vienna, 1949, PhD in Physics, 1951; PhD in Molecular Physics, U. Uppsala (Sweden), 1974. Phys. chemist Farbwerke Hoechst, Frankfurt, Fed. Republic Germany, 1952-55, plant mgr., 1956-59; sr. staff scientist Shockley Transistor Corp., Palo Alto, Calif., 1960-61; staff scientist Hewlett-Packard Co., Palo Alto, 1961-72; guest prof., docent Uppsala U., 1973-78; head lab. Swedish Agrl. U., Uppsala, 1980-88, prof., 1986-89; pres. HB Biocentra, Uppsala, 1988—; prof. emeritus, 1989—; bd. dirs. Scienta Instrument AB, Uppsala; Nobel Found.'s guest prof., 1975-77. Inventor microwave time domain spectroscopy; developer of electron spectroscopy with monochromatized X-rays; contbr. articles to profl. jours.; holder of various patents. Recipient Curzon award, 1975.

FELLOWS, ROBERT ELLIS, medical educator, medical scientist; b. Syracuse, N.Y., Aug. 4, 1933; s. Robert Ellis and Clara (Talmadge) F.; m. Karlen Kiger, July 2, 1983; children—Kara, Ari. A.B., Hamilton Coll., 1955; M.D., C.M., McGill U., 1959; Ph.D., Duke U., 1969. Intern N.Y. Hosp., N.Y.C., 1959-60; asst. resident N.Y. Hosp., 1960-61, Royal Victoria Hosp., Montreal, Que., Can., 1961-62; asst. prof. dept. medicine Duke U., Durham, N.C., 1966-76; asst. prof. dept. physiology and pharmacology Duke U., 1966-70, assoc. prof. dept. physiology and pharmacology, asso. dir. med. scientist tng. program, 1970-76; prof., chmn. dept. physiology and biophysics U. Iowa Coll. Medicine, dir. med. sci. tng. program, 1976—, dir. physician sci. program, 1984—, dir. neurosci. program, 1984—; mem. Nat. Pituitary Agy. Adv. Bd.; mem. NIH Population Rsch. Com., 1981-86, VA Career

Devel. Rev. Com., 1985-88; cons. NIH, NSF March of Dimes. Mem. editorial bd.: Endocrinology, Am. Jour. Physiology. Mem. AAAS, Am. Chem. Soc., Am. Fedn. Clin. Research, Am. Physiol. Soc., Am. Soc. Biol. Chemists, Am. Soc. Cell Biology, Assn. Chairmen Depts. Physiology, Biochem. Soc., Biophys. Soc., Endocrine Soc., Internat. Soc. Neuroendocrinology, N.Y. Acad. Scis., Soc. for Neurosci., Tissue Culture Assn., Sigma Xi, Alpha Omega Alpha. Home: 15 Prospect Pl Iowa City IA 52246-1932 Office: 5-660 Bowen Sci Bldg Iowa City IA 52242

FELSHER, MURRAY, geologist, publisher; b. Bronx, N.Y., Oct. 8, 1936; s. Harry and Ruth (Kleinman) F.; m. Natalie Yagodnick, June 10, 1961; children: Elyann Lee, Harry David, Joshua Jeremy. BS, CCNY, 1959; MS, U. Mass., 1963; postgrad., McMaster U., Hamilton, Ont., Can., 1967-68; PhD, U. Tex., 1971. Asst. prof. Syracuse (N.Y.) U., 1967-69; assoc. dir. Am. Geol. Inst., Washington, 1969-71; sr. scientist U.S. EPA, Washington, 1971-75; chief geologist energy applications, program scientist NASA, Washington, 1975-80; pres. Assoc. Tech. Cons., Germantown, Md., 1980—; pub. Washington Remote Sensing Letter, 1981—, Washington Fed. Sci. Newsletter, 1989—; trustee Syracuse Rsch. Corp., 1988—; mem. adv. bd. Spot Image Corp., Reston, Va., 1992—; mem. sci. adv. bd. Orbital Scis. Corp., Fairfax, Va., 1990; speaker in field. Fellow NRC, Shell Oil Co., Owen-Coates fellow. Fellow Geol. Soc. Am.; mem. Am. Astron. Soc. (sr.), AAAS, Am. Geophys. Union, Am. Soc. Photogrammetry and Remote Sensing, Armed Forces Comm. and Electronics Assn., D.C. Sci. Writers Assn., Md. Space Bus. Roundtable, Nat. Def. Preparedness Assn., Nat. Mil. Intelligence Assn., Security Affairs Support Assn., Soc. for Sedimentary Geology, Washington Space Bus. Roundtable, Assn. Old Crows, NASA Alumni League, Nat. Press Club, Nat. Press Found., Sigma Xi, others. Home: PO Box 20 Germantown MD 20875-0020 Office: 1057B National Press Bldg Washington DC 20045

FELTHOUS, ALAN ROBERT, psychiatrist; b. San Francisco, Oct. 16, 1944; s. Robert Alan and Agnetta Wilhelmena (Blindheim) F.; m. Mary Louise Wilkins, Aug. 6, 1971; children: Erik Alan, Emily Anna, Elizabeth Ashley. BS, U. Wash., 1967; MD, U. Louisville, 1971. Diplomate Nat. Bd. Med. Examiners, Am. Bd. Psychiatry and Neurology, Am. Bd. Forensic Psychiatry (v.p. 1992-93). Intern Roosevelt Hosp., N.Y.C., 1971-72; resident in psychiatry McLean Hosp./Harvard Med. Sch., Belmont, Mass., 1972-75; staff psychiatrist Naval Regional Med. Ctr., Oakland, Calif., 1975-77; psychiatrist, sect. chief The Menninger Found., Topeka, Kans., 1977-83, acting dir. adult div., 1993—; chief forensic svc. Dept. Psychiatry and Behavioral Scis., U. Tex., Galveston, 1984—, assoc. prof., 1984-89, prof., 1989—; cons., mem. expert panel on psychiat. disorders and comml. drivers U.S. Dept. Transp., Fed. Hwy. Adminstrn., Washington, 1990. Author: The Psychotherapist's Duty to Warn or Protect, 1989; newsletter editor Am. Acad. Psychiatry and the Law, 1988—; sect. editor: The Duty to Warn, 1989; contbr. articles to profl. jours. Capt. USNR, 1975-77, 92—. Recipient Wood-Prince award for sci. pubs. The Menninger Found., 1978, 79, 80, 81, 82, Outstanding Achievement award Gulf Coast Mental Health and Mental Retardation, Galvestion, 1991, Exemplary Psychiatrist award for 1993 Nat. Alliance for the Mentally Ill. Fellow Am. Acad. Forensic Scis., Am. Psychiat. Assn.; mem. German Soc. for Psychiatry, Naval Res. Assn. (life). Achievements include studies of abnormal aggressive behaviors produced findings, including the association between childhood cruelty to animals and other forms of untoward aggression. Office: Univ of Tex Med Br Dept Psychiatry Graves Bldg D29 Galveston TX 77555-0428

FELTZ, CHARLES HENDERSON, former mechanical engineer, consultant; b. Channing, Tex., Sept. 15, 1916; s. John Henderson and Josie Mary (Hamm) F.; m. Juanita May Donnell, July 23, 1941; children: Janis Lynn Zaharis, Jennifer Gayle Wortzman. BSME, Tex. Tech U., 1940. Design group engr. N.Am. Aviation, Inc., L.A., 1940-48, asst. project engr., 1948-56, chief engr., program mgr., 1956-62; cheif engr. Apollo Command and Svc. Module space divsn. N.Am. Rockwell Corp., Downey, Calif., 1962-64, asst. program mgr. Apollo Command and Svc. Module, 1964-69, asst. program mgr. Space Shuttle Orbiter Program, 1969-74, tech. asst. to pres. space divsn., 1974-76; v.p., asst. to pres. N.Am. Aerospace Ops. Rockwell Internat., El Segundo, Calif., 1976-80; pres. transp. system devel. prodn. divsn. Space Systems Group Rockwell Internat., Downey, 1980-81; cons. NASA. Recipient Personal Contbn. award Flight Rsch. Ctr., NASA, 1966, Disting. Engr. award Sch. Engring. Tex. Tech. Coll., 1967, Cert. Appreciation, Manned Spacecraft Ctr., NASA, 1969, Apollo Achievement award NASA, 1969, Pub. Svc. award NASA, 1969, Disting. Alumnus award Tau Beta Pi, 1972, Skylab Achievement award Skylab Program, NASA, 1974, Cert. Merit, 1974, Cert. Appreciation, NASA, 1977, Outstanding Engr. Merit award Inst. Advancement Engring., 1978, Disting. Pub. Svc. medal NASA, 1981, Group Achievement award NASA, 1988, Spirit of St. Louis medal ASME, 1990. Fellow AIAA, Am. Astronautical Soc., Inc. (W. Randolph Lovelace II award 1981); mem. Tex. Tech U. Mech. Engring. Acad. Achievements include design and development of the wing structure of the F-82 and F-86 series airplane, the Navy's FJ-2 airplane and production of F-86D airplane, X-15 hypersonic vehicle which was the forerunner to the development of the U.S. space programs; design and production of shuttle orbiter, space shuttle main engine, global position satellite, conceptual phase of Space Station. Home: 41800 Chaparral Dr Temecula CA 92592

FENDERSON, CAROLINE HOUSTON, psychotherapist; b. East Orange, N.J., June 17, 1932; d. George Cochran and Mary Bullard (Saunders) Houston; m. Kendrick Elwell Fenderson, Jr.; 1 child, Karen Sibley. BA, Vassar Coll., 1954; MA, U. So. Fla., 1973. Lic. mental health counselor, Fla.; diplomate Am. Bd. Cert. Managed Care Providers; cert. trainer, development of human capacities Found. for Mind Rsch.; ordained to ministry of edn. Unitarian Univeralist, 1981. Dir. of religious edn. Unitarian Universalist Ch., St. Petersburg, Fla., 1960-80; min. of religious edn. Unitarian Universalist Ch., Clearwater, Fla., 1981-83; counselor and staff devel. cons. Pinellas County (Fla.) Schools, 1973-83; pvt. practice Clearwater and Palm Harbor, Fla., 1983—. Author: Life Journey, 1988; (with Kendrick Fenderson Jr.) Magnets, 1961, Southern Shores, 1964; (with others) Man the Culture Builder, 1970, U.U. Identity, 1979; contbr. articles to profl. jours. Pub. affairs chmn. St. Petersburg Jr. League, 1960; founder Childbirth and Parent Edn. League of Pinellas County, 1960-70, pres., v.p., com. chair, tchr.; v.p. Child Guidance Clinic, St. Petersburg, 1960. Mem. AACD, Liberal Religious Edn. Dirs. Assn. (v.p. 1980-81), Assn. for Transpersonal Psychology, Assn. for Humanistic Psychology, Internat. Transpersonal Assn., Unitarian Universalist Assn. (com. mem. 1975-79), Phi Beta Kappa, Kappa Delta Pi. Home: 29 Freshwater Dr Palm Harbor FL 34684-1106 Office: Caroline H Fenderson MA PA 25 400 US 19 N Ste 172 Clearwater FL 34623

FENG, HSIEN WEN, biochemistry educator, researcher; b. Wu-Xi, China, May 16, 1928; came to U.S., 1963; s. Fei-Dong and Tse (Hong) F.; children: Helen I-Tzu, Mark I-Ming. BS in Pharmacy with honors, Nat. Def. Med. Ctr., 1948-53; MS, PhD in Biochemistry, U. Mich., 1963-68. From asst. to assoc. prof. Nat. Def. Med. Ctr., Taipei, China, 1954-74, prof. biochemistry, 1974-80; postdoctoral scholar, rsch. investigator U. Mich., Ann Arbor, 1982—; investigator molecular biology Veterans Gen. Hosp., Taipei, 1968-82; advisor Youth Monthly, Taipei, 1976-82. Contbr. articles to profl. jours. Col. Health Svc., 1948-80. Recipient Good Tchr. award Nat. Def. Med. Ctr., Taipei, 1960, Prize Medal Rsch. award premier, Taipei, 1982. Mem. Sigma Xi (assoc.). Home: 3445 Burbank Ann Arbor MI 48105

FENN, JIMMY O'NEIL, physicist; b. Brunswick, Ga., Nov. 18, 1937; s. Raymond Hume and Mae Elizabeth (Maxwell) F.; m. Nancy Sue Smith, Nov. 18, 1958 (div. 1974); children: Daniel Stewert, Nancy Anne, Margaret Elizabeth; m. Deborah Broadway, Sept. 2, 1989. BS, Lincoln Meml. U., 1963; MS, Emory U., 1967; PhD, Ga. Inst. Tech., 1980. Instr. dept. radiology Emory U., Atlanta, 1967, asst. prof., 1973-74; asst. prof. Med. U. S.C., Charleston, 1968-73, 74-82; assoc. prof. radiol. oncology Med. U. S.C., Charleston, ;, 1978-93; prof. radio oncology Med. U. S.C., Charleston, 1993—; med. bd. advisors radiation oncology Gen. Electric, 1992. Contbr. articles to So. Med. Jour., Cancer, Internat. Jour. Hyperthermia, Jour. Cell Biology, Jour. S.C. Med. Assn. Fellow Am. Coll. Med. Physics (chair, sec. 1984-91), Am. Coll. Radiology, Am. Assn. Med. Physicists (bd. dirs. 1984-88). Presbyterian. Office: Med U SC 171 Ashley Ave Charleston SC 29425

FENNEMA, OWEN RICHARD, food chemistry educator; b. Hinsdale, Ill., Jan. 23, 1929; s. Nick and Fern Alma (First) F.; m. Ann Elizabeth Hammer, Aug. 22, 1948; children: Linda Gail, Karen Elizabeth, Peter Scott. BS, Kans. State U., 1950; MS, U. Wis., 1951, PhD, 1960; PhD of Agrl. and Environ. Scis. (hon.), Wageningen Agrl. U., The Netherlands, 1993. Project leader for research and devel Pillsbury Co., Mpls., 1953-57; asst. prof. food sci. dept. U. Wis., Madison, 1960-64, assoc. prof., 1964-69, prof., 1969—, chmn. dept., 1977-81; cons. Pillsbury Co., Mpls., 1979—; pub. mem. Internat. Life Scis. Inst.-Nutrition Found., 1987-90. Author: Low Temperature Preservation of Foods, 1973; editor: Principles of Food Science, 2 vols., 1976, Proteins at Low Temperatures, 1979; Food Chemistry, 2d edit., 1985; mem. editorial bd. Cryobiology, 1966-82, Internat. Jour. Food Sci. and Nutrition, Jour. Food Sci., 1975-77, Jour. Food Processing Preservation, 1977—, Jour. Food Biochemistry, 1977-80, Nutrition Rsch. Newsletter, 1983—, Acta Alimentaria (Budapest, Hungary), 1990—, South African Jour. Food Sci. and Nutrition, 1991—. Served to 2d lt. U.S. Army, 1951-53. Recipient Excellence in Teaching award U. Wis., Madison, 1977; Fulbright disting. lectr., Spain, 1992. Fellow Am. Chem. Soc., Inst. Food Technologists (pres. 1982-83, Excellence in Teaching award 1978, Carl R. Fellers award 1988, Nicholas Appert award 1988); mem. Am. Inst. Nutrition, Soc. Cryobiology, Am. Dairy Sci. Assn., Internat. Union of Food Sci. and Tech. (del. 1983-88, exec. com. 1988—, v.p. 1992—). Home: 5010 Lake Mendota Dr Madison WI 53705-1305 Office: U Wis 1605 Linden Dr Madison WI 53706-1565

FENOGLIO-PREISER, CECILIA METTLER, pathologist, educator; b. N.Y.C., Nov. 28, 1943; d. Frederick Albert and Cecilia Charlotte (Asper) Mettler; m. John Fenoglio Jr., May 27, 1967 (div. 1977); children: Timothy, Johanna, Andreas, Nicholas; m. Wolfgang F.E. Preiser, Feb. 16, 1985. MD, Georgetown U., 1969. Diplomate Am. Bd. Pathology. Intern Presbyn. Hosp., N.Y.C., 1969-70; dir. Central Tissue Facility Columbia-Presbyn. Med. Ctr., N.Y.C., 1976-83; co-dir. div. surg. pathology Presbyn. Hosp., N.Y.C., 1978-82; div. div. surg. pathology Presbyn. Hosp., 1982-83; dir. Electron Microscop. Lab. Internat. Inst. Human Reprodn., 1978-85; assoc. prof. pathology Coll. Physicians and Surgeons, Columbia U., 1981-82, prof., 1982-83, attending pathologist, 1982-83; dir. lab. services Albuquerque VA Med. Ctr., 1983-90; prof. pathology U. N.Mex. Sch. Medicine, Albuquerque, 1983-90, also vice-chmn. dept. pathology; MacKenzie prof., chmn. dept. pathology and lab. medicine U. Cin. Sch. Medicine, 1990—. Author: General Pathology, 1983, Gastrointestinal Pathology, An Atlas and Text, 1989,, Tumors of the Large and Small Intestine, 1990; editor: Advances in Pathobiology Cell Membranes, 1988-92, Advances in Pathobiology: Aging and Neoplasia, 1976, Progress in Surgical Pathology, vols. I-XIV, 1980-87, Advances in Pathology, vols. I-V, 1988-89. Grantee NIH, 1973, 79-82, 94-87, 85-92, Casncer Rsch. Ctr., 1975-83, Population Coun., 1977-83, Nat. Ileitis and Colitis Found., 1979-80, Am. Cancer Soc., 1987-90. Mem. AAAS (life), Internat. Acad. Pathology (edn. com. 1980-85, coun. 1984-87, exec. com. 1987—, v.p. 1987, pres.-elect 1988, pres. 1989), Am. Assn. Pathologists, N.Y. Acad. Sci., N.Y. Acad. Medicine, Fedn. Am. Scientists for Exptl. Biology, Gastrointestinal Pathologist Group (founding mem., edn. com. 1983-85, sec.-treas. 1993), Arthur Purdy Stout Soc. (coun. 1987—). Office: U Cin Sch Medicine 231 Bethesda Ave Cincinnati OH 45267-0001*

FENSELAU, CATHERINE CLARKE, chemistry educator; b. York, Nebr., Apr. 15, 1939; d. Lee Keckley and Muriel (Thomas) Clarke; m. Allan Herman Fenselau, 1962 (div. 1980); children: Andrew Clarke, Thomas Stewart; m. Robert James Cotter, 1984. A.B., Bryn Mawr Coll., 1961; Ph.D., Stanford U., 1965. Research scientist U. Calif.-Berkeley, 1965-67; instr. to prof. Johns Hopkins U., Balt., 1967-87; prof., chmn. dept. chemistry, biochemistry U. Md., Balt. County, 1987—; cons. NIH, NSF, U.S. Dept. Agr., U.S. Army, FDA, others. Editor Biomed. Environ. Mass Spectrometry, 1973-89; assoc. editor Analytical Chemistry, 1990—; contbr. articles to profl. jours. Mem. Am. Soc. Mass Spectrometry (pres.), Am. Chem. Soc. (Garvan medal 1985, Md. Chemist award Md. sect. 1989), AAAS, Am. Soc. Pharmacology and Exptl. Therapeutics. Office: U Md Dept Chemistry & Biochemistry 5401 Wilkens Ave Baltimore MD 21228-5329

FENSTAD, JENS ERIK, mathematics educator; b. Trondheim, Norway, Apr. 15, 1935; s. Erik and Margit (Wullum) F.; m. Grete Usterud Hansen, Jan. 28, 1939; children—Anne Marie, Erik, Hakon. Mag. Scient., U. Oslo, 1959. Prof. math. U. Oslo, 1968—; chmn. Natural Sci. Research Council Norway, 1985-89; vice rector U. Oslo, 1989—; pres. Internat. Union of History and Philosophy of Sci., 1991—; mem. sci. com. NATO, 1992—. Author: General Recursion Theory, 1980; Nonstandard Methods in Stochastic Analysis and Mathematical Physics, 1986; Situations, Language and Logic, 1986. Mem. Norwegian Acad. Letters and Sci., Academia Europaea. Office: U Oslo Inst Math, PO Box 1053 Blindern, 0316 Oslo 3, Norway

FENTIMAN, AUDEEN WALTERS, nuclear engineer, educator; b. Athens, Ohio, Aug. 29, 1950; d. Rob Roy and Mary Frances (Bean) Walters; m. Allison F. Fentiman, June 2, 1984. BS in Math., Glenville (W.Va.) State Coll., 1972; MA in Math., W.Va. U., 1974; MS in Nuclear Engring., Ohio State U., 1977, PhD in Nuclear Engring., 1982. Tchr. math. St. Marys (W.Va.) High Sch., 1974-76; rsch. asst. Ohio State U., Columbus, 1976-79; rsch. scientist Battelle-Nuclear Systems Sect., Columbus, 1979-85; sr. engr. Battelle-Office Nuclear Waste Isolation, Columbus, 1985-87; assoc dept. mgr. Battelle-Ordnance Systems and Tech., Columbus, 1987-89; sr. specialist EG&G Mound Applied Techs., Miamisburg, Ohio, 1989-90; asst. prof. engring. graphics Ohio State U., Columbus, 1990—; cons. Sci. Applications Internat., Columbus, 1990—; mem. Dayton-Montgomery County Math. Collaborative, Ohio, 1989-90; mem. indsl. and profl. adv. coun. Pa. State Coll. Engring., 1989—. Author abstracts. Battelle Meml. Inst. fellow, 1982. Mem. AAAS, Am. Nuclear Soc. (local sect. pres. 1991-92), Ohio Acad. Sci., Am. Soc. for Engring. Edn., Sigma Xi. Office: Ohio State U Dept Engring Graphics 2070 Neil Ave Columbus OH 43210

FENTON, WAYNE S., psychiatrist; b. Mar. 24, 1953. BA in Exptl. Psychology, Bard Coll., 1975; MD, George Washington U., 1979. Cert. Am. Bd. Medical Examiners, 1980; cert. Md., Conn., Va.; Diplomate in Psychiatry. Rotating internship, dept. internal medicine Norwalk Hosp., Conn., 1979-80; resident, post doctoral fellow psychiatry Yale U., 1980-83; fellow Inst. Social and Policy Studies, Yale U., 1983-84; staff psychiatrist Yale Psychiat. Inst., Yale U., New Haven, Conn., 1983-84, Chestnut Lodge, Rockville, Md., 1984-85; rsch. assoc. Chestnut Lodge Rsch. Inst., Rockville, Md., 1984-90; clin. adminstrv. psychiatrist Chestnut Lodge Hosp., Rockville, Md., 1985-90; dir. rsch. Chestnut Lodge Rsch. Inst., Rockville, Md., 1990—; asst. clin. dir. Chestnut Lodge Hosp., Rockville, Md., 1990—; assoc. clin. prof., psychiatry and behavior scis. George Washington U., D.C., 1990—; facualty Washington Sch. Psychiatry, D.C., 1991—; cons. Montgomery County Pub. Defender, Md., 1984—, McAuliffe House, Md., 1990—. Editorial cons: Schizophrenia Bulletin, 1986—, Journal of Nervous and Mental Disease, 1986—, Am. Journal Psychiatry, 1989—; contbr. to profl. jours. Recipient Nat. Rsch. Svc. award USPHS, 1983-84, Young Investigator award NIH, 1989, Young Investigator award Nat. Alliance for Rsch. on Schizophrenia and Depression, 1989, Gralnick award Am. Suicide Found., 1992. Mem. Am. Psychiat. Assn., Wash. Psychiat. Soc., Nat. Alliance for Mentally Ill, NAPPH. Office: Chestnut Lodge Hospital 500 West Montgomery Ave Rockville MD 20850

FER, AHMET F., electrical engineer, educator; b. Ankara, Turkey, July 31, 1945; came to U.S., 1959; s. Muslih F. and Hayrunnisa (Gurkan) F.; m. Esther Elizabeth Horvath, Nov. 14, 1987; children: Danyal, Adam. BSEE, Mid. East Tech. U., 1967, MSEE, 1970; PhD, U. Birmingham, Birmingham, U.K., 1975. Asst. dept. chmn. elec. engring. dept. Mid. East Tech. U., Ankara, 1979-81, asst. prof., 1975-83, assoc. prof., 1983-84; exec. sec. Engring. Rsch. div. Sci. and Tech. Rsch. Coun., Ankara, 1984-85; assoc. prof. engring. and tech. Purdue U., Indpls., 1985-91; cons. Technalysis, Inc., Indpls., 1989—. Co-author: Microwave Techniques, 1978; contbr. articles to profl. jours. Fund raiser Multiple Sclerosis Soc., Indpls., 1989-90; mem. electromagnetic wave propagation panel AGARD/NATO. Mem. IEEE, Instn. Elec. Engrs. Eng., Scientech Club. Home: 5223 Mosswood Dr Indianapolis IN 46254-9796 Office: Technalysis Inc 7120 Waldemar Rd Indianapolis IN 46268-4193

FERDERBER-HERSONSKI, BORIS CONSTANTIN, process engineer; b. Craiova, Romania, May 17, 1943; came to U.S., 1980; s. Boris Modest and Anetta (Mihail) F.; m. Alexandra Ionescu; children: Boris Constantin Jr., Alexandru Vlad. MS in Process Engring., Poly. Inst., Bucharest, Romania, 1968; diploma fgn. trade, Romanian U., Bucharest, 1975. Registered profl. engr., Romania; engr.-in-tng., N.J. Plant engr. Pham. Complex, Bucharest, 1968-69, plant mgr., 1969-73; prin. engr. Indsl. Export Import, Bucharest, 1973-75, fgn. trade diplomate, 1975-80; sr. process engr. Foster Wheeler Corp., Livingston, N.J., 1980-85; projects mgr. CPC Internat./Best Foods, Fairfield, N.J., 1985-91; sr. process engr. Allied Signal Aquatech div., Morristown, N.J., 1991—; founder, pres. B.F.H. Design Corp., 1984—. Inventor in field. Mem. Rep. Nat. Com., Washington, 1981. Mem. Am. Inst. Chem. Engrs., Instrument Soc. Am., Am. Rowing Assn. Avocations: electronic applications, water and snow skiing. Office: PO Box 376 Hopatcong NJ 07843-0376

FERET, ADAM EDWARD, JR., dentist; b. Newark, Mar. 5, 1942; s. Adam Edward and Bronislawa Anne (Szorc) F. BA (athletic scholar), Seton Hall U., 1963; DMD, U. Medicine & Dentistry of N.J., 1967. Pvt. practice Westfield, N.J., 1972—. With USNR, 1967-70. Fellow Am. Acad. Gen. Dentistry; mem. ADA, N.J. Dental Assn., L.D. Pankey Study Club, Soc. Oral Physiology and Occlusion, Quest Study Club, Internat. Coll. Oral Implantologists, Am. Soc. Oral Implantology, Central Dental Soc., Balloon Fedn. Am., Polish-Am. Guardian Soc., Polish Falcons of Am., Copernicus Soc. Am., Toastmasters, Psi Omega. Roman Catholic. Home and Office: 440 E Broad St Westfield NJ 07090-2124

FERGASON, JAMES L., optical company executive; b. Wakenda, Mo., Jan. 12, 1934; s. Joshua E. and Sarah Margret (Cary) F.; m. Dora Delaine Barlish, June 10, 1956; children: Teresa Neal, Jeffrey, John, Susan. BS in Physics, U. Mo., 1956. Sr. engr. Westinghouse Electric, 1956-66; assoc. dir. Liquid Crystal Inst. Kent State (Ohio) U., 1966-70; pres. Internat. Liquid Xtal Co., Cleve., 1970-75, Am. Liquid Xtal Chem. Corp., Kent, 1975-83; pvt. cons. Menlo Park, Calif., 1983—; pres., CEO, chief scientist Optical Shields, Inc., Menlo Park, Calif., 1987—. Contbr. articles to Sci. Am., Applied Optics, Phys. Rev. Letters, Molecular Crystals and Liquid Crystals, SID Internat. Symposium Digest of Tech. Papers. Recipient IR 100 award Indsl. Rsch. mag., 1965, Laurels award Aviation Week and Space Tech., 1989; named Disting. Inventor of Yr., Intellectural Property Owners, 1989. Mem. IEEE (con. on intellectual property 1989—), Am. Phys. Soc. (dir. for info.), Soc. Info. Display (Francis Rice Darne Meml. award 1986), N.Y. Acad. Scis., Rotary. Achievements include patents for Thermal Imaging Device, Display Devices Utilizing Liquid Crystal Light Modulation, Encapsulated Liquid Crystal and Method, Modulated Retroreflector System. Home: 92 Adam Way Atherton CA 94027-3902 Office: Optical Shields Inc 1390 Willow Rd Menlo Park CA 94025

FERGUSON, BRUCE R., information systems consultant; b. Kroonstadt, Orange, July 2, 1959; came to U.S., 1986; s. John Robert and Elizabeth Joan (Wheals) F.; m. Nancy Marie O'Bannon, Sept. 27, 1987; children: Scott, David. BS, Rhodes U., Grahamstown, South Africa, 1982; MS, U. Calif., Irvine, 1991. Analyst Rhodes U., 1982-85; systems analyst Fridgitronics, Mpls., 1985-87; rsch. asst. Chevron Oil Fields Rsch., LaHabra, Calif., 1987-90; sr. cons. Digital Equipment Corp., Costa Mesa, Calif., 1990—. Contbr. articles to profl. jours. Chmn. in yard Balboa Yacht Club, 1989. Office: Digital Equipment Corp 3890 Harbor Blvd Costa Mesa CA 92626

FERGUSON, ERIK TILLMAN, transportation planning educator; b. Seattle, Apr. 3, 1957; s. Charles Harvey and Alice Eloise (Storaasli) F.; m. Elaine Ana Samayoa, May 17, 1985; children: Britnay Alexandra, Erik Tillman II. BA in History, U. So. Calif., L.A., 1979; PhD., U. So. Calif., 1988; M.C.R.P., Harvard U., 1982. Engring. aide Burmah Oil Co., Huntington Beach, Calif., 1975-77; roustabout Union Oil, Carpinteria, Calif., 1977-79; fabricator Lockheed Aircraft, Burbank, Calif., 1979-82; community planner Transp. Systems Ctr., Cambridge, Mass., 1982-84; projects specialist Commuter Transp. Svcs., L.A., 1984-86; assoc. planner Orange County Transit Dist., Garden Grove, Calif., 1986-88; asst. prof. Ga. Inst. Tech., Atlanta, 1988—; sr. assoc. Ekistic Mobility Cons., Gardena, 1986—. Editor newsletter/jour. Transp. Planning, 1990—; mem. editorial bd. Jour. Planning Edn. and Rsch., 1991—; contbr. articles to profl. jours. Mem. Task force on Regionally Important Resources, State of Ga., 1990. German Marshall Fund scholar, 1982083, Nat. Merit scholar, 1974; Fed. Transit Adminstrn. grantee, 1991, 92; Lilly Found. teaching fellow, 1990-91; recipient Rsch. Practitioner award Western Govtl. Rsch. Assn., 1988. Mem. Am. Planning Assn. (chair transp. planning div. 1993—), Transp. Rsch. Bd., Assn. for Commuter Transp. (bd. dirs. 1991—). Achievements include extensive research on the new field of Transportation Demand Management. Home: 2419 Leisure Lake Dr Atlanta GA 30338-5318 Office: Ga Inst Tech 245 W 4th St Atlanta GA 30332-0001

FERGUSON, JACKSON ROBERT, JR., astronautical engineer; b. Neptune, N.J., Aug. 18, 1942; s. Jackson Robert and Charlotte Carter (Rudewick) F.; m. Christina Mary Staley, Aug. 24, 1968; children: Jack Christopher, Joy Heather. BS in Engring. Sci., USAF Acad., 1965; MS in Astronautics, Air Force Inst. Tech., Dayton, Ohio, 1971; PhD in Aerospace Engring., U. Tex., 1983. Registered profl. engr., Tex. Astronautical engr. NORAD, Colorado Springs, Colo., 1972-76; asst. prof. USAF Acad., Colorado Springs, Colo., 1976-80, assoc. prof., 1982-84;, 1991-93; chief scientist European Office of Aerospace Rsch. & Devel., London, 1984-86; program mgr. Software Engring. Inst. Air Force Systems Command, Boston, 1986-88; detachment comdr. Air Force Systems Command, Colorado Springs, 1988-91; sr. mem. tech. staff Software Engring. Inst. Carnegie-Mellon U., Pitts., 1993—; ind. rev. team mem. USAF Data System Modernization Program, Washington, 1988; head ind. rev. team USAF System 1 Software Devel. Program, 1992; vis. prof. USAF Acad., 1991-93; program mgr. space systems, Software Engring. Inst., Carnegie Mellon U. Contbr. to reference book: Handbook of Engineering Fundamentals, 1984. Parish coun. pres. Our Lady of the Pines Cath. Ch., Black Forest, Colo., 1989. Col. USAF, 1965-91. Recipient USAF Rsch. and Devel. award, 1980. Mem. AIAA, Am. Soc. Engring. Edn. Roman Catholic. Achievements include research in Navstar global positioning system, spacecraft control. Office: Carnegi-Mellon U Software Engring Inst Pittsburgh PA 15213

FERGUSON, JAMES CLARKE, mathematician, algorithmist; b. Spokane, Wash., June 23, 1938; s. James Forsythe and Dorothy Eileen (Dillon) F. MS in Math., U. Wash., 1963; PhD in Math., U. N.Mex., 1984. Sci. programmer Boeing, Seattle, 1960-64; staff mem. GE Tech. Mil. Planning Office, Santa Barbara, Calif., 1964-66; mathematician TRW, Inc., Redondo Beach, Calif., 1966-71, Teledyne-Ryan Aero., San Diego, 1971-77; staff mem. Los Alamos (N.Mex.) Nat. Lab., 1977-85; sr. scientist Teletronix, Beaverton, Oreg., 1985-87, BBN Systems and Techs. Corp., Bellevue, Wash., 1987-92; with Point Control, Eugene, 1993—; cons. in field, 1975-87. Co-author: Key Works in Geometric Modeling, 1991, Fundamental Developments of Computer Aided Geometric Modeling, 1992; contbr. articles to profl. jours. Recipient advanced study fellowship, Los Alamos Nat. Lab., 1981. Mem. Assn. Computing Machinery, Soc. Indsl. and Applied Math. Achievements include introduction of parametric curve and surface techniques into computer aided geometric design field; complete classification of parametric planar cubics; application of parametric curve techniques to problem of shape preservation.

FERGUSON, JOHN BARCLAY, biology educator; b. Balt., July 5, 1947; s. John Miller and Helen (Sucro) F.; m. Jane Hough, June 28, 1970 (div. 1987); children: Hallam H., Gillian D.; m. Valeri J. Thomson, July 1, 1988. BS, Brown U., 1969; PhD, Yale U., 1973. Asst. prof. Bard Coll., Annandale, N.Y., 1977-83, assoc. prof., 1983-92, prof., 1992—. Contbr. to 1 book and articles to profl. jours. Bd. trustees Ch. St. John Evangelist, Barrytown, N.Y., 1988—. NIH Postdoctoral fellow, 1974-76. Mem. AAAS, Am. Soc. Microbiology, N.Y. Acad. Scis., Sigma Xi. Home: Rd 3 Box 305 Red Hook NY 12571 Office: Bard Coll Dept Biology Annandale NY 12504

FERGUSON, JOHN CARRUTHERS, biologist; b. Tuscaloosa, Ala., Mar. 2, 1937; s. John Howard and Rosalind Vera (Carruthers) F.; m. Rebecca Arletta Folsom, July 15, 1961; children: Joellyn, John. BA, Duke U., 1958; MA, Cornell U., 1961, PhD, 1963. Asst. prof. Fla. Presbyn. Coll., St. Petersburg, 1963-67, assoc. prof., 1967-72; assoc. prof. Eckerd Coll., St. Petersburg, 1972, prof., 1972—; coord. biology Eckerd Coll., 1985—; bd. Eckerd Coll. London Study Ctr., 1990. Contbr. articles to profl. jours. Mem. St. Petersburg Environ. Devel. Com., 1973-75, Citizen's Adv. com. for

Coastal Zone Mgmt., Pinellas County, Fla., 1976-77, Tampa Bay Area Regional Planning Coun. Adv. com. for Coastal Zone Mgmt., 1977-81. Rsch. grant NSF, 1964-74, 81, grant Rsch. Corp. of Am., 1977-79. Mem. Sea Edn. Assn. (acad. rev. bd. 1982—). Achievements include rsch. on studies on starfish nutrition, circulation, and water volume relationships. Home: 2127 Inner Circle S Saint Petersburg FL 33712 Office: Eckerd Coll Box 12560 Saint Petersburg FL 33733

FERGUSON, KENNETH LEE, nuclear engineer, engineering manager; b. Bayonne, N.J., Feb. 18, 1948; s. Harold L. and Mildred G. Ferguson; m. Linda Christine Lanz, Sept. 25, 1976. BS in Physics, U. Mich., 1969; PhD in Nuclear Engring., Carnegie Mellon U., 1973. Sr. engr. Westinghouse-Nuclear Tech., Monroeville, Pa., 1973-77, Westinghouse-Advanced Reactors, Madison, Pa., 1977-81; mgr. tech. mktg. Westinghouse-Nuclear Divs., Monroeville, 1981-87; mgr. transp. devel. Westinghouse-Waste Isolation Div., Carlsbad, N.Mex., 1987-89; mgr. Washington D.C. office Westinghouse-Savannah River Co., Washington, 1989-91; mgr. projects systems engring. Westinghouse-Savannah River Co., Aiken, S.C., 1991—. Contbr. articles to profl. jours. Mem. AAAS, Am. Nuclear Soc., Sigma Xi. Achievements include development in improved data interpretation for nuclear power reactors; successful licensing of environmental report for innovative reactor design; successful evolution for safety analysis techniques for reactor evaluations; certification of innovative design for waste container; plant life extensions; engineering assessments of nuclear, chemical and fossil fueled facilities; strategic planning for environmental technical alliance. Office: Westinghouse-Savannah River North Augusta SC 29841

FERGUSON, LLOYD ELBERT, manufacturing engineer; b. Denver, Mar. 5, 1942; s. Lloyd Elbert Ferguson and Ellen Jane (Schneider) Romero; m. Patricia Valine Hughes, May 25, 1963; children: Theresa Renee, Edwin Bateman. BS in Engring., Nova Internat. Coll., 1983. Cert. hypnotherapist, geometric tolerance instr. Crew leader FTS Corp., Denver, 1968-72; program engr. Sundstrand Corp., Denver, 1972-87, sr. assoc. project engr., 1987-90, sr. liaison engr., 1990—; v.p. Valine Corp. Team captain March of Dimes Team Walk, Denver, 1987; mem. AT&T Telephone Pioneer Clowns for Charity. Recipient recognition award AT&T Telephone Pioneers, 1990. Mem. Soc. Mfg. Engrs. (chmn. local chpt. 1988, zone chmn. 1989, achievement award 1984, 86, recognition award 1986, 90, appreciation award 1988), Nat. Mgmt. Assn. (cert., program instr. 1982—, honor award 1987, 90), Am. Indian Sci. and Engring. Soc., Colo. Clowns. Republican. Religious Science. Home: 10983 W 76th Dr Arvada CO 80005-3481 Office: Sundstrand Corp 2480 W 70th Ave Denver CO 80221-2500

FERGUSON, RICHARD PETER, project engineer; b. Chgo., Feb. 13, 1952; s. Peter and Clara Josephine (Anderson) F. BA in History, St. Olaf Coll., Northfield, Minn., 1974; BS in Engring., Ill. Inst. Tech., Chgo., 1990. Lic. land surveyor in tng., Ill.; registered profl. engr. in tng., Ill. Project engr. Amoco Chem. Co. Ctrl. Engring., Houston, 1991—. 1st lt. USAF, 1974-78. Mem. ASCE, NSPE, Am. Congress of Surveying and Mapping, Project Mgmt. Inst., Tex. Soc. Profl. Surveyors. Home: 900 Henderson # 1315 Houston TX 77058 Office: Amoco Chem Co 2525 Bay Area Blvd Ste 450 Houston TX 77058

FERGUSON, SHEILA ALEASE, psychologist, consultant, researcher; b. Cleve., Feb. 17, 1955; d. Harold Clayton and Thelma (Kibler) Lewis; m. Kenneth Duane Ferguson, July 16, 1977; children: Kenneth D. II, Justin K. BS in Psychology, John Carroll U., 1976; MA in Clin. Psychology, Cleve. State U., 1981; PhD in Organizational Behaviour, Case Western Res. U., 1991. Lic. profl. clin. counselor, Ohio. Mental health and substance abuse counselor City of Cleve./Office Mental Health and Substance Abuse, 1979-84; psychologist/employee assistance specialist City of Cleve./Dept. of Health, 1985-87; trainer and assoc. Ohio Corrective Counseling Inst., Cleve., 1983—; coordinating trainer and dir. of multicultral diversity Diocese of Cleve., 1987—; historian Ctr. for Urban Poverty and Social Change/Case Western Res., Cleve., 1989-90; assoc. dir. Econs. as a Second Lang. Found., Cleve., 1990-91, Phillis Wheatley Assn., Cleve., 1992; cons. Elsie Y. Cross Assocs., Phila., 1990-91. Author: (book) Making It in The Black Music Industry, 1990. Trustee Commn. on Cath. Community Action, Diocese of Cleve., 1988-92. Recipient Black Scholars award John Carroll U., Cleve., 1974-76; fellow The Inst. of Ednl. Leadership, Washington, 1984-85. Mem. Assn. Black Psychologists, Organizational Behavior Teaching Soc., Alpha Kappa Alpha. Avocations: textile art, writing, ethnomusicology. Home: 4147 E 147th St Cleveland OH 44128-1864

FERGUSON, SUSAN KATHARINE STOVER, nurse, psychotherapist, consultant; b. Warsaw, Ind., Mar. 11, 1944; d. Robert Eugene and Barbara Louise (Swaney) Stover; m. Philip Charles Ferguson, Oct. 2, 1965 (div.); children: Scott Duane, Shawn Alaine, Erin Kirsten. Diploma in nursing, Meth. Hosp., 1966; BA in Psychology, Purdue U., 1988; MSW, Smith Coll., 1991. Staff nurse, health hazard appraiser Meth. Hosp. of Ind., Indpls., 1966-68; staff nurse USPHS, Bethel, Alaska, 1968-70; instr. childbirth preparation Wabash, Ind., 1973-83; nurse Family Physicians Associated, Wabash, 1976-83; rsch. asst. Purdue U., Ft. Wayne, Ind., 1986-88; staff nurse, self-awareness seminar coord. Charter Beacon Hosp., Ft. Wayne, Ind., 1988-89; intern clin. social work Clifford Beers Guidance Ctr., New Haven, Conn., 1990-91; psychiat. nurse Yale-New Haven Hosp., 1990-91; pvt. practice Citadel Psychiat. Clinic, Ft. Wayne, Ind., 1991—. Bd. dirs. Hoosiers for Safety Belts, Indpls., 1987-88, Ind. Med. Pol. Action Com., Indpls., 1986-87; coordinator, founder Safe Start Infant Safety Seat Loan Program, Wabash, 1981-87; participant in leadership devel. com. Wabash County C. of C., 1983; workshop leader Wabash County Hosp. Stop Smoking Program, 1982-83. Mem. NAEW (family rels. coun.), Ind. Child Passenger Safety Assn. (pres. 1985-87), Am. Psychol. Assn. (student affiliate), Am. Assn. Marital and Family Therapists (student), Sierra Club, Nat. Audobon Soc., The Wilderness Soc., Nat. Family Rels. Coun., Nat. Wildlife Fedn., Smithsonian Assn., Kappa Kappa Kappa. Republican. Home: 2611 Neptunes Xing Fort Wayne IN 46815-8339 Office: Citadel Psychiat Clinic 2001 Reed Rd Fort Wayne IN 46815-7311

FERIC, GORDAN, mechanical engineer; b. Rijeka, Croatia, Oct. 25, 1961; s. Marin and Aleksandra (Mardesic) F.; m. Lena Zlatar, June 20, 1987; 1 child, Marina. BS in Mech. Engring., U. Zagreb, Croatia, 1986; MS in Mech. Engring., George Washington U., 1989. Engr. in tng., 1991. Faculty asst. U. Zagreb, 1987; sys. engr. Gilbert/Commonwealth, Inc., Gaithersburg, Md., 1989—. Recipient Best Paper award SEAM 30, 1992. Home: 14023 Jump Dr Germantown MD 20874 Office: Gilbert/Commonwealth Inc 19644 Club House Rd #820 Gaithersburg MD 20879

FERIN, MICHEL JACQUES, reproductive endocrinologist, educator; b. Louvain, Belgium, Apr. 23, 1939; came to U.S., 1965; s. Jacques and Paule (Goffaux) F. MD, U. Louvain, 1964. Rsch. fellow U. Geneva, Switzerland, 1964-65; trainee in reproductive biology Worcester Found., Shrewsbury, Mass., 1965-66; rsch. assoc. Columbia U., N.Y.C., 1966-73, asst. prof., then assoc. prof., 1973-89, prof. reproductive endocrinology, 1989—; population researcher NIH, Bethesda, Md., 1986-90, mem. study sect. on endocrinology, 1992—. Author: The Menstrual Cycle, 1992; contbr. articles to Endocrinology jour. Mem. Endocrine Soc., Soc. Study Reprodn., Soc. Neuroscis., Soc. Gynecol. Investigation. Achievements include research on neuroendocrine control of reproductive cycle in primates. Office: Columbia U Dept Ob-Gyn 630 W 168th St New York NY 10032

FERNALD, JAMES MICHAEL, engineer; b. Portsmouth, N.H., Oct. 12, 1964; s. R. Alden and Ruth Ann (Conlon) F. BS in Aerospace Engring., Syracuse U., 1986; BSME, U. N.H., 1992. Engring. technician Aquidneck Mgmt. Assn., Middleton, R.I., 1988; engr. Life Cycle Engring., Inc., Portsmouth, 1988—. Treas. troop 164 Boy Scouts Am., Portsmouth, 1989-91. Mem. AIAA, ASHRAE (assoc.), Assn. Energy Engrs., Syracuse U. Alumni Assn., Golden Key, Tau Beta Pi (pres. 1991-92), Theta Chi. Roman Catholic. Avocations: chess, golf, raquetball, skiing. Office: Life Cycle Engring Inc 500 Spaulding Tpke Ste 120N Portsmouth NH 03801-3162

FERNANDES, PRABHAVATHI BHAT, molecular biologist; b. Bangalore, Karnataka, India, Apr. 11, 1949; came to U.S., 1972; d. Govinda and Meenakshi (Rao) Bhat; m. Michael V. Fernandes, June 21, 1971; children: Meena, Sheila. MS, Madras (India) U., 1971; PhD, Thomas Jefferson U., 1974. Postdoctoral fellow Inst. Cancer Rsch., Fox Chase, Phila., 1976-78; asst. prof., postdoctoral fellow Temple U. Sch. Medicine, Phila., 1978-80; sr. rsch. microbiologist Squibb Inst. Med. Rsch., Princeton, N.J., 1980-82; project leader Abbott Labs., Abbott Park, Ill., 1983-85, sr. project leader, 1985-88; dir. Bristol-Myers Squibb, Princeton, N.J., 1988-90, exec. dir., 1990-92, v.p., 1993—. Editor: Quinolones: Mode of Action and Mechanisms of Resistance, 1987, Telesymposium on Fluorquinolones, 1989, New Approaches for Antifungal Drugs, 1992; mem. editorial bd. Diagnostic Microbiology and Infective Disease, 1989—, Drug News and Perspectives, 1989—, Jour. Antibiotics, 1991—. Trustee Princeton (N.J.) Day Sch., 1991—. Mem. AAAS, Am. Soc. Microbiology (editorial bd. Clin. Microbial Rev. 1990—, Soc. Gen. Microbiology, Soc. Indsl. Microbiology, Am. Chem. Soc. Hindu. Achievements include development of a novel macrolide antibiotic Clarithromycin; in vivo work for development of Aztreonam, the first monobactam antibiotic which is in clinical use; inventor of Tosufloxacin. Office: Bristol Myers Squibb Pharm Rsch Inst PO Box 4000 Princeton NJ 08543-4000

FERNANDEZ, FERNANDO LAWRENCE, research company executive, aeronautical engineer; b. N.Y.C., Dec. 31, 1938; s. Fernando and Luz Esther (Fortuno) F.; m. Carmen Dorothy Mays, Aug. 26, 1962; children: Lisa Marie, Christopher John. BSME, Stevens Inst. Tech., 1960, MS in Applied Mechanics, 1961; PhD in Aeronautics, Calif. Inst. Tech., 1969. Engr. Lockheed Missiles & Space Co., Sunnyvale, Calif., 1961-63; div. mgr. The Aerospace Corp., El Segundo, Calif., 1963-72; program mgr. R & D Assocs., Santa Monica, Calif., 1972-75; v.p. Phys. Dynamics, Inc., San Diego, 1975-76; pres. Arete Assocs., San Diego, 1976—; mem. Chief Naval Ops. Exec. Panel, Washington, 1983—. Editor Jour. AIAA, 1970; contbr. articles to Fluid Mechanics. Office: Arete Assocs PO Box 8050 La Jolla CA 92038

FERNANDEZ, RENÉ, aerospace engineer; b. Havana, Cuba, Oct. 2, 1961; came to U.S., 1967; s. Ramon and Emma (Fumero) F. Student, Broward Community Coll., Ft. Lauderdale, Fla., 1979-80; BS in Engring., Case Western Res. U., 1986, MSc in Engring., 1993. Rsch. fellow Univ. Space Rsch. Assn., Cleve., 1986; grad. teaching asst. Case Western Res. U., Cleve., 1986-87; rsch. engr. NASA Lewis Rsch. Ctr., Cleve., 1987—; mem. speaker's bur. NASA Lewis Rsch. Ctr., 1988—; sci. fair judge Ohio Acad. Sci., Columbus, Ohio, 1988, 90. Mem. AIAA (chmn. No. Ohio sect. 1991-92, dep. dir. for young mem. programs region III), ASTM, ASME, AAAS, IEEE, Optical Soc. Am., Soc. Photo-optical Instrumentation Engrs., NASA Ski Club, NASA Karate Club, NASA Chess Club. Roman Catholic. Avocations: scuba diving, skiing, dancing, astronomy, art. Office: NASA Lewis Rsch Ctr 21000 Brookpark Rd Cleveland OH 44135-3191

FERNANDEZ-MARTINEZ, JOSE, physician; b. San Juan, P.R., Apr. 2, 1930; s. Telesforo and Luisa (Martinez) Fernandez; m. Carmen Dolores Noya, Dec. 26, 1954. BS, Villanova U., 1951; MD, U. Pa., 1955. Diplomate Am. Bd. Internal Medicine, Sub-Bd. Cardiovascular Diseases. Intern U. Pa. Hosps., Phila., 1955-56, resident in internal medicine, 1956-59, fellow in hypertension and cardiovascular diseases, 1956-57; practice medicine specializing in cardiovascular diseases Santurce, P.R., 1961—; attending physician in internal medicine San Juan City Hosp., 1961—; assoc. prof. medicine U. P.R., 1978-88, prof. Sch. of Medicine, 1988—. Served to capt. U.S. Army, 1959-61. Fellow ACP, Am. Coll. Cardiology; mem. P.R. Med. Assn. (pres. sci. coun. 1968), Alpha Omega Alpha. Office: Ashford Med Ctr Ashford & Washington Sts Ste 208 Santurce PR 00907

FERNÁNDEZ MIRANDA, JORGE, physico-mathematician, researcher; b. Havana, Cuba, Nov. 18, 1950; s. David Fernández Merino and Emma Isabel Miranda Vargas; m. Maria Ofelia Prendes Vazquez, Sept. 15, 1973 (div.); children: Jorge Fernández Prendes, Joyma Fernández Prendes; m. Elsa Lidia Castillo Rodriguez, Dec. 25, 1992. Grad., Havana U., Cuba, 1972; MS, Moscow U., 1977; PhD, Bratislava (Czechoslovakia) U., 1980. Jr. researcher Nuclear Rsch. Inst., Havana, 1972-75, researcher, 1975-77, head of dept. solid state physics, 1977-80; sr. physicist Nat. Secretariat for Nuclear Problems, Havana, 1980-85; head of lab. irradiation techniques CENSA, Havana, 1985—; assoc. prof. Havana U., 1975-80; advisor for nuclear techniques applications Nat. Secretariat for Nuclear Problems, Havana, 1980-85. Contbr. numerous articles to profl. jours. Recipient 20th Anniversary Found. of Nat. Ctr. Animals and Plants Health (CENSA) medal Higher Edn. Min. of Cuba, Havana, 1990, 30th Anniversary Found. of Cuban Acad. Scis. medal, 1993. Mem. ASTM, Sci. Coun. CENSA. Achievements include patents for radiation processing related to chemicals dosimeters; development of radiation processing in Cuba. Office: Irradiation Tech Lab CENSA, San Jose de las Lajas PO Box 10, Havana Cuba

FERNANDEZ-MORAN, HUMBERTO, biophysicist; b. Maracaibo, Venezuela, Feb. 18, 1924; s. Luis and Elena (Villalobos) Fernandez-M.; m. Anna Browallius, Dec. 30, 1953; children: Brigida Elena, Veronica. M.D., U. Munich, Germany, 1944, U. Caracas, Venezuela, 1945; M.S., U. Stockholm, Sweden, 1951, Ph.D., 1952. Fellow neurology, neuropath. George Washington U., 1945-46; intern George Washington U. Hosp., 1945-46; resident Serafimerlasarettet, Stockholm, 1946-58; fgn. asst. Neurosurg. Clinic, Stockholm, 1946-48; research fellow Nobel Inst. Physics, Stockholm, 1947-49, Inst. Cell Research & Genetics, Karolinska Institutet, Stockholm, 1948-51; asst. prof. Inst. Cell Research & Genetics, Karolinska Institutet, 1952; prof., chmn. dept. biophysics U. Caracas, 1951-58; dir. Venezuelan Inst. Neurology and Brain Research, Caracas, 1954-58; asso. biophysicist neurosurg. service Mass. Gen. Hosp., Boston, 1958-62; vis. lectr. dept. biology Mass. Inst. Tech., 1958-62; research asso. neuropath. Harvard, 1958-62; prof. biophysics U. Chgo., 1962—, A.N. Pritzker prof. biophysics, now prof. emeritus; spl. sci. attache Embassy of Venezuela, Bern, Switzerland, 1989; sci. adv. Embassy of Venezuela, Portugal, 1991—; min.-counsellor Embassy of Venezuela, Stockholm; sci. and cultural attaché to Venezuelan legations, Sweden, Norway, Denmark, 1947-54; head Venezuelan commn. Atomic Energy Conf., Geneva, 1955; chmn. Venezuelan commn. 1st Inter-Am.-Symposium on Nuclear Energy, Brookhaven, N.Y., 1957; minister of edn., Venezuela, 1958; mem. Orgn. Am. States adv. commn. on sci. devel. in Latin Am., Nat. Acad. Scis., 1958; mem. U.S. Nat. Com. UNESCO, 1957. Author: The Submicroscopic Organization of Vertebrate Nerve Fibres, 1952, The Submicroscopic Organization of the Internode Portion of Vertebrate Myelinated Nerve Fibers, 1953, Cryoelectronmicroscopy; Superconductivity; Diamond Knife Ultramicrotomy, 1955-76; author series publs. in fields molecular biology, nerve ultrastructure, electron and cryo-electron microscopy, electron and x-ray diffraction, cell ultrastructure, neurobiology, superconducting lenses, superconductivity, others.; editorial bd. Jour. of Research & Development. Decorated Knight of Polar Star Sweden; Claude Bernard medal Canada; Medalla Andres Bello Venezuela, 1973; Recipient Gold medal City Maracaibo, 1968, John Scott award for invention of diamond knife, 1967; medal Bolivarian Soc. U.S., 1973. Fellow Am. Acad. Arts and Sci.; mem. Venezuelan Acad. Medicine (hon.), Academia Ciencias Fisicas y Matematicas (Caracas), Am. Acad. Neurology (corr. mem.), Internat. Soc. Cell Biology, Buenos Aires, Santiago, Lima, socs. Neurology, Buenos Aires, Santiago, Lima, Porto Alegre societies surgery, Electron Microscopy Soc. Am. (spl. citation), Am. Nuclear Soc., Pan Am. Med. Assn., Sociedad Bolivárianade Arquitectos (Venezuela) (hon.), Pan Am. Assn. of Anatomy (hon.). Office: Embassy Venezuela Portugal/Min-, Coun Stockholm Engelbrektsg 35, Stockholm 114 32, Sweden

FERNANDEZ-POL, BLANCA DORA, psychiatrist, researcher; b. Buenos Aires, Mar. 5, 1932; came to U.S., 1967; d. Balbino Fernandez and Maria Remedios van Pol. MD, U. Buenos Aires, 1958. Diplomate Am. Bd. Psychiatry and Neurology. Intern N.Y. Polyclinic Med. Sch., 1967-68; resident in psychiatry UCLA/Brentwood Hosp., 1968-69, NYU/Bellevue Hosp., 1969-71; gen. practitioner Hosp. Espanol, Buenos Aires, 1959-62; forensic psychiatrist Criminoloy Inst., Buenos Aires, 1963-65; clin. attending psychiatrist Bellevue Psychiat. Hosp., N.Y.C., 1971-75; pvt. practice St. Petersburg, Fla., 1976-78; chief psychiat. svcs. USAF Hosp. Yokota, Tokyo, 1980, USAF Hosp., Homestead, Fla., 1981; chief continuing treatment program dept. psychiatry Bronx-Lebanon Hosp., Bronx, 1983—; prof. psychology U. Moran, Buenos Aires, 1962-67; asst. prof. psychiatry N.Y. Med. Coll., N.Y.C., 1972-74; clin. asst. prof. psychiatry Albert Einstein Coll. Medicine, Bronx, 1982—. Contbr. articles to profl. jours. Maj. USAF, 1978-81. Mem. Am. Psychiat. Assn., N.Y. Acad. Scis., Am. Acad. Psychiatrists in Alcoholism and Addictions, Am. Soc. Addiction Medicine, Assn. Mil. Surgeons of U.S. Avocations: travel, painting, sculpture. Home: PO Box 21644 Brooklyn NY 11202-0036 Office: Bronx Lebanon Hosp 1285 Fulton Ave Bronx NY 10456-3401

FERNANDEZ-POL, JOSE ALBERTO, physician, radiology and nuclear medicine educator; b. Buenos Aires, Mar. 17, 1943; came to U.S., 1971; s. Manuel and Antonia (Pol) Fernandez; m. Maria Eugenia Mengual, June 11, 1971; children: Sebastian, Julia. MD, U. Buenos Aires, 1969. Diplomate Am. Bd. Nuclear Medicine. Physician Hosp. San Martin, U. Buenos Aires, 1969-70; rsch. assoc. Ctr. Nuclear Medicine, U. Buenos Aires, 1970-71; resident SUNY, Buffalo, 1971-72, rsch. fellow in nuclear medicine, 1972-75; rschl fellow molecular biology lab. Nat. Cancer Inst./NIH, Bethesda, Md., 1975-77; prof. medicine, radiology St. Louis U., VA Med. Ctr., 1977—. Contbr. articles to Molecular Oncololgy, Biotech., other publs. Mem. Am. Assn. Cancer Rsch., Soc. Biol. Chemistry and Molecular Biology. Achievements include original findings in area of growth regulation and cancer; discovery of a new human gene denoted metallopanstimulin. Office: VA Med Ctr 915 N Grand Blvd Saint Louis MO 63106

FERNANDEZ-REPOLLET, EMMA D., pharmacology educator; b. Ciales, P.R., 1951; d. Angel M. and Carmen M. (Repollet) Fernández; m. Abraham Schwartz, May 5, 1990. PhD, U. P.R., 1979. Postdoctoral fellow Duke U., Durham, N.C., 1979-80, U. N.C., Chapel Hill, 1980-82; with lab. VA Med. Ctr., San Juan, P.R., 1982-86; asst. prof. pharmacology U. P.R. Sch. Medicine, San Juan, 1982-86, assoc. prof. in pharmacology, 1986—; assoc. dir. Rsch. Ctrs. in Minority Insts., 1986—; advisor Interam. U., San Juan, 1990-92. Contbg. author: Acute Renal Failure, 1983; contbr. articles to Jour. Physiology, Am. Jour. Physiology, Jour. Clin. Investigation, Am. Jour. Med. Sci. Recipient medal Sci. Tchrs. Assn., San Juan, 1972, Young Investigator award P.R. Acad. Arts and Sci., 1992; travel grantee IX Internat. Congress Nephrology, Can., 1984, X, Eng., 1987. Mem. Am. Soc. Nephrology, Am. Diabetes Assn., N.Y. Acad. Scis., Sigma Xi. Achievements include patent on methods of use of non-fluorescent particles. Office: U PR Dept Pharmacology GPO Box 36-5067 San Juan PR 00936-5067

FERNANDEZ STIGLIANO, ARIEL, chemistry educator; b. Bahia Blanca, Argentina, Apr. 8, 1957; came to U.S., 1981; s. Domingo Fernandez and Haydee Erminia (Stigliano) Belinky. MPhil, Yale U., 1982, PhD, 1983. Rsch. assoc. Princeton (N.J.) U., 1984-86; vis. scientist Max-Planck-Institut fur Biophysikalische Chemie, Gottingen, Germany, 1986-88; assoc. prof. phys. chemistry/biochemistry/molecular biology U. Miami Sch. Medicine, Coral Gables, Fla., 1988—; permanent reviewer, cons. Math. Revs., Ann Arbor, 1990—; editor, organizer Miami Bio/Tech. Symposium, Miami Beach, Fla., 1993. Contbr. articles to profl. jours. Recipient medal Govt. of State of Buenos Aires, 1980, Disting. New Faculty award Camille & Henry Dreyfus Found., N.Y.C., 1989, Tchr. scholar award, 1991; Max Planck Soc. scholar, Bonn, Germany, 1988. Achievements include research in a novel theoretical and computational approach to elucidate the folding of biopolymers concurrent with the synthesis of the molecule; exptl. probes for this sequential folding approach. Home: 3246 Virginia St Coconut Grove FL 33133 Office: U Miami Med Sch Gautier Bldg Miami FL 33101-6129

FERNANDO, CECIL T., engineering executive; b. Mt. Lavinia, Sri Lanka, Nov. 10, 1924; s. Condegamage Theodore and Cornelia Henrietta (Fonseka) F.; m. Delicia Mary Senaratne, June 9, 1947; children: Ramya Surangani, Shreeni Damitha. BS of Engring. with honors, Coll. Engring., Poona, India, 1950. Irrigation engr. Dept. Irrigation, Sri Lanka, 1950-65; gen. mgr. River Valleys Devel. Bd., Sri Lanka, 1966-67; cons. Bank of Ceylon, Sri Lanka, 1968; mng. ptnr. Enging. Cons., Colombo, 1968-77, mng. dir., chmn., 1977-89; mng. ptnr. Engring. Cons. Internat., Colombo, 1990—; chmn. Samitar Ltd., Colombo, 1978-89, Engring. Cons. Ltd., Colombo, 1990—. Fellow Inst. Engrs. India, Inst. Engrs. Sri Lanka, ASCE, Assn. Cons. Engrs. Sri Lanka, Sri Lanka Assn. Advancement Sci. Buddhist. Avocations: philately, sports, classical music, nature. Home: 30/60 Longdon Pl, Colombo 7, Sri Lanka Office: Engring Cons Ltd, 60 Dharmapala Mawatha, Colombo 3, Sri Lanka

FERNANDO, HARINDRA JOSEPH, engineering educator; b. Colombo, Sri Lanka, Oct. 19, 1955; came to U.S. 1980; s. Joseph Raymond and Audrey (Phyllys) F.; m. Ravini Marina Cooray, Jan. 5, 1985; children: Ravi, Sahan. BS with honors, U. Sri Lanka, 1979; MA, PhD, Johns Hopkins U., 1980-83. Rsch. asst. Johns Hopkins U., Balt., 1980-83; rsch. assoc. Calif. Inst. Tech., Pasadena, 1983-84; prof. engring. Ariz. State U., Tempe, 1984—; panelist NRC, Washington, 1991—. Tech. assoc. editor Applied Mechanics Rev./ASME, 1990—; contbr. articles to profl. jours. Recipient Rotary Club award, 1980, UNESCO Gold Medal, 1980; Gilman fellow, 1980-83; NSF Presdl. Young Investigator award, 1987. Mem. ASME, Am. Phys. Soc., Am. Geophys. Union. Office: Ariz State Univ Dept Mech/Aerosapce Engring Tempe AZ 85287-6106

FERNICOLA, NILDA ALICIA GALLEGO GÁNDARA DE, pharmacist, biochemist; b. Bahia Blanca, B.A., Argentina, Jan. 9, 1931; d. Francisco and Alicia (Gándara) Gallego; m. Lucio Fernicola, Jan. 2, 1964; 1 child, Pablo Francisco. BS in Pharmacy, U. Buenos Aires, Argentina, 1955, BS in Biochemistry, 1959, PhD in Pharmacy and Biochemistry, 1962. Analyst Nat. Chemistry Office, Buenos Aires, Argentina, 1957-67; instr. Coll Pharmacy & Biochemistry, U. Buenos Aires, 1962-75; prof. U. Sao Paulo, Brazil, 1976-82; team leader Fundacentro, Sao Paulo, 1975-76; head toxicology divsn., agy. for environ. control. Companhia de Tecnologia de Saneamento Ambiental, Sao Paulo, 1976-82, head human toxicology, 1991—; toxicologist cons. PanAm. Health Orgn., Mexico, 1982-91; mem. rsch. group Nat. Coun. Science, Tech., Rsch., Buenos Aires, 1967. Co-author: Nociones Básicas de Toxicologia, 1985, Toxicologia Ocupacional , 1989; contbr. articles to profl. jours. Recipient award Argentina Congress, 1975. Mem. Argentinian Acad. Environ. Sci., Toxicology Soc. Panama (founder), Brazilian Toxicology Soc. (sec. Sao Paulo 1992-93). Home: Rua Joao Ramalho 586 Ap 31B, 05008 São Paulo SP, Brazil Office: Ave Frederico Herman 345, 05489 São Paulo SP, Brazil

FERRACUTI, STEFANO EUGENIO, forensic psychiatrist; b. Rome, June 2, 1958; s. Franco and Mirella (Garutti) F.; m. Lucia Fusco; 1 child, Giorgia. Cert. in teaching, Liceo Sperimentale, Rome, 1977; MD, U. Rome, 1985. Tchr. Istis Loiza Cordeiro for Blind, San Juan, P.R., 1977-78; resident dept. neurology U. Rome, 1985-90; asst. prof. dept. psychiatry and clin. psychology U. Rome La Sapienza, 1991—; active ednl. bd. Revista de Derecho Penal y Criminologia Quad. Psich. For. ; cons. in field, 1991—. Contbr. articles to profl. publs. Lt. Italian Army, 1986-87. Mem. Internat. Assn. for Study of Pain, Soc. Personality Assessment, N.Y. Acad. Sci., Italian Soc. Neurology, Am. Acad. Psychiatry and Law, Italian Soc. Criminology, Amnesty Internat. Avocations: chess, computers, travel. Home: Via Ugo Balzani 57, 00162 Rome Italy Office: U Rome Dept Psychiatry, Pzle A Moro 5, 00185 Rome Italy

FERRANDO, RAYMOND, animal nutrition scientist, educator; b. Constantine, Algeria, Mar. 3, 1912; s. Joseph and Etiennette (Dessens) F.; D.Vet.Sci., D.Sci., Vet. Nat. Sch. and Faculties Scis., Lyon, France, 1937, D.Scis., 1952; m. Raymonde Boulud, Dec. 3, 1943. Maitre-asst. Vet. Sch., Lyon, from 1945; prof. nutrition Ecole Veterinaire Alfort, Paris, 1955—, dir. sch., 1957-64, hon. dir., 1968; chmn. Commission interministerielle alimentation animale France; mem., past chmn. sci. com. Animal Nutrition Econ. Community Europe; expert WHO; cons. FAO. Decorated officer Legion of Honor. Mem. Am. Chem. Soc., N.Y. Acad. Scis., Acad. Nat. Medecine France (vice chmn. 1988, chmn. 1989), Acad. Agr. (vice chmn. France 1990, chmn. 1991), Société Française de Therapeutique, Acad. Veterinaire France (chmn. 1987, Acad. Royale Medecine Belgique. Roman Catholic. Club: Rotary (pres. club 1971-72, dist. gov. 1982-83) (Paris). Research, numerous publs. on vitamins, nutrition, antibiotics; editor books. Home: 20 Rue de Boulainvilliers, 75016 Paris France Office: 107 Rue de Reuilly, Bat 1, 75012 Paris France

FERRARO, VINCENT, mechanical engineer, civil engineer; b. Cleve., Oct. 22, 1947; s. Vincent and Virginia (Domiano) F.; m. Sarah Edwards, Aug. 29, 1970; children: Alexandra, Christina. BSME, Case Inst. Tech., 1970; MS in Civil Engring., Ohio State U., 1981. Registered profl. engr., Va. Commd. 2d lt. U.S. Army, 1977, advanced through grades to maj., 1991; co. comdr. Ft. Polk, La., 1977-79; asst. chief engring. div. N.Y.C., 1981-82; engr. plans officer The Netherlands, 1982-86, dir. engring. and housing, 1986-89; dep.

dist. engr. Rock Island, Ill., 1989-91; dir. constrn. Cuyahoga Metro Housing Authority, Cleve., 1991—. Decorated Legion of Merit. Mem. Soc. Am. Mil. Engrs. (past pres. 1990-91), Soc. Profl. Engrs., Cleve. Soc. Profl. Engrs. (dir.). Home: 31507 Carlton Dr Bay Village OH 44140

FERREE, DAVID CURTIS, horticultural researcher; b. Lock Haven, Pa., Feb. 9, 1943; s. George H. and Ruth O. (McClain) F.; m. Sandra J. Corman, Aug. 31, 1968; children: Curtis P., Thomas A. BS, Pa. State U., 1965; MS, U. Md., 1968, PhD, 1969. From asst. to assoc. prof. Ohio State U., Wooster, 1971-81, prof., 1981—. Editor Fruit Varieties Jour.; contbr. numerous articles to profl. jours. Capt. U.S. Army, 1969-71. Fellow Am. Soc. Hort. Sci. (assoc. editor 1983-86, v.p. 1988-89, J. H. Gouty award 1982, Stark award 1983), Am. Pomological Soc. (editor), Internat. Dwarf Fruit Tree Assn. (disting. researcher 1989), Gamma Sigma Delta (rsch. award 1981). Lutheran. Office: Ohio Agrl R & D Ctr Dept Horticulture Wooster OH 44691

FERREIRA, JAY MICHAEL, mechanical engineer; b. Allentown, Pa., July 21, 1967; s. Jacob and Margaret Louise (Frey) F. BSME, Drexel U., 1990; MSME, Ariz. State U., 1992, postgrad., 1992—. Engr. in tng. Draftsman Siebert Ferreira Assocs., Allentown, 1985-86; with nuclear licensing dept. coop. Pa. Power and Light Co., Allentown, nuclear lic. engring. asst. coop., nuclear maintenance engring. asst. coop.; rsch. engr. Ariz. State U., Tempe, 1992—; presenter in field. Contbr. articles to Composites Engring. Contbr. Statue of Liberty Ellis Island Found., 1985. 1st lt. Ordnance U.S. Army, 1990—. Mem. ASME (assoc.), AIAA, Am. Helicopter Soc., Pi Tau Sigma. Republican. Roman Catholic. Achievements include development of computer codes for design and optimization of composite structures; research in composite design, optimization, helicopter rotorblades and crashworthiness. Home: 1649 Washington Ave Northampton PA 18067 Office: Ariz State U Tempe AZ 85287

FERREIRA, PAULO ALEXANDRE, molecular biologist; b. Coimbra, Portugal, Sept. 19, 1964; came to U.S., 1986; s. Antonio Lopes Ferreira and Maria Celina (Baptista) M. Pinto; m. Mirna Farid El-Khoury Ghanem, Jan. 4, 1991. BS with distinction and honors, Jacksonville State U., 1988; Degree in Biology, U. Porto, Portugal, 1988. Vis. scholar U. Kiril I Metodiy Internat. Assn. of Exchange of Students for Tech. Experience, Skopje, Yugoslavia, 1985; acad. scholar Internat. House Program, Jacksonville (Ala.) State U., 1986-88; teaching, rsch. fellow Purdue U., West Lafayette, Ind., 1988—; NATO predoctoral fellow NATO, 1989-91; predoctoral fellow SCIENCE Program E.E.C., 1991—; vis. scientist Harvard Med. Sch., Boston, 1990; vis. fellow Karolinska Inst., Stockholm, 1990. Contbr. articles to profl. jours. including Procs. Nat. Acad. Sci. U.S.A. Past mem. Internat. House Program, Jacksonville State U., 1986-88. Recipient Assitantship Purdue U., 1988—, NATO fellowship, 1989-91, NATO postdoctoral fellowship, 1993—, Travel award Inst. Internat. Edn., 1989, Hon. Mention Prize of Med. Genetics, 1989, fellowship U. Tenn. Ctrs. of Excellence Program, 1987; named Hon. Citizen of Montgomery, Ala. Mem. AAAS, Fedn. Am. Scientists, Soc. Neurosci., Phi Eta Sigma, Tri Beta. Achievements include identification and molecular characterization of photoreceptor protein homologues of Drosphilia in mammalian retina; study of cell signal transduction in neuronal cells. Office: Purdue U Lilly Hall Biology Dept West Lafayette IN 47907

FERREIRA FALCON, MAGNO, economist; b. Bogado, Itapua, Paraguay, Sept. 3, 1936; s. Venancio and Juana (Falcon) Ferreira; m. Querubina Daniela Caballero, Aug. 19, 1963; children: Orlando Eugenio, Daysi Patricia, Carlos Enrique. MPA, U. Pitts., 1963; PhD in Econs., U. Asuncion, Paraguay, 1968. Acct., adminstrv. officer Customs Office, Asunción, Paraguay, 1954-58; sec. gen. Antelco-Telecommunications Adminstrn., Asunción, Paraguay, 1958-72; fin. mng. dir. City Adminstrn. of Asunción, 1972-73; exec. dir. Finan, Binational Entity, Asunción, 1974-82; mng. dir. Oga Rape, S.A., Asunción, 1983—; cons., economist in field; prof. adminstrn. Cath. and Nat. U. of Asunción, 1963—. Author: El Complejo Hidroelectric Yacyreta, 1990; contbr. articles to profl. jours. Mem. Club Centenario, Asunción, 1974—; pres. Centro de Economistas Colorados, Asunción, 1971-73. Recipient Gold medal Centro de Estudiantes de Ciencias Económicas, 1968, Colegio Grad. Ciencias Economicas, 1968, 85, Sch. of Econs., 1969, Ministerio Obras Publicas, 1970. Mem. Centro Ciencias Económicas, Colegio Graduados Ciencias Económicas, Club Cerro Porteño. Roman Catholic. Avocations: fishing, football, soccer. Home: Centenario 1545/47, Asunción Paraguay Office: Oga Rape SA Ahorro/Prestamo, Casilla de Correos #1815, Asunción Paraguay

FERRELL, REBECCA V., biology educator; b. Springfield, Mo., Aug. 29, 1955. BS in Biology and English, S.W. Mo. State U., 1978, MS in Biology, 1982, PhD in Molecular Microbiology, 1990. Teaching asst. dept. biology S.W. Mo. State U., Springfield, 1978-81; instr. in biology European div. U. Md., Schweinfurt, Fed. Republic Germany, 1982-84; teaching asst., rsch. molecular biology and immunology U. Mo.-Columbia Sch. Medicine, 1984-90; NIH postdoctoral rsch. assoc. dept. MCD biology U. Colo., Boulder, 1990-91; rsch. assoc. dept. med. microbiology and immunology U. Colo. Health Scis. Ctr., Denver, 1992—; asst. prof. Met. State Coll., Denver, 1991—; presenter in field. Contbr. articles to profl. publs., chpt. to book. Sec. Substance Abuse Adv. Commn., Columbia, 1986-90. Mem. AAAS, Am. Soc. Microbiology (treas. Rocky Mountain br. 1992—). Achievements include investigation of iron-regulated gene expression in Pseudomonas; discovery of 1st repetitive sequence in porcine mycoplasmas, existence of spleen-derived murine macrophage suppressor factor. Office: Met State Coll Campus Box 53 PO Box 173362 Denver CO 80217-3362

FERRELL, WILLIAM GARLAND, JR., industrial engineering educator; b. Lynchburg, Va., Feb. 24, 1955; s. William Garland and Mary Lucille (Jennings) F.; m. Henrietta Susan Fields, Nov. 14, 1981; 1 child, William Garland III. BA, Wake Forest U., 1977; MS, Va. Poly. Inst. and State U., 1979; PhD, N.C. State U., 1989. Registered profl. engr., S.C. Engr. Babcock & Wilcox, Lynchburg, 1977-81, project mgr., 1981-84; asst. prof. indsl. engring. Clemson (S.C.) U., 1988—; cons. Oak River Mills, Bennettsville, S.C., 1989—, Dixianna Mills, Dillon, S.C., 1991—, Gibbes and Clarkson, Greenville, S.C., 1991—, Sonoco Products, Hartsville, S.C. Contbr. articles to profl. jours. Grad. fellow Dept. Energy, 1978. Mem. Inst. Indsl. Engrs., Ops. Rsch. Soc. Am. (editor student newsletter 1986-88), Sigma Xi, Omega Rho. Achievements include implementation of results in optimal control of productive systems for improving quality. Office: Clemson U 104 Freeman Hall Clemson SC 29634-0920

FERRERO, THOMAS PAUL, engineering geologist; b. Santa Monica, Calif., May 21, 1951; s. Charles Fredrick Ferrero and Mary Elizabeth (McDonald) Stannard; m. Catherine Adams Kesler, June 21, 1980; children: Lia Rae, Andreu Joseph, Martin Thomas. BS, U. Oreg., 1983. Cert. engring. geologist, Oreg., Calif. Geologist various mining and engring. cos., western U.S., 1975-83; cons. geologist Ferrero Geologic, Ashland, Oreg., 1983—. Author publs. in field. Mem. Assn. Engring. Geologists. Achievements include structural and engineering geologic mapping of 305 square miles of Curry and Josephine Counties, Oregon; establishing and maintaining a successful small geologic consulting business for more than 10 years. Home and Office: 340 Avery St Ashland OR 97520

FERRIER, JOSEPH JOHN, atmospheric physicist; b. Weehawken, N.J., Jan. 28, 1959; s. Henry Pierre and Josephine (Logalbo) F. BS, Columbia U., 1980; MS, NYU, 1983. Sci. programmer Sigma Data Svcs. Corp., N.Y.C., 1980-81; programmer/analyst M/A-Com Info. Systems, Inc., N.Y.C., 1981-86; atmospheric physicist, planetary group mgr. Centel Fed. Svcs. Corp., N.Y.C., 1986-89, Hughes Aircraft Co., N.Y.C., 1989—. Mem. AAAS. Office: Hughes Aircraft 2880 Broadway New York NY 10025-7848

FERRIS, THEODORE VINCENT, chemical engineer, consulting technologist; b. Rochester, N.Y., Apr. 26, 1919; s. Theodore Clodoveo and Lucille T. (Pucci) F.; m. Doris Donaghue, June 26, 1943; children: William, Donald, Jean, Peter, Kathleen. BSChemE, MIT, 1941, MS in Chem. Engring. Practice, 1942. Registered profl. engr., Mass. Process engr. Allied Chem. Corp., Buffalo, 1942; devel. engr. GE Plastics, Pittsfield, Mass., 1943-44; process engr. Aspinook Corp., Lawrence, Mass., 1946-48; chief engr. Dehydrating Process Co., Boston, 1949-54; project engr., cons. Monsanto Co., Springfield, Mass., 1954-85; adj. prof. mech. engring. Western New England Coll., Springfield, 1986; cons. Ferris Tech. Svcs., Longmeadow, Mass., 1985—. Contbr. articles to profl. jours. Coach Little League, Longmeadow, 1955-65; cubmaster Boy Scouts Am., Longmeadow, 1955-65; co-author, mem. Bldg. Code Com., Longmeadow, 1955-56; chmn. Regional MIT Ednl. Coun., 1987—. Lt. USN Ordnance, 1944-46. Mem. Am. Inst. Chem. Engrs. (chmn. west Mass. sect. 1958), Assn. Cons. Chemists and Chem. Engrs., MIT Club of Conn. Valley (treas. 1987—). Roman Catholic. Achievements include 10 patents for organic chemical processing and distillation; development of spray drying processes for resin emulsions including equipment design; computerized process simulations of chemical manufacturing; design of rupture disks for formaldehyde converters; research on more efficient removal of methanol from formalin. Home and Office: 58 Clairmont St Longmeadow MA 01106-1002

FERRY, JOHN DOUGLASS, chemist; b. Dawson, Can., May 4, 1912; s. Douglass Hewitt and Eudora (Bundy) F.; m. Barbara Norton Mott, Mar. 25, 1944; children—Phyllis Leigh, John Mott. A.B., Stanford U., 1932, Ph.D., 1935; student, U. London, 1932-34. Pvt. asst. Hopkins Marine Sta., Stanford, Calif., 1935-36; instr. biochem. scis. Harvard U., 1936-38; mem. Soc. Fellows, 1938-41; assoc. chemist Woods Hole Oceanographic Inst., 1941-45; research assoc. Harvard U., Cambridge, Mass., 1942-45; asst. prof. chemistry U. Wis., 1946, assoc. prof., 1946-47, prof., 1947-82, prof. emeritus, 1982—, Farrington Daniels Research prof., 1973-82, chmn. dept., 1959-67; chmn. Internat. Com. on Rheology, 1963-68; vis. lectr. Kyoto U., Japan, 1968, Ecole d'Ete, U. Grenoble, France, 1973. Author: Viscoelastic Properties of Polymers, 1961, 2d edit., 1970, 3d edit., 1980; co-editor: Fortschritte der Hochpolymeren Forschung, 1958-85. Recipient Eli Lilly award Am. Chem. Soc., 1946, Bingham medal Soc. Rheology, 1953, Kendall Co. award Am. Chem. Soc., 1960, Witco award, 1974, Colwyn medal Instn. Rubber Industry, U.K., 1972, Tech. award Internat. Inst. Synthetic Rubber Producers, 1977. Fellow Am. Phys. Soc. (high polymer physics prize 1966), Am. Acad. Arts and Scis.; mem. Nat. Acad. Sci., NAE, Am. Chem. Soc. (Goodyear medal Rubber div. 1981, Polymer div. award 1984), Am. Soc. Biol. Chemists, Soc. Rheology (pres. 1961-63), Internat. Soc. Hematology, d'Honneur Groupe Francais Rheologie, Soc. Rheology Japan (hon.), Phi Beta Kappa, Sigma Xi, Phi Lambda Upsilon, Alpha Chi Sigma. Lodge: Rotary. Home: 137 N Prospect Ave Madison WI 53705-4069

FERSLEW, KENNETH EMIL, pharmacology and toxicology educator; b. Chgo., Sept. 26, 1953; s. Robert George and Phyllis Marie (Fialkowski) F.; m. Susan Marie Pobletts, Dec. 20, 1973; children: Matthew Ryan, Brian Christopher. BS, U. Fla., 1975, MS, 1976; PhD, La. State U., Shreveport, 1982. Lic. lab. dir., med. lab. dir. toxicology, Tenn. Lab. technologist Coll. Vet. Medicine, U. Fla., Gainesville, 1976-77; grad. rsch./teaching asst. dept. pharmacology La. State U. Med. Ctr., Shreveport, 1977-79, grad. rsch. asst., 1979-82; instr. dept. pharmacology East Tenn. State U., Johnson City, 1982-83, asst. prof., 1983-88, assoc. prof., 1988—, dir. toxicology, 1982—; ad hoc cons. Am. Assn. Acreditated Lab. Animal Care, Bethesda, Md., 1991-94; cons. toxicologist VA Med. Ctr., Mountain Home, Tenn., 1984—; forensic toxicologist State of Tenn. Forensic Program, Johnson City, 1984—. Contbr. articles to profl. jours. Mem. Soc. Toxicology (assoc., councilor S.E. chpt. 1983-84, pres. 1987-88), Am. Acad. Forensic Sci., Am. Coll. Clin. Pharmacology, S.W. Assn. Toxicologists, Sigma Xi. Achievements include copyright for Pursuit Meter II Program. Home: 2132 David Miller Rd Johnson City TN 37604 Office: East Tenn State U Sect Toxicology Box 70 422 W272 Meml Ctr Johnson City TN 37614-0422

FERTNER, ANTONI, electrical engineer; b. Krakow, Poland, Oct. 8, 1950; s. Antoni and Janina (Zaleska) F.; divorced; 1 child, Antoni; m. Jolanta Sekowska, 1993. MSc, U. Mining and Metallurgy, Krakow, 1973; PhD, Warsaw Polytech. Inst., 1978. Rsch. engr. Inst. Nuclear Physics and Techniques, Krakow, 1973-74, Inst. Radioelectronics and Warsaw Polytech., Warsaw, 1974-78; asst. prof. Polish Acad. Sci., Krakow, 1978-79; teaching asst. Royal Inst. Tech., Stockholm, Sweden, 1981-82; scientist Swedish Inst. Microelectronics, Stockholm, 1982-90, Catella Generics AB, Stockholm, 1990-91, Ericsson Telecom AB, Stockholm, 1991—. Office: Ericsson Telecom AB, Stockholm S-12625, Sweden

FESCEMYER, HOWARD WILLIAM, entomology educator, insect physiology reseacher; b. Glassport, Pa., Jan. 21, 1957; s. William Howard and Louise Anne (Fato) F.; m. Kathleen Ann Mann, Nov. 24, 1991. BS in Biology, Pa. State U., 1979, MS in Entomology, 1982; PhD in Entomology, La. State U., 1986. Rsch. assoc. in chem. ecology Pa. State U., University Park, 1979-82; rsch. assoc. in insect physiology La. State U., Baton Rouge, 1982-86; postdoctoral rsch. assoc. in chemosensory physiology U. Md. Balt. County Campus, Catonsville, 1986-88; postdoctoral rsch. affiliate in insect neurohormones U.S. Dept. Agr., Agrl. Rsch. Svc., Beltsville, Md., 1988-90; asst. prof. entomology Clemson (S.C.) U., 1990—; teaching asst. in insect physiology Pa. State U., University Park, 1980-81, La. State U., Baton Rouge, 1982-86; lectr. in neurobiology Young Scholars Program U. Md., Catonsville, 1987, 88. Contbr. 18 articles to profl. jours. Grantee La. State U., U.S. Dept. Agr., 1984-85, 1986-88, 1988-90, 1991—; Provost Rsch. award Clemson U., 1991, 92. Mem. Am. Assn. Advancement of Sci., Entomological Soc. Am., Entomological Soc. Pa., Md. Entomological Soc., Am. Soc. Zoologists, N. Ctrl. Regional Com. 148, S.C. Entomological Soc., Alliance for Aerobiology Rsch. Home: 113 Karen Dr Clemson SC 29631-1722 Office: Clemson University Dept Entomology 114 Long Hall Box 340365 Clemson SC 29634-0365

FESCHENKO, ALEXANDER, nuclear scientist. Researcher Inst. Nuclear Rsch., Moscow, Russia. Recipient Faraday Cup award Acad. Scis EUR, Russia, 1992. Office: Inst of Nuclear Research, Prospekt 60-Letiya Oktyabrya 7A, 117312 Moscow Russia*

FESHBACH, HERMAN, physicist, educator; b. N.Y.C., Feb. 2, 1917; s. David and Ida (Lapiner) F.; m. Sylvia Harris, Jan. 28, 1940; children: Carolyn Barbara, Theodore Philip, Mark Frederick. B.S., CCNY, 1937; Ph.D., MIT, 1942; DSc, Lowell Tech. Inst., 1975. Tutor CCNY, 1937-38; instr. MIT, Cambridge, 1941-45, asst. prof., 1945-47, assoc. prof., 1947-55, prof., 1955-87, Cecil and Ida Green prof. physics, 1976-83, inst. prof., 1983-87, inst. prof. emeritus, 1987—, dir. Ctr. for Theoretical Physics, 1967-73, head dept. physics, 1973-83; cons. AEC; chmn. nuclear sci. adv. com. of Dept. Energy and NSF, 1979-82. Author: (with P.M. Morse) Methods of Theoretical Physics, 1953, (with A. deShalit) Theoretical Nuclear Physics, Nuclear Structure, 1974, Theoretical Nuclear Physics; Nuclear Reactions, 1992; editor: Annals of Physics, Contemporary Concepts in Physics; contbr. articles to sci. jours. Trustee Associated Univs. Inc., 1974-87, 1990—. John Simon Guggenheim Meml. Found. fellow 1954-55; Ford fellow CERN, Geneva, Switzerland, 1962-63; recipient Harris medal CCNY, 1977, Nat. Medal of Sci., 1987. Mem. Am. Phys. Soc. (chmn. div. nuclear physics 1970-71, divisional councillor 1974-78, exec. com. 1974-78, chmn. panel on pub. affairs 1976-78, v.p. 1979-80, pres. 1980-81, Bonner prize 1973), Nat. Acad. Scis., NRC, Am. Acad. Arts and Scis. (v.p. Class I 1973-76, pres. 1982-86), AAAS (chmn. physics sect. 1987-88), Internat. Union Pure and Applied Physics (chmn. nuclear physics sect. 1984-90). Home: 5 Sedgwick Rd Cambridge MA 02138-2037

FESQ, LORRAINE MAE, aerospace and computer engineer; b. Pennsauken, N.J., June 26, 1957; d. John Fred Henry and Natalie Nicola (Nasuti) F.; m. Frank Tai, May 14, 1988. BA in Math., Rutgers U., 1979; MS in Computer Sci., UCLA, 1990, PhD in Computer Sci. and Astrophysics, 1993. Sci. programmer Systems and Applied Sci. Corp., Greenbelt, Md., 1979-81; computer engr./mgr. Ball Aerospace Systems Div., Boulder, Colo., 1981-86; systems engr. OAO, El Segundo, Calif., 1986-87; spacecraft systems engr. TRW, Redondo Beach, Calif., 1987—. Contbr. articles to IECEC Proceedings, AAS Proceedings, Diagnostic Workshop (DX-92) Proceedings, NASA Goddard Space Applications of Artificial Intelligence Proceedings. Mem. Playa Del Rey (Calif.) Network, 1988—. MS fellow TRW, 1988-89, PhD fellow, 1990-93. Mem. AIAA (sr., tech. com. mem. artificial intelligence tech. com. 1990—), Am. Astronautical Soc., Am. Assn. for Artificial Intelligence, Am. Astronomical Soc. Home: 6738 Esplanade Playa Del Rey CA 90293-7525 Office: TRW MS M2/2375 1 Space Pk Redondo Beach CA 90278

FETTER, STEVE, physicist; b. Sunbury, Pa., Oct. 2, 1959; s. Arthur and Betty (Wetzel) F.; m. Marie Redniss, May 24, 1980; children: Emily, Maxwell. BS, MIT, 1981; PhD, U. Calif., Berkeley, 1985. Assoc. prof. U. Md. Sch. Pub. Affairs, College Park, 1988—; internat. affairs fellow U.S. Dept. State Bur. Polit.-Mil. Affairs, Washington, 1992-93; cons. U.S. Office Tech. Assessment, Washington, 1992, Fedn. Am. Scientists, Washington, 1988-92, EG & Idaho, Idaho Falls, 1986-90. Author: Toward a Comprehensive Test Ban, 1988; contbr. articles to profl. jours. Postdoctoral fellow Lawrence U. Nat. Lab., Livermore, Calif., 1985-86, Harvard U. Ctr. Sci. & Internat. Affairs, Cambridge, Mass., 1986-88, Sloan Found. fellow, 1986-88, Regents fellow U. Calif., 1981-82. Mem. Am. Phys. Soc., Fedn. Am. Scientists, Coun. Fgn. Rels., Arms Control Assn. Democrat. Home: 7208 Hitching Post Ln University Hills MD 20783-1935 Office: U Md Sch Pub Affairs College Park MD 20742

FETTING, FRITZ, chemistry educator; b. Itzehoe, Holstein, Germany, June 28, 1926; s. Friedrich and Kaethe (Tempel) F. Dr., U. Goettingen, 1955; Habilitation, U. Hannover, 1962. Prof. chem. tech. Tech. Inst. Darmstadt (Fed. Republic Germany), 1966—. Mem. German Soc. Engrs., German Soc. Chemists, Rotary, Sigma Xi. Lutheran. Home: Auf dem Sand 3, D-6109 Muehltal Germany Office: Tech Inst Darmstadt, Petersenstrasse 20, D-64287 Darmstadt Germany

FETTIPLACE, ROBERT, neurophysiologist; b. Nottingham, Eng., Feb. 24, 1946; came to U.S. 1990; s. George Robert and Maisie (Rolson) F.; m. Merriel Cleone Kruse, Oct. 8, 1977; children: David George, Michael Robert. BA, Cambridge U., Eng., 1968, PhD, 1974. Postdoctoral fellow Stanford (Calif.) U., 1974-76; Elmore rsch. fellow Cambridge U., 1976-79, Howe sr. rsch. fellow of Royal Soc., 1979-90; Steenbock prof. behavioral neurosci. U. Wis., Madison, 1990—. Editorial bd. Jour. Physiology, Eng., 1982-89; contbr. articles to profl. jours. Recipient Claude Pepper award Nat. Inst. Deafness and Comm. Disorders, 1991. Fellow Royal Soc. Achievements include discovery of electrical tuning mechanisms in cochlear hair cells; mechanisms of transduction in cochlear hair cells; development of techniques for measuring small movements of hair cells. Office: Univ of Wis Dept Neurophysiology 1300 University Ave Madison WI 53706

FETTWEIS, GÜNTER BERNHARD LEO, mining engineering educator; b. Düsseldorf, Germany, Nov. 17, 1924; came to Austria, 1959; s. Ewald Ignaz Maria and Anna Maria (Leuschner-Fernandes) F.; m. Alice Yvonne Frieda Maria, Apr. 23, 1949; children: Astrid Maria Barbara, Raimund Ewald Rudolf, Annette Alice, Ursula Melanie. Diploma Engring., Tech. U., Aachen, Germany, 1950, D Engring., 1953, D. Engring. (hon.), 1980; D (hon.), Tech. U., Miskolc, Hungary, 1987. Sci. asst. Tech. U., Aachen, 1950-53; jr. mining inspector State of Nordrhein-Westfalia, Dortmund, Germany, 1953-55; mining engr. Mining Co. Neue Hoffnung, Oberhausen, Germany, 1955-57, prodn. mgr., 1957-59; prof. mining engring., head dept. mining and mineral econs. Montan U., Leoben, Austria, 1959—; v.p. Internat. Organizing Com. of World Mining Congress, 1974—. Contbr. articles to profl. jours., also several books. Recipient awards states of Austria, Germany and Poland. Mem. Austrian Acad. Scis., Polish Acad. Scis., Hungarian Acad. Scis. (hon.), Academia Scientiarum et Artium Europaea, Mining Assn. Austria (v.p. 1964—). Roman Catholic. Home: Gasteigergasse 5, A-8700 Leoben Austria Office: Montanuniversitat, Franz-Josef-Str 18, A-8700 Leoben Austria

FEUCHT, DONALD LEE, research institute executive; b. Akron, Ohio, Aug. 25, 1933; s. Henry George and Dorothy Fern (Kroeger) F.; m. Janet Wingerd, Aug. 18, 1958; children: Lynn Janet Feucht Malloy, Paul Henry. B.S., Valparaiso U., 1955; M.S., Carnegie Inst. Tech., 1956, Ph.D., 1961. Electronics engr. Convair, San Diego, 1956; instr. elec. engring. Carnegie-Mellon U. (formerly Carnegie Inst. Tech.), Pitts., 1958-61; asst. prof. Carnegie-Mellon U. (formerly Carnegie Inst. Tech.), 1961-65, assoc. prof., 1965-69, prof., assoc. head dept., 1969-73, prof. elec. engring., assoc. dean, 1973-77; chief advanced materials research and devel. br. Dept. Energy, Washington, 1977-78; mgr. photovoltaic program office Solar Energy Research Inst., Golden, Colo., 1978-80; mgr. photovoltaics div. Solar Energy Research Inst., 1980-81, acting dir. for R & D, 1981, dep. dir., 1981-89; mgr. process devel. program EG&G Rocky Flats Inc., Golden, Colo., 1990-92; dir. ops. Associated Western Universities, Inc., 1992—; research scientist IBM Research Center, Yorktown Heights, N.Y., 1961; cons. Power Components Inc., Scottsdale, Pa., 1965-67, PPG Industries, Pitts., 1967-69, Essex Internat., Pitts., 1967-74. Author: (with A.G. Milnes) Heterojunctions and Metal Semiconductor Junctions, 1972; contbr. numerous articles to sci. publs. Recipient Alumni award Valparaiso U., 1979, Francis Van Morris award Midwest Research Inst., 1987. Fellow IEEE (past treas., dir. Pitts.), Group Electron Devices (past chmn. Pitts., IEEE electron devices adminstrv. com., chmn. energy device tech. com. 1983-86, chmn. photovoltaics tech. com. 1986-88), Sigma Xi, Phi Kappa Phi, Tau Beta Pi, Theta Chi. Achievements include patents (with A.G. Milnes) method for making semicondrs. for solar cells. Home: 8881 S Tracy Dr Sandy UT 84093 Office: Associated Western Univs 4190 S Highland Dr Ste 211 Salt Lake City UT 84124

FEUILLET, CHRISTIAN PATRICE, botanist; b. Rouen, France, July 25, 1948; came to U.S., 1988; s. Pierre Hector and Huguette Marie (Lecourt) F.; m. Amy Yarnell Rossman, Sept. 4, 1988; 1 child, Diane Karen. MD in Plant Biology, U. P&M Curie, Paris, 1979, PhD in Tropical Botany, 1981. Botanist ORSTOM, Paris, 1981-82, Cayenne, French Guiana, 1982-88, Washington, 1988—. With French Army, 1972-73. Achievements include research in taxonomy of Gesneriaceae and Passifloraceae, and plant architecture. Office: Smithsonian Institution Botany Dept NHB-166 Washington DC 20560

FEULNER, EDWIN JOHN, JR., research foundation executive; b. Chgo., Aug. 12, 1941; s. Edwin John and Helen J. (Franzen) F.; m. Linda C. Leventhal, Mar. 8, 1969; children: Edwin John III, Emily V. BS, Regis Coll., 1963, MBA, U. Pa., 1964; PhD, U. Edinburgh, 1981; hon. degree, Nichols Coll., 1981, Universidad Francisco Marroquin, Guatemala City, 1982, Hanyang U., Seoul, Korea, 1982, Bellevue Coll., Nebr., 1987, Gonzaga U., 1992. Richard Weaver fellow London Sch. Econs., 1965; pub. affairs fellow Hoover Instn., 1965-67; confidential asst. to sec. def. Melvin Laird, 1969-70; adminstrv. asst. to U.S. Congressman Philip M. Crane, 1970-74; exec. dir. Rep. Study Com., Ho. of Reps., 1974-77; pres. Heritage Found., Washington, 1977—; chmn. Inst. European Def. and Strategic Studies, 1977—; mem. U.S. adv. com. pub. diplomacy USIA, 1982-93, chmn., 1982-91; vice chmn. bd. dirs. Fed. Capital Bank, N.A.; mem. nat. adv. bd. ctr. for Edn. and Rsch. in Free Enterprise, Tex. A&M U.; disting. fellow Mobilization Concepts, Devel. Ctr. Nat. Def. U., 1983-89; mem. Pres.'s Commn. on White House Fellows, 1981-83, mem. Exec. Com. of the Presdl. Transition, 1980-81; mem. U.S. Delegation to 1974, 75, 76 IMF World Bank; mem. Carlucci Commn. on Fgn. Assistance, 1983; pub. del. UN 2d Spl. Session on Disarmament, 1982; mem. U.S. Commn. Improving Effectiveness of UN, 1989—; White House cons. domestic policy, 1987. Author: Congress and the New International Economic Order, 1976, Looking Back, 1981, Conservatives Stalk the House, 1983; contbr. articles to profl. jours., newspapers, chpts. to books. Trustee Lehrman Inst., 1981-90, Sarah Scaife Found., 1988—, Regis U., 1991—, St. James Sch., 1990—; vice chmn. bd. Aequus Inst., Intercollegiate Studies, Inst., 1979—, chmn. 1989-93; vice chmn. bd. dirs. Roe Found.; bd. govs., mem. exec. com. Coun. Nat. Policy; trustee Am. Coun. Germany, N.Y., 1982-92; Found. Francisco Marroquin; trustee Inst. Rsch. Econs. Taxation, 1980-87; chmn. Citizens for Am. Edn. Found., 1985-89; vice chmn., trustee Manhattan Inst. Policy Studies, 1977-86; mem. coun. acad. advisors Bryce Harlow Found.; bd. trustees, vice chmn. Manhattan Inst. Policy Studies, 1977-86; bd. trustees Inst. Rsch. Economics of Taxation, 1980-87. Recipient Disting. Alumni award Regis Coll., 1985, Superior Pub. Svc. award Dept. of Navy, 1987, Presdl. Citizens medal, 1989, Dir.'s Svc. award USIA, 1992; named Free Enterprise Man of Yr., Tex. A&M U., 1985; decorated Order of Brilliant Star with Grand Cordon, Rep. of China, 1992. Mem. Am. Econs. Assn., Am. Polit. Sci. Assn., Internat. Inst. Strategic Studies, U.S. Strategies Inst., Inst. d'Etudes Politiques, Phila. Soc. (treas. 1964-79, pres. 1982-83), Mont Pelerin Soc. (treas.), Internat. Com. of the G.K. Chesterton Soc. (chmn. 1989-92), Belle Haven Country Club, Union League (N.Y.C.), Met. Club, Reform Club (London), Bohemian Club (San Francisco), Knights of Malta, Alpha Kappa Psi. Republican. Roman Catholic. Office: The Heritage Found 214 Massachusetts Ave NE Washington DC 20002-4999

FEUSNER, LEROY CARROLL, chemical engineer; b. Greybull, Wyo., Feb. 27, 1945; s. Wayne LeRoy and Freda Lucille (Niday) F.; m. Lynnette Adele Reichert, June 9, 1968; children: Kristi Ann, Katreena Lynn. B-SChemE, U. Wyo., 1968. Registered profl. engr., Wyo., S.D. Commd. 2d lt. USAF, 1968, advanced through grades to capt., 1970; bioenviron. engr. USAF, various mil. bases, 1968-78, released from active duty, 1978; engring. supr. S.E. dist. Wyo. Dept. Environ. Quality, Cheyenne, 1979-85, environ. emergency response supr., water quality div., 1985-89, supr. underground tank storage program, 1990—. Lt. col. Wyo. Air Nat. Guard, 1979—. Recipient Outstanding Leadership awards U.S. EPA, Denver, 1990, 92. Mem. AIChE, Am. Acad. Environ. Engrs. (diplomate, chmn. exam. devel. and upgrading com., mem. membership com., trustee, Wyo. membership chmn. 1984—), Am. Conf. Govt. Indsl. Hygienists (hazardous waste com. 1986—), Nat. Coun. Examiners for Engring. and Surveyors (environ. engring. profl. engr. exam. com.), Air and Waste Mgmt. Assn., Cheyenne Kiwanis. Republican. Lutheran. Home: 5461 Atlantic Dr Cheyenne WY 82001 Office: Wyo Dept Environ Quality Water Quality Div Herschler Bldg-4W Cheyenne WY 82002

FEYL, SUSAN, safety engineer, educator; b. Buffalo, Apr. 21, 1947; d. John George and Violet Ruth (Klees) F. BA in Physics, SUNY, Buffalo, 1969; MS in Physics, State Univ. Coll., Fredonia, N.Y., 1971. Materials engr. Aerojet Electrosystems, Azusa, Calif., 1977-79; instr. physics Calif. State Poly. U., Pomona, 1974-82; systems analytical engr. Lockheed, Ontario, Calif., 1979-80; project mgr. Dynalectron, Norco, Calif., 1981; systems engr. So. Calif. Rapid Transit Dist., L.A., 1983-85; quality engr. Gen. Dynamics, Ontario, 1986-90; sr. rapid transit control systems specialist State of Calif., L.A., 1990—. Mem. Am. Phys. Soc. Home: 7424 Hollaway Rd Rancho Cucamonga CA 91730 Office: CPUC 107 S Broadway Rm 5109 Los Angeles CA 90012

FICA, JUAN, endocrinologist; b. Santiago, Chile, Jan. 18, 1949; came to U.S., 1974, naturalized, 1977; s. Moises and Olga (Cisternas) F.; B.S., U. Chile, 1971, M.D., 1974; m. Margarita Martinez, Dec. 30, 1971; children—Michelle, Pamela, Vanessa Olga, Kristina Margarita. Intern, Booth Meml. Hosp., NYU, N.Y.C., 1974-75, resident, 1975-76; sr. resident Wellesley Hosp., U. Toronto (Ont., Can.), 1976-77; fellow in endocrinology U. Conn., Farmington, 1977-79; clin. instr., 1979—; practice medicine specializing in endocrinology, Waterbury, Conn., 1979—; chmn. endocrinology Waterbury Hosp., Yale New Haven; founder, dir. Diabetes Ctr. Bd. dirs. Am. Field Service Internat. scholar, 1965. Diplomate Am. Bd. Internal Medicine, Am. Bd. Endocrinology and Metabolism. Mem. AMA, New Haven County Med. Soc., Waterbury Med. Soc., Endocrine Soc., Conn. Endocrine Soc., Am. Diabetes Assn., N.Y. Acad. Scis., Assn. Insulin Pump Therapists, Met. Opera Assn. Lodge: Rotary. Home: PO Box 1102 159 Northridge Dr Middlebury CT 06762-1420 Office: 171 Grandview Ave Waterbury CT 06708-2517

FICALORA, JOSEPH PAUL, optical engineer; b. Flushing, N.Y., Apr. 7, 1957; s. Carmelo Augustus and Veronica Marie (Barbera) F.; m. Lynn MArie Gronberg, Sept. 26, 1982; children: Anne Marie, Katherine Theresa. BS, Rensselaer Polytech Inst., 1979; M of Engring., Stevens Inst. Tech., 1988. Assoc. engr. Sperry Gyroscope, Great Neck, N.Y., 1979-82; engr. Raytheon Equipment Div., Sudbury, Mass., 1982-84; sr. engr. KEarfott Guidance & NAvigation, Little Falls, N.J., 1985; staff engr. Bendix Guidance Systems Div., Teterboro, N.J., 1986-88; sr. staff engr. Bendix Guidance & Controls Systems, Teterboro, N.J., 1988—. Republican. Roman Catholic. Achievements include patents on ring laser cavity length control assembly, transverse discharge in an RLG, RLG with reduced sensitivity to magnetic effects. Office: Allied Signal Aerospace Rte 46 Teterboro NJ 07608

FICHMAN, MARK, industrial/organizational psychologist; b. Haifa, Israel, Mar. 14, 1952; came to U.S., 1957; s. Yehudi David and Pearl (Spiegel) F.; m. Ruth Ann Fauman, Mar. 22, 1980; children: Michael A., Jonathan S. BA in Psychology, Brown U., Providence, 1973; PhD in Psychology, U. Mich., 1982. Rsch. psychologist Ann Arbor (Mich.) Vets. Hosp., 1976-77; asst. study dir. Survey Rsch. Ctr., Inst. Social Rsch. U. Mich., Ann Arbor, 1975-80, teaching fellow Summer Inst., Inst. Social Rsch., 1978; asst. prof. orgnl. behavior Grad. Sch. Indsl. Adminstrn., Carnegie Mellon U., Pitts., 1980-87, assoc. prof., 1988—; NIMH trainee in social and orgnl. psychology U. Mich., 1974-75; cons. Exxon Corp., Planned Parenthood of U.S., Polaroid, Marriot Corp., McGraw-Hill, Dryden Press, Little Brown, Addison-Wesley. Assoc. editor Orgnl. Behavior Psycholoquy, 1990—; mem. editorial bd. Orgnl. Behavior and Human Decision Processes, Adminstrv. Sci. Quar.; contbr. articles and book revs. to profl. jours. Mem. APA, Acad. Mgmt., Am. Psychol. Soc., Am. Arbitration Assn. Democrat. Achievements include development of new methods to study absence and attendance; co-development of new methods for analyzing interorgn. rels. Home: 5715 Solway St Pittsburgh PA 15217-1203 Office: Carnegie Mellon U Grad Inst Indsl Adminstrn 5000 Forbes Ave Pittsburgh PA 15213-3890

FIDELL, LINDA SELZER, psychology educator, consultant; b. Hammond, Ind., Feb. 4, 1942; d. Maurice and Beatress (McMillin) Selzer; m. Sanford Fidell, Aug. 18, 1967 (div. Aug. 1990); children: Megan S., Tracy S. BS, Purdue U., 1964; MA, U. Mich., 1966, PhD, 1968. Prof. psychology Calif. State U., Northridge, 1968—; statis. cons. Med. Rsch. Consultation Assocs., Northridge, 1990—; cons. in field. Co-author: Using Multivariate Statistics, 1983, 2d edit., 1989; contbr. chpts. to Gender and Psychopathology, 1982, Sex Roles and Psychopathology, 1984, Gender in Transition, 1989. Pres. ParsArt, Northridge, 1991—. Mem. Am. Psychol. Soc. Democrat. Office: Calif State U 18111 Nordhoff St Northridge CA 91330

FIEDLER, ROBERT MAX, management consultant; b. Midland, Mich., June 19, 1945; s. Edward Louis and Lenora Margaret (Winterberg) F.; m. Carol Ann Raddatz, Nov. 28, 1981; children: Katy, Christa. BS in Packaging Sci., Mich. State U., 1967, MS in Packaging Sci., 1971, MBA, 1971. Grad. rsch. asst. Mich. State U., East Lansing, 1969-71; div. mgr. MTS Systems Corp., Mpls., 1971-80; cons. Robert Fiedler & Assocs., Mpls., 1980—; asst. prof. dept. engring. U. Minn., Mpls., 1990; mem. adv. bd. ad-hoc com. on packaging U. Minn., 1985—; speaker Arab Packaging Conf., Tunis, Tunisia, 1985; seminar lectr. in field, Brazil, Mex., China, Italy, The Netherlands. Contbg. editor: Fundamentals of Packaging Dynamics, 1985; contbr. articles to profl. jours. Lt. USNR, 1967-69. Recipient Diamond Wing award U.S. Parachute Assn., 1974. Mem. ASTM (chmn. subcom. 1986—), Inst. Packaging Profls. (cert. profl. in packaging, chmn. cons. coun., chmn. tech. com. 1993—). Office: Robert Fiedler & Assocs PO Box 24405 Minneapolis MN 55424-0405

FIELD, ALEXANDER JAMES, economics educator, dean; b. Boston, Apr. 17, 1949; s. Mark George and Anne (Murray) F.; m. Valerie Nan Wolk, Aug. 8, 1982; children: James Alexander, Emily Elena. AB, Harvard U., 1970; MS, London Sch. Econs., 1971; PhD, U. Calif., Berkeley, 1974. Asst. prof. econs. Stanford (Calif.) U., 1974-82; assoc. prof. Santa Clara (Calif.) U., 1982-88, acad. v.p., 1986-87, prof., chmn. dept. econs., 1988-93; mem. bd. trustees Santa Clara U., 1988-91. Author: Educational Reform and Manufacturing Development in Mid-Nineteenth Century Massachusetts, 1989; author, editor: The Future of Economic History, 1987; assoc. editor Jour. Econ. Lit., 1981—; editor: Research in Economic History, 1993—; mem. editorial bd. Explorations in Econ. History, 1983-89. Recipient Nevins prize Columbia U., 1975; NSF rsch. grantee, 1989. Mem. Phi Beta Kappa, Beta Gamma Sigma. Home: 3762 Redwood Cir Palo Alto CA 94306-4255 Office: Santa Clara Univ Dept Econs Santa Clara CA 95053

FIELD, DOROTHY, gerontologist, educator; b. Kansas City, Mo., Jan. 21, 1926; d. Edmund Mills and Dorothy Hazel (Browne) F.; m. Graham B. Moody, Jr., May 12, 1945 (div. Apr. 1970); children: Graham B. III, Stuart F., Katherine L., Carlin Mills. BA, U. Calif., Berkeley, 1954; MSc, U. London, 1972, PhD, 1975. Acting asst. prof. Humboldt State U., Arcata, Calif., 1975-76; rsch. psychologist U. Calif., Berkeley, 1976-90; Fulbright prof. U. Bonn, Germany, 1990-91; adj. prof., sr. rsch. scientist U. Ga., Athens, 1992—; adj. assoc. prof. Pa. State U., University Park, 1982-85. Editor: (with others) Family, Self, and Society, 1993; spl. issue editor Jour. Gerontology, 1991; editorial bd. Cognitive Devel. Abstracts, 1982-84; organizer Berkeley Gerontology Group, 1984-91; dir. Calif. Coun. Gerontology and Geriatrics, 1990-92; contbr. articles to profl. jours. Grantee Nat. Inst. Aging 1978-80; recipient Excellence in Rsch. award Mensa Edn. and Rsch. Found. 1988-89. Mem. APA, Am. Psychol. Soc., Brit. Psychol. Soc., Gerontol. Soc. Am., Internat. Soc. Study of Behavioral Devel., Jr. League Am., The Nature Conservancy, Sierra Club, Smith Coll. Club, League of Women Voters. Office: Univ Ga Dept Child & Family Devel Dawson Hall Athens GA 30602

FIELD, FRANCIS EDWARD, electrical engineer, educator; b. Casper, Wyo., Nov. 20, 1923; s. Jesse Harold and Persis Belle (St. John) F.; m. Margaret Jane O'Bryan, Oct. 13, 1945; children: Gregory A., Christopher B., Sheridan Diane. BSEE, U.S. Naval Acad., 1945; MA in Internat. Affairs, George Washington U., 1965; AMP, Harvard Bus. Sch., 1970. Master cert. graphoanalyst; comml. pilot. Owner Field Lumber Co., Lander, Wyo., 1948-50; commd. ensign, U.S. Navy, 1945, advanced through grades to capt. 1966, ret., 1975; rsch. engr., George Washington U., 1975-90, adj. faculty, 1977-90; program dir. NSF, Washington, 1982-90; pres. EXTANT, cons. firm, McLean, Va., 1981—. Author: Chronicle of a Workshop, 1977. Mem. Internat. Graphoanalysis Soc. (award of merit 1984), Sigma Xi. Republican. Lodges: Masons, Elks. Home: 8122 Dunsinane Ct Mc Lean VA 22102-2719

FIELD, GEORGE BROOKS, theoretical astrophysicist; b. Providence, R.I., Oct. 25, 1929; s. Winthrop Brooks and Pauline (Woodworth) F.; m. Sylvia Farrior Smith, June 23, 1956 (div. Oct. 1979); children: Christopher Lyman, Natasha Suzanne; m. Susan Alice Gebhart, Feb. 26, 1981. BS in Physics, MIT, 1951; PhD in Astronomy, Princeton U., 1955. Asst. prof., then assoc. prof. astronomy Princeton U., 1957-65; vis. prof. Calif. Inst. Tech., 1964; prof. astronomy U. Calif., Berkeley, 1965-72; chmn. dept. U. Calif., 1970-71; Phillips visitor Haverford (Pa.) Coll., 1965, 71; vis. prof. Cambridge (Eng.) U., 1969; prof. astronomy Harvard U., 1972—, now Willson prof. applied astronomy, dir. Harvard Coll. Obs., 1973-82; dir. Smithsonian Astrophys. Obs., 1973-82, now physicist; lectr. Ecole d'Ete de Physique Theorique, Les Houches, France, 1974; mem. Nat. Commn. on Space, 1985-86; mem. study group NRC-Space Sci. Bd.; chmn. NAS-NRC Coun. Astronomy Survey Commn., 1978-82. Recipient Disting. Pub. Svc. medal NASA, 1986, Joseph Henry medal Smithsonian Inst., 1983; Guggenheim fellow, 1960-61. Fellow AAAS, Am. Phys. Soc.; mem. NAS, Am. Acad. Arts and Scis., Am. Astron. Soc., Astron. Soc. Pacific, Internat. Astron. Union, Explorers Club, Harvard Club N.Y., Sigma Xi. Office: Harvard U Observatory 60 Garden St Cambridge MA 02138-1596

FIELD, JOHN DOUGLAS, marine biologist; b. Livonia, Mich., Aug. 17, 1964; s. David Barrows and Dolores Isabel (Richardson) F. BS, William & Mary U., 1987, MA, 1991. Rsch. asst. Va. Inst. Marine Sci., Gloucester Point, 1988-91; marine biologist Dept. Commerce/NOAA/NOS, Boothbay Harbor, Maine, 1991—. Contbr. articles to profl. publs. Mem. Am. Fisheries Soc., Nature Conservancy, Sigma Xi. Achievements include demonstration of efficacy of in situ silhouette photography in fishery-independent stock assessment programs for striped bass, compilation of nationwide inventory of estuarine living marine resources. Office: Dept Commerce/NOAA/NOS Marine Resources Lab Box 8 West Boothbay Harbor ME 04575

FIELD, MICHAEL STANLEY, information services company executive; b. London, Sept. 28, 1940; came to U.S., 1966; s. Stanley Frank Owen and Violet May (Collins) F.; m. Jenny Callen Chitwood, May 27, 1972; children: Shelley Callen, Heather Collins, Michael Randolph. B of Tech. Electrical Engring., U. Loughborough, 1962. Grad. engr. British Aerospace Corp., Luton, England, 1962; systems mgr. IBM (UK) Ltd., London, 1963-66; systems designer IBM Rsch. Ctr., Cambridge, Mass., 1966-68; dir., mgr. Nat. CSS, Wilton, Conn., 1969-74, v.p. data systems, 1974-80; v.p. applied tech. The Dun & Bradstreet Corp., N.Y.C., Conn., 1981-83; mng. dir. Dun & Bradstreet European Bus. Info. Ctr., Harefield, England, 1983-85; group exec. v.p. Dun & Bradstreet Info. Svcs., Murray Hill, N.J., 1989—; corp. v.p. The Dun & Bradstreet Corp., N.Y.C., 1981—. contbr. articles to profl. jours. and mags. Mem. Engring. Coun. (charter mem.), Brit. Computer Soc., Inst. Elec. Engrs. Avocations: squash, tennis, boating. Home: 73 Turning Mill Ln New Canaan CT 06840-3832 Office: Dun & Bradstreet Corp 299 Park Ave New York NY 10171

FIELD, ROBERT EUGENE, mechanical engineer, educator, consulant, researcher; b. Davenport, Iowa, July 14, 1946; s. Carl Ludgwig Field and Geraldine Gladis (Stahl) Weeks; m. Wanda M. Osborn, June 8, 1968; 1 child, Juliet M. BSME, Bradley U., 1968; MSME, Rensselaer Poly. Inst., 1972; PhD, Purdue U., 1989. Registered profl. engr., Ill., Fla. Sr. engr. Pratt & Whitney Comml. Products Div., East Hartford, Conn., 1968-74; mgr. rsch. and exploration Rockwell Internat., Admiral Appliance Group, Galesburg, Ill., 1974-77; project engr. govt. products div. Pratt & Whitney, West Palm Beach, Fla., 1977-91; asst. prof. No. Ill. U., DeKalb, Ill., 1991—; instr. Fla. Engring. Edn. Delivery System, Gainesville, 1983-85. Contbr. articles to profl. jours. Recipient David Ross summer grant Purdue U., 1986. Mem. ASME. Achievements include patents in field of gas turbine cooling, high temperature structures cooling, research in gas/steam turbine design, radiation heat transfer in semitransparent materials. Office: No Ill U Mechanical Engring Dept De Kalb IL 60115

FIELD, ROBERT WARREN, chemistry educator; b. Wilmington, Del., June 13, 1944; s. Edmund Kay Huebsch (Field). A.B., Amherst Coll., 1965; M.A., Harvard U., 1971, Ph.D., 1972. Adj. asst. prof. chemistry U. Calif.-Santa Barbara, 1974; asst. prof. chemistry M.I.T., Cambridge, 1974-78, assoc. prof. phys. chemistry, 1978-82, prof., 1982—. Mem. editorial bd. Jour. Molecular Spectroscopy, Chem. Physics Letters; contbr. articles to profl. jours. Alfred P. Sloan fellow, 1975-77. Fellow Am. Phys. Soc. (H.P. Broida prize 1980, E.K. Plyler prize 1988); mem. Am. Chem. Soc. (Nobel Laureate Signature award to Y. Chen, co-preceptor with J.L. Kinsey 1990), Optical Soc. Am. (E. Lippincott award 1990). Office: MIT Dept Chemistry Rm 6-219 Cambridge MA 02139

FIELD, STEVEN PHILIP, medical educator; b. Newark, Feb. 21, 1951; s. Irving and Florence (Engel) F. BA, Yale U., 1973; MD, NYU, 1977. Diplomate Am. Bd. Internal Medicine, Am. Bd. Gastroenterology. Intern in internal medicine Bellevue Hosp., N.Y.C., 1977-78, resident in internal medicine, 1978-81; instr. in medicine Mt. Sinai Hosp., N.Y.C., 1981-83; instr. in medicine NYU Sch. of Medicine, N.Y.C., 1983—, clin. asst. prof. medicine, 1991—. Contbr. articles to profl. jours. Recipient John Addison Porter Prize Yale U., 1973. Mem. Am. Gastroenterological Assn., N.Y. Acad. of Gastroenterology, N.Y. Soc. for Gastrointestinal Endoscopy (mem. exec. coun. 1991—), Yale Club of Cen. N.J. Office: 245 E 35th St New York NY 10016

FIELDHAMMER, EUGENE LOUIS, civil engineer; b. N.Y.C., Feb. 11, 1925; s. Louis and Agda Elvira (Anderson) F.; m. Genevieve Mullin, Aug. 26, 1950; children: Keith A., Michael D., Nancy H. BCE, Kans. State U., 1950. Registered profl. engr., Mo., Ill., N.Y. Design engr. Edwards and Kelcy, Newark, 1950, D.B. Steinman, N.Y.C., 1952; project engr. Goodkind and O'Dea, Inc., Hamden, Conn., 1956; project/office mgr. Goodkind and O'Dea, Inc., Chgo., 1959; from sect. chief to pres. Booker Assocs. Inc., St. Louis, 1963-85; pres. Fieldhammer Inc., St. Louis, 1985-92; cons. in field St. Louis, 1992—. Active Personnel Bd., Ferguson, Mo., 1965-72, Archtl. Rev. Bd., 1985—. With USNR, 1943-46, ATO. Inducted into Engring. Hall of Fame Kans. State U., 1990. Mem. ASCE (life), Nat. Soc. Profl. Engrs., Mo. Soc. Profl. Engrs., Engrs. Club St. Louis (life), Soc. Am. Mil. Engrs. (pres. St. Louis chpt. 1982-83). Achievements include design and design management of bridges over Mackinac Straits, Hudson, Mississippi, and Missouri Rivers. Home and Office: 153 S Clay Ave Ferguson MO 63135-2447

FIELDING, RONALD ROY, aeronautical engineer; b. Saskatoon, Sask., Can., July 24, 1961; s. Stanford R. and Helen A. (Rogers) F. BSc in Aero. Engring., Embry-Riddle Aero. U., 1986; BA, U. Sask., 1987. Cert. profl. engr. Assn. Profl. Engrs. of the Province of Manitoba. Material rev. bd. engr. Menasco Aerospace Ltd., Oakville, Ont., Can., 1987-89, test engr.-flight controls, 1989-90; target systems engr. Boeing Can. Tech. Ltd., Winnipeg, Manitoba, Can., 1990-93, engr. assembly tech. support unit, 1993—. Named to Hon. Citizenship, Daytona Beach (Fla.) C. of C., 1986; recipient Sask. Proficiency award Sask. Govt., 1979. Mem. AIAA (treas. Prescott, Ariz. sect. 1984-86), Can. Aero. and Space Inst. (treas. Winnipeg br. 1991-93), Am. Helicopter Soc., Boeing Engring. Social Club (activities dir. 1991-92), Alpha Eta Rho (treas. Pi Rho chpt. 1984-85). Home: 4310 193 Victor Lewis Dr, Winnipeg, MB Canada R3P 2A3 Office: Boeing Can Tech Ltd, Winnipeg Div 99 Murray Park, Winnipeg, MB Canada R3J 3M6

FIELDING, STUART, psychopharmacologist; b. Bronx, N.Y., Oct. 31, 1939; s. Harry and Ethel (Weisberg) Feinblatt; m. Maralyn J. Lowy, Aug. 26, 1962; children: Kimberly Ellen, Bradford Scott. BA, Monmouth Coll., 1962; MS, Howard U., 1964; PhD, U. Del., 1968. Mgr. psychopharmacology rsch. Ciba-Geigy Corp., Summit, N.J., 1967-75; assoc. dir. pharmacology Hoechst-Roussel Pharms., Inc., Somerville, N.J., 1975-76, assoc. dir. biol. sci., mgr. pharmacology, 1977-84, dir. pharmacology, 1984-86, dir. biol. rsch., 1987-89; v.p. R & D, dir. Interneuron Pharms., Inc., Lexington, Mass., 1989-92; chmn., CEO Bio-Enhancement Systems Corp., Morris Plaines, N.J., 1992—. Editor: (book) Psychopharmacology of Clonidine, 1981, (book series) Industrial Pharmacology: A Monograph Series, 1974-79, (jour.) Drug Devel. Rsch., 1980-92; contbr. articles to profl. publs. Fellow Am. Psychol. Assn.; mem. Am. Chem. Soc., Am. Soc. Pharmacology and Exptl. Therapeutics, Soc. Neurosci. Home and Office: 16 Bromleigh Way Morris Plains NJ 07950

FIELDS, BERNARD NATHAN, microbiologist, physician; b. Bklyn., Mar. 24, 1938; s. Julius and Martha F.; m. Ruth Peedin, Sept. 10, 1966; children—John, Edward, Michael, Daniel, Joshua. A.B., Brandeis U., 1958; M.D., N.Y. U., 1962; A.M. (hon.), Harvard U., 1976. Intern Beth Israel Hosp., Boston, 1962-63; resident in medicine Beth Israel Hosp., 1963-64; officer USPHS, Nat. Communicable Disease Center, Atlanta, 1965-67; fellow Albert Einstein Coll. Medicine, N.Y.C., 1967-68; asst. prof. medicine and cell biology Albert Einstein Coll. Medicine, 1968-71, asso. prof., 1971-75, chief infectious disease, 1971-75; prof. microbiology and molecular genetics Harvard Med. Sch., 1975-84, chmn. dept. microbiology and molecular genetics, 1982-87, Adele H. Lehman prof. microbiology and molecular genetics, 1984—; chief infectious diseases Peter Bent Brigham Hosp., Boston, 1975-80; chief infectious disease div. Brigham and Women's Hosp., Boston, 1980-87; mem. and chmn. exptl. virology study sect. NIH, 1977-81 ; mem. Multiple Sclerosis Adv. Commn. on Fundamental Research, 1976; mem. coun. Nat. Inst. Allergy and Infectious Disease, 1987—. Contbr. articles to profl. jours. Recipient Faculty Research Asso. award Am. Cancer Soc., 1969-74, Irma T. Hirschel scholar, 1974-76; Career Scientist award Health Research Council N.Y., 1974-75; 12th Ann. Redway medal N.Y. State Med. Soc., 1974; Dyer lecture award NIH, 1987; Bristol-Myers Suibb award Disting. Achievement Infectious Disease Rsch., 1993; grantee NIH, 1969—. Fellow AAAS; mem. Am. Soc. Microbiology, Am. Soc. Virology (pres. 1990-91), Am. Soc. Clin. Investigation, Harvey Soc., Am. Assn. Immunologists, Infectious Disease Soc. Am., Assn. Am. Physicians, Nat. Acad. Scis., Inst. Medicine, Am. Acad. Arts & Scis. Home: 281 Otis St Newton MA 02165-2531 Office: Harvard Med Sch Dept Microbiology and Molecular Genetics 25 Shattuck St Boston MA 02115-6092

FIELDS, ELLIS KIRBY, research chemist; b. Chgo., May 10, 1917; m. Jeanette Shames, Nov. 18, 1939; children—Jennifer Fields Grunschlag, Diana Carroll, Wendy Fields Abondolo. S.B., U. Chgo., 1936, Ph.D., 1938. Eli Lilly postdoctoral fellow U. Chgo., 1938-41; research dir. Research Corp., Chgo., 1941-50; sr. research chemist Amoco Chems. Corp. (formerly Standard Oil Co. Ind.), Naperville, Ill., 1950-92; Todd Prof. Ill. Inst. Tech., Wheaton, 1993—; vis. prof. King's Coll., U. London, 1962-63. Contbr. articles to profl. jours.; patentee in field. Chmn. com. Chgo. Sci. Ctr., 1972-74; mem. com. Unity Temple Music Bd., 1984-89. Recipient Almquist award U. Idaho, 1979, award Chgo. Tech. Socs. Council, 1979. Fellow AAAS, Can. Inst. Chemists (hon.); mem. Am. Chem. Soc. (pres. 1985, Petroleum award 1978), Chgo. Zool. Soc. Office: Amoco Chem Co PO Box 3011 Naperville IL 60566-7011

FIELDS, JOSEPH NEWTON, III, oncologist; b. Beaumont, Tex., Feb. 14, 1949; s. Joseph Newton Jr. and Charlie Maxine (Edwards) F.; m. Deborah Ann Larson, Dec. 24, 1972; 1 child, Emily Joan. PhD in Physics, Stanford U., 1977; MD, U. Miami, Fla., 1982. Diplomate Am. Bd. Radiology. Fellow Brookhaven Nat. Lab., Upton, N.Y., 1977-78; mem. tech. staff Hughes Rsch. Lab., Malibu, Calif., 1978-80; resident Washington U. Med. Sch., St. Louis, 1982-86, asst. prof. Oncology, 1986-88; clin. asst. prof. Medicine So. Ill. U., Springfield, Ill., 1988—; mem. brain com. Radiation Therapy Oncology Group, Phila., 1986—. Author (book chpt.) Physics of Fiber Optics, 1981; contbr. articles to profl. jours. W. Churchill fellow Churchill Coll., 1971-72. Mem. Am. Phys. Soc., Am. Coll. Radiology, Radiol. Soc. N.Am. Achievements include patent for Coupled Waveguide Acousto-optic Hydrophone. Office: Meml Med Ctr 800 N Rutledge Springfield IL 62781

FIELDS, WILLIAM ALEXANDER, naval officer, mechanical engineer; b. Queens, N.Y., June 21, 1959; s. William Gerard and Alice Mary (Byrne) F.; m. Katherine Elizabeth Rooney, Sept. 14, 1984; children: William Alexander Jr., John Joseph, Peter James. BS in Chem. Engring., Manhattan Coll., 1982; MSME, Naval Postgrad. Sch., 1990. Registered profl. engr., Calif. Commd. ensign USN, 1982, advanced through grades to lt., 1986; elec. officer, then engring. officer USS Bonefish USN, Charleston, S.C., 1984-87; ship supt. Portsmouth (N.H.) Naval Shipyard USN, 1991—, advanced through grades to lt. comdr., 1993, docking officer, 1993—. Mem. NSPE, Am. Soc. Naval Engrs., K.C. Roman Catholic. Home: 4 Hickory Ln South Berwick ME 03908-2118 Office: USN Portsmouth Naval Shipyard Portsmouth NH 03804-5000

FIERMAN, GERALD SHEA, electrical distribution company executive; b. Wilkes Barre, Pa., Dec. 11, 1924; s. Abe and Mary (Jacobs) F.; A.B. in Liberal Arts, Pa. State U. 1948; m. Bernice Perloff, June 12, 1949; children: Robert Alan, Lawrence David, Daniel Jon. Pres. Shea Realty Corp., Wilkes-Barre, 1959—, Barre Realty Corp., Wilkes-Barre, 1955—, Chase Wholesale Elec. Supply, Stroudsburg, Pa. 1960—, Tomberg Elec. Supply Co., Wilkes-Barre 1934—, ANESCO, Kingston, Pa. 1949—, v.p. L&R Elec. Supply Co., Scranton, Pa., Effco Inc., Scranton. Chmn. United Jewish Campaign, Wilkes-Barre, 1963; pres. Jewish Fed. of Wyoming Valley (Pa.), 1971-74. Served with 82d Airborne Div., AUS, 1942-46. Decorated Purple Heart. Mem. Temple Israel of Wilkes-Barre. Clubs: Westmoreland of Wilkes-Barre, Jockey of Miami, Huntsville Golf, Valley Tennis, Mason, Keystone Consistory, Huntsville Golf Club, Williams Island Club (Miami). Home: 76 James St Wilkes Barre PA 18704-4730 Office: 517 Pierce St Kingston PA 18704

FIFE, WILLIAM J., JR., metal products executive; b. 1938. BSBA, John Carroll U. With Inland Steel, 1958-66, Chase Brass & Cooper, 1966-75; with Combustion Engring., Inc., 1975-81, pres. C-E invalco measurement and control divsn., 1981-85; pres. Kearney & Trecker Corp., 1985-87; pres. machine tool divsn. Giddings & Lewis, Inc., 1987—, chmn. bd., pres., CEO, 1989—. With USN. Office: Giddings & Lewis Inc 142 Doty St Fond Du Lac WI 54935-3331

FIFE, WILMER KRAFFT, chemistry educator; b. Wellsville, Ohio, Oct. 19, 1933; s. Wilmer George and Lourene Elizabeth (Krafft) F.; m. Betsy Louise Jones, Dec. 26, 1959; children: Kimberly, Julia, Steven. B.Sc. in Chemistry, Case Inst. Tech., 1955; Ph.D. in Organic Chemistry, Ohio State U., 1960. Applications chemist Monsanto Chem. Co., Dayton, Ohio, summers 1955, 57; instr. Muskingum (Ohio) Coll., 1959-60, asst. prof., 1960-64, asso. prof., 1964-70, prof., 1970-71, chmn. dept. chemistry, 1966-71; prof. chemistry Ind. U.-Purdue U. at Indpls., 1971—, chmn. dept., 1971-80. NIH postdoctoral fellow Harvard U., 1965-66; NIH postdoctoral fellow Columbia U., 1968-69; NSF fellow, 1955-56; Sinclair Oil Co. fellow, 1958-59; DuPont fellow, 1960; Danforth assoc., 1969—; others. Mem. Am. Chem. Soc., AAAS, Sigma Xi, Tau Beta Pi, Phi Lambda Upsilon. Home: 7102 Dean Rd Indianapolis IN 46240-3626 Office: IUPUI Chemistry 402 N Blackford St Indianapolis IN 46202-3274

FIGEN, I. SEVKI, computer company executive; b. Istanbul, Turkey, Nov. 26, 1924; s. Osman Sevki and Atiye F.; student Sankt-Georg Austrian Coll., 1937-43, Robert Coll. Engring. Sch., 1944-48; BS in C.E., Ind. Tech. Coll., Ft. Wayne, 1953; M.S. in C.E., U. Tex., 1955; m. Evin Keseroglu, June 29, 1962 (dec. Aug. 1980); 1 child, Alp; m. Leyla Oksar, Sept. 7, 1983. Acting procurement mgr. 4th Div., Public Rds. Adminstrn., Ankara, Turkey, 1948-49; expediter Metcalfe-Hamilton-Grove, Inc., Ankara, 1950-52; design engr. Southwestern Engring., Los Angeles, 1955-58; design engr. Ralph M. Parsons Co., Los Angeles, 1958-59, Wheeler and Gray Cons. Engrs., Los Angeles, 1959; plant engr. Turkish Automotive Industries, Istanbul, 1960-61, asst. gen. mgr., 1961-63; aviation mgr. Mobil Oil Turk Co., Istanbul, 1963-65, area sales mgr., 1965-67; mktg. mgr. Mobil Gas Co., Istanbul, 1967-69, gen. mgr., 1969-73; mng. dir., gen. mgr. Turyag Co., Izmir, Turkey, 1973-85, mng. dir., 1985-86, vice chmn., 1985-89, advisor, 1986-89; chmn. bd. Turset Co., 1985-87; ptnr., chmn. bd. Eritenel Chem. Trading Co., Izmir, 1987—; ptnr., pres. FEKOM Computer Ctr. Ltd., 1990—; bd. dirs. EGEFREN Co., 1990-93. Trustee, Robert Coll. of Istanbul, 1980—; hon. mem. local exec. coun. Am. Collegiate Inst., Izmir, 1984-86; bd. dirs. Cimentas Edn. and Health Found., Consumer Protection Union Found. Author: Letters to My Friends, 1989, German version, Briefe an Meine Freunde, 1989; columnist Yeni Asir newspaper, 1985, 1992, Gunaydin-Izmir newspaper, 1989-90. With Turkish Army, 1949-50. Recipient awards Assn. Journalists of Izmir, 1988, Gov. of Izmir, 1986. Mem. Aegean Chamber of Industry (dir., chmn. vegetable oils profl. com. 1975-83), Turkish Industrialists and Businessmen's Assn. (adv. coun. Istanbul), Turkish-Am. Businessmen's Assn., Turkish-Am. Assn. Izmir (bd. govs.), Robert Coll. Alumni Assn. (past pres.), Robert Collegeians and Bosphorusian Alumni Assn. Istanbul (past pres.), Alumni of U.S. Univs. in Izmir (pres. 1988-90), Masons, Rotary (past pres. Izmir club). Home: 328/8 Ataturk Caddesi, Yeni Kordon Apt, 35220 Izmir Turkey Office: Fekom Computer Ctr Ltd, 1358 Sok. 9/b Kahramanlar, 35040 Izmir Turkey

FIGLEY, CHARLES RAY, psychology educator; b. Chgo., Oct. 6, 1944; s. John David and Geni (Bartley) F.; m. Marily G. Reeves, Feb. 8, 1983; children: Jessica, Laura. MS, Pa. State U., 1971, PhD, 1974. Instr. Bowling Green (Ohio) State U., 1971-72; grad. asst. Pa. State U., University Park, 1972-74; prof. Purdue U., West Lafayette, Ind., 1974-89, Fla. State U., Tallahassee, 1989—; founding pres. Soc. for Traumatic Stress Studies, Chgo., 1985-87. Author: Helping Traumatized Families, 1989; editor: Strangers at Home, 1980, Stress and the Family, vols. I & II, 1983, Trauma and its Wake, vol. I, 1985, vol. II, 1986, Treating Stress in Families, 1989; founding editor: Jour. Family Psychotherapy, 1982-87, Jour. Traumatic Stress, 1987-92. Sgt. USMC, 1963-67, Vietnam. Fellow APA, Am. Assn. for Marriage and Family Therapy, Am. Orthopsychiatric Assn., Am. Psychol. Soc. Achievements include establishment of post-traumatic stress disorder as a diagnosis, studies of Vietnam vets., studies of traumatized families, studies of secondary traumatic stress disorders/compassion fatigue. Office: Fla State U 103 Sandels Bldg Tallahassee FL 32306

FIGLIOLA, RICHARD STEPHEN, aerospace, mechanical engineer; b. Wilmington, Del., Sept. 14, 1952; s. Anthony A. and Ann Marie F.; m. Suzanne Speyer, Nov. 4, 1978; 1 child, Elizabeth Aline. BS in Aero. Engring., U. Notre Dame, 1974, PhD, 1979. Registered profl. engr. Test engr. Pratt & Whitney Aircraft, East Hartford, Conn., 1974; rsch. fellow von Karman Inst. Fluid Dynamics, Brussels, Belgium, 1979-80; prof. dept. mech. engring. Clemson (S.C.) U., 1980—; referee Jour. of Heat Transfer, 1980—, Internat. Jour. Heat and Mass Transfer, 1985—, Jour. Fluid Engring., 1982—. Author: Theory and Design for Mechanical Measurements, 1991; contbr. articles to profl. jours. Fellow in Sci., NATO, 1979. Mem. ASME (mem. heat transfer divsn. 1984—, tech. chair 1988-90), Sigma Xi (pres. Clemson chpt. 1990-91). Roman Catholic. Achievements include patent for Improved Atomizing Nozzle; rsch. on fluid metering, two-phase flows, basic fluid mechanics and heat transfer. Office: Clemson U PO Box 340921 Clemson SC 29634-0921

FIGUEROA, JUAN MANUEL, physicist; b. Morelia, Mexico, May 26, 1949; s. Plutarco and Carmen Figueroa; m. Olga Leticia Hernandez, Aug. 25, 1973; children: Claudia Yuritzi, Juan Carlos. BS in Physics, Escuela Superior Fisica, Mexico, 1973, MS, 1975; DSc, Cinvestav-IPN, Mexico, 1991. Prof. physics Escuela Superior de Fisica y Matematicas IPN, Mexico, 1975-91; dir. electric metrology Centro Nat. de Metrologia, Queretaro, Mexico, 1992—; guest worker Nat. Bur. Standards, Gaithersburg, Md., 1981-82. Contbr. articles to Jour. Applied Physics, Temperature. Mem. Am. Phys. Soc., AAAS, Soc. Nuclear Mexicana, Soc. Mexicana de Superficie y Vacio, Mexican Peace Orgn. (pres. 1988-90). Achievements include discovery of origin of photoluminescence emission; development of a new cryogenic technique for water degasing. Home: Matagalpa # 973, Mexico City 07300, Mexico Office: Centro Nacional de Metrolog, Periferico Sur 3449 3er Pis, Mexico City 10200, Mexico

FIGWER, JOZEF JACEK, acoustics consultant; b. Mielec, Poland, Mar. 16, 1928; s. Jozef M. and Zofia (Haladej) F.; m. Magda L. Rzadkowska, Sept. 15, 1955; children: Kai J., Ulla T. MSEE, Silesian Poly., Gliwice, Poland, 1951; PhD in Applied Acoustics, NIKFI, Moscow, 1958. Engr. rsch. and devel. Motion Picture Industry, Warsaw, 1951-60, Deutsche Grammophon GmbH, Hannover, Germany, 1960-62; supr., cons. Bolt Beranek and Newman, Cambridge, Mass., 1962-78; prin. cons. Jacek Figwer Assocs., Inc., Concord, Mass., 1978—. Fellow Acoustical Soc. Am., Audio Engring. Soc. Office: Jacek Figwer Assocs Inc 85 The Valley Rd Concord MA 01742

FILBIN, GERALD JOSEPH, ecologist; b. Boston, Mar. 23, 1951; s. John Thomas and Alice Marie (Burke) F. BS, Suffolk U., 1973; PhD, Wayne State U., 1980. Adj. asst. prof. Alma (Mich.) Coll., 1979-84; rsch. biologist U.S. Army Corps of Engrs., Alma, 1980-84; lab. coord. Lockheed Engring. and Mgmt., Las Vegas, 1984-86; lab. dir., quality assurance dir. Internat. Sci. and Tech., Reston, Va., 1986-88; program dir. Tech. Resources, Inc., Rockville, Md., 1988-91; environ. cons. Sci. Consulting Group, Rockville, Md., 1991-92; ecologist, policy analyst U.S. EPA, Washington, 1992—; working com. mem. Water Quality 2000, Washington, 1989-90; mem. editorial bd. Jour. of Freshwater Ecology, LaCrosse, Wis., 1985-88; session chair North Am. Lake Mgmt. Soc., 1988. Contbr. articles and reports to profl. publs. and chpts. to books. Chair, quality assurance com. Whitman-Walker Clinic, Washington, 1991—, mem. med. svcs. com., 1989-90, AIDS/HIV educator, 1988—. Recipient Wilhelmina Haley scholarship Wayne State U., 1978. Mem. AAAS, Am. Soc. Limnology and Oceanography, Internat. Limnology Soc., North Am. Lake Mgmt. Soc. Achievements include a quality assurance plan for case mgmt. for community base AIDS clinic. Office: US EPA PM223X 401 M St SW Washington DC 20460

FILE, JOSEPH, research physics engineer; b. Lecce, Italy, May 6, 1923; s. Carlo and Laura (Nuzzi) F.; m. Dorothy Richards, Sept. 2, 1944; children: Joseph C., Laurel M., Jeannette. BME, Cornell U., 1944; MS, Columbia U., 1958, PhD, 1967; Dr.Physics, U. Lecce, Italy, 1978. Design engr. Petro Chem. Devel. Co., N.Y.C., 1946-56; rsch. sr. Princeton (N.J.) U., 1956—. Contbr. articles to profl. jours. Col. USMCR, 1942-74; PTO, Korea. Fulbright fellow, 1978. Roman Catholic. Achievements include patent on bending free D, shaped magnetic coils for fusion reactors, and fabrication and operation of world's first sixth order superconducting magnet now used on MRI imaging devices. Office: PPPL Princeton U Princeton NJ 08543

FILIATRAULT, ANDRE, civil engineering educator. Prof. civil engring. U. Montreal, Can. Recipient Sir Casimir Gzowski medal Can. Soc. Civil Engring., 1990. Office: U Montreal Dept of Civil Engr, CP 6079, SUCC A, Montreal, PQ Canada H3C 3A7*

FILIPPONE STEINBRICK, GAY, pharmacist, educator; b. Long Branch, N.J., Jan. 28, 1960; d. Frederick James and Gloria (Ricciardi) F.; m. Mark Gerard Steinbrick, Dec. 18, 1987. BS, Mass. Coll. Pharmacy, 1983, PharmD, 1986. RPh., N.J. Poison info. specialist Mass. Poison Control Ctr., Boston, 1985-86; postdoctoral resident U. Pa., Phila., 1987; asst. prof. Rutgers Coll. Pharmacy, Piscataway, N.J., 1987—; assoc. dir. Schering Labs., Kenilworth, N.J., 1989-91, dir., 1991—. Contbr. articles to profl. jours. Rsch. grantee Frederick Korr Chem. Co., 1985, NIH, 1985; named one of Outstanding Young Women Am., 1985, 86; recipient Excellence in Clin. Pharmacy award Hoechst-Roussel Pharms., 1986. Mem. N.J. Soc. Hosp. Pharmacists (pres. cen. chpt. 1990—, pres. elect 1989-90), Rho Chi. Office: Schering Plough 2000 Galloping Hill Rd Kenilworth NJ 07033-1310

FILISKO, FRANK EDWARD, physicist, educator; b. Lorain, Ohio, Jan. 29, 1942; s. Joseph John and Mary Magdalene (Cherven) F.; m. Doris Faye Call, Aug. 8, 1970; children: Theresa Marie, Andrew William, Edward Anthony. BA, Colgate U., 1964; MS, Purdue U., 1966; PhD, Case Western Res. U., 1969. Post doctoral fellow Case Western Res. U., Cleve., 1968-70; prof. materials sci. engring. and macromolecular sci. U. Mich., Ann Arbor, 1970—, acting dir. macromolecular sci. and engring., 1987—; dir. Polymer Lab., U. Mich. Contbr. more than 75 articles to profl. jours. Mem. Am. Phys. Soc., Am. Chem. Soc., So. Rheology, Materials Rsch. Soc., KC. Roman Catholic. Achievements include patents for Electric field dependent fluids and Electric dependent fluids-CIP. Office: Materials Sci & Engring Univ of Mich Ann Arbor MI 48109

FINAISH, FATHI ALI, aeronautical engineering educator; b. Tripoli, Libya, July 22, 1954; came to U.S., 1981; s. Ali Finaish and Zuhra (Lamin) Mahfud; m. Deborah Lynn Demijohn, Dec. 28, 1984. BS in Aero. Engring., U. Al-Fateh, Tripoli, 1978; MS in Aerospace Engring., U. Colo., 1984, PhD in Aerospace Engring., 1987. Lic. pvt. pilot; FAA airframe and power plant cert. mechanic. Rsch. asst. U. Colo., Boulder, 1984-87, adj. asst. prof., 1987-88; asst. prof. aero. engring. U. Mo., Rolla, 1988—; airworthiness engr. Dept. Civil Aviation, Tripoli, 1979-81; ground sch. instr. Tripoli Flight Ctr., 1980-81; rsch. fellow Naval Under Water Systems Ctr., Newport, R.I., 1991, NASA Langley Rsch. Ctr., Hampton, Va., 1992; lectr. various univs.; advisor Licking High Sch., St. James High Sch.; summer rsch. fellow U.S. Navy-Am. Soc. Engring. Edn., 1991, NASA-Am. Soc. Engring. Edn., 1992. Head coach Rolla Soccer Club. Grantee U. Mo., Rolla, 1988-92, U. Mo. System, 1991-92, Office Naval Rsch., 1991, NASA, 1993-94. Mem. AIAA, ASEE, ASHRAE (grantee 1992-93). Achievements include development of several ednl. computer codes written in C for DOS environ.; design and bldg. an exptl. system that generates and visualizes impulsive and accelerating motions and other unsteady airflow histories; designed and developed several wind tunnels for steady and unsteady aerodynamic testing at the University of Missouri-Rolla. Office: U Mo Dept Mech Engring Rolla MO 65401

FINCHER, DARYL WAYNE, fire protection engineer; b. Arlington, Ga., Nov. 25, 1962; s. Gerald Wayne and Margaret (Watkins) F.; m. Janice Scarbrough, Feb. 28, 1987. BS in Mech. Engring., Auburn U., 1985. Licensed profl. engr., Ga., Fla. Loss prevention assoc. Indsl. Risk Insurers, Atlanta, 1986-87, loss prevention rep., 1987-89, loss prevention cons., 1989, supr. dist. loss prevention, 1989—. Mem. NSPE, ASME, Soc. Fire Protection Engrs., Nat. Fire Protection Assn., Profl. Alarm Soc. N.Am., Kappa Sigma (v.p. 1983-84). Republican. Home: 310 Hayes Rd Winter Springs FL 32708 Office: Indsl Risk Insurers 900 Winderley Pl Ste 222 Maitland FL 32751

FINDLAY, JOHN WILSON, retired physicist; b. Kineton, Eng., Oct. 22, 1915; came to U.S., 1956, naturalized, 1963; s. Alexander Wilson and Beatrice Margaret (Thornton) F.; m. Jean Melvin, Dec. 14, 1953; children—Stuart E.G., Richard A.J. B.A., Cambridge U., 1937, M.A., 1940, Ph.D., 1950. With British Air Ministry, 1939-40; fellow, lectr. in physics Queens' Coll., Cambridge, Eng., 1945-52; asst. dir. for electronics research Ministry of Supply, London, 1952-56; with Nat. Radio Astronomy Obs., Charlottesville, Va., 1956-85; dep. dir. Nat. Radio Astronomy Obs., 1961-65; dir. Arecibo Obs., 1965-66, sr. scientist, 1978-85; mem. space sci. bd. Nat. Acad. Scis., 1961-71; chmn. lunar and planetary missions bd. NASA, 1967-71; chmn. space sci. bd. study Scientific Uses of the Space Shuttle, 1973. Contbr. articles to profl. jours. Served with RAF, 1940-45. Decorated Order of Brit. Empire. Fellow IEEE, AAAS; mem. Internat. Sci. Radio Union, Internat. Astron. Union. Clubs: Cosmos, Farmington. Home: Millbank PO Box 317 Greenwood VA 22943

FINDLEY, JAMES SMITH, biology and zoology educator, museum director; b. Cleve., Dec. 28, 1926; s. Howard Nevin and Dorothy Georgine (Smith) D.; m. Muriel Thomson, June 18, 1949; children: Stuart Thomson, Heidi Ann, Douglas Smith, Joan Nevin. AB, Western Res. U., 1949; PhD, U. Kans., 1955. Asst. instr. zoology U. Kans., Lawrence, 1950-54; curatorial asst. U. Kans. Mus. Nat. History, 1953-54; instr. zoology U. S.D., Vermillion, 1954-55; from asst. prof. to prof. biology U. N.Mex., Albuquerque, 1955-92, chmn. dept. biology, 1978-82, dir. Mus. Southwestern Biology, 1982-92. Author: Mammals of New Mexico, 1976, Natural History of New Mexican Mammals, 1986, Bats: a community perspective, 1993; contbr. numerous articles on mammalian systematics, ecology and biogeography to profl. jours. Served with U.S. Army, 1945-46. Mem. Am. Soc. Mammalogists (pres. 1980-82, C. Hart Merriam award 1978), Ecol. Soc. Am., Am. Soc. Naturalists, Soc. for Study Evolution. Democrat. Avocation: farming. Home: PO Box 44 Corrales NM 87048-0044 Office: U NMex Mus SW Biology Biology Bldg Albuquerque NM 87131

FINDLING, DAVID MARTIN, application engineer; b. Hinsdale, Ill., Dec. 6, 1960; s. Martin J. and Carol A. (Rusch) F. Student, Washington U., 1979-81; BSCS, Ill. Benedictine Coll., Lisle, Ill., 1991. Systems engr. Intel Corp., Rolling Meadows, Ill., 1982-86; applications engr. Intel Corp., Schaumburg, Ill., 1986-88; software engr. Wizdom Systems, Naperville, Ill., 1988-90; applications engr. Pioneer-Standard Electronics, Addison, Ill., 1990-93, tech. mktg. mgr., 1993—. Disaster vol. ARC, Lombard, Ill., 1984—. Mem. IEEE, ACM, Am. Radio Relay League. Evangelical Ch. Office: Pioneer-Standard Elec 2171 Executive Dr #200 Addison IL 60101

FINE, MORTON SAMUEL, civil engineer; b. Worcester, Mass., June 3, 1916; s. Jacob and Mary (Savatsky) F.; m. Frances Diana Kaufman, Dec. 15, 1940; children: Philip J., Paula J. Ridge. BS with Distinction, Worcester Poly. Inst., 1937. REgistered profl. engr. Designer, draftsman Riley Stoker Corp., Worcester, 1937-38; office engr. U.S Engr.'s Office, Hartford, Conn., 1939-43; tool designer, adj. prof. Porter Sch. Trinity Coll., Hartford, 1943-52; cons. civil engr. Morton S. Fine and Assocs., Hartford, 1950-75; nat. exec. dir. nat. Coun. Examiners for Engrs., Clemson, S.C., 1976-81; program dir. Worcester Poly. Inst., 1982-83; freelance cons. engr. Bloomfield, Conn., 1983--; pres. Nat. Coun. Examiners for Engrs. and L.S., Clemson, 1974-75; nat. chmn. Nat. Soc. Prof. Engrs., Washington, 1970-71; state pres. Conn. Soc. Prof. Engrs., 1964-65. Author, editor, pub.: EIT-A Strategy for the Engineer in Training Exam, 1984; contbr. articles to profl. jours. Mem., bd. chmn. State Bd. Registration for PE&LS, Hartford, 1962-74; mem. Engr. Adv. Com. U. Hartford, 1971-76; mem., chmn. Town Planning and Zoning Commn., West Hartford, 1958-63. Recipient Archimedes Engring. Achievement award Calif. Soc. Profl. Engrs., 1978. Fellow Am. Soc. Civil Engrs. (life, Benjamin Wright award); mem. Nat. Soc. Profl. Engrs., Am. Soc. for Engring. Edn., Am. Arbitration Assn., Tau Beta Pi, Sigma Xi. Home: 14 E West Ln Bloomfield CT 06002 Office: 707 Bloomfield Ave Bloomfield CT 06002

FINE, SIDNEY GILBERT, chemist; b. N.Y.C., Apr. 7, 1954; s. Max and Florence (Silver) F.; m. Judy Beth Ratz, Nov. 10, 1985; children: Sara Elizabeth, Elena Stephanie, Aaron David. BS in Chemistry cum laude, Coll. Staten Island, 1977; MS in Chemisty, St. John's U., 1992. Analytical chemist Charles M. Shapiro and Sons, Bklyn., 1979-82; analytical chemist Forest Labs., Inc., Inwood, 1983-85, quality control supr., 1985--. Founding mem. Jewish Aid Com., Bklyn., 1971--. Mem. Mu Alpha Theta. Republican. Office: Forest Labs Inc 300 Prospect St Inwood NY 11696

FINE, STANLEY SIDNEY, pharmaceuticals and chemicals executive; b. N.Y.C., Sept. 26, 1927; s. Morris and Sophie (Brajer) F.; m. Eleanor D. Baker, July 21, 1955 (dec. 1972); children: Lauren Allison Caban, Stephen Sidney (dec.); m. Astrid E. Merget, June 8, 1984 (div. Apr. 1987); m. Li L. Yang, July 31, 1991. Student, NYU, 1944-45; B.S., U.S. Naval Acad., 1949; postgrad., Coll. William and Mary, 1955-56, U. Va., 1956-57; MBA, Am. U., 1959; postgrad., Harvard U., 1963-65. Commd. ensign U.S. Navy, 1949, advanced through grades to rear adm., 1972; comdg. officer USS Hawk, 1954-56, Polaris Program, 1956-59, USS Lowe, 1961-63; comdr. Escort Div. 33, 1963; comdg. officer USS Ingraham, 1965-67; br. head Navy Material Command, Washington, 1967-68; exec. asst., naval aide to asst. sec. Navy, 1968-70; study dir. Center for Naval Analysis Navy Dept., Washington, 1970; dep. dir. Navy Program Info. Center, 1970-71; br. head OPNAV, 1971; spl. asst. to dir. Navy Program Planning, Washington, 1971-72; dep. chief Programs and Fin. Mgmt.; comptr. Naval Ship Systems Command, Washington, 1972-73; dir. fiscal mgmt. div. Office Chief Naval Ops., Washington, 1973-78; dir. budget and reports Navy Dept., 1975-78; ret., 1978; sr. v.p. United-Guardian, Inc. (AMEX), Hauppauge, N.Y., 1979—, also bd. dirs.; v.p., bd. dirs. New Energy Leasing Corp., McLean, Va.; bd. dirs. Micron Products Inc.; cons. GAO. Co-author: The Federal Budget: Cost Based in the 1980's, 1979, The Military Budget on a New Plateau: Strategic Choices for the 1990's; contbr. articles to profl. jours. and other publs.; lectr., TV commentator on def. and fed. budget issues. Mem. Presdl. transition team Dept. Commerce, 1980-81; bd. dirs. Bronx High Sch. of Sci. Found., N.Y.C., 1987-91; dir. Com. for Nat. Security, 1987-92, Montgomery County Fiscal Affairs Com., 1987-88. Decorated D.S.M., Navy Commendation medal, Legion of Merit with gold star; recipient outstanding Mgmt. Analyst award Am. Soc. Mil. Comptrollers, 1971, cert. of recognition and appreciation Montgomery County, Md. Mem. Naval Inst., World Affairs Coun. D.C., Naval Acad. Alumni Assn., Harvard U. Bus. Sch. Alumni Assn. Democrat. Jewish. Avocation: collecting ancient Roman coins. Office: United-Guardian Inc 230 Marcus Blvd Hauppauge NY 11788-3751

FINEGOLD, SYDNEY MARTIN, microbiology and immunology educator; b. N.Y.C., Aug. 12, 1921; s. Samuel Joseph and Jennie (Stein) F.; m. Mary Louise Saunders, Feb. 8, 1947; children: Joseph, Patricia, Michael. A.B., UCLA, 1943; M.D., U. Tex., 1949. Diplomate: Am. Bd. Med. Microbiology (mem. bd. 1979-85), Am. Bd. Internal Medicine. Intern USPHS, Galveston, Tex., 1949-50; fellow in medicine U. Minn. Med. Sch., 1950-52, research fellow, 1951-52; resident medicine Wadsworth Hosp., VA Ctr., Los Angeles, 1953-54; instr. medicine U. Calif. Med. Ctr., Los Angeles, 1955-57, asst. clin. prof., 1957-59, asst. prof., 1959-62, assoc. prof., 1962-68, prof., 1968—, prof. microbiology and immunology, 1983—; chief chest and infectious disease sect. Wadsworth Hosp., 1957-61, chief infectious disease sect., 1961-86, assoc. chief staff for research and devel., 1986—; mem. pulmonary disease research program com. VA, 1961-62, infectious disease research program com., 1961-65, merit rev. bd. (infectious diseases), 1972-74, med. research program specialist, 1974-76, adv. com. on infectious disease, 1974-87; mem. NRC-Nat. Acad. Sci. Drug Efficacy Study Group, 1966-69; mem. subcom. on gram-negative anaerobic bacilli Internat. Com. on Nomenclature Bacteria, 1966—, chmn., 1972-78; mem. adv. panel U.S. Pharmacopoeia, 1970-75; chmn. working group on anaerobic susceptibility test methods Nat. Commn. Clin. Lab. Standards, 1987—. Mem. editorial bd. Calif. Medicine, 1966-73, Western Jour. Medicine, 1974-77, Applied Microbiology, 1973-74, Jour. Clin. Microbiology, 1975-85, Am. Rev. Respiratory Disease, 1974-76, Infection, 1976—, Antimicrobial Agts Chemotherapy, 1980-89, Jour. Infectious Disease, 1979-82, 84-85, Diagnostic Microbiology and Infectious Diseases, 1982-90; editor Revs. of Infectious Diseases, 1990-91, Clin. Infectious Diseases, 1992—; sect. editor: infectious disease vols. Clinical Medicine, 1978-82, Microbiol. Ecology in Health and Disease, 1987-90. Vice chmn. UCLA Acad. Senate, 1986-87, chair, 1987-88. Served with USMCR; Served with USNR, 1943-46; to 1st. lt. AUS, 1952-53. Co-recipient V.A. Williams S. Middleton award for biomed. research; recipient Profl. Achievement award UCLA, 1987, Mayo Soley award Western Soc. Clin. Investigation, 1988, Disting. Alumnus award U. Tex. Med. Br., 1988, UCLA Med. Alumni Assn Med. Scis. award, 1990, Hoechst Roussel award Am. Soc. Microbiology, 1992. Master ACP; fellow Am. Pub. Health Assn., Am. Acad. Microbiology, AAAS, Infectious Diseases Soc. Am. (councilor 1976-79, pres.-elect 1980-81, pres. 1981-82, exec. com. 1980-83, Bristol award 1987); mem. Assn. Am. Physicians, Am. Soc. Microbiology (chmn. subcom. on taxonomy of Bacteroidaceae 1971-74), Am. Thoracic Soc., Western Soc. Clin. Research, Western Assn. Physicians, Wadsworth Med. Alumni Assn. (past pres.), Anaerobe Soc. of the Ams. (interim pres. 1992-94), Soc. Intestinal Microbiology Ecology and Disease (interim pres. 1982-83, pres. 1983-87), Va. Soc. Physician in Infectious Diseases (pres. 1986-88), Am. Fedn. Clin. Rsch., Sigma Xi, Alpha Omega Alpha. Democrat. Jewish. Home: 421 23d St Santa Monica CA 90402 Office: UCLA Medical Ct Clinical Microbiology Room A2-250 Los Angeles CA 90024

FINERTY, MARTIN JOSEPH, JR., military officer, researcher; b. Wilmington, Del., July 22, 1936; s. Martin Joseph and Jane Morris (McClenaghan) F.; m. Joan Eddleman, Dec. 3, 1960; children: Nancy Jane, Laura Tourison. BSE, U.S. Naval Acad., 1959; MS in Phys. Oceanography, U. Miami, Coral Gables, Fla., 1966; MS in Indsl. Mgmt., Coll. of the Armed Forces, 1979. Commd. ensign USN, 1959, advanced though grades to capt., 1985; head, polar programs Office of Oceanographer of Navy, Alexanrdria, Va., 1975-76; spl. asst. submarines Office of Asst. Sec. of Navy, Washington, 1976-77; spl. asst. ocean environment Office of Chief of Naval Ops., Washington, 1977-78; commanding officer Naval Polar Oceanography Ctr., Washington, 1982-85; program officer Nat. Acad. Scis., Washington, 1985-87; asst. dir. research ASME, Washington, 1987-88; gen. mgr. Marine Tech. Soc., Washington, 1988—; expert in ocean and hydro survey ops., polar programs and assn. mgmt. Author/editor tech. publs. Mem. Marine Tech. Soc., Assn. of U.S. Naval Acad. Class of 1959 (sec. 1971-74). Lodge: Masons. Avocations: reading, gardening. Home: 1841 Northbridge Ln Annapolis MD 21401-6576 Office: Marine Tech Soc Ste 906 1828 L St NW Washington DC 20036-5104

FINGER, LARRY WAYNE, crystallographer, mineralogist; b. Terril, Iowa, May 22, 1940; s. Wayne W. and Lucille F. (Hewitt) F.; m. Denise J. Lanning, June 16, 1962; children: Cynthia D., Pamela J. B in Physics, U. Minn., 1962, PhD, 1967. Post-doctoral fellow Carnegie Inst. Washington, 1967-69, crystallographer, 1969—; vis. prof. SUNY, Stony Brook, 1975-76, Va. Poly. Inst. & State U., 1984-85; mem. Earth Sci. Bd. NRC/NAS, Washington, 1986-89. Co-author: Comparative Crystal Chemistry, 1982. Recipient Alan Berman Rsch. Publ. award Naval Rsch. Lab., 1991. Fellow Mineral. Soc. Am. (sec. 1974-78); mem. Am. Geophysical Union, Am. Crystallographic Assn. Office: Carnegie Inst Washington 5251 Broad Br Rd NW Washington DC 20015-1305

FINGER, STANLEY MELVIN, chemical engineer, consultant; b. N.Y.C., Feb. 16, 1947; s. Herman and Ruth (Goldenberg) F. BS in Chemistry, Pratt Inst., 1969; MSChemE, U. Md., 1972, PhDChemE, 1975. Registered environ. mgr. Sr. engr. David Taylor Rsch. Ctr., Annapolis, Md., 1969-79, br. chief, 1979-84; dir. environ. programs ECO, Inc., Annapolis, Md., 1984—; rsch. prof. chem. engring. Cath. U. Am., Washington, 1984—; pres. Environ. Cons. and Investigations, Inc., North Potomac, Md., 1993—; mem. coms. NATO Info. Exch. Program, Washington, 1979-84; mem. Navy Chem. Warfare Rsch., Devel., Test and Evaluation group, Washington, 1979-84; Navy rep. UN Internat. Maritime Orgn., Washington, 1971-79. Author: (chpt.) Mushroom Fermentation, 1979; chmn. conf. sessions, 1983, 92; contbr. articles to profl. jours. Bd. dirs. Juvenile Diabetes Found., Washington, 1983-86. Recipient rsch. grant U.S. EPA, 1989-91, spl. fellowship NIH, 1974-75. Fellow Am. Inst. Chemists; mem. AICE, Am. Chem. Soc., Tau Beta Pi, Pi Mu Epsilon (chpt. pres. 1969). Achievements include development of transportable emergency response monitoring module for counter-terrorist operations, porous glass lysimeter for groundwater monitoring, model for aerobic microbial growth in semi-solid matrices, integrated shipboard collective protection system for CB defense and firefighting, oil content monitors and oil/water separators for shipboard pollution control. Office: Environ Cons and Investigating Inc 14 Rolling Green Ct North Potomac MD 20878

FINGERSON, LEROY MALVIN, corporate executive, engineer; b. Rochester, Minn., July 1, 1932; s. Malvin Ferdinand and Corolla Racelia (Sundet) F.; m. Ruth Anne Johnson, Nov. 26, 1960; children: Mark, Karin, Laura. BSME, U. Minn., 1954, MS in Mech. Engring., 1955, PhD in Mech. Engring., 1961. Chief exec. officer TSI, Inc., St. Paul, 1961—. Contbr. articles to profl. jours. Lutheran. Office: TSI Inc 500 Cardigan Rd PO Box 64394 Saint Paul MN 55164-0394

FINK, CHARLES AUGUSTIN, behavioral systems scientist; b. McAllen, Tex., Jan. 1, 1929; s. Charles Adolph and Mary Nellie (Bonneau) F.; m. Ann Heslen, June 1, 1955 (dec. June 1981); children: Patricia A., Marianne E., Richard G., Gerard A. A.A., Pan-Am. U., 1948; B.S., Marquette U., 1950; postgrad., No. Va. Community Coll., 1973, George Mason U., 1974; M.A., Cath. U. Am., 1979. Journalist UP and Ft. Worth Star-Telegram, 1950-52; commd. 2d lt. U.S. Army, 1952, advanced through grades to lt. col., 1966, various positions telecommunications, 1952-56, instr., 1956-58, exec. project mgmt., 1958-62, def. analysis and rsch., 1962-65, fgn. mil. rels., 1965-67, def. telecommunications exec., 1967-69, chief planning, budget and program control office Def. Satellite Communications Program, Def. Communications Agy., 1969-72, ret., 1972; pvt. practice cons. managerial behavior Falls Church, Va., 1972-77; pres. Behavioral Systems Sci. Orgn. (and predecessor firms), Falls Church, 1978—; leader family group dynamics, 1958-67; pub (jour.) Circle, 1985—. Developer hierarchial theory of human behavior, 1967—, uses in behavioral, social and biol. sci. and their applications, 1972—, behavioral causal modeling research methodology, 1974—, computer-aided behavior systems coaching for persons and orgns., 1982—, telecoaching, 1989; microbiol. chromatographic profiling, 1989—; adv. for copyrighting computer graphics displays and multi-media communications in scis. Adv. bd. Holy Redeemer Roman Cath. Ch., Bangkok, Thailand, St. Philip's Ch., Falls Church, Va., 1971-73. Decorated Army Commendation medals, Joint Services Commendation medal; named to Fink Hall of Fame, 1982; recipient Behavior Modeling award Internat. Congress Applied Systems Research and Cybernetics, 1980. Mem. Internat. Soc. Systems Scis., Am. Soc. Cybernetics, Internat. Assn. Cybernetics, Internat. Network for Social Network Analysis, Assn. U.S. Army, Ret. Officers Assn., Finks Internat. (v.p. 1981—), K.C. Home: 3305 Brandy Ct Falls Church VA 22042-3705 Office: PO Box 2051 Falls Church VA 22042-0051

FINK, JAMES BREWSTER, geophysicist, consultant; b. Los Angeles, Jan. 12, 1943; s. Odra J. and Gertrude (Sloot) F. BS in Geophysics and Geochemistry, U. Ariz., 1969; MS in Geophysics cum laude, U. Witwatersrand, Johannesburg, Transvaal, Republic of South Africa, 1980; PhD in Geol. Engring., Geohydrology, U. Ariz, 1989. Registered profl. engr., Ariz., N.Mex.; registered land surveyor, Ariz.; registered profl. geologist, Wyo.; cert. environ. inspector. Geophysicist Geo-Comp Exploration, Inc., Tucson, 1969-70; geophys. cons. IFEX-Geotechnica, S.A., Hermosillo, Sonora, Mex., 1970; chief geophysicist Mining Geophys. Surveys, Tucson, 1971-72; research asst. U. Ariz., Tucson, 1973; cons. geophysics Tucson, 1974-76; sr. minerals geophysicist Esso Minerals Africa, Inc., Johannesburg, 1976-79; sr. research geophysicist Exxon Prodn. Research Co., Houston, 1979-80; pres. Geophynque Internat., Tucson, 1980-90, hydroGeophysics, Tucson, 1990—; cons. on NSF research U. Ariz., 1984-85, adj. lectr. geol. engring., 1985-86, assoc. instr. geophysics, 1986-87, supr. geophysicist, geohydrologist, 1986-88, bd. dirs. Lab. Advanced Subsurface Imaging, 1986—; v.p. R&D Alternative Energy Engring., Inc., Tucson, 1992—; lectr. South African Atomic Energy Bd., Pelindaba, 1979. Contbr. articles to profl. jours. Served as sgt. U.S. Air NG, 1965-70. Named Airman of Yr., U.S. Air NG, 1967. Mem. Soc. Exploration Geophysicists (co-chair internat. meetings 1980, 81, sr. editor monograph 1990, reviewer), Am. Geophys. Union, European Assn. Exploration Geophysicists, South African Geophys. Assn., Assn. Ground Water Scientists, Nat. Water Well Assn. (reviewer), Mineral and Geotech. Explorationists, Ariz. Geol. Soc., Ariz. Computer-Oriented Geol. Soc. (bd. dirs., v.p.), Soc. Engring. and Minerals Exploration Geophysicists. Republican. Avocations: reading, computers, natural sciences. Home and Office: Hydrogeophysics 5865 S Old Spanish Trl Tucson AZ 85747-9487

FINK, WILLIAM LEE, ichthyologist, systematist; b. Coleman, Tex., July 22, 1946; s. Fred William Fink and Anna L. (Cobb) Davis; m. Sara V. Haase, June 17, 1972; 1 child, William Coleman. BS, U. Miami, 1967; PhD, George Washington U., 1976. Asst. prof., assoc. prof. Harvard U., Cambridge, Mass., 1976-82; assoc. prof. U. Mich., Ann Arbor, 1982—. Office: Univ Mich Museum of Zoology Ann Arbor MI 48109

FINKEL, ROBERT WARREN, physicist; b. N.Y.C., Aug. 17, 1934; s. Abraham B. and Elizabeth (Michaels) F.; m. Carla E. Gamburg, Dec. 10, 1961; children: Julia and Ruth (twins), James. BA in Math., NYU, 1956, MS in Physics, 1963, PhD in Thoretical Physics, 1966. Programmer, math. analyst Space Tech. Labs., Hawthorne, Calif., 1956-60; asst. prof. St. John's U. Dept. Physics, Jamaica, N.Y., 1962, assoc. prof., 1973, chmn., 1981—; mem. adv. bd. Ednl. Design, Inc., N.Y.C., 1973—; mem. edit. bd. Jour. Biol. Physics, 1973-75. Author: The Brainbooster, 1983, The New Brainbooster, 1991; contbr. articles to profl. jours. NSF fellow, 1960. Mem. AAUP, Am. Phys. Soc., Am. Assn. Physics Tchrs., St. John's U. Faculty Assn. Achievements include development of a theoretical relationship between temperature and the average rate of change of molecular states; description of how chemical kinetic systems can exhibit collective quantum behavior. Office: St John's U Dept Physics Grand Ctrl & Utopia Pkys Jamaica NY 11439

FINKELSHTEIN, ANDREY MICHAILOVICH, astronomy educator; b. Tavda, Russia, Aug. 7, 1942; s. Michail Samoilovich Finkelshtein and Olga Aleksandrovna Kusnezova-Zachoder; m. Elvira Vladimirovna Morozova, Oct. 27, 1967; 1 child, Morozova Tanija. Grad., Leningrad State U., 1968; Cand. Sci., Ioffe Phys. Tech. Inst., Leningrad, 1973; DS, Main Astron. Observatory, Leningrad, 1990. Sci. researcher Inst. Theoretical Astronomy, Leningrad, 1968-73; sr. sci. researcher, vice-dir. Spl. Astrophys. Observatory Russian Acad. of Scis., Leningrad, 1973-85, dir., head Inst. for Applied Astrophys. Lab.-Astronomy, 1986—; dir. St. Petersburg br. Internat. Nongovtl. Orgn., 1989-92; gen. dir. Soviet-Switzerland Joint Venture, Leningrad, 1990-91. Author: Introduction in Radio Astrometry, 1984; contbr. more than 100 papers to sci. jours. Mem. Internat. Astron. Union (mem. commn. 1988), Am. Math. Soc., Einstein Fund (co-dir. 1985). Achievements include 6 patents. Home: Institute of Applied Astronomy, Trefoleva St 6/30 Apt 68, 197042 Saint Petersburg 198097, Russia Office: Inst for Applied Astronomy, Zdanovskaya St 8, Saint Petersburg 197042, Russia

FINKELSTEIN, RICHARD ALAN, microbiologist; b. N.Y.C., Mar. 5, 1930; s. Frank and Sylvia (Lemkin) F.; m. Helen Rosenberg, Nov. 30, 1952; children: Sheri, Mark, Laurie; m. Mary Boesman, June 20, 1976; 1 dau., Sarina Nicole. B.S., U. Okla., 1950; M.A., U. Tex., Austin, 1952, Ph.D., 1955. Teaching fellow, research scientist U. Tex., Austin, 1950-55; fellow, instr. U. Tex. Southwestern Med. Sch., Dallas, 1955-58; chief bioassay sect. Walter Reed Army Inst. Research, Washington, 1958-64; dep. chief, chief dept. bacteriology and mycology U.S. Army Med. Component, SEATO Med. Research Lab., Bangkok, Thailand, 1964-67; assoc. prof. dept. microbiology U. Tex. Southwestern Med. Sch., Dallas, 1967-73; prof. U. Tex. Southwestern Med. Sch., 1973-79; prof., chmn. dept. microbiology Sch. Medicine U. Mo., Columbia, 1979—, curators' prof., 1990—; Millsap Disting. Prof., 1985—; mem. Nat. Com. for Coordination of Cholera Research, Ministry of Pub. Health, Bangkok, 1965-67; cons. WHO, 1970—, to comdg. gen. U.S. Army Med. Research and Devel. Command, 1975-79, Schwarz-Mann Labs., 1974-79; vis. assoc. prof. U. Med. Scis., Bangkok, 1965-67; vis. prof. U. Chgo. Med. Sch., 1977; vis. scientist Japanese Sci. Council, 1976; Ciba-Geigy lectr. Waksman Inst., Rutgers U., 1975; guest lectr. FEMS/SGM Lab. Course, Trinity Coll., Dublin, Ireland, 1986. Contbr. articles on cholera, enterotoxins, gonorrhea, and role of iron in host-parasite interactions to profl. jours. Recipient Robert Koch prize Bonn, Fed. Republic Germany, 1976; Chancellor's award for outstanding faculty rsch. in biol. scis. U. Mo.-Columbia, 1985, Sigma Xi Rsch. award U. Mo.-Columbia, 1986. Fellow Am. Acad. Microbiology (hon. Tex. br., div. councilor, chmn. program com. 1979-82, sec.-treas, Mo. br. 1985-87, v.p. 1987-89, pres., 1989-91, councillor 1991-92, council policy com. 1992—, pres. Tex. br. 1974-75), Am. Assn. Immunologists, Soc. Gen. Microbiology, Pathol. Soc. Gt. Britain and Ireland, Sigma Xi. Achievements include first purification of cholera enterotoxin; first purification of heat-labile enterotoxin from Escherichia coli; patent for living attenuated candidate cholera vaccine. Home: 3207 Honeysuckle Dr Columbia MO 65203-0901 Office: U Mo-Columbia Sch Medicine Dept Molecular Microbiology and Immunology Columbia MO 65212

FINKL, CHARLES WILLIAM, II, geologist, educator; b. Chgo., Sept. 19, 1941; s. Charles William and Marian L. (Hamilton) F.; m. Charlene Bristol, May 16, 1965 (div.); children: Jonathan William Frederick, Amanda Marie. BSc, Oreg. State U., 1964, MSc, 1966; Ph.D., U. Western Australia, 1971. Instr. natural resources Oreg. State U., 1967; demonstrator U. Western Australia, Perth, 1968, staff geochemist for S.E. Asia, Internat. Nickel Australia Pty. Ltd., 1970-74; chief editor Ency. Earth Sci., N.Y.C., 1974-87; dir. Inst. Coastal Studies Nova. U., Port Everglades, Fla., 1979-83; pres. Resource Mgmt. & Mineral Exploration Cons., Inc., Ft. Lauderdale, Fla., 1974-85, Info. Mgmt. Corp. (IMCO), Ft. Lauderdale, 1985-87; exec. dir., v.p. Coastal Edn. and Rsch. Found., Charlottesville, Va., 1983-89; pres. Coastal Edn. and Rsch. Found., Charlottesville, 1990—; prof. dept. geology Fla. Atlantic U., Boca Raton, 1983-88; mem. survey and mapping and subcommn. on morphotectonics; exec. bd. dirs. Internat. Geol. Correlation Program project 174 on Quaternary Coastal Evolution, 1989-93; mem. exec. bd. Skagen (Denmark) Odde Project Mus. for Coastal Geomorphology; mem. marine adv. bd. com. Broward County Bd. Commrs., 1990—; radio and TV appearances. Author: Soil Classification, 1982; vol. editor, contbg. author: The Encyclopedia of Soil Science, Part I: Physics, Chemistr, Biology, Fertility and Technology, 1979; editor, contbg. author: The Encyclopedia of Applied Geology, 1983, The Encyclopedia of Field and General Geology, 1988; vol. editor The Encyclopedia of Soil Science and Technology, 1991—; editor in chief Jour. Coastal Rsch.: An Internat. Forum for the Littoral Scis., 1984—; series editor Benchmark Papers in Soil Sci., 1982-86; editor: Current Titles in Ocean, Coastal, Lake and Wateway Sciences, 1985-88. Mem. Am. Geophys. Union, Am. Geog. Soc., Am. Quaternary Assn. Am. Littoral Soc., Am. Soc. Photogrammetry and Remote Sensing, Am. Shore and Beach Preservation Assn., Australasian Inst. Mining and Metallurgy, Brit. Geomorphological Rsch. Group, Brit. Soc. Soil Sci., Can. Geophys. Union, Coastal Soc., Deutsche Bodenkundlichen Gesellschaft, Deutsche Geologische Vereinnung, European Assn. Earth Sci. Editors, Estuarine and Brackish-Water Scis. Assn., Geol. Assn. Can., Geol. Soc. Am., Geol. Soc. Australia, Geol. Soc. London, Geol. Soc. South Africa, Geologists Assn., Geosci. Info. Soc., Internat. Soc. Reef Studies, Soc. Wetland Sci., Internat. Soil Sci. Soc., Soc. Wetland Scientists, Internat. Geographical Union (mem. neotectohics, commn. on rapid geomorphological hazards working group, corr. mem. working group on paleosols, commn. on coastal systems), Internat. Union Geol. Scis. (mem. project 317, paleoweathering records and paleosurfaces), Mineral. Assn. Can., Nature Conservancy, Nat. Parks and Conservation Assn., Soil Sci. Soc. Am., Société de Belge de Pedologie, Soc. Econ. Paleontologists and Mineralogists, Soc. Scholarly Publ., Soc. Mining Engrs., Am. Inst. Profl. Geologists (cert. profl. geologist scientist), Am. Registry Cert. Profls. in Agronomy, Crops and Soils (cert. profl. soil scientist), Gamma Theta Upsilon. Republican. Presbyterian. Home: 4310 NE 25th Ave Fort Lauderdale FL 33308-4803 Office: Fla Atlantic U Dept Geology Boca Raton FL 33431

FINKLER, KAJA, anthropologist, educator; b. Poland, Jan. 4, 1935; came to U.S., 1946; d. Chaim and Golda (Taub) F. MA, Hunter Coll., 1968; PhD, CUNY, 1973. Prof. anthropology U. N.C., 1984—. Grantee NSF, 1977-79, 80-81, 86-87, 89, NIMH, 1974, 87-89, Health and Human Svcs., 1982, 84-86. Home: 1906 Overland Chapel Hill NC 27514 Office: U NC Chapel Hill NC 27514

FINKS, ROBERT MELVIN, paleontologist, educator; b. Portland, Maine, May 12, 1927; s. Abraham Joseph and Sarah (Bendette) F. B.S. magna cum laude, Queens Coll., 1947; M.A., Columbia U., 1954, Ph.D., 1959. Lectr. Bklyn. Coll., 1955-58, instr., 1959-61; lectr. Queens Coll., CUNY, 1961-62, asst. prof., 1962-65, acting chmn., 1963-64, assoc. prof. geology, 1966-70, prof., 1971—; geologist U.S. Geol. Survey, 1952-54, 63—; research assoc. Am. Mus. Natural History, 1961-77, Smithsonian Instn., 1968—; doctoral faculty CUNY, 1983—; cons. in field. Author: Late Paleozoic Sponge Faunas of the Texas Region, 1960; Editor: Guidebook to Field Excursions, 1968; Contbr. articles profl. jours. Queens Coll. Scholar, 1947. Fellow AAAS, Geol. Soc. Am., Explorers Club; mem. Paleontol. Soc. (vice chmn. Northeastern sect. 1977-78, chmn. 1978-79), Paleontol. Assn. Britain, Soc. Econ. Paleontologists and Mineralogists, Internat. Palaeontol. Assn., Geol. Soc. Vt. (charter mem.), Planetary Soc. (charter), Phi Beta Kappa (v.p. Sigma chpt. N.Y. 1993—), Sigma Xi. (exec. sec. Queens Coll. chpt. 1982-85). Office: Queens Coll CUNY Dept Geology Flushing NY 11367

FINLAN, MARTIN FRANCIS, physicist, consultant; b. Widnes, Lancs, Eng., Sept. 26, 1930; s. Martin and Lilian Rose (Haney) F.; m. Margaret Jean Passey, Apr. 9, 1955 (div. Aug. 1983); children: Christine Jean, Stephen Martin Francis; m. Corinne Sylvia Barker, Apr. 7, 1984. BS, Liverpool (Eng.) U., 1951. Chartered physicist, Eng. Exptl. officer Ministry of Def. Dept. Atomic Energy, Culcheth, Lancashire, Eng., 1951-56; from exptl. officer to sr. scientific officer U.K. Atomic Energy Authority, Dounreay, Caithness, Scotland, 1956-60; cyclotron group mgr. The Radiochem. Ctr., Amersham, Bucks, Eng., 1963-83; phys. scis. advisor Amersham Internat. PLC, Amersham, Bucks, Eng., 1983-88, 90—, asst. rsch. dir., 1988-90; cyclotron cons. Oxford (Eng.) Instruments, 1986—. Author: (with others) Metals Reference Book, 1976, 3rd edit., 1991, In Situ Hybridization, 1990; contbr. articles to profl. jours. Mem. Inst. Physics, British Med. Ultrasonic Soc. Achievements include patents for electrongun, superconducting compact cyclotron, surface plasmon resonance biosensors. Office: Amersham Internat, HP79LL Amersham England

FINLEY, RICHARD WADE, internist, educator; b. Amarillo, Tex., Aug. 11, 1949; s. Maurice Charles and Virginia Mae (Wade) F. MS, U. Denver, 1974; MD, Tulane U., 1976. Diplomate Am. Bd. Internal Medicine, Am. Bd. Infectious Diseases. Instr. in medicine Rush Prebyn. St. Lukes Hosp., Chgo., 1979; immunology rsch. fellow WHO, Geneva, 1980-82; infectious disease fellow U. Colo. Health Sci. Ctr., Denver, 1982-84; med. staff fellow NIH, Bethesda, Md., 1984-87; vis. scientist Inst. Oswaldo Cruz, Rio de Janiero, 1987-88; staff physician Proctor Emergency Dept., Peoria, Ill., 1988-89; asst. prof. medicine U. Miss., Jackson, 1989—. Author: (book chpt.) Infection as a Cause of Altered Mental State, 1992. Recipient Faculty Enhancement award, Oak Ridge Assn. Univs., 1992; grantee: U. Miss., NIH, 1990. Mem. AAAS, Am. Coll. Physicians, Infectious Disease Soc., Am. Soc. Micro Biology. Achievements include study of nucleoside transport processes in protoza. Office: U Miss Immunology Dept 2500 N State St Jackson MS 39216

FINLEY, WAYNE HOUSE, medical educator; b. Goodwater, Ala., Apr. 7, 1927; s. Byron Bruce and Lucille (House) F.; m. Sara Will Crews, July 6, 1952; children: Randall Wayne, Sara Jane. B.S., Jacksonville State U., 1948; M.A., U. Ala., 1950, M.S., 1955, Ph.D., 1958, M.D., 1960; postgrad., U. Uppsala, Sweden, 1961-62. Cert. clin. cytogenetics Am. Bd. Med. Genetics. Sci. tchr. High Sch., Tuscaloosa, Ala., 1949-51; intern U. Ala. Hosps. and Clinics, 1960-61; asst. prof. pediatrics U. Ala. Sch. Medicine, 1962-66, assoc. prof., 1966-70, asst. prof. biochemistry, 1965-75, assoc. prof., 1975-77, asst. prof. physiology and biophysics, 1968-75, assoc. prof., 1975, prof. epidemiology, pub. health and preventive medicine, 1975—, adj. prof. biology, 1980—, dir. Lab. Med. Genetics, 1966—, dir. med. genetics grad. program, 1983—; chmn. med. Student Rsch. Day, 1965-75, chmn. faculty coun. Sch. Medicine, 1977-78, 84-87; mem. nat. adv. rsch. resources coun. NIH-HEW, 1977-80; sr. scientist Comprehensive Cancer Ctr., Cystic Fibrosis Rsch. Ctr.; bd. dirs. Southeastern Regional Genetics Group, 1983—; dir. Ala. Med. Genetics Program, 1978—; chmn. steering com. Reynolds Hist. Libr. Assocs., 1981—, Carmichael Fund for Grad. Students, 1989—. Contbr. articles on human malformations and clin. cytogenetics to tech. jours. Served with AUS, 1945-46, 51-53; lt. col. Res.; ret. Recipient med. award Ala. Assn. Retarded Children, 1969, Disting. Alumni award U. Ala. Med. Sch. Alumni Assn., 1978, Tarlington award, 1982, Disting. Faculty lectr. award U. Ala. Med. Ctr., 1983, Alumnus of Yr. award Jacksonville State U., 1989; Wayne H. and Sara C. Finley chair in med. genetics established U. Ala., Birmingham, 1986; portrait in Reynolds Hist. Libr., 1991. Fellow Am. Coll. Med. Genetics (founder, edn. com. 1993—); mem. AAAS, N.Y. Acad. Scis., Soc. Exptl. Biology and Medicine, Am. Inst. Chemists, Am. Fedn. Clin. Rsch., Am. Soc. Human Genetics, So. Med. Assn., So. Soc. Pediatric Rsch., Med. Assn. Ala., Jefferson County Med. Soc. (maternal and child health com. 1975-79, chmn. 1976-77), (pres. 1983), Jefferson County Pediatrics Soc., U. Ala. Sch. Medicine Alumni Assn. (pres. 1974-75), Greater Birmingham Area C. of C. (bd. dirs. 1983-86), Newcomen Soc., Sigma Xi (pres. U. Ala. in Birmingham chpt. 1972-73), Kappa Delta Pi, Phi Delta Kappa, Alpha Omega Alpha, Phi Beta Pi (McBurney cup 1960), Omicron Delta Kappa. Baptist. Clubs: Caduceus, U. Ala. Sch. Medicine Faculty Coun. (pres. 1984-86), Kiwanis (pres. Shades Valley 1973-74), Rotary. Home: 3412 Brookwood Rd Birmingham AL 35223-2023 Office: U Ala Lab of Medical Genetics Univ Sta Birmingham AL 35294

FINNBERG, ELAINE AGNES, psychologist, editor; b. Bklyn., Mar. 2, 1948; d. Benjamin and Agnes Montgomery (Evans) F.; m. Rodney Lee Herndon, Mar. 1, 1981; 1 child, Andrew Marshal. BA in Psychology, L.I. U., 1969; MA in Psychology, New Sch. for Social Rsch., 1973; PhD in Psychology, Calif. Sch. Profl. Psychology, 1981. Lic. psychologist, Calif. Rsch. asst. in med. sociology Med. Coll. Cornell U., N.Y.C., 1969-70; med. abstractor USV Pharm. Corp., Tuckahoe, N.Y., 1970-71, Coun. for Tobacco Rsch., N.Y.C., 1971-77; editor, writer Found. of Thanatology Columbia U., N.Y.C., 1971-76, cons. family studies program cancer ctr. Coll. Physicians &Surgeons, 1973-74; dir. grief psychology and bereavement counseling San Francisco Coll. Mortuary Scis., 1977-81; rsch. assoc. dept. epidemiology and internat. health U. Calif., San Francisco, 1979-81, asst. clin. prof. dept. family and community medicine, 1985-93, assoc. clin. prof., dept. family and community medicine, 1993—; chief psychologist Natividad Med. Ctr., Salinas, Calif., 1984—; asst. chief psychiatry svc. Natividad Med. Ctr., 1985—, acting chief psychiatry, 1988-89, vice chair medicine dept., 1991-93, sec. treas. med. staff, 1992—. Editor: (newspaper) The California Psychologist, 1988—; editor Jour. of Thanatology, 1972-76, Cathexis, 1976-81. Mem. gov.'s adv. bd. Agnews Devel. Ctr., San Jose, Calif., 1988—, chair, 1989-90. Mem. APA, Nat. Register Health Svc. Providers in Psychology, Calif. Psychol. Assn. (Disting. Svc. award 1989), Soc. Behavioral Medicine, Mid-Coast Psychol. Assn. (sec. 1985, treas. 1986, pres. 1987). Office: Natividad Med Ctr 1330 Natividad Rd PO Box 81611 Salinas CA 93912-1611

FINNEY, CLIFTON DONALD, inventor, manufacturing executive; b. Dubuque, Iowa, Apr. 7, 1941; s. Clifton Monroe and Violet Irene (Snyder) F.; m. Kazuko Akiyama, Aug. 17, 1968; 1 child, Ann. BA in Chemistry, Austin Coll., 1964; PhD in Phys. Chemistry, Kans. State U., 1970. Postdoctoral fellow U. Toronto, Ont., Can., 1969-71; asst. prof. chemistry Drake U., Des Moines, 1971-75; pres. Natural Dynamics, Des Moines and Houston, 1975-86, Golf Physics Co., Baton Rouge, 1986—; assoc. Ames (Iowa) Lab., U.S. AEC, 1971-75; instr. computer sci. U. Houston, 1984-86. Contbr. articles to Phys. Chemistry, Sci., Computers and Edn. Recipient energy rsch. grant Iowa Energy Policy Coun., 1975, USERDA, 1976. Mem. Am. Chem. Soc., N.Y. Acad. Scis. Achievements include patents in Advanced Rail-Back and Corner-Back Inertial Weighting Systems for Golf Clubheads. Home and Office: Golf Physics Co 1057 Oak Hills Pky Baton Rouge LA 70810-4705

FINNEY, ESSEX EUGENE, JR., science executive; b. Powhatan, Va., May 16, 1937; s. Essex Eugene Sr. and Etta Francis (Burton) F.; m. Rosa Ellen Bradley, June 13, 1959; children: Essex Eugene III, Karen Renee Finney Shelton. BS, Va. Poly. Inst., 1959; MS, Pa. State U., 1960; PhD, Mich. State U., 1963. Agrl. engr. U.S. Dept. Agr., Beltsville, Md., 1965-77, asst. dir., 1977-83, assoc. area dir., 1983-87; assoc. area dir. U.S. Dept. Agr., Phila., 1987-89; Beltsville area dir. U.S. Dept. Agr., 1989-92; assoc. admin. Agrl. Rsch. Svc. U.S. Dept. Agr., Washington, 1992—; sr. policy analyst Office of Sci. & Tech. Policy, Washington, 1980-81. Author: Quality Control, 1973; editor: (CRC handbook) Transportation & Marketing, 1981. Councilman Town of Glenarden, Md., 1975; sec. County Fedn. of Civic Assns., Md., 1973. Lt. U.S. Army, 1963-65. Princeton fellow Princeton U., 1973-74; recipient Outstanding Engring. Alumni award Pa. State U., 1985, Adminstrn. award Gamma Sigma Delta, U. Md., 1985, Outstanding Alumni award Pa. State U. Coll. Agr., 1993. Fellow Am. Soc. Agrl. Engrs. (bd. dirs. 1970-72); mem. AAAS, Inst. Food Technologists. Office: Dept of Agriculture Science & Education 12th & Independence Ave SW Washington DC 20250

FINNIGAN, JAMES FRANCIS, civil engineer; b. Johnson City, N.Y., Jan. 9, 1957; s. James Francis and Betty Frances (Laskoski) F. AS, Broome Community Coll., 1979; BS, Clarkson U., 1980. Registered profl. engr., Calif. Civil engr. I L.A. County Flood Control Dist., 1980-85; dist. engr. Monterey Peninsula (Calif.) Water Mgmt. Dist., 1985-87; project engr. Sear-Brown Group, Rochester, N.Y., 1987-89, Metcalf & Eddy, Inc., Wakefield, Mass., 1989-91; project mgr. IT Corp., Rochester, 1991-92, tech. assoc., 1992-93; assoc. Erdman, Anthony & Assocs., Rochester, 1993—. Asst. capt. Va. Jets., Hockey N. Am., Vienna, Va., 1990-91. Mem. ASCE (environ. hazardous waste tech. com. 1992—), NSPE, Water Environ. Fedn., Trout Unltd. Roman Catholic. Office: Erdman Anthony & Assocs 259 Monroe Ave Rochester NY 14607

FINS, JOSEPH JACK, internist, medical ethicist; b. N.Y.C., Nov. 16, 1959; s. Herman and Ruth (Lovett) F.; m. Amy B. Ehrlich, July 2, 1989. BA with honors, Wesleyan U., 1982; MD, Cornell U., 1986. Diplomate Am. Bd. Internal Medicine. Intern in psychiatry N.Y. Hosp. Payne Whitney Clinic, N.Y.C., 1986-87; resident in medicine N.Y. Hosp., N.Y.C., 1987-89; instr. Cornell U. Med. Coll., N.Y.C., 1990; fellow in medicine N.Y. Hosp. Cornell Med. Ctr., N.Y.C., 1990-92; vis. assoc. for medicine Hastings Ctr., Briarcliff Manor, N.Y., 1990-92; instr. Cornell U. Med. Coll., N.Y.C., 1992-93, asst. prof. medicine, 1993—; assoc. for medicine Hastings Ctr., Briarcliff Manor, 1992—; asst. attending physician N.Y. Hosp., 1992—; vis. scholar Hastings Ctr., Briarcliff Manor, 1989; mem. ethics com. dept. medicine N.Y. Hosp., N.Y.C., 1991—. Assoc. editor for abstracts Jour. Am. Geriatrics Soc., 1991-92; contbr. articles to profl. jours. Fellow N.Y. Acad. Medicine; mem. ACP, Am. Geriatrics Soc. (ethics com. 1992—), Assn. of Bar of City of N.Y. (adj.). Office: NY Hosp Cornell Med Ctr Dept Medicine 525 E 68th St New York NY 10021

FINZEL, BARRY CRAIG, research scientist; b. Monroe, Mich., Dec. 19, 1956; s. Donald A. and Bonadine (Donnelly) F.; m. Muriel Ann Henry, Nov. 19, 1977; children: Kimberly, Torre, Callie, Helena. BS, Ea. Mich. U., 1979; PhD in Chemistry, U. Calif., San Diego, 1983. Rsch. scientist Genex Corp., Gaithersburg, Md., 1983-86; prin. scientist E.I. du Pont de Nemours & Co., Wilmington, Del., 1986-87; rsch. scientist The Upjohn Co., Kalamazoo, 1988—. Office: The Upjohn Co 301 Henrietta St Kalamazoo MI 49001

FIORI, MICHAEL J., pharmacist; b. Brunswick, Maine, Nov. 25, 1951; s. Columbus H. and Marie Alice (Pelletier) F.; m. Anna Marie Robinson, Dec. 25, 1980; 1 child, Michela. BA in Biology, Bowdoin Coll., 1974; BS in Pharmacy, Mass. Coll. Pharmacy, 1977; MBA, U. Maine, 1987; PhD in Bus. Adminstr., LaSalle U., 1993. Rsch. student Rsch. Inst. Gulf Maine, 1974-75; pharmacy intern Newton-Wellesley (Mass.) Hosp., 1977; pharmacist Allen Drug Store, Brunswick, 1977-78; cons. pharmacist Allen Drug Store, Bangor, Maine, 1977-84; pres., chief operating officer Downeast Pharmacy, Inc.; pres., cons. Pharmacists of New Eng. Downeast Pharmacy, Inc., various cities, Maine, 1984—; pres. Guardian Healthcare Downeast Pharmacy, Inc., 1990—; commr. Maine Commn. Pharmacy, 1985-90; pres. and chief exec. officer Vector Assocs., Inc. dba ODV, Inc. commr. Maine Commn. Pharmacy, 1985-90; pres., chief exec. officer Vector Assocs., Inc. dba ODV, Inc. Earle S. Thompson scholar, 1971-73, Charles Lowery scholar, 1974; named one of Outstanding Young Men Am., 1986. Fellow Am. Soc. Cons. Pharmacists; mem. Am. Soc. for Pharmacy Law, Narcotic Enforcement Officers Assn., Internat. Assn. for Identification, Internat. Assn. of Chiefs of Police, Nat. Assn. of Bds. of Pharmacy, Maine Pharmacy Assn., Health Care Providers, Inc., Nat. Assn. Retail Druggists, Mass. Coll. Pharmacy Alumni Assn., Bowdoin Coll. Alumni Assn., U. Maine Alumni Assn., Italian Heritage Soc., Gyro Internat., Maine Health Care Assn., NRA (life), KC, Elks, Beta Theta Pi (chpt. pres. 1972-73, ho. corp. pres. 1985—, ho. corp. dir. 1975—, dist. chief 1979-84, 87-90). Democrat. Roman Catholic. Home: 2079 Essex St Bangor ME 04401-2112 Office: Downeast Pharmacy Inc 185 Harlow St Bangor ME 04401-4933

FIRESTONE, DAVID, chemist; b. N.Y.C., Sept. 15, 1923; s. Harry and Anna (Koved) F.; m. Berdie Flegenheimer, May 26, 1946; children: Richard Ira, Michael Paul, Janice Celia. BS, CCNY, 1948; MS, Poly. Inst. N.Y., 1951; PhD, George Washington U., 1968. Chemist FDA, N.Y.C., 1948-54; chemist FDA, Washington, 1954-63, supervisory rsch. chemist, 1963-73, sr. rsch. chemist, 1973—; mem. ad hoc study group on pentachlorophenol contaminants EPA, 1977-79, dioxin/furan protocol rev. panel, 1987—. Editor: Am. Oil Chem. Soc. Ofcl. Methods and Recommended Practices, 1987—; contbr. chpts.: Chlorinated Dioxins and Dibenzofurans in Perspective, 1986, Advances in Lipid Methodology One, 1992. With U.S. Army, 1942-46, PTO. Recipient Fachini award Italian Soc. for Fat Rsch., 1986. fellow Assn. Ofcl. Analytical Chemists (gen. referee an oils and fats 1962—), Am. Oil Chemists Soc. (pres. 1979-80, chair uniform methods com. 1986-93), Internat. Union Pure and Applied Chemistry (pres. commn. on oils, fats and derivatives 1981-83), Am. Chem. Soc. (adv. bd. jour. 1976-79), Inst. Food Techs., B'nai B'rith (pres. met. lodge 1974-75, nat. capital assn. 1981-82, dir. philatelic svc. 1979—), Sigma Xi. Jewish. Achievements include development of FDA's initial methodology for determination of dioxins and related compounds in foods and biological tissues; research in composition and toxicity of frying fats, analysis of dioxins, food fats, and oils. Home: 906 Playford Ln Silver Spring MD 20901 Office: FDA 200 C St SW Washington DC 20204

FISCH, NATHANIEL JOSEPH, physicist; b. Montreal, Quebec, Can., Dec. 29, 1950; s. Mandel and Helene (Greenfield) F.; m. Tobe Michelle Mann, Aug. 12, 1984; children: Jacob, Benjamin, Adam. BS, MIT, 1972, MS, 1975, PhD, 1978. Researcher Princeton (N.J.) Plasma Physics Lab., 1978-91, assoc. dir. for acad. affairs, 1993—; dir. program in plasma physics Princeton U., 1991—, prof. astrophys. scis., 1991—; cons. Exxon Rsch. and Engring., Clinton, N.J., 1981-86; vis. scientist IBM, Yorktown Heights,

N.Y., 1986. Recipient fellowship Guggenheim Found., 1985, 1992 APS award for Excellence in Plasma Physics, Am. Phys. Soc., 1992. Fellow Am. Phys. Soc. Achievements include patents in new ways to produce current in plasmas. Office: Princeton U Forrestal Campus Box 451 Princeton NJ 08543

FISCH, RICHARD S., physicist, psychophysicist; b. N.Y.C., May 18, 1932; s. Abraham and Ruth (Leslie) F.; m. Lois Berber; children: Alan, Eric. BS in Physics, NYU, 1958; MS in Physics, Columbia U., 1961. With 3M Co., 1961—; supr. color physics 3M Co., St. Paul, 1963-72, mgr. material resources, 1972-82, sr. rsch. scientist unconventional imaging, 1982-86, div. scientist, 1986—. Editor graphic arts Jour. Imaging Tech.; exec. v.p. Tech. Assn. Graphic Arts Procs., 1989—; editor-in-chief Tech. Assn. Graphic Arts jour., 1991—; contbr. articles to profl. publs. Fellow Soc. Imaging Sci. and Tech., Royal Photographic Soc. (Eng.), Inst. Printing (Eng.); mem. Tech. Assn. Graphic Arts (v.p. tech. sect., bd. dirs., hons. 1992), Soc. Photographic Scientists and Engrs. (bd. dirs.). Achievements include 29 patents on photography and graphic arts materials and processes including color imaging. Home: 395 Woodlawn Ave Saint Paul MN 55105 Office: 3M Co 3M Ctr 235-1C-35 Saint Paul MN 55144

FISCHER, ALFRED GEORGE, geology educator; b. Rothenburg, Germany, Dec. 10, 1920; came to U.S., 1935; s. George Erwin and Thea (Freise) F.; m. Winnifred Varney, Sept. 26, 1939; children: Joseph Fred, George William, Lenore Ruth Fischer Walsh. Student, Northwestern Coll., Watertown, Wis., 1935-37; BA, U. Wis., 1939, MA, 1941; PhD, Columbia U., 1950. Instr. Va. Poly. Inst. and State U., Blacksburg, 1941-43; geologist Stanolind Oil & Gas Co., Kans. and Fla., 1943-46; instr. U. Rochester, N.Y., 1947-48; from instr. to asst. prof. U. Kans., Lawrence, 1948-51; sr. geologist Internat. Petroleum, Peru, 1951-56; prof. geology Princeton (N.J.) U., 1956-84, U. So. Calif., Los Angeles, 1984—. Co-Author: Invertebrate Fossils, 1952, The Permian Reef Complex, 1953, Electron Micrographs of Limestone, 1967; editor: Petroleum and Global Tectonics, 1975. Recipient Verrill medal Yale U. Fellow Geol. Soc. Am. (Arthur L. Day medal 1993), Soc. Econ. Paleontologists (hon., Twenhofel medal); mem. AAAS, Am. Assn. Petroleum Geologists, Paleontol. Soc., Deutsche Geologische Gesellschaft (Leopold von Buch medal), Geologische Vereinigung, Sigma Xi. Home: 1736 Perch St San Pedro CA 90732-4218 Office: U So Calif Dept Geol Scis University Park Dept Geological Sciences Los Angeles CA 90089-0741

FISCHER, BERNHARD FRANZ, physicist; b. Karlsruhe, Germany, Feb. 6, 1948; s. Emil Otto and Erika Maria (Mook) F.; m. Lieselotte Pleger, May 28, 1976; children: Antonia, Claudius. Diplom Physiker, U. Frankfurt, Germany, 1971; Doctor rer. nat., U. Stuttgart, Germany, 1973. Rsch. asst. Max-Planck Inst., Solid State Rsch., Stuttgart, Germany, 1972-81; postdoctoral fellow IBM Rsch. Lab., Yorktown Hgts., N.Y., 1974-75; vis. scientist Hewlett-Packard Labs., Palo Alto, Calif., 1981-83; tech. staff mem. Hewlett-Packard, Med. Div., Boeblingen, Germany, 1983-85, project mgr., 1985—. Patentee in fiber optic sensor technology; contbr. articles to profl. jours. and encys. Recipient Postdoctoral fellowship IBM Corp., 1974. Mem. Deutsche Physikalische Gesellschaft, Optical Soc. Am. Avocations: violin, woodworking, hiking, gardening. Office: Hewlett Packard GmbH, PO Box 1430, D-7030 Böblingen Germany

FISCHER, EDMOND HENRI, biochemistry educator; b. Shanghai, Republic of China, Apr. 6, 1920; came to U.S., 1953; s. Oscar and Renée (Tapernoux) F.; m. Beverley B. Bullock. Lic. es Sciences Chimiques et Biologiques, U. Geneva, 1943, Diplome d'Ingenieur Chimiste, 1944, PhD, 1947; D (hon.), U. Montpellier, France, 1985, U. Basel, Switzerland, 1988, Med. Coll. of Ohio, 1993. Pvt. docent biochemistry U. Geneva, 1950-53; research assoc. biology Calif. Inst. Tech., Pasadena, 1953; asst. prof. biochemistry U. Wash., Seattle, 1953-56, assoc. prof., 1956-61, prof., 1961-90, prof. emeritus, 1990—; mem. exec. com. Pacific Slope Biochem. Conf., 1958-59, pres. 1975; mem. biochemistry study sect. NIH, 1959-64; symposium co-chmn. Battelle Seattle Research Ctr., 1970, 73, 78; mem. sci. adv. bd. Biozentrum, U. Basel, Switzerland, 1982-86; sci. adv. bd. Friedrich Miescher Inst., Ciba-Geigy, Basel, 1976-84, chmn. 1981-84. Contbr. numerous articles to sci. jours. Mem. sci. council on basic sci. Am. Heart Assn., 1977-80, sci. adv. com. Muscular Dystrophy Assn., 1980-88. Recipient Lederle Med. Faculty award, 1956-59, Guggenheim Found. award, 1963-64, Disting. Lectr. award U. Wash., 1983, Laureate Passano Found. award, 1988, Steven C. Beering award, 1991, Nobel Prize in Physiology or Medicine, 1992; NIH spl. fellow, 1963-64. Mem. AAAS, AAUP, Am. Soc. Biol. Chemists (council mem. 1980-83), Am. Chem. Soc. (biochemistry div., mem. adv. bd. 1962, exec. com. div. biology 1969-72, monography adv. bd. 1971-73, editorial adv. bd. Biochemistry Jour., 1961-66, assoc. editor 1966-92), Swiss Chem. Soc. (Werner medal), Brit. Biochem. Soc., Am. Acad. Arts and Scis., Nat. Acad. Scis., Sigma Xi. Achievements include life-long study of the metabolism of glycogen and its ATP/ADP cycle. Office: U Washington Med Sch Dept Biochemistry SJ70 Seattle WA 98195

FISCHER, ERNST OTTO, chemist, educator; b. Munich, Germany, Nov. 10, 1918; s. Karl T. and Valentine (Danzer) F. Diplom, Munich Tech. U., 1949, Dr. rer. nat., 1952, Habilitation, 1954, Dr. rer. nat. h.c., 1972, D.Sc.h.c., 1975, Dr. rer. nat. h.c., 1977, Dr.h.c., 1983. Assoc. prof. inorganic chemistry U. Munich, 1957, prof., 1959; prof. inorganic chemistry inst. Munich Inst. Tech., 1964—. Author: (with H. Werner) Metall-pi-Komplexe mit di- und oligoolefinischen Liganden, 1963; transl. Complexes with di- and oligo-olefinic Ligands, 1966; Contbr. (with H. Werner) numerous articles in field to profl. jours. Recipient ann. prize Göttingen Acad. Scis., 1957, Alfred Stock Meml. prize Soc. German Chemists, 1959, Nobel Prize in Chemistry, 1973; Am. Chem. Soc. Centennial fellow, 1976. Mem. Bavarian Acad. Scis., Soc. German Chemists, German Acad. Scis. Leopoldina, Austrian Acad. Scis. (corr.), Accademia Nazionale dei Lincei, Italy (fgn.), Acad. Scis. Göttingen (corr.), Am. Acad. Arts and Scis. (fgn., hon.), Chem. Soc. (hon.). Spl. research in organometallic chemistry: metal pi complexes of arenes, olefins, carbene and carbyne complexes with metals, ferrocene type sandwich compounds, metal carbonyls. Home: 16 Sohnckestrasse, 8000 Munich 71, Germany

FISCHER, GREGORY ROBERT, geotechnical engineer; b. Quincy, Ill., Oct. 3, 1963; s. Lee Frank and Frances Janet (Schneider) F.; m. Lyn Josephine Sagaser, June 28, 1986; 1 child, Michael Gregory. BS, U. Ill., 1984, MS, 1986. Staff engr. Shannon & Wilson, Inc., St. Louis, 1986-87; prin. engr. Shannon & Wilson, Inc., Seattle, 1990—; project engr. Converse Cons. NW, Seattle, 1987-90. Co-author: Geotextile Filter Design Based on Pore Size Distribution, 1990, Principles, Practices and Problems in Geotextile Filter Design, 1991, Comparative Study of Different Geotextile Perometric Testing Methods, 1992, Research Needs in Geotextile Filter Design, 1992. Fellow U. Ill., 1985, Valle scholar U. Wash., 1989. Mem. Internat. Geotextile Soc., North Am. Geosynthetic Soc., ASCE, Soc. Am. Mil. Engrs. Office: Shannon & Wilson Inc 400 N 34th St Seattle WA 98103

FISCHER, HARRY WILLIAM, radiologist, educator; b. St. Louis, June 4, 1921; s. Harry William and Amy Babette (Gieselman) F.; m. Kay Fischer, 1943; 5 children. BS, U. Chgo., 1943, MD, 1945; MD (hon.), Goteborg U., 1991. Diplomate: Am. Bd. Radiology. Asst. prof., then assoc. prof. radiology U. Ia. Med. Sch., 1956-63, prof., head sect. diagnostic radiology, 1963-66; prof. radiology U. Mich. Med. Sch., 1966-71; dir. dept. radiology Wayne County Gen. Hosp., 1966-71; prof. radiology, chmn. dept. U. Rochester (N.Y.) Sch. Medicine and Dentistry, 1971-85. Mem. editorial bd. Investigative Radiology, 1966-91, Radiology, 1971-90. Served to lt. (j.g.) M.C. USNR, 1946-48. Award for Excellence established in his name by Contrast Media Rsch. Soc., 1989. Fellow Am. Coll. Radiology; mem. Radiol. Soc. N.Am., Assn. Univ. Radiologists (Gold medal), Am. Roentgen Ray Soc., Uroradiology Soc., U. Chgo. Med. Sch. Alumni Assn.(Disting. Svc. award), Sigma Xi. Home: 405 Paseo Del Canto Green Valley AZ 85614-1715

FISCHER, JOHN GREGORY, mechanical engineer; b. Houston, Aug. 2, 1956; s. John Francis and Velda Marie (Kuntz) F.; m. Alicia Kay Prime, Mar. 19, 1983 (div. Dec. 1987); 1 child, Shawna. BSME, U. So. Calif., 1979; postgrad., So. Meth. U., 1991—. Registered profl. engr., Tex. Indsl. engr. McDonnell Douglas, Long Beach, Calif., 1979-80; regional tech. svc. engr. Reed Rock Bit Co., Oklahoma City, 1980-82; cons. engr. West Coast Supply, Long Beach, 1982-84; v.p. mktg., field engr. Am. Bit Co., Ontario, Calif., 1984-85; mgr. environ. engring. and security div. Dresser Industries, Dallas, 1985—. Health/fitness spokesman at local day care and elem. schs., Dallas, 1988—; mem. adv. com. Valley Ranch Master Assn., Irving, Tex., 1992. Mem. ASME, ABA, IADC (mem. subcom. 1990-91), Nat. Soc. Profl. Engrs., Tex. Soc. Profl. Engrs., Dallas Assn. Young Lawyers, Am. Soc. Metals Internat., Am. Powder Metallurgy Inst., Am. Assn. Drilling Engrs., Phi Alpha Delta. Roman Catholic. Achievements include patents for cutter mounting for drag bits, drag bit with improved cutter mount, photoelectric mensuration device and method of use, PDC drill bit with facetted wings, publication of intersection solution method for drill bit design. Home: 1100 Stone Gate Dr Irving TX 75063 Office: Dresser Industries Security Div 3400 W Illinois Ave Dallas TX 75211

FISCHER, LAWRENCE JOSEPH, toxicologist, educator; b. Chgo., Sept. 2, 1937; s. Lawrence J. and Virginia H. (Dieker) F.; m. Elizabeth Ann Dunphy, Oct. 24, 1964; children—Julie Ann, Pamela Jean, Karen Sue. B.Sc., U. Ill.-Chgo., 1959, M.S., 1961; Ph.D., U. Calif.-San Francisco, 1965. NIH postdoctoral fellow St. Mary's Hosp. Med. Sch., London, 1965-66; sr. research pharmacologist Merck Sharp and Dohme, West Point, Pa., 1966-68; asst. prof. pharmacology U. Iowa, Iowa City, 1969-73, assoc. prof., 1974-76, prof., 1976-85; prof., dir. Ctr. Inst. for Environ. Toxicology Mich. State U., East Lansing, Mich., 1985—; cons. FDA Bur. Vet. Medicine, 1974-77; mem. bd. scientific counselors div. of cancer Etiology Nat. Cancer Inst., 1986-92. Mem. editorial adv. bd. Jour. Pharmacology and Exptl. Therapeutics, Toxicology and Applied Pharmacology, Drug Metabolism Revs. Recipient Faculty Scholar award Josiah Macy Found., U. Geneva, 1976. Mem. Am. Soc. for Pharmacology and Exptl. Therapeutics, Soc. Toxicology, AAAS, Soc. for Environ. Toxicology and Chemistry. Avocations: hunting upland birds, tennis. Home: 11630 Center Rd Bath MI 48808-9431 Office: Mich State U Inst for Environ Toxicology C231 Holden Hall East Lansing MI 48824

FISCHER, MARSHA LEIGH, civil engineer; b. San Antonio, May 9, 1955; d. Joe Henry and Ellen Joyce (Flake) F. BSCE, Tex. A&M U., 1977. Engring. asst. Tex. Dept. Hwys. and Transp., Dallas, 1977-79; outside plant engr. Southwestern Bell Telephone Co., Dallas, 1979-82, staff mgr. for budgets, 1982-84; area mgr., engring. design Southwestern Bell Telephone Co., Wichita Falls, Tex., 1984-86; area mgr. Southwestern Bell Telephone Co., Ft. Worth, 1986-88; dist. mgr., local provisioning application Bell Communications Rsch., Piscataway, N.J., 1988-91; dist. mgr. engring. Southwestern Bell Telephone, Ft. Worth, Tex., 1992—. Named one of Outstanding Women of Am., 1987. Mem. NSPE, Tex. Soc. Profl. Engrs., Tex. Soc. Civil Engrs., Profl. Engrs. in Industry, Tex. A&M Assn. Former Students. Republican. Avocations: tennis, travel, reading, cycling. Home: 6724 Johns Ct Arlington TX 76016 Office: Southwestern Bell Telephone 1116 Houston Rm 737 Fort Worth TX 76102

FISCHER, MICHAEL JOHN, ophthalmologist, physician; b. Norwood, Mass., July 28, 1948; s. John Edwin and Patricia (Murphy) F.; m. Linda K. Stertzer, Aug. 5, 1972; children: Jessica Lynn, Erica Leigh, Rebecca Lynn, Evan Michael. BS, Ohio State U., 1972; MD, Univ Autonoma de Guadalajara, Mex., 1976. Orthopedic resident St. Luke's Hosp., Cleveland, 1978-80; ophthalmology resident Mt. Sinai Med. Ctr., Cleveland, 1980-83; staff Carson-Tahoe Hosp., Carson City, Nev., 1983—; mem. Am. Acad. of Ophthalmology, San Francisco, 1985—; adminstr. Eye Surgery Ctr. of Nev., Carson City, Nev., 1985—; pres. Nev. Ophthalomogical Soc., Nev., 1988-92; clinical instr. U. Nev. Sch. of Med., Reno, 1989—; pres. Carson Douglas Med. Soc., Carson City, Nev., 1989-90; chmn. Nev. Peer Review Coun., Nev., 1990-91. Mem. Lion's Club Internat. Roman Catholic. Office: 3839 N Carson St Carson City NV 89706-1935

FISCHER, PETER, research immunologist; b. Frankfurt, Federal Republic Germany, July 26, 1959; came to U.S., 1991; s. Herbert and Anni F. Diploma, J.W. Goethe U., Frankfurt, Republic of Germany, 1984, PhD, 1990. Postdoctoral assoc. U. Frankfurt, 1990; rsch. fellow Scripps Rsch. Inst., La Jolla, Calif., 1991-92; rsch. immunologist U. Calif. at San Diego, La Jolla, 1993—. Contbr. articles to profl. jours. Liebig fellowship Stift. Stipendien Fonds, 1990. Mem. AACR, Ges für Biol. Chemie, Ges. Für Immunologie. Achievements include rsch. on monclonal antibody reactor with GGT from renal carcinomas. Office: U Calif at San Diego 9500 Gilman Dr La Jolla CA 92093

FISCHER, ROBERT LEE, electrical engineering executive, educator; b. Huntington, W.Va., Feb. 4, 1947; s. Charles Lee and Frances Louise (Pennington) F.; m. Mona Lynn Reeser, Oct. 27, 1966; children: Robert Lee Jr., Amy Lynn, Cory Brandon. Cert. in electronics tech., Huntington East Vocat. Tech., 1965; BA in Physics and Gen. Sci., Marshall U., 1970, MS in Vocat. Tech. Edn., 1976; PhD in Elec. Engring., Kennedy-Western U., 1993. Registered profl. electrical engr.; lic master electrician; cert. plant engr. Electrical engr. J.F. & M. Co., Huntington, 1970-71; electronics prodn. supr. polan ind. div. Wollensak, Inc., Huntington, 1971-72; electrical maintenance supr. ACF Industries, Inc., Huntington, 1972-76, electrical maintenance supt., 1976-78, sr. maintenance engr., 1978-80, plant engr., 1980-84, mgr. plant, prodn. and tooling engring., 1984-85; engr., prin. cons. Fischer Tech. Svcs., Huntington, 1979—; electrical, instrumentation and utilities mgr. Calgon Carbon Corp., Catlettsburg, Ky., 1985-93, maintenance mgr., 1993—; robotics instr. Marshall U. Community and Tech. Coll., Huntington, 1986—; instrumentation and control engring. curriculum adv. com. Shawnee State U., Portsmouth, Ohio, 1985—. patentee electronic height control device, robot safety mechanism. Elected to West Jr. High Sch. Hall of Fame, Huntington, 1988; recipient Sr.-Under Black Belt-Open 3d Place award United Fighting Arts Fedn. Nat. Karate Tournament, 1984; named W.Va. ambassador of sci. and engring. among all people, 1982. Mem. NSPE, W.Va. Soc. Profl. Engrs., W.Va., Acad. Sci., Ohio Valley Astron. Soc., Am. Radio Relay League, Six Meter Internat. Radio Club. Democrat. Avocations: ham radio, martial arts, amateur astronomy. Home: 3606 Route 75 Huntington WV 25704 Office: Calgon Carbon Corp PO Box 664 Catlettsburg KY 41129-0664

FISCHER LINDAHL, KIRSTEN, biologist, educator; b. Rønne, Denmark, Aug. 11, 1948; d. Poul Fischer and Bodil (Jørgensen) F.; m. Johann Diesenhofer, June 19, 1989. MSc, U. Copenhagen, 1973; PhD, U. Wis., 1975. Postdoctoral fellow Inst. Genetics U. Cologne, Fed. Republic Germany, 1976-78; mem. staff Basel (Switzerland) Inst. for Immunology, 1978-85; assoc. prof. microbiology and biochemistry U. Tex. Southwestern Med. Ctr., Dallas, 1985-92, prof. microbiology and biochemistry, 1992—; investigator Howard Hughes Med. Inst., Chevy Chase, Md., 1985—; reviewer immunobiology study sect. NIH, Bethesda, 1989—; reviewer eukaryotic genetics panel NSF, Washington, 1986-89. Mem. editorial bd. Trends in Genetics, Tissue Antigens, Internat. Immunology, Mamalian Genome, Immunogenetics, 1983—; contbr. over 90 articles to profl. publs. European Molecular Biol. Orgn. fellow, 1976-78. Mem. Am. Assn. Immunologists, Am. Soc. Cell Biology, Genetics Soc. Am., Brit. Soc. Immunology. Achievements include discovery and molecular definition of maternally transmitted antigen of mice, antigen presentation function of non-classical major histocompatibility antigens, recombinational hotspots in mice. Office: Howard Hughes Med Inst 5323 Harry Hines Blvd Dallas TX 75235-9050

FISCHETTE, MICHAEL THOMAS, engineering executive; b. Jersey City, N.J., Apr. 3, 1957; s. Carmen and Concetta (Tecchio) F.; m. Stephanie Balzano, Oct. 27, 1984. BSME, Rutgers U. Engr. Stone & Webster Engring. Corp., Cherry Hill, N.J., 1980-86; pres. MTF Inc., Sewell, N.J., 1986-89; co-pres. Concord Engring. Group Inc., Voorhees, N.J., 1989—. Mem. ASME, ASHRAE, Assn. Energy Engrs. Home: 6 Hamilton Ct Southampton NJ 08088 Office: Concord Engring 1200 Laurel Oak Rd Ste 195 Voorhees NJ 08043

FISCHHOFF, BARUCH, psychologist, educator; b. Detroit, Apr. 21, 1946; s. Henry and Shirley (Levine) F.; m. Andrea Marks, Dec. 22, 1968; children: Maya, Ilya, Noam. BS in Math., Wayne State U., 1967; MA in Psychology, Hebrew U., Jerusalem, 1972, PhD in Psychology, 1975. Rsch. assoc. Oreg. Rsch. Inst., Eugene, 1974-76, Decision Rsch., Eugene, 1976-85, Applied Psychology Unit Med. Rsch. Coun., Cambridge, Eng., 1981-82, Eugene Rsch. Inst., 1985-87; prof. Carnegie-Mellon U., Pitts., 1987—; vis. prof. U. Stockholm, 1982-83; mem. panels NRC; cons. in field. Author: Acceptable Risk, 1981; mem. editorial bd. jours. Cognitive Psychology, Policy Scis., Risk, others; contbr. numerous articles to profl. jours. Mem. Eugene Commn. on Rights of Women, 1975-81; pres. Eugene Human Rights Coun., 1979-81. Fellow Soc. for Risk Analysis (Disting. Achievement award 1991); mem. APA (Disting. Sci. award 1981, Psychology in Pub. Interest award 1991), Soc. Judgment and Decision-Making (coun. mem. 1988—, pres. 1990-91), Inst. of Medicine, Phi Beta Kappa. Home: 1437 Denniston Ave Pittsburgh PA 15217-1332 Office: Carnegie Mellon U Dept Engring & Pub Policy Pittsburgh PA 15213-3890

FISCHL, MYRON ARTHUR, psychologist; b. N.Y.C., Oct. 13, 1929; s. Louis Frank and Sally (Maeth) F.; m. Suzanne Dona Borowsky, June 21, 1958; children: Jeffrey, Sally, Amy. BA, NYU, 1950; MA, Hofstra U., 1954; PhD, Purdue U., 1956. Diplomate Am. Bd. Profl. Psychology. Sr. project dir. W.R. Simmons & Assocs., Inc., N.Y.C., 1956-57; sr. scientist The Matrix Corp., Arlington, Va., 1957-58, 59-63; human factors scientist GE Co., Phila., 1958-59; dir. pers. rsch. Nat. Analysts, Inc., Phila., 1963-65; lectr. Villanova (Pa.) U., 1966-71; rsch. assoc. Applied Psychol. Svcs., Wayne, Pa., 1965-71; various positions U.S. Army Rsch. Inst., Alexandria, Va., 1971-86; lectr. U. Md. U. Coll., College Park, 1975—; pers. psychologist U.S. Army Hdqrs., Pentagon, Washington, 1986—; co-chair Mil. Testing Assn. Ann. Meeting, Alexandria, 1981. Contbr. articles to profl. jours. including Jour. of Applied Psychology, Human Factors, Applied Psychol. Measurement, Armed Forces and Soc., Youth and Soc., Organizational Behavior and Human Performance. Pres. Brotherhood, Beth El Hebrew Cong., 1983-85, Tantallon South Homeowners Assn., Fort Washington, Md., 1981-83. With U.S. Army, 1951-52. Fellow APA (fellow, chair div. mil. psychology, mem. fellows com. 1989-93); mem. Soc. Indsl. and Organizational Psychology (membership com. 1981-82, chair membership com. 1983-85, mem. div. evaluation and measurement, div. applied exptl. and engring. psychologists), Am. Statis. Assn. Home: 317 Rexburg Ave Fort Washington MD 20744 Office: US Army Hdqrs - Office of Dep Chief of Staff Pers The Pentagon Washington DC 20310-0300

FISH, FALK, microbiologist, immunologist, researcher, inventor; b. Rehovoth, Israel, July 10, 1946; s. Aharon and Rivka (Halperin) F.; m. Tamar David, Aug. 18, 1975; children: Shlomi, Michal, Noah. MSc, Tel Aviv U., 1974, PhD, 1979. Faculty assoc. U. Tex. Health Sci. Ctr., Dallas, 1979-81; vis. fellow Bur. Biologics, FDA, Bethesda, Md., 1981-82; asst. prof. Tel Aviv U., 1982-88; co-founder, dir. R & D Orgenics Ltd., Yavne, Israel, 1985-89, v.p. tech., 1989—. Contbr. articles to profl. jours. Mem. Am. Assn. Immunologists, Am. Soc. for Microbiology, Israel Immunological Soc., Israel Microbiol. Soc. Achievements include patents for method for non-radioactive DNA labeling and its use in diagnosis, isolation of specific DNAs, enhanced enzyme immunoassay format and kit, rapid method and kit for detection of metabolic toxicity, novel molded strip format for blood tests. Home: 4 Dresner St Apt 25, Tel Aviv 69497, Israel Office: Orgenics Ltd, North Industrial Zone, Yavne 70650, Israel

FISH, JOHN PERRY, oceanographic company executive, historian; b. Boston, Jan. 13, 1949; s. Robert Story and Sylvia Colby (Draper) F.; m. Marjorie Ann Moore, May 15, 1982; children: Madelyn Moore, Colby Draper. AS, Boston U., 1970; BA in Biology, Windham Coll., 1972. Mktg. assoc. Benthos, Inc., N. Falmouth, Mass., 1975-81; ops. dir. Hydrolab Project/NOAA-FDU, St Croix, V.I., 1981-82; pres. Oceanstar Systems, Inc., Cataumet, Mass., 1983—; v.p. Am. Underwater Search and Survey, Ltd., Cataumet, 1986—; bd. dirs. Hist. Maritime Group New Eng., Cataumet, Draper Bros. Co., Canton, Mass.; archael. cons. Debraak Recovery Project, Lewes, Del., 1984; ambassador Nat. Assn. Underwater Instrs., Colton, Calif., 1983. Author: Unfinished Voyages, 1989, Sound Underwater Images: Guide to Generation of Side Scan Sonar Data, 1990, Discrete Object Recognition in Side Scan Sonar Data, 1992; contbr. articles to profl. jours. Recipient Recognition award Nat. Oceanic and Atmospheric Adminstrn., 1986. Mem. Marine Tech. Soc., Soc. Colonial Wars, U.S. Naval Inst., Woods Hole Oceanographic Inst.

FISHBINE, BRIAN HOWARD, physicist; b. Los Alamos, N.Mex., Aug. 11, 1948; s. Harold Louis and Blanche Helene (Torres) F. BS, U. N.Mex., 1970, PhD, 1984. Analyst programmer; anayst/programmer Quantum Systems, Inc., Albuquerque, 1979-80; rsch. asst. dept. chem. and nuclear engring. U. N.Mex., Albuquerque, 1980-84; rsch. asst. prof. Inst. for Accelerator and Plasma Beam Tech., U. N.Mex., Albuquerque, 1984-85; adj. rsch. prof. dept. civil engring. U. N.Mex., Albuquerque, 1988-91; rsch. physicist Intellisys Corp., Albuquerque, 1987-90; sr. rsch. assoc. NRC Phillips Lab., Albuquerque, N.Mex., 1993—; cons. Digital Biometrics, Inc., Minnetonka, Minn., 1985—; pres., founder Duke City Sci., Inc., Albuquerque, 1990—. Contbr. articles to profl. jours. Mem. AAAS, Am. Phys. Soc., Microbeam Analysis Soc., Nat. Space Soc., N.Y. Acad. Scis. Achievements include two patents on system for generating rolled fingerprint images; development of high framing rate scanning electron microscope; development in situ dynamic loading device for scanning electron microscope. Home: 2384 35th St Los Alamos NM 87544-2004

FISHE, GERALD RAYMOND AYLMER, engineering executive; b. Farnham Royal, Eng., Feb. 22, 1926; s. Daniel Hamilton and Dorothy Vida (Norton) F.; m. Patricia Ann Roach, Aug. 18, 1949; children: Martha Vida Bindshedler, Raymond Patrick Hamilton, G. Keith Hamilton. BS in Mech. Engring., Duke U., 1949. Registered profl. engr., Ala., Fla., Ga., Ill., Iowa, Mo.; Tenn. and W.Va. (inactive). Project engr. E.I. DuPont de Nemours & Co., Martinsville, Va., 1952-58; architect's staff engr. So. Ill. U., Carbondale, 1958-63; sec. Adair Brady & Fishe, Inc., Lake Worth, Fla., 1965-66; chief engr. Gamble Pownall & Gilroy, Ft. Lauderdale, Fla., 1963-65; cons. forensic engr. Ft. Lauderdale, 1966—; pres. Fishe & Kleeman, Inc., Ft. Lauderdale, 1974—, Fidelity Inspection and Svc. Co., Ft. Lauderdale, 1983-91, Farletot Found., Inc., Ft. Lauderdale, 1985-91. Patentee in field. With U.S. Army, 1944-45. Fellow Nat. Acad. Forensic Engrs., Am. Acad. Forensic Scis. (chmn. engring. sect. 1988-89, bd. dirs. 1990-93); mem. ASHRAE, Constrn. Specification Inst., Nat. Fire Protection Assn. Republican. Episcopalian. Home: 2031 SW 36th Ave Fort Lauderdale FL 33312-4208 Office: GRA Fishe Cons Engr 601 S Andrews Ave Fort Lauderdale FL 33301-2800

FISHER, ANNA LEE, physician, astronaut; b. St. Albans, N.Y., Aug. 24, 1949; m. William Frederick Fisher; children: Kristin Anne, Kara Lynne. B.S. in Chemistry, UCLA, 1971, M.D., 1976, M.S. in Chemistry, 1987. Physician, 1976-78; astronaut NASA Johnson Space Ctr., Houston, 1978—, mission specialist STS, 51-A, 1984. Mem. Sigma Xi. Office: NASA Johnson Space Ctr Astronaut Office Houston TX 77058

FISHER, ARON BAER, physiology and medicine educator; b. Phila., Apr. 20, 1936; m. Joan C. Fisher, 1957; children: Marc L., Steven A., Eric R., Mara E. BS in Chemistry summa cum laude, Dickinson Coll., 1956; MD, U. Pa., 1960. Diplomate Am. Bd. Internal Medicine; diplomate Nat. Bd. Med. Examiners. Intern and resident in medicine U. Hosps., Cleve., 1960-61, 64-65; resident in pulmonary medicine Hosp. U. Pa., 1965-66; fellow dept. physiology U. Pa., 1966-68, assoc. in medicine, assoc. in physiology, 1968-70, from asst. prof. to assoc. prof. medicine, 1970-80, prof. medicine, 1980—, from asst. prof. to assoc. prof. physiology, 1970-1980, prof. physiology, 1980—, prof. environmental medicine, 1986—; staff physician VA Hosp., Phila., 1968-73, clin. investigator, 1973-76, cons. in pulmonary medicine, 1976-82; mem. med. staff Hosp. U. Pa., 1976—, dir. hyperbaric medicine clin. practice, 1985—; dir. Inst. Environ. Medicine U. Pa., 1985—; mem. Am. Heart Assn. student rsch. fellowship adv. com. U. Pa., 1983—, mem. diabetes ctr. adv. com., 1985—, mem. teaching awards com., 1989—, chmn. animal care com. 1982-84, 87-89, chmn. com. for animal facility planning, 1985-86, chmn. transgenic mouse facility com., 1989, chmn. instnl. animal care and use com., 1989—, mem. bioengring. grad. group, 1988—, chmn. biochemistry grad. group rev. com., 1989-90, others, supr. grad. students; fellow dept. biophysics and phys. chemistry U. Pa., 1971-72; mem. study sect. Pa. Coal Worker's Respiratory Disease Program, 1976-78; mem. cardiovascular study sect. A NIH, 1979-81, mem. respiratory and applied physiology sect., 1981-83; mem. adv. panel U.S. Army Med. R&D Command, 1980-85. Editor: (with others) Handbook of Physiology: The Respiratory System (Section 3), vol. 1, 1980-85; mem. editorial bd. Exptl. Lung Rsch. 1979-88, Am. Rev. Respiratory Diseases, 1981-87, Jour. Applied Physiology, 1984-87, Am. Jour. Physiology, 1988—; guest editor Symposium on Lung Surfactant Apoproteins, 1984; contbr. numerous articles and revs. to profl. jours., chpts. to books. With USPHS, 1958, 59-61; capt. MC USAR, 1961-65. Grantee NIH, 1986-91, 1988—; recipient Clin. Investigator award VA Res. Svc., 1973-76, Established Investigator award Am. Heart

Assn., 1977-82, Christian R. and Mary F. Lindback Found. award for Disting. Teaching, 1984. Mem. AAAS, ACP, Am. Physiol. Soc. (chmn. respiration dinner 1991, councillor respiratory sect. 1991—), Am. Thoracic Soc. (sec. assembly on structure, function and metabolism 1973-74, chmn. 1981, sec. sect. on pulmonary circulation 1979, councillor ea. sect. 1973-77, chmn. ann. meeting program com. 1976, pres. 1983), Am. Fedn. Clin. Rsch., Am. Soc. Clin. Investigation, Am. Heart Assn. (cardiopulmonary coun.), Am. Soc. Cell Biology, Undersea and Hyperbaric Med. Soc., Oxygen Soc., Aerospace Med. Assn., John Morgan Soc. U. Pa., Laennec Soc. Phila., Pa. Thoracic Soc. (chmn. rsch. com. 1985-87), Phi Beta Kappa, Alpha Omega Alpha. Achievements include co-determination that lung lamellar bodies maintain an acidic internal pH, that phospholipids co-isolated with rat surfactant protein-C account for the apparent protein-enhanced uptake of liposomes into lung granular pneumocytes, that secretogues for lung surfactant increase lung uptake of alveolar phospholipids, that cAMP increases synthesis of surfactant-associated protein A by perfused rat lung; research on secretory granule calcium loss after isolation of rat alveolar type II cells, on alveolar uptake of lipid and protein components of surfactant, on oxygen-dependent peroxidation during lung ischemia, on choline transport by lung epithelium, and on role of acidic compartment in synthesis of disaturated phosphatidylcholine by rat granular pneumocytes. Home: 239 E Gowen Ave Philadelphia PA 19119-1021 Office: U Pa Inst Environ Medicine One John Morgan Bldg 36th St and Hamilton Walk Philadelphia PA 19104-6068

FISHER, BARBARA TURK, school psychologist; b. Bklyn., Feb. 21, 1940; d. Jack and Reva (Miller) Turk; m. Ronnie Herbert Fisher, Aug. 15, 1961; children: Sylvia Kay, Mark Lee. BA, Fla. State U., 1961; MS, Barry U., 1966, EdS, 1977; EdD, Nova U., 1989. Coord. of counseling Immaculate La Salle High Sch., Miami, Fla., 1966-69; rehab. counselor Miami Adult Tng. Ctr., 1969-70; sch. psychologist Dade County Sch. System, Miami, 1970—; instr. Miami Dade Community Coll., 1991; presenter at profl. assn. meetings and social sci. seminars, 1993. Fellow AAUW; mem. Dade County Asns. Sch. Psychologists. Republican. Jewish. Avocations: reading, classical music, needlepoint. Home: 234 Antiquera Ave Apt 3 Miami FL 33134-2914 Office: Dade County BPI Region I Office 733 E 57th St Hialeah FL 33013-1357

FISHER, BERNARD, surgeon, researcher, educator; b. Pitts., Aug. 23, 1918; s. Reuben and Anna (Miller) F.; m. Shirley Kruman, June 5, 1947; children: Beth, Joseph, Louisa. BS, U. Pitts., 1940, MD, 1943; DSc, Mt. Sinai Sch. Medicine CCNY, 1986. Diplomate Am. Bd. Surgery. Intern Mercy Hosp., Pitts., 1943-44, resident in surgery, 1944-48; fellow in surg. research, resident in gen. surgery Harrison Dept. Dept. Surg. Research U. Pa., Phila., 1950-52; fellow London Postgrad. Med. Sch. Hammersmith Hosp., 1955-56; teaching fellow in pathology U. Pitts., 1944-45, teaching fellow in surgery, 1945-47, assoc. prof., 1956-59, prof. surgery, 1959-86, Disting. Svc. prof., 1986—; med. surg. staff Presbyn.-Univ. Hosp.; mem. staff Children's Hosp., Pitts.; cons. staff Magee-Women's Hosp., VA Hosp., Pitts.; chmn. Nat. Surg. Adjuvant Breast and Bowel Project, 1967—, Adjuvant Therapy Ctr., 1973—, Breast Care and Diagnostic Ctr., 1980, Pitts. Cancer Inst., 1985—, Comprehensive Breast Care Ctr., 1992—; mem. spl. del. People's Republic of China, 1977; mem. Pres.'s Cancer Panel, 1979-82, Nat. Cancer Adv. Bd., 1986-92, Inst. of Medicine, NAS. Mem. editorial bd. Transplantation, 1966-71, Cancer, 1969-73, 75, Year Book of Cancer, 1973-85, Internat. Jour. Radiation Oncology Biology Physics, 1975-78, Cancer Clinical Trials, 1977, Invasion and Metastasis, 1981, Cancer Metastasis Revs., 1981-85, Jour. Clin. Oncology, 1982-87, Internat. Jour. Breast and Mammary Pathology, 1982-84, Cancer Research, 1976—, Breast Diseases, 1992—, Annals of Surgical Oncology, 1993—, Internat. Jour. Oncology, 1993—; mem. editorial adv. bd. Seminars in Oncology, 1979, Breast Cancer Research and Treatment, 1980—; contbr. 477 articles to profl. jours. Bd. dirs. Allegheny County Am. Cancer Soc. Recipient Man of Yr. award in medicine Pitts. Jr. C. of C., 1966, Philip Hench Disting. Alumnus award U. Pitts. Sch. Medicine, , 1976, McGraw medal Detroit Surg. Assn., 1978, Lucy Wortham James Clin. Research award, 1981, Heath Meml. award, 1982, Joseph H. Morton Meml. award, 1983, Julia Hudson Freund Meml. award. 1983, Albert Lasker Med. Research award, 1985, Hammer Cancer prize 1988, Am. Cancer Soc. Medal of Honor, 1986, Milken Med. Found. Ctr. Rsch. award, 1989, Assn. Commn. Cancer Ctrs. award, 1990, Nat. Health Couns. Med. Rsch. award, 1992, Brinker Internat. Breast Cancer award, 1992, Durham N.C. city of Medicine award, 1992, Dr. Josef Steiner Cancer Rsch. prize, 1992, GM Cancer Rsch. Found. Kettering prize, 1993, Susan Komen Found. Sci. Distinction award, Bristol-Myers Squib Co. Unrestricted Med. Grants Program, 1993; Markle scholar in med. sci. John and Mary Markle Found., 1953-58; Alpha Omega Alpha, 1989, Fisher Breast Cancer lectureship established in his honor U. Pitts., 1989, Bernard Fisher ICI-Pharma professorhip in surgery, 1989. Fellow AAAS; mem. AAUP, ACS, AMA, Assn. Cancer Edn., Am. Assn. Cancer Research (bd. dirs.), Am. Soc. Clin. Oncology (pres. 1992-93, bd. dirs., Karnofsky award 1980), Am. Physiol. Soc., Assn. Am. Med. Colls., Cell Kinetic Soc., Am. Surg. Soc., N.Y. Acad. Scis., Soc. Surg. Oncology, Soc. Univ. Surgeons, Am. Socs. for Exptl. Biology, Pa. Med. Soc., Allegheny County Med. Soc. (Man of Yr. award 1983), Pitts. Acad. Medicine, Pitts. Surg. Soc. (pres. 1979), Peruvian Acad. Surgery (hon.), Italian Surg. Research Assn., Assn. Italiana per la Divulgaxione Sci. della Cancerologia Clinica, Internat. Assn. Breast Cancer Research, Am. Italian Fedn. Cancer Research, Phi Beta Kappa. Office: U Pitts Sch Medicine Rm 914 Scaife Hall 3550 Terrace St Pittsburgh PA 15213-2500

FISHER, CHARLES HAROLD, chemistry educator, researcher; b. Hiawatha, W.Va., Nov. 20, 1906; s. Lawrence D. and Mary (Akers) F.; m. Elizabeth Dye, Nov. 4, 1933 (dec. 1967); m. Lois Carlin, July 1968 (dec. June 1990); m. Elizabeth Snyder Kiser, Nov. 29, 1991. BS in Chemistry, Roanoke Coll., 1928, ScD (hon.), 1963; MS in Chemistry, U. Ill., 1929, PhD, 1932; DSc (hon.), Tulane U., 1953. Teaching asst. in chemistry U. Ill., Urbana, 1928-32; instr. Harvard U., 1932-35; rsch. group leader U.S. Bur. Mines, Pitts., 1935-40; head carbohydrate div. Ea. Regional Rsch. Lab. Eastern Regional Rsch. Ctr. USDA, 1940-50; dir. So. mktg. and nutrition rsch. div. So. Regional Rsch. Ctr., USDA, New Orleans, 1950-72; adj. rsch. prof. Roanoke Coll., Salem, Va., 1972—; established The Elizabeth Snyder Fisher Scholarship, Roanoke Coll., 1992. Co-author: Profiles of Eminent American Chemists, 1988; contbr. over 200 articles to profl. jours. Co-inventor 72 patents. Pres. New Orleans Sci. Fair, 1967-69; bd. dirs. Salem Hist. Soc., 1982-85, Salem Ednl. Found., 1991—; established Lawrence D. and Mary A. Fisher Scholarship Roanoke Coll., 1978, Lois Carlin Fisher Scholarship, 1991. Recipient So. Chemists award, 1956, Herty medal, 1959, Chem. Pioneer award Am. Inst. Chemists, 1966; named Polymer Science Pioneer, 1981; The Charles H. Fisher Lecture established in his honor Roanoke Coll., 1990. Mem. AAAS, Am. Inst. Chemists (hon., pres. 1962-63, chmn. bd. dirs., Presdl. citation of merit, 1986), Oil Chem. Soc., Am. Chem. Soc. (dir. region IV 1969-71), Chemurgic Council (dir.), Am. Assn. Textile Chemists and Colorists, Sigma Xi, Alpha Chi Sigma, Gamma Alpha, Phi Lambda Upsilon. Club: Cosmos, (Washington); Internat. House, Round Table (New Orleans); Chemists (N.Y.C.). Achievements include co-invention of acrylic rubber. Office: Roanoke Coll Chemistry Dept Salem VA 24153

FISHER, CHARLES RAYMOND, JR., marine biologist; b. Ann Arbor, Mich., Sept. 6, 1954; s. Charles Raymond and Mary Jane (Fyke) F.; m. Deborah Diane Doca, Aug. 12, 1977; 1 child, Lisann Mary. BS, Mich. State U., 1976; MA, U. Calif., Santa Barbara, 1981, PhD, 1985. Rsch. asst. U. Calif., Santa Barbara, 1982-84, postdoctoral rsch. biologist, 1984-86, asst. rsch. biologist, 1987-90; asst. prof. Pa. State U., University Park, 1990—; mem. steering com. NSF RIDGE program, Washington, 1992—. Contbr. articles and revs. to profl. jours. Grantee Mineral Mgmt. Svc., 1991, NSF, 1991, Nat. Oceanographic Atmosphere Adminstrn., 1987-93, Office Naval Rsch., 1988; recipient NSF Presdl. Young Investigator award, 1991. Mem. AAAS, Am. Soc. Limnologists and Oceanographers, Am. Soc. Zoologists, Oceanography Soc. Achievements include rsch. in chemoautotrophic and methanotrophic symbioses; demonstration of symbioses and dual methanotrophic/chemoautotrophic symbioses, description of nutrient exch. in chemoautotrophic symbiosis. Office: Pa State U 208 Mueller Lab University Park PA 16802

FISHER, DALE DUNBAR, animal scientist, dairy nutritionist; b. Lewisburg, Pa., Feb. 13, 1945; s. Glenn Murray and Elsie May (Bryson) F.; divorced; children: Elsie Maria, Maria Vanessa. BS in Animal Sci., Pa. State U., 1967, MS in Animal Industry, 1978, PhD in Animal Industry, 1980. Vol. animal husbandry Peace Corps, Ciudad Quesada, Costa Rica, 1967-71; area animal husbandry-pasture specialist Costa Rican Ministry of Agr., Ciudad Quesada, 1971-73; vis. scientist Internat. Ctr. for Tropical Agr., Cali, Colombia, 1973-75; animal nutritionist Co-op. Feed Dealers, Inc., Chenango Bridge, N.Y., 1981—. Contbr. articles to profl. jours. Eva B. and G. Weidman Groff Meml. scholar Pa. State U., 1979. Mem. Am. Soc. Animal Sci., Am. Dairy Sci. Assn., Am. Soc. Agronomy, Am. Acad. Vet. Nutrition, N.Y. Acad. Scis., Sigma Xi, Phi Kappa Phi, Gamma Sigma Delta. Democrat. Avocations: jogging, reading. Home: 578 Chenango St Binghamton NY 13901-2134 Office: Coop Feed Dealers Inc PO Box 670 Chenango Bridge NY 13745-0610

FISHER, DANIEL ROBERT, consultant; b. Phila., May 23, 1925; s. John Schaefer and Dorothy May (Gear) F.; m. Jean Mahar, Mar. 20, 1924 (dec. Feb. 1957); m. Joan Francis Gauff, July 23, 1977; children: Derek, Lynne, Marc, Foster, Susan. BME, Rensselaer Poly. Inst., 1945, MME, 1949; MBA with distinction, Harvard U., 1955. Prodn. mgr. Raytheon Mfg. Co., Andover, Mass., 1955-61; v.p., gen. mgr. Sperry Gyroscope, Great Neck, N.Y., 1961-65; v.p., div. mgr. Magnavox Co., Ft. Wayne, Ind., 1965-69; group v.p. Sanders Assocs., Nashua, N.H., 1969-71; pres. Kollsman Instruments, Nashua, 1971-75, ECRM, Inc., Bedford, Mass., 1975-82; cons. DRF Assocs., Bald Peak and Washington, 1982—. Troop master Boy Scouts Am., Andover, 1957-69; v.p. North Essex Coun., Andover, 1975-62. Lt. (j.g.) USN, 1942-46, PTO. Baker scholar Harvard U. Mem. ASME (sr., Postgrad. award 1949). Republican. Congregationalist. Achievements include research on the effects of evaporative cooling on a compressible fluid. Home: Bald Peak Colony Club Melvin Village NH 03850 Office: DRFA Cons 1001 Wilson Blvd Rosslyn VA 22209

FISHER, DARRELL REED, medical physicist, researcher; b. Salt Lake City, Feb. 12, 1951; s. Wayne Ekman and Zina Vae (Moore) F.; m. Anna Jeanetta Thomas, June 23, 1988; children; Bryan, Jenny, Aaron, Brit, Jake, Kimberly. BA in Biology, U. Utah, 1975; MS, U. Fla., 1976, PhD in Nuclear Enging. Scis., 1978. Rsch. asst. Radiobiology Lab., U. Utah, Salt Lake City, 1973-75; grad. asst. Nuclear Enging, U. Fla., Gainesville, 1975-78; rsch. scientist Battelle, Pacific N.W. Labs., Richland, Wash., 1978-80; sr. rsch. scientist Battelle, Pacific N.W. Labs., Richland, 1980—, tech. group leader, 1991-93, staff scientist, 1993—; affiliate asst. prof. U. Wash. Sch. Medicine, Seattle, 1991—; adj. assoc. prof. Wash. State u., Richland, 1992—; affiliate investigator Fred Hutchinson Cancer Rsch. Ctr., Seattle, 1993—, cons., 1987—, Div. Nuclear Medicine, U. Wash., Seattle, 1986—, NeoRx Corp., Seattle, 1987—. Editor: Current Concepts in Lung Dosimetry, 1983, Inhaled Particles VI, 1988, Population Exposure from the Nuclear Fuel Cycle, 1988; assoc. editor Jour. Health Physics, 1985-91, Antibody Immunoconjugates and Pharmaceuticals, 1991—. Cert. arbitrator Better Bus. Bur. Ea. Wash., Yakima, 1981—. Mem. Am. Assn. Physicists in Medicine, Health Physics Soc. (Elda E. Anderson award 1986, bd. dirs. 1988-93, pres. Columbia chpt. 1983-84), Radiation Rsch. Soc. Republican. Mem. LDS Ch. Achievements include rsch. on dosimetry and biol. effects of internally deposited radionuclides, uranium toxicology, internal dosimetry of radiolabeled antibodies in cancer therapy, and dosimetry and effects of plutonium. Home: 229 Saint St Richland WA 99352 Office: Battelle Pacific NW Labs Health Physics Dept Battelle Blvd Richland WA 99352

FISHER, DAVID GEORGE, physics educator; b. Pottsville, Pa., Aug. 1, 1955; s. Harvey Kermit and Joan Leah (Kunkel) F. BS, Pa. State U., 1977; MS, U. Del., 1980, PhD, 1983. Postdoctoral fellow Bartol Rsch. Found. of Franklin Inst., Newark, Del., 1983-84; assoc. prof., chmn. dept. physics and astronomy Lycoming Coll., Williamsport, Pa., 1984—. Contbr. articles to profl. publs. and to book. Mem. Am. Phys. Soc., Nat. Space Soc., Planetary Soc., Soc. Vertebrate Paleontology, Phi Beta Kappa, Sigma Pi Sigma, Pi Mu Epsilon, Phi Kappa Phi. Office: Lycoming Coll Dept Physics Williamsport PA 17701

FISHER, DAVID MARC, software engineer; b. N.Y.C., Mar. 5, 1955; s. David and Bettie Lou (Merrill) F.; m. Karol Audrey Paltsios, Apr. 13, 1991; children: Hayden Merrill, McKenna Kay. Degree, Carnegie Mellon U., 1977. Software engr. CompuGuard, Pitts., 1975-78, dir. adv. product devel., 1978-79; co-founder, sr. v.p. Am. Automatrix, Export, Pa., 1979—. Contbr. chpt. to Open Protocols, 1989; contbr. articles to Specifying Engr., Engrs. Digest, Heating/Piping/Air Conditioning, Bldg. Operating Mgmt. Named Entrepreneur of Yr. Arthur Young/Venture Mag., 1987. Mem. ASHRAE (chmn. SPC135P working group 1987—), Cum Laude Soc. Achievements include development of first direct digital control system for commercial automation, first open protocol for building automation systems; patent for multidrop fiber optic transceiver; patents include FumeHood control. Office: Am Automatrix 1 Technology Dr Export PA 15632-8903

FISHER, DELBERT ARTHUR, physician, educator; b. Placerville, Calif., Aug. 12, 1928; s. Arthur Lloyd and Thelma (Johnson) F.; m. Beverly Carne Fisher, Jan. 28, 1951; children: David Arthur, Thomas Martin, Mary Kathryn. BA, U. Calif., Berkeley, 1950; MD, U. Calif., San Francisco, 1953. Diplomate Am. Bd. Pediatrics. Intern, resident in pediatrics U. Calif. Med. Center, San Francisco, 1953-55; resident in pediatrics U. Oreg. Hosp., Portland, 1957-58; intern, resident in pediatrics U. Calif. Med. Ctr., San Francisco, 1953-55; from asst. prof. to assoc. prof. pediatrics Med. Sch. U. Ark., Little Rock, 1960-67, prof. pediatrics, 1967-68; prof. Med. Sch. UCLA, 1968-73; prof. pediatrics and medicine Med. Sch., UCLA, 1973-91, prof. emeritus, 1991—; chief, pediatric endocrinology Harbor-UCLA Med. Ctr., 1968-73, rsch. prof. devel. and perinatal biology, 1975-85, chmn. pediatrics, 1985-89, sr. scientist Rsch. and Edn. Inst., 1991—, dir. Walter Martin Rsch. Ctr., 1986-91; pres. Nichols Inst. Reference Labs, 1991-93; chief sci. officer, pres. Nichols Acad. Assocs., 1993—; cons. genetic disease sect. Calif. Dept. Health Svcs., 1978—; mem. organizing com. Internat. Conf. Newborn Thyroid Screening, 1977-88; examiner Am. Bd. Pediatrics, 1971-80, mem. subcom. on pediatric endocrinology, 1976-79. Co-editor: Pediatric Thyroidology, 1985, four other books; editor-in-chief Jour. Clin. Endocrinology and Metabolism, 1978-83, Pediatric Rsch., 1984-89; contbr. chpts. to numerous books; contbr. over 400 articles to profl. jours. Capt. M.C., USAF, 1955-57. Recipient Career Devel. award NIH, 1964-68. Mem. Inst. Medicine NAS, Am. Acad. Pediatrics (Borden award 1981), Soc. Pediatric Rsch. (v.p. 1973-74), Am. Pediatric Soc. (pres. 1992-93), Endocrine Soc. (pres. 1983-84), Am. Thyroid Assn. (pres. 1988-89), Am. Soc. Clin. Investigation, Assn. Am. Physicians, Lawson Wilkins Pediatric Endocrine Soc. (pres. 1982-83), Western Soc. Pediatric Rsch. (pres. 1983-84), Phi Beta Kappa, Alpha Omega Alpha. Home: 24582 Santa Clara Ave Dana Point CA 92629 Office: Nichols Inst 33608 Ortega Hwy San Juan Capistrano CA 92690

FISHER, FARLEY, chemist, federal agency administrator; b. Cleve., Apr. 30, 1938; s. Benjamin and Esther Lea (Begun) F. SB, MIT, 1960; PhD, U. Ill., 1965. Asst. prof. Tex. A&M U., College Station, 1966-69; tech. dir. Office of Toxic Substances, EPA, Washington, 1972-77; program dir. NSF, Washington, 1977—; vis. lectr. Bucknell U., Lewisburg, Pa., 1969-70; vis. prof. U. Minn., Mpls., 1984-85. Editor: Hazard Assessment of Chemicals, 1981. Home: 1435 4th St SW Washington DC 20024 Office: NSF Div Chem & Thermal Systems Washington DC 20550

FISHER, GEORGE MYLES CORDELL, electronics equipment company executive, mathematician, engineer; b. Anna, Ill., Nov. 30, 1940; s. Ralph Myles and Catherine (Herbert) F.; m. Patricia Ann Wallace, June 18, 1965; children: Jennifer, Barcy, William. BS in Enging., U. Ill., 1962; MS in Enging., Brown U., 1964, PhD in Applied Maths., 1964-66. Mem. tech. staff Bell Telephone Labs., Murray Hill, N.J., 1965-67; supr. Bell Telephone Labs., Holmdel, N.J., 1967-71; dept. head Bell Telephone Labs., Indpls., 1971-76; dir. mfg. systems. Motorola Inc., Schaumberg, Ill., 1976-77; asst. dir. mobile ops. Motorola Inc., Ft. Worth, Tex., 1977-78; v.p. portable ops. Motorola Inc., Ft. Lauderdale, Fla., 1978-81, v.p. paging div., 1981-84; asst. gen. mgr. communications sector Motorola Inc., Schaumberg, 1984-86, sr. exec. v.p., 1986-88, pres., chief exec. officer, 1988-90, chmn., chief exec. officer, 1990—. Contbr. articles on continuum physics; 3 patents in optical wave guides and digital communications. Trustee Brown U.; bd. dirs. U. Ill. Found., Nat. Merit Scholarship Bd., Chgo., 1986—. Mem. IEEE. Office: Motorola Inc 1303 E Algonquin Rd Schaumburg IL 60196

FISHER, JOHN WILLIAM, civil engineering educator; b. Ancell, Mo., Feb. 15, 1931; s. Nevan August and Nettie (Miller) F.; m. Nelda Rae Adams, Oct. 11, 1952; children: John Timothy, Christopher Lee, Elizabeth Rene, Nevan Andrew. BSCE, Washington U., St. Louis, 1956; MS, Lehigh U., 1958, PhD, 1964; Dr. honoris causa, Swiss Fed. Inst. Tech., Lausanne, Switzerland, 1988. Registered profl. engr., Ill. Asst. bridge research engr. Nat. Acad. Scis., Ottawa, Ill., 1958-61; from research instr. to assoc. prof. Lehigh U., Bethlehem, Pa., 1961-69, prof. civil enging., 1969—, Joseph T. Stuart prof., 1988—, assoc. dir. Fritz Enging. Lab., 1972-84, co-chmn. civil enging., 1984-85, dir. advanced tech. for large structural systems NSF-Enging. Research Ctr., 1986—; cons. Washington Metro Area Transit Authority, 1979—, Bethlehem Steel Corp., 1964-69, 77-91, Conn. Dept. Transp., 1983-92, Md. Dept. Transp., Balt., 1981-88, N.Y.C. Dept. Transp., 1987-90, Triborough Bridge & Tunnel Auth., 1991—; civil col. eminent overseas speaker Inst. Engrs. Australia, 1983; vis. prof. Swiss Fed. Inst. Tech., Lausanne, Switzerland, 1982; sr. vis. scholar (lectr.) Peoples Republic of China, 1985; mem. Trans. Rsch. Bd. Author: Structural Steel Design, 1974, Guide to Design Criteria for Bolted Joints, 1974, 2d edit., 1987, Bridge Fatigue Guide, 1977, Fatigue and Fractures in Steel Bridges, 1984; contbr. over 190 articles to profl. jours. Mem. directory council Southside Ministries, Bethlehem, 1983—; bd. dirs. New Bethany Ministries, Bethlehem, 1985-90. 2d lt. U.S. Army, 1951-53. Recipient Alumni Achievement award Washington U., 1987, Frank P. Brown medal Franklin Inst., 1992; named Constrn. Man of Yr., ENR, 1987, Engr. of Yr. in Rsch., Inst. for Bridge Integrity and Safety, 1989. Mem. ASCE (hon. 1989, Huber Rsch. prize 1969, Ernest E. Howard award 1979, R.C. Reese Rsch. prize 1981), NAE (chmn. NAE/NRC com. internat. constrn. study 1987-88, mem. Internat. Affairs adv. com. 1988-92, mem. program adv. com. 1992—), NSPE, Am. Soc. Enging. Educators, Am. Ry. Enging. Assn. (steel structures com.), Am. Welding Soc. (Adams mem.), Internat. Assn. Bridge and Structural Engrs., Am. Inst. Steel Constrn. (T.R. Higgins lectr. 1977), Swiss Acad. Enging. Scis. Republican. United Methodist. Avocations: hiking, canoeing. Office: Lehigh U 117 Atlss Dr Bethlehem PA 18015-4729

FISHER, MARYE JILL, physical therapist, educator; b. Franklin, Ind., May 22, 1949; d. William Jackson and Bernice Janice (Schneider) F. BS, Washington U., 1972, Mich. State U., 1979; PhD, Mich. State U., 1984. Clin. instr. Washington U., Lansing (Mich.) Community Coll., St. Louis, 1972-75, 80-82; NSF pre-doctoral fellow Mich. State U., East Lansing, 1981-84, NIH post-doctoral fellow, 1985-87, asst. prof. radiology, 1987-91, clin. asst. prof. in radiology and physiology, 1991—; pediatric phys. therapist Ingham Intermediate Schs., Mason, Mich., 1991—; cons. education, phys. therapy Creative Cons., Clarkston, Mich., Southfield, Mich., 1991—. Contbr. articles and abstracts to Am. Jour. Physiology, Jour. Magnetic Resonance Medicine, Canadian Jour. Physiology, Jour. Applied Physiology, NMR in Biomedicine, Physiologist, Abstracts; contbr. chpt.: Muscle Energetics, 1988. Mem. Am. Phys. Therapy Assn., Assn. Univ. Radiologists (Stauffer award 1991), Soc. Magnetic Resonance in Medicine. Democrat. Roman Catholic. Home: 917 Huntington East Lansing MI 48823 Office: Ingham Intermediate Schs 2805 Howell Rd Mason MI 48823

FISHER, MICHAEL ALAN, air force officer, mechanical engineer; b. Monterey, Calif., Apr. 7, 1954; s. Thurman Danton and Edith (Morris) F. BSME, U. Fla., 1977; M of Aero. Sci., Embry-Riddle U., 1992. Field engr. Schlumberger Offshore Svcs., New Orleans, 1977-78; commd. 2d lt. USAF, 1977, advanced through grades to maj., 1988; engr. space shuttle systems Air Force Space Div., L.A., 1979-84; space shuttle payload officer Johnson Space Ctr., NASA, Houston, 1984-88; tech. officer Weapons Lab., USAF, Kirtland AFB, N.Mex., 1988-89; dir. space mission ops. Phillips Lab., USAF, Kirtland AFB, 1989-92, program mgr. relay mirror experiment, 1991-92; program mgr. Theatre Missile Def., Strategic Def. Initiative Orgn., Pentagon, 1992—. Mem. ASME, AIAA, Soc. Phot-Optical Instrumentation Engrs., Assn. Unmanned Vehicle Systems.

FISHER, MICHAEL ELLIS, mathematical physicist, chemist; b. Trinidad, W.I., Sept. 3, 1931; m. Sorrel Castillejo; children: Caricia J., Daniel S., Martin J., Matthew P.A. B.S. with 1st class honors in Physics, King's Coll., London, 1951, Ph.D., 1957; DSc (hon.), Yale U., 1987, Tel Aviv U., 1992. Lectr. math. RAF, 1952-53; lectr. theoretical physics King's Coll., 1958-62, reader physics, 1962-64; prof. physics U. London, 1965-66; prof. chemistry and math. Cornell U., 1966-73, Horace White prof. chemistry, physics and math., 1973-89, chmn. dept. chemistry, 1975-78; Wilson H. Elkins Disting. Prof. Inst. for Physical Sci. and Tech. U. Md., 1987—; guest investigator Rockefeller Inst., 1963-64; vis. prof. applied physics Stanford U., 1970-71; Buhl lectr. theoretical physics Carnegie-Mellon U., 1971; Richtmyer Meml. lectr. Am. Assn. Physics Tchrs., 1973; S.H. Klosk lectr. NYU, 1975; 17th F. London Meml. lectr. Duke U., 1975; Walker-Ames prof. U. Wash., Seattle, 1977; Loeb lectr. physics Harvard U., 1979; vis. prof. physics MIT, 1979; Welsh Found. lectr. in physics U. Toronto, Ont., Can., 1979; 21st Alpheas Smith lectr. Ohio State U., 1982; Fairchild scholar, Calif. Inst. Tech., 1984; Cherwell-Simon lectr., vis. prof. Oxford U., 1985; Schlapp lectr., Edinburgh U., 1987; Marker lectr., Penn. State U., 1988; Nat. Sci. Coun. lectr., Taiwan, 1989; Hamilton Meml. lectr., Princeton U., 1990; 65th J.W. Gibbs lectr., Am. Math. Soc., 1992. Author: (with D.M. MacKay) Analogue Computing at Ultra-High Speed, 1962, The Nature of Critical Points, 1964, The Theory of Equilibrium Critical Phenomena, 1967; assoc. editor: Jour. Math. Physics, 1965-68, 72-75, 86-89; adv. bd.: Jour. Theoretical Biology, 1969-82, Chem. Physics, 1972-84, Discrete Math., 1971-78, Jour. Statis. Physics, 1978-81; editorial bd. Communications Math. Phys., 1984—, Phys. Rev. A, 1987-92; contbr. 310 articles to profl. jours. Recipient award in phys. and math. scis. N.Y. Acad. Scis., 1978; Guthrie medal and prize Inst. Physics, London, 1980; Wolf prize in physics, 1980; Michelson-Morely award Case Western Res. U., 1982; Boltzmann medal IUPAP, 1983; Guggenheim fellow, 1970-71, 78-79. Fellow Am. Acad. Arts and Scis., Am. Assn. Adv. Sci., Royal Soc. (London) (Bakerian lectr. 1979, regional editor, 1989-93), Phys. Soc. London, Am. Phys. Soc. (Langmuir prize chem. physics 1970), Kings Coll. London, Royal Soc. Edinburgh (hon.); mem. Am. Chem. Soc., Am. Philos. Soc., Soc. Indsl. and Applied Math., Math. Assn. Am., Nat. Acad. Scis. (fgn. assoc., James Murray Luck award 1983), N.Y. Acad. Scis. Office: Inst Phys Sci & Tech U Md Inst Phys Sci and Tech College Park MD 20742

FISHER, PAUL DOUGLAS, psychologist, program director; b. Nashville, Aug. 6, 1956; s. John Clark and Iris (Pierce) F.; m. Donna Nichols, May 28, 1988. BS magna cum laude, Vanderbilt U., 1978; PhD, U. Ala., 1983. Psychologist, mental health dir. USAF, worldwide, 1982-85; resident Wilford Hall Med. Ctr., San Antonio, 1983; v.p. Profl. Performance Devel. Group, San Antonio, 1983-86; pvt. practice psychology Houston, various other locations, 1986-88; clin. dir., psychologist Addiction Recovery Corp., Harrison, Tenn., 1988-90; cons. dir. Tenn. Dept. Corrections-S.E. Region, 1988—; owner, dir. Tenn. Transitional Living Ctrs., Harrison, 1989—; examiner Tenn. Bd. Psychol. Examiners, 1988—; cons. Centurion Police Stress Program, 1990—; pres. Tenn. Transitional Living Ctrs., 1988-90. Author: Issues in Human Adjustment, 1980; contbr. articles to profl. publs. Rsch. edn. dir. Gov's. Alliance for a Drug-Free Tenn., 1988—; mem. Rep. Senatorial Com., Washington, 1982—, Tenn. Pub. Broadcasting, 1987—, Tenn. Friends of Folk Music, 1988—. Capt. USAF, 1982-86. Recipient Presdl. Order of Merit from Pres. George Bush, 1991; Smithsonian Instn. fellow, 1978-81, NIMH fellow, 1978-80. Mem. APA, Psychologists for Advancement of Conservative Thought (v.p. 1981—), Am. Police Assn. (life), Tenn. Sheriffs Assn., Tenn. Assn. for Child Care, Mensa. Avocations: martial arts, music, antique autos, antique firearms, animal rights. Office: TTLC PO Box 1006 Harrison TN 37341-1006

FISHER, ROBERT ALAN, laser physicist; b. Berkeley, Calif., Apr. 19, 1943; s. Leon Harold and Phyllis (Kahn) F.; m. Andrea Lapitski, Mar. 18, 1967; children—Andrew Leon, Derek Martin. A.B., U. Calif., Berkeley, 1965, M.A., 1967, Ph.D., 1971. Programmer Stanford Linear Accelerator, Stanford University, Calif., 1965; staff mem. Granger Assocs., Palo Alto, Calif., 1966; lectr. U. Calif., Davis, 1972-74; physicist Lawrence Livermore Lab., Calif., 1971-74; laser physicist Los Alamos Nat. Lab., N. Mex., 1974-86; cons. to R.A. Fisher Assocs., Santa Fe, N.Mex., 1986—; instr. Enging. Tech., Inc., 1982—; mem. Air Force ABCD Panel, 1982; program com. mem. Internat. Quantum Electronics Conf., 1982, 86; vice chmn. Gordon Conf. on Lasers and Non-linear Optics, 1981; chmn. Soc. Photo-Optical Instrumentation Engrs. conf. on Optical Phase Conjugation/Beam Combining/Diagnostics, 1987—; mem. Air Force Red Team for Space-Based Laser, 1983-86, HEDS II SDI Red Team, 1986; U.S. Ballistic Missile Office

Options Team, 1986; mem. secretariat SDI Red/Blue Sensor Teams, 1986, SDI GBL Red/Blue Team Interaction, 1987-88; mem. architecture panel SDI SDS Phase 1, 1990, Air Force Laser 21 Working Group, 1990. Assoc. editor Optics Letters, 1984-86, Applied Optics, 1984-91; editor: Optical Phase Conjugation, 1973; contbr. articles to profl. jours. Vol. coach elem. sch. chess team Pojoaque Elem. Sch. (winner nat. elem. championship 1984), Santa Fe, 1984. Fellow Optical Soc. Am. (guest editor Jour.'s spl. issue on optical phase conjugation); mem. IEEE (Sr.), Am. Phys. Soc. Avocations: restoring old houses; skiing; music. Home and Office: Route 5 Box 230 Santa Fe NM 87501-9309

FISHER, ROBERT AMOS, physical chemist; b. Honey Grove, Pa., Mar. 25, 1934; s. Robert A. and Edna M. (Maus) F.; m. Marianne C. Donadio, Aug. 30, 1958; children: Tracy A. Rice. BS, Juniata Coll., 1956; PhD, Pa. State U., 1960. Rsch. chemist U. Calif., Berkeley, 1960-91, Lawrence Berkeley Lab., 1991--. Contbr. articles to profl. jours. Union Carbide fellow, 1958-60. Mem. Am. Assn. for Advancement Sci., Am. Phys. Soc., Sigma Xi. Achievements include fundamental research in magnetic materials and superconducting materials particularly the high temperature superconductors and heavy fermions. Home: 1841 Yosemite Rd Berkeley CA 94707 Office: U Calif Low Temperature Lab Berkeley CA 94720

FISHER, SALLIE ANN, chemist; b. Green Bay, Wis., Sept. 10, 1923. BS in Chemistry, U. Wis., 1945, MS, 1946, PhD, 1949. Instr. Mt. Holyoke Coll., South Hadley, Mass., 1949-50; asst. prof. U. Minn., Duluth, 1950-51; group leader Rohm & Haas Co., Phila., 1951-60; assoc. dir. rsch. Robinette Rsch. Labs., Berwyn, Pa., 1960-72; v.p. Puricons, Inc., Malvern, Pa., 1972-76; pres. Puricons, Inc., Malvern, 1976—; mem. adv. bd. Internat. Water Conf., Pitts., 1976-91, Reactive Polymers, Netherlands, 1982-88. Contbr. chpts. to books and over 95 articles to profl. jours. Recipient award of merit Engring. Soc. Western Pa., Pitts., 1984. Fellow ASTM (vice-chmn. D-19 1972-78, award of merit 1974, Max Hecht award com. D-19 1975); mem. Soc. Chem. Industry, Am. Chem. Soc., Am. Waterworks Assn., Nat. Assn. Corrosion Engrs. Achievements include patent for regeneration of anion resins; research in process for the concentration and recovery of uranium; devel. of methodology for analysis of resins for nuclear industry. Office: Puricons Inc 101 Quaker Ave Malvern PA 19355

FISHER, STEVEN KAY, neurobiology eductor; b. Rochester, Ind., July 18, 1942; s. Stewart King and Hazel Madeline (Howell) F.; m. Dinah Dawn Marschall, May 2, 1971; children: Jenni Dawn, Ward, Brian Andrew, Steven William. BS, Purdue U., 1964, MS, 1966; postgrad., Johns Hopkins U., 1967-69; PhD, Purdue U., 1969. Postdoctoral fellow Johns Hopkins U., Balt., 1969-71; prof. U. Calif., Santa Barbara, 1971—, dir. Inst. Environ. Stress, 1985-88, dir. Neurosci. Rsch. Inst., 1989—; cons. Ultrastructure Tech., Goleta, Calif., 1984—. Contbr. numerous articles to profl. jours. Recipient Devel. award NIH, 1980-84, M.E.R.I.T. award NIH, 1989—; NIH grantee, 1971—. Mem. AAAS, Assn. Research in Vision and Ophthalmology (mem. program com. 1979-80), Soc. Neurosci., Am. Soc. Cell Biology, Electron Microscopy Soc. Am., NIH Visual Scis. A2 Study Sect. Avocations: music, gardening, literature, swing dancing. Home: 6890 Sabado Tarde Rd Santa Barbara CA 93117-4305 Office: U Calif Neuroscience Research Institute Santa Barbara CA 93106

FISHER, SUZANNE EILEEN, health scientist administrator; b. Mpls., Jan. 15, 1950; d. Arthur Clayton and Virginia Eileen (Rapacz) Ostlund; m. E. Marshall Fisher, Aug. 12, 1972; 1 child, James Marshall. BS, Mich. State U., 1972; PhD, U. Ill., 1978. Postdoctoral fellow Nat. Cancer Inst. NIH, Bethesda, Md., 1978-81, staff fellow Nat. Inst. Child Health and Human Devel., 1982-84, health sci. administr. Nat. Cancer Inst., 1984-87, health sci. administr. Nat. Eye Inst., 1987-89, asst. chief referral divn. rsch. grants, 1989—; chmn. Westwood Libr. Adv. Comm., Bethesda, 1991—; mem. NIH Adv. Com. Libr., Bethesda, 1992—. Contbr. articles to profl. jours. Chmn. Rockville (Md.) Libr. Adv. Com., 1991—; mem. Peerless Rockville Hist. Soc., 1986—. NSF pre-doctoral fellow, 1972-76. Mem. Am. Assn. for the Advancement Sci., Phi Kappa Phi, Phi Beta Kappa, Sigma Xi, Mortar Bd. Office: NIH Rm 248 Westwood 5333 Westbard Ave Bethesda MD 20892

FISHER, THORNTON ROBERTS, physicist; b. Santa Monica, Calif., Feb. 16, 1937; s. Vardis and Margaret (Trusler) F.; m. Yvonne Habib, June 22, 1968. AB, Wesleyan U., 1958; PhD, Calif. Inst. Technology, 1963. Rsch. assoc. Stanford (Calif.) U., 1963-68; rsch., cons. scientist Lockheed Palo Alto (Calif.) Rsch. Lab., 1968-87, program mgr., 1987-89, mgr. space payloads dept., 1989—. Contbr. articles to profl. jours. Recipient NSF fellowship, 1958-62. Mem. Am. Phys. Soc., Soc. Photo-Optical and Instrumentation Engrs., Sigma Xi, Soc. for Indsl. and Applied Math. Republican. Achievements include development of wire-shadow technique for measuring brightness in neutral particle beams; first measurement of pion cross section for heavy polarized nucleus. Home: 25603 Fernhill Dr Los Altos CA 94024-6335 Office: Lockheed Missiles and Space Dept 91-21/B255 3251 Hanover St Palo Alto CA 94304-1121

FISHER, WILLIAM LAWRENCE, geologist, educator; b. Marion, Ill., Sept. 16, 1932; s. Henry Adam and Madge Lenora (Moore) F.; m. Marilee Booth, Dec. 18, 1954; children: Leah, Karl, Peter. B.S., So. Ill. U., 1954, D.Sc., 1986; M.S., U. Kans., 1958, Ph.D. (Shell fellow), 1961. Research scientist Tex. Bur. Econ. Geology, Austin, 1960-68; assoc. dir. Tex. Bur. Econ. Geology, 1968-70, dir., 1970-75, 77—; asst. sec. for energy and minerals Dept. Interior, Washington, 1975-77; prof. dept. geol. scis. U. Tex., Austin, 1969—, Morgan J. Davis prof. petroleum geology, 1984-86, Leonidas T. Barrow chair in mineral resources, 1986—, chmn. dept. geol. scis., 1984-90; dir. Geol. Found. Tex., 1984—; mem. geology assoc. bd. U. Kans., 1972-74, 83—; mem. adv. coun. Gas Rsch. Inst.; mem. Tex. Sci. Adv. Coun.; mem. Gov.'s Energy Coun.; pres. Am. Geol. Inst., 1989-90; mem. White House Sci. Coun., Nat. Petroleum Coun. and Sec. Energy Adv. Bd. Author: Mineral Resources of East Texas, 1964, Depositional Systems in the Wilcox Group, 1969, Delta Systems in the Exploration for Oil and Gas, 1969, Environmental Geologic Atlas of Texas Coastal Zone, 1972, National Energy Policies, 1977, 78, 79, 80. Served with AUS, 1954-56. Recipient Hedberg award So. Methodist U. Fellow Geol. Soc. Am. (councillor); mem. NAS (bd. mineral and energy resources, U.S. nat. com. on geology, chmn. bd. on earth scis. and resources), Am. Inst. Profl. Geologists (pres. Tex. sect. 1979, pres. 1993, Pub. Svc. award), Am. Assn. Petroleum Geologists (pres. 1981), Am. Geol. Inst. (pres. 1991, Campbell medal), Austin Geol. Soc. (pres. 1973-74), Gulfcoast Assn. Geol. Scis. (hon.). Home: 8705 Ridgehill Dr Austin TX 78759-7342 Office: U Tex Bur Econ Geology Univ Sta Box X Austin TX 78712

FISHMAN, JACOB ROBERT, psychiatrist, educator, corporate executive, investor; b. N.Y.C., Aug. 6, 1930; s. Samuel and Francis (Goldin) F.; A.B., Columbia U., 1952; M.D., Boston U., 1956; m. Tamar Hendel, June 1, 1958; children: Marc Judah, Risa Esther, Zalman Schneur, Rebecca Anne. Intern in medicine Einstein Coll. Medicine, Bronx, N.Y., 1956-57, resident psychiatry, 1957-59; research psychiatrist NIMH, Washington, 1959-62; prof. psychiatry Howard U. Coll. Medicine, Washington, 1962-71; dir. Howard-D.C. Comprehensive Mental Health Center, 1966-68; chmn. bd., pres. Univ. Research Corp., Washington, 1968-78, Am. Health Services, Inc., 1971-78; pres. Ctr. for Human Services, 1968-74, Human Service Group, 1971-78, Horizon Mental Health Group, Inc., 1981-84, Cumberland Psychiat. Hosp., 1979-84; chmn. bd. dirs. Am. Mental Health Group, Inc., Am. Health Group Inc.; chmn. psychiatry So. Md. Hosp. Ctr., 1978-81; cons. fed. agys., U.S. Congress, numerous pvt. corps.; bd. dirs. Create Inc., Entertainment Concepts Inc., First Grafton Corp., Med. Services Corp., Inc., Am. Health Group, Inc., Md. Treatment Ctrs., Inc. Bd. dirs. Webster Coll., Washington, 1971-75, Ctr. for Human Services, 1967-75, DePaul Hosp., New Orleans, 1973-78, St. Elizabeth's Hosp., Richmond, Va., 1971-78, Cin. Mental Health Inst., 1971-78, Nat. Capital Day Care Assn., 1966-68; mem. D.C. Public Health Adv. Council, 1966-68; attending psychiatrist Freedman's Hosp., Washington Vets. Hosp., D.C. Gen. Hosp., 1962-68; dir. Potomac Psychiat. Assocs., 1978—, Am. Health Group, 1985—. Served with USPHS, 1959-61. Recipient Gold medal award Phi Lambda Kappa Med. Soc. Fellow Am. Public Health Assn., Am. Assn. Social Psychiatry; mem. Am. Psychiat. Assn., D.C. Psychiat. Soc., Potomac Psychiat. Assocs. (pres. 1978—), AAAS, D.C. Public Health Assn. (Disting. Service award), Am. Med. Soc. on Alcoholism and Other Addictive Drugs, Am. Council on Alcoholism, Nat. Assn. for New Careers in Human Svcs. (vice chmn. 1965-70), various others. Author numerous profl. articles and books. Bd. editors Nat. Jour. Research on Crime and Delinquency, 1965-71. Home: 1717 Poplar Ln NW Washington DC 20012-1135 Office: Am Health Group 3800 Frederick Ave Baltimore MD 21229-3618

FISHMAN, MARK BRIAN, computer scientist, educator; b. Phila., May 17, 1951; s. Morton Louis and Hilda (Kaplan) F.; m. Alice Faber, Feb. 20, 1977 (div. 1986); m. E. Alexandra Baehr, Apr. 13, 1992. AB summa cum laude, Temple U., 1974; postgrad. Northwestern U., 1974-76; MA, U. Tex., 1980. Bilingual tchr. Wilmette Pub. Schs., 1974; rsch. assoc., programmer, asst. instr. U. Tex., Austin, 1976-80; instr. computer and info. scis. U. Fla., Gainesville, 1980-85; asst. prof. computer sci. Eckerd Coll., St. Petersburg, Fla., 1985-90, dept. coord., 1988-90, 91—, assoc. prof. computer sci., 1991—; instrnl. cons. to IBM, 1980—; cons. artificial intelligence, Battelle Corp., 1987-89, USN Naval Tng. Systems Ctr., 1987—, Advanced Techs., Inc., 1988—, LBS Capital Mgmt., 1990—. Series editor: Advances in Artificial Intelligence Rsch., Vol. I, 1989; editor: Proc. of the First Florida Artificial Intelligence Rsch. Symposium, 1988, Proc. of the Second Florida Artificial Intelligence Rsch. Symposium, 1989, Advances in Artificial Intelligence Research, vol. I, 1989, vol. II, 1992, Proc. of the Third Florida Artificial Intelligence Rsch. Symposium, 1990, Proc. of the Fourth Florida Artificial Intelligence Rsch. Symposium, 1991, Proc. of the Fifth Artificial Intelligence Rsch. Symposium, 1992; guest editor: International Journal of Expert Systems, Vol. 5, no. 2; steering com. First Internat. Conf. Human and Machine Cognition; contbr. articles to profl. jours; presenter in field. U. Tex. fellow, 1978-80; F.C. Austin scholar, 1975; Nat. Def. Fgn. Lang. fellow, 1974. Mem. Assn. Computing Machinery (Tchr. of Yr. award U. Fla. 1984), IEEE Computer Soc., Am. Assn. Artificial Intelligence, Assn. Computational Linguistics, Fla. Artificial Intelligence Research Soc. (proc. chair 1988—, sec. 1988-89, v.p. 1989-91, pres. 1991—), Am. Soc. Engring. Edn. (faculty research fellow summer 1986, 91), Sigma Xi, Phi Beta Kappa, Phi Kappa Phi, Upsilon Pi Epsilon. Home: 6166 Lynn Lake Dr S Saint Petersburg FL 33712-6115 Office: Eckerd Coll Computer Sci Dept Saint Petersburg FL 33733

FISHMAN, ROBERT ALLEN, educator, neurologist; b. N.Y.C., May 30, 1924; s. Samuel Benjamin and Miriam (Brinkin) F.; m. Margery Ann Satz, Jan. 29, 1956 (dec. May 29, 1980); children: Mary Beth, Alice Ellen, Elizabeth Ann.; m. Mary Craig Wilson, Jan. 7, 1983. A.B., Columbia U., 1944; M.D., U. Pa., 1947. Mem. faculty Columbia Coll. Physicians and Surgeons, 1954-66, asso. prof. neurology, 1962-66; asst. attending neurologist N.Y. State Psychiat. Inst., 1955-66; asst. attending neurologist Neurol. Inst. Presbyn. Hosp., N.Y.C., 1955-61, asso., 1961-66; co-dir. Neurol. Clin. Research Center, Neurol. Inst., Columbia-Presbyn. Med. Ctr., 1961-66; prof. neurology U. Calif. Med. Ctr., San Francisco, 1966—, chmn. dept. neurology, 1966-92; cons. neurologist San Francisco Gen. Hosp., San Francisco VA Hosp., Letterman Gen. Hosp.; dir. Am. Bd. Psychiatry and Neurology, 1981-88, v.p., 1986, pres., 1987. Author: Cerebrospinal Fluid in Diseases of the Nervous System, 1992; Contbr. articles to profl. jours. Nat. Multiple Sclerosis Soc. fellow, 1956-57; John and Mary R. Markle scholar in med. sci., 1960-65. Mem. Am. Neurol. Assn. (pres. 1983-84), Am. Fedn. for Clin. Research, Assn. for Research in Nervous and Mental Diseases, Am. Acad. Neurology (v.p. 1971-73, pres. 1975-77), Am. Assn. Physicians, Am. Soc. for Neurochemistry, Soc. for Neurosci., N.Y. Neurol. Soc., Am. Assn. Univ. Profs. Neurology (pres. 1972-73), AAAS, Am. Epilepsy Soc., N.Y. Acad. Scis., AMA (sec. sect. on nervous and mental diseases 1964-67, v.p. 1967-68, pres. 1968-69), Alpha Omega Alpha (hon. faculty mem.). Home: 50 Summit Ave Mill Valley CA 94941-1819 Office: U Calif Med Center 794 Herbert C Moffitt Hosp San Francisco CA 94143

FISK, LENNARD AYRES, physicist, educator; b. Elizabeth, N.J., July 7, 1943; s. Lennard Ayres and Elinor (Fischer) F.; m. Patricia Elizabeth Leuba, Dec. 28, 1966; children: Ian, Justin, Nathan. AB, Cornell U., 1965; PhD, U. Calif., San Diego, 1969. Postdoctoral fellow NASA/Goddard Space Flight Ctr., Greenbelt, MD., 1969-71, astrophysicist, 1971-77; assoc. prof. U. N.H., Durham, 1977-81, prof., 1981-87, dir. rsch., 1982-83, interim v.p./fin. affairs, 1983-84, v.p. rsch. and fin., 1984-87; assoc. administr. space sci. and applications NASA Hdqrs., Washington, 1987-93; prof. U. Mich., 1993—; advisor NAS, NASA, 1980-87. Author more than 80 sci. publs., 1969-86. Fellow Am. Geophys. Union. Office: NASA Space Sci & Applications 300 E St SW Washington DC 20546-0001 Office: U Mich Dept Atmospheric Oceanic & Space Scis Ann Arbor MI 48109-2143

FISK, ZACHARY, physical scientist; b. N.Y., Sept. 3, 1941; s. James Brown and Cynthia Hoar F.; m. Mary Bayley, Dec. 28, 1964 (div. 1979); 1 child, Rebekah; m. Jehanne Teilhet, April 22, 1979; 1 child, Samantha. BA, Harvard, 1964; PhD, U. Calif., San Diego, 1969. Asst. prof. physics U. Chgo., 1970-71; from asst. researcher physics to researcher physics U. Calif., San Diego, 1971-81; staff mem. Los Alamos Nat. Lab., 1981—; prof. physics U. Calif., 1991—; cons. Bell Tel. Labs., Murray Hill, N.J., 1972-80; editor Physics B, 1990—. Contbr. over 400 articles to sci. jours. Recipient Ernest Orlando Lawrence Meml. award U.S. Dept. Energy, Washington, 1991. Fellow Am. Phys. Soc. (recipient Internat. prize for New Materials, 1990); mem. AAAS. Office: Los Alamos Nat Lab PO Box 1663 Los Alamos NM 87545*

FISKE, RICHARD SEWELL, geologist; b. Balt., Sept. 5, 1932; s. Franklin Shaw and Evelyn Louise (Sewell) F.; m. Patricia Powell Leach, Nov. 28, 1959; children: Anne Powell, Peter Sewell. BS in Geol. Engring., Princeton U., 1954, MS, 1955; PhD in Geology, Johns Hopkins U., 1960. With U.S. Geol. Survey, 1964-76; chief Office Geochemistry and Geophysics, Reston, Va., 1972-76; geologist, curator dept. mineral scis. Smithsonian Instn., Washington, 1976-80; dir. Nat. Mus. Natural History, 1980-85, geologist dept. mineral scis., 1985—; Phi Beta Kappa vis. scholar, 1990-91. Am. Chem. Soc. postdoctoral fellow U. Tokyo, 1960-61; recipient Meritorious Service award Dept. Interior, 1976. Fellow AAAS, Geol. Soc. Am.; mem. Am. Geophys. Union, Geol. Soc. Washington. Home: 4938 Western Ave Bethesda MD 20816-1714 Office: Smithsonian Instn Dept Mineral Scis NHB-119 Washington DC 20560

FISKE, SANDRA RAPPAPORT, psychologist, educator; b. Syracuse, N.Y., Sept. 25, 1946; d. Sidney Saul and Helen (Lapides) Rappaport; B.S., Cornell U., 1968; M.Ed., Tufts U., 1969; M.A., Columbia U., 1971, Ph.D., 1974; m. Jordan J. Fiske, June 22, 1974. Supervising sch. psychologist St. Elizabeth's Sch., N.Y.C., 1971-76; instr. clin. psychology Tchrs. Coll., Columbia, N.Y.C., 1973, clin. asst. dept. psychology, 1975-76; adj. prof. Syracuse U., 1976; sch. psychologist Syracuse Bd. Edn., 1976-77; asso. prof. Onondaga Community Coll., Syracuse, 1976-87, prof. 1988—; pvt. practice psychology, Syracuse, 1976—. NIMH fellow, 1969-72. Mem. Am. Psychol. Assn., Psychologists of Central N.Y., Am. Orthopsychiat. Assn., Sigma Xi, Psi Chi. Home: 2 Signal Hill Rd Fayetteville NY 13066-9674 Office: Onondaga Community Coll Syracuse NY 13215

FISZER-SZAFARZ, BERTA (BERTA SAFARS), research scientist; b. Wilno, Poland, Feb. 1, 1928; m. David Safars; children—Martine, Michel. M.S., U. Buenos Aires, 1955, Ph.D., 1956. Lab. chief Cancer Inst. Villejuif, France, 1961-67; vis. scientist Nat. Cancer Inst., Bethesda, Md., 1967-68; lab. chief Institut Curie, Orsay, France, 1969—; vis. scientist Inst. Applied Biochemistry, Mitake, Gifu, Japan, 1986. Contbr. articles to profl. jours. Mem. European Assn. Cancer Research, Am. Assn. Cancer Research (corres. mem.), N.Y. Acad. Scis., European Cell Biology Orgn., French Soc. Cell Biology. Office: Institut Curie-Biologie, Batiment 110, Orsay 91405, France

FITCH, GREGORY KENT, biologist; b. Enid, Okla., Apr. 10, 1955; s. Arthur William and Virginia Elizabeth (Newmann) F.; m. Barbara Gayle Anderson, Sept. 13, 1987. BS in Biology and Psychology, Kans. State U., 1977, MS in Neurophysiology, 1982. Instr. Old Trooper U., Fort Riley, Kans., 1983-85, Div. of Biology, Kans. State U., Manhattan, Kans., 1983—; mentor Project Choice of the Kaufman Found., Kansas City, Mo., 1991-92. Author: Understanding Human Anatomy and Physiology, 1993; contbr. articles to profl. jours. Recipient Stamey Outstanding Teaching award Kans. State U., 1991, Outstanding Prof. award, 1991-92. Mem. Sigma Xi. Home: 2606 Margot Ln Manhattan KS 66502 Office: Kansas State U Div of Biology Ackert Hall Manhattan KS 66506

FITCH, VAL LOGSDON, physics educator; b. Merriman, Nebr., Mar. 10, 1923; s. Fred B. and Frances Marion (Logsdon) F.; m. Elise Cunningham, June 11, 1949 (dec. 1972); children: John Craig (dec. 1987), Alan Peter; m. Daisy Harper Sharp, Aug. 14, 1976. B. of Engring., McGill U., 1948; Ph.D., Columbia U., 1954. Instr. Columbia, 1953; instr. physics Princeton, 1954-56, asst. prof., 1956-59, assoc. prof., 1959-60, prof., 1960—, Class 1909 prof. physics, 1968-76, Cyrus Fogg Bracket prof. physics, 1976-84, James S. McDonnell Distinguished Univ. prof. physics, 1984—; Mem. Pres.'s Sci. Adv. Com., 1970-73. Trustee Asso. Univ., Inc., 1961-67. Served with AUS, 1943-46. Recipient Research Corp. award, 1967; E.O. Lawrence award, 1968; Wetherill medal Franklin Inst., 1976; Nobel prize in physics, 1980; Nat. Medal of Sci., Nat. Sci. Found., 1993; Sloan fellow, 1960. Fellow Am. Phys. Soc. (pres. 1987-88); mem. Am. Acad. Arts and Scis., Nat. Acad. Sci. Office: Princeton U Dept Physics PO Box 708 Princeton NJ 08544-0708

FITCHETT, VERNON HAROLD, retired physician, surgeon, educator; b. Grover, Colo., May 14, 1927; s. Harold Leroy and Mazie (Bengston) F.; m. Kathryn Hellen Mullin, Aug. 3, 1963; children: Michael, Elizabeth, Benjamin. B.S., Buena Vista Coll., 1949; M.D., U. Iowa, 1953. Diplomate Am. Bd. Surgery. Commd. officer U.S. Navy, 1956, advanced through grades to capt., intern U.S. Naval Hosp., Bremerton, Wash., 1953-54; resident VA Hosp., Portland, Oreg., 1955-56, U.S. Naval Hosp., St. Albans, N.Y., 1956-59; med. officer-in-charge USPHS, DaNang, Vietnam, 1964-65; chief surgery U.S. Navy, DaNang, 1968-69; chmn. dept. surgery Naval Hosp., Oakland, Calif., 1970-76; ret., 1976; mem. staff Jamestown Hosp., N.D., 1976-93; asst. clin. prof. surgery N.D. Med. Sch., 1982-87, clin. assoc. prof. surgery, 1987-93; chmn. bd. Jamestown Clinic, 1979-85; surg. cons. State Hosp., Jamestown, Ann Carlson Sch., 1976-93. Author: War Surgery, 1971. Contbr. articles to profl. jours. Mem. exec. com. N.D. chpt. Am. Cancer Soc., 1979-84. Decorated Legion of Merit with Combat V. Fellow ACS (past pres. N.D. chpt.); mem. AMA, Pan Pacific Surg. Soc., Assn. Mil. Surgeons. Roman Catholic. Lodges: Eagles, Elks, K.C.

FITES, DONALD VESTER, tractor company executive; b. Tippecanoe, Ind., Jan. 20, 1934; s. Rex E. and Mary Irene (Sackville) F.; m. Sylvia Dempsey, June 25, 1960; children: Linda Marie. B.S. in Civil Engring., Valparaiso U., 1956; M.S., M.I.T, 1971. With Caterpillar Overseas S.A., Peoria, Ill., 1956-66; dir. internat. customer div. Caterpillar Overseas S.A., Geneva, 1966-67; asst. mgr. market devel. Caterpillar Tractor Co., Peoria, 1967-70; dir. Caterpillar Mitsubishi Ltd., Tokyo, 1971-75; dir. engine capacity expansion program Caterpillar Tractor Co., Peoria, 1975-76, mgr. products control dept., 1976-79; pres. Caterpillar Brasil S.A., 1979-81; v.p. products Caterpillar Tractor Co., Peoria, 1981-85, exec. v.p., 1985-89; pres., chief opd. officer Caterpillar Inc., Peoria, 1989-90, chmn., chief exec. officer, 1990—, also bd. dirs.; bd. dirs. Mobil Corp., First Chgo. Corp., Equip. Mfg. Inst., Ga.-Pacific Corp. Trustee Farm Found., 1985—, Meth. Med. Ctr., 1985—, Knox Coll., 1986—; mem. adv. bd. Salvation Army, 1985—, admintrv. bd. First United Meth. Ch., 1986—; bd. dirs. Valparaiso U., Keep Am. Beautiful. Mem. Agrl. Roundtable (chmn. 1985-87), SAE, ACTPN, Bus. Coun., Bus. Roundtable (policy com.). Republican. Clubs: Mt. Hawley Country, Creve Coeur. Office: Caterpillar Inc 100 NE Adams St 100 NE Adams St Peoria IL 61629-7210

FITTANTE, PHILIP RUSSELL, air force officer, pilot; b. Red Bank, N.J., May 27, 1965; s. Alfred Richard and Maryanne (Quattrocchi) F.; m. Anna Jane Andrews, June 27, 1987. BS in Math. Scis., U. N.C., Chapel Hill, 1987; MS in Computer Sci., Midwestern State U., 1992. Commd. 2d lt. USAF, 1987, advanced through grades to capt., 1991; instr. pilot 90th flying tng. squadron USAF, Sheppard AFB, Tex., 1988-90, pilot instr. trainer 88th flying tng. squadron, 1990-92, functional check flight test pilot 88th flying tng. squadron, 1992—, exec. officer 88th flying tng. squadron, 1991-92; B-1B aircraft comdr. Dyess AFB, Tex., 1993—. Speaker Ptnrs. in Edn., Wichita Falls, Tex., 1992. Mem. Air Force Assn., Phi Eta Sigma. Roman Catholic. Home: 108A Polaris St Sheppard AFB TX 76311 Office: 88th Flying Tng Squadron Bldg 2320 Sheppard AFB TX 76311

FITTON, GARY MICHAEL, electronics engineer; b. Trenton, N.J., Nov. 3, 1939; s. Arol Ashworth Fitton and Lillian Flora (Bayer) Huber; m. Carol Helen Gant, June 22, 1963; children: Gary Patrick, Stephen Paul, Laura Louise. B Engring., Stevens Inst. Tech., 1961. Assoc. engr. RCA Astrospace, Hightstown, N.J., 1961-68; sr. engr. Lockheed Electronics, Plainfield, N.J., 1968-70; staff engr. EMR Photoelectric, Princeton, N.J., 1970-77; chief engr. ORS Automation, Princeton, 1977-86; staff engr. GE Astrospace, Hightstown, 1986—. Contbr. articles to Soc. Photo-Optical Instrumentation Engrs. Jour, Machine Vision Assn. Soc. Mfg. Engrs. Mem. CAP. Achievements include research on optical data digitizer, real time pattern recognition, machine vision, Apollo and space shuttle TV systems. Home: 135 Reservoir Rd Hopewell NJ 08525 Office: GE Astrospace PO Box 800 Princeton NJ 08543

FITZGERALD, JOHN EDMUND, civil engineering educator, dean; b. Revere, Mass., Sept. 29, 1923; s. John Valentine and Gertrude Margaret (Doyle) F.; m. Elaine Louise Ohlson, Feb. 24, 1945; children: Deborah Lee, Christine Louise, David John, John Paul (dec.). Student, Tufts U., 1941-42, 46; MCE, Harvard U., 1947; MS in Math.-Physics, Nat. U. Ireland, Cork, 1970, DSc, 1972. Registered profl. engr., Utah, N.D.; chartered physicist, U.K. Regional constrn. engr. Liberty Mut. Ins. Co., Dallas, 1947-48; assoc. prof. N.D. State U., Fargo, 1948-51; supr. structures and dynamics Armour Research Found., Chgo., 1951-53; mgr. applied mechanics and med. physics Research div. Am. Machine & Foundry Corp., Chgo., 1953-56; mgr. applied math. and mechanics Borg-Warner Cen. Research Labs., Des Plaines, Ill., 1956-59; dir. devel. br. Lockheed Propulsion Co., Redlands, Calif., 1959-66; prof. civil engring., chmn. dept. U. Utah, Salt Lake City, 1966-74, prof., assoc. dean, 1973-74; prof., dir. Sch. Civil Engring. Ga. Inst. Tech., Atlanta, 1975-89, prof. emeritus, 1991—, assoc. dean, 1989-91; cons. numerous aerospace cos., govt. agys., 1966—; guest lectr. Trinity Coll., Dublin, Ireland, U. Bristol, U.K., U. Marseilles, France, NATO Advanced Study Inst., Italy, others., 1968—. Author: Engineering Structural Analysis of Solid Propellants, 1971; editor Structural Integrity Handbook, 1972; contbr. over 100 articles to profl. jours.; 27 patents. Served with submarine service USN, 1942-46, ETO. Recipient U.S. Sr. Scientist award for teaching and research Alexander von Humboldt Found., 1973-74. Fellow Inst. Physics U.K., ASCE, AIAA (assoc., Outstanding Achievement in Solid Propulsion award 1987); mem. Soc. Rheology, Am. Acad. Mechanics, Am. Phys. Soc. Roman Catholic. Club: Royal Cork Yacht (Crosshaven, Ireland). Avocations: swimming, bicycling, sailing, tennis. Home: 4252 Loch Highland Pkwy Roswell GA 30075-2042 Office: Ga Inst of Tech Coll Engring Atlanta GA 30332-0355

FITZGERALD, ROBERT HANNON, JR., orthopedic surgeon; b. Denver, Aug. 25, 1942; s. Robert Hannon and Alyene (Webber) Fitzgerald Anderson; m. Lynda Lee Lang, Apr. 27, 1968 (div. 1984); children—Robert III, Shannon, Dennis, Katherine, Kelly; m. Jamie Kathleen Dent, Mar. 9, 1985; children: Brian, Steven. B.S., U. Notre Dame, 1963; M.D., U. Kans., 1967; M.S., U. Minn., 1974. Instr. orthopedic surgery Mayo Med. Sch., Rochester, Minn., 1974-77, cons. orthopedic surgery, 1974-89, asst. prof., 1977-82, assoc. prof., 1982-86 , prof., 1986-89, chief adult reconstructive surgery, 1987-89, dir. orthopaedic rsch., 1988-89; prof. chmn. dept. orthopaedic surgery Wayne State U. Sch. Medicine, 1989—; chief orthopaedic surgery Hutzel Hosp., 1989—, Detroit Receiving Hosp., 1989—; orthopaedist-in-chief Detroit Med. Ctr., 1989—, chmn. coun., specialist-in-chief 1993—; cons. Ctr. Disease Control, Atlanta, 1981—, NIH, 1987, chmn. orthopaedic study sect., 1989-91. Assoc. editor Jour. Orthopedic and Traumatology, 1978—, Jour. Bone Joint Surgery, 1982-86, Clin. Orthopaedica and Related Research, 1988—; trustee Jour. Bone Joint Surgery, 1987-92, sec. 1988-92, Hutzel Hosp., 1989—. Mem. bd. edn. St. John's Grade Sch./Jr. High Sch., Rochester, 1983-87; mem. Bd. Devel. Mayo Clinic, 1984-87; mem. bd. devel. St. John's Ch., 1988-89; trustee Lourdes High Sch. Devel. Bd., Rochester, 1982-88. Served to capt. USAF, 1968-70. Decorated Air Commendation medal; recipient Kappa Delta award for musculoskeletal research, 1983. Fellow Am. Acad. Orthopedic Surgeons; mem. Orthopedic Research Soc., AMA, Assn. Bone and Joint Surgeons, Internat. Soc. Microbiology, Zumbro County Med. Soc., Min-Da-Man Orthopedic Soc., Minn. Orthopedic Soc., Am. Soc. Microbiology, N.Y. Acad. Scis., Am. Hip Soc. (Stinchfield award 1985, Charnley award 1986, pres. 1993—), Internat. Hip Soc., Am. Orthopedic Assn., Surg. Infection

Soc. (charter mem.), Clin. Orthopaedic Soc., Internat. Soc. Orthopaedic Surgery and Traumatology, Mid-Am. Orthopedic Soc. (bd. dirs. 1989—), Detroit Acad. Orthopedic Surgery, Mich. Orthopedic Soc., Mich. State Med. Soc., Detroit Acad. Medicine, Interurban Club, Sigma Xi, Kappa Delta, Alpha Epsilon Delta. Republican. Roman Catholic. Avocations: cross-country and downhill skiing; swimming; coaching children's sports. Home: 350 Provencal Rd Grosse Pointe MI 48236-2908 Office: Hutzel Hosp 4707 St Antoine St Detroit MI 48201-1498

FITZGERALD, WILLIAM F., chemical oceanographer, educator; b. Boston, Aug. 16, 1936; s. Thomas Francis and Julia Agnes (McDonough) F.; m. Jeanette Mae Dunlop, June 4,1960 (dec. 1971); 1 child, Julie Ann; m. Patricia Marie Kelly, May 28, 1972. BS in Chemistry, Boston Coll., 1960; MS in Chemistry, Coll. of Holy Cross, 1961; PhD, MIT/Woods Hole Oceanographic, 1970. Prof. chem. oceanography U. Conn., Groton, 1970—. Contbr. articles to sci. jours. Grantee NSF, 1970—. Home: Box 613 Old Lyme CT 06371 Office: Univ Conn Groton CT 06340

FITZMAURICE, MICHAEL WILLIAM, electrical and mechanical engineer; b. Washington, Mar. 29, 1939; s. Patrick Francis and Kathleen (Rogers) F.; m. Jeanne Ethel Miller, Nov. 10, 1962; children: Michael Jr., Karen. BSME, Md. U., 1964, MSEE, 1966, PhD, 1969. Engr. optical systems br. GSFC NASA, Greenbelt, Md., 1964-70, head space info. systems sect. GSFC, 1970-78, head electro-optical inst. br. GSFC, 1978-88, asst. divsn. chief GSFC, 1988-92, head inst. systems engr. office GSFC, 1992—; lectr. in field. Contbr. articles to Jour. Optical Soc. Am., Applied Optics. pres. ch. parish coun., Gambrills, Md., 1985, swim club, 1982; coach little league baseball, 1973. Mem. Optical Soc. Am. Catholic. Achievements include patents for Focus Spoiling Retrodirective Modulator, Polarization Compensator for Optical Communications, Channel Simulator for Optical Communication Systems, Multi-Access Laser Communications Transceiver System. Home: 927 Winterhaven Dr Gambrills MD 21054 Office: NASA GSFC Code 704 Greenbelt MD 20771

FITZPATRICK, THOMAS BERNARD, dermatologist, educator; b. Madison, Wis., Dec. 19, 1919; s. Joseph J. and Grace (Lawrence) F.; m. Beatrice Devaney, Dec. 27, 1944; children: Thomas B., Beatrice, John, L. Scott, Brian. BA with honors, U. Wis., 1941; MD, Harvard U., 1945; fellow, Mayo Found., 1948-51; PhD, U. Minn., 1952; fellow, Commonwealth Fund, Oxford, 1958-59; DSc (hon.), U. Mass., 1987. Intern 4th (Harvard) Med. Service, Boston City Hosp., 1945-46; biochemist Army Med. Ctr., Md., 1946-48; asst. prof. dermatology U. Mich. Med. Sch., 1951-52; prof., head div. dermatology U. Oreg. Med. Sch., 1952-58; Edward Wigglesworth prof. dermatology Harvard Med. Sch., 1959-90, prof. emeritus, 1990—, head dept., 1959-87; chief dermatology service Mass. Gen. Hosp., Boston, 1959-87; Prosser White orator St. John's Dermatol. Soc., London, 1964; Dohi Internat. exchange lectr. on dermatology, Japan, 1969; spl. cons. USPHS, NIH; cons. dermatology Peter Bent Brigham Hosp., Children's Hosp. Med. Ctr., Boston, 1962; mem. sci. adv. bd. EPA, 1985; mem. climatic impact com., chmn. health effects NAS; pres. Dermatology Found., 1971, Internat. Pigment Cell Soc., 1978-81, Assn. Profs. Dermatology, 1983. Chief editor: Dermatology in General Medicine, 1971, 3d edit., 1987, 4th edit., 1993; mem. editorial bd. New Eng. Jour. Medicine, 1961-69; editor Year Book Dermatology, 1984—. Decorated officer Order of Rising Gold Rays (Japan), 1986; recipient Mayo Found. Alumni Rsch. award, 1951, Outstanding Achievement award U. Minn. Bd. Regents, 1964, Myron Gordon award 6th Internat. Pigment Cell Conf., 1965, Disting. Svc. award Dermatology Found., 1989, U. Wis., 1983, award for discovery of PUVA photochemotherapy for psoriasis Nat Psoriasis Found., 1993; Outstanding Achievement in field of Cultaneous Melanoma World Health Orgn., 1993. Fellow Am. Acad. Dermatology (hon., master, past bd. dirs.); mem. Royal Soc. Medicine (hon.), Am. Acad. Arts and Scis., Assn. Am. Physicians, Soc. Investigative Dermatology (hon., pres. 1959-60, Stephen Rothman award, gold medal 1970, Outstanding Achievement award in Cutaneous Melanoma WHO Melanoma program 1993), Am. Soc. for Clin. Investigation (emeritus 1965), Brit. Assn. Dermatology (hon.), South African Dermatol. Soc. (hon.), Med. Assn. Israel Dermatol. Soc. (hon.), St. John's Hosp. Dermatol. Soc. (London, hon.), Argentina, Danish, Italian, Finnish, German, Polish, Austrian dermatol. socs. (hon.), Pacific Dermatologic Assn. (hon.), French Soc. Dermatology and Syphiligraphy (fgn. corr.), Australasian Coll. Dermatologists, Alpha Omega Alpha. Home: 209 Newton St Weston MA 02193-2338 Office: Mass Gen Hosp Dermatology Svc 32 Fruit St Boston MA 02114-2698

FITZSIMMONS, JOHN PATRICK, electrical engineer; b. Freeport, Ill., Aug. 2, 1964; s. Yvonne Jean (Hannan) Fitzsimmons; m. Vicki L. Mueller, Sept. 2, 1989. BSEE, Iowa State U., 1987. Application engr. microswitch div. Honeywell, Freeport, 1987-89, design engr. microswitch div., 1989—; mem. Tech. Adv. Com., Freeport, 1991—. Achievements include co-design of industrialized, on-line color sensing which approaches the capabilities of lab-based equipment; engineering support for high speed, high accuracy industrial machine vision products. Office: Honeywell 11 W Spring St B4-523 Freeport IL 61032

FITZSIMMONS, SOPHIE SONIA, interior designer; b. Paris, July 6, 1943; came to U.S., 1947; d. Oleg and Sophie (Ovsianico-Koulikovsky) Yadoff; m. J. Heath Fitzsimmons, Sept. 8, 1962; children: Gregory James, Raymond Heath, Douglas Paul. AAS with honors, Fashion Inst. Tech., N.Y.C., 1964. Design intern Euster Assocs., Inc., Armonk, N.Y., 1964; prin. Sophie Y. Fitzsimmons Interior Design, N.Y., Conn, 1964-77; co-owner Avon (Conn.), Interiors, Inc., 1977-89; prin. Sophie Fitzsimmons Interior Design, N.Y.C., 1989—; guest exhibitor Fashion Inst. Tech. Symposium, 1984. Chair ann. show Hope Benefit, Hartford, Conn., 1975; mem. Rep. Women's Club Conn., 1978-89; bd. dirs. Friends of Hartford Ballet, 1986-88; vol. N.Y. Commn. UN, Consular Corps and Internat. Bus., 1992—; vol. tchr. East Internat. Community Ctr., 1993—; pres., bd. dirs. Squadron Line PTAA, 1976; bd. dirs. Simsbury chpt. Federated Women's Club, 1976. Decorated Medal of Recognition, French Resistance Movement, World War II; recipient Award Edn. Civique Chevalier. Mem. Nat. Soc. Interior Designers (adv. panel 1967), Hartford Stage Co. Stagehands, World Affairs Coun. (exec. forum), Mark Twain Meml. Wadsworth Atheneum, Bushnell Meml., Simsbury Farms Golf Assn. (bd. dirs. 1989), Bamm Hollow Women's Golf Assn. (bd. dirs. 1992). Avocations: French and Russian languages, tennis, bridge, drawing, travel. Office: Sophie Fitzsimmons Interior 55 Liberty St New York NY 10005-1015

FITZSIMONS, CHRISTOPHER, design engineer; b. Oak Park, Ill., Aug. 17, 1964; s. Robert Christopher and Kathleen (Daugherty) F.; m. Stephanie Lorraine Riedel, Dec. 15, 1990. AS with honors, Elgin (Ill.) C.C., 1984; B of Engring., Iowa State U., 1988. Engr. scientist McDonnell Douglas Space Systems Co., Huntington Beach, Calif., 1989-92; design engr. Sr. Flexonics, Bartlett, Ill., 1992—; lead engr. Resistojet Thrusters, Huntington Beach, 1989-92, REM configuration mgr., 1992. Vol. Seal Beach (Calif.) Animal Care Ctr., 1989-92, bd. dirs. 1992; adv. Boy Scout Explorer Post 5, Elgin, 1983. Mem. AIAA. Roman Catholic. Office: Sr Flexonics 300 E Devon Ave Bartlett IL 60103

FITZSTEPHENS, DONNA MARIE, biologist; b. Phila., Nov. 28, 1966; d. Vincent James and Loretta Ann (Mingarino) Cicirello; m. Scott Edward Fitzstephens, Sept. 14, 1991. BS in Biology, Villanova U., 1988; PhD in Zoology, Mich. State U., 1994. Animal caretaker Villanova (Pa.) U., 1985-87, animal care supr., 1987-88, rsch. asst., 1987-90; teaching asst. Mich. State U., East Lansing, 1989; rsch. asst. Kellogg Biol. Sta., Hickory Corners, Mich., 1990, teaching asst., 1990, NSF grad. rsch. fellow, 1990-93; test supr. Ednl. Testing Svcs., Kalamazoo, 1991—; advisor undergrad. ind. rsch. project, 1990. Contbr. articles to profl. jours. Theodore Roosevelt Meml. grantee Am. Mus. Natural History, 1992. Mem. Sigma Xi (grantee 1990, 92), Phi Beta Kappa, Phi Kappa Phi, Phi Sigma. Achievements include research in age and status related color changes in male black-winged damselflies, calopteryx maculata; previous research includes mating and social behavior in white-footed mice, peromyscus leucopus and deer mice, peromyscus maniculatus. Home: 9952 N 40th St Hickory Corners MI 49060 Office: Kellogg Biol Sta 3700 E Gull Lake Dr Hickory Corners MI 49060

FIXMAN, MARSHALL, chemist, educator; b. St. Louis, Sept. 21, 1930; s. Benjamin and Dorothy (Finkel) F.; m. Marian Ruth Beatman, July 5, 1959 (dec. Sept. 1969); children—Laura Beth, Susan Ilene, Andrew Richard; m. Branka Ladanyi, Dec. 7, 1974. A.B., Washington U., St. Louis, 1950; Ph.D., MIT, 1954. Jewett postdoctoral fellow chemistry Yale U., 1953-54; instr. chemistry Harvard U., 1956-59; sr. fellow Mellon Inst., Pitts., 1959-61; prof. chemistry, dir. Inst. Theoretical Sci., U. Oreg., 1961-64, prof. chemistry, research asso. inst., 1964-65; prof. chemistry Yale U., New Haven, 1965-79; prof. chemistry and physics Colo. State U., Ft. Collins, 1979—. Assoc. editor Jour. Chem. Physics, 1962-64, Jour. Phys. Chemistry, 1970-74, Macromolecules, 1970-74, Accounts Chem. Research, 1982-85, Jour. Polymer Sci. B, 1991—. Served with AUS, 1954-56. Fellow Alfred P. Sloan Found., 1961-63; recipient Governor's award Oreg. Mus. Sci. and Industry, 1964. Mem. NAS, Am. Acad. Arts and Scis., Am. Chem. Soc. (award pure chemistry 1964, award polymer chemistry 1991), Am. Phys. Soc. (high polymer physics award 1980), Fedn. Am. Scientists. Office: Colo State U Dept Chemistry Fort Collins CO 80523

FJERDINGSTAD, EJNAR JULES, retired biological scientist and educator; b. Copenhagen, Jan. 28, 1937; s. Einer Svend Aage and Else Emilie Sofie (Andersen) F.; m. Karen Madsen, Aug. 10, 1963; children: Svend Jules, Else Juliette. PhD, U. Copenhagen, 1962, U. Bergen, Norway, 1975. Asst. dept. zoology U. Copenhagen, 1961-67, 71-72; rsch. assoc. dept. biochemistry Duke U., Durham, N.C., 1967-68; rsch. assoc. dept. anesthesiology Baylor Coll. Medicine, Houston, 1968-69; asst. prof. biochemistry U. Tenn., Memphis, 1969-71; assoc. prof. U. Aarhus, Denmark, 1972-80, ret., 1980. Author: Cell Biology Perspectives, 1980; contbr., editor: Chemical Transfer of Learned Information, 1971; contbr. sci. articles to profl. jours. Mem. AAAS, Am. Cetacean Soc., Internat. Brain Rsch. Orgn., Planetary Soc., Jane Goodall Inst., Gorilla Found. Avocations: history, classical music, photography. Home: Svendgaardsvej 52, DK-8330 Beder Denmark

FLACKS, LOUIS MICHAEL, consulting physician; b. Manchester, Lancashire, Eng., Mar. 24, 1937; s. Maurice and Lily Ruby (Shaffer) F.; m. Denise Dine (div. Feb. 1976); children: Sarah Louise, David Nathaniel; m. Barbara Jane Buttimer, Feb. 9, 1979. B in Medicine and Surgery, Leeds (Eng.) U., 1962. Sr. house officer North Devon (Eng.) Inf., 1964-65; med. registrar West Park Macclesfield, Eng., 1965-67, Stepping Hill Hosp., Stockport, Eng., 1967-69, Leeds Gen. (Eng.) Inf., 1969; med. adviser Glaxo Labs., London, 1969-71; cons. physician Whakatane (New Zealand) Hosp., 1972-77, Rotorua (New Zealand) Hosp., 1977-86; cons. physician, dir. community and geriatric medicine Fremantle (Western Australia) Hosp., 1987—; chmn. med. edn. Bay of Plenty (New Zealand) Hosp. Bd., 1972-75; divisional chmn. Nat. Kidney Found., New Zealand, 1983-84. Scrutineer Liberal Party, Perth, Western Australia, 1990. Mem. Rotorua-Taupo Post Grad. Med. Soc. (sec. New Zealand chpt.), Rotorua Multiple Sclerosis Soc. (hon.), N.Y. Acad. Scis., Canning Bridge Club, Melville Bridge Club. Jewish. Avocations: photography, bridge, chess, literature, music. Office: Fremantle Hosp, Alma St, Fremantle 6160, Australia

FLAGAN, RICHARD CHARLES, chemical engineering educator; b. Spokane, Wash., June 12, 1947; s. Robert George and Frances Cory (Arnold) F.; m. Aulikki Tellervo Pekkala, Aug. 4, 1979; children: Mikko, Suvi, Taru. BME, U. Mich., 1969; MME, MIT, 1971, PhDME, 1973. Research assoc. MIT, Cambridge, 1973-75; asst. prof. environ. engring. sci. Calif. Inst. Tech., Pasadena, 1975-81, assoc. prof., 1981-85, prof. environ. engring. sci. and mech. engring., 1986-90; prof. chem. engring. Calif. Inst. of Tech., Pasadena, 1990—. Assoc. editor Aerosol Sci. and Tech. Mem. AICE, Am. Assn. Aerosol Rsch. (Sinclair award 1993, treas.), Gesellschaft fur Aerosolforschung (Smoluchowski award 1990). Office: Calif Inst of Tech Dept Chem Engring Pasadena CA 91125

FLAGG, RAYMOND OSBOURN, biology executive; b. Martinsburg, W.Va., Jan. 31, 1933; s. Dorsey Slemons and Dorothy (Hobbs) F.; m. Ann Quinlan Birmingham, Oct. 3, 1956; children: Richard Matthew, Elizabeth Ann, Catherine Garnett. BA with honors, Shepherd Coll., 1957; PhD in Biology, U. Va., 1961. Math tchr. Boonsboro (Md.) High Sch., 1957; rsch. asst. Blandy Exptl. Farm, Boyce, Va., 1957-61; rsch. assoc. U. Va., Charlottesville, 1961-62; dir. Botany Carolina Biol. Supply Co., Burlington, N.C., 1962-80, v.p., 1980—; v.p. Wolfe Sales Corp., Burlington, 1985—; head Cabisco Biotech., Burlington, 1988-91; v.p. Found. for Ednl. Devel., Research Triangle Park, N.C., 1983-85; vice chmn. N.C. Plant Conservation Bd., Raleigh, 1984-88. Contbr. articles to profl. jours. Chmn. Beautification Commn., Burlington, 1976-80; chmn. Hist. Dist. Commn., Burlington, 1981-82; bd. dirs. United Way of Alamance County, Burlington, 1984-88. With U.S. Army, 1952-55. Rsch. grant Am. Cancer Soc., 1960, rsch. equipment grant Va. Acad. Sci., 1961; recipient Community Leadership award No. Piedmont Devel. Assn., 1977. Mem. AAAS, Assn. Southeastern Biologists (pres. 1978-79), N.C. Acad. Sci. (pres. 1983-84), Am. Inst. Biol. Scis., Va. Acad. Sci., Rotary (pres. Alamance A.M. 1988-89). Democrat. Presbyterian. Achievements include invention of instant drosophila medium, Carosafe, FlyNap, Sterigel, Planoslo, Vitachrome, Alga-Gro. Office: Carolina Biol Supply 2700 York Rd Burlington NC 27215-3398

FLAGG, ROBERT FINCH, research aerospace engineer; b. Somerville, Mass., Mar. 6, 1933; s. Donald Fairbanks and Helen Constance (Finch) F.; m. Lois-Ann Davis Laughton, June 14, 1958 (div. 1975); children: Scott, Susan, Marc. BS in Aero. Engring., MIT, 1959, MS in Aero. and Astronautical Engring., 1960; PhD in Engring. Physics, U. Toronto, Ont., Can., 1967. Teaching asst. MIT, Cambridge, 1959; rsch. assoc. U. Toronto, 1964-67; program mgr. Physics Internat. Co., San Leandro, Calif., 1967-68; dir. rsch. Holex Inc., Hollister, Calif., 1968-71; v.p. tech. ops. X-Demex Corp., Dublin, Calif., 1972-79; program mgr. Artec Assocs., Hayward, Calif., 1979-80; mgr. ordnance engring. Tracor Aerospace, San Ramon, Calif., 1980-84; tech. dir. Tracor Aerospace, Camden, Ark., 1986-90; dir. IR countermeasures Bermite div. Whittaker Corp., Saugus, Calif., 1984-85; R & D scientist Lockheed Advanced Aeros., Valencia, Calif., 1985-86; sr. engring. specialist Aerojet Ordnance, Downey, Calif., 1991—. Contbr. numerous articles to sci. and tech. jours. Staff sgt. USAF, 1950-54. Staff scholar MIT, 1954-59, N. Stewart Robinson scholar U. Toronto, 1964-67, U. Toronto Inst. Aerospace Studies scholar, rsch. fellow, Presdl. fellow U. Calif. Lawrence Berkeley Lab., 1971-72. Mem. AIAA, Am. Def. Preparedness Assn., Soc. Explosives Engrs., Internat. Pyrotechnic Soc., Sigma Gamma Tau. Republican. Achievements include patents for ordnance and ordnance instrumentation. Home: 5793 Greenridge Rd Castro Valley CA 94552 Office: Aerojet Ordnance 9236 E Hall Rd Downey CA 90241

FLAGLE, CHARLES LAWRENCE, pharmaceutical industry software firm executive; b. Balt., Feb. 27, 1949; s. Charles Denhard and Lois (Hagaman) F.; m. Judy Marie Riley, Aug. 17, 1976; children: Heather Marie, Charles David II. BSc, Towson State U., 1976. Systems programmer Md. State Colls. Info. Ctr., Towson, 1974-78, project leader, mgr., 1978-84; founder, prin. cons. Rsch. Info. Assocs., Inc., Towson, 1978-84, pres., chief exec. officer, 1984—. Contbr. articles to profl. jours. With U.S. Army, 1970-72, Vietnam. Grantee U.S. Dept. HHS Ctrs. for Disease Control, 1989-93. Mem. Assn. for Computing Machinery, IEEE Computer Soc., AAAS. Unitarian-Universalist. Office: Rsch Info Assocs Inc 809 Glen Eagles Ct Ste 111 Baltimore MD 21204-6210

FLAKNE, DAWN GAYLE, electronics engineer; b. Mpls., Jan. 29, 1959; d. John D. and Gail L. F. BA in Math., Computer Sci., Augustana Coll., 1981; BS in Computer Engring., U. Ill., 1981. Devel. engr. Universal Oil Products, Des Plaines, Ill., 1981-84, devel. coord., 1984-86, sr. devel. coord., 1986-89, group leader, 1989—. Winner IR-100 award R&D mag., 1986. Mem. IEEE. Achievements include design of software for hydrogen monitor. Office: UOP 175 W Oakton St Des Plaines IL 60018

FLANAGAN, FREDERICK JAMES, water systems engineer; b. Poughkeepsie, N.Y., Sept. 2, 1941; s. Fredrick and Jane (Poplawska) F.; m. Sandra Dianni, July 5, 1969; 1 chld, Kathleen. Student, Rensselaer Poly. Inst., Troy, N.Y., 1959-60; BA, SUNY, Oneonta, 1967; MA, SUNY, New Paltz, 1974. Tchr. math., sci. various pub. schs., N.Y., 1967-73; gen.mgr. WKIP Radio, Poughkeepsie, N.Y., 1969-72; pubr. Hudson Valley Mag., Poughkeepsie, N.Y., 1972-74; pres., owner Aqua King Internat., Poughkeepsie, N.Y., 1974—; reg. dir. R.E.T.A., Chgo., 1985-89. Author: Hexameron, 1975, Reflections, 1981, The Nature of Water, 1989. Sec. Arlington Relays, Poughkeepsie, 1973—, Sports Mus. Dutchess County, 1972—. With C.E., U.S. Army, 1960-63. Mem. Hall of Fame, Sports Mus. of Dutchess County, 1982, Dutchess County Slow Pitch Softball, 1986. Mem. New Eng. Ice Assn., Mid-Atlantic Ice Assn., Packaged Ice Assn. (stds. com. 1985-86), Cooling Tower Inst., Refrigerating Engrs., Am. Water Wks. Assn., Apple Valley Softball League (pres. 1973—), Bridge City Bowling League (pres. 1972—), Elks. Roman Catholic. Avocations: sports, photography, writing, travel. Home: 48 Mandalay Dr Poughkeepsie NY 12603-2633 Office: Aqua King Internat 22 Freedom Plains Rd Poughkeepsie NY 12603-2600

FLANAGAN, JAMES LOTON, electrical engineer, educator. B.S. in Elec. Engring., Miss. State U., 1948; S.M. in Elec. Engring., MIT, 1950; Sc.D. in Elec. Engring., M.I.T., 1955. Mem. elec. engring. faculty Miss. State U., 1950-52; mem. tech. staff Bell Labs., Murray Hill, N.J., 1957-61; head dept. speech and auditory research Bell Labs., 1961-67, head dept. acoustics research, 1967-85, dir. info. prins. research lab., 1985-90; dir. ctr. for computer aids for indsl. productivity Rutgers U., Piscataway, N.J., 1990—. Author: Speech Analysis, Synthesis and Perception, 1972; contbr. numerous articles to profl. jours. Mem. evaluation panel Nat. Bur. Standards/NRC, 1972-77; mem. adv. panel on White House tapes U.S. Dist. Ct. for D.C., 1973-74; bd. govs. Am. Inst. Physics, 1974-77; mem. sci. adv. bd. Callier Center, U. Tex., Dallas, 1974-76; mem. sci. adv. panel on voice communications Nat. Security Agy., 1975-77; mem. sci. adv. bd. div. communications research Inst. Def. Analyses, 1975-77. Recipient Disting. Svc. award in sci. Am. Speech and Hearing Assn., 1977, L.M. Ericsson Internat. prize in telecomms., 1985; Marconi Internat. fellow, 1992. Fellow IEEE (mem. fellow selection com. 1979-81, Edison medal 1986), Acoustical Soc. Am. (assoc. editor Speech Comm. 1959-62, exec. coun. 1970-73, v.p. 1976-77, pres. 1978-79, Gold Medal award 1986); mem. NAE, NAS, Acoustics, Speech and Signal Processing Soc. (v.p. 1967-68, pres. 1969-70, Achievement award 1970, Soc. award 1976). Achievements include U.S. and foreign patents in field. Office: Rutgers U Ctr Computer Aids for Indsl Productivity Piscataway NJ 08855

FLANIGEN, EDITH MARIE, materials scientist. Sr. rsch. fellow materials sci. VOP Tarrytown (N.Y.) Tech. Ctr. Recipient Perkin medal Am. Chem. Soc., 1992, Francis P. Garvan-John M. Olin medal Am. Chem. Soc., 1993. Office: UOP Tarrytown Tech Ctr 777 Old Saw Mill River Rd Rte 100 C Tarrytown NY 10591*

FLANNELLY, KEVIN J., psychologist, research analyst; b. Jersey City, Nov. 26, 1949; s. John J. and Mary C. (Walsh) F.; m. Laura T. Adams, Jan. 10, 1981. BA in Psychology, Jersey City State Coll., 1972; MS in Psychology, Rutgers U., 1975; PhD in Psychology, U. Hawaii, 1983. Rsch. asst. dept. psychology U. Ill., Champaign, 1972-73; rsch. intern Alcohol Behavior Rsch. Lab. Rutgers U., New Brunswick, N.J., 1973-75; rsch. scientist Edward R. Johnstone Tng. and Rsch. Ctr., Bordentown, N.J., 1975-78; teaching asst. dept. psychology U. Hawaii, Honolulu, 1980-81, rsch. asst. Pacific Biomed. Rsch. Ctr., 1981-83, asst. prof. Bekesy Lab. Neurobiology, 1983-85; rsch. statistician, statewide transp. planning office Hawaii Dept. Transportation, Honolulu, 1986-89; researcher Office of Lt. Gov., Honolulu, 1989—; statis cons. U. Hawaii Sch. Nursing, Honolulu, 1986, Hawaii Dept. Health, Honolulu, 1986; v.p., rsch. dir. Ctr. Psychosocial Rsch., Honolulu, 1987—; instr. dept. social scis. Honolulu Community Coll., 1981; ptnr. Flannelly Cons., 1991—; rsch. dir. Mktg. Rsch. Inst., 1992—. Editor: Biological Perspective on Aggression, 1984, Introduction to Psychology, 1987; reviewer 8 sci. and profl. jours., 1978—; grant reviewer NSF, 1984—; contbr. numerous articles to profl. jours. Polit. survey cons., Honolulu, 1988—; transp. cons., Honolulu, 1989—; mktg. cons., Honolulu, 1990—. Grantee NIH, 1984, Fed. Hwy. Adminstrn., 1987; N.J. State scholar N.J. Dept. Higher Edn., 1968-72. Fellow Internat. Soc. Rsch. on Aggression; mem. AAAS, Am. Psychol. Soc., Am. Statis. Assn., Internat. Soc. Comparative Psychology, N.Y. Acad. Scis., Pacific and Asian Affairs Coun., Psychonomic Soc., Sigma Xi. Achievements include research in social and emotional behavior, transportation planning, policy analysis, stochastic models of decision-making. Home: 445 Kaiolu St Apt 1207 Honolulu HI 96815-2255 Office: Office of Lt Gov Hawaii State Capitol Honolulu HI 96813

FLANNELLY, LAURA T., mental health nurse, nursing educator, researcher; b. Bklyn., Nov. 7, 1952; d. George A. Adams and Eleanor (Barragry) Mulhearn; m. Kevin J. Flannelly, Jan. 10, 1981. BS in Nursing, Hunter Coll., 1974; MSN, U. Hawaii, 1984, postgrad., 1988—. RN, N.Y., Hawaii. Psychiat. nurse Bellevue Hosp., N.Y.C., 1975, asst. head nurse, 1975-77; psychiat. nurse White Plains (N.Y.) Med. Ctr., 1978-79; community mental health nurse South Beach Psychiat. Ctr., N.Y.C., 1979-81; psychiat. nurse The Queen's Med. Ctr., Honolulu, 1981-83; crisis worker Crisis Response Systems Project, Honolulu, 1983-86; instr. nursing U. Hawaii, Honolulu, 1985-92, asst. prof., 1992—; adj. instr. nursing Hawaii Loa Coll., Honolulu, 1988, Am. Samoa Community Coll., Honolulu, 1987, 89, 90; mem. adv. bd., planning com. Psychiat. Day Hosp. of The Queen's Med. Ctr., Honolulu, 1981-82; program coord. Premenstrual Tension Syndrome Conf., Honolulu, 1984; dir. Ctr. Psychosocial Rsch., Honolulu, 1987—; program moderator 1st U.S-Japan Health Behavioral Conf., Honolulu, 1988; faculty Ctr. for Asia-Pacific Exch., Internat. Conf. on Transcultural Nursing, Honolulu, 1990. Contbr. articles to profl. jours. N.Y. State Bd. Regents scholar, 1970-74; NIH nursing trainee, 1983-84; grantee U. Hawaii, 1986, 91, Hawaii Dept. Health, 1990. Fellow Internat. Soc. Rsch. on Aggression; mem. AAAS, Am. Ednl. Rsch. Assn., Am. Psychol. Soc., Am. Statis. Assn., Nat. League for Nursing, N.Y. Acad. Scis., Pacific and Asian Affairs Coun., Sigma Theta Tau. Achievements include research in aggressive behavior, learning styles, problem-based learning, cross-cultural differences, statistical modeling. Home: 445 Kaiolu St Apt 1207 Honolulu HI 96815-2255 Office: U Hawaii Sch Nursing Webster Hall Honolulu HI 96822

FLANNERY, BRIAN PAUL, physicist, educator; b. Utica, N.Y., July 30, 1948; m. Sharon Ann Parkinson, May 23, 1970; children: Colleen Catherine, Paul Edward. AB, Princeton U., 1970; PhD, U. Calif., Santa Cruz, 1974. Postdoctoral fellow Inst. Advanced Study, Princeton, N.J., 1974-76; asst. prof. astronomy Harvard U., Cambridge, Mass., 1976-80, assoc. prof., 1980; physicist Exxon Rsch. and Engring. Co., Clinton, N.J., 1980—; mem. global climate change working group Internat. Petroleum Industry Environ. Conservation Assn., 1988—; tech. adviser panel World Coal Inst., 1992—. Author: Numerical Recipes, 1986, 2d edit., 1992; editor: Global Climate Change: A Petroleum Industry Perspective, 1992; editorial com. Ann. Revs. Energy and Environ., 1992—. Mem. Am. Phys. Soc., Internat. Astron. Union, Am. Petroleum Inst. Achievements include patents for x-ray microtomography. Office: Exxon Rsch and Engring Co Rte 22 E Annandale NJ 08801

FLANNERY, KENT V., anthropologist, educator. Prof. anthropology U. Mich., Ann Arbor. Recipient Alfred Vincent Kidder award Am. Anthrop. Assn., 1992. Office: Univ of Mich Dept of Anthropology Ann Arbor MI 48109*

FLANNERY, WILBUR EUGENE, health science association administrator, internist; b. New Castle, Pa., June 19, 1907; s. Charles Francis and Mary Catherine (McGrath) F.; m. Ruth Iva Donaldson, June 27, 1929; children: Charles, John, Richard, Harry. BA, Dartmouth Coll., 1929; MA, Oberlin Coll., 1930; MD, Harvard U., 1935. Diplomate Nat. Bd. Medical Examiners. Minister Meth. Ch., New Castle, 1930-31; intern Cleve. City Hosp., 1935-36; resident physician Jameson Meml. Hosp., New Castle, 1936-37; fellowship Cleve. Clinic Found., 1937-40; practice medicine specializing in internal medicine New Castle, 1940—; med. dir. Hospice of St. Francis Hosp., New Castle, 1987—; chmn. bd. Pa. Blue Shield, Harrisburg, Pa., 1975-80. Contbr. numerous articles to med. jours. Pres. Bd. of Edn., New Castle, 1947-53; former trustee Knoville (Tenn.) Coll.; former pres. or chmn. Lawrence County chpt. ARC, Lawrence County chpt. Pa. Assn. for Blind, Lawrence County Mental Health Clinic, New Castle Exec. Club, Greater New Castle C. of C. Recipient Disting. Citizens award Optimists Club, 1974; named Boss of Yr., Am. Bus. Women's Assoc., 1987. Mem. AMA (del. 1953-63), Pa. Med. Soc. (pres. 1963-64), Lawrence County Med. Soc. (sec. 1954-55, pres. 1955-56), Am. Soc. Internal Medicine, Am. Med. Writers Assn., Acad. Hospice Physicians (pres. 1990), Internat. Platform Assn., Pa. Soc., New Castle Country Club, Univ. Club (Pitts.), Lawrence Club, Youngstown (Ohio) Club, Elks, Lions (pres. New Castle 1943-44, Disting. Svc. award). Republican. Presbyterian. Home: 106 E Hazelcroft Ave New

Castle PA 16105-2133 Office: Hospice St Francis 1000 S Mercer St New Castle PA 16101-4673

FLANSBURGH, EARL ROBERT, architect; b. Ithaca, N.Y., Apr. 28, 1931; s. Earl Alvah and Elizabeth (Evans) F.; m. Louise Hospital, Aug. 27, 1955; children: Earl Schuyler, John Conant. B.Arch., Cornell U., 1954; M.Arch., MIT, 1957; S.C.M.P., Harvard U. Sch. Bus., 1982. Job capt., designer The Architects Collaborative, Cambridge, Mass., 1958-62; partner Freeman, Flansburgh & Assos., Cambridge, 1961-63; prin. Earl R. Flansburgh & Assocs., Cambridge, 1963-69, pres., dir. design, 1969—; bd. dirs. daka, Inc.; exec. v.p. Environment Systems Internat., Inc.; vis. prof. archtl. design Mass. Inst. Tech., 1965-66; instr. art Wellesley Coll., 1962-65, lectr. art, 1965-69; cons. Arthur D. Little, Inc., Cambridge, 1964-70. Archtl. works include Weston (Mass.) High Sch. Addition, 1965-67, Cornell U. Campus Store, 1967-70, Cumnock Hall, Harvard U. Bus. Sch, 1973-75, Acton (Mass.) Elementary schs, 1966-68, 69-71, Wilton (Conn.) High Sch, 1968-71, 14 Story St. Bldg, 1970, Boston Design Ctr., 1985-86, Glenwood Sch., Dallas, 1985-88, New Univ. No. B.C., Prince George, Can., 1991—; exhibited works Light Machine I, IBM Gallery, N.Y.C., 1958, Light Machine II, Carpenter Center, Harvard, 1965, 5 Cambridge Architects, Wellesley Coll., 1969, Work of Earl R. Flansburgh and Assos, Wellesley Coll., 1969, New Architecture in New Eng, DeCordova Mus., 1974-75, Residential Architecture, Mead Art Gallery, Amherst Coll., 1976, works represented in, 50 Ville del Nostro Tempo, 1970, Nuove Ville, New Villas, 1970, Vacation Houses, 1970, Vacation Houses, 2d edit., 1977, Interior Design, 1970, Drawings by American Architects, 1973, Interior Spaces Designed by Architects, 1974, New Architecture in New England, 1974, Great Houses, 1976, Architecture Boston, 1976, Presentation Drawings by American Architects, 1977, Architecture, 1970-1980, A Decade of Change, 1980, Old and New Architecture, A Design Relationship, 1980, 25 Years of Record Houses, 1981, School Ways: The Planning and Design of American Schools, 1992; Author: (with others) Techniques of Successful Practice, 1975. Chmn. architecture com. Boston Arts Festival, 1964, Downtown Boston Design adv. com.; bd. dirs. Cambridge Ctr. Adult Edn.; trustee Cornell U., 1972—; chmn. bldgs. and properties com., 1976-87; mem. exec. com. academic affairs com.; class sec. SCMP VII Harvard Bus. Sch., 1982-89. Served to 1st lt. USAF, 1954-56. Recipient design awards Progressive Architecture, design awards Record Houses, design awards AIA, design awards City of Boston, design awards Mass. Masonry Inst., spl. design citations Am. Assn. Sch. Adminstrs., spl. 1st prize Buffalo-Western N.Y. chpt. AIA Competition., Walter Taylor award Am. Assn. Sch. Adminstrs., 1986; Fulbright research grantee Bldg. Research Sta., Eng., 1957-58. Fellow AIA; mem. Royal Inst. Brit. Architects, Boston Soc. Architects (chmn. program com., 1969-71, commr. pub. affairs 1971-73, commr. design 1973-74, dir. 1971-74, pres. 1980-81), Boston Found. Architecture (treas. 1984-89), Cornell U. Coun., Quill and Dagger Soc., Tau Beta Pi. Home: 225 Old County Rd Lincoln MA 01773-4601 Office: 77 N Washington St Boston MA 02114-1908

FLATTÉ, MICHAEL EDWARD, physicist, researcher; b. Walnut Creek, Calif., Apr. 14, 1967; s. Stanley Martin and Renelde Marie (Demeure) F.; m. Jennifer Beatrice Kirsch, Aug. 20, 1989; 1 child, Devra Tamar. AB, Harvard U., 1988; PhD, U. Calif., Santa Barbara, 1992. Teaching asst., rsch. asst. dept. physics U. Calif., Santa Barbara, 1989-92, postdoctoral rsch. assoc. Inst. Theoretical Physics, 1992-93; postdoctoral rsch. fellow divsn. applied scis. Harvard U., Cambridge, Mass., 1993—. NSF fellow, 1988, Russell and Sigurd Varian fellow Am. Vacuum Soc., 1991. Mem. Am. Phys. Soc. Office: Harvard U Divsn Applied Scis Pierce Hall 204C Cambridge MA 02138

FLEENER, TERRY NOEL, marketing professional; b. Ottumwa, Iowa, May 26, 1939; s. Lowell F. and Freda B. (Sparks) F.; m. Jane A. Bacon, Dec. 9, 1969; children: Clinton Todd, Clayton Scott. BSME, U. Iowa, 1963. Engr. Bendix Corp., Davenport, Iowa, 1963-67, Ball Aerospace, Boulder, Colo., 1967-74; bus. mgr. Ball Aerospace, Boulder, 1974-78; v.p. gen. mgr. Entropy Ltd., Boulder, 1978-80; pres. Energy Bank, Inc., Golden, Colo., 1980-82; program mgr. Ball Aerospace, Boulder, 1982-84, dir. mktg., 1984—; pres. U.S. Rugby Assn., Colorado Springs, 1987-89, Pam-Am. Rugby Assn., Miami, 1991-93. Mem. ASME, AIAA, Am. Astron. Soc., Cryogenic Soc. Am. Office: Ball Aerospace PO Box 1062 Boulder CO 80306

FLEISCH, HERBERT ANDRÉ, pathophysiologist; b. Lausanne, Switzerland, July 22, 1933; s. Alfred and Ilse (Ullmann) F.; m. Mariapia Ronchetti, May 18, 1959; children: Marie-Gabrielle, Isabelle, Marie-Laure. MD, U. Lausanne, Switzerland, 1959. Asst. Dept. Physiology U. Lausanne, Switzerland, 1958-59, Dept. Surgery U. Lausanne, 1961-62; post doctoral fellow Dept. Radiation Biology, U. Rochester, N.Y., U.S., 1959-60; dir. Lab. Experimental Surgery, Davos, Switzerland, 1963-67; prof. and chmn. dept. pathophysiology of med. sch. U. Berne, Switzerland, 1967—, dean med. sch., 1981-83; pres. Union Swiss Socs. of Exptl. Biology, 1987-90. Recipient William F. Neuman award Am. Soc. Bone and Mineral Rsch., 1992. Achievements include research in physiology, pathophysiology and pharmacology of bone and calcium metabolism, especially the development of a new class of drugs for bone disease (the bisphosphonates). Home: Pourtalèsstr 10, 3074 Muri-Bern Switzerland Office: U Bern Dept Pathophysiology, Murtenstr 35, CH3010 Bern Switzerland

FLEISCHER, NORMAN, director of endocrinology, medical educator; b. Springfield, Tenn., Jan. 24, 1936; s. Paul and Eva (Cohen) F.; m. Eva Lessy, Apr. 7, 1966; children: Deborah, Arlene. AB, Vanderbilt U., 1958, MD, 1961. Med. resident Albert Einstein Coll. of Medicine, Bronx, 1961-64; fellow in endocrinology Vanderbilt U., Nashville, 1964-66; dir. endocrinology, dir. Diabetes Ctr. Albert Einstein Coll. of Medicine, Bronx, 1976—, prof., 1978—; fellow in endocrinology Sch. of Medicine Vanderbilt U., Nashville, 1964-66; asst. prof. Coll. of Medicine Baylor U., Houston, 1966-71, assoc. prof. Sch. of Medicine, 1971-73; assoc. prof. Albert Einstein Coll. of Medicine, Bronx, 1973-77. Author chpts. in books; contbr. numerous articles to profl. jours. NIH grantee, 1966—. Fellow ACP; mem. Am. Fedn. Clin. Rsch., Am. Soc. Clin. Investigation, Am. Assn. Physicians, Am. Diabetes Assn., Endocrine Soc. Office: Albert Einstein Coll of Medicine 1300 Morris Park Ave Bronx NY 10461-1924

FLEISCHER, ROBERT LOUIS, physics educator; b. Columbus, Ohio, July 8, 1930; s. Leo H. and Rosalie (Kahn) F.; m. Barbara L. Simons, June 10, 1954; children: Cathy Ann, Elizabeth Lee. A.B., Harvard U., 1952, A.M., 1953, Ph.D., 1956. Asst. prof. metallurgy MIT, 1956-60; physicist Gen. Elec. Rsch. Lab., Schenectady, 1960-92; rsch. prof. earth and environmental scis. Rensselaer Poly. Inst., Troy, N.Y., 1992—; sr. rsch. fellow physics Calif. Inst. Tech., 1965-66; adj. prof. physics and astronomy Rensselaer Poly. Inst., 1967-68; adj. prof. geol. sci. SUNY, Albany, 1982-87; cons. U.S. Geol. Survey, 1967-70, GE R & D Ctr., 1992-93; vis. scientist Nat. Ctr. for Atmospheric Rsch., Nat. Oceanic and Atmospheric Adminstrn., 1973-74; adj. prof. applied physics and mech. engring. Yale U., 1984. Author: Nuclear Tracks in Solids, 1975; editor: Intermetallic Compounds: Principles and Practice; assoc. editor: 1st-4th Lunar Sci. Conf. Procs., 1970-73. Pres. Zoller Sch. PTA, 1968-69; mem. com. on candidates Schenectady Citizens Conv. for Sch. Bd., 1969-72, 82-83, chmn., 1969-70, 71-72, vice chmn. conv., 1977-78, chmn., 1978-79; mem. com. on priorities Schenectady Sch. Bd., 1974-75; Bd. dirs. Schenectady Citizens' League, Freedom Forum, Inc. Recipient awards Indsl. Rsch., 1964, 65, 72, Spl. award Am. Nuclear Soc., 1964, Ernest O. Lawrence award AEC, 1971, Gen. Elec. Silver medallion Inventor's award, 1971, Gold Medallion Inventor's award, 1991, Golden Plate award Am. Acad. Achievement, 1972, Coolidge award Gen. Electric Rsch. and Devel. Ctr., 1972; NASA Exceptional Sci. Achievement award, 1973; spl. recognition, 1979. Fellow AAAS, Am. Acad. Arts and Scis., Nat. Acad. Engring., Am. Phys. Soc., Am. Geophys. Union; mem. AIME (Disting. Career award Hudson-Mohawk chpt. 1991), NAE, Am. Soc. Metals, Health Physics Soc., Sigma Xi. Achievements include research in charged particle tracks in solids and their use in several fields, including cosmic ray and meteorite sci., geochronology, nuclear physics, radiobiology, environmental radon, mineral exploration; defects in solids and their effects on mech. properties and superconducting properties, high temperature materials. Home: 1356 Waverly Pl Schenectady NY 12308-2629 Office: Rensselaer Poly Inst Dept Earth & Environ Sci West Hall G-17 Troy NY 12180-3590

FLEISCHMAN, MARVIN, chemical and environmental engineering educator; b. N.Y.C., May 19, 1937; s. Julius and Miriam (Kuropatva) F.; children: Sam, Steve, Richard. B.Ch.E., CCNY, 1959; M.S., U. Cin., 1965, Ph.D., 1968. Registered profl. engr. Research chemist Monsanto Research Corp., Miamisburg, Ohio, 1959-60; sr. asst. san. engr. USPHS, Washington and Cin., 1961-63; research engr. Exxon Co. U.S.A., Florham Park, N.J., 1968-70, Amoco Chems., Naperville, Ill., 1977-78; prof. dept. chem. and environ. engring. U. Louisville, 1970—, chmn. dept., 1980-85; engr., dir. USPHS, Cin., 1985; dir. Waste Minimization Assessment Ctr., 1988—; 3M McKnight vis. prof. U. Minn., Duluth, 1993. Contbr. articles to profl. jours. With U.S. Army Summer Associateship, 1986. 87; Served to lt. USPHS, 1961-63, col., 1985. Grantee NSF, 1971, Office Water Rsch. Tech., 1974, AID, 1975, EPA, 1989—. Fellow AAAS/EPA Environ. Soc., AICE (chmn. AID-LIFE com. 1975-77, chmn. continuing edn. com. 1985-87, speakers bur. 1988—, ABET evaluator 1989—), Am. Soc. Engring. Edn. Democrat. Jewish. Home: 6811 Greenlawn Rd Louisville KY 40222-6630 Office: U Louisville Dept Chem Engring Louisville KY 40292

FLEISCHMANN, MARTIN, chemistry educator; b. Carlsbad, Czechoslovakia, Mar. 29, 1927; s. Hans and Margarethe (Srb) F.; m. Sheila Flinn, 1950; 3 children. Student, Imperial Coll., London. ICI fellow U. Durham, Eng., 1952-57; lectr., then reader U. Newcastle-upon-Tyne, Eng., 1957-67; Electricity Coun. Faraday prof. electrochemistry U. Southampton, Eng., 1967-77, rsch. prof. chemistry, 1982—; pres. Internat. Soc. Electrochemistry, 1970-72; sr. fellow Sci. & English Rsch. Coun., 1977-82; rsch. prof. chemistry U. Utah, 1988—. Contbr. chpts. to books; pub. papers. Office: U of Southhampton-Dep of Physics, Highfield, Southhampton England S09 5NH*

FLEISHER, PAUL, elementary education educator. BA, Brandeis U., 1970; MEd, Va. Commonwealth U., 1975. Tchr., coord. Providence Free Sch., 1970-72; tchr. corps intern Va. Commonwealth U., Richmond, 1973-75; 6th grade tchr. Petersburg (Va.) Pub. Schs., 1975-76, Williamsburg (Va.) Pub. Schs., 1976-78; tchr. programs for gifted Richmond Pub. Schs., 1978—, trainer computer programming and applications, 1983-85; instr. div. continuing studies Va. Commonwealth U., 1981-86; adj. vaculty Ctr. for Talented Youth, Johns Hopkins U., Balt., 1989-90; leader workshops in field. Author: Secrets of the Universe, 1987, Understanding the Vocabulary of the Nuclear Arms Race, 1988, Write Now!, 1989, (with Patricia Keeler) Looking Inside, 1991, Changing the World: A Handbook for Young Activists, 1992, The Master Violinmaker, 1993; also computer software in field; contbr. articles to profl. jours.; editor: Va. Educators for Peace newsletter, 1982-86; mem. editorial bd. Va. Forum, 1990—. Mem. NEA (editor Peace Caucus News 1987-88), Va. Edn. Assn., Richmond Edn. Assn. (faculty rep., del. to convs., editor REAlworld and Actionline, 1980-85). Home: 2781 Beowulf Ct Richmond VA 23231

FLEJTERSKI, STANISLAW, economist, educator, consultant; b. Tomaszow, Poland, Sept. 10, 1948; s. Franciszek and Genowefa (Blazejewska) F.; m. Genowefa Maria Graj, Apr. 3, 1971; children: Tomasz, Ewa. Mgr. in Econ., Tech. U., Szczecin, Poland, 1970, D in Econ., 1974. Asst. lectr. Econ. Inst., Tech. U., 1970-85, dep. dir., 1978-82; sec. of organizing founder's com. U. Szczecin, 1985, sr. lectr. Econ. Inst., 1985-89, dir. Rsch. Ctr., 1987-89; sr. lectr. internat. econs. and bus. Agrl. Acad., Szczecin, 1989-93; sr. lectr. fin. and banking U. Szczecin, 1992—; lectr. Westpomeranian Bus. Sch., 1992—; bd. dirs. Szczecin br. Pekao Bank;mem. com. on econ. scis. Polish Acad. Scis., Warsaw, 1987-89; chmn. supervisory bd. Pomeranian Cons. Agy., 1990—; mem. supervisory bd. Westpomeranian Industry and Trade Chamber, Western Pomerania, 1991—; banking practice at Banca Nazionale del Lavoro, 1992; adviser Szczecin br. BIG Bank, 1993—, Mktg. Agy. IPH and Cons. Agy. Stetinum, 1993—. Co-editor: Polish Studies on Asian, African and Latin American Affairs, 1991—; contbr. articles to profl. jours. Dep. chmn. Social-Econ. Coun. Provincial Bd., Szczecin, 1987-89. Recipient 2 prizes Ministry of Edn.; faculty fellow U. Zurich, 1978, Free U. Amsterdam, 1987. Mem. Soc. Internat. Devel. (Rome), Polish-Dutch Assn. (co-founder, v.p. 1985-90), Club of Intellectuals (co-founder, dep. chmn. 1989-92). Avocations: family, The Netherlands, East Asia, popularization of market economy, jogging. Home: Jarogniewa 38-8, 71-664 Szczecin Poland

FLEMING, JULIAN DENVER, JR., lawyer; b. Rome, Ga., Jan. 12, 1934; s. Julian D. and MargaretMadison (Mangham) F.; m. Sidney Howell, June 28, 1960; 1 dau., Julie Adrianne. Student, U. Pa., 1951-53; B. Chem. Engring., Ga. Inst. Tech., 1955, Ph.D, 1959; J.D., Emory U., 1967. Bar: Ga. 1966, D.C. 1967; registered profl. engr., Ga., Calif. Research engr., prof. chem. engring. Ga. Inst. Tech., 1955-67; ptnr. Sutherland, Asbill & Brennan, Atlanta, 1967—. Contbr. articles to profl. jours.; patentee in field. Bd. dirs. Mental Health Assn. Ga., 1970-80; bd. dirs. Mental Health Assn. Met. Atlanta, 1970-80, pres., 1974-75; mem. council legal advisers Rep. Nat. Com., 1981-85. Fellow Am. Inst. Chemists, Am. Coll. Trial Lawyers; mem. AAAS, ABA (coun. sect. sci. and tech. 1980-82, vice chmn. 1982-84, chmn.-elect 1984-85, chmn. 1985-86, mem. ho. dels. 1990), AICE, Nat. Conf. Lawyers and Scientists (chmn. ABA del. 1988-90, ABA liaison 1990—, mem. standing com. on nat. conf. groups 1990, chmn. 1992—), Bleckley Inn of Ct. (master of bench). Achievements include patent for data apparatus. Home: 2238 Hill Park Ct Decatur GA 30033-2716 Office: Sutherland Asbill & Brennan 999 Peachtree St NE Ste 2300 Atlanta GA 30309-3964

FLEMING, LAWRENCE THOMAS, engineering executive; b. Tacoma, Sept. 26, 1913; s. Thomas Patrick and Dora Martha (Eichenhofer) F.; m. Frances Heaney, Nov. 5, 1937; 1 child, James Lawrence. BS, Calif. Inst. Tech., 1937. Registered profl. engr., Tex.; registered patent agt. Examiner U.S. Patent Office, Washington, 1937-41; engr. U.S. Naval Ordnance Lab., Washington, 1941-50; physicist Nat. Bur. Stds., Washington, 1950-54; sect. chief Diamond Ord. Fuze Labs., Washington, 1955-56; mgr. devel. Southwestern Indsl. Electronics Co., Houston, 1956-59; prin. rschr., engr. Bell and Howell Co., Pasadena, Calif., 1959-67; pres. Innes Instrument Co., Forest City, Pa., 1967—. Fellow Acoustical Soc. Am. Democrat. Achievements include 16 patents in field; development of definition for random nature of vibration in rockets; research in instrumentation for sensing vibration. Home: 7 Mountain Laurel Dr Forest City PA 18421

FLEMING, TIMOTHY PETER, molecular biologist; b. St. Louis, Jan. 18, 1954; s. Frank Peter and Betty L. (Clark) F.; m. Anne Cecelia Schrodt, Apr. 26, 1986; children: Jack, David. BA, St. Louis U., 1975; PhD, U. Mo., 1985. Postdoctoral fellow Cell and Molecular Biol. Lab. NIH, Bethesda, Md., 1985-87, sr. staff fellow Cell and Molecular Biol. Lab., 1987-91; asst. prof. Sch. of Medicine Ophthalmology and Genetics Wash. U., St. Louis, 1991—. Mem. AAAS, Assn. Rsch. Vision and Ophthalmology. Achievements include patent for efficient directional genetic cloning system; discoveries include isolated receptor for keratinocyte growth factor, role of platelet-derived growth factor in human malignancy; co-discoverer of novel oncogene ECT-2; reported that RAS effector, GTPase Activating Protein, was phosphorylated on thyrosine upon PDGF addition, coupling RAS activity to signal transduction pathway. Office: Wash U 660 S Euclid Box 8096 Saint Louis MO 63110

FLEMINGS, MERTON CORSON, engineer, materials scientist, educator; b. Syracuse, N.Y., Sept. 20, 1929; s. Merton C. and Marion (Dexter) F.; m. Elizabeth Goodridge, Sept. 7, 1956 (div. 1976); children: Anne, Peter; m. R. Elizabeth ten Grotenhuis, Feb. 20, 1977; children: Cecily, Elspeth. S.B., MIT, 1951, S.M., 1952, Sc.D., 1954. Mem. faculty MIT, Cambridge, Mass., 1956—, ABEX prof. Metallurgy, 1970-75, Ford prof. engring., 1975-81, dir. materials processing ctr., 1979-82, Toyota prof. materials processing, 1981—, dept. head materials sci. and engring., 1982—; mem. tech. adv. bd. Norton Co., Ampersand Splty. Materials Venture, Molten Metals Tech., Inc.; mem. adv. bd. New Ct. Ptnrs.; bd. dirs. Hitchiner Corp., Metal Casting Tech., Inc.; Hatfield Meml. lectr., 1989. Author: Foundry Engineering, 1959; Solidification Processing, 1974. Contbr. numerous articles on metallurgy to profl. jours. Recipient Simpson Gold medal Am. Foundrymen's Soc., 1961, Henri Sainte-Claire Deville medal Soc. Francaise de Metallurgie, 1977. Fellow TMS (Leadership award 1990, Bruce Chalmers award 1993), ASM Internat. (Henry Marion Howe medal 1973, 90, Edward DeMille Campbell Meml. lectr. 1990); mem. NAE, Am. Inst. Metall. Engrs. (Mathewson Gold medal 1969), Am. Acad. Arts and Scis., Japan Foundrymen's Soc. (hon.), Iron and Steel Inst. Japan (hon., Yukawa Meml. lectr. 1985), Italian Metall. Assn. (Luigi Losana Gold medal 1986). Home: 11 Hillside Ave Cambridge MA 02140-3615 Office: MIT Dept Materials Sci and Engring 8-309 Cambridge MA 02139

FLEMMING, STANLEY LALIT KUMAR, family practice physician, state legislator; b. Rosebud, S.D., Mar. 30, 1953; s. Homer W. and Evelyn C. (Misra) F.; m. Marth Susan Light, July 2, 1977; children: Emily Drisana, Drew Anil, Claire Elizabeth Misra. AAS, Ft. Steilacoom Coll., 1973; BS in Zoology, U. Wash., 1976; MA in Social Psychology, Pacific Luth. U., 1979; DO, Coll. Osteopathic Med. Pacific, 1985. Diplomate Am. Coll. Gen. Practice; cert. ATLS. Intern Pacific Hosp. Long Beach (Calif.), 1985-86; resident in family practice Pacific Hosp. Long Beach, 1986-88; fellow in adolescent medicine Children's Hosp. L.A., 1988-90; clin. preceptor Family Practice Residency Program Calif. Med. Ctr., U. So. Calif., L.A., 1989—; clin. instr. Sch. Medicine U. So. Calif., L.A., 1989-90; clin. instr. Coll. Osteopathic Medicine Pacific, Pomona, Calif., 1989-90; clin. asst. prof. Family Medicine Coll. Osteopathic Medicine Pacific, Pomona, 1987—; exam. commr., expert examiner Calif. Osteo. Med. Bd., 1987-89; med. dir. Community Health Care Delivery System Pierce County, Tacoma, Wash., 1990—; clin. instr. U. Wash. Sch. Medicine, 1990—; bd. dirs. Calif. State Bd. Osteo. Physicians Examiners, 1989—, cons., 1989. Lt. col. with med. corps U.S. Army, 1976—. Named Outstanding Young Man of Am. 1983, 1985; recipient Pumerantz-Weiss award 1985. Mem. Fedn. State Bds. Licensing, Am. Osteopathic Assn., Am. Acad. Family Practice, Soc. Adolescent Medicine, Assn. Military Surgeons U.S., Assn. U.S. Army (chpt. pres.), Soc. Am. Military Engrs. (chpt. v.p.), Calif. Med. Assn., Wash. Osteopathic Med. Assn., Calif. Family Practice Soc., Long Beach Med. Assn. (com. mem.), N.Y. Acad. Sci., Calif. Med. Review Inc., Sigma Sigma Phi, Am. Legion. Episcopalian. Home: 7619 Chambers Creek Rd W Tacoma WA 98467-2015 Office: Community Health Care Delivery System of Pierce County J428 J L O'Brien Bldg Olympia WA 98504

FLESCHER, ELIEZER, immunologist; b. Israel, Oct. 18, 1954; s. Menachem and Shoshana (Nemet) F.; m. Esther Pass, Feb. 27, 1984; 1 child, Asaf. MSc, Tel Aviv U., Israel, 1979, PhD, 1987. Rsch. asst., then instr. Tel Aviv U., 1976-86; postdoctoral rsch. fellow U. Tex. Health Sci. Ctr., San Antonio, 1986-89, instr., 1989-91, asst. prof., 1991-93; asst. prof. dept. environ. medicine NYU, 1993—. Contbr. articles to Jour. Nat. Cancer Inst., Jour. Clin. Investigation, other sci. publs. Mem. Am. Assn. Immunologists. Achievements include research on eicosanoid-independent effects of aspirin-like drugs on lymphocytes, effects of oxidative stress on lymphocyte signal transduction. Office: NYU Inst Environ Medicine Long Meadow Rd Tuxedo Park NY 10987

FLETCHER, CRAIG STEVEN, electronic technician; b. Phoenix, Dec. 10, 1967; s. Homer Woodrow Jr. and Virginia Sue (McBride) F.; m. Beatriz V. Arvizu, Apr. 28, 1990; 1 child, Savanna Laurén. AAS, Cochise Coll., 1989. Electronic technician, mfg. engr. Hutronix Inc., Douglas, Ariz., 1989-90; technician Harris Corp., Sierra Vista, Ariz., 1990—. Named Dec. Student of the Month, Morning Lions Club, 1985; scholar Morning Lions and Lioness, 1986-87, Fry Found., 1987-89, Standex Electronics, 1988. Avocations: computers, all types of sports, outdoor activities. Home: 810 Tacoma Pl Sierra Vista AZ 85635 Office: Harris Corp 1001 E Executive Dr Sierra Vista AZ 85635-4991

FLETCHER, DAVID QUENTIN, civil engineering educator; b. Brisbane, Queensland, Australia, May 16, 1946; came to U.S., 1951; s. Quentin Henderson and Muriel Mary (Beeston) F.; m. Donna Elaine Worley, Sept. 1, 1968; children: Duncan Edward, Meredith Elaine. BS, U. Calif., Davis, 1967, MS, 1970, PhD, 1973. Registered profl. engr., Calif. Rsch. engr. U.S. Bur. Mines, Denver, 1971-73; asst. prof. civil engring. dept. U. of Pacific, Stockton, Calif., 1973-79, assoc. prof., 1979-82, prof., 1982—, dept. chmn., 1988—; vis. lectr. U. Queensland, 1978, U. Calif., Davis, 1984. Author: Mechanics of Materials, 1985. Pres. Ctrl. Valley Youth Symphony Assn., Stockton, 1987-88; coach Valley Volleyball Club, Stockton, 1989, Stockton Volleyball Club, 1990-91. NDEA Title IV fellow, 1968. Mem. ASCE (pres. Ctrl. Valley 1978-80, nat. dept. heads coun. 1991—), Am. Soc. Engring. Edn. Office: U Pacific Dept Civil Engring 3601 Pacific Ave Stockton CA 95211

FLETCHER, JEFFREY EDWARD, biochemist, researcher; b. Toledo, Mar. 11, 1948; s. John Harper and Eleanore (Jackson) F.; m. Marcia Ruth Miller, Mar. 21, 1970 (div. Mar. 1977); m. Jeanne Claire Untied, Aug. 22, 1981; children: Katherine Ann, Lindsay Nicole, Sarah Jeanne. AA, Mohegan Community Coll., 1974; BA, Conn. Coll., 1976; PhD, U. Conn., 1981. Resident rsch. assoc. NRC, Washington, 1981-83; sr. instr. dept. anesthesia Hahnemann U., Phila., 1983-85, asst. prof. dept. anesthesia, 1985-90, assoc. prof. dept. anesthesia, 1990—; mem. editorial coun. sci. jour. Toxicon, 1991—. Contbr. articles to profl. jours. With USN, 1967-73, Vietnam. Conn. Coll. scholar, 1974-76. Mem. Am. Soc. for Pharmacology and Exptl. Therapeutics, Internat. Soc. Toxinology, Am. Soc. Anesthesiologists, Am. Soc. Biochemistry and Molecular Biology, Soc. for Exptl. Biology and Medicine, Soc. for Neurosci., Biophys. Soc., Phi Beta Kappa, Sigma Xi, Rho Chi. Episcopalian. Avocations: photography, tennis, fishing. Office: Hahnemann U Dept Anesthesia Broad and Vine Sts Philadelphia PA 19102-1192

FLETCHER, JOHN LYNN, psychology educator; b. Springdale, Ark., Apr. 18, 1925; s. Lynn Harrington and Elsie Irene (Jones) F.; m. Mary Lou Campbell, Aug. 21, 1949 (div. July 1974); children: Lynn Gray, Jana Lee. BA, U. Ark., 1950, MA, 1951; PhD, U. Ky., 1955. Commd. 2nd lt. U.S. Army, 1943, advanced through ranks to lt. col.; chief audition br. Med. Rsch. Lab. Ft. Knox, Ky., 1953-70; ret. U.S. Army, 1970; prof. psychology Memphis State U., 1970-75; prof., dir. rsch. dept. otolaryngology U. Tenn. Ctr. for Health Sci., Memphis, 1975-81; prof., chair psychology dept. U. Mo., Rolla, 1981-87; lectr. psychology S.W. Tex. State U., San Marcos, 1987—; cons. NASA Space Shuttle, Kennedy Space Ctr., 1972-76; mem. Commn. on Hearing and Bio Acoustics, 1956—. Editor: Effects of Noise on Animals, 1978; contbr. articles to profl. jours. Decorated Bronze Star. Fellow Acoustical Soc. Am., Am Speech, Lang. Hearing Soc.; mem. NAS, NRC, N.Y. Acad. Scis. (life), Human Factors Soc. (vice chmn. com. standards). Republican. Presbyterian. Achievements include patents for Acoustic Reflex Ear Defender. Home: PO Box 309 Martindale TX 78655-0309 Office: SW Tex State Univ Dept Psychology San Marcos TX 78666

FLETCHER, LEROY STEVENSON, mechanical engineer, educator; b. San Antonio, Oct. 10, 1936; s. Robert Holton and Jennie Lee (Adkins) F.; m. Nancy Louise McHenry, Aug. 14, 1966; children: Laura Malee, Daniel Alden. BS, Tex. A&M U., 1958; MS, Stanford U., 1963, Engr., 1964; PhD, Ariz. State U., 1968. Registered profl. engr., Ariz., N.J., Va., Tex., Australia; chartered engr., U.K. Research scientist Ames Research Ctr., NASA, Moffett Field, Calif., 1958-62; instr. Ariz. State U., Tempe, 1964-68; prof. aero., engring. Rutgers U., New Brunswick, N.J., 1968-75, assoc. dean, 1974-75; prof., chmn. dept. mech. and aero. engring. U. Va., Charlottesville, 1975-80; dir. Ctr. Energy Analysis, 1979-80; assoc. dean Tex. A&M U., College Station, 1980-88, assoc. dir. Tex. Engring. Expt. Sta., 1985-88, Dietz prof. mech. engring., 1988—; vis. prof. Tokyo Inst. Tech., 1993; hon. prof. Ruhr U.-Bochum, Germany, 1988—; cons. to various industries and univs. Author: Introduction to Engineering Including FORTRAN Programming, 1977, Introduction to Engineering Design with Graphics and Design Projects, 1979; editor: Aerodynamic Heating and Thermal Protection, 1978, Heat Transfer and Thermal Control Systems, 1978. Served to capt. USAF, 1958-61. Recipient Disting. Alumni award Ariz. State U., 1985. Fellow ASME (bd. govs. 1983-87, pres. 1985-86, Charles Russ Richards award), AAAS (chmn. sect. M-Engring. 1988-89), Accreditation Bd. Engring. and Tech. (dir. 1991—, 1979-89), Am. Astron. Soc., Inst. Engrs. Australia, Inst. Mech. Engrs. U.K., Am. Soc. Engring. Edn. (dir. 1978-80, George Westinghouse award 1982, Ralph Coats Roe award 1983, Donald E. Marlowe award 1986, Leighton W. Collins award 1993), AIAA (dir. 1981-84, 1991—, v.p. edn. 1991—, Aerospace Edn. Achievement award 1982, Energy Systems award 1984, Thermophysics award 1992), Internat. Acad. Astronautics; mem. Sigma Xi, Tau Beta Pi, Pi Tau Sigma, Sigma Gamma Tau, Phi Kappa Phi. Office: Tex A&M Univ Mech Engring College Station TX 77843

FLETTRICH, ALVIN SCHAAF, JR., civil engineer; b. New Orleans, Dec. 18, 1944; s. Alvin Schaaf and Ada Shelby (Flaspollar) F.; m. Carolyn Walls, June 10, 1966; children: Katharine, Kyle. BS, Tulane U., 1966, MS, 1967. Registered profl. engr., land surveyor, La. Engr. Boh Bros. Constrn. Co., Inc., New Orleans, 1966—. Lt. USPHS, 1967-69. Fellow USPHS, 1967. Mem. ASCE, NSPE, La. Associated Gen. Contractors (vice chmn. hwy. div. 1989, chmn. hwy. div. 1990-91, v.p. 1992-93, pres. 1994, La. Engring. Soc.

(chmn. constrn. div. 1993-94), La. Soc. Profl. Land Surveyors, Tchefuncta Country Club (pres. tennis assn. 1991), U.S. Tennis Assn., Elmwood Health Club. Republican. Episcopalian. Home: 12 Hummingbird Rd Covington LA 70433 Office: Boh Bros Constrn Co Inc 730 S Tonti St New Orleans LA 70119

FLETTRICH, CARL FLASPOLLER, structural engineer; b. New Orleans, Apr. 13, 1948; s. Alvin S. and Shelby (Flaspoller) F.; m. Doris Marie Brewer Flettrich, Mar. 12, 1977; children: Erica K., Annie L. BA in History, La. Tech. U., 1971; BS in Engrng., L.A. Tech. U., 1973. Project engr. Walk Haydel, New Orleans, 1973-75, W.S. Nelson, Inc., New Orleans, 1975-78, Denopulos & Ferguson, Inc., Shreveport, La., 1978-80, Petrocon, Inc., Sulphur, La., 1980-83, N.Y. Assoc., Inc., New Orleans, 1983—; mem. ASCE, New Orleans, 1973—. Designed City Park Driving Range, New Orleans, 1977, Greek Orthodox Cathedral, 1984, Ore Storage Facility, Gulfport, Miss., 1991, 13 story Hosp. Bldg., Shreveport, La., 1979. Pres. St. Paul's Sch. Bd., Pass Christian, Miss., 1985, St. Martin's Alumni Assn., New Orleans, 1988-90; chmn. St. Paul's Carnival Assn., Pass Christian (Miss.) Festival, 1984-86. Home: 905 Haring Rd Metaire LA 70001 Office: NY Assocs Inc 2750 Lake Villa Dr Metairie LA 70002

FLEURY, PAUL AIMÉ, physicist; b. Balt., July 20, 1939; m. Carol Anne Moss, Aug. 22, 1964; children: Ellen, Laura, Jennifer. BS in Physics, John Carroll U., 1960; PhD in Physics, MIT, 1965. Mem. tech. staff AT&T Bell Labs., Murray Hill, N.J., 1965-70, head condensed state physics rsch., 1970-79, dir. materials rsch., 1979-84, dir. phys. rsch., 1984-92; v.p. rsch. Sandia Nat. Lab., Albuquerque, 1992—. Editor: Coherence and Energy Transfer in Glasses, 1983; contbr. over 120 articles to Phys. Rev., Sci., others. Fellow AAAS, Am. Phys. Soc. (Frank Isakson prize Optical Effects in Solids 1992). Achievements include 5 patents for optical devices, lasers, optical fibers; research in laser spectroscopy. Office: Sandia Nat Lab Albuquerque NM 87185

FLICK, WILLIAM FREDRICK, surgeon; b. Lancaster, Pa., Aug. 18, 1940; s. William Joseph and Anna (Volkl) F.; m. Jacqueline Denise Phaneuf, May 21, 1966; children: William J., Karen E., Christopher R., Derrick W., Brian A. BS, Georgetown U., 1962, MD, 1966; MBA, U. Colo., 1990. Cert. Am. Bd. Surgeons, 1976. Self employed surgeon Cheyenne, Wyo., 1973-84; pres., surgeon Cheyenne Surgical Assocs., 1984—. Trustee Laramie County Sch. Dist. #1, Cheyenne, 1988-92. Maj., chief of surgery USAF, 1971-73. Fellow ACS, Southwestern Surg. Congress; mem. Am. Coll. Physician Execs., Rotary. Republican. Roman Catholic. Office: Cheyenne Surg Assocs 603 E 17th St Cheyenne WY 82001-4709

FLINN, DAVID R., federal agency research director; b. Jennings, Okla., Oct. 21, 1937; s. Roy Herbert and Clela Alice (Canfield) F.; m. Sarah Jane Denton, Dec. 13, 1957; children: Cynthia, Davetta, Dana. BS, East Tex. State Coll., 1960, MEd, 1961; PhD, North Tex. State U., 1968. Postdoctoral rsch. assoc. Nat. Rsch. Coun. U.S. Naval Rsch. Lab., Washington, 1968-70, rsch. chemist, 1970-75; rsch. chemist U.S. Bur. Mines, Coll. Park, Md., 1975-78; rsch. supr. U.S. Bur. Mines, Avondale, Md., 1978-87, Albany, Oreg., 1987-88; rsch. dir. U.S. Bur. Mines, Tuscaloosa, Ala., 1988—; instr. math, sci. Ranger (Tex.) Jr. Coll., 1961-64; part-time instr. North Tex. State U., Denton, 1964-68; chmn. subgroup U.S./Can. Transboundary Air Pollution Working Group, Washington, Toronto, 1980-83; chmn. task group Nat. Acid Precipitation Assessment Program, Washington, 1986-87. Contbr. articles to profl. jours.; mem. editorial bd. Surface and Coatings Tech. Jour., 1985-92; patentee. Chmn. combined fed. campaign Tuscaloosa County United Way, 1992-93. Fellow Washington Acad. Scis.; Nat. Assn. Corrosion Engrs. (chmn. T-3R com. 1984-86), Electrochem. Soc. (chmn. nat. capital sect. 1978-79), Am. Chem. Soc. Republican. Mem. Christian Ch. (Disciples of Christ). Avocations: jogging, reading. Office: US Bur Mines PO Box L Tuscaloosa AL 35486-9777

FLINN, PAUL ANTHONY, materials scientist; b. N.Y.C., Mar. 25, 1926; s. Richard A. and Anna M. (Weber) F.; m. Mary Ellen Hoffman, Aug. 20, 1949; children: Juliana, Margaret, Donald, Anthony, Patrick. AB, Columbia Coll., 1948, MA, 1949; ScD, MIT, 1952. Asst. prof. Wayne U., Detroit, 1953-54; research staff Westinghouse Research Lab., Pitts., 1954-63; prof. Carnegie-Mellon U., Pitts., 1964-78; sr. staff scientist Intel Corp., Santa Clara, Calif., 1978—; vis. prof. U. Nancy, France, 1967-68, U. Fed. do Rio Grand du Sol, Porto Allegro, Brazil, 1975, Argonne (Ill.) Nat. Lab., 1977-78, Stanford (Calif.) U., 1984-85; cons. prof., 1985—. Contbr. sci. articles to profl. jours. Served with USN, 1944-46, PTO. Fellow Am. Phys. Soc.; mem. AAAS, Metall. Soc., Materials Rsch. Soc., Phi Beta Kappa, Tau Beta Pi. Office: Intel Corp SC1-2 PO Box 58119 Santa Clara CA 95052-8119

FLINT, MARY LOUISE, entomologist; b. Montreal, Que., Can., May 21, 1949; came to U.S., 1953; d. John Barlow and Gyda (Sheppard) F.; m. Stephen James Meyer, Apr. 3, 1982; children: Nicholas Flint, William Flint. BS, U. Calif., Davis, 1972; PhD, U. Calif., Berkeley, 1979. Asst. dir. Environ. Assessment Team, Calif. Fed. Food and Agr., Sacramento, 1977-79; extension entomologist U. Calif., Davis, 1980—, dir. Integrated Pest Mgmt. Edn. and Pubs., 1980—; tech. com. mem. Statewide Integrated Pest Mgmt. Project U. Calif., Davis, 1983—, Sustainable Agr. Program, 1988-92; mem. policy adv. com. Calif. Agr., Oakland, Calif., 1991—. Co-author: Introduction to Integrated Pest Management, 1981; author: Pests of the Garden and Small Farm, 1990; tech. editor U. Calif. IPM Manuals, 1981—. Mem. AAAS, Entomol. Soc. Am., Assn. Applied Insect Ecologists. Achievements include research in biol. control, alternatives to pesticides, extension and adoption of Integrated Pest Mgmt. techs. Office: IPM Edn and Pubs U Calif Davis CA 95616

FLISS, ALBERT EDWARD, JR., chemical engineer; b. Harrisburg, Pa., Nov. 8, 1959; s. Albert Edward and Irene (Pierlioni) F.; m. Maria Filomena Ventura, Aug. 10, 1985. BS, U. Ctrl. Fla., 1982, MS, 1983; BSChemE, U. Fla., 1992. Rsch. assoc. USDA, Orlando, Fla., 1984-85; biol. scientist U. Fla., Gainesville, 1985-88, USDA, Gainesville, 1988-89; rsch. assoc. U. Fla., Gainesville, 1989, sr. biologist, 1989-92; sr. scientist BioNebraska, Inc., Lincoln, Nebr., 1992; rsch. assoc. Tex. A&M U., College Station, 1992—; CEO, pres. Designer Genes, Inc., Gainesville, 1990—. Contbr. articles to profl. jours. Recipient Superior Accomplishment award U. Fla., 1990. Mem. Lambda Chi Alpha, Omicron Delta Kappa. Democrat. Roman Catholic. Achievements include patent in Recombinant Uteroferrin. Home: 714 Inwood Rd Bryan TX 77802 Office: Tex A&M U Kleberg Ctr College Station TX 77843

FLITMAN, STEPHEN SAMUEL, neurologist; b. N.Y.C., Feb. 28, 1966; s. Robert Morris and Sheila Jill (Cohen) F; m. Mary Katherine Timm, Apr. 18, 1993. BS, Union Coll., 1988; MD, Albany Med. Coll., 1990. Diplomate Nat. Bd. Med. Examiners. Intern SUNY, Stony Brook, 1990-91; resident in neurology Barrows Neurologic Inst., Phoenix, 1991—, mem. house staff exec. com., 1992—; chief resident, 1993-94; pres., CEO Xenoscience, Inc., Phoenix, 1993—. Mem. AAAS, Am. Acad. Neurology, Phi Beta Kappa. Home and Office: Ste 202 525 W Earll Dr Phoenix AZ 85013

FLOCH, HERVE ALEXANDER, medical biologist; b. Lambezellec, France, Oct. 3, 1908; s. Herve Marie and Jeanne (Le Rouzic) F.; m. Lucie Henry; children: Therese, Herve Henri, Daniele. M.D., Faculté de Medecine de Bordeaux, 1932. Asst. Colonial Hosp., scholar Pasteur Inst., Paris, from 1938; mil. physician, medicin col., until 1956; dir., founder Pasteur Inst. Cayenne, French Guiana (br. Pasteur Inst., Paris), 1940-66; dir. Pasteur Inst. Pointe à Pitre, Guadeloupe, 1969-71; chief Anti-Mosquito Svc. and Leprosy Svc., French Guiana, 1940-66, Guadeloupe, 1969-71; chief lab. Inst. Pasteur, Paris, 1956-73; biologist chief Lab. Svc. Hosp., Morlaix, France, 1974-79; prof. microbiology Faculty Odontology, Brest, 1978-85; malariologist and pathologist WHO. Author over 918 publs. to profl. jours. Mem. French Acad. Medicine (6 prizes), French Acad. Sci. (Prix Muteau, grand prix Etancelin 1974), French Acad. Overseas Sci., other sci. socs. Rsch., publs. on leprology, promoter use of D.D.S. in treatment of leprosy, malariology, epidemiology, entomology, acarology parasitology, mycology, virology, bacteriology, biology and tropical pathology; studies on tropical alimentation-nutrition habitat. Home: 45 Ave Camille Desmoulins, 29200 Brest France

FLOCH-BAILLET, DANIELE LUCE, ophthalmologist; b. Brest, France, Jan. 9, 1948; d. Herve Alexandre and Lucie (Henry) Floch; m. Gilles Pierre Baillet, Dec. 6, 1980; 1 child, Victoire-Amelie. MD, Med. U. Brest, 1972. Cert. in ophthalmology, 1975. Med. cons. ophthalmology Brest Hosp., 1976-85; gen. practice ophthalmology Landivisiau, France, 1977—; researcher ophthalmic bacteriology, 1985—. Author: (with P. Francois) Nosological Outlines from Coats, 1975, Exsudation from Coats, 1976. Mem. French Ophthalmologist Soc., European Contact Lenses Soc. Ophthalmologists, Nat. Syndicat French Ophthalmology, Contact Lens Assn. Ophthalmologists, Assn. Ophthalmologic Improvement from East-Paris. Roman Catholic. Home: 11 Rue Creach Joly, 29600 Morlaix France Office: 7 Rue Georges Pompidou, 29400 Landivisiau France

FLOERSHEIM, ROBERT BRUCE, military engineer; b. Atlanta, Nov. 30, 1967; s. Robert Myron and Sandra (Kelton) F. BS in Aerospace Engring., U.S. Mil. Acad., Highland Falls, N.Y., 1989. Commd. 2d lt. U.S. Army, 1989, advanced through grades to capt., 1993; platoon leader 4th Engr. Bn., Ft. Carson, Colo., 1989-91, co. exec. officer, 1991-92; project engr. U.S. Army C.E., Diyarbakir, Turkey, 1992—. Mem. AIAA, Soc. Am. Mil. Engrs., U.S. Army Engr. Assn., Phi Kappa Phi. Home: 995 D Lamesa Ter Sunnyvale CA 94086

FLOOD, HAROLD WILLIAM, chemical engineer, educator; b. Alton, Ill., Oct. 7, 1922; s. Benjamin William and Hilda Elizabeth (Schmidt) F.; m. Jeanne Dietrich Bradfield, Nov. 2, 1946; children: Alan B., Gayle J., Anne E., Dona C., Betsy H., Mary K., Bryan W. BSChemE, U. Mo., Rolla, 1943, MS, 1974. Registered profl. engr., Mass. Chem. engr. Freeport (Tex.) Sulphur Co., 1943-47; mill and roast dept. head Mut. Chem. Co., Balt., 1947-52; plant engr. Dewey and Almy Chem. Co., Acton, Mass., 1952-56; sr. staff mem. Arthur D. Little, Inc., Cambridge, Mass., 1956-70; mgr. process engring. Kennecott Copper Corp., Lexington, Mass., 1970-81; mgr. bus. planning Resource Engring. Inc., Waltham, Mass., 1981-83; assoc. prof. Univ. Mass., Lowell, 1983—; adv. coun. mem. Coll. Engring., Univ. Mass., Lowell, 1979-83. Mem. Acton Planning Bd., 1955-60; chair Acton Bd. Appeals, 1964-87. With USAAF, 1945-46. Fellow AICE (nat. dir. 1978-80, Founder's award 1982); mem. AIME, Am. Chem. Soc., Assn. Energy Engrs. (dir. 1980-88), Am. Soc. Engring. Edn. Republican. Methodist. Achievements include patents in fluid bed processing methods, molten metal instrumentation and electrorefining. Home: 183 Main St Acton MA 01720-3616

FLOR, LOY LORENZ, chemist, corrosion engineer, consultant; b. Luther, Okla., Apr. 25, 1919; s. Alfred Charles and Nellie M. (Wilkinson) F.; BA in Chemistry, San Diego State Coll., 1941; m. Virginia Louise Pace, Oct. 1, 1946; children: Charles R., Scott R., Gerald C., Donna Jeanne, Cynthia Gail. With Helix Water Dist., La Mesa, Calif., 1947-84, chief chemist, 1963—, supr. water quality, 1963—, supr. corrosion control dept., 1956—. 1st. lt. USAAF, 1941-45. Registered profl. engr., Calif. Mem. Am. Chem. Soc. (chmn. San Diego sect. 1965—), Am. Water Works Assn. (chmn. water quality div. Calif. sect. 1965—), Nat. Assn. Corrosion Engrs. (chmn. western region 1970), Masons. Republican. Presbyterian.

FLORES-LOPEZ, AUREMIR, microbiologist; b. Mayaguez, P.R., Oct. 7, 1965; d. Jose Florencio Flores-Lebron and Luisa Lopez-Ithier. BS in Biology cum laude, U. P.R., Mayaquez, 1987; MSc in Microbiology, U. P.R., San Juan, 1992. Quality control inspector Baxter-Fenwal Div., Anasco, P.R., 1988; asst. researcher Med. Scis. Campus dept microbiology U. P.R., San Juan, 1989-92; quality control asst. Caribbean Microparticles Corp., San Juan, 1991, lab. coord., asst. researcher, 1992; quality ops. auditor Merck, Sharp & Dohme-Quimica de P.R. (MSDQ-Arecibo), Caguas, 1992—. Recipient V.P. Prodn. of Yr. Jr. Achievement, Mayaguez, 1982. Mem. AAAS, Am. Soc. Microbiology, Am. Soc. Quality Control, P.R. Am. Soc. Microbiology, N.Y. Acad. Scis., Coll. Agriculture and Mech. Arts-Alumni U. P.R. Roman Catholic. Achievements include research in germination of the Conidia to the yeast form in Sporothrix schenckii, molecular and cellular events during the Conidia to the yeast form of Sporothrix scheneckii.

FLORIN-CHRISTENSEN, JORGE, biologist; b. Buenos Aires, Jan. 25, 1951; came to the U.S., 1989; s. Vladimir and Nora (Christensen) Florin; m. Monica Ofelia Jacobsen, May 21, 1981; 1 child, Nicolas. Degree, U. Buenos Aires, 1979, PhD, 1988. Fellow Nat. Rsch. Coun. Argentina, Buenos Aires, 1981-82, 86-89, ind. researcher, 1992—; fellow DANIDA, Copenhagen, Odense, Denmark, 1982-83; rsch. assoc. Natural Scis. Rsch. Coun., Odense, 1983-84; fellow EMBO, Munster, Germany, 1984; NIH-Fogarty postdoctoral fellow Yale U., New Haven, Conn., 1989-91; postdoctoral assoc. U. Cin., 1991-92. Contbr. chpt. to book; contbr. articles to TIBS Trends Biochem. Sci., Microbial Ecology, Jour. Biol. Chemistry, Biochem. Jour. Recipient award Dansk-Argentinsk Kulturfond, 1984, 87, Erik of Martha Scheibel's Legat., 1984. Mem. Sigma Xi. Roman Catholic. Achievements include discovery of an extracellular phospholipase system in Tetrahymena, the biological role of exoenzymes in Tetrahymena, the mechanism of diacyglycerol signal termination, unique sterols in the pathogen Pneumocystis carinii. Home: Calle 63 No 5628, 1653 Chilavert Argentina Office: U Buenos Aires, Pab II Piso 4 Ciudad Universitaria, 1428 Buenos Aires Argentina

FLORY, CHARLES DAVID, retired psychologist, consultant; b. Nokesville, Va., Sept. 28, 1902; s. Samuel Henry and Lydia Frances (Kerlin) F.; m. Mary Amanda Grossnickle, June 4, 1925; children: Elaine Marie, Frances Ann. AB, Manchester Coll., North Manchester, Ind., 1924, LLD (hon.), 1968; AM, U. Chgo., 1928, PhD, 1933. Diplomate in clin. psychology Am. Bd. Examiners in Profl. Psychology. Instr. Park Coll., Parkville, Mo., 1928-30; rsch. assoc. U. Chgo., 1930-35; asst. prof., assoc. prof., prof. Lawrence Coll., Appleton, Wis., 1935-43; psychol. cons. Stevenson, Jordan, Harrison, Chgo. and N.Y.C., 1943-45; founding ptnr. Rohrer, Hibler & Replogle, Chgo., 1943-68; ret., 1968. Author: (with Frank N. Freeman) Growth in Intellectual Ability as Measured by Repeated Tests, 1937, (with McKenzie) The Credibility Gap in Management, 1971, (with Sherman and Sherman) Infant Behavior; editor: Managers for Tomorrow, 1965, Managing through Insight, 1968. Mem. adv. bd. White Plains (N.Y.) Hosp.; pres. Edgemont Sch. Bd., Scarsdale, N.Y. Recipient Alumni award Manchester Coll., 1962. Mem. AAAS, APA, Eastern Psychol. Assn., N.Y. State Psychol. Assn., N.Y. Assn. for Applied Psychology, N.Y. Met. Assn. for Applied Psychology, N.Y. Acad. Scis., Am. Assns. Ret. Persons, Green Hill Yacht and Country Club, Inst. for Ret. Persons, Phi Delta Kappa. Republican. Methodist. Home: 621 Douglas Rd Salisbury MD 21801-6701

FLORY, WALTER S., JR., geneticist, botanist, educator; b. Bridgewater, Va., Oct. 5, 1907; s. Walter Samuel and Ella May (Reherd) F.; m. Nellie Maude Thomas, Apr. 24, 1930 (dec. 1971); children: Kathryn Sue Flory Maier, Walter Samuel, Thomas Reherd; m. Gale Crews Gramley, June 25, 1975. A.B. with honors, Bridgewater Coll., 1928, Sc.D. (hon.), 1953; A.M. (Blandy fellow, 1928-31), U. Va., 1929, Ph.D., 1931; Nat. Rsch. fellow biol. scis. and rsch. assoc., Harvard U., 1935-36. In charge tech. work Shaver Bros., Inc., Jacksonville and Tampa, Fla., 1931-32; instr. in sci. Greenbrier Coll., Lewisburg, W.Va., 1932-34; prof. biology Bridgewater Coll., 1934-35; horticulturist Tex. Agrl. Expt. Sta., 1936-44, Va. Agrl. Expt. Sta., 1944-47; prof. exptl. horticulture U. Va., 1947-63; vice dir., mgr. Blandy Exptl. Farm, 1947-63, vis. prof., summer 1964; curator O.E. White Rsch. Arboretum, 1955-63; bd. dirs. Winston-Salem (N.C.) Nature Sci. Ctr., 1964-69; treas. Winston-Salem (N.C.) Nature Sci. Center, 1965-66; Babcock prof. botany Wake Forest U., Winston-Salem, 1963-80; Babcock prof. emeritus, biosystemic rschr. Wake Forest U., 1980—; dir. Reynolda Gardens, 1964-76; rsch. cons. Fairchild Tropical Garden, 1972; instnl. lectr. Piedmont U. Ctr., 1965-70; collaborator U.S. Dept. Agr., 1945-48; del. Internat. Botany Congress, Paris, 1954, Montreal, 1959, Edinburgh, 1964, Seattle, 1969, Sydney, 1981; mem. Internat. Genetics Congress, Montreal, 1958, Tokyo, 1968; mem. Internat. Hort. Congress, College Park, Md., 1966, Sydney, Australia, 1978; invited lectr. Internat. Chromosome Seminar, Calcutta, 1968, 76; mem. arboretum coun. Forsyth Counties, Tanglewood Park, 1985—. Contbr. articles on genetics, cytology and hort. subjects to profl. jours. and mags. Trustee, mem. exec. com. Highlands Biol. Sta.; pres., 1969-72; life trustee Bridgewater Coll. Recipient J. Shelton Horsley Rsch. award Va. Acad. Sci., 1949, Pres. and Visitors Rsch. prize U. Va., 1951, Bridgewater Coll. Alumni award, 1956, I.F. Lewis disting. svc. award Va. Acad. Sci., 1969, spl. citation Highlands Biol. Sta., 1973, Outstanding Svc. award Bridgewater Coll., 1981; Bicentennial hon. fellow Royal Hort. Soc. (Eng.). Fellow AAAS, Va. Acad. Sci. (pres. 1956, Hon. Life Mem. award 1981); mem. Genetics Soc. Am., Am. Genetics Assn., Soc. Study Evolution, Bot. Soc. Am. (chmn. southeastern sect. 1951-52), Assn. Southeastern Biologists (pres. 1962-63, rsch. award 1978), Am. Boxwood Soc. (hon. life mem., co-founder, treas., editor 1961-63, bd. dirs. 1982-84), Am. Assn. Bot. Gardens and Arboretums (editorial bd. 1962-64), Am. Begonia Soc. (hon.), Fairchild Tropical Garden (life), So. Appalachian Bot. Club (v.p. 1962), Am. Plant Life Soc. (Herbert medal 1978), La. Soc. Hort. Rsch. (hon.), Am. Magnolia Soc. (v.p. 1968-79), Friends of Va. State Arboretum (hon. life), Phi Beta Kappa (chpt. pres. 1974), Sigma Xi (chpt. pres. 1970), Tau Kappa Alpha, Phi Sigma. Democrat. Mem. Ch. of Brethren. Club: Torch (local pres. 1970). Home: 2025 Colonial Pl Winston Salem NC 27104-3128

FLOTTORP, GORDON, audiophysicist; b. Mandal, Norway, May 17, 1920; s. Knut and Agnes (Pettersen) F.; m. Randi Elise Hagnor, June 30, 1945; children: Anne-Marie, Brit Anette, Knut Dag. DSc, U. Oslo, 1947, PhD, 1976. Govt. grantee psycho-acoustic lab. Harvard U., Cambridge, Mass., 1950-51, rsch. asst. psycho-acoustic lab., 1950-51, rsch. fellow psycho-acoustic lab., 1951-52; audiophysicist Inst. of Audiology, Univ. Hosp., Rikshospitalet, Oslo, 1952-90, sr. cons., 1990—; chmn. Norwegian TC29 Internat. Electrotech. Commn., 1984—; internat. mem. com. hear. bioacoustical biomedicine NAS, 1960—; mem. Royal Norwegian Aircraft Noise Commn., Oslo, 1965-85; mem. adv. com. on noise problems Norwegian Labor Insp., 1979-82. Contbr. articles to profl. jours. Rsch. fellow Harvard U., 1956. Fellow Acoustical Soc. Am.; mem. Norwegian Phys. Soc., Internat. Soc. Audiology (exec. com. 1982-88), Inst. Noise Control Engring. (corr., peer com. 1983—), Norwegian Missionary Alliance (pres. 1963-87). Lutheran. Achievements include research on leisure-time noise such as discoteque music, rock and roll concerts, etc.; suceeded in inducing Norwegian Health Directorate to work on sound limits for leisure-time noise. Home: Grefsenveien 87, 0487 Oslo Norway Office: Inst Audiology Nat Hosp, Rikshospitalet, 0027 Oslo Norway

FLOURNOY, THOMAS HENRY, mechanical engineer; b. N.Y.C., July 4, 1958; s. John James and Ruth (Schumacher) F.; m. Martha Louise Wisenbaker, June 30, 1990. BSE, Duke U., 1980, PhD, 1991. Commd. ensign USN, 1980, advanced through grades to lt., 1984, resigned, 1988; mech. engr. FAA, Atlantic City, N.J., 1991—. Contbr. articles to profl. jours. Case-NASA Aero. Rsch. fellow, 1989. Mem. ASTM, Nat. Eagle Scout Assn., Sigma Xi. Republican. Home: 486 Chestnut Neck Rd Port Republic NJ 08241 Office: Fed Aviation Adminstrn ACD 220 Atlantic City NJ 08405

FLOWERS, HAROLD LEE, consulting aerospace engineer; b. Hickory, N.C., June 25, 1917; s. Edgar Lee and Olive Kathryn (Deal) F.; m. Doris Louis Hexamer, Apr. 18, 1941; children: Josselyn, Harold Jr. BSEE, Duke U., 1938; MS in Engring., U. Cin., 1941; DSc (hon.), Lenoir Rhyne Coll., Hickory, N.C., 1993. Rsch. scientist Naval Rsch. Lab., Washington, 1942-50; mgr. electronic div. Goodyear Aerospace, Akron, Ohio, 1950-61; dir. engring. div. AVCO Electronics, Cin., 1961-63; from dir. to mgr. of missiles div. McDonnell Douglas Astronautics Co., St. Louis, 1963-87; engring. cons. CP Assocs., Washington, 1987—; mem. adv. bd. Mark Twain Bank, Chesterfield, Mo. Contbr. articles to profl. jours. and mags. Mem. bd. visitors Duke U., Durham, N.C. Fellow IEEE; assoc. fellow AIAA; mem. Phi Beta Kappa, Tau Beta Pi. Presbyn. Achievements include patents for Radar Receiver Design, Atran Dead Reckoning Navigation System. Office: CP Assocs Ste 201 1755 Jefferson Davis Hwy Arlington VA 22202

FLURCHICK, KENNETH MICHAEL, computational physicist; b. Paterson, N.J., Aug. 21, 1951; s. John Flurchick and Lena Guerin Lynch. BA in Math., William Patterson Coll., 1975; MS in Physics, Colo. State U., 1979, PhD in Physics, 1987. Computational scientist ETA Systems, Inc., St. Paul, 1984-88, project leader, 1988-89; sr. computer performance analyst Amdahl Corp., Sunnyvale, Calif., 1989; rsch. cons. N.C. Supercomputing Ctr., Research Triangle Park, 1989-90, computational scientist, 1990, acting dir. rsch. inst., 1990, computational scientist of rsch. inst., 1990-91, asst. dir., 1992—. Contbr. articles to profl. jours. Mem. AAAS, Math. Soc. Am., Am. Phys. Soc. Achievements include research in theoretical/computational solid state physics, simulational physics, computational chemistry, both classical and quantum approaches. Office: NC Supercomputing Ctr 3021 W Cornwallis Rd Durham NC 27705-5205

FLUSS, HAROLD SHRAGE, engineer; b. Bklyn., Nov. 4, 1956; s. Hersh Isaac and Fannie (Kinderman) F.; m. Donna Michelle Grey, July 6, 1986; children: Joshua Dov, Gabriella Anne. MS, Columbia U., 1979, PhD, 1982. Mem. tech. staff AT&T Bell Labs., Whippany, N.J., 1982—. Contbr. articles to Connection Tech., 1990, Applied Math. Modeling, 1980 and various procs. Mem. ASME, Tau Beta Pi, Sigma Xi. Achievements include rsch. on modeling fabric reinforced composites, and analyses of connector products and designs.

FLYNN, PATRICK JOESPH, aerospace engineer; b. Phoenix, Mar. 22, 1965; s. Patrick Francis and Elizabeth Louise (Hentz) F. BS, Ariz. State U., 1989. Design engr./knowledge base engr. Garrett Aux. Power, Phoenix, 1989—. Mem. Soc. Automotive Engrs., AIAA, Exptl. Aircraft Assn.

FLYNN, THOMAS R., engineering executive; b. Fall River, Mass., Mar. 5, 1940; s. John J. and Rita R. (Thibault) F.; (div.); children: Dominic, Bridget. BS in Indsl. Engring., U. Mass., Dartmouth, 1962. Indsl. engr. Allied Chem. Corp., Chesterfield, Va., 1965-66; contract adminstr. United Technologies, Hamilton Standard Divsn., Windsor Locks, Conn., 1966-68; contracts mgr. Atkins & Merrill, Inc., Maynard, Mass., 1969-72, account exec. Merrill Lynch, Boston, 1972-75; mktg. mgr. Brunswick Corp., Def. Divsn., Hopkinton, Mass., 1975-80; dir. mktg. Brunswick Corp., Def. Divsn., Lincoln, Nebr., 1980-89; v.p. Brunswick Corp., Tech. Group, Arlington, Va., 1989—. Chmn. fin. com. Town of Hopkinton (Mass.), 1975-78; mem. Am. Legion Post 3, Lincoln, Nebr., 1980-87, Lions, East Windsor, Conn., 1967-79. Lt. USAF, 1962-65. Recipient Outstanding Svc. award Boston C. of C., 1972. Mem. AIAA, Am. Def. Preparedness Assn., Am. Helicopter Soc., Assn. of the U.S. Army, Navy League of the U.S., Quequechan Club. Avocations: fishing, boating. Home: 8244 Carrleigh Pky Springfield VA 22152 Office: Brunswick Technical Group 1745 Jeff Davis Hwy Ste 410 Arlington VA 22202

FLYNN, WILLIAM THOMAS, civil engineer; b. St. Louis, Oct. 7, 1916; s. William Thomas Sr. and Olympia Galacis (Biegler) F.; m. Charlotte Martha Louise Koehler, July 11, 1941; children: Elizabeth Ann Flynn Stigen, Mary Lou Flynn Dupart, W. Gregory. BSCE, Washington U., St. Louis, 1939, MSCE, 1941. Registered profl. engr., Tex., Mo., Wis., Calif. Asst. engr. Panama Canal, Ancon Engring. Dist., Diablo Heights, C.Z., 1941-43; design engr. Fruin-Colnon Contracting Co., St. Louis, 1946-48; project engr. Oscar Janssen Architects, Engrs., St. Louis, 1948-51; office engr. Vollmar Bros. Constrn. Co., St. Louis, 1951-53; staff constrn. engr. Joseph Schlitz Brewing Co., Milw., 1953-69; dir. engring. Seton Med. Ctr., Austin, Tex., 1969-77; engr. water and wastewater utility City of Austin, 1977—. Contbr. to profl. pubis. Trustee Theinsville (Wis.) Village Bd., 1959-67, pres., 1967-69; mem. Community-Police Rels. Coun., Austin, 1981-83. Lt. USNR, 1943-46, PTO. Decorated Silver Star medal. Mem. ASCE, ASTM, Am. Water Works Assn., Sigma Xi. Democrat. Roman Catholic. Office: Water & Wastewater Utility Ste 400 625 E 10th St Austin TX 78701

FLYNT, CLIFTON WILLIAM, computer programmer, software designer; b. Waterville, Maine, Feb. 7, 1953; s. Jerrold Miller and Alta Livia (Marasi) F.; m. Carol Ann Clapper, May 1, 1993. BS in Biochemistry, SUNY, Syracuse, 1975, postgrad., 1975. Chemist SUNY, Syracuse, 1975-77, Gelman Sci., Ann Arbor, Mich., 1978-79; programmer Applied Dynamics, Ann Arbor, 1979-81, GeoSpectra, Ann Arbor, 1981-83, BioImage, Ann Arbor, 1983-86; cons. Resource One, Chgo., 1986-88; sr. programmer Cimage, Ann Arbor, 1988-90; project leader, v.p. Veracity, Inc., Walled Lake, Mich., 1990-92; mem. tech. staff Computational Bioscis. Inc., 1992-93; sr. applications software engr. Applied Intelligent Systems, Inc., Ann Arbor, Maine, 1993—; mem. tech. resource com. Arbornet, Ann Arbor, 1986-90. Contbr. to Dept. Energy Small Bus. Innovative Rsch. Grant. Mem. Am. Chem. Soc., Assn. for Computing Machinery, U.S. Figure Skating Assn. Avocations: songwriting, woodworking, guitar, skating.

FOBBE, FRANZ CASPAR, radiologist; b. Thüngersheim, Bavaria, Fed. Republic Germany, Dec. 26, 1948; m. Doris Rummer-Löns, Mar. 10, 1989; children: Lukas, Lea. Student, Hansa Koll., Hamburg, Fed. Republic Germany, 1971-73, U. Glasgow, Scotland, 1979-80; MD, Med. Hochschule, U. Hannover, 1982; Habilitation, Freie U. Berlin, 1990. Asst. arzt. A.K. Barmbek, Chirurgie, Hamburg, 1981-83; asst. arzt. radiology U. Berlin, 1983-89, cons., 1990—. Contbr. numerous articles to profl. jours. Recipient Röntgenpreis Deutsche Röntgengesellschaft, 1991. Office: Freie U Berlin Klinikum Steglitz, Hindenburgdamm 30, 1000 Berlin 45, Germany

FOCKS, DANA ALAN, medical entomologist, epidemiologist; b. Salt Lake City, July 16, 1948; s. Fred Albert and Juanita Louise (Anderson) F.; m. Debby Darnell, Oct. 20, 1973; children: Rebekah M., Peter C., Kelly O. BS in Zoology, U. Fla., 1971, PhD in Med. Entomology, 1977. Rsch. entomologist Med. and Vet. Entomology Rsch. Lab. USDA, Gainesville, Fla., 1977—; adj. prof. U. Fla., Gainesville, 1977—, U. S.C., Columbia, 1985—; cons. WHO, Geneva, Switzerland, 1989, USAID, 1985—, Dept. Def., 1986—. Author: (pub. series) Biocontrol of Yellow Fever Mosquitoes, 1977-85, Population Dynamics and Control of Riceland Mosquitoes, 1987-91, Simulation Studies with Aedes Aegypti, Dengue Transmission, 1991-93. Grantee USDA, 1980. Mem. Am. Soc. Tropical Medicine and Hygiene, Entomol. Soc. Am., Soc. Vector Entomologists. Achievements include patent for sampling device for culicids; developed integrated control strategy for control of yellow fever mosquitoes for developing countries; identified key regulatory factors in dynamics of riceland mosquitoes. Home: 7409 NW 23 Ave Gainesville FL 32606 Office: USDA/ARS Med and Vet Entomol Rsch Lab 1600 SW 23 Dr Gainesville FL 32608

FODOR, GEORGE EMERIC, chemist; b. Mako, Hungary, Feb. 13, 1932; came to U.S., 1956; s. Imre and Ilona (Messinger) F.; m. Marjory A. Byrne, Feb. 13, 1965; children: Cara A., John E. Diploma chemistry, U. Scis., Szeged, Hungary, 1955; PhD, Rice U., 1965. Chemist Hungarian Oil Refinery, Petfurdo, Hungary, 1955; staff chemist Hungarian Oil and Gas Rsch. Inst., Veszprem, 1955-56; chemist Pontiac Refining Corp., Corpus Christi, Tex., 1957-60; rsch. chemist E.I. duPont, Parlin, N.J., 1965-66; chemist, staff scientist S.W. Rsch. Inst., San Antonio, 1966—. Contbr. articles to profl. jours. Mem. Am. Chem. Soc., Soc. Tribologists and Lubrication Engrs. Achievements include development of fire-resistant diesel fuel, analytical chemical methods using FTIR spectroscopy; rsch in synthesis of natural products, study of microemulsified fuels, study of the kinetics of hydrocarbon oxidation. Office: SW Rsch Inst PO Drawer 28510 6220 Culebra Rd San Antonio TX 78228-0510

FODSTAD, HARALD, neurosurgeon; b. Mo-i-rana, Norway, Sept. 4, 1940; came to the U.S., 1991; s. Reidar and Agathe Ursula (Jorfald) F.; m. Michiko Okayasu, Apr. 6, 1965; children: Tor, Henrik. MD, U. Berne, 1967; PhD, U. Umea, 1980. Intern Bodens Ctrl. Hosp., Sweden, 1968-71; resident Umea U. Hosp., Sweden, 1971-76; cons. neurosurgeon Umea Univ. Hosp., Sweden, 1976-88, Tawam Univ. Hosp., Alain, United Arab Emirates, 1988-90; chief surgeon Dobelle Inst., Glen Cove, N.Y., 1991-92; cons. neurosurgeon Health Ins. Plan Greater N.Y., N.Y.C., 1992—; expert witness Supreme Ct. South Australia, Adelaide, 1992; assoc. prof. U. Umea, 1981. Author: Untersuchung Zur Frage der Alkoholparanoia, 1968, Tranexamic Acid in Subarachnoid Hemorrhage, 1980, Antifibrinolytics in Subarachnoid Hemorrhage, 1990, Epidemiology of Aneurysmal SAH, 1990; cons. editor: Emirates Med. Jour., 1990' author, co-author 160 scientific publs. Lt. Norwegian Navy, 1970-71. Scholar Swedish-Japanese Found., 1981. Mem. World Confedn. Neuroscis. (pres. 1979-87), Study Group for Microphysiology in Stereotaxy, European Neurosurg. Standards Com. EANS. Lutheran. Achievements include development of phrenic nerve stimulators for respiratory paralysis of central nervous origin. Home: 415 C Bayville Ave Bayville NY 11709 Office: Health Ins Plan Greater NY 178 E 85th St New York NY 10028

FODY, EDWARD PAUL, pathologist; b. Balt., June 11, 1947; s. Edward Paul and Frances Dorothy (Schultz) F.; m. Nancy June Keipe, July 19, 1974. BS, Duke U., 1969; MS, U. Wis., 1971; MD, Vanderbilt U., 1975. Diplomate Am. Bd. Pathology. Resident in pathology Vanderbilt U. Hosp., Nashville, 1975-78; fellow in chemistry U. Tex. Med. Sch., Houston, 1979-80, asst. prof. pathology, 1980-81; chief lab. VA Hosp., Little Rock, 1981-87; assoc. prof. pathology U. Ark. Med. Sch., Little Rock, 1981-87; dir. pathology Bethesda Hosp., Cin., 1987—. Editor, author: Clinical Chemistry, 1984, chpt. to book. Mem. Cin.-Kharkov Sister City Project, 1990. Fellow Coll. Am. Pathologists, Am. Soc. Clin. Pathologists; mem. AMA, Am. Assn. for Clin. Chemistry, Am. Soc. for Microbiology, Ohio Med. Assn., Cin. Acad. Medicine. Republican. Lutheran. Avocations: boating, photography. Home: 7730 Coldstream Woods Dr Cincinnati OH 45255-5612 Office: Bethesda Hosp Dept Pathology 619 Oak St Cincinnati OH 45206

FOE, ELIZABETH, biologist, educator. Recipient MacArthur fellow John D. and Katherine T. MacArthur Found., Ill., 1993. Office: Univ of Washington Dept of Biology Seattle WA 98195*

FOERSTER, CONRAD LOUIS, project engineer; b. Balt., Jan. 19, 1938; s. George Leroy Sr. and Jane Ruth (Carson) F.; m. Tina M. Capone, Sept. 20, 1964; children: Christopher C., George A. AS in Engring. Sci., Nassau C.C., 1971; BS in Engring. Sci., L.I. U., 1975; postgrad., Poly. Inst., N.Y.C., 1978, U.S. Partical Accelerator Sch., 1992. Vacuum technician Veeco Instrument Corp., Plainview, N.Y., 1959-68; electronic technician Gen. Instrument Inc., Hicksville, N.Y., 1968, eqipment supr., 1973; mfg. engr. Deutsch Relays Inc., East Northport, N.Y., 1968-73; product engr. Deutsch Relays Inc., East Northport, 1973-79; project engr. Brookhaven Nat. Lab., Upton, N.Y., 1979-86, vacuum group leader, 1986—. Contbr. articles to profl. jours. With USAF, 1955-59. Achievements include research in vacuum science. Office: Brookhaven Nat Lab NSLS Bldg 725C Upton NY 11973

FOGEL, IRVING MARTIN, consulting engineer; b. Gloucester, Mass., Apr. 15, 1929; s. Jacob and Ethel (David) F.; children: Ethan, Ronit. BS, Ind. Inst. Tech., 1954, D of Engring. (hon.), 1982. Registered profl. engr., 22 states, D.C., Israel. Civil engr. Ill. Hwy. Dept., Peoria, 1954-55; field engr. Peter Kiewit Sons Co., East Gary, Ind., 1955, field engr., progress engr., cost engr., Ogdensburg, N.Y., 1955-56; supt. grading and paving Merritt, Chapman & Scott, Binghamton, N.Y., 1956; cost engr. Drake-Merritt, Goose Bay, Labrador, 1956-57; constrn. mgmt. engr. Mil. Estimating Corp., Madrid, Spain, also P.I., 1957-58; project mgr. Ministry of Def., State of Israel, 1958-59, Frederic R. Harris (Holland) N.V., The Hague, also Tehran, Iran, 1959-61; project mgr. Solel Boneh & Assocs., Addis Ababa, Ethiopia, 1961-63; asst. to tech. dir. Frederic R. Harris, Madrid, 1963-64; chief engr. McKee-Berger-Mansueto, Inc., N.Y.C., 1964-65, v.p. constrn. mgmt., 1965-69; pres. Fogel & Assocs., Inc., N.Y.C., and Ft. Lauderdale, Fla., 1969—; lectr. Fellow ASCE, Nat. Acad. Forensic Engrs.; mem. NSPE, Am. Arbitration Assn., Am. Assn. Cost Engrs., Am. Inst. Constructors, Constrn. Specifications Inst., Nat. Contract Mgmt. Assn., N.Y. Bldg. Congress, Project Mgmt. Inst., Soc. Am. Mil. Engrs. Author guides and handbooks on constrn. bus., latest being Construction Owner's Handbook of Property Development, 1992; contbr. articles to profl. jours. Home: 525 E 86th St New York NY 10028-7512 Office: 15 E 26th St Ste 1700 New York NY 10010-1505

FOGLE, HOMER WILLIAM, JR., electrical engineer, inventor; b. Harrisonburg, Va., Apr. 8, 1948; s. Homer William and Mary Elizabeth (Gaetano) F.; m. Irene Susan Pasternak, Feb. 11, 1978; children: Homer William III, Nikolaus Gerhardt Werner. BSEE, Cornell U., 1970; MSc, U. Cambridge, Eng., 1975. Sr. rsch. devel. engr. Scott Paper Co., Phila., 1976-79; sr. systems analyst ICI Americas Inc., Wilmington, Del., 1979-85, sr. electronics devel. engr., 1985-91; sr. rsch. scientist TRW Safety Systems Inc., Mesa, Ariz., 1992—. Maj. USMC, 1970-74, with USNR, 1966-70, with USMCR, 74-91. Mem. IEEE, Delta Kappa Epsilon (exec. bd. dirs., historian Delta Chi chpt. 1985—). Republican. Roman Catholic. Achievements include patents for automatic speed control of a rewinder, local-field type article removal alarm, currncy alarm pack, filtered electrical connection assembly using potted ferrite element, hermetically-sealed electrically absorptive low-pass radio frequency filters and materials for same; inventions include steam shower control system, programmable cam switch emulator, rotary kiln thermal insulation technique, cash-in-transit system AKA remote initiation of pyrotechnic devices, synchronous detection transceiver, electrically actuated fluid control valve with non-electrical manual override, electromagnetically lossy glass sealing material for low pass radio frequency interference suppression filters, photochemically activated indicator for identifying documents subjected to reproduction by xerographic processes, nondestructive electro-acoustic testing of bridgewire EEDs. Home: 3237 E Fox St Mesa AZ 85213

FOGLEMAN, GUY CARROLL, physicist, mathematician, educator; b. Lake Charles, La., Dec. 29, 1955; s. Louis Carroll and Peggy Joyce (Trahan) F. BS in Physics, La. State U., 1977; MS in Physics, Ind. U., 1979, MA in Math., 1981, PhD in Physics, 1982. Rsch. assoc. Tri Univ. Meson Facility U. B.C., Vancouver, Can., 1982-84; assoc. prof. San Francisco State U., 1984-87, adj. prof., 1987—; project scientist RCA Govt. Svcs., Moffett Field, Calif., 1987-88; prin. investigator Search for Extraterrestrial Intelligence Inst., Mountain View, Calif., 1988-89; mgr. advanced programs life scis. div. NASA Hdqrs., Washington, 1990—; vis. physicist Stanford (Calif.) Linear Accelerator Ctr., 1984-86. Contbr. articles to sci. jours. Travel grantee NSF and NATO, 1980; rsch. grantee NASA, 1988, 89. Mem. Am. Phys. Soc., Sigma Xi (assoc.), Sigma Pi Sigma. Achievements include research in physics of particles in microgravity, theoretical elementary particle physics, technologies for the collection of cosmic dust particles, the origins of life. Office: NASA Code ULF Washington DC 20546

FOGLESONG, PAUL DAVID, microbiology educator; b. Marion, Va., June 24, 1949; s. Everett Paul and Thelma Brouchelle (Conner) F.; m. Clare Maria Wright, July 1, 1978 (div. Jan. 1985). BS, Va. Poly. Inst. & State U., 1971; PhD, SUNY, Stony Brook, 1980. Rsch. asst. SUNY, Stony Brook, 1973-80; postdoctoral fellow Albert Einstein Coll. of Medicine, Bronx, N.Y., 1980-82, rsch. fellow, 1982; rsch. asst. St. Jude Children's Rsch. Hosp., Memphis, 1982-86; dir. biochemistry Biotherapeutics Inc., Memphis, 1986-88; asst. prof. Memphis State U., 1988-89, Rutgers U., Camden, N.J., 1989—; cons. So. Rsch. Inst., Birmingham, Ala., 1989—. Contbr. articles to Jour. Virol., Cancer Immunol. Immunother., Anal. Biochem. Treas. Am. Guild Organists, Memphis, 1988-89; trustee St. Mark's Ch., Phila., 1993—, treas., 1991—; asst. organist St. Mary's Cathedral, Memphis, 1984-89. Gov. Westmoreland Davis scholar, 1967-71. Mem. AAAS, Am. Soc. Microbiology, Am. Soc. Virology, Am. Assn. Cancer Rsch., Phi Kappa Phi. Achievements include characterization of the inhibition of type I DNA topoisomerase by ATP analogs, antibiotics, and antitumor drugs; characterization of DNA topoisomerase levels in tumor tissues by immunohistochemistry; development of an in vitro system for faithful transcription of vaccinia virus genes, map of Frog Virus 3 genome. Office: Biology Dept Rutgers Univ Camden NJ 08102-1401

FOLAN, LORCAN MICHAEL, physicist; b. Dublin, Ireland, Dec. 4, 1960; came to U.S., 1981; s. Aidan John Patrick and Mary (McShane) F.; m. Estelle Albala June 6, 1986; 1 child, Aidan Benjamin. BSc in Applied Sci., Trinity Coll., Dublin, 1981; PhD in Physics, Polytech. U., 1987. Asst. prof. Polytech U., Bklyn., 1987-93, assoc. prof., 1993—. Contbr. articles to profl. jours. Mem. Am. Phys. Soc., Optical Soc. Am., Sigma Xi (chpt. pres. 1990-92). Achievements include research in fluorescence spectroscopy. Office: Polytechnic Univ 6 Metrotech Ctr Brooklyn NY 11201

FOLCHETTI, J. ROBERT, water, wastewater engineer; b. Brewster, N.Y., Aug. 26, 1934; s. Innocenzo and Maragaret (Larkin) F.; m. Elisa Meloni, July 7, 1956; children: John E., Robert W., Melinda Anne, Gregory L. BS, St. Lawrence U., 1960; MS, Brown U., 1962. Petroleum engr. Humble Oil Co., New Orleans, 1962-66; engr. Bechtel Assocs., N.Y.C., 1966-68; project engr. Metcalf and Eddy, N.Y.C., 1968-70; project mgr. Malcolm Pirnie, Inc., White Plains, N.Y., 1970-77; dir. environ. svcs. Putnam County, Carmel, N.Y., 1977-84; prin. J. Robert Folchetti and Assocs., Brewster, N.Y., 1984—. With USMC, 1951-54, Korea. Mem. Am. Soc. Civil Engrs., Water Environ. Fedn., Am. Water Works Assn., Am. Inst. Profl. Geologists. Office: J Robert Folchetti Assocs 98 Mill Plain West Danbury CT 06811

FOLDES, FRANCIS FERENC, anesthesiologist; b. Budapest, Hungary, June 13, 1910; came to U.S., 1941; s. Leopold and Nellie (Friedman) F.; m. Edith, Oct. 9, 1938; children: Eva, Judith, Barbara. MD, U. Budapest, 1934; MD (hon.), U. Szeged Sch. Medicine, Hungary, 1982, Free U. Berlin, 1989, U. Vienna, Austria, 1990, U. Rostock, 1992. Diplomate Hungarian Bd. Internal Medicine, Nat. Bd. Med. Examiners, Am. Bd. Anesthesiology. Intern U. Clinic Hosp. of U. Budapest, 1933-34; resident physician Svabhegyi Sanatorium, Budapest, 1934-39; pvt. practice in medicine and rsch. in applied pharmacology Budapest, 1939-41; from rsch. fellow in anesthesiology to asst. anesthetist Mass. Gen. Hosp., Boston, 1941-47; asst. in medicine Harvard Med. Sch., Cambridge, Mass., 1943-46, asst. in anesthesia, 1946-47; dir. dept. anesthesia Mayview (Pa.) State Hosp., 1950-62; from assoc. prof. to clin. profl. anesthesiology U. Pitts. Med. Ctr., 1948-62; dir. dept. anesthesiology Mercy Hosp., Pitts., 1947-62; clin. prof. anesthesiology Coll. Physicians and Surgeons of Columbia U., N.Y.C., 1962-64; chmn. dept. anesthesia Montefiore Hosp. & Med. Ctr., Bronx, N.Y., 1962-76; prof. anesthesiology Albert Einstein Coll. Medicine, Bronx, 1964-75, prof. emeritus anesthesiology, 1975—; prof. emeritus anesthesiology U. Miami Sch. Medicine, 1976—; cons. in anesthesia Leech Farm Vets. Hosp., Pitts., 1952-62, Deshon Vets. Hosp., Burler, Pa., 1947-62, Montefiore Med. Ctr., Bronx, 1975—; pres. World Fedn. of Socs. of Anaesthesiologists, 1972-76; pres. sr. med. bd. Montefiore Med. Ctr., 1973-74. Recipient fellowship Royal Coll. Surgeons of Eng., 1971, 82, fellowship Royal Coll. Surgeons in Ireland, 1979, Semmelweiss award Am. Hungarian Med. Assn., 1966, George Washington award Am. Hungarian Studies Found., 1970, Gold medal Spanish Soc. Anesthesiologists, 1974, Ralph M. Waters award Internat. Anesthesia Award Commn., 1976, Issekutz medal Hungarian Pharmacologist Soc., 1979, Harold R. Griffith medal McGill U., 1986, Order of the Rising Sun, Emperor of Japan, 1990. Fellow Internat. Anaesthesia Rsch. Soc., Am. Coll. Anesthesiologists, N.Y. Acad. Medicine, N.Y. Acad. Scis., Royal Soc. Medicine; mem. Am. Soc. Anesthesiologists (Disting. Svc. award 1972, Excellence in Rsch. award 1988), N.Y. State Soc. Anesthesiologists, Assn. Univ. Anesthetists, Am. Soc. for Pharmacology and Exptl. Therapeutics, Internat. Soc. for Biomed. Pharmacology, Myasthenia Gravis Found. (med. adv. bd. 1959—), Am. Dental Soc. Anesthesiology (adv. bd. 1975—, Heidbrink award 1975). Office: Montefiore Med Ctr 111 E 210th St Bronx NY 10467

FOLDVARI, ISTVAN, physicist; b. Budapest, Hungary, Sept. 9, 1945; came to U.S., 1990; s. Aladar and Maria (Vogl) F.; m. Andrea Fekete, July 17, 1970; 1 child, Adam. MS in Chemistry, Eotvos Lorand U., Budapest, 1969, PhD in Chemistry, 1971. Postgrad. fellow Rsch. Lab. Chem. Structure, Budapest, 1969-70; asst. lect. Semmelweis Med. U., Budapest, 1970-72; researcher Rsch. Lab. Crystal Physics, Budapest, 1972-77, sr. reseracher, mem. adv. bd., 1977-90; vis. researcher Okla. State U., Stillwater, 1990—. Co-author: Properties of Lithium Niobate, 1989; contbr. articles to profl. jours. Mem. Internat. Orgn. Crystal Growth, Am. Phys. Soc., Eotvos Lorand Phys Soc. (bd. dirs.), Hungarian Chem. Soc. Achievements include interpretation of non-stoichiometric character of $LiNbO_3$, first growth of Bi_2TeO_5 in optical quality and showing photorefractive effects, improving quality of several oxide materials. Home: 21 Bem, H 1151 Budapest Hungary Office: Okla State Univ 411 Noble Rsch Ctr Stillwater OK 74078

FOLEY, ARVIL EUGENE, mechanical engineer; b. Williamsburg, Ky., Oct. 27, 1924; s. James Millard and Mary Maude (Prewitt) F.; m. Joan Teresa Urbanski, June 13, 1953; children: Roger Allen, Debra Ann. BSME, U. Toledo, 1959. Supr. gen. engr. U.S. Army Corps Engrs., Huntington, W.Va., 1959-86; retired U.S. Army Corps Engrs., 1986. Sgt. U.S. Army Air Force, 1945-46. Mem. ASME, NSPE, W.Va. Soc. Profl. Engrs. (chmn. 1983-92), Propeller Club of U.S. Baptist. Home: 315 W 10th Ave Huntington WV

FOLEY, JAMES DAVID, computer science educator, consultant; b. Palmerton, Pa., July 20, 1942; s. Marvin Winfield and Stella Elizabeth (Ziegler) F.; m. Mary Louise Herrman, Aug. 22, 1964; children: Heather, Jennifer. BSEE, Lehigh U., 1964; MSEE, U. Mich., 1965, PhD, 1969. Group mgr. Info. Control Systems, Ann Arbor, Mich., 1969-70; asst. prof. U. N.C., Chapel Hill, 1970-76; sr. systems analyst Bur. of Census, Washington, 1976-77; assoc. prof. George Washington U., Washington, 1977-81, prof., 1981-90, chmn. dept. elec. engring. and computer sci., 1988-90; prof. Ga. Inst. Tech., Atlanta, 1991—, dir. graphics and visualization ctr., 1991—; pres. Computer Graphics Cons., Washington, 1979—. Author: (with others) Fundamentals of Computer Graphics, 1982, (with others) Computer Graphics: Principles and Practice, 1990, (with others) Introduction to Computer Graphics; co-author (graphics standard) Core System, 1977. Fellow IEEE; mem. Human Factors Soc., Assn. Computing Machinery, Spl. Interest Group for Graphics (vice chair 1973-75), Nat. Computer Graphics Assn. (bd. dirs. 1982-84). Avocations: skiing, sailing, model railroading. Office: Ga Inst Tech GVU Ctr Atlanta GA 30332-0280

FOLEY, KATHLEEN M., neurologist, educator, researcher; b. Flushing, N.Y., Jan. 28, 1944; d. Joseph Cyril and Catherine (Cribbin) Maher; m. Charles Thomas Foley, Aug. 10, 1968; children: Fritz, David. BA in Biology magna cum laude, St. John's U., N.Y.C., 1965, DSc (hon.), 1992; MD, Cornell U., 1969. Diplomate Am. Bd. Psychiatry and Neurology (examiner 1980—); lic. physician, N.Y. Intern, then resident in neurology The N.Y. Hosp., N.Y.C., 1969-74; asst. attending neurologist, neuology dept. Meml. Sloan-Kettering Cancer Ctr., N.Y.C., 1974-79, assoc. attending neurologist, 1979-88, chief-pain svc., 1982—, attending neurologist, 1988—; attending neurologist Manhattan (N.Y.) Eye & Ear Hosp., 1974-83; instr. in neurology, Med. Coll. Cornell U., N.Y.C., 1974-75, asst. prof., 1975-79, assoc. prof., 1979-89, assoc. prof. pharmacology, 1979-89, prof. neurology and neuroscience, 1989—, prof. clin. pharmacology, 1990—; rsch. assoc. lab. neuro-oncology Sloan-Kettering Inst. Cancer Rsch., N.Y.C., 1981-84; vis. asst. physician, cons. in neurology Rockefeller U. Hosp., 1975-79, vis. assoc. physician, 1979—; cons. Calvery Hosp., 1982—; assoc. mem. Meml. Sloan-Kettering Cancer Ctr., 1985-88, mem. 1988—. Editor Clinical Jour. Pain, 1985-87, Jour. Pain and Symptom Mgmt., 1987—, Palliative Medicine Jour., 1993—. Patient Svcs. Adv. Group, Am. Cancer Soc. Genetic Training grant NIH, 1970-71, Program for Pain Rsch. grant Bristol-Myers, 1988-92; Neuro-Oncology spl. fellow Meml. Sloan-Kettering Cancer Ctr., 1975-78; recipient Jr. Faculty award Am. Cancer Soc., 1975-78, Disting. Svc. award, 1992, Nat. Bd. award The Med. Coll. Pa., 1986, Willaim M. Witter award U. Calif. San Francisco, 1987, Annie Blount Storrs award Calvery Hosp., 1988, Balfour M. Mount award Am. Jour. Hospice Care, 1988, Disting. Oncologist award Dayton Oncology Soc., 1990, Tenth Barbara Bohen Pfeifer award Am. Italian Found. for Cancer Rsch., 1993; named Outstanding Women Scientist Women in Sci. Met. N.Y. Chpt., 1987, A. Soriano Jr. Meml. Lectr. The Andres Soriano Cancer Rsch. Found. Inc., 1992. Mem. AAAS, AMA (ad hoc adv. panel mgmt. chronic pain, DATTA reference panel), Acad. Hospice Physicians, Am. Acad. Neurology (chmn. long range planning com. 1990—, scientific program com. 1990, and other coms.), Am. Fedn. Clin. Rsch., Am. Med. Womens Assn., Am. Neurological Assn. (mem. com. 1984-85, councilor 1984-78), Am. Pain Soc. (bd. dirs. 1980-82, pres. 1984-85, bylaws com. 1986-87 long range planning task force 1989—), Am. Soc. Clin. Onocology (program com. 1991-92, com. on care at the end of life 1993—, and other coms.), Am. Soc. Clin. Pharmacology and Therapeutics, Assn. Rsch. in Nervous and Mental Diseases, Children's Hospice, Children's Hospice Internat., Cornell U. Med. Coll. Alumni Assn. (bd. dirs., nominating com.), Eastern Pain Assn. (John J. Bonica award 1986), Harvey Soc., Internat. Assn. Study Pain (councilor 1984-90, edn. com. 1986-93, and various coms.), N.Y. Acad. Scis. (USP adv. panel on neurology 1990—), Soc. for Neuroscience, Alpha Omega Alpha. Office: Meml Sloan-Kettering Cancer Ctr 1275 York Ave New York NY 10021-6306

FOLK, GEORGE EDGAR, JR., environmental physiology educator; b. Natick, Mass., Nov. 14, 1914; s. George Edgar and Minnie May (Davis) F. AB, Harvard U., 1937, MA, 1940, PhD, 1947. Instr. New England Secondary Schs., 1937-39, 40-42; asst. prof. Bowdoin Coll., Brunswick, Maine, 1947-52; prof. environ. physiology U. Iowa, Iowa City, 1952—. Author: Textbook of Environmental Physiology, 1965, 3d edit., 1984; contbr. over 160 articles to scientific jours., chpts. to books. Mem. Internat. Hibernation Soc., Internat. Soc. Biometeorology, Internat. Soc. Zoologists, Am. Soc. Mammalogy, Am. Meteorol. Soc., Am. Soc. Circumpolar Health, Am. Physiological Soc., Explorers Club. Office: U Iowa Dept Physiology BSB Coll Medicine Iowa City IA 52242

FOLKENS, ALAN THEODORE, microbiologist; b. Graceville, Minn., Oct. 26, 1936; s. Martin and Catherine (Laman) F.; m. Pearl June Putnam, July 29, 1961; children: Lee Alan, Kimberly Mae Folkens Anderson, Shannon Lee Folkens Tobin, Eric Martin. BA, Omaha U., 1962; PhD, U. S.D., 1971. Acting dir., dir. allied health professions Ill. State U., Normal, 1971-73; chief clin. microbiologist Peoria (Ill.) Tazewell Pathology Group, 1973-84; lab. dir. Delta Med. Ctr., Greenville, Miss., 1984-85; R&D clin. microbiologist Alcon Labs., Inc., Fort Worth, Tex., 1985—; vis. faculty E. Tenn. State U., Johnson City, 1978-86; adj. faculty U. Ill. Peoria Sch. Medicine, 1980-84, Ill. State U., Normal, 1973-84, U. N. Tex., Denton, 1992—; presenter symposium in field. Contbr. articles to profl. jours., publs. Bd. edn., past pres. Blessed Sacrament Sch., Morton, Ill., 1978-81; chmn. sickle cell anemia screening Ill. State U., Normal, 1972; chmn. Tootsie Roll drive for retarded children, K.C., Morton, 1975. Trainee NIH, 1967. Mem. Am. Soc. Microbiology, Am. Soc. Clin. Pathology, Am. Acad. Microbiology (diplomate), Assn. Rsch. in Vision and Ophthalmology, N.Y. Acad. Scis., Phi Sigma, Sigma Xi. Independent. Roman Catholic. Achievements include work in FDA approval of Ciloxan for topical ophthalmic therapy.

FOLKERS, KARL AUGUST, chemistry educator; b. Decatur, Ill., Sept. 1, 1906; married, 1932; 2 children. BS with honors, U. Ill., 1928; PhD in Organic Chemistry, U. Wis., 1931; ScD (hon.), Phila. Coll. Pharmacy and Sci., 1962, U. Wis., 1969, U. Ill., 1973; PharmD (hon.), U. Uppsala, Sweden, 1969; degree in medicine and surgery (hon.), U. Bologna, Italy, 1989, MD (hon.), 1990. Squibb & Lilly research fellow Yale U., 1931-34; researcher Merck & Co. Inc., 1934-38, asst. dir. research, 1938-45, dir. organic and biol. chemistry research div., 1953-56, exec. dir. fundamental research, 1956-62, v.p. exploration research, 1962-63; pres., chief exec. officer Stanford Research Inst., 1963-68; prof. chemistry, dir. Inst. Biomed. Research U. Tex., Austin, 1968—, Ashbel Smith prof., 1973—; pres. Karl Folkers Found. for Biomed. and Clin. Research, 1990—; mem. div. 9, Nat. Def. Research Com., 1943-46; Baker nonresident lectr. Cornell U., 1953; lectr. med. faculty Lund, Stockholm Uppsala, Gothenburg univs., Sweden, 1954; chmn. conf. on vitamins and metabolism Gordon Research Confs., 1956, trustee, 1971; Strumer lectr., 1957; mem. sci. adv. com. Inst. of Microbiology, Rutgers U., 1957-60; chmn. adv. council dept. chemistry Princeton U., 1958-64; Regent's lectr. UCLA, 1960; guest lectr. Am.-Swiss Found. Sci. Exchange, 1961; Robert A. Welch Found. lectr., 1963; courtesy prof. Stanford U., 1963—, U. Calif.-Berkeley, 1963—; Marchon vis. lectr. U. Newcastle, 1964; F.F. Nord lectr. Fordham U., 1971; mem. rev. com. US Pharmacopoeia; chmn. Nat. Acad. Sci., 1975; mem. internat. adv. bd. 4th Intersci. World Conf. on Inflammation, Geneva, 1991; numerous invited lecture posts, 1953—. Mem. bd. editors Internat. Jour. Vitamin Research, 1968, Research Communications in Chem. and Pharmacological Pathology, 1969. Recipient Alexander von Humboldt-Stiftung award Fed. Republic Germany, 1977, Presdl. Cert. of Merit, 1948, Merck & Co. Inc. award, 1951, Spencer award, 1959, Perkin medal, 1960, President's Nat. Medal of Sci. NSF, 1990; co-recipient Mead Johnson and Co. award, 1940, 49, Van Meter prize Am. Thyroid Assn., 1969, Robert A. Welch Internat. award, 1972, Am. Pharm. Assn. Achievement award, 1974, Priestley medal Am. Chem. Soc., 1986. Fellow Am. Inst. Nutrition; fgn mem. Royal Swedish Acad. Engring. Scis.; mem. AAAS, Nat. Acad. Sci., Am. Soc. Biol. Chemistry, Am. Inst. Chemists, Am. Chem. Soc. (chmn. North Jersey sect., chmn. div. organic chemistry, pres. 1962, Harrison-Howe award and lectr. 1949, Spencer award Kansas City sect. 1959, Nichols medal N.Y. sect., 1967, Priestley medal 1986), Internat. Soc. Metabolic Therapy (bd. dirs 1988), Soc. Italiana di Scienze Pharmaceutiche (hon.), Phi Lambda Epsilon (hon.). Office: U Tex Inst Biomed Rsch Welch Hall 4.304 Austin TX 78712

FOLKERTS, DENNIS MICHAEL, telecommunications specialist; b. St. Louis, May 11, 1960; s. Jack Eugene and Ila Dolores (Mohr) F.; m. Cynthia Folkerts, Sept. 10, 1982; children: Deanna Marie, Angela Michelle. Student, Florrisant Valley C.C. Sr. data process coord. Venture Stores, O'Fallen, Mo., 1978-85; telecommunications specialist Kellwood Co., St. Louis, 1985-. Bd. dirs. St. John Bosco Men's Club, Maryland Heights, Mo.; mem. Tenn. Squires, Moore County, 1985—. Mem. KC (3rd degree). Achievements include development of high level data security for call in users to host system and development of sophisticated auto call feature for EDI application and built and maintained massive data base for on call security. Office: Kellwood Co 600 Kellwood Pkwy Saint Louis MO 63017

FOLKINS, JOHN WILLIAM, speech scientist, educator; b. Redlands, Calif., Mar. 19, 1948; s. Hugh Montgomery and Evelyn Louise (Smith) F.; m. Georgianna Marie Lewis, July 10, 1971; children: Pollyanna Erin, Claire Victoria. BA, U. Redlands, 1970, MS, 1971; PhD, U. Wash., 1976. Postdoctoral fellow U. Wash., Seattle, 1976-77; from asst. prof. to prof. Dept. Speech Pathology and Audiology, U. Wash., Seattle, 1977—; chair dept. speech pathology and audiology U. Iowa, Iowa City, 1986-93; assoc. provost U. Iowa, 1993—; adv. coun. Nat. Inst. Deafness and Other Communicative Disorders, NIH, 1989-92. Author: Atlas of Speech and Hearing Anatomy, 1984; contbr. sci. articles to profl. publs.; speech editor Jour. Acoustical Soc. Am., 1987-90. Grantee NIH, 1977—. Fellow Am. Speech-Lang.-Hearing Assn. (publs. bd. 1987-92), Acoustical Soc. Am. Democrat. Home: 21 East View Pl NE Iowa City IA 52240 Office: Univ Iowa Office of Provost 111 Jessup Hall Iowa City IA 52242

FOLKMAN, MOSES JUDAH, surgeon; b. Cleve., Feb. 24, 1933; s. Jerome D. and Bessie Folkman. B.A., Ohio State U., 1953; M.D., Harvard U., 1957. Intern, then asst. resident in surgery Mass. Gen. Hosp., Boston, 1957-60, sr. asst. resident in surgery, 1962-64, chief resident, 1964-65; chief resident in pediatric surgery Phila. Children's Hosp., 1969; instr. surgery Harvard U. Med. Sch., 1965-66, assoc. in surgery, 1967, prof. surgery, 1967—, Julia Dyckman Andrus prof. pediatric surgery, 1968—, prof. anatomy and cellular biology, 1989—; asst. surgeon Boston City Hosp., 1965-66; assoc. dir. Sears Surg. Lab., 1966-67; sr. surgeon Children's Hosp. Med. Ctr., Boston, 1968—. Served as officer M.C. USN, 1960-62. Recipient Career Devel. award NIH, 1966, Boylston Med. prize Harvard Med. Sch., 1957, Soma Weiss award, 1957, Lila Gruber award Am. Acad. Dermatology, 1974, Gairdner Found. internat. award, 1991, Christopher Columbus Commemorative Sci. medal, Wolf award (Israel), 1992. Fellow ACS (Sheen award 1989); mem. NAS, Am. Coll. Chest Physicians, Am. Surg. Assn., Assn. Acad. Surgery, Soc. U. Surgeons, Surg. Biology Club I, Am. Acad. Pediatrics, Allen O. Whipple Soc., Am. Pediatric Surg. Assn., N.Y. Acad. Scis., Mass. Med. Soc. Office: 300 Longwood Ave Boston MA 02115-5737

FOLKMAN, STEVEN LEE, engineering educator; b. Logan, Utah, Mar. 24, 1952; s. Basil Willmet and Rachel (Zollinger) F.; m. Marianne Boudrero, June 7, 1975; children: Melanie, Jennifer, Wendy, Ashley, Jason. BS in Nutrition & Food Sci., Utah State U., 1975, PhD in Mechanical Engring., 1990. Sr. engr. Thiokol Corp., Brigham City, Utah, 1978-80; rsch. engr. Utah State U., Logan, 1980-90, asst. prof., 1990—; cons. Space Dynamic Lab., Logan, 1990-92; energy conservation specialist State of Utah, Salt Lake CIty, 1980-80. Co-author: Food Engineering Fundamentals, 1983. Scoutmaster Boy Scouts Am., Logan, 1989-92. Mem. AIAA, ASME. Mem. LDS Ch. Office: Utah State U Mechanical Engring Dept Logan UT 84322-4130

FOLKS, F(RANCIS) NEIL, biologist, researcher; b. Ashland, Kans., Mar. 31, 1939; s. Francis Elmer and Velsie Pearl (Franks) F.; m. Merlene Folks, Mar. 15, 1963; children: Vala, Neilene, Courtland. BS, Ft. Hays (Kans.) U., 1961; MS, Utah State U., 1969. With Seacat Feedlot, Ashland, 1965-66; vet. technician Rangelands Vet. Clinic, Ashland, 1966-67; biologist, supt. Utah Div. Wildlife Resources, Salt Lake City, 1967—; cons. fed. agys., N.E. Utah, 1967—. Vol. Browns Hole Homemakers Club, Browns Park, Utah. Mem. Wildlife Soc., Utah chpt. Wildlife Soc., Sigma Xi. Avocations: photography, philosophy. Home: Daggett County Utah Rd 2606 PO Box 1 Maybell CO 81640-0001 Office: Utah Div Wildlife Resources 152 E 100 N Vernal UT 84078-2126

FONASH, STEPHEN JOSEPH, engineering educator; b. Phila., Oct. 28, 1941; s. Raymond Leo and Margaret Frances (Carney) F.; m. Joyce Maria Maurin, Dec. 7, 1968; children: Stephen Eugene, David Raymond. BS, Pa. State U., 1963; PhD, U. Pa., 1968. Postdoctoral fellow U. Pa., Phila., 1968; asst. prof. Pa. State U., University Pk., 1968-73, assoc. prof., 1973-76, prof., 1976—, Disting. prof., 1986—, dir. Electronic Materials and Processing Rsch. Lab, 1986—; vis. mem. tech. staff Bell Labs., Murray Hill, N.J., May-Sept. 1972; vis. faculty mem. Université de Lyon (France), Jan.-Aug. 1973; cons. Amoco Co., Chgo., 1983-88, Dept. Def., Washington, 1984—, Ametek Corp., Phila., 1983-90, Gen. Electric Corp., Schenectady, 1988-90, 3M Corp., Mpls., 1989—, Sematech, Austin, 1990—. Author: Solar Cell Device Physics, 1981; patentee in field. Fellow IEEE; mem. Electrochem. Soc. (chmn. electronic div. awards com. 1986—), Am. Vacuum Soc. Pa. State U. Alumni Assn. Roman Catholic. Avocations: skiing, hiking, carpentry. Office: Pa State U Electronic Material Processing Rsch Lab 113 Elec Engring W Bldg University Park PA 16802

FONCK, EUGENE JASON, electronics engineer, physicist; b. Taipei, Republic of China, Feb. 27, 1954; m. Ingrid Y. Liou, Apr. 12, 1991. MS in Physics, Ind. U., 1983; MSEE, U. N.M., 1989. Electronics engr. Uthe Tech., Inc., Milpitas, Calif., 1989—. Mem. IEEE, Am. Inst. Physics. Achievements include research on properties and application of ultrasonic transducer for wire bonding, on semiconductor device as a secondary. Home: 2809 Westberry Dr San Jose CA 95132-1776 Office: Uthe Tech Inc 455 Montague Expy Milpitas CA 95035-6800

FONDAHL, JOHN WALKER, civil engineering educator; b. Washington, Nov. 4, 1924; s. John Edmund and Mary (DeCourcy) F.; m. Doris Jane Plishker, Mar. 2, 1946; children: Lauren Valerie, Gail Andrea, Meredith Victoria, Dorian Beth. B.S., Thayer Sch. Engring., Dartmouth, 1947, M.S. in Civil Engring, 1948. Instr., then asst. prof. U. Hawaii, 1948-51; constrn. engr. Winston Bros. Co., Mpls., 1951-52; project engr. Nimbus Dam and Powerplant project, Sacramento, 1952-55; mem. faculty Stanford U., 1955—, prof. civil engring., 1966-90, Charles H. Leavell prof. civil engring., 1977-90, prof. emeritus, 1990—; bd. dirs. Caterpillar Inc., Peoria. Author reports in field. Served with USMCR, 1943-46. Recipient Golden Beaver award Heavy Constrn. Industry, 1976. Fellow ASCE (Constrn. Mgmt. award 1977, Peurifoy Constrn. Rsch. award 1990), Project Mgmt. Inst. (hon. life, Fellow award 1981); mem. Nat. Acad. Engring., Phi Beta Kappa. Republican. Achievements include patents in field. Home and Office: 12810 Viscaino Rd Los Altos CA 94022

FONG, WANG-FUN, biochemist; b. Hong Kong, Feb. 6, 1947; s. Chung-Chak and Wai-Kuen (cho) F.; m. Doris Lai-Han Man, Aug. 14, 1971; children: Emily Kate, Celine Soen-Si. BSc, Chinese U., Hong Kong, 1968; BSc with spl. honor, U. Hong Kong, 1969; PhD, Notre Dame U., Ind., 1974. Rsch. assoc. Yale U., New Haven, 1974-76; allied scientist Ottawa (Can.) Civic Hosp., 1978-80; lectr. U. Hong Kong, 1980-87; prin lectr. Hong Kong Bapt. Coll., 1987-90, reader, 1990-91; reader Hong Kong City Poly., 1991—; cons. biotechnology Chinese edit. Time-Life Books, 1990—; specialist Hong Kong Coun. for Acad. Accreditation, 1988—. Contbr. articles to profl. jours. Sci. devel. fellow NSF, 1971, A.J. Schmitt fellow U. Notre Dame, 1974, fellow Nat. Rsch. Coun., 1976-78. Mem. Biochem. Soc. U.K., Fedn. of Asian and Oceanian Biochemists (del. 1989-91), Hong Kong Biochem. Assn. (sec. 1987-89, chmn. 1989-91). Office: City Polytechnic Dept Biology and Chemistry, Tat Chee Ave Kowloon, Hong Kong Hong Kong

FONKEN, GERHARD JOSEPH, chemistry educator, university administrator; b. Krefeld, Fed. Republic of Germany, Aug. 3, 1928; came to U.S., 1930, naturalized, 1935; s. Henry A. and Wilhelmina Katerina (von Eyser) F.; m. Carolyn Lee Stay, Dec. 20, 1952; children—David, Katherine, Steven, Karen, Eric. B.S., U. Calif., Berkeley, 1954, Ph.D., 1957. Chemist Procter & Gamble Co., 1957-58; chemist Stanford (Calif.) Research Inst., 1958-59; instr. U. Tex., Austin, 1959-61; asst. prof. U. Tex., 1961-66, asso. prof., 1966-72, prof. chemistry, 1972—, asso. provost, 1972-75, acting v.p. acad. affairs, 1975-76, exec. asst. to pres., 1976-79, v.p. research, 1979-80, v.p. acad. affairs and research, 1980-85, exec. v.p., provost, 1985—. Contbr. articles to chemistry jours. Served with U.S. Army, 1946-49, 50-51, Korea. NIH grantee, 1961-64; Robert A. Welch Found. grantee, 1962-79. Mem. Am. Chem. Soc. Home: 6612 Lost Horizon Dr Austin TX 78759-6116 Office: U Tex Dept Chemistry Main Bldg Ste 201 Austin TX 78712

FONS, MICHAEL PATRICK, virologist; b. Pontiac, Mich., Aug. 17, 1959; s. Anton C. and Margaret A. (Colfer) F.; m. Susan A. George, Oct. 23, 1980; children: Michael George, Julie Ann. BA, Wabash Coll., Crawfordsville, Ind., 1981; PhD, U. Tex., Galveston, 1989. Postdoctoral fellow dept. microbiology U. Tex., Galveston, 1989-91, asst. prof., 1991—. Predoctoral fellow McLaughlin Found., Galveston, 1986, postdoctoral fellow, 1989. Mem. Soc. Exptl. Biology and Medicine, Am. Soc. Virology, N.Y. Acad. Scis. Achievements include description of importance of Na in the devel. and progression of Cytopathology and replication of human cytomegalovirus. Office: U Tex Med Br Dept Microbiology 14th and Mechanic Galveston TX 77555

FONTA, CAROLINE, biologist; b. Paris, Apr. 1, 1957; d. René and Jacqueline (Thiphonnet) F.; children: Pierre, Sébastien. Diploma, Lycee Racine, Paris, 1975; Ingenieur, Ecole Nat. Superieure Feminine d'Agronomie, Rennes, France, 1979. Postdoctoral fellow Brit. Coun., Cambridge, Eng., 1985; rschr. Ctr. Nat. Rsch. Sci., Bures sur Yvette, France, 1986-93; mem. faculté de médecine Rangueil, Centre de Recherche Cerveau et Cognition Université Paul Sabatier, Toulouse Cedex, France, 1993—. Contbr. articles to profl. jousrs. Office: Université Paul Sabatier, 133 Route de Narbonne, 31062 Toulouse Cedex, France

FONTAINE, ARNOLD ANTHONY, mechanical engineer; b. Pawtucket, R.I., Nov. 19, 1962; s. Arnold Rene Fontaine and Elizabeth Ann (Bois) Girard; m. Sharon Masters, Sept. 1, 1986; children: Eric, Elizabeth. BS, U. R.I., 1984, MS, 1986; PhD, Pa. State U., 1993. Grad. rsch. asst. U. R.I., Kingston, 1984-86, Pa. State U., State College, 1986-88; rsch. asst. Applied Rsch. Lab., State College, Pa., 1988-93; postdoctoral fellow Ga. Inst. Tech., Atlanta, 1993—. Contbr. articles to profl. jours. Mem. AIAA, ASME, Am. Phys. Soc., Tau Beta Pi, Pi Tau Sigma. Office: Ga Inst Tech Chem Engring Dept 778 Atlantic Dr Atlanta GA 30332-0100

FONTANA, ALESSANDRO, Italian minister of universities and research; b. Marcheno, Brescia, Italy, Aug. 15, 1936. Degree in history, Cath. U. Provincial vice-sec. Democrazia Cristiana, regional vice-sec., regional councillor, 1970, regional assessor culture and local authorities, v.p. regional coun.; pres. Democrazia Cristiana Lombardy; mem. cen. leadership Democrazia Cristiana, 1980—; now prof. Ministry of Universities and Rsch., Rome; nat. vice-sec. Democrazia Cristiana, 1985-86, dir. daily newspaper, Il Popolo; elected senator Fermo, 1987; mem. Com. Constl. Affairs; lectr. modern history U. Brescia; specialist in Cath. culture from 19th Century, student of contemporary ideologies. Author: Books on Lamennais, Leone XIII, Tommaseo, Cattaneo, Murri, Gobetti, Sturzo, Aldo Moro, 'La controrivoluzione cattolica 1820-30, 1968, "Oltre il riformismo", 1973, "I cattolici e l'unitá sindacale", 1978, "Antonomia della cultura, cultura delle autonomie", 1980, L'identitá minacciata", 1986, "Il fascismo e le autonomie locali", "La grande guerra: operai e contadini lombardi nel primo conflitto mondiale'; writer for Il Ponte, Humanitas, Il politico, Terza Fase, Il Giorno, Il Corriere della Sera; polit. editor Cittadino, Voce del Popolo, Eco di Brescia. Partito della Democrazia Cristiana. *

FONTANA, DAVID A., chemical engineer, military officer; b. Niagara Falls, N.Y., Nov. 19, 1947; s. Albert A. and Sophia M. (Tarasek) F.; m. Judith L. Glass, Apr. 3, 1970; children: Alexa L., David T., Katherine M., Christina B. BSChemE, SUNY, Buffalo, 1969; MBA, U. Utah, 1975. Commd. 2d. lt. USAF, 1970, advanced through grades to col., 1991; missile/explosives safety staff officer Air Force Inspection & Safety Ctr., Norton AFB, Calif., 1981-85; comdr. 509th Munitions Maintenance Squadron, Pease AFB, N.H., 1985-87, 400th Munitions Maintenance Squadron, Kadena Air Base, Okinawa, 1987-90; chief munitions & missiles, plans & policy divsn. USAF Hdqrs., The Pentagon, Washington, 1991—; chmn. Air Force Munitions Logistics Steering Group, Washington, 1991—. Author: U.S. Air Force Explosives Safety Standards, 1984. Office: HQ USAF/LGMW Pentagon Rm 4B259 Washington DC 20330-1030

FONTANA, MARIO H., nuclear engineer; b. West Springfield, Mass., Mar. 30, 1933; s. Remo and Sabina (De Angelis) F.; m. Sue Janeway, Apr. 12, 1958; children: Richard, Edward. BS, U. Mass., 1955; MS, MIT, 1957; PhD, Purdue U., 1968. Registered engr., Tenn. Instr. Purdue U., Oak Ridge, Tenn., 1964-65; dir. industry degraded core program Tech for Energy, Inc., Knoxville, Tenn., 1981-84; v.p. engring. Energex Oak Ridge, 1984-85; dir. nuclear safety tech. IT Corp. (not Tenera, L.P.), Knoxville, 1985-90; sr. scientist Avco Radiology, Wilmington, Mass., 1963-64; instr. Purdue U., 1964-65; mem. rsch. staff Oak Ridge Nat. Lab., 1957-63, 65-81, 90—, asst. dir. nuclear safety rsch., 1968-72, head advanced concepts devel. engring. tech. div., 1972-81, asst. dir. engring. tech. div., 1990—; cons. U.S. Dept. Energy, Washington, 1986-89, U.S. AEC, Washington, 1972-73, U.S. NRC, Washington, 1979-81, 91—. Author more than 100 reports and articles. Fellow Am. Nuclear Soc. (chmn. nuclear reactor safety div. 1972-73); mem. ASME, Am. Mgmt. Assn., Soc. Risk Analysis, Rotary Internat., Sigma Xi, Tau Beta Pi. Achievements include patents for method of arc synthesis of uranium carbide from UF6 and Graphite, others. Office: Oak Ridge Nat Lab PO Box 2009 Oak Ridge TN 37831-2009

FONTANA, PETER ROBERT, physics educator; b. Bern, Switzerland, Apr. 20, 1935; came to U.S., 1956; s. René Fontana and Mathilde Erika (Lanz) Sandoz; m. Sheryl Patricia Nelson, June 30, 1984; children: Erika Christine, Laura Suzanne. MS, Miami U., Oxford, Ohio, 1958; PhD, Yale U., 1960. Rsch. assoc. U. Chgo., 1960-62; asst. prof. U. Mich., Ann Arbor, 1962-67; assoc. prof. Oreg. State U., Corvallis, 1967-74, prof., 1974—; vis. prof. Swiss Inst. of Tech., Lausanne, Switzerland, 1976, U. Tubingen, Germany, 1982; cons. NSF, 1967-70. Author: Atomic Radiative Processes, 1982. Fellow Am. Phys. Soc. Achievements include patent in Multi-defector Intensity Interferometer. Office: Oregon Stat U 373 Weniger Hall Corvallis OR 97331

FOOTE, SIMON JAMES, molecular biologist; b. Kyneton, Victoria, Australia, Aug. 24, 1958; s. Owen James Foote and Nancy Lorna Price; m. Susan Margaret Robinson, Aug. 25, 1989; 1 child, Madeleine. B of Med. Sci., U. Melbourne, 1981, MBBS, 1984, PhD, 1990; student, Walter & Eliza Hall Inst., 1986-89. Resident Royal Melbourne Hosp., Victoria, 1985; postdoctoral fellow Whitehead Inst., Cambridge, Mass., 1990-93, rsch. scientist, 1993—; sr. rsch. fellow Walter & Eliza Hall Inst., Melbourne, Australia, 1993—. Recipient Postgrad. Rsch. award Nat. Health and Med. Rsch. Coun., Melbourne, 1984; Lucille P. Markey Australian postdoctoral fellow Lucille P. Markey Charitable Trust, 1989, Sr. Australian fellow Wellcome Trust, 1992. Achievements include work on drug resistance in malaria, mapping the human Y chromosome. Office: Walter & Eliza Hall Inst, PO Royal Melbourne Hosp, Parkville Victoria, Australia 3050

FOOTE, WARREN EDGAR, neuroscientist, psychologist, educator; b. Boston, Nov. 5, 1935; s. Warren Edgar and Edith Irene (Landry) F.; B.A., Hamilton Coll., 1958; M.A., Boston U., 1960; Ph.D., Tufts U., 1965; m. Cynthia Sue Hall, July 21, 1973; children: Pamela Fowler, Sarah Canby, Julia Landry, Christopher Warren. Research assoc. Harvard U. Med. Sch., 1966-67, vis. asst. prof. psychology, 1970-73, asst. prof., 1974-83, assoc. prof., 1983—; USPHS postdoctoral fellow Yale, 1967-69; research scientist Norwich (Conn.) State Hosp., 1969-70; sr. Fulbright scholar Max-Planck Inst., Munich, Germany, 1973-74; assoc. psychologist Mass. Gen. Hosp., Boston, 1974—, psychologist, 1984—; cons. Gen. Foods Corp., 1970-74, Neurotech Corp., 1987-88. Served with M.C., AUS, 1959-60. Recipient McCurdy prize Mass. Soc. Research in Psychiatry, 1962; sr. Fulbright fellow, 1973-74; Nat. Inst. Neurol. Disease and Stroke grantee, 1974-77; NIMH grantee, 1970-73; Nat. Eye Inst. grantee, 1979—; Wayland Pub. Sch. Found. advisor, 1982—; Nat. Inst. Communicative Disorders and Stroke grantee, 1983—. Mem. AAAS, N.Y. Acad. Scis., Soc. Neurosci., Am. Psychol. Assn., Sigma Xi. Club: Harvard (Boston). Contbr. articles, revs. to profl. jours. Home: 5 Hilltop Pk Wilbraham MA 01095-1753 Office: Mass Gen Hosp PO Box 70 Boston MA 02114

FORABOSCHI, FRANCO PAOLO, chemical engineering educator; b. Florence, Italy, Sept. 23, 1932; s. Ezio and Fedora (Ricci) F.; m. Maria Luisa Borgioli, Mar. 15, 1958; children: Paolo, Alberto. Laurea Ingegneria Chimica, U. Bologna, Italy, 1957, Libero Docente, 1963. Assoc. prof. Facoltà Ingegneria, Bologna, 1959-68, prof. chem. engring., 1968—, dean, 1974-77; mem. adminstrv. bd. U. Bologna, 1977-80, prof. chem. engring. principles and environ. chem. engring. principles; cons. Studio di Ingegneria Foraboschi, Bologna, 1960; pres. METIS srl, 1991—. Author: Principi di Ingegneria Chimica, 1973; contbr. articles to profl. jours. Mem. AICE, Soc. Chimica Italiana, Assn. Nat. Italiana Automazione, Accademia delle Scienze dell'Instituto di Bologna, Coun. Profl. Engrs. Bologna, Rotary Club Bologna Est. Office: Dept Ingegneria Chimica, Viale Risorgimento 2, I 40136 Bologna Italy

FORASTE, ROLAND, psychiatrist; b. N.Y.C., Mar. 1, 1938; s. Paul Foraste and Anita Schonbachler. AB honors cum laude, Coll. Holy Cross, 1960; cert. neurology, U. London, 1965; MD, SUNY, Downstate, 1965. clin. instr. psychiatry Cornell Med. Coll., 1967-73, clin. asst. prof., 1973-86; attending physician Gracie Sqq. Hosp., N.Y.C., 1969—, Rye (N.Y.) Psychiat. Ctr., 1969—. Med. intern Jefferson Hosp., Phila., 1965-66; resident in psychiatry N.Y. Hosp., 1966-69, resident in child psychiatry, 1968-71, chief resident in adolescent psychiatry, 1970-71, asst. attending psychiatrist, 1973-86; med. dir. psychiatry U.S. Healthcare and Total Health, N.Y.C., 1986-90, cons., 1990—; clin. instr. psychiatry Cornell Med. Coll., 1967-73, clin. asst. prof. psychiatry, 1973-86; attending physician Gracie Sq. Hosp., N.Y.C., 1969—, Rye (N.Y.) Psychiat. Ctr. Co-author: (audiotext) The Drug Syndrome and the Teacher, 1971; co-editor, contbr. Biology Jour. of Coll. of Holy Cross, 1957-58. Benefit chmn. Hosp. Audiences, Inc., N.Y.C., 1984; benefit com. Cultural Coun. Found., N.Y.C., 1984-86, Big Apple Circus, 1984-86, Five Men Named Moe, 1992; mem. Met. Mus. Art, Mus. Modern Art. Recipient Physicians Recognition award AMA. Mem. Am. Acad. Child and Adolescent Psychiatry. N.Y. Coun. Child and Adolescent Psychiatry, Am. Soc. for Adolescent Psychiatry (asst. continuing med. edn. officer 1989-90, continuing med. edn. officer 1990—), N.Y. Soc. for Adolescent Psychiatry (bd. dirs. 1987-88, sec. 1988-90, pres-elect 1990-91, pres. 1991-93), Am. Psychiat. Assn., Flying Physicians Assn. (instrumentrated), N.Y. Hosp.-Cornell Med. Ctr. Alumni Assn., Aircraft Owners and Pilots Assn., Greenwich Boat and Yacht Club, Holy Cross Club N.Y., Player's Club, Westchester Flying Club, Porsche Club Am., Westchester Country Club (mem. nominating com. 1987-88, membership com. 1987-88, 90-91, entertainment com. 1987-88, 90—), U.S. Ski Assn., Wintergreen Club (dir.-trustee 1990-93, chmn. entertainment com. 1990—), KC. Republican. Roman Catholic. Avocations: flying, riding, sailing, skiing, windsurfing. Home and Office: 623 Steamboat Rd Greenwich CT 06830-7140 Home and Office: 420 E 51st St New York NY 10022-8014

FORBES, BO CROSBY, clinical psychologist; b. Boston, Aug. 25, 1964; d. Allan Jr. Forbes. BA/MA, U. Chgo., 1986; D of Psychology, Ill. Sch. Profl. Psychology, 1990. Lic. psychologist, Ill. Rsch. asst. U. Chgo. Sleep Rsch. Lab., 1984-85; editor rsch. asst. Dept. Psychiatry, U. Chgo. Hosp., 1985-87; psychology extern Michael Reese Hosp., Chgo., 1987-88, Easting Disorders Clinic/Northwestern U., Chgo., 1988-89; staff psychologist Luth. Gen. Optifast Program, Park Ridge, Ill., 1988-89; psychology intern Hartgrove Hosp., Chgo., 1989-90; clin. psychologist The Touchstone Group, Westmont, Ill., 1990—. Mem. Health Adv. Bd., CEDA, Orland Park, Ill., 1991-92, Jesse White campaign, Chgo., 1990-92. Mem. APA (div. 35), Ill. Psychol. Assn., Nat. Assn. of Alcohol and Other Drug Abuse Counselors, Ill. Group Psychotherapy Assn. Office: The Touchstone Group Ste 140 999 Oakmont Plz Westmont IL 60559

FORBES, GILBERT BURNETT, physician, educator; b. Rochester, N.Y., Nov. 9, 1915; s. Gilbert DeLeverance and Lillian Augusta (Burnett) F.; m. Grace Moehlman, July 8, 1939; children: Constance Ann (Mrs. Joseph F. Citro), Susan Young (Mrs. William A. Martin). B.A., U. Rochester, 1936, M.D., 1940. Intern Strong Meml. Hosp., Rochester, 1940-41; resident St. Louis Children's Hosp., 1941-43; practice medicine, specializing in pediatrics Los Alamos, 1946-47; instr. pediatrics Sch. Medicine, Washington U., St. Louis, 1943-46; asst. prof. Sch. Medicine, Washington U., 1947-50; prof. pediatrics, chmn. dept. Southwestern Med. Sch., Dallas, 1950-53; assoc. prof. pediatrics Sch. Medicine, U. Rochester, 1953-57, prof., 1957-68, prof. pediatrics, prof. radiation biology, 1968—, Alumni Disting. Service prof. pediatrics, 1978—, chmn. faculty council, 1969-70, acting co-chmn. dept. pediatrics, 1974-76; cons. Nat. Inst. Child Health and Human Devel.; mem. sci. adv. com. Nutrition Found., 1963-66; mem. Nat. Council on Radiation Protection; mem. com. infant nutrition, com. dietary allowances NRC, 1960-63; vis. research fellow U. Oxford, Eng., 1970-71. Author: Human Body Composition, 1987; assoc. editor: Am. Jour. Diseases Childhood, 1964-72; chief editor, 1973-82; assoc. editor: Nutrition Revs, 1961-71; editor Pediatric Nutrition Handbook, 1985; contbr. numerous articles to profl. jours. Recipient Research Career award USPHS, NIH, 1962—, Borden award Am. Acad. Pediatrics, 1964, Alumni award to faculty U. Rochester, 1975; Albert David Kaiser award, Rochester Acad. Medicine, 1979. Mem. AAAS, AMA, Am. Pediatric Soc. (coun., v.p. 1975-76, John Howland award 1992), Soc. Pediatric Research (past pres.), Am. Acad. Pediatrics (com. on nutrition 1974-80), U. Rochester Med. Alumni Assn. (past pres., Gold medal 1982), Rotary, Sigma Xi, Alpha Omega Alpha, Theta Chi. Home: 2021 Westfall Rd Rochester NY 14618-3113 Office: Univ of Rochester Dept of Pediatrics 601 Elmwood Ave Rochester NY 14642-9999

FORBES, JERRY WAYNE, research physicist; b. Oquawka, Ill., July 12, 1941; s. Ernest Louis and Mabel Minetta (Jones) F.; m. Cynthia Joyce Schleifer, Dec. 2, 1961; children: Stephanie Ann, Jason Lloyd. BS in Physics, Western Ill. U., 1963; MS in Physics, U. Md., 1967; PhD in Physics, Wash. State U., 1976. Gen. sci. tchr. Edison Jr. High Sch., Macomb, Ill., 1962-63; rsch. physicist Naval Surface Warfare Ctr., Silver Spring, Md., 1963—. Co-editor: Shock Waves in Condesned Matter, 1991; mem. editorial bd. for Technical Digest, 1993. Fellow Am. Phys. Soc. (sec./treas. topical group on shock compression 1983-91, chmn.-elect 1991-92, chmn. 1992-93); mem. Sigma Pi Sigma (pres. chpt. 1962-63), Sigma Zeta, Kappa Delta Pi. Achievements include identification of cubic BN from shocked hexagonal BN, explosive Comp B3 as elastic-plastic; demonstration that incipient bcc nuclei always present in hcp iron could be frozen in by shock wave resulting in the phase transformation, that the bcc to hcp phase transformation in iron occurs at a rate greater than 10,000,000/sec. Home: 12509 Silverbirch Ln Laurel MD 20708 Office: Naval Surface Warfare Ctr 10901 New Hampshire Ave Silver Spring MD 20903-5640

FORBES, KENNETH ALBERT FAUCHER, urological surgeon; b. Waterford, N.Y., Apr. 28, 1922; s. Joseph Frederick and Adelle Frances (Robitaille) F.; m. Eileen Ruth Gibbons, Aug. 4, 1956; children: Michael, Diane, Kenneth E., Thomas, Maureen, Daniel. BS cum laude, U. Notre Dame, 1943; MD, St. Louis U., 1947. Diplomate Am. Bd. Urology. Intern St. Louis U. Hosp., 1947-48; resident in urol. surgery Barnes Hosp., VA Hosp., Washington U., St. Louis. U. schs. medicine, St. Louis, 1948-52; asst. chief urology Letterman Army Hosp., San Francisco, 1952-54; fellow West Roxbury (Harvard) VA Hosp., Boston, 1955; asst. chief urology VA Hosp., East Orange, N.J., 1955-58; practice medicine specializing in urology Green Bay, Wis., 1958-78, Long Beach, Calif., 1978-85; mem. cons. staff Fairview State Hosp. U. Calif. Med. Ctr., Irvine, VA Hosp., Long Beach; asst. clin. prof. surgery U. Calif., Irvine, 1978-85; cons. Vols. in Tech. Assistance, 1986—. Contbr. articles to profl. jours. Served with USNR, 1944-46; capt. U.S. Army, 1952-54. Named Outstanding Faculty Mem. by students, 1981. Fellow ACS, Royal Soc. Medicine, Internat. Coll. Surgeons; mem. AMA, AAAS, Calif. Med. Assn., Am. Urol. Assn. (exec. com. North Ctrl. sect. 1972-75, Western sect. 1980—), N.Y. Acad. Scis., Surg. Alumni Assn. U. Calif.-Irvine, Justin J. Cordonnier Soc. Washington U., Confedn. Americana Urologia, Urologists Corr. Club, Notre Dame Club (Man of Yr. award 1965), Union League Club, Phi Beta Pi. Republican. Roman Catholic. Home and Office: 11579 Sutters Mill Cir Gold River CA 95670-7214

FORD, BRIAN J., research biologist, author, broadcaster; b. Corsham, Wiltshire, Eng., May 13, 1939; m. Janice May Smith; 4 children, 2 foster children. Student, King's Sch., Peterborough, Cardiff U., Eng. Mem. ct. U. Cardiff, 1982-87, 91—, fellow, 1986—; lectr. Brit. Coun. divsn. Foreign and Commonwealth Office, 1978, Inter Micro, Chgo., 1984—. Author: Microbiology and Food, 1970, Nonscience, 1971, Revealing Lens, 1973, Microbe Power, 1976, Patterns of Sex, 1978, Cult of the Expert, 1982, Single Lens, 1985, Leeuwenhoek Legacy, 1990, Images of Science: First Encyclopedia of Science, 1993; co-author The Cardiff Book, 1971, Viral Pollution of the Environment, 1983, Sex and Health, 1989; founding editor: Biology History, 1988, Science Diary, 1967-78; mem. editorial bd., 1989-93, editor-in-chief, 1993; editor: Sci. and Tech. Authors' Newsletter, 1988—, Voice of British Industry, 1987-88; contbg. editor: European Biotechnology Newsletter, 1967-69; cons. sci. editor Guinness Book Records, 1993—; writer: The Human Body (Julie and Robert Brown), 1992; prodr., dir. (film) The Fund; creator radio and TV programs Science Now, Where Are You Taking Us?, It's Your Line, Food for Thought, Computer Challenge; contbr. The Times, Guardian, New Scientist, Nature, Microscopy, Private Eye, Brit. Med. Jour., reports in field. Fellow Royal Microscopical Soc., Inst. Biology (chmn. history com. 1989—, pub. policy com. 1993—), Cambridge Philos. Soc.; mem. Keynes Coll. U. Kent (hon. 1988—), Linnean Soc. London (surveyor sci. instruments 1985—, mem. council 1992—), Nat. Book League (mem. council 1983-86), Royal Soc. Health (vice-chmn. nutrition 1988-89), European Union Sci. Journalists Assns. (Brussels, pres. 1984-86), Soc. Authors (chmn. sci. and tech. writers 1986-88), Cambridge Soc. for Application Rsch., Cambridge U. Soc. for Cognitive Rsch. (life), Soc. for History Sci., Firend of Cambridge U. Libr. (life), Savage Club (dep. chmn., head entertainments), Old Petriburgians Assn. (chmn.), BBC Club, Architecture Club, Quekett Micros. Club. Achievements include pioneering research in theory of development of life from prebiotic molecules in outer space; research in interdisciplinary science, plant physiology, microbial ecology, haemostatic mechanisms, biohazard, science education, development of science, history of microscope. Address: Rothay House, Mayfield Rd, Eastrea Cambridgeshire PE7 2AY, England

FORD, BYRON MILTON, computer consultant; b. Hayden, Colo., Feb. 24, 1939; s. William Howard and Myrtle Oretta (Christian) F.; B.S., U. Colo., 1964; M.S. in Mgmt. Sci., Johns Hopkins U., 1971; m. Shirley Ann Edwards, Sept. 4, 1958; children—Gregory Scott, Barry Matthew. Sr. mathematician Applied Physics Lab., Johns Hopkins U., Laurel, Md., 1964-79; computer cons., Laurel, 1979—. Mem. Ops. Research Soc. Am., Nat. Assn. Self-Employed. Address: 6909 Redmiles Rd Laurel MD 20707

FORD, CLYDE GILPIN, chemistry educator; b. Augusta, Ark., Aug. 27, 1933; s. William Clyde and Mildred A. (Gilpin) F.; m. Virginia Lee McClung, Aug. 29, 1958. BA, Ark. State U., 1955; MS, U. Okla., 1960; PhD, U. Tex., 1976. Sr. rsch. chemist Internat. Paper, Mobile, Ala., 1960-68; engr. specialist LTV Corp., Dallas, 1968-76; mem. tech. staff Rockwell Internat., Tulsa, Okla., 1976-87; sr. engr. specialist E Systems, Greenville, Tex., 1987-90; telemarketer E. Blank Assoc., Eules, Tex., 1991-92; physics instr. Eastfield Coll., Mesquite, Tex., 1992; instr. chemistry Tex. State Tech. Coll., Marshall, 1992; cons. in field, 1971-76; mem. com. Tulsa Jr. Coll. Chem. Tech., Tulsa, 1985-86; adj. prof. chemistry Wiley Coll., Marshall, Tex., 1993. Contbr. articles to profl. jours. Lt. U.S. Army, 1955-57. Mem. Am. Chem. Soc. (chmn. 1982-83), Pi Lambda Upsilon, Sigma Xi, Pi Gamma Mu (Gold medal 1953). Republican. Baptist. Home: 706 Forest Trace Rockwall TX 75087 Office: Tex State Tech Coll 2615 East End Blvd S Marshall TX 75671

FORD, DENNIS HARCOURT, laser systems consultant; b. Bethel, Vt., July 15, 1954; s. Guy Raymond and Jean (McIntosh) F.; m. Cheryl Ann Ruvo, Mar. 17, 1981; children: Richard, Courtney, Justin. BS in Physics, Antioch U., 1980. Technologist LLNC, Livermore, Calif., 1980-84; rsch. assoc. STI Optronics, Bellevue, Wash., 1984-90; owner Lab Space, Issaquah, Wash., 1990—. Contbr. 20 articles in 15 different profl. jours. Coach Youth Soccer Am., Lake Washington Youth Svcs., 1990-92, Boys and Girls Club Basketball, Kirkland, 1984, Little League, Kirkland, 1987. Achievements include contribution to laser based instrumentation and techniques, created radially polarized laser beam. Office: Lab Space PO Box 195 Issaquah WA 98027

FORD, LEE ELLEN, scientist, educator, retired lawyer; b. Auburn, Ind., June 16, 1917; d. Arthur W. and Geneva (Muhn) Ford; BA, Wittenberg Coll., 1947; MS, U. Minn., 1949; PhD, Iowa State Coll., 1952; JD, U. Notre Dame, 1972. Bar : Ind. 1972. CPA auditing, 1934-44; assoc. prof. biology Gustavus Adolphus Coll., 1950-51; prof. and head biology dept. Anderson (Ind.) Coll., 1952-55; vis. prof. biology U. Alta. (Can.), Calgary, 1955-56; assoc. prof. biology Pacific Luth. U., Parkland, Wash., 1956-62; prof. biology and cytogenetics Miss. State Coll. for Women, 1962-64; chief cytogeneticist Pacific N.W. Rsch. Found., Seattle, 1964-65; founder, dir. Canine Genetics Cons. Svc., Parkland, 1963-69; pvt. practice, Ind., 1972-92. Founder, sponsor Companion Collies for the Adult, Jr. Blind, 1955-65; dir. Genetics Rsch. Lab., Butler, Ind., 1955-75, cons. cytogenetics, 1969-75; legis. cons., 1970-79; dir. chromosome lab. Inst. Basic Rsch. in Mental Retardation, S.I., 1968-69; founder, dir. Legis. Bur. U. Notre Dame Law Sch., founder, editor New Dimensions in Legis., 1969-72; editor Butler Record Herald, 1972-76; founder, dir. Ind. Interreligious Com. on Human Equality, 1976-80; exec. asst. to Gov. Otis R. Bowen, Ind., 1973-75; founder, bd. dirs. Ind. Commn. on Status Women, 1973-74; bd. dirs. Ind. Coun. Chs.; editor Ford Assocs. pubs., 1972-86; mem. Pres.'s Adv. Coun. on Drug Abuse, 1976-77. Admitted to Ind. bar, 1972. Adult counselor Girl Scouts U.S., 1934-40; bd. dirs. Ind. Task Force Women's Health, 1976-80; mem. exec. bd., bd. dirs. Ind.-Ky. Synod Lutheran Ch., 1972-78; bd. dirs., mem. coun. St. Marks Luth. Ch., Butler, 1970-76; mem. social svcs. pers. bd., 1970-76; mem. DeKalb County (Ind.) Sheriff's Merit Bd., 1983-87; founder, dir., pres. Ind. Caucus for Animal Legis. and Leadership, 1984-87. Mem. AAUW, AAAS, Genetics Soc. Am., Am. Human Genetics Soc., Am. Genetic Assn., Am. Inst. Biol. Scis., Am. Soc. Zoologists, La. Acad. Sci., Miss. Acad. Sci., Ind. Acad. Sci., Iowa Acad. Sci., Bot. Soc. Am., Ecol. Soc. Am., ABA (bd. dir.), Ind. Bar Assn. (bd. dir.), DeKalb County Bar Assn. (bd. dir.) Bar Assn., Humane Soc. U.S. (bd. dir. 1970-88), DeKalb County Humane Soc. (founder, bd. dir. 1970-86), Ind. Fedn. Humane Socs. (bd. dir. 1970-84), Nat. Assn. Women Lawyers (bd. dir.), Bus. and Profl. Women's Club, Nat. Assn. Rep. Women (bd. dir.), Women's Equity Action League (bd. dir.), Assn. So. Biologists, Phi Kappa Phi. Club: Altrusa. Author: Lee's 7 Lives, 1992; founder, editor: Breeder's Jour., 1958-63; numerous vols. on dog genetics and breeding, guide dogs for the blind. Contbr. over 4000 sci. and popular publs. on cytogenetics, dog breeding and legal topics; contbr. articles to Am. Kennel Club Gazette, 1970-81, also others; researcher in field. Home and Office: 336 Hickory St Butler IN 46721-1471

FORD, LORETTA C., retired university dean, nurse, educator; b. N.Y.C., Dec. 28, 1920; d. Joseph F. and Nellie A. (Williams) Pfingstel; R.N., Middlesex Gen. Hosp., New Brunswick, N.J., 1941; BS in Nursing, U. Colo., 1949, MS, 1951, EdD, 1961; DSc (hon.), Ohio State Med Coll.; LLD (hon.) U. Md., 1990; m. William J. Ford, May 2, 1947; 1 dau., Valerie. Staff nurse New Brunswick Vis. Nurse Service, 1941-42; supr., dir. Boulder County (Colo.) Health Dept., 1947-58; asst. prof., then prof. U. Colo. Sch. Nursing, 1960-72; dean Sch. Nursing, dir. nursing, prof. U. Rochester (N.Y.), 1972-86, acting dean Grad. Sch. Edn. & Human Devel., 1988-89; vis. prof. U. Fla., summer 1968, U. Wash., Seattle, 1974; mem. educators adv. panel GAO; dir. Security Trust Co., Rochester, Rochester Telephone Co.; internat. cons. in field. Bd. dirs. Threshold Alternative Youth Svcs., Easter Seal Soc., ARC, Monroe Community Hosp.; mem. adv. com. Commonwealth Fund Exec. Nurse Fellowship Program. Served with Nurse Corps, USAAF, 1942-46. Named Colo. Nurse of Year; recipient N.Y. State Gov.'s award for women in sci., medicine and nursing. Fellow Am. Acad. Nursing; mem. Nat. League Nursing (fellowship, Linda Richards award), Am. Coll. Health Assn. (Boynton award), Am. Nurses Assn., Am. Public Health Assn. (Ruth B. Freeman award), NAS Inst. Medicine (Gustav O. Leinhard award, 1990). Author articles in field, chpts. in books. Office: Univ Rochester Med Ctr 601 Elmwood Ave Box HWH Rochester NY 14642

FORD, MARY ELIZABETH (LIBBY FORD), environmental health engineer; b. Huntington, W.Va., Aug. 22, 1953; d. John Thomas and MAry Elizabeth (Gleason) F.; children: Emily, Caitlin. BS, U. Notre Dame, 1975, MS, 1976. Cert. environ. profl.; cert. hazardous materials mgr.; registered environ. assessor Calif. Engr. Philip J. Clark, P.E., Consulting Engrs., Rochester, N.Y., 1976-77; sr. environ. health engr. Nixon, Hargrave, Devans & Doyle, Rochester, 1977—; mem. adv. com. N.Y. State Dept. Environ. Conservation Divsn. Water, Albany, 1988—. Contbr. numerous articles to profl. jours.; presenter seminars. Vol. Girl Scouts Genesee Valley, Rochester, 1975—; vol. counselor Career Resources Ctr., Rochester, 1985—; eucharist min. St. Josephs of Rush, Rochester, 1985—. Recipient John Chester Brigham award N.Y. Water Pollution Control Assn., Syracuse, 1986. Mem. Water Environment Fedn. (chair govt. affairs com. 1991—), Nat. Assn. Environ. Profls., Inst. Hazardous Materials Mgmt., Soc. Women Engrs., Rochester Com. for Sci. Info. (past v.p. sci. info.). Democrat. Roman Catholic. Office: Nixon Hargrave et al Clinton Sq PO Box 1051 Rochester NY 14603

FORD, MARY SPENCER (JESSE), ecologist, writer; b. N.Y.C., June 11, 1948; d. Henry Rutgers and Mary Spencer (Blackford) F.; m. Mer Wiren, Apr. 10, 1993. BA with distinction, Swarthmore Coll., 1973; MS, Yale U., 1976; PhD, U. Minn., 1984. Rsch. assoc. Ecosytems Rsch. Ctr. Cornell U., Ithaca, N.Y., 1984-87; sr. scientist Nat. Coun. for Air and Stream Improvement US EPA Environ. Rsch. Lab., Corvallis, Oreg., 1987-90; assoc. prof. Oreg. State U., Corvallis, 1991—; tech. expert (U.S.) Arctic Monitoring and Assessment Workshop, Oslo, 1990; mem. steering com. Internat. Conf. on the Ecological Effects of Arctic Airborne Contaminants, 1993. Editorial bd.: Ecological Applications, 1993—. Recipient NSF Doctoral fellowships, 1973-76, 80. Mem. Assn. for Women in Sci. (co-founder Mpls.-St. Paul chpt.), Sigma Xi. Achievements include application of paleoecological techniques to elucidate process of natural tenestrial and aquatic ecosystem acidification; design of monitoring network to assess changes in surface water acidification in acid sensitive U.S. waters; design of network to elucidate status and extent of contamination of U.S. Arctic by long range atmospheric contaminants; research of air toxics, effects on ecostsems. Home: 24146 Old Peak Rd Philomath OR 97370 Office: Oregon State U Dept Fisheries & Wildlife care USEPA 200 SW 35th St Corvallis OR 97333

FORD, PETER C., chemistry educator; b. Salinas, Calif., July 10, 1941; s. Clifford and Thelma (Martin) F.; children: Vincent, Jonathan. BS with honor, Calif. Inst. Tech., 1962; MS, Yale U., 1963, PhD, 1966. Postdoctoral fellow Stanford U., 1966-67; asst. prof. chemistry U. Calif., Santa Barbara, 1967-72, assoc. prof. chemistry, 1972-77, prof. chemistry, 1977—; grad. advisor, Dept. Chem. Univ. Calif., 1980-81, co-grad. advisor, 1985-92; guest prof. H.C. Oersted Inst., Denmark, 1981; lecturer, Univ. Berne, Switzerland, 1989; lecturer MITI-ASTI, Japan, 1990. Contbr. to profl. jours. Nat. Inst. Health fellow, 1963-66, Sterling fellow Yale U., 1963, NSF fellow, 1966-67, Sr. Fulbright fellow, 1974, Vis. fellow Australian Nat. Univ., 1974; Dreyfus Found. Tchr. scholar, 1971-76; recipient Alexander van Humboldt-Stiftung Sr. Scientist Rsch award, 1992; recipient Richard C. Tolman medal Am. Chem. Soc., 1993. Achievements include research in the photochemical, photocatalytic and photophysical mechanisms of transition metal complexes and with homogeneous catalysis mechanisms as probed by modern kinetics techniques; the bioinorganic chemistry of metal nitrosyl complexes. Office: Univ of California Dept of Chemistry 552 University Ave Santa Barbara CA 93106

FORD, ROBERT ELDEN, natural resources educator, geographer, consultant; b. Puerto Castilla, Honduras, May 12, 1945; s. Robert Elden and Vanessa (Standish) F.; m. Karen Elaine Storz, June 15, 1969;children: Bryan A., Colby A. BA in History and Spanish, Pacific Union Coll., 1968; MA in Sociology, Loma Linda (Calif.) U., 1969, MPH, 1971; PhD in Geography, U. Calif., Riverside, 1982. Instr., asst. prof. Loma LInda U., 1973-83; acad. dean Adventist U. of Ctrl. Africa, Gisenyi, 1983-87; assoc. prof. Bringham Young U., Provo, Utah, 1988-92; asst. prof. Utah State U., Logan, 1992—; rsch. assoc. U. Calif., Riverside, 1975-77; cons. Adventist Devel. and Relief Agy. Internat., Silver Springs, Md., 1987-92; bd. dirs. Assn. Am. Geographers Africa Specialty Group, Washington, 1988-91; adj. assoc. prof. internat. health, Loma Linda U., 1990—. Contbr. articles to profl. jours. Elder Wasatch Hills Seventh-day Adventist Ch., Salt Lake CIty, 1991-92. Post-doct. fellow Ind. U., Bloomington, 1982; rsch. grantee David M. Kennedy Ctr. for Internat. Studies, Provo, 1991, USAID U., Washington, 1992. Mem. Assn. Am. Geographers (dir. 1988-91), Soc. Internat. Devel., African Studies Assn., Internat. Mountain Soc., Internat. Studies Assn. Office: Dept Geography & Earth Res CNR Utah State U Logan UT 84322-5240

FORD, SUE MARIE, toxicology educator, researcher. BS, Cornell U., 1975; MS, Mich. State U., 1978, PhD, 1985. Postdoctoral fellow SUNY, Buffalo, N.Y., 1984, Bristol-Myers, Syracuse, N.Y., 1985-87; asst. prof. St. John's U., Jamaica, N.Y., 1987-93, assoc. prof., 1993—. Grantee Am. Assn. Coll. Pharm., 1988. Mem. Soc. Toxicology, Am. Inst. Nutrition, Am. Chem. Soc., Tissue Culture Assn. Achievements include research in renal toxicology, vitro toxicology and cell physiology. Home: 45-11 168th St Flushing NY 11358-3238 Office: St Johns University Grand Central & Utopia Pkwy Jamaica NY 11439

FORD, VICTOR LAVANN, research scientist, forest biologist; b. Johnson City, Tenn., Oct. 18, 1955; s. Verlin Lavann and Shirley Mae (Hall) F.; m. Rhonda Darlene Bombailey, June 11, 1977; children: Christine Nicole, Nathan Zachary. BS, U. Tenn., 1977, MS, 1978; PhD, Va. Tech., 1982. Grad. teaching/rsch. asst. U. Tenn., Knoxville, 1977-78, Va. Polytechnic Inst. & State U., Blacksburg, 1978-82; ext. forester U. Ark., Hope, 1983-88; with Westvaco Central Forest Rsch., Wickliffe, Ky., 1988—; chmn. Ross Found. Stewardship Adv. Panel, Arkadelphia, Ark., 1986-88; mem. FORS Timber Inventory Software Evaluation, 1987. Contbr. articles to profl. jours. Mem. Soc. of Am. Foresters (sec. silviculture working group 1993-94). Republican. Baptist. Home: 114 Windmere Dr Paducah KY 42001 Office: Westvaco Central Forest Rsch PO Box 458 Wickliffe KY 42087

FORDTRAN, JOHN SATTERFIELD, physician; b. San Antonio, Nov. 15, 1931; s. William M. and Josephine (Bell) F.; m. Jewel Evans, July 25, 1953; children: William, Bess, Josephine, Amy. Student, U. Tex., 1949-52; M.D., Tulane U., 1956. Internal medicine intern Parkland Meml. Hosp., Dallas, 1956-57; asst. resident internal medicine Parkland Meml. Hosp., 1957-58; research fellow gastroenterology Mass. Meml. Hosp., Boston, 1960-62; instr. internal medicine U. Tex. Southwestern Med. Sch., Dallas, 1962-63; asst. prof. internal medicine U. Tex. Southwestern Med. Sch., 1963-67, assoc. prof. internal medicine, 1967-69, prof., 1969-79, chief sect. gastroenterology, 1963-79; chief dept. internal medicine Baylor U. Med. Center, Dallas, 1979—; mem. attending staff Parkland Meml. Hosp., Dallas, 1963—; cons. gastroenterology Dallas VA Hosp., 1963—. Contbr. articlkes to profl. jours.; editorial bd. Jour. Clin. Investigation, 1968-73; editor Gastroenterology, 1977-81; co-editor: Gastrointestinal Disease, 5th edit., 1993. Served with USPHS, 1958-60. Recipient King Faisal prize in medicine Saudi Arabia, 1984. Mem. ACP, Am. Soc. Clin. Investigation (past pres.), Am. Gastroent. Assn. (Disting. Achievement award 1971, Kirsner prize 1990, Disting. Educator award 1991, Friedenwald medal 1993), Assn. Am. Physicians. Home: 3508 Hanover Ave Dallas TX 75225-7434 Office: Baylor U Med Ctr 3500 Gaston Ave Dallas TX 75246-2088

FORDYCE, JAMES STUART, federal agency administrator; b. London, Dec. 10, 1931; came to U.S., 1947; s. James Wilfred and Doris Vera (Macrae) F.; m. Beverly Ann Arnold, June 12, 1954; children: Cameron James, Jean Margaret. AB, Dartmouth Coll., 1953; PhD in Phys. Chemistry, MIT, 1959. Rsch. scientist Parma (Ohio) rsch. lab. Union Carbide Corp., 1959-66; rsch. scientist Lewis rsch. ctr. NASA, Cleve., 1966-68, head electrchemical fundamentals, 1968-73, mgr. environ. monitoring office, 1973-76, chief electrochemistry br., 1976-80, dep. chief space power tech. divsn., 1980-81, chief, 1981-84, dep. dir. aerospace tech., 1984-85, dir., 1985-91, dep. ctr. dir., 1991—; spl. lectr. Internat. Space U.; disting. space tech. lectr. Columbia U., 1988; bd. dirs. Edison Polymer Innovation Corp., Akron, Ohio. Author: (with others) Solar Power Satellites, 1993; contbr. articles to profl. jours. Mem. spl. com. Mus. Natural History, Cleve., 1991—; active Leadership Cleve., 1992—; trustee, mem. Coun. Human Rels. Mem. AAAS, AIAA, Am. Chem. Soc., Fedn. Am. Scientists, Electrochemical Soc. (lectr. 106th meeting 1985), Sigma Xi. Democrat. Unitarian. Home: 21295 Cromwell Ave Fairview Park OH 44126 Office: NASA Lewis Rsch Ctr 21000 Brookpark Rd Cleveland OH 44135

FORDYCE, SAMUEL WESLEY, electrical engineer, communications company executive; b. Jackson, Miss., Feb. 28, 1927; s. Samuel Wesley and Polly Adams (White) F.; S.B., Harvard U., 1949; M.S., Washington U., St. Louis, 1953; m. Sally Gillespie, Apr. 9, 1970; children—Katherine Peake, Debbie Fordyce, Wesley, Polly. Project engr. Emerson Electric Co., St. Louis, 1949-58; mem. tech. staff Ramo Wooldridge, Los Angeles, 1958-60, Gen. Electric Tempo, Santa Barbara, Calif., 1960-62; chief engr. communications div. NASA, Washington, 1962-84; chief oper. officer Advanced Bus. Communications, Inc., McLean Va., 1986-88, cons. Cape York Space Agy., Brisbane, Australia, 1988—; pres. Riparian Research Corp., 1984—. Served with USN, 1944-46. Registered profl. engr., Mo. Assoc. fellow AIAA; sr. mem. IEEE. Clubs: St. Louis Country; Met. of Washington; Chevy Chase. Achievements include design and development of radio communications systems used on Apollo (manned lunar landing) Program. Home: 6716 Selkirk Ct Bethesda MD 20817-4936

FORE, CLAUDE HARVEL, III, scientist, information analyst; b. Columbus, Ohio, Jan. 2, 1957; s. Claude Harvel and Theresa Bertha (Vereb) F.; m. Carla Diane Snider, Jan. 19, 1980; children: Mariah Theresa, Charlotte Diane. BS in Physics, Va. Mil. Inst., 1979; MS in Nuclear Sci., Air Force Inst. Tech., 1987. Commd. 2d lt. U.S. Army, 1979, platoon leader, then ops. platoon leader 545th Ordnance Co., 1980-83, advanced through grades to capt., 1983; with 24th transp. bn. U.S. Army, Ft. Eustis, Va., 1983-85; physicist Def. Nuclear Agy. U.S. Army, Alexandria, Va., 1987-90; resigned U.S. Army, 1990; scientist, info. analyst Kaman Scis. Corp., 1990—. Mem. IEEE, Am. Soc. Testing Materials. Roman Catholic. Home: 4000 Forge Dr Woodbridge VA 22193 Office: Kaman Scis Corp Ste 500 2560 Huntington Ave Alexandria VA 22303-1410

FORER, BERTRAM ROBIN, psychologist, researcher; b. Springfield, Mass., Oct. 24, 1914; s. Maurice and Ida Edith (Robinson) F.; m. Lucille Kremith, Sept. 27, 1941; children: Stephen Keith, Robert William. BS in Premed., U. Mass., 1936; MA in Psychology, Exptl. Esthetics, UCLA, 1938, PhD in Experimental & Social Psychology, 1941. Test and measurement U.S. Civil Svc. Commn., Washington, 1941-42; psychologist U.S. Army, France, 1942-46; test specialist Office Sec. of War, Washington, 1946-47; clin. psychologist VA, L.A., 1947-59; pvt. practice L.A., 1959-83; retired, 1984; prof. UCLA, 1947-55; exec. bd. mem. Viewer Sponsored TV, L.A., 1966-70, Nat. Assn. Better Broadcasting, L.A., 1965—;. Contbr. articles to profl. jours. Exec. bd. Malibu (Calif.) Twp. Coun., 1983-85. Lt. U.S. Army, 1942-46. Recipient Second Prize Sculpture award Malibu Art Festival, 1979. Fellow Am. Psychol. Assn., Calif. State Psychol. Assn.; mem. L.A. Soc. Clin. Psychologists (pres. 1967), L.A. County Psychol. Assn. (pres. 1975-76), Soc. for Personality Assessment (pres. 1960-61, journal editor 1955-66). Achievements include work in 44 publications and Simplestat computer templates, Homework computer templates and structured sentence completion tests. Home: 19854 Pacific Coast Hwy Malibu CA 90265

FORESMAN, JAMES BUCKEY, geologist, geochemist, industrial hygienist; b. Neosho, Mo., Apr. 8, 1935; s. Frank James and Helen Blackburn (Buckey) F.; m. Barbara Ellen Runkle, Aug. 13, 1961; children: James Runkle, Robert Buckey. BSBA, BS, Kans. State U., 1962; MS, U. Tulsa, 1970. Cert. petroleum geologist; cert. insp., mgmt. planner, contractor, supr. and project designer, in asbestos control EPA. From geologist, geochemist to staff dir. geology N.Am.-S.Am. Phillips Petroleum Co., Denver, Midland, Tex., Bartlesville, Okla., 1962-83; petroleum cons. Bartlesville, 1983-84; v.p. Mopro, Inc., Lyons, Mich., 1985-87; indsl. hygienist Pittsburg (Kans.) State U., 1987—; geochemistry advisor Joint Oceanographic Instsn. for Deep Earth Sampling, 1974-75; ocean drilling advisor NSF, Washington, 1974-75; indsl. rep. for joint ventures with USSR, 1978; rep. Univ.-Indsl. Assoc. Programs, N.Y., Tex., Ariz., Mass., Calif., Cambridge (Eng.), 1981-83; citizen amb. programs Environ. Del. to Russia, Latvia, and Estonia, 1992. Contbr. articles to periodicals, jours., chpts. to books. Cm. mem. Boy Scouts Am., Bartlesville, 1975-82; bd. dirs. U.S. Little League, Bartlesville, 1975; smoke jumper U.S. Dept. Agr., Forest Svc. Sgt. USMC, 1954-57, Korea. Recipient Disting. Svc. award City of Bartlesville, 1977. Mem. Am. Assn. Petroleum Geologists (founding mem. energy minerals divsn., charter mem. divsn. environmental geologists), Am. Soc. Safety Engrs., Am. Conf. Govtl. Indsl. Hygienists, Am. Indsl. Hygiene Assn. (bd. dirs.), Kiwanis. Republican. Presbyterian. Avocations: cycling, reading, collecting rare and antique books, YMCA activities. Home: 1506 Woodland Ter Pittsburg KS 66762-5551

FOREST, CARL ANTHONY, lawyer; b. Tiffin, Ohio, May 9, 1940; s. John Thomas and Luella (Dehn) Baumgardner; m. Angela Leachman, Nov. 5, 1988. BS, U. Detroit, 1962; MS, Mich. State U., 1964, PhD, 1967; JD, Boston U., 1977. Bar: Idaho 1977, U.S. Patent and Trademark Office 1977, U.S. Ct. Appeals Fed. Cir., 1982, Ind. 1986, Colo. 1992, U.S. Dist. Ct. Colo. 1992. Asst. prof. physics U. Idaho, Moscow, 1967-72; clk. Wolf Greenfield & Sacks, Boston, 1976-77; assoc. Moffatt Thomas Barret & Blanton, Boise, Idaho, 1977-78; patent atty. Medtronic, Inc., Mpls., 1978-82; sr. patent atty. Bristol Myers-Squibb, N.Y.C., 1982-85; sr. group patent counsel Black & Decker Corp., Towson, Md., 1985-90; of counsel Dorr Carson Sloan & Petersen, Denver, 1991-92; chief patent counsel Symetrix Corp., Colorado Springs, Colo., 1992; dir., sec.-treas. Duft Graziano & Forest, Boulder, Colo., 1992—; bd. dirs. Fir Pub. Co., Boulder. Author: The Physics of Caribou Creek, 1991; contbr. articles to profl. publs. Bd. dirs. McGovern Presdl. Campaign, State of Idaho, Boise, 1972, Ctr. for Attitudinal Healing, Indpls., 1990-91. NSF fellow, 1963. Mem. Am. Phys. Soc., Am. Intellectual Property Law Assn., Colo. Bar Assn., Boulder Bar Assn. Democrat. Achievements include patents for first layered superlattice ferroelectrics; discovery of electrooptical effect in absorption spectrum of III-IV semiconductors. Office: Duft Graziano & Forest 1790 30th St Boulder CO 80301

FORESTER, DAVID ROGER, research scientist; b. Tyler, Tex., May 8, 1953; s. Jesse Roger and Dorothy Nell (Crow) F.; m. Joan Ann Smith, Nov. 19, 1977; children: Carrie Ann, Benjamin Roger, Susanna Faith. BA in Chemistry, Tex. A&M U., 1975. Chemist Texaco Rsch., Port Arthur, Tex., 1975-79; scientist Ethyl Petroleum Additives, St. Louis, 1979-81, Betz Labs., Woodlands, Tex., 1981-85; rsch. scientist Betz Process Chems., Inc., Woodlands, 1985—. Alt. del. Rep. Party of Tex., Fort Worth, 1990. Mem. ASTM (vice chmn. user group on catalysts 1990—), S.W. Catalysis Soc. (treas. 1991-93), Toastmasters Internat. (edn. v.p. 1991, competent toastmaster 1991). Baptist. Achievements include more than 28 patents in field, including advancements in refinery/petrochem. streams, contaminant passivation on FCC catalysts, and pyrolytic processing. Office: Betz Process Chems Inc 9669 Grogans Mill Rd Spring TX 77380-1096

FORGACS, OTTO LIONEL, forest products company executive; b. Berlin, Jan. 4, 1931; emigrated to Can., 1955; s. Joseph and Luise (Schick) Forgacs; m. Patricia Purdom Saunders, Sept. 24, 1960; children: Anthony, Stephen, Jonathan. B.Sc. in Tech., U. Manchester, Eng., 1955; Ph.D., McGill U., 1959. Pulp and paper research Inst. Can., Montreal, 1958-63; research mgr. Domtar, Ltd., Montreal, 1963-73; research dir. MacMillan & Bloedel, Ltd., Vancouver, B.C., Can., 1973-77, v.p. reserach and devel., 1977-79, sr. v.p. research and devel., 1979-85; dir. Forest Engring. Inst. of Can., 1985—; dir. Zellstoff, Papierfabrik Frantschach, Austria; mem. Sci. Council B.C., 1977-83. Contbr. numerous articles tech. jours. Fellow TAPPI (chmn. research and devel. div.), Chem. Inst. Can.; mem. Can. Pulp and Paper Assn. (councillor, exec. council 1977-80). Club: Vancouver Lawn Tennis. Home: 1843 Acadia Rd, Vancouver, BC Canada V6T 1R2 Office: Noranda Inc, 4225 Kincaid St, Burnaby, BC Canada V5G 4P5

FORKS, THOMAS PAUL, osteopath; b. Great Lakes Naval Station, Ill., Apr. 15, 1952; s. Louis John and Rhoda Joan (Miles) F.; m. Sharron Elizabeth Wells, Dec. 15, 1979; 1 child, Joseph Miles. BA, St. Mary's U., 1975, MS, U. Tex., San Antonio, 1977; PhD, U. So. Miss., 1981; DO, U. Health Scis., Kansas City, Mo., 1988. Teaching asst., instr., then asst. prof. biology U. So. Miss., Hattiesburg, 1977-82; asst. prof. biology Wilkes Coll., Wilkes-Barre, Pa., 1982-83; intern Corpus Christi (Tex.) Osteo. Hosp., 1988-89; resident, then chief resident in family practice U. Miss., Jackson, 1989-91; emergency rm. physician Rush Hosp., Newton, Miss., 1989-91, Lackey Meml. Hosp., Forest, Miss., 1991—; physician N.E. Family Practice, San Antonio, 1992-93; family physician Morton, Miss., 1993—; lectr. in emergency med. mgmt. of snakebites. Contbr. articles to profl. publs. Charity physician Stewpot Free Clinic, Jackson, 1990-91; adv. bd. Cath. Charities, Jackson, 1990-91. Mem. AMA, AAFP, Am. Osteo. Assn., Miss. Acad. Sci., So. Med. Assn., Soc. for Study Amphibians and Reptiles, So. Biol. Assn., Assn. Southeastern Biologists, Chgo. Herpetological Soc. Republican. Roman Catholic. Avocations: photography, hunting, jogging. Home: 4232 Oak Lake Dr Jackson MS 39212-5369

FORLINI, FRANK JOHN, JR., cardiologist; b. Newark, Mar. 30, 1941; s. Frank Sr. and Rose Theresa (Parussini) F.; m. Joanne Marie Horch, July 19, 1969; children: Anne Marie, Victoria, Frank III, Anthony. BS in Biology, Villanova (Pa.) U., 1963; MD, George Washington U., 1967. Diplomate Am. Bd. Internal Medicine, Am. Bd. Cardiovascular Disease. Intern Bklyn.-Cumberland Med. Ctr., N.Y., 1967-68, resident in internal medicine, 1968-70; fellow in cardiology Inst. Med. Sci. Pacific Med. Ctr., San Francisco, 1970-72; practice medicine specializing in cardiology Rock Island, Ill., 1974—; sr. ptnr. Forlini Med. Speciality Clinic, Rock Island, 1974—; owner Forlini Farm and Forlini Devel. Enterprises; assoc. prof. pharmacy L.I. U., Bklyn., 1969-70; pres., chief exec. officer U.S. Oil and Transp. Co., Inc., 1966-89; pres. Profl. and Execs. Ins. Assocs., 1973-89, Profl. Assocs., 1973-89; med. and exec. dir. Cardiovascular Inst. Northwestern Ill., 1984—; exec. dir., owner Franksoft Pub., 1988—; bd. dirs. Shelter for Abused Women and Children, Rock Island, Ill., 1992—, Christian Family Ctr. Shelter for Men,

Rock Island, Ill., 1992—. Contbr. articles to profl jours. Chmn. D.C. Young Reps., 1965-66; mem. exec. com. Rep. Cen. Com., Washington, 1965-66; mem. nat. com. Coll. Young Republicans, 1965-66; mem. exec. com. Young Rep. State Cen. Com., Washington, 1965-66; vice chmn. Rock Island Reps., 1985-90, precinct committeeman, 1985-90, 92-93; dep. registrar County of Rock Island, Ill., 1985—; trustee South Rock Island Twp., Rock Island County, 1987—; mem. exec. com. Rock Island County Republican Cen. Com., 1992—; del. Ill. State Rep. Conv., 1992; pres. parish coun., extraordinary minister. Served to maj. USAF, 1972-74. Nat. Inst. Heart Disease NIH-USPHS grantee, 1964-66, 70-72. Fellow Am. Coll. Cardiology, N.Am. Soc. Pacing and Electrophysiology; mem. Rock Island County Med. Soc. (chmn. com. on ins. 1990—), Ill. State Med. Soc., AMA. Roman Catholic. Office: 2701 17th St Rock Island IL 61201-5383

FORNATTO, ELIO JOSEPH, otolaryngologist, educator; b. Turin, Italy, July 2, 1928; came to U.S., 1953; s. Mario G. and Julia (Stabio) F.; m. Mary Elizabeth Pearson, Dec. 17, 1960; children: Susan, Robert, Daniel. MD, U. Turin, Italy, 1952. Diplomate Am. Bd. Otolaryngology. Intern Edgewater Hosp., Chgo., 1956-57; resident U. Ill., Chgo., 1953-56; chief otolaryngologist Elmhurst (Ill.) Clinic, 1958—; sr. otolaryngologist Elmhurst (Ill.) Meml. Hosp., 1964—; med. dir. Chgo. Eye Ear Nose Throat Hosp., 1966-69; clin. asst. prof. Loyola U., Chgo., 1967—; bd. dirs. Du Page County Am. Cancer Soc., 1977—; chmn. Elmhurst Clinic, 1980-89. Founder Centurion Club, Deafness Research Found., N.Y.C., 1960—. Mem. AMA, Ill. Med. Soc., Am. Acad. Facial Plastic and Reconstructive Surgery, Am. Acad. Otolaryngologic Allergy, Am. Acad. Otolaryngology and Head and Neck Surgery. Roman Catholic. Avocations: music, bicycling. Home: 200 W Jackson St Elmhurst IL 60126-4807 Office: Elmhurst Clinic 172 Schiller St Elmhurst IL 60126-2885

FORSEN, HAROLD KAY, engineering executive; b. Sept. 19, 1932; s. Allen Kay and Mabel Evelyn (Buehler) F.; m. Betty Ann Webb, May 25, 1952; children: John Allen, Ronald Karl, Sandra Kay. AA, Compton Jr. Coll., 1956; BS, Calif. Inst. Tech., 1958, MS, 1959; PhD, U. Calif., Berkeley, 1965. Research assoc. Gen. Atomic, San Diego, 1959-62; research assoc., elec. engr. U. Calif., Berkeley, 1962-65; assoc. prof. nuclear engring. U. Wis., Madison, 1965-69, prof., 1969-73, dir. Phys. Sci. Lab., 1970-72; v.p. Exxon Nuclear Co., Bellevue, Wash., 1973-75, v.p., bd. dirs., 1975-80, exec. in charge laser equipment, 1981; exec. v.p. Jersey-Avco Isotopes, Inc., 1975-80, pres., 1981, dir., 1975-81; mgr. engring. and materials Bechtel Group, Inc., San Francisco, 1981-83, dep. mgr. research and engring., 1983-84, mgr. advanced systems, 1984-85, mgr. R&D, 1986-91; sr. v.p. Bechtel Nat., San Francisco, 1986-91, exec. v.p., 1991; mgr. Bechtel Tech. Group, 1992—; mem. fusion power reactor sr. rev. com. Dept. Energy, 1977, magnetic fusion adv. com., 1982-86, fusion policy adv. com., 1990; chmn. U.S. del. of AEC on Ion Sources to Soviet Union, 1972; cons. Oak Ridge Nat. Lab., Tenn., 1969-72, Argonne Nat. Lab., Ill., 1970-72, Exxon Nuclear Co., 1970-73, Battelle N.W. Lab., 1971-72; mem. sci. and tech. adv. com. Argonne Nat. Lab., 1983-85; mem. fusion energy adv. com. Oak Ridge Nat. Lab., 1977-84; mem. vis. com. dept. nuclear energy Brookhaven Nat. Lab., 1992—. V.p., trustee Pacific Sci. Ctr. Found., 1977, pres., 1978-80, chmn., 1981; mem. dean's vis. com. Coll. Engring., U. Wash., 1981—, dept. nuclear engring. and engring. physics indsl. rels. coun. U. Wis., 1987-92; mem. com. magnetic fusion in energy policy Nat. Rsch. Coun., 1987; pres. bd. dirs. Bay Area Sci. Fair, Inc., 1988-89; bd. dirs. Plasma and Materials Techs., Inc., 1988-91; West Coast adv. bd. Inst. Internat. Edn., 1991—; mem. bd. overseers Superconducting Super Collider, 1991-92; mem.-at-large bd. on assessment Nat. Inst. Standards and Tech., 1991—; mem. adv. bd. to coll. Engring., U. Calif., Berkeley, 1993—. Served with USAF, 1951-55. Named San Francisco Bay Area Eminent Engr., 1990. Fellow Am. Phys. Soc., Am. Nuclear Soc. (Arthur H. Compton award 1972, chmn. tech. group controlled nuclear fusion 1973); mem. IEEE (sr.), NAE (bd. chancellors 1993—), Sigma Xi, Tau Beta Pi. Home: 255 Tim Ct Danville CA 94526-3240 Office: Bechtel Group Inc PO Box 193965 San Francisco CA 94119-3965

FORSSANDER, PAUL RICHARD, inventor, artist, entrepreneur; b. Chgo., Oct. 10, 1944; m. M. Andrea Peake, Dec. 30, 1967. BA in Econs., Marian Coll., Indpls., 1967. Sales and ops. adminstr. ITT Pub., Indpls., 1967-80; v.p., gen. mgr. Kutt Inc., Boulder, Colo., 1982-86; pres., chief exec. officer Skynasaur Inc., Boulder, 1986-89; pres., chief exec. officer, founder Zephyr Co. Inc., Boulder, 1989-91, Quillum Co., 1990—, Notetote Co., Boulder, 1990-91, PRF Designs, 1991—. Inventor flying and wind powered high-tech recreational products and parts, energy generation/conservation products, writing instrument designs, gift and office products; designer glass & metal sculptures, illuminaries & table art; developed forming and finishing process. Bd. dirs. PBS Sta. KGNU, Boulder, 1983-86, EMT Assn. Colo., Boulder, 1985-88; designer of fundraising strategies for non-profit orgns. Mem. Gift Assn. Am., Nat. Sporting Goods Assn., Toy Industry Am., Nat. Soc. Fundraising Execs., Nature Conservancy, Greenpeace, Sierra Club. Avocations: flying wind-powered aircraft, photography, bicycling, hiking, sculpting. Home: PO Box 1010 Boulder CO 80306-1010

FORSSELL, BÖRJE ANDREAS, electronics engineer, educator, consultant; b. Burträsk, Västerbotten, Sweden, Apr. 2, 1939; s. Emil Andreas and Helga Maria (Burman) Forssell; m. Anna Elisabet Norberg, May 19, 1963; children: Johan, Anders, Henrik, Elisabet. MSc in Elec. Engring., Royal Inst. Tech., Stockholm, 1965; PhD in Elec. Engring., Norwegian Inst. Tech., Tondheim, 1976. Lab. engr. Philips Teleindustri, Stockholm, 1965-69; scientist Electronics Rsch. Lab., Trondheim, Norway, 1969-83; prof. Norwegian Inst. Tech., Trondheim, 1983—; adv. Petroleum Rsch. Inst., Trondheim, 1985-93, Delab, 1993—. Author: Radionavigation Systems, 1991. Fellow Royal Inst. Navigation; mem. IEEE (sr. mem., section chmn. 1975-82), Instn. Elec. Engrs., Inst. Navigation, Deutsche Gesellschaft für Ortung und Navigation, Inst. Français de Navigation. Home: Dalhaugvegen 62, Trondheim N-7020, Norway Office: U Trondheim Norwegian Inst Tech, Divsn Telecom, Trondheim N-7034, Norway

FORSTER, BRUCE ALEXANDER, economics educator; b. Toronto, Ont., Can., Sept. 23, 1948; m. Margaret Jane Mackay, Dec. 28, 1968, (div. Dec. 1979); 1 child, Kelli Elissa; m. Valerie Dale Pendock, Dec. 8, 1979; children: Jeremy Bruce, Jessica Dale. BA in Math., Econs., U. Guelph, Ont., 1970; PhD in Econs., Australian Nat. U., Canberra, 1974. Asst. prof. U. Guelph, 1973-77, assoc. prof., 1977-83, prof. econs., 1983-88; vis. assoc. prof. U. B.C., Vancouver, 1979; vis. fellow U. Wyoming, 1979-80, vis. prof., 1983-84, 87, prof. econs., 1987—, dean Coll. Bus., 1991—; vis. prof. Profl. Tng. Ctr., Ministry of Econ. Affairs, Taiwan, 1990, 91, 92, 93; acad. assoc. The Atlantic Coun. of the U.S., cons. in field. Author: The Acid Rain Debate: Science and Special Interest in Policy Formation, 1993; co-author: Economics in Canadian Society, 1986; assoc. editor Jour. Applied Bus. Rsch., 1987, editorial adv. bd., 1987—; editorial coun. Jour. Environ. Econs. and Mgmt., 1989, assoc. editor, 1989-91; contbr. articles to profl. jours. Jayes-Qantas Vis. scholar U. Newcastle, Australia, 1983. Mem. Am. Econ. Assn., Assn. Environ. and Resource Economists, Faculty Club U. Guelph (treas. 1981-82, v.p. 1982-83, 85-86, pres. 1986-87). Avocations: weight lifting, swimming, skiing, scuba diving. Home: 3001 Sage Dr Laramie WY 82070-5751 Office: U Wyo Coll Bus Laramie WY 82071

FORSTER, MICHAEL LOCKWOOD (MIKE FORSTER), software engineer; b. Glendale, Calif., Feb. 20, 1950; s. William Lockwood and Elizabeth Irene (Hulse) F.; divorced; children: Eric, Shannon. BA in Math., U. So. Calif., 1971, MS in Computer Sci., 1974. Programmer TRW, Redondo Beach, Calif., 1974-76; sr. analyst SRI Internat., Menlo Park, Calif., 1976-80; sr. engr. Control Data Corp., Sunnyvale, Calif., 1980-81; CASE mgr. ESL, Inc., Sunnyvale, 1981-90; software rsch. and devel. mgr. Loral Western Devel. Labs., San Jose, Calif., 1990-; adv. bd. Indsl. and Systems Engring. Dept. San Jose State U., Calif., 1991-92. Mem. Assn. Computing Machinery, Alpha Phi Omega. Office: Loral WDL Po Box 49041 San Jose CA 95161-9041

FORSYTH, KEITH WILLIAM, optical engineer; b. Marion, Ohio, Mar. 15, 1950; s. William Richard and Twila June (Hickok) F.; m. Susan Grossinger, Aug. 29, 1981; children: Adam Glen, Micah Zachary. BS in Physics, SUNY, Albany, 1987; MSEE, Drexel U., 1992. Engr. Denton Vacuum Co., Cherry Hill, N.J., 1984-85; engring. scientist EG&G Princeton (N.J.) Applied Rsch., 1985-89; cons. Forsyth Electro-Optics, Phila., 1989—; chair sci. lab. com. Henry Elem. Sch. Home and Sch. Assn., Phila., 1990—. Contbr. articles to profl. jours. Mem. IEEE Laser and Electro-Optics Soc., AAAS, Internat. Soc. Optical Engring., Optical Soc. Am. Achievements include patent for electro-optic sampling system with dedicated electro-optic crystal and removable sample carrier. Home: 4402 Dexter St Philadelphia PA 19128-4823 Office: Forsyth Electro Optics 4402 Dexter St Philadelphia PA 19128

FORSYTHE, ROBERT ELLIOTT, economics educator; b. Pitts., Oct. 25, 1949; s. Robert Elliott and Dolores Jean (Davis) F.; m. Lynn Maureen Zollweg, June 17, 1970 (div. July 1978); m. Patricia Ann Hays, June 20, 1981; 1 child, Nathaniel Ryan. BS, Pa. State U., 1970; MS, Carnegie-Mellon U., Pitts., 1972, Carnegie-Mellon U., Pitts., 1974; PhD, Carnegie-Mellon U., Pitts., 1975. Ops. rsch. analyst PPG Industries Inc., Pitts., 1970-72; instr. Carnegie-Mellon U., Pitts., 1974-75; asst. prof. Calif. Inst. Tech., Pasadena, 1975-81; assoc. prof. U. Iowa, Iowa City, 1981-86, prof. econs., 1986-90, dept. chmn., 1990—; founder Iowa Polit. Stock Market; pres. Iowa Market Systems, Inc., 1993—. Author: Forecasting Presidential Elections: Polls, Markets, Models. Univ. faculty scholar U. Iowa, 1985-88. Mem. Econometric Soc., Am. Econ. Assn., Econ. Sci. Assn. (sect. head 1989-92, pres.-elect 1992—). Congregationalist. Home: 1806 E Court St Iowa City IA 52245-4643 Office: U Iowa Dept Econs Iowa City IA 52242

FORTENBERRY, JEFFREY KENTON, nuclear engineer; b. McComb, Miss., Dec. 25, 1957; s. Jasper Estes Jr. and Addie Lee (Adams) F.; m. Susan Fox Groome, Apr. 13, 1985; children: Susan Leigh, Elizabeth Ann, Margaret Elaine. BS in Nuclear Engring., Miss. State U., 1979. Registered profl. engr.; licensed sr. reactor operator. Thermo/hydraulic design engr. Knolls Atomic Power Lab., Schenectady, N.Y., 1979-83, core mech. design engr., 1983-84; nuclear fuels engr. Miss. Power and Light Co., Jackson, Miss., 1984-88; nuclear licensing engr. System Energy Resources, Inc., Jackson, 1988-89; shift tech. advisor Entergy Ops., Inc., Port Gibson, Miss., 1989-91, SRO shift supr., 1991-92; sr. staff engr. Def. Nuclear Facilities Safety Bd., Washington, 1992—. Mem. Miss. Engring. Soc., Am. Nuclear Soc. Home: 14601 Creek Valley Ct Centreville VA 22020 Office: Def Nuclear Facilities Safety Bd Ste 700 625 Indiana Ave NW Washington DC 20004

FORTIN, JOSEPH ANDRÉ, forestry educator, researcher; b. Quebec, Que., Can., Sept. 28, 1937; m. Monique Grenier; children: Christine, Brigitte, Elizabeth, René, Claude. B.Sc., Laval U., 1962, D.Sc., 1966; M.S., U. Wis., Madison, 1964. Prof. dept. forest scis. Laval U., Quebec, 1966—; mem. biotechnology com. Can. Nat. Research Council, 1983-84; organizer 5th North American Conf. on Mycorrhizae; chmn. Research Ctr. in Forest Biology, 1985-90, Rsch. Inst. Plant Biology. Assoc. editor: Canadian Botany Jour., 1980-90; contbr. numerous articles to profl. jours. Recipient Leo Parizeau prize Assn. Canadienne Française pour L'Avancement des Scis., 1982, Rene Pomerleau medal, 1982, Sci. Merit award Canadian Inst. Forestry, 1983. Mem. Royal Soc. Can., Can. Botany Assn. (v.p. 1973-74), Can. Assn. on Conifer Biotechnology (v.p. 1986-88), Internat. Found. for Sci. (trustee 1988-91). Office: U Montreal, Rsch Inst on Plant Biology, 4101 rue Sherbrooke E, Montreal, PQ Canada H1X 2B2

FORTNER, RICHARD J., physical scientist. Ph.D. physics, U Notre Dame, Ind., 1968. Rschr. Lawrence Livermore Nat. Lab., Livermore, Calif. Recipient Ernest Orlando Lawrence Meml. award U.S. Dept. Energy, 1991. Office: Lawrence Livermore National Lab PO Box 808 Livermore CA 94550*

FORTUIN, JOHANNES MARTINUS H., chemical engineer; b. Rotterdam, The Netherlands, Sept. 30, 1927; s. Hermanus Johannes Antonius and Anna (de Groot) F.; m. Wilhelmina J.T. van de Ven, Sept. 28, 1961; children: Anke, Karen, Wilke, Jan. Degree in Engring. cum laude, Tech. U., Delft, The Netherlands, 1953, D in Engring. cum laude, 1955. Cert. chem. engr. Rsch. asst. Tech. U., Delft, 1953-54; researcher Cen. Lab. Naamloze Vennootschap DSM, Geleen, 1954-61; head tech. dept. Cen. Lab. DSM, Geleen, 1961-82; coord. external rsch. in process tech. DSM Rsch. BV, Geleen, 1982-87; prof. chem. engring. U. Amsterdam, The Netherlands, 1977-92; sec. working party on chem. reaction engring. European Fedn. Chem. Engring., 1992-93. Editor: Physical and Chemical Aspects of Adsorbents and Catalysts, 1970, Chemical Reaction Engineering, 1972; contbr. articles to profl. jours. on chemistry and chem. engring. Lt. Netherlands Mil., 1954-56. Named Officer Order of Orange-Nassau, 1987; recipient Process Tech. award Hollandsche Maatschappij der Wetenschappen, 1978. Mem. Royal Netherlands Chem. Soc., Royal Inst. Engrs. Avocation: history of Europe. Office: U Amsterdam/Chem Engring, 166 Nieuwe Achtergracht, 1018 WV Amsterdam The Netherlands

FORTUNA, WILLIAM FRANK, architect, architectural engineer; b. Paris, Ill., Apr. 3, 1948; s. William F. Sr. and Mary O. (Komatz) F.; m. Gayle M. Meadors, June 11, 1983. BArch, U. Ill., 1972, MS in Archtl. Engring., 1973. Registered architect, Ill., Wis.; registered structural engr., Ill.; registered profl. engr., Wis.; registered archtl. engr. specializing in crisis mgmt. Designer Unteed Assocs. Ltd., Champaign, Ill., 1973-76; structural engr. Consoer Townsend, Chgo., 1976-79, Schmidt, Garden & Erikson, Chgo., 1979-83; sr. project structural engr. Skidmore Owings & Merrill, Chgo., 1983-87; pres. W.F. Fortuna Ltd., Lake Bluff, Ill., 1987—; project engr. World Trade Ctr., Cairo, Egypt; structural engr. exhibition ctr. McCormick Place Annex, Chgo., United Airlines terminal O'Hare, Bishop's Gate, London; contract adminstr. One and Two Prudential Plaza, Chgo., Sporting Club at Ill. Ctr., Chgo. Active mem. Illinois Emergency Mgmt. Agency. Mem. AIA (Chgo. chpt., Ill. chpt. nat.), Structural Engrs. Assn. Ill., Am. Concrete Inst., Am. Inst. Steel Constrn., Chgo. Hist. Soc., Nat. Trust Hist. Preservation. Home and Office: 530 E Prospect Ave Lake Bluff IL 60044-2616

FOSS, RALPH SCOT, mechanical engineer; b. Perth Amboy, N.J., Aug. 19, 1945; s. Frank Allen and Bessie Christine (Hopla) F.; m. Gail C. Fagre (div. 1983). Student, Pa. State U., 1963-65; BME, Parsons Coll., Fairfield, Iowa, 1966; MBA in Bus., London Sch. Econs., 1970; postgrad., Heidelberg U., Fed. Republic Germany, 1970-71. Mgr. advanced facilities, compressed air Vollkswagen GmbH, Wolfsburs, W.Ger., 1968-71; product mgr. Ingersoll Rand Co., East Brunswick, N.J., 1971-73; mgr. air compressor group Minn. Inst. Tech., Mpls., 1973-76; sales and mktg. mgr. Sullair Corp., Michigan City, Ind., 1976-81; sys. engring. mgr. Ingersoll Rand Co., Charlotte, N.C., 1982-88; pres. Plant Air Tech., Charlotte, 1988—, Winburn Assocs., Inc., Charlotte, 1989—; bd. dirs. Amethyst Found., Charlotte, 1987-90. Author: Compressed Air Systems, 1989; inventor in field. Mem. Presdl. Task Force, Washington, 1982-88; pres.'s coun. Fla. Hosp. Found., Orlando, 1990—. With U.S. Army, 1966-68. Named to Hon. Order Ky. Cols., 1990. Mem. Am. Inst. Plant Engrs. (pres., corp. mems. coun. 1985-89, dir. 1989-90), Assn. Energy Engrs., Instrument Soc. Am., Internat. Platform Assn. Republican. Presbyterian. Avocations: golf, writing. Office: Plant Air Tech 5121 Hunt Stand Ln Charlotte NC 28226-3215

FOSSUM, ROBERT MERLE, mathematician, educator; b. Northfield, Minn., May 1, 1938; s. Inge Martin and Tina Otelia (Gaudland) F.; m. Cynthia Carol Foss, Jan. 30, 1960 (div. 1979); children: Karen Jean, Kristin Ann; m. Barbara Joel Mason, Aug. 4, 1979 (div. 1993); children: Jonathan Robert, Erik Anton. B.A., St. Olaf Coll., 1959; A.M., U.Mich., 1961, Ph.D., 1965. Instr. U. Ill., Urbana, 1964-66; asst. prof. U. Ill., 1966-68, assoc. prof., 1968-72, prof. math., 1972—; lectr. Aarhus U., Denmark, 1971-73, Copenhagen U., Denmark, 1976-77; vis. prof. Université de Paris VI, 1978-79, Oslos U., 1968-69. Editor: Notices of American Mathematical Society, Abstracts of the Ameican Mathematical Society; contbr. numerous articles to profl. jours. Fulbright grantee Oslos U., 1967-68. Mem. AAAS, Nat. Assn. Mathematicians, Math. Assn. Am., Am. Math. Soc. (assoc. sec. cen. sect. 1983-87, sec.-designate 1988, sec. 1989—, editor Notices, Abstracts), Dansk Matematisk Forening, Inst. Algebraic Meditation (sec.), Swedish Math. Soc. Democrat. Lutheran. Club: Heimskringla (Urbana). Office: U Ill Dept Math 1409 W Green St Urbana IL 61801-2917

FOSTER, BARBARA MELANIE, microscopist, consultant; b. Los Alamos, N.Mex., Apr. 21, 1945; d. Lawrence Marvin and Evelyne Marilyn (Caro) Litz; m. John Michael Foster, Sept. 4, 1966 (div. Mar. 1984); m. Kenneth Martin Piel, Nov. 30, 1991. BSEd, Ohio U., 1967; MS in Chemistry, U. Mass., 1979; postgrad., Brunel U., Uxbridge, Eng., 1979-81. Tchr. sci. Athens (Ohio) High Sch., 1967-68, West Springfield (Mass.) High Sch., 1968-81; cons., owner Microscopy/Microscopy Edn., Springfield, Mass., 1981-84; dir. tech. applications group Unitron, Inc., Plainville, N.Y., 1984; field product specialist, adj. mem. mktg. mgmt. group Carl Zeiss, Inc., Thornwood, N.Y., 1984-86; applications mgr. rsch. microscopy Cambridge Instruments, Buffalo, 1986-87, mgr. product devel., 1988, mgr. ednl. mktg., 1988-89; mgr. tech. mktg. Sarastro, Inc., Phila. and Bethel, Conn., 1989-91; COO FM2, Bethel, Conn., 1990; cons., owner Microscopy/Mktg. & Edn., Springfield, 1989—; coord., prin. lectr. short courses ACS, 1982—. Author: Use & Care of the Microscope, 1986; editor The Quarterly newsletter, 1984; contbr. articles to profl. jours. and on-going column. Bldg. rep. West Springfield Edn. Assn., 1970, salary rep., negotiations com., 1971, chair com. 1972, 73, cons. com. 1974; founder, pres. N.E. Assn. Microscopists, 1981-85. Fellow Royal Microscopical Soc.; mem. Am. Soc. Materials, Microscopy Soc. of Am. Avocations: music, dance, swimming, sailing.

FOSTER, EDWARD JOHN, engineering physicist; b. N.Y.C., Aug. 10, 1938; s. John Paul and Mildred Julia (Hassiak) F.; m. Sandra Thornton Christie (div. 1989); children: Sandra Foster Swindler, Mary Elizabeth Foster. BS in Physics cum laude, Fordham U., 1959; MS in Physics, Syracuse (N.Y.) U., 1965; MBA, Iona U., 1973. Mgr. magnetics dept. Shephard Industries, Inc., Nutley, N.J., 1960-61; founder, CEO S.E.D. Memories, Inc., Rutherford, N.J., 1961-63; br. mgr. rsch. CBS Labs., Stamford, Conn., 1963-73; v.p. tech. ByWord Corp., Armonk, N.Y., 1973-76; pres. Diversified Sci. Labs., Redding, Conn., 1976—; cons. Electronics Industries Assn., Washington; del. U.S. Internat. Electrotech. Com., Geneva, Switzerland, 1982—. Author: Effects and Degrees of Error of Modulation-Demodulation, 1965; contbg. editor: Acquisition Reduction and Analysis of Acoustical Data, 1974; contbr. articles to profl. jours. Woodrow Wilson fellow, 1959, fellowship NSF, 1959-60. Fellow Audio Engring. Soc.; mem. IEEE, Sigma Xi, Delta Mu Delta. Achievements include patents for Automatic Recording Level Control, Directional Microphone Arrays. Home: 16 Drummer Ln West Redding CT 06896

FOSTER, EDWARD JOSEPH, structural engineer; b. Elmhurst, Ill., Oct. 9, 1967; s. Gerald Joseph and Nancy Ann (Banducci) F.; m. Rebecca Lyn Pfeifer, July 6, 1991. BS, Marquette U., 1990. Civil engr. Harza Engring. Co., Chgo., 1990-92; structural engr. Safway Steel Products, Milw., 1992—. Mem. ASCE, Chi Epsilon, Tau Beta Pi.

FOSTER, EDWARD PAUL (TED FOSTER), process industries executive; b. Pawtucket, R.I., Aug. 23, 1945; s. Edward Francis and Vivian Adrienne (Davagne) F.; m. Barbara Philomena Cook, Dec. 17, 1965 (div. Apr. 1978); children: Edward Robert, Gwendolyn Lucy; m. Johanna Helena Klaassen, June, 1985 (div. 1988); 2 children. BSChemE with distinction, U. R.I., 1967; MSChemE, Worcester Poly. Inst., 1970; MBA, Lehigh U., 1981. Mfg. melting engr. Corning Glass Works, Central Falls, R.I., 1966-67; group leader rsch. and devel. The Babcock & Wilcox Co., Alliance, Ohio, 1968-71; mgr. tampella process The Babcock & Wilcox Co., Barberton, Ohio, 1972-74; from commercial devel. engr. to dir. commercial devel. in gases, metallurgy, coal, chems. and polymers, and environ. areas Air Products and Chem., Inc., Allentown, Pa., 1974—; cons. U.S. Army Natick (Mass.) Lab., 1966-67. Contbr. articles to profl. jours.; patentee in field. Chmn. fin. Unitarian Ch., Bethlehem, Pa., 1985, chmn. social, 1983-84. NDEA fellow U.S. Dept. Health, Edn. and Welfare, 1967-69; ROTC scholar U.S. Army, 1965, Nat. Merit scholar, 1963. Mem. Commercial Devel. Assn., Am. Inst. Chem. Engrs., Am. Chem. Soc., Phi Kappa Phi, Tau Beta Pi. Avocations: tennis, downhill and cross-country skiing, biking, racquetball. Home: 6023 Fairway Ln Allentown PA 18106-9610 Office: Air Products and Chems 7201 Hamilton Blvd Allentown PA 18195-1501

FOSTER, EUGENE LEWIS, engineering executive; b. Clinton, Mass., Oct. 9, 1922; s. George Frank and Georgie Nina (Lewis) F.; m. Mavis Estelle Howard, July 30, 1944; children—Kaye Louise, Eugene Howard, Mark Edward, Carol Anne. B.S.M.E., U. N.H., 1944, M.S., 1951; Mech. E., M.I.T., 1953, Sc.D., 1954. Research engr. Procter and Gamble, Cin., 1946-47; instr. U. N.H., 1947-49; asst. prof. mech. engring. M.I.T., Cambridge, 1950-56; pres., chmn. Foster-Miller Assocs., Inc., Waltham, Mass., 1956-72; cons. Office of Sec. of Transp., Washington, 1972-73; chmn. Foster-Miller Assocs., 1974—; pres. UTD, Inc., Alexandria, Va., 1976—; mem. U.S. nat. com. for tunneling tech. Nat. Acad. Sci., 1975-79. Author: (with W.A. Wilson) Experimental Heat Power Engineering, 1956; also articles. Mem. Recreational and Environ. Com. Fairfax County, Va., 1976-78. Served with C.E. U.S. Army, 1943-46. Mem. ASME, ASCE, N.Y. Acad. Scis., AAAS, Nat. Soc. Profl. Engrs. Home: 3316 Wessynton Way Alexandria VA 22309-2229 Office: UTD Inc 8560 Cinder Bed Rd Ste 1300 Newington VA 22122-9999

FOSTER, JAMES JOSEPH, chemical engineer; b. Hammond, Ind., Jan. 18, 1957; s. Raymond Orville and Dorothy Ann (Schilling) F.; m. Susanna Marie Ams, Oct. 3, 1981; children: Karl, Peter. BS in Math., Loyola U., 1979, BS in Chemistry, 1979; MS, U. Ill., 1982, PhD, 1984. With Westvaco Corp., Covington, Va., 1984--, sr. tech. engr. Mem. Am. Inst. Chem. Engrs., Am. Chem. Soc. (high sch. award com. 1988, 89), CAnadian Pulp and Paper Assn., Tech. Assn. Pulp and Paper Industry. Achievements include 2 patents for Kappa Number Calibration Standard. Home: 72 Bath St Clifton Forge VA 24422 Office: Westvaco Bleached Bd Div 104 E Riverside Covington VA 24426-0950

FOSTER, JOHN STUART, JR., physicist, former defense industry executive; b. New Haven, Sept. 18, 1922; s. John Stuart and Flora (Curtis) F.; m. Frances Schnell, Dec. 28, 1978; children: Susan, Cathy, Bruce, Scott, John. BS, McGill U., 1948; PhD in Physics, U. Calif., Berkeley, 1952; DSc (hon.), U. Mon., 1979. Dir. Lawrence Livermore (Calif.) Lab., 1952-65; dir. def. rsch. and engring. Dept. Def., Washington, 1965-73; v.p. TRW Energy Systems Group, Redondo Beach, Calif., 1973-79; v.p. sci. and tech. TRW Inc., Cleve., 1979-88, also bd. dirs.; mem. nat. adv. bd. Am. Security Coun.; chmn. Def. Sci. Bd., 1989-93. Decorated knight Comdr.'s Cross, Badge and Star of Order of Merit (Federal Republic of Germany); comdr. Legion of Honor (France); recipient Ernst Orlando Lawrence Meml. award AEC, 1960, Disting. Pub. Svc. medal Dept. Def., 1969, 73, Enrico Fermi Award, U.S. Dept. of Energy, 1992. Mem. NAE (Founders award 1989), AIAA, Am. Def. Preparedness Assn., Nat. Security Indsl. Assn. Office: TRW Inc Bldg E1-5010 1 Space Pk Redondo Beach CA 90278

FOSTER, JOY VIA, library media specialist; b. Besoco, W.Va., Aug. 11, 1935; d. George Edward and Burgia Stafford (Earls) Via; m. Paul Harris Foster, Jr., Dec. 8, 1956 (dec. Dec. 1962); children: Elizabeth Lee, Michael Paul. BS, Radford Coll., 1971; MS, Radford U., 1979. Cert. pub. sch. libr., Va. Clk. Va. Tech. and State U., Blacksburg, 1955-57; clk. Christiansburg (Va.) Primary Sch., 1971-72, libr., 1972-85; libr. Auburn Mid. and High Sch., Riner, Va., 1985—. Meml. chmn. Am. Cancer Soc., Christiansburg, 1965-66; area chmn. Am. Heart Fund, Christiansburg, 1990-93, block worker, 1985-91. Mem. NEA, ALA, Am. Assn. Sch. Librs., Montgomery County Edn. Assn. (v.p. 1988-89, sec. 1989-91, bldg. rep. 1991—), Va. Ednl. Media Assn., Va. Ednl. Assn., Women of the Moose. Presbyterian. Avocations: reading, bowling, sailing, flea marketing, antique collecting. Office: Auburn Mid and High Sch 4163 Riner Rd Riner VA 24149

FOSTER, KENNITH EARL, life sciences educator; b. Lamesa, Tex., Jan. 20, 1945; s. John Hugh and Mamie (Hyatt) F.; children: Sherry, Kristi. BS, Tex. Tech. U., 1967; MS, U. Ariz., 1969, PhD, 1972. Prof. and dir. Office of Arid Lands Studies, U. Ariz., Tucson, 1983—. Contbr. articles to profl. jours. Grantee NASA, NSF, USDA, U.S. AID, industry, 1983—. Home: 651 Avenida Princesa Tucson AZ 85748 Office: Univ of Ariz Office of Arid Lands Studies 845 N Park Ave Tucson AZ 85719

FOSTER, KIRK ANTHONY, emergency medical service administrator, educator, consultant; b. Shreveport, La., June 4, 1959; s. William L. and Marzette P. (Stephens) F.; m. Eleanor Kelly Moore, June 28, 1984; children: Stephen, Shannon, Caitlin, Justin. BA, U. Southwestern La., 1983. Cert. paramedic, instr. ACLS. Resident supr. Acadiana Drug Elimiation Ctr., Lafayette, La., 1977-78; patrol dep. Lafayette Parish Sherrifs Dept., 1978-79; shift commdr. univ. police U. Southwestern La., Lafayette, 1979-81; flight paramedic, supr. safety offshore ops. Acadiana Ambulance Svc., Lafayette, 1980-88; instr., cons. emergency med. svc. REDCRES project U.S./Saudi Arabia Joint Commn. on Econ. Cooperation, Riyadh, Saudi Arabia, 1988-92, dir. REDCRES project, 1992—; cons. Saudi Arabian Red Crescent Soc., Riyadh, 1988-92. Mem. Am. Civil Def. Assn., Nat. Assn. Search and Rescue, Nat. Assn. Emergency Med. Tecnicians, Internat. Red Cross and Red Crescent Movement. Achievements include direction of emergency management, disaster preparation, and emergency medical services during the Iran-Iraq and Persian Gulf Wars. Home and Office: JECOR/RED-CRES Unit 61306 Box 103 APO AE 09803-1306

FOSTER, LANNY GORDON, writer, publisher; b. Harrisburg, Pa., Sept. 27, 1948; s. Gordon Eugene and Georgina Lillian (Kramer) F.; m. Carol Prescot McCoy, Nov. 29, 1975 (div. 1987); m. Denise Joy Freiman, Sept. 10, 1988. BA, Rutgers U., 1970, PhD, 1976. Postdoctoral fellow Bellevue and N.Y.U., N.Y.C., 1975-79; prodn. analyst, mgr. ops. Irving Trust Co., N.Y.C., 1979-82; med. writer Am. Home Prods., N.Y.C., 1982-83; writer various corps., N.Y.C., 1983-88; writer, owner Beta Books, Cragsmoor, N.Y., 1988—; mem. Mgmt. of HIV Disease Delegation to People's Republic of China, Citizen Ambassador Program, 1990. Author: The ABC of AIDS, 1990, The Third Epidemic, 1990; contbr. articles to Immunology Jour., 1973-79. Edn. vol. Gay Men's Health Crisis, N.Y.C., 1986-87; speaker on AIDS various radio and TV talk shows, 1987. Mem. N.Y. Acad. Scis., Am. Med. Writer's Assn., Com. Small Mag. Editors and Pubs., Internat. Platform Assn., The Harvey Soc., Writers Guild of Am. East. Avocations: drawing, music, hiking, camping. Office: Beta Books PO Box 40 Cragsmoor NY 12420-0040

FOSTER, LINDA ANN, biomaterials research scientist; b. Canal Zone, Panama, Jan. 19, 1956; d. Augustus Hunter and Catherine Minerva (Pierce) F. MS, North Tex. State U., 1985; PhD, U. N. Tex., 1991. Teaching asst. biological scis. U. N. Tex., Denton, 1982-85, electron microscope teaching fellow, 1985-91, rsch. scientist dept. Chemistry, 1991-1993; biomaterial rsch. scientist Med. Scis. Rsch. Inst., Herndon, Va., 1993—. Contbr. articles to profl. jours. Mem. Am. Soc. for Microbiology, Microscopy Soc. Am., Sigma Xi. Democrat. Seventh Day Adventist. Achievements include patent pending for embedding composition and associated method and method for tem analysis of eutectic solder copper joints. Office: Med Scis Rsch Inst 2190 Fox Mill Rd Herndon VA 22071

FOSTER, LYNN IRMA, physicist; b. Webster, Mass., Jan. 23, 1969; d. George E. and Carolyn M. (Sherman) F. BS in Physics, U. Mass., 1991. NASA cert. for Hubble Space Telescope's sci. instruments and thermal control subsystems. Satellite flight contr. Lockheed Tech. Ops. Co., GSFC, NASA, Greenbelt, Md., 1991—. Mem. Nat. Soc. Physics Students.

FOSTER, NANCY MARIE, natural resource management specialist, government official; b. Electra, Tex., Jan. 23, 1941; d. Evelyn Ann (Spurrier) F. BS, Tex. Women's U., 1963; MS, Tex. Christian U., 1965; PhD, George Washington U., 1969. Chmn. biology dept. Dunbarton Coll. for Women, Washington, 1969-73; environ. analyst U.S. Fish & Wildlife Service, Washington, 1973-74; sr. environ. analyst Dept. Interior, Washington, 1974-78; dir. U.S. Nat. Marine and Estuarine Sanctuary Programs, NOAA, Washington, 1978-87, dir. protected resources, 1987—; speaker to profl. and cultural groups on govt. involvement in hist. preservation of nationally significant maritime cultural resources, 1981—. Contbr. articles to profl. jours. Vol. programs for elderly, 1983—. Recipient Sustained Superior Performance award NOAA, 1983, Outstanding Performance award, 1979-85. Mem. AAAS, Womens Aquatic Network (founding trustee), Caribbean Island Directorate (UNESCO), Biosphere Res. Directorate (UNESCO). Avocations: ceramics, aerobics, English history. Office: Nat Marine Fisheries Svc 1335 East-West Hwy Silver Spring MD 20910

FOSTER, NORMAN HOLLAND, geologist; b. Iowa City, Oct. 2, 1934; s. Holland and Dora Lucinda (Ransom) F.; BA, U. Iowa, 1957, MS, 1960; PhD, U. Kans., 1963; m. Janet Lee Grecian, Mar. 25, 1956; children—Kimberly Ann, Stephen Norman. Instr. geology U. Iowa, 1958-60, U. Kans., 1960-62; sr. geologist, geol. specialist, exploration team supr. Sinclair Oil Corp. and Atlantic Richfield Co., Casper, Wyo. also Denver, 1962-69; dist. geologist Trend Exploration Ltd., 1969-72, v.p., 1972-74; v.p. exploration geology Filon Exploration, Denver, 1974-79, dir., 1977-79; ind. geologist, Denver, 1979—; guest lectr. geology Colo. Sch. Mines, 1972—, U. Colo., 1972—, U. Iowa, 1975—, U. Kans., 1975—; adv. bd. geology U. Kans., 1982—, U. Iowa, 1988—, U. Colo., 1990—; chmn. fin. com. 28th Internat. Geol. Congress, 1987-89, ofcl. U.S. rep., Washington, 1989. Served to capt. inf. AUS, 1957. Recipient Haworth Disting. Alumni award dept. geology U. Kans., 1977, Disting. Alumni award dept. geology U. Iowa, 1992. Fellow Geol. Soc. Am.; mem. Am. Assn. Petroleum Geologists (del. 1972-75, 79-82, disting. lectr. 1976-77, pres. Rocky Mountain sect. 1979-80, treas. 1982-84, adv. council 1985-88, chmn. astrogeology com. 1984-88, nat. pres. 1988-89, found. trustee 1979—, hon. mem. 1993—, Levorsen award 1980, Disting. Service award 1985), Rocky Mountain Assn. Geologists (sec. 1970, 1st v.p. 1974, pres. 1977, best paper award 1975, Explorer of Yr. award 1980, Disting. Service award 1981, hon. mem. 1983—), Soc. Econ. Paleontologists and Mineralogists, Am. Inst. Profl. Geologists, Soc. Ind. Profl. Earth Scientists, Soc. Exploration Geophysicists, Soc. Petroleum Engrs., Nat. Acad. Scis. (bd. earth scis. and resources, U.S. nat. com. geology, com. adv. to U.S. geol. survey 1989-92), Colo. Sci. Soc., Sigma Xi, Sigma Gamma Epsilon. Republican. Mem. Christian Ch. Assoc. editor Guidebook to Geology and Energy Resources of Piceance Basin, Colorado, 1974, Mountain Geologist, 1967-68, 71-85, editor, 1968-70; co-editor, compiler Treatise of Petroleum Geology, 1984—; contbr. papers on geology to profl. publs.

FOSTER, OTTIS CHARLES, civil engineer; b. San Antonio, Sept. 26, 1959; s. Raymond Edgar and Maria Elsa (Senior) F.; m. Heidi Lynne Burke, July 9, 1988; children: Jenna, Andrea. BCE, U. Tex., El Paso, 1982; MS in Urban Affairs, U. Tex., Arlington, 1989. Registered profl. engr., Tex. Geotech. engr. Trinity Engring. Testing Corp., Waco, Tex., 1983-87, 90-92; pres., owner Tejas Soils Engring. Co., Waco, 1992—. Active high speed rail task force City of Waco, 1992; chair mcpl. affairs Sanger Heights Neighborhood Assn., Waco, 1991—; coord. Adopt-A-Sch. Austin Ave. United Meth. Ch., Waco, 1992-93. Mem. ASCE (local v.p. 1992—, vice chmn. bldg. standard com. 1991—), Tex. Soc. Profl. Engrs., NSPE, Chi Epsilon, Tau Beta Pi. Office: Tejas Soils Engring Co 2204 Bosque Waco TX 76707

FOSTER, ROBERT EDWIN, energy efficiency engineer, consultant; b. Bloomington, Ind., Apr. 25, 1962; s. John Robert and Joanne Barbara (Ploski) F. BS in Mech. Engring., U. Tex., 1984. Solar technician Cole Solar Systems, Inc., Austin, Tex., 1984; project engr. S.W. Tech. Develop. Inst. N.Mex. State U., Las Cruces, 1989—; co-chmn. S.W. Border States Solar Conf., El Paso, 1991; coord. workshops on water pumping with solar and wind energy Sandia Nat. Labs., Mexico City and Guatemala City, 1992. Author conf. articles. Vol. appropriate tech. U.S. Peace Corps, Neyba, Dominican Republic, 1985-88. Acad. specialist grantee U.S. Info. Agy. (Chile), 1992. Mem. ASHRAE (bd. dirs. El Paso chpt. 1990—), Am. Solar Energy Soc. (assoc.), Nat. Speleological Soc. (Grotto chmn. 1990—), El Paso Solar Energy Assn. (pres. 1990-92), Evaporative Cooling Inst. Inc. (treas. 1990—). Episcopalian. Achievements include research on evaporative coolers and water consumption characteristics, Latin Am. solar and wind energy projects. Office: NMex State U SW Tech Devel Inst PO Box 30001 Dept 3 Solar Las Cruces NM 88003-0001

FOSTER, ROGER SHERMAN, JR., surgeon, educator, health facility administrator; b. Washington, Jan. 8, 1936; s. Roger Sherman and Genevieve Wakeman (Bartlett) F.; m. Joan Crile, June 25, 1960; children: Roger Sherman III, Charles Bartlett, Elizabeth Crile, Halle Crile. AB, Haverford Coll., 1957; MD, Case Western Res. U., 1961. Diplomate Am. Bd. Surgey, Nat. Bd. Med. Examiners; lic. Vt., Ga. Intern then resident in surgery Univ. Hosps., Cleve., 1961-66; research fellow Roswell Park Meml. Inst., Buffalo, 1966-68; asst. prof. surgery U. Vt., Burlington, 1970-73, assoc. prof. surgery, 1973-80, prof. surgery, 1980-92, dir. comprehensive cancer ctr., 1984-92; attending surgeon Med. Ctr. Hosp. of Vt., 1970-92; Wadley Glenn prof. surgery Emory U., Atlanta, 1992—; chief surgical svcs. Crawford Long Hosp. of Emory U., 1992—; mem. cancer clin. investigation rev. com. NIH, 1987-92, chmn., 1991-92, chmn. various coms.; cons. Am. Internat. Health Alliance for Tblisi, Georgia Hosp., 1992—. Assoc. editor: Clinical Surgery, 1987; co-editor: Essentials of Clinical Surgery, 1991; editor-in-chief: Breast Surgery: Index and Reviews, 1991; manuscript reviewer Jour. Am. Med. Assn., Jour. Trauma, and others; contbr. over 100 articles to profl. jours. Trustee Univ. Health Ctr., Burlington, 1986-89. Served to maj. U.S. Army, 1968-69. Grantee NIH, 1971-92; summer rsch. fellow Josiah Macy Jr, Found., 1958-59. Fellow Am. Surg. Assn., ACS (bd. regents 1991—, bd. govs. 1981-87, adv. coun. for gen. surgery 1989-92, sec./treas. Vt. chpt. 1979-80, v.p. 1980-81, pres. 1981-82); mem. AMA, AAAS, New Eng. Surg. Soc. (treas. 1986-89, exec. com. 1981-92), Soc. Univ. Surgeons, Soc. Surg. Oncology, Internat. Soc. Preventive Oncology, Ea. Surg. Soc., Am. Endocrine Surg. Soc. (coun. 1992—), Am. Soc. Clin. Oncology (pub. rels 1989-91 and pub. issues coms. 1989—), Am. Assn. Cancer Rsch., Transplantation Soc., New Eng. Cancer Soc. (treas. 1983-87, v.p. 1988-89, pres. 1989-90), Assn. Acad. Surgery, Nat. Surg. Adjuvant Breast Project (exec. com. 1978-81), Vt. State Med. Soc., Newfoundland Club Am. (bd. dirs. 1976-77, first v.p. 1979). Club: Newfoundland of Am. (trustee 1976-77, 1st v.p. 1979). Avocations: white water canoeing, breeding Newfoundland dogs, wilderness travel, chamber music. Home: 1750 Winterthur Close Atlanta GA 30318-1750 Office: Crawford Long Hosp Emory U 550 Peachtree St N E Atlanta GA 30365-2225

FOSTER, STEPHEN ROCH, civil engineer; b. New Rochelle, N.Y., Aug. 16, 1964; s. Stephen Peter and Charlynn Ann (Williams) F. AS in Bus. Acctg., Westchester C.C., Valhalla, N.Y., 1989; BE in Civil Engring., Stevens Inst. Tech., Hoboken, N.J., 1992. Commd. ensign U.S. Navy, 1992; asst. resident office in charge of constrn. C.E. Corps, U.S. Navy, Pearl Harbor, Hawaii, 1992—. Mem. ASCE (assoc.; Robert Ridgeway award 1992), Soc Am. Mil. Engrs., Tau Beta Pi. Republican. Mormon. Office: ROICC Pearl 4262 Radford Dr Honolulu HI 96818

FOTHERGILL, JOHN WESLEY, JR., systems engineering and design company executive; b. San Francisco, June 24, 1928; s. John Wesley and Madeline Mary (Frates) F.; m. Nancy F. Bacon, 1990; children from previous marriage: Nancy Gay, Wesley Daykin, Michael James, Karen Renee. BS, Calif. State U., Sacramento, 1955, postgrad., 1956-59; postgrad., Utah State U., 1960-63, SUNY, 1964-67; D of Environ. Engring. (hon.), World U., 1985. Mem. staff theoretical div. U. Calif. Radiation Lab., Livermore, 1955-56; mem. staff computer div. Aerojet Gen. Corp., Sacramento, Calif., 1957-59; computer div. Thiokol Chem. Corp., Brigham City, Utah, 1959-64; staff engr. Link group Gen. Precision Systems, Inc., Binghamton, N.Y., 1964-67; mgr. sci. dept., link Info. Sci. div. Gen. Precision Systems, Inc., Silver Spring, Md., 1967-69; gen. mgr. Washington div. Singer Info. Services Corp., Bethesda, Md., 1969-72; pres., chief scientist, chmn. bd. Integrated Systems, Inc., Brunswick, Md., 1972-86, Analytic Adv. Group, Inc., McLean, Va., 1980-86; project engr. firefighting tng. system Link Sim System div. CAE Corp., Silver Spring, Md., 1986-90, Contraves USA-SSI, Tampa, Fla., 1990-91; cons. engr. smoke control tech, firefighting tng. systems, facilities tech. and equipment, 1992—. Contbr. articles to profl. jours. Served with U.S. Army, 1945-46, with USAF Res., 1947-53, USNR, 1955-73. Mem. Smoke Control Assn., Nat. Fire Protection Assn., AAAS, N.Y. Acad. Scis., Naval Inst., ASHRAE. Achievements include patent pending for firefighting training system and components and ammunition processing device. Home and Office: 203 W Laurel Ave Sterling VA 20164

FOUASSIER, JEAN-PIERRE, chemist, educator; b. Chateau-Gontier, Mayenne, France, Mar. 5, 1947; s. Amedee Eugene and Daniele Marthe (Verneau) F.; m. Genevieve Therese Guillain, Dec. 28, 1970; children: Patrick, Laurence, Yann. PhD, U. Strasbourg, 1975. Prof. chemistry U. Haute Alsace, Mulhouse, France, 1978—; head Laboratoire de Photochimie Generale-Ura CNRS, Ecole Nationale Superieure de Chimie Université de Haute Alsace, Mulhouse, France, 1980—; mem. organizing com. of sci. internat. meetings; lectr. in field. Editor: Lasers in Polymer Science and Technology: Applications, 1990, Radiation Curing in Polymer Science and Technology, 1993; contbr. over 160 sci. papers to profl. jours. Recipient numerous rsch. grants from indsl. cos., univs. Mem. Am. Chem. Soc. (Polymer div.), European Photochemistry Assn., Societe Française de Chimie, Radtech Europe. Home: 11 rue des Campanules, Morschwiller Le Bas France 68790 Office: Lab de Photochimie Generale, 3 rue Alfred Werner, Mulhouse France 68093

FOUGHT, LORIANNE, chemist; b. Upper Darby, Pa., Oct. 5, 1962; d. Edwin Howard and Jeanette Marie Matthews; m. Daniel Lynn Fought, Jan. 22, 1990; children: Bethannie, Angelique, Daniel. BS, Pa. State U., 1985; MS, U. Ky., 1988, PhD, 1992. Rsch. aid Pa. State U., University Park, 1982-85; grad. rsch. asst. U. Ky., Lexington, 1985-91; chemist II Miles, Inc./Miles Rsch. Park, Stilwell, Kans., 1991—. Contbr. articles to profl. jours. Tchr. So. Hills United Meth. Ch., Lexington, 1989-90. Recipient dept. fellowship Dept. Plant Pathology, Univ. Ky., Lexington, 1985-91. Mem. Am. Phytopathol. Soc. (sec. grad. student com. 1990-91), AAAS, N.Y. Acad. Sci., Gamma Sigma Delta. Republican. Achievements include research in metabolism of xenobiotics in plants and animals, compound isolated from cucumber tissues which induces sytemic resistance to disease in cucumbers. Investigating metabolism of agricultural chemicals in plants. Office: Miles Inc/Miles Rsch Park 17745 Metcalf Ave Stilwell KS 66085-9104

FOUGHT, SHERYL KRISTINE, environmental scientist, engineer; b. Washington, Mo., Oct. 17, 1949; d. James Paul and Alice Marie (Kasper) McSpadden; m. Randy Bruce Stucki, Nov. 23, 1968 (div. 1974); children: Randy Bruce, Sherylynne Sue; m. Larry Donald Fought, July 31, 1980 (div. 1982); 1 child, Erin Marie. BS, N.Mex. State U., 1976, postgrad., 1977-79. Tchr. N.Mex. State U., Las Cruces, 1977-78; hydrologist U.S. Dept. Interior, Las Cruces, 1978-81; environ. scientist U.S. EPA, Dallas, 1981-84; hazardous waste inspector Ariz. Dept. Health Svc., Phoenix, 1984-85; environ. engr., technician Yuma Proving Ground U.S. Army, 1985-87, chief phys. scientist environment div. Yuma Proving Gound, 1987-88, chief hazardous waste mgmt br. Aberdeen Proving Ground, 1988—. Co-author: The Ghost Town Marcia, 1975, tng. manuals. With USMC, 1968-69. Recipient 2 Environ. Quality awards U.S. Army, 1986, Army Materiel Command, 1986. Mem. NAFE, Nat. Environ. Tng. Assn., Federally Employed Women, Fed. Women Ewngrs. and Scientists, Wildlife Soc., Air Pollution Control Assn., Dept. Def. Excellent Installations, Internat. Platform Assn. Democrat. Office: Aberdeen Proving Ground STEAP-SA-DSHE-E Aberdeen MD 21001

FOULKES, WILLIAM DAVID, psychologist, educator; b. East Orange, N.J., May 29, 1935; s. Paul Bergen and Alice (Hinson) F.; m. Nancy Helen Kerr, Apr. 19, 1978. BA, Swarthmore Coll., 1957; PhD, U. Chgo., 1960; MD (hon.), U. Ferrara, 1992. Instr. Lawrence Coll., Appleton, Wis., 1960-63; rsch. assoc. U. Chgo., 1963-64; from asst. prof. to prof. U. Wyo., Laramie, 1964-77; prof. psychiatry Emory U., Atlanta, 1977—. Author: The Psychology of Sleep, 1966, A Grammar of Dreams, 1978, Children's Dreams, 1982, Dreaming: A Cognitive Psychological Analysis, 1985; co-editor: Dreaming as Cognition, 1993. Fellow Ctr. for Advanced Study in Behavioral Scis., 1974-75; recipient Disting. Scientist award Sleep Rsch. Soc. Mem. Southeastern Psychol. Assn., Midwestern Psychol. Assn., Internat. Soc. for Study of Behavioral Devel., European Sleep Rsch. Soc.

FOUNTAIN, WILLIAM DAVID, laser/optics engineer; b. Moscow, Idaho, Aug. 21, 1940; s. Lloyd James and Gail E. (Moore) F.; m. Audrey Joan Delamater, Sept. 4, 1964; children: Jeffrey Scott, John Christopher, Joseph Chandler, Julian Andrew. BA in Physics, Reed Coll., 1961; MS in Physics, U. Pa., 1963. Sr. rsch. engr. Martin-Marietta Corp., Orlando, Fla., 1963-68; advanced R&D engr. GTE Sylvania, Mountain View, Calif., 1968-73; physicist Lawrence Livermore (Calif.) Nat. Lab., 1973-77; staff engr. Cooper LaserSonics Inc., Santa Clara, Calif., 1977-85; optics and solid state systems mgr. XMR, Inc., Santa Clara, 1985-90; optical engring. mgr. Phoenix Laser Systems, Inc., Fremont, Calif., 1990-92; sr. project engr. Med. Optics, Inc., Carlsbad, Calif., 1993—; cons. WDF Cons., Fremont, 1990—. Contbr. articles to profl. jours. Nat. Merit scholar, 1957-61. Mem. IEEE, Optical Soc. Am., SPIE-The Internat. Soc. for Optical Engring. Achievements include 4 patents and 4 patents pending in field. Office: Med Optics 2752 Loker Ave W Carlsbad CA 92008-6603

FOURNET, GERARD LUCIEN, physics educator; b. Paris, May 2, 1923; s. Paul Yves and Germaine (Tournaud) F.; m. Noele Valentine Tournier, Mar. 13, 1947 (dec. Mar. 1981); children: Jerome Denis, Christian François. Physicist Engr., Ecole de Physique et Chimie, Paris, 1944; PhD, Faculte des Scis., Paris, 1950. Engr. Ferisol, Paris, 1944-46; rsch. engr. Onera, Paris, 1946-52; chief engr. SACM, Paris, 1952-61; asst. prof. Ecole de Physique et Chimie, Paris, 1950-59, prof., 1959-81; prof. Ecole Superieure des Telecommunications, Paris, 1958-62, U. Pierre et Marie Curie, Paris, 1961—, Ecole Superieure d'Electricite, Gif-Sur-Yvette, France, 1961-91. Author: Small Angle Scattering of X-Rays, 1955, Physique Electronique des Solides, 1962, Solid State Electronics, 1968, Electronique, 1970, Electromagnetisme a partir des equations locales, 1979, 84; contbr. over 120 articles to profl. jours. Recipient Prix Ancel award Soc. Française de Physique, 1960, Medaille Blondel, 1967, Prix, Acad. des Scis., 1974. Mem. Soc. Française de Physique, Soc. Française des Electriciens, Am. Phys. Soc., Ctr. Nat. de la Recherche Scientifique. Avocation: walking. Home: 41 Ave Le Notre, 92420 Vaucresson France Office: Lab de Genie Electrique, ESE Plateau du Moulon, 91192 Gif sur Yvette France

FOURNIER, DONALD JOSEPH, JR., mechanical engineer; b. Norwich, Conn., July 27, 1962; s. Donald Joseph Sr. and Juanita L. (Malone) F.; m. Deborah D. Miller, Aug. 5, 1983; children: Catherine, Jacqueline, Evan. BSME, U. Fla., 1986, MSME, 1988. Rsch. engr. Combustion Lab. U. Fla., Gainesville, 1985-88; mech. engr. Envireco, Gainesville, 1987-88; project engr. Acurex Environ., Jefferson, Ark., 1988-93; rsch. engr. and cons. Gould, Lewis & Proctor, Gainesville, 1992—. Contbr. articles to profl. jours. U. Fla. grad. scholar's fund fellow, 1986. Mem. ASTM (subcom. sec. 1987-91, chmn. task group 1987-91), ASME, SAE, Air and Waste Mgmt. Assn., Tau Beta Pi, Pi Tau Sigma. Achievements include identification of metal behavior in hazardous waste incineration. Office: Gould Lewis & Proctor 6712 NW 18th Dr Gainesville FL 32606

FOURNIER, JEAN PIERRE, architect, real estate developer; b. Angers, Maine et Loire, France, Mar. 6, 1941; s. Fernand and Marie Anne (Chuche) F.; m. Marie Chantal Bazin, Oct. 1, 1966; children: Arnaud, Laurent. Baccalaureat, Lycee David D'Angers, 1961; diploma, Ecole Nationale des Beaux Arts de Paris, 1961, 68; Logiste du Prix de Rome (hon.), 1967. Architect B. Zehrfuss, Paris, 1968, Skidmore Owings & Merrill, N.Y.C., 1969-71; mng. ptnr. Chapus Claudon Herbez Fournier Semmel, Paris and Nice, 1971-84, Chapus Fournier et Associes, Paris and Nice, 1984—; mng. ptnr. B.E.R.B.A., Paris and Nice, 1978—, Realisations Argane, Paris and Nice, 1984—, Sarl Land, Les Trois Ilets, Martinique, 1988—. Author: (with B. Zehrfuss) Musee Archeologique de Lyon, (with F. Chapus) Direction Departementale de L'Equipement du lot et Garonne, A Paris Nombreaux Sieges Sociaux de Societes, 6000 Logements, Hotels et Bureaux an Afrique et au Moyen Orient. Recipient Franklin Delano Aldrich award, 1968. Mem. Cercle de la Voile D'Angers, Union Nat. de la Course Au Large, Am. C. of C. Avocations: voile, golf. Office: Chapus Fournier et Associes, 54 Bis Rue Michel Ange, 75016 Paris France also: Le Quadra, 455 Promenade des Anglais, 06200 Nice France

FOUTCH, GARY LYNN, chemical engineering educator; b. Poplar Bluff, Mo., Aug. 26, 1954; s. Cecil Foutch and Edith Frances Gardner Wood; m. Pamela Lynn Smith, May 28, 1977; children: Aaron Lloyd, Brendan Lee, Keely Anne. BS, U. Mo., Rolla, 1975, MS, 1977, PhD, 1980. Asst. prof. chem. engring. Okla. State U., Stillwater, 1980-85, assoc. prof., 1985-89, prof. chem. engring., 1989—; NASA-ASEE fellow Jet Propulsion Lab., Pasadena, Calif., summers 1981, 82; engr. Dow Chem. Co., Freeport, Tex., summer 1983; vis. scientist Smith Kline and French, Phila., summer 1985, Phillips Petroleum, Bartlesville, Okla., summer 1987, 92; vice chair faculty coun. Okla. State U., 1992—; lectr. in field. Contbr. articles to profl. jours. Witness Okla. State Senate Subcom., Oklahoma City, 1985. Halliburton Young faculty, Coll. Engring., Arch. & Tech., 1985, 90; named Outstanding Young Engr. Okla. Soc. Profl. Engrs., 1987; Fulbright scholar Loughborough U. of Tech., Eng., 1990-91. Mem. NSPE, Am. Chem. Soc., Am. Inst. Chem. Engrs. Achievements include 1 patent. Office: Okla State Univ Chemical Engring Dept 423EN Stillwater OK 74078

FOUTS, JAMES FREMONT, mining company executive; b. Port Arthur, Tex., June 3, 1918; s. Horace Arthur and Willie E. (Edwards) F.; m. Elizabeth Hanna Browne, June 19, 1948; children: Elizabeth, Donovan, Alan, James. B Chem Engring., Tex. A&M U., 1940. Div. supt. Baroid div. N.L. Industries, U.S. Rocky Mountain area and Can., 1948-60; pres. Riley-Utah Co., Salt Lake City, 1960-67, Fremont Corp., Monroe, La., 1967—, Auric Metals Corp., Salt Lake City, 1972—; bd. dirs. La Fonda Hotel, Santa Fe, N.Mex., High Plains Natural Gas Co., Canadian, Tex. Hon. asst. sec. of State of La. Served to lt. col. arty U.S. Army, 1942-46. Mem. Wyo. Geol. Assn. (v.p. 1958), Rocky Mountain Oil & Gas Assn. (bd. dirs. 1959), Res. Officers Assn. Wyo. (pres. 1948), Am. Assn. Petroleum Geologists, Internat. Geol. Assn., Mont. Geol. Assn., Ind. Petroleum Producers Assn. Republican. Episcopalian. Club: Univ. Lodge: Elks. Home: 4002 Bon Aire Dr Monroe LA 71203-3015 Office: Fremont Corp PO Box 7070 Monroe LA 71211-7070 also: Auric Metals Corp 443 E Adaley Ave Salt Lake City UT 84107

FOUTS, JAMES RALPH, pharmacologist, educator, clergyman; b. Macomb, Ill., Aug. 8, 1929; s. Ralph Butler and Mary May (Lingenfelter) F.; m. Joan Laverne Van Dyke, June 20, 1964; children: Mary, Jeffrey, Carolyn. BSc with highest honors, Northwestern U., 1951, PhD, 1954; MDiv summa cum laude, Duke U., 1984. Ordained priest Episcopal Ch., 1986. Instr. biochemistry Northwestern U. Med. Sch., Chgo., 1952-54; postdoctoral fellow Lab. Chem. Pharmacology Nat. Heart Inst., NIH, Bethesda, Md., 1954-56; sr. rsch. biochemist Wellcome Rsch. Labs., Burroughs Wellsome & Co., Tuckahoe, N.Y., 1956-57; from asst. prof. to prof. dept. pharmacology Coll. Medicine, U. Iowa, Iowa City, 1957-70, dir. Toxicology Ctr., 1968-70; Claude Bernard prof. Inst. Medicine, U. Montreal, Que., Can., 1970; chief pharmacology and toxicology br. Nat. Inst. Environ. Health Scis., Research Triangle Park, N.C., 1970-76, sci. dir., 1976-78, chief lab. pharmacology, 1978-82, sr. exec. svc., 1979—; sr. scientific adv. to dir. sr. sci. advisor to dir., Research Triangle Park, N.C., 1986—; cons. pharm. cos.; adj. prof. pharmacology Sch. Medicine, U. N.C., 1970—; adj. prof. toxicology N.C. State U., 1971—; cons. coms. NIH, FDA, NIDA, EPA, Nat. Inst. Occupational Safety and Health; cons. com. dept. commerce NOAA; chmn. Gordon Rsch. Conf. on Drug Metabolism, 1977-78; vis. prof. Swiss Fed. Inst. Tech. and U. Zurich, Switzerland, 1978-79. Editor Chemico-Biological Interactions, 1973-76, Ann. Report on Carcinogens, 1986—; assoc. editor Pharmacological Revs., 1977-88; mem. editorial bd. 13 sci. publs.; contbr. over 200 articles to profl. jours. Mem. coms. on AIDS, environ. and stewardship, secretariat of Cursillo Episcopal Diocese of N.C.; asst. to Episcopal chaplain N.C. State U., 1985-86; asst. to rector Ch. of the Holy Family, Chapel Hill, 1986-88; priest assoc. Chapel of the Cross, Chapel Hill, 1988—. Lt. USPHS, 1954-56. Recipient Marple-Schweitzer award Northwestern U., 1950, Superior Achievement award NIH, 1975, Spl. Teaching award U. N.C. Sch. Nursing, 1978, Spl. Achievement award NIH, 1987, Dirs. award, 1990, Spl. Achievement award Sr. Execs. Svc., HHS, 1990, 92. Mem. Am. Soc. Pharmacology and Exptl. Therapeutics (chmn. drug metabolism divsn. 1975-78, councilor 1980-81, Abel award 1964), Am. Assn. for Cancer Rsch., Soc. Toxicology, Mt. Desert Island Biol. Lab., Phi Beta Kappa, Sigma Xi, Phi Lambda Upsilon. Democrat. Episcopalian. Home: 212 Ridge Trl Chapel Hill NC 27516-1641 Office: Nat Inst Environ Health Sci MD WC-05 PO Box 12233 Durham NC 27709-2233

FOWLER, CECILE ANN, nurse, professional soloist; b. Paterson, N.J., Feb. 14, 1920; m. Chester A. Fowler, Mar. 9, 1942. Grad., Passaic (N.J.) Gen. Hosp. Nursing Program, 1941. Nurse Beth Israel Hosp., Newark, 1941-42, Orange (N.J.) Meml. Hosp., 1942-50; asst. receptionist Dr. Stokes, Urologist, East Orange, N.J., 1943-44; nurse Mountainside Hosp., Montclair, N.J., 1960-69, head nurse, premature and newborns, 1966-67; profl. soloist, 1952-69; part-time nurse Upper Three Hosps., 1950-60; co-founder The Oratorio Soc. of N.J., Montclair, 1952; mem. quartet First Baptist Ch., Montclair. Active various coms. PTA, 1951-62; co-counder CD, Little Falls, N.J., 1967; sponsor Met. Opera Guild N.Y., 1977—; child sponsor World Vision, 1983—; mem. Rep. Presdl. Task Force, 1987; founder Challenger Ctr. for Math., Space and Sci. Edn., 1990—, Ptnrs. in Hope: St. Jude's Rsch. Ctr., 1991—. Recipient Vocal Accomplishment award Griffith Music Found., 1944, 45, medal of Merit, Pres. Reagan, 1988, Pres. Bush, 1990, Rep. Presdl. Legion of Merit, 1992. Mem. Lincoln Ctr. for the Performing Arts, Friends of Carnegie Hall, Am. Biog. Inst. Am. (rsch. bd. advs. 1989—, dep. gov., life mem., fellowship, Commemorative medal of Honor 1991),

Heritage Found. (U.S. English mem. 1986—), U.S. Senatorial Club (preferred mem. 1988—), Little Falls Woman's Club (edn. chmn.), Montclair (gov. 1979-81), Montclair Operetta (various chmnships 1943—, gov. 1990—). Republican. Roman Catholic. Avocations: painting, needlepoint, knitting, reading. Home: 9 Lotz Hill Rd Clifton NJ 07013-2312

FOWLER, DAVID WAYNE, architectural engineering educator; b. Sabinal, Tex., Apr. 25, 1937; s. Otis Lindley and Sadie Gertrude (Cox) F.; m. Maxine Yvonne Thomson, Mar. 31, 1961; children: Teresa, Leah. BS in Archtl. Engring., U. Tex., Austin, 1960, MS, 1962; PhD in Civil Engring., U. Colo., 1965. Design engr. W.C. Cotten (Cons. Engr.), Austin, Tex., 1961-62; asst. prof. archtl. engring. U. Tex., Austin, 1964-69; asso. prof. U. Tex., 1969-75, prof., 1975—, Taylor prof., 1981—, dir. Ctr. Aggregates Rsch.; vis. prof. Nihon U., Japan, 1981; bd. dirs. Univ. Fed. Credit Union, 1976-84; pres. Internat. Congress on Polymers in Concrete, 1981-87; bd. dirs. Ctr. Aggregates Rsch. Editor procs. 2d Internat. Congress on Polymers in Concrete, 1978; contbr. articles to profl. jours. Ford Found. faculty devel. grantee, 1962-64; recipient Teaching award Gen. Dynamics, 1975, Teaching award Amoco Found., 1978, Disting. Engring. Alumnus award U. Colo., 1993; cited by Engring.-News Record, 1975. Fellow ASCE (pres. Austin br. 1976-77), Am. Concrete Inst. (Delmar L. Bloem award 1985, bd. dirs. 1993—); mem. Am. Soc. Engring. Edn. (chmn. archtl. engring. div. 1971-72), Tex. Soc. Profl. Engrs. (bd. dirs. Travis chpt. 1968), Russian Acad. Engring. (hon.), Tau Beta Pi, Chi Epsilon. Mem. Ch. of Christ. Home: 612 Brookhaven Trl Austin TX 78746-5455 Office: Univ of Tex Archtl Engring Group ECJ 5 208 Austin TX 78712

FOWLER, FLOYD EARL, national security consultant; b. Florence, Ala., June 20, 1937; s. Charles Owen and Tina Louise (Phillips) F.; m. Betty Inez Mathias, Feb. 14, 1959; children: Deborah Faye, Keith Douglas. BA, U. Md., 1978. Staff mem. (fellow) U.S. Ho. of Reps., Washington, 1982; sr. exec. Nat. Security Agy., Ft. Meade, Md., 1960-91; nat. security cons. Severn, Md., 1991—. Contbr. articles to profl. jours. Contbr. SpaceCause, Washington, 1992. With USAF, 1956-60. LEGIS fellow U.S. Office of Pers. Mgmt., 1982. Mem. AIAA, Armed Forces Communications Assn., OPSEC Profl. Soc., Security Affairs Support Assn. Ch. of Christ. Home and Office: Fowler Rsch 8254 Riviera Dr Severn MD 21144

FOWLER, GEORGE SELTON, JR., architect; b. Chgo., Jan. 20, 1920; s. George Selton and Mabel Helena (Overton) F.; m. Yvonne Fern Grammer, Nov. 25, 1945; 1 child, Kim Ellyn. Cert. Hamilton Coll., 1944; B.S., Ill. Inst. Tech., 1949, postgrad. city and regional planning, 1968; cert. Elec. Assn. Ill., 1976. Registered architect, Ill., Ohio. Co-founder, pres. The Modern Arts Press, Chgo., 1946; instr. archtl. and related engring. subjects Am. Sch. and Tech. Soc., Chgo., 1948-65; urban planner Chgo. Land Clearance Commn., 1949-50; liaison architect Chgo. Housing Authority, 1950-68, chief design-tech. div., 1968-80, dir. dept. engring., 1980-84; prin. George S. Fowler, Architect, Chgo., 1984—; treas., bd. dirs. Chgo. Housing Authority Credit Union, 1963-65; architect, community planner and cons. Interconco., 1965-66; cons. in field. Author: (text book study guide) Reinforced Concrete Design, 1959. Patentee. Subcommittee chmn. Mayor's Adv. Commn. to Revise the Bldg. Code, 1986—; founder, pres. EFCO, Chgo., 1988—. Served with C.E., U.S. Army, 1942-46. Recipient Citation for Residential Devel., Mayor Richard J. Daley, Chgo., 1960, Black Achievers of Industry Recognition award YMCA, Chgo., 1977; Kappa Alpha Psi grantee, 1936. Mem. Architects in Industry, Nat. Assn. Housing and Redevelopment Officials, Internat. Platform Assn., Inventors Coun. of Chgo. Home and Office: 8209 S Rhodes Ave Chicago IL 60619-5005

FOWLER, JOHN FRANCIS, radiobiologist; b. Bridport, Eng., Feb. 3, 1925; s. Norman Vaughan and Marjorie Vivian (White) F.; m. Kathleen Hardcastle Sutton (div.); children: Julie, Shirley, Robert, David, Esther, Helen, Jenny; m. Anna Edwards, June 20, 1992. BSc, London U., 1944; MSc, Londan U., 1946; PhD, U. London, 1955, DSc, 1974; MD, U. Helsinki, Finland, 1981; DS, Med. Coll. of Wis., 1989. Radiation physicist NHS, Newcastle-on-Tyne and London, 1950-59; sr. scientist Med. Rsch. Coun./Hammersmith Hosp., London, 1959-62; reader in physics applied to medicine St. Barts Med. Coll., London, 1962-63; prof. med. physics Roy Postgrad. Med. Sch., London, 1963-70; dir. Gray Lab of UK Cancer Rsch. Campaign, London, 1969-88; prof. human oncology U. Wis., Madison, 1991—; vis. prof. human oncology, U. Wis., Madison, 1988-91; mem. main commn. Internat. Commn. on Radiol. Units, Washington, 1962-65. Author: X-Ray Induced Conductivity, 1956, Nuclear Particles in Cancer Treatment, 1981; co-author: Theory of Fractionated Schedules in Radiotherapy, 1961-92, Linear Quadratic Model for Radiation Response, 1976; contbr. articles to profl. jours. Recipient Roentgen plaque Roentgen Mus., Remscheid, Germany, 1978, Marie Curie medal Polish Soc. Radiobiology, Krakow, 1986. Fellow Am. Coll. Radiology (hon.), Inst. Physics; mem. Hosp. Physicist's Assn. (pres. 1967-68), European Soc. Radiobiology (pres. 1974-75), Brit. Inst. Radiology (hon., pres. 1977-78, Roentgen prize 1964), Am. Assn. Med. Physicists (hon.), Royal Coll. Radiology (hon.). Achievements include early work on modeling dose-time factors in radiotherapy; experiments on normal tissue and tumor response to radiation; application of linear-quadratic theory to radiotherapy. Office: U Wis Dept Human Oncology and Med Physics K4/336 CSC 600 Highland Ave Madison WI 53792

FOWLER, NANCY CROWLEY, government economist; b. Newton, Mass., Aug. 8, 1922; d. Ralph Elmer and Margaret Bright (Tinkham) Crowley; m. Gordon Robert Fowler, Sept. 11, 1949; children: Gordon R., Nancy Pualani, Betty Kainani, Diane Kuulei. AB cum laude, Radcliffe Coll., 1943; grad. Bus. Admnstrn. Program, Harvard-Radcliffe U., 1946; postgrad., U. Hawaii, 1971-76. Econ. rsch. analyst Dept. Planning & Econ. Devel., Honolulu, 1963-69; assoc. chief rsch. Regional Med. Program, Honolulu, 1969-70; economist V and VI Dept. Planning and Econ. Devel., Honolulu, 1970-78, chief policy analysis br., 1978-83, tech. info. services officer, 1985-87; staff rep. State Energy Functional Plan Adv. Com., Honolulu, 1983-89, Hawaii Integrated Energy Assessment, 1978-81, Energy Resources Coord.'s Annual Report, 1988-92, Energy Fact Sheets, 1990. Contbr. articles to profl. jours. Recipient Employee of Yr. award Dept. Planning and Econ. Devel., Honolulu, 1977, others. Mem. Hawaii Econs. Assn. (various offices), Harvard Bus. Sch. Club, Radcliffe club of Hawaii, Propeller Club, Port of Honolulu (past pres., bd. govs.), Navy League Club. Democrat. Avocations: gardening, surfing.

FOWLER, THOMAS BENTON, JR., electrical engineering education, consultant; b. Balt., Jan. 30, 1947; s. Thomas Benton Sr. and Mabel Elizabeth (Corkran) F. MS, Columbia U., 1973; DSc, George Washington U., 1986. Mem. tech. staff MITRE Corp., McLean, Va., 1973-82, prin. engr., 1987—; ind. cons. Reston, Va., 1983-87; asst. prof. math. and physics Christendom Coll., Front Royal, Va., 1982-87; adj. prof. elec. engring. George Washington U., Washington, 1986-92; dir. project to design next generation telecom. sys. for fed. govt. Translator (book) Nature, History, God, 1981; presenter internat. confs. on physics, elec. engring. and ops. rsch.; contbr. articles to profl. jours. Mem. IEEE, AAAS, Am. Phys. Soc., Am. Math. Soc. Republican. Roman Catholic. Achievements include development of mathematical model indicating that a current U.S. drug interdiction policy will fail to achieve objectives; proved that stochastic methods can be applied to control of chaotic nonlinear systems; research on limitations of model reduction to nonlinear systems and evolution as a stability theory problem; research on genetic algorithms. Office: MITRE Corp 7525 Colshire Dr Mc Lean VA 22102-7500

FOWLER, THOMAS KENNETH, physicist; b. Thomaston, Ga., Mar. 27, 1931; s. Albert Grady and Susie (Glynn) F.; m. Carol Ellen Winter, Aug. 18, 1956; children—Kenneth, John, Ellen. B.S. in Engring, Vanderbilt U., 1953, M.S. in Physics, 1955; Ph.D. in Physics, U. Wis., 1957. Staff physicist Oak Ridge Nat. Lab., 1957-65, group leader plasma theory, 1961-65; staff physicist Gen. Atomic Co., San Diego, 1965-67, head plasma physics div., 1967; group leader plasma theory Lawrence Livermore Lab., Livermore, Calif., 1967-69, div. leader, 1969-70, assoc. dir. magnetic fusion, 1970-87; prof., chmn. dept. nuclear engring. U. Calif., Berkeley, 1988—. Fellow Am. Phys. Soc. (chmn. plasma physics div. 1970); mem. Nat. Acad. Scis., Sigma Xi, Sigma Nu. Home: 221 Grover Ln Walnut Creek CA 94596-6310 Office: U Calif 4155 Etcheverry Hall Berkeley CA 94720

FOWLER, WILLIAM ALFRED, retired physics educator; b. Pittsburgh, Penn., Aug. 9, 1911; s. John McLeod and Jennie Summers (Watson) F.; m. Ardiane Olmsted, Aug. 24, 1940 (dec. May 1988); children: Mary Emily Fowler Galowin, Martha Summers Fowler Schoenemann; m. Mary Dutcher, Dec. 14, 1989. B of Engring. Physics, Ohio State U., 1933, DSc (hon.), 1978; PhD, Calif. Inst. Tech., 1936; DSc (hon.), U. Chgo., 1976, Denison U., 1982, Ariz. State U., 1985, Georgetown U., 1986, U. Mass., 1987, Williams Coll., 1988; Doctorat honoris causa, U. Liège (Belgium), 1981, Observatoire de Paris, 1981. Asst. prof. physics Calif. Inst. Tech., 1939-42, asso. prof., 1942-46, prof. physics, 1946-70, Inst. prof. physics, 1970-82; prof. emeritus, 1982—; conducted rsch. on nuclear forces and reaction rates, nuclear spectroscopy, structure of light nuclei, thermonuclear sources of stellar energy and element synthesis in stars and supernovae and the early universe, including recently proposed inflationary model; study of gen. relativistic effects in quasar and pulsar models, nuclear cosmochronology; Fulbright lectr. Pembroke Coll. and Cavendish lab. U. Cambridge, 1954-55; Guggenheim fellow, 1954-55; Guggenheim fellow St. John's Coll. and dept. applied math. and theoretical physics U. Cambridge, 1961-62; vis. fellow Inst. Theoretical Astronomy, summers 1967-72; vis. scholar program Phi Beta Kappa, 1980-81; asst. dir. rsch. sect. L Nat. Defense Rsch. Com., 1941-45; tech. observer, office of field service OSRD, South Pacific Theatre, 1944; sci. dir., project VISTA, Dept. Def., 1951-52; mem. nat. sci. bd. NSF, 1968-74; mem. space sci. bd. Nat. Acad. Scis., 1970-73, 77-80; chmn. Office of Phys. Scis., 1981-84; mem. space program adv. council NASA, 1971-73; mem. nuclear sci. adv. com. Dept. Energy/Nat. Sci. Found., 1977-80; E.A. Milne Lectr. Milne Soc., 1986; named lectr. univs., colls.; hon. fellow Pembroke Coll., Cambridge U., 1992. Contbr. numerous articles to profl. jours. Bd. dirs. Am. Friends of Cambridge U., 1970-78. Rsch. fellow Calif. Inst. Tech., Pasadena, 1936-39; recipient Naval Ordnance Devel. award USN, 1945, Medal of Merit, 1948; Lammé medal Ohio State U., 1952; Liège medal U. Liège, 1955; Calif. Co-Scientist of Yr. award, 1958; Barnard medal for contbn. to sci. Columbia, 1965; Apollo Achievement award NASA, 1969; Vetlesen prize, 1973; Nat. medal of sci., 1974; Bruce gold medal Astron. Soc. Pacific, 1979; Nobel prize for physics, 1983; Légion d'Honneur, 1989; Benjamin Franklin fellow Royal Soc. Arts; named to Lima Ohio City Schs. Disting. Alumni Hall of Fame, Ohio Sci. and Tech. Hall of Fame; named hon. fellow Pembroke Coll., Cambridge U., 1992. Fellow Am. Phys. Soc. (Tom W. Bonner prize 1970, pres. 1976, 1st recipient William A. Fowler award for excellence in physics So. Ohio sect. 1986), Am. Acad. Arts and Scis., Royal Astron. Soc. (assoc., Eddington medal 1978); mem. NAS (council 1974-77), AAAS, Am. Astron. Soc., Am. Inst. Physics (governing bd. 1974-80), AAUP, Am. Philos. Soc., Soc. Royal Sci. Liège (corr. mem.), Brit. Assn. Advancement Sci., Soc. Am. Baseball Research, Mark Twain Soc. (hon.), Naturvetenskapliga Foreiningen (hon.), Sigma Xi, Tau Beta Pi, Tau Kappa Epsilon. Democrat. Clubs: Athenaeum (Pasadena); Cosmos (Washington). Office: Calif Inst Tech Kellogg 106-38 Pasadena CA 91125

FOWLER, WILLIAM MAYO, JR., rehabilitation medicine physician; b. Bklyn., June 16, 1926. BS, Springfield Coll., 1948, MEd, 1949; MD, U. So. Calif., L.A., 1957. Diplomate Am. Bd. Phys. Medicine and Rehab. Intern UCLA, 1958, resident in pediatrics, 1959, resident in phys. medicine, rehab., 1963; chmn. dept. phys. medicine, rehab. U. Calif., Davis, 1968-82, mem. faculty dept. phys. medicine, rehab., 1972-91, prof. emeritus, 1991—. Fellow Am. Acad. Phys. Medicine and Rehab. (pres. 1981), Am. Coll. Sports Medicine; mem. Assn. Acad. Physiatrists. Office: U Calif Davis Dept Phys Med & Rehab TB191 Davis CA 95616

FOX, BARRY HOWARD, nuclear engineer; b. Long Beach, N.Y., Dec. 18, 1957; s. Donald Irwin and Rosalind Audrey (Rubin) F.; m. Joan Renee Schindler, May 27, 1988; children: David Roy, Stephen Gregg. BS in Nuclear Engring., Rensselaer Poly. Inst., 1980. Prin. staff ORI, Inc., Alexandria, Va., 1984-85; sr. reliability engr. GPU Nuclear Corp., Middletown, Pa., 1985—; cons. Fox Computer Consulting, Harrisburg, Pa., 1990—. Author: (tech. paper) 17th Inter-RAM Conf., 1990, Inter-RAMQ Conf., 1992, ANS/ASME Nuclear Energy Conf., 1992. Lt. USN, 1980-84. Mem. Am. Nuclear Soc. Republican. Jewish. Home: 545 Cardinal Dr Harrisburg PA 17111-5010 Office: GPU Nuclear Corp Rte 441 S Middletown PA 17057

FOX, CARL ALAN, research institute executive; b. Waukesha, Wis., Nov. 24, 1950; s. Frank Edwin and Margaret Alvilda (Rasmussen) F.; m. Susan Jane Smith, June 18, 1977; children: Thomas Gordon, James David, Joseph Carl. BS, U. Wis., River Falls, 1973; MS, U. Minn., 1975; PhD, Ariz. State U., 1980. Lab. asst. dept. biology U. Wis., River Falls, 1971-73; rsch. asst. dept. agronomy and plant genetics U. Minn., St. Paul, 1973-75; tchr. high sch. Le Center (Minn.) Pub. Schs., 1975-76; rsch. fellow dept. botany Ariz. State U., Tempe, 1976-79; rsch. asst. Lab. Tree-Ring Rsch. U. Ariz., Tucson, 1978-79; rsch. scientist, then sr. rsch. scientist So. Calif. Edison Co., Rosemead, 1979-87; rsch. assoc. agrl. experiment sta. U. Calif., Riverside, 1986—; exec. dir. Desert Rsch. Inst., Reno, 1987—; rsch. adviser Electric Power Rsch. Inst., Palo Alto, Calif., 1983-87; liaison Utility Air Regulatory Group, Washington, 1983-87, cons., 1989-91; mem. peer rev. panel EPA, Corvallis, Oreg., 1986; invited reviewer air quality rsch. div. Nat. Park Svc., Denver, 1989. Contbr. numerous papers to profl. publs. Asst. troop leader Newport Beach (Calif.) area Boy Scouts Am., 1981-82, cub scout leader Reno area, 1990-91; bd. dirs. World Rainforest Found., Reno, 1989—, Internat. Visitors Coun. No. Nev., Reno, 1991—; coach YMCA, Reno, 1989-91; deacon Covenant Presbyn. Ch., 1989—; judge State of Nev. Odyssey of the Mind. NSF fellow, 1976-79; grantee EPA, 1978-79, 83-85, 89-92, NSF, 1987—, Dept. Def. and Energy, 1987. Mem. AAAS, Air Pollution Control Assn., Ecol. Soc. Am., Am. Soc. Agronomy, Greentree Gators Swim Team (pres. 1986-87), Beta Beta Beta. Republican. Presbyterian. Avocations: camping, canoeing, tennis, gardening, basketball. Office: Desert Rsch Inst PO Box 60220 7010 Dandini Blvd Reno NV 89512-3998

FOX, CHRISTOPHER, educational administrator; b. Yonkers, N.Y., Aug. 9, 1957; s. Edward John and Helen (Abate) Fuchs; m. Nancy Susan Shultz, July 11, 1992 (div.); children: Tristan Douglas, Erin Margaret. BA, Glassboro State Coll., 1980; MA, U. Md., 1983. Staff asst. U. Md., College Park, 1984-89, asst. dir. Ctr. for Global Change, 1989—; lectr. U. Md. Univ. Coll., College Park, 1985—. Mem. Lambda Alpha. Democrat. Mem. Soc. of Friends. Home: 2830 Nine Mile Cir Catonsville MD 21228 Office: Ctr for Global Change 7100 Baltimore Ave College Park MD 20740

FOX, DANIEL MICHAEL, foundation administrator, author; b. N.Y.C., Aug. 20, 1938; s. Alexander E. and Rose (Leitner) F.; m. Carol Anne Kemps, Sept. 8, 1963 (div. 1985); children: Aaron, Miriam, Joshua, Benjamin; m. Louise O. Vasvari, Dec. 26, 1988. AB, Harvard U., 1959, AM, 1961, PhD, 1964. Instr. Harvard U., Cambridge, Mass., 1964-65, asst. prof., 1967-72; dir. field ops. Applachian Vols., Berea, Ky., 1965-67; prof., v.p. SUNY, Stony Brook, 1972-89; assoc. dir. Nat. Ctr. for Health Svcs. Rsch., Rockville, Md., 1975-78; pres. Milbank Meml. Fund, N.Y.C., 1990—; cons. in field. Author: Engines of Culture, 1963, The Discovery of Abundance, 1967, Economists and Health Care, 1979, Health Policies, Health Politics, 1986, Photographing Medicine, 1988, AIDS: The Burdens of History, 1989, AIDS: The Marking of a Chronic Disease, 1992, Power and Illness: The Failure and Future of American Health Policy, 1993. Shaw travel fellow Harvard U., 1959-60, Sheldon travel fellow, 1962; also numerous grants. Mem. NAS, Inst. Medicine, Am. Hist. Assn. (Beveridge prize 1965), Am. Assn. for the History of Medicine, N.Y. Acad. Medicine, Harvard Club of N.Y. Jewish. Office: Milbank Meml Fund 1 E 75th St New York NY 10021-2601

FOX, DAVID PETER, psychologist, educator; b. Brockton, Mass., Nov. 5, 1953; s. George and Grace Fox. MA, Loyola Marymount U., 1976; PhD, U.S. Internat. U., 1980. Lic. psychologist, Md., Calif. Staff psychologist North Charles Gen. Hosp., Balt., 1980-82; asst. prof. Loyola Coll., Balt., 1980-82; dir. Phobia Clinic, Lutherville, Md., 1980-82; prof. psychology Ryokan Coll., L.A., 1983-88; CEO, founder Dr. David Fox, Inc. Psychol. Corp., Beverly Hills, Calif., 1982—; assoc. prof. Calif. Sch. Profl. Psychology, L.A., 1984—; assoc. clin. prof. Sch. Medicine Loma Linda (Calif.) U., 1988—; cons. Century City Hosp., L.A., 1990—, L.A. Police Dept., 1984—, Dept. Pub. and Social Svcs., L.A., 1986—; bd. dirs. Network of Therapist Abuse, L.A., 1989—; presenter in field. Contbr. articles to profl. jours. Mem. APA, Am. Psychol. Soc., Assn. Orthodox Scientists (co-chair behavioral scis.), Nat. Register Health Svcs. Providers. Achievements include development of training program on crosscultural psychology, national centralized network on dangerous patients. Office: 215 S Cienega Blvd Beverly Hills CA 90211

FOX, EMILE, physician; b. Luxembourg, Jan. 4, 1953; s. Nicolas and Lucie (Waltzing) F. MD, U. Nancy (France), 1980, diploma in pub. health, 1982; MS in Tropical Medicine, London Sch. Hygiene, 1983. Registrar Cen. Hosp., Luxembourg, 1980-83; specialist physician Nat. Health Lab., Luxembourg, 1983-84; sr. med. officer Internat. Ctr. Med. Rsch., Lahore, Pakistan, 1984-85; lectr. medicine U. Papua New Guinea, 1985-86; staff epidemiologist USN Med. Rsch. Unit # 3, Cairo, 1986-90; med. officer WHO, Kigali, Rwanda, 1990—; rsch. asst. prof. internat. health U. Md., Balt., 1984-90; hon. cons. Port Moresby Gen. Hosp., Papua New Guinea, 1985-86; team leader epidemiol. rsch. expeditons NAMRU-3, Republic Djibouti, 1987-90. Author: (with others) Tuberculosis in the Tropics, 1991; contbr. articles to profl. jours. Recipient Frederic Murgatroyd award London Sch. Hygiene, 1983. Fellow Royal Soc. Tropical Medicine and Hygiene; mem. Am. Soc. Microbiology, Am. Soc. Tropical Medicine and Hygiene, Am. Pub. Health Assn. Home: 16 rue Van Werveke, L-2725 Luxembourg Office: WHO, BP 1324, Kigali Rwanda

FOX, JAMES CARROLL, aerospace engineer, program manager; b. Saginaw, Mich., Jan. 15, 1932; s. Arthur Clarence Fox and Evelyn Irene (Scott) Goodrick; m. Lois Ann Pitsch, Sept. 22, 1952; children: Susan Carol, James Michael, Kevin Matthew. BSEE, U. Mich., 1960; MSEE, U. So. Calif., 1966. Mem. tech. staff TRW Systems Inc., Redondo Beach, Calif., 1960-67; project engr. Bendix Aerospace Systems, Ann Arbor, Mich., 1967-77, program mgr., 1977-82; div. mgr. KMS Fusion Space/Def. Systems, Ann Arbor, Mich., 1982-89, dir., 1989-91; pres., chief oper. officer Canopus Systems Inc., Ann Arbor, Mich., 1991—. Author: Six D.O.F. Simulation of Man Orbital Docking, 1963. With USN, 1951-55. Mem. AIAA, IEEE, Tech. Mktg. Soc. Am. Republican. Presbyterian. Achievements include co-design of Nimbus and Landsat satellites active attitude control system; management of first Landsat ground digital data processing system, first Space Shuttle Orbital Acceleration Rsch. Experiment. Home: 2010 Hogback Rd Dexter MI 48130 Office: Canopus Systems Inc 3601 Plymouth Rd Ann Arbor MI 48113-0319

FOX, JOAN PHYLLIS, environmental engineer; b. Rockledge, Fla., July 16, 1945; d. John A. and Nonie L. (Knutson) Fox. BS in Physics with high honors, U. Fla., 1971; PhD in Civil Engring., U. Calif., Berkeley, 1980. Engr. Bechtel, Inc., San Francisco, 1971-76; dir. program and prin. investigator Lawrence Berkeley Lab., 1977-81; prin. engr., pres. Fox Environ. Mgmt., Berkeley, 1981—; guest lectr. dept. conservation and resource studies U. Calif., Berkeley, 1980-84. Contbr. articles on oil shale, hazardous waste, water quality, water resources and air polution in San Francisco Bay area to profl. publs. Grantee Dept. Energy, 1976-81, EPA, 1978-81. Mem. Am. Chem. Soc., Am. Water Resources Assn., Air and Waste Mgmt. Assn., NAS (past mem. com. on surface mining and reclamation, mem. subcom. on QA/QC of com. irrigation-induced water quality problems 1986-90), Phi Beta Kappa, Sigma Pi Sigma. Office: 2526 Etna St Berkeley CA 94704-3115

FOX, JOSEPH CARL, veterinary medicine educator, researcher, parasitologist; b. Provo, Utah, Dec. 11, 1941; s. Joseph and Clara (Hatley) F.; m. Sharon Lael Larson, Nov. 16, 1962; children; Scott Joseph, Jeffrey Carl, Shalene Lael. BS, Brigham Young U., 1961, MS, 1970; PhD, Mont. State U., 1975. Rsch. assoc. veterinary medicine U. Ill., Urbana, 1975-78; prof. Okla. State U., Stillwater, 1978—. Author: Sewage Organisms- A Color Atlas, 1981; contbr. articles to profl. jours. Active Stillwater Sister Cities Coun., 1986-92. Republican. Mormon. Office: Okla State U Coll Veterinary Medicine Stillwater OK 74078-0353

FOX, KARL AUGUST, economist, eco-behavioral scientist; b. Salt Lake City, July 14, 1917; s. Feramorz Young and Anna Teresa (Wilcken) F.; m. Sylvia Olive Cate, July 29, 1940; children: Karl Richard, Karen Frances Anne. BA, U. Utah, 1937, MA, 1938; PhD, U. Calif., 1954. Economist USDA, 1942-54; head div. statis. and hist. rsch. Bur. Agrl. Econs., 1951-54; economist Coun. Econ. Advisers, Washington, 1954-55; head dept. econs. and sociology Iowa State U., Ames, 1955-66; head dept. econs. Iowa State U., 1966-72, disting. prof. scis. and humanities, 1968-87, prof. emeritus, 1987—; vis. prof. Harvard, 1960-61, U. Calif., Santa Barbara, 1971-72, 78, vis. scholar, Berkeley, 1972-73; William Evans vis. prof. U. Otago, N.Z., 1981; Bd. dirs. Social Sci. Rsch. Coun., 1963-67, mem. com. econ. stability, 1961-65, chmn. com. areas for social and econ. statistics, 1964-67; mem. Com. Reg. Accounts, 1963-68. Author: Econometric Analysis for Public Policy, 1958, (with M. Ezekiel) Methods of Correlation and Regression Analysis, 1959, (with others) The Theory of Quantitative Economic Policy, 1966, rev. edit., 1973, Intermediate Economic Statistics, 1968, rev. edit, (with T.K. Kaul), 1980, (with J. K. Sengupta) Economic Analysis and Operations Research, 1969, (with W.C. Merrill) Introduction to Economic Statistics, 1970, Social Indicators and Social Theory, 1974, Social System Accounts, 1985, The Eco-Behavioral Approach To Surveys and Social Accounts for Rural Communities, 1990, Demand Analysis, Econometrics and Policy Models, 1992, Urban-Regional Economics, Social System Accounts and Eco-Behavioral Science, 1993; author-editor: Economic Analysis for Educational Planning, 1972; co-editor: Readings in the Economics of Agriculture, 1969, Economic Models, Estimation and Risk Programming (essays in honor of Gerhard Tintner), 1969, Systems Economics, 1987; contbr. articles to profl. jours. Recipient superior service medal USDA, 1948, award for outstanding pub. research Am. Agrl. Econs. Assn., 1952, 54, 57, for outstanding doctoral dissertation, 1953. Fellow Econometric Soc., Am. Statis. Assn. (Census Research fellow 1980-81), Am. Agrl. Econs. Assn. (v.p. 1955-56, award for publ. of enduring quality 1977), AAAS; mem. Am. Econs. Assn. (research and publs. com. 1963-67), Regional Sci. Assn., Ops. Research Soc. Am., Am. Ednl. Research Assn., Phi Beta Kappa, Phi Kappa Phi. Home: 234 Parkridge Cir Ames IA 50014-3645 Office: Iowa State U Econs Dept Ames IA 50011

FOX, MARIAN CAVENDER, mathematics educator; b. Paris, Tenn., Dec. 12, 1947; d. G.W.F. and Marilee (Boden) Cavender; m. James H. Fox, Jan. 24, 1973; children: Brendon Dutch, Michael Garrett. BS in Math., Miss. U. for Women, 1969; MA in Edn., George Washington U., 1973; PhD in Math. Edn., Ga. State U., 1989. Secondary tchr. math., Mo., Va., 1970-77; instr. math. North Cen. Tech. Coll., Mansfield, Ohio, 1979-81; prof. Manatee Community Coll., Bradenton, Fla., 1981-86; grad. tchr., rsch. asst. Ga. State U., Atlanta, 1986-89; asst. prof. Math. Kenesaw State Coll., Marietta, Ga., 1989—; cons. Manatee County Schs., Bradenton, 1983-85, State of Ga. Dept. of Edn., Atlanta, 1987-90, Fulton County Schs., Atlanta, 1987-90. Mem. Assn. Tchr. Educators, Nat. Coun. Tchrs. Math., Math. Assn. Am., Ga. Coun. Tchrs. Math., Phi Delta Kappa. Office: Kennesaw State Coll PO Box 444 Marietta GA 30061-0444

FOX, MAURICE SANFORD, molecular biologist, educator; b. N.Y.C., Oct. 11, 1924; s. Albert and Ray F.; m. Sally Cherniavsky, Apr. 1, 1955; children: Jonathan, Gregory, Michael. B.S. in Meteorology, U. Chgo., 1944, M.S. in Chemistry, 1951, Ph.D., 1951. Instr. U. Chgo., 1951-53; asst. Rockefeller Inst., 1953-55, asst. prof., 1955-58, assoc. prof., 1958-62; assoc. prof. MIT, Cambridge, 1962-66, prof., 1966-79, Lester Wolfe prof. molecular biology, 1979—, head dept. biology, 1985-89. Served with USAAF, 1943-46. USPHS fellow, 1952-53; Nuffield Research fellow, 1957; Fogarty scholar, 1991. Fellow AAAS; mem. Inst. Medicine, NAS. Office: MIT Dept Biology 77 Massachusetts Ave Cambridge MA 02139

FOX, MICHAEL WILSON, veterinarian, animal behaviorist; b. Bolton, Eng., Aug. 13, 1937; came to U.S., 1962; s. Geoffrey and Elizabeth (Wilson) F.; m. Deanna L. Krantz, May 1989; children by previous marriage: Michael Wilson, Camilla, Mara. B. in Vet. Medicine, Royal Vet. Coll., London, 1962; Ph.D., U. London, 1967, D.Sc., 1975. Postdoctoral fellow Jackson Lab., Bar Harbor, Maine, 1962-64; med. research assoc. State Research Hosp., Galesburg, Ill., 1964-67; assoc. prof. psychology Washington U., St. Louis, 1967-76; v.p. Humane Soc. U.S., Washington. Contbg. editor: McCall's mag; author: syndicated newspaper column Ask Your Animal Doctor; author: Canine Behavior, 1965, Canine Pediatrics, 1966, Integrative Development of Brain and Behavior in the Dog, 1971, Behavior of Wolves, Dogs and Related Canids, 1971, Understanding Your Dog, 1972, Understanding Your Cat, 1974, Concepts in Ethology: Animal and Human Behavior, 1974, Between Animal and Man: The Key to The Kingdom, 1976,

The Dog, Domestication and Behavior, 1977, Wild Dogs Three, 1977, What is Your Cat Saying?, 1978; (juveniles), The Wolf, 1973 (Christopher award), Vixie, The Story of a Fox, 1973, Sundance Coyote, 1974, Ramu and Chennai, 1975 (Sci. Tchrs.' award); co-author: What Is Your Dog Saying?, 1977, Dr. Fox's Fables, 1980, The Touchlings, 1981, Understanding Your Pet, 1978, The Soul of the Wolf, 1980, One Earth One Mind, 1980, Returning to Eden: Animal Rights and Human Responsibility, 1980, How to be Your Pet's Best Friend, 1981, The Healing Touch, 1982, Love is a Happy Cat, 1982, Farm Animal Husbandry, Behavior and Veterinary Practice, 1983, The Whistling Hunters: Field Studies of the Asiatic Wild Dog (Cuon alpinus), 1984; The Animal Doctor's Answer Book, 1984; Laboratory Animal Care, Welfare and Experimental Variables, 1986, Agricide-The Hidden Crisis That Affects Us All, 1986, The New Animal Doctor's Answer Book, 1989, The New Eden, 1989, Superdog, 1990, Inhumane Society, The American Way of Animal Exploitation, 1990, Animals Have Rights Too, 1991, You Can Save The Animals 50 Things To Do Right Now, 1991, Supercat, 1991, Superpigs and Wondercorn: How the Brave New World of Biotechnology Will Affect Us All, 1992; editor: Abnormal Behavior in Animals, 1968, Readings in Ethology and Comparative Psychology, 1973, The Wild Canids, 1975, On the Fifth Day: Animal Rights and Human Ethics, 1978, Internat. Jour. for Study of Animal Problems, Advances in Animal Welfare Sci. Mem. AVMA, AAAS, Brit. Vet. Assn., Animal Behavior Soc.

FOX, RICHARD ROMAINE, geneticist, consultant; b. New Haven, Nov. 12, 1934; s. Clarence Romaine and Lydia Adella (Ellsworth) F.; m. Sally Ann Chudoba, Jan. 31, 1956 (dec. 1980); children: Sally, Susan, John; m. Barbara Helen Downing, Aug. 1, 1981; stepchildren: Lincoln, Daniel, Lorna. BS, U. Conn., 1956; MS, U. Minn., 1958, PhD, 1959. Lectr. U. Maine, Orono, 1961-90; postdoctoral fellow, then assoc. staff scientist Jackson Lab., Bar Harbor, Maine, 1959-65, staff scientist, 1965-80, sr. staff scientist, 1980-90, emeritus, 1990—; cons. Jackson Lab., 1990-91, FOXRUN Assocs., Bar Harbor, 1985—; bd. dirs. Bar Harbor Savs. and Loan. Contbr. to profl. publs. Mem. Rotary (past pres.), Arcady Music Soc. (bd. dirs. 1985—). Republican. Home: 3 Seely Rd Bar Harbor ME 04609-1506 Office: Jackson Lab 600 Main St Bar Harbor ME 04609

FOX, RONALD FORREST, physics educator; b. Berkeley, Calif., Oct. 1, 1943; s. Sidney Walter and Raia (Joffe) F.; children: Daniel, Lara. BS, Reed Coll., 1964; PhD, Rockefeller U., 1969. Postdoctoral fellow Miller Inst., U. Calif., Berkeley, 1969-71; asst. prof. Ga. Inst. Tech., Atlanta, 1971-74, assoc. prof., 1974-79, prof., 1979—, Regents prof. physics, 1991—, asst. dir. Sch. Physics, 1982-84, assoc. dir. Sch. Physics, 1986-89. Author: Biological Energy Transduction, 1982, Energy and the Evolution of Life, 1988; contbr. over 80 articles to sci. jours. Recipient W. Roane Beard Outstanding Tchr. award Ga. Inst. Tech., 1992; fellow Alfred P. Sloan Found., 1974-78, Guggenheim Fellow, 1985; grantee NSF, 1973—. Fellow Am. Phys. Soc.; mem. N.Y. Acad. Scis. Avocations: squash, raquetball, jazz piano. Office: Ga Inst Tech Dept Physics Atlanta GA 30332

FOX, STUART IRA, physiologist; b. Bklyn., June 21, 1945; s. Sam and Bess F.; m. Ellen Diane Berley; 1 child, Laura Elizabeth. BA, UCLA, 1967; MA, Calif. State U., L.A., 1967; postgrad., U. Calif., Santa Barbara, 1969; PhD, U. So. Calif., 1978. Rsch. assoc. Children's Hosp., L.A., 1972; prof. physiology L.A. City Coll., 1972-85, Calif. State U., Northridge, 1979-84, Pierce Coll., 1986—; cons. William C. Brown Co. Pubs., 1976—. Author: Computer-Assisted Instruction in Human Physiology, 1979, Laboratory Guide to Human Physiology, 2d edit., 1980, 3d edit., 1984, 4th edit., 1987, 5th edit., 1990, 6th edit., 1993, Textbook of Human Physiology, 1984, 3d edit., 1990, 4th edit., 1993, Concepts of Human Anatomy and Physiology, 1986, 2d edit., 1989, 3d edit., 1992, Laboratory Guide to Human Anatomy and Physiology, 1986, 2d edit., 1989, 3d edit., 1992, Perspectives on Human Biology, 1991. Mem. AAAS, So. Calif. Acad. Sci., Am. Physiol. Soc., Sigma Xi. Home: 5556 Forest Cove Ln Agoura Hills CA 91301-4047 Office: Pierce Coll 6201 Winnetka Ave Woodland Hills CA 91371-0002

FOXX, RICHARD MICHAEL, psychology educator; b. Denver, Oct. 28, 1944; s. James Martin and Marie Louise (Harris) F.; m. Susan J. Massee, Nov. 20, 1988; children: Alyssa, Marie. BA, U. Calif., Riverside, 1967; PhD, So. Ill. U., 1971. Lic. Psychologist Md., Ill., Pa. Rsch. scientist State of Ill., Anna, 1970-74; asst. prof. pediatrics U. Md. Med. Sch., Balt., 1974-80; assoc. prof. psychology U. Md. Baltimore County, Balt., 1976-80; dir. rsch. Anna State Hosp., 1980-91; adj. clin. prof. psychiatry So. Ill. Med. Sch., Springfield, Ill., 1987-91; prof. psychology Pa. State Harrisburg, Middletown, 1991--; adj. clin. prof. pediatrics Pa. State Med. Coll., Hersey, 1992--; pres. Help Svcs. Inc., Harrisburg, Pa., 1979—; bd. dirs. Soc. for Advancement Behavior Analysis, Kalamazoo, exec. coun., 1991. Author: Toilet Training in Less Than a Day, 1974, Increasing Behavior, 1982, Decreasing Behavior, 1982, Looking for the Words, 1986; producer, editor: (film) Harry: Treatment of Self Abusive Behavior, 1980. Recipient Disting. Alumni award Calif. State U. Fullerton, 1981, Rsch. award Am. Assn. Mental Deficiency, 1979, First Pl. award Interant. Rehab. Film Festival, 1981, First Pl. award Am. Film Festival, 1981; Erskine fellow U. Canterbury, 1987. Fellow APA (pres. divsn. 33 1991-92), Am. Psychol. Assn., Am. Psychol. Soc., Am. Assn. on Mental Retardation. Achievements include discovery of overcorrection; developer nicotine fading treatment for cigarette smoking and developer of time out ribbon program. Office: Pa State Harrisburg US Rt 230 Middletown PA 17057

FRADKOV, VALERY EUGENE, materials scientist; b. Kharkov, Ukraine, Sept. 9, 1954; came to U.S., 1991; s. Eugene and Svetlana (Tseitlina) F.; m. Anna Galina, June 10, 1983; 1 child, Elena. ME in Matallurgy, Moscow Inst. Steel & Alloys, Moscow, Russia, 21977; PhD in Physics, Inst. Solid State Physics, Chernogolovka, Russia, 1981. Researcher Acad. Scis. of the USSR, Chernogolorka, Russia, 1977-90; vis. scientist Tech. U. of Aachen, Germany, 1990-91, U. Pa., Phila., 1991; vis. scholar Rensselaer Poly. Inst., Troy, N.Y., 1991. Recipient Alexander von Humboldt fellowship, 1990, Exxon Rsch. fellowship, 1991, 92. Mem. Materials Rsch. Soc., Minerals, Metals and Materials Soc., Internat. Materials Info. Soc., Sigma Xi. Achievements include devel. of a set fo complementary computer models for microstructural evolution. Office: Rensselaer Poly Inst MRC-Bldg 275D Troy NY 12180

FRAHM, VERYL HARVEY, JR., laboratory manager; b. Lewellen, Nebr., Sept. 11, 1948; s. Veryl Harvey and Elaine Eloise (Cornelius) F.; m. Vicki Anne Olson, May 29, 1971; children: Errin Wilson, Megan Joy, Branden Corey. BA in Chemistry, U. Colo., 1971; MBA, U. Phoenix, 1989. Rsch. asst. Nat. Ctr. for Atmospheric Rsch., Boulder, Colo., 1968-72; support scientist Nat. Ctr. for Atmospheric Rsch., Boulder, 1976-79; rsch. metallurgist Cato Rsch. Corp., Wheatridge, Colo., 1972-75, 79-86; chemist scientist U.S. Geol. Survey, Denver, 1975-76; sr. radiochemist Pub. Svc. Co. of Colo., Denver, 1986-89; radiochemistry supr. Omaha Pub. Power Dist., 1989-92; lab. mgr. Scientech Inc. Environ. Labs., Gaithersburg, Md., 1992—. Treas. Alpha Chi Sigma Chemistry Frat., Boulder, 1967-71; guild mem. Nebr. Choral Arts Soc., Omaha, 1990-91. Mem. Am. Nuclear Soc., Mensa, Intertel. Office: Gaithersburg Lab Scientech Environ Labs 205 Perry Pky Ste 10 Gaithersburg MD 20877-2141

FRAIR, WAYNE FRANKLIN, biologist, educator; b. Pitts., May 23, 1926; s. Herbert E. and Elizabeth M. (Greenawald) Gfroerer. BA, Houghton Coll., 1950; BS, Wheaton (Ill.) Coll., 1951; MA, U. Mass., 1955; PhD, Rutgers U., 1962. Tchr. sci. Ben Lippen Sch., Asheville, N.C., 1951-52; mem. faculty King's Coll., Briarcliff, N.Y., 1955—, prof. biology, 1967—. Chmn. Heart Club, Phelps Meml. Hosp., Tarrytown, N.Y., 1979-80. With USN, 1944-46, PTO. Author: A Case for Creation, 3d edit., 1983; contbr. articles to profl. jours. Fellow AAAS, Am. Sci. Affiliation, Creation Rsch. Soc. (sec. 1974-84, v.p. 1985-86, pres. 1987—); mem. Am. Inst. Biol. Sci., Am. Soc. Zoology, Evang. Theol. Soc., Sigma Xi (life). Baptist. Club: Saw Mill River Audubon Soc. (bd. dirs. 1963-66). Home: 34 Piping Rock Dr Ossining NY 10562-2308 Office: The King's Coll Briarcliff Manor NY 10510

FRAITAG, LEONARD ALAN, mechanical and design engineer; b. N.Y.C., Dec. 23, 1961; s. David and Lucille Reneé (Jay) F.; m. Dorann Elizabeth Meecham, June 28, 1987; children: Shoshana Elizabeth, Aaron Joseph. BSME, San Diego State U., 1987; AA, Grossmont Coll., 1983. Registered profl. engr., Calif. Design engr. Restaurant Concepts, San Diego, 1987; mech. engr. Vantage Assocs., Inc., San Diego, 1988-89; design engr. Mainstream Engring. Co., Inc., San Diego, 1989, Pilkington Barnes Hind, San Diego, 1989—. Inventor safe product moving device for contact lens. Mem. Masons (officer 1988—), Shriners (noble 1989), Scottish Rite (class pres. 1989), Pi Tau Sigma. Avocations: computers, sports, camping. Office: Pilkington Barnes Hind 8006 Engineer Rd San Diego CA 92111-1975

FRAKES, LAWRENCE WRIGHT, program analyst, logistics engineer; b. Evanston, Ill., June 27, 1951; s. Linden Maurice and Dorothy Marie (Wright) F.; children: Laurenna Eileen, Darla Lamay. BA, Knox Coll., 1973; MS, Boston U., 1981; MBA, Fla. Inst. Tech., Melbourne, 1984. Instr. instrumental music Avon (Ill.) Community Unit Sch. Dist., 1973-74; commd. 2d lt. U.S. Army, 1974, advanced through grades to capt., 1978, ret., 1983; tng. instr. logistics U.S. Army Missile and Munitions Sch., Redstone Arsenal, Ala., 1983-85; logistics engr., program analyst CAS, Inc., Huntsville, Ala., 1985—. Condr. Ft. Knox Community Wind Ensemble, 1976-78, Fanfare '77 Summer Music Theatre, Radcliff, Ky., 1977; tenor sect. leader German-Am. Community Choir of Frankfurt, Fed. Republic Germany, 1980-82; tenor Huntsville Opera Theater, 1986—, soloist, 1989, 91, bd. dirs., 1989—, Huntsville Community Chorus, 1989—, soloist, 1991. Mem. North Ala. Trail Riders Club (sec.-treas. 1985-88), Brown's Creek Sailing Club. Presbyterian (choir dir. 1989-90). Avocations: opera, classical music, sailing, off-road motorcycling. Home: 202 Skylane Blvd Madison AL 35758

FRAKNOI, ANDREW, astronomy educator, astronomical society executive; b. Budapest, Hungary, Aug. 24, 1948; came to U.S., 1959; naturalized; s. Emery I. and Katherine H. (Schmidt) F.; m. Lola Goldstein, Aug. 16, 1992. B.A. in Astronomy, Harvard U., 1970; M.A. in Astrophysics, U. Calif.-Berkeley, 1972. Instr. astronomy and physics Cañada Coll., Redwood City, Calif., 1972-78; exec. dir. Astron. Soc. of Pacific, San Francisco, 1978-92; chmn. dept. astronomy Foothill Coll., Los Altos, Calif., 1992—; part-time prof. San Francisco State U., 1980-92; dir. Project ASTRO Astron. Soc. Pacific, 1992—; fellow Com. for Sci. Investigation of Claims of Paranormal, 1984—; bd. dirs. Search for Extra Terrestrial Intelligence Inst., Palo Alto, Calif., 1984—; host radio program Exploring the Universe Sta. KGO-FM, San Francisco, 1983-84; rev. panelist informal sci. edn. NSF, 1989-93; adv. com. Astrophysics div. NASA, 1992—. Author: Resource Book for the Teaching of Astronomy, 1978, (with others) Effective Astronomy Teaching and Student Reasoning Ability, 1978, Universe in the Classroom, 1985, (with R. Robert Robbins) The Universe at Your Fingertips, 1985, (with T. Robertson) Instructor's Guide to the Universe, 1991, (with Douglas Brown) Instructor's Manual to Discovering the Universe, 1993; editor: The Planets, 1985, Interdisciplinary Approaches to Astronomy, 1985, The Universe, 1987; editor Mercury Mag., 1978-92, The Universe in the Classroom Newsletter, 1985-92; assoc. editor: The Planetarian, 1986-88; columnist monthly column on astronomy San Francisco Examiner, 1986-87 and others. Bd. dirs. Bay Area Skeptics, San Francisco, 1982-91. Recipient award of merit Astron. Assn. No. Calif., 1980; Asteroid 4859 named Asteroid Fraknoi, 1992. Mem. AAAS (astronomy sect. com. 1988-92), Am. Astron. Soc. (astronomy edn. adv. bd. 1988—), Astron. Soc. Pacific (dir. ASTRO project), Am. Assn. Physics Tchrs., Nat. Assn. Sci. Writers, No. Calif. Sci. Writers Assn. (program chmn. 1983-85). Avocations: music, astronomy, sci., lit. Office: Foothill Coll Dept Astronomy 12345 El Monte Rd Los Altos CA 94022

FRANCE, SAMUEL EWING HILL, mechanical engineer; b. Fort Totten, N.Y., Aug. 22, 1925; s. Ewing Hill and Helen Gould (McKean) F.; m. Ethel Martin Albertson, June 25, 1949; children: Samuel William, Ethel Sanford, Susan Hill, Ann McKean. BS, U.S. Mil. Acad., 1946. Registered profl. engr., N.J. Sales engr. Carborundum Co., Niagara Falls, N.Y., 1947-53; mech. engr. Alvord & Swift, N.Y.C., 1953-55; sales engr. Am. Standard, Newark, 1955-57; project engr. Vogelbach and Baumann, Scotch Plains, N.J., 1957-64, 77-80; sales engr. Brangs and O'Brien, Springfield, N.J., 1964-72, Associated Air Products, West Orange, N.J., 1972-76; asst. v.p. EI Assocs., East Orange, N.J., 1981-93. Councilman Borough of Verona, N.J., 1962-73; active Verona Bd. Health, 1974-76. Mem. ASHRAE, NSPE, N.J. Soc. Profl. Engrs. (treas. North Cen. Jersey chpt. 1985—). Republican. Episcopalian. Home: 22 Mountain Rd Verona NJ 07044-1113

FRANCIS, CHARLES ANDREW, agronomy educator, consultant; b. Monterey, Calif., Apr. 12, 1940; s. James Frederick and G. Louise (Epperson) F.; m. Barbara Louise Hanson, June 23, 1962; children: Todd (dec.), Kevin, Andrea, Karen. BS, U. Calif., Davis, 1961; MS, Cornell U., 1967, PhD, 1970. Dir., maize breeder Internat. Ctr. for Tropical Agr., Cali, Colombia, 1970-72, dir., bean agronomist, 1973, dir. small farm systems, 1974-75, rsch. agronomist, 1976-77; prof. U. Nebr., Lincoln, 1977—, dir. Morocco project, 1982-84; dir. internat. program Rodale Inst., Emmaus, Pa., 1984-85; agronomist U.S. AID, Botswana, Liberia, 1978-90, World Bank, Colombia, S.Am., 1980; dir. Ctr. Sustainable Agr. Systems, 1990—; bd. dirs., sec. The Land Inst., Salina, Kans, 1990—; cons. OTA, Rockefeller Found., FAO/UN, 1978—. Editor: Multiple Cropping Systems, 1986, Breeding Crops for Sustainable Systems, 1993; co-editor: Sustainable Agriculture, 1990; contbr. chpts. to books and numerous articles to profl. jours. Cubmaster Cub Scout Pack 20, Lincoln, 1978-81; mem. ch. bd. Unitarian Universalist Ch., Lincoln, 1987-89; bd. dirs., v.p. sch. bd. Colegio Bolivar, Cali, 1973-77. 1st lt. U.S. Army, 1961-63. Fellow Am. Soc. Agronomy (divsn. chair 1968—, Robert E. Wagner award for Efficient Agriculture 1992), Crop Sci. Soc. Am.; mem. Phi Kappa Phi, Phi Beta Delta, Gamma Sigma Delta, Alpha Zeta. Democrat. Avocations: jogging, camping, biking, reading, travel. Office: U Nebr 219 Keim Hall Lincoln NE 68583

FRANCIS, EULALIE MARIE, psychologist; b. Holmdel, N.J.; d. Richard Erickson and Cora Mina (Patterson) F. BS, Newark State, N.J., 1945; EDM, Rutgers U., New Brunswick, 1957, MA, 1961; PhD, Rutgers U., Harvard U., N.J, Mass., 1971, 1973. Cert. Edn. Psychology. Tchr. Elem. Edn. Pub. Schs., Middletown, N.J., 1945-51; Elem. supr. Pub. Schs., Red Bank, N.J., 1951-63; dir. learning disability and psychologist Pub. Schs., East Brunswick, N.J., 1984; Cons. Nat. Assn. Mental Health N.Y.C., Family and childrens Sorriccs Natigna, N.Y.C., Lincoln Sch. Tchrs. Coll., Columbia U. 1981-89; Dir. Rsch. Div. NEA Assn. Trenton N.J. 1988-89. Author, editor: Kinesthetic Method of Reading, Theory and Techniques of Auditory Perception in Reading 1964-68. Adv. State Hist. Site Coun. Trenton N.J. 1986, Cultural and Heritage Com. Holmdel N.J. 1987-89. Mem. Arts Counc. of Princeton, Monmouth Mus. Lincroft, N.J., AAUW, Adv. Com. on Status of Women, Dir. Youth and Family Svcs., Princeton Child Devel. Inst., Rumson Country Club, Springlake Golf. Republican. Presbyterian. Avocations: flying private plane, swimming, photography, oil painting. Home: PO Box 43 Holmdel NJ 07733-0043

FRANCIS, FAITH ELLEN, biochemist; b. Batavia, N.Y., Dec. 28, 1929; d. Charles Lee and Julia (Maloney) F. BA, D'Youville Coll., 1952; PhD, St. Louis U., 1957. Rsch. assoc. St. Louis U., 1957-59, instr., 1959-64, asst. prof., 1964-72, assoc. prof., 1972—. Contbr. articles to Jour. Biol. Chemistry, Biochemistry, Jour. Clin. Endocrinology, Jour. Chromatoq, Soc. Exptl. Biology and Medicine. Fellow AAAS; mem. AAUP, Am. Chem. Soc., N.Y. Acad. Scis., Soc. Exptl. Biology and Medicine, Endocrine Soc., Sigma Xi. Roman Catholic. Home: Apt 902 40 Plz Sq Saint Louis MO 63103 Office: St Louis U Sch Medicine 3635 Vista Ave at Grand Blvd Saint Louis MO 63110-0250

FRANCIS, KENNON THOMPSON, physiologist; b. Camp LeJeune, N.C., July 8, 1945; m. Sheryl East, Dec. 29, 1966; children: Connie Jill, Wendy Kay. PhD, Auburn U., 1972; MRE, Birmingham Theol. Sem., 1992. Asst. prof. Troy State U., Montgomery, Ala., 1972-73; prof. physiology U. Ala., Birmingham, 1973—; chmn. Ala. Fitness Coalition, Birmingham, 1991—; mem. nat. adv. coun. for guidance on PhD program in phys. therapy at U. So. Calif., 1977; manuscript reviewer Phys. Therapy Jour., 1983—; vis. prof. S.W. Bapt. U., 1993. Author: Computer Essentials in Physical Therapy, 1986; assoc. editor Jour. Ala. Acad. Sci., 1983-84; contbr. articles to sci. publs. Mem. Sigma Xi, Alpha Eta, Gamma Sigma Delta. Home: 3851 Orleans Rd Birmingham AL 35243 Office: Univ Ala Birmingham Div Phys Therapy Birmingham AL 35294

FRANCIS, TIMOTHY DUANE, chiropractor; b. Chgo., Mar. 1, 1956; s. Joseph Duane and Barbara Jane (Sigwalt) F. Student, U. Nev., 1974-80, We. Nev. C.C., 1978; BS, L.A. Coll. Chiropractic, 1982, Dr. of Chiropractic magna cum laude, 1984; postgrad., Clark County Community Coll., 1986—; MS in Bio/Nutrition, U. Bridgeport, 1990. Diplomate Internat. Coll. Applied Kinesiology, Am. Acad. Pain Mgmt.; cert. kinesiologist; lic. chiropractor, Calif., Nev. Instr. dept. recreation and phys. edn. U. Nev., Reno, 1976-80; from tchng. asst. to lead instr. dept. principles & practice L.A. Coll. Chiropractic, 1983-85; pvt. practice Las Vegas, 1985—; asst. instr. Internat. Coll. Applied Kinesiology, 1990. Recipient Key award, 1990, Internat. Cultural Diploma of Hon., 1991. Fellow Internat. Acad. Clin. Acupuncture; mem. Am. Chiropractic Assn. (couns. on sports injuries, nutrition, roentgenology, technic, and mental health), Nev. State Chiropractic Assn., Nat. Strength and Conditioning Assn., Gonsted Clin. Studies Soc., Found. for Chiropractic Edn. and Rsch., Nat. Inst. Chiropractic Rsch., Nat. Acad. Rsch. Biochemists. Republican. Roman Catholic. Avocations: karate, weightlifting. Home: 3750 S Jones Las Vegas NV 89103

FRANCIS, WILLIAM KEVIN, civil engineer; b. East Orange, N.J., Apr. 2, 1965; s. William Herbert and Susan Louise (Gormly) F.; m. Amy Marie Winger, May 27, 1989; children: William Kevin Jr., Kristina Marie. BS in Civil Engring., Ariz. State U., 1988. Engr. in tng. Project engr. Eastcoast Testing and Engring., Ft. Lauderdale, Fla., 1989-90, Keith & Schnars, P.A., Ft. Lauderdale, 1990-91, Universal Engring. Scis., Orlando, Fla., 1991-92; dir. geotech. div. Keith & Schnars, P.A., Ft. Lauderdale, 1992—. Recipient Blue Cir. award Blue Cir. of Ariz., 1984. Mem. ASCE, NSPE, Fla. Engring. Soc., Peace River Engring. Soc. Republican. Roman Catholic. Home: 23104 SW 56th Ave Boca Raton FL 33433 Office: Keith & Schnars PA 324 SW 13th Ave Pompano Beach FL 33069

FRANCK, ARDATH AMOND, psychologist; b. Wehrum, Pa., May 5, 1925; d. Arthur and Helen Lucille (Sharp) Amond; m. Frederick M. Franck, Mar. 18, 1945; children—Sheldon, Candace. B.S. in Edn., Kent State U., 1946, M.A., 1947; Ph.D., Western Res. U., 1956. Cert. high sch. tchr., elem. supr., sch. psychologist, speech and hearing therapist. Instr., Western Res. U., Cleve., summer 1953, U. Akron, 1947-50; sch. psychologist Summit County Schs., Ohio, 1950-60; cons. psychologist Wadsworth Pub. Schs., Ohio, 1946-86; dir. Akron Speech & Reading Ctr., Ohio, 1950—; cons., dir. Hobbitts Pre-Sch., 1973-88. Author: Your Child Learns, 1976. Pres. Twirling Unltd., 1982—. Mem. Am. Speech and Hearing Assn., Internat. Reading Assn., Ohio Psychol. Assn., Mensa, Soroptomist (Akron). Home: 631 Ghent Rd Akron OH 44333-2629 Office: Akron Speech & Reading Ctr 700 Ghent Rd Akron OH 44333-2698

FRANCKLYN, CHRISTOPHER STEWARD, molecular biologist; b. Hartford, Conn., Nov. 13, 1957; s. Reginald Endicott and Phyllis (DeVeau) F. BA, U. Calif., Santa Barbara, 1979, MA, 1983, PhD, 1988. Rsch. chemist M&T Lab., Santa Barbara, 1979-82; rsch. & teaching asst. U. Calif., Santa Barbara, 1983-88; postdoctoral fellow MIT, Cambridge, 1988-91; asst. prof. U. Vt. Coll. Medicine, Burlington, 1991—. Contbr. articles to profl. jours. Mem. Am. Chem. Soc., N.Y. Acad. Scis. Achievements include a complete model of the arabinose operon in Escherichia Coli (with Nancy Lee), structural dissection of transfer RNA: partial transfer RNAs act as substrates for aninocyn tRNA synthetases. Office: U Vt Coll Medicine Dept Biochemistry Given Bldg Burlington VT 05405-0066

FRANDINA, PHILIP FRANK, civil engineer, consultant; b. Buffalo, N.Y., June 4, 1928; s. Frank Philip and Rose (Gugino) F.; m. Josephine Falsone, Apr. 15, 1950 (widowed Apr. 1983); children: Frank Philip, Joseph S., Rosanne; m. Mary Lou Klice, May 3, 1986. BS in Civil Engring., SUNY, 1964. Cert. Engring. Surveying. Asst. engr. Erie County, Buffalo, N.Y., 1958-60, hwy. maint. engr., 1960-64, sr. civil engr., 1964-72, chief engr., 1972-81, supt. of hwys., 1981-82, commr. pub. works, 1982-88; cons. pvt. practice, Buffalo, N.Y., 1988—; bd. mem. Erie C.C. Adv. Bd., Buffalo, N.Y., 1975—; arbitrator Am. Arbitrator Assn., 1985—; chmn. Transp. Adv. Bd., Buffalo, N.Y., 1988—; bd. mem. N.Y. State Hazardous Waste Citing Bd., Albany, N.Y., 1992—. Sec., treas. Erie County Credit Union, Buffalo, N.Y., 1974-81; founder, mgr., N. Buffalo Rockets, Buffalo, N.Y., 1976-80; chmn. Blessed Sacrament Parish Adminstr. Commn., Buffalo, N.Y., 1978-80. Named Engr. of Yr. N.Y. State Soc. Profl. Engrs., Buffalo, N.Y., 1987, Top Ten Pub. WOrks Officials in USA, Am. Pub. Works Assn., Albany, 1988; recipient Resolution of Commendation award Erie County Legis., Buffalo, N.Y., 1987, Nina award Nat. Columbus Day Commn., Buffalo. N.Y., 1989. Mem. Am. Soc. Civil Engrs., Assn. for Bridge Constrn. and Design. (founder), Am. Acad. Forensic Sci., Triple Nine Soc. Republican. Roman Catholic. Achievements include design in construction of the new Erie County Correctional Facility and the Erie County Holding Ctr.; development of energy saving concepts in Erie County Bldgs. with significant cost savings. Home: 75 Chatham Pkwy Buffalo NY 14216-3108 Office: 90 John Muir Dr Buffalo NY 14228

FRANE, JAMES THOMAS, mechanical engineer, author; b. Independence, Iowa, Nov. 27, 1942; s. Gerald E. and Alice M. (Crowell) F.; m. Carol Louise Trestrail, June 14, 1964; children: Sara Louise, Bryan Thomas. BS, Calif. Maritime Acad., 1963; postgrad. in Bus., Sacramento State U., 1963-65. Registered profl. engr., Calif.; registered environ. assessor, Calif. Test engr. Aerojet-Gen. Corp., Nimbus, Calif., 1963-65, McDonnell-Douglas Corp., Rancho Cordova, Calif., 1965-70; sr. engr. Bechtel Corp., San Francisco, 1970-76, engring. supr., 1976—. Author: Drywall Contracting, 1987; contbr. articles to profl. jours. Asst. scoutmaster, mem. Troop 303 com. Boy Scouts Am., Orinda, Calif., 1987-93. Mem. ASTM, Am. Waterworks Assn., Audio Engring. Soc. (assoc.). Office: Bechtel Corp 45 Fremont St San Francisco CA 94119

FRANETOVIC, VJEKOSLAV, physicist; b. Split, Croatia, Jan. 5, 1946; came to U.S., 1980; s. Vinko and Marija (Tasovac) F.; m. Snjezana Milat; children: Marija, Lucija. BSc in Physics, U. Zagreb, Croatia, 1971, MSc in Physics, 1978. Rsch. and teaching asst. Inst. of Physics/U. Zagreb, 1971-78; asst. prof. U. Zagreb, 1978-80; vis. scientist MIT, Cambridge, Mass., 1980-83; project scientist GM Rsch. Labs., Warren, Mich., 1983-88, sr. project scientist, 1988—; analytical electron microscopy expert GM Co., Warren, 1984-91. Author poetry; contbr. articles to Jour. Materials Sci., Metallography, Materials Sci. and Engring. Participant Pres. Bush Leader's Roundtable Discussion, Detroit, 1990; juror 6th Jud. Cir. Ct., County of Oakland, Mich., 1990; pres. Detroit Jr. Star Tamburitzans, 1987-89; chmn. The Coordinating Com. for Croatian Affairs. Mem. N.Y. Acad. Scis., Electron Microscopy Soc. Am., Croatian Acad. Am., Inc., Assn. Alumni Croatian U. (pres. 1991, 92, 93, 94), Metall. Soc., Materials Rsch. Soc. Almae Matris Alumni Croaticae (world com., pres. 1990—), Sigma Xi. Roman Catholic. Achievements include development of new splat cooling device; first to report measurements for stacking faulty energy in rapid quenched materials; discovery of a new metastable phase in Ag-In alloy; development of equipment for preparing electropolished specimens at controlled and low temperatures; investigation of nodular iron which provided an increased understanding of the austempering process and explained the high impact properties of this material; investigation of the influence of ion bombardment on the surface composition of alpha brass which provided new findings about the influence of ion beams on solid materials. Office: GM Rsch Lab 30500 Mound Rd Warren MI 48092-2054

FRANGAKIS, GERASSIMOS P., electronic engineer; b. Kalamata, Greece, Apr. 21, 1940; s. Panayotis and Pepi (Stavropoulos) F.; m. Bess Bastoulis, July 2, 1978; children: Penelope, Paraskevi. Diploma, Coll. Electronics, 1962; MSc, U. Southampton, Eng., 1972, PhD, 1985. Electronics engr. Nat. Cash Register Co., Athens, 1963; electronics engr. Demokritos div. NRC, Athens, 1964-71, rsch. scientist, 1971-85, sr. rsch. scientist, 1986—. Author: Logic Circuits, 1975; co-author: Digital Computers, 1978, Computer Architectures, 1987; contbr. articles to profl. jours. Mem. Greek Soc. Med. Informatics, Greek Soc. Artificial Intelligence. Avocations: gardening, swimming, chess, reading. Home: 14 Thessaloniki Str, Cholargos, Athens 15562, Greece Office: NCSR Demokritos, Aghia Paraskevi, Athens 15310, Greece

FRANK, CHRISTOPHER LYND, mechanical engineer; b. Chesterton, Ind., Dec. 26, 1949; s. Clarence Edward and Marie Caroline (Saylor) F.; m. Deborah Lynn Tanner, July 3, 1971; chileren: Erin Marie, Christopher David. BS in Engring., Calif. State U., Sacramento, 1983; cert. injection molding, U. Lowell, 1986. Plant mgr. Redelco Plastics, Clovis, Calif., 1975-79; owner, designer The Energy Factory, Fresno, Calif., 1977-79, Solar Utility Network, Yuba City, Calif., 1979-81; engr., designer Houston

Fearless 76, Carson, Calif., 1983-86; engr., designer Air Force Advanced Composites Program, Sacramento, 1986—, head thermoplastics devel. Contbr. articles to profl. jours. Served as sgt. USAF, 1970-74. Recipient Logistics Civilian Engr. of the year award U.S. Air Force. Mem. Soc. Automotive Engrs., ASME, Soc. Mfg. Engrs., Soc. Plastics Engrs. Achievements include introduction of injection molding into the Air Force; application of liquid crystal polymers to rocket motor cases. Office: SM-ALC/TIEC Bldg 243-E McClellan AFB CA 95652

FRANK, HELMAR GUNTER, educational cyberneticist; b. Waiblingen, Baden-Württemberg, Germany, Feb. 19, 1933; s. Manfred Helmut and Erna Hedwig (Glocker) F.; m. Brigitte Christine Böhringer, June 30, 1970 (dec. Mar., 1990); children: Ines Ute, Tilo Ingmar; Vera Barandovska, Dec. 20, 1991. M in Math., Tech. U., Stuttgart, Fed. Republic Germany, 1956, PhD, 1959; Universitätsdozent, Johann Kepler U., Linz, Austria, 1970; Prof. in Cybernetics (hon.), Pedagogical U., West Berlin, 1972. Tchr. math., physics gymanasiums Bad Württemberg, 1958-61; sci. collaborator Tech. U. Karslruhe, 1961-63; assoc. prof. cybernetics Pedagogical U., West Berlin, 1963-70, prof., 1971-72; prof. U. Paderborn, North Rhine Westphalia, Fed. Republic Germany, 1972—; prof. in comm. scis. (hon.) Tech. U., Berlin, 1992; Pres. Soc. Programmed Instruction, 1964-70, Internat. Acad. Scis., Republic of San Marino, 1987—; unlimited vis. prof. U. Nat. Rosario, Argentina, 1984, South China Normal U., Guangzhou, Peoples Republic China, 1986, U. Sibiu-Hermannstadt, Romania, 1990, Tech. Univ., Berlin, 1992, Universitas Carolina, Prague, 1992. Author: Cybernetic Foundation of Education Science, 1962, 2d edit., 1969, Cybernetics and Philosophy, 1966, Propedeutics to Prospective Education Science, 1984, 2d edit., 1991. Mem. Internat. Assn. Cybernetics Namur (bd. dirs.), World Assn. Cybernetics, Informatics and Systems Theory (v.p.). Mem. Free Dem. Party. Club: European (pres. 1978-80). Office: U Paderborn Cybernetics, KleinenbergerWeg 16B, D 33100 Paderborn Germany

FRANK, ILYA MIKHAILOVICH, physicist; b. Leningrad, Oct. 23, 1908. Ed., Moscow U. Asst. to prof. S. I. Vavilov, 1928; with Leningrad Optical Inst., 1930-34, Lebedev Inst. Physics, USSR Acad. Scis., 1934-70; prof. physics Moscow U., from 1944; head lab. of neutron physics, Joint Inst. for Nuclear Research, from 1957; corr. mem. USSR Acad. Scis., 1946-48, academician, from 1968. Author: Function of Excitement and Curve of Absorption in Optic Dissociation of Tallium Ioclate, 1933; Coherent Radiation of Fast Electron in a Medium, 1937; Pare Formation in Krypton under Gamma Rays, 1938; Doppler Effect in Refracting Medium, 1942; Radiation of a Uniformly Moving Electron Due to Its Transition from One Medium into Another, 1945; Neutron Multiplication in Uranium-Graphite System, 1955; On Group Velocity of Light in Radiation in Refracting Medium, 1958; Optics of Light Sources Moving in Refracting Media, 1960; On Some Peculiarities of Vavilov-Cherenkov Radiation, 1986. Recipient Nobel prize for physics (with Tamm and Cherenkov), 1958; State Prize, 1946, 54, 71; Order of Lenin (3); Order of Red Banner of Labor; Order of October Revolution 1978; Varilov Gold medal, 1979. Home: Moscow Russia *Died June 22, 1990.*

FRANK, KARL H., civil engineer, educator; b. San Francisco, Apr. 16, 1944; s. Heinz Benno and Jessie (Johnson) F.; m. Jeanne Doelp, June 28, 1969; children: Erik, Ilse. BCE, U. Calif., Davis, 1966; MCE, Lehigh U., 1969, PhD in Civil Engring., 1971. Registered profl. engr., Tex. Rsch. asst. Fritz Engring. Lab. Lehigh U., Bethlehem, Pa., 1967-71; asst. prof. civil engring. U. Tex., Austin, 1974-80, assoc. prof., 1980-88, prof., 1988—, dir. Ferguson Structural Engring. Lab., 1991—; structural rsch. engr. Fed. Hwy. Adminstrn., 1971-74; cons. in field. Named one of Outstanding Young Men Am., 1974. Mem. ASCE (assoc., mem. com. structural fatigue, 1971-77, flexural mems. com. 1974-77, chmn. subcom. on longitudinally stiffened plate girders, mem. com. structural connections 1980-83, sec., vice chmn., chmn. structures group Tex. sect. 1984-87, Raymond C. Reese award), Am. Welding Soc. Avocations: auto racing, golf. Home: 6005 Ivy Hills Dr Austin TX 78759-5522 Office: U Tex Dept Civil Engring Balcones Rsch Ctr 10100 Burnet Rd Austin TX 78758

FRANK, MARY LOU BRYANT, psychologist, educator; b. Denver, Nov. 27, 1952; d. W.D. and Blanche (Dean) Bryant; m. Kenneth Kerry Frank, Sept. 9, 1973; children: Kari Lou, Kendra Leah. BA, Colo. State U., 1974, MEd, 1983, MS, 1986, PhD, 1989. Tchr. Cherry Creek Schs., Littleton, Colo., 1974-80; from grad. asst. to dir. career devel. Colo. State U., Ft. Collins, 1980-86; intern U. Del., Newark, 1987-88; psychologist Ariz. State U., Tempe, 1988-93; assoc., lead prof. psychology Clinch Valley Coll. U. Va., Wise, 1992—; instr. Colo. State U., Ft. Collins, 1981-82, counselor 1984-85, 86-87; prof. Ariz. State U., Tempe, 1989-92; assoc. prof. psychology Clinch Valley Coll. U. Va., 1992—. Author: (program manual) Career Development, 1986; contbr. book chpts. on eating disorders and existential psychotherapy. Com. mem. Missions and Social Concerns, Gilbert, 1988—. Mem. APA, AACD, Am. Coll. Pers. Assn., Am. Assn. for Marriage and Family Therapy (assoc.), Phi Kappa Phi, Phi Beta Kappa, Pi Kappa Delta. Avocations: music, hiking, reading. Office: Clinch Valley Coll of U Va Psychology Dept Wise VA 24293

FRANK, MICHAEL VICTOR, risk assessment engineer; b. N.Y.C., Sept. 22, 1947; s. David and Bernice (Abrams) F.; m. Jane Griminger, Dec. 21, 1969; children: Jeffrey, Heidi, Heather. BS, UCLA, 1969; MS, Carnegie-Mellon U., 1972; PhD, UCLA, 1978. Registered profl. engr., Calif.; cert. profl. cons. to mgmt. Engr. Westinghouse Electric Corp., Pitts., 1970-72, Southern Calif. Edison, Los Angeles, 1972-74; lectr. U. Calif., Santa Barbara, 1976-77; task leader General Atomics, San Diego, 1977-81; sr. exec. engr. NUS Corp., San Diego, 1981-85; with Mgmt. Analysis Co., San Diego, 1985-86; sr. cons. PLG, Newport Beach, Calif., 1986-89; pres. Safety Factor Assocs., Inc., Encinitas, Calif., 1989—; tech. dir. risk and reliability studies of NASA facilities, space vehicles and nuclear facilities worldwide; risk assessment cons. U.S. Interagy. Nuclear Safety Rev. Panel, NASA hdqrs., NASA Ames Rsch. Ctr.; lectr. on risk assessment at various NASA ctrs.; qualified forensic cons. in product defects and hazards. Contbr. articles to profl. jours. Mem. AIAA, Soc. for Risk Analysis, Am. Assn. for Artificial Intelligence, Forensic Cons. Assn. (past pres.), Nat. Bur. Profl. Mgmt. Cons., Affiliation Profl. Cons. Orgns. (bd. govs.). Avocations: family activities, tennis, outdoor activities.

FRANK, PAUL SARDO, JR., forester, air force officer; b. Balt., Apr. 20, 1936; s. Paul Sardo and Amalie (Hafer) F.; m. Nancy Ann Parker, June 19, 1960 (div. Apr. 1976); children: Paul Sardo III, Karen Marie Frank Yoder, Kimberly Ann; m. Marlene Ann Fruehauf, Aug. 9, 1981. BS, U. Md., 1958, MEd, 1964; MS in Forestry, W.Va. U., 1978, PhD, 1981. Cert. forester, Ala., W.Va.; cert. tchr., Md. Asst. prof. Garrett C.C., McHenry, Md., 1972-76; fire staff officer Ala. Forestry Commn., Montgomery, 1982-83; tchr. earth sci. Frederick County Pub. Schs., Brunswick, Md., 1983-86; commd. capt. WVANG, 1974; advanced through grades to col. USAFR, 1992; chief promotion policy Hdqs. Air Res. Pers. Ctr., Lowry AFB, Denver, 1986-87; chief promotion secretariat Hdqrs. Air Res. Pers. Ctr., Lowry AFB, Denver, 1987-91; asst. for res. affairs USAF Air U., Maxwell AFB, Ala., 1992—; cons. on urban forestry dirctorate environ. planning Hdqs. Air Force Engring. and Svcs. Ctr., Tyndall AFB, Panama City, Fla., summers 1983-86; presenter 6th N.Am. Forest Soils Conf., 1983. Contbr. articles to Castanea, Can. Jour. Forest Rsch., Atlantic Naturalist, Forest Ecology Mgmt. Scoutmaster Boy Scouts Am., Accident, Md., 1966-76; radiol. def. officer CD Office, Garrett County, Oakland, Md., 1969-81; Colo. wing staff aerospace edn. officer CAP, Lowry AFB, 1987-89. Recipient Gill Rob Wilson award CAP, 1969, Wood Badge award Boy Scouts Am., 1973, 1st ann. award for Outstanding Rsch. Hardwood Rsch. Coun., 1985. Mem. NSTA (life), Am. Forestry Assn., Res. Officers Assn. (life), Sigma Xi. Office: Hdqrs AU/RPR 55 LeMey Plaza S Maxwell AFB AL 36112

FRANK, RICHARD STEPHEN, chemist; b. Teaneck, N.J., Sept. 7, 1940; s. Max and Viola (Ollert) F.; m. Nancy Sheridan Dempster, May 30, 1964; children: Karen, Richard Jr. BS, Washington Coll., 1962. Tchr. Kent County Bd. Edn., Chestertown, Md., 1962-63; chemist FDA, Balt., 1963-68, Arlington, Va., 1968; forensic chemist Drug Enforcement Adminstrn., Washington, 1968-70, supervising forensic chemist, assoc. dep. asst. administr., 1970—; mem. adv. task force local govt. police mgmt., Washington, 1989-91; mem. operating com. Crime Lab Info. Systems, Washington, 1976-78, Criminalistics Cert. Study Com., 1976-78; rep. U.S. govt. INTERPOL Forensic Scis. Symposia, St. Cloud, France, 1978, 80, 83, 86, Lyons, France, 1989, 92, mem. organizing com., 1989—, chmn., 1992—; rep. U.S. govt. UN Expert Group on Pre-Trial Destruction of Seized Narcotic Drugs, Psychotropic Substances, Precursors and Essential Chems., Bangkok, 1990. Contbr. articles to Jour. Forensic Scis., Analytical Chemistry, Internat. Microform Jour. Legal Medicine, Police Chief, Jour. Analytical Toxicology, Toxicology Talk; contbr. chpts.: Forensic Science Progress, Vol. 4, 1990. With U.S. ANG, 1963-67. Fellow Am. Acad. Forensic Scis. (pres. 1988-89), Am. Soc. Crime Lab Dirs. (sec. 1980-81), Fed. Exec. Inst. Alumni Assn., Nat. Assn. Ret. Fed. Employees. Office: Drug Enforcement Adminstrn 600-700 Army Navy Dr Washington DC 20537

FRANK, SANDERS THALHEIMER, physician, educator; b. Middletown, Conn., May 11, 1938; s. Harry S. and Pauline (Thalheimer) F.; B.A., Amherst Coll., 1959; M.D., N.Y. Med. Coll., 1963; children: Geoffrey Brooks, Susan Kimberly, Jonathan Blair. Intern, Sinai Hosp., Balt., 1963-64; resident Wilford Hall Med. Center, San Antonio, 1965-68; pvt. practice medicine, specializing in pulmonary disease, Monterey Park, Calif., 1971-92; dir. respiratory care Garfield Hosp., Monterey Park, 1974-92, Beverly Hosp., Montebello, Calif., 1975-78; assoc. prof. medicine U. So. Calif., L.A., 1972-90; pvt. practice, Zanesville, Ohio, 1992—. Served to maj. USAF, 1964-71. Decorated USAF Commendation medal; recipient Philip Hench award for demonstrating relationship of rheumatoid arthritis to lung disease, 1968; award of merit Los Angeles County Heart Assn., 1974. Fellow Royal Soc. Medicine (London), Am. Coll. Chest Physicians, A.C.P.; mem. Am. Thoracic Soc., Calif. Thoracic Soc., Nat. Assn. Dirs. Respiratory Care, Respiratory Care Assembly Calif., Alpha Omega Alpha. Contbr. articles in field to med. jours. Recorded relationship of ear-lobe crease to coronary artery disease, 1973.

FRANK, STEVEN NEIL, chemist; b. Red Oak, Iowa, Feb. 15, 1947; s. Robert Joseph and Joyce (Erickson) F.; m. Carol Bert Femmer, Jan. 4, 1975. BS, Colo. State U., 1969; PhD, Calif. Inst. Tech., 1974. Sr. mem. tech. staff, solar energy project Tex. Instruments, Dallas, mgr. fuel cell devel., 1980-83, mgr. charge coupled imagers, 1983-86, mgr. wafer fabrication, focal plane array, 1986-88, mfg. mgr., focal plane array, 1988-90, mgr. focal plane array assembly and testing, 1990-91, mgr. uncooled imaging, 1990—. Author (with others): Laboratory Techniques in Electro-Analytical Chem, 1984; referee Jour. Applied Physics, 1977—, Jour. Phys. Chemistry, 1977—; contbr. articles to profl. jours. Robert A. Welch fellow U. Tex., 1974-77. Fellow Am. Inst. Chemists; mem. Am. Chem. Soc., Electrochem. Soc. Achievements include 4 patents for Configuration for Gas Redox Fuel Cell employing an Ion Exchange Membrane; novel method for Catalyst Application to a Substrate for Fuel Cells; Graphite Flow Through Electrode for Fuel Cell Use; Tin Oxide CCD Imager. Home: 471 Hackberry Dr Mc Kinney TX 75069-9511 Office: Tex Instruments Inc PO Box 655012 Dallas TX 75265-5474

FRANKE, GEORGE EDWARD, highway engineer; b. Uniontown, Mo., Feb. 23, 1936; s. Albert Carl August and Leola Susanna (Kassel) F.; m. Luella Augusta Koenig, Feb. 21, 1959; children: David George, Matthew Paul, Sarah Ann. BSCE, Mo. Sch. Mines and Metallurgy, 1958. Registered profl. engr., Ky. Asst. resident engr. Ky. Dept. Hwys., Cadiz, Lietchfield, Ky., 1958-64; dist. traffic engr. Ky. Dept. Hwys., Paducah, Ky., 1964-67, dist. maintenance engr., 1967-70, dist. ops. engr., 1970-73; asst. dir. divsn. maintenance Ky. Dept. Hwys., Frankfort, Ky., 1973-78; chief dist. engr. Ky. Dept. Hwys., Lexington, Ky., 1978-80; transp. engr. Ky. Dept. Hwys., Frankfort, 1981—. Scoutmaster, troop leader Boy Scouts Am., Frankfort, 1975-81; com. chmn. elders, stewardship and evangelism Our Redeemer Luth. Ch., Lexington, 1980-90; Sunday sch. tchr. 5th-8th grades, 1990-93. col. Corps of Engrs., USAR, 1958-90. Recipient Dist. Award of Merit, Boy Scouts Am., 1980; decorated Meritorious Svc. medal (2), Army Commendation medal with oak leaf cluster, Armed Forces Res. medal wiht 10 yr. device, Army Res. Component Achievment medal with 2 oak leaf clusters. Mem. NSPE, Ky. Soc. Profl. Engrs. (state scholarship chmn. 1990-92, Pres.'s award 1992-93), Ky. Assn. Transp. Engrs., Res. Officers Assn. (state resolutions chmn. 1980-93, Cert. of Appreciation 1992). Lutheran. Home: 1208 Miami Trail Frankfort KY 40601 Office: Ky Dept Hwys State Office Bldg 7th Fl Frankfort KY 40622

FRANKE, HILMAR, physics educator; b. Meinerzhagen, Germany, June 26, 1946; s. Josef and Hedwig (Bredenbröker) F.; m. Ruth Sigrid Kuprat, July 21, 1978; children: Heidrun, Joern-Holger. Diploma, U. Bochum (Germany), 1974, D. in Natural Scis., 1979. Researcher Dornier System, Friedrichshafen, 1979-80; acad. lectr. U. Osnabrück, 1980-89; prof. U. Duisburg (Germany), 1989—; vis. scientist IBM, Yorktown Heights, 1984-85. Contbr. over 50 applied physics articles to profl. jours. SPD. Lutheran. Achievements include patents in the application of photopolymers and polymers for optoelectronics and optical sensors. Home: Rossinistr 11, 4550 Bramsche Germany Office: U Duisburg, Lotharstr 1, 4100 Duisburg Germany

FRANKEL, BARBARA BROWN, cultural anthropologist; b. Phila., Dec. 24, 1928; d. Paul and Sarah (Magil) Brown; m. Herbert L. Frankel, Feb. 27, 1949 (dec. Sept. 1976); children: Claire R. Sholes, Joan L. Frankel, David S. Frankel; m. Donald T. Campbell, Mar. 19, 1983. PhB, U. Chgo., 1947; BA, Goddard Coll., 1966; MA in Anthropology, Temple U., 1970; PhD, Princeton (N.J.) U., 1974. Asst. prof. Lehigh U., Bethlehem, Pa., 1973-77, assoc. prof., 1977-85, assoc. dean arts and sci., 1981-83, prof. anthropology, 1985—; rsch. assoc. prof. Boston U., 1980-81. Author: Childbirth in the Ghetto, 1977, Transforming Identities, 1989; contbr. articles to profl. jours. Bd. dirs. Pinebrook Svcs. for Children and Youth, Whitehall, Pa., 1987-93. Grad. fellowship for Women Danforth Found., Princeton U., 1969-73; predoctoral fellowship AAUW, 1971-72; rsch. grant Mellon Faculty Devel. Grant, Boston U., 1980-81, Provost's Rsch. award Lehigh U., 1987. Fellow Am. Anthropol. Assn., Royal Anthrolpol. Soc. (U.K.), Soc. for Applied Anthropology (chair ethics com. 1986-88), mem. AAAS, Soc. for Humanistic, Urban, Med. Anthropology, Social Studies of Sci., Phila. Anthropol. Soc. (pres. 1988), Phi Beta Kappa (pres. Beta chpt. 1989-90). Democrat. Agnostic Jewish. Achievements include rsch. on utopian/therapeutic communities, urban society, epistemology of anthropology. Home: 637 N New St Bethlehem PA 18018 Office: Lehigh U Sociology & Anthropol 681 Taylor St Bethlehem PA 18015

FRANKEL, IRWIN, chemical engineer; b. New Orleans, Nov. 25, 1919; s. Louis A. and Renee (Levy) F.; m. Saidee Kanoff, Dec. 1945 (div. Nov. 1976); children: Raymond E., Laurence J.; m. Barbara Lazier, Dec. 11, 1976. B in Engring., Tulane U., 1942; MS in Chem. Engring., Case Inst. Tech., 1948; D in Chem. Engring., Rensselaer Poly. Inst., 1951. Profl. engr. Va., N.Y. Mgr. process devel. Olin Mattheison Corp., Niagara Falls, N.Y., 1957-62; mgr. pilot plants Allied Chem. Corp., Richmond, Va., 1962-73; project mgr. Versar Inc., Springfield, Va., 1974-78; staff mem. MITRE Corp., McLean, Va., 1978-82; sr. engr. U.S. Synthetic Fuels Corp., Washington, 1982-86; sr. staff engr. Versar Inc., Springfield, 1986-91, Sci. Applications Internat. Corp., Falls Church, Va., 1991--. Contbr. articles to profl. jours. Chmn. Commn. on Pub. Edn., Oak Lawn, Ill., 1953-55; v.p., Richmond Band Soc., 1968-72; mgr. Beacon Concert Orch., N.Y., 1947-49. Lt. col. USAF, 1942-79, CBI. Fellow Am. Inst. Chem. Engrs. (chmn. Va. tidewater sect. 1972-73, nat. capital sect. 1979-80, gov. programs steering com. 1979-81), Tau Beta Pi, Sigma Xi, Phi Lambda Upsilon. Achievements include patents. Home: 8423 Briar Creek Dr Annandale VA 22003 Office: Sci Applications Internat 7600-A Leesburg Pike Falls Church VA 22043

FRANKEL, KENNETH MARK, thoracic surgeon; b. Bklyn., July 29, 1940; s. Clarence Bernard and Ruth (Rutes) F.; m. Felice Cala Oringel, Dec. 10, 1967; children: Matthew David, Michael Jacob. B.A., Cornell U., 1961; M.D., SUNY, Bkyln., 1965. Diplomate Am. Bd. Surgery, Am. Bd. Thoracic Surgery. Intern in surgery Yale New Haven Hosp., 1965-66; resident in surgery Kings County-SUNY Med. Ctr., Bklyn., 1966-67, 69-71, chief resident in gen. surgery, 1971-72, resident in thoracic surgery, 1972-73, chief resident thoracic and cardiovascular surgery, 1973-74; attending thoracic surgeon Mercy Hosp., Springfield, Mass., 1974—, Holyoke (Mass.) Hosp., 1974—, Providence Hosp., Holyoke, 1974—; pvt. practice medicine specializing in thoracic surgery Springfield, 1974—; chief thoracic surgery Baystate Med. Ctr., Springfield, 1977—; assoc. clin. prof. cardiothoracic surgery Tufts U. Sch. Medicine, 1978—; cons. in thoracic surgery Ludlow (Mass.) Hosp., Noble Hosp., Westfield, Mass., 1976—; cons. Shriners Hosp. for Crippled Children. Contbr. articles to profl. jours. Corporator Springfield (Mass.) Symphony Orch., Stage West, Springfield. Served to capt. U.S. Army, 1967-69. Decorated Bronze Star, Gallantry Cross (Republic of Vietnam). Fellow ACS, Am. Coll. Chest Physicians; mem. AMA, ACLU, Soc. Thoracic Surgeons, Am. Thoracic Soc., Springfield Acad. Medicine (past pres), Mass. Med. Soc. (councilor 1981-83), Hampden Dist. Med. Soc. (exec. com. 1990—), Physicians for Social Responsibility, Amnesty Internat., Internat. Physicians for Prevention Nuclear War, Union Concerned Scientists. , Cornell Club of Western Mass., Porsche Club of Am., Maimonides Med. Club (past pres.). Democrat. Jewish. Home: 202 Ellington Rd Longmeadow MA 01106 Office: Baystate Med Ctr 2 Medical Center Dr Springfield MA 01107-1270

FRANKEL, MICHAEL S., telecommunication and automation sciences executive; b. L.A., Sept. 22, 1946; s. Eugene D. and Susana R. (Mutal) F.; m. Shayne M. Larson; children: Jeffrey D. Barrom, Mahriya A. BS in Elec. Engring., Stanford U., 1968, MS in Elec. Engring., 1970, PhD in Elec. Engring., 1973. Instr., rsch. assoc. Stanford U., Palto Alto, Calif., 1968-73, faculty lectr., 1973-74, rsch. engr., 1974-77, sr. rsch. engr., 1978-79; asst. dir. SRI Internat., RPL, Menlo Park, Calif., 1978-80; dept. dir. SRI Internat., AITAD, Menlo Park, 1980-82; ctr. dir. SRI Internat., ITSC, Menlo Park, 1982-87; v.p., divsn. dir. SRI Internat., ITSTD, Menlo Park, 1987—; bd. dirs. Cisco System, Composities Automation Consortium; v.p., dir. ITAD div. SRI Internat., Menlo Park, 1987—; mem. Army Sci. bd. Washington, 1992—. Contbr. articles to profl. jours. adv. bd., corp. affiliate Divsn. Computer Sci., U. Calif., Davis. Recipient Bausch and Lomb Science award. Mem. IEEE, NAS (Radio elec. battle mgmt. panel,Navy 21 study, Naval Studies bd. 1987-88, Radio elec. and acoustic battle mgmt. panel Future Aircraft Carrier study, 1990-91), Am. Defense Preparedness Assn., Armed Forces Communications and Electronics Assn., Cosmo Club, Sigma Xi, Tau Beta Pi. Achievements include patent for passive, high gain, freequency steerable satellite repeater, patent for advanced Teleeducation concept. Avocations: woodworking, antique restoration, camping.

FRANKEL, SHERMAN, physicist; b. N.Y.C., Nov. 15, 1922; s. Harry and Rose F.; m. Ruzena Bajcsy, Oct. 22, 1981; 1 son by previous marriage, Walter. B.A., Bklyn. Coll., 1943; M.S., U. Ill., 1947, Ph.D., 1949. Mem. staff radiation lab M.I.T., 1943-46; instr. U. Pa., Phila., 1950-52; asst. prof. physics U. Pa., 1952-56, asso. prof., 1956-60, prof., 1960—; vis. scientist Niels Bohr Inst., Denmark, 1968, C.E.R.N. Geneva, 1975, C.E.N. de Saclay, France, 1979; guest fellow Stanford U. Ctr. for Internat. Security Arms Control, 1987; guest scholar Brookings Inst., 1987. Assoc. editor Rev. of Sci. Instruments, 1952-53. Guggenheim fellow, 1957, 79. Fellow Am. Phys. Soc.; mem. AAUP, AAAS, Sigma Xi, N.Y. Acad. Sci., Pi Mu Epsilon. Home: 2320 Delancey Pl Philadelphia PA 19103-6407 Office: U Pa Physics Dept 33d and Walnut Sts Philadelphia PA 19104

FRANKENBERG, DIRK, marine scientist; b. Woodsville, N.H., Nov. 25, 1937; s. Charles Henry and Patricia Edith (Smith) F.; m. Susan Alice Campbell, June 25, 1960; children—Elizabeth Alice, Eben Whitfield. A.B. in Biology, Dartmouth Coll., 1959; M.S. in Biology, Emory U., 1960, Ph.D., 1962. Asst. prof. zoology, research asso. Marine Inst., U. Ga., 1962-66; adj. lectr. Dartmouth Coll., 1965-69; asst. prof. dept. biol. scis. U. Del., 1966-67; asso. prof. zoology, research asso. Marine Inst., U. Ga., 1967-72, prof. zoology, research asso., 1972-74; dir. ocean scis. div. NSF, Washington, 1978-80; dir. marine sci. program, prof. U. N.C., Chapel Hill, 1974—; vice chmn. Univ. Nat. Oceanographic Lab. System, 1981-84; mem. Duke U.-U. N.C. Oceanographic Consortium Policy Bd., 1980—. Contbr. articles to profl. jours. Mem. AAAS, Assn. Southeastern Biologists, Am. Soc. Limnology and Oceanography, Ecological Soc. Am., Estuarine and Brackish Water Scis. Assn. Office: Univ of NC 3407 Arendell St Chapel Hill NC 18557

FRANKENBERGER, GLENN F(RANCES), lawyer; b. Commack, N.Y., Dec. 22, 1968; s. Clifford and JoAnn (Romagnano) F. BS, Fordham U., 1991. Teaching asst. Fordham U., Bronx, 1989-90, lab. aid, 1990-91; legal rschr. Fordham U., N.Y.C., 1991—. Recipient Victor Hess award Fordham Physics Dept., 1991. Mem. Am. Physics Soc. (v.p. 1991), Circle K Internat., Phi Beta Kappa, Alpha Sigma Nu. Home: 7 Map Ln Commack NY 11725 Office: Fordham Univ Dept Physics 7 Map Ln Commack NY 11725-3610

FRANKENHAEUSER, MARIANNE, psychology educator; b. Helsinki, Finland, Sept. 30, 1925; d. Tor and Ragni (Alfthan) von Wright; m. Bernhard Frankenhaeuser, Mar. 26; 1 child, Carola. Dipl. in Psychology, Oxford U., 1947; BA, U. Helsinki, 1950; MA, U. Stockholm, 1954; PhD, U. Uppsala, 1959; D in Polit. Sci (hon.), U. Turku, Finland, 1990. Clin. psychologist Seraphimer Hosp., Stockholm, 1951-55; rsch. psychologist Karolinska Inst., Stockholm, 1955-60; asst. prof. U. Stockholm, 1960-63; rsch. fellow Swedish Coun. for Social Sci. Rsch., 1963-65; assoc. prof. Swedish Med. Rsch. Coun., 1965-69, prof., 1969-80; prof. psychology Karolinska Inst., Stockholm, 1980-92, head psychology div., 1982-92; mem. sci. adv. panel Swedish Nat. Bd. Health and Welfare, 1968-91; med. adv. bd. Swedish Tobacco Co., 1971-83; bd. dirs. Bank of Sweden Tercentenary Found., 1974-80, rev. com. on behavioral scis., 1974-80; vis. prof. psychiat. Stanford U., 1976; cons. U.S. Social Sci. Rsch. Coun. Com. on Life Course Perspectives in Middle and Old Age, 1978—; bd. dirs. Swedish Coun. for Mgmt. and Work Life Issues, 1981-91, Swedish Soc. for Mems. Parliament and Scientists, 1981—, many others in past. Editorial bd. Acta Psychologica, European Jour. Psychonomic Sci., 1970-72, Jour. Human Stress, 1975-84, Motivation and Emotion, 1975-89, Internat. Jour. Psychology, 1976-80, Human Neurobiology, 1981-84, Plenum Series on Stress and Coping, 1985—, Stress Medicine, 1985—, New Trends in Exptl. and Clin. Psychiatry, 1986—, Jour. Orgnl. Behavior, 1987—, Internat. Jour. Behavioral Medicine, 1992—; contbr. articles to profl. jours.; author: Stress - en del av livet, 1983, 4th edit. 1992, Women, Work and Health, 1991, Kvinnligt, Manligt, Stressigt, 1993, others. Sci. coun. Swedish Nat. Ency., 1987—; mem. com. on human rights Nat. Acad. Scis., 1989—; mem. productivity commn. Swedish Govt., 1989-91; mem. work environ. coun. Swedish Govt., Dept. Labor, 1989-91; mem. Swedish Work Environment Fund Adv. Com. on Long Range Rsch. Investments, 1988-90; sci. coun. Swedish Inst. for Opinion Rsch. SIFO/Holen Network AB, 1988-90; com. on jobs in future soc. Swedish Employment Security Coun., 1986-89, others. Resident scholar Rockefeller Found. Study and Conf. Ctr., Bellagio, Como, Italy, 1980; recipient Royal award The King of Sweden's Medal of the 8th dimension with the ribbon of the Seraphimer Order, 1985, Swedish Nat. award for zealous and devoted svc., 1986, Gold leaf award and hon. mem. Swedish Forum for Psychosocial Worklife Issues, 1988, Internat. Women's Forum award, 1989, others. Mem. Swedish Psychol. Assn. (chmn. sci. coun. 1970-73), European Brain and Behaviour Soc. (pres.), Academie Internationale de Philosophie des Sciences (corres. mem.), NAS (fgn. assoc.), Academia Europaea, N.Y. Acad. Scis. Home: Vargardsvagen 75, Saltsjobaden Sweden 133 36 Office: U of Stockholm Psychology, Frescati Hagvag 14, 106 91 Stockholm Sweden 106 91

FRANKL, WILLIAM STEWART, cardiologist, educator; b. Phila., July 15, 1928; s. Louis and Vera (Simkin) F.; m. Razelle Sherr, June 17, 1951; children: Victor S. (dec.), Brian A. B.A. in Biology, Temple U., 1951, M.D., 1955, M.S. in Medicine, 1961. Diplomate: Am. Bd. Internal Medicine, Am. Bd. Cardiovascular Disease. Intern Buffalo Gen. Hosp., 1955-56; resident in medicine Temple U., Phila., 1956-57, 59-61; mem. faculty Temple U. (Sch. Medicine), 1962-68, dir. EKG sect. dept. cardiology, 1966-68, dir. cardiac care unit, 1967-68; prof. medicine, dir. div. cardiology Med. Coll. Pa., Phila., 1970-79; prof. medicine, assoc. dir. cardiology div. Thomas Jefferson U., Phila., 1979-84; physician-in-chief Springfield (Mass.) Hosp., 1968-70; pvt. practice medicine specializing in cardiology Phila., 1962-68, 70—; prof. medicine, co-dir. William Likoff Cardiovascular Inst. Hahnemann U., Phila., 1984-86, dir. William Likoff Cardiovascular Inst., dir. div. cardiology, 1986-92, Thomas J. Vischer Prof. medicine, chmn. dept. medicine, 1987-92; prof. medicine, dir. cardiovascular regional programs for the Med. Coll. Hosps. and the Med. Coll. Pa., 1992—; cons. cardiology Phila. Va Hosp., 1970-79; Fogarty Sr. Internat. fellow Cardiothoracic Inst., U. London, 1978-79; pres. Pa. affiliate Am. Heart Assn., 1985-86. Contbr. articles to profl. jours. Capt. Med. Corps, U.S. Army, 1957-59. Cardiovascular rsch. fellow U. Pa., Phila., 1961-62; recipient Golden Apple award Temple U. Sch. Medicine, 1967; award Med. Coll. Pa., 1972; Lindback award for distinguished teaching, 1975. Fellow ACP, Am. Coll. Cardiology (gov. Ea. Pa. 1986-89), Phila. Coll. Physicians, Am. Coll. Clin Pharmacology (regent 1980-85), Coun. Clin. Cardiology of Am. Heart Assn. (coun. on arteriosclerosis); mem. AAUP, AAAS, N.Y. Acad. Scis., Am. Fedn. Clin. Rsch., Assn. Am. Med. Colls., Am. Heart Assn. (bd. govs. S.E. Pa. chpt. 1972-84, pres. 1976), Am. Soc. Clin. Pharmacology and Exptl. Therapeutics, Philadelphia County Med. Soc. (pres. 1993-94). Home: 536 Moreno Rd Wynnewood PA 19096-1121 Office: Med Coll Hosps Main Clin Campus 3300 Henry Ave Philadelphia PA 19129

FRANKLIN, DAVID LEE, economist; b. Yuma, Ariz., Sept. 28, 1943; s. Damon and Dolores (Carmona) F.; m. Sydney Lantz, Aug. 22, 1960 (div. 1983); children: Ralph Louis, Heather Lee, Holly Lyn, Kathleen Lantz; m. Marie Louise Worden, Sept. 15, 1990. BS in Systems Engring., U. Ariz., 1965, MS in Systems Engring., 1966; PhD in Econs., N.C. State U., Raleigh, 1978. Pres. Markstems Inc., Tucson, 1965-69; sr. scientist Franklin Inst., Phila., 1967-68; Fulbright prof. U. Sonora, Mexico, 1968-69; assoc. prof., systems engr. U. Ariz., Tucson, 1969-70; rsch. assoc. Harvard U., Boston, 1970-73; dir. sys. dept. Internat. Ctr. for Tropical Agr., Cali, Colombia, 1973-78; sr. economist Rsch. Triangle Inst., Research Triangle Park, N.C., 1978-81; pres. Sigma One Corp., Research Triangle Park, 1981—. Contbr. articles to profl. jours.; patentee in field. Mem. Ariz. 100 Coun., Tucson, 1989—, Nat. Acad. Sci. Com., Washington, 1981. Rockefeller Found. fellow, 1976; NSF fellow, 1965; named N.C. Minority Entrepreneur of Yr., 1991. Republican. Roman Catholic. Avocations: history, Latin Am. literature; pioneer in design of computer based point of sale systems for data acquisition and management control in supermarkets. Office: Sigma One Corp PO Box 12836 Research Triangle Park NC 27709-2836

FRANKLIN, GENE FARTHING, electrical engineering educator, consultant; b. Banner Elk, N.C., July 25, 1927; s. Burnie D. and Delia (Farthing) F.; m. Gertrude Stritch, Jan. 1952; children: David M., Carole Lea. BSEE, Ga. Inst. Tech., 1950; MSEE, MIT, 1952; DEngSc, Columbia U., 1955. Asst. prof. Columbia U., N.Y.C., 1955-57; prof. elec. engring. Stanford (Calif.) U., 1957—, chmn. elec. engring. dept/, 1993—; cons. IBM, Rochester, Minn., 1982—. Author: Sampled-Data Control, 1958, Digital Control, 1980, 2d edit., 1990, Feedback Control, 1986, 2d edit., 1991. With USN, 1945-47. Recipient Edn. award Am. Automatic Control Coun., 1985. Fellow IEEE (life), Control Soc. of IEEE. Democrat. Office: Stanford U Dept Elec Engring Stanford CA 94305

FRANKLIN, HOWARD DAVID, chemical engineer; b. Montreal, Que., Can., Oct. 16, 1953; s. Karl and Mary (Maislin) F.; m. Malka Grinkorn, June 29, 1979. B in Engring., McGill U., 1975; MS, U. Mich., 1976; PhD, MIT, 1980. Dynamic simulation engr. Exxon Rsch. Engring., Florham Park, N.J., 1981-84, team leader dynamic simulation, 1984-89, team leader advanced control user support, 1989-. Contbr. articles to profl. jours. Mem. Am. Inst. Chem. Engrs., Am. Chem. Soc., Sigma Xi. Achievements include development of several major process dynamic simulations and several major advanced computer process control application.

FRANKLIN, KEITH BARRY, entrepreneur, technical consultant, former military officer; b. Paris, Tenn., Apr. 12, 1954; s. Jesse Earl and Ina Mae (Sykes) F.; m. Lucille Maude Espey, Jan. 3, 1987 (div. Jan. 1990). BS in Agri-Bus., U. Tenn., Martin, 1977; postgrad., U. Okla., 1978-79, U. Md., 1987-88. Commd. U.S. Army, 1977, active duty, 1977-89, telephone systems officer, community rels. officer, radio systems officer, wideband radio tech. evaluation program team chief., tng. devel. officer; comm.-electronic officer 101st Airborne Divsn Artillery U.S. Army, Ft. Campbell, Ky., 1985-87; chief material fielding team Comm.-ELectronics Command-Europe U.S. Army, Seckenheim, Germany, 1987-89; comm. staff officer Directorate Ops., U.S. European Command U.S. Army, Stuttgart, Germany, 1989-90; comm. ops. officer Dep. Chief Staff Ops., Hdqrs., U.S. Army Europe and 7th Army U.S. Army, Heidelberg, Germany, 1990; officer U.S. Army Signal Corps, Europe, Korea, U.S.A, 1977—; staff announcer, engr. Sta. WALR-FM, Union City, Tenn., 1975-77; owner, mgr. Franklin Homes, 1980—; mobilization augmentee USAR, 1989—; adv. to Saudi Arabian Nat. Guard Vinnell Corp., Riyadh, Saudi Arabia, 1989—; sr. doctrine developer Saudi Arabian Nat. Guard, 1989-90; sr. adv. dir. intelligence Saudi Arabian Nat. Guard Signal Corps, 1990; prin. adv. Saudi Arabian Nat. Guard Mil. Schs., 1991—. Mem. 101st Airborne Divsn. Assn. (life), Assn. U.S. Army (Disting. Svc. award 1984), Armed Forces Comm.-Electronics Assn., U. Tenn. Alumni Assn. Century CLub., Riyadh HHH Running Club (exec. com. 1990-92), Alpha Phi Omega (Torchbearer award 1988, 90, 92). Address: 306 Bancroft Ct Clarksville TN 37042

FRANKLIN, KEITH JEROME, electrical engineer; b. Pine Bluff, Ark., June 27, 1963; s. Sarah Mae (Lang) F.; m. Merline Armstrong, Sept. 19, 1981; children: Keith II, Caressa. BSEE, So. Ill. U., 1988; BSEd, Kensington U., 1990; MBA, City Univ., Bellevue, Wash., 1992. Cert. quality engr., quality auditor. Project mgr. Johnson & Johnson, Randolph, Mass., 1988-90; sr. quality engr. Warner-Lambert, Lititz, Pa., 1990-91; dept. mgr. Intertech, Ft. Myers, Fla., 1991—; adj. instr. computers, bus. and engring. Edison Coll., Fort Myers, Fla., 1991—. Contbr. articles to profl. jours. With USN, 1981-87. Recipient Malcolm Baldrige Nat. Quality award. Mem. Am. Soc. Quality Control (sr. mem., guest editor, tech. advisor 1987—), Am. Mgmt. Assn. (reviewer 1991—). Home: 2509 High St Pine Bluff AR 71601 Office: Franklin End Franklin 1060 SE 20th Ct Cape Coral FL 33990

FRANKLIN, NANCY JO, psychology educator; b. Tacoma, Wash., Nov. 14, 1963; d. Stanley and Sylvia (Fox) Newman; separated. BA, U. Calif., Santa Barbara, 1985; PhD, Stanford U., 1989. Asst. prof. psychology SUNY, Stony Brook, 1989—. Mem. Psychonomic Soc. Achievements include research in switching points of view in spatial mental models, spatial representation for described environs., qualitative physics, and reality monitoring. Office: SUNY Stony Brook Dept Psychology Stony Brook NY 11794-2500

FRANKLIN, THOMAS DOYAL, JR., medical research administrator; b. Morganton, N.C., Sept. 25, 1941; s. Thomas Doyal Sr. and Mabel (Smith) F.; m. Annie Faye Jones, Nov. 8, 1964; children: Michael, Jonathan. BS, Wake Forest U., 1963; MS, Bowman Gray Sch. Medicine, Winston-Salem, N.C., 1967; PhD, U. Ill., 1972. Assoc. physiologist McDonnel-Douglas Corp., St. Louis, 1967-69; rsch. scientist Intersci. Rsch. Inst., Champaign, Ill., 1969-72; asst. prof. radiology Ind. U., Indpls., 1972-79, assoc. prof. radiology, 1979-90; pres. Tex. Back Inst. Rsch. Found., Plano, Tex., 1990—; asst. dir. ultrasound Indpls. Ctr. for Advanced Rsch., 1974-80, assoc. dir. ultrasound, 1980-81, dir., 1981-82, acting exec. dir., 1982-83, exec. dir., 1983-90, pres., 1985-90. Contbr. chpts. to books and articles to profl. jours. Bd. dirs. Consortium for Urban Edn., Indpls., 1982-86, Rose Hulman Inst. Tech., Terre Haute, Ind., 1985-90, Walther Cancer Inst., Indpls., 1986-90, Sci. Edn. Found. Ind., Indpls., 1988-90 Crossroads of Am. Coun. Boy Scouts Am., Indpls., 1990; trustee Children's Mus. Indpls., 1985-90. Karl R. Ruddell scholar, 1975; grantee NIH, 1978-85, Showalter Residuary Trust, 1982-90. Mem. Am. Inst. Ultrasound in Medicine, Am. Soc. Echocardiography (bd. dirs. 1979-82), Radiol. Soc. N.Am., N.Y. Acad. Scis., Indpls. Athletic Club, Signature Athletic Club, Scientech Club Indpls., Rotary (Indpls. and Dallas/dir. Indpls. chpt. 1988-90), Sigma Xi. Presbyterian. Achievements include rsch. in cardiovascular, med. ultrasound, echocardiographic, microvascular, tissue ablation. Home: # 614 5330 Bent Tree Forest Dr Dallas TX 75248 Office: Tex Back Inst Rsch Found 3801 W 15th # 375 Plano TX 75075-7788

FRANKS, PAUL TODD, laboratory manager; b. Malvern, Ark., Feb. 4, 1962; s. Jim Paul and Clyda Joyce (Blevins) F.; m. Janet Gwynne Immel, Oct. 4, 1986; children: Benjamin, Laura, Daniel. BS in Chemistry, Henderson State U., 1984. Assoc. chemist Siplast Inc., Arkadelphia, Ark., 1984-86, lab. mgr., 1992—; chemist II Ark. Dept. Health, Little Rock, 1986-91; supr. chromatography sect. Entek, Little Rock, 1991-92. Deacon, Sunday Sch. tchr. 1st Bapt. Ch., Arkadelphia, 1992. Mem. Am. Chem. Soc., Single Ply Roofing Inst.

FRANKUM, RONALD BRUCE, communications executive, entrepreneur; b. Winfield, Kans., Nov. 17, 1935; s. Bruce Edward and Hildreth Bessie (Spencer) F.; m. Virginia Karam, Aug. 4, 1962; children: Katherine Jamalle, Ronald Bruce Jr. BA in History, Govt., U. Tex., 1958; JD, U. San Diego, 1965. Bar: Calif. 1972. Asst. to gov. Gov. of Calif., Sacramento, 1966-72; pres. Ctr. for Tech. Svcs., San Diego, 1973-81; dep. dir. office policy devel. The White House, Washington, 1981-82, dep. sci. advisor to pres. U.S., 1982-83; chmn. TFI Ltd., Washington, 1984-87; mng. gen. ptnr. Cellular Fund One, McLean, Va., 1988-93; dir., founder Pakistan Mobile Comms. Ltd., Islamabad, Pakistan, 1987—; pres. SAIF Telecom, 1993—; cons., advisor, lawyer, 1983—. Author: A Tax Policy Analysis of the Transient Occupancy Tax, 1980, Emerging Issues in Local Government Law, 1976, Politics of Change in Local Government, 1973; editor, mgr., pub. over 50 rsch. projects. With USNR, 1958-63. Named Chevalier, Sovereign Mil. Order of the Temple of Jerusalem, London, 1988. Republican. Office: Cellular Fund One 1350 Beverly Rd Ste 115-294 Mc Lean VA 22101

FRANSSON, TORSTEN HENRY, mechanical engineering educator, researcher; b. Malmo, Sweden, June 27, 1949; s. Henry E. and Elsie A. I. (Olsson) F.; m. Eva M. Berling, May 31, 1979; children: Patrick, Anna, Lena. Fil. Mag., Fil. Kand., U. of Lund, Sweden, 1974; PhD, Swiss Inst. of Tech., Lausanne, Switzerland, 1986. NRC assoc. Naval Postgrad. Sch., Monterery, Calif., 1986-87; lectr. Swiss Fed. Inst. of Tech., Switzerland, 1987-91; prof. Royal Inst. of Tech., Stockholm, 1990—; dir. chair heat and power technology Swiss Fed. Inst. of Tech., Switzerland. Contbr. over 30 articles to profl. jours. Recipient award Cray Supercomputer Contest, 1988. Mem. AIAA, ASME (cogen-turbo European program chmn. 1988, 89, 90, 91, 94, vice chmn. structural dynamics com. internat. gas turbine inst. 1992—), Sigma Xi (v.p. Swiss chpt. 1990-91). Office: Royal Inst Tech, Thermal Engring, S-100 44 Stockholm Sweden

FRANTZ, DEAN LESLIE, psychotherapist; b. Beatrice, Nebr., Mar. 27, 1919; s. Oscar C. and Flora Mae (Gish) F.; m. Marie Flory, Aug. 31, 1940; children: Marilyn, Shirley, Paul. BA, Manchester (Ind.) Coll., 1942; MDIV, Bethany Theol. Sem., Oak Brook, Ill., 1945; Diploma, C.G. Jung Inst. Zurich, 1977. Dir. ch. rels. Manchester Coll., North Manchester, Ind., 1964-72; assoc. prof. Bethany Theol. Sem., 1957-64; pvt. practice Jungian analyst Ft. Wayne, Ind., 1977—. Author: Meaning for Modern Man in the Paintings of Peter Birkhauser; editor: Barbara Hannah: The Cat, Dog, and Horse Lectures, and the Beyond, 1992. Mem. Internat. Assn. Analytical Psychology, Assn. Grad Analytical Psychologists. Home: 3831 Evergreen Ln Fort Wayne IN 46815-4707

FRANTZIDES, CONSTANTINE THEMIS, general surgeon; b. Limassol, Cyprus, Nov. 6, 1950; came to U.S., 1983; s. Themistokles and Christothea (Papageorgeou) F.; m. Eleni Kasapi, May 8, 1981; children: Alexander, Marlena. MD, U. Athens, Greece, 1976; PhD, Med. Sch. Athens U., Greece, 1987. Chief resident Athens U., Greece, 1981-82; rsch. fellow Med. Coll. Wis., Milw., 1983-84, asst. clin. prof., 1984-85, vis. asst. prof., 1986-88, asst. prof., 1989—; staff surgeon Milw. Regional Med. Ctr., Milw., 1989—, Froedtert Meml. Hosp., Milw., 1989—; com. mem. Laser Safety Com., Milw., 1990—, Adv. Com., Clin. Rsch., Milw., 1990—; sr. advidor Clin. Edn., Milw., 1990—. Author: Postcholecystectomy Syndrome, 1991. Recipient Shipley medal So. Surgical Assn., Hot Springs, Va., 1985; Physician's Recognition award AMA, Chgo., 1990; 1st Ind. Rsch. award NIH, 1986. Mem. Am. Gastroenterology Assn., Collegium Internat. Chrurgiae Digestivae, N.Y. Acad. Scis., Soc. for Surgery of Alimentary Tract. Achievements include discovery of a method for intrinsic denervation of the small intestine to study the physiology of this organ, the first laparoscopic highly selective vagotomy in the USA. Home: 3690 Emberwood Dr Brookfield WI 53005 Office: Medical Coll Wisconsin 9200 W Wisconsin Ave Milwaukee WI 53226

FRANZ, CRAIG JOSEPH, biology educator; b. Balt., Apr. 12, 1953; s. Harry Joseph and Vera Lee (Garrett) F. BA in Biology, Bucknell U., Lewisburg, Pa., 1975; MSc in Environ. Engring. & Sci., Drexel U., Phila., 1977; PhD in Biology, U. R.I., Kingston, 1988. Tchr. biology LaSalle Coll. High Sch., Phila., 1977-79; instr. biology St. John's Coll., Washington, 1980-84; teaching asst. U. R.I., Kingston, 1984-86; malacological researcher Estacion de Investigaciones, Margarita, Venezuela, 1986-87; univ. fellow U. R.I., Kingston, 1987-88; asst. prof. biology LaSalle U., Phila., 1988—. Author: Invertebrate Zoological Investigations, 1988; co-author: The Cornerstone, 1989. Bd. mem. IRB, Einstein Med. Ctr., Phila., 1989—; bd. trustees Calvert Hall Coll., Balt., 1982-84. Univ. fellow U. R.I., Kingston, 1987. Mem. AAAS, Am. Soc. Zoologists, Am. Malacological Union, Delta Upsilon (bd. dirs. 1975, 1989—), Phi Kappa Phi, The Demosthenean Club (bd. dirs. 1987-89). Democrat. Roman Catholic. Achievements include discovery of new species of chiton, new genus of copepod. Office: LaSalle U 20th St at Olney Ave Philadelphia PA 19126

FRANZ, JOHN E., bio-organic chemist, researcher; b. Springfield, Ill., Dec. 21, 1929; m. Elinor Thielken, Aug. 7, 1951; children—Judith, Mary, John, Gary. B.S., U. Ill., 1951; Ph.D., U. Minn., 1955. Sr. research chemist Monsanto Agrl. Co., St. Louis, 1955-60, research group leader, 1960-63, fellow, 1963-75, sr. fellow, 1975-80, disting. fellow, 1980—. Inventor roundup herbicide; holder 840 U.S. and fgn. patents; contbr. articles to sci. pubis. Recipient Indsl. Rsch. Mag., award, 1977, Indsl. Research Inst. Achievement award, Washington, 1985, J.F. Queeny award Monsanto Co., 1981, Inventor of Yr. award St. Louis Bar Assn., 1986; The Nat. Medal of Tech., Washington, 1987, Outstanding Achievement award, U. Minn., 1988, The Mo. award, Gov. of Mo., 1988, Perkin medal Am. Sect. Soc. Chem. Industry, 1990. Mem. AAAS, Am. Chem. Soc. (Carothers award Del. sect. 1989). Office: Monsanto Agri Co Ste 640 17600 Chesterfield Airport Rd # 104 Chesterfield MO 63005-1246

FRANZEN, HUGO FRIEDRICH, chemistry educator, researcher; b. N.Y.C., Aug. 27, 1934; s. Raymond Hugo and Louisa Hubbard (Blaine) F.; children: Stefan Hugo, Alice Anne, Kurt Eric. BS, U. Calif., 1957; PhD, U. Kans., 1962. Postdoctoral fellow U. Stockholm, 1962-63; instr. Iowa State U., Ames, 1963-66, asst. prof., 1966-69, assoc. prof., 1969-74, prof., 1974—; vis. scientist U. Groningen, Netherlands, 1973-74, Max Planck Inst., Stuttgart, Germany, 1981, Ariz. State U., 1982, Aachen Tech. Hochschule, Germany, 1989; mem. editorial adv. bd. Jour. Solid State Chemistry, 1970—; chmn. Gordon Conf. on Solid State Chemistry, 1984. Author: The Physical Chemistry of Inorganic Crystalline Solids, 1986; editor Jour. of Alloys and Compounds, 1985—; contbr. 120 articles to sci. and profl. jours., 1970—. Bd. dirs. Iowa Civil Liberties Union, 1972. Outstanding Rsch. in Materials Chemistry recognition, U.S. Dept. Energy. Mem. Am. Chem. Soc (chmn. solid state chemistry subsect, 1986). Democrat. Unitarian. Achievements include discovery of new metal-rich refractory compounds; application of symmetry theory to phase transitions; innovations in teaching chem. thermodynamics; contbns. to understanding of role of metal-metal bonds in solid compounds. Home: 1216 Scott Ave Ames IA 50010 Office: Iowa State U Chemistry Dept Ames Lab Ames IA 50011

FRANZKE, HANS-HERMANN, engineering scientist, educator; b. Clausthal, Germany, Mar. 18, 1927; s. Karl Hermann and Elisabeth Agnes (Richter) F.; m. Herta Leonore Schomer, Sept 27, 1952; children: Wolfgang, Klaus, Sigrid. Diploma in Mining Engring, Tech. U., Clausthal, 1951, D. in Engring. Scis., 1957. Cert. in engring. Asst. instr. Tech. U., Clausthal, 1951-56; head sect. Allgem. Elektricitaets-Ges., Berlin, 1957-62; head dept. Allgem. Elektricitaets-Ges., Essen, Fed. Republic Germany, 1962-67; mem. mng. bd. Projekta, Duesseldorf, Fed. Republic Germany, 1968-75; prof. Tech. Coll., Cologne, Fed. Republic Germany, 1975-77, Tech. U., Berlin, 1977—; dept. dir. Inst. Tech. U., Berlin, 1978—. Author: Maschinen-u. Anlagentechn., 1986, 90; inventor in field. Fellow Berliner Wissenschaftliche Gesellschaft; mem. Franzke'sche Stiftung (chmn. bd.), Verein Deutscher Ingenieure, Verband Deutscher Wirtschaftsingenieure, Deutscher Kaeltetechnischer Verein, Internat. Solar Energy Soc. Home: Hochbaumstr 36, D 1000 Berlin 37, Germany Office: Tech U Berlin, Ernst Reuter Platz 1, D 1000 Berlin 12, Germany

FRASER, JANE MARIAN, operations research specialist, educator; b. Glen Ridge, N.J., Sept. 3, 1950; d. John McBride and Jean Carol (Hardman) F. BA in Math. with honors, Swarthmore Coll., 1971; MS in Ops. Rsch., U. Calif., Berkeley, 1972, PhD, 1979. Rsch. and teaching asst. dept. indsl. engring. ops. rsch. U. Calif., Berkeley, 1972-75, rsch. asst. Sch. Edn., 1976-78; rsch. asst. Sch. Pub. Health U. Wash., Seattle, 1979; assoc. rsch. scholar Inst. for Interdisciplinary Engring. Purdue U., West Lafayette, Ind., 1979-82, asst. prof. Sch. Indsl. Engring., 1981-86; vis. asst. prof. dept. indsl. and systems engring. Ohio State U., Columbus, 1986, asst. prof., 1986-90, assoc. prof., 1990—; from assoc. to co-dir. Ctr. for Advanced Study Telecommunications Ohio State U., Columbus, 1988—; lectr. on bus. uses of telecommunications Sino-Ohio Ctr., Columbus, 1989—; cons. Sci. Applications Internat. Corp., Columbus, 1987-88, Kaiser Found. Health Plan, Oakland, 1972-78. Assoc. editor: Engring. Econ., 1989-91; contbr. articles to Legis. Studies Quar., Jour. Politics, Polity, Polit. Methodology, European Jour. Operational Rsch., Engring. Econ., Immunohematology, Transfusion. Vol. Morris Udall for Pres. Campaign, 1976. Kellogg Found. fellow, 1982-85, Ameritech fellow, 1991-92. Mem. AAAS, Am. Soc. for Engring. Edn. (newsletter editor engring. econ. div. 1987-88, sec./treas. 1988-89, program chmn. 1989-90, chmn. 1990-91), Inst. Indsl. Engrds., Ops. Rsch. Soc. Am., IEEE Systems, Man and Cybernetics Soc. (adminstrv. com. 1989-92), Inst. Mgmt. Sci., Columbus Area C. of C. (info. svcs. com. 1990—), Phi Sigma Rho (faculty advisor 1988—), Phi Beta Kappa, Sigma Xi. Office: Ohio State U 1971 Neil Ave 210 Baker Systems Columbus OH 43210

FRASER, MALCOLM JAMES, JR., biological sciences educator; b. Troy, N.Y., Oct. 20, 1952; s. Malcolm James and Rose-Marie Evelyn (Jordan) F.; Tresa Marie Strauss; children: Steven James, Nicholas Alan, Mark Evan. BS, Wheeling (W.Va.) Coll., 1975; MS, Ohio State U., 1979, PhD, 1981. Postdoctoral fellow Pa. State U., State College, 1981; postdoctoral assoc. Tex. A&M U., College Station, 1981-83; asst. prof. U. Notre Dame, Ind., 1983-89, assoc. prof., 1989—; cons. Am. Biogenetic Scis., Notre Dame, 1985-92, scientific adv. bd., 1986-92. Contbr. articles to Jour. of Virological Methods, Virology, Gene, Biochemistry and In Vitro. Asst. coach Irish Youth Hockey League, South Bend, Ind., 1990-91; den leader Boy Scouts Am., South Bend, 1989—. Recipient Rsch. Career Devel. award NIH, 1991-96, rsch. grantee, 1985-89, 90-95, U.S. Dept. Agriculture, 1984-87, 88-91. Mem. AAAS, Am. Soc. for Virology, Tissue Culture Assn. (chmn. invertebrate div. 1990-92). Achievements include patent on recombinant baculovirus occlusion bodies in vaccines and biological insecticides; co-development of baculovirus expression vector systems; research on baculovirus molecular biology and genetics, with emphasis on transposon mediated mutagenesis of baculovirus genomes by host cell elements. Office: U Notre Dame Biol Scis Notre Dame IN 46556

FRASER-REID, BERTRAM OLIVER, chemistry educator; b. Coleyville, Jamaica, Feb. 23, 1934; Jamaican and Can. citizen; married, 1963. BS, Queen's U., 1959, MS, 1961; PhD in Chemistry, U. Alta., 1964. Postdoctoral fellow Imperial Coll., U. London, 1964-66; prof. U. Waterloo, Can., 1966-80; prof. chemistry U. Md., 1980-82; prof. chemistry Duke U., Durham, N.C., 1982—, James B. Duke prof. chemistry, 1985—; mem. med. chem. study sect. A, NIH, 1979-83, pharmcol. sci. rev. panel, 1984-88; cons. Burroughs Wellcome Co., Glaxo, Glycomed. Mem. Chem. Inst. Can. (Merck, Sharp & Dohme award 1977), Am. Chem. Soc., Brit. Chem. Soc. Rsch. in synthesis organic compounds of pharmacological, biological, and theoretical importance from readily available carbohydrate derivatives; influence of electronic effects on reactivities of carbohydrates and carbocycles. Office: Duke U Dept Chemistry Durham NC 27706

FRASSINELLI, GUIDO JOSEPH, aerospace engineer; b. Summit Hill, Pa., Dec. 4, 1927; s. Joseph and Maria (Grosso) F.; m. Antoinette Pauline Clemente, Sept. 26, 1953; children: Lisa, Erica, Laura, Joanne, Mark. BS, MS, MIT, 1949; MBA, Harvard U., 1956. Treas. AviDyne Rsch., Inc., Burlington, Mass., 1958-64; asst. gen. mgr. Kaman AviDyne div. Kaman Scis., Burlington, 1964-66; asst. dir. strategic planning N. Am. ACFT OPNS, Rockwell Internat., L.A., 1966-69; mgr. program planning Rockwell Space Systems Div., Downey, Calif., 1970-76; project leader R&D Rockwell Space Systems Div., Downey, 1976-79, chief analyst bus. planning, 1980-85, project mgr. advanced programs, 1986—. Mem. Town Hall of Calif., L.A., 1970—; treas. Ecology Devel. and Implementation Commitment Team Found., Huntington Beach, Calif., 1971-75; founding com. mem. St. John Fisher Parish Coun., Rancho Palos Verdes, Calif., 1978-85. Recipient Tech. Utilization award, NASA, 1971, Astronaut Personal Achievement award, 1985. Fellow AIAA (assoc., tech. com. on econs. 1983-87, exec. com. L.A. sect. 1987-91); mem. Sigma Xi, Tau Beta Pi. Roman Catholic. Achievements include determination of aircraft damage limits and atomic-weapon-delivery capabilities of aircraft; development of cost models to account for advances in engineering state of art, of cost prioritization techniques for space shuttle improvements, of software to produce business plans. Home: 29521 Quailwood Dr Palos Verdes Peninsula CA 90274-4930 Office: Rockwell Internat Space Systems Divsn 12214 Lakewood Blvd Downey CA 90241

FRATI, LUIGI, oncologist, pathologist; b. Siena, Tuscany, Italy, Apr. 10, 1943; s. Tosco and Anna (Spediacci) F.; m. Luciana Rita Angeletti; children: Paola, Giacomo. MD, Cath. U., Rome, 1967; postgrad., State U., Rome, 1975. Intern Cath. U., 1965-67; resident Med. Sch. Hosp., Perugia, Italy, 1969; resident clin. endocrinology br. Nat. Inst. Arthritis and Metabolic Diseases, U.S., 1970; asst. prof. faculty medicine U. Perugia, 1967-71, acting prof. pathology faculty scis., 1971-72; acting prof. pathology faculty pharmacy U. Rome, 1973—, acting prof. pathology faculty medicine, 1974-79, prof. pathology faculty medicine, 1980—, chmn. faculty medicine, 1983—; pres. Nat. Conf. Faculty Medicine, Rome, 1986—, Nat. Univs. Coun., Rome, 1980—; Nat. Sci. and Tech. Coun., 1988—; chmn. Internat. Inst. Molecular Medicine, Luzern. Author: General Pathology, 1980, Molecular Biology Cancer, 1988, 89; contbr. articles to profl. jours. Grantee NIH, Bethesda, Md., 1971, 78, EEC, Brussels, 1985, 88, 90, 91, 92. Mem. Am. Assn. Cancer Rsch., EB Virus Assn., European Assn. Against Cancer. Office: Dept Exptl Medicine, Vle Regina Elena 324, 00161 Rome Italy

FRAUENFELDER, HANS, physicist, educator; b. Neuhausen, Switzerland, July 28, 1922; came to U.S., 1952, naturalized, 1958; s. Otto and Emma (Ziegler) F.; m. Verena Anna Hassler, May 16, 1950; children: Ulrich Hans, Kätterli Anne, Anne Verena. Diploma, Swiss Fed. Inst. Tech., 1947, Ph.D. in Physics, 1950. Asst. Swiss Fed. Inst. Tech., 1946-52; asst. prof. physics U. Ill. at Urbana, 1952-56, assoc. prof., 1956-58, prof., 1958-92, prof. emeritus, 1992—; mem. staff Los Alamos (N.Mex.) Nat. Labs., 1992—; Guggenheim fellow, 1958-59, 73; vis. scientist CERN, Geneva, Switzerland, 1958-59, 63, 73. Author: The Mossbauer Effect, 1962, (with E.M. Henley) Subatomic Physics, 1974, 2d edit., 1991, Nuclear and Particle Physics, 1975; contbr. articles to profl. jours. Recipient Humboldt award, 1987-88. Fellow AAAS, Am. Phys. Soc. (Biol. Physics prize 1992), N.Y. Acad. Sci.; mem. NAS, Am. Inst. Physics (chmn. governing bd. 1986-93), Am. Acad. Arts and Sci., Am. Philos. Soc., Acad. Leopoldina. Home: PO Box 449 Tesuque NM 87574-0449 Office: Los Alamos Nat Lab P-DO MS M715 Los Alamos NM 87515

FRAUMENI, JOSEPH F., JR., scientific researcher, medical educator, physician, military officer; b. Boston, Apr. 1, 1933; s. Joseph Francis and Pauline (Malta) F.; m. Patricia Welch D'Arcy, Apr. 23, 1977. AB, Harvard U., 1954; MD, Duke, 1958; ScM, Harvard U., 1965. Diplomate Am. Bd. Internal Medicine. Commd. lt. USPHS, 1962, advanced through grades to Capt., 1968; med. intern, resident Johns Hopkins Hosp., Balt., 1958-60; med. resident, chief resident Meml. Sloan-Kettering Cancer Ctr., N.Y.C., 1960-62; staff assoc. Nat. Cancer Inst., Bethesda, Md., 1962-65, assoc. chief, 1966-75, chief environ. epidemiology br., 1975-82, dir. epidemiology & biostats. program, 1979—; attending physician Clin. Ctr. NIH, Bethesda, 1966—; prof. epidemiology uniformed svcs. U. Health Scis., Bethesda, 1985—. Editorial Bds. Jour. Nat. Cancer Inst., 1966-69, Teratology, 1974-78, Med. and Pediatric Oncology, 1974-78, Cancer Rsch., 1974—, Am. Jour. Indsl. Medicine, 1979—, Oncology, 1980—, Cancer Investigation, 1980—, Preventive Medicine, 1982—, Genetic Epidemology, 1984—, Cancer Causes and Control, 1989—, Cancer Epidemiology, Biomarkers and Prevention, 1990—, Cancer, 1991—, Internat. Jour. Oncology, 1992—; contbr. more than 500 articles to profl. jours., books. Recipient Gorgas medal Assn. Mil. Surgeons of U.S., 1989, vis prof. GM Cancer Rsch. Found. Internat. Agy. Rsch. Cancer, 1990, W.W. Sutow award U. Tex. M.D. Anderson Cancer Ctr., 1992, Disting. Alumnus award Duke U. Med. Ctr., 1992, Alumni award Merit Harvard Sch. Pub. Health, 1992. Fellow AAAS, Am. Coll. Physicians, Am. Coll. Epidemiology (Lilienfeld award 1993, bd. dirs. 1985-89), Am. Coll. Preventive Medicine; mem. Inst. Medicine NAS, Am Soc. Preventive Oncology (Disting. Achievement award 1993, pres. 1981-83), Am. Assn. Cancer Rsch. (bd. dirs. 1983-87, Am. Cancer Soc. award rsch. excellence epidemiology, prevention 1993). Achievements include research in environmental and genetic determinants of cancer. Office: Nat Cancer Inst EPN/543 Div of Cancer Etiology 6130 Executive Blvd Bethesda MD 20892

FRAUTSCHI, STEVEN CLARK, physicist, educator; b. Madison, Wis., Dec. 6, 1933; s. Lowell Emil and Grace (Clark) F.; m. Mie Okamura, Feb. 16, 1967; children—Laura, Jennifer. B.A., Harvard U., 1954; Ph.D., Stanford U., 1958. Research fellow Kyoto U., Japan, 1958-59, U. Calif.-Berkeley, 1959-61; mem. faculty Cornell U., 1961-62, Calif. Inst. Tech., Pasadena, 1962—; prof. theoretical physics Calif. Inst. Tech., 1966—; vis. prof. U. Paris, Orsay, 1977-78. Author: Regge Poles and S-Matrix Theory, 1963, The Mechanical Universe, 1986. Guggenheim fellow, 1971-72. Mem. Am. Phys. Soc. Research, publs. on Regge poles, bootstrap theory, cosmology. Home: 1561 Crest Dr Altadena CA 91001-1838 Office: 1201 E California Blvd Pasadena CA 91125-0001

FRAYSER, MICHAEL KEITH, electrical engineer; b. Clarksville, Tenn., Dec. 31, 1966; s. Charles A. and Carolyn Sue (Simmons) F. AS, Southeastern Ill. Coll., Harrisburg, 1987; BS, Murray State U., 1990. Jr. engr. Lafarge Corp., Joppa, Ill., 1990-92; elec. engr. Lane Erectors, Inc., Harrisburg, 1992; test tech./engr. Bendix Auto. N.Am., Clarksville, 1992—. Mem. IEEE Controls Soc. Avocations: tennis, basketball. Home: 2190 Memorial Dr A18 Clarksville TN 37043 Office: Bendix Auto N Am 780 Arcata Blvd Clarksville TN 37041

FRAZIER, KIMBERLEE GONTERMAN, veterinarian; b. St. Louis, Mar. 5, 1953; d. Joseph Wilbur Jr. and Melody (Engleman) Gonterman; m. Burk Ralph Frazier, Oct. 11, 1985; 1 child, Weston James. DVM magna cum laude, U. Mo., 1979. Vet. intern U. Mo. Coll. Vet. Medicine, Columbia, 1979-80; relief vet. St. Louis and Kansas City, Mo., 1980-84; account exec. Merrill Lynch, Clayton, Mo., 1984-85; owner, dir., small animal practitioner VET STOP Animal Clinics, St. Charles, St. Peters, Florissant, Kirkwood, St. Louis, Manchester, Mo., 1985—; owner, dir. HealthyPet Vet. Svcs. in PetsMart, St. Charles, Mo., and O'Fallon, Ill., 1993—; advisor Math. Sci. Network, St. Louis, 1987. Frank Wells scholar U. Mo., 1978; First Pl. in Exhibn. Sport for synchronized swimming, Munich Olympics, 1972. Mem. AVMA, Mo. Vet. Med. Assn., St. Louis Vet. Med. Assn., Mo. Bot. Garden, Friends of Zoo, Gamma Sigma Delta, Phi Zeta. Republican. Avocations: commercial hot air balloon pilot, scuba diving, hobie cat sailing, skiing. Home and Office: 4601 Maryland Ave Saint Louis MO 63108-1912

FRAZIER, RONALD GERALD, JR., mechanical engineer; b. Winchester, Tenn., May 5, 1965; s. Ronald Gerald and Paulette (Stephens) F.; m. Catheryn Elaine Watts, NOv. 4, 1989 (div. 1991); 1 child, Adam Lee. BS in Mech. Engring., Miss. State U., 1990. Engr. in tng. Waterways Experiment Sta., Vicksburg, Miss., 1985-90; spreader engr. Fontaine Body & Hoist Co., Collins, Miss., 1991-93, Warren Inc., Collins, Miss., 1993—. Mem. ASME, SAE, Nat. Soc. Profl. Engrs. Home: 103 Walnut Ave Seminary MS 39479

FRECH, BRUCE, mathematician; b. Norristown, Pa., Dec. 11, 1956; s. Henry and Ruth (Derstein) F.; m. Toni Fredette, May 19, 1991. BS, Rensselaer Poly. Inst., 1978; PhD, U. Va., 1983. Asst. prof. math. Lehigh U., Bethlehem, Pa., 1984, U. Scranton, Pa., 1986-88, Colby Coll., Waterville, Maine, 1988-90; mathematician Pacer Systems, Horsham, Pa., 1985; tech. advisor Tailwind Bicycles, Schwenksville, Pa., 1990-92; engr. Am. Elec. Lab., Lansdale, Pa., 1992—. Coach bicycle racing team Colby Coll., 1988-90. Mem. Am. Math. Soc., U.S. Cycling Found. Avocation: bicycling. Home: 160 Main St Schwenksville PA 19473

FRECHETTE, VAN DERCK, ceramic engineer; b. Ottawa, Ont., Can., Jan. 5, 1916; s. Howells and Lena D. (Derick) F.; m. Sarah W. Houghton, Apr. 4, 1940; children: William G.H., Howells Van Derck, Christopher J., Margaret Kathleen, Judith L. Student, U. Toronto, 1934-36; BS, Alfred U., 1939, DSc honoris causa, 1991; MS, U. Ill., 1940, PhD., 1942. Registered profl. engr., N.Y. Research physicist Corning Glass Works, N.Y., 1942-44; prof. ceramic sci. State U. N.Y. Coll. Ceramics, Alfred U., 1944—; guest prof. U. Göttingen, Germany, 1955-56, Max Planck Inst., 1965-66, U. Erlangen-Nurnberg, 1973; cons. on fractology, ceramic problems and microscopy. Pres. Alfred Delta Sig Corp., 1971-76. Author: Microscopy of Ceramics, 1955, Failure Analysis of Brittle Materials, 1990; editor: Noncrystalline Solids, 1960, Kinetics of Reactions in Ionic Systems, 1970, Surfaces and Interfaces of Glass and Ceramics, 1974, Ceramic Engineering and Science-Emerging Priorities, 1974, Borate Glasses, 1977, Quality Assurance in Ceramic Industries, 1979, Natural Glasses, 1984, Fractography of Glasses and Ceramics, 1988, Applied Mineralogy Series, 1971-80. Recipient Gordon Rsch. Conf. award 1955, Western Elec. award, 1969, Merit award Alfred U., 1985; named Outstanding Educator, Ceramic Edn]. Coun., 1983, A.V. Bleininger medal, 1991; Fulbright fellow, 1955. Fellow Am. Ceramic Soc. (Albert Victor Bleininger award 1991); mem. Swedish Royal Acad. Scis., N.Y. Acad. Sci., Acad. of Ceramics, Sigma Xi, Delta Sigma Phi, Phi Kappa Phi. Home: 22 S Main St Alfred NY 14802-1317

FREDERICK, CLAY BRUCE, toxicologist, researcher; b. Hamlin, Tex., Oct. 29, 1948; s. Billy Bob and Mildred Lenora (Kemplin) F.; m. Anne Patricia Jones, Apr. 14, 1973; children: Scott Christopher, Erin Elizabeth. BS in Chemistry and Biology, Tulane U., 1971; PhD in Organic Chemistry, U. Tex., 1979. Diplomate Am. Bd. Toxicology. Chemist, asst. tng. coord. ICI U.S., Inc., Wilmington, Del., 1971-73; postdoctoral rsch. scientist Nat. Ctr. for Toxicol. Rsch., Jefferson, Ark., 1979-82; rsch. scientist Uniroyal Chem., Naugatuck, Conn., 1982-85; rsch. fellow in biochem. toxicology Rohm and Haas Co., Spring House, Pa., 1985—; adj. assoc. prof. pathobiology dept. U. Pa. Sch. Vet. Medicine, Phila., 1992—. Contbr. articles to profl. jours. Mentor Project LABS, Spring House, 1989; outside expert reviewer U.S. EPA, Washington, 1990, 93. Mem. Am. Assn. Cancer Rsch., Am. Coll. Toxicology, Soc. of Toxicology, Soc. for Risk Analysis, Phi Kappa Phi. Achievements include patent for Method for Reducing the Carboxylester Content in an Emulsion Polymer; research in toxicological risk assessment and biological modeling. Office: Rohm and Haas Co 727 Norristown Rd Spring House PA 19477

FREDERICK, EDWARD RUSSELL, chemical engineering consultant; b. Pitts., Oct. 20, 1913; s. Edward Ferdinand and Edna Anna (Mall) F.; m. Marie Margaret Busch, Sept. 29, 1939; children: E. Russell, Marlene G. Frederick Lurtwieler, Kenneth C. BS in Chemistry, Carnegie Inst. Tech., 1936; BSChemE, U. Pitts., 1951. Chemist Westinghouse Electric Co., Wilmerding, Pa., 1936-37; chemist, asst. supt. Liberty Powder Co., Uniontown, Pa., 1937-39; rsch. asst., sr. fellow Mellon Inst., Pitts., 1939-83, tech. dir., 1971-83; part-time cons. in field. Author annotated bibiography Electrical Effects, 1986-91; contbr. articles to profl. jours. Mem. Sigma Xi, Phi Lambda Upsilon. Republican. Lutheran. Achievements include patents for life preserver, dielectric fabrication, fibrous filters and processes, insulating materials and production methods, chemical modifications, triboelectric evaluation and triboelectric adjustment of fibrous materials, triboelectric property modification and selection of fabrics for filtration applications. Home: 294 Sunset Rd Pittsburgh PA 15237

FREDERICK, JAMES PAUL, chemical engineer; b. Billings, Mont., July 22, 1943; s. John William and Alice Murtle (Avery) F.; m. Karen Ann Roth, Aug. 3, 1962; children: Rand John, Ardith Jean Frederick Flight. BSChemE, Mont. State U., 1968. Registered profl. engr., Wyo., Colo., Ohio. Chem. engr. Shell Oil Co., Houston, 1968-77; project mgr. Shell Mining Co., Houston, 1977-92; tech. mgr. Zeigler Coal Holding Co., 1992—. Contbr. articles to profl. jours. Office: Encoal Corp PO Box 3038 Gillette WY 82717

FREDERICK, NORMAN L., JR., electrical engineer; b. Hopkinsville, Ky., Feb. 7, 1965; s. Norman L. and Nancy A. (Bass) F. ASES, Hudson Valley C.C., 1985; BSEE, Union Coll., Schenectady, 1987; MSEE, Syracuse U., 1990. Comm. engr. GE, Schenectady, 1986; systems engr. Rome Air Force devel. ctr. MITRE, Griffiss AFB, Rome, N.Y., 1987-89; researcher, teaching asst. Syracuse (N.Y.) U., 1989-90; R&D elec. engr. EESOF, Inc., Westlake Village, Calif., 1991—. Mem. IEEE, Tau Beta Pi, Eta Kappa Nu, Sigma Xi. Achievements include research on one to three phase converter circuits, near fields for phased array antennas; development of T-Matrix method for relation of current distbn. to near field.

FREDERICK, RONALD DAVID, aerospace engineer; b. Dayton, Ohio, Feb. 17, 1966; s. Ronald Richard and Judith Anne (Echle) F. BS in Aerospace Engring., U. Cin., 1989; MS in Aerospace Engring., U. Dayton, 1992. Aerospace engr. 4950 Test Wing, USAF, Wright Patterson AFB, Ohio, 1989—. Mem. AIAA. Republican. Roman Catholic.

FREDERICKSON, ARTHUR ROBB, physicist; b. Rahway, N.J., July 5, 1941; s. Arthur Raymond and Bertine Lavinia (Beecher) F.; m. Christine Magnuson, June 6, 1970; children: Timothy R., Nathan B., Julie H. BSc, Rensselaer Poly. Inst., 1965; PhD, U. Mass., Lowell, 1991. Physicist Cambridge Rsch. Labs., Bedford, Mass., 1967-80, Rome Air Devel. Ctr., Bedford, 1980-87, Air Force Geophysics Lab., Hanscom AFB, Mass., 1987—; mem. spl. topics rev. groups, panels Dept. Def., 1980—. Author: Spacecraft Dielectric Material Properties, 1986; contbr. articles to Jour. Applied Physics, Jour. Elec. Materials, numerous other jours., conf. procs. Chmn. Town Ctr. Com. Planning Bd., Stow, Mass., 1974-75; adult leader Carlisle (Mass.) area Boy Scouts Am., 1981-90. Mem. IEEE (sr., chmn. Boston sect. nuclear and plasma soc. 1983-92), Am. Phys. Soc., Sigma Xi. Achievements include patent on device to aid centering of high-energy beams, method and system for secondary emission system, charge accumulation gamma radiation detector, process for prevention of spontaneous discharging in irradiated insulators. Home: 488 West St Carlisle MA 01741 Office: Space Physics Div PL/GPSP Hanscom AFB Bedford MA 01731

FREDINE, C(LARENCE) GORDON, biologist, former government agency official; b. St. Paul, Aug. 15, 1909; s. Andrew Clarence and Hulda (Anderson) F.; m. Edith Louise Handy, June 7, 1934; children: John Gordon, Patricia Ann Narrowe. B.S. in Biology, Hamline U., 1932; postgrad. in zoology, U. Minn., 1932-35. Assoc. biologist Minn. Emergency Conservation Work, St. Paul, 1935-36; chief biologist game and fish div. Minn. Conservation Dept., 1936-41; asst. prof. dept. forestry and conservation Purdue U., 1941-47; with fish and wildlife service Dept. Interior, 1947-55; prin. biologist Nat. Park Service, 1955-60, park planner Mission 66, 1960-62, chief div. extension services, 1962-64, chief div. internat. affairs, 1964-71, staff dir. 2d World Conf. on Nat. Parks, 1972-73, ret. from, 1973, vol. asst. editor Parks Mag., 1975-80; exec. dir. Renewable Natural Resources Found., 1980-81; charter mem. The Wildlife Soc., 1937, exec. sec., 1960-63, hon. mem., 1963. Author govt. bulls., articles. Served to lt. USNR, 1943-46, PTO. Recipient Disting. Service award Dept. Interior, 1967, Conservation award Gulf Oil Corp., 1984. Mem. Am. Fisheries Soc. (Disting. Svc. award 1983, vol. staff coord. 1982-91), Washington Biologists Field Club (pres. 1973-76), Internat. Assn. Fish and Wildlife Agys., Student Conservation Assn., George Wright Soc.

FREE, ALFRED HENRY, clinical chemist, consultant; b. Bainbridge, Ohio, Apr. 11, 1913; s. Alfred Harry and Alice Virginia (Clymer) F.; m. Dorothy Hoffmeister, June 20, 1934 (div. Mar. 1947); children: Charles Alfred, Jane Alison, Barbara Beth; m. Helen Mae Murray, Oct. 18, 1947; children: Eric Scot, Penny Alene, Kurt Allen, James Jacob, Bonnie Anne, Nina Joann. AB, Miami U., Oxford, Ohio, 1934; MA, PhD, Western Res. U., 1939. Diplomate Am. Bd. Clin. Chemistry. Lab. asst. Cleve. Clinic Rsch. Found., 1934-35; teaching fellow biochemistry dept. Western Res. U. Sch. Medicine, Cleve., 1935-39, instr., 1939-40, sr. instr., 1940-42, assoc. prof., 1942-46; numerous sci. positions to v.p. sci. rels. Ames div. Miles Inc., Elkhart, Ind., 1946-78, sr. sci. cons. diagnostics div., 1978—; bd. dirs. Nat. Com. on Clin. Lab. Standards, 1976-83. Author: (with Helen Free) Urodynamics, 1972, Urinalysis in Clinical Laboratory Practice, 1974; editor blood and body fluids Biol. Abstracts, 1940-78; contbr. over 200 articles to sci. jours. Bd. dirs. Elkhart County chpt. Am. Cancer Soc., 1972; mem. blood collection com. ARC. Recipient award for 40 yrs. outstanding med. sci. contbns. Med. Econs., 1986. Fellow AAAS, Assn. Clin. Scientists (nat. pres. 1972, Diploma of Honor 1973); mem. Am. Chem. Soc. (past chmn. nat. coms., chmn. St. Joseph Valley sect. 1962-63, 75-76, Mosher award 1984), Am. Assn. Clin. Chemistry (chmn., councilor Chgo. sect. 1974, 75-77, Honor Scroll 1967), Am. Inst. Chemistry (Chgo. chpt.), Lions (a founder Elkhart, 1st chmn. Ind. Lions Eye Bank, bd. dirs., Citizen of Yr. award Chgo. 1983, Melvin Jones Fellow award 1990), Elks (exalted ruler), Phi Beta Kappa, Sigma Xi, Sigma Alpha Epsilon. Presbyterian. Achievements include several patents on clinical laboratory methodology; development of dipstick method of urinalysis through expansion of a series of dry chemical reagents. Home: 3752 E Jackson Blvd Elkhart IN 46516-5205 Office: Miles Inc Diagnostics Divsn PO Box 70 Elkhart IN 46515-0070

FREE, HELEN M., chemist, consultant; b. Pitts., Feb. 20, 1923; d. James Summerville and Daisy (Piper) Murray; m. Alfred H. Free, Oct. 18, 1947; children: Eric, Penny, Kurt, Jake, Bonnie, Nina. B.A. in Chemistry, Coll. of Wooster, Ohio, 1944; DSc (hon.), Coll. of Wooster, 1992; M.A. in Clin. Lab. Mgmt., Central Mich. U. Cert. clin. chemist Nat. Registry Clin. Chemistry. Chemist Miles Labs., Elkhart, Ind., 1944-78, dir. mktg. services research products div., 1978-82, chemist, mgr., cons. diagnostics div. Miles Inc., 1982—; mem. adj. faculty Ind. U., South Bend, 1975—. Author: (with others) Urodynamics and Urinalysis in Clinical Laboratory Practice, 1972, 76. Contbr. articles to profl. jours. Patentee in field. Women's chmn. Centennial of Elkhart, 1958. Recipient Disting. Alumni award Coll. of Wooster, 1980, award Medi Econ. Press, 1986; named to Hall of Excellence, Ohio Found. Ind. Colls., 1992; named Woman of Yr. YWCA, 1993. Fellow AAAS, Am. Inst. Chemists (co-recipient Chicago award 1967), Royal Soc. Chemistry; mem. Am. Chem. Soc. (pres. 1993, bd. dirs., chmn. women chemists com. internat. activities com., grants and awards com., profl. and member relations com., nominating com., council policy pub. affairs and budget, Service award local chpt. 1981, councilor; Garvan medal 1980, co-recipient Mosher award, 1983), Am. Assn. for Clin. Chemistry (council, bd. dirs., nominating com. and pub. relations com., nat. membership chmn., profl. affairs coordinator, pres.), Assn. Clin. Scientists (diploma of honor 1992), Am. Soc. Med. Tech. (chmn. assembly, Achievement award 1976), Nat. Com. Clin. Lab. Standards (bd. dirs.), Iota Sigma Pi (hon.), Sigma Delta Epsilon (hon.), Presbyterian. Lodge: Altrusa (pres. 1982-83, bd. dirs.). Home: 3752 E Jackson Blvd Elkhart IN 46516-5205 Office: Miles Inc Diagnostics Divsn PO Box 70 Elkhart IN 46515-0070

FREE, MARY MOORE, anthropologist; b. Paris, Tex., Mar. 6, 1933; d. Dudley Crawford and Margie Lou (Moore) Hubbard; m. Dwight Allen Free, Jr., June 26, 1954; children: Hardy (dec.), Dudley (dec.), Margery, Caroline. BS, So. Meth. U., 1954, MLA, 1981, MA, 1987, PhD, 1989. Instr. So. Meth. U., Dallas, 1982-89, prof. continuing edn., 1989-90, prof. Dedman Coll., 1990—, adj. asst. prof. dept. anthropology, 1990—; prof. Richland Community Coll., Dallas, 1986; house anthropologist Baylor U. Med. Ctr., Dallas, 1990—; adv. bd. geriatrics Vis. Nurse Assn., Dallas, 1984-91. Contbr. articles to Anthropology Newsletter, Am. Jour. Cardiology., Cahiers de Sociologie Economique et Culturelle-Ethnopsychologie, Jour. Heart Failure. Named one of Notable Women of Tex., 1984. Fellow Am. Anthrop. Assn., Inst. for Study of Earth and Man; mem. Dallas Women's Club, Pi Beta Phi. Methodist. Achievements include development of position of house anthropologist in non-academic medical center, community medicine program. Home: 4356 Edmondson Ave Dallas TX 75205 Office: Baylor U Med Ctr 3500 Gaston Ave Dallas TX 75246

FREE, ROSS VINCENT, federal official; b. Bathurst, Australia, Mar. 7, 1943. BScin with honors, U. New South Wales, Australia; diploma in edn., Sydney. Mem. standing com. for publs. Australia House Rep., 1980-83, mem. for expenditure com., 1983-87, mem. for employment, edn., and tng. com., 1987—; now Minister for Sci. Australian Labor Party. Office: Dept of Industry Tech & Commerce, 51 Allara St, Canberra ACT 2601, Australia*

FREED, EDMOND LEE, podiatrist; b. Phila., Sept. 7, 1935; s. Frank and Jean D. (Schultz) F.; m. Judith Hope Falk (div. 1982); children: David Scott, Eric Corey. D of Podiatric Medicine, Temple U., 1960. Diplomate Am. Bd. Podiatric Surgery, Am. Bd. Podiatric Orthopedics. Pvt. practice Phila., 1960—; chmn. dept. podiatric surgery Met. Hosp., Phila., 1985-89, dir. podiatric residency program, 1983-89, co-chmn. limb salvage team, 1983—; mem. clin. faculty Pa. Coll. Podiatric Medicine, Phila., 1968—. Co-author booklets: Limb Salvage Concepts, 1984, Lower Extremity Ulcerations, 1985, Neurological Manifestations of Diabetes Mellitus, 1986. Fellow Am. Coll. Foot Surgeons (Ea. div. pres. 1985-88), Am. Coll. Foot Orthopedics, Am. Assn. Hosp. Podiatrists; mem. Am. Acad. Podiatric Sports Medicine, Phila. County Podiatry Soc., Am. Podiatric Med. Assn. Avocations: tennis, biking, model trains, numismatics. Office: Graduate Hosp Med Bldg Ste 2 N 520 S 19th St Philadelphia PA 19146

FREED, KARL FREDERICK, chemistry educator; b. Bklyn., Sept. 25, 1942; s. Nathan and Pauline (Wolodarsky) F.; m. Gina P. Goldstein, June 14, 1964; children: Nicole Yvette, Michele Suzanne. B.S., Columbia U., 1963; A.M., Harvard U., 1965, Ph.D., 1967. NATO postdoctoral fellow U. Manchester (Eng.), 1967-68; asst. prof. U. Chgo., 1968-73, assoc. prof., 1973-76, prof. chemistry, 1976—, dir. James Frank Inst., 1983-86. Author: Renormalization Group Theory of Macromolecules, 1987; editorial bd. Jour. Statis. Physics, 1976-78, Advances in Chem. Physics, 1985—; adv. editor Chem. Physics, 1979-92, Chem. Revs., 1981-83; assoc. editor Jour. Chem. Physics, 1982-84; contbr. articles to profl. jours. Recipient Marlow medal Faraday div. Chem. Soc. London, 1973; recipient Pure Chemistry award Am. Chem. Soc., 1976; fellow Sloan Found., 1969-71; Guggenheim fellow, 1972-73; fellow Dreyfus Found., 1972-77. Fellow Am. Phys. Soc.; mem. Royal Soc. Chemistry (London). Office: U Chgo 5640 S Ellis Ave Chicago IL 60637

FREEDMAN, ALFRED MORDECAI, pscyhiatrist, educator; b. Albany, N.Y., Jan. 7, 1917; s. Jacob Abraham and Pauline Rebecca (Hoffman) F.; m. Marcia Irene Kohl, Mar. 24, 1943; children: Paul Harris, Daniel Sholom. AB, Cornell U., 1937; MD, U. Minn., 1941. Diplomate Am. Bd. Psychiatry and Neurology. Intern Harlem Hosp., N.Y.C., 1941-42; resident and fellow Bellevue Hosp., N.Y.C., 1948-51, sr. psychiatrist, 1951-54; asst. pediatrician Babies Hosp.-Columbia, N.Y.C., 1953-60; assoc. prof. psychiatry SUNY Downstate Med. Sch., Bklyn., 1955-60; prof. and chair psychiatry N.Y. Med. Coll., Valhalla, 1960-89, prof. psychiatry emeritus, 1989—; vis. prof. Harvard Med. Sch., Boston, 1988—; dir. psychiatry Westchester Med. Ctr., Valhalla, 1979-89; cons. WHO, Geneva, 1984, 89—; S.Y. Mak vis. prof. U. Hong Kong, 1989; mem. awards jury Anna Monika Stiftung, Dortmund, Germany, 1983—; mem. internat. com. Prevention and Treatment of Depression, 1983—; sec.-treas. Ctr. for Comprehensive Health Practice Svc., N.Y., 1990—. Sr. editor textbook: Comprehensive Psychiatry, 1967-80; sr. editor book: Issues in Psychiatric Classification, 1986; editor-in-chief Polit. Psychology, 1981-90, Integrative Psychiatry, 1981—; contbr. articles to profl. jours. Mem. N.Y. State Commn. to Evaluate Drug Laws, Albany, 1970-73; founding trustee Ctr. for Urban Edn., N.Y.C., 1965-70; dir. Upper Park Ave. Boys Club of Am., N.Y.C., 1970-80; NGO rep. UN for World Psychiat. Assn., 1985-90, NGO rep. UN for World Assn. Psychosocial Rehabilitation, 1989—. Recipient Henry Wismer Miller award Manhattan Soc. Mental Health, 1964, Terence Cardinal Cooke medal N.Y. Med. Coll., 1985, Lapinlahti medal U. Helsinki, 1990, Wyeth Ayerst award World Psychiat. Assn., Athens, 1989, A.M. Freedman Ann. award Internat. for Polit. Psychology, 1990. Fellow Am. Psychiat. Assn. (pres. 1973-74, Rush medal 1974), Am. Psychopathol. Assn. (pres. 1971-72, Hamilton medal 1972), Am. Coll. Neuropsychopharmacology (pres. 1972-73), Am. orthopsychiat. Assn. (dir. 1962-64), Academia Medicinae et Psychiatricae (founding fellow, pres. 1990—); mem. N.Y. Psychiat. Soc. (pres. 1986-87), Nat. Com. on Confidentiality of health Records (pres. 1976—). Avocations: music, travel, gardening, sailing. Home and Office: 1148 Fifth Ave New York NY 10128-0807

FREEDMAN, LAURENCE STUART, statistician; b. London, Feb. 6, 1948; came to U.S., 1988; s. Hyman and Winnie (Faibis) F.; m. Nanette Maria Theresa Marmorstein, July 1, 1971; children: Tamara, Lewis, Shira, Rachel. BA, Cambridge U., 1966-69, MA, 1977; Diploma in Statistics, Univ. Coll., London U., 1971. Statistician MRC Statis. Rsch. Unit, London, 1971-75; rsch. fellow Wirral Area Health Authority, Liverpool, U.K., 1975-77; chief statistician MRC Cancer Trials Office, Cambridge, 1977-78; vis. scientist biometry br., divsn. cancer prevention Nat. Cancer Inst., Bethesda, Md., 1988-93, acting br. chief, 1993—; examiner in statistics Royal Coll. Radiologists, London, 1978-83; mem. cancer therapy com. Med. Rsch. Coun., London, 1979-88; mem. program com. Soc. for Clin. Trials, Balt., 1989-90, 91-92. Editor (jour.) Statistics in Medicine, 1981—; contbr. articles to Cancer Rsch., Jour. Nat. Cancer Inst. Recipient NIH Merit award, 1992. Fellow Inst. Statisticians, Internat. Statis. Inst.; mem. Royal Statis. Soc. (med. sect. com. mem. 1975—), Am. Statis. Assn., Am. Soc. for Preventive Oncology. Jewish. Achievements include development of new methodology for design and analysis of clinical trials; developed statistical methods of synthesizing results of laboratory experiments. Office: NCI Cancer Prevention & Control 6130 Executive Blvd Bethesda MD 20892

FREEDMAN, LOUIS MARTIN, dentist; b. Newark, Mar. 19, 1947; s. Morris and Sylvia (Swimmer) F.; m. Elizabeth Norine Palmer, June 17, 1978; children: Steven, Julie, Brian. Student, Emory U., 1963-66, DDS, 1970. Gen. dentist Freedman, Freedman & Weitman DDS, P.C., Atlanta, 1970—; clin. instr. Emory U. Dental Sch., Atlanta, 1970-77; team dentist Atlanta Hawks Basketball Team, 1971—, Atlanta Flame Hockey Team, 1979-80, Atlanta Knights Hockey Team, 1992—. Mem. Exch. Club, Atlanta, 1970-73; mgr. Sandy Springs Youth Sports Little League Baseball, 1979—; head coach Sandy Springs United Meth. Ch. basketball program, 1991—. Mem. Alpha Epsilon Delta, Omicron Kappa Upsilon. Jewish. Avocations: softball, little league managing, gardening, snow skiing, water skiing, swimming. Office: Freedman Freedman & Weitman 3111 Piedmont Rd Atlanta GA 30305

FREEDMAN, MICHAEL HARTLEY, mathematician, educator; b. Los Angeles, Apr. 21, 1951; s. Benedict and Nancy (Mars) F.; 1 child by previous marriage, Benedict C.; m. Leslie Blair Howland, Sept. 18, 1983; children: Hartley, Whitney, Jake. Ph.D., Princeton U., 1973. Lectr. U. Calif., 1973-75; mem. Inst. Advanced Study, Princeton, N.J., 1975-76; prof. U. Calif., San Diego, 1976—; Charles Lee Powell chair math. U. Calif., 1985—. Author: Classification of Four Dimensional Spaces, 1982; assoc. editor Jour. Differential Geometry, 1982—, Annals of Math., 1984-91, Jour. Am. Math. Soc., 1987—. MacArthur Found. fellow, 1984-89; named Calif. Scientist of Yr., Calif. Mus. Assn., 1984; recipient Veblen prize Am. Math. Soc., 1986, Fields medal Internat. Congress of Mathematicians, 1986, Nat. Medal of Sci., 1987, Humboldt award, 1988. Mem. Nat. Acad. Scis., Am. Assn. Arts and Scis., N.Y. Acad. Scis. Avocation: technical rock climber (soloed Northeast ridge Mt. Williamson 1970, Great Western boulder climbing champion 1979). Office: U Calif San Diego Dept Math 9500 Gilman Dr La Jolla CA 92093-0112

FREELAND, JOHN CHESTER, III, neuropsychologist; b. Wilmington, N.C., May 11, 1950; s. John Chester and Mary Frances (Stephens) F.; m. Jeanne M. Roy, Feb. 1, 1955. M in Psychology, U. Miss., 1983, PhD in Psychology, 1985. Lic. psychologist, Calif. Neuropsychologist VA Med. Ctr., Martinez, Calif., 1984-86, Transitional Learning Community, Galveston, Tex., 1986-88; program dir. Neurocare, San Diego, 1988-90; clin. dir. Casa Colina, Pomona, Calif., 1991—. Contbr. articles to profl. jours. Univ. fellow U. Miss., 1980. Mem. APA, Internat. Neuropsychol. Soc., Nat. Head Injury Found. Home: 296 E Lincoln Ave Pomona CA 91767 Office: Casa Colina 255 E Bonita Ave Pomona CA 91767

FREEMAN, ALBERT E., agricultural science educator; b. Lewisburg, W.Va., Mar. 16, 1931; s. James A. and Grace Vivian (Neal) F.; m. Christine Ellen Lewis, Dec. 23, 1950; children: Patricia Ellen, Lynn Elizabeth, Ann Marie. BS, W.Va. U., Morgantown, 1952, MS, 1954; PhD, Cornell U., 1957. Grad. asst. W.Va. U., Morgantown, 1952-54; grad. asst. Cornell U., Ithaca, N.Y., 1955-57; asst. prof animal sci Iowa State U., Ames, 1957-61, assoc. prof. animal sci., 1961-65, prof. animal sci., 1965-78, Charles F. Curtiss Disting. prof. agriculture, 1978—. Contbr. numerous articles to profl. jours. Active Collegiate Presbyterian Ch., Ames. Recipient 1975, Sr. Fulbright-Hays award, 1975, First Miss. Corp. award, 1979, award of appreciation for contbns. to Dairy Cattle Breeding 21st Century Genetics, 1984, Disting. Alumni award W.Va. U., 1985, faculty citation Iowa State U., 1987; named Charles F. Curtiss Disting. Prof. Agr., 1978. Fellow Am. Soc. Animal Sci. (Rockefeller Prentice Meml. award 1979, award of Honor 1987); mem. Am. Diary Sci. Assn. (bd. dirs. 1981-83, Nat. Assn. Animal Breeders Research award 1975, Borden award, 1982, J.L. Lush award 1984), Biometrics Soc., Am. Dairy Sci. Assn., First Acad. Disting. Alumni W.Va. U., Gamma Sigma Delta (award of Merit). Office: Iowa State Univ 239 Kildee Hall Ames IA 50010

FREEMAN, BRIAN S., electrical engineer; b. Naha, Okinawa, June 15, 1967; s. Max C. and Mary A. (Stonham) F.; m. Lynette A. Jankowski, July 8, 1989. BSEE, Worcester Poly. Inst., 1989. Engr.-in-tng., Wash. Commd. 2d lt. USAF, 1989, advanced through grades to capt., 1993; design elec. engr. 92d Civil Engring. Squadron USAF, Fairchild AFB, Wash., 1989-90; chief project mgmt. 20th Civil Engring. Squadron USAF, RAF Upper Heyford, U.K., 1990-93, wing energy conservation officer 20th Fighter Wing, 1990—; chief environ. flight 20th Civil Engring. Squadron USAF, 1993—. Tech. dir. Bicester (U.K.) Drama Group, 1991—. Recipient Gen. Rawling award Air Force Assn., 1992. Mem. IEEE (assoc.), Nat. Soc. Profl. Engrs., Soc. Am. Mil. Engrs. (sec. 1991—). Home: PSC 43 Box 7034 APO AE 09466 Office: 20 CES/CEV RAF Upper Heyford APO AE 09466

FREEMAN, DANIEL HERBERT, JR., biostatistician; b. Annapolis, Md., July 7, 1945; s. Daniel Herbert and Mary Virginia (Fiske) F.; m. Jean Louise Otis, May 26, 1971; 1 child, Elizabeth Grace. BA, Boston U., 1968, MA, 1970; PhD, U. N.C., 1975. Asst. prof. Yale U. Sch. Medicine, New Haven, 1975-81, assoc. prof., 1981-85; prof. Dartmouth Med. Sch., Hanover, N.H., 1985-92; dir. N.H. Cancer Registry Norris Cotton Cancer Ctr., Lebanon, 1986-92; prof. U. Tex. Med. Br., Galveston, 1992—, dir. Office of Biostatistics, 1992—. Author: Applied Categorical Data Analysis, 1989; contbr. articles to profl. jours. Mem. vestry St. John's Ch., New Haven, 1982-83; mem. Hanover (N.H.) Conservation Coun., 1987-88, Hanover Planning Bd., 1990-92. Indo-Am. fellow CIES, 1983. Mem. Am. Statis. Assn., Biometric Soc., Am. Pub. Health Assn., Soc. Epidemiol. Rsch. Democrat. Episcopalian. Office: Univ of Tex Med Br Office of Biostatistics 1 134 Ewing Hall Galveston TX 77535

FREEMAN, DAVID LAURENCE, chemist, educator; b. L.A., Mar. 10, 1946; s. Abe Mordechai and Edith (Cohen) F.; m. Donna Beth Meister, June 27, 1971; 1 child, Mark S. BS, U. Calif., Berkeley, 1967; PhD, Harvard U., 1973. Prof. chemistry U. R.I., Kingston, 1976—; collaborator Los Alamos (N.Mex.) Nat. Lab., 1982—. Co-author: Algebraic and Diagrammatic Methods in Many Fermion Theory; contbr. to sci. publs. Mem. Am. Phys. Soc., Am. Chem. Soc., Phi Beta Kappa. Office: Univ RI Dept Chemistry Pastore Chem Lab Kingston RI 02881

FREEMAN, EUGENE EDWARD, electronics company executive; b. L.A., Jan. 10, 1952; s. Louis Harold and Bella Ida (Yellin) F.; m. Claire Elaine Weinstein, Feb. 14, 1977; children: Benjamin, Hani. BA, Occidental Coll., 1973; BSEE, U. Houston, 1984. Layout engr. Tex. Instruments, Houston, 1979-84; cons. engr. NCR, Colorado Springs, Colo., 1984-91, sr. product mgr., 1991—; lectr. local schs. Contbr. articles to profl. jours. Mem. IEEE. Achievements include patent for latch with feedback, tolerant SCSI pads, electronic checkbook. Office: NCR 1635 Aeroplaza Dr Colorado Springs CO 80916

FREEMAN, GORDON RUSSEL, chemistry educator; b. Hoffer, Sask., Can., Aug. 27, 1930; s. Winston Spencer Churchill and Aquila Maud (Chapman) F.; m. Phyllis Joan Elson, July 9, 1927; children: Mark Russel, Michèle Leslie. B.A., U. Sask., 1952, M.A., 1953; Ph.D., McGill U., 1957; D.Phil., Oxford (Eng.) U., 1957. Postdoctoral fellow Centre D'Etudes Nucleaires, Saclay, France, 1957-58; asst. prof., then assoc. prof. chemistry U. Alta. (Can.), Edmonton, 1958-65; prof. U. Alta. (Can.), 1965—, chmn. div. phys. and theoretical chemistry, 1965-75, dir. radiation rsch. ctr., 1968—; exec. Chem. Inst. Can., 1974-80, chmn. phys. chemistry div., 1976-78, councillor, 1978-80. Contbr. articles to jours., chpts. to books. Research grantee Nat. Research Council Can., 1959-78; research grantee Natural Scis. and Engring. Research Council Can., 1978—, Def. Research Bd. Can., 1965-72. Mem. Chem. Inst. Can., Am. Phys. Soc., Can. Assn. Physicists, Epigraphic Soc., Can. Assn. Archaeologists. Office: University of Alberta, Radiation Research Center, Edmonton, AB Canada T6G 2G2

FREEMAN, MARJORIE KLER, interior designer; b. Phila., June 30, 1929; d. Joseph H. and Elizabeth VanHoesen (Vaughan) Kler; m. John Martin Hale, Dec. 26, 1953 (div. 1974); children: John Marshall, David Maclain; m. Bruce George Freeman, Dec. 17, 1983. Cert. Interior Design, Pratt Inst., 1951, BFA, 1952; MA, U. Mich., 1954. Dir. design studio Handicraft Furniture Co., Ann Arbor, Mich., 1953-63; design cons. dorms U. Mich., Ann Arbor, 1955-62; design cons. U. Del., Newark, 1963-67; bldg. and maint. designer and studio mgr. Vallery Miller Interiors, Woodland Hills, Calif., 1969-74; office mgr. Joseph H. Kler, M.D., New Brunswick, N.J., 1974-83; pres. Marjorie Kler Interiors Inc., Bound Brook, N.J., 1980—, Jewel Box of Princeton, Inc., N.J., 1988—; design cons. East Jersey Olde Towne, Inc., Piscataway, 1974—; buyer EJOT Gift Shop, 1977—. Author/editor cookbooks: Educated Palate, 1969, Grand Slam, 1990, Indian Queen Tavern, 1991. Pres. Bucceleuch Mansion Found., New Brunswick, 1983—; past pres., v.p. East Jersey Olde Towne, Inc., 1983-90, 91—. Mem. DAR (Jersey Blue chpt.), N.J. Assn. Bus. Women Owners, Princeton C. of C., Penn Hall Alumnae Assn. (pres., dir. 1989—), The Trowel Club of New Brunswick (pres. 1993). Republican. Presbyterian. Avocations: bridge, flower arranging, stamp collecting. Home and Office: 6 Mimosa Ct Princeton NJ 08540

FREHLICH, RODNEY GEORGE, engineer, researcher; b. Wilkie, Sask., Can., Aug. 7, 1952; came to U.S., 1977; s. Mathew and Katherine (Fenrich) F. PhD in Elec. Engring./Applied Physics, U. Calif., San Diego, 1982; BSc in Physics, U. Sask., 1974, MSc in Physics, 1977. Rsch. scientist La Jolla (Calif.) Inst., 1983-85; rsch. assoc. U. Colo., Boulder, 1986—. Topical editor Jour. Optical Soc. Am., 1989—; contbr. articles to Applied Optics, Modern Optics, Optics Letters, Physics Rev. Letters, others. Bd. dirs. Village Arts Coalition, Boulder, 1991—. NSF grantee, 1989-91, 91-92; NASA grantee, 1992—. Mem. IEEE, Optical Soc. Am., Acoustical Soc. Am. Achievements include research findings in the areas of lidar performance, lidar design, laser propagation, imaging, measurements of atmospheric turbulence using laser scintillation, theory of wave propagation in random media. Office: U Colo CIRES Campus Box 449 Boulder CO 80309

FREIBERG, JEFFREY JOSEPH, civil engineer; b. Casper, Wyo., Mar. 24, 1960; s. Patrick Joseph and Ferne Joan (Horton) F.; m. Celeste Ketterman, May 25, 1991. BSCE, U. Wyo., 1984. Registered profl. engr., Nev.; cert. water-rights surveyor, Nev. Designer Pulte Home Corp., Denver, 1985-86; engr. G.C. Wallace, Inc., Las Vegas, 1986—. Fellow mem. NSPE; mem. Nat. Assn. Indsl. and Office Pks. Republican. Roman Catholic. Home: 4044 Laurel Hill Dr Las Vegas NV 89030 Office: G C Wallace Inc 1555 S Rainbow Blvd Las Vegas NV 89102

FREIBERGER, WALTER FREDERICK, mathematics educator, actuarial science consultant, educator; b. Vienna, Austria, Feb. 20, 1924; came to U.S., 1955, naturalized, 1962; s. Felix and Irene (Tagany) F.; m. Christine Mildred Holmberg, Oct. 6, 1956; children: Christopher Allan, Andrew James, Nils H. B.A., U. Melbourne, 1947, M.A., 1949; Ph.D., U. Cambridge, Eng., 1953. Rsch. officer Aero. Rsch. Lab. Australian Dept. Supply, 1947-49, sr. sci. rsch. officer, 1953-55; tutor U. Melbourne, 1947-49, 53-55; asst. prof. div. applied math. Brown U., 1956-58, assoc. prof., 1958-64, prof., 1964—, dir. Computing Center, 1963-69, dir. Ctr. for Computer and Info. Scis., 1969-76, chmn. div. applied math., 1976-82, chmn. grad. com., 1985-88, assoc. chmn. div. applied math., 1988-91, chmn. univ. com. on statis. sci., 1991—; lectr., cons. program in applied actuarial sci., Bryant Coll., 1986—; mem. fellowship selection panel NSF, Fulbright fellowship selection panel. Author: (with U. Grenander) A Short Course in Computational Probability and Statistics, 1971; editor: The International Dictionary of Applied Mathematics, 1960, (with others) Applications of Digital Computers, 1963, Advances in Computers, Volume 10, 1970, Statistical Computer Performance Evaluation, 1972; mng. editor: Quarterly of Applied Mathematics, 1965—; Contbr. numerous articles to profl. jours. Served with Australian Army, 1943-45. Fulbright fellow, 1955-56; Guggenheim fellow, 1962-63; NSF Office Naval Research grantee in field. Mem. Am. Math. Soc. (assoc. editor Math. Reviews 1957-62), Soc. for Indsl. and Applied Math., Am. Statis. Assn., Inst. Math. Stats., Assn. Computing Machinery. Republican. Episcopalian. Club: Univ. (Providence). Home: 24 Alumni Ave Providence RI 02906-2310 Office: Brown U 182 George St Providence RI 02912-0001

FREIBOTT, GEORGE AUGUST, physician, chemist, priest; b. Bridgeport, Conn., Oct. 6, 1954; s. George August and Barbara Mary (Schreiber) F.; m. Jennifer Noble, July 12, 1980 (div.); children: Jessica, Heather, George; m. Arlene Ann Steiner, Aug. 1, 1982. BD, Am. Bible Coll., Pineland, Fla., 1977; BS, Nat. Coll. NHA, International Falls, Minn., 1978; ThM, Clarksville (Tenn.) Sch. Theology, 1979; MD, Western U., Phoenix, 1982; ND, Am. Coll., 1979; MsT, Fla. Sch. Massage, 1977. Diplomate Nat. Bd. Naturopathic Examiners; ordained priest Ea. Orthodox Ch., 1983. Chief mfg. cons. in oxidative chemistry Am. Soc. Med. Missionaries, Priest River, Idaho, 1976-88; mfg. cons. Oxidation Products Internat. div. ASMM, Priest River, 1974—; chemist/oxidative chemistry Internat. Assn. Oxygen Therapy, Priest River, 1985—; oxidative chemist, scientist, priest A.S. Med. Missionaries, Priest River, 1982—; massage therapist Fla. Dept. Profl. Registration, Tallahassee, 1977-91; cons. Benedict Lust Sch. Naturopathy; lectr. in field. Author: Nicola Tesla and the Implementation of His Discoveries in Modern Science, 1984, Warburg, Blass and Koch: Men With A Message, 1990, Free Radicals and Their Relationship to Complex Oxidative Compounds, 1991; contbr. articles to profl. jours. Recipient Tesla medal of Scientific Merit, Benedict Lust Sch. Natural Scis., 1992. Mem. Tesla Meml. Soc., Tesla Coil Builder's Assn., Internat. Bio-Oxidative Med. Found., British Guild Drugless Practitioners, Am. Colon Therapy Assn., Am. Massage Therapy Assn., Am. Naturopathic Med. Assn., Am. Soc. Med Missionaries, Am. Coll. Clinic Adminstrs., Nat. Assn. Naturopathic Physicians, Am. Psychotherapy Asns., Am. Soc. Metals, Am. Naturopathic Assn. (trustee, pres.), Eagles. Achievements include research conducted in organic and inorganic oxidative chemistry, thermoelectric/thermionic materials in relation to oxygen, oxygen as related to superconductivity and molecular makeup, energy studies, material science, archaeology, ancient Biblical and medical studies; developer and co-designer advanced oxidative equipment and testing apparatus of oxidation and oxidative studies. Home: PO Box 1360 Priest River ID 83856-1360 Office: Am Soc Med Missionary PO Box 1360 Priest River ID 83856-1360

FREID, JAMES MARTIN, mechanical engineer; b. Far Rockaway, N.Y., Nov. 3, 1965; s. Robert Charles and Renate (Hiller) F.; m. Michelle Francis Mahon, June 24, 1989 (div. Apr. 1991). BS, Southwest Tex. U., 1989. Engring. designer BDM Ford Aerospace, Austin, Tex., 1987-88, Carroll Touch Technologist, Round Rock, Tex., 1988-89; mech. engr. IDM Corp., Austin, 1989-91, U.S. Med. Products, Austin, 1991—; med. designer U.S. Med. Adv. Bd., Austin, 1991—. Mem. Soc. Mfg. Engrs., Nat. Assn. Cad Cam Opers. Jewish. Achievements include development of medical femoral broach holder, medical acetabular shell holder, industrial catscan used in super collider magnet inspection. Office: US Medical Products 912 Capital of Tex Hwy 100 Austin TX 78746

FREIDENBERGS, INGRID, psychologist; b. Latvia, Aug. 6, 1944; came to U.S, 1951; d. Olgerts and Marta (Purvins) F.; m. Jack Feder, June 21, 1980; 1 child, Paul. BA, CCNY, 1966, MS, 1970; MA, L.I. U., 1973, PhD, 1975; cert. in psychoanalysis, NYU, 1983. Lic. psychologist, N.Y. Sch. psychologist Bur. of Guidance N.Y.C. Bd. Edn., 1971-73; intern in clin. psychology Bellevue Psychiat. Hosp., N.Y.C., 1973-74; with Inst. Rehab. Medicine NYU, N.Y.C., 1974—, dir. psychology intern program Inst. Rehab. Medicine, 1983-85, dir. psychol. svcs. Cancer Rehab. Svc., 1979—; adj. asst. prof. dept. counselor edn. NYU, 1978-82, clin. instr. dept. psychiatry NYU Med. Ctr., 1981—; presenter in field. Contbr. numerous articles to profl. jours. Mem. med. adv. bd. Skin Cancer Found. NSF fellow Yeshiva U., 1966, L.I. U. fellow, 1971-72. Mem. Am. Psychol. Assn., N.Y. State Psychol. Assn., Psychoanalytic Soc. of NYU, Assn. for the Advancement of Psychology. Avocation: art. Office: 29 W 9th St New York NY 10011-8942

FREIDKIN, EVGENII S., physicist; b. Kishinev, Russia, Aug. 2, 1948; came to the U.S., 1983; s. Solomon A. Freidkin and Pearl Ya; 1 child, David. BS, Kishinev U., 1970; PhD, Poly. U., 1987. Part-time lectr. Rutgers U., New Brunswick, N.J., 1992—. Contbr. articles to profl. jours. With Israeli Army, 1979-80. Mem. Am. Phys. Soc., N.Y. Acad. Scis. Office: Rutgers U Dept Physics Piscataway NJ 08855

FREITAG, PETER ROY, transportation specialist; b. L.A., Dec. 19, 1943; s. Victor Hugo and Helen Veronica (Burnes) F. Student, U. Fla., 1961-63, George Washington U., 1964-65. Chief supr. Eastern Airlines, L.A., 1965-77; tariff analyst, instr. United Airlines, San Francisco, 1977-84; mng. ptnr. Bentdahl, Freitag & Assoc., San Francisco, 1984-86; v.p. ops. PAD Travel, Inc., Mountain View, Calif., 1985-86; travel mgr. Ford Aerospace, San Jose, Calif., 1986—. Co-editor: (textbook) International Air Tariff and Ticketing, 1983. Vol. San Francisco Bay chpt. Oceanic Soc., 1984-87. Mem. Silicon Valley Bus. Travel Assn., Bay Area Bus. Travel ASsn. Episcopalian. Avocations: travel, cooking, oenology, hiking.

FREITAG, ROBERT FREDERICK, government official; b. Jackson, Mich., Jan. 20, 1920; s. Fred J. and Beatrice (Paradise) F.; m. Maxine Pryer, Apr. 13, 1941; children—Nancy Marie (Mrs. Stephen Sprague), Janet Louise (Mrs. Richard Wasserstrom), Fred John II, Paul Robert. B.S.E. in Aero. Engring., U. Mich., 1941; postgrad., MIT, 1941-42. Commd. ensign USNR, 1941; lt. comdr. U.S. Navy, 1946, advanced through grades to capt., 1960; various guided missile programs, 1941-55; project officer Jupiter and Polaris intermediate range ballistic missiles (Chief Naval Operations), 1955-57; range planning officer, also spl. asst. to comdr. (Pacific Missile Range), Point Mugu, Calif., 1957-59; astronautics officer (Bur. Naval Weapons), 1959-63; ret., 1963; dir. launch vehicles and propulsion NASA, 1963; dir. Manned Space Flight Field Center Devel., 1963-72, dir. manned space flight advanced programs, 1973-82; dep. dir. NASA Space Sta. Task Force, 1982-85, dir. Space Sta. Office of Policy, 1985, dep. assoc. adminstr. for Space Sta., 1985—; Mem. NACA Com. Propellers, 1944-46, Sec. Def. Spl. Com. Adequacy Range Facilities, 1956-58, Joint Army-Navy Ballistic Missile Com., 1955-57, NACA Spl. Com. Space Tech., 1958-59; re-adv. com. missile and spacecraft aerodynamics NASA, 1960-63; joint Def. Dept.-NASA-Astronautics Coordinating Bd. (on launch vehicles panel), 1960-64. Author tech. papers. Decorated Legion of Merit, 1959; recipient Spl. Commendation from Comdr.-in-Chief U.S. Pacific Fleet, 1953, Spl. Commendation from Sec. Def., 1958, Sec. Navy Commendation medal, 1959, Disting. Alumnus award U. Mich., 1957, Sesquicentennial medal and cert., 1967, NASA Exceptional Service medal, 1969, 81, Outstanding Leadership medal, 1985, Bronze medal Brit. Interplanetary Soc., 1979; named Meritorious Exec., 1987; elected to Mich. Aviation Hall Fame, 1991. Fellow AIAA (pres. Cen. Calif. sect. 1958-59, dir. Washington sect. 1964-65, 69, Internat. Cooperation award 1990), Am. Astronautical Soc., Royal Aero. Soc.; mem. Internat. Acad. Astronautics (Allen D. Emil award 1986), Deutsche Gesellschaft für Luft-und Raumfahrt (hon.). Home: 4110 Mason Ridge Dr Annandale VA 22003-2034 Office: NASA Space Sta Task Force 600 Independence Ave SW Washington DC 20546-0002

FRENCH, ALFRED DEXTER, chemist; b. Boston, June 27, 1943; s. Dexter and Mary Catherine (Martin) F.; m. Sandra Sue Alleman, Aug. 7, 1965 (div. 1984); m. Mary An Holcombe, Jan. 22, 1991. BS, Iowa State U., 1965; PhD, Ariz. State U., 1971. Chemist So. Regional Rsch. Ctr., New Orleans, 1971—. Editor: Fiber Diffraction Methods, 1980, Computer Modeling of Carbohydrate Molecules, 1990; contbr. articles to profl. jours. Mem. Am. Crystallographic Assn., Am. Chem. Soc. (exec. sec. carbohydrate chemistry 1991-92, 93—, chmn. La. sect. 1992), Sigma Xi. Home: 1513 Madison St Metairie LA 70001 Office: USDA PO Box 19687 New Orleans LA 70179

FRENCH, ANTHONY PHILIP, physicist, educator; b. Brighton, Eng., Nov. 19, 1920; came to U.S., 1955; s. Sydney James and Elizabeth Margaret (Hart) F.; m. Naomi Mary Livesay, Oct. 6, 1945; children—Martin Charles, Gillian Ruth. BA with honors, Cambridge (Eng.) U., 1942, MA, 1946, PhD, 1948; ScD (hon.), Allegheny Coll., 1989. Mem. atomic bomb projects Tube Alloys and Manhattan Project, 1942-46; demonstrator, lectr. physics Cambridge U., 1948-55; fellow Pembroke Coll., 1950-55; prof. physics U. S.C., 1955-63, chmn. dept., 1956-62; vis. prof. MIT, 1962-64, prof., 1964-91, prof. emeritus, 1991—; vis. fellow Pembroke Coll., Cambridge, 1975; chmn. Internat. Commn. on Physics Edn., 1975-81. Author: Principles of Modern Physics, 1958, Special Relativity, 1966, Newtonian Mechanics, 1971, Vibrations and Waves, 1971, (with Edwin F. Taylor) Introduction to Quantum Physics, 1978, (with M.G. Ebison) Introduction to Classical Mechanics, 1986; editor: Einstein: A Centenary Volume, 1979, Physics in a Technological World, 1988; co-editor : Niels Bohr: A Centenary Volume, 1985; contbr. articles to profl. jours. Recipient Univ. medal Charles U., Prague, 1980, Bragg medal Inst. Physics, U.K., 1988, Oersted medal Am. Assn. Physics Tchrs., 1989. Fellow Am. Phys. Soc.; mem. Am. Assn. Physics Tchrs. (pres. 1985-86, Oersted medal 1989, Melba Newell Phillips award 1993), Sigma Xi, Sigma Pi Sigma. Office: Mass Inst Tech Rm 6-101 Cambridge MA 02139

FRENCH, EDWARD RONALD, plant pathologist; b. Buenos Aires, Apr. 28, 1937; s. Daniel Argentino and Federica Romana (Tonizzo) F.; m. Delia G. Monar-Peralta, Mar. 9, 1968; children: Vivian Marie, Ronald David, Sandra Janice. BS, U. R.I., 1960; MSc, U. Minn., 1963; PhD, N.C. State U., 1965. Plant pathologist agrl. mission to Peru N.C. State U., Raleigh, 1965-71, asst. prof. dept. plant pathology, 1971-72; head pathology dept. Internat. Potato Ctr., Lima, Peru, 1972-91, leader disease mgmt. program, 1992—; vis. prof. U. Agraria La Molina, Lima, 1967—; vis. plant pathologist Cen. Agrl. Rsch. Inst., Gannoruwa, Sri Lanka, 1980-81; vis. scientist Sta. de Pathologie Vegetale, Rennes, France, 1990; mem. Jakob Eriksson Prize Com., Stockholm, 1980—. Author, editor: Prospects For The Potato in the Developing World, 1972; author: (with T.T. Hebert) Metodos de Investigacion Fitopatologica, 1980; editor: (with G. Galvez) Plant Pathologists in Latin America, 1990. Pres. Tuqui Urco Housing Devel., Monterrico, Lima, 1977-78, 84-85, 93—; pres., bd. dirs. F.D. Roosevelt Am. Sch. Lima, Comacho, 1987-89; assoc. bd. dirs. F.D. Roosevelt Ednl. Inst., Lima, 1986—; pres. Am. Sch. Lima Found., Wilmington, Del., 1990—. Named Hon. Citizen City of Huanuco, Peru, 1985; E.R. French Bd. Rm. named in his honor FDR Am. Sch. Lima, Camacho, 1989. Mem. Internat. Soc. for Plant Pathology (v.p. 1983-88, coun. mem. 1973—), Assn. Latin Am. Fitopatologia (pres. 1970-74, 85-87, exec. sec. 1974-80, 92—), Rinconada Country Club, Club Tenis Terrazas Miraflores, Sigma Chi (magister 1958-59, Freshman award 1957, Found. award 1960), Alpha Zeta, Sigma Xi (hon. mem.), Phi Sigma (hon. mem.), Phi Kappa Phi (hon. mem.). Roman Catholic. Avocations: tennis, swimming. Office: Internat Potato Center, Apartado 5969, Lima 100, Peru

FRENCH, JUDSON CULL, government official; b. Washington, Sept. 30, 1922; s. Morrison Brady and Ethel Haviland (Cull) F.; m. Julia A. McAllister, Aug. 1, 1951; 1 child, Judson Cull. B.S. cum laude, Am. U., 1943; M.S., Harvard U., 1949; postgrad., Bus. Sch., 1968, Johns Hopkins U., 1943-44, George Washington U., 1944-45, M.I.T., 1951. Instr. physics Johns Hopkins U., Balt., 1943-44, George Washington U., Washington, 1944-47; sec., dir. Home Title Ins. Co., Washington, 1956-71; with Nat. Bur. Standards (now Nat. Inst. Standards and Tech.), Commerce Dept., Washington, 1948—; asst. chief electron devices sect. Nat. Bur. Standards (now Nat. Inst. Standards and Tech.), Commerce Dept., 1964-68, chief electron devices sect., 1968-73, chief electronic tech. div., 1973-78, dir. Ctr. for Electronics and Elec. Engring., 1978-91; dir. Electronics and Elec. Engring. Lab., Nat. Inst. Standards and Tech., Gaithersburg, Md., 1991—. Contbr. articles to profl. jours. Recipient Silver medal for meritorious service Commerce Dept., 1964; Gold medal for exceptional service, 1978; Edward Bennett Rosa award Nat. Bur. Standards, 1971; presdl. rank of Meritorious Exec., Sr. Exec. Service, 1980, Disting. Exec., 1984. Fellow IEEE; mem. Am. Phys. Soc., ASTM, Nat. Acad. Engring., Sigma Pi Sigma, Pi Delta Epsilon, Alpha Kappa Pi. Office: Nat Inst Standards and Tech Electronics and Elec Engring Lab Bld 220 Rte 270 Gaithersburg MD 20899

FRENCH, WILLIAM J., cardiologist, educator; b. Lawrence, Mass., Sept. 14, 1942; s. Harry and Catherine (McCoole) F.; m. Ninoska French; children: Michael, Natascha, Christopher. BS, U. N.H., 1964; MD, U. Vt., 1968. Diplomate Am. Bd. Internal Medicine, Am. Bd. Cardiology. Intern Harlem Hosp.-Columbia, N.Y.C., 1968-69, resident, 1969-71; resident Grad. Hosp.-U. Pa., 1971-72; critical care fellow U. So. Calif., 1972-73; dir. cardiac catheterization lab. Harbor-UCLA Med. Ctr., Torrance, 1980—; prof. medicine UCLA, 1990—. Office: Harbor UCLA Med Ctr 1000 W Carson St Torrance CA 90509-2059

FRENGER, PAUL FRED, medical computer consultant, physician; b. Houston, May 9, 1946; s. Fred Paul and Frances Mae (Mitchell) F.; m. Sandra Lee Van Schreeven, Aug. 17, 1979; 1 child, Kirk Austin. BA in Biology, Rice U., 1968; MD, U. Tex.-San Antonio, 1974. Lic. physician, Tex., Colo. Pediatric intern Keesler USAF Med. Ctr., Biloxi, Miss., 1974-75; course dir. U.S. Air Force Physician Assistant Sch., Sheppard AFB, Tex., 1976-78; spl. projects cons. Med. Networks, Inc., Houston, 1979-81; dir. med. products Microprocessor Labs., Inc., Houston, 1983; chief med. officer, dir. Mediclinic, Inc., Houston, 1984-85; pres., cons. Working Hypothesis, Inc., Houston, 1981-83, 85-91; project leader Telescan, Inc., Houston, 1987-89; med. dir. McCarty Clinic, 1989-90, Ft. Bend Family Health Ctr., 1990-91, Doctors at the Galleria, 1991-92; chief med. officer Health Testing, Inc., 1991; med. dir. Houston Pro Med., 1992-93. Contbr. over 70 articles to profl. jours.; patentee life raft test device, 3 patents. Mem. Rocky Mountain Bioengring. Symposium. Served to lt. col. USAF, 1969-78. Decorated Air Force Commendation medal. Mem. IEEE, Am. Assn. Med. Dirs., Am. Assn. Med. Systems and Informatics, Assn. for Computing Machinery (editor SIG Forth Newsletter, ACM News Houston chpt.), Internat. Neural Network Soc., Mensa. Episcopalian. Avocations: model engineering, railroading. Home: 814 Silvergate Houston TX 77079

FRENKEL, EUGENE PHILLIP, physician; b. Detroit, Aug. 27, 1929; s. David Eugene and Eva (Antin) F.; m. Rhoda Beth Smilay, Dec. 21, 1958; children: Lisa Michelle, Peter Alan. B.S., Wayne State U., 1949; M.D., U. Mich., 1953. Diplomate Am. Bd. Internal Medicine (hematology, med. oncology; bd. govs. 1980-87, chmn. subspecialty com. hematology 1980-85). Intern Wayne County Gen. Hosp., Eloise, Mich., 1953-54; resident in internal medicine Boston City Hosp., 1954-55; resident in internal medicine, then instr. U. Mich. Med. Center, 1957-62; mem. faculty U. Tex. Southwestern Med. Ctr., Dallas, 1962—, prof. internal medicine and radiology, 1969—, chief div. hematology-oncology, 1962-91, Patsy R. and Raymond D. Nasher Disting. chair in cancer rsch., 1990—; chief nuclear medicine, cons. hematology-oncology VA Med. Center, Dallas, 1962-80; cons. com. on evaluation research hematology: nutrition Nat. Inst. Arthritis and Metabolic Diseases, 1979-82. Author numerous research papers in field. Served as officer M.C. USAF, 1955-57. Fellow ACP, Internat. Soc. Hematology; mem. Am. Soc. Hematology (treas. 1976-84), Am. Soc. Clin. Oncology (chmn. membership com. 1982—), Am. Cancer Soc. (pres. Dallas unit 1970-71, dir. Tex. div. 1978—, sci. adv. com. on clin. investigations II—chemotherapy and hematology 1978-82, Emma Freeman prof. 1981-91, nat. clin. fellowship com. 1978-87, internat. rsch. grants com. 1988-90, sci. adv. coun. 1991—), Assn. Am. Physicians, Am. Assn. Cancer Research, Am. Assn. Cancer Edn., Am. Soc. Biol. Chemists, Am. Soc. Clin. Investigation, So. Soc. Clin. Investigation, Soc. Nuclear Medicine, Am. Fedn. Clin. Research, Internat. Soc. Hematology (elected councillor 1992—), Internat. Assn. Study Lung Cancer, Alpha Omega Alpha. Office: U Tex Southwestern Med Ctr Dallas TX 75235-8852

FRENKIEL, RICHARD HENRY, electronics company research and development executive; b. N.Y.C., N.Y., Mar. 4, 1943; s. Lucjan and Stefani (Komorowska) F.; kkm. Annamae Mary Rollason, Dec. 28, 1963; children: Scott Thomas, Kathleen Ann. BSME, Tufts U., 1963; MS Engring. Mechanics, Rutgers U., 1965. Mem. tech. staff Bell Labs., Holmdel, N.J., 1963-71, supr., 1973-77, dept. head, 1977—; asst. engring. mgr. AT&T, N.Y.C., 1971-73; mem. com. on cellular standards Electronics Industry Assn., Washington, 1980; session chmn. Nat. Electronics Conf., Chgo., 1974. Patentee in field. Recipient Vice Chmn.'s award AT&T, 1987; Bell Labs. fellow, 1990, Achievement award Indsl. Rsch. Inst., 1992. Fellow IEEE (speaker Outstanding Lecture Tour, 1975-76, Alexander Graham Bell medal, 1987). Republican. Office: AT&T Bell Labs Crawford Corner Rd Holmdel NJ 07733

FRENZ, BERTRAM ANTON, crystallographer; b. Port Washington, Wis., Sept. 23, 1945; s. Wilbert A. and Alice C. (Ernst) F.; m. Sharon Kath, Aug. 27, 1966; 1 child, Melissa. BS, U. Wis., 1967; PhD, Northwestern U., 1971. Postdoctoral fellow Tex. A&M U., College Station, Tex., 1971-74; owner, pres. Molecular Structure Corp., College Station, Tex., 1973-81, B.A. Frenz & Assocs., Inc., College Station, Tex., 1981—; owner, v.p. Computalytics, Inc., College Station, Tex., 1982—; owner, pres. List Run, Inc., Huntsville, Tex., 1985—; cons. IBM Corp., 1983-92; adv. bd. mem. ComputerLand Corp., Pleasanton, Calif., 1983-91. Contbr. articles to profl. publs. and chpts. to books. Mem. Am. Crystallographic Assn. (program chmn. 1981), Am. Chem. Soc. (local treas. 1979-80), Rio Brazos Audubon Soc. (dir. 1985-

92), Holy Cross Luth. Ch. (elder 1992—). Lutheran. Achievements include authoring and distributing of SDP, a leading software package for x-ray crystallography. Office: B A Frenz & Assocs Inc 209 University Dr E College Station TX 77840

FRENZ, DOROTHY ANN, cell and developmental biologist; b. New Rochelle, N.Y., Jan. 17, 1954; d. Anthony Joseph and Angelina Marie (Guida) Chiodo; m. Michael Richard Frenz, Sept. 15, 1974; children: Christopher, Elizabeth. BA summa cum laude, Iona Coll., 1978; MS, N.Y. Med. Coll., 1986, PhD, 1988. Postdoctoral fellow Albert Einstein Coll. Medicine, Bronx, N.Y., 1988-91; asst. prof. dept. otolaryngology Albert Einstein Coll. Medicine, Bronx, 1991—, asst. dir. rsch., 1993—, anatomy instr., 1991-92, asst. prof. anatomy and structural biology, 1993—; chairperson resident rsch. com. Albert Einstein Coll. Medicine, 1991—, senator faculty senate, 1991—. Contbr. chpts. in books and articles to profl. jours. Bd. dirs. New Rochelle YMCA; pres. Isaac E. Young Mid. Sch. PTA, New Rochelle, 1988-90; rec. sec. New Rochelle PTA Coun., 1992—; tchr., lector Blessed Sacrament Ch., New Rochelle, 1986—, parish bull. editor 1988—, parish coun. rec. sec. 1990—. Mem. Am. Assn. Anatomists, Cell Biology Soc., Assn. for Rsch. on Otolaryngology, N.Y. Acad. Scis., Soc. Devel. Biology. Roman Catholic. Office: A Einstein Coll Medicine 1300 Morris Park Ave Bronx NY 10461

FRERE, MAURICE HERBERT, soil scientist; b. Sheridan, Wyo., Sept. 8, 1932; s. Jules James and Eunice Marie (Sage) F.; m. Margaret Etta Rutherford, June 8, 1957; children: Ann Marie, James Michael, William Patrick. BS, U. Wyo., 1954, MS, 1958; PhD, U. Md., 1962. Soil scientist USDA Agrl. Rsch. Svc., Beltsville, Md., 1958-71, Durant, Okla., 1971-75; rsch. leader USDA Agrl. Rsch. Svc., Chichasha, Okla., 1975-79; rsch. adminstr. USDA Agrl. Rsch. Svc., New Orleans, 1979-85; rsch. leader USDA Agrl. Rsch. Svc., Watkinsville, Ga., 1985—; vis. scientist Agr. U., Wageningen, Netherlands, 1966-67. Author over 60 rsch. jour. articles, abstracts, conf. proceedings and book chpts. Mem. AAAS, Am. Chem. Soc., Grazing Lands Forum, Internat. Soil Sci. Soc., Soil Sci. Soc. Am., Soil and Water Conservation Soc., Am. Soc. Agronomy, Rotary, Sigma Xi, Alpha Zeta.

FRESHWATER, MICHAEL FELIX, hand surgeon; b. N.Y.C., Feb. 4, 1948; s. Jack and Rhoda Freshwater; m. Shawn M. Porter; 2 children. BS magna cum laude, Brooklyn Coll., 1968; MD, Yale U., 1972. Diplomate Am. Bd. Med. Examiners (cert. in surgery of hand), Am. Bd. Plastic Surgery. Asst. resident in surgery Yale New Haven Hosp., 1972-74; fellow in plastic surgery Med. Sch. Johns Hopkins U., Balt., 1974-77; resident, then chief resident in plastic surgery Jackson Meml. Hosp., 1977-78; Kleinert fellow hand and microsurgery U. Louisville, 1979; pvt. practice medicine specializing in hand surgery Miami, 1979—; pres., dir. Miami (Fla.) Inst. Hand and Microsurgery, 1980—; dir. hand and micro surgery Cedars Med. Ctr., 1985—, Deering Hosp., Miami, 1993—; chief surgery Cedars Med. Ctr., 1988-90, bd. dirs., 1990-92; vis. prof. Javeriana U., Bogota, Colombia, 1980-82, Hosp. Militar, Bogota, 1983-85, Centro Medico de los Andes, 1983-86; cons. Fla. Children's Med. Svc., Tallahassee, 1979—, Fla. Elks Crippled Children Soc., Orlando, 1983—, Fla. Dept. Profl. Regulation, Tallahassee, 1984—, League Against Cancer, 1983—, Scientists Inst. for Pub. Info., 1985—. Contbr. chpts. to books and articles to profl. jours.; mem. bd. reviewers Plastic and Reconstructive Surgery, 1976—, Internat. Abstracts of Plastic Surgery, 1987—. Trustee Yale U. Med. Libr., New Haven, 1972-77, D.R. Millard Found., 1987—; bd. dirs. V and A Gildred Found., Miami, 1980-86; bd. dirs. Yale Sch. of Medicine Alumni Fund, 1991-94; active nat. campaign com. Yale Sch. Medicine, 1993—. Recipient Letter Commendation Gov. Bob Graham, 1984; Weinberger fellow NIH, 1974-76; Jonas Salk scholar CUNY, 1968-72. Fellow Internat. Coll. Surgeons; mem. AMA (Physicians Recognition award 1976, 79, 82, 85, 88, 90, 93), Am. Assn. Hand Surgery, Am. Burn Assn., Am. Soc. Reconstructive Microsurgery, Internat. Soc. Reconstructive Microsurgery, Royal Soc. Medicine, Greater Miami Soc. Plastic and Reconstructive Surgeons (sec.-treas. 1987-88, pres.-elect 1988-89, pres. 1989-90), Am. Soc. of Peripheral Nerve, Miami Assn. for Surgery of Hand (dir. 1991—), Yale Club (Miami, N.Y.), Phi Beta Kappa. Avocation: skiing. Office: Miami Inst Hand & Micro Surgery 1150 NW 14th St Ste 713 Miami FL 33136-2118

FREUDENBURG, WILLIAM R., sociology educator; b. Norfolk, Nebr., Nov. 2, 1951; s. Eldon G. and Betty D. Freudenburg. BA, U. Nebr., 1974; MA, Yale U., 1976, MPhil, 1977, PdD, 1979. Research assoc. Yale U., New Haven, 1975-77; asst. prof. sociology and rural sociology Wash. State U., Pullman, 1978-83; assoc. prof. rural sociology Wash. State U., 1983-86, U. Wis., Madison, 1986-91; prof. rural sociology U. Wis., 1991—; mem. sci. com. U.S. Dept. Interior, minerals mgmt. svc., 1982-91, chair socioecon. subcom., 1986-91; researcher, cons. in field. Author: Public Reactions to Nuclear Power: Are There Critical Masses?, 1984, Paradoxes of Western Energy Development, 1984; contbr. articles to profl. jours. Hawksworth scholar, 1970-72, Nat. Merit scholar, 1970-74; NSF grad. fellow, 1975-79. Fellow Soc. for Applied Anthropology; mem. Am. Sociol. Assn. (congl. fellow 1983-84, coun., sect. on environ. sociology 1980-83, chair-elect 1987-89, chair sect. on enivron. and tech. 1989-91), Internat. Assn. for Impact Assessment, NAS (panelist, adv. com. on future nuclear power 1984, com. Alaska outer continental shelf oil and gas program 1992—), Rural Sociol. Soc. (v.p. 1993—, chmn. natural resources rsch. group 1982-83, program chmn. 1983-84, mem. various coms., local arrangements chmn., 1986-87, award of merit, natural resources rsch. group 1991), AAAS (life, Rural Sociol. Soc. rep. 1979-86, sec., sect. on social, econ. and polit. scis. 1986—), Soc. for Risk Analysis, coun. for Agrl. Sci. and Tech., Wis. Sociol. Assn., Law and Soc. Assn. (life), Midwest Sociol. Assn., Phi Beta Kappa, Phi Eta Sigma. Office: Univ Wis Dept Rural Sociology 350 Agriculture Hall 1450 Linden Dr Madison WI 53706-1562

FREUDENHEIM, JO L., social and preventive medicine educator; b. Akron, Ohio, Sept. 6, 1952; s. Milton B. and Elizabeth (Ege) F.; m. Michael Frisch. MS in Preventive Medicine, U. Wis., 1980, PhD in Nutrition Sci., 1986. Registered dietitian. Asst. to assoc. prof. social and preventive medicine SUNY, Buffalo, 1988—. Recipient Rsch. Career Devel. award NIH, 1992. Achievements include research in diet and nutrition in the epidemiology of chronic disease especially breast and colon cancer. Office: SUNY Dept Social and Preventive Medicine 270 Farber Hall Buffalo NY 14214

FREUDENSTEIN, FERDINAND, mechanical engineering educator; b. Frankfurt, Germany, May 12, 1926; came to U.S., 1942, naturalized, 1945; s. George Gerson and Charlotte (Rosenberg) F.; m. Leah Schwarzschild, July 5, 1959 (dec. May 1970); children: David George, Joan Merle; m. Lydia Gersten, 1980. Student, N.Y.U., 1942-44; MS, Harvard U., 1948; PhD, Columbia U., 1954. Devel. engr. instrument div. Am. Optical Co., 1948-50; mem. tech. staff Bell Telephone Labs., 1954; mem. faculty Columbia U., 1954—, prof. mech. engring., 1959—, Stevens prof. mech. engring., 1981—, Higgins prof. mech. engring., 1985—, chmn. dept., 1958-64; cons. to industry, 1954—. Served inf. AUS, 1944-46. Recipient Gt. Tchr. award Soc. Older Grads., Columbia U., 1966, Applied Mechanisms Conf. award, 1989, Egleston medal for disting. engring. achievement, 1992; Guggenheim fellow, 1961-62, 67-68; guest of honor at conf. Tribute to Work of Ferdinand Freudenstein, Brainard, Minn., 1991. Fellow ASME (hon. life 1992, Jr. award 1955, Machine Design award 1972, Mechanisms com. award 1978, Charles russ Richards Meml. award 1984), N.Y. Acad. Scis.; mem. Harvard Soc. Engrs. and Scientists, Columbia Engring. Soc., N.Y. Acad. Scis., Nat. Acad. Engring., Sigma Xi. Achievements include research in kinematics, dynamics, mechanisms, engring. design. Home: 435 W 259th St Bronx NY 10471-1617 Office: Columbia U S W Mudd Bldg 116th St and Broadway New York NY 10027

FREUDENTHAL, RALPH IRA, toxicology consultant; b. N.Y.C., Aug. 27, 1940; m. Susan E. Loy; children: Judith, Jennifer, Ralph D. BS, NYU, 1963; PhD, SUNY, Buffalo, 1969. Biochem. pharmacologist Rsch. Triangle Inst., Research Triangle Park, N.C., 1969-73; assoc. mgr. Battelle Meml. Inst., Columbus, Ohio, 1973-77; dir. toxicology Stauffer Chem. Co., Farmington, Conn., 1977-84; dir. health, safety and regulatory affairs Stauffer Chem. Co., Westport, Conn., 1984-88; cons. Toxicology Consultancy, West Palm Beach, Fla., 1988—; steering com. Rene Dubos Ctr. for Environ., N.Y.C., 1982-89. Contbr. articles to profl. jours. Mem. Soc. of Toxicology, Am. Soc. Pharmacology and Exptl. Therapeutics, Am. Assn. for Cancer Rsch., Soc. for Environ. Toxicology and Chemistry. Home and Office: 8737 Estate Dr West Palm Beach FL 33411

FREUND, ECKHARD, electrical engineering educator; b. Düsseldorf, Germany, Feb. 28, 1940; s. Karl and Margret (Meya) F.; m. Brigitte Keudel; children: Viviane, Ariane. Diploma in engring., Tech. Sch. Darmstadt, Fed. Republic Germany, 1965; D Engring., Tech. U. Berlin, 1968. Scientist U. Raumfahrt, Oberpfaffenhofen, Fed. Republic Germany, 1965-70; guest prof. aero. engring. U. So. Calif., L.A., 1972-76, 83; guest scientist European Space Ops. Ctr., Darmstadt, Fed. Republic Germany, 1970-71; sci. coord. Fraunhofer Inst., Karlsruhe, Fed. Republic Germany, 1976-78; prof. dept. elec. engring. Fernuniversität, Hagen, Fed. Republic Germany, 1978-84; prof. dept. elec. engring., dir. Inst. Robotics Rsch. U. Dortmund, Fed. Republic Germany, 1985—; sci. adviser Jet Propulsion Lab., NASA, Pasadena, Calif., 1983. Author: Zeitvariable Mehrgrössensysteme, 1971, Regelungssysteme im Zustandsraum, I/II, 1986, 87; contbr. numerous articles on robotics to tech. publs. Office: Inst Robotics Rsch Dortmund, Otto Hahn Strasse 8, D 44221 Dortmund 50, Germany

FREUND, EMMA FRANCES, medical technologist; b. Washington; d. Walter R. and Mabel W. (Loveland) Ervin; m. Frederic Reinert Freund, Mar. 4, 1953; children: Frances, Daphne, Fern, Frederic. BS, Wilson Tchrs. Coll., Washington, 1944; MS in Biology, Catholic U., Washington, 1953; MEd in Adult Edn., Va. Commonwealth U., 1988; cert. in mgmt. devel. Va. Commonwealth U., 1975, MEd, 1988; student SUNY, New Paltz, 1977, J. Sargeant Reynolds Community Coll., 1978. Cert. Nat. Cert. Agy. for Clin. Lab. Pers. Tchr. math. and sci. D.C. Sch. System, Washington, 1944-45; technician in parasitology lab., zool. div., U.S. Dept. Agr., Beltsville, Md., 1945-48; histologic technician dept. pathology Georgetown U. Med. Sch., Washington, 1948-49; clin. lab. technician Kent and Queen Anne's County Gen. Hosp., Chestertown, Md., 1949-51; histotechnologist surg. pathology dept. Med. Coll. Va. Hosp., Richmond, 1951—, supr. histology lab., 1970—; mem. exam. coun. Nat. Cert. Agy. Med. Lab. Pers. Asst. cub scout den leader Robert E. Lee coun. Boy Scouts Am., 1967-68, den leader, 1968-70. Co-author: (mini-course) Instrumentation in Cytology and Histology, 1985. Mem. AAAS, Am. Soc. Med. Technology (rep. to sci. assembly histology sect. 1977-78, chmn. histology sect. 1983-85, 89-92), Va. Soc. Med. Technology, Richmond Soc. Med. Technologists (corr. sec. 1977-78, dir. 1981-82, pres. 1984-85), Va. Soc. Histology Technicians (dir. 1979—, pres. 1982-88), Nat. Certification Agy. (clin. lab. specialist in histotech., clin. lab. supr., clin. lab. dir.), N.Y. Acad. Scis., Am. Assn. Clin. Chemistry (assoc.), Am. Soc. Clin. Pathologists (cert. histology technician), Nat. Geog. Soc., Va. Govtl. Employees Assn., Nat. Soc. Histotech. (by-laws com. 1981—; C.E.U. com. 1981—, program com. regional meeting 1984, 85, chmn. regional meeting 1987), Am. Mus. Natural History, Smithsonian Instn., Am. Mgmt. Assn., Clin. Lab. Mgmt. Assn., Nat. Soc. Historic Preservation, Sigma Xi, Phi Beta Rho, Kappa Delta Pi, Phi Lambda Theta. Home: 1315 Asbury Rd Richmond VA 23229-5305 Office: Med Coll VA Hosp Surg Pathology Dept PO Box 240 Richmond VA 23202-0240

FREUND, LAMBERT BEN, engineering educator, researcher, consultant; b. McHenry, Ill., Nov. 23, 1942; s. Bernard and Anita (Schaeffer) F.; m. Colleen Jean Hehl, Aug. 21, 1965; children: Jonathan Ben, Jeffrey Alan, Stephen Neil. B.S., U. Ill., 1964, M.S., 1965; Ph.D., Northwestern U., 1967. Postdoctoral fellow Brown U., Providence, 1967-69; asst. prof. Brown U., 1969-73, assoc. prof., 1973-75, prof. engring., 1975—, Henry Ledyard Goddard prof., 1988—, chmn. div., 1979-83; vis. prof. Stanford (Calif.) U., 1974-75; cons. Aberdeen Proving Ground, U.S. Steel Corp., vis. scholar Harvard U., 1983-84; mem.-at-large U.S. Nat. Com. for Theoretical and Applied Mechanics, NRC, 1985—; mem. IUTAM Gen. Assembly, 1987—. Author: Dynamic Fracture Mechanics, 1990; editor in chief: ASME Jour. Applied Mechanics, 1983-88, editor Cambridge monographs on Mechanics and Applied Mathematics, 1989—, Jour. Mechanics and Physics of Solids, 1992—; mem. editorial adv. bd. Acta Mechanica Sinica, 1990—; contbr. articles to tech. jours. NSF trainee, 1964-67; grantee NSF, Office Naval Research, Army Research Office, Nat. Bur. Standards. Fellow ASME (Henry Hess award 1974, mem. Applied. Mech. Div. Exec. Com. 1989--), Am. Acad. Mechanics, Am. Acad. Arts and Scis.; mem. Am. Geophys. Union, ASTM (George R. Irwin medal 1987). Home: 3 Palisade Ln Barrington RI 02806-3921 Office: Brown U 79 Waterman St Providence RI 02912

FREVERT, DONALD KENT, hydraulic engineer; b. Des Moines, Mar. 23, 1950; s. Richard Keller and Corine (Twetley) F.; m. Maria Carmen Tarazon, Mar. 16, 1973; children: Richard Paul, Erica Lynn. BS in Hydrology, U. Ariz., 1972; MS in Hydrology and Water Resources, Colo. State U., 1974, PhD in Irrigation and Drainage, 1983. Registered profl. engr., Colo. Engring. aid USDA Agrl. Rsch. Svc., Tucson, 1970-72; grad. rsch. asst. Colo. State U., Fort Collins, 1972-74; hydrologist, water rights engr. Woodward-Clyde Cons., Denver, 1975-76; grad. rsch. asst. Colo. State U., Fort Collins, Colo., 1977-80; hydraulic engr. U.S. Bur. Reclamation, Lakewood, Colo., 1980—; faculty affiliate Colo. State U. civil engring. dept., Ft. Collins, 1986—; trustee Rocky Mountain Hydrologic Rsch. Ctr., 1992—. Co-author: (manuals) Comparison of Equations Used for Estimating Agricultural Crop Evapotranspiration with Field Research, 1983, Applied Stochastic Techniques Users Manual, 1990; contbr. articles to profl. jours. Age group coord. Lakewood Swim Club, 1988—. Recipient of Paul Elliott Ullman scholarship U. Ariz., Tucson, 1969, Pima Mining Co. scholarship, U. Ariz., 1970. Mem. ASCE (chmn. surface water com. 1988-90, mem. exec. com. irrigation and drainage div. 1991—), Phi Kappa Phi, Alpha Epsilon. Home: 2034 S Xenon Ct Lakewood CO 80228-4355 Office: US Bur Reclamation D-5755 PO Box 25007 Denver CO 80225-0007

FREYBERG, DALE WAYNE, technical trainer; b. Fargo, N.D., Dec. 20, 1927; s. Carl William and Ruth Eloise (Cram) F.; m. Jane Ann Owen, Sept. 3, 1949; children: Leah, Lynn, Owen. BSBA, Gustavus Adolphus C.C., 1951. Collector small loans 1st Nat. Bank, Mankato, Minn., 1951-52; sales tr. mgr. Tubular Micrometer Co., St. James, Minn., 1952-59; mng. dir. Scherr-Tumico Israel Ltd., Lod, Israel, 1959-61; product mgr. Engis Equipm Co., Morton Grove, Ill., 1962-70; sales mgr. Mahr Gage Co., N.Y.C., 1971-76; regional mgr., dir. MTI-Mitutoyo Inc., Paramus, N.J., 1976—. 2d lt. U.S. Army, 1946. Mem. Soc. Mfg. Engrs., Am. Soc. Quality Control, Masons. Republican. Methodist. Home: 1242 Wintergreen Batavia IL 60510 Office: MTI Mitutoyo Corp Edn Dept 945 Corporate Dr Aurora IL 60504

FREYERMUTH, CLIFFORD L., structural engineering consultant. BS in Civil Engring., State U. Iowa, 1956, MS in Structural Engring., 1958. Registered structural engr., Ariz. Consulting engr. structural design Ned L. Ashton, 1955-57; grad. teaching asst. structural mechanics State U. Iowa, 1957-58; with bridge divsn. Ariz. State Hwy. Dept., 1958-64; with Portland Cement Assn., Chgo., Skokie, Ill., 1964-71; dir. post-tensioning divsn. Prestessed Concrete Inst., 1971-76; mgr. Post-Tensioning Inst., 1976-88; pres. Clifford L. Freyermuth, Inc., 1988—; mem. cable-stayed bridges com. Post-Tensioning Inst, editor various publs.; prin. investigator Nat. Coop. Hwy Rsch. Project, Washington, 1988. Contbr. articles to profl. jours. Recipient Martin P. Korn award Prestressed Concrete Inst., 1969. Fellow Am. Concrete Inst. (prestressed concrete com., standard bldg. code com., bd. dirs. 1991—, Henry C. Turner medal 1991); mem. ASCE (prestressed concrete com.), Internat. Assn. Bridge and Structural Engrs., Structural Engrs. Assn. Ariz., Chi Epsilon. Office: Clifford L Freyermuth Inc 9201 N 25th Ave #1508 Phoenix AZ 85021*

FREYMANN, RAYMOND FLORENT, aeronautical engineer; b. Esch/Alzette, Luxembourg, May 30, 1952; s. Arthur and Catherine (Graas) F. Engring. Degree, Tech. U., Braunschweig, Fed. Republic of Germany, 1976, DEng Sci., 1981. Scientist DFVLR, Göttingen, Fed. Republic of Germany, 1976-86; vis. scientist Air Force Flight Dynamics Lab., Dayton, Ohio, 1983-84; lectr. Inst. Sup. de Tech., Luxembourg, 1985-90; div. head BMW, Munich, Fed. Republic of Germany, 1986-90, head dept., 1990—; panel mem. Adv. Group for Aero. R&D-Structures and Materials Panel, NATO, 1981—, chmn. subcom. on acoustic loads, 1990—, chmn. work group on structures, 1991—, subcom. landing gear design, 1993—. Contbr. articles to jours. AGARD, AIAA. Recipient ASME award 1987. Mem. AIAA, Deutsche Gesellschaft fuer Luft- und Raumfahrt. Achievements include patents in servohydraulics, active controls, structural dynamics and acoustics; research in nonlinear aeroservoelastic analyses and structural-acoustics. Home: Bahnhofstrasse 27, 85386 Eching Germany Office: BMW AG, PO Box 40 02 40, 80788 Munich Germany

FRIBOURGH, JAMES HENRY, university administrator; b. Sioux City, Iowa, June 10, 1926; s. Johan Gunder and Edith Katherine (James) F.; m. Cairdenia Minge, Jan. 29, 1955; children: Cynthia Kaye, Rebecca Jo, Abbie Lynn. Student, Morningside Coll., 1944-47; BA, U. Iowa, 1949, MA, 1949, PhD, 1957; LHD (hon.), Morningside Coll., 1989, DHL (hon.), 1989. Instr. Little Rock Jr. Coll., 1949-56; assoc. prof. biology Little Rock U., 1957-60, prof., chmn. div. life scis., 1960-69; vice chancellor U. Ark., Little Rock, 1969-72, interim chancellor, 1972-73, exec. vice chancellor for acad. affairs, 1973-82, interim chancellor, exec. vice chancellor for acad. affairs, 1982, provost, exec. vice chancellor, 1983—, disting. prof., 1984—; cons. in field; assoc. Marine Biol. Lab., Woods Hole, Mass. Contbr. articles to profl. jours. Mem. Ark. Gov.'s Com. on Sci. and Tech., 1969-71; bd. dirs., mem. nat. adv. bd. Nat. Back Found., 1979; vice chmn. NCCJ, 1981-82; div. rep. United Way of Pulaski County, 1980-82; bd. dirs. Ark. Dance Theatre, Little Rock, 1980-82; vestryman Good Shepherd Episcopal Ch.; del. Episcopal Diocese of Ark.; fellow Ark. Mus. Sci. and History, 1987. NSF fellow Hist. of Sci. Inst., 1959-60. Fellow AAAS, Coll. Preceptors (London), Am. Inst. Fishery Rsch. Biologists, Ark. Mus. Sci. and History; mem. Am. Fisheries Soc. (chmn. com. on internationalism cert. fisheries scientist), AAUP (pres. Ark. conf.), Electron Microscopy Soc. Am., Am. Soc. Swedish Engrs. (corr. mem.), Ark. Acad. Sci. (pres. 1966), Ark. Dean's Assn. (pres. 1982), Am. Assn. State Colls. and Univs., Am. Swedish Inst., Swedish Club (Chgo.), Rotary (Paul Harris fellow), Vasa Order Am. Lodge, Sigma Xi, Phi Kappa Phi. Democrat. Clubs: Swedish, Vasa Order Am. Lodge: Rotary (Paul Harris fellow). Office: U Ark 33d and University Ave Little Rock AR 72204

FRIDAY, ELBERT WALTER, JR., federal agency administrator, meteorologist; b. DeQueen, Ark., July 13, 1939; s. Elbert Walter and Mary Elizabeth (Ward) F.; m. Karen Ann Hauschild, Nov. 14, 1959; children: Kristine Ann, Kelly Sue. BS in Engring. Physics, U. Okla., Norman, 1961, MS in Meteorology, 1967, PhD in Meteorology, 1969. Commd. 2d lt. USAF, 1961, advanced through grades to col., weather officer, 1961-81, dir. environ. and life scis., Dept. Def., 1978-81, ret., 1981; dep. dir. Nat. Weather Svc., Silver Spring, Md., 1981-87, dir., 1987—; mem. com. on low level wind shear NAS, Washington, 1985-86. Contbr. articles to prof. jours. Elder Calvary Christian Ch., Burke, Va., 1985-89, trustee, 1989—. Decorated Bronze Star; recipient Def. Superior Svc. medal Dept. Def., 1981, Presdl. Rank award, 1988, Disting. achievement award U. Okla., 1992. Fellow Am. Meteorol. Soc. (councilor 1988-90); mem. Nat. Weather Assn., Sigma Xi. Office: Dept Commerce Nat Weather Svc 1325 East West Hwy Silver Spring MD 20910

FRIDLEY, ROBERT BRUCE, agricultural engineering educator, academic administrator; b. Burns, Oreg., June 6, 1934; s. Gerald Wayne and Gladys Winona (Smith) F.; m. Jean Marie Griggs, June 12, 1955; children: James Lee, Michael Wayne, Kenneth Jon. BSME, U. Calif., Berkeley, 1956; MS in Agrl. Engring., U. Calif., Davis, 1960; PhD in Agrl. Engring., Mich. State U., East Lansing, 1973; D honoris causa, Universidad Polytecnica de Madrid, 1988. Asst. specialist U. Calif., Davis, 1956-60, prof. agrl. engring., 1961-78, acting assoc. dean engring., 1972, chmn. dept. agrl. engring., 1974-76, dir. aquaculture and fisheries program, 1985-90, exec. assoc. dean agrl. and environ. scis., 1990—; dept. mgr. R & D Weyerhaeuser Co., Tacoma, 1977-85; vis. prof. Mich. State U., East Lansing, 1970-71; NATO vis. prof. U. Bologna, Italy, 1975. Co-author: Principles and Practices for Harvesting and Handling Fruits and Nuts, 1973; contbr. articles to profl. jours.; patentee in field. Recipient Charles G. Woodbury award Am. Soc. Hort. Sci., 1966, Alumni citation Calif. Aggie Alumni Assn., 1990. Fellow Am. Soc. Agrl. Engrs. (Young Researchers award 1971, Concept of Yr. award 1976, Outstanding Paper awards 1966, 68, 69, 76, 86, Disting. Svc. award 1988, v.p. Found. 1989-93, pres. Found. 1993—); mem. NAE, Am. Soc. Agrl. Engrs., Am. Soc. Engring. Edn., World Aquaculture Soc., Am. Fisheries Soc. Office: U Calif 228 Mrak Hall Davis CA 95616

FRIED, JOEL ROBERT, chemical engineering educator; b. Memphis, Dec. 9, 1946; s. Samuel J. and Mathilda (Kleinman) F.; m. Ava S. Krinick, June 8, 1969; children: Marc S., Aaron M. BS, Rensselaer Poly. Inst., 1968, 71, ME, 1972; MS, U. Mass., 1975, PhD, 1976. Mem. assoc. rsch. staff GE, Schenectady, N.Y., 1972-73; sr. rsch. engr. Monsanto Co., St. Louis, 1976-78; asst. prof. chem. engring. U. Cin., 1978-83, assoc. prof. chem. engring., 1983-90, dir. grad. studies, 1986-90, dir. polymer rsch. ctr., 1989-92, prof. chem. engring., 1990—; pres. Polymer Rsch. Assocs., Inc., Cin., 1984—; vis. fellow Sci. and Engring. Rsch. Coun., U.K., 1986. Patentee permeation modified separation membranes; author: Polymer Science and Technology, 1993; contbr. articles to sci. jours. Jr. Morrow rsch. chair U. Cin., 1980; USAF summer faculty rsch. fellow, 1981, 93. Mem. Am. Chem. Soc., Am. Inst. Chem. Engrs., Soc. Plastics Engrs. Office: U Cin Dept Chemical Engineering Cincinnati OH 45221-0171

FRIED, JOHN H., chemist; b. Leipzig, Fed. Republic Germany, Oct. 7, 1929; s. Abraham and Frieda F.; m. Heléne Gellen, June 29, 1955; children: David, Linda, Deborah. AB, Cornell U., 1951, PhD, 1955. Steroid chemist, research assoc. Merck and Co., Rahway, N.J., 1956-64; with Syntex Research, Palo Alto, Calif., 1964-92, dir. inst. organic chemistry, 1967-74, exec. v.p., 1974-76, pres., 1976-92; sr. v.p. Syntex Corp., 1981-86, vice chmn., 1986-92; dir. Corvas Internat., Inc., 1992—; chmn. bd. dirs. Alexion Pharms., Inc., 1992—; bd. dirs. Syntex Corp. Mem. Am. Chem. Soc. Office: 1238 Martin Ave Palo Alto CA 94301

FRIED, JOSEF, chemist, educator; b. Przemysl, Poland, July 21, 1914; came to U.S., 1938, naturalized, 1944; s. Abraham and Frieda (Fried) F.; m. Erna Werner, Sept. 18, 1939 (dec. Nov. 1986); 1 dau., Carol Frances. Student, U. Leipzig, 1934-37, U. Zurich, 1937-38; Ph.D., Columbia U., 1941. Eli Lilly fellow Columbia U., 1941-43; research chemist Givaudan, N.Y., 1943; head dept. antibiotics and steroids Squibb Inst. Med. Research, New Brunswick, N.J., 1944-59; dir. sect. organic chemistry Squibb Inst. Med. Research, 1959-63; prof. chemistry, biochemistry and Ben May Lab. Cancer Research, U. Chgo., 1963—, Louis Block prof., 1973—, chmn. dept. chemistry, 1977-79; mem. med. chem. study sect. NIH, 1963-67, 68-72, chmn., 1971; mem. com. arrangements Laurentian Hormone Conf., 1964-71; Knapp Meml. lectr. U. Wis., 1958. Mem. bd. editors: Jour. Organic Chemistry, 1964-69, Steroids, 1966-86, Jour. Biol. Chemistry, 1975-81, 83-88; contbr. articles to profl. jours. Recipient N.J. Patent award, 1968. Fellow AAAS, N.Y. Acad. Scis.; mem. Am. Chem. Soc. (award in medicinal chemistry 1974, Roussel prize 1992), Nat. Acad. Scis., Am. Acad. Arts and Scis., Am. Soc. Biol. Chemists, Swiss Chem. Socs., Brit. Chem. Socs., Sigma Xi. Patentee in field. Home: 5715 S Kenwood Ave Chicago IL 60637-1742

FRIED, ROBERT, psychology educator; b. Linz, Austria, July 27, 1935; s. Georg and Alice (Schwartz) F.; m. Virginia Lynn Cutchin, Nov. 16, 1991; children: Paul M., Steven G., Dennis A. AB, CCNY, 1959; PhD, Rutgers U., 1964. Lic. psychologist, N.Y.; cert. psychotherapy. Prof. CUNY, N.Y.C., 1964—; dir. biofeedback clinic Inst. for Rational Emotive Therapy, N.Y.C., 1985—; dir. rehab. rsch. Internat. Ctr. for the Disabled, N.Y.C., 1981-84. Author: The Hyperventilation Syndrome, 1987, The Breath Connection, 1990, The Psychology and Physiology of Breathing in Behavioral Medicine, 1993; contbr. over 50 articles to profl. jours. Cpl. U.S. Army, 1953-55. Recipient Superior Accomplishment award USN, 1965, Citation for Outstanding Contbn. to Profession, N.Y. State Biofeedback Soc., 1988. Fellow N.Y. Acad. Scis.; mem. APA, Assn. for Applied Psychophysiology and Biofeedback. Achievements include patents for Cardiac Computer, Muscle Activity Recorder and Brain Wave Computer. Home: 1040 Park Ave New York NY 10028 Office: Hunter Coll 695 Park Ave New York NY 10021

FRIEDBERG, SIMEON ADLOW, physicist, educator; b. Pitts., July 7, 1925; s. Emanuel B. and Lillian (Adlow) F.; m. Joan Brest, Sept. 4, 1950; children: Elizabeth B., Aaron L., Susan A. A.B., Harvard, 1947; M.S., Carnegie Inst. Tech., 1948, D.Sc., 1951. Fulbright grantee U. Leiden, Netherlands, 1951-52; research physicist Carnegie Inst. Tech., Pitts., 1952-53; mem. faculty Carnegie Inst. Tech., 1953-67, prof. dept. physics, 1962-67; prof. physics Carnegie-Mellon U., Pitts., 1967—; chmn. dept. physics Carnegie-Mellon U., 1973-80. Westinghouse fellow, 1950-51; Alfred P. Sloan Found. research fellow, 1957-61; Guggenheim fellow Imperial Coll., London,

Eng., 1965-66. Fellow Am. Phys. Soc., AAAS; mem. Sigma Xi, Tau Beta Pi, Phi Kappa Phi, Pi Mu Epsilon. Achievements include studies of magnetic ordering in many compounds of transition series and rare earth metals by magnetic, thermal and neutron scattering methods, among them quasi-one- and two-dimensional spin systems, which verify predictions of microscopic theory; of magnetic critical phenomena and phase diagrams, of the role of magnons in heat transport, and of special properties of singlet ground-state systems. Home: 1220 S Negley Ave Pittsburgh PA 15217-1219

FRIEDHOFF, ARNOLD J., psychiatrist, medical scientist; b. Johnstown, Pa., Dec. 26, 1923; s. Abraham M. and Stella (Beerman) F.; m. Frances Wolfe, Feb. 24, 1946; children: Lawrence, Nancy, Richard. B.A., U. Pa., 1944, M.D., 1947. Diplomate: Am. Bd. Psychiatry and Neurology. Intern Western Pa. Hosp., 1947-48; resident psychiatry U.S. Army, 1952-53, Bellevue Hosp., N.Y.C., 1953-55; instr., to Menas S. Gregory prof. psychiatry Sch. Medicine, NYU, N.Y.C., 1956—, head psychopharmacology rsch. unit, 1956-63, co-dir. Ctr. for Study Psychotic Disorders, 1963-69, dir., 1970—, dir. Millhauser Labs., 1970—; mem. clin. projects rsch. rev. com. NIMH, 1970-74, mem., 1977-81; dir. MV/NIMH Mental Health Clin. Rsch. Ctr., 1981—; mem. Mayor's Com. on Prescription Drugs, N.Y.C.; mem. sci. coun. Nat. Alliance for Rsch. in Schizophrenia and Depression, 1986—; mem. rsch. scientist devel. award rev. com., NIMH, 1983-85, mem. nat. adv. mental health coun., 1987—; hon. prof. Basque U., Bilbao, Spain, U. Seoul, Republic of Korea. Co-editor: Yearbook of Psychiatry and Applied Mental Health, 1968-80; assoc. editor Biol. Psychiatry, 1989—, mem. adv. bd., 1969—; contbr. numerous reports on biochem. psychiatry, psychopharmacology. Served to 1st lt. M.C. U.S. Army, 1951-53. Recipient Research Scientist award NIMH, 1967—. Fellow Am. Coll. Neuropsychopharmacology (past councillor and past pres. 1978-79), Am. Psychiat. Assn., Am. Soc. Clin. Pharmacology and Therapeutics, Royal Coll. Psychiatrists (Gt. Britain); mem. Am. Chem. Soc., Internat. Soc. Neurochemistry, Assn. for Research in Nervous and Mental Diseases (past asst. sec.-treas.), Am. Psychopath. Assn. (past pres., Samuel B. Hamilton award), Soc. Biol. Psychiatry (past pres., Gold medal 1989). Office: NYU Med Ctr Millhauser Labs 560 1st Ave New York NY 10016-6402

FRIEDHOFF, LAWRENCE TIM, internist, bio-medical researcher; b. Pitts., Nov. 12, 1948; s. Arnold Jerome and Frances (Wolfe) F.; m. Beth Ann Mazur, Aug. 12, 1981; children: Emily, Jane, Sarah. BA, NYU, 1970, MD, 1977; PhD in Chemistry, Columbia U., 1977. Diplomate Am. Bd. Internal Medicine. Intern Beth Israel Med. Ctr., N.Y.C., 1977-78; rsch. fellow endocrine div. Meml. Hosp., N.Y.C., 1978-81; instr. medicine Cornell Grad. Sch. Med. Sci., N.Y.C., 1978-81; resident U. Medicine and Dentistry of N.J., New Brunswick, 1981-82; asst. assoc., dir. E.R. Squibb Clin. Pharm., Princeton, N.J., 1982-88; sr. v.p. R&D EISAI Am., Inc., Teaneck, N.J., 1988—. Contbr. articles to profl jours. Mem. Am. Soc. Microbiology, N.Y. Acad. Sci., Phila. Coll. Physicians. Office: EISAI Am Inc Glen Pointe Ctr E Teaneck NJ 07666

FRIEDHOFF, RICHARD MARK, computer scientist, entrepreneur; b. N.Y.C., Dec. 2, 1953; s. Arnold Jerome and Frances (Galanter) F.; m. Livia R. Antola, May 5, 1988. BA, Columbia U., 1976; MA, Yale U., 1978. Sci. cons. PBS's The Brain, N.Y.C., 1978-80; industry adviser Polaroid Corp., Cambridge, Mass., 1981-82; v.p. Internat. Sci. Exch., N.Y.C., 1982-85; pres. Visicom Corp., L.A., 1986—; cons. U. Calif., 1990-91, Silicon Graphics Inc.; speaker Smithsonian Instn., also various corps. and sci. socs., 1989—. Author: Visualization: The 2nd Computer Revolution, 1989, 2d edit., 1991; contbr. articles to profl. jours. Dir. S & A Friedhoff Found. Fellow AAAS; mem. IEEE, Soc. for Photo Optical Instrumentation, Assn. for Computing Machinery, Authors Guild of Am., N.Y. Acad. Scis., Phi Beta Kappa. Office: Visicom Corp 1100 Glendon Ave 12th Flr Los Angeles CA 90024

FRIEDLAENDER, GARY ELLIOTT, orthopedist, educator; b. Detroit, May 15, 1945; s. Alex Seymour and Eileen Adrianne (Berman) F.; m. Linda Beth Krohner, Mar. 16, 1969; children: Eron Yael, Ari Seth. BS, U. Mich., 1967, MD, 1969; MA (hon.), Yale U., 1984. Diplomate Am. Bd. Orthopaedic Surgery. Intern, resident in surgery U. Mich., Ann Arbor, 1969-71; resident orthopaedics Yale New Haven Hosp., 1971-74; dir. tissue bank Naval Med. Research Inst., Bethesda, Md., 1974-76; instr. surgery Yale U., New Haven, 1974, asst. prof., 1976-79, assoc. prof., 1979-84, prof., chief orthopaedics, 1984-86, prof. chmn. dept. orthopaedics and rehab., 1986—. Bd. cons. editors: Jour. Bone and Joint Surgery, Boston, 1981—; bd. assoc. editors: Clin. Orthopaedics and Related Rsch., 1986—, Modern Medicine, 1988—; editor: Orthopaedic Rheumatology Digest, 1986—; mem. editorial bd. Transplantation Scis., 1991—; contbr. articles to profl. jours., chpts. to books. Served to lt. comdr. USN, 1974-76. Recipient Kappa Delta Outstanding Research award, 1982. Fellow ACS, Am. Acad. Orthopaedic Surgeons (chmn. com. biol. implants 1987-93); mem. AMA, NIH (orthopaedics and musculoskeletal study sect. 1986-89, Nat. Adv. Bd. arthritis and musculoskeletal and skin diseases 1991—, chmn. 1993—),Am. Tissue Banks (pres. 1983-85), Orthopaedic Rsch. Soc., Transplantation Soc., Musculoskeletal Tumor Soc., Am. Coun. on Transplantation (pres. 1983-85), Soc. for Surg. Oncology, Am. Soc. Transplant Surgeons, Am. Orthopaedic Assn., Assn. Bone and Joint Surgeons, Acad. Orthopaedic Soc. Jewish. Home: 15 Old Still Rd Woodbridge CT 06525-1101 Office: Yale U Dept Orthopedics & Rehab PO Box 208071 New Haven CT 06520-8071

FRIEDLAND, BERNARD, engineer, educator; b. Bklyn., May 25, 1930; s. Irving and Beckie (Kissen) F.; m. Zita Isa Silverman, Aug. 16, 1959; children: Barbara, Irene, Rochelle. AB, Columbia U., 1952, BSEE, 1953, MSEE, 1954, PhD, 1957. Registered profl. engr., Calif. Instr. Columbia U., N.Y.C., 1953-57, asst. prof., 1957-61; head control lab. Melpar, Inc., Watertown, Mass., 1961-62; prin. scientist Kearfott Guidance and Navigation Corp. (formerly The Singer Co.), Little Falls, N.J., 1962-90; disting. prof. N.J. Inst. Tech., Newark, 1990—; adj. prof. Columbia U., 1965-72, NYU, 1970-73, Poly. U. (formerly Poly. Inst. N.Y.), Bklyn., 1974-90. Author: Control System Design, 1986; co-author: Principles of Linear Networks, 1961, Linear Systems, 1965; contbr. over 80 articles to profl. jours. Chmn. The Hilary Sch., Newark, 1965. Fellow ASME (various offices, Oldenburger medal 1982), IEEE (disting. mem., various offices), AIAA (assoc.; assoc. editor jour.). Democrat. Jewish. Avocations: skiing, swimming, tennis, reading, sculpture. Office: NJ Inst Tech Dept Elec and Computer Engring Newark NJ 07102

FRIEDLAND, BETH RENA, ophthalmologist; b. Ft. Campbell, Ky., July 23, 1954; d. Bernard and Shirley (Denmark) Friedland; m. Robert Rosenthal, Aug. 30, 1981; 1 child, Alexander. BSc, Fla. Atlantic U., 1974; postgrad., U. Miami, 1974-75; MD, U. Fla., 1989. Predoctoral fellow Fla. Lions Eyebank, Miami, 1974-75; postdoctoral fellow Dept. Pharmacology, U. Fla., Gainesville, 1979-80; intern internal medicine Duke U., Durham, N.C., 1980-81; resdient in ophthalmology U. Miami, 1981-84; clin. asst. prof. ophthalmology U. N.C., Chapel Hill, 1984—; med. dir. Park Ophthalmology, Research Triangle Park, N.C., 1984—; cons. Cilco Labs., Sanford, N.C., 1985-88, Burroughs Wellcome, 1985—, Glaxo Rsch., 1980—, Merck Rsch. Labs., 1991—. Editor teaching packages: Basics of Human Neurosciences Vision, 1978; contbr. articles to profl. jours. Donor chmn. Hadassah, Raleigh, N.C., 1988. Recipient Arvo Travel award, 1978, Watson Clinic award for outstanding rsch., 1979, Nat. Rsch. Svc. awardee, 1979. Fellow Am. Acad. Ophthalmology, Contact Lens Assn. Am.; mem. Assn. for Rsch. in Vision and Ophthalmology, Am. Women's Med. Assn. (life), Alpha Omega Alpha. Democrat. Jewish. Achievements include discovery of role of lens carbonic anhydrase in lens metabolism; involved in development of topical carbonic anhydrase inhibitor for glaucoma; research in beta blocker research for glaucoma treatment. Office: Park Ophthalmology PO #12765 10 Park Plaza Research Triangle Park NC 27709-2765

FRIEDLANDER, GERHART, nuclear chemist; b. Munich, Germany, July 28, 1916; came to U.S., 1936, naturalized, 1943; s. Max O. and Bella (Forchheimer) F.; m. Gertrude Maas, Feb. 6, 1941 (dec. 1966); children: Ruth Ann F. Huart, Joan Claire F. Hurley; m. Barbara Strongin, 1983. BS, U. Calif., Berkeley, 1939, PhD, 1942; hon. docorate, Clark U., 1991; hon. doctorate, U. Mainz, Germany, 1992. Instr. U. Idaho, Moscow, 1942-43; staff Los Alamos Sci. Lab., 1943-46; research assoc. Gen. Electric Co. Research Lab., Schenectady, 1946-48; vis. lectr. Washington U., St. Louis, 1948; chemist Brookhaven Nat. Lab., Upton, N.Y., 1948-52; sr. chemist Brookhaven Nat. Lab., 1952-81, 89-91, cons., 1981-89, 91-93, chmn. chemistry dept., 1968-77; chmn. Gordon Research Conf. on Nuclear Chemistry, 1954; mem. adv. com. for chemistry Oak Ridge Nat. Lab., 1966-70; mem. program adv. com. Los Alamos Meson Physics Facility, 1971-75; chmn. vis. com. nuclear chemistry Lawrence Berkeley Lab., 1974-75; exec. sec. basic energy scis. lab. program panel Dept. Energy, 1976-80; mem. adv. com. for nuclear chemistry div. Lawrence Livermore Lab., 1977-83; mem. sci. adv. com. Gesellscheft für Schwerionenforschung, Darmstadt, Fed. Republic of Germany, 1980-85; mem. adv. com. for chem. tech. div. Oak Ridge Nat. Lab, 1988-91; mem. adv. com. nuclear sci. divsn. Lawrence Berkeley Lab., 1992—. Author: (with J.W. Kennedy) Introduction to Radiochemistry, 1949, Nuclear and Radiochemistry, 1955, (with J.M. Miller), 1964, (with E.S. Macias), 1981; also articles; assoc. editor: Ann. Rev. Nuclear Sci., 1958-67; editor: Radiochimica Acta, 1972-73. Recipient Alexander von Humboldt award Institut für Kernchemie, Mainz, Fed. Republic of Germany, 1978-79, 87, 92, 93. Fellow Am. Phys. Soc., AAAS; mem. NAS (mem. assembly math. and phys. scis. 1981-82, mem. commn. phys. scis., math. and resources 1982-90, chmn. ad hoc panel on future nuclear sci. 1975-76, chmn. com. on recommendations for U.S. Army basic sci. research 1977-81, mem. com. on postdoctoral and doctoral research staff 1977-80), Am. Acad. Arts and Scis., Am. Chem. Soc. (chmn. div. nuclear chemistry and tech. 1967, award for nuclear applications in chemistry 1967). Achievements include research on chem. effects of nuclear transformations, properties of radioactive isotopes, mechanisms of nuclear reactions, especially those induced by protons of very high energies, solar neutrino detection, cluster impact phenomena. Home: 5 Lorraine Ct Smithtown NY 11787-1633 Office: Brookhaven Nat Lab Upton NY 11973

FRIEDLANDER, MICHAEL J., neuroscientist, animal physiologist, medical educator; b. Miami, Fla., Jan. 30, 1950; 3 children. BS, Fla. State U., 1972; MS, U. Ill., 1974, PhD, 1977. NIH fellow physiology U. Ill., 1974-77, U. Va., 1977-79; from rsch. asst. prof. anatomy to asst. prof. neurobiology SUNY, Stony Brook, 1979-80; from asst. prof. to assoc. prof. U. Ala., Birmingham, 1980-87, prof. physiology and biophysics, 1987—, dir. Neurobiology Rsch. Ctr., 1987—; co-investigator rsch. project Nat. Eye Inst., 1979-80, Sloan Found. Computer Modeling Award, 1982; prin. investigator Develop. Structure & Function Vis. System, 1981-84, NATO Collaborative Rsch., 1983-87, NSF Devel. Vis. System, 1985—, Effects of Vis. Deprivation on Genicolocortical Pathway, 1984—; sr. inst. rsch. fellow for Australia, 1988. Recipient Sloan Young Neurosci. award. Mem. AAAS, Soc. Neurosci., Assn. Rsch. Vis. Ophthalmologists, Sigma Xi. Achievements include rsch. in structural basis of function of individual mammalian brain cells involved in processing visual information in the normal adult brain and during postnatal devel. chem. communication. Office: U Ala Neurobiology Rsch Ctr Volker Hall G82B UAB Sta Birmingham AL 35294*

FRIEDLANDER, RALPH, thoracic and vascular surgeon; b. N.Y.C., Oct. 2, 1913; s. Samuel and Mollie (Drimmer) F.; m. Sybil Rainsbury, Apr. 10, 1950; children: Andrea Lynn, Beth Caryn. BA, Columbia Coll., 1934; MD, U. Chgo., 1938. Diplomate Am. Bd. Surgery, Am. Bd. Thoracic Surgery. Intern Bellevue Hosp., N.Y.C., 1938; intern surg. Michael Reese Hosp., Chgo., 1939, resident in surgery, 1940; resident in surgery Mt. Sinai Hosp., N.Y.C., 1941; sr. resident in surgery Michael Reese Hosp., Chgo., 1942; adj. surgeon Mt. Sinai Hosp., N.Y.C., 1946-50; chief surgery and thoracic surgery VA Hosp., Castle Point, N.Y., 1947-50, Ft. Hamilton, N.Y., 1950-53; dir. surgery Bronx-Lebanon Hosp. Ctr., N.Y.C., 1953-64, attending surgeon, cons. thoracic, vascular, gen. surgery, 1964—; attending surgeon, cons. thoracic, vascular and gen. surgery Union Hosp., N.Y.C., 1976—; attending surgeon thoracic and vascular surgery Beth Israel Hosp. North, N.Y.C., 1977—; cons. gen. and thoracic surgery Dept. of Health N.Y. State, 1963—; cons. thoracic and cardiovascular surgery Hebrew Hosp. for the Chronic Sick, N.Y.C., 1955-89, Hebrew Home for the Aged, Riverdale, N.Y., 1956—, Health Ins. Plan, N.Y.C., 1954—, Bur. Disability Determinations, State of N.Y., 1960—, City of N.Y. Med. Assistance Program, 1966—, Bronx County Supreme Ct., 1958-91; instr. anatomy Sch. Medicine NYU, 1946-49; assoc. clin. prof. surgery Albert Einstein Coll. Medicine, 1959-90. Contbr. articles to Diseases of the Chest, Tuberculosis, N.Y. State Jour. Medicine, Am. Jour. Gastroenterology, Clin. Rsch. Mem. med. bd. Bronx-Lebanon Hosp. Ctr., sec., v.p., pres., 1975-78. Major Med. Corps, U.S. Army, 1942-46, ETO. Alfred Moritz Michaelis fellow in physics Columbia U., 1934; grantee USPHS, 1960. Fellow ACS, N.Y. Acad. Scis., N.Y. Acad. Medicine; mem. AMA, Am. Heart Assn. (coun. on cardiovascular surgery), N.Y. Heart Assn., N.Y. County Med. Soc., N.Y. State Med. Soc., N.Y. Soc. for Thoracic Surgery, Am. Assn. for Thoracic Surgery, Harvey Soc., N.Y. Gastroent. Assn., Phi Beta Kappa, Alpha Omega Alpha. Home: 535 E86th St New York NY 10028-7533

FRIEDMAN, ARNOLD CARL, diagnostic radiologist; b. Bronx, N.Y., Nov. 17, 1951; s. Isidore and Helen and (Lowenthal) F.; m. Wendy Sue Corn, June 8, 1975; children: Jeffrey Jonathan. BA in Chemistry, Cornell U., 1972; MD, Albert Einstein Coll., 1975. Intern Mt. Sinai Hosp., Hartford, Conn., 1975-76; resident Montefiore Hosp., Bronx, N.Y., 1976-79; asst. prof. Uniformed Svcs. U., Bethesda, Md., 1979-83; assoc. prof. George Washington U., Washington, 1983-84; assoc. prof. Temple U., Phila., 1984-88, prof. radiology, 1989-92; prof. and acting chmn. dept. radiology scis. Med. Coll. of Pa., Phila., 1992—. Editor: Radiology of Liver, Spleen, Pancreas, Biliary Tract, 1987, Clinical Pelvic Imaging, 1990. Mem. Radiologic Soc. N.Am., Am. Roentgen Ray Soc., Assn. Univ. Radiologists, Assn. Ultrasound in Medicine, Soc. Gastrointestinal Radiology. Avocations: tennis, basketball, ice skating, skiing, fitness. Home: 524 Hoffman Dr Bryn Mawr PA 19010-1745

FRIEDMAN, AVNER, mathematician, educator; b. Petah-Tikva, Israel, Nov. 19, 1932; came to U.S., 1956; s. Moshe and Hanna (Rosenthal) F.; m. Lillia Lynn, June 7, 1959; children—Alissa, Joel, Naomi, Tamara. M.Sc., Hebrew U., Jerusalem, 1954, Ph.D., 1956. Prof. math. Northwestern U., Evanston, Ill., 1962-86; prof. math. Purdue U., West Lafayette, Ind., 1984-87, dir. Ctr. Applied Math., 1984-87; prof. math., dir. Inst. Math. and Its Applications U. Minn., 1987—. Author: Generalized Functions and Partial Differential Equations, 1963, Partial Differential Equations of Parabolic Type, 1964, Partial Differential Equations, 1969, Foundations of Modern Analysis, 1970, Advanced Calculus, 1971, Differential Games, 1971, Stochastic Differential Equations and Applications, vol. 1, 1975, vol. 2, 1976, Variational Principles and Free Boundary Problems, 1983, Mathematics in Industrial Problems, 6 vols., 1988-93; contbr. articles to profl. publs. Fellow Sloan Found., 1962-65, Guggenheim, 1966-67; recipient Creativity award NSF, 1983-85, 90-92. Mem. AAAS, NAS, Am. Math. Soc., Soc. Indsl. Applied Math. (pres. 1993-94). Office: U Minn IMA 206 Church St SE Minneapolis MN 55455

FRIEDMAN, GARY DAVID, epidemiologist, health facility administrator; b. Cleve., Mar. 8, 1934; s. Howard N. and Cema C. F.; m. Ruth Helen Schleien, June 22, 1958; children: Emily, Justin, Richard. Student, Antioch Coll., 1951-53; BS in Biol. Sci., U. Chgo., 1956, MD with honors, 1959; MS in Biostatics, Harvard Sch. Pub. Health, 1965. Diplomate Am. Bd. Internal Medicine. Intern, resident Harvard Med. Svcs., Boston City Hosp., 1959-61; 2d yr. resident Univ. Hosps. Cleve., 1961-62; med. officer heart disease epidemiology study Nat. Heart Inst., Framingham, Mass., 1962-66; chief epidemiology unit, field and tng. sta., heart disease ctrl. program USPHS, San Francisco, 1966-68; sr. epidemiologist divsn. rsch. Kaiser Permanente Med. Care Program, Oakland, Calif., 1968-76, asst. dir. epidemiology and biostatics, 1976-91, dir., 1991—; rsch. fellow, then rsch. assoc. preventive medicine Harvard Med. Sch., 1962-66; lectr. dept. biomedical and environ. health scis., sch. pub. health U. Calif. Berkeley, 1968—; asst. prof. dept. epidemiology and biostatics U. Calif. Sch. Medicine, San Francisco, 1967-75, assoc. prof., 1975-92, lectr., 1980—, now clin. prof. depts. medicine and family and community medicine; mem. U.S.-USSR working group sudden cardiac death Nat. Heart, Lung and Blood Inst., 1975-82, com. on epidemiology and veterans follow-up studies Nat. Rsch. Coun., 1980-85, subcommittee on twins, 1980—, epidemiology and disease ctrl. study sect. NIH, 1982-86, U.S. Preventive Svcs. Task Force, 1984-88, scientific rev. panel on toxic air contaminants State of Calif., 1988—, adv. com. Merck Found./Soc. Epidemiol.Rsch., Clin. Epidemiology Fellowships, 1990—; sr. advisor expert panel on preventive svcs. USPHS, 1991—. Author: Primer of Epidemiology, 1974, 2d edit., 1980, 3d edit., 1987; assoc. editor, then mem. editorial bd. Am. Jour. Epidemiology, 1988—; mem. editorial bd. HMO Practice, 1991—. Oboist San Francisco Recreation Symphony, 1990—; bd. dirs. Chamber Musicians No. Calif., Oakland, 1991—. Sr. surgeon USPHS, 1962-68. Recipient Merit award Nat. Cancer Inst., 1987, Outstanding Investigator grantee, 1989; named to Disting. Alumni Hall of Fame Cleve. Heights High Sch., 1991. Fellow Am. Heart Assn. (chmn. com. on criteria and methods 1969-71, chmn. program com. 1973-76), Am. Coll. Physicians, Coun. on Epidemiology; mem. APHA, Am. Epidemiol. Soc. (mem. com. 1982-86), Am. Soc. Preventive Oncology, Internat. Epidemiol. Assn., Internat. Soc. Twin Studies, Soc. Epidemiologic Rsch., Phi Betta Kappa, Alpha Omega Alpha (Roche award for Outstanding Performance as Med. Student), Delta Omega. Achievements include research on cancer, cardiovascular disease, gall bladder disease, effects of smoking, alcohol and medicinal drugs, evaluation of health screening tests. Office: Kaiser Permanente Med Care Program Divsn Rsch 3451 Piedmont Ave Oakland CA 94611

FRIEDMAN, GERALD MANFRED, geologist, educator; b. Berlin, July 23, 1921; came to U.S., 1946, naturalized, 1950; s. Martin and Frieda (Cohn) F.; m. Sue Tyler Theilheimer, June 27, 1948; children: Judith Fay Friedman Rosen, Sharon Mira Friedman Azaria, Devorah Paula Friedman Zweibach, Eva Jane Friedman Scholle, Wendy Tamar Friedman Spanier. Student, U. Cambridge, Eng., 1938-39; BSc, U. London, Eng., 1945, DSc, 1977; student, U. Wyo., 1949; MA, Columbia U., 1950, PhD, 1952; Dr rer nat (hon.), U. Heidelberg, Fed. Republic Germany, 1986. Lectr. Chelsea Coll., London, 1944-45; analytical chemist E.R. Squibb & Sons, New Brunswick, N.J., also J. Lyons & Co., London, 1945-48; asst. geology Columbia U., 1950; temporary geologist N.Y. State Geol. Survey, 1950; instr., then asst. prof. geology U. Cin., 1950-54; cons. geologist Sault Ste. Marie, Ont., Can., 1954-56; mem. rsch. dept. Pan Am. Petroleum Corp. (Amoco), Tulsa, 1956-64; sr. rsch. scientist Pan Am. Petroleum Corp. (Amoco), 1956-60, rsch. assoc., 1960-62, supr. sedimentary geology rsch., 1962-64; Fulbright vis. prof. geology Hebrew U., Jerusalem, Israel, 1964; prof. geology Rensselaer Poly. Inst., 1964-84, prof. emeritus, 1984—; prof. geology Bklyn. Coll., 1985-88, Disting. prof. geology, 1988—; prof. earth and environ. scis. Grad. Sch. CUNY, 1985-88, disting. prof. earth and environ. scis., 1988—, dep. exec. officer, 1992—; pres. Gerry Exploration Inc., 1982—; rsch. scientist Hudson Labs., Columbia, 1965, 66-69, rsch. assoc. dept. geology, 1968-73; vis. prof. U. Heidelberg, 1967; cons. scientist Inst. Petroleum Rsch. and Geophysics, Israel, 1967-71; lectr. Oil & Gas Cons. Internat., 1968—; pres. Northeastern Sci. Found. Inc., 1979—; vis. scientist Geol. Survey of Israel, 1970-73, 78; mem. Com. Sci. Soc. Pres., 1974-76; Gerald M. Friedman post-doctoral fellowship, Inst. Earth Scis., Hebrew U., Israel, 1990. Co-author: Principles of Sedimentology (Outstanding Acad. Books, Choice, 1979), 1978, Exploration for Carbonate Petroleum Reservoirs, 1982, Exercises in Sedimentology, 1982, Principles of Sedimentary Deposits: Stratigraphy and Sedimentology, 1992; pub. Northeastern Environ. Sci., 1982-90; editor: Jour. Sedimentary Petrology, 1964-70 (Best Paper award 1961, hon. mention 1964, 66), Northeastern Geology, 1979—, Earth Scis. History, 1982—, Carbonates and Evaporites, 1986—, 10th Internat. Congress on Sedimentology, 1978; sect. co-editor: Chem. Abstracts, 1962-69, abstractor, 1952-69; editorial bd. Jour. Geol. Edn., 1951-55, Sedimentary Geology, 1967—, Israel Jour. Earth Scis, 1971-76, Coral Reef Newsletter, 1973-75, Jour. Geology, 1977—, GeoJour., 1977-83, Facies, 1987—; mng. editor Earth Sci. Revs., 1992—; contbg. co-editor: Carbonate Sedimentology in Central Europe, 1968, Hypersaline Ecosystems: The Gavish Sabkha, 1985; editor, contbr.: Depositional Environments in Carbonate Rocks, 1969; co-editor: Modern Carbonate Environments, 1983, Lecture Notes in Earth Scis., 1985—; contbr. articles to profl. jours; patentee in field. Mem. phys. edn. com., judo instr. Tulsa YMCA, 1958-64; adviser, instr. Judo Club, Rensselaer Poly. Inst., 1964-84; bd. dirs. Troy Jewish Community Coun., 1966-72, 74-77; v.p. Temple Beth El, 1986-89, pres., 1989-91, bd. dirs., 1965-76; bd. dirs. Leo Baeck Inst., N.Y.C., 1986—. Named hon. alumnus dept. geology Bklyn. Coll., 1989; grantee Office Naval Rsch., AEC, Dept. Energy, Petroleum Rsch. Fund, N.Y. Gas Assn. Fellow AAAS (chmn. geology and geography 1978-79, councillor 1979-80), Mineral. Soc. Am. (mem. nominating com. for fellows 1967-69, awards com. 1977-78), Geol. Soc. Am. (sr., chmn. sect. program com. 1969, candidate sect. chmn. 1969, publs. com. 1980-82), Geol. Soc. London (life, chartered geologist), Geol. Assn. Can., Soc. Econ. Geologists (sr.); mem. Am. Inst. Profl. Geologists (cert.), Am. Chem. Soc. (group leader 1962-63), Mineral. Soc. of Gt. Brit. (abstractor mineralogical abstracts 1963-64), Am. Assn. Petroleum Geologists (nat. hon. mem. 1990, Nat. Did 1988, chmn. carbonate rock com. 1965-69, mem. rsch. com. 1965-71, 76-82, lectr. continuing edn. program 1967—, adv. coun. 1974-75, Disting. lectr. 1972-73, mem. disting. lectr. com. 1975-78, membership com. 1982-86, ho. of dels. 1977-80, 83-87, 91—, alt. del. 1980-83, 87-90, sect. sec. 1979-80, sect. treas. 1980-81, sect. v.p. 1981-82, sect. pres 1982-83, div. profl. affairs rep. from Eastern sect. 1983-84, nat. v.p. 1984-85, cert. petroleum geologist, h0n. mem. Eastern sect. 1984, chmn. sect. awards com., 1989-92), Soc. for Sedimentary Geology (nat. v.p. 1970-71, pres. 1974-75, sect. pres. pro tem 1966-67, sect. pres. 1967-68, chmn. Shepard award selection com. 1966-67, Best Paper award Gulf Coast sect. 1974, hon. mention to Outstanding Paper award Jour. Paleontology 1971, hon. mem. 1984), Capital Dist. Geologists Assn. (chmn. program 1966-73), New Eng. Intercollegiate Geol. Conf. (program chmn. 1979), Am. Geol. Inst. (governing bd. 1971-72, 74-75), Geologists' Assn. (life), Internat. Assn. Sedimentologists (v.p. 1971-75, pres. 1975-78, nat. corr. U.S.A. 1971-73, hon. mem. 1982), Geol. Soc. Israel (hon.), Indian Assn. Sedimentologists (mem. governing coun. 1978-82), Geol. Vereinigung, Deutsche Geol. Gesellschaft, Soc. Venezolana Historia Geociencias (internat. corr. mem.), Nat. Assn. Geology Tchrs. (nat. treas. 1951-55, chmn. organizing com. establish east-ctrl. sect. 1952-53, pres. Okla. 1962-63, pres. Ea. sect. 1983-84), Assn. Earth Sci. Editors (v.p. 1970-71, pres. 1971-72, host 1991), N.Y. Acad. Scis. (vice chair. geol. scis. section 1993—), N.Y. State Geol. Assn. (pres. 1978-79, bd. dirs. 1979-84), Cin. Mineral Soc. (v.p., program chmn. 1953-54), U.S. Judo Fedn. (San Dan), Okla. Judo Fedn. (pres. 1959-60, v.p. 1961-64), Empire State Judo Assn. (v.p. 1975-77, dir. coll. devel. 1972-82), Kodokan (Japan), Sigma Gamma Epsilon (nat. v.p. 1978-82, nat. pres 1982-86, nat. hon. mem. 1986), Sigma Xi (v.p. Rensselaer chpt. 1969-70). Home: 32 24th St Troy NY 12180-1915 Office: CUNY Bklyn Coll Dept Geology Brooklyn NY 11210

FRIEDMAN, HERBERT, physicist; b. N.Y.C., N.Y., June 21, 1916; s. Samuel and Rebecca (Seligson) F.; m. Gertrude Miller, 1940; children—Paul, Jon. BA, Bklyn. Coll., 1936; PhD in Physics, Johns Hopkins U., 1940; DSc (hon.), U. Tübingen, Fed. Republic Germany, 1977, U. Mich., 1979. With U.S. Naval Rsch. Lab., Washington, 1940—, supt. atmosphere and astrophysics div., 1958-63, supt. space sci. div., 1963-80; chief scientist E. O. Hulburt Ctr. Space Rsch. U.S. Naval Rsch. Lab., 1963-80, Emeritus, 1980—; adj. prof. physics U. Md., 1960-80, U. Pa., 1974—; vis. prof. Yale U., 1966-68; Martin-Marietta fellow in Space Sci., NASM, 1986-87; mem. space sci. bd. Nat. Acad. Scis.-NRC, 1962-75, 86—, chmn. com. on solar-terrestrial rsch., 1968-71; mem. com. sci. and pub. policy Nat. Acad. Scis., 1967-71, mem. geophysics rsch. bd., 1969-71, chmn., 1976-79, mem. adv. com. internat. orgns. and programs, 1969-77; pres. Interunion Com. on Solar-Terrestrial Physics, 1967-74; pres. spl. com. on solar-terrestrial physics ICSU, 1975-80; chmn. COSPAR working group II, Internat. Quiet Sun Yr.; v.p. COSPAR, 1970-75, 86—; mem. Gen. Adv. Com. on Atomic Energy, 1968-73; mem. Pres.'s Sci. Adv. Com., 1970-73, Advisory Bd. Fermi Lab., 1986-88; chmn. commn. on phys. scis., math and resources NRC, 1984-86. Recipient Disting. Svc. award Dept. Navy, 1945, 80; medal Soc. Applied Spectroscopy, 1957; Disting. Civilian Svc. award Dept. Def., 1959; Disting. Achievement in Sci. award, 1962; Janssen medal French Photog. Soc., 1962; Presdl. medal for disting. fed. svc., 1964; Eddington medal Royal Astron. Soc., 1964; R.D. Conrad medal Dept. Navy, 1964; Rockefeller Pub. Svc. award, 1967; Nat. Medal Sci., 1969; medal for exceptional sci. achievement NASA, 1970, 78; Michelson medal Franklin Inst., 1972; Dryden Rsch. award, 1973; Wolf Found. prize in physics, 1987; Russell award Am. Astron. Soc., 1980; Sci. award Nat. Space Club, 1990; Janssen medal French Astron. Soc., 1990; Massey medal, Royal Soc. London, 1992. Fellow AIAA (hon., Space Sci. award 1963, Internat. Cooperation in Space Sci. medal 1991), Am. Phys. Soc., Am. Optical Soc., Am. Geophys. Union (pres. sect. on solar-planetrary relationships 1967-70, Bowie medal 1981), Am. Astronautical Soc. (Lovelace award 1973), AAAS (v.p. 1972); mem. NAS (council 1979-82, chmn. assembly of math. and phys. scis. 1980-83), Am. Acad. Arts and Scis., Internat. Acad. Astronautics, Am. Philos. Soc. (coun. 1992—); hon. mem. Spl. Com. on Solar-Terrestrial Physics, 1984. Club: Cosmos. Achievements include discovery of solar x-ray emission and its role in control of the ionosphere; demonstration of connection between solar flare x-rays and radio fadeout; identification of x-ray star, the crab pulsar and x-ray galaxy. Home: 2643 N Upshur St Arlington VA 22207-4025 Office: Naval Rsch Lab Code 7690 Washington DC 20375

FRIEDMAN, HOWARD SAMUEL, cardiologist, educator; b. N.Y.C., Dec. 27, 1940; s. Harry and Bella Esther (Israel) F.; m. Maud Tanowitz, June 18, 1961; children: Shawn Marcus, Saroya Danielle, Heather Eve. BA, Bklyn. Coll., 1962; MD, SUNY, Buffalo, 1966. Intern St. Louis City Hosp., 1966-67; asst. resident Barnes Hosp., St. Louis, 1967-68; asst. resident Mt. Sinai Hosp., N.Y., 1968-69, resident cardiology, 1969-70, 72-73; acting chief cardiology VA Hosp., Bronx, N.Y., 1974-76; chief cardiology Bklyn. Hosp. Ctr., 1977-90; chmn. dept. medicine L.I. Coll. Hosp., Bklyn., 1990—; prof. medicine SUNY Health Sci. Ctr. at Bklyn., 1987—. Contbr. articles to profl. jours. Dir. Am. Heart Assn. N.Y.C. Affiliate Bd., 1985-88; chmn. Coronary Care Com., 1982-88. Fellow Am. Coll. Physicians, Am. Heart Assn., Am. Coll. Cardiology; mem. Am. Physiology Soc., Soc. for Exptl. Biology and Medicine. Democrat. Jewish. Home: 401 E 84th St New York NY 10028-6268 Office: L I Coll Hosp 340 Henry St Brooklyn NY 11201-5525

FRIEDMAN, JEROME ISAAC, physics educator, researcher; b. Chicago, Ill., Mar. 28, 1930; married, 1956; 4 children. A.B., U. Chgo., 1950, M.S., 1953, Ph.D. in Physics, 1956. Research assoc. in physics U. Chgo., 1956-57; research assoc. in physics Stanford U., Calif., 1957-60; from asst. prof. to assoc. prof. MIT, Cambridge, 1960-67, prof. physics, 1967—, dir. lab. nuclear sci., 1980-83, head dept. physics, 1983-88. Recipient Nobel prize in physics, 1990. Fellow Am. Phys. Soc., AAAS (co-recipient W.H.K. Panofsky prize 1989); mem. NAS. Office: MIT Dept of Physics Cambridge MA 02139

FRIEDMAN, KENNETH MICHAEL, energy policy analyst; b. Queens, N.Y., July 15, 1945; s. Harry and Rose (Stoloff) F.; m. Janet Eve Bader, June 21, 1968; children: Robert C., Stuart B. BA, Queens Coll., 1966; MA, Mich. State U., 1968, PhD, 1973. Asst. prof. polit. sci. Purdue U., West Lafayette, Ind., 1971-77; dir. policy coordination and legis. support U.S. Dept. Energy, Washington, 1977—, spl. asst. to dep. asst. sec. for conservation, 1979-81, spl. asst. to depl asst. sec. indsl. techs., 1986-89; dir. tech. studies div. Internat. Energy Energy, Paris, 1989-93; book series editor Marcel Dekker Inc., N.Y.C., 1975-82. Author: Cigarette Smoking and Public Policy, 1974, Towards a National Health Policy, Public, 1975. Mem. Telecomms. Adv. Bd., Montgomery County, Md., 1981. Achievements include research on conservation multi-year plans. Home: 11321 Dunleith Pl Gaithersburg MD 20878 Office: Internat Energy Agy, 2 rue Andre-Pascal, 75775 Paris Cedex 16, France

FRIEDMAN, LAWRENCE SAMUEL, gastroenterologist, educator; b. Newark, May 11, 1953; s. Maurice and Esther (Slansky) F.; m. Mary Jo Cappuccilli, Apr. 12, 1981; 1 child, Matthew Jacob. Student, Princeton U., 1971-73; BA, Johns Hopkins U., 1975, MD, 1978. Intern dept. medicine Johns Hopkins Hosp., Balt., 1978-79, resident dept. medicine, 1979-81; fellow Mass. Gen. Hosp. & Harvard Med. Sch., Boston, 1981-84; asst. prof. Jefferson Med. Coll., Phila., 1984-87, assoc. prof., 1987-93, vice chmn., 1987-92; assoc. prof. Mass. Gen. Hosp. & Harvard Med. Sch., Boston, 1993—. Editor: Gastrointesinal Disorders in the Elderly, 1990; contbr. articles to profl. jours. Med. adv. com. mem. (Phila. chpt.) Crohn's & Colitis Found. Am., 1985; nat. mem. Am. Liver Found., 1986. Fellow ACP, Am. Coll. Gastroenterology, Coll. Physicians of Phila.; mem. Am. Assn. for Study of Liver Diseases, Am. Fedn. for Clin. Rsch., Am. Soc. Gastrointestinal Endoscopy, Am. Gastroent. Assn. Jewish. Avocations: Am. history, woodwind instruments, travel, basketball. Office: Mass Gen Hosp GI Unit Bulfinch 127 Boston MA 02114

FRIEDMAN, MARION, internist, family physician, medical administrator; b. Onley, Va., Aug. 15, 1918; s. Jacob and Bertha (Bernstein) F.; m. Esther Lerner, May 29, 1941; 1 son, Barry Howard. BS, U. Md., 1938, MD, 1942. Diplomate Am. Bd. Family Practice (charter). Rotating intern U.S. Marine Hosp., Norfolk, Va., 1942-43; asst. health officer Montgomery County (Kans.), 1943-44; health officer Cherokee County (Kans.), 1944-45; asst. health commr. St. Louis County (Mo.), 1945-46; resident internal medicine U.S. Marine Hosp., Balt., 1946-49; fellow medicine Johns Hopkins Sch. Medicine, Balt., 1948-49; individual practice medicine, specializing in family practice internal medicine, Balt., 1949-84; asst. medicine U. Md., Balt., 1954-72; chief dept. gen. practice Doctors Hosp., 1952-54; chief dept. family practice N. Charles Gen. Hosp., Balt., 1972-75, med. dir. ambulatory svcs., 1972-86, assoc. chief medicine, 1975-88, pres. med. staff, 1964, 68, chmn. med. exec. com., 1984-85, trustee, 1984-85, physician advisor, 1984-91; med. dir. Chesapeake Health Plan, 1991-92. Contbr. numerous articles to sci. jours. Chmn. cultural com. Liberty Jewish Ctr., 1960-62; mem. Md. High Blood Pressure Coordinating Coun., 1980-82; mem. task force on family physicians Md. Health Resources Planning Commn., Md. State Legislature, 1983-84; trustee Jimmie Swartz Found., 1982—. Served with USPHS, 1942-49. Fellow Am. Acad. Family Physicians (charter), Md. Acad. Family Physicians (pres. 1983-84, prodn. editor 1984-86, editor jour. 1986—); mem. AMA, Am. Acad. Family Physicians, Balt. City Med. Soc. (alt. del. 1978-82, 93—, del. 1982-88, profl. edn. com. 1985-87, chmn. 1987-90, nominating com. 1992), Med. and Chirurg. Faculty Md. (lectr. 1983, legis. com. 1985-87, 90—, pro com. 1989—), World Med. Assn., Pan-Am. Med. Assn., Md. Heart Assn., Md. Thoracic Soc., Am. Thoracic Soc., Am. Heart Assn., Phi Kappa Phi. Democrat. First to suggest use of steroid in subacute deltoid bursitis in world lit., 1952. Home: 7906 Terrapin Ct Baltimore MD 21208-3126

FRIEDMAN, MARK, physician, consultant; b. Bklyn., Mar. 25, 1932; s. Samuel and Ann (Sapan) F.; m. Myrna Cohen, Nov. 1, 1959; children: Suzanne, Melanie, Barbra. BA in Physics, Adelphi U., 1953; MD, Wake Forest U., 1958. Diplomate Am. Bd. Phys. Medicine and Rehab. Rotating intern Mamimodes Hosp., Bklyn., 1959-61; resident VA Hosp., Coral Gables, Fla., 1961-62, East Orange, N.J., 1962-64; attending physician John F. Kennedy Hosp., Edison, N.J., 1973—; cons. physician VA Hosp., Lyons, N.J., 1978-82; med. dir., pres. Middlesex Rehab. Hosp., North Brunswick, N.J., 1965-73, med. dir. phys. medicine Muhlenberg Hosp., Plainfield, N.J., 1965-88; cons. physician South Amboy (N.J.) Hosp., 1978—. Capt. USAF, 1959-61, CBI. Fellow Am. Acad. Phys. Medicine and Rehab., Am. Coll. Sports Medicine, Am. Acad. Disability Evaluating Physicians; mem. Am. Holistic Med. Assn. (founder), Am. Assn. Orthopedic Medicine. Office: 2509 Park Ave South Plainfield NJ 07080-5370

FRIEDMAN, MOSHE, research physicist; b. Tel Aviv, Apr. 12, 1936; came to U.S., 1969; s. Itzhak and Rivka (Azulay) F.; m. Sarah Landau, June 11, 1967; children, Daphne, Ruth. MSc in Physics, Hebrew U. Jerusalem, 1961, PhD in Physics, 1964. Rsch. physicist Culham Lab., Eng., 1964-69, Cornell U., Ithaca, N.Y., 1969-71; Naval Rsch. Lab., Washington, 1971—. Contbr. articles to Rev. Sci. Instrumentation, Applied Physics Letters, Jour. Applied Physics, Phys. Rev. Letters. Recipient rsch. publ. award Naval Rsch. Lab., 1973, 75, 85, 89, 91. Fellow Washington Acad. Sci.; mem. Am. Phys. Soc., Sigma Xi. Achievements include co-patent for Compact High Power Accelerator. Office: Naval Rsch Lab Code 4632 Plasma Physics Washington DC 20375

FRIEDMAN, PHILIP HARVEY, psychologist; b. N.Y.C., Oct. 4, 1941; s. Leonard and Miriam Rosalyn (Solomon) F.; m. Teresa Jean Molinaro, Dec. 22, 1965; 1 son, Mathew Alan. BA, Columbia Coll., 1963; MA, U. Wis., 1965, PhD, 1968. NIMH postdoctoral rsch. fellow Temple U. Med. Sch., 1968-69; clin. psychologist ea. Pa. Psychiat. Inst., Phila., 1969-73; sr. family therapist dept. psychiatry Jefferson U. and Community Mental Health Ctr., Phila., 1973, instr., 1975-77, asst. prof., 1977—, program adminstr., 1978—; exec. dir. Found. for Well-Being, Plymouth Meeting, Pa., 1987—; coord. tng. child and family divsn. CATCH CMHC, 1979-81, coord. tng. in marital and family therapy, 1981-82; referring specialist HMO of Pa., 1979-81; dir. Friedman Family Circle Assocs., 1980-81; mem. staff Ea. Cons. Assocs., 1980—; sr. supr. Masters in Marital and Family Therapy program Hahnemann Med. Sch., 1981, asst. prof., 1982—; dir. Ctr. for Integrative Psychotherapy and Tng., 1983-87; exec. dir. Attitudinal Healing Ctr. Delaware Valley, Plymouth Meeting, Pa., 1983-87; cons. Phila. Bd. Edn., 1978—, Ctr. for Study of Adult Devel., 1982. Author: Creating Well-Being: The Healing Path to Love, Peace, Self-Esteem and Happiness, 1989. Guest various radio and TV programs, 1980—. Fellow Pa. Psychol. Assn.; mem. APA, Assn. Transp. Psychologists, Am. Assn. Marriage and Family Therapists, Am. Family Therapy Acad., Phila. Soc. Clin. Psychologists, Internat. Assn. for Study Subtle Energy and Energy Medicine, Soc. for Profl. Well Being, Toastmasters. Achievements include research in psychological, emotional, interpersonal and spiritual correlates of well-being. Home: 46 Red Rowen Ln Plymouth Meeting PA 19462 Office: Found for Well-Being PO Box 627 Plymouth Meeting PA 19462

FRIEDMAN, ROBERT JAY, physician; b. Queens, N.Y., Mar. 20, 1948; s. Josef and Evelyn (Promisloff) F.; m. Nancy L. Weiner, June 14, 1969; children: Scott, Seth. BA, Adelphi U., 1969; MSc in Medicine, McMaster U., 1972; MD, Albert Einstein Coll. Medicine, 1978. Rsch. asst. McMaster Univ. Sch. of Medicine, Hamilton, Ont., 1972-74; rsch. assoc. Albert Einstein Coll. of Medicine, Bronx, N.Y., 1975-76; NIH fellow dept hematology Albert Einstein Coll. Medicine, Montefiore Hosp. and Med. Ctr., Bronx, 1975-76, Albert Einstein Coll. Medicine, Bronx, 1976-78; intern anatomic pathology NYU Med. Ctr., N.Y.C., 1978-80, resident dermatopathology, 1979-80, dermatology melanoma fellow, 1980-82, clin. attending physician, 1982—; clin. instr. NYU Sch. of Medicine, N.Y.C., 1982—; clin. asst. prof. NYU Sch. Medicine, N.Y.C., 1989—. Editor: Melanoma, 1985, Pigmented Lesions of the Skin, 1989; chief editor: Cancer of the Skin, 1991. Program chmn. Westchester Acad. Medicine Dermatology, N.Y.C., 1983-86. Recipient Mead Johnson award for excellent med. rsch. AMA, 1977, Husik award NYU Dept. Dermatology; fellow NIH, 1980-82. Mem. Am. Acad. Dermatology (skin cancer com. 1982-84, adv. coun. 1985—), Internat. Soc. Dermatopathology, Am. Cancer Soc. (bd. dirs. 1985—, profl. edn. com. 1985—, Jr. Faculty fellowship 1981-81) Skin Cancer Found. (med. edn. com. 1982—). Achievements include research on malignant melanoma and pathology of malignant melanoma. Office: 350 5th Ave Ste 7805 New York NY 10118-0189

FRIEDMAN, ROBERT MORRIS, pathologist, molecular biologist; b. N.Y.C., Nov. 21, 1932; s. Jack and Rose M. (Weiss) F.; m. Ina Reichler, Dec. 15, 1989 (dec. Aug. 1992); children: Thomas, Deborah, Antony. BA with honors, Cornell U., 1954; MD, NYU, 1958. Intern Mt. Sinai Hosp., NYC, 1958-59; resident The Clinical Ctr., NIN, Bethesda, Md., 1961-63; pathologist NIH, Bethesda, Md., 1963-73, lab. chief, 1973-80; chmn. dept. pathology Uniformed Svcs. U., Bethesda, Md., 1981—; vis. scientist Nat. Inst. for Med. Rsch., Mill Hill, London, 1963-64, 71-73; vis. prof. Warwick U., Coventry, U.K., 1981; adj. prof. pathology Georgetown U. Med. Sch., 1992—. Author: The Interferons: A Primer, 1981, Interferons, 1982, Interferons as Cell Growth Inhibitors & Anti-tumor Factors, 1986. Capt. USPHS, 1979-90. Mem. Am. Soc. for Investigative Pathology, Am. Soc. for Microbiology, Internat. Acad. Pathology, Phi Beta Kappa, Phi Kappa Phi, Alpha Omega Alpha. Democrat. Jewish. Achievements include research in replication of arboviruses, mechanism of antiviral activity of interferons, mechanism of action of tumor suppressor genes. Office: Uniformed Services Univ Dept Pathology 4301 Jones Bridge Rd Bethesda MD 20814

FRIEDMAN, RONALD MARVIN, cellular biologist; b. Brooklyn, N.Y., Apr. 26, 1930; s. Joseph and Helen (Plotkin) F. B.S., Columbia U., 1960; M.S., NYU, 1967, Ph.D., 1976. Predoctoral fellow Inst. Microbiology, 1968-72, NYU, 1972-76; postdoctoral fellow Columbia U., 1976, Yale U., 1977-78; vis. fellow Princeton U., 1978-79; vis. scientist N.Y. State Inst. Basic Rsch., 1979-81; NIH fellow Albert Einstein Coll. Medicine, 1981-82; sci. advisor Royal Arch Med. Rsch. Found., N.Y.C., 1982-88; research fellow meml. Sloan-Kettering Cancer Center, N.Y.C., N.Y., 1984-85; sr.research assoc. dept. pathology Cath. Med. Ctr., 1983-84; research assoc. dept. biochemistry U. Medicine and Dentistry N.J., Newark, 1984-85; sr. research assoc. in hematology CUNY, 1985-86; research assoc. dept. immunology and biochemistry Roswell Park Meml. Inst., Buffalo, 1986-87; research assoc., infectious disease, Channing Laboratory, Harvard Medical School, Boston, 1987-88; spl. asst. Sec. Gen. U.N. (promoting the philosophy of human dignity and its impact on world peace), N.Y., 1987-88; asst. research prof. CCNY, 1988-89; vis. scientist in molecular biology, Lewis Thomas Labs, Princeton U.; vis. prof. Kasetsart U. Bangkok, Thailand, 1992—, vis. scholar Boston U. 1992-93; scientific cons. U. Rangoon, 1993—, inspector of Cell Biology Research in Poland and Russia, Citizens Ambassador Program; advisor curriculum research and cell/molecular biology, lectr. cellular and molecular biology. Asst. to Sec. of Agr., 1970-71 serving as spl. liaison to Congress; conducted survey of emergency med. home call service, Bronx County, NY., 1971-72. Knights Templar fellow, 1973-87; NIH fellow, 1981-82. Mem. Harvey Soc., Fedn. Am. Soc. for Exptl. Biology, Am. Soc. Cell Biologists, N.Y. Acad. Scis., Am. Chemical Soc., Soc. of Environ. Toxicology and Chemistry, AAAS, Sigma Xi. Lodges: Masons, Shriners, K.T. Home: 315 W 232d St Riverdale NY 10463

FRIEDMAN, SANDER BERL, engineering educator; b. Chgo., Oct. 28, 1927; s. Albert J. and Rae (Lazar) F.; m. Jeanne Himmel, Oct. 13, 1951; divorced; children: Elise, Marc, Victoria; m. Luella Wilson Knotts, June 1, 1989. MS, No. Ill. U., 1971; PhD, Waterloo U., 1974. Profl. engr. Ill., N.Y., Ohio. Engr. various, 1949-52; chief engr. Universal Circuit Controls, Skokie, Ill., 1952-64; engr. mfg. Shure Bros., Evanston, Ill., 1964-69; prof. Wm. Rainey Harper Coll., Palatine, Ill., 1969-74; prof. mfg. engring. Miami U., Oxford, Ohio, 1974-87; prin. SBF Profl. Engrs., Oxford and Boulder, Colo., 1978—. Author: Logical Design of Automation Systems, 1990; contrb. articles to profl. jours. Lt. U.S. Army. Achievements include patents for laminar flow liquid fluidic device, liquid fluidic control device. Home: 2331 Larkspur Ave Estes Park CO 80517-7117

FRIEDMAN, SHARON MAE, science journalism educator; b. Phila., Apr. 28, 1943; d. Thomas and Evelyn Eva (Gordon) Berschler; m. Kenneth A. Friedman, July 12, 1963; children: Melissa, Michael. BA in Biology, Temple U., 1964; MA in Journalism, Pa. State U., 1974. Sci. writer/editor Pa. State U., University Park, 1966-67; assoc. info. officer Nat. Acad. Sci., Washington, 1967-70; editor Ctr. for Study of Higher Edn., University Park, 1970-71; adminstrv. and info. officer U.S. Com. for Internat. Biol. Program, State College, Pa., 1971-74; asst. and assoc. prof. journalism Lehigh U., Bethlehem, Pa., 1974-86, dir. sci. writing program, 1977—, prof. journalism, chmn. dept. journalism, 1986—, Iacocca prof., 1992—; cons. Pres.'s Commn. on the Accident at Three Mile Island, Washington, 1979, Clement Internat. Corp., Washington, 1988-90; cons. Environ. Unit, UN Econ. and Social Commn. for Asia/Pacific, Bangkok, 1987-89; mem. adv. bd. Environmental Reporting Forum, Radio-TV News Dirs. Found., 1991—, environ. journalism program Found. Am. Comm., 1992—; mem. bd. trustees Internat. Food Info. Coun. Found., 1992—; Fulbright Disting. lectr., Brazil, 1982, Bosch Found. lectr., Fed. Republic Germany, 1984, 92; cons. in field. Co-author: Reporting on the Environment - Handbook for Journalists, 1988; sr. editor: Scientists and Journalists: Reporting Science as News, 1986; contbr. articles to profl. jours., chpts. to books. U.S. EPA grantee, 1990, 92, Gen. Mtrs. Corp. grantee, 1986—. Fellow AAAS (sect. Y officer 1991—); mem. Sci. Writing Educators Group (chmn. 1985-91), Assn. for Edn. in Journalism and Mass Communications (mag. div. officer 1988-91), Nat. Assn. Sci. Writers, Soc. Environ. Journalists. Achievements include research on mass media coverage of science, environment and technology issues, environmental risk communication, coverage of Alar, radon, Chernobyl and Three Mile Island radiation issues, and international environmental journalism and training. Office: Lehigh Univ Dept Journalism & Comm 29 Trembley Dr Bethlehem PA 18015

FRIEDMAN, SHELLY ARNOLD, cosmetic surgeon; b. Providence, Jan. 1, 1949; s. Saul and Estelle (Moverman) F.; m. Andrea Leslie Falchook, Aug. 30, 1975; children: Bethany Erin, Kimberly Rebecca, Brent David, Jennifer Ashley. BA, Providence Coll., 1971; DO, Mich. State U., 1982. Diplomate Nat. Bd. Med. Examiners, Am. Bd. Dermatology. Intern Pontiac (Mich.) Hosp., 1982-83, resident in dermatology, 1983-86; assoc. clin. prof. dept. internal med. Mich. State U., 1984-89, adj. clin. prof., 1989—; med. dir. Inst. Cosmetic Dermatology, Scottsdale, Ariz., 1986—. Contbr. articles to profl. jours. Mem. B'nai B'rith Men's Council, 1973, Jewish Welfare Fund, 1973. Am. Physicians fellow for medicine, 1982. Mem. AMA, Am. Osteopathic Assn., Am. Assn. Cosmetic Surgeons, Am. Acad. Cosmetic Surgery, Internat. Soc. Dermatologic Surgery, Internat. Acad. Cosmetic Surgery, Am. Acad. Dermatology, Am. Soc. Dermatol. Surgery, Frat. Order Police, Sigma Sigma Phi. Jewish. Avocations: karate, horseback riding. Office: Scottsdale Inst Cosmetic Dermatology 5206 N Scottsdale Rd Scottsdale AZ 85253

FRIEDMANN, E(MERICH) IMRE, biologist, educator; b. Budapest, Hungary, Dec. 20, 1921; came to U.S., 1965; s. Hugo and Gisella (Singer) Friedmann; 1 dau., Daphna; m. Roseli Ocampo, July 22, 1974. Ph.D., U. Vienna, 1951. Instr., lectr. Hebrew U., Jerusalem, 1952-66; assoc. prof. Queens U., Kingston, Ont., Can., 1967-68; assoc. prof. Fla. State U., Tallahassee, 1968-76, prof., 1976—, Robert Lawton Disting. prof., 1991—, dir. Polar Desert Rsch. Ctr., 1985—; concurrent prof. Nanjing U., People's Republic of China, 1987—; vis. prof. Fla. State U., Tallahassee, 1966-67, U. Vienna, 1975. Contbr. articles to profl. jours. Recipient Congl. Antarctic Service medal NSF, 1979, Alexander v. Humboldt award, 1987, resolution of commendation Gov. of Fla., 1978. Fellow Linnean Soc. London, Royal Micros. Soc.; mem. Am. Soc. Microbiology, Brit. Phycol. Soc., Indian Phycol. Soc., Am. Phycol. Soc., Internat. Phycol. Soc., Soc. Phycol. France, Internat. Soc. for Study Origins of Life, Hungarian Algological Soc. (hon.). Jewish. Discoverer micro-organism (cryptoendolithic lichens) living in Antarctic rocks, 1976. Home: 692 Duparc Circle Tallahassee FL 32312 Office: Fla State U Dept Biol Sci Tallahassee FL 32306

FRIEDMANN, PAUL, surgeon, educator; b. Vienna, Austria, Dec. 2, 1933; came to U.S., 1938; s. Erich and Rochelle (Behar) F.; m. Janee Armstrong, Apr. 24, 1962; children: Pamela, Cynthia. BA, U. Pa., 1955; MD, Harvard U., 1959. Diplomate Am. Bd. Surgery, Am. Bd. Vascular Surgery. Chmn. dept. surgery Baystate Med. Ctr., Springfield, Mass., 1971—; prof. surgery Tufts U. Sch. Medicine, Boston, 1978—; mem. residency rev. com., 1985-91, chmn., 1989-91; chmn. RRC Coun., Accreditation Coun. for Grad. Med. Edn., 1989-91. Contbr. articles to profl. jours. Served to capt. USAF, 1961-63. Fellow ACS (bd. govs. 1978-84, pres. Mass. chpt. 1987); mem. Am. Surg. Assn., Assn. Program Dirs. in Surgery (sec. 1985-87, pres. 1987-89), New Eng. Soc. Vascular Surgery (recorder 1988-91, pres.-elect 1991-92, pres. 1992—), New Eng. Surg. Soc. (treas. 1991—, Harvard Club (Boston), Colony Club (Springfield, Mass.). Office: Baystate Med Ctr 759 Chestnut St Springfield MA 01199-0001

FRIEDRICH, CHRISTOPHER ANDREW, internist, geneticist; b. Abington, Pa., July 26, 1956; s. Charles Adam Jr. and Mary Nadine (Donegan) F. PhD, U. Tex. Health Sci. Ctr., 1986; MD, Rutgers Med. Sch., 1988. Diplomate Am. Bd. Internal Medicine. Intern, resident U. Minn. Hosp. and Clinics, Mpls., 1988-91; fellow med. genetics Johns Hopkins Hosp., Balt., 1991—. Contbr. articles to profl. jours. Pres. Grad. Student Assn., U. Tex., Houston, 1983; co-chmn. High Sch. Tchrs. Workshop, U. Tex., Houston, 1984, Grad. Student Rsch. Symposium, U. Tex., Houston, 1983. Named winner Clin. Rsch. Competition for Assocs., Am. Coll. Physicians, 1989, finalist Clin. Rsch. Competition for Assocs., Am. Coll. Physicians, 1990. Mem. AAAS, Nat. Coun. Against Health Fraud, Am. Soc. for Human Genetics, Am. Coll. Physicians, Sigma Xi. Achievements include determination of the correct enzyme involved in the metabolism of aromatic alpha-keto acids; discovery of synergistic effect of storage pool disease in patient with autosomal dominant exudatine vitreoretinopathy. Office: Johns Hopkins U Ctr for Med Genetics 600 N Wolfe St Baltimore MD 21287-4922

FRIEND, ALEXANDER LLOYD, forester educator; b. N.Y.C., Mar. 16, 1960; s. Henry Parker Friend and Caryl (Daly) Johnson; m. Julia Downs, June 5, 1982. MS, N.C. State U., 1984; PhD, U. Wash., 1988. Post doctoral assoc. Environ. Scis., Oak Ridge, Tenn., 1988-89; asst. prof. forestry Miss. State U., Starkville, 1989—; cons. Winrock Internat. Inst. for Agrl. Devel., New Delhi, India, 1990, 91; reviewer Environ. Protection Agy., Corvallis, Oreg., 1992. Contbr. articles to profl. jours. Mem. Ecol. Soc. Am., Soil Sci. Soc. Am., Sigma Xi. Achievements include research on root physiology and rhizosphere relations; anthropogenic effects on tree and ecosystem function, particularly ozone and carbon dioxide; plant carbon allocation, especially as influenced by environmental stress. Office: Miss State U Dept Forestry PO Drawer FR Starkville MS 39762

FRIEND, CYNTHIA M., chemist, educator; b. Hastings, Nebr., Mar. 16, 1955; d. Matthew Charles and Elise Germaine Friend; children: Ayse K., Kurt Y. BS, U. Calif., Davis, 1977; PhD, U. Calif., Berkeley, 1981. Postdoctoral assoc. Stanford (Calif.) U., 1981-82; asst. prof. Harvard U., Cambridge, Mass., 1982-86; assoc. prof. Harvard U., Cambridge, 1986-89, prof., 1989—; rsch. collaborator Nat. Synchrotron Light Source/Brookhaven Nat. Labs.; Lucy Pickett lectr. Mt. Holyoke Coll., 1991; Cargill lectr. U. South Fla., 1992; Robert Welch lectr., Bernhard vis. fellow Williams Coll., 1992; Procter & Gamble lectr. U. Cin., 1993. Recipient Presdl. Young Investigator award NSF, 1985, Am. Chem. Soc. Garvan medal, 1990, Iota Sigma Pi Agnes Fay Morgan award, 1991. Mem. Am. Phys. Soc., Am. Chem. Soc. (Francis P. Garvan-John M. Olin medal 1990), Am. Vacuum Soc., Phi Beta Kappa (hon., Iota chpt.). Avocations: golf, swimming, weightlifting. Office: Harvard U Dept Chemistry 12 Oxford St Cambridge MA 02138-2900

FRIESEN, HENRY GEORGE, endocrinologist, educator; b. Morden, Man., Can., July 31, 1934; s. Frank Henry and Agnes (Unger) F.; m. Joyce Marylin Mackinnon, Oct. 12, 1967; children: Mark Henry, Janet Elizabeth. BSc, U. Man., 1958, MD, 1958. Diplomate: Am. Bd. Internal Medicine. Intern Winnipeg (Man.) Gen. Hosp., 1958-60; resident Royal Victoria Hosp., Montreal, Que., 1961-62; rsch. assoc. New Eng. Centre Hosp., Boston, 1962-65; prof. exptl. medicine McGill U., Montreal, 1965-73; prof. physiology and medicine U. Man., 1973—, head dept. physiology, 1973—; pres. Med. Rsch. Coun. Can., 1991—; chmn. exec. com. Med. Rsch. Coun. Can., mem. exec. com., 1984—, active, 1981—; pres. Nat. Cancer Inst. Can., 1990-92. Contbr. numerous articles to profl. jours. Decorated Officer Order of Can.; recipient Gairdner award 1977, Killam scholar, 1979. Fellow Royal Soc. Can. (McLaughlin medal 1987), Royal Coll. Physicians and Surgeons; mem. AAAS, Am. Physiol. Soc., Endocrine Soc. (Koch award 1987), Can. Soc. Clin. Investigation (pres. 1974, G. Malcolm Meml. award 1982, Disting. Sci. award 1987), Can. Physiol. Soc., Am. Fedn. Clin. Research, Am. Soc. Clin. Investigation, Can. Soc. Endocrinology and Metabolism (past pres.), Internat. Soc. Neuroendocrinology. Mennonite. Office: Med Rsch Coun Can, 1600 Scott St Tower B Rm 500, Ottawa, ON Canada K1A 0W9*

FRIEZE, ALAN MICHAEL, mathematician, educator; b. London, Oct. 25, 1945; came to U.S., 1987; s. Sidney and Esther (Dayan) F.; m. Carol Mayfield, June 27, 1969; children: Nancy, Adam. BA, Oxford U., Eng., 1966; PhD, London U., 1975. Rsch. officer Brit. Rail, London, 1968-69; programmer ICL, Reading, 1969-70; lectr. Poly. North London, 1970-71, Queen Mary Coll., London, 1972-87; prof. math. Carnegie Mellon U., Pitts., 1987—. Contbr. articles to profl. jours. Recipient Delbert Ray Fulkerson prize Am. Math. Soc./Math. Programming Soc., 1991. Office: Carnegie-Mellon University Dept of Math 5000 Forbes Ave Pittsburgh PA 15213*

FRINAK, SHEILA JO, engineer; b. Flatwoods, Ky., May 12, 1963; d. David Lester and Donna (Sigler) Sherman; m. Paul Eric Frinak, June 18, 1988; children: Eric Nathan, Kyle Wesley. BS, Univ. Ky., 1985; MS, Univ. Cin., 1993. Materials engr. Robins Air Force Base, Warner Robins, Ga., 1985-86; materials devel. engr. Westinghouse, Cin., 1986-88; lead engr. General Elec. Aircraft Engines, Cin., 1988—; partnership in edn. Westinghouse, 1986-88; engr. minority females General Elec., 1990. Recipient General Elec. Achievement award, 1990. Mem. Am. Soc. Metals, Am. Ceramic Soc. Baptist. Office: General Elec Aircraft Engines One Newmann Way MD D95 Cincinnati OH 45215

FRISCH, HARRY LLOYD, chemist, educator; b. Vienna, Austria, Nov. 13, 1928; s. Jacob J. and Clara F. (Spondre) F.; children—Benjamin, Michael. B.A., Williams Coll., 1947; Ph.D., Poly. Inst. Bklyn., 1952. Research asso. physics Syracuse U., 1952-54; instr. U. So. Calif., 1954-55, asst. prof., 1955-56; mem. tech. staff Bell Telephone Labs., Inc., Murray Hill, N.J., 1956-67; prof. chemistry SUNY, Albany, 1967-78, disting. prof. chemistry, 1978—; assoc. dean Coll. Arts and Sci., 1969-71; vis. assoc. prof. physics Yeshiva U., 1963-65, Inst. Study Metals, U. Chgo., 1960; asst. to dean Belfer Grad. Sch. Yeshiva U., 1963-65; cons. in field. Editor: (with J. Lebowitz) The Equilibrium Theory of Classical Fluids, 1964, (with Z. Salsburg) Simple Dense Fluids, 1968; assoc. editor: Jour. Chem. Physics, 1964-66, Jour. Statis. Physics, 1970-75; mem. editorial bd.: Jour. Phys. Chemistry, 1976-80, Jour. Polymer Sci. (Physics edit.), 1976—, Jour. Membrane Sci, 1976-80, Jour. Colloid and Interface Sci., 1978-81, Jour. Adhesion, 1970-75; contbr. articles to profl. jours. NSF grantee, 1968—; recipient Sr. U.S. Scientist Humboldt award, 1987-89. Fellow Am. Phys. Soc.; mem. Am. Chem. Soc., Cosmos Club, Williams Club, Sigma Xi. Democrat. Jewish. Office: 1400 Washington Ave Albany NY 12222-0001

FRISCH, IVAN THOMAS, computer and communications company executive; b. Budapest, Hungary, Sept. 21, 1937; came to U.S., 1939, naturalized, 1941; s. Laszlo and Rose (Balog) F.; m. Vivian Scelzo, June 6, 1962; children: Brian, Bruce. B.S., Queens Coll., N.Y., 1958, Columbia U., 1958; M.S., Columbia U., 1958, Ph.D., 1962. Asst. prof. elec. engring. and computer sci. U. Calif., Berkeley, 1962-65; asso. prof. U. Calif., 1965-69; Ford Found. resident engring. practice Bell Labs., Holmdel, N.J., 1965-66; founding mem. Network Analysis Corp., Great Neck, N.Y., 1969—; sr. v.p. Network Analysis Corp., 1971—, gen. mgr., 1978-85; v.p. Control Bus. Networks, 1985-87; dir. Ctr. on Advanced Tech. in Telecommunications, prof. Poly. U., Bklyn., 1987—; provost Polytech. U., 1992—; adj. prof. computer sci. SUNY, Stony Brook, 1975—, Columbia U., N.Y.C., 1977—; cons. in field. Author: (with Howard Frank) Communication, Transmission and Transportation Networks, 1971; Founding editor-in-chief: Networks, 1971—; contbr. articles to profl. publs. Guggenheim fellow, 1969. Fellow IEEE; mem. N.Y. Acad. Scis., Cable TV Assn. Am., Phi Beta Kappa, Tau Beta Pi, Eta Kappa Nu. Office: Poly U 333 Jay St Rm 321 Brooklyn NY 11201-2990

FRISCH, KURT CHARLES, educator, administrator; b. Vienna, Austria, Jan. 15, 1918; came to U.S., 1939; s. Jacob J. and Clara F. (Spondre) F.; m. Sally Sisson, Sept. 14, 1946; children: Leslie Frisch Nickerson, Kurt C. Jr., Robert J. MA, U. Vienna, 1938; candidate Sc. Chim., U. Brussels, 1939; MA, Columbia U., 1941, PhD, 1944; DSc (hon.) U. Detroit, 1989. Project leader Gen. Electric Co., Pittsfield, Mass., 1944-52; acting mgr. rsch. E.F. Houghton & Co., Phila., 1952-56; dir. polymer research and devel. Wyandotte Chems. Corp., Mich., 1956-68; prof., dir. Polymer Inst., U. Detroit Mercy, 1968-86; pres., dir. rsch. Polymer Techs. subs. U. Detroit Mercy, 1986-90, Polymer Techs., Inc., U. Detroit Mercy, 1990—; pres. Kurt C. Frisch, Inc., Grosse Ile, Mich., 1982—; cons. various corps. Patentee in field. (55); author, co-author, editor 29 books. Contbr. articles to profl. jours. Recipient medal of merit German Foam Soc., 1981, medal of merit Brit. Rubber and Plastics Group, 1982, Gold medal U. Tuzla, Yugoslavia, 1987; named to Polyurethane Hall of Fame, 1984; IR-100 award Indsl. Research Inst. Fellow Am. Inst. Chemists; mem. Soc. Plastics Industry (div. chmn.), Soc. Plastics Engrs.(Outstanding Achievement in Plastics Edn. award 1986), Am. Chem. Soc., Soc. Coating Tech. Republican. Episcopalian. Home: 17986 Parke Ln Grosse Ile MI 48138-1042

FRISQUE, GILLES, forestry engineer; b. Brussels, June 18, 1943; s. Joseph-Jean Frisque and Georgette Versele; m. Cécile Ugeux, Oct. 17, 1967; children: Catherine, Véronique. Degree in agrl. engring., U. Louvain, Belgium, 1965, MS in Forestry, 1967; PhD in Forestry, Laval U., Que., Can., 1977. Rsch. scientist Laurentian Forest Rsch. Ctr., Quebec City, Can., 1968-72; project leader Can. Forestry Svc., Quebec City, 1972-78, program mgr. rsch. program, 1978-82, dir., 1982-85; forest rsch. dir. Univ. du Québec, Quebec City, 1985-92, Inst. Armand-Frappier, Mont., 1992—; mem. James Bay Adv. Com. on Environment, Quebec City, 1986—; vice-chmn. Conseil de la Recherche forestière du Québec, Quebec City, 1990-92, Forest Rsch. Adv. Coun. Can., Ottawa, 1990—. Assoc. editor: Can. Jour. Forest Rsch., 1988; mem. editorial bd. (sci. jour.) Agricultures, 1990; contbr. more than 60 sci. and tech. publs. on forest mgmt., silviculture, ecology, forest and rsch. policy to profl. jours. Mem. Internat. Union of Forest Rsch. Orgn., Can. Pulp and Paper Assn., Can. Inst. Forestry, Ordre des Ingénieurs forestiers du Québec. Home: University of Quebec, 2196 Chemin du Foulon, Sillery, PQ Canada G1T 1X4 Office: Inst Armand-Frappier, 531 Blvd des Prairies, Laval, PQ Canada H7V 1B7

FRISSE, RONALD JOSEPH, telecommunications engineer; b. Highland, Ill., Aug. 22, 1966; s. Norbert Francis and JoAnn Ruth (Vohradsky) F. BSEE, U. Mo., Rolla, 1990. Telecomm. engr. Union Pacific R.R., Omaha, 1990—. Mem. Omaha Jaycees. Office: Union Pacific RR 1416 Dodge St Rm 232 Omaha NE 68179

FRITCHER, EARL EDWIN, civil engineer, consultant; b. St. Ansgar, Iowa, Nov. 24, 1923; s. Lee and Mamie Marie (Ogden) F.; m. Dorsille Ellen Simpson, Aug. 24, 1946; 1 child, Teresa. BS, Iowa State U., 1950. Registered civil engr., Calif. Project devel. engr. dept. transp. State of Calif., Los Angeles, 1950-74, traffic engr. dept. transp., 1974-87; pvt. practice cons. engr. Sunland, Calif., 1987—; consulting prin. traffic engr. Parsons DeLeuw Inc., 1990—. Co-author: Overhead Signs and Contract Sign Plans, 1989; patentee in field. Served to 2d lt. USAF, 1942-46, 50-51. Mem. Iowa State Alumni Assn. Republican. Methodist. Clubs: Verdugo Hills Numismatic (Sunland), Glendale Numismatic. Avocations: collecting coins, stamps, and Indian artifacts.

FRITTS, HAROLD CLARK, dendrochronology educator, researcher; b. Rochester, N.Y., Dec. 17, 1928; s. Edwin Coulthard and Ava Lee (Washburn) F.; m. Barbara Smith, June 11, 1955 (dec.); children: Marcia L., Paul T.; m. Miriam Colson, July 19, 1982. AB, Oberlin Coll., 1951; MS, Ohio State U., 1953, PhD in Botany, 1956. Asst. prof. botany Eastern Ill. U., Charleston, 1956-60; asst. prof. dendrochronology U. Ariz., Tucson, 1960-64, assoc., 1964-69, prof., 1969-92, emeritus, 1992—; adj. prof. in rsch. Desert Rsch. Inst., U. Nev.; dir., founder Internat. Tree-Ring Data Bank, 1975-90; adj. prof. Desert Rsch. Inst., Reno, 1992—; NSF faculty, mem. Task Group 3 adv. com. on paleoclimatology, Climate Dynamics Program, 1978—; lectr. NATO Advanced Study Inst. on Climatic Variability, Sicily, 1980; vis. dr. U. Wyo. Summer Sci. Camp, summer 1956; mem. U. Ariz. del. to People's Republic of China, 1976; participant Nat. Def. Univ., 1978-79; mem. organizing group internat. conf. on dendroclimatology, England, 1980. Author: Tree Rings and Climate, 1976, Reconstructing Large-Scale Climate Patterns from Tree-Ring Data, 1991; mem. editorial adv. bd. Quaternary Rsch., 1977-82; contbr. articles to profl. jours. Mem. local sch. bd., 1971-72. Recipient Dendrochronological award of Appreciation Sci. Community, Lund, Sweden, 1990; Grad. fellow Ohio State U., 1954-56, NSF fellow Oreg. Inst. Marine Biology, summer 1957, Guggenheim fellow, 1968-69; grantee NSF 1971-87, U. Calif. Lawrence Livermore Lab., 1978-79, State of Calif., 1979-80, 85-86. Fellow AAAS; mem. Am. Assn. Quaternary Environment (council 1978-82, adv. com. paleoclimatology), Ecol. Soc. Am. (edit. bd. 1964-66, council rep., chmn. paleoecology sect 1984), Am. Inst. Biol. Scis., Tree-Ring Soc., Am. Meteorol. Soc. (Oustanding Achievement in Bioclimatology award 1982), Am. Quaternary Assn. (mem. coun. 1978-80), Internat. Assn. for Ecology, Internat. Soc. Ecol. Modeling, Internat. Union Quaternary Research, Sigma Xi. Home: 5703 N Lady Ln Tucson AZ 85704-3905 Office: U Ariz Lab of Tree-Ring Rsch Bldg 58 Tucson AZ 85721

FRITZ, EDWARD WILLIAM, mechanical engineer; b. Kenosha, Wis., Nov. 30, 1953; s. Edward Wilgar and Marilyn Jean (Pflug) F.; m. Cindy Susan Smith, Nov. 25, 1977; children: Edward W. Jr., Michael V. BS, U.S. Merchant Marine Acad., 1975. Registered profl. engr., N.C., S.C. Test engr. Ingalls Shipbuilding, Pascagoula, Miss., 1975-77, shift test engr., 1977-78; assoc. engr. Duke Power Co., Charlotte, N.C., 1978-80, design engr., 1980-87, sr. engr., 1987-91; engring. supr., 1991—; York, S.C. Mem. ASME, Am. Nuclear Soc. Presbyterian. Office: Duke Power Co Catawba Nuclear Sta 4800 Concord Rd York SC 29745-9635

FRITZSCHE, HELLMUT, physics educator; b. Berlin, Feb. 20, 1927; came to U.S., 1952; s. Carl Hellmut and Anna (Jordan) F.; m. Sybille Charlotte Lauffer, July 5, 1952; children: Peter Andreas, Thomas Alexander, Susanne Charlotte, Katharina Sabine. Diploma in Physics, U. Göttingen, Fed. Republic Germany, 1952; PhD in Physics, Purdue U., 1954, DSc (hon.), 1988. Instr. physics Purdue U., Lafayette, Ind., 1954-55, asst. prof., 1955-56; asst. prof. U. Chgo., 1957-61, assoc. prof., 1961-63, prof., 1963—, dir. Materials Rsch. Lab., 1973-77, chmn. dept., 1977-86, Louis Block prof. physics, 1989—; v.p., bd. dirs. Energy Conversion Devices, Inc., Troy, Mich.; mem. adv. com. Encyclopaedia Britannica, 1969—. Editor 7 sci. books; assoc. editor Jour. Applied Physics, 1975-80; regional editor Jour. Non-Crystalline Solids, 1987—; contbr. 220 articles to profl. jours. Named hon. prof. Shanghai Inst. Ceramics, 1985, Nanjing U., 1987, Beijing U. Astronautics, 1988. Fellow Am. Physical Soc. (Oliver Buckley Condensed Matter Physics prize 1989), N.Y. Acad. Scis. (chmn. div. condensed matter physics 1979-80). Avocations: the violin, sailing, skiing. Home: 5801 S Blackstone Ave Chicago IL 60637-1855 Office: U Chgo James Franck Inst 5640 S Ellis Ave Chicago IL 60637-1467

FROEHLICH, FRITZ EDGAR, telecommunications educator and scientist; b. Worms am Rhine, Hesse, Fed. Republic Germany, Nov. 12, 1925; came to U.S., 1938; s. Julius and Ida (Heilborn) F.; m. Eileen Karch, Dec. 25, 1949; children: Laurence Alan, Georgine K. Froehlich Scharff, Philip Marc. BS in Physics magna cum laude, Syracuse U., 1950, MS in Physics, 1952, PhD in Physics, 1955. Rsch. asst. Syracuse (N.Y.) U., 1950-54; asst. instr. Utica (N.Y.) Coll., 1952-54; with AT&T Bell Labs., 1954-87; tech. staff Whippany, N.J., 1954-56; supr. data transmission div. Murray Hill, N.J., 1956-63; head data theory dept. Holmdel, N.J., 1963-68, head telecommunications and data systems dept., 1968-83; head univ. relations AT&T Info. Systems and Communications, Lincroft, N.J., 1983-87; ret. AT&T, Holmdel, 1987; prof. telecommunications U. Pitts., 1987—; mem. adv. bd. Ctr. for Info. and Communication Scis. Ball State U., Muncie, Ind., 1987—; nat. telecommunications adv. coun. U. Pitts., 1992—. Editor-in-chief Ency. of Telecommunications, 1988—; sr. editor IEEE Trans. on Communications, 1988-93; contbr. articles to profl. jours.; holder 7 patents. Trustee Cong. B'nai Israel, Rumson, N.J., 1970-84, v.p. cong., 1974-76. With U.S. Army, 1944-46. Recipient Hon. Alumnus award Pitts. U., 1992; Ann. Fritz Froehlich award established in his honor Pitts. U. Sch. Libr. and Info. Sci., 1992. Fellow IEEE (Data Transmission and New Telephone Svcs. award, chmn. tellers com. 1972), Comm. Soc. IEEE (chmn. N.J. Coast sect. 1970, mem. data com., trans. system com. 1960—, chmn. comms. terminal com. 1981-84, mem. awards bd. 1992—), Phi Beta Kappa, Sigma Xi, Sigma Pi Sigma (pres. Syracuse U. chpt. 1949), Pi Mu Epsilon. Home: 10621 NW 71st Ct Tamarac FL 33321 Office: U Pitts 135 N Bellefield St 743 SLIS Bldg Pittsburgh PA 15260

FROESSL, HORST WALDEMAR, business executive, data processing developer; b. Mannheim, Baden-Württemberg, Germany, Apr. 12, 1929; s. Otto and Friederike (Wieder) F.; m. Waltraut Kühnreich, Apr. 26, 1963 (div. Sept. 1971); m. Monika Morgener, Nov. 3, 1972. Student, pvt. schl., Shanghai, People's Republic China, 1945-50, pvt. schl., Mannheim, 1958-60. Interpreter, sect. chief Ordnance Procurement Ctr. U.S. Army, Mannheim, 1951-57; system analyst, mgr. data processing U.S. Army Indsl. Ctr. Europe, Mannheim, 1961-65; systems deliverer AEG-Telefunken, Konstanz, Fed. Republic Germany, 1966-68; mgr. orgn. and data processing Pakistan Machine Tool Factory, Karachi, 1969-71; researcher, inventor Hemsbach, Fed. Republic Germany, 1972-78; inventor, mgr., co-owner Froessl GmbH, Hemsbach, 1979—. Author 17 patents various data processing systems. Avocations: chess, writing poetry. Home and Office: Froessl GmbH, Gutenberg Strasse 2-4, D 69502 Hemsbach Germany

FRÖHLICH, JÜRG MARTIN, physicist, educator; b. Schaffhausen, Switzerland, July 4, 1946; s. Walter Werner and Annemarie (Roth) F.; m. Eva Daniela Schubert, Aug. 31, 1972; children: Judith Monica, Sonja Gabriela. Diploma, Eidgenössische Technische Hochschule, Zurich, 1969, PhD in Physics, 1972. Research asst. U. Geneva, 1972-73; research fellow Harvard U., Cambridge, Mass., 1973-74; asst. prof. Princeton (N.J.) U., 1974-77; prof. Inst. Hautes Études Sci., Paris, 1978-82, vis. prof., 1987; full prof. Eidgenössische Technische Hochschule, 1982—, head Ctr. Theoretical Studies, 1985—; vis. mem. Inst. Advanced Study, Princeton, N.J., 1984-85; speaker in field. Author: Progress in Physics, 1983; editor, contbr. Scaling and Self-Similarity in Physics; contbr. more than 100 articles to profl. jours. Recipient Latsis prize Swiss Nat. Sci. Found., 1984; Alfred P. Sloan fellow, 1976; numerous research grants from U.S. and Swiss Nat. Sci. Founds. Mem. Am. Math. Soc. (Dannie N. Heineman Math. Physics prize 1991), Naturforschende Gesellschaft, Swiss Phys. Soc. Avocations: drawing, painting, psychology, hiking, skiing. Home: Neuhausstrasse 10, 8044 Zurich Switzerland Office: ETH-Z Theoretical Physics, ETH-Honggerberg, 8093 Zurich Switzerland

FROLOV, KONSTANTIN VASILIEVITCH, mechanical engineer, science administrator; b. Kirov, Kaluga, Russia, July 22, 1932; s. Vasiliy and Alexandra F.; m. Galina Alekseevna; 1 child, Konstantin Konstantinovitch. Dipl.Mech.Eng., Transport. Machine Inst., Briansk, 1956; PhD, VAK, Moscow, 1961, DSc, 1970, prof., 1971. Engr., researcher Metal Working Plant, Leningrad, Russia, 1956-58; sr. researcher Rsch. Inst., Moscow, 1961-76, head of lab., 1961-76, head of dept., 1961-76, dir. inst., 1976—; v.p. Russian Acad. Sci., Moscow, 1985—; head dept., prof. Bauman Tech. U., Moscow, 1978; dir., academician Blagonravov Inst. Theoretical Engring. Author: Metody Sovershenstvovania Mashin i Problemy Mashinovendeniya, 1984, Teoria Vibrastsionnoi Tekniki i Technologii, 1981; editor-in-chief Soviet-Hungarian Jour., 1980. Mem. USSR Supreme Soviet, Moscow, 1989, The Cen. Com. of the CPSU, 1989. Recipient Lenin Prize Govt. of USSR, 1980, State prize, 1986. Mem. ASME (life), Internat. Soc. Biomechanics, IFTOMM (exec. coun.), Fellowship of Engring., Swedish Royal Engring. Acad. (fgn. mem.), Nat. Engring. Acad. (fgn. mem.). Home: 4 Griboedov St, Moscow Russia 101830 Office: Blagonravov Inst, Moscow Ctr Ulitsa Griboyedova 4, Moscow Russia 101830*

FROMHAGEN, CARL, JR., obstetrician/gynecologist; b. Tampa, Fla., 1926; s. Carl Frederick and Minnette Gertrude (Douglass) Von Fromhagen; children: Dana Lynn, Carol Leslie, Carl Scott. BS, U. Miami, 1950; student U. Utah, 1949; grad. mil. pilot tng. USAF, 1951; MS, U. Colo., 1952; MD, Emory U., 1955. Diplomate Am. Bd. Ob-Gyn. Intern, Baylor U., 1955-56, resident in ob-gyn, 1956-59; instr. Sch. Medicine U. Miami, Coral Gables, Fla., 1959-62, asso. prof., 1975—; obstetrician, gynecologist, specialist in aviation medicine, FAA sr. med. examiner, Clearwater, Fla., 1960—; pres. Fromhagen Aviation Inc., 1969—; chmn. bd. Navigate Inc., 1970-73; med. cons. Planned Parenthood, 1963-67; chief of staff Clearwater Community Hosp., 1991-92. Mem. Fla. State Aviation Coun., 1966-67; mem. Com. of 100 Pinellas County, pres. Honduras Relief Soc., 1970; bd. dirs. Am. Cancer Soc., 1962-68; bd. dirs. Interprofl. Family Coun., 1967-68. Served to col. USAFR. Named outstanding resident Baylor U. Med. Sch., 1959; recipient award merit Res. Officers Assn., 1964; Silver Wings Frat. award of honor, 1981. Fellow ACOG, ACS, Am. Coll. Abdominal Surgeons, Internat. Coll. Surgeons; mem. Pan Am. Med. Assn., Fla. Soc. for Preventive Medicine (pres. 1968), Aerospace Med. Assn., Civil Aviation Med. Assn., Flying Physicians (v.p. 1967-68, dir., 1968-74, state pres. 1966-74), Res. Officers Assn. Fla. (Clearwater chpt. pres. 1963-67, state surgeon 1964), N.Y. Acad. Sci., Confederate Air Force, Clan Douglas Soc., Aviation Maintenance Found., U.S. Power Squadron (fleet surgeon), Iron Arrow, Omicron Delta Kappa, Pi Kappa Alpha, Beta Beta Beta. Clubs: Carlouel Yacht. Home: 1666 Robinhood Ln Clearwater FL 34624-6431 Office: 1745 S Highland Ave Clearwater FL 34616

FROMLET, K. HUBERT, banking economist; b. Stuttgart, Fed. Republic Germany, May 22, 1947; arrived in Sweden, 1975; s. Kurt and Marianne (Schnitzler) F.; m. Cristina Lindqvist, June 1, 1979; children: Camilla, Pia. Diploma in bus., U. Würzburg, Fed. Republic Germany, 1971, D. in Polit. Sci., 1975. Researcher Saab-Scania, Södertälje, Sweden, 1975-81; researcher Swedish Coop. Banks, Stockholm, 1981-83, chief economist, 1983-84; chief economist Swed Bank, Stockholm, 1984—. Author: Das schwedische Bankensystem, 1975; contbr. articles to profl. jours. Avocations: sports, art.

FROMM, ELI, engineering educator; b. Niedaltdorf, Germany, May 7, 1939; s. Siegfried and Helen (Lucas) F.; m. Dorothy Mildred Gold, Dec. 23, 1962; children: Stephen Arthur, Larry Brian, Richard Michael. BSEE, Drexel U., 1962, MSE, 1964; PhD, Jefferson Med. Coll., 1967. Engr. missile and space div. GE Co., Phila., 1962; engr. Applied Physics Lab. E.I. DuPont Co., Wilmington, Del., 1963; from asst. prof. to prof. biomed. sci. Drexel U., Phila., 1967-80, prof. elec. and computer engring., 1980—, acting head dept. biol. sci., 1984-85, asst. head dept. elec. and computer engring., 1987-89, assoc. dean. Coll. Engring., 1988-89, interim dean, 1989-90, vice provost for rsch. and grad. studies, 1990—; mem. staff, congl. fellow com. sci. and tech. U.S. Ho. of Reps., 1980-81; program dir. NSF, Washington, 1983-84; vis. scientist Legis. Rsch. Office Pa. Ho. Reps., Harrisburg, 1986-87. Contbr. over 50 articles to profl. jours. Recipient Centennial medal Drexel U.; Spl. fellow NIH, 1964-67; grantee NIH, 1969-78, NSF, 1969-71, 79, 84, 88—. Fellow IEEE (bd. dirs. 1983-84, mem. coms., Centennial medal 1984), Am. Inst. of Med. and Biologic Engring.; mem. Am. Soc. Engring. Edn., Sigma Xi. Jewish. Home: 2604 Selwyn Dr Broomall PA 19008-1632 Office: Drexel U Vice Provost for Rsch Philadelphia PA 19104

FROMMHOLD, WALTER, radiologist; b. Geringswalde, Germany, Aug. 28, 1921; s. Arno and Welly (Thalheim) F.; MD, U. Würzburg, 1944; Dr. h.c., U. Bordeaux, 1981, U. Pécs, 1985, U. Poznán, 1986; m. Gabriele Körner, Mar. 17, 1951; children: Anke, Uwe. Resident radiology Karlsruhe Hosp., 1946-50; teaching fellow radiology Harvard Med. Sch., 1957; asst. prof. radiology Free U. Berlin, 1952; radiologist-in-chief Auguste Viktoria Hosp., Berlin, 1956-68; prof. radiology, dir. dept. U. Tübingen Med. Sch., 1968-88. Served with German Air Force, 1939-45. Recipient medal Slovakian Med. Soc., Radiol. Soc. Netherlands; C. Wegelius medal Finnish Soc. Radiology; Boris-Rajewsky medal European Soc. Radiology; Röntgen medal City of Würzburg, Bundesverdienstkreuz I Kl. M. Mem. Acad. Leopoldina, German Soc. Radiology (pres. 1971-75, H. Rieder medal), IV European Congress Radiology (pres. 1979); hon. mem. Columbian, Indonesian, French. Hungarian, Yugoslavian, Luxembourg, Egyptian, Czechoslovakian, Belgian, Polish, Italian, Austrian, Berlin, Finnish, Swiss, German radiol. socs.; corr. mem. Swedish Soc. Radiology. Lodge: Lions. Editor: Klin-radiol. Seminar, Vols. 1-18, Röntgen-Wie Wann?, Vols. 4-11; co-editor: Fortschritte auf dem Gebiete der Röntgenstrahlen, 1969-90; Lehrbuch der Röntgendiagnostik, 7th edit. Home: 10 Schwabstr, D 72074 Tübingen Germany

FROSCH, ROBERT ALAN, retired automobile manufacturing executive, physicist; b. N.Y.C., May 22, 1928; s. Herman Louis and Rose (Bernfeld) F.; m. Jessica Rachael Denerstein, Dec. 22, 1957; children: Elizabeth Ann, Margery Ellen. A.B., Columbia U., 1947, A.M., 1949, Ph.D., 1952; DEng (hon.), U. Miami, 1982, Mich. Technol. U., 1983. Scientist Hudson Labs. Columbia U., 1951-53, asst. dir. theoretical div., 1953-54, asso. dir., 1954-56, dir., 1956-63; dir. nuclear test detection Advanced Research Projects Agy., Office Sec. Def., 1963-65; dep. dir. Advanced Research Projects Agy., 1965-66; asst. sec. navy for research and devel. Washington, 1966-73; asst. exec. dir. UN Environment Programme, 1973-75; asso. dir. for applied oceanography Woods Hole (Mass.) Oceanographic Instn., 1975-77; adminstr. NASA, Washington, 1977-81; pres. Am. Assn. Engring. Socs., N.Y.C., 1981-82; v.p. in charge Research Labs. Gen. Motors Corp., Warren, Mich., 1982-93; sr. rsch. fellow Harvard U. John F. Kennedy Sch. Govt., Cambridge, Mass., 1993—; chmn. U.S. del. to Intergovtl. Oceanographic Commn. meetings UNESCO, Paris, 1967, 70. Research and publs. numerous sci. and tech. articles. Recipient Arthur S. Flemming award, 1966, NASA Disting. Service award, 1981. Fellow AAAS, AIAA, NAE (sr.), IEEE, Acoustical Soc. Am., Am. Astronautical Soc. (John F. Kennedy Astronautics award 1981); mem. Am. Geophys. Union, Seismol. Soc. Am., Am. Acad. Arts and Scis., Soc. Exploration Geophysicists (spl. commendation 1981), Marine Tech. Soc., Nat. Acad. Engring. (sr. fellow), Am. Phys. Soc., Soc. Naval Architects and Marine Engrs., Soc. Automotive Engrs., Engring. Soc. Detroit. Office: Harvard U John F Kennedy Sch Govt CSIA 79 John F Kennedy St Cambridge MA 02138 also: GM Gen Motors Bldg 3044 W Grand Blvd Detroit MI 48202

FROSSI, PAOLO, engineer, consultant; b. Verona, Italy, Mar. 30, 1921; s. Luigi and Pina (Bellavite) F.; dottore in ingegneria indsl., Poly. Milan, 1945; M.B.A. (spl. student), Harvard U., 1955; m. Gabriella Crespi, Oct. 27, 1962. With Società Edison, Milan, 1956-59; mng. dir. Edison Page, Rome, 1960-67; dir. Montedel, Rome, 1968-72; bus. cons., Verona and Rome, 1973—. Mem. Assn. Elettrotecnica Italy, Assn. Internat. Ingegneri Telecommunicazioni, Union Cristiana Impreditori e Dirigenti, Roman Catholic. Clubs: Harvard Bus. Italy, MIT Alumni of Italy. Lodge: Verona East Rotary. Author papers in field. Address: Lungadige Matteotti 1, 37126 Verona Italy

FROST, JACK MARTIN, civil engineer; b. Denver, Oct. 12, 1928; s. Harold Orville and Nena Anne (Rhodes) F.; m. Mary Jane Hamilton, Sept. 17, 1955; children: Roxanna, Scott, Joel, Jack. BSCE, BSBA, U. Colo., 1953. Registered profl. engr., Ohio, N.Y., N.C. Design engr. Jeffrey Mfg. Co., Columbus, Ohio, 1953-62; dist. engr. Clay Sewer Pipe Assn., Columbus, Ohio, 1962-68, Nat. Clay Pipe Inst., Crystal Lake, Ill., 1968-76; engr. Logan (Ohio) Clay Products Co., 1977—. Staff sgt. U.S. Army, 1946-58, Korea. Mem. NSPE (PEI gov. 1989-91), Ohio Soc. Profl. Engrs. (v.p. 1992—), Water Environ. Fedn. Home: 267 Galloway Rd Galloway OH 43119 Office: Logan Clay Products Co 201 Bowen St Logan OH 43138

FROST, JOHN ELLIOTT, minerals company executive; b. Winchester, Mass., May 20, 1924; s. Elliott Putnam and Hazel Lavera (Carley) F.; m. Carolyn Catlin, July 12, 1945 (div. 1969); children: John Crocker, Jeffrey Putnam, Teresa Baird, Virginia Nicholl; m. Martha Hicks, June 6, 1969 (div. 1984); m. Catherine Kearns, July 27, 1985; 1 stepchild, Colleen Depew. BS, Stanford U., 1949, MS, 1950, PhD, 1965. Geologist Asarco, Salt Lake City, 1951-54; chief geologist, surface mines supt. Philippine Iron Mines Inc., Larap, Camarines Norte, 1954-60; chief geologist Duval Corp. (Pennzoil Corp.), Tucson, 1961-67; minerals exploration mgr. Exxon Corp., Houston, 1967-71; minerals mgr. Esso Eastern Inc. div., 1971-80; sr. v.p. Exxon Minerals Co. div., Houston, 1980-86; pres., Frost Minerals Internat., 1986—; bd. dirs., United Engring. Trustees, N.Y.C., 1981—, chmn. real estate com., 1986-89, v.p., 1989-91, pres., 1991-93. Mem. adv. bd. Sch. Earth Scis., Stanford (Calif.) U., 1983-85; pres. SEG Found., 1984. Served to 1st lt. USAAF, 1943-45, PTO. Fellow Geol. Soc. Am., Soc. Econ. Geologists (pres. 1989-90, councilor 1982-84, program com., nominating com. chmn. 1982); mem. AIME (chmn. edn. com. Soc. Mining Engrs. 1971; Charles F. Rand medal 1984, Disting. Mem. award 1984, Dist. Svc. award 1991), Australian Inst. Mining and Metallurgy; mem. Am. Inst. Profl. Geologists, Mining Club of Southwest (Tucson), Hearthstone Country Club, Sigma Xi. Republican. Methodist. Home and Office: 602 Sandy Port St Houston TX 77079-2419

FROST, SUSAN COOKE, biochemistry educator; b. Providence, Feb. 12, 1949; m. Meryll M. Frost Jr., July 11, 1970; 1 child, Christopher. BS, U. R.I., 1971; PhD, U. Ariz., 1979. Adj. instr. U. Ariz., Tucson, 1979-82; rsch. assoc. Johns Hopkins U., Balt., 1982-85; asst. prof. U. Fla., Gainesville, 1985-91, assoc. prof., 1991—. Mem. rev. bd. Am. Heart Assn., Dallas, 1992-95. Recipient rsch. award NIH, 1985-93, 92-95. Office: U Fla Box 100 245 JNMHC Gainesville FL 32610

FROVA, ANDREA FAUSTO, physicist, author; b. Venice, Italy, Dec. 11, 1936; s. Carlo and Evelina (Schenardi) F.; m. Mariapiera Marenzana, Nov. 16, 1960; children: Elena, Luisa. Laurea, Univ. Pavia, Italy, 1959; docenza, Univ. Roma, 1967. Lectr. Univ. Pavia, Italy, 1959-62; asst. prof. Univ. Messina, Italy, 1962-63; rsch. assoc. Univ. of Ill., Urbana, 1963-65; mem. tech. staff Bell Labs., Murray Hill, N.J., 1965-67; assoc. prof. Univ. Rome, 1967-76; prof. Univ. Modena, 1976-78, Univ. Rome, 1978—; cons. Bell Labs., 1968, 74; vis. Ecole Polytech. Fed., 1977, 78, 79, Univ. Calif., 1984, 92, Univ. Stuttgart, Germany, 1978, 89. Author: Semiconduttori, 1977, La Rivoluzione Elettronica, 1981, Luce Colore Visione, 1984, Bravo, Sebastian, 1989; editor Semiconductor Light Emitters and Detectors, 1976; contbr. articles to profl. jours; mem. bd. of editors Semiconductor Physics and Technology, Solar Energy Materials and Solar Cells, Jour. De Physique III. Recipient Premio Fondazione Della Riccia, Italy, 1979. Mem. Internat. Union of Pure & Applied Physics (v.p. 1990—), Italian Physical Soc., European Physical Soc. Achievements include development of electroabsorption spectroscopy, explanation of Kerr-Pockels electrooptic effect in terms of bandstructure changes, observation of exciton-plasma transition in III-V semiconductors, production of amorphous semicon. Office: Università di Roma, P Aldo Moro 2, Rome Italy 00185

FRUCHT, HAROLD, physician; b. N.Y.C., June 3, 1953; s. Sam and Sara (Jagoda) F. BA, SUNY, Albany, 1975; MD, SUNY, Syracuse, 1982. Cert. Am. Bd. Internal Medicine, Nat. Bd. Med. Examiners. Resident SUNY, Syracuse, 1982-85, fellow, 1985-86; med. staff fellow NIH, Bethesda, Md., 1986-89, sr. staff fellow, 1989-90; dir. of gastroenterology Fox Chase Cancer Ctr., Phila., 1991—. Contbr. articles to profl. jours. Recipient Glaxo Fellow Travel award, 1988. Fellow Am. Coll. Physicians; mem. Am. Gastroenterological Assn., Am. Assn. for Cancer Rsch., Am. Assn. for the Advancement Sci., Am. Soc. for Gastrointestinal Endoscopy, Phila. GI Tng. Group, Am. Soc. Clin. Oncology. Office: Fox Chase Cancer Ctr 7701 Burholme Ave Philadelphia PA 19111

FRUZZETTI, ORESTE GIORGIO, geologist; b. Rome, Italy, Aug. 23, 1938; s. Italo Vittorio and Marianna (Conti) F.; m. Anna Maria Manna, Oct. 3 1968; children: Viviene, Lara Jayne. Dr. in Geol. Scis., U. Rome, 1964;

diploma in English, British Sch., 1967; postgrad., Australian Mineral Found., 1973. Registered Geologist, Italy; registered profl. Engr., Kenya. Cons. geologist Macerata, Italy, 1964-68; resident geologist Bur. Mineral Resources Geology and Geophysics, Canberra, Australia, 1968-69, No. Terr. Geol. Survey, Darwin, Australia, 1969-76; mgr., tech. dir. Beacon Svcs. Ltd., Kano, Nigeria, 1976-83; tech. dir. Cons. Internat., Cons. Engrs., Rome, 1978-92; project mgr. Ismes Spa, Bergamo, Italy, 1988-90; area mgr. (South-East Asia and East Africa) Censulint Internat., Cons. Engrs., Rome, 1990-92. Contbr. articles to profl. jours. Ofcl. consular rep. Italian Embassy, Canberra, Italian Consulate Brisbane, Alice Springs, Australia, 1973-75. Lt. Mining Engrs. Corps, Italian Army, 1965-66. Decorated knight Sovereign Order of Hospitallers of St. John of Jerusalem. Fellow Australasian Inst. Mining and Metallurgy; mem. AAAS, Italian Assn. Profl. Geologists, N.Y. Acad. Scis., Geol. Soc. Australia, Italian Verdi Club (sec. 1973-75). Roman Catholic. Avocations: motoring, hi-fi, video, reading. Home: Via P Capuzi 49, Macerata Italy 62100

FRYBERGER, THEODORE KEVIN, mechanical and ocean engineer; b. Harrisburg, Pa., Jan. 21, 1950; s. Wilbert Witmer and Grace Elizabeth (Cochran) F. BSME, Pa. State U., 1972; MS in Mech./Ocean Engring., U. Calif., Berkeley, 1980. Registered profl. engr., Pa., Md. Engr. AMP, Inc., Harrisburg, Pa., 1973-75, Am. Chain and Cable Co., York, Pa., 1975-77; rsch. asst. U. Calif., Berkeley, 1978-80; engr. Applied Physics Lab., Johns Hopkins U., Laurel, Md., 1980-86; pres. TKF Systems, Columbia, Md., 1986—; personal computer tech. com. UNIX task force Applied Physics Lab., Johns Hopkins U., 1983-86. Mem. ASME. Lutheran. Achievements include design and construction of a diver propulsion vehicle and a microprocessor based decompression computer for a diver, of custom microprocessor based single board computers; design of wire terminating machinery; research of stress and thermal analysis on machinery and missile systems; development of software for engineering applications. Office: TKF Systems 5478-A3 Harpers Farm Rd Columbia MD 21044

FRYCZKOWSKI, ANDRZEJ WITOLD, ophthalmologist, educator, business executive; b. Mstyczow, Poland, Oct. 10, 1939; came to U.S. 1981; s. Jan and Anna (Kugler) F.; m. Hanna B. Bruszewska, Dec. 27, 1962; children: Krzysztof J., Piotr T. MD, U. Lodz Mil. Med. Acad., 1962; PhD, U. Lodz, 1971. Intern U. Lodz, 1962-64, resident in gen. ophthalmology, 1964-66, fellow in ophthalmology, 1966-70; fellow in cornea and external eye disease Dept. Ophthalmology, U. Gent, Belgium, 1975-76; instr., asst. prof. anatomy U. Lodz (Poland), 1958-64, asst. prof., assoc. prof. ophthalmology, 1964-73; cons. in ophthalmology Inst. Hygiene and Epidemiology, Warsaw, Poland, 1973-76; assoc. prof. Mil. Med. Acad., Warsaw, 1976-80; rsch. fellow dept. ophthalmology U. N.C., Chapel Hill, 1981-84; chief ophthalmology Kino Hosp., Tucson, 1984-85; rsch. assoc. prof. U. Ariz., Tucson, 1984-87; vis. assoc. prof. ophthalmology Ohio State U., Columbus, 1987-88, asst. prof., 1988-91, assoc. prof., 1991—; vis. assoc. prof. U. Ghent (Belgium), 1975-76; pres. Al-Bio-Cosmetics, Inc., Tucson, 1986-88, Frysko Enterprises, Inc., 1990—, Friendly Help, Inc., 1991—; lectr. in field. Contbr. 200 articles to med. jours., also chpts. to books and poems to publ.; reviewer scientific jours.; inventor refractive sutures and vaccum corneal trephanon. V.p. Chopin Assn., Tucson, 1985-87. Recipient bronze medal Francois Assn., Ghent, 1980, lst photography award Assn. Scanning Electron Microscopy Microbeam Analysis, 1984. Fellow AAUP, Assn. for Rsch. in Vision and Ophthalmology, Ohio Ophthal. Soc.; mem. AMA, Am. Acad. Ophthalmology. Avocations: reading, swimming, chess, dancing. Home: 895 Dennison Ave Columbus OH 43215-1321 Office: Ohio State U Dept Ophthal 456 W 10th Ave Columbus OH 43210-1240

FRYD, DAVID STEVEN, biostatistician, consultant; b. Bklyn., Oct. 25, 1950; s. Benjamin and Ida Fryd; m. Dolores M. Lippert, June 18, 1977. MS, U. Minn., 1974, PhD, 1976. Assoc. prof., dir. statis. and computer studies U. Minn., Mpls., 1976-89; dir. clin. rsch. Schneider Inc., Mpls., 1989—; cons. to various univs., corps., and hosps. Contbr. over 145 articles to books, profl. jours. Mem. Am. Statis. Assn., Biometric Soc. Office: Schneider Inc 5905 Nathan Ln Minneapolis MN 55442

FRYE, RAYMOND EUGENE, geotechnical engineer; b. Shelbyville, Ky., Jan. 1, 1961; s. Raymond Eugene and Mary Lee (Kelly) F.; m. Theresa Lynn Veasey, June 20, 1987 (div. 1992). BS, U. Louisville, 1983, M Engring., 1987. Registered profl. engr., Ky., Tenn. Staff engr. Law Engring., Inc., Nashville, 1985-87, project engr., 1987-91, constrn. svcs. mgr., 1988-89; project engr. Law Engring., Inc., Louisville, 1990-91, sr. engr., 1991—. Mem. ASTM, ASCE (editor Nashville sect. 1988-90), Harrods Creek Field and Stream Club. Democrat. Home: 908 1/2 Blankenbaker Ln Louisville KY 40207 Office: Law Engring Inc 11125 Decimal Dr Louisville KY 40299

FRYREAR, DONALD WILLIAM, agricultural engineer; b. Haxtun, Colo., Dec. 8, 1936; s. William Alfred and Majorie (Adams) F.; m. Sherry Janice Watson, Sept. 16, 1956; children: Debra Lou, Kenneth William. BSAE, Colo. State U., 1959; MSAE, Kans. State U., 1962. Registered profl. engr., Tex. Engr. USDA-Agrl. Rsch. Svc., Akron, Colo., 1959-60, Manhattan, Kans., 1960-62; rsch. engr. USDA-Agrl. Rsch. Svc., Temple, Tex., 1962-65; rsch. leader USDA-Agrl. Rsch. Svc., Big Spring, Tex., 1965—; erosion cons. UNESCO, Medmine, Tunisia, 1983, Pretoria, South Africa, 1985. Contbr. over 75 articles to profl. jours. Recipient Appreciation award Howard Coll., 1977; Soil Conservation Soc. Am. fellow, 1982. Mem. Am. Soc. Agrl. Engrs. (assoc. editor 1974), Soil and Water Conservation Soc. (charter pres. 1972), Am. Soc. Agronomy (state pres. 1977), N.Y. Acad. Sci. Baptist. Achievements include development of graded furrow concept for controlling water erosion, techniques for analyzing field erosion data; design and construction of five wind tunnels; design of first field equipment for measuring wind erosion. Home: RR 1 Box 319 Big Spring TX 79720-9128 Office: USDA-ARS PO Box 909 Big Spring TX 79721-0909

FRYT, MONTE STANISLAUS, petroleum company executive, speaker, advisor; b. Jackson, Mich., Aug. 3, 1949; s. Marion E. and Dorothy A. (Fischman) F.; m. Pollyanna Hayes, May 26, 1990. BS in Aerospace Engring., U. Colo., Boulder, 1971; MBA in Mgmt., U. Colo., Denver, 1988. Field engr. Schlumberger Well Svcs., Bakersfield, Calif., 1971-75; computer R & D engr. Schlumberger Well Svcs., Houston, 1975-77; account devel. engr. Schlumberger Well Svcs., L.A., 1977-78; dist. mgr. Schlumberger Well Svcs., Abilene, Tex., 1978-80, Williston, N.D., 1980-81; v.p. ops. Logmate Svcs. Inc., Calgary, Alta., Can., 1981-84; pres. Fryt Petroleum Inc., Denver, 1984-91; mgr. petrophysics Am. Hunter Exploration, Ltd., Denver, 1991-92; prin. Reservoir Evaluations Group, Denver, 1992—. Mem. Colo. Rep. Com., 1990—, Rep. Nat. Com., Colo. Rep. Leadership Program, 1992-93; mem. exec. com. Colo. Rep. Bus. Coalition, 1993—. Mem. Am. Mgmt. Assn., Am. Assn. Petroleum Geologists, Brit. Am. Bus. Assn., German Am. C. of C., Rocky Mountain Assn. Geologists, Greater Denver C. of C., Elks, Rockies Venture Club. Roman Catholic. Avocations: mountain climbing, skiing, running, biking, cultural and political reading. Home: 24245 Choke Cherry Ln Golden CO 80401-9203 Office: 410 17th St Ste 1220 Denver CO 80202-4425

FRY-WENDT, SHERRI DIANE, psychologist; b. Clinton, Mo., Mar. 30, 1958; d. Charles Pierce and Norma Geraldine (Croft) Fry; m. Joseph Otto Wendt, May 24, 1980; children: Benjamin, Ethan. BSE, Cen. Mo. State U., 1979, MS, 1981; PhD, U. Mo., 1989. Lic. psychologist, Mo. Mental health therapist Wyandot Mental Health Ctr., Kansas City, Kans., 1981-88; EAP contract psychologist Menninger Found., Topeka, 1988-89; contract psychologist Tri-County Mental Health Ctr., Kansas City, 1988-89; pvt. practice Kansas City, 1988—; trainer various workshops Wyandot Tng. Inst., 1985-88; expert witness State of Kans., 1985—. Youth group sponsor Hillside Christian Ch., Kansas City, 1982-86, children's choir dir., 1983-87, deaconess, 1983—, dir. vacation bible sch., 1992. Mem. APA, Internat. Soc. Study of Multiple Personality and Dissociation, Greater Kansas City Psychol. Assn., Phi Kappa Phi, Psi Chi. Avocations: piano, guitar, crafts, traveling. Office: 4901 Main St Ste 408 Kansas City MO 64112-2635

FRYZUK, MICHAEL DANIEL, chemistry educator; b. Sarnia, Ont., Can., Mar. 15, 1952; s. Michael and Yvette (Benoit) F.; m. Alice Lee Hanlan, Aug. 26, 1978; children: Jeremy Thomas, Brett Andrew. B.Sc., U. Toronto, Ont., Can., 1974, Ph.D., 1978. Postdoctoral fellow Calif. Inst. Tech., Pasadena, 1978-79; asst. prof. U. B.C., Vancouver, Can., 1979-84, assoc. prof., 1984-89, prof., 1989—; cons. Ballard Techs., North Vancouver, 1986—, E.I. du Pont de Nemours and Co., Wilmington, Del., 1990—. Recipient Alfred P. Sloan Found. fellow, 1984-87, Alexander von Humboldt Found. fellow, 1987-88, E.W.R. Steacie fellow Natural Scis. and Engring. Rsch. Coun., 1990-92. Mem. Royal Canadian Soc. (Rutherford medal in chemistry, 1990), Canadian Soc. of Chemistry (ALCAN Lecture award 1992), Chem. Inst. Can., Am. Chem. Soc. Avocations: golf, softball, hockey, skiing. Home: 3590 W 19th Ave, Vancouver, BC Canada V6S 1C4 Office: Univ of BC Dept Chemistry, 2036 Main Mall, Vancouver, BC Canada V6T 1Z1

FTHENAKIS, VASILIS, chemical engineer, consultant, educator; b. Chania, Crete, Greece, July 21, 1951; came to U.S., 1976; naturalized, 1986; s. Menelaos and Antonia Korkidis; m. Christina Georgakopoulos, Feb. 6, 1982; children: Antonia, Menelaos. Diploma in Chemistry, U. Athens, 1975; MS in Chem. Engring., Columbia U., 1978; PhD in Fluid Dynamics & Atmospheric Sci., NYU, 1990. Rsch. analyst Columbia U., N.Y.C., project engr.; rsch. engr. Brookhaven Nat. Lab., Upton, N.Y., 1980—; cons. in chem. engring., 1986—, semiconductor and photovoltaic cons., 1987—, petroleum and petrochemical cons., 1989—; founder, pres. Environ Inc., Upton, N.Y., 1991; chmn. confs.; adj. prof. environ. engring. CCNY, 1992—, Columbia U., 1993—. Author Prevention & Control of Accidental Releases of Hazardous Gases, 1993; editor: (newsletter) Fossil Energy and the Environment, 1991—; contbr. articles to sci. jours., chpts. to books. Mem. ACGIH, Am. Inst. Chem. Engrs., Semiconductor Safety Assn. Home: 88 Ledgewood Dr Smithtown NY 11787-4247 Office: Brookhaven Nat Lab Eviron Assessment Group Bldg 490A Upton NY 11973

FU, ALBERT JOSEPH, astrophysicist; b. Chgo., Mar. 21, 1959; s. John Evangelist and Frances (Lee) F. BS in Physics summa cum laude, U. Ill., 1981; MS in Astronomy and Astrophysics, U. Chgo., 1983, PhD in Astronomy and Astrophysics, 1987. Rsch. assoc. Northwestern U., Evanston, Ill., 1987-89, U. Calif., Santa Cruz, 1989-90; software engr. JVC Lab. Am., Santa Clara, Calif., 1990-91, The Santa Cruz (Calif.) Operation, 1991—. Contbr. articles to profl. jours. Student advisor, music dir. Newman Cath. Community, Santa Cruz, 1989-91; discussion leader Holy Name Young Adult Ministry, Chgo., 1988-89. Robert R. McCormick fellow U. Chgo., 1981; recipient Marc Perry Galler prize U. Chgo., 1988, Lincoln Meml. award State of Ill., 1981. Mem. Am. Astron. Soc., Phi Beta Kappa, Phi Kappa Phi. Achievements include design of analytic mathematical model of Supernova 1987A, the observed optical light over the first 100 days of evolution; research on general relativistic treatment of radiation from the prominent x-ray source A0620-00 that provided new and independent support for its status as the most likely black hole yet discovered.

FU, GANG, nuclear engineer; b. Tianjin, China, Dec. 11, 1956; came to U.S., 1986; s. Hongen Fu and Erxia Cao; m. Han Ni, Mar. 26, 1985; 1 child, Anqi Fu. BS, Tsinghua U., Beijing, 1982, MS, 1984; ScD, MIT, 1991. Asst. prof. Tsinghua U., Beijing, 1984-86; engr. 53 Techs., Columbia, Md., 1991-92; sr. engr. 53 Techs., 1993—. Contbr. articles to profl. jours. Mem. Sigma Xi. Office: S3 Techs 8930 Stanford Blvd Columbia MD 21045

FU, SHOU-CHENG JOSEPH, biomedicine educator; b. Peking, China, Mar. 19, 1924; came to U.S., 1946, naturalized, 1957; s. W.C. Joseph and W.C. (Tsai) F.; m. Susan B. Guthrie, June 21, 1951; children: Robert W.G., Joseph H.G., James B.G. BS, MS, Catholic U., Peking, 1944; PhD, Johns Hopkins U., 1949. Postdoctoral fellow Nat. Insts. of Health, Bethesda, Md., 1949-51, scientist, 1951-55; Gustav Bissing fellow Johns Hopkins U., at Univ. Coll., London, 1955-56; chief Enzyme and Bioorganic Chemistry Lab. Children's Cancer Research Found. (now Dana Farber Cancer Ctr.), 1956-66; research assoc. Harvard U. Med. Sch., Boston, 1956-66, Univ. prof., chmn. bd. chemistry Chinese U., Hong Kong, 1966-70, univ. dean sci. faculty, 1967-69; vis. prof. Coll. Physicians and Surgeons, Columbia U., N.Y.C., 1970-71; prof. biochemistry and molecular biology U. Medicine & Dentistry of N.J., Newark, 1971—, asst. dean, 1975-77, acting dean Grad. Sch. Biomed. Scis., 1977-78, prof. ophthamology, 1989—. Contbr. articles to profl. jours. Capt. USPHS Res., 1959—. Named Hon. Disting. Prof. and Academic Advisor Inner Mongolia Med. Univ., Huthot, Peoples Republic China, 1988—. Fellow AAAS, Royal Soc. Chemistry (London); mem. Royal Hong Kong Jockey Club, American Club Hong Kong, Sigma Xi (chpt. pres. 1976-80, sec. 1974-76, 81-82). Home: 693 Prospect St Maplewood NJ 07040-3105 Office: U of Medicine and Dentistry NJ Med Sch Med Sci Bldg 185 S Orange Ave Newark NJ 07103-2714

FU, YUAN CHIN, chemical engineering educator; b. Ta-hsi, Taiwan, China, Feb. 16, 1930; s. Tsu-chien and Ahmay (Yu) F.; children: Eugene, Steven. BS, Nat. Taiwan U., Taipei, 1953; PhD, U. Utah, 1961. Rsch. engr. Union Indsl. Rsch. Inst., Sinchu, Taiwan, China, 1953-56; rsch. assoc. U. So. Calif., L.A., 1964-65; rsch. engr. Phillips Petroleum Corp., Bartlesville, Okla., 1961-64; rsch. chemist U.S. Bur. of Mines, Pitts., 1965-75; rsch. engr. Energy Rsch. and Devel. Adminstrn., Pitts., 1975-77, Pitts. Energy Tech. Ctr. DOE, Pitts., 1977-89; prof. Muroran Inst. Technol., Japan, 1989—. Patentee in field; author: Processes for Direct Coal Liquefaction, 1981, Coal Liquefaction, 1984. Recipient Bituminous Coal Rsch. award Am. Chem. Soc., 1968, Grant-in-Aid of Sci. Rsch., 1990. Mem. Am. Chem. Soc., Chem. Soc. Japan, Soc. Chem. Engring. Japan, Japan Inst. Energy. Avocation: golfing. Office: Muroran Inst Technol, 27-1 Mizumoto-cho, Muroran 050, Japan

FUCHIGAMI, LESLIE HIRAO, horticulturist, researcher; b. Lanai, Hawaii, June 11, 1942; s. Susumi and Shigeko (Sakamura) F.; m. Elaine Rei Kisaba, June 1, 1963; children: Sheila Shigeko, Michelle Michiko, Tammy Tamiko, Summer Sachiko, Shane Satomi. MS, U. Minn., St. Paul, 1964, PhD, 1970. From asst. to assoc. prof. Oreg. State U., Corvallis, 1970-80, prof., 1980—; chairperson dept. horticulture U. Hawaii, Honolulu, 1981-82, acting asst. dir. Coll. Tropical Agr. and Human Resources, 1990. Assoc. editor Environ. Horticulture Jour., 1984—; contbr. over 50 articles to profl. jours. Grantee USDA, 1986-88, 87-91, 89-92, Binat. Agrl. R & D Fund, 1988-91. Fellow Am. Soc. for Hort. Sci. (v.p. rsch. 1991-93, assoc. editor 1983-86, Ornamentals Publ. award 1987, Cross-Commodity Publ. award 1990). Office: Oreg State U Dept Horticulture Corvallis OR 97331-2911

FUCHS, BETH ANN, research technician; b. Moberly, Mo., July 22, 1963; d. Larry Dale and Marilyn Sue (Summers) Williams; m. Fred Albano Fuchs Jr., Sept. 30, 1989. AA, Cottey Coll., 1983; BS in Engring., U. N.Mex., 1987. Bookkeeper, chemistry technician U. N.Mex., Albuquerque, 1984-88; rsch. technician Sandia Nat. Labs., Albuquerque, 1988—. Mem. Am. Vacuum Soc. Republican. Avocations: reading, cooking, crocheting, bowling. Home: 909 Carlisle Blvd SE Apt D Albuquerque NM 87106-1646

FUCHS, LASZLO JEHOSHUA, mathematician; b. Nyirbator, Hungary, May 19, 1949; s. Sandor and Roza (Klein) F.; m. Ilona Lewensohn, Aug. 22, 1984; children: Alexander, Robert-Gabriel. BSc, U. London; Dr.Engring., Royal Inst. Tech., Stockholm; BSc in math., U. London, 1972. Rsch. assoc. Royal Inst. Tech., Stockholm, 1977-83, assoc. prof., 1984-88, prof. applied computational fluid dynamics, 1989—. Contbr. articles to profl. jours. Mem. AIAA. Achievements include development of computational methods in fluid dynamics; applications to combustion and haemodynamical problems. Office: Royal Inst Tech, Teknikringen 8, Stockholm Sweden 100 44

FUCHS, OWEN GEORGE, chemist; b. Austin, Tex., June 22, 1951; s. Emil George and Hazel June (Johnson) F.; children from previous marriage: Ginny Lynn, William Oberholz, Owen George; m. Caroline S. Crook, Dec. 15, 1990. AA, Lee Jr. Coll., 1970, AS, 1973; BS, U. Houston, 1972. Chemist, Merichem Co., Houston, 1972-73; lab. mgr. Superintendence Co., Inc., Houston, 1973-78; dir. labs. and hydrocarbon research Chas. Martin Internat., Pasadena, Tex., 1978-79; pres., chief exec. officer Alpha-Omega Labs., Inc., Houston, Tex., 1979-88; bd. dirs. A.O.L. Inc., Houston, 1988—; pres., chief exec. officer Owen G. Fuchs & Assocs., Houston, 1988—, Texas City Testing Inc., 1989—, Environ. Testing Enterprises, Inc., 1991—, La. Testing Labs., Inc., 1992—. Mem. ASTM, Am. Chem. Soc. Home: PO Box 613 Highlands TX 77562-0613 Office: PO Box 3921 Texas City TX 77592-3921

FUCHS, ROLAND JOHN, geography educator, university administrator; b. Yonkers, N.Y., Jan. 15, 1933; s. Alois L. and Elizabeth (Weigand) F.; m. Gaynell Ruth McAuliffe, June 15, 1957; children: Peter K., Christopher K., Andrew K. BA., Columbia U., 1954, postgrad., 1956-57; postgrad., Moscow State U., 1960-61; MA, Clark U., 1957, PhD, 1959. Asst. prof. to prof. emeritus U. Hawaii, Honolulu, 1958—; chmn. dept. geography U. Hawaii, 1964-86, asst. dean to assoc. dean coll. arts and scis., 1965-67, dir. Asian Studies Lang. and Area Ctr., 1965-67, adj. rsch. assoc. East West Ctr., 1980—, spl. asst. to pres., 1986; vice rector UN U., Tokyo, 1987—; vis. prof. Clark U., 1963-64, Nat. Taiwan U., 1974; mem. bd. internat. orgns. and programs Nat. Acad. Scis., 1976-81, chmn., 1980-81, mem. bd. sci. and tech. in devel., 1980-85; mem. U.S. Nat. Commn. for Pacific Basin Econ. Coop., 1985-87; sr. advisor United Nations U., 1986. Author, editor: Geographical Perspectives on the Soviet Union, 1974, Theoretical Problems of Geography, 1977, Population Distribution Policies in Development Planning, 1981, Urbanization and Urban Policies in the Pacific-Asia Region, 1987; asst. editor Econ. Geography, 1963-64; mem. editorial adv. com. Soviet Geography: Review and Translation, 1966—, Geoforum, 1988—, African Urban Quar., 1987, Global Environmental Change, 1990. Ford Found. fellow, 1956-57; Fulbright Rsch. scholar, 1966-67. Mem. AAAS, Internat. Geog. Union (v.p. 1980-84, 1st v.p. 1984-88, pres. 1988-92, past pres. and v.p. 1992—), Assn. Am. Geographers (honors award 1982), Am. Assn. Advancement Slavic Studies (bd. dirs. 1976-81), Pacific Sci. Assn. (coun. 1978—, exec. com. 1986—, sec. gen., treas. 1991—). Home: 5136 Maunalani Cir Honolulu HI 96816-4020

FUCHS, SHELDON JAMES, plant engineer; b. Bklyn.; s. Louis edward and Matilda (Klar) F.; m. Myrna Faith Korchin, Jan. 30, 1956 (dec. 1987); children: Linda Fuchs Ivans, Laura Fuchs Schector. BCE, CCNY, 1950, MBA, 1956. Asst. plant engr. Kollsman Instrument Corp., Elmhurst, N.Y., 1954-61; maint. control supr. Republic Aviation, Farmingdale, N.Y., 1961-62; dir. maint. and constrn. Bklyn. Pub. Libr. System, 1962-64; supt. bldgs. and grounds Hofstra U., Hempstead, N.Y., 1964-78; plant/facilities mgr. Baldwin (N.Y.) Sch. Dist., 1978-86; cons. engr. S. Fuchs - Cons. Engr., Merrick, N.Y., 1986—. Contbg. author: Plant Engineers Manual and Guide, 1973, Encyclopedia of Professional Management, 1977; author/editor: Complete Building Equipment Maintenance Desk Book, 1982, 2nd edit. 1992, Supplement to Complete Building Equipment Maintenance Desk Book, 1993. Com. mem. Town of Hempstead Firematics, 1986-93, Nassau County Youth Bd., Mineola, N.Y., 1992-93;/ founder, chmn. Hofstra U. Plant Maint. Seminar and Exhibit. Recipient Merit citation County Exec.. (Nassau County), 1985, Nat. Educators award Old Timers Assn. of Oil Industry, 1983, Cert. of Appreciation, Hofstra U. Conf., 1974. Mem. Am. Inst. Plant Engrs. (chpt. pres. 1963-64, 64-65, 65-66), Nassau County Dirs. of Sch. Facilities and Mgmt. (pres. 1981-82). Home: 1705 James St Merrick NY 11566 Office: SUNY Coll of Tech Farmingdale NY 11735

FUENO, TAKAYUKI, chemistry educator; b. Osaka, Japan, Sept. 12, 1931; s. Takaichi and Chiyoko (Taniuchi) F.; m. Michiko Nonomura, Nov. 4, 1956; children: Yumiko, Hiroyuki. BA, Kyoto (Japan) U., 1953, MA, 1955, PhD, 1958. Instr. Kyoto U., 1958-63, assoc. prof., 1963-66; prof. of chemistry Osaka U., 1966—; vis. prof. Free U. Berlin, 1980-81. Recipient award Japan Chem. Soc., 1990. Home: Seiwadai-Nishi 4-4-60, Kawanishi, Hyogo 666-01, Japan Office: Osaka U, Toyonaka Osaka 560, Japan

FUENTEVILLA, MANUEL EDWARD, chemical engineer; b. Havana, Cuba, Feb. 17, 1923; s. Fernando and Edith Agnes (Pira) F.; B.Ch.E., Poly. Inst. Bklyn., 1947; M.S., Drexel U., 1954; m. May Belle Tutwiler, Oct. 18, 1945; children—William F., Diane G., Austin D., Eve J., Inez M. Sr. engr. Catalyic Inc., Phila., 1951-60; chief engr. Stokes Equipment div. Pennwalt Corp., 1960-67; asst. mgr. mfg. Esso Eastern, Tokyo, 1967-69, tech. supt., Okinawa, Japan, 1969-72; project mgr. Jacobs Engring. Co., Cherry Hill, N.J., 1972-75; project mgr. Stauffer Japan Ltd., Tokyo, 1975-77; dir. process devel. Alfa Laval Process, Mt. Laurel, N.J., 1977-79; tech. dir., sr. project mgr. Synergo Inc., Phila., 1979-82; chief mech. engring. Kling/Lindquist Inc., Phila., 1982—; pres. Cerus, Inc., Cherry Hill, N.J., 1986—; process and tech cons. pharm. and chem. applicators. Served with USNR, 1943-46. Mem. Am. Inst. Chem. Engrs., Soc. History of Tech., Phi Lambda Upsilon. Club: Cooper River Yacht (Collingswood). Patentee in indsl. processes. Home: 314 Tearose Ln Cherry Hill NJ 08003-3524

FUHRMANN, HORST, science administrator; b. Kreuzburg, Germany, June 22, 1926; s. Karl and Susanna F.; m. Ingrid Winkler-Lippoldt, 1954; 2 children. Dr.jur. h.c., U. Tübingen; Dr.phil. h.c., Bologna, Columbia, New York. Collaborator Monumenta Germaniae Historica, 1954-56; asst. Monumenta Germaniae Historica, Rome, 1957, asst., lectr., 1957-62; pres. Monumenta Germaniae Historica, Munich, 1971-93; prof. U. Tübingen, 1962-71, U. Regensburg, 1971-93; pres. Bavarian Acad. Humanities & Sci., Munich, 1992—. Author: The Donation of Constantine, 1968, Influence and Circulation of the Pseudoisidorian Forgeries (3 vols.), 1972-74, Germany in the High Middle Ages, 1978, From Petrus to John Paul II: The Papacy, 1980, Invitation to the Middle Ages, 1987, Far from Cultured People: An Upper Silesian Town around 1870, 1989Pour le Mérite: On Making Merit Visible, 1992. Recipient Premio Spoleto, 1962, Cultore di Roma, 1981, Upper Silesian Culture prize, 1989, Premio Ascoli Piceno, 1990, Ordre Pour le Mérite, Grosses Bundesverdienstkreuz mit Stern, Bayerische Verdienstorden. Office: Bavarian Academy of Sciences, Marstallplatz 8, Munich 22, Germany*

FUHS, G(EORG) WOLFGANG, medical research scientist; b. Cologne, Germany, May 19, 1932; came to U.S., 1964; s. Friedrich Karl and Lisette I. (Stayen) F.; children: Lisette Fuhs Mallary, H. Georg, Dagmar Fuhs Haswell. Diploma in biology, D in Nat. Scis., U. Bonn, Germany, 1956; postdoctoral, Tech. U. Delft, The Netherlands, 1956-57. Sci. employee dept. botany U. Frankfurt, Fed. Republic Germany, 1957-58, research assoc. dept. hygiene U. Bonn Sch. Medicine, 1958-63; fellow dept. genetics U. Cologne, 1963-64; sr., prin. research scientist div. labs. and research N.Y. State Dept. Health, Albany, 1964-72, dir. environ. health labs., 1973-85; chief div. labs. Calif. Dept. Health Services, Berkeley, 1985-89; rsch. scientist div. labs. Calif. Dept. Health Svcs., Berkeley, 1989—; vis. prof. U. Wis., Milw., 1973; rsch. assoc. U. Minn. Sch. Pub. Health, Mpls., 1970-71; adj. prof. dept. biology SUNY, Albany, 1984-86; mem. exptl. com. on human health effects of Great Lakes water quality U.S./Can. Internat. Joint Commn., 1978—; tech. adv. com. San Francisco Estuary Project, 1987-92; mem. Calif. Environ. Technol. Partnership, Calif. Comparative Risk Project, 1993—. Contbr. articles to profl. jours. (Inst. Sci. Info. award 1969); mem. editorial bd. Jour. Phycology, 1972-74, Limnology and Oceanography, 1973-76, Microbial Ecology, 1974-89. Mem. AAAS, Am. Soc. Microbiol. (past chmn. Eastern N.Y. br.), Internat. Assn. Theoretical Applied Limnology, Am. Pub. Health Assn., Water Pollution Control Fedn. Office: Calif-EPA Dept Toxic Substance Control Lab 2151 Berkeley Way Berkeley CA 94704-1011

FUJII, AKIRA, pharmacology educator; b. Ueda, Nagano, Japan, Oct. 29, 1942; s. Yunosuke and Kane (Kashino) F.; m. Yoko Maezawa, June 21, 1969; children: Nobumitsu, Utako. BS, Shinshu U., Nagano, Japan, 1965; MS, St. Thomas Inst., Cin., 1969, PhD, 1970; DDS, Nihon U., Tokyo, 1983. Postdoctoral fellow St. Thomas Inst., 1970-72, asst. prof., 1972-74, assoc. prof., 1974-76; asst. prof. Nihon U., 1976-91, assoc. prof. pharmacology, 1991—; cons. Sperti Drug Corp., Covington, Ky., 1973-76. Author: Experimental Method in Pharmacology, 1977, Systemic and Local Drugs in Dentistry, 1982. Mem. Am. Chem. Soc., Am. Pharm. Assn., N.Y. Acad. Sci., Japanese Pharmacological Soc. (councilor), Japanese Soc. Oral Therapeutics and Pharmacology (councilor), Internat. Assn. Dental Rsch., Internat. Conf. Calcium Regulating Hormones, Japanese Soc. Clin. Pharmacology and Therapeutics (councilor), Japanese Soc. Toxicological Sci. (councilor), Sigma Chi. Avocation: fishing. Home: 1465-5 Kamihongo, Matsudo Chiba 271, Japan Office: Nihon U Sch Dent at Matsudo, 2-870-1 Sakaecho-Nishi, Matsudo Chiba 271, Japan

FUJII, HIRONORI ALIGA, aerospace engineer, educator; b. Himeji, Hyogo, Japan, Apr. 6, 1944; s. Tokuichi and Chizuko (Ariga) F.; m. Naomi Matsumoto, Dec. 24, 1985; 1 child, Tomonori Fujii. M Engring., Kyoto (Japan) U., 1969, D Engring., 1975. Asst. prof., assoc., then prof. Tokyo Met. Coll. Tech., 1972-86; prof., chmn. dept. aero. engring. Tokyo Met. Inst. Tech., 1986—; adv. bd. Tokyo Met. Govt., 1991—; chmn. Japan Rsch. Group on Control of Space Structures, 1985—, Working Group for Space Robot Forum, Tokyo, 1988-90; mem. Com. for Utilization of Tokyo Area, 1991—. Author: Handbook of Aerospace Engineering, 1992; contbr. articles to profl. publs. Mem. Japan Soc. Japanese Chess. Achievements include research on mechanics in control analysis with application to aerospace engineering. Home: 1-24-9, Lamiere Fussa 409, Musashinodai, Fussa, Tokyo 197, Japan Office: Tokyo Met Inst Tech, 6-6 Asahigaoka, Hino Tokyo 191, Japan

FUJII, KOZO, engineering educator; b. Kofu, Yamanashi, Japan, Oct. 17, 1951; s. Yoshio and Tamie (Odagiri) F.; m. Yoshimi Hirai, July 8, 1974; children: Kotaro, Kenneth, Fujii. MS, U. Tokyo, 1977, PhD, 1980. Nat. Rsch. Coun. rsch. assoc. NASA Ames Rsch. Ctr., Moffett Field, Calif., 1981-83; rsch. scientist Nat. Aerospace Lab., Tokyo, 1984-86, sr. rsch. scientist, 1986-88; Nat. Rsch. Coun. sr. rsch. assoc. NASA Ames Rsch. Ctr., Moffett Field, Calif., 1986-87; assoc. prof. Inst. of Space and Astronautical Sci., Kanagawa, Japan, 1988—. Contbr. articles to AIAA Jour., Jour. of Aircraft. Mem. AIAA (Aerospace Image award 1987), Japan Soc. for Aeronautics and Astronautical Sci. (dir. for gen. affairs 1990-91), Japan Soc. for Mech. Engring. Office: Inst Space/Astronautical Sci, Yoshinodai 3-1-1, Sagamihara 229, Japan

FUJII, MASAYUKI, chemistry educator; b. Nishiwaki, Hyogo, Japan, Oct. 18, 1958; s. Masaichi and Kazuko Fujii; m. Emi Fujimoto, May 23, 1987; children: Shun, Eri. MS, Kyoto U., 1985, DSc, 1990. Rsch. asst. Kinki U., Higashiosaka, 1987; rsch. asst. Kinki U., Iizuka, 1987-90, lectr., 1990-92, asst. prof., 1992—. Contbr. articles to profl. jours. Jour. of Organic Chemistry, Jour. of the Chem. Soc., Jour. of Heterocyclic Chemistry, Bull. of the Chem. Soc. of Japan and Tetrahedron Letters. Grant for Scientific Rsch. Ministry of Edn., 1989, 91, Itoh Sic. Found., 1991, Ichikizaki Fund for Young Chemist, 1992, Sasagawa Scientific Rsch. grant Japan Sci. Soc., 1991. Mem. Chem. Soc. of Japan, Assn. of Synthetic Organic Chemistry (Promotion award 1991), Am. Chem. Soc. Achievements include rsch. on chemoselective reduction of olefins, photochemical desulfurization for ketone synthesis, stereoselective reductions with axially chiral coenzyme NAD(P)H model, synthesis of intelligent host molecule using axially chiral heteroaromatics. Office: Kinku U, 11-6 Kayanomori, Iizuka 820, Japan

FUJIME, YUKIHIRO, horticultural science educator, researcher; b. Hiroshima, Japan, Jan. 5, 1945; s. Shigeo and Shizue F.; m. Yasuko Monju, Apr. 29, 1970; children: Yuichi, Satoshi. BS, Kyoto (Japan) U., 1967, MS, 1969, PhD, 1982. Asst. prof. horticulture Kyoto U. Faculty of Agr., 1969-72; asst. prof. Kagawa (Japan) U. Faculty of Agr., 1972-84, assoc. prof., 1984-86, prof., 1986—; prof. Ehime (Japan) U. Faculty of Agr., 1988—; hort. expert Japan Internat. Coop. Agy., Nairobi, Kenya, 1985-86, 91; tech. advisor Tianjin (People's Republic of China) Acad. Agrl. Scis., 1988—;. Author: (book) Cauliflower & Broccoli, 1988. Expert in com. Kagawa Prefecture, Takamatsu, 1986—. Mem. Internat. Soc. Hort. Sci., Japanese Soc. Hort. Sci. (editor jour. 1991—, award 1983), Am. Soc. Hort. Sci. Avocations: travel, classical music, reading, painting. Office: Kagawa U Faculty Agr, Miki-cho kida-gun, Miki-lyo Kagawa 761-07, Japan

FUJIOKA, ROGER SADAO, research microbiology educator; b. Pearl City, Hawaii, May 11, 1938; s. Nobuichi and Hisayo (Iboshi) F.; m. Ruby Nanaye Yamashita, July 2, 1966; 1 child, Ryan Makoto. BS in Med. Tech., U. Hawaii, 1960, MS in Microbiology, 1966; PhD in Virology, U. Mich., 1970. Assoc. research U. Hawaii, Honolulu, 1963-66, research microbiologist, 1972—; predoctoral fellow U. Mich., Ann Arbor, 1966-70; postdoctoral fellow Baylor Coll. Medicine, Houston, 1970-71; sec.-treas. Hawaii Water Pollution Control, Honolulu, 1986—; mem. com. Standard Methods, Washington, 1986—. Contbr. articles to profl. jours. Served to capt. USAR, 1960-66. Grantee Sea Grant Coll. Program, 1976-84, Office Water United States Geol. Survey, 1978—, Office Naval Research, 1985—, Dept. Health, 1986—, all in Honolulu. Mem. AAAS, Am. Soc. Microbiology, Water Pollution Control Fedn., Am. Water Works Assn., Internat. Assn. Advancement Aquatic Mammals. Avocation: tennis. Office: U Hawaii Water Resources Research Ctr Holmes Hall 283 2540 Dole St Honolulu HI 96822-2333

FUJITA, EIICHI, educator emeritus; b. Osaka, Japan, Feb. 2, 1922; s. Shinji and Etsuko Fujita; m. Michiko Miyamoto Fujita, Nov. 17, 1953. B in Pharm. Sci., Kyoto U., 1943, PhD, 1952. Assoc. prof. U. Tokushima, 1951-54, prof., 1954-62; postdoctoral rsch. assoc. U. Wis., Madison, 1960-62; prof. Kyoto U., 1962-85, dir. Inst. for Chem. Rsch., 1982-84, prof. emeritus, 1985—; pres. Osaka U. of Pharm. Scis., Matsubara, 1985-91; councillor U. Coun., Kyoto U., 1982-84, Coun., Inst. for Molecular Sci., Okazaki, Aichi, 1987—. Contbr. articles to profl. jours. including Internat. Rev. of Sci., Alkaloids, Progress in the Chemistry of Organic Natural Products, Advances in Heterocyclic Chemistry, Jour. Chem. Soc., Jour. Am. Chem. Soc. Fulbright grantee, 1960-62. Mem. Pharm. Soc. of Japan (Scientific prize 1978), Chem. Soc. of Japan, Am. Chem. Soc., The Royal Soc. of Chemistry, Rotary Club of Kyoto-East. Achievements include rsch. on the chemistry of physiologically active natural products including alkaloids and terpenoids as well as organic synthesis especially asymmetric synthesis. Home: 5-52 Nanmeicho Fukakusa, Fushimi-ku, Kyoto 612, Japan

FUJITA, TETSUYA THEODORE, educator, meteorologist; b. Kitakyushu City, Japan, Oct. 23, 1920; came to U.S., 1953, naturalized, 1968; s. Tomojiro and Yoshie (Kanesue) F.; m. Sumiko Yamamoto, June 13, 1969; 1 son, Kazuya. B.S.Eq. in Mech. Engring, Meiji Coll. Tech., Kitakyushu City, 1943; Dr.Sci., Tokyo U., 1953. Asst. prof. Meiji Coll. Tech., Kitakyushu, 1943-49; asst. prof. Kyushu Inst. Tech., Kitakyushu, 1949-53; sr. meteorologist U. Chgo., 1953-62, assoc. prof., 1962-65, prof., 1965-89, Charles E. Merriam Disting. Svc. prof., 1989-90, Charles E. Merriam Disting. Svc. prof. emeritus, 1991—. Recipient Kamura award Kyushu Inst. Tech., 1965, Aviation Week and Space Tech. Disting. Svc. award Flight Safety Found., 1977, Adm. Luis de Florez Flight Safety award, 1977, Ann. award Nat. Weather Assn., 1978, Disting. Pub. Svc. award NASA, 1979, Losey Atmospheric Sci. award AIAA, 1982, 25th yr. Weather Satellite medal Dept. Commerce, 1985, Vermeil Gold medal French Nat. Acad. Air and Space, 1989, Second Order of the Sacred Treasure, Gold and Silver Star Govt. Japan, 1991. Mem. Am. Meteorol. Soc. (Meisenger award 1967, Applied Meteorology award 1988), Japan Meteorol. Soc. (Okada award 1959, Fujiwara award 1990). Specialized research on tornadoes, microburst-related aircraft accidents and satellite meteorology. Home: 5727 S Maryland Ave Chicago IL 60637-1425 Office: U Chicago Dept Geophys Scis 5734 S Ellis Ave Chicago IL 60637-1472

FUJITA, TSUNEO, systems analysis educator; b. Hiroshima, Japan, May 26, 1933; s. Tatsuo and Sakae (Takemura) F.; m. Sachiko Fujita, Apr. 29, 1961; children: Masami, Junko. BEngring., Tokyo U., 1956; MS, Rensselaer Poly. Inst., 1971; DCS, Chuo U., 1989. Mining engr. Hokkaido Tanko-Kisen Co., Japan, 1956-65; researcher Sanno Inst. Bus. Adminstrn., Tokyo, 1965-72; asst. prof., then prof. Sanno Jr. Coll., Tokyo, 1972-80; prof. Sanno Coll., Isehara, Japan, 1980—; dean, faculty mgmt. and informatics Sanno Coll., 1989—; asst. mgr. systems devel. lab., Sanno Inst., Tokyo, 1973-80, mgr. regional scis. lab., Isehara, 1980-84; paper referee, All Japan Fedn. Mgmt. Orgns., Tokyo, 1976—. Author: Elementary Systems Analysis, 1975, 2d edit., 1983, Organization Systems Theory, 1989, Introductory Decision Analysis (with others), 1989. Chmn. Adminstrv. Reform Coun. Isehara, 1985-88; councilor, Comprehensive City Planning Coun. Isehara, 1986—. Mem. Japan Indsl. Mgmt. Assn. (councilor), Ops. Rsch. Soc. Japan, Japan Soc. Planning and Adminstrn., Project Mgmt. Inst., U.S. Japan Soc. Ednl. Info. Scis. (dir.), Decision Sci. Inst. U.S., Japan Soc. Mgmt. Info. (dir.). Avocations: Noh plays, industrial archaeology, swimming. Home: 56-22 Higashi-kamigo, Yokohama 247, Japan Office: Sanno Coll, 1573 Kamikasuya, Isehara 259-11, Japan

FUJITANI, MARTIN TOMIO, software quality engineer; b. Sanger, Calif., May 3, 1968; s. Matsuo and Hasuko Fujitani. BS in Indsl. and Systems Engring., U. So. Calif., 1990. Sec. Kelly Svcs., Inc., Sacramento, 1987; receptionist Coudert Bros., L.A., 1988; rsch. asst. U. So. Calif., L.A., 1988-89; math. aide Navy Pers. Rsch. and Devel. Ctr., San Diego, 1989; quality assurance test technician Retix, Santa Monica, Calif., 1989-90; software engr. Quality Med. Adjudication, Inc., Rancho Cordova, Calif., 1990-92; test engr. Worldtalk Corp., Los Gatos, Calif., 1993—. Assemblyman Am. Legion Calif. Boys State, 1985. Recipient Service Above Self award East Sacramento Rotary, 1986. Mem. Ops. Rsch. Soc. Am. (assoc.), Sacramento Sr. Young Buddhist Assn. (treas. 1990-91), Gen. Alumni Assn. U. So. Calif. (life). Avocations: studying Japanese, watching ballet, listening to jazz music. Home: 205 Milbrae Ln Apt 2 Los Gatos CA 95030 Office: Worldtalk Corp 475 Alberto Way Los Gatos CA 95032

FUKAI, YUH, physics educator; b. Chiba, Japan, Nov. 1, 1934; s. Eiichi and Shima (Iida) F.; m. Reiko Nakakimura, Jan. 9, 1960; children: Sumi, Moto. BA, U. Tokyo, 1958, DSc, 1963. Asst. prof. Chuo U., Tokyo, 1963-71, prof., 1971—, dean Grad. Sch. of Sci. and Engring., 1991—; assoc. prof. U. Grenoble, France, 1980; vis. prof. U. Ill., Urbana, 1967-69, Hiroshima U., 1978-79, Niigata U., 1980-81, Tohoku U., Sendai/Miyagi, 1981-82, Nagoya Inst. Tech., 1985, Inst. Study of Earth's Interior, Misasa/Tottori, 1985-86, 88, Hokkaido U., 1990, U. of the Air, 1991—. Editor-in-cheif: Japanese Jour. of Applied Physics, Tokyo, 1985-87; author: Physics of Diffusion Phenomena, 1988, The Metal-Hydrogen System, 1993. Mem. Phys. Soc. Japan, Japan Inst. Metals, Japan Soc. Applied Physics, Japan Soc. High Pressure Sci. and Engring. Home: 897-7 Mogusa, Hino 191, Japan Office: Chuo Univ Dept Physics, 1-13-27 Kasuga Bunyo-ku, Tokyo 112, Japan

FUKUDA, ICHIRO, astronomer, researcher; b. Kumamoto, Japan, July 7, 1940; s. Katsuki and Toki (Uesaka) F.; m. Tomoko Koishi. DSc, Kyoto U., 1982. Lectr. Kanazawa Inst. of Tech., Ishikawa, Japan, 1970-74; assoc. prof. Kanazawa Inst. of Tech., Ishikawa, 1974-93, prof., 1993—. Author: Lectures in Modern Astronomy, 1981; contbr. Astronomical Herald Japan, Publs. of Astron. Soc. of Pacific, Memoirs of Kanazawa Inst. of Tech., 1989. Mem. Internat. Astron. Union, Am. Astron. Soc., Astron. Soc. of the Pacific, Astron. Soc. Japan. Home: 1-108 Awada, Nonoichi 921, Japan Office: Kanazawa Inst Tech Dept Physics, 7-1 Ogigaoka, Nonoichi Ishikawa 921, Japan

FUKUDA, MORIMICHI, medical educator; b. Sapporo, Hokkaido, Japan, June 20, 1929; s. Yoneichi and Kiyo (Kikuchi) F.; m. Morimichi Ryoko, Nov. 1957; children: Hiroyuku, Toru. MD, Hokkaido U., Sapporo, Japan, 1953. Lectr., asst. prof. Sapporo Med. Coll., 1955-58, assoc. prof., 1968-80, assoc. prof. medicine, 1980-84, prof., chmn., 1984—; rsch. immunochemist UCLA, 1958-68; rsch. immunochemist UCLA, 1958-60. Author, editor Ultrasonic Differential Diagnosis of Tumors, 1984, Ultrasound Diagnosis (Japanese) 1989. Mem. Japanese Soc. Ultrasonics Medicine (councilor 1979-85), World Fedn. Ultrasound Med. Biology (councilor 1982-85, v.p. 1985-88, pres.-elect 1988-91, pres. 1991—). Home: S-17 W-6 Chou-ku, Sapporo Hokkaido 060, Japan Office: Sapporo Med Coll, S-1 W-16 Chuo-ku, Sapporo Hokkaido 060, Japan

FUKUDA, STEVEN KEN, materials science educator; b. Honolulu, Sept. 14, 1952; s. Kenneth Masaru and Naoe (Toyofuku) F. BS in Engring., Case Western Res., 1974; MS in Nuclear Engring., U. Md., 1976; PhD in Material Sci., U. Washington, 1988. Engr. Westinghouse-Handford, Richland, Wash., 1976-78, 80-82, EG&G Idaho, Idaho Falls, 1979-80; asst. prof. N.Y. State Coll. Ceramics Alfred (N.Y.) U., 1988—. Mem. Am. Phys. Soc., Am. Chem. Soc., Am. Soc. Engring. Educators, Am. Ceramic Soc. Office: Alfred U N Y State Coll Ceramics Alfred NY 14802

FUKUI, GEORGE MASAAKI, microbiology consultant; b. San Francisco, May 25, 1921; s. Tsunejiro and Kimiko (Wada) F.; m. Yuri Lillyn Kenmotsu, Sept. 23, 1944; children: Lisa Jo, Tenley Kay. BS, U. Conn., 1945, MS, 1948; PhD, Cornell U., 1952. Instr. bacteriology U. Conn., Storrs, 1948-49; lab. instr. Cornell U., Ithaca, N.Y., 1949-52, mem. adv. bd. microbiology dept., 1985-88; asst. br. chief U.S. Army, Frederick, Md., 1952-60; dir. microbiology and immunology Wallace Labs., Cranbury, N.J., 1960-77, Hazelton Labs., Vienna, Va., 1977-78; dir. microbiology Abbott Labs., North Chicago, Ill., 1978-79; rsch. microbiologist Abbott Labs., Irving, Tex., 1979-86; pres. Internat. Cons. in Microbiology, Irving, 1986—. Contbr. articles to sci. jours. Asst. scoutmaster troop 712, Boy Scouts Am., Topaz, Utah, 1943; recruiter Cornell U., Princeton, N.J., 1964-69. With U.S. Army, 1945-46. Recipient commendation Rsch. Soc. Am., 1959, medal for sci. achievement Hiroshima (Japan) U., 1973, Gran Amigo de Mex. commendation Nat. U. Mex., 1982, commendation Tohoku U., Sendai, Japan, 1983. Fellow Am. Acad. Microbiology (charter, diplomate); mem. Am. Soc. for Microbiology, Rutgers Soc. Japan (hon.), Phi Beta Kappa, Sigma Xi. Republican. Episcopalian. Achievements include patents for Non-Allergenic Penicillin; Phenoxypropanediols on Reduction of Penicillin Allergy; Synthetic Penicillin, Non-Allergenic; Suppression of Histamine Release; Salicylates for Quantification of Antibiotics in Sera. Home and Office: 3813 E Greenhills Ct Irving TX 75038-4819

FUKUI, HATSUAKI, electrical engineer, art historian; b. Yokohama, Japan, Dec. 14, 1927; came to U.S., 1962, naturalized, 1973; s. Ushinosuke and Yoshi (Saito) F.; m. Atsuko Inamoto, Apr. 1, 1954 (dec. 1973); children: Mayumi, Naoki; m. Kiku Kato, Dec. 12, 1975. Diploma, Miyakojima Tech. Coll. (now Osaka City U.), 1949; BS, Tokyo Coll. Sci.; D.Eng., Osaka U., 1961. Rsch. assoc. Osaka City U., 1949-54; engr. Shimada Phys. and Chem. Indsl. Co., Tokyo, 1954-55; sr. engr. to mgr. Sony Corp. semi-condr. div., Tokyo, 1955-61; mgr. engring. div. Sony Corp., 1961-62; mem. tech. staff Bell Telephone Labs., Murray Hill, N.J., 1962-69, supr., 1969-73; v.p. Sony Corp. Am., N.Y.C., 1973; asst. to chmn. Sony Corp., Tokyo, 1973; staff mem. Bell Labs., Murray Hill, N.J., 1973-81; supr. Bell Labs., 1981-83, AT&T Bell Labs., 1984-89; lectr. Tokyo Met. U. (part-time), 1962. Author: Esaki Diodes, 1963, Solid-State FM Receivers, 1968; contbr. to: Semiconductors Handbook, 1963, GaAs FET Principles and Technologies, 1982; editor: Low-Noise Microwave Transistors and Amplifiers, 1981; contbr. articles to profl. jours.; patentee in field. Fellow IEEE (standardization com. 1976-82, editorial bd. IEEE Transactions on Microwave Theory and Techniques 1980—, com. on U.S. competitiveness 1988—); mem. Inst. Electronics, Info. and Communication Engrs. Japan (Inada award 1959), IEEE Communications Soc., IEEE Electron Devices Soc., IEEE Lasers and Electro-Optics Soc., IEEE Microwave Theory and Techniques Soc. (Microwave prize 1980, Pioneer award 1990), Electromagnetics Acad., Japan Soc. Applied Physics, Inst. TV Engrs. Japan (tech. steering com. 1973-74), Am. Assn. Museums, Gakushikai, Internat. House Japan. Home: 53 Drum Hill Dr Summit NJ 07901-3141 also: 1-21-16-802 Nakane Meguro, Tokyo 152, Japan

FUKUI, KENICHI, chemist; b. Nara, Japan, Oct. 4, 1918; s. Ryokichi and Chie Fukui; m. Tomoe Horie, 1947; 2 children. Student, Kyoto Imperial U. Researcher synthetic fuel chemistry Army Fuel Lab., 1941-45; lectr. in fuel chemistry Kyoto Imperial U., 1943-45; asst. prof. Kyoto U., 1945-51, prof., 1951-82; pres. Kyoto Inst. Tech. (formerly Kyoto U. Indsl. Arts and Textile Fibres), 1982-88; dir. Inst. for Fundamental Chemistry, 1988—; councillor Kyoto U., 1970-73, dean faculty engring., 1971-73 ; chemist U.S.-Japan Eminent Scientist Exchange Programme, 1973; counselor Inst. Molecular Sci., 1976—;. Contbr. articles to profl. jours. Chmn. exec. com. 3d Internat. Congress Quantum Chemistry, Kyoto, 1979. Sr. Fgn. Scientist fellow NSF, 1970; fgn. assoc. NAS; recipient Japan Acad. medal, 1962, Nobel Prize for chemistry, 1981, Order of Culture award, 1981; named Person of Cultural Merits, 1981. Mem. Am. Acad. Arts, Scis. and Humanities, Japan Acad., Pontifical Acad. Scis., Chem. Soc. Japan (v.p. 1978-79, pres. 1983-84), Royal Soc. (London, fgn. mem.), Royal Instn. Gt. Britain (hon.), Academia Europaea (fgn. mem.). Home: 23 Kitashirakawa-Hiraicho, Sakyo-ku, Kyoto 606, Japan Office: Inst Fundamental Chemistry, 34-4 Takano-Nishihirakicho, Sakyo-ku, Kyoto 606, Japan

FUKUI, YASUO, astronomer; b. Osaka, Japan, Sept. 29, 1951; s. Takeshi and Takako (Nishi) F.; m. Yuriko Saito, Mar. 2, 1974; children: Eriko, Ririko. MS, U. Tokyo, 1976, DSc, 1979. Rsch. assoc. Nagoya (Japan) Univ., 1980-87, assoc. prof., 1987—; Humboldt fellow Cologne Univ., 1984-85. Author: High Energy Astrophysics, 1990; contbr. articles to Nature, Astrophys. Jour. Recipient Vainu Bappu Meml. Gold medal Astron. Soc. of India, 1987, Inoue Sci. award Inoue Found. for Sci., Tokyo, 1991, Grant-in-Aid for Specially Promoted Rsch., Ministry of Edn., Sci. and Culture, Tokyo, 1989. Mem. Internat. Astron. Union Paris, Astron. Soc. Japan (councillor 1990—), Am. Astron. Soc. Achievements include development of the 4m millimeter-wave telescope facility equipped with the most sensitive superconducting receiver at 3mm wavelength, at Nagoya Univ., and its usage for an extensive survey of star formation regions in the galaxy. Home: Higashisakura 2-18-24-1105, Naka-ku, Nagoya 460, Japan Office: Nagoya Univ Dept Astrophysics, Chikusa-ku, Nagoya 464-01, Japan

FUKUI, YASUYUKI, psychology educator; b. Kyoto, Japan, July 8, 1934; s. Terutaro and Umeno (Okumura) F.; m. Terue Nakamura, Apr. 30, 1961; children: Miyuki, Hitoshi, Kaori. BS, Osaka (Japan) City U., 1959; MS, Kyoto (Japan) U., 1967, postgrad., 1967-69; researcher, Gestalt Inst., L.A., 1986. Student asst. Shinshyu U., Matsumoto, Nagano, Japan, 1959-62, Osaka (Japan) Ednl. Coll., 1962-65; tchr. Osaka High Sch., 1965-69; counselor Kanazawa U., Kanazawa, Ishikawa, Japan, 1969-76; assoc. prof. Ehime U., Matsuyama, Ehime, Japan, 1976-81; prof. Ehime U., Matsuyama, Ehime, 1981—; counselor Shikoku Postal Svc. Matsuyama, Ehime, 1989—; dir. Shikoku Counseling Ctr., Matsuyama, Ehime, 1991—. Author: Anxiety and Growth in Adolescence, 1980, Psychology of Gazing, 1984, Psychology of Emotions, 1990; editor: Development of Group Approach, 1981; translator: Counseling Workbook, 1992. Com. mem. Japan Life Line Fedn., Tokyo, 1989—. Mem. Japanese Soc. Psychologist, Japanese Soc. Personality Psychology, Japanese Soc. Rsch. on Emotions, Japan Clin. Psychologist Assn. Avocations: fishing, badminton, tennis, mystery books. Home: 13-11 Tani Machi, Matsuyama Ehime 791, Japan Office: Ehime U, 3 Bunkyo-Chyo, Matsuyama Ehime 790, Japan

FUKUMOTO, NEAL SUSUMU, civil engineer; b. Honolulu, Oct. 11, 1958; s. Harold Tatsumi and Mary Umeyo (Narahara) F. BS in Civil Engring., U. Hawaii, 1980. Registered profl. engr., Hawaii. Engr. Charles Pankow Assocs., Honolulu, 1981-84, S&M Sakamoto, Honolulu, 1984; civil engr., project mgr. Kennedy/Jenks Cons., Honolulu, 1984-91; project mgr., asst. office mgr. Parametrix, Inc., Honolulu, 1991—. Mem. ASCE (assoc. mem.), U. Hawaii Engring. Alumni (dir. 1988—), Chi Epsilon Alumni Assn. (pres. 1984-88). Office: Parametrix Inc Ste 1600 1164 Bishop St Honolulu HI 96813

FUKUYAMA, TOHRU, organic chemistry educator; b. Anjo, Japan, Aug. 9, 1948; married; 3 children. BA, Nagoya U., 1971, MA, 1973; PhD in Chemistry, Harvard U., 1977. Fellow organic chemistry Harvard U., 1977-78; from asst. prof. to assoc. prof. Rice U., Houston, 1978-88, prof. organic chemistry, 1988—. Recipient Disting. Investigator award Abbott Labs. pharm. divsn., 1993. Mem. Am. Chem. Soc. (Arthur C. Cope Scholar award 1993). Achievements include research in total synthesis of complex natural products of biological importance. Office: Rice University PO Box 1892 6100 South Main Houston TX 77251*

FUKUYAMA, YUKIO, child neurologist, pediatrics educator; b. Takachihomachi, Miyazaki, Japan, May 28, 1928; s. Masaharu and Kiku Fukuyama; m. Ayako Arai, Nov. 6, 1954. MD, U. Tokyo, 1952, postgrad., 1953-56, PhD, 1959. Intern U. Tokyo Hosp., 1952-53; asst. prof. pediatrics U. Tokyo Faculty Medicine, 1960-64, assoc. prof., 1964-65; dir. div. neurology Nat. Children's Hosp., Tokyo, 1965-67; prof. pediatrics Tokyo Women's Med. Coll., 1967—, chmn. dept., 1967—. Editor: (monographs) Epilepsy Bibliography, 6th edit., 1989, Child Neurology Atlas, 1986, Modern Perspectives of Child Neurology, 1991, Fetal and Neonatal Neurology, 1992; editor-in-chief No To Hattatsu, 1969-87, Brain and Devel., 1979—. Mem. Internat. Child Neurology Assn. (pres. 1982-86, v.p. 1986-90), Asian and Oceanian Assn. Child Neurology (pres. 1983-90), Japanese Soc. Child Neurology (chmn. bd. trustees 1968—), Am. Neurol. Assn. (corr.), Child Neurology Soc. (hon.), Am. Acad. Neurology (hon.), Canadian Child Neurology Soc. (hon.), Czechoslovakian Neurol. Soc. (hon.). Avocation: philately. Home: 6-12-16 Minami-Shinagawa, Shinagawa-ku, Tokyo 140, Japan Office: Tokyo Women's Med Coll Pd, 8-l Kawadacho Shinjuku-ku, Tokyo 162, Japan

FUKUZUMI, NAOYOSHI (HAI-CHIN CHEN), pathology educator; b. Taipei, China, Aug. 7, 1924; arrived in Japan, 1972; s. Chin-Sun and Chen-Lin (Tien) Chen; m. Noriko Fukuzumi, Nov. 15, 1960; children: Yayoi, Eiko. MD, Nat. Taiwan U., Taipei, 1947; PhD, Chiba U., Chiba, Japan, 1955. Intern Nat. Taiwan U., Taipei, 1947-48, resident in pathology, 1948-53, Prof. pathology coll. medicine, 1964-72; cons. in pathology U.S. Naval Med. Rsch. Unit, Taipei, San Francisco, 1967-72; rsch. fellow inst. for rsch. med. sci. Tokyo U., 1972-78, Tokyo Cancer Rsch. Inst., 1978-79; prof. pathology, surg. pathology, clin. pathology Coll. Medicine, Kyorin U., Mitaka, Japan, 1979—; Contbr. articles to profl. jours. Panelist, IX Internat. Congress Pediatrics, Tokyo, 1965, internat. symposium on nasopharyngeal carcinoma WHO, Kyoto, 1977; mem. Internat. Symposium on Human Tumor Virology and Immunology, Tokyo, 1970; hon. v.p. Internat. Congress Histochemistry and Cytochemistry, Kyoto, 1972. Rsch. fellow coll. physicians and surgeons Columbia U., 1957-59, Armed Forces Inst. Pathology, Washington, 1959. Fellow Internat. Coll. Surgeons; mem. AAAS, Japanese Path. Soc., Japanese Cancer Assn., Japan Soc. for Cancer Therapy, N.Y. Acad. Scis. Office: Kyorin U Coll Medicine, Shinkawa 6 20 2, Mitaka 181, Japan

FULBRIGHT, DENNIS WAYNE, plant pathologist, educator; b. Lynwood, Calif., Aug. 20, 1952; s. Ross Edward and Beulah Mae (Hollingshead) F.; m. Joanne Shutt, Jan. 7, 1978 (div. May 1992); children: Kimberly Ann, Scott Paul; m. Jane Schneider, June 19, 1993. AB, Whittier Coll., 1974; PhD, U. Calif., Riverside, 1979. Asst. prof. Mich. State U., East Lansing, 1979-85, assoc. prof., 1985-90, prof., 1990—; bd. dirs. Am. Chestnut Found., Morgantown, W.Va., 1985—. Contbr. over 25 articles to profl. jours. Office: Mich State U Dept Botany/Plant Pathology East Lansing MI 48824-1312

FULDA, MICHAEL, political science educator, space policy researcher; b. Liverpool, Eng., Apr. 21, 1939; came to U.S., 1962, naturalized, 1966; s. Boris and Catherine (Von Dehn) F.; m. Rosa Bongiorno, July 19, 1970; children: Robert, George. Student Polytechnique, Grenoble, France, 1956-57, Tech. U., West Berlin, Germany, 1957-58, Karl Eberhardt U., Tubingen, Germany, 1963-66; MA, Am. U., 1968, PhD in Internat. Studies, 1970. Prof. polit. sci. Fairmont State Coll., W.Va., 1971—; internat. rls. specialist NASA, Washington, 1979; fellow NASA Marshall Ctr., Huntsville, Ala., summer 1977, Langley Ctr., Hampton, Va., 1976. Author: Oil and International Relations, 1979; (with others) United States Space Policy, 1985. Contbr. articles to profl. jours. Bd. dirs. Fairmont Chamber Music Soc., 1983—; W.Va. state com. chmn., dir. space policy Nat. Unity Campaign for John Anderson, 1980; mem. nat. adv. com. John Glenn Presdl. Com., 1984, space policy group Dukakis/Bentsen Com., 1988. With U.S. Army, 1962-66. Woodrow Wilson Found. fellow, 1969-70; Humanities Found. W.Va. grantee, 1978-80, NASA W.Va. Space Grant Consortium grantee, 1991—. Fellow AIAA (assoc.); mem. Am. Astronautical Soc., Nat. Space Soc. (dir. 1991—), World Future Soc. (pres. W.Va. chpt. 1977-80), Nat. Space Club, Inst. for the Social Sci. Study of Space (pres. 1988—). Home: 1 Timothy Ln Fairmont WV 26554-1331

FULERO, SOLOMON M., psychologist, educator, lawyer; b. Norfolk, Va., Oct. 18, 1950; s. Leon and Fani Fulero; m. Kris Blom, 1973 (div. 1977); children: Joshua, Asher; m. Lynda Olsen, Jan. 1, 1982; 1 child, David. JD, U. Oreg., 1979, PhD, 1979. Lic. psychologist, Ohio. Pvt. practice law Dayton, Ohio, 1980—; prof. psychology Sinclair Coll., Dayton, 1980—; pvt. practice psychology Dayton, 1986—; clin. dir. Ctr. for Forensic Psychiatry, Hamilton, Ohio, 1992—. Author: Ohio Law and Psychology Handbook, 1988; contbr. articles to profl. jours. Mem. APA, Am. Psychol. Soc., Am. Psychology-Law Soc., Ohio State Bar Assn. Office: Ctr for Forensic Psychiatry 222 High St Ste 225 Hamilton OH 45011

FULLER, KATHRYN SCOTT, environmental association executive, lawyer; b. N.Y.C., July 8, 1946; d. Delbert Orison and Carol Scott (Gilbert) F.; m. Stephen Paul Doyle, May 29, 1977; children: Sarah Elizabeth Taylor, Michael Stephen Doyle, Matthew Scott Doyle. BA English, Am. Lit., Brown U., 1968, LHD (hon.), 1992; JD with honors, U. Tex., 1976; postgrad., U. Md., 1980-82; DSci. (hon.), Wheaton Coll., 1990; LLD (hon.), Knox Coll., 1992. Bar: Tex. 1977, D.C. 1979. Rsch. asst. Yale U., New Haven, Conn., 1968-69, Am. Chem. Soc., 1970-71, Harvard U. Mus. Comparative Zoology, Cambridge, Mass., 1971-73; law clerk Dewey, Ballantine, Bushby, Palmer & Wood and Vinson & Elkins, N.Y.C., Houston, 1974-76, U.S. Dist. Ct. (so. dist.), Tex., 1976-77; atty., advisor Office Legal Counsel Dept. Justice, Washington, 1977-79, atty. Wildlife and Marine Resources sect., 1979-80, chief Wildlife and Marine Resources sect., 1981-82; exec. v.p., dir. Traffic USA, pub. policy, gen. counsel World Wildlife Fund, Washington, 1982-89, pres., CEO, 1989—. Contbr. articles to profl. jours. Adv. com. Trade Policy and Negotiations; Pres'. Commn. Environ. Quality; bd. dirs. Brown U. Recipient William Rogers Outstanding Grad. award Brown

U., 1990, UNEP Global 500 award, 1990; outstanding woman law student Tex. scholar, 1975. Mem. State Tex. Bar, D.C. Bar (coun. fgn. rels., internat. coun. environ. law, overseas devel. coun.), Zonta Internat. (hon.). Avocations: squash, trekking, scuba diving, gardening, fishing. Office: World Wildlife Fund 1250 24th St NW Washington DC 20037

FULLER, RICHARD MILTON, physics educator; b. Crawfordsville, Ind., July 23, 1933; s. Harold Q. and Charlotte Mae (Gohl) F.; m. Judith Wheaton, Aug. 28, 1955; children: Cynthia, Christopher, Janet. BA, DePaul U., 1955; MA, U. Minn., 1960; PhD, Mich. State U., 1965. From instr. to asst. prof. Alma (Mich.) Coll., 1958-68; from assoc. prof. to prof. Gustavus Adolphus Coll., St. Peter, Minn., 1968—; vis. prof. U. Mo., Rolla, 1966, 67, Howline U., St. Paul, 1982, 85, Jilin U. Tech., Chaugchun, People's Rep. China, 1988-89. Author: (text book) Physics: Including Human Applications, 1978, (lab. book), 1978; contbr. articles to profl. jours. Recipient Edgar M. Carlson award for Teaching, 1971. Mem. AAAS, Am. Assn. Physics Tchrs., Am. Physical Soc., Fedn. Am. Scientists, Sigma Xi. Methodist. Home: 723 Upper Johnson Cir Saint Peter MN 56082 Office: Gustavus Adolphus Coll Saint Peter MN 56082

FULLERTON, JESSE WILSON, applications specialist; b. Norfolk, Va., Apr. 21, 1947; s. James A. and Betty (James) F.; m. Susan Richter, Aug. 21, 1970; children: Jennifer Anne, Kimberly Marie. AA in Indsl. Engring., Del. Tech. Coll., 1981; BS in Mktg., Goldey Beacom Coll., 1985. Pressman W.Va. Pulp and Paper, Newark, Del., 1965-66; rsch. asst. E.I. DuPont de Nemours, Wilmington, Del., 1967-78; specialist ICI Americas, Wilmington, 1978-88; applications coord. ICI Americas, Wilmington, Del., 1988—. Contbr. to profl. publs. Vice pres. Talleyville Softball Assn., 1985-88, mgr., 1988—. mem. ASTM, Am. Soc. Testing Methods, Soc. Plastics Engrs., Soc. Quality Control. Democrat. Methodist. Achievements include U.S. and foreign patents in polymer additives. Home: Pierson Farm 3347 Pierson Dr Wilmington DE 19810 Office: ICI Americas Rt 202 Murphy Rd Wilmington DE 19898

FULMER, KEVIN MICHAEL, environmental engineer; b. Birmingham, Ala., Sept. 14, 1968; s. Cecil Harold and Pauline Fulmer. BSCE, U. Ala., Tuscaloosa, 1991. Cert. engr. in tng. Environ. engr. Ala. Dep. Environ. Mgmt., Montgomery, 1991—. Named Eagle Scout, Boy Scouts Am. Mem. ASCE (assoc.). Office: Ala Dept Environ Mgmt 1751 Cong Dickinson Dr Montgomery AL 36130

FULTON, ALICE BORDWELL, biochemist, educator; b. Nyack, N.Y., July 26, 1952; d. Paul Dent and Charlotte Joyce (Pluckhahn) Bordwell; m. Thomas Harald Haugen, Dec. 26, 1982; children: Frances Bordwell, Peter Thomas. BSc magna cum laude, Brown U., 1973, PhD, 1977. Asst. prof. U. Iowa, Iowa City, 1981-87, assoc. prof., 1987-93, prof., 1993—. Author: The Cytoskeleton: Cellular Architecture and Choreography, 1984; contbr. numerous articles and abstracts to profl. jours. Predoctoral fellow NSF, 1974, postdoctoral fellow, 1977; postdoctoral fellow Muscular Dystrophy Assn., 1978. Mem. AAAS, Am. Soc. for Cell Biology. Achievements include discovery of cotranslational assembly of cytoskeletal proteins in vivo. Home: 1483 Grand Ave Iowa City IA 52246 Office: U Iowa Coll Medicine 4-170 Bowen Sci Bldg Iowa City IA 52242-1109

FULTON, DARRELL NELSON, infosystems specialist; b. Urbana, Ill., Nov. 11, 1946; s. Arthur Nelson and Mabel Rose (Felix) F.; m. Janet Marie Arndt, Dec. 28, 1968; children: Michael Nelson, Kevin James, Steven Lloyd. BBA, U. Iowa, 1968; MS, Air Force Inst. Tech., 1969; MA, U. Pa., 1975, PhD, 1975. CPA, Iowa. Commd. 2d lt. USAF, 1968; logistics officer Air Force Inst. Tech., Wright-Patterson AFB, Ohio, 1968-69; advanced through ranks to maj. Air Force Inst. Tech., 1978; logistics programmer U.S. Air Force Security Svc., Kelly AFB, Tex., 1969-72; assoc. prof. Air Force Inst. Tech., Wright-Patterson AFB, 1975-78, dep. dir. resource mgmt., 1978-80; computer systems mgr. C.H. Dean & Assocs., Dayton, Ohio, 1980-84; assoc. prof. acctg. U. Dayton, 1984-87; dir. info. systems C.H. Dean & Assocs., 1987-89, asst. v.p. info. systems, 1989-92; v.p. C.H. Dean & Assocs., Dayton, 1992—; cons. D.N. Fulton Cons., Huber Heights, Ohio, 1975-80, 1984-87. Author: Defense Resource Management Systems, 1976-78; contbr. articles to The Basic-2 Report. Treas. Huber Heights Amateur Baseball Assn.; chief umpire Huber Heights Umpire Assn.; umpire Amateur Softball Assn.; commr. Huber Heights Amateur Baseball; mgr. Huber Heights Little League. Decorated AF Commendation medal, 1972; recipient AF Inst. Tech. Pride in Excellence award, 1969, 71, AF Inst. of Tech. Silver Scroll award, 1976, 77. Mem. Macintosh Computer Users Group, Am. Acctg. Assn., Rotary, Am. Legion, Beta Alpha Psi. Avocations: bowling, softball, baseball. Office: CH Dean & Assocs 2480 Kettering Tower Dayton OH 45423

FUNABA, MASATOMI, political economy educator; b. Wakayama-Ken, Japan, Feb. 20, 1938; s. Masao and Fujiko (Taki) F.; m. Kimiko Komuro, Oct. 7, 1968; children: Hisamichi, Chisumi, Yuki. BA, Kyoto U., 1960, LittM, 1962, M in Econs., 1966, D in Econs., 1974. Assoc. prof. Ryukoku U., Kyoto, Japan, 1969-76; assoc. prof. Hiroshima (Japan) U., 1976-80, prof., 1980-92, prof. Grad. Sch. Regional Devel. and Environ. Study, 1978-92; prof. Kobe (Japan) U. of Commerce, 1992—; mem. Com. Local Fin., Osaka Prefecture, Japan, 1975—, Com. Housing, Osaka, 1980—, Com. Indsl. Policy, Ministry of Internat. Trade and Industry of Japan at Hiroshima, 1980; chmn. Com. on Taxing, Kyoto-shi, Japan, 1976—; vis. prof. Boston U., 1989, U. Wis., Madison, 1992. Author: History of British Public Credit, 1971, Local Finance in Japan, 1974, Choice of Environment, 1986; editor: Life of Information Era, 1985, Strategy towards Information Society, 1991 Mem. Japan Inst. Pub. Fin. (bd. dirs.). Buddhist. Home: 29-9 Ansyu Babahigashi Yamashina, Kyoto-shi 607, Japan Office: Kobe U of Commerce, Gakuen-Nishimachi, Nishi-ku, Kobe 651-21, Japan

FUNABIKI, RYUHEI, nutritional biochemist; b. Tokyo, Oct. 23, 1931; s. Masayuki and Kou (Yamashita) F.; m. Atsuko Tanaka, Dec. 28, 1963; children: Kohei, Yusuke, Naoko. BS, Tokyo Noko U., 1955; MS, Tokyo U., 1959, PhD, 1964. Asst. prof. agrl. Iwate U., Morioka, 1963-64, lectr., 1964-65, assoc. prof., 1965-71; assoc. prof. Tokyo Noko U., Fuchu, 1972-78, full prof., 1978—; vis. prof. U. Wis., Madison, 1970-71; mem. Rsch. Com. of Essential Amino Acids, Tokyo, 1983—. Co-author: Nutrition: Proteins and Amino Acids, 1990. Member Adminstrn. Com. of Pub. Utility Trust, Tokyo, 1988—; dir. Agrl. Chemistry Soc. of Japan, Tokyo, 1973-74. Recipient Nogeikagakusho Agrl. Chemistry Soc. Japan, 1966. Mem. Japanese Soc. Nutrition and Food Sci. (councilor Tokyo chpt. 1982—). Avocations: sketching, hiking. Home: 4-18-15 Higashimotomachi, Kokubunji Tokyo 185, Japan Office: Tokyo Noko U, 3-5-8 Saiwaicho, Fuchu Tokyo 183, Japan

FUNAHASHI, AKIRA, physician, educator; b. Chingtao, China, Mar. 5, 1928; came to U.S., 1969; s. Shikanosuke and Masu (Yoshida) F.; m. Masako Kinukawa, Oct. 21, 1956; children: Yuri, Tadashi, Kenji. MD, Kyushu U., Fukuoka, Japan, 1954, PhD, 1959. Diplomate Am. Bd. Internal Medicine. Intern Kyushu U. Hosp., Fukuoka, Japan, 1954-55, resident, 1955-59; resident U. Cin. Hosps., 1964-66; chief of medicine Japan Bapt. Hosp., Kyoto, 1960-67, chief of staff, 1967-69; assoc. prof. medicine Med. Coll. Wis., Milw., 1977-83, fellow, 1969-70; chief pulmonary function lab. VA Med. Ctr., 1972—; prof. Med. Coll. Wis., Milw., 1983—; cons. in field. Contbr. articles to profl. jours. Fulbright Found. scholar, 1964-66. Fellow Am. Coll. Chest Physicians (pres. Northland chpt. 1977-78). Home: 2160 Possum Ct Brookfield WI 53045-4723 Office: VA Med Ctr 5000 W National Ave Milwaukee WI 53295-0002

FUNDENBERG, HERMAN HUGH, research scientist; b. N.Y.C., Oct. 24, 1928; s. Nathan and Frances (Chachowi) F.; m. Betty Roof (div.); children: Drew, Brools, David, Haskell. BA, UCLA, 1949; DMS, U. Chgo., 1956; MA, Boston U., 1958; postgrad., U. Kupio, Finland, 1982, U. Clauxe Bernard, Lyon, France, 1984. Diplomate Am. Bd. Lab. Med. Immunology. Instr. Rockefeller U., N.Y.C., 1958-60; from instr. to assoc. prof. of medicine U. Calif., San Francisco, Berkeley, 1960-69, prof. of medicine, 1969-75; prof., chmn. dept. immunology U. S.C., Charleston, 1975-91; dir. rsch. NeuroImmunotherapeutics Rsch. Found., 1991—; lectr. U. Messina, Sicily; mem. nat. adv. coun. Nat. Inst. Allergies and Infectious Desease, 1981-85, sci. adv. bd. Pasteur Inst., 1974-90, South African Med. Rsch. Clinic, Chinese Inst. Immunology, 1989—, Ponce Sch. Medicine and various sci. jour. editorial bds.; hon. prof. U. Beijing, 1993; plenary lectr. 9th Internat. Symposium Transfer Factor, Beijing, 1993. Author: Basic and Clinincal Immunology, 2d edit., 1988; editor: Clinical Immunology and Immunopathology, 1974-90; translator 6 books; contbr. numerous publs. Recipient Prof. Honoris Causa awards France, Italy, Finland and China, various soc. awards Russia, Hungary, Denmark, Mexico, Brazil, award Pasteur Inst., 1962, Cooke medal Am. Acad. Allergy, Contbns. to Quality of Life award Pres. of Italy, Internat. First Med. Sci. prize Italian Acad. Arts and Scis., 1972. Fellow Am. Assn. Advancement Sci.; mem. Am. Cancer Soc., Am. Soc. Human Genetics Internat. (former mem. exec. coun.), IMMUNOTECH (bd. dirs.). Achievements include patents in field. Office: Neuroimmunotherapeutics Rsch Found 145 N Church St Spartanburg SC 29301

FUNG, HENRY CHONG, microbiologist, educator, administrator; b. San Francisco, Feb. 5, 1939; s. Henry G. and Gay Kee (Wong) F.; m. Janet Yee, June 18, 1961; children: Brian, Bruce, Brent. BA, U. Calif., Berkeley, 1959; MA, San Francisco State U., 1962; PhD, Wash. State U., 1966. Cert. lab. technologist, Calif. Asst. prof. microbiology Calif. State U., Long Beach, 1966-70, assoc. prof. microbiology, 1970-75, prof., 1975—, acting assoc. dean nat. scis., 1991—; chair undergrad. edn. Am. Soc. for Microbiology-Bd. Edn. and Tng., Washington, 1981-90; mem. Emerson Elem. Adv. Coun., chair, 1972-74; mem. curriculum com. Dist. Sr. High Sch., 1984-86; external reviewer microbiology U. Wis., 1988, Calif. Poly. U., 1990. Contbng. author College Board Guide to Popular College Majors, 1992; contbr. Clin. Infectious Diseases, Cancer Rsch., Reticulendothelial Soc. Jour. Pres. Millikan Orch-A-Band Assn., 1980-81, 86-87; active Millikan PTA. Mem. APHA (rep. to Am. Bd. Med. Microbiology 1987-92), Calif. Assn. Pub. Health Lab. Dirs. (chair academicians 1986-88), Chinese Assn. Orange County (pres. 1980-82). Achievements include research in cancer and immunology. Office: Calif State U Office Dean Coll Natural Scis and Math Long Beach CA 90840

FUNG, KEE-YING, engineer, educator, researcher; b. Hunan, Peoples Republic of China, Dec. 20, 1948; came to U.S., 1972; s. Yuk-Kowng and Nam-Hing (Tam) F.; m. E.E. Ho, Jan. 18, 1976 (div. Apr. 1985); m. Ivy Wang, Mar. 15, 1987; children: Elysia, E-Dean. BS, Nat. Taiwan (Republic of China) U., 1972; PhD, Cornell U., 1976. Research assoc. aerospace and mech. engring. dept. U. Ariz., Tucson, 1976-79, asst. prof., 1979-85, assoc. prof., 1985-93; prof. mech. engring. U. Miami, Coral Gables, 1993—; vis. scientist DFVLR, Göttingen, Fed. Republic of Germany, 1981; cons. Ariz. State U., Phoenix, 1985—, Zonatech., 1985—. Contbr. articles to profl. jours. A.V. Humboldt fellow, 1981. Mem. AIAA, Tucson Soaring Club. Home: 9020 SW 68th Ave Miami FL 33156 Office: Dept Mech Engring PO Box 248294 Coral Gables FL 33124-0624

FUNG, SUN-YIU SAMUEL, physics educator; b. Hong Kong, Dec. 27, 1932; came to U.S., 1953; s. Lok-Chi and Lai-Lan Fung; m. Helen Wu, Feb. 9, 1964; children: Eric, Linette. BS, U. San Francisco, 1957; PhD, U. Calif., 1964. Rsch. physicist Rutgers U., New Brunswick, N.J., 1964-66; asst. prof. physics U. Calif., Riverside, 1966-70, assoc. prof., 1970-76, prof., 1976—, chmn. physics dept., 1980-85, 90-91. Chmn. Chinese Meml. Pavilian Com., Riverside, 1985-88. Mem. AAAS, Am. Phys. Soc., Overseas Chinese Physicist Assn., Chinese Am. Faculty Assn. (pres. 1988-89, 90-92). Office: U Calif Riverside CA 92521

FUNG, YUAN-CHENG BERTRAM, bioengineering educator, author; b. Yuhong, Changchow, Kiangsu, China, Sept. 15, 1919; came to U.S., 1945, naturalized, 1957; s. Chung-Kwang and Lien (Hu) F.; m. Luna Hsien-Shih Yu, Dec. 22, 1949; children: Conrad Antung, Brenda Pingsi. BS, Nat. Central U., Chungking, China, 1941, MS, 1943; PhD, Calif. Inst. Tech., 1948. Research fellow Bur. Aero. Research China, 1943-45; research asst., then research fellow Calif. Inst. Tech., 1946-51, mem. faculty, 1951-66, prof. aeros., 1959-66; prof. bioengring. and applied mechanics U. Calif., San Diego, 1966—; cons. aerospace indsl. firms, 1949—. Author: The Theory of Aeroelasticity, 1985, 69, 93, Foundations of Solid Mechanics, 1965, A First Course in Continuum Mechanics, 1969, 77, 92, Biomechanics, 1972, Biomechanics: Mechanical Properties of Living Tissues, 1980, 92, Biodynamics: Circulation, 1984, Biomechanics: Motion, Flow, Stress and Growth, 1990; also papers,; Editor: Jour. Biorheology, Jour. Biomech. Engring. Recipient Achievement award Chinese Inst. Engrs., 1965, 68; Landis award Microcirculatory Soc., 1975, Poiseville medal Internat. Soc. Biorheology, 1986, Engr. of Yr. award San Diego Engring. Soc., 1986; von Karman medal ASCE, 1976; ALZA award Biomedical Engring. Soc., 1989, Borelli award Am. Soc. Biomechanics; Guggenheim fellow, 1958-59. Fellow AIAA, ASME (Lissner award 1978, Centennial medal 1978, Worcester Reed Warner medal 1984, Timoshenko medal 1991); mem. NAS, NAE, Inst. Medicine, Soc. Engring. Sci., Microcirculatory Soc., Am. Physiol. Soc., Nat. Heart Assn., Basic Sci. Coun., Sigma Xi. Home: 2660 Greentree Ln La Jolla CA 92037-1148 Office: Dept of Ames Bioengineering 9500 Gilman Dr La Jolla CA 92093

FUNKEN, KARL-HEINZ, chemist; b. St. Tönis, Germany, June 17, 1953; s. Wilhelm Joseph and Auguste Josepha (Dörper) F.; m. Elisabeth Maria Strohscheidt, Apr. 5, 1988. Dipl. Chem., Tech. U., Aachen, Germany, 1985, Dr. rer. nat., 1988. Rschr. Deutsche Forschungsanstalt, Köln, Germany, 1988-91; br. head Deutsche Forschungsanstalt, Köln, 1991—. Co-editor: Solarchemische Technik, vol. 1 and 2, 1989, Solar Thermal Energy Utilization, vol. IV, V, VI, VII, 1991-92, Solares Testzentrum Almería, 1993; contbr. articles to profl. jours. Mem. Gesellschaft Deutscher Chemiker. Achievements include patents in field. Office: Deutsche Forschungsaustalt, f Luft-U Raumfahrt Linder Höhe, 51140 Köln Germany

FUNK ORSINI, PAULA ANN, pharmaceutical administration educator; b. Marietta, Ohio, May 16, 1956; d. James Corwin and Inez Aline (Smith) Funk; m. Michael Joseph Orsini, May 4, 1991. BS, Ohio State U., 1979, MS, 1986, PhD, 1992. Pharmacy resident Hosp. of U. Pa., Phila., 1979-80; staff pharmacist Nursing Ctr. Svcs., Hilliard, Ohio, 1984-85; mktg. rsch. assoc. Strategic Mktg. Corp., Bala Cynwyd, Pa., 1986; staff pharmacist Hahnemann U. Hosp., Phila., 1986-87, Mt. Carmel East Drugstore, Columbus, Ohio, 1987-91, Ohio State U. Hosps., Columbus, 1980-86, 87-91; asst. prof. U. Md., Balt., 1991—; presenter in field. Author: (with others) Modern Medicine Dermatology Pocket Guide, 1982; reviewer Am. Jour. Hosp. Pharmacy, 1987; contbr. articles to Jour. Pharm. Mktg. & Mgmt., Am. Jour. Hosp. Pharmacy, Hosp. Pharmacy. Scholar in pharm. scis. IAPS-Am. Found. Pharm. Edn., 1988; Nat. Assn. Bds. Pharmacy-Am. Found. Pharm. Edn. fellow in pharmacy adminstrn., 1990, Albert B. Fisher Jr. PhD Citation fellow Am. Found. Pharm. Edn., 1989. Mem. Coun. Grad. Students in Pharm. Scis. (pres. 1990-91), Am. Assn. Pharm. Scientists, Am. Soc. Hosp. Pharmacists, Am. Pharm. Assn., Assn. for Consumer Rsch., Rho Chi. Office: Pharm Practice & Sci Allied Health Bldg 100 Penn St Ste 240 Baltimore MD 21201-1082

FUNSTEN, HERBERT OLIVER, III, physicist; b. Princeton, N.J., May 29, 1962; s. Herbert O. and Edythe R. (Reed) F.; m. Ann W., Aug. 20, 1988; 1 child, H. Oliver. BS in Physics, Washington and Lee U., 1984; PhD in Engring. Physics, U. Va., 1990. Postdoctoral fellow Los Alamos (N.Mex.) Nat. Lab., 1990-93, staff mem., 1993—. Contbr. articles to profl. jours. Mem. Am. Phys. Soc., Am. Geophys. Union, Phi Beta Kappa. Home: 875 Kristi Ln Los Alamos NM 87544-2881 Office: Los Alamos Nat Lab MS D466 Los Alamos NM 87545

FURCHTGOTT, DAVID GROVER, computer engineer; b. Knoxville, Tenn., May 30, 1955; s. Ernest and Mary Alma (Wilkes) F.; m. Aviva Kahn, June 10, 1979; children: Michael Max, Lisa Rachael. BSEE with highest honors, U. Ill., 1976; PhD in Computer Engring., U. Mich., 1984. Mem. tech. staff AT&T Bell Labs., Naperville, Ill., 1983-89, supr., 1989-92, disting. mem. tech. staff, 1992—. Contbr. articles to profl. publs. Mem. AAAS, IEEE, Assn. Computing Machinery, Tau Beta Pi, Eta Kappa Nu, Phi Eta Sigma. Achievements include patent for method and apparatus for providing variable reliability in telecommunications switching system, information transfer method and arrangement. Home: 456 Lenox St Oak Park IL 60302 Office: AT&T Bell Labs 2000 N Naperville Rd Naperville IL 60566

FUREY, DEBORAH ANN, aerospace engineer, naval architect; b. Fairfax, Va., Apr. 19, 1966; d. Roger Joseph Furey and Mary Nancy (Bean) Stone. BS in Aero. Engring., Va. Poly. Inst. and State U., 1989, MS in Engring. Mechanics, 1991. Grad. rsch. asst. dept. engring. Va. Poly. Inst. and State U., Blacksburg, 1989-91; mech. engr., naval architect David Taylor Rsch. Ctr. Dept. of Navy, Bethesda, Md., 1991—. Mem. AIAA, Am. Soc. Naval Engrs., Tau Beta Pi, Sigma Gamma Tau. Home: 822 Villa Ridge Rd Falls Church VA 22046 Office: Dept Navy David Taylor Rsch Ctr Bethesda MD 20084

FUREY, ROBERT L., research chemist; b. Canton, Ohio, June 25, 1941. BS, Kent State U., 1963, PhD, 1967. Assoc. sr. rsch. chemist GM Rsch. Labs., Warren, Mich., 1969-73, sr. rsch. chemist, 1973-77, sr. rsch. scientist, 1977-80, staff rsch. scientist, 1980-85, sr. staff rsch. scientist, 1985—. Contbr. articles on automotive fuels and emissions to profl. jours. Capt. U.S. Army, 1967-69. Mem. Am. Chem. Soc., Soc. Automotive Engrs. (disting. speaker). Home: 1196 Maple Leaf Dr Rochester Hills MI 48309 Office: GM Rsch & Devel Ctr PO Box 9055 30500 Mound Rd Warren MI 48090-9055

FURIGA, RICHARD DANIEL, government official; b. New Eagle, Pa., June 20, 1935; s. Joseph and Edith (Cain) F. B.S., U.S. Naval Acad., 1957; M.S., U. Kans., 1970. Logistics officer Strategic Petroleum Res. Dept. Energy, Washington, 1977-79, dep. dir., 1979-81, dep. asst. sec., 1982—. Served to comdr. USN, 1957-77. Mem. Soc. Petroleum Engrs. Republican. Lutheran. Home: 7823 Water Valley Ct Springfield VA 22153-3017 Office: US Dept Energy 1000 Independence Ave SE Washington DC 20585-0001

FURNAS, DAVID WILLIAM, plastic surgeon; b. Caldwell, Idaho, Apr. 1, 1931; s. John Doan and Esther Bradbury (Hare) F.; m. Mary Lou Heatherly, Feb. 11, 1956; children: Heather Jean, Brent David, Craig Jonathan. AB, U. Calif.-Berkeley, 1952, MS, 1957, MD, 1955. Diplomate: Am. Bd. Surgery, Am. Bd. Plastic Surgery (dir. 1979-85). Intern U. Calif. Hosp., San Francisco, 1955-56, asst. resident in surgery, 1956-57; asst. resident in psychiatry, NIMH fellow Langley Porter Neuropsychiat. Inst. U. Calif., San Francisco, 1959-60; resident in gen. surgery Gorgas Hosp., C.Z., 1960-61; asst. resident in plastic surgery N.Y. Hosp., Cornell Med. Center, N.Y.C., 1961-62; chief resident in plastic surgery VA Hosp., Bronx, N.Y., 1962-63; registrar Royal Infirmary and Affiliated Hosps., Glasgow, Scotland, 1963-64; assoc. in hand surgery U. Iowa, 1965-68, asst. prof. surgery, 1966-68, assoc. prof., 1968-69; assoc. prof. surgery, chief div. plastic surgery U. Calif., Irvine, 1969-74, prof., chief div. plastic surgery, 1974-80, clin. prof., chief div. plastic surgery, 1980—; surgeon East Africa Flying Doctors Service, African Med. and Research Found., Nairobi, Kenya, 1972-90; plastic surgeon S.S. Hope, Nicaragua, 1966, Sri Lanka, 1968; mem. Balakbayan med. mission, Mindanao and Sulu, Philippines, 1980, 81, 82. Contbr. chpts. to textbooks, articles to med. jours.; author/editor 5 textbooks; assoc. editor Jour. Hand Surgery, Annals of Plastic Surgery, Jour. Craniofacial Surgery. Expedition leader Explorer's Club Flag 171 Skull Surgeons of the Kisii Tribe, Kenya, Flag 44 Skull Surgeons of the Marakwet Tribes, Kenya, 1987. Capt. Med. Corps, USAF, 1957-59; col. Med. Corps., USAR, 1989-92. Recipient Golden Apple award for teaching excellence U. Calif.-Irvine Sch. Medicine, 1980, Kaiser-Permanente award U. Calif.-Irvine Sch. Medicine, 1981, Humanitarian Service award Black Med. Students, U. Calif. Irvine, 1987, Sr. Research award (Basic Sci.) Plastic Surgery Ednl. Found., 1987; named Orange County Press Club Headliner of Yr., 1982. Fellow ACS, Royal Coll. Surgeons Can., Royal Soc. Medicine, Explorers Club, Royal Geog. Soc.; mem. AMA, Calif., Orange County med. assns., Am. Soc. Plastic and Reconstructive Surgeons, Am. Soc. Reconstructive Microsurgery, Soc. Head and Neck Surgeons, Am. Cleft Palate Assn., Am. Soc. Surgery of Hand, Soc. Univ. Surgeons, Am. Assn. Plastic Surgeons (trustee 1983-86, treas. 1988-91, v.p. 1993—), Am. Soc. Aesthetic Plastic Surgery, Am. Soc. Maxillofacial Surgeons, Assn. Acad. Chmn. Plastic Surgery (bd. dirs. 1986-89), Assn. Surgeons East Africa, Pacific Coast Surg. Assn., Internat. Soc. Aesthetic Plastic Surgery, Internat. Soc. Reconstructive Microsurgery, Internat. Soc. Craniomaxillofacial Surgery, Assn. Mil. Surgeons U.S., NG Assn. Calif., Pan African Assn. Neurol. Scis., African Med. and Research Found. (bd. dirs. U.S.A. 1987—), Muthaiga Club, Ctr. Club, Club 33, Phi Beta Kappa, Alpha Omega Alpha. Office: U Calif Div Plastic Surgery Irvine Med Ctr 101 City Dr W Orange CA 92668-2901

FURNEAUX, HENRY MORRICE, biochemist, educator; b. Aberdeen, Scotland, Jan. 2, 1954; came to U.S. 1979; s. Henry Cox and Flora (Robson) F.; m. Jean Watson, Sept. 28, 1978 (div. 1985); m. Betsy Joyce Kaufman, Aug. 16, 1987. BSc with hons., Aberdeen U., 1975, PhD, 1978. Postdoctoral fellow Albert Einstein Coll. Medicine, N.Y.C., 1979-84; rsch. assoc. Meml. Sloan Kettering Cancer Ctr., N.Y.C., 1984-87, asst. lab. mem., 1987-89, asst. mem., 1989—; asst., prof. neurosci. Cornell U. Med. Sch., N.Y.C., 1989—. Mem. N.Y. Acad. Scis., Harvey Soc., Soc. for Neursci. Achievements include several patents on cloning paraneoplastic neural antigens. Office: Meml Sloan Kettering Ctr 1275 York Ave New York NY 10021

FURSIKOV, ANDREI VLADIMIROVICH, mathematics educator; b. Saratov, Russia, June 12, 1945; s. Vladimir Avksentievich and Vera Leontievna (Kniazeva) F.; m. Irina Viktorovna Fadeeva, Dec. 14, 1968; 1 child, Tatiana. PhD in Physics and Math., Moscow State U., 1986. Jr. research in math. Moscow State U., 1971-76, lectr., 1976-81, asst. prof. math., 1981-90, prof. math., 1990—. Author: Mathematical Problems of Statistical Hydromechanics, 1980; contbr. articles to profl. jours. Recipient First award in USSR competition of student research works Ministry of High Edn. of USSR, 1969. Mem. Am. Math. Soc., Moscow Math. Soc. Home: Krupskoy 19 108, Moscow Russia 117331 Office: Moscow State U, Lenin Hills, Moscow Russia 119889

FURST, GEORGE, forensic dentist; b. Bklyn., May 1, 1918; s. Isidore Jacob and Regina (Moskowitz) F.; m. Mildred Berman, Aug. 22, 1948 (div., June 1969); children: Andrea Carole, Francine. Gen. practice dentistry, 1946-78, forensic dentist, 1971—; DDS, U. Pa., 1942. Diplomate Am. Bd. Forensic Odontology. Dental cons. Chief Med. Examiner, N.Y.C., 1978-84, 88—. Editor (newsletter) Am. Soc. Forensic Odontology. Lt. USPHS, 1942-46. Recipient Sect. award Am. Acad. Forensic Scis., 1988. Home: 75-06 197th St Flushing NY 11366-1817

FURSTE, WESLEY LEONARD, II, surgeon, educator; b. Cin., Apr. 19, 1915; s. Wesley Leonard and Alma (Deckebach) F.; m. Leone James, Mar. 28, 1942; children: Nancy Dianne, Susan Deanne, Wesley Leonard III. A.B. cum laude (Julius Dexter scholar 1933-34); Harvard Club scholar 1934-35), Harvard U., 1937, M.D., 1941. Diplomate: Am. Bd. Surgery. Intern Ohio State U. Hosp., Columbus, 1941-42; fellow surgery U. Cin., 1945-46; asst. surg. resident Cin. Gen. Hosp., 1946-49; sr. asst. surg. resident Ohio State U. Hosps., 1949-50, chief surg. resident, 1950-51; limited practice medicine specializing in surgery Columbus, 1951—; instr. Ohio State U., 1951-54, clin. asst. prof. surgery, 1954-66, clin. assoc. prof., 1966-74, clin. prof. surgery, 1974-85, clin. prof. emeritus, 1985—; mem. surg. staff Mt. Carmel Med. Center, chmn. dept. surgery, 1981-85, dir. surgery program, 1981-82; mem. surg. staff Children's, Grant Med. Ctr., Univ., Riverside, Meth. Hosps., St. Anthony Med. Ctr., Park Med. Ctr. (all Columbus); surg. cons. Dayton (Ohio) VA Hosp., Columbus State Sch., Ohio State Penitentiary, Mercy Hosp., Benjamin Franklin Hosp., Columbus; regional adv. com. nat. blood program ARC, 1951-68, chmn., 1958-68; invited participant 2d Internat. Conf. on Tetanus, WHO, Bern, Switzerland, 1966, 3d, São, Paulo, Brazil, 1970, 4th, Dakar, Sénégal, 1975, 5th, Ronneby Brunn, Sweden, 1978, 6th, Lyon, France, 1981, 7th, Copanello, Italy, 1984, 8th, Leningrad, USSR, 1987, 9th, Granada, Spain, 1991; invited rapporteur 4th Internat. Conf. on Tetanus, Dakar, Sénégal, 1975; mem. med. adv. com. Medic Alert Found. Internat., 1971-73, 76—, bd. dirs., 1973-76; Douglas lectr. Med. Coll. of Ohio, Toledo; founder Digestive Disease Found. Prime author: Tétanos; Tetanus: A Team Disease; contbg. author: Advances in Military Medicine, 1948, Management of the Injured Patient, Immediate Care of the Acutely Ill and Injured, 1978, Anaerobic Infections, 1989, Procs. of Internat. Confs. in Switzerland, Brazil, Sweden, Sénégal, France, Italy, USSR, Spain, Current Therapy in Emergency Medicine, Surgical Infectious Diseases (3 edits.), Currenty Emergency Therapy, Surgical Infections, Current Diagnosis (multiple edits.), Current Therapy (multiple edits.), Surgical Infections, 5 Minute Clinical Consult, Medical Microbiology and Infectious Diseases, editor Surgical Monthly Review; contbr. articles to profl. jours. Mem. Ohio Motor Vehicle Med. Rev. Bd., 1965-67; bd. dirs. Am. Cancer Soc. Franklin County, pres., 1964-66. Served to maj., M.C. AUS, 1942-46, CBI, 1951-53. Recipient China Liberation medal, 2 commendations for surg. service in China U.S.

Army; cert. of merit Am. Cancer Soc.; award for outstanding achievement in field clostridial infection dept. surgery Ohio State U. Coll. Medicine, 1984, Outstanding Service award, 1985; award for outstanding and dedicated service Mt. Carmel Med. Ctr., 1985; award for over 25 yrs. service St. Anthony Med. Ctr. Mem. AMA, AAAS, Cen. Surg. Assn., Surgical Infection Soc., Internat. Biliary Assn., Shock Soc., Soc. Am. Gastrointestinal Endoscopic Surgeons (com. on standards of practice, resident and fellow edn., mem. com. legis. review), Soc. Surgery of Alimentary Tract, A.C.S. (gov.-at-large, chmn. Ohio com. trauma; nat. subcom. prophylaxis against tetanus in wound mgmt., Ohio chapter Disting. Service award 1987; regional credentials com.), Am. Assn. Surgery of Trauma, Ohio Surg. Assn., Columbus Surg. Assn. (hon. mem.; pres. 1983), Am. Trauma Soc. (founding mem., dir.), Ohio Med. Assn., Acad. Medicine Columbus and Franklin County (Award of Merit for 17 yrs. service, chmn. blood transfusion com.), Acad. Medicine Cin., Am. Public Health Assn., Am. Med. Writers Assn., Grad. Surg. Soc. U. Cin., Robert M. Zollinger Surg. Ohio State U. Surg. Soc., Mont Reid Grad. Surg. Soc., Am. Geriatrics Soc., N.Y. Acad. Scis., Assn. Program Dirs. in Surgery, Assn. Physicians State of Ohio, Collegium Internationale Chirurgiae Digestivae, Assn. Am. Med. Colls., Internat. Soc. Colon and Rectal Surgeons, Soc. Internat. de Chirurgie, Am. Assn. Sr. Physicians, Société Internationale sur le Tétanos, Am. Physicians Art Assn., China-Burma-India Vets., Assn. Columbus Basha (vice comdr. 1992-93, comdr. 1993-94), Am. Med. Golfing Assn., Internat. Brotherhood Magicians, Soc. Am. Magicians, N.Y. Cen. System Hist. Soc., U.S. Squash Racquets Assn. Presbyterian. Clubs: Scioto Country, Ohio State U. Golf, Ohio State Faculty, Capital (Columbus); University (Cin.); Harvard (Boston). Home and Office: 3125 Bembridge Rd Columbus OH 43221-2203

FURTH, HAROLD PAUL, physicist, educator; b. Vienna, Austria, Jan. 13, 1930; came to U.S., 1941, naturalized, 1947; s. Otto and Gertrude (Harteck) F.; m. Alice May Lander, June 19, 1959 (div. Dec. 1977); 1 son, John Frederick. Grad., Hill Sch., 1947; A.B., Harvard U., 1951; Ph.D., Harvard, 1960; postgrad., Cornell U., 1951-52. Physicist U. Calif. Lawrence Radiation Lab., Livermore, 1956-67; prof. astrophys. scis. Princeton U., 1967—; dir. Plasma Physics Lab., 1981-90. Bd. editors: Physics of Fluids, 1965-67, Nuclear Fusion, 1964-88, Revs. Modern Physics, 1975-80, Plasma Physics and Controlled Fusion, 1984-88; contbr. articles to profl. jours. Recipient E.O. Lawrence award AEC, 1974, Joseph Priestley award Dickinson Coll., 1985, Delmer S. Fahrney medal Franklin Inst., 1992. Fellow Am. Phys. Soc. (J. C. Maxwell prize 1983); mem. NAS, Am. Acad. Arts and Scis. Achievements include patents in field. Home: 36 Lake Ln Princeton NJ 08540-7212 Office: Princeton U Plasma Physics Lab PO Box 451 Princeton NJ 08543-0451

FURUICHI, SUSUMU, physics researcher; b. Kurashiki, Okayamaken, Japan, Apr. 23, 1931; s. Eitaroh and Taka (Takatori) F.; m. Tsuneko Shiogami, Apr. 12, 1959; children: Naoko, Masafumi. Bs, U. Hiroshima, Japan, 1953, MS, 1955, PhD in Sci., 1959. Lectr. Dept. Physics, Rikkyo U., Tokyo, 1958-63, asst. prof., 1963-68, prof., 1968—; dean faculty of sci. Rikkyo U., Tokyo, 1975-79, registrar, 1983-86. Contbr. sci. papers to profl. jours. Mem. The Phys. Soc. Japan, Japanese Alpine Club. Avocations: mountaineering, bowling. Office: Rikkyo Univ, Faculty of Sci, 3-34-1 Nishi-Ikebukuro, Toshima-Ku Tokyo, Japan

FURUYA, TSUTOMU, plant chemist and biochemist, educator; b. Nyuzen, Toyama, Japan, Nov. 17, 1928; s. Tomojiro Kosugi and Tomi Furuya; m. Yayoi Yoshida, Mar. 25, 1951; children: Tatsumi, Takashi, Shu. BSc in Pharm. Sci., U. Tokyo, 1951, PhD in Pharm. Sci., 1959. Asst. prof. faculty of pharm. sci. U. Tokyo, 1955-64; postdoctoral Sch. Pharmacy U. Calif., San Francisco, 1963-64; prof. Sch. Pharm. Sci. Kitasato U., Tokyo, 1965—. Author, editor: (books) Thin-layer Chromatography, 1963, Plant Tissue Culture, 1972, New Plant Tissue Culture, 1979, Naturally Occurring Medicine, 1987. Rsch. grantee Mitsubishi Found., 1973, Naito Found., 1971. Mem. Japan Pharm. Libr. Assn. (pres.), Japan Info. Ctr. Sci. and Tech. (med. mem.), Japan Bioindustry Assn. (head bio-project), Petroleum Energy Ctr. (head bio-project), Pharm. Soc. Japan (Promotion award 1973, award 1988), Japanese Soc. Pharmacognosy (councilor), Japanese Assn. for Plant Tissue Culture (councilor). Liberal Democrat. Buddhist. Avocations: Yookyoku, golf, billiards, skiing. Home: 4-20-33 Minami-ooizumi, Nerima-ku Tokyo 178, Japan Office: Kitasato U Sch Pharm Scis, 5-9-1 Shirokane, Minato-ku Tokyo 108, Japan

FUSCO, PENNY PLUMMER, mechanical engineer; b. Ft. Worth, Feb. 28, 1968; d. Jerald M. and Anita F. Plummer; m. Andrew Malcolm Fusco, Aug. 15, 1992. BS, MIT, 1990; MS, Stanford U., 1992. Product design engr. Apple Computer, Inc., Cupertino, Calif., 1991; tech. staff mem. Los Alamos (N.Mex.) Nat. Lab., 1992—. Pres. Christian Youth Fellowship Ridglea Christian Ch., Ft. Worth, 1984-86. Recipient Bronze award Lincoln Arc Welding Found., Luis DeFlorez Engring. Design award MIT, 1990; scholar Tex. Soc. Profl. Engrs., 1986, scholar Eugene McDermott Found., 1986-90. Mem. Sigma Xi, Alpha Phi. Republican. Achievements include design of Stand-Up Mobility Device for Paraplegics; research in linear accelerator equipment & support. Home: 136 Longview Dr # 5 White Rock NM 87544 Office: Los Alamos Nat Lab PO Box 1663 MSH826 Los Alamos NM 87545

FUSSELL, PAUL STEPHEN, mechanical engineer; b. Havre, Mont., Jan. 5, 1958; s. J. Aubrey and Nancy E. Fussell; m. Caroline A. Haessly, Dec. 27, 1989; 1 child, Alexandra. BS in Mech. Engring., Mont. State U., 1980; ME, Carnegie Mellon U., 1983, PhD in Mech. Engring., 1993. Mem. tech. staff Advanced Robotics Corp., Columbus, Ohio, 1983-85; sr. engr. Alcoa Tech. Ctr., Pitts., 1985-86, staff engr., 1987-92, tech. specialist, 1992—. Contbr. articles to profl. jours. Mem. Am. Soc. Materials, Sigma Xi, Tau Beta Pi, Phi Eta Sigma, Pi Tau Sigma. Achievements include research in sprayed metal shells for tooling, microstructure, properties and methods; sprayed tooling for rapid prototyping; controlled microstructure of arc sprayed metal shells; sprayed steel tool for permanent mold casting. Office: Alcoa Tech Ctr 100 Technical Dr Alcoa Center PA 15069-0001

FUSSICHEN, KENNETH, computer scientist; b. Bklyn., Aug. 3, 1950; s. Lorenzo Anthony and Sue (Treppiedi) F.; m. Bobbie J. Ezra, May 18, 1974; children: Matthew, David, Vanessa, Natalie. AS in Data Processing, San Antonio Coll., 1975; BS in Bus., Ind. U., Indpls., 1980; MS in Mgmt., Ind. Wesleyan U., 1991. Programmer, analyst Computer Mgmt. Systems, Indpls., 1976-81; sr. programmer, analyst Jefferson Nat. Life, Indpls., 1981-84; project leader HAS (Healthcare Adminstrv. Systems), Inc., Indpls., 1984-87; sr. computer scientist Computer Scis. Corp., Indpls., 1987—; assoc. prof. computer scis. Ind. U./Purdue U., Indpls., 1981-88. Info. Systems mgr. Cerebral Palsy Support Group, Indpls., 1987-89; computer com. United Cerebral Palsy Cen. Ind., 1984-85; participant Ada 9X Lang. Rev., 1990—. Mem. Assn. Computing Machinery, Indpls. Computer Soc. (pres. 1989-90). Home: 130 Massie Dr Xenia OH 45385

FUTAI, MASAMITSU, biochemistry researcher, educator; b. Tokyo, Nakano, Japan, May 13, 1940; s. Shonosuke and Yoshiko F.; m. Taeko Hamamoto; children: Kensuke, Eugene. BS, U. Tokyo, Japan, 1964, MS, 1966, PhD, 1969. Rsch. assoc. U. Tokyo, Japan, 1969-70; postdoctoral fellow U. Wis., Madison, 1970-72; rsch. assoc. Cornell U., Ithaca, N.Y., 1972-74, U. Tokyo, 1974-77; prof. biochemistry Okayama U., 1977-85, Osaka U., 1985—. Mem. editorial bd. Jour. Bacterial, 1982-88, Jour. Biochemistry, 1988-90, 92—, Jour. Bioengring. Biomed., 1985—, Archives Biochem. Biophysics, 1992—; contbr. articles to profl. jours. Recipient Eleanor Roosevelt fellowship Am. Cancer Soc., 1977. Mem. Japanese Biochem. Soc., Am. Soc. for Microbiology, Am. Chem. Soc. Office: Osaka U ISIR, 8-1 Mihogaoka, Ibaraki Osaka 567, Japan

FUTERNICK, KENNETH DAVID, education educator, software developer; b. San Francisco, Jan. 20, 1953; s. Joseph Israel and Mary Lee (Guynn) F.; m. Joy Louise Pelton, Nov. 6, 1982; 1 child, Emily. BS, U. Calif., Davis, 1975; MA, U. Calif., Berkeley, 1981, PhD, 1988. Tchr. Rescue (Calif.) Union Sch. Dist., 1976-80; pres. Bear Rock Techs. Corp., Shingle Springs, Calif., 1984-93; asst. prof. edn. Calif. State U., Sacramento, 1988—; pres. Calif. Assn. Philosophy Edn., Sacramento, 1991—. Author: A Problem-Censered Approach to Computer Education, 1990, Computer Programming as Education Experience, 1988. Mem. Philosophy Edn. Soc., Calif. Assn. Philosophy Edn. (pres. 1991—). Achievements include development of software programs: Print Bar I, 1984, Print Bar Softfonts, 1985, Print Bar Mac, 1988, Bear Rock Labeler for Windows, 1991. Office: Calif State Univ Sacramento Dept Tchr Edn 6000 J St Sacramento CA 95819

FUTRELL, NANCY NIELSON, neurologist; b. Salt Lake City, Nov. 8, 1947; d. John Willard Jr. and Dorothy Jay (Clark) Nielson; m. Clark H. Millikan, Dec. 28, 1987. MusB, U. Utah, 1971, MD, 1981. Diplomate Am. Bd. of Psychiatry and Neurology. Resident U. Utah, Salt Lake City, 1981-86; resident, fellow U. Miami, Fla., 1986-88; sr. staff mem. Henry Ford Hosp., Detroit, 1988-92; dir. Stroke Unit, asst. prof. dept. neurolgoy Creighton U. Sch. Medicine, Omaha, 1992—; Author rsch. papers. Asst. editor Jour. Stroke Cerebrovase Disease; reviewer Neurology, Stroke and Rhematology. Fulbright scholar in piano, Germany, 1971-72; grantee NIH, 1988-89, 92—, Am. Fedn. Aging Rsch., 1990-91, Mich. Heart Assn., 1989-92. Mem. ACP, AMA, Am. Acad. Neurology (working group on edn. for women and minorities 1992—, legis. affairs com. 1993—, program and accreditation subcom. 1993—), Soc. Exptl. Neuropathology (exec. com. 1991—), Assn. Women in Sci. (chair edn. com. 1992—), Am. Geriatrics Soc., Nat. Stroke Assn. (rsch. com. 1993—). Achievements include development of successful model of stroke in geriatric rats, research defining pathology of multi-infarct dementia, development of educational program for high school females and minorities. Office: Creighton U Sch Medicine 601 N 30th St Omaha NE 68131

FUTUYMA, DOUGLAS JOEL, ecology educator; b. New York, Apr. 24, 1942; s. Joseph and Eleanor (Haessler) F. BS, Cornell U., 1963; MS, U. Mich., 1966, PhD, 1969. Asst. prof. SUNY, Stony Brook, 1969-76, assoc. prof., 1976-83, prof. ecology, 1983—; pres. Soc. for Study of Evolution, 1987; mem. task force for the 90's Am. Inst. Biol. Scis., 1990—. Author: Evolutionary Biology, 1979, 2d edit., 1986, Science on Trial: The Case for Evolution, 1983; co-editor: Coevolution, 1983, Oxford Surveys in Evolutionary Biology Vol. 7, 1991, Vol. 8, 1992, Vol. 9, 1993; editor Evolution, 1981-83; assoc. editor Annual Review of Ecology and Systematics, 1992—. Fellow J. S. Guggenheim Meml. Found., 1992; rsch. grantee NSF, 1974—. Office: State Univ New York Dept Ecology and Evolution Stony Brook NY 11794-5245

FYFE, RICHARD WARREN, electro-optics executive; b. Hawthorn, Nev., Feb. 6, 1942; s. James Robert Jr. and Irene Evelyn (Hinrichson) F.; m. Edna Gay Swain, Sept. 4, 1965; children: Steven Patrick, Stephanie Ann. Grad. in Electronics Tech. (radar), Navy ETA Sch., San Francisco, 1962. Electro-optics supr. EG&G/EM, Las Vegas, 1966—. Donor United Blood Svcs., Las Vegas, 1992; mem. internship program Area Tech. Tng. Ctr., Las Vegas, 1991, 92; ptnr. in edn. Clark County Sch. Dist., Las Vegas, 1990, 91. With USNR, 1962-64. Republican. Achievements include patent for optical fiber stripper positioning apparatus; designed and built parts cleaner; helped design and build multipin connector continuity and hipot tester. Home: 3820 Asbury Ct Las Vegas NV 89130

FYFE, WILLIAM SEFTON, geochemist, educator; b. New Zealand, June 4, 1927; s. Colin Alexander and Isabella Fyfe; m. Patricia Walker, Feb. 27, 1981; children: Christopher, Catherine, Stefan. BSc, U. Otago, New Zealand, 1948, MS, 1949, PhD, 1952; DSc (hon.), Meml. U., Lisbon, Portugal, Lakehead U. Prof. chemistry in N.Z., 1955-58; prof. geology U. Calif., Berkeley, 1958-66; research prof. Manchester Coll. and Imperial Coll., London, 1966-72; chmn. dept. geology Western Ont. U., 1972-84, prof. dept. geology, 1984—, dean faculty sci., 1986-90. Decorated companion Order of Can.; Commemorative medal (New Zealand), Commemorative medal (Canada); recipient Logan medal Geol. Assn. Can., Arthur Holmes medal European Union of Geoscis., Can. Gold medal for Sci. and Engring., 1991; Guggenheim fellow, 1964, 83. Fellow Geol. Soc. London (hon.), Royal Soc. London, Geol. Soc. Am. (hon. life, Day medal), Mineral Soc. Am.; mem. Internat. Union Geoscis. (pres.), Nat. Sci. and Engring. Rsch. Council Can., Royal Soc. Can., Acad. Sci. Brazil, Brit. Chem. Soc., Explorers Club. Home: 1197 Richmond, London, ON Canada N6A 3L3 Office: U Western Ont, Dept Geology, London, ON Canada N6A 5B7

FYFFE, LES, earth scientist. Recipient Julian Boldy Meml. award Can. Inst. Mining and Metallurgy, 1992. Office: care Xerox Tower Ste 1210, 3400 de Maisonneuve Blvd W, Montreal, PQ Canada H3Z 3B8*

GAA, PETER CHARLES, organic chemist, researcher; b. Springfield, Ill., Mar. 30, 1955; s. Peter Carl and Patricia (Stewart) G.; m. Sheila Bourque, Aug. 2, 1980; children: Lily, Charlie. BS in Chemistry, U. Notre Dame, 1977; PhD in Organic Chemistry, Ga. Tech. Inst., 1982. Rsch. chemist PPG Industries, Pitts., 1982-84, sr. rsch. chemist, 1984-87, rsch. project chemist, 1987-91, group leader, 1987-91, rsch. assoc., 1989-91; dir. R&D Premark Internat., Temple, Tex., 1991—. Contbr. articles to profl. jours. Mem. TAPPI (vice chmn. div. indsl. nonwoven materials 1989—, chmn. tech. program div. nonwovens 1991-92), Am. Chem. Soc. (PPG Indsl. sponsor and rep. div. polymers 1987—), Product Devel. Mgmt. Assn., Adhesion Soc. Republican. Roman Catholic. Achievements include 6 patents on chemical coatings for fiber glass, including first patent on a silylated polyurethane emulsion; development of 7 new chemical coatings for fiber glass plastic applications. Home: 131 Maple Dr Wexford PA 15090 Office: Ralph Wilson Plastics 600 S General Bruce Dr Temple TX 76503

GAAB, MICHAEL ROBERT, neurosurgery educator, consultant; b. Landau, Germany, Mar. 11, 1947; s. Erich and Margarete (Hollinger) G.; m. Hannelore Sommer (div. 1972); 1 child, Marcus; m. Katharina Maria Weller, Sept. 28, 1974; children: Oliver, Jasmin, Florian. BS, U. Würzburg and U. Kiel, Germany, 1972; MD, PhD, U. Würzburg, 1973, U. Vienna., Austria, 1982, 83. House officer Knappschafts Hosp., Bochum, Germany, 1973; registrar, asst. neurosurgeon U. Hosp. Würzburg, Germany, 1974-79, sr. neurosurgeon, 1979-81, assoc. prof. neurosurgery, 1981-82; assoc. prof., head dept. pediatric neurosurgery U. Hosp. Vienna, 1982-84; assoc. prof. Med. Sch., U. Hannover, Germany, 1984-87, prof., 1987-92; full prof., head dept. neurosurgery Ernst Moritz Arndt U., U. Hosp. Greifswald, Germany, 1992—; cons. neurosurgeon City Hosp. Braunschweig, Germany, 1984-92. Author: Registration of Intracranial Pressure, 1981 (E.K. Frey award 1981); mem. editorial adv. bd. Brit. Jour. Neurosurgery, 1987—; patentee device for intracranial pressure monitoring. Mem. Christian Dem. Union, Hannover, 1973—. German Rsch. Community grantee, 1981, 83, German Ministry Rsch. and Tech. grantee, 1981, 82, German Cen. Nervous System bd. trustees grantee, 1991. Mem. German Neurosurg. Soc., Austrian Soc. Physicians, European Assn. for Pediatric Neurosurgery (sec. congress 1984), World Fedn. Neurosurg. Socs. (asst. treas. congress 1981), Bavarian Soc. Emergency Medicine, German Soc. Neuroendoscopy. Roman Catholic. Avocations: tennis, flying, flight instructor. Home: Senator-Bauer-Strasse 36, D-30625 Hannover Germany Office: U Hosp Ernst Moritz Arndt U, Löfflerstr 23 Dept Neurosurgery, D-17487 Greifswald Germany

GABBAI, ALBERTO ALAIN, neurology educator, researcher; b. Cairo, May 19, 1953; arrived in Brazil, 1958; s. Maurizio and Farida (Sasson) G.; m. Miriam Benasayag Birmann, Sept. 21, 1978; children: Carolina, Lisa. Physician, Paulista Sch. Medicine, Sao Paulo, Brazil, 1976, MS, 1981, D in Medicine, 1986. Med. diplomate. Intern Paulista Sch. Medicine, 1977, resident in clin. neurology, 1978-80; rsch. fellow Tufts-New Eng. Med. Ctr., Boston, 1981-82; clin. and rsch. fellow Mass. Gen. Hosp.-Harvard U., Boston, 1987-89; asst. prof. Paulista Sch. Medicine, 1982-86, assoc. prof. medicine, 1989—; cons. Rsch. Found. State of Sao Paulo, 1991—. Contbr. articles on AIDS and neuro-oncology to med. jours. Fellow Brazilian Acad. Neurology; mem. Am. Acad. Neurology (clin. assoc.), N.Y. Acad. Scis. Home: 203 Afonso Braz St, 04511 São Paulo Brazil Office: Paulista Sch Medicine, PO Box 20212, 04034 São Paulo Brazil

GABELNICK, HENRY LEWIS, medical research director; b. Boston, May 10, 1940; s. Murray and Lillian Gabelnick; m. Faith Schectman, June 17, 1962; children: Deborah Anne, Tamar Miriam; m. Clare Ann Donaher, May 22, 1987. BS, MIT, 1961, MS, 1962; PhD, Princeton U., 1966. Sr. chem. engr. Monsanto Co., Springfield, Mass., 1966-68; biomed. engr. NIH, Bethesda, Md., 1968-1986; dir. extramural rsch. CONRAD Program Ea. Va. Med. Sch., Arlington, 1986-89, dep. dir. CONRAD Program, 1989-90, dir. CONRAD Program, 1990—; tech. advisor World Health Orgn., Geneva, Switzerland, 1977—; tech. expert UN Devel. Program, Haifa, Israel, 1973. Editor: Rheology of Biological Systems, 1973, Drug Delivery Systems, 1976. Fellow Textile Research Inst.; mem. N.Y. Acad. Scis., Am. Chem. Soc., Controlled Release Soc., Sigma Xi. Avocation: nature photography. Home: 11612 Danville Dr Rockville MD 20852-3716

GABER, ROBERT, psychologist; b. N.Y.C., Nov. 5, 1923; s. William and Freda (Harris) G.; m. Heidi Walters, Apr. 3, 1967 (div. Jan. 5, 1976); 1 child, Nathan. BA, NYU, 1949, MA, 1951; PhD, Columbia Pacific U., San Rafael, Calif., 1982. Psychotherapist Nat. Hosp. for Speech Disorders, N.Y.C., 1954-57; psychologist Indsl. Home for the Blind, N.Y.C., 1957-58; sch. psychologist Roosevelt Sch., Stamford, Conn., 1958-60; sr. clin. psychologist N.Y. State Dept. Mental Hygiene, Thiells, 1960-64; staff psychologist N.Y. Med. Coll., N.Y.C., 1965-66; cons. psychologist The Salvation Army, Phila., 1971-72; psychologist Md. Dept. Mental Hygiene, 1975-76, Dept. Corrections, Balt., 1979-80; chief exec. officer Axxiom De-Stress Ctrs., Balt., 1980—; cons. Family Crisis Ctr. of Balt., 1973-74, Gov., Pa. Dept. Corrections, 1971; dir. mental health, nursery div. Dept. Welfare, N.Y.C., 1953-56. Author: Federal Prisoners' Attitudes Toward Crime and Confinement, 1982, The Experience of Enlightenment, 1980; author booklet: Comprehensive Therapy Questionnaire, 1978; author articles, pamphlets on crime, human behavior and higher states of consciousness. With USAAF, 1942-46; PTO. Mem. Am. Psychol. Assn. Democrat. Avocations: golf, horseback riding, snow and water skiing. Office: Axxiom De-Stress Ctrs PO Box 22115 Baltimore MD 21203-4115

GABLE, ROBERT S., psychology educator; b. Canton, Ohio, Mar. 21, 1934; s. Harry C. and Mary (Blackburn) Schwitzgebel. EdD, Harvard U., 1964; PhD, Brandeis U., 1964; LLD (hon.), San Fernando Valley Coll. Law, 1982. Asst. prof. psychology UCLA, 1964-70; assoc. prof. Claremont (Calif.) Grad. Sch., 1970-86, prof., 1986—; bd. dirs. Human Interaction Rsch. Inst., L.A., 1987-92; CEO Life Sci. Rsch. Group, Inc., Claremont, 1975—; mem. rev. panel NSFA, 1979. Co-author: Law and Psychological Practice, 1980, (text) Computer Aptitude and Literacy, 1984; contbr. articles to profl. publs. Grantee NIMH, 1970, 74-77, John Randolf & Dora Haynes Found., 1990. Mem. APA, Am. Psychology-Law Soc. Achievements include patent for radio controlled audio-resonators. Office: Claremont Grad Sch Psychology Dept Claremont CA 91711-6175

GABLE, ROBERT WILLIAM, JR., aerospace engineer; b. Clarinda, Iowa, Nov. 20, 1939; s. Robert William and Elsie Pearl (Stone) G.; m. Karen Elaine Clay, Feb. 4, 1961; children: Susan, Barbara, Robert. BS, Iowa State U., 1963. Engr. Boeing, Seattle, 1963; engr. GM, Indpls., 1964-71, engr., project mgr., 1973—; engr. AiRsch., Phoenix, 1971-73. Event chmn. Avon (Ind.) Jaycees, 1982; bd. dirs. Chapel Hill United, Indpls., 1972. Recipient Internal Program of Yr. award Avon Jaycees, 1982. Mem. AIAA. Home: 36 N County Rd 450 E Danville IN 46122-9412 Office: Allison Gas Turbine/GM PO Box 420 Mail Stop S49 Indianapolis IN 46206

GABUTTI, ALBERTO, physicist; b. Turin, Italy, Nov. 15, 1960; s. Vittorio and Franca (Chianale) G.; m. Carmen Marquez, Jan. 13, 1990. PhD in Physics, U. Turin, 1986. Rsch. fellow CSELT, Turin, 1986-87; researcher Inst. Elettrotecnico Nat. IEN-GF, Turin, 1987-88; postdoctoral researcher Argonne (Ill.) Nat. Lab., 1988-90; researcher in physics U. Bern, Switzerland, 1991—. Contbr. sci. articles to profl. jours. Mem. Am. Phys. Soc. Avocations: rock climbing, skiing, hiking. Office: U Bern, Sidlerstrasse 5, CH 3012 Bern Switzerland

GADDIS, M. FRANCIS, mechanical and marine engineer, environmental scientist; b. Boston, July 27, 1920; s. Michael Joseph and Catherine Agnes (Lavelle) G.; m. Marie B. Leen, Nov. 22, 1946 (dec. Feb. 1979); children: Robert L., Paul L.; m. Jeanne Bowen Crites, Oct. 27, 1990. BS, U. Ala., 1945; BA, Adelphi U., 1977, MSc, 1979, cert. environ. mgmt., 1979; MPhil, Columbia U., 1981, PhD, 1988. Chief marine engr. U.S. Army, Port of N.Y., 1944-45; svc. engr. Garlock Inc., Boston, 1946-47; ter. mgr. Garlock Inc., N.Y.C., 1947-61; pres., chief engr. Gaddis Engring. Co., Port Washington, N.Y., 1961—; mem. seals edn. workshop com. Dept. Energy, Office Naval Rsch., Am. Soc. Lubrication Engrs., ASME, 1979-80; mem. naval arch. and marine engring. com. People to People Delegation to China, 1986, Delegation to Bicentennial Maritime Symposium, Australia, 1988, ASME Delegation to S.Am., 1989. Author: Awareness of Environmental Hazards in Risk Management, 1979, Siting Criteria in Hazardous Waste Disposal, 1987, The Politics of Waste Disposal, 1989, Environmental Awareness, 1990. Recipient Disting. Alumni medal Adelphi U., 1984. Mem. ASME, Am. Soc. Tribologists and Lubrication Engrs. (emissions com. 1980-84), Nat. Assn. of Environ. Profls., Environ. Law Inst., Pacific Basin Consortium for Hazardous Waste Rsch., East-West Inst., Soc. Naval Architects and Marine Engrs., Assn. Environ. Profls. (Calif.), Marine Tech. Soc., N.Y. Acad. Scis., John Henry Newman Hon. Soc., Columbia (N.Y.C.) Club, North Shore Yacht (Port Washington) Club, Delta Tau Delta (MIT chpt. adviser 1946-47). Achievements include research on hydrogeological environmental considerations in toxic waste disposal facility siting; research on materials and systems in mechanical sealing and containment; co-development of cryogenic vapor barrier. Home: PO Box 411 Locust Valley NY 11560 Office: Gaddis Engring Co PO Box 689 Port Washington NY 11050-0215

GADDY, JAMES LEOMA, chemical engineer, educator; b. Jacksonville, Fla., Aug. 16, 1932; s. Leoma Ithama and Mary Elizabeth (Edwards) G.; m. Betty Maricella, Sept. 7, 1952; children: James, Teresa. B.S. in Chem. Engring., La. Poly. U., 1955; M.S. in Chem. Engring., U. Ark., 1968; Ph.D. in Chem. Engring., U. Tenn., 1972. Registered profl. engr., Ark. Process engr. Ethyl Corp., Baton Rouge, 1955-60; project mgr., engring. supr. Ark.-La. Gas, Shreveport, La., 1960-66; assoc. prof. chem engring. U. Mo.-Rolla, 1972-79, prof., 1979, dir. research ctr., 1980; prof., head chem. engring. U. Ark., Fayetteville, 1980-88, disting. prof., 1988—; pres. Engring. Resources, Fayetteville; cons. to 15 orgns. including TVA and UN; tchr. 85 short courses in chem. engring. for industry. Mem. editorial bd. Biomass and Biofuels, Chem. Engring. Rsch. and Devel.; contbr. 450 articles to tech. jours. Faculty fellow Swiss Fed. Inst. Tech. Zurich, 1978. Mem. Am. Inst. Chem. Engrs. (speakers bur.), Am. Chem. Soc., Am. Soc. Engring. Edn., AAAS, Tau Beta Pi (Eminent Engr. 1976), Alpha Chi Sigma, Omega Chi Epsilon. Baptist. Home: 2207 Tall Oaks Dr Fayetteville AR 72703-6126 Office: U Ark Bell Engring Ctr Fayetteville AR 72701

GAD-EL-HAK, MOHAMED, aerospace and mechanical engineering educator, scientist; b. Tanta, El-Gharbia, Egypt, Feb. 11, 1945; came to U.S., 1968; s. Mohamed Gadelhak and Samira (Hosni) Ibrahim; m. Dilek Karaca, July 19, 1976; children: Kamal, Yasemin. BSc in Mech. Engring. summa cum laude, Ain Shams U., Cairo, 1966; PhD in Fluid Mechanics, Johns Hopkins U., 1973. Instr. Air Shams U., Cairo, 1966-68; postdoctoral fellow Johns Hopkins U., Balt., 1973, U. So. Calif., L.A., 1973-74; asst. prof. engring. sci. & systems U. Va., Charlottesville, 1974-76; program mgr. Flow Rsch. Co., Seattle, 1976-86; prof. aerospace & mech. engring. U. Notre Dame, Ind., 1986—; cons. USN, Washington, 1990-91, UN, N.Y.C., 1991, many others; lectr. in field. Assoc. tech. editor AIAA Jour., 1988-91; assoc. editor Applied Mechanics Revs., 1988—; contbg. editor: Springer Verlag's Lecture Notes in Engineering, 1988—; reviewer Jour. Fluid Mechanics, Physics of Fluids, AIAA Jour., Jour. of Aircraft, many others; editor: Advances in Fluid Mechanics Measurements, 1989, Frontiers in Experimental Fluid Mechanics, 1989; contbr. over 160 articles to profl. jours. Whitehead fellow Johns Hopkins U., Balt., 1968-73; professeur invite Univ. de Grenoble, France, 1991; sr. guest NATO, Paris, 1991; rsch. grantee USN, 1976-80, USCG, 1976-78, NASA-Ames, 1981, NASA-Langley, 1985-87, 86, ONR, 1981-85, AFOSR, 1982-85, 85, Boeing Co., 1984, NSF, 1986, Flow Industries, Inc., 1986-88, Cortana Corp., 1989-90, ONR, 1991, DARPA, 1991, Bourse de Haut Niveau Ministere de la Recherche et de la Technologie, Paris, 1991-92, NATO, 1991-92, others. Fellow AIAA (assoc.); mem. AAAS, ASME, Am. Phys. Soc. (life), Am. Chem. Soc. Achievements include patents on method and apparatus for controlling bound vortices in the vicinity of lifting surfaces, for reducing turbulent skin friction, for controlling turbulent boundary layers. Office: U Notre Dame Dept Aero & Mech Engring Notre Dame IN 46556

GAERLAN, PUREZA FLOR MONZON, pediatrician; b. Imus, Cavite, Philippines, Oct. 16, 1933; came to U.S., 1958; d. Luis and Paz (Monzon) G. MD, Far Ea. U., Manila, 1957; postgrad., U. Pa., 1962-63. Diplomate Am. Bd. Pediatrics. Rotating intern Bapt. Meml. Hosp., Jacksonville, Fla., 1958-59; resident in pediatrics Albert Einstein Med. Ctr., Phila., 1959-61;

pediatric fellow M.D. Anderson & Tumor Inst., Houston, 1961-62; trainee in clin. investigation Oak Ridge (Tenn.) Inst. Nuclear Studies, 1963-64; house pediatrician Queens Hosp. & Med. Ctr., Honolulu, 1964-66; pediatric cons. Delgado Clin. & St. Rita Hosp., Manila, 1966-68; fellow in pediatric pulmonary diseases Babies Hosp. Columbia Presbyn. Med. Ctr., N.Y.C., 1968-70, from assoc. to asst. prof. in clin. pediatrics Babies Hosp., 1971-77; asst. prof. in clin. pediatrics Sch. Medicine NYU, 1977-81; assoc. attending pediatrician to attending pediatrician St. Vincents Hosp. & Med. Ctr. N.Y., N.Y.C., 1979—; assoc. dir. Cystic Fibrosis Pediatric Pulmonary Ctr., N.Y.C., 1977—; assoc. prof. clin. pediatrics N.Y. Med. Coll., Valhalla, 1981—. Contbr. articles to profl. jours. Fellow Am. Acad. Pediatrics, Royal Soc. for the Promotion of Health; mem. Am. Thoracic Soc., Pan Am. Med. Assn. (diplomate), N.Y. Acad. Scis., N.Y. Trudeau Soc., Womens Med. Assn. Roman Catholic. Achievements include research in cystic fibrosis and use of influenza vaccines. Office: 36 7th Ave Ste 509 New York NY 10011

GAERTNER, ALFRED LUDWIG, biochemist; b. Wildberg, Germany, Apr. 20, 1953; came to U.S., 1988; s. Georg and Hildegard (Clapier) G.; m. Debbra Linda Hodgson, Apr. 30, 1987; children: Alexander, Joanna, Emma, Hannah. MS in Biochemistry, U. Tuebingen, Germany, 1981, PhD in Biochemistry, 1985. Asst. U. Tuebingen, 1984-85; Univ. Grants Com. fellow Massey U., Palmerstown, N.Z., 1985-86; postdoctoral fellow U. B.C., Vancouver, Can., 1986-88; vis. scientist Genencor, Inc., South San Francisco, Calif., 1988-91; sr. scientist Genencor Internat., South San Francisco, 1991—. Contbr. to profl. publs. Coord. youth affairs Flying Club, Germany, 1972-76; animateur French-German Exch. Programs, 1975-77. Cpl. German armed forces, 1973-74. Mem. AAAS, TAPPI, Am. Chem. Soc., Internat. Union of Pure and Applied Chemistry (affiliate), Gesellschaft Biologische Chemie. Achievements include research in metalloproteins and oxidations, downstream processing and product formulations, pulp and paper biotechnology. Office: Genencor Internat 180 Kimball Way South San Francisco CA 94080

GAERTNER, RICHARD FRANCIS, manufacturing research center executive; b. Pitts., Aug. 10, 1933; s. John William and Alma Louise (Heimbuecher) G.; m. Nancy Lawlor Keary, Sept. 29, 1962; children—Barbara, Richard, Linda, Catherine. BSChemE, W.Va. U., 1955, MSChemE, 1957; PhD, U. Ill., 1959. With Gen. Electric Co., 1959-77, mgr. laminated products dept., 1971-74, mgt. tech. resources planning chem. and metall. div., 1974-77; research dir. Tech. Center Owens-Corning Fiberglas Corp., 1977-79, dir. strategic tech. planning, 1979-87; dir. Ctr. for Advanced Tech. Devel. Iowa State U., Ames, 1987—. Dow Chem. Co. fellow. Mem. Am. Inst. Chem. Engrs., Sigma Xi, Tau Beta Pi, Phi Lambda Upsilon. Achievements include patents in field. Office: Iowa State U CATD/IPRT 153 A ASC II Ames IA 50011-3041

GAETA, VINCENT ETTORE, laboratory technologist; b. N.Y.C., May 4, 1963; s. Anthony Arnold and Raffaela (Simeone) G. BA, Hofstra U., 1985; MS, NYU, 1990; postgrad., Rutgers U., Univ. Medicine & Dentistry N.J., 1992—. Cert. lab. technologist, N.Y. Sr. rsch. technologist N.Y. Hosp./Cornell U. Med. Coll., N.Y.C., 1986-88; lab. technologist Nat. Health Labs., Plainview, N.Y., 1990-92. Mem. N.Y. Acad. Scis. Achievements include discovery of new ways to isolate steroids. Office: U Medicine Dentistry NJ 675 Hoes Ln Piscataway NJ 08854-5635

GAFFAR, ABDUL, research scientist, administrator; b. Rangoon, Burma, Dec. 10, 1940; came to U.S., 1964; s. Ismmail and Khatija (Mohamed) Darji; m. Maria C. Gaffar, May 23, 1970; 1 child, Yousuf A. MS, Brigham Young U., 1965; PhD, Ohio State U., 1967. Rsch. chemist CSIR, Karachi, 1962-63; rsch. assoc. Brigham Young U., Provo, Utah, 1964-65, Ohio State U., Columbus, 1965-67; sr. scientist Colgate Tech Ctr., Piscataway, N.J., 1967-72, rsch. mgr., 1975-80, v.p. adv. tech., 1982—. Author over 100 publs. in field. Mem. AAAS, Am. Soc. Microbiology, Am. Chem. Soc. Achievements include 86 patents in field. Home: 89 Carter Rd Princeton NJ 08540 Office: Colgate Tech Ctr 909 River Rd Piscataway NJ 08854

GAFFNEY, PATRICK MICHAEL, marine biologist, aquacultural geneticist; b. New Orleans, Dec. 3, 1951; s. Peter Charles and Mary Jane (Ballina) G.; m. Perrin Dunlap Smith, Apr. 2, 1978; children: Alison, Conor, Nathaniel. AB in Zoology, U. Calif., Berkeley, 1973; PhD in Biol. Sci., SUNY, Stony Brook, 1986. Environ. cons. Jones & Stokes, Assocs., Sacramento, 1976-78; postdoctoral fellow SUNY, Stony Brook, 1986-87; asst. prof. Coll. Marine Studies U. Del., Lewes, 1987-93, assoc. prof., 1993—. Assoc. editor Estuaries, 1992—; contbr. articles to Marine Biology, Genetics, Aquaculture, others. Fulbright fellow, Cambridge, Eng., 1973-74; NOAA Sea Grant scholar, 1983-86; grad. fellow SUNY Grad. Coun., 1974. Mem. Soc. Study Evolution, Genetics Soc. Am., Phi Beta Kappa, Sigma Xi. Office: U Del Coll Marine Studies Lewes DE 19958

GAGGIOLI, NESTOR GUSTAVO, physicist, researcher, educator; b. Argentina, Mar. 7, 1940; s. Gustavo Adolfo and Clelia Petronila (Costa Frugoni) G.; m. Delia Edelmira Rodriguez, Oct. 8, 1965; children: Naymé, Natalia. M of Physics, U. Buenos Aires, 1966, PhD in Physics, 1976; Diploma d'Etudes Approfondis, U. Paris, 1971. Assoc. researcher Mil. Rsch. Inst., Buenos Aires, 1966-70; vis. researcher GE Labs., Paris, 1968-70, Inst. d'Optique, Paris, 1970-72; asst. prof. U. Buenos Aires, 1972-74; prof. U. La Pampa, Santa Rosa, 1974-77; prin. researcher Nat. Inst Indsl. Tech., Buenos Aires, 1977-86, NRC, Buenos Aires, 1976—; head optical group Nat. Agy. Atomic Energy, Buenos Aires, 1986—; vis. prof. U. Paris, 1979-80. Contbr. 60 articles to profl. jours. Mc. Adv. Coun. Nat. Nat. Pedagogic Congress, Buenos Aires, 1985-87; ministry edn. advisor Ministry Edn., Buenos Aires, 1989; sec. of sci. and tech. advisor NRC, Buenos Aires, 1991-93. Mem. Argentine Physics Assn. (pres. 1989-94), Latinoam. Fedn. Physics Assn. (pres. 1992—), Optical Soc. Am., Soc. Photo-optical Instrumentation Engrs., Societé Francaise d'Optique, European Optical Soc. Home: Cafayate 4369, 1439 Buenos Aires Argentina Office: Dept Ensayos No Destructivos, Av del Libertador 8250, 1429 Buenos Aires Argentina

GAGLIARDI, RAYMOND ALFRED, physician; b. New Haven, Nov. 20, 1922; s. Carl Albert and Carmela (Esposito) G.; m. Patricia DeTuncq, Apr. 6, 1946; children: Laura E. Bucci, John Bell. BS, Yale U., 1943, MD, 1945. Pvt. practice radiology Pontiac, Mich., 1951-92; chmn. dept. radiology St. Joseph Mercy Hosp., Pontiac, 1976-91, chmn. emeritus, 1991—. Contbr. articles to profl. jours. Capt. U.S. Army, 1946-48; PTO. Fellow Am. Coll. Radiology; mem. Am. Roentgen Ray Soc. (pres. 1987-88, recipient Gold Medal award 1989), Mich. Radiol. Soc. (pres. 1972), Mich. Med. Soc. (Disting. Svc. award, 1988), Oakland Hills Country Club, Royal Palm Yacht and Country Club. Republican. Avocation: golf. Home: 2100 Queen Palm Rd Boca Raton FL 33432

GAGLIARDI, UGO OSCAR, systems software architect, educator; b. Naples, Italy, July 23, 1931; came to U.S., 1956; s. Edgardo and Lina (Valenzuela) G.; m. Anna Josephine Italiano, July 7, 1954 (div. May 1972); children: Oscar Marco, Alex Piero. Diploma in Math. and Physics, U. Naples, Italy, 1951; DEng in Elec. Engring., U. Naples, 1954. Chief scientist U.S. Air Force, Hanscom AFB, Mass., 1966-67; v.p. tech. ops. Interactive Scis., Inc., Braintree, Mass., 1968-70; dir. engring. Honeywell Info. Systems, Waltham, Mass., 1970-75; lectr. Harvard U., Cambridge, Mass., 1967-74, prof. practice computer engring., 1974-83, Gordon McKay prof. practice computer engring., 1983—; pres. Gen. Systems Group, Inc., Salem, N.H., 1975—; chmn. Ctr. for Software Tech., Inc.; mem. NAS rsch. coun. panel Nat. Computer Systems Lab. (formerly Inst. Computer Scis. and Tech.), Nat. Inst. Standards and Tech. (formerly Nat. Bur. Standards), 1985-91, chmn., 1988-91; mem. Centerline Software, Inc., bd. dirs. 1988—. Fulbright scholar, 1955-56. Home: 5 Manor Pky Salem NH 03079-2842 Office: Harvard U 33 Oxford St Cambridge MA 02138-2901

GAGLIARDO, VICTOR ARTHUR, environmental engineer; b. Springfield, Mass., Apr. 2, 1957. BS in Civil Engring., U. Mass., 1979; MSEE, U. Fla., Gainesville, 1986. Profl. Engr., Fla. Enforcement supr. SW Fla. Water Mgmt. Dist., Tampa, 1988-93; hydrologic cons., 1993—. Contbr. Stormwater Management Model, 1986. Mem. ASCE. Home: 501 Carolyne St Temple Terrace FL 33617-3713

GAGNON, JOHN HARVEY, psychotherapist, educator; b. Derby, Conn., Dec. 16, 1946; s. Ernest John and Pauline Stella (Dziedulonis) G.; m. Carolyn Ingersoll, Oct. 4, 1980; 1 child, Isabelle Eleanor. BS, Fairfield U., 1969; MS, Western Conn. State Coll., 1976; PhD, Union Inst., 1982. Diplomate Am. Bd. Med. Psychotherapists (fellow), Am. Bd. Psychotherapy; cert. marriage and family therapist, EMT, Conn.; cert. family life educator. Counselor in tng. Conn. Valley Hosp., 1972-73; counselor Whiting Forensic Inst., 1973; coord., dir. Danbury Hosp. Day Treatment Program, 1973-77; pvt. practice, 1977-80; psychotherapy intern Counseling Ctr. and N.Y. Inst. for Gestalt Therapy, 1981-83; pvt. practice, 1983—; rsch. cons. Newtown Counseling Ctr., 1987-89; instr. N.Y. Inst. for Gestalt Therapy, 1983-89; lectr. Yale U., 1983; adj. prof. Western Conn. State U., 1983-86, U. Bridgeport, 1988-90; adj. lectr. U. Conn., Torrington, 1990-93. Author: Gagnon's Directory, 1986, Wounded Healer, 1993; contbr. articles to profl. jours. Officer emergency sta. Am. Radio Relay League, 1992—; chmn. adult program Unitarian-Universalist Soc. North Fairfield County, West Redding, Conn., 1990-91, tchr. religious edn., 1984-86; judge sr. div. Conn. State Fair, 1986—. Fellow Internat. Coun. for Sex Edn. and Parenthood; mem. ACA, AAUP, Assn. for Counselor Edn. and Supervision, Am. Soc. for Group Psychotherapy and Psychodrama, Am. Acad. Psychotherapists, Am. Assn. for Marriage and Family Therapy, Assn. for Humanistic Psychology, Internat. Assn. Marriage and Family Counselors, Nat. Coun. Family Rels., Phi Delta Kappa. Democrat. Home: 99 Stadley Rough Rd Danbury CT 06811-3230

GAHAGAN, THOMAS GAIL, obstetrician/gynecologist; b. Brush Valley, Pa., Apr. 14, 1938; s. Ben D. and Zula C. (Brown) G.; m. Mary A. Miller, Dec. 23, 1960; children: David, Diane, Kevin, Keith. BA, Washington and Jefferson Coll., 1960; MD, U. Pa., Phila., 1964. Diplomate Am. Bd. Ob/Gyn. Intern U. Ky., Lexington, 1964-65, resident in ob/gyn., 1965-68; group practice Dr. Jones and Kelch P.A., Newark, Ohio, 1970-71, Naples (Fla.) Ob/Gyn., 1971-85; pvt. practice Naples, 1985—. Capt. USAF, 1968-70. Fellow Am. Coll. Ob/Gyn., Fla. Ob/Gyn. Soc.; mem. AMA, Am. Cancer Soc. (life, bd. dirs. Collier unit 1973—, bd. dirs. Fla. div. 1976-91, pres. 1986-87, St. George medal 1990), Fla. Med. Assn., Collier County Med. Soc. (exec. com. 1989—, pres.-elect 1991-92, pres. 1992-93). Republican. Presbyterian. Avocations: scuba diving, flying, golf, snow skiing, fishing. Office: 700 2d Ave N # 305 Naples FL 33940

GAHAN, PETER BRIAN, cell biologist, researcher; b. London, Aug. 23, 1933; s. Desmond Edgeworth and Lily (Pam) G.; m. Viviane Maggi, Feb. 1963 (dec. 1974); m. Danielle France Carmignac; 1 child, Jonathan David Edgeworth. BS, U. London, 1960, PhD, 1964. Head biology Thames Poly., London, 1966-70; prof. Meml. U., Newfoundland, Can., 1970-74, Queen Elizabeth Coll., London, 1974-85; lectr. King's Coll. London, 1962-66, prof., 1985—; assoc. prof. U. Geneva, Switzerland, 1982-90. Author: Plant Histochemistry and Cytochemistry, 1984; co-author: Vascular Differentiation, 1988; editor: Autoradiography for Biologists, 1972. Fellow Inst. Biology (v.p. 1990-92), Royal Soc. Arts, Royal Microscopical Soc. (chmn. histochemistry 1976-80, coun. 1977-80); mem. Biochem. Soc., Soc. for Exptl. Biology (coun. 1984-86), Brit. Soc. of cell Biology. Avocations: opera, theatre, cricket, lit., walking. Office: King's Coll London, Campden Hill Rd, London W8 7AH, England

GAILLARD, MARY KATHARINE, physics educator; b. New Brunswick, N.J., Apr. 1, 1939; d. Philip Lee and Marion Catharine (Wiedemayer) Ralph; children: Alain, Dominique, Bruno. BA, Hollins (Va.) Coll., 1960; MA, Columbia U., 1961; Dr du Troiseme Cycle, U. Paris, Orsay, France, 1964, Dr-es-Sciences d'Etat, 1968. With Centre National de Recherche Scientifique, Orsay and Annecy-le-Vieux, France, 1964-84; maitre de recherches Centre National de Recherche Scientifique, Orsay, 1973-80; maitre de recherches Centre National de Recherche Scientifique, Annecy-le-Vieux, 1979-80, dir. research, 1980-84; prof. physics, sr. faculty staff Lawrence Berkeley lab. U. Calif., Berkeley, 1981—; Morris Loeb lectr. Harvard U., Cambridge, Mass., 1980; Chancellor's Disting. lectr., U. Calif., Berkeley, 1981; Warner-Lambert lectr. U. Mich., Ann Arbor, 1984; vis. scientist Fermi Nat. Accelerator Lab., Batavia, Ill., 1973-74, Inst. for Advanced Studies, Santa Barbara, Calif., 1984, U. Calif., Santa Barbara, 1985; group leader L.A.P.P., Theory Group, France, 1979-81, Theory Physics div. LBL, Berkeley, 1985-87; sci. dir. Les Houches (France) Summer Sch., 1981; cons., mem. adv. panels U.S. Dept. Energy, Washington, and various nat. labs. C0-editor: Weak Interactions, 1977, Gauge Theories in High Energy Physics, 1983; author or co-author 140 articles, papers to profl. jours., books, conf. proceedings. Recipient Thibaux prize U. Lyons (France) Acad. Art & Sci., 1977, E.O. Lawrence award, 1988, J.J. Sakurai prize for theoretical particle physics, APS, 1993; Guggenheim fellow, 1989-90. Fellow Am. Acad. Arts and Scis., Am. Physics Soc. (mem. various coms., chairperson com. on women, J.J. Saburai prize 1993); mem. AAAS, NAS. Office: U Calif Dept Physics Berkeley CA 94720

GAINTNER, J(OHN) RICHARD, health facility executive, medical educator; b. Lancaster, Pa., Feb. 18, 1936; s. Joseph Richard and Sarah K. (Long) G.; m. Suzanne Butler, July 29, 1961; children—Wendy, Sally, Jenny. B.A. in Philosophy magna cum laude, Lehigh U., 1958; M.D., Johns Hopkins U., 1962. Diplomate Am. Bd. Internal Medicine. Intern Univ. Hosp. Cleve., 1962-63, resident, 1963-64; mem. staff John Dempsey Hosp. U. Conn., Farmington, 1967-77, Hartford Hosp., 1970-77, Johns Hopkins Hosp., Balt., 1978-83, Albany Med. Ctr. Hosp., N.Y., 1983—; assoc. dean for clin. affairs U. Conn. Sch. Medicine, Farmington, 1970-75, asst. prof. dept. clin. medicine and health care, 1967-72, assoc. prof. depts. medicine and community medicine, 1972-77; v.p. med. affairs New Britain Gen. Hosp., Conn., 1975-77, mem. staff; assoc. dean adminstrn. Johns Hopkins U. Sch. Medicine, Balt., 1977-80, assoc. prof. medicine, 1977-83; v.p., dep. dir. Johns Hopkins Hosp., Balt., 1981-83; prof. medicine Albany Med. Coll. of Union U.; pres., CEO Albany Med. Ctr., Cancer Rsch. Inst., Boston; dir. program for internat. edn. in gynecology/obstetrics Johns Hopkins U. Contbr. articles to profl. jours. Mem. Tricentennial Com., Albany, 1985-86, Mayor's Strategic Planning com., Albany, 1983—, Health Commn. Bus. Council of N.Y. State, Albany, 1984 , Fifty Group, Albany, 1983—. Served to capt. U.S. Army, 1964-66; Vietnam. Henry Strong Denison scholar Johns Hopkins U., 1960-62; Borden Undergrad. Research awardee in medicine Johns Hopkins U., 1960. Fellow Am. Coll. Physician Execs., ACP; mem. AMA, N.Y. State Med. Soc., Albany County Med. Soc., Am. Hosp. Assn. (com. on med. edn., 1984—, ho. of dels.), Med. Adminstrv. Conf., Soc. Med. Adminstrs., Assn. for Biomed. Research (exec. com.), Phi Beta Kappa. Clubs: Ft. Orange (Albany); Schuyler Meadows (Loudonville, N.Y.). Avocations: golf; sailing. Home: 14 Birch Hill Rd Albany NY 12211-2003 Office: Cancer Rsch Inst New England Deaconess Hosp 185 Pilgrim Boston MA 02215*

GAITONDE, SUNIL SHARADCHANDRA, computer engineer; b. Nagpur, India, Dec. 12, 1960; came to U.S., 1983; s. Sharadchandra Shantaram and Sumati Sharadchandra (Prabhu) G.; 1 child, Shweta. BTech in Elec. Engring., Indian Inst. Tech., Kharagpur, 1983; MS in Computer Engring., Iowa State U., 1985, PhD in Computer Engring., 1988. Architect distributed dab mgmt. architecture IBM, Rochester, Minn., 1988-90, designer distributed computing environ. on AS/400, 1990—; rsch. contact Western Mich. U., Kalamazoo, 1989—. Contbr. articles to Comm. of ACM, IBM Systems Jour., IEEE Macro, IEEE Network mag. Nat. merit scholar Govt. of India, 1976. Mem. IEEE, Sigma Xi, Tau Beta Pi, Eta Kappa Nu. Achievements include patents pending on a data language of data description, multi-dimensional windows and editors.

GAJDUSEK, DANIEL CARLETON, pediatrician, research virologist; b. Yonkers, N.Y., Sept. 9, 1923; s. Karl A. and Ottilia D. (Dobroczki) G.; children: Ivan Mbagintao, Josede Figirliyong, Jesus Raglmar, Jesus Mororui, Mathias Maradol, Jesus Tamel, Jesus Salalu, John Paul Runman, Yavine Borima, Arthur Yolwa, Joe Yongorimah Kintoki, Thomas Youmog, Toni Wanevi, Toname Ikabala, Magame Prima, Senavayo Anua, Igitava Yoviga, Luwi Ikavara, Iram'bin'ai Undae'mai, Susanna Undapmaina, Steven Malrui, John Fasug Raglmar, Launako Wate, Louise Buwana, Regina Etangthaw Raglmar, Vincent Ayin, Daniel Sumal, Iyo Fanechigiy Raglmar, John Clayton Harongsemal, Peter Paul Ffiran, Jason Sohorang, Edwina Wes Mugunbey, Brenda Gillippin, Carleton Kalikaipapadaua Mbagintao, Basil Talonu, Gideon Waiwaime, OKovi Yarao. BS, U. Rochester, 1943; MD, Harvard U., 1946, DSc (hon.), 1987; NRC fellow, Calif. Inst. Tech., 1948-49; DSc (hon.), U. Rochester, 1977, Med. Coll. Ohio, 1977, Washington & Jefferson Coll., 1980, Hahneman Med. Coll., 1983, Med. and Dental Coll. of N.J., 1987; DHL (hon.), Hamilton Coll., 1977, U. Hawaii, 1986; LL.D. (hon.), U. Aberdeen, Scotland, 1980; Dr. honoris causa, U. Aix-Marseille, France, 1977, U. Lisbon, Portugal, 1991, U. Milan, Italy, 1992. Diplomate Am. Bd. Pediatrics. Intern, resident Babies Hosp., Columbia Presbyn. Med. Center, N.Y.C., 1946-47; resident pediatrics Children's Hosp., Cin., 1947-48; pediatric med. mission Germany, 1948; resident, clin. and research fellow Childrens Hosp., Boston, 1949-51; research fellow pediatrics and infectious diseases Harvard U., 1949-52; rsch. virologist Walter Reed AMSGS, Washington, 1952-53; with Institut Pasteur, Teheran, Iran, 1954-55; vis. investigator Nat. Found. Infantile Paralysis, Walter and Eliza Hall Inst. Med. Research, Melbourne, Australia, 1955-57; dir. program for study child growth and devel. and disease patterns in primitive cultures and lab. slow, latent and temperate virus infections NINDS, NIH, Bethesda, Md., 1958—; chief Central Nervous System Studies Lab., 1970—; chief scientist rsch. vessel Alpha Helix expdn. to Banks and Torres Islands, New Hebrides, South Solomon Islands, 1972; hon.prof. virology Hupei Med. Coll., Wuhan, Peoples Rep. of China, 1986; hon. prof. neurology Beijing Med. U., Peoples Republic of China, 1987, Las Palmas de Gran Canaria, Spain, 1993; hon. faculty Med. Sch. U. of Papua New Guinea, 1980; vis. prof. Royal Soc. of Medicine, London, 1987. Author: Hemorrhagic Fevers and Mucotoxicoses in the USSR, 1951, Jours., 45 vols., 1954-92, Slow Latent and Temperate Virus Infections, 1965, Correspondence on the Discovery of Kuru, 1976, (with Judith Farquhar) Kuru, 1980. Recipient E. Meade Johnson award Am. Acad. Pediatrics, 1963, Superior Service award NIH, HEW, 1970, Disting. Service award HEW, 1975, Prof. Lucian Dautrebande prize in pathophysiology Belgium, 1976, Nobel prize in physiology and medicine, 1976, Cotzias prize Am. Neurol. Assn., 1978, Huxley medal Royal Anthrop. Inst. Gt. Britain and Ireland, 1989, Gold medal Czechoslovak Med. Soc., 1989, Mudd award Internat. Union Microbiol. Socs., 1990, Gold medal, prize 3d. Internat. Congress on Alzheimers Disease, 1992, 2nd Pacific Rim Biotech. award, 1992, Gold medal Basque Acad. Medicine, 1993; Dyer lectr. NIH, 1974; Heath Clark lectr. U. London, 1974; B.K. Rachford lectr. Children's Hosp. Research Found., Cin., 1975; Langmuir lectr. CDC, Atlanta, 1975; Withering lectr. U. Birmingham, Eng., 1976; Cannon Elie lectr. Boston Children's Med. Center, 1976; Zale lectr. U. Tex., Dallas, 1976; Bayne-Jones lectr. Johns Hopkins Med. Sch., Balt., 1976; Harvey lectr. N.Y. Acad. Medicine, 1977; J.E. Smadel lectr. Infectious Disease Soc. Am., 1977; Burnet lectr. Australasian Soc. Infectious Disease, 1978; Mapother lectr. U. London, 1978; Disting. lectr. in medicine Mayo Clinic, 1978; Kaiser Meml. lectr. U. Hawaii, 1979; Eli Lilly lectr. U. Toronto, 1979; Payne lectr. Children's Hosp. D.C., 1981; Ray C. Moon lectr. Angelo State U., Tex., 1981; Silman lectr. Yale U., 1981; Blackfan lectr. Children's Hosp. Med. Ctr., Boston, 1981; Hitchcock Meml. lectr. U. Calif.-Berkeley, 1982; Nelson lectr. U. Calif.-Davis, 1982; Derick-MacKerres lectr. Queensland Inst. Med. Research, 1982; Bicentennial lectr. Harvard U. Sch. Medicine, 1982; Cartwright lectr. Columbia U., 1982; lectr. Chinese Acad. Med. Sci., 1983; Michelson lectr., prof., U. Tenn., Memphis, 1986; plenary lectr., Chinese Assn. Med. Virology, Yentai, 1986; returned Nobel Laureate, Karolinska Inst., Stockholm and U. Tromsö, Norway, 1986; Nobel Jubilee lectr. Karolinska U., Uppsala U., U. Trondheim, Norway, 1991; Rubbo Orator Australian Soc. Microbiology, 1992. Mem. NAS, Am. Acad. Arts and Scis., Am. Philos. Soc., Deutsche Akademie Naturförscher Leopoldina, Russian Acad. Med. Sci., Australian Acad. Sci., World Acad. Art and Sci., Royal Acads. Medicine Belgium (Antwerp, Brussels), Royal Coll. Physicians, Royal Anthrop. Inst. Gt. Britain and Ireland, Soc. Pediatric Rsch., Am. Pediatric Soc., Am. Soc. Human Genetics, Am. Acad. Neurology (Cotzias prize 1979), Soc. Neurosci., Am. Epidemiol. Soc., Infectious Diseases Soc., Am. Soc. des Oceanistes, Paris, Papua and New Guinea Sci. Soc., Academia Nacional de Medicina Mexico and Columbia, Czech and Slovak Acad. Scis., third Wordl Acad. Scis., Internat. Acad. Scis., Phi Beta Kappa, Sigma Xi.

GAL, AARON, electronics engineer; b. Petach Tikva, Israel, Aug. 15, 1955; came to the U.S., 1989; s. Chanoch and Hana (Cohen) Grinbaum; m. Olga Zevulukov Gal, July 9, 1981; children: Rickie, Assaf. Degree in electronics, Amal Petch Tikva, 1974. Mgr. final test dept. Laser Industries, Israel, 1982-83; R&D electronics engr. Laser Industries, Tel Aviv, 1983-89; tech. support mgr. Sharplan Lasers, Allendale, N.J., 1989—. Achievements include first to design computerized control system for 100W laser. Home: 28 Vivian Ct Fair Lawn NJ 07410

GAL, DAVID, gynecologic oncologist, obstetrician, gynecologist; b. Debrecen, Hungary, June 4, 1946; came to U.S., 1975; s. Sandor and Sophie (Kestenbaum) G.; m. Judy Haron, Jan. 7, 1969; children: Ori, Eldad, Daniel. MD, Technion, Haifa, Israel, 1972. Residency ob/gyn Rambam Hosp. Medical Ctr., Haifa, 1972-75, Brookdale Hosp., Bklyn., 1975-79; rsch. scientist Southwestern U.T., Dallas, 1979-81, fellow gynecologic oncologist, 1981-83; dir. gynecologic oncologist Rambam Hosp. Medical Ctr., Haifa, 1983-89, Maimonides Medical Ctr., Bklyn., 1989-90; co-dir. gyecologic oncologist North Shore Hosp. Cornell Medical Coll., Manhasset, N.Y., 1990—; pres. Israeli Soc. Coloposcopy, Israel, 1987-89; bd. dirs. Region Care North Shore Hosp., 1991—. Contbr. articles to profl. jours. With U.S. Navy, 1972—. Recipient fellow Am. Cancer Soc., 1979-83; rsch. grant UTHSC, 1982, Upjohn, 1992. Fellow Am. Coll. Surgeon, Am. Coll. Obstetrics and Gynecology; mem. Am. Soc. Clinical Oncologist, Soc. Gynecologic Oncologists. Office: Ob/Gyn North Shore Univ Hosp 300 Community Dr Manhasset NY 11020

GALAMBOS, THEODORE VICTOR, civil engineer, educator; b. Budapest, Hungary, Apr. 17, 1929; s. Paul and Magdalena (Potzner) G.; m. Barbara Ann Asp, June 25, 1957; children: Paul, Ruth, Ronald, John. BSCE, U. N.D., 1953, MSCE, 1954; PhD in CE, Lehigh U., 1959; Dr. Honoris Causa, Tech. U., Budapest, 1982. Registered profl. engr., Pa., Minn., Mo. From asst. to assoc. prof. civil engring. Lehigh U., Bethlehem, Pa., 1959-65; prof. Washington U., St. Louis, 1965-81, head dept., 1970-78; prof. U. Minn., Mpls., 1981—; cons. engr. Steel Joist Inst., Myrtle Beach, S.C., 1965—; vis. prof. U.S. Mil. Acad., West Point, 1990. Author, co-author 4 books in field; editor 1 book; contbr. over 80 articles to profl. jours. Served with U.S. Army, 1954-56. Recipient T.R. Higgins award Am. Inst. Steel Constrn., 1981. Mem. ASCE (hon., Norman medal 1983, Shortridge Hardesty award 1988, E.E. Howard award 1992), NAE, Am. Soc. Engring. Educators, Internat. Assn. Bridge and Structural Engrs. Democrat. Baptist. Avocation: photography. Home: 4375 Wooddale Ave Minneapolis MN 55424-1060 Office: Univ of Minn CME Dept Minneapolis MN 55455

GALANOPOULOS, KELLY, biomedical engineer; b. Athens, Greece, Jan. 4, 1952; came to U.S. 1970, naturalized, 1976; d. Panayotis and Catherine (Calas) G.; m. Dale S. Kruchten, Sept. 4, 1982; children: Catherine Roberta Kruchten, Stephanie Diane Kruchten. BA, CUNY, 1974; MS, Poly. Inst. N.Y., 1978, postgrad., 1982—; postgrad. L.I. U., 1982—. Dir. bio-med. engring. Wyckoff Heights Hosp., Bklyn., 1980-83, Bronx Lebanon Hosp., N.Y.C., 1983-89, Mt. Sinai Med. Ctr., N.Y.C., 1991—; cons. Environ. Co., N.Y.C., 1980-85, Joint Purchasing, N.Y.C., 1980-87; lectr. in field. Mem. Am. Soc. for Hosp. Engring. of Am. Hosp. Assn., Am. Coll. Clin. Engrs., Assn. Advancement Med. Instrumentation, IEEE, Soc. Women Engrs., N.Y. Acad. Scis. Office: Mt Sinai Med Ctr 1 Gustave L Levy Pl New York NY 10029-6504

GALANTE, JOSEPH ANTHONY, JR., computer programmer; b. Yonkers, N.Y., July 15, 1947; s. Joseph Anthony Sr. and Lavinia (Brue) G. BS in secondary edn., physics, U. Md., 1971; MS in Tech. Mgmt., Am. U., 1974. Programmer UNISYS, Green Belt, Md., 1971-76; programmer analyst N.Y. Tel., White Plains, 1976-79, Telic Corp., Darien, Conn., 1978-81; assoc. mgr. N.Y. Tel., N.Y.C., 1981-85; sr. systems specialist Telesector Resources Group, Pearl River, N.Y., 1985--. Pres. Communicators Westchester, Hasting, N.Y., 1979; chmn. by law com. Masthope Rapids Property Owners Coun., Lackawaxen, Pa., 1977-79. Recipient Apollo 11 TEam Medallion NASA, 1971, Apollo 17 Personal Contribution plaque, 1975; N.Y. State scholar, 1965. Mem. Masthope Rapids Prop. Owners Coun. (com.chmn. 1977-79), Kappa Delta Pi. Roman Catholic. Achievements include development of data conversion techniques for information interchange between UNISYS and IBM using ASCII COBOL; conversion of real time basic assembly lang. program complex from SVS to MVS. Office: NYNEX TRG 2 Blue Hill Plz 4th Flr Pearl River NY 10965

GALAS, DAVID JOHN, molecular biology educator, researcher; b. St. Petersburg, Fla., Feb. 25, 1944; s. David Emanuel and Catherine Elizabeth (Filan) G.; m. Linda Elaine Hubbard, July 1, 1967; children: David John Jr., John Ryan. BA in Physics, U. Calif. Berkeley, 1967; MA in Physics, U. Calif. Davis, 1968, PhD in Physics, 1972. Sr. scientist Lawrence Livermore Nat. Labs., Livermore, Calif., 1974-77; chargé de recherche dept. molecular biology U. Geneva (Switzerland), 1977-81; asst. prof. U. So. Calif., L.A., 1981-83, assoc. prof., 1983-86, prof., 1986—, dir. dept. molecular biology, 1986-90; dir. Health and environ. rsch. Dept. Energy, Washington, 1990-93; v.p. rsch. Darwin Molecular Corp., Seattle, 1993—. Contbr. articles and revs. to profl. jours., chpts. to books. Capt. USAF, 1972-74.

GALBRAITH, LISSA RUTH, industrial engineer; b. Balt., Mar. 9, 1954; d. James Charles and Margaret Esther (Marley) G. MS, Va. Polytech., 1984; PhD, U. South Fla., 1990. Sr. indsl. engr. BMC Industries Interconics Div., Tampa, Fla., 1984-85, E-Systmes ECI Div., St. Petersburg, Fla., 1985-86; asst. prof. Fla. State U., Tallahassee, 1990—; cons. Honeywill Defense Comm. Products Div., Tampa, 1987-88; investigator LORAL Am. Beryllium, Tallevast, Fla., 1989-90; proposal reviewer NSF, Washington, 1991, 92. Co-author: Industrial Safety in the Age of High Technology, 1992; reviewer: Irwin Publishing, 1992, (jour.) Computers and Industrial Engineering, 1989—, Journal of Production Planning and Control, 1993—. Judge Leon County Sci. Fair, Tallahassee, 1991—; mem. Fla. High Tech. and Indsl. Coun., Tallahassee, 1991—; coord. Tec-Net Fla. Indsl. Edn. Coalition, Tallahassee, 1991—; mem. planning com. Enterprise Florida. Mem. IIE (sr.), Soc. Mgf. Engrs. (sr.), Am. Soc. Engring. Edn. (Nat. Dean's List 1987), Alpha Pi Mu. Office: Fla State Univ Dept Indsl Engring 2525 Pottsdamer St Tallahassee FL 32310

GALBRAITH, NANETTE ELAINE GERKS, forensic and management sciences company executive; b. Chgo., June 15, 1928; d. Harold William and Maybelle Ellen (Little) Gerks; m. Oliver Galbraith III, Dec. 18, 1948; children: Craig Scott, Diane Frances Galbraith Ketcham. BS with high honors with distinction, San Diego State U., 1978. Diplomate Am. Bd. Forensic Document Examiners. Examiner of questioned documents San Diego County Sheriff's Dept. Crime Lab., San Diego, 1975-80; sole prop. Nanette G. Galbraith, Examiner of Questioned Documents, San Diego, 1980-82; examiner of questioned documents Galbraith Forensic & Mgmt. Scis., Ltd., San Diego, 1982—; keynote speaker Internat. Assn. Forensic Scis., Adelaide, South Australia, 1990. Contbr. articles to profl. jours. Fellow Am. Acad. Forensic Scis. (questioned documents section, del. to Peoples Rep. of China 1986, USSR, 1988); mem. Am. Soc. Questioned Document Examiners, Southwestern Assn. Forensic Document Examiners (charter), U. Club Atop Symphony Towers, Phi Kappa Phi. Republican. Episcopalian. Office: Galbraith Forensic & Mgmt Scis Ltd 701 B St Ste 1300 San Diego CA 92101-8194

GALBREATH, GARY JOHN, biology educator, researcher; b. Dallas, Mar. 20, 1950; s. Donald Howard and Ruth Ellen (Weber) G.; m. Sharon Kay Walker, Aug. 3, 1985. BS in Zoology cum laude, U. Louisville, 1972; PhD in Evolutionary Biology, U. Chgo., 1980. Vis. asst. curator mammals Field Mus. Natural History, Chgo., 1981-82, curatorial assoc. dept. geology, 1982-85, rsch. assoc. dept. geology, 1986—; vis. asst. prof. Purdue U., Hammond, Ind., 1982-85, U. Ill., Chgo., 1985-87; assoc. prof. Coll. of DuPage, Glen Ellyn, Ill., 1987-88; vis. assoc. prof. Northwestern U., Evanston, Ill., 1988-89, assoc. dir. biol. sci., 1989—. Contbr. chpts. to books, articles to sci. jours. Pres. Kennicott Club, Chgo., 1983-85, 88-90, v.p., 1986-88; mem. Chgo. Rainforest Action Group, 1988—, pres., 1991—. Mem. AAAS, Am. Soc. Mammalogists, Paleontological Soc., Orgn. Tropical Studies, Sigma Xi. Achievements include research on genetic drift as the primary cause of karyotypic evolution. Office: Northwestern U Program Biol Sci Evanston IL 60208

GAL-CHEN, TZVI, geophysicist; b. Bett Oved, Israel, Sept. 1, 1941; s. Abraham Moses and Rebecca Miriam (Wilsker) Gal-Chen; m. Josepha, Feb. 28, 1968; children: Oren, Rebecca Rikki. M.Sch. in Math., Tel-Aviv U., 1970; PhD in Geology, Columbia U., 1973. Postdoctoral fellow NCAR, Boulder, Colo., 1973-74, CIRES, Boulder, Colo., 1974-75; asst. prof. U Toronto, Ont., 1975-79; vis. scientist NCAR, Boulder, 1979-80; rsch. assoc. NASA, Greenbelt, Md., 1980-82; assoc. prof. U. Okla., Norman, 1982-86, prof. meteorology, 1986—; dir. summer sch. NATO, Bonas, France, 1982; adv. bd. NCAR Rsch. Application Program, Boulder, 1991—; program chair Internat. Conf. Radar Meteorology, 1992-93. Co-chief, editor Jour. Atmospheric Scis., 1993—; assoc. editor Jour. Applied Meteorology, 1982-88; editor: Meso-Scale Meteorology, 1983; contbr. articles to profl. jours. Rsch. grantee NSF, NASA, NOAA, AAAS, Dept. of Def. Mem. AAAS (mem. nominating com. 1986-90), Am. Geophys. Union, Am. Meteorol. Soc. Jewish. Achievements include research in computational fluid dynamics, numerical weather predictions, atmospheric turbulence and waves, data analysis, remote sensing of the atmosphere by radars and lidars, and climate dynamics. Office: U Okla Dept Meteorology Norman OK 73019

GALDIKAS, BIRUTE, primatologist; m. Rod Brindamour; 1 child, Binti; m. Pak Bohap bin Jalan, 1981; children: Frederick, Jane. Studied at, UCLA. Investigator of orangutans Borneo, 1971—; lectr. Simon Fraser Univ., Vancouver, B.C., Can.; founder Orangutan Found. Contbr. articles to profl. journals. Address: Camp Leakey, Pangkalanbun Borneo, Indonesia

GALE, GEORGE DANIEL, JR., philosophy of science educator, researcher; b. Granite City, Ill., Aug. 10, 1943; s. George D. and Virginia Lee (Bruch) G.; m. Carol Jean Hammes, Aug. 6, 1966. BA, U. Santa Clara, 1965; MA, San Francisco State U., 1967; PhD, U. Calif., Davis, 1971. Engring. asst. Aerojet-Gen. Corp., Sacramento, 1963-65; adminstr., tchr., Lux Lab. San Francisco Unified Sch. Dist., 1965-67; winemaker, owner Midi Vineyard Ltd., Lone Jack, Mo., 1973-85; prof. philosophy and phys. sci. U. Mo., Kansas City, 1971—; vis. prof., fellow U. Pitts. Ctr. for Philosophy of Sci., 1983; vis. prof. Oxford (Eng.) U., 1984, Wuhan (China) U., 1986. Author: Theory of Science, 1979; contbr. articles to Sci. Am., Nature, Am. Jour. Physics. Capt. U.S. Army, 1970-73. Mem. AAAS, Philosophy of Sci. Assn., History of Sci. Soc. Home: 5210 W 77th Ter Prairie Village KS 66208 Office: U Mo Dept Philosophy Kansas City MO 64110

GALE, STEPHAN MARC, civil engineer; b. Cleve., Jan. 29, 1952; s. Irwin and Bernice (Raim) G.; m. Suzanne S. Napolitan, May 5, 1979; children: Kelly D., Michael J., Joseph J. BSCE, Ohio State U., 1974, MS, 1976. Registered profl. engr., Minn., Ill., Wis., Ohio, S.D. Researcher NSF, Prince of Wales Island, Alaska, 1974; field engr. Herron Testing Labs., Inc., Cleve., 1974; project engr. Soil Testing Svcs., Inc., Chgo., 1976-80; sr. engr. STS Cons. Ltd., Mpls., 1980-83, dir. engring., 1984—; internat. adv. coun. Geotech. Fabrics Report, St. Paul, 1990—. Contbr. papers to profl. publs. Planning commr. City of Brooklyn Park, Minn., 1980-87. Fellow ASCE (Young Engr. of Yr. award 1988); mem. ASTM, Minn. Soc. Profl. Engrs. (bd. dirs. 1987-89, sec. 1988-89, dir. pub. rels. 1987, Young Engr. of Yr. award 1986), Transp. Rsch. Bd. Achievements include research in field of geosynthetic engineering. Home: 14325 44th Ave Minneapolis MN 55446 Office: STS Cons Ltd 3650 Annapolis Ln Minneapolis MN 55447

GALEAZZA, MARC THOMAS, neuroscientist; b. Syracuse, N.Y., May 20, 1962; s. Wayne Harry and JoAnne (Bonadonna) Ferris. BS, William and Mary Coll., 1984; M Engring., U. Va., 1987. Lab. technician Uniformed Svcs. U. of Health Scis., Bethesda, Md., 1987-88; predoctoral fellow U. Minn., Mpls., 1988—. Del. Dem. State Senate Dist. Conv., Roseville, Minn., 1992; vol. scientist Sci.-by-Mail Sci. Mus. Minn., St. Paul, 1992-93. Mem. AAAS, Soc. Neurosci. (sec. Voyageurs chpt. 1991—), World Fedn. Neuroscientists, N.Y. Acad. Scis. Democrat. Achievements include research in peptides and brain research. Office: U Minn Dept Cell Biology 4-135 Jackson Hall 321 Church St SE Minneapolis MN 55455

GALEYEV, ALBERT ABUBAKIROVICH, physicist; b. Oct. 19, 1940. Univ. Novosibirski. With Inst. Nuclear Physics, U.S.S.R. Acad. Scis., 1961-70; sr. researcher Inst. High Temperatures, U.S.S.R. Acad. Scis., 1970-73; head of sect. Inst. Space Rsch., U.S.S.R. Acad. Scis., 1973—, prof., 1980. Recipient Lenin prize, 1984. Achievements include research on physics of plasma and cosmic physics. Office: Institute of Space Research, Ulitsa Profsoyuznaya 84/32, 117810 Moscow Russia*

GALIATSATOS, VASSILIOS, chemist, educator; b. Athens, Greece, Apr. 1, 1958; came to U.S., 1982; s. Gerassimos and Paraskevi (Lenoutsou) G.; m. Kleoniki Stathis, Nov. 4, 1985. BS, U. Thessaloniki, Greece, 1982; MS, PhD, U. Cin., 1986. Rsch. assoc. U. Wash., Seattle, 1986-87; sr. rsch. physicist Goodyear Tire and Rubber Co., Akron, Ohio, 1987-90; asst. prof. U. Akron, 1990—. Contbr. articles to Macromolecules, Jour. Computational Chemistry, Rubber Chem. Tech., Jour. Polymer Sci., Polymer Bull. Mem. Amnesty Internat., N.Y.C., 1990—. Mem. Am. Chem. Soc., Am. Phys. Soc., AAAS, Soc. Plastic Engrs. Office: U Akron Inst Polymer Sci Akron OH 44325-3909

GALINDO, MIGUEL ANGEL, economics educator; b. Madrid, June 7, 1960; s. Angel and Teresa (Martin) G.; m. Maria Isabel Abradelo, Jan. 24, 1986; children: Miguel Angel, Isabel. D. in Econs., U. Complutense, Madrid, 1988. Lectr. Colegio Universitario San Pablo Ceu, Madrid, 1984-91; assoc. prof. U. Complutense, Madrid, 1987-91; mgr. econs. sect. Colegio Universitario San Pablo Ceu, Madrid, 1988-91, prof. econs., 1991—; prof. econs. U. Complutense, Madrid, 1991—, Colegio Cardenal Cisneros, Madrid, 1991—; collaborator in rsch. Fiscal Studies Inst., Madrid, 1988—, Ceseden (Def. Ministry), Madrid, 1990—. Author: Fiscal Policy Theory, 1990, Development Policy, 1989, Monetary Policy Course, 1991, Fiscal Policy: New Approaches, 1991, Applied Economics Controversies, 1992. Scholar Consejeria de Economia y Hacienda, Toledo, Spain, 1989-90, Fin. Ministry, Spain, 1984-85. Mem. Sci. Orgn. European Applied Econs., Royal Econ. Soc., Am. Econ. Assn., Assn. for Evolutionary Econs., Western Econ. Assn. Internat., Soc. for Econ. Devel. Roman Catholic. Avocations: classical music, chess, reading. Home: Avenida de America 50, 28028 Madrid Spain Office: Coll U San Pablo CEU, c/ Julian Romea, 23, 28003 Madrid Spain also: U Complutense Madrid, Campus de Somosaguas, 28023 Madrid Spain

GALL, DONALD ALAN, data processing executive; b. Reddick, Ill., Sept. 13, 1934; s. Clarence Oliver and Evelyn Louise (McCumber) G.; m. Elizabeth Olmstead, June 25, 1960 (div. 1972); children: Christopher, Keith, Elizabeth; m. Kathleen Marie Insognia, Oct. 13, 1973; 1 child, Kelly Marie. BSME, U. Ill., 1956; SM, MIT, 1958, ME, 1960, ScD, 1964. Research engr. Gen. Motors, Detroit, 1956-57; staff engr. Dynatech Corp., Cambridge, Mass., 1959-60; mgr. ctr. systems Dynatech Corp., Cambridge, 1962-63; asst., assoc. prof. Carnegie-Mellon U., Pitts., 1964-69; assoc. prof. surgery and anesthesiology U. Pitts. Sch. Medicine, 1969-73; vis. fellow IBM Research Lab., Rueschlikon, Switzerland, 1970-71; pres. Omega Computer Systems, Inc., Scottsdale, Ariz., 1973—. Contbr. articles to profl. jours.; inventor fuel injection system. Bd. dirs. Scottsdale Boys and Girls Club, 1982—; mem. Scottsdale Head Honchos, 1978-87; mem. Verde Vaqueros, 1987—. Recipient Taylor medal Internat. Conf. on Prodn. Rsch. Mem. AAAS, ASME, Sigma Xi, Pi Tau Sigma, Tau Beta Pi, Phi Kappa Phi. Avocations: horseback riding, skiing, golf. Home: 9833 E Cortez St Scottsdale AZ 85260-6012 Office: Omega Computer Systems Inc 4300 N Miller Rd Ste 136 Scottsdale AZ 85251-3620

GALL, ERIC PAPINEAU, physician educator; b. Boston, May 24, 1940; s. Edward Alfred and Phyllis Hortense (Rivard) G.; m. Katherine Theiss, Apr. 20, 1968; children: Gretchen Theiss Gall, Michael Edward. AB, U. Pa., 1962, MD, 1966. Asst. instr. U. Pa., Phila., 1970-71, post doctoral trainee, fellow, 1971-73; asst. prof. U. Ariz., Tucscon, 1973-78, assoc. prof., 1978-83, prof. internal medicine, 1983—, prof. surgery, 1983—, prof. family/community medicine, 1983—, chief rheumatology allergy and immunology, 1983—, dir. arthritis ctr., 1986—. Author, editor: Rheumatoid Arthritis: Illustrated Guide to Path DX and Management of Rheumatoid Arthritis, 1988, Rheumatic Disease: Rehabilitation and Management, 1984, Primary Care, 1984; contbr. numerous articles to profl. jours. Chmn. med. and scientific com. Arthritis Found., Tucson, 1979-81. Maj. M.C., U.S. Army; Vietnam. Decorated Bronze Star; recipient Addie Thomas Nat. Svc. award Arthritis Found., 1988. Fellow ACP, Am. Coll. Rheumatology (founding chair ednl. materials com. 1986-89, bd. dirs. 1992-95, chmn. rehab. section 1992-95); mem. Arthritis Health Professions Assn. (nat. pres. 1982-83), Am. Assn. Med. Colls., Am. Fedn. Clin. Rsch., Arthritis Found. (vice chmn. nat. 1982-83, profl. edn. com. 1991—, chmn. ednl. materials com. 1991—), Alpha Omega Alpha, Alpha Epsilon Delta. Avocations: photography, fishing. Office: U Ariz Athritis Ctr 1501 N Campbell Ave Tucson AZ 85724-0001

GALLAGER, ROBERT GRAY, electrical engineering educator; b. Phila., May 29, 1931; s. Jacob Boon and May (Gray) G.; m. Ruth Atwood, Oct. 19, 1957 (div. July 1981); children: Douglas, Ann, Rebecca; m. Marie Tarnowski, July 18, 1981. BEE, U. Pa., 1953; MEE, MIT, 1957, ScD, 1960. Mem. tech. staff Bell Telephone Labs., Murray Hill, N.J., 1953-54; rsch. asst. MIT, Cambridge, Mass., 1956-60, asst. prof., 1960-64, assoc. prof., 1964-67, prof., 1967—; co-dir. Lab. Info. and Decision Systems, 1986—; chmn. adv. com. NSF Div. on Networking and Comm. Rsch. and Intrastructure, Washington, 1989-92; mem. adv. coun. Elec. Engring. Dept., U. Pa., 1991—. Author: Information Theory and Reliable Communication, 1968; co-author Data Networks, 1987, 2d edit. 1992; patentee in field. Recipient Gold medal Moore Sch., U. Pa., 1973; Guggenheim fellow, 1978. Fellow IEEE (Baker prize 1966, Medal of Honor 1990); mem. NAS, NAE, Infor. theory Soc. of IEEE (bd. govs. 1965-72, 79-88, pres. 1971). Avocations: piano, skiing, windsurfing. Home: 7 Wainwright Rd # 31 Winchester MA 01890 Office: MIT Rm 35-206 Dept Elec Engring & Computer Sci Cambridge MA 02139

GALLAGHER, EDWARD JOSEPH, scientific marketing executive; b. Passaic, N.J., June 10, 1952; s. Edward Joseph and Elizabeth Alice (Lennon) G.; m. Patricia Ellen Connolly, June 4, 1977 (div. Dec. 1982); m. Carole Mary Andresko, Dec. 1, 1984; children: Erin, John. BS, M in Mgmt. Sci., Stevens Inst. Tech., 1976. Sr. cons. Chase Econometrics/Interactive Data Corp., West Orange, N.J., 1978-80; mgmt. scientist Ethicon Inc., Somerville, N.J., 1980-82, mgr. mktg. info., forecasting, 1982-85; forecasting analyst Glaxo Inc., Research Triangle Park, N.C., 1985-87, mgr. forecasting, 1987-89, group mgr. forcasting, price, 1989-91, dir. mktg. econs. and rsch., 1991-93; dir. new product planning Glaxo Inc., Reseach Triangle Park, N.C., 1993—. Sgt. USMC, 1971-72. Mem. Pharm. Mgmt. Sci. Assn. (pres. 1989-90), Mgmt. Sci. Roundtable. Roman Catholic. Office: Glaxo Inc 5 Moore Dr Research Triangle Park NC 27709

GALLAGHER, JOAN SHODDER, research immunologist; b. Camden, N.J., June 19, 1941; d. Emanuel S. and Marie T. (Danella) Shodder; m. Joseph P. Gallagher, Jan. 30, 1965; children: Joseph, Timothy. BS, Drexel U., 1964; postgrad., Stanford U., 1964-65; MS, U. Ill., 1967, PhD, 1970. Rsch. asst. U. Pa., Phila., 1962-64; asst. prof. immunology U. Ill., Champaign-Urbana, 1970-74; asst. prof., assoc. prof. U. Cin., 1980-89, rsch. assoc., 1974-80, 89—; reviewer Jour. Allergy and Immunology, 1986-90; mem. grant rev. com. for small bus. innovative rsch. NIH, 1986-87. Contbr. articles to profl. jours. Foster parent Montgomery County Children's Svcs., Dayton, Ohio, 1973-79; editor history book Bellbrook (Ohio) Hist. Soc., 1981; active sch. bd. campaigns Sugarcreek Sch. System, Bellbrook, 1985-90, pres. Athletic Boosters, 1981-83. Recipient Adolph G. Kammer merit award in authorship Am. Occupational Med. Assn., 1983. Mem. Am. Assn. Immunologists, Am. Acad. Allergy and Immunology, Sigma Xi. Democrat. Roman Catholic. Home: 3340 Ferry Rd Bellbrook OH 45305 Office: U Cin Med Ctr Pathology Dept 231 Bethesda Ave Cincinnati OH 45267

GALLAGHER, NEIL PAUL, metallurgical engineer; b. Eureka, Calif., Apr. 12, 1962; s. William Anthony Gallagher and Peggy Jean (Patterson) Rose; m. LeeAnn Colegrove, Aug. 23, 1980; children: Jacob, Nicholas, Daniel, Maggie. BS in Chem. Engring., U. Nev., 1984, MS in Metall. Engring., 1987. Chem. engr. U.S. Dept. Interior Bur. Mines, Reno, Nev., 1985-87; metallurgist Golden Sunlight Mines Inc., Whitehall, Mont., 1987—. Pres. Whitehall AAU Wrestling Club, 1991, treas., 1991—; alderman Whitehall Town Coun., 1992—. Mem. AIME, Am. Inst. Chem. Engrs., KC (trustee, dept. grand knight). Achievements include research on effect of electrochem. potential on the absorptive properties of activated carbon for gold complexes. Home: PO Box 262 112 N Division St Whitehall MT 59759 Office: Golden Sunlight Mines Inc 453 Mt Hwy 2 E Whitehall MT 59759

GALLAGHER, RICHARD HUGO, university official, engineer; b. N.Y.C., N.Y., Nov. 17, 1927; s. Richard Anthony and Anna (Langer) G.; m. Terese Marylyn Doyle, May 17, 1952; children: Marylee, Richard, William, Dennis, John. BCE, NYU, 1950, MCE, 1955; PhD, SUNY, Buffalo, 1966; Dr. Tech. (honoris causa), Tech. U. Vienna, 1987; Hon. F., U. Coll. Swansea, U. Wales, 1987; PhD (hon.), Shanghai U. Tech., 1992. Field engr. CAA, Dept. Commerce, Jamaica, N.Y., 1950-52; structural designer Texaco, N.Y.C., 1952-55; asst. chief engr. Bell Aerospace Co., Buffalo, 1955-67; prof. civil engring. Cornell U., 1967-78, chmn. dept. structural engring., 1969-78; dean Coll. Engring., U. Ariz., 1978-84; provost, v.p. acad. affairs Worcester (Mass.) Poly. Inst., 1984-88; pres. Clarkson U., Potsdam, N.Y., 1988—; cons. in field. Author: Finite Element Analysis, 1975, Matrix Structural Analysis, 1979; editor: Internat. Jour. Numerical Methods in Engring, 1969—. Served with USNR, 1945-47. Fulbright fellow Australia, 1973, Sci. Rsch. Coun. fellow U. Wales, 1974; recipient Clifford C. Furnas Meml. award SUNY-Buffalo, 1991. Fellow ASCE, ASME (Worcester Reed Warner medal 1985, ASME medal 1993), AIAA (Structures, Structural Dynamics and Materials medal 1990), Am. Soc. Engring. Edn. (Lamme medal 1990, Centennial medallion 1993, Hall of Fame 1993), Sigma Xi, Chi Epsilon, Tau Beta Pi; mem. IACM (Congress medal 1991), Nat. Acad. Engring., Soc. Exptl. Stress Analysis. Roman Catholic. Home: 71 Pierrepont Ave Potsdam NY 13676-2109 Office: Clarkson U Office of Pres Potsdam NY 13699-5500

GALLAGHER, TANYA MARIE, speech pathologist, educator; b. Rockford, Ill., Aug. 19, 1945; m. Kenneth L. Watkin; 1 child, Laura. BS summa cum laude, U. Ill., 1967, MS in Speech-Lang. Pathology, 1969, PhD in Speech and Lang. Sci., 1971. Asst. prof. dept. speech and dept. phys. medicine & rehab. U. Mich./U. Mich. Med. Sch., Ann Arbor, 1972-77, assoc. prof., 1977-78, assoc. prof. speech & hearing scis. program, 1978-87; prof. & dir. sch. human communication disorders McGill U. Faculty of Medicine, Montreal, Que., 1987—; assoc. dean allied health scis. McGill U. Faculty of Medicine, Montreal, 1991—. Editor: Pragmatics of Language: Clinical Practice Issues, 1991; contbr. chpts. to books; contbr. numerous articles to profl. publs.; editor Jour. Speech & Hearing Rsch., 1981-83; assoc. editor Linguistics and Edn.: An Internat. Rsch. Jour., 1987-92, Human Comm. Can., 1987-89, Jour. Speech-Lang. Pathologists and Audiologists, 1989—, Jour. Speech & Hearing Rsch., 1977-80; editorial cons. Jour. Speech & Hearing Rsch., 1974-77; reviewer Jour. Applied Psycholinguistics, 1980—. Grantee U. Mich., 1981-82, 89-91, Spencer Found., 1982-83, IBM Corp., 1983-86, 86-87, Nat. Inst. Edn., 1986-88, Fonds pour la Formation de Chercheurs et l'Aide a la Recherche, Que., 1989-91, 93—, Coopération Que.-Provinces Canadiennes Enseignement Superieur et Recherce, Que., 1992-93, MacKay Ctr./McGill U., 1992—. Fellow Am. Speech-Lang.-Hearing Assn. (v.p. rsch. & tech. 1992—, chair publs. bd. 1988-91, chair coun. editors 1988-91, trustee Am. Speech-Lang.-Hearing Found. Bd. 1992—, mem. publs. bd. 1981-83, 85-88, chair lang. scis. program com. Nat. Conv. 1986, others); mem. Can. Assn. Speech-Lang. Pathologists & Audiologists (publs. bd. 1992—), Phi Beta Kappa, Zeta Phi Eta. Office: McGill University, 1266 Pine Ave W, Montreal, PQ Canada H3G 1A8

GALLAHER, WILLIAM MARSHALL, dental laboratory technician; b. Philipsburg, Pa., June 10, 1952; s. Marshall William and Florence Marie (Milner) G. Degree in Dental Tech., Hiram G. Andrews Ctr., 1971; BS, Rutgers U., 1979. Cert. dental technician in full dentures. Dental lab. technician to pvt. practice dentist Osceola Mill, Pa., 1971-72; dental lab. technician Profl. Dental Lab., South Amboy, N.J., 1972-79; instr. dental lab. tech. Union Tech. Inst., Neptune, N.J., 1979-84, Hiram G. Andrews Ctr., Johnstown, Pa., 1980-91; owner Gallaher's Dental Lab., Asbury Park, N.J., 1982-90; sr. dental lab. technician Denture Walk-In Ctr., Harrisburg, Pa., 1991—; adv. bd. Union Tech. Inst., 1984-90, Hiram G. Andrews Ctr., 1991-92; founder, pres. Person Enjoying New and Innovative Software User Group, Asbury Park, 1985-90. Author instrnl. manuals. Vol. deaf svcs. Monmouth County Deaf Group, Asbury Park, 1976-77; publicity chmn. Neighbor Preservation Program, Asbury Park, 1979-82. Mem. Nat. Dental Lab. Assn., Nat. Denturist Soc., N.J. Denturist Soc., Internat. Brotherhood Magicians, Masons (sr. master of ceremonies 1982—). Achievements include research on low-cost denture procedures, cleft palate and post cancerous intra-oral appliances. Home: PO Box 2767 Harrisburg PA 17105-2767 Office: Denture Walk In Ctr 2023 N 2d St Harrisburg PA 17102

GALLAWAY, BOB M., consulting engineer; b. Kosciusko, Miss., Oct. 14, 1916; s. William Franklin and Lieura (Williams) G.; m. Emily Susan Gillehay, Feb. 1, 1941; children: Robert Michael, Suzanne, Mary Cay. B-SChemE, Tex. A&M U., 1943, MSChemE, 1946, MSCE, 1956. Registered profl. engr., Tex. Teaching asst. Tex. A&M U., College Station, 1941-43, instr. drafting and descriptive geometry, 1945-46, asst. rsch. engr. Tex. Transp. Inst., 1947-57, asst. prof. civil engring., 1947-57, assoc. rsch. engr. Tex. Transp. Inst., 1957-59, assoc. prof. civil engring., 1957-59, rsch. engr. Tex. Tranp. Inst., 1959-85, prof. civil engring., 1959-85, prof. emeritus, 1985—; pres. Cons. and Rsch. Svcs., Inc., Bryan, Tex., 1965—; bd. cons. Transp. Ctr., U. P.R., U. Ill.. Contbr. over 250 articles to profl. publs., chpts. to books. Recipient Engr. of Yr. award Brazos chpt. Tex. Soc. Profl. Engrs., 1969-70. Fellow ASCE (life, publs. com. 1985-88, tech. papers com. 1970—); mem. Assn. Asphalt Paving Techs. (Svc. award 1975, 2d v.p. 1973, 1st v.p. 1974, pres. 1975, dir.-at-large 1972-74), Nat. Asphalt Pavement Assn. (Hall of Fame 1988), Phi Kappa Phi. Methodist. Achievements include 4 patents in field of asphalt pavement technology. Home: 2904 Par Dr Bryan TX 77802 Office: Tex A&M U Civil Engring Dept College Station TX 77843

GALLAWAY, LOWELL EUGENE, economist, educator; b. Toledo, Jan. 9, 1930; s. Leroy and Bessie Marguerite (Hiteshew) G. Means; m. Gladys Elinor McGhee, Dec. 19, 1953; children: Kathleen Elizabeth Gallaway Searles, Michael Scott, Ellen Jane Gallaway Kroutel. B.S., Northwestern U., 1951; M.A., Ohio State U., 1955, Ph.D., 1959. Asst. prof. Colo. State U., Fort Collins, 1957-59; asst. prof. San Fernando Valley State Coll., Northridge, Calif., 1959-62; vis. asso. prof. U. Minn., Mpls., 1962-63; chief analytic studies sect. Social Security adminstrn., Balt., 1963-64; asso. prof. U. Pa., Phila., 1964-67; prof. econs. Ohio U., Athens, 1967-74; disting. prof. Ohio U., 1974—; vis. prof. U. Lund, Sweden, 1973, U. Tex., Arlington, 1976, U. New South Wales, Australia, 1978, U. N.C., Chapel Hill, 1980, Mara Inst. Tech., Kuala Lumpur, Malaysia, 1987; staff economist Joint Econ. Com. U.S. Congress, 1982. Author: The Retirement Decison, 1965, Inter-industry Labor Mobility in the United States 1957-1960, 1967, Geographic Labor Mobility in the United States 1957-1960, 1969, Manpower Economics, 1971, Poverty in America, 1973, The "Natural Rate" of Unemployment, 1982, Paying People to Be Poor, 1986, Poverty, Income Distribution, The Family and Public Policy, 1986, Doing More with Less, 1991, Out of Work, 1992; contbr. articles to profl. jours. Served with USN, 1951-54. Ford Found. faculty fellow, 1960, fellow Gen. Electric Found., 1962, Pacific Inst. for Pub. Policy Research, Liberty Fund fellow Inst Humane Studies, 1983; Ford Rockefeller Population policy research grantee, 1974-75; Fulbright-Hays sr. scholar Australia, 1978. Mem. Manhattan Inst. (assoc.), European Acad. Arts and Scis. and Humanities (corr. mem.), Phi Beta Kappa, Beta Gamma Sigma. Home: 33 Longview Heights Rd Athens OH 45701-3335

GALLEGO-JUAREZ, JUAN ANTONIO, ultrasonics research scientist; b. Porcuna, Jaen, Spain, July 10, 1941; s. Francisco and Providencia (Juarez-Villa) Gallego-Casado; m. Julia Garrido-Sebastian, Apr. 7, 1969; children: Juan, Javier. Licenciatura' in Sci., U. Madrid, 1966, PhD in Phys. Sci., 1971; D of Physics, U. Rome, 1970. Grad. fellow Spanish Rsch. Coun., Madrid, 1965-66, postgrad. fellow, 1966-67, Sci. collaborator, 1970-72, sci. researcher, 1972-88, rsch. prof., 1988—; sci. collaborator Italian Rsch. Coun., Rome, 1967-70; head ultrasonics unit Inst. AcusticaCSIC, Madrid, 1978-91; dir. Inst. Acustica CSIC, Madrid, 1991—; invited lectr. Internat. Soc. Phys. Acoustics, Erice, Italy, 1985, U. Santiago, Chile, 1986, Spanish Inst., Tech. U., Denmark, 1986, 1st and 2d Workshop Power Sonic Ultrasonic Transducers, Lille, France, 1988, Toulon, 1990, Ultrasonics Internat., Madrid, 1989, ASME, Atlanta, 1991. Contbr. articles to profl. jours., chpts. to books; patentee in field. Recipient Sci. Rsch. prize Savings Bank Cordoba, 1971; grantee Royal Soc. London, Bristol, Eng., 1981, Danish Ministry Edn., Lyngby, Denmark, 1982. Fellow Inst. Acoustics (Eng.); mem. IEEE, Ultrasonics Internat. (jour. adv. bd. 1979—, internat. conf. session pres., chmn. 1983-91, organizer 1987-89, panel mem. 1990-91), Acoustical Soc. Am., Spanish Acoustical Soc. (bd. dirs. 1991—), N.Y. Acad. Scis. Avocations: skiing, Andalusian folklore, theatre, literature, classical music. Office: Inst Acustica CSIC, Serrano 144, 28006 Madrid Spain

GALLETTI, PIERRE MARIE, artificial organ scientist, medical science educator; b. Monthey, Switzerland, June 11, 1927; s. Henri and Yvonne (Chamorel) G.; m. Sonia Aidan, Dec. 31, 1959; 1 son, Marc-Henri. BA in Classics, St. Maurice Coll., Switzerland, 1945; MD, U. Lausanne, Switzerland, 1951; PhD in Physiology and Biophysics, U. Lausanne, 1954; ScD (hon.), Roger Williams Coll., U. Nancy, France, U. Ghent, Belgium. Asst. prof. physiology Emory U., 1958-62, assoc. prof., 1962-66, prof., 1966-67, vis. prof., 1967-68; prof. med. sci. Brown U., 1967—, Univ. prof., 1991—, chmn. div. biol. sci., 1968-72, v.p. biology and medicine, 1972-91; pres. Found. for Biotech., Turin, Italy, 1991—; sci. adv. com. I-Stat, Princeton, N.J., 1984-90, Sorin Biomedica S.P.A., Turin, Italy, 1985—, Cardiopulmonics, Salt Lake City, 1988—, Cytotherapeutics, Inc., Providence, 1989—; bd. dirs. Sorin Biomedica s.p.a., Turin, 1985—, chmn. bd., 1987-90; bd. dirs. Sorin Biomedical, Inc., Irvine, Calif., 1993—; chmn. Consensus Devel. Conf. NIH, chmn. devices and tech. br. task force; Hastings lectr. NIH, 1979; plenary lectr. World Biomaterials Conf., 1980, 88; McNeil Pharm. Spring Sci. lectr., 1982; lectr. Whittaker; lectr. German Surg. Soc., 1987, Japan Soc. Artificial Organs, 1987, 92, Soc. Cardiac Anesthesiologists, 1992, Bio-engineering Soc., 1992; trustee Morehouse Sch. Medicine, 1975—, overseer Tufts U. Sch. Medicine, 1984—. Author: Heart-Lung Bypass: Principles and Techniques of Extracorporeal Circulation, 1962; contbr. chpts. to books, articles, abstracts to profl. jours. R.I. Philharmonic, 1988—. Recipient John H. Gibbon award Am. Soc. Extracorporeal Technology, 1980, R.I. Gov.'s Sci. and Tech. award, 1987, Runzi prize, Switzerland, 1988, R.I. Commodore award, 1987; grantee NIH, 1962-91. Fellow Am. Coll. Cardiology, Am. Inst. Med. Biol. Engring. (v.p. 1991-93, pres.-elect 1993-94); mem. AAAS, Am. Physiol. Soc., Swiss Physiol. Soc., Royal Acad. Medicine (Brussels, fgn. corr.). Office: Brown U Box G B-393 Providence RI 02912

GALLI, PAOLO, chemical company executive; b. Bassano del Grappa, Italy, Aug. 29, 1936; came to U.S., 1985; s. Carlo and Vittoria (Nanni) G.; m. Anna Maria Marcon; children: Giulia, Stefano, Massimo. PhD in Indsl. Chemistry, U. Padua, Italy, 1962. Permanent adj. prof. phys. chemistry U. Bologna and Ferrara, Italy, 1962-84; permanent prof. macromolecular sci. U. Ferrara, Italy, 1984-85; dir. rsch. Giulio Natta Rsch. Ctr., Ferrara, 1977-83; gen. mgr. Montedison SpA, Milan, 1983-85; v.p. HIMONT Inc. HIMONT Inc., Wilmington, Del., 1985—. Holder of 30 patents in field; contbr. aricles to profl. jours. Office: Himont Inc 3 Little Falls Ctr 2801 Centerville Rd PO Box 15432 Wilmington DE 19850-5439

GALLIAN, JOSEPH ANTHONY, mathematics educator; b. New Kennington, Pa., Jan. 5, 1942; s. Joseph Anthony Gallian and Alvira Helen (Gardner) Strauss; m. Charlene Toy, May 29, 1965; children: William, Ronald, Kristin. BA, Slippery Rock State U., 1966; PhD, Notre Dame U., 1971. Vis. asst. prof. Notre Dame (Ind.) U., 1971-72; asst. prof. U. Minn., Duluth, 1972-76, assoc. prof., 1976-80, prof., 1980—. Author: Contemporary Abstract Algebra, 1990, 2d edit. Recipient Allendoerfer award Math. Assn. Am., 1976, Disting. Teaching award, 1993. Home: 1522 Triggs Ave Duluth MN 55811 Office: Univ of Minnesota Duluth MN 55812

GALLIN, JOHN ISAAC, health science association administrator; b. N.Y.C., Mar. 25, 1943; s. Nathaniel Mitchell and Hellen (Cohen) G.; m. Elaine Barbara Klimerman, June 23, 1966; children: Alice Jennifer, Michael Louis. BA (cum laude), Amherst Coll., 1965; MD, Cornell U. Med. Sch., 1969. Diplomate Nat. Bd. Med. Examiners. Intern medicine Bellevue Hosp., N.Y.C., 1969-70, asst. resident, 1970-71; teaching asst., instr. in medicine NYU Sch. Medicine, 1970-74, 74-81; clin. assoc. lab. clin. investigation Nat. Inst. Allergy and Infectious Diseases, NIH, Bethesda, Md., 1971-74; sr. investigator lab. clin. investigation Nat. Inst. Allergy and Infectious Diseases, NIH, Bethesda, 1975—; cons. infectious diseases U.S. Navy Med. Ctr., Bethesda, 1971-74; cons. Allergy/Clin. Immunology Svc., Walter Reed Army Med. Ctr., Washington, 1982—; dir. div. intramural rsch. Nat. Inst. Allergy and Infectious Diseases, NIH, 1985—, chief lab. of host defenses, 1991—; asst. surgeon gen.-rear admiral USPHS; guest lectr. in field and speaker in field. Co-editor: Inflammation, Basic Principles and Clinical Correlates, 1988, 2nd edit. 1992, Advances in Host Defenses, 1981—; contbr. over 240 publs to profl. jours. Recipient Dean William Mecklenburg Polk Meml. prize in Rsch., 1969, Anthony Seth Werner Meml. prize in Infectious Diseases, 1969, Rsch. award Am. Fedn. for Clin. Rsch., 1984, Squibb award Infectious Diseases Soc. Am., 1987, The Jeffrey Modell Found. Life Time Achievement award, 1990, The Commendation medal Pub. Health Svc., 1980, Outstanding Svc. medal Pub. Health Svc., 1985, Meritorious Svc. medal Pub. Health Svc., 1988, Dist. Svc. medal Pub. Health Svc., 1992. Fellow Infectious Diseases Soc. of Am.; mem. Assn. Am. Physicians, Am. Soc. for Clin. Investigation, Am. Fedn. for Clin. Rsch., Internat. Immunocompromised Host Soc. (pres. 1992—), Am. Assn. Immunologists, The Am. Soc. for Cell Biology, Sigma Xi, Soc. for Leukocyte Biology. Office: NIH Bldg 10/4A31 9000 Rockville Pike Bethesda MD 20892-0001

GALLINA, CHARLES ONOFRIO, nuclear regulatory official; b. New Brunswick, N.J., Oct. 10, 1943; s. Matthew Salvatore and Mary (Piazza) G.; m. Ellen Mary Romano, Oct. 10, 1976; children: Mary Catharine, Matthew Charles, Maria Christine. BS, Fordham U., 1965; MS, Rutgers U., 1967, PhD, 1971. Environ. radiation specialist Consol. Edison N.Y., N.Y.C., 1971-72; radiation specialist AEC, Newark, 1972-73; sr. radiation specialist Nuclear Regulatory Commn., King of Prussia, Pa., 1973-76, sr. duty officer, 1973-82, investigation specialist, 1976-80, coord. emergency preparedness, 1980-82; sr. emergency preparedness engr. Tera Corp., King of Prussia, 1982; sr. radiol. engr. Hydro Nuclear Svcs., Marlton, N.J., 1982-84, dir. tech. mktg., 1984-85; mgr. bus. devel. Westinghouse Electric Corp., Moorestown, N.J., 1985-87; mgr. tech. program devel. Westinghouse Radiological Svcs., Moorestown, N.J., 1987-89; sr. nuclear scientist Dept. Nuclear Safety State of Ill., Springfield, 1990—; exec. cons. Profl. Nuclear Assocs., Springfield, Ill., 1990—; mem. bd. sci. and policy advisors Am. Coun. on Sci. and Health, 1991—. Tech. reviewer, contbr. Radiation Protection Management Mag., Health Physics Soc. Jour.; contbr. articles to Health Physics Soc. Jour. Pres. Providence Force Condominium Assn., 1973-77. AEC fellow, 1968-70, USPHS fellow, 1967, fellow Fed. Water Pollution Control Assn., 1971. Mem. Am. Nuclear Soc. (vice chmn., chmn.-elect Midwest Ill. chpt. 1991-93), Delaware Valley Soc. Radiation Protections, Health Physics Soc. (charter mem. Prairie State chpt. 1990—, bd. dirs. Prairie State chpt. 1993—), Am. Coun. of Sci. and Health (bd. sci. and policy advisors 1991—). Home: 3505 Bluff Rd Springfield IL 62707-9674 Office: 1035 Outer Park Dr Springfield IL 62704-4462

GALLINAT, MICHAEL PAUL, fisheries biologist; b. Flint, Mich., Nov. 1, 1962; s. Paul John Richard and Myrna Mae (Dingman) G.; m. Carol Ann Koshko, Sept. 8, 1989; 1 child, Nathan Michael. BS in Fisheries and Wildlife Mgmt., Lake Superior State U., Sault Ste. Marie, Mich., 1985; MS in Fisheries Biology, Ball State U., 1987. Grad. rsch. asst. Ball State U., Muncie, Ind., 1985-87; pvt. aquatic contractor, Flushing, Mich., 1987-88; rsch. asst. U. Mich., Ann Arbor, 1988; fisheries biologist, program adminstr. Red Cliff Band of Lake Superior Chippewa, Bayfield, Wis., 1988—; mem. Wis. Coastal Mgmt. Coun., Madison, 1991—, Native Peoples Fisheries Com., 1990—; adj. mem. Lake Superior Tech. Com., 1988—. Mem. Am. Fisheries Soc., Sigma Xi (assoc.). Home: 1012 3d Ave W Ashland WI 54806 Office: Red Cliff Fisheries Dept PO Box 529 Bayfield WI 54814

GALLO, ROBERT CHARLES, research scientist; b. Waterbury, Conn., Mar. 23, 1937; s. Francis Anton and Louise Mary (Ciancuilli) G.; m. Mary Jane Hayes, July 1, 1961; children: Robert Charles, Marcus. BA, Providence Coll., 1959, DSc (hon.), 1974; MD, Jefferson Med. Coll., 1963; 11 hon. degrees from univs. in U.S., Belgium, Italy, Israel. Intern, resident medicine U. Chgo., 1963-65; clin. assoc. med. br. Nat. Cancer Inst. NIH, Bethesda, Md., 1965-68, sr. investigator human tumor cell biology br., 1968-69, head sect. cellular control mechanisms, 1969-72, chief lab. tumor cell biology, 1972—; adj. prof. genetics George Washington U.; U.S. rep. to world com. Internat. Comparative Leukemia and Lymphoma Assn., 1981—; hon. prof. biology Johns Hopkins U., 1985—; bd. govs. Franco Am. AIDS Found., 1987, World AIDS Found., 1987. With USPHS, 1965-68. Recipient Dameshek award Am. Hematol. Soc., 1974, CIBA-GEIGY award in biomed. sci., 1977, 88, Superior Service award USPHS, 1979, Meritorious Service medal, 1983, Stitt award, 1983, Disting. Service medal, 1984, F. Stohlman lecture award, 1979, Lasker award for basic biomed. research, 1982, 86, Abraham white award in biochemistry George Washington U., 1983, 1st Otto Herz award for cancer research Tel Aviv U., 1982, Griffuel prize Assn. for Cancer Research, France, 1983, Gen. Motors award in cancer research, 1984, Gruber prize Am. Soc. Investigative Dermatology, 1984, Lucy Wortham prize in cancer research Soc. for Surg. Oncology, 1984, Gold medal Am. Cancer Soc., 1984, Birla Internat Sci. prize, India, 1985, Hammer prize for Cancer Rsch., 1985, Gairdner prize for Biomed. Research from Can., 1987, spl. award Am. Soc. Infectious Disease, 1986, Gold Plate award Am. Acad. Achievement, Lions Humanitarian award, 1987, Japan prize in Preventative Medicine, 1988, Ciba Corning award, 1993. Mem. NAS, Inst. Medicine, Internat. Soc. Hematology, Am. Soc. Clin. Investigation, Am. Soc. Biol. Chemists, Am. Microbiology Soc., Biochem Soc., Am. Assn. Cancer Rsch. (Rosenthal award 1983), Am. Fedn. Clin. Rsch., Fedn. for Advanced Edn. in Scis., Royal Soc. Physicians of Scotland (hon.), Alpha Omega Alpha (hon.). Achievements include research on viruses, AIDS, and Leukemia. Office: Nat Cancer Inst Tumor Cell Biol Lab 9000 Rockville Pike Bethesda MD 20892-0001

GALLOPOULOS, NICHOLAS EFSTRATIOS, chemical engineer; b. Athens, Apr. 5, 1936; came to U.S., 1953; s. Efstratios C. and Lucia N. (Romanides) G.; m. Mary Frances Veale, Oct. 25, 1958; children: Gregory S., Lucia Anne. BS in Chem. Engring., Tex. A&M U., 1958; MS in Chem. Engring., Pa. State U., 1959. Tech. specialist Humble Oil & Refining Co. (Exxon), Houston, 1967-68; rsch. engr. Gen. Motors Rsch. Labs., Warren, Mich., 1959-67, 68-75, asst. dept. head fuels and lubricants, 1975-85, head dept. environ. sci., 1985-89, head dept. engine rsch., 1989—; mem. Coordinating Rsch. Coun., Atlanta, 1974-89. Author: Future Automotive Fuels, 1977; contbr. chpts. to books, articles to Sci. American, Indsl. and Engring. Chemistry. Mem. Econ. Devel. Corp., Rochester Hills, Mich., 1978-91; mem., chmn. Planning Commn., Rochester Hills, 1982-92. Mem. AAAS, Soc. Automotive Engrs., Am. Chem. Soc., Sigma Xi. Achievements include reseach on the chemical mechanism of action of various lubricating oil additives, alternative fuels and their role in automotive transportation. Home: 1565 Hampstead Ln Rochester Hills MI 48309 Office: Gen Motors Rsch Box 9055 30500 Mound Rd Warren MI 48090-9055

GALSTER, RICHARD W., civil engineer. Recipient Claire P. Holdredge award Assn. Engring. Geologists, 1991. Home: 18233 13th NW Seattle WA 98177*

GALTERIO, LOUIS, healthcare information executive; b. N.Y.C., Apr. 20, 1951; s. Elio and Angelina (Mattina) G.; m. Elizabeth Anne Coddington, May 2, 1971; children: Jason, Heather. Student, CCNY, 1969-70, Baruch Coll., 1970-75; BS in Mgmt. summa cum laude, Mercy Coll., 1978; MBA in Fin., L.I. U., 1980. Asst. mgr. Mfrs. Hanover Trust, N.Y.C., 1971-82; v.p. Bankers Trust Co., N.Y.C., 1982-87; dir., tech. mgr. Mortgage Backed Securities Clearing Corp., N.Y.C., 1987-88; mgr. capital markets and mktg. Digital Equipment Corp., N.Y.C., 1988-90, integration exec., 1990-91; chief info. officer healthcare Health and Hosps. Corp. NYC, 1991—; pres., mgmt. cons. Galterio Cons., N.Y.C., 1987—. Mem., sect. capt. Throgs Neck (N.Y.) Estates, 1988. Mem. IEEE (assoc.), Bankers Trust Alumni Orgn., Electronic mail and Messaging Assn., Am. Hosp. Assn., Coll. Healthcare Info. Mgmt. Execs., Healthcare Info. & Mgmt. Systems Soc., Alpha Chi. Republican. Home: 1417 Shore Dr Bronx NY 10465-1560

GALVIN, ROBERT W., electronics executive; b. Marshfield, Wis., Oct. 9, 1922. Student, U. Notre Dame, U. Chgo.; LL.D. (hon.), Quincy Coll., St. Ambrose Coll., DePaul U., Ariz. State U. With Motorola, Inc., Chgo., 1940—, exec. v.p., 1948-56, former pres., from 1956, chmn. bd., 1964-90, now chmn. exec. com., 1990—, chief exec. officer, 1964-86, also dir. Former mem. Pres.'s Commn. on Internat. Trade and Investment.; chmn. industry policy adv. com. to U.S. Trade Rep.; mem. Pres.'s Pvt. Sector Survey; chmn. Pres.'s Adv. Council on Pvt. Sector Initiatives; chmn. Ill. Inst. Tech., U. Notre Dame; bd. dirs. Jr. Achievement Chgo. Served with Signal Corps, AUS, World War II. Recipient Nat. medal Tech. U.S. Dept. Commerce Tech. Adminstrn.; named Decision Maker of Yr. Chgo. Assn. Commerce and Industry-Am. Statis. Assn., 1973; Sword of Loyola award Loyola U., Chgo.; Washington award Western Soc. Engrs., 1984. Mem. Electronic Industries Assn. (pres. 1966, dir., Medal of Honor 1970, Golden Omega award 1981). Office: Motorola Inc 1303 E Algonquin Rd Schaumburg IL 60196

GALYSH, ROBERT ALAN, information systems analyst; b. Cleve., Apr. 4, 1954; s. Fred Theodore and Jennie Catherine (Masiglowa) G.; m. Nanette Kappus, Mar. 3, 1984; 1 child, Joanna Marie. BA in Econs., Cleve. State U., 1976, MA in Econs., 1982. Savs. officer Cleve. Fed. Savs., 1977-79; asst. v.p. systems, procedures analyst Continental Fed. Savs. (formerly Cleve. Fed. Savs.), Cleve, 1979-84; data processing officer, mgr. systems and procedures Continental div. Dollar Bank FSB, Cleve, 1984-86; systems analyst Cleve. Met. Gen. Hosp., 1986-87, sr. systems analyst, 1987-90; project leader info. systems MetroHealth System (formerly Cleve. Met. Gen. Hosp.), 1990—; cons. on microcomputer installations and applications, Cleve., 1984—. Mem. Gt. Lakes Hist. Soc., Omicron Delta Epsilon. Presbyterian. Avocations: home computing, photography, travel, military and aviation history. Home: 26602 Sudbury Dr North Olmsted OH 44070-1844

GAMARNIK, MOISEY YANKELEVICH, solid state physicist; b. Khmelnizky, Ukraine, USSR, Nov. 3, 1936; s. Yankel Khaymovich and Polya Iserovna (Gendelman) G.; m. Evgeniya Adolfovna Lubomirskaya, Nov. 3, 1965; children: Yan, Alexander. Candidate of Scis. Phys.-Math., U. Kharkov, USSR, 1984, DSc Phys.-Math., 1992. Tchr. Pilyava (USSR) secondary sch., 1959-60, Kiev (USSR) Secondary Sch. N96, 1960-62; researcher, engr. Inst. of Geol. Scis. Acad. Sci., Kiev, 1962-69; sr. researcher Inst. Geochemistry and Physics Minerals Acad. Sci., Kiev, 1969-85, scientist, 1985-89, sr. scientist, 1989—. Contbr. articles to Phys. State Sollids. Mem. Internat. Union Crystallography, Assn. for Aerosol Rsch. Achievements include research in problem of character and nature of size-related changes structure of lattice parameters in small crystal particles. Home: Vernadsky 85 Apt 101, Kiev 252142, Ukraine Office: Inst Geochem Phys Miner, Acad Sci Ukr Palladin's 34, Kiev 252680, Ukraine

GAMBARINI, GRAZIA LAVINIA, engineering educator; b. Milan, Italy, Aug. 3, 1942; parents Giuseppe and Anna (Gasparinetti) G. Laurea in fisica, U. degli Studi, Milan, 1966; professore associato degree, Facoltà di Ingegneria, Milan, 1982. Assistente volontario Facoltà di Scienze, U. Milan, 1966-70; tecnico laureato Facoltà di Ingegneria, Politecnico Milan, 1967-70, assistente di ruolo, 1970-82; professore incaricato Facoltà di Farmacia, U. Milan, 1971-75; professore incaricato Facoltà di Ingegneria, Politecnico Milan, 1974-82, professore di ruolo, 1982—. Contbr. articles to profl. jours. and confs. Scholar Ministry of Edn., Italy, 1967. Mem. Soc. Italiana di Fisica, Gruppo Nazionale di Struttura della Materia, Istituto Nazionale di Fisica Nucleare, Assn. Italiana di Fisica Biomedica. Avocations: music, rock climbing, alpine skiing. Office: Dipartimento di Fisica, via Celoria 16, 20133 Milan Italy

GAMBLE, FRANCOISE YOKO, structural engineer; b. Nashville, Aug. 28, 1962; d. Flem Brown and Yoshiko (Sasagawa) O.; m. Robert Curtis Gamble, Dec. 7, 1985. BS, U.S. Mil. Acad., 1985; MCE, Auburn U., 1990. Registered profl. engr., Ga. Commd. 2d lt. U.S. Army, 1985, advanced through grades to capt., served with C.E., 1985-88, resigned, 1988; grad. teaching asst. Auburn (Miss.) U., 1988-89; engr. Fluor Daniel, Greenville, S.C., 1989—. Office: Fluor Daniel 100 Fluor Daniel Dr Greenville SC 29615

GAMBLING, WILLIAM ALEXANDER, optoelectronics research center director; b. Port Talbot, Glam, Wales, Oct. 11, 1926; s. George Alexander and Muriel Clara (Bray) G.; m. Margaret Pooley, July 26, 1952 (separated 1987); children: Paul Maitland, Alison Jill, Vivien Ruth. BSc, Bristol U., Eng., 1947, DSc, 1968; PhD, Liverpool U., 1955; DSc (hon.), Eurotech. Rsch. U., 1984. Registered electronic engr. NRC fellow U. B.C., Can., 1955-57; lectr., reader Southampton U., Eng., 1958-64, prof., 1964—, head, electronics dept., 1976-79, dean engring., 1972-75, dir. optoelectronics rsch. ctr., 1989—; vis. prof. Colo. U., 1966-67, Osaka U., Japan, 1977, Capetown U., South Africa, 1979; bd. dirs. York Tech. Ltd., Chandersford, Eng., York Ltd., Chandlersford; cons. in field; hon. prof. Huazhung U. Sci., Wuhan, China, 1986, Beijing Inst. Telecommunications, 1987, Shanghai U. Sci. Tech., 1991. Contbr. articles to profl. jours. Freeman City of London, 1988. Selby fellow Australian Acad. Sci., 1982; named Hon. dir. Beijing Glass Rsch. Inst., 1987. Fellow Royal Soc. London (fellowship 1983), Institution of Elec. Engrs. (hon.), Royal Acad. Engring. (coun. mem. 1989-92); mem. Polish Acad. Scis. (fgn.). Avocations: walking, reading, music. Office: U Southampton Optoelectronics Rsch Ctr, University Rd, Southampton England S09 5NH

GAMBRELL, RICHARD DONALD, JR., endocrinologist, educator; b. St. George, S.C., Oct. 28, 1931; s. Richard Donald and Nettie Anzo (Ellenburg) G.; m. Mary Caroline Stone, Dec. 22, 1956; children—Deborah Christina, Juliet Denise. B.S., Furman U., 1953; M.D., Med. U. S.C., 1957. Diplomate Am. Bd. Obstetrics and Gynecology, Diplomate Div. Reproductive Endocrinology. Intern Greenville Gen. Hosp., S.C., 1957-58, resident, 1961-64; commd. USAF, 1958, advanced through grades to col.; chmn. dept. ob-gyn, cons. to surgeon gen. USAF Hosp. USAF, Wiesbaden, Germany, 1966-69; chief gynecologic endocrinology Wilford Hall USAF Med. Ctr. USAF, Lackland AFB, Tex., 1971-78; ret. USAF, 1978; clin. prof. ob-gyn and endocrinology Med. Coll. Ga., Augusta, 1978—; practice medicine specializing in reproductive endocrinology Augusta, 1978—; fellow in endocrinology Med. Coll. Ga., 1969-71; mem. staff Westlawn Bapt. Mission Med. Clinic, San Antonio, 1972-78; assoc. clin. prof. U. Tex. Health Sci. Ctr., San Antonio, 1971-78; internat. lectr.; mem. ob-gyn. adv. panel U.S. Pharmacopeial Conv., 1986-90, sci. adv. bd. Nat. Osteoporosis Found., 1988-91. Co-author: The Menopause: Indications for Estrogen Therapy, 1979, Sex Steroid Hormones and Cancer, 1984, Unwanted Hair: Its Cause and Treatment, 1985, Estrogen Replacement Therapy, 1987, Hormone Replacement Therapy, 3d edit., 1992, Estrogen Replacement Therapy Users Guide, 1989; mem. editorial bd. Jour. Reproductive Medicine, 1982-85, Maturitas, 1982—; mem. editorial bd. Internat. Jour. Fertility, 1986-91, assoc. editor, 1988-91; contbr. articles to med. jours., chpts. to books. Deacon, Sunday sch. tchr. Baptist Ch., 1971—; mem. sci. adv. bd. Nat. Osteoporosis Found., 1988-91. Recipient Chmn.'s Best Paper in Clin. Research from Teaching Hosp. award Armed Forces Dist. Am. Coll. Ob-Gyn, 1972, 88, Host award, 1977, Chmn.'s award, 1978, Purdue-Frederick award, 1979, Outstanding Exhibit award Am. Fertility Soc., 1983, Am. Coll. Obstetricians and Gynecologists award, 1983, Thesis award South Atlantic Assn. Ob.-Gyn., Winthrop award Internat. Soc. Reproductive Med., 1985, Chmn.'s Best Paper award Pan Am. Soc. for Fertility, 1986, Outstanding Sci. Exhibit award Am. Acad. Family Practitioners, 1986, 87, 92. Fellow Am. Coll. Obstetricians and Gynecologists (subcom. on endocrinology and infertility 1983-86); mem. Pacific N.W. Ob-gyn Soc. (hon.), So. Med. Assn. (2d place Sci. Exhibit award 1992), Am. Fertility Soc., Tex. Assn. Ob-Gyn, San Antonio Ob-Gyn. Soc. (v.p. 1975-76), Chilean Soc. Ob-Gyn. (hon.), Soc. Obstetricians and Gynecologists of Can. (hon.), Internat. Family Planning Research Assn., Internat. Menopause Soc. (exec. com. 1981-84), Internat. Soc. for Reproductive Medicine (program chmn. 1980, pres. 1986-88), Am. Geriatric Soc. (editorial bd. 1981-83), Nat. Geog. Soc., Phi Chi, Am. Philatelic Soc., Alpha Epsilon Delta. Home: 3542 National Ct Augusta GA 30907-9517 Office: 903 15th St Augusta GA 30910-0192

GAMELLI, RICHARD L., surgeon, educator; b. Springfield, Mass., Jan. 18, 1949; married; 3 children. AB in Chemistry magna cum laude, St. Michael's Coll., Colchester, Vt., 1970; MD, U. Vt., 1974. Diplomate Nat. Bd. Med. Examiners, Am. Bd. Surgery (examination cons. 1988—, guest examiner 1993); lic. surgeon, Vt., Ill. Straight surg. intern. Med. Ctr. Hosp. Vt., 1974-75, surg. resident PG-II, PG-III, PG-IV, 1975-79; asst. prof. surgery U. Vt. Coll. Medicine, 1979-85, assoc. prof., 1985-89, prof., 1989-90, dir. surg. rsch. labs. dept. surgery, 1985-90, dir. house staff tng. program., 1985-89, vice chmn. dept. surgery, 1985-90, chmn. sect. gen. surgery, 1989; attending surgeon Med. Ctr. Hosp. Vt., 1979-90, dir., founder burn program, 1980-90, dir. nutritional support svcs., 1980-88, dir. resident teaching conf., 1983-90, assoc. surgeon-in-chief, 1985-89, dir. burn-shock-trauma svc., 1988-90; prof. surgery Strich Sch. Medicine, dir. Shock-Trauma Inst., chied burn ctr. Foster G. McGaw Hosp. Loyola U. Med. Ctr., Maywood, Ill., 1990—; chmn. quailty assurance com. burn ctr. Loyola U. Med. Ctr., 1990—, infection control com., 1990—, rsch. com. coun., 1990—, surg. rsch. com., 1991—, intensive care unit com., 1991—, EMS bldg. com., 1991—, med. chmn. nutrition com., 1992—, managed care tsck force, 1993—, commitment to teaching task force, 1993—; Dr John C. Hartnett lectr. St. Michael's Coll., 1983; mem. spl. study sect. NIH, 1991; mem. physicians adv. coun. Marianjoy Rehab. Hosps. and Clinics, 1993. Co-author: Trauma 2000, 1992, A Compendium of Slides on Surgical Infections, 1992; co-editor: Clinical Surgery, 1987, Early Care of the Injured Patient, 1990, Essentials of Clinical Surgery, 1991; mem. editorial bd., reviewer Jour. Trauma, 1984—, Essentials Clin. Surgery, 1988—, Clin. Surgery, 1990—, Shock, 1993—; reviewer Circulatory Shock, Surgery, Jour. Surg. Rsch.; contbr. 93 articles to profl. jours., 16 chpts. to books. Recipient Dr. James E. DeMeules 1st Annual Rsch. award U. Vt. Dept. Surgery, 1990; grantee NIH, 1981-84, 89-93, Ethicon, Inc., 1988-90, Genetech., Inc., 1988-89, Amgen, Inc., 1989-90, U. Ill., Chgo., 1991. Fellow ACS (vice chmn. Vt. state com. on trauma 1984-86, chmn. Vt. state com. on trauma 1986-91, sec.-treas. Vt. state chpt. 1987-90, subcom. on publs. 1987-90, exec. com. 1991-93, 93, fed. com. on trauma verification/consultaion program for hosps., 1991, 92, 93 chmn. audit com. 1992, 93, bd. dirs. 1992, cons., beta test site NTRACS, 1993); mem. Am. Burn Assn. (instr., dir. advanced burn life suuport course 1988—, regionalization com. 1992—, chair region V. 1992—, beta test site registry 1993), N.Am. Burn Soc. (pres. 1991), Shock Soc., Soc. for Leukocyte Biology, Soc. Univ. Surgeons (chmn. com. on social and legis. issues 1990-93, exec. com. 1990-93), Surg. Infection Soc. (edn. com. 1988—, chmn. fellowship com. 1990—), Surg. Biology Club III, Ea. Assn. for Surgery Trauma (exec. com. 1991-93, bd. dirs. 1992, chmn. audit com. 1992, 93), Internat. Soc. for Burn Injuries, John H. Davis Soc. (founding., bd. dirs. 1988—, coun. 1988-90, sec.-treas. 1990-92, pres. 1993—), New England Surg. Soc. Office: Loyola U Chgo Shock Trauma Institute 2160 S 1st Ave Maywood IL 60133

GAMERITH, GERNOT, chemist, researcher; b. Villach, Austria, Nov. 20, 1953; s. Hermann and Gertraud (Vanino) G.; m. Helga Wallner, July 8, 1978; children: Gabriele, Clemens, Caroline. Diploma in engring., Tech. U. Graz, Austria, 1978, D of Tech., 1982. Rsch. biochemist Inst. F. Med. Biology/Human Genetics, Graz, 1979-85, Institute fuer Biotech., Graz, 1985-86; rsch. chemist Lenzing (Austria) AG, 1986-88, project leader, 1988-91, rsch. mgr., 1991—; instr. Inst. F. Med. Biology/Human Genetics, Graz, 1980-85, head chemistry group students' presentation, 1975, mem. faculty group students' representation, 1975-77, mem. med. faculty assts.' representation, 1980-85. Author: (with others) Analysis of Amino Acids by Gas Chromatography, 1987; contbr. articles to profl. jours. Begabtenstipendium scholar Ministry for Edn., Graz, 1977. Mem. Österr. Gesellschaft für Biotechnologie, Österreichische Biochemische Gesellschaft, Österreichische Gesellschaft für Bioprozesstechnik. Achievements include patent for production of xylanase for use in pulp industry; patent pending for use of xylanase for pulp industry; research on rapid amino acid analysis by gas chromatography and application of method in errors of metabolism. Office: Lenzing AG, A 4860 Lenzing Austria

GAMLEN, JAMES ELI, JR., corrosion engineer; b. Pitts., Jan. 3, 1950; s. James Eli Gamlen and Alys Ann (Simons) Simons; m. Catherine Paula Turner, Dec. 21, 1976; children: Allison, Matthew, Emily, Tyler. BA in Biology, Chemistry, San Francisco State U., 1974. Registered profl. engr., Calif. Engring. mgr. Western Polymer Corp., Burlingame, Calif., 1970-73, pres. (acting), 1973-76; field engr. Garratt-Callahan Co., Millbrae, Calif., 1976-89, corrosion engr., 1989-91, mktg. dir., 1991—. Mem. Nat. Assn. Corrosion Engrs., NSPE, Nat. Assn. Energy Engrs., No. Calif. Cogeneration Assn. Achievements include discovery that Chelation treatment can be used successfully in high pressure steam boilers up to at least 1800 psig. Office: Garratt Callahan Co 111 Rollins Rd Millbrae CA 94030-3114

GAMMANS, JAMES PATRICK, ceramic engineer; b. Atlanta, Sept. 18, 1952; s. Raymond Francis Jr. and Alice Jane (Rhodes) G.; m. Brenda Diane Klingenberg, Dec. 10, 1983. BS magna cum laude, U. Ga., 1976; B Ceramic Engring. with highest honors, Ga. Inst. Tech., 1976; MBA, Ga. State U., 1985. Reg. profl. engr. Ceramic engr. Ga. Sanitary Pottery Inc., Atlanta, 1976-80, Universal Rundle Corp., Monroe, Ga., 1980—. TRI-FIRE scholar The Refractories Inst., Ga. Inst. Tech., 1975-76. Mem. Am. Ceramic Soc. (sec. S.E. sect. 1988-89, vice chair 1990, chair 1992). Roman Catholic. Office: Universal Rundle Corp PO Box 828 Vine St Monroe GA 30655

GAMOTA, DANIEL ROMAN, materials engineer; b. Ann Arbor, Mich., Oct. 11, 1965; s. George and Christina Stephanie (Dawydowycs) G. BSE in Chem. Engring., U. Mich., 1987, PhD in Materials Engring., 1992. Student engr. KMS Fusion, Ann Arbor, 1983-85, Siemens RTL, Princeton, N.J., 1986, Thermo Electron, Waltham, Mass., 1987; rsch. asst. U. Mich., Ann Arbor, 1988-91, rsch. fellow, 1992—. Contbr. articles to profl. jours. Mem. Am. Phys. Soc., Am. Ceramic Soc., Soc. Rheology, Ukrainian Students Orgn. (treas. 1985-87), Alpha Sigma Mu (hon. mem.), Phi Lambda Upsilon (hon. mem.). Achievements include research on characterization of the dynamic mechanical properties of an electrorheological material; observation of a transition in the mode of deformation: from linear viscoelastic to nonlinear viscoelastic response. Office: Univ of Mich 2300 Hayward St Ann Arbor MI 48109-2136

GAN, FELISA SO, physician; b. Manila, June 29, 1943; d. Victor Ang So and Siok Gee Tan; m. David Jr. Lo Gan, June 30, 1968; children: Jason, Johann, Tanya. BA, U. Santo Tomas, Manila, 1963; MD, 1968; postgrad., U. Santo Tomas, U. East, Manila, 1977—. Asst. to med. dir. Met. Gen. Hosp., Manila, 1968—, perceptorship, 1968-79; gen. practice internal medicine Manila, 1968—. Contbr. articles to profl. jours. Pres. Greenhills Christian Women's Fellowship, Manila, 1985-86, v.p., 87-88; speaker for Campus Crusade for Christ at Asian Christian Med. Congress, Los Banos, Rizal Laguna, Philippines, 1988, 89. Fellow Philippines Soc. Gastroenterology (assoc.); mem. Philippines Med. Assn. Office: Met Gen Hosp, 1357 G Masangkay St, Binondo Manila, The Philippines

GAN, LEONG-HUAT, chemist, educator; b. Klang, Malaysia, June 20, 1945; arrived in Singapore, 1991; s. Kok-Yong and Yoke-Mea (Ng) G.; m. Yik-Yuen Yap, Dec. 17, 1976; children: Julia Sze-Ling, Chee-Chun. BS, Victoria U., New Zealand, 1968, MS, 1969; PhD, Queen's U., Kingston, Canada, 1972. Postdoctoral fellow Nat. Rsch. Coun. Canada, Ottawa, Ontario, 1972-74; lectr. U. Malaya, Kuala Lumpur, 1974-83, assoc. prof., 1983-91; sr. lectr. Nanyang Tech. U., Singapore, 1991—. Contbr. articles to Jour. Am. Oil Chemist's Soc., Jour. Applied Polymer Sci., Jour. Colloid and Interface Sci., Rubber Chem. and Tech., Australian Jour. Chemistry, Jour. Am. Chem. Soc., European Polymer Jour., Canadian Jour. Chem., Jour. Phys. Chemistry, Jour. Polymer Sci., Jour. Polymer Letter, Jour. Chemistry. Grantee Govt. of Malaysia, 1989-91. Mem. Malaysian Inst. Chemistry (assoc.), Am. Chem. Soc., Singapore Nat. Inst. Chemistry. Achievements include research in surfactant chemistry, chemical kinetics, catalysis, polymer chemistry. Home: 199 Eng Kong Garden, Singapore 2159, Singapore Office: Nanyang Tech U Div Chemistry, 469 Bukit Timah Rd, Singapore 1025, Singapore

GANDHI, OM PARKASH, electrical engineer; b. Multan, Pakistan, Sept. 23, 1934; came to U.S., 1967, naturalized, 1975; s. Gopal Das and Devi Bai (Patney) G.; m. Santosh Nayar, Oct. 28, 1963; children: Rajesh Timmy, Monica, Lena. BS with honors, Delhi U., India, 1952; MSE, U. Mich., 1957, Sc.D., 1961. Rsch. specialist Philco Corp., Blue Bell, Pa., 1960-62; asst. dir. Cen. Electronics Engring. Rsch. Inst., Pilani, Rajasthan, India, 1962-65, dep. dir., 1965-67; prof. elec. engring., rsch. prof. bioengring. U. Utah, Salt Lake City, 1967—; chmn. elec. engring., 1992—; cons. U.S. Army Med. Rsch. and Devel. Command, Washington, 1973-77; cons. to industry and govtl. orgns.; mem. Internat. URSI Commn. B. and K; mem. study sect. on diagnostic radiology NIH, 1978-81. Author: Microwave Engineering and Applications, 1981; editor: Engineering in Medicine and Biology mag., 1987, Electromagnetic Biointeraction, 1989, Biological Effects and Medical Applications of Electromagnetic Energy, 1990; contbr. over 200 articles to profl. jours. Recipient Disting. Rsch. award U. Utah, 1979-80; grantee NSF, NIH, EPA, USAF, U.S. Army, USN, N.Y. State Dept. Health, others. Fellow IEEE (editor Procs. of IEEE Spl. Issue, 1980, co-chmn. com. on RF safety standards 1988—, Tech. Achievement award Utah sect. 1975); mem. Electromagnetics Acad., Bioelectromagnetics Soc. (bd. dirs. 1979-82, 87-90, v.p., pres. 1991—). Office: U Utah Elec Engring Dept 4516 Merrill Engring Salt Lake City UT 84112

GANDHI, SHAILESH RAMESH, biotechnologist; b. Bombay, July 21, 1960; came to U.S., 1986; s. Ramesh Moreshwar and Shakuntala Shankar (Shetye) G.; m. Medha Phrabhakar Manchekar, Dec. 27, 1991. BSc, Bombay U., 1982, MSc, 1985. Rsch. fellow dept. microbiology Bhavan's Coll., Bombay, 1985-86; rsch. asst. dept. botany and microbiology Auburn (Ala.) U., 1986-89, teaching asst., 1989-92, postdoctoral fellow, 1992—. Contbr. chpt.: Handbook of Applied Mycology, 1992, Industrial Applications of Single Cell Oils, 1992. Mem. Wilderness Soc., World Wildlife Fund, Sigma Xi. Hindu. Office: Auburn Univ Dept Botany and Microbiology 129 Funchess Hall Auburn AL 36849

GANDHI, SUKETU RAMESH, chemist; b. Jhagadia, India, Aug. 5, 1959; came to the U.S., 1969; s. Ramesh N. and Sharmi R. (Shah) G. BS in Chemistry, U. Ill., 1981; PhD, UCLA, 1988. Postdoctoral rsch. assoc. Princeton (N.J.) U., 1988-91, Max Planck Inst., Göttingen, Germany, 1991—. Contbr. numerous articles to profl. jours. Alexander von Humboldt rsch. fellow, 1992-93. Mem. Am. Phys. Soc., Am. Chem. Soc. Achievements include research in orientation measurements of molecules, production of molecules in a pure rotational state. Office: Max Planck Inst, Bunsenstrasse 10, W-3400 Göttingen Germany

GANDSEY, LOUIS JOHN, petroleum and environmental consultant; b. Greybull, Wyo., May 19, 1921; s. John Wellington and Leonora (McLaughlin) G.; m. Mary Louise Alviso, Nov. 10, 1945; children: Mary M., Catherine K., John P., Michael J., Laurie A. AA, Compton Jr. Coll., 1941; BS, U. Calif. Berkeley, 1943; M in Engring., UCLA, 1958. Registered profl. engr., Calif., environ. assessor. With Richfield Oil Corp., L.A., 1943-65, process engr., processing foreman, sr. foreman, mfg. coord., 1943-61, project leader process computer control, 1961-63, light oil oper. supt., 1963-64, asst. refinery supt., 1964-65; mgr. planning Richfield div. Atlantic Richfield Co., L.A., 1966-68, mgr. evaluation products div., L.A., 1968-69, mgr. supply and transp., Chgo., 1969-71, mgr. planning and mgmt. sci., N.Y.C., 1971, mgr. supply and transp., L.A., 1971-72, mgr. coordination and supply, 1972-75, mgr. domestic crude, 1975-77; v.p. refining Lunday-Thagard Oil Co., South Gate, Calif., 1977-82; petroleum cons. World Oil Corp., L.A., 1982-85; gen. cons., 1986—; instr. chem. and petroleum tech. L.A. Harbor Coll., 1960-65; cons. on oil crops, Austria, 1991; U.S. del. in environ. affairs to Joint Inter-Govtl. Com. for Environ. Protection, USSR, 1991, asphalt tech. to Joint Inter-Govtl. Com. for Highway Design CWS, 1992 . Contbr. articles to profl. jours. Active Boy Scouts of Am. Served with C.E., AUS, 1944-45. Mem. AAAS, AICE, Am. Chem. Soc., Pacific Energy Assn., Calif. Soc. Profl. Engrs., Environ. Assessment Assn.. Home: 2340 Neal Spring Rd Templeton CA 93465-9610

GANDY, GERALD LARMON, rehabilitation counseling educator; b. Thomasville, Ga., Feb. 9, 1941; s. Larmon Brinkley and Ruby Wylene (Vickers) G.; m. Patricia Kay Haltiwanger, Jan. 22, 1966. BA, Fla. State U., 1963; MA, U. S.C., 1968, PhD, 1971. Lic. profl. counselor, Va.; lic. counseling psychologist, S.C.; nat. cert. rehab. counselor; nat. cert. counselor; nat. registered psychologist. Profl. counselor U. S.C. Counseling Ctr., Columbia, 1968-70; counseling psychologist VA Regional Office, Columbia, 1970-75, chief counseling psychologist, 1974-75; prof., program dir. Va. Commonwealth U., Richmond, 1975—; chair nat. com. on undergrad. rehab. edn. Nat. Coun. on Rehab. Edn., 1984-89; mem. numerous state and govt. adv. coms., 1970—. Co-author: Rehabilitation and Disability, 1990; co-author/editor: Rehabilitation Counseling and Services, 1987; co-editor: International Rehabilitation, 1980; contbr. numerous articles to profl. jours. Faculty pres. Sch. of Community and Pub. Affairs, VA Commonwealth U., 1989-93. Capt. U.S. Army, 1963-66. Recipient Disting. Svc. award Sch. of Community and Pub. Affairs, 1988, Schooland U. Leadership award, 1993. Fellow Internat. Acad. of Behavioral Medicine, Counseling and Psychotherapy (diplomate); mem. APA, Am. Counseling Assn., Internat. Assn. for Applied Psychology, World Fedn. for Mental Health. Democrat. Home: Highland Springs 300 Southern Ct Richmond VA 23075 Office: Va Commonwealth U PO Box 2030 921 W Franklin St Richmond VA 23284

GANDY, JAMES THOMAS, meteorologist; b. Memphis, Tenn., Nov. 25, 1952; s. Thomas Marion and Sible Christine (McBride) G.; m. Ann Cuppia, Apr. 12, 1986. BS, Fla. State U., 1974; postgrad., U. S.C. Meteorologist WREG-TV (CBS affiliate), Memphis, 1975-77; staff meteorologist KTVY-TV (NBC affiliate), Oklahoma City, 1977-82; dir. ops. Weather Data, Inc., Wichita, Kans., 1982-84; meteorologist Kans. State Network (NBC affiliate), Wichita, 1982-84; chief meteorologist WIS TV (NBC affiliate), Columbia, S.C., 1984—; guest lectr. U. S.C., Columbia, 1991. Mem. Am. Meteorol. Soc. (TV Seal of Approval 1985), Nat. Weather Assn., AAAS, Planetary Soc. (charter mem.). Home: 133 Outrigger Ln Columbia SC 29212-8056 Office: WIS Television 1111 Bull St Columbia SC 29201-3775

GANGOPADHYAY, SUNITA BHARDWAJ, physicist, researcher; b. New Delhi, India, Nov. 14, 1964; came to U.S., 1987; d. Dwarka Das and Raj (Vaid) Bhardwaj; m. Subhagat Gangopadhyay, Aug. 8, 1987. BSc in Physics with honors, Delhi U., 1985; MSc in Physics, Indian Inst. Tech., 1987. Rsch. asst. dept. physics and astronomy U. Del., Newark, 1987—. Mem. Am. Physical Soc. Achievements include research in high magnetization and coercivity in ultrafine iron particles. Office: U Del Dept Physics & Astronomy Newark DE 19716

GANGULY, ASHIT KUMAR, organic chemist; b. New Delhi, Aug. 9, 1934; came to U.S., 1967; s. Apurba Kumar and Protiva (Chatterji) G.; m. Jean Currie Gowans, Sept. 10, 1966; 1 child, Nomita. PhD, U. Delhi (India), 1959, Imperial Coll., London, 1962. Sr. scientist Schering-Plough Corp., Bloomfield, N.J., rsch. fellow, dir., presdl. fellow, v.p.; permanent mem. medicinal chemistry study sect. NIH, Bethesda, Md., 1986—; Khaira Disting. profl. Indian Assn. Cultivation Scis., Calcutta, 1975; Charles Sabat lectr. Rutgers U., N.J., 1987. Contbr. articles to profl. jours. Recipient Seshadri Meml. award Delhi U., 1982, Outstanding Scientist award Assn. Scientists Indian Origins in Am., 1991. Fellow Royal Soc. Chemistry; mem. ACS, N.Y. Acad. Scis. Achievements include patents on Oligosaccharide Antibiotics, Penems, Macrolide Antibiotics, PAF Antagonists, 5-Lipoxygenase Inhibitors. Office: Schering Plough Rsch Inst 2015 Galloping Hill Rd Kenilworth NJ 07033-0539

GANNON, JANE FRANCES, information scientist, researcher; b. N.Y.C., Feb. 25, 1964; d. John Jogues and Mary Elizabeth (Fay) G. Student, Brn Mawr Coll., 1983-85; BA in English, BS in Biochemistry, Niagara U., 1991. Med. researcher Med. Found. Buffalo, 1990-91; dir. N.Am. Rsch. Systems Egon Zehnder Internat. Inc., N.Y.C., 1991—. Mem. AAAS, Am. Chem. Soc., Am. Inst. Biol. Scis. Office: Egon Zehnder Internat Inc 55 E 59 St New York NY 10022

GANNON, MICHAEL ROBERT, biology educator; b. Suffren, N.Y., Sept. 11, 1958; s. Robert P. and Rose E. (Gaeta) G. AAS, SUNY, Suffren, 1978; BS, SUNY, Oswego, 1980; MS, SUNY, Brockport, 1984; PhD, Tex. Tech. U., 1991. Microbiology tech. Lederle Labs., Pearl River, N.Y., 1979, pharm. inspector, 1981, chemistry tech., 1982; teaching asst. dept. biology SUNY, Brockport, 1982-83; rsch. asst. Tex. Tech. U., Junction, 1986, Ctr. Environ. & Energy Rsch., Luquillo Mtns., P.R., 1987-90; instr. Tex. Tech. U., Lubbock, 1985-90, postdoctoral rsch. assoc., 1991; asst. prof. dept. biology Ga. Southern U., Statesboro, 1991-92, Pa. State U., Altoona, 1992—; ecological cons. Environ. Impact State of Tex., 1989; speaker in field. Contbr. articles to profl. jours. Mem. AAAS, Am. Soc. Mammalogists, Ecological Soc. Am., Southwest Assn. Naturalists, Tex. Soc. Mammalogists, Tex. Acad. Sci., Soc. Systematic Biology, Assn. Tropical Biology, Animal Behavior. Office: Pa State U Dept Biology 3000 Ivyside Pk Altoona PA 16601-3760

GANOE, GEORGE GRANT, electrical engineer; b. Williamsport, Pa., Dec. 21, 1950; s. Blair David and Beatrice Ruth (Wurster) G.; m. Thea Ellen Morgan, July 31, 1982; children: Michelle Claire, Rene Elise. BSEE, U. Fla., 1980. Design engr. Facility Engring. div. NASA Langley Rsch. Ctr., Hampton, Va., 1980-87, opts. analyst space sta. freedom office, 1987—. With USAF, 1970-74. Mem. IEEE, Toastmasters (sec. 1992—). Home: 300 Beechwood Ln Yorktown VA 23693 Office: NASA Langley Rsch Ctr MS288 Hampton VA 23665

GANONG, WILLIAM FRANCIS, III, speech sciences research executive; b. Boston, Nov. 16, 1951; s. William F. and Ruth (Jackson) G.; m. Marilyn Newman, July 31, 1977; children: Peter, Anna. AB in Math., Harvard U., 1973; PhD in Psychology, MIT, 1977. Postdoctoral fellow Brown U., Providence, 1977-79; asst. prof. U. Pa., Phila., 1979-82; v.p. rsch. Kurzweil Applied Intelligence, Waltham, Mass., 1982—. Contbr. articles to profl. jours.; patentee speech recognition devices. Mem. worship-study cong. Harvard U. Hillel. NSF fellow, 1973, NIMH fellow NSF, 1977. Mem. Fedn. Am. Scientists, Psychonomics Soc., Acoustical Soc. Am. Office: Kurzweil Applied Intelligence Inc 411 Waverly Oaks Rd Waltham MA 02154-8414

GANZ, WILLIAM I., radiology educator; b. Munich, Jan. 2, 1951; s. Lazar and Jean Ganz; m. Susan Rebecca Sirota, June 22, 1980; children: Tova, Debora, Harry. BA, Adelphi U., 1972; MS, Albert Einstein Coll. of, Medicine, 1979, MD, 1979. Diplomate Am. Bd. Nuclear Medicine. NIH med. scientist trainee Albert Einstein Coll. Medicine, Bronx, N.Y., 1972-78, pharmacology rsch. fellow, 1978-79, NIH cardiovascular fellow, 1979-80, radiology resident, 1980-83; radiology/nuclear medicine fellow Barnes Hosp./Inst. Radiology, St. Louis, 1983-85; asst. prof. U. Miami, 1985-90, assoc. prof., 1990—; prof. panel Pfizer Pharms., Miami, 1986-91. Exhibitor in field. Recipient NIH Svc. awards 1975-78, NSF award 1976, others. Mem. Am. Coll. Cardiology, Radiol. Soc. Mid Am., Soc. Nuclear Medicine, Soc. Computer Resonance Imaging. Democrat. Jewish. Home: 4333 Adams Ave Miami FL 33140-2927

GAO, HONG WEN, chemical engineer; b. Taipei, Taiwan, Jan. 31, 1945; came to U.S., 1970; s. Don Ching and Mahn (Chen) G.; m. Ellen Ling Tsuei, Apr. 7, 1973; children: Karen G., Judy G., Stanley G. BSChemE, Nat. Taiwan U., 1968; PhD in Chem. Engring., U. Utah, 1979. Postdoctoral rsch. assoc. U. Utah, Salt Lake City, 1979-83; sr. engr. Nat. Inst. for Petroleum and Energy Rsch., Bartlesville, Okla., 1984—. Contbr. articles to profl. jours. including Jour. of Rheology, Metall. Trans B, Macromolecules, Ency. of Engring. Materials, SPE Reservoir Engring. Fund raiser Am. Heart Assn., 1991. Mem. AICE, Am. Chem. Soc., Soc. of Petroleum Engrs., Sigma Xi. Republican. Achievements include patent for improved method for mobility control and permeability modification in subterranean formations. Office: Nat Inst Petroleum and Energy Rsch PO Box 2128 Bartlesville OK 74005

GAO, HONG-BO, physics educator, researcher; b. Guanghan, Sichuan, Peoples Republic of China, June 17, 1961; s. Shi-Long and Zhong-Su (Ho) G.; m. Pei Wu, Dec. 25, 1985; 1 child, Jing-Xuan. BS, Lanzhou U., Peoples Republic China, 1982; PhD, U. Sci. and Tech. of China, Hefei, Peoples Republic of China, 1989. Nuclear engr. Southwestern Inst. Nuclear Physics, Chengdu, China, 1982-85; bus. negotiator Sichuan Provincial Fgn. Trade Co., Chengdu, 1985-86; assoc. prof. physics, rschr. Zhejiang U., Hangzhou, China, 1989—; vis. scientist U. Rome, 1987-88, Internat. Ctr. for Theoretical Physics, Trieste, Italy, 1988; Humboldt rsch. fellow U. Frieburg, Germany, 1992-93; Royal Soc. K C Wong rsch. fellow U. Durham, Eng., 1993—. Contbr. rsch. papers to profl. publs. Mem. Chinese Phys. Soc. World Fedn. Scientists, Internat. Ctr. for Theoretical Physics (assoc.). Avocations: travel, sports, music, fine arts. Office: Univ Freiburg Fakultät Physik, Hermann-Herder-Strasse 3, W-79117 Freiburg Germany

GAO, YI-TIAN, physicist, educator, astronomer; b. Beijing, China, May 6, 1959. BSc, Nankai U., China, 1982; MA in Physics, CUNY, 1986; MS in Astronomy, UCLA, 1988, PhD in Astronomy, 1991. Teaching asst. CUNY, 1983-87; teaching assoc. UCLA, 1987-88, rschr. in astrophysics, 1987-91; postdoctoral rsch. fellow U. Mich., 1991-93; guest scientist Fermi Nat. Accelerator Lab., 1991-93; prof. physics Lanzhou U., China, 1993—. Recipient Julius Goodman Meml. Prize in Theoretical Physics, 1991; Ou Jou Yi scholar, 1990. Mem. Am. Physical Soc. Office: Lanzhou U, Lanzhou 730000, China

GAPONOV-GREKHOV, ANDREY VIKTOROVICH, physicist; b. Moscow, Russia, June 7, 1926. Gorky State U., Russia, Postgrad., 1949-1952. Instr. Gorky Polytech Inst., 1952-55; sr. sci. assoc., head dept. radio physics Applied Physics Inst., Gorky State U., 1955—. Recipient U.S.S.R. People's Deputy award, 1989, Hero of Socialist Labour award, 1986, State prize, 1967, 83. Mem. Russian Acad. Scis. Achievements include research in numerous theoretical and experimental works in the field of inducted cyclotronic radiation, which led to development of a new class of electronic instruments maser with cyclotronic resonance. Office: Institute of Applied Physics, Ulitsa Ulyanova 46, 603600 Nizhniy Novgorod Russia*

GARABEDIAN, CHARLES, JR., mathematics educator; b. Whitinsville, Mass., July 16, 1943; s. Charles and Sadie (Madanjian) G.; m. Manoushag Manougian. BS, Worcester State Coll., 1965; MEd, Framingham State Coll., 1970; PhD, U. Conn., 1981. Cert. secondary tchr., Mass. Math. tchr. Holliston (Mass.) High Sch., 1965—; assoc. prof. math. Framingham (Mass.) State Coll., 1971-75, 84—; math. tchr. Ea. Conn. State Coll., Willimantic, 1976. Recipient Presdl. Disting. Tchr. award , Harvard U. Practitioner award, 1988. Mem. Nat. Coun. Tchrs. of Math., ASCD, N.E. Assn. Tchrs. of Math., Mass. Assn. Tchrs. of Math., Mass. Assn. Supervision and Curriculum Devel., Mass. Assn. RR Passengers, Nat. Assn. RR Passengers, Knights of Vartan, Phi Delta Kappa. Mem. Armenian Evangelical Ch. Avocations: music, photography, model railroads, cooking, reading. Home: PO Box 452 MO Shrewsbury MA 01545

GARBERS, DAVID LORN, biochemist; b. La Crosse, Wisc., Mar. 17, 1944. BSc, U Wisconsin, Madison, 1966; MSc, 1970, Ph.D., biochemistry, 1972. Assoc. physiologist Vanderbilt U., Nashville, Tenn., 1972-74; asst. prof., 1974-76, assoc. prof., 1977—; investigator Howard Hughes Med. Inst., Nashville, Tenn., 1976—; visiting prof. Johns Hopkins Med. Sch., Baltimore, Md., 1984-85; NIH, 1984-87. mem., Am. Soc. Biol. Chemists. Achievements include research in the molecular biology of fertilization. Office: Univ of Texas SW Med Ct Howard Hughes Med Inst 5323 Harry Hines Blvd Dallas TX 75235

GARCELON, JOHN HERRICK, structural engineer, researcher; b. Winchester, Mass., Oct. 30, 1957; s. William Stetson Garcelon and Rosemary (Frend) Dunn. BSCE, Tulane U., 1980; MSCE, U. Calif., Berkeley, 1984; PhD in Engring. Mechanics, U. Fla., 1993. Design engr. Petro Marine Engring., Inc., Gretna, La., 1980-82; engr. Structural Software Devel., Inc., Berkeley, 1983-85; software engr. DATEC, Inc., Gretna, 1985-86; cons. Fla. Dept. Transp., Talahassee, 1987-88; rsch. engr. U. Fla., Gainesville, 1991—. Contbr. articles to profl. jours. Mem. ASCE, ASME. Office: Univ Fla 325 Aerospace Gainesville FL 32611

GARCIA, ANTONIO AGUSTIN, chemical engineer, bioengineer, educator; b. Placetas, Cuba, June 13, 1959; s. Antonio Agustin and Isabel (Vergara) G.; m. Beatriz Castañer, Aug. 29, 1981; children: Rebecca Marie, Jessica Christina. BS, Rutgers U., 1981; PhD, U. Calif., Berkeley, 1988. Project engr. Exxon Rsch. and Engring. Co., Florham Park, N.J., 1981-83; rsch. asst. U. Calif., Berkeley, 1983-88; rsch. engr. Eastman Kodak Co., Rochester, N.Y., 1988-89; asst. prof. chem. and bioengring. Ariz. State U., Tempe, 1989—; vice-chmn. adv. bd. Rutgers U. Minority Engring. Students, 1988-89; cons. Genencor Internat., Rochester, 1991—. Contbr. articles to Biotech. Progress, other profl. jours. Adv. bd. Community Documentation Program, Tempe, 1991—. Recipient Rsch. Initiation award NSF, 1989. Mem. AICE, Am. Chem. Soc., Sigma Xi. Achievements include imaging of bacterial and yeast cells with scanning tunneling and atomic force microscopes, invention of method for recovery of polar organic compounds using adsorbents. Office: Ariz State U Tempe AZ 85287-6006

GARCÍA, DOMINGO, mathematics educator; b. Lorquí, Murcia, Spain, Jan. 7, 1958; s. Jesús and Máxima (Rodriguez) G.; m. Amparo Carbonell, Apr. 1, 1990. BS in Math. magna cum laude, U. Valencia, Spain, 1979, PhD in Math. magna cum laude, 1984. Fellow U. Valencia, 1980-81, asst. prof., 1981-87, lectr., 1987—, dep. head, sec. dept. math. analysis, 1990—; vis. prof. U. Coll. Dublin, Ireland, 1985, U. Estadual de Campinas, Brazil, 1993. Contbr. articles to profl. jours. Grantee Ministry Edn. and Sci., Madrid, 1974-79, Nat. Programme Tng. of Rsch. Pers., Ministry Edn. and Sci., Madrid, 1980, Juan March Found., Madrid, 1981, Comisión Asesora para la Investigación Científica y Técnica, Ministry Edn. and Sci., 1983-87, Dirección Gen. de Investigación Científica y Técnica, Ministry Edn. and Sci., Madrid, 1988-92, 92—, Mobility Programme of Reseacher Pers., 1993, Commn. European Communities, Tempus Joint European Project, Brussels, 1991-92, Erasmus Program, Brussels, 1992-94. Mem. Math. Assn. Am., Am. Math. Soc. Avocations: reading, listening to music, playing golf. Office: U Valencia, Doctor Moliner 50, 46100 Burjasot Valencia, Spain

GARCIA, ERNEST VICTOR, medical physicist; b. Havana, Cuba, Sept. 14, 1948; came to U.S., 1960; s. Antonio Librado Garcia and Norma Victoria (Giro) Triana; m. Terri Marylin Spiegel, Nov. 16, 1980; children: Meredith Sara, Evan Randall. MS, U. Miami, 1972, PhD, 1974. Asst. prof. radiology U. Miami (Fla.)/Jackson Meml. Hosp., 1976-79, assoc. prof. radiology, 1979-80; dir. nuclear med. computer sci. Cedars-Sinai Med. Ctr., L.A., 1980-85; prof. radiology Emory U. Sch. Medicine, Atlanta, 1990—, dir. P.E.T. Ctr., 1990—; co-dir. Emory U./Ga. Tech. Biomed. Tech. Rsch. Ctr., Atlanta, 1990-93; adj. assoc. prof. Sch. Medicine UCLA, 1983-85; mem. diagnostic radiology study sect. div. rsch. grants NIH, Washington, 1986-90; mem. med. imaging drugs adv. com. FDA, Bethesda, Md., 1990—. Mem. editorial bd. Internat. Jour. Cardiovascular Imaging, 1985—; contbr. articles to profl. publs. Grantee NIH, 1987—, 1988—, 1990—. Mem. IEEE, Soc. Nuclear Medicine (4th Tetalman Meml. award 1984, 1st Winfield Evans Meml. Lectr. award S.W. chpt. 1991, pres. cardiovascular coun. 1990—), Am. Heart Assn. (mem. coun. on cardiovascular radiology 1987—), Am. Assn. Physicists in Medicine. Achievements include development of computer interface and 3D reconstruction software for multiplane tomographic scanner, real-time computer acquisition for 2D echocardiography, software to quantify thallium-201 myocardial SPECT, software to quantify Tc99m Sestamibi myocardial SPECT distributions. Office: Emory U Dept Radiology 1364 Clifton Rd NE Atlanta GA 30322

GARCIA, PEDRO IVAN, psychologist; b. Humacao, P.R., Aug. 18, 1947; s. Jose Adelaido Garcia and Josefine Delgado Sanafria; m. Isabel Lopez Marcano, Nov. 19, 1946 (div. 1976); children: Yamir Ivan, Nilka Idalis; m. Sandra Maria Sanchez, June 15, 1983; children: Pedro Ivan, Sandra MAria, Ivan Alejandro. MS with honors, Carribean Ctr. Advance Studies, San Juan, P.R., 1972, PhD, 1978. Pvt. practice psychology Manati, P.R.; cons. Pers. Performance Cons., P.R., 1991—, Managed health Network, Inc., 1990—, Disability Determination Svc., P.R., 1985—, Preventive Health Programs, Inc., Washington, 1980-83; psychol. assessment dir. State sychiat. Hosp. of P.R., 1984-89; dir. psychology dept. Cuban Resettlement Programs, Ft. Smith, Ark., 1980-83; dir. psycho/social svcs. St. Elizabeth Hosp., Washington, 1982-83; mem. com. social behavioral scientists Harvard Med. Sch., Boston, 1973-74. Contbr. articles to profl. jours. Cons. to Mayor of the Municipality of Manati, 1986—. Boston U. Teaching fellow, 1973, Harvard U. CLin. fellow, 1973; recipient Meritorious Cert. for Colaboration in selection of ofcls. in P.R. Penal System, 1986, Meritorious Svc. in Cuban Resettlement Programs, 1980-83. Mem. APA, Am. Psychol. Assn., P.R. Psychol. Assn. Roman Catholic. Home: 221 Paseo Real Montejo Manati PR 00674-5710 Office: Borreteaga St # 26 Manati PR 00674

GARCIA, RAFAEL JORGE, chemical engineer; b. Havana, Cuba, July 2, 1933; came to U.S., 1962; s. Rafael and Martha Teresa (Suarez) G.; m. Amelia Fernandez, Feb. 23, 1958; children: Amelia Maria, Rafael Jorge Jr. BA, Columbia Coll., 1954; BS in Chem. Engring., La. State U., 1957; MS in Environ. Engring., Johns Hopkins U., 1975. Registered profl. engr., Ind., Ky., La., Md.; registered environ. mgr. Chem. engr. Freeport Sulphur Co., New Orleans, 1957-58; prodn. supt. Litografia Garcia Muniz, Havana, 1958-62; chem. engr. The Am. Sugar Refining Co., Balt., 1962-63, The House of Seagram, Balt., 1963-80; chief ecology engr. The House of Seagram, Louisville, 1981—. Mem. Am. Inst. Chem. Engrs., Instrument Soc. Am., Assn. Energy Engrs., Engring. Soc. Balt., St. Matthews Lions (pres. 1986-87). Republican. Roman Catholic. Home: 912 Lake Forest Pky Louisville KY 40245 Office: The House of Seagram Barkley Bldg Ste 200 12700 Shelbyville Rd Louisville KY 40243

GARCIA, RAYMOND LLOYD, dermatologist; b. Paterson, N.J., Jan. 24, 1942; s. Raymond and Ruth Elaine (De Graff) G.; m. Cynthia Ruth Towne (div.); m. Toy Ping Woo, Dec. 22, 1984; 1 child, Christopher Drew. BA cum laude, Drew U., 1963; MD, Temple U., 1967. Diplomate Am. Bd. Dermatology, Am. Bd. Dermatology-Pathology. Commd. Col. USAF, 1966; intern Wilford Hall USAF Med. Ctr., San Antonio, 1967-68; dermatology resident, 1969-72; vice-chmn. residency tng. program Wilford Hall USAF Med. Ctr., San Antonio, 1972-82; asst. chief aerospace medicine USAF Acad., Colorado Springs, Colo., 1968-69; chief dermatology Carswell USAF Hosp., Ft. Worth, 1982-86; pvt. practice Irving, Tex., 1986—; asst. prof. U. Tex. Med. Sch., San Antonio, 1972-82; assoc. prof. Tex. Coll. Osteo. Medicine, Ft. Worth, 1982—; cons. to surgeon gen. USAF, 1979-86. Editor: Jour. the Assn. Mil. Dermatologists, 1978-86, Handbook of Dermatology, 1980; contbr. over 50 articles to profl. jours. Decorated Nat. Defense medal USAF, 1972, Meritorious Svc. medal 1982. Fellow Am. Acad. Dermatology (legis. liaison com. 1972-78); mem. Assn. Mil. Dermatologists (sec., treas. 1973-75, v.p. 1977-78, pres. 1980), Tex. Med. Assn., Tarrant County Med. Soc., Babcock Surg. Soc. of Temple U. Med. Sch., Tex. Dermatol. Soc., Biol. Honor Soc. of Drew U., Alpha Kappa Kappa, Beta Beta Beta. Republican. Baptist. Avocations: collecting coins, sports memorabilia, antique gun and flag collecting. Home: 1110 San Juan Ct Arlington TX 76012-2750 Office: Dermatology Center 2015 W Park Dr Irving TX 75061-2141

GARCIA, RICHARD LOUIS, plant physiologist/micrometeorologist, researcher; b. Horton, Kans., Jan. 15, 1952; s. Jack Anthony and Irene (Kloepper) G.; m. Paula S. Marten, June 17, 1978; children: Maria, David, Daniel, Anna. BS, Kans. State U., 1978, MS, 1986; PhD, U. Nebr., 1991. Mgr. Dryland Crop Demonstration Farm, Palapye, Bostwana, 1978-84; rsch. asst. Evapotranspiration Lab., Manhattan, Kans., 1984-86; rsch. asst. dept. agrl. meteorology U. Nebr., Lincoln, 1986-91; plant physiologist U.S. Water Conservation Lab., Phoenix, 1991—; mem. Arable Rsch. Priorities Com., Botswana, 1980-83. Contbr. chpt. to book, articles to profl. jours. Mem. Am. Soc. Agronomy, Crop Sci. Soc. Am., Am. Meteorol. Soc., Phi Kappa Phi. Achievements include research in gas exchange and light use efficiency of plants and the influence of global climate change on crop productivity. Office: US Water Conservation Lab 4331 E Broadway Phoenix AZ 85040

GARCIA MARTINEZ, HERNANDO, agrochemical company executive, consultant; b. Bogota, Colombia, Mar. 5, 1942; s. Carlos Garcia and Maria Martinez; m. Elsa Arenas, Sept. 9, 1967; children: Juan Camilo, Maria Paola, Natalia, Katty. Grad. in agronomy, Nat. U., Bogota, 1965; grad. in bus. adminstrn., ESAM, Lima, Peru, 1970. Agronomist Inst. Fomento Algodonero, Armero, Colombia, 1966; with devel. and sales depts. Hoechst Colombiana S.A., Bogota, 1967-69, sales chief, 1970-73, mktg. mgr., 1974-76; mgr. Rhone-Poulenc Agro, Colombia, 1977-81, Colombia, Ecuador, 1982-84; dep. dir. for Latin Am. Rhone-Poulenc Agro, Lyon, France, 1985-87; dir. for Andean Pact Rhone-Poulenc Agro, Bogota, 1988-89, dir. for S.Am., 1990—; dir. for Adean Pact Rhone-Poulenc Agro, Carribean and Cen. Am., 1991—. Author: Manual on Weeds, 1976. Recipient award XIII Agronomists Congress, Cúcuta, Colombia, 1989. Mem. Soc. Colombiana Malezas (founder, pres. 1968-69, award 1976), Entomology Soc. Colombia (founder), Phytopathology Soc. Colombia, Latin Am. Weed Soc., Agronomists Assn. Colombia, Agronomists Assn. Valle (life). Avocations: golf, tennis. Home: Diagonal 128C No 19-17, Bogota Colombia Office: Rhone-Poulenc Agro, Calle 100 8A-55, Of 801/802, Bogota Colombia

GARCIA MARTINEZ, RICARDO JAVIER, architect; b. Monterrey, Nuevo Leon, Mexico, Apr. 3, 1943; s. Jose Antonio Garciaelizondo and Ramona (Martinez) Garcia; m. Jackeline Julieta Santos Luna, Dec. 21, 1985; children: Ricardo, Esau; children by previous marriage: Caleb, Aura, Estefania. Architect, U. Nuevo Leon, Monterrey, 1964; student, U. Nat. Mexico, Mexico City, 1974-76, Inst. Internat. D'Urbanisme, Brussels, 1970-71, U. Perpignan, France, 1980. Registered architect, urban planning, solar energy. Designer Junta de Mejoras Materiales, Monterrey, 1964-65, Ricardo Garcia Y Asociados, Monterrey, 1965-69, Architecture Plus Urbanisme, Brussels, 1970; advisor direction Fideicomiso de Ciudades Industriales, Mexico City, 1972-77; dir. design and urban planning Fondo Nacional de Turismo en Nacional Financiera, Mexico City, 1977-80; mgr. sub-metropoli ctrl. Secretaria de Asentamientos Humanos, Monterrey, 1980-85; dir. design and devel. Ricardo Garcia Y Asociados, Monterrey, 1985—; sec. Fideicomiso de Ciudad Indsl., Morelia, Michoacan, 1975-77; advisor Direccion Aguas Salinas y Energia Solar Secretaria Asentamientos Humanos, Mexico City, 1979-80; sponsor solar energy Study and Devel. Solar Energy Civil Assn., Monterrey, 1988-92. Co-editor: Planeacion y Operacion de Ciudades Industriales, 1975, Boletin Tecnico, 1974-77; editor Folleto Tecnico Consultivo, 1973-74. Grantee Office Belge du Commerce Exterieure, Brussels, 1970, Ente Nazionale Idrocarburi, Urbino, Italy, 1979, Ambassade of France, Perpignan, 1980. Mem. Solar Energy Civil Assn. (pres. 1988—), Sociedad Mexicana de Planification, Sociedad Interamericana de Planificacion, Colegio de Arquitectos de Nuevo Leon. Achievements include urbanistic design of neighborhood in Can Cun, Quintana Roo, design of turistic areas in Can Cun, Quintana Roo, Ixtapa, Zihuatanejo, San Jose Del Cabo and Loreto-Nopolo Baja Calif. Sur, design of houses and other bldgs. the use solar energy in Monterray, Mexico, Distrito Federal and Cuauhtemoc, Chihuahua. Home: Bruselas 800 Penthouse, Monterrey Nuevo Leon 64070, Mexico Office: Solar Energy Assn Civil, Bruselas 800 Fl One, Monterrey Nuevo Leon 64070, Mexico

GARCIA-MORAN, MANUEL, surgeon; b. Oviedo, Asturias, Spain, Feb. 17, 1935; s. Joaquin and Cecilia (Lopez) G.; m. Beatriz Bezares, Aug. 30, 1962; children: Beatriz, Alfredo, Elvira, Fernando. Lic. Med. Surgery, Med. Complutense U., Madrid, Spain, 1959. Intern Jimenez Diaz Found., Madrid, Spain, 1955-60; fgn. asst., resident Paris U., 1960-64; fellow in surgery NYU Med. Ctr., N.Y.C., 1968-70; asst. chief surgery Hosp. Gen. de Asturias, Oviedo, Asturias, Spain, 1964-68, 70-74, chief surgery, 1974—, head surgery, 1975-78; full prof. Oviedo U. Faculty of Medicine, Oviedo, Asturias, Spain, 1984—; pres. Asturias Fedn. U. Sports, Oviedo, 1973-74; pres. Acad. Med. Quirurgica, Asturias, Oviedo, 1981-84. Author: Liver Transplantation, 1973; contbr. articles to profl. jours. Competitor Spanish Olympic Ski Team, Squaw Valley, Calif., 1960; mem. Medicus Mundo, Asturias, 1975—. Recipient Scholarship Colegio de Espana, 1961-63, Premio San Nicolas, Real Acad. Nac. Medicine, 1974. Mem. Acad. Medico Quirurgica Asturiana, Soc. Espanola de Patologia Digestiva, Assn. Espanola de Cirujanos, Assn. Française de Chirurgie, AAAS, N.Y. Acad. Scis. Roman Catholic. Avocations: ski, tennis, golf, bridge. Home: Cervantes 4, 33004 Oviedo Asturias, Spain Office: Uria 18, 33003 Oviedo Asturias, Spain

GARCIA-RILL, EDGAR ENRIQUE, neuroscientist; b. Caracas, Venezuela, Oct. 31, 1948; came to U.S., 1973; s. Juan Garcia and Aracelis (Rill) Ramirez; m. Sherrie Hunt, Oct. 2, 1978 (div.); children: Sarah Thais; m. Susan Gene Ebel, May 13, 1984. BA, Loyola of Montreal, 1968; PhD, McGill U., 1973. Grad. student dept. physiology McGill U., Montreal, Canada, 1969-73; rsch. asst. dept. psychiatry McGill U., Montreal, 1972-73; postdoctoral fellow UCLA, 1973-78; asst. prof. anatomy U. Ark. for Med. Scis., Little Rock, 1978-82, assoc. prof., 1982-87, prof., 1987—, prof. psychiatry, 1990—; mem. biomed. rsch. study sect. NIAAA, Washington, 1983-87, biopsychology study sect. NIH, Washington, 1988-93; chmn. biopsychology rev. com. NIH, 1991-93; reviewer several neurosci. jours., 1979—, small bus. innovative rsch. study sect. NIH, 1988-92. Editor: (videotape) The Basal Ganglia and the Locomotor Regions, 1986; patentee in field. V.p., bd. dirs. Morris Found., Little Rock, 1985—. Postdoctoral fellow Que. Med. Rsch. Coun., 1973; grantee NSF, 1985-85, 88—, NIH, 1983—. Mem. Soc. for Neurosci. (chpt. com. 1991-93), Am. Assn. Anatomists. Avocation: sailboat racing. Office: U Ark for Med Scis 4301 W Markham St Little Rock AR 72205-7101

GARCZEWSKI, RONALD JAMES, transportation engineer; b. Warren, Ohio, Apr. 14, 1969; s. Joseph Bernard and Mary Elizabeth (Michalsky) G. BS in Civil Engring., U. Akron, 1992. Engr. in tng. Coop. technician Finkbeiner, Pettis and Strout, Ltd., Akron, Ohio, 1990-91, transp. engr., 1992—. Mem. ASCE, Am. Soc. Hwy. Engrs. Office: Finkbeiner Pettis & Strout 1725 Merriman Rd Akron OH 44313

GARDE, ANAND MADHAV, materials scientist; b. Sangli, India, Jan. 1, 1945; came to U.S., 1968; s. Madhav Moreshwar and Malati Madhav (Javadekar) G.; m. Vandana Mukund Joshi, Jan. 22, 1972; children: Vinaya, Preeti. B in Tech., Indian Inst. Tech., Bombay, 1967; MS, Syracuse U., 1970; PhD, U. Fla., 1973. Asst. metallurgist Argonne (Ill.) Nat. Lab., 1974-79; prin. engr. Combustion Engring., Windsor, Conn., 1979-88; consulting engr. ABB Combustion Engring., Windsor, 1989—; adj. lectr. Hartford (Conn.) Grad. Ctr., 1989—; symposium chmn. 10th Internat. Symposium on Zirconium in the Nuclear Industry, Balt., 1993. Contbr. over 26 articles, 32 tech. reports and 20 abstracts to profl. jours. Pres. India Assn. of Greater Hartford, 1982; program coord. India Festival of Sci., West Hartford, 1988-89. Mem. AIME (Nuclear Metallurgy com. 1984—), ASTM (B 10, G1 coms. 1985—, tech. editor spl. tech. publ. 1132 1991), Am. Soc. Metals Internat., Indian Inst. of Metals. Republican. Hindu. Achievements include patents for Ductile Irradiated Zirconium Alloys and Corrosion Resistant Zirconium Alloys, patents pending. Office: ABB Combustion Engring 1000 Prospect Hill Rd Windsor CT 06095-1564

GARDNER, AUDREY V., chemist, researcher; b. Newark, N.Y., Mar. 20, 1946; d. Marinus and Elizabeth (VanDerlinde) VanKoeveringe; m. Thomas C. Gardner, Apr. 16, 1966 (div. 1977); children: Lance T., Stephen J. AAS, Agril. & Tech. Coll., Alfred, N.Y., 1966; BS, Rochester Inst. Tech., 1975. Med. technologist Newark-Wayne Community Hosp., 1966-69, Newark Med. Ctr., 1969-75; rsch. technician Cornell U., Geneva, N.Y., 1975-83, sr. rsch. support specialist, 1983—. Author: Laboratory Methods Manual, 1984; (with others) Laboratory Methods Manual, 1979. Mem. AOAC Internat. (chair constn. com. 1986-89, chair regional sect. com. 1992—, N.E. regional sect. exec. com. , pres. 1982-83, treas. 1991—), Assn. Feed Microscopists. Achievements include research in snap bean response to sources and rate of N of K. Office: Cornell U NY Agrl Exptl Sta Food Rsch Lab Geneva NY 14456-0462

GARDNER, BRUCE LYNN, agricultural economist; b. Solon Mills, Ill., Aug. 31, 1942; s. Robert W. and Jeannette (Hopper) G.; m. Mary Agacinski, Sept. 5, 1964. BS, U. Ill., 1964; PhD, U. Chgo., 1968. Asst. prof. econs. N.C. State U., Raleigh, 1968-75; sr. staff economist Pres. Council of Econ. Advisers, Washington, 1975-77; prof. Tex. A&M U., College Station, 1977-80, U. Md., College Park, Md., from 1981; asst. sec. U.S. Dept. Agriculture, Washington, 1989—. Author: Optimal Stockpiling of Grain, 1979 (Outstanding Research award Am. Assn. Agrl. Econs. 1980), The Governing of Agriculture, 1981, The Economics of Agricultural Policies, 1987. Fellow Am. Assn. Agrl. Econs. (bd. dirs. 1984-87). Office: Dept Agriculture 14th & Independence Ave SW Washington DC 20250

GARDNER, CLYDE EDWARD, health care executive, consultant, educator; b. Steubenville, Ohio, Oct. 8, 1931; s. Peter D. and Louella Mary (Gillespie) G.; m. Patricia Jackson, Oct. 4, 1953 (div. Dec. 1977); 1 child, Bruce Stephen. BA, San Francisco State U., 1969, MS, 1971. Adminstr. Gardner Convalescent Hosp., Napa, Calif., 1955-68; exec. dir. Haight Ashbury Free Med. Clinic, San Francisco, 1970-71; lectr. San Francisco State U., 1969-71; dir. planning and rsch. div. N. Country Com. on Area Wide Health Planning, Canton, N.Y., 1971-77; prof. Gov.'s State U., University Park, Ill., 1977-83; sr. ptnr. Health Care Cons., Park Forest, Ill., 1983-86; exec. dir. Mahoning Shenango Area Health Edn. Network, Youngstown, Ohio, 1986-90; pres., chief exec. officer Mahoney Edn. and Tng. Network, Youngstown, Ohio, 1990-92; pres., CEO Health Sci. Assocs., Tucson, 1992—; bd. dirs. rec. sect. Mahoning Shenango Area Health Edn. Network, Youngstown, 1986-90; adj. prof. SUNY, Canton, 1975-76, Youngstown State U., 1987—. Author: Data Book for Health and Institutional Planning, 1981; author of numerous pub. health planning, health edn. studies and funded pvt., state and fed. health care grants, 1971-90. Pres. Found. I Ctr. for Human Devel., Harvey, Ill., 1978-83, U. Profls. of Ill., Chgo., 1982-83; bd. dirs. Blue Cross/Blue Shield Drug and Alcohol Benefit Study, Chgo., 1980-83. Recipient Recognition award Ill. Dangerous Drugs Commn., 1980, 81, Outstanding Svc. award U. Profls. Ill., 1983-84, Outstanding Svc. award Ill. Fedn. Tchrs., 1983. Democrat. Avocations: painting, writing.

GARDNER, GREGORY ALLEN, industrial engineer; b. Abilene, Tex., July 20, 1958; s. Richard Allen and Iva Grace (Nash) G.; m. Jeanne Marie Mason, Jan. 7, 1980; 1 hild, Keir Grey. BS in Indsl. Tech., So. Ill. U., 1982; MS in Aero. Sci., Embry-Riddle U., 1987. Engr., mgr. McDonnell Douglas Corp., St. Louis, 1982—; mng. dir. Bridge Assocs., St. Louis, 1990—. Contbr. articles to profl. jours. Capt. USAF, 1974-82, ANG, 1984—. Decorated Army Achievement medal, Air Force Commendation medal. Mem. AIAA (chmn. subcom. on aerospace maintenance 1989—), Inst. Indsl. Engrs., Soc. Logistics Engrs. (Frank Winship award 1989). Republican. Achievements include patent for air-driven mechanical agitation device for chemical tanks. Home: 727 Shenandoah Ave Saint Louis MO 63104 Office: McDonnell Douglas Missile M/C 106 4297 PO Box 516 Saint Louis MO 63166-0516

GARDNER, HOWARD EARL, psychologist, author; b. Scranton, Pa., July 11, 1943; s. Ralph and Hilde (Weilheimer) G.; m. Ellen Winner; children: Kerith, Jay, Andrew, Benjamin. AB summa cum laude, Harvard U., 1965, PhD, 1971; hon. degree, Curry Coll., 1992, New Eng. Conservatory of Music, 1993. Lectr. edn. Harvard U., 1971-86, co-dir. Project Zero, 1972—, prof. edn., 1986—, affiliated prof. psychology, 1987—; prof. neurology Boston U. Sch. Medicine, 1984-87, adj. prof. neurology, 1987—; rsch. psychologist Boston VA Med. Ctr., 1978—. Author: The Shattered Mind, 1975, Art, Mind and Brain, 1982, Frames of Mind, 1983 (Best Book award Am. Psychol. Assn. 1984), The Mind's New Science, 1985 (William James award 1988), To Open Minds, 1989, The Unschooled Mind, 1991, Multiple Intelligences, 1993, Creating Minds, 1993. MacArthur Prize fellow, 1981, Grawemeyer award in edn., 1990; rsch. grantee numerous govtl. and pvt. founds. Fellow AAAS; mem. Nat. Acad. Edn. (v.p.), Phi Beta Kappa. Office: Harvard U Grad Sch Edn Project Zero Cambridge MA 02138

GARDNER, MICHAEL LEOPOLD GEORGE, biochemistry and physiology researcher, educator; b. Perth, Scotland, May 7, 1946; s. George Gordon Wrangles and Grace Eirene (Critchley) G.; m. Marjorie Minshull, Aug. 22, 1987. BSc in Biochemistry with honors, U. Edinburgh, Scotland, 1969, PhD, 1971, DSc, 1984. Rsch. scholar Med. Sch. U. Edinburgh, 1969-71, univ. demonstrator in biochemistry, 1971-76, rsch. fellow, 1976-85, lectr. in biochemistry, 1985; lectr. in med. biochemistry U. Bradford, Eng., 1985-86, reader in physiol. biochemistry, 1986—; mem. editorial bd. Clin. Sci., London, 1981-87, referee, reviewer numerous sci. jours. Author: Medical Acid-Base Balance, 1978; contbr. articles to sci. jours. Fellow Inst. Biology; mem. Physiol. Soc., Biochem. Soc., Nutrition Soc. Avocations: fishing, photography, wines, pre-war cars. Home: Hazelhurst Brow Farm, Malvern Rd, Bradford BD9 6AR, England Office: U Bradford, Dept Biomed Scis, Bradford BD7 1DP, England

GARDNER, ROBIN PIERCE, engineering educator; b. Charlotte, N.C., Aug. 17, 1934; s. Robin Brem and Margaret (Pierce) G.; m. Linda Jean Gardner, Oct. 21, 1976. B.Ch.E., N.C. State U., 1956, M.S., 1958; Ph.D., Pa. State U., 1961. Scientist Oak Ridge Inst. Nuclear Studies, 1961-63; research engr., asst. dir. measurement and controls lab. Research Triangle Inst., Research Triangle Park, N.C., 1963-67; research prof. nuclear engring. and chem. engring., dir. Center Engring. Applications of Radioisotopes, N.C. State U., 1967—; cons. Oak Ridge Inst. Nuclear Studies, Research Triangle Inst., Oak Ridge Nat. Lab., Internat. Atomic Energy Agy., NASA, AEC, TVA, Alcoa. Author: (with Ralph L. Ely, Jr.) Radioisotope Measurement Applications in Engineering, 1967; Contbr. articles to sci. jours. Served to 1st lt. AUS, 1956. Recipient Alcoa Found. Dist. Research award N.C. State U. Sch. Engring., 1986;. Fellow Am. Nuclear Soc. (Radiation Industry award isotopes and radiation div. 1984), Am. Inst. Chem. Engrs., Sigma Xi, Phi Kappa Phi, Phi Lambda Upsilon. Home: 3005 Randolph Dr Raleigh NC 27609-6941 Office: NC State U Dept Nuclear Engring Raleigh NC 27695-7909

GARDNER, STANLEY, forensic engineer, expert witness; b. Birmingham, Eng., Apr. 8, 1934; s. Albert and Beatrice (May) G.; m. Beverly H. Heiberg, Aug. 1985; three children. Higher Nat. Cert., Chance Tech. Coll., Birmingham, 1955. Registered profl. engr., Calif. Design engr., mfg. engr. Tappan, L.A., 1961-64; engr. Walt Disney, Anaheim, Calif., 1964-66; project engr. Fender Musical Instruments, Fullerton, Calif., 1966-68, Ling Electronics, Anaheim, 1968-72; chief engr. Ormond, Santa Fe Springs, Calif., 1972-74; sr. project engr. United Concrete Pipe, Irwindale, Calif., 1974-85; v.p., forensic engr. Snyder Rsch. Labs, Pico Rivera, Calif., 1985—. Author publs. in field. Mem. ASME, ASCE, Internat. Assn. Arson Investigators, Nat. Fire Prevention Assn., Soc. Automotive Engrs., Am. Soc. Safety Engrs., Nat. Assn. Fire Investigators. Home: 8181 Malloy Dr Huntington Beach CA 92646 Office: Snyder Rsch Labs Inc 4740 Durfee Ave Pico Rivera CA 90660

GARDNER, WILFORD ROBERT, physicist, educator; b. Logan, Utah, Oct. 19, 1925; s. Robert and Nellie (Barker) G.; m. Marjorie Louise Cole, June 9, 1949; children: Patricia, Robert, Caroline. B.S., Utah State U., 1949; M.S., Iowa State U., 1951, Ph.D., 1953. Physicist U.S. Salinity Lab., Riverside, Calif., 1953-66; prof. U. Wis., Madison, 1966-80; physicist, prof., head dept. soil and water sci. U. Ariz., Tucson, 1980-87; dean coll. natural resources U. Calif., Berkeley, 1987—. Author: Soil Physics, 1972. Served with U.S. Army, 1943-46. NSF sr. fellow, 1959; Fulbright fellow, 1971-72. Fellow AAAS, Am. Soc. Agronomy; mem. Internat. Soil Sci. Soc. (pres. physics comm. 1968-74), Soil Sci. Soc. Am. (pres. 1990, Rsch. award 1962), Nat. Acad. Scis. Office: U Calif Coll Natural Resources Berkeley CA 94720

GARDNER, WILLIAM ALLEN, electrical engineering educator; b. Palo Alto, Calif., Nov. 4, 1942; s. Allen Frances McLean and Francis Anne Demma; m. Nancy Susan Lenhart Hall, June 19, 1966. MS, Stanford U., 1967; PhD, U. Mass., Amherst, 1972. Engr. Bell Telephone Labs., North Andover, Mass., 1967-69; asst. prof. U. Calif., Davis, 1972-77, assoc. prof., 1977-82, prof. elect. engring., 1982—; pres. Statis. Signal Processing Inc., 1982—; chmn., organizer workshop on Cyclostationary Signals, NSF, Air Force Office of Sci. Rsch., Army Rsch. Office, Office Naval Rsch. Author: Introduction to Random Processes with Applications to Signals and Systems, 1985, 2d edit., 1989, Statistical Spectral Analysis: A Nonprobabilistic Theory, 1987, Cyclostationarity in Communications and Signal Processing, 1993; contbr. over 100 articles to profl. jours.; patentee in field. Recipient Disting. Engring. Alumnus award, 1987; grantee Air Force Office Sci. Rsch., 1979-82, 92-93, NSF, 1983-84, 89—, Electromagnetic Systems Labs., 1984-92, Army Rsch. Office, 1989—, Office of Naval Rsch., 1991-93. Fellow IEEE (S.O. Rice Prize Paper award in Comm. Theory, 1988); mem. AAAS, European Assn. Signal Processing (Best Paper award 1986), Sigma Xi, Eta Kappa Nu, Tau Beta Pi. Office: U Calif Dept Elec and Computer Engring Davis CA 95616

GARDON, JOHN LESLIE, paint company research executive; b. Budapest, Hungary, June 5, 1928; came to U.S., 1958; s. Louis and Clara (Popper) G.; m. Berta Rost, Dec. 26, 1951; children: Jessica Joan, Frederic Paul. B.S., Swiss Fed. Inst. Tech., 1951; Ph.D., McGill U., 1955. Chemist Can. Internat. Paper Co., Hawkesbury, Ont., Can., 1955-58; sr. chemist, group leader, research assoc., mgr. Rohm and Haas Co., Springhouse, Pa., 1958-68; dir. research and devel. M & T Chems., Southfield (Mich.) and Rahway (N.J.), 1969-80; v.p. corp. research and devel. Sherwin Williams Co., Chgo., 1981-85; v.p. research and devel. Akzo Coatings, Inc., Troy, Mich., 1985—; trustee Paint Research Inst., Kent, Ohio, 1972-81; chmn. Gordon Research Conf. of Adhesion, New Hampton, N.H., 1976. Editor Non-Polluting Coatings and Processes, 1973, Emulsion Polymerization, 1976; contbr. articles to profl. jours., patentee in field. Mem. Am. Chem. Soc. (chem. organic coatings and plastics div. 1980), Fedn. of Soc. for Paint Tech. (Roon award 1966, Mattiello lectr. 1992), Chem. Inst. Can., Soc. plastics Engrs., Soc. Mfg. Engrs., N.Y. Acad. Scis., Sigma Xi. Office: Akzo Coatings Inc PO Box 7062 Troy MI 48007-7062

GAREGG, PER JOHAN, chemist, educator; b. Oslo, Norway, July 11, 1933; arrived in Sweden, 1958; s. Syver and Hjørdis (Ingebretsen) G.; m. Sheila Barnet Hamilton, June 25, 1955 (div. 1986); children: Inez Hamilton, Irene Carter; m. Margaret Alice Clarke, May 25, 1991. PhD, Stockholm U., 1965. Chemist Bakelite Ltd., Birmingham, England, 1955-58; rsch. grantee Swedish Forest Products Lab., Stockholm, 1958-59, rsch. asst., 1959-64; lectr. Stockholm U., 1964-68, assoc. prof., 1969-84, prof., 1985—; rsch. assoc. U. Alberta, Edmonton, Canada, 1968-69; prof. Uppsala (Sweden) U., 1984-85; mem. scientific adv. bd. Glycomed, Inc., Alameda, Calif., 1990—. Contbr. more than 200 articles to profl. jours. Recipient Lindblom award Royal Swedish Sci. Acad., 1974, Norblad-Ekstrand medal Swedish Chem. Soc., 1979, Claude S. Hudson award Am. Chem. Soc. Achievements include research in synthetic aspects of carbohydrate chemistry, polysaccharide chemistry, and nucleotide synthesis. Home: Rorstrandsg 16 NB OG, S-11340 Stockholm Sweden Office: Stockholm U, Arrhenius Lab, Dept Organic Chemistry, S-10691 Stockholm Sweden

GAREY, DONALD LEE, pipeline and oil company executive; b. Ft. Worth, Sept. 9, 1931; s. Leo James and Jessie (McNatt) G.; BS in Geol. Engring., Tex. A&M U., 1953; m. Elizabeth Patricia Martin, Aug. 1, 1953; children: Deborah Anne, Elizabeth Laird. Reservoir geologist Gulf Oil Corp., 1953-54, sr. geologist, 1956-65; v.p., mng. dir. Indsl. Devel. Corp. Lea County, Hobbs, N.Mex., 1965-72, dir., 1972-86, pres., 1978-86; v.p., dir. Minerals, Inc., Hobbs, 1966-72, pres., dir., 1972-86, chief exec. officer, 1978-82; mng. dir. Hobbs Indsl. Found. Corp., 1965-72, dir., 1965-76; v.p. Llano, Inc., 1972-74, exec. v.p., chief operating officer, 1974-75, pres., 1975-86, chief exec. officer, 1978-82, also dir.; pres., chief exec. officer, Pollution Control, Inc., 1969-81; pres. NMESCO Fuels, Inc., 1982-86; chmn., pres., chief exec. officer Estacado Inc., 1986—, Natgas Inc., 1987—; pres. Llano Co2, Inc., 1984-86; cons. geologist, geol. engr., Hobbs, 1965-72. Chmn., Hobbs Manpower Devel. Tng. Adv. Com., 1965-72; mem. Hobbs Adv. Com. for Mental Health, 1965-67; chmn. N.Mex. Mapping Adv. Com., 1968-69; mem. Hobbs adv. bd. Salvation Army, 1967-78, chmn., 1970-72; mem. exec. bd. Conquistador coun. Boy Scouts Am., Hobbs, 1965-75; vice chmn. N.Mex. Gov.'s Com. for Econ. Devel., 1968-70; bd. regents Coll. Southwest, 1982-85. Capt. USAF, 1954-56. Registered profl. engr., Tex. Mem. Am. Inst. Profl. Geologists, Am. Assn. Petroleum Geologists, AIME, N.Mex. Geol. Soc., Roswell Geol. Soc., N.Mex. Amigos Club, Rotary. Home: 315 E Alto Dr Hobbs NM 88240-3905 Office: Broadmoor Tower PO Box 5587 Hobbs NM 88241-5587

GARFINKEL, HARMON MARK, specialty chemicals company executive; b. Bklyn., May 20, 1933; s. Samuel and Elsie (Schwartz) G.; m. Lorraine Plawsky, Mar. 4, 1956; children—Elyse, Michelle. B.A., Bklyn. Coll., 1957; Ph.D., Iowa State U., 1960; postgrad. program for mgmt. devel., Harvard U. Bus. Sch., 1973. Dir. bio-organic tech. Corning Inc., N.Y., 1973-74, dir. applied chemistry and biology, 1974-75, dir. biomed. and chem. tech., 1975-78, dir. research, 1978-85; v.p. R&D Engelhard Corp., Edison, N.J., 1985—; bd. dirs. NECC Co.; instr. math. Elmira Coll., 1964. Patents and publs. in field. Mem. Am. Chem. Soc., Am. Phys. Soc., Am. Inst. Chemists, Am. Ceramics Soc. Republican. Jewish. Home: 1584 Mountain Top Rd Bridgewater NJ 08807-2320 Office: Engelhard Corp 101 Wood Ave Iselin NJ 08830-3503

GARFINKLE, DEVRA, mathematician, educator; b. Northampton, Mass., Aug. 21, 1956; d. Norton and Vivienne (Feigenbaum) G.; m. Joseph Francis Johnson, Dec. 19, 1981; 1 child, Robert Francis Johnson. BSc, Princeton U., 1978; PhD, MIT, 1982. Fellow NSF, Cambridge, Mass., 1982-83; instr. dept. math. MIT, Cambridge, 1982-83, U. Utah, Salt Lake City, 1983-87; vis. researcher U. Paris VII, 1987; fellow Math. Scis. Rsch. Inst., Berkeley, Calif., 1988; asst. prof. Coll. Arts and Scis., Rutgers U., Newark, 1988—; fellow Inst. for Advanced Study, Princeton, N.J., 1992—. Contbr. chpts. to books. Mem. Am. Math. Soc. Democrat. Roman Catholic. Home: 2037 Spruce St Philadelphia PA 19103 Office: Rutgers Univ Dept Math NCAS Newark NJ 07102

GARG, ANUPAM K., physicist; b. Amritsar, Punjab, India, Aug. 17, 1956; came to U.S., 1977.; s. Hem Raj and Sukarma (Gupta) G.; m. Neerja Gupta, Aug. 16, 1986; children: Gaurav, Arjun. MS in Physics, Indian Inst. Tech., Delhi, India, 1977; PhD, Cornell U., 1983. MacArthur rsch. assoc. U. Ill., Champaign, 1983-85; rsch. assoc. Inst. Theoretical Physics, U. Calif., Santa Barbara, 1985-88; asst. prof. Northwestern U., Evanston, Ill., 1988—. Contbr. articles to profl. jours. Mem. Am. Phys. Soc. Achievements include rsch. on macroscopic quantum phenomena in Josephson junctions and in magnets, foundations of quantum mechanics. Office: Northwestern U Dept Physics 2145 Sheridan Rd Evanston IL 60208

GARG, DEVENDRA PRAKASH, mechanical engineer, educator; b. Roorkee, India, Mar. 22, 1934; came to U.S., 1965; s. Chandra Gopal and Godawari (Devi) G.; m. Prabha Govil, Nov. 19, 1961; children—Nisha, Seema. B.Sc., Agra (India) U., 1954; B.S. in Mech. Engring, U. Roorkee, 1957; M.S. (Tech. Coop. Mission Merit scholar), U. Wis.-Madison, 1960; Ph.D., N.Y. U., 1969. Lectr. mech. engring. U. Roorkee, 1957-62, reader, 1962-65, vis. prof., 1978; instr. NYU, 1965-69; asst. prof. MIT, Cambridge, 1969-71, assoc. prof., 1971-72, chmn. engring. projects lab., 1971-72, lectr., 1972-75, vis. prof., 1976-80; prof. Duke U., Durham, N.C., 1972—, dir.

undergrad. studies dept. mech. engring. and materials sci., 1977-86; vis. prof. dept. automatic control Georgian Poly. Inst., Tbilisi, USSR, 1988; program dir. dynamic systems and control program civil and mech. systems div. NSF, Washington, 1992-95. Author: An Introduction to the Theory and Use of the Analog Computer, 1963, A Textbook of Descriptive Geometry, 1964; asso. editor Jour. Dynamic Systems, Measurement, and Control, 1971-73, Jour. Interdisciplinary Modeling and Simulation, 1978-80; contbr. numerous articles to profl. jours. Recipient Founder's Day award NYU, 1969, Fulbright Sr. Scholar award, 1987-88, cert. of commendation Acoustical Soc. Am., 1983; U.S. Dept. Transp. faculty fellow, 1980-81, NASA/ASEE faculty fellow, 1986-87, U.S. Army faculty fellow, 1988. Fellow ASME (reviewer, co-guest editor spl. issues on ground transp. 1974, socioecon. and ecol. systems 1976, sec. dynamic systems and control div. 1980-83, vice chmn. 1984-85, chmn. div. 1985-86, cert. of appreciation tech. and society div. 1986-87); mem. IEEE (reviewer), Instrument Soc. Am. (reviewer, chmn. adv. panel 1987-88, 89—, chmn. honors com. 1987-88, nat. nominating com. 1989—, system and design group operating bd. 1989—), Sigma Xi (chpt. sect. 1970-72). Home: 2815 Dekalb St Durham NC 27705-5601 Office: Duke U Sch Engring Box 90300 Durham NC 27708-0300

GARGIULO, GERALD JOHN, psychoanalyst, writer; b. N.Y.C., Nov. 12, 1934; s. Fred Nunzio and Fanny Joy (Tarantino) G.; m. Julia Caldiero, Apr. 12, 1964; children: Paul Gerald, Connie Joy. BA in Philosophy, St. Bonaventure U., Orlean, N.Y., 1958; MA in Religious Studies, Washington Theol. Union, 1962. Mem. faculty Manhattan Coll., N.Y.C., Riverdale, 1962-70; pvt. practice N.Y.C., Greenwich, Conn., 1970—; dir. The Inst. N.P.A.P., N.Y.C., 1988-92, pres. tng. inst., 1992—; assoc. editor The Psychoanalytic Review, N.Y.C., 1987—; pres. Coun. of Psychoanalytic Psychotherapists, N.Y.C., 1981-83; bd. dirs. N.Y. Ctr. for Psychoanalysis, 1973-75; mem. faculty Nat. Psychol. Assn. for Psychoanalysis, 1972—. Contbr. articles to profl. pubs. Bd. dirs. Psychoanalysts Against Nuclear Weapons, 1982-85; com. mem. Greenwich Nuclear Freeze Orgn., 1987-90. Fellow Inst. for Psychoanalytic Tng. & Rsch. (bd. dirs. 1976), Coun. of Psychoanalytic Psychotherapists; mem. Internat. Fedn. for Psychoanalytical Edn. (bd. dirs. 1992—), Internat. Psychoanalytical Assn., Nat. Psychol. Assn. for Psychoanalysis. Avocations: poetry, piano. Home: 11 Lafayette Ct Apt 1D Greenwich CT 06830 Office: 158 Greenwich Ave Greenwich CT 06830-6548

GARIBALDI, LOUIS, aquarium administrator. With New Eng. Aquarium, Steinhart Aquarium; assoc. dir. N.Y. Aquarium, Bklyn., 1983-88, dir., 1988—; v.p. N.Y. Zoological Soc./Wildlife Conservation Soc., 1993—. Recipient Edward H. Bean Meml. award Am. Assn. Zoological Parks & Aquariums, 1992. Office: NY Aquarium Boardwalk & W 8th St Brooklyn NY 11224*

GARIBAY, JOSEPH MICHAEL, mechanical engineer; b. Burlington, N.C., Sept. 14, 1960; s. Rafael and Doris Faye (Pugh) G. BS in Mech. Engring., N.C. State U., 1985. Registered profl. engr., N.C. HVAC tech. Glosson Svc., Burlington, 1978-86; project mgr. AC Corp., Greensboro, N.C., 1986—. Mem. Nat. Soc. Profl. Engrs., ASHRAE. Office: AC Corp 301 Creek Ridge Rd Greensboro NC 27416-0367

GARIMELLA, SURESH VENKATA, mechanical engineering educator; b. Nedunuru, Andhra, India, July 8, 1963; came to U.S. 1985; s. Sastry Viswanadha and Radha Devi (Gorti) G. BTECH, Indian Inst. Tech., Madras, India, 1985; MS, Ohio State U., 1986; PhD, U. Calif., Berkeley, 1989. Instr. U. Calif., Berkeley, 1990; Cray Rsch. asst. prof. U. Wis., Milw., 1990--; cons. Cray Rsch., Inc., Chippewa Falls, Wis., 1991--. Contbr. articles to profl. jours. Faculty mentor NIH, 1991; instr. Coll. for Kids, Milw., 1991. Recipient Rsch. award Nat. Sci. Found., 1992--, Outstanding Tchr. award Coll. Engring. Applied Sci., 1992; Engring. Achievement award Ohio State U., 1986. Mem. Soc. Automotive Engrs. (Teetor Ednl. award 1992), Am. Soc. Mechanical Engrs., Am. Soc. Engring. Edns., Sigma Xi. Office: U Wis 3200 N Cramer St Milwaukee WI 53211

GARLAND, HARRY THOMAS, research administrator; b. Detroit, Jan. 18, 1947; s. Harry George and Rose (Bonn) G.; m. Roberta Joy Siciliano; children: Eva, Harry, Brad, Ken. BA, Kalamazoo Coll., 1968; PhD, Stanford U., 1972. Lectr. Stanford (Calif.) U., 1972-73, asst. dept. chmn., 1973-76; pres. Cromemco, Inc., Mountain View, Calif., 1976-89; v.p. Canon Rsch. Ctr., Palo Alto, Calif., 1990—; trustee Kalamazoo (Mich.) Coll., 1986—. Author: Introduction to Microprocessor System Design, 1979; co-author: Understanding IC Operational Amplifiers, 1971, Understanding CMOS Integrated Circuits, 1975; contbr. articles to profl. jours. Recipient NIH traineeship, Disting. Alumni award Kalamazoo Coll., 1986. Office: Canon Rsch Ctr 4009 Miranda Ave Palo Alto CA 94304

GARMIRE, ELSA MEINTS, electrical engineering educator, consultant; b. Buffalo, Nov. 9, 1939; d. Ralph E. and Nelle (Gubser) Meints; m. Gordon P. Garmire, June 11, 1961 (div. 1975); children: Lisa, Marla; m. Robert Heathcote Russell, Feb. 4, 1979. AB in Physics, Harvard U., 1961; PhD in Physics, MIT, 1965. Rsch. scientist NASA Electronics Rsch. Ctr., Cambridge, Mass., 1965-66; rsch. fellow Calif. Inst. Tech., Pasadena, 1966-73; sr. rsch. scientist U. So. Calif. Ctr. for Laser Studies, L.A., 1974-78, assoc. prof. elec. engring. and physics, 1981-92, assoc. dir. Ctr. for Laser Studies, 1978-83, dir., 1984—, William Hogue prof. of engring., 1992—; vis. fellow Standard Telecommunication Labs., Eng., 1973-74; cons. Aerospace Corp., L.A., 1975-91, sci. adv. bd. Air Force, Washington, 1985-89, TRW, L.A., 1988-89, McDonnell Douglas, St. Louis, 1990—. Contbr. over 160 sci. papers and articles to profl. publs.; patentee in field. Recipient Soroptimist Achievement award Soroptimist Club, L.A., 1970, K. C. Black award N.E. Electronics Rsch. and Engring. Meeting, 1972; named Mademoiselle Woman of Yr. Mademoiselle Mag., 1970. Fellow IEEE (bd. dirs. 1985-89), Optical Soc. Am. (bd. dirs. 1983-86, pres. elect 1992, pres. 1993); mem. NAE, Am. Phys. Soc. (rep. 1962—), Soc. Women Engrs. (sr.), Harvard Radcliffe Club (v.p. 1984-86). Democrat. Avocation: music. Office: U So Calif Ctr for Laser Studies DRB17 Los Angeles CA 90089-1112

GARNER, DOUGLAS RUSSELL, science writer; b. Orange, Tex., Aug. 7, 1953; s. Jim Buck and Ruthie Delores (Seastrunk) G. BA in Biology, U. Tex., 1975; postgrad. in medicine, Creighton U., 1976-77. Pub. health investigator Houston Health Dept., 1979-84; freelance tech. writer, translator Houston, 1984-86; tech. translations editor McElroy Translation Co., Austin, Tex., 1986-89; freelance sci. writer, editor L.A., 1989—; sci. researcher Calif. Afro-Am. Mus., L.A., 1991—. Author: The Adventures of Teddy Wallace, 1991. Recipient award Charles Palmer Davis Found., 1970. Mem. AAAS. Home and Office: PO Box 7026 Austin TX 78713

GARNER, JASPER HENRY BARKDOLL, ecologist; b. Bulsar, Gujarat, India, Nov. 7, 1921; s. Holly Pearl and Kathryn Elizabeth (Barkdoll) G.; m. Lois Marie Harman, Aug. 30, 1954; 1 child, Eric Lynn. BA in History, Manchester Coll., Ind., 1948; MA in Botany, Ind. U., 1953; PhD in Botany, U. Iowa, 1955. Asst. prof. botany U. Nebr., Lincoln, 1955-56; instr. botany to asst. prof. botany U. Ky., Lexington, 1956-67; vis. assoc. prof. botany Coll. Agr. and Vet. Medicine U. Indonesia, Bogor, 1957-61; assoc. prof. biology East Tenn. State U., Johnson City, 1967-70; rsch. assoc. plant pathology N.C. State U., Raleigh, 1969-71; botanist Office of Air Programs, EPA, Research Triangle Park, N.C., 1971-72; botanist, spl. studies staff EPA, Research Triangle Park, N.C., 1972-78, ecologist environ. criteria and assessment office, 1978—; advisor Dean, Coll. Agr., U. Indonesia, Ky.-AID Program, 1957-61, acting chief of party, 1960-61. Contbr. articles to profl. jours. Rsch. grantee U. Ky. Faculty Rsch. Fund, 1962, NSF, summer 1964, U. Ky., summer 1965, Nat. Air Pollution Control Adminstrn., 1969-71; recipient Manchester Coll. Alumni Honor award, 1980, U.S. EPA Bronze medal, 1980, 82, 87. Mem. Ecol. Soc. Am., Mycol. Soc. Am., Sigma Xi. Office: EPA Environ Criteria & Assessment Office MD-52 Research Triangle Park NC 27711

GARNER, PATRICK LYNN, nuclear engineer; b. Lynchburg, Va., Sept. 23, 1950; s. Curtis Lee and Helen Frances (Vaughn) G.; m. Deborah Gail Alstedt, Nov. 21, 1981. BS, U. Va., 1972, MS, 1974, PhD, 1977. Registered profl. engr., Ill. Asst. nuclear engr. Argonne (Ill.) Nat. Lab., 1977-81, nuclear engr., 1981--. Contbr. articles to profl. jours. Deacon Community Presbyn. Ch., Clarendon Hills, Ill., 1981-83, elder, 1984-86; videographer Village of Downers Grove, Ill., 1990. Mem. Am. Nuclear Soc. (program com. 1984-86), Sigma Xi (asst. sec. local sect. 1982-84, sec. 1984-86). Home: 6650 Barrett St Downers Grove IL 60516 Office: Argonne Nat Lab 9700 S Cass Ave Argonne IL 60438

GARNES, DELBERT FRANKLIN, clinical and consulting psychologist, educator; b. Lorain, Ohio, Jan. 13, 1943; s. Delbert Chauncey and Virginia (Scott) G.; m. Bertha J. Smith (div.); m. Joyce M. Roberts; children: Franklin Chauncey, Charles Deltre. BA, Ohio State U., 1969; MA, Xavier U., 1974; PhD, St. Louis U., 1980. Lic. clin. psychologist. Counselor Fairfield Sch. for Boys, Lancaster, Ohio, 1970-71; instr. Met. Coll., St. Louis, 1974-75; psychologist Narcotics Svc. Coun., St. Louis, 1974-75, Roxbury Ct. Clinic, Boston, 1975-77, Fuller Mental Health Ctr., Boston, 1976-77; asst. prof. psychology Tex. So. U., Houston, 1980-86, assoc. prof., 1986—; chmn. dept. counseling and psychology, 1990—; cons. Mass. Bar Assn., Boston, 1976-77, Mass. Parole Bd., Boston, 1977, Harris County Juvenile Probation, Houston, 1981—, Harris County Dept. Edn., Houston, 1988—. Judge, Internat. Sci. and Engring. Fair, 1982, Houston Sci. and Engring. Fair, 1983-85. Cpl. USMC, 1961-65. Recipient Disting. Svc. award Nat. Tech. Assn., 1985, Svc. award Student Nat. Pharm. Assn., 1987, Sam Houston Area coun. Boy Scouts Am., 1987; NIMH fellow, 1977; Boston U. Sch. Medicine teaching fellow, 1975. Mem. Houston Assn. Black Psychologists (pres. 1981-82), Am. Psychol. Assn., Nat. Assn. Black Psychologists, Am. Fedn. Tchr., Tex. State Tchrs. Assn. Avocations: travel, golf, fishing. Office: Tex So U 3100 Cleburne St Houston TX 77004-4501

GARON, CLAUDE FRANCIS, laboratory administrator, researcher; b. Baton Rouge, Nov. 5, 1942; s. Ivy Joseph and Janith (Latil) G.; m. Sally Sheffield; children: Michele, Anne, Julie. BS, La. State U., 1964, MS, 1966; PhD, Georgetown U., 1970. Predoctoral fellowship La. State U., Baton Rouge, 1964-66; predoctoral traineeship Georgetown U., Washington, 1966-69; postdoctoral fellowship Nat. Inst. Allergy and Inf. Diseases, Bethesda, Md., 1971-73, staff fellowship, 1971-73, sr. staff fellowship, 1973-74, rsch. microbiologist, 1974-81; head electron microscopy Rocky Mountain Labs, Hamilton, Mont., 1981-85, chief, pathobiology, 1985-89, chief, lab. vectors and pathogens, 1989—; bd. govs. Ctr. Excellence in Biotech., Missoula, Mont., 1988—; faculty affiliate U. Mont., 1989—. Mem. editorial bd. Jour. Clin. Microbiology, 1993. Recipient award of merit NIH, Dirs. award, 1988. Mem. Am. Soc. for Microbiology, Am. Soc. Biochemistry and Molecular Biology, Microscopy Soc. Am., Am. Soc. Rickettsiology, Pacific N.W. Electron Microscopy Soc., Lions (pres. Hamilton 1989-90). Office: Rocky Mountain Labs Lab Vectors & Pathogens 903 S 4th St Hamilton MT 59840-2999

GARRATTY, GEORGE, immunohematologist; b. London, July 2, 1935; came to the U.S., 1968; s. George H. and Marjory Garratty; m. Eileen Margaret Carroll, 1969. PhD, Columbia Pacific U., 1985. Chief rsch. lab. scientist Royal Postgrad. Med. Sch. London, 1963-68; rsch. assoc. U. Calif., San Francisco, 1968-78; scientific dir. ARC, L.A., 1978—; clin. prof. pathology UCLA, 1985—. Co-author: Handbook of Haematology and Blood Transfusion, 1969, Acquired Immune Hemolytic Anemias, 1980; editor 5 books; contbr. articles to profl. jours. Recipient Lyndall Molthan Meml. award Pa. Assn. Blood Banks, 1990, Wiener Meml. award N.Y. Blood Ctr., 1991. Fellow Inst. Med. Lab. Sci., Royal Coll. Pathologists; mem. Am. Assn. Blood Banks (Ivor Dunsford Meml. award 1978, Emily Cooley Meml. award 1989), Am. Soc. Hematology, Am. Assn. Immunologists, Calif. Blood Bank Soc. (pres. 1985-86, Owen Thomas Meml. award 1987), Internat. Soc. Blood Transfusion (councillor 1991—). Office: ARC Blood Svcs 1130 S Vermont Ave Los Angeles CA 90006

GARRELICK, JOEL MARC, acoustical scientist, consultant; b. N.Y.C., May 20, 1941; s. Samuel J. Garrelick and Phyllis Weidenbaum; m. Renee Brosell, Dec. 23, 1963; chilren: Kevin, Jenine, Daniel. BCE, CCNY, 1963, ME, 1965; PhD, CUNY, 1969. Lectr. CCNY, 1968-69; scientist Cambridge (Mass.) Acoustical Assocs., Inc., 1969-75, prin. scientist, 1976—, pres., 1990—. Contbr. articles to profl. jours. Fellow Acoustical Soc. Am.; mem. ASCE. Office: CAA Inc 200 Boston Ave Medford MA 02155

GARRETSON, HENRY DAVID, neurosurgeon; b. Woodbury, N.J., June 8, 1929; s. O.K. and Mary Marjorie (Davis) G.; m. Marianna Schantz, July 4, 1964; children: John, Steven. B.S., U. Ariz., 1950; M.D., Harvard U., 1954; Ph.D., McGill U., 1968. Diplomate: Am. Bd. Neurol. Surgery (mem. 1981-87, vice chmn. 1985-86, chmn. 1986-87. Surg. intern Royal Victoria Hosp., Montreal, 1954-55; resident Montreal Neurol. Inst., 1959-63; asst. prof. neurosurgery McGill U., Montreal, 1966-71; prof., dir. neurol. surgery Sch. Medicine U. Louisville, 1971—, assoc. dean clin. affairs Sch. Medicine, 1975-79, dir. neuroscis. programs Sch. Medicine, 1979-82; individual practice medicine, specializing in neurosurgery, Montreal, 1963-71; asso. with Dr. William Feindel, Montreal, 1963-71; with Granthan & Garretson, Louisville, 1971—; mem. staff Humana Hosp. U., Norton Children's, VA, Surburban, Ky. Baptist hosps., all Louisville; staff Inst. Phys. Medicine and Rehab. Contbr. numerous articles, abstracts, editorials, presentations in field. Served with USNR, 1955-58. Fellow ACS; mem. AAAS, AMA, Am. Assn. Neurol. Surgeons (bd. dirs. 1983-85, sec. 1985-86, pres. elect 1986-87, pres. 1987-88), Am. Acad. Neurol. Surgery (pres. 1991-92), Congress Neurol. Surgeons, Ky. Neurosurg. Soc., Ky. Surg. Soc., Louisville Surg. Soc., Ky. Med. Assn., Soc. Neurol. Surgeons, Soc. U. Neurosurgeons (pres. 1983-84), So. Neurosurg. Soc. (pres. 1986-87), Jefferson County Med. Soc., Phi Beta Kappa, Phi Kappa Phi, Sigma Xi. Home: 517 Tiffany Ln Louisville KY 40207-1438 Office: U Louisville Health Scis Ctr Louisville KY 40292

GARRETT, CHARLES GEOFFREY BLYTHE, engineering consultant; b. Ashford, Kent, Eng., Sept. 15, 1925; came to U.S., 1950, naturalized, 1989; s. Charles Alfred Blythe and Laura Mary (Lotinga) G. B.A. in Natural Scis., Trinity Coll., Cambridge U., Eng., 1946; M.A. in Natural Scis., Ph.D. in Physics, Cambridge U., 1950. Instr. physics Harvard U., 1950-52; mem. tech. staff Bell Labs., Murray Hill, N.J., 1952-54; supr. Bell Labs., 1955-56, dept. head, 1960-69; dir. AT&T Bell Labs., Murray Hill-Morristown, N.J., 1969-87; chmn. Gordon Conf. on non-linear optics, 1964. Author: Magnetic Cooling, 1954, Gas Lasers, 1963; contbr. articles to profl. jours.; patentee in field. Named knight of Sovereign Order St. John of Jerusalem (Orthodox). Fellow Am. Phys. Soc., IEEE; mem. Guild of Carillonneurs in N.Am. Episcopalian. Avocations: piano, harpsichord, carillon, restoring 18th century houses and older Rolls-Royce cars. Home: 7 Fithian Ln East Hampton NY 11937-2605

GARRETT, JAMES HENRY, JR., civil engineering educator; b. Pitts., Jan. 12, 1961; s. James Henry and Marie Angela (Tommasino) G.; m. Ruth Ann Killmeyer, June 2, 1984; children: Ellen Grace, Patrick James. BS in Civil Engring., Carnegie-Mellon U., 1982, MS in Civil Engring., 1983, PhD in Civil Engring., 1986. Project engr. Schlumberger Well Svcs., Houston, 1986-87; asst. prof. civil engring. U. Ill., Urbana, 1990-93; asst. prof. civil engring. Carnegie-Mellon U., Pitts., 1990—, assoc. prof. civil engring., 1993—; presdl. young investigator NSF, 1989—. Mem. Zoning Hearing Bd., Shaler Twp., 1991. Mem. ASCE (assoc., Moisseiff award 1990), IEEE, Internat. Assn. Bridge and Structural Engrs. (prize 1992), Internat. Neural Network Soc. Achievements include research in computing in civil engring., structural engring., engring. desing, and analysis. Home: 307 S Highlander Heights Dr Glenshaw PA 15116 Office: Carnegie Mellon U 5000 Forbes Ave Pittsburgh PA 15213

GARRETT, JOSEPH EDWARD, aerospace engineer; b. Hendersonville, N.C., Mar. 4, 1943; s. Kenneth Pace and Anna Lou (Lytle) G.; m. Aurelia Jane Pryor, Aug. 7, 1971. BS in Aerospace Engring., N.C. State U., 1966; MS in Aerospace Engring., Ga. Inst. Tech., 1978. Registered profl. engr., Ga. Basic and fatigue loads assoc. aircraft engr. LASC-Ga. (formerly Lockheed-Ga.), Marietta, 1966-67, basic and fatigue loads structures engr., 1967-75, fatigue and fracture mechanics sr. structures engr., 1975-80, company planning, 1980-82, fracture mechanics structures engr., 1982-91, advanced structures sr. engr., 1991—. Loaned exec. United Way, Atlanta, 1984, Cobb County chmn. for Individual Gifts, Marietta, 1985, chmn. Cobb County Adv. Com., Marietta, 1987-88, bd. dirs., Atlanta, 1987-88. Assoc. fellow AIAA (dir. Region II 1990—, Mem. of Yr. Atlanta sect. 1986, Booster of Yr. 1988, 1992); mem. Inst. Cert. Mgrs., Lockheed Ga. Mgmt. Assn. (v.p. mem. achievement 1988-89, v.p. adminstrn. 1989-90, Booster of Month 1980). Republican. Baptist. Avocations: landscaping, woodworking. Home: 2291 Goodrum Ln Marietta GA 30066-5200 Office: LASC-Ga Dept 73-C2 Zone 0160 86 S Cobb Dr Marietta GA 30063-0160

GARRETT, SUZANNE THORNTON, management educator; b. Charlottesville, Va., Jan. 2, 1960; d. Stafford E. and Frances Caroline (Umberger) Thornton; m. John Robert Garrett, Aug. 17, 1985. BSCE, W.Va. Inst. Tech., 1982; MBA, John F. Kennedy U., Orinda, Calif., 1989. Engr. Dominion Resources, Richmond, Va., 1982-86; dir. Computer Ctr. John F. Kennedy U., Orinda, Calif., 1988-92, inst. Sch. Mgmt., 1989-90; chair John F. Kennedy U., Walnut Creek, Calif., 1990—. Assoc. mem. ASCE (treas. Va. sect. 1984-85). Home: 2312 Benham Ct Walnut Creek CA 94596 Office: John F Kennedy U 1250 Arroyo Way Walnut Creek CA 94596

GARRETT-PERRY, NANETTE DAWN, chemical engineer; b. Galesburg, Ill., Aug. 21, 1968; d. Noel Dwight and Mary Catherine (Hasselbacher) Garrett; m. Jonathan Paul Perry, Sept. 19, 1992. BSChemE, U. Ill., 1991; degree in chemistry, Western Ill. U., 1991. Chem. engr. tech. svc. Ill. refining div. Marathon Oil Co., 1991—; mem. Safety Adv. Group. Mem. AICE. Office: Marathon Oil Co Ill Refining Div Box 1200 Robinson IL 62454

GARRIDO PEREZ, MERCEDES, chemist; b. Madrid, July 10, 1956; s. Jose and Carmen (Perez) Garrido; m. Victor Garcia Pidal, July 19, 1980; children: Victor J., Javier. Degree in Chemistry, Complutense U., Madrid, 1979, PhD, 1985. Prof. Complutense U., Madrid, 1980-89, Autonoma U., Madrid, 1989-90; postdoctoral Ecole de Chemie, Strasbourg, France, 1989-90; rschr. Centro Invest Justesa Imagen SA, Madrid, 1990—. Contbr. articles to profl. jours. Home: Alcala 231 3oA, 28028 Madrid Spain Office: Centro Inves Justesa Imagen, Roma 19, 28028 Madrid Spain

GARRY, VINCENT FERRER, environmental toxicology researcher, educator; b. Larchmont, N.Y., Nov. 6, 1937; s. Vincent Ferrer and Lucy Enes (Galasso) G.; m. Kathleen Anne Myers, Nov. 23, 1964; children: Colleen A., Michael J., Sara A. BA, Providence Coll., 1959; MS, Fordham U., 1964; MD, U. Mich., 1967. Diplomate Am. Bd. Toxicology. Resident in pathology U. Kans., Kansas City, 1967-68, postdoctoral fellow in toxicology, 1969-71; instr. U. Minn., Mpls., 1974-76; asst med. dir. Minn. Mut. Life, St. Paul, 1976-78; assoc. prof. lab. medicine and pathology, dir. Lab. of Environ. Medicine and Pathology U. Minn., Mpls., 1978—; reviewer NIH, Environ. Medicine, Washington, 1990—, ATSDR, Washington, 1990—. Contbr. articles to Sci., Proceedings NAS, Environ. Mutagenesis, others. Dist. leader Dem. Party, Mpls., 1980—; mem. citizen panel Mpls. Park Bd., 1982—; ch. vol. Basilica of St. Mary, Mpls., 1980—. Maj. U.S. Army, 1971-74, col. Res. Grantee NIH, 1992, U.S. EPA, 1986—, USDA, 1990—, others. Mem. AAAS, Soc. of Toxicology, Environ. Mutagen Soc., Am. Legion. Roman Catholic. Achievements include first demonstration of toxicity of formaldehyde in the home, of human genotoxicity of ethylene oxide, of linkage between chromosome effects and tumor development in humans, of human genotoxicity of phosphne. Office: U Minn 421 29th Ave SE Minneapolis MN 55414

GARSHELIS, DAVID LANCE, wildlife biologist; b. N.Y.C., Jan. 16, 1953; s. Ivan Jules Garshelis and Rhoda (Wolf) Lechenger; m. Judith Ann Swain, June 25, 1977; children: Daren Swain, Brian David. BA, U. Vt., 1975; MS, U. Tenn., 1978; PhD, U. Minn., 1983. Bear project leader Minn. Dept. Natural Resources, Grand Rapids, Minn., 1983--; adj. prof. wildlife conservation U. Minn., St.Paul, 1986--; cons. Valdez oil spill Exxon USA, Houston, 1989--; coun. mem. Internat. Assn. Bear Rsch. and Mgt., 1989--. Contbr. chpts. to books, articles to profl. jours. Chmn. Planning Evaluation and Reporting Com., Grand Rapids, Minn., 1991-92. Mem. Am. Soc. Mammalogists, Soc. Marine Mammalogists, Soc. Conservation Biology, The Wildlife Soc., Sigma Xi. Achievements include development of new method of estimating statewide bear population using tetracyline marked baits, method to reduce bias in estimates of large mammal populations on small study areas using radio telemetry; evaluation of techniques used to monitor bear populations; Species Action Plan for the worldwide conservation of sloth bears (in India, Nepal, Sri Lanka); assessment of effect of Valdez oil spill on sea otters in Prince William Sound. Home: 57 Hanna Rd Bass Brook MN 55721 Office: Dept Natural Resources 1201 E Highway 2 Grand Rapids MN 55744

GARSKE, JAY TORING, geologist, oil and minerals consultant; b. Fargo, N.D., Jan. 5, 1936; s. Vincent Walter and Margaret Anna (Toring) G.; m. Margo Joan Galloway, Aug. 31, 1957 (div. Mar. 1989); children: Mara Jayne, Brett Andrew; m. Carol Jean Apker, Apr. 24, 1993. BS in Geology, U. N.D., 1957. Geologist Superior Oil Co., Rocky Mountain Region, 1959-62; pvt. practice cons. geologist Denver, 1962—; pres., bd. dirs. Kudu Oil Corp., Denver, 1982—, Garske Energy Corp., Denver, 1984—, Frontier Gold Resources, Inc., Denver, 1987-92, Omega Oil Corp., Denver, 1992—. 1st lt. U.S. Army, 1958-59. Mem. Am. Assn. Petroleum Geologists, Am. Soc. Photogrammetry and Remote Sensing, Rocky Mountain Assn. Geologists, Colo. Mining Assn., Sigma Gamma Epsilon. Home: 1583 S Spruce St Denver CO 80231-2615 Office: Omega Oil Corp 1616 Glenarm Pl Ste 2970 Denver CO 80202-4304

GARSTEN, JOEL JAY, gastroenterologist; b. N.Y.C., Jan. 10, 1948; s. Richard Maxwell and Gertrude Ann (Perlberg) G.; m. Marion Susan Moscovitz, July 10, 1971; children: Bryan David, Lauren Roberta. BA in Biology, CUNY, 1968; MD, Georgetown U., 1973. Resident in internal medicine Cornell-Coop. Hosps. Program, N.Y.C., 1973-76; fellow gastroenterology Yale Affiliated Gastroenterology Program, New Haven and Waterbury, Conn., 1976-78; gastroenterologist Gastroenterology Assocs. of Waterbury, 1978-80; physican, mng. ptnr. Digestive Disease Ctr. of Conn., 1980—; dir. sect. of gastroenterology Waterbury Hosp. Health Ctr., 1990—; assoc. dir. Yale Affiliated GI fellowship program Waterbury Hosp. and Hosp. of St. Raphael, New Haven and Waterbury, 1990—; clin. instr. internal medicine Yale U. Sch. Medicine, New Haven, 1978, asst. clin. prof., 1981, assoc. clin. prof., 1987—; med. dir. Liberty Health Plan, Naugatuck, Conn., 1987-89, Physicians Health Plan, Trumbull, Conn., 1989-90, med. adv. bd., 1990—. Contbr. articles to profl. jours. Med. adv. chmn. Crohn's and Colitis Found., WTBY Satelite, Waterbury, 1990—; resource speaker Waterbury Celiac Group, Thomaston, Conn., 1990—, Am. Cancer Soc., 1991—; prin. investigator multiple drug trials. Fellow ACP; mem. Am. Soc. for Liver Disease, Conn. Soc. of Internal Medicine (exec. bd. sect. gastroenterology), Am. Soc. Internal Medicine, Am. Gastroenterology Assn., Am. Soc. Parenteral and Enteral Nutrition, others. Achievements include introduction of home parenteral nutrition to Waterbury, of sclerotherapy, esophageal stenting, percutaneous gastrostomy, other endoscopic techniques to Waterbury; prin. investigator in drug rsch. trials. Home: 47 Harvest Ct Cheshire CT 06410-1844 Office: Digestive Disease Ctr Conn 60 Westwood Ave Waterbury CT 06708-2460

GARTLING, DAVID KEITH, mechanical engineer; b. Balt., June 16, 1947; s. Richard William and Josefina Andrea (Gonzalez) G.;m . Laura Jane Garcia, Jan. 1, 1981; 1 child, Soña Juliana. BS in Aero. Engring., U. Tex., 1969, MS in Aero. Engring., 1971, PhD in Aero. Engring., 1975. Mem. tech. staff Sandia Nat. Labs., Albuquerque, 1974-85, disting. mem. tech. staff, 1985-86, 91—, supr., 1986-91. Adv. editor: Internat. Jour. Numerical Methods in Fluids, 1981—, Communications in Applied Numerical Methods, 1985—; contbr. articles to refereed jours. Recipient von Karman prize von Karman Inst., brussels, 1971; sr. scholar Fulbright-Hays fellowship, U. Sydney, Australia, 1981. Mem. ASME, Soc. Rheology, Tau Beta Pi, Sigma Gamma Tau. Home: 1308 Sierra Larga St NE Albuquerque NM 87112 Office: Sandia Nat Labs Dept 1511 PO Box 5800 Albuquerque NM 87185

GARTZ, PAUL EBNER, systems engineer; b. Chgo., July 17, 1946; s. Friedrich Samuel and Lillian Louise (Koroschetz) G. BSEE, Ill. Inst. Tech., 1969; MSEE, Stanford U., 1970. Engring. co-op Western Electric, Chgo., 1965-69; mem. tech. staff Bell Telephone Labs., Whippany, N.J., 1969-74; sales mgr. Evelyn Wood Reading Dynamics, N.Y.C., 1975-78; owner Gartz Design, Montclair, N.J., 1976-79; mktg. rep. United Computing Systems, Seattle, 1979; sr. prin. engr. Boeing, Seattle, 1980—; bd. dirs. Walla Walla (Wash.) Coll. of Engring., CASE Outlook, Inc., Portland, Oreg.; chmn. bd., pres. SDF, Inc., L.A.; educator Seattle U., 1987—, Walla Walla Coll., 1989, U. Wash., 1992—. Contbr. articles to profl. publs. Recipient Nat. Hist.

Preservation award Nat. Hist. Preservation Soc., N.J., 1980. Mem. Structured Devel. Forum (pres. 1987—), Nat. Coun. Systems Engr., Am. Inst. Aerospace and Astronautics, IEEE (Harry Rowe Mimno award 1987). Achievements include advances in systems engring., scis., methods and tools on aerospace and computing software systems; rsch. in state-of-the-art application of general systems theory to man-made systems, human engring. and mgmt. orgnl. structures. Home: 9912 Arrowsmith Ave S Seattle WA 98118-5907 Office: Boeing PO Box 3707 MS 7X-MR Seattle WA 98124-2207

GARVEY, DANIEL CYRIL, mechanical engineer; b. Chgo., Nov. 25, 1940; s. Cyril and Genei Marie (McCarthy) G; children: Michael Daniel, Erin T. BSME, Marquette U., Milw., 1963; MS in Mech. Engring., IIT, Chgo., 1965. With Kearney & Trecker Corp., Milw., 1960-63, A C Electronics div. Gen. Mtrs. Corp., Milw., 1965-68; vibration and control sys. engr. Woodward Governor Co., Ft. Collins, Colo., 1970—. Reviewer tech. papers IEEE, 1980—; contbr. articles to profl. jours.; patentee in field. Recipient Arch T. Colwell Merit award, SAE, 1984, Internal Combustion Engine award ASME, 1990. Mem. IEEE, SAE. Home: 5205 Mail Creek Ln Fort Collins CO 80525-3876 Office: Woodward Governor Co 1000 E Drake Rd Fort Collins CO 80525-1800

GARVEY, JAMES FRANCIS, physical chemist; b. Passaic, N.J., Feb. 6, 1957; s. James and Christina (Shields) G.; m. Pamela Lee, Aug. 15, 1987; 1 child, Patricia Anne. BS, MS, Georgetown U., 1978; PhD, Caltech, 1985. Postdoctoral fellow UCLA, 1985-87; asst. prof. SUNY, Buffalo, 1987-91, assoc. prof., 1991—; vis. assoc. prof. Rice U., 1993-94; cons. Hughes Rsch., Malibu, Calif., 1985-87. Contbr. numerous articles to profl. jours. Recipient ACS-PRF award, 1987-89, Am. Inst. Chemist award, ARCS fellowship, First prize in 1975, 76 for Undergrad. Rsch.; named Alfred P. Sloan fellow, 1991-93; Fulbright scholar U. Sussex, U.K., 1993—. Mem. Am. Chem. Soc., Am. Phys. Soc., Am. Inst. Chemistry, Am. Soc. Mass Spectroscopy, Soc. for Applied Spectroscopy. Democrat. Roman Catholic. Achievements include patent in Method and Apparatus for the Prodn. of Superconducting Films via cluster deposition. Office: SUNY at Buffalo Dept Chemistry Acheson Hall Buffalo NY 14214

GARVEY, JUSTINE SPRING, immunochemistry educator, biology educator; b. Wellsville, Ohio, Mar. 14, 1922; d. John Sherman and Lydia Kathryn (Johnsten) Spring; m. James Emmett Garvey, June 15, 1946; children: Johanna Xandra Kathryn, Michaela Garvey-Hayes. BS, Ohio State U., 1944, MS, 1948, PhD, 1950. Analytical chemist Sun Oil Refinery Lab., Toledo, 1944-46; Office of Naval Rsch. predoctoral fellow in microbiology U. Rochester, N.Y., 1946-47; AEC predoctoral fellow microbiology Ohio State U., Columbus, 1948-50; rsch. fellow chemistry Caltech, Pasadena, Calif., 1951-57, sr. rsch. fellow chemistry, 1957-73, rsch. assoc. chemistry, 1973-74; assoc. prof. biology Syracuse (N.Y.) U., 1974-78, prof. immunochemistry, 1978-89, emeritus, 1990—; vis. assoc. biology Caltech, 1990—; bd. sci. counselors Nat. Inst. Dental Rsch., NIH, Bethesda, Md., 1979-82; ad hoc study sects. NIH, Bethesda, 1979-88. Co-author: (textbook) Methods in Immunology, 1963, 2d edit., 1970, 3d edit., 1977; editorial bd. Immunochemistry Jour., 1964-71, Immunological Methods Jour., 1971-77; contbr. 125 articles to profl. jours. Grantee NIAID, 1951-72, NSF, 1977-79, Nat. Inst. on Aging, 1978-87, Nat. Inst. Environ. Health Scis., 1980-88. Mem. AAAS, Am. Assn. Immunologists, N.Y. Acad. Scis., Sigma Xi. Avocations: painting, hiking.

GARVEY, MICHAEL STEVEN, veterinarian, educator; b. Chgo., Dec. 5, 1950; s. Charles Anthony and Jane O. G. BS in Vet. Medicine, U. Ill., 1972, DVM, 1974; cert. internship, The Animal Med. Ctr., N.Y.C., 1976, cert. med. residency, 1978; cert. advanced mgmt. program, Wharton Sch., U. Pa., 1992. Diplomate Am. Coll. Vet. Internal Medicine, Am. Coll Vet. Emergency and Critical Care. Staff veterinarian Bevlab Vet. Hosp., Blue Island, Ill., 1974-75; intern in medicine and surgery The Animal Med. Ctr., N.Y.C., 1975-76, resident in medicine, 1976-78; staff internist Bevlab Vet. Hosp., Blue Island, 1978-81; dir. medicine The Animal Med. Ctr., N.Y.C., 1981-83, chmn. dept. medicine, 1983—, vice-chief of staff, 1993—; cons. Office of Animal Care, U. Chgo. Sch. Medicine, 1979—, Nat. Bd. Vet. Examiners, Schaumburg, Ill., 1981—, Mercy Coll. Animal Health Tech., Dobbs Ferry, N.Y., 1984—, Reader's Digest and Good Housekeeping mags., N.Y.C., 1984—; tech. cons. Sesame St., N.Y.C., 1990—; adv. bd. Profl. Examination Svc., N.Y.C., 1981—, vice-chmn., 1991—; vet. adv. panel Alpo Pet Foods, Inc., Allentown, Pa.,1981—; adj. prof. vet. medicine Tex. A&M U., 1986—; faculty assoc. U. Maine Animal Health Tech., Orono, 1976-78; chmn. vet. adv. panel Scheing Plough Inc., Madison, N.J., 1993—. Author: Animal Medical Center Hospital Formulary, 1990; author: (with others) Keeping Your Dog Healthy, 1985, Canine Emergencies, 1985, Symptoms of Illness in Dogs, 1985, Infectious and Contagious Diseases of Dogs, 1985, Feline Emergencies, 1985, Symptoms of Illness in Cats, 1985, Infectious and Contagious Diseases of Cats, 1985, Feeding the Sick Cat, 1989; editor: Canine Allergic Inhalant Dermatitis, 1982; cons. editor Small Animal Medicine, 1990; editorial review bd. Jour. of Am. Animal Hosp. Assn., 1985-93, Jour. Vet. Emergency Critical Care, 1992—; contbr. articles to profl. jours. Bd. dirs. Blue Island Community Theatre, 1974-75, 78-80; treas. 440 E 62d St. Owners' Corp., N.Y.C., 1985—. Recipient Disting. Leadership award Am. Biog. Inst., 1993. Mem. AVMA (del. 1990-91), Am. Animal Hosp. Assn. (Friskie's award for excellence in feline medicine 1993), Am. Assn. Vet. Clinicians (pres. 1988-89, Pres. Gavel award 1989, Faculty Achievement award 1993), Am. Coll. Vet. Emergency Critical Care, Am. Coll. Vet. Internal Medicine, Acad. Vet. Cardiology, N.Y. State Vet. Med. Soc., Soc. Comparative Endocrinology, Comparative Gastroenterology Soc., Vet. Emergency Critical Care Soc., Vet. Med. Assn. N.Y.C. (Outstanding Svc. award 1984), Vet. Endoscopy Soc., Soc. Internat. Vet. Symposia (bd. dirs. 1990-92, pres. 1992—). Republican. Roman Catholic. Avocations: golf, swimming, snorkeling, personal computers. Home: 440 E 62d St Apt 2B New York NY 10021 Office: The Animal Med Ctr 510 E 62d St New York NY 10021

GARWOOD, WILLIAM EVERETT, chemist researcher; b. Kirkwood, N.J., Oct. 25, 1919; s. Everett and Ethel Mary (Horner) G.; m. Betty Marie Spangberg, June 19, 1946; children: John Ernest, Christine Louise, Deborah Ann. BA in Chemistry, U. N.C., 1942; postgrad., Temple U., 1947-54. Rsch. scientist Mobil R & D Corp., Paulsboro, N.J., 1942-87; cons. Mobil R & D Corp., Paulsboro, 1987—; vis. scientist U. Ill., 1969; adj. prof. Rowan Coll. of N.J., 1990—. Co-author book; 117 patents in field; contbr. 30 articles to profl. jours. With USN, 1944-46. Mem. Am. Chem. Soc. (chmn. South Jersey sect. 1960), Phila. Catalysis Club, Rotary. Republican. Methodist. Avocations: violin, harpsichord bldg., jogging, swimming. Office: Mobil R & D Corp Billingsport Rd Paulsboro NJ 08066-1003

GARY, WALTER J(OSEPH), entomologist, educator; b. Flower Hill, N.Y., June 8, 1944; s. Walter J. and Mary E. G.; m. Margaret Frances, June 30, 1971; children: Ryan, Sean. BA, BS, Oreg. State U., 1968; MS, U. Nebr., 1973, PhD, 1978. Integrated pest mgmt. specialist U. Nebr., Lincoln, 1973-78; educator agriculture Wash. State U., Pullman, 1978—. Bd. dirs. Walla Walla (Wash.) Valley Pioneer and Hist. Soc.; senator Wash. State U. Faculty Senate, Pullman; com. chair Walla Walla C. of C. With U.S. Army, 1968-70. Recipient Disting. Svc. award Nat. Assn. County Agrl. Agts., 1985. Mem. Entomol. Soc. of Am. (bd. cert.). Home: 834 Wauna Vista Dr Walla Walla WA 99362-4260

GARZIONE, JOHN EDWARD, physical therapist; b. Newburgh, N.Y., Jan. 3, 1950; s. John Edward and Della Elizabeth (Gentila) G.; m. Anita Louise Hirschman, Sept. 21, 1974; children: Adriana, Katrina. AAS Orange County Community Coll., Middletown, N.Y., 1970; BS, Ithaca Coll., 1973. Mem. staff phys. therapy Chenango Meml. Hosp., Norwich, N.Y., 1973-74; sr. phys. therapist N.Y. State Vets. Home, Oxford, N.Y., 1974-86; CEO Chenango Therapeutics, Norwich, 1975--; cons. phys. therapy Broome Devel. Ctr., Binghamton, N.Y., 1985--, Upstate Home for Children, Milford, N.Y., 1986-88, Hospice Chenango County, Norwich, 1991--; examer N.Y. State Phy. Thearapy Bd., Albany, 1976-90; adj. instr. Cazenovia Coll., N.Y., 1985-92, Ithaca Coll., 1993. Contbr. artilce sto profl. jours. Mem. Am. Phys. Therapy Assn., Am. Acad. Pain Mgmt. (clin. assoc.), N.Y. Acad. Scis., Lions (v.p. 1990). Home: Box 451 Cunningham Hill Rd Sherburne NY 13460 Office: Chenango Therapeutics 50 W Main St Norwich NY 13815

GASCO, ALBERTO, medicinal chemistry educator; b. Jan. 12, 1938; s. Gian Mario and Ida (Penati) G.; m. Maria Luisa Sartoris, July 14, 1965; children: Annalisa, Gianmario. D Chemistry, U. Torino, Italy, 1962. Asst. U. Torino, 1965-75, prof. medicinal chemistry, 1975—, dean pharmacy faculty, 1983-92. Home: Strada Gorree n 83, 10024 Moncalieri Italy Office: Facoltá Farmacia Via, Pietro Giuria 9, 10125 Torino Italy

GASKELL, ROBERT WEYAND, physicist; b. Providence, July 9, 1945; s. Robert Eugene and Jane Ardith (Weyand) G.; m. Virginia Ann Dibble, Aug. 26, 1967; children: Lisa Ellen, Robert-Eric, Peter Edward. ScB, Brown U., 1967; PhD, McGill U., 1972. Postdoctoral fellow Carleton U., Ottawa, Ontario, Can., 1972-74, U. Toronto, Ontario, Can., 1974-75; rsch. assoc. McGill U., Montreal, Quebec, Can., 1975-78; asst. prof. Lafayette Coll., Easton, Pa., 1978-84; mem. tech. staff Jet Propulsion Lab., Pasadena, Calif., 1984—. Contbr. articles to profl. jours. Asst. dir AYSO Region 13, Pasadena, Calif., 1992—, bd. dirs., 1987—. Mem. Am. Phys. Soc., Am. Geophys. Union. Office: 301/125 L JPL/Caltech 4800 Oak Grove Dr Pasadena CA 91109

GASKIN, FELICIA, biochemist, educator; b. Carlisle, Pa., Jan. 17, 1943; d. Joseph A. and Wanda J. (Rakowski) G.; m. Shu Man Fu, Nov. 29, 1969; children: Kai-Ming, Kai-Mei. AB in Chemistry, Dickinson Coll., Carlisle, Pa., 1965; MA in Organic Chemistry, Bryn Mawr Coll., 1967; PhD in Biochemistry, U. Calif., San Francisco, 1969. Postdoctoral fellow Stanford U., Palo Alto, Calif., 1969-71; rsch. assoc. Rockefeller U., N.Y.C., 1971-72, Columbia U., N.Y.C., 1972-74; asst. prof., then assoc. prof. Albert Einstein Coll. Medicine, N.Y.C., 1974-82; prof. U. Okla., Oklahoma City, 1982-88; prof. Sch. medicine U. Va., Charlottesville, 1988—. Contbr. articles to profl. jours. Recipient rsch. career devel. award NIH, 1975-80; Nat. Inst. Neurol. Diseases and Stroke spl. fellow, 1972-74. Mem. Am. Soc. Biochemistry and Molecular Biology, Am. Soc. for Cell Biology, Soc. Neurosci., N.Y. Acad. Sci., Okla. Med. Rsch. Found. Office: U Va Sch Medicine Charlottesville VA 22908

GASKINS, H. REX, animal sciences educator; b. Morehead City, N.C., Jan. 30, 1958; s. Harvey Rexford Jr. and Ann (Oglesby) G.; m. Laurie Ann Rund, Nov. 21, 1987. BS, N.C. State U., 1981, MS, 1986; PhD, U. Ga., 1989. Postdoctoral fellow Jackson Lab., Bar Harbor, Maine, 1989-92; asst. prof. dept. animal scis. Div. Nutritional Scis., U. Ill., Urbana, 1992—. Contbr. articles to profl. jours. including Sci., Jour. Clin. Investigation and Proc. Nat. Acad. Scis. Internat. postdoctoral fellow Juvenile Diabetes Found., 1991-92, postdoctoral fellow Am. Diabetes Assn., 1989-91; recipient Grad. Student Rsch. award Am. Inst. Nutrition, 1988, Future Leaders award Internat. Life Scis. Inst., 1992. Mem. AAAS, The Endocrine Soc., Am. Inst. of Nutrition, Am. Diabetes Assn., Am. Assn. of Immunologists, Soc. for Exptl. Biology and Medicine, Gamma Sigma Delta, Sigma xi. Democrat. Methodist. Achievements include creation of a mouse pancreatic beta cell line for diabetes research, efforts to identify genes encoding susceptibility to autoimmune diabetes. Home: 780 County Rd 100 E Ivesdale IL 61851 Office: U Ill 1207 W Gregory Dr Urbana IL 61801

GASPAROTTO, RENSO, civil engineer; b. Detroit, June 26, 1952; s. Giuseppe and Nives (Minatel) G.; m. Joanne Talis, Mar. 20, 1993. BSCE, Mich. State U., 1974. Registered profl. engr., Ill., Wis. Project mgr. Greeley and Hansen, Engring., Chgo., 1974—. Contbr. articles to profl. jours. Leader bible study Willow Creek Community Ch., South Barrington, Ill., 1986-92. Mem. ASCE, Chi Epsilon. Achievements include design of river water oxygenation facilities utilizing innovative sidestream elevated pool aeration concept. Office: Greeley and Hansen Engrs 100 S Wacker Dr Chicago IL 60606

GASPER-GALVIN, LEE DELONG, chemical engineer; b. Des Moines, July 13, 1956; d. Wayne Lee and Norma Jean (Schreiner) Gasper; m. Gary James Galvin, July 27, 1985. BS in Chemistry, U. Iowa, 1978, BSChemE, 1979; MSChemE, Cornell U., 1982; PhD in Chem. Engring., U. Iowa, 1989. Teaching asst. Cornell U., Ithaca, N.Y., 1979-81; assoc. rsch. engr. Internat. Mineral and Chem., Terre Haute, Ind., 1981-83; rsch. teaching asst. U. Iowa, Iowa City, 1983-87; chem. engr. Morgantown (W.Va.) Energy Tech. Ctr., 1988—. Contbr. articles to one book and profl. jours. Shell Oil Co. scholar, 1978; recipient Monsanto award, U. Iowa, 1976. Mem. Am. Inst. Chem. Engrs. (sec. 1984-85, Student award 1977), Nat. Soc. Prof. Engrs. (reg. profl. engr.), Am. Chem. Soc., Alpha Chi Sigma, Tau Beta Pi, Omicron Delta Kappa, Sigma Xi. Republican. Avocations: weight lifting, racquetball, barbershop singing, trumpet, keyboards. Office: Morgantown Energy Tech Ctr 3610 Collins Ferry Rd Morgantown WV 26507

GASS, ARTHUR EDWARD, chemist; b. Dallas, July 23, 1931; s. Arthur E. and Alice Elizabeth (Fooshee) G.; m. Gloria Jean Carter, Apr. 18, 1954; children: Kathleen, Mark E., Laura, Amy, Andrew C., Susan G. BS in Biology, So. Meth. U., 1953; MS in Biology, Trinity U., 1965. Civilian rsch. chemist USAF Sch. Aerospace Medicine, San Antonio, 1959-75; indsl. hygienist OSHA, U.S. Dept. Labor, Washington, 1975-89; mgr. health and safety Gassco Petroleum Corp., Seguin, Tex., 1989—; indsl. hygiene cons. Alamo Risk Mgmt. Svc., San Antonio, 1992—; instr. biology grad. sch. Trinity U., San Antonio, 1963-66; chair, bd. advisors Satellite Safety TV Network, San Antonio, 1992—. Contbr. over 22 articles to tech. sci. jours. Grantee USAF, Dept. Def., 1968-69; recipient Rsch. Chemist Outstanding Civilian award USAF, 1967, Spl. Achievement awards U.S. Dept. Labor, 1977, 79, 87. Mem. Am. Chem. Soc., Am. Indsl. Hygiene Assn., Am. Conf. Govt. Indsl. Hygienists, Sigma Xi. Presbyterian. Achievements include establishment of passive gut transport of chloride; development of federal field sanitation health standard; publication that promotes or protects health of 600,000 migrant workers on a yearly basis.

GASS, TYLER EVAN, hydrogeologist; b. N.Y.C., Nov. 20, 1948; s. Saul and Manya (Greenberg) G.; m. Madeline S. Steiner, July 8, 1971; children: Brianna Lynn, Jenna Layne. BA in Geology, SUNY, Buffalo, 1970; MS in Geoscis., U. Ariz., 1977. Geologist City of Tucson, 1972-73; hydrogeologist Holzmarker, McLendon & Murrell, Melville, N.Y., 1973-75; dir. rsch. Nat. Water Well Assn., Worthington, Ohio, 1975-81; pres. Bennett, Gass & Williams, Westerville, Ohio, 1981-85; exec. v.p. Blasland, Bouck & Lee, Syracuse, N.Y., 1985—; cons. EPA, Washington, 1985-86. Author: (textbook) Manual of Water Well Maintenance and Rehabilitation Technology, 1980; co-author: (textbook) Domestic Water Treatment, 1980. Recipient Cert. Appreciation Ohio Environ. Health Assn., Columbus, 1984, 85, Chevron's Mgr. award, 1991. Mem. ASTM, Nat. Water Well Assn. (chair aquifier protection com. 1985-89, rsch. and edn. com. 1990—), Assn. Ground Water Scientists and Engrs. (bd. dirs. 1985-89, sec.-treas. 1989-91, chmn. 1992—). Home: 8232 Verbeck Dr Manlius NY 13104-9320 Office: Blasland Bouck & Lee 6723 Towpath Rd Syracuse NY 13214

GAST, ALICE P., chemical engineering educator. Prof. dept. chem. engring. Stanford (Calif.) U. Recipient Initiatives in Rsch. award NAS, 1992. Office: Stanford U Dept Chem Engring Stanford CA 94305*

GASTON, HUGH PHILIP, marriage counselor, educator; b. St. Paul, Sept. 12, 1910; s. Hugh Philander and Gertrude (Heine) G.; BA, U. Mich., 1937, MA, 1941; postgrad. summers Northwestern U., 1938, Yale U., 1959; m. Charlotte E. Clarke, Oct. 1, 1945 (dec. 1960); children: Trudy E. Gaston Crippen, George Hugh. Counselor, U. Mich., Ann Arbor, 1936; tchr., counselor W. K. Kellogg Found., Battle Creek, Mich., 1937-41; tchr. spl. edn., Detroit, 1941; instr. airplane wing constrn. Briggs Mfrs. Co., Detroit, 1942 (rep. Mich. Indstrl. Tng. Coun.); psychologist VA, 1946-51; chief VA guidance ctr. U. Mich., 1949-51; with Mich. State VA Guidance Ctr, 1951; sr. staff asso. Sci. Rsch. Asso., Chgo., 1951-55; marriage counselor Cir. Ct., Ann Arbor, 1955-60; pvt. practice marriage counseling, Ann Arbor, 1955—; educator 14 different courses Ea. Mich. U., 1963-80; former chief Guidance Ctr., V.A. Grad. Ctr. U. Mich. and Mich. State U.; lectr., Ea. Mich. U., Ypsilanti, 1964-67, asst. prof., 1967-81; mem. Study Group for Health Care of Elderly, China, USSR, 1983, Profl. Study Group on Family Affairs, USSR, 1986. Acting postmaster, Ann Arbor, 1960-61. Chief insp. U.S. Army Engrs. Civil Svc., 1930-35, Monroe, Mich. and Toledo, Ohio, 1930-35, insp. Livingston Channel, Detroit; chmn. Wolverine Boys State, Am. Legion, 1957-86; chmn. com. on Christian marriage Presbyn. So. Mich., 1962-69; mem. exec. com., legis. agt., chmn. legis. com. Mich. Coun. Family Rels., 1972-74; bd. dirs. Internat. Parents Without Partners, 1968-69, 1st pres. Mich. chpt. 38, 1961; bd. dirs. Ann Arbor Sr. Citizens, 1982-85, Washtenaw County Coun. Alcoholism, 1982-84. Served with U.S. Army, 1943-46. Decorated Purple Heart (2), Bronze Star; Medallion of Nice (France); named Citizen of Year, Am. Legion, 1968, PWP, 1938, Single Parent of Yr. Parents Without Ptnrs. chpt. 38, 1978, Patriot of Yr. State of Mich., Mil. Order of Purple Heart, 1987-88. Mem. Am. Assn. Marriage Counselors, Circumnavigators Club, Am. Personnel and Guidance Assn., Nat. Vocat. Guidance Assn., D.A.V. (past comdr. local chpt. 13), Am. Soc. Tng. Dirs., Mich. Indsl. Tng. Coun. (charter), SAR (past pres.), U. Mich. Band Alumni Assn. (pres. 1957-58), Mil. Order Purple Heart (nat. exec. com. 1977-82, 1st comdr. chpt. 459 Mich., state comdr. Mich. 1984-85, nat. historian 1981-85), Rotary (Paul Harris fellow 1989), Phi Delta Kappa (past pres. U. Mich., 50 yr. mem.). Address: 513 4th St Ann Arbor MI 48103

GASTON, JERRY COLLINS, sociology educator; b. Trinidad, Tex., Oct. 16, 1940; s. James Elmore and Alice Audrey (Airheart) G.; m. Mary Frank Ballow, Sept. 16, 1961; 1 child, Jeremy. BA, East Tex. State U., 1962, MA, 1963; MPhil, Yale U., 1967, PhD, 1969. Asst. prof. So. Ill. U., Carbondale, 1969-81, assoc. dean liberal arts, 1974-75, chair dept. sociology, 1975-81; prof. Tex. A&M U., College Station, 1981—, head dept. sociology, 1981-86, assoc. provost, 1986-91, exec. assoc. prof., 1991—. Author: Originality and Competition in Science, 1973, The Reward System in British and American Science, 1978; editor: The Sociology of Science, 1978; co-editor: The Sociology of Science in Europe, 1977; editor jour. Sociol. Quarterly, 1978-81; cons. editor Am. Jour. Sociology, 1979-81. Faculty fellow Yale U., 1965-69; grantee Nat. Inst. Edn., 1978-79, NSF, 1971-73, 78-80. Mem. AAAS, Am. Sociol. Assn., Midwest Sociol. Soc., Soc. for Social Studies of Sci., Rotary (pres. College Station club 1986-87), Sigma Xi. Home: 611 E 29th St Bryan TX 77802 Office: Tex A&M U College Station TX 77843-1248

GASTWIRT, LAWRENCE E., chemical engineer; b. N.Y.C., Mar. 7, 1936; s. Morris and Regina (Wurm) G.; widowed, 1990; children: Leslie, Richard, Allison. PhD in Chem. Engring., Princeton U., 1962; MBA, NYU, 1968. Tech. mgr. Exxon Chem. Co., Linden, N.J., 1970-75; ops. and planning mgr. Exxon Chem. Europe, Brussels, Belgium, 1975-79; product exec. Exxon Chem. Co., Darien, Conn., 1979-85; mgr. new bus. devel. polymers group Exxon Chem. Co., Darien, Conn., 1986-92; prof. mgmt. and engring. mgmt., exec. dir. Stevens Alliance for Tech. Mgmt., Stevens Inst. Tech., Hoboken, N.J., 1992—; adv. bd. mem. Stevens Tech. Ventures Incubator, Hoboken, 1992—. Co-author: Turning Research and Development into Profits - A Systematic Approach, 1979. Mem. Am. Chem. Soc. (rubber div., chem. mktg. & econs. div.), N.Y. Acad. Sci., Sigma Xi. Office: Stevens Inst Tech Castle Point on the Hudson Hoboken NJ 07030

GATELL, JOSE MARIA, physician; b. Brafim, Tarragona, Spain, Jan. 14, 1951; s. Jose and Maria (Artigas) G.; m. Rosa Gatell, Sept. 29; children: Mariano, Griselda, Violeta. MD, Med. Sch., Barcelona, 1976, PhD, 1981. From intern to resident Hosp. Clinic, Barcelona, 1976-80; rsch. fellow Mass. Gen. Hosp., Boston, 1982; chief infectious disease unit Hosp. Clinic, Barcelona, 1990—; assoc. prof. medicine Faculty of Medicine, Barcelona, 1982—. Author and editor of numerous books and articles on infectious diseases, AIDS. Mem. Am. Soc. Microbiology, Spanish Infectious Diseases Soc. (bd. dirs. 1988-92), Nat. Geographic Soc., Spanish AIDS Soc. (vice chmn. 1990—). Avocations: tennis, skiing. Home: Santapau 62, Barcelona 08016, Spain Office: Hosp Clinic, Villarroel 170, Barcelona 08036, Spain

GATELY, MAURICE KENT, research immunologist; b. Omaha, Feb. 3, 1946; s. Harold Stephen and Alys Marie (Witt) G.; m. Celia Lin, July 8, 1972; children: Lynn Christine, Mark Stephen. BA, Johns Hopkins U., 1968, PhD in Microbiology, 1974, MD, 1975. Resident in pediatrics St. Louis Children's Hosp., 1975-76; rsch. fellow in pathology Harvard Med. Sch., Boston, 1976-79; sr. staff fellow NIH, Bethesda, Md., 1979-83; sr. scientist Hoffmann-La Roche, Inc., Nutley, N.J., 1983-85, rsch. investigator, 1985-88, rsch. leader, 1988—; ad hoc mem. contracts rev. com. Nat. Cancer Inst., Bethesda, 1982-83. Contbr. chpts. to books; contbr. articles to Sci., Jour. Immunology, Cell Immunology, Procs. Nat. Acad. Sci., others. Pres. Montville (N.J.) Soccer Assn., 1988-89. Helen Hay Whitney Found. fellow, 1976-79. Mem. AAAS, Am. Assn. Immunologists, N.Y. Acad. Scis. Achievements include patent for vitamin D analogs as immunosuppressive agents; co-discovery of cytotoxic lymphocyte maturation factor (IL-12). Office: Hoffmann La Roche Inc Inflammation/Autoimmune Diseases 340 Kingsland St Nutley NJ 07110-1199

GATES, BRUCE CLARK, chemical engineer, educator; b. Richmond, Calif., July 5, 1940; s. George Lawrence and Frances Genevieve (Wilson) G.; m. J. Margarete Reichert, July 17, 1967; children: Robert Clark, Andrea Margarete. BS, U. Calif., Berkeley, 1961; PhD in Chem. Engring., U. Wash., 1966. Rsch. engr. Chevron Rsch. Co., Richmond, Calif., 1967-69; from asst. prof. to assoc. prof. U. Del., Newark, 1969-77, prof. chem. engring., 1977-85, H. Rodney Sharp prof., 1985-93, assoc. dir. Ctr. Catalytic Sci. & Tech., 1977-81, dir. Catalytic Ctr. Sci. & Tech., 1981—; prof. chem. engring. U. Calif., Davis, 1992—. Author: Catalytic Chemistry, 1992; co-author Chemistry of Catalytis Processes, 1979; co-editor: Metal Clusters in Catalysis, 1986, Surface Organometallic Chemistry, 1988. Fulbright Rsch. grantee Inst. Phys. Chemistry U. Munich, 1966-67, 75-76, 83-84. Mem. Am. Inst. Chem. Engrs. (Alpha Chi Sigma award 1989), Am. Chem. Soc. (Del. sect. award 1985, Petroleum Chemistry award 1993). Achievements include research in catalysis, surface chemistry and reaction kinetics, chemical reaction engineering, petroleum and petrochemical proccesses, catalysis by superacids, zeolites, soluble and supported transition-metal complexes and clusters, catalytic hydroprocessing. Office: Univ of California Dept of Chemical Engineering Davis CA 95616

GATES, MARSHALL DEMOTTE, JR., chemistry educator; b. Boyne City, Mich., Sept. 25, 1915; s. Marshall DeMotte and Virginia (Orton) G.; m. Martha Louise Meyer, Sept. 9, 1941; children—Christopher David, Catharine Louise, Marshall DeMotte III, Virginia Alice. B.S., Rice Inst., 1936, M.S., 1938, Ph.D., Harvard, 1941, D.Sc. (hon.), MacMurray Coll., 1963. Asst. prof. chemistry Bryn Mawr Coll., 1941-43; vis. prof. Harvard, 1946; assoc. prof., 1947-49, Max Tishler lectr., 1953; tech. aid NDRC, 1943-46; lectr. chemistry U. Rochester, 1949-52, part-time prof., 1952-60, prof., 1960-68, Charles Frederick Houghton prof. chemistry, 1968-81, prof. emeritus, 1981—; Welch Found. lectr., 1960; adv. bd. Chem. Abstracts Services, 1974-76; vis. prof. Dartmouth Coll., 1982, 84, 85, 86; charter fellow Coll. Problems Drug Dependence, 1992—. Mem. com. on drug addiction and narcotics, div. med. scis. NRC, 1956-70, also com. on organic nomenclature div. of chemistry; mem. Pres.'s Com. on Nat. Medal of Sci., 1968-70. Recipient Edward Peck Curtis award for excellence in undergrad. teaching, 1967; Armed Services cert. of appreciation, 1946; Disting. Alumnus award Rice U., 1986. Fellow Am. Acad. Arts and Scis., N.Y. Acad. Scis.; mem. Am. Chem. Soc. (editor Jour. 1963-69), Nat. Acad. Scis. Office: U Rochester Rochester NY 14627

GATES, MARVIN, construction engineer; b. N.Y.C., Oct. 29, 1929; s. Sam and Pauline (Struver) G.; m. Norma Friedlander, June 24, 1956; children: Julia, Paula. Lic. profl. engr., landscape architect, Conn.; cert. cost analyst. V.p. Gates-Scarpa and Assocs. Inc. Cons. Engrs., Newington, Conn., 1961—; profl. lectr. Worcester (Mass.) Poly. Inst., 1974-80; disting. vis. prof. U. Hartford, Conn., 1981-83; lectr. in field, 1980-90; mem. Conn. Bd. Registration for Profl. Engrs. and Land Surveyors. Contbr. numerous rsch. papers to Jour. Transp. Engring., Jour. Constrn. Engring. and Mgmt., Cost Engring. Active various offices Town of Newington. Fellow ASCE (past chmn. exec. com. constrn. div., com. on estimating and cost coontrol, James Laurie prize, Walter Huber Rsch. prize); mem. Nat. Soc. Profl. Engrs., Conn. Soc. Civil Engrs. (past dir.), Inst. Cost Analysis, Tau Beta Pi, Chi Epsilon. Achievements include devel. of Gates bidding model, Gates formula for pile bearing capacity, preliminary cash flow analysis, other. Home: 32 Dover Rd Newington CT 06111

GATES, THOMAS EDWARD, civil engineer, waste management administrator; b. Tachikawa AFB, Japan, June 25, 1953; came to U.S., 1954; s. Harold Charles and Masako (Endo) G.; m. LeAnn Faye Eakins, Aug. 19, 1981 (div. 1986); m. Nancy Neer Adams, July 24, 1993. BS, Kans. State U., 1979, MS, 1981. Registered profl. engr., Wash., Kans., Alaska. Advt. salesman Junction City (Kans.) Daily Union, 1972-74; co-op student Burns & McDonnell, Kansas City, Mo., 1975-76; state insp. Riley County Pub.

Works, Manhattan, Kans., 1977-78, field supr., 1978, cons., 1979; grad. rsch. asst. Kans. State U., Manhattan, 1979-81; engr. Battelle Pacific N.W. Labs., Richland, Wash., 1981-83, rsch. engr., 1983-85; sr. rsch. engr. Battelle Pacific N.W. Labs., Richland, 1985-86; mgr. waste package projects BWIP, 1986-88, acting mgr. support projects, 1988; mgr. for def. programs Westinghouse Hanford Co., Richland, 1988-89, staff mgr. engring. and devel. divs., 1990, mgr. tech. assessment and application, 1990-91; mgr. tech. demonstration program ops. Westinghouse Hanford Co., Washington, 1991—; cons. Elec. Power Rsch. Inst., Washington, Atomic Energy of Can., Ltd. Rsch. Co., Ottawa, Can.; lead judge Wash. State Sci. Talent Search, Richland, 1985-90. Contbr. 7 articles to profl. jours., 12 tech. reports; session works, obtaining accelerated data on concrete degradation, 1981, concrete durability and degradation processes, 1986. Councilman City of Richland, 1988—, mayor, 1990; mem. Phys. Planning Com., Richland, 1982-87, vice chmn. 1983-84, chmn. 1984-87; precinct. chmn. Rep. Cen., Richland, 1984-90, dep. registrar 1985—, state del. 1988, county del. 1986, 88; instr. christian catechism doctrine Christ The King Ch., Richland, 1981-82; chmn. Sausage Festival Vol., 1984-86; bd. dirs. Salvation Army Adv. Coun., Richland, 1987-89, chmn., 1988-89; vice chmn., program chmn. Benton-Franklin Community Action Com., Pasco, Wash., 1988-89, bd. dirs., 1992—; mem. March of Dimes, Junction City, Kans., 1976-80, Walk-A-Thon, 1973-80, campaign chmn., 1978-79, chmn. bd., 1979-80; chmn. dept. campaign United Way, Richland, 1984; Cen. Bus. Dist. Master Plan Team, Richland C. of C., 1987; bd. dirs. Assn. of Washington Cities, 1990—, mem. resolution com., 1989-91, mem. legis. com., 1989-92, mem. energy adv. com. 1990—, mem. local govt. adv. com., 1990—, mem. mcpl. rsch. coun., 1991—; mem. Benton County Solid Waste Adv. Com., 1990—, chmn., 1990—; mem. hazardous materials mgmt. tech. adv. com. Columbia Basin Coll., 1990—, chmn., 1991—. Named one of Outstanding Young Men of Am., Montgomery, Ala., 1987. Mem. Am. Concrete Inst. (tech com. 118 computers 1983—, 227 radioactive waste mgmt. 1983—, E801 student concrete projects 1984—, Harry F. Thomson scholarship 1980), ASCE (tech. coun. on computer practices pub. com. 1986—), Kiwanis, KC (Sir Knight of the Yr. 1992). Roman Catholic. Avocations: gardening, woodworking, reading. Home: 1937 Forest Ave Richland WA 99352-2155 Office: Westinghouse Hanford Co MSIN T3-01 PO Box 1970 Richland WA 99352-2155

GATES, WILLIAM HENRY, III, software company executive; b. Seattle, Wash, Oct. 28, 1955; s. William H. and Mary M. (Maxwell) G. Grad. high sch., Seattle, 1973; student, Harvard U., 1975. With MITS, from 1975; founder, chmn. bd. Microsoft Corp., Redmond, Wash., 1976—, now chief exec. officer. Recipient Howard Vollum award, Reed Coll., Portland, Oreg., 1984, Nat. medal Tech. U.S. Dept. Commerce Tech. Adminstrn., 1992. Office: Microsoft Corp 1 Microsoft Way Redmond WA 98052-6399

GATEWOOD, BUFORD ECHOLS, retired educator, aeronautical and astronautical engineer; b. Byhalia, Miss., Aug. 23, 1913; s. Robert P. and Irene (Echols) G.; m. Margaret Murphy, June 28, 1939; 1 dau., Marianne. B.S. in Mech. Engring, La. Poly. Inst., 1935; M.S., U. Wis., 1937, Ph.D., 1939. Faculty La. Poly. Inst., 1939-42, Air Force Inst. Tech., 1947-60; with McDonnell Aircraft Corp., 1942-46, Beech Aircraft Corp., 1946-47; prof. aero. and astronautical engring. Ohio State U., Columbus, 1960-78; Cons. on structural design and analysis, structural fatigue, problems in dynamics, thermal Problems to various cos., 1949—. Author: Thermal Stresses, 1957, Virtual Principles in Aircraft Structures, 2 vols., 1989. Mem. Am. Inst. Aeros. and Astronautics, Soc. Exptl. Stress Analysis, Math. Assn. Am., ASME, Am. Soc. Engring. Edn., Sigma Xi. Achievements include research and publications on thermal stresses and inelastic structures for flight vehicle structures. Home: 2150 Waltham Rd Columbus OH 43221-4150

GATLEY, DONALD PERKINS, mechanical engineer, building scientist; b. Pueblo, Colo., Jan. 24, 1932; s. William P. and Lenore (Brown) G.; m. Jane Douglass, Dec. 7, 1954; 1 child, Eric D. BSME, Vanderbilt U., 1954. Registered profl. engr., 21 states. Dir. engring. The Trane Co., Atlanta, 1957-70; ea. regional mgr., v.p. Sam P. Wallace Co., Dallas and Atlanta, 1970-76; exec. v.p. McKenney's Inc., Atlanta, 1976-78; pres. Gatley & Assocs. Inc., Atlanta, 1978—; also moisture cons., bldg. scientist. Co-author: AH&MA Mold and Mildew Handbook, 1991, Moisture and Mildew Control, 1992, Cool Storage Ethylene Glycol Design Guide; contbr. articles to profl. jours. Chmn. Ga. Bldg. Energy Code Com., Atlanta, 1991—. Lt. USN, 1954-59. Named Engr. of Yr. in Constrn. Ga. Soc. Profl. Engrs., 1975. Fellow ASHRAE (pres. Atlanta chpt. 1964, chmn. U.S. and Can. energy conservation com. 1975-76, Internat. Energy award 1981, 84, Disting. Svc. award 1982). Office: Gatley & Assocs Inc 489 Westover Dr NW Atlanta GA 30305

GATLIN, LARRY ALAN, pharmaceutical professional; b. Salem, Oreg., Aug. 16, 1950; s. Norman Lee and Evelyn Marie (Sikorra) G.; m. Carol Ann Brister, Sept. 2, 1972; children: Jason Trevor, Brandon Travis, Courtney Alison. BS in Pharmacy, Oreg. State U., 1973; PhD in Pharmaceutics, U. Ky., 1981. Pharmacist NIH Clin. Ctr., Bethesda, Md., 1973-75; scientist Upjohn Co., Kalamazoo, Mich., 1979-82, Genentech, South San Francisco, Calif., 1982-88; sect. head Glaxo, Research Triangle Park, N.C., 1988—. Author: Pharmaceutical Dosage Forms, 1992; contbr. articles to profl. publs. Co-pres. local PTA, Chapel Hill, N.C., 1991. 2d lt. USPHS, 1973-75. Mem. AAAS, Am. Assn. Pharm. Scientists (co-chmn. 1993), Parenteral Drug Assn., World Future Soc., Phi Kappa Phi, Rho Chi. Achievements include patent for tissue plasminogen activator.

GAULT, DONALD EIKER, planetary geologist; b. Chgo., Feb. 12, 1923; s. Frank Leonard and Arabella Branan (Genge) G.; m. Mary Lucy Deaver, Sept. 7, 1947; children: Kathleen Ann, Frank Leonard, Jan Marie. BS in Aero. Engring., Purdue U., 1944. Rsch. scientist Nat. Adv. Com. for Aeronautics, Moffett Field, Calif., 1944-59; rsch. scientist NASA, Moffett Field, 1959-64, chief planetology br., 1964-71, sr. staff scientist, 1972-76; chief scientist Murphys (Calif.) Ctr. Planetology, 1976—; adj. prof. U. Ariz., Tucson, 1977-85; presenter papers Royal Soc., London, NATO, COSPAR, Joint Soviet-Am. Conf., others. Editorial bd.: The Moon and Planets, Modern Geology; contbr. numerous articles to sci. jours., chpts. to books. Lt. (j.g.) USN, 1944-46. Guggenheim fellow Max Planck Inst., Heidelberg, Germany, 1971-72; named Fairchild Disting. scholar, Calif. Inst. Tech., 1978. Mem. Am. Geophys. Union, Geol. Soc. Am. (G.K. Gilbert award 1987), Meteoritical Soc. (Barringer medal 1986), Tau Beta Pi, Pi Tau Sigma. Achievements include participation in lunar sample analysis planning for allocation of samples from Apollo 14-17, Mariner 10 imaging team.

GAUNAURD, GUILLERMO C., physicist, engineer, researcher; b. Havana, Cuba, July 19, 1940; came to U.S., 1961; s. Celestino Carlos and Ana Marie (Herrera) G.; m. Marlene Jane Johnson, June 10, 1967. AB in Math., Cath. U. Am., Washington, 1964; BSME, Cath. U. Am., 1966, MS, 1967, PhD, 1971. Cons. engr. Ocean Systems Inc. (div. Union Carbide), Arlington, Va., 1966-68; sr. cons. engr. Litton Industries Inc., College Park, Md., 1968-71; rsch. physicist, group leader Rsch. Dept. Naval Surface Warfare Ctr., Silver Spring, Md., 1971—; lectr. U. Md. Sch. Engring., College Park, 1983-92, Cath. U. Am. Sch. Engring., Washington, 1974-78. Contbr. over 100 articles to profl. jours., chpts. to books; patentee in field. Mem. Randolph Hills Civic Com., Rockville, Md., 1971—. Recipient various publ. awards and sci. excellence medals; grantee Office Naval Rsch., 1967—; fellow NDEA, 1967-70. Fellow Acoustical Soc. Am. (various offices), Washington Acad. Scis.; mem. IEEE (sr. mem., editor Jour. Oceanic Engring. 1987—, assoc. editor Jour. of Ultrasonics, Ferroelectrics and Frequency Control 1992—), ASME, Philos. Soc. Washington, Am. Phys. Soc., Am. Acad. Mechanics, Washington Soc. Engrs., N.Y. Acad. Scis., Sigma Xi, Tau Beta Pi. Avocations: photography, classical music. Home: 4807 Macon Rd Rockville MD 20852-2348 Office: Naval Surface Warfare Ctr Code R-14 White Oak Detachment Silver Spring MD 20903-5000

GAUTHIER, JON LAWRENCE, telecommunications engineer; b. Lafayette, La., Mar. 3, 1962; s. Lawrence Joseph and Shirley Mae (Poché) G.; m. Shawna Renea Holly, Nov. 26, 1992. BSEE, U. Southwestern La., 1985; MS, So. Meth. U., 1991. Inside plant engr. Contel of Tex., Inc., Dallas, 1985-87; plant engr. Ericsson Network Systems, Inc., Richardson, Tex., 1987-91, network systems programmer, 1991—. V.p. Holly Hill Townhomes Homeowners Assn., Dallas, 1990. Scholar faculty La. U., 1980-81, La. Telephone Assn., 1983-85. Mem. IEEE, Dallas/Ft. Worth Heath Computer User's Group (pres. 1990), U. Southwestern La. Alumni Assn. Office: Ericsson Network Sytems 730 International Pky Richardson TX 75081

GAUVENET, ANDRÉ JEAN, engineering educator; b. Nuits-St.-Georges, Côte d'Or, France, Mar. 31, 1920; s. Emile Félix and Louise Adrienne (Tixier) G.; m. Hélene Frédérique Gras-Gauvenet, May 15, 1948; children: Christian, Françoise, Anne. Cert., Ecole Normale Supérieure, Saint-Cloud, 1942; Diplôme d'Etudes Supérieures, U. Sorbonne, Paris, 1944. Prof. Coll. Turgot, Paris, 1942-43; engr. Lab. Cen. d'Electricité, Paris, 1943-45; researcher Soc. Alsacienne Constrns. Mech., Paris, 1945-48; prof. Ecole Normale Supérieure, Saint-Cloud, 1948-54; sci. attache French Embassy, N.Y.C., 1954-56; engr. nuclear safety and radiation protection Commissariat à l'Energie Atomique, Paris, 1956-82; inspector gen. nuclear safety and security Electricité France, Paris, 1982-85; prof. safety engr. Ecole Centrale, Paris, 1988—; supr. d'electricite Ecole Poly. Féminine, Sceaux, 1990—. Author: (with others) Space Techniques, 1965, Images de la science, 1984; contbr. articles to profl. jours. Decorated Légion d'honneur French govt., 1972, Mérite Commandeur, Palm Academiques. Fellow Am. Nuclear Soc.; mem. Soc. Française Nucléaire, Am. Phys. Soc., European Phys. Soc., Soc. Française Physique, Cercle d'Union Interalliée, N.Y. Acad. Scis. Home: 31 Rue Censier, 75005 Paris France Office: Electricité France, 9 Rue d'Aguesseau, 75008 Paris France

GAVALER, JUDITH ANN STOHR VAN THIEL, epidemiologist; b. Pitts., Aug. 5; d. Frank Howell and Nancy Helen (Hoovler) Stohr; m. John Raymond Gavaler, Nov. 17, 1962 (div. Apr. 1974); children: Joan Susan, Christopher Paul; m. David Hoffman Van Thiel, May 13, 1978. BS, Hood Coll., 1961; PhD, U. Pitts., 1986. Jr. engr. Westinghouse Rsch., Pitts., 1961-63; rsch. asst. U. Pitts. Sch. Medicine, 1974-78, rsch. assoc., 1978-86, asst. prof., 1986-88, assoc. prof., 1988-92, prof., 1992-93; assoc. prof. dept. epidemiology U. Pitts. Grad. Sch. Pub. Health, 1988-93; sr. scientist women's rsch. Okla. Transplantation Inst., Bapt. Med. Ctr. Okla., 1993—; mem. Inst. Medicine, Conf. Com., 1991—, Okla. Med. Rsch. Found., 1993—. Editorial bd. Alcoholism, Clinical and Experimental Research jour., 1989—, Digestive Diseases and Sciences jour., 1990—; contbr. articles to profl. jours. Mem. LWV, Pitts., 1965—. Grantee Nat. Inst. Alcohol Abuse and Alcoholism, 1985—; recipient Young Investigator award Rsch. Soc. Alcoholism, 1990. Fellow Am. Coll. Nutrition; mem. Internat. Assn. Study of the Liver, Rsch. Soc. on Alcoholism (Young Investigator award 1990), Internat. Soc. Biomed. Rsch. on Alcoholism, Am. Assn. for Study of Liver Diseases, Am. Gastroenterol. Assn. Democrat. Achievements include research on beneficial effects of moderate alcoholic beverage consumption in normal postmenopausal women, presence of plant estrogens (phyto estrogens) in alcoholic beverages, deleterious effects of ethanol on reproductive parameters in both male and female experimental animals. Home: 7441 Country Club Dr Oklahoma City OK 73116 Office: Baptist Med Ctr Okla Transplantation Inst 3300 NW Expressway Oklahoma City OK 73112

GAVANDE, SAMPAT ANAND, agricultural engineer, soil scientist; b. Nasik, India, Mar. 1, 1936; came to U.S., 1960; s. Ananda Bala and Saraja G. Gavande; m. Shaila Sawant, Feb. 25, 1968; children: Neil, Vikram. MS in Agrl. Engring., Kans. State U., 1962; PhD in Soils, Irrigation and Drainage, Utah State U., 1966. Registered profl. engr., Tex.; cert. profl. soil scientist; cert. profl. agronomist. Tech. officer FAO, UN, Turrialba, Costa Rica, 1966-69, Chapingo, Mex., 1969-72, Saltillo, Mex., 1973-77; chief tech. advisor FAO, UN, Rome and Asuncion, Paraguay, 1987-89; sr. scientist/engr. Radian Corp., Austin, Tex., 1977-82; hydrologist/sr. engr. Tex. Water Commn., Austin, 1983-87; chief tech. support br. Tex. Dept. Health, Austin, 1989-92; team leader Tex. Natural Resources Commn., Austin, 1992—; cons. soil/water FAO, UN, Kenya, 1985, watershed cons., Chile, 1987, India, 1989, watershed mgmt. cons., Iran, 1989, Indonesia, 1990. Author: (textbook in Spanish lang.) Soil Physics and Its Applications, 1972; contbr. over 60 articles to tech. publs., 1968-77. Mem. Rep. Presdl. Task Force, Austin and Washington, 1989-91; bd. dirs. India Community Ctr., Inc., Austin, 1991—. Mem. ASTM, Am. Soc. Agrl. Engrs., Am. Soc. Soil Sci., Am. Soc. Agronomy. Achievements include design of cost-effective, renovative tillage, drainage land reclamation, solid and liquid waste management, soil and water conservation systems for arid, semi-arid and humid areas. Home: 4501 Upvalley Ct Austin TX 78731 Office: Tex Water Commn PO Box 13087 Austin TX 78711

GAVELIS, JONAS RIMVYDAS, dentist, educator; b. Boston, Jan. 11, 1950; s. Mykolas and Janina (Povydis) G.; m. Bonnie Sylvester; children: Gregory, Nikolas. B.S., U. Mass., Amherst, 1971; D.M.D., U. Conn., 1975. Resident in dentistry Cabrini Health Care Center, N.Y.C., 1975-76; fellow in prosthetic dentistry Harvard U. Sch. Dental Medicine, Boston, 1976-78, instr., 1978-79; asst. prof. U. Conn. Sch. Dental Medicine, Farmington, 1979-82; practice dentistry specializing in prosthodontics Harvard Community Health Plan, Boston, 1982-92, Rockport, Mass., 1991—; asst. prof. Harvard Sch. Dental Medicine, 1982—. Contbr. articles on prosthetic dentistry to profl. jours. Fellow Acad. Gen. Dentistry (Vernon S. Johnson award 1981). Recipient Diamond award Harvard Community Health Plan, 1988. Mem. ADA, Northeast Prosthodontic Soc., Harvard Odontological Soc., Am. Acad. Crown and Bridge Prosthodontics, Am. Coll. Prosthodontists, Omicron Kappa Upsilon. Roman Catholic. Clubs: Southboro (Mass.), Rod and Gun, New Eng. Aquarium Dive (Boston), Southboro Rod and Gun, Cape Ann Sportsman's, Rotary. Home: 1238 Washington St Gloucester MA 01930-1056 Office: 23 Main St Rockport MA 01966-1512

GAVEZZOTTI, ANGELO, chemistry educator; b. Albizzate, Italy, Oct. 14, 1944; m. Alessandra Gara, Oct. 2, 1971. Chemistry Degree, U. Milano, 1968. Reader U. Milano, 1972-83, assoc. prof., 1983-86, prof. chemistry, 1986—. Co-editor Acta Crystallographica, Chester, U.K., 1988-91; contbr. over 90 articles to profl. jours. Mem. Am. Crystallographic Assn. Achievements include research on the structure of molecular crystals. Office: Dept Chimica Fisica, Via Golgi 19, Milan Italy 20133

GAWRYLOWICZ, HENRY THADDEUS, aerospace engineer; b. Passaic, N.J., Sept. 9, 1928; s. Stanley and Eva (Gilarek) G.; m. Alice Jean Sargent, Oct. 11, 1952; children: David Stanley, Amy. BS in Chem. Engring., Purdue U., 1955; MS in Mgmt. Engring., L.I. U., 1967. Registered profl. engr., Tex. Chief test engr. Lia Prop Lab., Picatinna Arsenal, Dover, N.J., 1960-63; mgr. propulsion system (Apollo) NASA, Bethpage, N.Y., 1963-70; with Apollo program and Space Shuttle systems & integration engring. NASA, Bethpage, 1970-83; mgr. adv. tech. propulsion Propulsion div. NASA, Washington, 1983-86; mgr. tech. resources Space Sta. program Grumman, Reston, Va., 1989-92; chmn. bd. Quantum Tech. Corp., Clifton, Va., 1988—, exec. v.p., 1990—. Contbr. articles to profl. jours. Mem. Ferguson Knolls Assn., Clifton, 1980. With USAF, 1951-52. Recipient 1st Moon Landing award NASA, 1969, LM-7 Snoopy award, 1970, Apollo Mission J award, 1971, many others. Fellow AIAA (assoc.). Achievements include patents for variable thrust rocket injection. Home: 12510 Knollbrook Dr Clifton VA 22024 Office: Quantec 12510 Knollbrook Dr Clifton VA 22024

GAY, DAVID HOLDEN, project technician; b. Boston, Jan. 5, 1954; s. Ernest and Mary (Holden) G.; m. Ann Marie Grieve, June 19, 1982; 1 child, David C. Student, Northeastern U., Boston, 1972, 75, 80, 81, 83, 84, Lowell U., 1987-92, Lesley Coll., 1992—. Floor mgr. Harvard Coop. Soc., Cambridge, Mass., 1972-76; R&D tech. Polariod Corp., Cambridge, 1976-85; project tech. Draper Lab., Cambridge, 1985—. Pres. R&D and Tech. Employee's Union, Belmont, Mass., 1990—; legis. agt. State of Mass., 1991—, legis agt. federal, 1991—. Roman Catholic. Avocations: baseball, basketball, skiing, swimming. Home: 5 Decarolis Dr Tewksbury MA 01876 Office: Draper Lab 555 Technology Dr Cambridge MA 02139

GAY-BRYANT, CLAUDINE MOSS, physician; b. Alma, Ga., Nov. 30, 1915; d. Fred and Rosa (Mercer) Moss; B.S., Coll. William and Mary, 1935; M.D., U. Va., 1939; m. Lendall C. Gay, June 29, 1940 (dec. 1971); children—Gordon B., Spencer B.; m. J. Marion Bryant, 1974 (dec. 1986). Intern, Gallinger Mcpl. Hosp., Washington; practice medicine specializing in family practice, Washington, 1940-91 ; mem. staff, exec. bd. Sibley Meml. and Capitol Hill Hosp., Washington; mem. Pres.'s Council on Malpractice, 1965; mem. health adv. commn. HEW, 1971-78; U.S. del. Med. Women's Internat. Congress, 5 times; del. Pres.'s Workshop on Non-Govtl. Orgn. Trustee Moss Charity Trust Fund, 1966—; adv. bd. Med. Coll. Pa., 1977; mem. president's council Coll. William and Mary. Recipient Capitol Hill Community Achievement award, 1986; Claudine Moss Radiological Apatheatre donated to U. Va. Med. Sch., 1991. Fellow Am. Acad. Family Practice (del. 1971-81; alt. del. to ho. dels. 1964-71); mem. Assn. Med. Women Internat. (del. 1966-72, councillor 1978-84), Royal Acad. Medicine, Pan Am. Med. Soc., D.C. Acad. Gen. Practice (pres.), Am. Med. Women's Assn. (councilor orgn. and mgmt. 1972-73, v.p. 1974, nat. pres. 1977, Blackwell medal 1988), D.C. Med. Women's Assn. (pres.), AMA, D.C. Med. Soc. (dir., exec. bd., past v.p., mem. nominating com. 1970, 81, relative value study com. 1970-72, constn. and constn. bylaws com., sec. family practice sect. 1966, 69, 78), DAR Regaret. Clubs: Women's Roundtable for Health Issues, Washington Forum (pres. 1987-88), Zonta (dir.). Home: 5030 Loughboro Rd NW Washington DC 20016-2613 Office: 5000 Macomb St NW Washington DC 20016-2610

GAYLE, JOSEPH CENTRAL, JR., computer information professional; b. N.Y.C., July 31, 1942; s. Joseph Sr. and Margaret Louemma (Smith) G.; m. Xenia Patricia Cockburn, June 22, 1968; children: Sean C., Melanie D. AS, CUNY, 1964; cert. with honors, Monroe Sch. Bus., 1963; cert., IBM Edn. Ctr., N.Y.C., 1974, Burroughs Edn. Ctr., Oakbrook, Ill., 1982, AT&T Edn. Ctr., Cin., 1983, AT&T Edn. Ctr., Atlanta, 1986. Treasury dept. clk. Tidewater Oil Co. subs. Getty Co., N.Y.C., 1962-64; computer operator analyst Chase Manhattan Bank, N.Y.C., 1966-68; sr. computer operator analyst Great Am. Ins. Co., N.Y.C., 1968-70; lead computer ops. analyst C.P.C. Internat. Co., Englewood Cliffs, N.J., 1970-71; asst. computer ops. supr. Irving Trust Bank, N.Y.C., 1971-77; computer ops. network supr. ABC, Hackensack, N.J., 1977-79, Warner Communications, N.Y.C., 1979-80; sr. system network supr. City Fed. Savs. and Loan, Somerset, N.J., 1980-82; sr. telecommunications cons. CIBA Geigy Corp., Ardsley, N.Y., 1982—; sr. computer specialist U.S. Army Data Processing Office Hdqrs., Hampton, Va., 1965-66; county supr. Bergen County Data Processing, Hackensack, 1975-80; computer sales cons. Computerland Micro Sales, Union, N.J., 1985-87; with AT&T Ednl. Ctr., Cin. and Fla., 1989. Contbr. articles to profl. jours. Town mem. Bd. Elections, Teaneck, N.J., 1978; sr. mem. Com. to Elect Bernie Brooks Mayor, Teaneck, 1980; county supr. Bergen County Civil and Criminal Justice System, Hackensack, 1980; state del. U.S. Congl. Adv. Bd., Washington, 1986. Served with U.S. Army, 1964-66. Recipient Nat. Peace award U.S. Army, 1966; Speaking of People honoree Ebony Mag. Publ., Chgo., 1981. Mem. Communications Mgrs. Assn., Sunguard Computer Disaster Group, Nat. Assn. Advancement Sci., N.Y. Acad. Scis., Nat. Space Soc., Am. Legion. Democrat. Club: Afro Civic League (Teaneck) (computer cons. 1983-86). Avocation: music. Home: 628 George St Teaneck NJ 07666-5356 Office: CIBA-Geigy Corp 444 Saw Mill River Rd Ardsley NY 10502-2600

GAYLOR, DONALD HUGHES, surgeon, educator; b. Bklyn., Apr. 17, 1926; s. Norman Hunter and Frances (Hughes) G.; m. Joan Winifred Power, Apr. 3, 1948; children: David, Christopher, Steven, Susan, Timothy. AB, U. Rochester, 1946, MD, 1949. Diplomate Am. Bd. Surgery, Am. Bd. Thoracic Surgery. Commd. lt. (j.g.) USN, 1949, advanced through grades to capt. M.C., 1966; intern U.S. Naval Hosp., Phila., 1949-50; student flight surgeon Sch. Aviation Medicine, Pensacola, Fla., 1950-51; flight surgeon U.S. Naval Sta., Trinidad, B.W.I., 1951-53; resident gen. surgery U.S. Naval Hosp., St. Albans, N.Y., 1953-57; postgrad. fellow surgery Royal Victoria Hosp., McGill U., Montreal, Can., 1957; resident thoracic surgery U.S. Naval Hosp., St. Albans, N.Y., 1957-59; resident cardiovascular surgery St. Francis Hosp., Roslyn, N.Y., 1958; staff thoracic surgeon U.S. Naval Hosp., Portsmouth, Va., 1959-64; surgeon U.S.S. Enterprise, 1964; staff thoracic surgeon U.S. Naval Hosp., Nat. Naval Med. Ctr., Bethesda, Md., 1964-65, chief thoracic and cardiovascular surgery, 1965-68; chief surgery, exec. officer U.S.S. Repose, 1968-69; exec. officer Naval Med. Sch., Bethesda, Md., 1969-72; ret., 1972; clin. assoc. surgery U. Pa. Sch. Medicine, 1976—; prof. clin. surgery Hahnemann U. Sch. Medicine, 1986—; chief surgery Allentown (Pa.) Hosp., 1972-90, Sacred Heart Hosp., 1973-76, Lehigh Valley Hosp. Ctr., 1974-90. Contbr. articles to profl. jours. Fellow ACS; mem. AMA, Am. Thoracic Soc., Am. Trauma Soc. (pres. Pa. div. 1979-83, treas. 1985-91), Soc. Thoracic Surgeons (founding), Pa. Assn. Med. Edn. (pres. 1983-84), Pa. Med. Soc., Assn. for Hosp. Med. Edn., Pa. Assn. for Thoracic Surgery, Assn. Mil. Surgeons U.S., Am. Trauma Soc. (founding mem., Assoc. program dirs. surgery 1982-90). Roman Catholic. Home and Office: 3761 Devonshire Rd Allentown PA 18103-9628

GAYLOR, JAMES LEROY, biomedical research director; b. Waterloo, Iowa, Oct. 1, 1934; s. David P. and Lena (Livingston) G.; m. Marilyn Louise Gibson, Mar. 25, 1956; children—Douglas, Ann, Robert, Kenneth. B.S., Iowa State U., 1956; M.S., U. Wis., 1958, Ph.D., 1960. From asst. prof. to prof. biochemistry Cornell U., Ithaca, N.Y., 1960-77, chmn. biochemistry, molecular and cell biology sect., 1970-76; prof., chmn. dept. biochemistry U. Mo., Columbia, 1977-80; assoc. dir. life scis. rsch. E.I. duPont Cen. Rsch., Wilmington, Del., 1981-83, dir. health sci. rsch., 1984-85; dir. biol. rsch. E.I. duPont Pharms., Wilmington, Del., 1986-87; corp. dir. sci. and technology Johnson & Johnson, New Brunswick, N.J., 1987—; vis. prof. U. Ill., summers 1964-65; sabbatical leave U. Oreg. Sch. Medicine, 1966-67, U. Osaka, Japan, 1973-74; vis. lectr. La Molina, Peru, summer 1962; nutrition cons. Pew Found., Phila., 1986-92; mem. bd. sci. counselors div. cancer prevention Nat. Cancer Inst., NIH, Bethesda, Md., 1987-91. Mem. various editorial bds.; contbr. over 150 rsch. articles to sci. jours. NIH fellow, 1958-60; spl. fellow, 1966-67; Guggenheim fellow, 1973-74. Mem. AAAS, Am. Chem. Soc., Am. Soc. Biochemistry and Molecular Biology, Am. Inst. Nutrition, Am. Heart Assn., Am. Assn. Pharm. Scientists. Achievements include patents for specific synthetic inhibitors of cholesterol synthesis; research on biosynthesis of cholesterol and other membrane-bound enzymes. Office: Cosat Johnson & Johnson 410 George St New Brunswick NJ 08901-2021

GAYLORD, EDSON I., manufacturing company executive. Chmn., pres. Ingersoll Milling Machine Co., Rockford, Ill. Recipient M. Eugene Merchant Mfg. medal ASME/SME, 1991. Office: Ingersoll Milling Machine Co 707 Fulton Ave Rockford IL 61103-4092*

GAYLORD, ROBERT STEPHEN, aerospace engineering manager; b. Santa Ana, Calif., Apr. 8, 1933; s. George Tucker and Alice (Francis) G.; m. Marva Rose Engbaum, Nov. 25, 1953; children: Karen, Kevin, David. BS, UCLA, 1956, MS, 1961. Registered profl. engr., automatic control, Calif. Group leader TRW Systems Group, Redondo Beach, Calif., 1956-62; system engring. dir. The Aerospace Corp., El Segundo, Calif., 1962-69; devel. program mgr. The Boeing Co., Kent, Wash., 1969-84; gen. mgr. systems planning and devel. Aerospace Corp., El Segundo, 1984—; chmn. Laser Crosslink Blue Ribbon Rev. Team, 1984, Titan Four Ind. Readiness Rev., 1989. Author: Advances in Astronautics, 1962, Advances in Automatic Control, 1966; contbr. articles to profl. jours. Councilor, com. chair Boy Scouts Am., Tacoma, Wash., 1973-80; com. mem. Rep. Party, Puyallup, Wash., 1980. Mem. AIAA, IEEE (chpt. pres. 1983), Armed Forces Communications & Electronics Assn., Nat. Coun. Systems Engrs. (charter). Presbyterian. Home: 13603 Marina Pointe Dr C339 Marina Del Rey CA 90292 Office: The Aerospace Corp 2350 E El Segundo Blvd El Segundo CA 90245

GAYNOR, JOSEPH, chemical engineering consultant; b. N.Y.C., Nov. 15, 1925; s. Morris and Rebecca (Schnapper) G.; m. Elaine Bauer, Aug. 19, 1951; children—Barbara Lynne, Martin Scott, Paul David, Andrew Douglas. B.Ch.E., Polytechnic Inst. Bklyn., 1950; M.S., Case-Western Res. U., 1952, Ph.D., 1955. Research asst. Case Inst., Cleve., 1952-55; with Gen. Engring. Labs. Gen. Electric Co., Schenectady, N.Y., 1955-66, sect. mgr. research and devel., 1962-66; group v.p. research Bell & Howell Co., 1966-72; mgr. comml. devel. group, mem. pres.' office Horizons Research, Inc., Cleve., 1972-73; pres. Innovative Tech. Assos., Ventura, Calif., 1973—; mem. nat. materials adv. bd. com. NAS; chmn. conf. com. 2d internat. conf. on bus. graphics, 1979, program chmn. 1st internat. congress on advances in non-impact printing techs., 1981, mem. adv. com. 2d internat. congress on advances in non-impact printing techs., 1984, chmn. publs. com. 3rd internat. congress on advances in non-impact printing techs., 1986, chmn. internat. conf. on hard copy media, materials and processes, 1990. Editor: Electronic Imaging, 1991, Procs. Advances in Non-Impact Printing Technologies, Vol. I, 1983, Vol. II, 1988, 3 spl. issues Jour. Imaging Tech., Proc. Hard Copy Materials Media and Processes Internat. Conf., 1990; patentee in field. Served with U.S. Army, 1944-46. Fellow AAAS, AICE, Imaging Sci. and Tech. Soc.; mem. Am Chem. Soc., Soc. PHotographic Scientists and Engrs. (sr., gen. chmn. 2d internat. conf. on electrophotography 1973, chmn. bus. graphics tech. sect. 1976—, chmn. edn. com. L.A. chpt. 1978—), Am. Soc. Photobiology, Sigma Xi, Tau Beta Pi, Phi Lambda Upsilon, Alpha Chi Sigma. Home: 108 La Brea St Oxnard CA 93035-3928 Office: Innovative Tech Assocs 3639 E Harbor Blvd # 203E Ventura CA 93001

GAYTHWAITE, JOHN WILLIAM, civil engineer; b. Boston, Jan. 15, 1948; s. John Ingham and Doris (Rich) G.; m. Michele Rabot, July 7, 1979. BS in Civil Engring., Northeastern U., Boston, 1971. Registered profl. engr., several states. Project engr. Crandall Dry Dock Engrs. Inc., Dedham, Mass., 1971-79, Parson, Brinkerhoff, Quade & Dougals, Boston, 1979-80, Fay, Spofford & Thorndike, Inc., Boston, 1980-81; project mgr. The Maguire Group, Inc., Foxborough, Mass., 1981-82; cons. engr. Manchester, Mass., 1982-88; pres. Maritime Engring. Cons. Inc., Manchester, 1988—; guest lectr. univs. including U. Main, U. Wis., U. R.I., U. N.H. Author: The Marine Environment and Structural Design, 1981, Design of Marine Facilities, 1990. Mem. ASCE (exec. com. coastal engring. tech. com. 1989—), Marine Tech. Soc. (councilor, exec. com. 1978-81), Boston Soc. Civil Engrs. (chmn. waterway, port and coastal tech. group and com. to develop coastal zone bldg. code 1989-91, editorial bd. Jour. Civil Engring. Practice 1990—, Tech. Paper award 1984), Soc. Naval Architects and Marine Engrs. (Paper award 19710, Permanent Internat. Assn. Navigation Congresses (invited paper 1988), Am. Shore and Beach Preservation Assn. Achievements include project management of design of many large scale marine civil engring. works including floating dry docks, marine terminals, seawall and breakwaters. Home: 155 Pine St Manchester by Sea MA 01944 Office: Maritime Engring Cons Inc 155 Pine St Manchester MA 01944

GAZE, NIGEL RAYMOND, plastic surgeon; b. Leamington Spa, Warwick, Eng., Nov. 2, 1943; s. Raymond Ernest and Beatrice Maud (Caswell) G.; m. Heather Winifred Richardson, Aug. 6, 1966; children: Julia, Celia, Richard, Thomas, Mary, Harry. MB, ChB, Liverpool U., 1966; BMus, London U., 1986. House officer Whiston Hosp., Prescot, Lancashire, Eng., 1966-67, sr. house officer orthopaedics, 1967-68; sr. casualty officer Royal So. Hosp., Liverpool, Eng., 1969-70; surg. registrar Liverpool Regional Hosp. Bd., 1970-72; gen. surgery registrar Chester (Eng.) Royal Infirmary, 1972-73; registrar in plastic surgery Wordsley Hosp., Stourbridge, Worcs, Eng., 1973-75; sr. registrar plastic surgery Yorks Region Health Authority, Leeds, 1975-79; cons. plastic surgery Royal Preston (Lancashire) Hosp., 1980—. Contbr. med. articles to profl. jours.; composer choirs, organs, and solos. Condr. Elizabethan Singers, Preston; accompanist County Hall Singers, Preston, 1980—, Clitheroe Assn. Ch. Choirs, Preston, 1984—; assoc. organist Preston Parish Ch., Lancashire, 1980—. Fellow Trinity Coll. Music, Royal Coll. Organists. Fellow Royal Coll. Surgeons Edinburgh, Royal Coll. Surgeons; mem. Royal Acad. Music (licentiate), Brit. Inst. Organ Studies, Brit. Assn. Plastic Surgeons, Brit. Assn. Aesthetic Plastic Surgeons, Victorian Soc., Select Vestry Club, Assn. British Choral Conductors. Mem. Conservative party; mem. Ch. of Eng. Avocations: collecting books and antiques, walking. Home: Priory House, 35 Priory Ln Penwortham Lancashire, Preston PR1OAR, England Office: Fulwood Hall Hosp, Midgery Ln, Lancashire, Preston PR1 OAR, England

GAZINSKI, BENON, agricultural economics educator; b. Gebice, Bydgoszcz, Poland, Mar. 11, 1953; s. Kazimierz and Janina (Jercha) Z.; m. Barbara Smyk, Oct. 16, 1988; 1 child, Aleksandra. MSc, U. Econs., Poznan, 1976; PhD, Agrl. U., Olsztyn, 1980. Postdoctoral fellow Agrl. U., Olsztyn, 1976-79, jr. asst., 1979-80, sr. asst., 1980, lectr. agrl. econs., 1980—; guest lectr. Papal Acad. of Teology, Krakow, Br. of Olsztyn, 1986—; cons. AGrl. Extension Ctr., Olsztyn, 1991—. Editorial bd. Nowe Rolnictwo, 1985-90, Rolnictwo na Swiecie, 1983-89; translator 50 profl. articles from English to Polish; contbr. over 70 articles to profl. jours.; author/editor newsletter: Solidarnosc Kortowska. Mem. Presidium, Solidarity Trade Union, Olsztyn, 1980-81, chmn revision com., 1989—. Recipient Honor Medal of Agrl. Faculty, Agrl. U., 1990, Award of the Minister of Nat. Edn., 1989. Mem. European Assn. Agrl. Econs., European Distant Edn. Network, Internat. Assn. Agrl. Economists, Polish Assn. Economists (chmn. Olsztyn br. 1986—). Roman Catholic. Avocations: table tennis, swimming, biking, chess. Home: Kortowo 45 B/23, Olsztyn Poland PL10718 Office: Agrl Univ, Kortowo 41, Olsztyn Poland PL10718

GE, GUANG PING, mathematics educator, statistician; b. Nanjing, Jiangsu, China, Sept. 13, 1934; s. Zhao Xun and Pei Su (Li) G.; m. Yue Er Chen, Feb. 15, 1961; children: Qian Hong, Wan Zi. Grad., Beijing Normal U., 1957. Asst. Hebei Normal U., Shijiazhuang, Peoples Republic China, 1957-58; asst. and lectr. Hebei Normal U., Shijiazhuang, 1960-82, assoc. prof., 1983-86, prof., 1986-90, chmn. dept. math., 1984-88; trainer Beijing U., 1958-60; sr. vis. scholar CUNY, U.S., 1982-83, U. Calif., Berkeley, 1983-83; prof., dean dept. sci. Shanghai (Peoples Republic China) U. of Tech., 1990—; cons. Hebei Cadre Inst. Econ. Mgmt., Shijiazhuang, 1984—, Ordnance Engring. Inst., 1984—; reviewer Math. Revs., Ann Arbor, Mich., 1989—; referee IEEE Transactions on Reliability, 1991—. Contbr. articles to profl. jours., chpts. to books. Recipient Sci. award Edn. Com. Hebei Province, Shijiazhuang, 1989, Sci. and Tech. Progress award Com. Sci. and Tech. Hebei Province, 1992; named Outstanding Expert Hebei Province, 1988. Mem. Shanghai Math. Soc. (dir. 1992—), Chinese Math. Soc., Chinese Assn. for Applied Stats. (v.p. 1989—), Internat. Chinese Stats. Assn., Am. Math. Soc. Avocations: swimming, bicycling, cooking. Office: Shanghai U Tech Dept Sci, 149 Yanchang Rd, Shanghai 200072, China

GE, LI-FENG, engineer; b. Chuzhou, Anhui, People's Republic of China, Mar. 18, 1947; s. Tian-Min and Ji-Hua (Fan) G.;m. Jing-Ping Shao, Mar. 24, 1982; 1 child, Zhong-Qi. BS, Hefei (Peoples Rep. of China) U. Tech., 1970; MS, U. Sci. & Tech. of China, Beijing, 1980; postgrad., Nat. Inst. Metrology, Beijing, 1980-81. Registered profl. engr., People's Republic of China. Engr. Wuhu (People's Republic of China) Diesel Engine Factory, 1970-77, Wuhu Bur. of Standards and Metrology, 1978, Nat. Inst. Metrology, Beijing, 1981-82; rsch. engr. Anhui Bur. of Standards and Metrology, Hefei, People's Rep. of China, 1982-86; sr. rsch. engr., chmn. Acoustic Lab. Anhui Bur. of Standards and Metrology, Hefei, 1988-92; dep. dir. Inst. Measurement and Testing Tech. Anhui Bur. Tech. Supervision, 1992—; guest scientist Nat. Inst. Standards & Tech., Gaithersburg, Md., 1986-88; vis. scholar PCB Piezotronics Inc., Buffalo, 1986. Inventor piezoelectric reciprocity method and apparatus for absolute calibration of high-frequency primary vibration standards. Mem. Acoustical Soc. China, Acoustical Soc. Am., On-Line Measurement Soc. China (vice dir. Anhui chpt. 1988—). Avocations: music, traveling. Office: Anhui Inst Measurement and Testing Tech, 5 Taihu Rd, Hefei 230022, China

GE, WEIKUN, physicist, educator; b. Beijing, China, Mar. 25, 1942; came to U.S., 1988; s. Li Ge and Jinlan Wang; m. Fuxing Hou, June 9, 1967; children: Cheng, Qiong. BSc, Peking U., Beijing, 1965; PhD, U. Manchester, Eng., 1983. Engr. Inst. Non-Ferrous Metals, Beijing, 1965-78; teaching asst. U. Manchester Inst. Sci. and Tech., 1981-83, postdoctoral fellow, 1983; rsch. assoc. Inst. Semiconductors, Beijing, 1984-85, assoc. prof., 1985-88; rsch. assoc. Dartmouth Coll., Hanover, N.H., 1988-91, rsch. assoc. prof., 1991-93; sr. lectr. Hong Kong U. Sci. and Tech., Hong Kong, 1993—; dir. Fibernet Rsch. Inc., Nashua, N.H., 1992—. contbr. articles to profl. publs. Recipient 3d degree award Chinese Acad. Scis., 1987, 2d degree award, 1989. Mem. Am. Phys. Soc., Chinese Profl. Club USA (chmn. bd. dirs.). Achievements include contributions to the discovery of Ga-O-Ga local vibrational mode absorption, its model and its relationship with EL2 levels in GaAs, and identifying electron states in short period GaAs/AIAs superlattices. Office: Hong Kong U Sci & Tech, Dept Physics, Hong Kong Hong Kong

GEBBIE, KATHARINE BLODGETT, astrophysicist; b. Cambridge, Mass., July 4, 1932. BA, Bryn Mayr Coll., 1957; BSc, U. London, 1960, PhD, 1965. Rsch. assoc. astrophysics Joint Inst. Lab. Astrophysics, U. Colo., 1967-68, lectr. physics and astrophysics, 1974-77; astrophysicist Nat. Bur. Standards, 1968-85, supervisory physicist, 1985-89; dir. physics lab. Nat. Bur. Standards & Tech., 1989—; adj. prof. astro-geophysics U. Colo., 1977-89. Editor: The Observatory, 1965-67. Fellow Joint Inst. Lab. Astrophysics; mem. Internat. Astron. Union, Am. Astron. Soc., Am. Phys. Soc., Royal Astron. Soc. Achievements include rsch. in planetary nebulae, stellar atmospheres; physics of solar atmosphere. Office: Joint Inst Lab Astrophysics Nat Inst Standards & Tech Commerce Ctr Boulder CO 80303 also: Natl Inst of Standards & Tech Physics Lab Bldg 221, Rte 270 Gaithersburg MD 20899*

GEBBIE, KRISTINE MOORE, health official; b. Sioux City, Iowa, June 26, 1943; d. Thomas Carson and Gladys Irene (Stewart) Moore; divorced; children: Anna, Sharon, Eric. BSN, St. Olaf Coll., 1965; MSN, UCLA, 1968. Project dir. USPHS tng. grant, St. Louis, 1972-77; coord. nursing St. Louis U., 1974-76, asst. dir. nursing, 1976-78, clin. prof., 1977-78; adminstr. Oreg. Health Div., Portland, 1978-89; sec. Wash. State Dept. Health, Olympia, 1989-93; coord. Nat. AIDS Policy, Washington, 1993—; assoc. prof. Oreg. Health Scis. U. Portland, 1980—; chair, U.S. dept. energy secretarial panel on Evaluation of Epidemiologic Rsch. Activities, 1989-90; mem. Presdl. Commn. on Human Imunodeficiency Virus Epidemic, 1987-88. Author: (with Deloughery and Neuman) Consultation and Community Orgn., 1971, (with Deloughery) Political Dynamics: Impact on Nurses, 1975; (with Scheer) Creative Teaching in Clinical Nursing, 1976. Bd. dirs. Luth. Family Svcs. Oreg. and S.W. Wash., 1979-84; bd. dirs. Oreg. Psychoanalytic Found., 1983-87. Recipient Disting. Alumna award St. Olaf Coll., 1979; Disting. scholar Am. Nurses Found., 1989. Fellow Am. Acad. Nursing; mem. Assn. State & Territorial Health Ofcls., 1988 (pres. 1984-85, exec. com. 1980-87, McCormick award 1988), Am. Pub. Health Assn. (exec. bd.), Inst. Medicine, Hastings Ctr., N.Am. Nursing Diagnosis Assn. (treas. 1983-87), Oreg. Pub. Health Assn., Am. Soc. Pub. Adminstrn. (adminstrn. award II 1983), City Club of Portland. Office: 750 17th NW Ste 1060 Washington DC 20503

GEBHARD, DAVID FAIRCHILD, aeronautical engineer, consultant; b. Mt. Vernon, N.Y., Nov. 16, 1925; s. John Gabriel and Helen Louise (Fairchild) G.; m. Shirley Hodges Wallace, Sept. 8, 1951; children: David F. Jr., Jennifer L., Douglas H. BS in Aero. Engring., Princeton U., 1948, MS in Aero. Engring., 1949. Head design analysis Gyrodyne Co. Am., St. James, N.Y., 1950-52; rsch. assoc. Forestal Rsch. Ctr., Princeton, N.J., 1952-56; chief preliminary design Kellett Aircraft Corp., Willow Grove, Pa., 1956-59; project engr. for advanced aircraft Grumman Aerospace Corp., Bethpage, N.Y., 1959-76, prin. engr. for advanced aircraft, 1976-84, cons., 1984-91; pres. Task and Analysis Corp., Northport, N.Y., 1991—. Contbr. articles to profl. jours. With USAAF, 1944-45. Mem. Am. Helicopter Soc., Northport Yacht Club (sailing champion 1969, 73), Sigma Xi. Achievements include patent on short take-off jump mode for airplane landing gear struts and 2 aircraft design patents; director design and test effort leading to successful jump mode flight demonstrations; manager Apollo lunar module test requirements program fisrt defining and then completing the design flight certification tests at both assembly and vehicle levels. Home and Office: 96 Markan Dr Northport NY 11768

GECZIK, RONALD JOSEPH, pharmaceutical researcher; b. N.Y.C., Mar. 22, 1933; s. Joseph Michael and Marie (Kirby) G.; m. Olive Marie Shanley, Apr. 15, 1961; children: Catherine Mary, Michael Joseph, Paul James. BS, Fordham Coll., 1954, MS, 1957, PhD, 1959; JD, Seton Hall U., 1971. Bar: N.J. Sr. rsch. scientist Colgate Palmolive Co., New Brunswick, 1959-64; dir. tech. adminstrn. Squibb Inst. Med. Rsch., New Brunswick, 1964-73; corp. v.p. licensing and prodn. acquistion Carter Wallace, Inc., N.Y.C., 1973-84; v.p. SNW, Inc., Montclair, N.J., 1984-89; pres. Paul Michael Assoc., Inc., South River, N.J., 1989—. Mem. ASPET, AAAS, ABA, Licensing Execs. Soc., N.Y. Acad. Scis., N.J. Bar Assn., Sigma Xi. Office: Paul Michael Assocs Inc PO Box 595 South River NJ 08882

GEERLINGS, PETER JOHANNES, psychiatrist, psychoanalyst; b. Jakarta, Indonesia, Nov. 13, 1939; arrived in Netherlands, 1951; s. Johannes J. and Anna W. (Bauer) G.; m. Eugenie A. Oosterhuis, Feb. 24, 1965 (div. May 1985); children: Suzanne, Mirjam, Paulien. MD, U. Amsterdam, Netherlands, 1965. Chmn. dept. psychiatry U. Amsterdam, 1984-91, assoc. prof. dept. psychiatry, 1984—; med. dir. Jellinekcentrum for Addictions, Amsterdam, 1991—; coord. ednl. program, U. Amsterdam. Contbr. numerous articles to med. jours. Chmn. com. on addictions, Nat. Coun. Health in the Netherlands, 1988-90; mem. com WHO, Div. Mental Health, Geneva, 1989. Mem. Dutch Assn. for Group Psychotherapy (supr.), Dutch Assn. for Psychoanalytic Psychotherapy (training analyst), Dutch Psychoanalytic Assn. Home: C Krusemanstraat 8, 1075 NL Amsterdam The Netherlands Office: Jellinekcentrum, Jacob Obrechtstraat 92, 1071 KR Amsterdam The Netherlands

GEHLERT, DONALD RICHARD, pharmacologist; b. Milw., June 27, 1958; s. William Richard and Barbara Elaine (Gescheidle) G.; m. Susan Nestle, June 19, 1982. BS in Pharmacy, Purdue U., 1981; PhD in Pharmacology, U. Utah, 1985. Rsch. assoc. dept. psychiatry U. Utah, Salt Lake City, 1985; pharmacology rsch. assoc. tng. staff fellow Nat. Inst. Gen. Med. Scis./NIH, Bethesda, Md., 1985-87; unit chief exptl. therapeutics br. Nat. Inst. Neurol. Disorders and Stroke/NIH, Bethesda, Md., 1987-89; sr. pharmacologist Lilly Rsch. Labs., Indpls., 1989-92; rsch. scientist, 1992—. Contbr. numerous articles to profl. jours. Organizer Together Fest, Indpls., 1991; mem. Old Northside Hist. Found., Indpls., 1991-92, Nat. Found. for Hist. Preservation, Indpls., 1992. Recipient Travel award Am. Coll. Neuropsychopharmacology, Maui, Hawaii, 1985, Tng. award-PRAT, NIH, Bethesda, 1985-87, Travel award Am. Soc. for Pharmacology and Toxicology, Sydney, Australia, 1987, Travel award Danish Med. Rsch. Coun., Copenhagen, 1988. Mem. Soc. for Neuroscience, Ind. Acad. Sci., Sigma Xi. Achievements include patent in novel analogs of tomoxetine with high affinity and selectivity for the norepinephrine transporter. Office: Lilly Rsch Labs CNS Pharmacology Mail Code 0815 Indianapolis IN 47285

GEHLING, MICHAEL PAUL, engineering executive. BSME, Iowa State U. V.p. engring. Kahler Corp., Rochester, Minn., 1982—. Mem. NSPE, Am. Hotel and Motel Assn. Exec. Engrs. Office: Kahler Corp 20 SW 2d Ave Rochester MN 55902

GEHMAN, BRUCE LAWRENCE, materials scientist; b. Akron, Ohio, Oct. 22, 1937; s. Samuel D. and Helen (McCaughey) G.; m. Judith Lawson, Dec. 11, 1960; children: Jean, Jeffrey, James. BS in Engring., U. Mich., 1959; PhD in Physics, U. Calif., San Diego, 1970. Mgr. R&D Cominco Electronic Materials, Spokane, 1971-86; tech. dir. Deposition Tech., Inc., San Diego, 1986-88; v.p. for bus. devel. ISM Techs., San Diego, 1988-89; v.p. R&D Leybold Materials, Inc., Morgan Hill, Calif., 1989—. Contbr. articles to profl. publs. Mem. ASTM (chmn. subcom. on sputtered thin films 1990—, vice chmn. com. on electronics 1990—), ASM, IEEE. Achievements include 2 patents in field of electronic materials and manufacture. Office: Leybold Materials Inc 16035 Vineyard Blvd Morgan Hill CA 95037

GEHO, WALTER BLAIR, biomedical research executive; b. Wheeling, W.Va., May 18, 1939; s. Blair Roy and Susan (Yonko) G.; m. Marjorie Cooper, Aug. 25, 1962; children: Hans, Alison, Robert, David, Daniel. BS, Bethany Coll., 1960; PhD in Pharmacology, Western Res. U., 1964, MD, 1966. Instr. pharmacology Sch. of Medicine Western Res. U., Cleve., 1966-67; pres. SDG Tech. Inc., Wooster, 1993—; staff researcher Procter & Gamble Co., Cin., 1968-74, head pharmaceutical rsch. sect., 1974-81; v.p., dir. rsch. Tech. Unltd., Inc., Wooster, Ohio, 1981-89, pres., 1989-93. Contbr. articles to Phamacology of Disphosphates, Clin. Pharmacology of Didronel, Genetics of Myositis Ossificans. Recipient two Ohio Innovator awards Edison Fund Ohio, 1987,. Mem. AMA, Am. Chem. Soc. Achievements include patents in pharmaceuticals; contributions to development of osteoscan and didronel, and targeted drug delivery systems. Office: SDG Tech Inc PO Box 723 Wooster OH 44691-0723

GEHRING, DAVID AUSTIN, physician, adminstrator, cardiologist; b. Bryn Mawr, Pa., Dec. 6, 1930; s. Harry Rittenhouse and Anne Gardiner (Bozarth) G.; m. Joan Helen Lotz, June 7, 1953 (div. Aug. 1982); children: David, Paul, Peter, Sue, Barbara, Eric; m. Victoria Marie Damiano, Sept. 2, 1982; children: Theresa, Judy Lynne, Michael Austin. BA magna cum laude, U. Pitts., 1952, MD, 1956. Diplomate Am. Bd. Internal Medicine. Commd. USN, 1956, advanced through grades to lt. comdr.; intern, then resident in internal medicine U.S. Naval Hosp. USN, Phila., 1956-60, mem. staff internal medicine U.S. Naval Hosp., 1960-61; chief internal medicine heart sta. U.S. Naval Hosp. USN, Annapolis, Md., 1961-63; resigned USN, 1963; cardiologist K.G.E. Med. Group, Woodbury, N.J., 1963-82; cardiologist, pres. Hobbs Cardiology, P.A., Hobbs, N.Mex., 1982-86; med. dir. Polk (Pa.) Ctr., 1986-91; physician, chief grade VA Med Ctr., Coatesville, Pa., 1991—, assoc. chief of staff for ambulatory care, 1993—; testing cardiologist Anthropometrics United Med. Group, Cherry Hill, N.J., 1974-82; clin. asst. prof. medicine Temple U. Hosp., Phila., 1975-82; adj. asst. prof. medicine Jefferson Meml. Coll., Phila., 1981-82; chief cardiac rehab. unit Lea Regional Hosp., Hobbs, 1982-86; chief med. svcs. 829th Sta. Hosp. USAR, Lubbock, Tex., 1984-86; cons. cardiology Oil City, Pa., 1986-91; staff Franklin (Pa.) Regional Med. Ctr., 1986-90, Oil City Area Health Ctr., 1986-91; teaching staff St. Joseph Hosp., Lancaster, Pa., 1991—; clinical preceptor U. Pa. Sch. Nursing, 1993—. Author: EKG Workbook, 1972, EKG Workbook I, 1978; contbr. articles to profl. jours. Project dir. 23 Greater Del. Valley Reg. Med. Prog., Pa., 1971-75; mem. ACLS Inst. and affiliated faculty Pa. Heart Assn., 1986—, bd. dirs. N.W. chpt. 1988-90; bd. dirs. adv. com., chmn. personnel com. med. health, rehab., drugs and alcohol Venango County, Franklin, Pa., 1986-90, pres., 1988-89; mem. Health Care Adv. Com. to Congressman William F. Clinger, Jr., 23d Dist., Pa., 1989-91; lector St. Joseph Ch., Oil City, 1987-91, eucharistic min.. 1990-92; eucharistic min. St. Joseph Ch., Swedesboro, N.J., 1992—; mem. Pitts. Opera Soc. Lt. col. USAR, 1983-90. Recipient Outstanding Svc. award Am. Cancer Soc. N.J., 1967, Benjamin Berkowitz award N.J. Heart Assn., 1975, Nat. Def. Svc. medal 1975, USAR Components Achievement medal, 1988, Letter of Commendation, USAR, 1988, 90, Pres.'s medal of Merit, Rep. Task Force, 1984; Cert. of Appreciation, Sec. of State N.Mex., 1982, Venango County Commr.'s, 1987, 88, 89, 90, Polk Ctr. award of merit, 1991. Fellow Am. Coll. Cardiology, Am. Coll. Chest Physicians, Coll. Physicians of Phila., Am. Coll. Clin. Pharmacology, ACP (life, Recognition awards 1967-70); mem. AMA, Am. Geriatrics Soc., St. Jude Soc., Holy Name Soc., Assn. Miraculous Medal (promoter 1987—), Venango County Med. Soc. (pres. 1989-91), Franklin Club, Assn. Mil. Surgeons, Am. Coll. Physician Execs., Am. Legion. Democrat. Roman Catholic. Avocations: stamp collecting, reading, walking, swimming, opera. Home: 138 Harvest Rd 865 W Red Bank Ave West Deptford NJ 08096 Office: VA Med Ctr 1400 Blackhorse Hill Rd Coatesville PA 19320-2097

GEHRING, GEORGE JOSEPH, JR., dentist; b. Kenosha, Wis., May 24, 1931; s. George J. and Lucille (Martin) G.; m. Ann D. Carrigan, Aug. 2, 1982; children: Michael, Scott. DDS, Marquette U., 1955. Pvt. practice dentistry, Long Beach, Calif., 1958—. Author: The Happy Flosser. Chmn. bd. Long Beach affiliate Calif. Heart Assn.; mem. Long Beach Grand Prix com. of 300; ind. candidate for pres. of the U.S., 1988, 92. Served with USNR, 1955-58. Fellow Internat. Coll. of Denists, Am. Coll. Dentists; mem. Harbor Dental Soc. (dir.), Pierre Fauchard Acad., Delta Sigma Delta. Club: Rotary. Home: 1230E Ocean Blvd # 603 Long Beach CA 90802-6909 Office: 532 E 29th St Long Beach CA 90806-1645

GEHRING, RICHARD WEBSTER, structural engineer; b. Rockford, Ill., Mar. 16, 1927; s. John Gottlieb and Lucia Mae (Webster) G.; m. Ellen Elizabeth Hansen, Sept. 4, 1948; children: Katherine Louise, John Webster, Richard Mentor. BS in Aerospace Engr., Tri-State Coll., 1949; postgrad., Pa. State U., 1965, 66. Structural test engr. Glenn L. Martin Co., Balt., 1950-52; structures engr. N.Am. Aviation Inc., Columbus, Ohio, 1952-68; sr. engr. specialist Rockwell Internat., Columbus, 1968-89; cons. Snow Aviation Internat., Columbus, 1990—; mem. MIL-HOBK-5 com. Rep. from N.Am. Aviation, Columbus, 1964-68; reviewer USAF Design Guide for Advanced Composites, USAF/Rockwell Internat., Dayton and Columbus, 1971; guest lectr. airframe design Ohio State U., Columbus, 1982, 84. With USCG, 1944-46. Recipient Letter of Appreciation, U.S. Naval Air Engring. Ctr., Phila., 1967. Mem. AIAA (vice-chmn. local chpt. 1954), U.S. Naval Inst., Lions Internat. (dir. local chpt. 1970-72). Republican. Lutheran. Achievements include development of structural analysis method for inelastic structural joints with temperature and mixed materials, methods for defining elevated temperature strength of airframes using room temperature tests. Home: 2310 Middlesex Rd Columbus OH 43220

GEHRLEIN, MICHAEL TIMOTHY, air force officer; b. Nicosia, Cyprus, Jan. 28, 1966; s. Richard Charles and Carolyn Elizabeth (Zamaria) G. BS in Aerospace Engring., Pa. State U., 1988; MS in Aerospace Engring., Northrop U., 1991. Commd. 1st lt. USAF, 1989—; project engr. Space Surveillance Program Office, L.A., 1989-91; project mgr. Brilliant Eyes Program Office, 1991—. Vol. Ft. MacArthur Open House Com., L.A., 1990; judge Elem. Sch. Sci. Fair, L.A., 1992. Mem. AIAA, Air Force Assn., Co. Grade Officers Coun., Mensa, Tau Beta Pi, Sigma Gamma Tau. Roman Catholic. Achievements include development of sensor and cryogenic cooler requirements for space shuttle experiments; led development and demonstration of critical technologies; developed cryocooler technology development program; research on orbital debris. Home: 800 Meyer Ln Apt # 8 Redondo Beach CA 90278

GEIER, GERHARD, chemistry educator; b. Schaffhausen, Switzerland, Mar. 19, 1935; s. Gottlieb and Marie (Sieber) G.; m. Marthe Lis Baechtold, Nov. 26, 1966; children: Christian Florian, Eva Ruth. Chemistry diploma, Eidgenoessische Technische Hochschule, 1958, Dr. Sci. Tech., 1962. Rsch. assoc. Max-Planck Inst., Göttingen, Germany, 1963-64, Mich. State U., East Lansing, 1964-65; rsch. est. chemistry dept. Swiss Federal Inst. Tech., Zurich, 1966-69, asst. prof., 1970-75, assoc. prof., 1976—. Recipient Ruzicka prize Schweizer Schulrat, 1970. Evangelical. Achievements include research in field of kinetics and thermodynamics of reactions of metal complexes in solutions, proton-transfer reactions. Home: Hasenweg 6, CH-8606 Greifensee Switzerland Office: ETH Zentrum, Universitaetsstr 6, CH-8092 Zurich Switzerland

GEIGER, DANIEL JAY, mining engineer; b. Canton, Ohio, Oct. 11, 1949; s. Harold Hoover and Helen (Musser) G.; m. Robin Tannenbaum, Dec. 24, 1971; children: Stuart, Michael. BSCE, Ohio U., 1972. Registered profl. engr., Ky., W.Va., Tenn. Hwy. engr. W.Va. Hwy. Dept., Charleston, 1972-78; chief engr. Cedar Coal Co., Charleston, 1978-82; v.p. engring. Transco Coal Co., London, Ky., 1982—. Asst. scout leader Boy Scouts of Am., Corbin, Ky., 1985—. Mem. NSPE (chpt. pres. 1984-92, Achievement in Mining award 1992), Soc. Mining Engrs. Republican. Home: 186 Whirlaway Trail Corbin KY 40701 8518 Office: Interstate Coal Co 100 Coal Dr London KY 40741

GEIGER, LOUIS CHARLES, electrical engineer, retired; b. Fillmore City, Minn., Feb. 6, 1921; s. Charles Frederick and Lorena A. (Irle) G.; m. Mary E. Wade, Aug. 12, 1967. BEE, Marquette U., Milw., 1946. Tech. staff person Bell Tel. Labs., Murray Hill, N.J., 1951-52; engring. asst., asst. engr. Wis. Tel., Milw., 1946-50, engr., 1953-82. Scoutmaster Boy Scouts Am., Summit, N.J., 1952. With USN, 1942-45, PTO. Mem. NSPE, Am. Interprofl. Inst. (Milw. chpt. pres. 1966), Inst. Electric and Electron Engrs., Engrs. and Scientists. of Milw. (dir. 1978-81), Wis. Soc. Profl. Engrs. (pres. 1968). Presbyterian.

GEIJO, FERNANDO ANTONIO, chemist; b. Barcelona, Spain, Jan. 10, 1955; s. Juan J. and M. Dolores (Caballero) G.; m. Luisa Lopez, June 26, 1982; 1 child, Alda. BS, Barcelona U., 1977; PhD in Organic Chemistry, 1985. Investigation fellowship Barcelona U., 1979-82, asst. prof., 1981-84; investigation scientist Bosch-Gimpera Found., Barcelona, 1984-85; chemist S.A. Lasa Labs., Barcelona, 1986-87, head of organic synthesis, 1988-91, quality assurance and gen. mgr., 1991—; scis. educator Liceo Ortega, Barcelona, 1988-89; scientific collaborator Barcelona U., 1987-90. Contbr. articles to profl. jours. including Israel Jour. Chem., Tetrahedron Letters, Jour Heterocyclic Chem., Hetrocycles. Treas. St. Martin Chess Club, Barcelona, 1977-82. Grantee Nat. Inst. of Social Security, 1969-77, Edn. and Svcs. Ministry, 1979-82. Mem. Official Chemists Coll., Spanish Assn. of Therapeutic Chemistry, Am. Chem. Soc., Spanish Quality Assurance Soc. Achievements include 20 patents for new phamaceutical products; discover of new reactions; rsch. on theoretical chemistry. Home: Maladeta 47, Vallirana E-08759 Barcelona Spain Office: SA Lasa Labs, Ctra-Laurea Miro 395, Sant Feliu Llobregat E-08980 Barcelona Spain

GEISE, HARRY FREMONT, retired meteorologist; b. Oak Park, Ill., Jan. 8, 1920; student U. Chgo., 1938-39, Meterorol. Service Sch., Lakehurst, N.J., 1943-44; m. Juanita Calmer, 1974; children: Barry, Gary, Harry (triplets); children by previous marriage: Marian Frances, Gloria Tara. Pioneered in extending pvt. weather svcs. in Chgo., 1937; chief meteorologist Kingsbury Ordnance, 1943; meteorologist radio sta. WLS, and Prairie Farmer Newspaper, 1941, 42, 46; asso. Dr. Irving P. Krick, metorol. cons., 1947-49; Army Air Corps research, 1948-49, developed new temperature forecasting technique; condr. weather and travel shows WBKB-TV, Chgo., also radio sta. WOPA, Oak Park, 1950-51; developed radio and television shows, San Francisco and San Jose, Cal., 1954-55; dir. media div. Irving P. Krick Assos., 1955-59; produced, appeared on weather programs Columbia Pacific Radio and TV Networks, also weatherman KNXT, Hollywood, Calif., 1957-58; comml. weather svc., 1962-80; instr. meteorology Santa Rosa Jr. Coll., 1964-66, Sonoma State Coll., 1967-68; weather dir. WCBS-TV, 1966-67, established weather ctr. for CBS, N.Y., 1966-67; produced 1 million weather forecasts, numerous programs for radio and TV. Research relationship between specified solar emission and major change in earth's weather patterns, tornado forecasting and long-range forecasting up to 4 years in advance. Meteorologist, Nat. Def. Exec. Res., 1968-74. Served with USMC, 1944-45. Recipient 1st Calif. Teaching Credential for Eminence in Meteorology, 1964. Mem. Royal Meterol. Soc. (life fgn. mem.). Author articles in field, contbr. to newspapers and mags. Contbr. long range forecasts. Home: 4585 Brighton Pl Santa Maria CA 93455-4252

GEISE, RICHARD ALLEN, medical physicist; b. Watertown, Wis., Nov. 27, 1945; s. Walter George and Verona (Zier) G.; m. Suzanne Arlene Johnson, Jan. 25, 1969; children: Peter Matthew, Caroline Suzanne. BEd, U. Wis., Whitewater, 1969; MS, U. Wis., 1973, U. Wis., 1976; PhD, U. Minn., 1991. Diplomate Am. Bd. Radiology in Radiol. Physics. Specialist, tech. instr. physics dept. U. Wis., Madison, 1969-76; med. physicist Rocky Mountain Med. Physics, Inc., Littleton, Colo., 1976-77, Meth. Hosp., Saint Louis Park, Minn., 1977-82, Midwest Radiation Consultants, Inc., Shoreview, Minn., 1978-85; med. physicist, instr. radiology dept. U. Minn., Mpls., 1985-92, asst. prof., med. physicist radiology dept., 1992—. Contbr. articles to profl. jours. Mem. Am. Assn. Physicists in Medicine, Am. Coll. Radiology, Health Physics Soc. Achievements include development of metal foil dosimeters for testing of shock wave lithotripters. Office: Radiology Box 292 UMHC 420 Delaware St SE Minneapolis MN 55455

GEISEL, CHARLES EDWARD, industrial engineer, consultant; b. St. Louis, Dec. 14, 1927; s. Gustav George and Marie (Dresch) G.; m. Marlene Helen Brom, Feb. 14, 1954; children: Laurie, Jane, Karl. BS in Indsl. Engring., Washington U., St. Louis, 1950. Registered profl. engr., Mo. Supr., then plant supt. Union Camp Corp., St. Louis, 1954-63; gen. mgr., then regional indsl. engr. Container Corp. Am., St. Louis, 1963-71; corp. mgr. indsl. engring. Container Corp. Am., Chgo., 1971-86; sr. staff engr. Jefferson Smurfit Corp., Carol Stream, Ill., 1986-90; pres. Simplified Systems, Inc., Naperville, Ill., 1990—. Author: Statistical Process Control, 1988; contbr. to profl. publs. including Nat. Safety News. 1st lt. U.S. Army, 1951-53. Mem. TAPPI (chair indsl. engring. 1983-85, Beloit Engring. award 1992), Inst. Indsl. Engring. (sr. mem., local v.p. 1965-66), Sigma Xi. Achievements include patent for caulking cartridge filling and seaming machine, development of approach to application of biomechanics, development of method of teaching statistical process control, development of equipment reliability system. Office: Simplified Systems Inc 1713 Towpath St Naperville IL 60565

GEISLER, LINUS SEBASTIAN, physician, educator; b. Vyskovce, Oct. 7, 1934; s. Linus and Hilde (Schaeffer) G.; children: Michael, Claudius. MD, Univ., Heidelberg, Germany, 1959. Asst. Univ., Heidelberg, Germany, 1962-67; prof. Univ., Giessen, Germany, 1967-73, Bonn, Germany, 1973-76; med. supt. St. Barbara Hosp., Gladbeck, 1976—; founder German League Against Respiratory Diseases, 1979. Author: Arztund Patient ImGesprach, 1987, others; contbr. about 200 articles to profl. jours. Avocations: communication research, writing, photography. Office: Saint Barbara Hosp, Barbarastr 1, 45964 Gladbeck Germany

GEIWITZ, (PETER) JAMES, psychologist, writer, researcher; b. Minneota, Minn., June 9, 1938; s. Peter H. and Hansina T. (Johanson) G.; m. Judith Haefele, 1964 (div. 1968); 1 child, Charles Paul; m. Roberta Klatzky, Dec. 1, 1972. BA, St. Olaf Coll., 1960; PhD, U. Mich., 1964. Rsch. assoc. U. Mich., Ann Arbor, 1964-65; asst. prof. Stanford U., Palo Alto, Calif., 1965-69; writer pvt. practice, Santa Barbara, Calif., 1969-85; rschr. Anacapa Scis., Santa Barbara, 1985-93; v.p. Advanced Scientific Concepts, Pittsburgh, 1993—; dir. Santa Barbara Breast Cancer Inst.. 1992—. Author: Approaches to Personality, 1979, Psychology: Looking at Ourselves, 1980, Adult Development and Aging, 1982, Guidebook for Maintenance Proficiency Testing, 1989, Knowledge Acquisition Guidebook, 1992. Mem. ACLU, Zero Population Growth, Drug Policy Found. Mem. Am. Psychol. Soc., Human Factors and Ergonomics Soc. Home: 5477 Aylesboro Ave Pittsburgh PA 15217

GELB, ARTHUR, business executive, electrical and systems engineer; b. N.Y.C., Sept. 20, 1937; m. Linda Lewis; children: Ronald, Caren, Laurie. BEE, CUNY, 1958; MS in Applied Math., Harvard U., 1959; ScD in Systems Engring., MIT, 1961. Engr. Aviation Gas Turbine div. Westinghouse Electric Corp., Kansas City, Mo., 1956, Am. Dist. Telegraph Co., N.Y.C., 1957-58, Draper Lab., Cambridge, Mass., 1959; dept. mgr. Dynamics Research Corp., Stoneham, Mass., 1961-66; pres., chief exec. officer TASC (The Analytic Sciences Corp.), Reading, Mass., 1966-93, chmn., 1993—; co-chmn. tech. and policy program rev. com. MIT, 1987; chmn. adv. bd. Ctr. for Tech., Policy and Indsl. Devel., MIT, 1987. Co-author: Multiple-Input Describing Fns., 1968, Applied Optimal Estimation, 1974; contbr. articles to profl. jours. Bd. dirs. Massport, Boston, 1977-85; bd. regents Higher Edn., Mass., 1989-90; mem. Higher Edn. Coord. Coun., Mass., 1990—. Named Outstanding Young Engr. CUNY, 1969. Fellow AIAA, IEEE (bd. editors Control Systems Mag. 1981—); mem. Mensa. Avocations: music, tennis, golf, microcomputing, math. Office: TASC 55 Walkers Brook Dr Reading MA 01867-3238

GELB, MICHAEL H., chemistry educator. Prof. chemistry U. Wash., Seattle. Recipient Pfizer Enzyme Chemistry award Am. Chem. Soc., 1993. Office: Univ of Washington Dept of Chemistry Seattle WA 98195*

GELB, RICHARD LEE, pharmaceutical corporation executive; b. N.Y.C., June 8, 1924; s. Lawrence M. and Joan F. (Bove) G.; m. Phyllis L. Nason, May 5, 1951; children: Lawrence N., Lucy G., Jane E., James M. Student, Phillips Acad., 1938-41; B.A., Yale, 1945; M.B.A. with Distinction, Harvard U., 1950. Joined Clairol, Inc., N.Y.C., 1950, pres., 1959-64; exec. v.p. Bristol-Myers Co., 1965-67, pres., 1967-76, chief exec. officer, 1972—, chmn. bd., 1976—; bd. dirs. N.Y. Times Co., N.Y. Life Ins. Co.; policy com. Bus. Roundtable; mem. Bus. Coun.; trustee Com. for Econ. Devel.; mem. Conf. Bd.; ptnr. N.Y.C. Partnerships, Inc. Charter trustee Phillips Acad., Andover; dir. Lincoln Ctr. for Performing Arts; mem., former dir. Coun. Fgn. Rels.; vice-chmn. bd. overseers, bd. mgrs. Meml. Sloan-Kettering Cancer Ctr.; chmn. bd. mgrs. Sloan Kettering Inst. Cancer Rsch.; vice-chmn., trustee N.Y.C. Police Found.; trustee N.Y. Racing Assn. Home: 1060 5th Ave New York NY 10128-0104 Office: Bristol-Myers Squibb Co 345 Park Ave New York NY 10154-0004

GELBART, ABE, mathematician, educator; b. Paterson, N.J., Dec. 22, 1911; s. Wolf and Pauline (Landau) G.; m. Sara Goodman, July 2, 1939 (dec. Nov. 23, 1988); children: Carol Marie (Mrs. Ivan P. Auer), Judith Sylvia (dec.), William Michael, Stephen Samuel; m. Mona Siegel, Mar. 4, 1990. B.Sc., Dalhousie U., 1938, LL.D. honoris causa, 1972; Ph.D. in Math, MIT, 1940; D.Sc. (h.c.), Bar-Ilan U., Israel, 1985. Asst. MIT, 1938-40; instr. math. N.C. State Coll., 1940-42; research asso. Brown U., 1942; asso. physicist NACA, Langley Field, Va., 1942-43; asst. prof. to prof. math. Syracuse U., 1943-58; dir. Inst. Math., Yeshiva U., 1958-59; dean Belfer Grad. Sch. Sci., 1959-70, dean emeritus, 1970—, disting. univ. prof. math., 1968—; vis. disting. prof. math. Bard Coll. and fellow Bard Coll. Center, 1979—, David and Rosalie Rose Disting. prof. natural sci. and math., 1983—; lectr., Sorbonne, Paris, 1949; vis. prof. U. So. Calif., 1951; mem. Inst. Advanced Study, Princeton, 1947-48, 77-81; Fulbright lectr., Norway, 1951-52; mem. directorate math. scis. USAF Office Sci. Research.; vice chmn. bd. dirs., chmn. sci. adv. bd. Daltex Med. Scis., Inc., 1983—; mem. adv. bd. Inst. for Thinking and Learning, Pace U., 1982—; founding dir. series, lectures Bard Coll. Scis., 1979—. Editor: Scripta Mathematica, 1957—; co-developer theory of pseudo-analytic functions. Trustee, chmn. acad. sci. com. Bar-Ilan U., Israel, 1982—. Recipient Bard medal, 1981; spl. award of recognition U. Pa. Sch. Nursing; chair in math. named in his honor Bar-Ilan U., 1983; Internat. Rsch. Inst. Math. Scis. Bar-Ilan U. renamed Gelbart Internat. Rsch. Inst. Math. Scis., 1990; appointed fgn. mem. Acad. Tech. Scis. Russian Fedn., 1992. Mem. Am. Math. Soc., Math. Assn. Am., Acad. Ind. Scholars (trustee 1982—), Russian Acad. Tech. Sci. (fgn.), City Athletic Club, Cosmos Club, Sigma Xi. Home and Office: 242 E 72nd St New York NY 10021

GELBOIN, HARRY VICTOR, biochemistry educator, researcher; b. Chgo., Dec. 21, 1929; s. Herman and Eva (Jurkowsky) G.; m. Marlena Maisels, Apr. 1, 1962; children: Michele Ida, Lisa Rebecca, Sharon Anna, Tamara Rachel. BA in Chemistry, U. Ill., 1951; MS in Chemistry and Oncology, U. Wis., 1956, PhD in Chemistry and Oncology, 1958; PhD (hon.), U. Inonu, Molaty, Turkey. Devel. chemist U.S. Rubber Co., Chgo., 1952-54; rsch. asst. McArdle Meml. Lab. for Cancer Rsch., U. Wis., 1954-58; biochemist lab. cellular pharmacology NIMH, 1958-60, biochemist lab. clin. sci., 1960-61; supervisory biochemist chemistry sect., diagnostic rsch. br. Nat. Cancer Inst., 1962-64, head chemistry sect., carcinogenesis studies br., 1964-66, chief lab. molecular carcinogenesis, div. cancer etiology, 1966—; adj. prof. Georgetown U., 1974-78; vis. prof. Hebrew U., Jerusalem, 1985-86; Keynote speaker Carcinogenesis, Gordon Res. Conf., 1965; Franz Bielschowsky Meml. lectr., Dunedin, New Zealand, 1966; Smith Kline French hon. lectr. U. Fla., 1974, U. Mich., 1976; hon. lectr. Israel Cancer Soc. and. U. Tel Aviv, Israel, 1983; Keynote lectr. Internat. Conf. Carcinogenesis, Alghero, Italy, 1986; Nakasone hon. lectr. Japan Found. Promotion Sci., Tokyo, Osako, 1989;keynote speaker U.S organizer and co-chmn. Princess Takamatsu Cancer Symposium, Tokyo, 1990; also speaker, lectr. various internat. sci. meetings, confs., symposiums in N.Am., Eng., Italy, Israel, Turkey, Brazil, Japan, China, New Zealand, etc. Editor 8 profl. books; assoc. editor. Cancer Rsch., 1968-79, 83-87, mem. editorial adv. bd., 1965-67; assoc. editor Biochem. Toxicology, 1984—; mem. editorial bd. Chemico-Biol. Interactions, 1969-75, Archives Biochemistry and Biophysics, 1969-76, Life Scis., 1976, Environ. Health Scis., 1976-78; contbr. and co-contbr. over 285 sci. papers to med. publs. Recipient Superior Svc. award NIH, 1970, Claude Bernard award U. Montreal, 1970, New Horizons award Radiol. Soc. N.Am., 1970, Merit awards Sr. Sci. Svc. NIH, 1983, 85, EEO award NIH, 1989. Mem. AAAS, Am. Assn. for Cancer Rsch., Am. Cancer Soc. (adv. com. on carcinogenesis, mem. coun. 1975—), Am. Soc. Biol. Chemists, Am. Soc. for Pharmacology and Exptl. Therapeutics, Internat. Soc. for Preventive Oncology, Internat. Soc. for Study Xenobiotics. Achievements include rsch. on molecular mechanism chem. carcinogenesis, enzyme regulation, genetics of metabolism and activation of drugs and environ. agts., toxicology of xenobiotics, biochem. individuality in carcinogenesis and drug metabolism. Office: Nat Insts Health Bethesda MD 20014

GELEHRTER, THOMAS DAVID, medical and genetics educator, physician; b. Liberec, Czechoslovakia, Mar. 11, 1936; married 1959; 2 children. BA, Oberlin Coll., 1957; MA, U. Oxford, Eng., 1959; MD, Harvard U., 1963. Intern, then asst. resident in internal medicine Mass. Gen. Hosp., Boston, 1963-65; rsch. assoc. in molecular biology NIAMD NIH, Bethesda, Md., 1965-69; fellow in med. genetics U. Wash., 1969-70; asst. prof. human genetics, internal medicine and pediatrics Sch. Medicine Yale U., 1970-73, assoc. prof., 1973-74; assoc. prof. U. Mich., Ann Arbor, 1974-76; prof. internal medicine and human genetics U. Mich., 1976-87, dir. div. med. genetics, 1977-87, chmn. dept. human genetics, prof. human genetics and internal medicine, 1987—; Josiah Macy Jr. Found. faculty scholar and vis. scientist Imperial Cancer Rsch. Fund Labs., London, 1979-80. Trustee Oberlin Coll., 1970-75. Rhodes scholar, 1957-59. Fellow Am. Coll. Med. Genetics; mem. Am. Soc. Human Genetics, Am. Soc. Clin. Investigation, Am. Soc. Biochemistry and Molecular Biology. Office: U Mich Med Sch Dept Human Genetics Box 0618 1500 E Medical Center Dr Ann Arbor MI 48109-0618

GELERNT, IRWIN M., surgeon, educator; b. N.Y.C., Sept. 27, 1935; s. Lipman and Ray (Samuels) G.; married, June 11, 1960; children: Lee, Alicia, Michelle. BS, CCNY, 1957; MD, SUNY, N.Y.C., 1961. Diplomate Am. Bd. Surgery. Intern Bellevue Hosp. Cornel Med. Svc.; attending surgeon Mt. Sinai Hosp., N.Y.C., 1962-67, pres. attending staff, 1985-87; clin. prof. of surgery Mt. Sinai Sch. Medicine, N.Y.C., 1987—. Contbr. articles to profl. publs., chpts. to books. Trustee Manhattan Country Sch., N.Y.C., 1978-84. Named Physician of Yr., Nurses Assn. of Mt. Sinai Hosp., 1991. Fellow ACS, Am. Coll. Gastroenterology; mem. Found. Ileitis and Colitis (Man of Yr. 1983), Phi Beta Kappa, Alpha Omega Alpha. Office: 25 E 69th St New York NY 10021-4925

GELINAS, PAUL JOSEPH, psychologist, author; b. Woonosket, R.I., July 17, 1914; s. Edmund J. and Marianne (Desaultnier) G.; m. Eva J. MacFarlane, 1935; 1 child, Robert P. B.A., Acadia U., 1933; M.A., Columbia U., 1954; M.S., CCNY, 1953; Ed.D., NYU, 1955. Supt. schs., Setauket, N.Y., 1950-70; cons., Halifax, N.S., 1967-70; pvt. practice psychology, Setauket, N.Y., 1970—; cons. Tax Action Group, Setauket, 1970-74. Author: History for Young Readers, 1967 (Book of the Month bonus selection), So You Want to Be a Teacher, 1967, Teenagers Can Get Good Jobs, 1970, 71, Coping With Anger, 1972, Coping With Fears, 1972, Coping With Loneliness, 1980, Coping With Emotions, 1981. Receiver taxes Town of Brookhaven, County of Suffolk, 1978-82; v.p. Setauket Civic Assn., 1970-72; mem. bd. edn., Setauket, 1982-84; bd. dirs. Coram Mediation, N.Y., 1982—. Served with USNG, 1930. Recipient Citations, C. of C., Setauket, 1966, 67. Mem. Am. Psychol. Assn., Am. Assn. Marriage and Family Therapists, N.Y. State Psychol. Assn., Suffolk County Psychol. Assn. Presbyterian. Lodge: Lions (pres. 1960-62). Avocation: business writing. Home: 31 W Meadow Rd Setauket NY 11733-2228

GELLER, GARY NEIL, systems engineer; b. Brookline, Mass., Oct. 14, 1954; s. Louis Maxwell and Harmona Fairlith (Jones) G. BS, Union Coll., 1976; MS, U. Wyoming, 1980; PhD, UCLA, 1986. Programmer/analyst Jet Propulsion Lab., Pasadena, Calif., 1986-88, subsystem lead, 1988-92; ASTER system engr. Jet Propulsion Lab., 1992—. Contbr. to profl. publs. Co-founder, pres. Equestrian Safety Com., Glendale, Calif., 1987—. Grantee NSF, 1984. Mem. Ecol. Soc. Am., Am. Inst. Biol. Sci. Achievements include advancements in computer modelling of plant architecture and soils, development of accurate model of light interception by cacti. Home: 1550 Randall St Glendale CA 91201 Office: Jet Propulsion Lab MS 169-315 4800 Oak Grove Dr Pasadena CA 91109

GELLER, HAROLD ARTHUR, earth and space sciences executive; b. Bklyn., June 14, 1954; s. Morris and Minnie (Kaplan) G. BS, SUNY, Albany, 1983; MA, George Mason Univ., 1992. Rsch. asst. SUNY at Downstate Med., Bklyn., 1972-74; rsch. asst. CUNY at Bklyn. Coll., 1974-75; engring. aide FBI, Washington, 1977-78; lab. supr. ENSCO Inc., Springfield, Va., 1978-80; assoc. engr. Def. Systems Inc., McLean, Va., 1980-83; staff scientist/systems engr. Sci. Applications Internat. Corp., McLean, 1983-87; systems engr. Grumman Aerospace, Reston, Va., 1987-88, Sci. Applications Internat. Corp., McLean, 1988-90; rsch. asst. Naval Rsch. Lab., George Mason U., 1990-91; project mgr. Rsch. and Data Systems Corp., Greenbelt, Md., 1991-92; dep. dir. Washington ops. Consortium Internat. Earth Sci. Info. Network, Washington, 1992—; instr. physics and astronomy George Mason U., 1993—; computer cons., Burke, Va., 1986-87. Commonwealth fellow, 1992-93. Mem. AIAA (chmn. corp. liaison com. 1989-90, chmn. pub. affairs com. 1990-91), Am. Astron. Soc., Am. Geophys. Union, AAAS. Democrat. Jewish. Office: Consortium Internat Earth Sci Info Network 1825 K St N W Washington DC 20006

GELLER, MARGARET JOAN, astrophysicist, educator; b. Ithaca, N.Y., Dec. 8, 1947; d. Seymour and Sarah (Levine) Geller. A.B., U. Calif.-Berkeley, 1970; M.A., Princeton U., 1972, Ph.D., 1975. Research fellow Center for Astrophysics, Cambridge, Mass., 1974-78; research assoc. Harvard Coll. Obs., Cambridge, 1978-80; sr. vis. fellow Inst. of Astronomy, Cambridge, Eng., 1978-82; asst. prof. Harvard U., 1980-83; prof. astronomy Harvard U., 1988—; astrophysicist Smithsonian Astrophys. Obs., Cambridge, 1983—. Contbr. articles to profl. jours. Bd. reviewing editors Science, 1991—. NSF fellow, 1970-73, MacArthur Found. fellow, 1990—; recipient Newcomb-Cleve. prize, 1989-90. Mem. NAS, Am. Acad. Art & Scis., Am. Astron. Soc. (councillor), Assoc. Univs. for Research in Astronomy (dir-at-large), AAAS, Internat. Astron. Union. Office: Harvard U Ctr for Astrophysics 60 Garden St Cambridge MA 02138-1596

GELLER, RONALD GENE, health administrator; b. Peoria, Ill., Jan. 15, 1943; s. Harold H. and Rose G.; m. Lois S. Geller, Sept. 5, 1971; children—Andrea, Steven, Lauren. B.S. in Zoology, U. Wis., 1964, Ph.D. in Physiology, 1969. Spl. research fellow Nat. Heart Inst. NIH, Bethesda, Md., 1969-71, sr. staff fellow Nat. Heart, Lung and Blood Inst., 1971-72, grants assoc., 1972-73, asst. chief, chief hypertension and kidney diseases br. Nat. Heart, Lung and Blood Inst., 1973-78, assoc. dir. extramural and collaborative programs Nat. Eye Inst., 1978-86, acting dir. div. program analysis Office Program Planning and Eval., 1986, dir. div. program analysis, 1987-89, dir. div. extramural affairs Nat. Heart, Lung and Blood Inst., 1989—; instr. Found. for Advanced Edn. in Sci.; USPHS trainee, 1966-67. Contbr. articles to profl. jours. Wis. Heart Assn. fellow, 1967-69. Mem. Am. Heart Assn. (mem. med. adv. bd. council for high blood pressure research), Am. Physiol. Soc., Soc. Exptl. Biology and Medicine. Home: 14960 Dufief Dr North Potomac MD 20878 Office: NIH Westwood Bldg Rm 7A17 Bethesda MD 20892

GELLMAN, ISAIAH, association executive; b. Akron, Ohio, Feb. 19, 1928; s. Meyer and Pearl (Milker) F.; m. Lola Malkis, Dec. 27, 1947; children: Paula, Judith. B in Chem. Engring., CCNY, 1947; MS, Rutgers U., 1950, PhD, 1952. Rsch. assoc. Rutgers U., 1948-52; process engr. Abbott Labs., 1952-56; with Nat. Coun. Paper Industry for Air and Stream Improvement Inc., N.Y.C., 1956—, tech. dir., 1969-77, exec. v.p., 1977-87, pres., 1987—; lectr. Johns Hopkins U., 1961-65. NIH fellow, 1948-52. Fellow TAPPI; mem. Water Pollution Control Fedn., Air Pollution Control Assn., Am. Inst. Chem. Engrs., N.Y. Acad. Scis., Sigma Xi. Office: Nat Coun Paper Industry Air and Stream Improvement 260 Madison Ave New York NY 10016-2401

GELL-MANN, MURRAY, theoretical physicist, educator; b. N.Y.C., Sept. 15, 1929; s. Arthur and Pauline (Reichstein) Gell-M.; m. J. Margaret Dow, Apr. 19, 1955 (dec. 1981); children: Elizabeth, Nicholas; m. Marcia Southwick, June 20, 1992; 1 stepson, Nicholas Levis. BS, Yale U., 1948; PhD, Mass. Inst. Tech., 1951; ScD (hon.), Yale U., 1959, U. Chgo., 1967, U. Ill., 1968, Wesleyan U., 1968, U. Turin, Italy, 1969, U. Utah, 1970, Columbia U., 1977, Cambridge U., 1980; D (hon.), Oxford (Eng.) U., 1992. Mem. Inst. for Advanced Study, 1951, 55, 67-68; instr. U. Chgo., 1952-53, asst. prof., 1953-54, assoc. prof., 1954; assoc. prof. Calif. Inst. Tech., Pasadena, 1955-56; prof. Calif. Inst. Tech., 1956—, now R.A. Millikan prof. physics; vis. prof. MIT, spring 1963, CERN, Geneva, 1971-72, 79-80; Mem. Pres.'s Sci. Adv. Com., 1969-72; mem. sci. and grants com. Leakey Found., 1977—; chmn. bd. trustees Aspen Ctr. for Physics, 1973-79; founding trustee Santa Fe Inst., 1982, chmn. bd. trustees, 1982-85, co-chmn. sci. bd. 1985—. Author: (with Y. Ne'eman) Eightfold Way. Regent Smithsonian Instn., 1974-88; bd. dirs. J.D. and C.T. MacArthur Found., 1979—. NSF post doctoral fellow, vis. prof. Coll. de France and U. Paris, 1959-60; recipient Dannie Heineman prize Am. Phys. Soc., 1959; E.O. Lawrence Meml. award AEC, 1966; Overseas fellow Churchill Coll., Cambridge, Eng., 1966; Franklin medal, 1967; Carty medal Nat. Acad. Scis., 1968; Research Corp. award, 1969; named to UN Environ. Program Roll of Honor for Environ. Achievement, 1988; Nobel prize in physics, 1969. Fellow Am. Phys. Soc.; mem. NAS, Royal Soc. (fgn.), Am. Acad. Arts and Scis. (v.p., chmn. Western ctr. 1970-76), Council on Fgn. Relations, French Phys. Soc. (hon.). Clubs: Cosmos (Washington); Century Assn., Explorers (N.Y.C.); Athenaeum (Pasadena). Office: Calif Inst Tech Dept Physics Pasadena CA 91125

GELSTON, JOHN HERBERT, electrical engineer; b. Mineola, N.Y., Mar. 12, 1949; s. Herbert John and Constance Joyce (Barley) G.; m. Jane Blair, June 2, 1973; children: Jennifer Barley, Kerry Anne. BSEE, Northeastern U., 1972. Sr. elec. engr. Bechtel Power Corp., Gaithersburg, Md., 1972-75; train control engr. Bechtel Assocs. Profl. Corp., Washington, 1975-77; sr. elec. engr. R.M. Parsons Co., Washington, 1977-79; prin. elec. engr. Stone and Webster Engring. Corp., Cherry Hill, N.J., 1979-85; dir. engring. svcs. Nutech Engrs., Inc., Chgo., 1985-88; project mgr. Fluor Daniel, Inc., Chgo., 1988—. Mem. IEEE, Am. Nuclear Soc. Episcopalian. Home: 1125 Barcroft Ct Naperville IL 60540 Office: Fluor Daniel Inc 200 W Monroe St Chicago IL 60606

GEMINN, WALTER LAWRENCE, JR., computer programmer, analyst; b. Atlanta, Dec. 12, 1959; s. Walter Lawrence and Mildred E. (Walter) G.; m. Alice Marie Hotter, June 6, 1981; children: Wesley Louis, Adam Mitchell, Shelby Marie. AS, Belleville (Ill.) Area Coll., 1979; BS, So. Ill. U., 1981. Programmer/analyst Computer Graphics Labs., Inc., Glen Cove, N.Y., 1981-85; sr. programmer Crosfield Electronics, Inc., Glen Rock, N.J., 1985-87, Dicomed, Inc., Burnsville, Minn., 1987-90; prin. engr. Dicaned Inc., Burnsville, Minn., 1990-92, engring. mgr. workstations, 1992—. Contbr. articles to profl. jours. Mem. ACM. Achievements include video, prepress, graphic arts digital workstation design and development. Office: Dicomed Inc 1401 Rupp Dr Burnsville MN 55337

GEMSA, DIETHARD, immunologist; b. Berlin, Aug. 9, 1937; s. Hans and Hedwig (Reinhard) G.; m. Inken Fischer, Apr. 11, 1968; children: Jan Ulrich, Kerstin Friederike, Meike Charlotte. MD, U. Freiburg, 1968; Pvt. Dozent, U. Heidelberg, 1974. Rsch. fellow U. Wash., Seattle, 1965-67; resident U. Mainz, Germany, 1968-70; rsch. assoc. in internal medicine U. Calif., San Francisco, 1970-73; asst. prof. Inst. Immunology, U. Heidelberg, 1974-82; assoc. prof. Med. Sch. Hannover, 1983-84; prof. head Inst. Immunology, Philipps U., Marburg, 1985—. Editor German Textbook of Immunology; contbr. over 180 articles to med. jours.; editor-in-chief Immunobiology, 1978—. Mem. Am. Assn. Immunologists, German Soc. Immunology (councillor 1980—), German Soc. Cancer Rsch. Liberal Party. Roman Catholic. Avocations: painting, ecology, long-distance running. Office: Inst Immunology, U Marburg, Robert-Koch-Str 17, Marburg D-35037, Germany

GENARO, DONALD MICHAEL, industrial designer; b. Hoboken, N.J., Feb. 22, 1932; s. Gustav G. and Margaret (DeMave) G.; m. Margaret Hermes, June 23, 1956; children: Susan, Karen. BID, Pratt Inst., 1957. Archtl. designer F.W. Fisher-Architects, N.J. and N.Y., 1951-52; indsl. designer Henry Dreyfuss Assocs., N.Y.C., 1957-63, assoc., 1963-68, ptnr., 1968-82, sr. ptnr., 1982—; cons. AT&T, Bell Labs., John Deere, Polaroid, and various others; lectr. on design, 1962—. Designer of Trimline Phone; holder over 200 patents; contbr. numerous articles to profl. jours. Trustee, vice chmn., bd. dirs. Pascack Valley Hosp. Represented in permanent collection at Mus. of Modern Art and Cooper-Hewitt (Smithsonian) Museum; recipient Contemporary Achievement award Pratt Inst., 1970, Best Product Design 1983 Time Mag., several design awards from Indsl. Designers Soc. of Am. and Indsl. Design Mag.; named one of 25 Best Designed Products Fortune Mag., 1977. Mem. Indsl. Designers Soc. Am. Office: Henry Dreyfuss Assocs 423 W 55th St New York NY 10019-4460

GENDELMAN, HOWARD ELIOT, biomedical researcher, physician; b. Phila., Mar. 18, 1954; s. Seymour and Soffia (Raphael) G.; m. Bonnie Rae Bloch, June 15, 1980; children: Lesley, Sierra, Adam. BS, Muhlenberg Coll., 1975; MD, Pa. State U., 1979. Rsch. assoc. Pa. State U., Hershey, 1978-79; med. resident in internal medicine Montefiore Hosp. Ctr. Albert Einstein Coll. Medicine, Bronx, 1979-82; clin. and rsch. fellow depts. neurology and medicine Med. Ctr. Johns Hopkin's U., Balt., 1982-85, asst. prof. div. infectious diseases Med. Sch., 1985-89; staff physician in infectious diseases Walter Reed Army Inst. Rsch., Washington, 1993—; prof., chief Lab. Viral Pathogenesis Univ. Nebr. Med. Ctr., Omaha, Nebr., 1985-87; spl. expert Lab. Molecular Microbiology Sect. Biochem. Virology, NIH, Bethesda, 1985-87; prin. scientist Henry M. Jackson Found. for Advancement of Mil. Medicine Uniformed Svcs. Univ. Health Sci. Ctr., 1988-92, rsch. assoc. prof. dept. pathology, 1990-92; lectr. dept. infectious diseases the Johns. Hopkins U. Sch. Pub. Health, Balt., 1985-92; cons. Bethesda Rsch. Labs., Inc., Rockville, Md., 1987, Schering-Plough, Kenilworth, N.J., Applied Biotechs. Beltsville, Md., 1990, Viragen, 1990, others. Editorial reviewer Jour. Histochemistry and Cytochemistry, Jour. Virology, Am. Jour. Pathology, Jour. Clin. Investigation, Jour. Neuroimmunology, AIDS, Jour. Immunology, Gastroenterology, Lab Investigation, Jour. Infectious Disease, Reviews of Infectious Disease Sci.; contbr. articles to Jour. Immunology, Jour. Virology, Jour. Exptl. Medicine, Sci., Proc. Nat. Acad. Sci. (US) and numerous others. Capt. USAR, 1984-87, maj., 1987—. Named Carter Wallace fellow 1987—; grantee NIH, Amfar, and others. Mem. ACP, AMA, Am. Soc. Virology, Am. Soc. Microbiology, Reticuloendothelial Soc. Achievements include rsch. in cellular tropism and pathogenesis for the Human Immunodeficiency virus, the etiologic agt. of AIDS. Home: 125 S 127th St Omaha NE 68154 Office: Univ Nebr Med Ctr 600 So 42d St Omaha NE 68198

GENEGA, STANLEY G., career officer, federal agency administrator; married; children: Beth, Stan Jr. Grad., U.S. Mil. Acad., 1965; MS, MIT; postgrad., U. Bonn, Germany. Commd. 2d lt. U.S. Army, 1965, advanced through grades to major gen.; comdr. southwestern divsn. U.S. Army Corps of Engrs., Dallas; comdr. South Atlantic divsn. U.S. Army Corps of Engrs., Atlanta; dir. Civil Works Hdqs. U.S. Army Corps of Engrs., Washington, 1992—; prin. advisor nat. security affairs to Chmn. Joint Chiefs of Staff; comdr. Army Engr. Dist., Savannah, Ga., Engr. Battalion, Fort Polk, La., engr. cos. Vietnam and Fort Meade, Md. Olmsted scholar. Office: Dept of the Army Civil Works 20 Massachusetts Ave NW Washington DC 20314*

GENEL, MYRON, pediatrician, educator; b. York, Pa., Jan. 6, 1936; s. Victor and Florence (Mowitz) G.; m. Phyllis Norma Berkman, Aug. 25, 1968; children: Elizabeth, Jennifer, Abby. Grad., Moravian Coll., 1957; M.D., U. Pa., 1961; M.A. hon., Yale U., 1983. Diplomate: Am. Bd. Pediatrics. Intern Mt. Sinai Hosp., N.Y.C., 1961-62; resident in pediatrics Children's Hosp. Phila., 1962-64; trainee pediatric endocrinology Johns Hopkins Hosp., Balt., 1966-67; instr. pediatrics U. Pa. Sch. Medicine, 1967-69, assoc. in pediatrics, 1969-71; trainee in genetics, inherited metabolic diseases Children's Hosp. Phila., 1967-69, assoc. physician, 1969-71; attending physician Yale-New Haven Hosp., 1971—; mem. faculty Yale U. Sch. Medicine, New Haven, 1971—, dir. pediatric endocrinology, 1971-85, program dir. Children's Clin. Rsch. Ctr., 1971-86, prof., 1981—, assoc. dean, 1985—, dir. Office Govt. and Community Affairs, 1985—; mem. genetic adv. bd. State of Conn., 1979-82; cons. subcom. investigations, oversight com. sci. and tech. U.S. Ho. of Reps., 1982-84; mem. adv. bd. New Eng. Congenital Hypothyroidism Collaborative: cons. Newington Children's Hosp., Hosp. St Raphael, Milford Hosp., Norwalk Hosp., Stamford Hosp., Danbury Hosp.; chmn. trnsplant adv. com. Office of Commr., Conn. Dept. Income Maintenance, 1984-92; health policy fellowship bd. Inst. Medicine, 1989—. Contbr. articles to profl. jours. Served as capt. U.S. Army, 1964-66. Robert Wood Johnson Health Policy fellow Inst. Medicine NAS, Washington, 1982-83; recipient ann. award Conn. Campaign Against Cooley's Anemia, 1979, Ann. Comenius Alumni award Moravian Coll., 1990. Mem. AAAS, AMA, Am. Acad. Pediatrics (task force organ transplants, coun. on govt. affairs), Am. Coll. Nutrition, Am. Diabetes Assn. (co-recipient Jonathan May award 1979), Am. Fedn. Clin. Rsch., Am. Pub. Health Assn., Am. Pediatric Soc., Am. Soc. Bone and Mineral Rsch., Assn. Am. Med. Colls. (mem. adminstrv. bd. coun. acad. socs. 1987-92, chmn.-elect coun. acad. socs. 1989-90, chmn. coun. acad. socs. 1990-91, mem. exec. coun. 1989-92), Assn. Health Svc. Rsch., New Haven County Med. Assn. (bd. govs. 1990—), Assn. Program Dirs. (pres. elect 1980-81, pres. 1981-82), Nat. Assn. Biomed. Rsch. (bd. dirs. 1990-93, exec. com. 1991-93), Conn. Endocrine Soc., Conn. United for Rsch. Excellence (chmn. steering com. 1989-90, pres. 1990—), Endocrine Soc., Lawson Wilkins Pediatric Endocrine Soc., Soc. Pediatric Rsch., Conn. Acad. Sci. and Engring, Sigma Xi. Jewish. Home: 30 Richard Sweet Dr Woodbridge CT 06525-1126 Office: PO Box 3333 New Haven CT 06510-0333

GENET, JEAN PIERRE, chemistry educator, researcher; b. Tulle, France, May 3, 1942; s. Pierre and Anna (Teilhac) G.; m. Nelly Genet Keyser, July 15, 1967; children: Olivier, Remi. BS, Lycee Claude Bernard, 1962; MS, U. Paris, 1966, PhD, 1972. Asst. prof. U. Paris, 1966-79, prof., 1980-87; prof. Ecole Nat. Superieure de Chimie de Paris, 1988—. Contbr. 70 scientific papers and articles to profl. jours. Mem. Am. Chem. Soc., French Chem. Soc. (v.p. 1990, award 1990). Achievements include 12 patents in field. Office: École Nationale Superieure, Pierre Marie Curie, 75231 Paris France

GENETET, BERNARD, hematologist, immunologist, educator; b. Chaumont, Champagne, France, Dec. 17, 1931; s. René Emile and Marie Louise (Henry) G.; m. Francine Belleville, May 4, 1957 (div. Mar., 1971); children: Anne, Isabelle; m. Noëlle Pirron Apr. 25, 1972; 1 child, Julien. MD, U. Nancy, France, 1964, D in U. Scis., 1973. Chief of lab. Regional Blood Ctr., Nancy, France, 1962-73; dir. of lab. Regional Blood Ctr., Nancy, 1973; tchr. Medicine U. Nancy, 1964-73; tchr. Medicine U. Rennes, France, 1973-86, prof. Immunology, 1986; expert Cour Cassation (Paris); mem. Nat. Coun. Univs., French Ministry of Edn., Paris, 1987; expert in human biology, genetics, immunology French Ministry Health, Paris, 1989; auditor IHEDN; cons. Internat. Fed. Blood Donors, 1971; pres. XIV congress of French Transfusion, 1988. Author: La Transfusion, 1978 (French, Italian, spanish), Aide Memoire de transfusion, 1984 (French, Portuguese, Italian), 2d edit., 1991, Glossaire de Transfusion, 1987 (French, Italian, Spanish, Arabic, Neederland), Immunology, 1989 (French), Hematology, 1989 (French, Spanish), Transfusion in Europe, 1990. Active Vol. Donors Assn. Col. French Health Svc., 1959-61. Decorated with Legion d' Honneur, 1984; recipient Silver medal French Red Cross, 1971, Nat. Order of Merit, France, 1972, Nat. Order of Health, Tunisia. Mem. N.Y. Acad. Sci., Transplantation Soc., French Soc. Transfusion (v.p. 1986), French Sopc. Immunology, French Soc. Hematology, Assn. Immunologies Univs., Grand Orient of France (pres. of the convent, Paris 1971), J. Callot (venerable master Nancy 1969-71). Mem. Socialist Party. Avocations: climbing, flying, classical music. Home: Rue Baudelaire, Bretagne 35700 Rennes France Office: Regional Ctr Blood Transfusion, Pierre Jean Gineste, Bretagne 35016 Rennes France

GENIS, VLADIMIR I., electrical engineer; b. Kiev, Ukraine, June 24, 1946; came to U.S. 1989; s. Ivan Y. and Yevgenia P. (Lukatskaya) G.; m. Larisa M. Shukhman, Sept. 14, 1968; 1 child, Alexander Genis. MSEE, Kiev Poly. Inst., 1969. Project mgr. Kiev Poly. Inst., 1969-78; chief specialist Design Co., Kiev, 1978-88; rsch. engr. Drexel U., Phila., 1989--; councillor Design Co., Kiev, 1987-88. Mem. Acoustical Soc. Am. Achievements include invention certs. in electronics. Office: Drexel U 32 Chestnut St Philadelphia PA 19104

GENOVA, VINCENT JOSEPH, physicist; b. Johnson City, N.Y., Sept. 28, 1959; s. Benjamin Vincent and Rose Marie (Giallo) G. BS in Physics, SUNY, Binghamton, 1981, MEng, Cornell U., 1983. Assoc. engring. scientist IBM, East Fishkill, N.Y., 1984-86; sr. assoc. engring. scientist IBM, Owego, N.Y., 1987-89, staff engring. scientist, 1990—; interdivisional tech. liaison for thin films IBM Corp., Owego, 1990—. Contbr. articles to profl. jours. Cornell Engring. fellow, 1981. Mem. Am. Inst. Physics, Am. Vacuum Soc., Cornell Soc. Engrs., Sigma Pi Sigma. Republican. Roman Catholic. Achievements include patent for platinum arsenide resistor formation, reliable capacitor fabrication process; research on GaAs device physics processing, surface physics characterization. Office: IBM Corp MS 0302 RT 17C Owego NY 13827

GENOWAYS, HUGH HOWARD, museum director; b. Scottsbluff, Nebr., Dec. 24, 1940; s. Theodore Thompson and Sarah Louise (Beales) G.; m. Joyce Elaine Cox, July 28, 1963; children: Margaret Louise, Theodore Howard. AB, Hastings Coll., 1963; postgrad. U. Western Australia, 1964; PhD, U. Kans., 1971. Curator The Mus., Tex. Tech U., Lubbock, 1972-76, lectr. Mus. Sci. Program, 1974-76; curator Carnegie Mus. Natural History, Pitts., 1976-86; dir. U. Nebr. State Mus., Lincoln, 1986—; chair mus. studies program U. Nebr., 1989—, prof. state mus., 1986—, prof. biol. scis., 1987—, prof. mus. studies, 1990—. Author, editor: Mammalian Biology in South America, 1982; Natural History of the Dog, 1984; Contributions in Vertebrate Paleontology, 1984; Species of Special Concern in Pennsylvania, 1985; Current Mammalogy, 1987, 90. Packmaster, Allegheny Trails coun. Boy Scouts Am., 1981-83, asst. scoutmaster, 1983-86. Grantee, Fulbright, 1964, NSF, 1977-86, R.K. Mellon Found., 1981-86, Smithsonian Fgn. Currency Program, 1983-84, Inst. Mus. Svcs., 1989-93. Mem. Am. Soc. Mammalogists (pres. 1984-86, C. Hart Merriam award 1987), Southwestern Assn. Naturalists (pres. 1984-85), Am. Assn. Museums, Nebr. Mus.. Assn. (pres. 1990-92), Assn. Systematics Collections (bd. dirs. 1993—), Lincoln Attractions and Mus. Assn. (chair 1987—), Coun. Biology Editors, Soc. Systematic Zoologists, Rotary (bd. dirs. Lincoln N.E. club, 1990-92). Office: U Nebr State Mus 307 Morrill Hall 14th & U Sts Lincoln NE 68588-0338

GENSHEIMER, KATHLEEN FRIEND, epidemiologist; b. Sussex, N.J., Oct. 27, 1950; d. Douglas William and Catherine (Decker) Friend; m. Gregory George Gensheimer, Mar. 5, 1977; children: katherine, William, Maryl, James. BA, Pa. State U., 1973; MD, U. Rochester, 1977; MPH, Harvard U., 1991. Intern Thomas Jefferson U. Hosp., Phila., 1977-78; resident in pub. health N.J. Dept. Health, Trenton, 1978-81; mem. epidemic intelligence svc. Ctrs. for Disease Control, Atlanta, 1981-83; state epidemiologist Maine Dept. Human Svcs., Augusta, 1981—; mem. gov.'s com. on biotech. Maine Dept. Agr., Augusta, 1989—; v.p. Coun. State and Territorial Epidemiologists, Atlanta, 1992-93; mem. Maine tuberculosis elimination com. Am. Lung Assn. Maine, Augusta, 1992—. Recipient Charles C. Lund award ARC Blood Svcs., Dedham, Mass., 1989. Fellow Am. Coll. Preventive Medicine; mem. Am. Soc. Microbiology, Phi Beta Kappa. Home: 164 Main St Yarmouth ME 04096 Office: Maine Bur Health State House Sta # 1 Augusta ME 04333

GENSKOW, JOHN ROBERT, civil engineer, consultant; b. Hinsdale, Ill., Mar. 21, 1955; s. Roy Donald and Helen (Nauman) G.; m. Deborah Ann Freed, July 24, 1959; 1 child, Matthew Robert. BS in Civil Engring., Iowa State U., 1978. Registered profl. engr., Wash. Student engr. City of Ames (Iowa) Pub. Works, 1975-77; jr. engr. Gray & Osborne Inc., P.S., Seattle, 1978-83; assoc. engr. Group Four Inc., Lynnwood, Wash., 1983-85; project engr. URS Cons., Seattle, 1985-90; sr. project mgr. HDR Engring., Inc., Bellevue, Wash., 1990—; mem. drainage manual adv. com. King County Surface Water Mgmt. Div. Mem. Univ. Presbyn. Ch., Seattle, 1980—. Mem. ASCE (Local Civil Engring. Outstanding Project award 1992), Am. Pub. Works Assn. (surface water mgr. com. 1988—). Office: HDR Engring Inc 500 108th NE Ste 1200 Bellevue WA 98004-5538

GENTER, ROBERT BRIAN, biology educator; b. Detroit, June 17, 1956; s. Ralph Robert Genter and Gwendolyn Ann (Martin) Balance. BS in Oceanography, U. Mich., 1978; MS in Biology, Bowling Green State U., 1983; PhD in Biology, Va. Poly. Inst., 1986. Rsch. asst. Great Lakes Rsch. Div., Ann Arbor, Mich., 1979-81; teaching asst. Bowling Green (Ohio) State U., 1981-83; teaching asst. Va. Poly. Inst. and State U., Blacksburg, 1983-86, rsch. asst. Ctr. for Environ. Studies, 1984; teaching asst. U. Mich. Biol. Sta., Pellston, 1984; assoc. prof. Johnson (Vt.) State Coll., 1986—; project counselor NASA-Kennedy Space Ctr., Cape Canaveral, Fla., 1987; ecol. cons. Langrock, Sperry, Parker & Wool, Middlebury, Vt., 1988. Contbr. articles to profl. jours. Reviewer NSF, Washington, 1988, 90, Environ. Toxicology & Chemistry, Pensacola, Fla., 1992; mem. exec. bd. Lake Champlain Rsch. Consortium, Burlington, Vt., 1991—. Rsch. grantee Bowling Green State U., 1982, U. Mich. Biol. Sta., 1984, Sigma Xi, 1984, Johnson State Coll., 1986, 89, 90, NSF, 1988, 89. Mem. Soc. of Environ. Toxicology and Chemistry, N.Am. Benthological Soc., Am. Soc. Limnology and Oceanography, Phycological Soc. Am. Achievements include research in freshwater algae useful as biological monitors of environmental pollution. Office: Johnson State Coll Johnson VT 05656

GENTLE, KENNETH WILLIAM, physicist; b. Oak Park, Ill., Oct. 27, 1940; s. William and Cathryn Mary (Spence) G. B.S., MIT, 1962, Ph.D., 1966. Asst. prof. dept. physics U. Tex., Austin, 1966-69, assoc. prof., 1970-75, prof. physics, 1976—. Sloan fellow, 1973-75. Mem. Am. Phys. Soc. Home: 212 Buckeye Trl Austin TX 78746-4420 Office: Univ Tex Dept Physics Austin TX 78712

GENTNER, PAUL LEFOE, architect, consultant; b. Seattle, Feb. 24, 1944; s. Edward George and Opal Eloise (Davis) G.; m. Glenda Frank Hoy, May 25, 1975; 1 stepchild, Robert Michael Hurd. AA in Architecture, Anne Arundel Community Coll., Arnold, Md., 1970; BS in Engring., Century U., 1984. Cert. Constrn. Specifier. Project rep. RTKL Assocs., Inc., Balt., 1970-73; staff architect James R. Grieves Assocs., Balt., 1973-77; sr. engr. Morrison-Knudsen (MKSAC), Columbia, Md., 1977-79; staff engr. Morrison-Knudsen (MKSAC), Saudi Arabia, 1979-81; planning mgr. Morrison-Knudsen Internat. Inc., Barranquilla, Colombia, S.Am., 1981-86; staff architect RTKL Assocs., Inc., Balt., 1986-92; specifications writer Sverdrup Cpr., Arlington, Va., 1992—. With USNR, 1965-68, Vietnam; Persian Gulf, 1990-91. Mem. AIA, Constrn. Specifications Inst. (bd. dirs. Balt. chpt. 1991-92, 1st v.p. 1993), Soc. Am. Mil. Engrs., Bricklayers Local No. One, NAval Res. Mobile COnstrn. Bn. Home: 2028 Park Ave Baltimore MD 21217-4816 Office: Sverdrup Facilities Inc 1001 19th St N Ste 700 Arlington VA 22209

GENTRY, DAVID RAYMOND, engineer; b. Easley, S.C., Sept. 26, 1933; s. Thomas Herbert and Rosalie (Howard) G.; m. Mary Lynn White, June 5, 1955; children: David R. Jr., Mary Diane Gentry Farley. BS, Clemson Coll., 1955; MS, Inst. Textile Tech., Charlottesville, Va., 1957; PhD, Clemson U., 1972. Rsch. engr. WestPoint (Ga.) Mfg. Co., 1957-60; asst. prof. Clemson (S.C.) U., 1960-67; mgr. testing and evaluation Phillips Fibers Corp., Greenville, S.C., 1967-73; assoc. prof. Ga. Inst. Tech., Atlanta, 1973-78; sr. devel. engr. Amoco Fabrics & Fibers Co., Atlanta, 1978-80, mgr. fibers devel., 1980-84, dir. fibers devel., 1985-90; rsch. assoc. Amoco Fabrics & Fibes Co., Atlanta, 1990-92, sr. rsch. assoc., 1992—. fellow NSF, 1966, Sirrine Found., 1965-67, Inst. Textile Tech., 1955-57. Mem. Fiber Soc., ASTM (sec. com. D-13 textiles 1974-80), Am. Assn. Textile Technologists (sec. Piedmont chpt. 1963-65), Phi Psi (sec., pres. Iota chpt. 1953-55, faculty adviser 1961-65), Phi Kappa Phi. Home: 3456 Embry Cir Atlanta GA 30341-5612 Office: Amoco Fabrics and Fibers Co 260 The Bluffs Atlanta GA 30336

GENTRY, JAMES FREDERICK, chemical engineer, consultant; b. Charleston, W.Va., Apr. 21, 1954; s. Gene Allen and Patricia Katherine (McDonie) G.; m. Deborah May Brown, June 9, 1979. BSChemE, La. State U., 1976. Devel. engr. Ethyl Corp., Baton Rouge, 1976-78; process engring. mgr. Borden Chems. and Plastics, Geismar, La., 1978-89; engring. mgr. Borden Inc., Calabasas, Calif., 1989-92; cons. Gentry Engring., Thousand Oaks, Calif., 1992-93; sr. project engr. Unocal, L.A., 1993—. Recipient Nat. Energy Innovation award Dept. Energy, 1988, Indsl. Energy Innovation award Indsl. Energy Tech. Conf., 1988. Mem. Tau Beta Pi, Phi Kappa Phi, Phi Eta Sigma. Home: 267 Scarborough St Thousand Oaks CA 91361

GEOFFRION, ARTHUR MINOT, management scientist; b. N.Y.C., Sept. 19, 1937; s. Arthur Joseph and Dorothy Arlene (Senter) G.; m. Helen Mathilda Hamer, Dec. 22, 1962; children: Susan, Deborah. BME, Cornell U., 1960, M Indsl. Engring., 1961; PhD, Stanford U., 1965. Asst. prof. in ops.rsch. UCLA, 1965-67, assoc. prof., 1968-70, prof. Grad. Sch. Mgmt., 1971—; dir. Insight, Inc., Alexandria, Va. Author: Perspectives on Optimization, 1972; contbr. 46 articles to profl. jours., 3 chpts. to books. Ford Found. faculty rsch. fellow, 1967-68; rsch. grantee NSF, 1968-91, Ford Found., 1969-72, Office Naval Rsch., 1972-90; recipient System Sci. Prize NATO, Brussels, 1976. Mem. Inst. Mgmt. Scis. (pres. 1981-82, Disting. Svc. medal 1992), Ops. Rsch. Soc. Am., Omega Rho (hon.). Achievements include research on optimization theory (parametric concave programming, integer programming, multi-criterion optimization, large-scale, decomposition, duality theory), optimization applications (to logistics, prodn., fin.), aggregation. Home: 322 24th St Santa Monica CA 90402 Office: U Calif Sch Mgmt Los Angeles CA 90024

GEOFFROY, DONALD NOEL, civil engineer, consultant; b. Lowell, Vt., Dec. 25, 1937; s. Paul E. and Lillian (Comtois) G.; m. Shirley A. Dusablon, Aug. 20, 1960; children: Sandra, Michael, Catherine, Richard. BSCE, U. Vt., 1960; M Engring., Rensselaer Poly. Inst., 1966. Registered profl. engr., N.Y. Prin. civil engr. N.Y. State Dept. Transp., Albany, 1973-77, 79-82, regional dir., 1977-79, 82-86, dep. chief engr., 1986-89, asst. commr., 1989-92; intelligent vehicle hwy. system contract mgr. Transcom, Jersey City, 1993—; mem. joint task force on pavements Am. Assn. State Hwy. and Transp. Ofcls., Washington, 1986-90, standing com. on rsch., 1988-92; chair project panel Nat. Coop. Hwy. Rsch. Program, Washington, 1990—; exec. com. Strategic Hwy. Rsch. Program, Washington, 1990-92. Author tech. reports. Fellow ASCE; mem. Inst. Transp. Engrs., Tau Beta Pi, Chi Epsilon. Home and Office: 22 Northgate Dr Albany NY 12203

GEOKAS, MICHAEL C., gastroenterologist; b. Villia-Attiha, Greece, Aug. 25, 1924. MD, Athens U., 1951; MSc, McGill U., 1964; PhD in Invvest. Medicine. Chief med. svc. VA Med. Ctr., Martinez, Calif., 1974—. Fellow ACP, Am. Gastroenterol. Assn.; mem. Am. Physiol. Soc., N.Y. Acad. Sci. Office: Dept VA Med Ctr Enzymology Rsch Lab 150 Muir Rd Dept Medicine Martinez CA 94553*

GEORGAKIS, CHRISTOS, chemical engineer educator, consultant, researcher; b. Patra, Greece, Aug. 13, 1947; came to U.S., 1970; s. Theofilaktos and Penelope (Rompoti) G.; m. Konstantina Hinou, July 12, 1970; children: Alexander, Natalie. Chem. engring. diploma, Nat. Tech. U., Athens, Greece, 1970; MSChemE, U. Ill., 1972; PhDChemE, U. Minn., 1975. Cert. chem. engr., Greece. Asst. prof. MIT, Cambridge, 1975-79, assoc. prof., 1979-83, Du Pont prof., 1975-76, Edgerton prof., 1977-79; prof. U. Thessaloniki, Greece, 1979-83; assoc. prof. chem. engring. Lehigh U., Bethlehem, Pa., 1983-87, prof., 1987—, dir. PMC Rsch. Ctr., 1985—; dir. Am. Automatic Control Coun., 1993—. Dreyfus Found. tchr.-scholar, 1979-81. Mem. AICE (chmn. process control area 10b 1990-92, vice chmn. 1988-90). Achievements include elucidation of interaction between process modeling and process control with several applications in different types of continuous and batch reactors; development of overall approach to the optimization and control of chemical processes based on approximate models called tendency models; research on the use of thermodynamic variables in the systematic understanding of process dynamics nonlinear model predictive control. Office: Lehigh U PMC Rsch Ctr Iacocca Hall 111 Research Dr Bethlehem PA 18015-4732

GEORGANAS, NICOLAS D., electrical engineering educator; b. Athens, Greece, June 15, 1943; s. Demetrios N. and Athanasia (Kotsovou) G.; m. Jacynthe Savard, June 17, 1972; children: Nikita, Emmanuel. Diploma in Engring., Nat. Tech. U. Athens, 1966; PhD summa cum laude, U. Ottawa, Ont., Can., 1970. Registered profl. engr., Ont. Lectr., elec. engring. U. Ottawa, 1970-71, asst. prof., 1971-76, assoc. prof., 1976-80, prof., 1980—, chmn., 1981-84, dean engring., 1986-93; researcher Bell-No. Rsch., Ottawa, 1993-94; vis. prof. IBM, LaGaude, France, 1977-78, INRIA/Bull-Transac, Paris, 1984-85, BNR, Ottawa, 1993-94. Author: Queueing Networks - Exact Computational Algorithms: A Unified Theory by Decomposition and Aggregation, 1989; contbr. over 60 articles to profl. jours., over 90 conf. articles. Fellow IEEE. Home: 1915 Montereau Ave, Gloucester, ON Canada Office: U Ottawa Faculty Engring, 161 Pasteur St, Ottawa, ON Canada K1N 6N5

GEORGE, DONALD JAMES, architecture educator, administrator; b. Canton, Ohio, Dec. 1, 1938; s. Frank and Mary Elizabeth (Bradley) G.; m. Norma Nadine Miller, May 26, 1984. A in theology, St. Leo's Sem., 1961; postgrad., U. Md., 1970, N.Y. Coll., 1973. Tech. engr. Weber Dental Mfg. Co., Canton, 1966-73; classification and coding officer Hoover Co., North Canton, Ohio, 1973-89; head architecture and mech. applied sci. E.T.I. Tech. Coll., North Canton, 1990—; mem. indsl. rels. adv. bd. E.T.I. Tech. Coll., 1992—; bar coding advisor U.S. Army Ordnance Dept., Washington, 1988-89. Chmn. Elks blood bank ARC, North Canton, 1992; adv. bd. United Ch. of Christ, Paris, Ohio, 1992. With U.S. Army, 1962-65. Recipient Legion of Honor, Nat. Chaplains Assn., 1977. Mem. Am. Design Drafting Assn. (advisor 1990—), Arion Singing Soc., Ohio Welsh (v.p. 1987-88), Elks (editor Elk News North Canton chpt. 1982-92). Democrat. Home: 6663 Columbus Rd NE Louisville OH 44641 Office: ETI Tech Coll 1320 W Maple St NW North Canton OH 44720

GEORGE, JOHN MICHAEL, chemical engineer; b. New Kensington, Pa., Jan. 1, 1969; s. John James George and Catherine Pauline Kotermanski Kento. BS in Chem. Engring., Pa. State U., 1991. Cert. EIT, hazardous materials tech. Lab. tech. Kurt J. Lesker Co., Pitts., 1989-91, lab. supr., 1991-92, lab. mgr., 1993—. Mem. Am. Inst. Chem. Engrs., Pa. State U. Alumni Assn. Home: 1203 Liberty Ave Natrona Heights PA 15065 Office: Kurt J Lesker Co 255 William Pitt Way Pittsburgh PA 15238

GEORGE, KATTUNILATHU OOMMEN, homoeopathic physician, educator. Practicing homoeopathic physician; developer L.A. Internat. U. divsn. Samuel Hahnemann Sch. Homoeopathic Medicine, founding pres., dir. Hahnemann Rsch. Ctr., Inc., tchr. clin. practice; conductor seminars, trainer in field. Author: Twelve Energy Medicine, How to Balance your Body Dynamically, Ways to Change the Health Care Crisis in America, Comprehensive Therapeutics in Homoeopathy, Twelve Energy Supplements for your Health with Nutrition. Mem. Homoeopathic Med. Assn. Am. (founding pres.). Office: Hahnemann Research Ctr Inc 2232 SE Bristol 102 Santa Ana Heights CA 92707*

GEORGE, RUSSELL JOSEPH, design engineer; b. Buffalo, Dec. 1, 1958; s. Russell Joseph and Anita (Uli) G.; m. Carol Ann Dittenhauser, Sept. 13, 1980; children: Elisa Anne, Alex Joseph. BSME, SUNY, Buffalo, 1986. Mech. technician Viatran Corp., Grand Island, N.Y., 1978-79, Moore Rsch. Ctr., Grand Island, N.Y., 1979-80, Gaymar Industries, Orchard Park, N.Y., 1980-81; pneumatic technician Carleton Controls Corp., East Aurora, N.Y., 1981-83; quality engring. technician Moog Inc., East Aurora, N.Y., 1983-87; sr. design engr. Thiokol Corp., Brigham City, Utah, 1987—; adj. faculty Weber State U., Ogden, Utah, 1990-91, 92-93; cons. AIAA Utah Sect. Sounding Rocket Project, Ogden, 1992. Candidate City Coun., Pleasant View, Utah, 1991. Recipient Group Achievement award NASA, 1988. Mem. AIAA, K.C. Achievements include development of numerous nozzle design concepts for performance improvements of shuttle-class solid rocket motors and hybrid rocket motors, organized the nozzle post-flight evaluation team for shuttle solid rocket motor nozzles. Office: Thiokol Corp PO Box 707 MS L62A Brigham City UT 84302-0707

GEORGIANA, JOHN THOMAS, electrical engineer; b. Gibbstown, N.J., June 5, 1942; s. Andrew Michael and Loretta Shirley (Cassidy) G.; m. Dorothy A. Cameron, June 30, 1963 (div. 1983); children: Lori Alice, Maria JoAnn, Andrew Michael. BS in Elec. Engring., Drexel U., Phila., 1966; MBA, Webster U., Webster Groves, Mo., 1989, postgrad., 1989—. Registered profl. engr., Tex. Power engr. Texaco, Inc., Westville, N.J., 1966-68; project engr. Owens Corning Fiberglass, Barrington, N.J., 1968-77; elec. engr., project engr. Pepperidge Farm/Campbell Soup, Norwalk, Conn., 1970-77, asst. mgr. new plant devel., 1975-76, mgr. maintenance engring., 1976-77; mgr. project engring. The Great Atlantic and Pacific Co., Montvale, N.J., 1977-82; major projedct engr. Interstate Brands Corp., Kansas City, Mo., 1982-83; project engr. Anheuser Busch/Campbell Taggart, St. Louis, 1983-89, Anheuser Busch/Metal Container, St. Louis, 1989-91; equip. engr., packaging specialist The Pritchard Corp., Kansas City, Kans., 1991—. Pres. PTA, Gibbstown, N.J., 1971-72. Mem. IEEE. Avocations: tennis, bowling, education. Home: 13305 W 78th Pl Lenexa KS 66216-3025 Office: The Pritchard Corp 10950 Grandview St Overland Park KS 66210-1505

GERALD, NASH OGDEN, environmental engineer; b. Birmingham, Ala., Aug. 9, 1946; s. Nash Ogden III and Elsie Louise (Emken) G. BChemE, Ga. Inst. Tech., 1968, MS in Indsl. Mgmt., 1970. Registered profl. engr., N.C. Supr./analyst Taylor/Sybron Corp., Arden, N.C., 1971-73; with Dept. Natural Resources and Community Devel., State of N.C., Asheville, environ. engr./supr., 1984-86, asst. chief air quality sect., 1973-84, chief air quality sect., 1986-89; chief monitoring sect. U.S. EPA, Durham, N.C., 1989-91, sr. environ. engr., 1989—; state rep. State and Territorial Air Pollution Program Adminstrs., Raleigh, N.C., 1986-89; mem. Standing Air Simulation Work Group, Raleigh, 1987-89; chmn. Standing Air Monitoring Work Group, Durham, 1989-91. With fire/rescue svc. Skyland (N.C.) Vol. Fire Dept., 1973-81; SCUBA instr. YMCA, Profl. Assn. Diving Instrs., Asheville, N.C., 1973-84. Capt. USAR. Named Disting. Mil. Grad., U.S. Dept. of Army, 1968, Outstanding Officer-3COBC, 1971; recipient Disting. Svc. award Dept. Environ., Health and Natural Resources, State of N.C., 1989, Bronze medal U.S. EPA, 1992; named EPA Co-Engr. of Yr., 1993, top ten finalist for Federal Engr. of Yr., 1993. Mem. NSPE, Soc. Am. Mil. Engrs., Profl. Engrs. N.C., Carolinas Air Pollution Assn. (bd. dirs. 1987-89), Air and Waste Mgmt. Assn. Avocation: martial arts (T'ai Chi). Office: US EPA MD-14 Research Triangle Park NC 27701

GERARD, MARK EDWARD, mechanical engineer; b. Feb. 4, 1952; s. Donald Edward and Helen Lorraine (Gans) G. BSME, Kans. State U., 1975. Lic. profl. engr., Kans. Plant engr. Beech Aircraft Corp., Wichita, Kans., 1975-78; sr. engr. Kans. Power & Light Co., Topeka, 1978-91; environ. engr. Kans. Dept. of Health & Environ., Topeka, 1991—. Mem. ASME, Kans. Engring. Soc. (pres. 1992—). Home: 610 Hillcrest Rd Wamego KS 66547 Office: Kans Dept Health & Environ Bldg 740 Forbes Field Topeka KS 66620

GERARDI, ROY G., JR., mechanical engineer; b. Bklyn., May 8, 1948; s. Roy G. and Mildred (Jacobs) G.; m. Margaret H. Muir, Dec. 1, 1974; children: Jane Elizabeth, Roy Andrew, Bradley Muir. BS in Mech. Engring., Manhattan Coll., 1969. Applications engr. Dean Products, Bklyn., 1969-72;

br. mgr. ITT Grinnell, Lincoln Park, N.J., 1972-81; dir. sales Lightnin' Mixers, Rochester, N.Y., 1981-88; pres. Repco Inc., Metuchen, N.J., 1988—. Commr. St. Anselm Little League, Bklyn., 1990; dir. St. Anselm T-Ball, Bklyn., 1990. Office: Repco Inc 54 Bridge St Metuchen NJ 08840

GERASCH, THOMAS ERNEST, computer scientist; b. New Ulm, Minn., Jan. 6, 1949; s. Earl Joseph and Doris L. (Weymann) G.; m. Pearl Yun Wang, Aug. 9, 1980. MA in Math., U. Wis., Milw., 1971, PhD in Math., 1977. Asst. prof. computer sci. George Mason U., Fairfax, Va., 1980-86; mem. tech. staff Perkin-Elmer Advanced Devel. Ctr., MRJ Inc., Oakton, Va., 1986-88; prin. investigator Sparta, Inc., McLean, Va., 1988-89; mem. tech. staff Mitre Corp., McLean, 1990—. Contbr. articles on computer sci. and math. to profl. publs. Mem. AAAS, IEEE Computer Soc., Assn. for Computing Machinery, European Assn. for Theoretical Computer Sci. Office: The Mitre Corp 7525 Colshire Dr Mc Lean VA 22102

GERBA, CHARLES PETER, microbiologist, educator; b. Blue Island, Ill., Sept. 10, 1945; s. Peter and Virginia (Roulo) G.; m. Peggy Louise Scheitlin, June 6, 1970; children: Peter, Phillip. BS in Microbiology, Ariz. State U., 1969; PhD in Microbiology, U. Miami, 1973. Postdoctoral fellow Baylor Coll. Medicine, Houston, 1973-74, asst. prof. microbiology, 1974-81; assoc. prof. U. Ariz., Tucson, 1981-85, prof., 1985—; cons. EPA, Tucson, 1980—, World Health Orgn., Pan Am. Health Orgn., 1989—; advisor CRC Press, Boca Raton, Fla., 1981—. Editor: Methods in Environmental Virology, 1982, Groundwater Pollution Microbiology, 1984, Phage Ecology, 1987; contbr. numerous articles to profl. and sci. jours. Mem. Pima County Bd. Health, 1986-92; mem. sci. adv. bd. EPA, 1987-89. Named Outstanding Research Scientist U. Ariz., 1984, 92; environ. sci. and engring. fellow AAAS, 1984. Mem. Am. Soc. Microbiology (div. chmn. 1982-83, 87-88, pres. Ariz. chpt. 1984-85, councilor 1985-88), Internat. Assn. Water Pollution Rsch. (sr. del. 1985-91), Am. Water Works Assn. Achievements include rsch. in environmental microbiology, colloid transport in ground water, wastewater reuse. Home: 1980 W Paseo Monserrat Tucson AZ 85704-1329 Office: U Ariz Dept Microbiology and Immunology Tucson AZ 85721

GERBER, ARTHUR MITCHELL, chemist; b. N.Y.C., Feb. 22, 1940; s. Phillip and Mollie (Glick) G. BS, Fla. State U., 1961. Chemist rsch. div. Merck and Co., Inc., Rahway, N.J., 1961-67; sr. chemist nuclear reactor Union Carbide, Sterling Forest, N.Y., 1967-69; sr. rsch. group leader Polaroid Corp., Cambridge, Mass., 1969-83; chief chemist Analogic Corp., Wakefield, Mass., 1983-84; tech. cons. Cipher Data, Inc., San Diego, 1984-86; v.p. Mesa Systems, San Diego, 1986—; cons. IBM, White Plains, N.Y., 1991. Mem. Am. Chem. Soc. Achievements include 22 patents in photographic emulsions, optical date storage systems, and optical data storage media.

GERBER, MICHAEL ALBERT, pathologist, researcher; b. Kassel, Germany, Oct. 18, 1939; came to U.S., 1966; s. Bruno and Luise (Kramer) G.; m. Luviminda Gerber, Oct. 8, 1971; 1 child, Elisa. MS, Gutenberg Gymnasium, Wiesbaden, Fed. Republic Germany, 1960; MD, Gutenberg U., Mainz, Fed. Republic Germany, 1966. Chief electron microscopy VA Hosp., Bronx, N.Y., 1973-74; from asst. prof. to assoc. prof. pathology Mt. Sinai Sch. Medicine, N.Y.C., 1973-79, prof. pathology, 1980-87; co-dir. cellular molecular pathology City Hosp. Ctr. at Elmhurst, N.Y., 1982-87; prof., chmn. pathology Tulane U. Sch. Medicine, New Orleans, 1987—, dir. grad. program in molecular & cellular biology, 1991; assoc. editor Hepatology. Contbr. over 200 articles to profl. jours. Mem. adv. com. FDA, Washington, 1985. Recipient Rsch. Career Devel. award NIH, 1980. Mem. Am. Assn. Pathologists, Am. Assn. Study Liver Diseases, U.S. and Can. Acad. Pathology, Am. Coll. Gastroenterology. Office: Tulane U Sch Medicine 1430 Tulane Ave New Orleans LA 70112-2699

GERDING, THOMAS GRAHAM, medical products company executive; b. Evanston, Ill., Feb. 11, 1930; s. Louis Henry and Helen Frances (Graham) G.; m. Beverly Ann Starnes, June 18, 1955; children: Mark, David, Gail, Gene Ann. Student, U. Notre Dame, 1948-49; BS in Pharmacy, Purdue U., 1952, MS, 1954, PhD, 1960. Registered pharmacist, Ind. From instr. to asst. prof. Purdue U., West Lafayette, Ind., 1956-62; dir. product devel. Pitman-Moore divsn. Dow Chem., Indpls., 1961-64; tech. dir. new products Glenbrook Labs., N.Y.C., 1964-66; dir. product devel. Sterling-Winthrop Rsch. Inst., Rensselaer, N.Y., 1966-70; v.p. rsch. and devel. Calgon Consumer Products, Rahway, N.J., 1970-77; v.p., dir. rsch. and devel., quality assurance, consumer affairs, engring. Johnson & Johnson Products Inc., New Brunswick, N.J., 1977-88; pres. Thomas G. Gerding, Inc., Georgetown, Tex., 1988—; dir. Drug Dynamics Inst. U. Tex., Austin, 1988—; pres., CEO Newform Devel. Labs., Inc., Austin, 1993—. Sgt. U.S. Army Med. Svc. Corp, 1954-56. Recipient Disting. Alumni award Purdue U., 1984. Mem. Am. Chem. Soc., Am. Assn. Pharm. Scientists, Nonprescription Drug Mfrs. Assn. (assoc., mem. assoc. mem. adv. com.), Union League Club (Chgo.), Berry Creek Country Club. Republican. Roman Catholic. Clubs: Union League (Chgo.); Shrewsbury River Yacht, Channel. Achievements include research in pharmaceutics and wound care and healing. Home: 355 Logan Ranch Rd Georgetown TX 78628 Office: Newform Devel Labs Inc PO Box 52 Georgetown TX 78627

GERGELY, JOHN, biochemistry educator; b. Budapest, Hungary, May 15, 1919; naturalized U.S.; married; 8 children. MD, U. Budapest, 1942, PhD in Phys. Chemistry, 1948. From asst. prof. pharmacology to asst. prof. biochemistry U. Budapest, 1942-48; asst. prof. biochemistry New Sch. Social Rsch., 1948-50; Nat. Heart Inst. sr. trainee U. Wis., 1950-51; rsch. assoc. medicine Harvard Med. Sch., 1951-62, asst. prof., 1962-71, assoc. prof., 1971-80, prof. biol. chemistry, 1980—; biochemist Mass. Gen. Hosp., 1969—; dir. dept. muscle rsch. Boston Biomed. Rsch. Inst., 1970—; NIH spl. rsch. fellow, 1948-50; established investigator Am. Heart Assn., 1951-58; from asst. biochemist to assoc. biochemist Mass. Gen. Hosp., 1954-69; tutor biochemistry sci. Harvard Med. Sch., 1957-72; dir. dept. muscle rsch. Retina Found., 1961-70. Mem. Am. Soc. Biol. Chemistry, Am. Chem. Soc., N.Y. Acad. Sci., Brit. Biochemistry Soc., Biophys. Soc. Achievements include research in biochemistry of muscle contraction, enzymes, nuclear magnetic rsonance, physical chemistry of proteins, electron spin resonance. Office: Boston Biomedical Research Inst 20 Staniford St Boston MA 02114*

GERGELY, PETER, structural engineering educator; b. Budapest, Feb. 12, 1936; m. Kinga M. Mecs, Dec. 26, 1964; children: Zoltan, Illa. B Engring., McGill U., 1960; PhD, U. Ill., 1963; Dr Honoris Causa, Tech. U. of Budapest, 1992. Structural engr. Dominion Bridge Co., Montreal, Que., Can., 1959-60; rsch. asst. U. Ill., Champaign-Urbana, 1960-63; prof. Cornell U., Ithaca, N.Y., 1963—, chmn., dept. structural engring., 1983-88, dir., sch. civil engring., 1985-88; cons. Westinghouse Savannah River Co., Aiken, S.C., 1990—. Co-author: Structural Engineering, Vol. 1,2,3, 1974. Fellow ASCE (State of Art of Civil Engring. prize 1974), Am. Concrete Inst. (Raymond C. Reese Rsch. prize 1976, D.L. Bloem Disting. Svc. award 1981, Wason Medal for Most Meritorious Paper 1992); mem. Earthquake Engring. Rsch. Inst., Am. Soc. Engring. Edn. Home: 106 Juniper Dr Ithaca NY 14850 Office: Cornell U Hollister Hall Ithaca NY 14853

GERGELY, TOMAS, astronomer; b. Budapest, Hungary, Oct. 14, 1943; came to U.S., 1976, naturalized, 1982; s. Tibor and Magda (Szilasi) G.; m. Ana Lajmanovich, Mar. 6, 1970; children—Gabriela S., Esteban A., Daniel M. Licenciado in Physics, U. Buenos Aires, Argentina, 1967; Ph.D. in Astronomy, U. Md., 1974. Asst. prof. Nat. Tech. U., Buenos Aires, 1974, researcher, 1975; research assoc. U. Md., College Park, 1976-81, sr. research assoc., 1981-82, assoc. research scientist, 1982-85; astrophysicist NASA Hdqrs., Washington, 1985-86; mgr. electromagnetic spectrum, NSF, 1986—. Editor: (with others) Radio Physics of the Sun. Contbr. articles to profl. jours. Recipient Young Scientist award French Govt., 1976. Mem. Internat. Astron. Union, Am. Astron. Soc. (solar physics div.), Internat. Radio Physics Union. Home: 8217 Windsor View Ter Rockville MD 20854-4028 Office: NSF Divsn Astron Scis 1800 G St NW Washington DC 20550-0002

GERGESS, ANTOINE NICOLAS, civil engineer; b. Beirut, Oct. 31, 1965; s. Nicolas and Victoria Emily (Nasr) G. BSCE, Am. U., Beirut, 1987; MSCE, U. S. Fla., 1988, postgrad., 1992. Registered profl. engr., Fla. Project engr. Conseil Executif Des Grands Projets, Beirut, 1986-87; grad. researcher U. S. Fla., Tampa, 1987-88; sr. bridge engr. Post, Buckley, Schuh & Jernigan, Miami, 1988-93; profl. engr. II concrete and movable bridge tech. group (CMBT) Parsons Brinckerhoff Quade & Douglas, Tampa, 1993—; adj. prof. Fla. Internat. U., Miami, 1990-93. Contbr. articles to profl. jours. Mem. ASCE, (v.p. 1992—, Young Engr. of Yr. 1990-91), Can. Soc. Civil Engrs., NSPE, Fla. Engring. Soc., Phi Kappa Phi. Greek Orthodox. Achievements include rsch. of cast-in-place post-tensioned voided slab bridges for Fla. Dept. Transp. Home: 2706 Acorn Ct 19 C Tampa FL 33613 Office: Parsons Brinckerhoff Quade & Douglas 4200 W Cypress St Ste 700 Tampa FL 33607

GERHARD, LEE CLARENCE, geologist, educator; b. Albion, N.Y., May 30, 1937; s. Carl Clarence and Helen Mary (Lahmer) G.; m. Darcy LaFollette, July 22, 1964; 1 dau., Tracy Leigh. B.S., Syracuse U., 1958; M.S., U. Kans., 1961, Ph.D., 1964. Exploration geologist, region stratigrapher Sinclair Oil & Gas Co., Midland, Tex. and Roswell, N.Mex., 1964-66; asst. prof. geology U. So. Colo., Pueblo, 1966-69, assoc. prof., 1969-72; assoc. prof., asst. dir. West Indies Lab. Fairleigh Dickinson U., Rutherford, N.J., 1972-75; asst. geologist State of N.D., Grand Forks, 1975-77, geologist, 1977-81; prof., chmn. dept. geology U. N.D., Grand Forks, 1977-81; mgr. Rocky Mountain div. Supron Energy Corp., Denver, 1981-82; owner, pres. Gerhard & Assocs., Englewood, Colo., 1982-87; prof. petroleum geology Colo. Sch. Mines, Denver, 1982—, Getty prof., 1984-87; state geologist, dir. geol. survey State of Kans., Lawrence, 1987—; co-dir. Energy Rsch. Ctr. U. Kans., 1990—; presdl. appointee Nat. Adv. Com. on Oceans and Atmosphere, 1984-87. Contbr. articles to profl. jours. Served to 1st lt. U.S. Army, 1958-60. Danforth fellow, 1970-72. Mem. Am. Assn. Petroleum Geologists (Disting. Svc. award 1989), Geol. Soc. Am., Am. Inst. Profl. Geologists, Rocky Mountain Assn. Geologists, Colo. Sci. Soc., Kans. Geol. Soc., Sigma Xi, Sigma Gamma Epsilon. Home: 1628 Alvamar Dr Lawrence KS 66047-1714 Office: Kans Geol Survey 1930 Constant Ave Lawrence KS 66047-3726

GERHARDINGER, PETER F., engineer; b. Toledo, Feb. 6, 1957; s. William J. and Marcella M. (Eggerstorfer) G.; m. Susan L., July 14, 1984; children: Eric, Robert, Laura. BEE, U. Toledo, 1984. Solar energy technologist Libbey-Owens-Ford Co., Toledo, 1977-80; sr. product devel. technologist Phton Power Inc., El Paso, Tex., 1980-82; devel. engr. Libbey-Owens-Ford Co., Toledo, 1982-84, sr. engring. archtl. svcs., 1984-89, mgr. new product tech., 1989-93; rsch. assoc., 1993—; bd. dirs. Am. Coun. Energy Efficient Economy, Washington, 1992—, Internat. Inst. Energy Conservation, Washington, 1992—; tech. chmn. Primary Glass Mfg. Coun., Topeka, 1988-90; expert testimony 2 Congl. subcoms. energy efificency. Contbr. articles to profl. pubs. Achievements include participation in Dept. of Energy Bldg. Energy Efficiency Program Rev.; devel. of proprietary coated glass product for emerging thin film phtovoltaic market; influence of addendum provisions of ASHRAE 90.1. Office: Libbey-Owens-Ford Co 811 Madison Ave Toledo OH 43695-0799

GERHARDT, DOUGLAS L., computer scientist. BS, Carnegie-Mellon U., 1973. Software engr. E-Systems Inc., Garland, Tex., 1975-80; pres. Gerhardt Engring., Richardson, Tex., 1980—; cons. DARPA, Washington, 1989-90, U.S. Govt., Washington, 1990-91, U.S. Mil., Washington, 1992—. Mem. IEEE. Achievements include research in software engineering techniques, operating system architecture, programming language design. Office: PO Box 831267 Richardson TX 75083

GERHARDT, JON STUART, mechanical engineer, engineering educator; b. Springfield, Ohio, June 5, 1943; s. Robert William and Mary Josephine (Jones) G.; m. Claudia Jay Sadler, Feb. 7, 1970; children: Kirsten Lea, Benjamin Luke. BS in Mech. Engring., U. Cin., 1966, MS in Mech. Engring., 1968, PhD, 1971. Registered profl. engr., Ohio. Asst. prof. U. N.C., Charlotte, 1971-73; project engr. Duff-Norton, Charlotte, 1973; sr. devel. engr. Gen. Tire and Rubber Co., Akron, Ohio, 1973-77, group leader, 1977-79, mgr. tech. staff devel., 1979-84, mgr. product engring., rsch. ctr. adminstrn., 1984-87, dir. rsch., Gen. Tire, Inc., 1987-92, dir. rsch. and tire testing, 1992—; instr. U. Akron, 1976—. Mem. ASME (bd. dirs. 1981-85), Akron Rubber Group, Soc. Automotive Engrs., Sigma Xi.

GERHARDT, ROSARIO ALEJANDRINA, materials scientist; b. Lima, Peru, May 20, 1953; d. Jacob K. and Tarcila (La Cruz) G.; m. Michael Paul Anderson, Sept. 27, 1980; children: Heidi Margaret, Kathleen Elizabeth. BA, Carroll Coll., 1976; MS, Columbia U., 1979, D Engring. Sci., 1983. Teaching asst. Columbia U., N.Y.C., 1978-79, grad. asst., 1979-83, research assoc., 1983-84; postdoctoral fellow Rutgers U., Piscataway, N.J., 1984-86, asst. rsch. prof., 1986-90; assoc. prof. Ga. Inst. Tech., 1990—; cons. in field. Contbr. articles to profl. jours. Mem. Am. Ceramic Soc., Am. Phys. Soc., N.Y. Acad. Sci., Electron Microscopy Soc. Am., Materials Rsch. Soc., Sigma Xi, Materials Sci. Club N.Y. (sec. 1988-90). Roman Catholic. Avocations: walking, stamp collecting, travel, language. Home: 124 Infantry Way Marietta GA 30064-5000

GERKEN, GEORGE MANZ, neuroscientist, educator; b. Hackensack, N.J., July 12, 1933; s. Henry Luhrs and Margaret Barbara (Manz) G.; m. Dora Vernon Udall, Aug. 11, 1956; children: Christopher, Cynthia. BS, MIT, 1955; PhD, U. Chgo., 1959. Asst. prof. U. Va., Charlottesville, 1959-67; scientist Callier Ctr., Dallas, 1967-74; assoc. prof. U. Tex., Dallas, 1974-82, prof., 1982—; vis. prof. U. Tex. Southwestern Med. Ctr., Dallas, 1988-90. Contbr. sci. articles to profl. publs. Mem. Acoustical Soc. Am., Assn. for Rsch. in Otolaryngology, Sigma Xi. Achievements include research on brain mechanisms of hearing. Office: Univ Tex at Dallas 1966 Inwood Rd Dallas TX 75235

GERLACH, ROBERT LOUIS, research and development executive, physicist; b. Guthrie, Okla., Nov. 16, 1940; s. Charles Frederick and Avalie Genevive (d'Avignon) G.; m. Carol Louise McKee; children: Robert Scott, Brett Cornell, Avalie Lynn, Cheri Louise. BS, Northwestern U., 1964; PhD, Cornell U., 1969. Staff scientist Sandia Corp., Albuquerque, N.Mex., 1968-73; sr. engr. Varian Assocs., Palo Alto, Calif., 1973-74; project scientist Physical Electronics div. Perkin Elmer Corp., Eden Prairie, Minn., 1974-85, dir. R&D, 1985-90; v.p. FEI Co., Beaverton, 1990—. Contbr. numerous articles to profl. jours.; patentee in field. Troop com. chmn. Boy Scouts Am., Mpls., 1979-89. Recipient IR 100 award R & D Mag., 1980, 81. Mem. Am. Phys. Soc., Am. Vacuum Soc., Microbeam Analysis Soc. Republican. Mormon. Avocation: water skiing. Office: FEI Corp 19500 NW Gibbs Dr Ste 100 Beaverton OR 97006-6900

GERLACH, THURLO THOMPSON, electrical engineer; b. Sparta, Ill., Oct. 30, 1916; s. Kenneth Frederick and Golda M. (Thompson) G.; m. Ellen Marie Kuhn, July 14, 1946. BEE, Tri-State U., 1937; grad., Air Force Command and Staff, 1952. Registered profl. engr., Ill., Mont. Dist. engr. Ill. Power Co., Centralia, 1937-40; area engr. Ill. Power Co., Sparta, 1940-41, Granite City, Ill., 1946-48; engr. U.S. Bur. Standards, Washington, 1948-50, Fed. Power Commn., Washington, 1953-56, Bur. of Reclamation, Billings, Mont., 1950-51, 56-77; cons. Billings, 1977—; del. heavy engr. constrn. program Citizen Amb. Program, Peoples Republic of China, 1990, USSR, 1991. Active People to People Internat., Kansas City, Mo., 1989—. Major USAF, 1941-46, 51-53. Mem. NSPE, IEEE, U.S. Com. Large Dams, Elks, Masons. Methodist. Achievements include participation in development of Missouri River Basin power system. Home and Office: 533 Park Ln Billings MT 59102

GERLING, GERARD MICHAEL, neurologist; b. St. Louis, July 19, 1939. BS, St. Louis U., 1961, MD, 1966. Diplomate Am. Bd. Psychiatry and Neurology, Am. Bd. Electroencephalography, Am. Bd. Electrodiagnostic Medicine. Intern in medicine St. Louis U. Group of Hosps., 1966-67; resident U. Mich. Med. Ctr. Hosp., Ann Arbor, 1967-70; clin. asst. Flagler Hosp., Inc., St. Augustine, Fla., 1979—; neurologist St. Luke's Hosp., Jacksonville, Fla., 1990—, Meml. Hosp., Jacksonville, Fla., 1991—. Recipient fellowship U. Minn., 1975-76. Fellow Am. Acad. Neurology; mem. AMA, Am. Assn. Electromyography and Electro-diagnosis, Am. Soc. Neuro Imaging, Am. Med. EEG Assn., Am. Electroencephalographic Soc. Office: Neurology Specialist Clinic 1955 US 1 S Ste A-1 Saint Augustine FL 32086

GERMAN, MARJORIE DACOSTA, mechanical engineer; b. N.Y.C., Feb. 1, 1932; d. Austin and Lily (Elliott) DaCosta; m. Donald German, Apr. 7, 1973 (dec. 1991); children: Preston, Vicki. MS in Math., Union Coll., 1958, MS in Engring., 1982. Systems analyst Gen. Electric Knolls Atomic Power Lab., Schenectady, N.Y., 1953-59; specialist in numerical methods Gen. Electric Rsch. Lab., Schenectady, 1959-79; mech. engr. Gen. Electric, Schenectady, 1979—. Contbr. articles to profl. jours. Mem. ASME, AIAA, Soc. Women Engrs. Office: Gen Electric Corp R&D PO Box 8 Schenectady NY 12301

GERMAN, NORTON ISAIAH, pathologist, educator; b. N.Y.C., Mar. 6, 1933; s. Harry Aaron and Dora (Seldin) G.; m. Judith Ellen Stolov, June 12, 1960; children: Linda G. Ackerman, Mark A. AB, NYU, 1956; MD, Albert Einstein Coll., 1960. Diplomate Am. Bd. Pathology. Dir. U.S. Army Med. Lab., Ft. Meade, Md., 1973-74; vice chmn. pathology dept. Walter Reed Army Med. Ctr., Washington, 1974-77; dep. dir. U.S. Armed Med. Rsch. Inst. Infectious Disease, Frederick, Md., 1977-78; hematopathology staff Armed Forces Inst Pathology, Washington, 1978-79; dir. dept. pathology St. Thomas Hosp. Med. Ctr., Akron, Ohio, 1979-83; chmn. dept. pathology St. Elizabeth Hosp. Med. Ctr., Youngstown, Ohio, 1983-90, assoc. dir. dept. pathology, 1990-93; prof. clin. pathology Northeastern Ohio U. Coll. Medicine, Rootstown, 1980—; bd. dirs. Nat. Accrediting Agy. for Clin. Lab. Scis., Chgo., Clin. Lab. Mgmt. Assn., Ohio chpt., Independence; pathology cons. Hillside Hosp., Warren, Ohio, 1984—, St. Joseph Hosp., Warren, 1988-93. Col. U.S. Army, 1952-79. U.S. Pub. Health fellow USPHS, 1957, 58. Fellow Am. Soc. Clin. Pathologists, Coll. Am. Pathologists (commr. No. Ohio Lab. Accreditation Program, del.); mem. Ohio Soc. Pathologists (pres. 1991-93). Home: 980 Royal Arms Dr Girard OH 44420-1652

GERMAN, RANDALL MICHAEL, materials science educator, consultant; b. Bainbridge, Md., Nov. 12, 1946; s. Eugene Knox and Helen (Schrufer) G.; m. Carol Jean Hosmer, Dec. 21, 1968; children: Eric, Garth. BS in Materials Sci., San Jose State U., 1968; MS in Metall. Engring., Ohio State U., 1971; PhD in Materials Sci., U. Calif., Davis, 1975; cert. mgmt. devel., Hartford Grad. Ctr., 1979. Materials scientist Battelle Columbus Labs., Columbus, Ohio, 1968-69; tech. staff Sandia Nat. Lab., Livermore, Calif., 1969-77; dir. R&D Mott Metall. Corp., Farmington, Conn., 1977-78; dir. rsch. J.M. Ney Co., Bloomfield, Conn., 1978-80; Hunt prof. Rensselaer Poly. Inst., Troy, N.Y., 1980-91; Brush chair materials Pa. State U., University Park, 1991—; founder Xform, Inc., Troy, 1989—; dir. PIM Symposium, San Francisco, 1990—. Author: Powder Metallurgy Science, 1984, Liquid Phase Sintering, 1985, Powder Packing, 1989, Injection Mold, 1990; contbr. over 320 articles to profl. jours.; patentee in field. Named Hon. Prof. N.E. U. Tech., 1985, Disting. Alumni U. Calif., 1990. Fellow ASM Internat. (chmn., Geissler award 1983); mem. Am. Powder Metallurgy (speaker, organizer), Minerals, Metals, Materials Soc. (chmn. 1983-85), Am. Ceramic Soc., Materials Rsch. Soc., Alpha Sigma Mu (hon.). Avocations: running, bicycling. Office: Brush Chair in Materials Pa State U 118 Rsch Bldg West University Park PA 16802-6809

GERMINO, FELIX JOSEPH, chemist, research-development company executive; b. N.Y.C., July 14, 1930; s. Thomas J. and Philamena (Di Clemente) G.; m. Faith E. Kramer, Oct. 11, 1952; children: Tom, Bill, Joe, Greg, Mary, Kevin, Faith. BS, Fordham U., 1952; MBA, U. Chgo., 1977; postgrad. advanced mgmt. program, Harvard U., 1977. Chemist Gen. Foods Corp., Tarrytown, N.Y., 1954-59; exploratory rsch. chemist Am. Machine & Foundry, Springdale, N.Y., 1959-64; asst. mgr. for starch CPC Internat., Argo, Ill., 1964-72; dir. pet foods Quaker Oats Co., Barrington, Ill., 1972-76, v.p. R & D for pet foods, 1976-78, v.p. R & D human foods, 1978-82; pres. F. Germino & Assocs., Inc., Orland Park, Ill., 1982—; bd. dirs. Knechtel Rsch. Scis. Inc.; bd. dirs. food update Food and Drug Law Inst., 1986-89, program chmn., 1986-87, chmn. bd. govs., 1987-88, chmn. long range planning com., 1988-89. Contbr. articles to Jour. Polymer Sci. Bd. dirs. Chgo. Opera Theatre, 1983-88. With U.S. Army, 1952-54. Mem. Am. Chem. Soc., Am. Inst. Chemists, Am. Assn. Cereal Chemists, Inst. Food Technologists (chmn. 50th anniversary subcom. for membership 1984-89), N.Y. Acad. Scis. Achievements include numerous patents in polysaccharide chemistry. Office: 9763 W 143d St Orland Park IL 60462

GERNER, FRANK MATTHEW, mechanical engineering educator; b. Ft. Thomas, Ky., Dec. 3, 1961; s. Edward Carl and Virginia Mary Margaret (Zier) G.; m. Deborah June Hamilton, Mar. 22, 1992. BS, U. Ky., 1984; MS, U. Calif., Berkeley, 1986, PhD, 1988. Engring. technician Eastman Kodak, Kingsport, Tenn., 1982-83; rsch. asst. U. Ky., Lexington, 1984, U. Calif., Berkeley, 1984-88; UES fellow Wright-Patterson AFB, Dayton, Ohio, 1989; asst. prof. mech. engring. U. Cin., 1988-93, assoc. prof., 1993—. Contbr. articles to profl. pubs.; editor: Heat Transfer on the Microscale, 1992. Treas. Campbell County (Ky.) Jaycees, 1991, bd. dirs., 1992. Eastman Kodak scholar, 1981-84; NSF fellow, 1984-87. Mem. AIAA, ASME, Sigma Xi, Tau Beta Pi, Pi Tau Sigma. Home: 302 Bluestone Ct Cold Spring KY 41076 Office: U Cin 683 Rhodes Hall Cincinnati OH 45221-0072

GERSCH, HAROLD ARTHUR, physics educator; b. N.Y.C., Jan. 8, 1922; S. Adolph and Marie (Reder) G.; m. Thelma Lee Gardner, Mar. 21, 1947; children: Lee, Harold Jr., Robert. BS in Physics, Ga. Inst. Tech., 1949; PhD, John Hopkins U. Asst. prof. Ga. Inst. Tech., Atlanta, 1953-56, assoc. prof., 1956-62, prof., 1962-70, regent's prof., 1970-87, regent's prof. emeritus, 1987—; vis. lectr. Johns Hopkins U., Balt., 1956-57; vis. prof. U. New Orleans, 1979-80, Oglethorpe U., 1987-88, U.S. Mil. Acad., 1989-90; cons. physics and solid state div. Oak Ridge (Tenn.) Nat. Lab., 1956—, Ford Found., Caracas, Venezuela, 1966. Contbr. over 50 articles to profl. jours. Served in USN, 1940-46. Named NATO Sr. fellow NSF, 1973. Fellow Am. Physical Soc. Republican. Presbyterian. Avocation: sailing. Home: 40 Wilson Rd Apt H West Point NY 10996-1914 Office: Ga Inst of Tech Dept of Physics Atlanta GA 30332-0183

GERSHENGORN, MARVIN CARL, physician, educator; b. N.Y.C.. M.D., NYU, 1971. Diplomate Am. Bd. Internal Medicine. Intern Strong Meml. Hosp., Rochester, N.Y., 1971-72, asst. resident in medicine, 1972-73; asst. prof. medicine NYU Sch. Med., 1976-80, assoc. prof., 1980-83; prof. medicine Cornell U. Med Coll., N.Y.C., 1983—; Abby Rockefeller Mauze disting. prof. Cornell U. Med Coll. Office: Cornell Univ Med Coll 1300 York Ave New York NY 10021-4896

GERST, STEVEN RICHARD, healthcare director, physician; b. N.Y.C., Oct. 20, 1958; s. Paul Howard and Elizabeth (Carlsen) G.; m. Isabelle Sylvie Meier, Apr. 21, 1987 (div.); 1 child, Chantal Elizabeth. BA, Columbia U., 1981, MD, 1986, MPH, 1987. Lic. ins. broker, N.C. Med. affairs coord. Sun Health Care Plans, Charlotte, N.C., 1987-88, cons., 1988-90; dir. Prefered Provider Orgns. Crawford and Co., Atlanta, 1990-93; v.p. Imaginative Devices Inc., Atlanta, 1993—; interviewer Columbia Coll., N.C., 1987. Editor-in-chief: Handbook Coll. Physicians and Surgeons (Alumni award), 1983, Columbian (Robert Shellow Gerdy award), 1981. Vol. Presbyn. Hosp., N.Y.C., 1979-81, St. Lukes Hosp., N.Y.C., 1978-79. Mem. AMA, Am. Acad. Med. Dirs., Am. Coll. Med. Staff Affairs, Am. Coll. Physician Execs., Am. Coll. Health Care Execs., Andover Alumni Soc. N.Y. (dir. 1986-87), Alliance Francaise (v.p. Charlotte, N.C. chpt. 1987-88). Avocations: bicycling, tennis, golf, dance, gardening. Home: 5450 Glenridge Dr # 372 Atlanta GA 30342 Office: 5620 Glenridge Dr NE Atlanta GA 30342-1399

GERSTEIN, MARK BENDER, biophysicist; b. N.Y.C., Feb. 23, 1966; s. David Brown and Jane Ellen (Bender) G. AB summa cum laude, Harvard U., 1989; PhD, Cambridge U., 1992. Mem. staff Thermwell Products Co., Paterson, N.J., 1987-88; rsch. scientist MRC Lab Molecular Biology & Univ. Chem. Labs., Cambridge, Eng., 1992—. Contbr. articles to sci. jours. Mem. Phi Beta Kappa, Sigma Pi Sigma. Home: 432 Long Hill Dr Short Hills NJ 07078

GERSTNER, ROBERT WILLIAM, structural engineering educator, consultant; b. Chgo., Nov. 10, 1934; s. Robert Berty and Martha (Tuchelt) G.; m. Elizabeth Willard, Feb. 8, 1958; children: Charles Willard, William Mark. B.S., Northwestern U., 1956, M.S., 1957, Ph.D., 1960. Registered structural and profl. engr., Ill. Instr. Northwestern U., Evanston, Ill., 1957-59; research fellow Northwestern U., 1959-60; asst. prof. U. Ill., Chgo., 1960-63; asso. prof. U. Ill., 1963-69, prof. structural engring., architecture, 1969-92, prof. emeritus, 1992—; structural engr. cons., 1959—. Contbr. articles to profl. jours. Pres. Riverside Improvement Assn., 1973-77, 79-82. Fellow

ASCE; mem. AAUP, ACLU, Am. Concrete Inst., Am. Soc. Engring. Edn., Structural Engrs. Assn. Ill. (bd. dirs. 1986-89, 92—, sec. 1989-91, pres. 1991-92). Home: 2628 W Agatite Ave Chicago IL 60625-3011

GERTIS, NEILL ALLAN, writer; b. Buffalo, Mar. 24, 1943; s. Alfred Charles and Gertrude Charlotte (Hurst) G.; m. Gail C. Morgan, Oct. 3, 1966 (div. Aug. 1982); m. Alma Ann Sullivan, Sept. 15, 1984; children: Charlotte Ann, Joseph Alfred, Daniel Andrew, Martin Alexander. Community planner Alaska State Housing Authority, Anchorage, 1968-72, libr. dir., 1968-72; real estate appraiser Gertis Assocs., Buffalo, 1972-76; ops. mgr. Chem. Equipment Labs., Phila., 1979-83; tech. writer Gen. Dynamics, Groton, Conn., 1983-84; sr. tech. writer communication products MTS Systems Corp., Mpls., 1984—. Author: Student Housing Demand in Anchorage, 1972; editor: Guide to Periodical Holdings in the Anchorage Area, 1970, Storm Drainage for Chester Creek, Anchorage, 1969; editor Engring. Graphics, Anchorage, 1968-72. Served with U.S. Army, 1965-69. Mem. Alaska Library Assn. Republican. Avocations: computers, antiques, restoration, writing, reading. Home: 1157 Tyler St Shakopee MN 55379-2070 Office: MTS Systems Corp 14000 Technology Dr Eden Prairie MN 55344-2290

GERTLER, MENARD M., physician, educator; b. Saskatoon, Sask., Can., May 21, 1919; came to U.S., 1947, naturalized, 1953; s. Frank and Clara (Handelman) G.; m. Anna Paull, Sept. 4, 1943; children—Barbara Lynn, Stephanie Jocelyn, Jonathan Paull. BA, U. Sask., 1940; MD, McGill U., 1943, MS, 1946; DSc, NYU, 1960. Intern Royal Victoria Hosp., Montreal, Que., Can., 1943-44; resident Mass. Gen. Hosp., Boston, 1947-50; also research fellow in medicine Mass. Gen. Hosp., Harvard Med. Sch., 1947-50; dir. cardiology Francis Delafield div. Columbia Presbyn. Med. Ctr., N.Y.C., 1950-54; spl. research fellow NIH, NYU Dept. Biochemistry, 1954-56; prof. Sch. Medicine, dir. cardiovascular research Rusk Inst. NYU Med. Ctr., 1958-71; sr. med. examiner FAA, 1975; med. dir. Sinclair Oil Corp., 1958-68; dir. Washington Fed. Savs. & Loan Assn., 1972-83; internat. cons. cardiovascular diseases, social and rehab. services HEW, Washington, 1968—. Author: Coronary Heart Disease in Young Adults, 1954, Coronary Heart Disease, 1974; Contbr. articles to profl. jours. Pres. Friends of McGill U.; mem. dean's com. McGill U. Med. Sch. With M.C., Royal Can. Army, 1940-43. Recipient Founders Day award NYU, 1959. Mem. Gallatin Assocs. NYU, Cosmos Club (Washington), Harvard Club (Boston), Univ. Club. Home: 1000 Park Ave Apt 2C New York NY 10028-0934 Office: NYU Med Ctr Rusk Inst 400 E 34th St New York NY 10016-4998

GERVASI, ANNE, language professional, English language educator; b. Rochester, N.Y., Jan. 6, 1947; d. Francis Charles and Mildred Inez (Quillin) G.; m. John David Van Camp, Jan. 28, 1968 (div.); 1 child, Kierstin Laine. BA in Theater, Baylor U., 1968; MLS in Info. Sci., U. Tex., 1971; PhD in English, Tex. Woman's U., 1992. Div. chief Austin (Tex.) Pub. Libr., 1971-74; grants adminstr. Tex. State Libr., Austin, 1974-76; systems dir. South Tex. Libr. System, Corpus Christi, 1976-80; county libr. Wise County, Decatur, Tex., 1982-85; teaching fellow Tex. Woman's U., Denton, 1984-91; instr. North Lake Coll., Irving, Tex., 1991—; pres. Wordhill Cons., Sanger, Tex., 1985—. Editor: Out of Chaos: Semiotics, 1990; co-author: Handbook for Small, Rural & Emerging Public Libraries, 1988. Recipient Toulouse fellowship Fedn. of North Tex., 1988. Mem. MLA, Nat. Coun. Tchrs. English, Coll. English Assn., Tex. Coun. Tchrs. English. Democrat. Episcopalian. Home: 109 Culp Branch Rd Sanger TX 76266

GESELOWITZ, DAVID BERYL, bioengineering educator; b. Phila., May 18, 1930; s. Sidney W. and Fannie (Charny) G.; m. Lola Wood, June 21, 1953; children: Daniel, Michael, Ari. B.S. in E.E, U. Pa., 1951, M.S. in E.E, 1954, Ph.D., 1958. Asst. prof., asso. prof. U. Pa., Phila., 1951-71; prof., head biomed. engring. Pa. State U., University Park, 1971-88, prof. medicine, 1982—, Alumni Disting. prof. biomed. engring., 1985—; vis. asso. prof. elec. engring. M.I.T., Cambridge, 1965-66; cons. to med. dir. Provident Mut. Life Ins. Co., Phila., 1959-71; vis. prof. biomed. engring. Duke U., Durham, N.C., 1978-79; vis. prof. medicine U. Okla., Oklahoma City, 1987-88. Editor: (with C. V. Nelson) Theoretical Basis of Electrocardiography, 1976, IEEE Transactions on Biomed. Engring, 1967-72; mem. editorial bd.: Jour. Electrocardiology, 1974—, CRC Critical Revs., 1979-88; contbr. articles to various publs. Chmn. com. electrocardiography Am. Heart Assn., 1976-81. J.S. Guggenheim fellow, 1978-79. Fellow Am. Coll. Cardiology, IEEE (Career Achievement award 1985); mem. ISE (founding), AAAS, Am. Inst. Engring. Medicine and Biology (founding), Biomed. Engring. Soc., Am. Assn. Physics Tchrs., Nat. Acad. Engring. Office: Pa State Univ University Park PA 16802

GESKIN, ERNEST S(AMUEL), science administrator, consultant; b. Dnepropetrovsk, Ukraine, USSR, June 4, 1935; came to U.S., 1977; s. Samuel A. and Rosa M. (Raskin) G.; m. Doris M. Osherenko, June 12, 1964; 1 child, Ellen. M in MetE, Inst. Mettalurgy, Dnepropetrovsk, 1957; PhD in ME, Inst. Steel and Alloys, Moscow, 1967. Engr. Inst. Automation, Dnepropetrovsk, 1957-67, mgr. lab., 1967-74; assoc. rsch. prof. George Washington U., Washington, 1977-78; assoc. prof. Clarkson Coll. Tech., Potsdam, N.Y., 1979-80; rsch. scientist, lab. mgr. Revere Rsch. Inc., Edison, N.J., 1981-83; dir. waterjet cutting lab. Revere Rsch. Inc., Edison, 1986—; spl. lectr. N.J. Inst. Tech., Newark, 1984-85, assoc. prof., 1986-90, prof. 1991—. Author/co-author over 80 papers and presentations; editor various symposia (Cert. Recognition, 1984, 89); 22 U.S. and USSR patents, 1969—. Mem. Iron & Steel Soc. AIME, Waterjet Tech. Assn., Sigma Xi. Avocations: swimming, gardening. Office: NJ Inst Tech 323 King Blvd Newark NJ 07102

GESS, ALBIN HORST, lawyer; b. Lithuania, Apr. 22, 1942; came to U.S. 1956; s. Albin and Emily (Block) G.; m. Brenda Martha Massaroni, Dec. 30, 1966; children: Lisa, Brent. BEE, U. Detroit, 1966; JD, Am. U., 1971. Bar: Calif. 1972, U.S. Dist. Ct. (cen. dist.) Calif. 1972, U.S. Ct. Appeals (9th cir.) 1972, U.S. Supreme Ct. 1977, U.S. Ct. Appeals (1st and 10th cirs.) 1979, U.S. Ct. Appeals (fed. cir.) 1982, U.S. Dist. Ct. (so. and no. dists.) Calif. 1985. Student engr. Detroit Edison Co., 1964-66; patent examiner U.S. Patent Office, Washington, 1966-68; patent agt. Office of Naval Rsch., Washington, 1968-69, Burroughs Corp., Washington, 1969-71; sr. patent atty. Burroughs Corp., Pasadena, Calif., 1971-74; patent atty. Jackson & Jones, Tustin, Calif., 1974-85, Price, Gess & Ubell, Irvine, Calif., 1985—. Fellow Inst. for Advancement of Engring.; mem. ABA (patent, trademark, copyright and litigation sect.), Fed. Bar Assn., Orange County Bar Assn., Orange County Patent Law Assn. (bd. dirs. 1985—, pres. 1990—), IEEE (sec. Orange County sect. 1992, 93), Am. Electronics Assn., Am. Intellectual Property Law Assn., L.A. Intellectual Property Law Assn., Licensing Execs. Soc., Lions (v.p. 1988, bd. dirs.). Avocations: enduro motorcycle riding, racquetball, skiing. Office: Price Gess & Ubell 2100 Main St Ste 250 Irvine CA 92714-6238

GEST, HOWARD, microbiologist, educator; b. London, Oct. 15, 1921; m. Janet Olin, Sept. 8, 1941; children: Theodore Olin, Michael Henry, Donald Evan. B.A. in Bacteriology, UCLA, 1942; postgrad. in biology (Univ. fellow), Vanderbilt U., 1942; Ph.D. in Microbiology (Am. Cancer Soc. fellow), Washington U., St. Louis, 1949. Instr. microbiology Western Res. U. Sch. Medicine, 1949-51, asst. prof. microbiology, 1951-53, asso. prof., 1953-59; USPHS spl. research fellow in biology Calif. Inst. Tech., 1956-57; prof. Henry Shaw Sch. Botany, Washington U., 1959-64, dept. zoology, 1964-66; prof. Ind. U., Bloomington, 1966-78, Disting. prof. microbiology, 1978—, Disting. prof. emeritus microbiology, 1987—, adj. prof. history and philosophy of sci., 1983—, chmn. dept. microbiology, 1966-70; NSF sr. postdoctoral fellow Nat. Inst. Med. Rsch., London, 1965-66; Guggenheim fellow Imperial Coll., London, U. Stockholm, U. Tokyo; vis. prof. dept. biophysics and biochemistry U. Tokyo and Japan Soc. Promotion Sci., 1970; mem. study sect. bacteriology and mycology NIH, 1966-68, chmn. study sect. microbial chemistry, 1968-69, mem. study sect. microbial physiology and genetics, 1988-90; mem. com. microbiol. problems of man in extended space flight Nat. Acad. Scis.-NRC, 1967-69. Guggenheim fellow Imperial Coll., London, UCLA, 1979-80; recipient Disting. Faculty Research Lecture award, Ind. U., 1987. Office: Ind U Dept Biology Bloomington IN 47405

GEST, ROBBIE DALE, aerospace engineer; b. Austin, Tex., Apr. 13, 1964; s. Arbie James and Betty Gene (Adkins) G. BS in Aerospace Engring., Tex. A&M U., 1986, postgrad. in aerospace engring., 1993. Registered profl. engr., Tex. Rsch. asst. Ctr. for Strategic Tech., College Station, Tex., 1987-88; propulsion engr. Rockwell Space Ops. Co., Houston, 1988—. Contbr. articles to profl. publs. Recipient Group Achievement award NASA, 1989. Mem. AIAA, Nat. Mgmt. Assn. (cert. mgr.). Achievements include development of a reaction cross-section theoretical model for hydrazine reaction in a rarified environment. Office: Rockwell Space Ops Co Johnson Space Ctr Mail Code DF63 Houston TX 77058

GETTIG, MARTIN WINTHROP, retired mechanical engineer; b. South Bend, Ind., Nov. 8, 1939; s. Joseph H. and Esther (Scheppele) G.; m. Nancy Caroline Buchannan, June 25, 1960 (dec. 1965). Student, Pa. State U., 1957-60, 89—. Process engr. Gettig Tech. Inc., Spring Mills, Pa., 1960-88. Inventor ultralight non-solid state miniature ignition systems for model aircraft employing small two cycle spark ignition engines. Staff sgt. Pa. N.G., 1961-67. Mem. NRA, Model Engine Collectors Assn., Soc. Antique Modelers and Model Airplanes, Acad. Model Awronautics, Univ. Club Pa. State U., Delta Phi. Republican. Lutheran. Home: PO Box 85 Boalsburg PA 16827-0085

GETZ, MORTON ERNEST, internist, gastroenterologist; b. Bklyn., May 22, 1930; s. Jacob Michael and Regina (Kohn) G.; m. Carol Washer, Aug. 12, 1956; children: Jacob Michael, Deborah Etta. AB, Emory U., 1950; MS, Purdue U., 1952; MD, Wake Forest U., 1956. Intern Jackson Meml. Hosp., Miami, Fla., 1956-57; resident in medicine Jackson Meml. Hosp., 1957-58; sr. surgeon NIH, Atlanta and Bethesda, Md., 1958-60; chief resident in medicine Jackson Meml. Hosp., 1960; NIH fellow in gastroenterology U. Miami, 1960-61; pvt. practice internal medicine and gastroenterology Coral Gables, Fla.; mem. courtesy staff South Miami Hosp. Contbr. articles to profl. jours. With USPHS, 1958-60. Mem. Miami Fla. Gastroenterologic Soc., Dade County Soc. Internal Medicine, Am. Soc. Internal Medicine, Am. Geriatric Soc., So. Med. Assn., Fla. Med. Assn., Dade County Med. Assn., AMA, Ind. Acad. Scis., N.C. Acad. Sci., Phi Rho Sigma. Democrat. Jewish. Avocations: art collecting, fishing. Office: 4675 Ponce De Leon Blvd Miami FL 33146-2113

GEUSIC, JOSEPH EDWARD, physicist; b. Nesquehoning, Pa., Nov. 21, 1931; s. Joseph John and Mary Martha (Kosch) G.; m. Irene Jean Hosak, July 18, 1953; children: Patricia, Mark, Michael, Mary Ellen, Robert, Joseph. BS in Physics, Lehigh U., 1953; MS in Physics, Ohio State U., 1955, PhD in Physics, 1958. Rsch. assoc. physics dept. Ohio State U., Columbus, 1955-58; mem. tech. staff AT&T Bell Labs., Murray Hill, N.J., 1958-62, supr. solid state maser group, 1962-66, head solid state optical device dept., 1966-70, head magnetics dept., 1970-84, head semiconductor laser dept., 1984—. Recipient R.W. Wood Prize Optical Soc. Am., 1993, Patent trophy AT&T, 1993. Fellow IEEE (Quantum Electronics award 1992); mem. Am. Inst. Physics, Sigma Xi. Achievements include first report of paramagnetic spectra of Cr 3 in Ruby; invention and devel. of Nd/YAG laser, barium sodium niobate nonlinear optical material; first demonstation of continuous operating optical parametric oscillator; devel. of semiconductor lasers for terrestrial and undersea lightwave communication systems, magnetic bubble materials and devices; over 30 patents in field. Home: 261 Lorrainne Dr Berkeley Heights NJ 07922 Office: AT&T Bell Labs 600 Mountain Ave Rm 2D-354 Murray Hill NJ 07922

GEZELTER, ROBERT L., computer systems consultant; b. N.Y.C., Apr. 17, 1959; s. Bertram and Francine (Waltzman) G.; m. Edna Shoshani, Aug. 26, 1990; 1 child, Edmund Abraham. BA, NYU, 1981, MS, 1983. Rsch. staff Courant Inst. NYU, 1977-82; prin. Robert Gezelter Software Cons., Flushing, N.Y., 1978—. Contbr. articles to profl. jours.; contbg. editor: Hardcopy Mag, 1987-89, Digital News, 1989-92, Computer Purchasing Update, 1992—. Office: Robert Gezelter Software Cons 35-20 167th St Ste 215 Flushing NY 11358-1731

GFELLER, DONNA KVINGE, clinical psychologist; b. Chgo., Jan. 15, 1959; d. Milton Melvin and Doris Ann (Chapman) Kvinge; m. Jeffrey Donald Gfeller, Aug. 2, 1986. BS in Biol. Scis., Ill. State U., 1980, MS in Clin. Psychology, 1984; PhD in Clin. Psychology, Ohio U., 1987. Lic. psychologist. Staff psychologist Cardinal Glennon Children's Hosp., St. Louis, 1986-87, sr. psychologist, 1988-89, dir. dept. psychology, 1990—. Mem. APA (divsn. clin. psychology, sect. on clin. child psychology), Soc. Pediatric Psychology, World Wildlife Fund, Nat. Audubon Soc. Avocations: travel, horseback riding. Office: Cardinal Glennon Children's Hosp 1465 S Grand Blvd Saint Louis MO 63104-1095

GHADIA, SURESH KANTILAL, chemical engineer; b. Ahmedabad, India, Sept. 12, 1948; came to the U.S., 1970; s. Kantilal B. and Chandrakanta K. J. Shah; m. Kokila Suresh Ghadia, Jan. 18, 1974; 1 child, Birva S. BS in Chemistry, Gujarat U., 1969; BSChE, Tex. A&I U., 1973; MSChE, Lamar U., 1981. Registered profl. engr., Tex., La. Rsch. engr. Salastomer Chgo. Inc., Bensenville, Ill., 1973-75, City of Houston, 1975-77; chem. engr. Gilbert/Commonwealth Inc., Jackson, Mich., 1977-79; lead planning engr. Gulf States Utilities, Beaumont, Tex., 1979-92; supr. process utilities and offsites SAMAREC-Riyadh (Saudi Arabia) Refinery, 1992—. Mem. Indian Assn. S.E. Tex. (exec. com. 1990-91). Home: 2115 Somerset Dr Beaumont TX 77707 Office: Samarec Riyadh Refinery, c/o Tech Svcs Dept, PO Box 2946, Riyadh 11194, Saudi Arabia

GHALI, AMIN, civil engineering educator. Prof. civil engring. U. Calgary, Alta., Can. Recipient Le Prix A.B. Sanderson award Can. Soc. Civil Engring., 1992. Office: Univ of Calgary, Dept of Civil Engring, Calgary, AB Canada T2N 1N4*

GHALI, ANWAR YOUSSEF, psychiatrist, educator; b. Cairo, May 30, 1944; came to U.S., 1974, naturalized, 1980; s. Youssef and Insaf Wahba (Soliman) G.; m. Violette Fouad Saleh, May 23, 1968; 1 child, Susie. MD, Cairo U., 1966, DPM, 1970, DM, 1971. Diplomate Am. Bd. Psychiatry and Neurology; cert. adminstrv. psychiatry. Registrar in psychiatry Woodilee Hosp., Glasgow, Scotland, 1973-74; resident in psychiatry N.J. Med. Sch., Newark, 1974-77, instr., 1977-78, clin. asst. prof., 1978-79, asst. prof., 1979-83, clin. assoc. prof., 1983—; chief Outpatient Dept.-Community Mental Health Ctr., N.J. Med. Sch., Newark, 1978-86; dir. Emergency Psychiat. Svcs. Univ. Hosp., U. Medicine and Dentistry of N.J., Newark, 1986-87; med. dir. Profl. Counsel Ctr., Westfield, N.J., 1984-87; med. dir. chief ambulatory psychiat. svcs. Elizabeth (N.J.) Gen. Hosp., 1987-89; dir. psychiat. tng. VA Med. Ctr., East Orange, N.J., 1989—, asst. chief psychiatry, 1990-91, assoc. chief psychiatry, 1991—. Contbr. articles to profl. jours. Recipient Exceptional Merit award Coll. Medicine & Dentistry, Newark, 1981. Mem. AMA, Christian Med. Soc., Am. Psychiat. Assn., N.J. Psychiat. Assn., N.Y. Acad. Scis. Republican. Presbyterian. Home: 22 Benvenue Ave West Orange NJ 07052-3202

GHANDHI, SORAB KHUSHRO, electrical engineering educator; b. Allahabad, India, Jan. 1, 1928; came to U.S., 1947, naturalized, 1960; s. Khushro S. and Dina (Amroliwalla) G.; m. Cecilia M. Ghandhi; children: Khushro, Rustom, Behram. B.Sc. in Elec. and Mech. Engring, Benares (India) Hindu U., 1947; M.S., U. Ill., 1948, Ph.D., 1951. Mem. electronics lab. Gen. Electric Co., 1951-60; mgr. electronic components and functions lab., research div. Philco Corp., 1960-63; prof. elec. engring. Rensselaer Poly. Inst., Troy, N.Y., 1963—, chmn. electrophysics and electronic engring. div., 1968-75, prof. electrophysics, elec., computer and systems engring. dept., 1975—; cons. to industry, 1963—. Co-author: (with R. F. Shea editor) Principles of Transistor Circuits, 1953, Transistor Circuit Engineering, 1957, Amplifier Handbook, 1966; author: The Theory and Practice of Microelectronics, 1968, Semiconductor Power Devices, 1977, VLSI Fabrication Principles: Silicon and Gallium Arsenide, 1983. J.N. Tata fellow, 1947-50. Fellow IEEE; mem. Electrochem. Soc., Am. Standards Assn., Sigma Xi, Eta Kappa Nu, Pi Mu Epsilon, Phi Kappa Pi. Home: 7 Linda Ln Schenectady NY 12309-1911 Office: Rensselaer Poly Inst Troy NY 12181

GHANI, ASHRAF MUHAMMAD, mechanical engineer, business and engineering consultant; b. Wazirabad, Pakistan, Oct. 12, 1931; emigrated to Saudi Arabia, 1967; s. Abdul and Alam (Bibi) G.; m. Yasmeen Elahi, Nov. 15, 1964; children: Faiza, Saad, Farha. BS in Mech. Engring., Ind. Inst. Tech., 1962; MS in Mech. Engring., Columbia U., 1963; PhD in Mgmt., Franklin U., 1977. Tech. dir. Engring. Controls, Karachi, Pakistan, 1963-67; asst. prof. Riyadh U., Saudi Arabia, 1967-76; tech. expert Saudi Fund for Devel., Riyadh, 1976-86; cons. dir. Poly Engring. Co., Riyadh, 1979-87; chmn. Polyconsult Internat., Inc., Riyadh, 1979—, Inter-Services Corp., Metuchen, N.J., 1982—, chmn. Vols. Orgn. for Tech. Assistance to Underdeveloped Countries, U.K., 1985—; convener Internat. Solidarity for Peace, U.K., 1986—; sec. Internat. Vols. for Human Relief, Vienna, 1986—; chmn. Pakistion Biographical Inst., 1991—. Author: Management of Complex Development Projects, 1979. Hon. sec. Pakistan Red Crescent Soc., Lahore, 1965. Recipient William Henry Caswell award Ind. Inst. Tech., 1961. Fellow Pakistan Assn. Mgmt. Cons. (Achievement award 1981); mem. Soc. Am. Mil. Engrs., mem. Brit. Inst. Mgmt. (assoc.), Soc. Internat. Devel., Internat. Journalists Assn. London, Am. Soc. for Tng. and Devel. Moslem. Club: Gymkhana (Lahore, Pakistan). Home: 11-H Gulberg Three, Lahore 54660, Pakistan

GHAUSI, MOHAMMED SHUAIB, electrical engineering educator, university dean; b. Kabul, Afghanistan, Feb. 16, 1930; came to U.S., 1951, naturalized, 1963; s. Mohammed Omar and Homaira G.; m. Marilyn Buchwold, June 12, 1961; children: Nadjya, Simine. B.S. summa cum laude, U. Calif., Berkeley, 1956, M.S., 1957, Ph.D., 1960. Prof. elec. engring. NYU, 1960-72; head elec. scis. sect. NSF, Washington, 1972-74; prof., chmn. elec. engring. dept. Wayne State U., Detroit, 1974-77; John F. Dodge prof. Oakland U., Rochester, Mich., 1978-83; dean Sch. Engring. and Computer Sci., Oakland U., 1978-83; dean Coll. Engring., U. Calif., Davis, 1983—; mem. adv. panel NSF, 1989. Author: Principles and Design of Linear Active Circuits, 1965, Introduction to Distributed-Parameter Networks, 1968, Electronic Circuits, 1971, Modern Filter Design: Active RC and Switched Capacitor, 1981, Electronic Devices and Circuits: Discrete and Integrated, 1985, Design of Analog Filters, 1990, also numerous articles.; cons. editor Van Nostrand Rinehold Pub. Co., 1968-71. Mem. disting. alumni rev. panel Elec. Engring. and Computer Sci. programs U. Calif., Berkeley, 1973; mem. external bd. visitors U. Pa., 1974. Fellow IEEE (chmn. edn. medal com. 1990—, Centennial medal, Alexander von Humboldt prize, circuits and systems soc. edn. award); mem. Circuits and System Soc. (v.p. 1970-72, pres. 1976), N.Y. Acad. Scis., Engring. Soc. Detroit, Sigma Xi, Phi Beta Kappa, Tau Beta Pi, Eta Kappa Nu. Office: U Calif Office of Dean Coll Engring Davis CA 95616

GHAZALI, SALEM, linguist, educator; b. Ghoumrassen, Tunisia, Jan. 17, 1944; m. Zohra Zaghbani, Sept. 3, 1971 (dec. 1983); children: Anis, Nizar; m. Maria Martinez, May 11, 1985. BA, UCLA, 1969, MA, 1973; PhD in Linguistics, U. Tex., 1977. Lectr. U. Tunis, 1971-73, asst. prof., 1977-80; assoc. prof. Bourguiba Inst. Modern Langs., Tunis, 1980-87, prof., 1984—; rsch. dir. Inst. Regional des Scis. Informatiques et Telecommunications Speech Communication Lab., 1986—. Contbr. articles to profl. jours. Grantee Ford Found., 1974, Fulbright Found., 1980; scholar U.S. Agy. for Internat. Devel., 1971-73, African Am. Inst., 1975-77. Mem. Linguistic Soc. Tunisia, Soc. Francaise d'Acoustique, Acoustical Soc. Am., Found. Nat. de la Rsch., 1991. Achievements include research in phonetic theory, speech timing, coarticulation phenomena in production of Arabic, development of Arabic text-to-speech system operational on a Sun workstation, English-to-Arabic computer assisted translation system. Office: IRSIT, PO Box 212, 1082 Mahrajane Tunisia

GHAZARIAN, ROUBEN, structural engineer; b. Abadan, Khoozestan, Iran, Jan. 5, 1956; came to U.S., 1983; s. Samson and Astkhik (Abramian) G.; m. Ida Kehish, Dec. 11, 1982; children: Armen A., Arminen L. A. Crawley (Eng.) U., 1979; BSc, Southampton (Eng.) U., 1981, MSc, 1982. Registered profl. engr., D.C., Md. Engr. Woodward Engring., Palm Beach, Fla., 1983, Causway Lumber Co., Ft. Lauderdale, Fla., 1983-85, Carson Mok Engring., Silver Spring, Md., 1985; project engr. Neubauer-Sonn Consulting Engrs., Ptomac, Md., 1985-88, Tadjer-Cohen-Edelson & Assoc., Silver Spring, 1988-90; sr. engr. Alpha Corp., Sterling, Va., 1991—. Mem. ASCE. Home: 14853 Hammersmith Cir Silver Spring MD 20906 Office: Alpha Corp 45665 Willow Pond Plz Sterling VA 20164

GHENT, ROBERT MAYNARD, JR., clinical audiologist, engineer, consultant; b. Compton, Calif., Apr. 5, 1956; s. Robert Maynard Sr. and Dorothy Francis (Sharpnack) G.; m. Judith Irene Davis, Oct. 1, 1977; children: Sarah, Holly, Gabriel, Jacqueline. BS in Communicative Disorders, Brigham Young U., 1991, MS in Audiology, 1993. Audio engr. Tychobrahe Sound Co., Hermosa Beach, Calif., 1974-77; product engr. Internat. Rectifier Corp., El Segundo, Calif., 1980-86; owner Associated Electronic Artists, Redondo Beach, Calif., 1978—; cons. electronic music applications for gifted and talented elem. edn. Alpine Sch. Dist., Lindon, Utah, 987—; audio/acoustics engr., Orem, Utah, 1987—; operator computer music studio; presenter in field. Contbr. articles to profl. jours. Choir dir. various LDS congregations, 1976—. Recipient Alonzo & Eloise Morley award for academic excellence in communicative disorders, Nat. Collegiate Edn. award USAA All-Am. Scholar At Large; audiology rsch. grantee Brigham Young U., 1992. Mem. Acoustical Soc. Am., Am. Speech-Lang.-Hearing Assn., Am. Auditory Soc. Democrat. Mem. Ch. LDS. Achievements include development of automated semiconductor handling and manufacturing processes and systems for wafers and finished product, automated semiconductor test program development and new test and quality philosophies pioneered for start up of International Rectifier's Hexfet America Facility, rsch. in devel. of technical standards for recorded speech audiometry and central auditory test materials. Office: AEA PO Box 1536 Orem UT 84059-1536

GHIARA, PAOLO, immunopharmacologist; b. Naples, Italy, Apr. 18, 1958; s. Gianfranco and Tina (Lolli) G.; m. Francesca Righi-Parenti, Dec. 8, 1984; children: Silvia, Guido. Diploma in Music, Conservatory Mascagni, Livorno, Italy, 1980; laurea in Farmacia cum laude, U. Naples, 1982. Diplomate in pharmacy. Student intern Inst. Exptl. Pharmacology, Univ. Naples, 1976-82; fellowship pharmacology Inst. Exptl. Pharmacology U. Naples, 1982-83; rsch. asst. Sclavo Rsch. Ctr., Siena, Italy, 1983-85, researcher, 1985-88, sr. investigator Sclavo Rsch. Ctr., Siena, Eng., 1988-90, chief lab. immunopharmacology Sclavo Rsch. Ctr., Siena, 1991; sr. staff scientist dep. immunology Immunobiol. Rsch. Inst., Siena, 1992, sr. staff scientist dept. pharmacology, 1993—; visiting scientist Wellcome Rsch. Labs., Beckenham, Eng., 1985, Ludwig Inst. Cancer Rsch., Epalinges, Lausaune, C.H., 1988. Mem. Gruppo Di Coop. Immunology, ESO Alumni Group, Am. Assn. Immunologist. Roman Catholic. Avocations: classical music, informatics. Home: Via Lorenzo Maitani, Siena 53100, Italy Office: Immunobiol Rsch Inst Dept Pharmacology, Via Fiorentina 1, Siena 53100, Italy

GHIZONI, CÉSAR CELESTE, electrical engineer; b. San Joaquim, Brazil, Feb. 16, 1945; s. Estevão and Ruth B. (Brasil) G.; m. Ielva Fátima Costa, Mar. 21, 1970; children: Enrico, César, Leonardo. Diploma in engring., Fed. U., Rio Grandesul, Brazil, 1969; PhD, Cornell U., 1976. Head microwave lab. Nat. Inst. Space, St. Campos, Brazil, 1970-72, head sensor div., 1976-81, dir. engring., 1986-91; rsch. asst. Cornell U., Ithaca, N.Y., 1973-76; head laser div. Air Force Tech. Ctr., St. Campos, 1981-86; mgr. space systems ESCA Engring., Barueri, Brazil, 1991—. Contbr. articles to profl. publs. Office: ESCA, Al Araguaia, 1142 Alphaville, 06460 Barueri Brazil

GHORPADE, AJIT KISANRAO, chemical engineer; b. Wai, Maharashtra, India, Mar. 18, 1954; s. Kisanrao R. and Nalini K. (Sumati) G.; m. Rashmi A. Leele Behere, Dec. 20, 1982; children: Kaustubh A., Divya A. BS, U. Mysore, India, 1977; MChemE, U. Bombay, India, 1980; MS, U. Bombay, 1985; PhD, U. Louisville, 1989. Rsch. engr. Hindustan Lever Ltd., Bombay, 1980-82; devel. engr. Menon and Menon, Kolhapur, India, 1982-83; project engr. Eckenfelder, Inc., Nashville, Tenn., 1989-91; sr. engr. Merck and Co., Elkton, Va., 1991—. Recipient Grad. award Rotary Internat. 1983. Mem. Am. Inst. Chem. Engrs., Am. Chem. Soc., Water Environ. Fedn. Hindu. Home: 2214 Lonicera Way Charlottesville VA 22901 Office: Merck and Co PO Box 7 Elkton VA 22827-0007

GHOSH, ARUN KUMAR, economics, social sciences and accounting educator; b. Burdwan, West Bengal, India, Feb. 1, 1930; came to U.S., 1986; s. Ashu Tosh and Indu Prova (Roy Mitter) G.; m. Krishna Datta, Dec. 10, 1986. BA with Honors in Econs. and Polit. Sci., U. Calcutta, India, 1948, MA in Econs., 1950. Asst. tchr. Burdwan Town Sch., 1950-51; rsch. fellow Dept. Econs. U. Calcutta, 1952-55, examiner, re-examiner, scrutineer BA and

B of Commerce exams., 1952-66, asst. prof., 1955-56, rsch. asst. in indsl. fin., 1956-66; tutor Inst. Cost and Works Accts. India, Calcutta, 1966-69, asst. dir. rsch., 1970-85, faculty mem. exec. and profl. devel. programs, 1970-86, head rsch. directorate, 1981-84, asst. dir. exams., 1985-88; socio-polit. commentator, analyst The Radical Humanist, 1950-62; vis. prof. Indian Inst. Mgmt., Calcutta, 1973-74; chmn. exams com. Internat. Inst. Mgmt. Scis., Calcutta, 1984-86, papersetter MBA exams, 1985; cons. U.S. AID, Indonesia, 1992-93; researcher and cons. in field. Author: The Collective Economy and the Cooperative Economy, 1954, Individual Freedom, Economic Planning and Cooperation, 1956, Government and Private Enterprise: Their Place in the Economy, 1957, Economic Growth and Integral Humanism, 1957, Fiscal Problem of Growth with Stability, 1959, Fiscal Policy and Economic Growth I and II, 1962 and 1963, Inflation and Price Control, 1975, (with C.R. Sengupta) Bank Finance Criteria and the Tandon Committee Report, 1975, Cost Accounting in Commercial Banking Industry, 1979, Introduction to Cost Accounting in Commercial Banking Industry, 1983, Cost Accounting and Farm Product Costing, 1990, Fiscal Policy, Stability and Growth: Experience and Problems of the Underdeveloped Economies, 1929-39, 1945-65, 1990, Fiscal Debt Management, Monetary-Credit Policy, and Growth-with-Stability, 1963, Management Accountants' Role in Monitoring Bank Finance, 1982; founder, editor rsch. bull. Inst. Cost and Works Accts. India, 1982-84; contbr. over 30 articles to profl. jours. Active Radical Humanist Movement in India, Indian Renaissance Inst., 1950-62. Mem. Am. Econ. assn., Cine Club Calcutta, Am. Univ. Ctr. Jazz Club. Avocations: painting, sculpture, architecture, performing arts, international and Indian history and culture. Mailing: 11500 Bucknell Dr # 3 Wheaton MD 20902 Home: Punascha 72/1 BC Rd, Burdwan 713 101, India

GHOSH, ASISH, control engineer, consultant; b. Calcutta, India, Sept. 2, 1935; came to U.S., 1978; s. Sudhangsu Kumar and Lotika (Roy) G.; m. Aparna, Sept. 20, 1968; children: Annapurna, Ashapurna. BSc, Delhi (India) U., 1954; diploma in advanced studies, Cambridge (Eng.) U., 1968. Chartered engr., U. K. Rsch. scientist Imperial Chem. Industries, Runcorn, Eng., 1968-74; systems engr. The Foxboro Can. Inc., Montreal, Que., 1974-80; project engr. The Foxboro (Mass.) Co., 1980-82, sr. engr., 1982-87, prin. engr., 1987-88, cons., 1989—; course dir. Ctr. for Profl. Advancement, East Brunswick, N.J., 1988—; mem. product rsch. panel Chem. Engring., N.Y.C., 1992-93. Co-author: Batch Process Automation, 1987; also articles. Mem. Instrument Soc. Am., Inst. Elec. Engrs. (U.K.). Hindu. Achievements include pioneering work in automating fluid batch manufacturing processes. Home: 45 Warren Dr Wrentham MA 02093 Office: The Foxboro Co MC B52-2J 33 Commercial St Foxboro MA 02035

GHOSH, CHUNI LAL, physicist; b. Bilsara, Bengal, India, Jan. 3, 1948; came to the U.S., 1979; s. Arabinda and Bimala (Ghosh) G.; m. Malathi Ghosh, Feb. 19, 1973; children: Shukti, Suchira. MS in Physics, U. Burdwan, India, 1969; PhD, U. Bombay, 1976. Scientist Bhabha Atomic Rsch. Ctr., Bombay, 1969-79; vis. prof. Oreg. State U., Corvallis, 1979-80; dept. mgr. ITT Corp., Roanoke, Va., 1980-85; pres. Tachonics Corp., Plainsboro, N.J., 1985-90; lab. dir. David Sarnoff Rsch. Ctr., Princeton, N.J., 1990—. Author: Thin Solid Alums, 1983; contbr. articles to profl. jours. Named Young Scientist of Yr. Indian Nat. Sci. Acad., 1978. Mem. IEEE, Am. Phys. Soc., Indian Physics Assn. Achievements include patents for a new type of night vision device, a novel high speed electronic transistor, and a new technique for lithography of fine lines. Office: David Sarnoff Rsch Ctr CN 5300 Princeton NJ 08543

GHOSH, DEEPAK RANJAN, chemical/environmental engineer; b. Calcutta, India, July 17, 1965; s. Salil Kumar and Ashoka G.; m. Madhumanjari Dutta, June 6, 1990. BChemE, Jadavpur U., Calcutta, 1987; MS in Environ. Engr., Clemson U., 1990. Project engr. Indian Petrochems. Corp. Ltd., Baroda, India, 1986, Balmer Lawrie & Co. Ltd., Calcutta, 1987-88; software engr. Data Stream Systems, Greenville, S.C., 1989-90; sr. engr. Engring. Sci., Inc., Atlanta, 1990-93; environ. engr. Law Environ., Inc., Kennesaw, Ga., 1993—. Contbr. articles to profl. jours. Recipient Merit Cert. Govt. of India. Mem. AICHE, Water Environ. Fedn. (toxic substances com. and program com. 1991-93), Ga. Water and Pollution Control Assn., Internat. Assn. on Water Quality. Home: 1176 Francis St NW 6640 Akers Mill Rd #30T4 Atlanta GA 30339 Office: Law Environ Inc Ste 590 57 Executive Park S 114 Townpark Dr Ste 400 Kennesaw GA 30144

GHOSH, MALATHI, physicist; b. Mysore, Karnataka, India, Aug. 19, 1947; came to U.S., 1979; d. Yedetore G. and Najamma Rao.; m. Feb. 19, 1973; children: Shukti, Suchira. MS, Mysore U., India, 1969. Sci. officer Bhabha Atomic Rsch. Ctr., Bombay, India, 1969-79; sect. mgr. ITT Corp., Roanoke, Va., 1982-86; R & D mgr. Epitaxx Corp., Princeton, N.J., 1986-87; sr. scientist Advanced Photovoltaic Systems, Inc., Lawrenceville, N.J., 1990—. Contbr. articles to profl. jours. Mem. IEEE. Achievements include patents in field on Stabilization of Amorphous Photovoltaic Solar Cells. Office: Advanced Photovoltaic PO Box 7093 Princeton NJ 08543

GHOSH, SID, telecommunications engineer; b. Calcutta, India, June 16, 1934; came to U.S., 1969; s. Bibhuti and Sarala Bala (Bose) G.; m. Joan Mary Moyes, June 30, 1966. BSc, Calcutta U., 1956, London U., 1962. Prin. engr. ITT Telecomm., Raleigh, N.C., 1969-76; supr. TRW-Vidar, Mountain View, Calif., 1976-78; engring. mgr. GTE Fiber Optics, Mountain View, 1978-80, Karkar Electronics, Santa Clara, Calif., 1980-82, DSC Corp., Santa Clara, 1982-86; dir. engring. Fujitsu Am., San Jose, Calif., 1986-90; prin. Design Assistance, Crescent City, Calif., 1990—. Co-author: Designers Handbook, 1984; contbr. articles to profl. publs. Mem. IEEE (sr.). Achievements include 5 patents for electronics designs. Home and Office: 1692 Del Mar Crescent City CA 95531

GHOSH, SUBIR, statistician; b. Calcutta, India, Aug. 26, 1950; came to the U.S., 1972; s. Subimal and Padma Renu (Guha) G.; m. Susnata Roy, Apr. 27, 1978; 1 child, Malancha. MS in Stats., Calcutta U., 1970; PhD in Stats., Colo. State U., 1976. Grad. rsch. asst. Colo. State U., Ft. Collins, 1972-75, postdoctoral fellow, 1976; asst. prof. Indian Statis. Inst., Calcutta, 1976-80; from asst. to assoc. prof. to prof. U. Calif., Riverside, 1980—. Editor: Statistical Design and Analysis of Industrial Experiment, 1990; assoc. editor Jour. Statis. Planning of Inference, 1989—, Comms. in Stats., 1992—; contbr. over 40 articles to profl. jours. Rsch. grantee USAF, 1985-92. Mem. Am. Statis. Assn. (pres.-elect So. Calif. chpt. 1992—), Am. Soc. Quality Control. Inst. Math. Stats. Hindu. Achievements include rsch. in robustness of designs against the unavailability of data, development of statis. design of experiments. Home: 6873 Ranch Grove Rd Riverside CA 92506 Office: U Calif Dept Statis Riverside CA 92521-0138

GIACCO, ALEXANDER FORTUNATUS, chemical industry executive; b. San Giovanni di Gerace, Italy, Aug. 24, 1919; naturalized in 1927; s. Salvatore J. and Maria Concetta (de Maria) G.; m. Edith Brown, Feb. 16, 1946; children: Alexander Fortunatus Jr., Richard John, Mary P. Giacco Walsh, Elizabeth B. Giacco Brown, Marissa A. Giacco Rath. BSChemE, Va. Poly. Inst., 1942; postgrad. in mgmt., Harvard Grad. Sch. Bus., 1965; DBA (hon.), William Carey Coll., 1980; D.Bus. (hon.), Goldey Beacom Coll., 1984; LLD (hon.), Widener U., 1984, Cath. U. of Am., 1990; LHD (hon.), Mt. Saint Mary's Coll., 1988. With Hercules Inc., Wilmington, Del., 1942-87, gen. mgr. polymers dept., 1968-73, dir., 1970-87, gen. mgr. operating dept. (Hercules Europe), 1973, v.p. parent co., 1974-76, mem. exec. com., 1974-87, exec. v.p., 1976-77, pres., chief exec. officer, chmn. exec. com., 1977-87, chmn. bd., 1980-87; chmn. bd. HIMONT Inc., Wilmington, Del., 1983-91, chief exec. officer, 1987-91; dep. chmn. Montedison SpA, Milan, 1988-89; mng. dir. Axess Corp., Palm Beach, Fla., 1991—; bd. dirs. China Trust Bank, N.Y.C., 1988—, Carlisle Plastics, Boston, 1991-93, Erbamont, N.V., 1983-91, Feruzzi Finanziaria, Milan, Italy, 1988-91; mem. adv. bd. Marvin & Palmer Assocs., Wilmington, Del., 1987—. Trustee, bd. dirs., mem. exec. com. Med. Ctr. of Del., 1975-88; trustee, bd. visitors Va. Poly. Inst. and State U., 1980-87, rector, 1984-87; chmn. bd. dirs. Grand Opera House, Wilmington, 1980—; pres. Italian Am. Heritage Ctr. at Cath. U. of Am., Washington, 1991—, bd. dirs., 1987—; adv. bd. New Ctr. for History of Chemistry, 1982-87; exec. com. Pres. Private Sector Survey, Grace Commn., Washington, 1982-86; mem. U.S. Com. on New Initiatives in East-West Cooperation, 1976-86, Adv. Coun. on Japan-U.S. Econ. Rels., 1982-88; dep. chmn. Propulsion Com. for Guided Missiles and JATO, 1960; mem. Junior Achievement of Del., 1976-84, pres., chmn., 1971-78, hon. chmn., 1979-84; mem. Del. Roundtable Inc., 1980-90, Nat. Conf. Christians and Jews, assoc. chmn, 1986. Decorated commendatore Order of Merit (Italy); recipient Disting. Achievement award Va. Poly. Inst. and State U., 1989, Bergamotto d'Oro award Lions, 1988, Honor award Comml. Devel. Assn., 1987, Disting. Citizens award Del-Mar-Va Coun. Boy Scouts Am., 1987; named One of Ten Outstanding Chief Exec. Officers Fin. World, 1980, 87, Best Chief Exec. Officer in Chem. Industry Fin. World, 1984, Outstanding Chief Exec. Officer in the Chem. Industry Wall Street Transcript, 1983, 84, 85, 87, Excellence in Mgmt. award Administrv. Mgmt. Soc. (Del. chpt.), 1986, Ann. Indsl. award Nat. Italian Am. Found., 1980, Outstanding Svc. to Community awardCath. Charities, 1992; named to Del. Bus. Leaders Hall of Fame by Jr. Achievement of Del. and Hagley Mus. & Libr., 1992. Mem. Am. Assn. of the Sovereign Mil. Order of Malta, Chem. Manufacterers Assn. (bd. dirs.), Manmade Fiber Producers Assn. Clubs: Wilmington, Wilmington Country (bd. dirs.), Vicmead Hunt, Palm Beach Yacht Club. Achievements include Can. patent for Propellant Charge and Method of Manufacture (relates to propellant charges and, more particularly, to thrust unit or gas-producing device charges which permit a comparatively wide choice of propellant web and surface and a high-loading density). Office: Axess Corp Phillips Point W Tower 777 S Flagler Dr Ste 1112 West Palm Beach FL 33401

GIACCONI, RICCARDO, astrophysicist, educator; b. Genoa, Italy, Oct. 6, 1931; came to U.S., 1956, naturalized, 1967; s. Antonio and Elsa (Canni) G.; m. Mirella Manaira, Feb. 15, 1957; children: Guia Giacconi Chmiel, Anna Lee, Marc A. Ph.D., U. Milan, Italy, 1954; Sc.D. (hon.), U. Chgo., 1983; laurea ad honorem in astronomy, U. Padua, 1984. Asst. prof. physics U. Milan, 1954-56; research assoc. Ind. U., 1956-58, Princeton U., 1958-59; exec. v.p., dir. Am. Sci. & Engring. Co., Cambridge, Mass., 1959-73; prof. astronomy Harvard U.; also assoc. dir. high energy astrophysics div. Center Astrophysics, Smithsonian Astrophys. Obs./Harvard Coll. Obs., Cambridge, 1973-81; dir. Space Telescope Sci. Inst., Balt., 1981-92; prof. astrophysics Johns Hopkins U.; dir. general European So Observatory, Germany, 1993—; mem. space sci. adv. com. NASA, 1978-79, mem. adv. com. innovation study, 1979—; mem. NASA Astrophysics Council; mem. adv. com. innovation study astronomy adv. com., 1979—; mem. high energy astronomy survey panel Nat. Acad. Scis., 1979-80, mem. Space Sci. Bd., 1980-84, 89—; mem. adv. com. Max-Planck Inst. für Physik und Astrophysik; chmn. bd. dirs. Instituto Guido Donegani, Gruppo Montedison, 1987-89; mem. vis. com. to div. of phys. scis. U. Chgo., U. Padova. Co-editor: X-ray Astronomy, 1974, The X-Ray Universe, 1985; author numerous articles, papers in field; inventor x-ray telescope, discovered x-ray starsmen. Fulbright fellow, 1956-58; recipient Röntgen prize astrophysics Physikalish-Medizinische Gesellschaft, Wurzburg, Germany, 1971; Exceptional Sci. Achievement medal NASA, 1971, 80; Disting. Public Service award, 1972; Space Sci. award AIAA, 1976; Elliott Cresson medal Franklin Inst., 1980; Gold medal Royal Astron. Soc., 1982; A. Cressy Morrison award N.Y. Acad. Sci., 1982; Bruce medal; Heinneman award, Wolf Prize in Physics, 1987; Russell lectr. Mem. Am. Astron. Soc. (Helen B. Warner award 1966, chmn. hign energy astrophysics dept. 1976-77, councilor 1979-82, task group on directions in space sci. 1995-2015), Italian Phys. Soc. (Como prize 1967), AAAS, Internat. Astron. Union (nat. Acad. Scis. rep. 1979-82), Nat. Acad. Scis., Am. Acad. Arts and Scis., Md. Acad. Sci. (sci. coun. 1982—), Accademia Nazionale dei Lincei (fgn.), Royal Astronomical Soc., Am. Phys. Soc. Club: Cosmos (Washington). Office: European So Observatory, Karl-Schwarzschild-Strasse 2, D-8046 Garching bei Munich Germany

GIAEVER, IVAR, physicist; b. Bergen, Norway, Apr. 5, 1929; came to U.S., 1957, naturalized, 1963; s. John A. and Gudrun (Skaarud) G.; m. Inger Skramstad, Nov. 8, 1952; children: John, Anne Kari, Guri, Trine. Siv. Ing., Norwegian Inst. Tech., Trondheim, 1952; Ph.D., Rensselaer Poly. Inst., 1964. Patent examiner Norwegian Patent Office, Oslo, 1953-54; mech. engr. Can. Gen. Electric Co., Peterborough, Ont., 1954-56; applied mathematician Gen. Electric Co., Schenectady, 1956-58, physicist Research and Devel. Ctr., 1958-88; Inst. prof. Rensselaer Poly. Inst., Troy, N.Y., 1988—; also prof. U. Oslo, 1988—. Served with Norwegian Army, 1952-53. Recipient Nobel Prize for Physics, 1973; Guggenheim fellow, 1970. Fellow Am. Phys. Soc. (Oliver E. Buckley prize 1965); mem. IEEE, Norwegian Profl. Engrs., Nat. Acad. Sci., Nat. Acad. Engring. (V.K. Zworykin award 1974), Am. Acad. Arts and Scis., Norwegian Acad. Sci., Norwegian Acad. Tech. Office: Rensselaer Poly Ins Physics Dept Troy NY 12180-3590

GIALLORENZI, THOMAS GAETANO, optical engineer; b. N.Y.C., Feb. 28, 1943; s. Amedeo and Eleanor (Spica) G.; m. Margaret Mary Marrin, Sept. 6, 1966; children: Thomas R., Kathy. BS in Engring. Physics, Cornell U., 1965, MS in Engring. Physics, 1966, PhD, 1969. Tech. staff Gen. Telephone & Electronics Lab., Bayside, N.Y., 1969-70; sect. head, optical techniques br. Naval Rsch. Lab., Washington, 1970-76, head optical techniques br., 1976-79, supt. optical scis. div., 1979—; lectr. in field and at profl. soc. confs. Editor Jour. Lightwave Tech., 1983-88; contbr. over 80 articles to profl. jours.; over 30 patents in field. Mem. adv. bd. U. Va., 1986-92. Recipient Applied Sci. award Rsch. Soc. Am., 1973; Meritorious Civilian Svc. award USN, 1978, Conrad award, 1985; Meritorious award Pres. of U.S., 1983, Disting. Exec. award, 1981, 90, Meritorious Exec. Rank award, 1986; Disting. Civilian Svc. award Dept. Def., 1987. Fellow IEEE (assoc. editor Procs. 1990-93, Lightware Comms. 1989-92, Harry Diamond award 1986, John Tyndell award 1990), Optical Soc. Am. (editor Jour. Lightware Tech. 1983-89, assoc. editor Applied Optics 1991—, John Tyndall award 1990); mem. NAE. Home: 8704 Side Saddle Rd Springfield VA 22152-2731 Office: Naval Rsch Lab Optical Scis Divsn Washington DC 20375-5000

GIAM, CHOO-SENG, marine science educator; b. Singapore, Apr. 2, 1931; came to U.S., 1964, naturalized, 1970; s. Chong-Hing and Eng-Keow (Tan) G.; m. Mun-Yung Ng, Feb. 25, 1956; children: Benny Y.B., Patrick Y.Y., Michael Y.K. M.Sc., U. Sask, 1961, Ph.D., 1963. Research chemist Imperial Oil, Sarnia, Can., 1963-64; postdoctoral fellow Pa. State U., State College, 1964-65; research assoc. U. Calif.-Irvine, 1965-66; asst. prof. Tex. A&M U., College Station, 1966-70, assoc. prof., 1970-72, prof. dept. chemistry, 1972-81, prof. chemistry/oceanography, chmn. organic div. chemistry, 1972-82; dean Coll. Sci., prof. chemistry and geol. scis. U. Tex., El Paso, 1981-82; prof. chemistry U. Pitts., 1983-87; prof. marine sci., dir. Coastal Zone Lab. Tex. A&M U., Galveston, 1987—; adj. prof. dept. preventive medicine and community health U. Tex. Med. Br. Contbr. articles to profl. jours.; patentee in field. Mem. Am. Chem. Soc., Can. Chem. Soc., Royal Inst. Chemistry, N.Y. Acad. Scis., Sigma Xi, Phi Lambda Upsilon. Home: 9 Quintana Pl Galveston TX 77554-9302 Office: Tex A&M U at Galveston PO Box 1675 Galveston TX 77553-1675

GIAMBRA, LEONARD MICHAEL, psychologist; b. Jamestown, N.Y., Mar. 21, 1941; s. Mario and Frances E. (D'Angelo) G.; m. Nancy Ann Sullivan, Aug. 24, 1963. BA in Psychology, U. Rochester, 1963; PhD in Psychology, Ohio State U., 1968. Asst. prof. Miami U., Oxford, Ohio, 1967-72; rsch. psychologist Nat. Inst. on Aging, Balt., 1972—. Fellow APA, Gerontol. Soc. Am., Am. Psychol. Soc.; mem. Psychonomic Soc., Orgn. Sons of Italy in Am. (treas. 1982-92). Achievements include research in life span course of cognition and attention in normal adults, especially in forgetting likelihood of task-unrelated thought intrusions, sustained attention and mind wandering. Office: Nat Inst Aging Gerontol Rsch Ctr 4940 Eastern Ave Baltimore MD 21224

GIANAKOS, NICHOLAS, engineering executive, consultant; b. New London, Conn., Mar. 13, 1947; s. Louis and Chrysula (Verenes) G.; m. Marilyn Louise Streeter, Oct. 11, 1976; children: Jacqueline, Stephanie. BSME, U. Conn., 1970; MBA, U. Hartford, 1983. Registered engr., Conn., Mass., R.I., N.H., Vt., Maine. Consulting engr. JP Legnos Assocs., Hartford, Conn., 1970-72; product mgr. Dunham Bush Inc., West Hartford, 1972-83; project mgr. BG Mech. Engrs., Holyoke, Mass., 1983-89; sr. project mgr. CN Flagg Co., Meriden, Conn., 1989-90; gen. mgr. The Fleming Group, Windsor, Conn., 1990—; energy cons. Town of Cromwell (Conn.) Schs., 1991. Chmn. Cromwell Town Bldg. com., 1984, Sch. Bldg. com., 1989. Mem. ASHRAE, Assoc. Energy Engrs., Assn. Demand Side Mgmt. Planning. Achievements include development of measurement procedures for evaluation of utility demand-side mgmt. program in the Northeast. Home: 21 Greendale Ave Cromwell CT 06416 Office: The Fleming Group 427 Hayden Station Rd # A Windsor CT 06095-1335

GIANCOTTI, FRANCESCA ROMANA, immunologist, cancer research scientist; b. N.Y.C., June 20, 1956; d. Antonio Giancotti and Anna (Conte) Malamood. MA, Columbia U., 1987, PhD, 1989. Research asst. Meml. Sloan-Kettering Inst. Cancer Rsch., N.Y.C., 1978-80; rsch. assoc. Columbia U., N.Y.C., 1981-88; sr. rsch. assoc. Lenox Hill Hosp./Cornell U., N.Y.C., 1988—; cons. health issues; cons. New Woman Mag., N.Y.C., 1992—. Contbr. articles to profl. jours. Grantee Am. Cancer Soc., 1983. Mem. AAAS, Soc. Exptl. Biology, Soc. Analytical Cytology, Am. Assn. Cancer Rsch. Democrat. Achievements include development of monoclonal antibody against cancer of ovary. Home: 155 W 68th St New York NY 10023 Office: Lenox Hill Hosp 100 E 77th St New York NY 10021

GIANNAROS, DEMETRIOS SPIROS, economist, educator; b. Karlovasi, Samos, Greece, Oct. 4, 1949; came to U.S., 1964; s. Spiridon Demetrios and Irene (Kiriakou) G.; m. Elizabeth Sampson, June 5, 1977; children: Edward, Spiros Jason. BA in Econs., U. Mass., 1972; MA in Econ. Devel., Boston U., 1976, MAPE in Polit. Econ., 1977, PhD in Econs., 1981. Mgr. Samos Imex Corp., Boston, 1974-77; asst. prof. econs. Suffolk U., Boston, 1977-79; prof. U. Hartford, West Hartford, Conn., 1980—, dir. internat. programs, 1993—; dir. exec. MPA program U. Hartford, West Hartford, 1986-88, assoc. to sr. v.p., dir. internat. studies, 1988-91; mem. Bd. Edn., Farmington, Conn., 1993—; spl. asst. to pres. George Washington U., Washington, 1988-89; cons. to pub. and pbt. orgns., 1977—; bd. dirs. Coll. Southeastern Europe, 1992—. Mem. Bd. Edn., Farmington, Conn., 1993—. NSF grantee, 1983-84, U. Hartford Coffin grantee, 1983-8, Mellon Found. grantee, 1991-92; Am. Coun. on Edn. fellow, 1988-89. Fellow Am. Coun. on Edn. (mem. exec. bd. coun.); mem. Am. Econ. Assn., Internat. Econ. Assn., N.E. Bus. and Econs. Assn. (pres. 1990-92, bd. dirs. 1989—), Helicon Soc. (pres., bd. dirs. 1975-78), Hellenic Soc., Paideia, World Affairs Coun. Greek Orthodox. Avocations: travel, water sports, museums, political activities. Home: 56 Basswood Rd Farmington CT 06032-1142 Office: U Hartford Econs Dept 200 Bloomfield Ave West Hartford CT 06117-1500

GIANNINI, A. JAMES, psychiatrist, educator, researcher; b. Youngstown, Ohio, June 11, 1947; s. Matthew and Grace Carla (Nistri) G.; m. Judith Ludvik, Apr. 26, 1975; children: Juliette Nicole, Jocelyn Danielle. B.S., Youngstown State U., Ohio, 1970; M.D., U. Pitts., 1974; postgrad., Yale U., 1974-78. Diplomate Nat. Bd. Med. Examiners. Intern St. Elizabeth Med. Ctr., Youngstown, 1974; resident dept. psychiatry Yale U., New Haven, 1975-78, chief resident, 1977-78; assoc. psychiatrist Elmcrest Psychiat. Inst., Portland, Conn., 1976-78; acting ward chief Conn. Mental Health Ctr., New Haven, 1977; assoc. dir. family medicine, psychiatry St. Elizabeth Med. Ctr., Youngstown, 1978-80; asst. prof. N.E. Ohio Med. Coll., 1978-80, program dir., 1980-88, assoc. prof. dept. psychiatry, 1980-84, prof., 1984-90, vice chmn., 1985-89; assoc. clin. prof. dept psychiatry Ohio State U., 1983-89, clin. prof., 1989—; chmn. depts. psychiatry and toxicology Western Res. Care System Hosp., 1985-87, med. dir. toxicology, 1987; examiner in psychology LaTrobe U., Bundoora, Australia, 1988-89; sr. cons. Fair Oaks Hosp., Summit, N.J., 1979, Regent Hosp., N.Y.C., 1981—, chmn. Nat. Adv. Council Prevention and Control of Rape, NIMH, Rockville, Md., 1983-86; mem. drug abuse clin., behavioral and rsch. rev. com. Nat. Inst. Drug Abuse, Rockville, Md., 1987-88; chief forensic psychiatrist Mahoning County Prosecutor, 1989—; Am. Participant USIA Drug Abuse program to Cyprus, Italy, Can., Barbados, St. Lucia and Yugoslavia, 1990—; cons. Smith-Kline Labs., McNeil Labs., Excerpta Medica Pubs., Amino Labs., Fund for Am. Renaissance; dir. clin. rsch. Princeton Diagnostic Labs., South Plainfield, N.J., 1987-89; med. dir. med. adv. bd. Neurodata Inc., 1987-89, pres., 1989—, med. dir. Chem. Abuse Ctrs. Inc., 1987. Author: (with Henry Black) Psychiatric, Psychogenic, Somatopsychic Disorders, 1978; (with Robert Gilliland) Neurologi and Neuropsychiatric Disorders, 1983; (with Andrew Slaby) Overdose and Detoxification Emergencies, 1983; Biological Foundation of Clinical Psychiatry, 1988, (with Andrew Slaby and Mark Gold) Drugs of Abuse, 1989, Comprehensive Laboratory Services in Psychiatry, 1986, (with Philip Jose Farmer) Red Orc's Rage, 1991, (with Andrew Slaby) The Eating Disorders, 1993; contbr. numerous articles to profl. jours. Vice chmn. Mahoning County (Ohio) Mental Health Bd., 1982-84, chmn. 1984-86; councilor Nat. Italian Am. Found. Recipient James Earley award U. Pitts., 1974, Upjohn Research prize Upjohn Co., 1974; recipient Fair Oaks Research award Fair Oaks Hosp., 1979, Bronze award Brit. Med. Assn, 1983, Outstanding Leadership award Mahoning County Mental Health Bd., 1986; Entrepreuner of Yr. nominee Inc. Mag., 1989, Silver Rose award Assn. Italiano Donati d'organo, Milan, Italy, 1990. Recipient Physician's Recognition award, 1978—. Fellow Royal Acad. Medicine (Eng.), Acad. Medicine, Am. Coll. Clin. Pharmacology (sec.-treas. Ohio chpt. 1990—), Am. Psychiat. Assn.; mem. Soc. Neurosci., Brit. Brain Soc., European Neurosci., Royal Coll. Medicine, N.Y. Acad. Scis., Am. Psychiat. Assn., Acad. Clin. Psychiatry, Youngstown C. of C. (vice chmn. health com. 1986-89, chmn. 1989—), Youngstown Club, Atrium Club (Warren, Ohio), Poland Club (Ohio), Swim and Racquet Club, Sigma Xi. Republican. Roman Catholic.

GIANNIS, ATHANASSIOS, chemist, physician; b. Drama, Greece, Jan. 22, 1954; s. Nikolaos and Ariadne (Petridis) G.; m. Sabine Kleine-Rueschkamp, Apr. 11, 1980; children: Melina, Alexander, Lisa, Eleni, Aristotelis. Diploma in chemistry, U. Bonn, Fed. Republic Germany, 1980, PhD, 1986, MD, 1988. Postdoctoral researcher Inst. Organic Chemistry/Biochemistry, U. Bonn, 1986, asst., 1988—; habil., 1992; lectr. in organic chemistry and biochemistry. Contbr. articles to profl. jours.; patentee in field. Mem. Am. Chem. Soc., Gen. Med. Coun. Germany. Orthodox. Home: Londoner str 13, 5300 Bonn 1, Germany Office: Inst Organic Chemistry, Gerhard Domagk Str 1, 5300 Bonn 1, Germany

GIANNOPOULOS, JOANNE, pharmacist, consultant; b. Chgo.; m. James Giannopoulos, July 16, 1972; children: Alexandra, Andronike. BS in Pharmacy, U. Ill., Chgo., 1967, PharmD, 1988; MBA, Rosary Coll., River Forest, Ill., 1985. Asst. dir. pharmacy NW Hosp., Chgo., 1969-85; dir. pharm. svcs. and lab. Forest Health Systems, Des Plaines, Ill., 1990-91; dir. pharm. Rehab. Inst. Chgo., Northwestern U., 1991—. Mem. Plato Sch. Bd., Chgo., 1988-93. Mem. Ill. Pharmacists Assn. (hosp. ednl. com. mem.). Office: Rehab Inst Chgo 345 E Superior St Chicago IL 60611-4496

GIBBONS, EDWARD FRANCIS, psychobiologist; b. Bronx, N.Y., Dec. 25, 1949; s. Edward Francis and Mary Theresa (Westervelt) G. BS, SUNY, Stony Brook, 1977, PhD, 1986. Dir. ctr. for sci. and tech. Briarcliffe Coll., Woodbury, N.Y., 1992—. Series editor: SUNY Press Series on Endangered Species, 1986—; editor: Naturalistic Environments in Captivity for Animal Behavior Research, 1993. Sgt. USAF, 1970-74. Grantee Inst. Mus. Svcs., 1988-90, N.Y. State Dept. Edn., 1993—. Mem. Animal Behavior Soc., Soc. Behavioral Medicine, Sigma Xi. Roman Catholic. Home: 33 Warren St Brentwood NY 11717 Office: Briarcliffe Coll 250 Crossways Park Dr Woodbury NY 11797-2015

GIBBONS, JOHN HOWARD (JACK), physicist, government official; b. Harrisonburg, Va., Jan. 15, 1929; s. Howard K. and Jessie C. G.; m. Mary Ann Hobart, May 21, 1955; children: Virginia Neil, Diana Conrad, Mary Marshall. BS in Math. and Chemistry, Randolph-Macon Coll., 1949, ScD, 1977; PhD in Physics, Duke U., 1954. Physicist and group leader nuclear geophysics Oak Ridge Nat. Lab., 1954-69, dir. environ. program, 1969-73; dir. Office Energy Conservation, Fed. Energy Adminstrn., Washington, 1973-74; prof. physics, dir. Energy, Environ. and Resources Center, U. Tenn., Knoxville, 1974-79; dir. Office of Tech. Assessment, U.S. Congress, 1979-92; chmn. panel on demand and conservation CONAES Study Nat. Acad. Sci., 1975-79; asst. to Pres. for sci. and tech., 1993—; dir. Office of Sci. and Tech. Policy Exec. Office of Pres., Washington, 1993—; Mem. sr. adv. panel Energy Modeling Forum, 1980-92; mem. steering com. Symposium Series on Tech. and Soc., NAE, 1984-92; mem. energy rsch. adv. bd. U.S. Dept. Energy, 1978-79; bd. dirs. Resources for the Future, 1983-92 bd. dirs., mem. bd. sci. and tech. internat. devel. NAS, 1979-88; mem. energy and resources com. Aspen Inst., 1979-92; mem. adv. coun. Stanford U. Sch. Engring., 1984-87, Electric Power Rsch. Inst., 1986-92; mem. internat. environ. tech. transfer adv. bd. EPA, 1989-92; mem. Carnegie Commn. Sci., Tech. and Govt. Task Force on Long Term Goals and Priorities, 1990-92. Mem. editorial bd. Forum for Applied Research and Pub. Policy Jour., TVA, 1985—; contbr. articles to profl. jours. Chmn. bd. assocs. Randolph-Macon Coll., Ashland, Va., 1980-83, trustee, 1977-79; bd. dirs. Knoxville Energy Expo '82, 1978-79, chmn. energy com. and nat. adv. coun., 1979-82;

bd. dirs. State of Tenn. Energy Authority, 1977-79; adv. bd. Trinity Inst., N.Y.C., 1987-92. Recipient Disting. Svc. award Fed. Energy Adminstrv., 1974, Pub. Svc. award Fedn. Am. Scientists, 1990, Officer's Cross Order of Merit, Fed. Rep. Germany, Disting. Alumni award James Madison U., 1993. Fellow AAAS (bd. dirs. 1988-90, Philip Hauge Abelson prize 1993), Am. Phys. Soc. (Leo Szilard award for physics in pub. interest 1991); mem. Coun. Fgn. Rels., Cosmos Club, Sigma Xi (long-range planning com. 1989-92, nat. lectr. 1978-79), Phi Beta Kappa, Pi Gamma Mu, Omicron Delta Kappa, Pi Mu Epsilon. Episcopalian. Home: PO Box 497 The Plains VA 22171-0497 Office: Office Sci and Tech Policy Exec Office of Pres Exec Office Bldg Washington DC 20500

GIBBS, ELIZABETH DOROTHEA, developmental psychologist; b. Ithaca, N.Y., July 4, 1955; d. Robert Henry and Sarah Preble (Bowker) G. AB, Cornell U., 1976; MA, U. Vt., 1981, PhD, 1984. Lic. psychologist, Vt.; cert. N.H. Postdoctoral fellow Brown U. Child Devel. Ctr., R.I. Hosp., Providence, 1984-85; asst. prof. Rutgers U., New Brunswick, 1986-87; assoc. dir. rsch., asst. prof. Clin. Genetics and Child Devel. Ctr., Dartmouth Med. Sch., Hanover, N.H., 1987-92; owner Positive Devel. Cons., Newport, N.H., 1992—; Presenter in field. Editor: Interdisciplinary Assessment of Infants: A Guide for Early Intervention Professionals, 1990; contbr. articles to profl. jours.; videotape producer: Early Use of Total Communication: Parents' Perspectives on Using Sign Language with Young Children with Down Syndrome. Bd. dirs. Early Intervention Network of N.H., Concord, 1988—. Grantee Dartmouth Med. Sch., 1988-89, N.H. Dept. Edn., 1989-92, Dept. Edn. Office of Spl. Edn. Programs, Handicapped Children's Early Edn. Programs, 1988-91, 88-91, Dartmouth Med. Sch., 1988-89. Mem. Soc. for Rsch. in Child Devel., Nat. Ctr. for Clin. Infant Programs, Coun. for Exceptional Children (div. early childhood), Am. Psychol. Assn. (developmental psychology div.), N.H. Early Intervention Network. Avocations: knitting, sailing, gardening, cross country skiing, birdwatching. Office: 4 Fletcher Rd Newport NH 03773-9723

GIBBS, JAMES ALANSON, geologist; b. Wichita Falls, Tex., June 18, 1935; s. James Ford and Clovis (Robinson) G.; m. Judith Walker, June 18, 1966; children: Ford W., John A. BS, U. Okla., 1957, MS, 1962. Cert. profl. geologist. Geologist Calif. Co., New Orleans, 1961-63, Lafayette, La., 1963-64; cons. geologist, oil producer, Dallas, 1964—; owner, chief exec. officer Five States Energy Co., 1984—. Author: Finding work as a Petroleum Geologist: Hints to the Jobseeker, 1984. Trustee Inst. for Study Earth and Man, So. Meth. U. Lt. USNR, 1957-59. Mem. AAAS, Geol. Soc. Am., Dallas Geol. Soc. (hon., pres. 1975-76), Am. Assn. Petroleum Geologists (sec. 1983-85, pres. 1990-91, Disting. Svc. award 1987), Am. Inst. Profl. Geol., Ind. Petroleum Assn. Am., Nat. Petroleum Coun., Tex. Ind. Producers and Royalty Owners Assn., Dallas Geophys. Soc., Houston Geol. Soc., Lafayette, La. Geol. Soc., Soc. Ind. Profl. Earth Scientists (past chmn. Dallas chpt.), Petroleum Engrs. Club, Dallas Country Club, Dallas Petroleum, Energy Club of Dallas, Explorers Club, Sigma Xi, Sigma Gamma Epsilon, Phi Delta Theta. Republican. Methodist. Home: 3514 Caruth Blvd Dallas TX 75225-5001 Office: 1220 One Energy Sq Dallas TX 75206

GIBBS, MARTIN, biologist, educator; b. Philadelphia, Penn., Nov. 11, 1922; s. Samuel and Rose (Sugarman) G.; m. Svanhild Karen Kvale, Oct. 11, 1950; children—Janet Helene, Laura Jean, Steven Joseph, Michael Seland, Robert Kvale. B.S., Phila. Coll. Pharmacy, 1943; Ph.D., U. Ill., 1947. Scientist Brookhaven Nat. Lab., 1947-56; prof. biochemistry Cornell U., 1957-64; prof. biology, chmn. dept. Brandeis U., Waltham, Mass., 1965-93; cons. NSF, 1961-64, 69-72, NIH, 1966-69; mem. corp. Marine Biol. Lab., Woods Hole, Mass., 1970, RESA lectr., 1969; NATO cons. fellowship bd., 1968-70; mem. Coun. Internat. Exch. of Scholars, 1976-82; chmn. adv. com. selection Fulbright Scholars for Eastern Europe; adj. prof. Bot. Inst., U. Munster, Fed. Republic of Germany, 1978, 80, 87; adj. prof. dept. botany U. Calif., Riverside, 1979-89. Author: Structure and Function of Chloroplasts, Crop Productivity-Research Imperative, Revisited, Hungarian-USA Binational Symposium on Photosynthesis; editor in chief Plant Physiology, 1963-92; assoc. editor: Physiologie Vegetale, 1966-76, Ann. Rev. Plant Physiology, 1966-71. Recipient Adolph E Gude, Jr. award Am. Soc. Plant Physiologists, 1992; Alexander von Humboldt fellow, 1987. Mem. AAUP, Am., Japanese socs. plant physiologists, Am. Acad. Arts and Scis., Am. Soc. Biol. Chemists, Can. Soc. Plant Physiologist (hon. life mem., Charles Reid Barnes award 1984, Adolph E. Gude award 1993, Martin Gibbs medal 1993), Coun. Biology Editors, Nat. Acad. Scis., Sigma Xi. Home: 32 Slocum Rd Lexington MA 02173

GIBBY, MABEL ENID KUNCE, psychologist; b. St. Louis, Mar. 30, 1926; d. Ralph Waldo and Mabel Enid (Warren) Kunce; student Washington U., St. Louis, 1943-44, postgrad., 1955-56; B.A., Park Coll., 1945; M.A., McCormick Theol. Sem., 1947; postgrad. Columbia U., 1948, U. Kansas City, 1949, George Washington U., 1953; M.Ed., U. Mo., 1951, Ed.D., 1952; m. John Francis Gibby, Aug. 27, 1948; children—Janet Marie (Mrs. Kim Williams), Harold Steven, Helen Elizabeth, Diane Louise (Mrs. Roderick Rohrich), John Andrew, Keith Sherridan, Daniel Jay. Dir. religious edn. Westport Presbyn. Ch. Kansas City, Mo., 1947-49; tchr. elementary schs., Kansas City, 1949-50; high sch. counselor Arlington (Va.) Pub. Schs., 1952-54; counselor adult counseling services Washington U., 1955-56; counseling psychologist Coral Gables (Fla.) VA Hosp., 1956—; counseling psychologist Miami (Fla.) VA Hosp., 1956—, chief counseling psychology sect., 1982-86; sr. psychologist Office Disability Determination Fla. Hdqrs., 1987—. Sec. bd. dirs. Fla. Vocat. Rehab. Found. Recipient Meritorious Service citation Fla. C. of C., 1965, President's Com. on Employment of Handicapped, 1965; commendation for meritorious service Com. on Employment of Physically Handicapped Dade County, 1965, named Outstanding Rehab. Profl., 1966, 81; named Profl. Fed. Employee of Year, Greater Miami Fed. Exec. Council, 1966; Outstanding Fed. Service award Greater Miami Fed. Exec. Council, 1966; Fed. Woman's award U.S. Civil Service Commn., 1968, Community Headliner award Theta Sigma Phi, 1968, Outstanding Alumni award Park Coll., 1968, Freedom award The Chosen Few, Korean War Vets. Assn. 1986; certificate of appreciation Bur. Customs, U.S. Treasury Dept., 1969, Fla. Dept. Health and Rehab. Services, 1970. Mem. Am., Dade County (past sec.) psychol. assns., Nat., Fla. (past dir. Dade County chpt.) rehab. assns., Nat. Rehab. Counseling Assn. (past sec.). Patentee in field. Home: 10260 SW 56th St Miami FL 33165-7099

GIBERSON, KARL WILLARD, physics and philosophy educator; b. Bath, N.B., Can., May 13, 1957; s. Philip Sidney and Ursula Margaret (Steeves) G.; m. Myrna Loree Fuller, July 28, 1979; children: Sara Jacqueline, Laura Louise. BA in Philosophy/BS in Physics and Math., Ea. Nazarene Coll., 1979; MA/PhD in Physics, Rice U., 1985. Prof. Physics Ea. Nazarene Coll., Quincy, Mass., 1984—. Author: Worlds Apart: The Unholy War Between Religion and Science, 1993; contbr. 10 articles to profl. jours. Recipient Wilson award Rice U., 1984. Mem. Nat. Sci. Tchrs. Assn. Mem. Ch. of the Nazarene. Home: 329 Whiting Hingham MA 02043 Office: Ea Nazarene Coll 23 E Elm St Quincy MA 02170

GIBSON, CLIFFORD WILLIAM, military officer; b. Brunswick, Ga., Nov. 27, 1933; s. Leroy Lewis Lee Gibson and Lady Bird Paulk; m. Merian Elaine Jackson, Dec. 23, 1956; children: Valencia Yvette, Clifford William II, Frederick Leroy, Raymond Patrick. BS in Chemistry, Morris Brown Coll., 1955; BS in Aerospace Engring., NAvy Postgrad. Sch., 1963; MS in Mgmt. Engring., George Washington U., 1973, postgrad., 1983-85. Commd. ensign USN, 1955, advanced through grades to commdr., 1973, retired, 1982; prof. naval sci. N.C. Ctrl. U., Durham, 1973-76; instr. naval scis. H.D. Woodson High Sch., Washington, 1982-89. Mem. AIAA, Nat. Naval Officers Assn, Armed Forces Communications and Electronics Assn., Mil. Retirees Tri-State Area, Vets. Fgn. Wars, U.S. Naval Inst., Alpha Phi Alpha. Home: 4801 Acme Cove Memphis TN 38128

GIBSON, DAVID ALLEN, civil engineer, career officer; b. Neon, Ky., Sept. 5, 1957; s. Hubert and Ramona Blanche (Stallard) G.; m. Alice Marie Clarkston, May 28, 1977; children: Mary Elizabeth, Douglas Lee. BS in Civil Engring., Ohio State U., 1984; MS in Bus. Mgmt., U. LaVerne, 1990. Registered profl. engring. intern, Okla. Commd. 2d lt. USAF, 1984, advanced through grades to capt., 1988; contract planning engr. 2854 CES/DEEX USAF, Tinker AFB, Oka., 1984-85, chief requirements 2854 CES/DEMR, 1985-86, design civil engr. 2854 CES/DEM-2, 1986-87; lead program engr. 343 CES/DEEP USAF, Eielson AFB, Alaska, 1987-90, chief civil engr. 343 CES/DEEE, 1990-92; program devel. engr. HQ AMC USAF, Scott AFB, Ill., 1992—. Mem. ASCE (assoc.), Soc. Am. Mil. Engrs. Home: 404 David Dr Fairview Heights IL 62208 Office: HQ AMC/CEPD 507 A St Scott AFB IL 62225-5022

GIBSON, ELEANOR JACK (MRS. JAMES J. GIBSON), psychology educator; b. Peoria, Ill., Dec. 7, 1910; d. William A. and Isabel (Grier) Jack; m. James J. Gibson, Sept. 17, 1932; children: James J., Jean Grier. BA, Smith Coll., 1931, MA, 1933, DSc (hon.), 1972; PhD, Yale U., 1938; DSc (hon.), Rutgers U., 1973, Trinity Coll., 1982, Bates Coll., 1985, U. S.C., 1987, Emory U., 1990, Middlebury Coll., 1993; LHD (hon.), SUNY, Albany, 1984, Miami U., 1989. Asst., instr., asst. prof. Smith Coll., 1931-49; research assoc. psychology Cornell U., Ithaca, N.Y., 1949-66; prof. Cornell U., 1972—, Susan Linn Sage prof. psychology, 1972—; fellow Inst. for Advanced Study, Princeton, 1959-60, Inst. for Advanced Study in Behavioral Scis., Stanford, Calif., 1963-64, Inst. for Advanced Study, Ind. U., fall 1990; vis. prof. Mass. Inst. Tech., 1973, Inst. Child Devel., U. Minn., 1980; Disting. vis. prof. U. Calif., Davis, 1978; vis. scientist Salk Inst., La Jolla, Calif., 1979; vis. prof. U. Pa., 1984; Montgomery fellow Dartmouth Coll., 1986; Woodruff vis. prof. psychology Emory U., 1988-90. Author: Principles of Perceptual Learning and Development, 1967 (Century award), (with H. Levin) The Psychology of Reading, 1975, Odyssey in Learning and Perception, 1991. Recipient Wilbur Cross medal Yale U., 1973, Howard Crosby Warren medal, 1977, medal for disting. svc. Tchrs. Coll., Columbia U., 1983, Nat. Medal Sci., 1992; Guggenheim fellow, 1972-73, William James fellow Am. Psychol. Soc., 1989. Fellow AAAS (div. chairperson 1983), Am. Psychol. Assn. (Disting. Scientist award 1968, G. Stanley Hall award 1970, pres. div. 3 1977, Gold medal award 1986); mem. NAS, Eastern Psychol. Assn. (pres. 1968), Soc. Exptl. Psychologists, Nat. Acad. Edn., Psychonomic Soc., Soc. Rsch. in Child Devel. (Disting. Sci. Contbn. award 1981), Am. Acad. Arts and Scis., Brit. Psychol. Soc. (hon.), N.Y. Acad. Scis. (hon.), Italian Soc. Rsch. in Child Devel. (hon.), Phi Beta Kappa, Sigma Xi. Home: RR 1 Box 265A Middlebury VT 05753-9705

GIBSON, GERALD JOHN, physician; b. Wetherby, Yorkshire, Eng., Apr. 3, 1944; s. Maurice and Margaret (Cronin) G.; m. Mary Teresa Cunningham, Feb. 12, 1977; children: Paul Daniel, Michael John, David James. Student, St. Michaels Coll., Leeds, Eng., 1954-61; BSc with honors, U. London, 1965, MBBS with honors, 1968, MD, 1976. House physician Guys Hosp., London, 1968; house surgeon neurosurgery Leeds Gen. Infirmary, 1969; house physician Hammersmith Hosp., London, 1969-70; casualty med. officer Middlesex Hosp., London, 1970; resident physician McMaster U., Hamilton, Can., 1970-71; registrar respiratory and gen. medicine Hammersmith Hosp., London, 1971-73, sr. registrar respiratory and gen. medicine, 1973-77; cons. physician, hon. sr. lectr. Dept. Medicine U. Newcastle Upon Tyne, 1978—. Author: Clinical Tests of Respiratory Function, 1984; contbr. articles to profl. jours; co-editor Respiratory Medicine, 1990. Fellow Royal Coll. Physicians; mem. Brit. Thoracic Soc. (hon. sec. 1986-88), Am. Thoracic Soc., Med. Rsch. Soc., European Respiratory Soc. (chmn. sci. com. 1991—), Assn. Physicians of Gt. Britain and Ireland. Roman Catholic. Avocations: photography, opera. Home: 36 High St Gosforth, Newcastle-upon-Tyne NE3 1LX, England Office: Freeman Hosp, Newcastle upon Tyne NE7 7DN, England

GIBSON, GORDON RONALD, chemist; b. Buffalo, Sept. 14, 1929; s. Sandy Wellington and Geneva Lucy (Hill) G.; m. Janet Long, Feb. 10, 1954 (div. 1961); children: Andrew, Robert, Douglas; m. Marilyn Jean Kirkendoll, Oct. 20, 1966; children: Nicholas John, Holli Rae. BA in Chemistry, U. Buffalo, 1957. Process devel. chemist Dunlop Tire & Rubber Co., Buffalo, 1957-59; sr. process engr. Hercules Inc., Salt Lake City, 1961-68, Radford, Va., 1969-76; analytical chemist Biomed. Test Lab. U. Utah, Salt Lake City, 1976-77; analytical chemist OSHA U.S. Dept. Labor, Salt Lake City, 1977-81; chemistry specialist Aerojet Propulsion div. GenCorp, Sacramento, 1981-92. Active Reps., Sacramento. With U.S. Army, 1951, Korea. Achievements include patent for Polyurethane Molding; development of solid propellant having highest delivered specific impulse in the world. Home: 6634 Quanah Way Orangevale CA 95662-3332

GIBSON, KATHLEEN RITA, anatomy and anthropology educator; b. Phila., Oct. 9, 1942; d. Keath Pope and Rita Irene (Shewell) G. BA, U. Mich., 1963; MA, U. Calif., Berkeley, 1969, PhD, 1970. Teaching assoc. U. Calif., Berkeley, 1965-69; lectr., adj. assoc. prof. Rice U., Houston, 1973—; asst. prof. U. Tex. Health Sci. Ctr., Houston, 1970-73, assoc. prof., 1973-80, prof., 1980—; mem. com. on parenting behavior Social Sci. Rsch. Coun., N.Y.C., 1980-89; mem. fellowship rev. panel NSF, 1992. Editor: (with S. Parker) Language and Intelligence in Monkeys and Apes, 1990, (with A. Petersen) Brain Maturation and Cognitive Development, 1991, (with Tim. Ingold) Tools, Language and Intelligence in Human Evolution, 1993; author: (with M. Thames and K. Molokon) Genealogy and Demography of the West Main Crees, 1989; contbg. editor Anthropology Newsletter, 1990—; contbr. articles, commentaries and abstracts to profl. jours. Conf. grantee Wenner Gren Found., 1990, Sloan Found., 1985, travel grantee NSF, 1984, 86, Brit. Soc. Devel. Biology, 1982. Fellow Am. Assn. Phys. Anthropologists, Am. Assn. Anthropologists; mem. AAAS, Am. Assn. Anatomists, Internat. Primatological Assn., Lang. Origins Soc., Am. Assn. Primatologists (publs. com. 1987-89), Am. Assn. Dental Schs. (chair sect. on anatomical scis.). Office: U Tex Health Sci Ctr Dept Anatomical Sciences Houston TX 77225

GIBSON, KATHY HALVEY, nuclear reactor technology educator; b. Pitts., June 13, 1956; d. John J. and Mildred (Lauer) Halvey; m. Raymond D. Gibson Jr., Aug. 9, 1980; children: Jennifer Ashley, James Joseph, Megan Elizabeth. BS, Indiana U. of Pa., 1978; postgrad., Carnegie-Mellon U., Pitts., 1983. Chemistry lab. teaching asst. Indiana U. of Pa., 1976-78; chem. analyst, chemist-nuclear, radiochemist Duquesne Light Co., Pitts., 1979-83, chemistry supr., 1983-84, quality assurance engr., 1984-85, sr. quality assurance specialist, 1985-86; reactor engr. U.S. Nuclear Regulatory Commn., King of Prussia, Pa., 1986, resident inspector, 1986-88, sr. resident insp., 1988-90; reactor tech. instr. U.S. Nuclear Regulatory Commn., Chattanooga, 1990-92, section chief, 1992—; lectr. in field. Editor, co-author USNRC Tech. Tng. Manuals, 1990—; contbr. articles to profl. jours. Recipient Disting. Alumni award, Indiana U. of Pa., 1989. Mem. Am. Nuclear Soc. Roman Catholic. Office: US NRC Chattanooga TN 37411-4017

GIBSON, MICHAEL ADDISON, chemical engineering company executive; b. Dallas, Sept. 7, 1943; s. Horace Foster and Inez (Farmer) G.; m. Rose Mary Rodgers, Aug. 21, 1971. BS, Cornell U., 1965, MChemE, 1966; PhD, Rice U., 1973. Registered profl. engr., Tex. Engr. Humble Oil & Refinery Co., Baytown, Tex., 1966-67; research assoc. Exxon Research & Engring. Co., Houston, Baytown, 1973-81; cons. engr. Gibson Research & Cons., Inc., Houston, 1981-84; chmn. Carbotek, Inc., Houston, 1984—; research contractor NASA/Johnson Space Ctr., Houston, 1985—, dept. energy Morgantown (W.Va.) Entech., 1986—. Contbr. articles to profl. jours. 1st lt. Corps Engrs., U.S. Army, 1969-73. Recipient (with others) Presdl. Medal of Freedom for recovery effort during Apollo 13 mission. Mem. Am. Inst. Chem. Engrs., Am. Chem. Soc., Petroleum Club, Tau Beta Pi, Sigma Xi. Baptist. Achievements include 6 patents for in situ coal and heavy oil recovery, and steamflooding and materials processing in space; co-discovery of 2 proprietary catalytic petrochemical processes now under patent application, of extensively-used mathematical models for fluidized bed and in situ coal gasification; first experimental demonstration of fluidized bed reactors at lunar gravity; rsch. on leading candidate processes for prodn. of oxygen and hydrogen for lunar bases from lunar minerals. Avocations: hunting, fishing, hiking, jogging, weight-lifting. Office: Carbotek Inc 16223 Park Row Ste 100 Houston TX 77084-5137

GIBSON, RAYMOND NOVARRO, computer programmer, analyst; b. Chgo., July 11, 1961; s. Novarro and Virginia (Gipson) G.; m. Sonya Dee Harper, Aug. 25, 1984; children: Brittany Chantel, Raymond Novarro II. A in Data Processing, U. Toledo, 1989, BM, 1993. Computer operator U. Toledo, Ohio, 1983-87, programmer/analyst, 1987-91, LAN server adminstr., 1991—. With USNG, 1979-85. Mem. Soc. Black Profls. (faculty advisor), Assn. Black Faculty and Staff (chmn. pub. rels.). Office: U Toledo 2801 W Bancroft Toledo OH 43606

GIBSON, ROBERT LEE, astronaut; b. Cooperstown, N.Y., Oct. 30, 1946; s. Paul A. Gibson; m. M. Rhea Seddon; children: Paul, Julie, Dann. BS in Aero. Engring., Calif. Poly. State U., 1969. Commd. ensign USN, 1969, advanced through grades to capt., 1990; served in Vietnam; astronaut NASA, Houston, 1978—; pilot Shuttle Mission 41-B, 1984; spacecraft comdr. Shuttle Mission 61-C, 1986, STS-27, 1988; chief astronaut office NASA, Houston, 1992—. Office: Johnson Space Ctr Astronaut Office Houston TX 77058

GIBSON, SAM THOMPSON, internist, educator; b. Covington, Ga., Jan. 1, 1916; s. Count Dillon and Julia (Thompson) G.; m. Alice Chase, Oct. 31, 1942 (dec. Jan. 1971); children: Lena S., Stephen C., Judith Gibson Hammer, Lucy F.; m. Madge L. Crouch, Sept. 20, 1986. B.S. in Chemistry, Ga. Inst. Tech., 1936; M.D., Emory U., 1940. Diplomate: Am. Bd. Internal Medicine. Med. house officer Peter Bent Brigham Hosp., Boston, 1940-41, asst. resident medicine, 1946-47, asst. medicine, 1947-49; rsch. fellow medicine Harvard Med. Sch., 1941-42, spl. rsch. assoc., 1943, Milton fellow medicine, 1947-49; assoc. medicine George Washington U. Med. Sch., also George Washington U. Hosp., 1949-63, asst. clin. prof. medicine, 1963—, clin. asst. prof. medicine, Uniformed Svcs., Univ. Health Scis., 1980—; asst. med. dir. ARC Blood Program, 1949-51, assoc. med. dir., 1951-53, assoc. dir., 1953-56, dir., 1956-66; sr. med. officer ARC, 1957-67; asst. dir. div. biologics standards NIH, 1967-72; asst. dir. Bur. of Biologics, FDA, Bethesda, Md., 1972-74; asst. to dir. Bur. Biologics, FDA, Bethesda, Md., 1974-77, dir. div. biologics evaluation, 1977-83; dir. div. biol. product compliance Ctr. for Drugs and Biologics, 1983-85; assoc. dir. sci. and tech. Office of Compliance, Ctr. Drugs and Biologics, FDA, 1985-88; dir. sci. and tech. Office of Health Affairs FDA, Rockville, Md., 1988-89; cons. blood Naval Med. Sch., Nat. Naval Med. Center, Bethesda, 1950-63; mem. med. adv. bd. CARE-Medico, 1962-70, cons., 1970-89; chmn. U.S. com. for tranfusion equipment for med. use Am. Standards Assn., 1954-66, tech. adv. group transfusion equipment for med. use Nat. Commn. Clin. Lab. Standards/Am. Nat. Standards Inst., 1975-89; adviser orgn. blood transfusion services League Red Cross Socs., 1955-66. Contbg. editor: Vox Sanguinis Jour. Blood Transfusion, 1956-65; mem. adv. bd., 1965-76. Served from lt. (j.g.) to comdr., M.C. USNR, 1941-46; capt. Res. ret. Mem. AMA, AAAS, Internat., Am. socs. hematology, Nat. Health Coun. (dir. 1957-60, 61-64), Internat. Soc. Blood Transfusion (regional counselor 1962-66), Am. Fedn. Clin. Rsch., N.Y. Acad. Scis., Delta Tau Delta, Alpha Kappa Kappa, Alpha Chi Sigma, Tau Beta Pi, Phi Kappa Phi, Omicron Delta Kappa, Alpha Omega Alpha. Home: 5801 Rossmore Dr Bethesda MD 20814-2229

GIBSON, WILLIAM CHARLES, neuropsychologist; b. Beaver Falls, Pa., June 26, 1959; s. Charles Harvey and Margaret (Gordon) G.; m. Mary Anne Della Santa, May 19, 1984. BA in Psychology, Gannon U., Erie, Pa., 1981; MA in Urban Affairs and Policy Analysis, New Sch. for Social Rsch., N.Y.C., 1984; PhD in Clin. Psychology, St. John's U., Jamaica, N.Y., 1990. Lic. psychologist, Pa. Psychology intern VA Med. Ctr., Coatesville, Pa., 1989-90; fellow in neuropsychology Bryn Mawr Rehab. Hosp., Malvern, Pa., 1990-91; instr. dept. psychology Immaculata (Pa.) Coll., 1991—; neuropsychologist Jefferson Med. Coll., Phila., 1991—. Contbr. articles to profl. jours. St. John's U. fellow, 1985-86; Erwin S. Wolfson fellow, 1982-84. Mem. APA, Internat. Neuropsychol. Soc., Nat. Acad. Neuropsychology, Phila. Neuropsychol. Soc. Democrat. Home: 236 E Greenwood Ave Lansdowne PA 19050-1708 Office: Jefferson Med Coll Dept Psychiatry/Human Behavior 1025 Walnut St Ste 301 Philadelphia PA 19107-5083

GIDDENS, DON PEYTON, engineering educator, researcher; b. Augusta, Ga., Oct. 24, 1940. BS in Engring., Ga. Inst. Tech., 1963, MS in Aerospace Engring., 1965, PhD, 1967. Assoc. aircraft engr. Lockheed-Ga. Co., Atlanta, 1963; mem. tech. staff Aerospace Corp., San Bernardino, Calif., 1966-67; asst. prof. Ga. Inst. Tech., Atlanta, 1968-70, assoc. prof., 1970-77, prof., 1977-82, Regents prof., 1982-92, dir. Sch. Aerospace Engring., 1988-92; dean engring. Johns Hopkins U., Balt., 1992—. Contbr. numerous articles to profl. jours. Fellow ASME (chmn. bioengring. div. 1986-87); mem. Soc. Sigma Xi (nat. lectr. 1983-87). Avocation: whitewater canoeing. Office: Johns Hopkins U 3400 N Charles St Baltimore MD 21218-2694*

GIDDENS, JOHN MADISON, JR., nuclear engineer; b. Childersburg, Ala., Sept. 4, 1962; s. John Madison and Lyla Joyce (Burchell) G.; m. Angela Faye McDonough, Sept. 22, 1990. BSESE engr, Auburn U., 1987; postgrad. law, Kensington U., 1988-89. Nuclear licensing engr. So. Co. Svcs., Birmingham, Ala., 1987-91; sr. nuclear licensing engr. So. Nuclear Operating Co., Birmingham, Ala., 1991—; vice chmn. Westinghouse Owners Group Lic. Renewal Com., Pitts., 1993—; mem. Nuclear Utilities Mgmt. and Resources Coun. licensing subcom., Washington, 1989-91, Boiling Water Reactor Owner's Group Plant Optimization and License Renewal com., San Jose, Calif., 1991—, Elec. Power Rsch. Inst. Life Cycle Mgmt. Subcom., Palo Alto, Calif., 1991—. Named Young Engr. of the Yr., Ala. Soc. Profl. Engrs., 1990, 89. Mem. NSPE, Engring. Coun. of Birmingham, So. Nuclear Nat. Mgmt. Assn., Am. Nuclear Soc. Birmingham Chpt. (pres. 1990), Ala. Soc. Profl. Engrs., Nat. Inst. for Engring. Ethics, Nat. Am. Nuclear Soc. Home: 111 Cedar Cove Dr Pelham AL 35124-1659 Office: So Nuclear Operating Co 40 Inverness Ctr Pky Birmingham AL 35201-1295

GIDDINGS, J. CALVIN, chemistry educator. Prof. dept. chemistry U. Utah, Salt Lake City. Recipient William H. Nichols medal Am. Chem. Soc., 1991. Office: Univ of Utah Dept of Chemistry Salt Lake City UT 84112*

GIDH, KEDAR KESHAV, chemical engineer; b. Bombay, Sept. 1, 1967; came to the U.S., 1989; s. Keshav Hiraji and Kishori (Murkute) G.; m. Sharmila Arvind Nadkarni, Aug. 23, 1991. BSChE, Banaras Hindu U., 1988; MSChE, Bucknell U., 1991. Mktg. exec. Hindustan Computers Ltd., New Delhi, 1988-89; applications programmer Bucknell U., Lewisburg, Pa., 1991—. Mem. AICHE (assoc.), Sigma Xi (assoc.). Home: 140 N 3d St Apt # 2 Lewisburg PA 17837 Office: Bucknell U 230 Dana Engring Bldg Lewisburg PA 17837

GIER, AUDRA MAY CALHOON, environmental chemist; b. Bella Vista, Peru, Aug. 21, 1940; came to U.S., 1944; d. Nathan Moore and Olivia Cleo (Hite) Calhoon; m. Delta Warren Gier, Apr. 4, 1968. BA, Austin Coll., 1962; MS in Chemistry, Kans. State Coll., 1964; MA in History of Sci., U. Wis., 1974; postgrad., York U., Toronto, Can., 1974-79. Food technologist Midwest Rsch. Inst., Kansas City, Mo., 1963-64; chemist Mobay (formerly Chemagro), Kansas City, 1964-67; instr. chemistry St. Andrews Presbyn. Coll., Laurinburg, N.C., 1967-68; chemist Cardinal Chem. Co., Columbia, S.C., 1968; asst. prof. chemistry Lea Coll., Albert Lea, Minn., 1969-72; psychology intern emergency unit Thistletown Regional Centre for Children & Adolescents, Toronto, Ont., Can., 1975-77; assoc. prof. chemistry Cleveland Chiropractic Coll., Kansas City, 1979-84; adj. faculty Pk. Coll., Parkville, Mo., 1982-92; environ. chemist, quality assurance specialist Ecology & Environ., Inc., Overland Park, Kans., 1987—; pres. Delta and Assocs., Inc., Kansas City, 1988—; co-founder, v.p. Midwest Sci. Found., Kansas City, 1990—; adj. faculty Donnelly Coll., 1992—, dean adminstrn. health scis. program, 1992—. Author: Highlights of Organic Chemistry, 1985; co-editor: (with D.W. Gier) History and Directory of Chemical Education, 1974, (with D.W. Gier) Peace is Something Speshl; co-inventor, co-patentee acetylenic ketones as herbicides. Mem. adv. bd. Kansas City Interfaith Peace Alliance, 1980—, bd. dirs., 1982-85, pres., 1985-86; bd. dirs. Prairie Star Dist./Unitarian-Universalist Midwest (Upper), 1985-91; co-chair Bragg Symposium on Humanism, Kansas City, 1980-90; chair Social Responsibility Com., Prairie Star Dist. UUA, 1986-91; mem. N.Am. Com. for Humanism and Fellowship of Religious Humanists. Recipient Social Justice award Social Justice Com. Prairie Star Dist, 1985; named Woman of Yr., 1982, Humanist of Yr., 1987, All Souls Unitarian Ch., Kansas City. Mem ACLU, DAR, Am. Chem. Soc., Am Soc. Quality Control (cert.), Inst. for Soc., Ethics & Life Scis., Midwest Bioethics Ctr., Planned Parenthood, NARAL, Hazardous Materials Control Rsch. Inst., Assn. for Quality and Participation, Alpha Chi. Democrat. Avocations: bio-med. ethics, peace & social justice activities, needlepoint, knitting, movies. Home: 5828 Cherry St Kansas City MO 64110-3024

GIER, KARAN HANCOCK, counseling psychologist; b. Sedalia, Mo., Dec. 7, 1947; d. Ioda Clyde and Lorna (Campbell) Hancock; m. Thomas Robert Gier, Sept. 28, 1968. BA in Edn., U. Mo., Kansas City, 1971; MA Teaching in Math/Sci. Edn., Webster U., 1974; MA in Counseling Psychology, Western Colo. U., 1981; MEd Guidance and Counseling, U. Alaska, 1981; PhD in Counseling, Pacific Western U., 1989. Nat. cert. counselor. Instr. grades

5-8 Kansas City-St. Joseph Archdiocese, 1969-73; ednl. cons. Pan-Ednl. Inst., Kansas City, 1973-75; instr., counselor Bethel (Alaska) Regional High Sch., 1975-80; ednl. program coord. Western Regional Resource Ctr., Anchorage, 1980-81; counselor U. Alaska, Anchorage, 1982-83; coll. prep. instr. Alaska Native Found., Anchorage, 1982; counselor USAF, Anchorage, 1985-86; prof. U. Alaska, Anchorage, 1982—; dir. Omni Counseling Svcs., Anchorage, 1984—; prof. Chapman Coll., Anchorage, 1988—; workshop facilitator over 100 workshops on the topics of counseling techs., value clarification, non-traditional teaching approaches, peer-tutor tng. Co-author: Coping with College, 1984, Helping Others Learn, 1985; editor, co-author: A Student's Guide, 1983; contbg. author developmental Yup'ik lang. program, 1981; contbr. photographs to Wolves and Related Canids, 1990, 91; contbr. articles to profl. jours. Mem. Am. Bus. Women's Assn., Blue Springs, Mo., 1972-75, Ctr. for Environ. Edn., World Wildlife Fund, Beta Sigma Phi, Bethel, Alaska, 1976-81. Recipient 3d place color photo award Yukon-Kuskokwim State Fair, Bethel, 1978, Notable Achievement award USAF, 1986, Meritorious Svc. award Anchorage Community Coll., 1984-88. Mem. Coll. Reading and Learning Assn. (editor, peer tutor sig leader 1988—, Cert. of Appreciation, 1986-93, bd. dirs. Alaska state, coord. internat. tutor program), AACD, Alaska Assn. Counseling and Devel. (pres. 1989-90), Alaska Career Devel. Assn. (pres.-elect 1989-90), Nat. Rehab. Assn., Nat. Rehab. Counselors, Greenpeace, Human Soc. of the U.S. Wolf Haven Am., Wolf Song of Alaska. Avocations: raising and training hybrid wolves, wolf preservation, photography, classical music, British mysteries. Home and Office: Omni Counseling Svcs 8102 Harvest Cir Anchorage AK 99502-4682

GIESE, CLAYTON FREDERICK, physics educator, researcher; b. Mpls., July 19, 1931; s. Theodor F. and Emma R. (Bechtold) G.; m. Joyce A. Woods, Mar. 20, 1985. BS, U. Minn., 1953, PhD, 1957. Asst. prof. U. Chgo., 1957-65; assoc. prof. U. Minn., Mpls., 1965-73, prof., 1973—. Office: Univ of Minn Sch of Physics Minneapolis MN 55455

GIESEN, HERMAN MILLS, engineering executive, consultant, mechanical forensic engineer; b. San Antonio, Sept. 22, 1928; s. Herman Iglehart and Emeline Barbara (Frey) G.; m. Linda B. Williams, Aug. 9, 1979; 1 child, Jonathan; children by previous marriage: John Herman, David Douglas, Amy Lynn. Student Tex. A&M U., 1946-47; BS in Engring., U.S. Naval Acad., 1951; MSEE, USAF Inst. Tech., 1960; MS in Ops. Mgmt., U. So. Calif., 1966. Commd. 2d lt. USAF, 1951, advanced through grades to maj., 1966; served as aircraft maintenance mgr., 1954-56, flight instr., 1957-59, rsch. and devel. program officer, 1960-63, aircraft, flight commdr., 1963-64, elec. engr.-analyst, 1964-66, resigned, 1966, now col. Res., ret.; exec. adviser in program control McDonnell-Douglas Corp., Huntington Beach, Calif., 1966-68; sr. bus. planner E-Systems, Inc., Greenville, Tex., 1968-71; pres. Giesen & Assos., Inc., indsl. mgmt. engring. cons., Dallas, 1971-72, 78—; plant engr. Dixie Metals of Tex., Dallas, 1972-73; plant engr. Murph Metals Div., R.S.R. Corp., Dallas, 1973-74, ops. maintenance/engring. mgr., 1974-76; mfg. mgr. Ferguson Industries, Dallas, 1976-78; self-employed cons. design engr., 1978-84; mech. forensic engr. AID Cons. Engrs., Inc., Dallas, 1985-90; pres., mech. forensic engr. Environ. Issues Support, Inc., 1990—. Decorated Air medal, USAF Commendation medal, Air Force Meritorious Service medal; registered profl. engr., Tex., environ. assessor; cert. flight instr., advanced instrument ground aircraft instr. FAA. Mem. NSPE, Tex. Soc. Profl. Engrs. Contbr. articles to profl. jours. Home: 3636 Shenandoah St Dallas TX 75205-2119

GIFFORD, ERNEST MILTON, biologist, educator; b. Riverside, Calif., Jan. 17, 1920; s. Ernest Milton and Mildred Wade (Campbell) G.; m. Jean Duncan, July 15, 1942; 1 child, Jeanette. A.B., U. Calif., Berkeley, 1942, Ph.D., 1950; grad., U.S. Army Command and Gen. Staff Sch., 1965. Asst. prof. botany, asst. botanist expt. sta. U. Calif.-Davis, 1950-56, assoc. prof. botany, assoc. botanist, 1957-61, prof. botany, botanist, 1962-87, prof. emeritus, 1988—, chmn. dept. botany and agrl. botany, 1963-67, 74-78. Author: (with A. S. Foster) Morphology and Evolution of Vascular Plants, 3d edit., 1989, (with T. L. Rost) Mechanisms and control of Cell Division, 1977; editor in chief Am. Jour. Botany, 1975-79; advisor to editor Ency. Brit.; contbr. articles on anatomy, ultrastructure and morphogenesis of higher plants to profl. jours. Served to maj. U.S. Army, 1942-46; ETO; to col. USAR, 1946-73. Decorated Bronze Star medal; named disting. contbr. Ency. Brit., 1964; NRC fellow Harvard U., 1956; Fulbright research scholar, France, 1966; John Simon Guggenheim Found. fellow, France, 1966; NATO sr. postdoctoral fellow, France, 1974; recipient Acad. Senate Disting. Teaching award U. Calif.-Davis, 1986. Fellow Linnean Soc. (London); mem. Bot. Soc. Am. (v.p. 1981, pres. 1982, merit award 1981), Internat. Soc. Plant Morphologists (v.p. 1980-84), Am. Inst. Biol. Scis., Calif. Bot. Soc., Sigma Xi. Office: U Calif Dept Botany Robbins Hall Davis CA 95616

GIFFORD, GERALD FREDERIC, environmental program director; b. Chanute, Kans., Oct. 24, 1939; s. Gerald Leo and Marion Lou (Browne) G.; m. Cinda Jean Lowman, June 26, 1982. Student, Kans. U., 1957-60; BS in Range Mgmt., Utah State U., 1962, MS in Watershed Mgmt., 1964, PhD in Watershed Sci., 1968. Asst. prof. watershed sci. Utah State U., Logan, 1967-72, assoc. prof., 1972-80, prof., 1980-84, chmn. watershed sci. unit, 1967-84, dir. Inst. Land Reclamation, 1982-84; head range, wildlife and forestry U. Nev., Reno, 1984-92, chmn. environ. and resource sci. dept., 1992—; exchange scientist NSF, Canberra, Australia, 1974; cons. Smithsonian Inst., Nat. Park Service, Office of Tech. Assessment, Tex. Tech U., U. Minn., Bur. Land Mgmt. AMAX Coal Co., Nat. Commn. Water Quality, 1967—. Author: Rangeland Hydrology, 1981; assoc. editor Arid Soil Rsch. and Rehab., 1985-90; contbr. papers to profl. pubs. Mem. Am. Water Resources Assn., Soc. Range Mgmt. (assoc. editor 1982-86, 91—), Soil and Water Conservation Soc., Soc. Wetland Scientists. Avocations: racquetball, antiques, garage sales. Home: 3880 Squaw Valley Cir Reno NV 89509-5663 Office: U Nev Environ and Resource Scis 1000 Valley Rd Reno NV 89512-2899

GIFKINS, ROBERT CECIL, materials engineer; b. London, May 30, 1918; s. Cecil and Ida (Gill) G.; m. Elizabeth Emily Glaisker, Feb. 14, 1942; children: Kenneth John, Pamela, Roger. BS, London U., 1941; DSc, Melbourne U., 1961. Cert. engr., Eng.; cert. profl. engr., Australia. Tech. asst. Brit. Indsl. Solvents, Carshalton, Surrey, 1936-37; exptl. officer Nat. Phys. Lab., Teddington, Middlesex, 1937-46, Brit. Atomic Energy Authority, Harwell, 1946-48; from sci. officer to chief rsch. officer Commonwealth Sci. and Indsl. Rsch. Orgn., Melbourne, Australia, 1948-78; sr. rsch. fellow Imperial Coll., London, 1962-63, Southampton (Eng.) U., 1978-79; bd. dirs. Acta Metallurgica, Inc., Mass., 1980-83, 86-89, lectr., 1990-92. Author: Optical Microscopy of Metals; contbr. over 100 rsch. papers to profl. jours. Pres. High Sch. Coun., Blackburn, 1975-78.80. Recipient Hofmann prize, 1974. Fellow Inst. Materials, Inst. Metals and Materials Australia (hon., bd. dirs.), Inst. Engrs. Australia (hon.); mem. Australian Inst. Metals (pres. 1965, editor jour. 1956-78, annual lectr. 1974, Silver medal 1965), Iron and Steel Inst. Japan (hon.). Home: 31 Kennedy Rd, Somers Victoria, Australia 3927

GIFT, JAMES J., toxicologist. BA in Biology, Harvard U, 1964; MA in Environ. Sci., Rutgers U., 1968, PhD in Environ. Sci., 1970. Sr. toxicologist, dir. risk assessment and mgmt. bus. EA Engring., Sci. & Tech. Inc., Md.; Contbr. articles to profl jours. Mem. Am. Fisheries Soc., Soc. Environ. Toxicology and Chemistry, Water Environ. Fed. Achievements include direction of a multimedia assessment contrasting ocean disposal of sewage sludge with various land based waste management options; direction of ocean site designation studies for New York City and other municipalities; preparation of the first Special Permit Application for ocean disposal of sewage sluge; conducted research on the physiological effects of thermal gradients of numerous marine, estuarine, and freshwater fish species. Office: EA Engring Sci & Tech Inc 11019 McCormick Rd Cockeysville Hunt Valley MD 21031

GIGA, YOSHIKAZU, mathematician; b. Musashino, Japan, Nov. 4, 1955; s. Kenjiro and Kazuyo (Hasumi) G. BS, U. Tokyo, 1979, MS, 1981, DSc, 1985. Rsch. assoc. Nagoya (Japan) U., 1981-86; asst. prof. math. Hokkaido U., Sapporo, Japan, 1986-88, assoc. prof., 1988-92, prof., 1992—; vis. mem. Courant Inst., N.Y.C., 1982-84, Inst. for Math. and its Applications, U. Minn., Mpls., winters 1985, 90; vis. asst. prof. U. Md., College Park, fall 1984; vis. prof. U. Paderborn, Fed. Republic Germany, winter 1987. Editor Mathematische Annalen. Mem. Math. Soc. Japan, Am. Math. Soc. Home: 2-20-402 North, Nango Ave 19, Hokkaido, Shiroishi, Sapporo 003, Japan Office: Hokkaido U, Dept Math, Sapporo 060, Japan

GIGER, PETER, engineer; b. Zurich, Switzerland, Mar. 3, 1945; s. Paul and Claudie (Irniger) G.; m. Judith Schwitzer, Jan. 26, 1985. Dipl.-Ing, Dr sc. techn., Swiss Fed. Inst. Tech., Zurich, 1971. Project mgr. Swiss Fed. Rys., Zurich, 1973-78; rsch. asst. Swiss Fed. Inst. Tech., 1972-73, head sect. ry. constrn., 1978—. Contbr. numerous articles on math. methods in transport planning to profl. jours. Office: ETH, Hoenggerberg, CH-8093 Zurich Switzerland

GIGLI, IRMA, physician, educator, academic administrator; b. Cordoba, Argentina, Dec. 22, 1931; d. Irineo and Esperanza Francisca (Pons de Gigli) G.; m. Hans J. Muller-Eberhard, June 29, 1985. B.A., Liceo Nacional Manuel Belgrano, Cordoba, 1950; M.D., Universidad Nacional de Cordoba, 1957. Intern Cook County Hosp., Chgo., 1957-58; resident in dermatology Cook County Hosp., 1958-60; fellow in dermatology NYU, 1960-61; mem. faculty Harvard Med. Sch., 1967-75, asso. prof. dermatology, 1972-75; chief dermatology service Peter Bent Brigham Hosp., Robert B. Brigham Hosp., 1971-75; prof. dermatology and exptl. medicine N.Y. U. Med. Center, N.Y.C., 1976-82; mem. Irvington Houst Inst. N.Y. U. Med. Center, mem. faculty N.Y. Grad. Sch. Med. Scis., dir. Asthma and Allergic Disease Center for Immunodermatology Studies, 1980-91; prof. medicine, chief div. dermatology U. Calif.-San Diego, 1983—; chmn. study sect. Allergy and Immunology Inst., NIH; Guggenheim Found. Western Hemisphere and Phillippines Com. of Selection. Contbr. articles to profl. jours. Recipient research award Am. Cancer Soc., 1970-72, research award NIH, 1972-76; Guggenheim Found. grantee, 1974-75. Mem. Am. Soc. Clin. Investigation, Am. Assn. Immunologists, Am. Acad. Dermatology, Soc. Investigative Dermatology, Am. Acad. Allergy, Assn. Am. Physicians, Am. Dermatol. Assn., Soc. for Investigative Dermatology (pres. 1990-91), Inst. Medicine. Office: U Calif-San Diego Med Ctr 200 W Arbor Dr San Diego CA 92103-8420

GIHWALA, DHERENDRA ISVER, chemist; b. Cape Town, South Africa, Dec. 26, 1951; s. Isver and Padma (Jeena) G. MSc, U. Durban, Natal, South Africa, 1979; PhD, U. Cape Town, 1982. Cert. natural scientist. Rsch. asst. Nat. Accelerator Centre, Cape Town, 1977-82; sr. lectr. Peninsula Technikon, Bellville, South Africa, 1983-84, head chemistry dept., 1984-88, dir. sci. sch., 1988—; disting. vis. prof. Macalester Coll., St. Paul, Minn., 1990, 91. Author: (with others) Physiologic Foundations of Perinatal Care, 1985, Elemental Analysis By Particle Accelerators, 1992; contbr. articles to profl. jours. Mem. Fedn. Cape Civic Assns., Cape Province, South Africa, 1978—; sec. and publs. officer New Unity Movement, South Africa, 1985-90; mem. adv. bd. Found. for Rsch. Devel., Pretoria, South Africa, 1990—; trustee Gateway Sci. Exporatorium, Cape Town, 1991—. Coun. for Sci. and Indsl. Rsch. scholar, 1977-83. Mem. Am. Chem. Soc. (Chem. Edn. div.), South Africa Spectroscopic Soc., N.Y. Acad. Scis., Assn. So. Inst. Tech. (exec. com. 1990—, rsch. rep. com. Technikon principals 1992—). New Unity Movement. Achievements include research in charged particle prompt photon spectrometry; application of prompt photon spectrometry in geology, archaeology, medicine, industry and mining. Home: 18 Doreen Rd Rylands Estate, Cape Town 7764, South Africa Office: Peninsula Tech, Modderdam Rd, Bellville 7535, South Africa

GIL, JANUSZ ANDRZEJ, physics educator; b. Bielsko, Poland, Nov. 24, 1951; s. Teodor and Anna (Sztwiorok) G.; m. Emilia Antonina Wabia, July 3, 1975; children: Michal, Jaroslaw. MS, Pedagogical U., Rzeszow, Poland, 1976; PhD in Physics, Warsaw U., 1983. Rsch. assoc. Pedagogical U., Rzeszow, 1975-83; vis. rsch. assoc. Nat. Radio Astronomy Lab., Charlottesville, Va., 1984-85; vis. asst. prof. Ky. State U., Lexington, 1985-87; prof. physics Pedagogical U., Zielona Gora, Poland, 1988-91; Humboldt rsch. fellow Max Planck Inst. for Radio Astronomy, Bonn, 1991—. Editor Proceedings of IAU Colloquium 128, Lagow, Poland, 1991. Fellow Am. Astron. Soc., Royal Astron. Soc. Office: The Astron Ctr, Lubuska 2, Zielona Gora Poland 65001

GILBERT, ARTHUR CHARLES, aerospace engineer, consulting engineer; b. N.Y.C., Sept. 23, 1926; s. Phillip Saul and Annie (Taishoff) G.; m. Suzanne Debra Teperson, June 18, 1953 (div. 1986); children: Pamela Stephanie Gilbert Remis, Randi Ilene Gilbert Cutler. B Aero. Engring., NYU, 1946, M Aero. Engring., 1947, ScD in Engring., 1956. Registered profl. engr., N.Y., Mich., D.C. Rsch. engr., designer, educator various orgns., 1951-67; v.p., mng. ptnr. Systems Technology Lab., Inc., 1968-70; founder Auto-Train Corp., 1968-79; chief scientist Chief Naval Ops. Exec. Panel, 1970-75; v.p., dir. engring. R & D Data Solutions Corp., 1975-77; v.p. Unified Industries, 1977-78; with OAO Corp., 1978-81; pres. Arthur C. Gilbert, SCD, PE, Arlington, Va., 1981—; consulting engr. DAR, Washington, 1982-85, sr. engr R&D, mgmt. major aerospace corps.; consulting engr. naval systems div. FMC Corp., Mpls., 1987—; special asst. to USN-sec. Nav R&D, Washington, 1987—; vis. prof. Navy War Coll., Newport, R.I., 1979. Mem. Spitfire Soc. U.K., Cosmos Club Washington. Achievements include 2 patents for high speed machinery, helicopter propulsion systems; contributing to numerous articles and books on structures, acoustics, vibration and photo-elasticity. Home and Office: Arthur C Gilbert SCD PE 1201 S Eads St Apt 910 Arlington VA 22202-2840

GILBERT, CHARLES D., neurobiologist; b. N.Y.C., Jan. 15, 1949; s. Gustave M. and Matilda S. (Safran) G. BA in Biophysics, Amherst U., 1971; MD, Harvard U., 1977, PhD in Neurobiology, 1977. Teaching fellow Harvard Med. Sch., Cambridge, Mass., 1977-79, prin. rsch. assoc., 1979-81, asst. prof., 1981-83; asst. prof. Rockefeller U., N.Y.C., 1983-85, assoc. prof., 1985-91, prof. neurobiology, 1991—; mem. adv. bd. Klingenstein Fund, N.Y.C., 1983—; mem. NSF Adv. Panel, Washington, 1984-87. Contbr. articles to profl. jours. Exec. sec. Pew Charitable Trust Latin Am. Scholars, N.Y.C., 1989-91. Recipient Weill-Caulier award, 1984, Rita Allen Found. award, 1986, Presdl. Young Investigator award NSF, 1984, Devel. award McKnight Found., 1991, Cortical Discoverer award Cajal Club, 1993; fellow Danforth Found., 1971-75, Med. Found., 1978-80. Mem. Assn. for Rsch. in Vision and Ophthalmology (editorial bd. Jour. Neurophysiology 1989—), Soc. for Neurosci.

GILBERT, DAVID WALLACE, aerospace engineer; b. Berkeley, Calif., June 20, 1923; s. Wallace William and Elizabeth Wilson (Findlay) G.; m. Jeanne Wilson Gillette, June 27, 1948; children: Laurence R., LeeAnn, Dean A., Barbara L., John L., Mary H. BSME, U. Calif., Berkeley, 1948. Flight test analyst Convair, San Diego, 1948-51, dynamics engr., 1951-55, design specialist, 1955-59, chief GN&C systems, 1959-62; mgr. GN&C JSC Apollo Project Office NASA, Houston, 1962-64, br. chief GN&C Systems div., 1964-89. With U.S. Army, 1942-45, ETO. Mem. AIAA (assoc. fellow). Achievements include technical development for numerous aspects of F-102, F-106, CV-880 and 990 aircraft and Apollo and space shuttle systems. Home: 10019 Cedarhurst Dr Houston TX 77096-5102

GILBERT, FRED IVAN, JR., physician, researcher; b. Newark, Mar. 5, 1920; s. Fred I. and Gertrude Olga (Lund) G.; m. Helen Ruth Odell, Sept. 21, 1943 (div. Jan. 1974); children—Rondi, Kristin, Galen, Gerald, Fred I. III, Lisa, Cara; m. Gayle Yamashiro, Sept. 16, 1978; children—Heidi, John. Jr. cert., U. Hawaii, 1940; B.S., U. Calif.-Berkeley, 1942; M.D., Stanford U., 1946. Diplomate Nat. Bd. Med. Examiners, Am. Bd. Internal Medicine. Intern Stanford-Lane Hosp., San Francisco, 1945-46; assoc. clin. prof. medicine Stanford U., Calif., 1948-51; internist Straub Clinic and Hosp., Honolulu, 1951—; resident in nuclear medicine Straub Clinic and Hosp., 1979-80; resident in neurology U. London, 1960-61; chief medicine Queen's Hosp., Honolulu, 1965-69; now staff Queen's Hosp.; prof. pub. health and medicine U. Hawaii, Honolulu, 1969—; med. dir. Pacific Health Research Inst., Honolulu, 1960—; resident in nuclear medicine U. Hawaii and U. Calif.-Sacramento, 1978-80; mem. staff St. Francis, Kuakini, Castle hosps., Honolulu; pres. Hawaii Acad. Sci., 1959-60, Hawaii Heart Assn., 1959; bd. dirs. Cancer Control Cancer Ctr. Hawaii, 1985-86. Contbr. numerous articles to profl. jours. Served to capt. U.S. Army, 1943-48. Recipient Disting. Physician award Nat. Acad. Practice, 1985. Fellow AAAS, ACP; mem. Am. Soc. Nuclear Medicine (pres. Hawaii chpt. 1980-82), Am. Coll. Nuclear Physicians, Internat. Health Evaluation Assn. (pres. 1971-72), AMA, Inst. Medicine of Nat. Acad. Sci., Social Sci. Club, Alpha Omega Alpha. Avocations: surfing, diving, tennis. Home: 2112 Mott Smith Dr Honolulu HI 96822-2511 Office: Straub Clinic and Hosp 846 S Hotel St Ste 303 Honolulu HI 96813-3083

GILBERT, JAMES FREEMAN, geophysics educator; b. Vincennes, Ind., Aug. 9, 1931; s. James Freeman and Gladys (Paugh) G.; m. Sally Bonney, June 19, 1959; children: Cynthia, Sarah, James. BS, MIT, 1953, PhD, 1956. Research assoc. MIT, Cambridge, 1956-57; asst. research geophysicist Inst. Geophysics and Planetary Physics at UCLA, 1957, asst. prof. geophysics, 1958-59; sr. research geophysicist Tex. Instruments, Dallas, 1960-61; prof. Inst. Geophysics and Planetary Physics at U. Calif. San Diego, La Jolla, 1961—, assoc. dir., 1976-88; chmn. grad. dept. Scripps Inst. Oceanography, La Jolla, 1988-91; chmn. steering com. San Diego Supercomputer, 1984-86. Contbr. numerous articles to profl. jours. Recipient Arthur L. Day medal Geol. Soc. Am., 1985, Internat. Balzan prize , 1990; Fairchild scholar Calif. Inst. Tech., Pasadena, 1987; fellow NSF, 1956, Guggenheim, 1964-65, 72-73, Overseas fellow Churchill Coll. U. Cambridge, Eng., 1972-73. Fellow AAAS, Am. Geophys. Union; Nat. Acad. Scis., European Union Geoscis. (hon.); mem. Seismology Soc. Am., Am. Math. Soc., Royal Astron. Soc. (recipient Gold medal 1981), Sigma Xi. Home: 780 Kalamath Dr Del Mar CA 92014-2630 Office: U Calif Inst Geophysics and Planetary Physics 0225 La Jolla CA 92093

GILBERT, JO, psychologist; b. L.A., July 25, 1949; d. Joseph Raymond and Rochelle Rose (Burdman) G.; divorced; 1 child, Branden Christopher Smale. BA in Psychology cum laude, UCLA, 1972; postgrad., U. Houston, 1971-72, William Marsh Rice U., 1972-77; PhD in Clin. Psychology, Calif. Sch. Profl. Psychology, 1980. Lic. psychologist, Calif. Psychol. intern, researcher, then counselor Olive St. Bridge, Fresno, Calif., 1978-80; registered psychologist FCEOC Project Pride, Fresno, 1980-82; psychologist Fox, Pick and Assocs., Napa, Calif., 1982-85; pvt. practice Napa, 1985—; adj. faculty in forensic psychology Calif. Sch. Profl. Psychology, Berkeley, 1987; faculty U. San Francisco, 1987-88; presenter at profl. confs.; mem. Sacramento County panel ct.-appointed psychologists, Yolo County panel ct.-appointed psychologists, Solano County panel ct.-appointed psychologists; cons. Bd. Prison Terms. Contbr. articles to profl. publs. Mem. APA, Calif. Psychol. Assn. (bd. dirs.), Napa Valley Psychol. Assn. (past pres.), Soc. Personality Assessment. Democrat. Jewish. Avocations: music, literature, fitness training.

GILBERT, JOHN JOUETT, aquatic ecologist, educator; b. Southampton, N.Y., July 18, 1937; s. Seymour Parker Gilbert and Louise Ross (Todd) Stanley; m. Caroline Spalding Colburn, June 16, 1959; children: John Spalding, Anne Menefee. BA, Williams Coll., 1959; PhD, Yale U., 1963. Asst. prof. Princeton (N.J.) U., 1964-66; asst. prof. dept. biol. scis. Dartmouth Coll., Hanover, N.H., 1966-69, assoc. prof., 1969-74, prof., 1974—. Contbr. over 100 articles to profl. jours. Recipient Career Devel. award, 1973-78; NSF, NIH, EPA grantee, 1965—. Mem. Ecol. Soc. Am., Am. Soc. Limnology and Oceanography, Internat. Soc. Theoretical and Applied Limnology (nat. rep. 1971-83). Avocation: fly fishing. Office: Dartmouth Coll Dept Biol Scis Hanover NH 03755

GILBERT, KEITH DUNCAN, electronics executive; b. Boyertown, Pa., Nov. 3, 1941; s. Harry Irman and Wilma Marie (Hudson) G.; m. Susan Jane Kelley, June 26, 1971; 1 child, Amanda Kelley. BSEE, MIT, 1963, SMEE, 1965. With Watkins-Johnson Co., Palo Alto, Calif., 1966—, tech. staff, 1966-74, sect. head, 1974-76, dept. mgr., 1976-78, div. mgr., 1978-86, group v.p., 1986—. NSF fellow U.S. Govt., 1964-65. Mem. Assn. Old Crows, University Club. Avocations: tennis, bridge. Office: Watkins-Johnson Co 3333 Hillview Ave Palo Alto CA 94304-1204

GILBERT, NATHAN, mechanical engineer; b. N.Y.C., Nov. 8, 1916; s. Joseph and Eva (Post) G.; m. Helen Dragilow, Oct. 31, 1942; children: Roger William, Jane Ellen, Nancy G. McMullen-Knapp. BME, Coll. of the City of N.Y., 1941; MME, NYU, 1947. Registered profl. engr., Tex., N.J., N.Y. Mech. engr. N.Y. Naval Shipyard, Brooklyn, 1941-46; mech. engr. M.W. Kellogg Co., N.Y.C., 1946-57, sr. engring. cons., 1971-81; rsch. engr., cons. Kellogg Rsch. and Engring. Devel. Lab., Piscataway, N.J., 1957-71; engring. cons. Foster Wheeler USA Corp., Houston, 1981-86; pvt. practice in engring. cons. Houston, 1986—; lectr. U. Houston, 1975-86; instr. Fairleigh Dickenson U., Teaneck, N.J., 1958-71, Rutgers U., New Brunswick, N.J., 1966-71. Conbr. articles to profl. jours. including Chem. Engring., Jour. of Pressure Vessel Tech., Chem. Engring. Progress, ASME Transactions. Mem. ASME (mem. pressure vessel and pressure piping code coms., Dean W.R. Woolrich Engr. of Yr. award 1985), Sigma Xi. Achievements include patents for sensor apparatus for primary battery system and external fluid catalytic cracking unit regulator plenum manifold. Home and Office: Lake Conroe Hills 299 Bunker Hill Dr Willis TX 77378-8620

GILBERT, PAUL H., engineer, consultant; b. Healdsburg, Calif., Apr. 23, 1936; s. Lindley D. and Beatrice G.; m. Elizabeth A. Gilbert, July 13, 1963; children: Christopher, Gregory, Kevin. BSCE, U. Calif., Berkley, 1959, MSCE, 1960. Registered profl. engr. in 17 states. Project mgr. Calif. State Water Project, Sacramento, 1959-68; officer U.S. Army Corp Engrs., Heidleberg, Germany, 1960-61, capt., 1961-68; project mgr. Parsons Brinckerhoff, N.Y.C., 1969-73; regional mgr./ptnr. Parsons Brinckerhoff, San Francisco, 1973-85; deputy COO, sr. v.p. Parsons Brinckerhoff, N.Y.C., 1985-90, chmn. of bd., 1990—; project dir. supercollider design and constrn. Parsons Brinckerhoff, Dallas, 1990—; prin.-in-charge of award projects, Glenwood Canyon I70 tunnels, San Francisco Ocean Outfall, Seattle Bus. Tunnel, Hood Canal Floating Bridge and West Seattle High Level and Low Level Swing Bridges, others; reviewer NSF, Washington, D.C., 1992. Contbr. articles to profl. jours. Recipient Lincoln Art Welding award, 1966. Fellow ASCE (Rickey medal 1969); mem. Project Mgmt. Inst., Soc. Am. Mil. Engrs., Moles. Republican. Roman Catholic. Office: The PB/MK Team 5510 S Westmoreland Rd Dallas TX 75237

GILBERT, RALPH WHITMEL, JR., engineering company executive; b. Columbia, Miss., Dec. 19, 1939; s. Ralph Whitmel and Jack Pearl (Anderson) G.; m. Dorothy Kresevich, Feb. 19, 1977; children: Randall, Richard. BSCE, U. Miss., 1962; MSCE, Princton U., 1969. Registered profl. engr., La., Pa. Commd. 2d lt. U.S. Army, 1962, advanced through grades to lt. col., 1980, ret., 1982, career officer, 1962-82; project mgr. Michael Baker, Jr., Inc., Beaver, Pa., 1982-83, dept. mgr., 1983-84; sr. engr. Parsons Brinckerhoff, Inc., Pitts., 1984-89; program mgr. HDR-Richardson, Gordon, Inc., Pitts., 1989-90; dept. mgr., sr. v.p. HDR Engring., Inc., Pitts., 1990—; bd. dirs. and officer Consulting Engrs. Coun., Pa., chmn. transp. com., 1992-93. Prodr. (video prodn.) Wabash Bridge and Tunnel Project, 1992. Asst. cubmaster Pack 62 Cub Scouts of Am., 1987-92; coach Penn Hills (Pa.) Soccer Assn., 1985-92; mem. Crescent Hills Civic Assn., Penn Hills, 1979-92; elder, deacon Beulah United Presbyn. Ch., Churchill, Pa., 1984—; coach and referee Oakmont Athletic Assn., 1991-93. Decorated Bronze Star, Legion of Merit; named to Hall of Fame, U. Miss., 1962. Fellow Soc. Am. Mil. Engrs. (pres. 1981-82); mem. ASCE, Pa. Soc. Profl. Engrs., Am. Soc. Hwy. Engrs., Engrs. Soc. of We. Pa. Republican. Office: HDR Engring Inc 3 Gateway Ctr Pittsburgh PA 15222

GILBERT, RICHARD MICHAEL, physician, educator; b. N.Y.C., Dec. 27, 1942; s. Joseph George and Gertrude (Pincus) G.; m. Diana Wasserman, Jan. 3, 1970; children: Alexander Jon, Darren Todd, Elizabeth Merril. BS, U. Mich., 1964; MD, NYU, 1968. Diplomate Am. Bd. Internal Medicine, splty. in medicine, 1972, 80, splty. in nephrology, 1976. Intern NYU Bellevue Med. Ctr., N.Y.C., 1968-69, resident, 1969-71, chief resident, 1971-72; nephrology fellow Albert Einstein Med. Ctr., Bronx, N.Y., 1974-76; nephrology rsch. fellow Montifiore Hosp., Bronx, 1976-77; attending physician NYU Hosp., N.Y.C., 1977—; asst. prof. NYU Sch. Medicine, N.Y.C., 1977—. Contbr. articles to Mutation Rsch., Jour. Clin. Investigation. Major U.S. Army, 1972-74. Recipient cert. of merit for rsch. U.S. Army, 1974, Nat. Rsch. Svc. award NIH, 1975-76. Fellow ACP. Achievements include rsch. on involvement of separate pathways in the repair of mutational and lethal lesions induced by a monofunctional sulfur mustard; intrarenal recycling of urea in the rat with chronic exptl. pyelonephritis. Office: 530 First Ave New York NY 10016

GILBERT, ROBERT OWEN, veterinary educator, researcher; b. Pretoria, Transvaal, South Africa, Dec. 22, 1954; came to U.S., 1988; s. Raymond Cyril and Sheila (Bewick) G.; m. Kathleen Clarke, Nov. 26, 1977; children:

David Robert, Mary-Anne. B.V.Sc., U. Pretoria, 1977, M.Med.Vet. cum laude, 1985. Diplomate Am. Bd. Theriogenologists; registered veterinarian, veterinary specialist in Genesiology, South Africa. Asst. veterinarian with pvt. practice, Ladysmith, Natal, South Africa, 1977-78, Alberton, Transvaal, South Africa, 1978; vet. supt. Taurus A.I. Coop., Irene, South Africa, 1978-81; state veterinarianA.I. sect. Vet. Rsch. Inst., Onderstepoort, South Africa, 1981; sr. lectr. dept. genesiology U. Pretoria, Onderstepoort, 1981-84, assoc. prof. theriogenology, 1986-88; resident in Theriogenology U. Wis. Sch. Vet. Medicine, Madison, 1984-85; asst. prof. theriogenology Cornell U. Coll. Vet. Medicine, Ithaca, N.Y., 1988-92, chief theriogenology sect., 1990—; dir. Cornell U. Bovine Rsch. Ctr., Ithaca, N.Y., 1991—; head theriogenology clinic, U. Pretoria, 1986-88; mem. various coms. and subcoms. U. Pretoria, 1987, 88, Cornell U., 1989—; ad hoc reviewer various grant applications; cons.; speaker in field. Author: (with others) Small Animal Endocrinology, 1987, Veterinary Clinics of North America, 1992, Foreign Animal Diseases: Their Prevention, Diagnosis and Control, 1992; editor Vet. Student Jour., 1976-77; mem. editorial & rev. com. Jour. South African Vet. Assn., 1987, 88; mem. editorial com. Animal Health Newsletter, Cornell U., 1990, 91, 92; ad hoc reviewer Vet. Medicine, 1989, Theriogenology, 1990, 91, 92, Cornell Veterinarian, 1990, Vet. Pathology, 1992, Lab. Animal Sci., 1993; contbr. articles to profl. jours. Active Tugela Basin Agrl. Show Com., 1977-80; parish councillor St. Martin's Anglican Ch., Irene, 1979-80, 80-81, St. Francis' Anglican Ch., Waterkloof, Pretoria, 1987-88; mem. exec. com. KwaNdebele Mission, 1987-88. Achievement scholar U. Pretoria, 1973, 74, 75, 76; scholar Witwatersrand Agrl. Soc., 1977; grantee Dorper Sheep Breeders' Assn., 1982, AVMA Found., 1985, U. Pretoria, 1986-87, Hoechst, Germany, 1987-88, Cornell U. Coll. Vet. Medicine, 1989, 90, Eastern Artificial Insemination Coop., 1991, Nat. Assn. Animal Breeders, 1991, NRSA, 1991, Cornell U. Coll. Vet. Medicine, 1991, 92. Mem. South African Vet. Assn. (exec. com. No. Transvaal Branch 1982-83, 83-84, exec. com. sec. Embryo Transfer Group, 1982-83, 83-84, Production and Reproduction Group), Rural Practitioners' Group, Veterinarians in Industry Group (organizer 1980-84), Royal Coll. Vet. Surgeons, Soc. Theriogenology, N.Y. State Vet. Med. Soc. Office: Cornell U Coll Vet Medicine Bovine Rsch Ctr Ithaca NY 14853-6401

GILBERT, WALTER, molecular biologist, educator; b. Boston, Mass., Mar. 21, 1932; s. Richard V. and Emma (Cohen) G.; m. Celia Stone, Dec. 29, 1953; children: John Richard, Kate. AB, Harvard U., 1953, AM, 1954; PhD, Cambridge U., 1957; DSc (hon.), U. Chgo., 1978, Columbia U., 1978, U. Rochester, 1979, Yeshiva U., 1981. NSF postdoctoral fellow Harvard U., Cambridge, Mass., 1957-58, lectr. physics, 1958-59, asst. prof. physics, 1959-64, assoc. prof. biophysics, 1964-68, prof. biochemistry, 1968-72, Am. Cancer Soc. prof. molecular biology, 1972-81, prof. biology, 1985-86, H.H. Timken prof. sci., 1986-87, Carl M. Loeb Univ. prof., 1987—, chair dept. cellular and devel. biology, 1987-93; chmn. sci. bd. Biogen N.V., Dutch Antilles, 1978-83, co-chmn., supervisory bd., 1979-81, chmn. supervisory bd., chief exec. officer, 1981-84; V.D. Mattia lectr. Roche Inst. Molecular Biology, 1976. Recipient U.S. Steel Found. NAS, 1968, Ledlie prize Harvard U., 1969, Warren triennal prize Mass. Gen. Hosp., 1977, Louis and Bert Freedman Found. N.Y. Acad. Scis., 1977, Prix Charles-Leopold Mayer Academie des Scis., Inst. de France, 1977, Nobel prize in chemistry, 1980, New Eng. Entrepreneur of Yr. award, 1991; co-winner Louisa Gross Horwitz prize Columbia U., 1979, Gairdner prize, 1979, Albert Lasker Basic Sci. award, 1979; Guggenheim fellow, 1968-69; hon. fellow Trinity Coll., Cambridge, U.K., 1991. Mem. Am. Phys. Soc., Nat. Acad. Scis., Am. Soc. Biol. Chemists, Am. Acad. Arts and Scis.; fgn. mem. Royal Soc. Office: The Biol Labs 16 Divinity Ave Cambridge MA 02138-2097

GILBREATH, WILLIAM POLLOCK, federal agency administrator; b. Portland, Oreg., Nov. 10, 1936; s. Chester Edson and Naomi (Coffield) G.; m. Vibeke Kjalke, Oct. 18, 1965 (div. 1992); children: Joan K., Susan V.; m. Anita von Koor, Jan. 23, 1993. BA, Reed Coll., 1958; PhD, U. Wash., 1962. Aerospace engr. NASA-AMES, Moffett Field, Calif., 1962-75; program mgr. NASA-Hdqs., Washington, 1975-76; rsch. scientist NASA-AMES, 1976-79, staff engr., 1979-83, project mgr., 1983-86; program mgr. NASA-Hdqrs., Washington, 1986-93, br. chief, 1993—. Editor: Advanced Automation for Space Missions, 1982, Space Research and Space Settlements, 1979; contbr. articles to profl. jours. Mem. Hist. Soc., Washington, 1989—. Mem. AIAA, Am. Chem. Soc. Achievements include patent for Electrical conductivity cell and method for fabricating same. Office: NASA Washington DC 20546

GILBY, STEVE, metallurgical engineering researcher; b. Dayton, Ohio, Sept. 22, 1939. BS, U. Cin., 1962; PhD in Metall. Engring., Ohio State U., 1966. Rsch. engr. steelmaking Youngstown Steel Co., 1966-76; rsch. engr. Armco Steel Co., 1967-69, sr. rsch. engr., 1969-72, rsch. assoc., 1972-75, mgr. steelmaking rsch., 1975-82, dir. process rsch., 1982—; chmn. external adv. commn. mat. sci. & engring. dept. Ohio State U., 1988—. Mem. Am. Iron & Steel Soc., Am. Soc. Metals Internat. Achievements include research in steelmaking and continuous casting process development. Office: Armco Inc 300 Interpace Pkwy Parsippany NJ 07054*

GILES, GLENN ERNEST, JR., nuclear engineer; b. Newport News, Va., Dec. 26, 1943; s. Glenn Ernest and Irene (Caldwell) G.; m. Nancy Stuart Powell, Oct. 12, 1964 (div. Apr. 1974); children: Russell Stuart, Douglas Chandler; m. Carol Scott, Aug. 16, 1975. BSCE, Va. Poly. Inst. and State U., 1967. Registered profl. engr., Va. Design engr. Newport News Shipbuilding, 1967-69, 71-72, mech. test engr., 1969-70, 72-73, shift test engr., 1973-84, supr. test engring., 1984-88, mgr. nuclear engring. tng., 1988-93, mgr. nuclear engring svcs., 1993—. Mem. Keel Club, United Way of Va. Peninsula, Newport News, 1988-90; judge Tidewater Regional Sci. Fair, Newport News, 1984, 85. Mem. Am. Nuclear Soc., Soc. Naval Architects and Marine Engrs., Nautilus Soc. Reactor Test Engrs. (pres. 1984-88), Navy League U.S. Republican. Episcopalian. Home: 1307 Chesapeake Ave Hampton VA 23661-3121 Office: Newport News Shipbuilding 4101 Washington Ave Newport News VA 23607-2734

GILL, AJIT SINGH, civil engineer; b. Mullanpur, Ludhiana, India, Sept. 21, 1933; came to U.S., 1955; s. Chanan Singh and Indkaur Gill; m. Sharon Rose, Nov. 24, 1960 (div.); 1 child, Daniel P.; m. Sandra Chloe, Jan. 21, 1972; 1 child, Mira A. BSc in Econs., Brigham Young U., 1961, BSc in Engring. Scis., 1962. Registered profl. engr., Utah. Roadway designer Dept. Transp., Salt Lake City, 1963-66, hydraulic and civil engr., 1966-80; water resources engr. Div. Water Resources State of Utah, Natural Resources & Energy, Salt Lake City, 1980—. Mem. Instrumentation Soc. Am. Achievements include patents in Presssure Regulating Valves. Home: 4169 Bennion Rd Salt Lake City UT 84119-5467

GILL, ANNA MARGHERITA ANYA, application specialist; b. Seattle, Mar. 28, 1963; d. Paul Mark and Dagmar Tatiana (Lebedevs) Lorenz; m. Eric Justin Gill, May 27, 1984; children: Alexandra Marie, Peter Michael. BSc in Mineral Engring., Colo. Sch. Mines, 1984. Chemist Tex. Air Control Bd., Austin, 1984, Drug Dynamics Inst., Austin, 1984-85; internat. sales adminstr. Tracor, Austin, 1985-88; sales engr. Applied Automation Hartmann & Braun, Bartlesville, Okla., 1988-90; product mgr. Applied Automation Hartmann & Braun, Bartlesville, 1990—. Contbr. articles to profl. jours. Children's choir dir. First United Meth. Ch., Bartlesville, 1991—; troop leader Daisy Girl Scouts Am.; instr. Greek folk dance. St. Elias, Austin, 1985-87, Holy Trinity, Tulsa, 1992—. Mem. IEEE (assoc., cement industry com., first woman member, first woman presenter), Air and Waste Mgmt. Assn., Instrument Soc. Am. Eastern Orthodox. Home: 722 Sooner Park Dr Bartlesville OK 74006 Office: Hartmann & Braun Applied Automation Pawhuska Rd Bartlesville OK 74006

GILL, HENRY LEONARD, civil engineer, consultant; b. Flora, Ill., Jan. 16, 1939; s. Delbert A. and Bessie V. (Rogers) G.; m. Barbara Joan Warren, Aug. 27, 1941; children: Debra, Julia, Karen, Gary. BSCE, U. Ill., 1960, MSCE, 1966. Civil Engr., Ill. Civil engr. Ill. Dept. Transp., Effingham, 1960-61; rsch. engr. Naval Civil Engring. Lab., Port Hueneme, Calif., 1961-68, div. dir., 1968-74, dept. head, 1974-79; gen. mgr. Clay Electric Coop., Flora, Ill., 1980-82; mgmt. cons. pvt. practice, Flora, Ill., 1983—; bd. mem. Ill. Mcpl. Electric Agy., Springfield, Ill., 1989—, Cin-Made Corp., Cin., 1990—. Mem. Bd. Dirs. Soyland Power Coop., Springfield, Ill., 1980-82, Flora Airport Auth., 1985—. recipient Collingwood prize Am. Soc. Civil Engrs., N.Y.C., 1964, Young Engr. award Calif. Soc. Civil Engrs., Ventura, Calif., 1972; fellow Navy Civil Engring. lab., Port Huenume, Calif., 1965. Mem. Am. SOc. Civil Engrs. Achievements include patents on soil testing equipment. Home and Office: 35 Stone Forge Pike Flora IL 62839

GILL, MOHAMMAD AKRAM, civil engineer; b. India, June 1, 1935; came to U.S. 1983, naturalized.; s. Rukn Uddin and Barkat Bibi; m. Fauzia Rasheed, Sept. 26, 1970; children: Salman, Faheem, Shuaib, Farheen. BSc in Civil Engring., Punjab U., Lahore, Pakistan, 1956; PhD in Hydraulics, London U., 1970. Registered profl. engr., Mich., Iowa. Asst. engr. Warsak (Pakistan) Dam Authority, 1956-60; asst. exec. engr. Water & Power Devel. Authority, Lahore, 1960-61; exec. engr. Ministry of Wks. of N. Nigeria, Maiduguri, 1961-67; sr. irrigation engr. Ministry of Agr., Sokoto, Nigeria, 1970-72; prof. Ahmadu Bello U., Zaria, Nigeria, 1972-83; vis. prof. U. Iowa, Iowa City, 1985; engr. water systems Detroit Water and Sewerage Dept., 1985—; advisor Kaduna (Nigeria) State Water Bd., 1980-83. Contbr. articles to profl. jours. Mem. ASCE (J.C. Stevens award 1990), Internat. Assn. for Hydraulic Rsch., Am. Water Wks. Assn. (project adv. com. rsch. found.). Achievements include development of theoretical formulation of shape of sediment stable channels and aggradation/degradation in alluvial channels. Home: 24259 Leewin Detroit MI 48219-9999 Office: Detroit Water & Sewerage Dept 735 Randolph St Detroit MI 48226-2830

GILL, WILLIAM NELSON, chemical engineering educator; b. N.Y.C., Sept. 13, 1928; s. William Nelson and Frances (Murphy) G.; m. Chandlee Stevens, Aug. 13, 1982; children: Alison Louise, Christine Marie, Douglas Max, Max William. BSChemE, Syracuse U., 1951, MA, 1955, PhD, 1960. Field engr. Am. Blower Corp., 1951-55; mem. faculty Syracuse U., 1957-65, assoc. prof., 1963-65; prof. chem. engring., chmn. dept. Clarkson Coll. Tech., 1965-71; provost engring. and applied sci. SUNY, Buffalo, 1971-78, prof. chem. engring., 1982-87; Glenn Murphy Disting. prof. engring. Iowa State U., Ames, 1980-82; Russell Sage prof., chmn. chem. engring. Rensselaer Polytech. Inst., Troy, N.Y., 1987—; cons. in field. Editor: Chem. Engring. Communications, 1979—; mem. editorial adv. bd. Fuel, Processing Tech.; mem. bd. cons. editors Elsevier Texts in Engring.; editor Chem. Engring. series Elsevier Sci. Pub. Co.; author numerous articles in field. Fulbright-Hays sr. research scholar U. Coll., London, 1977-78, Fulbright sr. research scholar U. Old, Australia, 1986-87. Fellow AICE; mem. AAAS, Am. Chem. Soc., Am. Soc. Engring. Edn. (Chem. Engring. Divsn. Lectureship award 1992), AAUP, N.Y. Acad. Sci., Sigma Xi. Office: Rensselaer Polytech Inst Chem Engring-Ricketts Troy NY 12180

GILL, WILLIAM ROBERT, soil scientist; b. McDonald, Pa., July 21, 1920; s. William Merle and Mary Della (Leiden) G.; m. Irene Victoria Majorkiewicz, July 10, 1947; children: William Robert, John Philip, David C., Michael J., Elaine N. BS, Pa. State U., 1942; MS, U. Hawaii, 1949; PhD, Cornell U., 1955. Asst. soil scientist Pineapple Rsch. Inst., Honolulu, 1949-50; rsch. soil scientist USDA-ARS, Auburn, Ala., 1955-80, dir., 1971-80; adj. prof. agrl. engring. Auburn U., 1957-88; collaborator Nat. Soil Dynamics Lab. (formerly Nat. Tillage Machinery Lab.), Auburn, 1980—; exch. scientist in Soviet Union, 1970. Author: Soil Dynamics in Tillage and Traction, 1967, History of the National Tillage Machinery Laboratory, 1990; contbr. articles to profl. jours. Col. AUS ret., 1943-47, 51-52. Recipient Recognition award Internat. Soil Dynamics Conf., 1985; inducted into Officer's Candidate Sch. Hall of Fame, 1993. Mem. Am. Soc. Agrl. Engrs. (Peer Recognition award 1985, Disting. Engr. award 1988, John Deere medal 1990), Soil Sci. Soc. Am. Achievements include translation of Russian soil dynamics articles into English, fgn. analysis and technology transfer for internat. audience. Office: Nat Soil Dynamics Lab USDA-ARS PO Box 3439 N Donahue Dr Auburn AL 36831-3439

GILLANI, NOOR VELSHI, mechanical engineer, researcher, educator; b. Arusha, Tanzania, Mar. 8, 1944; came to U.S., 1963, naturalized, 1976; s. Noormohamed Velshi and Sherbanu (Kassam) G.; m. Mira Teresa Pershe, Aug. 13, 1971; children: Michael, Michelle, Nicole. GCE (Ordinary Level Div. I), U. Cambridge, 1960, (Advanced Level), U. London, 1963; AB cum laude, Harvard U., 1967; MS in Mech. Engring., Washington U., St. Louis, 1969, DSc, 1974. Vis. scientist Stockholm U., 1977; research assoc. Washington U., 1975-76, research scientist, 1976-77, asst. prof., 1977-80, assoc. prof. 1981-84, prof. mech. engring., 1985-91, faculty assoc. CAPITA, 1979-91, dir. air quality spl. studies data ctr., 1981-88, mech. engring research computing facility, 1988-90; vis. scientist Brookhaven Nat. Lab., 1990-91; pres. N.V. Gillani & Assocs., Inc., 1991—; vis. scientist EPA/RTP, 1992; prof atmospheric sci. N.C. State U., 1993—; organizer NATO CCMS 15th internat. tech. meeting on air pollution modeling and its applications, St. Louis, Apr. 1985; mem. Sci. Bd. NATO/CCMS Air Pollution Pilot Study, 1986-94; mem. tech. adv. bds. U.S. EPA, DOE and others, 1980—; hon. mem. Aga Khan Bd. Edn. for the u.S.A., 1987-90. Author 2 chpts. in EPA Critical Assessment Document on Acid Deposition, 1984; editor: Air Pollution Modeling and Its Applications V, vol. 10, 1986; contbr. articles on superconductivity, bioengring., atmospheric scis. and air pollution to nat. and internat. profl. jours. Dir. Nat. Aga Khan Bd. Edn. Program for Parental Involvement in Children's Edn., USA; Aga Khan scholar and travel grantee, 1961-63; Harvard Coll. scholar, 1963-67; Washington U. Grad. Engring. fellow, 1967-69; research assistantships NIH, EPA, 1971-74; EPA Research grantee, 1978—. Mem. N.Y. Acad. Scis., Air Pollution Control Assn., Am. Meteorol. Soc., Am. Chem. Soc., ASME, Nat. Assn. for Edn. Young Children. Avocations: religious studies, music, tennis, early childhood education, computers. Office: NC State U Dept Marine Earth Atmospheric Scis Raleigh NC 27695

GILLASPIE, ATHEY GRAVES, JR., pathologist, researcher; b. Asheville, N.C., July 30, 1938; s. Athey Graves Sr. and Virginia Graves (Shackelford) G.; m. Margaret Ellen Fleming, Aug. 14, 1965; children: Timothy Graves, Jonathan Todd. BA, Miami U., 1960; MS, Purdue U., 1962, PhD, 1965. Rsch. plant pathologist Agr. Rsch. Svc., USDA, Houma, La., 1965-72; rsch. plant pathologist Agr. Rsch. Svc., USDA, Beltsville, Md., 1972-80, supr. rsch. plant pathologist, 1980-86; supr. rsch. plant pathologist Agr. Rsch. Svc., USDA, Griffin, Ga., 1986-93, rsch. plant pathologist, 1993—; adj. prof. plant pathology U. Ga., Griffin, 1988—. Editor: Plant Pathogenic Bacteria, 1987; editor, author: chpt. Diseases of Sugarcane, 1989. Rsch. Orgn. grantee, 1980, 82, 85. Fellow Wash. Acad. of Sci.; mem. Am. Phytopathological Soc., So. Div. Am. Phytopathological Soc. Baptist. Avocation: trap shooting. Office: USDA So Regional Plant Intro Sta 1109 Experiment St Griffin GA 30223-1797

GILLEN, HOWARD WILLIAM, neurologist, medical historian; b. Chgo., Nov. 25, 1923; s. John Howard and Emily Elizabeth (Bayley) G.; m. Corinne V. Neese, July 24, 1948. BS, U. Ill., 1947; MD, U. Ill., Chgo., 1949. Hon. active neurologist New Hanover Regional Med. Ctr., Wilmington, N.C., 1973—; cons. neurologist Cape Fear Meml. Hosp., Wilmington, 1973—; clin. prof. neurology U. N.C., Chapel Hill, 1973—; adj. prof. biol. sci. U. N.C., Wilmington, 1986—; rsch. assoc. I.R.I.S.C., Wilmington, 1989—. Capt. USN, ret. Home: 2038 Trinity Ave Wilmington NC 28405 Office: 1301 Cypress Grove Rd Wilmington NC 28401

GILLESPIE, GARY DON, physician; b. Jackson, Mich., Apr. 23, 1943; s. Harold Don and Marion Estella (Diemer) G.; m. Nancy Bliven Hinkle, June 29, 1969 (div. July 1980; children: Brian James, Julie Elizabeth; m. Elaine Marie Beard, July 25, 1984. BS, U. Mich., 1966, D of Medicine, 1971. Diplomate Am. Bd. Family Practice. Intern Edward W. Sparrow Hosp., Lansing, Mich., 1971-72, resident in family practice, 1971-74; physician Dept. Family Practice, USN Med. Corps., Orlando, Fla., 1974-76; pvt. practice Okemos, Mich., 1976—; chmn. continuing edn., dept. family practice Edward W. Sparrow Hosp., 1976-91; asst. clin. prof. dept. family practice Mich. State U. Coll. Medicine, East Lansing, 1981—. Lt. comdr. USN, 1974-76. Mem. AMA, Am. Acad. Family Physicians, Am. Bd. Family Practice, Mich. Acad. Family Physicians (treas. Capitol chpt. 1982-92). Republican. Avocations: reading, music, photography, travel, golf. Office: 1745 Hamilton Rd # 340 Okemos MI 48864-1810

GILLESPIE, ROBERT BRUCE, biology educator; b. Cin., June 23, 1953; s. Charles Halliwell and Ruth Ann (Anderson) G.; m. Maria Helena Kranjc, June 25, 1982; children: Sara Ann, Sean Michael. MS, U. Akron, 1981; PhD, Ohio State U., 1985. Biologist Ichthyological Assocs., Forked River, N.J., 1976-79; researcher Battelle Meml. Inst., Columbus, Ohio, 1983-86; postdoctoral fellow Miami U., Oxford, Ohio, 1986-88; rsch. asst. prof. SUNY, Fredonia, 1988-91; asst. prof. biology Ind.-Purdue U., Ft. Wayne, 1991—. Contbr. articles to profl. jours. Purdue Rsch. Found. fellow, 1992. Mem. Am. Fisheries Soc., Soc. Environ. Toxicology and Chemistry (coord. panel discussion 1992). Achievements include research in teratogenic effects of selenium in fishes. Office: Ind-Purdue U Biology Dept Fort Wayne IN 46805-1499

GILLESPIE, SHANE PATRICK, chassis engineer; b. Dayton, Ohio, Dec. 14, 1968; s. Vernon Paul and Judith Ann (Moen) G. BS in Aerospace Engring., U. Notre Dame, 1991. Chassis engr. Toyota Tech. Ctr., Ann Arbor, 1991—. Mem. AIAA, Soc. Automotive Engrs. Home: 5111 Prairie View Brighton MI 48116

GILLESPIE, THOMAS DAVID, mechanical engineer, researcher; b. Beaver Falls, Pa., Dec. 3, 1939; s. Thomas Benjamin and Faith Lavern (Garvin) G.; m. Susan Louise Lampus, Sept. 27, 1963; children: David, Darren, Devin, Jessica. BSME, Carnegie Inst. Tech., 1961; MS, Pa. State U., 1965, PhD, 1970. Registered profl. engr., Pa. Engr. PPG Industries, Pitts., 1963-64; rsch. assoc. Pa. State U., University Park, 1970-73; group leader Ford Motor Co., Dearborn, Mich., 1973-76; rsch. scientist U. Mich., Ann Arbor, 1976—; dir. Great Lakes Ctr. for Truck and Transit Rsch., 1988—; pvt. cons. T.D. Gillespie Inc., Ann Arbor, 1976—; commr. Mich. Truck Safety Commn., 1989—. Editor: Measuring Road Roughness and It's Effects on User Cost and Comfort, 1985; author: Fundamentals of Vehicle Dynamics, 1992. Sr. policy analyst Office of Sci. and Tech. Policy, White House, Washington, 1987-88. Capt. C.E., U.S. Army, 1964-66. Mem. Soc. Automotive Engrs. (faculty advisor 1991—, L. Ray Buckendale award 1985), Phi Kappa Phi (pres. elect Mich. chpt. 1991—). Achievements include development of national action plan on advanced superconductivity research and development; co-development of International Roughness Index for road measurement. Home: 1083 Bandera St Ann Arbor MI 48103-9703 Office: U Mich Transp Rsch Inst 2901 Baxter Rd Ann Arbor MI 48109-2150

GILLET, ROLAND, financial economist; b. Vielsalm, Luxembourg, Belgium, June 25, 1962; s. Ernest Gillet and Ghislaine Machuraux. Lic. in Econs., Liege (Belgium) U., 1985; M. in Econs., Cath. U. Louvain (Belgium), 1986, PhD in Econs., 1991. Asst. prof. Cath. U. Louvain, 1986-88; researcher Nat. Fund of Sci. Rsch., Belgium, 1989-90; prof. fin. Cath. U. Lille, France, 1991—, Cath. U. Louvain, Belgium, 1991—; vis. fellow Harvard U., Boston, 1989-90, MIT, Boston, 1989-90; cons. Brussels Stock Exch., Belgian Banks and Brokers, 1989—; sci. collaborator Liege U., 1989—. Contbr. fin. articles to profl. jours.; patentee in field. With Belgian Mil., 1981. Recipient 1st prize Belgian Ministry of French Community, 1986, Laureat/Winner prize Belgian Ministry of French Community, 1986. Mem. Inst. Recherches Economiques et Social, Am. Econ. Assn., European Econ. Assn., Ctr. for Econ. Policy Rsch., Nat. Bur. Econ. Rsch. Roman Catholic. Avocations: tennis, sport fishing, skiing, chess, philately. Home: 16 Rue Chamont, 6980 La Roche Belgium Office: Cath U Lille, 60 Blvd Vauban, 59016 Lille France

GILLETT, JOHN BLEDSOE, civil engineer; b. Statesville, N.C., Oct. 14, 1927; s. Rupert and Mary Lina (Bledsoe) G.; m. Minnie Louise Rountree, June 11, 1949; children: John, David, Warren, Mary, Mark. BSCE, N.C. State U., 1948. Registered profl. engr., Md. Field office engr. Duke Power Co., Charlotte, N.C., 1948-51; engr. Whitman, Requardt & Assocs., Balt., 1951-58, project engr., 1959-63, assoc. engr., 1964-75, ptnr., 1976-90; civil engring. cons. Balt., 1991—. Contbr. articles to profl. jours. Vol. Boy Scouts Am., 1948-92; mem. Mayor's Sch. Constrn. Com., Balt., 1965-66; vice chmn. Bd. of Health, Baltimore County, 1962-69. Recipient Silver Beaver Boy Scouts Am., 1969; named Md. Outstanding Young Man Md. Jaycees, 1963. Fellow ASCE (life); mem. Am. Water Works Assn. (life), Am. Acad. Environ. Engrs. (diplomate), Engring. Soc. Balt. (pres. 1984-85), Engrs. Coun. Md. (sec. 1991-93), Md. Cons. Engrs. Coun. (pres. 1979-80). Episcopalian.

GILLETTE, FRANK C., JR., aeronautical engineer; m. Jane Gillette; 3 children. BS in Mech. Engring., U. Fla. Mech. designer Pratt & Whitney, 1962-77, chief of structures, 1977-80, engring. mgr. YF119 program, dir. engring. programs F119 engine projects for Govt. Engines and Space Propulsion, 1980—; presenter numerous papers to profl. socs. Recipient Disting. Svc. award U. Fla. Coll. Engring., Laurels award Aviation Week, 1991. Mem. AIAA (Nat. Engr. of Yr. award 1991), ASME. Achievements include design of the RL10 rocket chamber, the turbine section of the J58; management of the overall structural engineering effort of the J52, TF30, F100 rockets and preliminary design; patents in field. Office: Pratt & Whitney Aircraft Govt Prod Division PO Box 2691 West Palm Beach FL 33402*

GILLETTE, (PHILIP) ROGER, physicist, systems engineer; b. Mt. Vernon, Iowa, May 12, 1917; s. Clinton Edgar and Celia (Rogers) G.; m. Bettelaine Dunbar, April 26, 1947 (dec. Mar. 1986); children: Kenneth Lee, Sandra Jo. B.A., in Physics, Cornell Coll., 1937; B.S. in Engring. Physics, U. Ill., 1938, M.S. in Physics, 1939, Ph.D. in Physics, 1942. Staff mem. Radiation Lab. MIT, Cambridge, Mass., 1942-45; research engr. Sperry Gyroscope Co., Great Neck, N.Y., 1945-48; physicist Hanford Works Gen. Electric Co., Richland, Wash., 1948-50; sr. research physicist SRI Internat. Menlo Park, Calif., 1950-92. retired, SRI Internat., 1992. Co-author: Pulse Generators, 1948. Bd. dirs. West Bay Opera Assn., Palo Alto, Calif., 1959-64, 1977-79. Mem. AAAS, IEEE (life mem.), Am. Phys. Soc. (life), Sigma Xi, Phi Beta Kappa, Tau Beta Pi, Phi Kappa Phi. Achievements include development of Pulse Transformer Theory, of system design concepts for command, control, communications, and intelligence systems, electronic combat systems, and air combat training systems. Home: 2385 Crestview Dr S Salem OR 97302

GILLIAM, M(ELVIN) RANDOLPH, urologist, educator; b. Elliott County, Ky., Jan. 5, 1921; s. Adolphus and Grace (Thornsberry) G.; m. Sara Dee Rainey, May 15, 1948; children: Elizabeth Neal, Virginia Dee, Bryan Randolph, Frank Stuart, Grace Carroll. Student Centre Coll. of Ky., 1938-41; MD, U. Louisville, 1944. Diplomate Am. Bd. Urology. Intern Norfolk (Va.) Marine Hosp., 1944-45; resident in urology Nichols VA Hosp., Louisville, 1947-50; pvt. practice medicine specializing in urology, Lexington, Ky., 1950—; ptnr. Commonwealth Urology, P.S.C., Lexington, 1971—; clin. prof. urology, U. Ky. Med. Sch., 1964—; chief of urology Good Samaritan Hosp.; mem. staff Central Baptist Hosp., St. Joseph's Hosp. Capt. U.S. Army, 1945-47. Mem. AMA, Ky. Med. Assn., Fayette County Med. Soc. (past pres.), Am. Urology Assn. Republican. Methodist. Home: 1244 Summitt Dr Lexington KY 40502-2273 Office: 1760 Nicholasville Rd Lexington KY 40503-2518

GILLIE, MICHELLE FRANCOISE, industrial hygienist; b. Phila., Oct. 24, 1956; d. Marino and Marcelle Jeannine (Boyer) Lazarich; m. Alan Deane Gillie, May 22, 1982; children: Patrick Alan, Caroline Elizabeth. BS, Pa. State U., 1977; MS, Drexel U., 1981. Diplomate Am. Bd. Indsl. Hygiene. Clin. chemist Pa. Hosp., Phila., 1976-80; indsl. hygienist Stewart-Todd Assocs., Wayne, Pa., 1981-82, S.W. Occupational Health Svcs., Houston, 1982-84; clin. toxicologist Smith-Kline-Beckman Labs., Houston, 1984-85; sr. indsl. hygienist Am. Analytical Labs., Akron, Ohio, 1985-88; indsl. hygiene cons. AMP Technical Svcs., Cleve., 1988-91; sr. indsl. hygienist Environ. Mgmt., Inc., Anchorage, 1991-92; indsl. hygiene cons. AMP Tech. Svcs., Bakersfield, Calif., 1992—. Mem. Am. Indsl. Hygiene Assn. (pub. rels. com. chair Midnight Sun chpt. 1991—). Avocations: reading, cross-country skiing. Home: 11310 Cedarhaven Ave Bakersfield CA 93312-3653 Office: AMP Tech Svcs 11310 Cedarhaven Ave Bakersfield CA 93312-3653

GILLIGAN, JOHN GERARD, nuclear engineer, educator; b. Beech Grove, Ind., Jan. 17, 1949; s. John Bernard and Mary Ann (Prieshoff) G.; m. Barbara Ann Bertolami, Dec. 8, 1979; children: Theresa, Lisa, Michael. BS in Engring. Sci., Purdue U., 1971; MS, U. Mich., 1974, PhD, 1977. Asst. prof. U. Ill., Champaign, 1977-83; assoc. prof. nuclear engring. N.C. State U., Raleigh, 1983-90, prof., 1990—, dir. grad programs nuclear engring., 1986—. Editor: Nuclear Engineering Education Sourcebook, 1986—; contbr. articles on plasma physics to Jour. Applied Physics, Physics of Fluids, others. Mem. IEEE (exec. com. plasma scis. div. 1989-92), Am. Nuclear Soc. (chair edn. div. 1990-91). Achievements include discovery of magnetic plasma shielding effect for materials under extremely high heat flux. Office: NC State U Dept Nuclear Engring Box 7909 Raleigh NC 27695

GILLIN, JOHN F., quality/test engineer; b. Agana, Guam, Sept. 8, 1956; s. George F. and Betsey (Berg) G. BS, SUNY, Stony Brook, 1978. Materials lab. asst. Unisys Corp., Great Neck, N.Y., 1979-81, matls. engr., 1981-90, quality/test engr., 1990—. Vol. Christ the King Youth Coun., Commack, N.Y., 1971-87. Recipient Joseph Furman Meml. award Christ The King Youth Coun., 1976, Cert. of Appreciation, 1982, 84; letter of appreciation Program Mgmt. office, USN., 1993. Mem. Am. Vacuum Soc. Achievements include patent in Target Source for Ion Beam Sputter Deposition. Office: Unisys Corp Mail Sta I-9 365 Lakeville Rd Great Neck NY 11020

GILLIS, JOHN SIMON, psychologist, educator; b. Washington, Mar. 21, 1937; s. Simon John and Rita Veronica (Moran) G.; m. Mary Ann Wesolowski, Aug. 29, 1959; children: Holly Ann, Mark, Scott. B.A., Stanford U., 1959; M.S. (fellow), Cornell U., 1961; Ph.D. (NIMH fellow), U. Colo., 1965. Lectr. dept. psychology Australian Nat. U., Canberra, 1968-70; sr. psychologist Mendocino (Calif.) State Hosp., 1971-72; asso. prof. dept. psychology Tex. Tech U., Lubbock, 1972-76; prof. psychology Oreg. State U., Corvallis, 1976—, chmn. dept. psychology, 1976-84; cons. VA, Ciba-Geigy Pharms., USIA, UN High Commn. for Refugees; commentator Oreg. Ednl. and Pub. Broadcasting System, 1978-79; Fulbright lectr., India, 1982-83, Greece, 1992; vis. prof. U. Karachi, 1984, 86, U. Punjab, Pakistan, 1985, Am. U., Cairo, 1984-86. Contbr. articles to profl. jours. Served with USAF, 1968-72. Ciba-Geigy Pharms. grantee, 1971-82. Mem. Am. Psychol. Assn., Western Psychol. Assn., Oreg. Psychol. Assn. Roman Catholic. Home: 7520 NW Mountain View Dr Corvallis OR 97330-9106 Office: Oreg State U Dept Psychology Corvallis OR 97331

GILLIS, STEVEN, biotechnology company executive; b. Phila., Apr. 25, 1953; s. Herbert and Rosalie Henrietta (Segal) G.; m. Anne Cynthia Edgar, June 26, 1976; children: Sarah Milne, Bradley Stirling. BA cum laude, Williams Coll., 1975; PhD, Dartmouth Coll., 1978. Lectr. in biology Dartmouth Coll., Hanover, N.H., 1977-78; research assoc. Dartmouth Med. Sch., Hanover, 1978-79; assoc. researcher Meml. Sloan-Kettering, N.Y.C., 1979-80; asst. prof. U. Wash., Seattle, 1980-83; exec. v.p., dir. R & D Immunex Corp., Seattle, 1982—, pres., chief oper. officer, 1988—, chief exec. officer, 1990, also bd. dirs.; adj. assoc. prof. U. Wash., 1982-90, adj. prof., 1990—; bd. dirs. Seattle Biomed. Research Found. Editor: Lymphokines, 1985, Recombinant Lymphokines and Their Receptors, 1987; contbr. articles to profl. jours. Asst. mem. Fred Hutchinson Cancer Ctr., Seattle, 1980-82, affiliate investigator, 1982—. Recipient Internat. Immunopharmacology award, 1983. Mem. Am. Assn. Immunologists, N.Y. Acad. Scis., Am. Assn. Arts and Scis. Club: Columbia Tower (Seattle). Avocations: golf, swimming. Office: Immunex Corp 51 University St Seattle WA 98101-2936

GILLMAN, LEONARD, mathematician, educator; b. Cleve., Jan. 8, 1917; s. Joseph Moses and Etta Judith (Cohen) G.; m. Reba Parks Marcus, Dec. 24, 1938; children: Jonathan Webb, Michal Judith. Diploma (fellow in piano 1933-38), Juilliard Grad. Sch. Music, 1938; BS, Columbia U., 1941, MA (Carnegie fellow math. statistics 1942-43), 1945, PhD, 1953. Asst. in math. dept. Columbia U., 1941-42, lectr., 1942-43; ops. analyst Tufts Coll., MIT, 1943-51; from instr. to assoc. prof. math. Purdue U., 1952-60; prof. math., chmn. dept. U. Rochester, 1960-69; prof. math. U. Tex., Austin, 1969-87, prof. emeritus, 1987, chmn. dept., 1969-73; mem. Inst. Advanced Study, Princeton, 1958-60; cons. editor W.W. Norton Co., Inc., 1967-80. Author: (with Meyer Jerison) Rings of Continuous Functions, 1960, 76, You'll Need Math, 1967, (with Robert H. McDowell) Calculus, 1973, 78, Writing Mathematics Well, 1987; editorial bd.: Topology and Its Applications, 1971—. Guggenheim fellow, 1958-59; NSF sr. post-doctoral fellow, 1959-60. Mem. Am. Math. Soc. (assoc. sec. 1969-71, mem. com. to monitor problems in communication 1972-77), Nat. Council Tchrs. Math., Math. Assn. Am. (bd. govs. 1973—, treas. 1973-85, pres.-elect 1986, pres. 1987-88, past pres. 1989). Home: 1606 The High Rd Austin TX 78746

GILLMER, THOMAS CHARLES, naval architect; b. Warren, Ohio, July 17, 1911; s. Derr Oscar and Hazel May (Voit) G.; m. Anna May Derge, June 5, 1937; children: Christina Gesell Gillmer Erdmann, Charles Voit. BS, U.S. Naval Acad., 1935; postgrad., Case Western Res. U., Johns Hopkins U., 1946. Commd. ensign USN, 1935, advanced through grades to lt. (j.g.), 1939; lt. comdr. USNR, 1944-46; mem. faculty U.S. Naval Acad., Annapolis, Md., 1946-68, prof., dir. Naval Architecture, 1961-68; chmn. naval engring. dept., 1963-68; pvt. practice naval architecture Annapolis, 1968—; mem. panel experts FAO, UN, Rome, 1963-66. Author: Construction and Stability of Naval Ships, 1959, Modern Ship Design, 1970, 2d edit., 1975, Working Water Craft, 1972, Chesapeake Sloops, 1981, Introduction to Naval Architecture, 1982, The Story of Baltimore Clippers, 1991, Old Ironsides, Rise, Decline, and Resurection of Uss Constitution, 1993; designer PRIDE of Balt., 1976-77, Lady Maryland, 1985, PRIDE of Balt. II, 1987-88; contbr. articles to profl. jours. and papers to profl. confs.; patentee in field. Mem. bd. govs. Chesapeake Bay Maritime Mus., 1981-91, curaturial chmn., 1989-91. Recipient 1st Maritime Preservation award Chesapeake Bay Maritime Mus., 1992. Mem. Soc. Naval Architects and Marine Engrs. (award for individual efforts to further body of knowledge 1988), Am. Soc. Naval Engrs., Soc. Nautical Research, AAUP, PRIDE of Balt. (ops. com. 1979—), Hellenic Inst. Marine Archaeology Athens (Bronze medal 1985). Clubs: Annapolis Yacht; de Voile (France). Office: 300 State St Annapolis MD 21403

GIL-LOYZAGA, PABLO ENRIQUE, neurobiologist, researcher, educator; b. Madrid, Apr. 13, 1954; s. Enrique Gil and Maria Elvira Loyzaga; m. Ana Perez, May 22, 1980; children: David, Ana, Pablo. Bachelors degree, Sagrada Familia, Madrid, 1970; MD, U. Complutense Madrid, 1981; degree in pathology, Hosp. C., San Carlos, Madrid, 1982; PhD, U. S.T.L., France, 1990. Asst. prof. Sch. Medicine, Univ. Complutense, Madrid, 1978-82, assoc. prof., 1982-85, titular prof., 1985—; sci. pharm. cons. Juste SAQF Labs., Madrid, 1983-90; cons. to Scientific Evaluation Nat. Agy., 1991—; dir. Spanish commintment to Franco-Hispano Inner Ear Biology Agreement, 1986—. Contbr. articles to 76 profl. pubs.; presenter at numerous sci. meetings. Pres. 15th Commn. for Sci. Evaluation, Fondo de investigaciones Sanitarias de la Seguridad Social (Ministry of Health). Recipient Conde de Cartagena award, Royal Acad. Medicine, Madrid, 1984, Amel Albarrar award, Med. Coun. Badajoz, 1984, Sureste award Med. Coun. Madrid, 1986; grantee from 20 pub. and private funds. Avocation: classical guitar. Office: Sch Medicine, Apdo Correos 60075, Madrid 28080, Spain

GILMAN, ALFRED GOODMAN, pharmacologist, educator; b. New Haven, July 1, 1941; s. Alfred and Mabel (Schmidt) G.; m. Kathryn Hedlund, Sept. 21, 1963; children: Amy, Anne, Edward. BS, Yale U., 1962; MD, PhD, Case Western Res. U., 1969. Pharmacology research assoc. NIH, Bethesda, Md., 1969-71; from asst. prof. to assoc. prof. pharmacology U. Va., Charlottesville, 1971-77, prof., 1977-81, dir. med. sci. tng. program, 1979-81; prof. pharmacology, chmn. dept. U. Tex. Southwestern Med. Ctr., Dallas, 1981—, Raymond and Ellen Willie prof. molecular neuropharmacology, 1987—; mem. pharmacology study sect. NIH, 1977-81; bd. sci. counselors Nat. Heart, Lung & Blood Inst. NIH, 1982-86; sci. adv. com. Am. Cancer Soc., N.Y.C., 1982-86; adv. com. Lucille P. Markey Charitable Trust, Miami, Fla., 1984—; sci. rev. bd. Howard Hughes Med. Inst., Bethesda, 1986—. Editor: The Pharmacological Basis of Therapeutics, 1975, 80, 85; contbr. over 100 articles to profl. jours. Recipient Poul Edvard Poulsson award Norwegian Pharmacology Soc., 1982, GairdnerFound. Internat. award, Can., 1984, Albert Lasker Basic Med. Rsch. award, 1989, Passano Sr. award Passano Found., 1990, Waterford Biomedical Sci. award Scripps Clinic and Rsch. Found. 1990. Mem. Am. Soc. Pharmacology & Exptl. Therapeutics (John J. Abel award in pharmacology 1975), Am. Soc. Biol. Chemistry, Nat. Acad. Scis. (Richard Lounsbery award 1987), Am. Acad. Arts and Scis., Inst. Medicine of NAS. Office: U Tex Southwestern Med Ctr Dept Pharmacology 5323 Harry Hines Blvd Dallas TX 75235-7200

GILMAN, JAMES RUSSELL, petroleum engineer; b. Sheridan, Mont., Apr. 13, 1956; s. Lowell Lincoln and Laura Isabelle (Moore) G.; m. Anita Marie Gasperino, Sept. 24, 1956; children: Stephanie, Michelle. BS, Mont. State U., 1978; MS in Chem. Engring., Colo. Sch. Mines, 1983. Registered profl. engr., Colo. Assoc. engr. Marathon Oil Co., Littleton, Colo., 1978-89, sr. engr., 1989—. Contbr. chpt. to book, articles to Petroleum Engring. Mag. Mem. Soc. Petroleum Engrs., Am. Inst. Chem. Engrs., Sigma Xi. Achievements include developments in numerical simulation of naturally fractured reservoirs. Office: Marathon Oil Co PO Box 269 7400 S Broadway Littleton CO 80160

GILMARTIN, MALVERN, oceanographer; b. L.A., Nov. 14, 1926; s. Malvern Sr. and Gladys Edna (Hall) G.; m. Amy Jean Finch, June, 1954 (div. 1973); children: Malvern III, Dale Moana, Sheila Ann, Ian Harvey; m. Noelia Revelante, Dec., 1974; 1 child, Darren. BA, Pomona Coll., 1954; MS, U. Hawaii, 1956; PhD, U. B.C., Vancouver, 1960. Sr. scientist Inter-Amer. Tropical Tuna Commn., Ecuador, 1960-64; assoc. prof. oceanography U. Hawaii, 1964-66; prof., dir. oceanography program Stanford U., Calif., 1966-74; founding dir. Australian Inst. Marine Sci., Townsville, 1974-78; founding dir. Ctr. Marine Studies U. Maine, 1978-82, prof. oceanography, 1982—; cons. Nat. Fish Inst. Ecuador, Chile, 1961-64, Empresa Puertos Colombia, 1962-64, C.Am. Regional Fishery Devel., El Salvador, 1964-66; mem. Internat. Com. Sci. Exploration Mediterranean Sea, 1977—, Znanstveni Savjenik Ctr. Marine rsch., Rovinj Croatia, 1972—. Contbr. over 74 articles to sci. jours. Lt. USN, 1943-47, 50-53, PTO, Korea. Recipient Fulbright scholar, 1983-84, 90. Mem. Am. Soc. Limnology Oceanography, Am. Fisheries Soc., Sci. Rsch. Soc. Am.

GILMER, ROBERT, mathematics educator; b. Pontotoc, Miss., July 3, 1938; s. Robert William and Lucy Marie (Jernigan) G.; m. Rachel Grace Colson, Aug. 24, 1963; children: David Patrick, Stephen Douglas. Student, Itawamba Jr. Coll., 1955-56; B.S., Miss. State U., 1958; M.S., La. State U., 1960, Ph.D., 1961. Instr., Miss. State U., Starkville, 1958; vis. prof. Miss. State U., 1962; research instr. La. State U., Baton Rouge, 1961-62; vis. lectr. U. Wis., Madison, 1962-63; mem. faculty Fla. State U., Tallahassee, 1963—; prof. math. Fla. State U., 1968—, Robert O. Lawton Disting. prof., 1981—; vis. prof. Latrobe U., Bundoora, Victoria, Australia, 1974, U. Tex., Austin, 1976-77; vis. rsch. prof. U. Conn., Storrs, 1982; visitor Inst. for Advanced Study, 1990. Author: Multiplicative Ideal Theory, 1967, 72, 92, Commutative Semigroup Rings, 1984; also articles; assoc. editor Am. Math. Mo., 1971-73; editorial bd. Jour. Communications in Algebra, 1974-85. Office Naval Research fellow, 1962-63; Alfred P. Sloan Found. fellow, 1965-67; NSF grantee, 1965-89; Fulbright sr. scholar to Australia, 1974. Mem. Am. Math. Soc., Math. Assn. Am. (gov. Fla. sect. 1986-89, cert. meritorious svc. 1992). Baptist. Home: 2414 Perez Ave Tallahassee FL 32304-1329

GILMORE, ALLAN EMORY, forensic consultant; b. Riverside, Calif., Apr. 7, 1924; s. Elvin E. and Emma Martha (Picker) G.; m. Betty Jo Hoskinson, Jan. 2, 1952; children: William Allan, Robert Donald, David Ernest. Student, The Citadel, 1943-44, Loughborough Tech. Coll., England, 1945-46; AB, U. Calif., Berkeley, 1948. Chief chemist Western Gulf Oil Co., Bakersfield, Calif., 1948-53; narcotic chemist, inspector State of Calif., Sacramento, 1953-56; criminalist Calif. State Crime Lab., Sacramento, 1956-67; dir. crime lab. Sacramento County, 1967-81; cons. Calif. Forensic Lab., Foster City, 1981-85; pvt. practice criminalist, cons. Sacramento, 1985—; part-time instr. Sierra Coll., Rockland, Calif., 1972-76, Sacramento City Coll., 1981-86, Sacramento State Coll., 1969. Contbr. articles to profl. jours.; co-author publ. in field. Sgt. U.S. Army, 1943-46, ETO. Mem. Calif. Assn. Criminalists (life, treas. 1970-71), Internat. Assn. Toxicologists, Internat. Assn. Arson Investigators, Calif. Peace Officers Assn. (life), N.W. Assn. Forensic Scientists, Nat. Assn. Watch and Clock Collectors, Masons. Republican. Home and Office: 7635 Bar Du Lane Sacramento CA 95829

GILMORE, MAURICE EUGENE, mathematics educator; b. N.Y.C., Jan. 2, 1938; s. Maurice Eugene and Mary Wells (Barnes) G.; m. Julie Anne Rogers, June 20, 1964 (div. 1988); children: Peter Barnes, Christopher Alan, Jessica Lynn; m. Cathi Leslie Sonneborn, Sept. 1, 1991. BA, Georgetown U., 1959; MS, Syracuse U., 1961; PhD, U. Calif., Berkeley, 1966. Instr. Northeastern U., Boston, 1966-68, asst. prof., 1968-72, assoc. prof., 1972-78, prof., 1978—, chmn. math. dept., 1975-88; vis. prof. U. Tecnica Del Estado, Santiago, Chile, 1968, U. of Sussex, Falmer, U.K., 1989. NSF grantee, 1979, 92. Mem. Math. Assn. Am., Am. Math. Soc. Office: Northeastern U 360 Huntington Ave Boston MA 02115-5096

GILMORE, THOMAS DAVID, biologist; b. Uniontown, Pa., Dec. 28, 1952; s. Thomas Hardy and Marian (Madeline) G. AB in English, Princeton U., 1974; PhD in Zoology, U. Calif., Berkeley, 1984. Grad. student U. Calif., Berkeley, 1978-84; postdoctoral fellow U. Wis., Madison, 1984-87; asst. prof. Boston U., 1987-93, assoc. prof., 1993—; grant reviewer NIH, Bethesda, Md., 1992, Am. Cancer Soc., Atlanta, 1991-93. Contbr. articles to profl. jours. Recipient Jr. Faculty Rsch. award Am. Cancer Soc., 1988-91, 92-96, NIH First award, 1988-93, 93-97, Postdoctoral fellowship Jane Coffin Childs, 1984-87. Mem. AAAS, Am. Soc. Microbiology. Democrat. Roman Catholic. Achievements include characterization of the rel oncogene. Office: Boston U Biology Dept 5 Cummington St Boston MA 02215

GILRUTH, ROBERT ROWE, aerospace consultant; b. Nashwauk, Minn., Oct. 8, 1913; s. Henry Augustus and Frances Marion (Rowe) G.; m. E. Jean Barnhill, Apr. 24, 1937 (dec. 1972); 1 dau., Barbara Jean (Mrs. John Wyatt); m. Georgene Hubbard Evans, July 14, 1973. B.S. in Aero. Engring, U. Minn., 1935, M.S., 1936, D.Sc., 1962; D.Sc., George Washington U., 1962, Ind. Inst. Tech., 1962; D.Eng., Mich. Tech. U., 1963; LL.D., N.Mex. State U., 1970. Flight research engr. Langley Aero. Lab., NACA, Langley Field, Va., 1937-45; chief pilotless aircraft research div. Langley Aero. Lab., NACA, 1945-50, asst. dir., 1950-58; dir. NASA Project Mercury, 1958-61, NASA Manned Spacecraft Ctr., Houston, 1961-72; dir. key personnel devel. NASA Manned Spacecraft Ctr., 1972-73, ret., 1973; cons. to adminstr. NASA, 1974—; dir. Bunker Ramo Corp. Ind. experimenter and cons. hydrofoil craft, 1938-58; advisor on guided missiles, aeros. and structures, high temperature facilities U.S. Dept. Def., 1947-58; mem. com. space systems NASA Space Adv. Council, 1972—; chmn. mgmt. devel. edn. panel NASA, 1972-73; mem. ad hoc com. fire safety aspects of polymeric materials Nat. Materials Adv. Bd., 1973-74. Recipient Outstanding Achievement award U. Minn., 1954, Great Living Am. award U.S. C. of C., 1962, Disting. Fed. Civilian Service award Pres. U.S., 1962, Americanism award CBI Vets. Assn., 1965, Spirit of St. Louis medal, 1965, Internat. Astronautics award Daniel and Florence Guggenheim, 1966, Disting. Service medal NASA, spring 1969, fall 1969, Pub. Service at Large award Rockefeller Found., 1969, ASME medal, 1970, James Watt Internat. medal, 1971, Achievement award Nat. Aviation Club, 1971, Robert J. Collier trophy with Nat Aero. Assn., 1972, Space Transp. award Louis W. Hill, Disting. Service medal NASA, medal of honor N.Y.C., Robert H. Goddard Meml. trophy Nat. Rocket Club, Nat. Air and Space Mus. trophy, 1985; named to Nat. Space Hall of Fame, 1969, Internat. Space Hall of Fame, 1976. Mem. Nat. Acad. Engring. (aeros. and space bd. 1974—), Nat. Acad. Scis. Home: RR 1 Box 1486 Kilmarnock VA 22482-9769

GILSON, ARNOLD LESLIE, engineering executive; b. Perrysburg, Ohio, Apr. 10, 1931; s. Leslie Clair and Velma Lillian (Hennen) G.; B.S. in Mech. Engring., U. Toledo, 1962; m. Phyllis Mary Seiling, Sept. 15, 1951 (dec. May 1982); children—David, Jeffrey, Luann, Suzanne. Engr., Miller, Tillman & Zamis engrs., Toledo, 1962-67, regional mgr., Phoenix br., 1967-69; owner, mgr. A B S Tech. Services, Phoenix, 1969—. Served with U.S. Army, 1952; Korea. Decorated Bronze Star. Mem. Nat. Mil. Intelligence Assn. Republican. Roman Catholic. Commd. extraordinary minister, 1975. Patentee in several fields. Home: 8226 E Meadowbrook Ave Scottsdale AZ 85251-1739 Office: PO Box 2440 Scottsdale AZ 85252

GIMPEL, RODNEY FREDERICK, chemical engineer; b. Idaho Falls, Idaho, Apr. 19, 1953; s. Frederick Isaac Gimpel and Sheila Leora (Lords) Young; m. Victoria Lynn Pabst; children: Tonya Lynn, David Tyson, Tara Lacey. Student, Idaho State U., 1971-73; BSChemE, U. Idaho, 1975. Process engr. Atlantic Richfield, Richland, Wash., 1975-76; devel. engr. Rockwell Internat., Richland, 1976-77; feasibility engr. Allied Chem., Idaho Falls, Idaho, 1977-78; project engr. Exxon, ENICO, Idaho Falls, 1978-83, Westinghouse, WINCO, Idaho Falls, 1984-85; project engr., mgr. Westinghouse WMCO, Cin., 1986-88; environ. clean-up project mgr. to devel. program mgr. for making glass of radioactive and hazardous wastes Westinghouse, WEMCO, Cin., 1989-93; devel. program mgr. Federal Environ. Restoration Mgnt. Corp., Cin., 1993—; Contbr. articles to profl. jours. Pres. Head Start, Idaho Falls, 1985; leader Boy Scouts Am., Cin., 1987-88. Mem. Haztech, Am. Ceramic Soc. Republican. Mem. LDS Church. Achievements include copyrighting and computer programs; designing and building of clock and clockworks completely out of wood; first to make holograms in Idaho; developing processes for turning hazardous wastes into glass. Home: 5920 Coachmont Dr Fairfield OH 45014 Office: FERMCO PO Box 398704 Cincinnati OH 45239-8704

GINDER, JOHN MATTHEW, physicist; b. Galion, Ohio, Aug. 8, 1961; s. John Ogle and Eileen Marie (Hoffman) G.; m. Jill Adair Flick, June 25, 1983. BS in Physics, Rensselaer Poly. Inst., 1983; MS in Physics, Ohio State U., 1985, PhD in Physics, 1988. Postdoctoral rsch. assoc. Ohio State U., Columbus, 1988-90; rsch. scientist sr. Ford Motor Co., Dearborn, Mich., 1990—; mem. Dept. Energy/Nat. Renewable Energy Lab. Surface Processing Workshop, Dearborn, 1991-92. Contbr: Conducting Polymers, 1987, Spectroscopy of Advanced Materials, 1991; contbr. articles to profl. jours. Recipient Presdl. fellowship Ohio State U., 1987, Univ. fellowship Ohio State U., 1983, Alumni scholarship Rensselaer Poly. Inst., 1979-83. Mem. Am. Phys. Soc., Materials Rsch. Soc., Phi Kappa Theta, Sigma Xi, Sigma Pi Sigma. Achievements include patents in optical and electrical properties of conducting polymers. Office: Ford Motor Co Rsch Lab PO Box 2053 MD 3028 Dearborn MI 48121

GINGERICH, PHILIP DERSTINE, paleontologist, evolutionary biologist, educator; b. Goshen, Ind., Mar. 23, 1946; s. Orie Jacob and Miriam (Derstine) G.; m. B. Holly Smith, 1982. A.B., Princeton U., 1968; Ph.D., Yale U., 1974. Prof. U. Mich., Ann Arbor, 1974—, dir. Mus. Paleontology, 1981-87, 1989—. Contbr. articles to sci. jours. Recipient Henry Russel award U. Mich., 1980; Shadle fellow Am. Soc. Mammalogists, 1973-74, NATO fellow, 1975, Guggenheim fellow, 1983-84. Fellow AAAS, Geol. Soc. Am.; mem. Am. Assn. Phys. Anthropologists, Paleontol. Soc. (Schuchert award 1981), Soc. Study Evolution, Am. Geophys. Union, Am. Soc.Naturalists. Office: U Mich Mus Paleontology 1109 Geddes Rd Ann Arbor MI 48109

GINIECKI, KATHLEEN ANNE, environmental engineer; b. Pompton Plains, N.J., Apr. 5, 1966; d. William Vincent and Lois Mae (Hazel) G. BS in Mech. Engring., U. Rochester, 1988, MS in Biomed. Engring., 1989. Cert. in biomed. engring. Mech. technician U. Rochester (N.Y.) Lab. for Laser Energetics, 1986-87; teaching asst. dept. mech. engring. U. Rochester, 1988-89; nuclear/environ. engr. U.S. Dept. Energy, Schenectady, 1989-92; environ. engr. Dames & Moore Cons., Dallas, 1992—; Job Fair rep. U.S. Dept. Energy, 1989. Folk music soloist St. Mary's Ch., Schenectady, 1985—; active Big Bros./Big Sisters, Rochester, 1985; vol. Ellis Hosp., Schenectady, 1992. Rochester scholar, 1984-88. Mem. ASME, Soc. Women Engrs. (activity chairperson 1986-87), U. Rochester Alumni Assn. (com. mem. 1991—). Republican. Roman Catholic. Avocations: tennis, musician (flute, saxophone), fitness walking, fishing, golfing. Home: 18800 Lina St # 301 Dallas TX 75287 Office: Dames & Moore Cons 5151 Beltline Rd # 700 Dallas TX 75240

GINNETT, ROBERT CHARLES, organizational psychologist; b. Washington, Aug. 17, 1947; s. Hubert W. and Eleanor (Walker) G.; m. Sherrill Absher, Aug. 7, 1971; children: Laura, Bradley. MBA, U. Utah, 1976; MA, Yale U., 1982, MPhil, 1983, PhD, 1985. Commd. 2d lt. USAF, 1970, advanced through grades to lt. col., 1986, retired, 1990; prof. USAF Acad., Colo. Springs, Colo., 1980-90; dir. rsch. Ctr. for Creative Leadership, Colo. Springs, 1990—; bd. dirs. Leadership 2000, Colo. Springs, 1986—; cons. major airlines, Colo. Springs, 1987—. Author: Leadership: Enhancing the Lessons of Experience, 1993; contbr. chpts. to books: Groups that Work, 1990, Cockpit Resource Mgmt., 1993. Decorated Air medal and Bronze Star; rsch. grantee NASA, Calif., 1984-93. Mem. Am. Psychol. Soc. (charter), Acad. Mgmt. Presbyterian. Achievements include research that discovered behavior used by effective leaders in developing high performance work teams (airline crews, surgical teams, total quality teams, space shuttle teams). Office: Ctr for Creative Leadership 850 Leader Way Colorado Springs CO 80906

GINOZA, WILLIAM, retired biophysics educator; b. L.A., Feb. 7, 1914; s. Shinkichi and Kame (Yamashiro) G.; m. Midori Sugita, Oct. 4, 1944 (dec. May 1987); children: Lillian, Donn. BA, U. Calif., Berkeley, 1937, MA, 1939; PhD, UCLA, 1952. Asst. rsch. biochemist dept. botany UCLA, 1952-55, rsch. scientist atomic energy commn., 1956-61; assoc. prof. dept. biophysics Pa. State U., University Park, 1961-67, prof., 1967-79, prof. emeritus, 1979—; invited speaker ednl. insts. and scientific confs. including Internat. Congress Biochemistry, Vienna, 1958, Farraday Soc. meeting on nucleic acids, Birmingham, England, 1958; vis. prof. U. Kyoto, Japan, 1974. Co-author: Methods in Virology, Vol IV, 1968.; contbr. reviews to Ann. Reviews Nuclear Sci., Ann. Reviews Microbiology; contbr. articles to profl. jours. Fellow AAAS; mem. Biophys. Soc., Sigma Xi, Phi Lambda Upsilon. Achievements include illucidation of molecular structure of Tobacco Mosaic Virus and its RNA, mechanisms by which heat or high energy radiations destroy the biological functions of nucleic acids of viruses and bacteria. Home: 962 Mccormick Ave State College PA 16801-6529

GINSBERG, DONALD MAURICE, educator, physicist; b. Chgo., Nov. 19, 1933; s. Maurice J. and Zelda (Robbins) G.; m. Joli D. Lasker, June 10, 1957; children: Mark D., Dana L. BA, U. Chgo., 1952, BS, 1955, MS (NSF fellow), 1956; PhD (NSF fellow), U. Calif. at Berkeley, 1960. Mem. faculty U. Ill. at Urbana, 1959—, prof. physics, 1966—; vis. scientist in physics Am. Assn. Physics Tchrs.-Am. Inst. Physics, 1965-71; vis. scientist IBM, 1976; mem. evaluation com. for Nat. High-Field Magnet Lab., NSF, 1977-79, 85, 91; mem. rev. com. for solid state sci. div. Argonne Nat. Lab., 1977-83, chmn., 1980; mem. rev. panel for basic energy scis. div. Dept. Energy, 1981. Editor: Physical Properties of High Temperature Superconductors, Vols. 1, 2 and 3, 1989, 90, 92; contbr. to Ency. Britannica, 1971, 82, 88, Concise Ency. of Magnetic and Superconducting Materials, 1992. Alfred P. Sloan research fellow, 1960-64; NSF fellow, 1966-67. Fellow Am. Phys. Soc.; mem. AAAS, Phi Beta Kappa, Sigma Xi. Research and publs. on low temperature physics, superconductivity, cryogenic instrumentation. Home: 2208 Grange Cir Urbana IL 61801-6607

GINSBERG, MYRON, research scientist; b. Brockton, Mass., May 3, 1943; s. Frank and Evelyn Hazel (Spekin) G.; m. Judith Beverly Rosenbaum, Nov. 19, 1989; 1 stepchild, Ellen Joy Schoenfeld. BA in Math., Boston U., 1965; MA in Math., Clark U., 1967; PhD in Computer Sci., U. Iowa, 1972. Instr. dept. computer sci. U. Iowa, Iowa City, 1969-72; from asst. prof. to assoc. prof. computer sci. So. Meth. U., Dallas, 1972-77, 77-79; NASA/ASEE rsch. fellow NASA Langley Rsch. Ctr, Hampton, Va., summer 1979; assoc. sr. rsch. scientist GM, Warren, Mich., 1979-81; sr. rsch. scientist GM Rsch. Labs, Warren, Mich., 1981-82, staff rsch. scientist, 1982-92; cons. systems engr. EDS Advanced Computing Ctr., Warren, 1992—; mathematician U.S. Army Ballistics Rsch. Lab., Aberdeen Proving Ground, Md., summers, 1964-67; data systems analyst NASA Electronics Rsch. Ctr., Cambridge, Mass., summers, 1968-69; adj. assoc. prof. U. Mich., Ann Arbor, 1990; editorial bd. ComputinSystems in Engring., 1988-93; adv. bd. Cray Rsch. Fortran, 1991-92; grant review panelist NSF, 1992; GM/EDS rep. to Nat. Consortium for Advancement of Automotive Engring., 1992—. Editor: Supercomputers in the Auto Industry, 1985, Automotive Applications of Supercomputers, 1988, High-Speed and Large-Scale Computing: A Panoramic View, 1988, Automotive Applications of Vector/Parallel Computers: State-of-the-Art, 1992; contbr. articles to profl. jours. Grantee Mobil Oil Found., 1975, U.S. Army C.E., 1977-78, NSF, 1983-84, 77-79, Alfred P. Sloan Found., 1975-78; recipient award for excellence in oral presentation SAE, 1985, 86, 87, Disting. Speaker plaque SAE, 1988. Mem. IEEE, IEEE Computer Soc. (lectr.), ASME (lectr.), Soc. for Indsl. and Applied Math. (lectr., spl. group on super computing), Assn. for Computing Machinery (lectr., bd. dirs. SIGNUM 1976-80, editor-in-chief SIGNUM newsletter 1976-80), Sigma Xi (lectr.). Avocations: playing alto sax, tenor sax, clarinet and flute; listening to jazz and classical music. Office: EDS Advanced Computing Ctr GM NAO R & D Ctr PO Box 9055 30500 Mound Rd Warren MI 48090-9055

GINSBERG, MYRON DAVID, neurologist; b. Denver, Aug. 26, 1939; s. Morris Seymour and Evelyn (Fishman) G.; m. Bryna Greenberg, June 22, 1969; children: Deborah Mara, Emily Michelle. BA, Wesleyan U., 1961; MD, Harvard U., 1966. Intern, resident Harvard Med. Svc., Boston City Hosp., 1966-68; neurology resident, fellow Mass. Gen. Hosp., Boston, 1968-70, 72-73; staff assoc. Lab. Perinatal Physiology, NIH, Bethesda, Md., 1970-72; asst. prof., assoc. prof. dept. neurology U. Pa., Phila., 1973-79; assoc. prof. neurology U. Miami (Fla.) Sch. Medicine, 1979-81, prof. neurology,

1981—, dir. cerebral vascular disease rsch. ctr., 1981—, dir. neurotrauma clin. rsch. ctr., 1991—, Peritz Scheinberg endowed chair of neurology, 1992—; mem. study sect. NIH, Bethesda, 1982-86; nat. rsch. com. Am. Heart Assn., Dallas, 1986-91. Editor: Cerebrovascular Diseases, 16th Princeton Conf., 1989; editor: Jour. Cerebral Blood Flow and Metabolism, 1992—; contbr. over 170 articles to profl. jours. Lt. comdr. USPHS, 1970-72. Recipient Fulbright scholarship U.S. Govt., 1961-62, Jacob Javits Neuroscience Investigator award NIH, 1985-92. Fellow Am. Acad. Neurology; mem. Am. Neurol. Assn. (membership com. 1990-91), Am. Physiol. Soc., Internat. Soc. Cerebral Blood Flow & Metabolism (dir. 1985-89), Phi Beta Kappa, Alpha Omega Alpha. Office: U Miami Sch Medicine PO Box 016960 Miami FL 33101-6960

GINSBERG-FELLNER, FREDDA, pediatric endocrinologist, researcher; b. N.Y.C., Apr. 21, 1937; d. Nathaniel and Bertha (Jagendorf) Ginsberg; m. Michael J. Fellner, Aug. 27, 1961; children: Jonathan R., Melinda B. AB, Cornell U., 1957; MD, NYU, 1961. Diplomate Am. Bd. Pediatrics, Am. Bd. Pediatric Endocrinology. Intern Albert Einstein Coll. Medicine, N.Y.C., 1961-62, fellow in pediatrics, 1962-63, 64-65, 66-67, resident in pediatrics, 1963-64, 65-66, clin. instr. pediatrics, 1967; assoc. in pediatrics Mt. Sinai Sch. Medicine, N.Y.C., 1967-69, asst. prof., 1969-75, assoc. prof., 1975-81, dir. div. pediatric endocrinology, 1977—, prof. pediatrics, 1981—. Mem. med. scis. rev. com. Juvenile Diabetes Found., 1985-88, mem. scis. adv. bd., 1991—; mem. N.Y. State Coun. on Diabetes, Albany, 1988-89; chmn. Camp NYDA for Diabetic Children, Burlington, 1977-89. Grantee NIH, 1977—, Am. Diabetes Assn., 1978, March of Dimes, 1983-87, Juvenile Diabetes Found., 1982-88, 93—, Wm. T. Grant Found., 1985-89. Fellow Am. Acad. Pediatrics; mem. Am. Diabetes Assn. (chmn. cody 1992—, Outstanding Contbns. award 1991), Soc. Pediatric Rsch., Am. Pediatric Soc., Endocrine Soc., Lawson Wilkins Pediatric Endocrine Soc., N.Y. Diabetes Assn. (pres.-elect 1985-87, pres. 1987-89, Svc. award Camp NYDA 1989, Max Ellenberg Profl. Svc. award 1993). Office: Mt Sinai Med Ctr 1 Gustave Levy Pl Box 1198 New York NY 10029

GINTAUTAS, JONAS, physician, scientist, administrator; b. Justinava, Lithuania, Oct. 3, 1938; came to U.S., 1967; s. Jonas and Elena (Zaveckaitė) Sinsinas; m. Kristina Zebrauskaite, June 13, 1970; children: Pasaka, Vadas. PhD, Northwestern U., 1976; MD, U. Juarez, Mex., 1984. Assoc. prof. Tex. Tech. U., Lubbock, 1975-77; assoc. prof. and dir. rsch. Tex. Tech. U. Health Scis. Ctr., Lubbock, 1979-82; dir. basic and clin. rsch., prof. neurology Brookdale Hosp. Med. Ctr., N.Y.C., 1985—; cons. Amtorg Corp., N.Y.C., 1987—, Raley Internat. Co., Boston, 1988—, Arrow Biomed Inc., Metuchen, N.J., 1988—. Editorial cons. Jour. Aphasia Agnosia Apraxia, 1979—; contbr. articles to profl. jours. Charter mem. Rep. Presdl. Task Force, Washington, 1982— (medal of honor 1982). Grantee more than $1 million for rsch. pvt. and govt. agys. Fellow Internat. Coll. Physicians and Surgeons (hon.); mem. Am. Biog. Inst. (dep. gov. 1987—), U.S. Senatorial Club (preferred). Roman Catholic. Avocations: wordworking, camping, fishing, reading. Home: 84-19 107 St Richmond Hill NY 11418 Office: Brookdale Hosp Med Ctr Linden and Rockaway Brooklyn NY 11212

GIOLITTI, ALESSANDRO, chemist; b. Rome, Feb. 28, 1955; s. Ugo and Anna Maria (Giurlani) G.; m. Elisabetta Nice, Dec. 2, 1978; children: Giovanni, Bianca Maria. Laurea in Chimica, U. Degli Studi, Firenze, Italy, 1978. Researcher Montedison, Porto Marghera, Italy, 1979-80; researcher Menarini Pharm., Firenze, 1980-86, group leader, 1987-91, head dept. drug design, 1992—. Contbr. articles to profl. jours. Mem. Ordine dei Chimici, Am. Chem. Soc. Roman Catholic. Achievements include numerous patents. Home: Via Fabroni 45, 50134 Florence Italy Office: Menarini Pharms, Via Sette Santi 3, 50131 Florence Italy

GIOMINI, MARCELLO, chemistry educator; b. Rome, June 9, 1939; s. Raffaele and Caterina (Schwoerer) G.; m. Antonella Tesei, Sept. 18, 1965; children: Bruno, Claudia. PhD, U. Rome, 1963. Cert. chemistry. Scholar Nat. Coun. Rsch., Rome, 1965-67; rsch. assoc. U. Rome, 1968-85, prof. chemistry, 1986—. Author: Principles of General Chemistry 1980, Origins of Life on the Earth, 1985; contbr. about 40 articles to profl. jours. including Biochimica et Biophysica Acta, Biophys. Chemistry, Chem. Physics Letters. Mem. Am. Chem. Soc., Italian Chem. Soc., Italian Biochem. Soc. Achievements include patent for Anthracycline Gels and Their Use in Therapy. Office: U La Sapienza, P Le Aldo Moro 5, 00185 Rome Italy

GIONFRIDDO, MAURICE PAUL, research and development manager, aeronautical engineer; b. Medford, Mass., Feb. 19, 1931; s. Santo and Germaine Camille (Gaillard) G.; m. Joan Marie Powers, Apr. 26, 1956; children: Marianne E., Linda. BS in Aero. Engring., MIT, 1953, MS in Aero. Engring., 1969. Rsch. asst. Aeroelastic and Structures Rsch. Lab., MIT, Cambridge, Mass., 1953-54; aero. rsch. engr. Air Force Cambridge Rsch. Ctr., Bedford, Mass., 1956-57; aero. engr. Army Natick (Mass.) Rsch., Devel. and Engring. Ctr., 1957—; mem. Nat. Parachute Tech. Coun., 1991—. Class agt. MIT Class of 1953, 1968-78. 1st lt. USAF, 1954-56. Fellow AIAA (assoc., charter, aerodyn. decelerator tech. com. 1964-67, Aerodyn. Decelerator award 1990); mem. Sigma Xi. Roman Catholic. Home: 20 Westminster Way Westborough MA 01581 Office: US Army Natick RD&E Ctr Kansas St Natick MA 01760

GIORDAN, ANDRE JEAN PIERRE HENRI, biologist researcher; b. Nice, France, Nov. 7, 1946; s. Francois and Laurence (Abbo) G.; 1 child, Severine. D of Physiology, U. Nice, France, 1970; Postmaster of History of Scis., U. Paris, 1974; D on Didactic and Epistemology of Scis., U. Paris V "Sorbonne", 1976. Physiologist researcher U. Nice, France, 1968-71; tchr. Ministry of Edn., France, 1971-80, sci. didactic researcher Nat. Inst. Edn., Paris, 1972-76, rsch. dir., 1976-80; extraordinary prof. U. Geneva, Switzerland, 1980-83; rsch. dir. U. Paris VII, 1978—; prof. U. Geneva, 1983—; LDES dir.; pres. ednl. scis. sect. 1992—; rsch. dir. Nat. Ctr. for Sci. Rsch., Paris, 1976-80; environ. cons. UNESCO and UNEP, 1976-87; CECSI network dir., 1980—; sci. and tech. communication cons. OCDE, 1993—, CCE, 1993—. Author: Une pedagogie pour les sciences experimentales, 1978, Quelle education scientifique pour quelle societe, 1978, L'education relative a l'environnement, 1986, Histoire de la biologie, 1987, Les origines du savoir scientifique, 1987, L'enseignement scientifique, 1989, Psychologie genetique et didactique des sciences, 1989, Maitriser l'information scientifique et medicale, 1990, L'education pour l'environnement, 1991; contbr. articles to profl. jours. Recipient Grameyer award U. Louisville, 1988, Cage a mouche award, Nice, 1989. Mem. Swiss Soc. Edn., INISTE Network, Internat. Union Biol. Scis., Commn. Biol. Edn., Assn. for Devel. Rsch. on Didactic Scis., Pansemiotic Assn., European Soc. didactic Biology (v.p. 1990—), CECSI Network (pres. 1988—). Avocations: theater, plastic arts, cuisine, sports. Home: Benoit Bunico 3, F-06300 Nice France Office: LDES, FPSE, 9 Rt de Drize, 1227 Carouge Geneva Switzerland

GIORGIO, TODD DONALD, chemical engineering educator; b. Passaic, N.J., Dec. 23, 1960; s. Donald Henry and Carole Barbara (Singlary) G. BS, Lehigh U., 1982; PhD, Rice U., 1987. Asst. prof. chem. engring. Vanderbilt U., Nashville, 1987-93, assoc. prof. chem. engring., 1993—. Contbr. articles to profl. jours. Recipient 1st award NIH, Biomed. Engring. award Whitaker Found. Mem. AICE, AAAS, Biomed. Engring. Soc., Am. Chem. Soc., Am. Soc. for Engring. Edn., N.Am. Soc. Biorheology, Am. Heart Assn., Sigma Xi, Tau Beta Pi. Office: Vanderbilt U Box 60 Station B Nashville TN 37235

GIOVACCHINI, PETER LOUIS, psychoanalyst; b. N.Y.C., Apr. 12, 1922; s. Alex and Therese (Chicca) G.; m. Louise Post, Sept. 29, 1945; children: Philip, Sandra, Daniel. BS, U. Chgo., 1941, MD, 1944; postgrad., Columbia U., 1939; cert., Chgo. Inst. Psychoanalysis, 1954. Diplomate Am. Bd. Psychiatry and Neurology. Intern Fordham Hosp., N.Y.C., 1944-45; resident U. Chgo. Clinics, 1945-46, resident and research fellow, 1948-50; candidate Chgo. Inst. Psychoanalysis, 1949-54, clin. assoc., 1957—; clin. prof. U. Ill. Coll. Medicine, 1961-92, prof. emeritus, 1992—; pvt. practice, Chgo., 1950—; chief cons. psychodynamic unit Barclay Hosp., Chgo., 1979-81; cons. Wilmette (Ill.) Family Svc. Bur. and United Charities, Boyer-Marin Lodge, Marin County, Calif., 1986—, Mario Martin Inst. for Psychotherapy, 1989—, Psychoanalytic Ctr. Calif., L.A., 1990—. Author: (with L.B. Boyer) Psychoanalytic Treatment of Schizophrenia and Characterological Disorders, 1967, Psychoanalytic Treatment, 1971, also several books on character structure, primitive mental states, psychopathology and psychoanalytic technique, psychoanalysis; also articles.; Co-editor: Annals of Adolescent Psychiatry, 1972-80. Capt. M.C. AUS, 1946-48. Fellow Am. Psychiat. Assn., Am. Orthopsychiat. Assn. (bd. dirs. 1979-83), Am. Coll. Psychiatrists; mem. Am. Soc. Adolscent Psychiatry, Chgo. Soc. Adolscent Psychiatry (pres. 1972-73), Internat. Psychianalytic Soc., Am. Psychoanalytic Assn., Chgo. Psychoanalytic Soc. Home: 270 Locust Rd Winnetka IL 60093-3609 Office: 505 N Lake Shore Dr Chicago IL 60611-3427

GIOVANETTI, KEVIN LOUIS, physicist, educator; b. Pittsfield, Mass., Jan. 5, 1953; s. Louis J. and Elizabeth Ann (Byrne) G.; m. Lisa F. Goldstein, July 11, 1982; children: Graham, Eli, Simone. BS in Physics, Lowell Tech. Inst., 1974; MS in Physics, Coll. of William and Mary, 1977, PhD in Physics, 1982. Rsch., teaching asst. Coll. of William and Mary, Williamsburg, Va., 1974-82; physics instr. Fed. Tech. Inst., Zurich, Switzerland, 1982-85; rsch. assoc. medium energy physics SIN Lab., Switzerland, 1982-85, U. Va., Charlottesville, 1985-88; asst. prof. James Madison U., Harrisonburg, Va., 1989—. Contbr. numerous rsch. articles to profl. jours. Achievements include electron scattering studies of the few nucleon system at MIT Bates Lab.; work with high resolution pionic x-ray crystal spectrometer. Office: James Madison U Physics Dept Harrisonburg VA 22807

GIRAUDET, MICHELE, mathematics educator, researcher; b. Cauderan, France, July 28, 1945; d. Gaston and Jacqueline-Marguerite (Landreau) G.; m. Jambu Michel, June 24, 1968 (div. 1984). PhD in Math., U. Paris 7, 1979. Asst. U. Nanterre, 1971-83; maitre de conf. U. Le Mans, France, 1983—; vis. asst. prof. Bowling Green (Ohio) State U., 1980; mem. equipe de logique U. Paris 7; organizer day in ordered algebra, 1987, meeting in ordered groups and permutation groups, 1990, 93. Editor Actes Journee Algebre Ordonnee, 1987; contbr. articles to profl. jours. Mem. Math. Soc. France, Am. Math. Soc., French Assn. for Ordered Algebra (pres.). Home: 32 Rue de La Reunion, 75020 Paris France

GIRONDA, A. JOHN, III, engineer; b. Dayton, Ohio, Nov. 26, 1956; s. A. John Jr. and Rose Marie (Dibiase) G.; m. Florence Ann Boyd, Sept. 8, 1990; children: Jeffrey Brown, Ryan Brown. BSME, U. Calif., Irvine, 1978; MA in Mgmt., Webster U., 1990. registered profl. engr.N.C., Calif. Commd. 2d lt. USAF, 1978, advanced through grades to capt., retired, 1990; pres. Criterium-Gironda Engrs., Seal Beach, Calif., 1991—; scholar chmn. Profl. Engrs. N.C., South Ctrl. chpt., 1991. Chmn. Local Keep Am. Beautiful Chpt., Fayetteville, N.C., 1988-89. Named U.S. Jaycees Outstanding Young Man Am., 1989, Eagle Scout. Mem. NSPE (sec. Orange County chpt. 1992, 1st v.p. 1993—), Soc. Am. Mil. Engrs. (membership chmn. 1987-89, Silver medal 1989), Calif. Soc. Profl. Engrs. (sec. 1st v.p. Orange County chpt. 1993-94), K.C. (sr. knight, treas. 1991-92, cert. merit 1991), Calif. Real Estate Inspectors Assn. (sec. Orange County chpt. 1993-94). Republican. Roman Catholic. Office: Criterium-Gironda Engrs PO Box 2664 Seal Beach CA 90740

GIROUARD, KENNETH, civil engineer; b. Providence, May 23, 1955; s. Fernand Honree and Mildred (Cowsill) G.; m. Diane Marie Adamson, Dec. 24, 1988. BS in Computer Sci., Calif. State Hayward, 1985; MSCE, San Jose State Coll., 1987; BSCE, U. Ariz., 1977. Cert. profl. engr., Calif. Vol. Peace Corps, Guatemala, 1978-80; civil engr. Bala & Strandgaard, Mill Valley, Calif., 1980-83; jr. asst. engr. Santa Clara Valley Water Dist., San Jose, Calif., 1987-88; assoc. engr. City of Watsonville, Calif., 1988-91; dist. engr. San Lorenzo Valley Water Dist., Boulder Creek, Calif., 1991—. Mem. Latin Am. Task Force, San Francisco, 1981-85, Chiapas (Mex.) Relief and Encouragement, 1986. Ariz. Gen. Rsch. scholar, 1977. Mem. Am. Water Works Assn., Water Pollution Control Fedn. Achievements include research in recycling, portable water projects.

GIRTMAN, GREGORY IVERSON, psychologist; b. Oakland, Calif., May 10, 1956; s. Louie Franklin and Opal (Bowen) G. BA, U. Calif., Berkeley, 1978; MA, Ariz. State U., 1986; EdD, No. Ariz. U., 1991. Lic. psychologist Ariz. Psychology intern U. So. Calif., L.A., 1989-90; psychologist Ariz. State Prison, Florence, 1990—. Mem. APA. Office: Ariz State Prison Spl Programs Unit Florence AZ 85232

GITLITZ, MELVIN HYMAN, chemist, researcher; b. Montreal, Quebec, Can., Feb. 28, 1940; came to U.S., 1965; s. Saul and Ruth (Heit) G.; m. Margaret E. Macdonald, Aug. 22, 1964; children: David, Karin. BS, McGill U., Montreal, 1961; PhD, U. Western Ont., London, 1965. Rsch. chemist M&T Chems. Inc., Rahway, N.J., 1965-68, sr. rsch. chemist, 1968-73, rsch. assoc., 1973-75, rsch. mgr., 1975-89; sr. staff chemist Atochem Inc., King of Prussia, Pa., 1989—; recognized global expert on bioactivity, toxicology of organotin compounds; expert in rsch. mgmt., comml. devel. on new products. Author: (chpt.) Disinfection, Sterilization & Preservation, 1991; author (monograph) Ency. of Chem. Tech., 1984. Mem. Am. Chem. Soc., N.Y. Acad. Scis. Jewish. Achievements include patents on agrl. pesticides, catalysts, marine antifoulants, transparent conductive coatings, metal alkoxides. Home: 92 Daylesford Blvd Berwyn PA 19312 Office: Elf Atochem N Am Inc 900 First Ave King Of Prussia PA 19406

GITLOW, STANLEY EDWARD, internist, educator; b. N.Y.C., Jan. 28, 1926; s. Max and Ida (Birnbaum) G.; children: William, Stuart. MD, L.I. Coll., 1948. Diplomate Am. Bd. Internal Medicine. Intern Mt. Sinai Hosp., N.Y.C., 1948-49; resident in medicine U.S. VA Hosp., Bronx, N.Y., 1949-50; rsch. fellow in pharmacology SUNY Coll. Medicine, N.Y.C., 1950-51; asst. resident in medicine Mt. Sinai Hosp., N.Y.C., 1951-52; sr. asst. surgeon USPHS, 1952-54; from sr. clin. asst. physician to attending physician Mt. Sinai Hosp., N.Y.C., 1957—; from asst. physician to chief 1st med. svc. Metro. Hosp., N.Y.C., 1957-68; from instr. in medicine to vis. clin. prof. medicine N.Y. Med. Coll., N.Y.C., 1961—; clin. prof. medicine Mt. Sinai Sch. Medicine, N.Y.C., 1972—; bd. dirs. Am. Soc. Addiction Medicine, Washington. Mem. editorial adv. bd. The Alcoholism Digest, 1981-87; mem. editorial bd. Alcoholism, Clin. & Exptl. Rsch., 1978-84, Jour. of Substance Abuse Treatment, 1984, Alcohol and Alcoholism, 1984—, Advances in Alcohol & Substance Abuse, 1985—; contbr. over 200 articles to profl. jours. Mem., former chmn. Bd. for Profl. Med. Conduct, Dept. Health, N.Y. State, Albany, 1975—; mem. med. adv. bd. N.Y.C. Dept. Correction; councilor Alumni Assn., SUNY Coll. Medicine, Bklyn.; mem. community adv. coun. Jr. League of City of N.Y., 1976-79. Recipient Annual award Am. Soc. Addiction Medicine, 1990, Citation of Merit in Humanities, Malvern Inst., 1980, Ford prize L.I. Coll. Medicine, 1948, grant NIH, 1961-69, grant Am. Heart Assn., 1965-68, grant Am. Cancer Soc., 1965-73, grant Lic. Beverage Industries, 1965-66, grant Nat. Heart and Lung Inst., 1969-79, grant Genetics Ctr., 1972-76. Fellow Am. Coll. Physicians, N.Y. Acad. Medicine, N.Y. Acad. Scis.; mem. AAAS, AMA (regional cons., residency review com. internal medicine), N.Y. County Med. Soc., N.Y. State Med. Soc., N.Y. Heart Assn., Am. Fedn. for Clin. Rsch., N.Y. Med. Soc. on Alcoholism (pres. 1961), Am. Soc. for Pharmacology and Exptl. Therapeutics, Am. Soc. for Clin. Pharmacology and Therapeutics, Am. Heart Assn. (coun. for high blood pressure rsch., med. adv. bd., rsch. study com.), Med. Soc. of State of N.Y. (chmn. com. addiction to alcohol and narcotics 1972-90), Am. Med. Soc. on Alcoholism (pres. 1970-72), Am. Soc. for Neurochemistry. Achievements include research in field of hypertension and addiction medicine. Home and Office: 1136 Fifth Ave New York NY 10128

GITTES, RUBEN FOSTER, urological surgeon; b. Mallorca, Spain, Aug. 4, 1934; s. Archie and and Cicely Mary (Foster) G.; m. K.S. Zipf, June 10, 1955; m. Rita R. Drum, Feb. 21, 1976; children: Julia S., Frederick T., George K., Melissa S., Robert F. Grad., Phillips Acad., Andover, Mass., 1952; A.B., Harvard U., 1956, M.D., 1960. Intern, then resident in surgery and urology Mass. Gen. Hosp., Boston, 1960-67; asst. prof. UCLA Med. Sch., 1968-69; assoc. prof., then prof., chief urology U. Calif. at San Diego Med. Sch., 1969-75; prof. urol. surgery Harvard U. Med. Sch., chmn. Harvard program urology Longwood area, 1975-87; chmn. dept. surgery Scripps Clinic and Rsch. Found., La Jolla, Calif., 1987—; mem. study sects., task forces NIH, 1973—. Author, editor pubs. in field. Served with USPHS, 1963-65. NIH grantee, 1969—. Mem. AAAS, Endocrine Soc., Soc. Univ. Surgeons, Soc. Univ. Urologists, Am. Assn. Genito-Urinary Surgeons, Clin. Soc. Genito-Urinary Surgeons, A.C.S., Am. Surg. Assn., Am. Urol. Assn., Am. Soc. Transplant Surgeons, Soc. Ancient Numismatics, Phi Beta Kappa, Alpha Omega Alpha. Office: Scripps Clinic & Rsch Found 10666 N Torrey Pines Rd La Jolla CA 92037-1027

GITTLEMAN, MORRIS, consultant, metallurgist; b. Zhidkovitz, Minsk, Russia, Nov. 2, 1912; came to U.S., 1920, naturalized; s. Louis and Ida (Gorodietsky) G.; B.S. cum laude, Bklyn. Coll., 1934; postgrad. Manhattan Coll., 1941, Pratt Inst., 1943, Bklyn. Poly. Inst., 1946-47; m. Clara Konefsky, Apr. 7, 1937; children—Arthur Paul, Michael Jay. Metall. engr. N.Y. Naval Shipyard, 1942-47; chief metallurgist, chemist Pacific Cast Iron Pipe & Fitting Co., South Gate, Calif., 1948-54, tech. mgr., 1954-57, tech. and prodn. mgr., 1957-58; cons. Valley Brass, Inc., El Monte Calif., 1958-61, Vulcan Foundry, Ltd., Haifa, Israel, 1958-65, Anaheim Foundry Co. (Calif.), 1958-63, Hollywood Alloy Casting Co. (Calif.), 1960-70, Spartan Casting Co., El Monte, 1961-62; Overton Foundry, South Gate, Calif., 1962-70, cons., gen. mgr., 1970-71; cons. Familian Pipe & Supply Co., Van Nuys, Calif., 1962-72, Comml. Enameling Co., Los Angeles, 1963-68, Universal Cast Iron Mfg. Co., South Gate, 1965-71; pres. MG Coupling Co., 1972-79; instr. physics Los Angeles Harbor Coll., 1958-59; instr. chemistry Western States Coll. Engring., Inglewood, Calif., 1961-68. Registered profl. engr., Calif. Mem. Am. Foundrymen's Soc., Am. Foundrymen's Soc. So. Calif. (dir. 1955-57), AAAS, Am. Soc. Metals, N.Y. Acad. Scis., Internat. Solar Energy Soc. (Am. sect.). Contbr. to tech. jours.; inventor MG timesaver coupling, patents worldwide. Home: 17635 San Diego Circle Fountain Valley CA 92708-5243 Office: 17044 Montanero Ave Carson CA 90746-1311

GIULIANI, ELEANOR REGINA, biology educator; b. Bayonne, N.J., Aug. 10, 1949; d. Edward and Eleanor (Weth) McMahon; m. Dennis Charles Giuliani, Mar. 31, 1973; children: Megan McMahon, Denielle Alicia, David Charles, Anne Marie. BA, Cardinal Cushing Coll., Brookline, Mass., 1971; MS, Rutgers U., 1979, PhD, 1979. Rsch. asst. U. Medicine & Dentistry N.J., Newark, 1971-73; teaching asst. Rutgers U., Newark, 1973-77, grad. asst., 1977-78; instr. Seton Hall U., South Orange, N.J., 1978-79; asst. prof. St. Peter's Coll., Jersey City, 1979-84, assoc. prof., 1984-90, prof. biology, 1990—, natural sci. coord. Inst. for Advancement Urban Edn., 1990-92, dir. high sch. and coll. rels., 1989-92. Contbr. articles to profl. jours. Faculty fellow St. Peter's Coll., Jersey City, 1988. Mem. AAUP, AAAS, Sigma Xi. Roman Catholic. Office: St Peters Coll Dept Biology Jersey City NJ 07306

GIURGIUTIU, VICTOR, aeronautical engineer, educator; b. Bucharest, Romania, May 9, 1949; s. Stefan and Constantza (Alecsandrescu) G.; m. Dana Mihaela Teodoru, Feb. 9, 1979; 1 child, Dan-Victor. BS in Engring., Imperial Coll., London, 1972, PhD, 1977. Group leader Aviation Inst., Bucharest, 1977-82, sect. leader, 1982-88; tech. dir. Bucharest Aircraft Factory, 1988-90; sci. sec. Aviation Inst., Bucharest, 1990-91; pres., dir. gen. Straero Inst., Bucharest, 1991—; assoc. prof. Bucharest Poly. Inst. and Tech. U., 1977—; vis. prof. Va. Poly. Inst. and State U., Blacksburg. Author: Helicopter Aeroelasticity-Blade Studies, 1982, Stability of Aircraft Structure, 1990; translator: (3 books) English Translation of Romanian Maths, 1982-90. Recipient Herici medal Imperial Coll., 1971, Aurel Vlaicu award Romanian Acad., 1984. Mem. AIAA (sr.), Royal Aero. Soc. (Finsbury medal 1972), City and Guild Inst. (assoc.), Old Centralian Assn. (life), Instn. Engrs. (U.K.). Achievements include research in semi-analytic method for rotating Timoshenko beams, vibrations and aeroelasticity of rotor blades, theoretical and exptl. studies of aero-servo-elasticity and group vibrations, studies flight flutter testing, elasto-plastic stress determination using strain gauges, failure of random composites, active control with smart materials, NDE of aging aircraft. Home: 1200 Hunt Club Rd # 6800D Blacksburg VA 24060 Office: Va Poly State U Tech Inst ESM Dept Blacksburg VA 24061-0219

GIVENS, STEPHEN BRUCE, medical physicist; b. Owensboro, Ky., Mar. 13, 1952; s. David Gayle and Doris Ann (Lowe) G.; m. Janet Sue Glikes, Jan. 7, 1984 (div. May 1989); children: Erica Lane; m. Donna Rose Edenfield, July 29, 1991. BS, U. Ky., 1974, MS, 1976. Diplomate Am. Bd. Radiology. Instr. in radiology La. State U. in Shreveport, 1976-77; asst. in physics M.D. Anderson Hosp., Houston, 1977-80; radiol. physicist West Coast Cancer Found., San Francisco, 1980-83; cons. physicist Oakland, 1983-85; instr. U. San Francisco, 1985; med. physicist Meth. Hosp. of Ind., Indpls., 1985-88, Mayo Clinic Jacksonville, 1988—; assoc. mem. Ctrs. for Radiol. Physics, Radiation Therapy Task Group, 1980-83; mem. ultrasound task group, 1981-83, community hosp. oncology project, 1982-83. Author FORTRAN computer programs used in radiation oncology for patient dose calculation; artist: exhibits include Gallery 88, Ponte Vedra Beach, Fla., 1991, 92, Arts Mania, N.E. Fla. Competition, Jacksonville, 1991, Mayor's Invitational Show, City Hall, Jacksonville, 1992; contbr. articles to profl. jours. Mem. Jacksonville Coalition for the Visual Arts, Jacksonville, Fla. 1991—. Mem. Am. Assn. of Physicists in Medicine (mem. Fla. chpt.), Jacksonville Art Mus. Achievements include rsch. on advanced techniques in Brachytherapy and stereotactic I-125 brain implant. Office: Mayo Clinic Jacksonville 4500 San Pablo Rd Jacksonville FL 32224

GJEDDE, ALBERT HELLMUT, neuroscientist, neurology educator; b. Copenhagen, Denmark, Jan. 10, 1946; arrived in Can., 1986; s. Albert Stoll and Elisabeth Gjedde; m. Susanne Borum Andreasen, May 5, 1972 (div. 1981); 1 child, Nanna Louise; m. Suzan Eva Dyve, June 4, 1983; children: Laura Sophie, Nikolaj Kristian. BA, Rungsted Acad., Denmark, 1964; MD, Copenhagen U., 1973, PhD, 1983. Postdoctoral fellow N.Y. Hosp.-Cornell Med. Ctr., N.Y.C., 1973-76; rsch. assoc. Copenhagen U., 1976-79; asst. prof. Panum Inst., Copenhagen U., 1979-81, assoc. prof., 1981-86; assoc. prof. McGill U., Montreal, Que., 1986-89, prof., 1989—; dir. McConnell Brain Imaging Ctr., Montreal, 1993—. Author: Life of P.L. Panum, 1971; contbr. articles to profl. jours. Served with Royal Life Guards (Army), 1965-66. Grantee MRC, Can., 1992. Mem. Am. Physiol. Soc., Soc. for Cerebral Blood Flow & Metabolism (founding mem., dir. 1987-91), Univ. Club Montreal. Achievements include research on kinetics of blood-brain glucose transport, DOPA decarboxylase activity of the living human brain. Office: McGill University, 3801 University St, Montreal, PQ Canada H3A 2B4

GJERDE, ANDREA JO, entomologist; b. Whittier, Calif., Aug. 25, 1955; d. Erwin Gerald and Mary (Steinberg) G. BS, Calif. State U., Long Beach, 1978; MS, U. Calif., Riverside, 1985. Lic. pest control advisor, Calif. Field tech. Limoniera Co., Santa Paula, Calif., 1983, Pest Mgmt. Assocs., Exeter, Calif., 1984; lab. tech. in entomology U. Calif., Riverside, 1985; entomologist, field rschr. Entomol. Svcs., Inc., Corona, Calif., 1985—. Mem. Entomol. Soc. Am., Calif. Agrl. Prodn. Cons. Assn., Assn. Applied Insect Ecologists, Calif. Women in Agr. Democrat. Home: 1656 E Magill Fresno CA 93710 Office: Entomol Svcs 510 1/2 Chase Dr Corona CA 91720

GJOVIG, BRUCE QUENTIN, manufacturing consultant; b. Crosby, N.D., Mar. 24, 1951; s. Ronald David and Agnes (Smedberg) G.; children: Mike Mohn, Todd Chaffee. BA, BS, U. N.D., 1974. Rsch. chemist Man-in-the-Sea Project, Grand Forks, N.D., 1975-76; campaign advisor Elkin for Gov. Com., Bismarck, N.D., 1976; exec. officer Grand Forks Bd. Realtors, 1977-81; devel. officer U. N.D. Found., 1981-84; founder, dir. Ctr. for Innovation & Bus. Devel., Grand Forks, 1984—; dir. 1st Seed Capital Co., Grand Forks, Innovative Systems, Inc., Brooklyn Park, Minn., Microdose Internat., Grand Forks; founder, dir., vice chmn. Ask-Me Info. Sys., Inc., Fargo, 1985-88; founder, chmn. N.D. Entrepreneur Hall of Fame, 1985—. Editor: The Business Plan: Step-by-Step, 1988, The Marketing Plan: Step-by-Step, 1990; author, editor: Boxcar of Peaches: Nash Finch Co., 1990, Pardon Me, Your Manners are Showing!, 1992; contbr. articles to profl. jours. Founder, sponsor 67th Patent & Trademark Depository Libr., 1991—. Named Friend of Sml. Bus., Fargo C. of C., 1988; named U. N.D. Outstanding Greek Alumnus, 1990, Outstanding Svc. award, U. N.D. Alumni Assn., 1984, others. Mem. Assn. Univ. Tech. Mgrs., Assn. Univ. Related Rsch. Pks., Univ. Small Bus. Tech. Consortium (state dir. 1986-90), Alumni Inter-Fraternity Coun. (chmn. 1982-86, 90—, Outstanding Alumnus 1990), Rotary, Delta Tau Delta. Republican. Episcopalian. Avocations: reading, politics, art collector, fund raising, entrepreneur history collector. Home: Condo # 31 2501 26th Ave S Grand Forks ND 58201-6483 Office: Ctr Innovation & Bus PO Box 8372 100 Harrington Grand Forks ND 58202-8372

GLAD, DAIN STURGIS, aerospace engineer; b. Santa Monica, Calif., Sept. 17, 1932; s. Alma Emanuel and Maude La Verne (Morby) G.; BS in Engring., UCLA, 1954; MS in Elec. Engring., U. So. Calif., 1963. Registered profl. engr., Calif. m. Betty Alexandra Shainoff, Sept. 12, 1954 (dec. 1973); 1 child, Dana Elizabeth; m. Carolyn Elizabeth Giffen, June 8, 1979. Electronic engr. Clary Corp., San Gabriel, Calif., 1957-58; with Aerojet Electro Systems

Co., Azusa, Calif., 1958-72; with missile systems div. Rockwell Internat., Anaheim, Calif., 1973-75; with Aerojet Electrosystems, Azusa, 1975-84; with support systems div. Hughes Aircraft Co., 1984-90; with Electro-Optical Ctr. Rockwell Internat. Corp., 1990—. Contbr. articles to profl. jours. Ensign, U.S. Navy, 1954-56; lt. j.g. Res., 1956-57. Mem. IEEE, Calif. Soc. Profl. Engrs., Soc. Info. Display. Home: 1701 Marengo Ave South Pasadena CA 91030-4818 Office: Rockwell Internat Corp Electro-Optical Ctr 3370 E Miraloma Ave Anaheim CA 92803-3105

GLADER, MATS LENNART, economics educator, researcher; b. Harnosand, Sweden, Nov. 20, 1945; s. Gustaf L. and Inga-Marta (Svanholm) G.; m. Berit Margareta Kallin, June 23, 1967; children: Christine, Eva-Lotta, Anna-Lena. MBA, Umea U., 1969, PhD, 1971. Asst. prof. Umea U., Sweden, 1969-74, assoc. prof., 1974-80; assoc. prof. Sch. of Forestry/Swedish Agr. U., Umea, 1980-81, Stockholm Sch. Econs., 1985—; vis. assoc. prof. Swedish Sch. Econs., Helsinki, Finland; vis. prof. U. Tampere, Finland; vis. sr. lectr. Norwegian Sch. Econs., Bergen, Norway; mng. dir. Ind. Mgmt. Tng. AB, Stockholm, 1981-83; bd. dirs. various cos., Sweden; lectr. in field. Software developer; columnist: Datavarden, 1984-92; contbr. numerous articles to profl. jours., books. Home: Rodhakevagen 48A, 90651 Umea Sweden Office: Stockholm Sch of Econs, PO Box 6501, Sveavagen 65, 11383 Stockholm Sweden

GLADFELTER, WILBERT EUGENE, physiology educator; b. York, Pa., Apr. 29, 1928; s. Paul John and Marea Bernadette (Miller) G.; m. Ruth Isabelle Ballantyne, Jan. 26, 1952; children: James W., Charles D., Mary A. AB magna cum laude, Gettysburg (Pa.) Coll., 1952; PhD, U. Pa., 1960. NSF fellow U. Pa., Phila., 1956-58, NIH fellow, 1958-59, asst. instr., 1954-56; instr. physiology W.Va. U., Morgantown, 1959-61, asst. prof., 1961-69, assoc. prof., 1969—. Contbr. articles to profl. jours. Treas., Monongalia County chpt. W. Va. Heart Assn., 1976—. With USN, 1946-48. NSF fellow, 1956-58. Mem. Am. Physiol. Soc., Soc. Neurosci., Am. Soc. Zoologists, Sigma Xi, Phi Beta Kappa, Beta Beta Beta. Lutheran. Home: RR 7 Box 528 Morgantown WV 26505-9118 Office: WVa U Health Sci Ctr Dept Physiology Morgantown WV 26506

GLADKI, HANNA ZOFIA, civil engineer, hydraulic mixer specialist; b. Krakow, Poland, Dec. 30, 1933; came to U.S., 1984; d. Stanislaw Wojtanowski and Maria (Ekiert) Wojtanowska; m. Jozef Gladki, July 2, 1955 (dec. 1982); 1 child, Ania. ScD, Tech. U., Warsaw, Poland, 1966; postgrad. degree, Agrl. U., Wroclaw, Poland, 1977. Asst. prof. Agrl. Acad. Krakow, 1966-70, assoc. prof., 1970-81, chair dept., 1973-83, dean of faculty, 1977-81, prof., 1981-85; hydraulic mixer specialist ITT Flygt Corp., Norwalk, Conn., 1985—; presenter at profl. confs. Contbr. articles to profl. publs. Mem. ASCE, AICE, Internat. Assn. Hydraulic Rsch. Roman Catholic. Achievements include expertise in hydraulics, flow velocity, pressure, mixing slurry and viscous fluid in tanks; designing mixers for biological and sludge treatment; development of method for sizing mixers in oxidation ditches with clarifier. Home: 79 Melville St Stratford CT 06497 Office: ITT Flygt Corp 35 Nutmeg Drive Trumbull CT 06611

GLADSTEIN, MARTIN KEITH, electrical engineer; b. New Britain, Conn., Nov. 30, 1957; s. Philip and Jean B. G.; m. Shelly R. Field, Nov. 2, 1980; children: Katie Jill, Jaclyn Lisa. BSEE, U. Conn., 1980; MSEE, U. Lowell, 1988. Analog coax systems engr. AT&T, North Andover, Mass., 1980-82, lightwave engr., 1982—. AT&T sci. and engring. fellow, 1986. Mem. IEEE, AT&T Engring. Excellence Soc. Office: AT&T Dept 47520 1600 Osgood St North Andover MA 01845

GLAENZER, RICHARD HOWARD, electrical engineer; b. St. Louis, Nov. 29, 1933; s. Frederick Robert and Evelyn Amy (Harker) G.; Jeanette Audrey Cornelius, June 29, 1963; children: Bryan Kent, James Martin, Justin Charles, Tanya Marie. BS in Engring. Physics, Washington U., St. Louis, 1960, MSEE, 1964; PhD, Carnegie Mellon U., 1968. Rsch. scientist McDonnell Douglas Rsch. Labs., St. Louis, 1967-71; engr. McDonnell Douglas Astronautics, St. Louis, 1971-76; sect. mgr. McDonnell Douglas Electronics Co., St. Charles, Mo., 1976-83; sr. rsch. scientist Monsanto, St. Louis, 1984-89; fellow MEMC Electronic Materials, St. Peters, Mo., 1989—; vis. prof. Inst. Nat. Astrofisica Optica y Electronic., Puebla, Mex., 1983-84. Chmn. IEEE Combined Group, St. Louis, 1976-78. With U.S. Army, 1954-56. Lutheran. Home: 7112 Westmoreland Dr University City MO 63130

GLANZER, MURRAY, psychology educator; b. N.Y.C., Nov. 18, 1922; s. Max and Norma (Reichenthal) G.; m. Mona Naomi Sorcher, Sept. 20, 1953; children: Michael, Marla, James. BA, City Coll., N.Y.C., 1943; MA, U. Mich., 1948, PhD, 1952. Instr. Bklyn. Coll., 1949-53; project dir. to program dir. Am. Inst. Rsch., 1954-58; lectr. U. Pitts., 1955-58; rsch. assoc. Walter Reed Army Inst. Rsch. U. Md. Sch. Medicine, 1958-63; prof. N.Y.U., 1963—. Numerous publications; contbr. articles to profl. jours. Fellow Ford Found. U. Chgo., 1953-54, Guggenheim, Hebrew U., Jerusalem, 1969-70. Mem. Am. Psychol. Assn., Am. Psychol. Soc., Ea. Psychol. Assn., Psychonomic Soc. Home: 17 Weston Pl Lawrence NY 11559 Office: NYU 6 Washington Pl New York NY 10003

GLASER, DONALD A(RTHUR), physicist; b. Cleveland, Ohio, Sept. 21, 1926; s. William Joseph Glaser. B.S., Case Inst. Tech., 1946, Sc.D., 1959; Ph.D., Cal. Inst. Tech., 1949. Prof. physics U. Mich., 1949-59; prof. physics U. Calif., Berkeley, 1959—; prof. physics, molecular and cell biology, and neurobiology U. Calif., 1964—. Recipient Henry Russel award U. Mich., 1955, Charles V. Boys prize Phys. Soc., London, 1958, Nobel prize in physics, 1960, Gold Medal award Case Inst. Tech., 1967, Golden Plate award Am. Acad. of Achievement, 1989; NSF fellow, 1961, Guggenheim fellow, 1961-62, fellow Smith-Kettlewell Inst. for Vision Rsch, 1983-84. Fellow AAAS, Fedn. Am. Scientists, The Exploratorium (bd. dirs.), Royal Soc. Sci., Royal Swedish Acad. Sci., Assn. Rsch. Vision and Ophthalmology, Neurosci. Inst., Am. Physics Soc. (prize 1959); mem. Nat. Acad. Scis., Internat. Acad. Sci., Sigma Xi, Tau Kappa Alpha, Theta Tau. Office: U Calif Dept Molecular & Cell Biology Neurobiology Divsn Stanley Hall Berkeley CA 94720

GLASGOW, J. C. (JOHN CARL GLASGOW), systems engineering executive; b. Des Moines, Jan. 26, 1947; s. John Norman and Alice Bertha (Bruning) G.; m. Janet Gay Slothower, June 26, 1971; children: Jonathan Lucas, Sarah Elizabeth, Ruth Renate, Lorna Jane, Jeanette Carrie, Eric Oliver, Bernadette Fiona, Karl Dietrich. BA, San Jose State U., 1976; MS in Constrn. Mgmt., Boston U., 1980. Master planner Lockheed Missle & Space Co., Sunnyvale, Calif., 1980-81; program rep. Martin Marietta Orlando (Fla.) Aerospace, 1981-87; pres., chief systems engr. Hi-Tech Rsch. Co., San Jose, Calif., 1988—. Elder, Latter Day Saints Ch. With U.S. Army, 1966-68, 76-80. Achievements include design of semiautomatic screw gun for drywall application of expert system shell Quantum-DECISION, of knowledge base manage system and intelligent spreadsheet and database Quantum-FOCUS, of general intelligence software and hardware. Office: Hi-Tech Rsch Co 888 N First St Ste 311 San Jose CA 95112

GLASHOW, SHELDON LEE, physicist, educator; b. N.Y.C., Dec. 5, 1932; s. Lewis and Bella (Rubin) G.; m. Joan Glashow; children: Jason David, Jordan, Brian Lewis, Rebecca Lee. AB, Cornell U., 1954; AM, Harvard U., 1955, PhD, 1958; DSc (hon.), Yeshiva U., 1978, U. Marseille, 1982, Adelphi U., 1989, Bar Ilan U., 1989, Gustave Adolphus Coll., 1989. NSF fellow U. Copenhagen, Denmark, 1958-60; rsch. fellow Calif. Inst. Tech., 1960-61; asst. prof. Stanford U., 1961-62; asst. prof., assoc. prof. U. Calif. at Berkeley, 1962-66; mem. faculty Harvard U., 1966—, prof. physics, 1967—, Higgins prof. physics, 1979—, Mellon prof. scis., 1988—; disting. vis. scientist Boston U., 1984—; cons. Brookhaven Nat. Lab., 1966-73, 75—; mem. sci. policy com. CERN, 1979-84; vis. prof. U. Marseille, 1971, MIT, 1974-80, Boston U., 1983; affiliated sr. scientist U. Houston, 1983—; univ. scholar Tex. A&M U., 1983-86. Author: (with Ben Bova) Interactions, 1988; contbr. articles to profl. jours. and popular mags.; editor-in-chief for physics Quantum mag., 1989—. Pres. Andrei Sakharov Inst., 1980-85, Nat. Com. for Excellence in Edn., 1985-88. Recipient J.R. Oppenheimer Meml. prize, 1977, George Ledlie prize, 1978, Nobel prize in physics, 1979, Castiglione di Sicilia prize, 1983; NSF fellow, 1955-60, Sloan fellow, 1962-66, CERN vis. fellow, 1968. Fellow Am. Phys. Soc., AAAS; mem. Am. Acad. Arts and Scis., Nat. Acad. Scis., Sigma Xi.

GLASMAN, MICHAEL MORRIS, metallurgical engineer; b. Des Moines, Iowa, May 18, 1956; s. Harry and Carol (Ryngermacher) G.; m. Harriet Mottsman, Oct. 28, 1984; 1 child, Jonathan. BS in Metallurgical Engring. and Psychology, Iowa State U., 1981. Quality engr. Sandvik Steel Co., Clarks Summit, Pa., 1985-87; reactor inspector U.S. Nuclear Regulatory Commn., Atlanta, 1987-92, resident inspector, 1992—. Office: US NRC Region II 101 Marietta St Ste 2900 Atlanta GA 30323

GLASS, ARNOLD LEWIS, psychology educator; b. Newark, Mar. 9, 1951; s. Julius and Beatrice (Derchin) G.; m. Lynne Barbara Cohen, Sept. 5, 1973; 1 child, Brian. BA, SUNY, Buffalo, 1971; PhD, Stanford U., 1975. Asst. prof. Rutgers U., New Brunswick, N.J., 1975-78, assoc. prof., 1980—. James McKeen Cattell Fund fellow, 1981-82. Achievements include algorithm for parsing context-free languages in linear time patent (pending). Home: 944 Kensington Ave Plainfield NJ 07060 Office: Rutgers U Psychology Dept New Brunswick NJ 08903

GLASS, DAVID EUGENE, mechanical engineer, educator; b. Roanoke Rapids, N.C., Sept. 27, 1956; s. Ernest Wilson and Charlotte Marjorie (Magruder) G.; m. Rebecca Latrell Evans, Nov. 22, 1986. BS, Wake Forest U., 1978; MS in Pub. Health, U. N.C., Chapel Hill, 1980; MME, N.C. State U., Raleigh, 1982; PhD, N.C. State U., 1986. Vis. instr. N.C. State U., 1986-87; aerospace technologist NASA-Kennedy Space Ctr., Fla., 1987-88; rsch. engr. Analytical Svcs. and Materials-NASA, Hampton, Va., 1988—; ajd. asst. prof. Old Dominion U., Norfolk, Va., 1990—. Mem. ASME, AIAA (sr.). Bapt. Achievements include design, fabrication and testing of carbon-carbon/heat-pipe-cooled wing leading edge for Nat. Aero-Space Plane. Office: NASA-Langley Mail Stop 396 Hampton VA 23681

GLASS, THOMAS GRAHAM, JR., retired general surgeon, educator; b. Marlin, Tex., Dec. 28, 1926; s. Thomas Graham and Dorothy (Marshall) G.; m. Helen Kathryn Burk, June 9, 1951; children: Thomas Graham III, Phillip K., Dean B., Helen Kathryn Glass Gough. Student, U. Tex., Austin, 1944, 46-49; MD, U. Tex., Galveston, 1953. Diplomate Am. Bd. Surgery. Intern VA Hosp., Oklahoma City, 1953-54; resident in gen. pathology Postgrad. Sch. Medicine, U. Tex., San Antonio, 1954-55, resident in gen. surgery, 1955-58; chief resident Postgrad. Sch. Medicine, U. Tex., San Antonio, $D, $D, 1957-58; Am. Cancer Soc. fellow in surgery M.D. Anderson Hosp., Houston, 1958-59; pvt. practice, San Antonio, 1959-91, ret., 1991; prof. surgery U. Tex. Health Sci. Ctr., San Antonio; pub. Glass Pub. Co., San Antonio; chief of staff N.E. Bapt. Hosp., San Antonio, 1980-81; cons. Aerospace Sch. Medicine, Brooks AFB, Tex., 1965-81, Tex. Medicine, 4th Continental Army U.S., 1957-63; examiner, cons. Armed Forces Exam. Sta., 1958-73; surg. and orthopedic cons. VA Regional Office, 1959-63; chmn. bd. dirs. Venereal Disease Action Coun. Tex.; mem. Gov.'s Steering Com. for Immunization, State of Tex.; med.-surg. expert witness, cons. Author: Latin and Greek Stems, 1957, Snakebite First Aid, 1974, rev., 1993, The Management of Poisonous Snakebite, 1976, rev., 1993, The Art of Medicine, 1990, Mah Jongg 2000, 1992, A Russian Soldier in the German Army 1941-45, 1991, I, Witness, 1991; contbr. articles to med. jours. Precinct chmn. Bexar County Rep. Com., Castle Hills Rep. Com.; mem. Bexar County Rep. Exec. Com. With USN, 1944-46, USNR, 1950-54. Recipient awards for sci. exhibits on western diamondback rattlesnake bite treatment, 1971, 72, appreciation award Vet. Med. Assn. Bexar County, 1973. Fellow ACS; mem. AMA, AAAS, Tex. Med. Assn., Bexar County Med. Soc., San Antonio Surg. Soc. (pres. 1975-76), Southwestern Surg. Congress, William S. McCombs Soc. (founding), Am. Phys. Soc., Am. Inst. Physics, Am. Radio Relay League, Acad. Model Aeros., Book Pubs. Tex., Alamo Radio Control Soc., San Antonio Yacht Club (past pres.), San Antonio Radio Club, Kiwanis, Masons, Shriners, Sigma Pi Sigma, Alpha Epsilon Delta, Phi Kappa Sigma, Osteon. Episcopalian. Avocations: writing, travel, ham radio, minicomputer construction, radio controlled model planes.

GLASSER, JOSHUA DAVID, computer scientist; b. Pitts., Mar. 19, 1961; s. Mervin Lawrence and Judith Jay (Sensibar) G.; m. Lisa Marie Bianchi, Sept. 25, 1987; children: Eliezra Ariel, Sacha Rose. BS, Clarkson Coll. Tech., 1983; MS, Clarkson U., 1985, postgrad., 1992—. Surveyor Constrn. Aggregates Corp., Miami Beach, Fla., 1978-79; engr. Standard Sand, Ferrysburg, Mich., 1980; rsch. scientist Reasoning Systems, Inc., Palo Alto, Calif., 1985-86, Honeywell Systems and Rsch. Ctr., Mpls., 1986-89; sr. software engr. Artificial Intelligence Techs. Inc., Hawthorne, N.Y., 1990-91; teaching asst. Clarkson U., Potsdam, N.Y., 1992—; cons. Bus. and Acctg. Software Systems, Vail, Colo., 1985. Democrat. Jewish. Achievements include research in cryptography and automated reasoning. Home: 7 1/2 Pleasant St Potsdam NY 13676 Office: Clarkson U Dept Math Potsdam NY 13676

GLASSER, WOLFGANG GERHARD, forest products and chemical engineering educator; b. Zwickau, Ger., Oct. 9, 1941; came to U.S., 1969; s. Joachim and Charlotte (Syjatz) G.; m. Heidemarie Reinecke, Mar. 18, 1969; children—Christine M., Stephan A. Degree in Wood Tech., U. Hamburg (W.Ger.), 1966, Ph.D. in Wood Chemistry, 1969. Research assoc. U. Wash., Seattle, 1969-70, research asst. prof., 1970-71; asst. prof. Va. Poly. Inst. and State U., Blacksburg, 1972-75, assoc. prof., 1975-80, prof. wood chemistry, 1980—; dir. Pulp and Paper Research Inst., Sao Paulo, Brazil, 1976, Biobased Materials Ctr., 1988-91; chmn. panel Nat. Acad. Scis., 1974-76; mem. editorial adv. group Holzforschung, Braunschweig, Ger., 1984—, Cellulose Chemistry and Tech. (Romania), 1986—, Jour. of Applied Polymer Sci., 1990— . Patentee; contbr. articles to profl. jours., book editor. Co-recipient George Olmsted award Am. Paper Inst., 1974; recipient Sci. Achievement award Internat. Union Forest Rsch. Orgns., 1986. Mem. Am. Chem. Soc. (alt. councilor 1983-85, pub. chmn. 1985-88, chmn. elect 1989, chmn. 1990, councilor 1991—, program chmn. 1993—), Soc. Wood Sci. Tech., Soc. Plastics Engrs., Sigma Xi, TAPPI. Lutheran. Office: Va Tech Thomas M Brooks Forest Products Blacksburg VA 24061

GLASSMAN, ARMAND BARRY, physician, pathologist, scientist, educator, administrator; b. Paterson, N.J., Sept. 9, 1938; s. Paul and Rosa (Ackerman) G.; m. Alberta C. Macri, Aug. 30, 1958; children: Armand P., Steven B., Brian A. BA, Rutgers U., N.J., 1960; MD magna cum laude, Georgetown U., Washington, 1964. Diplomate Am. Bd. Pathology, Am. Bd. Nuclear Medicine. Intern Georgetown U. Hosp., Washington, 1964-65; resident Yale-New Haven Hosp., West Haven VA Hosp., 1965-69; asst. prof. pathology U. Fla. Coll. Medicine; chief radioimmunoassary lab. Gainesville VA Hosp.; practice lab. and nuclear medicine, 1969-71; dir. clin. labs., assoc. prof., prof. pathology, cellular and molecular biology Med. Coll. Ga., Augusta, 1971-76; cons. physician in pathology VA Hosp., Augusta, 1973-76; cons. physician in nuclear medicine Univ. Hosp., Augusta, 1973-76; med. dir. clin. labs. Med. U. S.C. Hosp., Charleston, 1976-87; attending physician in lab. and nuclear medicine Med. U. S.C., Charleston, 1976-87, assoc. med. dir. Med. U. Hosp. and Clinics, 1982-86; med. dir. clin. labs. Charleston Meml. Hosp., S.C., 1976-87; cons. VA Hosp., Charleston, 1976-87; prof., chmn. dept. lab. medicine Med. U. S.C., 1976-87, med. dir. MT and MLT programs, 1976-87, clin. prof. pathology, lab. medicine, and radiology, 1987—, acting chmn. dept. immunology and microbiology, 1985-87, assoc. dean Coll. Medicine, 1979-85, asst. and assoc. dean Coll. Allied Health Sci., 1984-87, chmn. hosp. exec. com., 1985-86, acting med. dir. Univ. Hosp. and Clinics, 1985-86; clin. prof. pathology Med. U. S.C., Charleston, 1987—; sr. v.p. med. affairs, prof. lab. medicine and nuclear medicine Montefiore Med. Ctr. and Albert Einstein Coll. Medicine, Bronx, N.Y., 1987-89; v.p., lab. dir. Nat. Reference Lab., Nashville, 1989-92; cons., 1992—; clin. prof. Dept. Pathology Vanderbilt U., Nashville, 1990-92, prof. pathology, 1992—; dir. Vanderbilt Pathology Lab. Svcs., 1992—; dir. clinical labs. Vanderbilt U. Med. Ctr., 1993—; mem. adv. coun. Trident Tech. Coll., 1976-87; bd. dirs. Fetter Family Health Ctr.; founding dir., bd. dirs. Sealite, Inc.; mem. med. adv. com. Nashville Red Cross Blood Ctr., 1991—, acting med. dir., 1991-92; mem. Bd. Sci. Advisors NHL/NRL, 1992—. Editor, co-editor 4 books; contbr. over 90 refereed articles to profl. jours., 30 chpts. to books. Trustee Coll. Prep. Sch., 1979-84, chmn. bd., 1983-84; trustee, bd. dirs., v.p. Mason Prep. Sch., 1984-87; bd. dirs. United Way, 1983-87, Am. Cancer Soc., 1984-87. Served with USMCR, 1956-64. Johnson and Avalon Found. scholar Georgetown U., 1961-64; State scholar Rutgers U., 1956-60. Fellow Coll. Am. Pathologists (numerous coms. including edn. com. 1983-89), ACP, Assn. Clin. Scientists (Diploma of Honor 1987, pres. 1990-91, exec. com. 1990—, Clin. Scientist of Yr. 1993), Am. Soc. Clin. Pathology (coun. immunohematology and blood banking 1983-89, Commr.'s award for CCE 1989), Am. Bd. Pathology (blood bank test com. 1984-88), Am. Coll. Nuclear Medicine, N.Y. Acad. Medicine; mem. Internat. Acad. Pathology, Am. Assn. Pathologists, Soc. Nuclear Medicine (chmn. edn. com. 1973-77, acad. council 1979—), AMA (Physician's Recognition award, instnl. rep. to sect. on med. schs.), So. Med. Assn., Am. Geriatric Soc. (founding fellow So. div.), Am. Soc. Microbiology, Am. Assn. Blood Banks (chmn. cryobiology com. 1974-83, edn. com. 1978-85, sci. program com. 1981-84, autologous transfusion com. 1979-83, bd. dirs. 1984-87, transfusion transmitted diseases com. 1992—), Assn. Schs. Allied Health Professions (bd. editors jour. 1979-83), Soc. Cryobiology (treas., bd. dirs. 1978-80), AAAS, N.Y. Acad. Scis., Acad. Clin. Lab. Physicians and Scientists (exec. council 1978-85, pres. 1982-83), S.E. Area Blood Bankers (pres. 1979-81, exec. council 1980-85), Tenn. Assn. Blood Banks (treas. 1993—), Am. Coll. Physician Execs., Sigma Xi, Alpha Eta, Alpha Omega Alpha. Avocations: jogging, tennis, community svc.

GLASSMAN, GEORGE MORTON, dermatologist; b. N.Y.C., Sept. 7, 1935; s. Oscar and Jeanette (Bitterbaum) G.; m. Carol Beth Frankford, July 10, 1960; children: Keith F., Laurie C. BA cum laude, Brown U., 1957; MD, NYU, 1962. Diplomate Nat. Bd. Med. Examiners. Rotating intern Greenwich (Conn.) Hosp., 1962-63; resident in dermatology NYU Med. Ctr., N.Y.C., 1963-66; chief dermatology U.S. Navy, St. Albans, N.Y., 1966-68; pvt. practice White Plains, N.Y., 1968—; clinical asst. prof. Albert Einstein Coll. Medicine, Bronx, N.Y., 1970-75, N.Y. Med. Coll., Valhalla, N.Y., 1975-87; assoc. attending dermatologist Westchester County Med. Ctr., Valhalla, 1974-87, St. Agnes Hosp., White Plains, 1978—; attending dermatologist White Plains Hosp., 1977—. Contbr. articles to profl. jours. Lt. comdr. USN, 1966-68. Mem. Am. Acad. Dermatology (Continuing Med. Edn. award, 1980—), Internat. Soc. Tropical Dermatology, Dermatology Found., Westchester County Med. Soc., Westchester Acad. Medicine (pres. dermatology sect., 1990-91), N.Y. State Soc. Dermatology, AMA (Physician's Recognition award, 1980—), World Med. Assn., N.Y. Acad. Scis., Am. Soc. Dermatologic Surgery, Soc. for Pediatric Dermatology, Dermatologic Radiotherapy Soc. Home: 268 Stuart Dr New Rochelle NY 10804-1423 Office: George M Glassman MD PC 1 Old Mamaroneck Rd White Plains NY 10605-1703

GLASSMAN, IRVIN, mechanical and aeronautical engineering educator, consultant; b. Balt., Sept. 19, 1923; s. Abraham and Bessie (Snyder) G.; m. Beverly Wolfe, June 17, 1951; children: Shari Powell, Diane Geinger, Barbara Ann. B.E., Johns Hopkins U., 1943, D.Eng., 1950. Research asst. Manhattan Project, Columbia U., N.Y.C., 1943-46; mem. faculty Princeton U., N.J., 1950—, prof. mech. and aero. engring., 1964—, Robert H. Goddard prof. mech. and aero. engring., 1988—, dir. Ctr. for Energy and Environ. Studies, 1972-79; cons. to industry; vis. prof. U. Naples, Italy, 1966-67, 78-79, Stanford U., 1975; mem. adv. coms. Sibley Sch., Cornell U., Cifer Colo. Sch. Mines, United Tech. Research Ctr., Chrysler Corp., CNR Istituto Motori, Italy, NSF, Ctr. for Fire Research, Nat. Inst. Standards and Tech. Author: (with R.F. Sawyer) Performance of Chemical Propellants, 1971, Combustion, 1977, 2d edit., 1987; editor Combustion Sci. & Tech. Jour., also 3 books; contbr. articles to tech. jours. Served with U.S. Army, 1944-46. NSF fellow, 1966-67. Fellow AIAA; mem. Combustion Inst. (Sir Alfred Edgerton Gold medal 1982), Am. Soc. Engring. Edn. (Roe award 1984), Am. Chem. Soc., AAUP, Tau Beta Pi. Patentee 2 rocket propellants. Home: PO Box 14 Princeton NJ 08542-0014 Office: Princeton U Princeton NJ 08544

GLASSMAN, LAWRENCE S., plastic surgeon; b. June 20, 1953. BA, Johns Hopkins U., 1975, MD, 1978. Diplomate Am. Bd. Surgery, Am. Bd. Plastic Surgery. Surgeon Columbia Presbyn. Med. Ctr., N.Y.C., 1978-83; plastic surgeon Montefiore Med. Ctr. and Albert Einstein Coll. of Medicine, Bronx, N.Y., 1983-85; asst. clin. prof. plastic surgery Albert Einstein Coll. Medicine, 1985—; plastic surgeon, dir. Inst. Aesthetic and Reconstructive Surgery, Pomona, N.Y., 1985—; plastic surgeon Good Samaritan Hosp., Suffern, N.Y., 1985—, Nyack (N.Y.) Hosp., 1985—, St. Anthony Community Hosp., Warwick, N.Y., 1985—, Chilton Meml. Hosp., Pompton Plains, N.J., 1985-89. Contbr. articles to profl. jours. Fellow ACS; mem. AMA, Am. Soc. Plastic and Reconstructive Surgery, Am. Assn. Surgery of the Hand, Med. Soc. Rockland County, Phi Beta Kappa, Alpha Omega Alpha. Office: Northside Plz Rt 45 Pomona NY 10970

GLASSMEYER, JAMES MILTON, aerospace and electronics engineer; b. Cin., Mar. 31, 1928; s. Howard Jerome and Ethel Marie (Nieman) G.; m. Anita Mary Tschida, Apr. 21, 1979. Student, U. Cin., 1947-49; BSEE with spl. honors, U. Colo., Boulder, 1958; MS in Aeronautics and Astronautics, MIT, 1960. Commd. 2d lt. USAF, 1950, advanced through grades to lt. col., 1971; astron. engr. Air Force Space Systems Div. Hdqrs. USAF, L.A., 1960-64; astronautical engr. and astronautics intelligence analyst Air Force Rocket Propulsion Lab USAF, Edwards AFB, Calif., 1967-73; ret. USAF, 1973; pvt. practice aerospace and electronics rsch. and analysis, 1973—. Contbr. articles to jours. in field. Recipient Air Force Inst. Tech. scholarship, U. Colo., 1956-58, MIT, 1958-60, Am. Rocket Soc. Grad. Student Nat. 1st Pl. award, MIT, 1960, USAF Master Missileman badge, Air Force Rocket Propulsion Lab., 1970. Mem. AIAA, IEEE, Air Force Assn., Planetary Soc., Ret. Officers Assn., Tau Beta Pi (1st grand prize Greater Interest in Govt. Nat. Essay Contest 1957), Eta Kappa Nu, Sigma Tau, Sigma Gamma Tau, Sigma Xi. Roman Catholic. Home: 61 Brookhill Woods Ln Tipp City OH 45371-1951 Office: PO Box 84 Tipp City OH 45371-0084

GLASSOCK, RICHARD JAMES, nephrologist; b. San Bernardino, Calif., Feb. 4, 1934; s. Richard James and Merne Rosalin (Wickham) G.; m. Jo-Anne Theresa Bourke, May 21, 1977; 1 child, Mark Andrew; children by previous marriage—Ellen Virginia, Scott Laurance, Sharon Elde. B.S. in Pharmacy, U. Ariz., 1956; postgrad., Duke U., 1956-57; M.D., UCLA, 1960. Diplomate Am. Bd. Internal Medicine (chmn. 1990-91); lic. pharmacist, Calif. Intern UCLA Center for Health Scis., 1960-61, resident in internal medicine, 1961-63; rsch. assoc. Scripps Clinic and Rsch. Found., 1965-66; jr. assoc. medicine, coord. USPHS transplantation and immunology Peter Bent Brigham Hosp., 1966-67; chief div. nephrology Harbor-UCLA Med. Center, 1967-80, chmn. dept. medicine, 1980-92; prof., chmn. dept. medicine U. Ky., Lexington, 1992—; instr. medicine Harvard U., 1966-67; asst. prof. medicine UCLA, 1967-71, assoc. prof., 1971-75, prof., 1975-92; cons. NIH, 1980—; chmn. sci. adv. bd. Nat. Kidney Found., 1977-78, pres., 1986-88; cons. career devel. program VA; NIH mem. Pathology Study Sect. A. Author: (with S. Massry) Textbook of Nephrology, 1992; contbr. numerous articles to profl. jours., chpts. in books. Served with USNR, 1951-61. USPHS fellow, 1963-65; NIH grantee, 1967-85; Am Heart Assn. grantee, 1967-69, recipient Hume Memorial award, Nat. Kidney Found., 1990. Fellow ACP; mem. AAAS, Assn. Am. Physicians, Am. Fedn. Clin. Research, Western Soc. Clin. Investigation, Western Assn. Physicians, Australasian Soc. Nephrology (hon.), Italian Soc. nephrology, German Soc. Nephrology, Chinese Soc. Nephrology, Am. Lupus Soc. (adv. bd.), Am. Soc. Artificial Internal Organs, Am. Assn. Pathology, Am. Soc. Nephrology (council, pres. 1984), Internat. Soc. Nephrology (council), Transplant Soc., N.Y. Acad. Scis., Kidney Found. So. Calif., Alpha Omega Alpha, Rho Chi.

GLAUSER, MARK NELSON, mechanical and aeronautical engineering educator; b. Buffalo, Mar. 10, 1957; s. Wilbert Gordon and Jennie Bernice (Wiggins) G.; m. Gina Jiho Lee, July 29, 1984; children: Stephen Jinho, Kristen Unok. BS in Mech. Engring., SUNY, Buffalo, 1982, PhD in Fluid Dynamics, 1987. Jr. mech. fellow Alcan Aluminum Co., Oswego, N.Y., 1981; rsch. asst. turbulence rsch. lab. SUNY, Buffalo, 1981-82, 83-84, teaching asst., 1982-83; rsch. fellow NASA, Buffalo, 1984-87; asst. prof. Clarkson U., Potsdam, N.Y., 1987-93, assoc. prof., 1993—; panelist NSF, Washington, 1991; sen. Clarkson U. Faculty Senate, 1988—, vice-chmn., 1992-93. Co-editor: (book) Eddy Structure Identification in Free Turbulent Shear Flow; contbr. chpts. to books: Studies in Turbulence, 1992, Turbulence and Coherent Structures, 1992, articles to Exptl. Thermal Fluid Sci., Physics of Fluids. Lectr. on aerodynamics to Tiger Cubs, Boy Scouts, local high schs. Grantee NSF, 1988-90, 90-92, 92-93, Pratt and Whitney/UTC, 1989, 1991-94, NASA Lewis 1991-94, NASA Ames/Dryden, 1991-94, NASA Langley, 1991-96. Mem. AIAA (faculty advisor, contbr. articles to jour.), Am. Phys. Soc., Am. Soc. Mech. Engrs. Democrat. Presbyterian. Office: Clarkson U Dept Mech & Aero Engring Potsdam NY 13699-5725

GLAVOPOULOS, CHRISTOS DIMITRIOS, mechanical engineering consultant; b. Thessaloniki, Greece, Mar. 10, 1958; s. Dimitrios C. and Miranta A. Glavopoulos; m. Xanthippi Krikelli, Sept. 17, 1988; children: Dimitrios, Labros. BSME, U. Sussex, Eng., 1982; MSc in Automotive Engring., U. Southampton, Eng., 1983. Mng. dir. C.D. Glavopoulos & Ptnrs. Cons. Engrs., Thessaloniki, 1985—; export sales engr. Pyramis Metall. S.A., Thessaloniki, 1988-89, export mgr., 1989-91. Mem. Inst. Bus. Adminstrn. Greece, Inst. Mech. Engrs. U.K., Inst. Fire Engrs. U.K., Nat. Fire Protection Assn. U.S.A., Nat. Assn. Fire Investigators U.S.A., Amorc Philipos Thessaloniki Lodge # 38, Tech. Chamber Greece, Engring. Coun. U.K. (chartered engr.), Feani Europe Engr., Soc. Mech. Engrs. (gen. sec. 1991-93), Assn. Mech. Engrs. No. Greece (gen. sec. 1991—), Hellenic Soc. (gen. sec. 1978, pres. 1983). Avocations: reading, fishing, swimming. Office: C D Glavopoulos Cons Engrs, PO Box 18269, GR540 08 Thessaloniki Greece

GLAWE, LLOYD NEIL, geology educator; b. Des Moines, Aug. 21, 1932; s. John Franklin and Irene Elizabeth (Royer) G.; m. Nancy Louise Gholson, Jan. 24, 1970; children: Lori Michelle, John Dana. BS, U. Ill., 1954; MS, La. State U., 1960, PhD, 1966. Rsch. asst. La. Geol. Survey, Baton Rouge, 1961; rsch. investigator Ala. Geol. Survey, Tuscaloosa, 1964; asst. prof. N.E. La. U., Monroe, 1964-69, assoc. prof., 1969-77, prof. geology, 1977—; pvt. cons., Monroe, 1974-84. Author: Pecten perplanus stock, 1969, Physical Geology Laboratory Manual, 1978. Fellow Pan Am. Petroleum Corp., Baton Rouge, 1962-64; faculty rsch. grantee N.E. La. U., Monroe, 1977. Mem. Paleontol. Rsch. Instn., Soc. Econ. Paleontologists and Mineralogists, La. Acad. Scis., Sigma Xi. Avocation: photography. Office: NE La U Geoscis Dept Monroe LA 71209-0550

GLAZER, GARY MARK, radiology educator; b. Feb. 13, 1950; children: Daniel I., David A. AB, U. Mich., 1972; MD, Case Western Res. U., 1976. Intern in internal medicine U. Calif., San Francisco, 1976-77, resident in diagnostic radiology, 1977-80, clin. instr., fellow in diagnostic radiology, 1980-81; asst. prof. radiology, dir. div. body computed tomography U. Mich., Ann Arbor, 1981-84, assoc. prof. radiology, 1984-87, dir. divs. magnet resonance imaging and body computed tomography, 1984-89, assoc. prof. cancer ctr., 1986-87, prof. radiology, prof. cancer ctr., 1987-89; prof., chmn. dept. radiology Stanford (Calif.) Sch. Medicine, 1989—. Cons., assoc. editor, reviewer Radiology; cons., reviewer Jour. Computer Assisted Tomogrphy; cons., chmn., reviewer, mem. editorial bd. Radiographics; contbr. articles to profl. publs. Fellow Am. Cancer Soc., 1980-82, Clarence Heller Found., 1980-81. Mem. Am. Roentgen Ray Soc., Radiology Soc. N.Am., Soc. Magnetic Resonance in Medicine, Fred Jenner Hodges Soc., Soc. Magnetic Resonance Imaging, Alpha Omega Alpha. Office: Stanford Sch Med Dept Radiology Rm S-078 MC 5105 Rt 1 Stanford CA 94305-5105

GLAZKO, ANTHONY J(OACHIM), pharmaceutical consultant; b. San Francisco, Aug. 15, 1914; s. Joachim Nicholas and Josephine (Rutkovska) G.; m. Margaret Jean Hemans, Dec. 17, 1959; children: John Nicholas, Susan Jean, Mary Janet. PhD, U. Calif., 1939. Asst. prof. Emory U., Atlanta, 1946-47; rschr. Parke-Davis & Co., Detroit and Ann Arbor, Mich., 1947-70, Warner-Lambert/Parke Davis, Ann Arbor, 1970-80; cons. new drug devel. Ann Arbor, 1980—. Contbr. chpts. to books and contbr. over 200 articles to profl. jours. Lt. USNR, 1941-46. Grantee NIH. Mem. N.Y. Acad. Scis., Am. Soc. Pharmacology and Exptl. Therapeutics, Am. Physiol. Soc., Am. Chem. Soc., Sigma Xi. Republican. Presbyterian. Achievements include pioneering efforts in pharmacokinetics; first analytical procedures for many drugs in blood; rsch. on cause of grey syndrome in infants receiving chloramphenicol; absorption differences in generic drugs, pediatric dosage monogram. Home: 1245 Fair Oaks Pky Ann Arbor MI 48104

GLEASON, THOMAS CLIFFORD, university official; b. Elizabeth, N.J., Aug. 30, 1953; s. Thomas Patrick and R. Lois (Nilsen) G.; 1 child, Brianna Dillon. BFA, Tex. Tech U., 1981; postgrad., U. So. Maine. Graphics specialist Furrs' Supermarkets, Inc., Lubbock, Tex., 1981-82; graphic artist Duncan Press., Lubbock, 1982-83, PenMor Printers, Inc., Lewiston, Maine, 1983-84; microcomputer ops. mgr. U. So. Maine, Portland/Lewiston, 1986—; mem. staff devel. com. U. So. Maine, Gorham, 1988-91, mem. campus ctr. com., Portland, 1988-91, mem. recycling com., 1990-91, mem. art com. LAC, Lewiston, 1991. Exhibited in over 20 group shows, 1976-92. Treas. MountainView Condominium Owners Assn., Lewiston, 1992-93. Recipient various awards juried art shows, 1978-82; Meml. art scholar Tex. Tech U., 1979, state tuition scholar, 1977-80. Mem. Maine Tchrs. Assn., Phi Kappa Phi, Phi Eta Sigma. Office: U So Maine Lewiston-Auburn Coll 51 Westminster St # 118-C Lewiston ME 04240-3534

GLEICH, GERALD JOSEPH, immunologist, medical scientist; b. Escanaba, Mich., May 14, 1931; s. Gordon Joseph and Agnes (Ederer) G.; m. Elizabeth Louise Hearn, Aug. 16, 1955 (div. 1976); children: Elizabeth Genevieve, Martin Christopher, Julia Katherine; m. Kristin Marie Leiferman, Sept. 25, 1976; children: Stephen Joseph, David Francis, Caroline Louise, William Gerald. B.A., U. Mich., 1953, M.D., 1956. Diplomate: Am. Bd. Internal Medicine. Intern Phila. Gen. Hosp., 1956-57; resident Jackson Meml. Hosp., Miami, Fla., 1959-61; instr. in medicine and microbiology U. Rochester, N.Y., 1961-65; cons. in medicine, prof. immunology and medicine Mayo Clinic-Med. Sch., Rochester, Minn., 1965—; chmn. dept. immunology Mayo Clinic, Rochester, Minn., 1982-90; mem. bd. sci. counselors Nat. Inst. Allergy and Infectious Disease, 1981-83; chmn. subcom. on standardization allergens WHO, Geneva, 1974-75; lectr. Am. Acad. Allergy, 1976, 82; mem., chmn. immunological scis. study sect. NIH, 1984-87; John M. Sheldon Meml. lectr., 1976, 82, 88; Steve Lang Meml. Lectureship, 1980, Stoll-Stunkard lectr. Am. Soc. Parasitologists, 1986, David Talmage Meml. lectureship, 1987, Disting. lectr. Med. Scis. Mayo Clinic, 1988. Contbr. articles on eosinophilic leukocyte to profl. jours. Served to capt. USAF, 1957-59. Recipient Landmark in Allergy award, 1990; grantee Nat. Inst. Allergy and Infectious Disease, 1970—. Fellow ACP, Am. Acad. Allergy and Immunology (hon. fellow award 1992), AAAS; mem. Am. Soc. Clin. Investigation, Am. Assn. Immunologists, Assn. Am. Physicians, Phi Beta Kappa, Phi Kappa Phi, Alpha Omega Alpha. Roman Catholic. Home: 799 SW 3d St Rochester MN 55902 Office: Mayo Clinic Mayo Found 200 1st St SW Rochester MN 55905-0001

GLEICHAUF, JOHN GEORGE, ophthalmologist; b. Rochester, N.Y., Mar. 21, 1933; s. George William and Cecelia Frieda (ehner) G.; m. Barbara Helen Warm, Aug. 20, 1960 (div. 1980); children: Kurt John, Karin Marie; m. Jacqueline Kay Thompson, June 20, 1985. AB, U. Rochester, 1955; MD, SUNY, Buffalo, 1962. Diplomate Nat. Bd. Med. Examiners, Am. Bd. Ophthalmology. Resident in ophthalmology Thomas Jefferson U., Phila., 1964-67; pvt. practice Santa Fe, N.Mex., 1967-79; ophthalmologist, chief of svc. Littleton Clinic, Denver, 1981-85; pres. Aiken (S.C.) Ophthalmology, 1985—; cons. State of N.Mex., Santa Fe, 1973-77, Moonshot Program, Holloman AFB, N.Mex., 1968-69, Med-Tech. Inc., Aiken, 1986-87. Active Aiken Choral Soc., Aiken Arts Coun. 1st lt. USMC, 1955-58. Fellow ACS, Internat. Coll. Surgery, Am. Acad. Ophthalmology; mem. AMA, N.Y. Acad. Scis., S.C. Med. Soc., Cataract and Refractive Surgery Soc. Republican. Roman Catholic. Avocations: skiing, travel, symphonic music, sailing. Home: 111 Northwood Dr Aiken SC 29803-5281

GLEICHMAN, JOHN ALAN, safety and loss control executive; b. Anthoney, Kans., Feb. 11, 1944; s. Charles William and Caroline Elizabeth (Emch) G.; m. Martha Jean Cannon, July 1, 1966; 1 son, John Alan Jr. BS in Bus. Mgmt., Kans. State Tchrs. Coll., 1966. Cert. hazard control mgr.; cert. safety profl.; cert. safety exec. Office mgr. to asst. supt. Barton-Malow Co., Detroit, 1967-72, safety coord., 1972-76, corp. mgr. safety and security, 1976-89, dir. corp. safety and loss control, 1989—; instr. U. Mich., Wayne State U., 1977-81; mem. constrn. safety standards commn. adv. com. for concrete constrn. and steel erection Bur. of Safety and Regulations, Mich. Dept. Labor, 1977—. Author: (with others) You, The National Safety Council, and Voluntary Standards, 1981, Construction Accident Analysis: The Inductive Learning Approach, 1991. Instr. multi media first aid ARC, 1976-89; past trustee Apostolic Christian Ch., Livonia, Mich. Recipient Safety Achievement awards Mich. Mut. Ins. Co., 1979-83; Cameron award Constrn. sect. Indsl. div. Nat. Safety Conf., 1982, 1987. Mem. Mich. Safety Coun. (pres. 1984-85), Am. Soc. Safety Engrs. (pres. Detroit chpt. 1982, nat. adminstr. constrn div. 1988-89, bd. dirs. 1988-90, Safety Prof. of Yr. 1984) Nat. Safety Coun. (chmn. tech. rev. constrn. sect. indsl. div. 1980-84, chmn. standards com. indsl. div. 1983-85, chmn. assn. com. indsl. div. 1985-86, dir. tech. support com. indsl. div. 1986-87, dir. sects. group indsl. div. 1987-89, chmn. elect indsl. div. 1989-90, chmn. 1990-91, bd. dirs. 1987-92, rep. Am. Nat. Standards Inst., Disting. Svcs. to Safety award, 1993), Am. Arbitration Assn. (panel arbitrators 1985). Office: PO Box 5200 Detroit MI 48235-0995

GLEIM, JEFFREY EUGENE, research chemist; b. Harrisburg, Pa., June 20, 1956; s. James Earl and Eldora E. (Fansler). BS in Chemistry, Lebanon Valley Coll., Annville, Pa., 1978; MS in Chemistry, Pa. State U., 1981, M in Environ. Pollution Control, 1993. Chem. technician Andrew S. McCreath & Son, Harrisburg, 1974; rsch. chemist P.H. Glatfelter & Co., Spring Grove, Pa., 1981—. Contbr. articles to Jour. Phys. Chemistry, Internat. Jour. Chem. Kinetics. Mem. Aircraft Owners and Pilots Assn., White Rose Flying Club, Phi Kappa Phi, Phi Lambda Upsilon. Republican. Office: PH Glatfelter co 228 S Main St Spring Grove PA 17362

GLEIXNER, RICHARD ANTHONY, materials engineer; b. Milw., July 28, 1955. BS in Materials Engring./Mfg. Engring., U Wis., Milw., 1980; MS in Metall. Engring., U Wis., Madison, 1983, PhD in Metall. Engring., 1985. Rsch. scientist Battelle Columbus (Ohio) Labs., 1985-87, prin. rsch. scientist, 1987-90; group supr. Babcock & Wilcox Rsch. Div., Alliance, Ohio, 1990—. Contbr. chpts. to books, chpt. to handbook, articles to profl. jours. Forging Found. scholar, 1978. Mem. Am. Soc. for Materials (scholar 1979), Tau Beta Pi, Alpha Sigma Mu. Achievements include a patent; research in advanced methods of process control for use in manufacturing operations. Office: Babcock & Wilcox Rsch Ctr 1562 Beeson St Alliance OH 44601

GLENCER, SUZANNE THOMSON, science educator; b. Monongahela, Pa., Feb. 7, 1942; d. John Cuddy and Sue Elizabeth (DeForrest) Thomson; m. May 9, 1970 (div.). BS in Zoology, Pa. State U., 1964; MEd in Biology, Calif. (Pa.) State U., 1968; postgrad., U. Pitts., 1970-84. Biology, health instr. Allegheny Community Coll., Pitts., 1969-85; instr. Pa. State U., New Kensington, 1978-84; sci. tchr. Northgate Sch. Dist., Bellevue, Pa., 1967—, also drug/alcohol coord., 1978—; cons. area sch. dists., Pitts., 1990—; speaker numerous local orgns., Pitts., 1985—. Author: Adventures of Atom, 1985, Concepts in Kindness, 1987; contbr. articles to local newspaper, 1980—; writer and presenter numerous grants. Bd. dirs. Animal Friends, Pitts., 1974-84, Am. Cancer Soc., Pitts., 1979-83, Teen Recreation Fedn., Bellevue, 1989—; mem. adv. bd. Citizens Against Substance Abuse, Bellevue, 1990—. Named Outstanding Young Educator, North Hill's Jaycees, 1979, Pa. Outstanding Young Educator, State of Pa. Jaycees, 1979, Pa. State Tchr. of Yr., Dept. Edn., 1982, Citizen of Yr., Pitts. City & Suburban Life, 1982, Citizen of Yr., Pa. Police, 1989; recipient Nat. Drug Free Sch. award Pres. Bush, 1989. Mem. NEA, Pa. State Edn. Assn. Republican. Roman Catholic. Avocation: boarding kennel. Home: The Ter Box 617 RR # 1 Fombell PA 16123 Office: Northgate Sch Dist 589 Union Ave Pittsburgh PA 15202-2999

GLENISTER, BRIAN FREDERICK, geologist, educator; b. Albany, Western Australia, Sept. 28, 1928; came to U.S., 1959, naturalized, 1967; s. Frederick and Mabel (Frusher) G.; m. Anne Marie Treloar, Feb. 16, 1956; children: Alan Edward, Linda Marie, Kathryn Grace. BSc, U. Western Australia, Perth, 1949; MSc, U. Melbourne, Australia, 1953; PhD, U. Iowa, 1956. Lectr., then sr. lectr. geology U. Western Australia, 1956-59; asst. prof. U. Iowa, Iowa City, 1959-62, assoc. prof., 1962-66, prof., 1966-74, chmn. geology dept., 1968-74, A.K. Miller prof. geology, 1974—. Mem. AAAS, Paleontol. Soc. (pres. 1988-89), Paleontol. Assn., Palaontologischen Gesellschaft, Soc. Econ. Paleontologists and Mineralogists, Geol. Soc. Am., Geol. Soc. Iowa (pres. 1991), Paleontol. Rsch. Inst., Internat. Paleontol. Union, Am. Assn. Petroleum Geologists, Sigma Xi. Home: 2015 Scales Bend Rd NE North Liberty IA 52317-9331

GLENN, JOHN HERSCHEL, JR., senator; b. Cambridge, Ohio, July 18, 1921; s. John Herschel and Clara (Sproat) G.; m. Anna Margaret Castor, Apr. 1943; children: Carolyn Ann, John David. Student, Muskingum Coll., 1939-42, B.Sc., 1962; naval aviation cadet, U. Iowa, 1942; grad. flight sch., Naval Air Tng. Center, Corpus Christi, Tex., 1943, Navy Test Pilot Tng. Sch., Patuxent River, Md., 1954. Commd. 2d lt. USMC, 1943, assigned 4th Marine Aircraft Wing, Marshall Islands campaign, 1944, assigned 9th Marine Aircraft Wing, 1945-46; with 1st Marine Aircraft Wing, North China Patrol, also Guam, 1947-48; flight instr. advanced flight tng. Corpus Christi, 1949-51; asst. G-2/G-3 Amphibious Warfare Sch., Quantico, Va., 1951; with Marine Fighter Squadron 311, exchange pilot 25th Fighter Interceptor Squadron USAF, Korea, 1953; project officer fighter design br. Navy Bur. Aero. Washington, 1956-58; astronaut Project Mercury, Manned Spacecraft Center NASA, 1959-65; pilot Mercury-Atlas 6, 1st orbital space flight launched from Cape Canaveral, Fla., Feb. 1962; ret. as col., 1965; v.p. corp. devel. and dir. Royal Crown Cola Co., 1966-74; pres. Royal Crown Internat.; U.S. senator from Ohio, 1975—. Co-author: We Seven, 1962; author: P.S., I Listened to Your Heart Beat. Made first supersonic transcontinental flight, July 16, 1957; trustee Muskingum Coll. Decorated D.F.C. (six), Air medal (18); recipient Astronaut medal USMC, Navy unit commendation, Korean Presdl. unit citation, Disting. Merit award Muskingum Coll., Medal of Honor N.Y.C., Congl. Space Medal of Honor, 1978, Centennial awd., Nat. Geographic Soc., 1988, other decorations, awards and hon. degrees. Mem. Soc. Exptl. Test Pilots, Internat. Acad. of Astronautics (hon.). Democrat. Presbyterian. Office: US Senate 503 Hart Senate Bldg Washington DC 20510-3501

GLENN, MARK WILLIAM, operations research analyst; b. Washington, Feb. 22, 1953; s. Edmund Stanislaw and Marjorie (Rugg) G.; m. Jean Louise Mulcahy, Aug. 17, 1974; 1 child, Melody Jean. BA in Econs., U. Del., 1976, M Econs., 1979. Supply mgmt. analyst U.S. Army Aviation Systems Command, St. Louis, 1976-79, ops. rsch. analyst, 1980-85; ops. rsch. analyst Hdqrs. Dept. of Army, Washington, 1985-86, Hdqrs. U.S. Army Materiel Command, Alexandria, Va., 1986-88, U.S. Army Missile Command, Huntsville, Ala., 1989—. Mem. Am. Soc. Mil. Contrs. (nat. award for analysis and evaluation 1991), Soc. Cost Estimating and Analysis (treas. Greater Midwestern chpt. 1985, cert. 1985), Redstone MacIntosh Users Group (pres. 1991). Achievements include numerous weapon system cost estimates; research in definitional problems and their large cost impact in the area of initial spare and repair parts; redesign of inflation methods used by Dept. of Army. Home: 5004 Sunset Bluff Huntsville AL 35803 Office: US Army Missile Command AMSMI-OR Redstone Arsenal AL 35803

GLENN, ROGERS, psychologist, student advisor, consultant; b. Tallahassee, Jan. 22, 1930; s. James Burdock Glenn and Carl (Howard) Lucas; m. Ora Lee Goodwin, Oct. 9, 1946 (div. 1982); children: Frances C., Angela M. MS in Art Edn., Edn. Adminstrn., Fla. A&M U., 1961; MFA in Creative Arts, Fla. State U., 1973; MS in Instrn. System Tech., Ind. U., 1973, MS in Edn. Psychology, 1977. Art instr., dept. chmn. Fla. A&M U., Tallahassee, 1957-69, acad. adminstrv. coord. instructional media ctr., 1969-70; instr. art Fla. State U., Tallahassee, 1970-71; resident grad. asst. Ind. U., Bloomington, 1971-73, assoc. instl. devel., 1974-75; adminstv. program asst. Ind. U., Indpls., 1976-77; teaching asst. Purdue U., West Lafayette, Ind., 1978-80; media cons. to v.p. Fla. A&M U., 1980-81, student advisor, 1981—; race rels. cons. Ind. U., 1974-78, Purdue U., 1978-80, Tri-State Area Law Enforcement, Tallahassee, 1986-87, County Pub. Sch. System, Tallahassee, 1987-88, Fla. A&M U., 1981—, Fla. State U., 1981—; expert witness racial bias Fla. Supreme Ct. Tallahassee, 1990; cons. seminar racial bias, Fla. Corrections Acad., North Oulstee, 1991; presenter M.P.A. Psychol. Assn. St. Louis, 1980, Detroit, 1981, S.E. Psychol. Assn. Atlanta, 1981, New Orleans, 1982, Chgo., 1984, Washington, 1989 and numerous others. With U.S. Army, 1951-52. Recipient Black & Minorities fellowship, Purdue U., 1978, rsch. grant Fla. A&M U., 1985, Profl. Devel. award, Fla. A&M U., 1989. Mem. APA, S.E. Psychol. Assn. (rsch. award 1982), North Fla. Black Psychol. Assn., Midwestern Psychol. Assn., Phi Delta Kappa (Achievement award 1984). Democrat. Methodist. Avocations: computer programming, reading, artistic creations, skating, fishing. Home: 1820 Saxon St Tallahassee FL 32310-5349 Office: Fla A&M Univ Tallahassee FL 32304

GLENN, ROLAND DOUGLAS, chemical engineer; b. Somerville, Mass.; s. Charles Rathford and Anna Amanda (Card) G.; m. Eleanor Norwood Greene, June 19, 1939; children: Mary Eleanor, Nancy Anne Hansen, Sara Elisabeth Baker, Rolene Douglas Ramsey. BSChemE, MIT, MSChemE, postgrad. Registered profl. engr., N.Y., Conn., Va. Prodn. supr. Union Carbide Corp., South Charleston, W.Va.; devel. group leader Union Carbide Corp., South Charleston, plant mgr., 1953-56; div. v.p. Union Carbide Corp., N.Y.C., 1957-68; v.p. Pope, Evans & Robbins, N.Y.C., Alexandria, Va., 1969-71; pres. Combustion Processes, Inc., N.Y.C., 1972-90, Darien, Conn., 1991—. Editor: (directory) Consulting Services, 1978-88; contbr. numerous reports and papers to profl. jours. Sloan fellow, MIT. Mem. Am. Inst. Chem. Engrs., Am. Chem. Soc., Assn. Cons. Chemists & Chem. Engrs. (dir. 1974-92). Office: Combustion Processes Inc 53 Goodwives River Rd Darien CT 06820-5919

GLICK, JOHN H., oncologist, medical educator; b. N.Y.C., May 9, 1943; s. Arthur W. and Sybil (Goldman) G.; m. Jane Mills, May 25, 1968; children: Katherine, Sarah. AB magna cum laude, Princeton U., 1965; MD, Columbia U., 1969. Diplomate Am. Bd. Med. Oncology, Am. Bd. Internal Medicine (sec. subspecialty com. med. oncology 1976-83, mem. subspecialty bd. med. oncology 1983-87, chmn. 1987-89, cert. examination com. 1986-88, bd. govs. 1987-89). Intern in medicine Presbyn. Hosp., N.Y.C., 1969-70, asst. resident in medicine, 1970-71; commd. surgeon, clin. assoc. medicine br. Nat. Cancer Inst., USPHS, Bethesda, Md., 1971-73; postdoctoral fellow in med. oncology Stanford (Calif.) U., 1973-74; asst. prof. medicine U. Pa., Phila., 1974-79, Ann B. Young asst. prof. cancer rsch., 1974, assoc. prof., 1979-83, prof., 1983—, Madlyn and Leonard Abramson prof. clin. oncology, 1988—; dir. clin. trials U. Pa. Cancer Ctr., Phila., 1977-79, assoc. dir. for clin. rsch., 1980-85, dir. Cancer Ctr., 1985—; mem. numerous acad. coms., dept. medicine coms., hosp. coms., 1974—; attending physician Hosp. of U. Pa., 1974—, dir. Hematology-Oncology Clinic, 1974-76; cons. Phila. VA Hosp., 1974—; mem. NIH clin. trials rev. com., 1980-83, radiosensitizer/radioprotector working group, radiotherapy devel. br., 1980-85, chmn. consensus devel. panel conf. adjuvant therapy for breast cancer, 1985, all Nat Cancer Inst., NIH; mem. com. accreditation med. oncology tng. progams, 1983—, mem. appeals panel, 1987—, Accreditation Coun. Grad. Med. Edn.; prin. investigator Ea. Coop. Oncology Group, U. Pa. Mem. editorial bd. Am. Jour Clin. Oncology, 1983-89, Blood jour., 1983-86, Jour. Clin. Oncology, 1986-88, Jour. Cancer Rsch. and Clin. Oncology, 1987—; bd. editors Internat. Jour. Radiation Oncology, Biology and Physics; assoc. editor Cancer Rsch. jour., 1984-88; contbr. over 100 original papers to profl. jours. Recipient Am. Cancer Soc. Faculty Rsch. award, 1982-86; rsch. grantee Nat. Cancer Inst., Eastern Coop. Oncology Group, Am. Cancer Soc., others. Fellow ACP (mem. various specialty coms. 1983-85), Coll. Physician and Surgeons; mem. Am. Soc. Clin. Oncology (chmn. program com. 1983-84, nominating com. 1983-84, pub. issue com. 1984-85, bd. dirs. 1988—, other coms.), Am. Assn. Cancer Edn., Am. Assn. Cancer Rsch., Am. Radium Soc. (exec. com. 1986—), Am. Soc. Hematology, Am. Fedn. Clin. Rsch., John Morgan Soc. U. Pa., Phi Beta Kappa, Alpha Omega Alpha. Office: U Pa Cancer Ctr 3400 Spruce St Philadelphia PA 19104-4220

GLICK, MICHAEL ANDREW, aerospace engineer; b. Harrisonburg, Va., Oct. 17, 1958; s. Phillip Andrew and Shirley (Forren) G.; m. Deborah Robertson, Aug. 23, 1980; children: Matthew Aaron, Jon-Michael. BS in Engring. Sci., Old Dominion U., 1981. Test engr. Gen. Electric, Cin., 1982-84; devel. engr. Garret Turbine, Phoenix, Ariz., 1984-86; sr. systems engr. Atlantic Rsch., Gainesville, Va., 1987-90; sr. project engr. Hughes Aircraft Missiles, Tucson, Ariz., 1991—; mem. JANNAF Airbreathing Propulsion Com., 1992—. Mem. AIAA (sr.). Home: 1361 S Barbara Dr Tucson AZ 85748

GLICKSMAN, MARTIN EDEN, materials engineering educator; b. N.Y.C., Apr. 4, 1937; s. Nathan Henry and Ruth Elaine (Rosensaft) G.; m. Lucinda Jeanette Mulder, May 7, 1967. B in Metall. Engring., Rensselaer Poly. Inst., 1957, PhD, 1961. Metall. engr. Procter & Gamble Co., Cin., 1957-58; research metallurgist Naval Research Lab., Washington, 1961-75, assoc. supt. materials sci. divsn., 1974-75; chmn. materials engr. dept. Rensselaer Poly. Inst., Troy, N.Y., 1975-86, prof. 1975-86, prof., 1986—; prof. materials engring., chmn. dept. materials engring. Rensselaer Poly. Inst., Troy, N.Y., 1975-86; John Tod Horton prof. materials engring. Rensselaer Poly. Inst., 1986—; Van Horn lectr. Case Western Res. U., 1984; cons. in field. Contbr. in articles to profl. jours. Recipient Pure Sci. Rsch. award Rsch. Soc. of Am., 1968, Arthur Flemming award Washington Jr. C. of C.; Minerals Metals and Materials Soc. fellow AIME, 1994. Fellow Am. Soc. Metals (M.E. Grossman award 1971); mem. AIME (Bruce Chalmers award 1992), U. Space Rsch. Assn. (chmn. bd. trustees 1986, dir. microgravity divsn., 1986—). Home: 22 Schuyler Hills Rd Albany NY 12211-1445 Office: Rensselaer Poly Inst MRC 104 Troy NY 12180-3590

GLIDDEN, BRUCE, structural engineer, construction consultant; b. Cleve., Mar. 20, 1928; s. Herbert Wiley and Ruth (Winters) G.; m. Marilou Seigler, Aug. 27, 1960. BS in Engring., U. Calif., Berkeley, 1950. Registered structural engr., Calif., civil engr., Calif., profl. engr., Pa. Various engring. positions Consol. Western Steel Div. U.S. Steel, L.A., 1956-64; dist. engr. Am. Bridge div. U.S. Steel, L.A., 1964-67; v.p. engring. Am. Bridge div. U.S. Steel, Pitts., 1969-74; v.p. western Am. Bridge div. U.S. Steel, L.A., 1974-79; gen. mgr. design U.S. Steel Corp., Pitts., 1979-81; pres. Am. Bridge div. U.S. Steel, Pitts., 1981-85; prin. Glidden & Co., Ltd., Pitts., 1985—; mem. design specification com. Am. Inst. Steel Constrn., Chgo., 1969-74, code std. practice com., 1969-79. Econs. instr. Bethal Park (Pa.) Sch. Dist. Jr. Achievement, 1988-89; dist. dir. Boy Scouts Am., Pitts., 1973-74; advisor Jr. Achievement, Pitts., 1971-72. Lt. USNR, 1950-53. Fellow ASCE; mem. Structural Engrs. Assn. Calif. code com. 1965-67), Am. Welding Soc. (code com. 1969-74), Senickley Heights Golf Club (pres. 1987-90). Office: Glidden and Co Ltd PO Box 34 Bridgeville PA 15017

GLIMCHER, MELVIN JACOB, orthopedic surgeon; b. Brookline, Mass., June 2, 1925; s. Aaron and Clara (Fink) G.; m. Geraldine Lee Bogolub, June 22, 1946; children: Susan Deborah, Laurie Hollis, Nancy Blair. Student, Duke U., 1943-44; B.S. in Mech. Engring. with highest distinction; B.S. in Physics with highest distinction, Purdue U., 1946; M.D. magna cum laude, Harvard, 1950; postgrad., Mass. Inst. Tech., 1956-59. Intern surgery Strong Meml. Hosp., Rochester, N.Y., 1950-51; 3d asst. resident surgery Mass. Gen. Hosp., Boston, 1951-52; 2d asst. resident Mass. Gen. Hosp., 1952-53, asst. resident orthopedic surgery, 1954-55, chief resident, 1956, chief orthopedic service, 1965-71, chmn. dept. orthopedic surgery, 1968-71; asst. resident orthopedic surgery Children's Med. Center, Boston, 1953-54; jr. resident Children's Med. Center, 1955-56; mem. faculty Harvard Med. Sch., 1956—, Edith M. Ashley prof. orphopedic surgery, 1965-71, Harriet M. Peabody prof., 1971—; also chmn. dept.; orthopedic surgeon-in-chief Children's Hosp. Med. Center, Boston, 1971-81, dir. Lab. for Study of Skeletal Disorders and Rehab., 1980—. Trustee Forsyth Dental Infirmary. Served with USMCR, World War II. Recipient Soma Weiss award Harvard Med. Sch., 1950, Borden Research award, 1950; Kappa Delta award, 1959; Internat. Assn. Dental Research award, 1964; Ralph Pemberton award Am. Rheumatism Soc., 1969; Bristol-Meyers/Zimmer instl. grant for excellence; Disting. Achievement in Orthopaedic Research award Orthopaedic Research Edn. Found. Fellow Am. Acad. Arts and Scis., Am. Acad. Orthopaedic Surgeons (Silver anniversary Kappa Delta prize 1974), Am. Orthopedic Assn.; mem. Orthopedic Research Soc. (past pres.), Assn. Bone and Joint Surgeons (Nicholas Andry award 1978), Internat. Soc. for Study Lumbar Spine (Volvo award 1983), Societe Internationale de Chirurgie Orthopedique et de Traumatologie. Office: 300 Longwood Ave Boston MA 02115-5737

GLIMM, JAMES GILBERT, mathematician; b. Peoria, Ill., Mar. 24, 1934; s. William Frederick and Barbara Gilbert (Hooper) G.; m. Adele Strauss, June 30, 1957; 1 dau., Alison. A.B., Columbia U., 1956, A.M., 1957, Ph.D., 1959. From asst. prof. to prof. math. MIT, 1960-69; prof. Courant Inst., NYU, 1969-74; prof. math. Rockefeller U., N.Y.C., 1974-82; prof. Courant Inst., NYU, N.Y.C., 1982-89; disting. prof., chair dept. applied math. and statis. SUNY, Stony Brook, 1989—, dir. Inst. for Math. Modeling, 1989—, dir. Ctr. for Excellence for Nonlinear Math., 1991—; dir. Advanced Mfg. Initiative, 1993—; bd. dirs. Advanced Mfg. Initiative. Co-author: Quantum Physics, 1981; Collected Papers, Vols. I and II, 1985; mem. editorial bds. profl. jours.; contbr. articles to sci. publs. Recipient Dannie Heineman prize in math. physics, 1980; Guggenheim fellow, 1963, 65. Mem. NAS, Internat. Assn. Math. Physicists, Am. Phys. Soc., Am. Math. Soc. (Leroy P. Steele prize 1992), Soc. Indsl. and Applied Math., Math. Assn. Am., Am. Acad. Arts and Scis., Soc. Petroleum Engrs., N.Y. Acad. Scis. (award in phys. and math. scis. 1979). Office: SUNY Dept Applied Math & Stats Math Tower Stony Brook NY 11794-3600

GLINER, ERAST BORIS, theoretical physicist; b. Kiev, USSR, Feb. 3, 1923; came to U.S., 1980; s. Boris Moses Gliner and Bella Boris (Pauckman) Rubinstein; m. Galina Ilchenko, Dec. 12, 1944; children: Bella, Arkady. MS in Physics, Leningrad U., USSR, 1963; PhD in Physics, Tartu U., Estonia, 1972. Head theoretical dept. Spl. Design Office, Leningrad, 1954-63; sr. scientist A Ioffe Inst. of Soviet Acad. Scis., Leningrad, 1963-80; rsch. assoc. McDonnel Ctr. Space Sci. Washington U., St. Louis, 1983-86; vis. fellow Inst. Lab. Astrophysics Colo. U., Boulder, 1982-83; vis. scientist Stanford U., SLAC, Palo Alto, Calif. Co-author: Differential Equation of Mathematical Physics (English, Russian, Japenese edits.), 1962; contbr. articles to profl. jours. Polit. prisoner USSR, 1945-54. Sgt. field arty. Soviet Army, 1942-44. Decorated Russian Orders Red Star. Mem. Am. Phys. Soc., N.Y. Acad. Scis. Jewish. Achievements include patents in field (USSR); research in Einstein gravitational theory, introduction of vaccumlike state of matter, covaiant energy description in general relativity, foundation of general relativity on the basis of Sakharov's concept of gravity. Office: Stanford U SLAC Bin 81 PO Box 4349 Stanford CA 94309

GLINES, STEPHEN RAMEY, software industry executive; b. N.Y.C., Mar. 3, 1952; s. Earl Stanley and Catherine Van Arman (Stevenson) G.; m. Susan Leigh Collings; children: Elizabeth, Catherine. Pres. Brahman Publ., Cambridge, Mass., 1973-77; system designer Wakefield Software System, Woburn, Mass., 1973-74; sr. product mgr. Atex, Lexington, Mass., 1983; pres. S.R. Glines & Co., Belmont, Mass., 1983-91; rsch. assoc. J. Forrester, MIT, Cambridge, 1978-79; cons. Logistics Systems, Newton, Mass., 1979-82; fellow Ctr. for Advanced Profl. Studies, Cambridge, 1991—. Author 4 books; columnist Altos World, 1988-90; contbr. articles to profl. jours. Mem. IEEE, IUSR group, Mt. Auburn Club (Watertown, Mass.). Avocation: sailing. Home: 62 Tobey Rd Belmont MA 02178

GLISKY, ELIZABETH LOUISE, psychology educator; b. Toronto, Ont., Can., Oct. 28, 1941; d. Robert Laird Borden and Ruby Lillian (Jeeves) Joynt; m. Eugene Peter Glisky, Sept. 15, 1962; children: William Peter, Martha Louise, Sarah Marie, Linda Natalia. BA, U. Toronto, 1962, PhD, 1983. Rsch. fellow unit for memory disorders U. Toronto, 1983-86, rsch. assoc. unit for memory disorders, 1986-87; vis. asst. prof. dept. psychology U. Ariz., Tucson, 1987-89, asst. prof. psychology, 1989—; gov. bd. Ariz. Head Injury Found., 1990-92. Mem. editorial rev. bd. Neuropsychol. Rehab., 1990—, Neuropsychology, 1992—. Contbr. articles to profl. jours., chpts. to books; author instr.'s manual Methods Toward a Science of Behavior and Experience, 1993. Scholar Natural Scis. and Engring. Rsch. Coun., 1980-81; fellow Govt. of Ontario, 1982-83; grantee Ariz. Disease Control Rsch. Commn., 1989-90, U. Ariz. Found., Social and Behavioral Scis. Rsch. Inst., 1990, Nat. Inst. on Aging, 1991—. Mem. APS, Internat. Neuropsychol. Soc., Memory Disorders Rsch. Soc., Psychonomic Soc. Achievements include demonstration of acquisition of domain-specific knowledge in amnesic patients by use of method of vanishing cues, beginning development of scientific foundation or basis for rehabilitation of patients with memory disorders. Office: U Ariz Dept Psychology Tucson AZ 85721

GLOVER, ANTHONY RICHARD, engineer; b. London, Dec. 3, 1944; came to U.S. 1970.; s. Herbert William and Evelyn May Elizabeth (Lucas) G.; m. Jeanne Elizabeth Cleary, Feb. 6, 1971; children: Sarah Elizabeth, Wendy Louise. BSc in Engring. with hons., Manchester U., Eng., 1966. Profl. engr. 1974, lic. surveyor 1979. Hwy. engr. Wiltshire County Coun., Eng., 1966-69; field engr. Eastern Construction Co., Toronto, Can., 1969-71; site engr. Sear Brown and Assocs., Rochester, N.Y., 1971-75; land surveyor George Nechwort PE LS, Sangerfield, N.Y., 1975-79; asst. city engr., water supr. City of Oneida, N.Y., 1979-88; free lance consulting engr., land surveyor Vernon, N.Y., 1988--; town engr. Town of Vernon, 1989--; village engr. Village of Wampsville, N.Y., 1991--; v.p. Cent. N.Y. Waterworks Conf., 1985-88. Mem. Planning Bd. Honeoye Falls, n.Y., 1972-75. Mem. Mohawk Valley Land Surveyors Assn., N.Y. Assn. Profl. Land Surveyors. Home: PO Box 61 Vernon NY 13476 Office: Churton Rd Vernon NY 13476

GLOVER, EVERETT WILLIAM, JR., environmental engineer; b. Fairmont, W.Va., Apr. 4, 1948; s. Everett William and Pearl Irene (Bollman) G.; m. Joyce Ann Linville, May 24, 1969 (div.); 1 child, Alison Renee. BSCE with high honors, W.Va. U., 1970, MSCE, 1975. Registered profl. engr., Md., Va., N.C., S.C., Ga., Tenn., Fla., Ala., Miss., La. Ky., N.Y. Project engr. Law Engring. Testing Co., Atlanta, 1972-78; project mgr. Soil & Material Engrs. Inc. (S&ME), Atlanta and Raleigh, N.C., 1978-87; br. mgr. Westinghouse Environ. & Geotech. Svcs. (formerly S&ME), Atlanta, 1987-90, project dir., 1990-92; project dir. Rust Environment & Infrastructure (formerly SEC Donohue, Inc.), 1992—. Mem. ASCE, ASTM, Air and Waste Mgmt. Assn., Project Mgmt. Inst., Chi Epsilon, Tau Beta Pi.

GLOVER, FRED WILLIAM, artificial intelligence and optimization research director, educator; b. Kansas City, Mo., Mar. 8, 1937; s. William Cane and Mary Ruth (Baxter) G.; m. Diane Tatham, June 4, 1988; children from previous marriage: Dana Reynolds, Paul Glover. BBA, U. Mo., 1960; PhD, Carnegie-Mellon U., 1965. Asst. prof. U. Calif., Berkeley, 1965-66; assoc. prof. U. Tex., Austin, 1966-69; prof. U. Minn., Mpls., 1969-70; John King Prof. U. Colo., Boulder, 1970-87, US West Chair in System Sci., 1987—; research dir. Artificial Intelligence Ctr., Boulder, 1984—; invited disting. lectr. Swiss Fed. Inst. Tech., Lausanne, 1990—, IMAG Labs., U. Genoble, France, 1991; vis. Regents Chair in Engring, U. Tex., Austin, 1989; cons. U.S. Congress, 1984, Nat. Bur. Standards, 1986, also over 70 U.S. corps. and govt. agys., 1965—; lectr. NATO, France, Italy, Germany, Denmark, 1970, 78, 80, 82, 89, Inst. Decision Scis., 1984; bd. dirs. Mgmt. Systems, Boulder, Decision Analysis Inst., Boulder, 1974-82; chmn. bd. Analysis, Research & Computation, Austin, 1971-83; head, rsch. assoc. Global Optimization Space Constrn. Ctr., Boulder, 1988—; rsch. prin. U. Colo.-U.S. West Joint Rsch. Initiative, 1990—; prin. investigator Air Force Office Sci. Rsch, Office Naval Rsch., 1990—. Author: Network Decision Models, 1983 (DIS award 1984), Tabu Search I, 1989, Tabu Search II, 1990, Tabu Search (special vol.) 1993, Ghost Image Processes for Neural Networks, 1993, Linkages with Artificial Intelligence, 1990, Network Models in Optimization and Their Application in Practice, 1992, also others; contbr. over 200 articles on math. optimization and artificial intelligence to profl. jours. Participant Host Vis. Exchange, Nat. Acad. Scis., 1981; mem. grants com. Queen Elizabeth II fellowships, Australia and U.S., 1984; mem. U.S. nat. adv. bd. Univ. Rsch. Initiative on Combinatorial Optimization. Recipient Internat. Achievement award Inst. Mgmt. Scis., 1982, energy Rsch. award Energy rsch. Inst., 1983, Univ. Disting. Rsch. Lectr. award U. Colo., 1988, rsch. Excellence prize Ops. Rsch. Soc., 1989; named first US West Disting. fellow, 1987. Fellow AAAS, Am. Inst. Decision Scis (lectr. 1984, Outstanding Achievement award 1984), Am. Assn. Collegiate Schs. Bus., ICC Inst.; mem. Alpha Iota Delta. Achievements include design of software systems used throughout the U.S. and Europe. Office: U Colo Coll Bus Box 419 Boulder CO 80309-0419

GLOYNA, EARNEST FREDERICK, environmental engineer, educator; b. Vernon, Tex., June 30, 1921; s. Herman Ernst and Johanna Bertha (Reithmayer) G.; m. Agnes Mary Lehman, Feb. 17, 1946; children: David Frederick, Lisa Anna (Mrs. Jack Grosskopf). BS in Civil Engring., Tex. Technol. U., 1946; MS in Civil Engring., U. Tex., 1949; Dr. Engring., Johns Hopkins U., 1952. Registered profl. engr. Jr. engr. Tex. Hwy. Dept., 1945-46; office engr. Magnolia Petroleum Co., 1946-47; instr. civil engring. U. Tex., Austin, 1947-49, asst. prof., 1949-53, assoc. prof., 1953-59, prof., 1959-70, Joe J. King prof. engring., 1970-82, dir. Environ. Health Engring. Labs., 1954-70, dir. Ctr. for Research in Water Resources, 1963-73, dean Coll. Engring., 1970-87, dir. Bur. Engring. Research, 1970-87, Bettie Margaret Smith chair in environ. engring., 1987—; cons. on water and wastewater treatment and water resources, 1947—; dir. Parker Drilling Co.; cons. numerous industries, WHO, World Bank, U.S. Air Force, U.S. Army, U.S. Senate, fgn. cities and govts., UN, 1952—; mem., past chmn. sci. adv. bd. EPA; chmn. various coms. NRC, Nat. Acad. Sci., Nat. Acad. Engring.; chmn. Tex. State Bd. Registration Profl. Engrs., 1992-93. Author: Waste Stabilization Ponds, 1971 (also French and Spanish edits), (with Joe O. Ledbetter) Principles of Radiological Health, 1969; Editor: (with W. Wesley Eckenfelder, Jr.) Advances in Water Quality Improvement, 1968, Water Quality Improvement by Physical and Chemical Processes, 1970, (with William S. Butcher) Conflicts in Water Resources Planning, 1972, (with Woodson and Drew) Water Management by Electric Power Industry, 1975, (with Malina and Davis) Ponds as a Wastewater Treatment Alternative, 1976, (with Richard B. McCaslin) Commitment to Excellence, 1990; contbr. numerous articles to profl. jours. Served with Corps Engrs. AUS, 1942-46, ETO, lt. col. Ret. Named Disting. Engr. Grad. Tex. Tech. U., 1971, Disting. Alumnus, 1973, Disting. Engring. Grad. U. Tex., Austin, 1982, Disting. Alumnus, 1992, Disting. Alumnus Johns Hopkins U., 1993; recipient Joe J. King award U. Tex., Austin, 1982; EPA regional environ. educator award, 1977; Nat. Environ. Devel. award, 1983, Sci. award Nat. Wildlife Fedn., 1986; Order of Henri Pittier, Nat. Conservation medal Venezuela, 1983. Fellow ASCE (Hon. Mem. award, Meritorious Paper award Tex. sect. 1968, Award of Honor Tex. sect. 1985), Simon W. Freese environ. engr. award 1986), Pub. Health Engrs. Pakistan (hon.); mem. NAE, NSPE (past dir.), Am. Inst. Chem. Engrs., Assn. Environ. Engring. Profs. (past pres.), Am. Soc. for Engring. Edn. (Centennial Medallion), Am. Water Works Assn. (water resources divsn. award 1959), Am. Acad. Environ. Engrs. (diplomate, past pres., Gordon Maskew Fair award 1982, Arthur Bedell award 1991), Water Pollution Control Fedn. (Harrison Prescott Eddy medal 1959, past pres., hon. mem.), Tex. Soc. Profl. Engrs. (Engr. of Year award Travis chpt. 1972, Award of Honor 1985, pres. 1986), Southwestern Soc. Nuclear Medicine (hon.), Nat. Acad. Engring. Mex. (fgn. corr. mem.), Nat. Acad. Scis. Venezuela (fgn. corr.), Sociedad Mexicana de Aguas (Jack Huppert award), Sigma Xi, Tau Beta Pi, Chi Epsilon, Phi Kappa Phi, Pi Epsilon Tau (hon.), Omicron Delta Kappa. Clubs: Cosmos (Washington); Headliners, Faculty Center, Rotary (Austin). Office: U Tex Coll of Engring Austin TX 78712

GLUSHKO, VICTOR, medical products executive, biochemist; b. Kiel, Germany, June 26, 1946; came to U.S., 1951; s. George and Maria Glushko; 1 child, Sarah Rachel Alexandra. AB, Earlham Coll., 1968; PhD, Ind. U., 1972. Rsch. assoc. Meml. Sloan Kettering, N.Y.C., 1973-77; asst. prof. Temple U. Med. Sch., Phila., 1977-83; dir. product devel. Helitrex, Plainsboro, N.J., 1982-85; exec. dir. Am. Biomaterials, Plainsboro, 1985-88; v.p. Vitaphore, Menlo Park, Calif., 1988-92; pres. SRA Assocs., Burlingame, Calif., 1992—. Developer drug delivery systems, collagen dressings, perio dressing; contbr. articles to profl. jours. Sloan-Kettering fellow, 1972-73; rsch. grantee (9), 1973-82. Mem. Am. Chem. Soc. (chmn. continuing edn. Phila. sect. 1981-82), Am. Assn. Pharm. Sci., N.Y. Acad. Scis., Biophys. Soc., Tissue Culture Assn., Controlled Release Soc., Wound Healing Soc. Home: 2504 Hale Dr Burlingame CA 94010 Office: Pharmetrix 1330 O'Brien Dr Menlo Park CA 94025

GOBER, HANS JOACHIM, physics educator; b. Berlin, Oct. 30, 1931; m. Caecilia Bantle, Oct. 2, 1964; children: Peter, Maria. D Engring., Tech. U. Berlin, 1964. Asst. prof. Tulane U., New Orleans, 1965-69; prof. physics and acoustics Fachhochschule Luebeck, Germany, 1970—; acoustic cons. IHK-Lübeck, Luebeck, 1973—. Mem. Acoustical Soc. Am., Deutsche Physikalische Gesellschaft, Verein Deutscher Ingenieure. Home: Junoring 43, Luebeck Germany Office: Fachhochschule Lübeck, Stephensonstr 1, Luebeck Germany

GOBRECHT, HEINRICH FRIEDRICH, physicist, educator; b. Bremen, Germany, July 20, 1909; s. Heinrich Ludwig and Caroline Friederike (Oesterhelweg) G.; m. Christa Johanna Schubbe, Aug. 25, 1938; children: Klaus, Juergen, Jens. Student, U. Hanover, 1929-32, U. Göttingen, 1932-33, U. Marburg, 1933-34, U. Dresden, 1934-36; diploma in engring., Tech. High Sch., Dresden, 1936, D in Engring., 1937, D Engring. Habilitation, 1939. Chief engr. TV dept. Loewe-Radio, Berlin, 1938-45, Siemens-Radio, Arnstadt, 1946-47; prof. physics and dir. Phys. Inst. U. Berlin, 1948-77, prof. emeritus, 1977—; dir. Phys. Inst. U. Berlin, 1952-77. Author, editor: Bergmann-Schaefer, Experimentalphysics, (5 vols.); contbr. over 200 articles to spl. periodicals. Burgomaster City of Oberlungwitz, Saxonia, Germany, 1945. Mem. German Phys. Soc., Gesellschaft Deutscher Naturforscher und Ärzte. Home: Rheinbabenallee 17 a, D-1000 Berlin 33, Germany Office: Tech U Berlin, Hardenbergstrasse 36, D-10623 Berlin Germany

GODARA, LAL CHAND, electrical engineering educator; b. Sri Ganganagar, Rajasthan, India, Jan. 2, 1952; arrived in Australia, 1979; s. Nand Ram G.; m. Saroj, July 17, 1975; children: Pankaj, Vikas. BE with honors, B.I.T.S., Pilani, India, 1975; M of Tech., I.I.Sc., Banglore, 1977; PhD, Newcastle (Australia) U., 1982. Asst. lectr. B.I.T.S., 1977-79; rsch. scientist Defence Rsch. Lab., Hyderabad, India, 1983; lectr. U. Canberra, Australia, 1985-87; lectr., sr. lectr. U. New South Wales, Canberra, 1988—; vis. fellow Stanford U., Polo Alto, Calif., 1988-89; rsch. affiliate Yale U., New Haven, Conn., 1991. Contbr. articles to profl. jours. Postdoctoral fellow Australian Nat. U., Canberra, 1984-88. Mem. IEEE (sr. mem., chmn. ACT sect. 1988-91), Acoustical Soc. Am. Home: 4 Brinkley Circuit, Palmerston 2913 Australian Capital Territory, Australia Office: Australian Defence Force Acad, Dept Electrical Engring, Canberra 2600 Australian Capital Territory, Australia

GODBOLD, JAMES HOMER, JR., biostatistician, educator; b. Natchez, Miss., Nov. 1, 1947; s. James Homer Godbold and Josephine (Lessley) Williams; m. Cheryl Jean Anderson, June 23, 1984; children: Ryan James, Anna Margaret. BA, Millsaps Coll., 1969; MS, Va. Poly. Inst. and State U., 1970; PhD, Johns Hopkins U., 1976. Biostatistician USPHS, Rockville, Md., 1970-72; assoc. scientist Oak Ridge (Tenn.) Associated Univs., 1976-78; sr. biostatistician Johnson and Johnson, New Brunswick, N.J., 1978-80; asst. mem. Meml. Sloan-Kettering Cancer Ctr., N.Y.C., 1980-87; rsch. assoc. prof. Mt. Sinai Sch. Medicine, N.Y.C., 1987—; adj. assoc. prof. Cornell U. Med. Coll., N.Y.C., 1987—; cons. Coll. S.I., N.Y., 1987-88. Assoc. editor: Environ. Rsch., 1987—, Mt. Sinai Med. Jour., 1990—, Am. Jour. Indsl. Medicine, 1991—; contbr. articles to Lancet, Am. Jour. Medicine, New Eng. Jour. Medicine, others. Elder, trustee 1st Presbyn. Ch., Hightstown, N.J., 1991-93. Lt. (s.g.) USPHS, 1970-72. Grantee NIH, 1992. Mem. Am. Statis. Assn., Biometric Soc., Am. Coll. Epidemiology. Achievements include statistical collaboration on studies of cancer, AIDS and occupational/environmental disease. Home: 107 Maple Stream Rd East Windsor NJ 08520 Office: Mount Sinai Med Ctr Box 1057 1 Gustave Levy Pl New York NY 10029

GODDARD, DAVID BENJAMIN, physician assistant, clinical perfusionist; b. Panguitch, Utah, Dec. 6, 1947; s. Edward Pershing and Emma Louise (Stander) G.; m. Ann Braodbent, Dec. 22, 1975; children: Cecilee, Yorke B., Chelsea. BS, Brigham Young U., 1974; B. Health Sci., Duke U., 1976, Physician Assoc. Cert., 1976; cert., Stanford U., 1985; postgrad., U. Nebr. Cert. physician asst.-surgery/primary care Nat. Commn. Cert. Physician Assists., cert. clin. perfusionist Am. Bd. Cardiovascular Perfusion. Rsch. asst. Dept. Parasitology Brigham Young U., Provo, Utah, 1974; physician asst. Peace Corps, Kingdom of Tonga, 1976-78, Peace Corps Health Svcs., Washington, 1979, Cardiovascular & Chest Surg. Assocs., Boise, Idaho, 1979-81; physician asst., clin. perfusionist Starr-Wood-Chapman-Ahmad, PC, Portland, Oreg., 1981-85, Cardiovascular Surgery Assocs., Las Vegas, Nev., 1985-88; program dir., clin. perfusion sci. U. Nebr. Med. Ctr., Omaha, 1989-91; pvt. practice cons., 1992—. Contbr. articles to profl. jours. Panel mem. Medicare Adv. Panel, Portland, Oreg., 1985. Fellow Am. Assn. Surgeon's Assts. (dir.-at-large 1987-88, pres. 1989-90), Am. Acad. Physician Assts. (reimbursement subcom. 1984), Assn. Physicians Assts. in Cardiovascular Surgery, Nev. Acad. Physician Assts. (chmn. legis. affairs 1987-88), Nebr. Acad. Physician Assts. (alt. del. 1989, legis. affairs com. 1990-92), Am. Soc. Extra-Corporeal Tech.; mem. N. Am. Transplant Coords. Orgn. (assoc.), Idaho' Acad. Physician Assts. (CME chmn. 1980-81), Oreg. Soc. Physicians Assts. (pres. 1983-84). Mormon. Avocations: bicycling, amateur radio, snow skiing, boating, photography. Office: Good Samaritan Hosp PO Box 2361 Kearney NE 68848

GODDARD, JEROME, medical entomologist; b. Booneville, Miss., Apr. 12, 1957; s. E.J. Jr. and Clarene (Stevens) G.; m. Rosella May Blackman, May 19, 1979; children: Jerome II, Joseph. MS, U. Miss., 1981; PhD, Miss. State U., 1984. State med. entomologist Miss. Dept. Health, Jackson, 1989—; clin. asst. prof. U. Miss., Jackson, 1991—. Author: Ticks and Tickborne Disease Affecting the Military, 1989, Physician's Guide to Arthropods of Medical Importance, 1993; contbr. articles to profl. jours. Capt. USAF, 1985-89. Capt. USAF, 1985-89. Mem. Am. Soc. Rickettsiology, Miss. Entomological Assn., Sigma Xi. Methodist. Office: Miss Dept Health PO Box 1700 Jackson MS 39215

GODDARD, KENNETH WILLIAM, forensic scientist; b. San Diego, July 6, 1946; s. Joseph William and Bernice Elizabeth (Cahoon) G.; m. Georgene Shitara, June 22, 1968; 1 child, Michelle Suni. BS in Biochemistry, U. Calif., Riverside, 1968; MS in Criminalistics, Calif. State U., L.A., 1971. Criminalist Riverside County Sheriff's Office, 1968-69; criminalist, dep. sheriff San Bernadino County (Calif.) Sheriff's Dept., 1969-72; chief criminalist Huntington Beach (Calif.) Police Dept., 1972-79; chief forensics br. U.S. Fish and Wildlife Svc., Washington, 1979-87; lab. dir. Nat. Fish and Wildlife Forensics Lab., Ashland, Oreg., 1987—; adj. prof. So. Oreg. State Coll., Ashland, 1991—, mem. sci. adv. bd., 1990—. Author: Crime Scene Investigation, 1972, Balefire, 1983, The Alchemist, 1985, Digger, 1990, Prey, 1992. Fellow Am. Acad. Forensic Scis.; mem. Am. Assn. Crime Lab Dirs. (charter) Authors Guild, Mystery Writers Am. Achievements include development of first national and international wildlife crime laboratory in the world. Office: Nat Fish Wildlife Forensics 1490 E Main St Ashland OR 97520

GODDARD, RALPH EDWARD, electrical engineer; b. East Liverpool, Ohio, Sept. 28, 1951; s. Ralph and Charlene June (Stephens) G. BSEE cum laude, Ohio State U., 1979, MSEE, 1981, PhD, 1990. Rsch. assoc. Ohio State U., Columbus, 1979-80; mem. tech. staff GE Aerospace, Binghamton, N.Y., 1980-81; subsystem engr. TRW, Redondo Beach, Calif., 1982-86; teaching assoc. Ohio State U., Columbus, 1986-90, postdoctoral researcher, 1990-91; work unit mgr. Jet Propulsion Lab., Pasadena, Calif., 1991—. Contbg. author: Micro Processors in Robotics and Manufacturing, 1992; contbr. articles to profl. jours. Grantee GE, 1988, 89. Mem. IEEE, AIAA, Tau Beta Pi, Eta Kappa Nu. Achievements include design and constrn. of a digital image processor. Home: 5003 Halison St Torrance CA 90503 Office: The Jet Propulsion Lab 4800 Oak Grove Dr Pasadena CA 91109

GODDARD, WILLIAM ANDREW, III, chemist, applied physicist, educator; b. El Centro, Calif., Mar. 29, 1937; s. William Andrew and Barbara Worth (Bright) G.; m. Yvonne Amelia Correy, Oct. 27, 1957; children: William Andrew, Susan Yvonne, Cecelia Monique, Lisa Sharéll. B.S. in Engring. with highest honors, UCLA, 1960; Ph.D. in Engring. Sci, Calif. Inst. Tech., 1964. Mem. faculty Calif. Inst. Tech., Pasadena, 1964—; asso. prof. theoretical chemistry Calif. Inst. Tech., 1971-75, prof. theoretical chemistry, 1975-78, prof. chemistry and applied physics, 1978-84, Charles and Mary Ferkel prof. chemistry and applied physics, 1984—, dir. Caltech-NSF materials research group, 1985-91; dir. materials/molecular simulation ctr. Beckman Inst., Calif. Inst. Tech., 1990—; co-founder Molecular Simulations, Inc., 1984, Schrodinger Inc., 1990; vis. staff mem. Los Alamos Sci. Lab., 1973-92; cons. GM Rsch. Labs., 1978—, Argonne Nat. Lab., 1978-82, GE Rsch. and Devel. Labs., 1982—, Failure Analysis Assocs., 1988-90; mem. adv. com. for chemistry NSF, 1983-86, chmn., 1985-86; mem. coun. Gordon Rsch. Confs., 1985-87, trustee, 1987-93. Mem. adv. editorial bd. Chem. Physics, 1972-92, Catalysis Letters, 1988—, Computer Methods in Materials Sci. & Engring, 1991—. Mem. bd. on chem. scis. and tech. Nat. Rsch. Coun., 1985-88, mem. nat. materials adv. bd. com. on tribology of ceramics, 1985-88, mem. U.S. nat. com. for internat. union of pure and applied chemistry, commn. on physical scis., math. and resources, 1985-88, mem. nat. materials adv. bd. com. on computer simulation and analysis on complex materials phenomena, 1986-88, mem. bd. on computer sci. and tech., 1988-91, mem. chemistry panel Nat. Sci. Found. Grad. Fellowship program, 1990-93; mem. adv. coun. Dept. Chemistry Princeton U., 1987—; mem. vis. com. for chemistry dept. Weiss Sch. Natural Scis., Rice U., Houston, 1989—. Recipient Buck-Whitney medal for major contbns. in chemistry, 1978; NSF fellow, 1960-61, 62-64; Shell Found. fellow, 1961-62; Alfred P. Sloan Found. fellow, 1967-69. Fellow Am. Phys. Soc., Am. Assn. Advance Sci.; mem. Nat. Acad. Scis., Materials Rsch. Soc., Am. Chem. Soc. (award for computers in chemistry 1988), Am. Vacuum Soc., Calif. Catalysis Soc., Sigma Xi, Tau Beta Pi. Home: 955 Avondale Rd San Marino CA 91108-1133 Office: Calif Inst Tech Beckman Inst Mail Code 139-74 Pasadena CA 91125

GODET, MAURICE, mechanical engineer. With Lab de Mechanique des Contacts, Villeurbanne Cedex, France. Recipient Mayo D. Hersey award ASME, 1992. Office: Lab de Mechanique des Contacts, Bat 113-INSA, 69621 Villeurbanne Cedex, France*

GODETTE, STANLEY RICKFORD, microbiologist; b. Berbice, Guyana, Nov. 27, 1940; s. Wilfred Arnold and Dorothy Charles (Jordan) G.; m. Desiree Carole Hazlewood, Sept. 9, 1967; 1 child, Dionne Carole. BS, Howard U., 1965, MS, 1972. Microbiologist Children's Nat. Med. Ctr., Washington, 1965-69, lab. mgr., 1969-79; health care cons. DIMPEX Assocs., Washington, 1979-85; dir. Met. Health Svcs. Lab., Washington, 1985-90; asst. adminstr. Al-Amal Hosp., Jeddah, Saudi Arabia, 1990-92; pres. Fredol Assocs., Ft. Washington, Md., 1992—. Mem. Am. Soc. for Med. Tech. (bd. dirs. 1980—), N.Y. Acad. Scis., Clin. Lab. Mgmt. Assn., Gyana Berbice Assn. (treas. 1985-90, pres. 1990—, award of merit 1990). Home: 9601 Wedgewood Dr Fort Washington MD 20744-5715

GODFRAIND, THEOPHILE JOSEPH, pharmacologist educator; b. Bande, Belgium, Feb. 18, 1931; m. De Becker Anne, July 12, 1957; children: Pierre, Catherine. MB, U. Libre de Bruxelles, Belgium, 1951; MD, U. Catholique de Louvain, Belgium, 1955; Cert., Inst. de Med. Tropicale, Anvers, Belgium, 1958; PhD, U. Catholique de Louvain, Belgium, 1958. Prof. U. Lovanium, Leopoldville, Congo, 1958-65, Université Catholique de Louvain, Brussels, 1965—; fellow Royal Acad. Medicine, Brussels, 1974-88, v.p., 1988-91, pres., 1991—; sec. gen. Internat. Union Pharmacology, 1987—. Recipient Lauréat du Concours des Bourses de Voyage, 1955, Lauréat du Prix Spécia, 1955, Lauréat du Priz J.F. Heymans, 1967, Lauréat du prix Quinquennal des Sciences Thérapeutiques, 1973, Lauréat du Prix Smith Kline, 1982, Peter Debye prize U. Limburg, 1987, Lauréat du Prix de la Fondation de Physiopathologie Prof. Lucien Dautrebande, 1988, ASPET award, 1991. lMem. Acad. Royale de Médecine de Belgique, Acad. Nat. de Médecine de France, Acad. Nat. de Pharmacie de France, Acad. Europaea, Assn. des Physiologist, Deutsche Pharmakologische Gesellschaft, Biochem. Soc., Brit. Pharmacol. Soc., Physiol. Soc., N.Y. Acad. Scis., Am. Soc. for Pharmacology and Exptl. Therapeutics. Home: Rue du Bémel 19, B 1150 Brussels Belgium Office: Lab de Pharmacologie, UCL 5410, Av Hippocrate 54, B 1200 Brussels Belgium

GODFREY, BRENDAN BERRY, physicist; b. Norfolk, Va., July 17, 1945; s. Wehrle Berry and Lillian Josephine (Hauan) G.; m. Kathryn Dorothy Burke, June 17, 1972; children: Helen Kathryn, Brendan Berry Jr., Elizabeth Dorothy. BS, U. Minn., 1967; PhD, Princeton U., 1970. Project officer Air Force Weapons Lab., Albuquerque, 1970-72, chief scientist, 1989-90; group leader Los Alamos (N.Mex.) Nat. Lab., 1972-79; v.p. Mission Rsch. Corp., Albuquerque, 1979-89; dir. advanced weapons Phillips Lab., Albuquerque, 1990—; adj. prof. Univ. N.Mex., Albuquerque, 1982—; reviewer several physics jour., 1974—. Author several multi-dimensional plasma simulation computer programs; contbr. articles to profl. jours., publs. 1st lt. USAF, 1970-72. Fellow Am. Phys. Soc. (exec. bd. 1989-91); mem. IEEE (sr. mem.), U. S. Chess Fedn. (master). Achievements include patent for "E-Beam ionized channel guiding of a relationistic electron beam"; rsch. in relativistic electron beam applications, high-power microwave sources, particle accelerators and numerical methods. Office: Phillips Lab PL/WS Kirtland AFB NM 87117-6008

GODFREY, DAVID WILFRED HOLLAND, aeronautical engineer, communication educator; b. Whitstable, Eng., May 28, 1928; arrived in Can., 1959; s. Wallace Frederick and May Florence (Copping) G.; m. Joy Carole Lewis, Sept. 1, 1962; children: Michael, Mark, Daniel. Grad., Royal Aero. Soc., London, 1948. Registered profl. engr., Ont., Can.; chartered engr., London. Tech. asst. De Havilland Propellers Ltd., Hatfield, Eng., 1949-54; asst. tech. editor The Aeroplane, Temple Press, London, 1954-59; infor. svcs. specialist Canadair Ltd., Montreal, Que., Can., 1959-66; sr. info. rsch. specialist Lockheed-Calif. Co., Burbank, 1966-71; prof. Ryerson Poly. U., Toronto, Ont., Can., 1971-93; engring. editor Wings and Helicopters, Calgary; Can. correspondant Interavia Air Letter, Revista Aerea, N.Y.C.; freelance editorial cons. Author: Modern Technical Communication, 1983; contbr. articles to profl. jours. Fellow AIAA (assoc.); mem. Can. Aviation Hist. Soc., Royal Aero. Soc., Aviation Space Writers Assn. (excellence award Can. region 1977, 90). Home and Office: 653 Village Pky # 35, Unionville, ON Canada L3R 2R2

GODFREY, GEORGE CHEESEMAN, II, surgeon; b. Atlantic City, Oct. 15, 1926; s. William M. and Elizabeth (Uzzell) G.; m. Evelyn Fry, Sept. 20, 1952; children: Cheryl Lynn, George Cheeseman III. Student St. Bonaventure Coll., 1944, U. Ky., 1945; AB, Colgate U., 1948; MD, Jefferson Med. Coll., 1952. Intern, Atlantic City Hosp., 1952-53; resident in gen. surgery U.S. VA Hosp., Ft. Howard, Balt., 1953-57; practice medicine specializing in surgery, Somers Point, N.J., 1957—; chief gen. and trauma surgery, dir. dept. surgery Shore Meml. Hosp., Somers Point, 1973-76, dir. surgery, 1982-87; instr. surgery Jefferson Med. Coll., Phila., 1958-84; cons. in orthopedics and neurology N.J. Div. Disability Determinations, N.J. Rehab. Program, 1960—; physician FAA Tech. Center, part-time, 1977—, med. mgr., 1982-85; pres. Shore Surg. P.A., Atlantic Indsl. Med. Assocs. Contbr. article to profl. jours. Pres. Linwood (N.J.) Bd. Edn., 1972-73; mem. Atlantic County United Way, Atlantic County YMCA, Atlantic Performing Arts. Served with U.S. Army, 1944-46. Recipient Disting. Service award N.J. Jr. C. of C., 1960, Maroon Citation, Colgate U. Diplomate Am. Bd. Surgery, Nat. Bd. Med. Examiners. Fellow ACS; mem. AMA, Am. Trauma Soc., Am. Soc. Abdominal Surgeons, Aerospace Med. Assn., Am. Coll. Occupational and Environ. Medicine, N.J., Atlantic County Med. Socs., Atlantic Indsl. Med. Physicians (pres.), Chainede Rotesseurs, Atlantic City Country Club, Marriott Seaview Country Club, Resorts Internat. Racquet Club, Kiwanis, Masons, Shriners, KT, Phi Kappa Tau, Phi Beta Pi. Methodist. Home: 112 Glenside Ave Linwood NJ 08221-2424 also: 5550 N Ocean Dr West Palm Beach FL 33404 also: 674 Shore Rd Somers Point NJ 08244 also: 1616 Pacific Ave Atlantic City NJ 08401 also: 705 White Horse Pike Absecon NJ 08201

GODFREY, JOHN CARL, medicinal chemist; b. Cornelius, Oreg., Mar. 11, 1929; s. Carl H. and Ruth Emma (James) G.; m. Nancy Jane Williams, June 12, 1954; children: Laura Alexis, Helen Rebecca, Sabrina Lee. BA in Chemistry, Pomona Coll., Claremont, Calif., 1951; PhD in Organic Chemistry, U. Rochester, 1954. Rsch. chemist Shell Devel. Co., Emeryville, Calif., 1954-55; instr. chemistry Rutgers U., New Brunswick, N.J., 1955-59; asst. dir. clin. rsch. Bristol Labs., Syracuse, N.Y., 1959-79, Revlon Health Care, Tuckahoe, N.Y., 1979-86; assoc. dir. clin. rsch. Rorer Pharm. Corp., Horsham, Pa., 1986-90; pres. Godfrey Sci. & Design, Inc., Huntington Valley, Pa., 1979—, cons., 1990—; dir. Hill Abrasives, Inc., Ontario, N.Y., 1990—; mem. sci. adv. bd. Quigley Corp., Doylestown, Pa., 1992—, adv. coun. Preconception Care Found., Inc., Quakertown, Pa., 1993—. Contbr. over 53 articles to profl. jours. NSF fellow, 1951; DuPont fellow, 1952-53. Fellow Am. Inst. Chemists; mem. AAAS, Am. Soc. Microbiology, Am. Chem. Soc. Achievements include patents for formulation to deliver active zinc in treatment of common cold (U.S., U.K., Can.), 23 others; invention of Godfrey Stereomodels which uniquely demonstrate mechanisms of formation, properties and reactions of new class of carbon compounds known as Fullerenes. Home: 1649 Old Welsh Rd Huntingdon Valley PA 19006-5835 Office: Godfrey Sci & Design 1649 Old Welsh Rd Huntingdon Valley PA 19006-5835

GODFREY, MAURICE, biomedical scientist; b. Addis Ababa, Ethiopia, June 11, 1956; s. Robert and Liliana (Gandolfi) G.; m. Matilde Elena Almeida, July 5, 1985; children: C. Maximilian, R. Alessandro, D. Guillermo. BS, Monmouth Coll., 1977; MS, Columbia U., 1980, M in Philosophy, 1983, PhD, 1986. Postdoctoral fellow Oreg. Health Sci. U., Shrine Hosp., Portland, 1986-89; asst. prof. pediatrics, dir. connective tissue lab. U. Nebr. Med. Ctr., Omaha, 1990—. Author: (with others) McKusick's Heritable Disorders of Connective Tissue, 1992; contbr. articles to profl. jours. Grant-in-aid Am. Heart Assn., 1989, 93; Basil O'Connor scholar March of Dimes, 1991; recipient First award NIH, 1992. Mem. AAAS, Am. Soc. of Human Genetics, Am. Fedn. for Clin. Rsch., Basic Sci. Coun. of the Am. Heart Assn. Achievements include co-discovery of fibrillin gene the cause of the Marfan syndrome. Office: U Nebr Med Ctr 600 S 42d St Omaha NE 68198

GODFREY, PAUL JOSEPH, science foundation director; b. Brockton, Mass., Sept. 29, 1944; s. Joseph and Jeannette Aldora (Paul) G.; m. Laurie Ann Rohde, June 13, 1968; children: Darren, Mollie. BS, Tufts U., 1966; MS, U. Mass., 1970; PhD, Cornell U., 1977. Staff asst. Water Resources Rsch. Ctr. U. Mass., Amherst, 1978-80, dir. Water Resources Rsch. Ctr., 1980—. Editor: Ecological Considerations in Wetland Treatment of Municipal Wastewater, 1985. Recipient Cert. of Recognition Gov. Michael Dukakis, 1983, Silver Trout award Trout Unltd., 1984, Conservation award Gulf Oil, 1985, Searching for Success award Renew Am., 1990, Environ. Programs award Friends of the UN, 1990. Mem. Am. Soc. Limnological Oceanography, Nat. Insts. Water Resources (sec.-treas. 1987-93, treas. 1993—), Soc. for Internat. Limnologists. Office: U Mass Blaisdell House Amherst MA 01003

GODO, EINAR, computer engineer; b. Aalesund, Möre, Norway, May 31, 1926; came to U.S. 1953; s. Lars and Oline (Blindheim) G.; m. Betty Jane Graba, 1955; children: Kjell Einar, Greta Anne, Erik Lars. BS in Aero. Engring., U. Wash., 1956, BSEE, 1958, MS, 1964. Electronic designer Boeing Aerospace, Seattle, 1959-82; prime investigator Computer Devel., Bellevue, Wash., 1982—. Achievements include patent for bating machine (Norway), Word Recognition System(USA, Can., Japan, Brit., France, Ger.); contributor engineering to most or all programs for putting man/hardware on the moon including lunar orbiter, lunar rover, Saturn 5 booster.

GODSEY, WILLIAM COLE, physician; b. Memphis, Dec. 11, 1933; s. Monroe Dowe and Margaret Pauline (Cole) G.; m. Norma Jean Wilkinson, June 18, 1958; children: William Cole, John Edward, Robert Dowe. B.S., Rhodes Coll., 1955; M.D., U. Tenn., 1958. Diplomate Am. Bd. Psychiatry and Neurology. Intern John Gaston Hosp., Memphis, 1958-59; resident in psychiatry Gailor Meml. Hosp., Memphis, 1960-63; practice medicine specializing in psychiatry and neurology; asst. supt. Memphis Mental Health Inst., 1965-74; supt. Cen. State Hosp., Nashville, 1974-75; med. dir. Whitehaven Mental Health Ctr., Memphis, 1975-84, St. Joseph Hosp. Life Ctr., 1984-88; dir. West Tenn. Civilian Material Assistance, Memphis, 1988—; mem. staff Bapt. Meml. Hosp., Lakeside Hosp., St. Joseph Hosp., Eastwood Hosp.; asst. prof. U. Tenn. Coll. Medicine, 1965-74, Coll. Pharmacy, 1972-75; chief of staff Lakeside Hosp., Memphis, 1976-77; songwriter, pub.; pres. Memphis Country Music, Inc. Dir. West Tenn. sect. Civilian Material Assistance, past pres. Memphis chpt. Fellow Am. Psychiatric Assn. (past pres. West Tenn. chpt.); mem. Tenn. Psychiat. Assn. (exec. coun., past pres.), Tenn. Med. Assn., Memphis and Shelby County Med. Soc., Nat. Rifle Assn., Moose. Methodist. Office: 5118 Park Ave Ste 323 Memphis TN 38117-5711

GODSON, WARREN LEHMAN, meteorologist; b. Victoria, B.C., Can., May 4, 1920; s. Walter Ernest Henry and Mary Edna (Lehman) G.; m. Harriet Burke, Dec. 28, 1977; children: Elliott, Marilyn, Murray, Ralph, Ellen; stepchildren—Alan, Alison, Stephen Bloom. BA, U. B.C., Vancouver, 1939, MA, 1941; MA, U. Toronto, Ont., 1944, PhD, 1948; DSc (hon.), U. Victoria, B.C., 1992. Lab. demonstrator U. B.C., 1939-41; lab. demonstrator U. Toronto, 1941-42, spl. lectr. physics dept., 1948-61, hon. prof., 1975; meteorologist Can. Meteorol. Service (name changed to Atmospheric Environ. Service 1971), Toronto, 1942-51; supt. atmospheric research sect. Can. Meteorol. Service (name changed to Atmospheric Environ. Service 1971), 1951-72, dir. atmospheric processes research br., 1972-73; dir. gen. Atmospheric Research Directorate, 1973-84, sr. sci. advisor, 1984-91, research scientist emeritus, 1991—. Author: (with J.V. Iribarne) Atmospheric Thermodynamics, 1974, 81; Contbr. (over 100) articles to profl. jours. Recipient Gov.-Gen.'s medal, 1935, Lefevre Gold medal, 1939, Patterson medal, 1968, IMO prize, 1975, Ministerial Gold medal, 1992. Fellow Royal Soc. Can., Am. Meteorol. Soc. (councillor 1967-70, assoc. editor Jour. Atmospheric Scis. 1962-70); mem. World Meteorol. Orgn. (v.p. Commn. for Atmospheric Scis. 1957-61, pres. 1973-77, alt. permanent rep. of Can. 1977-84, chmn. six working groups), Internat. Assn. Meteorology and Atmospheric Physics (sec., bur. dir. 1960-75, v.p. 1975-79, pres. 1979-83, convenor com. on meteorol. data for research 1960-64), Can. Meteorol. Soc. (exec. com. 1955-61, pres. 1957-59, Pres.'s prize), Can. Assn. Physicists (councillor 1955-57, chmn. earth physics div. 1967-68), Royal Meteorol. Soc. (v.p. Can. 1959-61, Can. Darton prize, Buchan prize 1964). Pioneer Arctic stratospheric jet stream and final warming process in polar winter stratosphere, Curtis-Godson approximation technique, Ozonagram diagram used for ozone representation. Home: 39 Dove Hawkway, North York, Willowdale, ON Canada M2R 3M8 Office: Atmospheric Environ Svc, 4905 Dufferin St, North York, ON Canada M3H 5T4

GODWIN, LARS DUVALL, civil engineer; b. Kinston, N.C., Mar. 22, 1953; s. Roland Duvall and Jackie Geraldine (Oliver) G.; m. Deborah Anne Credle, May 16, 1987. BSCE, N.C. State U., 1975. Community devel. engr. City of Sanford, N.C., 1976-77; environ. engr. N.C. Dept. Natural Resources, Raleigh, 1977-84; county engr. Wake County, Raleigh, 1984—. Mem. ASCE, N.C. Assn. Floodplain Mgrs. (treas. 1990-92), Am. Water Works Assn., Water Pollution Control Assn., Nat. Assn. County Engrs. Office: Wake County 312 Fayetteville St Mall Raleigh NC 27602

GODWIN, STEPHEN ROUNTREE, not-for-profit organization administrator; b. Morehead City, N.C., Sept. 17, 1953; s. James Leroy and Elizabeth (Rountree) G.; m. Margaret Louise Hilton, Apr. 7, 1983; children: Miriam Elizabeth Godwin, James Rountree Godwin. BA, U. N.C., 1975, M in Regional Planning, 1980. Program analyst HUD, Washington, 1979-81; rsch. assoc. Urban Inst., Washington, 1981-83; study dir. Transp. Rsch. Bd. Nat. Rsch. Coun., Washington, 1983—. Contbr. articles, reports to profl. jours. Mem. Transp. Rsch. Forum, Phi Beta Kappa. Office: Transp Rsch Bd 20001 Wisconsin Ave NW Washington DC 20007

GOEDDE, JOSEF, physiologist; b. Ahlen, Germany, Nov. 20, 1953; s. Bernhard and Maria (Pollmeier) Gödde; m. Dorothea C. Neugebauer, Aug. 4, 1984; (div. Sept. 1986). Diploma, U. Muenster, 1978, PhD, 1983. Univ. asst. U. Regensburg, Germany, 1984-91; developer, cons. Haag Elektronische Messgerate, Waldbrunn. Roman Catholic. Avocation: solar energy. Home: Steinklepperweg 1, Greifenstein-Nenderoth 35753, Germany Office: Haag Elektronische Gerate, Emil-Hurm-Str 18-20, 65620 Waldbrunn 2, Germany

GOEHMAN, M. CONWAY, mechanical engineer; b. Jackson, Mo., June 15, 1938; s. Marvin C. and Helen L. (Chapman) G.; children: Tanya L. Goehman Busch, Zane L., Melissa A.; m. Alice L. Mayfield. BSME, Mo. Sch. Mines and Metallurgy, 1962. Registered profl. engr., Mo.; cert. plant engr. Dep. crew comdr. USAF Titan II ICBM, Tucson, 1961-65; process engr. Ramsey Corp., Div. of TRW, Sullivan, Mo., 1966-67; asst. plant engr. ASARCO Inc., Glover, Mo., 1967-79, plant engr., 1979-88, engring. mgr., 1988—. Author: Lead Blast Furnace Design & Construction, Belt Conveyors and Drives, 1983. 1st lt. USAF, 1961-65. Mem. NRA (life), Am. Inst. Plant Engrs., S.E. Mo. Civil War Roundtable. Baptist. Home: Rte 1 Box 121 Arcadia MO 63621 Office: ASARCO Inc Rte 1 Box 60 Annapolis MO 63620

GOEI, BERNARD THWAN-POO (BERT GOEI), architectural firm executive; b. Semarang, Indonesia, Jan. 27, 1938; came to U.S., 1969; naturalized, 1976; s. Ignatius Ing-Khien Goei and Nicolette Giok-Nio Tjioe; m. Sioe-Tien Liem, May 26, 1966; children: Kimberley Hendrika, Gregory Fitzgerald. BA in Fine Arts, Bandung Inst. Tech. State U. Indonesia, 1961, MA in Archtl. Space Planning, 1964; postgrad., U. Heidelberg, Germany, 1967-68. Co-owner, chief designer Pondok Mungil Interiors Inc., Bandung, 1962-64; dept. mgr., fin. advisor Gumarna Architects, Engrs. and Planners, Inc., Bandung, Jakarta, Indonesia, 1964-67; shop supr., model maker Davan Scale Models, Toronto, Ont., Can., 1968-69; chief archtl. designer George T. Nowak Architects and Assocs., Westchester, Calif., 1969-72; sr. archtl. designer Krisel & Shapiro Architects and Assocs., L.A., 1972-74; sr. supervising archtl. designer The Ralph M. Parsons A/E Co., Pasadena, Calif., 1974—; v.p. United Gruno U.S.A. Corp. Import/Export, Monterey Park, Calif., 1980-89. Mem. Rep. Presdl. Task Force, Washington, 1982—, Nat. Rep. Senatorial Com., Washington, 1983—, Nat. Rep. Congrl. Com., Washington, 1981—, Rep. Nat. Com., 1982—. Recipient Excellent Design Achievement commendation Strategic Def. Initiative "Star Wars" Program, 1988, USAF Space Shuttle Program, West Coast Space-Port, 1984; scholar U. Heidelberg, 1967-68. Mem. NRA, Indonesian Am. Soc., Dutch Am. Soc., Second Amendment Found., The Right to Keep and Bear Arms Com. Republican. Roman Catholic. Avocations: fire arms and daggers, photography, hi-tech electronics, stamps and coins, world travel. Office: Ralph M Parsons A/E Co 100 W Walnut St Pasadena CA 91124-0001

GOEL, ADITYA PRASAD, engineering educator, consultant; b. Columbia, S.C., Jan. 14, 1966; s. Kanti and Uma Rani (Goel) Prasad. BSEE, U. Lowell, 1987, MSEE, 1990. Engr. Hybricon Corp., Ayer, Mass., 1989-90; asst. prof. Merrimack Coll., North Andover, Mass., 1990—. Contbr. articles to profl. jours. Mem. IEEE, Sigma Xi, Tau Beta Pi, Eta Kappa Nu. Hindu. Home: 25 Selfridge Rd Bedford MA 01730 Office: 9 Redcoat Rd Bedford MA 01730

GOEL, AMRIT LAL, engineering educator; b. Meerut Cantt, India, Mar. 4, 1938; came to U.S., 1962; s. Gujjar Mal and Shanti Devi (Mittal) G.; m. Norma Lynn Currie, Mar. 27, 1967; children: Alok, Nandita, Neha. BS, Agra U., India, 1957; BE, U. Roorkee, India, 1961; MS, U. Wis., 1963, PhD, 1968. Asst. mech. engr. Atomic Energy Commn., Bombay, 1961-62; vis. prof. U. Md., College Park, 1962-63, Vienna (Austria) Tech. U., summer 1963; instr. Peace Corp U. Wis., Milw., 1964; instr. engring. U. Wis., Madison, 1965-66; from asst. prof. to prof. Syracuse (N.Y.) U., 1968—; pres. Software Modeling & Analysis, Inc., Fayetteville, N.Y., 1985—, CSMA, Inc., Fayetteville, 1991—. Editor: Internat. Jour. Reliability, Charity and Safety, 1992—; guest editor: Jour. Systems and Software, 1984; contbr. articles to profl. jours. Rsch. grantee NSF, U.S. Army, USAF, NASA, 1969-92. Fellow IEEE (guest editor jour. 1985, 86, P.K. McElroy Best Tech. Paper award Reliability Soc. 1979, Disting. Visitor award Computer Soc. 1988-91), AAAS, Am. Statis. Assn., Sigma Xi (pres. Syracuse U. chpt. 1981-82). Home: 5011 Woodside Rd Fayetteville NY 13066 Office: Syracuse U 2-183 Ctr for Sci & Tech Syracuse NY 13244

GOEL, VIJAY KUMAR, biomedical engineer; b. Sangrur, India, Mar. 5, 1945; came to U.S., 1979; s. Radha K. and Laxmi D. (Mittal) G.; m. Shashi Rani Gupta, June 5, 1972; children: Anuj, Anish. BME with honors, Panjabi (India) U., 1966; MME, Roorkee (India) U., 1971; PhD, U. New South Wales, Australia, 1977. Asst. prof. mech. engring. Thapar Coll. Engring., Patiala, India, 1966-74; teaching fellow U. New South Wales, Kensignton, Australia, 1974-77; asst. prof. biomed. engring. Indian Inst. Tech., Hauz Khas, India, 1978-79; rsch. assoc. orthopaedics Yale U., 1979-82; asst. prof. biomed. engring. U. Iowa, Iowa City, 1982-86, assoc. prof., 1986-90, prof. orthopaedics and biomed. engring., 1990—, dir. biomechanics lab. II, 1988—, chmn. dept. biomed. engring., 1990—; lectr. in field. Author: Mechanical Machine Design, India, 1974, 2d edit., 1982; editor: (with J.N. Weinstein) Biomechanics of the Spine: Clinical and Surgical Perspective, 1989, 2d edit., 1990; contbr. numerous articles to Jour. Ins. Engrs., Jour. Biomechanics, Complete Biol. Medicine, Jour. Biomech. Engring., and others. Recipient Volvo award Internat. Soc. Study of Lumbar Spine, 1981, 90; Old Gold fellow, 1983-84, various drs. fellow, 1985-92; grantee NIH, 1983—, AcroMed Corp., 1987—, Acsulap Inc., 1991-93, and others. Mem. ASME, Am. Soc. Biomechanics, Am. Soc. Engring. Edn., Am. Inst. Med. and Bio. Engring., Internat. Soc. for Study of Lumbar Spine, Orthopedic Rsch. Soc., Cervical Spine Rsch. Soc., Coun. of Chairs Biomed. and Bioengr. Undergrad. Programs, Rotary Internat. Achievements include research in lumbar spine, orthopaedics; head, neck and spine injuries. Office: U Iowa 1202 Engring Bldg Iowa City IA 52242

GOENKA, PAWAN KUMAR, mechanical engineer; b. Harpalpur, India, Sept. 23, 1954; s. Ram Kishore and Savitri Devi (Saraf) G.; m. Mamta Ruia, Jan. 16, 1981; children: Pooja, Puneet. B in Tech., Indian Inst. Tech., Kanpur, India, 1975; MS, Cornell U., 1978, PhD, 1979. Registered profl. engr. Mich. Sr. rsch. engr. GMR, Warren, Mich., 1979-83, staff rsch. engr., 1983-88, sr. staff rsch. engr., 1988-92, sect. mgr. engine structure and components, 1992—. Assoc. editor ASME Jour. Tribology, 1991—. Mem. Soc. Automotive Engrs., Soc. Tribologists and Lubrication Engrs., Am. Soc. Mech. Engrs. (Best Paper award, 1985, Burt L. Newkirk award 1987). Home: 4540 Whisper Way Troy MI 48098 Office: GM Rsch 30500 Mound Rd Warren MI 48090

GOERSS, JAMES MALCOLM, statistician, electrical engineer; b. North Tonawanda, N.Y., Apr. 30, 1964; s. James Malcolm and Carol Ann (Setlik) G.; m. Lynnette Allis, Nov. 5, 1988. BSEE, Rochester Inst. Tech., 1987; MA, U. Pitts., 1990. Jr. engr. IBM, Endicott, N.Y. and Boca Raton, Fla., 1985-87; elec. engr. Westinghouse Elec. Corp., Monroeville, Pa., 1987-89; rsch. asst. U. Pitts., 1990, teaching asst., 1989-90; statis. process control engr. Compaq Computer Corp., Houston, 1991—. Recipient Statis. Rsch. grant U. Pitts., 1990. Mem. Am. Statis. Assn., Am. Soc. for Quality Control, Tau Beta Pi, Eta Kappa Nu. Office: Compaq Computer Corp 20555 SH 249 Houston TX 77070-2698

GOESCH, WILLIAM HOLBROOK, aeronautical engineer; b. Mpls., Mar. 15, 1937; s. William Extrand and Gladys (Holbrook) G.; m. Lorraine Hazel Hanson, July 23, 1961; children: William David, Nancy A. BS in Aero. Engring., U. Minn., 1961; MS in Aerospace Engring., U. Dayton, 1975; grad., Air War Coll., 1983. Civilian, gen. mgr. Dept. of Air Force, 1961; aero. engr. Air Force Flight Dynamics Lab., Wright-Patterson AFB, Ohio, 1961-67, group leader, 1971-74; dep. program mgr. Air Force Materials Lab., Wright-Patterson AFB, Ohio, 1974-75; program mgr. Air Force Flight Dynamics Lab., Wright-Patterson AFB, 1975-77, dep. program element mgr., 1977-79, br. chief, 1979-89; div. chief Air Force Wright Lab., Wright-Patterson AFB, 1989—; chmn. MIL-HDBK-23, Wright-Patterson AFB, 1967-74; Air Force mem. NASA Space Shuttle Working Group, Johnson Space Ctr., 1971-73; Air Force mem. adv. composite structures Tech. Coordinating Panel-H-Action Group-5, DOD, 1979-83. Mem. AIAA, Sci. Rsch. Soc. Republican. Lutheran. Home: 4421 Powder Horn Dr Beavercreek OH 45432-4029 Office: Wright Lab XPT Wright Patterson AFB OH 45433

GOETHALS, ERIC JOZEF, chemistry educator; b. Ghent, Belgium, Dec. 6, 1936; s. Gaston and Marguerite (Mortier) G.; m. Eliane Baudewyns, May 25, 1960 (div. 1984); children: Ann, Tom; m. Yvette Vanderstraeten, Sept. 19, 1986; children: Gunther, Fabienne. Lic. chemistry, U. Ghent, 1958, PhD in Chemistry, 1963. Asst. prof. U. Ghent, Belgium, 1958-70; assoc. prof. U. Ghent, 1970-80, prof., 1980—. Editor: Polymeric Amines, 1980, Cationic Polymerization & Related Processes, 1984, Telechelic Polymers, 1989; editorial bd.: Makromol. Chemie, Polymer Internat., Receuil Trav. Chim. Pays-Bas, New Polymeric Materials, Macromolecular Reports; congbt. over 200 articles to profl. jours. Mem. Am. Chem. Soc. (affiliate mem., polymer div.), Royal Flemish Chem. Soc. Achievements include 10 patents in the field of polymer chemistry. Home: Huisgaverstraat 52, 9750 Ouwegem-Zingem Belgium Office: Univ Ghent, Lab Org Chem, Krijgslaan 281 S-4, 9000 Ghent Belgium

GOETTL, BARRY PATRICK, personnel research psychologist; b. Mankato, Minn., May 23, 1960; s. Ronald Alfonse and Mary Isabelle (Kallberg) G.; m. Pamela Lynne Harrison, July 1, 1989; 1 stepchild, Kimberly Shawn Townsend; 1 child, Meghan Aileen. BS in Psychology, U. Dayton, 1981; MA, PhD in Psychology, U. Ill., 1987. Asst. prof. Clemson (S.C.) U., 1987-92; pers. rsch. psychologist Air Force Armstrong Lab., Brooks AFB, Tex., 1992—; reviewer Human Factors, 1991—. Contbr. articles to Human Factors, Ergonomics, Perception and Psychophysics. Recipient Office Univ. Rsch. award Clemson U., 1989; Rsch. Initiation Program grantee USAF, 1989, travel grantee NSF, 1990, sci. affairs grantee NATO, 1990. Mem. Human Factors and Ergonomics Soc. (editor visual performance tech. group newsletter 1992-93), Am. Psychol. Soc. (charter). Home: 15034 Digger Dr San Antonio TX 78247 Office: Armstrong Lab/HRTI Bldg 578 Brooks AFB TX 78235

GOFF, HAROLD MILTON, chemistry educator; b. St. Louis, Sept. 24, 1947; s. Olden Milton and Veola Faye (Oliver) G.; m. Barbara Jean Kessell, June 19, 1971; children: Jason Edward, Justin Allen. BS in Chemistry, U. Mo., 1969, MA in Chemistry, 1971; PhD in Chemistry, U. Tex., 1976. Lab. technician River Cement Co., Festus, Mo., 1965-66; med. lab. instr. U.S. Army, San Antonio, 1971-73; postdoctoral fellow U. Calif., Davis, Calif., 1976; prof. U. Iowa, Iowa City, 1976—; metallobiochemistry study sect. NIH, Washington, 1985-89; mem. editorial bd. Inorganic Chemistry, Washington, 1993—. Contbr. articles to profl. jours. Named Univ. Faculty scholar U. Iowa, 1982-85; grantee NIH, 1986, NSF, 1993, Petroleum Rsch. Fund, 1991. Mem. Am. Chem. Soc., Iowa Acad. Sci., Alpha Chi Sigma Chemistry Frat. Achievements include approximately 150 presentations at acad. institutions and indsl. labs. Home: 445 Kimball Rd Iowa City IA 52245 Office: U Iowa Dept Chemistry Iowa City IA 52242

GOFF, KENNETH ALAN, engineering executive; b. Providence, June 16, 1941; s. Maurice A. and Florence E. (Gillett) G.; m. Cristel Noe, Sept. 7, 1963; children: David, Karl. BSCE, U. Mass., 1963; MSCE, Northwestern U., 1969. Registered profl. engr., Mass. Chief engr. water resources Stone and Webster Engring. Corp., Boston, 1984-86, asst. chief engr. civil/structural engring., 1987-89, project mgr., dir., 1987-90; v.p., mgr. Stone and Webster Engring. Corp., Ft. Lauderdale, Fla., 1990-91, N.Y.C., 1991—; bd. dirs. N.Y.C. Bldg. Congress, N.Y.C. Transit Mus. Mem. Am. Water Works Assn., Water Environ. Fedn., New Eng. Water Works Assn., New Eng. Water Pollution Control Fedn.,. Office: Stone and Webster Engring 1 Penn Plz 250 W 34th St New York NY 10119

GOFFI, RICHARD JAMES, nuclear engineer; b. Denver, Mar. 28, 1963; s. Henry Joseph and Marilyn Helen (Burns) G.; m. Angela J. Eckerd, Sept. 23, 1989. BS, U.S. Naval Acad., 1985; student, U. Md., 1991—. Officer U.S. Navy, various cities, 1985-90; tech. analyst NASA Johnson Space Ctr., Houston, 1990-91; nuclear engr. Jacob's Engring. Group, Washington, 1991—. Lt. USN, 1981-90. Mem. AIAA (Individual Student Design award 1985), Am. Nuclear Soc. Home: L'Enfant Plz Box 23219 Washington DC 20026 Office: Weston Civilian Radioactive Waste Mgmt Support Team 955 L'Enfant Plz Ste 8000 Washington DC 20024

GOFFINET, SERGE, neuropsychiatrist, researcher; b. Brussels, Feb. 14, 1959; s. Gilbert Goffinet; children: Hubert, Hadrian Laurent. Degree in neuropsychiatry, U. Louvain, Belgium, 1985. Neuroscientist Pet Lab., Louvain-la-Neuve, Belgium, 1988—; psychoanalyst Brussels, 1990, family therapist, 1991; researcher Psychiatry Svc.; prof. psychiatric nursing. Contbr. articles to profl. jours. Mem. Assn. Freudienne, Ctr. Etudy Famille, Imaging Neuropsychiatry, Soc. Royale Medicine Mentale, Soc. Circulation et Metabolisme du Cerveau, Soc. Spiritualite Ethique et Hygiene Mentale. Home: 60 Ave des Volontaires, B 1040 Brussels Belgium Office: Clin U St Luc, Ave Hippocrate 10/1360, B-1180 Brussels Belgium also: Clin La Ramee, Ave Boetendael 34, B-1180 Brussels Belgium

GOFFMAN, THOMAS EDWARD, radiobiologist, researcher; b. Chgo., Apr. 16, 1953; s. Erving and Angelica (Choate) G.; m. Pamela Howard, June 28, 1987; 1 child, James Edward. BA, Yale U., 1975; MD, Hahnemann U., 1979. Diplomate Am. Bd. Radiology, Am. Bd. Internal Medicine. Intern, resident Georgetown U. Hosp., Washington, 1979-82; med. staff fellow, epidemiology tng. program Nat. Cancer Inst., NIH, Bethesda, Md., 1982-83; resident in radiotherapy, Joint Ctr. for Radiation Therapy Harvard U. Med. Sch., Boston, 1983-86; instr. in radiation oncology Columbia U., N.Y.C., 1986-87, asst. prof. of radiation oncology, 1987; attending in radiation oncology Washington Hosp. Ctr., 1987-89, vice chmn. dept. radiation oncology, 1988-89; asst. dir. radiation oncology Sibley Meml. Hosp., 1989; asst. clin. prof. radiation medicine Georgetown U., 1989—; head clin. therapy sect., radiation oncology br., Nat. Cancer Inst., Bethesda, 1989—; asst. prof. radiology USUHS, Bethesda, 1989-91; dir. radiation oncology tng. USUHS, Bethesda, 1989—; dir. radiation oncology tng. Nat. Cancer Inst., USUHS, Bethesda, 1990—; assoc. prof. radiology USUHS, 1991—; dir. radiation oncology St. Agnes Hosp., Balt., 1992—. Contbr. articles to profl. jours. Recipient Mosby scholarship for acad. achievement, 1979, Excellence in Medicine award, 1979, Blue Ribbon award, 1979, Nat. Rsch. Svc. award, 1983, Epidemiology Tng. fellowship Nat. Cancer Inst.-NIH, 1983. Mem. AMA, AAAS, ACP, N.Y. Acad. Sci., Com. on Physicians Assn., D.C. Med. Soc. (legis. com.), Nat. Cancer Inst. (internal review bd. 1989-90, biol. operating com. 1991—), Balt. City Med. Soc.

GOFFMAN, WILLIAM, mathematician, educator; b. Cleve., Jan. 28, 1924; s. Sam and Mollie (Stein) G.; m. Patricia McLoughlin, Feb. 7, 1964. B.S., U. Mich., 1950, Ph.D., 1954. Math. cons., 1954-59; research asso. prof. Case Western Res. U., Cleve., 1959-71; dean Case Western Res. U. (Sch. Library Sci.), 1971-77; dir. Case Western Res. U. (Complex Systems Inst.), 1972-75. Contbr. numerous publs. to sci. jours. Served with USAAF, 1943-46. Recipient research grants NSF, research grants NIH, research grants USAF, research grants others. Fellow AAAS. Home: II Bratenahl Pl Bratenahl OH 44108 Office: Case Western Res Univ Cleveland OH 44106

GOFRON, KAZIMIERZ JAN, physicist; b. Bochnia, Tarnow, Poland, Feb. 3, 1962; s. Franciszek and Maria (Przybylo) G.; m. Urszula Gofron. BS, U. Ill., Chgo., 1987, MS, 1988, postgrad., 1988—. Teaching asst. U. Ill., Chgo., 1987-90; lab. grad. Argonne (Ill.) Nat. Lab., 1990—. Contbr. articles to profl. jours. V.p. Grad. Student Coun., U. Ill., Chgo., 1991-92. Mem. Am. Phys. Soc. Office: U Ill 829 W Taylor Rm 2236 SES 801 W Taylor Rm 2344 SES Chicago IL 60607-7059

GOGAN, CATHERINE MARY, dental educator; b. Buffalo, Feb. 9, 1959; d. John Francis and Mary Louise (Solomon) G. BA, SUNY, Buffalo, 1981, DDS, 1985. Resident Erie County Med. Ctr., Buffalo, 1985-86, attending dentist, 1986—, dental residency coord., 1987—, dental dir. skilled nursing facility, 1989—; pvt. practice Buffalo, 1986—; clin. instr. SUNY, Buffalo, 1987-88, asst. prof., 1988—. Editor mag. UB Dental Report, 1989— (Golden Scroll award); contbr. articles to profl. jours. Fellow Am. Assn. Hosp. Dentists; mem. ADA, Am. Assn. Dental Schs., Orgn. Tchrs. Oral Diagnosis, Acad. Dentistry for Handicapped, Am. Soc. Geriatric Dentistry, U. Buffalo Dental Alumni Assn. (sec. 1988-89, v.p. 1989-90, pres. 1990-91), Mt. Mercy Acad. Alumni Assn. (bd. dirs. 1990-91), Omicron Kappa Upsilon. Roman Catholic. Office: SUNY at Buffalo Dept Oral Medicine 315 Squire Hall Buffalo NY 14214

GOGATE, KAMALAKAR CHINTAMAN, metallurgist; b. Pune, India, Dec. 17, 1940; came to U.S., 1990; s. Chintaman Gopal and Saraswati (Chintaman) G.; m. Urmila Madhuri Bhave, May 21, 1971; children: Bhagyasri, Chaitrali, Deepali. B. Engring. (Metallurgy), U. Poona, Pune, 1962, M. Engring. (Metallurgy), 1964. Chartered engr. Inst. of Engrs. (India). Lectr. Coll. of Engring., Aurangabad, India, 1963-65; supt. heat treatment Bharat Forge Ltd., Pune, 1965-79; mng. ptnr. Trinity Udyog, Pune, 1980-90; mng. dir. Trinity Thermal Pvt. Ltd., Pune, 1991—; metallurgy cons. Trinity Die Forgers Pvt. Ltd., Pune, 1980—, Trinity Forge Ltd., Ahmednagar, 1980—. Author scholar papers. Mem. Inst. of Metals India (life), Inst. of Engrs. (India), ASTM, Am. Soc. Materials Internat. USA. Home: Chiman Baag, 1551 Sadashiv Peth, Pune 411 030, India Office: Trinity Thermal Pvt Ltd, Plot No 151 S Block Bhosari, Pune 411 026, India

GOGGIN, NOREEN LOUISE, kinesiology educator; b. Cleve., Oct. 10, 1954; d. James Martin and Carla Elizabeth (Mannu) G. BS, Bowling Green U., 1976; MS, Pa. State U., 1978; PhD, U. Wis., 1989. Instr. Marietta (Ohio) Coll., 1978-79; adj. asst. prof. U. Wis.-Parkside, Kenosha, 1979-84; asst. prof. U. North Tex., Denton, 1989—. Contbr. articles to profl. publs. Vol. ARC, Denton, 1989—. Recipient Young Scholar award So. Assn. Phys. Edn. of Coll. Women, 1990, Travel award NSF, 1990. Mem. AAHPERD, N.Am. Soc. for Psychology of Sport and Physical Activity, Gerontol. Soc. Am. Office: U North Tex PO Box 13857 Denton TX 76203

GOGULSKI, PAUL, construction consultant; b. South Bend, Ind., Jan. 13, 1938; s. Cassimer and Bernice (Korecei) G.; m. Nuala Margaret O'Shea, Sept. 2, 1967; children: David, Stephen, Scott, Alaxander, Joseph. BSCE, U. Mich., 1960. Dir. constrn. Lanning Corp., Boston, 1973-77; devel. mgr. Kaiser Engrs., Boston, 1977-79; dir. facilities Teradyne Corp., Boston, 1979-81; project mgr., cons. Univ. Hosp., Ralph M. Parsons Corp., Boston, 1981-83, Saudi Arabia, 1981-83; v.p., dir. constrn. Drexel Properties, Dallas and Princeton, 1983-88; project mgr. NICO Constrn. Co., N.Y.C., 1988-91; prin. Gogulski & Co., Princeton, 1991—; instr. CPM Scheduling, Orlando (Fla.) Jr. Coll., 1992—. Author: Using Guaranteed Maximum Contracts to Reduce Contruction Cost and Improve Performance, Consulting Today, 1993. 2d lt. U.S. Army, 1960-61. Mem. ASCE, NSPE, Am. Arbitration Assn. (panel mem. 1992—), Facilities Devel. Corp. State N.Y. (chmn. panel for selection of contractors and cons., 1993—). Home: 24 Bayberry Ln Belle Mead NJ 08502 Office: Gogulski & Co Ste # 121 CN 5330 1330 Hwy 206 Princeton NJ 08543

GOHAGAN, JOHN KENNETH, medical institute administrator, educator; b. Barrington, N.J., July 24, 1939; s. John K. Gohagan and Marion A. Happ; m. Dorothy L. Childs. BS in Physics, LaSalle Coll., 1964; MS in Physics, Temple U., 1969; D, MIT/Harvard Sch. Pub. Health, 1973. Tchr. basic electonics and KO-6 digital electronic cryptographic equipment U.S. Army Signal Corps Sch., Ft. Monmouth, 1958-60; mem. exec. staff, project engr., rschr., educator U.S. Dept. Defense Armaments Commd./Frankford Arsenal, Phila., 1964-74; tchr. probability, stats., decision analysis, calculus ctr. for advanced engring. study divsn. sponsored rsch. MIT, 1971-73; acting dir. divsn. healthcare, prof. preventive medicine, dir. program indsl. prodn. mgmt., rschr., investigator schs. medicine and engring. Washington U., St. Louis, 1974-91; dir., project officer prostate, lung, colon-rectum and ovary cancer screening trial divsn. cancer prevention and control br. early detection Nat. Cancer Inst., Bethesda, Md., 1990—, chief br. early detection, 1992—; mem. sci. working group to explore issues in breast self-examination Nat. Cancer Inst., 1982, DRCCA Bd. Sci. Counselors. subcommittee on screening, 1982, breast cancer task force meeting, 1983, program project site visit team, 1987; mem. tech. advisors task forces for plan devel. on acute care, new. tech., and long term care, 1977; mem. long term care project review task force St. Louis Health Systems Agy, 1977-82, chmn. tech. innovative equipment task force emission computed tomography, 1978, radiation therapy tech. advisors task force, 1979, project review com., 1979-82, acute care project review task force, 1979-82; resource profl. Mo. Cancer Programs Planning, 1978-79; mem. biostatistics and epidemiology com. Mo. Cancer Programs, 1978-79; vis. rschr. divsn. cancer prevention and control Nat. Cancer Inst., 1985; sr. fellow dept. finance Wharton Sch. Business, vis. profl. divsn. clin. epidemiology dept. medicine U. Pa., Phila., 1989; proposal reviewer NSF, 1982; mem. sci. bd. advisors Somanetics Corp., 1985-88; cons. Fed. Hwy. Adminstrn., 1975, Mo. Dept. Revenue, 1975, Exec. Tng. Workshops, 1976-80, So. Ill. U., 1978, Interagency Inst. for Fed. Healthcare Execs., 1980, Pentax Corp., 1984, Boatman's Bank N.A., 1988-90; invited lectr. in field. Author: Quantitative Analysis for Public Policy, 1980, Solved Problems and Case Teaching Notes (companion to Quantitative Analysis for Public Policy), 1980, (with others) Early Detection of Breast Cancer: Risk, Detection Protocols, and Therapeutic Implications, 1982; co-contbr. chpts. to Anesthetic Side Effects and Complications: Seeking, Finding and Treating, 1980, Public Policy Programs: The First Ten Years, 1982, Magnetic Resonance Imaging, 1987, 2nd. edit., 1992, Magnetic Resonance in Experimental and Clinical Oncology, 1990; manuscript reviewer Science, 1978, Mgmt. Sci., 1978, Jour. Am. Planning Assn., 1982, Health Care, 1983, 89, Jour. Nat/ Cancer Inst., 1990-92; contbr., co-contbr. over 30 papers and reports in field; contbr. articles to profl. and popular jours. Chmn. St. Louis region MIT Alumni Fund Drives, 1977-82; active MIT Alumni Fund Bd., 1979-82. Rsch. grantee NASA, 1974-76, Am. Cancer Soc., 1976-82, Rand Grad. Inst., 1977, U.S. Congress Office Tech. Assessment, 1979-80, Washington U., 1979-80, 81-82, Henry J. Kaiser Found., 1983-84, Mallinckrodt Inst. Radiology, 1984, U.S. Dept. Health and Human Svcs., 1984-89, NIH, 1989-90; Dissertation grantee NSF; Systems Analysis fellow U.S. Civic Svc. Fellow Am. Coll. Epidemiology; mem. Sigma Xi. Office: Nat Cancer Inst Br Early Detection Divsn Cancer Prevention & Control 9000 Rockville Pike EPN 305 Bethesda MD 20892

GOKARN, VIJAY MURLIDHAR, pharmacist, consultant; b. Bombay, Sept. 16, 1953; came to U.S., 1977; s. Murli B. and Sunanda (Chandavarkar) G.; m. Aparna V. Burde, Mar. 28, 1979; children: Niketa, Sarita. BS in Pharmacy, Coll. Pharmacy, Manipal, India, 1975. With Am. Home Products, Bombay, India, 1975-77; prodn. mgr. Pvt. Formulations, Hempstead, N.Y., 1977-78, Tishon Corp., Westbury, N.Y., 1978-79; v.p. Excel Coating, Inc., Westbury, 1979-81; gen. mgr. Nat. Nutritional Labs., Huntington, N.Y., 1981-86; pres. VANS, Inc., S.I., N.Y., 1986-92, MaxPharma U.S.A., Inc., S.I., 1993—; cons. Nat. Nutritional Labs., Huntington, 1986-89. Office: MaxPharm USA Inc PO Box 061205 Staten Island NY 10306-0012

GOLAB, WLODZIMIERZ ANDRZEJ, biologist, geographer, librarian; b. Kozmin, Poland, Mar. 10, 1938; s. Andrzej and Irena (Borowska) G.; m. Maria Malgorzata Grygiel, June 18, 1969; children: Dagmara, Filip. Degree, U. Mickiewicza, Poznan, Poland, 1968. Libr. Polish Acad. Sci., Poznan, 1968-75; dir. reading rm. U. Poly, Poznan, 1975-81; dir. main libr. U. Agr., Poznan, 1981—; cons. Green Libr., Poznan, 1990-92; mem. com. info. sci. Polish Acad. Sci., 1972—t. Contbr. to profl. publs. Roman Catholic. Home: Osiedle Przyjazni 11/91, Poznan 61685, Poland Office: Biblioteka Glowna AR, Ul Witosa 45, Poznan Poland

GOLAND, MARTIN, research institute executive; b. N.Y.C., July 12, 1919; s. Herman and Josephine (Bloch) G.; m. Charlotte Nelson, Oct. 16, 1948; children—Claudia, Lawrence, Nelson. M.E., Cornell U., 1940; LL.D. (hon.), St. Mary's U., San Antonio. Instr. mech. engring. Cornell U., 1940-42; sect. head structures dept. research lab., airplane div. Curtiss-Wright Corp., Buffalo, 1942-46; chmn. div. engring. Midwest Research Inst., Kansas City, Mo., 1946-50; dir. for engring. scis. Midwest Research Inst., 1950-55; v.p. Southwest Research Inst., San Antonio, 1955-57; dir. Southwest Research Inst., 1957-59, pres., 1959—; pres. S.W. Found. Research and Edn., San Antonio, 1972-82; dir. Nat. Bancshares Corp. Tex.; chmn. subcom. vibration and flutter NACA, 1952-60; chmn. research adv. com. on aircraft structures NASA, 1960-68, chmn. materials and structures group, aeros. adv. com., 1979-82; sci. adv. com. Harry Diamond Labs., U.S. Army Materiel Command, 1955-75; adv. panel com. sci. and astronautics Ho. of Reps., 1960-73; mem. adv. bd. on undersea warfare Dept. Navy, 1968-70, chmn., 1970-73; mem. spl. aviation fire reduction com. FAA, 1979-80; sci. adv. panel Dept. Army, 1966-77; chmn. U.S. Army Weapons Command Adv. Group, 1966-72; mem. materiels adv. bd. NRC, 1969-74; vice-chmn. Naval Research Adv. Com., 1974-77, chmn., 1977; dir. Nat. Bank Commerce, San Antonio; dir. Engrs. Joint Council, 1966-69; mem. adv. group U.S. Armament Command, 1972-76; mem. sci. adv. com. Gen. Motors, 1971-81; mem. Nat. Commn. on Libraries and Info. Scis., 1971-78; chmn. NRC Bd. Army Sci. and Tech., 1982-89; chmn. Commn. Engring. and Tech. Systems, NRC, 1980-86. Editor: Applied Mechanics Review, 1952-59; editorial adviser, 1959-84. Bd. govs. St. Mary's U., San Antonio, 1970-76, 85—; pres. San Antonio Symphony, 1968-70, chmn. bd., 1970-71; bd. dirs. So. Meth. U. Found. Sci. and Engring., Dallas, 1979-90; trustee Univs. Research Assocs., Inc., 1979-84; mem. Tex. Nat. Rsch. Lab. Commn., 1986-91. Recipient Spirit of St. Louis Jr. award ASME, 1945, Jr. award, 1946, Alfred E. Noble prize ASCE, 1947, Outstanding Civilian Svc. award U.S. Army, 1972, 88, Nat. Engring. award, 1985, W.W. McAllister Patriotism award, 1986, Herbert Hoover medal, 1987; named Employer of Yr. Nat. Employee Svcs. and Recreation Assn., 1993. Fellow AAAS, Am. Inst. Aeros. and Astronautics (pres. 1971); hon. mem. ASME (dir., mem. bd. tech., mem. tech. devel. com., v.p. communications); mem. C. of C. (dir.), Nat. Acad. Engring., Sigma Xi, Tau Beta Pi. Home: 306 Country Ln San Antonio TX 78209-2319 Office: Southwest Rsch Inst PO Drawer 28510 San Antonio TX 78228-0510 also: 6220 Culebra Rd San Antonio TX 78238-5166

GOLANT, VICTOR EVGEN'EVICH, physicist, researcher, educator; b. St. Petersburg, Russia, Jan. 14, 1928; s. Evgenij Yakovlevich and Ida Borisovna (Berkovskaya) G.; m. Natalija Georgievna Edelgauz, Mar. 5, 1966; children: Galina, Evgeniya. Physicist-engr., Poly. Inst., St. Petersburg, Russia, 1944-50, Cand. Sci., 1955; Dr.sc., A.F. Ioffe Physico-Tech. Inst., St. Petersburg, Russia, 1964. Engr., head of lab. factory Svetlana, St. Petersburg, 1950-56; lectr. Poly. Inst., St. Petersburg, 1956-61; head of lab., dir. of div. A. F. Ioffe Physico-Tech. Inst., St. Petersburg, 1961—; prof., dept. head Poly. Inst., Tech. U., St. Petersburg, 1961—; mem. bur. Plasma Physics Coun., 1968—. Author: Microwave Methods of Plasma Investigation, 1968; co-author: Fundamentals of Plasma Physics, 1977, English transl., 1980, RF Plasma Heating in Toroidal Fusion Devices, 1986, English transl., 1989; chief editor: Jour. Tech. Physics, 1987—; editorial bd. mem. Nuclear Fusion, 1980—. Recipient State prize, USSR, 1983. Mem. Russian Acad. Scis. Achievements include research in pulse microwave discharges, plasma diffusion in magnetic field, microwave plasma diagnostics, wave-plasma interaction, plasma heating in tokamaks and improved confinement in tokamaks. Office: A F Ioffe Physico-Tech Inst, Polytechnicheskaya 26, 194021 Saint Petersburg Russia

GOLASKI, NICHOLAS JOHN, information scientist; b. Washington, Pa., Dec. 29, 1954; s. John L. and Ann (Balog) G.; m. Kathleen S. Davis, June 30, 1978; children: Matthew, Steven, Jordan. Assoc. Econs., Washington & Jefferson Coll., Washington, Pa., 1980; BS in Bus. Adminstrn., Sacred Heart Coll., 1984. Data processing support Camalloy, Inc., Washington, Pa., 1973-78; project leader Pitney Bowes, Stamford, Conn., 1978-85; project mgr. Ga. Pacific, Westport, Conn., 1985-87; info. svcs. mgr. Sealed Air Corp., Saddlebrook, N.J., 1987—. Office: Sealed Air Corp 19-01 State Hwy 208 Fair Lawn NJ 07410

GOLBRAYKH, ISAAK GERMAN, geologist; b. Beshenkovichi, Belorussia, USSR, July 15, 1935; s. German and Rachil G.; m. Shulamith Bromberg, Nov. 26, 1958 (wid. Dec. 1986); 1 child, Victoria; m. Alla Mitskevich; 1 child, Maxim. MS, Leningrad Mining Inst., 1957; PhD, All-Union Scientific-Rsch. Petroleum Exploration Inst., 1970. Cert. petroleum geologist. Geologist All-Union Geol. Inst., St. Petersburg, U.S.S.R., 1957-60; geologist to sr. geologist All-Union Scientific-Rsch. Petroleum Exploration Inst., St. Petersburg, U.S.S.R., 1960-86; cons. N.Y., 1987-90, Exxon Co., Internat., Houston, 1991—. Author: Tectonic Fracture Analysis of Covered Platform Oil-Bearing Territories/NEDRA, USSR, 1968; contbr. articles to profl. jours.; patentee in field. Mem. N.Y. Acad. Scis., Am. Assn. Petroleum Geologists, Houston Geol. Soc. Home: 20018 Telegraph Square Ln Katy TX 77449 Office: Exxon Co Internat 820 Gessner Houston TX 77024

GOLD, DANIEL HOWARD, ophthalmologist, educator; b. N.Y.C., Sept. 21, 1942; s. Isadore and Leona (Cotton) G.; m. Joann Aaron, Oct. 22, 1966 (div. Sept. 1985); m. Barbara Wood, June 19, 1988; children: David, Abigail, Michael. Student, U. Mich., 1959-66. Diplomate Am. Bd. Ophthalmology. Asst. chief dept. ophthalmology Walter Reed Army Med. Ctr., Washington, 1972-74; asst. prof. dept. ophthalmology Montefiore Hosp. Med. Ctr., Bronx, N.Y., 1974-76; asst. clin. prof. med. branch U. Tex., Galveston, 1977-85, assoc. clin. prof. med. branch, 1986-91; physician, ophthalmologist Eye Clinic of Tex., Galveston, 1977—; clin. prof. ophthalmology med. branch U. Tex., Galveston, 1991—; exec. com. mem. St. Mary's Hosp. Med. Staff, Galveston, 1989-90; mem. self assessment com. Am. Acad. Ophthalmology, San Francisco, 1989-92. Editor: (textbook) The Eye in Systemic Disease, 1990; sec. editor Duanes Clinical Ophthalmology, 1974—, Current Opinion in Ophthalmology, 1991—; contbr. articles to profl. jours. Maj. U.S. Army, 1972-74. Fellow Am. Acad. Ophthalmology (Honor award 1985), Coll. Ophthalmologists Great Britain, N.Y. Acad. Medicine; mem. Macula Soc., Assn. for Rsch. in Vision and Ophthalmology, Pan Am. Assn. Ophthalmology, Galveston Physicians Svc. Assn. (bd. dirs. 1985—, pres. 1993—). Jewish. Office: Eye Clinic Tex 2302 Avenue P Galveston TX 77550-7992

GOLD, JOSEPH, medical researcher; b. Binghamton, N.Y., Jan. 17, 1930; s. Leon and Gertrude J. G.; m. Judith Barbara Taylor, June 12, 1955; children: Shannon Gabriel, Skye Raphael. AB, Cornell U., 1952; MD, SUNY Health Sci. Ctr., Syracuse, 1956. Fellow dept. pharmacology SUNY Health Sci. Ctr., Syracuse, 1961-62, rsch. asst. prof., 1962-64, asst. prof. pathology, 1964-65; dir. Syracuse Cancer Rsch. Inst., 1965—, trustee, 1965—. Contbr. chpts. to med. textbooks on heat stress and numerous articles on cancer rsch. and therapy to profl. publs. Served with USAF, 1958-61. Recipient Presdl. citation for work in Mercury Astronaut Selection Program, 1960; USPHS postdoctoral research fellow U. Calif. Sch. Medicine, Berkeley, 1956-58. Mem. Am. Assn. Cancer Research, N.Y. Acad. Scis., Onondaga County Med. Soc., Med. Soc. State N.Y. Achievements include pioneering work in proposing gluconeogenesis as a biochemical mechanism of cancer cachexia, 1968; development of hydrazine sulfate, 1st specific anti-cachexia drug to be used in human cancer; invention of process for the synthesis and prodn. of DL-Glyceraldehyde-3-phosphate, process for the synthesis and prodn. of DL-Glyceraldehyde-3-Phosphate in a pure and stable form; patentee in field. Home: 127 Edgemont Dr Syracuse NY 13214-2010 Office: 600 E Genesee St Syracuse NY 13202-3108

GOLD, LORNE W., Canadian government official; b. Saskatoon, Sask., Can., June 7, 1928; s. Alexander Stewart and Grace Dora (Davis) G.; m. Elizabeth Joan L'Ami, Sept. 8, 1951; children: Catherine Anne, Patricia Ellen, Judith Sharon, Kenneth Robert. B.Sc., U. Sask., 1950; M.Sc. in Physics, McGill U., 1952, Ph.D., 1970. Research officer div. bldg. research Nat. Research Coucil Can., Ottawa, Ont., 1950-52, head snow and ice sect., 1953-69, head geotech. sect., 1969-74, asst. dir. div., 1974-79, assoc. dir. div., 1979-86, chmn., assoc. com. geotech. research, 1976-83, guest worker inst. research on constrn., 1987, researcher emeritus, 1988—; Canadian del. to Intern. Union of Testing and Research Labs. for Materials and Structures, 1982-87, bd. dirs. Coun. Internat. du Batiment, 1983-86; sr. visiting scientist Ctr. for Cold Oceans Resources Engring., Meml. U. of Newfoundland, 1987-88. Fellow Royal Soc. Can., Engring. Inst. Can., Can. Soc. Civil Engrs. (Horst Leipholz medal 1991); mem. Internat. Glaciol. Soc. (pres. 1978-81), Assn. Profl. Engrs. Ont., Engring. Inst. Can., Can. (hon. treas 1991), Geotech. Soc., Arctic Inst. N.Am. Mem. United Ch. of Canada. Home: 1903 Illinois Ave, Ottawa, ON Canada K1H 6W5 Office: Nat Rsch Coun of Can, Inst for Rsch in Constrn, Ottawa, ON Canada K1A 0R6

GOLD, PHIL, physician, educator; b. Montreal, Que., Can., Sept. 17, 1936; m. Evelyn Katz; 3 children. BSc in Physiology with honors, McGill U., Montreal, 1957, MSc, M.D., 1961, PhD in Physiology, 1965. Licentiate Med. Council Can. Jr. rotating intern Montreal Gen. Hosp., 1961-62, jr. asst. resident in medicine, 1962-63, sr. resident in medicine, 1965-66, jr. asst. physician, asst. and assoc. physician, 1967-73, sr. physician, 1973—, physician-in-chief, 1980—, dir. div. clin. immunology and allergy, 1977-80, dir. McGill U. Med. Clinic, 1980—, also sr. investigator Research Ins.; faculty dept. physiology McGill U., 1964—, mem. faculty of medicine, 1965—, prof. medicine and clin. medicine, 1973—, chmn. dept. medicine and clin. medicine, 1985-90, prof. physiology, 1974—, prof. oncology, 1989—, mem. faculty of medicine exec. com. representing clin. depts., 1985—, D. G. Cameron prof. medicine (inauguaral), 1987—; vis. scientist Pub. Health Research Inst. N.Y.C., 1967-68; Chester M. Jones Meml. lectr. Mass. Gen. Hosp., 1974; vis. prof. U. Caracas, Venezuela, 1974; Squires Club vis. prof. Wellesley Hosp., Toronto, 1983; Cecil H. and Ida Green vis. prof., 1984 autumn lectures U. Brit. Columbia; cons. in allergy and immunology Mt. Sinai Hosp., St. Agathe des Monts, Quebec, 1975—; hon. cons. dept. medicine Royal Victoria Hosp., Montreal; cons. dept. internal medicine Douglas Hosp. Ctr., Montreal; vice chmn. med. adv. com. Council of Physicians, Dentists and Pharmacists, 1985-90; mem. Conseil d'Adminstrn., Found. Quebecoise du Cancer, 1986-88; speaker at convocations and invited lectr. numerous univs. Mem. editorial bd. Clin. Immunology and Immunopathology, 1972—, Immunopharmacology, 1978—, Diagnostic Gynecology and Obstetrics, 1978-83, Oncodevelopmental Biology and Medicine, 1979—, Modern Medicine of Can., 1984-90, Current Therapeutic Rsch., 1992—; editorial cons. Jour. Chronic Diseases, 1981-84; mem. editorial adv. bd. Cancer Research, 1971-73, assoc. editor 1973-80; contbr. over 140 articles to med. jours. External referee Can. Red Cross Soc. Recipient Hiram Mills Gold medal, Mosby Scholarship Book award, Wood Gold medal, E.W.R. Steacie prize Nat. Research Council Can., 1973, Can. Silver Jubilee medal, 1977, Johann-Georg-Zemmerman prize for cancer research Medizinische Hochschule, Hannover, Fed. Republic Germany, 1978, Gold medal award of merit Grad. Soc. McGill U., 1979, Internat. award Gardner Found., Ernest C. Manning prize, F.N.G. Starr award Izzak Walton Killam Prize Can. Council, 1985, Canadian Soc. for Clin. Investigation Disting. Svc. award, 1992, R.M. Taylor medal Nat. Cancer Inst, Can., 1992, Internat. Soc.Oncodevelop. Biol. Medicine Internat. Abbott award; decorated companion Order of Can., 1986; Great Montrealer, 1986; Knight Comdr., Sovereign Order St. John Jerusalem, Knights of Malta, 1986; MacDonald scholar, J. Francis Williams scholar, Univ. scholar. Fellow Royal Coll. Physicians and Surgeons Can. (cert. internal medicine, medal 1965, chmn. examing bd., 1975-77, mem. research com. 1986-90), Royal Soc. Can., ACP ; mem. AAAS, Am. Acad. Allergy and Immunology, Am. Assn. Cancer Research, Am. Assn. Immunologists, Am. Fedn. Clin. Research, Am. Soc. Clin. Investigation, Assn. Am. Physicians, Am. Bd. Med. Lab. Immunology (mem. adv. panel 1978—), Assn. Medicale du Que., Can. Assn. Radiologists (hon.), Can. Fedn. Biologic Socs., Can. Oncology Soc. (founding mem., mem. sci. com. 1979—), Can. Soc. Allergy and Clin. Immunology, Can. Soc. Clin. Investigation, Corp. Profl. des Medecins Que., Fedn. des Medecins Specialists Que., Montreal Physiol. Soc., N.Y. Acad. Scis., Reticuloendothelial Soc., Alzheimer Soc. Montreal (patron), Can. Soc. for Immunology (pres. 1975-77), Internat. Soc. for Oncodevel. Biology and Medicine (editorial bd. jour. 1979—, mem. constn. com. 1976—, pres. 9th ann. meeting 1981, Inaugural Outstanding Scientist award 1976), Can. Med. Assn. (F.N.G.), Med. Research Council Can. (chmn. panel 1972-77, chm grants com. 1981-83, council mem. 1986-92, com. membership of coms. 1989—), Nat. Cancer Inst. Can. (mem. cancer grants panel B 1977-79, mem. research adv. com. 1984—), Med-Chi Soc. (pres. 1986-88), Sigma Xi, Alpha Omega Alpha, others. Office: McGill Univ Med Clinic, 1650 Cedar Ave, Montreal, PQ Canada H3G 1A4

GOLD, PHILIP WILLIAM, neurobiologist; b. Newport News, Va., Sept. 23, 1944; s. Jonah and Miriam (Novileasky) G.; m. Carol Dornself, Jan. 23, 1971; children: Jonathan, Leah, Sarah. AB, Duke U., 1966, MD, 1970. Intern in medicine Boston City Hosp., 1970-71; resident in psychiatry Harvard Med. Sch., Boston, 1971-73; asst. in medicine Peter Bent Brigham Hosp., Boston, 1973-74; clin. assoc. NIMH, Bethesda, Md., 1974-77, endocrinology fellow, 1977-80, sr. investigator, 1980-86, lab. chief, 1986—; mem. med. bd. NIH Clin. Ctr., 1977-81; vis. prof. Duke U. Sch. Medicine, 1988—. Mem. editorial bd. Jour. Neuroendocrinoology (London), 1989—; contbr. articles to jours. in field; patentee in field. Trustee Nat. Child Rsch. Ctr., Washington, 1981-84. Capt. USPHS, 1984—. Recipient C. Richter prize for rsch. Inst. Soc. Psychoneuroendocrinology, Vienna, Austria, 1984, Outstanding Svc. medal USPHS, 1985, Walters prize for rsch. U. Toronto Sch. Medicine, 1987, Found. prize for rsch. APA, 1988. Mem. Am. Coll. Neuropharmacology, Endocrine Soc., Libr. Congress Coun. Scholars. Democrat. Jewish. Avocations: music, creative writing. Office: NIH Clin Neuroendocrinology Clin Ctr Rm 35231 9000 Rockville Pike Bethesda MD 20892-0001

GOLDANSKII, VITALII IOSIFOVICH, chemist, physicist; b. Vitebsk, USSR, June 18, 1923; s. Iosif Efimovich and Yudif' Iosifovna (Melamed) G.; m. Lyudmila Nikolaevna Semenova; children: Dmitrii, Andrei. Grad. in Chemistry, Moscow U., 1944, M of Chemistry, 1947, DSc in Physics, 1954. Scientist Inst. Chem. Physics-USSR Acad. Scis., Moscow, 1942-52, 1961—, from div. head to dir., 1988—; sr. scientist P.N. Lebedev Phys. Inst.-USSR Acad. Scis., Moscow, 1952-61; asst. prof. Phys.-Tech. Inst., Moscow, 1947-51; asst. prof., then prof. Inst. Phys. Engring., Moscow, 1951—. Author: Kinematics of Nuclear Reactions, 1959, Mössbauer Effect and its Applications in Chemistry, 1963, Physical Chemistry of Positron and Positronium, 1968, Tunneling Phenomena in Chemical Physics, 1986, many others; contbr. numerous articles and revs. to profl. jours.; patentee (numerous) in field. Chmn. Russian Pugwash Com., Moscow, 1987—; people's dep. of USSR; mem. com. fgn. affairs Supreme Soviet of USSR, 1989-92. Decorated Lenin Order, Order of October Revolution, numerous other orders and medals; recipient Lenin prize, 1980; Golden Mendeleev medal USSR Acad. Scis., 1975, Karpinsky prize Friedrich von Schiller Found., Hamburg, Germany, 1983, Boris Pregel award N.Y. Acad. Scis., 1990, Alexander von Humboldt award, Germany, 1991. Fellow Am. Chem. Soc. (hon.), Am. Phys. Soc., Am. Acad. Arts and Scis., Am. Philos. Soc., Acad. Scis. German Dem. Republic, Royal Swedish Acad. Scis., Royal Danish Acad. Scis. and Lettrs, Deutsche Akademie der Naturforscher Leopoldina, Russian Acad. Scis., Finnish Acad. Scis., Acad. Europaea. Avocations: writing humor and aphorisms, record collecting, movies, CDs, videos. Home: Ulitsa Kosygina 6, Apt 49, Moscow 117334, Russia Office: Russian Acad of Scis, Inst Chem Physics, Ulitsa Kosygina 4, Moscow 117334, Russia

GOLDBART, PAUL MARK, theoretical physicist, educator; b. Hertfordshire, Eng., Aug. 17, 1960; came to U.S., 1985; s. Colin Kenneth and Carole Jean (Turk) G.; m. Jenny Lee Singleton, Sept. 4, 1988. BA, Gonville and Caius Coll., Eng., 1978-81; PhD, Imperial Coll. U. London, 1982-85. Postdoct. rsch. assoc. dept. physics U. Ill., Urbana, 1985-87, asst. prof. physics, 1987—; proposal reviewer NSF, Washington, 1991—. Article reviewer profl. publs., including Phys. Rev., 1987—; contbr. articles to profl. jours. Recipient Myerscough prize U. London, 1984, Beckman award U. Ill., 1988, Presdl. Young Investigator award NSF, Washington, 1991, Xerox award U. Ill., 1992. Mem. Am. Phys. Soc., Internat. Liquid Crystal Soc., U.K. Inst. Physics. Achievements include research on condensed matter theory. Office: Univ Ill Dept Physics 1110 W Green St Urbana IL 61801-3080

GOLDBAUM, MICHAEL HENRY, ophthalmologist; b. Bklyn., Apr. 17, 1939; s. Samuel Zolman and Sarah (Kramer) G.; m. Brenda Scott Leggio, July 7, 1964; children: David, Stephen, Rachel. BA in Physics, Syracuse U., 1961; MD, Tulane U., 1965; MS in Med. Informatics, Stanford U., 1988. Fellow retina Cornell U. Med. Sch., N.Y.C., 1972-73; asst. prof. opthalmology U. Ill. Sch. Med., Chgo., 1973-76, U. Calif. San Diego, La Jolla, Calif., 1977—. Contbr. articles to profl. jours. With USN, 1967-72. Fellow Am. Acad. Ophthalmology (cert. 1972, honor 1989); mem. Am.

Assn. Artificial Intelligence, Retina Soc., Macula Soc. Achievements include research in digital image processing in opthalmology, retina images, extracellular matrix of the optical nerve, retinal depressions, choroidal ischemia. Office: U Calif San Diego Shiley Eye Ctr 9415 Campus Point Dr La Jolla CA 92093-0946

GOLDBECK, ROBERT ARTHUR, JR., physical chemist; b. Evanston, Ill., July 25, 1950; s. Robert Arthur Sr. and Ruth Marilyn (Nordwall) G.; m. Jennifer Jane Tollkuhn, Aug. 19, 1989; stepchildren: Jessica Kathleen Tollkuhn, Brenna Maurin Tollkuhn. BS, U. Calif., Berkeley, 1974; PhD, U. Calif., Santa Cruz, 1982. Postdoctoral fellow Stanford (Calif.) U., 1983-84, rsch. assoc., 1984-87; rsch. chemist U. Calif., Santa Cruz, 1987—; lectr. in chemistry U. Calif., 1980, 84, 86. Contbr. articles to Biophys. Jour.; contbr. articles to profl. jours. Mem. AAAS, Am. Chem. Soc., Biophys. Soc. Achievements include development of nanosecond time-resolved magnetic circular dichroism spectroscopy; research in the time-resolved MCD and natural CD of photolyzed hemeprotein-ligand complexes. Office: U Calif Dept Chemistry/Biochemistry Santa Cruz CA 95064

GOLDBERG, ARNOLD IRVING, psychoanalyst, educator; b. Chgo., May 21, 1929; s. Morris Henry and Rose (Auerbach) G.; m. Constance Obenhaus; children: Andrew, Sarah. BS, U. Ill., 1949; MD, U. Ill., Chgo., 1953. Diplomate Am. Bd. Psychiatry and Neurology; cert. psychoanalyst. Tng. and supervising analyst Chgo. Inst. for Psychoanalysis, 1970—, dir., 1990-92; assoc. psychiatrist Rush Presbyterian St. Lukes Hosp., Chgo., 1982—; prof. psychiatry Rush Med. Coll., Chgo., 1982—. Author: Models of the Mind, 1973, A Fresh Look at Psychoanalysis, 1988, The Prisonhouse of Psychoanalysis, 1990; editor: Future of Psychoanalysis: Progress in Self Psychology, Vols. 1-9, 1976-92; contbr. numerous articles to profl. jours. Capt. U.S. Army, 1955-57. Fellow Am. Psychiat. Assn. (life); mem. Am. Psychoanalytic Assn. Home: 844 W Chalmers Pl Chicago IL 60614-3223 Office: Institute for Psychoanalysis Chicago Institute 180 N Michigan Ave Chicago IL 60601-7401

GOLDBERG, DAVID BRYAN, biomedical researcher; b. San Bernardino, Calif., Mar. 29, 1954; s. Gus and Rose (Goldrich) G.; m. Dianne Rae, Dec. 19, 1976; children: Jason, Mark, Eric, Ashley. BA, UCLA, 1976, PhD, 1987. Rsch. asst. Calif. State U., L.A., 1976-79; rsch. assoc. UCLA, 1979-82; sci. project mgr. Alpha Therapeutic Corp., L.A., 1989—; adj. prof. Chaffey Coll., Alta Loma, Calif., 1990—. Contbr. articles to N.Y. Acad. Scis., Jour. Clin. Apheresis, Proceedings of ASCO, FASEB Jour., Fedn. Preceedings, Nat. Hemophelia Found. Mem. PTA, Alta Loma, 1991. Basic Rsch. grantee, Cancer Rsch. Ctr., 1987, 88, Cancer Seed grantee 1989; Teaching fellow, UCLA, 1982-87, Rsch. fellow II, City of Hope, Duarte, Calif., 1987-89. Mem. Fedn. Am. Socs. Experimental Biology. Achievements include patents; development of IL-2/LAK immunotherapy for the treatment of malignant melanoma; formulation chemistry; pharmaceutical product and device development. Office: Alpha Therapeutic Corp 1213 S John Reed Ct La Puente CA 91745-2455

GOLDBERG, ERWIN, biochemistry educator; b. Waterbury, Conn., Jan. 14, 1930; m. Geraldine Bloom, Aug. 26, 1951 (div. Sept. 1983); m. Pauline Bentley, May 12, 1985; children: Samuel, Larry, Jeffrey, Thomas, Katherine. BA, Harpur Coll., 1951; PhD, U. Iowa, 1956. Asst. prof. W.Va. U., Morgantown, 1958-61; from asst. prof. to assoc. prof. N.D. State U., Fargo, 1961-63; from asst. prof. to prof. dept. biochem., molecular biol., cell biol. Northwestern U., Evanston, Ill., 1963—. Contbr. over 100 articles to profl. jours. NSF, NIH Rsch. grantee, 1958—. Fellow AAAS; mem. Am. Soc. Biochem. Molecular Biology, Am. Soc. Andrologists, Soc. Study Reproduction, Protein Soc. Office: Northwestern U Dept BMBCB 2153 Sheridan Rd Evanston IL 60208

GOLDBERG, HAROLD HOWARD, psychologist, educator; b. N.Y.C., Apr. 30, 1924; s. Julius and Fannie (Somers) G.; m. Roslyn Jacobowitz, June 26, 1948; children: Barbara Balsam, Susan Pitcher, Lisa. BA, NYU, 1948, MA, 1949; PhD, AMW U., Tulsa, 1970. Cert. psychologist, sch. psychologist, sch. adminstr., therapist, psychol. diagnostician; approved fellow in marital and family therapy. Psychotherapist in pvt. practice N.Y.C., 1949—; chief psychologist Queens-Island Reading Ctrs., N.Y.C., 1959-62; sch. adminstr. League Sch./Rsch. Ctr. for Seriously Emotionally Disturbed, Bklyn., 1967-78; acting dir. Community Guidance Svc., N.Y.C., 1977-78; acting dir. Am. Inst. for Psychotherapy and Psychoanalysis, N.Y.C., 1977-78, mem. faculty, 1974-78; mem. faculty Greenwich Inst. for Psychotherapy and Psychoanalysis, N.Y.C., 1978—; supr. of psychotherapists 5th Ave. Ctr. for Psychotherapy, Nat. Inst. for the Psychotherapies. Founder, editor Jour. Clin. Issues in Psychology, 1969-78, Profl. Digest of N.Y. Soc. Clin. Psychology, 1965-69, News and Notes, 1963-65; editor Newsletter for the N.Y. Soc. Clin. Psychologists, 1965-69. With USAF, 1942-45. Mem. APA, N.Y. Soc. Clin. Psychologists (pres. 1972-73), N.Y. Psychol. Assn., N.Y. Acad. Scis., Am. Assn. Marriage and Family Therapy, Psi Chi, Phi Delta Kappa, Kappa Delta Pi. Avocations: writing, swimming, tennis. Office: 105 E 63d St 3-A New York NY 10021

GOLDBERG, HAROLD SEYMOUR, electrical engineer, academic administrator; b. Bklyn., Jan. 22, 1925; s. David and Rose (Maslow) G.; m. Florence Meyerson, May 29, 1949; children: Lawrence, Irene. B.E.E. (Schweinberg scholar), Cooper Union, 1944; M.E.E., Poly. Inst. Bklyn., 1949; student, Columbia U. Engring. draftsman Cole Electric Products Co., 1944-45; radio engr. Press Wireless, Inc., 1945-47; asst. project engr. Radio Receptor Co., 1947-48; project engr. No. Radio Co., 1948-50; mgr. prodn. test, test equipment design sects. Allen B. DuMont Labs., Inc., 1950-56; mgr. engring. fabrication dept. Emerson Radio & Phonograph Corp., 1956-57; chief devel. engr. Consol. Avionics Corp., Westbury, N.Y., 1957-59; engring. mgr. data systems EPSCO, Inc., Cambridge, Mass., 1959-62; v.p. research Lexington Instruments Corp., Waltham, Mass., 1962- 66; prin. research engr. AVCO Research div., Everett, Mass., 1966-68; ops. mgr. Orion Research Inc., 1968-70; v.p. applications Analogic Corp., Wakefield, Mass., 1970-71; ops. mgr. Data Precision Corp., Danvers, Mass., 1971-72; pres. Data Precision Corp., 1972-82; v.p. Analogic Corp., 1979-85; pres. Acrosystems Corp., Beverly, Mass., 1985-88; assoc. dean Gordon Inst., Wakefield, Mass., 1988-93; cons. Wakefield, 1988-92; assoc. dean Gordon Inst. Tufts U., Medford, Mass., 1988-93; lectr. Tufts U., 1993—; cons. Analogic Corp., 1988—. Served with AUS, 1945-47. Recipient award of distinction Poly. Inst. N.Y., 1980, John Fluke Sr. Pioneer award, 1989, Haraden Pratt award, IEEE, 1993, Allen Ploss award Electro, 1992; N.Y. State Vets scholar, 1957. Fellow IEEE (chmn. Boston group on medicine and biology 1965-66, mem. exec. com. Boston sect. 1967-69, vice chmn. Boston 1969-70, chmn. Boston 1970-71, internat. bd. dirs. 1971-75, 89-90, v.p. 1975, dir. Electro 1975-89, treas. tech. activities bd. 1991, citation of honor U.S. Activities Bd. 1978); mem. Instrumentation and Measurement Soc. of IEEE (sec.-treas. 1983, pres. 1985-87), Tau Beta Pi. Home: 10 Alcott Rd Lexington MA 02173-1950 Office: Audubon Rd Wakefield MA 01880-1203

GOLDBERG, KIRSTEN BOYD, science journalist; b. San Bernardino, Calif., Oct. 29, 1963; d. Jerry Dock and Jewel Marie (Purkiss) Boyd; m. Paul Boris Goldberg, Aug. 25, 1985; 1 child, Katherine. BA, U. Calif., Berkeley, 1984. News editor Reston (Va.) Connection, 1985-86; reporter Edn. Week, Washington, 1986-88; assoc. editor Cancer Letter Inc., Washington, 1989-90, editor, pub., 1990—. Editor newsletter The Clin. Cancer Letter. Mem. AAAS, Newsletter Pubs. Assn., Amnesty Internat., Soc. Profl. Journalists. Democrat. Jewish. Office: Cancer Letter Inc PO Box 15189 Washington DC 20003

GOLDBERG, LEWIS ROBERT, psychology educator, researcher; b. Chgo., Jan. 28, 1932; s. Max Frederick and Gertrude (Lewis) G.; m. Robin Montgomery; children: Timothy Duncan, Holly Lynn, Randall Monte; m. Janice C. Goldberg. AB in Social Rels., Harvard U., 1953; PhD in Psychology, U. Mich., 1958. Acting asst. prof. Stanford U., 1958-60; asst. prof. psychology U. Oreg., Eugene, 1960-65, assoc. prof., 1965-68, prof., 1968—; rsch. scientist Oreg. Rsch. Inst., Eugene, 1961—; Fulbright prof. Istanbul U., 1974-75, U. Nijmegen, Netherlands, 1966-67; vis. prof. U. Calif., Berkeley, 1970-71; field selection officer U.S. Peace Corps, 1962-66; mem. intelligence div. U.S. Secret Svc., 1980-86; mem. rsch. coms. NIMH, 1973-88. Mem. editorial bd. Applied Psychol. Measurement, 1976—, Rev. of Personality and Social Psychology, 1980—, Jour. Personality Assessment, 1985—, European Jour. Personality, 1986—, Personality and Individual Differences, 1988—, Jour. Personality and Social Psychology, 1989—, Psychol. Inquiry, 1990—, others. Netherlands Inst. for Advanced Study fellow, 1981-82. Mem. APA, AAAS, Am. Psychol. Soc., Soc. Personality Assessment; mem. Soc. Multivariate Exptl. Psychology (pres. 1974-75), Psychometric Soc. Office: Oreg Rsch Inst 1899 Willamette St Eugene OR 97401

GOLDBERG, MORTON FALK, ophthalmologist, educator; b. Lawrence, Mass., June 8, 1937; s. Maurice and Helen Janet (Falk) G.; m. Myrna Davidov, Apr. 6, 1968; children—Matthew Falk, Michael Falk. A.B. magna cum laude, Harvard U., 1958, M.D. cum laude, 1962. Diplomate Am. Bd. Ophthalmology. Intern Peter Bent Brigham Hosp., Boston, 1962-63; resident John Hopkins Hosp, Balt., 1963-67; prof. and head ophthalmology Eye and Ear Infirmary U. Ill. Hosp, Chgo., 1970-89; head dept., dir. Wilmer Ophthalmol. Inst. Johns Hopkins, Balt., 1989—. Author: (with D. Paton) Injuries of the Eye, the Lids and the Orbit: Diagnosis and Management, 1968, Management of Ocular Injuries, 1976; editor: Genetic and Metabolic Eye Disease, 1974, (with G.A. Peyman and D.R. Sanders) Principles and Practice of Ophthalmology (3 vols.), 1984; editor-in-chief Archives of Ophthalmology, Chgo., 1984—; contbr. articles to profl. jours. Served to lt. comdr. USPHS, 1967-69. Recipient award for outstanding contbns. in the field of vision research Alcon Research Inst., 1987, Univ. Scholar award U. Ill.-Chgo., 1987. Fellow Am. Acad. Ophthalmology (sr. honor award 1985); mem. Am. Ophthal. Soc., Chgo. Ophthal. Soc. (pres. 1985-86), Assn. Research in Vision and Ophthalmology (trustee 1985-90, pres. 1989-90), Assn. Univ. Profs. Ophthalmology (trustee 1985-91, pres. 1990-91), Macula Soc. (pres. 1980-82). Avocation: snorkelling. Home: 3607 Anton Farms Rd Baltimore MD 21208-1705 Office: Johns Hopkins Med Insts Wilmer Eye Inst 600 N Wolfe St Baltimore MD 21287

GOLDBERG, MYRON ALLEN, physician, psychiatrist; b. Bronx, June 4, 1942; s. Marcus and Rose (Spiegel) G. AB, Hunter Coll., 1965; MD, Universidad del Noreste, Tampico, Mexico, 1979. Staff psychiatrist, forensic unit Bronx-Lebanon Hosp., Bronx, N.Y., 1985-86, team leader inpatient unit, 1986-88, chief physician geriatric svcs., dept. psychiatry, 1986—; clin. instr. Albert Einstein Coll. of Medicine, Bronx, 1986—, team leader inpatient unit, 1991-92, unit chief inpatient svc., 1992—. Inventor fluid pressure relief valve, 1973. Judicial del. Dem. Coop City Club, Bronx, 1986; exec. com. Dem. Club, Bronx, 1987-88 and other offices. Recipient recognition Crotona Park Community Mental Health Ctr., 1986. Mem. Am. Psychiatric Assn., AMA. Democrat. Hebrew. Avocations: tropical fish, science fiction, woodworking. Office: Bronx-Lebanon Hosp Franklin Pavilion 8th Flr 1276 Fulton Ave Bronx NY 10456

GOLDBERG, RAY ALLAN, agribusiness educator; b. Fargo, N.D., Oct. 19, 1926; s. Max and Anne G.; m. Thelma R. Englander, May 20, 1956; children: Marc E., Jennifer E., Jeffrey L. AB, Harvard U., 1948, MBA, 1950; PhD, U. Minn., 1952. Officer, dir. Moorhead (Minn.) Seed & Grain Co., 1952-62; dir. Experience, Inc., Mpls., 1963-78, Arbor Acres Farm, Inc., N.Y.C., H.K. Webster Co.; mem. faculty Harvard U. Grad. Bus. Sch., 1955—, Moffett prof. agr. and bus., 1970—, also dir. continuing edn. programs, participant seminars; bd. dirs. Pioneer Hi-Bred Internat., Inc., Archer Daniels Midland, Vigoro Corp., Eco Sci. Labs. Inc., All-Flow, Inc.; vis. prof. U. Minn. Grad. Sch., summer 1960; adv. council Foods Multinat., Inc., 1972-77; mem. agrl. investment com. John Hancock Ins. Co., 1971—; cons. in field, 1955—; adviser Instituto Centroamericano de Administracion de Empresa, Managua, Nicaragua, 1973—, Instituto Panamericano de Alta Direccion de Empressa, Mexico City, 1973—, U.S. Comptroller of Currency, 1975—, Food and Agr. Policy Project, Ctr. Nat. Policy, 1984—; mem. study team, subgroup chmn. world food and nutrition study NRC, 1975—; mem. com. tech. factor contbg. to nation's fgn. trade positions Nat. Acad. Engring., 1976—; chmn. agribus. adv. com. on Caribbean Basin USDA, 1982—; mem. com. on indsl. policy for developing countries Commn. on Engring. and Tech. Systems, NRC, 1982—; mem. task force on agr. Fowler-McCracken Commn., 1984—; adv. bd. The First Mercantile Currency Fund Inc., 1985—; internat. adv. bd. Atlantic Exchange Program, 1987—; mem. V.I. Lenin All-Union Acad. of Agrl. Scis., 1988—; mem. U.S. Presdl. Econ. Del. to Poland, Nov., 1989; vice chmn. Bd. Spoleto Festival U.S.A., 1993. Author numerous books, 1948—including Agribusiness Management for Developing Countries-Latin America, 1974, (with Lee F. Schrader) Farmers' Cooperatives and Federal Income Taxes, 1974, (with John T. Dunlop et. al.) The Lessons of Wage and Price Controls—The Food Sector, 1977, (with Richard C. McGinity et. al.) Agribusiness Management for Developing Countries—Southeast Asian Corn Study, 1979; editor: Research in Domestic and International Agribusiness Management, Vol. 1, 1980, Vol. 2, 1981, Vol. 3, 1982, Vol. 4, 1983, Vol. 5, 1984, Vol. 6, 1986, Vol. 7, 1987, Vol. 8, 1988, Vol. 9, 1989, Vol. 10, 1991; co-editor (with Gerald E. Gaul) New Technologies and the Future of Food and Nutrition, 1991, The Emerging Global Food System: Public and Private Sector Issues, 1993; contbr. numerous articles to profl. jours.; chmn. editorial adv. bd. Agribus.: An Internat. Jour., 1983—. Bd. govs. Internat. Devel. Rsch. Ctr., Govt. of Can., 1978—; trustee Roxbury Latin Sch., Boston, 1973-76, Beth Israel Hosp., Boston, 1978—, mem. com. on patents and tech. transfer, 1982—, chmn. gerontology com., 1991—; mem. adv. com. to prep. sch. New Eng. Conservatory Music, 1974—, assoc. trustee, 1978—; vice chmn. bd. Spoleto Festival U.S.A., 1993. Recipient Outstanding Alumni award, Dept. Agrl. Econs. U. Minn., 1992. Mem. V.I. Lenin All-Union Acad. Agrl. Scis. (fgn.), Am. Agrl. Econs. Assn. (editorial coun. 1974-78, nat. agribus. edn. comn. 1988—), Internat. Agribus. Mgmt. Assn. (pres. 1990-92), Agribus. Inst. Cambridge (chmn. bd., treas. 1991—), Am. Mktg. Assn., Am. Dairy Sci. Assn., Food Distbn. Rsch. Soc., Harvard Club (Boston and N.Y.C.). Address: 975 Meml Dr Apt 701 Cambridge MA 02138

GOLDBERG, WILLIAM K., chemist; b. Scranton, Pa., 1954; s. Henry and Ann (Levy) G. BS in Chemistry, Lebanon Valley Coll., 1976; MA in Biochemistry, U. Scranton, 1981. Project engr. Gen. Battery, Reading, Pa., 1981-82; product engr. SAFT, Valdosta, Ga., 1983-85; chemist, supr. Associated Minerals, Green Cove Springs, Fla., 1986, chemist Envirolab, Ormond Beach, Fla., 1987-88; chemist II HRS, Jacksonville, Fla., 1988; lab. officer USAR, Jacksonville, 1987—; chemist, supr. Tech. Svcs., Jacksonville, 1989—. Capt. U.S. Army, 1990-91. Mem. Am. Chem. Soc., Am. Inst. Chemists, Soc. Applied Spectroscopy. Office: Tech Svcs PO Box 52329 Jacksonville FL 32201-2329

GOLDBERGER, ARTHUR EARL, JR., industrial engineer, consultant. BS in Systems Engring., U. Ariz., 1974, BS in Indsl. Engring., 1975; MS in Indsl. Engring., Tex. A&M U., 1977, postgrad., 1991—. Registered novell engr., profl. engr., Ky., Tex., Mo., Ariz. Gen. engr. DARCOM/RRAD, Texarkana, Tex., 1975-77; mgr. DARCOM/AVSCOM, St. Louis, 1977-81; div. dir. prodn. improvement McDonnell Douglas, St. Louis, 1981-90; pres. Spectrum Techs., Inc., St. Louis, 1990—; chmn. CAD/Expert System Tool Design, Seattle, 1991. Author: Real Leadership, 1993; contbr. articles to profl. jours. Bd. dirs. Engrs. Club St. Louis, 1978, Nat. Com. on U.S. Competitiveness, Washington, 1989—; mem. Scientific Olympiad, Mo., 1989. Recipient Quality Leadership award McDonnell Douglas Corp., 1988. Mem. IEEE (chmn. 1987-88, vice chmn. vehicle tech. soc. conf. 1991, Leadership award 1988), Inst. Indsl. Engrs., Soc. Mfrg. Engrs., Alpha Pi Mu. Achievements include research in manufacturing technology, RF and Network Communications operations integration, and proactive quality/process improvement.

GOLDEMBERG, JOSE, educator. Prof. Universidad de Sao Paulo, Brazil. Rosenblith lectr. NAS, 1993. Office: Universidad de São Paulo, Ciudade Universaria CP 8191, São Paulo SP, Brazil*

GOLDEN, CAROLE ANN, immunologist, microbiologist; b. L.A., Sept. 23, 1942; d. Floyd Winfred and Betty Lee (Cantland) G. AB, Okla. Coll. Liberal Arts, 1963; MS in Microbiology, Miami U., Oxford, Ohio, 1969, PhD in Immunology, 1973. Research asst. prof. medicine U. Utah Med. Sch., Salt Lake City, 1973-79; sr. scientist Utah Biomed. Test Lab., Salt Lake City, 1976-82; v.p., sci. dir. Microbiol. Research Corp., Bountiful, Utah, 1978-87; v.p. research and devel. Editek, Inc., Burlington, N.C., 1987—; chmn. R&D mgmt. com. Assn. Biotech. Cos., 1990-91. Contbr. articles to profl. jours. Recipient citation Tech. Commn., 1983. Mem. AAAS, Am. Soc. Microbiology, N.Y. Acad. Sci., Soc. Leukocyte Biology, Soc. Analytical Cytology, Am. Mensa, Assn. Ofcl. Analytical Chemists. Republican. Home: 2 White Oak St Elon College NC 27244-9106 Office: Environ Diagnostics Inc PO Box 908 1238 Anthony Rd Burlington NC 27215

GOLDEN, JOHN JOSEPH, JR., manufacturing company executive; b. New Milford, Conn., Jan. 13, 1943; s. John Joseph and Anne Munroe (Hope) G.; m. Carolyn Joan Pachesa, May 29, 1965 (div. July 1984); children: Elizabeth Susan, Jennifer Leigh, John Joseph III, Matthew Benjamin; m. Ethel M. (Piercy) O'Neill, June 8, 1991; 1 child, Michael Joseph. BS, MIT, 1966. V.p. systems devel. Quantum Computing Corp., Newton, Mass., 1968-70; mgr. computer ops. Polaroid Corp., Cambridge, Mass., 1970-75; dir. info. processing Schering-Plough Corp., Kenilworth, N.J., 1975-78; dir. info. systems Compugraphic Corp., Wilmington, Mass., 1978-80; dir. info. systems electro-optics div. Honeywell, Lexington, Mass., 1981-83; dir. adminstrn. electro-optics div. Honeywell, Wilmington, Mass., 1983-87; dir. materials electo-optics div. Honeywell, Marlboro, Mass., 1987-90; gen. mgr. Micracor, Acton, Mass., 1990—. With USAR, 1964-70. Mem. Assn. for Computing Machinery, IEEE, MIT Alumni Orgn., Mass. Iota Tau Assn. (treas. 1970—), Optical Soc. Am. Roman Catholic. Home: PO Box 761 North Andover MA 01845-0761 Office: Micracor Acton MA 01720

GOLDEN, LESLIE MORRIS, software development company executive; b. Chgo., Dec. 4, 1943; s. Irving R. and Anne K. (Eisenberg) G. Student, U. Ill., 1963; BS, Cornell U., 1965, M of Engring. Physics, 1966; MA, U. Calif., Berkeley, 1971, PhD, 1977; postgrad., Northwestern U., 1980. Rsch. assoc. Jet Propulsion Lab./NASA, Pasadena, Calif., 1977-79; astronomer Aerospace Corp., El Segundo, Calif., 1979-82; prof. Rosary Coll., River Forest, Ill., 1982-84; prof. Heller Coll. Bus. Adminstrn. Roosevelt U., Chgo., 1985-87, 93—; prof. Northeastern Ill. U., Chgo., 1985-87; pres. Edn. Software Cons., Oak Park, Ill., 1987—; prof. U. Ill., Chgo., 1993—; feature editor Cornell Engr. Mag., 1964-65, editor-in-chief, 1965-66. Author: Basic Composer, 1989, Scientific Approach to Creativity: The Improvisational Techniques of the Chicago School of Comedy; contbr. articles to sci. jours. Pres. Citizens Active for a Responsible Electorate, Oak Park, 1989-92; coord. Allianc of County Taxpayers, Cook County, Ill., 1992. Recipient editing and composition awards Engring. Coll. Mags. Assn., 1966, Sr. Divsn. Copernicus Essay Contest award Am. Coun. Polish Cultural Clubs, 1973; Ill. State scholar, 1961, McMullen scholar Cornell U., 1962-66; Interfound. Com. fellow Am. Inst. Econ. Rsch., 1961-65; Lili Fabilli-Eric Hoffer Essay Contest winner, 1972. Fellow Pi Delta Epsilon, Tau Beta Pi, Phi Beta Kappa; mem. SAG, AFTRA. Jewish. Achievements include rsch. in radio source populations distributed isotropically with position over the sky, quasar optical and radio luminosities evolve similarly with time, evolution of quasar optical and radio luminosity, nature of the subsurface of the planet Mercury. Home: 934 Forest Ave Oak Park IL 60302

GOLDEN, REYNOLD STEPHEN, family practice physician, educator; b. Herkimer, N.Y., Jan. 11, 1937; s. Harold Theodore and Ethel Anne (Myers) G.; m. Gale Holtz, Nov. 26, 1959 (div. May 1978); children: Nathan Myers, Jennifer Lynn (dec.), Laura Beth (Lieba); m. Ellen Jean Moore, Sept. 9, 1978; children: Melissa Nan, Benjamin Harold. AB cum laude, Harvard Coll., 1958; MD, SUNY, Syracuse, 1962. Diplomate Am. Bd. Family Practice, Am. Bd. Internal Medicine. Intern Lankenau Hosp., Phila., 1962-63; resident in internal medicine SUNY, Syracuse, 1963-66; pvt. practice Utica, N.Y., 1966-78; dir. family practice residency St. Elizabeth Hosp., Utica, 1978-92, St. Francis Hosp., Poughkeepsie, N.Y., 1992—; clin. assoc. prof. dept. family medicine SUNY, Syracuse, 1991—; cons. residency assistance program, Kansas City, Mo., 1988—; charter mem. N.Y. State Coun. on Grad. Med. Edn., N.Y.C., 1987-89. Editor N.Y. Family Physician, 1987-92. Mem. N.Y. State Acad. Family Physicians (chmn. bd. dirs. 1988-89, pres. 1992-93, Presdl. Citation 1985, 86, 88), Cen. N.Y. Acad. Medicine (pres. 1977-79, Golden Torch award 1989). Jewish. Avocations: travel, computers, music, theater, skiing. Office: Family Health Ctr 4 Jefferson Plaza Poughkeepsie NY 12601

GOLDEN, TIMOTHY CHRISTOPHER, chemist, researcher; b. Spokane, Wash., Aug. 18, 1956; s. William Edwin and Anna Marie (Palladino) G.; m. Audrey Jean Saul, July 25, 1991. BS in Chemistry, Wilkes Coll., 1978; PhD of Fuel Sci., Pa. State U., 1983. Rsch. asst. Pa. State U., University Park, 1983-84; from sr. rsch. chemist to prin. rsch. chemist Air Products and Chem., Inc., Allentown, Pa., 1984-91; sr. prin. rsch. chemist Air Products and Chem., Inc., Allentown, 1991—. Contbr. articles to profl. jours. Coach Parkland Wrestling, Allentown, 1991—. Mem. Am. Chem. Soc., Sigma Xi, Phi Lambda Upsilon. Achievements include patents for Highly Dispersed Cuprous Compositions 1992, Gas Separation Adsorbent Membranes 1992, Oxygen Selective Desiccants 1992, and others. Home: 4104 Hampshire Ct Allentown PA 18104 Office: Air Products & Chem Inc PO Box 25780 Lehigh Valley PA 18002

GOLDENBERG, DAVID MILTON, experimental pathologist, oncologist; b. N.Y.C., Aug. 2, 1938; s. Leo and Lillie (Spivak) G.; m. Hildegard Gruenbaum, Apr. 28, 1961; children: Eva, Deborah, Denis, Neil, Lee. Student, Shimer Coll., 1954-56; BS, U. Chgo., 1958; ScD, U. Erlangen-Nuremberg, Fed. Republic of Germany, 1965; MD, U. Heidelberg, Fed. Republic of Germany, 1966. Assoc. rsch. prof. pathology U. Pitts. Med. Sch., 1968-70; assoc. prof. pathology Temple U. Med. Sch., Phila., 1970-72, U. Ky. Med. Ctr., Lexington, 1972-73; prof., dir. div. exptl. pathology U. Ky., Lexington, 1973-83; pres. Ctr. for Molecular Medicine and Immunology, Newark, 1983—; adj. prof. medicine and surgery N.J. Med. Sch., U. of Medicine and Dentistry of N.J., Newark, 1983-93; adj. prof. microbiology N.Y. Med. Coll., Valhalla, 1993—; mem. VA Merit Rev. Bd. for Oncology, Washington, 1974-77; exec. dir. Ephraim McDowell Community Cancer Network, Lexington, 1975-80; pres. Ephraim McDowell Cancer Rsch. Found., 1978-80; sec., treas. Ky. Cancer Commn., Frankfort, 1978-80; mem. sci. adv. bd. German Fund for Cancer Rsch., Bonn, 1980—; chmn. bd. Immunomedics, Inc., Morris Plains, N.J., 1983—; mem. exptl. immunology study sect. NIH, Bethesda, Md., 1980-83. Author more than 700 articles, book chpts., abstracts, 1962—; mem. editorial bd. Tumor Biology, Antibody, Immunoconjugates and Radiopharms., Jour. Nuclear Medicine and Allied Scis. Recipient Rsch. Found. award U. Ky., 1978, Outstanding Investigator grant Nat. Cancer Inst., 1985, 92, N.J. Pride award in sci. and tech. N.J. Monthly, 1986, Excellence in Cancer Rsch. award N.J. Senate and Assembly, 1985-86, Otto Herz Meml. lectureship Tel Aviv U., 1991, 3M/Mayneord Meml. lectureship Brit. Inst. Radiology, 1991. Hon. mem. Argentine Cancer Assn. Jewish. Achievements include more than 20 patents in field. Office: Ctr Molecular Medicine & Immunology 1 Bruce St Newark NJ 07103-2709

GOLDFARB, MARVIN AL, civil engineer; b. Memphis, Tenn., Dec. 12, 1928; s. Al Bohne and Melba (Pollock) G.; m. Lorene Shelton, June 13, 1965 (div. 1974); children: David Al, Julie Lin. BCE, U. Tenn., 1950; MS in Engring. Mgmt., U. Mo., 1971. Registered profl. engr., Ala., Mo. Field engr. Inter-Am. Geodetic Survey, Panama, 1950-53; engr., supr. Rust Engring. Co., Birmingham, Ala., 1956-64; engr., prin. Monsanto Co., St. Louis, 1964-90, ret., 1990. Author: An Owner's Approach to Project Scheduling, 1975. Councilman City of Maryland Heights, Mo., 1985-86; commr. Planning and Zoning Commn., chmn. 1986-93. Mem. Mo. Soc. Profl. Engrs. (pres. 1985-86), Am. Legion (comdr. post 213 1991-93). Jewish. Home: 1474 Glenmeade Dr Maryland Heights MO 63043

GOLDFARB, RICHARD CHARLES, radiologist; b. N.Y.C., Feb. 22, 1946; s. Harold and Lenore (Goldenheim) G.; m. Linda Markovitz; children: Adina, Akiva, Aviva, Aliza, Atara. AB, Columbia U., 1966; MD, N.Y. Med. Coll., 1970. Diplomate Am. Bd. Radiology, Am. Bd. Nuclear Medicine. Intern Met. Hosp. Ctr., N.Y.C., 1970-71; resident in diagnostic radiology St. Luke's Med. Ctr., N.Y.C., 1971-74, fellow in nuclear medicine and ultrasound, 1974-75; dir. nuclear medicine Nassau County Med. Ctr., East Meadow, N.Y., 1976-79; chief nuclear medicine Beth Israel Med. Ctr., N.Y.C., 1979—; assoc. prof. radiology Mt. Sinai Sch. Medicine, N.Y.C., 1988—. Editor Practical Reviews in Nuclear Medicine, 1992—; contbr. articles to profl. jours. Maj. U.S. Army, 1971-84. Named Tchr. of the Yr. Radiology Dept., Nassau County Med. Ctr., 1979. Mem. N.Y. Acad. Sci., Radiol. Soc. N.Am., Soc. Nuclear Medicine, Am. Coll. Radiology. Jewish. Home: 490 W End Ave New York NY 10024 Office: Beth Israel Med Ctr 1st Ave and 16th St New York NY 10003

GOLDFARB, RONALD B., research physicist. BA, Rice U., 1973, MS, 1975; MS, Colo. State U., 1976, PhD, 1979; MBA, U. Colo., 1991. NRC postdoctoral rsch. assoc. Nat. Inst. Standards and Tech. (former Nat. Bur. Standards), Boulder, Colo., 1979-81, sr. rsch. physicist, 1981—; lectr. physics dept. U. Colo., Boulder, 1989—. Contbr. over 50 articles to profl. jours. Mem. IEEE (sr.), ASTM, Am. Phys. Soc. Achievements include research in magnetics, superconductivity and instrumentation. Office: Nat Inst Standards and Tech 325 Broadway Boulder CO 80303-3328

GOLDFIELD, JOSEPH, environmental engineer; b. N.Y.C., Sept. 11, 1918; s. Abraham and Mollie (Levitt) G.; m. Roslyn Rose, Jan. 6, 1940; children: Michael H., Steven A. B in Chem. Engring., CCNY, 1939. Registered profl. engr., N.J., Colo., Miss., Mass. Pvt. practice N.Y.C., 1939-42; chem. engr. Chem. Warfare Svcs., Edgewood, Md., 1942-43, 45-46; rsch. assoc. MIT, Cambridge, Mass., 1943-45; dir. tng. Air Conditioning & Refrigeration Inst., Balt., 1946-51; mgr. environ. engring. Manville Corp., N.J., Denver, 1951-83; cons. Denver, 1983—. Contbr. articles to profl. jours. Pres. North Plainfield, N.J. High Sch. PTA, 1959-60; cubmaster Cub Scouts North Plainfield Boy Scouts Am., 1957; mem. Air Pollution Control com., Somerset County, N.J. Named Disting. Engr. in Industry Profl. Engrs. of Colo., 1980-81. Mem. NSPE, ASHRAE, Am. Acad. Environ. Engrs. (diplomate), Am. Chem. Soc., Am. Waste Mgmt. Assn. Achievements include design and installation of largest fabric filter ever (5,000,000 CFM) at asbestos mine and mill in Can., patent for high efficiency air filter using ordinary fiber glass matts to filter sub-micron particulates. Home and Office: 129 Elm St Denver CO 80220

GOLDHABER, GERSON, physicist, educator; b. Chemnitz, Germany, Feb. 20, 1924; came to U.S., 1948, naturalized, 1953; s. Charles and Ethel (Frisch) G.; m. Judith Margoshes, May 30, 1969; children—Amos Nathaniel, Michaela Shally, Shaya Alexandra. M.Sc., Hebrew U., Jerusalem, 1947; Ph.D., U. Wis., 1950; PhD honoris causus, U. Stockholm, 1986. Instr. Columbia U., N.Y.C., 1950-53; acting asst. prof. physics U. Calif., Berkeley, 1953-54, asst. prof., 1954-58, assoc. prof., 1958-63, prof. physics, 1963-92, prof. physics emeritus, 1992—; Miller research prof. Miller Inst. Basic Sci. U. Calif.-Berkeley, 1958-59, 75-76, 84-85, group leader Lawrence Berkeley Lab., 1962—; Morris Loeb lectr. in physics Harvard U., 1976-77. Named Calif. Scientist of Yr., 1977, Sci. Assoc., CERN, 1986; Ford Found. fellow CERN, 1960-61; Guggenheim fellow CERN, 1972-73. Fellow AAAS, Am. Phys. Soc. (Panofsky prize 1991), Sigma Xi; mem. Royal Swedish Acad. Sci. (fgn.), Nat. Acad. Sci. Office: U Calif-Berkeley Lawrence Berkeley Lab Berkeley CA 94720

GOLDHABER, GERTRUDE SCHARFF, physicist; b. Mannheim, Fed. Republic of Germany, July 14, 1911; came to U.S., 1939, naturalized, 1944; d. Otto and Nelly (Steinharter) Scharff; m. Maurice Goldhaber, May 24, 1939; children: Alfred Scharff, Michael Henry. Student, univs. Freiburg, Zurich, Berlin; Ph.D., U. Munich, 1935. Research assoc. Imperial Coll., London, Eng., 1935-39; research physicist U. Ill., 1939-48, asst. prof., 1948-50; assoc. physicist Brookhaven Nat. Lab., Upton, N.Y., 1950-58; physicist Brookhaven Nat. Lab., 1958-62, sr. physicist, 1962—; cons. nuclear data group NRC, Nat. Acad. Scis., AEC Labs. ACDA, 1974-77; adj. prof. Cornell U., 1980-82, Johns Hopkins U., 1983-86; Phi Beta Kappa vis. scholar, 1984-85; mem. sci. conf. Yamada Found., Japan, 1983; discovered pseudomagic nuclei, 1986, researched their level schemes and theoretical interpretation. Mem. editorial com. Ann. Rev. Nuclear Sci, 1973-77; N. Am. rep. bd. editors Jour. Physics G (Europhysics Jour.), 1978-80. Trustee-at-large Univ. Research Assn. governing Fermi Nat. Accelerator Lab., 1972-77; ednl. adv. com. N.Y. Acad. Scis., 1982—; Nat. Adv. Com. on Pre-Coll. Material Devel., 1984-88. Fellow Am. Phys. Soc. (council 1979-82, chmn. panel on improvement pre-coll. physics literacy 1979-82, chmn. audit com. 1980, mem. com. on profl. opportunities 1979-81, com. on history of physics, exec. com. 1983-84), AAAS (mem.-at-large sect. B physics com. 1986-88); mem. Nat. Acad. Scis. (mem. report rev. com. 1973-81, mem. acad. forum adv. com. 1974-81, mem. com. on edn. and employment of women in sci. and engring. 1978-83, commn. on human rights 1984-87), Sigma Xi. Achievements include discovery of pseudomagic nuclei, 1986, rsch. on their level schemes and theoretical interpretation, 1986—. Home: 91 S Gillette Ave Bayport NY 11705-2226 Office: Brookhaven Nat Lab #510A Upton NY 11973

GOLDHABER, MAURICE, physicist; b. Lemberg, Austria, Apr. 18, 1911; came to U.S., 1938, naturalized, 1944; s. Charles and Ethel (Frisch) G.; m. Gertrude Scharff, May 24, 1939; children: Alfred S., Michael H. Ph.D., Cambridge U., Eng., 1936; Ph.D. honoris causa, Tel-Aviv U., Israel, 1974; Dr. honoris causa, U. Louvain-La-Neuve, Belgium, 1982; D.Sc. honoris causa, SUNY, Stony Brook, 1983. Bye fellow Magdalene Coll., Cambridge, 1936-38; asst. prof. physics U. Ill., 1938-43, assoc. prof., 1943-45, prof., 1945-50; sr. sci. Brookhaven Nat. Lab., 1950-60, chmn. dept. physics, 1960-61, dir., 1961-73, AUI distinguished scientist, 1973—; cons. labs. AEC; Morris Loeb lectr Harvard U., 1955; adj. prof. physics SUNY, Stony Brook, 1965—, Royal Soc. Rutherford Meml. lectr., Can., 1987; Mem. nuclear sci. com. NRC. Assoc. editor: Phys. Rev, 1951-53; Contbr. articles on nuclear physics to sci. jours. Mem. bd. govs. Weizmann Inst. Sci., Rehovoth, Israel, Tel Aviv U.; trustee Univs. Research Assn. Recipient citation for meritorious contbns. U.S. AEC, 1973, J. Robert Oppenheimer meml. prize, 1982, Nat. Medal Sci., 1983, Am. Acad. Achievement award, 1985, Wolf Found. prize in physics (Jerusalem), 1991; co-recipient Rossi prize Am. Astron. Soc., High Energy Astrophysics div., 1989. Fellow Am. Phys. Soc. (pres. 1982), Am. Acad. Arts and Scis., AAAS; mem. Nat. Acad. Sci., Am. Philos. Soc. (Tom W. Bonner prize in nuclear physics 1971). Office: Brookhaven Nat Lab Bldg 510 Upton NY 11973

GOLDIE, PETER, electronic publisher, computer consultant; b. Manchester, Conn., Aug. 15, 1955; s. Mark Goldie and Marguerite (Garges) White. BS, Lehigh U., 1980; MS, Cornell U., 1982, NYU, 1987; PhD, NYU, 1988. Researcher St. Luke's Roosevelt Hosp. Ctr., N.Y.C., 1982-85; founder, pres. PMC Rsch., Inc., N.Y.C., 1983—, Lightbinders, Inc., San Francisco, 1989—. Author, editor: Darwin Multimedia CD-ROM, 1992; contbr. articles to sci. publs. Mem. AAAS, Am. Med. Informatics Assn., Am. Soc. Tropical Medicine and Hygiene, Internat. Assn. Law Enforcement Intelligence Analysts. Achievements include publication of first complete basic research journal and first major collection of works of Charles Darwin on a multimedia CD-ROM. Office: Lightbinders Inc 2325 3d St Ste 320 San Francisco CA 94107-3138

GOLDIN, BARRY RALPH, biochemist, researcher, educator; b. Bklyn., Feb. 25, 1942; s. Morris and Florence (Levy) G.; married; children: David, Rachel. BS, Bklyn. Coll., 1963; MS, U. Mass., 1967, PhD, 1968. Postdoctoral fellow Washington U. Sch. Medicine, St. Louis, 1968-71; staff fellow NIH, Bethesda, Md., 1972; rsch. assoc. Harvard U., Boston, 1972-74; research biochemist UCLA, 1974-75; mem. sci. staff Tufts U., Boston, 1975-77, asst. prof., 1977-83, assoc. prof., 1984—; mem. cadre Nat. Cancer Ist. Large Bowel Cancer Project, Houston, 1981-84; chmn. Inst Animal Care and Use Commn., Tufts U., Boston, 1985—. Contbr. articles to profl. jours., chpts. to books. Mem. AAAS, Am. Assn. Cancer Rsch., Am. Chem. Soc., N.Y. Acad. Scis., Sigma Xi. Achievements include patents in isolation technique and clinical applications of Lactobacillus. Office: Tufts U Sch Medicine 136 Harrison St Rm 330 Boston MA 02111

GOLDIN, DANIEL S., government agency administrator; b. N.Y.C., July 23, 1940; m. Judith Linda Kramer; children: Aerial, Laura. BS in Mech. Engring., City Coll. N.Y., 1962. Rsch. scientist Lewis Rsch. Ctr., NASA, Cleve., 1962-67; with TRW, from 1967, mem. tech. staff, 1967; then v.p., gen. mgr. space and tech. group TRW, Redondo, Calif., 1987-92; adminstr. NASA, Washington, 1992—. Office: NASA Office of Adminstr Washington DC 20546

GOLDMAN, ALLEN MARSHALL, physics educator; b. N.Y.C., Oct. 18, 1937; s. Louis and Mildred (Kohn) G.; m. Katherine Virginia Darnell, July 31, 1960; children—Matthew, Rachel, Benjamin. A.B., Harvard U., 1958; Ph.D., Stanford U., 1965. Research asst. Stanford U., Calif., 1960-65, research assoc., 1965; asst. prof. physics U. Minn., Mpls., 1965-67, assoc. prof., 1967-73, prof., 1974—, inst. tech. prof., 1992—, dir. Ctr. for Sci. and Application of Superconductivity, 1989—; co-chmn. Gordon Conf. on Quantum Liquids and Solids, 1981; dir. NATO Advanced Study Inst., 1983; mem. materials rsch. adv. com. NSF, 1985-88; mem. vis. com. Francis Butter Nat. Magnet Lab., 1986-89, chmn., 1987-89; mem. vis. com. Nat. Nanofabrication Facility at Cornell, 1988-90; mem. vis com. U. Chgo. Materials Program of Argonne Nat. Lab., 1992—. Contbr. numerous articles to profl. jours. Alfred P. Sloan Found. fellow, 1966-70. Fellow AAAS, Am. Phys. Soc. (divisional councillor divsn. condensed matter physics 1994—); mem. Israeli Phys. Soc. Jewish. Club: Harvard. Home: 1015 James Ct Mendota Heights MN 55118-3640 Office: U Minn Sch Physics and Astronomy 116 Church St SE Minneapolis MN 55455-0149

GOLDMAN, ARNOLD IRA, biophysicist, statistical analyst; b. Chgo., Mar. 13, 1945; s. Morton Irving and Rita Mae (Satten) G.; m. Sandra Gail Lipman, Aug. 2, 1971; children: Jennifer Lauren, Lesley Ann. MS in Radiol. Health, U. Pitts., 1968; PhD in Biophysics, Med. Coll. Va., 1974. Postdoctoral fellow Jules Stein Eye Inst., UCLA, 1974-75; staff fellow Nat. Eye Inst., NIH, Bethesda, Md., 1975-79; asst. prof. Med. Coll. Wis., Milw., 1979-82; dir. rsch. ophthalmol. products and dir. M.I.S. Biomatrix, Inc., Ridgefield, N.J., 1983—; ind. statis. cons., Milw., 1980-83. Contbr. to profl. publs. including Science. Lt. (j.g.) USPHS, 1968-70. Mem. Assn. Rsch. in Vision and Ophthalmology. Home: 15 Stockbridge Ave Suffern NY 10901 Office: Biomatrix Inc 65 Railroad Ave Ridgefield NJ 07657

GOLDMAN, CHARLES REMINGTON, environmental scientist, educator; b. Urbana, Ill., Nov. 9, 1930; s. Marcus Selden and Olive (Remington) G.; m. Shirley Ann Aldous, Apr. 4, 1953 (div. June 1975); children: Christopher Selden (dec.), Margaret Blanche, Olivia Remington, Ann Aldous; m. Evelyne de Amezaga, May 12, 1977. B.A., U. Ill., 1952, M.S., 1955; Ph.D., U. Mich., 1958. Asst. aquatic biologist Ill. Natural History Survey, 1954-55; teaching fellow fisheries U. Mich., 1955-58; fishery research biologist U.S. Fish and Wildlife Service, Alaska, 1957-58; mem. faculty U. Calif. at Davis, 1958—, prof. zoology, 1966-71, dir. Inst. Ecology, 1966-69, prof. limnology, div. environ. studies, 1971—, chairperson div. environ. studies, 1988—; dir. Tahoe Research Group, 1973—; chairperson Man and Biosphere Freshwater program (MAB-5), 1988—; cons. hydroelectric and water pollution to govt. and industry, U.S., Africa, S. Am., Cen. Am., Australia and N.Z., 1959—; cons. UN Purari River Dam project, Pupua New Guinea, 1974, Niger River Dam Project, Nigeria, Africa, 1977-78, Parana River Flood Control, Argentina, 1979, El Cajon dam project, Honduras, 1979-84; mem. Calif. Assembly Sci. and Tech. Adv. Council, 1970-73, Calif. Solid Waste Mgmt. Bd., 1973-77; U. Calif. rep. Orgn. Tropical Studies, 1977—, mem.-at-large, 1985-86; NSF and Nat. Acad. Scis. on coastal pollutions problems, Taiwan, 1974; mem. Sci. Com. on Problems of Environ., Nat. Research Council coms., 1979-87; chmn. fresh water resources directorate U.S. Nat. Com. for Man and the Biosphere. Co-author: Limnology, 1983; editor: Primary Productivity in Aquatic Environments, 1966, Freshwater Crayfish V, 1983; Co-editor: Environmental Quality and Water Development, 1973. Served to capt. USAF, 1952-54. Guggenheim fellow, 1965; NSF sr. fellow, 1964; Goldman Glacier named in Antarctica, 1967; recipient Antartic Service medal, 1968; Fulbright Disting. prof., 1985. Fellow AAAS, Calif. Acad. Scis. (fed. del. to USSR on water pollution 1973); mem. Am. Soc. Limnology and Oceanography (editorial bd. 1964-67, nat. pres. 1967-68, pres. Western sect. 1966-67), Ecol. Soc. Am. (editorial bd. 1966-68, mem.-at-large 1972-73, v.p. 1973-74), Internat. Soc. Theoretical and Applied Limnology (nat. rep.); hon. mem. Culver chpt. Cum Laude Soc. Club: Explorers. Achievements include discovery of trace element limiting factors in N.Am. and N.Z. lakes.

GOLDMAN, ERNEST HAROLD, computer engineering educator; b. Lynn, Mass., Oct. 4, 1922; s. Samuel Abraham and Julia (Portnoy) G.; m. Muriel Lyons, Dec. 19, 1948; children: Robyn S., Wayne A. BS, USCG Acad., 1943; postgrad., Princeton U., MIT, 1943; MS, Harvard U., 1949, DSc, 1952. Instr. in elec. engring. Harvard U., Cambridge, Mass., 1950-51; sr. engr. IBM, Poughkeepsie, N.Y., 1951-57; rsch. dir. IBM, Yorktown Hgts., N.Y., 1958-64; lab. mgr. IBM, Fishkill, N.Y., 1965-74; program mgr. IBM Corp. Hdqrs., Armonk, N.Y., 1975-81; adj. full prof. Pace U., Pleasantville, N.Y., 1974-81, prof. info. systems, 1981-82; vis. prof. Rutgers U., New Brunswick, N.J., 1990; Bannow-Wahlstrom prof. computer engring. U. Bridgeport, Conn., 1982-92; cons. in field. Contbr. articles to profl. jours. Lt. USCG, 1943-47. Recipient Bowdoin prize Harvard U., 1950. Mem. IEEE, Computer Soc., Assn. for Computer Machinery, Spl. Interest Group Computer Communications, Harvard Club of Fairfield County, Rsch. Soc. Am., Sigma Xi. Avocations: skiing, tennis. Home: 4 Regulation Rd West Redding CT 06896-1225

GOLDMAN, ISRAEL DAVID, hematologist, oncologist; b. N.J., Nov. 17, 1936; married; 3 children. BA, NYU, 1958; MD, U. Chgo., 1962. Diplomate Am. Bd. Internal Medicine; lic. physician, Ill., N.C., Va. Intern internal medicine U. Chgo. Hosps., 1962-63, jr. and sr. asst. resident internal medicine, 1963-65; postdoctoral fellow biophysics Biophys. Lab. Harvard Med. Sch., Boston, 1965-66; asst. prof. medicine U. N.C., Chapel Hill, 1969-72, assoc. prof. medicine and pharmacology, 1972-74; assoc. prof. medicine, vice-chmn. dept. medicine Med. Coll. Va., Richmond, 1974-83, prof. medicine and pharmacology, 1979-81, prof. medicine pharmacology, chmn. div. oncology, 1982—; mem. bd. sci. counselors div. cancer treatment Nat. Cancer Inst., 1982-86, exptl. therapeutics study sect., 1986-80; mem. sci. adv. com. Damon Runyon-Waler Winchell Cancer Fund, 1986-90; mem. tobacco related disease study sect. U. Calif., 1992—; dir. Massey Cancer Ctr., 1988—. Mem. editorial adv. bd. Biochem. Pharmacology, 1984—, Cancer Comm., 1989-92, Oncology Rsch., 1992—. With USPHS, Nat. Cancer Inst., NIH, Bethesda, Md., 1966-69. Recipient Faculty Salary awar Pharm. Mfrs. Foun., 1972, Rsch. Career Devel. award Nat. Cancer Inst., 1973, Outstanding Investigator award, 1985-92, 92—. Fellow ACP; mem. Am. Soc. Clin. Investigation, Assn. Am. Physicians. Home: 1812 Grove Ave Richmond VA 23220 Office: Med Coll Va Div Hematology/Oncology Box 230 MCV Sta Richmond VA 23298

GOLDMAN, LEONARD MANUEL, physicist, engineering educator; b. N.Y.C., Mar. 22, 1925; s. Robert and Edith G.; m. Dovie Lee McSwain, June 15, 1952; children: Douglas Alan, Ellen Rebecca, Judith Andrea. Student, CCNY, 1941-44; A.B., Cornell U., 1945; M.Sc., McGill U., Can., 1948; Ph.D., U. Rochester, 1952. Rsch. assoc. Princeton U., 1952-56; rsch. physicist GE Rsch. and Devel. Ctr., Schenectady, 1956-75; prof. mech. and aerospace sci. U. Rochester, 1975-87, prof. emeritus, 1988—, assoc. dir. lab. for laser energetics, 1977-87; acting mgr. applied physics, R & D div. Bechtel Corp., San Francisco, 1985-86, mgr. applied physics, 1988-92; vis. fellow Culham (Eng.) Lab., U.K. Atomic Energy Assn., 1965-66; cons. in field. Contbr. articles to physics jours. Mem. Schenectady Sch. Bd., 1968-72. Served with USNR, 1944-46. Fellow Am. Phys. Soc.; mem. Am. Nuclear Soc., AAAS. Home: 2307 Ridgefield Dr Chapel Hill NC 27514

GOLDMAN, PETER, health science, chemistry, molecular pharmacology educator; b. N.Y.C., May 23, 1929; married, 1959; 2 children. BEngPhys, Cornell U., 1952; MA, Harvard U., 1953; MD, Johns Hopkins U., 1957. Rsch. assoc. biochemistry Nat. Heart Inst., Washington, 1959-63; sr. investigator biochem. pharmacology Nat. Inst. Arthritis, Metabolism and Digestive Diseases, Washington, 1963-72; prof. clin. pharmacology Harvard Sch. Pub. Health, Cambridge, Mass., 1972—, prof. Health Sci., dept. nutrition, 1982—, acting chmn., from 1984, now Maxwell Finland prof. biol. chemistry and molecular pharmacology. Office: Harvard U Dept Biol Chemistry Cambridge MA 02138

GOLDMAN, YALE E., physiologist, educator. MD, U. Pa., 1975, U. Pa., 1976. Assoc. prof. physiology U. Pa. Sch. Medicine, Phila., 1985—. Achievements include research in optics. Office: U Pa Muscle Inst D701 Richards Bldg Philadelphia PA 19104-6083*

GOLDMANN, NAHUM, product development executive; b. Odessa, Ukraine, June 24, 1948; s. Solomon and Brunia (Zilberberg) G.; m. Natalya Shteinberg, Apr. 30, 1976; 1 child, Alice. M of Engring., Electrotech. Inst., Leningrad, 1972. Registered profl. engr. Scientist Med. Rsch. Inst., Leningrad, 1970-73, Cybernetics Inst./Ukraine Acad. of Sci., Odessa, USSR, 1973-74; prin. scientist, lectr. Lomonosov Inst. Tech., Odessa, 1974-78; acoustical cons. Rome, 1978; engr. WCB of BC, Vancouver, Can., 1979-80; scientist Can. Ctr. Occupational Health and Safety, Hamilton, Can., 1980-83; mgr. acoustic dept. Bell-No. Rsch., Ottawa, Can., 1983-93; v.p. R & D Array Devel., Inc., Ottawa, 1993—; participant numerous scientific confs. Author: Online Information Hunting, 1992, Online Research and Retrieval, 1985; contbr. articles to Western and Russian publs.; patentee in field. Mem. Assn. of Profl. Engrs. of Ont., Acoustical Soc. Am., Internat. Electrotech. Com., Internat. Telecommunications Com. Home: 51 Wallford Way, Nepean, ON Canada K2E 6B7 Office: Array Devel Inc, PO Box 5145 Sta F, Ottawa, ON Canada K2C 3H3

GOLDREICH, JOSEPH DANIEL, consulting structural engineer; b. N.Y.C., Oct. 8, 1925; s. Pincus and Frieda (Weinstein) G. m. Vivian Cherie Jaffe, Oct. 22, 1950; children: Peter Jay, Andrew Steven, Matthew Howard. Student, CCNY, 1942-43; BS in Civil Engring., Union Coll., 1945; MCE, NYU, 1952. Registered profl. engr. 14 states. Surveyor Starret Bros. and Eken, Builders, 1946-47; sructural designer Howard, Needles, Tammen & Bergendoff cons. egnrs., 1947-48, Fred V. Severud, cons. engr., 1948-49, Eipel Engring., 1950-54; structural engr., engr.-in-charge, supr. engr. Seelye, Stevenson, Value & Knecht, 1955-60; cons. engr. N.Y.C., 1960-61; ptnr. Goldreich, Page & Thropp Cons. Engrs., N.Y.C., 1961—. Mem. sch. bldg. adv. com. Greenburgh Sch. Bd., 1955-60; athletic coach Recreation Commn., Greenburgh, 1960-66; trustee Hartdale Lawns Civic Assn.; vol. leader Boy Scouts Am., 1965—, dist. commr., 1979-81, mem. exec. bd. Westchester Putnam Coun., 1987—; trustee, chmn. bldg. com. Temple Beth Abraham, Tarrytown, N.Y., trustee, mem. bldg. com. Temple Beth El, Chappaqua; marshal Meml. Day Parade, Town of Newcastle, N.Y., 1978-81, 1983—; mem. adv. bd. Green Chimeys Childrens Svcs. Lt. j.g., CEC, USNR, World War II. Recipient Shofar award Boy Scouts Am., 1973, Silver Beaver award, 1977, Good Scout award, 1985; award of Merit, N.Y. Concrete Industry Bd., 1979. Mem. ASCE, Am. Concrete Inst., Concrete Industry Bd. (dir. 1987-90, sec., treas., v.p. 1991-93), Am. Cons. Engrs. Coun., N.Y. Assn. Cons. Engrs. (past. dir., sec., v.p.), N.Y. Bldg. Congress, Soc. Am. Mil. Engrs., Coalition Am. Structural Engrs. (vice chmn., sec.-treas., chmn.-elect 1990-92, chmn. 1993), N.Y. Acad. Scis., Am. Arbitration Assn. (nat. panel arbitrators), Am. Soc. Travel Agts., Assoc. Retail Travel Agts, Tau Beta Pi (Eminent Engr. 1992). Home: Brandon Dr Mount Kisco NY 10549 Office: Goldreich Page & Thropp 45 E 20th St New York NY 10003-1308

GOLDREICH, PETER MARTIN, astrophysics and planetary physics educator; b. N.Y.C., July 14, 1939; s. Paul and Edith (Rosenfield) G.; m. Susan Kroll, June 14, 1960; children: Eric, Daniel. BS in Physics, Cornell U., 1960, PhD in Physics, 1963. Instr. Cornell U., summers 1961-63; postdoctoral fellow Cambridge U., 1963-64; asst. prof. astronomy and geophysics UCLA, 1964-66, assoc. prof., 1966; assoc. prof. planetary sci. and astronomy Calif. Inst. Tech., 1966-69, prof. planetary sci. and astronomy, 1969—, Lee DuBridge prof. astrophysics and planetary physics, 1981—. Recipient Chapman medal Royal Astron. Soc., 1985, Gold medal, 199; named Calif. Scientist of Yr. 1981; Woodrow Wilson hon. fellow, 1960-61, fellow NSF, 1961-63, Sloan Found., 1968-70. Mem. NAS (fellow 1972), Am. Acad. Arts and Scis. (fellow 1973), Am. Astron. Soc. (Henry Norris Russell lectr., Dick Brouwer award 1986, George P. Kuiper prize divsn. planetary sci. 1992). Office: Calif Inst Tech 1201 E California Blvd Pasadena CA 91125-0001

GOLDSBERRY, RICHARD EUGENE, mobile intensive care paramedic, registered nurse; b. Colorado Springs, Feb. 17, 1956; s. Eugene Theodore and Martha Blanche (Geiser) G. Cert. paramedic, U. Calif., San Diego, 1979; AS in Nursing, Imperial Valley Coll., 1992. Cert. mobile intensive care paramedic. With Circle K Corp., El Centro, Calif., 1970-75, La Junta (Colo.) Med. Ctr., 1975-76, Pioneers Meml. Hosp., Brawley, Calif., 1976-77; emergency rm. technician Pioneers Meml. Hosp., Brawley, 1977-79; mobile intensive care paramedic Gold Cross Ambulance Co., El Centro, 1979—, mgr., 1984-88; instr., cons. Imperial Valley Coll., 1984—; cons., advisor Imperial Valley Reg. Occupational Program, 1984. Rep. Emergency Med. Care Com., El Centro, Calif., 1987-88. Named Advanced Life Support Provider of Yr. Imperial County, 1991. Mem. Calif. Rescue and Paramedic Assn., Cousteau Soc., Greenpeace USA, Sierra Club. Avocations: scuba diving, rock climbing, reading, teaching. Home: 371 West C St 585 W H St #12 Brawley CA 92227 Office: Gold Cross Ambulance Co PO Box 1834 El Centro CA 92244-1834

GOLDSBY, RICHARD ALLEN, biochemistry educator; b. Kansas City, Mo., Dec. 19, 1934. BS, U. Kans., 1957; PhD in Chemistry, U. Calif., Berkeley, 1961; AM (hon.), Amherst Coll., 1985. Jr. organic chemist Monsanto Inc., 1957-58; biochemist, virologist E.I. du Pont de Nemours & Co. Inc., 1961-66; from asst. prof. to assoc. prof. biology Yale U., 1966-72; prof. chemistry U. Md., College Park, from 1972; now Amanda and Lisa Cross prof. biology Amherst (Mass.) Coll.; master Pierson Coll., 1971-72. Bd. dirs. Carver Found., Tuskegee Inst., 1973—. Morse fellow Yale U., 1970-71; Nat. Research Council sr. fellow Ames Research Ctr., 1975—. Mem. Am. Sociology. Office: Amherst Coll Dept of Biology Amherst MA 01002

GOLDSCHMIDT, BERND, mathematics educator; b. Helbra, Eisleben, Germany, Apr. 13, 1950; s. Ewald and Ruth (Schwarz) G.; m. Ilona Goldschmidt, Sept. 7, 1974; children: Anja, Ines, Katy. Diploma, Martin-Luther U., Halle, Germany, 1973, D Natural Sci., 1980. Lectr. Martin Luther U., Halle, 1984-87, prof. math., modelling simulation & optimization in biotech., 1987—. Office: Martin Luther U Inst Biotechnologie, Weinbergweg 16A, 06099 Halle Germany

GOLDSMITH, LOWELL ALAN, medical educator; b. Bklyn., Mar. 29, 1938; s. Isidore Alexander and Ida (Kaplan) G.; m. Carol Amrein, June 11, 1960; children: Meredith, Eileen. AB, Columbia Coll., 1959; MD, SUNY, Bklyn., 1963. Diplomate Am. Bd. Dermatology. Intern, then resident in medicine UCLA Med. Ctr., 1963-65; resident in dermatology Harvard Med. Sch., Boston, 1967-69; asst. prof. dermatology Harvard U. Med. Sch., Boston, 1970-73; asst. in dermatology Mass. Gen. Hosp., Boston, 1970-71, asst. dermatologist, 1971-73; assoc. prof. medicine Duke U. Med. Ctr., Durham, N.C., 1973-78, prof., 1978-81; James H. Sterner prof. dermatology Sch. Medicine and Dentistry, U. Rochester, N.Y., 1981—, chief dermatology unit, 1981-87, acting chmn. dept. medicine, 1985-87, chmn. dept. dermatology, 1987—; mem. dermatology adv. com. FDA, 1983-87; chmn. Gordon Rsch. Conf. on Epithelial Differentiation and Keratinization, 1987; mem. gen. medicine A study sect. USPHS, NIH, 1988-92, chmn. 1990-92; chmn. med. adv. bd. Nat. Ichthyosis Found., 1981-85, Nat. Alopecia Areata Found., 1981-87, 90—, bd. dirs.; bd. dirs. Monroe Community Hosp., Rochester, Ctr. Alternatives Animal Testing, Balt.; chmn. NIH Consensus Conf. on Diagnosis and Treatment of Early Melanoma, Bethesda, 1992. Author, editor: Biochemistry and Physiology of the Skin, 1983, 2d edit., 1991; Physiology, Biochemistry and Molecular Biology of the Skin, 1991; mem. editorial bd. Archives Dermatology, 1982-92, Clinics in Dermatology, 1982, Jour. Investigative Dermatology, 1987—, Seminars in Dermatology, 1991—; also numerous articles. With USPHS, 1965-67. Recipient Rsch. Career Devel. award USPHS, 1975-80; Macy Found. fellow, 1978-79. Mem. Assn. Am. Physicians, Am. Soc. Clin. Investigation, Soc. Investigative Dermatology (bd. dirs., pres.-elect 1993—), N.Y. State Soc. Dermatology (pres. 1985-89), Buffalo-Rochester Dermatology Soc. (pres. 1987), Assn. Profs. Dermatology (bd. dirs. 1984-87, pres. 1992—), Am. Bd. Dermatology (bd. dirs. 1993—), Nat. Ichtyhosis Found. (chmn. adv. bd. 1981-85), Rochester Dermatology Soc., Rochester Acad. Medicine, Polish Dermatol. Assn. (hon.), Alpha Omega Alpha. Office: U Rochester Dept Dermatology 601 Elmwood Ave # 697 Rochester NY 14642-9999

GOLDSMITH, MARY HELEN M., biology educator; b. Boston, May 2, 1933; d. Monroe H. and Virginia (Parker) Martin; m. Timothy H. Goldsmith, Aug. 20, 1955; children: Kenneth Martin, Margaret Parker. BA with distinction, Cornell U., 1955; AM, Radcliff Coll., 1956, PhD, 1959. NIH postdoctoral fellow Harvard U., 1959-60, Kings Coll., U. London, 1960-61; rsch. assoc. Yale U., New Haven, 1961-73, lectr. dept. biology, 1963-73; assoc. prof. dept. biology Yale U., 1974-84, prof. biology, 1984—, dir. Marsh Bot. Garden, 1986—; master Silliman Coll., Yale U., 1987—; vis. assoc. prof. dept. biol. scis. Sanford U., Calif., 1971. Contbr. articles to profl. jours.; editorial bd. Plant Physiology, 1978-92, Planta, 1981-90. Carnegie fellow, 1978, Guggenheim fellow, 1987, Brenda Ryman vis. fellow, 1987. Mem. Am. Soc. Plant Physiologists (exec. com. 1985-88, pres. 1991-92). Home: 555 Forest Rd Northford CT 06511-6606 Office: Yale U Kline Biology Tower Dept Biology PO Box 6666 New Haven CT 06511

GOLDSMITH, MICHAEL ALLEN, medical oncologist, educator; b. Bronx, N.Y., Jan. 28, 1946; s. Walter and Bertha (Tannenberg) G.; m. Judith Harriet Plaut, June 6, 1971; children: Sharon, Esther, Eva, Steven. BA,

Yeshiva U., 1967; MD, Albert Einstein Coll. Medicine, 1971. Diplomate Am. Bd. Internal Medicine. Intern Bronx Mcpl. Hosp. Ctr., 1971-72; staff assoc. Nat. Cancer Inst., Bethesda, Md., 1972-74; resident in medicine Mt. Sinai Hosp., N.Y.C., 1974-75, fellow in neoplastic diseases, 1975-77, asst. clin. prof. medicine and neoplastic diseases, 1977—; attending physician Oncology Consultants, P.C., N.Y.C., 1977—; reviewer Jour. AMA, 1988-90. Contbr. articles to med. jours. Vice-pres. Congregation Orach Chaim, N.Y.C., 1978-83. Lt. comdr. USPHS, 1972-74. Fellow ACP; mem. Am. Soc. Clin. Oncology, Am. Assn. Cancer Rsch. Achievements include research in new anticancer drugs. Office: Oncology Cons PC 1045 5th Ave New York NY 10028-0138

GOLDSON, ALFRED LLOYD, oncologist educator. BS, Hampton Inst., 1968; MD, Howard U., Washington, 1972. Diplomate Am. Bd. Therapeutic Radiology; med. lic. D.C.; cert. Ga. state med. bd. Resident in radiation therapy Meml. Sloan-Kettering Cancer Ctr., N.Y.C., 1972-75, fellow, 1975-76; clin. instr. radiation therapy coll. medicine Howard U., 1976, from asst. prof. to assoc. prof., 1977-79, chmn. dept. radiotherapy, 1979—, prof., 1984; clin. assoc. prof. radiation oncology coll. medicine, Georgetown U., Washington, 1979; chmn. radiation therapy Greater Southeastern Community Hosp., 1991; chmn. Howard U. Cancer Com., 1985—; interim dir. Howard U. Cancer Ctr., 1991, exec. com. 1979—; chmn. adv. bd. Howard U. Coll. Allied Health Scis. Radtation Therapy Tech., 1977—. Contbr. articles to profl. jours. Chmn. D.C. Cancer Consortium; chmn. trial com., Nat. Cancer Inst., 1984-86, patient data query editorial bd., 1984-86; program com. nat. conf., Am. Cancer Soc., 1984, trustee 1979—. Jr. Faculty Clin. fellow Am. Cancer Soc., 1977-79. Mem. Am. Soc. Therapeutic Radiology (scientific program com. 1982-85), Am. Coll. Radiology (com. radiotherapy rsch. and devel. 1982-85), Am. Soc. Clin. Onocology, Nat. Med. Assn., Radiologic Soc. North Am., Mid-Atlantic Soc. Radiation Oncologists, N.Y. Acad. Scis., Meml. Sloan-Kettering Radiation Therapy Dept. Alumni Assn. (pres. 1989). Home: 4015 28th Pl NW Washington DC 20008 Office: Howard U Dept Radiotherapy 2041 Georgia Ave NW Washington DC 20060*

GOLDSTEIN, ALLAN LEONARD, biochemist, educator; b. Bronx, N.Y., Nov. 8, 1937; s. Morris and Miriam (Siegel) G.; m. Linda Jo Tish, Dec. 23, 1975; children: Jennifer Joy, Dawn Eden, Adam Lee. B.S., Wagner Coll., 1959; M.S., Rutgers U., 1961, Ph.D., 1964. Teaching asst. Rutgers U., New Brunswick, N.J., 1959-61; asst. instr. biology, 1961-63, instr. physiology, 1963-64; research fellow Albert Einstein Coll. Medicine, 1964-66, instr. biochemistry, 1966-67, asst. prof., 1967-71, asso. prof., 1971-72; prof., dir. div. biochemistry U. Tex. Med. Br., Galveston, 1972-78; acting dir. multidisciplinary research program in mental health U. Tex. Med. Br., 1973-78; prof., chmn. dept. biochemistry and molecular biology George Washington U. Sch. Medicine, Washington, 1978—, pres., sci. dir. Inst. for Advanced Studies in Immunology and Aging,, 1985—; chmn. bd. Alpha 1 Biomeds., 1982—; cons. Syntex Research, 1972-74, Hoffmann-LaRoche, 1974-82; spl. cons. bd. sci. counselors Nat. Inst. Allergy and Infectious Diseases, 1975-80; mem. med. research service rev. bd. in oncology VA, 1977-80; cons. mem. decisive network com. Biol. Response Modifiers Program, Div. Cancer Treatment, Nat. Cancer Inst., 1982—; mem. sci. adv. com. to pres. Papanicolaou Cancer Research Inst. Miami, Inc., 1981-84 ; mem. AIDS task force adv. com. Nat. Cancer Inst., 1983-84, sci. bd. Alliance for Aging Research, 1986—. Discoverer (with Abraham White) Thymosins, hormones of thymus gland and HGP-30 a "core" based p17 AIDS Vaccine currently undergoing Phase I human testing. Decorated Chevalier des Palmes Academiques (France); recipient Career Scientist award N.Y.C. Health Rsch. Coun., 1967, Alumni Achievement award Wagner Coll., 1974, Gordon Wilson medal Am. Clin. and Climatol Soc., 1976, Disting. Faculty Rsch. award U. Tex. Sch. Biomed. Scis., 1976, Van Dyke award in pharmacology Columbia Coll. Physicians and Surgeons, 1984; vis. prof. award Burroughs Wellcome Found., FASEB, 1986, Fermandez-Cruz award, 1989, Martin Rubin award Am. Coll. Advancement in Medicine, 1990, Michele Fodera Internat. prize for Biomed. Rsch., 1990. Mem. AAAS, Endocrine Soc., Am. Soc. Biol. Chemists and Molecular Biology, Am. Assn. Immunologists, Internat. Soc. Immunopharmacology (council mem. 1985—), Assn. Med. Sch. Chmn. of Depts. Biochemistry, AAUP, Sigma Xi. Club: Toastmasters Internat. (pres. N.Y. chpt. 1971). Home: 6407 Bradley Blvd Bethesda MD 20817 Office: George Washington U Med Ctr Dept Biochemistry/Molecular Biology 2300 I St NW Washington DC 20037

GOLDSTEIN, DAVID ARTHUR, biophysicist, educator; b. Rochester, N.Y., Nov. 8, 1934; s. Jacob David and Elizabeth Maude (Brown) G.; m. Marie Elaine Nardone, May 25, 1969; 1 child, David James. AB in Physics, Harvard U., 1956, MD, 1960. Rsch. fellow biophysical lab. Harvard Med. Sch., Cambridge, Mass., 1960-62; rsch. assoc. biophysical lab. Harvard Med. Sch., Cambridge, 1964-65; asst. prof. radiation biology & biophysics Rochester (N.Y.) Med. Coll., 1965-68, assoc. prof. biophysics, 1968—, assoc. prof. biomath., 1969-74, assoc. prof. med. informatics, 1988—; dir. Med. Ctr. Computing U. Rochester Med. Sch., 1975-77, assoc. chmn. dept. radiation biology & biophysics, 1980-85, dir. div. med. informatics, 1988—; consulting mathematician, NIMH, Bethesda, Md., 1963-64. Contbr. articles to profl. jours. Treas. Stormers Soccer Club, Rochester, 1983-93; bd. dirs. Monroe County Girls Soccer League, Rochester, 1988-93. Surgeon, USPHS, 1963-64. Grantee AEC, NIH, NSF, ERDA, DOE, 1965-91. Mem. AAAS, Biophysical Soc., Harvey Soc., Protein Soc., N.Y. Acad. Scis., Am. Med. Informatics Assn. (edn. working group), Nat. Edn. Med. Sch. Consortium. Home: 75 Deer Creek Rd Pittsford NY 14534-4147 Office: U Rochester Dept Biophysics Med Ctr Box BPHYS Rochester NY 14642

GOLDSTEIN, IRVING ROBERT, mechanical and industrial engineer, educator, consultant; b. Jersey City, N.J., Apr. 28, 1916; s. David and Anna (Krug) G.; m. Natalie E. Glattstein, Jan. 30, 1949; children: Barbara Joy, David Lee. BSME, Newark Coll. Engring.; MSME, Stevens Inst. Tech., 1947. Registered profl. engr., N.J., Calif. Field worker, N.J. Dept. Edn., 1938-39, indsl. engr., Maidenform Co., 1939-40, cost analyst, William Bal Corp., 1940-41, resident insp. N.Y. Ordnance Dist., War Dept., U.S. Army, 1941-43, sales rep., Eagle Hosiery Co., 1946-47, instr. dept. indsl. and mgmt. engring. N.J. Inst. Tech., 1947-50, asst. prof., 1950-55, assoc. prof., 1955-70, prof., 1970-81, prof. emeritus, 1981—, prof. dept. info. sci. and systems, Fairleigh Dickinson U., 1992; cons. engr. Irving R. Goldstein, P.E., Springfield, N.J., 1967—; lectr. in field; examiner profl. engring. exam State of N.J., 1967-82; rep. Am. Nat. Standards Inst., 1970-83, Engr. Joint Coun. Com. for Am. Bicentennial, 1975-78; vice-chmn. N.J. Engrs. Com. for Student Guidance, 1981-83, state meetings coord., 1974-81, treas., 1983-86. Contbr. articles to profl. jours. Served with U.S. Army, 1943-46, ETO. Fellow Inst. Indsl. Engrs. (dir. work measurement and method engring div. 1970-73, conf. chmn. 1973-81, publs. chmn. 1966-70, Phil Carroll Achievement award 1975, pres. Met. N.J. chpt. 1977-78, v.p. rsch. and edn. 1968-73, 75-76, 79-81, chmn. bd. gov.'s Metro N.J. chpt. 1966-68, 81-82, faculty advisor N.J. Inst. Tech. U. chpt. 1962-77, Disting. Svc. award Met. N.J. chpt. 1970, 76, 85, Walter Salabun award 1989, author, historian Metro N.J. chpt. 1982-92, dir. student affairs Dist. 2, 1989-90); mem. IEE (life), ASME (life), Ops. Rsch. Soc. Am., Inst. Mgmt. Scis., Nat. Soc. Profl. Engrs., Prodn. Ops. Mgmt. Soc., N.Y. Acad. Scis., Order of Engr., Alpha Pi Mu, Pi Tau Sigma. Home and Office: 21 Janet Ln Springfield NJ 07081

GOLDSTEIN, IRWIN JOSEPH, medical research executive; b. Newark, N.J., Sept. 8, 1929; 2 children. BA, Syracuse U., 1951; PhD in Biochemistry, U. Minn., 1956. Rsch. fellow, dept. agrl. biochemistry U. Minn., St. Paul, 1956-59; asst. prof., dept. biochemistry State U. N.Y., Buffalo, 1961-65; assoc. prof., dept. biological chemistry U. Mich., Ann Arbor, 1965-72, prof., dept. biological chemistry, 1972—; cons. Ann Arbor Community, 1968-71, Procter & Gamnbe Co., 1968—; assoc. dean Rsch. and Grad. Studies U. Mich., 1986—; mem. rsch com. Henry Ford Hosp., 1983—. Editorial bd.: Journal of Biological Chemistry, 1983-88, 1991—, Plant Physiology, 1983-86, Carbohydrate Research, 1984-87, Glycoconjugate Journal, 1985-88, Archive Biochemistry and Biophysics, 1989—. Rsch. bd. United Fund, Buffalo, N.Y., 1963-65; bd. dirs. Guild House U. Mich., 1975-80. Recipient Kaiser Permanente Pre Clinical Teaching award, 1980, Claude S. Hudson Carbohydrate Chemistry award Am. Chem. Soc., 1993; Guggenheim fellow, 1959-60, NIH fellow, 1960-61. Mem. Am. Heart Assn., Biochemical Soc., Chem. Soc., Am. Soc. Biological Chemists, Am. Chem. Soc., Soc. Complex Carbohydrates (exec. com. 1987—), Sigma Xi, Phi Lambda Upsilon. Achievements include research in carbohydrate-protein interactions; isolation, purification and characterization of lectins (carbohydrate-binding proteins); use of lectins to study cell-surface phenomena; studies on the structure and biosynthesis of glycoproteins; immunochemistry of carbohydrates. Office: U Mich Cancer Ctr Cancer Rsch Com 101 Simpson Dr Box 0752 Ann Arbor MI 48109*

GOLDSTEIN, JOEL, management science educator, researcher; b. N.Y.C., Mar. 29, 1938; s. Jack and Regina (Gross) G.; m. Marcia Rosen, Sept. 5, 1966; children: Jennifer Ann, Carol Lynn. BME, CCNY, 1967; MS, N.Y.U., 1971; PhD, Polytech. Inst. N.Y., 1980. Analyst Allied Corp., N.Y.C., 1963-67; automation engr. Ebasco Svcs., N.Y.C., 1967-68; mgr. Bunker Ramo Corp., Trumbull, Conn., 1969-74; sr. analyst Getty Oil Co., N.Y.C., 1974-77; dir. Am. Express Co., N.Y.C., 1978-83; v.p. Citicorp, NA, N.Y.C., 1983-86; assoc. prof. Western Conn. State Univ., Danbury, 1987—. Author: (with R. Montague) Lotus 1-2-3 The Easy Way, 1989; contbr. articles to profl. jours. Mem. IEEE, Am. Statistical Assn., Assn. Computing Machinery, Ops. Rsch. Soc. Am. Office: Western Conn State Univ 181 White St Danbury CT 06810

GOLDSTEIN, JOSEPH LEONARD, physician, medical educator, molecular genetics scientist; b. Sumter, SC, Apr. 18, 1940; s. Isadore E. and Fannie A. Goldstein. BS, Washington and Lee U., Lexington, Va., 1962; MD, U. Tex., Dallas, 1966; DSc (hon.), U. Chgo., 1982, Rensselaer Poly. Inst., 1982, Washington and Lee U., 1986, U. Paris, 1988, U. Buenas Aires, 1990; So. Meth. U., 1993. Intern, then resident in medicine Mass. Gen. Hosp., Boston, 1966-68; clin. assoc. NIH, 1968-70; postdoctoral fellow U. Wash., Seattle, 1970-72; mem. faculty U. Tex. Southwestern Med. Ctr., Dallas, 1972—, Paul J. Thomas prof. medicine, chmn. dept. molecular genetics, 1977—; regental prof. U. Tex. Health Scis. Ctr., Dallas, 1985—; Harvey Soc. lectr., 1977; mem. sci. rev. bd. Howard Hughes Med. Inst., 1978-84, med. adv. bd., 1985-90; non-resident fellow The Salk Inst., 1983—. Co-author: The Metabolic Basis of Inherited Disease, 5th edit., 1983; editorial bd. Jour. Biol. Chemistry, 1981-85, Cell, 1983—, Jour. Clin. Investigation, 1977-82, Ann. Rev. Genetics, 1980-85, Arteriosclerosis, 1981-87, Sci. 1985—. Sci. adv. bd. Welch Found., 1986—; bd. dirs. Passano Found., 1985—. Recipient Heinrich-Wieland prize, 1974, Pfizer award in enzyme chemistry Am. Chem. Soc., 1976; Passano award Johns Hopkins U., 1978; Gairdner Found. award, 1981; award in biol. and med. scis. N.Y. Acad. Scis., 1981, Lita Annenberg Hazen award, 1982; Rsch. Achievement reward Am. Heart Assn., 1984; Louisa Gross Horwitz award, 1984; 3M Life Scis. award, 1984, Albert Lasker award in Basic Med. Rsch., 1985; Nobel Prize in Physiology or Medicine, 1985, Trustees' medal Mass. Gen. Hosp., 1986, U.S. Nat. medal of Sci., 1988. Mem. NAS (Lounsbery award 1979, coun. 1991—), ACP (award 1986), Assn. Am. Physicians, Am. Soc. Clin. Investigation (pres. 1985-86), Am. Soc. Human Genetics (William Allan award 1985), Amer. Acad. Arts and Scis., Am. Soc. Biol. Chemists, Am. Fedn. Clin. Research, Am. Philos. Soc., Inst. Medicine, Royal Soc. London (fgn. mem.), Phi Beta Kappa, Alpha Omega Alpha. Home: 3831 Turtle Creek Blvd # 22B Dallas TX 75219-4417 Office: U Tex Southwestern Med Ctr 5323 Harry Hines Blvd Dallas TX 75235-7200

GOLDSTEIN, MARK KINGSTON LEVIN, high technology company executive, researcher; b. Burlington, Vt., Aug. 22, 1941; s. Harold Meyer Levin and Roberta (Butterfield) G.; m. Kyoko Matsubara, Mar. 8, 1985. B.S. in Chemistry, U. Vt., 1964, Ph.D., U. Miami-Coral Gables, 1971. Pres. IBR, Inc., Coral Gables, Fla., 1970-74; group leader Brookhaven Nat. Lab., Upton, N.Y., 1974-77; sr. researcher East-West Ctr., Honolulu, 1977-79; sr. tech. advisor JGC Corp., Tokyo, 1979-81; pres., chmn. bd. Quantum Group, Inc., La Jolla, Calif., 1981—; exec. dir. Magnatek, Inc., Brotas, Brazil, 1982—. Contbr. articles to profl. jours.; contbr. poetry to mag. NSF fellow, 1964, 65. Mem. AAAS, Am. Chem. Soc., Hawaii Yacht Club (Honolulu). Patentee, inventor devices including biomimetic carbon monoxide sensor, thaser co-generators, supermitters, thermphotovoltaics self powered gas appliance, photon control systems and gas safety valve. Home: 2500 Torrey Pines Rd Apt 805 La Jolla CA 92037-3430 Office: Quantum Group Inc 11211 Sorrento Valley Rd San Diego CA 92121-1323

GOLDSTEIN, MARVIN EMANUEL, aerospace scientist, research center administrator; b. Cambridge, Mass., Oct. 11, 1938; s. David and Evelyn (Wilner) G.; m. Priscilla Ann Beresh, July 5, 1965; children: Deborah, Judy. BS in Mech. Engring., Northeastern U., 1961; MS in Mech. Engring., MIT, 1962; PhD in Mech. Engring., U. Mich., 1965. Engr. Arthur D. Little, Inc., Cambridge, 1958-61; rsch. asst. MIT, Cambridge, 1961-63, rsch. assoc., 1965-67; aerospace engr. Lewis Rsch. Ctr., NASA, Cleve., 1967-79, chief scientist, 1980—. Author: Aeroacoustics, 1976; contbr. articles to profl. jours. Fellow AIAA (assoc. editor jour. 1977-79, chmn. aeroacoustics tech. com. 1979-81, mem. publs. com. 1980-83, Aeroacoustics award 1983, Pendray award 1983), Am. Phys. Soc. (exec. com. div. fluid dynamics 1991—); mem. NAE. Avocation: automobile racing and rebuilding. Office: NASA Lewis Rsch Ctr MS 3-17 21000 Brookpark Rd Cleveland OH 44135

GOLDSTEIN, MURRAY, health organization official; b. N.Y.C., Oct. 13, 1925; s. Israel and Yetta (Zeigen) G.; m. Sue Mary Michael, June 13, 1957; children: Patricia Sue Robertson, Barbara Jean Warner. BA, NYU, 1947; DO, Des Moines Still Coll. Osteo. Medicine, 1950; MPH, U. Calif., 1959; DSc (hon.), Kirksville Coll. Osteo. Medicine, 1970, U. New Eng., 1984, Ohio U., 1986, U. Osteo. Medicine and Health Scis., 1990; LLD (hon.), N.Y. Inst. Tech., 1982; Dr. honoris causa, Med. Univ. Pecs, Hungary, 1985; LHD (hon.), Coll. Osteo. Medicine Pacific, 1988. Diplomate Am. Osteo. Bd. Internal Medicine, Am. Osteo. Bd. Preventative Medicine (sec.-treas. 1987-88, vice chmn. 1988-92). Rotating intern Still Coll. Osteo. Hosp., Des Moines, 1950-51, resident internal medicine, 1951-53; commd. corps USPHS, 1953, advanced through grades to asst. surgeon gen., 1980, resigned, 1993; asst. to chief, then asst. chief, grants and tng. br., Nat. Heart Inst. NIH, Bethesda, Md., 1953-58, dir. epidemiology and biometry tng. grant program, divsn. rsch. grants, 1956-58, asst. chief rsch. grants rev. br., divsn. rsch. grants, 1959-60; chief spl. projects br. Nat. Inst. Neurol. Diseases and Blindness, NIH, Bethesda, Md., 1960-61; exec. sec. joint coun. subcom. cerebrovascular disease Nat. Inst. Neurol. Diseases and Stroke and Nat. Heart and Lung Inst., NIH, Bethesda, Md., 1961-67, 69-75; dir. extramural programs Nat. Inst. Neurol. and Communicative Disorders and Stroke, NIH, Bethesda, Md., 1961-76, dir. stroke and trauma program, 1976-78, dep. dir., 1978-81, acting dir., 1981-82, dir., 1982-93; pub. health trainee epidemiology Calif. State Dept. Pub. Health, Berkeley, 1958, acting chief sect. virus diseases ctrl. nervous system, Bur. Acute Communicable Disease, 1958; med. dir. United Cerebral Palsy Rsch. and Edn. Found., Washington, 1993—, bd. dirs., 1972—; clin. prof. neurol. medicine N.Y. Coll. Osteo. Medicine, 1977—; sr. lectr. dept. neurology Uniformed Svcs. U. Health Scis., 1986—; chmn. Commd. Corps Adv. Com. to NIH dir., 1990-93, WHO Task Force on stroke and other vascular cerebral disorders, 1986-89; dir. WHO Neurosci. Collaborating Ctr., Bethesda, 1981-93; liaison, mem. sci. adv. bd. Kent Waldrep Nat. Paralysis Found., 1989—; vis. prof. med. rsch. Semmelweis Med. U., Budapest, Hungary, 1975; vis. sci. sect. neurology Mayo Clinic and grad. sch., Rochester, Minn., 1967-68; vis. scholar Henry Ford Hosp., 1979, 80; v.p. Eisenhower Inst. Stroke Rsch., 1975-88; lectr., cons. in field. Assoc. editor Stroke: A Journal of Cerebral Circulation, 1976-91, consulting editor, 1992—; mem. editorial bd. Osteo. Annals, 1973-85, 87-88, Internat. Jour. Neurology, 1980—, Jour. Neuroepidemiology, 1981-90, Hosp. and Community Psychiatry, 1980—, Alzheimer Disease: An Internat. Jour., 1985—, Cerebralvascular and Brain Metabolism Revs., 1985—; contbr. articles to profl. jours. With U.S. Army, 1943-45. Decorated D.S.M., Purple Heart, Silver Star; recipient Founders Day medal U. Osteo. Medicine and Pub. Health, 1983, Patenge Pub. Svc. medal Mich. State U., 1987, Marjorie Guthrie award The Huntington's Disease Soc. Am., 1988, Burke award Burke Found., 1988, Spl. Leadership award United Cerebral Palsy Rsch. & Ednl. Found., 1989, Phillips Pub. Svc. medal Ohio U., 1990, Alumnus of Yr. award U. Osteo. Medicine and Pub. Health Scis., 1990, Pioneer award The Am. Speech-Lang.-Hearing Assn., 1992, The Surgeon Gen.'s Exemplary Svc. medal, 1993, Profl. Support award Nat. Headache Found., 1993; named Educator of Yr. by Nat. Osteo. Found./Am. Osteo. Assn., 1993. Fellow Am. Acad. Neurology (mem. long range planning com. 1972-75, mem. manpower com. 1979-85, mem. neurology in govtl. svcs. and insts. com. 1979-85, chmn. 1981-83, 93—, mem. internat. affairs com. 1981-90, chmn. 1981-83, mem. com. govtl. rels. 1983-85, ANA-AAN del. to World Fedn. Neurology 1983-85, mem. AAN com. on pub. comm. and legislation 1983-85, mem. ad hoc com. for soc. neurology liaison 1987-89, sr. advisor uniformed svcs. orgn.ts com. 1987-93), APHA (mem. epidemiology sect.), Am. Osteo. Coll. Preventative Medicine (trustee 1977-92, pres. 1989-90), Pan Am. Med. Assn. (mem. coun. neurol. sect.), Am. Coll. Osteo. Internists (hon.), Am. Coll. Neuropsychiatry (hon.), N.Am. Spine Soc. (hon.); mem. AAAS, Am. Osteo. Assn. (cons. bur. rsch. 1985—), Am. Soc. Neurorehabilitation (bd. dirs.), Am. Neurol. Assn. (hon. mem., 2nd v.p. 1982-83, ANA-AAN del. to World Fedn. Neurology 1981-87, chmn. adv. com. hon. membership 1980-82, mem. com. constitution and bylaws 1979-80, mem. exec. com. joint com. for stroke resources 1970-78), Am. Assn. Neurol. Surgeons (hon.), Am. Soc. Neurol. Investigation (hon.), Internat. Stroke Soc. (mem. exec. com. 1989-92, counselor exec. com. 1992—), Assn. Rsch. in Nervous and Mental Disease, Soc. Neurosci., Neurotrauma Soc., World Fedn. Neurology (mem. rsch. com. neuroepidemiology, mem. rsch. com. cerebrovascular disorders), Internat. Brain Rsch. Orgn., Nat. Acad. Practice (Disting. Practitioner in Osteo. Medicine), Fulton Soc. (hon.), Italian Soc. Neurology (corr.), Peruvian Assn. Neurology, Psychiatry and Neurosurgery (corr.), Beta Alpha Epsilon, Psi Chi, Sigma Alpha, Delta Omega. Avocations: gardening, golf, swimming. Home: 6210 Swords Way Bethesda MD 20817 Office: Nat Inst Neuro Disorders Stroke NIH Bldg 31 Rm 8A52 9000 Rockville Pike Bethesda MD 20892

GOLDSTEIN, RICHARD JAY, mechanical engineer, educator; b. N.Y.C., Mar. 27, 1928; s. Henry and Rose (Steierman) G.; m. Barbara Goldstein; children: Arthur Sander, Jonathan Jacob, Benjamin Samuel, Naomi Sarith. BME, Cornell U., 1948; MS in Mech. Engring., U. Minn., 1950, MS in Physics, 1951, PhD in Mech. Engring., 1959. Instr. U. Minn., Mpls., 1948-51, instr., rsch. fellow, 1956-58, mem. faculty, 1961—, prof. mech. engring., 1965—, head dept., 1977—, James J. Ryan prof., 1989—, Regents' prof., 1990—; devel. rsch. engr. Oak Ridge Nat. Lab., 1951-54; sr. engr. Lockheed Aircraft, 1956; asst. prof. Brown U., 1959-61; vis. prof. Technion, Israel, 1976, Imperial Coll., Eng., 1984; cons. in field, 1956—; chmn. Midwest U. Energy Consortium; chmn. Council Energy Engring. Research; NSF sr. postdoctoral fellow, vis. prof. Cambridge (Eng.) U., 1971-72; Prince lectr., 1983, William Gurley lectr., 1988, Hawkins Meml. lectr., 1991; disting. lectr. Pa. State U., 1992. Editorial adv. bd. Experiments in Fluids, Heat Transfer-Japanese Rsch., Heat Transfer-Soviet Rsch., Bull of the Internat. Centre for Heat andMass Transfer, Internat. Archives of Heat and Mass Transfer; hon. editorial adv. bd. Internat. J. Heat and Mass Transfer, Internat. Communications in Heat and Mass Transfer. 1st U.S. Army lt. AUS, 1954-55. NATO fellow Paris, 1960-61; Lady Davis fellow Technion, Israel, 1976; recipient NASA award for tech. innovation, 1977, MUEC dist. svc. award, 1986, George Taylor Alumni Soc. award, 1988, A.V. Lykov medal, 1990, Max Jakob Meml. award ASME/AIChE, 1990, Nusselt Reynolds prize, 1993. Fellow ASME (Heat Transfer Meml. award 1978, Svc. award 1978, Centennial medallion 1980, BEG v.p. 1984—, 50th anniv. award of heat transfer div. 1988, sr. v.p. 1989-93, hon. mem. 1992), Assembly for Internat. Heat Transfer Confs. (pres. 1986-90), Internat. Ctr. for Heat and Mass Transfer (exec. com. 1985—, chmn. 1992), AAAS, Am. Phys. Soc., Japan Soc. Promotion of Sci.; mem. Am. Phys. Soc., Am. Soc. Engring. Edn., Minn. Acad. Sci., Nat. Acad. Engring., Nat. Acad. Engring.-Mex. (corr. 1991), Golden Key Nat. Honor Soc., Sigma Xi, Tau Beta Pi, Pi Tau Sigma. Achievements include research and publications in thermodynamics, fluid mechanics, heat transfer, optical measuring techniques. Home: 520 Janalyn Cir Minneapolis MN 55416-3327 Office: U Minn Dept Mech Engring 111 Church St SE Minneapolis MN 55455-0150

GOLDSTEIN, RUBIN, consulting physicist; b. N.Y.C., Mar. 29, 1933; s. Sam and Ray (Faier) G.; m. Sylvia Fay Galitzer, June 3, 1956; children: Lori Jean, Susan May, Pamela Joy. AB in Physics, Princeton U., 1955; AM in Physics, Harvard U., 1956, PhD in Physics, 1960. Rsch. assoc. in applied physics and instr. applied math. Harvard U., Cambridge, Mass., 1960; asst. prof. nuclear engring. U. Calif. Berkeley, Berkeley, 1960-65; physicist Brookhaven Nat. Lab., Upton, N.Y., 1965-71; cons. staff physicist Combustion Engring., Windsor, Conn., 1971-73, sr. cons. physicist and area mgr., 1973-79, prin. cons. scientist and mgr., 1980-82, sr. prin. cons. scientist, 1983-86; cons. scientist Goldstein Cons., West Hartford, Conn., 1987—; cons. Aetna Life & Casualty Co., Hartford, Asea Brown Boveri, Inc., Windsor, GE, San Jose and Vallecitos, Calif., N.E. Util., Berlin, Conn., Reactor Safety Rsch. dir. NRC, Washington. Contbr. over 55 articles to tech. jours., books. Princeton U. scholar, 1951-55, Harvard U. fellow, 1955-57, NSF fellow, 1958-60. Mem. Am. Nuclear Soc. (exec. com. of math. and computation div., honors and awards com.), Am. Phys. Soc., Phi Beta Kappa, Sigma Xi. Avocations: chess, basketball, softball, biking. Home and Office: 8 E Normandy Dr West Hartford CT 06107-1405

GOLDSTEIN, STEVEN ALAN, medical educator, engineering educator; b. Reading, Pa., Sept. 15, 1954; m. Nancy Ellen Gehr, Aug. 22, 1976; children: Aaron Michael, Jonathan David. BS in Mech. Engring., Tufts U., 1976; MS in Bioengring., U. Mich., 1977, PhD in Bioengring, 1981. Rsch. investigator dept. surgery U. Mich., Ann Arbor, 1981-83, asst. prof. surgery, 1983-88, assoc. prof. surgery, 1988-92, prof. surgery, 1992—; co-dir. orthopaedic biomechanics lab. U. Mich., 1981-82, dir. orthopaedic rsch. labs. U. Mich., 1982—; adj. asst. prof. mech. engring. and applied mechanics U. Mich., 1985-88, adj. assoc. prof. mech. engring. and applied mechanics, 1988-92, prof. mech. engring. and applied mechanics, 1992—, rsch. dir., advisor, 1981—, mem. faculty bioengring. program, 1982—, interim chmn., 1985-89, rsch. scientist inst. gerontology, 1993—, asst. dean rsch. & grad. studies U. Mich. Med. Sch., 1993—; rsch. asst. bioengring. ctr. Tufts New England Med. Ctr., 1974-76; mem. calcium homeostasis adv. group NASA, 1987-89; cons. Libbey-Owens Ill., Gen. Tire & Rubber, Upjohn, Ethyl Corp., Norwich Eaton, KMS Fusion, Whitby Pharmaceuticals, Norian Corp., Genetics Inst.; dean's rep. rsch. and devel. com. Ann Arbor Vets. Adminstrn. Med. Ctr., 1984-87; mem. various univ. coms.; lectr. in field. Author: Advances in Engineering, 1991; author (with others) Biomechanics of Diathrodial Joints, 1990, Molecular Biology of the Cardiovascular System, 1991, Surgery: Scientific Principles and Practice, 1993, Limb Development and Regeneration, 1993, Accidental Injury: Biomechanics and Prevention, 1993; reviewer Math. Bioscis., 1982—, Annals of Biomed. Engring., 1983—, Clin. Orthopaedics and Related Rsch., 1983—, Jour. Rehab. Rsch. and Devel., 1987—; reviewer Jour. Biomechanics, 1982—, editorial cons., 1992—; reviewer Jour. Biomech. Engring., 1982—, assoc. editor, 1991—; reviewer Jour. Orthopaedic Rsch., 1984—, mem. bd. assoc. editors , 1992—, reviewer Jour. Bone and Joint Surgery, 1987—, mem. bd. assoc. editors for rsch., 1989—; reviewer, mem. study section NIH, NSF, NASA, Nat. Inst. Occupational Health & Safety, 1983—. Recipient Young Rsch. Investigator award 3M Corp., 1984, Nicolas Andre award Assn. Bone & Joint Surgeons, 1987-88. Mem. ASME (chair program com. 1989-92, sec.-elect 1993, exec. com. bioengring. divsn. 1989—, Y.C. Fung Young Investigator award 1987), Am. Soc. Biomechanics (exec. bd. 1984-85), Am. Acad. Orthpaedic Surgeons (com. biomed. engring. 1991—, Kappa Delta award 1989-90), Orthopaedic Rsch. Soc. (adj. program com. 1990-91, program com. 1992), Biomed. Engring. Soc., Engring. Soc. Detroit (Sigma Xi Young Engr. of Yr. award 1987), The Knee Soc. Achievements include patents (with other) for Intracone Reamer, Instacone Prosthetic Surface, Flexible Connecting Shaft for Intramedullary Reamer, Tissue Pressure Measurement Transducer System, Continuous Flow Tissue Pressure Measurement Transducer System, Prosthesis Interface Surface and Method of Implanting. Office: U Mich Orthopaedic Rsch Labs 400 N Ingalls Bldg Rm G161 Ann Arbor MI 48109

GOLDSTEIN, WALTER ELLIOTT, biotechnology executive; b. Chgo., Nov. 28, 1940; s. Henry Harold and Dorothy (Davidson) G.; m. Paula G. Copen, Feb. 18, 1962; children: Susan, Marc. BS in Chem. Engring., Ill. Inst. Tech., 1961; MBA, Mich. State U., 1968; MSChemE, U. Notre Dame, 1971, PhDChemE, 1973. Registered profl. engr., Ind. Process devel. engr. Linde div. Union Carbide, Tonawanda, N.Y., 1961-64; assoc. project engr. Miles Labs., Elkhart, Ind., 1964-67, assoc. rsch. scientist, 1967-72, rsch. scientist, 1972-73, rsch. supr., 1973-76, mgr. Chem. Engring. Rsch. & Pilot Svcs., 1976-78, dir., 1978-82, v.p. R & D Chem. Engring. Rsch. & Pilot Svcs., 1982-87; v.p. process svcs. Phytopharms., Inc. and ESCAgenetics Corp., San Carlos, Calif., 1987-93, also v.p. R&D, 1993—; adj. asst. prof. chem. engring. U. Notre Dame, 1974-75; cons. Bernard Wolnak, Chog., 1987. Contbr. chpts. to books; inventions and publs. in chem. engring. and biotech. field. Vice-pres. B'nai B'rith, South Bend., Ind., 1978-89. Mem. AAAS, Am. Inst. Chem. Engrs., N.Y. Acad. Scis., Inst. Food Technologists, Sigma Xi. Jewish. Avocations: reading, computers, outdoor sports.

GOLDSTONE, JEFFREY, physicist; b. Manchester, Eng, Sept. 3, 1933; came to U.S., 1977; m. Roberta Gordon; 1 child, Andrew. B.A., Cambridge U. Eng., 1954, Ph.D., 1958. Fellow Trinity Coll., Cambridge, Eng., 1956-60, 62-82; lectr., reader U. Cambridge, Cambridge, 1961-76; MIT, Cambridge,

1977—, Cecil and Ida Green prof. physics, 1983—. Recipient Dannie Heineman prize Am. Phys. Soc., 1981, Guthrie medal Inst. Physics, 1983, Dirac prize Internat. Centre for Theoretical Physics, 1991. Mem. Royal Soc., Am. Acad. Arts and Scis. Office: MIT 6-313 77 Massachusetts Ave Cambridge MA 02139

GOLDSTONE, PHILIP DAVID, physicist; b. Bklyn., Mar. 5, 1950; s. Herman and Carolyn Martha (VanGelderen) G.; m. Joyce Ann Roberts, Aug. 26, 1972 (div. Jan. 1989); m. Carol Lee Sorensen, Apr. 28, 1990 (div. Feb. 1993). BS in Physics, Polytechnic Inst. of Bklyn, 1971, MS in Physics, 1972; PhD, SUNY, Stony Brook, N.Y., 1975. Nuclear physics group postdoctoral appointee Los Alamos (N.Mex.) Nat. Lab., 1976-77, laser ops. group staff mem., 1977-78, high energy density physics group staff mem., 1978-81, laser matter interaction and fusion physics group leader, 1981-89, inertial fusion expts. program mgr., 1986-89, staff mem. R & D tech. program, 1989-92, chief scientist inertial confinement fusion and high energy density physics, 1992—. Contbr. articles to profl. jours. Chief adminstrv. officer La Cueva (N.Mex.) Vol. Fire Dept., 1984-87; mem. planning com. 1991 N.Mex. Men's Wellness Conf.; bd. dirs. Santa Fe County div. Am. Heart Assn., 1992—, chmn. community programs, 1993—. Fellow Woodrow Wilson Found.; mem. AAAS, Am. Phys. Soc. Div. Plasma Physics (exec. com. 1984-85), Optical Soc. Am. Avocations: skiing, photography, hiking. Office: Los Alamos Nat lab NWT/ICF MS-E527 Los Alamos NM 87545

GOLECKI, ILAN, physicist, researcher, educator; b. Haifa, Israel; came to U.S., 1978; s. Moshe and Rebecca (Lazarovici) G. BS cum laude in Physics, Technion, Israel Inst. Tech., Haifa, Israel, 1970, MS in Physics, 1974; PhD in Physics, U. Neuchâtel, Neuchâtel, Switzerland, 1978. Rsch. fellow Calif. Inst. Tech., Pasadena, Calif., 1978-79, vis. assoc., 1979-86; mem. tech. staff Rockwell Internat. Corp., Anaheim, Calif., 1979-85, Thousand Oaks, Calif., 1985-86; ind. cons. Thousand Oaks, Calif., 1986-87; sr. rsch. physicist Allied-Signal, Inc., Morristown, N.J., 1987-93, rsch. scientist, 1993—; organizer, chmn. of session on silicon-on-insulators, SPIE Conf., L.A., 1986; referee of jour. articles in field. Contbr. 45 articles to sci. jours. Mem. AAAS, IEEE, Am. Assn. for Crystal Growth, Am. Phys. Soc., Am. Vacuum Soc., Böhmische Phys. Soc., Electrochem. Soc., Materials Rsch. Soc., Mineral, Metals and Materials Soc. Achievements include patents concerning the processing of silicon-on-sapphire, SiC epitaxial growth by chemical vapor deposition; devel. of new silicon-on-insulator technologies; co-discovery of ion beam induced epitaxial regrowth effect in silicon; devel. of apparatus enabling ion channeling measures; research interests include analytical characterization and growth of thin films and porous composites by chemical vapor deposition and infiltration; ion beam analysis; vacuum science and technology. Home: 100 Vail Rd #N-5 Parsippany NJ 07054 Office: Allied Signal Inc 101 Columbia Rd MS CTC-3 Morristown NJ 07962

GOLERKANSKY, PETER JOSEPH, electrical engineer; b. Kishinev, USSR, Jan. 24, 1950; came to the U.S., 1975; s. Joseph Michael and Berta (Langer) G.; m. Linda Diane Collins, Mar. 18, 1979; children: Larissa Danielle, Bryan Michael. BEE, U. So. Calif., 1984; MBA, Pepperdine U., 1990. Cert. plant engr. Elec. engr. Cedars-Sinai Med. Ctr., L.A., 1978-86; dir. engring. Maxicare Med. Ctr., L.A., 1986-87, Centinela Hosp. Med. Ctr., Inglewood, Calif., 1987—. Mem. IEEE, Am. Inst. Plant Engrs., Am. Soc. for Hosp. Engring. Home: 18016 Raymer St Northridge CA 91325

GOLIGHTLY, DANOLD WAYNE, chemist; b. Cape Girardeau, Mo., Apr. 12, 1941; s. J. C. and Dorothy Rosina (Koch) G.; m. Marilyn Victoria Brodine, May 2, 1969; 1 child, Yvonne Marie. AB, S.E. Mo. State U., 1962; MS, Iowa State U., 1965, PhD, 1967. Teaching/rsch. asst. dept. chemistry Iowa State U., Ames, 1962-67, rsch. assoc. Ame Lab., 1970-71; rsch. assoc. materials rsch. lab. U. Ill., Urbana, 1967-69; stagier etrange Centre d'Etudes Nucléaires, Grenoble, France, 1969-70; rsch. chemist Nat. Bur. Standards, Gaithersburg, Md., 1971-73, U.S. Geol. Survey, Reston, Va., 1973-88; sr. rsch. scientist Ross Labs., Columbus, Ohio, 1988—; cons. Docegeo, Belem, Brazil, 1983. Editor: Inductively Coupled Plasmas in Analytical Atomic Spectrometry, 1987, 2d rev. edit., 1992; contbr. articles to profl. publs., chpts. to books. Mem. Am. Chem. Soc., Soc. for Applied Spectroscopy, Internat. Assn. Ofcl. Analytical Chemists. Achievements include patent for image dissector photomultiplier for direct reading spectrometry. Office: Ross Labs 625 Cleveland Ave Columbus OH 43215

GOLIJANIN, DANILO M., materials scientist, physics educator; b. Belgrade, Serbia, Yugoslavia, May 28, 1952; came to U.S., 1983; s. Milan B. and Zora (Ristic) G.; m. Liliana Golijanin, Dec. 14, 1987; 1 child, Gala. BSEE, U. Belgrade, 1976; MSEE, U. So. Calif., L.A., 1983, PhD in Material Sci., 1988. Rsch. assoc. electron microanalysis lab. U. So. Calif., 1976-82, rsch. asst., 1983-88, lectr., 1989—; sr. scientist Fisons Instruments, Valencia, Calif., 1992—; sr. failure analyst Xerox Corp., El Segundo, Calif., 1985-92. Contbr. articles to jour. Applied Physics; confs. Internat. Congress on X-Ray Optics, Microbeam Analysis Soc. Recipient Birks award Microbeam Analysis Soc., 1988, 89; Fulbright fellow, 1983. Mem IEEE, Sigma Xi. Mem. Serbian Orthodox Ch. Achievements include research in fabrication of doubly bent crystal diffractors; patents pending for focusing of x-ray beams, scanning x-ray microprobe. Office: Fisions Instruments X Ray Div 24911 Stanford Ave Valencia CA 91355

GOLITSYN, GEORGIY, research institute director; b. Moscow, Jan. 23, 1935; s. Serguey M. and Claudia (Bavykina) G.; m. Ludmila V. Lisitskaya, Sept. 12, 1933; children: Anna, Maria. Masters, Moscow U., 1958, postgrad., 1961, DS, 1971. Jr. rsch. scientist Inst. Atmospheric Physics, Moscow, 1958-64, sr. rsch. assoc., 1964-75, head lab., 1975—, head sect., 1982—, dir., 1990—; adj. prof. Phys.-Tech. Inst., Moscow, 1975—; coun. chmn. Internat. Inst. Applied System Analysis. Author: Introduction to Dynamics of Planetary Atmospheres, 1973, Global Climate Catastrophes, 1986. Mem. Acad. Scis. of USSR (corr. mem. 1979—, full mem. 1987—, mem. presidium 1988-90, 1992—), Moscow House of Scientists. Avocations: poetry, history, canoeing. Office: Inst Atmospheric Physics, 3 Pyzhevsky, 109017 Moscow Russia

GOLOMB, FREDERICK MARTIN, surgeon, educator; b. N.Y.C., Dec. 18, 1924; s. Jacob J. and Hannah (Loewy) G.; m. Joan E. Schneider, Nov. 28, 1954; children: James Bradley, Susan Lynn. B.S., Yale U., 1945; M.D., U. Rochester, 1949. Diplomate: Am. Bd. Surgery. Intern Johns Hopkins Hosp., 1949-50; resident NYU Hosp., 1950-56; pvt. practice specializing in surgery N.Y.C.; mem. staff NYU Med. Center, 1950—, dir. chemoimmunotherapy div. tumor service dept. surgery, 1967—; dep. div. dir., chief patient research unit div. II, clin. research div., chief chemotherapy unit div. IV NYU Cancer Center, 1975-79; attending surgeon Beth Israel North Hosp.; cons. in gen. surgery Manhattan VA Hosp.; cons. surgeon Cabrini Health Care Center; vis. surgeon Bellevue Hosp.; mem. faculty NYU Sch. Medicine, 1956—, prof., clin. surgery, 1977—; cons. N.Y.C. div. Am. Cancer Soc., 1968—; mem. clin. trials rev. com. Nat. Cancer Inst., 1976-79; chmn. melanoma com. Eastern Coop. Oncology Group, 1978-80; prin. investigator Central Oncology Group, 1969-77, exec. com., 1976-77; mem. met. med. com. Chemotherapy Found.; co-prin. investigator Eastern Coop. Oncology Group NYU, 1978—. Editorial adv. bd., contbg. editor Oncology News; contbr. articles to profl. jours. Served with M.C. AUS, 1953-54, Korea. Fellow A.C.S.; mem. Soc. Head and Neck Surgeons, Soc. Surgery Alimentary Tract, Am. Assn. Cancer Research, Am. Soc. Clin. Oncology (a founder), AMA, N.Y. Cancer Soc. (pres. 1974-75), N.Y. Surg. Soc., N.Y. State, N.Y. County med. socs., Soc. Surg. Oncology, George Hoyt Whipple Soc., Brit. Assn. Surg. Oncology (editorial adv. panel 1980-85), Pan Am. Med. Soc., Sigma Xi. Clubs: Am. Alpine, Explorers. Office: NYU Sch Medicine 530 1st Ave New York NY 10016-6402

GOLOVCHENKO, JENE ANDREW, physics and applied physics educator; b. N.Y.C., July 14, 1946; s. Boris Andrew and Marilyn (Lee) G.; m. Elizabeth Marie Catricala; children: Peter, Eric, Katya. BEE, Rensselaer Poly. Inst., 1967, MEE, 1968, PhD, 1972; MA (hon.), Harvard U., 1987. Lectr. Inst. Physics Aarhus U., Denmark, 1974-76; mem. tech. staff Bell Labs., Murray Hill, N.J., 1976-87, Disting. mem. tech. staff, 1984; prof. physics Harvard U., Cambridge, Mass., 1987—, Gordon McKay prof. applied physics, div. applied scis., 1987—; mem. Rowland Inst. Sci., Cambridge, 1989—. Contbr. numerous articles to profl. publs. Office: Harvard U Physics Dept Cambridge MA 02138

GOLUB, GENE HOWARD, computer science educator, researcher; b. Chgo., Feb. 29, 1932; s. Nathan and Bernice (Gelman) G. BS in Math., U. Ill., 1953, MA in Math.-Stats., 1954, PhD in Math., 1959; Tech. Dr. (hon.), Linköping U., 1984; Dr. honoris causa, U. Grenoble, 1986, U. Waterloo, 1987, U. of Dundee, 1987, Katholieke Univ. Leuven, Belgium, 1992; hon. degree, U. Ill., 1991. Mem. tech. staff Space Tech. Labs., Inc., 1961-62; vis. asst. prof. computer sci. Stanford U., 1962-64, assoc. prof., 1966-70, prof., 1970—, Fletcher Jones prof. computer sci., 1991, chmn. dept. computer sci., 1980-84, dir. sci. computing/computational math.; adj. asst. prof. Courant Inst. Math. Scis. NYU, N.Y.C., 1965-66; Forsythe lectr. Assn. Computing Machinery, 1978; mem. adv. com. on computer sci. NSF, 1982-84; cons. in field. Author: (with Gerard Meurant) Resolution Numerique des Grandes Systems Lineaires, 1983, (with Charles Van Loan) Advanced Matrix Computations, 1984, rev. edit., 1989, (with James M. Ortega) Scientific Computing and Differential Equations: An Introduction to Numerical Methods, 1992, Scientific Computing: An Introduction with Parallel Computing, 1993; contbr. articles to profl. jours.; assoc. editor various jours., 1967-79; editor: Numerische Mathematik, 1978; founding editor SIAM Jour. on Matrix Analysis and Applications. Recipient Alumni Honor award U. Ill., Urbana, 1984; hon. fellow St. Catherine's Coll., Oxford U., Eng., 1983; Guggenheim fellow, 1987. Fellow AAAS (hon.); mem. NAS, NAE, Soc. for Indsl. and Applied Math. (mem. coun. 1975-77, trustee 1982—, pres. 1985-87, vis. lectr. 1976-77, founder, editor Jour. Sci. and Statis. Computing 1980—), Spl. Interest Group in Numerical Analysis (bd. dirs. 1976-78), Gesellschaft fuer angewandte Mathematik und Mechanik (governing coun. 1982—), U.S. Nat. Com. for Math., Royal Swedish Acad. Engring. Scis. Home: 576 Constanzo St Palo Alto CA 94305-8418 Office: Stanford U Dept U Bldg 460 Dept Computer Sci Stanford CA 94305

GOLUBITSKY, MARTIN AARON, mathematician, educator; b. Phila., Apr. 5, 1945; s. Isaac and Rose (Sarvetnick) G.; m. Barbara Lee Keyfitz, May 30, 1976; children: Elizabeth Ann, Alexander. AB, U. Pa., 1966, AM, 1966; PhD, MIT, 1970. Vis. lectr. UCLA, 1970-71; lectr. MIT, Cambridge, 1971-73; asst. prof., then assoc. prof. Queens Coll., CUNY, N.Y.C., 1973-79; prof. math. Ariz. State U., Tempe, 1979-83, U. Houston, 1983—; editorial bd. Jour. Math. Analysis, 1981-92, Archives Rational Mechanics and Analysis, 1984—, Jour. Nonlinear Sci., 1990—. Co-author: Stable Mappings and Their Singularities, 1978, Singularities and Groups in Bifurcation, vols. I and II, 1985, 88, Fearful Symmetry, 1992, Symmetry in Chaos, 1992. Cullen prof. U. Houston, 1989. Fellow AAAS; mem. Am. Math Soc., Soc. Indsl. and Applied Math. Home: 6419 Sewanee St Houston TX 77005 Office: Univ Houston Dept Math 4800 Calhoun St Houston TX 77204-3476

GOLUBSKI, JOSEPH FRANK, pathologist, physician; b. Cleve., Apr. 30, 1953; s. Joseph John and Rita Dolores (Krysinski) G.; m. Wanda Beth Kalencki, Nov. 11, 1983; children: Anne Elise, Joseph Edward. BA, Ohio Wesleyan U., 1975; MS, Cleve. State U., 1976; DO, U. Health Scis., Kansas City, Mo., 1980. Diplomate Am. Bd. Pathology. Intern Brentwood Hosp., Warrensville Heights, Ohio, 1980-81, chmn. pathology dept., 1987-88; resident in pathology Naval Regional Med. Ctr., Portsmouth, Va., 1981-85, mem. staff, head of autopsy svc. and clin. chem., 1985-87; assoc. pathologist Sheboygan (Wisc.) Meml. Med. Ctr., 1988—; cons. pathology Naval Hosp. Guantanamo Bay, Cuba, 1985-87; asst. clin. prof. pathology Ea. Va. Med. Sch., Norfolk, 1985-87. Comdr. USNR-IRR (Ind. Ready Res.), 1990—. Hall undergrad. fellow in chem. Ohio Wesleyan U, 1974. Fellow Am. Coll. Pathologists, Am. Soc. Clin. Pathologists; mem. AMA, Am. Osteo. Assn., Wis. Med. Assn., Sheboygan County Med. Soc. (sec.-treas. 1992—), Farmer's and Sportsman's Conservation Club (bd. dirs. 1990), Sheboygan Falls Conservation Club, Sheboygan Yacht Club. Avocations: sailing, flying, skeet and trap shooting, personal computers. Home: 2232 N 7th St Sheboygan WI 53083-4923 Office: Sheboygan Meml Med Ctr 2629 N 7th St Sheboygan WI 53083-4998

GOMBLER, WILLY HANS, chemistry educator; b. Hostenbach, Saarland, Germany, May 29, 1941; s. Peter and Katharina (Burbach) G.; m. Monika Stutz, July 30, 1965; 1 child, Melanie. Diploma in chemisry, U. Saarland, Saarbrucken, 1974, D in Natural Sci., 1975; D in Habilitation, Ruhr U, Bochum, Fed. Republic of Germany, 1982. Chem. engr. Chemische Werke Offenburg, Fed. Republic of Germany, 1964; chem. engr. U. Saarland, 1965-70, asst. scientist, 1975-76; asst. scientist Ruhr U., 1976-82, lectr. Inorganic Chemistry, 1983; prof. Instrumental Analysis Fachhochschule Ostfriesland, Emden, Fed. Republic of Germany, 1984—; dean faculty Naturwissenschaftliche Technik Fahhochschule Ostfriesland, Emden, 1993—. Contbr. numerous articles to profl. jours.; patentor vacuum technique. Mem. Gesellschaft Deutscher Chemiker, Am. Chem. Soc. (fluorine divsn.). Avocations: tennis, sailing. Home: Kopersand 34, 26723 Emden Germany Office: Fachhochschule Ostfriesland, Constantiaplatz 4, 26723 Emden Germany

GOMER, ROBERT, chemistry educator; b. Vienna, Mar. 24, 1924; m. Anne Olah, 1955; children: Richard, Maria. B.A., Pomona Coll., 1944; Ph.D. in Chemistry, U. Rochester, 1949; AEC fellow chemistry, Harvard, 1949-50. Instr. dept. chemistry James Franck Inst. U. Chgo., 1950-51, asst. prof., 1951-54, assoc. prof., 1954-58, prof., 1958—, Carl William Eisendrath Disting. Service prof., 1984—; dir. James Franck Inst. U. Chgo., 1977-83. Bd. dirs. Bull. Atomic Scientists, 1960-84. Served with AUS, 1944-46. Recipient Kendall award in surface chemistry Am. Chem. Soc., 1975; Davisson-Germer prize Am. Phys. Soc., 1981, Medard W. Welch award Am. Vacuum Soc., 1989; Sloan fellow, 1958-62; Guggenheim fellow, 1969-70; Bourke lectr. Eng., 1959. Mem. Leopoldina Acad. Scis., Nat. Acad. Scis., Am. Acad. Arts and Sci. Home: 4824 S Kimbark Ave Chicago IL 60615-1916 Office: 5640 S Ellis Ave Chicago IL 60637-1467

GOMES, JOÃO FERNANDO PEREIRA, chemical engineer; b. Lisbon, Portugal, Mar. 26, 1960; s. João Pedro Antunes and Maria Fernanda (Almeida Pereira) G.; m. Maria Paula Nunes, Sept. 20, 1988. BSc in Chem. Engring., U. Lisbon, 1983; degree in chem. reactor tech., U. West Ont., London, Can., 1985; degree in coal combustion, U. Sheffield, Eng., 1986. Asst. lectr. chem. engring. Tech. U. Lisbon, 1981-84; rsch. asst. Ctr. for Chem. Processes, Lisbon, 1981-84; rsch. fellow Nat. Labs. for Engring., Lisbon, 1983-87; head div. Inst. Welding and Quality, Oeiras, Portugal, 1987—; invited tchr. Tech. U. Lisbon, 1988—; tchr. Inst. de Soldadura e Qualidade, Lisbon, 1988—; cons. CETEL, Lisbon, 1989—. Author: (in Portuguese) Simulations of Unit Operations of CPI, 1984, Coal Combustion, 1986, Wood Gasification, 1987. Rsch. grantee Nat. Body for Environment, Lisbon, 1987; recipient Conservation prize Nat. Parks Body, Lisbon, 1990. Fellow Internat. Union Pure Applied Chemistry, Portuguese Instn. Engrs.; mem. Air Wastes Mgmt. Assn., Combustion Inst., Portuguese Chem. Soc. Avocations: music, tennis, swimming, collecting stamps and coins. Home: R Prof Mario Chico 9-7-H, 1600 Lisbon Portugal Office: Inst Soldadura e Qualidade, R Francisco A Silva, 2780 Oeiras Portugal

GOMES, NORMAN VINCENT, retired industrial engineer; b. New Bedford, Mass., Nov. 7, 1914; s. John Vincent and Georgianna (Sylvia) G.; grad. U.S. Army Command and Gen. Staff Coll., 1944; B.S. in Indsl. Engring. and Mgmt., Okla. State U., 1950; M.B.A. in Mgmt., Xavier U., 1955; m. Carolyn Moore, June 6, 1942 (dec. Apr. 1983). Asst. chief engr. Leschen div. H.K. Porter Co., St. Louis, 1950-52; staff mfg. cons. Gen. Electric Co., Cin., 1952-57: lectr. indsl. mgmt. U. Cin., 1955-56; vis. lectr. indsl. mgmt. Xavier U. Grad. Sch. Bus. Adminstrn., 1956-57; staff indsl. engr. Gen. Dynamics, Ft. Worth, 1957-60; chief ops. analysis Ryan Electronics, San Diego, 1960-64; sr. engr., jet propulsion lab. Calif. Inst. Tech., Pasadena, 1964-67, mem. tech. staff, 1967, mgr. mgmt. systems, 1967-71; industry rep. and cons. U.S. Commn. on Govt. Procurement, Washington, 1970-72; adminstrv. officer GSA, Washington, 1973-78, program dir., 1979; vis. lectr. mgmt. San Antonio Coll., 1982-85. Active Serra, Internat.; mem. Drug and Alcohol Adv. Coun. N.E. Ind. Sch. Dist., San Antonio. 2d lt. to maj. C.E., AUS, 1941-46; engring. adviser to War Manpower Bd., 1945. Decorated Army Commendation medal, Armed Svcs. Res. medal; recipient Apollo Achievement award, 1969; Outstanding Performance award GSA, 1974- 75, 76, 77, 79. Mem. Am. Inst. Indsl. Engrs. (nat. chmn. prodn. control research com., 1951-57; bd. dirs. Cin., Fort Worth, San Diego, Los Angeles, San Antonio chpts. 1954-84, pres. Cin. chpt. 1956-57, pres. Los Angeles 1970-71, nat. dir. community services 1969-73), Ret. Officers Assn. U.S. (chpt. pres. 1968-69, recipient Nat. Pres. certificate Merit 1969), Nat. Security Indsl. Assn. (mgmt. systems subcom. 1967-69), Vis. Nurse Assn. of San Antonio (mem. adv. coun. 1988—), Freedoms Found. at Valley Forge (v.p. edn. and youth leadership programs San Antonio chpt. 1987-89), Old Dartmouth Hist. Soc., Equestrian Order of the Holy Sepulchre of Jerusalem (knight comdr.). Republican. Roman Catholic. Club: K.C. (4th deg.). Home: 2719 Knoll Tree St San Antonio TX 78247-3915

GOMEZ, JAIME G., neurosurgeon; b. Bucaramana, Colombia, Apr. 3, 1932; came to U.S., 1957:; s. Heriberto and Anita (Gonzalez) G.; m. Lucy Gonzalez, Oct. 28, 1956; children: Claudia, Robert, Maurice, Richard, Phillip. MD, Nat. U., Bogota, Colombia, 1956. Asst. resident neurology San Juan deDios U. Hosp., Bogota, 1956; fellow neuropathology Columbia U., N.Y.C., 1957; asst. resident neurology N.Y.U., Bellevue Hosp., 1958; resident St. Vincent's Hosp., N.Y.C., 1959-60; attending neurosurgeon Children Hosp., Bogota, 1961, Mil. Hosp., Bogota, 1962-66; founder, dir. Neurol. Inst., Bogota, 1966-87; pvt. practice Ft. Lauderdale, Fla., 1987—; cons. Pan Am. Health Orgn., Washington, 1980-87; interim bd. mem. Ptnrs. of Alliance, Washington, 1973-75; vis. prof. U. Cin., 1980, U. Fla., Gainesville, 1981, 92; assoc. prof. orthopedics U. Miami, 1989. Author: Last MRI of the Nervous System, 1990 (4 others); editor: Neurology in Colombia; contbr. articles in profl. jours. Lt. Colombia Corp. Engrs., 1956. Recipient Best Jr. Exec. Bogota C. of C. 1966, Disting. Svcs. award Govt. Colombia 1956, Order of City of Bogota 1973, Medal of Nat. Acad. Medicine Bogota 1976, Medal of Univ. of Valencia 1980. Mem. Am. Acad. Neurol. Surgery, Congress of Neurol. Surgery, Broward Med. Soc. Roman Catholic. Office: 5353 N Federal Hwy 210 Fort Lauderdale FL 33308

GOMEZ, ROMEL DEL ROSARIO, physicist; b. Manila, Jan. 6, 1960; s. Rafael Dela Merced and Julita (Del Rosario) G.; m. Aurean Magalit Castillo, Jan. 6, 1986; children: Ryan Rafael, Rainier Adrian. BS in Physics, U. Philippines, 1980; MS in Physics, Wayne State U., 1984, U. Md., 1987; PhD in Physics, U. Md., 1990. Instr. U. Philippines, Diliman, 1980-81; rsch. asst. Wayne State U., Detroit, 1981-84, U. Md., College Park, 1984-90; faculty rsch. assoc. Dept. Elec. Engring. U. Md., College Park, 1990-91, rsch. scientist, 1991—; sci. cons., College Park, 1990—. Contbr. articles to profl. jours. Mem. AAAS, IEEE, Am. Phys. Soc., Sigma Xi. Achievements include development of surface microscopy capable of imaging real space magnetic field distributions with high resolution; this is used in imaging patterns of computer hard disks to study recording problems (like increasing storage densities) and retrieve erased data, a novel technique to map individual Cartesian components of surface magnetic fields. Office: U Md Dept Elec Engring and Lab Phys Scis 8050 Greenmead Dr College Park MD 20740

GOMIDE, FERNANDO DE MELLO, physics educator, researcher; b. Rio de Janeiro, June 4, 1927; s. José G. Jr. and Erycina Marés (Dias) G.; m. Luiza Yamashita, July 13, 1990. BS, Fed. U. Rio de Janeiro, 1952. Scholar Ctr. Brasileiro Pesquisas Físicas, Rio de Janeiro, 1950-53; instr. Nat. Inst. Tech., Rio de Janeiro, 1953-54; instr. physics Tech. Inst. Aeros., São José dos Campos, Brazil, 1955-62, asst. prof., 1963-77, assoc. prof., 1977-88; fellow Inst. Physics, U. Bologna (Italy), 1962-63; vis. prof. physics Cath. U. Petrópolis (Brazil), 1989—; cons. Fundação Amparo à Pesquisa Estado São Paulo (Brazil), 1980-85. Author: (with M.S. Berman) Cálculo Tensorial e Relatividade Geral uma Introdução, 1986, 2d edit., 1987, Introdução à Cosmologia Relativistica, 1986, 2d edit., 1988; Filosofia do Conhecimento Cientí, Hipóteses e Aprioris, 1988, Diálogo entre Filosofia e Ciência, 1990; contbr. articles to sci. and philos. jours. Mem. AAAS, Soc. Brasileira para o Progresso Cie, Inst. Brasileiro Filosofia, Soc. Brasileira Fiscica, Soc. Astronomica Brasileira, Soc. Brasileira Filósofos Católicos, Assn. Católica Interam. Filosofia, Internat. Astron. Union, Planetary Soc., N.Y. Acad. Scis., Am. Phys. Soc. Roman Catholic. Avocation: classical music. Home: R Angélica Lopes de Castro, 71, 25655-430 Petrópolis Brazil Office: Cath U Petropolis, Postgrad Ctr, R Monte Caseros 576, 25600 Petrópolis Brazil

GOMORY, RALPH EDWARD, mathematician, manufacturing company executive, foundation executive; b. Brooklyn Heights, N.Y., May 7, 1929; s. Andrew L. and Marian (Schellenberg) G.; m. Laura Dumper, 1954 (div. 1968); children: Andrew C., Susan S., Stephen H. BA, Williams Coll., 1950, ScD (hon.), 1973; postgrad., Kings Coll., 1950-51, Cambridge U., Eng., 1950-51; PhD, Princeton U., 1954; LHD (hon.), Pace U., 1986; DSc (hon.), Poly. U., 1987, Syracuse U., 1989, Worcester Poly. U., 1989, Carnegie-Mellon U., 1989. Rsch. assoc. Princeton U., 1951-54, asst. prof. math., Higgins lectr., 1957-59; with IBM, Yorktown Heights, N.Y., 1959-86; dir. math. scis., rsch. div. IBM, Armonk, 1968-70, dir. rsch., 1970-86, v.p., 1973-84, sr. v.p., 1985-89, sr. v.p. for sci. and tech., 1986-89, also mem. corp. mgmt. bd., 1983-89, dir. Asia Pacific Group, 1982-88; pres. Alfred P. Sloan Found., N.Y.C., 1989—; Andrew D. White prof.-at-large Cornell U., 1970-76; bd. dir. Bank of N.Y., Lexmark Internat., Inc., Washington Post Co., Ashland Oil, Inc.; mem. adv. coun. dept. math. Princeton, 1982-85, chmn., 1984-85; mem. adv. coun. Sch. Engring. Stanford U., 1978-85; chmn. vis. com. div. applied scis. Harvard U., 1987-91; mem. White House sci. coun., 1986-89, Coun. on Fgn. Rels.; chmn. adv. com. to Pres. on High Temperature Superconductivity, 1987-88; mem. coun. on grad. sch. Yale U., 1988—; mem. vis. com. elec. engring. and computer sci. MIT, 1988-90; mem. Pres.' Coun. Advisors on Sci. and Tech., 1990-93; researcher in integer and linear programming, non-linear differential equations. Trustee Hampshire Coll., 197-86, Princeton U., 1985-89, Alfred P. Sloan Found., 1988-89; mem. governing bd. Nat. Rsch. Coun., 1980-83; chmn. com. on mandatory retirement in higher edn. Nat. Rsch. Coun., 1989-91. With USN, 1954-57. Recipient Lanchester prize Ops. Rsch. Soc. Am., 1964, Harry Goode Meml. award Am. Fedn. Info. Processing Socs., 1984, John Von Neumann Theory prize Ops. Rsch. Soc. Am. and Inst. Mgmt. Scis., 1984, IRI medal Indsl. Rsch. Inst., 1985, Engring. Leadership Recognition award IEEE, 1988, Nat. Medal of Sci., 1988, Arthur M. Bueche award NAE, 1993; IBM fellow, 1964. Fellow Econometric Soc., Am. Acad. Arts and Scis.; mem. NAS (coun. 1977-78, 80-83, com. sci. engring. and pub. policy 1985—), Nat. Acad. Engring. (coun. 1986-92, Arthur M. Bueche award 1993), Am. Philos. Soc. (coun. 1986-92). Home: 260 Douglas Rd Chappaqua NY 10514-3100 Office: Alfred P Sloan Found 630 Fifth Ave New York NY 10111-0242

GONÇALVES DA SILVA, CYLON EUDÓXIO TRICOT, physics educator; b. Ijuí, Rio Grande do Sul, Brazil, Nov. 14, 1946; s. Solon and Jorgelina (Tricot) Gonçalves da S.; m. Jennifer Isabel Harris, Apr. 26, 1972; children: Anders, Per. BS in Physics, Rio Grande do Sul Fed. U., 1967; MA in Physics, U. Calif., Berkeley, 1971, PhD in Physics, 1972. Asst. prof. Unicamp, Campinas, SP, Brazil, 1974-78, prof., 1982—; maître asst. U. Lausanne, Switzerland, 1978-79, vis. prof., 1979-82; vis. prof. Ecole Normale Superieure, Paris, 1985; rsch. fellow IBM Rsch. Lab., Yorktown Heights, N.Y., 1985-86; dir. Lab. Nac. Luz Síncrotron, Campinas, 1986—. Editor Solid State Communications, 1987—, 5 sci. conf. proceedings, 1983-91; contbr. numerous articles to profl. jours., papers to sci. meetings. Fellow Guggenheim Found., 1978. Mem. Brazilian Acad. Sci. Office: Lab Nac Luz Síncrotron, Rua Lauro Vannucci 1020, 13087-410 Campinas SP, Brazil

GONCAROVS, GUNTI, radio chemist; b. Seneca Falls, N.Y., Sept. 5, 1956; s. Olegs and Skaidrite (Krebs) G.; m. Joan Margaret Sherman, Dec. 29, 1979; children: Kristina, Rebekah, Nina. BS in Chemistry, Charter Oak Coll., 1983. Chemistry technician Conn. Yankee Atomic Power, Haddam Neck, 1981-82, chemistry supr., 1982-83, chemist, 1983-91, sr. chemist, 1991-92, acting chemistry dept. mgr., 1992—; chmn. Sorrento Electronics Radiation Monitoring Systems Users Group, 1991-93. Contbr. articles to Nuclear Technology, 1993. Mem. Electric Coun. New Eng. (chemistry com. 1986—), Am. Nuclear Soc., Am. Chem. Soc., Assn. Official Analytical Chemists. Achievements include derivation of new technique to evaluate radiochemical data for nuclear fuel evaluation using combination of radioactive noble gas and radioiodines. Office: Conn Yankee Atomic Power Co PO Box 127E East Hampton CT 06424

GONCZ, DOUGLAS DANA, information broker; b. Brighton, Mass., Dec. 18, 1959. Student, MIT, 1977-79, No. Va. Community Coll., 1981-90; AAS in Mech. Engring. Tech., No. Va. Community Coll., 1990. Owner Universal Quantum Antientropics, Annandale, Va., 1980-85; info. broker Replikon Rsch., Seven Corners, Va., 1985—. Achievements include discovery that amplification ratio of Vortec's #903 Transvector decreases with increasing primary fluid pressure instead of remaining constant as indicated in some of Vortec's catalogs. Home: 6187 Greenwood Dr #102 Seven Corners VA 22044-2514

GONDA, IGOR, pharmaceutical scientist; b. Bratislava, Slovak Republic, Dec. 9, 1947; came to U.S., 1992; s. Iko and Bianca (Szende) G.; m. Daniela Rose Kantor, Aug. 1969; children: Ellen Danielle, Abigail. BSc, U. Leeds, Eng., 1971, PhD, 1974. Lectr. Aston U., Birmingham, Eng., 1975-83; sr. lectr. U. Sydney, Australia, 1983-92; sr. scientist Genentech Inc., South San Francisco, Calif., 1992—; bd. mem. Internat. Soc. Aerosols in Medicine, 1991—; pres. Australasian Pharm. Sci. Assn., 1991-92. Contbr. chpts. to books. Fellow Royal Soc. Chemistry; mem. Am. Assn. Aerosol Rsch., N.Y. Acad. Sci., Sigma Xi. Achievements include research in pharmaceuticals, drug delivery, therapeutic drug carrier systems and pharmaceutical inhalation aerosols technology. Office: Genentech Inc 460 Point San Bruno Blvd South San Francisco CA 94080

GONDEK, JOHN RICHARD, quality assurance professional; b. Heidelberg, Fed. Republic Germany, July 11, 1959; s. John Richard and Alice Emma (Kump) G.; m. Nancy Ann Poznick, Aug. 14, 1983; 1 child, Heather Michelle. Tech. draftsman Elizabethtown Gas Co., Iselin, N.J., 1977-79, corrosion technician, 1979-82, sr. corrosion technician, 1982-86, engring. technician, 1986-87, sr. engring. technician, 1987-89, quality assurance coord., 1989—; mem. standard practice com. Elizabethtown Gas Co., Iselin, 1986—, chmn., 1991—; mem. N.Y. Metro Gas Materials and Systems Group, 1986—. Fireman Clark (N.J.) Fire Dept., 1979-81; first aider Iselin First Aid Squad, 1982; notary pub. N.J. State, Middlesex County, 1990-95; loan officer EGCO Fed. Credit Union, Elizabethtown, N.J., 1989—; mem. adminstrv. bd., co-chair newsletter United Meth. Ch., Woodbridge, 1992—. Mem. ASTM (voting mem. G-01 and F-17 coms. 1986—), Internat. Mgmt. Coun. (bd. of control, keyperson chmn. 1989-90). Home: 12 Wedgewood Ave Woodbridge NJ 07095-3829 Office: Elizabethtown Gas Co Brown Ave Iselin NJ 08830

GONG, YITAI, information specialist; b. Wuhan, Hubei, Peoples Republic of China, Aug. 5, 1941; s. Wenli and Muxiang (Xia) G.; m. Liping Fu, Oct. 1, 1970; 1 child, Yi. BA, Shanghai Fgn. Langs. Inst., 1965; M in Philosophy, City U., London, 1986. Reference libr. Shanghai Libr. Acad. Sinica, 1965-82, dep. dir., 1982-84, dir., 1987—; sr. researcher dept. info. sci. City U., London, 1984-86; hon. prof. dept. libr. and info. studies East China Normal U., Shanghai, 1987—. Editor in chief: Chinese Biol. Abstracts, 1990—; contbr. articles to profl. jours. Recipient Second award for advancement sci. and tech. Chinese Acad. Scis., 1991, 93. Mem. Chinese Libr. Soc. (coun. 1987—, Best Articles award 1989), Chinese Acad. Scis. (vice chmn. rsch. consortium sci. and tech. info. 1987), Shanghai Libr. Soc. (vice chmn.). Achievements include development of Shanghai documentation and information center information system, Chinese biological database, classification of classic laws in bibliometrics, dual nature of citations, aging of biochemical literature. Office: Shanghai Libr Acad Sinica, 319 Yueyang Rd, Shanghai 200031, China

GONGORA-TREJOS, ENRIQUE, mathematician, educator; b. San José, Costa Rica, June 4, 1931; s. Enrique Gongora Umaña and Clara Rosa Trejos Matamoros; divorced; children: Tania, Anjte, Federico; m. Helena Ramirez, 1991. Licentiate, Chem. U. Costa Rica, San José, 1961; Diploma in Math., Georgia Augusta U., Göttingen, Fed. Republic Germany, 1966. Chair math. and math. logic U. Costa Rica, Nat. U., 1966-76; dean faculty sci. Nat. U., Heredia, Costa Rica, 1974-76, head math. dept., 1976-77; acad. vice-rector U. Estatal a Distancia, San José, 1977-80, rsch. vice-rector, 1980-89; lectr. philosophy and history of music U. Costa Rica, San José, 1980-89; alt. amb. Internat. Atomic Energy Agy., Vienna, Austria, 1989—; mem. founder com., Nat. U., 1974, U. Estatal a Distancia, 1976; counsellor energy Oficina de Planificaciōn Nat. y Politica Econōmica, San Josē, 1975-81. Author: (books) Introdución al Pensamiento Logico Matemātico, 1979 (award 1980), Falso-Verdadero, 1983, Que son los Reactores Nucleares?, 1989 (award 1992), Sobre la Afinación de los Instrumentos de Tecla, 1992, several textbooks; contbr. articles to profl. jours. Sec. Com. Interamericano de Edn. Math., 1972-82; bd. dirs. AEC, Costa Rica, 1984—, pres. AEC, 1986—; founder Collegium Musicum, 1974. Mem. Inst. Costarricense de Cultura Hispānica. Avocations: early music, fencing. Home: PO Box 2717, 1000 San José Costa Rica Office: Atomic Energy Commn Costa Rica, PO Box 6681, San Jose 1000, Costa Rica*

GONNERING, RUSSELL STEPHEN, ophthalmic plastic surgeon; b. Milw., Nov. 21, 1949; s. Russell Richard and Virginia Mary (Mlinar) G.; m. Sandra Lynne Brubaker, Aug. 6, 1971; children: Julie Kathleen, Stephen Russell, Scott Duncan. Student, Univ. Wien, Vienna, Austria, 1969-70; AB in History cum laude, Boston Coll., 1971; MD, The Med. Coll. of Wis., 1975. Cert. Am. Bd. Ophthalmology; med. lic. Wis. Intern St. Luke's Hosp., Milw., 1975-76; resident in ophthalmology The Med. Coll. of Wis., Milw., 1977-80; fellow in ophthalmic plastic and reconstructive surgery U. Wis., Madison, 1980-81, asst. clin. prof. dept. ophthalmology, 1981-92, assoc. clin. prof. dept. ophthalmology, 1992—; asst. clin. prof. dept. ophthalmology The Med. Coll. Wis., Milw., 1985—; ophthalmologist Children's Hosp. Wis., Milw.; assoc. St. Joseph's Hosp., Milw., Milw. County Med. Complex; chief of ophthalmology St. Luke's Hosp., Milw., 1983-92; pvt. practice Ophthalmic Plastic & Reconstructive Surgery, 1981—; rsch. assoc. in corneal physiology The Med. Coll. Wis., Milw., 1976-77; rsch. advisor to fellowship in ophthalmic plastic and reconstructive surgery U. Wis., Madison, 1983—. Author: (with others) Infections of the Eye and Ocular Adnexa, 1986, Oculoplastic, Orbital and Reconstructive Surgery, 1988, Oculoplastic and Orbital Emergencies, 1990; contbr. numerous articles to profl. jours.; presenter papers. Fellow ACS, Am. Acad. Ophthalmology (basic and clin. sci. course com. 1986-92, chmn. 1988-92, Honor Award 1990), Am. Soc. Ophthalmic Plastic and Reconstructive Surgery (edit. bd. 1987—, edn. com. 1988—, Marvin H. Quickert Award 1982, Rsch. Award 1982); mem. AMA, Internat. Soc. for Orbital Disorders, Assn. for Rsch. in Vision and Ophthalmology, Med. Soc. Wis., Milw. County Med. Soc. (del. to state med. soc. 1987-90, bd. dirs. 1989—), Milw. Acad. Medicine, Milw. Ophthalmological Soc. (treas. 1989-90, sec. 1990-91, v.p. 1991-92, pres. 1992—), Am. Soc. Ocularists (med. adv. bd. 1987—), Nat. Soc. to Prevent Blindness (med. adv. bd. Wis. chpt. 1987-88). Avocations: sailing, skiing. Office: Oculoplastic & Orbital Cons 2600 N Mayfair Rd Ste 950 Milwaukee WI 53226-1307

GONSALVES, ROBERT ARTHUR, electrical engineering educator, consultant; b. Woburn, Mass., Apr. 17, 1935; s. John Henry and Rowena Mary (Nolan) G.; m. Patricia Liberge, Apr. 28, 1962; children: Maria, Robert, Richard, Joanna, Paul. BS, Tufts U., 1956; MS, Northeastern U., 1961, PhD, 1965. Mem. staff RCA, Burlington, Mass., 1959-62; rsch. assoc. Northeastern U., Boston, 1962-65, prof., 1965-85; prof. Tufts U., Medford, Mass., 1985—, dir. Electro-Optics Tech. Ctr., 1985—; cons. Instrument Devel. Labs., Attleboro, Mass., Itek Corp., Lexington, Mass., Bell Telephone Labs., North Andover, Mass., Sanders Assocs., Nashua, N.H., Raytheon Co., Norwood, Mass., Mitre Corp., Bedford, Mass., Eikonix, Bedford, Boston U., Fairchild Camera, Syosett, N.Y., Lockheed Co., Palo Alto, Calif., Martin Marietta Co., Denver, ERIM, Ann Arbor, Mich., Ektron Applied Imaging, Bedford, Betagen Corp., Waltham, Mass., Phase One Devel. Corp., Newton, Mass., Mission Rsch. Corp., Nashua, A.J. Devaney Assocs., Boston. Contbr. articles to profl. jours. Elected mem. Woburn Sch. Com., 1969-75; elected del. Dem. Nat. Conv., Miami, Fla., 1972. Lt. USN, 1956-59. Warren Acad. scholar, 1952. Fellow Internat. Soc. Optical Engineers; mem. AAAS, IEEE, Sigma Xi, Eta Kappa Nu, Phi Kappa Phi, Tau Beta Pi. Achievements include patents (with others) for wavefront sensing by phase retrieval, and for achromatic volume holography apparatus. Office: Tufts U Electro Optics Tech Ctr 25 Crosby Dr Chestnut Hill MA 02167-1172

GONTIER, JEAN ROGER, internist, physiology educator, consultant; b. Lens, France, Mar. 8, 1927; s. Paul Maurice and Marie Jeanne (Tricoche) G.; m. Sylviane Prevost, Dec. 8, 1968; children: Sylviane, Yannick, Jean-Yves, Yann. AB, Coll. d'Etampes, France, 1945; MS, Coll. Scis., Paris, 1947; MD with honors, Sch. Medicine, Paris, 1965. Prof. physiology UGSEL, Paris, 1957-62; instr. in medicine Sch. Medicine, Paris, 1960-65; intern Hop Cochin, Paris, 1965; resident Hop Bicetre, Paris, 1966; dir. physiology Sch. Medicine, Reims, 1966-68; prof. physiology U. Montreal, 1970-78; cons. in internal medicine Paris, 1979—; cons. in physiology Bicetre U. Hosp., Paris, 1967-68; cons. editor various pubs., N.Y.C., 1975-78, Paris, 1969-73, Montreal, 1986-89. Author: Hormones, Nervous System and Digestion, 1968, Respiration, 1971, 77, Digestion, 1969, 82, Textbook of Medical Physiology, 1980, Human Physiology, 1989. Recipient Man of Yr. award Internat. Biog. Ctr., 1991-92. Mem. AAAS, Am. Physiol. Soc. (teaching physiology sect.), Can. Physiol. Soc., N.Y. Acad. Scis., Assn. Physiologists, Cercle de l'Etrier Club, La Baule Country Club. Roman Catholic. Avocation: sailing. Home and Office: 133 Rue Michel Ange, F75016 Paris France

GONZALEZ, DAVID ALFONSO, environmental engineer; b. Kingsville, Tex., Mar. 24, 1962; s. Alfonso Mario and Grace (Ramirez) G. BSChemE, U. Tex., 1985; MBA, U. St. Thomas, 1992. Registered profl. engr., Tex. Environ. engr. U.S. EPA, Dallas, 1986-89; sr. environ. engr. Destec Energy, Houston, 1989—. Mem. AICE, NSPE, MBA Profl. Soc., Nat. Air and Waste Assn. Office: Destec Energy 2500 City West Blvd Houston TX 77042

GONZALEZ, MICHAEL JOHN, nutrition educator, nutriologist; b. N.Y.C., July 5, 1962; s. R. Miguel and Daisy (Guzman) G.; m. Enid J. Bauza, Mar. 28, 1987; 1 child, Michael John Jr. BS in Biology, Cath. U., 1983; MS in Cell Biology, Nova Coll., 1985; MNS in Nutrition and Biochemistry, U. P.R., 1986; NMD in Nutrition, John F Kennedy, 1988; DSc in Health Sci, Lafayette U., 1989; postgrad. in tumor biology, Mich. State U., 1987—. Rsch. asst. dept. chemistry Cath. U., Ponce, P.R., 1982-83; lab. instr. dept. biology U. P.R., Mayaguez, 1983-85; rsch. asst. dept. biochemistry U. P.R., Rio Piedras, 1985-86; mem. dept. biology faculty Cath. U., Ponce, 1986-87; rsch. asst. dept. human nutrition Mich. State U., East Lansing, 1987-90, sci. instr. dept. Upward Bound, 1990-91, lab. instr., rsch. asst. dept. food sci. and pharmacology, 1991—; mem. adv. bd. Internat. Health Coun., East Lansing, 1988—. Contbr. articles to profl. jours. Fellow Am. Nutritional Med. Assn. (v.p. 1991—); mem. Am. Inst. Nutrition, Am. Assn. Cancer Rsch., Soc. for Exptl. Biology, N.Y. Acad. Sci., Am. Assn. Police, United Farmers. Democrat. Roman Catholic. Home: Cherry Ln # 919 C East Lansing MI 48823-5502 Office: Mich State U Dept Pharmacology Giltner Hall 349 East Lansing MI 48824

GONZALEZ, WILLIAM JOSEPH, sales executive, geosynthetic engineer; b. Dayton, Ohio, Jan. 11, 1965; s. Cesar William and Juanita Louise (Thoma) G. BSCE, U. Cin., 1990. Engring. tech. Westinghouse Environ. & Geotech. Engring., Cin., 1987-89; sales engr. Contech Constrn. Products, Inc., Middletown, Ohio, 1990—. Cons. Jr. Achievement of N.E. Kans., Topeka, 1993. Mem. ASCE (v.p. Young Mems. Forum 1992-93, sec.-treas. 1991-92), Internat. Erosion Control Assn., Toastmasters (v.p. 1992-93). Roman Catholic. Office: Contech Constrn Products 5883 SW 29th St Topeka KS 66614

GONZÁLEZ FLORES, AGUSTÍN EDUARDO, physicist; b. Orizaba, Veracruz, Mexico, Dec. 27, 1948; s. Guillermo González Correa and Angela (Flores) González. BSc in Physics, Nat. U. Mexico, Mexico City, 1973, MSc in Physics, 1977; PhD in Physics, Boston U., 1981. Assoc. prof. Met. Autonomous U., Mexico City, 1974-77; adj. asst. prof. UCLA, 1981-82; vis. rsch. physicist Inst. Theoretical Physics U. Calif., Santa Barbara, 1982-83; investigador titular Inst. Physics Nat. U. Mexico, Mexico City, 1983—. Contbr. articles to profl. jours. Nat. System of Investigators fellow Secretaria Edn. Pública, Mexico City, 1984—. Mem. Mexican Acad. Scis., Am. Phys. Soc., Am. Chem. Soc. (polymer div.). Roman Catholic. Home: Cerro del Quetzal 289, 04200 Mexico City DF, Mexico Office: Inst Physics Nat U Mexico, Apdo Postal 20-364, 01000 Mexico City Mexico

GONZÁLEZ UREÑA, ANGEL, chemistry researcher and educator; b. Roquetas de Mar, Almeria, Spain, Aug. 20, 1947; s. Francisco Gonzalez Vizcaino and Gloria Ureña Crespo; m. Dolores González Baena; children: Gloria, Alicia. MS, Granada U., 1968; PhD with honors, Madrid Complutense U., 1972. Teaching asst. Complutense U. Madrid, 1968-69, 69-70, lectr., 1969-71, asst. prof., lectr. phys. chemistry, 1971-72, assoc. prof., 1974-75, permanent assoc. prof., 1975-83, prof., 1983-92, dir. rsch. group on molecular beams and lasers; Juan March Found. fellow U. Madison, Wis., 1972-73; postdoctoral fellow U. Wis., Madison, 1972-73; postdoctoral rsch. assoc. Tex. U., Austin, 1973-74; Brit. Coun. fellow U. Cambridge, Eng., 1981; vis. rsch. scholar Johns Hopkins U., Balt., 1982; vis. prof. chemistry dept. Toronto U., Can., 1982; NATO vis. prof. chemistry dept., Manchester, Eng.; SERC vis. rschr. chemistry dept. Nottingham (Eng.) U., 1987; vis. prof. U. Swansea, Wales, 1990; vis. prof. U. Paris-Sud Orsay, France, 1992. Author: Cinética y dinámica molecular de las reacciones quimicas elementales, 1985, Cinética y dinámica química, 1991; co-author: Advanced in Chemical Physics, 1987, Selectivity in Chemical Reactions, 1988, Nuevas tendencaise en Física Atómica y Molecular, 1988, Láseres y rayos moleculares, 1990, Láser and Chemical Reactions, 1990; contbr. over 100 articles to profl. publs. Recipient 1st Rsch. award Complutense U. of Madrid, 1991. Mem. Am. Inst. Physics, Royal Soc. (divsn. molecular beams), Real Sociedad Española de Fisica y Química, Atomic and Molecular Physics, Internat. Com. Molecular Beams, Spanish Group of Física Atómica y Molecular. Home: Jose Fentanes 89C-2D, 28035 Madrid Spain Office: U Complutense Madrid, Faculty Quimicas, Dept Quimica Fisica, 28040 Madrid Spain also: U Complutense Spain, Inst Pluridisciplinar Juan XXIII-1, 28040 Madrid Spain

GONZO, ELIO EMILIO, chemical engineer, educator; b. Rosario de Lerma, Argentina, Sept. 3, 1945; s. Antonio and Gilda (Caregnato) G.; m. Gladys Ana Maria Monasterio, Oct. 27, 1973; children: Luis Federico, Andres Sebastian, Aneli Beatriz, Ivan Alejandro. Degree in chem. engring., Nat. U. Tucuman, Argentina, 1969; MSChemE, Stanford U., 1975. Teaching asst. Nat. U. Tucuman, 1965-69; asst. prof. Nat. U. Salta, 1969-73; rsch. asst. Stanford U., Palo Alto, Calif., 1974-75; assoc. prof. Nat. U. Salta, 1976-80, prof., 1981-90, plenarium prof., 1990—; researcher Nat. Coun. Sci. and Technol. Rsch., Argentina, 1978—; dean, Tech. Sci. Dept. Nat. U. Salta, 1992—; dir., Rsch. Inst. INIQUI, 1992—. Contbr. chpts. to books, articles to profl. jours. Mem. Heat and Mass Transfer Assn. Argentina, Catalysis Nat. Com. Argentina. Roman Catholic. Avocations: sports, soccer, paddle tennis, philately. Home: Urquiza 62 Dept 4, Salta 4400, Argentina Office: Nat U Salta INIQUI, Buenos Aires 177, Salta 4400, Argentina

GOOCH, LAWRENCE LEE, astronautical engineering executive; b. Reno, Dec. 6, 1940; s. Ira Lee and Mary Geraldine (Hill) G.; m. Penelope Jane McKay, June 7, 1962; children: Jennifer Lynn, Mary Winifred, Laura Jane. BS in Basic Scis., USAF Acad., 1962; MS in Astronautics, Air Force Inst. Tech., 1964; PhD in Mech. Engring., U. Tex., 1972. Commd. 2d lt. USAF, 1962, advanced through grades to col., 1981; program mgr. USAF Office Spl. Projects, L.A., 1976-83; commdr. USAF Western Space & Missile Ctr., Vandenberg AFB, Calif., 1983-86, USAF Space Tech. Ctr., Kirtland AFB, N.Mex., 1986-87, USAF Ea. Space & Missile Ctr., Patrick AFB, Fla., 1987-89; ret. USAF, 1989; mgr. advanced solid rocket motor devel. Aerojet Solid Rocket Co., Sacramento, 1989-90; mgr. bus. devel. EG&G Inc., Wellesley, Mass., 1990-92; pres. Analex Corp., Brook Park, Ohio, 1992—; adv. bd. mem. Tex. A&M Space Rsch. Ctr., College Station, 1987-89; cons. Def. Advanced Rsch. Project Agy., Washington, 1989-90. Contbr. articles to Banking Jour., Credit Jour. Elder Presbyn. Ch., Austin, Tex., 1970-92. Named Outstanding Young Men of Am., 1971; decorated Legion of Merit, USAF, 1987, 89; recipient Exceptional Svc. award NASA, Kennedy Space Ctr., 1989. Mem. AIAA (sr.), Phi Kappa Phi. Achievements include program management of the only space booster with 100% success rate. Office: Analex Corp 3001 Aerospace Pkwy Brookpark OH 44142

GOOD, DAVID MICHAEL, engineer; b. Hamilton, Bermuda, July 18, 1961; s. Ronald Dean and Sandy Lynn (Freil) G.; m. Kathy L. Lindman, June 28, 1984; 1 child, Grace L. BS, Ohio State U., 1984. Registered profl. engr., Tex.; lic. risk mgr., Tex. Dist. supr. loss prevention Indsl. Risk Insurers, Dallas, 1985-88; account engr. Am. Internat. Group, Dallas, 1988-90; loss prevention specialist CNA Ins. Co., Dallas, 1990-92, Cigna Property and Casualty Corp., Dallas, 1992—. Mem. environ. liaison com. City of Dallas, 1990-91. Mem. NSPE (govtl. affairs com.), Am. Soc. Safety Engrs. Home: 2200 Daybreak Plano TX 75093

GOOD, HARVEY FREDERICK, biologist, educator; b. Detroit, June 4, 1938; s. Jacob Benjamin and Violet Marie Zimmel (Weimar) G.; m. Constance Lee Forror, June 16, 1961; children: Sandra Jolene, Randall Frederick, Ronald Harvey. BA in Biology, LaVerne Coll., 1960, EdD, 1981; MS in Biology, Purdue U., 1968. Artisan Conrad Divine Artisans, LaVerne, Calif., 1960-61; instr. LaVerne Coll., 1961-65; tchr. Chaffey Union High Sch. Dist., Ontario, Calif., 1965-69; prof. Biology U. LaVerne, 1969—. Bd. dirs. Pomona (Calif.) Valley Humane Soc. and SPCA, pres. 1991-93. Mem. AAUP, Nat. Assn. Biology Tchrs. Mem. Brethren Ch. Home: PO Box 364 28 Oak Grove Ave Mount Baldy CA 91759 Office: U LaVerne Dept Biology 1950 Third St La Verne CA 91750

GOOD, MARY LOWE (MRS. BILLY JEWEL GOOD), government official; b. Grapevine, Tex., June 20, 1931; d. John W. and Winnie (Mercer) Lowe; m. Billy Jewel Good, May 17, 1952; children: Billy, James. BS, Ark. State Tchrs. Coll., 1950; MS, U. Ark., 1953, PhD, 1955, LLD (hon.), 1979; DSc (hon.), U. Ill., Chgo., 1983, Clarkson U., 1984, Ea. Mich. U., 1986, Duke U., 1987; hon. degree, St. Mary's Coll., 1987, Kenyon Coll., 1988, Stevens Inst. Tech., 1989, Lehigh U., 1989, Northeastern Ill. U., 1989, U. S.C., 1989, N.J. Inst. Tech., 1989; hon. law degree, Newcomb Coll. of Tulane U., 1991; DSc (hon.), Manhattan Coll., 1992. Instr. Ark. State Tchrs. Coll., Conway, summer 1949; instr. La. State U., Baton Rouge, 1954-56; asst. prof. La. State U., 1956-58; asso. prof. La. State U., New Orleans, 1958-63; prof. La. State U., 1963-80; Boyd prof. materials sci., div. engring. research La. State U., Baton Rouge, 1979-80; v.p., dir. research UOP, Inc., Des Plaines, Ill., 1980-84; pres. Signal Research Ctr. Inc., 1985-87; pres. engineered materials research div. Allied-Signal Inc., Des Plaines, Ill., 1986-88; sr. v.p.-tech., Allied-Signal Inc., Morristown, N.J., 1988-93; under sec. of commerce for technology Dept. of Commerce, Washington, DC, 1993—; chmn. Pres.'s Com. for Nat. Medal Sci., 1979-82; mem. Nat. Sci. Bd., 1980-91 (chmn. 1988-91), chmn., 1988-90; mem. adv. bd. NSF Chemistry Sect., 1972-76; mem. com. medicinal chemistry NIH, 1972-76, Office of USAF Rsch., 1974-78, chemist div. Brookhaven and Oak Ridge Nat. Labs., 1973-83, chem. tech. div. Oak Ridge Nat. Lab., catalysis program Lawrence-Berkeley Lab.; catalysis program coll. engring. La. State U.; vice chmn. Nat. Sci. Bd., 1984, chmn., 1988-90; bd. dirs. Cin. Milacron Inc., bd. dirs. Ameritech. Contbr. articles to profl. jours. Mem. Nat. Sci. Bd., 1980-91; mem. Pres.' Coun. Advisors for Sci. and Tech., 1991-93. Recipient Agnes Faye Morgan rsch. award, 1969, Disting. Alumni citation U. Ark, 1973, Scientist of Yr. award Indsl. R&D Mag., 1982, Delmer S. Fahrney medal Franklin Inst., 1988, N.J. Women of Achievement award Douglass Coll., Rutgers U., 1990, Indsl. Rsch. Inst. medal, 1991; AEC tng. grantee, 1967, NSF Internat. travel grantee, 1968, NSF rsch. grantee, 1969-80. Fellow AAAS, Am. Inst. Chemistry (Gold medal 1983), Chem. Soc. London; mem. NAE, Am. Chem. Soc. (1st woman dir. 1971-74, regional dir. 1972-80, chmn. bd. 1978, 80, pres. 1987, Garvan medal 1973, Herty medal 1975, award Fla. sect. 1979, Charles Lathrop Parsons award 1991), Internat. Union Pure and Applied Chmistry (pres. inorganic div. 1980-85), Zonta (past pres. New Orleans club, chmn. dist. status of women com. and nominating com., chmn. internat. Amelia Earhart schoolarship com. 1978-88, pres. internat. Found. 1988-93, mem. internat. bd. 1988-90), Phi Beta Kappa, Sigma Xi, Iota Sigma Pi (regional dir. 1967-93, hon. mem. 1983). Home: 3321 O St NW Washington DC 20007 Office: Dept of Commerce Rm 4824 14 and Constitution Ave NW Washington DC

GOOD, ROBERT ALAN, physician, educator; b. Crosby, Minn., May 21, 1922; s. Roy Homer and Ethel Gay (Whitcomb) G.; m. Noorbibi K. Day, 1986; children from previous marriage: Robert Michael, Mark Thomas, Alan Maclyn, Margaret Eugenia, Mary Elizabeth. BA, U. Minn., 1944, MB, 1946, PhD, 1947, MD, 1947, DSc (hon.), 1989; MD (hon.), U. Uppsala, Sweden, 1966; DSc (hon.), N.Y. Med. Coll., 1973, Med. Coll. Ohio, 1973, Coll. Medicine and Dentistry N.J., 1974, Hahnemann Med. Coll., 1974, U. Chgo., 1974, St. John's U., 1977, U. Health Scis., Chgo. Med. Sch., 1978, Miami Children's Hosp., 1986, Med. Sch., U. Minn., 1989. Teaching asst. dept. anatomy U. Minn., Mpls., 1944-45; instr. pediatrics U. Minn. (Med. Sch.), 1950-51, asst. prof., 1951-53, asso. prof., 1953-54, Am. Legion Meml. research prof. pediatrics, 1954-73, prof. microbiology, 1962-72, Regents prof. pediatrics and microbiology, 1969-73, prof., head dept. pathology, 1970-72; intern U. Minn. Hosps., 1947, asst. resident pediatrics, 1948-49; pres., dir. Sloan-Kettering Inst. for Cancer Research, 1973-80, mem., 1973-81; prof. pathology Sloan-Kettering div. Grad. Sch. Med. Scis. Cornell U., 1973-81, dir., 1973-80; adj. prof., vis. physician Rockefeller U., 1973-81; prof. medicine and pediatrics Cornell U. Med. Coll., 1973-81; dir. research Meml. Sloan-Kettering Cancer Ctr., v.p., 1980-81; dir. research Meml. Hosp. for Cancer and Allied Diseases, 1973-80, also attending physician depts. medicine and pediatrics; attending pediatrician N.Y. Hosp., 1973-81; mem., head cancer research program Okla. Med. Research Found., 1982-85; prof. pediatrics, research prof. medicine, Okla. Med. Rsch. Found. prof. microbiology and immunology U. Okla. Health Scis. Ctr., 1982-85; attending physician, head div. immunology Okla. Children's Meml. Hosp., 1982-85; attending physician in internal medicine Okla. Meml. Hosp., 1983-85; physician-in-chief All Children's Hosp., St. Petersburg, Fla., 1985—; prof., chmn. dept. pediatrics U. South Fla., St. Petersburg, 1985-91, prof. depts. pediatrics, microbiology, immunology and medicine, 1985—, head allergy and clinical immunology, dept. pediatrics, 1985—, disting. grad. rsch. prof., 1989—; vis. investigator Rockefeller Inst. for Med. Research, N.Y.C., 1949-50, asst. physician to Hosp., 1949-50; attending pediatrician Hennepin County Gen. Hosp., 1950-73, cons., 1960-73; Mem. Unitarian Service Commn. Med. Exchange Team to, France, Germany, Switzerland and Czechoslovakia, 1958; cons. VA Hosp., Mpls., 1959-60; cons., sci. adviser Nat. Jewish Hosp., Denver and Childrens Asthma Research Inst. and Hosp., Denver, 1964-69; mem. study sects. USPHS, 1952-69; mem. expert adv. panel on immunology WHO, 1967—; cons. Merck & Co., N.J., 1968—, Nat. Cancer Inst., 1973-74; mem. ad hoc com. President's Sci. Adv. Council on Biol. and Med. Sci., 1970, Pres.'s Cancer Panel, 1972; mem. Lyndon B. Johnson Found. awards com., 1972; mem. adv. com. Bone Marrow Transplant Registry, 1973—; chmn. Internat. Bone Marrow Registry, 1977-79; bd. dirs. Nat. Marrow Donor Program, 1987—; fgn. adv. Acad. Med. Scis., People's Republic of China, 1980—; chmn. Fla. Gov.'s Task Force on AIDS, 1985-87, mem., 1988—. Author, editor numerous books; contbr. many articles to profl. jours. Mem. adv. council Childrens Hosp. Research Found., Cin., 1954-58; bd. dirs. Allergy Found. Am., 1973; bd. sci. advisers Jane Coffin Childs Meml. Fund Med. Research, 1972-74, Merck Inst. Therapeutic Research, 1972-76; trustee Eleanor Naylor Dana Charitable Trust, 1982—. Recipient Borden Undergrad. Research award U. Minn. Med. Sch., 1946, E. Mead Johnson First award, 1955, Theobald Smith award, 1955, Parke-Davis 6th Ann. award, 1962, Rectors medal U. Helsinki, 1963-64, Pemberton Lectureship award, 1966, Gordon Wilson Gold medal, 1967, R.E. Dyer Lectureship award, 1967, Clemens Von Pirquet Gold medal 9th Ann. Forum on Allergy, 1968, Presidents medal U. Padua, Italy, 1968, Robert A. Cooke Gold medal Am. Acad. Allergy, 1968, John Stewart I. award Dalhousie U., 1969, Borden award Assn. Am. Med. Colls., 1970, Howard Taylor Ricketts award U. Chgo., 1970, Gairdner Found. award, 1970, City of Hope award, 1970, Am. Acad. Achievement golden plate award, 1970, Albert Lasker award for clin. and med. research, 1970, ACP award, 1972, Am. Coll. Chest Physicians award, 1974, Lila Gruber award Am. Acad. Dermatology, 1974, award in cancer immunology Cancer Research Inst. N.Y., 1975, Outstanding Achievement award U. Minn., 1978, award Am. Dermatological Soc. Allergy and Immunology, 1978, 1st Sarasota Med. award, 1979, sect. on mil. pediatrics award Am. Acad. Pediatrics, 1980, recipient Univ. medal Hacettepe U., Ankara, Turkey, 1982, Pres.' medal U. Lyon, France, 1986, Merieux Found. award Internat. Soc. Preventive Oncology, 1987, Claude Bernard prize World Med. Communications, Inc., 1987, Disting. Med. Sci. of Fla. award Fla. Soc. Pathologists, 1989, Gold-Headed Cane award Dept. Pediatrics U. Minn., 1990, Askounces-Ashford Disting. Scholar award, USF, 1990, Excellence award Ronald McDonald Children's Charities, 1991, Councill C. Rudolph award, All Children's Hosp., 1991, Merit award for outstanding and superior res., Dept. Health and Human Services, Md., 1992, Lifetime Achievement award Immune Deficiency Found., Md., 1992, Asthma and Allergy Found. Am. award 1993, Univ. Rsch. medallion award Disting. Rsch. Prof. U. South Fla., 1993, award Asthma and Allergy Found. Am., 1993, numerous others; Fellow Nat. Found. for Infantile Paralysis, 1947, Helen Hay Whitney Found. fellow, 1948-50; Markle Found. scholar, 1950-55. Fellow AAAS, Am. Coll. Physicians, 1993, Royal Soc. Medicine, 1992, Acad. Multidisciplinary Research, N.Y. Acad. Sci., Am. Acad. Arts and Scis., Am. Coll. of Allergy and Immunology (hon.); mem. NAS, Am. Soc. Transplant Surgeons (hon.), Am. Assn. History of Medicine, Am. Fedn. Clin. Research, Am. Assn. Anatomists, Am. Assn. Immunologists (past pres.), AAUP, Am. Pediatric Soc. (John Howland award 1987), Mpls. Pediatric Soc., Northwestern Pediatric Soc., Am. Rheumatism Assn., Am. Soc. Clin. Investigation (past pres.), Am. Soc. Exptl. Pathology (past pres.), Am. Soc. Microbiology, Assn. Am. Physicians, Central Soc. Clin. Research (past pres.), Harvey Soc., Infectious Disease Soc. Am. (Squibb award 1968), Internat. Soc. Nephrology, Internat. Acad. Pathology, Internat. Soc. for Transplantation Biology, Minn. State Med. Assn., Nat. Acad. Sci. Inst. Medicine (charter), Reticuloendothelial

Soc. (past pres.), Soc. for Exptl. Biology and Medicine (past pres.), Soc. for Pediatric Research, Am. Clin. and Climatol. Assn. (Gordon Wilson Gold medal 1967), Detroit Surg. Assn. (McGraw medal 1969), Internat. Soc. Blood, Transfusion, Practitioners' Soc., Am. Assn. Pathologists, Internat. Soc. Exptl. Hematology, Transplant Soc., Western Assn. Immunologists, Internat. Soc. Immunopharmacology (founding mem.), Am. Soc. Transplant Surgeons, Pioneer, Internat. Bone Marrow Transplant Registry (charter), Phi Beta Kappa, Sigma Xi, Alpha Omega Alpha. Office: All Children's Hosp 801 6th St S Saint Petersburg FL 33701-4899

GOODALL, FRANCES LOUISE, nurse, production company assistant; b. Gove, Kans., Apr. 30, 1915; d. Francis Mitchell and Ella Aurelia (Brown) Sutcliffe; m. Richard Fred Goodall, Feb. 22, 1946; children: Roy Richard, Gary Frederick. Student, U. Kans., 1932-33, Ft. Hays State Coll., 1933-34; BS in Nursing, U. Wash., 1939. RN, Wash. Nurse King County Hosp. System, Seattle, 1939-41; office nurse Dr. Cassius Hofrictor, Seattle, 1941-42; founder Goodall Prodns., Seattle, 1971—. Pres. Hawthorne Elem. Sch. PTA, Seattle, 1960-61, Caspar Sharples Jr. High Sch. PTA, Seattle, 1967-68; historian Seattle Coun. PTAs, 1964-65, 68-69; den mother Boy Scouts Am., Seattle, 1963-67; active United Good Neighbors, Seattle, 1964-68; treas. Women's Overseas Svc. League, Seattle, 1970-74, treas., 1987-91. 1st lt. Nurses Corps, AUS, 1942-46, PTO. Recipient vol. award King County Hosp. System, 1964, Acorn award Franklin High Sch. PTA, 1965, Woman Achievement Cert. award Past Pres. Assembly, 1992. Mem. U. Wash. Alumni Assn. (v.p. 1966-70), U. Wash. Nursing Alumni Assn., Seattle Mus. Art Soc. (assoc., social com., bd. dirs.), Pres's. Forum, Seattle Fedn. Women's Clubs (chmn. community improvement program 1990—), Seattle Geneal. Soc., Lake City Emblem Club, Order Eastern Star, Seattle Sorosis (pres. 1990—), Sigma Sigma Sigma, Kappa Delta (pres. Seattle alumni 1954-55, sec. alumnae coop. bd. 1963-82), Nat. Assn. Parliamentarians (pres. parliamentary law unit 1989—, treas. 1960-61, 64-66, 75-89), Am. Legion (life mem. Fred Hancock post #19 Renton, Wash.). Republican. Presbyterian. Avocations: photography, snow shoeing, travel, camping, cooking. Home: 4111 51st St S Seattle WA 98118

GOODALL, JANE, ethnologist; b. London, Apr. 3, 1934; d. Mortimer Herbert and Vanne (Joseph) Morris-Goodall; m. Hugo Van Lawick, 1964 (div. 1974); one child, Hugo Eric Louis; m. Derek Bryceson, 1975 (dec. 1980). PhD, Cambridge U., Eng., 1965. Asst., sec. to Dr. Louis S. B. Leakey Coryndon Meml. Mus. Nat. History, Olduvai Gorge, Tanzania; rschr. in animal behavior, sci. dir. Gombe Stream Rsch. Ctr., Tanzania, 1960—; vis. prof. psychiatry, human biology Stanford U., 1970-75; hon. vis. prof. zoology U. Dar Es Salaam, Tanzania; lectr. Yale U., 1973. Author: In the Shadow of Man, 1961, My Friends, the Wild Chimpanzies, 1967, The Chimpanzees of Gombe, 1986, Through a Window, 1990; contbg. author: Primate Behavior, 1965, Primate Ethology, 1967. Recipient conservation award N.Y. Zool. Soc., Franklin Burr award Nat. Geographic Soc. (2), Centennial award, 1988. Mem. Am. Acad. Arts. and Scis. (hon. fgn.). Office: Jane Goodall Inst PO Box 41720 Tucson AZ 85717*

GOODCHILD, MICHAEL, geographer, educator. Prof. geography U. Calif., Santa Barbara. Recipient Scholarly Distinction in Geography award Can. Assn. Geographers, 1990. Office: U Calif Dept Geography Santa Barbara CA 93106-4060*

GOODELL, JOHN DEWITTE, electromechanical engineer; b. Omaha, Nebr., Sept. 20, 1909; s. Edwin Dewitte and Vera May (Watts) G.; m. Bernadette Michel, Apr. 27, 1943; children: Mary, Greg, Thomas, Caroline, Daniel. Cons. engr. N.Y.C., 1931-41; tech. dir. U.S. Army Detroit Signal Lab., 1941-43; dir. engring. Minn. Electronics, St. Paul, 1946-57; mgr. new product design CBS Lab., Stamford, Conn., 1957-60; dir. engring. Robodyne, U.S.Industries, Silver Spring, Md., 1960-61; corp. tech. dir. U.S. Industries, N.Y.C., 1962-63; producer Goodell Motion Pictures, St. Paul, 1964-75; cons. engr. New Product Design, St. Paul, 1976-90; exhibit prototyper Sci. Mus. of Minn., St. Paul, 1990—. Author: The World of Ki, 1967; writer, dir. (motion picture) Always a New Beginning, 1973, (acad. nominated best documentary 1973), (TV documentary) Wisdom and Change, 1992; dir. Challenge for Tomorrow, 1964 (indsl. Oscar); inventor: automatic mail handler, automatic manipulator, magnetic pulse controlling device, conditioned reflex teaching machines and others, 1954—; editor Jour. of Computing Systems, 1965-70. With U.S. Navy, 1943-46, S. Pacific. Recipient Master Design award, Product Engring., 1962. Mem. IEEE (sr.), Soc. Motion Picture and TV Engrs. Avocations: Oriental Game of Go (capt. U.S team 2nd place winners Internat. Go Cong., 1964). Home: 751 Mt Curve Blvd Apt 5 Saint Paul MN 55116-1113

GOODEN, ROBERT, chemist; b. Shreveport, La., Jan. 4, 1949; s. Arthur and Geneva (Knighten) G.; div.; children: Nakida, Michele. BS in Chemistry, So. U., 1974; PhD in Chemistry, Stanford U., 1979. Mem. tech. staff AT&T Bell Labs., Murray Hill, N.J., 1978-89; rsch. leader Dow Chem. Co., Plaquemine, La., 1989-91; assoc. prof. dept. chem. U. Baton Rouge, 1992—; vis. prof. dept. chemistry So. U., Baton Rouge, 1987-88. Contbr. over 28 articles to tech. jours., chpts. to books. With USMC, 1969-72. Mem. AAAS, Nat. Orgn. Profl. Advancement Black Chemists, Soc. Plastics Engrs., Am. Chem. Soc. Home: 12074 Newcastle Ave Apt 1707 Baton Rouge LA 70816-8993 Office: So U Dept Chemistry Baton Rouge LA 70813

GOODES, MELVIN RUSSELL, manufacturing company executive; b. Hamilton, Ont., Can., Apr. 11, 1935; s. Cedric Percy and Mary Melba (Lewis) G.; m. Arlene Marie Bourne, Feb. 23, 1963; children: Melanie, Michelle, David. B in Commerce, Queen's U., Kingston, Ont., Can., 1957; MBA, U. Chgo., 1960. Rsch. assoc. Can. Econ. Rsch. Assocs., Toronto, Ont., 1957-58; market planning coord. Ford Motor Co. Can., Oakville, Ont., 1960-64; asst. to v.p. O'Keefe Breweries, Toronto, 1964-65; mgr. new product devel. Adams Brands div. Warner-Lambert Can., Scarborough, Ont., 1965-68; area mgr. Warner-Lambert Internat., Toronto, 1968-69; regional dir. confectionary ops. Warner-Lambert Europe, Brussels, 1969-70; pres. Warner-Lambert Mex., 1970-76; pres. Pan-Am. zone Warner-Lambert Internat., Morris Plains, N.J., 1976-77, pres. Pan-Am. and Asian zone, 1977-79; pres. consumer products div. Warner-Lambert Co., Morris Plains, N.J., 1979-81, sr. v.p., pres. consumer products group, 1981-83, exec. v.p., pres. U.S. ops., 1984-85, pres., chief operating officer, 1985-91, chmn., chief exec. officer, 1991—, also bd. dirs., chmn. bd., chief exec. officer, 1991—; bd. dirs. Chem. Banking Corp., Chem. Bank, Unisys; mem. exec. adv. council Nat. Ctr. Ind. Retail Pharmacy, 1984-85. Bd. dirs. Coun. on Family Health, N.Y.C., 1981-86, Advt. Edn. Found., N.Y.C., 1989-91; mem. fin. com. Nat. Coun. on Econ. Edn., 1984—, mem. exec. com., 1986—; mem. Internat. Exec. Svc. Corps., 1989—; mem. adv. coun. Sch. of Bus. Queen's U., Kingston, Ont., Can., 1980-84; trustee Drew U., Madison, N.J., 1985-88, Queen's U., 1988—. Fellow Ford Found., 1958, Sears, Roebuck Found., 1959. Mem. nat. Wholesale Druggists Assn. (assoc. adv. com.), Nat. Assn. Retail Druggists (exec. adv. coun. 1983-85), Pharm. Mfrs. Assn. (bd. dirs. 1989-91), Proprietary Assn. (v.p. 1983-88, bd. dirs., mem. exec. com. 1981-88), Nat. Alliance Bus. (bd. dirs. 1984-86), Plainfield Country Club (N.J.), Econ. (N.Y.C.). Unitarian. Avocations: golf, tennis. Office: Warner-Lambert Co 201 Tabor Rd Morris Plains NJ 07950-2693

GOODHUE, PETER AMES, obstetrician/gynecologist, educator; b. Ft. Fairfield, Maine, Feb. 26, 1931; s. Lawrence and Zylpha (Ames) G.; m. Edith Ann Helfenstein, June 21, 1958; children: Lisa Grace, Scott Ames. BA, Amherst Coll., 1954; MD, U. Vt. 1958. Diplomate Am. Bd. Ob-Gyn. Intern Bellevue Hosp., N.Y.C., 1958-59; resident Yale-New Haven Med. Center, 1959-62; practice medicine specializing in ob-gyn, Stamford, Conn., 1964—; assoc. clin. prof. ob-gyn N.Y. Med. Coll. Contbr. articles to profl. jours. Served to capt. USAF, 1962-64. Recipient Carbee prize, U. Vt., 1958. Fellow ACS, Am. Fertility Soc., Am. Coll. Obstetricians and Gynecologists, Am. Soc. For Colposcopy and Cervical Pathology, Am. Assn. Of Gynecologic Laproscopists; mem. Conn. Med. Soc., Conn. Soc. Am. Bd. Obstetricians and Gynecologists, Fairfield County Med. Soc., Stamford Med. Soc. Republican. Episcopalian. Office: Stamford Gynecology PC 70 Mill River St Stamford CT 06902-3725

GOODING, CHARLES THOMAS, psychology educator, college dean; b. Tampa, Fla., Nov. 18, 1931; s. Charles T. and Gladys (Bingman) G.; m. Shirley Ann Puckett, June 7, 1953; children: Steven Thomas, Carol Ann, David Lee, Mark Charles. B.A., U. Fla., 1954, M.Ed., 1962, Ed.D., 1964; postgrad., U. Tampa, 1956-58. Tchr. Meml. Sch., Tampa, 1956-58; asst. prin., then prin. St. Mary's Sch., Tampa, 1958-62; grad. fellow U. Fla., Gainesville, 1962-63, instr., 1963-64; assoc. prof., then prof. SUNY, Oswego, 1964-79, prof. psychology, 1980—, assoc. dean, dean grad. studiesand rsch., 1989-89, dean grad. studies and rsch., 1989—; vis. prof. U. Liverpool, Eng., 1979-80; mem. SUNY Chancellor's Task Force on Tchr. Edn., 1984; grad. fellow U. Fla., Gainesville, 1962-63. Author: Learning Theories in Educational Practice, 1971; contbg. author: Questioning and Discussion: A Multidisciplinary Study, 1988, Research Matters to the Science Teacher, 1992; contbr. articles to profl. jours. Bd. dirs. Oswego County unit Am. Cancer Soc., N.Y., 1972-74, 82-84; mem. commn. on ordination Episcopal Diocese Central N.Y., 1980—; bd. dirs. Lancaster Career Devel. Ctr., 1984—. Served to 1st lt. USAR, 1954-56. SUNY Rsch. Found. grantee, 1966, 69, 70; N.Y. State Dept. Edn. grantee, 1971-72, 88-94; NSF grantee, 1980-81, 85-88, 90—. Mem. AAAS, Ea. Ednl. Rsch. Assn. (v.p. 1979-81, treas., dir. 1983-85, pres. elect 1987-88, pres. 1989-91, editorial bd. 1991-93), APA, Brit. Ednl. Rsch. Assn., Am. Ednl. Rsch. Assn., Nat. Assn. for Rsch. in Sci. Teaching. Avocation: antique and classic automobiles-specialist in Jaquar sports cars. Home: 4169 W River Rd PO Box 231 Minetto NY 13115 Office: SUNY Grad Office Culkin Hall Oswego NY 13126

GOODKIND, JOHN MORTON, physics educator; b. N.Y.C., Aug. 27, 1934; s. Herbert Knolton and Mabel (Goldhammer) G.; m. Alice Anderson, Sept. 7, 1963; children: Hilary Mabel, Lane Charles. BA, Amherst Coll., 1956; PhD, Duke U., 1960. Postdoctoral Stanford U., Palo Alto, Calif., 1960-62; asst. prof. U. Calif., San Diego, 1962-67, assoc. prof., 1967-75, prof., 1975—; ptnr. GWR Instruments, San Diego, 1979-92. Mem. fin. com. City of Del Mar, Calif., 1988—. Recipient fellowship Sloan Found., 1962, 64. Fellow Am. Phys. Soc. Achievements include discovery of low temperature magnetically ordered phase of solid 3He; development of practical application of nuclear cooling for submilli Kelvin measurements; of superconducting gravimeter applied to geophysical and fundamental physics. Office: U Calif 9500 Gilman Dr La Jolla CA 92093

GOODMAN, ALAN NOEL, physician; b. N.Y.C., May 5, 1934; s. Jack and Beatrice (Rosenblum) G.; m. Yoshiko Oishi, Aug. 28, 1968. BA in Zoology, U. Pa., 1956; MD, Boston U., 1960. Bd. cert. Am. Bd. Internal Medicine 1968, 74, Am. Bd. Pathology 1969. Intern Mt. Sinai Hosp., N.Y.C., 1960-61, resident pathology, medicine, 1961-63; resident medicine, hematology, chief resident Maimonides Hosp., Bklyn., 1963-65; NIH fellow in immunohematology Mt. Sinai Hosp., 1965-66; asst. prof. clin. pathology NYU Med. Ctr., 1968-70, assoc. prof. clin. pathology, 1970-71; asst. attending pathologist NYU Hosp., 1968-71; attending physician VA Hosp., Manhattan, N.Y., 1968-71; asst. attending hematologist Maimonides Hosp., 1968-71; chief hematology Booth Meml. Med. Ctr., N.Y.C., 1969-71; with Boca Raton (Fla.) Community Hosp., 1971—. With U.S. Army, 1966-68, Vietnam. Mem. Am. Coll. Physicians, Am. Soc. Hematology, Am. Assn. Blood Banks, Am. Soc. Clin. Pathologists, Am. Soc. Clin. Oncology, Internat. Soc. Blood Transfusion, N.Y. Acad. Scis., Coll. Am. Pathologist.

GOODMAN, COREY SCOTT, neurobiology educator, researcher; b. Chgo., June 29, 1951; s. Arnold Harold and Florence (Friedman) G.; m. Marcia M. Barinaga, Dec. 8, 1984. BS, Stanford U., 1972; PhD, U. Calif., Berkeley, 1977. Postdoctoral fellow U. Calif., San Diego, 1979; asst. prof. dept. biol. scis. Stanford (Calif.) U., 1979-82, assoc. prof., 1982-87; prof. neurobiology and genetics U. Calif., Berkeley, 1987—; investigator Howard Hughes Med. Inst., 1988—, head neurobiology div., 1992—. Mem. editorial bd. Sci., Neuron, Jour. of Neurosci.; contbr. more than 100 articles to profl. jours. Recipient Charles Judson Herrick award, 1982, Alan T. Waterman award Nat. Sci. Bd., 1983, Javits Neurosci. Investigator award NIH, 1985, 92. Office: Univ Calif Howard Hughes Med Inst Dept Molecular and Cell Biology Life Sci Addition Rm 519 Berkeley CA 94720

GOODMAN, DANIEL, electrical engineer, mathematical physicist; b. Island Park, N.Y., Oct. 25, 1957; s. Charles and Marlene (Albert) G. BSEE, Polytechnic U., 1979, MSEE, 1986; BS in Math., Regents Coll., Univ. State N.Y., Albany, 1988; MS in Physics, Adelphi U., 1992, MS in Math., 1993, postgrad. studies in math., 1993—. Registered profl. engr., N.Y., gen. radio and TV lic., FCC, amateur radio lic. technician class, FCC; EMT. With sales staff Lafayette Radio-Electronics Corp., Melville, N.Y., 1975-79; technician ABL Gen. Instruments, Freeport, N.Y., 1977; elec. engr. Rome Air Devel. Ctr., Hanscom AFB, Bedford, Mass., 1979-83; assoc. radar systems engr. Unisys Corp. (formerly Sperry Corp.), Great Neck, N.Y., 1984-85; radar systems engr. Grumman Aerospace Corp., Bethpage, N.Y., 1985-88; tech. editor Radio-Electronics mag. Gernsback Publs., Inc., Farmingdale, N.Y., 1989-90; lab. technician chemistry dept., grad. asst. physics dept. Adelphi U., Garden City, N.Y., 1990—; adj. faculty mem. Coll. Tech. SUNY, Farmingdale, 1986; grad. asst. chemistry dept. SUNY, Stony Brook, 1987-88. 1st. lt. USAF, 1979-83; capt. USAFR, 1983-92. Recipient Mil. History award Assn. of U.S. Army, 1978; U.S. Army ROTC scholar Polytechnic U., 1976-79; Challenger fellow for Tchrs., N.Y. State Dept. Edn., 1991-93. Mem. IEEE, Res. Officers Assn., Air Force Assn. (life), Civil Air Patrol, N.Y. State Soc. Profl. Engrs., Internat. Soc. Cert. Electronic Technicians (cert.). Jewish. Home: 204 Traymore Blvd Island Park NY 11558-1900 Office: Adelphi U Chemistry Dept Garden City NY 11530

GOODMAN, DAVID WAYNE, research chemist; b. Glen Allen, Miss., Dec. 14, 1945; s. Henry G. and Anniebelle G.; m. Sandra Faye Hewitt, June 9, 1967; 1 child, Jac Hewitt. BS, Miss. Coll., 1968; PhD, U. Tex., 1974. NATO postdoctoral fellow Tech. Hochschule, Darmstadt, Fed. Republic of Germany, 1974-75; NRC postdoctoral fellow NBS, Washington, 1975-76, mem. research staff, 1976-80; mem. research staff Sandia Labs., Albuquerque, 1980-85, head surface sci. div., 1985-88; prof. chemistry, Tex. A&M U., College Station, 1988—. Mem. Am. Chem. Soc. (treas. div. colloid and surface sci. 1980-83, vice chair 1983, chmn. 1984, Colloid or Surface Chemistry award 1993), Am. Vacuum Soc. (mem. exec. council 1981, 85-87). Home: 8707 Appomattox Dr College Station TX 77845-5590 Office: TX A&M University Dept of Chemistry College Station TX 77843

GOODMAN, DONALD JOSEPH, dentist; b. Cleve., Aug. 14, 1922; s. Joseph Henry and Henrietta Inez (Mandel) G.; BS, Adelbert Coll., 1943; DDS, Case-Western Reserve U., 1945; m. Dora May Hirsh, Sept. 18, 1947; children: Lynda (Mrs. Barry Allen Levin), Keith, Bruce; m. Ruth Jeanette Weber, May 1, 1974. Pvt. practice dentistry, Cleve., 1949-86; lectr. in field. With Dental Corps, USNR, 1946-48. Mem. Am. Acad. Gen. Dentistry, ADA Ohio State Dental Assn., Cleve. Dental Soc., Fedn. Dentaire Internationale, Cleve. Council on World Affairs, Greater Cleve. Growth Assn., Council of Smaller Enterprises, Phi Sigma Delta, Zeta Beta Tau, Alpha Omega. Clubs: Masons (32 deg.), Shriners, Travelers' Century (Gold award), Circumnavigators. Home: 29099 Shaker Blvd Cleveland OH 44124-5022

GOODMAN, ERIK DAVID, engineering educator; b. Palo Alto, Calif., Feb. 14, 1944; s. Harold Orbeck and Shirley Mae (Lillie) G.; m. Denise Rowand Dyktor, Aug. 10, 1968 (div. 1976); m. Cheryl Diane Barris, Aug. 27, 1978; 1 child, David Richard. BS in Math., Mich. State U., 1966, MS in Systems Sci., 1968; PhD in Computer Communication Sci., U. Mich., 1972. Asst. prof. elec. engring. Mich. State U., East Lansing, 1972-77, assoc. prof. elec. engring., 1977-84, dir. case ctr. for computer aided engring. and mfg., 1983—, prof. elec. engring., dir., 1984—, prof. mech. engring., 1992—; dir. Mich. State U. Mfg. Rsch. Consortium, 1993—; pres. Tech. Gateway, Inc., East Lansing; cons. Chinese Computer Comms., Inc., Lansing, 1988—; bd. dirs. Mid-Mich. Flight, Inc., East Lansing; gen. co-chmn. Internat. Computer Graphics Conf., Detroit, 1986. Author: (with others) SYSKIT: Linear Systems Toolkit, 1986; patentee in field. Academician, Internat. Informatization Acad. (Russia), 1993—. Mem. AIAA (chair rsch. and future dirs., subcom. CAD/CAM tech. com. 1987-89, Outstanding Svc. 1990), IEEE Computer Soc., Soc. Mfg. Engrs., Internat. Acad. Informatics (fgn. academician Russia 1993—), Aircraft Owners and Pilots Assn. Avocations: musician, ski racing, tennis, studying Chinese. Office: Mich State U Case Ctr Coll Engring 112 Engineering Bldg East Lansing MI 48824

GOODMAN, HAROLD, pediatrician, consultant; b. Chgo., Oct. 25, 1926; s. Morris and Dorothy (Rosenberg) G.; m. L. Jane Lippincott, Jan. 16, 1962; children: Anne M., David J. Student, Loyola U., Chgo., 1943-45; B.S., M.D., U. Ill. Coll. Medicine, 1949. Diplomate Am. Bd. Pediatrics. Attending physician Children's Meml. Hosp., Chgo., 1956—, Northwestern Meml. Hosp., Chgo., 1957—; clin. assoc. prof. pediatrics Northwestern U., Chgo., 1957—; clin. asst. prof. dept. pediatrics Rush U. Med. Sch., Chgo., 1976—; chmn dept. pediatrics Grant Hosp., Chgo., 1963-86, attending physician dept. pediatrics, 1956—. Capt. USAF, 1952-55. Mem. Chgo. Med. Soc. (del. Ill. State Med. Soc. 1985—, pres. n. side br., 1991-92), Am. Acad. Pediatrics (life), Alpha Omega Alpha, Phi Delta Epsilon. Home: 1020 Forest Ave Evanston IL 60202 Office: 4753 Broadway Chicago IL 60640

GOODMAN, HERBERT IRWIN, petroleum company executive; b. Pitts., Mar. 11, 1923; s. Meyer Irwin and Bessie (Crossof) G.; m. Mary Katherine Schilken, Aug. 12, 1978; children: Michael Christopher, Anne Katheryn, Nancy Hjortshoj, Sara Elizabeth, Mary Ellen. BS, U. Pitts., 1943; cert., U. Besancon, 1945; MBA, Harvard U., 1949, AM, 1950. Commd. officer U.S. Fgn. Service, 1951; served in U.S. Fgn. Service, Copenhagen, 1951-53, Vietnam, 1953-54, Cambodia, 1954-55; intelligence research officer Dept. State, 1956-57; with Gulf Oil Corp., 1957-84; coordinator European sales Gulf Oil Corp., London, 1957-59; gen. mgr. Pacific Gulf Oil, Tokyo, 1960-64; coordinator crude oil dept. Pacific Gulf Oil, Pitts., 1964-66, coordinator Far East, 1966-70; pres. Gulf Oil Co. South Asia, Singapore, 1970-72, Gulf Oil Trading Co., Pitts., 1972-80, Gulf Trading and Transp. Co., Houston, 1980-84, GOTCO USA, Inc., Houston, 1984-87, SARMAR Corp., Houston, 1987—; bd. chmn. Applied Trading Ssystems, Houston, 1988—; bd. dirs. Houston Livestock Show and Rodeo. Mem. U.S. Nat. Com. for Pacific Econ. Cooperation; bd. dirs. Inst. Pacific Asia, Tex. A&M U., U. Houston Coll. Bus.; bd. dirs. Asia Soc. Houston. 1st lt. U.S. Army, 1943-46. Decorated Bronze Star; médaille de la Réconnaissance (France). Mem. Am. Petroleum Inst., Am. Inst. Mgmt., Am. Mgmt. Assn., Coun. on Fgn. Rels., Assn. Asian Studies, Mid East Inst., Asia Soc. N.Y. (corp. coun.), Japan Soc., U.S. Korea Soc., Harvard Club (N.Y.C.), Lakeside Country Club, Racquet Club, Univ. Club, Petroleum Club. Office: SARMAR Corp 1800 W Loop S Ste 1510 Houston TX 77027

GOODMAN, HOWARD ALAN, human factors engineer, air force officer; b. Phila., Oct. 29, 1955; s. Irving Goodman and Sarah Elizabeth (Bancoff) Krassnoff; m. Karen Denise Farah, Mar. 15, 1977; children: Brandi Lynn, Erica Corinne. BA, Calif. State U., Carson, 1982; MA, Incarnate Word Coll., San Antonio, 1986. Enlisted USAF, 1974, commd. 2d lt., 1983, advanced through grades to capt., 1986; occupational analyst USAF Occupational Mgmt. Ctr., Randolph AFB, Tex., 1983-86; human factors engr. electronic systems div. Systems Program Office, Hanscom AFB, Mass., 1986-88; human factors engr. in artificial intelligence and virtual reality Human Systems Ctr., Brooks AFB, Tex., 1990—; chmn. human factors engring. tech. subgroup on human systems integration Dept. Def., Wright-Patterson AFB, Ohio, 1992—. Mem. APA (affiliate), Human Factors Soc. Republican. Achievements include conception of artificial intelligence applications to virtual reality environments, research on spatial disorientation in virtual reality, attempting to construct an authoring environment for virtual reality applications. Home: 10019 Windburn Trail Converse TX 78109 Office: AL/HRTI Bldg 578 Rm 106 Brooks AFB TX 78235

GOODMAN, HUBERT THORMAN, psychiatrist, consultant; b. Oklahoma City, Mar. 5, 1933; s. Hubert Thorman and Belle (Wilkonson) G.; m. Doris Alene Knight, Feb. 1, 1957 (div. Feb. 1975); children: Mark, Martha Harris, Mary Downs, Carmen Lugo, Valerie Weiner; m. Paulette Sue Freeman, Oct. 28, 1988. MD, Ind. U., 1957. Intern Riverside Hosp., Toledo, 1957-58; resident in pub. health Miss. Dept. Health, Jackson, 1958-60; resident in psychiatry Cen. Ohio Psychiat. Hosp., Columbus, 1960-63;; officer Pub. Health Svc., Jackson, 1958-60; pvt. practice Columbus, 1960—; clin. asst. prof. psychiatry Ohio State U., 1962-91; cons. Dept. Youth Svc., Columbus, 1983—, Ohio Dept. Mental Health, 1990—, Peer Rev. Systems Ohio, 1990—. Contbr. articles to profl. jours. Capt. USPHS, 1958-60. Recipient Felix Underwood award Miss. Med. Assoc., 1963. Mem. AMA, Cen. Ohio Med. Assn., Ohio State Med. Assn., Am. Psychiat. Assn., Ohio State Psychiat. Assn., Cen. Ohio Psychiat. Assn. Avocations: theater, music, dancing. Home: 4770 Dierker Rd Columbus OH 43220-2985 Office: 4700 Reed Rd Columbus OH 43220-3074

GOODMAN, JOSEPH WILFRED, electrical engineering educator; b. Boston, Feb. 8, 1936; s. Joseph and Doris (Ryan) G.; m. Hon Mai Lam, Dec. 5, 1962; 1 dau., Michele Ann. B.A., Harvard U., 1958; M.S. in E.E., Stanford U., 1960, Ph.D., 1963. Postdoctoral fellow Norwegian Def. Rsch. Establishment, Oslo, 1962-63; rsch. assoc. Stanford U., 1963-67, asst. prof., 1967-69, assoc. prof., 1969-72, prof. elec. engring., 1972—; vis. prof. Univ. Paris XI, Orsay, France, 1973-74; dir. Info. Systems lab., dept. elec. engring., Stanford U., 1981-83, chmn., 1988—; William E. Ayer prof. elec. engring. Stanford U., 1988—; cons. to govt. and industry, 1965—; v.p. Internat. Comm. for Optics, 1985-87, pres., 1988-90, past pres., 1991-93. Author: Introduction to Fourier Optics, 1968, Statistical Optics, 1985; editor: International Trends in Optics, 1991; contbr. articles to profl. jours. Recipient F.E. Terman award Am. Soc. Engring. Edn., 1971. Fellow Optical Soc. Am. (dir. 1977-83, editor jour. 1978-83, Max Born award 1983, Frederick Ives award 1990, v.p. 1990, pres. elect 1991, pres. 1992, past pres. 1993), IEEE (edn. medal 1987), Soc. Photo-optical Instrumentation Engrs. (bd. govs. 1979-82, 88-90, Dennis Gabor award 1987); mem. Nat. Acad. Engring., Electromagnetics Acad. Home: 570 University Ter Los Altos CA 94022-3523 Office: Stanford U Dept Elec Engring McCullogh 152 Stanford CA 94305

GOODMAN, LAWRENCE EUGENE, structural analyst, educator; b. N.Y.C., Mar. 12, 1920; s. Joseph John and Dorothy (Goldberger) G.; m. Katherine Cecilia Lewis, Sept. 16, 1951; children: Jennifer Robin, Jeanne, Alice. AB, Columbia U., 1939, BSCE, 1940; MS in Applied Mechanics, U. Ill., 1942; PhD, Columbia U., 1948. Registered profl. engr. Lectr. Columbia U., N.Y.C., 1946-48; assoc. prof. civil engring. U. Ill., Urbana, 1948-54; prof. applied mechanics U. Minn., Mpls., 1954-65, dept. head civil engring., 1965-72, Record prof. civil engring., 1975-88, Record prof. emeritus, 1988—; cons. Xerxes Corp., Mpls., 1988; vis., prof. Tex. A&M U., College Station, 1991-92. Co-author: Statics, 1963, Dynamics, 1963; contbr. articles to profl. jours. Lt. U.S. Navy, 1943-46, PTO. Recipient Disting. Teaching award Regents of U. Minn. NSF sr. fellow, 1962-63. Fellow ASCE (Newmark Gold medal 1990), Am. Soc. of ME, 1970. Jewish. Home: 1589 Vincent Ave Saint Paul MN 55108-1324

GOODMAN, MATTHEW SAMUEL, physicist; b. Oak Park, Ill., Oct. 22, 1947; s. Jacob and Lillian (Rozsa) G. BS with honors, Ind. U., 1969; MA, Johns Hopkins U., 1971, PhD, 1977. Rsch. fellow Harvard U., Cambridge, Mass., 1977-80, asst. prof. physics, 1980-85; mem. tech. staff Bellcore, Morristown, N.J., 1985—; vis. scientist CERN, Geneva, 1983-85, Fermilab, Batavia, Ill., 1977-83, Oxford (England) U., 1977-78, Oak Ridge (Tenn.) Nat. Labs., 1981-85. Contbr. articles to Phys. Rev. Letter, Electronics Letters, Nuclear Instrn. and Methods; editor: IEEE Transactions on Communications. Recipient Thomas T. Hoopes III Teaching award, 1984. Mem. IEEE (area editor for optical comm. in publ. Transactions on Communications), Am. Phys. Soc., Sigma Pi Sigma, Phi Eta Sigma. Achievements include 4 patents; research into rearrangable optical networks and technology, demonstration of high capacity multiwavelength optical networks, Generalized Heisenberg Uncertainty Relation, physics of liquid Argon calorimetry, shadowing of nuclei by virtual photons, and measurement of large mass diphoton cross section. Office: Bellcore 445 South St Rm 2L149 Morristown NJ 07960

GOODMAN, WILLIAM ALFRED, materials engineer; b. Bronx, N.Y., Feb. 19, 1961; s. Irwin F. and Alice Christine (Kraus) G.; m. Bobbie Sheri Anderson, Oct. 8, 1988; children: Britney Michel, Brandon Michael. BS in Chem. Engring., U. N.Mex., 1982; MS in Materials Sci. & Engring., UCLA, 1991. Chemist Materials Rsch. Corp., Orangeburg, N.Y., 1980; operator Ford Utilities Plant, Albuquerque, 1981; rsch. asst. dept. chem. engring. U. N.Mex., Albuquerque, 1981-82; physics technician Air Force Weapons Lab., Albuquerque, 1982; materials/systems engr. W.J. Schafer Assocs., Calabasas, Calif., 1983—. Sunday Sch. tchr. St. Mark's Ch., Glendale, Calif., 1987-89; grad. liaison Netherlands Industria U.S.A. tour UCLA, 1991; lifegiver Unied Blood Svcs., Thousand Oaks, Calif., 1992. Mem. AIAA, Am. Soc. Metals Internat., Am. Ceramic Soc. Roman Catholic. Achievements include authoring of advanced transp. analysis code; research in modeling of laser absorption calorimetry for uncooled silicon optics, an elegant computer program for metal forming problems, liquid propellant rocket engine cost model. Office: W J Schafer Assocs 26565 West Agoura Rd Calabasas CA 91302

GOODNER, HOMER WADE, safety risk analysis specialist, industrial process system failure risk consultant; b. Birmingham, Ala., May 28, 1929; s. Robert Wade and Sadie Mae (Daniel) G.; m. Kathleen Annette Holland, Oct. 20, 1950; children: Kathleen Cary, Jacquelyn Ruth, Keith Wade, Lela Suzanne. BSME, U. Ala., 1959. Registered profl. engr., Okla. Asst. supt. Lake Shore Pipeline Co., Ashtabula, Ohio, 1952-54; constrn. engr. Ala. Gas Corp., Anniston, 1954-58; sr. devel. engr. Monsant Corp., Pensacola, Fla., 1959-65, Decatur, Ala., 1965-67; sr. engr. Dow Badische Co., Anderson, S.C., 1967-73, Phillips Fibers Corp., Greenville, S.C., 1973-75, Phillips Petroleum Co., London, 1975-80; sr. engring. specialist Phillips Petroleum Co., Bartlesville, Okla., 1980-92; process system failure risk analyst. Author: Risk Analysis: HARA Technique; inventor in field; contbr. articles to ency. and profl. jours. With U.S. Army, 1946-48, PTO. Mem. Reliability Soc. of IEEE, Inst. Soc. Am., Nat. Soc. Profl. Engrs., Okla. Soc. Profl. Engrs., Internat. Platform Assn., Soc. for Risk Analysis, Pi Tau Sigma. Republican. Episcopalian. Avocations: photography, historical and biographical reading, woodworking, peregrinating. Home and Office: 106 Brittany Pk Anderson SC 29621

GOODNICK, PAUL JOEL, psychiatrist; b. Phila., Sept. 29, 1950. BA magna cum laude, U. Pa.; MD with honors, SUNY Downstate Med. Ctr., Bklyn. Diplomate Am. Bd. Psychiatry and Neurology. Resident Washington U., St. Louis, Mo., Columbia U., N.Y.C.; fellow Mt. Sinai Hosp., N.Y.C.; asst. prof. psychiatry Wayne State U., Detroit, 1980-81, U. Chgo., 1981-84, Columbia U., N.Y.C., 1984-87; asst. prof. psychiatry U. Miami, Fla., 1987-89, clin. assoc. prof. psychiatry, 1989-90, assoc. prof., 1990-93, prof., 1993—, dir. mood disorders program, dept. psychiatry, 1989—; dir. outpatient svcs. and affective disorders program Fair Oaks Hosp., Boca/Delray, Fla., 1987-90; cons. APA, 1991. Assoc. editor jour. Lithium, 1989—; editor: Chronic Fatigue and Related Immune Deficiency Syndromes, 1992. Mem. nat. adv. bd. Jerusalem Health Ctr. Recipient Clin. Excellence award N.Y. Alliance for Mentally Ill, 1987. Fellow Am. Psychopathological Assn., Am. Psychiat. Assn.; mem. Soc. Biol. Psychiatry, N.Y. Acad. Sci., AAAS, Am. Acad. Clin. Psychiatry, KP. Office: U Miami Dept Psychiatry 1400 NW 10th Ave Ste 304 Miami FL 33136-1032

GOODNICK, STEPHEN MARSHALL, electrical engineer, educator; b. Mt. Vernon, Ill., Aug. 20, 1955; s. Duane Simon and Betty Jo (Cleek) G.; m. Sara Lynn Holtke, Aug. 20, 1977. BS in Engring. Sci., Trinity U., San Antonio, 1977; MSEE, Colo. State U., 1979, PhD in Elec. Engring., 1983. Rsch. assoc., then asst. prof. Colo. State U., Ft. Collins, 1983-85; vis. scientist Solar Energy Rsch. Inst., Golden, Colo., 1985, U. Modena, Italy, 1985, Tech. U. Munich, 1986; asst. prof. Oreg. State U., Corvallis, 1986-90, assoc. prof. elec. engring., 1990-93, prof. elec. engring., 1993—; adv. bd. Nat. Ctr. Computational Electronics, Urbana, Ill., 1988—. Contbr. chpt. to Physics and Chemistry of Compound Semiconductor Interfaces, 1986, Hot Carriers in Semiconductor Microstructures, 1992; contbr. articles to Phys. Rev., other jours. Alexander von Humboldt regional fellow Fed. Republic Germany, Munich, 1986; Melchor vis. chair U. Notre Dame, South Bend, 1991. Mem. IEEE (sr. mem.), Am. Phys. Soc., Blue Key, Sigma Xi. Democrat. Achievements include quantitative measure of surface roughness degradation in MOSFETs, model for electron and hole relaxation in semiconductors, model for transport in nanostructures, development of multiterminal negative conductance device (with M.N. Wybourne and J. Wu). Office: Oreg State U Dept Elec/Computer Engring Corvallis OR 97331

GOODRICH, CRAIG ROBERT, business analyst; b. Lynwood, Calif., Mar. 10, 1949; s. Gordon Llewellyn and Dora Jeannette (Shannon) G. BA in Bus. Adminstrn., Calif. State U., Long Beach, 1972, BA in Psychology, 1986. Salesman Pacific Stereo, Torrance, Calif., 1973; mgr. Radio Shack, Long Beach, 1975-76; inventory contr. L.B. Ball & Co., Long Beach, 1979-84; sales, acctg. and edn. cons. Long Beach, 1984-85; adminstrv. analyst Rockwell Internat., Cypress, Calif., 1986; surcharge tax asst. L.A. County, Calif., 1987; sr. engring. bus. mgmt. analyst McDonnell Douglas, Long Beach, 1987-89, human resources adminstr., 1989-90, budget analyst, 1990—. Mem. Acacia Frat. Avocations: sports, science, bicycling, astronomy, music. Home: 1685 Loma Ave Apt 4 Long Beach CA 90804-2751 Office: McDonnell Douglas Aerospace 5301 Bolsa Ave Huntington Beach CA 92649

GOODRICH, DAVID CHARLES, management psychologist; b. Plymouth, Ind., Feb. 3, 1926; s. Clifford Oscar and Mae (Manuwal) G.; m. Bette L. Taubeneck, Oct. 22, 1953 (div. Nov., 1978); children: Charles Allen, Carol Ann; m. Eleanor Griener Gabriel, Jan. 18, 1983. BS, Purdue U., 1948; PhD, U. Rochester, 1953. Diplomate Am. Bd. Profl. Psychology. Chief psychologist Ryuukyus Army Hosp., Okinawa, 1954-56; sr. psychologist CIA, Washington, 1956-62; cons. Rohrer, Hibler, Repogle, N.Y.C., 1962-64, Rohrer, Hobler, Repogle, Phila., 1964-70; mgr. Rohrer, Hobler, Repogle, Washington, 1970-80, Roher, Hobler, Repogle, Stamford, Conn., 1980-83; pres., prin. David C. Goodrich Assocs., Fairfield, Conn., 1983—. Chmn., bd. dirs. Tredyfnin Pub. Libr., Paoli, Pa., 1965-70; bd. dirs. Upper Main Line YMCA, Berwyn, Pa., 1967-70; chmn. bd. of music First Congregationalist Ch., Fairfield, Conn., 1989-92, mem. parish coun., 1990-92. Mem. Am. Psychol. Assn. Achievements include rsch. on CEO succession. Home and Office: 125 Field Point Dr Fairfield CT 06430

GOODRICH, JAMES TAIT, neuroscientist, pediatric neurosurgeon; b. Portland, Ore., Apr. 16, 1946; s. Richard and Gail (Josselyn) G.; m. Judy Loudin, Dec. 27, 1970. Student, Golden West Coll., 1971-72, A.A., Orange Coast Coll., 1972; B.S. cum laude, U. Calif.-Irvine, 1974; M.Phil., Columbia U., 1979, Ph.D., 1970 M.D., 1980; Diplomate Am. Bd. Neurological Surgery. Neuroscientist, pediatric neurosurgeon N.Y. Neurol. Inst., N.Y.C., 1981-86; dir. div. pediatric neurosurgery Albert Einstein Coll. Medicine, N.Y.C., 1986—, assoc. prof. neurological surgery, pediatrics, plastics and reconstructive surgery, 1992—. Contbr. articles to profl. jours. Recipient Roche Labs. award in neuroscis., 1978, Mead-Johnson award, 1978, Bronze medal Alumni Assn. Coll. Physicians and Surgeons, 1980, Sandoz award for outstanding research, 1980; Willamette Industries scholar; NIH grantee. Fellow Royal Soc. Medicine (London); mem. Internat. Soc. Pediatric Neuro-Surgeons, Worshipful Soc. Apothecaries (London), N.Y. Acad. Medicine (Melicow award 1980), Am. Assn. History of Medicine (Sir William Osler medal 1977-78), AMA, Brit. Brain Research Assn., European Brain Research Assn., Friends of Columbia U. Libraries, Friends of Osler Library of McGill U., N.Y. Acad. Scis., Am. Assn. Neurol. Surgeons, Am. Assn. Neurological Surgery (chmn. sect. on history of neurological surgery), Congress Neurol. Surgeons, Med. History Soc. N.J., ISIS History of Sci. Soc., Soc. for Bibliography of Natural History (London), Columbia Presbyn. Med. Soc., U. Calif. Alumni Assn., Soc. Ancient Medicine, AAAS, Am. Osler Soc., Les Amis du Vin, South Coast Wine Explorers Club (past chmn.), Friends of Bacchus Wine Club (past chmn.), Dionysius Council of Presbyn. Hosp. of N.Y.C., Sigma Xi, Alpha Gamma Sigma. Research on neuronal regeneration, brain reconstruction and craniofacial reconstruction. Home: 214 Everett Pl Englewood NJ 07631-1650 Office: Albert Einstein Coll Medicine Montefiore Med Ctr Div Pediatric Neurosurgery New York NY 10467

GOODSON, RICHARD CARLE, JR., chemist, hazardous waste management consultant; b. Toledo, June 22, 1945; s. Richard Carle Goodson Sr. and Norma (Buehler) Robinson; m. Deborah Ann Hart, Mar. 29, 1969 (div. Feb. 1978); 1 child, Geoffrey Carle; m. Thelma Agnes Matthews, Nov. 22, 1978. BS in Chemistry, Union Coll., 1967; MS in Inorganic Chemistry, U. Conn., 1970. Dist. engr. Drew Chem. Corp., Boonton, N.J., 1972-74; product supr. Drew Chem. Corp., Boonton, 1974-75, regional tech. supr., 1975-76; chief chemist, tech. dir. Environ. Waste Removal, Waterbury, Conn., 1976-79; gen. mgr., dir. tech. lab. Conn. Treatment Corp., Bristol, Conn., 1979-82; pres., owner Goodson Assocs., Avon, Conn., 1982—; dir. ops., corp. dir. waste mgmt. and regulatory compliance Hampden Mathieu Chem. Co., Springfield, Mass., 1990—. Mem. Am. Chem. Soc. Republican. Avocations: boating, hiking, skiing, cycling. Home and Office: 51 Anvil Dr Avon CT 06001-3218

GOODWIN, DAVID GEORGE, mechanical engineering educator; b. Sacramento, Calif., Oct. 15, 1957; s. George Raymond and Verma Bell (Ledbetter) G. BS, Harvey Mudd Coll., 1979; MS, Stanford (Calif.) U., 1980, PhD, 1986. Rsch. fellow Max Planck Inst. for Quantum Optics, Garching, Germany, 1986-88; asst. prof. Calif. Inst. of Tech., Pasadena, 1988—. Contbr. articles to profl. jours. Named Presdl. Young Investigator NSF, 1990. Mem. Materials Rsch. Soc., Optical Soc. Am. Office: Calif Inst Tech MC104-44 Pasadena CA 91125

GOODWIN, FRANK ERIK, materials engineer; b. Bethlehem, Pa., Jan. 6, 1954; s. Francis Black and Grethe Julie (Andresen) G.; m. Rosalind Ann Volpe, May 30, 1987; 1 child, Adrian Edmond. BS, Cornell U., 1975; ScD, MIT, 1979. Plant engr. Chambersburg (Pa.) Engring. Co., 1979-80; devel. dir. Chromalloy Rsch. & Tech., Orangeburg, N.Y., 1980-82; mgr. devel. Internat. Lead Zinc Rsch. Orgn., Research Triangle Park, N.C., 1982-84, mgr. metallurgy, 1984-86, v.p. materials sci., 1986—; mem. peer review com. on lead Dept. Energy, Washington, 1987-89. Author: Galfan Galvanizing Alloy & Technology, 1984; editor: Stress Calculations for Zinc Die Castings, 1988, Engineering Properties of Zinc Alloys, 1988; contbr. articles to profl. jours., chpts. to books. Mem. ASM, N. Am. Die Casting Assn. (rsch. com.). Republican. Episcopalian. Achievements include patents (with other) for new aluminum alloy, new lead alloy for batteries. Office: Internat Lead Zinc Rsch Org 2525 Meridian Pky # 100 Durham NC 27713-2260

GOODWIN, FREDERICK KING, psychiatrist; b. Cin., Apr. 21, 1936; s. Robert Clifford and Marion Cronin (Schmadel) G.; m. Rosemary Powers, Oct. 19, 1963; children: Kathleen Kelly, Frederick King, Daniel Clifford. B.S., Georgetown U., 1958; philosophy fellow, St. Louis U., 1958-59, M.D., 1963. Intern medicine and psychiatry SUNY, Syracuse, 1963-64; resident psychiatry U. N.C., Chapel Hill, 1964-65; commd. med. officer USPHS, 1965; clin. assoc. adult psychiatry br. NIMH, 1965-67; research fellow Lab. Biochemistry, Nat. Heart Inst., NIH, Bethesda, Md., 1967-68; chief sect. on psychiatry Lab. Clin. Sci., NIMH, Bethesda, 1970-77; chief clin. psychobiology br. Lab. Clin. Sci., NIMH, 1970-81, sci. dir., 1981-88; adminstr. Alcohol, Drug Abuse and Mental Health Adminstrn., Washington, 1988-92; pvt. practice medicine, specializing in psychiatry Bethesda, 1967—; dir. NIMH, Rockville, Md., 1992—; faculty George Washington U. Sch. Medicine, Washington Sch. Psychiatry, Uniformed U. Sch. Health Scis.; vis. prof. U. Calif., Irvine, U. Wis., Boston U., U. So. Calif., Duke U.; cons. AMA Council on Drugs; AIDS coordinator Alcohol, Drug Abuse and Mental Health Adminstrn., 1986—; participant pub. edn. programs on local and network television and radio. Author: (with K.R. Jamison) Manic-Depressive Illness, 1990 (Best Med. Book award 1990 Assn. Am. Pubsd.); editor in chief Psychiatry Research, 1979—; mem. editorial bd. Archives of Gen. Psychiatry, 1977—, Psychopharmacology, 1976-79; Contbr. articles to med. jours. Mem. adv. bd. Max Planck Inst., Munich, W. Ger. Recipient Psychopharmacology Research prize Am. Psychol. Assn., 1971, Internat. Anna Monica prize for research in depression, 1971, Taylor Manor award, 1976, Adminstrs. award HEW, 1977, Superior Service award USPHS, 1980, Strecker award, 1983, Sr. Exec. Service Presdl. Meritorious Rank award, 1982, Disting. Rank award, 1986, Disting. Exec. Service award Sr. Exec. Assn. Profl. Devel. League, 1986, Best Tchr. in Am. Psychiatry award CME Inc., 1989, Svc. to Sci. award Nat. Assn. for Biomed. Rsch., 1990, Pub. Svc. award. Fed. Am. Socs. for Exptl. BNiologhy, 1990, 1st recipient of Fawcett Humanitarian award NDMDA, 1990, McAlpin award NMHA, 1991; NIMH Spl. fellow, 1967-68. Fellow Am. Psychiat. Assn. (chmn. com. on protection of human subjects, task force on research tng., Hofheimer prize for research 1971, chmn. task force on future of psychiat. research), Am. Coll. Neuropsychopharmacology (chmn. com. on problems of public concern); mem. Inst. Medicine, Nat. Acad. Scis., AAAS, Am. Psychosomatic Soc., Soc. Biol. Psychiatry (A.E. Bennett award 1970), Am. Acad. Psychoanalysis, Soc. for Neuroscience, Psychiat. Research Soc., Washington Psychiat. Soc. (peer rev. com.). Club: Cosmos (Washington). Home: 5712 Warwick Pl Bethesda MD 20815-5502 Office: HHS NIMH 5600 Fishers Ln Rockville MD 20857

GOODWIN, IRWIN, magazine editor; b. Chgo., Aug. 19, 1929; s. Albert and Sarah Esther (Wallen) G.; m. Mary Margaret Revell, Apr. 21, 1966 (div. 1987). AB, Roosevelt U., Chgo., 1948; MA, U. Mich., 1949. Reporter City News Bur., Chgo., 1949-50; reporter, asst. editor Newsweek, Chgo. and N.Y.C., 1952-58; dir. pub. info. Sci. Rsch. Assocs., Chgo., 1958-60; corr. Newsweek, London, 1960-70; Caribbean corr. Washington Post, San Juan, P.R., 1970-72; spl. asst. to dir. Smithsonian Instn., Washington, 1972-73; sr. editor Nat. Acad. Scis., Washington, 1973-82; editor Washington bur. Physics Today, Washington, 1983-93, sr. editor Washington bur., 1993—. Co-author: Physics and Nuclear Arms Today, 1991; editor: Paying for America's Health Care, 1973, Energy and Environment: Collision of Crises, 1974; contbr. articles to profl. jours. Sgt. maj. U.S. Army, 1950-52. Recipient News Writing award Overseas Press Club, 1971, 72, Pub. Svc. Group Achievement award NASA, 1981. Mem. Nat. Press Club, AAAS, Nat. Assn. Sci. Writers, Fedn. Am. Scientists, D.C. Sci. Writers Assn., N.Y. Acad. Scis., Phi Beta Kappa. Office: Physics Today 1050 National Press Bldg 529 14 th St NW Washington DC 20045

GOODWIN, RICHARD CLARKE, military analyst; b. Hancock, Mich., Mar. 24, 1949; s. Robert Clement and Jean (Gibson) G.; m. Linda Wells, Oct. 30, 1971; children: Katherine E., James G. BS in Nuclear Engring., U.S. Mil. Acad., 1971; MS in Systems Mgmt., U. So. Calif., 1977; MS in Nuclear Engring., Air Force Inst. Technology, 1978; D of Pub. Adminstrn., U. Ala., 1991. Asst. prof., physicist Dept. Physics, U.S. Mil. Acad., West Point, N.Y., 1978-81; asst. flight comdr. 441 Bomb Squadron, KI Sawyer AFB, Mich., 1981-84; rsch. fellow CADRE/ACSC, Montgomery, Ala., 1984-85; dep. chief, combat ops. 2nd Bomb Wing/DOXX, Bossier City, La., 1986-87; rsch. fellow CADRE, Montgomery, 1987-88; chief strategy & policy bd. Hdqrs. USAF, Studies & Analysis, Washington, 1988-90; mil. advisor for theater affairs Arms Control & Disarmament Agy., Washington, 1990-91; ret. USAF, 1991; sr. rsch. specialist Logicon/RDA, Hdqrs. Shape, Mons, Belguim, 1991—; vis. asst. prof. U. So. Ill., Carbondale, 1987; presenter studies at profl. confs. Contbr. articles to profl. jours. Mem. Am. Nuclear Soc., Soc. Am. Mil. Engrs. (past pres. 1978), Am. Soc. Pub. Adminstrs., Mil. Ops. Rsch. Soc., Pi Sigma Alpha. Home: Chaussée de Grammont 6 Bis, 7860 Lessines Belgium Office: DNA Field Office Belgium CMR 451 APO AE 09708

GOODYEAR, JACK DALE, electronic educator; b. Tonopah, Nev., June 8, 1935; s. George William Goodyear and Alice Nevada Steinman; m. Shirley Esther Bartell, Oct. 4, 1957; children: Barbara, Barry, Burtan, Edward, George. BSVTE, Idaho State U., 1987, BSCT, 1987. Gen. mgr., CEO Intermountain Power Industry Personel, Blackfoot, Idaho, 1979-82; corp. recruiter Energy Inc., Idaho Falls, 1982-86; cons. Electronic Svcs., Blackfoot, 1986-88; rsch. electronic tech. U. Idaho, Aberdeen, 1988—; computer cons. Energy Utility Ops. Svcs., Ann Arbor, Mich., 1987; cons. Nuclear Svcs., Gaithersville, Md., 1987. Author software program: Resume/Applicant Tracking System, 1986; contbr. articles to profl. jours. With USN, 1957-79. Mem. IEEE, Early Day Gas Engine & Tractor Assn. #7 (v.p. 1990-91). Achievements include development of system for automatic-realtime control of sprinkler irrigation machines; developed real time electro-hydraulic system to control loading boom on potato harvestor; developed system for precious metal recovery from scrap metal. Home: 767 W 100 St S Blackfoot ID 83221

GOOKIN, WILLIAM SCUDDER, hydrologist, consultant; b. Atlanta, Ga., Sept. 8, 1914; s. William Cleveland and Susie (Jaudon) G.; m. Mildred Hartman, Sept. 4, 1937; children: William Scudder Jr., Thomas Allen Jaudon. BSCE, Pa. State U., 1937. Registered profl. engr. and hydrologist. Engr. U.S. Geol. Survey, Tucson, 1937-38; inspector City of Tucson, 1938-39; steel designer Allison Steel Mfg. Co., Phoenix, 1939-40; engr. Bur. Reclamation, various locations, 1940-53; chief engr. San Carlos Irrigation and Drainage Dist., Coolidge, Ariz., 1953-58; chief engr. Ariz. Interstate Stream Commn., Phoenix, 1956-62, state water engr., 1962-68; adminstr. Ariz. Power Authority, Phoenix, 1958-60; cons. engr. Scottsdale, Ariz., 1968—; mem. exec. com. Cen. Ariz. Project Assn., Phoenix, 1985—. Contbr. articles to profl. jours. Dem. committeeman State of Ariz., 1979-84; Ariz. mem. Com. of 14, Western States Water Coun.; episcopal lay reader. Served to 2d lt. C.E., U.S. Army, 1938-42. Fellow Am. Soc. Civil Engrs.; mem. NSPE (outstanding engr. project 1988), Nat. Water Resources Assn. (small projects com.), Colo. River Water Users' Assn., State Bar Ariz. (assoc.), Assn. Western State Engr. (pres.), Am. Legion, Culver Legion, Order of the Engr., Mason, Chi Epsilon. Home: 9 Casa Blanca Estates Paradise Valley AZ 85253-6919

GOOS, ROGER DELMON, mycologist; b. Beaman, Iowa, Oct. 29, 1924; s. Gus and Georgiana Bertha (Witt) G.; m. Mary Lee Engel, Sept. 21, 1946; children: Marinda Lee, Suzanne Maurine. B.A., U. Iowa, 1950, Ph.D., 1958. Mycologist United Fruit Co., Norwood, Mass., 1958-62; scientist USPHS, NIH, Bethesda, Md., 1962-64; curator of fungi Am. Type Culture Collection, Rockville, Md., 1964-68; assoc. researcher, vis. assoc. prof. botany U. Hawaii, Honolulu, 1968-70; assoc. prof. botany U. R.I., Kingston, 1970-72, prof. botany, 1972—, chmn. dept. botany, 1971-86; trustee Am. Type Culture Collection, Rockville, Md., 1977-82; vis. rschr. U. B.C., 1977, U. Hawaii, 1977, U. Exeter, U.K., 1984, Bishop Mus., 1990. Served with U.S. Army, 1944-46, 50-51. Decorated Bronze Star, Purple Heart; Indo-Am. fellow, U. Madras, India, 1981; Fulbright scholar U. Lisbon, 1993. Mem. AAAS, Mycol. Soc. Am. (sec.-treas. 1980-83, v.p. 1983-84, pres.-elect 1984-85, pres. 1985-86), Bot. Soc. Am., Am. Soc. Microbiology, Am. Phytopath. Soc., Mycol. Soc. Japan, Brit. Mycol. Soc., Mycol. Soc. India. Home: 4 Tanglewood Trl Narragansett RI 02882-1034 Office: U RI Dept Botany Kingston RI 02881-0812

GOOSKENS, ROBERT HENRICUS JOHANNUS, pediatric neurologist; b. Eindhoven, Brabant, The Netherlands, Nov. 4, 1948; s. Theo M. and Anne Elisabeth (Gimbrère) G. MD, U. Utrecht, The Netherlands, 1975, PhD, 1988. Resident in neurology Univ. Hosp. Utrecht, 1975-81, resident in clin. neurophysiology, 1981-82; rsch. child neurologist Univ. Children's Hosp., Utrecht, 1982-84, cons. child neurologist, 1984—; cons. child neurologist Psychiat. Child Clinic, Vught, The Netherlands, 1984—. Contbr. articles to med. jours. Rsch. grantee Prevention Fund The Netherlands, 1987. Mem. Dutch Child Neurology Soc. (bd. dirs. 1990—), Dutch Spina Bifida Soc. (sec. 1990—), Internat. Child Neurology Assn., Internat. Soc. Rsch. into Hydrocephalus and Spina Bifida, Internat. Cerebral Palsy Soc., European Ultrasound Soc. Roman Catholic. Avocations: skiing, golf, hockey, sculpture, travel. Home: Voorstraat 62B, 3512AR Utrecht The Netherlands Office: U Childrens Hosp Child Neur, Heidelberglaan 100, 3584CX Utrecht The Netherlands

GOPAL, PRADIP GOOLAB, engineer, environmental consultant; b. Bethal, Transvaal, South Africa, Mar. 18, 1957; came to U.S., 1980; s. Goolab and Ratan (Deyar) G.; m. Dana Jill, June 8, 1982; children: Devran, Kaila. BS, U. London, 1980; MS, U. Wis., Milw., 1983. Teaching asst. U. Wis., Milw., 1980-83; chemist Afrox, Johannesburg, South Africa, 1983-84; sr. engr. Internat. Paper Co., Atlanta, 1985—; bd. dirs. Imaging Sci. and Tech., Ga. Environ. Coun., both Atlanta; mem. PIA Environ. Com., Atlanta, 1992—. Author: Directory of Free Engineering Journals, 1991; contbr. to sci. and tech. jours., books and mags. Mem. air pollution task force Atlanta C. of C., 1993; mem. Atlanta Regional Devel. Coun. Grantee Lab. for Surface Studies, Milw., 1981, 82. Mem. Imaging Sci. and Tech. Assn., Ga. Environ. Coun. (bd. dirs. 1992—). Achievements include design of imaging products and pollution reduction procedures for imaging industry.

GOPAL, RAJ, energy systems engineer; b. Bangalore, Mysore, India, Feb. 1, 1942; came to U.S., 1969; s. Muthuswamy and Saraswathi (Swaminathan) Ramamoorthy; m. Parvathi Bhaskar, June 11, 1967; children: Mira, Neil. BSME, Coll. Engring., Bangalore, 1963; MSME with distinction, Indian Inst. of Tech., 1965; PhDME, U. Akron, 1973. Registered profl. engr. Wis. Mgr. prodn. Hindustan Ferodo Ltd., Bombay, 1965-69; sr. rsch. engr. Johnson Controls Inc., Milw., 1973-81, sr. rsch. scientist, 1981-84; pres., owner Engring Cons. Svcs., Milw., 1984-87; sr. project engr. The Anco Cons. Group, Milw., 1987-88, mgr. R&D, 1988—; cons. Johnson Controls Inc., Milw., 1984-85, MCC Powers, Northbrook, Ill., 1985-86; lectr. mech. engring. U. Wis., Milw., 1986-88. Contbr. articles to profl. jours. Vol. YMCA, Brown Deer, Wis., 1989-90, Wis. Electrical, Milw., 1988. Named Competent Toastmaster Toastmasters Internat., 1982. Mem. ASHRAE (vice chmn. TC4-6 bldg. operation dynamics), Soc. Am. Mech. Engrs. (K-6 com. 1976-80), Assn. Energy Engrs. (session chmn. 1990-92), Instrument Soc. Am., Pi Mu Epsilon, Sigma Xi. Achievements include patents for solar energy control system, thermal energy storage apparatus; research in work on heating and air condidtioning system self dianoser helps detect faults well before unacceptable failure mode, sensor diagnostics energy optimization routines run only when reliable data available, cool-storage studies—potential to shift over 5 megawatt peak demand to Wisconsin Electric in off-peak hours, energy management control systems application software. Home: 6267 W Silver Brook Ln Brown Deer WI 53223 Office: Anco Consulting Group Inc 231 W Michigan P141 Milwaukee WI 53203

GOPALAKRISHNAN, BHASKARAN, manufacturing engineer; b. Tiruchendurai, Tamil Nadu, South India, Aug. 27, 1960; came to U.S., 1983; s. Gopala and Lalitha (Ramanathan) Bhaskaran; m. Uma Gururajan, Aug. 22, 1986. B in Engring. with honors, U. Madras, TN, India, 1983; MS, So. Meth. U., 1985; PhD, Va. Tech, 1988. Asst. prof. dept. indsl. engring. W.Va. U., Morgantown, 1988—. Contbr. articles to profl. jours. Recipient TAG award for Outstanding Undergrad. Student, 1983, Jawaharlal Nehru award for Outstanding Undergrad. Student, J.N. Meml. Fund, 1983. Mem. Sigma Xi, Omega Rho, Alpha Pi Mu. Hindu. Achievements include definition of CNC Machinability; devel. of machining advisor for concurrent engring.; devel. expert systems and methods for design for mfg. Office: WVa U Dept Indsl Engring PO Box 6101 Engring Scis Morgantown WV 26506-6101

GOPALAN, MUHUNDAN, software engineer; b. Karaikudi, India, Mar. 15, 1960; came to U.S., 1984; s. Srinivasan and Lakshmi G.; m. Subhashini Muhundan, Oct. 31, 1988. B of Tech., Indian Inst. Tech., 1982; M of Engring., Asian Inst. Tech., 1984; MS, Ariz. State U., 1985. Sr. engr. Auto-Trol Tech., Denver, 1986-88; advanced engring. systems engr. Elec. Data Systems, Troy, Mich., 1988—. Ariz. Bd. Regents scholar, 1984-85, Govt. of India Merit scholar, 1977-82; Govt. of Australia fellow, 1983-84. Mem. Soc. Mfg. Engrs., Phi Kappa Phi. Achievements include new surface triangulation, triming techniques, feature based creation techniques, surface modeling and algorithms. Office: Elec Data Systems 750 Tower Dr Troy MI 48007

GOPPELT, JOHN WALTER, physician, psychiatrist; b. Saginaw, Mich., Jan. 20, 1924; s. Paul Gustave and Marion LeRoy (Payne) G.; m. Martha Keller Rowland, Mar. 31, 1956; 1 child, Edmund H. S.B., MIT, 1949; M.D., U. Pa., 1955. Diplomate Am. Bd. Psychiatry and Neurology. Rotating intern Bryn Mawr Hosp., Pa., 1955-56; resident in psychiatry Inst. of Pa. Hosp., Phila., 1956-59; practice medicine, specializing in psychiatry, Haverford, Pa., 1959—; from instr. to assoc. dept. psychiatry Sch. Medicine, U. Pa., Phila., 1960-74. Contbr. articles to profl. jours. Chmn. Drug and Alcohol Council of Delaware County, Media, Pa., 1979-83; committeeman Republican Party, Haverford Twp., Pa., 1980. Served with U.S. Army, 1943-46. Recipient Legion of Honor award Chapel of Four Chaplains. Mem. AMA, Am. Psychiat. Assn., N.Y. Acad. Scis., Math. Assn. Am., Sigma Xi. Avocation: mathematics. Address: 369 Exeter Rd Haverford PA 19041

GORANSON, HARVEY EDWARD, fire protection engineer; b. Nashville, Oct. 27, 1952; s. Harvey Edward and Alice Isabelle (Dealy) G.; m. Linda Kay Bennett, May 19, 1984; children: Matthew Alan, Laura Kacyn; 1 stepchild, Ronald Edward Pruitt. BS in Fire Protection Engring., Ill. Inst. Tech., 1974; MBA, U. Tenn., 1979. Registered profl. engr., Tenn. Engring. rep. Ins. Svcs. Office of Nebr., Omaha, 1974-76, Ins. Svcs. Office of Tenn., Nashville, 1976-80; sr. fire protection cons. Wausau Ins. Cos., Nashville, 1980-84; sr. fire protection engr. Profl. Loss Control, Inc., Kingston, Tenn., 1984—. Author: (with others) Fire Protection Management for Hazardous Materials, 1991. Mem. NSPE, Soc. Fire Protection Engrs. (chpt. pres. 1989-91), Nat. Fire Protection Assn. Office: Profl Loss Control Inc 1 Locomotive Dr Kingston TN 37763

GORBATY, MARTIN LEO, chemist, researcher; b. Bklyn., Nov. 17, 1942; s. Julius and Florence (Birnbach) G.; m. Dianne Morse, June 30, 1968; children: Howard M., Matthew J., Lisa R. BS in Chemistry with honors, City Coll. of N.Y., 1964; PhD in Organic Chemistry, Purdue U., 1969. Rsch. chemist Esso Agrl. Products Lab. Esso Rsch. and Engring. Co., 1969-70; sr. rsch. chemist Corp. Rsch. Lab. Exxon Rsch. and Engring. Co., 1970-73, sr. rsch. chemist Baytown R & D divsn., 1973-75, group head Corp. Rsch. Labs., 1975-78, lab. dir. corp. rsch., 1978-84, sr. rsch. assoc. Corp.

Rsch.-Resource Chemistry Lab., 1984—; mem. internat. editorial bd. Fuel, 1983—; chmn. Gordon Conf. Fuel Sci., 1988. Editor, 5 books on synthetic crudes and coal sci.; contbr. 50 articles to profl. jours. Recipient R.A. Glenn award Bituminous Coal Rsch., Inc., 1990, Disting. Alumnus award Sch. of Sci. Purdue U., 1993. Mem. AAAS, Am. Chem. Soc. (chmn. divsn. petroleum chemistry 1983-84, program com. 1978—, councilor 1988—, divsn. fuel chemistry, adv. bd. ACS books 1984-87, editorial bd. Chemtech 1986—, Henry H. Storch award 1993), N.Y. Acad. Scis., Soc. Sigma Xi, Phi Lambda Upsilon. Achievements include invention of 4 new classes of highly active broad spectrum insecticides, inexpensive cosolvent system for producing high diene butyl rubber economically, new class of low molecular weight high diene isobutylene based polymers and sulfonated derivatives useful for coatings and thermoplastic elastomers, first catalyst recovery scheme for Exxon's catalytic coal gasification process; evaluation of potential environmental issues for a coal gasification process; invention and development of solution for critical problem during coal liquefaction; development and implementation of strategic research plan that led to several new propriety approaches for utilization of heavy oils, oil shale and coal; research on building structure-reactivity relationships for coals, oil shales and heavy petroleum; 30 patents, 5 patents pending. Office: Exxon Rsch & Engring Co PO Box 998 Annandale NJ 08801-0998

GORDIS, JOSHUA HAIM, mechanical engineer, researcher; b. N.Y.C., Dec. 7, 1960; s. Enoch and Lucille (Sapirstein) G. BSME, U. Vt., 1983; MS and PhD in Mech. Engring., Rensselaer Poly. Inst., 1990. Support engr. Stone & Webster Engring. Corp., Boston, 1984-85; project engr. Structural Dynamics Rsch. Corp., San Diego, 1990-91; mem. faculty Miracosta Community Coll., Oceanside, Calif., 1992, San Diego State U., 1992; asst. prof. mech. engring. Naval Postgrad. Sch., Monterey, Calif., 1992—; cons. Lord Corp., Cary, N.C., 1990, Dr. Larry Bank, Troy, N.C., 1990. Contbr. articles to Jour. Vibration, Acoustics, Stress and Reliability, Jour. Sound and Vibration, Proceedings 10th and 11th Internat. Modal Analysis Conf., Proceedings 34th AIAA/ASME/ASCE/RHS/ASC/SDM Conf., Jour. Am. Helicopter Soc. Disting. fellow Army Rsch. Office, 1988-90. Mem. AIAA, ASME, Am. Helicopter Soc. (Robert L. Lichten award 1990), Sigma Xi. Achievements include design of exact frequency domain theory for linear system identification with frequency dependent properties, unified and generalized theory of frequency domain structural synthesis in structural dynamics. Office: Naval Postgrad Sch Monterey CA 93943-5100

GORDON, CRAIG JEFFREY, oncologist; b. Detroit, Feb. 10, 1953; s. Maury Allen and Shirley Phoebe (Jacoby) G.; m. Susan Ann Blase, Aug. 3, 1980; children: Sari, Scott, Brittany. BS, Oakland U., 1978; DO, U. Osteopathic Medicine and, Health Scis./Des Moines, 1983. Diplomate Am. Bd. Internal Medicine, Am. Bd. Med. Oncology. Intern-chief Botsford Gen. Hosp., Farmington Hills, Mich., 1983-84, resident, 1984-87; fellow in hematology and oncology Wayne State Univ. (affiliated Hosp.'s Prog.), Detroit, 1987-90, fellow-chief, 1989-90; clin. asst. prof. dept. medicine Wayne State U., Detroit, 1990—; dir. divsn. hematology and oncology Botsford Hosp., Livonia, Mich., 1992—. Contbr. articles to profl. jours. Named Intern of the Yr. Botsford Hosp. Staff, 1984, Resident of the Yr., 1985-87; clin. fellow Am. Cancer Soc., 1987-90. Mem. Am. Osteopathic Assn., Am. Coll. Osteopathic Internists, Mich. Assn. Osteopathic Physicians and Surgeons, So. Med. Assn., Southwest Oncology Group, Am. Soc. Clin. Oncologists. Avocations: sports, popular music, astronomy, electronics. Office: Botsford Gen Hosp 28711 W Eight Mile Rd Ste E Livonia MI 48152

GORDON, EDMUND WYATT, psychologist, educator; b. Goldsboro, N.C., June 13, 1921; s. Edmund Tayloe and Mabel (Ellison) G.; m. Susan Elizabeth Gitt, Nov. 6, 1948; children: Edmund T., Christopher W., Jessica G., Johanna S. BS, Howard U., 1942; MA, Am. U., 1950; EdD, Columbia U., 1957; MA (hon.), Yale U., 1979; LHD (hon.), Yeshiva U., N.Y.C., 1986, Brown U., 1989, Bank St. Coll., 1992. Asst. dean men Howard U., Washington, 1946-50; from assoc. prof. to prof. Yeshiva U., N.Y.C., 1961-68; prof., chmn. dept. guidance Columbia U., N.Y.C., 1968-78, Richard March Hoe prof. psychology and edn., 1978-79; John M. Musser prof. emeritus of psychology Yale U., New Haven, 1979-91; prof. ednl. psychology CUNY, N.Y.C., 1992—. Author: Compensatory Education for the Disadvantaged: Programs and Practices, 1966, Education and Social Justice, 1994; editor: Equality of Educational Opportunity: Handbook for Research, 1974, Human Diversity and Pedagogy, 1989; editor Am. Jour. of Orthopsychiatry, 1978-83, Rev. of Rsch. in Edn., 1983-85; contbr. articles to N.Y. State Jour. of Medicine, Am. Jour. of Mental Deficiency, Am. Zoologist, Jour. of Genetic Psychology, Am. Child, and others. Pres. Rockland County NAACP. Fellow AAAS, APA, Am. Orthopsychiatric Assn., Am. Psychol. Soc.; mem. Am. Ednl. Rsch. Assn., Nat. Assn. Black Psychologists, Nat. Acad. Edn., N.Y. Acad. Sci. Democrat. Achievements include research in human diversity, cultural hegemony, and integrity, culture and cognitive development and education of low status populations; responsible for founding and directing research for Project Head Start, 1965-67. Home and Office: 3 Cooper Morris Dr Pomona NY 10970

GORDON, GEORGE STANWOOD, JR., physicist; b. Lynn, Mass., Jan. 8, 1935; s. George Stanwood and Alma (Langell) G.; m. Clarice Barrett, June 18, 1960; children: Kent, Kari, Glen, Wayne, Maurine. BA, Columbia U., 1959; PhD, MIT, 1962. Staff MIT, Cambridge, 1962—. 1st lt. U.S. Army, 1962-63. Home: 22 Irving St Arlington MA 02174 Office: MIT 77 Massachusetts Ave Cambridge MA 02139

GORDON, HELMUT ALBERT, biomedical researcher, pharmacology educator; b. Malinska, Austria, May 5, 1908; came to U.S., 1946; s. Albert John and Cornelia Leopoldina (de Adamich) Gordon-Königes; m. Irene Julianna Rontskevits, Nov. 28, 1942; children: Iretta Celta, Brent Helmut. MD, Med. Schs., Budapest, Hungary, Rome, 1932; Habilitation, Med. Sch., Budapest, 1944. Med. diplomate. Physiologist Hungarian Air Force, Budapest, 1934-44; from asst. prof. to assoc. prof. dept. physiology Med. Sch., U. Budapest, 1934-44; med. officer U.S. Mil. Govt., Wolfratshausen, Germany, 1945-46; assoc. prof. U. Notre Dame (Ind.), 1946-62; prof. pharmacology U. Ky., Lexington, 1962-78, emeritus prof., 1978—. Author: Clinical and Experimental Gnotobiotics, 1979, Recent Advances in Germfree Research, 1981, The Germfree Animal in Biomedical Reasearch, 1984; contbr. articles to profl. jours. Capt. Med. Corps, Hungarian mil., 1940-44. Recipient Eszterhazy award, Budapest, 1932, N.Y. Acad. Scis. Cressey-Morrison award, 1964. Mem. Assn. for Gnotobiotics, Am. Physiol. Soc., N.Y. Acad. Scis., Tutukaka South Pacific Yacht Club (Whangarei, New Zealand, charter mem.). Lutheran. Achievements include research in aviation medicine, host-microbial flora relationships, and marine environment.

GORDON, JACQUELINE REGINA, pharmacist; b. N.Y.C., Jan. 14, 1964; d. Reginald Augustas and Helen Isadora (Cohen-Pearl) G. BS in Pharmacy, Long Island U., 1987. Lic. Pharmacist, N.Y., N.J. Oncology pharmacist Dept. Vets. Affairs, N.Y.C., 1989-90. Mem. Student Nat. Pharm. Soc., Am. Pharm. Soc. Roman Catholic.

GORDON, JAMES POWER, optics scientist; b. N.Y.C., Mar. 20, 1928. BS, MIT, 1949; MA, Columbia U., 1951, PhD in Physics, 1955. Asst. physics dept. Columbia U., 1953-55; mem. tech. staff electronics rsch. AT&T Bell Labs., Murray Hill, N.J., 1955-59, head quantum electronics rsch. dept., 1959-80, sr. tech. staff cons., 1980—. Recipient Max Born award Optical Soc. Am., 1991. Fellow Am. Phys. Soc.; mem. IEEE (sr.), Nat. Acad. Engring. Achievements include research in quantum electronics, interaction of electromagnetic waves with matter, communication theory. Office: AT&T Bell Labs Research Lab 600 Mountain Ave Murray Hill NJ 07974*

GORDON, JOSEPH WALLACE, nuclear engineer; b. Parkersburg, W.VA., Dec. 30, 1957; s. Joseph Wallace and Elizabeth (Carroll) G. BS in Engring. Physics, U. Colo., 1981. Asst. engr. Gibbs & Hill, Inc., N.Y.C., 1981-83; cons. engr. Three Mile Island, Middletown, Pa., 1983-85; project engr. Dames & Moore, Pearl River, N.Y., 1985-88; sr. engr. Nuclear Energy Svcs., Danbury, Conn., 1988-90; mgr. safety analysis and risk assessment Dames & Moore, Denver, 1990—. Contbr. articles to profl. publs. Mem. ASTM (decommissioning standards com.), Am. Nuclear Soc., Health Physics Soc., Soc. for Risk Analysis. Home: 916 Pine St Boulder CO 80302 Office: Dames & Moore 1125 17th St Denver CO 80202

GORDON, MARK ELLIOTT, environmental engineer; b. Janesville, Wis., Mar. 18, 1956; s. Robert Elliott and Mary Eileen (Carolan) G.; m. Debra Bystrom, June 8, 1985; children: Martin Patrick Gordon, Erin Elizabeth Gordon. BS, U. Wis., 1979. Environ. engr. Wis. Dept. Natural Resources, Madison, 1979-84, engring. supr., 1984—. Contbr. articles to profl. jours. Named Employee of Yr. Solid and Hazardous Waste Program, 1992. Mem. ASCE. Home: 7836 Caribou Ct Verona WI 53593 Office: Dept Natural Resources 101 S Webster St Madison WI 53707-7921

GORDON, RICHARD EDWARDS, psychiatrist; b. N.Y.C., July 15, 1922; s. Richard and Virginia (Ryan) G.; m. Katherine Lowman Kline, Nov. 12, 1949; children: Richard Edwards, Katherine Lowman Gordon Reed, Virginia Lamborn Gordon Ford, Laurie Lloyd Gordon. B.S., Yale U., 1943; M.D., U. Mich., 1945; M.A., Columbia U., 1956, Ph.D., 1961. Diplomate: Am. Bd. Psychiatry and Neurology. Intern City Hosp., N.Y.C., 1945-46; resident in neurology N.Y. Postgrad Hosp., N.Y.C., 1946-47; resident in psychiatry N.Y. Psychiat. Inst., N.Y.C., 1947-48, Manhattan (N.Y.) State Hosp., 1948-49; fellow in psychosomatic medicine and child psychiatry Mt. Sinai Hosp., N.Y.C., 1949-51; practice medicine specializing in psychiatry N.Y.C., 1950-51, Englewood, N.J., 1953-67; mem. staffs Univ. Settlement House, 1950-51, Englewood Hosp., 1953-67, Shands Teaching Hosp., Gainesville, Fla., 1967—, Gainesville VA Hosp., 1967-76; sr. research psychiatrist, EEG cons. Rockland State Hosp., Orangeburg, N.Y., 1953-54; founder, dir. EEG Clinic Englewood Hosp., 1953, dir. rsch. unit, 1954-60; prof. psychology, cons. psychiatrist Wagner Coll., S.I., N.Y., 1960-67; prof. psychiatry and psychology, dir. Fla. Mental Health Inst., Tampa, 1974-79; assoc. prof. psychiatry, rsch. dir. Multiphasic Health Testing Ctr., U. Fla., Gainesville, 1967-87, emeritus prof. psychiatry Coll. Medicine, 1987—; adj. prof. clin. psychology U. South Fla., Tampa, 1977—; founder Mental Health Consultation Center, Hackensack, N.Y., 1956, trustee, 1956-57; founder Community Multiphasic Health Testing Center, Gainesville; mem. N.J. Mental Health Commn., 1957-61; bd. dirs. Biosystems, Inc., Reed Curve, Inc., Applied Digital Tech. Author: Prevention of Postpartum Emotional Difficulties, 1961, (with K.K. Gordon, M. Gunther) The Split-Level Trap, 1961, (with K.K. Gordon) The Blight on the Ivy, 1963, Systems of Treatment for the Mentally Ill: Filling the Gaps, 1981, (with B. Franklin et al) Towards Better Mental Health in New Jersey, 1961, (with C.J. Hursch, K.K. Gordon), Introduction to Psychiatric Research, 1988; contbr. numerous articles to profl. jours. Pres. Kirkwood Environ. Improvement Assn., Gainesville, 1970-75; cons./surveyor Joint Commn. on Accreditation of Hosps., 1980-84. Served to capt. AUS, 1943-45, 51-53. Grantee in field. Fellow Am. Psychiat. Assn. (del. to assembly), Soc. Advanced Med. Systems; mem. AAAS, Fla. Psychiat. Soc. (pres. 1978-79, newsletter editor), Sigma Xi. Club: Yale of Gainesville.

GORDON, RICHARD WARNER, naval propulsion engineer; b. Winchester, Va., May 20, 1969; s. James Carlyle and Abbye Caroline (Davis) G. BS in Ocean Engring., U.S. Naval Acad., 1991. Commd. ensign U.S. Navy, Annapolis, 1991; coastal engring. asst. U.S. Army Corps. Engrs., Duck, N.C., 1990; coastal engring researcher Naval Systems Div. USN, Annapolis, 1991, naval radiation dosimetry researcher, 1991; naval propulsion engr. U.S. Navy, Charleston, S.C., 1992—. Mem. Ruritan Club, Tau Beta Pi, Sigma Xi. Republican. Achievements include investigation of previous rsch. on offshore rubble mound breakwaters with examination of orientation effects on steep sloped beaches with predominating longshore current; reported previous work did not allow for extreme condition. Home: 524 Orchard Ln Winchester VA 22602

GORDON, RONNIE ROSLYN, pediatrics educator, consultant; b. N.Y.C., June 29, 1923; d. Maurice and Margaret (Leizer) Klein; divorced; children: Lyn Leslie, Barbara Margaret Gordon Simmons. BA In Math., Hunter Coll., 1944; MS in Child Devel., Bank St. Coll., 1961; postgrad., NYU. Mathematician Manhattan Project, Los Alamos, 1944-51; project dir. Rusk Inst. Rehab. Medicine NYU Med. Ctr. HEW Dept. Edn., 1969-78; dir. presch. and infant devel. programs Katherine Lilly Conroy Learning Lab. Infant Sch. Therapeutic Playground, 1962-85; asst. prof. clin. rehab. medicine Sch. Medicine NYU, 1970-74, assoc. prof. clin. rehab. medicine, 1974-86; dir. presch. and infant svcs. and rsch. dept. Inst. Rehab. Medicine NYU Med. Ctr., 1962-86; cons. pvt. practice N.Y.C., 1986—; presented at confs., seminars, and workshops at univs., annual meetings of nat. and internat. related assns. in Eng., Wales, France, Norway, Germany, Austria, Nova Scotia, Venezuela, Trinidad, Denmark, Sweden, Israel, Can., Netherlands, India, China, Hong Kong, Hawaii, Yugoslavia, Japan. Filmmaker: Special Children, Special Needs, 1972 (1st prize award), Special Children, Different Needs-Growing Up Handicapped, 1980 (award), Special Children 20 Years Later; author: (with others) Intervention Strategies for High Risk Infants and Young Children, 1974, Educational Programming for the Severely and Profoundly Handicaped, 1977; contbr. articles to Developmental Medicine and Child Neurology, Jour. Spl. Edn., Am. Sch. and Univ., Archtl. Forum, others. Bd. dirs. Louise Wise Svcs. for Children, 1984—, subcom. preventive svcs. Recipient Contbn. award Hong Kong Rehab. Ctr., 1982, award State of N.Y. Dept. of Parks, 1973-74; grantee Evan F. Lilly Found., 1966-68, HEW, 1970-72, 72-75, 75-78. Mem. World Orgn. for Presch. Edn. (N.Y.C. liaison for the U.S. Nat. Com.), Assn. for Children with Learning Disabilities, Assn. for Childhood Edn. Internat., Am. Soc. of Law and Medicine. Achievements include patents for custom design of adapted equipment for three therapeutic learning labs.; rsch. in interaction of handicapped children in mainstreamed setting and evaluation of behavioral change of children and parents. Home and Office: 3 Washington Sq Village 14T New York NY 10012

GORDON, ROY GERALD, chemistry educator; b. Akron, Ohio, Jan. 11, 1940; s. Nathan Gold and Frances (Teitel) G.; m. Myra Sheila Miller, Dec. 24, 1961; children: Avra Karen, Emily Francine, Steven Eric. A.B. summa cum laude, Harvard, 1961, A.M. in Physics, 1962, Ph.D. in Chem. Physics, 1964. Jr. fellow Soc. of Fellows, Harvard, 1964-66, mem. faculty, 1966—, prof., 1969—. Sloan Found. fellow, 1966-69, Einstein fellow, Israel, 1985. Fellow Am. Phys. Soc.; mem. Am. Chem. Soc. (award in pure chemistry 1972, Baekeland award 1979) R & D award 1991, Faraday Soc., Union of Concerned Scientists, NAS, Am. Acad. Arts and Scis., Phi Beta Kappa, Sigma Xi. Achievements include inventions in solar energy, energy conservation and microelectronics, theoretical research discovering forms of forces between molecules, the way molecules collide with each other, motion of molecules in liquids and solids. Office: Harvard U Dept Chemistry Cambridge MA 02138

GORDON, STEVEN B., chemist; b. Brookline, Mass., Nov. 10, 1948; s. Samuel N. and Dora I. (Resnick) G.; m. Cathy J. Hopewell, Oct. 20, 1973; children: Robin N., Jeffrey A. AB, Boston U., 1970; MA, Clark U., 1973. Chemist Reed Plastics Corp., Holden, Mass., 1972-74, adminstr. tech. svcs., 1974-80, corp. tech. mgr., 1980-84, tech. dir., 1984-88; dir. tech. ReedSpectrum div. Sandoz Chem. Corp., Holden, Mass., 1988—. Mem. Am. Chem. Soc., Am. Soc. of Textile Chemists and Colorists, Soc. Plastics Engrs. (tech. program com. 1976-79). Achievements include contributions to plastics coloring technology. Office: Sandoz Chem Corp ReedSpectrum Divsn Holden Indsl Park Holden MA 01520

GORDON, STEVEN JEFFREY, mechanical engineer; b. Phila., Feb. 1, 1958; s. Samuel Leopold and Phyllis (Gudis) G.; m. Mary Theresa Zuccarini, Dec. 2, 1989. BS, MIT, 1979, MSME, 1982, PhD, 1986. Engr. Digital Equip. Corp., Maynard, Mass., 1979-81; cons. GE Co., Somersworth, N.H., 1984, McKinsey & Co., Cleve., 1987; pres. Intelligent Automation Systems, Cambridge, Mass., 1987—. Contbr. articles to profl. jours. Achievements include 3 patents in field. Office: Intelligent Automation Sys 142 Rogers St Cambridge MA 02142-1024

GORE, JAMES WILLIAM, civil engineer; b. Cairo, Ill., Mar. 20, 1953; s. James Fletcher and Cora Irene (Matson) G.; m. Sandra Lee King, May 20, 1978; children: Brandon James, Lindsay Lee. BA in Econs., U. of the South, 1975; BCE, Ga. Tech., 1975; MBA, U. Ark., 1978. Registered profl. engr., Ala., Ark., N.C., S.C., Tenn., Mo., Va., W.Va., Ky., Ohio. Dir. civil and geotech. engr. Cooper Cons. divsn. Cooper Comm., Bella Vista, Ark., 1978-86; asst. office mgr. MCI Consulting Engrs., Huntsville, Ala., 1987-88; office mgr. The Edge Group, Charlotte, N.C., 1988-89; mgr. engring. svcs. Woolpert Cons., Charlotte, 1989-91; prin. engr. Huntsville (Ala.) Engring. and Land Surveying, Inc., 1991—. Prin. works include Tellico Village, Tenn., Lake Balboa Dam, Hot Springs Village, Ark.; prin. engr. storm drainage infrastructure, Rsch. Park West, Huntsville. Vestry St. Andrews episcopal Ch., Rogers, Ark., 1972.; mem. stormwater mgmt. bd. City of Huntsville, 1993; mem. gov. affairs com., HMCBA, 1993—. Named Eagle Scout Boy Scouts Am., 1967; recipient award for Excellence in Engring. Design and Constrn. Soc. Am. Mil. Engrs., 1984. Mem. ASCE (coun. chmn. dist. 14 1982-83, pres. Ark. sect. 1983-84), Consulting Engring. Coun. (Ala. and N.C. chpts., transp. com. 1990), Tau Beta Pi, Chi Epsilon. Home: 227 Pin Oak Dr Madison AL 35758 Office: Huntsville Engring and Land Surveying Inc 1218-B Church St Huntsville AL 35801

GORELICK, JEFFREY BRUCE, physician, educator; b. N.Y.C., Aug. 24, 1955; s. George and Harriet (Ferber) G.; m. Lisa Hope Greenfield, Sept. 5, 1982; children: Ariel Lynn, Lindsay Elyse. BA in Chemistry, SUNY, Binghamton, 1977; MD, Med. Coll. Wis., 1981. Diplomate Am. Bd. Phys. Medicine and Rehab. Asst. prof. Med. Coll. Wis., Milw., 1981—; med. dir. brain injury program N.W. Rehab. Ctr., Milw., 1990—; med. dir. outpatient physical med. Affiliated Health, Milw., Wis.; surveyor Commn. Accreditation Rehab. Facilities, Tucson, 1992—. Home: 7360 N Seneca Rd Milwaukee WI 53217 Office: Affiliated Health Outpatient Physical Med 2626 N 76th St Milwaukee WI 53213

GOREN, HOWARD JOSEPH, biochemistry educator; b. Bialocerkwe, Ukraine, Apr. 9, 1941; came to Can., 1940's; s. Morris Mordechai and Bracha (Nissenbaum) G.; m. Frances Claire, Sept. 18, 1965; children—Robyn Pearl, Jeffrey Michael. B.Sc., U. Toronto, Ont., Can.; Ph.D., SUNY, Buffalo. Postdoctoral fellow Weizman Inst, Rehovot, Israel, 1968-70; asst. prof. U. Calgary, Alta., Can., 1970-75, assoc. prof., 1975-82, prof., 1982—, chmn. univ. biochemistry group, 1987-91; vis. scientist NIH, Bethesda, Md., 1977-78; vis. prof. Harvard Med. Sch., Boston, 1984-85, U. Calif. San Francisco, 1992-93. Asst. editor Molecular Pharmacology, 1974-77; contbr. articles to profl. jours. Mem. Am. Soc. Biochemistry and Molecular Biology, Am. Soc. Pharmacology and Exptl. Therapeutics, Can. Biochem. Soc., Am. Diabetes Assn., The Protein Soc. Avocations: running, basketball, sports. Office: U Calgary, 3330 Hospital Dr, Calgary, AB Canada T2N 4N1

GORENBERG, NORMAN BERNARD, aeronautical engineer, consultant; b. St. Louis, May 18, 1923; s. Isadore and Ethel Gorenberg; m. Lucille Richmond, June 10, 1947; children: Judith Allyn Gorenberg Stein, Carol Ann, Gershom. BSME, Washington U., St. Louis, 1949. Registered profl. engr., Mo. Aero. engr. USAF Wright Air Devel. Ctr., Dayton, Ohio, 1949-51; aerodynamicist McDonnell Aircraft Corp., St. Louis, 1951-59; supervisory engr. Boeing Co., Vertol Div., Phila., 1959-62; R & D engr. Lockheed Corp., Burbank, Calif., 1962-89; vertical takeoff and landing aircraft cons. Dana Point, Calif., 1989—. Contbr. articles to profl. reports. With USAAF, 1943-46. Mem. AIAA, ASME, Am. Helicopter Soc. (chmn. St. Louis sect. 1955-56, nat. aerodyns. com. 1969-70, tech. dir. western region 1969-70), Nat. Mgmt. Assn. (life). Jewish.

GORENSTEIN, DAVID G., chemistry educator; b. Oct. 6, 1945; s. Ben and Shirley (Adelberg) G.; m. Deborah H. Joseph, June 11, 1967; 1 child, Jennifer. BS in Chemistry, M.I.T., 1966; MA in Chemistry, Harvard U., 1967, PhD in Chemistry, 1969. Asst. prof. U. Ill., Chgo., 1969-73, assoc. prof., 1973-76, prof., 1976-85; prof. chemistry Purdue Univ., West Lafayette, Ind., 1985—; dir. Purdue Biochem. MRI Lab., West Lafayette, Ind., 1985—, NSF Nat. Biol. Facilities Ctr., West Lafayette, 1987—, NMR and Structural Biology Cores, West Lafayette, 1988—; dep. dir. NIH Designated AIDS Rsch. Ctr., West Lafayette, 1993—; vis. assoc. prof. U. Wis., Madison, 1975; vis. prof. Oxford U., 1977-78, U. Calif., San Francisco, 1986; cons. Baxter Travenol, 1985—, Merck and Co., 1988, Eli Lilly, 1987-89, Ill. Tool Works, 1973-85, Chronomatic Inc., 1973-85, U.S. Dept. of Labor, 1975, Continental Group, Inc., 1982-84; active numerous univ. coms.; lectr. in field. Editor Bull. of Magnetic Resonance, 1982—; mem. editorial bd. Magnetic Resonance Revs., 1983-93, Jour. Magnetic Resonance, 1992—, Biophys. Jour., 1992—; pub. abstracts; contbr. articles to profl. jours. Grantee NSF, 1987-93, NIH, 1988—, Eli Lilly, 1988—, Genta Corp, 1989—, and numerous others; Teaching fellow Harvard U., 1966-69, Trainee Summer fellow NSF, 1966, Predoctoral fellow NIH, 1967-69, Alfred P. Sloan fellow, 1975-79, Sr. Rsch. fellow Fulbright, 1977-78, Guggenheim fellow, 1986; recipient Internat. Lectr. award Fulbright, 1978. Mem. Am. Soc. for Biochemistry and Molecular Biology, Am. Chemical Soc. (program chmn. divsn. biol. chemistry 1985-87, vice chmn. Purdue sect. 1990-91, chmn. 1991-92), Midwest NMR Group, Internat. Soc. of Magnetic Resonance (organizing com. VIIIth meeting 1983, exec. bd. XIth meeting 1992), Biophysical Soc., Protein Soc., Sigma Xi, Phi Lambda Upsilon. Achievements include patent in Process for Preparing Dithiophosphate Oligonucleotide Analogs via Nucleoside Thiophosphoramidite Intermediates; research in applications of NMR spectroscopy and other physical techniques to biological systems, theoretical organic and bio-organic chemistry, biomolecular design; development of high quality refined NMR solution structures of biopolymers, NMR method for understanding the detailed structure of proteins bound as a monolayer to a solid support and the detailed structure of proteins in non-aqueous media. Office: Purdue U Dept Chemistry West Lafayette IN 47907

GORHAM, EVILLE, scientist, educator; b. Halifax, N.S., Can., Oct. 15, 1925; s. Ralph Arthur and Shirley Agatha (Eville) G.; m. Ada Verne MacLeod, Sept. 29, 1948; children: Kerstin, Vivien, Jocelyn, James. BSc in Biology with distinction, Dalhousie U., 1945, MSc in Zoology, 1947, LLD (hon.), 1991; PhD in Botany, U. London, Eng., 1951; DSc (hon.), McGill U., 1993. Lectr. botany U. Coll., London, Eng., 1951-54; sr. sci. officer Freshwater Biol. Assn., Ambleside, Eng., 1954-58; lectr., asst. prof. botany U. Toronto, 1958-62; assoc. prof. botany U. Minn., Mpls., 1962-65; prof. U. Minn., 1966-84, head dept., 1967-71, prof. ecology, 1975-84, Regents prof. ecology and botany, 1984—; prof., head dept. biology U. Calgary, Alta., Can., 1965-66; Mem. for Can. Internat. Commn. on Atmospheric Chemistry and Radioactivity, 1959-62; mem. Scientists Inst. for Pub. Info., 1971—, fellow, 1972—; mem. vis. panel to review toxicology programs Nat. Acad. Scis.-NRC, 1974-75; mem. coordinating com. for sci. and tech. assessment of environ. pollutants Environ. Studies Bd., 1975-78; mem. com. on med. and biologic effects of environ. pollutants Assembly Life Scis., 1976-77; mem. com. on atmosphere and biosphere Bd. Agr. and Renewable Resources, 1979-81; mem. panel on environ. impact Diesel Impact Study Com., Nat. Acad. Engring.-NRC, 1980-81; mem. U.S./Can./Mex. Joint Sci. Com. on acid precipitation Environ. Studies Bd., Nat. Acad. Scis.-NRC-Royal Soc. Can.-Mex. Acad. Scis., 1981-84. Mem. editorial bd. Ecology, 1965-67, Limnology and Oceanography, 1970-72, Conservation Biology, 1987-88, Ecological Applications, 1989-92, Environmental Reviews, 1992—; contbr. articles on limnology, ecology and biochemistry to profl. jours. Bd. dirs. Acid Rain Found., 1982-87, sec. treas. 1982-84. Named Royal Soc. Can. rsch. fellow State Forest Rsch. Inst., Stockholm, Sweden, 1950-51; grantee NSF, AEC, NIH, ERDA, NASA, Dept. of Energy, NRC Can., Environment Can., Office Water Resources Rsch, Dept. Interior, Andrew W. Mellon Found., N.Y.C.; recipient Regents medal U. Minn., 1984. Fellow AAAS, Royal Soc. Can.; mem. Am. Soc. Limnology and Oceanography (G. Evelyn Hutchinson medal 1986), Ecol. Soc. Am., Brit. Ecol. Soc., Internat. Assn. Theoretical and Applied Limnology, Soc. Wetland Scientists, Internat. Ecol. Assn., Soc. Conservation Biology, Swedish Phytogeographical Soc. (hon.). Home: 1933 E River Ter Minneapolis MN 55414-3673

GORMAN, ROBERT SAUL, architect; b. N.Y.C., June 28, 1933; s. Philip and Lillian (Weiss) G.; B.Arch., M.Arch., Yale U., 1966; m. Judith Alice Albaum, July 2, 1965; children—Melissa, Sasha William Shannon. Apprentice to Frank Lloyd Wright, 1953-56; designer Eero Saarinen, Hamden, Conn., 1961-67; architect, planner Victor Gruen Assos., N.Y.C., 1967-69, Juster/Pope, Architects, Shelburne Falls, Mass., 1977-78; architect Robert Gorman Assos., Architects, Planners, Solar Energy Cons., Richmond, N.H., 1969-80; founder, prin. Rawson Place Architects, 1980-89; founder, prin. Green River Architects, 1989—; cons. Bklyn. Coll., 1967-69. Served with AUS, 1956-58. Frank Lloyd Wright Found. fellow, 1953-56. Mem. AIA (Design award 1972). Pioneer in solar energy archtl. applications; architect, planner many projects. Home: Richmond Rd Richmond NH 03470 Office: Green River Architects Box 817 RR4 Brattleboro VT 05301

GORSKI, JACK, biochemistry educator; b. Green Bay, Wis., Mar. 14, 1931; s. John R. and Martha (Kenney) G.; m. Harriet M. Fischer, Sept. 9, 1955; children: Michael, Jo Anne. Student, Calif. Poly. Coll., 1949-50; B.S., U. Wis., 1953; postgrad., U. Utah, 1957; M.S., Wash. State U., 1956, Ph.D., 1958. NIH postdoctoral fellow U. Wis., 1958-61; asst. prof., asso. prof. physiology U. Ill., Urbana, 1961-66; prof. physiology U. Ill., 1967—, prof. biochemistry, 1969—; prof. biochemistry and animal scis. U. Wis., Madison, 1973—; Wis. Alumni Research Found. prof. U. Wis., 1985; NSF research fellow Princeton, 1966-67; mem. endocrinology study sect. NIH, 1966-70, molecular biology study sect., 1977-81; mem. biochemistry adv. com. Am. Cancer Soc., 1973-76, mem. personnel for research com., 1983—. Contbr. articles to profl. jours. Recipient NIH Merit award, 1986. Fellow Am. Acad. Arts and Sci.; mem. NAS, Am. Soc. Biol. Chemists, Endocrine Soc. (Oppenheimer award 1971, Disting. Leadership award 1987, pres. 1990-91). Democrat. Unitarian. Office: U Wis Dept Biochemistry 420 Henry Mall Madison WI 53706-1569

GORSKI, ROBERT ALEXANDER, chemist, consultant; b. Passaic, N.J., Nov. 24, 1922; s. Stephen T. and Wanda P. (Amlicke) G.; m. Helen Marie Thompson, Aug. 19, 1944; children: Robert J., Mary Ann B., Mark G., Stephen J., Paul F. BA in Sci., La Salle U., 1947; MS in Chemistry, U. Pa., 1948, PhD in Phys. Chemistry, 1951. Chemist DuPont, Wilmington, Del., 1951-53; rsch. chemist DuPont Freon Products Lab., Wilmington, Del., 1953-60, sr. rsch. chemist, 1960-70, rsch. assoc., 1970-78, tech. assoc., 1978-85; cons. DuPont Fluorochemicals Lab., Wilmington, Del., 1985-91; ret., 1991—. Author book chpt.; contbr. articles to profl. jours.; patentee solvents. With U.S. Army, 1943-46, ETO. Mem. Am. Chem. Soc., ASTM, Nat. Geog. Soc., KC, Sigma Xi. Republican. Roman Catholic. Avocations: sports, reading. Home: 735 Harvard Ln Newark DE 19711-3134

GORSUCH, RICHARD LEE, psychologist, educator; b. Wayne, Mich., May 14, 1937; s. Culver C. and Velma L. (Poe) G.; m. Sylvia S. Coalson, Aug. 18, 1961; children: Eric, Kay. AB, Tex. Christian U., 1959; MA, U. Ill., 1962, PhD, 1965; MDiv, Vanderbilt U., 1968. Lic. psychologist, Calif. Asst. prof., then assoc. prof. psychology George Peabody Coll. for Tchrs., 1968-73; assoc. prof. Inst. Behavioral Rsch. Tex. Christian U., 1973-75; assoc. prof., then prof. psychology U. Tex., Arlington, 1975-79; prof. psychology Fuller Theol. Sem., Pasadena, 1979—. Author: Factor Analysis, 2d edit., 1983, Psychology of Religion, 1983; editor Jour. For. Sci. Study of Religion, 1975-78; cons. editor Ednl. and Psychol. Measurement, Multivariate Behavioral Rsch.; contbr. article to Ann. Rev. Psychology, 1988. Fellow APA (div. pres. 1990-91, William James award 1986), Soc. Sci. Study Religion; mem. Religious Rsch. Assn. (bd. dirs.). Mem. Disciples of Christ Ch. Achievements include development of UniMult computer program. Office: Fuller Theol Sem Grad Sch Psychology 180 N Oakland Ave Pasadena CA 91101

GORTNER, SUSAN REICHERT, nursing educator; b. San Francisco, Dec. 23, 1932; d. Frederick Leet and Erida Louise (Leuschner) R.; m. Willis Alway Gortner, Aug. 25, 1960; children: Catherine Willis, Frederick Aiken. AB, Stanford U., 1953; M Nursing, Western Res. U., 1957; PhD, U. Calif., Berkeley, 1964; postgrad., Stanford U., 1983. Staff nurse, instr., supr. Johns Hopkins Hosp. Sch. Nursing, Balt., 1957-58; instr. to asst. prof. Sch. Nursing U. Hawaii, Honolulu, 1958-64; staff scientist, rsch. adminstr. div. nursing USPHS, Bethesda, Md., 1966-78; assoc. dean rsch. Sch. Nursing U. Calif., San Francisco, 1978-86, acting chmn. dept. family health, 1982, prof. dept. family health care nursing, 1978—; fellow, assoc. faculty mem. Inst. Health Policy U. Calif., San Francisco, 1979—, affiliated faculty mem. Inst. for Aging and Health, 1981—, adj. prof. internal medicine dept. gen. medicine Sch. Medicine, 1989—, dir. cardiac recovery lab. Sch. Nursing, 1987—, spl. asst. to dean, 1993—; Fulbright lectr., rsch. scholar Norwegian Fulbright Commn., Oslo, 1988. Contbr. articles, papers to profl. publs., chpts. to books. Health advisor N. Fork Assn., Soda Springs, Calif., 1981—. Disting. scholar Nat. Ctr. Nursing Rsch., 1990; named Disting. Alumna Frances Payne Bolton Sch. Nursing, 1983. Fellow Am. Acad. Nursing; mem. ANA (chair exec. com., coun. nurse rsch. com. 1976-80, cabinet on nursing rsch. 1984-86), Am. Heart Assn. (coun. cardiovascular nursing exec. com. 1987-91, coun. epidemiology 1989—, Katharine A. Lembright award 1991, fellow in cardiovascular nursing coun. 1992), Soc. for Behavioral Medicine (program com. 1986). Home: 470 Cervantes Rd Portola Valley CA 94028-7660 Office: U Calif N411Y 4th and Parnassus San Francisco CA 94143-0606

GORUM, VICTORIA, computer engineer; b. Tucson, Ariz., Dec. 30, 1951; d. Alvin E. and Virginia L. (Don Carlos) G. BS in Zoology, U. Mass., 1974; MS in Math., Elec. Engring., U. Nev., 1982. Rsch. asst. Worcester Found. for Exptl. Biology, Shrewsbury, Mass., 1974-76, Stanford Rsch. Inst., Menlo Park, Calif., 1976-77; sr. rsch. asst. Lawrence Berkeley (Calif.) Labs., 1977-79; tech. writer Lynch Communications, Reno, 1981-82; cons. CPL Inc., Sunnyvale, Calif., 1982-83; mgr. software support Zilog, Inc., Campbell, Calif., 1983-85; mgr. network support Sun Microsystems, Mountain View, Calif., 1986-90; mgr. net. info. systems Next Computer, Inc., Fremont, Calif., 1990-93; mgr. network ops. Adobe Systems, Mountain View, Calif., 1993—. Mem. Assn. for Computing Machinery, IEEE, Am. Needlepoint Guild, Embroidery Guild Am. Avocations: needlework, scuba diving, camping, reading. Home: 14737 Clayton Rd San Jose CA 95127-5213

GOSS, PATRICIA LYNN, information specialist; b. Cumberland, Md., Nov. 25, 1968; d. William Franklin Sr. and Mary Marion Susie (Reynolds) G. BA, Frostburg State U., 1990. Info. specialist I Biospherics Inc., Cumberland, Md., 1990-92, info. specialist II, 1992—. Recipient Acad. Merit scholarship Frostburg State U., 1986, CSX Transp. scholarship, 1986. Mem. Pi Delta Phi, Delta Omicron. Democrat. Methodist. Home: 1406 Virginia Ave Cumberland MD 21502-4745

GOSSAGE, THOMAS LAYTON, chemical company executive; b. Nashville, May 7, 1934; s. Walker E. and Mildred (Davis) G.; m. Virginia Eastman, July 27, 1957; children: Laura Eastman, Virginia Lowry. BS, Ga. Inst. Tech., 1956, MS, 1957. Process engr. Humble Oil Co., 1957; asst. dir. govt. rels. Monsanto Rsch. Corp., Dayton, Ohio, 1961-66; dir. rsch. and devel. mktg. Monsanto Rsch. Corp., Dayton, 1966-68; group mktg. dir. Monsanto Co. New Enterprises div., St. Louis, 1968-70; mktg. dir. Monsanto Co. Splty. Products, St. Louis, 1970-75; dir. results mgmt. Monsanto Indsl. Chems. Co., St. Louis, 1975-77, asst. gen. mgr. plasticizers div., 1977, gen. mgr. plasticizers div., 1977-79, gen. mgr. detergents and phosphates, 1979-80, asst. mng. dir., 1980-81, v.p., mng. dir., 1981-83; group v.p., mng. dir. Monsanto Internat., St. Louis, 1983-86; group v.p., sr. v.p. Monsanto Chem. Co. div. Monsanto Co., St. Louis, 1986-88; pres. Hercules Splty. Chems. Co., a unit of Hercules, Inc., 1988-89; pres., CEO The Aqualon Group, a unit of Hercules, 1989-91; chmn., CEO Hercules Inc., 1991—; also bd. dirs. Hercules, Inc.; mem. adv. coun. Law Engring. Co., Atlanta. 1st lt. USAF, 1957-60. Mem. Chem. Mfrs. Assn. (bd. dirs.), Conf. Bd., Bus. Roundtable, Ga. Tech. Adv. Bd. Home: 8 Wood Rd Wilmington DE 19806-2022 Office: Hercules Inc Hercules Plz 1313 N Market St Wilmington DE 19894-0001*

GOSSELIN, ROBERT EDMOND, pharmacologist, educator; b. Springfield, Mass., Sept. 2, 1919; s. A. Edmond and Grace (Pettengill) G.; m. Ruth L. Smith, June 26, 1948 (dec. Dec. 1977); children—Peter Gordon, Andrea Lee; m. Patricia S. Whitaker, July 25, 1981. A.B., Brown U., 1941; Ph.D., U. Rochester, 1945, M.D., 1947. Med. intern Yale service Grace-New Haven Hosp., 1947-48; instr. pharmacology U. Rochester, 1948-52, asst. prof. pharmacology, scientist atomic energy project, 1954-56; prof. dept. pharmacology and toxicology Dartmouth Med. Sch., 1956-89, prof. emeritus, 1989—, chmn. dept., 1956-75; dir. poison info. center Hitchcock Hosp., 1957-82; cons. USPHS, 1959-63, U.S. Army Chem. Corps, 1954-59; mem. toxicology study sect. USPHS, 1964-68; mem. toxicology adv. bd. U.S. Consumer Product Safety Commn., 1978-85. Contbr. articles to profl. jours. Pres. Norwich Devel. Assn., 1965. Served with AUS, 1944-46; to capt., M.C. AUS, 1952-54. Mem. Am. Physiol. Soc., Am. Soc. Exptl. Pharmacology and Therapeutics, AAAS, Toxicology Soc., Phi Beta Kappa, Sigma Xi. Home: The Holm Hanover NH 03755 Office: Dartmouth Med Sch Hanover NH 03756

GOTHARD, MICHAEL EUGENE, industrial engineer; b. Regensburg, Bavaria, Germany, Feb. 22, 1948; s. William McKinley and Elfrieda (Schober) G.; m. Carolyn Rebecca Howton, June 11, 1970; 1 child, Michael David. BS in Indsl. Engring., U. Tenn., Chattanooga, 1970. Cert. Level III in Ultrasonic Liquid Penetrant, Magnetic Particle and Visual Examination. Rsch. engr. Combustion Engring. Inc., Chattanooga, Tenn., 1970-79; engring. mgr. Tenn. Valley Authority, Chattanooga, Tenn., 1979-90; NDE instr. EPRI NDE Ctr., Charlotte, N.C., 1990—; adv. bd. Electric Power Rsch. Inst., Palo Alto, Calif., 1985-90. Editor: (reference book) Ultrasonic Technical Skills-Series, 1991, NDE For Engineers, 1992. Capt. USAR, 1971-80. Mem. ASME, Am. Soc. for Nondestructive Testing. Achievements include development of qualified ultrasonic system for monitoring water level in scram discharge headers; certification as an instr. in detection and sizing of intergranular stress corrosion cracking. Office: EPRI NDE Ctr 1300 Harris Blvd Charlotte NC 28262

GOTO, KEN, chronobiologist, educator; b. Yokosuka, Kanagawa, Japan, Dec. 13, 1952; s. Tatsuwo and Kimiye (Shinohara) G.; m. Kazuko Takahashi, Nov. 18, 1975; children: Miki, Aki. BS, Kyoto U., 1975; MS, Nagoya (Japan) U., 1977, DSc, 1980. Postdoctoral rsch. assoc. Nat. Inst. for Basic Biology, Okazaki, Aichi, Japan, 1980-82, SUNY, Stony Brook, 1982-83, Inst. Applied Biochemistry, Mitake, Gifu, Japan, 1984-86; lectr. Obihiro (Hokkaido, Japan) U. Agr. and Vet. Medicine, 1986-87, assoc. prof., 1987—. Contbr. to books Models in Plant Physiology and Biochemistry, 1987, Endocytobiology III, 1987, Circadian Clocks and Ecology, 1989, Handbook for Chronobiology (in Japanese), 1991. Japan Soc. for Promotion of Sci. fellow, 1980-82, NSF fellow, 1982-83, Inst. Applied Biochemistry fellow, 1984-86; Ministry of Edn., Sci. and Culture rsch. grantee, 1987-89, Joint Internat. Sci. Rsch. Program rsch. grantee, 1990-92. Mem. AAAS, Bot. Soc. Japan, Japanese Biochem. Soc., Japan Soc. Plant Physiologists, Biophys. Soc. Japan, Internat. Soc. Chronobiology, Soc. for Rsch. on Biol. Rhythms, Am. Soc. Plant Physiologists, N.Y. Acad. Scis. Achievements include research in molecular mechanisms of circadian clocks; circadian clock control of cell division cycle; temporal organization in the unicell. Home: Nishi 2-13 Inada, 080 Obihiro Japan Office: Obihiro U Agr and Vet Med, Nishi 2-11 Inada, 080 Obihiro Japan

GOTO, NORIHIRO, aeronautical engineering educator; b. Sasebo, Japan, Dec. 9, 1943; came to U.S., 1975; s. Tatsumi and Miyuki (Koide) G.; m. Yoshie Yonebayashi, Mar. 21, 1972; children: Natsuko, Eiko. M. Engring., U. Tokyo, 1968, D. Engring., 1972; postgrad., MIT, 1968-69. Lectr. dept. aeronautical engring. Kyushu U., Fukuoka, Japan, 1972-73, assoc. prof. aeronautical engring., 1973-88; NRC assoc. NASA Ames Rsch. Ctr., Mountain View, Calif., 1975-76; prof. dept. aeronautics and astronautics Kyushu U., Fukuoka, 1988—. Reviewer Applied Mechanics Revs., 1979-83; contbr. articles AIAA Jour., Jour. Guidance and Control, Jour. Guidance, Control and Dynamics, Jour. Japan Soc. for Aeronautical & Space Scis. Fellow AIAA (assoc.), Japan Soc. for Aeronautical and Space Scis. Buddhist. Achievements include development of methods to identify pilot control behavior in practical multi-input and multi-output aircraft control systems. Home: 1-12-9 Takamidai Higashi-ku, Fukuoka 811-02, Japan Office: Kyushu U Dept Aeronautics Astronautics, 6-10-1 Hakozaki Higashi-ku, Fukuoka 812, Japan

GOTOVCHITS, GEORGY OLEXANDROVICH, electrical engineer, educator; b. Olgopol Village, Vinnitsa, Ukraine, Sept. 17, 1935; s. Olexander Ivanovich and Mariya Mikchailovna (Pomyatyna) G.; m. Larica Yakovlevna Vislouh; children: Alla Georgievna, Alexandra Georgievna. Grad., Poly. Inst., Odessa, Ukraine, 1958. Engr. electric equipment and automatics, head electro-tech. lab. Progress Plant Chem. Machine Engring., Berdichev, Zhitomir, 1958-60, head electro-supt., 1960-62, chief power engr., 1962-71; exec. for social and econ. devel. Town of Berdichev, 1971-78; dept. chmn. exec. power Zhitomir Region, 1978-90; first dep. chmn. exec. power Zhitomir Power, 1990; chmn. State Com. Ukraine for Population Protection from Consequences of Accident at Chernobyl Nuclear Power Plant, Kiev, 1990-91, min., 1991—. Contbr. chpt. to Chernobyl: Five Hard Years, 1992; contbr. articles to profl. jours. Sec. party com. orgn. Mcpl. Com. Communist Party Ukraine, Berdichev, 1970-78. Recipient 2 orders, 5 medals Govt. USSR, diploma Presidium Supreme Soviet Ukraine. Office: Min Ukraine Affairs for Pop Protection, from Chernobyl 8 Lvovskaya Sq, 254655 Kiev Ukraine

GOTSCHLICH, EMIL CLAUS, physician, educator; b. Bangkok, Thailand, Jan. 17, 1935; came to U.S., 1950, naturalized, 1955; s. Emil Clemens and Magdalene (Holst) G.; m. Kathleen-Ann Haines, May 24, 1975; children—Emil Christofer, Hilda Christina, Emil Chandler, Emily Claire. B.A., N.Y. U., 1955, M.D., 1959. Intern Bellevue Hosp., N.Y.C., 1959-60; mem. faculty Rockefeller U., N.Y.C., 1960—; prof. microbiology Rockefeller U., 1978—, sr. physician, 1978—. Served as capt. M.C. U.S. Army, 1966-68. Decorated Army Commendation medal; recipient Squibb award Am. Soc. Infectious Disease, 1974; Lasker award Albert and Mary Lasker Found., 1978. Mem. Am. Assn. Immunologists, Nat. Acad. of Scis., Assn. Am. Physicians, Am. Soc. for Clin. Investigation, Sigma Xi, Alpha Omega Alpha. Club: Peripatetic. Office: Rockefeller U Dept Immunology 1230 York Ave New York NY 10021-6341

GOTT, J. RICHARD, III, astrophysicist; b. Louisville, Feb. 8, 1947; s. J. Richard and Marjorie (Crosby) G.; m. Lucy Jennifer Pollard, June 10, 1978; 1 child, Elizabeth. AB, Harvard U., 1969; PhD, Princeton U., 1973. Postdoctoral fellow Calif. Inst. Tech., Pasadena, 1973-74; vis. fellow Cambridge U., Eng., 1975; asst. prof. astrophysics Princeton U., 1976-80, assoc. prof., 1980-87, prof., 1987—; chmn. of the judges Nat. Westinghouse Sci. Talent Search, Washington, 1986—; chmn. Hayden Planetarium vis. com., 1992-93. Contbr. articles to profl. jours. Recipient R. J. Trumpler award, Astron. Soc. of Pacific, 1975; Alfred P. Sloan fellow, 1977-81. Mem. Am. Astron. Soc., Internat. Astron. Union, Phi Beta Kappa. Achievements include rsch. in theory that the universe will continue to expand forever; solution to Einstein's field equations for the gravitational field around cosmic strings. Office: Princeton U Dept Astrophys Sci Princeton NJ 08544

GOTTA, ALEXANDER WALTER, anesthesiologist, educator; b. Bklyn., Apr. 10, 1935; s. A. Walter and Helen C. (Bruskewic) G.; m. Colleen A. Sullivan, July 17, 1965; 1 child, Nancy C. B.S. summa cum laude, St. John's U., 1956; M.D., NYU, 1960. Diplomate Am. Bd. Anesthesiology, Am. Bd. Med. Examiners. Intern, U. Chgo., 1960-61; resident Boston City Hosp., 1961-62, N.Y. Hosp.-Cornell U., N.Y.C., 1962-64; instr. anesthesiology Cornell U., 1964-66, asst. prof., 1978-79; dir. anesthesia St. Mary's Hosp., Bklyn., 1966-78; asst. prof. SUNY-Bklyn., 1966-78, assoc. prof., 1978-85, prof., 1985—; dir. anesthesia L.I. Coll. Hosp., Bklyn., 1983-90; dir. anesthesia Kings County Hosp. Ctr., 1990—; speaker in field. Contbr. articles to profl. jours. Served to capt. U.S. Army, 1966-68, Vietnam. Fellow N.Y. Acad. Medicine (chmn. anesthesia sect. 1990), Am. Coll. Anesthesiologists, Am. Soc. Anesthesiologists (ho. of del. 1986—); mem. N.Y. Soc. Anesthesiologists (bd. dirs. 1983—, chmn. sci. program com.), N.Y. Soc. Critical Care Medicine (pres. 1985), Assn. Univ. Anesthesiologists, Acad. Anesthesia. Republican. Roman Catholic. Club: Brooklyn. Avocation: History. Home: 29 Ascot Ridge Rd Great Neck NY 11021-2912 Office: Kings County Hosp Ctr 451 Clarkson Ave Brooklyn NY 11203-2097

GOTTESMAN, IRVING ISADORE, psychiatric genetics educator, consultant; b. Cleve., Dec. 29, 1930; s. Bernard and Virginia (Weitzner) G.; m. Carol Applen, Dec. 23, 1970; children—Adam M., David B. B.S., Ill. Inst. Tech., 1953; Ph.D., U. Minn., 1960. Diplomate in clin. psychology; lic. psychologist Minn., Calif., Va. Intern clin. psychology VA Hosp., Mpls., 1959-60; lectr. dept. social relations Harvard U., 1960-63; fellow psychiat. genetics Inst. Psychiatry, London, 1963-64; assoc. prof. psychiat. genetics, dept. psychiatry U. N.C., 1964-66; prof. dept. psychology U. Minn., 1966-80; prof. dept psychiatry Washington U., St. Louis, 1980-85; Commonwealth prof. psychology U. Va., Charlottesville, 1985—; cons. NIMH, Washington, 1975-79, 92—, NIMH Nat. Plan for Schizophrenia, 1988-89; mem. Pres.'s Commn. on Huntington Disease, 1977; tng. cons. VA, Washington, 1968-85; fellow Ctr. for Advanced Studies in the Behavioral Scis., Stanford, Calif., 1987-88; inst. of medicine com. cons. Viet Nam War Experience Study, 1987-88. Author: Schizophrenia and Genetics, 1972 (Hofheimer prize), Schizophrenia—The Epigenetic Puzzle, 1982, Schizophrenia Genesis: The Origins of Madness, 1991 (transl. into Japanese, German and Portuguese, William James book award, Phi Beta Kappa U. Va. Book award 1992); editor: Man, Mind and Heredity, 1971, Vital Statistics, Demography and Schizophrenia, 1989. Served with USN, 1953-56. Guggenheim fellow U. Copenhagen, 1972; recipient R. Thornton Wilson prize Eastern Psychiat. Rsch. Assn., 1965, Stanley Dean award Am. Coll. Psychiatrists, 1988, Eric Stromgren medal Danish Psychiat. Soc., 1991, Kurt Schneider prize, Bonn, Germany, 1992, Alexander Gralnick prize Am. Assn. Suicidology, 1992, U. Va. Pres.'s Report award, 1992; David C. Wilson lectr. U. Va. Sch. Medicine, 1967; Parker lectr. Ohio State U. Sch. Medicine, 1983, 93. Fellow APA, Am. Psychopathol. Assn., Royal Coll. Psychiatrists (hon.), Am. Psychol. Soc.; mem. Minn. Human Genetics League (v.p. 1969-71), Soc. Study Social Biology (v.p. 1976-80), Behavior Genetics Assn. (pres. 1976-77, T. Dobzhansky award 1990), Am Soc. Human Genetics (editorial bd. 1967-72), Soc. Rsch. in Psychopathology (pres. 1993). Home: 260 Terrell Rd W Charlottesville VA 22901-2167 Office: Univ Va Gilmer Hall Charlottesville VA 22903

GOTTESMAN, MICHAEL MARC, biomedical researcher; b. Jersey City, N.J., Oct. 7, 1946; s. Jacob Joseph and Frieda (Shapiro) G.; m. Susan Kemelhor, Feb. 5, 1966; children: Daniel Eric, Rebecca Fran. AB, Harvard Coll., 1966; MD, Harvard Med. Sch., 1970. Diplomate Am. Bd. Internal Medicine. Med. intern then resident Peter Bent Brigham Hosp., Boston, 1970-71, 74-75; rsch. assoc. NIH, Bethesda, Md., 1971-74; asst. prof. Harvard Med. Sch., Boston, 1975-76; sr. investigator Nat. Cancer Inst., Bethesda, 1976-80, sect. head, 1980-90, lab. chief, 1990—; acting dir. Nat. Ctr. for Human Genome Rsch., 1992-93. Author and editor: Molecular Cell Genetics, Molecular Genetics of Mammalian Cells, The Role of Proteases in Cancer. Capt. USPHS, 1971—. Recipient Milken Family award for cancer rsch., 1990. Fellow AAAS; mem. Am. Soc. Biochemistry Molecular Biology, Genetics Soc. Am., Am. Soc. Cell Biology, Am. Assn. Cancer Rsch. (Richard and Hinda Rosenthal Found. award 1992). Achievements include rsch. on molecular basis of resistance to anti-cancer drugs. Office: Nat Cancer Inst Natl Ctr for Human Genome Rsch 9000 Rockville Pike Bldg 38A Bethesda MD 20892

GOTTFREDSON, GARY DON, psychologist; b. Sonora, Calif., Sept. 4, 1947; s. Don Martin and Betty Jane (Hunt) G.; m. Denise Claire Ruff, Dec. 31, 1979; 1 child, Nisha Claire. BA in Psychology, U. Calif., Berkeley, 1969; PhD in Psychology, Johns Hopkins U., 1976. Assoc. adminstrv. officer APA, Washington, 1976-77; rsch. scientist Johns Hopkins U., Balt., 1977-85, prin. rsch. scientist, 1985—; mem. panel on rsch. on rehab. techniques Nat. Rsch. Coun., Washington, 1978-81, Com. on Occupational Classification and Analysis, 1978-81. Author: Victimization in Schools, 1985, Dictionary of Holland Occupational Codes, 1989; author psychol. tests. Vol. Peace Corps, Penang, West Malaysia, 1969-72. Fellow APA (J.L. Holland award 1989); mem. Am. Psychol. Soc., Am. Soc. Criminology, Am. Ednl. Rsch. Assn. Democrat. Achievements include development of occupational classification widely used in career counseling, development of job analysis procedures to implement a psychological taxonomy of work environments. Home: 11444 Old Frederick Rd Marriottsville MD 21104 Office: Johns Hopkins U 3505 N Charles St Baltimore MD 21218

GOTTFRIED, DAVID SCOTT, chemist researcher; b. Detroit, Oct. 25, 1962; s. Roger Michael Gottfried and Vivian Sharon (Liber) Friedman; m. Pamela (Jay) Gottfried, Jan. 19, 1992. BS, U. Mich., 1984; PhD, Stanford U., 1991. Researcher dept. chemistry U. Mich., Ann Arbor, 1983-84; teaching asst. Stanford (Calif.) U., 1984-87; post doctoral fellow Weizmann Inst. of Sci., Rehovot, Israel, 1990-91; assoc. Albert Einstein Coll. of Medicine, Bronx, N.Y., 1991—. Contbr. 9 articles to profl. jours. including Jour. of Physical Chemistry. Recipient Otto Graf scholarship U. Mich., 1983-84, Am. Inst. of Chemists award, 1984, pre-doctoral fellowship NSF, 1985-88, post-doctoral fellowship, Euopean Molecular Biology Orgn., 1990-91. Mem. AAAS, Am. Chem. Soc., Biophys. Soc., Protein Soc. Office: Albert Einstein Coll Medicine 1300 Morris Park Ave Bronx NY 10461

GOTTFRIED, EUGENE LESLIE, physician, educator; b. Passaic, N.J., Feb. 26, 1929; s. David Robert and Rose (Chill) G.; m. Phyllis Doris Swain, Aug. 16, 1957. AB, Columbia U., 1950, MD, 1954. Cert. Nat. Bd. Med. Examiners, Am. Bd. Internal Medicine. Intern Presbyn. Hosp., N.Y.C., 1954-55, asst. resident in medicine, 1957-58; resident Bronx (N.Y.) Mcpl. Hosp. Ctr., 1958-59, fellow in medicine, 1959-60; asst. instr. medicine Albert Einstein Coll. Medicine Yeshiva U., N.Y.C., 1959-60, instr., 1960-61, assoc., 1961-65, asst. prof., 1965-69; assoc. prof. medicine Cornell U. Med. Coll., N.Y.C., 1969-81, assoc. prof. pathology, 1975-81; clin. prof. dept. lab. medicine U. Calif., San Francisco, 1981—, vice chmn. dept. lab. medicine, 1981—; hosp. appointments include asst. vis. physician Bronx Mcpl. Hosp. Ctr., 1960-66, assoc. attending physician, 1966-69; assoc. attending physician N.Y. Hosp., N.Y.C., 1969-81, assoc. attending pathologist, 1975-81, dir. lab. clin. hematology, 1969-81; chief lab. medicine San Francisco Gen. Hosp. Med. Ctr., 1981—, dir. clin. labs., 1981—. Assoc. editor Jour. Lipid Research, 1971-72, 75-77; mem. editorial bd. Jour. Lipid Research, 1972-77. Served to lt. comdr. USNR, 1955-57. Recipient Career Scientist award Health Research Council City of N.Y., 1964-72. Fellow Am. Soc. Hematology, Internat. Soc. Hematology, ACP, Acad. Clin. Lab. Physicians and Scientists; mem. AAAS, Phi Beta Kappa, Alpha Omega Alpha. Office: San Francisco Gen Hosp Clin Labs 1001 Potrero Ave San Francisco CA 94110-3594

GOTTFRIED, KURT, physicist, educator; b. Vienna, Austria, May 17, 1929; came to U.S., 1952, naturalized, 1965; s. Salomon and Augusta (Werner) G.; m. Sorel B. Dickstein, June 26, 1955; children: David M., Laura S. B.Eng., McGill U., 1951, M.S., 1952; Ph.D., MIT, 1955. Jr. fellow Soc. Fellows, Harvard, 1955-58; research fellow Inst. Theoretical Physics, Copenhagen, 1958-59; research fellow Harvard, 1959-60, asst. prof. physics, 1960-64; assoc. prof. physics Cornell U., Ithaca, N.Y., 1964-68, prof. physics, 1968—, chmn. dept., 1991—; staff mem. European Orgn. for Nuclear Research, Geneva, Switzerland, 1970-73. Author: Quantum Mechanics, 1966, Concepts of Particle Physics, Vol. 1, 1984, Vol. 2, 1986; co-editor: Crisis Stability and Nuclear War, 1988. Fellow AAAS, Am. Acad. Arts and Scis., Am. Phys. Soc. (chmn. div. particles and fields 1981, councillor 1990—); mem. Union Concerned Scientists (bd. dirs. 1978—), Coun. on Fgn. Rels. Office: Cornell U Clark Hall 109 Ithaca NY 14853

GOTTLANDER, ROBERT JAN LARS, dental company executive; b. Bohuslan, Sweden, Sept. 5, 1956; came to U.S., 1986; s. Jan H. K. and Ragnhild S.E. (Rutgerson) G.; m. Eva L.M. Svenson, July 4, 1987; children: Daniel J.R., Magdalena A.E. Student, Kongahalla Coll., Sweden, 1975; candidate of odontology, U. Gothenburg, Sweden, 1976, DDS, 1980. Dentist Swedish Health Care, Trollhattan, Sweden, 1980-82; asst. prof. dept. orthodontics Community Dentistry, Trollhattan, 1982-84; mgr. tng. and edn. Nobelpharma AB, Gothenburg, 1984-85, product mgr., 1985; v.p., mgr. edn. and product Nobelpharma USA Inc., Waltham, Mass., 1986-87, v.p. profl. affairs, 1987-88; v.p., gen. mgr. Nobelpharma USA Inc., Chgo., 1988—; pres. V-Dal Union of Dentists, Trollhattan, Sweden, 1982-84; chmn. V-Dal Dental Soc., Sweden, 1983-84, sec. 1981-82. Lt. Swedish Royal Navy, 1976-79. Mem. AMA, Swedish Dental Soc., Swedish Orthodontic Soc.; affiliate mem. ADA, Acad. of Osseointegration. Lutheran. Avocations: sailing, skiing, tennis, reading. Home: 5101 S Keeler Ave Chicago IL 60632-4205 Office: Nobelpharma USA Inc 5101 S Keeler Ave Chicago IL 60632-4205

GOTTLIEB, A(BRAHAM) ARTHUR, medical educator, biotechnology corporate executive; b. Dec. 14, 1937; m. Marise S. Gottlieb, 1958; children: Mindy, Joanne. AB summa cum laude, Columbia Coll., 1957; MD, NYU, 1961. Med. house officer Peter Bent Brigham Hosp., Boston, 1961-62, asst. resident, 1962-63; rsch. fellow in chemistry, tutor in chemistry Harvard U., Peter Bent Brigham Hosp., Boston, 1965-67; assoc. in medicine Harvard U., Peter Bent Brigham Hosp., 1968; asst. prof. medicine Harvard Med. Sch., 1969; assoc. prof. microbiology Inst. Microbiology, Rutgers U., New Brunswick, N.J., 1969-72, prof. microbiology, 1972-75; prof., chmn. microbiology and immunology Tulane U. Sch. Medicine, New Orleans, 1975—, prof. medicine, 1975—; pres., sci. dir. IMREG, Inc., New Orleans; cons. in field; vis. prof. Walter and Eliza Hall Inst. Med. Rsch., Melbourne, Australia, 1979, Wakayama Med. Coll. and Gunma Med. Coll., Japan, 1980; vis. prof. medicine and pharmacology Shanghai Med. U., People's Republic of China, 1991-92. Mem. editorial adv. bd. several profl. publs. Mem. sci. adv. bd. Cancer Assn. Greater New Orleans, 1979-84; mem. tech. rev. bd. U. New Orleans, 1984-90. Recipient Frances Stone Bunrs award Am. Cancer Soc., 1968, Rsch. Career Devel. award Nat. Inst. Gen. Med. Scis., 1968-69. Fellow ACP, Am. Assn. Cancer Rsch., Am. Assn. Immunologists, Am.

Chem. Soc., Am. Soc. Biol. Chemists, Am. Soc. Cell Biology, Am. Soc. Clin. Investigation, Am. Soc. Microbiology, Am. Acad. Microbiology, Internat. Assn. Comparative Rsch. on Leukemia and Related Diseases, N.Y. Acad. Scis., Reticulendothelial Soc. (chmn. publs. com.), Assn. Med. Sch. Microbiology (chmn.), AAAS, Phi Beta Kappa, Sigma Xi, Alpha Omega Alpha. Office: IMREG Inc 144 Elk Pl Ste 1400 New Orleans LA 70112-2647 also: Tulane Med Sch 1430 Tulane Ave New Orleans LA 70112

GOTTLIEB, ARNOLD, dentist; b. N.Y.C., Jan. 8, 1926; s. Samuel J. and Matilda (Gross) G.; m. Joan F. Rigler, June 27, 1947; children: Susan, Jeffrey, Jane. Student, CCNY, 1943-45; DDS cum laude, U. Pitts., 1949. Pvt. practice reconstructive dentistry N.Y.C., 1949-92; ret., 1992; fin. cons., advisor, investor networking; vis. dentist Heart Ho. Hosp., 1948-49; attending dental surgeon Harlem Eye and Ear Hosp., 1955-65, chief dental cons., 1966-68; lectr. Fedn. Dentair-Austria, Australia, Brazil, Mex., France, Israel, AIMS, China, Russia; presenter at profl. confs.; bd. dirs. Haptentech, Sysdyne Corp. Contbr. articles to profl. jours. Past v.p. community synagogue, Rye, N.Y.; founder, pres. Homeowners Assn., Pine Ridge, Town of Rye, 1959-60; past pres. Hudson Investment Club; patron Met. Opera. Capt. U.S. Army, 1951-53. Recipient Doubles Champion, Conn. Sr. Olympics, 1993. Mem. ADA, Fedn. Dentaire Internat., Myodentronic Soc. N.Y., N.E. Prosthetic Study Club (past sec., treas.), Fedn. Dentair, Am. Internat. Med. Soc., Sea Pines Country Club (Hilton Head Island, S.C.), Pound Ridge Tennis Club, B'nai Brith (past pres. Henry Hudson Lodge), Dental Honor Soc., Omicron Kappa Upsilon, Nat. Dental Honor Soc. Jewish. Avocations: song writing, writing poetry, tennis, writing novels. Home: 211 Ferris Hill Rd New Canaan CT 06840

GOTTLIEB, H. DAVID, podiatrist; b. Washington, Mar. 2, 1956; s. Julius J. and Charlotte (Papernik) G.; m. Wendy Ilene Weisbard, June 17, 1979; children: Jason, Cheryl. BA, Cornell U., 1978; DPM, Pa. Coll. Podiatric Medicine, 1982. Diplomate Nat. Bd. Podiatry Examiners, Am. Acad. Pain Mgmt., Am. Coun. Cert. Podiatric Physicians and Surgeons. Podiatrist Dr. Julius J. Gottlieb, P.C., Washington, 1982-91; prin. H. David Gottlieb, DPM, P.C., Washington, 1991—. Author (book chpt.) Laser Surgery of the Foot, 1988. Mem. Chevy Chase (D.C.) Citizens Assn., 1982—; den master Cub Scouts Am., Gaithersburg, Md., 1991-93; pres. Young Couples Club Gaithersburg Hebrew Congregation, 1982-85. Felow Acad. Ambulatory Foot Surgery; mem. APHA, Am. Podiatric Med. Assn., Am. Assn. Podiatric Physicians and Surgeons, Am. Running and Fitness Assn., Internat. Soc. Podiatric Laser Surgery (assoc.). Avocations: Karate, gardening. Office: 3900 McKinley St NW Washington DC 20015-2993

GOTTLIEB, JULIUS JUDAH, podiatrist; b. Jersey City, May 27, 1919; s. Joseph Uziel and Gussie (Farber) G.; m. Charlotte Papernik, Oct. 18, 1942; children: Sheldon, Cynthia, Lorinda, David, Jonathan. Student, NYU, 1938-39, Ill. Coll. Podiatric Medicine, 1940-42; DPM, Ohio Coll. Podiatric Medicine, 1943. Diplomate Am. Podiatric Med. Specialties Bd. Pvt. practice podiatric medicine Washington, 1943-92; pres. Chevy Chase Profl. Cons., 1993—; past cons. Army Footwear Clinic. Co-inventor fiberglass foot prosthetics and plastic shoe lasts. Chmn. com. Nat. Capital area coun. Boy Scouts Am., 1969-73; pres. Franklin Knolls Citizens Assn., 1963; podiatry dir. Greater Washington Hewbrew Home for the Aged, 1963. Recipient Shofar award Boy Scouts Am. Fellow Acad. Ambulatory Foot Surgeons (region 8 sci. chmn. 1987-88), Nat. Coll. Foot Surgeons (founding); mem. Am. Podiatric Med. Assn., Am. Pub. Health Assn., Am. Coll. Podopediatrics, Am. Podiatric Circulatory Soc., Am. Bd. Foot Surgeons (founding diplomate), D.C. Podiatric Med. Soc. (past pres.), Am. Assn. Foot Specialists (past pres., Foot Specialist of the Yr. 1973), Am. Assn. Individual Investors, Internat. Platform Assn., Am. Physicians Fellowship Inc. for Medicine in Israel, Columbia Heights Bus. Men's Assn. (past pres., Man of Yr. 1964), Parents Assn. U. Md. (co v.p. parents fund 1980-81, co-recipient Outstanding Svc. Award), Chevy Chase Citizens Assn., B'nai B'rith. Republican. Jewish. Home and Office: 15812 Ancient Oak Dr Darnestown MD 20878-2110

GOTTLIEB, LEONARD SOLOMON, pathology educator; b. Boston, May 26, 1927; s. Julius and Jeanette (Miller) G.; m. Dorothy Helen Apt, Mar. 23, 1952; children: Julie Ann, William Apt, Andrew Richard. A.B., Bowdoin Coll., 1946; M.D., Tufts U., 1950; M.P.H., Harvard U., 1969. Diplomate: Am. Bd. Anatomic Pathology. Intern and resident in pathology Boston City Hosp., 1950-55; asst. chief pathology U.S. Naval Hosp., Chelsea, Mass., 1955-57; assoc. pathologist Mallory Inst. Pathology, Boston, 1957-66, assoc. dir., 1966-72, dir., 1972—; chief pathology dept. Boston U. Med. Ctr.-Univ. Hosp., 1973—; prof. pathology Boston U. Sch. Medicine, 1970—, chmn. dept., 1980—; chief pathology dept. Boston U. Med. Ctr./The U. Hosp., 1973—; dir. Mallory Inst. Pathology Found., 1980—; lectr. Harvard U., 1963—; dir. student faculty exch. program Boston U. and Hebrew U., Hadassah Med. Sch., 1988—. Gen. editor Biopsy Pathology Series, Chapman and Hall; mem. editorial bd. Am. Jour. Surgical Pathology; contbr. over 125 articles on exptl. and human gastrointestinal and liver diseases. Bd. govs. Hebrew U. Jerusalem, 1991—; mem. sci. adv. bd. Boston chpt. Israel Cancer Rsch. Fund, 1991—; mem. Greater Boston chpt. State of Israel Bonds Cabinet, 1991—; pres. Am. Physicians Fellowship, Inc. for Medicine in Israel, 1990—. Lt. comdr. M.C., USNR, 1955-57, 60-63. Named hon. mem. faculty medicine Hebrew U., 1987; recipient Stanley L. Robbins faculty award for excellence in teaching Boston U. Sch. Medicine, 1986; named Acad. Physician of Yr., recipient Jerusalem City of Peace award Boston chpt. State of Israel Bonds, 1992; James Bowdoin scholar, 1945. Mem. Am. Assn. Pathologists, Am. Assn. for Study Liver Disease, U.S.-Can. Acad. Pathology, Am. Soc. for Cell Biology, Am. Gastroent. Assn., New Eng. Soc. Pathologists (pres. 1968-69), Am. Soc. Clin. Pathologists, Am. Friends Hebrew U. Jerusalem (pres. New Eng. region 1989—). Office: Boston U Sch Medicine Dept Path and Lab Medicine 80 E Concord St Boston MA 02118

GOTTLIEB, MOSHE, chemical engineer, educator; b. Haifa, Israel, Oct. 29, 1948; s. Jacob and Eva (Weiss) Gotal; m. Ilana Urbaitel, Nov. 21, 1971; 1 child, Sigal. BSc, Technion, Haifa, 1973; PhD, U. Wis., 1978. Vis. asst. prof. U. Minn., Mpls., 1978-79; lectr. Ben-Gurion U., Beer Sheva, Israel, 1979-82, sr. lectr., 1982-85, assoc. prof. chem. engring., 1985-91, prof. chem. engring., 1991—; vis. prof. U. Mass., Amherst, 1986-87; cons. Los Alamos (N.Mex.) Nat. Labs., 1985—, Exxon Chems., Linden, N.J., 1991—; mem. adv. bd. Jour. Network and Gels, London, 1992—. Editor spl. issue Jour. Stat. Physics, 1990, 91. Maj. Israeli armed forces, 1966-69; Reserves, 1969—. Mem. Am. Chem. Soc., Am. Phys. Soc., Soc. Rheology (polymer networks group com. 1988). Office: Ben-Gurion U, Chem Engring Dept, Beer Sheva 84105, Israel

GOTTSCHALK, KURT WILLIAM, research forester; b. Bloomington, Ill., Apr. 30, 1952; s. Donald Eugene and Ruth Virginia (Ahrends) G.; m. Janice Jo Johnston, June 28, 1980; children: Stephanie Ruth, Katherine Marie. BS in Forestry, Iowa State U., 1974; MS in Forestry, Mich. State U., 1976, PhD in Forestry, 1984. Rsch. forester USDA Forest Svc. Northeastern Experiment Sta., Warren, Pa., 1979-83; rsch. forester USDA Forest Svc. Northeastern Experiment Sta., Mortantown, W.Va., 1983-87, rsch. forester, project leader, 1987—; adj. assoc. prof. U. W.Va., Morgantown, 1991—; vis. prof. Va. Poly. Inst. and State U., Blacksburg, 1992. Author, editor proc., articles in field. Mem. Am. Inst. Biol. Scis., Am. Forestry Assn., Ecol. Soc. Am., Soc. Am. Foresters (mem. exec. com. 1992—), Sigma Xi, Alpha Zeta, Xi Sigma Pi, Gamma Sigma Delta. Office: USDA Forest Svc Northeastern Forest Exp Sta 180 Canfield St Morgantown WV 26505-3101

GOTTSCHALK, WALTER HELBIG, mathematician, educator; b. Lynchburg, Va., Nov. 3, 1918; s. Carl and Lula (Helbig) G.; m. Margaret Hemsworth, Aug. 27, 1952; children: Heather, Steven. B.S., U. Va., 1939, M.A., 1942, Ph.D. in Math, 1944; M.A. (hon.), Wesleyan U., Middletown, Conn., 1964. From instr. to prof. math. U. Pa., 1944-63, chmn. dept., 1955-58; prof. math. Wesleyan U., 1963-82, prof. emeritus, 1982—, chmn. dept., 1964-69, 70-71; Mem. Inst. Advanced Study, Princeton, 1947-48; research assoc. Yale U., 1960-61. Author: (with G.A. Hedlund) Topological Dynamics, 1955; Mem. editorial bd.: Math. Systems Theory, 1967-75; Contbr. articles to profl. jours. Mem. Am. Math. Soc. (asso. editor proc. 1954-56, asso. sec. for East 1971-76), Math. Assn. Am., Soc. Indsl. and Applied Math., AAUP, Phi Beta Kappa, Sigma Xi. Democrat. Unitarian. Home: 500 Angell St Apt 414 Providence RI 02906-4455

GOTTZEIN, EVELINE, aerospace engineer; b. Leipzig, Germany, Sept. 30, 1931; d. Bruno and Charlotte G. Diploma in math., Tech. U. Darmstadt, Germany, 1959; D Engring., Tech. U. Munich, 1984. Mem. staff space transp. and propulsion systems Deutsche Aero. AG, Munich, 1959—; now head control dynamics and simulation dept. Deutsche Aero. AG Space Transp. and Propulsion Systems, Munich; tchr. Tech. U. Stuttgart, Germany, 1989—. Author: Magnetic Levitation of High Speed Vehicles, 1984; contbr. articles on magnetic levitation and control problems of aerospace vehicles to profl. publs.; editor: Automatic Control in Space, 1976, Automatic Control in Aerospace, 1992. Mem. AIAA (tech. com. 1990—), Internat. Fedn. Automatic Control (aerospace control tech. com. 1972—). Achievements include 5 patents in magnetic levitation and control of aerospace vehicles. Office: Deutsche Aerospace AG, PO Box 801169, D-8000 Munich Germany

GOTZOYANNIS, STAVROS ELEUTHERIOS, cardiologist; b. Piraeus, Greece, June 18, 1933; s. Eleutherios G. and Irene Stavros (Nikitaki) G.; M.D. (Greek Govt. scholar), U. Salonica, 1957; Doctorate, U. Athens, 1968; m. Ourania Cavoulacou, Nov. 20, 1969. Intern, 401 Army Gen. Hosp., Athens, 1957-58; resident in internal medicine Army Pansion Share Hosp., Athens, 1960-63; fellow in cardiology Hellenic Red Cross Hosp., Athens, 1964-65; commd. 2d lt. M.C., Greek Army, 1957, advanced through grades to brig. gen., 1984; dir. internal medicine 403 Army Gen. Hosp., Kozani, 1963-64; research asso., div. cardiology Phila. Gen. Hosp., 1970-71; asst. attending physician Georgetown U. Hosp., Washington, 1971-72; asso. CCU, 401 Army Gen. Hosp., Athens, 1972-75, dir. cardiac catheterization lab., 1975-76; dir. cardiology dept. 409 Army Gen. Hosp., Patras, 1976-78; cardiology cons. to chief Hellenic Nat. Def. Gen. Staff, 1978-80; mem. staff cardiology dept. Army Pansion Share Hosp., Athens, 1980-81; dir. div. cardiology dept. 401st Army Gen. Hosp., Athens, 1981-83; cons. cardiology Hellenic Red Cross Hosp., Athens, 1965-69; instr. Army Nursing Sch., Athens, 1976. Fellow Fedn. Internat. Sport Medicine, Am. Coll. Cardiology; mem. Athens Med. Assn., Hellenic Cardiologic Soc., Greek Assn. Sports. Greek Orthodox. Contbr. papers, abstracts to med. books, jours. Home: 3 Evrou St, GR 115 28 Athens Greece

GOUDY, JAMES JOSEPH RALPH, electronics executive, educator; b. Bloomfield, Iowa, Nov. 3, 1952; s. Charles Jacob and Marjorie Ethel (Morten) G.; m. Diane Marie Guenther, Nov. 24, 1978; children: Megan Joanne, Monica Victoria. BS, Wayne State Coll., 1976; AAS, Indian Hills C.C., Ottumwa, Iowa, 1978; MA, N.E. Mo. State U., 1980; BA, Iowa Wesleyan Coll., 1986. Cert. engring. technician Nat. Inst. Certification Engring. Technicians. Sr. electronic comm. cons. ANR Pipeline Co., Fairfield, Iowa, 1978—, instr. high tech., 1987—; sr. communications technician ANR Pipeline Co., Birmingham, Iowa, 1991—; owner Advanced Tech. Cons., 1993—; advanced tech. cons., 1993; temp. instr. Wayne (Nebr.) State Coll., 1976-77; instr. VA program Indian Hills C.C., Ottumwa, 1978, mem. high tech. programs adv. com., 1992—; instr. Iowa Wesleyan Coll., Mt. Pleasant, 1986. Bd. dirs. Wapello County Agrl. Fair, Eldon, 1988—; participant Nat. Runners Health Study, U. Calif. Mem. Masons, Shriners, Order Ea. Star. Avocations: amateur radio, computers, running. Home: RR 2 Box 81 Fairfield IA 52556-9550 Office: ANR Pipeline Co PO Box 9 Birmingham IA 52535-0009

GOUESBET, GERARD, systems and process engineering educator; b. Compiegne, France, Nov. 26, 1947; s. Maurice and Denise (Paillot) G.; m. Monique Levistre, July 26, 1969; children: Ludovic, Nicolas. Doctor, U. Rouen, France, 1973, D. of State, 1977. Lectr. Rouen U., 1974-75; researcher Nat. Ctr. Scientific Rsch., Rouen, 1975-93; prof. system and process engring. Nat. Inst. Applied Scis., Rouen, 1993—; cons. various French industries. Co-editor: Optical Particle Sizing: Theory and Practice, 1988; patentee in field; contbr. over 200 papers to proceedings, profl. jours. Grantee Kyoto U., 1988. Mem. AIAA, Am. Physica. Soc., Optical Soc. Am., French Soc. Physics, European Soc. Physics, Optical Soc. France. Avocations: hiking, music, gymnastics. Office: LESP INSA-Rouen VA CNRS 230, BP-08, 76131 Mont-Saint-Aignan France

GOUGÉ, SUSAN CORNELIA JONES, microbiologist; b. Chgo., Apr. 18, 1924; d. Harry LeRoy and Gladys (Moon) Jones; student Am. U., Washington, 1942-43, La. Coll., 1944-45; BS, George Washington U., 1948; postgrad. Georgetown U., 1956-58, 66-69, Vt. Coll. of Norwich U., M.A. in Pub. Health, 1984; m. John Oscar Gougé, Aug. 7, 1943; children: John Ronald, Richard Michael (dec.), Claudia Renée Gougé Carr. Med. technician Children's Hosp. Research Lab., Washington, 1948-49; bacteriologist George Washington U. Research Lab., D.C. Gen. Hosp., 1950-53; med. microbiologist Walter Reed Army Inst. Research, Washington, 1953-61; research asst. Dental Research, Walter Reed Army Med. Ctr., 1961-62; microbiologist antibiotics div. FDA, 1962-63; supr. quality control John D. Copanos Co., Pharms., Balt., 1963-64; research tng. asst. infectious diseases and tropical medicine Howard U. Med. Sch., 1964-65; research assoc. Georgetown U. Lab. Infectious Diseases, D.C. Gen. Hosp., 1966-69; mycologist Georgetown U. Hosp. Lab., 1969-70; microbiologist Research Found. of Washington Hosp. Ctr., 1971-73; dir. quality control Bio-Medium Corp., Silver Spring, Md., 1973-76; microbiologist Alcolac, Inc., Balt., 1976-77; microbiologist div. labs., dept. human resources Community Health and Hosps. Adminstrn., Washington, 1978-79; microbiologist div. ophthalmic devices, Office Device Evaluation Ctr. for Devices and Radiol. Health, FDA, Rockville, Md., 1979—. Sec. to exec. bd. Bethesda Project Awareness, 1970-71; vol. lead poisoning detection testing project, D.C. Office Vols. Internat. Tech. Assistance, 1970-71; vol. Zacchaeus Free Clinic, Washington, 1979-84. Mem. Nat. Capital Harp Ensemble, 1941-65; mem. parish social concerns com. Roman Cath. Ch., 1972-84. Recipient medal community service; registered microbiologist Nat. Registry Microbiologists; specialist microbiologist Am. Acad. Microbiology. Mem. AAAS, VITA, Am. Soc. for Microbiology, Am. Inst. Biol. Scis., Am. Chem. Soc., Internat. Union Pure and Applied Chemistry, N.Y. Acad. Scis., Am. Pub. Heath Assn., Bus. and Profl. Women (Capital Club, rec. sec. 1973-74, 1st v.p. 1974-75, pres. 1975-76), Winchester Bus. and Profl. Women, World Affairs Council of Washington D.C., Winchester-Frederick County Hist. Soc., Toastmasters Internat.(charter sec. BMD Club #3941 1979-80), Pi Kappa Delta, Sigma Xi. Methodist. Office: FDA Div Ophthalmic Devices Office Device Evaluation 1390 Piccard Dr Rockville MD 20850-4308

GOUGH, HARRISON GOULD, psychologist, educator; b. Buffalo, Minn., Feb. 25, 1921; s. Harry B. and Aelfreda (Gould) G.; m. Kathryn H. Whittier, Jan. 23, 1943; 1 child, Jane Kathryn Gough Rhodes. AB summa cum laude, U. Minn., 1942, AM (Social Sci. Research Council fellow 1946-47), 1947, PhD, 1949. Asst. prof. psychology U. Minn., 1948-49; asst. prof. U. Calif.-Berkeley, 1949-54, assoc. prof., 1954-60, prof., 1960-86, prof. emeritus, 1986—, assoc. dir. Inst. Personality Assessment and Research, 1964-67, dir., 1973-83, chmn. dept. psychology, 1967-72; cons. VA, 1951—; dir. cons. Psychologists Press, Inc., 1956—; mem. research adv. com. Calif. Dept. Corrections, 1958-64, Calif. Dept. Mental Hygiene, 1964-69, Gov.'s Calif. Adv. Com. Mental Health, 1968-74, citizens adv. council Calif. Dept. Mental Hygiene, 1968-71; clin. projects research review com. NIMH, 1968-72. Served to 1st lt. AUS, 1942-46, 1986. Recipient U. Calif. the Berkeley citation, 1986, Bruno Klopfer Disting. Contbn. award Soc. Personality Assessment, 1987; Fulbright research scholar, Italy, 1958-59, 65-66; Guggenheim fellow, 1965-66. Mem. Am., Western psychol. assns., Soc. Personality Assessment, Internat. Assn. Cross-Cultural Psychology, Académie National de Psychologie, Soc. Mayflower Desc., Phi Beta Kappa. Clubs: Commonwealth (San Francisco), Capitol Hill (Washington). Author: Adjective Check List, California Psychological Inventory, other psychol. tests; chmn. bd. editors U. Calif. Publs. in Psychology, 1956-58; cons. editor Assessment, 1993—, Jour. Cons. and Clin. Psychology, 1956-74, 77-84, Jour. Abnormal Psychology, 1964-74, Jour. Personality and Social Psychology, 1981-84, Med. Tchr., 1978-84, Cahiers d'Anthropologie, 1978-84, Population and Environment: Behavioral and Social Issues, 1977-80; Current Psychol. Research and Revs., 1985-93, Pakistan Jour. Psychol. Research, 1985—, Jour. Personality Assessment, 1986—, Psychological Assesment, 1991-92, Psychopathology and Behavioral Assessment, 1992—; assoc. editor Jour. Cross-Cultural Psychology, 1969-81. Home: PO Box 909 Pebble Beach CA 93953-0909 Office: U Calif Inst Personality and Social Rsch Berkeley CA 94720

GOUIN, WARNER PETER, electrical engineer; b. Internat. Falls, Minn., Sept. 14, 1954; s. Joseph Andre and Rose Marie (Grandaw) G.; m. Judith Ann Nelson, Aug. 25, 1979; 1 child, Nicole Renee. AA, Rainy River Community Coll., 1974; BS Mgmt., St. Cloud State U., 1979; BSEE, N.D. State U., 1985, MS in Indsl. Engring. and Mgmt., 1987. Purchasing/prodn. contr. Plastech Rsch., Inc., Rush City, Minn., 1979-80; inventory supr. Aero Systems Engring., St. Paul, 1980-81; grad. asst. N.D. State U., 1985-87; elec. engr. Marvin Windows, Warroad, Minn., 1987-93; systems integrator MIS dept., 1993—; project mgmt. trainer Process Re-Engring., TQM. Scoutmaster Boy Scouts Am., Warroad, 1989-91. Mem. IEEE, Office Automation Soc. Internat. (editor 1989—), Project Mgmt. Inst., Computer Soc. of IEEE, Inst. Indsl. Engrs. Avocations: computer integrated mfg. rsch., fishing, hunting, walking, guitar. Office: Marvin Windows PO Box 100 Warroad MN 56763

GOULD, DONALD EVERETT, chemical company executive; b. Concord, N.H., May 19, 1932; s. Everett Luther and Gladys (Wilcox) G.; B.S. in Chem. Engring., U. N.H. 1954; postgrad. math. Rutgers U., 1955-59; m. Marilyn Bachelder, June 13, 1953; children—Barbara, Allen, Douglas. Devel. chem. engr. plastics div. Union Carbide Co., Bound Brook, N.J., 1954-59, tech. service engr., Bound Brook and Wayne, N.J. 1959-64, mgr. tech. service indsl. bag dept., Wayne, 1964-66, mgr. tech. services indsl. fabricated products dept. 1966-67, mktg., mgr. indsl. bags, 1967-69, sr. packaging engr., 1969-72, mgr. packaging, 1972-74, mgr. distbn. safety and regulations, 1974-79, staff engr. for packaging, 1980-85, sr. staff engr. packaging, labeling, 1985-91, prin. engr. packaging, labeling, and regulations, 1991—. Mem. Inst. Packaging Profls. (vice chmn. films, foils and laminations com. 1962-64, chmn. 1964-66, sect. leader bottle containers, chmn. bag com. 1975-78, 85-88, exec. com. chem. packaging 1985—, hon. life mem., 1992), Am. Soc. Quality Control, Chem. Mfrs. Assn. (chmn. distbn. work group), Am. Council for Chem. Labeling, Alpha Chi Sigma. Club: Packanack Lake Country. Contbr. articles profl. jours., also to Ency. Engring. Materials and Processes. Home: 98 Lake Dr E Wayne NJ 07470-4253 Office: River Rd PO Box 670 Bound Brook NJ 08805-0670

GOULD, GORDON, physicist, retired optical communications executive; b. N.Y.C., July 17, 1920; s. Kenneth Miller and Helen Vaughn (Rue) G. B.S. in Physics, Union Coll., 1941, D.Sc., 1978; M.S. in Physics, Yale U., 1943, Columbia U., 1952. Physicist Western Electric Co., Kearny, N.J., 1941; instr. Yale U., 1941-43; physicist Manhattan Project, 1943-45; engr. Semon Bache Co., N.Y.C., 1945-50; instr. CCNY, 1947-54; research asst. Columbia U., 1954-57; research dir. TRG, Inc./Control Data Corp., Melville, N.Y., 1958-67; prof. electrophysics Bklyn. Poly. Inst., 1967-74; v.p. engring./mktg., dir. Optelecom, Inc., Gaithersburg, Md., 1974-85; dir. Optelecom, Inc., 1979—; bd. dirs. Polygon Inc., Dillon, Colo. Contbr. sci. articles to profl. jours. Recipient 63 research grants and contracts, 1958—; named Inventor of Year for laser amplifier Patent Office Soc., 1978; John Scott award for laser Phila. City Trust, 1983. Mem. Am. Inst. Physics, Optical Soc. Am., IEEE, AAAS, Fiber Optic Communications Soc., Laser Inst. Am. (pres. 1971-73, dir. 1971—). Patentee in field. Home and Office: RR 1 Box 112 Kinsale VA 22488-9720

GOULD, PHILLIP L., civil engineering educator, consultant; b. Chgo., May 24, 1937; s. David J. and Belle (Blair) G.; m. Deborah Paula Rothholtz, Feb. 5, 1961; children—Elizabeth, Nathan, Rebecca, Joshua. B.S., U. Ill., 1959, M.S., 1960; Ph.D., Northwestern U., 1966. Structural designer Skidmore, Owings & Merrill, Chgo., 1960-63; prin. structural engr. Westenhoff & Novick, Chgo., 1963-64; NASA trainee Northwestern U., Evanston, Ill., 1964-66; asst. prof. civil engring. Washington U., St. Louis, 1966-68, assoc. prof., 1968-74, prof., 1974—, chmn. dept. civil engring., 1978—, Harold D. Jolly prof. civil engring., 1981—; vis. prof. Ruhr U., Fed. Republic Germany, 1974-75, U. Sydney, Australia, 1981, Shanghai Inst. Tech., Peoples Republic of China, 1986; dir. Earthquake Engring. Rsch. Inst., exec. coun. Internat. Assn. for Shelland Spatial Structures, pres. Great Lakes chpt. Earthquake Engring. Rsch. Inst. Author: Static Analysis of Shells: A Unified Development of Surface Structures, 1977, Introduction to Linear Elasticity, 1984, Finite Element Analysis of Shells of Revolution, 1985, Analysis of Shells and Plates, 1987; co-author: Dynamic Response of Structures to Wind and Earthquake Loading, 1980; co-editor: Environmental Forces on Engineering Structures, 1979, Natural Draught Cooling Towers, 1985; editor: Engineering Structures, 1979—. Served to 1st lt. U.S. Army, 1959-61. Recipient Sr. Scientist award Alexander von Humboldt Found., Fed. Republic Germany, 1974-75. Fellow ASCE (bd. dirs. St. Louis sect. 1985-87); mem. Am. Soc. Engring. Edn., Internat. Assn. Shell Structures, Am. Acad. Mechanics, Structural Engrs. Assn. Ill., Civil Engring. Alumni Assn. U. Ill., Urbana-Champaign (disting. alumnus award), Sigma Xi. Home: 102 Lake Forest Saint Louis MO 63117 Office: Washington U Dept Civil Engring Box 1130 Saint Louis MO 63130

GOULDING, MERRILL KEITH, engineer, consultant; b. Erie, Pa., Jan. 21, 1933; s. Forest Clute and Felicita Clara (Johnson) G.; BS, UCLA, 1968, PhD, 1979; children: Merrill, Robert, Nida, Gina, Asst. to v.p. Internat. Controls Corp., 1963-69; chmn. bd. Village Verde Corp., 1963-64; pres. Merrill K. Goulding & Assocs., Inc., Los Angeles, 1974—; chief exec. officer Coin Cop Electronics Co., 1975-88; bd. dirs. Mid City Travel Industry; cons. FAA, DOT, DNA. Bd. dirs. Rio Hondo Area Action Com., 1970; guiding counselor Inst. Cultural Affairs; past pres. Request Computer Users Group. Served with USMC, 1953. Registered profl. engr., N.Y., Calif. Mem. ASME, IEEE, AIAA, NSPE, Calif. Soc. Profl. Engrs., Am. Soc. Metals, Constrn. Specifications Inst., Soc. Material and Process Engrs., Vols. in Tech. Assistance, Mensa, Am. Legion, Blue and Gold Circle Alumni Assn. UCLA. Republican. Clubs: Calif. Yacht. Lodges: Shriners, Los Angeles Consistory. Address: 8616 La Tijera Blvd Ste 300 Los Angeles CA 90045

GOULIANOS, KONSTANTIN, physics educator; b. Salonica, Greece, Nov. 9, 1935; came to U.S., 1958. naturalized, 1967; s. Achilles and Olga (Nakopoulou) G. Student, U. Salonica, 1953-58; Ph.D., Columbia U., 1963. Research assoc. Columbia U., N.Y.C., 1963-64; instr. physics Princeton U., N.J., 1964-67, asst. prof., 1967-71; assoc. prof. physics Rockefeller U., N.Y.C., 1971-81, prof., 1981—. Patentee electronic device of analysis of radioactivitively labeled gel electrophoretograms. Fulbright scholar, 1958-59. Home: 11 W 69th St Apt 4A New York NY 10023-4742 Office: Rockefeller U Lab Expt High-Energy Physics 1230 York Ave New York NY 10021-6341

GOUPY, JACQUES LOUIS, chemiometrics engineer; b. Paris, France, June 14, 1934; s. Paul Louis Goupy and Henriette Emilie (Augustine) Group Pouyet; m. Chantal Balut, Dec. 21, 1957 (div. 1982); children: Renaud, Sylvaine, Anne; m. Nicole Guillemain, Mar. 9, 1990. B, Lycee Henri IV, Paris, 1953; Ingenieur, Ecole de Physique Chimie, Paris, 1958. Registered profl. engr., Paris. Rsch. engr. Quartz et Silice, Boulogne, France, 1961-66; devel. indstl. engr. Quartz et Silice, Nemours, France, 1966-69; programme de recherche Total, paris, 1969-72; chief lab. Total, Courbevoie, France, 1972-88; conseiller Sci. Total, Paris, 1988—; cons. in exptl. designs. Author: La Methode des Plans d'Experiences, 1988, Methods for Experimantal Design, 1993; contbr. articles to profl. jours. Counselor Enseignement Technologique, Paris, 1970-92. With French Mil., 1958-61. Mem. French Chem. Soc. (pres. analyt. chemistry chpt. 1993). Home: 1 Rue Mignet, 75016 Paris France

GOURLEY, PAUL LEE, physicist; b. Fargo, N.D., Mar. 15, 1952; s. Harry Lee and Phyllis Niome (Cuty) G.; m. Gail Maureen Goforth, July 11, 1981; children: Cheryl Rose, Brett Andrew. BS with honors, U. N.D., 1974; MS, U. Ill., 1976, PhD, 1980. Rsch. asst. U. N.D., Grand Forks, 1972-73, U.S. Bur. Mines, Grand Forks, 1973-74; teaching asst., then rsch. asst. U. Ill., Urbana, 1974-80; mem. tech. staff Sandia Nat. Labs., Albuquerque, 1980-87, supr., 1987—, contract mgr., 1987—. Author: Protect Your Life in the Sun, 1993; contbr. sci. articles on solids, semiconductors and lasers, optical properties of materials to profl. publs. Discussion leader Bible Study Fellowship, Albuquerque, 1991-93; co-founder High Light Pub., Albuquerque, 1993. Recipient awards U.S. Dept. Energy, 1985, 91. Mem. Am. Phys. Soc. (symposium organizer), Optical Soc. Am. (presider), Norski Ski Club, U. N.D. Hockey, One Hundred Club U. N.D. Achievements include discovery of biocavity laser, invention of epitaxial surface-emitting laser wafer, discovery of excitonic molecules in silicon; research on semiconductor photonic lattice and surface-emitting lasers and physics of operation, critical layer thickness in III-V lattice-mismatched epitaxy. Home: 12508 Loyola St NE Albuquerque NM 87112 Office: Sandia Nat Labs Semiconductor Physics Albuquerque NM 87185

GOVIER, GEORGE WHEELER, petroleum engineer; b. Nanton, Alta., Can., June 15, 1917; married; 3 children. BASc, U. B.C., 1939; MSc, U. Alta., 1945; ScD in Chem. Engring., U. Mich., 1949; LLD (hon.), U. Calgary. Plant operator Standard Oil Co., B.C., 1939-40; from lectr. to asst. prof. chem. engring. U. Alta., 1940-48, prof. chem. engring., head dept. chem. and petroleum engring., 1948-59, dean faculty engring., 1959-63; chmn. Energy Resources Conservation Bd., 1962-78; pres. Govier Consulting Svcs. Ltd., 1978—; mem. permanent coun. World Petroleum Congress, 1960—; part time prof. U. Alta., U. Calgary, 1963—; v.p., bd. dirs. Petroleum Recovery Rsch. Inst., 1966—; chmn. sci. program com. World Petroleum Congress, 1973-85; dir. Can. Foremost Ltd. & Texaco, Inc., 1979—, Can.-Mont. Gas Co. Ltd., Can.-Mont. Pipe Line Co., Roan Resources Ltd., C-E Combustion Engine Ltd. & Stoned Webster Ltd., 1980—, Bow Valley Resources Ltd., 1981—, Coop. Energy Devel. Corp., 1982—. Recipient Anthony F. Lucas Gold medal Soc. Petroleum Engrs., 1990, Am. Inst. Mech. Engrs., 1989. Fellow Chem. Inst. Can. (R. S. Jane Meml. award 1966, Selwyn G. Blaylock medal 1971); mem. AICE, Nat. Acad. Engring. Achievements include research in pipeline flow of complex mixtures, especially non-Newtonians, gas-liquid, liquid-liquid and solid-liquid mixtures. *

GOVINDJEE, biophysics and biology educator; b. Allahabad, India, Oct. 24, 1933; came to U.S., 1956, naturalized, 1972; s. Visheshvar Prasad and Savitri Devi Asthana; m. Rajni Varma, Oct. 24, 1957; children: Anita Govindjee, Sanjay Govindjee. B.Sc., U. Allahabad, 1952, M.Sc., 1954; Ph.D., U. Ill., 1960. Lectr. botany U. Allahabad, 1954-56; grad. fellow U. Ill., Urbana, 1956-58; research asst. U. Ill., 1958-60, USPHS postdoctoral trainee biophysics, 1960-61, mem. faculty, 1961—, assoc. prof. botany and biophysics, 1965-69, prof. biophysics and plant biology, 1969—, disting. lectr. Sch. Life Scis., 1978. Author: Photosynthesis, 1969; editor: Bioenergetics of Photosynthesis, 1975, Photosynthesis: Carbon Assimilation and Plant Productivity; Energy Conversion by Plants and Bacteria, 2 vols, 1982 (Russian translation 1987); Light Emission by Plants and Bacteria, 1986; Excitation Energy and Electron Transfer in Photosynthesis, 1986, Molecular Biology of Photosynthesis, 1989, Photosynthesis: From Photoreactions to Productivity, 1993; guest editor spl. issue: Biophys. Jour., 1972; spl. issue: Photochemistry and Photobiology, 1978; editor in chief Photosynthesis Research, 1985-88, Advances in Photosynthesis, 1992—; editor Historical Corner: Photosynthesis Research, 1989—; co-author articles in Scientific American, 1965, 74, 90; contbr. articles to profl. jours. Fulbright scholar, 1956-61. Fellow AAAS, Nat. Acad. Scis. (India); mem. Am. Soc. Plant Physiologists, Biophys. Soc. Am., Am. Soc. Photobiology (council 1976, pres. 1981), Sigma Xi. Home: 2401 Boudreau Dr Urbana IL 61801-6655

GOWEN, RICHARD JOSEPH, electrical engineering educator, college president; b. New Brunswick, N.J., July 6, 1935; s. Charles David and Esther Ann (Hughes) G.; m. Nancy A. Applegate, Dec. 28, 1955; children: Jeff, Cindy, Betsy, Susan, Kerry. B.S. in Elec. Engring., Rutgers U., 1957; M.S., Iowa State U., 1961, Ph.D., 1962. Registered profl. engr., Colo. Research engr. RCA Labs., Princeton, N.J., 1957; commd. USAF; ground electronics officer Yaak AFB, Mont., 1957-59; instr. USAF Acad., 1962-63, research assoc., 1963-64, asst. prof., 1964-65, assoc. prof., 1965-66, tenured assoc. prof. elec. engring., 1966-70, tenured prof., 1971-77, dir., prin. investigator NASA instrumentation group for cardiovascular studies, 1968-77; mem. launch and recovery med. team Johnson Space Ctr., NASA, 1971-77; v.p., dean engring., prof. S.D. Sch. Mines and Tech., Rapid City, 1977-84, pres., 1987—; pres. Dakota State U., Madison, 1984-87; prin. investigator program in support space cardiovascular studies NASA, 1977-81; co-chmn. Joint Industry, Nuclear Regulatory IEEE, Am. Nuclear Soc. Probabilistic Risk Assessment Guidelines for Nuclear Power Plants Project, 1980-83; mem. Dept. Def. Software Engring. Inst. Panel, 1983; bd. dirs. ETA Systems, Inc., St. Paul, Minn., 1983-89, Data Max, Inc. Contbr. articles to profl. jours.; patentee in field. Bd. dirs. St. Martins Acad., Rapid City, S.D. Bd. dirs. St. Martins Acad., Rapid City, S.D., Greater Rapid City Econ. Devel. Partnership, 1991—, Data Max Inc. Fellow IEEE (Centennial Internat. pres. 1984, USAB/IEEE Disting. Contbns. to Engring. Professionalism award 1986); mem. Am. Assn. Engring. Socs. (chmn. 1988), Rotary, Sigma Xi, Phi Kappa Phi, Tau Beta Phi, Eta Kappa Nu, Pi Mu Epsilon. Roman Catholic. Home: 1609 Palo Verde Dr Rapid City SD 57701-4461 Office: SD Sch Mines & Tech Office of Pres Rapid City SD 57701

GOYLE, RAJINDER KUMAR, civil engineer, consultant; b. New Delhi, India, Aug. 1, 1945; came to U.S., 1971; s. Vidya Sagar and Bhagirthi G.; m. Madhur Bhashini, Jan. 28, 1973; children: Ashu K., Charu. MSCE, U. Wis., Milw., 1977. Registered profl. engr., Ohio, W.Va., Ky., Wis., Ind., Pa. Prin. engr. MEGT, Inc., Marietta, Ohio, 1977-91; owner, cons. Goyle Engring., Inc., Marietta, 1991—. Mem. ASCE. Home: 115 Rauch Dr Marietta OH 45750

GOZANI, TSAHI, nuclear physicist; b. Tel Aviv, Nov. 25, 1934; came to U.S., 1965; s. Arieh and Rivcca (Meiri) G.; m. Adit Soffer, Oct. 14, 1958; children: Mor, Shai N., Or P., Tal. BSc, Technion-IIT, Haifa, Israel, 1956, MSc, 1958; DSc, Swiss Fed. Inst. Tech. (ETH), Zurich, Switzerland, 1962. Registered profl. nuclear engr., Calif.; accredited nuclear material mgr. Rsch. physicist Israel Atomic Energy Commn., Beer-Sheva, 1962-65; rsch. assoc. nuclear engring. dept. Rensselaer Poly. Inst., Troy, N.Y., 1965-66; sr. staff scientist General-Atomic & IRT, San Diego, 1966-70, 71-75; prof. applied physics Tel Aviv U., 1971; chief scientist, div. mgr. Sci. Applications Internat. Corp., Palo Alto and Sunnyvale, Calif., 1975-84; v.p., chief scientist Sci. Applications Internat. Corp., Sunnyvale, 1984-87; corp. v.p. Sci. Applications Internat. Corp., Santa Clara, Calif., 1987—. Author: Active Non-Destructive Assay of Nuclear Materials, 1981; co-author: Handbook of Nuclear Safeguards Measurement Methods, 1983; contbr. over 150 articles to profl. jours. Recipient 1989 Laurel award Aviation Week Jour., R&D 100 awd, 1988, Most Innovative New Products, nominee for the Safe Skies award Conway Data Inc., 1991, 92, 93. Fellow Am. Nuclear Soc.; mem. Am. Phys. Soc., Inst. Nuclear Material Mgrs., Indsl. Liaison-Fermi Lab. Achievements include 6 patents in Explosive Detection, Explosive Detection System Using an Artificial Neural System, Multi Sensor Explosive Detection System, Apparatus and Method for Detecting Contraband Using Fast Neutron Activation, Contraband Detection System Using Direct Imaging Pulsed Fast Neutrons, 1989-92; invented method to measure nuclear reactor's reactivity, 1963. Office: Sci Applications Internat Corp 2950 Patrick Henry Dr Santa Clara CA 95054-1813

GOZUM, MARVIN ENRIQUEZ, internist; b. Phila., July 19, 1960; s. Filemon Tizon and Teresita Ver G. BS in Biology, Ateneo de Manila, The Philippines, 1980; MD, Fatima Coll. of Medicine, The Philippines, 1984. Intern internal medicine The Bklyn.-Caledonian Hosp. div. Downstate Med. Ctr., N.Y., 1984-85, resident internal medicine, 1985-87; attending physician Thomas Jefferson U. Hosp., Phila., 1987—; clin. instr. Jefferson Med. Coll., Phila., 1987-89, rsch. assoc. Ctr. for Rsch., 1989—, clin. asst. prof. medicine, 1989—, chief med. informatics div. internal medicine, 1989—; med. cons. Wills Eye Hosp., Phila., 1987—, chief med. cons., 1990—; mem. adv. bd. computers in medicine com. Thomas Jefferson U. Hosp., 1987—; mem. adv. bd. computer com. Wills Eye Hosp., 1987; mem. adv. bd. curriculum devel. com. Thomas Jefferson U., Phila., 1988; mem. adv. bd. Continuing Med. Edn. Com. Internal Medicine, 1991—. Developer: (computer programs) Diagnosticon Computer Assisted Diagnosis, 1982, Fluid/Electrolyte Calculator, 1984, Preoperative Evaluation, 1987; co-developer: (computer program) VACAD Image Processor, 1987. Named to Osteoporosis Project, Health Sci. Inst., 1990. Mem. AAAS, Am. Med. Informatics Assn., Soc. Gen. Internal Medicine. Achievements include development of computer assisted preoperative evaluation, automated report generation for preoperative evaluations, automated medical diagnosis, pocket intensive care calculator. Office: Jefferson Med Coll 1025 Walnut St Philadelphia PA 19107-5083

GRABOWSKY, CRAIG, aerospace engineer; b. Waltham, Mass., Feb. 16, 1957; s. Fred and Laila (Williams) G.; m. Tami Gay Macri, June 29, 1979; 1 child, Steven. BS in Aerospace Engring., U.S. Naval Acad., 1979. Registered profl. engr., Va. Officer, maj. USMC, Camp Lejeune, N.C., 1979-84; plant engr. Rochester (N.Y.) Products, Div. of GM, 1984; lead engr. Sverdrup Tech. Inc., Cleve., 1984-88; sr. staff mgr. BDM Internat., Columbia, Md., 1988—. Mem. AIAA, NSPE, Marine Corps Res. Officers Assn. Achievements include principle design and test engineering for the solid surface combustion experiment which flew on several space shuttle missions. Home: 2678 Walston Rd Mount Airy MD 21771 Office: BDM Internat 9705 Patuxent Woods Dr Columbia MD 21046

GRACE, GEORGE WILLIAM, linguistics educator; b. Corinth, Miss., Sept. 8, 1921; s. Herbert Landrum and Florence Estelle (Kennedy) G.; m. Peggy Anne Plummer, May 30, 1960 (div. Jan. 1970); children: Mark Lawrence, Erika Deanna; m. Elizabeth Matson Foster, Mar. 8, 1992. Licence, U. de Genève, Switzerland, 1948; PhD, Columbia U., 1958. Asst. prof. U. N.C., Greensboro, 1958-59; vis. asst. prof. Northwestern U., Evanston, Ill., 1959-60; asst. prof. So. Ill. U., Carbondale, 1960-63, assoc. prof., 1963-64; prof. U. Hawaii, Honolulu, 1964—. Editor-in-chief Oceanic Linguistics, 1961-91; author: (monograph) The Position of the Polynesian Languages, 1959, An Essay on Language, 1981, The Linguistic Construction of Reality, 1987; co-author: (monograph) The Sparkman grammar of Luiseño, 1960. 1st lt. USAAF, 1942-46. Fellow AAAS, Am. Anthrop. Assn.; mem. Linguistic Soc. Am. (life). Home: 2333 Kapiolani Blvd # 1914 Honolulu HI 96826

GRACE, MICHAEL JUDD, immunologist; b. Chambley, France, Oct. 20, 1957; came to U.S., 1958; s. Judd Harper and Laura Belle (Davidson) G.; m. Jo Ann Fox, June 2, 1979; children: Christina Marie, Stephanie Ann, David Michael. BS, U. Nebr., 1979, PhD, 1984. Scientist Procter and Gamble Corp., Cin., 1984-88; sr. scientist Schering-Plough Corp., Bloomfield, N.J., 1988-90; prin. scientist Schering, Plough Rsch. Inst., Kenilworth, N.J., 1990—; adv. bd. dept. chemistry U. Nebr., Lincoln, 1991—. Contbr. articles to profl. jours. Miessner Minerva grantee U. Nebr., 1976; Avery fellow U. Nebr., 1979. Mem. Am. Soc. Microbiologists, Reticuloendothelial Soc., N.Y. Acad. Scis., Phi Lambda Upsilon. Achievements include patent for method for increasing numbers of neutrophils and for treating wounds with IL-4, patent for method for increasing and activating monocytes and neutrophils and for inducing maturation of myeloid cells with IL-5. Home: 22 Bonnie Rae Dr Yardville NJ 08620

GRACIA, JUAN MIGUEL, mathematics educator; b. Tudela, Navarra, Spain, Mar. 15, 1945; s. Jorge and Carmen (Melero) G.; m. Maria del Carmen Vergara, July 8, 1972; 1 child, Maria. Lic. Math., U. Complutense, Madrid, 1970; D Math., U. Bilbao, Spain, 1978. Asst. prof. U. Navarra, Pamplona, Spain, 1970-74; adj. prof. U. Valladolid, Vitoria, Spain, 1974-78; adj. prof. U. Pais Vasco, Vitoria, 1978-84, titular prof. math., 1984—. Editor: Algebra Lineal y Aplicaciones, 1984; reviewer Jour. Math. Revs., 1989; contbr. articles to profl. jours. V.p. Assn. for Defence of U. Campus of Alava, Vitoria, 1987-91. Mem. Soc. Indsl. and Applied Math., Internat. Linear Algebra Soc. Avocations: reading, listening to music. Home: Rioja 27 - 2o izda, E-01005 Vitoria Alava, Spain Office: U Pais Vasco, Ftad de Farmacia Apdo 450, E-01080 Vitoria Alava, Spain

GRADIJAN, JACK ROBERTSON, software engineer; b. Worcester, Mass., May 12, 1932; s. John and Viola May (Watt) G.;m. Christine Malin, Apr. 21, 1963; children: Paul Timothy, Stephen John, Mark David, David Andrew. BS, MS in Elec. Engring., MIT, 1955. USPHS trainee neurobiology U. Rochester, 1963-66; computer cons. Rockland Rsch. Inst., Orangeburg, N.Y., 1966-75; computer scientist Lederle Labs., Pearl River, N.Y., 1975-77; software engr. GTE-Sylvania, Needham, Mass., 1977-79; mgr. software devel. Datamedix, Sharon, Mass., 1979-82, Instrumentation Lab., Lexington, Mass., 1982-86; sr. staff engr. CIBA Corning Diagnostics, Medfield, Mass., 1986—. Contbr. articles to profl. jours. Mem. IEEE. Lutheran. Home: 52 Walnut St Needham MA 02192 Office: CIBA Corning Diagnostics 63 North St Medfield MA 02052

GRADY, LEE TIMOTHY, pharmaceutical chemist; b. Chgo., Mar. 21, 1937; s. Thomas Aloysius and Lentella Kathryn (Eibel) G.; m. Ann Marie Gill, Aug. 8, 1964; children: Patricia Ann, Meghan Elizabeth. BS in Pharmacy with high honors, U. Ill., 1959, PhD in Chemistry, 1963. Registered pharmacist, Ill., Va., Md. Sr. rsch. pharmacologist Merck Inst. Therapeutic Rsch., West Point, Pa., 1965-68; dir. drug standards lab. Am. Pharm. Assn. Found., Washington, 1968-74; dir. drug rsch. and testing lab. U.S. Pharmacopeia, Rockville, Md., 1975-78, dir. drug standards div., 1979—; mem. expert coms. WHO, Geneva, Switzerland, 1980-87; temporary advisor Pan Am. Health Orgn., Washington, 1984; foreign adv. Chinese Pharmacopoeia, 1992; mem Pharmacopoeia Discussion Group (Internat), 1992. Contbr. articles to sci. jours.; sci. editor U.S. Pharmacopeia National Formulary, 1980—. Recipient Rsch. award Am. Soc. Hosp. Pharmacists, 1982. Fellow AAAS, Am. Assn. Pharm. Scientists; mem. Am. Pharm. Assn. (J.L. Powers Rsch. Achievement award 1990), Am. Chem. Soc., Internat. Pharm. Fedn., Rho Chi, Phi Kappa Phi. Roman Catholic. Avocations: swimming, hiking. Office: US Pharmacopoeia 12601 Twinbrook Pky Rockville MD 20852-1790

GRAEN, GEORGE B., psychologist, researcher; b. Mpls., Aug. 7, 1937; s. Clarence J. and Mary E. (Kuhn) G.; m. Joan A. Graen, Mar. 7, 1958; children: Michael R., Martin G. BA, U. Minn., 1961, MA, 1963, PhD, 1967. From rsch. asst. to assoc. U. Minn., Mpls., 1962-67; prof. psychology U. Ill., Urbana, 1967-77; prof. orgnl. behavior U. Cin., 1977-92; prof. orgnl. behavior Keio U., Tokyo, 1971-72, Nagoya (Japan) U., 1985-86. Author: Unwritten Rules, 1989, Cross-Cultural Leadership, 1992. Fellow Am. Psychol. Soc.; mem. Assn. Japanese Bus. Studies (pres. 1992-94). Home: 7412 Towerview Ln Cincinnati OH 45255 2101

GRAF, EDWARD DUTTON, grouting consultant; b. L.A., Dec. 31, 1924; s. John Edward and Florence Claire (Dutton) G.; m. Verna M. Greenfield (div.); children: Teri, Thomas, Eric; m. Joyce Main, Sept. 12, 1981. BSBA, UCLA, 1948. Field engr., estimator Bechtel Corp., 1950-52; engr., estimator EMSCO of San Francisco, 1952-55; founder, owner, pres. Pressure Grout Co., Foster City, Calif., 1955-86; ind. cons. in pressure grouting, 1986—; guest lectr. civil engring. Stanford (Calif.) U., U. Calif.-Berkeley, UCLA, Ga. Inst. Tech., Purdue U., Northwestern U., others; presenter seminars in geotech. grouting. Contbr. articles to profl. publs. Lt. USNR, World War II, PTO. Fellow Am. Concrete Inst. (past chmn. com. on geotech. cement grouting); mem. ASCE (chmn. nat. com. on grouting, past pres. San Francisco br., Martin S. Kapp Found. Engring. award 1990), Internat. Soc. Rock Mechanics Commn. on Rock Grouting, Soc. Mining Engrs., Assn. Engring. Geologists, Structural Engrs. Assn. Calif. Home and Office: 715 San Miguel Ln Foster City CA 94404

GRAF, JEFFREY HOWARD, cardiologist; b. N.Y.C., Apr. 6, 1955; s. Rudolf F. and Bettina L. (Knisbacher) G.; m. Roberta Ruth Rubin, June 26, 1982; children: Allison, David, Russell. BA magna cum laude, NYU, 1976, MD, 1980. Diplomate Am. Bd. Cardiology, Am. Bd. Internal Medicine. Cardiology fellow Mt. Sinai Med. Ctr., N.Y.C., 1983-86; cardiologist educator Mt. Sinai Sch. Medicine, Beth Israel Med. Ctr., N.Y.C., 1986—; cons., 1986—. Fellow Am. Coll. Cardiology; mem. Phi Beta Kappa, Alpha Omega Alpha. Office: 1111 Park Ave New York NY 10128

GRAF, KARL ROCKWELL, nuclear engineer; b. San Diego, Apr. 19, 1940; s. Frederic August and Beatrice (Rockwell) G.; m. Nancy Ann Scott, June 9, 1962; children: Robin Elizabeth, Scott Frederic. BS, U. S. Naval Acad., 1962. Submarine officer USN, 1962-84; sr. mgmt. cons. Advanced Sci. and Tech. Assn., Solana Beach, Calif., 1984; dir. nuclear support Ill. Power Co., Decatur, 1985, dir. ops. monitoring, 1986-89, dir. quality assurance, 1990-92, dir. engring. projects, 1992—; dep. comdr., readiness and tng. officer, Submarine Squadron One, USN, Pearl Harbor, Hawaii, 1982-84. Author: Monitoring Manual, 1986. Chmn. zoning bd. Village of Forsyth, Ill., 1988—, chmn. long-range plan com., 1989—. Mem. Am. Nuclear Soc., Am. Soc. for Quality Control, U.S. Submarine League, Ret. Officers Assn. Achievements include the development and implementation of an innovative monitoring program at Illinois Power Company's Clinton Nuclear Power Station to monitor, evaluate and trend such things as individual responsibility and professionalism and develop actions to improve performance standards relating to the nuclear reactor, steam turbine and electrical generating systems. Home: 736 Weaver Rd Decatur IL 62526-9777 Office: Ill Power Co 500 S 27th St Decatur IL 62525

GRAF, TIMOTHY L., mechanical engineer; b. Mpls., Aug. 11, 1963; s. Richard A. and Marilyn J. (Score) G. BSME, Rensselaer Poly. Inst., 1985. Sr. methods devel. engr. 3M Abrasive Systems Divsn., St. Paul, 1985—. Contbr. articles to profl. jours. Achievements include development and refinement of methods which allow robots to control forces accurately, allowing them to deburr, grind and finish parts accurately and quickly, used throughout industry. Office: 3M Abrasive Systems 240-01 3M Ctr Saint Paul MN 55144

GRAFTON, ROBERT BRUCE, science foundation official; b. Rochester, N.Y., May 15, 1935; s. Corydon Melvin and Beatrice (Hawes) G.; m. Carolyn Kolb Grafton, July 8, 1967 (div. Dec. 1990); children: Geoffrey Backus, Benjamin Robert. ScB, Brown U., Providence, R.I., 1958, PhD, 1967. Prof. U. Mo., 1967-71; visiting lectr. Leicester (England) U., 1970-71; prof. Trinity Coll., Hartford, Conn., 1971-75; program mgr. Office Naval Rsch., N.Y.C., 1975-78, Arlington, Va., 1978-86; program dir. NSF, Washington, 1986—; program com. COMPSAC Conf., Chgo., 1983-86. Facilitator New Beginnings, Washington, 1990—; Lt. USN, 1958-62. Recipient Svc. award IEEE Computer Soc., 1985. Avocations: music, hiking, reading, computational number theory. Office: Nat Sci Found 1800 G St NW Washington DC 20550-0002

GRAHAM, DONALD JAMES, food technologist; b. York, N.Y., Sept. 24, 1932; s. Howard Alexander Graham and Naomi Irene (Fletcher) Graham Horgan; m. Dorothy Jane Schroeder, Jan. 1, 1965; children: Christopher Howard, Jonathan Edward. AAS, N.Y. State Agrl. Tech. Inst., 1952; BS with honors, Mich. State U., 1958, MS, 1959; postgrad., Oreg. State U., 1959-62. Profit planning dir. Green Giant Co., LeSueur, Minn., 1962-67; dir. tech. svc. Green Giant of Can., Windsor, Ont., 1967-77; dir. quality assurance William Underwood Co., Westwood, Mass., 1977-83; internat. tech. dir. Pet, Inc., St. Louis, 1983-87; sr. food technologist, food sanitation cons., lectr. Sverdrup Corp., St. Louis, 1988—; faculty, com. mem. Food Processors Inst., Washington, 1980-92. Contbr. articles to tech. publs. Troop com. chmn. Boy Scouts Am., Medfield, Mass., 1979-82, treas., Chesterfield, Mo., 1984-89; mem. Minn. Rep. Com., 1965-67. Sgt. U.S. Army, 1952-54, Korea. Sverdrup fellow, 1993. Mem. Inst. Food Technologists, Internat. Assn. Milk, Food and Environ. Sanitarians, Inst. Thermal Processing Specialists (bd. dirs. 1980-82), Mo. Food Processors Assn. (bd. dirs. 1992—), Am. Soc. Quality Control, Alpha Zeta (chancellor Kedzie chpt. 1957-58). Avocations: photography, videotaping, genealogy. Home: 14318 Aitken Hill Ct Chesterfield MO 63017-2820 Office: Sverdrup Corp 801 N 11th St Saint Louis MO 63101-1000

GRAHAM, DOROTHY RUTH, software engineering consultant; b. Grand Rapids, Mich., June 11, 1944; d. Anthony Andrew and Ruth A. (Brink) Hoekema; m. Roger Graham, Aug. 16, 1969; children: Sarah Anne, James Stephen. Student, U. Coll., London, 1965-66; BA in Math., Calvin Coll., 1967; MS in Math., Purdue U., 1969. ATMS Bell Labs., Whippany, N.J., 1970-72; software engr. Ferranti Computer Systems, Manchester, Eng., 1973-79; sr. cons. The Nat. Computing Ctr., Manchester, 1979-80; cons. Cheadle Hulme, Eng., 1980-88; mng. dir. Grove Software Engring. Cons., Macclesfield, Eng., 1988—; program chair 1st European internat. conf. on software testing analysis and rev. Eurostar '93. Author: The Cast Report: Computer Aided Software Testing, 1991, (with Tom Gilb) Software Inspection, 1993; editorial bd. Jour. Software Testing, Verification and Reliability; contbr. articles to profl. jours. Mem. Brit. Computer Soc. (com. mem. spl. interest group on software testing, speaker, organizer 1988-93). Mem. Ch. Eng. Avocations: choral and madrigal singing. Home: Grove House 40 Ryles Pk Rd, SK11 8AH Macclesfield England Office: Grove Cons, Grove House 40 Ryles Pk Rd, SK11 8AH Macclesfield England

GRAHAM, JAMES, mineralogist; b. Perth, Australia, Oct. 13, 1929; s. James Mitchell and Dorothea Isobel (Thrum) G.; m. Valerie Lois Elms, Mar. 22, 1958; children: Carolyn, Dorothy, Margaret. MSc in Physics, U. Western Australia, 1953; PhD in Phys. Metallurgy, U. Birmingham, U.K., 1956. Rsch. fellow U. Birmingham, 1955-56; rsch. officer, sr. rsch. officer CSIRO, Melbourne, Victoria, Australia, 1956-63; prin. rsch. scientist CSIRO, Perth, Western Australia, 1964—; adv. bd. physics Curtin U., Perth, 1970—. Editor Australian Physicist, Perth, 1980-85; contbr. articles to profl. jours. Deacon Claremont Bapt. Ch., 1970—. Fellow Australian Inst. Physics, Brit. Inst. Physics; mem. Mineralogical Assn. Can., Mineralogical Soc. Am. (v.p. internat. coun. for applied mineralogy 1993—). Achievements include development of technique for trace analysis by electron microprobe analyses; development pyrolysis method for recovery of refractory gold. Home: 16 Marimba Crescent, City Beach Australia 6015 Office: CSIRO, Div Mineral Products, Private Bag PO, Wembley Australia 6014

GRAHAM, JAMES MILLER, physiology researcher; b. St. Louis, Oct. 16, 1945; s. Alvin Rudd and Edrie (Miller) G.; m. Linda Kay Edwards, May 3, 1969; children: Michael Edwards, Melissa Edwards. MA, U. Mich., 1968, PhD, 1979. Postdoctoral scholar environ. engring. U. Mich., Ann Arbor, 1979-80; lectr. zoology U. Wis., Madison, 1981-82, rsch. assoc. physiology, 1983-88, lectr. botany, 1987-88, physiology researcher, 1988—; reviewer Phycological Soc. Am., Jour. Great Lakes Rsch., Microbial Ecology. Contbr. chpt. to Periphyton of Freshwater Ecosystems, 1983; contbr. articles to Jour. Great Lakes Rsch., Microbial Ecology, Jour. Phycology, Aquatic Botany, Jour. Protozoology. With U.S. Army, 1969-72. Mem. Am. Soc. Limnology and Oceanography, Am. Inst. Biol. Scis., Phycological Soc. Am., Soc. Protozoologists, Phi Beta Kappa, Sigma Xi. Office: U Wis Dept Physiology 1300 University Ave Madison WI 53706-1532

GRAHAM, KENNETH ROBERT, psychologist, educator; b. Phila., June 5, 1943; s. Edgar and Margit (Leafgreen) Graham; m. Michele Carolyn Monroe, Aug. 10, 1968; children: Mark Andrew, Richard Alan. BA, U. Pa., 1964; PhD, Stanford U., 1969. Lic. psychologist, Pa. Asst. prof. Muhlenberg Coll., Allentown, 1970-77, assoc. prof., 1977-84, prof., 1984—, head psychology dept., 1984-93; rsch. psychologist Unit for Exptl. Psychiatry Inst. of Pa. Hosp., Phila., 1969-70; adjunct asst. prof. U. Pa., Phila., 1969-70; cons. smoking cessation various hosps., 1985—. Author: (text) Psychological Research, 1977; contbr. over 30 articles to profl. and sci. jours. Bd. dirs., pres. Lehigh Valley Child Care, Allentown, 1979-85; advisor Pathways (Conf. of Chs.), Allentown, 1989—, N.E. Pa. Synod Luth. Ch. in Am., Wescosville, Pa., 1989—. Mem. Am. Psychol. Assn. (pres. div. psychol. hypnosis 1980-81), Am. Soc. Clin. Hypnosis (asst. editor Jour. Clin. Hypnosis 1974—), Soc. for Clin. and Exptl. Hypnosis, Kiwanis Club (pres. Allentown, 1991-92). Democrat. Avocations: swimming, collecting glass paperweights and signatures of 19th century explorers. Office: Muhlenberg Coll Psychol Dept Allentown PA 18104

GRAHAM, KIRSTEN R., information service executive; b. Inglewood, Calif., July 20, 1946; d. Ray Selmer and Ella Louise (Carter) Newbury; m. Frank Sellers Graham, July 31, 1981. BS, U. Wis., Oshkosh, 1971; MS, U. Colo., 1980; postgrad., Army War Coll., 1987. Cert. Flight instr. Chief info. svc. Mont. State Dept. Labor and Industry, Helena, Mont.; dir., personal property and bus. lic. div. County of Fairfax, Va.; analyst officer U.S. Army Pentagon, Washington; battalion commdr. U.S. Army, Frankfurt, West Germany; assoc. prof. U.S. Army, West Point, N.Y.; del. People-to-People Women Computer Sci. Profls. program., China. Vice chair, bd. dirs. Helena Industries for Vocationally Disabled; del. to People's Republic of China Citizen's Amb. Program, 1993. LTC U.S. Army, 1964-88. Mem. Data Processing Mgr.'s Assn., Nat. Assn. State Info. Resource Execs. Achievements include selection as first woman in Signal Corps to command an operational battalion and command overseas.

GRAHAM, RICHARD DOUGLAS, computer company executive, consultant; b. Pascagoula, Miss., July 19, 1947; s. Robert A. and Lois Mary (Dillman) G.; m. Margaret Jean Laprade (div. May 1991); 1 child, Spence D. BS in Computer Sci., La. Tech. Coll., 1968. Cert. data processor. Programmer Nat. Shawmut Bank, Boston, 1968-69; mgr. First Nat. Bank, Mobile, Ala., 1969-75, Nat. Bank Commerce, Dallas, 1976-81; regional mgr. Apple Computer, Inc., Carrollton, Tex., 1981-83; pres. Graham Cons., Inc., Carrollton, 1983—; bd. dirs First Colony Bank, The Colony, Tex. Com. mem. Jaycees, Waltham, Mass., 1969; troop leader Boy Scouts Am., Carrollton, 1985, 86. Mem. Inst. for Cert. Computer Profls., Data Processing Mgmt. Assn. Republican. Methodist. Avocations: fishing, golfing, diving.

Office: Graham Cons Inc 1930 E Rosemeade Pky Ste 4 Carrollton TX 75007-2468

GRAHAM, ROBERT KLARK, lens manufacturer; b. Harbor Springs, Mich., June 9, 1906; s. Frank A. and Ellen Fern (Klark) G.; A.B., Mich. State U., 1933; B.Sc. in Optics, Ohio State U., 1937; O.D. (hon.), 1987; hon. Dr. Ocular Sci., So. Calif. Coll. Optometry, 1988; children (by previous marriage)-David, Gregory, Robin, Robert K., Janis, Wesley; m. Marta Ve Everton; children: Marcia, Christie. With Bausch & Lomb, 1937-40; Western mgr. Univis Lens Co., 1940-44, sales mgr., 1945-46; v.p., dir. research Plastic Optics Co., 1946-47; pres., chmn. bd. Armorlite, Inc., 1947-78; lectr. optics Loma Linda U.; assoc. prof. So. Calif. Coll. Optometry, 1948-60. Co-founder (with Hermann J. Muller) Repository for Germinal Choice; trustee Found. Advancement of Man; bd. dirs. Inst. for Research on Morality; bd. dirs. Intra-Sci. Research Found., v.p., 1980. Recipient Herschel Gold medal Germany, 1972, Feinbloom award Am. Acad. Optometry, 1987, Glenn Fry medal Physiol. Optics, 1992; named Disting. Alumnus, Ohio State U., 1987.. Fellow AAAS; mem. Am. Inst. Physics Profs., Optical Soc. Am., Am. Acad. Optometry, Rotary Club, Mensa, Sigma Xi. Republican. Author: The Evolution of Corneal Contact Lenses; The Future of Man; also articles in sci. publs. Inventor variable focus lens, hybrid corneal lens; directed devel. hard resin lenses. Home: 3024 Sycamore Ln Escondido CA 92025-7433 Office: Graham Internat Plz Ste 300 2141 Palomar Airport Rd Carlsbad CA 92009-1423

GRAHAM, RONALD LEWIS, mathematician; b. Taft, Calif., Oct. 31, 1935; s. Leo Lewis and Margaret Jane (Anderson) G.; children: Cheryl, Marc. Student, U. Chgo., 1951-54; BS, U. Alaska, 1958; MA, U. Calif., Berkeley, 1961, PhD, 1962; LLD (hon.), Western Mich. U., 1984; DSc, St. Olaf Coll., 1985, U. Alaska, 1988. Mem. tech. staff Bell Labs., Murray Hill, N.J., 1962—, head dept. discrete math., 1968—, dir. Math. Scis. Research Ctr., 1983—, adj. dir. research, info. scis. div., 1987—; prof. Rutgers U., 1987—; Regents' prof. UCLA, 1975; vis. prof. computer sci. Stanford U., 1979, 81, Princeton (N.J.) U., 1987, 89. Author: Ramsey Theory, 1980, Concrete Mathematics, 1989. Served with USAF, 1955-59. Recipient Polya prize, 1975; named Scientist of Yr., World Book Encyclopedia, 1981; scholar Ford Found., 1958, Fairchild Found. Disting. scholar Calif. Inst. Tech., 1983; fellow NSF, 1961, Woodrow Wilson Found., 1962. Fellow AAAS, N.Y. Acad. Scis.; mem. NAS, Am. Math. Soc. (pres. 1993-94), Math. Assn. Am., Soc. Indsl. and Applied Math., Assn. Computing Machinery, Internat. Jugglers Assn. (past pres.). Office: AT&T Bell Labs Murray Hill NJ 07974

GRAHAM, SUSAN LOIS, computer science educator, consultant; b. Cleve, Sept. 16, 1942; m., 1971. A.B. in Math., Harvard U., 1964; M.S., Stanford U., 1966, Ph.D. in Computer Sci., 1971. Assoc. research scientist, adj. asst. prof. computer sci. Courant Inst. Math. Sci., NYU, 1969-71; asst. prof. computer sci. U. Calif., Berkeley, 1971-76, assoc. prof., 1976-81, prof., 1981—; vis. scientist, Stanford U., 1981; adv. com. NSF div. computer and computation rsch., 1987-92, program for sci. and tech. ctrs., 1987-91; mem. MIT vis. com. for elec. engring. and computer sci., 1989—; mem. Nat. Rsch. Coun. div. on physical sci., math., and applications, 1992—. Co-editor Communications, 1975-79; editor Transactions on Programming Languages and Systems, 1978-92. NSF grantee. Fellow AAAS; mem. IEEE, Assn. Computing Machinery, Nat. Acad. Engring. Office: U Calif-Berkeley Computer Sci Div EECS Berkeley CA 94720

GRAMANN, RICHARD ANTHONY, aerospace engineer; b. Elgin, Ill., Feb. 14, 1960; s. Richard H. and Beatrice J. (Katsulos) G. BS, U. Va., 1982; PhD, U. Tex., 1989. System analyst BDM Corp., McLean, Va., 1981-83; engr. sci. assoc. McDonnell Douglas, Long Beach, Calif., 1983-84; grad. rsch. asst. U. Tex., Austin, 1984-86, grad. fellow, 1986-90, rsch. engring. assoc., 1990-91, rsch. engring. assoc. Applied Rsch. Labs., 1991—. Contbr. articles to profl. jours. Grad. fellow Army Res. Office, 1986-89. Mem. AIAA, Am. Soc. for Engring. Edn. Office: U Tex Applied Rsch Labs PO Box 8029 Austin TX 78713-8029

GRAMINES, VERSA RICE, telecommunications professional, retired; b. Dorset, Ohio, Sept. 24, 1907; d. Joel August and Cora Pearl (Day) Rice; m. Chris Thomas Gramines (wid.); children: Theofun Joel, Theodora. Student, Cleve. Coll. Sr. draftswoman Ohio Bell. Vol. Alzheimer Rsch., Fairhill Hosp., Cleve. Mem. IEEE, N.Y. Acad. Sci., Heights Civic Club. Methodist-Episcopalian.

GRANATSTEIN, VICTOR LAWRENCE, electrical engineer, educator; b. Toronto, Feb. 8, 1935; s. Charles Samuel and Bella (Godfrey) G.; m. Bethie Mills, Sept. 4, 1955; children—Rebecca Miriam, Abraham Solomon, Annie Sara Khaya. B.S., Columbia U., 1960, M.S., 1961, Ph.D., 1963. Research staff physicist Bell Telephone Labs. Murray Hill, N.J., 1964-72; head high power electromagnetic radiation br. Naval Research Lab., Washington, 1972-83; prof. elec. engring. U. Md., College Park, 1983—; acting dir. Lab. for Plasma Research, 1986-88; dir. 1988—. Vis. lectr. Hebrew U., Jerusalem, 1969-70; cons. BDM Corp., McLean, Va., 1981-83, Sci. Applications Corp., McLean, 1983—, Omega-P Inc., New Haven, 1983—, Pulse Scis. Inc., San Leandro, 1985-88, Jet Propulsion Lab., Pasadena, 1987—. Patentee microwave devices; contbr. articles to profl. jours.; editor Wave Heating and Current Drive in Magnetic Plasmas, 1985, High Power Microwaves, 1987. Pres., Bethesda Chevy Chase Jewish Community Group, 1983-84. Recipient R.D. Conrad Award Sec. Navy, 1981, Superior Civilian Service award, Office Naval Research, 1980, E.O. Hulbert award Naval Research Lab., 1980; Fulbright sr. scholar award, 1993-94. Fellow Am. Phys. Soc., IEEE (vice chmn. plasma sci. com. 1984-85, Plasma Sci. and applications award 1991). Democrat. Avocations: folk dancing; swimming. Home: 13508 Rippling Brook Dr Silver Spring MD 20906-3177 Office: U Md Lab Plasma Rsch College Park MD 20742

GRANBERRY, EDWIN PHILLIPS, JR., safety engineer, consultant; b. Orange, N.J., Aug. 20, 1926; s. Edwin Phillips Sr. and Mabel (Leflar) G.; children: Melissa, Edwin Phillips III, James, Jennifer. BS, Rollins Coll., 1950; MBA, Embry Riddle Aero. U., 1985. Cert. profl. chemist. Weapons system engr. Martin Co., Orlando, Fla., 1958-62; supt. indsl. safety Guided Missiles Range div. Pan Am. World Airways, Cape Canaveral, Fla., 1962-72; mgr. indsl. hygiene/safety engring. Pratt & Whitney, West Palm Beach, Fla., 1972-88; mgr. indsl. and systems safety engring. Chem. Systems div. United Techs. Corp., San Jose, Calif., 1988-89; pres. Granberry & Assocs., Winter Park, Fla., 1989—; adj. faculty mem. Valencia C.C., Orlando, Fla.; mem. Fla. State Toxic Substances Adv. Council, 1984-88, Fla. State Emergency Response Commn., 1988, chmn. local emergency response planning com. region 6, 1991-92. Scoutmaster Boy Scouts Am., 1946-74, dist. chmn. Wekiwa dist. Central Fla. council, 1946-74, also council commr. Served with USNR, 1944-46, PTO. Recipient Silver Beaver award, 1960. Fellow Am. Inst. Chemists; mem. Am. Welding Soc., Nat. Fire Protection Assn., Rollins Coll. Alumni Assn. (bd. dirs. 1958-61), Associated Industries Fla., Am. Chem. Soc., Am. Soc. Safety Engrs. (chmn. Gold Coast chpt. 1979-80, pres. 1981-84; regional v.p. 1984-88, v.p. divs. 1988-90, adminstr.-environ. div. 1990-92, nat. bd. dirs. 1984-90, Safety Profl. of Yr. Fla., Ga. and P.R. chpts. 1985, Safety Profl. Yr. divs. 1991), Safety Council Palm Beach County (pres. 1981-82, chmn. bd. 1983, treas. 1984), Am. Nat. Standards Inst., Am. Indsl. Hygiene Assn. Home: 521 Langholm Dr Winter Park FL 32789-5251 Office: Granberry & Assocs 2431 Aloma Ave Ste 276 Winter Park FL 32792-2566

GRANDI, ATTILIO, engineering consultant; b. La Spezia, Italy, Sept. 24, 1929; s. Luigi and Egle (Canese) G.; m. Maria Teresa Berti, Apr. 23, 1962; 1 child, Giovanni. Maturita scientifica, Liceo Scientifico Pacinotti, La Spezia, 1949; univ. degree in aero. engring., U. Pisa (Italy), 1958. Project engr. S.p.A. Piaggio, Pontedera, Italy, 1959-60; project engr. Termomeccanica Italiana, La Spezia, 1960-71, tech. mgr., 1971-85, rsch. and mktg. mgr., 1985-88; cons. hydraulic machinery and marine propulsive systems, 1988—. Patentee in field. Mem. Italian Standard Hydraulic Machinery. Roman Catholic. Avocations: mathematics, old languages, fishing.

GRANDON, RAYMOND CHARLES, physician, educator; b. Carlisle, Pa., Nov. 27, 1919; s. Frank Leonard and Teresa (Libonat) G.; m. Doris May New, June 8, 1945; children: Raymond, S. Suzanne, David. BS, Dickinson Coll., 1942; MD, Jefferson Med. Coll., 1945. Intern St. Lukes Hosp., Bethlehem, Pa., 1945-46; resident internal medicine Harrisburg (Pa.) Hosp., 1948-50; dir. Hahneman Affiliate Faculty, Harrisburg, 1958-64; assoc. prof. div. medicine Hahneman affiliate faculty Harrisburg Hosp., 1958-64; preceptor Physician Asst. program Hahneman Med. Coll. Physician's, 1975-76; clin. asst., prof. medicine Pa. State U. Milton S. Hershey Med. Ctr., 1979-80, clin. assoc. prof. medicine, 1980-89; mem. Corp. Pa. Blue Shield, 1975-92, Bd. Health Rsch. Inst. Pa., 1971-83, Bd. Health Corp. Pa. Skilled Nursing Home, 1970. Contbr. articles to Jour. AMA. Mem. Govnrs. Commn. Alcoholism, Pa., 1954-56; clinician Alcohol Counseling Ctr., Pa. Dept. Health, 1954-59. Recipient Founders award, Pa. Med. Soc. Auxiliary, 1984, Seilbert Meml. award, Harrisburg Acad. Medicine, 1957. Mem. ACP (life), AMA (del. 1972—), Am. Soc. Internal Medicine (trustee 1983-86, mem., pres. socio-econ. rsch. edn. found. 1985-86), Pa. Med. Soc. (pres. 1981-82), Pa. Soc. Internal Medicine (pres. 1974-76), Dauphin County Med. Soc., Med. Bur. Harrisburg (pres. 1971-76), Jefferson Med. Coll. Alumni Assn. (exec. com. 1960-85), Am. Heart Assn. (coun. clin. cardiology 1945—), Am. Diabetes Assn., N.Y. Acad. Scis., World Med. Assn., Assn. Life Ins. Med. Dirs. Am., Forum for Med. Affairs (pres. 1991), Pa. State Bd. Med. Edn. and Licensure, Phi Kappa Sigma, Phi Chi. Office: 131 State St Harrisburg PA 17101-1087

GRANDY, WALTER THOMAS, JR., physicist; b. Phila., June 1, 1933; s. Walter Thomas and Margaret Mary (Hayes) G.; m. Patricia Josephine Langan, Dec. 27, 1955; children: Christopher, Neal, Mary, Jeanne. BS, U. Colo., 1960, PhD, 1964. Physicist Nat. Bur. Standards, Boulder, Colo., 1958-63; mem. faculty U. Wyo., Laramie, 1963—; prof. physics U. Wyo., 1969—, head dept., 1971-78; Fulbright lectr. U. Sao Paulo, Brazil, 1966-67, vis. prof., 1982; vis. prof. U. Tubingen, W. Germany, 1978-79, U. Sydney, Australia, 1988. Author: Introduction to Electrodynamics and Radiation, 1970, Foundations of Statistical Mechanics: Volume I, Equilibrium Theory, 1987, Vol. II, Nonequilibrium Phenomena, 1988, Relativistic Quantum Mechanics of Leptons and Fields, 1991. Served with USNR, 1953-57. Fellow AAAS; mem. Am. Phys. Soc., Brasilian Phys. Soc., Am. Assn. Physics Tchrs., Sigma Xi, Sigma Pi Sigma. Rsch. on statis. mechanics, electrodynamics, quantum theory. Home: 604 S 18th St Laramie WY 82070-4304

GRANGER, HARRIS JOSEPH, physiologist, educator; b. Erath, La., Aug. 26, 1944; s. Willis Gabriel and Edith Ann (Hebert) G.; m. Ramona Ann Vice; children—Ashley, Jarrod, Brent. B.S., U. S.W. La., Lafayette, 1966; Ph.D., U. Miss., Jackson, 1970. Asst. prof. physiology U. Miss. Med. Ctr., Jackson, 1970-74, assoc. prof. physiology, 1974-76; vis. assoc. prof. U. Calif.-San Diego, LaJolla, 1975-76; assoc. prof. dept. med. physiology Tex. A&M U., College Station, 1976-78, prof. med. physiology, 1978—, head dept. med. physiology, 1982—; dir. Microcirculation Research Inst., Tex. A&M U., 1981; mem. study sect. NIH Exptl. Cardiovascular Scis. Study Sect., 1981-86, 88—; chmn. Gordon Conf., 1982. Co-author: Circulatory Physiology II, 1975; mem. editorial bd.: Circulation Research, 1982—, Microvascular Research, 1979-85; mem. editorial bd. Am. Jour. Physiology, 1986—, assoc. editor, 1987-93, editor, 1993—; contbr. chpts. to books, articles to profl. jours. Recipient Research Career Devel. award Nat. Heart Lung & Blood Inst., 1978-83; Disting. Achievement award in research Tex. A&M U., 1982, Merit award Nat. Heat, Lung and Blood Inst., 1987. Mem. Microcirculatory Soc. (council 1978-81, pres.-elect 1988), Am. Physiol. Soc. (Harold Lamport award 1978). Home: 1403 Mill Creek Court College Station TX 77842 Office: Tex A&M U Microcirculation Rsch Instit Coll Medicine College Station TX 77843

GRANGER, WESLEY MILES, medical educator; b. Tampa, Fla., Jan. 12, 1951; s. Reeves C. and Mary Jane (Parker) G.; m. Elizabeth Dianne Dunaway, June 14, 1975; children: Darah Elizabeth, John Wesley. BA, U. S. Fla., 1972; A. Deg. by Examination, U. Chgo. Hosps. and Clinics, 1972; PhD in Physiology, Med. Coll. Ga., 1983. Registered respiratory therapist; cert. respiratory therapy technician; lic. registered respiratory therapist, La. Trainee to CRTT and registry eligible therapist Univ. Community Hosp., Tampa, 1970-74; part-time staff therapist Ga. Bapt. Med. Ctr., Atlanta, 1974-76; staff therapist Emory U. Hosp., Atlanta, 1976-77; shift supr., insvc. instr. Aiken (S.C.) Community Hosp., 1977-79; grad. teaching asst. Med. Coll. Ga., 1979-81; part-time staff therapist Doctors Hosp. of Augusta (Ga.), 1979-83; instr. respiratory therapy program Univ. Ala., Birmingham, 1983-84; asst. prof. cardiopulmonary sci. La. State U. Med. Ctr., 1984-89, assoc. prof., 1989—; lectr. in field; conductor seminars in field. Contbr. articles to profl. jours. Mem. Am. Assn. Respiratory Care, La. Soc. Respiratory Care, Soc. for Computer Simulation. Avocations: photography, computer programming. Office: La State U Med Ctr 1900 Gravier St New Orleans LA 70112-2262

GRANLUND, THOMAS ARTHUR, engineering executive, consultant; b. Spokane, Wash., Mar. 1, 1951; s. William Arthur and Louise (Urie) G.; m. Jean MacRae Melvin, May 25, 1974 (div. Feb. 1991). BS, Wash. State U., 1973, BA, 1973; MBA, Gonzaga U., 1982. Engring. adminstr. Lockheed Aeronautical Systems Co., Burbank, Calif., 1978-91; mgmt. cons., 1991—. Co-author: (screenplay) Identities, 1988, Flash, 1989. 1st lt. USAF, 1973-78. Mem. Wash. State U. Alumni Assn. Avocations: skiing, golf, tennis. Home: 20924 Ben Ct Santa Clarita CA 91350-1418

GRANNAN, WILLIAM STEPHEN, safety engineer, consultant; b. Detroit, Nov. 10, 1929; s. William Stephen and Rose Marie (Gebel) G.; m. Mary Suzanne Malasky, Apr. 8, 1961; children: William Bernard, Douglas Andrew, John Charles. BS in Fire Protection and Safety Engring., Ill. Inst. Tech., 1952. P.E. Cert. Safety Engrs., Calif. Fire ins. rater Ins. Svcs. Office, Detroit, 1952-54 1956-60; sales field rep. Springfield Fire Ins., Detroit, 1960-63, Aetna Life & Casualty Co., Detroit, 1963-66; mgr. property loss control Amerisure Co., Detroit, 1966-92; sr. cons. Crawford Risk Control Svcs., Southfield, Mich., 1993—; industry ambassador comm. Engr. Soc. of Detroit 1970—; cons. pub. safety Commn. Arson Prevention, Greater Detroit C. of C. 1986—. With U.S. Army 1954-56, Germany. Mem. Engring. Soc. Detroit, Nat. Fire Protection Assn., Soc. Fire Protection Engrs., Cert. Safety Profls., Safety Com.'s Associated Gen. Contrs. Am., Am. Legion (Livonia, Mich.), Am. Assn. Ret. Persons. Unitarian. Avocation: photography. Home: 117 N Holbrook St Plymouth MI 48170-1441

GRANT, ANGUS JOHN, immunologist; b. London, Jan. 19, 1960; came to U.S., 1969; s. Alistair McKenzie and Hilary Anne (Beck) G. BA, U. Richmond, 1983; PhD, Med. Coll. Va., 1988. Biotech. fellow Nat. Cancer Inst., Bethesda, Md., 1988-90, staff fellow, 1990-91, sr. staff fellow, 1991-93; sr. staff fellow FDA, Bethesda, 1993—; cancer info. specialist Biospherics Inc., Beltsville, Md., 1989-90. Contbr. articles to Immunology, proceedings of NAS. Pres. Christopher Ct. Condominium Assn., Gaithersburg, Md., 1992. Mem. U.S. Masters Swimming Assn., Sigma Xi. Republican. Achievements include rsch. in devel. and appliation of immunologically based therapeutics for treatment of cancer from both the bench rsch. and bus./legal standpoint. Home: 18472 Bishopstone Court Gaithersburg MD 20879 Office: FDA Ctr Biologics Evaluation and Rsch HFM-518 Bldg 29A Rm 2B24 8800 Rockville Pike Bethesda MD 20892

GRANT, ANTHONY VICTOR, biology educator; b. London, Eng., Nov. 30, 1950; s. Joan Evelyn (Brann) G.; m. Kathryn Ann Schussler, July 15, 1978; children: Jordana Brann, Jennifer Kathryn. AA with honors, Santa Monica Community Coll., 1985; BS in Biology Edn., SUNY, Old Westbury, 1993. Teaching asst. Santa Monica (Calif.) Community Coll., 1983-85; design coord., cons., 1985-88; rsch. asst. SUNY, Old Westbury, 1990—. Mem. N.Y. Acad. of Sci., Soc. for Neurosci., Sigma Tau Sigma. Home: 2B Hendricks Ave E Glen Cove NY 11542

GRANT, JOHN ALEXANDER, JR., engineering consultant; b. Crockett, Tex., July 13, 1923; s. John Alexander and Anne Blackburn)Lentz) G.; m. Joan Marilyn Keith, July 2, 1946; children: Linda, John, W. Keith. AS, No. Tex. Agrl. Coll., Arlington, 1942; BS, Tex. A. and M. U., 1947. Registered profl. engr., Fla., Tex., Ky., Va., Del., Gal., Md., N.C., S.C.; registered land surveyor Fla., Md., Ky. Sr. hwy. engr. Del. State Hwy. Dept., Dover, 1947-49; hwy. engr. GS9 U.S. Bur. Pub. Rds., Richmond, Va., 1949-54; asst. engring. mgr. Michael Baker, Jr., Inc., College Park, Md., 1954-55; project engr. Michael Baker, Jr., Inc., Ft. Lauderdale, Fla., 1955-58; v.p. engring. Arvida Corp., Boca Raton, Fla., 1958-61; cons. engr. John A. Grant, Jr., Inc., Boca Raton, 1961—. Bd. dirs. Boca Forum, 1988-89. Capt. U.S. Army, 1943-46. Mem. Fla. Engring. Soc. (past pres., nat. bd. dirs.), Fla. Inst. Cons. Engrs., Fla. Soc. Profl. Land Surveyors, Nat. Soc. Profl. Engrs. (hons. com. 1988), Boca Raton C. of C. (pres. 1985-86), Elks, Exch. Club of Boca Raton, 100 Club of Broward County, Royal Palm Yacht and Country Club. Home: 3211 NE 27th Ave Pompano Beach FL 33064-8109 Office: 3333 N Federal Hwy Boca Raton FL 33431-6003

GRANT, MICHAEL JOSEPH, electrical engineering educator, consultant; b. New Albany, Ind., Aug. 4, 1944; s. Leslie Joseph and Evelyn (Ethel) G.; m. Lillian Doris Grant, June 8, 1966 (div. June 1984); children: Michael Jr., HEather; m. Carol Bonner, Mar. 9, 1985; stepchildren: Amanda, Ryan. BS, Ind. State U., 1980, MS, 1981. Electronics technician IBM, San Jose, Calif., 1966-71, Singer Bus. Machines, Louisville, 1973-77; field svc. rep. MSI Data Corp., Louisville, 1977-80; asst. prof. Ind. State U., Terre Haute, 1980—; cons. various electronics cos., Terre Haute, 1980—. Home: PO Box 1935 Terre Haute IN 47808

GRANT, MICHAEL PETER, electrical engineer; b. Oshkosh, Wis., Feb. 26, 1936; s. Robert J. and Ione (Michelson) G.; m. Mary Susan Corcoran, Sept. 2, 1961; children: James, Steven, Laura. B.S., Purdue U., 1957, M.S., 1958, Ph.D., 1964. With Westinghouse Research Labs., Pitts., summers 1953-57; mem. tech. staff Aerospace Corp., El Segundo, Calif., 1961; instr. elec. engring. Purdue U., West Lafayette, Ind., 1958-64; sr. engr. Combustion Engring. Corp., Columbus, Ohio, 1964-67, mgr. advanced devel. and control systems, 1967-72, mgr. control and info. scis. div., 1972-74, asst. gen. mgr. indsl. systems div., 1974-76, mgr. system design, 1976-87; v.p., chief scientist SynGenics Corp., Columbus, 1987—; dir. Nat. Ctr. for Mfg. Scis., Ann Arbor, MIch., 1987—. Contbr. articles to profl. jours.; holder 8 patents in field of automation. Mem. IEEE, Sigma Xi, Eta Kappa Nu, Pi Mu Epsilon, Tau Beta Pi. Home: 4461 Sussex Dr Columbus OH 43220-3857 Office: Nat Ctr Mfg Scis 3025 Boardwalk Ann Arbor MI 48108-1779

GRANT, PETER MICHAEL, biologist, educator; b. Erie, Pa., Feb. 23, 1953; s. Matthew Richard and Elizabeth Jane (Scheppner) G.; m. Marcia Lu McCord, Aug. 7, 1976; children: Emily Katherine, Nicholas Jacob. BS in Biology, Pa. State U., 1975; MS in Biology, North Tex. State U., 1978; PhD in Biology, Fla. State U., 1985. Asst. prof. biology Morris Coll., Sumter, S.C., 1985-88, Southwestern Okla. State U., Weatherford, 1988—; adj. asst. prof. entomology Clemson (S.C.) U., 1987—. Editor: The Mayfly Newsletter, 1990—; contbr. articles to profl. jours. Chmn. Environ. Recycling and Solid Waste Task Force, Weatherford, 1991—; vol. Vol. in Pub. Schs. program, Weatherford, 1991—. Mem. AAAS, N.Am. Benthological Soc., Ecol. Soc. Am., Am. Inst. Biol. Scis., Sigma Xi. Achievements include research in systematics and ecology of North American Ephemeroptera (mayflies). Office: Southwestern Okla State U Dept Biol Scis Weatherford OK 73096

GRANT, PETER RAYMOND, biologist, researcher, educator; b. London, Oct. 26, 1936; came to U.S., 1978; m. B. Rosemary Matchett, Jan. 4, 1962; children: Nicola, Thalia. BA with honors, Cambridge U., England, 1960; PhD, U. B.C., Vancouver, Can., 1964; PhD (hon.), U. Uppsala, 1986. Prof. McGill U., Montreal, 1965-78, U. Mich., Ann Arbor, 1978-85, Princeton (N.J.) U., 1985—. Author: Ecology and Evolution of Darwin's Finches, 1986; co-author: Evolutionary Dynamics of a National Population, 1989; co-editor: Molecules, Molds and Metazoa, 1992. Fellow AAAS, Royal Soc. London; mem. Am. Philosophical Soc. Office: Princeton U Dept Ecol & Evol Biology Princeton NJ 08544

GRANT, RHODA, biomedical researcher, educator, medical physiologist; b. Hopewell, N.S., Can., Jan. 12, 1902; d. James William and Marjorie Madelein (Cruickshank) G. BA in Biology and Chemistry with honors, McGill U., Montreal, 1924, MA in Biochemistry, 1930, PhD in Exptl. Medicine cum laude, 1932. Biochemistry rsch. technician Med. Lab., Royal Victoria Hosp., McGill U., Montreal, 1925-29; demonstrator in physiology Banting & Best Med. Rsch. Inst., U. Toronto, Ont., Can., 1933-35; researcher on hearing Physiology Dept., McGill U., Montreal, 1936; asst. prof. physiology Med. Coll., Dalhousie U., Halifax, N.S., Can., 1937-38; teaching and rsch. faculty dept. physiology McGill U., 1939-47; rsch. assoc. clin. sci. med. Coll. U. Ill., Chgo., 1948-61; rsch. assoc. pathology med. Coll. U. Ill., 1961-66; participant in Internat. Symposium on Gastric Cancer, Nat. Cancer Inst./NSF, Japan, 1969. Contbr. articles to profl. jours., med. texts, physiol. handbook. Fellow Royal Soc. Medicine (Eng.); mem. Can. Physiol. Soc., Am. Physiol. Soc., Sigma Xi. Avocation: research and writing on biblical truth in the context of science and history. Home: 525 University Dr East Lansing MI 48823-3046

GRANT, RODERICK MCLELLAN, JR., physicist, educator; b. Chgo., July 9, 1935; s. Roderick McLellan Grant and Elizabeth (Meriam) Hill; m. Susan Jane Fisher, Aug. 31, 1957; children: Carol, Patty, Sally. BS, Denison U., 1957; MS, U. Wis., 1959, PhD, 1965. Instr. U. Wis., Wausau, 1959-61; asst. prof. Denison U., Granville, Ohio, 1965-70, assoc. prof., 1970-77, prof., Henry Chisholm chair physics, 1977—; John R. Cameron lectr. U. Wis., Madison, 1985. Co-author: Physics of the Body, 1992. Mem. Am. Inst. Physics (sec. 1982—, sec. governing bd., 1978-79, 82—), Am. Assn. Physics Tchrs. (exec. bd. 1976-82, sec. 1976-82, Disting. Svc. award 1985). Office: Denison U Dept Physics Granville OH 43023

GRANT, WENDELL CARVER, II, civil engineer; b. Nassau, Bahamas, Apr. 21, 1967; s. Wendell Carver I and Veronica Deloris (Turnquest) G. BSCE, Howard U., 1990. Staff engr. asst. W. Carver Grant & Co. Ltd., Freeport, Bahamas, 1984-89; sr. staff engr. Thomas L. Brown Assocs., P.C., Washington, 1989—; jr. high sch. sci. fair judge ASCE, Washington, 1991, 92, 93; future city competition engring. advisor Am. Assn. Engr. Soc., Washington, 1992. Engring. advisor Meridian House Internat., Washington, 1992. Recipient Pres.'s award ASCE, 1991. Mem. ASCE (conducted contract document interpretation seminars for nat. capital sect. 1993). Anglican. Achievements include participation in the monitoring of subsurface movement during first application of soft ground tunneling using the "New Austrian Tunneling Method" in North America. Home: 19 Santa Maria Ave, Freeport PO Box F-1703, Bahamas Office: Thomas L Brown Assocs PC Ste 230 1010 Massachusetts Ave NW Washington DC 20001

GRANTHAM, CHARLES EDWARD, broadcast engineer; b. Andalusia, Ala., Mar. 15, 1950; s. J.C. and Geraldine (Brooks) G. Student, Enterprise State Jr. Coll., 1968-69; AA, Lurleen B. Wallace Coll., 1979; m. Sandra J. Mosley, Mar. 9, 1973; 1 child, Christopher Charles. Sales engr., draftsman S.E. Ala. Gas Co., Andalusia, 1968-70; asst. mgr., engr. Sta. WAAO, Andalusia, 1972-78; engr. Ala. Public TV, WDIQ-TV, Dozier, Ala., also chief technician Sta. WAAO, Andalusia, 1978—; South Ala. microwave engr. APTV, 1980-93; asst. dir. broadcasting ops., APTV, 1993—. Notary pub., Ala.; bd. dirs. Carolina Vol. Fire Dept., sec./treas., 1985-91; pres. Andalusia Men's Ch. Softball, 1985-86; youth dir. Cedar Grove Ch., 1987-89; pres. Andalusia High Sch. Band Boosters, 1990-91; coach Andalusia Little League, 1982-83. With inf. U.S. Army, 1970-72. Named Civitan Outstanding Young Am., 1967. Mem. I.E.E.E., I.S.C.E.T., Assn. Cert. NABER Technicians (sr. mem.), N.A.R.T.E. (master endorsement), Internat. Soc. Cert. Electronic Technicians, Am. Film Inst., Nat. Rifle Assn., Ala. State Employees Assn. (bd. dirs., pres. local chpt. 1991—), Country Music Assn., Nat. Assn. Bus. and Ednl. Radio, Soc. Broadcast Engrs., Country Music Disc Jockey Assn., Phi Theta Kappa. Mem. Ch. of Christ. Home: RR 5 Box 48-W Andalusia AL 36420-9203 Office: Sta WDIQ-TV RR 2 Dozier AL 36028

GRANTHAM, JARED JAMES, nephrologist, educator; b. Dodge City, Kans., May 19, 1936; married, 1958; 4 children. AB, Baker U., 1958; MD, U. Kans., 1962. Assoc. prof. med. U. Kans., Kansas City, 1969-76, head nephrology sect., 1970—, prof., 1976—. Fellow NIH, 1964-66; grantee Kaw Valley Heart Assn., 1969-70, Nat. Inst. Arthritis and Metabolism Diseases, 1969-83. Mem. Am. Soc. Nephrology, Am. Soc. Clin. Investigation, Am. Physiol. Soc., Am. Fedn. Clin. Rsch., Assn. Am. Phys. Achievements include research in fluid and electrolyte metabolism, electrolyte transport, mechanism of action of antidiuretic hormone and polycystic kidney disease. Office: U Kans Kidney & Urology Rsch Ctr 39th & Rainbow Blvd Kansas City KS 66103*

GRASS, JUDITH ELLEN, computer scientist; b. Hartford, Conn., Sept. 16, 1953; d. Luther Martin and Glorious Kathleen (Plourde) G. BS in Modern

Lans., Georgetown U., 1975; AM in Slavic Linguistics, U. Ill., 1977, MA in Computer Sci., 1982, PhD, 1986. Rsch. asst. U. Ill., Urbana, 1981-86; mem. tech. staff AT&T Bell Labs., Murray Hill, N.J., 1986-93; sr. mem. tech. staff Corp. for Nat. Rsch. Initiatives, Reston, Va., 1993—; mem. program com. Usenix Conf., San Antonio, 1992, SUUG Free Software Workshop, Moscow, 1992-93. Contbr. articles to IEEE Software, Computing Systems. Mem. Assn. for Computing Machinery, IEEE, Usenix Assn., U.S. Combined Tng. Assn., U.S. Dressage Fedn., Sigma Xi. Achievements include design of C information abstractor system. Office: Corp Nat Rsch Initiatives MH 3C-532C 1895 Preston White Dr Reston VA 22124

GRASSE, JOHN M., JR., physician, missionary; b. Chalfont, Pa., Sept. 7, 1927; s. John and Blanche (Landis) G.; m. Hannah Elizabeth Stover, Sept. 12, 1953; children: Elizabeth Ann, Linda Sue Siemens, Sandra Ruth Mckeon, Martha Jane Young, John Michael. BS in Biology, Juniata, Huntingdon, Pa., 1948; MD, Jefferson U., Phila., 1952; student, Menninger Sch. Psychiatry, Topeka, 1972. Med. missionary Mennonite Bd. Missions, Elkhart, Ind., 1954-63; asst. prof. psychiatry Tex. U. Health & Sci. Ctr., San Antonio, 1972-73; pvt. practice Akron, Pa., 1974-86; med. missionary Africa Inter-Mennonite Mission, Elkhart, Ind., 1986-90; pvt. practice Calico Rock, Ark., 1964-69, 91—; asst. prof., Psychiatry Hershey Med. Ctr., Pa., 1975-80; Psychiatric cons. Philhaven Hosp., Lebanon, Pa., 1974-86. Pres. Lions Club, Calico Rock, Ark., 1964-69, Fellows Assn. Menninger Sch. Psychiatry, Topeka, Kans., 1969-72.; with U.S. Army, 1946-47. Fellow Psychiatric Fed. Govt., Topeka, 1969-72. Mem. Am. Psychiatric Assn., Acad. Family Practice, Ark. Med. Assn., Tri County Med. Assn. Democrat. Avocations: travel, gardening, photography. Home: HC79 Box 192 Calico Rock AR 72519 Office: Med Ctr of CR Calico Rock AR 72519

GRASSELLI, JEANETTE GECSY, university official; b. Cleve., Aug. 4, 1928; d. Nicholas W. and Veronica (Varga) Gecsy; m. Glenn R. Brown, Aug. 1, 1987. BS summa cum laude, Ohio U., 1950, DSc (hon.), 1978; MS, Western Res. U., 1958; DSc (hon.), Clarkson U., 1986; D Engring. (hon.), Mich. Tech. U., 1989. Project leader, assoc. Infrared Spectroscopist, Cleve., 1950-78; mgr. analytical sci. lab. Standard Oil (name changed to BP Am., Inc. 1985), Cleve., 1978-83, dir. technol. support dept., 1983-85, dir. corp. rsch. and analytical scis., 1985-88; Disting. vis. prof., dir. rsch. enhancement Ohio U., Athens, 1989—; bd. dirs. Nicolet Instrument Co., Madison, Wis., 1982-92, B.F. Goodrich Co., Akron, Ohio, AGA Gas, Inc., USX Corp.; mem. bd. on chem. sci. and tech. NRC, 1986-91; chmn. U.S. Nat. Com. to Internat. Union of Pure & Applied Chemistry, 1992-94. Author, editor 8 books; editor: Vibrational Spectroscopy; contbr. numerous articles on molecular spectroscopy to profl. jours.; patentee naphthalene extraction process. Bd. dirs. N.E. Ohio Sci. and Engring. Fair, Cleve., 1977—; trustee Ohio U., Athens, 1985—, chmn., 1991-92; trustee Holden Arboretum, Cleve., 1988—, Edison Biotech. Ctr., Cleve., 1988—, Cleve. Playhouse, 1990—, Cleve. Scholarship Program, 1992—, Garden Ctr. Greater Cleve., 1990—, Mus. Arts Assn., 1991—, Gt. Lake Sci. Mus., 1991—, Cleve. Edn. Fund. Sci. Collaborative, 1991—, Nat. Inventor's Hall of Fame, 1993—, Rainbow Babies and Children's Hosp., 1992—; chmn. Ohio Coun. on Rsch. and Econ. Devel., 1992—. Recipient Disting. Svc. award Cleve. Tech. Soc. Coun., 1985; named Woman of Yr. YWCA, 1980; named to Ohio Women's Hall of Fame State of Ohio, 1989, Ohio Sci. & Tech. Hall of Fame, 1991. Mem. Am. Chem. Soc. (chair analytical div. 1990-91, Garvan medal 1986, Analytical Chem. award 1993), Soc. for Applied Spectroscopy (pres. 1970, Dist. Svc. award 1983), Coblentz Soc. (bd. govs. 1968-71, William Wright award 1980), Phi Beta Kappa, Iota Sigma Pi (pres. fluorine chpt. 1957-60, nat. hon. mem. 1987). Republican. Roman Catholic. Avocations: swimming, dance, music. Home: 150 Greentree Rd Chagrin Falls OH 44022-2424

GRASSIAN, VICKI HELENE, chemistry educator; b. N.Y.C., Nov. 29, 1959; d. Solomon and Marilyn G.; m. Mark Alan Young, May 29, 1988. BS, SUNY, Albany, 1981; MS, Rensselaer Poly. Inst., 1982; PhD, U. Calif., Berkeley, 1987. Postdoctoral fellow Colo. State U., Ft. Collins, 1988-89; rsch. assoc. U. Calif., Berkeley, 1990—; chemistry prof. Univ. Iowa, Iowa City, 1990—. Contbr. articles to Jour. Chem. Physics, Chemistry of Materials, Jour. of the Am. Chem. Soc. Recipient Iowa Career Devel. award in laser chemistry U. Iowa, Iowa City, 1992, GE Found. Faculty fellowship. Mem. Am. Chem. Soc., Am. Phys. Soc., Am. Vacuum Soc., Assn. for Women in Sci. (Ednl. award 1986), Sigma Xi, Sierra Club, NOW. Achievements include research in surface science, catalysis and materials chemistry. Home: 322 Beldon Ave Iowa City IA 52246 Office: U Iowa Dept Chemistry Iowa City IA 52242

GRASSO, PATRICIA GAETANA, biochemist, educator; b. Albany, N.Y., Mar. 19, 1940; d. Leonard Nicholas and Olga Helen (Krause) G. MS, Coll. of St. Rose, Albany, 1967; PhD, Georgetown U., 1984. Tchr. biology St. Patrick's Cen. Cath. High Sch., Catskill, N.Y., 1965-66; instr. biology Coll. of St. Rose, Albany, 1966-68; tchr. biology Cath. Cen. High Sch., Troy, N.Y., 1968-78; instr. anatomy George Washington U., 1983-84; asst. prof. biology Coll. St. Rose, Albany, 1985-86; asst. prof. biochemistry Albany Med. Coll., 1988-91, assoc. prof. biochemistry, 1991—; cons./author Globe Pub. Co., N.Y.C., 1981—, Modern Curriculum Press, Cleve., 1981—, Cambridge Book Co., N.Y.C., 1974-81. Contbr. articles to profl. jours. Recipient Thomas Edison Citation GE Corp., 1974, Sci. Tchr. Achievement Recognition Nat. Sci. Tchrs. Assn., 1970. Mem. Endocrine Soc., Soc. for Study of Reprodn., Sigma Xi (steering com.), Delta Epsilon Sigma. Office: Albany Med Coll Dept Biochemistry/Molec Bio Albany NY 12208

GRATCH, SERGE, mechanical engineering educator; b. Monte San Pietro, Italy, May 2, 1921; s. Isaak F. and Tatiana (Dermaner) G.; m. Rosemary Delay, June 30, 1931; children: Susan, Mary, Lucia, Karen, Elizabeth, Ann, Barbara, Amy, Ellen, Thomas Charles. BSChemE, U. Pa., 1943, MS, ME, 1945, PhD, ME, 1950. Instr., U. Pa., 1943-45, asst. prof., 1945-50, assoc. prof., 1950-51; research scientist Rohm & Haas Co., Phila., 1951-59; assoc. prof. mech. engring. Northwestern U., Evanston, Ill., 1959-61; supr. processes and devices Ford Motor Co., Dearborn, Mich., 1961-62, mgr. chem. processes and devices Ford Motor Co., 1963-69, asst. dir. engring. sci., 1969-72, dir. chem. sci. lab., 1972-85, dir. vehicles and component research lab., 1985-86; prof. mech. engring. GMI Inst., Flint, Mich., 1986—; mem. adv. bd. Coll. Engring. U. Iowa, 1969-73, Coll. Engring. U. Detroit, 1971-88; adv. bd. dept. mech. engring. U. Pa., 1973-88; chmn. air pollution rsch. adv. com. Coord. Rsch. Coun., 1983-85; mem. Nat. Alcohol Fuels Commn., 1979-81. Regional editor Internat. Jour. Fracture, 1965-91; contbr. articles to profl. jours. Mem. ASME (hon., past v.p. rsch., past pres., John Fritz medal 1992), NAE, AAAS, Am. Soc. Engring. Edn., Am. Chem. Soc., Engring. Soc. Detroit (past pres.), Soc. Automotive Engrs. (chmn. lubricant rev. bd. 1982-83), Sigma Xi, Tau Beta Pi, Sigma Tau. Roman Catholic. Home: 32475 Bingham Rd Bingham Farms MI 48025-2427 Office: GMI Engring & Mgmt Inst Flint MI 48504

GRAU, RAPHAEL ANTHONY, reliability engineer; b. Balt., Nov. 9, 1963; s. Raphael Manuel and Zoraide (Scanoni) G. BSEE, Tex. A&M U., 1986; MBA, U. Houston, 1992. Software engr. Ferranti Internat. Controls, Sugarland, Tex., 1986-90; computer engr. Ford Aerospace, Houston, 1990; reliability engr. NASA Johnson Space Ctr., Houston, 1990—. Mem. IEEE, Nat. Soc. Profl. Engrs., Nat. Mgmt. Assn. Home: 8402 Leader St Houston TX 77036-5523 Office: NASA Johnson Space Ctr NASA Rd 1 Houston TX 77058-3696

GRAUER, MANFRED, computer engineering educator, researcher; b. Herzberg/Elster, Brandenburg, Germany, Aug. 17, 1945; s. Erich and Lieselotte (Wilkniss) G.; m. Monika Schaarschmidt, Feb. 1, 1969; children: Juliane, Thomas. Diploma in Engring., Moscow Inst. Tech., USSR, 1970; DEng, U. Merseburg, Fed. Republic of Germany, 1975, DSc, 1979. Univ. asst. U. Merseburg, 1970-75, scientist, 1975-81; scientist Internat. Inst. Applied Systems Analysis, Laxenburg, Austria, 1981-84; prof. Inst. of Informatics, Berlin, 1984-86, U. Dortmund, Fed. Republic of Germany, 1987-88, U. Siegen, Fed. Republic of Germany, 1988—. Editor: (book) Interactive Decision Analysis, 1984, Plural Rationality, 1985, Large-Scale Modelling, 1986, Parallel Computing, 1991. Avocation: tennis. Office: U Siegen, Hölderlinstr 3, D-57068 Siegen Nordrhein Westfalia, Germany

GRAULE, RAYMOND S(IEGFRIED), metallurgical engineer; b. Phila., Feb. 7, 1932; s. Oscar P. and Elizabeth Keim (Merkle) G.; m. Beatrice D. Miller, Sept. 4, 1954 (div. Nov. 1982); children: Melissa, Jon; m. Marlys Ann Sunkle, Sept. 21, 1985; children: Troy, Tara, Tiffany. BSChemE, N.J. Inst. Tech., Newark, 1955; MS in Metallurgy, Stevens Inst. of Tech., Hoboken, N.J., 1961. Process engr. Wilbur B. Driver Co., Newark, 1954-62; supr. of prod. engring. G.T.E. Corp., Newark, 1962-77; engring. mgr. Amax Corp., Parsippany, N.J., 1977-84; sr. mfg. engr. Carpenter Tech. Corp., Orangeburg, S.C., 1984—; adj. instr. Essex County Coll., Newark, 1979-81, Orangeburg Calhoun Tech. Coll., 1984-87. County committeeman Rep. Party, Parsippany, 1965-81; advisor Bd. of Edn., 1969-73. With U.S. Army, 1956-58. Mem. Am. Soc. of Metals, Republican Club (Parsippany), Goodyear Blimp Club, Experimental Aviation Assn. Avocations: woodworking, flying, boating. Home: 2045 Longwood Rd NE Orangeburg SC 29115 Office: Carpenter Tech Corp Rt 33 & Cameron Rd Orangeburg SC 29115

GRAVELY, JANE CANDACE, computer company executive; b. Rocky Mount, N.C., Dec. 1, 1952; d. Edmund Keen and Janice Eleanor (Beavon) G.; m. Barney Ben Linthicum, July 13, 1985 (div. 1991). BS, N.C. Wesleyan Coll., 1974; MEd, Coll. William and Mary, 1980. Circulation and promotion mgr. Va. Gazette, Williamsburg, 1975-80; computer analyst, chief exec. officer Affordable Computer Systems, Rocky Mount, 1982-85, Goldsboro, N.C., 1985—; instr. bus., math., computers Nash Tech. Coll., Rocky Mount, 1980-83; instr. math., computers N.C. Wesleyan Coll., Rocky Mount, 1983-85, instr. computers, 1985-89. Mem. NAFE, United Meth. Womens Circle (pres. 1990-91), Rocky Mount C. of C., Goldsboro C. of C. (Chamber Amb., com. of 100), Kiwanis Internat., Goldsboro Club, Omicron Delta Kappa. Republican. Avocations: golf, tennis, music. Office: 1406 E Ash St Goldsboro NC 27530

GRAVER, JACK EDWARD, mathematics educator; b. Cin., Apr. 13, 1935; s. Harold John and Rose Lucille (Miller) G.; m. Yana Regina Hanus, June 3, 1961; children: Juliet Rose, Yana-Maria, Paul Christopher. BA in Math., Miami U., Oxford, Ohio, 1958; MA in Math., Ind. U., 1961, PhD in Math., 1964. Instr. Ind. U., Bloomington, 1964; John Wesley Young Rsch. instr. Dartmouth Coll., Hanover, N.H., 1964-66; asst. prof. math. Syracuse (N.Y.) U., 1966-69, assoc. prof., 1969-76; vis. prof. U. Nottingham (Eng.), 1971-72; prof. math. Syracuse U., 1976—, chmn. dept. math., 1979-82. Co-author book (with M. Watkins): Combinatorics with Emphasis on Graph Theory, 1977, (with J. Baglivo) Incidence and Symmetry in Design and Architecture, 1982; contbr. articles to profl. jours. With USN, 1953-55. Fellow Inst. Combinatories and its Applications; mem. Soc. Indsl. and Applied Math., Nat. Coun. Tchrs. Math., Assn. Math. Tchrs. N.Y. State, Math. Assn. Am. (bd. govs. 1985-88), Am. Math. Soc. Home: 871 Livingston Ave Syracuse NY 13210-2935 Office: Syracuse Univ Dept Math Syracuse NY 13244-1150

GRAVES, HARRIS BREINER, physician, hospital administrator; b. Lincoln, Nebr., Apr. 29, 1928; s. Fred T. and Marie (Breiner) G.; m. Marilyn J. Eidam, Apr. 6, 1950; children: John W., Stephen B. AB, U. Nebr., Lincoln, 1948; MD, U. Nebr., Omaha, 1952. Intern Kansas City Gen. Hosp., Omaha, 1952-53; with Physicians Clinic, Omaha, 1953-54, 56-67; physician Emergency Med. Svcs. Group, Omaha, 1967-75; dir. emergency services Midwest Physicians Svcs., Inc. (formerly Med. Svc. Group), Omaha, 1975-92, Meth. and Childrens Meml. Hosps., Omaha, 1981-92; assoc. prof. dept. family practice U. Nebr. Med. Ctr., 1988—; pres. Midwest Physicians Svcs., Inc., Omaha, 1975-90, Emergency Enterprises, Inc., Omaha, 1980; med. dir. Physicians Resources, Inc., Omaha, 1985-89; specialty site surveyor Accreditation Coun. on Grad. Med. Edn., 1991—, field rep., 1992—. Assoc. editor: Emergency Department Organization and Management, 1975; contbr. articles to profl. jours. Pres. Waterloo (Nebr.) Sch. Bd., 1972-76, Booster Club, Waterloo, 1971-73; v.p. Midlands Emergency Med. Service Council, Omaha, 1978-82. Capt. USAF, 1954-56. Mem. Am. Bd. Emergency Medicine (bd. dirs. 1980-87, sec., treas. 1983-84, pres. 1985-86, chmn. recert. com. 1987-89), Am. Coll. Emergency Physicians (bd. dirs. 1968-77, pres. 1975-76, chmn. standards com. 1989—, Meritorious Svc. award 1978), Alpha Omega Alpha, Mayor's Trophy 1984). Republican. Methodist. Avocations: golf, computers. Home: 820 Branding Iron Dr Elkhorn NE 68022-2136

GRAVES, JOAN PAGE, biologist; b. Richmond, Va., Mar. 4, 1959; d. Edward Merwin and Carol Joyce (Clarke) G. BS in Biology, Va. Commonwealth U., 1980; MS in Biology, U. Va., 1986. Lab. technician Philip Morris Inc., Richmond, Va., 1981; lab. technician physiology dept. Med. Coll. of Va., Richmond, Va., 1982-83; teaching asst. biology dept. U. Va., Charlottesville, 1983-85; biologist Nat. Inst. Environ. Health Scis., Research Triangle Park, N.C., 1986—; poster presenter 33d Drosophila Rsch. Conf., Phila., 1992. Contbr. articles to profl. jours. Achievements include research in the structure, function and regulation of eukaryotic organism Drosophila melanogaster and the molecular analysis of specific genes suppressor of sable and Minute (1) 1B at the levels of DNA and RNA. Office: Nat Inst Environ Health Sci Mail Drop D3-03 PO Box 12233 Research Triangle Park NC 27709

GRAVES, JOHN FRED, III, microelectronics engineer; b. Indpls., June 11, 1945; s. John Fred Jr. and Mildred (Tatchell) G.;m. Annette Marie Gigone, June 4, 1967; children: James, Hope. BS in Chemistry, Rose-Hulman Inst. Tech., 1967; MS in Engring., Materials Sci., Cath. U. Am., 1984. Engr. Westinghouse Electric Corp., Balt., 1967-79; sr. staff engr. Allied-Signal Corp., Balt., 1979—; gen. chair Internat. Microelectronics Symposium, Balt., 1989. Author, editor, pub.: The Micro-Tech Index, 1986, Bibliography of Microelectronics, Electronics Packaging and Interconnection References (1986-89), 1990; contbr. articles to tech. publs. Mem. Internat. Soc. Hybrid Microelectronics (chpt. officer 1977-85). Home: 20005 Graves Run Rd Hampstead MD 21074 Office: Allied Signal Communication Systems 1300 E Joppa Rd Baltimore MD 21286-5999

GRAVES, JOSEPH LEWIS, JR., evolutionary biologist, educator; b. Westfield, N.J., Apr. 27, 1955; s. Joseph L. and Helen (Tucker) G.; m. Suekyung Joe, Mar. 5, 1984; 1 child, Joseph L. III. AB, Oberlin Coll., 1977; PhD, Wayne State U., 1988. Pres.'s postdoctoral fellow U. Calif., Irvine, 1988-90, asst. prof. evolutionary biology, African Am. studies, 1990—, faculty mentor Howard Hughes biomed. fellowship, 1990—; dir. Calif. Alliance for Minority Participation Transfer Acad. Contr. chpt. to: Genetic Effects on Aging II, 1990, Insect Life Cycles, 1990; contbr. articles to Sci., Genetica, Physiological Zoology, Functional Ecology, Jour. Insect Physiology, Life Sci. Advances: Fundamental Genetics, Jour. Gerontology, Race Relations Abstracts; TV appearance Sta. KCET-TV, 1993. Josiah P. Macy fellow Marine Biol. Lab., Woods Hole, Mass., 1978; NSF fellow, 1979; Thomas Rumble grad. fellow Wayne State U., Detroit, 1985. Mem. Soc. Study Evolution, Genetics Soc. Am., Am. Soc. Naturalists, Gerontol. Soc. Am. Achievements include pioneering physiological research of Drosophila Melanogaster frutifly used in genetic research on postponed aging and comparative biology of aging in the genus Drosophila. Office: U Calif Irvine Dept Evolutionary Biology Irvine CA 92717

GRAVES, MICHAEL LEON, II, aeronautical engineer; b. Laredo, Tex., Feb. 1, 1970; s. Michael L. and Sherry A. (Moser) G. BS in Aero. Engring., Embry-Riddle Aero. U., 1992. Rsch. asst. ERAU's ctr. for Aviation/Aerospace Rsch., Daytona Beach, Fla., 1990-92; asst. staff mem. Galaxy Sci. Corp., Alexandria, Va., 1992—. Graphics editor (book) Automation and Systems Issues in ATC, 1990. Eagle Scout Boy Scouts Am., Ft. Ord, Calif., 1984. Mem. AIAA, NATO Advanced Study Inst. Home: 114 Almond Ct Sterling VA 20164-2838 Office: Galaxy Sci Corp 4900 Seminary Rd Ste 400 Alexandria VA 22311-1860

GRAVES, ROBERT JOHN, industrial engineering educator; b. Buffalo, Sept. 25, 1945; s. Paul Frederick and Ann (Mayer) G.; m. Virginia Jane Burry, June 8, 1968; children: Peter F., Anna K., Christopher J. BS Indsl. Engring., Syracuse U., 1967; MS Indsl. Engring., SUNY, Buffalo, 1969, PhD, 1974. Instr. indsl. engring. SUNY, Buffalo, 1973-74; asst. prof. sch. indsl. & systems engring. Ga. Tech., Atlanta, 1974-79; assoc. prof. indsl. engring. U. Mass., Amherst, 1979-80, prof. indsl. engring., 1988-91; prof. indsl. engring. Rensselaer Poly. Inst., Troy, N.Y., 1991—; pres. Coll. Industry Coun. on Material Handling Edn., Charlotte, N.C., 1990-92. Editor: Material Handling of the 90's, 1991, Progress in Material Handling Research, 1992; U.S. editor Internat. Jour. Prodn. Planning and Control, 1992; contbr. articles to profl. jours. Mem. sch. com. Town of Pelham, Mass., 1985-86, planning bd., 1987-92. Grantee Mass. Ctrs. Excellence Corp., 1987-90, NSF, 1989-91. Fellow Inst. Indsl. Engrs. (sr., rsch. chair 1980-81, program chair 1981-82, editor newsletter 1990-91, dir. div. 1991-92, Spl. Citation award 1985). Achievements include research in flexible assembly systems scheduling, printed circuit board assembly. Office: Rensselaer Poly Inst Dept Decision Scis and Engring Systems Troy NY 12180-3590

GRAY, CAROL HICKSON, chemical engineer; b. Atlanta, Jan. 3, 1958; d. Ronald Allen and Charlotte Patricia (Blitch) Hickson; m. Randy Lee Gray, June 25, 1983; children: Amanda Christine, Stephanie Lee. BSChemE, Ga. Inst. Tech., 1979. Process engr. Air Products and Chems., Inc., Calvert City, Ky., 1979-83, sr. process engr., 1983-86, sr. prodn. engr., 1986-87, prin. prodn. engr., 1987-89; engring. supr. Air Products and Chems., Inc., Pasadena, Tex., 1990-92; lead engr. Air Products and Chems., Inc., Calvert City, Ky., 1992-93, area supr., 1993—. Mem. NAFE, Internat. Platform Assn. Avocation: bicycling. Office: Air Products and Chems Inc PO Box 97 Calvert City KY 42029

GRAY, CHRISTOPHER DONALD, software researcher, author, consultant; b. Brookville, Pa., May 18, 1951; s. Donald Garrison and Patricia Lee (Huffman) G.; m. Allison Selby Farragher, Oct. 12, 1974 (div. 1989); children: Patrick Xanthe, Colin Christopher. BA in Math., Washington and Jefferson Coll., 1973; MS in Math., Carnegie-Mellon U., 1975. Mfg. systems analyst Ohaus Scale Corp., Florham Park, N.J., 1974-76; systems rep. Software Internat. Corp., Florham Park, N.J., 1976-77, cons. mfg. systems, 1977-78, mktg. rep., 1978-79; v.p. Mfg. Software Systems, Inc., Essex Junction, Vt., 1979-85, pres., 1985; pres. Oliver Wight Software Research, Inc., Essex Junction, Vt., 1985-88, Gray Rsch., Exeter, N.H., 1988—; pres. Gray Media, Inc., 1992—, Monochrome Press, Inc., 1992—; bd. dirs. Partners for Excellence, Inc., 1993—; assoc. R.D. Garwood, Inc., 1988-92, Oliver Wight Edn. Assocs., Newbury, 1982-88; cons. Oliver Wight Video Prodns., Essex Junction, Vt., 1980-84; advisor mfg. applications, Software News, Sentry Pub. Co., Hudson, Mass., 1983-89. Author: The Right Choice: The Complete Guide to Evaluating, Selecting and Installing MRP II Software, 1987; co-author MRP II Standard System, 1983 (rsch. report), MRP II Standard System: A Handbook for Manufacturing System Survival, 1989, MRP II Standard System Workbook, 1989; contbr. rsch. reports, articles and conf. papers to tech. lit. Fellow Am. Prodn. and Inventory Control Soc. (cert. fellow prodn. and inventory mgmt.; chpt. program chmn. 1978-79, v.p. 1979-80, pres. 1980-81); mem. Phi Beta Kappa. Republican. Presbyterian. Avocations: gardening, landscaping, house restoration, furniture building. Home: 4 Juniper Ridge Rd Exeter NH 03833 Office: Gray Rsch PO Box 424 Exeter NH 03833

GRAY, DAVID MARSHALL, chemical engineer; b. Fostoria, Ohio, Oct. 30, 1947. BSChE, Case Western Reserve U., 1969. Product engr., engring. specialist Leeds & Northrup, North Wales, Pa., 1969-88, sr. applications specialist, 1988—. Contbr. chpts. books, articles to profl. jours. Mem. Am. Soc. Testing and Materials (task group chmn. 1991—), Instrument Soc. Am. (sr. mem.). Achievements include original design and product development of state of the artpH control instrumentation introducing pH characterization to enable convenient correction for non-linearity of pH vs. reagent demand. Office: Leeds & Northrup 351 Sumneytown Pike North Wales PA 19454

GRAY, DONALD LEE, chemist; b. Dayton, Ohio, July 20, 1948; s. Donald Edward and Phyllis Jean (Nail) G.; m. Sheryl Gay Spacek, Feb. 1970 (div.); children: Brett, Sara. BS in Microbiology, Kans. State U., 1970; BS in Chemistry, Rockhurst Coll., 1975. With Bd. Pub. Utilities, Kansas City, Kans., 1970—, supt., 1985-89, dir., 1989—; presenter at profl. confs. Contbr. articles to profl. publs. Mem. Am. Waterworks Assn. (chmn. Kans. sect. 1991—, Oper. Meritorious award 1989),Eagles, Tau Kappa Epsilon. Achievements include development of pilot plant testing program to help meet changes in EPA Safe Drinking Water act. Office: Bd Pub Utilities 3601 N 12th St Kansas City KS 66104

GRAY, FESTUS GAIL, electrical engineer, educator, researcher; b. Moundsville, W.Va., Aug. 16, 1943; s. Festus P. and Elsie V. (Rine) G.; m. Caryl Evelyn Anderson, Aug. 24, 1968; children: David, Andrew, Daniel. BSEE, W.Va. U., 1965, MSEE, 1967; Ph.D., U. Mich., 1971. Instr. W.Va. U., Morgantown, 1966-67; teaching fellow U. Mich., 1967-70; asst. prof. Va. Poly. Inst. and State U., Blacksburg, 1971-77, assoc. prof., 1977-82, prof., 1983—; vis. scientist Rsch. Triangle Inst., N.C., 1984-85; faculty fellow NASA, 1975; cons. Inland Motors, Radford, Va., 1980, Rsch. Triangle Inst., 1987—; researcher Rome Air Devel. Ctr., N.Y., 1980-81, Naval Surface Weapons Ctr., Dahlgren, Va., 1982-83, Army Rsch. Office, 1983-86, NSF, 1991—; publs. chmn. Internat. Symposium on Fault Tolerant Computing, Ann Arbor, Mich., 1985; assoc. treas., Northside Presbyn. Ch., Blacksburg, 1986—. Contbr. articles to sci. jours. Bd. deacons Northside Presbyn. Ch., Blacksburg, 1980-83; coach S.W. Va. Soccer Assn., Blacksburg, 1980-86. Grantee NSF, Office Naval Research, NASA. Mem. IEEE (chpt. chmn. 1979-80), Computer Soc. of IEEE, Sigma Xi. Democrat. Research on fault tolerance, diagnosis, testing, and reliability issues for VLSI; distributed and multiprocessor computer architectures. Home: 304 Fincastle Dr Blacksburg VA 24060-5036 Office: Va Poly Inst and State U Blacksburg VA 24061-0111

GRAY, GARY GENE, ecologist, educator; b. El Dorado, Kans., Jan. 21, 1940; s. Cecil Tipton and Neva Nancy (Rounds) G.; m. Gertrude Rowena Wallace, Sept. 1961 (div.); 1 child, Nike Dawn; m. Sally Worthington Smith, June 1968 (div.); m. Elda Darlene Dove, Dec. 1979 (div.); m. Francine Estelle Dickinson, Sept. 1987. Student, U. Colo., 1957-63; BS cum laude, U. Mass., 1973, MS, 1975; PhD, Tex. Tech. U., 1980. Cert. wildlife biologist, 1983. Asst. ecologist Argonne (Ill.) Nat. Lab., 1980-82; asst. prof. biol. sci. No. Ill. U., DeKalb, 1981-88; vis. asst. prof. biol. sci. U. Ill., Chgo., 1989-92; writer, editor, cons. Lanark, Ill., 1993—. Author: Wildlife and People: The Human Dimensions of Wildlife Ecology, 1993; contbr. articles to profl. jours. Mem. Am. Soc. Mammalogists, Internat. Soc. Environ. Ethics, Soc. Conservation Biology, Wildlife Soc., Sigma Xi, Phi Kappa Phi, Xi Sigma Pi, Alpha Zeta. Home and Office: RR 1 Box 79 A Lanark IL 61046

GRAY, HARRY BARKUS, chemistry educator; b. Woodburn, Ky., Nov. 14, 1935; s. Barkus and Ruby (Hopper) G.; m. Shirley Barnes, June 2, 1957; children: Victoria Lynn, Andrew Thomas, Noah Harry Barkus. BS, Western Ky. U., 1957; PhD, Northwestern U., 1960, DSc (hon.), 1984; DSc (hon.), U. Chgo., 1987, U. Rochester, 1987, U. Paul Sabatier, 1991, U. Göteborg, 1991, U. Firenze, 1993. Postdoctoral fellow U. Copenhagen, 1960-61; faculty Columbia U., 1961-66, prof., 1965-66; prof. chemistry Calif. Inst. Tech., Pasadena, 1966—; now Arnold O. Beckman prof. chemistry and dir. Beckman Inst. Calif. Inst. Tech.; vis. prof. Rockefeller U., Harvard U., U. Iowa, Pa. State U., Yeshiva U., U. Copenhagen, U. Witwatersrand, Johannesburg, South Africa, U. Canterbury, Christchurch, New Zealand; cons. govt., industry. Author: Electrons and Chemical Bonding, 1965, Molecular Orbital Theory, 1965, Ligand Substitution Processes, 1966, Basic Principles of Chemistry, 1967, Chemical Dynamics, 1968, Chemical Principles, 1970, Models in Chemical Science, 1971, Chemical Bonds, 1973, Chemical Structure and Bonding, 1980, Molecular Electronic Structures, 1980. Recipient Franklin Meml. award Stanford U., 1967, Fresenius award Phi Lambda Upsilon, 1970, Shoemaker award U. Louisville, 1970, award for excellence in teaching Mfg. Chemists Assn., 1972, Centenary medal of Royal Soc. Chemistry, 1985, Nat. Medal of Sci., 1986, Alfred Bader Bioinorganic Chemistry award, 1990, Gold medal Am. Inst. Chemists, 1990, Linderstrom-Lang Prize, 1992, Priestley award Dickinson Coll., 1991; named Calif. Scientist of Yr., 1988, Achievement Rewards for Coll. Scis. Man of Sci., 1990; Guggenheim fellow, 1972-73; Phi Beta Kappa scholar, 1973-74. Fellow AAAS; mem. NAS, Am. Chem. Soc. (award pure chemistry 1970, award inorganic chemistry 1978, award for disting. service in advancement of inorganic chemistry 1984, Harrison Howe award 1972, Remsen Meml. award 1979, Tolman medal 1979, Pauling medal 1986, Priestley medal 1991, Willard Gibbs medal 1992), Royal Danish Acad. Scis. and Letters, Alpha Chi Sigma, Phi Lambda Upsilon. Home: 1415 E California Blvd Pasadena CA 91106-4101 Office: Calif Inst Tech Chemistry 127-72 1201 East California Blvd Pasadena CA 91125

GRAY, JAMES LEE, systems analyst; b. Burlington, Kans., Dec. 30, 1943; s. Irvin L. and Kathryn (Robe) G.; m. M. Kathryn Konefal, July 20, 1968; children: Matthew Konefal, Eric Zachary. BS, Kans. State U., 1966; PhD, SUNY, Stony Brook, 1974. Asst. prof. psychology U. Alaska, Fairbanks, 1974-79; postdoctoral fellow SUNY, Stony Brook, 1979-80, programmer analyst, 1980-82; systems analyst Grumman Data Systems, Woodbury, N.Y.,

1982-84, Dept. Vet. Affairs Med. Ctr., Northport, N.Y., 1984—. Mem. AAAS, Internat. Soc. for Human Ethology, M Tech. Assn., Masons. Home: 207 Thompson St Port Jefferson NY 11777 Office: Dept Vet Affairs Med Ctr Middlevill Rd Northport NY 11768

GRAY, JAMES RANDOLPH, III, air force officer; b. Monticello, Ky., Sept. 23, 1965; s. James Randolph Jr. and Gloria Mae (Kendrick) G. BSEE, USAF Acad., 1987, BS in Engring. Sci., 1987. Commd. 2nd lt. USAF, 1983—, advanced through grades to capt., 1991; missile design engr. AFIC/FASTC, Wright-Patterson AFB, Ohio, 1987-92; lead flight controls engr. ASC/VJE, Wright-Patterson AFB, Ohio, 1992—; USAF rep. Dept. of Def. Intel Missile Tech. Working Group, Huntsville, Ala., 1990-92. Creator AFIC/FASTC Angel Tree campaign, Dayton, 1988—. Recipient Martha Drew Shields award AFSC/FTD, 1989, Theodore Von Karman award Falcon Found., 1982. Mem. AIAA. Office: ASC/VJE Wright-Patterson AFB Bldg 14 Dayton OH 45433-6503

GRAY, JENNIFER EMILY, biology educator; b. Akron, Ohio, May 5, 1946; d. Edward Raynes and Marjorie (Ball) Eaton; m. Shelton R. Gray, May 17, 1969; children: Shelton R. Jr., Christopher E. BS in Med. Tech., U. Mass., 1967; MA in Biology, Calif. State U., Fresno, 1989; student, U. Calif. Berkeley, 1993. Lic. med. technologist. Med. technologist Leonard Morse Hosp., Natich, Mass., 1968, Ctrl. Calif. Blood Bank, Fresno, Calif., 1971-85; biology educator State Coll. Community System, Fresno, Calif., 1988—. Recipient Nat. Fenwal scholarship Am. Assn. Blood Banks, 1981. Mem. AAAS, Am. Soc. Clin. Pathology, Am. Biology Tchr. Home: 1527 W San Bruno Fresno CA 93711

GRAY, KENNETH WAYNE, chemist, researcher; b. DeRidder, La., Apr. 2, 1944; s. James L. and Nancy J. (Brown) Yarbrough; m. Patti Ann Arnold, Aug. 7, 1965; children: Dawn M., Tara L. BS in Chemistry, Ouachita U., Arkadelphia, Ark., 1966. Rsch. scientist Continental Can Co., Chgo., 1968-71, The Polymer Corp., Reading, Pa., 1971-77; product mgr. The Polymer Corp., Reading, 1977-80; v.p., gen. mgr. ICO-Spincote Plastic Coatings, Odessa, Tex., 1980-89; mgr. technical svcs. ICO, Inc., Odessa, 1989-91; cons. Cloud-Cote, Inc. Odessa, 1987—; tech. dir. The Arabian Pipecoating Co., Jubail, Saudi Arabia. Patentee in field; contbr. articles to profl. jours. 1st lt. U.S. Army Chem. Corp., 1966-68; Korea. Recipient Army Commendation Medal U.S. Army, 1967. Mem. Nat. Assn. Corrosion Engrs. Republican. Baptist. Avocations: computer science, hunting, fishing, art. Home: PO Box 834, 31951 Jubail Saudi Arabia

GRAY, MARY WHEAT, statistician, lawyer; b. Hastings, Nebr., 1939; d. Neil C. and Lillie W. (Alves) Wheat; m. Alfred Gray, Aug. 20, 1964. AB summa cum laude, Hastings Coll., 1959; postgrad., J.W. Goethe U., Frankfurt, Fed. Republic Germany, 1959-60; M.A., U. Kans., 1962, Ph.D., 1964; J.D. summa cum laude, Am. U., 1979; LLD (hon.), U. Nebr., 1993. Bar: D.C. 1979, U.S. Supreme Ct. 1983, U.S. Dist. Ct., D.C. 1980. Physicist Nat. Bur. Standards, Washington, summers 1959-63; asst. instr. U. Kans., Lawrence, 1963-64; instr. dept. math. U. Calif., Berkeley, 1965; asst. prof. Calif. State U., Hayward, 1965-67; assoc. prof. Calif. State U., 1967-68; assoc. prof. dept. math., stats. and computer sci. Am. U., 1968-71, prof., 1971—, chmn. dept., 1977-79, 80-81, 83—; statis. cons. for govt. agys., univs. and pvt. firms, 1976—. Author: A Radical Approach to Algebra, 1970; Calculus with Finite Mathematics for Social Sciences, 1972; contbr. numerous articles to profl. jours. Nat. treas., dir. Women's Equity Action League, from 1981, pres., from 1982; bd. dirs. treas. ACLU, Montgomery County, Md.; mem. adv. com. D.C. Dept. Employment Services, 1983—; dir. Amnesty Internat. USA, 1985—, treas., 1988—; mem. Commn. on Coll. Retirement, 1984-86; dir. Am.-Middle East Edn. Found., 1983—. Fulbright grantee, 1959-60; NSF fellow, 1963-64, NDEA fellow, 1960-63. Fellow AAAS (chmn. com. on women, com. on investments, com. on sci. freedom and responsibility); mem. AAUP (regional counsel 1984—, com. on acad. freedom 1978—, dir. Legal Def. Fund 1974-78, bd. dirs. Exxon Project on Salary Discrimination 1974-76, com. on status of women 1972-78, Georgina Smith award), Am. Math. Soc. (v.p. 1976-78, council 1973-78), Conf. Bd. Math. Scis. (chmn. com. on affirmative action 1977-78), Math. Assn. Am. (chmn. com. on sch. lectrs. 1973-75, vis. lectr. 1974—), Assn. for Women in Math (founding pres. 1971-74, exec. com. 1974-80, gen. counsel 1980—), D.C. Bar Assn., ABA, Am. Soc. Internat. Law, London Math. Soc., Societe de Mathematique de France, Brit. Soc. History of Math., Can. Soc. History of Math., Assn. Computing Machinery, N.Y. Acad. Scis., Am. Statis. Assn., Phi Beta Kappa, Sigma Xi, Phi Kappa Phi, Alpha Chi, Pi Mu Epsilon. Home: 6807 Connecticut Ave Chevy Chase MD 20815-4937 Office: American U Math & Stats Dept Washington DC 20016

GRAY, PHILIP HOWARD, psychologist, educator; b. Cape Rosier, Maine, July 4, 1926; s. Asa and Bernice (Lawrence) G.; m. Iris McKinney, Dec. 31, 1954; children: Cindelyn Gray Eberts, Howard. M.A., U. Chgo., 1958; Ph.D., U. Wash., 1960. Asst. prof. dept. psychology Mont. State U., Bozeman, 1960-65; assoc. prof. Mont. State U., 1965-75, prof., 1975-92; ret., 1992; vis. prof. U. Man., Winnipeg, Can., 1968-70; chmn. Mont. Bd. Psychologist Examiners, 1972-74; speaker sci. and geneal. meetings on ancestry of U.S. presidents. Organizer folk art exhbns. Mont. and Maine, 1972-79; author: The Comparative Analysis of Behavior, 1966, (with F.L. Ruch and N. Warren) Working with Psychology, 1963, A Directory of Eskimo Artists in Sculpture and Prints, 1974, The Science That Lost Its Mind, 1985, Penobscot Pioneers vol. 1, 1992, vol 2, 1992, vol. 3, 1993; contbr. numerous articles on behavior to psychol. jurs.; contbr. poetry to lit. jours. With U.S. Army, 1944-46. Recipient Am. and Can. research grants. Fellow AAAS, APA, Am. Psychol. Soc., Internat. Soc. Rsch. on Aggression; mem. SAR (v.p. Sourdough chpt. 1990, pres. 1991-93, trustee 1989), Nat. Geneal. Soc., New Eng. His. Geneal. Soc., Gallatin County Geneal. Soc. (charter, pres. 1991-93), Deer Isle-Stonington Hist. Soc., Internat. Soc. Human Ethology, Descs. Illegitimate Sons and Daus. of Kings of Britain, Piscataqua Pioneers, Order Desc. Colonial Psysicians and Chirugiens, Flagon and Trencher. Republican. Avocations: collecting folk art, first and signed editions of novels. Home: 1207 S Black Ave Bozeman MT 59715-5633

GRAY, ROBERT BECKWITH, engineer; b. Johnstown, Pa., Apr. 9, 1912; s. Edward Townsend and Sarah Jean (Lomison) G.; m. Mary Elizabeth Ann Lynch, Jan. 17, 1942; children: William E., Rebecca Jean Fluegel, Robert F., Thomas O., Barry J. Student, Cornell U., 1934; EE, U. Pitts., 1941. Registered profl. engr., N.Y. Lab asst. Westinghouse Rsch. Magnetics, 1935-38; chief physicist Erie (Pa.) Resistor Corp., 1941-46; supr. rsch. Am. Meter Co., Erie, 1946-54; rsch. engr. Erie Resistor/Erie Tech. Products, Erie, 1954-65, Acme Electric Corp., Cuba, N.Y., 1965-67; engr. Hartman Metal Fabricators/Hartman Material Handling Systems, Victor, N.Y., 1967-85; cons. Tech. Svcs., Erie, 1981—. Editor Erie IEEE newsletter, 1988—; patentee in field; contbr. articles to profl. jours. Initiator, sec. textbook com. Erie Engring. Socs. Coun., 1957-58. Mem. IEEE, Am. Assn. of Physics Tchrs., Pa. Inventors Assn. (sec. 1989—), Sigma Xi, Sigma Pi Sigma. Office: Tech Svcs PO Box 1967 Erie PA 16507-0967

GRAY, STANLEY RANDOLPH, JR., systems engineer; b. Macon, Miss., Oct. 13, 1958; s. Stanley and Margaret (Ewing) G.; m. Debby Diane Howard, Dec. 17, 1982; children: Brandy, Carrie. BS, Miss. U. for Women, 1984; BSEE, Miss. State U., 1987. Engring. aide U.S. Army C.E., Columbus, Miss., 1984-86; systems engr. Computer Scis. Corp., Edwards AFB, Calif., 1987-89; control systems engr. Weyerhaeuser Pulp and Paper Co., Columbus, Miss., 1989—. With USAF, 1976-81. Mem. IEEE (treas. Miss. State U. chpt. 1986-87), TAPPI, New Hope Exch. Club. Republican. Baptist. Home: 4989 New Hope Rd Columbus MS 39702 Office: Weyerhaeuser Pulp and Paper PO Box 1830 Columbus MS 39701

GRAY, T(HEODORE) FLINT, JR., chemist; b. Anniston, Ala., Feb. 25, 1939; s. Theodore Flint and Alice (Roebuck) G.; m. Dottye Benkert, Aug. 13, 1960; children: Theodore Flint III, Brett C., A. Holly. BA, Centre Coll., 1960; PhD, U. Fla., 1964. Rsch. chemist Eastman Chem. Co. divsn. Eastman Kodak Co., Kingsport, Tenn., 1965-66; sr. rsch. chemist Eastman Chem. Co. div. Eastman Kodak Co., Kingsport, Tenn., 1966-72, rsch. assoc., 1972-79, sr. rsch. assoc., 1979-84, staff asst. to v.p. R&D, 1984-85, mgr. tech. svc. and devel. animal nutrition, 1985-86, dir. polymer rsch. divsn., 1986-91, dir. spl. projects, 1991-92, dir. emerging tech., 1993—; mem. chemistry indsl. adv. bd. U. Fla., Gainesville, 1985—, chmn., 1992. Contbr. articles to Jour. Macromolecular Sci.; contbr. entries to Ency. Polymer Sci. and Tech., 1976; contbg. author: Ultra-High Modulus Polymers, 1979. Elder, clk. of session 1st Presbyn. Ch., Kingsport; chmn. troop com. Boy Scouts Am., Kingsport. Mem. Am. Chem. Soc. (div. polymer sci.), Soc. Plastics Engrs., AAAS, Comml. Devel. Assn., Lions (pres. 1984-85), Kingsport C. of C. Achievements include several patents; research in kinetics and stereochemistry of cyclopolymerization, polymer structure, processing and property relations, effects of polymer blends and reinforcements on properties, thermoplastic polyester. Home: 2153 Westwind Dr Kingsport TN 37660 Office: Eastman Chem Co Kingsport TN 37662

GRAYBEAL, JACK DANIEL, chemist, educator; b. Detroit, May 16, 1930; s. Paul Herman and Polly Dale (McClintic) G.; m. Evelyn Alice Nicolai, June 13, 1954; children: Daniel Lee, David Eugene, Dale Kevin. BS in Chemistry, W.Va. U., 1951; MS in Chemistry, U. Wis., 1953, PhD in Chemistry, 1955. Mem. tech. staff Bell Telephone Labs., Holmdel, N.J., 1955-57; asst. prof. chemistry W.Va. U., Morgantown, 1957-63, assoc. prof., 1963-68; assoc. prof. chemistry Va. Poly. Inst. and State U., Blacksburg, 1968-69, prof., 1969-75, prof. and assoc. head dept. chemistry, 1975—. Author: Molecular Spectroscopy, 1988; contbr. articles to profl. jours. Mem. Am. Chem. Soc., Am. Phys. Soc., Sigma Xi, Phi Lambda Upsilon (editor 1981-87, nat. sec. 1987—). Home: 312 Apperson Dr Blacksburg VA 24060 Office: Va Poly Inst and State U Dept Chemistry Blacksburg VA 24061-0212

GRAYSON, RICHARD ANDREW, aerospace engineer; b. Silver Spring, Md., Aug. 5, 1966; s. Benson Lee and Helen Marie (Donovan) G. BS in Aerospace Engring., U. Va., 1988. Engr. Army Rsch. Lab., Aberdeen Proving Ground, Md., 1989—; mem. Joint Tech. Coordinating Group Air Systems, Wright-Patterson AFB, Ohio, 1990—; leader PATRIOT Assessment Team, Riyadh, Saudi Arabia, 1991-92. Advisor youth group Episcopal Ch., McLean, Va., 1988—. Recipient comdr.'s award for civilian svc. Dept. Army, 1991. Republican. Achievements included on-site investigations in Saudi Arabia and at Ballistic Research Laboratory to assist in evaluating PATRIOT lethality against SCUD-B tactical ballistic missiles fired in Operation Desert Storm. Home: 7006 Capitol View Dr McLean VA 22101 Office: SLAD-BVLD-ASB Army Rsch Lab Aberdeen Proving Ground MD 21005-5068

GRAZIADEI, WILLIAM DANIEL, III, biology educator, researcher; b. Hartford, Conn., Mar. 12, 1943; s. William Daniel Jr. and Rose Eleanor (Viggiano) G.; children—William D., IV, Mark C., Michael J., Keith N., Kevin M. B.S. in Biology, Fairfield U., 1965; M.S. in Radiation Biology, Boston Coll., 1967, Ph.D. in Microbiology and Molecular Biology, 1970. NIH postdoctoral fellow Yale U., New Haven, Conn., 1970-73; asst. prof. SUNY, Plattsburgh, 1973-76, assoc. prof., 1976-80, prof. biology, 1980—; coordinator in vitro cell biology W. Alton Jones Cell Ctr., Plattsburgh, 1980-83; dir. in vitro cell biology and biotech. Miner Inst., Plattsburgh, 1982—. Editor and author lab. protocols in vitro cell biology and biotech., 1982; contbr. articles to profl. jours. Recipient Chancellor's Excellence in Teaching award SUNY, 1976, Disting. Teaching Professorship award, 1992; grantee NIH, 1972, 74, Am. Cancer Soc., 1977, SUNY Rsch. Found., 1973, 81, 82, Fitzpatrick Oncology Ctr., 1989. Mem. Tissue Culture Assn., Inc. (chmn. edn. com. 1982-84), AAAS, Sigma Xi (v.p., pres. Plattsburgh chpt. 1974-76). Democrat. Roman Catholic. Avocations: boating, fishing, snow skiing, windsurfing, golf. Home: 6 Laurel Ct Plattsburgh NY 12901-4213 Office: SUNY Plattsburgh Miner Inst In Vitro Cell Biology & Biotech Plattsburgh NY 12921

GREANEY, JOHN PATRICK, civil engineer; b. Hartford, Conn., Feb. 20, 1956; s. John Joseph and Bridget (McCarthy) G. BCE, Worcester Poly. Inst., 1977; MBA, Syracuse U., 1981. Registered profl. engr., N.Y., Tex. Engr. batch facilities Corning (N.Y.), Inc., 1978-81; engring. specialist Exxon Prodn. Rsch. Co., Houston, 1981-86; mem. tech. staff ballistic missiles divsn. TRW, San Bernadino, Calif., 1986-87; sr. cost estimator Ogden Environ. Svcs., San Diego, 1987-89; supr. estimating chems. group Ethyl Corp., Baton Rouge, 1989—. Mem. NSPE, Am. Assn. Cost Engrs., Tau Beta Pi. Achievements include development of INTERCOST facilities cost estimating system. Office: Ethyl Corp 451 Florida St 17th Fl Baton Rouge LA 70801

GREATBATCH, WILSON, biomedical engineer; b. Buffalo, Sept. 6, 1919; married; 5 children. BEE, Cornell U., 1950; MSEE, U. Buffalo, 1957; ScD (hon.), Houghton Coll., 1971, SUNY, Buffalo, 1984, Clarkson U., 1987, Roberts Wesleyan Coll., 1988. Project engr. Cornell Aeronaut Lab. Inc., 1950-52; asst. prof. elec. engring. U. Buffalo, 1952-57; mgr. electronics div. Taber Instrument Corp., 1957-60; v.p. Mennen Greatbatch Electronics Inc., 1962-78; adj. prof. elec. engring. SUNY, Buffalo, 1981—; adj. prof. engring. Cornell U., Ithaca, N.Y., 1989—; adj. prof. physical scis. Houghton (N.Y.) Coll., 1978—. Contbr. over 100 articles to sci. jours; holder over 150 U.S. and fgn. patents. Recipient Holley medal ASME, 1986, Chancellor Morton medal U. Buffalo, 1990, Disting. Svc. award NSPE, 1984, Pacemaker award Prince Rainier of Monaco, 1988, Nat. Medal of Tech. Pres. Bush, 1990, Vladimir Karapetoff award Eta Kappa Nu, 1992; named to Am. Inventors Hall of Fame, 1986, U.S. Space Tech. Hall of Fame, 1993; Paul Harris fellow Rotary Internat., 1993. Fellow AAAS, IEEE, Am. Coll. Cardiology, Royal Soc. Health, Am. Soc. Angiology, Am. Inst. Med. and Biol. Engring. (founder); mem. NAE, Am. Soc. Mech. Engring., Assn. Advancement Med. Instrumentation (Laufman award 1982), N.Y. Acad. Sci., Sigma Xi. Achievements include invention of implantable cardiac pacemaker; rsch. in implantable power supplies for medical uses, biomass energy, genetic engring. Office: Greatbatch Gen Aid Ltd 10871 Main St Clarence NY 14031-1707

GREAVER, JOANNE HUTCHINS, mathematics educator, author; b. Louisville, Aug. 9, 1939; d. Alphonso Victor and Mary Louise (Sage) Hutchins; 1 child, Mary Elizabeth. BS in Chemistry, U. Louisville, 1961, MEd, 1971; MAT in Math., Purdue U., 1973. Cert. tchr. secondary edn. Specialist math Jefferson County (Ky.) pub. schs., 1962—; part-time faculty Bellarmine Coll., Louisville, 1982—, U. Louisville, 1985—; project reviewer NSF, 1983—; advisor Council on Higher Edn., Frankfort, Ky., 1983-86; active regional and nat. summit on assessment in math., 1991, state task force on math., assessment adv. com., Nat. Assessment Ednl. Progress standards com.; lectr. in field. Author: (workbook) Down Algebra Alley, 1984; co-author curriculum guides. Charter mem. Commonwealth Tchrs. Inst., 1984—; mem. Nat. Forum for Excellence in Edn., Indpls., 1983; metric edn. leader Fed. Metric Project, Louisville, 1979-82; mem. Ky. Ednl. Reform Task Force, Assessment Com., Math. Framework, Nat. Nat. Assessment Ednl. Progress Rev. Com. Recipient Presdl. award for excellence in math. teaching, 1983; named Outstanding Citizen, SAR, 1984, mem. Hon. Order Ky. Cols.; grantee NSF, 1983, Louisville Community Found., 1984-86. Mem. Greater Louisville Council Tchrs. of Math. (pres. 1977-78, Outstanding Educator award 1987), Nat. Council Tchrs. of Math. (reviewer 1981—), Ky. Coun. Tchrs. of Math. (pres. 1990-91, Jeff County Tchr. of Yr. award 1985), Math. Assn. Am., Kappa Delta Pi, Zeta Tau Alpha. Republican. Presbyterian. Avocations: tropical fish; gardening; handicrafts; travel; tennis. Home: 11513 Tazwell Dr Louisville KY 40241 Office: Van Hoose Edn Ctr 3332 Newburg Rd PO Box 34020 Louisville KY 40232-4020

GREAVES, IAN ALEXANDER, occupational physician; b. Darwin, Australia, June 1, 1947; came to U.S., 1981; s. George A. and Norma J. (Blair) G.; m. Alison Mary Street, Nov. 23, 1972 (div. 1985); m. Melana Dee Catlett, June 19, 1987. BSc in Medicine, Monash U., 1969, MB BS, 1971. Med. resident Alfred Hosp., Melbourne, Australia, 1972-75; med. registrar Prince Henry Hosp., Sydney, Australia, 1976-77; rsch. fellow U. NSW, Sydney, 1978-80; vis. fellow Harvard Sch. Pub. Health, Boston, 1981-83, asst. prof., then assoc. prof., 1983-89; assoc. prof., div. head environ. and occupational health U. Minn. Sch. Pub. Health, Mpls., 1989—; mem. com. on toxicology Nat. Acad. Sci., Washington, 1989—; dir. Midwest Ctr. Occupational Health and Safety, Mpls., 1990—, Minn. Ctr. Rsch. in Agrl. Safety and Health, Mpls., 1991—. Contbr. articles to profl. pubs. Recipient Outstanding Svc. award Am. Lung Assn. Mass., 1989. Mem. Assn. Univ. Programs in Occupational Health and Safety (pres. 1990—). Office: U Minn Box 807 420 Delaware St SE Minneapolis MN 55455

GREAVES, WILLIAM WEBSTER, chemist, patent analyst; b. Queenstown, Md., Jan. 10, 1951; s. William Emory and Mary Elizabeth (Wood) G. BS in Chemistry, Bucknell U., 1973; PhD in Inorganic Chemistry, Iowa State U., 1978. Tech. publ. editor Standard Oil of Ind., Naperville, Ill., 1978-81, rsch. info. scientist, 1981-84; assoc. editor Science mag., Washington, 1984-86; supr. chem. data systems SK&F Labs., Upper Merion, Pa., 1986-88; sr. patent searcher Abbott Labs., Abbott Park, Ill., 1988-90; patent analyst Amoco Corp., Chgo., 1990—. Contbr. articles to profl. publs. Active Frontrunners Chgo., 1988—, sec., 1991, v.p., 1992, pres., 1993; active D.C. Front Runners, Washington, 1984—. Mem. AAAS, Am. Chem. Soc. (edn. com. Chgo. chpt. 1981-84, mgr. Chgo. chpt. student symposium 1982), Soc. Tech. Communication (sr., sec. Chgo. chpt. 1983), Sigma Xi. Republican. Roman Catholic. Office: Amoco Corp Mail Code 1904 200 E Randolph St Chicago IL 60601-7125

GREBENC, JOSEPH D., mechanical engineer, consultant; b. Cleve., Sept. 29, 1958; s. Frank Henry and Christine (Stih) G.; m. Patricia Kay Johanneck, May 11, 1991. BSME, U. Toledo, 1981. Application engr. Production Machinery Corp., Mentor, Ohio, 1982—; cons. Marquis Diamond Fabricator, Mentor, 1990—. Recipient coll. grant Foresters, 1976. Mem. AISI, Kirtland Hills Country Club. Republican. Roman Catholic.

GREBNER, EUGENE ERNEST, biochemist, educator; b. Pitts., Feb. 6, 1931; s. Eugene Eduard and Carolina (Fuerst) G.; m. Mary Alice Wagner, Nov. 26, 1955; children: Lisa J., Matthew E. AB, Hiram Coll., 1952; PhD, U. Pitts., 1964. Postdoctoral fellow NIH, Bethesda, Md., 1964-67; asst. mem. Albert Einstein Med. Ctr., Phila., 1967-75; assoc. prof. Thomas Jefferson U., Phila., 1975—; dir. Tay-Sachs Prevention Program of Delaware Valley. Contbr. articles to sci. jours. Bd. dirs. Upper Moreland Pub. Libr., Willow Grove, Pa., 1976-81. Sgt. U.S. Army, 1952-54, Korea. Mem. Nat. Tay-Sachs and Allied Diseases Assn. (mem. sci. adv. com. 1987—). Democrat. Office: Thomas Jefferson U 1100 Walnut St Philadelphia PA 19107

GRECO, RALPH STEVEN, surgeon, researcher, medical educator; b. N.Y.C., May 25, 1942; s. Charles Mario and Lydia Antoinette (Barone) G.; m. Irene Leonar Wapnir, Feb. 23, 1991; children: Justin Michael, Eric Matthew. BS, Fordham U., 1964; MD, Yale U., 1968. Instr. Yale U., New Haven, 1972-73; asst. prof. Med. Sch. Rutgers U., Piscataway, N.J., 1975-79, assoc. prof. Med. Sch., 1979-83; chief of surgery Robert Wood Johnson Med. Sch. U. Medicine & Dentistry of N.J., New Brunswick, 1982—, prof. Robert Wood Johnson Med. Sch., 1983—; cons. Nat. Heart, Lung and Blood Inst.-NSF, Bethesda, Md., 1991. Contbr. articles to profl. jours. Maj. U.S. Army, 1973-75. NHLBI grantee, 1980-84. Fellow Am. Surg. Assn.; mem. Soc. Univ. Surgeons. Achievements include research in antibiotic bonding, treatment of prosthetic infection, bonding of TPA to prostheses, drug delivery; patents in field. Home: 2 Quail Run Warren NJ 07059 Office: U Medicine & Dentistry NJ Robert Wood Johnson Med Sch 1 Robert Wood Johnson Pl # 19cn New Brunswick NJ 08901-1928

GRECO, THOMAS G., chemist, educator; b. Phila., Dec. 22, 1944; s. Louis and Jenny (Rago) G.; m. Catherine G. Greco, June 22, 1973; children: Susan, Jeffrey. BS, St. Joseph's U., 1966; PhD, U. Del., 1973. Resources chemist Del. Dept. Natural Resources and Environ. Control, Dover, 1973-76; from asst. prof. to prof. chemistry Millersville (Pa.) U., 1976—; cons. TMI, Middletown, Pa., 1976-78, NOAA, V.I., 1986, Smithsonian Inst., Washington, 1982. Author lab. manual; contbr. articles to profl. publs. Mem. Am. Chem. Soc., Delaware Valley Chromatography Forum, Optimists Club, Sigma Xi. Office: Millersville U Dept Chemistry PO Box 1002 Millersville PA 17551-0302

GREDEN, JOHN FRANCIS, psychiatrist, educator; b. Winona, Minn., July 24, 1942; m. Renee Mary Kalmes; children: Daniel John, Sarah Renee, Leigh Raymond. BS, U. Minn., 1965, MD, 1967. Diplomate Am. Bd. Psychiatry and Neurology. Assoc. dir. psychiat. research Walter Reed Army Med. Ctr., Washington, 1972-74; asst. prof. Dept. Psychiatry U. Mich., Ann Arbor, 1974-77, assoc. prof., 1977-81, dir. clin. studies unit for affective disorders, 1980-85, prof., 1981—, chmn., research scientist, 1985—. Contbr. 188 articles to profl. jours., 28 chpts. to books. Served to maj. U.S. Army, 1969-74. Recipient A.E. Bennett research award Cen. Neuropsychiat. Found., 1974, Ralph Patterson Meml. award Ohio State U., 1980, Nolan D.C. Lewis Vis. Scholar award Carrier Found., 1982. Fellow Am. Psychiat. Assn.; mem. AAAS, Soc. Biol. Psychiatry (past pres.), Am. Coll. Neuropsychopharmacology, Psychiat. Rsch. Soc. (past pres.). Office: U Mich Psychiat Ctr 1500 E Med Ctr Dr Ann Arbor MI 48109-0704

GREELEY, RICHARD FOLSOM, retired sanitary engineer; b. Hudson, Mass., Apr. 29, 1915; s. Willis Samuel and Eleanor (Stratton) G.; m. Evelyn Davenport, Sept. 7, 1946; children: Roger, Richard, Donald. BS in Civil Engring., Tufts U., 1936; MS in Engring., Harvard U., 1939. Registered profl. engr., Mass. Jr. sanitary engring. aid Dept. Pub. Health, Mass., 1939-41; sanitary engr. Camp Edwards U.S. War Dept., Falmouth, Mass., 1941-43; sanitary engr. N.E. regional office VA, Boston, 1946; sanitary engr. Inst. Inter-Am. Affairs, Mexico, Peru, Nicaragua, 1946-53, Metcalf & Eddy, Mass., 1953-55; chief const. mass divsn. W.P.C., 1955-83; commr. Ea. Water Dist., Grantham, N.H., 1984-88; health officer Town of Grantham, 1985-91. Author: Greeley Genealogy, 1989, Stratton Genealogy, 1990. Track ofcl. Dartmouth Coll., Hanover, N.H., 1983-90. Capt. U.S. Army, 1943-46. Mem. ASCE (life), Am. Acad. Environ. Engrs., Water Pollution Control Fedn. (life), N.E. Water Pollution Control Assn. (life, sec.-treas. 1960-67, Bedell award 1969), N.E. Water Works Assn. (life), Tau Beta Pi. Methodist. Home: Villa 279 Carolina Meadows Chapel Hill NC 27514

GREEN, DAVID MARVIN, psychology educator, researcher, consultant; b. Jackson, Mich., June 7, 1932; s. George Elmer and Carrie Ruth (Crawford) F.; m. Clara Loftstrom, Feb. 2, 1953 (dec. Dec. 1978); children: Allen, Phillip, Katherine, George; m. Marian Heinzman, June 7, 1980. BA, U. Mich., 1954, PhD, 1958; MA (hon.), U. Cambridge, 1973, Harvard U., 1973. Rsch. assoc. electronic def. group U. Mich., Ann Arbor, 1954-58; asst. prof. psychology MIT, Cambridge, Mass., 1958-63; assoc. prof. U. Pa., Phila., 1963-66; prof. U. Calif., San Diego, 1966-73; prof. psychophysics Harvard U., Cambridge, 1973-85; grad. rsch. prof. psychology U. Fla., Gainesville, 1985—; acoustic cons. Bolt, Beranek & Newman, Conoga Park, Calif., 1958—. Author: An Introduction to Hearing, 1976, Profile Analysis: Auditory Intensity Discrimination, 1988; co-author Signal Detection Theory and Psychophysics, 1966. Guggenheim fellow, 1973-74. Fellow NAS, AAAS, Acoustical Soc. Am. (pres. 1985, silver medal in psychol. and physiol. acoustics 1990), Com. on Hearing and Bio-Acoustics (chmn. 1970-71, 78-79, exec. com. 1968-71). Office: U Fla Dept Psychology Rm 078 Gainesville FL 32611

GREEN, DAVID RICHARD, development chemist; b. Bismarck, N.D., Feb. 11, 1964; s. Richard Vern and Dolores Jean (Schatz) G. BS in Chemistry, N.D. State U., 1987. Assoc. chemist Container Coatings div. BASF Corp., Cin., 1987-90, Hazmat technician, 1990—, chemist, 1990—. Achievements include contbn. in development of first prodn. novar white basecoat for 2 piece beer/beverage cans in U.S. and Europe, ultra-low V.O.C. water based white basecoats for 2 piece beer/beverage cans. Home: 5825 Monassas Run Rd Milford OH 45150 Office: Container Coatings Div BASF Corp 500 Technecenter Dr Milford OH 45150

GREEN, DON WESLEY, chemical and petroleum engineering educator; b. Tulsa, July 8, 1932; s. Earl Leslie and Erma Pansy (Brackins) G.; m. Patricia Louise Polston, Nov. 26, 1954; children—Guy Leslie, Don Michael, Charles Patrick. B.S. in Petroleum Engring., U. Tulsa, 1955; M.S. in Chem. Engring., U. Okla., 1959, Ph.D. in Chem. Engring., 1963. Research scientist Continental Oil Co., Ponca City, Okla., 1962-64; asst. to assoc. prof. U. Kans., Lawrence, 1964-71, prof. chem. and petroleum engring., 1971-82, chmn. dept. chem. and petroleum engring., 1970-74, co-dir. Tertiary Oil Recovery project, 1974—, Conger-Gabel Disting. prof., 1982—. Editor: Perry's Chemical Engineers' Handbook, 1984; contbr. articles to profl. jours. Served to 1st lt. USAF, 1955-57. Fellow Am. Inst. Chem. Engrs.; mem. Soc. Petroleum Engrs. (Disting. Achievement award 1983, chmn. edn. and accreditation com. 1980-81, Disting. mem. 1986, Disting. lectr. 1986). Democrat. Avocations: handball; baseball; mountain hiking. Home: 1020 Sunset Dr Lawrence KS 66044-4546 Office: U Kans Dept Chem and Petroleum Engring 4008 Learned Hall Lawrence KS 66045-0001

GREEN, DONALD ROSS, research psychologist; b. Holyoke, Mass., Aug. 12, 1924; s. Donald Ross and Constance (McLaughlin) G.; m. Mary Reese,

June 16, 1950; children: Alice Angell, Mitchell Reese. BA in Math., Yale U., 1948; MA in Edn., U. Calif., Berkeley, 1954, PhD in Edn., 1958. Instr. math. George Sch., Pa., 1948-50; statistician med. sch. cancer rsch. inst. U. Calif., Berkeley, 1953-56, asst., assoc. in edn., 1956-57; asso. prof. edn. and psychology Emory U., Atlanta, 1957-67; sr. rsch. psychologist CTB McGraw-Hill, Monterey, Calif., 1967—, dir. rsch., 1968-92. Author: Educational Psychology, 1964, Biased Tests, 1971, Racial and Ethnic Bias in Test Construction, 1972, Racial and Ethnic Bias in Achievement Tests and What to Do About It, 1973, the Aptitude-Achievemnt Distinction, 1974; co-author: Reading for Meaning in the Elementary School, 1969, Measurement and Piaget, 1971, Achievement Testing of Disadvantaged and Minority Students for Educational Program Evaluation, 1978; contbr. articles to profl. jours.; presenter in field. With U.S. Army, 1943-45. Fellow Am. Psychol. Assn., Am. Psychol. Soc.; mem. AAAS, ASUP, ASCD, Am. Ednl. Rsch. Assn., Assn. Measurement and Evalutaion in Counseling and Devel., Internat. Assn. Applied Psychology, Internat. Coun. Psychologists, Internat. Reading Assn., Nat. Soc. Study Edn., Psychometric Soc., Phi Delta Kappa.

GREEN, EDWARD CROCKER, health consulting firm executive; b. Washington, Nov. 29, 1944; s. Marshall and Lispenard Seabury (Crocker) G.; m. K. Shannon McCaffray, Sept. 22, 1967 (div. 1977); 1 child, Timothy A. BA, George Washington U., 1967; MA, Northwestern U., 1968; PhD, Cath. U. Am., 1974; postdoctoral, Vanderbilt U., 1978-79. Asst. prof. W.Va. U., Morgantown, 1976-78; devel. cons. various orgns., Washington, 1979-86; mgr. internat. programs John Short & Assocs., Columbia, Md., 1986-88; mgr. and researcher The Futures Group, Washington, 1988—; social scientist Acad. for Ednl. Devel., Swaziland, 1981-84; personal services contractor U.S. AID, Swaziland, 1984-85. Author: Planning Psychiatric Services for Southern Africa, 1979; editor: Practicing Development Anthropology, 1986; contbr. 55 articles to profl. jours. Mem. adv. council Health Communications for Child Survival Acad. for Ednl. Devel., Washington, 1985—. Recipient Praxis award Washington Assn. Profl. Anthropologists, 1982, 83; NIMH postdoctoral fellow, 1978-79; research grantee Sigma Xi, 1971. Fellow Am. Anthrop. Assn.; mem. N.Y. Acad. Scis., Soc. Med. Anthropology, Soc. Applied Anthropology, Nat. Council for Internat. Health, Sigma Xi Soc. Club: Swaziland Theatre (Mbabane) (social dir. 1982-83). Avocations: folk music, folk dancing, collecting primitive art, photography. Home and Office: 2807 38th St NW Washington DC 20007

GREEN, JAMES ANTON, systems engineer; b. May 13, 1949. Student, Friedrich Wilhelms U., Bonn, Fed. Republic Germany, 1971, U. Kans., 1973; MSEE, Wichita State U., 1977; postgrad., U. South Fla., 1988. Project engr. Mycro-Tek, Inc., Wichita, 1977-83; design engr. Kreonite, Wichita, 1983; chief engr. AlphaType/Berthold, Tampa, 1983-84, 85; engr. Gen. Def., Pinellas Park, Fla., 1984; prin. design engr. Honeywell Def. Co., Wichita, 1986-87; software engr. Compro, Tampa, Fla., 1988-90; mgr. systems design Watkins, Inc., Wichita, Kans., 1990; systems engr. Telos Fed. Systems, Lawton, Okla., 1990—. Author: Inside A Chambered Nautilus, Optimal Edge Detection and Digital Picture Processing, 1977, The Effects of Thermonuclear Weapons, 1991; designer computer hardware and software. Mem. IEEE, Sigma Pi Sigma, Pi Mu Epsilon. Achievements include work in field theory and physics. Home and Office: 1036 Murray Ct Wichita KS 67212

GREEN, JOHN DAVID, engineering executive; b. Newark, Ohio, July 20, 1948; s. John Robert and Ruth Ella (Lugenbeal) G.; m. Clare Marie McHenry, Dec. 22, 1970; children: Jennifer L., David E. BSEE, Case Inst. of Tech., 1970; MBA, Ohio U., 1983. Registered profl. engr., Ohio; cert. lighting efficiency profl. Engrin. mgr., sr. rsch. engr. Holophane Co., Inc., Newark, 1970—. Sgt. USAF, 1970-73. Recipient A award Manville Corp., 1979. Mem. Illuminating Engring. Soc. (cive chmn. 1989-91, chmn. 1991-93, author publ. 1980), Nat. Soc. Profl. Engrs. (Top 10 Achievement award 1975), Assn. Energy Engring., Soc. Plastic Engrs. Achievements include patents in Method and Assembly for Measuring Equivalent Sphere Illumination, Optical Design for Poster Panel Luminaire. Office: Holophane Co Inc 214 Oakwood Ave Newark OH 43055-6700

GREEN, JOSEPH, chemist; b. N.Y.C., Oct. 5, 1928; s. Sol and Rose G.; m. Phyllis Rothstein, Mar. 25, 1951; children: Howard S., Mitchell S., Robert K. BS, CCNY, 1950; MS, U. Kans., 1952. Chemist U.S. Rubber Reclaiming, Buffalo, 1951-55; sect. mgr. Thiokol Chem., Reaction Motors Div., Rockaway, N.J., 1955-67; dept. mgr. Cities Svc. Co., Cranbury, N.J., 1967-77; v.p. mktg. Saytech Inc., East Brunswick, N.J., 1977-80; prin. scientist FMC Corp., Princeton, N.J., 1980—. Contbr. over 175 articles to profl. jours. Fellow Am. Inst. Chemists; mem. Fire Retardant Chems. Assn. (pres. 1989-91, bd. dirs. 1991—), Soc. Plastics Engrs. (seminar staff 1975—). Achievements include over 65 patents in field. Home: 3 New Dover Rd East Brunswick NJ 08816-2746 Office: FMC Corp PO Box 8 Princeton NJ 08543-0008

GREEN, JOSEPH BARNET, neurologist; b. Phila., Aug. 2, 1928; s. Charles and Bella (Hurwitz) B.; married; children—Charna Alice Green Evans, Robert I. B.S., St. Joseph's Coll., Phila., 1950; M.D., Jefferson Med. Coll., Phila., 1954. Intern Wilkes-Barre (Pa.) Gen. Hosp., 1954-55; resident in neurology Georgetown U. Med. Center, 1955-58; asst. neurologist Pa. Hosp., Phila., 1960-64; asst. prof., then prof. neurology Ind. U. Med. Sch., 1964-72; prof. neurology and pediatrics, chmn. dept. neurology Med. Coll. Ga., Augusta, 1972-82; prof., chmn. dept. psychiatry and neurology Tulane U. Sch. Medicine, New Orleans, 1982-87, prof. neurology, clin. prof. pediatrics, 1986-87; dir. neurology VA Cen. Office, Washington, 1987-88; chmn. dept. med. and surg. neurology Tex. Tech U. Health Scis. Ctr., Lubbock, 1988—; chmn. profl. adv. bd. Epilepsy Assn.; mem. profl. adv. bd. Assn. Children with Learning Disabilities, Ga. chpt. Nat. Multiple Sclerosis Soc.; dir. NIH project Ga. Comprehensive Epilepsy Program, 1976; Fulbright lectr., Denmark, 1969; cons. VA Med. Ctr., Augusta. Author articles in field. Served with M.C. USNR, 1958-60. Fogarty Internat. Research fellow Israel, 1981. Mem. AMA, Am. Acad. Neurology, Child Neurology Soc., Am. Neurol. Assn., Am. EEG Soc., Am. Epilepsy Soc. (sec. 1972), Assn. U. Profs. Neurology (v.p. 1981). Democrat. Jewish. Club: B'nai B'rith. Home: 4619 7th St Lubbock TX 79416-4723 Office: Tex Tech U Health Sci Ctr Dept Med/Surg Neurology 3601 4th St Lubbock TX 79430

GREEN, LAWRENCE, neurologist, educator; b. Atlantic City, N.J., Oct. 23, 1938; s. Martin and Lillian (Spector) G.; m. Ann Etta Buchberg, Aug. 21, 1970; 1 child, Louis Aaron. BS in Chemistry cum laude, Dickinson Coll., 1960; MD, Jefferson Med. Coll., 1964. Diplomate Am. Bd. Psychiatry and Neurology, Am. Bd. Clin. Neurophysiology; cert. in neurology with added qualifications in neurophysiology. Intern Lakenau Hosp., Phila., 1964-65; resident in neurology Jefferson Med. Coll. Hosp., Phila., 1965-68; fellow in clin. neurophysiology Boston Va. Hosp., 1970-71; attending neurologist Phila. Gen. Hosp., 1971-72; chief, div. neurology and clin. neurophysiology Crozer-Chester (Pa.) Med. Ctr., 1971—, med. dir., Sch. of EEG Tech., 1974—; chief EEG Lab. Hahnemann U. Hosp., Phila., 1972-78, 87-91, electroencephalographer, 1978-87; asst. prof. neurology to clin. prof. neurology Hahnemann Univ., Phila., 1971—; instr. in neurology Boston U. Sch. Medicine, 1970-71; adj. prof. bioengring. Widener Coll., Chester, Pa., 1975-77; cons. in neurology and EEG Community Hcsp. Chester, 1971—, Taylor Hosp., Ridley Park, Pa., 1971—, St. Agnes Med. Ctr., Phila., 1971—, Hahnemann U. Hosp., 1972—; assoc. examiner Am. Bd. Clin. Neurophysiology, Am. Bd. Registered EEG Tech. Contbr. articles to Neurology, Experientia, Electroencephalography and Clin. Neurophysiology, Low Back Pain, Annals of Neurology, Electroencephalographer Clin. Neurophysiology. Lt. comdr. USNR, 1968-70. Fellow Am. EEG Soc. (coun., EEG lab. accreditation bd., rep. 1988-92, Am. EEG Soc./Joint Rev. Com. cons. to Com. on Allied Health Accreditation 1988—, program com. 1991); mem. AMA, ACP, Am. Acad. Neurology, Am. Epilepsy Soc., Am. Assn. EMG & ED, Pa. Med. Soc., Alpers Soc. for Clin. Neurology (past pres., sec., treas.), Am. Med. Tennis Assn. Office: EEG Lab Crozer Chester Med Med Ctr Blvd Upland PA 19013

GREEN, MAURICE, molecular biologist, virologist, educator; b. N.Y.C., May 5, 1926; s. David and Bessie (Lipschitz) G.; m. Marilyn Glick, Aug. 20, 1950; children—Michael Richard, Wendy Allison Green Lee, Eric Douglas. B.S. in Chemistry, U. Mich., 1949; M.S. in Biochemistry and Chemistry, U. Wis.-Madison, 1952, Ph.D. in Biochemistry and Chemistry, 1954. Instr. biochemistry U. Pa. Med. Sch., Phila., 1955-56; asst. prof. St. Louis U. Health Scis. Ctr., 1956-60; assoc. prof. St. Louis U. Med. Ctr., 1960-63, prof. microbiology, 1963-77; prof., chmn. Inst. for Molecular Virology, 1964—. Office: St Louis U Health Scis Ctr Inst Molecular Virology 3681 Park Ave Saint Louis MO 63110-2592

GREEN, MAURICE BERKELEY, agrochemical research consultant; b. London, Sept. 25, 1920; s. Walter Michael and Margaret Grace (MCNab) G.; m. Catherine Evelyn Nayler, Aug. 15, 1942 (dec. 1979); children: Sally, Patrick, Richard, Andrew; m. Renee Maria Karpati, Jan. 4, 1986. BSc, Imperial Coll., 1940; PhD, Queen Mary Coll., London, 1955. Cert. chemistry. Rsch. officer Wellcome Found., London, 1941-46, May and Baker Ltd., London, 1946-57; rsch. mgr. Imperial Chem. Industries Ltd., Cheshire, Eng., 1957-79; cons. Washington, 1979-83, London, 1983—; vis. prof. U. Stirling (U.K.), 1968-79, U. Aston, Birmingham, Eng., 1970-79, U. Fla., Gainesville, 1979-83. Pesticides: Boon or Bane, 1976, Chemicals for Crop Improvement and Pest Management, 1987; contbr. articles to profl. jours. Pres. Assn. Mgmt. and Profl. Staffs, London, 1974-79, exec. com. mem., 1983—. Hinchley med. fellow. Fellow Royal Chem. Soc., Am. Chem. Soc. (agrochem. div., exec. com. 1976-83); mem. Soc. Chem. Industry Pesticides Group (sec. 1964-70, chmn. 1970-72, exec. com. 1972—, Lampitt medal). Labour. Mem. Soc. of Friends. Avocations: chess, bridge, reading, writing. Home and Office: 1 Dinorben 79/81 Woodcote Rd, Wallington SM6 0PZ, England

GREEN, MAURICE RICHARD, neuropsychiatrist; b. Chgo., Oct. 28, 1922; divorced; children: Melissa, Suzanne, Constance. BS, Northwestern U., 1942; BM, Northwestern U. Med. Sch., 1945, MD, 1946; cert. in Psychoanalytic Tng., William Alanson White Inst., N.Y.C., 1954. Diplomate Am. Bd. Psychiatry and Neurology. Intern Passavant Hosp., Chgo., 1945-46; resident in psychiatry Bronx (N.Y.) VA Hosp., 1948-51; cons. psychiatrist Brookwood Hall, East Islip, L.I., N.Y., 1955-58; staff psychiatrist Psychiatric Clinic Ct. Spl. Sessions, 1956-60; cons. psychiatrist Bleuler Psychotherapy Ctr., Queens, N.Y., 1956-68; research psychiatrist, mem. psychiat. epidemiology sect. William Alanson White Inst., N.Y.C., 1968-72; attending geriatric psychiatrist Albert Einstein Med. Sch., 1974-76; attending child and adolescent psychiatry Harlem Hosp. of Columbia Presbyn. Med. Ctr., N.Y.C., 1974-75; med. dir. geriatric and family psychiatry Lincoln Hosp., 1974-76; chief psychiatrist Family Ct. Services div. South Beach Psychiat. Ctr., S.I., N.Y., 1976-80; sr. attending pyschiatrist Columbia-Presbyn. at St. Luke's-Roosevelt Hosp., N.Y.C., 1978—; cons. psychiatrist Liaison-Consultation Service NYU Med. Ctr., N.Y.C., 1985-86; psychiatrist spl. evaluation and treatment unit Rockland Psychiat. Ctr., 1985-87; mem. faculty William Alanson White Inst., N.Y.C., 1957—; cons. Goddard Coll., 1961-68; assoc. attending psychiatrist Bellevue Hosp., 1962-85, presently attending physician; supervisory and tng. analyst William Alanson White Inst., 1962—; clin. prof. psychiatry NYU Med. Sch., 1964—; mem. med. bd. Roosevelt Hosp., 1965-76; prin. investigator Diamox-Thiamine Research Unit Nathan S. Kline Research Inst., 1987; project dir. Brain Chemistry of Schizophrenia at Nathan Kline Inst., 1988-93; med. dir. Neurologic Systems, Inc., 1987; presidium Inst. for Brain Function Research, Inc., 1987; mem. Treatment Innovations Task Force-Soc. for Traumatic Stress Studies, 1987. Author: Interpersonal Psychoanalysis: Selected Papers of Clara Thompson, 1971, Psicoanalisi interpersonale, 1972, L'Esperiencze Prelogica, 1972, Violence and the Family, 1980; (with Edward S. Tauber) Prelogical Experience, 1959; assoc. editor Contemporary Psychoanalysis jour., 1968—; contbr. articles to profl. jours. Project dir. Nathan Kline Rsch. Inst., 1988—. Fellow Am. Psychiat. Assn. (com. on aging N.Y. Dist. br.), Am. Acad. Psychoanalysis, Am. Orthopsychiat. Assn. (publs. com. anniversary Vol. 1968-71), Am. Acad. of Child and Adolescent Psychiatry (com. on hospitalization of children, nat. legis. network 1982-86), N.Y. Acad. Medicine; mem. AMA, N.Y. Coun. on Child Psychiatry, N.Y. Soc. Clin. Psychiatry, William Alanson White Psychoanalytic Soc., Physicians for Social Responsibility, Nat. Assn. Patients Rights and Advocacy, Inst. Brain Function Rsch., Soc. Biol. Psychiatry, World Assn. for Psychosocial Rehab (v.p. USA br.). Home and Office: 275 Central Park W 15-D New York NY 10024

GREEN, MAYER ALBERT, physician; b. Pitts., May 29, 1909; s. Oscar and Elizabeth (Rosenbloom) G.; m. Phyllis Blumenfeld, Feb. 6, 1938; children: Patricia, Richard, Nancy. BS, U. Pitts., 1929, MD, 1932. Diplomate Am. Bd. of Internal Medicine and Allergy, Am. Bd. of Allergy and Immunology. Rotating intern Montefiore Hosp., Pitts., 1932-33, resident in roentgenology, 1933-34; resident in roentgenology Mt. Sinai Hosp., N.Y.C., 1937-38, asst. med. dir., 1938-39; pvt. practice in gen. medicine and internal medicine Pitts., 1933-37, pvt. practice in allergy and clin. immunology, 1939—; instr. in allergy Dept. Medicine, U. Pitts., 1933-37, Sch. Medicine, U. Pitts., 1953—; sr. physician, chief of allergy Columbia Hosp., now Forbes Health System, 1955—; cons. in allergy Shadyside Hosp., Braddock Med. Ctr., Jewish Home and Hosp. for the Aged, South Hills Health System, Ohio Valley Gen. Hosp.; emeritus physician, allergy cons. St. John's Gen. Hosp.; lectr. in field. Contbr. numerous articles to profl. jours. including Jour. of Allergy, Jour. of Investigative Dermatology, Hyggia, Annals of Internal Medicine, Annals of Allergy, Archieves of Dermatology, Pa. Med. Jour., Jour. of Clin. Medicine, Med. Times, Clin. Medicine, W.Va. Med. Jour. and others. Fellow ACP (life), AMA, Am. Coll. Allergists (first v.p. 1953-54, bd. regents 1962-63, pres. 1962-63, others), Am. Acad. of Psychosomatic Medicine, Am. Acad. Allergy, Am. Coll. of Chest Physicians, Internat. Assn. of Allergists and Allergology (hon.), Am. Med. Writer's Assn.; mem. AAAS, Assn. of Am. Med. Colls., Pa. Med. Soc., Pitts. Allergy Soc., Allegheny Med. Soc., Pa. Thoracic Soc., Assn. of Convalescent Homes and Hosp. for Asthmatic Children (various coms.), and many others. Home: Dithridge Ho Apt 1104 220 N Dithridge St Pittsburgh PA 15213 Office: Gateway Towers Ste 380 Pittsburgh PA 15222

GREEN, OTIS MICHAEL, electrical engineer, consultant; b. Sayre, Okla., Feb. 12, 1945; s. Raymond W. and Mattie L. (Griffen) G.; m. Janie L. Segraves, Mar. 3, 1973 (div. Nov. 1986); children: Don M., Phil J., Ken D.; m. Carol Ann Mitchell, Sept. 9, 1989; 1 stepchild, Marda Hatcher. AA, Sayre Jr. Coll., 1965; BSEE, Okla. State U., 1968. Design engr. Tex. Instruments Inc., Dallas, 1968-71, project engr., 1971-72, project mgr., 1972-74, control system cons., 1974-76; chief engr. Warren D. Segraves & Assocs., Architects & Engrs., Fayetteville, Ark., 1976-78; pres., CEO Warren D. Segraves & Assocs., Architects & Engrs., Fayetteville, 1978-81, O. Michael Green Engring., Inc., Fayetteville, 1981—. Asst. mayor Fayetteville, 1992; chmn. Advt. and Promotion Commn. Fayetteville, 1992; mem. Washington County Energy Adv. Com., 1977-79; chmn. Pollution Control and Energy Com. Fayetteville, 1978-80. Named Outstanding Young Man Am., 1979. Mem. NSPE, Illuminating Engrs. Soc., Nat. Coun. Engring. Examiners, Rotary (bd. dirs. Fayetteville club 1984). Office: O Michael Green Engring Inc 112 W Center St Ste 610 Fayetteville AR 72701

GREEN, PAUL ELIOT, JR., communications scientist; b. Durham, N.C., Jan. 14, 1924; s. Paul Eliot and Elizabeth Atkinson (Lay) G.; m. Dorrit L. Gegan, Oct. 30, 1948; children: Dorrit Green Rodemeyer, Nancy E., Judith Green Godin, Paul M., Gordon M. A.B., U. N.C., Chapel Hill, 1943; M.S., N.C. State U., 1948; Sc.D., MIT, 1953. Group leader MIT Lincoln Lab., Lexington, 1951-69; sr. mgr. IBM Research Div., Yorktown Heights, N.Y., 1969-81; mem. corp. tech. com. IBM Research Div., Armonk, N.Y., 1981-83; staff IBM Research Div., Yorktown Heights, N.Y., 1983—; mem. radio engring. adv. com. USIA, 1984-93; panel on survivable communications NRC, 1982-89. Author: Fiber Optic Networks, 1992; co-editor: Computer Communications, 1974; editor: Computer Network Architectures and Protocols, 1982, Network Interconnection and Protocol Conversion, 1988. Served to lt. comdr. USNR, 1943-60; ret. Named Disting. Engring. Alumnus N.C. State U., 1983. Fellow IEEE (chmn. info. theory group 1960, pres. Comm. Soc. 1992—, Aerospace Pioneer award 1981, E.H. Armstrong award 1989, Simon Ramo medal 1991); mem. NAE. Home: Roseholm Pl Mount Kisco NY 10549 Office: IBM PO Box 704 Yorktown Heights NY 10598-0704

GREEN, RICHARD, psychiatrist, lawyer, educator; b. Bklyn., June 6, 1936; s. Leo Harry and Rose (Ingber) G.; 1 child, Adam Hines-Green; life ptnr. Melissa Hines. AB, Syracuse U., 1957; MD, Johns Hopkins U., 1961; JD, Yale U., 1987. Diplomate Am. Bd. Psychiatry and Neurology; bar: Calif. 1987, D.C. 1989. Intern Kings County Hosp., Bklyn., 1962-64; resident in psychiatry Nat. Inst. Mental Health, Bethesda, Md., 1965-66; asst. prof., assoc. prof., then prof. dept. psychiatry UCLA, 1968-74; prof. psychiatry and psychology SUNY, Stony Brook, 1974-85; prof. psychiatry UCLA, 1986—, prof. law, 1988-90; part-time faculty mem. law sch. UCLA, 1991-92. Author: Sexual Identity Conflict in Children and Adults, 1974, Impotence, 1981, The Sissy Boy Syndrome and the Development of Homosexuality, 1987, Sexual Science and The Law, 1992; co-author, co-editor: Transsexualism and Sex Reassignment, 1969, Human Sexuality: A Health Practitioner's Text, 1975, 2d edit., 1979; editor Jour. Archives of Sexual Behavior, 1981—. Vol. atty., ACLU, LA. Vis. scholar, U. Cambridge, Eng., 1980-81; fellow, Ctr. Advanced Study in Behavioral Scis., Stanford, Calif., 1982-83; Fulbright scholar King's Coll., London, and Univ. of Cambridge, 1992. Fellow Soc. Sci. Study of Sex (past pres.), Internat. Acad. Sex Rsch. (founding pres. 1973); mem. Calif. Bar Assn., D.C. Bar Assn. Avocations: photography, traveling, antiques. Office: 760 Westwood Plz Los Angeles CA 90024-1759

GREEN, ROBERT EDWARD, JR., physicist, educator; b. Clifton Forge, Va., Jan. 17, 1932; s. Robert Edward and Hazle Hall (Smith) G.; m. Sydney Sue Truitt, Feb. 1, 1962; children: Kirsten Adair, Heather Scott. BS, Coll. William and Mary, 1953; PhD, Brown U., 1959; postgrad., Aachen (Germany) Technische Hochschule, 1959-60. Physicist underwater explosions rsch. divsn. Norfolk Naval Shipyard, Va., 1959; asst. prof. mechanics Johns Hopkins U., Balt., 1960-65, assoc. prof., 1965-70, prof., 1970—, chmn. mechanics dept., 1970-72, chmn. mechanics and materials sci. dept., 1972-73, chmn. civil engring./materials sci. and engring. dept., 1979-82, chmn. materials sci. and engring. dept., 1982-85, 91-93, dir. ctr. for nondestructive evaluation, 1985—; Ford Found. resident sr. engr. RCA, Lancaster, Pa., 1966-67; cons. U.S. Army Ballistic Research Labs., Aberdeen Proving Ground, Md., 1973-74; physicist Ctr. for Materials Sci., U.S. Nat. Bur. Standards, Washington, 1974-81; program mgr. Def. Advanced Research Projects Agy., 1981-82; mem. nat. materials adv. bd. Author: Ultrasonic Investigation of Mechanical Properties (Treatise on Materials Science and Technology, Vol. 3), 1973; co-editor 6 books; also articles. Fulbright grantee. Mem. ASM Internat., Am. Phys. Soc., Acoustical Soc. Am., Met. Soc. AIME, Am. Soc. Nondestructive Testing, Soc. for the Advancement of Material and Process Engring., Sigma Xi, Tau Beta Pi, Alpha Sigma Mu, Sigma Nu. Methodist. Achievements include research in recovery, recrystallization, elasticity, plasticity, crystal growth and orientation, X-ray diffraction, electro-optical systems, linear and non-linear elastic wave propagation, light-sound interactions, high-power ultrasonics, ultrasonic attenuation, dislocation damping, fatigue, acoustic emission, non-destructive testing, polymers, biomaterials, synchrotron radiation, composites, sensors and process control. Home: 936 Ellendale Dr Baltimore MD 21286 Office: Johns Hopkins U Ctr Nondestructive Evaluation 102 Maryland Hall Baltimore MD 21218

GREEN, ROBERT FREDERICK, physician, photographer; b. Newark, N.J., Oct. 13, 1923; s. Robert Aloysious (Green) and Sarah Wallington Schenck; m. June 20, 1960 (div.); children: Robert Daniel, Daniel Richard; m. Mary Ann Green, Oct. 5, 1982. BS, Seton Hall Coll., 1948; MD, SUNY, Bklyn., 1952. Diplomate Am. Bd. Psychiatry and Neurology, Nat. Bd. Med. Examiners. Intern Bklyn. Hosp., 1952-53; resident Peter Bent Brigham Hosp., Boston, 1953-54, Worcester (Mass.) State Hosp., 1954-57; asst. in psychiatry Tufts U., 1955-57; pvt. practice psychiatry Worcester, 1957-61, Ft. Wayne, Ind., 1961-82; chief mental hygiene clinic VA Med. Ctr., Ft. Wayne, 1982, chief of staff, 1983-88; med. dir. Grant-Blackford Mental Hygiene Clinic, Marion, Ind., 1988—; psychiatry cons. VA Med. Ctr., Ft. Wayne, 1961-82, Ft. Wayne Occupational Health Ctr., 1984—; cons. disability determintion sect. Dept. Vocat. Rehab., State of Ind., 1961—; teaching cons. Ind. Acad. Gen. Practice, 1971-75. Presenter weekly live TV segment Coping, 1980-81; contbr. articles to profl. jours.; photography published in The Rangefinder, Petersons Photographica, Modern Photography, Popular Photography, Darkroom, Camera 35, the Photo, Spec, Wall Street Jour.; numerous commissions by individuals, art galleries, museums and corps. Cpl. USAF, 1943-46. Fellow Acad. Psychosomatic Medicine; mem. AAAS, Pan Am. Med. Assn. (diplomate, life mem.), Ind. State Med. Assn., N.Y. Acad. Sci., Ft. Wayne Med. Soc. Roman Catholic. Home: 350 County Rd 20 Corunna IN 46730-9766 Office: Grant-Blackford Ctr Health Ctr 505 N Wabash Rd Marion IN 46952-9737

GREEN, ROBERT LAMAR, consulting agricultural engineer; b. Moultrie, Ga., Nov. 15, 1914; s. Louis Pinkney and Bessie (Tillman) G.; m. Frances Cowan, June 7, 1940; 1 son, Robert Lamar. B.S., U. Ga., 1934; M.S., Iowa State Coll., 1939; grad., Command and Gen. Staff Coll., 1944; Ph.D.; fellow Gen. Edn. Bd., Mich. State U., 1953. Registered profl. engr., Ga., Md. Terracing foreman Soil Erosion Service, Athens, Ga., 1934; camp engr. Civilian Conservation Corps., Bartow County, Ga., 1935; jr. agrl. engr. Soil Conservation Service, Lawrenceville and Americus, Ga., 1936-38; work unit conservationist Soil Conservation Service, Lawrenceville, 1939-47; asst. prof. agrl. engring. La. State U., 1947-50, 53-54; agrl. engr. U.S. Spl. Tech. and Econ. Mission to Indonesia (ECA, MSA, TCA, FOA), Djakarta, 1951-53; supt., agrl. engr. S.E. Tidewater Expt. Sta., Dept. Agr., 1954-58; state drainage engr. Md., 1958-73; prof., head dept. agrl. engring. U. Md., 1958-73; coordinator Water Resources Research Center, 1965-79, prof. emeritus agrl. engring., 1979; acting dir. Md. Agrl. Expt. Sta., 1972-76; cons. agrl. engr., Central African Republic, 1979, Guyana, 1981. Contbr. articles to profl. and trade jours. Chmn. Spl. Gov.'s Com. to study shore erosion Md., 1960-66, Spl. Gov's. Com. for Conservation and Devel. Natural Resources, 1960-66; mem. Md. Water Resources Commn., 1964-85, chmn., 1975-85; chmn. Md. Water Scis. Adv. Bd., 1968-73; bd. suprs. Dorchester County (Md.) Soil Conservation Dist., 1979. Served from 1st lt., cav. to maj., armor AUS, 1941-46; col. Res., ret. Life fellow Am. Soc. Agrl. Engrs. (chmn. D.C.-Md. sect. 1961-62, rep. to NRC 1959-66, dir. 1969-71, rep. Agrl. Research Inst., Nat. Acad. Scis.-NRC 1974-76, del. Pan Am. Union Engrs., hon. v.p. 1972, rep. XII-XVI Congresses, chmn. tech. com. on agrl. and food engring. 1982-84, organizing 1st Pan Am. Congress on Agrl. and Food Engring., XVII conv. in Caracas, 1984, John Deere gold medalist 1988); mem. Sigma Xi, Tau Beta Pi, Phi Kappa Phi, Epsilon Sigma Phi. Episcopalian. Home and Office: Bedford Ct Apt 621 3701 International Dr Silver Spring MD 20906-1309 Address: 312 Country Club Dr Rehoboth Beach DE 19971

GREEN, SAMUEL ISAAC, optoelectronic engineer; b. Chgo., Aug. 15, 1942; s. Joseph and Belle (Lepkovsky) G.; m. Marilyn Brandhorst, Aug. 11, 1989. BSEE, Northwestern U., 1964; PhD, U. Ill., 1969. Registered profl. engr., Mo. Specialist McDonnell Douglas Aerospace, St. Louis, 1969—. Contbr. articles to profl. jours. Mem. IEEE (sr. mem.), Soc. Photooptical Instrumentation Engrs. Achievements include five patents in the field of visible and near infrared photodetection. Office: McDonnell Douglas MC 1067160 PO Box 516 Saint Louis MO 63166-0516

GREEN, SHARON JORDAN, interior decorator; b. Mansfield, Ohio, Dec. 14, 1948; d. Garnet and L. Wynell (Baxley) Fraley; m. Trice Leroy Jordan Jr., Mar. 30, 1968 (dec. 1973); children: Trice Leroy III, Caerin Danielle, Christopher Robin; m. Joe Leonard Green, Mar. 13, 1978. Student, Ohio State U., 1966-67, 75-76. Typist FBI, Washington, 1968; ward clk. Means Hall, Ohio State U. Hosp., Columbus, 1970; x-ray clk. Riverside Hosp., Columbus, 1971; contr., owner T&D Mold & Die, Houston, 1988—; interior decorator, franchise owner Decorating Den, Houston, 1989-91; franchise owner T&D Interior Decorator, Houston, 1992—. Tchr. aide Bedford Sch., Mansfield, Ohio, 1976-77, Yeager Sch., 1981-82; pres. N.W. Welcome Wagon, Houston, 1980-81, Welcome Club, El Paso, 1986-87; active North Houston Symphony, 1992—, North Houston Performing Arts, 1993—, Mus. Fine Arts, Houston, 1993—, Edn. and Design Resouce Network, 1993—. Republican. Home: 16247 Morningbrook Dr Spring TX 77379-7158

GREEN, TODD LACHLAN, physiologist, educator; b. Memphis, Dec. 7, 1956; s. Charles Albert and Betty Ruth (McDonald) G. BS, Fla. State U., 1978; PhD, U. Va., 1986. Postdoctoral fellow Rutgers U., Newark, 1985-89, Washington U., St. Louis, 1989-91; asst. prof. physiology Marshall U., Huntington, W.Va., 1991—. NIH postdoctoral fellowship grantee Washington U., 1990. Mem. AAAS, Soc. Neurosci., Am. Soc. Microbiology, Sigma Xi. Democrat. Presbyterian. Achievements include rsch. on S-laminin. Office: Marshall U Sch Medicine 1542 Spring Valley Dr Huntington WV 25755

GREEN, WALTER VERNEY, materials scientist; b. Schenectady, N.Y., Sept. 20, 1929; s. Walter Verney and Rita (McGrath) G.; m. Marilyn

Johnson, 1954 (div. 1975); children: Wendy S., Scott K., Walter Verney; m. Sherron Louise Lockwood, Mar. 17, 1977. BS, U. Wis., 1952, MS, 1953, PhD, 1956. Mem. staff Los Alamos (N.Mex.) Nat. Lab., 1956-79; vis. scientist Paul Scherrer Inst., Villigen, Switzerland, 1979-91; cons. Purple Mountain Sci., Crested Butte, Colo., 1991—; vis. prof. Centro Atomico, Bariloche, Argentina, 1964-65; Swiss rep. to Internat. Energy Agy. Working Group on Materials for Fusion Reactors, 1980-91. Contbr. articles to profl. jours. Mem. Nat. Ski Patrol, Los Alamos, 1965-79. Mem. AIME. Achievements include establishment of limits on use of graphite in fusion reactors through workshops under International Energy Agency sponsorship. Home: PO Box 897 Crested Butte CO 81224

GREENBERG, ARNOLD H., pediatrics educator, cell biologist; b. Winnipeg, Man., Can., Sept. 29, 1941; s. Samuel H. and Bertha (Segal) G.; m. Sharon Domey, 1967 (div. 1983); children: David E., Juliet R.; m. Faye L. Hellner, 1981; children: Marni L. Hellner, Rachel D. Hellner, Katherine R. Hellner. BS in Med. Genetics, U. Man., 1965, MD, 1965; PhD in Immunology, U. London, 1974. Rotating intern Winnipeg Gen. Hosp., 1965-66; jr. resident in pediatrics Winnipeg Children's Hosp., 1966-67; asst. resident in pediatrics Johns Hopkins Hosp., Balt., 1967-68, rsch. fellow in pediatric endocrinology, 1968-70; sabbatical dept. immunology Nat. Inst. for Med. Rsch., Mill Hill, London, 1978-89; sabbatical Lab. of Chemoprevention Nat. Cancer Inst./NIH, Bethesda, 1988; from asst. prof. to assoc. prof. pediatrics and immunology U. Man., Winnipeg, 1974-85, prof. pediatrics, immunology and human genetics, 1986—; dir. Man. Inst. Cell Biology, Winnipeg, 1988—; mem. dept. pediatrics exec. coun. U. Man., 1978-81, dept. pediatrics continuing med. edn. com., 1981-85, chmn. dept. pediatrics GFT exec. com., 1983-88, animal care com., 1979-89, FACS mgmt. com., 1979-89, chmn. instnl. self study rsch. com. faculty of medicine, 1989-90, dept. pediatrics promotion com., 1984-91, chmn. biology cancer course, 1986-92, chmn. dept. immunology rev. com., 1992, dept. pediatrics exec. com., 1988-92, chmn. faculty of medicine rsch. com., 1991-93, inst. and interdisciplinary rsch. group com., 1991—; mem. grants panel B Nat. Cancer Inst., 1977-82, 84-85, fellowship panel, 1984, Terry Fox expansion programs grants panel, 1987, chmn., 1989-90, awards panel, 1990-91; adv. com. on rsch. Alberta Cancer Bd.; mem. immunology and transplantation grants panel Med. Rsch. Coun., 1986-90; med. adv. com. Children's Rsch. Found., 1980-90; lectr. WHO, Clin. Rsch. Ctr., Harrow, England, Middlesex Hosp. Med. Sch., London, Univ. Coll., London, U. Saskatchewan, Queen's U., U. Pa., Jackson Labs., Bar Harbor, Me., McGill U., U. Alberta, Can. Fed. Biol. Socs., Karolinska Inst., Stockholm, Weizmann Inst., Rehovot, Israel, Deutches Krebsforschungszentrum, Heidelberg, Germany, U. Rochester, NIH, Soc. for Behavioral Medicine, New Orleans, Bristol-Myers Co., McMaster U., Mt. Sinai Hosp. Rsch. Inst., Toronto, Nat. Inst. Med. Rsch., London, Charing Cross Rsch. Inst., London, UCLA, Salk Inst., RJR Nabisco Co., Colby-Sawyer Coll., Rutgers U., Quebec Cancer Inst., Brown U., U. Western Ontario, Ontario Cancer Inst., Hosp. for Sick Children, Toronto, Henry Ford Hosp., Detroit, Wayne State U., CIBA Found., Australian Behavioral Immunology Group, St. Vincent's Hosp., Sydney, Australia, Walter and Eliza Hall Inst. for Med. Rsch., Melbourne, Australia, NRC Biotechnology Rsch. Inst., Montreal, U. Minn., U. Toronto, Reichmann Rsch. Inst., Toronto, also various symposiums, confs., and workshops. Assoc. editor: Cancer and Metastasis Revs.; guest editor: Nature, Jour. Immunology, Molecular and Cellular Biology, Cancer Rsch., Molecular Carcinogenesis, Invasion and Metastasis, Immunology Letters; contbr. numerous articles to sci. jours. including Jour. Clin. Endocrinology, Jour. Pediatrics, Nature New Biology, Johns Hopkins Med. Bull., European Jour. Immunology, Nature. Recipient Am. Coll. Chest Physicians prize for student rsch., 1964, Terry Fox Career award Nat. Cancer Inst., 1982-85, Children's Hosp. Rsch. Found. Teddy award, 1991; named Nat. Cancer Inst. Terry Fox Cancer Rsch. Scientist, 1985—; fellow Med. Rsch. Coun., 1968-71, 71-74, Schering travel fellow, 1977, Nuffield Found. travel fellow, 1978, Dept. of External Affairs travel fellow, 1978; Med. Rsch. Coun. scholar, 1974-79, Nat. Cancer Inst. scholar, 1979-82. Mem. Am. Assn. for Cancer Rsch., Am. Assn. of Immunologists, Can. Soc. for Immunology (exec. coun. 1979-82, 89—), Soc. for Psychoneuroimmunology, Can. Med. Assn., Can. Inst. of Acad. Medicine, Nat. Cancer Inst. (mem. panel A 1991-93, assoc. editor jour. 1975—), Can. Fedn. of Biol. Studies (sci. policy com. 1979 Office: Man-Inst Cell Biology, 100 Olivia St, Winnipeg, MB Canada R3E OV9

GREENBERG, BERNARD, pediatrician; b. N.Y.C., Jan. 21, 1913; s. Joseph and Gussie (Gans) G.; m. Bernice Robbins, Nov. 1, 1942; children: Judith, Steven. BS, Columbia U., 1933; MD, U. Chgo., 1937. Intern (rotating) Israel Zion, Bklyn., 1937-39; resident pediatrics Home for Hebrew Infants, Bronx, N.Y., 1940; resident Morrisania City Hosp., Bronx, 1941; attending physician pediatrics Maimonides Med. Ctr., Bklyn., 1947-85, cons. pediatrician, 1985—; asst. prof. pediatrics Down State Med. Ctr., SUNY, Bklyn., 1965—; staff pres. Maimonides Md. Ctr., 1966-68. Contbr. articles to profl. jours. Major U.S. Army, 1941-46. Fellow Am. Acad. Pediatrics; mem. AMA, Bklyn. Pediatric Soc., Kings County Acad. Med. (past pres. pediatric sect.). Jewish. Office: 755 Ocean Ave Brooklyn NY 11226-4979

GREENBERG, FRANK, clinical geneticist, educator, academic administrator; b. Perth Amboy, N.J., Aug. 24, 1948. BA cum laude in Zoology, U. Mich., 1970; MMS, Rutgers U., 1972; MD, U. Pa., 1974. Cert. in pediatrics, medical genetics, Pa., N.J., Ga., Tex.; diplomate Am. Bd. Pediatrics, Am. Bd. Med. Genetics. Pediatric resident Children's Hosp. Pitts., 1974-76; pediatric resident St. Christopher's Hosp. for Children, Phila., 1976-77, fellow med. genetics, 1977-79; epidemic intelligence officer bur. epidemiology Ctrs. Disease Ctrl., Birth Defects Br., Chronic Diseases Divsn., Atlanta, 1979-81; dir. birth defects-genetics clinic, inst. molecular genetics Baylor Coll. Medicine, Houston, 1981—, dir. MSAFP screening program, 1990—; clin. asst. prof. divsn. med. genetics Emory U. and Grady Meml. Hosp., Atlanta, 1979-81; asst. prof. dept. pediatrics Baylor Coll. Medicine, 1981-88; clin. dir. MSAFP screening program, 1981-90, asst. prof. dept. ob-gyn., 1984-88, inst. molecular genetics, 1985-88; adj. asst. prof. dept. epidemiology U. Tex., 1985—, assoc. prof. clin. genetics, inst. molecular genetics, 1988—, assoc. prof. pediatrics, 1988—; mem. profl. adv. bd. Spina Bifida Assn. Tex., 1986—, Chromosome 18 Support Group, 1991—; mem. med. adv. bd. Williams Syndrome Assn., 1986—; numerous coms. and consultations in field. Assoc. editor Birth Defects Ency., sect. editor spl. syndromes sect.; peer reviewer Agy. Toxic Substances and Disease Registry; contbr. numerous articles to profl. jours. and book chapters; reviewer numerous profl. jours. With USPHS, 1979-81. March of Dimes Med. Svc. grantee 1981-85; Biomed. Rsch. support grantee 1983-85; grantee Tex. Genetics Network, 1988—. Mem. AAAS, APHA (genetics com. maternal and child health sect. 1983—), AMA, Am. Soc. Human Genetics (social issues com. 1983-88, task force maternal serum alpha fetoprotein screening 1986), Am. Acad. Pediatrics (perinatal pediatrics sect. 1981—, genetics sect. 1991—), Am. Fedn. Clin. Rsch., Am. Coll. Med. Genetics, So. Soc. Pediatric Rsch., Tex. Genetics Soc., N.Y. Acad. Scis., Tex. Med. Assn., Tex. Pediatric Soc. (environ. health and accident com. 1986-90), Harris County Med. Soc., Houston Pediatric Soc., Teratology Soc. Achievements include applied research on maternal serum alpha fetoprotein screening (e.g. effects of race, weight, diabetes, and smoking and in the interperetation of low MSAFP levels), on human chorionic gonadotropin and unconjugated estriol screening for Downes Syndrome; research on Williams Syndrome (estiology, genetics descriptive features, natural history), early detection of, descriptive epidemiology of, sleep studies in, growth hormone treatment of, and body composition in Prader-Willi Syndrome; research in etiology, pathogenesis and molecular genetics of DiGeorge Anomaly and clin. and molecular studies of Smith-Magenis Syndrome. Office: Baylor Coll Medicine Inst Molecular Genetics MC3-3370 6621 Fannin St Houston TX 77030

GREENBERG, JOSEPH H., anthropologist; b. Bklyn., May 28, 1915; s. Jacob and Florence (Pilzer) G.; m. Selma Berkowitz, Nov. 23, 1940. A.B., Columbia, 1936; Ph.D. in Anthropology, Northwestern U., 1940, D.Sc. (hon.), 1982. Faculty U. Minn., 1946-48; asst. prof. Columbia, 1948-53, asso. prof., 1953-57, prof. anthropology, 1957-62; prof. Stanford, 1962-85, Ray Lyman Wilbur prof. social scis. in anthropology, 1971; dir. Nat. Def. Edn. Act. African Lang. and Area Center, 1967-78; Vis. prof. Summer Linguistic Inst., Mich. U., 1957, U. Minn., 1960; mem. panel anthropology and philosophy and history of sci. NSF, 1959-61; vis. prof. summer inst. U. Colo., 1961; dir. West African Langs. Survey, 1959-66; Linguistic Soc. Am. prof. Summer Linguistic Inst., Oswego, N.Y., 1976; Collitz prof. Summer Linguistic Inst., Stanford, 1987, coord. Stanford Project on Language Universals. Author: Languages of Africa, 1963, Essays in Linguistics, 1957, Universals of Language, 1963, Influence of Islam on a Sudanese Religion, 1946, Anthropological Linguistics: An Introduction, 1968, Language, Culture and Communication: Essays by Joseph H. Greenberg, 1971, Language Typology, 1974, A New Invitation to Linguistics, 1977, Universals of Human Language, 4 vols, 1978, Language in the Americas, 1987, On Language: The Selected Writings of Joseph H. Greenberg, 1990; co-editor: Word, 1950-54. Served with Signal Intelligence Corp. AUS, 1940-45. Social Sci. Research Council fellow Northwestern U., 1940; Stanford humanities fellow, 1958-59, 82-83; Ford Found. grantee, 1952, 57-62; recipient Demobilization award Social Sci. Research Council, 1945-46; Guggenheim award, 1954-55, 58-59, 82-83; Haile Selassie award for African research, 1967; award in behavioral scis. N.Y. Acad. Scis., 1980. Mem. Am. Anthrop. Assn. (rep. to gov. bd. Internat. Inst. 1955—, 1st distinguished lectr. 1970), Linguistic Soc. Am. (exec. com. 1953-55, v.p. 1976, pres. 1977), West African Linguistics Soc. (chmn. 1965-66), African Studies Assn. (exec. com., also com. on langs. and linguistics 1959—, pres. 1964-65), Nat. Acad. Scis., Am. Acad. Arts and Scis., Am. Philos. Soc., Phi Beta Kappa. Home: 860 Mayfield Ave Stanford CA 94305-1051

GREENBERG, PAUL ERNEST, economics consultant; b. Montreal, Que., Can., Oct. 23, 1962; came to U.S., 1986; s. Arthur Peter and Ruth Roslyn (Bernstein) G.; m. Marla Choslovsky, July 2, 1989; 1 child, Talia. BA, Vassar Coll., 1983; MA, U. Western Ont., 1984; MS, MIT, 1988. Fin. economist Dept. of Fin. Govt. of Can., Ottawa, 1984-86; market analyst Alcan Aluminum Corp., Cambridge, Mass., 1987; litigation & health care economist, v.p. Analysis Group, Inc., Cambridge, Mass., 1988—; mem. editorial bd. Jour. Neglected Ideas, Poughkeepsie, N.Y., 1983—. Author: Canadian Energy Demand, 1989, Practice of Econometrics Using MicroTSP, 1990; contbr. articles to profl. econs. jours. Recipient U. Western Ont. scholarship, 1983, Social Sci. and Humanities Rsch. Coun. of Can. award, 1983, Govt. Can., Emilie Louise Wells fellowship, Wall St. Jour. awards, Vassar, 1983. Mem. ABA, Am. Econ. Assn., Nat. Assn. Bus. Economists, Phi Beta Kappa, Omicron Delta Epsilon. Jewish. Avocations: humor, comedy, writing. Home: 80 Pleasant St Apt 66 Brookline MA 02146-3461 Office: Analysis Group Inc 1 Brattle Square 5th Flr Cambridge MA 02138

GREENBERG, RAYMOND SETH, dean, educator; b. Chapel Hill, N.C., Aug. 10, 1955; s. Bernard George and Ruth Esther (Marck) G.; m. Leah Daniella Dacus, Oct. 23, 1988. BA with highest honors, U. N.C., 1976, PhD, 1983; MD, Duke U., 1979; MPH, Harvard U., 1980. Asst. prof. sch. medicine Emory U., Atlanta, 1983-86, assoc. prof., 1986-90, dep. dir. Winship Cancer Ctr., 1985-90, chair epidemiology/ biostat., 1988-90, dean sch. pub. health, 1990—; chair preventative medicine Nat. Bd. Med. Examiners, Phila., 1991-93; chair epidemiology study sect. NIH, Bethesda, Md., 1992—. Author: Medical Epidemiology, 1993; contbr. articles to profl. jours. Bd. dirs. Am. Cancer Soc. Ga. Divsn., 1987—. Fellow Am. Coll. Epidemiology (pres. 1990-91); mem. APHA, Am. Statis. Assn., Am. Epidemiology Soc., Soc. Epidemiol. Res., Assn. Tchrs. Preventative Medicine. Democrat. Jewish. Office: Emory U Sch Pub Health 1599 Clifton Rd NE Atlanta GA 30329

GREENBERG, RICHARD ALAN, psychiatrist, educator; b. Phila., Mar. 28, 1946; s. Benjamin and Natalie (Kaplan) G.; m. Laudelina Lahom, June 23, 1979; children: Eric Michael, Jason Paul. AB, Albright Coll., 1968; MD, Thomas Jefferson U., 1973. Diplomate Am. Bd. Psychiatry and Neurology. Resident in psychiatry Jefferson-Del. State Hosp., 1973-76; pvt. practice Bethesda, Md., 1979-86, Washington, 1986—; staff psychiatrist Psychiat. Inst. D.C., Washington, 1976—, pres. med. staff, 1985—; lectr. psychiatry George Washington U., Washington, 1979—, assoc. clin. prof., 1981—; chmn. Ann. Greenbrier Psychopharmacologic Conf., White Sulphur Springs, W.Va.; chief physician, dir. evaluation svcs. Clin. Neurosci. Program, Washington, 1987—. Fellow Am. Pschiat. Assn.; mem. AMA (med. staff sect.), Washington Psychiat. Soc., Med. Soc. D.C. (health svcs. facilities com. 1984—). Democrat. Roman Catholic. Avocations: baseball, reading, travel. Home: 10544 Hunters Way Laurel MD 20723-5724 Office: 2112 F St NW Ste 406 Washington DC 20037-2715

GREENBERG, SHARON GAIL, neuroscientist, researcher; b. Wareham, Mass., Nov. 2, 1956; d. Arthur Esau and Evelyn Bernice (Cutler) G. BS cum laude, U Mass., 1978; PhD, Case Western Res. U., 1986. Rsch. asst. U. Mass., Boston, 1978-81; asst. rsch. scientist NYU Med. Sch., N.Y.C., 1986-87; rsch. assoc. Albert Einstein Coll. Medicine, Bronx, N.Y., 1987-91; rsch. scientist Burke Med. Rsch. Inst., White Plains, N.Y., 1991—; asst. prof. Cornell Med. Coll., N.Y.C., 1993—. Contbr. articles to Jour. Biol. Chemistry, Jour. Neurosci., others. NIH trainee Case Western Res. U., 1981-86; Belfor fellow Albert Einstein Coll. Medicine, 1990; faculty scholar Alzheimers Disease and Related Disorders Assn., 1992—; recipient First award Nat. Inst. Neurol. Disorders and Strokes, 1993—. Mem. N.Y. Acad. Sci., Soc. Neurosci. Achievements include development of method to isolate soluble populations of paired helical filaments characterization of PHF proteins and isolation of the monoclonal antibody PHF-1. Office: Burke Med Rsch Inst 785 Mamaroneck Ave White Plains NY 10605

GREENBERG, SHELDON BURT, plastic and reconstructive surgeon; b. Bklyn., July 8, 1948; s. Morris and Lillian (Liss) G.; m. Andrea R. Levy, Feb. 10, 1991. BS, Muhlenberg Coll., 1970; MD, Chgo. Med. Sch., 1974. Diplomate Am. Bd. Otolaryngology, Plastic Surgery. Resident in surgery Lenox Hill Hosp., N.Y.C., 1974-75; resident in otolaryngology Met. Hosp., Manhattan Eye and Ear Hosp., N.Y.C., 1978; resident in plastic surgery Akron (Ohio) City Hosp., 1978-80, fellow in hand surgery, 1980; pvt. practice, Norwalk, Conn., 1981—. Fellow Am. Coll. Surgeons, Am. Soc. Plastic and Reconstructive Surgeons; mem. Conn. Med. Soc., Fairfield County Med. Soc., Fairfield Men's Club. Republican. Jewish. Avocations: tennis, American history, gardening. Office: 40 Cross St Norwalk CT 06851-4647

GREENBERG, WILLIAM MICHAEL, psychiatrist; b. Bklyn., Oct. 19, 1946; s. Benjamin Greenberg and Marilyn (Berger) Hamberg; m. Wendy Faith Megerman, June 14, 1992. BA, Queens Coll., 1968; postgrad., U. Medicine & Dentistry N.J., 1974-76; MD, Albert Einstein Coll. Medicine, 1978. Diplomate Am. Bd. Psychiatry and Neurology. Computer programmer Western Electric Co., N.Y.C., 1970-73; rsch. asst. Bklyn. Jewish Hosp., 1973-74; resident in psychiatry Bronx (N.Y.) Mcpl. Hosp. Ctr., 1978-83, house staff pres., 1981-82; acting med. dir. Met. Ctr. for Mental Health, N.Y.C., 1983; staff psychiatrist Bronx Psychiat. Ctr., 1983-84; dir. psychiatry clinic North Cen. Bronx Hosp., 1984-88; psychiatrist, cons. Montefiore Mental Health Svcs. at Rikers Island, East Elmhurst, N.Y., 1985-86; pvt. practice Bronx, 1985-88; chief psychiatrist, attending staff mem. Bergen Pines County Hosp., Paramus, N.J., 1988-93, dir. of psychiat. rsch., 1993—; asst. clin. prof. Albert Einstein Coll. Medicine, Bronx, 1988-90; vis. asst. prof. Med. Coll. Pa., 1990—. Asst. editor Community Psychiatrist, 1985-89; mem. editorial bd. Einstein Quar. Jour. Biology and Medicine, 1987—; contbr. articles to profl. jours.; reviewer profl. jours., books. Union rep. Com. Interns and Residents, N.Y.C., 1979-81; speaker's bur. Physicians for Social Responsibility, N.Y.C., 1982-84, Bergen Pines County Hosp., 1988—. Rock Sleyster Meml. scholar AMA, 1977; recipient Bergen Pines Psychiatry Residency Teaching award, 1991. Mem. AAAS, APHA, Am. Psychiat. Assn., Am. Assn. Community Psychiatrists, Assn. for Advancement of Philosophy and Psychiatry. Avocations: analytic philosophy, meditation, computers, photography. Office: Bergen Pines County Hosp Div Psychiatry Paramus NJ 07652

GREENE, CHARLES ARTHUR, materials engineer; b. Balt., Nov. 11, 1962; s. David Graham Silliman and Anne Constance (Williams) G.; m. Gretchen Raschelle Beal, Dec. 31, 1987. MS, U. Md., 1992, postgrad., 1992—. Sr. rsch. technician biophysics Johns Hopkins Med. Sch., Balt., 1988-89; rsch. asst. materials engring. U. Md., College Park, 1989—. Pres., treas., bd. dirs. Stillmeadows One Bd. Condo. Owners, Severn, Md., 1989-93. Nat. Def. Sci. and Engring. grad. fellow, 1991—. Mem. Am. Soc. Materials Internat., Materials Rsch. Soc., Minerals, Metals and Materials Soc., Sigma Pi Sigma, Phi Kappa Phi. Democrat. Mem. Soc. of Friends. Home: 8204 Coatsbridge Ct Severn MD 21144-2908 Office: U Md Dept Materials and Nuclear Engring College Park MD 20742-2115

GREENE, CHRIS H., physicist, educator. Prof. dept. physics U. Colo., Boulder. Recipient I.I. Rabi prize Am. Phys. Soc., 1991. Office: U Colo Dept Physics University Of Colorado CO 80309*

GREENE, CLIFFORD, psychologist; b. Bklyn., Feb. 3, 1941; s. Clifford and Ada (Chavis) G.; m. Sharon Johnston, July 11, 1981; children: Jasmin, Forrest, Greer, Julie, Meri. MA, NYU, 1978, PhD, 1984. Lic. psychologist, N.J. Staff psychologist Barnert Hosp., Paterson, N.J., 1980-85; sr. psychologist Woodhull Med. Ctr., Bklyn., 1985-87; dir. dept. counseling Passaic County Coll., Paterson, 1987-89; dir. psychol. svcs. Rutgers U., Newark, N.J., 1989—; pvt. practice, Englewood, N.J., 1984—; cons., clin. supr. Passaic County Coll., 1990—. Bd. dirs. Urban League Bergen County, Englewood, 1989-91. NIMH fellow, 1976-79. Mem. Am. Psychol. Assn., N.J. Psychol. Assn., N.J. Assn. Black Psychologists, N.Y. Acad. Sci.ž. Home: 329 Murray Ave Englewood NJ 07631-1418 Office: Rutgers U 249 University Ave Newark NJ 07102-1896

GREENE, DAVID LLOYD, transportation energy researcher; b. N.Y.C., Nov. 18, 1949; s. Donald and Alice (Lloyd) G.; m. Janet Margaret Lane, Jan. 2, 1971; children: Jennifer, Michael. BA, Columbia U., 1971; MA, U. Oreg., 1973; PhD, Johns Hopkins U., 1978. Rsch. assoc. Oak Ridge (Tenn.) Nat. Lab., 1977-80, leader transp. energy group, 1980-82, rsch. staff mem., 1982-84, sr. rsch. staff mem. I, 1984-87, head transp. rsch. sect., 1987-88; sr. rsch. analyst Office of Policy Integration U.S. Dept. of Energy (on assignment Oak Ridge Nat. Lab.), Washington, 1988-89; sr. rsch. staff mem. II, mgr. energy policy rsch. programs Ctr. for Trnsp. Analysis U.S. Dept. of Energy (on assignment Oak Ridge Nat. Lab.), Oak Ridge, 1989—; chmn. organizing com. Conf. on Transp. and Global Climate Change, Asilomar, Calif., 1991, 93; cons. Eno Transp. Found., 1991-92; mem. numerous coms. transp. rsch. bd. Nat. Rsch. Coun., 1982—. Author: (with others) Systems and Models for Energy and Environmental Analysis, 1983, Advances in Energy Systems and Technology, 1982, Changing Energy Use Futures, 1979; mem. editorial adv. bd. Transp. Rsch., 1986—; contbr. over 50 articles to profl. jours. Planning Commn. Town Farragut, TN, 1988; Mission coun. St. Elizabeth's Episcopal Ch., 1986-88. Recipient Disting. Svc. cert. Transp. Rsch. Bd., 1989, Tech. Achievement award Martin Marietta Energy Systems, 1985, Pyke Johnson award NAS, 1984, Johns Hopkins U. grad. fellowship, 1976-77, Ford Found. fellowship, 1973-75; Columbia Coll. faculty scholar, 1967-71, N.Y. State Regents scholar, 1967-71. Mem. AAAS, Assn. Am. Geographers (Energy Specialty Group Paper award 1986), Soc. Automotive Engrs., Sigma Xi. Achievements include research in vehicle stock modeling of transportation energy use, techniques for analyzing trends in transportation energy use, understanding the effects of fuel economy regulation, quantification of efficiency rebound effect on motor vehicle use, contributions to estimating the costs and benefits and technological potential for vehicle fuel economy improvement. Office: Oak Ridge Nat Lab Ctr Transp Analysis PO Box 2008 Oak Ridge TN 37831-6207

GREENE, DON HOWARD, product designer; b. Norcross, Ga., July 9, 1958; m. s. Paul Howard and Patricia Anne (Knox) G.; Cheryl Jeanne Garner, June 9, 1984; 1 child, April Elizabeth. B of Indsl. Engring., Ga. Inst. Tech., 1980; MBA, Ga. State U., 1988. Registered profl. engr., Ga. Mfg. engr. Sci.-Atlanta, Inc., Atlanta, 1981-82; indsl. engr. Sci.-Atlantic, Inc., Atlanta, 1982-84; staff indstrl. engr. Inst. Indsl. Engrs., Atlanta, 1984-88, tech. ops. mgr., 1988-91, product devel. mgr., 1991—, total quality mgmt. facilitator, 1991—. Mem. NSPE, Inst. Indsl. Engrs., Tau Beta Pi, Alpha Pi Mu, Beta Sigma Gamma. Methodist. Office: Inst Indsl Engrs 25 Technology Park Norcross GA 30092

GREENE, DOUGLAS A., internist, educator. AB in Biol. Scis., Princeton U., 1966; MD, Johns Hopkins U., 1970. Intern Johns Hopkins U. Hosp., Balt., 1970-71, asst. resident, 1671-72; postdoctoral rsch. fellow George S. Cox Med. Rsch. Inst. U. Pa. Hosp., Phila., 1972-75; asst. prof. medicine U. Pa., Phila., 1975-80; assoc. prof. medicine, dir. clin. rsch. unit and diabetes rsch. labs. U. Pitts., 1980-86; prof. internal medicine, dir. Mich. Diabetes Rsch. and Tng. Ctr. U. Mich., Ann Arbor, 1986—, mem. faculty neuroscience, 1988—, chief divsn. endocrinology and metabolism, 1991—; chmn. endocrinologic and metabolic adv. com. U.S. FDA, 1991—. Contbr. over 29 articles to Jour. Clin. Investigation, Frontiers in Diabetes, Diabetic Neuropathy, Diabetes, Diabetes Care, Am. Jour. Physiology, others. Office: U Mich Hosp-Diabetes Rsch & Tng Inst 3920 Taurman Ctr Box 0354 1331 E Medical Center Dr Ann Arbor MI 48109-0580

GREENE, ERIC, naval architect, marine engineer; b. Mineola, N.Y., Mar. 16, 1956; s. Stewart and Iris (Katz) G.; m. Lynn Ellen Morris, Apr. 13, 1985; children: Rebecca Morris, Cole Morris. BS in Naval Architecture/Marine Engring., MIT, 1979. Naval architect Kiwi Boats, Largo, Fla., 1979-80, Bainbridge Island, Wash., 1983; mgr. marine systems Severn Cos., Annapolis, Md., 1984-85; naval architect Giannotti & Assocs., Inc., Annapolis, 1985-87, Structural Composites, Inc., Melbourne, Fla., 1990—; east coast ops. mgr. DLI Engring. Corp., Inc., Bainbridge Island, 1981-82, 91—; pres. Eric Greene Assocs., Inc., Annapolis, 1987—; publ. cons. Ship Structure Com., Washington, 1991—. Author: Marine Composites, 1991; author/ editor: Ship Vibration Design Guide, 1990; contbr. articles to profl. jours. Mem. Soc. Naval Architects and Marine Engring. (materials and small craft com. 1990—), Soc. of Advanced Material Process Engrs., Am. Soc. of Naval Engrs. Achievements include development of methodology for determining thermo-mechanical response of compsite structures, application of PC's to measure physical phenomena in shipboard environments. Office: Eric Greene Assocs Inc 18 Cushing Ave Annapolis MD 21403

GREENE, ERNEST RINALDO, JR., anesthesiologist, chemical engineer; b. Mobile, Ala., Jan. 26, 1941; s. Ernest Rinaldo and Dorris Rolinha (Lassiter) G.; m. Lois Ellen Laura Zullig, Sept. 23, 1967; children: Laura Rolinha, Ernest Rinaldo III, Ellen Victoria, Max McKeen. BA, Rice U., 1962, BS, 1963; MA, Princeton U., 1966, PhD, 1968; MD, Washington U., St. Louis, 1981. Diplomate Am. Bd. Anesthesiology; diplomate, Nat. Bd. Med. Examiners; registered profl. engr., Ala. Tenured asst. prof. engring. U. Ala., Birmingham, 1970-84, asst. prof. anesthesiology, 1986-88; chief anesthesiology Cooper Green Hosp., Birmingham, 1986-90, VA Med. Ctr., Birmingham, 1987-90; assoc. prof. anesthesiology U. Ala., Birmingham, 1988-90; chief anesthesiology Vaughan Regional Med. Ctr., Selma, Ala., 1990-92; adjunct assoc. prof. biomed. engring. U. Ala., Birmingham, 1990—; founder, CEO Hivex, Inc.; with East Montgomery Anesthesia, Montgomery, Ala., 1993—; reviewer (bioengring.) NSF, Washington, 1981—; guest reviewer Anesthesiology (jour.), Phila., 1988-90. Author: (books) Homogenous Enzyme Kinetics, 1984, Immobilized Enzyme Kinetics, 1984; (with others) New Anesthetic Agents, Devices and Monitoring Techniques, 1984, Pain Management of Aids Patients, 1991. Mem. Am. Inst. Chem. Engrs., Internat. Anesthesia Rsch. Soc., N.Y. Acad. Sci., SAR, S.R., Soc. Colonial Wars, Sigma Xi (assoc.), Tau Beta Pi, Phi Lambda Upsilon, Sigma Tau. Republican. Methodist. Office: PO Box 241427 Montgomery AL 36124-1427

GREENE, FRANK SULLIVAN, JR., business executive; b. Washington, Oct. 19, 1938; s. Frank S. Sr. and Irma O. Greene; m. Phyllis Davison, Jan. 1958 (dec. 1984); children: Angela, Frank, Ronald; m. Carolyn W. Greene, Sept. 1990. BS, Washington U., St. Louis, 1961; MS, Purdue U., 1962; PhD, U. Santa Clara (Calif.), 1970. Part-time lectr. Washington U., Howard U., Am. U., 1959-65; dir. Tech. Devel. Corp., Arlington, Tex., 1985-92; pres. Zero One Systems, Inc. (formerly Tech. Devel. of Calif.) Santa Clara, Calif., 1971-87, Zero One Systems Group subs. Sterling Software Inc., 1987-89; asst. chmn., lectr. Stanford U., 1972-74; bd. dirs. Networked Picture Systems Inc., 1986, pres., 1989-91, chmn., 1991—. Author two indsl. textbooks; also articles; patentee in field. Bd. dirs. NCCJ, Santa Clara, 1980—, NAACP, San Jose chpt., 1986-89; bd. regents Santa Clara U., 1983-90, trustee, 1990—; mem. adv. bd. Urban League, Santa Clara County, 1986-89, East Side Union High Sch., 1985-88. Capt. USAF, 1961-65. Mem. IEEE, IEEE Computer Soc. (governing bd. 1973-75), Assn. Black Mfrs. (bd. dir., 1974-80), Am. Electric Assn. (indsl. adv. bd., 1975-76), Fairchild Rsch. and Devel. (tech. staff, 1965-71), Bay Area Purchasing Coun. (bd. dir. 1978-84), Security Affairs Support Assn. (bd. dir. 1980-83), Sigma Xi, Eta Kappa Nu, Sigma Pi Phi. Office: Network Picture Systems Inc 2041 Mission Coll Blvd Ste 255 Santa Clara CA 95054-1155

GREENE, GEORGE CHESTER, III, chemical engineer; b. Ocala, Fla., Feb. 8, 1944; s. George C. Jr. and Madeline Greene; m. Molly Feemster; children: Jennifer, George IV. BSChemE, U. Fla., 1967; MSChemE, Columbia U., 1969; PhDChemE, Tulane U., 1973. Registered profl. engr., S.C., N.C. With Lurgi Constrn. Co., Frankfurt, SC, Germany, 1973-77;

mngr. rsch. & engring. process devel. Exxon, N.J., 1977-81; v.p. Gen. Engring. Labs., Charleston, S.C., 1981—. Mem. editorial staff Am. Environ. Mag.; environ. editor S.C. Bus. Jour.; contbr. articles to profl. jours. Sr. warden St. Philip's Episcopal Ch., Charleston; bd. dirs. Bank of Charleston. Recipient Best Paper of Yr., So. Rubber Group, 1989. Mem. AICE, ASTM, Water Pollution Control Assn. S.C., Am. Coun. Ind. Labs., S.C. Lab. Mgmt. Soc. Office: Gen Engring Labs 2040 Savage Rd Charleston SC 29414

GREENE, JOHN M., physicist; b. Pitts., Sept. 22, 1928; s. John W. and Frances M. Greene; m. Alice Andrews; 1 child, Emily. BS, Calif. Inst. Tech., 1950; PhD, U. Rochester, 1956. Physicist Princeton (N.J.) Plasma Physics Lab., 1956-82, Gen. Atomics Co., San Diego, 1982—. Recipient James Clerk Maxwell prize Am. Phys. Soc., 1992, Plasma Physics Rsch. Excellence award Am. Phys. Soc., 1992. Office: General Atomics Co POB 85608 San Diego CA 92186

GREENE, JOYCE MARIE, biology educator; b. Pitts., Mar. 10, 1936; d. Hiram Lloyd and Catherine (Nelson) Green. AB, Bryn Mawr Coll., 1957, PhD, 1968; MA, Wesleyan U., Middletown, Conn., 1960. Asst. prof. Smith Coll., Northampton, Mass., 1969-75, SUNY, Fredonia, 1975-76; postdoctoral fellow Temple U. Med. Sch., Phila., 1986-88; asst. prof. C.C. of Phila., 1992—. Contbr. articles to profl. jours. Mem. AAAS, Am. Soc. Microbiology, Sigma Xi. Episcopalian. Office: CC of Phila 1700 Spring Garden St Philadelphia PA 19000

GREENE, LAURA HELEN, physicist; b. Cleve., June 12, 1952; d. Sam and Frances (Kain) G.; m. Russell W. Giannetta, Nov. 11, 1989; children: Max Greene Giannetta, Leo Greene Giannetta. BS, Ohio State U., 1974, MS, 1977; MS, Cornell U., 1980, PhD, 1984. Mem. tech. staff Hughes Aircraft Co., Torrence, Calif., 1974-75; teaching asst. Ohio State U., Columbus, 1975-76, rsch. asst., 1976-77; teaching asst. Cornell U., Ithaca, N.Y., 1977-79, rsch. asst., 1979-83; mem. tech. staff Bellcore, Red Bank, N.J., 1983-92; prof. dept. physics U. Ill., Urbana, 1992—. Editor Superconductivity Rev.; contbr. over 100 articles to profl. jours. Rsch. grantee NSF, 1991, 92. Mem. Am. Phys. Soc. (gen. coun. 1992-96), Materials Rsch. Soc. Achievements include rsch. in novel materials, in particular, high temperature superconductivity. Office: U Ill Loomis Lab Physics 1110 W Green St Urbana IL 61801

GREENE, LAURENCE WHITRIDGE, JR., surgical educator; b. Denver, Jan. 18, 1924; s. Laurence Whitridge Sr. and Freda (Schmitt) G.; m. Frances Steger, Sept. 16, 1950 (dec. Dec. 1977); children: Charlotte Greene Kerr, Mary Whitridge Greene, Laurence Whitridge III; m. Nancy Kay Bennett, Dec. 7, 1984. BA, Colo. Coll., 1945; MD, U. Colo., 1947; postgrad., U. Chgo., 1948-50. Diplomate Am. Bd. of Surgery. Intern St. Lukes Hosp., Denver, 1947-48; sr. intern in ob./gyn. U. Chgo. Lying-In Hosp., 1948-49; surg. resident U. Cin. Gen. Hosp., 1952-55, sr. surg. resident, 1955-57, chief surgery resident, 1957-58; clin. surgery asst. Sch. of Medicine U. Colo., Denver, 1958-61, clin. instr. Sch. of Medicine, 1961-67, asst. clin. prof. Sch. of Medicine, 1967-75, assoc. clin. prof. Sch. of Medicine, 1975-87, clin. prof. Sch. of Medicine, 1987—; adj. prof. zoology and physiology U. Wyo., Laramie, 1970—; mem. staff Ivinson Meml. Hosp., Laramie, Wyo., 1958—; chmn. Wyo. chpt. Com. on Trauma, 1973-89; tchr., mem. adv. staff U. Colo. Med. Sch., Denver, 1958-83; med. advisor, surgeon U. Wyo. Athletics, Laramie, 1975-80, Wyo. Hwy. Patrol, 1950—. Contbr. numerous articles to profl. jours. Lt. M.C. (s.g.) USN, 1950-52, Korea. Fellow ACS; mem. Am. Assn. for Surgery of Trauma, Southwestern Surgery Congress, Western Surg. Assn., Mont Reed Soc., Masons, Shriners, Sigma Xi. Republican. Episcopalian. Avocations: golf, sports, camping, hunting, fishing.

GREENE, ROLAND, chiropractor; s. Roland A. and Kathleen M. G.; m. Barbara Burchanowski, Apr. 2, 1985; children: Patrick, Thomas. BS in Biology and Chemistry, Fordham U., 1964; MA in Biology, SUNY, Buffalo, 1970, PhD, 1970; cert., Northwestern Coll. Chiropractic, 1984. Cert. chiropractor, Mass., R.I. Post-doctoral fellow dept. anatomy and cell biology U. Pitts. Med. Sch., 1970-71; rsch. assoc. dept. biol. and med. scis. Brown U. Med. Sch., 1971-75; asst. prof. dept. anatomy Med. Coll. of Penn., 1975-81; pvt. practice Townsend, Mass., 1985—. Am. Cancer Soc. Rsch. grantee, NIH Rsch. grantee; Gonstead scholar. Mem. Am. Chiropractic Assn., Nat. Inst. Chiropractic Rsch., Mass. Chiropractic Assn., Found. Chiropractic Edn. and Rsch., Sigma Xi. Office: Townsend Chiropractic Office 208 Main St Townsend MA 01469

GREENEBAUM, BEN, physicist; b. Chgo., Nov. 30, 1937; s. Ben I. Jr. and Maxine M. (Oberndorf) G.; m. Nancy Jung, June 16, 1963; children: Daniel, David, Edward. BA, Oberlin Coll., 1959; MA, Harvard U., 1961, PhD, 1965. Asst. prof. to prof. physics U. Wis. - Parkside, Kenosha, 1972—, acting vice chancellor, 1983-84, 85-86, assoc. dean faculty, 1977-90, dean of sci. and tech., 1990—; Vice-pres. First World Congress on Electricity and Magnetisim in Medicine and Biology, 1990—. Editor: Bioeletromagnetics jour., 1992—, assoc. editor 1990-92, editoral bd. 1989-90; contbr. articles to profl. jours., publs. Bd. dirs. Careers, Inc., Racine, Wis., 1985-91. Mem. Bioelectromagnetic Soc. (bd. dirs. 1989-91, 92—), Am. Phys. Soc., Am. Assn. Physics Tchrs., Bioelec. Repair and Growth Soc. Office: Univ Wis-Parkside Wood Rd Box 2000 Kenosha WI 53141-2000

GREENER, JEHUDA, polymer scientist; b. Cracow, Poland, Nov. 29, 1948; came to U.S., 1973; s. Solomon and Regina (Kauffman) Gruner; m. Helena Temkin, July 27, 1975; children: Avital, Daniel. BSCE cum laude, Technion (Israel), 1974; PhD in Polymer Sci., U. Mass., 1978. Rsch. scientist Eastman Kodak Rsch. Labs., Rochester, N.Y., 1977-81, sr. rsch. scientist, 1981-86, rsch. assoc., 1986—; bd. dirs. Engring. Properties and Structure Divsn. Soc. Plastics Engrs., treas., 1991-92, chmn. elect, 1992-93, chmn., 1993-94. Contbr. 1 chpt. to 1 book and articles to profl. jours. Lt. Israeli Army, 1966-69. Recipient Mees award Kodak. Mem. Am. Chem. Soc., Soc. Rheology, Soc. Plastics Engrs. Achievements include 1 patent. Office: Eastman Kodak Co Rochester NY 14650-2158

GREENEWALT, DAVID, geophysicist; b. Wilmington, Del., Mar. 26, 1931; s. Crawford Hallock and Margaretta Lammot (Dupont) G.; m. Charlotte Kemble Simonds, May 26, 1962; children: Gavin, David, Barbara, Henry, Charlotte, Frederick. BS in Physics, Williams Coll., 1953; PhD in Geophysics, MIT, 1960. Lectr. MIT, Cambridge, 1960-66; geophysicist Naval Rsch. Lab., Washington, 1966—. Mem. AAAS, Am. Geophys. Union. Republican. Home: 2509 Foxhall Rd Washington DC 20007 Office: Naval Rsch Lab Code 7340 4555 Overlook Ave Washington DC 29375

GREENFIELD, BRUCE PAUL, investment analyst, biology researcher; b. Bklyn., May 17, 1951; s. Stanley Samuel and Selma Margaret (Woolfson) G.; m. Linda Baldassini, May 15, 1988; 1 child, Skye Woolfson. BA in Biology and Religion, NYU, 1973, MS in Biology and Phys. Anthropology, 1977, cert. in commodities trading, 1991. Rsch. assoc. Interfaith Hosp., Bklyn., 1972-73; rsch. asst. lab. phys. anthropology NYU, N.Y.C., 1975-79, teaching asst. dept. anthropology, 1976-79, teaching fellow dept. biology, 1981-82, 82-83; rsch. fellow bioengring. lab. Hosp. for Joint Diseases, N.Y.C., 1983; ind. market analyst N.Y.C., 1987—; lectr. genetics dept. edn. S.I. (N.Y.) Zoo, 1977. Contbr. articles to Am. Jour. Phys. Anthropology, Lab. Primate Newsletter. Mem. Am. Assn. Phys. Anthropologists, N.Y. Acad. Scis., N.Y. Neuropsychology Group. Jewish. Achievements include development of method of identification of Awash Park baboons in Ethiopia using palm prints; development of model for genetic pre-analysis of endangered species Macaca silenus for maximization of breeding potential. Home: 425 W 57th St New York NY 10019-1764 Office: 446 W 55th St New York NY 10019-4437

GREENFIELD, JAMES DWIGHT, technical writer; b. Rochester, N.Y., Oct. 21, 1935; s. G. Dwight and Dorothy Laney (Fillingham) G.; m. Sarah Kathryn Curtice, June 3, 1961. BA, U. Rochester, 1957; MA, Ariz. State U., 1970. Publs. engr. Sperry Flight Systems, Phoenix, 1963-68; tech. writer Unidynamics/Phoenix, Phoenix, 1968-70, Arcata Data Mgmt., Phoenix, 1970-71; publs. mgr. E-H Rsch. Labs., Oakland, Calif., 1975-76; sr. tech. writer Fairchild Systems Tech. Divsn., San Jose, Calif., 1976-77; specification writer Motorola Semiconductor Sector, Phoenix, 1971-75, 77-78; sr. tech. writer/editor Motorola Computer Group, Tempe, Ariz., 1978—. Contbr. articles to Technical Communication. Vol. reader Recording for the Blind, Phoenix, 1980-91. Sp/5 U.S. Army, 1958-62. Mem. AAAS, IEEE, Soc. Tech. Communication (sr.). Republican. Presbyterian. Office: Motorola Computer Group 2900 S Diablo Way Tempe AZ 85282

GREENFIELD, JOHN CHARLES, bio-organic chemist; b. Dayton, Ohio, 1945; s. Ivan Ralph and Mildred Louise (House) G.; m. Liga Miervaldis, aug. 2, 1980; children: John Hollen, Mark Richard. B.S. cum laude, Ohio U., 1967; Ph.D., U. Ill., 1974. High sch. sci. instr. Dayton, 1968-71; grad. research asst. U. Ill., 1971-74; postdoctoral research fellow Swiss Fed. Inst. Tech., Zurich, 1975-76; research chemist infectious diseases research Upjohn Co., Kalamazoo, 1976-82, sr. rsch. scientist drug metabolism rsch., 1982-93; sr. project mgr. Upjohn Labs., Kalamazoo, 1993—; lectr. in field. Contbr. articles to sci. jours.; patentee in field. Am.-Swiss Found. for Sci. Exchange fellow, 1975; NSF-NATO postdoctoral fellow, 1975-76. Mem. Am. Chem. Soc., AAAS, N.Y. Acad. Sci., Internat. Soc. Study of Xenobiotics, Internat. Isotope Soc., Am. Assn. Pharm. Scientists, Sigma Xi. Achievements include mgmt. and evaluation of worldwide devel. projects for new pharm. agts. Home: 6695 E E Ave Richland MI 49083-9729 Office: Upjohn Co 301 Henrietta St Kalamazoo MI 49007

GREENFIELD, MOSES A., medical physicist, educator; b. N.Y.C., Mar. 8, 1915; s. Benjamin and Goldie (Seewald) G.; m. Sylvia Sorkin, June 13 (dec. 1982); children: Richard, Carolyn Greenfield Sargeant; m. Bella Manel. Sept. 17, 1984. BS, CCNY, 1935; MS, NYU, 1937, PhD, 1941. Diplomate Am. Bd. Radiology. Tutor dept. physics CCNY, N.Y.C., 1937-39; jr. asst. scientist Coast and Geodetic Survey, Washington, 1939-41; scientist D.W. Taylor Model Basin/USN, Carderock, Md., 1941-46, North Am. Aviation, L.A., 1946-48; prof. dept. radiological scis. sch. medicine UCLA, 1948-82, prof. emeritus, 1982—; cons. Atomic Energy Commn., 1960s, North Am. Aviation, Rockwell Internat., 1948-80, Environ. Evaluation Group, Albuquerque., 1979—; lectr. radiological physics Vets. Hosp., USN Hosp., San Diego, 1953-82. Contbr. over 150 papers to reviewed jours. Fellow Am. Coll. Radiology, Am. Assn. Physics in Medicine (Coolidge award 1991); mem. Inst. Phys. Scis. in Medicine, Radiological Soc. North Am. Achievements include 2 patents based on work in nuclear medicine. Home: 10824 Wilshire Blvd # 212 Los Angeles CA 90024 Office: UCLA Dept Radiological Scis Los Angeles CA 90024

GREENFIELD, VAL SHEA, ophthalmologist; b. N.Y.C., Apr. 20, 1932; s. Frank Lynne and Helen (Meyers) G. Student, Brown U., 1948-49, 50-51, St. John's U., 1949; BA cum laude, Bklyn. Coll., 1952; MD, Yale U., 1956. Diplomate Am. Bd. Ophthalmology; lic. to practice med. Pa., N.Y., and N.J., 1959—. Intern Walter Reed Army Hosp., Washington, 1956-57; asst. chief U.S. Army Dispensary, Phila., 1957-59, chief, 1959-60; postgrad. preceptorship in ophthal. under co-chief ophthal. Presbyn.-U. Pa. Med. Ctr., Phila., 1963-66; practice medicine specializing in obstetrics Phila., Riverside, N.J., 1960-63; practice medicine specializing in ophthalmology Phila., 1966—; assoc. dir., lectr. in neuro-ophthalmology Hahneman U., Phila., 1978—, assoc. clin. prof. Robert Wood Johnson Med. Sch. of N.J. U. of Medicine and Dentistry, 1988—; attending surgeon in ophthalmology Frankford and Rolling Hills Hosps., Phila, 1970—; asst. to assoc. prof. Ophthalmology Hahneman U. Sc. Medicine, 1977-88; lectr. Biblical topics U.S., Israel, Europe, New Zealand, USSR.; guest speaker TV stas., clubs. Contbr. articles to profl. jours., chpts. to textbooks. Bd. deacons Community Ch., Mt. Laurel Chapel, 1974—. Served to capt., M.C., U.S. Army, 1955-60. Industed into Chapel of 4 Chaplains, Temple U., 1981; inducted Hon. Brave Cherokee Indians by Chief Rising Sun, Chief and High Priest of all N. and S. Am. Indian Tribe and Couns., 1947; recipient AMA Physicians Recognition award in continuing med. edn., 1974— (award tri-annually). Fellow ACS, Phila. Coll. Physicians; mem. AMA, Pa. Med. Soc., Phila. County Med. Soc., Am. Acad. Ophthalmology, Pa. Acad. Ophthalmology, Pan-Am. Soc. Ophthalmology, Soc. Contemporary Ophthalmology, Christian Med. Soc., Am. Soc. Cataract and Refracture Surgery, Internat. Platform Soc., Am. Judeo-Christian Fellowship, Alpha Kappa Kappa. Democrat. Avocation: book collecting, Bible lectures and writings. Office: 4500 Arthur Kill Rd Staten Island NY 10309-1318

GREENGARD, PAUL, neuroscientist; b. N.Y.C., Dec. 11, 1925; married, 1954; two children. AB, Hamilton Coll., 1948; PhD, Johns Hopkins U., 1953. NSF fellow in neurochemistry U. London (Eng.)Inst. Psychiatry, 1953-54; Nat. Found. Infantile Paralysis fellow U. Cambridge (Eng.) Molteno Inst., 1954-55; Paraplegia Found. fellow Nat. Inst. Med. Rsch., Eng., 1955-56; fellow Nat. Inst. Neurological Diseases and Blindness, 1956-58; dir. biochemistry dept. Ciba-Geigy Rsch. Labs., 1958-67; prof. pharmacology Yale U. Sch. Medicine, New Haven, 1968-83; Andrew D. White prof.-at-large Cornell U., Ithaca, N.Y., from 1981; prof. Rockefeller U., N.Y.C., 1983—; vis. scientist Nat. Heart Inst., 1958-59; vis. assoc. prof. Albert Einstein Coll. Medicine, 1961-65, vis. prof., from 1968; vis. prof. Vanderbilt U., 1967-68. Recipient Ciba-Geigy Drew award, 1979, Biol. and Med. Scis. award N.Y. Acad. Scis., 1980, 3M Life Scis. award Fedn. Am. Socs. Exptl. Biology, 1987, Bristol-Myers award for disting. achievement in neurosci. rsch., 1989, Goodman and Gilman award in receptor pharmacology, 1992. Mem. NAS (award in neuroscis. 1991), Am. Acad. Arts and Scis., Soc. for Neurosci. (Grass lectr. 1986). Office: Rockefeller U 1230 York Ave New York NY 10021-6341

GREENGRASS, ROY MYRON, plant engineer; b. Hazelton, Pa., Oct. 12, 1953. BSME, Northeastern U., 1976; Ms, U. Conn., 1978; MBA, Calif. State U., Stainislaus, 1992. Registered profl. engr., Calif. Plant engr. Kendall Co., Merced, Calif., 1980-82; project engr. Heublein Wines, Madera, Calif., 1982; corp. engr. Foster Farms, Livingston, Calif., 1983; sr. engr. Ragu Foods/Unilever, Merced, 1984-89; plant engr. LePrino Foods, Tracy, Calif., 1989—. Home: 92 Shell Rd Byron CA 94514 1110

GREENLEAF, JOHN EDWARD, research physiologist; b. Joliet, Ill., Sept. 18, 1932; s. John Simon and Julia Clara (Flint) G.; m. Carol Lou Johnson, Aug. 28, 1960. MA, N.Mex. Highlands U., 1956; BA in Phys. Edn., U. Ill., 1955, MS, 1962, PhD in Physiol., 1963. Teaching asst. N.Mex. Highlands U., Las Vegas, Nev., 1955-56; engring. draftsman Allis-Chalmers Mfg. Co., Springfield, Ill., 1956-57; teaching asst. in phys. edn. U. Ill., Urbana, 1957-58, rsch. asst. in phys. edn., 1958-59, teaching asst. in human anatomy and physiology, 1959-62; summer fellow NSF, 1962; pre-doctoral fellow NIH, 1962-63; rsch. physiologist Life Scis. Directorate, NASA, Ames Rsch. Ctr., Moffett Field, Calif., 1963-66, 67—; postdoctoral fellowship Karolinska Inst., Stockholm, 1966-67. Mem. editorial bd. Jour. Applied Physiology, 1989—; contbr. articles to profl. jours. Recipient George Huff award for Scholarship, U. Ill. 1954-55, NASA Spl. Achievement award, 1973. Served with U.S. Army, 1952-53. Exch. fellow Nat. Acad. Scis., 1973-74, 77, 89, NIH, 1980; recipient Disting. Alumni award N.Mex. Highlands U., 1990. Fellow Am. Coll. Sports Medicine (trustee 1984-87), Aerospace Med. Assn. (Harold Ellingson award 1981, 82, Eric Liljencrantz award 1990); mem. Am. Physiol. Soc. (mem. com. on coms. 1984-87), Shooting Sports Rsch. Coun. (internat. shooters devel. fund 1984), Sigma Xi. Achievements include patents in field. Home: 12391 Farr Ranch Ct Saratoga CA 95070-6527 Office: NASA Ames Rsch Ctr Life Sci Div MS 239-11 Moffett Field CA 94035-1000

GREENLEAF, MARCIA DIANE, psychologist, writer, educator; b. N.Y.C., Feb. 15, 1944; d. Selig and Irene (Gardner) Finkelstein; m. William Greenleaf, Nov. 26, 1972 (div. 1977); m. Herbert Spiegel, Jan. 29, 1989. BA, Conn. Coll., 1965; MA in Psychology, New Sch. for Social Rsch., 1979; PhD, Yeshiva U., 1986. Lic. psychologist, N.Y. Instr. modern dance Pine Manor Jr. Coll., Chestnut Hill, Mass., 1969-72; dir. human support svc. dept. nursing Hosp. of Albert Einstein Coll. Medicine, Bronx, N.Y., 1983-91, asst. prof. psychiatry, 1986—; asst. clin. attending psychologist Montefiore Med. Ctr., Bronx, 1986—; pvt. practice health psychology N.Y.C., 1986—; NIH rsch. fellow Meml. Sloan-Kettering Cancer Ctr., N.Y.C., 1992, psychologist, 1992—. Contbr. articles to profl. publs., chpts. to books. Recipient Milton H. Erickson award for Excellence in Sci. Writing, Am. Jour. Clin. Hypnosis., Nat. Svc. Rsch. award Nat. Cancer Inst., 1992-93. Fellow Am. Soc. Clin. Hypnosis (sec. 1988, 90, mem.-at-large 1987, chair speakers bur. 1992—, mem. faculty 1986—); mem. APA, Soc. Clin. and Exptl. Hypnosis. Office: 19 E 88th St New York NY 10128

GREENLEE, KENNETH WILLIAM, chemical consultant; b. Leon, W.Va., Jan. 23, 1916; s. Roy Elsworth and Lola Ethel (Woodall) G.; m. Mary Elizabeth Boord, Dec. 14, 1947; children: Sarah Anne, William John, Mark Leroy. AA, U. Charleston, 1935; BS, Antioch Coll., 1938; PhD, Ohio State U., 1942. Rsch. assoc. dept. chemistry Ohio State U., Columbus, 1942-49; rsch. supr. Ohio State U. Rsch. Found., Columbus, 1949-63; lectr. dept. chemistry Ohio State U., Columbus, 1959-63; pres. Chem. Samples Co., Inc., Columbus, 1963-78, Chemsampco Inc./A.I. Corp., Columbus, 1978-80; v.p. Albany Internat. Chem. Div., Columbus, 1981-84; chem. cons. Columbus, 1984—; chem. cons. Goodyear Rsch. Dept., Akron, Ohio, 1957-81, Ohio Indsl. Commn., Columbus, 1950-52. Contbr. articles to profl. jours. Mem. Community Svcs. Coun. of Columbus and Franklin County, Columbus, 1956-66, Christian Bus. Men's Com., Columbus, 1989—. Mem. Am. Chem. Soc. (co-founder small chem. businesses div. 1978, editor Small Chem Biz News 1978—, Award for Outstanding Achievement in Promotion of Chem. Scis. 1988, chmn. Columbus sect. 1955-56), Ohio Acad. Sci., Soc. of the Sigma Xi. Republican. Presbyterian. Avocations: woodsmanship, reading, meetings with other U.S. Civil War buffs. Home: 98 Blenheim Rd Columbus OH 43214-3230 Office: ACS Divsn Small Chem Bus Editorial Office PO Box 14373 Columbus OH 43214-0373

GREENLER, ROBERT GEORGE, physics educator, researcher; b. Kenton, Ohio, Oct. 24, 1929; s. Dallas George and Ruth Edna (Mallett) G.; m. Barbara Stacy, May 30, 1954; children: Leland S., Karen R., Robin A. BS in Physics, U. Rochester, 1951; PhD in Physics, Johns Hopkins U., 1957. Research scientist Allis-Chalmers Mfg. Co., Milw., 1957-62; assoc. prof. physics U. Wis., Milw., 1962-67, prof., 1967—; sr. vis. fellow U. East Anglia, Norwich, Eng., 1971-72; traveling lectr. Optical Soc. Am., 1973-74; lectr. Coop. Edn. Program, Malaysia, 1990-91. Author: Rainbows, Halos and Glories, 1980; contbr. articles to profl. jours. Sr. Fulbright scholar Fritz Haber Inst. of Max Planck Soc., West Berlin, 1983; grantee NSF, Petroleum Research Fund, Am. Chem. Soc. Fellow AAAS, Optical Soc. Am. (v.p. 1985, pres.-elect 1986, pres. 1987); mem. Am. Assn. Physics Tchrs. (Millikan Lecture award 1988). Research in surface science, infrared spectroscopy of adsorbed molecules, meteorological optics, iridescent colors in biological systems. Home: 1901 W Pioneer Rd Mequon WI 53092-8330 Office: U Wis Milw Dept Physics Milwaukee WI 53201

GREENSTEIN, JEROME, mechanical engineer; b. Chgo., May 27, 1925; s. Dennis and Helen (Davis) G.; m. Lillian Sporer, June 30, 1946; children: Joyce Lynn, Marcia Rose, Robert Steven. BS in Mech. Engring., Ill. Inst. Tech., 1949. Registered profl. engr. Ill., Fla.; cert. mech. contractor Fla.; cert. energy mgr. Mech. engr. E.R. Gritschke & Assocs., Inc., Chgo., 1949-58; exec. v.p. ThermoDynamics, Inc., Chgo., 1958-77; mech. and energy engr. Gritschke & Cloke, Inc., Chgo., 1977-78; mech. engr. R.L. Anderson, Inc., Venice, Fla., 1978-81, pres., 1981-88; ret. Sarasota, Fla., 1988-91; part-time cons. Sarasota, 1991—; chmn. County Licensing and Examination Bd., Sarasota, 1979—; vol., guest lectr. County Tech. Inst., Sarasota, 1988—, mem. adv. com., 1983—. 2d lt. USAF, 1943-45, ETO. Mem. ASHRAE (life), Assn. Energy Engrs. (sr. charter), Tau Beta Pi (life). Jewish. Home and Office: 8776 Midnight Pass Rd Sarasota FL 34242-2837

GREENWALD, ANTHONY GALT, psychology educator; b. N.Y.C., Jan. 30, 1939. BA magna cum laude, Yale U., 1959; MA in Social Psychology, Harvard U., 1961, PhD, 1963. Rsch. asst. Yale U., 1959-63; postdoctoral rsch. fellow Ednl. Testing Svc., Princeton, N.J., 1963-65; from asst. prof. to prof. psychology Ohio State U., Columbus, 1965-86; prof. psychology U. Wash., Seattle, 1986—; vis. scholar Stanford U., 1978-79; presenter in field. Assoc. editor: Jour. Personality and Social Psychology, 1972-76, editor, 1977-79; editorial bd. Psychonomic Sci., 1971-72, Jour. Personality and Social Psychology, 1971-72, Memory & Cognition, 1972—, Psychol. Rev., 1985-89, Jour. Exptl. Psychology: Gen., 1990—; contbr. articles to Psychol. Found. Attitudes, Psychol. Rev., Jour. Exptl. Psychology, Psychol. Perspectives on the Self, Contemporary Psychology, Jour. Applied Psychology, Memory and Cognition, Jour. Personality and Social Psychology. Achievements include rsch. on double-blind tests of subliminal self-help audiotapes, differences between backward and simultaneous masking, defining attitude and attitude theories, motivational facets of the self, explorations in social psychology. Office: U Washington Dept Psychology NI 25 Seattle WA 98195

GREENWALD, ANTON CARL, physicist, researcher; b. Bklyn., Apr. 7, 1947; s. Leslie and Grace Greenwald; m. Sherrill Joy Vine, Jan. 15, 1978; 1 child, Lindsay Joanna. BS, Cooper Union, 1968; MS, U. Md., 1971, PhD, 1975. Staff scientist Spire Corp., Bedford, Mass., 1975-81, sr. scientist, 1982—. Mem. Am. Phys. Soc., Materials Rsch. Soc. Achievements include patents for chemical vapor deposition techniques for ceramics; development of pulsed annealing of ion implantation for large scale manufacture of solar cells. Office: Spire Corp 1 Patriots Park Bedford MA 01730

GREENWALD, PETER, physician, government medical research director; b. Newburgh, N.Y., Nov. 7, 1936; s. Louis and Pearl (Reingold) G.; m. Harriet Reif, Sept. 6, 1968; children—Rebecca, Laura, Daniel. BA, Colgate U., 1957; MD, SUNY Coll. Medicine, 1961; MPH, Harvard U., 1967, DrPH, 1974. Intern Los Angeles County Hosp., 1961-62; resident in internal medicine Boston City Hosp., 1964-66; asst. in medicine Peter Bent Brigham Hosp., 1967-68; mem. epidemiology and disease control study sect. NIH, 1974-78; mem. N.Y. State Gov.'s Breast Task Force, 1976-78; with N.Y. State Dept. Health, Albany, 1968-81; dir. N.Y. State Dept. Health, 1968-76, dir. epidemiology, 1976-81; prof. medicine Albany Med. Coll., 1976-81; attending physician Albany Med. Ctr. Hosp., 1968-81; adj. prof. biomed. engring. Rensselaer Poly. Inst., Troy, N.Y., 1976-81; assoc. scientist Sloan-Kettering Inst. for Cancer Research, N.Y.C., 1977-81; dir. Div. Cancer Prevention and Control, Nat. Cancer Inst., NIH, 1981—; mem. VA Merit Rev. Bd. Med. Oncology, Washington, 1972-74. Editor-in-chief Jour. Nat. Cancer Inst., NIH, 1981-87; contbr. articles to profl. jours. Served with USPHS, 1962-64, 81—. Recipient Disting. Service award N.Y. State Dept. Health, 1975; Redway medal and award for med. writing N.Y. State Jour. Medicine, 1977; N.Y. State Gov.'s Citation for pub. health achievement, 1981, PHS commendation, 1983, 88. Fellow ACP, Am. Coll. Preventive Medicine, Am. Pub. Health Assn. (epidemiology sect. chmn. 1981); mem. Am. Assn. Cancer Research, Am. Soc. Clin. Oncology, Am. Coll. Epidemiology (bd. dirs. 1981-82), Am. Cancer Soc., Am. Soc. Preventive Oncology, Internat. Cancer Registry Assn., Internat. Epidemiology Soc., Nat. Acad. Scis. (food and nutrition bd. 1982-88). Office: NIH Bldg 31 Rm 10A52 9000 Rockville Pike Bethesda MD 20892-3100

GREENWOOD, FRANK, information scientist; b. Rio de Janeiro, Mar. 6, 1924; came to U.S., 1935; s. Heman Charles and Evelyn (Heyns) G.; m. Mary Mallas, Oct. 24, 1972; children: Margaret, Ernest, Nicholas. BA, Bucknell U., 1950; MBA, U. So. Calif., 1959; PhD, UCLA, 1963. Cert. systems profl. Various positions The Tex. Co., U.S., Africa and Can., 1950-60; assoc. prof. U. Ga., Athens, 1961-65; chmn. dept. computer systems Ohio U., Athens, 1966-76; dir. computer ctr. U. Mont., Missoula, 1977-84; prof. mgmt. info. systems Southeastern Mass. U., North Dartmouth, 1985-89, Ctrl. Mich. U., Mt. Pleasant, 1990-93; assoc. Asia Pacific Inst. for Mgmt. Devel., Singapore, Singapore, 1993—; bd. dirs. Ctr. for Productivity, Inc., Mt. Pleasant, 1981. Author: Casebook for Management and Business Policy: A Systems Approach, 1968, Managing the Systems Analysis Function, 1968; (with Nicolai Siemens and C.H. Marting Jr.) Operations Research: Planning, Operating and Information Systems, 1973; (with Mary Greenwood) Information Resources in the Office Tomorrow, 1980, Profitable Small Business Computing, 1982, Office Technology: Principles of Automation, 1984, Business Telecommunications: Data Communications in the Information Age, 1988, Introduction to Computer-Integrated Manufacturing, 1990, How to Raise Office Productivity, 1991; columnist: Computerworld mag., 1972-73, The Daily Record, 1982-83, (with Mary Greenwood) Herald News, 1986, The Beacon, 1986, Morning Sun, 1990-93; contbr. monographs, articles to profl. jours. and chpts. to books. Sgt. AUS, 1943-45. UCLA Alumni scholar, 1961; Ford Found. fellow, 1962-63. Mem. Assn. for Systems Mgmt., Wamsutta Club (New Bedford, Mass.). Presbyterian. Avocation: exercise.

GREENWOOD, M. R. C., college dean, biologist, nutrition educator; b. Gainesville, Fla., Apr. 11, 1943; d. Stanley James and Mary Rita (Schmeltz) Cooke; m. (div. 1968); 1 child, James Robert. AB summa cum laude, Vassar Coll., 1968; PhD, Rockefeller U., 1973; LHD (hon.), Mt. St. Mary Coll., 1989. Rsch. assoc. Inst. of Human Nutrition, Columbia U., N.Y.C., 1974-

75, adj. asst. prof., 1975-76, asst. prof., 1976-78; assoc. prof. dept. biology Vassar Coll., Poughkeepsie, N.Y., 1978-81, prof. biology, 1981-86, dir. animal model, CORE Lab. of Obesity Rsch. Ctr., 1985-89, dir. undergrad. rsch. summer inst., 1986-88, dir. Howard Hughes biol. scis. network program, 1988, chmn. of biology dept., John Guy Vassar prof. natural scis., 1986-89; prof. nutrition and internal medicine, dean grad. studies U. Calif., Davis, 1989—; mem. nutrition study sect. NIH, 1983-87; mem. NRC. Editor: Obesity, Vol. 4, 1983; contbr. over 250 articles and abstracts to profl. jours., 1974-89. Recipient Rsch. Career Devel. award NIH, 1978-83; Mellon scholar-in-residence St. Olaf Coll., Northfield, Minn., 1978; N.Y. State Regents fellow, 1968. Mem. Inst. Medicine of Nat. Acad. Scis. (chair food and nutrition bd., diet and health subcom. 1986—), N.Am. Soc. for Study of Obesity (pres. 1987-88), Am. Inst. Nutrition (BioServ 1982), Am. Physiol. Soc., The Harvey Soc., Am. Diabetes Assn., Internat. Assn. for Study of Obesity (treas. 1991—). Home: 5033 El Cemonte Ave Davis CA 95616-4413 Office: U Calif Grad Studies Dept 252 Mrak Hall Davis CA 95616

GREENWOOD, RICHARD P., wastewater superintendent; b. St. Joseph, Mo., Sept. 17, 1941; s. Victor K. and Agnes (Bing) G.; m. Jacqueline Greenwood, Oct. 1, 1962 (div. Oct. 1976); m. Cheryl A. Greenwood, Feb. 15, 1980; children: Philip Joanie, Christopher, Brandee, Ricelle. Power plant operator City Fairbury, Nebr., 1962-73, wastewater plant supr., 1973--. Democrat. Lutheran. Office: City of Fairbury 612 D St Fairbury NE 68352-2320

GREER, CLAYTON ANDREW, electrical engineer; b. Clinton, Iowa, Nov. 25, 1966; s. Ronald Edward and Tanya Jo (Begley) G.; m. Janna Leigh Winkelmann, June 14, 1989; 1 child, Haleigh Lauren. BSEE, U. Tulsa, 1989. Engr. asst. Pub. Svc. Co. of Okla., Tulsa, 1988-89, power system planning engr., 1989-92, substa. design engr., 1992—. Mem. NSPE, Tau Beta Pi, Eta Kappa Nu. Republican. Home: 9130 S 91st East Ave Tulsa OK 74133

GREER, DAVID LLEWELLYN, electrical engineer; b. N.Y.C., Mar. 26, 1930; s. Thomas L. and Gladys (Pearson) G.; m. Jane Ann Solan, Nov. 8, 1958; children: Laurie Anne, Ellen Frances, Sara Louise, Amy Lynn. BEE, Cornell U., 1954. Devel. engr. GE Co., Ithaca, Syracuse, N.Y., 1956-69; cons. engr. GE Co., Syracuse, 1969-82, mgr., 1982-84; sr. cons. GE Co., Charlottesville, Va., 1984-87; dir. Interlogic Inc., Charlottesville, 1987—. Contbr. articles to IEEE Trans. on Computers, Internat. Solid State Cirs. Conf., IEEE Conf. on Computer Design, IEEE Jour. Solid State Cirs. Achievements include patents for Magnetic Second Harmonic Analog Devices, Multiple Level Associative Logic Circuits, Electrically Programmable Logic Circuits, Segmented Associative Logic Circuits; in 1971 developed first working ultraviolet light erasable, electrically programmable logic device; produced programmable logic arrays in 1974 incorporating new concepts of OR-AND plane signal feedback and line folding. Home and Office: Interlogic Inc 2500 Thrush Rd Charlottesville VA 22901-8813

GREER, EDWARD COOPER, chemist; b. Chapel Hill, N.C., Feb. 27, 1957; s. Paul Shryock and Sylvia (Morene) G.; m. Melinda Sue Wolff, July 9, 1988; children: Douglas Russell, Timothy David. BS with honors, U. N.C., 1977; PhD in Analytical Chemistry, U. Wis., 1982. Scientist Rohm and Haas Co., Spring House, Pa., 1982-84, rsch. sect. mgr., 1984—. Mem. Am. Chem. Soc., Sigma Xi. Office: Rohm and Haas Co PO Box 904 727 Norristown Rd Spring House PA 19477-0904

GREER, R. DOUGLAS, behaviorology and education educator; b. Elizabethton, Tenn., June 21, 1942; s. Robert Lee and Cleo Virginia (Miller) G.; divorced; children: Angela Rene, John Boone, Laura Melissa. BME, Fla. State U., 1963, MME, 1966; PhD, U. Mich., 1969. Instr. Tallahasse (Fla.) Schs., 1963-66, U. Mich., Ann Arbor, 1966-69; prof. Columbia U. Tchrs. Coll., N.Y.C., 1969—; sr. cons. Comprehensive Application Behavior Analysis to Schooling, U.S., Italy, 1980—. Author: A Science of Teaching: Learner Driven Professional Expertise for Superior Schools, The Teacher as Strategic Scientist: A Solution to Our Educational Crisis, 1992. Pres. bd. dirs. The Fred S. Keller Sch., Yonkers, N.Y., 1986—. Recipient Disting. Vis. Prof. award Ohio State U., 1991. Mem. Assn. for Behavior Analysis (chmn. various task forces 1980—), The Internat. Behaviorology Assn. (assoc.; editor Experimental Analysis 1990). Achievements include development of the CABAS model of schooling, and R&Dof 1 of 5 behavioral interventions for feeding disorders with young children; indentified and researched peer mediating establishing operation. Office: Columbia U Tchrs Coll 525 W 120th St New York NY 10027

GREGERSEN, MAX A., structural, earthquake and civil engineer; b. Blackfoot, Idaho, Apr. 6, 1951; s. Garth Clifford and Ella Lavere (Adamson) G.; m. Berneitta Ann Pieper, July 24, 1982; children: Dusty Rae, Molly Malinda. BS in Civil Engring., U. Utah, 1976. Registered profl. engr. 34 states, Can., P.R. Civil/structural engr. Kellogg-Rust Engring., Salt Lake City, 1976-83; mgr. civil/structural engring. dept. Ford, Bacon & Davis Techs., Inc., Salt Lake City, 1983—; corr. mem. 1994 Fed. Emergency Mgmt. Agy. Nat. Earthquake Hazards Reduction Program, seismic provisions update com. for steel structures Bldg. Seismic Safety Coun., Washington, 1992—; corr. com. mem. applied tech. coun. Fed. Emergency Mgmt. Agy.-sponsored ATC-33 Project devel. guidelines for seismic rehab. existing bldgs., 1993—; mem. curriculum adv. bd. dept. civil engring. U. Utah, 1993—. Mem. ASCE, Am. Concrete Inst. (com. 369 Seismic repair and rehab. 1992—), Earthquake Engring. Rsch. Inst., Constrn. Specifications Inst., Structural Engrs. Assn. Utah, Steel Structures Painting Coun., North Am. Geosynthetics Soc., Internat. Geotextile Soc., Internat. Conf. Bldg. Ofcls., Bldg. Ofcls. and Code Adminstrs. Internat., So. Bldg. Code Congress Internat., Assn. Profl. Engrs., Geologists, and Geophysicists Alta. Achievements include seismic evaluation and retrofit of existing heavy indsl. facilities; engring. designs and constrn. of major petroleum, chem., hazardous waste incineration, indsl. wastewater treatment, mining, milling, smelting, coal handling, power plant, slipformed concrete silos, solid fuel rocket motor production, roadway, railroad and geosynthetic-lined waste containment facilities. Office: Ford Bacon and Davis Techs Inc 375 Chipeta Way PO Box 58009 Salt Lake City UT 84158-0009

GREGOR, CLUNIE BRYAN, geology educator; b. Edinburgh, Scotland, Mar. 5, 1929; came to U.S., 1968; s. David Clunie Gregor and Barbara Mary Moller-Beilby; m. Suzanne Assir, Apr. 24, 1955 (div. Apr. 1969); 1 child, Andrew James; m. Anna Bramanti, Apr. 15, 1969; children: Thomas James, Matthew James. BA, Cambridge (Eng.) U., 1951, MA, 1954; DSc, U. Utrecht, The Netherlands, 1967. Instr. Am. U. Beirut, 1958-64; rsch. asst. Delft (The Netherlands) Inst. Tech., 1964-65, dir. Crystallographic Lab., 1965-67; vis. prof. Case Western Res. U., Cleve., 1968-69; prof. West Ga. Coll., Carrollton, 1969-72, Wright State U., Dayton, Ohio, 1972—; vice chmn. panel on geochem. cycles NAS, 1988-90. Author: (monograph) Geochemical Behaviour of Sodium, 1967; editor: Chemical Cycles in the Evolution of the Earth, 1988. Grantee NSF, 1977-82, Sicily, 1978-80. Fellow Geol. Soc. (London); mem. Am. Geol. Soc., Am. Geophys. Union, Geochem. Soc. (sec. 1983-89). Home: 136 W North College St Yellow Springs OH 45387-1563 Office: Wright State U Dept Geol Scis Dayton OH 45435

GREGOR, EDUARD, laser physicist; b. Dnepropetrovsk, Ukraine, Jan. 9, 1936; came to U.S., 1955; s. Waldemar and Concordia (Teschke) G.; m. Marie L. Carlin, June 29, 1968; 1 child, Eduard Joseph. BS in Physics, Calif. State U., 1964, MS in Physics, 1966. Instr. Calif. State U., L.A., 1963-66; optical physicist TRW Instruments, El Segundo, Calif., 1966-68; laser physicist Union Carbide (Korad), Santa Monica, Calif., 1968-72; prodn. mgr. holography Quantrad Corp., El Segundo, Calif., 1972-75, ops. mgr., 1975-79; sr. project physicist Hughes Aircraft Co., El Segundo, 1979-82, dept. mgr., 1982—. Contbr. over 20 tech. articles on laser tech., coherent optics and holography to profl. jours. Sgt. U.S. Army, 1959-61. Recipient IR 100 award Indsl. Rsch. Mag., 1975. Mem. Optical Soc. Am., Soc. Photo-optical Instrumentation Engrs. Achievements include 4 patents for Mobile Laser Holocamera, Two Cavety Laser, Varible Lens and Birefringence Compensator, Phase Conjugate Laser with a Temporal Square Pulse. Home: 820 Las Lomas Ave Pacific Palisades CA 90272-2428

GREGORIOU, GREGOR GEORG, aeronautical engineer; b. Athens, Greece, Feb. 5, 1937; s. Georg and Athina (Koulia) G.; m. Josephine Nievelstein, Aug. 31, 1962; children: Lauretta, Katja. Dipl. Ing., Tech. U., Aachen, Germany, 1962; Dr.-Ing., Tech. U., Munich, 1973. Aerodynamicist Vereinigte Flugtechn Werke, Bremen, Germany, 1963-64, Messerschmitt-Bolkow-Blohm, Ottobrunn, Germany, 1964-71; mgr. aerodynamics Messerschmitt-Bolkow-Blohm, Ottobrunn, 1971-84; pres. ENVIRO, Putzbrunn, Germany, 1985—. Contbr. articles to profl. jours. Mem. AIAA, Gesellschaft fur Angewandte Mathematick und Mechanik, Deutsche Gesellschaft für Luft und Raumfahrt. Office: Oedenstockacher Strasse 5, 8011 Putzbrunn Germany

GREGORY, BRUCE NICHOLAS, astrophysicist, educator; b. N.Y.C., July 12, 1938; s. Nicholas and Florence (Hoagland) G.; m. Jane Gray Jacobik, Jan. 4, 1983; children: David, Christopher, James, Christianne. AB, U. Mass., 1960, MS, 1963. Astronomer U.S. Naval Rsch. Lab., Washington, 1963-65; staff dir. NAS, Washington, 1966-82; assoc. dir. Harvard-Smithsonian Ctr. for Astrophysics, Cambridge, Mass., 1983—. Author: Inventing Reality, 1988; co-author: Project STAR, 1993. Home: 86 Fox Hill Rd Pomfret CT 06259 Office: Smithsonian Astrophysical Obs 60 Garden St Cambridge MA 02138

GREGORY, PATRICIA JEANNE, corporate relations director; b. St. Louis, Feb. 13, 1951; d. Kenneth Robert and Mary Jane (Gibbs) Reilly; m. Mark Hitchcock Gregory, Apr. 20, 1985; children: Martin Hitchcock Gregory, Michael Wilford Gregory. BA, U. Mo., St. Louis, 1975; MS, U. Mo., 1978; PhD, U. Vt., 1986. Rsch. technician Washington U. Sch. Medicine, St. Louis, 1972-79; rsch. assoc. Univ. Coll., London, 1979-81; postdoctoral fellow U. Chgo., 1986-88; asst. chmn. Biochemistry Northwestern U., Evanston, Ill., 1988-92; dir. corp. and found. rels. Washington U. Sch. Medicine, St. Louis, 1992—; dir. minority recruitment for the life scis., Northwestern U., Evanston, 1988-92; assoc. dir. Ctr. for Biotech., Northwestern U., Evanston, 1990-92. Contbr. articles to profl. jours. Recipient Curator's Scholar award U. Mo. Bd. Curators, 1969, fellow Acad. Leadership Program, Com. on Instl. Cooperation, Urbana, Ill., 1990. Home: 404 Edgewood Dr Clayton MO 63105 Office: Washington U Sch Medicine Box 8049 660 S Euclid Saint Louis MO 63110

GREGORY, RICHARD LEE, immunologist; b. Elmhurst, Ill., Oct. 31, 1954; s. Walter Leonard Jr. and Jetta Mae (Jensen) G.; m. Rebecca Jo Brown, Aug. 22, 1981; children: Amanda Kristin, Emily Sara, Eric Richard. BS, Ea. Ill. U., 1976; PhD, So. Ill. U., 1982. Cert. med. technologist. Rsch. assoc. U. Ala., Birmingham, 1982-84; asst. prof. Emory U., Atlanta, 1984-90, assoc. prof., 1990-91; assoc. prof. Ind. U., Indpls., 1991—; vis. instr. U. Ala., Tuscaloosa, 1984; ad hoc grant reviewer NIH, Bethesda, 1987—; review bd. Clin. Lab. Sci. jour., Atlanta, 1989—. Contbr. articles to profl. jours., publs.; editorial bd. Clin. and Diagnostic Lab. Immunology Jour., 1993—. Vol. tchr. Woodbrook Elem. Sch., Carmel, Ind., 1991—. Grantee NIH, Birmingham, Atlanta and Indpls., 1984—; rsch. grantee Smokeless Tobacco Rsch. Coun., N.Y., 1986—. Fellow Am. Soc. Microbiology (branch newsletter editor 1986-91, ednl. rep., v.p., pres.), Am. Assn. Immunologists, Am. Soc. Med. Techs. Achievements include identification of role of salivary antibodies in protection from tooth decay. Office: Ind U 1121 W Michigan St Indianapolis IN 46202

GREGORY, ROBERT SCOTT, pharmacist; b. Binghamton, N.Y., May 30, 1954; s. Floyd A. and Betty J. (Jenson) G.; m. Karen J. Shislo, Sept. 20, 1980. AA, Broome Community Coll., 1973; BS in Pharmacy, Mass. Coll. Pharmacy, 1977, MS in Pharmacy, 1983; MBA, U. Houston, 1992. Registered pharmacist in Tex., Mass., N.H. Staff pharmacist Colonial Pharmacy, New London, N.H., 1977-78, Darmouth-Hitchcock Med. Ctr., Hanover, N.H., 1978-80; pharmacy supr. Quincy (Mass.) City Hosp., 1980-83; pharmacy dir. Hosp. Corp. Am., Coronado Hosp., Pampa, Tex., 1983-86, Diagnostic Ctr. Hosp., Houston, 1986-93, Group Health Assn. Washington, 1993—; cons. HCA Div. Support Network, Dallas and Fort Worth, Tex., 1986—. Mem. Am. Soc. Hosp. Pharmacists, Assn. Managed Care Pharmacists, Tex. Soc. Hosp. Pharmacists (bd. dirs. 1984-86), Panhandel Soc. Hosp. Pharmacists (pres. 1985-86, President's award 1986), Houston-Galveston Soc.. Hosp. Pharmacists, Rho Chi. Methodist. Avocations: softball, racquetball, photography, collecting pharmacy antiques. Office: Group Health Assn Washington 4301 Connecticut Ave NW Washington DC 20008

GREGORY, THOMAS BRADFORD, mathematics educator; b. Traverse City, Mich., Dec. 13, 1944; s. Philip Henry and Rhoda Winslow (Hathaway) G. BA, Oberlin (Ohio) Coll., 1967; MA, Yale U., 1969, M of Philosophy, 1975, PhD, 1977. Lectr. Ohio State U., Mansfield, 1977-78, asst. prof. math., 1978-84, assoc. prof. math., 1984—. Reviewer: Math. Revs., 1984—; contbr. articles to profl. jours. Active Mansfield (Ohio) Symphony Chorus, 1977—, Presbytery Youth Ministries Com., New Philadelphia, Ohio, 1980-87, Ohio State U. Community Singers, Mansfield, 1985—. Comdr. USNR, 1969—. Fellow NSF, Washington, 1967; hon. fellow U. Wis., Madison, 1987-88, 92. Mem. Am. Math. Soc. (translator 1974-82), Ohio Coun. Tchrs. Math., Am. Soc. Naval Engrs., Naval Inst., Res. Officers Assn., Naval Res. Assn., Navy League, Phi Beta Kappa, Sigma Xi. Republican. Avocations: classical piano, singing, amateur radio, volleyball, jogging. Home: 930 Maumee Ave Mansfield OH 44906-2909 Office: Ohio State U 1680 University Dr 0-15 Mansfield OH 44906-1547

GREGORY, VANCE PETER, JR., chemist; b. Jacksonville, Fla., Sept. 7, 1943; s. Vance Peter and Roberta Marie (White) G.; m. Sharon Keilman, Apr. 27, 1990. BA, Vanderbilt U., 1965. Tech. svc. chemist CPC Internat., Argo, Ill., 1966-74; project chemist Uarco, Inc., Barrington, Ill., 1974-86; mgr. R & D Weber Marking Systems, Arlington Heights, Ill., 1986; plant chemist Wallace Computer Svcs., Streetsboro, Ohio, 1986-88; sr. project chemist Wallace Computer Svcs., Bellwood, Ill., 1988—. Mem. Historic Sites Commn., Glen Ellyn, Ill., 1985. Mem. Am. Chem. Soc., Tech. Assn. Pulp and Paper Industries. Avocations: travel, flying, fishing, gardening. Home: 2286 Durham Dr Wheaton IL 60187-8818 Office: Wallace Computer Svcs 10 Davis Dr Bellwood IL 60104-1093

GREGORY, W. LARRY, social psychology educator; b. Alamogordo, N.Mex., Oct. 18, 1950; s. William Vernon and Nina (Holladay) G.; m. Janice Margaret Russell, Jan. 24, 1971 (div. 1986); children: Siusla, Ariel. BA, Ariz. State U., 1973, PhD, 1981. Asst. prof. N.Mex. State U., Las Cruces, 1980-86, assoc. prof., 1986—; cons. ARC, Las Cruces, 1987-89, CSAP-funded substance abuse prevention programs, N.Mex., various law firms, Tex. and N.Mex. Editor: Introduction to Applied Psychology, 1989; contbr. articles to profl. jours. Mem. Am. Psychol. Soc. (charter). Office: NMex State U Psychology Dept 3452 Las Cruces NM 88003

GREINER, THOMAS MOSELEY, physical anthropologist, archaeologist, consultant; b. Huntington, N.Y., July 24, 1961; s. Ronald Moseley and Rosemary (Piatt) G. AB in Anthropology, U. Chgo., 1983; MA in Anthropology, SUNY, Binghamton, 1988, PhD in Anthropology, 1993. Archaeologist Pub. Archaeology Facility, Binghamton, 1985-87; instr. SUNY, Binghamton, 1987; rsch. anthropologist U.S. Army Natick (Mass.) R&D Ctr., 1988-92; instr. dept. anthropology SUNY, Albany; specialist reviewer Appraisal, Boston, 1990-92; organizer symposium on race, ethnicity and applied bioanthropology, 1991. Author: Hand Anthropometry of U.S. Army Personnel, 1991; contbr. articles, papers, tech. reports to profl. publs. Mem. AAAS, Am. Assn. Phys. Anthropologists, Am. Anthropol. Assn., Sigma Xi. Achievements include collection and analysis of major anthropometric data base on human hand, creation of predictive models of anthropometric change used by U.S. Army in clothing and equipment design, creation of hindlimb computer program to model the evolution of human bipedalism. Office: SUNY Binghamton Dept Anthropology Binghamton NY 13902

GREINER, WALTER ALBIN ERHARD, physicist; b. Neuenbau, Germany, Oct. 29, 1935; s. Albin and Elsa (Fischer) G.; m. Barbara Chun; children: Martin, Carsten. MS, U. Darmstadt, Fed. Republic Germany, 1959; PhD, U. Freiburg, Fed. Republic Germany, 1961; DSci (hon.), U. Witwatersrand, South Africa, 1982, U. Beijing, 1990, U. Tel Aviv, 1991, U. Louis Pasteur, Strasbourg, France, 1991, U. Bucharest, 1992. Rsch. asst. U. Freiburg, 1961-62; asst. prof. U. Md., 1962-64; prof. theoretical physics U. Frankfurt, Fed. Republic Germany, 1965—; dir. Inst. Theoretical Physics, 1965—; guest prof. at numerous univs.; adj. prof. Vanderbilt U.-Oak Ridge Nat. Lab., 1975—; hon. prof. U. Beijing, 1990; permanent sci. cons. Gesellschaft fur Schwerionenforschung, Darmstadt. Author: (with others) Nuclear Theory, Nuclear Models Vol. 1, 1970, Excitation Mechanism of Nuclei Vol. 2, 1970, Theory of the Nucleus Vol. 3, 1972, 3d edition, 1987-89, Theoretische Physik Vols. 1-10, 1974-89; editor: Jour. of Physics, 1975-89. Recipient Max Born prize Inst. Physics, 1974, Otto Hahn prize, 1982. Fellow Royal Soc. Arts and Scis.; mem. European Physics Soc., Am. Physics Soc., Sci. Soc. Johann Wolfgang Goethe U., Eötvös Lorand Soc. Hungary (hon.), Acad. Sci. Romania (hon.), Lions. Home: 44 Gundelhardstrasse, Taunus, D-6233 Kelkheim Federal Republic of Germany Office: 8-10 Robert Mayer Strasse, D-6000 Frankfurt Germany

GREITZER, EDWARD MARC, aeronautical engineering educator, consultant; b. N.Y.C., May 8, 1941; s. Arthur O. and Harriet G.; m. Helen Moulton, Nov. 24, 1966; children: Mary Lee, Jennifer Elizabeth. B.A., Harvard U., 1962, M.S., 1964, Ph.D., 1970. Asst. project engr. Pratt & Whitney Aircraft div. United Techs., East Hartford, Conn., 1969-76; indsl. fellow commoner Churchill Coll., Cambridge U., Eng., 1975-76; sr. research engr. United Techs. Research Ctr., East Hartford, 1976-77; asst. prof. MIT, Cambridge, 1977-79, assoc. prof., 1979-84, prof., dir. Gas Turbine Lab., 1984—, H.N. Slater prof. aero. and astronautics, 1988—; Royal Soc. guest fellow, SERC vis. fellow, overseas fellow Churchill Coll., Cambridge U., 1983-84; vis. fellow Japan Soc. for Promotion of Sci., 1987, Peterhouse, Cambridge U., 1990-91; mem. aeronautics adv. com. NASA, 1990—; mem. sci. adv. bd. USAF, 1992—. Contbr. articles to profl. jours., handbooks. Recipient T. Bernard Hall prize Instn. Mech. Engrs., London, 1978, Air Breathing Propulsion Best Paper award AIAA, 1987. Fellow AIAA, ASME (gas turbine power award 1977, '79, Freeman scholar in fluids engring. 1980, bd. dirs. internat. Gas Turbine Inst. 1993—, chmn. turbomachinery com. 1989-91, chmn. gas turbine scholar selection com. 1989-93, turbometremy com., Best Paper award 1991, '92); mem. Sigma Xi. Avocations: jogging, photography, rock climbing. Home: 77 Woodridge Rd Wayland MA 01778-3611 Office: MIT Dept Aeronautics & Astronautics Bldg 31-264 Cambridge MA 02139

GREMILLION, CURTIS LIONEL, JR., psychologist, hospital administrator, musician; b. Slaughter, La., Feb. 26, 1924; s. Curtis Lionel and Beatrice (Watson) G.; m. Rosemary Duhon, Dec. 8, 1951; children: Suzanne Lynelle Gremillion (Rader), Curtis Lionel III, Monique Angele Gremillion Smith. BA in Psychology and Music, U. Southwestern La., 1948; postgrad. in psychology, La. State U., 1948-49, 53. Profl. musician, 1940-43, 46-53; staff psychologist East La. State Hosp., Jackson, 1949-83; dir. psychology and social svc. depts. East La. State Hosp., 1953- 57, asst. supt., 1957-62, adminstr., 1961, 62-64, acting supt., 1964-66, assoc. adminstr., 1966-81, patient advocate, 1981-83; cons. psychology, 1983—; clin. dir. Pace Ctr., Baton Rouge, 1983-88; notary pub., 1954—; dir. regional coun. Alcoholism and Drug Abuse, 1972-76; initiated numerous modern treatment programs for mentally ill. Author: History of The East Louisiana State Hospital; contbr. to publs. on New Orleans jazz. Bd. dirs. La. State Credit Union, 1962-68; chmn. East Feliciana Parish United Givers Fund, 1960; regional chmn. Am. Heart Assn., 1968-76, ARC, 1954-55, 62, Boy Scouts Am., 1964-69, Am. Cancer Soc., 1963-64; bd. dirs. So. Behavioral Rsch. Found., 1970-76. With USNR, 1943-46. Recipient Outstanding Leadership and Svc. award La. Dept. Hosps., 1966. Mem. La. Psychol. Assn. (charter), So. Sociol. Assn., La. Music Therapy Assn. (dir. 1966-74), Am. Legion, Internat. Platform Assn., Psi Chi, Sinfonia, Pi Gamma Mu, Kappa Delta Pi. Democrat. Baptist. Clubs: New Orleans Jazz, Lions. Address: PO Box 306 Slaughter LA 70777

GRENIER, FERNAND, geographer, consultant; b. East Broughton, Que., Can., Mar. 30, 1927; m. Nilma St. Gelais, 1984; children: Mira, Chloé. BA, BPh, Laval U., Quebec City, Que., 1948, MA, 1950; DES, U. Paris, 1955; D honoris causa, Athabasca U., Edmonton, Alta., Can., 1979. Prof. geography Laval U., 1955-73, dean Faculty Letters, 1967-73; dir. gen. Télé U. Que., Quebec City, Montreal, 1973-81; sr. adminstr. U. Que. Presses, Quebec City, 1983-88. Author: Papiers Contrecoeur, 1953 (Prix Casgrain 1953), Atlas du monde contemporain, 1967, (art book) De Ker-Is à Québec Légendes, 1990; founder, editor rev. Cahiers de géographie, 1952-65; contbr. to Dictionnaire biographique du Canada, 1960-67, Ency. Americana, 1960-83, Dictionnaire canadien des noms propres, 1988-89, Dictionnaire toponymique du Québec, 1989-93. Pres. Salon du livre, Quebec, 1968, Festival d'été, Quebec City, 1980; mem. Commn. des Biens culturels, Quebec province, 1972-79, Commn. de toponymie, Quebec province, 1976-84. Mem. Can. Assn. Geographers (pres. 1964, Svc. to Profession of Geography award 1993), Can. Assn. for Study Names, Que. Geog. Soc. (life). Home: 1035 Route Laurier, Sainte-Croix, PQ Canada G0S 2H0

GRENLEY, PHILIP, urologist; b. N.Y.C., Dec. 21, 1912; s. Robert and Sara (Schrader) G. BS, NYU, 1932, MD, 1936. Diplomate Am. Bd. Urology; m. Dorothy Sarney, Dec. 11, 1938; children: Laurie (Mrs. John Hallen), Neal, Jane (Mrs. Eldridge C. Hanes), Robert. Intern, Kings County Hosp., Bklyn., 1936-38, resident, 1939; resident in urology L.I. Coll. Hosp., Bklyn., 1939-41; pvt. practice medicine specializing in urology, Tacoma, Wash., 1946-90, ret., 1990; urologist Tacoma Gen. Hosp., St. Joseph Hosp., Tacoma, Good Samaritan Hosp., Puyallup, Wash.; pres. med. staff St. Joseph Hosp., Tacoma, 1968-69, mem. exec. bd., 1950-54, 67-68; cons. urologist to Surgeon Gen., Madigan Army Med. Ctr., Tacoma, 1954-87, VA Hosp. at American Lake, 1953-80, USPHS McNeil Island Penitentiary, 1955-82, Good Samaritan Rehab. Ctr., Puyallup, 1960-90; med. cons. State of Wash. Dept. Soc. and Health Svcs., 1990—; chief of urology 210th Gen. Hosp., Ft. Jackson Regional Hosp., Ft. McClellan Regional Hosp., 1941-46; lectr. in sociology U. Puget Sound, Tacoma, 1960—. Trustee Wash. Children's Home Soc., 1951-60, Charles Wright Acad., 1961-69, Wash. State Masonic Home, 1984—; trustee Pierce County Med. Bur., 1949-51, 59-61, 71-73, pres., 1973-74, mem. exec. bd., 1975-77; mem. Lakewood adv. commn. TAE Pierce County Coun., 1992—. With AUS, 1941-46. Fellow ACS; mem. Am. Urol. Assn., AMA, Wash., Pan Am. med. assns., Pierce County Med. Soc., Masons, Shriners (med. dir. 1965-78, imperial coun. 1982-85, potentate 1983), Royal Order Jesters (dir. 1986, 87), Lions, Elks, Red Cross of Constantine (knight). Home: 40 Loch Ln SW Tacoma WA 98499-1432

GRENZEBACH, WILLIAM SOUTHWOOD, nuclear engineer; b. Chgo., Sept. 5, 1945; s. William Southwood Sr. and Edla (Edin) G.; m. Judith Samuels, June 16, 1968 (div. Feb. 1978). BA, Grinnell Coll., 1967; MA, Brandeis U., 1970, PhD in Comparative History, 1978; MS in Engring., Boston U., 1988; MS in Indsl. Engring., Northeastern U., Boston, 1988. Cert offshore drilling rig supt., U.S. Geol. Survey; cert. in offshore rescue, Govt. of Newfoundland. Mgr. Greyhound Lines East, Boston, 1970-77; engring. technician Sylvester Assocs., Rockland, Mass., 1977-78; subsea engr. ODECO, Inc., New Orleans, 1978-81; project mgr. SEDCO, Inc., Dallas, 1981-85; rsch. assoc. dept. energy Northeastern U., 1987-89; with Applied Mgmt. Cons., Assonet, Mass., 1989-90; sr. nuclear engr. Yankee Atomic Electric Co., Bolton, Mass., 1990-92; applied mgmt. cons. Assonet, Mass., 1992-93; Cons. Palisades Nuclear Power Sta., Covert, Mich., 1989, Peach Bottom/Limerick (Pa.) Power Sta., Delta and Limerick, 1989-90, Fitzpatrick Station, Lycoming, N.Y., 1992, Indian Point III, Buchanan, N.Y., 1993, New Brunswick Power Commn., Fredricton, 1990; mem. Conf. Group on Cen. European History. Contbr. articles to profl. publs.; author (computer software) Reactor Coolant Expert System, 1988 (Copywrite award 1990). Supporter Friends of Cohasset (Mass.) Libr., 1978-83; mem. Cohasset Hist. Soc., 1978-83. Fellow, U. Calif. San Diego, 1967, Brandeis U., Waltham, Mass., 1968-71. Mem. AAAS, Am. Nuclear Soc., Am. Soc. Quality Control, Inst. Indsl. Engrs., Soc. for Risk Analysis, Marine Tech. Soc., Soc. for History of Tech., Statis. Process Control Soc. Avocations: historical writing, reading, SCUBA diving, horseback riding. Office: Box 5 325 Huntington Ave Boston MA 02115

GRESSER, HERBERT DAVID, physicist; b. Pitts., Aug. 2, 1930; s. Samuel Marcus and Marion (Bialo) G.; m. Adele Davidson, Mar. 25, 1956; children: Mark Geoffrey, Nina Suddenly, Daniel Stuart. BS in Physics, Queens Coll., 1957; postgrad., Bklyn. Polytech. Sr. engr. T.R.G. Inc., Melville, N.Y., 1955-69; v.p. engring. Mallinkrodt/Serosonics, Bohemia, N.Y., 1969-71; chief engr. Holobeam Laser, Ridgefield, N.J., 1971-73; sr. engr. Quantronix Corp., Hauppauge, N.Y., 1973-75; pres. Group II Mfg., Plainview, N.Y., 1975—. Contbr. articles to profl. jours. With U.S. Army, 1953-55. Mem. Jewish War Vet. (comdr. 1992-94). Democrat. Achievements include patents for

Photo-Cauterizer with Coherent Light Source, Method and Apparatus for Placing Identifying Indicia on the Surface of Precious Stones Including Diamonds, Creating of a Parting Zone in a Crystal Structure. Home: 3 Gainsville Dr Plainview NY 11803

GRESSER, MARK GEOFFREY, podiatrist; b. Flushing, N.Y., Feb. 28, 1958; s. Herbert David and Adele (Davidson) G. BS, BA, SUNY, Stony Brook, 1980; D of Podiatric Medicine, N.Y. Coll. Podiatric Medicine, 1984. Diplomate Am. Bd. Podiatric Orthopedics, Nat. Bd. Podiatry Examiners. Resident in podiatry Foot Clinics N.Y., N.Y.C., 1985; podiatrist North Country Podiatry, Miller Place, N.Y., 1986—, Ctr. Moriches, N.Y., 1988—; Assoc. Am. Coll. Foot Surgeons, 1987—. Sec. Suffolk County Handicapped Adv. Bd., Hauppage, N.Y., 1991. Mem. Am. Podiatric Med. Assn. (dir.), Miller Pl.-Mt. Sinai C. of C. (v.p.). Democrat. Jewish. Home: 20 Bell Ave Blue Point NY 11715 Office: N Country Podiatry 765-3 Route 25A Miller Place NY 11764-2738

GRESSHOFF, PETER MICHAEL, molecular geneticist, educator; b. Berlin, Nov. 26, 1948; came to U.S., 1988; s. Horst B. and Margot (Graeper) G.; m. Rosalyn Marie Williams, Nov. 24, 1973; children: Michael, Nikolas. BSc, U. Alta., Edmonton, Can., 1970; PhD, Australian Nat. U., Canberra, 1973, DSc, 1989. Alexander von Humboldt fellow U. Hohenheim, Stuttgart, Germany, 1974-75, U. Bielefeld, Germany, 1985-86; rsch. fellow Australian Nat. U., Canberra, 1975-79, sr. lectr. botany, 1979-87; prof., chair of excellence molecular genetics U. Tenn., Knoxville, 1988—; cons. Internat. Atomic Energy Agy., Vienna, Austria, 1991—, European Community Commn., 1992—. Editor: Molecular Biology of Symbiotic Nitrogen Fixation, 1990, Nitrogen Fixation (with others), 1990, Plant Biotechnology and Development, 1992, Plant Responses to the Environment, 1993; editorial bd. Physiologia Plantarum, 1992—, Jour. Plant Physiology, 1984—, Plant Physiology, 1988-92; contbr. articles to profl. publs. Mem. Knoxville C. of C., 1988-91. Grantee Agrigenetics Corp., Madison, Wis., 1981-86, 91-92, Am. Soybean Assn., 1992—, Human Frontier Sci. Program, Strasbourg, France, 1992. Mem. Internat. Soc. Plant Microbe Interaction, Internat. Soc. Plant Molecular Biology, Sigma Xi, Phi Kappa Phi. Achievements include patent for nitrate-tolerant soybeans; co-invention of DNA amplification fingerprinting, DNA silver staining. Home: 801 Kempton Rd Knoxville TN 37909 Office: U Tenn Inst Agr Knoxville TN 37901-1071

GRETZINGER, JAMES, engineering consultant; b. Pitts., Nov. 7, 1925; s. William J. and Flora Louise (Hubner) G.; m. Nancy Jane Nason, Dec. 20, 1954; children: Susan, Anne, Betsy, Kathryn. BS, Kans. State U., 1949; PhD, U. Wis., 1956. Engr. Eastman Kodak Co., Rochester, N.Y., 1949-51; sr. rsch. assoc. E.T. DuPont Nemours, Inc., Wilmington, Del., 1956-90; cons. Gretzinger, Inc., High Point, N.C., 1991—. Fin. com. Tar Heel Triad Girl Scout Coun., High Point, 1991—. Sgt. U.S. Army, 1943-46. Mem. AICE, Sigma Xi. Achievements include patent for seating support systems.

GREVE, JOHN HENRY, veterinary parasitologist, educator; b. Pitts., Aug. 11, 1934; s. John Welch and Edna Viola (Thuenen) G.; m. Sally Jeanette Doane, June 21, 1956; children—John Haven, Suzanne Carol, Pamela Jean. B.S., Mich. State U., East Lansing, 1956, D.V.M., 1958, M.S., 1959; Ph.D., Purdue U., West Lafayette, Ind., 1963. Assoc. instr. Mich. State U., East Lansing, 1958-59; instr. Purdue U., West Lafayette, 1959-63; asst. prof. Iowa State U., Ames, 1963-64, assoc. prof., 1964-68, prof. dept. vet. pathology, 1968—, interim chair dept. vet. pathology, 1992—, counselor acad. and student affairs, 1991-92; cons. parasitologist various zoos. Contbr. chpts. to books, articles to profl. jours; mem. editorial bd. Lan. Animal Sci., 1971-83, Vet. Rsch. Communications, 1977-84, Vet. Parasitology, 1984—. Dist. chmn. Broken Arrow Dist., Boy Scouts Am., Ames, Iowa, 1975-77. Named Disting. Tchr. Norden Labs., 1965; Outstanding Tchr. Amoco Oil, Iowa State U., 1972; recipient Faculty Citation Iowa State U. Alumni Assn., 1978. Mem. AVMA (mem. editorial bd. jour. 1975—, Excellence in Teaching award student chpt. 1990), Iowa Vet. Med. Assn., Am. Soc. Parasitologists, Midwestern Conf. Parasitologists (sec.-treas. 1967-75, presiding officer 1975-76), Am. Assn. Vet. Parasitologists (pres. 1968-70), Helminthological Soc. Washington, World Assn. for Advancement Vet. Parasitology, Am. Assn. Vet. Med. Colls., Izaak Walton League (bd. dirs. Iowa 1968-70), Honor Soc. Cardinal Key, Gamma Sigma Delta, Phi Eta Sigma, Phi Kappa Phi, Phi Zeta. Republican. Lodges: Kiwanis (Town and Country-Ames pres. 1967, Nebr.-Iowa lt. gov. 1972-73). Avocations: philately, camping, gardening. Office: Iowa State U Dept Vet Pathology Ames IA 50011

GREW, PRISCILLA CROSWELL, academic administrator, geology educator; b. Glens Falls, N.Y., Oct. 26, 1940; d. James Croswell and Evangeline Pearl (Beougher) Perkins; m. Edward Sturgis Grew, June 14, 1975. BA magna cum laude, Bryn Mawr Coll., 1962; PhD, U. Calif., Berkeley, 1967. Instr. dept. geology Boston Coll., 1967-68, asst. prof., 1968-72; asst. research geologist Inst. Geophysics UCLA, 1972-77, adj. asst. prof. environ. sci. and engring., 1975-76; dir. Calif. Dept. Conservation, 1977-81; commr. Calif. Pub. Utilities Commn., San Francisco, 1981-86; dir. Minn. Geol. Survey, St. Paul, 1986-93; prof. dept. geology U. Minn., Mpls., 1986-93; vice chancellor for rsch. U. Nebr., Lincoln, 1993—, prof. dept. geology, prof. conservation and survey divsn., 1993—, prof. conservation/survey divsn. Inst. Agr., 1993—; vis. asst. prof. dept. geology U. Calif., Davis, 1973-74; chmn. Calif. State Mining and Geology Bd., Sacramento, 1976-77; exec. sec., editor Lake Powell Research Project, 1971-77; cons., vis. staff mem. Los Alamos (N.Mex.) Nat. Lab., 1972-77; mem. com. minority participation in earth sci. and mineral engring. U.S. Dept. Interior, 1972-75; chmn. Calif. Geothermal Resources Task Force, 1977, Calif. Geothermal Resources Bd., 1977-81; mem. earthquake studies adv. panel U.S. Geol. Survey, 1979-83; mem. adv. com. U.S. Geol. Survey, 1982-86; mem. adv. council Gas Research Inst., 1982-86, research coordination coun., 1987—; mem. bd. on global change rsch. NRC, 1992—, subcom. earthquake rsch. NRC, 1985-88, bd. on mineral and energy resources Nat. Acad. Scis., 1982-88, Minn. Minerals Coordinating Com., 1986-93, bd. on earth scis. and resources NRC, 1988-91, adv. bd. earth scis. Stanford U., 1989—. Contbr. articles to profl. jours. Fellow NSF, 1962-66. Fellow AAAS, (chmn. electorate nominating com. sect. E 1980-84, mem. at large 1987-91), Geol. Soc. Am. (chmn. com. on geology and pub. policy 1981-84, com. on coms. 1986-87, 91-92, councilor 1987-91), Mineral. Soc. Am., Geol. Assn. Can.; mem. Am. Geophys. Union (chmn. com. pub. affairs 1984-89), Soc. Mayflower Descs., Nat. Parks and Conservation Assn. (trustee 1982-86), Nat. Assn. Regulatory Utility Commrs. (com. on gas 1982-86, exec. com. 1984-86, com. on energy conservation 1983-84), U.S. Nat. Com. Geology (mem. at large 1985—), NSF (com. equal opportunities in sci. and tech. 1985-86, adv. com. on earth scis. 1987-91, adv. com. on sci. and tech. ctrs. devel. 1987-91), Cosmos Club. Office: U Nebr Vice Chancellor for Rsch 302 Adminstrn Bldg Lincoln NE 68588-0433

GREW, RAYMOND EDWARD, mechanical engineer; b. Metamora, Ohio, Jan. 11, 1923; s. Edward F. and Coletta (Crassel) G.; children: Elizabeth, Mary, Janet, John. BS in Mech. Engring., U. Mich., 1948. Registered profl. engr., N.J., Calif. Prin. engr. Hoffmann La Roche, Nutley, N.J., 1957-83. Mem. ASHRAE, Am. Assn. Energy Engrs. (specialist in rsch. lab. design) , Am. Soc. Profl. Engrs., English Speaking Union, Pilgrims of U.S. Achievements include patent for chromatographic device. Home: 28124 Hamden Ln Escondido CA 92026

GREY, ROBERT DEAN, biology educator; b. Liberal, Kans., Sept. 5, 1939; s. McHenry Wesley and Kathryn (Brown) G.; m. Alice Kathleen Archer, June 11, 1961; children: Erin Kathleen, Joel Michael. BA, Phillips U., 1961; PhD, Washington U., 1966. Asst. prof. Washington U., St. Louis, 1966-67; from asst. prof. to full prof. zoology U. Calif., Davis, 1967—, chmn. dept., 1979-83, dean biol. scis., 1985—, interim exec. vice chancellor, 1993—. Author: (with others) A Laboratory Text for Developmental Biology, 1980; contbr. articles to profl. jours. Recipient Disting. Teaching award Acad. Senate U. Calif., Davis, 1977, Magnar Ronning award for teaching Associated Students U. Calif., Davis, 1978. Mem. Am. Soc. Cell Biology, Soc. Developmental Biology, Phi Sigma. Lodge: Rotary. Avocations: music, hiking, gardening. Office: U Calif Davis Div of Biol Scis Davis CA 95616

GREYWALL, MAHESH INDER-SINGH, mechanical engineer; b. Patiala, India, Oct. 15, 1934; came to U.S., 1954; s. Harichand Singh and Sahib (Kaur) G.; m. Hermine Fischer, Apr., 1960; children: Shaun, Paul. BS, U. Calif., Berkeley, 1957, MS, 1959, PhD, 1962. Tech. staff Aerospace Corp., L.A., 1963-65; physicist Lawrence Radiation Lab., Livermore, Calif., 1965-69; prof. Wichita (Kans.) State U., 1969—. Home: 2707 Rushwood Ct Wichita KS 67226 Office: Wichita State U Dept Mechanical Engring Wichita KS 67208

GRIBOV, VLADIMIR N., physicist. Senior scientist Landau Inst. Theoretical Physics, Moscow. Recipient J.J. Sakurai prize Am. Phys. Soc., 1991. Office: Landau Inst Theoretical Physics, UL A N Kosyeina 2, 117940 Moscow V-234, Russia*

GRIEGER, GÜNTER, physicist; b. Berlin, Feb. 26, 1931; s. Kurt Willy Heinrich and Marie-Elisabeth Schmidt; m. Ursula Erika Dreissig, Apr. 21, 1960; 1 child, Martina. Degree, Free U. Berlin, 1954; PhD, Ludwig-Maximilian U., Munich, 1959. Rsch. physicist inst. physics and astrophysics Max-Planck-Inst., Munich, 1959-64; group leader Max-Planck-Inst. for Plasmaphysics, Garching, Fed. Republic Germany, 1965-70; div. head Max-Planck-Inst., Garching, Fed. Republic Germany, 1971—; mem. numerous sci. orgns. Contbr. articles to profl. publs. Mem. European Phys. Soc. (officer 1985), German Phys. Soc., Am. Phys. Soc., German Nuclear Soc. Home: Am Mühlbach 28, D-85748 Garching Germany Office: Max-Planck Inst Plasmaphys, Boltzmannstr 2, D-85748 Garching Germany

GRIEM, HANS RUDOLF, physicist, educator; b. Kiel, Schleswig-Holstein, Fed. Republic Germany, Oct. 7, 1928; came to U.S., 1954; s. Rudolf H. and Paula D. (Schwarz) G.; m. Irmgard H. Hôhling, May 11, 1957; children: Jens, Torsten, Rowena, Bridget. Abitur, Max-Planck Sch. Kiel, 1949; PhD, U. Kiel, 1954; PhD (hon.), Ruhr U., Bochum, Fed. Republic Germany, 1990. Rsch. asst. U. Md., College Park, 1954-55, asst. prof., 1957-61, assoc. prof., 1961-63, prof., 1963—; Wissenschaftlicher asst. U. Kiel, 1955-57; dir. Lab. for Plasma Rsch. U. Md., 1980-87; cons. Naval Rsch. Lab., Washington, 1957—, Lawrence Livermore (Calif.) Nat. Lab., 1979—, Los Alamos (N.Mex.) Nat. Lab., 1980—. Author: Plasma Spectroscopy, 1964, Spectral Line Broadening by Plasmas, 1974, editor: Methods of Experimental Physics, Vol. 9A, 1970; contbr. articles to profl. jours. and chpts. to books. NSF sr. postdoctoral fellow, 1963; Guggenheim Found. fellow, 1968; European Space Rsch. Orgn. fellow, 1971; recipient Humboldt prize, 1978, William F. Meggers award Optical Soc. Am., 1987. Fellow Am. Phys. Soc. (councilor 1983-87, J.C. Maxwell prize 1991). Achievements include devel. of quantitative spectroscopic methods for high temperature plasma diagnostics. Office: Lab Plasma Rsch U Md College Park MD 20742-3511

GRIEMAN, FRED JOSEPH, chemist, educator; b. Long Beach, Calif., May 19, 1952; s. Fred J. and Virginia (Elwood) G.; m. Janet Marie Bloom, Aug. 28, 1976; children: Zachary Paul, Mackenzie Marie. BA, U. Calif., Irvine, 1974; PhD, U. Calif., Berkeley, 1979. Postdoctoral rsch. assoc. dept. physics and chemistry U. Oreg., Eugene, 1980-82, rsch. assoc., 1981-82; asst. prof. chemistry Pomona Coll., Claremont, Calif., 1982-88, assoc. prof., 1988-93, chmn. chemistry dept., 1993—; vis. scientist U. Paris-Sud, Orsay, France, 1980, Oxford (Eng.) U., 1988-89. Contbr. articles to Jour. Chem. Physics, other profl. publs. Mem. Am. Chem. Soc. (grantee 1983, 90, 92). Home: 524 W 10th St Claremont CA 91711 Office: Pomona Coll Seaver Chemistry Lab 645 N College Ave Claremont CA 91711

GRIER, NATHANIEL, chemist; b. N.Y.C., Mar. 27, 1918; s. Max and Esther Annie (Zucker) G.; m. Roslyn Levine, Dec. 18, 1941; children: Paul Carl, Eli, Diane. BS, L.I. U., 1937; MS, PhD, U. Mich., 1938, 43. Sr. chemist Hoffman LaRoche, Inc., Nutley, N.J., 1942-46; v.p., dir. rsch. Metalsalts Corp., Hawthorne, N.J., 1946-66; sr. investigator Merck, Sharp & Dohme Rsch. Labs. (div. Merck & Co. Inc.), Rahway, N.J., 1966-80. Contbg. author: Disinfection, Sterilization and Preservation, 3rd edit., 1983, The Ency. of the Chemical Elements, 1968, The Ency. of Chemistry, 1973; contbr. articles to profl. jours. Past bd. dirs. United Jewish Appeal, Englewood, N.J., Joint Distbn. Com., N.Y.C., 1974, Inci, Jewish Nat. Fund, Bergen County, N.J., 1981. F.S. Stearns fellow U. Mich., Ann Arbor, 1942. Mem. Am. Chem. Soc. (emeritus mem.), AAAS (emeritus fellow), Sigma Xi. Hebrew. Achievements include more than 50 U.S. patents; development of novel intestinal antiseptics, cholesterol-lowering polymers; first U.S. comml. prodn. of B-hydroxyquinoline from coal-tar quinoline, and widely used biocides. Home: 153 Morse Pl Englewood NJ 07631

GRIERSON, WILLIAM, retired agriculture educator, consultant; b. Boscombe, Eng., Dec. 15, 1917; came to U.S., 1952; s. Edward James and Winifred (Burridge) Grierson-Jackson; m. Agnes Cray; children: Peter Robert, John Patrick. BSc in Agr., Ont. Agrl. Coll., Guelph, Can., 1938; PhD, Cornell U., 1951. Asst. prof. U. B.C., Vancouver, Can., 1945-51; asst. prof. U. Fla., Lake Alfred, 1952-60, prof., 1964-82; assoc. dir. Food Industries Rsch. and Engring., Yakima, Wash., 1960-64; prof. emeritus, cons. Winter Haven, Fla., 1983—. Author, editor: (textbook) Fresh Citrus Fruits, 1986; author 4 manuals; contbr. over 200 articles to sci. jours. Maj. RCAF, 1940-45. Fla. Citrus Packers grad. fellow, 1982; named Researcher of Yr. Fla. Fruit and Vegetable Assn., 1972. Fellow Am. Soc. Hort. Sci. (assoc. editor 1970-74); mem. Fla. State Hort. Soc. (hon., pres. 1981-82, editor 1972-79, Gold medal 1969). Achievements include devel. of designs for citrus degreening rooms now used world wide, of methods for the marketing of Florida lemons; first identification of two physiological diseases of citrus ("zebraskin" and "sloughing"). Home: 18 Golf View Cir NE Winter Haven FL 33881-4302

GRIESÉ, JOHN WILLIAM, III, astronomer; b. Norwalk, Conn., Sept. 27, 1955; s. John William Jr. and Celia (Bolté) G. Diploma, Morse Sch. Bus., Hartford, 1986; student, U. Conn., 1991—. Asst. dir. Stamford (Conn.) Obs., 1978—; observer Van Vleck Obs., Middletown, Conn., 1986, asst. astrometry program, 1992—; user Perkin-Elmer PDS, New Haven, Conn., 1992—; lectr. Stamford Mus., 1985—; presenter in field; lectr. in field. Contbr. articles to Jour. Am. Assn. Variable Star Obs., Deep Sky Mag.; observations of variable stars pub. on circulars of Cen. Bur. for Astron. Telegrams, Internat. Astron. Union, Smithsonian Astrophys. Obs. Named Outstanding Young Man of Am., 1987. Mem. Am. Assn. Variable Star Observers (preliminary charts com., supernova search com.), Am. Astron. Soc., Am. Assn. Variable Star Obs. (coun. 1985-90, liaison and rep. to mems. in Hungary, contbr. Variable Star Atlas, edits. I and II), Royal Astron. Soc. Canada, Hungarian Astron. Assn., Astron. Soc. Pacific, Mt. Wilson Observatory Assn., Western Observatorium, Western Amateur Astronomers (pub. info. coord. 1989-90, Caroline Herschel Astronomy project award 1988, v.p., acting pres. 1992, v.p. 1992—), L.A. Astron. Soc., Fairfield County Astron. Soc. (treas. 1985-88, pres. 1988—), Astron. Soc. Greater Hartford (pres. 1992—), Aston. League (long range planning com. 1992—). Democrat. Home: 965 Elms Common Dr Rocky Hill CT 06067-1821

GRIESSER, JAMES ALBERT (JAMIE), computer scientist; b. Cleve., Nov. 25, 1952; s. Robert Charles and Doris Evelyn (Willberg) G.; m. Kathryn Ann Stevens, Apr. 20, 1974; children: Beth Ann, Laura. BS in Microbiology, Ohio U., 1974. Systems engr. IBM, Cleve., 1974-78; cons. IBM, Atlanta, 1978-82, mfg. systems designer, 1982-92; mfg. systems designer IBM/Marcam Alliance, Atlanta, 1992—; MAPICS Inc., Atlanta, Ga.; mem. Mapics Ops. Bd., Atlanta, 1986-90, cons., 1990—. Contbr. articles to profl. jours. Mem. Lindley Soc. of Ohio U. Achievements include design/development of IBM's capacity requirements planning module for the manufacturing accounting production information control system MAPICS/DB data base. Office: MAPICS Inc. 5775-D Glenridge Dr Ste 300 Atlanta GA 30328

GRIEVE, GRAHAM ROBERT, civil engineering executive; b. Durban, Natal, Republic of South Africa, Nov. 1, 1946; s. Colin Nicol Grieve and Anne Caroline (Petrie) Consani; m. Katharine Wyche Taylor, Mar. 9, 1974; children: Andrew James, Julian Stuart. M. Ing., Pretoria U., Republic of South Africa, 1984; PhD, U. Witwatersrand, Johannesburg, Republic of South Africa, 1991. Tech. asst. Rijkswatersstaat, Den Haag, the Netherlands, 1971-72; rsch. officer Coastal Rsch. Unit, CSIR, Stellenbosch, Republic of South Africa, 1972-74; resident engr. B.S. Bergman & Ptnrs., Tzaneen, Republic of South Africa, 1974-78; materials engr. B.S. Bergman & Ptnrs., Pretoria, Republic of South Africa, 1978-81; researcher Pretoria U., 1981; dir. lab. svcs. Portland Cement Inst., Midrand, Republic of South Africa, 1982-90, exec. dir., 1990—; chmn. adv. bd. Concrete Durability Rsch. Program, Johannesburg, 1992—. Mem. ASTM, South African Cement Prodrs. Assn. (chmn. tech. com. 1990—), South African Instn. Civil Engrs., Instn. Concrete Tech., Concrete Soc. (UK), Am. Concrete Inst., Concrete Soc. South Africa (nominated rep. from PCI group membership), RILEM (titular), South African Coal Ash Assn., South African Inst. Bldg. Home: Waterkloof Ridge, 344 Delphinus St, Pretoria 0181, South Africa Office: Portland Cement Inst, PO Box 168, Halfway House 1685, South Africa

GRIFF, IRENE CAROL, cell biology researcher; b. Phila., Aug. 31, 1965; d. Leonard and Lillian Lena (Golub) G. BS, MIT, 1987; MA, Princeton U., 1989, PhD in Molecular Biology, 1993. Vet. med. asst. Rau Animal Hosp., Glenside, Pa., 1979-83; lab. mgr. Wistar Inst., Phila., 1987; vis. researcher Sloan Kettering Inst., N.Y.C., 1991-93; biology and chemistry text recorder Recording for the Blind, Princeton, N.J., 1989-91. Contbr. articles to jour. Biol. Chemistry, Cell, Nature. Tutor N.Y. pub. schs., 1992. Mem. AAAS, MIT Alumni Princeton (asst. treas. 1990-91). Office: Sloan Kettering Inst 1275 York Ave PO Box 251 New York NY 10021

GRIFFIN, DEWITT JAMES, architect, real estate developer; b. Los Angeles, Aug. 26, 1914; s. DeWitt Clinton and Ada Gay (Miller) G.; m. Jeanmarie Donald, Aug. 19, 1940 (dec. Sept. 1985); children: Barbara Jean Griffin Holst, John Donald, Cornelia Caulfield Claudius, James DeWitt; m. Vivienne Dod Kievenaar, May 6, 1989. Student, UCLA, 1936-38; B.A., U. Calif., 1942. Designer Kaiser Engrs., Richmond, Calif., 1941; architect CF Braun & Co., Alhambra, Calif., 1946-48; pvt. practice architecture Pasadena, Calif., 1948-50; prin. Goudie & Griffin Architects, San Jose, Calif., 1959-64, Griffin & Murray, 1964-66, DeWitt J. Griffin & Assocs., 1966-69; pres. Griffin/Joyce Assocs., Architects, 1969-80; chmn. Griffin Balzhiser Affiliates (Architects), 1974-80; founder, pres. Griffin Cos. Internat., 1980—; founder, dir. San Jose Savs. and Loan Assn., 1965-75, Capitol Services Co., 1964-77, Esandel Corp., 1965-77. Pub. Sea Power mag, 1975-77; archtl. works include U.S. Post Office, San Jose, 1966, VA Hosp, Portland, 1976, Bn. Barracks Complex, Ft. Ord, Calif., 1978. bd. dirs. San Jose Symphony Assn., 1973-84, v.p. 1977-79, pres. 1979-81; active San Jose Symphony Found., 1981-86, v.p. 1988-90; bd. dirs. Coast Guard Acad. Found., 1974-87, Coast Guard Found., 1987-90; founder, bd. dirs. U.S. Navy Meml. Found., 1978-80, trustee, 1980—; trustee Montalvo Ctr. for Arts, 1982-88. Served to comdr. USNR, 1942-46, 50-57. Recipient Navy Meritorious Pub. Svc. medal, 1971, Disting. Service medal Navy League of U.S., 1973; Coast Guard Meritorious Pub. Svc. medal, 1975; Navy Disting. Pub. Svc. medal, 1977; Coast Guard Disting. Pub. Svcs. medal, 1977. Fellow Soc. Am. Mil. Engrs.; mem. AIA (emeritus), U.S. Naval Inst., Navy League U.S. (pres. Santa Clara Valley coun. 1963-66, Calif. state pres. 1966-69, nat. dir. 1967—, exec. com. 1968—, pres. 12th region 1969-71, nat. v.p. 1973-75, nat. pres. 1975-77, chmn. 1977-79), U.S. Naval Sailing Assn., Naval Order of U.S., Wash. Athletic Club (Seattle), Marin Yacht Club, St. Francis Yacht Club, Commonwealth of San Francisco Club, Phi Gamma Delta. Republican. Congregationalist. Home and Office: 8005 NE Hunt Club Ln Hansville WA 98340-0124

GRIFFIN, DONALD S., nuclear engineer, consultant. BME, Cornell U., 1952; MS in Engring. Mechanics, Stanford U., 1953, PhD in Engring. Mechanics, 1959. From sr. engr. to mgr. structural mechanics Bettis Atomic Lab. Westinghouse, Pitts., 1959-72, with advance energy systems divsn., 1974-91; ind. cons. Pitts., 1972-74, 91—; ad hoc visitor Accreditation Bd. Engring. and Tech., 1977-85. Assoc. editor Jour. Applied Mathematics, 1973-80; former mem. editorial bd. International Jour. Computers and Structures, Jour. Strucural Mechanics Software; contbr. articles, papers to profl. jours. Officer Civil Engring. Corps, USN, 1953-56. Recipient Literature award PVP, 1987. Fellow ASME (life, divsn. applied mechanics, chmn. divsn. pressure vessels adn piping, publs. com., mem. com. computer tech., com. computing in applied mechanics, op. group materials and structures, subcom. bioler and pressure vellel code, com. solar energy standards codes, policy bds. comm. and rsch., Pressure Vessel and Piping award 1992), Nat. Rsch. Coun. (computational math. com.), Welding Rsch. Coun. (pressure vessel rsch. com.). Achievements include research in design methods, design criteria and software for structural software analysis and computer operations for design of advanced e nergy systems. Home: 208 Oakcrest Lane Pittsburgh PA 15236

GRIFFIN, EDWIN H., JR. (HANK GRIFFIN), chemist; b. Hanover, N.H., Sept. 23, 1935; s. Edwin H. and Rosanne (Gore) G.; m. Joyce M. Krekich, Jan. 27, 1962; children: Linda Griffin Geraghty, Debra, Kelly Griffin Crawford, Edwin. BS in Chemistry, U. N.H., 1959; MS in Chemistry, Lehigh U., 1962. Chemist Gulf Rsch. Devel. Co., Pitts., 1962-66, Hunt Chem. Co., Lincoln, R.I., 1966-68; chemist Tex. Instruments, Attleboro, Mass., 1968-69, chemist, lab. supr., 1969-80, mgr. tech. svc. lab., 1980-86, sr. mem. tech. staff, 1986—; dir. Nat. Conf. on Spectrochem. Excitation and Analysis, Edgartown, Mass., 1981-85. Contbr. over 80 articles to profl. jours. Leader Boy Scouts Am., Troop 6, Barrington, R.I., 1983-85. Mem. Am. Chem. Soc., Soc. for Applied Spectroscopy, New Eng. Thermal Forum. Achievements include development of wavelength scanner, automatic sample introduction, and computer interface with software for a direct current plasma spectrometer. Home: 46 Bluff Rd Barrington RI 02806-4314 Office: Tex Instruments 34 Forest St MS10-16 Attleboro MA 02703

GRIFFIN, JOHN JOSEPH, JR., chemist, video producer; b. Chgo., Sept. 11, 1946; s. John Joseph, Sr. and Louise (Griswold) G.; m. Ramona Rodriguez, Apr. 19, 1969; 1 child, Marcus Allan. BS, Tex. A&M U., 1972, MS, 1974. Lab. technician Johns-Manville, Chgo., 1964-66; chemist, rsch. chemist Dow Chemical USA, Tex. Divsn., Freeport, 1974-78; sr. chemist Soltex Polymers, Deer Pk., Tex., 1978-80; plant chemist Air Products & Chemicals, Pasadena, Tex., 1980-84; quality assurance supr. Core Lab., Chromaspec Divsn., Houston, 1984-88; plant chemist and quality assurance supr. Ga. Gulf Corp., Pasadena, Tex., 1988—; propr. and owner JJ's Quality Custom Video, Houston, 1986-88; propr. and producer Pro-Star Video Productions, Houston, 1988—. pres. Kirkwood Civic Club, Houston, 1991-93; bd. dirs. Southbelt Security Alliance, Houston, 1988-93. Served in USAF, 1966-70. Mem. ASTM, Am. Chem. Soc. Roman Catholic. Office: Ga Gulf Corp 3503 Pasadena Fwy Pasadena TX 77503

GRIFFIS, FLETCHER HUGHES, civil engineering educator, engineering executive; b. Wauchula, Fla., Apr. 22, 1938; s. Fletcher Hughes and Eva (Murphy) G.; m. Nancy Inch, Oct. 16, 1960; children: Hugh, Greg. BS, U.S. Mil. Acad., 1960; MSCE, Okla. State U., 1965, PhD, 1970, MS in Indsl. Engring., 1971. Registered profl. engr., N.Y., Okla. Commd. 2d lt. U.S. Army Corps of Engrs., 1960, advance through grades to colonel; program mgr. Waterways Experiment Sta. U.S. Army Corps of Engrs., Vicksburg, Miss., 1972-76; comdr. 79th engring. bn. U.S. Army, Karlsruhe, Federal Republic of Germany, 1976-79; student U.S. Army War Coll., Carlisle, Pa., 1979-80; area engr. Ramon Air Base, Ramon Air Base, Israel, 1980-82; dist. engr. U.S. Army War Coll., N.Y.C., 1983-86; ret. U.S. Army War Coll., 1986—; prin. Robbins, Pope and Griffis, P.C., N.Y.C., 1989—; prof. Columbia U., N.Y.C., 1986—; cons. N.Y.C. Dept. Transp., 1987—, Tyger Constrn. Co., Spartansburg, S.C., 1986-88, Guy F. Atkinson, Inc., San Francisco, 1986-88. Contbr. papers and articles to profl. jours. and conf. procs. Dir. N.Y. Dist. Explorer, Boy Scouts Am., N.Y.C., 1990—. NSF grantee; recipient Bronze Star, 1968, Legion of Merit, Vinh Long, Viet Nam, 1969, Karlsruhe, 1979, Ramon, Israel, 1982, N.Y.C., 1986. Fellow ASCE (pres. Met. sect. 1988-89, chmn. profl. publs., exec. com. constrn. div., named Civil Engr. of Yr. 1993); mem. Soc. Am. Mil. Engrs. (past pres., dir., Gold medal 1985), Project Mgmt. Inst., Chi Epsilon, Sigma Tau, Sigma Xi. Republican. Home: 25 Claremont Ave # 3B New York NY 10027

GRIFFITH, B(EZALEEL) HEROLD, physician, educator, plastic surgeon; b. N.Y.C., Aug. 24, 1925; s. Bezaleel Davies and Henrietta (Herold) G.; m. Jeanne B. Lethbridge, 1948; children: Susan, Tristan. BA, Johns Hopkins U., 1992; M.D., Yale U., 1948. Diplomate: Am. Bd. Plastic Surgery (dir. 1976-82, chmn. 1981-82). Intern Grace New Haven Community Hosp.-Yale U., 1948-49; resident in surgery VA Hosp., Newington, Conn., 1949-50; asst. resident in surgery 2d (Cornell) Surg. Div., Bellevue Hosp., N.Y.C., 1952-53; resident in plastic surgery VA Hosp., Bronx, 1953-55, U. Glasgow, Scotland, 1955, N.Y. Hosp. Cornell Med. Center, N.Y.C., 1956; rsch. fellow in plastic surgery Cornell U. Med. Coll., 1956-57; pvt. practice specializing in plastic surgery Chgo., 1957—; attending plastic surgeon Northwestern Meml., Children's Meml., VA Lakeside hosps., Rehab. Inst. Chgo.; instr. surgery Northwestern U., 1957-59, assoc. in surgery, 1959-62, asst. prof. surgery, 1962-67, assoc. prof., 1967-71, prof., 1971—, chief div. plastic surgery, 1970-91. Assoc. editor: Plastic and Reconstructive Surgery, 1972-78; contbr.

articles to profl. jours. Served to lt., M.C. USNR, 1950-52. Fellow ACS, Am. Assn. Plastic Surgeons, Chgo. Surg. Soc., Royal Soc. Medicine; mem. AAAS, AMA, Am. Soc. Plastic and Reconstructive Surgeons (sec. 1972-74), Brit. Assn. Plastic Surgeons, Plastic Surgery Research Council (chmn. 1969), Am. Cleft Palate Assn., N.Y. Acad. Scis., Ill., Chgo. med. socs., Midwestern Assn. Plastic Surgeons, Soc. Head and Neck Surgeons, Ill., Chgo. hist. socs., Civil War Round Table, Evanston Hist. Soc. (trustee 1974-78), Sigma Xi (pres. Northwestern U. 1986-87). Club: Yale (Chgo.). Lodge: Masons. Achievements include research in transplantation, skin tumors, cleft palate, paraplegia. Office: 251 E Chicago Ave Chicago IL 60611-2614

GRIFFITH, CARL DAVID, civil engineer; b. Hill City, Kans., Mar. 1, 1937; s. Wilfred Eugene and Veda May (Jackson) G.; m. Mariana Segall, Mar. 26, 1988; stepchildren: Laurie Ann Segall, Allen Segall. BSCE summa cum laude, West Coast U., 1978; MSCE in Water Resources, U. So. Calif., 1980, MS in Engring. Mgmt., 1983. Profl. engr., Calif. Chief draftsman Bear Creek Mining Co., Spokane, Wash., 1959-64; right-of-way technician So. Calif. Edison Co., Los Angeles, 1964-65; engr. treatment plant design of spl. projects br. Metropolitan Water Dist. So. Calif., Los Angeles, 1965—, com. chmn. employees assn.; assoc. prof. Sch. Engring., West Coast U. Sustaining mem. Calif. Republican party. Served with USAF, 1957-58. Mem. ASME, ASCE, NSPE, Am. Water Works Assn., Nat. Mgmt. Assn., Metropolitan Water Dist. Mgmt. Club. Lodge: Masons. Home: PO Box 923122 Sylmar CA 91392-3122 Office: PO Box 54153 Los Angeles CA 90054

GRIFFITH, CARL DEAN, electronics engineer; b. Mammoth Springs, Ark., Sept. 13, 1935; s. Carl Dan and Doris Maxine (Dubois) G.; m. Anna Marilyn Dupre, July 4, 1959; children: John Carl, Curtis Wayne, Clark Daniel. BSEE, U. Mo., Rolla, 1958; MSEE, Tex. Tech U., 1961. Staff engr. Sperry Flight Systems, Phoenix, 1961-76; engring. sect. head Sperry Avionics Div., Glendale, Ariz., 1976-81, engring. dept. head, 1981-83; engring. dept. head Sperry Def. Systems Div., Albuquerque, 1983-87; sr. staff engr. Honeywell Def. Avionics, Albuquerque, 1987-88; staff engr. Bell Helicopter Textron, Ft. Worth, 1988-90; dir. design engring. Am. Eurocopter, Grand Prairie, Tex., 1990—; mem. shuttle cert. assessment staff NASA, L.A., 1980. Contbr. articles to profl. jours. Mem. chorus Phoenix Opera, 1978-82, N.Mex. Symphony Orch., Albuquerque, 1983-88. Mem. Am. Helicopter Soc. Mem. Ch. of Christ. Office: Am Eurocopter Corp 2701 Forum Dr Grand Prairie TX 75051-7099

GRIFFITH, CHARLES RUSSELL, nuclear operations consultant; b. Pontiac, Mich., Sept. 12, 1953; s. Barclay Caldwell and Flora (Harris) G.; m. Patricia Manuel, Mar. 23, 1985; children: Eric Tyler, Brian Michael. Grad. high sch., Russellville, Ky. Lic. sr. reactor operator Nuclear Regulatory Commn. Reactor operator, nuclear watch engr. Fla. Power and Light, Ft. Pierce, 1980-83; start-up test engr. So. Calif. edison, San Clemente, 1983-85; ops. tng. qualifier Ga. Power Co., Waynesboro, 1985-89; ops. shift adviser Savannah River Site, Aiken, S.C., 1989—. With USN, 1974-80. Republican. Methodist.

GRIFFITH, JERRY DICE, government official, nuclear engineer; b. Sturgis, Mich., Sept. 8, 1933; s. Levi Robert and Vivian Marie (LeVeck) G.; m. Gloria Louise Hessie, June 25, 1965; children—Jennifer Lynn, Bradley Jerome. BS summa cum laude, Mich. State U., 1955, MS, 1957; ME, Calif. Inst. Tech., 1959; PFPA, Princeton U., 1967. Dir. nuclear safety C.E., U.S. Army, Washington, 1967-72; chief research and devel. br. AEC and ERDA, Washington, 1972-76; asst. dir. for reactor safety Dept. Energy, Washington, 1976-79, dir. div. nuclear power devel., 1979-80, dir. office light water reactors, 1980-85, assoc. dept. asst. sec. reactor systems devel. and tech., 1985—, acting asst. sec. for nuclear energy, 1989, acting prin. dept. asst. sec. for nuclear energy, 1990-92; U.S. rep. to OECD Nuclear Energy Agy., Paris, 1976-86, 89—. Contbr. articles to profl. jours., 1967—; patentee inherent reactor control concept, small reaction turbine. Served to capt. U.S. Army, 1959-62. Recipient Meritorious Civilian Service award U.S. Army, 1970; Congl. fellow, 1969. Mem. Am. Nuclear Soc. Home: 14711 Bauer Dr Rockville MD 20853-3621 Office: Dept Energy NE 40 Washington DC 20545

GRIFFITH, PATRICK THEODORE, systems engineer; b. Pierre, S.D., Apr. 30, 1962; s. Charles James and Patricia Louise (Raymond) G.; m. Barbara Mary Rokoski, June 15, 1985. AS in Refrigeration, Air Conditioning, Dunwoody Indsl. Inst., Mpls., 1982. Rsch. technician McQuay Inc., Plymouth, Minn., 1982; product devel. technician Litton Microwave Products Inc., Plymouth, Minn., 1982-84; engring. technician Thermo-King Inc., Bloomington, Minn., 1984-85, Honeywell Residential Divsn., Golden Valley, Minn., 1985-89; application engr. Johnson Controls Inc., Mpls., 1989-92, Protective Systems Group, Mlps., 1992—. Home: 8171 Lawndale Ln N Maple Grove MN 55311-1719

GRIFFITH, ROBERT CHARLES, allergist, educator; b. Shreveport, La., Jan. 9, 1939; s. Charles Parsons and Madelon (Jenkins) G.; m. Loretta Dean Secrist, July 15, 1969; children: Charles Randall, Cameron Stuart, Ann Marie. BS, Centenary Coll., 1961; MD, La. State U., 1965. Intern, Confederate Meml. Med. Ctr., Shreveport, 1965-66, resident in internal medicine, 1966-68; fellow in allergy and chest disease, instr. Va. Med. Sch. Hosp., Charlottesville, 1968-70; practice medicine specializing in allergies, Alexandria, La., 1970-72, The Allergy Clinic, Shreveport, 1972; pres. Griffith Allergy Clinic, Shreveport, 1973—; faculty internal medicine La. State U., 1972—. Bd. dirs. Caddo-Bossier Assn. Retarded Citizens, 1977-84, Access (fomerly Child Devel. Ctr.), Shreveport, 1979-85; mem. med. adv. com., spl. edn. adv. com. Caddo Parish Sch. Bd., 1977—; mem. commission on missions and social concerns First Methodist Ch., 1981-84, mem. adminstrv. bd., 1981-84; mem. med. panel for transfer Caddo Parish Sch. Bd., 1974—. Served to maj. M.C., U.S. Army, 1965-71. Recipient Physician of the Yr. award Shreveport-Bossier Med. Assts., 1984. Fellow Am. Coll. Allergy and Immunology, Am. Coll. Chest Physicians (assoc.), Am. Thoracic Soc.; mem. AMA, SAR, Am. Acad. Allergy, Internat. Platform Assn., Jamestowne Soc., So. Med. Assn., La. Med. Soc., Shreveport Med. Soc. (allergy sposkesman 1984—), La. Allergy Soc. (charter; past pres.), U. Va. Med. Alumni Assn. (life), Pace Soc. Am., La. State U. Med. Alumni Assn., SCV, Mil. Order Stars and Bars, Order of So. Cross and Pub. Solicitation Review Coun., Shreveport C. of C., Kappa Alpha, Methodist. Lodges: Masons (32 degree), Shriners. Clubs: Shreveport Country, Petroleum of Shreveport, Shreveport, Ambs., Cotillion, Royal, Plantation, Jesters, Les Bon Temps., The Order of the Southern Cross, Demoiselle Club. Home: 7112 E Ridge Dr Shreveport LA 71106-4749 Office: 2751 Virginia Ave Shreveport LA 71103-3940

GRIFFITH, STEVEN LEE, research physiologist; b. Prosser, Wash., Apr. 1, 1960; s. Donald Robert and Joyce Elvirta (Polman) G.; m. Debra Lynne Schantz, Aug. 28, 1982; children: Amanda Faith, Samantha Joy. BS, Grace Coll., 1982; MSc, Ball State U., 1986; PhD, Ind. U., Indpls., 1991. Asst. ultrasound researcher Indpls. Ctr. Advanced Rsch., 1984-86, rsch. assoc., 1986-91, asst. dir. life scis., 1990-91; clin. rsch. assoc. St. Francis Hosp. Ctr., Beech Grove, Ind., 1989-90; dir. rsch. Tex. Back Inst. Rsch. Found., Plano, 1991—. Contbr. articles to profl. jours. Paul A. Nicoll fellow Ind. affiliate Am. Heart Assn., 1989. Mem. AAAS, Am. Coll. Sports Medicine, Am. Inst. Ultrasound in Medicine, Am. Physiol. Soc., Am. Soc. Bone and Mineral Rsch. Republican. Achievements include patent for one-punch catheter, patent for tissue ablation system using ultrasound.

GRIFFITHS, PHILLIP A., mathematician, academic administrator; b. Raleigh, N.C., Oct. 18, 1938. BS, Wake Forest U., 1959; PhD, Princeton U., 1963; D Honoris Causa, Angers U., France, 1979; Hon. degree, Wake Forest U., 1973, U. Peking, Peoples Republic China, 1983. Prof. math. Princeton U., N.J., 1968-72; prof. math. Harvard U., 1972-83, Dwight Parker Robinson prof. math., 1983; provost, James B. Duke prof. math. Duke U., Durham, N.C., 1983-91; dir. Inst. for Advanced Study, Princeton, N.J., 1991—; mem. faculty U. Calif. Berkeley, 1964-67; vis. prof. Princeton (N.J.) U., 1967-68; mem. Inst. Advanced Study, 1968-70; former chmn. bd. on math scis., NRC, chmn. Commn. on Phys. Scis., Math. and Resources; chair def. conversion and community assistance program State of N.J.; active Nat. Sci. Bd. Editor: Jour. Differential Geometry, 1980-90, Composito Mathematica, 1980-92, Duke Mathematical Jour. Bd. trustees Woodward Acad., N.C. Sch. Sci. and Math. Recipient LeRoy P. Steel prize Am. Math. Soc., 1971, Dannie Heineman Preis, Acad. Scis. Gottingen, 1979; Miller fellow U. Calif. Berkeley, 1962-64, 1975-76, Guggenheim fellow, 1980-82. Mem. NAS (mem. Coord. Coun. Edn., chair sci., engring. and pub. policy), Am. Acad. Arts and Scis., Am. Philos. Soc., Nat. Sci. Bd. Office: Inst Advanced Study Office of Dir Olden Ln Princeton NJ 08540-4920

GRIFFITHS, ROBERT BUDINGTON, physics educator; b. Etah, India, Feb. 25, 1937; s. Walter Denison and Margaret (Hamilton) H. A.B., Princeton U., 1957; M.S., Stanford U., 1958, Ph.D., 1962. Postdoctoral fellow U. Calif. at San Diego, 1962-64; asst. prof. Carnegie-Mellon U., Pitts., 1964-67; assoc. prof. Carnegie-Mellon U., 1967-69, prof. physics, 1969—, Otto Stern prof., 1979—. NSF postdoctoral fellow, 1962-64; Alfred P. Sloan research fellow, 1966-68; J.S. Guggenheim fellow, 1973; recipient Sr. Scientist award Humboldt Found., 1973, A. Cressy Morrison award Acad. Scis., N.Y., 1981, Dannie Heineman prize for math. physics, 1984. Mem. Am. Phys. Soc., Am. Sci. Affiliation, U.S. Nat. Acad. Scis., Phi Beta Kappa, Sigma Xi. Presbyterian. Achievements include research in statistical and quantum mechanics. Office: Carnegie-Mellon U Physics Dept Pittsburgh PA 15213

GRIFFITTS, JAMES JOHN, physician; b. Springfield, Ill., Dec. 13, 1912; s. Thomas Houston and Elizabeth (Glynn) G.; m. Leola Horton, June 13, 1940 (dec. 1985); children: Susan, Sally, Sharon, Shelley. BS, U. Va., 1933, MD, 1937. Intern Univ. Hosp., Cleve., 1937-40; dir. Blood Bank of Dade County, Miami, Fla., 1949-54; pres. Dade Reagents, Miami, 1954-71; dir. Am. Hosp. Corp., Evanston, Ill., 1956-71; ret. Author: Call Me A Doctor, 1990; author books on immunology, 1940-49, books on blood transfusion, 1950-60. With USPHS, 1940-49. Mem. Am. Assn. Blood Banks (pres. 1955-56, John Elliott award 1960). Episcopalian.

GRIGGER, DAVID JOHN, chemical engineer; b. Cleve., May 5, 1960; m. Michael Steven Grigger. B Chem. Engring., Cleve. State U., 1982; MS in Chem. Engring., Ohio State U., 1983; MBA, John Carroll U., 1993. Assoc. mem. tech. staff The RCA Corp., Findlay, Ohio, 1983-87; chem. engr. Life Systems Inc., Cleve., 1987—. Mem. AICE, Internat. Assn. Hydrogen Energy, Am. Soc. Metals Internat., Soc. Automotive Engrs. Internat. Achievements include patents in field.

GRIGGS, GARY BRUCE, earth sciences educator, oceanographer, geologist, consultant; b. Pasadena, Calif., Sept. 25, 1943; s. Dean Brayton and Barbara Jayne (Farmer) G.; m. Venetia Bradfield, Jan. 11, 1980; children: Joel, Amy, Shannon, Callie, Cody. BA in Geology, U. Calif., Santa Barbara, 1965; PhD in Oceanography, Oreg. State U., 1968. Registered geologist, Calif.; cert. engr. geologist. Fulbright fellow Inst. for Ocean & Fishing Rsch., Athens, Greece, 1974-75; oceanographer Joint U.S.A.-N.Z. Rsch. Program, Calif., 1980-81; prof. U. Calif., Santa Cruz, 1968—, chair earth scis., 1981-84; dir. Inst. of Marine Scis., 1991—. Author: (with others) Geologic Hazards, Resources & Environmental Planning, 1983, Living With The California Coast, 1985, Coastal Protection Structures, 1986, California's Coastal Hazards, 1992; editor Jour. of Coastal Rsch. Fellow Geol. Soc. Am.; mem. Am. Geophys. Union, Am. Geol. Inst., Coastal Found. Office: U Calif Div Natural Scis Applied Sciences Rm 269 Santa Cruz CA 95064

GRILLO, MARIA ANGELICA, biochemist; b. Rovereto, Trento, Italy, May 11, 1928; d. Virginio and Giuseppina (Chimelli) G. PhD in Chemistry, U. Torino, 1952. Asst. prof. biochemistry U. Torino, Italy, 1952-59, 61-68, prof. of biochemistry, 1975—; rsch. assoc. U. Ill., 1959-61; prof. biochemistry U. Sassari, Italy, 1968-75, U. Torino, Italy, 1975—. Contbr. articles on biochemistry to profl. jours. Office: Via Michelangelo 27, 10126 Torino Italy

GRIMES, CRAIG ALAN, research scientist; b. Ann Arbor, Mich., Nov. 6, 1956; s. Dale Mills and Janet LaVonne (Moore) G.; m. Jean Cardenas, Aug. 18, 1984. BS in Physics, Pa. State U., 1984, BSEE, 1984; MS, U. Tex., 1985, PhD, 1990. Engr. Applied Rsch. Labs., Austin, Tex., 1981-83; chief scientist Crale, Inc., Austin, 1985-90; rsch. scientist Lockeed Rsch. Labs., Palo Alto, Calif., 1990-92; dir. advanced materials lab. Southwall Techs., Palo Alto, Calif., 1992—; rsch. asst. U. Tex., Austin, 1985-88; teaching asst., 1987-90; ; cons. Eastman Kodak, San Diego, 1989, Storage Tech., Boulder, Colo., 1989. Co-author: Essays on the Formal Aspects of E&M Theory, 1992; contbr. articles to profl. jours. Active Nature Conservancy, New Braunfels, Tex., 1988-90, Austin Triathletes, 1987-90. Mem. AAAS, IEEE, Mountain View Masters. Achievements include 1 patent, 2 patents pending in field; development and manufacture of permeameters, magnetic measurement tools for high frequency permeability measurements; development of size independent antennae. Home: 736 LoLa Ln Mountain View CA 94040

GRIMES, JAMES GORDON, geologist; b. Kenosha, Wis., Mar. 18, 1951; s. James Gordon Bennett Jr. and Alyce Louise (Gannaway) G. BS in Earth Sci., U. Wis., Parkside, 1974; MS in Geology, Mich. Tech. U., 1977. Geologist nat. uranium resource evaluation project Union Carbide Corp. Nuclear Div., Oak Ridge, Tenn., 1977-84; geologist Martin Marietta Energy Systems Inc., Oak Ridge, Tenn., 1984—; geol. cons. UCC-ND Mercury Task Force, Oak Ridge, 1983. Mem. AAAS, Am. Statis. Assn., Am. Meteorol. Soc., Am. Mgmt. Assn., Am. Water Resources Assn., Nat. Weather Assn., Geol. Soc. Am., Computer Oriented Geol. Soc., Air and Waste Mgmt. Assn. Achievements include technical management of Y-2 plant mateorological information support system (including emergency atmospheric dispersion modeling). Office: Martin Marietta Energy Systems Inc PO Box 2009 MS 8219 Oak Ridge TN 37831-8219

GRIMES, RICHARD ALLEN, economics educator; b. Toledo, Ohio, Apr. 24, 1929; s. Robert Howell and Mary Mildred (Hatcher) G.; m. Helen Ann Schaeffer, Aug. 25, 1951; children: Gregory Allen, Julianne, Frank Edwin, Mary Ann. BS in Chemistry, U. Ga., 1951; MS in Indsl. Mgmt., Ga. Inst. Tech., 1959; postgrad., Ga. State U., 1979. Commd. lt. U.S. Army, 1951, advanced through grades to lt. col., ret. 1971; asst. prof. econs. Clayton State Coll., Morrow, Ga., 1971-74; assoc. prof. econs. DeKalb Coll., Decatur, Ga., 1974—; adj. prof. Jacksonville State U., 1959-63, Va. Commonwealth U., 1964-67; ednl. cons.; real estate broker. Reviewer: (textbook) Economics, 1979, 93. Umpire Atlanta Area Football Ofcls. Assn., treas. 1971—; active Spl. Olympics, Atlanta, 1971—; founding pres. Rex Civic Assn., 1973. Decorated Solider's medal for valor, Vietnam, 1963; named Rotarian of Yr. 1976, Football Ofcl. of Yr., Atlanta area, 1980. Mem. soc. Econ. Assn., Am. Acctg. Assn., AAUP (pres. DeKalb chpt. 1987-93), Ga. Assn. Econs. and Fin. (pres. 1992-93), Ga. Assn. Acctg. Profs. (past pres.), Nat. Soc. Pub. Accts., Delta Pi Epsilon, So. Metro Ga. Tech Alumni Club, Atlanta UGA Alumni Club (scholarship chmn.). Republican. Presbyterian (elder). Avocations: football, golfing, camping, swimming. Home: Eagles Landing 118 Carron Ln Stockbridge GA 30281 Office: DeKalb Coll 555 N Indiana Creek Dr Clarkston GA 30021-2396

GRIMM, CURT DAVID, anthropologist; b. St. Louis, Sept. 15, 1957; s. David C. and Lucille (Davis) G.; m. Tamara Lynn Bray, May 26, 1990. BA, U. N.H., 1979; MA, SUNY, Binghamton, 1985, PhD, 1991. Vol. Peace Corps, Burkina Faso (formerly Upper Volta), West Africa, 1979-81; sr. researcher Inst. for Devel. Anthropology, Binghamton, 1982-92; AAAS Sci. and Diplomacy Fellow USAID, Washington, 1992—; long-term researcher Manantali Resettlement project, USAID, Mali, West Africa, 1985-87, project dir. Tunisia Rural Water User Assn., USAID, N. Africa, 1990-92. Contbr. articles to profl. jours. Organizer Jamie Malley Election Campaign, Broome County, N.Y., 1990. Mem. Am. Anthrop. Assn., Washington Assn. Practicing Anthropologists. Office: USAID AFR/DP/PSE Rm 2495 NS Washington DC 20523

GRIMMEISS, HERMANN GEORG, physics educator, researcher; b. Hamburg, Germany, Aug. 19, 1930; arrived in Sweden, 1966; s. Georg and Franziska (Marz) G.; m. Hildegard Maria Weizmann, Nov. 17, 1956; children: Bernd, Daniela. PhD in Phys. Chemistry, U. Munich, 1957. Researcher Philips Rsch. Lab., Aachen, Fed. Republic of Germany, 1957-65; prof. solid state physics U. Lund, Sweden, 1966-73, 74-81, 83—; prof. physics U. Frankfurt, Fed. Republic of Germany, 1973-74; dir. Inst. Semiconductor Physics, Frankfurt/Oder, Germany, 1991-93; cons. Philips Rsch. Lab., Eindhoven, Holland, 1966-70, ASEA, Vasteras, Sweden, 1981-82, Rifa, Stockholm, 1984-86. Contbr. articles to profl. jours. Fellow Am. Phys. Soc.; mem. Royal Swedish Acad. Engring. Scis., Royal Swedish Acad. Scis., Societas Scintarum Sennica of Finland, Royal Physiographic Soc. of Lunds, N.Y. Acad. Scis. Roman Catholic. Avocation: tennis. Home: Malsmansvagen 5, 22367 Lund Sweden Office: U Lund Solid State Physics, Box 118, 22100 Lund Sweden

GRIMSBO, RAYMOND ALLEN, forensic scientist; b. Portland, Oreg., Apr. 25, 1948; s. LeRoy Allen and Irene Bernice (Surgen) G.; m. Barbara Suzanne Favreau, Apr. 26, 1969 (div. 1979); children: John Allen, Kimberly Suzanne; m. Charlotte Alice Miller, July 25, 1981; children: Sarah Marie, Benjamin Allen. BS, Portland State U., 1972; D of Philosophy, Union for Experimenting Colls. & Univs., Cin., 1987. Cert. of profl. competency in criminalistics DEA Researcher Registration. Med. technician United Med. Labs., Inc., Portland, 1969-74; criminalist Oreg. State Police Crime Lab., Portland, 1975-85; pvt. practice forensic science Portland, 1985-87; pres. Intermountain Forensic Labs., Inc., Portland, 1987—; adj. instr. Oreg. Health Scis. U., Portland, 1987—; adj. prof. Portland State U., 1986-88, adj. asst. prof., 1988—; clin. dir. Intermountain Forensic Labs., Inc., 1988-92, Western Health Lab., Portland; adj. faculty mem. Union Inst.; mem. substance abuse methods panel Oreg. Health Div. Contbr. articles to profl. jours. Fellow Royal Microscopical Soc.; mem. STM, Am. Acad. Forensic Scientist, Soc. Forensic Haemogenetics, N.W. Assn. Forensic Scientists, Internat. Assn. Bloodstain Pattern Analysts, Internat. Electrophoresis Soc., Internat. Assn. Identification, Internat. Assn. of Forensic Toxicologists, Pacific N.W. Forensic Study Club, New Horizons Investment Club. Avocations: gardening, camping, photography, study of ritualistic crime. Home: 16936 NE Davis St Portland OR 97230-6239 Office: Intermountain Forensic Labs Inc 11715 NE Glisan St Portland OR 97220-2141

GRINDALL, EMERSON JON, civil engineer; b. Jackson, Mich., May 13, 1938; s. Emerson Leroy and Eva Jane (Stone) G.; divorced; children: Timothy Jon, Lara Jean Brilla; m. Helen Marie Close, Dec. 12, 1987; 1 child, Theresa Ann Long. BSCE, Pa. State U., 1964, M Engring., 1976. Registered profl. engr., land surveyor, Pa. Field engr. Balt. Contractors, Inc., 1964-66; chief dam safety sect. Pa. Fish and Boat Commn., Bellefonte, Pa., 1966—. Contbr. articles to profl. publs. With U.S. Army, 1958-60. Mem. ASCE, Assn. State Dam Safety Ofcls. (assoc.). Mem. Ch. of the Brethren. Home: 721 E 15th St Tyrone PA 16686-2008 Office: Pa Fish and Boat Commn 450 Robinson Ln Bellefonte PA 16823

GRINDEA, DANIEL, international economist; b. Galatz, Romania, Feb. 23, 1924; came to U.S., 1975; s. Samy and Liza (Kaufman) Grünberg; M.Econs. and M.Law, Inst. Econ. Scis. and Faculty of Law, Bucharest, Romania, 1948; Ph.D. in Econs., Inst. Fin. and Planning, St. Petersburg, Russia, 1953; m. Lidia Bunaciu; 1 child, Sorin. Asso. prof. econs. various univs. in Bucharest, Romania, 1953-69; full prof. econs., 1969-75; cons. State Planning Com., 1953-56, Ministry of Fin., 1956-68; mem. Sci. Council of the Cen. Statis. Office, 1956-68; internat. economist Republic Nat. Bank of N.Y., N.Y.C., 1976-78, sr. internat. economist, dept. head, 1978-79, v.p., sr. internat. economist, 1979-84, sr. v.p., chief economist, 1984-89, sr. cons., 1990—; pres. Romanian-Am. C. of C., N.Y.C., 1990-92; sr. advisor U.S. Congl. Adv. Bd., 1988; prof., elected mem. sci. coun. L'Ecole Superioure des Sciences Commercieles d'Angers, France, 1989; mem. econ. adv. bd. Inst. Internat. Fin., Washington, 1988; invited vis. prof. l'Institut International de la Planification de l'Education (UNESCO), Paris, 1973; mem. adv. group Com. on Asian Econ. Studies, 1983. Recipient first prize in econ. research Ministry of Edn., Romania, 1969. Mem. Nat. Assn. Bus. Economists. Achievements include correcting predictions on world economy and individual countries; special research regarding the transition period to a free market economy in the ex-communist European countries; contbr. articles on forecasts in field U.S. and internat. to publs.; papers presented to profl. confs. U.S., France, Sweden, Ireland, Bulgaria, Romania. Office: Republic Nat Bank of NY 452 Fifth Ave New York NY 10018-2706

GRINDLAY, JONATHAN ELLIS, astrophysics educator; b. Richmond, Va., Nov. 9, 1944; s. John Happer and Elizabeth (Ellis) G.; m. Sandra Kay Smyrski, Oct. 10, 1970; children: Graham Charles, Kathryn Jane. A.B., Dartmouth Coll., 1966; M.A., Harvard U., 1969, Ph.D., 1971. Jr. fellow Harvard U., Cambridge, Mass., 1971-74, asst. prof., 1976-81, prof. astronomy, 1981—; chmn. dept. astronomy Harvard U., 1985-90; astrophysicist Smithsonian Obs., 1974-76; cons. MIT Lincoln Lab., Bedford, Mass., 1982—; mem. vis. com. astronomy U. Chgo., 1983, Astrophys. Lab. Saclay, France, 1988—; mem. ONR Panel on Rsch. in Astronomy, 1988—; mem. users com. Cerro Tololo Interam. Obs., La Serena, Chile, 1981-84; mem. astrophysics program com. Aspen Ctr. for Physics, Colo., 1983—, trustee, 1989-90; chmn. high energy astrophysics mgmt. ops. group NASA, 1986-88, Compton Gamma Ray Obs. users com., 1992—; mem. Space Sci. Bd., NAS, 1986-89; mem. Astronomy and Astrophysics commn. NRC, 1992—; mem. Space and Telescopic Inst. Coun., 1993—; chmn. Space Sci. Working Group. Contbr. articles to profl. jours. and books. Recipient Bart J. Bok prize dept. astronomy, Harvard U., 1976; NSF, NASA rsch. grantee, 1978—; Guggenheim fellow, 1993—; Sloan fellow, 1981-84. Fellow AAAS, Am. Phys. Soc., Am. Astron. Soc. (councilor, nat. sec.-treas. high energy astrophysics div. 1982-84); mem. Internat. Astron. Union (pres., commn. 6, 1991-94). Home: 195 Lincoln Rd Lincoln MA 01773-4102 Office: Harvard Coll Obs 60 Garden St Cambridge MA 02138-1596

GRINER, DONALD BURKE, engineer; b. Walnut Cove, N.C., July 7, 1941; s. Alonzo A. and Nealie Ann (Williams) G.; m. Carolyn Lee Spencer, Aug. 13, 1967; children: Kimberly, Stacy, David. BS, U. Fla., 1966. Engr. NASA, Ala., 1967-86, Teledyne Brown Co., Huntsville, Ala., 1986-87, Control Dynamics Corp., Huntsville, 1987-88; engr. GE Aerospace, Washington, 1988-90, Huntsville, 1990--; mem. optics panel for Hubble telescope NASA, 1984-85, chmn. AXAR optics panel, 1985-86. Contbr. articles to profl. jours. Mem. SPIE, OSA (chmn.), NCOSE. Achievements include patents for Ultra Low Light Level Measurement device, and Laser Furnace for Material Processing. Home: 248 Eastview Dr Madison AL 35758 Office: GE Aerospace 4040 S Memorial Pkwy S Huntsville AL 35812

GRINNELL, ALAN DALE, neurobiologist, educator, researcher; b. Mpls., Nov. 11, 1936; s. John Erle and Swanhild Constance (Friswold) G.; m. Verity Rich, Sept. 30, 1962 (div. 1975). BA, Harvard U., 1958, PhD, 1962. Jr. fellow Harvard U., 1959-62; research assoc. biophysics dept. Univ. Coll. London, 1962-64; asst. research zoologist UCLA, 1964-65, from asst. prof. to prof. dept. biology, 1965-78, prof. physiology, 1972—; dir. Jerry Lewis Neuromuscular Research Ctr. UCLA Sch. Medicine, 1978—; head Ahmanson Lab. Cellular Neurobiology UCLA Brain Research Inst, 1977—; dir. tng. grant in cellular neurobiology UCLA, 1968—, rsch. assoc. Fowler Mus. Cultural History, 1990—. Author: Calcium and Ion Channel Modulation, 1988, Physiology of Excitable Cells, 1983, Regulation of Muscle Contraction, 1981, Introduction to Nervous Systems, 1977, others; contbr. editorial revs. to profl. jours., pub. houses, fed. granting agys. Guggenheim fellow, 1986; recipient Sr. Scientist award Alexander von Humboldt Stiftung, 1975, 79, Jacob Javits award NIH, 1986. Mem. Muscular Dystrophy Assn. (mem. med. adv. com. L.A. chpt. 1980-92), Soc. for Neurosci. (councilor 1982-86), Am. Physiol. Soc. (mem. neurophysiol. steering com. 1981-84), Soc. Fellow, Phi Beta Kappa, Sigma Xi, others. Avocations: music, anthropology, archaeology, travel, sports. Home: 510 E Rustic Rd Santa Monica CA 90402-1116 Office: UCLA Sch Medicine Jerry Lewis Neuromuscular Rsch Ctr Los Angeles CA 90024

GRISAR, JOHANN MARTIN, research chemist; b. Görlitz, Germany, July 10, 1929; came to U.S., 1955, naturalized, 1962; s. Charles Martin and Dora (Stoess) G.; m. Carol Lee Hanson, Jan. 2, 1960 (div. Aug. 1981); children: Caia, Margot, Paul; m. Gabriele L. von Oettingen, June 26, 1983. Diploma in chemistry, Swiss Fed. Inst. Tech., Zurich, 1954; PhD, MIT, 1959. Rsch. chemist Marion Merrell Dow Inc. (formerly Wm. S. Merrell Co.), Cin., 1963-81, Strasbourg, France, 1981—. Contbr. articles to profl. jours.; inventor, patentee in field drug rsch. Mem. Am. Chem. Soc. Home: 7 Rue de Mulhouse, 67160 Wissembourg France

GRISCHKOWSKY, DANIEL RICHARD, research scientist; b. St. Helens, Oreg., Apr. 17, 1940; s. Oscar Edward and Christine Hazel (Olsen) G.; m. Frieda Rosa Bachmann; children: Timothy and Stephanie (twins), Daniela. BS, Oreg. State U., 1962; AM in Physics, Columbia U., 1965, PhD in Physics, 1968. Postdoctoral studies Columbia U., N.Y.C., 1968-69; mem. rsch. staff IBM Watson Rsch. Ctr., Yorktown Heights, N.Y., 1969-77; sci. advisor to dir. rsch. div. IBM, Yorktown Heights, 1978; mgr. atomic physics with lasers group IBM Watson Rsch. Ctr., Yorktown Heights, 1979-83, mgr.

ultra-fast sci. with lasers group, 1983-93; Bellmon chair optoelectronics Sch. Elec. and Computer Engring. Okla. State U., Stillwater, 1993—; chmn. Internat. Coun. on Quantum Electronics, 1989-93, Am. Phys. Soc./Optical Soc. Am./IEEE Joint Coun. on Quantum Electronics, 1989-93. Contbr. articles to profl. jours.; patentee in field. Recipient Boris Pregel award N.Y. Acad. of Sci., 1985. Fellow IEEE, Am. Phys. Soc., Optical Soc. Am. (R.W. Wood prize 1989). Office: Okla State Univ Sch Elec Computer Engring 215 Engineering S Stillwater OK 74078-0545

GRISEWOOD, NORMAN CURTIS, chemical engineer; b. Bklyn., Apr. 4, 1929; s. Edgar Norman and Dorothy Elizabeth (Sharpe) G.; m. Shirley Alice Robinson, Oct. 21, 1950; children: Sheryll, Bonnie, Jeffery, Kenneth, Sandra. BSChemE, Yale U., 1950. Process engr. The Am. Sugar Refining Co., N.Y.C., 1953-62; process engr. The F&M Schaefer Brew Co., N.Y.C., 1962-69, asst. tech. dir., 1969-73, tech. dir., 1973-77, v.p., 1977-81; dir. prodn. The F.X. Matt Brewing Co., Utica, N.Y., 1982—. Lt. USN, 1950-53. Mem. Master Brewers Assn. Home: 251 Old Turnpike Rd Califon NJ 07830 Office: F X Matt Brewing Co 811 Edward St Utica NY 13502

GRISHAM, JOE WHEELER, pathologist, educator; b. Smith County, Tenn., Dec. 5, 1931; s. William Wince and Grace (Allen) G.; m. Jean Evelyn Malone, July 2, 1955. B.A., Vanderbilt U., 1953, M.D., 1957. Intern Washington U.-Barnes Hosp., St. Louis, 1957-58; resident in pathology Washington U.-Barnes Hosp., 1958-60; mem. faculty Washington U., Med. Sch., 1960-73, prof. pathology and anatomy, 1969-73; assoc. pathologist Barnes Hosp., 1969-73; vis. instr. Makerere Med. Coll., Kampala, Uganda, 1961; Kenan prof. pathology, chmn. dept. U. N.C. Med. Sch., Chapel Hill, 1973-91, Kenan prof., chair dept. pathology, 1992—; also pathologist-in-chief U. N.C. Hosp., 1973—; mem. pathology study sect. A NIH, 1969-73, chmn., 1970-73, chmn. pathology study sect. B, 1979-83; Kenan prof. U. N.C. Med. Sch., Chapel Hill, 1992—; bd. sci. counsellors Nat. Inst. Environ. Health Scis., 1974-78; mem. sci. advisory panel Chem. Industry Inst. Toxicology, 1977-88, chmn., 1980-88; adv. bd. Given Inst. Pathobiology, 1983-87. Contbr. articles to med. jours. Served to lt. comdr. USNR, 1961-63. John and Mary R. Markle scholar acad. medicine, 1964-69; fellow Life Ins. Med. Research Fund, 1959-61; fellow Nat. Cancer Inst., 1958-59. Mem. Am. Assn. Pathologists (pres. 1984-85), Am. Assn. Cancer Research, Fedn. Am. Soc. Exptl. Biology (pres., chmn. bd. 1984-85), Am. Assn. Study Liver Diseases, Am. Soc. Cell Biology, Univ. Assn. Research and Edn. in Pathology (v.p. 1985-86), Tissue Culture Assn., Internat. Acad. Pathology, Cell Kinetics Soc., AMA, AAAS. Home: 1703 Curtis Rd Chapel Hill NC 27514-7614 Office: Univ NC Med Sch Dept Pathology CB # 7525 Chapel Hill NC 27599

GRISOLIA, SANTIAGO, biochemistry educator; b. Valencia, Spain, Jan. 6, 1923; s. Santiago and Concepcion (Garcia) G.; m. Frances Lena Thompson, Aug. 16, 1949; children: James S., William F. BA, Nat. Inst., Cuenca, Spain, 1939; postgrad., Med. Sch., Madrid, 1940-41; MD, Med. Sch., Valencia, 1944; D Medicine and Surgery, U. Madrid, 1949; Hon. Degree in Medicine, U. Salamanca, Spain, 1968, U. Valencia, 1973; Hon. Degree in Chemistry, U. Barcelona, Spain, 1972, U. Madrid, 1973; Hon. Degree in Med. Surgery, U. Siena, Italy, 1980; Hon. Degree in Biol. Sci., U. LéOn, Spain, 1982; Hon. Degree in Med. Oncology, U. Basque Country, Bilbao, Spain, 1988; Hon. Degree in Med. Surgery, U. Florence, Italy, 1988; Hon. Degree in Politenic, U. Valencia, Valencia, 1991. Asst. prof. Med. Sch., U. Valencia, 1944-45; fellow in pharmacology NYU, N.Y.C., 1946; rsch. assoc., asst. prof. physiology and chemistry Med. Sch., U. Wis., Madison, 1947-54; assoc. prof., then prof. medicine and biochemistry Med. Sch., U. Kans., Kansas City, 1954-62, chmn. biochemistry dept., 1962-73, disting. prof. biochemistry, 1973—; Inst. Investigation Citology, Valencia, 1977-92, disting. prof., 1992—; disting. prof. U. Los Palmas, Canary Islands; vis. asst. prof. U. Chgo., 1946-47; sec., pres. sci. com. Found. for Advanced Studies, Valencia, 1978—; pres. coord. com. human genome project UNESCO, Paris, 1988—; cons. GLAXO Labs., Madrid, 1980—, Sigma Tau Labs., Madrid, 1988—; organizer symposia in field; dir. summer course on human genetic map Complutense U., El Escorial, Spain, 1989, Menendez y Pelayo U., Santander, Spain, 1990; disting. prof. U. Las Palmas, Canary Islands; researcher in field. Contbr. articles to profl. publs.; co-editor various books in field. Pres. bd. dirs. Multiple Sclerosis Found., Madrid, 1990. Recipient Disting. Svc. citation U. Kans., 1991, Grand Crosses in Edn., Health, Agriculture and Civil Merit, Spain. Mem. Am. Soc. Biol. Chemists (mem. various coms.), Spanish Soc. Biochemistry (hon.), Spanish Soc. Physiology (hon.), Internat. Soc. Neurochemistry, Royal Acad. Medicine Belgium (fgn. hon. mem.), Royal Acad. Pharmacy (corr. academician), Internat. Soc. Clin. Enzymology (hon.), Academia Patavina (Italy, corr. mem.), Royal Acad. Medicine Valencia and Rome (hon. mem.), Coll. of Physicians (hon. mem.), Royal Acad. Scis. (Madrid, corr. mem.), Academia Galesga de Ciencias (Santiago de Compostela, corr. mem.), Sigma Xi, Alpha Omega Alpha (hon.). Office: U Kans Med Ctr Deans Ofce 39th and Rainbow Kansas City KS 66103 also: Inst Investigation Citol, Amadeo de Saboya 4, 46010 Valencia Spain

GRISSOM, CHARLES BUELL, chemistry educator; b. San Diego, July 1, 1959; s. Zadie Buell and Stella (Carrithers) G.; m. Janet Wisniewski, July 10, 1982. BA, U. Calif., Riverside, 1981; PhD, U. Wis., 1985. NIH postdoctoral scientist U. Wis., Madison, 1985-87, U. Calif., Berkeley, 1987-89; asst. prof. chemistry U. Utah, Salt Lake City, 1989—, adj. asst. prof. biochemistry, 1989—. Contbr. articles to profl. jours.; producer videotape: Chemical Career Counseling, 1985-90. Asst. scoutmaster Boy Scouts Am., Madison, 1983-87; sci. fair judge Greater San Diego Sci. Fair, 1977-79. NIH postdoctoral fellow, 1985-88, grantee, 1993—. Mem. AAAS, Am. Chem. Soc. (younger chemists com. 1985-94), Am. Soc. Biochemistry and Molecular Biology, Utah Cancer Ctr. Achievements include research in enzyme kinetics, enzyme mechanisms, isotope effects and photochemistry. Office: U Utah Dept Chemistry Salt Lake City UT 84112-1194

GRISSOM, RAYMOND EARL, JR., toxicologist; b. Raleigh, N.C., Oct. 27, 1943; s. Raymond Earl and Edith (Ballance) G.; m. Lorraine Rankin, Aug. 11, 1974; children: Kelly Daniel, James Earl, Mary Elizabeth. 2 BS degrees with high honors, N.C. State U., 1976, PhD, 1982. Scientist N.C. Bd. Sci. and Tech., Raleigh, 1982-83; postdoctoral rsch. assoc. N.C. State U., Raleigh, 1983-87; toxicologist Environ Monitoring and Svcs., Inc., Chapel Hill, N.C., 1987-88; toxicologist Agy. for Toxic Substances and Disease Registry, Atlanta, 1988-90, sr. toxicologist, 1990—; cons. Becton Dickinson Rsch. Ctr., Research Triangle Park, N.C., 1986-87; speaker internat. symposiums, Eng., Malaysia, and Can., 1993. Author book chpts., govt. documents; contbr. articles to profl. jours. Coach South Guinnett League Basketball, Atlanta, 1990-93. Recipient Citation for Disting. Svc. State of N.C., 1983, Atlanta, 1990-92. Mem. AAAS, Soc. Toxicology, Sigma Xi, Gamma Sigma Delta. Presbyterian. Home: 4959 Joy Ln SW Lilburn GA 30247-5119 Office: Agy Toxic Substances & Disease Registry 1600 Clifton Rd NE # E-57 Atlanta GA 30333

GRIST, CLARENCE RICHARD, chemist, precious metals investor; b. High Point, N.C., Dec. 17, 1932; s. James Wiley and Margaret Hazel (Ewell) G. BA, Bridgewater Coll., 1957; postgrad., U. Richmond, 1961, U. W.Va., 1963, U. Tenn., 1963-64. Chemist FDA, Washington, 1957-58; pvt. practice Rockville, Md., 1967-68; clk. letter sorting machine operator City Post Office U.S. Postal Svc., Washington, 1968-92. With U.S. Army, 1954-56, Germany. Mem. Am. Chem. Soc. (assoc.), N.Y. Acad. Scis., Chem. Soc. Washington, D.C., Washington Tennis Found. (top seed 1968-91, one trophy), Masons (master the veils 1962-65). Democrat. Methodist. Achievements include product development for future manufacturing. Home and Office: 301 Seth Pl Rockville MD 20850-1547

GRITZNER, JEFFREY ALLMAN, geographer, educator; b. Newaygo, Mich., Jan. 10, 1941; s. Charles Frederick II and Laura Elizabeth (Chamberlain) G.; m. Yvonne Gastineau, Jan. 4, 1969; children: Jason Montague, Ingeborg, Justus Gallatin. Student, Ariz. State U., 1959-61, Earlham Coll., 1961-62; AB, U. Calif., Berkeley, 1966; postgrad., Royal Univ. Lund, Sweden, 1966, U. Neuchâtl, Switzerland, 1970, École Supéerieure de Commerce, Neuchâtel, Switzerland, 1970; AM, U. Chgo., 1974, PhD, 1986. Agronomist Peace Corps, Ahwaz, Iran, 1962-64; tech. dir. Agricola du Tchad, N'Djamena, Chad, 1974-75; instr., chmn. dept. geography, geology and anthropology Trinidad (Colo.) State Jr. Coll., 1975-78, instr. dept. phys. sci., curator coll. mus., 1975-78, dir. program for investigation of natural hazards, 1977; sr. program officer NAS, Washington, 1978-88; sr. assoc. World Resources Inst., Washington, 1988-89; rsch. prof. U. Mont., Missoula, 1989—, dir. Pub. Policy Rsch. Inst., 1989—; mem. commn. World Conservation Union, Gland, Switzerland, 1980—; cons. UN, N.Y.C., 1981—; sessional lectr. Fgn. Fvc. Inst. Dept. State, 1983-89. Author: Environmental Degradation in Mauritania, 1981, The West African Sahel, 1988; contbg. author: Integrated Resource Management, 1992; contbr. over 50 chpts. to books, articles, revs. and reports to jours. Pres. Hollin Meadows Elem. Sch. PTA, Alexandria, Va., 1984-85; mem. citizen's task force Fairfax County (Va.) Pub. Schs., 1984-85; coord. com. Ctr. for U.S.-USSR Initiatives, San Francisco, 1988-89. Fellow Ford Found., 1967-68, NDEA, 1968-71, Fulbright-Hays, 1972-73; rsch. grantee U.S. AID, Australian Devel. Assistance Bur., Found. Microbiology, NRC, World Wildlife Fund, Nat. Geographic Soc.; travel grantee Swedish Internat. Devel. Authority, Andrew W. Mellon Found., Nitrogen Fixing Tree Assn., Cervia Amviente, NSF. Mem. AAAS, Am. Soc. Environ. History, Assn. Am. Geographers, Hakluyt Soc., Gamma Theta Upsilon. Democrat. Mem. Soc. of Friends. Home: 378 One Horse Creek Rd Florence MT 59833 Office: U Mont Pub Policy Rsch Inst Missoula MT 59812

GRIVNA, EDWARD LEWIS, electronics engineer, consultant; b. Mpls., June 19, 1956; s. Lawrence John and Emma (Tischler) G.; m. Sieglinde Yvette Nelson, May 29, 1982; children: Margaret Eileen, Mara Lynn, Meridith Mae. B in Electronic Engring., DeVry Inst., Chgo., 1977. Product engr. CIRCOM Inc., Bensenville, Ill., 1977; design engr. Magnetic Peripherals Inc., Edina, Minn., 1977-82; project engr. Magnetic Peripherals Inc., Edina, 1982-84, prin. engr., 1984-88; consulting engr. Imprimis Tech., Eden Prairie, Minn., 1988-89; design ctr. mgr. Cypress Semiconductor, Minnetonka, Minn., 1989—. Contbr. articles to profl. jours. Cantor St. Gerard Cath. Ch., Brooklyn Park, Minn., 1982—; trumpet St. Louis Park (Minn.) Community Band, 1979—. Recipient Great Performer award Control Data Corp., 1985. Mem. Am. Nat. Standards Inst. Achievements include patent for metastable prevention. Developed first 10MB/s IPI controller, first 25MB/s enhanced IPI controller. Home: 8913 Montegue Terr Brooklyn Park MN 55443 Office: Cypress Semiconductor 14525 Hwy 7 Ste 360 Minnetonka MN 55345

GROCE, JAMES FREELAN, petroleum engineer; b. Lubbock, Tex., Nov. 24, 1948; s. Wayne Dee and Betty Jo (Rice) G.; m. Patricia Kay Rogers; 1 child, Jason Eric. BS cum laude, Tex. Tech U., 1971. Registered profl. engr., Tex. Petroleum engr. Texaco, Inc., Sweetwater, Tex., 1971-74; drilling and prodn. engr. Texaco, Inc., Wichita Falls, Tex., 1974-77; asst. dist. engr. Texaco, Inc., Midland, Tex., 1977-78; sr. prodn. engr. Bass Enterprises Prodn., Midland, 1978-81; petroleum engr. Murphy H. Baxter Co., Midland, 1981-82, Henry Engring., Midland, 1982-87; petroleum engr. Fasken Oil and Ranch Interests, Midland, 1987, mgr. engring./ops., 1987—. Scoutmaster Boy Scouts Am., Midland, 1980-83, merit badge counselor, 1987; mem. Community Bible Study, Midland, 1987-93. Mem. Soc. Petroleum Engrs. (sect. chmn. 1987), Soc. Petroleum Evaluation Engrs., Nat. Assn. Corrosion Engrs., Mensa, Tau Beta Pi. Presbyterian. Avocations: individual investments, real estate, gardening. Home: 2117 Bradford Ct Midland TX 79705

GROCE, WILLIAM HENRY, III, environmental engineer, consultant; b. Greer, S.C., Feb. 9, 1940; s. William Henry Jr. and Mary Alvis (Williams) G.; 1 child, William H. IV. BS in Chemistry, Newberry Coll., 1964. Registered profl. engr., S.C.; diplomate environ. engring.; master hazardous material mgmt. Chemist FDA, Atlanta, 1966-68; chemist, engr. Celanese Corp., Greer, 1968-74; prin. engr. Groce Labs., Greer, 1974-86; dir. rsch. Aqua-Tech Corp., Port Washington, Wis., 1986-89; cons. Chemotech Corp., Greer, 1989-90; sr. scientist Savannah River Site U.S. Dept. of Energy, Aiken, S.C., 1990—; cons. Environ. Resource Tech., Greer, 1990. Recipient Silver Beaver award Boy Scouts Am., 1990. Fellow Am. Inst. Chemists; mem. NSPE, Am. Inst. Chem. Engrs. Achievements include research in process and treatment methods for hazardous waste including reclaimation, detoxification, in situ treatment, and deactivation of highly reactive materials. Office: Savannah River Site Bldg 755-11A Aiken SC 29801

GROCOTT, STEPHEN CHARLES, industrial research chemist; b. Perth, W.A., Australia, Nov. 8, 1957; s. Charles Harold and Beatrice Audrey (Neal) G.; m. Dianne Julie Mueller, Nov. 29, 1980. BS with Honors, U. WA, Australia, 1978, PhD, 1982. Devel. chemist Alcoa of Australia Ltd., WA, Australia, 1981-83, rsch. chemist, 1983-85, sr. rsch. chemist, 1985-88, sr. cons. chemistry, 1988—. Mem. Royal Australian Chem. Inst., Am. Chem. Soc. Church of Christ. Achievements include research on the impact, removal, chemistry and analysis of organic impurities in alumnia refining. Office: Alcoa of Australia Ltd, PO Box 161, Kwinana 6167, Australia

GRODBERG, MARCUS GORDON, drug research consultant; b. Worcester, Mass., Jan. 27, 1923; s. Isaac and Rosalie (Hirsch) G.; m. Shirley Florence Merkle, Apr. 15, 1951; children: Joel David, Kim Gordon, Jeremy Daniel. AB, Clark U., Worcester, 1944; MS, U. Ill., 1948. Jr. research chemist Schenley Labs Inc., Lawrenceburg, Ind., 1944-47; research and devel. chemist Marine Products Co., Boston, 1948-50, Brewer and Co. Inc., Worcester, 1950-55; tech. dir. Gray Pharm. Co. Inc., Newton, Mass., 1955-58; dir. research and devel. Colgate Hoyt Labs., Canton, Mass., 1958-89; cons. Newton, 1989—. Author: Fluorides; patentee in field. Asst. leader Cub Scouts, Newton, 1960-62, fund raiser United Fund, Newton, 1960-62. Mem.Internat. Assn. Dental Research, Am. Chem. Soc., Acad. Pharm. Scis., Am. Pharm. Assn.,N.Y. Acad. Scis., others. Jewish. Avocations: tennis, opera, traditional jazz. Home: 4091 Hearthstone Dr Sarasota FL 34238

GROENEVELD, DAVID PAUL, plant ecologist; b. Harvey, Ill., Mar. 20, 1952; s. Robert D. and Ruth M. (Terranova) G. BA, U. Colo., 1975, MA, 1977; PhD, Colo. State U., 1985. Rsch. asst. Inst. Arctic and Alpine Rsch., Boulder, Colo., 1972-76; cons., pres. GEOS, Inc., Telluride, Colo., 1977-81; plant ecologist Inyo County Water Dept., Bishop, Calif., 1981—; cons. Resource Mgmt. Cons., Bishop, 1983—; mem. tech. group Inyo City/City of L.A., 1983—; tech. advisor Great Basin Air Pollution Control Dist., Bishop, 1985—; mem. mined land reclamation expert's panel, Denver, 1980. Contbr. articles to profl. jours. Mem. Ecol. Soc. Am., Range Soc. Am. (life). Achievements include development of management tools to monitor and protect vegetation from effects of groundwater pumping; research in physiology and evapotranspiration of Owens Valley floor vegetation that requires shallow groundwater. Home: 618 Keough St Bishop CA 93514 Office: Inyo County Water Dept 163 May St Bishop CA 93514

GROESCH, JOHN WILLIAM, JR., marketing research consultant; b. Seattle, Nov. 22, 1923; s. John William and Jeanette Morrison (Gilmur) G.; B.S. in Chem. Engring., U. Wash., 1944; m. Joyce Eugenia Schauble, Apr. 25, 1948; children—Sara, Mary, Andrew. Engr., Union Oil Co., L.A., 1944-48, corp. economist, L.A., 1948-56, chief statistician, 1956-62, mgr., 1962-68, mgr., Schaumburg, Ill., 1968-90; exec. v.p. Performance Systems, Inc., Barrington, Ill., 1990—. Bd. dirs. Arlington Heights (Ill.) Boy Scouts Am., 1977-88, adv. bd., 1989—, v.p., 1979, 82-85; commr. 1980-81, mem. Oak-Brook (Ill.) East Cen. Region, 1984—; treas. Scout Cabin Found., Barrington, 1977—. Served with USN, 1944-47. Mem. West Coast Mktg. Research Council (chmn. 1969), Am. Petroleum Inst. (chmn. com. 1970-72). Lodge: Mason. Home: 17 Shady Ln Deer Park Barrington IL 60010 Office: PO Box 56 Barrington IL 60011-0056

GROGAN, EDWARD JOSEPH, clinical engineer; b. Camden, N.J., Nov. 22, 1959; s. Edward Charles and Elizabeth Gertrude (Shields) G.; m. Margaret Mary Reisinger, Aug. 16, 1986, children: Joseph Michael, John Paul. BS in Engring., Duke U., 1981. Cert. clin. engr. Dir. biomed. engring. and safety Alexandria (Va.) Hosp., 1981—. Contbr. articles to profl. jours., chpt. to textbook related to med. lasers. S.E. regional coord. Nat. Youth Pro-Life Coalition. Recipient Svc. Excellence award, Work Improvement Initiative award Alexandria Hosp. Mem. Assn. for Advancement of Med. Instrumentation, Am. Soc. Hosp. Engring., Va. Soc. for Human Life (ednl. co-chair), Phi Beta Kappa, Tau Beta Pi, Phi Eta Sigma. Roman Catholic. Home: 25 Dudley Ct Sterling VA 20165 Office: Alexandria Hosp 4320 Seminary Rd Alexandria VA 22304-1500

GROH, GABOR GYULA, mathematician, educator; b. Budapest, Hungary, Apr. 26, 1948; came to Switzerland, 1957, naturalized, 1970; s. Julius and Margit (Molnar) G.; m. Keiko Fuse, July 21, 1979; children: Christoph Ken, Miya Valentine. Diploma in physics Eidgenössische Technische Hochschule, Zurich, 1974, Dr. Sci. Math., 1982. Research assoc. Eidgenössiche Technische Hochschule, 1975-82, 85—; research fellow, research assoc. Swiss Nat. Sci. Found.-Lawrence Berkeley Lab., Berkeley, Calif., 1982-83; vis. lectr. dept. math. U. Calif.-Berkeley, 1984; assoc., cons. Medichem S.A., Fribourg, Switzerland, 1984-85. Mem. Soc. Indsl. and Applied Math., AIAA, N.Y. Acad. Sci., Swiss Math. Soc. Office: Inst Energy Tech Supercomputing, ETH-Zentrum, 8092 Zurich Switzerland

GRONBECK, CHRISTOPHER ELLIOTT, energy engineer; b. Ann Arbor, Mich., July 31, 1969; s. Bruce Elliott and Wendy Lee (Gilbert) G. BS, Stanford U., 1991. Programmer and technician Shuttle Electrodynamic Tether System Project, Stanford U., Calif., 1988-91; rsch. assoc. Internat. Inst. for Energy Conservation, Washington, 1991—. Mem. ASHRAE, Assn. Energy Engrs., Illuminating Engring. Soc. N.Am. (assoc.). Office: Internat Inst Energy Cons Ste 940 750 First St NE Washington DC 20002

GRONER, PAUL S(TEPHEN), electrical engineer; b. Binghamton, N.Y., May 23, 1937; s. David and Ruth Groner; m. Millie Ayscue, Sept. 30, 1967; children: Carl, Daniel. Head circuit design group Hughes Aircraft Co., Fullerton, Calif., 1969-73; mgr. circuit design Sperry Univac, Irvine, Calif., 1973-82; dir. tech. Computer Consoles Inc., Irvine, 1982-88; dir. hardware AST Rsch., Irvine, 1988-89; pres. Creative Silicon, Irvine, 1989—. Contbr. articles to profl. jours. Achievements include patents in field. Home: 2139 N Ross St Santa Ana CA 92706

GRONLUND, SCOTT DOUGLAS, psychology educator, researcher; b. Hibbing, Minn., Mar. 24, 1959; s. Gerald Wayne and Kathleen Fern (Fritcher) G.; m. Christine Norrell Cox, May 19, 1990; 1 child, Taylor Christine. BA, U. Calif., Irvine, 1981; PhD, Ind. U., 1986. Postdoctoral researcher Northwestern U., Evanston, Ill., 1986-89; prof. U. Okla., Norman, 1989—. Contbr. articles to Jour. Exptl. Psychology, Jour. Math. Psychology, Psychol. Rev., Applied Cognitive Psychology. Active Big Bros./Little Bros., Chgo., 1987-89. Mem. APA, Am. Psychol. Soc., Soc. for Math. Psychology, Southwestern Psychol. Assn. (Okla. rep. 1990-92), Psychonomics Soc. (assoc.). Democrat. Achievements include rsch. in modularity in automation, initial availability of item information precedes the initial availability of associative info. in recognition memory. Office: U Okla Dept Psychology 455 W Lindsey Norman OK 73019

GROOMS, JOHN MERRIL, research and development engineer; b. Goshen, Ind., May 24, 1957; s. Laroy Brook and Barbara Ann (Culp) G.; m. Donna Sue Weaver, Aug. 18, 1979; children: Ryan Laroy, Shawn Paul. BSME, Tri-State U., 1980. Cert. engr.-in-tng. Asst. engr. AIRVAC, Rochester, Ind., 1980-87, R&D engr., 1987-91, R&D mgr., 1991—. Co-author, editor 1989 AIRVAC Design Manual, 1989. Mem. ASTM, Soc. Plastic Engrs. Mem. Brethren Ch. Achievements include patents for vacuum sewage transport system, vacuum sewerage system with in pit breather, inlet vacuum valve with quick release mounting apparatus, electric air admission control, vacuum system with electric air admission control, sewerage system with non-jamming valve with tapered plunger, non-jamming valve with tapered plunger. Home: 2621 E Hickory Dr Rochester IN 46975 Office: AIRVAC Old US 31 N Rochester IN 46975

GROSS, A. CHRISTOPHER, environmental manager; b. S.I., N.Y., Oct. 7, 1942; s. Herbert Carl and Clara Nancy (Hage) G.; m. Barbara Ann Rippen, Sept. 2, 1967; children: Amy Milbrook, Alexander Carl. AB, Wabash Coll., 1964; MA, Conn. Coll., 1966. Dir. biol. rsch. Wapora, Inc., Bethesda, Md., 1970-74; mgr. environ. compliance L.I. Lighting Co., Hicksville, N.Y., 1974—; environ. cons., Centerport, N.Y., 1980—; bd. dirs. Harbor Adv. Coun., Huntington, N.Y. Contbr. articles to profl. jours. Pres. Urbana (Md.) Civic Assn., 1973-74, Little Neck Peninsula Civic Assn., Centerport, 1976-78. With U.S. Army, 1967-70. Recipient Innovators award Elec. Power Rsch. Inst., 1991. Mem. Ecol. Soc. Am., Am. Fisheries Soc., Am. Inst. Biol. Scis., Commodore Centerport Yacht Club (pres. 1993-94). Achievements include demonstration of use of non-toxic coatings to prevent or control fouling of power plant cooling water intake systems; research in reducing environmental impacts of power generation, transmission and use. Home: 12 Harbor Ridge Dr Centerport NY 11721 Office: LI Lighting Co 175 E Old Country Rd Hicksville NY 11801

GROSS, DENNIS MICHAEL, pharmacologist; b. L.A., Sept. 12, 1947; s. Edward and Norma (Berman) G.; m. Rosemarie Ganley, Nov. 25, 1973; children: Kevin Michael, Alexis Diane. BA, Calif. State U., 1969, MS, 1970; PhD, UCLA, 1974. Instr. Tulane Med. Sch., New Orleans, 1976-77; sr. rsch. pharmacologist Merck Sharp & Dohme Rsch. Labs., West Point, Pa., 1977-80, rsch. fellow, 1981-86; mgr. Merck Sharp & Dohme Rsch. Labs., Rohway, N.J., 1987-89; assoc. dir. Merck Sharp & Dohme Rsch. Labs., West Point, 1990-91; dir. Merck Rsch. Labs., West Point, 1992—; adj. asst. prof. Jefferson Med. Coll., Phila., 1977-85, adj. assoc. prof., 1986—. Mem. Am. Soc. Cell Biology, Am. Heart Assn., N.Y. Acad. Scis., Sigma Xi. Home: 1621 Clearview Rd Blue Bell PA 19422 Office: Merck Rsch Labs WP42-300 West Point PA 19486

GROSS, HERBERT GERALD, space physicist; b. Chgo., Dec. 9, 1916; s. William Theodore and Lucille Eleanor (Powalski) G.; m. Regina Marie Dwyer, Sept. 22, 1951 (div. Dec. 1962); children: Regina Marie, Gerald, Paul, Robert; m. Alice Marie Molway, June 13, 1964; stepchildren: Susan, Margaret, Judith, Joseph. BA in Philosophy, Cath. U. Am., 1940, MS in Physics, 1950; postgrad., Mass. Inst. Tech., 1952-53, Northeastern U., 1954-55, UCLA, 1968, Colo. State U., 1968. Registered profl. engr., Mass. Physicist, project engr. Raytheon Co., Newton, Mass., 1951-58, Edgerton, Germeshausen & Grier Co., Boston, 1958-59; staff scientist Geophysics Corp. Am., Bedford, Mass., 1959-64; prin. scientist McDonnell Douglas Astronautics Co., Huntington Beach, Calif., 1964-75; dir. advanced devel. H Koch & Sons, Anaheim, Calif., 1977-93; pres. Gross Emergency Lighting Co., Santa Ana, Calif., 1993—; cons. Harvard U. Obs., Cambridge, Mass., 1962-63, Sci. and Applications, Inc., La Jolla, Calif., 1975-76, Planning Rsch., Inc., Newport Beach, Calif., 1977-78, Domar, Ltd., 1990-93; mem. task force Nat. Fire Protection Assn.; co-prin. investigator Apollo lunar rocks for luminescence; prin. investigator laser-induced luminescence remote sensing of oil and hazardous materials spills in L.A. Harbor. Inventor in field; contbr. articles to profl. jours. Pres. Cedarwood Assn., Waltham, Mass., 1951-54, Waltham Mus. Club, 1954-55; founder Waltham Community Symphony, 1955; dist. dir. Waltham Hosp. Fund Drive, 1956, Citizens Edn. Adv. Group, Tustin, Calif., 1967; treas. South Coast Homeowners Assn., Santa Ana, Calif., 1974-76, pres., 1976-78; mem. Laguna Beach Arts Festival. Mem. Planetary Soc., Survival and Flight Equipment Assn., Illuminating Engring. Soc., Soc. Automotive Engrs. (exec. com., chmn. A-20C aircraft interior lighting subcom., mem. A-20 aircraft lighting com.), Elks. Roman Catholic. Achievements include breakthrough science and invention of Emergency Egress Signage and Marker Lighting Systems for Smoke and Adverse Optical Conditions; co-invention of Helicopter Emergency Egress Lighting System used on U.S. Navy Helicopters; successful demonstration of 2d worst case tests for shipboard application by Royal Navy, U.K.; design and implementation of complete LED Emergency Egress System in building research establishment in U.K. and demonstrating breakthrough superiority for all buildings; co-invention of Machine Vision Fire Detection System eliminating false alarms, new concept for assessing whether potential sources can and will cause false alarms in fire detection systems in Air Force hangars. Home and Office: 2011 W Summer Wind Santa Ana CA 92704-7133

GROSS, KAUFMAN KENNARD, applications engineer, consultant; b. Phila., Nov. 28, 1916; s. Maurice H. and Esther (Manko) G.; m. Hilda N. Harris, Oct. 16, 1949; children: Louis J., Henry C., Hannah M., Martin R. BEE, Drexel U., 1941. Elec. engr. Leeds and Northrup, Phila., 1935-45, various cos., 1946-65; sr. applications engr. Barber-Colman, Rockford, Ill., 1965-86; cons. Phila., 1986—. Bd. dirs. Camp Coun. Inc., Phila., 1940—. With USAF, 1945-46. Home and Office: 6503 N 12th St Philadelphia PA 19126-3641

GROSS, MICHAEL LAWRENCE, chemistry educator; b. St. Cloud, Minn., Nov. 6, 1940; s. Ralph J. and Margaret T. (Iten) G.; m. Kathleen M. Trammer, June 13, 1966 (div. 1981); children: Matthew R. and Michele R. (twins). BA, St. John's U., St. Cloud, 1962; PhD, U. Minn., 1966.

Postdoctoral fellow U. Pa., Phila., 1966-67, Purdue U., Lafayette, Ind., 1967-68; asst. prof. chemistry U. Nebr., Lincoln, 1968-72, assoc. prof., 1972-78, prof., 1978-83, 3M alumni prof., 1983-88, C. Petrus Peterson prof., 1988—; dir. Midwest Ctr. for Mass Spectrometry, Lincoln, 1978—; mem. metallobiochemistry study sect. NIH, Washington, 1985-88; mem. bd. on chem. scis. and technology Nat. Rsch. Coun., 1986-91; vis. prof. Internat. Grad. Sch., U. Amsterdam, Netherlands, 1990. Author: High Performance Mass Spectrometry, 1978; editor Mass. Spectrometry Revs., 1982-90, Jour. Am. Soc. Mass. Spectrometry, 1990—; contbr. over 250 articles to profl. jours. Mem. instnl. rev. bd. St. Elizabeth, Lincoln Gen. and Bryan Meml. hosps., 1982-90. Recipient award for disting. teaching U. Nebr., 1978, Pioneer award Commonwealth of Mass., 1987; named one of Top 50 Cited Chemists in World, 1984-91. Mem. Am. Chem. Soc., Am. Soc. for Mass. Spectrometry, Union Concerned Scientists, Sigma Xi, Phi Lambda Upsilon. Democrat. Roman Catholic. Home: 3135 S 30th St Lincoln NE 68502 Office: U Nebr Dept Chemistry Lincoln NE 68588

GROSS, PAUL HANS, chemistry educator; b. Berlin, Apr. 17, 1931; came to U.S., 1962; s. Paul Karl Friedrich and Olga Frieda (Saacks) G.; m. Uta Maria Freudiger, June 8, 1957; children: Thomas, Klaus, Michael, Eva. Diploma, F.U., Berlin, 1958, Dr.rer.nat., 1961. Rsch. chemist Schering A.G., Berlin, 1961-62; postdoctoral fellow U. Pacific, Stockton, Calif., 1962-64; rsch. fellow, instr. Harvard Med. Sch., Boston, 1965-66, Mass. Gen. Hosp., Boston, 1965-66; assoc. prof. U. Pacific, Stockton, Calif., 1966-70, prof., 1970—; vis. prof., vis. scientist Tech. Hochschule, Munich, 1973, Freie U., Berlin, 1978, U. Autonoma, Baja Calif., Mex., 1978, Med. Hochschule Hannover, Germany, 1983, U. de Sevilla, Spain, 1988; cons. Cell Pathways, Inc., Tucson, 1990—, United Pharms., 1984—, Dupont Merck, 1992—. Contbr. numerous articles to profl. jours. Rsch. grantee NSF, 1968, 69-71, Rsch. Corp., 1972-74, Med. Sch. Hann., 1983, NIH, 1987-89. Mem. Gesellschaft Deutscher Chemiker, Am. Chem. Soc., Alpha Chi Sigma, Sigma Xi, Phi Kappa Phi. Achievements include 6 patents; rsch. in rheology, carbohydrate chemistry, peptide chemistry and organic chemistry. Office: U of the Pacific Dept Chemistry Stockton CA 95211

GROSS, PETER ALAN, epidemiologist, researcher; b. Newark, Nov. 18, 1938; s. Meyer P. and Nathalie (Bass) Denburg) G.; m. Regina Teri Gittlin, May 30, 1964; children: Deborah Karen, Michael Philip, Daniel Brian. BA cum laude, Amherst Coll., 1960; MD, Yale U., 1964. Diplomate Am. Bd. Internal Medicine. Intern Yale-New Haven Hosp., 1964-65, jr. resident, 1965-66; sr. resident Peter Bent Brigham Hosp., Boston, 1968-69; research and edn. assoc. Va Hosp., West Haven, Conn., 1971-73, acting chief infectious disease sect., 1972-73; chief infectious disease sect. VA Hosp., West Haven, Conn., 1973-74; chief infectious disease sect. Hackensack (N.J.) Med. Ctr., 1974—, chmn. dept. medicine, 1980—, chmn. med. bd., 1986; prof. medicine N.J. Med. Sch., Newark, 1981—, vice chmn. dept. medicine, 1989-90; assoc. clin. prof. medicine Columbia U. Coll. Physicians and Surgeons, N.Y.C., 1971-81, asst. clin. prof., 1974-77; asst. prof. medicine Yale U. Shc. Medicine, New Haven, 1971-74; ad hoc reviewer rsch. grants NIH, Nat. Inst. Allergy and Infectious Diseases; investigator Ctr. for Biologic Evaluation and Rsch. FDA, 1974—; mem. clin. indicators task force Joint Commn. on Accreditation of Healthcare Orgns., 1987-89. Author: Gram Strain Recognition, 1975, 2d edit., 1980; assoc. editor: Clinical Performance and Quality Health Care; mem. editorial bd. Jour. Clin. Microbiology, 1980—, Infection Control, 1980-90. Served to lt. comdr. USPHS, CDC, 1966-68. NIH fellow Yale U., 1969-71. Mem. AAAS, ACP (task force on adult immunization), Infectious Disease Soc. Am. (fellow clin. affairs com.), Am. Acad. Microbiology, Am. Soc. Virology, Am. Soc. Microbiology, Soc. Hosp. Epidemiologists Am. (councillor 1986-88, v.p. 1992, pres.-elect 1993). Republican. Jewish. Office: Hackensack Med Ctr Hackensack NJ 07601

GROSS, SAMSON RICHARD, geneticist, biochemist, educator; b. N.Y.C., July 27, 1926; s. Isidor and Ethel (Mermelestein) G.; m. Helen Hudi Steinmetz, Sept. 16, 1952; children—Deborah Ann, Michael Robert, Eva Elizabeth. B.A., NYU, 1949; A.M., Columbia, 1951, Ph.D. (USPHS fellow), 1953. Asst. prof. genetics Stanford U., 1956-57; asst. prof. genetics Rockefeller U., N.Y.C., 1957-60; assoc. prof. dept. microbiology and immunology Duke, Durham, N.C., 1960-65, prof. genetics and biochemistry, 1965-91, prof. emeritus genetics and biochemistry, 1991—; dir. div. genetics dept. biochemistry Duke, 1965-77, dir. univ. program in genetics, 1967-77; bd. dirs. Cold Spring Harbor Lab. Quantitative Biology, N.Y., 1967-72. USPHS Spl. fellow Weizmann Inst., 1969-70; Josiah Macy Found. fellow Hebrew U., 1977-78; John Simon Guggenheim fellow Hebrew U., 1985-86. Mem. Genetic Soc. Am., AAAS, Am. Soc. Microbiology, Am. Soc. Biol. Chemists, Phi Beta Kappa. Home: 2411 Prince St Durham NC 27707-1432

GROSS, SEYMOUR PAUL, retired civil engineer; b. N.Y.C., Sept. 26, 1930; s. Bernard and Sarah (Pildes) G.; m. Dolores B. Adelson, Jan. 7, 1956; children: George M., Robert D. BSCE, CCNY, 1952; MSCE, NYU, 1960. Diplomate Am. Acad. Environ. Engrs.; Registered profl. engr. N.J. Engr. Amman and Whitney, N.Y.C., 1954-55; hwy. project engr. King and Gavaris, N.Y.C., 1956-60, Frederic R. Harris, Assocs., Norwalk, Conn., 1960-61; drainage project engr. Levitt and Sons, Willingboro, N.J., 1961-62; water resources engr. Delaware River Basin Commn., West Trenton, N.J., 1962-74, supervising engr., 1975-92; retired, 1993; mem. rsch. com. Water Environ. Fedn., 1970-77; publ. com., N.J. Effluents, 1968-85. Fellow ASCE; mem. Water Environ. Fedn. Achievements include contributions to development and implementation of regulatory framework for wasteload allocations for upgrading treatment of municipal and industrial wastewater discharge to Delaware River resulting in substantial recovery of quality of river.

GROSS, SHARON RUTH, psychology educator, researcher; b. L.A., Mar. 21, 1940; d. Louis and Sylvia Marion (Freedman) Lackman; m. Zoltan Gross, Mar. 1969 (div.); 1 child, Andrew Ryan. BA, UCLA, 1983; MA, U. So. Calif., L.A., 1985, PhD, 1991. Tech. Rytron, Van Nuys, Calif., 1958-60; computress on tetrahedral satellite Space Tech. Labs., Redondo Beach, Calif., 1960-62; owner Wayfarer Yacht Corp., Costa Mesa, Calif., 1962-64; electronics draftsperson, designer stroke-writer characters Tasker Industries, Van Nuys, 1964-65; pvt. practice cons. Sherman Oaks, Calif., 1965-75, 77-80; printed circuit bd. designer Systron-Donner, Van Nuys, Calif., 1975-76; design checker Vector Gen., Woodland Hills, CAlif., 1976-77; undergrad. adv. U. So. Calif., L.A., 1987-89, rsch. asst. prof., rsch. assoc. social psychology, 1991—. Contbr. chpts. to books. Mem. ACLU, L.A., 1991. Recipient Haynes Found. Dissertation fellowship U. So. Calif., 1990. Mem. APA (student dissertation rsch. award 1991), AAAS, Computer Graphics Pioneers, Am. Psychol. Soc., Western Psychol. Assn. Democrat. Jewish. Office: U So Calif Dept Psychology Los Angeles CA 90089-1061

GROSS, STANISLAW, environmental sciences educator, activist; b. Lodz, Poland, Nov. 27, 1924; came to U.S., 1962; s. Oskar and Janina (Gundelach) G.; children: Krzysztof, Zbigniew, Richard. BChemE, Tech. U., Lodz, 1947, MChemE, 1949; PhD in organic chemistry, U. London, 1961. Dep. mgr. Boruta, Zgierz, Poland, 1946-50; assoc. prof. Inst. for Indsl. Medicine, Lodz, 1950-58; researcher Chester Beatty Rsch. Inst., London, 1958-62; sr. scientist Boyce Thompson Inst., Yonkers, N.Y., 1962-66; assoc. mem., head divsn. Inst. for Muscle Diseases, N.Y.C., 1966-74; dir. environ. health projects, adj. prof. N.Y. Med. Coll. Grad. Sch. Health Scis., Valhalla, 1973—. Contbr. over 45 articles to profl. jours. Tech. advisor to Congressman from N.Y., 1985—. Mem. N.Y. Acad. Scis., Biophys. Soc., Polish Chem. Soc., Polish Inst. of Arts and Scis. of Am., Radiation Rsch. Soc. Home: PO Box 422 Tarrytown NY 10591-0422 Office: NY Med Coll Munger Pav Rm 525 Valhalla NY 10595

GROSS, THOMAS PAUL, lawyer; b. Cleve., Sept. 4, 1949; m. Eleanor J. Hill, Apr. 7, 1990. BS, Case Inst. Tech., 1971; MS, Stanford U., 1973; JD, Case Western Res. U., 1977. Bar: Ohio 1977, D.C. 1979, Va. 1988. Process engr. Nat. Semiconductor Corp., Santa Clara, Calif., 1973-74; patent atty. U.S. Dept. Energy, Washington, 1977-79, atty., dep. asst. gen. counsel, 1979-87; atty. McGuire Woods Battle Boothe, Washington, 1987-92; ptnr. Katten, Muchin, Zavis and Dombroff, Washington, 1992—. Mem. D.C. Bar Assn., Va. State Bar Assn., Ohio State Bar Assn. Office: Katten Muchin Zavis and Dombroff Ste 700 East 1025 Thomas Jefferson NW Washington DC 20007

GROSS, WILLIAM SPARGO, civil engineer; b. Corpus Christi, Tex., Aug. 30, 1946; s. Lewis and Jean (Spargo) G.; m. Barbara Ruth Harland, Mar. 19, 1983; 1 child, David Robinson. BS, Tex. A&M U., 1969, M. Engring., 1974. Engr. Brown & Root Inc., Houston, 1974-79; project engr. Stubbs, Overbeck & Assocs., Houston, 1979-83; structural engr. Sargent & Lundy, Chgo., 1983-84; civil engr. City of Tex., Dallas, 1984—. Capt. U.S. Army, 1969-72, Vietnam. Decorated Bronze Star, Meritorious Svc. medal. Mem. ASCE, NSPE. Office: Office Emergency Preparedness 1500 Marilla L2AN Dallas TX 75201

GROSSBERG, DAVID BURTON, cardiologist; b. Bronx, N.Y., Oct. 28, 1956; s. Jules Harold and Florence (Greenbaum) G.; m. Karen Leslie Sonin, Apr. 17, 1988; children: Samuel Benjamin, Hannah Rachel. BA, SUNY, Binghamton, 1977; MD, SUNY, Syracuse, 1981. Diplomate Am. Bd. Internal Medicine, Am. Bd. Cardiology. Resident in internal medicine Overlook Hosp., Coll. Physicians and Surgeons, Columbia U., Summit, N.J., 1981-84; asst. clin. prof. medicine George Washington U., Washington; adj. asst. prof. medicine Baylor U. Sch. Medicine; staff physician St. Mary's Hosp., East Orange, N.J., 1982; internist Sumter County Pub. Health, Wildwood, Fla., 1984-86; cardiology fellow Albany (N.Y.) Med. Ctr. Hosp., 1986-88; cardiologist Md. Cardiology Assoc., Silver Spring, 1988-91; pvt. practice Silver Spring, Rockville, Md., 1991—; mem., dir. Cen. Fla. Ambulance Svcs., Sumterville, 1984-85; active attending staff Washington Adventist Hosp., Shady Grove Adventist Hosp., Holy Cross Hosp., Suburban Hosp., Laurel Hosp.; co-investigator gusto trial-thrombolytic therapy post myocardial infarction. Recipient Elsbeth Kroeber Meml. award N.Y. Biology Tchrs. Assn., 1973, Regents scholar, 1973. Fellow Am. Coll. Cardiology, Am. Coll. Chest Physicians; mem. ACP, Physicians for Social Responsibility, Md. Med. Soc. (alternate del. 1992-93), Sierra Club (vol. physician Wilderness Project 1982), Audubon Soc., Md. Soc. Cardiology, Montgomery County Med. Soc. Avocations: karate, hiking, philately, numismatics. Office: 2415 Musgrove Rd Ste 307 Silver Spring MD 20904

GROSSBERG, JOHN MORRIS, psychologist, educator; b. Bklyn., Oct. 11, 1927; s. Percy and Edith (Garcee) G.; m. Carolyn Waldkoetter, July 2, 1955; children: Paul, Michael, Andrew. AB, Bklyn. Coll., 1950; MA, Ind. U., 1955, PhD, 1956. Asst. prof. psychology U. Calif., Riverside, 1955-58; chief psychologist Mound Bldrs. Guidance Ctr., Newark, Ohio, 1958-62; lectr. in psychology Denison U., Granville, Ohio, 1958-62; asst. prof. San Diego State U., 1962-68, prof. psychology, 1969—, chmn. dept. psychology, 1976-81, co-founder joint PhD program, 1978-80; vis. prof. psychology Stanford U., 1968-69; cons. energy usage Navy Ocean Systems Ctr., San Diego, 1977-78; co-prin. investigator U.S. Dept. HEW, Vocat. Rehab. Adminstrn., San Diego, 1963-64; prin. investigator NIMH, San Diego, 1966-68; faculty participant NSF Undergrad. Rsch. Participant Program, 1964-65. Contbr. numerous articles to profl. jours. With USN, 1945-46; PTO. Mem. AAAS (Pacific div. v.p. 1981-82, pres. 1982-83), APA, Am. Psychol. Soc., Western Psychol. Assn. (co-mgr. ann. conv. 1979, bd. dirs. 1978-80), Sigma Xi, Phi Beta Kappa. Home: 6760 Bonnie View Dr San Diego CA 92119 Office: San Diego State U Dept Psychology San Diego CA 92182-0350

GROSSFELD, MICHAEL L., hospital administrator; b. N.Y.C., Jan. 26, 1955; s. Irving and Mildred (Eigen) G.; m. Cathleen Ann Del Giorno, Aug. 21, 1977; 1 child, Matthew Brandon. BA, Queens Coll., 1978, MA, 1980. Coord. computer svcs. rehab. medicine Med. Ctr. NYU, N.Y.C., 1980-83, dir. computer svcs. rehab. medicine, 1983-84; dir. hosp. computer svcs. Goldwater Meml. Hosp. N.Y.C. Health and Hosps. Corp., 1984-85; asst. adminstr. Goldwater Meml. Hosp., 1985-87, assoc. adminstr., 1987-88; asst. v.p. Huntington Hosp., 1988—; cons. Computer Cons. Svcs., Old Bethpage, n.Y., 1989--; adj. prof. Adelphi U., Garden City, N.Y., 1985; bd. dirs. Spectrum Healthcare Solutions Users Assn., Hauppaugae, N.Y., 1990--, IBAX Healthcare Corp. User Assn., Longwood, Fla., 1992--. Editor: Microcomputer Applications in the Rehabilitation of Communication Disorders, 1986. Recipient Hosp. Info. Systems award of Excellence Shared Data Rsch. Hudson, 1989. Mem. Am. Hosp. Assn., Healthcare Fin. Mgmt. Assn., Healthcare Info. and Mgmt. Systems Soc., Disabled Vets. Comdrs. Club. Democrat. Jewish. Avocations: long distance running, golf, scuba diving, tennis. Home: 204 Haypath Rd Old Bethpage NY 11804-1434 Office: Huntington Hosp Assn 270 Park Ave Huntington NY 11743-2799

GROSSFELD, ROBERT MICHAEL, physiologist, zoologist, educator, neurobiologist; b. N.Y.C., Oct. 15, 1943. BS, U. Wis., 1963; PhD, Stanford U., 1968. Instr. Harvard Med. Sch., Boston, 1968-70; asst. prof. Cornell U., Ithaca, N.Y., 1970-74; rsch. scientist VA Hosp., Sepulveda, Calif., 1974-76; vis. asst. prof. U. Tex., Austin, 1976-79; asst./assoc. prof. zoology/physiology N.C. State U., Raleigh, 1979—. Contbr. articles to profl. jours. Office: NC State U Zoology Dept Raleigh NC 27695-7617

GROSSKREUTZ, JOSEPH CHARLES, physicist, engineering researcher, educator; b. Springfield, Mo., Jan. 5, 1922; s. Joseph Charles and Helen (Mobley) G.; m. Mary Catherine Schubel, Sept. 7, 1949; children—Cynthia Lee, Barbara Helen. B.S. in Math., Drury Coll., 1943; postgrad., U. Calif.-Berkeley, 1946-47; M.S., Washington U., St. Louis, 1948, Ph.D. in Physics, 1950. Research physicist Calif. Research Corp., La Habra, 1950-52; asst. prof. physics U. Tex.-Austin, 1952-56; research scientist Nuclear Physics Lab., Austin, 1952-56; sr. physicist Midwest Research Inst., Kansas City, Mo., 1956-59, prin. physicist, 1959-63, sr. adviser, 1963-67; prin. advisor Midwest Research Inst., Kansas City, 1967-71; chief mech. properties sect. Nat. Bur. Standards, Washington, 1971-72; mgr. solar programs Black & Veatch Cons. Engrs., Kansas City, Mo., 1972-77, mgr. advanced tech. projects, 1979-88, project engr. design/constrn. solar thermal test facility, 1975-77; ret., 1988; research prof. physics U. Mo., Kansas City, 1989—; 1st dir. rsch. Solar Energy Rsch. Inst., Golden, Colo., 1977-79; spl. cons. NATO, 1967. Contbr. articles to profl. jours. Served to lt. USN, 1943-46. Recipient Disting. Service award Drury Coll., 1959; Washington U. fellow, 1948-49. Fellow Am. Phys. Soc., ASTM (dir. 1977-80, Merit award 1972); mem. Sigma Xi, Sigma Pi Sigma. Methodist. Achievements include patent for detecting defects by distribution of positron lifetimes in crystalline materials. Home: 4306 W 111th Ter Shawnee Mission KS 66211-1702 Office: U Mo Physics Dept Kansas City MO 64110

GROSSMAN, GARY DAVID, ecological educator; b. Rochester, N.Y., May 29, 1954; s. Joseph and Sylvia (Seigel) G.; m. Barbara June Mullen, May 24, 1980; 1 child, Rachel. BS in Conservation, Resource Studies, U. Calif., Berkeley, 1975; PhD in Animal Ecology, Limnology, U. Calif., Davis, 1979. Curator fish collection Calif. Poly. State U., San Luis Obispo, 1975-76; lectr. dept. biol. scis. Calif. State U., Sacramento, 1977-78, 80; asst. prof. Warnell Sch. Forest Resources, U. Ga., Athens, 1981-87, assoc. prof., 1987-92, prof., 1992—; invited lectr. U. Salzburg, Austria, 1983, Limnological Inst. of Austrian Acad. Scis., Mondsee, 1983, U. Innsbruck, Austria, 1983, Vanderbilt U., Nashville, 1984, Tex. A&M U., College Station, 1985, U. Lund, Sweden, 1985, U. So. Miss., Hattiesburg, 1987, U. Md., Frostburg, 1987, Va. Commonwealth U., 1988, U. Maine, Orono, 1988, Appalachian State U., Boone, N.C., 1988, U. Vienna, Austria, 1988, EPA, Duluth, Minn., 1988, Redwood Sci. Lab, USDA, Arcata, Calif., 1991; presenter at profl. confs. Contbr. to refereed jours., symposia. Grantee U.S. Dept. Commerce-NOAA, 1982-83, 83-84, 84-86, USDA, 1982-87, 87-88, 87-92, 88-89, U.S.-Spanish Com. for Sci. and Technol. Cooperation, 1985-88, NSF, 1985-90, 91—. Mem. Am. Fisheries Soc., Am. Soc. Ichthyologists and Herpetologists, Ecol. Soc. Am., European Ichthyologists Union, Ga. Fishery Workers Assn., N.Am. Benthological Soc., Assn. Southeastern Biologists, Southeastern Fishes Coun. Home: 237 Highland Ave Athens GA 30606 Office: U Ga Warnell Sch Forest Resource Athens GA 30602

GROSSMAN, JACOB S., structural engineer; b. Jerusalem, Apr. 29, 1930; came to U.S., 1950; s. Joseph and Rachel F. (Naviasky) G.; m. Wanda R. Brodsky, Aug. 3, 1952; children: Iris, Arnon, Lilit, Elana B. BS in Engring., UCLA, 1954; MSCE, U. So. Calif., 1956. Registered profl. engr., N.Y. Engr. several structural firms L.A., 1954-57; ptnr. Rosenwasser/Grossman Cons. Engrs., P.C., N.Y.C., 1957—; pres. WWES, Inc., West Hempstead, N.Y., 1970—; cons. ATC, NIST, NCEER, Seismic Com., N.Y.C.; lectr., researcher in field. Designer most slender structures and three from amongst the tallest 100 structures in the world; author: (with others) Building Structural Design Handbook, 1987; contb. papers on design and rsch. in steel and concrete structures to profl. jours. Sgt. Israeli Army, 1948-50. Fellow Am. Concrete Inst. (mem. concrete code com. 1972—, bd. dirs. 1983-86, Maurice P. van Buren award 1987, Alfred E. Lindau award 1989); mem. NSPE, ASCE. Achievements include design of first diagonally braced concrete tube structure, of most slender structures in the world, 3 of the tallest 100; advancement of design of concrete, tall buildings by incorporating innovative techniques. Office: Rosenwasser/Grossman Cons Engrs PC 1040 Ave of the Americas New York NY 10018-3703

GROSSMAN, LAWRENCE, biochemist, educator; b. Bklyn., Jan. 23, 1924; s. Isidor Harry and Anna (Lipkin) G.; m. Barbara Meta Mishen, June 24, 1949; children—Jon David, Carl Henry, Ilene Rebecca. Student, Coll. City N.Y., 1946-47; B.A., Hofstra U., 1949; Ph.D., U. So. Calif., 1954. Scientist NIH, Bethesda, Md., 1956-57; asst. prof. biochemistry Brandeis U., Waltham, Mass., 1957-62; assoc. prof. Brandeis U., 1962-67, prof., 1967-75; E.V. McCollum prof. and chmn. dept. biochemistry Johns Hopkins Sch. Hygiene and Pub. Health, Balt., 1975-89, Univ. Disting. prof., 1989—; mem. sci. adv. com. Am. Cancer Soc.; adviser in biochemistry NIH; U.S. rep. to Internat. Union of Biochemistry, NRC. Author: Method in Nucleic Acids, 12 vols., since 1968; assoc. editor: Cancer Research; editorial bd. Jour. Biological Chemistry; contbr. articles to profl. jours. Trustee Brandeis U. Served to lt. USNR, 1942-45. Decorated DFC, Air medal; Dept. Energy Rsch. grantee; Commonwealth fellow, 1963, Guggenheim fellow, 1973, Burrough-Wellcome fellow, 1989; recipient USPHS Career Devel. award, 1964-74, Merit award NIH, 1989—, Stebbins medal Pub. Health Edn. Mem. Am. Soc. Biol. Chemistry and Molecular Biology, Am. Chem. Soc., Am. Soc. Photobiology. Home: 5828 Pimlico Rd Baltimore MD 21209 Office: 615 N Wolfe St Baltimore MD 21205-2103

GROSSMAN, MARTIN BERNARD, data systems consultant; b. Phila., Oct. 9, 1961; s. Melvin Stephen and Anita Elfriede (Fuller) G.; m. Marlynn Susan (O'Haire) Grossman, May 26, 1984. BS, Rutgers U., 1983; MS, Cen. Mich. U., 1991. Integrated logistics support mgr. NATIONS, Inc., Shrewsbury, N.J., 1989-93; health svcs. planner, data systems cons. New Solutions, Inc., Rochelle Park, N.J., 1993—; health svcs. administr. 78th Divsn., Ft. Dix, N.J., 1992—. Capt. U.S. Army, 1983-89, ETO. Mem. N.Y. Acad. Scis., Soc. Logistics Engrs., N.J. Health Mktg. & Planning Assn. Jewish. Achievements include research on role of integrated logistics support management teams and their influence on favorable management value to supportability issues in the development of new and/or modified communications-electronics products; research on application of statistical and analytical components utilizing computer/automation tools to health planning and marketing. Home: 2020 Fanwood St Oakhurst NJ 07755-1115 Office: New Solutions Inc Rochelle Park NJ

GROSSMAN, NATHAN, physician; b. Milw., Jan. 31, 1914; s. Abraham Grossman and Anna Mekel; m. Feb. 10, 1944; children: Peter Grossman, Robin Bell. BS, U. Wis., 1935; MD, U. Ga., 1938. Diplomate Am. Bd. Internal Medicine 1947. Intern Milw. County Hosp., Wauwatosa, Wis., 1938-39; resident Muirdale Sanatorium, Wauwatosa, Wis., 1939-41, Montefiore Hosp., N.Y.C., 1941-43; cardiology fellow Michael Reese Hosp., Chgo., 1944-45, instr. cardiology, 1945-50; dir. cardiology catheterization lab. Marquette U. Med. Sch., Milw., 1950-55; pvt. practice Milw. Democrat. Jewish. Achievements include visualization of the coronary arteries in a living dog. Office: 411 D St SE Washington DC 20003

GROSSMANN, IGNACIO EMILIO, chemical engineering educator; b. Mexico City, Nov. 12, 1949; s. Donat and Marie-Louise (Epper) G.; m. Ignacio E. Blanca Espinal, Nov. 26, 1977; children: Claudia, Andrew, Thomas. BSc ChemE, U. Iberoamericana, 1974; MSc ChemE, Imperial Coll., 1975, hon. diploma, 1975, PhD ChemE, 1977. Research and devel. engr. Inst. Mexicano del Petroleo, Mexico City, 1978; asst. prof. chem. engring. Carnegie Mellon U., Pitts., 1979-83, assoc. prof., 1983-86, prof., 1986-90, Rudolph R. and Florence Dean prof. chem. engring., 1990—; Robert W. Vaughan lectr. Calif. Inst. Tech., Pasadena, 1986; Mary Upson vis. prof. engring., Cornell U., Ithaca, 1986-87; acad. trustee, v.p. Computer Aids for Chem. Engring. Edn. (CACHE), Austin, Tex., 1983—. Contbr. over 100 tech. publs.; editor: CACHE Design Case Studies; mem. editorial bd. Computers and Chem. Engring. Jour., 1987—, Jour. Global Optimization, 1991—. Recipient Presdl. Young Investigator award NSF, Washington, 1984. Mem. Am. Inst. Chem. Engrs. (chmn. computing and systems tech. div. 1992), Ops. Rsch. Soc. Am., Am. Chem. Soc., Sigma Xi. Roman Catholic. Avocations: classical music. Home: 6385 Douglas St Pittsburgh PA 15217-1821 Office: Carnegie Mellon Univ Dept of Chem Engring Pittsburgh PA 15213

GROSVENOR, GILBERT MELVILLE, journalist, educator, business executive; b. Washington, May 5, 1931; s. Melville Bell and Helen (Rowland) G.; m. Donna C. Kerkam, June 16, 1961 (div.); children: Gilbert Hovey II, Alexandra Rowland; m. Wiley Jarman, June 1, 1979; 1 child, Graham Dabney. BA, Yale U., 1954; D Pub. Service (hon.), George Washington U., 1983; LHD (hon.), U. Colo., 1983, Curry Coll., 1984; LLD (hon.), Coll. Wooster (Ohio), 1983; LHD (hon.), William and Mary Coll., 1987, Miami U., Oxford, Ohio, 1988, Syracuse U., 1989, R.I. Coll., 1991. With Nat. Geog. Soc., 1954—, trustee, 1966—, v.p., 1966-80, assoc. editor, 1967-70, editor, 1970-80, pres., 1980—, chmn. bd., 1987—; bd. dirs. White House Hist. Assn., Conservation Fund, Chevy Chase Bank, FSB, Chesapeake & Potomac Telephone Co., Charles Allmon Trust, Inc., Marriott Corp., Ethyl Corp.; former fellow Yale Corp. Trustee Nat. Wildflower Rsch. Ctr., Fed. City Coun., B.F. Saul Real Estate Trust, N.Y. Zool. Soc.; past vice chmn. Pres.'s Commn. Ams. Outdoors.; chmn. emeritus, found. bd. Alexander Graham Bell Assn. for Deaf; bd. dirs. Am. Farmland Trust; former bd. visitors Coll. William and Mary; ann. corp. mem. Children's Hosp.; former mem. Pres.'s Commn. Environ. Quality, Washington Cathedral Bldg. Com. Recipient Editor of Year award Nat. Press Photographers Assn., 1975; Disting. Achievement award U. So. Calif. Sch. Journalism and Alumni Assn., 1977. Mem. Assn. Am. Geographers, Explorers Club, Newcomen Soc., Alfalfa Club, Alibi Club, Cosmos Club, Chevy Chase Club (Md.). Office: Nat Geographic Soc 17th & M Sts NW Washington DC 20036

GROSZ, BARBARA JEAN, computer science educator; b. Phila., July 21, 1948; d. Joseph Eugene and Judith Phyllis (Zander) Gross. AB in Math., Cornell U., 1969; MA in Computer Sci., U. Calif., Berkeley, 1971, PhD in Computer Sci., 1977. Rsch. mathematician Artificial Intelligence Ctr., SRI Internat., Stanford, Calif., 1973-77, computer scientist, 1981-82, sr. computer scientist, 1981-82, program dir. nat. lang. and representation, 1982-83, sr. staff scientist, 1983-86; co-founder, mem. exec. com., prin. researcher Ctr. for Study of Lang. and Info. Stanford U. and SRI Internat., 1983-86; Gordon McKay prof. computer sci. Div. Applied Scis. Harvard U., Cambridge, Mass., 1986—; vis. faculty dept. computer sci. Stanford U., fall 1982, cons. assoc. prof. computer sci. and linguistics, 1984-85, computer sci., 1985-87; vis. scholar dept. computer and info. sci. U. Pa., Jan.-June 1982; conf. chair Internat. Joint Conf. on Artificial Intelligence (IJCAI-91), chair bd. trustees IJCAI Inc., 1989-91; invited speaker numerous nat. and internat. profl. assns., confs., symposia; participant Sloan workshops Computational Aspects of Linguistic Structure and Discourse Setting, U. Pa., Indefinite Reference, U. Mass., Amherst, 1978 ; organizer, chair panel on discourse, Workshop on Theoretical Issue in Nat. Lang. Processing, U. Ill., Champaign, 1978; reviewer program proposals NSF; participant adv. meetings for rsch. and funding various govtl. agys. Author: (with others) Elements of Discourse Understanding, 1982, Understanding Spoken Language, 1982, Foundations of Cognitive Science, 1988, Intentions in Communications, 1988; assoc. editor Ann. Rev. Computer Sci.; editorial bd. Artificial Intelligence, Am. Jour. Computational Linguistics, 1981-83, contbr. articles and papers to profl. jours., workshops and conf. procs. Fellow AAAS, Am. Asn. Artificial Intelligence (exec. coun. 1981-84, 86-89, pres.-elect 1991-93, pres. 1993—); mem. Assn. Computational Linguistics (exec. com. 1986-88), Spl. Interest Group on Artificial Intelligence of Assn. for Computing Machinery (vice chair 1979-81, chair 1981-83), Internat. Joint Conf. on Artificial Intelligence (program com. 1982). Avocations: hiking, wildflower photography, snorkeling. Office: Harvard U Aiken Computer Lab 33 Oxford St Cambridge MA 02138-2901

GROTTEROD, KNUT, retired paper company executive; b. Sarpsborg, Norway, Feb. 12, 1922; emigrated to Can., 1945, naturalized, 1954; s. Klaus and Maria Magdalena (Thoresen) G.; m. Isabel Edwina MacMaster, Feb. 25, 1950; children: Ingrid, Christopher, Karen. Grad., Tech. Coll., Horten, Norway, 1945; BME, McGill U., Can., 1949, postgrad, 1951; DS (hon.), U. Maine, 1987; Exec. in Residence (hon.), U. New Brunswick, 1989. With Consol. Bathurst Ltd. Que., 1951-70; v.p. prodn., gen. mgr. N.S. Forest

Industries, Port Hawkesbury, 1970-73; v.p. mfg. Fraser Cos. Ltd., Edmundston, N.B., Can., 1973-75, sr. v.p. ops., 1975-80; exec. v.p. Fraser Inc., Edmundston, 1980-85, pres., chief operating officer, 1985-87, chmn., chief exec. officer, 1987-90, ret., 1990; chmn. bd. Atlantic Waferboard, Chatham, N.B., 1985-87, Island Paper Mills, Vancouver, B.C., Can., 1985-87, Alta. Newsprint Co. Ltd., 1988-90, dir., 1988—, Rsch. and Productivity Coun., N.B., 1986; chmn. Potato Devel. and Mktg. Coun., N.B., 1989-90; pres., chmn. Incotech N.B., 1988 ; hon. exec. in residence U. N.B., 1989. Bd. dirs. Canadian-Scandinavian Found., Montreal, 1974-75, v.p., 1975-77, pres., 1978—; mem. bd. govs. U. New Brunswick. With Norwegian Underground Army, 1941-45. Mem. N.B. Forest Products Assn. (dir. 1983-88, pres. 1985-88), Canadian Pulp and Paper Assn., Corp. of Profl. Engrs. of Province of Que. and N.B., Tech. Assn. Pulp and Paper Industry. Lodge: Rotary. Home: 67 Castleton Ct, Fredericton, NB Canada E3B 6H3 Office: Rsch & Productivity Coun, 921 College Hill Rd, Fredericton, NB Canada E3B 5HI

GROTZINGER, TIMOTHY LEE, electronics engineer; b. Renovo, Pa., Jan. 4, 1951; s. Joseph I. and Barbara J. (Miller) G.; m. Kelly Anne Murphy, Oct. 16, 1981. BSEE Tech., Pa. State U., Middletown, 1985, MSEE, 1990. Registered profl. engr., Pa. Technician Syntonic Tech., Highspire, Pa., 1976-86; sr. engr. Laser Communications Inc., Lancaster, Pa., 1986-88, v.p. engring., 1988—. Staff sgt. USAF, 1971-75. Mem. IEEE, NSPE, Pa. Soc. Profl. Engrs. Achievements include patents for laser beam data verification system, laser signal mixer circuit; patent pending for alignment apparatus for laser communication stas.

GROUNES, MIKAEL, metallurgist; b. Warsaw, Poland, Sept. 5, 1933; s. Jacob and Dorota (Tylbor) G.; m. Birgitta Rind Hellkvist, Aug. 30, 1955; children: Jan, Lena, Eva Frida. M Met. Engring., Royal Inst. of Tech., Stockholm, 1956; SM, MIT, Cambridge, 1964; postgrad., Royal Inst. Tech., Stockholm, 1987. Rsch. metallurgist Swedish Ironmasters' Assn., Stockholm, 1956, 58, AB Atomenergi, Studsvik, Sweden, 1958-64; lab head, 1964-69; rsch. mgr. AB Åkers Styckebruk, Sweden, 1969-73; materials rsch. mgr. AB Bofors Åkers, 1973-82, mktg. mgr., 1982-83; product mgr. Studsvik AB, 1983-91; product mgr. fuel testing Studsvik Nuclear AB, Nykoping, Sweden, 1992—. Contbr. articles to profl. jours. Lt. Royal Swedish Navy, 1957. Grantee Swedish Solid State Physics Com., MIT, 1962-63. Mem. Am. Soc. for Materials, Materials Soc. (London), Swedish Materials Soc., Swedish Nuclear Soc., Rotary Club (pres. 1976). Avocations: collecting warship models and photographs. Home: Kvarnvillan, S 64060 Akers Styckebruk Sweden Office: Studsvik Nuclear AB, S 61182 Nykoping Sweden

GROUS, JOHN JOSEPH, hematologist, oncologist, educator; b. New Haven, July 22, 1955; s. John Joseph Sr. and Irene Olga (Vershish) G. BS in Biology, Trinity Coll., Hartford, Conn., 1977; MD, Tufts U., 1981. Intern U. Hosps. of Cleve., 1981-82; resident Case Western U., Cleve., 1982-84; fellow in med. oncology Meml. Sloan Kettering Cancer Ctr., N.Y.C., 1985-87; mem. staff div. hematology/oncology St. Elizabeth's Hosp., Boston, 1987—; asst. prof. medicine Tufts U., Boston, 1987—; lectr. in field. Contbr. articles to New Eng. Jour. Medicine. Mem. Am. Soc. Clin. Oncology. Office: St Elizabeth's Hosp 736 Cambridge St Boston MA 02135

GROVE, JACK STEIN, naturalist, marine biologist; b. York, Pa., Oct. 29, 1951; s. Samuel Hersner and Myrtle Elenor (Stein) G. AS, Fla. Keys Coll., 1972; BS, U. West Fla., 1976; postgrad., Pacific Western U. Chief naturalist Galapagos Tourist Corp., Guayaquil, Ecuador, 1977-84; rsch. assoc. Sea World Rsch. Inst., San Diego, 1981—, L.A. County Mus. Natural History, L.A., 1982—; expedition leader Soc. Expeditions Cruises, Seattle, 1985-91; marine biologist, underwater photographer Eye on the World, Inc., L.A., 1986-91; assoc. investigator Nat. Fisheries Inst., Guayaquil, 1982-85; park naturalist Galapagosd Nat. Park, Ecuador, 1977-85; founder Zegrahm Expdns., Seattle. Editor: Voyage to Adventure/Antarctica, 1985; photographer film documentaries. NSF grantee, 1986. Mem. AAAS, Acad. Underwater Scis., Am. Soc. Mag. Photographers, Am. Bamboo Soc., Profl. Assn. Diving Instrs. (divemaster), U.S. Nat. Recreation and Parks Assn. (supr.). Avocations: sailing, scuba diving, surfing. Office: 146 N Sunrise Dr Tavernier FL 33070

GROVER, JAMES ROBB, retired chemist, editor; b. Klamath Falls, Oreg., Sept. 16, 1928; s. James Richard and Marjorie Alida (Van Groos) G.; m. Barbara Jean Ton, Apr. 14, 1957; children: Jonathan Robb, Patricia Jean. BS summa cum laude, valedictorian, U. Wash., Seattle, 1952; PhD, U. Calif., Berkeley, 1958. Rsch. assoc. Brookhaven Nat. Lab., Upton, N.Y., 1957-59, assoc. chemist, 1959-63, chemist, 1963-67, chemist with tenure, 1967-77, sr. chemist, 1978-93; cons. Lawrence Livermore (Calif.) Nat. Lab., 1962; assoc. editor Ann. Revs., Inc., Palo Alto, Calif., 1967-77; vis. prof. Inst. for Molecular Sci., Okazaki, Japan, 1986-87; vis. scientist Max-Planck Inst. für Strömungsforschung, Göttingen, Fed. Republic Germany, 1975-76. Contbr. numerous articles to profl. jours. With USN, 1946-48. Mem. Am. Chem. Soc. (chmn. nuclear chemistry and tech. 1989), Am. Phys. Soc., Triple Nine Soc., Sigma Xi, Phi Beta Kappa, Phi Lambda Upsilon, Zeta Mu Tau, Pi Mu Epsilon. Libertarian. Presbyterian. Achievements include research in dynamics of chemical reactions, photoionization of van der Waals complexes, nuclear reactions. Home and Office: 1536 Pinecrest Terr Ashland OR 97520

GROVER, MARK DONALD, computer scientist; b. Augusta, Maine, July 12, 1955; s. Donald William and Aletha D. (Wells) G. BA, U. Fla., 1976; MS, Northwestern U., 1978, PhD, 1982. Instr. Northwestern U., Evanston, Ill., 1978-81; mem. tech. staff TRW Def. Systms, Redondo Beach, Calif., Fairfax, Va., 1981-85; sr. computer scientist Advanced Decision Systems, Arlington, Va., 1985-89; prin. software engr. Oberon Software Inc., Cambridge, Mass., 1990—; program chmn. Nat. Symbolics User Group Conf., Washington, 1986; presenter to confs. in field. Contbr. articles to sci. jours. Vol. cert. emergency med. technician. Mem. NRA (life), Phi Beta Kappa, Tau Beta Pi. Avocations: travel, rare books, drama, marksmanship, history. Office: Oberon Software Inc One Cambridge Ctr Cambridge MA 02142

GROVER, ROSALIND REDFERN, oil and gas company executive; b. Midland, Tex., Sept. 5, 1941; d. John Joseph and Rosalind (Kapps) Redfern; m. Arden Roy Grover, Apr. 10, 1982; 1 child, Rosson. BA in Edn. magna cum laude, U. Ariz., 1966, MA in History, 1982; postgrad. in law, So. Meth. U., Dallas. Libr. Gahr High Sch., Cerritos, Calif., 1969; pres. The Redfern Found., Midland, 1982—; ptnr. Redfern & Grover, Midland, 1986—; pres. Redfern Enterprises Inc., Midland, 1989—; chmn. bd. dirs. Flag-Redfern Oil Co., Midland. Sec. park and recreation commn. City of Midland, 1969-71, del. Objectives for Convocation, 1980; mem., past pres. women's aux. Midland Community Theatre, 1970, chmn. challenge grant bldg. fund, 1980, chmn. Tex. Yucca Hist. Landmark Renovation Project, 1983, trustee, 1983-88; chmn. publicity com. Midland Jr. League Midland, Inc., 1972, chmn. edn. com., 1976, corr. sec., 1978; 1st v.p. Midland Symphony Assn., 1975; chmn. Midland Charity Horse Show, 1975-76; mem. Midland Am. Revolution Bicentennial Commn., 1976; trustee Mus. S.W., 1977-80, pres. bd. dirs., 1979-80; co-chmn. Gov. Clements Fin. Com., Midland, 1978; mem. dist. com. State Bd. Law Examiners; trustee Midland Meml. Hosp., 1978-80, Permian Basin Petroleum Mus., Libr. and Hall of Fame, 1989—. Recipient HamHock award Midland Community Theatre, 1978. Mem. Ind. Petroleum Assn. Am., Tex. Ind. Producers and Royalty Owners Assn., Petroleum Club, Racquet Club (Midland), Horseshoe Bay (Tex.) Country Club, Phi Kappa Phi, Pi Lambda Theta. Republican. Home: 1906 Crescent Pl Midland TX 79705-6407 Office: PO Box 2127 Midland TX 79702-2127

GROVES, JOHN TAYLOR, III, chemist, educator; b. New Rochelle, N.Y., Mar. 27, 1943; s. John Taylor and Frances (Gaylor) G.; m. Karen Joan Morrison, Apr. 15, 1967; children—Jay, Kevin. B.S., M.I.T., 1965; Ph.D., Columbia U., 1969. Asst. prof. U. Mich., Ann Arbor, 1969-76, assoc. prof., 1976-79, prof. organic chemistry, 1979-85; prof. organic and inorganic chemistry Princeton (N.J.) U., 1985—, chmn. dept. chemistry, 1988—, Hugh Stott Taylor prof. chemistry, 1991—; Morris S. Kharasch Vis. Prof. U. Chgo., 1993; cons. in field; dir. Mich. Center for Catalytic and Surface Scis., Ann Arbor, 1981-85. Bd. editors Bioorganic Chemistry, 1984—; mem. editorial bd. Reaction Kinetics and Catalysis Letters, 1989—; contbr. articles to profl. jours. Recipient Phi Lambda Upsilon award for outstanding teaching and leadership, 1978, NSF Extension award, 1990-92. Fellow AAAS, Am. Acad. of Arts and Scis.; mem. Am. Chem. Soc. (Arthur C. Cope scholar award 1991), N.Y. Acad. Sci., Sigma Xi. Office: Princeton U Dept Chemistry 203 Hoyt Lab Princeton NJ 08544

GROVES, SHERIDON HALE, orthopedic surgeon; b. Denver, Mar. 5, 1947; s. Harry Edward Groves and Dolores Ruth (Hale) Finley; m. Deborah Rita Threadgill, Mar. 29, 1970 (div. Apr. 1980); children: Jason, Tiffany; m. Nanely Marie Lamont, July 1, 1980 (div. Dec. 1987); 1 child, Dolores; m. Elaine Robbins, Feb. 7, 1991. BS, U.S. Mil. Acad., 1969; MD, U. Va., Charlottesville, 1976. Commd. 2nd lt. US Army, 1969, advanced through grades to maj.; surg. intern US Army, El Paso, Tex., 1976-77, orthopedic surgery resident, 1977-80; staff orthopedic surgeon US Army, Killeen, Tex., 1980-83; resigned US Army, 1983; staff emergency physician various emergency depts. State of Tex., 1983-84, 87; emergency dept. dir. Victoria (Tex.) Regional Med. Ctr., 1984-86; med. dir. First Walk-In Clinic Victoria, 1986-87; tchr. U. Tex. Med. Br., Galveston, 1986-90; emergency dept. dir. Gulf Coast Med. Ctr., 1988-89; with Amerimed Corp., 1990-92, Primedex Corp., 1992—; lectr. Speakers Bur., Victoria, 1984-86, Cato Inst., Ludwig Von Mises Inst. Contbr. articles to profl. jours. Mem. Victoria Interagy. Council Sexual Abuse, 1984-86; treas. bd. dirs. Youth Home Victoria, 1986-90. Recipient Physician's Recognition award AMA, 1980, 83, 86, 89, 92. Fellow Am. Acad. Neurologic and Orthopedic Surgeons, Internat. Coll. Surgeons (U.S. sect.); mem. Am. Acad. Neurologic and Orthopedic Surgeons, Soc. Mil. Orthopedic Surgeons, Am. Coll. Emergency Physicians, Tex. Med. Found., Assn. Grads. of U.S. Mil. Acad. (life), Am. Assn. Disability Evaluation Physicians, Coalition of Med. Providers, Am. Coll. Sports Medicine, Am. Running and Fitness Assn. (cert. recognition 1987), Internat. Martial Arts Assn., Hurricane Sports Club of Houston, Smithsonian Assocs., So. Calif. Striders Track Club. Avocations: track and field masters (3-time nat. champion), martial arts.

GROW, ROBERT THEODORE, economist, association executive; b. Newton, Mass., Aug. 14, 1948; s. William and Lempi (Kangas) G.; m. Anita L. Capps, Nov. 20, 1982; 1 child, Margaret Celia. BS magna cum laude, U. Mass., 1970, MS, 1973. Regional economist Southeastern Va. Planning Dist. Commn., Norfolk, 1973-80; dir. met. coord. Met. Washington (D.C.) Coun. Govts., 1980-85; exec. dir. Washington/Balt. Regional Assn., Washington, 1985—; chmn. met. com. Capital Area chpt. Am. Planning Assn., Washington, 1988-89. Fellow Am. Ctr. for Internat. Leadership; mem. Am. Soc. Assn. Execs., Am. Econ. Devel. Council, Nat. Econimists Club, Southern Industrial Devel. Council, Md. Distbn. Coun., Md. Indsl. Developers Assn., Md. Internat. Trade Assn., Va. Econs. Devel. Assn., Balt. Econ. Soc. (mem. steering com. 1992), Phi Kappa Phi. Avocations: skiing, sailing, golf, trapshooting.

GRUA, CHARLES, government official; b. N.Y.C., Sept. 30, 1934; s. Peter and Lydia E. (Fieno) G.; m. Bonnie Jean Groetsch, June 27, 1959; children: Phyllis J., Sandra J., Charles P., Steven W. BS in Marine Engring., U.S. Merchant Marine Acad., Kings Point, L.I., N.Y., 1957. Maintenance engr. NIH, Bethesda, Md., 1966-67; resident engr. U.S. Dept. Interior, Office Saline Water, Freeport, Tex., 1967-69; resident mgr. U.S. Dept. Interior, Office Saline Water, Roswell, Fountain Valley, Calif., 1969-72; chief program mgmt. and plant engring. div. U.S. Dept. Interior, Office Saline Water, Washington, 1972-74; program mgr. coal gasification U.S. Dept. Interior, Office Coach Rsch., Washington, 1976-82; program mgr. environ. control tech. Energy Rsch. and Devel. Adminstrn. and U.S. Dept. Energy, Germantown, Md., 1982-88; sr. quality engr. safety programs U.S. Dept. Energy, Germantown, 1988-89, sr. quality engr. nuclear energy, 1989-91, tech. safety appraisal team leader, 1991-92; acting dep. dir. Office Spl. Projects U.S. Dept. Energy, Washington, 1992—. Lt. (j.g.) USNR, 1957-58. Mem. ASME, Am. Soc. Quality Control. Achievements include rsch. in desalination of sea and brackish waters; safety of nuclear production facilities; constrn. and operation of the vertical tube evaporated/ multistage flash prototype desalination plant; feasibility of desalination and electric power expansion for Beirut, Lebanon. Home: 11399 Bantry Ter Fairfax VA 22030 Office: US Dept Energy 1000 Independence Ave SW Washington DC 20585

GRUBA-MCCALLISTER, FRANK PETER, psychologist; b. Chgo., Sept. 22, 1952; s. John and Anna Marie (Zonzo) Gruba; m. Sandra Loretta McCallister, Dec. 28, 1974; children: Deirdre Joanna, Brian Francis. MS, Purdue U., 1976, PhD, 1978. Program dir. Epilepsy Rehab. Inst., Cicero, Ill., 1978-79; pvt. practice Park Ridge, Ill., 1979—; couseling psychologist U. Ill. at Chgo., 1979-83; teaching faculty Chgo. Sch. of Profl. Psychology, Chgo., 1981-83; clin. neuropsychologist Alexian Bros. Med. Ctr., Elk Grove Village, Ill., 1983-88; core faculty Ill. Sch. Profl. Psychology, 1984—, assoc. dean, 1990—. Contbr. articles to Jour. of the History of the Behavioral Sciences, Jour. of Humanistic Psychology, Advanced Development. Mem. APA, Ill. Psychol. Assn. Democrat. Roman Catholic. Office: Ill Sch of Profl Psychology 220 S State St Chicago IL 60604

GRUBB, DAVID CONWAY, mechanical engineer; b. Allentown, Pa., Oct. 12, 1944; s. John B. and Margaret (Snow) G.; m. Carol Ann Thane, June 10, 1967; children: Eric, John, Matthew. BS in Mech. Engring., Tri-State U., 1969. Project adminstr. Sanders & Thomas, Inc., Pottstown, Pa., 1970-74; pres. Grubb Engrs. & Constructors, Pottstown, 1974-78; v.p. Kirk Bros, Inc., Glenside, Pa., 1978-82; pres. David C. Grubb Assocs., Pennsburg, Pa., 1982—; v.p. Knoll Group Div. of Westinghouse, East Greenville, Pa., 1984—. Adv. bd. Villa Maria Retreat Ctr., Wernersville, Pa., 1983-87; presenting mem. Mrie Wald Retreat Ctr., Shillington, Pa., 1988—; bldg. com. Upper Perkiomen Libr., Pennsburg, 1989—, Most Blessed Sacrament Ch., Bally, Pa., 1990—; bd. dirs. Upper Perkiomen Valley YMCA, 1992—. Mem. ASHRAE, Am. Soc. Plant Engrs., Internat. Facility Mgrs. Assn., Nat. Soc. Plumbing Engrs. Republican. Roman Catholic. Home: PO Box 292B Pennsburg PA 18073-0221 Office: The Knoll Group PO Box 157 East Greenville PA 18041-0157

GRUBB, ROBERT LYNN, computer system designer; b. Knoxville, Tenn., Nov. 23, 1927; s. William Henry and DoLores Alfisi (Pierucci) Hollinshead; m. Donna Jean Chicado, May 28, 1973; children: Barbara, Robert Lynn, Paul, Werner, Luke, Jubal. BS, Central State Coll., Edmond, Okla., 1972. Air traffic contr. FAA, Fort Worth, 1955-62; engr. Philco-Ford Corp., Oklahoma City, 1962-65; svc. co. exec. Lear-Siegler Inc., Oklahoma City, 1965-67; computer specialist U.S. Navy, Corpus Christi, Tex., 1967-71, U.S. Army, Petersburg, Va., 1971-77, U.S. CSC, Washington, 1977-79, U.S. Dept. Justice, San Antonio, 1979-89, Defense Mapping Agy., Acad. of Health Scis., 1989—; chief exec. officer Tex. Office Systems Co., Inc., Wetmore 1980—; cons. Durham Bus. Coll., Corpus Christi; cons. Corpus Christi Pub. Sch. Bd. Author: Conversion and Implementation of CS3 Computer System, 1973; Economic Analysis of Automated System-TOPS, 1977. Contbr. articles and stories on Western history to various periodicals. Committeeman, Boy Scouts Am., 1963-64; bd. dirs., athletic coach Southside Youth League, 1970. With USNR, 1945-46, PTO. Mem. Western Writers Am. Home and Office: 7926 Broadway Ste 707 San Antonio TX 78209

GRUBBS, JEFFREY THOMAS, environmental biologist; b. Fulton, Ky., Dec. 12, 1957; s. James Thomas and Martha Jane (Johnson) G. BS in Agrl., Murray State U., 1982, MS in Plant Taxonomy, 1989. Teaching asst. Murray (Ky.) State U., 1986-89, herbarium asst., 1988-89; field botanist div. of water Nat. Resources Environ. Protection Cabinet, Frankfort, Ky., 1989-90; environ. biologist sr. Nat. Resources Environ. Protection Cabinet, 1990-91; environ. biologist prin., 1991—. Contbr. articles to Trans. Ky. Acad. Sci., 1989, '92, Castanea, 1991. Home: 901 Crosshill # 10 Frankfort KY 40601 Office: Water Div Natural Resources and Environ Protection 14 Reilly Rd Frankfort KY 40601

GRUBBS, ROBERT H., chemistry educator; b. Calvert City, Ky., Feb. 27, 1942; s. Henry Howard and Faye (Atwood) G.; m. Helen Matilda O'Kane; children—Robert B., Brendan H., Kathleen M. B.S., U. Fla., 1963, M.S., 1965; Ph.D., Columbia U., 1968. NIH postdoctoral fellow Stanford U., Calif., 1968-69; asst. prof. Mich. State U., East Lansing, 1969-73, assoc. prof., 1973-78; prof. chemistry Calif. Inst. Tech., Pasadena, 1978—, Victor and Elizabeth Atkins prof., 1989. Contbr. articles to profl. publs.; patentee in field. Fellow Sloan Found., 1974-76, Alexander von Humboldt Found., 1975; Dreyfus Found. scholar, 1975-78. Mem. Am. Chem. Soc. (award in organic chemistry 1989), NAS. Democrat. Home: 1700 Spruce St South Pasadena CA 91030-4721 Office: Calif Inst Tech Dept Chemistry 164-30 Pasadena CA 91125

GRUBBSTRÖM, KARL ROBERT WILLIAM, economist; b. Malmö, Sweden, Nov. 12, 1941; s. Jarl Ingemar and Dorothy Mary (Taylor) G.; m. Anne-Marie Behm; children: Henrik, Carl, Cecilia. MEE, Royal Inst. Tech., Stockholm, 1965, PhD in Indsl. Econs. and Mgmt., 1969; MBA, Stockholm Sch. Econs., 1968; D in Econs., Gothenburg Sch. Econs., 1973. Asst. Royal Inst. Tech., 1965-67, asst. lectr., 1967-68, acting prof. indsl. econs. and mgmt., 1968-69; rsch. asst. Gothenburg (Sweden) Sch. Econs. and Bus. Adminstrn., 1970; assoc. prof. bus. adminstrn. U. Stockholm, 1970-71, acting prof. bus. adminstrn., 1971-72; acting prof. prodn. econs. Linköping Inst. Tech., 1972-74, prof. prodn. econs., 1974—, head dept. prodn. econs.; vis. prof. Naval Postgrad. Sch., Monterey, Calif., 1978-79, 90-91; external examiner Cranfield Inst. Tech., Bedford, Great Britain, 1983-90; hon. rsch. fellow U. Lancaster, Great Britain, 1988—; cons. to numerous Swedish cos., banks, ednl. instrns., govt. authorities and mil. agys.; bd. dirs. Argentus Aktiesparfondforvaltnings AB, Ostgota Holding Aktiebolag, 1983-85, Scandinavian-Asian Venture Mgmt., Aktiebolag, 1986, others. Author: What is Operations Research, 1969, Economic Decisions in Space and Time, 1973, Decision and Game Theory with Applications, 1978, Production Economics: Trends and Issues, 1985, Recent Developments in Production Economics, 1987, Production Economics: State of the Art and Perspectives, 1989, Production Economics--Issues and Challenge for the 90's, 1991, others; prin. editor Internat. Jour. Prodn. Econs., 1979—, Studies in Production and Engineering Econs., 1979—; assoc. editor Opsearch, 1982-89, Managerial and Decision Econs., 1988—; mem. editorial bd. Kybernetes, 1971—. Bd. dirs. Swedish Young Enterprise Movement, 1980—. Mem. Am. Inst. Indsl. Engrs. (founder, sr. 1980—), Mgmt. Profls. Assn. India (hon. 1983—), Royal Acad. Engring. Scis., Internat. Soc. for Inventory Rsch. (v.p. 1982-90, pres. 1990-92), European Fedn. Young Enterprises (v.p. 1991—), World Orgn. Gen. Systems and Cybernetics, Swedish Inst. Indsl. Engrs. (pres. 1974-78, bd. dirs. 1974-81), Operational Rsch. Soc. Sweden, Inst. Mgmt. Scis. (bd. dirs. coll. prodn. and ops. mgmt. 1988-90), Internat. Found. Prodn. Rsch. (bd. dirs. 1993—), Rotary. Home: Stigbergsgatan 30, S-58245 Linkoping Sweden Office: Linkoping Inst Tech, S-581 83 Linköping Sweden

GRUBE, JOEL WILLIAM, psychologist, researcher; b. Oakland, Calif., May 19, 1949; s. Bruce William Grube and Katharine Dean (Hunting) Gannon; m. Kathleen Ann Kearney, July 14, 1973. AB, U. Calif., Berkeley, 1971; MS, Wash. State U., 1975, PhD, 1979. Asst. dir. Social Rsch. Ctr., Wash. State U., Pullman, 1978-81; sr. rsch. officer Econ. and Social Rsch. Inst., Dublin, Ireland, 1981-84; rsch. scientist Social Rsch. Ctr., Wash. State U., Pullman, 1984-85; postdoctoral fellow Sch. Pub. Health, U. Calif., Berkeley, 1985-86; sr. rsch. scientist Prevention Rsch. Ctr., Pacific Inst. Rsch. and Evaluation, Berkeley, 1986—. Author: Great American Values Test, 1984; author 4 monographs; contbr. over 25 articles to profl. jours. Recipient Prin. Investigator grant Nat. Inst. on Alcohol Abuse and Alcoholism, 1990—. Mem. APA (soc. for personality and social psychology div., evaluation and measurement div., health psychology div.), Am. Psychol. Soc., Rsch. Soc. on Alcoholism, Western Psychol. Assn., Kettil Bruun Soc. Office: Prevention Rsch Ctr 2532 Durant Ave Berkeley CA 94704-1769

GRUBER, JACK, medical virologist, biomedical research administrator; b. Bklyn., Apr. 18, 1931; s. Harry and Rose (Kramer) G.; m. Patricia Ann Mason, June 28, 1964; 1 child, Harry Mason. BS, CUNY, Bklyn., 1954; PhD, U. Ky., 1963. Rsch. asst., lab. instr. dept. microbiology U. Ky., Lexington, 1955-61; rsch. bacteriologist U.S. Army Biol. Labs., Ft. Detrick, Frederick, Md., 1962-63; bacterial immunology microbiologist Med. Scis. Lab., Ft. Detrick, 1963-67, viral immunology microbiologist, 1967-70; microbiologist, rsch. program adminstr. viral biology br. Nat. Cancer Inst., NIH, 1970-72, chief office of program resources and logistics, viral oncology program, 1972-78, asst. chief biol. carcinogenesis br., divsn. cancer etiology, 1978-80, chief, 1984—; dep. chief biol. carcinogenesis br., divsn. cancer etiology Nat. Cancer Inst., 1980-84. Editor: (with others) Primates and Human Cancer, 1979; contbr. articles to profl. jours. Achievements include research in attempts to produce rheumatic fever and in vitro leukocytic hypersensitivity with group A streptococci, purification, concentration and inactivation of Venezuelan equine encephalitis virus, the relationship of DNA viruses and cervical carcinoma, the biology of SV40 and polyomavirus transformations, prospects for human papillomavirus vaccines and immunotherapies, AIDS progress and prospects for vaccine development, pathogenic diversity of Epstein-Barr virus, viral T-antigen interactions with cellular proto-oncogenes and anti-oncogene products, the role of human immunodeficiency virus type 1 and other viruses in malignancies associated with AIDS, progress and prospects for human cancer vaccines. Office: National Cancer Institute NIH Exec Plz N # 540 Bethesda MD 20892

GRUBER, JACK ALAN, chemist; b. Sept. 10, 1943; s. Martin and Meriam (Lewis) G.; m. Dorothy Jean Ruggiero, Aug. 22, 1964; children: Neil Robert, Kelli Anne, Brian Scott. BS in Chemistry and Metallurgy, Youngstown (Ohio) State U., 1972, MS in Analytical Chemistry, 1977; postgrad., Pa. State U., Harvard U., U. Mich. Lab. technician Wheatland (Pa.) Tube Co., 1966-68, chief technician, 1968-70, staff chemist, 1970-72, chief chemist, 1972-79, chief chemist, mgr. quality control, 1980-92, dir. tech. svcs., 1992—; pres. J.A. Gruber & Assocs., Girard, Ohio, 1988—; mem. cadamium task group Am. Iron and Steel Inst., Washington, 1990—; instr. Pa. State U., Sharon, 1982-84. Fellow Am. Inst. Chemists (accredited profl. chemist 1980); mem. ASTM (com. A1 and F14), Am. Soc. Metals (exec. com. 1990—, vice chmn. Mahoning Valley chpt. 1992—), Assn. Iron and Steel Engrs., Am. Chem. Soc., Am. Soc. Quality Control, Am. Soc. Non-Destructive Testing, Nat. Fire Protection Assn., Nat. Elec. Mfrs. Assn. (vice chmn. tech. com.). Office: Wheatland Tube Co Council Ave Wheatland PA 16161-9999

GRUBER, JOHN BALSBAUGH, physics educator, university administrator; b. Hershey, Pa., Feb. 10, 1935; s. Irvin John and Erla R. (Balsbaugh) G.; m. Judith Anne Higer, June 20, 1961; children: David Powell, Karen Leigh, Mark Balsbaugh. B.S., Haverford (Pa.) Coll., 1957; Ph.D., U. Calif. at Berkeley, 1961. NATO postdoctoral fellow Inst. Tech. Physics, Tech. U. Darmstadt, Germany, 1961-62; gastdozent Inst. Tech. Physics, Tech. U. Darmstadt, 1961-62; asst. prof. physics U. Calif. at Los Angeles, 1962-66; assoc. prof. physics Wash. State U., Pullman, 1966-71; prof. chem. physics Wash. State U., 1971-75; asst. dean Wash. State U. (Grad. Sch.), 1968-70, assoc. dean, 1970-72; prof. physics, dean Coll. Sci. and Math., N.D. State U., Fargo, 1975-80; prof. physics and chemistry, v.p. for acad. affairs Portland (Oreg.) State U., 1980-84; prof. physics San Jose State U., 1984—, acad. v.p., 1984-86, v.p. devel., 1986, dir. Inst. for Modern Optics, 1992—; vis. prof. Joint Ctr. Grad. Study, Richland, Wash., 1964, 65, 66, Ames Lab., Dept. of Energy, Iowa State U., 1976-80; Disting. vis. prof. U.S. Navy Naval Weapons Ctr., China Lake, Calif., 1984-93, Stanford U., 1993—; invited lectr., U.S., Can., Europe, 1966—; cons. in solid state physics and spectroscopy Aerospace Corp., El Segundo, Calif., 1962-65, Douglas Aircraft and McDonnell Douglas Astronautics Co., Santa Monica, Calif., 1963-69, N.Am. Aviation, Space and Info. Systems, Downey, Calif., 1964-66, Battelle-Northwest, Richland, Wash., 1964-69, Los Alamos (N.Mex.) Sci. Lab., 1969-71, 73-74; mem. task force lunar exploration sci. Apollo, NASA, 1964-69, 71-73; cons. Harry Diamond Labs. U.S. Army, 1981—, IBM, 1985-90, GTE, 1986-89, Lasergenics, 1986—, Night Vision Lab. U.S. Army, Ft. Belvoir, 1988—, Deltron, 1990-91; mem. Rare Earth Research Conf. Com., 1976-83, exec. com., 1977-83, sec. bd. dirs., 1979-84; gen. conf. chmn. XIV Internat. Rare Earth Research Conf., 1979; exec. sec. Internat. Frank H. Spedding Award, 1979, 83, Willing award, 1986, Internat. Spencer prize for outstanding contbrn. to sci., 1987, Nom. U.S. Asst. Sec. Def. (Spl. Ops.), 1986-87; chmn. U.S. Office Naval Tech. Postdoctoral Selection Bd., 1988—, U.S. Nat. Inst. Sci. and Tech. Postdoctoral Selection Bd., 1989-91; mem. rev. panel Office Naval Grad. Fellowship Program, 1990—. Contbr. articles to profl. jours., chpts. to books; holder numerous patents in laser sci. and tech. Trustee Symphony Bd. Fargo-Moorhead Symphony Orch., 1978-80; pres. Franklin Elementary Sch. PTA, Pullman, 1973-74; pres. elect PTA coordinating council, City of Pullman, 1974-75; v.p. Horace Mann Elementary Sch. PTA, Fargo, 1975-76, pres., 1976-77; mem. PTA coordinating council, City of Fargo, 1976-77, N.D. State Bd. PTA; chmn. Univ., Coll. and Pub. Sch. Relations Bd., 1979-80; active Boy Scouts Am.; trustee Pullman Pub. Library, 1973-75, N.D. Symphony Orchestras Assn., 1978-80; mem. planning commn. City of Pullman, 1972-75; bd. dirs. Westminster Found., 1982-84. Disting. fellow in edn. Am. Soc. for Engring. Edn., 1988—; recipient Outstanding Merit and Performance award San Jose State U., 1990; grantee AEC-ERDA, 1963-75, NSF, 1966-72, 76-78, 79-80, 89—, NASA, 1966-72, 76-78, U.S. Army Research Office, Durham, 1979-80, Am. Chem. Soc. Petroleum Research Funds, 1979-80, Dept. Energy, 1979-84, Dept. Def., 1984—, Office Naval Rsch., 1987—, Office Naval Tech., 1988—; fellow

NASA Ames Lab., 1993—; vis. scholar Stanford U., 1993—. Fellow Am. Soc. Engring. Edn. (Disting.), Am. Phys. Soc. (chmn. nat. mtg. sessions); mem. AAAS, N.Y. Acad. Scis., N.D. Acad. Sci., Oreg. Acad. Sci., Coun. Colls. Arts and Scis., Optical Soc. No. Calif. (v.p. 1992, pres. 1993), Phi Beta Kappa, Sigma Xi, Phi Kappa Phi, Sigma Pi Sigma, Phi Sigma Iota. Presbyterian (ruling elder). Clubs: Mason (Shriner), Kiwanian. Home: 5870 Meander Dr San Jose CA 95120-3839

GRUBER, KENNETH ALLEN, health scientist, administrator; b. N.Y.C., June 21, 1948; s. Isidore and Frances (Schauben) G.; m. Gail Frances Miller, Dec. 1, 1982 (div. Aug. 1989); children: Vanessa Gail, Sarah Alexis. BA, NYU, 1971, PhD, 1974. Predoctoral fellow NYU, N.Y.C., 1971-74; adj. assoc. prof. L.I. U., Bklyn., 1974-75; postdoctoral fellow Roche Inst. Molecular Biology, Nutley, N.J., 1974-76; asst. prof. Wake Forest U. Sch. Medicine, Winston-Salem, N.C., 1976-80, assoc. prof., 1981-89; prof. physiology U. P.R. Sch. Medicine, San Juan, 1989-92; program adminstr. NIH, Rockville, Md., 1992—; cons. Thereplix Pharms., Paris, 1988, Ipsen Pharms., Paris, 1992. Capt. M.S. USAR, 1970—. Individual rsch. grantee NSF, 1978, NIH, 1983-88; recipient Internat. Travel award NRC, 1980, Rsch. Career Devel. award Nat. Heart, Lung and Blood Inst., 1980. Mem. Am. Physiol. Soc., Assn. Pour Les Exch. Sci. Internat. (hon.). Jewish. Achievements include research in humoral regulation of renal and cardiovascular systems. Home: 106 Olney Sandy Spring Rd Ashton MD 20861 Office: Nat Inst Deafness & Comm Disorders EPS 400B 6120 Executive Blvd Rockville MD 20892

GRUBIN, HAROLD LEWIS, physicist, researcher; b. Bklyn., Mar. 1, 1939; s. Saul Solomon and Sallye Grubin; m. Ruth Lena Levine, Aug. 26, 1961; children: Scott Geoffrey, Rachael Alyssa. BS, Bklyn. Coll., 1960; MS, Poly. Inst. Bklyn., 1962, PhD, 1967. Sr. theoretical physicist United Techs. Rsch. Ctr., East Hartford, Conn., 1966-81; v.p. Scientific Rsch. Assocs., Glastonbury, Conn., 1981—; adj. prof. N.C. State U., Raleigh, 1992—; internat. editor John Wiley & Sons, Great Britain, 1989—. Co-author: The Gunn Hilsum Effect, 1978, The Physics of Instabilities in Solid State Electron Devices, 1993; co-editor: The Physics of Submicron Structures, 1984, The Physics of Submicron Semiconductor Devices, 1988. Mem. IEEE, Am. Phys. Soc. Office: Scientific Rsch Assocs PO Box 1058 Glastonbury CT 06003

GRUCHALLA, MICHAEL EMERIC, electronics engineer; b. Houston, Feb. 2, 1946; s. Emeric Edwin and Myrtle (Priebe) G.; m. Elizabeth Tyson, June 14, 1969; children: Kenny, Katie. BSEE, U. Houston, 1968; MSEE, U. N.Mex., 1980. Registered profl. engr., Tex. Project engr. Tex. Instruments Corp., Houston, 1967-68; group leader EG&G Washington Analytical Services Ctr., Albuquerque, 1974-88; engring. specialist EG&G Energy Measurements Inc., Albuquerque, 1988—; cons. engring., Albuquerque; lectr. in field, 1978—. Contbr. articles to tech. jours.; patentee in field. Judge local sci. fairs, Albuquerque, 1983—. Served to capt. USAF, 1968-74. Mem. IEEE, Instrumentation Soc. Am., Planetary Soc., N.Mex. Tex. Instruments Computer Group (pres. 1984-85), Sigma Xi, Tau Beta Pi, Eta Kappa Nu. Avocations: electro-optics, photography, woodworking. Office: EG&G Energy Measurements Inc Kirtland Ops PO Box 4339 Albuquerque NM 87196-4339

GRUEBELE, MARTIN, chemistry educator; b. Stuttgart, Fed. Republic Germany, Jan. 10, 1964; came to U.S., 1980; s. Helmut and Edith Victoria (Berner) G.; m. Nancy Makri, July 10, 1992. BS in Chemistry, U. Calif., Berkeley, 1984, PhD in Chemistry, 1988. Rsch. fellow Calif. Inst. Tech., Pasadena, 1989-92; asst. prof. U. Ill., Urbana, 1992—; cons. Quantum Design Corp., San Diego, 1991. Fellow IBM, 1986-87, Dow Chem. Co., 1987-88; recipient New Faculty award Dreyfus Found., 1992. Mem. AAAS, Am. Phys. Soc., Am. Chem. Soc., Sigma Xi. Achievements include theoretical and experimental studies of previously unobserved transient molecular species, studies in laser-control chemical reactions, observation of unperturbed bimolecular reatcion using femtosecond lasers. Office: U Ill Dept Chemistry 505 S Mathews Ave Urbana IL 61801

GRUEN, ARMIN, photogrammetry educator; b. Berneck, Bavaria, Germany, Apr. 27, 1944; s. Walter and Hedwig (Ammon) G.; m. Gudrun Poehner, Feb. 12, 1969; children: Gunilla, Gillian. Diploma in engring., Tech. U. Munich, 1968, D of Engring., 1974. Researcher Bavarian Acad. Sci., Munich, 1968-69; asst. prof. Tech. U. Munich, 1969-74, assoc. prof., 1974-80; assoc. prof. Ohio State U., Columbus, 1980-84; prof. photogrammetry and remote sensing Swiss Fed. Inst. Tech., Zurich, 1984—; cons. to govt. agys. and cos., Germany, U.S., Switzerland, 1969—; v.p. Internat. Soc. for Photogrammetry and Remote Sensing. Editor: Optical 3-D Measurement Techniques, 1989, (proc.) Close-Range Photogrammetry Meets Machine Vision, 1990. Recipient Otto v. Gruber gold medal Internat. Soc. Photogrammetry and Remote Sensing, 1980. Mem. German Soc. Photogrammetry, Am. Soc. Photogrammetry and Remote Sension (Talbert Abrams award 1985), Swiss Soc. Photogrammetry. Avocations: reading, music, sports. Office: Inst Geodesy & Photog, ETH Hoenggerberg, CH-8093 Zurich Switzerland

GRUGEL, RICHARD NELSON, materials scientist; b. Milw., Dec. 22, 1951; s. Harry Robert and Corinne Linnea (Nelson) G.; m. Cynthia Carla Conwill, Dec. 21, 1991. MS in Metallurgy, U. Wis., Milw., 1980; PhD in Metallurgy, Mich. Tech. U., 1983. Postdoctoral rsch. assoc. Swiss Fed. Ins. Tech., Lausanne, 1983-87, rsch. assoc., 1987; vis. researcher Northwestern Poly. U., Xian, China, 1985-87; rsch. asst. prof. Vanderbilt U., Nashville, 1987-92, rsch. assoc. prof., 1992—. Rsch. grantee NASA, 1991. Mem. AIAA, Internat. Metallographic Soc., The Metall. Soc. Achievements include microstructural development in alloys as a function of processing parameters; novel dendrite growth procedures; novel processing of hypermonotectic alloys. Home: 907 Winston Pl Nashville TN 37204 Office: Vanderbilt U Box 6079 Sta B Nashville TN 37235

GRUGER, EDWARD HART, JR., retired chemist; b. Murfreesboro, Tenn., Jan. 21, 1928; s. Edward Hart and Edith (Sundin) G.; m. Audrey Ruth Lindgren, June 27, 1952; children: Sherri Jeanette, Lawrence Hart, Linda Gay. BS, U. Wash., 1953, MS, 1956; PhD, U. Calif., Davis, 1968. Chemist U.S. Bur. Comml. Fisheries, Seattle, 1953-54, organic chemist, 1955-59, supervisory chemist, 1959-65, rsch. chemist, 1965-70; rsch. assoc. Agrl. Exptl. Sta. U. Calif., Davis, 1965-68; supervisory rsch. chemist NOAA, Seattle, 1970-83; ret., 1983; rsch. prof. Seattle U., 1977-82; coord., rsch. advisor NRC resident rsch. associateship postdoctoral program NOAA, Seattle, 1977-83. Contbr. papers to profl. publs., chpts. to books. Mem. adv. com. Planned Parenthood of Seattle-Puget Sound, 1983-87; precinct com. officer Legis. Dist. Seattle, 1984—. With USN, 1946-48. Mem. AAAS (emeritus), Am. Oil Chemist' Soc. (emeritus), Elks (Wash. State Emerald award 1992), Sigma Xi. Democrat. Achievements include research to support vitamin E and similar molecules as biological antioxidants; polychlorinated biphenyls in fish activating enzymic systems for biotransformations of petroleum hydrocarbons; analysis of fish fatty acids. Home: 3727 NE 193d St Seattle WA 98155-2750

GRUMET, MARTIN, biomedical researcher; b. N.Y.C., June 30, 1954; s. Ephraim and Harriet (Blank) G.; m. Judith Brener, Feb. 17, 1986; children: Avi, Alexandra. BS in Physics, Cooper Union, 1976; PhD in Biophysics, Johns Hopkins U., 1980. Postdoctoral fellow Rockefeller U., N.Y.C., 1980-84, asst. prof., 1984-91; assoc. prof. NYU Med. Ctr., N.Y.C., 1991—. Contbr. articlees to JOur. Neurosci. Rsch., Current Opinion in Neurobiology. NIH grantee; recipient Career Scientist award Irma T. Hirschl Trust, 1986-91. Mem. AAAS, Am. Soc. for Cell Biology, Soc. for Neurosci. Democrat. Jewish. Achievements include discovery of several neural cell adhesion molecules including Ng-CAM, Nr-CAM and cytotactin; cloned cDNAs and determined structure of Ng-CAM and Nr-CAM. Office: NYU Med Ctr 550 1st Ave New York NY 10016

GRUNDY, RICHARD DAVID, engineer; b. San Mateo, Calif., Mar. 17, 1937; s. John Richard and Violette (Morris) G.; m. Claudia Copeland, Sept. 4, 1977 (div. 1992). BSEE, Stanford (Calif.) U., 1958; MS, U. Calif., 1963, postgrad., 1964; postgrad., George Wash. U., 1965-67, Harvard U., 1980. Profl. staff engr. subcom. on air and water pollution Com. on Environ. and Pub. Works, U.S. Senate, Washington, 1967-71; exec. sec., profl. staff mem. Senate's Nat. Fuels and Energy Policy Study, U.S. Senate, Washington, 1971-76; sr. profl. staff mem. Com. on Environ. and Pub. Works, U.S. Senate, Washington, 1971-76; sr. prof. staff mem. energy Com. on Energy and Natural Resources, U.S. Senate, Washington, 1977-86, minority sr. prof. staff mem., 1987—; chmn. protocol com. 2d Internat. Clear Air Congress, Internat. Union of Air Prevention Assns., 1970; mem. steering com. Aspen Inst. Energy Forum, 1985-91; observer White Ho. Conf. on Global Climate Change, Washington, 1990, 93, UN Negotiations on Climate Change, Geneva, 1993; participant UN Conf. on Clean Coal Tech. in Devel. Countries, Beijing, 1991. Author: (with others) Air Pollution and Industry, 1972, Consumer Health and Product Hazards/Cosmetics, 1974, Radiation Exposures from Consumer Radiologic Services, Toxic Substances: Possible Regulatory Policies; co-editor: Consumer Health and Product Hazards, 1974; contbr. numerous articles to profl. jours. Mem. adminstrv. bd. Foundry United Meth. Ch., Washington, 1960-62, 66-70, mem. coun. of mins., 1967-70, chmn. membership commn., 1968-70; mem. nat. planning com. Nat. Youth Govs. Conf., YMCA and Readers Digest Found., Washington, 1975-80; mem. Air Pollution Control Assn., 1967-82; pres. Nat. Capital Orchid Soc., Washington, 1989-90. Comdr. USPHS, 1959-67. Fellow AAAS; mem. IEEE, NSPE, Assn. of Energy Engrs., D.C. Soc. of Profl. Engrs. (Young Engr. of the Yr. 1970). Methodist. Home: 8905 Linton Ln Alexandria VA 22308-2731 Office: Com Energy & Nat Resources US Senate Washington DC 20510

GRUNES, ROBERT LEWIS, engineering consulting firm executive; b. Bklyn., Aug. 15, 1941; s. Abe and Doris (Dicker) G.; m. Eleonora Grasselli, Oct. 14, 1972; children: Natalie, Daniel, Ian. BS in Engring., Poly. Inst. Bklyn., 1963, MS in Engring., 1965, PhD in Phys. Metallurgy, 1970. Registered profl. engr., N.Y., N.J., Pa. Engr. Pratt & Whitney div. United Aircraft Corp., East Hartford, Conn., 1963; rsch. fellow Poly. Inst. Bklyn., 1963-64, rsch. assoc., 1966-70; rsch. engr. Lewis Rsch. Ctr. NASA, Cleve., 1965; pres. R. L. Grunes & Assocs., Inc., N.Y.C., 1970—; mem. adj. faculty N.J. Inst. Tech., Newark, 1974-78. Author: Pollution Control Market and Industries, 1971; contbr. articles to profl. jours. 1st lt. CE U.S. Army, 1964-66. Mem. ASME, ASCE, ASTM, Metall. Soc., Soc. Automotive Engrs., Nat. Fire Protection Assn., Am. Boat & Yacht Coun. Office: R L Grunes & Assocs Inc 521 Fifth Ave New York NY 10175

GRÜNEWALD, MICHAEL, physics researcher, research program coordinator; b. Bielefeld, Fed. Republic of Germany, June 24, 1954; s. Aloys and Helga (Kemper) G.; m. Ruth Brockmeier, June 13, 1985. Pre-Dipl. Math. and Physics, Philipps U., Marburg, Fed. Republic of Germany, 1974, Dipl.phys., 1979, Dr.rer.nat. in Physics, 1984. Sci. asst. in math. and physics Philipps U., Marburg, Fed. Republic of Germany, 1974-79, rsch. staff dept. physics, 1979-84; rsch. grantee German Rsch. Orgn., Marburg, Fed. Republic of Germany, 1984-85; rsch. scientist Dornier Med. Systems, Munich, 1985-90, sr. scientist, project leader, developer novel medical devices; program coord. Joint European Submicron Silicon Project, Munich, 1990—; internat. lectr. in field; referee for sci. papers. Patentee in field; contbr. numerous articles to profl. jours. Mem. Deutsche Physikalische Ges., Acoustical Soc. Am. Avocations: tennis, sailing, mountain hiking. Home: An der Biberwiese 15, D-8034 Germering Germany Office: JESSI Office, Elektrastr 6A, D-8000 Munich 81, Germany

GRUNINGER, ROBERT MARTIN, civil engineer; b. Paterson, N.J., Aug. 20, 1937; s. Martin A. and Henrietta (Van Decker) G.; m. Margaret M. Cooke, Aug. 29, 1959; children: Robert M., John M. BS in Civil Engring., N.J. Inst. Tech., 1960, MS in Civil Engring., 1968. Registered profl. engr., Md., N.J., N.Y., Pa., Fla.; diplomate environ. engr. Engr. N.Y. Telephone Co., Bronx, 1960-66; engr., project engr. Malcolm Pirnie, INc., White Plains, N.Y., 1966-68; project mgr. Malcolm Pirnie, INc., Paramus, N.J., 1968-79; v.p. C.C. Johnson & Malhotra, P.C., Silver Spring, Md., 1979-90, Lewis & Zimmerman Assocs., Inc., Rockville, Md., 1990-91, Gannett Fleming Inc., Balt., 1991—; chmn. Engring. Found. Conf. on Solid Waste Mgmt., N.Y.C., 1983. Editorial dir. Indsl. Water Engring., 1980-85. Mem. ASCE (com. chair Md. and D.C., 1986—), AAEE, Bergen County Soc. Profl. Engrs. (pres. 1975-76), Am. Water Works Assn., Water Environ. Fedn. Office: Gannett Fleming Inc Village of Cross Keys 200 E Quadrangle Baltimore MD 21210

GRUNKE, ANDREW FREDERICK, astronomy educator, instrument designer; b. West Palm Beach, Fla., May 20, 1946; s. Frederick Andrew and Elizabeth jane (Huntington) G.; m. Marsha Leonora Oliver, Jan. 5, 1968, (div. 1973); 1 child, Erika Corrinne; m. Jacqueline Christine Talerico. BA, Western State Coll., 1983; MS, Bowling Green State U., 1985. Computer operator IBM Svc. Bur., Charlotte, N.C., 1966, Sealtest Foods, Charlotte, Tarcon, Inc., Charlotte; with electronics staff Tesco, Inc., Charlotte, S.E. Film and Sch. Supply, Columbia, S.C., Boulder (Colo.) Pub. Libr., 1975; electronics instr. Phoenix Inst. Tech., 1987-90; astronomy instr. Yavapai Community Coll., Prescott, Ariz., 1992—; instrument designer, founder Aspect Instruments, Wickenburg, Ariz., 1991—. Contbr. articles to profl. jours. With communications Ariz. Wing Civil Air Patrol, Wickenburg, 1992. Recipient Achievement Cert. John Hopkins U., 1991; Wester State Found grantee, 1979-80. Achievements include design and construction of Czerny-Turner spectroscope for small aperture telescope users; design of pentatent, phloth, evolving baseline interferometry, machine controls EEG input, gravity telescope, universal symbol processor. Home and Office: Aspect Instruments PO Box 423 2451 Las Canoas Rd Santa Barbara CA 93105

GRUNTMAN, MICHAEL A., physicist, researcher, educator; b. Tashkent, USSR, Dec. 1, 1954; came to U.S., 1990; s. Alexander Yu and Raisa Z. (Donskaya) G. MSc in Physics, Moscow Phys.-Tech. Inst., 1977; PhD in Physics, Space Rsch. Inst., Moscow, 1984. Rsch. scientist Space Rsch. Inst., USSR Acad. Sci., Moscow, 1977-86, Inst. for Problems in Mechanics, USSR Acad. Sci., Moscow, 1987-90, U. So. Calif., L.A., 1990—; assoc. prof. dept. aerospace engring. U. So. Calif., 1993—; reporter Internat. Assn. Geomagnetism and Aeronomy, 1983-87; mem. com. on astronomical optical detectors Acad. of Scis., USSR, 1988-90; sci. sec. Internat. Cosmic Gas Dynamic Conf., Moscow, 1988. Contbr. articles to profl. jours. Mem. AIAA, Am. Phys. Soc., Am. Geophysical Union. Office: U So Calif Dept Aerospace Engring MC-1191 Los Angeles CA 90089

GRUNWALD, ARNOLD PAUL, communications executive, engineer; b. Berlin, Dec. 7, 1910; came to U.S., 1952, naturalized, 1957; s. Richard Michael and Hedwig (Bamann) G.; m. Grete Marie Gwinner, Dec. 29, 1945; children: Eva Dubowski, Peter. Degree in physics and math., Univ., Munich, 1933; degree in engring., Tech. Univ., Munich, 1945. Chief engr., gen. mgr. Wehoba GmbH, Weilheim, Germany, 1946-49; engr. Capital Engring. Co., Chgo., 1952-58; assoc. engr. Argonne (Ill.) Nat. Lab., 1958-77, chmn. Senate, 1971-72; cons., pres. Rsch. for Braille Communication, Chgo., 1977—; cons. Am. Found. for Blind, N.Y.C., 1973-76, divsn. for blind Libr. Congress, Washington, 1970; cons. engr. Chisholm, Boyd & White, Chgo., Ethicon Inc., Chgo.; participant internat. confs. on engring., social and ethical issues. Contbr. articles to profl. jours. Vice pres., edn. chmn. Parents of the Blind, Chgo., 1957-67; group discussion leader World Federalists, Chgo.; lectr. Union of Concerned Scientists, Argonne. Recipient Letter of Commendation, Pres. U.S., 1976, One of 100 Most Significant Products award Indsl. Rsch. mag., 1969; Hew grantee U.S. Dept. Health, Edn. and Welfare, 1969-75. Mem. ACLU, Fedn. Am. Scientists, Nat. Fedn. Blind, Sigma Xi. Achievements include U.S. and foreign patents in the field. Home: 18135 Martin Ave Homewood IL 60430

GRUPPEN, LARRY DALE, psychologist, educational researcher; b. Zeeland, Mich., Jan. 27, 1955; s. Howard Melvin and Gertrude Jean (Huizenga) G.; m. Mary Louise Shell, May 27, 1978; children: Timothy Andrew, Matthew Scott. MA, U. Mich., 1984, PhD, 1987. Rsch. investigator U. Mich. Med. Sch., Ann Arbor, 1987-88, asst. rsch. scientist, 1988—. Contbr. articles to Jour. AMA, Acad. Medicine, other profl. publs. Grantee Agy. Health Care Policy and Rsch., 1991—, Mich. Alzheimer's Disease Rsch. Ctr., 1992—, NIH, 1993—. Mem. APA, Am. Ednl. Rsch. Assn., Soc. Med. Decision Making, Soc. Judgement and Decision Making. Achievements include investigation of foundation and development of medical expertise, clinical reasoning and expert judgment, development of computer-based educational methods in medical education, exploration of process of innovation dissemination. Office: U Mich Med Sch G1211 Towsley Ctr Ann Arbor MI 48109-0201

GRUSKY, DAVID BRYAN, sociology educator; b. L.A., Apr. 14, 1958; s. Oscar Grusky and Jean (Roethlisberger) Richon; m. Szonja Szelényi, Sept. 3, 1988. BA, Reed Coll., 1980; MS, U. Wis., 1983, PhD, 1987. Asst. prof. U. Chgo., 1986-88; asst. prof. Stanford (Calif.) U., 1988-92, assoc. prof., 1992—. Editor: Social Stratification: Class, Race and Gender in Sociological Perspective, 1993; contbr. articles to profl. jours.; co-editor Westview Press Series on Social Inequality, 1987—; assoc. editor Am. Jour. Sociology, 1987-88; mem. editorial bd. Stanford U. Press, 1990—. Presdl. Young Investigator NSF, Washington, 1988-93; Spencer rsch. fellow Nat. Acad. Edn., 1989; fellow Ctr. for Advanced Study in Behavioral Scis., Stanford, 1991-92. Mem. AAAS, Am. Sociol. Assn., Population Assn. Am., Internat. Sociol. Assn., Phi Beta Kappa.

GRYCHOWSKI, JERRY RICHARD, mechanical engineer; b. Katowice, Silesia, Poland, Feb. 24, 1940; came to U.S., 1980; s. Eugeniusz Jerzy and Magdalena Maria (Ruranska) G.; m. Helen Maria Blaska, Oct. 27, 1967; children: Albert Jerry, Joanne Margaret. MSME, Silesian Tech. U., Gliwice, Poland, 1962, PhD in Tech. Sci., 1969. Asst. to the head of dept. Silesian Tech. U., Gliwice, 1962-80; new product engr. The Lockformer Co., Lisle, Ill., 1980-86; mgr. mech-engring. Northgate Rsch. Inc., Arlington Heights, Ill., 1986-92; mgr. rsch. and devel. aerosol applications in medicine Trudell Med. Co. Ltd., London, Ont., Can., 1992—; cons. TRAMEC, Inc., Iola, Kans., 1980-84, Barnant Co., Barrington, Ill., 1990—; pres. Engring. Soc. of Energy Dept. of Silesian Tech. U., Gilwice, 1975-80. Contbr. 27 articles to profl. jours. Achievements include 2 U.S. patents and 5 Polish patents for Centrifugal Pumps, Instrumentation Medical Devices. Home and Office: 535 Waterford Dr Lake Zurich IL 60047-2995 Office: Trudell Med Co Ltd 100 N Gordon St Elk Grove Village IL 60007

GSCHWINDT DE GYOR, PETER GEORGE, economist; b. Budapest, Hungary, Jan. 1, 1945; came to U.S., 1975; s. George and Marie Henrietta (Haggenmacher) G.; m. Michele Herman, Oct. 14, 1972; children: Henrik, Marie. MA in Econs., Brussels U., Belgium, 1967. Grad. trainee Samuel Montagu and Co., London, 1967-69; dep. mgr. Europe Chase Manhattan Bank, London, 1970; credit officer Banque de Commerce (Chase), Antwerp and Brussels, 1971-75; ops. officer IMF, Washington, 1975-80, economist, 1980—. Dir. Hosp. Relief Fund for Caribbean, Chevy Chase, Md., 1982—. Decorated knight Magistalis Grace Sovereign Mil. Order Malta, Rome, 1975. Mem. Am. Conf. on Religious Movements, European Conf. on Religious Movements (dir. Brussels 1991—). Roman Catholic. Avocation: golf. Home: 10108 Fleming Ave Bethesda MD 20814

GU, CLAIRE XIANG-GUANG, physicist; came to U.S., 1985; BS, Fudan U., Shanghai, China, 1985; PhD, Calif. Inst. Tech., 1989. Rsch. asst. Calif. Inst. Tech., Pasadena, 1986-89, vis. assoc., 1989-90; mem. tech. staff Rockwell Sci. Ctr., Thousand Oaks, Calif., 1989-92; asst. prof. Pa. State U., 1992—. Author: (with others) Optical Processing and Computing, 1989, An Introduction to Neural and Electronic Network, 1990; contbr. articles to Nature, Jour. Applied Physics, Optics Letters. Recipient Young Investigator award NSF, 1993; Calif. Inst. Tech. fellow, 1985, Fudan U. scholar, 1984, 85. Mem. Optical Soc. Am. Achievements include research in optical information processing, nonlinear optics, volume holography and optical neural networks. Office: Pa State U Dept Elec Engring 121 Elec Engring East University Park PA 16802

GU, JIANG, biomedical scientist; b. Beijing, China, Apr. 19, 1949; came to U.S., 1985; s. Liang and Jingli (Chang) G.; m. Bei Li, Dec. 25, 1984; 1 child, Ting L. MD, Jilin Med. U., 1977; PhD, Royal Postgrad. Med. Sch., London, Eng., 1984. Rsch. assoc. Royal Postgrad. Med. Sch., London, 1983-84; sr. lectr. Beijing Med. U., 1984-85; vis. fellow NIH, Bethesda, Md., 1985-86; head immunopathology Evanston (Ill.) Hosp., Northwestern U., 1986-87; sr. scientist Deborah Rsch. Inst., Browns Mills, N.J., 1987-90, chmn. sci. affairs, 1990—; assoc. prof. biochemistry Robert Wood Johnsom Med. Sch. U. Medicine and Dentistry N.J., North New Brunswick, 1991—; prof. pathology Health Sci. Ctr. SUNY, Brooklyn, 1992—. Editor: Modern Analytic Methods in Histochemistry, 1993; editor-in-chief Internat. Jour. of Analytical Morphology, 1993—; contbr. over 70 articles to profl. jours. Grantee NIH, 1990, 92, Am. Heart Assn., 1988. Fellow Am. Coll. Angiology, Am. Assn. for Lab. Animal Sci., Am. Histochemistry Soc. Achievements include first report of neuropeptide tyrosin in heart, of vasoactive intestinal polypeptide in male and female genital organs and linkage of them to sexual functions and dysfunctions; research in atrial natriuretic peptide mediated coronary perfusion regulatory system, peptide containing nerves in heart, viscera, urinary bladder and genital organs, in cardiac neuroendocrinology, and in AIDS pathogenesis. Office: Deborah Rsch Inst 1 Trenton Rd Browns Mills NJ 08015-3205

GU, YOUFAN, cryogenic researcher; b. Changshu, Jiangsu, China, May 19, 1963; came to U.S., 1988; s. Zukang and Yingnan (Ling) G.; m. Xiaoyan Shi, May 4, 1986; 1 child, Shuonan. BS, Xian (China) Jiaotong U., 1982, MS, 1985; PhD in Chem. Engring., U. Colo., 1993. Lectr. Xian Jiaotong U., 1985-88. Contbr. articles to Internat. Jour. of Refrigeration, Cryogenics, Advances in Cryogenic Engring., Gas Separation and Purification. Mem. AICE, Cryogenic Soc. Am., Sigma Xi. Achievements include research in multicomponent gas coadsorption breakthrough curve in molecular sieve adsorption system,; modified stability theory for thermal acoustic oscillation in cryogenic systems; development of damping thermal acoustic oscillations. Office: U Colo Dept Chem Engring Boulder CO 80309

GU, YUANCHAO, biomedical engineer, geneticist; b. Shanghai, People's Republic of China, Nov. 21, 1952; came to U.S., 1985; s. Zhi Fang and Yiming (Zhu) G.; m. Li Hao, July 30, 1980; 1 child, Liang. MD, Tongji Med. U., Wuhan, People's Republic of China, 1977. Rsch. assoc. Med. Coll. Ga., Augusta, 1985-92; biochemistry & genetic technologist II Emory U., Atlanta, 1993—. Mem. AAAS. Achievements include research on two new and different quadruplicated globin gene arrangements; detection of a new hybrid globin gene among African Americans; research on the gene expression and the promoter function from different promoter point mutation in globin gene and sickle cell disease, the function and structure of Guanylate cyclase-A/Atrial natriuretic factor receptor gene, mental retardation and Down's syndrome. Home: 603 Trafalgar Ln Augusta GA 30909-3331 Office: Emory U Sch Medicine Dept Genetics & Molecular Medicine Atlanta GA 30322

GUADAGNO, MARY ANN NOECKER, social scientist, consultant; b. Springville, N.Y., Sept. 21, 1952; d. Francis Casimer and Josephine Lucille (Fricano) Noecker; m. Robert George Guadagno, Aug. 29, 1970 (div. Mar. 1981). BS in Edn. cum laude, SUNY, Buffalo, 1974; MS, Ohio State U., 1977, PhD, 1978. Grad. teaching assoc. Ohio State U., Columbus, 1974-77, grad. rsch. assoc., 1977-78; asst. prof. U. Minn., St. Paul, 1978-83; cons. Nationwide Ins. Co., Columbus, 1982-83, rsch. assoc. Corp. Rsch., 1983-86, product devel. assoc., Office of Mktg., 1986-89; adjunct prof. Coll. Bus. & Pub. Adminstrn. Franklin U., Columbus, Ohio, 1985-89; lectr. Coll. Bus. Adminstrn. and Econ. Ohio Dominican Coll., Columbus, 1986-89; sci. Family Econ. Rsch. Group U.S. Dept. Agr., Washington, 1989—; com. mem. Fed. Women's Program, U.S. Dept. Agr., Beltsville, Md., 1991—; mem. Women in Sci., U.S. Dept. Agr., Beltsville, Md., 1991—. Author: Family Inventory of Money Management, 1982, Family Inventory, 1982; contbr. articles to profl. jours., 1978—. Com. mem. United Way, Mkt. Rsch. Info. Exchange, Columbus, Ohio. Recipient Spl. Recognition award Ohio House Reps., 1987, Cert. Grad. award Columbus Area Leadership Program, 1987, Cert. Appreciation award Am. Mktg. Assn., 1987, Cert. Merit award U.S. Dept. Agr., 1991. Mem. Columbus Area Leadership Program, Ohio State U. Coll. Human Ecology Alumni. Republican. Roman Catholic. Avocations: horseback riding, classical music, eastern philosophy, gardening. Home: 3401 Hampton Hollow Dr # M Silver Spring MD 20904-6179 Office: US Dept Agriculture Family Economics Rsch Group 6505 Belcrest Rd Rm 439A Hyattsville MD 20782-2011

GUANGZHAO, ZHOU, theoretical physicist; b. Changsha, Hunan, China, 1929. Grad., Qinghua U., 1951; grad. theoretical physics, Beijing U., 1954. Rschr. Dubra Joint Inst. Nuclear Rsch., Moscow, 1957-60; dir. 9th Rsch. Inst., 2d Ministry of Machine Building; rschr. and dir. Inst. Theoretical Physics, Chinese Acad. Scis.; v.p. Chinese Acad. Scis., pres., 1987—; mem. divsn. maths. and physics Chinese Acad. Scis.; vice-chair China Physics Soc. Co-recipient 1st Prize award for Natural Scis., issued by the state, 1964.

Achievements include pioneering achievements in putting forward theory of spiral amplitude of particle and established relevant mathematical method; advisor to research in detonation physics, radiation hydrodynamics and computing method in mechanics; contributed to designing theoretically China's first atomic and hydrogen bomb. Office: Academia Sinica, 52 San Li He Rd, Beijing 100864, China*

GUARDIA, DAVID KING, obstetrics/gynecology educator; b. New Orleans, June 13, 1952; s. Charles Edward and Dorothy (Tangye) G. BA in Biology, Pittsburg (Kans.) State U., 1974; MD, U. Kans., Kansas City, 1977. Diplomate Am. Bd. Ob-Gyn. Intern U. Kans. Sch. Medicine, Kansas City, 1978-79, resident in ob-gyn, 1979-81; asst. prof. ob-gyn. U. Kans. Med. Ctr., Kansas City, 1982-88; divsn. dir. Ob/Gyn Emergeny Svcs., Kansas City, 1982-88; clin. cons. VA Hosp., Kansas City, 1982-88; liaison dir. Kaiser-Permante, Shawnee Mission, Kans., 1984-88; clin. cons. Wyeth-Ayerst Labs., Ciba Geigy Pharms., Merck-Sharp-Dohme, Upjohn Labs., Ethicon Endosuture. Named one of Outstanding Young Men of Am., 1982, 85. Fellow Am. Coll. Ob-gyn.; mem. AMA, Internat. Coll. Surgeons, Assn. Profs. of Gyn-Ob. (undergrad. edn. com. 1983-88), Kansas City Round Table Endocrinology, Kermit E. Krantz Soc. Democrat. Episocpalian. Home: 5141 Mesa Del Oso Rd NE Albuquerque NM 87111-3706 Office: U NMex Sch Medicine Dept Ob/Gyn 2211 Lomas Blvd NE Albuquerque NM 87131

GUARNACCIA, DAVID GUY, mechanical engineer; b. Winchester, Mass., Jan. 21, 1961; s. Guy and Antoinette (DeSomonie) G.; m. Kathleen Ann Ellison, Nov. 11, 1989. MBA, Suffolk U., 1992, postgrad., 1990—. Devel. engr. Varian Assocs., Lexington, Mass., 1984-90; sr. engr. BTU Internat., North Billerica, Mass., 1990-92; engr. Norton Co., Northboro, Mass., 1992—. Contbr. articles to Am. Vacuum Jour. Mem. ASME, Am. Vacuum Soc. Home: 391 East St Carlisle MA 01741-1102

GUBBINS, KEITH EDMUND, chemical engineering educator; b. Southampton, Eng., Jan. 27, 1937; came to U.S., 1962; m. Pauline Margaret Payne, June 28, 1960; children: Nick, Vanessa. B.Sc. in Chemistry, Queen Mary Coll., U. London, 1958; Diploma in Chem. Engring., King's Coll., U. London, 1959, Ph.D. in Chem. Engring., 1962. Vis. lectr. U. London, Eng., 1960-62; postdoctoral fellow U. Fla., Gainesville, 1962-64, asst. prof., 1964-68, assoc. prof., 1968-72, prof., 1972-76; T.R. Briggs prof. engring. Cornell U., Ithaca, N.Y., 1976—, dir. Sch. Engring., 1983-90; vis. cons. theoretical physics div., U.K. Atomic Energy Authority, Harwell, U.K., 1971; vis. prof. physics dept., U. Guelph, 1971-73, 76, U. Kent, Canterbury, Eng., 1975; vis. prof. chemistry, U. Oxford, 1979-80, 86-87; vis. prof. chem. engring., U. Calif., Berkeley, 1982; Reilly lectr. U. Notre Dame, London, 1978, McCabe lectr. N.C. State U., 1986, Lindsay lectr. Tex. A&M U., 1989, Katz lectr. U. Mich., 1991, Wohl lectr. U. Del., 1991, Merck lectr. Rutgers U., 1992, Olaf Hougen vis. prof. chem. engring., U. Wis., 1993; cons. Mobil Oil, 1979, 80, Exxon Engring., 1980-81, Union Carbide Corp., 1981, Process Simulation Internat., 1982, Nat. Bur. Standards, 1983, BP Rsch., U.K., 1985, 89, Exxon Rsch. and Engring. Co, Clinton, N.J., 1985—, Unilever Rsch., Port Sunlight, U.K., 1985, Linde Div., Union Carbide Corp., 1988, Mobile Rsch., Princeton, 1991, Exxon Chem. Co., 1991; mem. NAS com. to study formation of Nat. Resource Ctr. for Computing in Chemistry, 1976-77, NRC Assessment Bd. to review NIST programs, 1988-91. Mem. editorial bd. Molecular Physics, 1978-87; mem. editorial bd. Molecular Simulation, 1986—, assoc. editor, 1990—; assoc. editor Am. Int. Chem. Engrs. Jour., 1988-91; editor: Topics in Chem. Engring., Oxford U. Press, 1991—. Recipient ann. award best paper Can. Soc. Chem. Engring., 1973; Eppley Found. fellow Imperial Coll., London, 1970-71, Guggenheim fellow, 1986-87, sr. vis. fellow (SERC award) U. Oxford, 1986-87. Mem. NAE, AAAS, Am. Chem. Soc., Am. Inst. Chem. Engrs. (program com. 1974-81, Alpha Chi Sigma award 1986), Am. Inst. Physics, Chem. Soc. (London). Office: Cornell U Sch Chem Engring 270 Olin Hall Ithaca NY 14853

GUBLER, DUANE J., research scientist, administrator; b. Santa Clara, Utah, June 4, 1939; s. June and Thelma (Whipple) G.; m. Bobbie J. Carroll, Mar. 1, 1958; children: Justin Chase, Stuart Jefferson. BS, Utah State U., 1963; MS, U. Hawaii, 1965; ScD, Johns Hopkins U., 1969; AS, So. Utah State Coll., 1962, DSc (hon.), 1988. Asst. prof. of pathobiology Sch. of Hygiene Johns Hopking U., Balt. and Calcutta, 1969-71; assoc. prof. tropical medicine Sch. Medicine U. Hawaii, Honolulu, 1971-75; head virology dept. Naval Med. Rsch. Unit Number 2, Jakarta, Indonesia, 1975-78; assoc. prof. U. Ill., Urbana, 1978-79; rsch. microbiologist div. vector-borne viral diseases Ctrs. for Disease Control, Fort Collins, Colo., 1980-81; dir. San Juan (P.R.) Labs. Ctrs. for Disease Control, 1981-89; dir. Div. Vector-Borne Infectious Diseases Ctrs. for Disease Control, Ft. Collins, Colo., 1989—; cons. South Pacific Commn., 1972-76, WHO, Geneva, 1974—, AID, Washington, 1977—, Pan Am. Health Orgn., 1981—, Internat. Devel. Rsch. Ctr., Ottawa, Can., 1977—. Contbr. numerous articles to profl. jours. Lt. USN, 1975-77; capt. USPHS. Recipient Commendation medal, 1984, Outstanding Svc. medal, 1988, Meritorious Svc. medal, 1991. Mem. AAAS, Am. Soc. Tropical Medicine (Charles Franklin Craig lectr. 1988), Am. Soc. Parasitologists, Am. Mosquito Control Assn., Entomol. Soc. Am. (highlights in med. entomol. lecture 1979), Rotary (Rotarian of Yr. San Juan chpt. 1986, meritorious svc. award Rotary Found., Evanston, Ill., 1990). Home: 717 Dartmouth Trl Fort Collins CO 80525-1522 Office: Ctrs for Disease Control USPHS PO Box 2087 Fort Collins CO 80522

GUCKENHEIMER, JOHN, mathematician; b. Baton Rouge, Sept. 26, 1945. BA, Harvard U., 1966; PhD in Math., U. Calif., Berkeley, 1970. Vis. lectr. IMPA, Rio de Janeiro, 1969; sr. rsch. fellow U. Warwick, 1969-70; mem. Inst. Advanced Study, 1970-72; lectr. Mass. Inst. Technol., 1972-73; from asst. prof. to prof. U. Calif., Santa Cruz, 1973-85; prof. Cornell U., 1985—; dir. Ctr. Appl. Math., 1989—; chmn. U. Calif. Coord. Com. Nonlinear Sci., 1983-85; bd. dirs. Math. Sci. Rsch. Inst., 1982-85. Mem. Am. Math Soc., Soc. Indsl. and Applied Math. Achievements include research in bifurcation theory. Office: Cornell U Ctr Applied Math 305 Sage Hall Ithaca NY 14853-6201*

GUDE, WILLIAM D., retired biologist; b. Balt., Feb. 27, 1914; s. William D. and Mary Cecilia (Mullikin) G.; m. Mary Lillian Stebbins, Feb. 14, 1942; children: Patricia Lucille Perkins, Katie Lee Dripps. AB, Tulane U., 1940; MS in Zoology, U. Tenn., 1952, MS in Indsl. Mgmt., 1959. Clin. lab. tech. Hotel Dieu Hosp., New Orleans, 1940-42; tech. dept. anatomy Tulane Sch. Medicine, New Orleans, 1946-48; supr. biology divsn. Oak Ridge (Tenn.) Nat. Lab., 1948-82, ret., 1982. Author: Autoradiographic Techniques, 1968, (with others) Histological Atlas Laboratory Mouse, 1982. Reader Recording for Blind, Oak Ridge, 1950-88; instr. in reading Tri-County Literacy Unit, Oak Ridge, 1988—. Lt. col. U.S. Army, 1942-46, ETO. Home: 128 Pembroke Rd Oak Ridge TN 37830

GUDEMA, NORMAN H., civil engineer; b. Bronx, N.Y., Nov. 7, 1936; s. Daniel and Theresa Gudema; m. Madeline Eisner, Sept. 28, 1992 (div.); children: Michelle, Daniel, Jonathan. BCE, CUNY, 1959; MBA, Fairleigh Dickinson U., 1969. Cert. prof. engr., N.J., N.J., Pa., Ill., Ariz. Project engr. Exxon Rsch. & Engring., Florham Park, N.J., 1966-68, Witco Chem., Oakland, N.J., 1968-70; project mgr. Warner Lambert, Morris Plains, N.J., 1970-81, Lehrer-Mcgovern, N.Y.C., 1985; dir. engring. Cosmair, inc., Clark, N.J., 1985-88, Revlon, Inc., N.Y.C., 1988-89, 81-85; site mgr. Liberty Sci. Ctr., Jersey City, N.J., 1989—. 1st lt. U.S. Army, 1959. Mem. ASCE, Chi Epsilon. Democrat. Jewish. Home: 27 Coddington Terr Livingston NJ 07039 Office: Liberty Sci Ctr 251 Phillip St Jersey City NJ 07305

GUDIN, SERGE, plant breeder; b. Paris, Oct. 18, 1959; s. Claude Romain Pierre and Jacqueline Anne-Marie (Chieusse) G.; m. Elisabeth Jeanne Juliette Vernhes, Sept. 1, 1979; children: Camille Lise Justine, Bastien Octave Clément. MS in Plant Devel. and Improvement, U. Marseille, France, 1984; PhD, U. Nice, France, 1989. Edn. asst. U. Nice, 1984; head applied rsch. Sélection Meilland, Antibes, France, 1985-88, rsch. dir., 1988—, head of selection, 1990—, production mgr., 1993—. Contbr. articles to sci. publs. Mem. Internat. Soc. Hort. Sci., French Soc. Vegetal Physiology. Roman Catholic. Avocations: gardening, fishing, reading, bull fights. Home: Domaine St André, 83340 Le Cannet des Maures France Office: Sélection Meilland, 83340 Le Cannet des Maures France

GUDJONSSON, BIRGIR, physician; b. Akureyri, Iceland, Nov. 8, 1938; s. Gudjon and Kristjana (Jakobsdottir) Vigfusson; m. Heidur Anna Vigfusdottir, Oct. 21, 1961; children: Asdis, Gunnar, Sigrun. MD, U. Iceland, 1965; postgrad. fellow Yale U., 1970-72. Diplomate internal medicine and gastroenterology; recert. in internal medicine. Intern, Stamford Hosp., Conn., 1966-67, resident, 1967-68; resident in medicine Yale New Haven Hosp., 1968-70; asst. prof. medicine Yale U. Med. Sch., 1972-73, 77-78, 82; practice medicine specializing in internal medicine and gastroenterology, Reykjavik, 1974—; cons. City Hosp., Reykjavik, 1974-77, Med. Clinic, 1974—. Author: (with H.M. Spiro) Controversies in Internal Medicine, 1980; contbr. articles to profl. jours. Mem. council Athletic Union Iceland, 1981—. Fellow ACP, Royal Coll. Physicians London, Royal Soc. Medicine; mem. Am. Gastroenterol. Assn., Brit. Soc. Gastroenterology, Brit. Assn. Sport and Medicine, Am. Coll. Sports-Medicine, World Assn. Hepato-Pancreato-Biliary Surgery, N.Y. Acad. Sci. Lutheran. Club: Reykjavik Athletic. Avocation: international judge in athletics and gymnastics. Home: Alftamyri 51, 108 Reykjavik Iceland Office: Medical Clinic, Alfheimum 74, 108 Reykjavik Iceland

GUDMUNDSON, BARBARA ROHRKE, ecologist; b. Chgo.; d. Lloyd Ernest and Helen (Bullard) Rohrke; m. Valtyr Emil Gudmundson, June 14, 1951 (dec. Dec. 1982); children: Holly Mekkin, Martha Rannveig. BA, U. Tenn., 1950; MA, Mankato State Coll., 1965; PhD, Iowa State U., 1969. pvt. practice cons. ecologist, Des Moines, Mpls., 1968-78; mem. adv. com. Mpls. Lakes Water Quality, Mpls., 1974-75; field ecologist Mississippi River Canoe Expedition, Coll. of the Atlantic, Bar Harbor, 1979. Microbiologist Hektoen Inst. & Ill. Ctr. Hosp., Chgo., 1950-52; immunologist Jackson Meml. Lab., Bar Harbor, Maine, 1952-54; dist. ecologist Corps of Engrs., St. Paul, 1971-72; sr. ecologist North Star Rsch. Inst., Mpls., 1972-76; staff engr. Met. Waste Control Commn., St. Paul, 1976-77; pres., prin. ecologist Ecosystem Rsch. Svc./Upper Midwest, Mpls., 1978—. Editor-in-chief The Icelandic Unitarian Connection, 1984; contbr. articles to profl. jours. Mem. from 61st dist. Dem.-Farmer-Labor Cen. Com., Minn. 1978-80; mgr. Minnehaha Creek Watershed Dist., Hennepin & Carver Counties, Minn., 1979-83; mem. Capital Long-Range Improvements com., Mpls., 1981. Recipient River Basin Ecology grantee Iowa Acad. Scis., Cedar Falls, 1976, Mississippi River Ecology grantee Freshwater Biol. Rsch. Found., Navarre, Minn., 1979, Fulbright Sr. Rsch. grantee USA/Iceland Fulbright Commns., Washington, Reykjavik, 1986, 92. Mem. Ecol. Soc. Am. (pres. Minn. chpt. 1971-75), Minn. Acad. Sci. (sec.-treas. 1973-75), Geol. Soc. Minn. (pres. 1981), Phycological Soc. Am., Sigma Delta Epsilon (nat. membership com. 1990—, chair 1991—), Phi Kappa Phi (chair 1991-93), Sigma Xi. Unitarian Universalist. Achievements include discovery of diatom genus Biddulphia in the state of Iowa; establishment of Diatom Herbarium of Iceland. Home: 5505 28th Ave S Minneapolis MN 55417-1957 Office: Ecosystem Rsch Svc/Upper Midwest PO Box 17102 Minneapolis MN 55417-0102

GUDRY, FREDERICK E., JR., aerospace medical researcher. Recipient Eric Liljencrantz award Aerospace Med. Assn., 1991. Home: 3861 Ochuse Drive Pensacola FL 32503*

GUEDES-SILVA, ANTÓNIO ALBERTO MATOS, economist; b. Porto, Douro Litoral, Portugal, Apr. 18, 1948; s. António Augusto Aires and M. Perpétua Rodrigues (Lucas) Guedes-S.; m. Maria Elisa Correia Gomes Costa, Feb. 4, 1967; children: Margarida Maria, Maria Celeste, Ivo Miguel. Degree in econ., U. Porto, Portugal, 1978. Econ. tchr. Rodrigues Freitas Secondary Sch., Porto, 1980-82; tchr. Perpétuo Socorro Pvt. Sch., Porto, 1982-83; econ. cons. Perpétuo Socorro Sch., Porto, 1988—; concillor Nat. Fedn. Edn. Unions, Porto, 1983, pres. auditing com., 1990—; concillor Union Geral de Trabalhadores, Lisboa, 1984; bd. dirs. North Region Tchrs. Union, econ. and unionism lectr., 1985, Pub. Svc. Union Fedn., Lisboa, Portugal, v.p. 1989; bd. dirs. Inst. Superior de Edn. Trabalho; tchr. Rodrigues de Freitas Secondary Sch., Porto, 1985—. Contbr. articles to profl. jours. Regional Com. SocialDem. Youth, Porto, 1976-77; v.p. tchrs. dept. Social Dem. Party, Porto, 1984-90; social judge Children Ct., Porto, 1988—; particiant edn. and tng. mtgs. European Community, European Union Confedn., OECD. Social Democrat Party. Roman Catholic. Home: Oliveira Monteiro 791-1 E, 4000 Porto Douro Litoral, Portugal Office: Sindicato dos Prof Zona N, D João IV 610, 4000 Porto Douro Litoral, Portugal

GUELL, DAVID CHARLES, chemical engineer. BSChemE, Iowa State U., 1984; MSChemE, MIT, 1987, PhD in Chem. Engring., 1990. Postdoctoral fellow Los Alamos (N.Mex.) Nat. Lab., 1990-92, tech. staff mem., 1992—. NSF grad. fellow. Mem. AICE. Office: Los Alamos Nat Lab MS G736 Los Alamos NM 87545

GUELLER, SAMUEL, civil and environmental engineer; b. Casares, Argentina, Feb. 9, 1935; came to U.S., 1985; s. Abraham and Paulina (Bilik) G.; m. Sofia Rebeca Rendler, Jan. 5, 1961 (dec. 1980); children: Daniel Horacio, Eduardo Javier; m. Berta Fanny Awruch, Feb. 12, 1981. MSc, U. Cin., 1987, PhD, 1991. Registered profl. engr., Argentina. Asst. prof. Northeast U., Resistencia, Argentina, 1968-81; sr. engr. Ministry Pub. Works, Corrientes, Argentina, 1970-82, Elscint Ltd., Sao Paulo, Brazil, 1982-85, Farlow, Inc., Indpls., 1989-90, Rumpke, Inc., Cin., 1990-92, Inter-Am. Devel. Bank, Washington, 1992—; founding mem. Congress Big Hydroelectric Works, Buenos Aires, 1970—. Author: Frontiers of Physics, 1987; author tech. reports and sci. rsch. papers. Recipient Internat. Cooperation award Delft (The Netherlands) U., 1971-72, Technion U., Haifa, Israel, 1973-74, U. Graz, Austria, 1976. Fellow ASCE; mem. Am. Water Works Assn., Internat. Assn. Hydraulic Rsch., Inter-Am. Soc. Sanitary Engrs., Sigma Xi. Jewish. Achievements include development of software; research in computer math models. Home: 1232 W Kemper Rd # 323 Cincinnati OH 45240-9999 Office: Inter-Am Devel Bank 1300 New York Ave NW Washington DC 20577-9999

GUENGERICH, FREDERICK PETER, biochemistry educator, toxicologist, researcher; b. Pekin, Ill., Jan. 1, 1949; married, 1973; 3 children. BS, U. Ill., 1970; PhD in Biochemistry, Vanderbilt U., 1973. Rsch. assoc. biochemistry Sch. Medicine U. Mich., 1973-75; prof. biochemistry Sch. Medicine Vanderbilt U., Nashville, 1975—, rschr. Ctr. Molecular Toxicology; fellow Nat. Inst. Gen. Med. Sci., 1974-75; prin. investor Nat. Cancer Inst., 1978—. Recipient Rsch. Career Devel. award Nat. Inst. Health Sci., 1978—, J.J. Abel award Am. Soc. Pharm. Exptl. Therapeutics, 1984, Founder's award Chem. Industry Inst. Toxicology, 1991. Mem. Am. Soc. Biol. Chemists, Am. Chem. Soc., Sigma Xi. Achievements include research in enzymic activation and detoxification of foreign chemicals of environmental interest. Office: Vanderbilt Univ Ctr in Molecular Toxicology 21st Ave S & Garland Nashville TN 37232*

GUENTER, THOMAS EDWARD, retired chemical engineer; b. Eveleth, Minn., May 10, 1927; s. William Theodore and Selma (Gliem) G.; m. Phyllis Jeanne Nuoffer, Aug. 20, 1949; children: Kathryn, William, Robert, Carol. BS in Engring., U. Mich., 1949. Process devel. engr. E.I. DuPont Co., Niagara Falls, N.Y., 1950-52; process ops. engr. E.I. DuPont Co., Memphis, 1952-62; rsch. engr. E.I. DuPont Co., Niagara Falls, 1963-75; tech. transfer engr. E.I. DuPont Co., Taiwan, Korea, Can., N.Z., 1976-92. Lutheran. Achievements include patent for mfg. of hydrogen peroxide. Home: 6796 Briarmeadows Dr Memphis TN 38120

GUERIN, PATRICK GERARD, veterinary surgeon, consultant; b. Paris, Oct. 7, 1965; s. Hubert and Marie-Claire (Tardy) G. DVM, Ecole Vet. de Nantes, France, 1987; Certificat d'Aptitude, à l'Administration des Ent., France, 1990; grad. in Law, Faculté de Droit de Nantes, France, 1991. Chartered acct. Vet. surgeon Sante Animale, Nantes, 1989—; chmn. bd. SA Lab. Celtipharm vet. lab., GFO fin. co.; cons. on vet. mkt. Contbr. articles to profl. jours. With Mil. Vet. Svcs. French Army, 1989-90, capt. Frency Army Res., 1990—. Mem. Nat. Coun. of Union pour la Democratie Francaise. Avocations: sailing, diving, chess. Home: Nisnizan, 22130 Corseul France

GUÉRITÉE, NICOLAS, endocrinologist; b. Bucharest, Romania, Dec. 29, 1920; s. Virgile-Georges and Marie-Antoinette (Gebhardt) G.; m. Gabriela Rizescu, Dec. 6, 1944 (div. 1954); 1 child, Jean-Claude; m. Lucienne Suzanne Taillebois, July 24, 1954; children: Catherine, Virginie. MD, U. Med. Sch., Bucharest, 1944, Faculté de Médecine, Paris, 1952. Pvt. practice medicine specializing in endocrinology Paris, 1957-85; cons. in endocrinology Hosp. of Nanterre, Paris, 1958-66, Endocrine Dept. Faculty of Medicine La Pitié-Salpétriére, Paris, 1960-85; head med. dept. French Sub. of Schering A.G. Berlin, Paris, 1949-58; head rsch. dept. Lab. Théramex SA, Monaco and Paris, 1958-75; bd. dirs. Laboratoire Théramex SA, 1975—; expert WHO, 1958; mem. French Nat. Com. of Qualification of the Endocrinologists, 1976-89; founder, exec. ann. Journees Francaises d'Endocrinolgie Clinique, 1980—; pres. EEC Monospecialist Sect. of Endocrinology, 1989—. founder and chief editor La Revue Francaise d'Endocrinologie Clinique, 1960—; inventor steroid compounds. Mem. French Nat. Union of Endocrinologists (founder, exec. pres. 1960-91), Societe Francaise d'Endocrinologie, Am. Soc. Bone and Mineral Rsch., Soc. for Study of Reprodn., Endocrine Soc. (Am.). Christian Orthodox. Avocations: skiing, fly fishing.

GUEST, GERALD BENTLEY, veterinarian; b. Hart County, Ga., July 20, 1936; s. Ottis Clayton and Esther Alma (Tankersley) G.; D.V.M., U. Ga., 1960; m. Anne Frances Monaghan, Sept. 1, 1962; children—Gerald, John. Research veterinarian U.S. Dept. Agr., U.Pa., Kennett Sq., 1963-70; spl. asst. to dir. Bur. Vet. Medicine, FDA, Rockville, Md., 1972-77, asso. dir. for food safety, 1978-82, dep. dir. bur., 1982-86, dir., 1986—; practice vet. medicine, Rockville, 1974—. Served with USAF, 1960-63. Mem. AVMA, Nat. Assn. Fed. Veterinarians, Assn. Food Hygiene Veterinarians, D.C. Vet. Med. Assn., Phi Zeta. Office: Food & Drug Adminstrn Ctr for Vet Medicine 5600 Fishers Ln Rockville MD 20857-0001

GUEST, WELDON S., biomedical products and services executive; b. Wichita Falls, Tex., Nov. 16, 1947; m. Cynthia Ann Terrill; children: Courtney, Emily. BA, U. Tex., Austin, 1970; MD, U. Tex., Dallas, 1974. Internat. biomed. cons., 1974-85; chmn. Biodynamics Internat., Tampa, Fla., 1985-90; chmn., pres. Xynet Labs., Houston, 1991—. Office: Xynet Labs Inc 4545 Bissonnet # 285 Bellaire TX 77401

GUETZKOW, DANIEL STEERE, computer company executive; b. Ann Arbor, Mich., May 19, 1949; s. Harold S. and Lauris (Steere) G.; m. Diana Gulbinowicz, April, 1979. Student, Columbia U., 1967-70; BSBA in Accountancy, Thomas Edison State Coll., 1989; MS in Bus. and Mgmt., Acctg. Systems, U. Md., 1991. CPA, D.C.; cert. mgmt. acct. Prodn. mgr. Rehrig-Pacific, Inc., L.A., 1975-78; maintenance mgr. Setco, Inc., Culver City, Calif., 1978-79; plant mgr. Veneer Tech., Inc., L.A., 1979; co-founder, chief fin. officer, chief ops. officer, exec. v.p. Netword, Inc., Riverdale, Md., 1981—; also dir. Netword, Inc., Riverdale. Author: (book) Indemnification of Officers and Directors, 1988, (software) Telemarketing Database Mgr., 1984-86, Systems Accounting Control, 1985, Electronic Mail Switcher, 1982, 84, Compute Marginal IRS Tax Rate Using Linear Programming Sensitivity Analysis, 1989, Working Capital Liquidity Mgmt. Simulation, 1990, Use of Information Theory to Determine When to Post Audit Capital Budgeting Decisions, 1991, Leading/Lagging Paradigm for Classifying Performance Indicators for Total Quality Management, 1991; contbr. articles to profl. jours. Mem. AICPA, D.C. Inst. CPAs (chief fin. officers and mgmt. cons. svcs. coms.), Ops. Rsch. Soc. Am., Inst. Mgmt. Accts., Md. Assn. CPAs, Nat. Assn. Corp. Dirs. Office: Netword Inc PO Box 840 Riverdale MD 20738-0840

GUEZENNEC, YANN GUILLAUME, mechanical engineering educator; b. Saint Brieuc, France, Oct. 2, 1956; came to U.S., 1977; s. Alain G. and Genevieve (Dauguet) G.; m. Colleen C. McGuinness, July 13, 1985; children: Alexandra M., Patrick M. Diplome d'Ingenieur cum laude, Inst. Nat. Scis. Appliquees, Lyon, France, 1979; PhD in Mech. and Aerospace Engring., Ill. Inst. Tech., Chgo., 1985. Rsch. asst., instr. dept. mech. and aerospace engring. Ill. Inst. Tech., Chgo., 1980-85, summer vis. prof., 1986, cons., 1986; vis. scientist Lulea U. Tech., Sweden, 1987, Ctr. Turbulence Rsch., Stanford U./NASA, Ames, Calif., 1987, 1990; vis. scientist Ford Motor Co., 1993; asst. prof. dept. mech. engring. Ohio State U., Columbus, 1986-91, assoc. prof., 1991—; founder, pres. Customsoft, Inc., 1992—. Contbr. articles to profl. jours. 2d lt. U.S. Army, 1979-80. Lilly Teaching fellow Ohio State U., 1988-89. Mem. ASME, AIAA, Am. Phys. Soc., SAE, LIA, Sigma Xi, Pi Tau Sigma, Tau Beta Pi. Roman Catholic. Achievements include research in turbulent flows and turbulence control for heat and drag modification, particle image velocimetry and related techniques, and advanced signal processing and statistical techniques. Home: 1999 Arlington Ave Upper Arlington OH 43212 Office: Ohio State U Dept Mech Engring Columbus OH 43210

GUIDA, JAMES JOHN, manufacturing engineer, consultant; b. N.Y.C., Apr. 6, 1957; s. James and Carmela (Maggiore) G.; m. Linda Ann Blier, May 29, 1983; children: James, Christopher. BS in Aero. Engring., Poly. U. Bklyn., 1979; MS in Engring. Mgmt., Northeastern U., 1985. Mfg. engr. GE Aircraft Engines, Lynn, Mass., 1979-88; dir. mfg. Sverdrup Tech., Inc., Eglin AFB, Fla., 1988-91; mgr. advanced mfg. engring. Sverdrup Corp., Seattle, 1991-93; mem. Aviation Week Adv. Bd., N.Y.C., 1989-91. Contbr. articles to profl. publs. Achievements include study of major Air Force weapons manufacture production operation and recommendations for improvement, statistical process control program for key manufacturing suppliers. Home: 1390 Windward Ln Niceville FL 32578 Office: Sverdrup Tech Inc PO Box 1935 Eglin A F B FL 32542

GUIDI, JOHN NEIL, scientific software engineer; b. Chgo., July 28, 1954. BSChE, Purdue U., 1976. Scientific application programmer/analyst The Jackson Lab., Bar Harbor, Maine, 1977-82, systems programmer/analyst, 1982-88, sr. applications programmer/analyst, 1988-89, scientific software engr., 1989—; vis. prof. math. Coll. of the Atlantic, Bar Harbor, 1990; faculty mem. Short Course in Med. & Experimental Mammalian Genetics, Bar Harbor, 1985-88, cons., 1989, 91-93; mem. People to People Computer Software Del., China, 1983. Contbr. chpts. to books and articles to profl. jours. Mem. IEEE Computer Soc., Am. Assn. Artificial Intelligence, Assn. Computing Machinery. Achievements include software releases MATRIX an on line strain, locus database for the mouse, khconduit providing MEDLINE access via TCP/IP. Office: The Jackson Lab 600 Main St Bar Harbor ME 04609-1500

GUILAK, FARSHID, biomedical engineering researcher, educator; b. Tehran, Iran, Sept. 17, 1964; came to U.S., 1970; s. Hooshang and Nahid (Toufigh) G. BS, Rensselaer Poly. Inst., 1985, MS in Biomed. Engring., 1987; MPhil, Columbia U., 1990, PhD in Mech. Engring., 1991. Computer programmer U. Houston, 1982; rsch. asst. Rensselaer Poly. Inst., Troy, N.Y., 1983, grad. rsch. asst., 1985-86; computer programmer Houston Mus. Natural Sci., 1983-84; rsch. fellow Columbia U., N.Y.C., 1986-91; asst. prof. orthopaedic surgery and mech. engring. SUNY, Stony Brook, 1991—. Reviewer Jour. Biomechanics, 1991—, NSF, Washington, 1991—, Jour. Biomedical Engring., 1992—, Jour. Orthopedic Rsch., 1993—, Bone, 1993—; contbr. articles to profl. jours., chpts. to books. Nat. Merit scholar, 1982; Frank E. Stinchfield fellow, 1990; recipient Young Investigator's award World Congress on Med. Physics, 1991. Mem. ASME (Best Paper award 1990), Orthopaedic Rsch. Soc. (Young Investigators award 1991), Am. Soc. for Cellular Stress Biology. Achievements include patent for apparatus for induction of high frequency strains into the axial skeleton to promote growth and repair; development in cell mechanics and cellular engineering using confocal microscopy and finite element modeling. Office: Musculo-Skeletal Rsch Lab Dept Orthopaedics HSC T18 030 Stony Brook NY 11794-8181

GUILBEAU, ERIC J., biomedical engineer, electrical engineer, educator; b. Tullos, La., June 5, 1944. BS, La. Tech. U., 1967, MS, 1968, PhD in Chem. Engring., 1971. Rsch. assoc. chem. engring. La. Tech. U., 1971-72, rsch. assoc. biomed. engring., 1972-73, from asst. to assoc. prof., 1973-77; prof. chem. and biomed. engring. Ariz. State U., Tempe, 1973-77, assoc. prof., 1977-81, prof. chem. and biomed. engring., 1981—, dir. bioengring., 1990—; affiliate med. staff St. Joseph's Hosp., Phoenix, 1977. m. AICE, Am. Chem. Soc., Internat. Soc. Study Oxygen Transport to Tissue, Biomed. Engring. Soc., Soc. Biomat. Achievements include research in biomedical engineering, development of transducers for measurement of cellular biological parameters, research in transport phenomena in physiological systems, investigation of myocardial protection techniques, development of pericardial substitutes. Office: Ariz State U Coll Engring COB B-338 Tempe AZ 85287*

GUILLAUMONT, ROBERT, chemist, educator; b. Lyon, France, Feb. 26, 1933; s. Jean and Alexandrine (Masbou) G.; m. Nicole Parent, Dec. 15, 1958; children: Cyrille, Elisabeth. Licence es scis., Faculty of Scis., Paris, 1957, PhD, 1966. Asst. in radiochemistry Faculty of Scis., 1958-67; asst.

prof. chemistry Faculty of Scis., Orsay, France, 1967-72, prof., 1972—; head radiochemistry group Nuclear Physics Inst., Orsay, 1979-90; head standing group radwaste mgmt. Ministry of Industry, 1986—; mem. numerous French nat. coms., 1973—. Author: Protactinium 1974, Fundamentals of Radiochemistry, 1993; contbr. articles to chemistry and radiochemistry mags. With French Army, 1960-62. Recipient Ordre Nat. du Merite Ministry of Industry, 1989. Home: 7 E Branly, F91 120 Palaiseau France Office: IPN Radiochemistry Group, BP no 1, F91 406 Orsay France

GUILLEMIN, MICHEL PIERRE, occupational hygienist; b. Neuchâtel, Switzerland, Nov. 23, 1943; s. Henri Joseph and Jacqueline (Rödel) G.; children: Pierre, Antoine. MS, U. Neuchatel, 1966, PhD, 1969. Head of lab. Inst. of Social and Preventive Medicine, Lausanne, 1970-75, head of div., 1976-78; dir., prof. Inst. Occupational Health Scis., Lausanne, 1979—. Contbr. numerous articles to profl. jours. Mem. Internat. Occupational Hygiene Assn. (pres. 1990). Office: Inst Occupational Health, Scis, 19 Rue Du Bugnon, CH-1005 Lausanne Switzerland

GUILLEMIN, ROGER C. L., physiologist; b. Dijon, France, Jan. 11, 1924; came to U.S., 1953, naturalized, 1963; s. Raymond and Blanche (Rigollot) G.; m. Lucienne Jeanne Billard, Mar. 22, 1951; children—Chantal, Francois, Claire, Helene, Elizabeth, Cecile. B.A., U. Dijon, 1941, B.Sc., 1942; M.D., Faculty of Medicine, Lyons, France, 1949; Ph.D., U. Montreal, 1953; Ph.D. (hon.), U. Rochester, 1976, U. Chgo., 1977, Baylor Coll. Medicine, 1978, U. Ulm, Germany, 1978, U. Dijon, France, 1978, Free U. Brussels, 1979, U. Montreal, 1979, U. Man., Can, 1984, U. Turin, Italy, 1985, Kyung Hee U., Korea, 1986, U. Paris, Paris, 1986, U. Barcelona, Spain, 1988, U. Madrid, 1988, McGill U., Montreal, Can., 1988, U. Claude Bernard, Lyon, France, 1989. Intern, resident univs. hosps. Dijon, 1949-51; asso. dir., asst. prof. Inst. Exptl. Medicine and Surgery, U. Montreal, 1951-53; asso. dir. dept. exptl. endocrinology Coll. de France, Paris, 1960-63; asst. prof. physiology Baylor Coll. Medicine, 1953-57, assoc. prof., 1957-63, prof., dir. labs. neuroendocrinology, 1963-70, adj. prof., 1970—; adj. prof. medicine U. Calif. at San Diego, 1970—; resident fellow, chmn. labs. neuroendocrinology Salk Inst., La Jolla, Calif., 1970-89, adj. rsch. prof., 1989—; Salk Inst.; Disting. Sci. prof. Whittier Inst., 1989—, dir., 1993—, also bd. dirs. Decorated chevalier Legion d'Honneur (France), 1974, officer, 1984; recipient Gairdner Internat. award, 1974; U.S. Nat. Medal of Sci., 1977; co-recipient Nobel prize for medicine, 1977; recipient Lasker Found. award, 1975; Dickson prize in medicine, 1976; Passano award sci., 1976; Schmitt medal neurosci., 1977; Barren Gold medal, 1979; Dale medal Soc. for Endocrinology U.K., 1980, Ellen Browning Scripps Soc. medal Scripps Meml. Hosps. Found., 1988. Fellow AAAS; mem. NAS, Am. Physiol. Soc., Am. Peptide Soc. (hon.), Assn. Am.Physicians, Endocrine Soc. (pres. 1986), Soc. Exptl. Biology and Medicine, Internat. Brain Rsch. Orgn., Internat. Soc. Rsch. Biology Reprodn., Soc. Neuro-scis., Am. Acad. Arts and Scis., French Acad. Scis. (fgn. assoc.), Academie Internationale de Medecine (fgn. assoc.), Swedish Soc. Med. Scis. (hon.), Academie des Scis. (fgn. assoc.), Academie Royale de Medecine de Belgique (corr. fgn.), Internat. Soc. Neurosci. (charter), Western Soc. Clin. Rsch., Can. Soc. Endocrinal Metabolism, (hon.), Club of Rome. Office: Whittier Inst 9894 Genesee Ave La Jolla CA 92037-1296

GUILLEN, MICHAEL ARTHUR, mathematical physicist, educator, writer, television journalist; b. L.A.; s. Marin Arthur and Betty Guillen; m. Laurel Lucas, Sept. 7, 1991. BS in Physics with distinction, UCLA; MS in Physics, Math. and Astronomy, Cornell U., PhD in Physics, Math. and Astronomy, 1982. Tchr. physics and math. Core Curriculum Program Harvard U., Cambridge, Mass., 1985—; sci. editor Sta. WCVB-TV, Boston, 1985—, ABC-TV program Good Morning Am., N.Y.C., 1988—; sci. correspondent ABC News, N.Y.C., 1990—; tech. advisor Metro Goldwyn Mayer; participant numerous ednl. improvement programs; sci. cons. MGM/VA; mem. adv. bd. AIP. Author: Bridges to Infinity: The Human Side of Mathematics, 1984; contbr. articles to numerous newspapers and mags. including N.Y. Times, Washington Post, Sci. Digest, Sci. News, Psychology Today, Esquire; chief cons. NOVA TV show A Mathematical Mystery Tour; host/writer (TV spls.) Time, Tides and Tuning Forks (Emmy award, Ohio State award), To Be or Not to Be: Endangered Species of New England (Ohio State award), Heads or Tails: Predicting the Unpredictable, 1987, Greenland Polar Ice Cap, 1991 (Ohio State award), War in the Gulf: Answering Children's Questions, 1991 (ACT award, Nat. TV Critics award, Dupont-Columbia U. award), Monteverde Cloud Forest, 1991 (TEDDY award), What are the Differences Between Men and Women?, 1991 (EMMA award), U.S. Disabled Ski Team, 1992 (EDI award), Russian Space Program-ABC News Nightline, 1992 (Aviation/Space Writers Assn. award); formerly sci., tech. contbr. CBS Morning News; TV spls. include Laetrile: The Last Chance. Recipient Danforth award for disting. teaching Harvard U., 1989, 90, Broadcast Media award for overall excellence AIAA, 1987. Mem. AAAS (chmn. sci. and math. edn. symposium), NAS (chmn. scis. and humanities conf., com. rsch. in math., sci. and tech. edn.), Leonardo Da Vinci Soc. (founder), Phi Eta Sigma, Sigma Pi Sigma, Pi Mu Epsilon. Achievements include research in theoretical plasma physics, liquid physics and astrophysics. Office: ABC-TV 7 W 66th St New York NY 10023

GUILLORY, JACK PAUL, chemist, researcher; b. Alwcandria, La., Feb. 28, 1938; s. Mayo P. and Gladys (Roy) G.; m. Gloria T. Bossier, Jan. 25, 1960; children: Theresa, Debra, Michael. BS in Chemistry, La. State U., 1960; PhD in Phys. Chemsitry, Iowa State U., 1965. Chemsit Phillips Petroleum Co. R & D, Bartlesville, Okla., 1965-70, sect. supr., 1970-80, br. mgr., 1980-90, div. mgr., 1990—. Mem. Am. Chem. Soc., Catalyst Soc. Achievements include patents in petroleum refining and catalysis. Office: Phillips Petroleum Co R & D 252 Rsch Forum Bartlesville OK 74004

GUIMOND, RICHARD JOSEPH, federal agency executive, environmental scientist; b. Massena, N.Y., Oct. 28, 1947; s. Lionel Emory and Agnes Mary (Tyo) G.; m. Sherry Lynn Masis, Sept. 11, 1971; 1 child, Nicole Angele. B-SMechE, U. Notre Dame, 1969; M Engring., Rensselaer Poly. Inst., 1970; MS in Environ. Health, Harvard U., 1973. Registered profl. engr., D.C. Commd. officer USPHS, 1970, advanced through grades to rear adm., 1989; with EPA, Washington, 1971—; spl. asst. Office Radiation Programs, 1971-74, environ. project leader, Criteria and Standards div., 1974-78; chief engr. Office of Chem. Control, 1978-79; chief spl. reports br. Office of Toxic Substances, 1981, chief chem. control br., 1981-82; dir. Criteria & Standards div. Office Radiation Programs, 1982-86, dir. Radon div., 1986-88, dir. Office Radiation Protection, 1988-91; dep. asst. adminstr. Office Solid Waste and Emergency Response, Washington, 1991-93, acting asst. adminstr., 1993—. Mem. Mem. Health Physics Soc., Commd. Officers Assn. USPHS, Res. Officers Assn. Roman Catholic. Avocations: boating, woodworking. Home: 12305 Firth Of Tae Dr Fort Washington MD 20744-7007 Office: EPA Office Solid Waste and Emergency Response SE360 401 M St SW Washington DC 20460-0002

GUINN, DAVID CRITTENDEN, petroleum engineer, drilling and exploration company executive; b. Port Arthur, Tex., Nov. 29, 1926; s. Leland Lee and Corrie Andrews (Avery) G.; AA, Lamar Inst. Tech., 1948; BS in Petroleum Engring., U. Tex., Austin, 1951; m. Marguerite V. Guinn, Oct. 7, 1966; children: Susan, David, Jay, Jeffrey. Engr. trainee Dowell, Inc., Alice, Tex., 1949; petroleum engr., exploration drilling eng., Calif. Co. (now Chevron USA Inc.), Lafayette, La., 1951-52, area prodn. and drilling engr., Venice, La., 1953-54, evaluation engr., New Orleans, 1954-55; dist. engr. Republic Natural Gas Co., 1955-56, div. drilling engr., 1956-57; div. engr. Shaffer Tool Works, Inc., 1957-63, sales mgr., Midcontinent div., Beaumont, 1963-65; div. mgr. Mid-Continent-Gulf Coast div., 1965-66; pvt. practice as cons. petroleum engr., Houston, 1966—; cons. petroleum engr., horizontal drilling and completion tech. Austin Chalk Trend in Tex.; cons. petroleum engr. Dept. Econ. and Social Devel., Sci., Tech., environment and Resources Dvsn., UN; owner Guinn and Assos., Engrs., Guinn Resources Co.; bd. chmn.; founder Internat. Offshore Operators, Tropic Drilling and Exploration Co., Consol. Offshore Corp., 1978, mining div. Guinn Resources Co., Henderson, Nev., Quo Vadis Ltd. Ptnrshp. Div., 1974-79. Designed, constructed, contracted world's 2nd largest drill ship "Tainaron", Hong Kong, Philippines, Southeast Asia. Served from cadet to sgt. USAAF, 1943-46. Life mem. Pres. Assn. U. Tex.; mem. Tejas Found. Registered profl. engr., La., Tex. ASME, AIME-SPE (sr.), SAR (life), Mem. Nat. Soc. Profl. Engrs., Tex. Soc. Profl. Engrs., Soc. Petroleum Engrs., Internat. Assn. Drilling Contractors, Am. Soc. Oceanology, Mid Continent Oil and Gas Assn., Am. Soc. Inventors, Life US Naval Inst., Internat. Oceanographic Found., Tex. State Rifle Assn. (life), Explorers Club, U. Tex. Ex-Students Assn. (life), 100 Club (Houston) (life). Contbr. articles to profl. jours.; pioneer, patentee in domestic and fgn. offshore and floating vessel drilling, offshore petroleum subsea prodn. systems. Office: PO Box 1126 Houston TX 77251-1126

GUINN, JANET MARTIN, psychologist, consultant; b. Rapid City, S.D., Aug. 16, 1942; d. Verne Oliver and Carolyn Yetta (Clark) Martin; m. David Lee Guinn, Oct. 27, 1962 (div. June 1988); children: Cynthia Gail, Kevin Scott, Garrett Lee. BS in Psychology, U. Alaska, 1980, MS in Counseling Psychology, 1983; PhD in Clin. Psychology, Calif. Sch. Profl. Psychology, 1988. Lic. psychologist, Alaska, Nev. Pvt. practice Anchorage, 1988—; clinician Behavior Medicine Cons., 1983-84; pvt. practice clinician, 1983-84; supr. Southcentral Counseling Ctr., Anchorage, 1984-85; cons. City/Borough of Juneau, Alaska, 1988; psychologist youth treatment program Alaska Psychiat. Inst., Anchorage, 1989-90; cons. in field; cons. Alaska Small Bus. Coalition, Anchorage, 1990-92; reviewer Blors Corp. Contbr. articles to profl. jours. Active in politics. Mem. Am. Psychol. Assn., Alaska Psychol. Assn., Internat. Neuropsychol. Soc., Rotary, Psi Chi. Republican. Avocations: skiing, gourmet cooking, dancing. Office: 9191 Old Seward Hwy # 14 Anchorage AK 99515-2032

GUINNESS, KENELM L., civil engineer; b. London, Dec. 13, 1928; came to U.S., 1948; s. K. Lee and Josephine (Strangman) G.; m. Jane Nevin, June 3, 1961; children: Kenelm, Sean. BSc, MIT, 1953. Sr. engr. World Bank (Internat. Bank for Reconstrn. and Devel.), Washington, 1954-75; ind. engring. cons., 1975—. Lt. Royal Horse Guards, 1946-48. Mem. ASCE, ICID. Home and Office: Rich Neck Claiborne MD 21624

GUINZBURG, ADIEL, mechanical engineer; b. Johannesburg, Republic of South Africa, Apr. 30, 1966; arrived in Switzerland, 1992; d. Israel Lucio and Leonor Rita (Bercovich) G. BSc in Engring., U. Witwatersrand, Republic of South Africa, 1985; MS, Calif. Inst. Tech., 1986, PhD, 1992. Rsch. asst. Calif. Inst. Tech., Pasadena, 1986-91; researcher Ecole Poly. Federale de Lausanne, Switzerland, 1992-93, Ingersoll Dresser Pump Co., Phillipsburg, N.J., 1993—. Mem. AIAA, ASME, Royal Aero. Soc. Democrat. Jewish. Office: Ingersoll Dresser Pump Co 942 Memorial Pkwy Phillipsburg NJ 08865

GUION, ROBERT MORGAN, psychologist, educator; b. Indpls., Sept. 14, 1924; s. Leroy Herbert and Carolyn (Morgan) G.; m. Mary Emily Firestone, June 8, 1947; children: David Michael, Diana Lynn, Keith Douglas, Pamela Sue, Judith Elaine. B.A., State U. Iowa, 1948; M.S., Purdue U., 1950, Ph.D., 1952. Vocat. counselor Purdue U., 1948-51, research fellow, 1951-52; mem. faculty Bowling Green (Ohio) State U., 1952—, prof. psychology, 1964—, univ. prof., 1983-85, univ. prof. emeritus, 1985—, chmn. dept., 1966-71; vis. prof. U. Calif. at Berkeley, 1963-64, U. N.Mex., summer 1965; tech. adviser Dept. Personnel Services, State Hawaii, summer 1970; vis. research psychologist Ednl. Testing Service, 1971-72; cons. in field, 1954—. Author: Personnel Testing, 1965; editor Jour. Applied Psychology, 1983-88. Served with AUS, 1943-46. Mem. Am. Psychol. Assn. (pres. div. 14 1972-73, pres. div. 5 1982-83, James McKeen Cattell award div. 14 1965, 81, Disting. Sci. Contbn. award div. 14 1987, Disting. Svc. award div. 14 1993), Midwestern Psychol. Assn., Internat. Assn. Applied Psychology, Am. Edn. Rsch. Assn., Nat. Coun. on Measurement in Edn., Am. Psychol. Soc., Sigma Xi, Psi Chi. Methodist. Home: 632 Haskins Rd Bowling Green OH 43402-1615

GUIVENS, NORMAN ROY, JR., mathematician, engineer; b. Brockton, Mass., May 8, 1957; s. Norman Roy and Lula Elizabeth (Wager) G. SB in Math., MIT, 1979, SM in Meteorology, 1979; MTS in Pastoral Ministry, St. Meinrad Sch. Theology, 1992. Teaching asst. MIT Dept. Ocean Engring., Cambridge, Mass., 1979; cons. to Lincoln Lab. Mass. Tech. Lab., West Newton, 1984; sr. engr. SPARTA, Inc., Lexington, Mass., 1984—. Contbr. over 15 articles to profl. jours. Lt. USN, 1979-84. Mem. IEEE (chpt. chmn. 1990-93), Soc. Photo-Optical Instrumentation Engrs., U.S. Naval Inst. Roman Catholic. Achievements include development of first successful simulation coherent laser radars, developed comprehensive model for optical detection systems; key contbr. to develop defense laser/target signatures (DELTAS) code. Office: SPARTA Inc 24 Hartwell Ave Lexington MA 02173-3157

GULATI, SURESH THAKURDAS, glass scientist, researcher; b. Kot Adu, Punjab, West Pakistan, Nov. 13, 1936; came to U.S., 1958, naturalized, 1975; s. Thakur Das and Vishan Devi (Kathuria) G.; m. Teresa Antoinette Davids, Aug. 19, 1961; children: Raj, Prem, Sonya. BSME, U. Bombay, 1957; MSME, Ill. Inst. Tech., 1959; PhD, U. Colo., 1966. Registered profl. engr., N.Y. Stress analyst Continental Can Co., Chgo., 1959-62; instr. U. Colo., Boulder, 1966-67; rsch. fellow Corning (Ind.) Inc., 1967—; adj. prof. Cornell U., Ithaca, N.Y., 1968-70. Contbr. numerous articles to tech. jours. Pres. Internat. Club Finger Lakes, Corning, 1975, Am. Field Svc., Elmira, N.Y., 1971. Mem. ASME, Am. Ceramic Soc., Rotary, Sigma Xi. Democrat. Hindu. Achievements include patents in optical Waveguide Preform, Lightweight Photchromic Lenses, Ceramic Honeycomb Substrates, Metal Shaping Die; research on fracture mechanics, brittle material behavior, fiber optics, composite materials, porous honeycomb materials, strong glasses, theoretical and applied mechanics. Home: 1001 W Water St Elmira NY 14905 Office: Corning Inc Sullivan Park RB4 Corning NY 14831

GULLAPALLI, PRATAP, chemist, researcher, educator; b. Hanamakonda, Andhra Pradesh, India, Feb. 1, 1959; came to U.S., 1988; s. Sekharaiah and Setyavati G.; m. Hae Ok, June 29, 1992. MSc in Chemistry, Kakatiya U., Warangal, Andhra Pradesh, India, 1981; PhD in Chemistry, Osmania U., Hyderabad, Andhra Pradesh, India, 1988. Jr. rsch. fellow Coun. Sci. and Indsl. Rsch., Hyderabad, 1983-85, sr. rsch. fellow, 1985-88; postdoctoral rsch. assoc. Petroleum Recovery Rsch. Ctr./N.Mex. Inst. Mining Tech, Socorro, 1988-90; rsch. assoc. Petroleum Recovery Rsch. Inst./N.Mex. Inst. Mining Tech, Socorro, 1990—. Contbr. articles to sci. publs. Rsch. grantee Amoco, 1990-91, JAPEX, 1990-92. Mem. Am. Chem. Soc., Sigma Xi. Achievements include patent pending for one-pot synthesis of living sulfonated polyisobutylene telechelics; research in methods of living carbocationic and free radical polymerizations, kinetics of polymerization, methods of sulfonation, ionomer syntheses; petrochemicals; silicone chemistry; syntheses and applications of new initiators and new monomers; elastomers, thermoplastics, urethanes, and engineering plastics; characterization by spectroscopic methods. Office: N Mex Petroleum Recovery Rsch Ctr Kelley Bldg Campus Station Socorro NM 87801

GULRAJANI, RAMESH MULCHAND, biomedical engineer, educator; b. Patna, Bihar, India, Dec. 12, 1944; arrived in Can., 1973; s. Mulchand T. and Saraswati Mirchandani G.; m. Lily Tikamdas Mani, Feb. 21, 1976; children: Nilima Ramesh, Rohan Ramesh. BEE, U. Bombay, 1964; MEE, Ill. Inst. Tech., 1965; MS in Physics, Syracuse U., 1972, DEng, 1973. Postdoctoral fellow dept. physiology U. Montreal, Can., 1973-75, rsch. assoc. dept. medicine, 1976-79, rsch. assoc. Inst. of Biomedical Engring., 1979-87, assoc. prof., 1987-90, prof., 1990—; vis. prof. dept. engring. sci. Auckland U., New Zealand, 1992-93; mem biomedical engring. grants com. Med. Rsch. Coun. of Can., Ottawa, 1988-91. Contbr. articles to profl. jours. and chpts. to books. Recipient Postdoctoral fellow Med. Rsch. Coun. of Can., 1973-75; rsch. scholarship Canadian Heart Found., 1978-81, sr. rsch. scholarship Fonds de la recherche en santé du Québec, 1986-89. Mem. IEEE. Office: U Montreal Inst of BioMed Eng, CP 6128 Succursale A, Montreal, PQ Canada H3C 3J7

GUMBS, GODFREY ANTHONY, physicist, educator, researcher; b. Georgetown, Guyana, Sept. 17, 1948; came to U.S., 1990; s. Charles Alexander and Mary Teresa (Jansen) G.; m. Jean Maydalyne Sickander, Oct. 2, 1971; children: Anthony, Alexander, Andrew. BA, Cambridge U., 1971; PhD, U. Toronto, 1978. Rsch. assoc. Nat. Rsch. Coun., Ottawa, Ont., 1978-82; asst. prof. Dalhousie U., Halifax, N.S., 1982-86; prof. U. Lethbridge, Alta., 1986-91; guest scientist MIT, Cambridge, Mass., 1990-91; prof. Hunter Coll./CUNY, N.Y.C., 1992—. Rsch. fellow NSERC of Can., Dalhousie U. & Lethbridge U., 1986-91; Humboldt fellow Humboldt Found., Julich, Germany, 1987. Mem. Am. Phys. Soc., Sigma Xi. Achievements include research on nonlinear effects in quasiperiodic systems, transport phenomena in electronic microstructures. Office: Hunter Coll 695 Park Ave New York NY 10021

GUMERMAN, GEORGE JOHN, archaeologist; b. Milw., Feb. 6, 1936; s. George John and Josephine (Berdoll) G.; m. Sheila Jane Gumerman, Dec. 27, 1959; children: George John, Steven James. BS, Columbia U., 1960; MA, U. Ariz., 1967, PhD, 1969. Asst. prof. Prescott (Ariz.) Coll., 1969-73; assoc. prof. So. Ill. U., Carbondale, 1973-74, prof., 1974-91, rsch. prof. archaeology, 1991—, dir. Ctr. for Archaeol. Investigations, 1981-91; mem. sci. bd. Santa Fe (N.Mex.) Inst., 1991—. Editor: The Anasazi in a Changing Environment, 1988, Dynamics of Southwestern Prehistory, 1989, Exploring the Hohokan, 1990; author: A View From Black Mesa, 1989, People of the Mesa, 1990. Bd. trustees Amerind Found., Dragoon, Ariz., 1989—, S.W. Parks & Monuments Assn., Tucson, 1992—. Recipient Emil W. Haury award S.W. Parks and Monuments Assn., Tucson, 1992. Mem. AAAS, Soc. Profl. Archaeologists, Am. Anthropol. Assn., Soc. Am. Archaeology (Disting. Svc. award 1992), Sch. Am. Rsch. Home: 597 Monte Alto Santa Fe NM 87501 Office: So Ill U Ctr Archaeol Investigations Carbondale IL 62901

GUMNICK, JAMES LOUIS, research institute executive; b. Balt., Oct. 5, 1930; s. Michael and Mary Rose (Schap) G.; m. Jean Kathleen Lawler, Oct. 30, 1959; children: John, Anne, Edward, Mary, Jane, Elizabeth. B.S., Loyola Coll., Md., 1953; Ph.D., U. Notre Dame, 1958. Prof. physics Martin-Loyola Coll., Balt., 1957-64; exec. dir. Nat. Council Energy Co., Phila., 1973-76; chmn. govt. program com. Pa. Gov's. Energy Council, Phila., 1973-74; dir. research devel. U. Houston, 1976-80; gen. mgr. Gulf Univs. Research Co., Houston, 1980-82, pres., 1982-84; univ. relations dir. Oak Ridge Assoc. Univs., 1984-89; exec. dir. of rsch. St. Francis Regional Med. Ctr., Wichita; pres. St. Francis Rsch. Inst. Contbr. articles to profl. jours.; inventor multiple reflection photocathode; discoverer cyclic migration of CS on refractory metals. Danforth Found. fellow, 1953; hon. mention Outstanding Young Scientist in Md., Md. Acad. Sci., 1959. Mem. Am. Physics Soc., Nat. Council Univ. Adminstrs., AAAS, Soc. Research Adminstrs., Sigma Xi, Alpha Sigma Nu. Democrat. Roman Catholic. Avocations: gardening, bonsai, antiques. Home: 2433 Benjamin St Wichita KS 67204-5519 Office: St Francis Regional Med Ctr 929 N St Francis Wichita KS 67214

GUMPER, LINDELL LEWIS, psychologist; b. Mo., Oct. 6, 1947; s. August David and Laura Charlotte (Waldecker) G.; m. Dianne Flesh, June 30, 1973; children: Emily Sara, Bethany Brooke. BA, Westminster Coll., 1968; JD, Yale U., 1972; PhD, Purdue U., 1979. Lic. clin. psychologist, Mont.; bar: Colo., Ind., Mich. Assoc. Holland & Hart, Denver and Aspen, Colo., 1972-74; dep. pros. atty. Tippecanoe County, Lafayette, Ind., 1975-78; clin. psychologist Psychol. Inst. Mich., Birmingham, 1979-83, Psychol. Studies and Consultation Program, Birmingham, 1979-83, Deaconess Behavioral Health Clinic, Billings, Mont., 1991—; pvt. practice clin. psychology Billings, 1983-91; clin. psychologist, co-founder Profl. Health Resources, Inc., Billings, 1988-91; forensic cons., expert witness, 1983—; lectr. in field, 1979—; psychol. cons. Yellowstone Treatment Ctrs., 1984—, dir. assessment, 1985-88. Reviewer Jour. Marital and Family Therapy, 1990—; author: Legal Issues in the Practice of Ministry, 1981; contbr. articles to profl. jours. Mem. Billings Symphony Chorale, 1987—. With USAR, 1970-76. Mem. APA, ABA, Mont. Psychol. Assn., Mental Health Assn. Mont. Mem. United Ch. of Christ. Office: Deaconess Behavioral Health Clinic 550 N 31st St Billings MT 59101

GUMPRIGHT, HERBERT LAWRENCE, JR., dentist; b. Newport, R.I., Apr. 23, 1946; s. Herbert Lawrence and Helen Marie (Broderick) G.; m. Cynthia Randolph Williams, June 16, 1972 (div. Apr. 1987); children: Broderick J., Tyler H. BS, U. R.I., 1968; DDS, NYU, 1972. Lic. dentist, Mass. Intern Waterbury (Conn.) Hosp. Affiliate Yale U. Sch. Medicine, 1972-73; assoc. Ronald J. Dowgiallo, DMD, Harwichport, Mass., 1973-78; pvt. practice Brewster, Mass., 1978—; producer, host Your Dental Health, Community Access TV, 1978-91. Fellow Acad. Gen. Dentistry; mem. Am. Dental Soc., Mass. Dental Soc., Cape Cod Dist. Dental Soc. (treas. 1988-90, v.p. 1989, pub. rels. com. chmn. 1988-91). Avocations: bicycle racing, riding, sailing. Office: 2452 Main St Box 1108 Brewster MA 02631

GUNDERSEN, MARTIN A., electrical engineering and physics educator; b. May 19, 1940. BA in Physics, U. Calif., Berkeley, 1965; PhD in Physics, U. So. Calif., 1972. Asst. prof. elec. engring. Tex. Tech. U., Lubbock, 1973-77, assoc. prof. elec. engring. and physics, 1977-80; assoc. prof. elec. engring. and physics U. So. Calif., L.A., 1980-83, prof., 1983—; vis. prof. UCLA, 1986-87; vis. scientist MIT, 1986-87, 89, C.E.R.N., 1987; presenter in field to numerous workshops and confs. Editor: (with G. Schaefer) The Physics and Applications of Pseudosparks, 1990; editor: (chpt.) Advances in Pulsed Power Technology Volume II: Gas Discharge Closing Switches, 1990; author numerous book chpts.; reviewer, contbr. articles to profl. jours. Fellow IEEE, Optical Soc. Am. Achievements include patents in light initiated high power electronic switch; in back-lighted thyratron Marx bank. Office: U So Calif Seaver Science Ctr 420 Los Angeles CA 90089-0484

GUNDERSON, DONALD RAYMOND, fisheries educator and researcher; b. La Jolla, Calif., Jan. 3, 1942; s. David Elling Gunderson and Elizabeth Topping Duke; m. Maure Dunn; children: David John, Dean Stransham. BS, Mont. State U., 1963, MS, 1966; PhD, U. Wash., 1976. Biologist Wash. State Dept. Fisheries, Seattle, 1967-75; rsch. biologist Nat. Marine Fisheries Svc., Seattle, 1975-78; asst. prof. U. Wash., Seattle, 1978-81, assoc. prof., 1981—; cons. various govt. agys. and pvt. corps., 1978—. Author: Surveys of Fisheries Resources, 1993; mem. editorial bd. Fisheries Rsch., 1984-89; contbr. articles to Can. Jour. Fisheries and Aquatic Sci., Fisheries Rsch., Marine Fisheries Rev., Estuaries, others. Scoutmaster Boy Scouts Am., Seattle, 1980-83, cubmaster, 1977. Mem. Am. Fisheries Soc., Am. Inst. Fisheries Rsch. Biologists, Sigma Xi. Achievements include research in population dynamics and recruitment processes in marine fish and shellfish; surveys of fisheries resources; biology of rockfish and other marine fish. Home: 123 Raft Island Gig Harbor WA 98335 Office: U Wash Fisheries Rsch Inst WH-10 Seattle WA 98195

GUNDERSON, EDWARD LYNN, environmental engineer; b. Houston, Sept. 4, 1952; s. Ernst Fischer and Doris Evelyn (Harris) G.; m. Nancy Jo Smith, Nov. 17, 1978; 1 child, Mary Evelyn. BSChemE, Lamar U., 1978. Registered profl. engr., Tex. Utilities engr. Chevron Chem. Co., Baytown, Tex., 1978-80, environ. engr., 1980-87; environ. engr. Stepan Co., Elwood, Ill., 1987-91; corp. environ. air and water adminstr. Stepan Co., Northfield, Ill., 1991—; mem. air subcom. Tex. Chem. Coun., Houston, 1980-87, fugitive emissions regulation devel. com., 1985-87; mem. air and water subcoms. Chem. Industries Coun. Ill, 1993—. Tech. reviewer: AWMA Air Pollution Control Manual, 1992. Mem., acting chmn. planning commn. Village of Channahon, Ill., 1988-92, zoning bd. appeals, 1988-92; pres. Optimists Internat., Crosby, Tex., 1986. With U.S. Army, 1972-74. Mem. NSPE, Phi Kappa Phi, Tau Beta Pi, Omega Chi Epsilon (pres. 1978, Outstanding Student 1978). Achievements include development of prototype for plastics solids removal from water; of fugitive emissions estimation guidelines for cooling towers; assisted in development of air emissions factors for publication in air pollution control manual. Office: Stepan Co 22 W Frontage Rd Northfield IL 60093

GUNTER, GORDON, zoologist; b. Goldonna, La., Aug. 18, 1909; s. John O. and Joanna (Pennington) G.; B.A., La. State Normal Coll., 1929; M.A., U. Tex., 1931, Ph.D., 1945; m. Frances M. Hudgins, Sept. 7, 1957; children—Edmund Osbon, Harry Allen; children by previous marriage—Charlotte A. Gunter Evans, Miles G., Forrest P. Biologist, U.S. Bur. Fisheries, intermittently, 1931-38; marine biologist Tex. Game, Fish and Oyster Commn., 1939-45; research scientist Inst. Marine Sci., U. Tex., 1945-49, dir., 1949-55; prof. zoology Marine Lab., U. Miami (Fla.), 1946-47; sr. marine biologist Scripps Instn. Oceanography, U. Calif., La Jolla, 1948-49; dir. Gulf Coast Research Lab., Ocean Springs, Miss., 1955-71, dir. emeritus, 1971—; also prof. biology U. So. Miss., prof. zoology Miss. State U., 1956-78; adj. prof. emeritus biology U. Miss., 1979-85; area cons. Tex. Office Coordinator Fisheries, 1942-45; adv. panel comml. seafoods div. La. Commn. Wild Life and Fisheries, 1953-54; vice chmn. biology, com. treatise on marine ecology NRC, 1942-57; sci. adv. panel Gulf State Marine Fisheries Commn., mem. bd. advisors Fla. Bd. Conservation, 1956-68; prin. investigator plankton studies OTEC program, Gulf of Mex., 1978-82; mem. standing com. Gulf of Mexico Fishery Mgmt. Council. Fellow La. Acad. Scis., Internat. Oceanographic Found., Internat. Acad. Fisheries Scientists, Am. Inst. Fisheries Research Biologists, La. Acad. Scis., Explorers Club; mem. Am. Fisheries Soc. (hon.), Am. Ornithologists Union, Am. Soc. Ichthyologists and Herpetologists, Am. Soc. Limnology and Oceanography, Am. Soc. Mammalogists, Am. Soc. Naturalists, Am. Soc. Zoologists, Ecol. Soc. Am., Miss., New Orleans acads. scis., Nat. Shellfisheries Assn. (hon.), Wildlife Soc., World Mariculture Soc. (pres. 1974, hon.), Miss. Acad. Scis. (pres. 1964-65), Sigma Xi, Phi Kappa Phi. Founder editor: Gulf Research Reports, 1961-74. Author: Gunter's Archives No. 1, 1984, Gunter's Archives No. 4, 1987, No. 5, 1988, No. 6, 1989, No. 7, 1990, No. 8, 1991; contbr. over 435 articles on marine biology to profl. and popular publs. in U.S. and fgn. countries. Address: 127 Halstead Rd Ocean Springs MS 39564

GUNTER, WILLIAM DAYLE, JR., physicist; b. Mitchell, S.D., Jan. 10, 1932; s. William Dayle and Lamerta Berniece (Hockensmith) G.; m. Shirley Marie Teshera, Oct. 24, 1955; children—Maria Jo, Robert Paul. B.S. in Physics with distinction, Stanford U., 1957, M.S., 1959. Physicist Ames Research Ctr. NASA, Moffett Field, Calif., 1960-81, asst. br. chief electronic optical engring., 1981-85; pvt. practice cons. Photon Applications, San Jose, Calif., 1985—. Patentee in field. Contbr. articles to profl. jours. Served with U.S. Army, 1953-55. Recipient Westinghouse Sci. Talent Search award, 1950; various awards NASA; Stanford U. scholar, 1950. Mem. Am. Assn. Profl. Cons., Optical Soc. Am., IEEE (sr.), Am. Phys. Soc., Soc. Photo-Optical Instrumentation Engrs., Planetary Soc., Nat. Space Soc., NASA Alumni League. Office: Photon Applications 5290 Dellwood Way San Jose CA 95118-2904

GÜNTHER, MARIAN W(ACLAW) J(AN), theoretical physicist; b. Warsaw, Poland, Nov. 27, 1923; came to U.S., 1960; s. Waclaw Henryk and Janina Leona (Wilke) G.; m. Marion Jeanette Koch, Mar. 22, 1976. MS, Jagiellonian U., Cracow, Poland, 1946; PhD, U. Warsaw, 1948, Veniam Legendi, 1951. Postdoctoral fellow U. Leiden, Holland, 1948-50, U. Birmingham, Eng., 1950; asst. prof. U. Warsaw, 1950-53, assoc. prof., 1956-60; assoc. prof. U. Wroclaw, Poland, 1953-56; vis. mem. Inst. for Advanced Study, Princeton, N.J., 1960-61; mem. sci. staff Boeing Sci. Rsch. Labs., Seattle, 1961-64; assoc. prof. of physics U. Cin., 1964-66, prof., 1966-74, prof. emeritus, 1984—. Contbr. articles to Phys. Rev., Jour. Math. Physics. Recipient cert. of achievement Am. Men and Women of Sci., 1987. Mem. AAAS, N.Y. Acad. Scis. Roman Catholic. Achievements include research on the simple exact solution of relativistic Bethe-Salpeter equation in g squared approximation of kernel for non-feynman propagators, on extending Gupta-Bleuler method of indefinite metric justifying general use of such non-feynman propagators in relativistic quantum field theory. Home: 310 Bryant Ave Apt 15 Cincinnati OH 45220-1682

GUNTHER, WILLIAM EDWARD, electrical engineer, researcher; b. N.Y.C., Apr. 6, 1948; s. William Edward and Joan (Liccardi) G.; m. Nancy Louise Hastings, Aug. 29, 1970; children: Brian Dana, Emily Elizabeth. BSEE, Northeastern U., 1970, MSEE, 1971. Registered profl. engr., N.Y. Plant engr. L.I. Lighting Co., Hicksville, N.Y., 1969-74; instrumentation and controls engr. L.I. Lighting Co., Shoreham, N.Y., 1975-82, ops. mgr., 1982-84; rsch. engr. Brookhaven Nat. Lab., Upton, N.Y., 1984-89, sr. rsch. engr., 1989—; chmn. nuclear power plant instrument engrs. Instrument Soc. Am., 1979-81; mem. reactor safety com. Brookhaven Nat. Lab., Upton, 1984-90. Contbr. articles to profl. jours. Leader Boy Scouts Am., Wading River, N.Y., 1984-90. Mem. IEEE. Achievements include research on the susceptibility to aging degradation of selected components used in nuclear power plants. Office: Brookhaven Nat Lab Bldg 130 Upton NY 11973

GUNTLY, LEON ARNOLD, mechanical engineer; b. Wayne Twp., Wis., Oct. 8, 1944; s. Arnold Reuben and Marie (Dorothea (Habeck) G.; m. Luan Margaret LaFave, June 1, 1974; children: Rebecca Lynn, Rachel Joy, Ruth Erin. BSME, U. Wis., 1966; MBA, U. Wis.-Parkside, Kenosha, 1986. Registered profl. engr., Wis. Product engr. Modine Mfg. Co., Racine, Wis., 1966-78, project engr., 1978-80, sr. project engr., 1980-90, sr. rsch. engr., 1990—; mem. heating/refrigeration/air conditioning adv. bd. Gateway Tech. Coll., Racine, 1990—. Author tech. papers in field. Mem. ASHRAE (mem. thermodynamics and psychometrics com. 1984-88, 92—), SAE (mem. cold weather truck and bus heating com. 1990—), TI Computer Club (pres. 1987, 90—). Achievements include patents for evaporator with improved condensate drainage, condenser with small hydraulic diameter flow path. Office: Modine Mfg Co 1500 DeKoven Ave Racine WI 53403-2540

GUNZO, IZAWA, chemistry educator; b. Japan, Mar. 28, 1940; s. Youjiro and Chikako I.; m. Yuuko Izawa, Dec. 16, 1972; 1 child, Sachiko. DSc in Chemistry, Tohoku U., Sendai, Japan, 1969. Rsch. asst. dept. chemistry Tohoku U., Seandai, 1969. 1970-88; assoc. prof. dept. chemistry Tohoku U., T, 1988; prof. Utsunomiya Bunsei Jr. Coll., Utsunomiya, Japan, 1988—. Office: Utsunomiya Bunsei Jr Coll, 4-8-15 Kamitomatsuri, Utsunomiya 320, Japan

GUO, CHU, chemistry educator; b. Fengyang, Shanxi, China, Aug. 25, 1933; came to U.S. 1986; s. FengTian and Fengyi G.; m. Shihua Wang, July 1, 1962; 1 child, Huizhong. PhD, Moscow State U., 1962. Sr. rsch. fellow Inst. Chem. Physics, Academia Sinica, Dalian, China, 1963-75; full rsch. prof. Inst. Chemistry, Academia Sinica, Beijing, China, 1975-86; vis. prof. Royal Instn. of G.B., London, 1986; vis. prof. chemistry CCNY, 1986, CASI, Dept. Chemistry, CCNY, 1990—. Editorial bd. Jour. Chinese Lasers, 1986—, Chinese Jour. Luminescence, 1988—; author: The Primary Process of Photosynthesis, 1986; contbr. articles to profl. jours. Mem. Chinese Optical Soc. (adv. bd. 1982-88), Sigma Xi. Achievements include research in photoinitiated electron transfer on membranes, interplay between electron transfer and conformetional change of surrounding protein. Office: CCNY Dept Chemistry W 130th St at Convent Ave New York NY 10031

GUO, DONGYAO, crystallographer, researcher; b. Liaozhong, Liaoning, People's Republic of China, Sept. 20, 1935; came to U.S., 1989; s. Chunrong Guo and Sukun Sun; m. Changfeng Ji, Jan. 31, 1966; children: Shishan, Shilin, Shilei. BS in Physics, Jilin U., 1961, PhD in Physics, 1966. Asst. prof. theoretical chemistry Jilin U., People's Republic of China, 1966-82, assoc. prof. theoretical chemistry, 1983-89; vis. scientiest Med. Found., Buffalo, 1984-86, sr. rsch. scientist, 1989—. Reviewer Math. Reviews, 1984—; contbr. over 35 articles to profl. jours. Recipient Sci. and Tech. Advancement award Nat. Edn. Com., People's Republic of China, 1988, Nat. Sci. and Technol. Achievement award Nat. Sc. and Tech. Com., People's Republic of China, 1991. Mem. Am. Crystallography Assn. Achievements include research in direct methods in crystallography and symmetry groups. Office: Med Found Buffalo 73 High St Buffalo NY 14203

GUO, HUA, chemist, educator; b. Sichuan, Peoples Republic China, Aug. 20, 1962; s. Xian-Jian Guo and Yu-Xian Zeng; m. Qian-Yun Zhang, Sept. 11, 1985; 1 child, Mindy. BS, Chengdu Inst. Elec. Engring., China, 1982; MS, Sichuan U., Chengdu, 1985; PhD, U. Sussex, Brighton, Eng., 1988. Rsch. assoc. Northwestern U., Evanston, Ill., 1988-90; asst. prof. dept. chemistry U. Toledo, 1990—. Contbr. articles Jour. Chem. Physics, Jour. Phys. Chemistry, others. Grantee NSF, 1992. Mem. Am. Chem. Soc., Am. Phys. Soc., Sigma Xi. Office: U Toledo Dept Chemistry Toledo OH 43606

GUO, XIN KANG, mathematics educator; b. Shanghai, People's Republic of China, Feb. 10, 1938; s. Zai and Li Juan (Wang) G.; m. Ying Zhu Chen, Nov. 28, 1967; 1 child, Jin. Grad., Fudan U., Shanghai, 1961. Asst. prof. dept. math. Guangxi U., Nanning, Peoples Republic of China, 1961-77, lectr., 1978-84, assoc. prof., 1985-89, prof., 1990—, vice-dir., 1981-84, vice-dir. dept. sci. rsch., 1984-88, dir. Inst. Math. and Computer Software, 1989—; Reviewer Math. Revs., 1988—. Contbr. articles on boundary value problems of partial differential equations to profl. jours. Recipient Sci. Progressive Prize, Guangxi Province Govt., Nanning, 1985, Brilliant Sci. Article Diploma, Guangxi Province Fedn. Sci., Nanning, 1988, Excellent Textbook Prize, Guangxi U., Nanning, 1988, Excellent Teaching Prize, Guangxi U., Nanning, 1989. Mem. Chinese Soc. Indsl. and Applied Math. (councillor 1990—), Am. Math. Soc., Math. Soc. People's Republic of China. Home and Office: Guangxi U, Dept Math, Nanning Guangxi, China

GUPTA, ANAND, mechanical engineer; b. Jabalpur, India, July 15, 1960; came to U.S., 1972; s. Mohandas and Tulsa (Bhoria) G.; m. Anita Nayak, Dec. 14, 1985; 1 child, Anurag. B Chem. Engring., U. Minn., 1980, MSME, 1986, MBA, 1990, PhD, 1991. Process engr. Tex. Instruments, Dallas, 1980-81; rsch. asst. U. Minn., Mpls., 1982-89; microcontamination engr. Applied Materials, Santa Clara, Calif., 1989—. Contbr. articles to profl. jours. Mem. AICE, Am. Assn. Aerosol Rsch., Gesellschaft Aerosolforschung. Hindu. Achievements include patent for particle monitor system and method, patents pending for particle reduction in semiconductor processing equipments, developments related to particle and contamination field. Home: 1270 Briarcreek Ct San Jose CA 95131 Office: Applied Materials 3050 Bowers Ave Santa Clara CA 95054

GUPTA, DHARAM V., chemical engineer; b. New Delhi, Sept. 4, 1945; came to U.S., 1970; s. Hari Charan Das and Kala Vati (Garg) G.; m. Shobha Karwan, Feb. 10, 1981; children: Mona, Tina, Neil. BTech with honors, Indian Inst. Tech., Kharagapur, 1968; PhD, Worcester (Mass.) Poly. Inst., 1975. Rsch. engr. Am. Cyanamid Co., Bound Brook, N.J., 1974-78, group leader, 1981-83; tech. dir. Am. Cyanamid Co., Charlotte, N.C., 1978-81; sr. staff engr. Am. Cyanamid Co., Stamford, Conn., 1983-89, sr. rsch. scientist, 1989—. Mem. AICE, Am. Mgmt. Assn. Republican. Hindu. Office: Am Cyanamid Co 1937 W Main St Ste 275 Stamford CT 06904

GUPTA, GOPAL DAS, mechanical engineer, researcher; b. Meerut, India, Aug. 1, 1946; came to U.S., 1967; s. Chandra R. and Hem D. (Agarwal) G.; m. Susham Gupta, Jan. 20, 1978; children: Shawn, Michelle. BS, Indian Inst. Tech., Kanpur, 1967; MS, Lehigh U., 1968, PhD, 1970. Asst. prof. Foster Wheeler, Livingston, N.J., 1970-73, devel. engr., 1973-76, system analysis sect. head, 1976-80, mgr. engring. sci. & tech. dept., 1980-87, program mgr., 1987-89, dir. govt. systems ops., 1989-90, v.p., 1990—; chmn. internat. environ. com. Foster Wheeler, 1992—. Contbr. over 80 articles to profl. jours. Chmn. East Hanover (N.J.) Environ. Commn., 1990—. Fellow ASME (chmn. computers in engring. divsn. 1989-90, Henry Hess award 1979). Home: 50 Timberhill Dr East Hanover NJ 07936 Office: Foster Wheeler 12 Peach Tree Hill Rd Livingston NJ 07039

GUPTA, HEM CHANDER, mechanical and electrical engineer; b. Bawal, India, June 17, 1931; came to U.S., 1953; s. Chooni Lal and Kalawati (Garg) G.; m. Asha Hem Sampson, Sept. 1, 1957; children: Reeta H. Gupta Brendamour., Raj P., David A., Mark C. BSEE, Delhi (India) Poly., 1952; MSME, U. Ill., 1954. Registered profl. engr., Ala., Calif., Conn., D.C., Fla., Ga., Ill., Ind., Iowa, Ky., La., Maine, Md., Mich., Minn., N.Y., N.C., Ohio, Pa., Tex., Vt., Va., W.Va., Wis. Project engr. Skidmore, Owings & Merrill, Chgo., 1954-59, A. Epstein & Sons, Chgo., 1959-61; project engr. Perkins & Will, Chgo., 1961-62, assoc., 1962-63, sr. assoc., 1963-66, chief engr., v.p., 1966-67; pres. Environ. Systems Design, Inc., Chgo., 1967-92; CEO Systems Design, Inc., Chgo., 1993—. Contbr. articles on mech. engring., induction units, heat transfer, air conditioning to profl. jours. Mem. exec. com. Boy Scouts Am., Chgo., 1988—; mem. Mayor's Adv. Com. on Bldg. Code Amendments, Chgo., 1977-88; mem. com. on standards and tests Mayor's Office of Intergovtl. Affairs, 1992—. Fellow ASHRAE; mem. NSPE, Chgo. Com. on High Rise Bldgs. (former officer), Chgo. Architecture Found. (former trustee), Ill. Soc. Profl. Engrs., Tavern Club, Univ. Club, Econ. Club, Flossmoor Country Club, The Club at Pelican Boy. Achievements include research on the heat transfer characteristics of a cellular steel radiant floor and performance of induction unit system with pyramidal type of ceiling. Home: 180 E Pearson Unit 7207 Chicago IL 60611 Office: Environ Systems Design Inc 55 E Monroe Ste 1660 Chicago IL 60603

GUPTA, KULDIP CHAND, electrical and computer engineering educator, researcher; b. Risalpur, India, Oct. 6, 1940; came to U.S., 1982; s. Chiranjiva Lal and Gauran (Agarwal) G.; m. Usha Agarwal, Apr. 4, 1971; children: Parul, Sandeep, Anjula. BSc, Punjab U., Chandigarh, India, 1958; BE, Indian Inst. Sci., Bangalore, India, 1961, ME, 1962; PhD, Birla Inst. Tech. Sci., Pilani, India, 1969. Asst. prof. Punjab Engring. Coll., Chandigarh, 1964-65, Birla Inst. Tech. and Sci., Pilani, 1968-69; asst. prof., then prof. Indian Inst. Tech., Kanpur, India, 1969-84; prof. U. Colo., Boulder, 1983—; vis. assoc. prof. U. Waterloo (Ont., Can.), 1975-76; vis. prof. Swiss Fed. Tech. Inst., Lausanne, 1976, Zurich, 1979, Tech. U. Denmark, Lynby, 1976-77, U. Kans., Lawrence, 1982-83; advisor, cons. UN Devel. Programme, People's Republic China, 1987, India, 1990. Author: CAD of Microwave Circuits, 1981, Chinese transl., 1986, Russian transl., 1987, Microstrip Lines and Slotlines, 1979, Microwaves, 1979, Spanish transl., 1983; editor, author: Microwave Integrated Circuits, 1974, Microstrip Antenna Design, 1988; founding editor Internat. Jour. Microwave Millimeter-Wave Computer Aided Engring., 1991—; contbr. numerous articles to profl. jours. and chpts. to books; patentee in field. Fellow IEEE (guest editor spl. issue IEEE Transactions on Microwave Theory and Tech. 1988), Instn. Electronics and Telecommunication Engrs. India (guest editor jour. July 1982). Hindu. Office: U Colo Campus Box 425 Boulder CO 80309

GUPTA, PRABHAT KUMAR, materials scientist. MS in Metallurgy, Case Western Res. U., 1969, MS in Physics, 1970, PhD, 1972. Postdoctoral fellow Yeshiva U., N.Y.C., 1971-72; vis. prof. Cath. U., Washington, 1972-77; sr. scientist Owens Corning Fiberglass Corp, Granville, Ohio, 1977-82; assoc. prof. Case Western Res. U., Cleve., 1982-83; rsch. assoc. Owens Corning Fiberglass Corp., Granville, 1983-86; prof. Ohio State U., Columbus, 1986—. Recipient Otto Schott Rsch. award 1993. Mem. Am. Ceramic Soc. (assoc. editor 1987-92, chmn. glass div. 1992-93), Materials Rsch. Soc., Sigma Xi. Achievements include patents for Optical Fibers with Compression Surface Layer and Method for Making Composite Glass Articles. Office: Ohio State U 2041 College Rd Columbus OH 43210

GUPTA, RAJESH, industrial engineer, quality assurance specialist; b. Calcutta, India, Oct. 27, 1961; came to U.S., 1986; s. Chandra Kumar and Indira (Karnani) G.; m. Amita Negi, Jan. 26, 1989. BSME, Bangalore (India) U., 1984; postgrad. in Indsl. Systems, San José State U., 1988. Engr. trainee Incoducts Pvt. Ltd., Bangalore, 1983-84, quality and prodn. control supr., 1984-86; quality assurance engring. supr. Gen. Signals Semiconductor Systems, Fremont, Calif., 1988-89, quality assurance engring. mgr., 1989-90; quality assurance engring. mgr. Semiconductor Systems, Inc., Fremont, 1990—. Mem. NSPE, Soc. Mfg. Engrs., Inst. Indsl. Engrs., Am. Soc. Quality Control. Office: Semiconductor Systems Inc 47003 Mission Falls Ct Fremont CA 94539

GUPTA, RAMESH CHANDRA, geotechnical engineer, consultant; b. Khurja, India, Jan. 1, 1941; came to U.S., 1978; s. Chhitar Mal and Gunvati Devi G.; m. Radha Devi, Feb. 23, 1966; children: Vineet, Suneet, Nidhi. B in Tech. with honors, Indian Inst. Tech., Kharagpur, 1962; M Engring., U. Fla., 1980, PhD, 1983. Registered profl. engr., Fla., Md., Va., Del., D.C. Asst. engr. U.P. Irrigation Dept., Lucknow, India, 1963-74; exec. engr. U.P. Irrigation Dept., Lucknow, 1974-78; geotechnical engr. Hayward Baker Co., Odenton, Md., 1983-87; assoc. Kidde Cons., Inc., Jessup, Md., 1987-89; sr. geotechnical engr. Greiner, Inc., Timonium, Md., 1989—; mem. Hydraulic Structures com., Indian Standards Inst., New Delhi, 1975-78. Contbr. articles to profl. jours. including Jour. Geotech. Engring., Jour. Soils and Found., Jour. Instl. Engrs., Indian Geotech. Jour. Mem. ASCE, U.S. Nat. Soc., Internat. Soc. Soil Mechanics and Found. Engrs., Tau Beta Pi. Republican. Hindu. Achievements include development of new test setup for in situ shear tests, of new method for determining consolidation characteristics of saturated clays, and of new methodology for analyzing strains around cone penetrometers; derivation of formulas for finite strain components in spherical and cylindrical coordinates. Home: 14914 Falconwood Dr Burtonsville MD 20866-1349 Office: Greiner Inc 2219 York Rd Ste 200 Timonium MD 21093-3189

GUPTA, RISHAB KUMAR, medical association administrator, educator, researcher; b. Nagina, Utter Pradesh, India, Apr. 18, 1943; came to U.S., 1965; s. Sahu Harbans Lal and Chandravati (Devi) G.; m. Mridula Gupta, May 2, 1972; children: Arvind, Anita. MSc, G.B. Plant U., Pantnagar, India, 1965; MS, PhD, Rutgers U., 1968. Asst. rsch. oncologist UCLA Sch. Medicine, 1972-75, assoc. rsch. oncologist., 1975-79, asst. prof., 1979-81, assoc. prof., 1981-85, prof., 1985-91; v.p., dir immunodiagnosis John Wayne Cancer Inst., Santa Monica, Calif., 1991—; mem. study sect. NIH, Bethesda, Md., 1989-92; spl. grant reviewer Med. Rsch. Coun. Can., 1991. Editorial bd. Contemporary Oncology, 1991—; Contbr. articles to med. jours. Pres.'s fellow Am. Soc. for Microbiology, 1967; Rsch. grantee Calif. Inst. for Cancer Rsch., UCLA, 1973-75, U. Calif. Cancer Rsch. Com., 1973, 74, 81, Nat. Cancer Inst/NIH, 1981-90. Mem. Am. Assn. for Cancer Rsch., Am. Assn. Immunologists, Am. Soc. for Clin. Oncology, Am. Acad. Microbiology. Achievements include definition, isolation, and characterization of tumor associated antigens that are immunogenic in cancer host from cultured and biopsy tumor cells; development of monoclonal antibodies to these antigens and untilization of these in immunoassays to detect the antigens in body fluids of cancer patients and determine their clinical significance in terms of immunodiagnosis and immunoprognosis. Home: 7118 Costello Ave Van Nuys CA 91405 Office: John Wayne Cancer Inst 2200 Santa Monica Blvd Santa Monica CA 90404

GUPTA, SUDHIR, immunologist, educator; b. Bijnor, India, Apr. 14, 1944; came to U.S., 1971; s. Tej S. and Jagdishwari Gupta; m. Abha, Jan. 28, 1980; children: Ankmalika Abha, Saurabh Sudhir. MD, King George's Med. Coll., Lucknow, India, 1966, PhD, 1970. Diplomate Am. Bd. Allergy and Immunology, Am. Bd. Diagnostic Lab. Immunology, Clin. Immunology Bd., Royal Coll. Physicians and Surgeons Can. Intern King George's Med. Coll., Lucknow, 1966, resident in medicine, 1967-70; teaching faculty fellow dept. medicine Tufts U. Med. Sch., Boston, 1971-72; fellow in allergy and immunology R. A. Cooke Inst., N.Y.C., 1972-74; rsch. fellow Sloan-Kettering Inst. Cancer Rsch., N.Y.C., 1974-76, asst. prof., 1976-78, assoc. prof., 1978-82; instr. Cornell U., N.Y.C., 1976-77, asst. prof., 1977-79, assoc. prof., 1979-82; prof. medicine U. Calif., Irvine, 1982—; mem. adv. panel FDA, Washington, 1989—; sci. advisor Inst. Immunopathology, Kohn, Germany, 1990—; mem. allergy-immunology subcom. NIH, Bethesda, Md., 1985-89; vis. prof. Hematologic Rsch. Found., Roslyn, N.Y., 1992. Editor in chief Jour. Clin. Immunology, 1980—; editor: Immunology of Clinical and Experimental Diabetes, 1984, Immunology of Rheumatic Diseases, 1985, Mechanisms of Lymphocyte Activities and Immune Regulation I-IV, 1985-92. Pres. Nargis Dutt Meml. Found., So. Calif., 1990; vice-chair AIDS Task Force, Orange County (Calif.) Med. Assn., 1987—; mem. Indo-Am. Republican Club, Orange County, 1991—. Recipient Arthur Manzel Rsch. award R.A. Cooke Inst., N.Y.C., 1976, Outstanding Achievement award in med. scis. Nat. Fedn. Asian Indians in N.Am., 1986, Lifetime Achievement award Jeffrey Modell Found., N.Y.c., 1990. Fellow ACP, Royal Coll. Physicians and Surgeons Can., Am. Soc. Medicine (London); mem. Am. Assn. Immunologists. Achievements include description of the presence of K channels in human T cells, their role in T cell function and assn. with exptl. autoimmune diseases, reversal of multidrug resistance of cancer cells by cyclosporin A both in vitro and in vivo, described a new human intracisternal retrovirus associated with CD4 cell deficiency without HIV infection. Office: U Calif Dept Medicine C240 Med Sci I Irvine CA 92717

GUPTA, SURAJ NARAYAN, physicist, educator; b. Haryana, India, Dec. 1, 1924; came to U.S., 1953, naturalized, 1963; s. Lakshmi N. and Devi (Goyal) G.; m. Letty J.R. Paine, July 14, 1948; children: Paul, Ranee. M.S., St. Stephen's Coll., India, 1946; Ph.D., U. Cambridge, Eng., 1951. Imperial Chem. Industries fellow U. Manchester, Eng., 1951-53; vis. prof. physics Purdue U., 1953-56; prof. physics Wayne State U., Detroit, 1956-61, Distinguished prof. physics, 1961—; researcher on high energy physics, nuclear physics, relativity and gravitation. Author: Quantum Electrodynamics, 1977. Fellow Am. Phys. Soc., Nat. Acad. Scis. of India. Achievements include generalization of the formalism of quantum mechanics by introducing negative probability and quantization of the electromagnetic field; introduction of flat-space interpretation of Einstein's theory of gravitation and quantization of the gravitational field; formulation of regularization by auxiliary fields and renormalization by counter terms for the interaction of elementary particles; development of theory of bound fermion-antifermion systems in quantum electrodynamics and quantum chromodynamics. Home: 30001 Hickory Ln Franklin MI 48025-1566 Office: Wayne State U Dept Physics Detroit MI 48202

GUPTA, UMESH CHANDRA, agriculturist, soil scientist; b. Kanpur, Uttar Pradesh, India, Oct. 25, 1937; arrived in Can., 1961;; m. Sharda Kumari, June 8, 1957; children Sharad, Kamal, Subhas. BS, Agara U., India, 1955, MS, 1957; PhD, Purdue U., 1961. Rsch. asst. Agrl. Coll., Kanpur, 1957-58; rsch. assoc. Purdue U., Lafayette, Ind., 1961; post doctoral fellow Soil Rsch. Inst. Agriculture Can., Ottawa, Ontario, 1961-63; rsch. scientist Agriculture Can., Charlottetown, P.E.I., 1963—; lectr. in field. Editor: Boron and Its Role in Crop Production, 1993; chief editor Can. Jour. Soil Sci., 1984-86; editorial policy bd. Agrl. Inst. Can. Jours., 1983-86; contbr. articles to profl. jours. chpts. to books. Fellow Am. Soc. Agronomy, Soil Soc. Am., Can. Soc. Soil Sci., Agrl. Inst. Can. (AIC Fellowship award, 1991; mem. P.E.I. Inst. Agrologists (pres. 1990-91). Achievements include research in micronutrients and selenium nutrition of crops and livestock. Office: Agriculture Can Rsch Sta, PO Box 1210, Charlottetown, PE Canada C1A 7M8

GUPTILL, STEPHEN CHARLES, physical scientist; b. Bryn Mawr, Pa., Aug. 17, 1950; s. James Ernest and Majorie Duncan (Healey) G.; m. Regina McLean, June 24, 1972; children: Daniel, Ellen. BA, Bucknell U., Lewisburg, Pa., 1972; MA, U. Mich., 1974, PhD, 1975. Geographer U.S. Geol. Survey, Reston, Va., 1975-80, chief br. of analysis, 1980-85, rsch. phys. scientist, 1985-90, sci. advisor, 1990—. Co-author: Elements of Cartography, 6th edit., 1993; editor Internat. Jour. Geog. Info. Systems, 1990-92, editorial bd., 1992—. Ford Found. fellow, 1974, U. Mich. Rackham fellow, 1974. Mem. Am. Cartographic Assn. (bd. dirs. 1991-93, v.p. 1993-94, pres. 1994-95), Am. Congress on Surveying and Mapping, Am. Soc. Photogrammetry and Remote Sensing, Assn. Am. Georgraphers (editorial bd. 1984-87). Achievements include lead scientist in design of a nat. geog. info. infrastructure; lead scientist in design of advanced data structures for digital spatial info.; design and devel. of geog. info. system and computer cartography software. Home: 3431 Lyrac St Oakton VA 22124 Office: US Geol Survey 519 Nat Ctr Reston VA 22092

GURALNICK, SIDNEY AARON, civil engineering educator; b. Phila., Apr. 25, 1929; s. Philip and Kenia (Dudnik) G.; m. Eleanor Alban, Mar. 10, 1951; children—Sara Dian, Jeremy. B.Sc., Drexel Inst. Tech., Phila., 1952; M.S., Cornell U., 1955, Ph.D., 1958. Registered profl. engr., Pa.; lic. structural engr., Ill. Instr., then asst. prof. Cornell U., 1952-58, mgr. structural research lab., 1956-58; mem. faculty Ill. Inst. Tech., Chgo., 1958—; prof. civil engring. Ill. Inst. Tech., 1967—, disting. prof. engring., 1982—, dir. structural engring. labs., 1968-71, dean Grad. Sch., 1971-75, exec. v.p., provost, 1975-82, trustee, 1976-82, dir. advanced bldg. materials and systems ctr., 1987—; devel. engr. Portland Cement Assn., Skokie, Ill., 1959-61; participant internat. confs.; cons. to govt. and industry. Author numerous papers in field. Trustee Inst. Gas Tech., 1976-81, Rsch. Inst. of Ill. Inst. Tech., 1976-82; commr.-at-large North Central Assn. Schs. and Colls., 1985-89, cons., evaluator, 1989—. With Corps of engrs. AUS, 1950-51. McGraw fellow, 1952-53; Faculty Research fellow Ill. Inst. Tech., 1960; European travel grantee, 1961. Fellow ASCE (Collingwood prize 1961); mem. Am. Concrete Inst., Am. Soc. for Engring. Edn., Soc. Exptl. Mechanics, Structural Engrs. Assn. Ill. (bd. dirs., pres.-elect 1989-90, pres. 1990-91, John F. Farmer award 1993), Transp. Rsch. Bd., Ill. Univs. Transp. Rsch. Consortium (adminstrv. com. 1983—), Sigma Xi, Phi Kappa Phi, Tau Beta Pi, Chi Epsilon. Office: Ill Inst Tech 3300 S Federal St Chicago IL 60616-3793

GURGIN, VONNIE ANN, social scientist; b. Toledo, Nov. 20, 1940. B.A., Ohio State U., 1962; M.A., U. Calif., Berkeley, 1966; D.Criminology, 1969. Research asst. Calif. Dept. Mental Hygiene, San Francisco, 1962-64; research sociologist U. Calif., Berkeley, 1964-66; dir. cons. services Survey Research Center, 1967-68, asst. prof. criminology, 1968-71; research sociologist Social Sci. Research and Devel. Corp., Berkeley, 1966-67; sr. research criminologist Stanford Research Inst. (now SRI Internat.), Menlo Park, Calif., 1971-72; research dir. Inst. Study Social Concerns, Berkeley, 1972—; asst. chief resource for cancer prevention and epidemiology sect. Calif. Tumor Registry, Calif. Dept. Health Services, Emeryville, 1981-86; dir. survey research No. Calif. Cancer Ctr., Belmont, 1982-86; dir. SEER programs No. Calif. Cancer Ctr., 1986; cons.; mgr. dept. family and community health Calif. Med. Assn., San Francisco, 1993—; pres. bd. dirs. Inst. Study Social Concerns, Berkeley; bd. dirs. Elmwood Coll., Berkeley, 1992—. Author reports, monographs, articles. Bd. dirs. Rsch. Guild, Sacramento, 1983-86. Mem. AAAS, Am. Sociol. Assn. Address: 1099 Sterling Ave Berkeley CA 94708

GURIAN, MARTIN EDWARD, textile engineer; b. N.Y.C., Nov. 9, 1943; s. Nathan and Elsie G. BS in Textile Engring., Phila. Coll. Textile & Sci., 1965; MS in Textile Engring., Ga. Inst. Technology, 1968; MBA in Mktg., U. Del., 1970. Rsch. engr. E.I. DuPont Co., Wilmington, Del., 1967-70;

prodn. devel. mgr. Cone Mills Mktg. Co., N.Y.C., 1970-71; mgr. market and product devel. Burlington Industries, Rockleigh, N.J., 1971-82; instr. Fashion Inst. Technology, N.Y.C., 1982-84; sales mgr. Stirling (N.J.) Textile & Chems., 1984-86; dir. StretchWall Fabrics, Long Island City, N.Y., 1986-92; mgr. tech. info. svcs. DesignTex Fabrics, Woodside, N.Y., 1992—; cons. in field. Contbr. articles to Fabrics & Architecture, Contract Mag., Indsl. Fabric Products Rev., Knitting Times. Exec. commn. Hi-Rise, Fort Lee, N.J., 1990-92; writer Horizon House Newsletter, Fort Lee, 1984—; nat. bd. dirs. Phila. Coll. Textiles & Sci. Alumni Assn., Phila., 1970-76, exec. dir., 1973-74. Recipient Bronze award Inst. Bus. Designs, 1989, 1st Prize Textile Engring., Phila. Coll. Textiles & Sci., 1965, A.A.T.T. award Am. Assn. Textiles Technology, 1965; grantee N.Y. State Dept. Edn., 1983-84. Mem. Inst. Bus. Designers (N.Y. chpt.), ASTM, Interior Designers for Legislation in N.Y. Democrat. Jewish. Achievements include patent in soft luggage construction; development of new fabrics for wallcovering; research in acoustical fabric. Home: 4 Horizon Rd Fort Lee NJ 07024-6533 Office: DesignTex Fabrics 56-08 37th Ave Woodside NY 11377

GURNIS, MICHAEL, geological sciences educator; b. Boston, Oct. 22, 1959; s. George Albert and Barbara (Dempsey) G. BS, U. Ariz., 1982; PhD, Australian Nat. U., Canberra, 1987. Rsch. fellow in geophysics Calif. Inst. Tech., Pasadena, 1986-88; asst. prof. geol. scis. U. Mich., Ann Arbor, 1988-93, assoc. prof., 1993—. Recipient Presdl. Young Investigator award NSF, 1989, fellowship David and Lucile Packard Found., 1991. Fellow Am. Geophys. Union (Macelwance medal 1993), Geol. Soc. Am. (Donath medal 1993). Achievements include research in the linkage of sedimentary rocks deposited in the interiors of continents to geodynamic processes within the earth; global dynamics, mantle convection, plate tectonics, sea level changes, evolution of mantle and crust; computational and visual fluid mechanics. Office: Univ Mich Dept Geology Ann Arbor MI 48109

GURR, CLIFTON LEE, engineering executive; b. Jacksonville, Fla., Feb. 24, 1932; s. George Walter and Mary Louise (Woodell) G.; m. Barbara Ann Wynn, June 13, 1951 (div. 1979); children: Barbara Wynne, Lee Ann, Dicksie Lynn, Kippy Louise; m. Sheila Diane McCormick, Aug. 31, 1984; children: Danielle, Kristi. BME, U. Fla., 1956, MME, 1967. Test engr. Lockheed Corp., Marietta, Ga., 1956-57; propulsion engr. Gen. Dynamics Corp., Cape Canaveral, Fla., 1957-59; mech. engr. Martin Mariella Corp., Cape Canaveral, Fla., 1959-71; program mgr. Martin Merietta Corp., Cape Canaveral, Fla., 1971-74, engring. mgr., 1974-88, activations mgr., 1988—; cons. engring. Clifton L. Gurr PE, Merritt Island, Fla., 1967—; gen. contractor Clift L. Gurr, Merritt Island, 1976—. Fellow AIAA (assoc.); mem. Canaveral Coun. Tech. Socs. (officer 1984-87). Democrat. Baptist. Achievements include participation in launching over 100 missiles, design of the largest self propelled structure for land use, buildings up to seven stories. Home: 1642 S Banana River Blvd Merritt Island FL 32952 Office: Martin Mareitta Corp PO Box 1399 Cocoa Beach FL 32951

GURUSIDDAIAH, SARANGAMAT, plant scientist; b. Chitradurga, Karnataka, India, Mar. 18, 1937; came to U.S., 1964; s. Chandrasekharaiah and Hampamma Sarangamat; m. Gowra Gurusiddaiah, Jan. 26, 1964; children: Manjesh, Sudha. BSc, Mysore U., Karnataka State, India, 1958, BSc with honors, 1960, MSc, 1961; PhD, Wash. State U., 1970. Mgr. Bioanalytical Ctr., Wash. State U., Pullman, 1971-76, asst. scientist, 1976-81, assoc. dir., assoc. scientist, 1986-81, assoc., dir., scientist, 1986—; vis. scientist Hoffmann-LaRoche, Inc., Nutley, N.J, 1988-89. Democrat. Hindu. Achievements include patents for antibiotics pantomycins, grahaminycins, treponimycin, chandramycin, wassumycin, also others. Home: NW 1230 Clifford Pullman WA 99163 Office: Bioanalytical Ctr Wash State U Pullman WA 99164-4235

GUSDON, JOHN PAUL, JR., obstetrics/gynecology educator, physician; b. Cleve., Feb. 13, 1931; s. John and Pauline (Malencek) G.; m. Marcelle Deiber, June 6, 1956 (dec. 1979); children: Marguerite, John Phillip, Veronique; m. R. Carolyn Gallager Aycock, July, 1989. BA, U. Va., 1952, MD, 1959. Diplomate Am. Bd. Ob-Gyn. Rotating intern U. Hosps. Cleve., 1959-60, resident, 1960-64; instr. ob-gyn. Sch. Medicine, Case Western Res. U., Cleve., 1964-66, asst. prof., 1967; asst. prof. ob-gyn. Bowman Gray Sch. Medicine, Wake Forest U., Winston-Salem, N.C., 1967-70, assoc. prof., 1970-74, prof., 1974-90, prof. emeritus, 1990—; Contbr. articles to sci. jours., chpts. to books. Lt. USN, 1952-55, Korea. Recipient John Horsley Meml. award U. Va., Charlottesville, 1968, Pres. award South Atlantic Assn. Ob-Gyn., 1973. Fellow ACOG (Pres. award 1970, 72), Am. Assn. Immunology; mem. Am. Soc. Immunology of Reproduction (founder, pres. 1981-84), Am. Gynecol. and Obstet. Soc. Republican. Roman Catholic. Avocations: reading, fishing.

GUSEK, TODD WALTER, food scientist; b. St. Louis Park, Minn., Oct. 24, 1959; s. Walter Thomas and Mary Virginia (Dustin) G.; m. Christine LaVerne Olson, Aug. 1, 1987; 1 child, Lyndsay Christine. BS, U. Minn., 1983; PhD, Cornell U., 1990. Sr. rsch. scientist Gen. Mills, Inc., Mpls., 1990—; cons., expert Teltech, Inc., Bloomington, Minn., 1988—. Contbr. articles to profl. jours.; patentee in field. Fulbright scholarship Fulbright Found., Mysore, India, 1991; recipient Excellence in Rsch. award Am. Chem. Soc., 1988, Biotech. Rsch. award Inst. Food Techs., 1990, Albert Flegenheimer Rsch. award Cornell U., 1989, Grad. Rsch. award Procter & Gamble, 1987. Mem. Inst. Food Techs., Sigma Xi, Phi Kappa Phi. Mem. Christian Ch. Achievements include a patent for discovery of a new enzyme for use in laundry detergents; invention of an edible composition which functions as a "susceptor" in microwaveable food products; invention of process of treating flour to eliminate need for chlorination. Home: 3240 Yates Ave N Crystal MN 55422 Office: General Mills Inc 9000 Plymouth Ave N Minneapolis MN 55427

GUSS, PAUL PHILLIP, physicist; b. Danville, Pa., Jan. 4, 1956; s. Jerome Vincent and Kathryn Mary (Orner) G. BA, Gettysburg Coll., 1977; PhD, Duke U., 1982. Rsch. prof. Coll. William and Mary, Williamsburg, Va., 1982-87; postdoctoral fellow Duke U., Durham, N.C., 1982; sect. head EG&G Energy Measurements, Suitland, Md., 1987—. Mem. ARC, Clinton, Md., 1989—. Mem. Am. Nuclear Soc., Am. Phys. Soc., Duke U. Alumni Assn., Gettysburg Alumni Assn., Oxon Hill Jaycees (pres. 1993—), Mensa, Sigma Xi. Democrat. Lutheran. Office: EG&G Energy Measurements PO Box 380 Suitland MD 20752-0380

GUSSIN, ROBERT ZALMON, health care company executive; b. Pitts., Jan. 5, 1938; s. Carl and Yetta G. B.S. in Pharmacy, Duquesne U., 1959, M.S. in Pharmacology, 1961; Ph.D. in Pharmacology, U. Mich., 1965. Rsch. fellow dept. pharmacology SUNY, 1965-67; rsch. pharmacologist Lederle Labs., N.Y.C., 1967-69; group leader dept. cardiovascular renal pharmacology Lederle Labs., 1969-73, dir. cardiovascular renal disease therapy sect., 1973-74; exec. dir. rsch. McNeil Labs., Ft. Washington, Pa., 1974-78; v.p. rsch. div. McNeil Labs., 1978, v.p. R & D, 1978-79; v.p. sci. affairs McNeil Pharm., Pa., 1979-86; corp. v.p., sci. and tech. Johnson & Johnson, New Brunswick, N.J., 1986—. Author: Introduction to Cardiovascular Pharmacology, 1976; mem. editorial bd. New Drug Evaluations, Drug Devel, Pharmaco Therapy; contbr. in field. Mem. Am. Soc. Clin. Pharmacology and Therapeutics, Am. Soc. Nephrology, Am. Soc. Pharmacology and Exptl. Therapeutics, Am. Fedn. Clin. Research, AAAS, N.Y. Acad. Scis., Am. Heart Assn. Office: Johnson & Johnson 410 George St New Brunswick NJ 08901-2020

GUSTAFSON, DAVID HAROLD, industrial engineering and preventive medicine educator; b. Kane, Pa., Sept. 11, 1940; s. Harold Edward and Olive Albertina (McKalip) G.; m. Rea Corina Anagnos, June 23, 1962; children—Laura Lynn, Michelle Elaine, David Harold. B.S. in Indsl. Engring., U. Mich., 1962, M.S. in Indsl. Engring, 1963, Ph.D., 1966. Dir. hosp. div. Community Systems Found., Ann Arbor, Mich., 1961-64; asst. prof. indsl. engring. U. Wis.-Madison, 1966-70, assoc. prof., 1970-74, prof., 1974—, dir., founder Ctr. for Health Systems and Analysis, 1974—, chmn. dept. indsl. engring., 1984-88; sr. analyst Dec. and Designs Inc., McLean, Va., 1974; dir. rsch. Govt. Health Policy Task Force, State of Wis., 1969-71; prin. cons. Medicaid Mgmt. Study Team, 1977-78; prin. investigator Nursing Home Quality Assurance System, 1979, Computer System for Adolscent Health Promotion, 1983, Computer System to Support Breast Cancer and People with AIDS, 1993; vis. prof. London Sch. Econs., 1983; developer computer-based support to measure and improve health care quality. Author: Group Techniques, 1975, Health Policy Analysis, 1992; contbr. articles to profl. jours. Adviser conflict resolution Luth. Ch., 1973-79; active numerous civic orgns. Recipient numerous grants, 1968—. Mem. Inst. Indsl. Engring., Ops. Research Soc., Med. Decision Making. Avocations: jogging, guitar, water sports, cross country skiing, parenting. Office: U Wis Ctr Health Systems 1300 University Ave Madison WI 53706-1572

GUSTAVINO, STEPHEN RAY, aerospace engineer; b. Lynwood, Calif., Mar. 27, 1962; s. Guido Remo and Patricia Ann (Kobey) G. BS in Aerospace Engring., San Diego State U., 1986. Engr. Douglas Aircraft Co., Long Beach, Calif., 1987-88; engr., scientist McDonnell Douglas Space Systems Co., Huntington Beach, Calif., 1988-92; rschr. spl. rsch. assignment NASA Ames Rsch. Ctr., Moffett Field, Calif., 1991—. Author, co-author tech. papers. Mem. AIAA, Cousteau Soc., Nature Conservancy, Laguna Festival of Arts. Achievements include development of a computer program designed to simulate plant behavior, of artificial environments to support human life on the moon or Mars. Office: McDonnell Douglas Space Sys 5301 Bolsa Ave Huntington Beach CA 92647

GUSTAVSON, KARL-HENRIK, physician; b. East Orange, N.Y., Mar. 31, 1930; s. Karl Helmer and Margit (von Delwig) G.; m. Gerd Margarete Thornberg, July 25, 1954; children: Karin, Lars. B Dental Sci., Royal Sch. Dentistry, Stockholm, 1950; MD, Karolinska Inst., 1958; PhD, U. Uppsala, Sweden, 1964. Asst. physician Univ. Hosp., Stockholm, 1958-60; rsch. physician dept. med. genetics U. Uppsala, 1960-64, lectr. in pediatrics, 1965-74; prof. pediatrics U. Umeå, Sweden, 1975-80; chmn. dept. pediatrics U. Umeå, 1975-80; head physician, chmn. dept. clin. genetics Univ. Hosp., Uppsala, 1980—; mem. adv. bd. on perinatal medicine Nat. Bd. Health and Medicine, Stockholm, 1989—; mem. planning groups for rsch. Swedish Med. Rsch. Coun., 1977—; cons. in med. genetics UN Devel. Fund, 1987-89. Contbr. articles to profl. jours.; author textbooks in field. Sub.-lt. Swedish Navy, 1949-64. Mem. Royal Soc. Sci. Sweden. Home: Döbelnsgat 30E, S-752 37 Uppsala Sweden Office: Univ Hosp, Dept Clin Genetics, S-751 85 Uppsala Sweden

GUSTIN, ANN WINIFRED, psychologist; b. Winchester, Mass., 1941; d. Bertram Pettingill and Ruth Lillian (Weller) G.; B.A. with honors in Psychology, U. Mass., 1963; M.S. (USPHS fellow), Syracuse U., 1966, Ph.D., 1969. Registered psychologist, Sask.; lic. psychologist, Ga.; Diplomate Am. Bd. Med. Psychotherapists. Research asst., psychology trainee U. Mass., Tufts U., Harvard U., Syracuse U., 1961-66; psychology intern VA, Canandaigua, N.Y., 1967-68; asst. prof. psychology U. Regina (Sask., Can.), 1969-74, assoc. prof. psychology, dir. counseling services, head clin. tng., 1974-78; pvt. practice psychology, Carrollton, Ga., 1978—, Atlanta, 19—; staff tng. cons. Frobisher Bay Dept. Social Services, N.W. Territories, Can., 1979-80; cons. staff Tanner Hosp.; ancillary staff West Paces Ferry Hosp.; psychiat. cons. Social Security Adminstrn., Ga. Dept. Human Resources, 1980—. Membership chmn. Carroll County Mental Health Assn., 1979-81. Fellow Ga. Psychol Assn. (exec. divsn. lic. psychologists 1986-91, disaster response team 1991—); mem. Am. Psychol. Assn., Can. Psychol. Assn., Sask. Psychol. Assn. (mem. exec. council 1971-72, registrar 1972-73), Nat. Assn. Disability Examiners, Ga. Assn. Disability Examiners. Office: 107 College St Carrollton GA 30117-3136 also: One Decatur Town Ctr 150 E Ponce de Leon Ave Ste 460 Decatur GA 30030

GUTH, ALAN HARVEY, physicist, educator; b. New Brunswick, N.J., Feb. 27, 1947; s. Hyman and Elaine (Cheiten) G.; m. Susan Tisch, Mar. 28, 1971; children: Lawrence David, Jennifer Lynn. SB and SM, MIT, 1969, PhD in Physics, 1972. Instr. Princeton U., 1971-74; research assoc. Columbia U., N.Y.C., 1974-77, Cornell U., Ithaca, N.Y., 1977-79, Stanford Linear Accelerator Ctr., Calif., 1979-80; assoc. prof. Physics MIT, Cambridge, 1980-86, prof., 1986-89, Jerrold Zacharias prof. physics, 1989-91, Victor F. Weisskopf prof. physics, 1992—; physicist Harvard-Smithsonian Ctr. for Astrophysics, 1984-89, vis. scientist, 1990-91. Alfred P. Sloan fellow, 1981; on Sci. Digest's list of America's 100 Brightest Scientists Under 40, 1984; on Esquire Mag.'s list of Men and Women Under 40 Who Are Changing the Nation, 1985; on Newsweek's list of 25 Top Am. Innovators, 1989. Fellow AAAS, Am. Phys. Soc. (mem. exec. com. astrophysics div. 1986-88, vice chmn. astrophysics div. 1988-89, chmn. div. 1989-90, recipient Lilienfeld Prize 1992), Am. Acad. Arts and Scis.; mem. NAS, Am. Astron. Soc. Achievements include being originator of inflationary model of early universe. Office: MIT Ctr Theoretical Physics 77 Massachusetts Ave Cambridge MA 02139

GUTHRIE, FRANK ALBERT, chemistry educator; b. Madison, Ind., Feb. 16, 1927; s. Ned and Gladys (Glick) G.; m. Marcella Glee Farrar, June 12, 1955; children: Mark Alan, Bruce Bradford, Kent Andrew, Lee Farrar. A.B., Hanover Coll., 1950; M.S., Purdue U., 1952; Ph.D., Ind. U., 1962. Mem. faculty Rose-Hulman Inst. Tech., Terre Haute, Ind., 1952—, assoc. prof., 1962-67, prof. chemistry, 1967—, chmn. dept., 1969-72, chief health professions adviser, 1975—; Kettering vis. lectr. U. Ill., Urbana, 1961-62; vis. prof. chemistry U.S. Mil. Acad., West Point, N.Y., 1987-88, 93—. Mem. exec. bd. Wabash Valley council Boy Scouts Am., 1971-87, adv. bd., 1988—, v.p. for scouting, 1976; selection chmn. Leadership Terre Haute, 1978-80. Served with AUS, 1945-46. Recipient Silver Beaver award Boy Scouts Am., 1980. Fellow Ind. Acad. Sci. (pres. 1970, chmn. acad. found. trustees 1986—); mem. Am. Chem. Soc. (sec. 1973-77, editor directory 1965-77, chmn. div. analytical chemistry 1979-80, chmn. 1958, counselor Wabash Valley sect. 1980—, local sect. activities com. 1982-86, nominations and elections com. 1988—, sec. 1992—, steering com. for Joint Cen.-Great Lakes Regional Meetings, Indpls., 1978, 91, vis. assoc. com. profl. tng. 1984—), Coblentz Soc., Midwest Univs. Analytical Chemistry Conf., Nat. Assn. Advs. Health Professions, Hanover Coll. Alumni Assn. (pres. 1974, Alumni Achievement award 1977), Sigma Xi, Phi Lambda Upsilon, Phi Gamma Delta, Alpha Chi Sigma (E.E. Dunlap scholarship selection com., 1986—, chmn. 1990—). Presbyterian. Club: Masons (32 deg.). Home: 19 S 21st St Terre Haute IN 47803-1819 Office: Rose Hulman Inst Tech 5500 Wabash Ave Terre Haute IN 47803-3999

GUTHRIE, HUGH DELMAR, chemical engineer; b. Murdo, S.D., May 11, 1919; s. John Arlington and Farol Venus (Smith) G.; m. Elizabeth Anne Harris, Mar. 4, 1950; children: Katherine Farol, Gretchen, Mary Melissa, Elizabeth Lenore, Emily Jo. BSChemE with highest distinction, State U. Iowa, 1943. Jr. engr., engr., group leader Shell Devel. Co., San Francisco, 1943-52; technologist, sr. technologist, asst. dept. mgr. Shell Oil Co., Wood River, Ill., 1952-56; staff engr., group leader Shell Oil Co., N.Y.C., 1956-60; dept. mgr. Shell Oil Co., Wood River, 1960-62; asst. mgr. to mgr. mktg. Shell Oil Co., N.Y.C., 1962-70; from dept. mgr. to sr. staff Shell Oil Co., Houston, 1970-76; div. dir. ERDA, Dept. Energy, Washington, 1976-78; dir. Energy Ctr., Stanford Rsch. Inst., Menlo Park, Calif., 1978-80; v.p. licensing, mgr. tech. assessment Occidental Rsch. Corp., Irvine, Calif., 1980-83; v.p. licensing, mgr. rsch. planning Cities Svc., Tulsa, 1983-86; dir. extraction divsn. Morgantown (W.Va.) Energy Tech. Ctr. Dept. Energy, 1987-92, gen. engr. products tech. mgmt., mgr. gas products, 1992—; cons. Hugh D. Guthrie & Assocs., Tulsa, 1986-87; mem. adv. bd. U. Iowa, U. Calif., Berkeley, Tulsa U., U. Tex., U. Pitts., W.Va. U. Former sr. warden Episcopal chs., Conn., Ill., Tex. Fellow Am. Inst. Chem. Engrs. (pres. 1969, chair Assembly of Fellows 1990—, chair mgmt. div. 1991, Founder's award 1974, F.J. Van Antwerpen award 1986, Robert L. Jacks Meml. award 1992, chair mem. campaign found.); mem. Am. Chem. Soc., Soc. Petroleum Engrs., Am. Assn. Arts and Scis., N.Y. Acad. Scis., Sigma Xi, Tau Beta Pi, Phi Lambda Upsilon, Omicron Delta Kappa. Republican. Achievements include patents on distillation equipment. Home: 901 Stewart Pl Morgantown WV 26505-3688 Office: Dept Energy Morgantown Energy Tech Ctr 3610 Collins Ferry Rd Morgantown WV 26505-1301

GUTHRIE, JOHN ROBERT, physician, health science facility administrator; b. Spartanburg, S.C., Mar. 8, 1942; s. Clarence L. and Rosa Jane (Thackston) G.; children: Jason, Elizabeth; m. Natasha K. Guthrie, Sept. 18, 1993; children from a previous marriage: Luke, Asia. BS, Mercer U., 1969; MS, Duke U., 1971; DO, Chgo. Coll. Osteo. Medicine, 1979. Diplomate Am. Bd. Osteopathic Medicine and Surgery. Instr. sci. N.C. Community Coll., 1971-75; intern Dr.'s Hosp., Atlanta, 1979-80; physician with family practice assoc. Spartanburg, S.C., 1980-82; med. dir. Guthrie Family Practice Clinic, Spartanburg, 1982—; med. cons. CBS-TV, Spartanburg, 1985-88; CEO, med. dir. The Agy. for Internat. Understanding. Contbr. articles to profl. jours., poetry and prose to various publs. Bd. dirs. Spartanburg Parents Who Care, 1982-87, Spartanburg Teen Ctr., 1984-85. Served to USMC, 1961-65, lt. comdr. USNR, 1983-87. Mem. AMA, Am. Osteo. Assn., S.C. Osteo Med. Assn. (past pres., past v.p.), S.C. Med. Assn., Spartanburg County Med. Assn., Acad. Neuromuscular Thermography, Am. Acad. Family Physicians. Democrat. Baptist. Club: Caroline Country (Spartanburg). Avocations: shooting, writing, flying. Home: PO Box 8493 Spartanburg SC 29305-8493 Office: Guthrie Family Practice Clinic 703 N Pine St Spartanburg SC 29303-3768

GUTHRIE, MICHAEL STEELE, magnetic circuit design engineer; b. Murray, Ky., Nov. 22, 1954; s. Steele G. and Lunelle (Holmes) G. BS in Physics, Murray State U., 1976. Engr. quality control & mfg. Allegheny Ludlum, Princeton, Ky., 1977-79; engr. applications & design Hitachi Magnetics Corp., Edmore, Mich., 1979-86; engr. applications & design Delco Remy div. GM, Anderson, Ind., 1986-91; regional mgr. applications engring. Stackpole Magnet div. Stackpole Carbon Co., Kane, Pa., 1991—, Carbone of Am., Farmville, Va., 1991—. Co-author: Rapidly Solidified Alloys, 1993. Mem. IEEE, Magnetics Soc., Ky. Cols. Home and Office: 9055 Ravinewood Ln South Lyon MI 48178

GUTHRIE, ROBERT VAL, psychologist; b. Chgo., Feb. 14, 1930; s. Paul Lawrence and Lerlene Yvette (Cartwright) G.; m. Elodia S. Guthrie, Sept. 15, 1952; children: Robert S., Paul L., Michael V., Ricardo A., Sheila E., Mario A. B.S., Fla. A&M U., 1955; M.A., U. Ky., 1960; Ph.D., U.S. Internat. U., 1970. Tchr. San Diego City Schs., 1960-63; instr. psychology San Diego Mesa Coll., 1963-68, chmn. dept., 1968-70; asso. prof. U. Pitts., 1971-73; sr. research psychologist Nat. Inst. Edn., Washington, 1973-74; asso. dir. organizational effectiveness and psychol. scis. Office of Naval Research, Arlington, Va., 1975; supervising research psychologist Naval Personnel, Research and Devel. Center, San Diego, 1975-82; pvt. practice psychology San Diego, 1982-90; prof. psychology So. Ill. U., Carbondale, 1991—; Adj. assoc. prof. George Washington U., Washington, 1975; lectr. Georgetown U., 1975; adj. assoc. prof. U. Pitts., 1977, adj. prof. San Diego State U., 1989. Author: Psychology in the World Today, 1968, 2d edit., 1971, Encounter, 1970, Black Perspectives, 1970, Man and Society, 1972, Psychology and Psychologists, 1975, Even the Rat Was White, 1976. Served with USAF, 1950-59, Korea. Mem. AAAS, Am., Western, Calif. psychol. assns., Fedn. Am. Scientists, Am. Acad. Polit. and Social Scis., Kappa Alpha Psi. Achievements include research on social psychology, organizational and personnel psychology variables in small groups.

GUTHRIE, RUSSELL DALE, vertebrate paleontologist; b. Nebo, Ill., Oct. 27, 1936; married; two children. BS, U. Ill., 1958, MS, 1959; PhD in Zoology, U. Chgo., 1963. Assoc. prof. U. Alaska, Fairbanks, 1963-74, prof. zoology, 1974—. NSF grantee, 1963-65, 68, 70-74, 80-81. Mem. AAAS, Soc. Study Evolution, Soc. Vertebrate Paleontology. Office: U Alaska Dept Zoology 116 Bunnell Fairbanks AK 99701-1520

GÜTLICH, PHILIPP, chemistry educator; b. Rüsselsheim, Hessen, Fed. Republic Germany, Aug. 5, 1934; s. Philipp and Eva Dorothea (Rabenstein) G.; m. Angelika Stoeck, Oct. 11, 1969; children: Katja, Daniel. Diplomingenieur, Tech. Hochschule, Darmstadt, 1961, PhD, 1963, habilitation, 1969. Prof. inorganic chemistry Tech. Hochschule, 1972-73, prof. of theoretical inorganic chemistry, 1973-75; prof. inorganic chemistry U. Mainz, Fed. Republic Germany, 1975—; Author: (with others) Mössbauer Spectroscopy and Transition Metal Chemistry, 1978; editor Comments on Inorganic Chemistry, 1981-90; contbr. numerous articles to profl. jours. Rsch. award Japan Soc. for the Promotion Sci., 1989. Mem. Gesellschaft Deutscher Chemiker, Deutsche Bunsengesellschaft Für Physik Chemie, Am. Chem. Soc., Royal Chem. Soc. (London), Deutsche Gesellschaft für Metallkunde. Achievements include patent for light induced excited spin state trapping. Home: Georg Büchnerstr 9, 64380 Rossdorf Hessen, Federal Republic Germany Office: U Mainz, Staudingerweg 9, 55099 Mainz Germany

GUTOFF, EDGAR BENJAMIN, chemical engineer; b. N.Y.C., June 2, 1930; s. Boris and Miriam Deborah (Tumarkin) G.; m. Hinda Oler, July 22, 1956; children: Joshua Jared, Jonathan Michael. B Chem. Engring., CCNY, 1951; SM, MIT, 1952, ScD, 1954. Registered profl. engr., Mass., N.Y. Sr. process engr. Brown Co. (now James River), Berlin, N.H., 1954-58; sr. chem. engr. Ionics, Inc., Watertown, Mass., 1958-60; scientist, sr. scientist, sr. prin. engr. Polaroid Corp., Waltham, Mass., 1960-88; ind. cons. chem. engr. Brookline, Mass., 1988—; part-time lectr., adj. prof. Northeastern U., Boston, 1981—, vis. rsch. prof., 1991. Co-author: The Application of Statistical Process Control to Roll Products, 1990, 91; co-editor: Modern Coating and Drying Technology, 1992; contbr. articles on coating, drying, crystallization, colloid chemistry to sci. publs. Fellow AICE; mem. Am. Chem. Soc., Soc. Imaging Sci. and Tech. (chpt. v.p. 1963-64, pres. 1980-82, Svc. award 1979). Jewish. Achievements include patent for a process for continuous formation of photographic emulsions, development of method to reliably scale-up photographic emulsions, simplified method for designing coating die internals for uniform flow across width, modeling the drying of coated webs. Home and Office: 194 Clark Rd Brookline MA 02146

GUTOWICZ, MATTHEW FRANCIS, JR., radiologist; b. Camden, N.J., Feb. 23, 1945; s. Matthew F. and A. Patricia (Walczak) G.; m. Alice Mary Bell, June 27, 1977; 1 child, Melissa. BA, Temple U., 1968; DO, Phila. Coll. Osteo. Medicine, 1972. Diplomate Am. Bd. Radiology, Am. Bd. Nuclear Medicine. Intern Mercy Hosp., Denver, 1972-73; resident in diagnostic radiology Hosp. of U. Pa., Phila., 1973-76, fellow in nuclear medicine, 1976-77; chief dept. radiology and nuclear medicine Fisher Titus Med. Ctr., Norwalk, Ohio, 1977—; pres. Firelands Radiology, Inc., Norwalk, 1977—. Republican. Roman Catholic. Avocations: photography, tennis, scuba diving. Home: 23 Patrician Dr Norwalk OH 44857-2463

GUTOWSKY, H. S., chemistry educator; b. Bridgman, Mich., Nov. 8, 1919; s. Otto and Hattie (Meyer) G.; m. Barbara Stuart, June 22, 1949 (div. Sept. 1981); children: Daniel Kurt (dec.), Robb Edward, Christopher Carl.; m. Virginia Warner, Aug. 1982. AB, Ind. U., 1940, DSc (hon.), 1983; MS, U. Calif.-Berkeley, 1946; PhD, Harvard U., 1949. Mem. faculty U. Ill., Urbana, 1948—, prof. chemistry, 1956—, head div. phys. chemistry, 1956-63, head dept. chemistry and chem. engring., 1967-70, dir. Sch. Chem. Scis., head dept. chemistry, 1970-83, mem. Ctr. for Advanced Study, 1983—; mem. chemistry panel NSF, 1963-66, chmn. panel, 1965-66, mem. adv. com. on planning, 1971-74; mem. Ill. Bd. Natural Resources and Conservation, 1973—; G.N. Lewis Meml. lectr., 1976, G.B. Kistiakowsky lectr., 1980. Mem. adv. bd. Petroleum Research Fund, 1959-61; mem. selection and scheduling com. Gordon Research Conf., 1959-64, 68-72, trustee, 1969-72, chmn. bd. trustees, 1971-72. Served to capt., chem. warfare service AUS, 1941-45. Recipient 1966 Irving Langmuir award Am. Chem. Soc., Midwest award St. Louis sect., 1973, prize Internat. Soc. Magnetic Resonance, 1974, Peter Debye award in phys. chemistry Am. Chem. Soc., 1975, Nat. Medal of Sci., 1977, Wolf prize in Chemistry, 1983, Chem. Pioneer award Am. Inst. Chemists, 1991, Pitts. Spectroscopy award, 1992; Guggenheim fellow, 1954-55. Fellow AAAS, Am. Phys. Soc. (chmn. div. chem. physics 1973-74), AAAS, Am. Acad. Arts and Scis.; mem. AAUP, NAS (mem. com. sci. and pub. policy 1972-75, chmn. panel on atmospheric chemistry 1975-77, mem. com. impacts of stratospheric change 1975-77), Am. Philos. Soc., Am. Chem. Soc. (chmn. div. phys. chemistry 1966-67, com. on profl. tng. 1969-77, chmn. 1974-77), Phi Beta Kappa, Sigma Xi. Home: 202 W Delaware Ave Urbana IL 61801-4905 Office: U Ill Noyes Lab 505 S Mathews Ave Urbana IL 61801-3664

GUTSCHE, CARL DAVID, chemistry educator; b. LaGrange, Ill., Mar. 21, 1921; s. Frank Carl and Vera (Mutchler) G.; m. Alice Eugenia Carr, June 4, 1944; children: Clara Jean, Betha Lynn, Christopher Glenn. B.A., Oberlin Coll., 1943; Ph.D., U. Wis., 1947. With U.S. Dept. Agr. Office Sci. Devel., 1943-44; instr. chemistry Washington U., St. Louis, 1947-48; asst. prof. Washington U., 1948-51, assoc. prof., 1951-59, prof., 1959—, chmn. dept., 1970-76; Robert A. Welch prof. chemistry Tex. Christian U., Fort Worth, 1989—; cons. to industry; mem. adv. bd. Petroleum Research Fund, 1971-74; chmn. medicinal chemistry study sect. NIH, 1978-81; Bd. dirs. St. Louis Conservatory and Schs. for Arts, 1978-82. Author: The Chemistry of Carbonyl Compounds, 1967, Carbocyclic Ring Expansion Reactions, 1968, Fundamentals of Organic Chemistry, 1975, Calixarenes, 1989; mem. adv. bd.: Jour. Organic Chemistry, 1979-83; mem. editorial bd.: Organic Preparations and Procedures Internat., 1968—; contbr. articles to profl. jours. Guggenheim fellow, 1981. Fellow AAAS; mem. Am. Chem. Soc. (chmn. St. Louis sect. 1959, mem. pub. com. 1974-77, mem. com. on coms. 1977-80, mem. com. on profl. tng. 1980-89, cons. to com. 1990—, councilor and dir., St. Louis sect. award 1971, Midwest award 1988), Chem. Soc. (London), AAUP, Phi Beta Kappa (mem. qualifications com. 1992—), Sigma Xi. Home: 3521 Arborlawn Dr Fort Worth TX 76109-2533 Office: Tex Christian U Dept Chemistry Fort Worth TX 76129

GUTSCHE, STEVEN LYLE, physicist; b. St. Paul, Nov. 10, 1946; s. Lyle David and Phyllis Jane (Stubstad) G.; divorced; children: Kristina, Angela; m. Marilyn D. Maloney, Oct. 4, 1980; children: Taylor Steven, Daniel Mark. BS, U. Colo., 1968; MS, U. Calif., Santa Barbara, 1970. Physicist USN Pacific Missile Range, Point Mugu, Calif., 1968-71; staff scientist Mission Rsch. Corp., Santa Barbara, 1971-76, group leader, 1977-79, div. leader, 1979—, v.p., 1987—; pres., 1989—; also bd. dirs. Mission Rsch. Corp., Santa Barbara. Contbr. articles to tech. publs. Presbyterian. Avocations: collecting oriental rugs, soccer, long distance running, reading. Office: Mission Rsch Corp 735 State St # 719 Santa Barbara CA 93102-0719

GUTSCHICK, RAYMOND CHARLES, geology educator, researcher, micropaleontologist; b. Chgo., Oct. 3, 1913; s. Anthony William and Elizabeth (Bessie) (Kostka) G.; m. Alice Edna Augusta Lude, July 2, 1939; children: Alice Antonette, Raal Emily. A.A., Morton Jr. Coll., 1934; B.S. in Engring. Physics, Univ. Ill., Urbana, 1938, M.S. in Geology, 1939, Ph.D. in Geology, 1942. Geologist, MOBIL-Magnolia Petroleum Co., Oklahoma City, 1943-45, ALCOA Alum. Ore Co., Rosiclare, Ill., 1942, 1946-47, Gulf Oil Co., Oklahoma City, 1947; asst. prof. dept. geology U. Notre Dame, Ind., 1947-50, assoc. prof., 1950-54, prof., 1954-79, dept. chmn., 1956-70, emeritus prof., cons., 1979—; geologist U.S. Geol. Survey, Denver, 1975—; cons. geologist Rogers Group, Inc., 1978—, Environment Service EIS, 1983—. Co-author geol. research: Redwall Limestone of N. Arizona, 1969; contbr., co-contbr. papers, articles to profl. jours. NSF grantee, 1954, 60; recipient ann. Teaching award U. Notre Dame, 1964, Neil A. Miner award Nat. Assn. Geology Tchrs., 1977, Raymond C. Moore Paleontology medal Soc. Sedimetary Geology, 1993. Fellow Geol. Soc. Am.; mem. Assn. Geol. Tchrs. (pres. 1952, 58), Am. Assn. Petroleum Geologists, Paleontol. Soc., Sigma Xi (pres. Notre Dame chpt. 1953). Independent Republican. Lutheran. Avocations: photography, golf, nature, history. Home: 2901 Leonard Medford OR 97504 Office: U Notre Dame Dept Earth Sciences Notre Dame IN 46556

GUTSTEIN, SIDNEY, gastroenterologist; b. Dortmund, Germany, Apr. 6, 1934; came to U.S. 1939; s. David and Mary (Goetz) G.; children: Jennifer, Benjamin Daniel. BA, Washington Sq. Coll., N.Y.C., 1955; MD, NYU, 1958. Intern Bronx Mcpl. Hosp., 1958-59, resident, 1959-60, 62-63, fellow in gastroenterology, 1963-65; instr. medicine Albert Einstein Coll. Medicine, Bronx, 1965-66, assoc. in medicine, 1966-67, asst. prof. medicine, 1967-72, assoc. prof. medicine, 1972-78, assoc. clin. prof. medicine, 1978—; dir. Klau Med. Svc. Montefiore Hosp. and Med. Ctr., Bronx, 1972-78; attending physician Montefiore Hosp. & Med. Ctr., 1976—, Hosp. of Albert Einstein Coll. Medicine, 1966—. Capt. U.S. Army, 1960-62. Fellow ACP, Am. Coll. Gastroenterology; mem. AAAS, Am. Gastroenterol. Assn. Democrat. Jewish. Office: 101 S Bedford Rd Mount Kisco NY 10549

GUTTERMAN, MILTON M., operations research analyst; b. N.Y.C., Nov. 5, 1927; s. Benjamin and Gussie (Rothchild) G.; m. Joan Helen Levey, Nov. 30, 1952; children: Gail Rosemary, Allen Bernard. BS, CCNY, 1948; MS, U. Chgo., 1949. Researcher Inst. for Air Weapons Rsch., Chgo., 1952-54; rsch. engr. IIT Rsch. Inst., Chgo., 1954-66; sr. ops. rsch. cons. Amoco Corp., Chgo., 1966-92; pvt. practice cons., 1992-93; rsch. cons. in field. Editor: (book) Computer Applications 1962, 1964; assoc. editor Transactions on Math. Software, 1975-80; asst. editor ORSA Jour. on Computing, 1987-91. With U.S. Army, 1946-47. Mem. Ops. Rsch. Soc. Am. (chmn. computer sci. tech. sect. 1989-91), Math. Programming Soc., Soc. for Indsl. and Applied Math., Assn. for Computing Machinery, Am. Rose Soc., No. Chicagoland Rose Soc., Am. Contact Bridge League (life master). Jewish. Home and Office: 5049 Lee St Skokie IL 60077-2336

GUTTORMSEN, MAGNE SVEEN, nuclear physicist; b. Halden, Norway, Nov. 4, 1950; s. Georg and Kari (Sveen) G.; m. Christa Diex, June 22, 1974; 1 child, Christian. Candidatus Realium, U. Oslo, Norway, 1975, PhD, 1980. Cert. physicist. Advisor Norwegian Railway, Oslo, 1974-76; rsch. asst. U. Oslo, 1976-80; rsch. fellow U. Bonn, Germany, 1979-82; postdoctorate Norwegian Rsch. Coun., Oslo, 1983-85; assoc. prof. U. Oslo, 1985—; bd. dirs. Oslo Cyclotron Lab., Oslo. Referee Nuclear Physics jour., 1987, Physica Scripta jour., 1988; contbr. articles to profl. jours. Recipient Price of Physics award F. Nansen Fund, Oslo, 1982. Mem. Nuclear Physics Com. of Norwegian Phys. Soc., Nat. Com. for Nuclear Rsch., Am. Phys. Soc. Avocation: fishing. Office: U Oslo Dept Physics, PO Box 1048 Blindern, Oslo 0316, Norway

GUTZWILLER, MARTIN CHARLES, theoretical physicist, research scientist; b. Basel, Switzerland, Oct. 12, 1925; married; 2 children. BS, Swiss Fed. Inst. Tech., 1947, MS, 1950; PhD in Physics, U. Kans., 1953. Physicist Brown, Boveri & Co., Baden, Switzerland, 1950-51; with exploration and production divsn. Shell Devel. Co., Tex., 1953-60; with rsch. divsn. Internat. Bus. Machines, Zurich, 1960-63; dir. gen. svc. dept. Watson Lab. T.J. Watson Rsch. Ctr. IBM Corp., Yorktown Heights, N.Y., 1963-70, 74-77, rsch. scientist, physicist T.J. Watson Rsch. Ctr., 1970—; adj. prof. Columbia U. Mem. Am. Phys. Soc. (Dannie N. Heineman prize for math. physics 1992). Achievements include research in propagation of waves, solid state physics, quantum and classical mechanics especially the chaotic phenomenum, ergodic theory, applied mathematics. Office: IBM T J Watson Rsch Ctr PO Box 218 Yorktown Heights NY 10598*

GUY, MATTHEW JOEL, gastroenterologist; b. Bklyn., Aug. 23, 1945; s. Rubin and Gertrude (Feinberg) Guy; B.A. summa cum laude, Bklyn. Coll., 1966; M.D., Columbia U., 1970, Med.Sc.D., 1974; m. Barbara Mae Sachartof, Oct. 21, 1979; children—Reuven Maxwell, Judah Philip, Alan Louis, David Charles, Goldie Hannah-Cheryl. Intern, Maimonides Hosp., Bklyn., 1970-71, resident, 1971-72; resident St. Luke's Hosp., N.Y.C., 1972-73, Columbia-Presbyn. Med. Center, N.Y.C., 1973-74; practice medicine specializing in gastroenterology, Bklyn., 1974—; mem. staff Kings Hwy. Hosp., 1974—, dir. gastro-intestinal endoscopy, 1976—; mem. staff Maimonides Hosp., SUNY Downstate Med. Center (all Bklyn.); asst. prof. medicine SUNY Downstate Med Center. Trustee, mem. bd. edn. Yeshiva of Flatbush. Diplomate Am. Bd. Internal Medicine. Fellow Am. Coll. Gastroenterology, Am. Soc. Gastrointestinal Endoscopy; mem. ACP, AMA, Kings County Med. Soc., Phi Beta Kappa, Sigma Xi. Office: 1401 Ocean Ave Brooklyn NY 11230-3917

GUY, TERESA ANN, aerospace engineer; b. Dayton, Ohio, Apr. 12, 1961; d. John Joseph and Charlotte Jean (Ninneman) Sollars; m. James Kevan Guy, June 10, 1989. BSME, U. Wash., 1983; MSAA, MIT, 1991. Registered profl. engr., Calif. With NASA Johnson Space Ctr., Houston, 1980-81; assoc. engr. Gen. Dynamics Space Systems Div., San Diego, 1983, engr., 1984-87, sr. engr., 1988-91; tech. specialist McDonnell Douglas Technologies, Inc., San Diego, 1991—; conf. presenter in field. Contbr. articles to profl. jours. Mem. USA Track and Field Assn., San Diego, 1991—; coach track high sch./coll., Seattle, San Diego, Cambridge, 1982—. Recipient First Shuttle Flight Achievement award NASA, JSC, 1981. Mem. ASME (assoc., Structures and Materials award 1988), Sigma Xi (assoc.). Achievements include rsch. on minimum compressive residual strength after impact of graphite/epoxy laminates was ind. of impact method; principal investigator on multiple composite structures related independent research and development (IRAD) programs. Emphasis on low cost fabrication damage tolerance, repair and structural stability. Office: McDonnell Douglas Tech Inc 16761 Via Del Campo Ct San Diego CA 92127

GUYER, J. PAUL, civil engineer, architect, consultant; b. Sacramento, Feb. 12, 1941; s. Paul Marline and Vivian Ruth (Mosher) G.; m. Judith Mae Overholser, June 28, 1968; children: J. Paul Jr., Christopher Meador. BS, Stanford U., 1962; postgrad., McGeorge Law Sch., 1962-65. Registered profl. engr., architect, Calif. Engr. State of Calif., Sacramento, 1962-67; ptnr. Guyer and Santin, Sacramento, 1967-75; pres. Guyer Santin, Inc., Sacramento, 1975—. Author: Planning and Development Manual, Calif. Dept. Parks and Recreation, 1984; editor: Recreation Planning and Devel., 1983, Infrastructure for Urban Growth, 1987; author tech. reports. Bd. dirs. Davis (Calif.) Sci. Ctr., 1991-92. Fellow ASCE (Harland Bartholemew award for contbns. to field of urban planning 1991); mem. ASME, Sacramento Symphony Assn., Sacramento Opera Assn., Sacramento Met. C. of C., El Macero Country Club, Capital Club, Comstock Club. Home: 44240 Clubhouse Dr El Macero CA 95618 Office: Guyer Santin Inc 917 7th St Sacramento CA 95814-2509

GUZEK, JAN WOJCIECH, physiology educator; b. Lublin, Poland, Mar. 28, 1924; s. Józef and Maria (Pelczarska) G.; m. Barbara Moskalewska, Oct. 24, 1952 (dec. Nov. 1980); children: Anna Maria, Maria Magdalena, Wojciech Józef. Grad. in medicine, Jagellonian U., Kraków, Poland, 1951; MD, U. Łódź, Poland, 1962; habilitation in human physiology, U. Lódz, Poland, 1968. Asst. lect. dept. gen. pathology, faculty of medicine U. Kraków, 1949-60; sr. lectr., reader, asst. prof. dept. physiology U. Łódź, 1960-74, vice dean Sch. Medicine, 1972, dep. dir. Inst. Physiology and Biochemistry, Sch. Medicine, 1973-74, prof.-in-ordinary, head dept. pathophysiology, 1974—; mem. com. for physiol. scis. Polish Acad. Scis., 1984—, mem. com. for basic med. scis., 1987—, com. for clin. pathophysiology, 1984-87, com. for cell pathophysiology, 1987-90. Editor, co-author textbooks in field, 1970, 80, 85, 90, 92; translator Hans Selye: The Stress of Life, Polish edit., 1960; contbr. articles to profl. publs. Mem. Internat. Parliament for Safety and Peace, Palermo, Italy, 1991—; mem. sci. coun. Ministry of Health, Warsaw, Poland, 1987-90. Rsch. fellow Free U., Brussels, 1963, U. Copenhagen, 1972, hon. rsch. fellow Univ. Coll. London, 1988; recipient Tchr. of Merit award State Coun. Poland, 1973, Chevalier of Polonia Restituta, State Coun. Poland, 1980; named Knight Sovereign Mil. Templar Order, Jerusalem, 1991. Mem. Polish Physiol. Soc. (pres. 1984-90), Polish Endocrinol. Soc., Internat. Soc. Neuroendocrinology, Internat. Brain Rsch. Orgn., European Neuroendocrine Soc., European Pineal Soc., Internat. Soc. Pathophysiology (chmn. ednl. commn., mem. coun. 1991—), Soc. Pathol. Physiology (corr.), Gen. Sikorski's Inst. Polish History. Roman Catholic. Avocations: science in history, classical painting, old Polish maps, walking. Home: ul Narutowicza 120 m 2, 90-145 Łódź Poland Office: Sch Medicine Dept Pathophys, ul Narutowicza 60, 90-136 Lodz Poland

GUZIK, MICHAEL ANTHONY, electrical engineer, consultant; b. Sharon, Pa., Sept. 25, 1950; s. John Stanley and Stephanie Mary (Vanecek) G.; m. Linda Liane Hassan, June 10, 1970 (dec. Aug. 1989); 1 child, Suzanne Marie. B of Engring., Youngstown State U., 1973, MS, 1977. Registered profl. engr., Tex., La., N.Mex., Ga., Tenn., Ky. Jr. engr. Michael Baker Jr., Inc., Beaver, Pa., 1973-77; elec. engr. Killebrew, Rucker and Assoc., Inc., Wichita Falls, Tex., 1977-78, Hammer Engrs., Inc., Austin, Tex., 1978-82; assoc. Espey, Huston and Assocs., Inc., Austin, Tex., 1982—. Mem. bldg. com. St. Paul's Ch., Austin, 1991—. Mem. NSPE, Tex. Soc. Profl. Engrs., K.C. (warden 1989-91, trustee 1992—). Republican. Roman Catholic. Home: 7316 Gaines Mill Ln Austin TX 78745-6016 Office: Espey Huston and Assocs Inc 916 Capital of Texas Hwy S Austin TX 78746

GWAL, AJIT KUMAR, electrical engineer, federal government official; b. Varanasi, India, Jan. 2, 1939; came to U.S., 1970; s. Shiv Nath and Sudah G.; m. Kadambari Singh, Nov. 21, 1965; children: Anuradha, Anita, Kriti. BTech with honors, Indian Inst. Tech., 1962; MSc in Engring., B.H.U., India, 1964, U. Pa., 1978. Registered profl. engr., N.Y., N.J., Pa. Tech. mgr. Electric Constrn. & Equipment Co., Calcutta, India, 1964-70; supr. elec. dept. T.B. Mac Cabe, Phila., 1970-72; project engr. Stone & Weber Engring. Co., Cherry Hill, N.J., 1972-88; technical specialist Defense Nuclear Facilities Safety Bd. Fed. Govt., Washington, 1990—; mem. bd. TVA Watts Bar Program Team, Knoxville, 1988-90. Mem. IEEE (chmn. IEEE-ICCP 848). Republican. Hindu. Achievements include development of new transformer design to reduce material and electrical losses, identification of actions to license nuclear power plant-Watts Bar; helped set standards for electrical industries-nuclear power plants and others. Home: 102 Kilburn Dr Cherry Hill NJ 08003 Office: Defense Nuclear Facilities Safety Bd 625 Indiana Ave Washington DC 20004

GWALTNEY, JACK MERRIT, JR., physician, educator, scientist; b. Norfolk, Va., Dec. 24, 1930; s. Jack Merrit and Mary Gordon (Weck) G.; m. Sarah Bulloch Parrott, June 26, 1954; children: Elizabeth Cromwall, Jack Merrit III. BA, U. Va., 1952, MD, 1956. Diplomate Am. Bd. Internal Medicine. Rotating intern Univ. Hosps., Cleve., 1956-57, resident in internal medicine, 1957-59; chief resident internal medicine U. Va. Hosp., Charlottesville, 1959-60; asst. respiratory virus rsch. U. Va. Sch. Medicine, Charlottesville, 1962-63; Nat. Inst. Allergy and Infectious Diseases, NIH rsch. postdoctoral fellow preventive medicine and medicine U. Va. Sch. Medicine, 1963-64, instr. preventive medicine and medicine, 1964-66, asst. prof., 1966-70, assoc. prof. internal medicine, 1970-75, Wade Hampton Frost prof., 1975—, head div. epidemiology and virology, 1970—, dir. Ctr. for Prevention of Disease and Injury, 1984—; assoc. mem. Commn. Acute Respiratory Diseases, Armed Forces Epidemiol. Bd., 1968-73; mem. adv. panel infectious disease therapy U.S. Pharmacopeia, 1970—; cons. NSF, 1976-79; trustee Am. Type Culture Collection, 1972, chmn. bd., 1976-78. Mem. editorial bd. Antimicrobial Agents and Chemotherapy, 1971-86, editor, 1985-90; contbr. numerous articles to profl. jours. Capt. U.S. Army, 1960-62. Recipient Rsch. Career Devel. award NIH, 1969-73. Fellow ACP; mem. AAUP, AAAS, Am. Epidemiol. Soc., Med. Soc. Va., Albemarle County Med. Soc., Am. Fedn. Clin. Rsch., Am. Soc. Microbiology, Am. Thoracic Soc. (Edward Livingston Trudeau fellow 1964-67), Va. Thoracic Soc. (sec.-treas. governing coun. 1973-75, v.p. 1975-76, pres. 1977—), Infectious Diseases Soc. Am. (Joseph E. Smadel award 1987), So. Soc. Clin. Investigation, Am. Clin. and Climatol. Assn. (Jermiah Metzgar lectr. 1984), Soc. Epidemiologic Rsch., Raven Soc., Sigma Xi, Alpha Omega Alpha. Home: RR 1 Box 208AA Free Union VA 22940-9801 Office: U Va Sch Medicine PO Box 473 Charlottesville VA 22908-0001

GWINN, WILLIAM DULANEY, physical chemist, educator, consultant; b. Bloomington, Ill., Sept. 28, 1916; s. Walter E. and Allyne (Dulaney) G.; m. Margaret Boothby, July 11, 1953; children—Robert B., Ellen, Kathleen. A.B., U. Mo., 1937, M.A., 1939; Ph.D., U. Calif. at Berkeley, 1942. Teaching asst. U. Calif., Berkeley, 1939-42, mem. faculty, 1942—, prof. phys. chemistry, 1955-79, prof. emeritus, 1979—; rsch. prof. Miller Rsch. Inst., 1961-62; pres. Environ. Conversion Tech., Inc., El Cerrito, Calif., 1992—; vis. prof. chemistry U. Minn., Mpls., 1969-70; cons. several energy-related fields. Assoc. editor: Jour. Chem. Physics, 1962-64. Guggenheim fellow, 1954; Sloan fellow, 1955-59; recipient citation merit U. Mo., 1964. Fellow Am. Phys. Soc.; mem. Am. Chem. Soc., Phi Beta Kappa, Sigma Xi, Pi Mu Epsilon. Achievements include spl. rsch. on molecular structure, microwave spectroscopy, quantum mechanics, direct digital control, rsch. and consulting in several energy-related fields. Home and Office: 8506 Terrace Dr El Cerrito CA 94530-2721

GWÓŹDŹ, BOLESŁAW MICHAEL, physiologist; b. Radzionków, Poland, Sept. 27, 1928; s. Peter and Maria (Cieśla) G.; m. Urszula Anna Krzeszowiak, June 12, 1954; children: Jolanta, Mark. MD, Silesian Med. Acad., Zabrze, Poland, 1961; PhD, Silesian Med. Acad., Katowice, Poland, 1966. Asst. dept. anatomy Silesian Med. Acad., Zabrze, 1951-53, asst. dept. physiology, 1954-56, sr. asst. dept. physiology, 1956-61, asst. prof., 1961-66; extraordinary prof. Silesian Med. Acad., Katowice, 1979, ordinary prof., 1990—, decanus, 1972-75, rector, 1975-78; prof. Inst. Physiology, Zabrze, 1967-72, dir., 1972—; cons. Therapeutic Occupational Ctr. Rehab., Repty, Poland, 1971-81. Author: Human Physiology, 1980; patentee walking apparatus for patients with hemiplegia. Founder Polish Red Cross, Zabrze, 1951—. Recipient award Polish Ministry of Mining, 1962, Ministry of Health and Social Welfare, 1976, 78, 79, 81. Fellow Polish Med. Soc.; mem. Polish Acad. Sci., Polish Soc. Ergonomics, Polish Physiol. Soc. (regional pres. 1982-90, award 1963). Avocations: historiography, political science, skiing. Home: 1 Szkolna, 42-600 Tarnowskie Góry Poland Office: Silesian Med Acad Physiol, 19 H Jordana, 41-808 Zabrze 8, Poland

GYONGYOSSY, LESLIE LASZLO, mechanical engineer, consultant; b. Szikszo, Hungary, June 29, 1932; came to U.S., 1978; s. Ferenc and Susanna (Erhardt) G.; m. Christina Hunyady, Apr. 1, 1960 (div. Feb. 1984); children: Laszlo, Ferenc. BSME, U. Miskolc, Hungary, 1955. Registered profl. engr. Project engr. Beker Industries, N.Y.C., 1975-78; sr. engr. Bechtel Engring. Co., San Francisco, 1978-91; engring. mgr. Yemen Hunt Oil Co., Dallas, 1991—. Mem. Soc. Petroleum Engrs. Achievements include patents in valve design. Home: 11140 Westheimer Rd # 358 Houston TX 77042 Office: Yemen Hunt Oil Co Ste 1700 1445 Ross Ave Dallas TX 75202

HA, SAM BONG, molecular geneticist; b. Pusan, Korea, Jan. 16, 1954; s. Man Jeung and Chan Bun (Oh) H.; m. Young Ok Um, Dec. 5, 1982. BS, Seoul Nat. U., 1980; PhD, Okla. State U., 1986. Rsch. assoc. Wash. State U., Pullman, 1986-89; asst. prof. Va. Poly. Inst. and State U., Blacksburg, 1989—. Contbr. articles to profl. publs., chpt. to book. Mem. AAAS, Am. Soc. Plant Physiologists, Am. Soc. Agronomy, Crop Sci. Soc. Am., Internat. Soc. Plant Molecular Biology, Phi Kappa Phi, Gamma Sigma Delta. Achievements include development of the first genetically engineered turfgrass. Office: Va Poly Inst and State U Smyth 365 Blacksburg VA 24061-0404

HA, SUNG KYU, mechanical engineer, educator; b. Seoul, Korea, June 18, 1960; s. Hee Y. and Jae K. (Lee) H. MS, Stanford U., 1985, PhD, 1989. Postdoctoral fellow Stanford (Calif.) U., 1989-91; prof. Hanyang U., Seoul, 1991—; cons. Am. Cup, 1990-91. Contbr. articles to Jour. Composite Materials. Mem. AIAA. Home: 502-8 Mock-2 Dong Yangcheon, 150-02 Seoul Korea Office: Hanyang U Dept Mech Engring, 17 Haengdang-Dong Sungdong-Gu, 133-791 Seoul Republic of Korea

HAACK, ROBERT ALLEN, forest entomologist; b. Madison, Wis., Apr. 28, 1952; s. Harold Richard William and Dorothy Elaine (Chase) H.; m. Sheridan Louise Kidd, June 6, 1983. BS in Forestry, U. Wis., 1974, MS in Entomology, 1980; PhD in Entomology, U. Fla., 1984. Rsch. asst. U. Wis., Madison, 1978-80, U. Fla., Gainesville, 1980-84; rsch. assoc. Mich. State U., East Lansing, 1984-86; forest entomologist North Cen. Forest Experiment Sta. USDA Forest Svc., East Lansing, 1986-88, project leader forest insect project, 1988—; adj. assoc. prof. dept. entomology, Pesticide Rsch. Ctr. Mich. State U., 1986—, dept. forestry, 1990—. Contbr. chpts. to books, articles to profl. jours. including Jour. Chem. Ecology. Mem. Entomological Soc. Am., Internat. Union Forestry Rsch. Orgns., Fla. Entomol. Soc. (membership com. 1982-84), Mich. Entomol. Soc. (newsletter editor 1989—), Sigma Xi, Phi Kappa Phi. Achievements include extension of theory on drought stress provoking insect outbreaks, study of ultrasonic acoustic emissions from drought-stressed trees changing insect host-finding behavior, study of bark beetle pheromone, pine shoot beetle. Home: 1280 Clark Rd Dansville MI 48819 Office: USDA Forest Svc 1407 S Harrison Rd East Lansing MI 48823

HAAG, JOEL EDWARD, architect; b. Wayne, Nebr., June 30, 1962; s. Robert James and Shirley Ann (Krutz) H. BS, N.E. Mo. State U., 1984; BArch, U. Kans., 1986. Architect Hollis & Miller Group, Prairie Village, Kans., 1986-90, Tognascioli, Gross, Kautz Architects, Inc., Kansas City, Mo., 1991-92; engr., structural design draftsman Borton Inc., Hutchinson, Kans., 1992-93; architect Mann & Co., Hutchinson, Kans., 1993—; co-chmn. Archtl. Explorer Post, Kansas City, Mo., 1989-92. Trustee Redeemer Luth Ch., Lawrence, Kans., 1987-88; mem. choir Bethany Luth. Ch., Overland Park, Kans., 1989-92. Mem. Reno Choral Soc., Kans. Mennonite Men's Choral. Home: 1010 N Washington St Hutchinson KS 67501 Office: Mann & Co 335 N Washington Ste 110 Hutchinson KS

HAAN, CHARLES THOMAS, agricultural engineering educator; b. Randolph County, Ind., July 10, 1941; s. Charles Leo and Dorothy Mae (Smith) H.; m. Janice Kay Johnson, June 3, 1967; children: Patricia Kay, Christopher Thomas, Pamela Lynn. B.S. in Agrl. Engring, Purdue U., 1963, M.S., 1965; Ph.D. in Agrl. Engring, Iowa State U., 1967. Grad. asst. Purdue U., W. Lafayette, Ind., 1963-64; research asso. Iowa State U., Ames, 1964-67; asst. prof., asso. prof., prof. U. Ky., Lexington, 1967-78; prof. agrl. engring. Okla. State U., Stillwater, 1978—, head dept., 1978-84, regents prof., 1986—, Sarkeys Disting. prof. agrl. engring., 1989—; cons. in area of hydrology various firms and govtl. orgns. Author: Statistical Methods in Hydrology, 1977, Hydrology and Sedimentology of Surface Mined Lands, 1978, Hydrology and Sedimentology of Disturbed Areas, 1981; editor: Hydrologic Modeling of Small Watersheds, 1981; contbr. tech. papers and reports to publs. and confs. Recipient and or adminstr. various research grants. Fellow Am. Soc. Agrl. Engrs. (Young Researcher of 1975, rsch. paper award 1969, Hancor Soil and Water Engring. award 1990), Am. Inst. Hydrology, Sigma Xi, Tau Beta Pi, Alpha Epsilon, Gamma Sigma Delta, Phi Kappa Phi. Roman Catholic. Home: 720 W Lakeshore Dr Stillwater OK 74075-1335 Office: Okla State U Agr Engring Dept Stillwater OK 74078

HAAR, ANDREW JOHN, environmental engineer; b. Hackensack, N.J., Dec. 23, 1963; s. Daniel S. and Emma F. (Metz) H.; m. Lea Ann Swanson, Jan. 16, 1988; 1 child, Alex J. BSChemE, U. Mo., 1987. Process engr. Harcross Chems., Kansas City, Kans., 1988-89; project engr. Farmland Industries, Coffeyville, Kans., 1989-91; environ. engr. Farmland Industries, Kansas City, Mo., 1991-93, sr. environ. engr., 1993—. Mem. Red Cross, Kansas City, Mo., 1983—; mem. Lions Club Internat., 1989—. Mem. AICE. Home: 1308 NW 67 St Kansas City MO 64118 Office: Farmland Industries PO Box 7305 Dept 141 Kansas City MO 64116

HAAS, DAVID COLTON, neurologist, educator; b. N.Y.C., Sept. 14, 1931; s. Harry Colton and Doris (Richman) H.; m. Barbara Ann Kaplan, Apr. 5, 1955; 1 child, Andrew. BA, Ohio Wesleyan U., 1953; MD, NYU, 1957. Diplomate Am. Bd. Psychiatry and Neurology. Intern Cin. Gen. Hosp., 1957-58; resident in neurology U. Mich., Ann Arbor, 1960-63; instr., asst. prof., then assoc. prof. SUNY Health Sci. Ctr., Syracuse, 1963-90, prof. neurology, 1990—. Contbr. articles to profl. jours. Capt. M.C. U.S. Army, 1958-60. Mem. AAAS, Am. Assn. Study of Headache, Internat. Headache Soc., N.Y. Acad. Scis. Achievements include identification of naturally-occurring canine model of Guillain-Barre Syndrome, identification of migraine attacks triggered by head trauma, identification of transient global amnesia attacks triggered by head trauma. Office: SUNY Health Sci Ctr 750 E Adams St Syracuse NY 13210

HAAS, JEFFREY EDWARD, aerospace engineer; b. West Reading, Pa., Jan. 8, 1949; s. Willis James and Thelma Mae (Edwards) H.; m. Jeanne Marie Kovalsky, Apr. 15, 1972; children: Shannon, Kevin, Todd. BS in Aerospace Engring., U. Pitts., 1970; MS in Mech. Engring., U. Toledo, 1975. Rsch. engr. U.S. Army Propulsion Lab., Cleve., 1970-85; facility mgr. NASA-Lewis Rsch. Ctr., Cleve., 1985-89, br. chief, 1989—. Author 30 NASA tech reports. Pres. St. Mary's Athletic Commn., Berea, Ohio, 1992—; bd. dirs. Berea Baseball Assn., 1992—. Mem. AIAA (vice-chmn. ground test tech. com. 1993—). Roman Catholic. Office: NASA-LERC MS 6-9 21000 Brookpark Rd Cleveland OH 44135

HAAS, ROBERT JOHN, aerospace engineer; b. Dayton, Ohio, Apr. 14, 1930; s. Robert J. Haas and Harriett (Longstreth) Bevan; m. Florence A. Eldred, June 6, 1952 (div. June 1984); adopted children: Jeffrey (dec.), Lisa Haas Cappucio; m. Gayle F. Byrne, Dec. 14, 1984; stepchildren: Patrick Barton, Marissa Barton; children: Amber Haas, Robert J. Haas III. Student, U.S. Mil. Acad., 1948-51; BS in Petroleum Engring., U. Tulsa, 1954. Petroleum engr. Skelly Oil Co., Tulsa, 1953-54; propulsion engr., supr. Marquardt, Van Nuys, Calif., 1957-64; mgr. rocket programs Marquardt, Van Nuys, 1964-69, dir. test and facilities, 1969-72, gen. mgr. environ. systems, 1972-75; plant gen. mgr. Williams Internat., Ogden, Utah, 1975-79; sr. v.p. engring. Williams Internat., Walled Lake, Mich., 1979-86; sr. v.p. product planning and mktg. Williams Internat., Walled Lake, 1986-90; sr. advisor, cons. Las Vegas, Nev., 1990—; cons. Marquardt, Van Nuys, 1961-75. Author: Approach to Aerospace Plane Propulsion, 1960. Lectr. and advisor Weber State Coll., U. Utah and various high schs. and clubs., 1975-79; pres. Marquardt Mgmt. Club, 1971. 1st lt. USAF, 1954-56. Mem. AIAA, Navy League (lifetime). Republican. Roman Catholic. Achievements include contribution to devel. and prodn. of world's smallest turbofan for cruise missiles; discoveries in the field of integrated propulsion modules for missiles, economical methods of testing ramjets, turbines and rocket engines. Home: PO Box 33126 Las Vegas NE 89133 Office: Haas Enterprize PO Box 33126 Las Vegas NE 89133

HAAS, THOMAS JOSEPH, coast guard officer; b. Staten Island, N.Y., Mar. 5, 1951; s. Joseph Walter and JoAnne (Pawloski) H.; m. Marcia Jane Knapp, Jan. 12, 1974; children: Eric, Gregory, Sarah. BS with honors,

USCG Acad., New London, Conn., 1973; MS in Chemistry, U. Mich., 1976, MS in Environ. Health Sci., 1977; MS in Human Rsch. Mgmt., Rensselaer Poly. Inst., 1981. Cert. indsl. hygienist. Commd. ensign USCG, 1973, advanced through grades to comdr., 1989; ops. officer USCG Cutter Acacia, Port Huron, Mich., 1973-75; mem. staff USCG Hdqrs., Washington, 1977-80, br. chief, 1980-81; from section chief to assoc. dean acads. USCG Acad., New London, 1981—; mem. group experts UN, Geneva, 1980; mem. Chem. Transport Adv. Com., Washington, 1977-81, 87-92; data mgr. USCG Valdez (Alaska) Oil Spill, 1989. Editor: Descriptions of Selected Hazardous Materials, 1991. Chair Ledyard (Conn.) Congregation Ch. Session, 1983-90; chair Scholarship Com., Ledyard, 1987-90; pres. Parsonage Hill Homeowners Assn., Ledyard, 1987-91. Yale fellow, 1991-92, Am. Coun. on Edn. fellow, 1992-93. Mem. Am. Chem. Soc., Am. Conf. Govtl. Indsl. Hygienists, N.Y. Acad. Scis. Achievements include research on investigation of synthetic properties. Home: 15 Seabury Ave Ledyard CT 06339 Office: USCG Acad New London CT 06320

HAAS, TRICE WALTER, researcher; b. Dallas, July 24, 1932; s. Rolf Oscar and Viola Jeannette (Shadday) H.; m. Joyce Marie Tauzin, Sept. 2, 1955; children: Ann Marie, Mark Anthony, Michael Joseph, Matthew Walter, Andrea Jean. BS in Chemistry, St. Mary's U., 1954; PhD in Phys. Chemistry/Chem. Engring., Iowa State U., 1960. Sr. rsch. technologist Mobil Oil Co., Dallas, 1960-62; rsch. scientist Nat. Cash Register Corp., Dayton, Ohio, 1962-64; rsch. group leader Aerospace Rsch. Lab., Wright Patterson AFB, Ohio, 1964-74, Wright Lab./Materials Directorar, Wright Patterson AFB, 1974—. Contbr. articles to profl. jours. With U.S. Army, 1954-56. Rsch. fellow USAF, 1991. Mem. Materials Rsch. Soc., Am. Phys. Soc. Roman Catholic. Achievements include discovery of chemical effects in Auger electron spectroscopy; development of model of dispenser cathodes, models for mechanisms of activation of ohmic contacts to compound semiconductors; patent on use of Pulse Laser Deposition to Deposit Solid Lubricant and Hard Coat Films; contributions to use of electron diffraction and optical probes for in-situ monitoring of molecular beam epitaxial growth of compound semiconductors. Home: 2412 Westlawn Dr Kettering OH 45440-2017

HAASE, ASHLEY THOMSON, microbiology educator, scientist; b. Chgo., Dec. 8, 1939; s. Milton Conrad and Mary Elizabeth Minter (Thomson) H.; m. Ann DeLong, 1962; children: Elizabeth, Stephanie, Harris. BA, Lawrence Coll., 1961; MD, Columbia U., 1965. Intern Johns Hopkins Hosp., Balt., 1965-67; clin. assoc. Nat. Inst. Allergy and Infectious Disease, Bethesda, Md., 1967-70; vis. scientist Nat. Inst. Med. Research, London, 1970-71; chief infectious disease sect. VA Med. Ctr., San Francisco, 1971-84, med. investigator, 1978-83; prof. microbiology U. Minn., Mpls., 1984—, head dept., 1984—; mem. fellowship screening com. Am. Cancer Soc., San Francisco, 1978-81; mem. UNESCO Internat. Cell Rsch. Orgn., India, 1978; mem. nat. adv. coun. Nat. Inst. Allergy and Infectious Diseases, 1986-91, mem. task force on microbiology and infectious diseases, 1991, merit investigator, 1989, mem. vaccine subcom. AIDS rsch. adv. com., 1993—; Javits neurosci. investigator Nat. Inst. Neurol. and Communicative Disorders and Stroke, 1988—; chmn. panel on AIDS, U.S.-Japan Coop. Med. Sci. Program, 1988—. Editor-in-chief Microbial Pathogenesis, 1988—; contbr. articles on neurovirology to profl. jours. Recipient Lucia R. Briggs Disting. Achievement award Lawrence Coll., 1990. Mem. Am. Soc. Microbiology, Assn. Am. Physicians, Am. Soc. Clin. Investigation, Am. Soc. Virology, Assn. Med. Schs. Microbiology (chmn.), Infectious Diseases Soc. Am., Nat. Multiple Sclerosis Soc. (adv. com. 1978-84), Phi Beta Kappa, Alpha Omega Alpha. Democrat. Home: 14 Buffalo Rd Saint Paul MN 55127-2136 Office: U Minn Dept Microbiology 420 Delaware St SE Minneapolis MN 55455-0374

HABBEN, DAVID MARSHALL, state official; b. Hammond, Ind., Aug. 22, 1952; s. Robert Woodward Habben and Elizabeth Ruth (Baxter) Spanos; m. Joanne Dorothy Gombert, May 25, 1974; children: David Marshall II, Rachel G. AS in Law Enforcement, Calumet Coll., East Chicago, Ind., 1973; student, Chemeketa Community Coll., Salem, Oreg., 1978-79, 83, Lane Community Coll., Eugene, Oreg., 1981, Fed. Emergency Mgmt. Agy., 1986—. Cert. nursing asst., EMT-4/Paramedic, Oreg., hazardous material response instr., ACLS affilliate faculty, BCLS affilliate faculty, EMT/paramedic instr., emergency response driving instr., critical incident stress debriefing coord.; cert. paramedic Nat. Registry Emergency Med. Technicians. Merchandising specialist, photo-sound mgr. Fred Meyer, Inc., Portland and Salem, Oreg., 1974-77; staff photographer Steimont's Studio, Salem, 1977-78; dept. mgr. Emporium, Inc., Salem, 1978-79; retail store mgr. Photo Factory, Inc., Salem, 1979; intern ambulance EMT Willamette Ambulance Svc., Salem, 1979; paramedic Med. Svcs., Inc., Salem, 1980-81; flight paramedic Mercy Flights Air Ambulance, Medford, Oreg., 1985-86; sr. paramedic, community rels. officer Community 1 Ambulance/Josephine Meml. Hosp., Grants Pass, Oreg., 1981-86; regional tng. specialist State of Idaho Emergency Med. Svc. Bur., Twin Falls, Idaho, 1986-90; statewide tng. coord. State of Idaho Emergency Med. Svc. Bur., Boise, 1990—; state Emergency Med. Svc. team coord., mem. Critical Incident Stress Debriefing Team, Boise, 1990—; disaster coord. State Health and Welfare Dept., Boise, 1992; CPR and first aid instr. Boise Community Edn., 1990—. Article reviewer JEMS mag.; manuscript reviewer Brady Pub., AAOS Pub., W.B. Saunders Pub. Emergency Med. Svc. advisor, coach Spl. Olympics, Idaho, Jerome, 1986-90; med. coord. Josephine County Disaster Svcs., Grants Pass, Oreg., 1981-86, Emergency Med. Svcs. advisor Josephine County Vol. Svcs., 1981-86; chmn. Speaker's Bur. S.A.F.E.T.Y. Com., Grants Pass, 1985-86; med. care div. coord. Josephine County Disaster Commn., Grants Pass, 1985-86; first aid instr., merit badge counselor Boy Scouts Am., 1982—; mem., Emergency Med. Svc. rep. Idaho Emergency Response Commn., 1987-92. Mem. Am. Critical Incident Stress Found., Nat. Coun. State Emergency Med. Svc. Tng. Coords., Associated Pub. Safety Communications Officers. Achievements include rsch. in night driving, continuing edn. and farm tractor rescue. Office: State of Idaho EMS Bur 450 W State St Boise ID 83720

HABER, PAUL, health psychologist, educator; b. Yonkers, N.Y., Jan. 30, 1936; s. Herbert Hubert and Sylvia Martha (Kliger) H.; m. Marsha Last, June 29, 1957; children: Kara Edin Haber Stevens, Robert Jacobs. BA in Psychology, St. Thomas U., Miami, Fla., 1978; MA in Counseling, Goddard Coll., Plainfield, Vt., 1979; PhD in Health Psychology, The Union Inst., Cin., 1981. Pres., dir. Stress Inst., Inc., Miami, 1978—; pvt. practice Miami, 1979—; v.p., project dir. Child Assault Prevention Project of South Fla., Inc., Miami, 1984—; adj. prof. various ednl. instns., including St. Thomas U., U. Miami, Miami-Dade C.C., Union Inst., 1978-89. Writer, moderator 11-part TV series on child abuse, 1985-86; author: Health, Stress and Type A Behavior, 1984-86; co-author: The Stress Reduction Workbook, 1987, Protecting Your Child, 1988. Active Dade County Task Force on Child Abuse, Miami, 1984—, Fla. Com. for Prevention of Child Abuse, 1989—, Fla. Ctr. for Children and Youth, 1989—, The Stop Smoking Program, Refocusing. Recipient Nat. Sony Innovator award, 1969. Fellow Am. Inst. Stress; mem. ACA, Am. Orthopsychiat. Assn., Assn. Transpersonal Psychology, Assn. Psychol. Type, Inst. Noetic Sci., Phi Theta Kappa. Avocations: computers, photography, writing, behavioral medicine.

HABER, PAUL ADRIAN LIFE, geriatrician; b. N.Y.C., Feb. 14, 1920; s. Benjamin Walter and Gussie Esther (Schnur) H.; m. Mary Agatha Crolley, Oct. 25, 1959; 1 child, Peter William. BA, U. Tex., 1941; MA, Columbia U., 1942; MD, U. Tex. Med. Br., Galveston, 1949; MS, George Washington U., 1967. Diplomate Am. Bd. Internal Medicine. Instr. in chemistry Hofstra U., Huntington, N.Y., 1941-42; rsch. chemist So. Alkali Corp., Corpus Christi, Tex., 1942-45; intern L.A. County Gen. Hosp., 1949-50; resident VA Med. Ctr., L.A., 1950-52; asst. clin. prof. medicine UCLA, 1959-64, George Washington U., Washington, 1965-80; asst. chief med. dir. U.S. Dept. Vets. Affairs, Washington, 1975-82, regional coord. for aging, 1986—; clin. prof. medicine Stanford U., Palo Alto, Calif., 1983-85; bd. dirs. Nat. Coun. Aging, Washington, 1975-79; mem. Presdl. Coun. on Aging, Washington, 1978-82; mem. Nat. Adv. Com. on Aging for Nat. Inst. Aging, Bethesda, Md., 1979-85. Contbr. chpt. to book Handbook on Aging, 1988, Merck Manual of Geriatrics, 1989. Capt. USAF, 1953-55. Fellow ACP (gov. 1980-82); mem. AMA (alternate del. 1980-82). Home: 7501 Honeywell Ln Bethesda MD 20814-1027 Office: US Dept Vets Affairs 50 Irving St NW Washington DC 20422-0002

HABERMAN, CHARLES MORRIS, mechanical engineer, educator; b. Bakersfield, Calif., Dec. 10, 1927; s. Carl Morris and Rose Marie (Braun) H. BS, UCLA, 1951; MS in Mech. Engring., U. So. Calif., 1954, ME, 1957. Lead, sr. and group engr. Northrop Aircraft, Hawthorne, Calif., 1951-59, cons., 1959-61; asst. prof. to prof. mech. engring. Calif. State U., L.A., 1959-91; cons. Royal McBee Corp., 1960-61. Author: Engineering Systems Analysis, 1965, Use of Computers for Engineering Applications, 1966, Vibration Analysis, 1968, Basic Aerodynamics, 1971. Served with AUS, 1946-47. Mem. Am. Acad. Mechanics, Am. Soc. Engring. Edn., AIAA, AAUP. Democrat. Roman Catholic.

HABERMANN, DAVID ANDREW, chemical engineer; b. Ames, Iowa, Dec. 21, 1957; s. Clarence Edward and Marlene Ann (Messerli) H.; m. Marilyn Martha Lopez, Oct. 19, 1992; 1 child, Rafael. BSChemE, Mich. State U., 1980. Rsch. leader Dow Chem. Co., Midland, Mich., 1980—; pres. and incorporating officer BBN Users Group Corp., Cambridge, Mass., 1991-92. Office: Dow Chem Co 438 Bldg Midland MI 48667

HABERMANN, HELEN MARGARET, plant physiologist, educator; b. Bklyn., Sept. 13, 1927. AB, SUNY, Albany, 1949; MS, U. Conn., 1951; PhD, U. Minn., 1956. Asst. botanist U. Conn., Storrs, 1949-51; asst. U. Minn., Mpls., 1951-53, asst. plant physiologist, 1953-55, head residence counselor, 1955-56; rsch. assoc. U. Chgo., 1956-57; rsch. fellow Hopkins Marine Sta. Stanford (Calif.) U., 1957-58; from asst. prof. to assoc. prof. biol. scis. Goucher Coll., Towson, 1958-70, chmn. dept. biology, 1963-66, 68, 78-79, prof., 1970-92, Lilian Welsh prof. biol. scis., 1982-92; prof. emeritus, 1992—. Co-author Biology: A Full Spectrum, 1973, Mainstreams of Biology, 1977. NIH spl. rsch. fellow Rsch. Inst. Advanced Study, Balt., 1966-67. Fellow AAAS; mem. Phytochem. Soc. N.Am. (sec. 1987-93), Am. Soc. Plant Physiologists, Am. Soc. Hort. Sci., Soc. Devel. Biology, Am. Soc. Photobiology, Am. Inst. Biol. Scis., Scandinavian Soc. Plant Physiology, Internat. Soc. Plant Molecular Biology, Japanese Soc. Plant Physiology, Soc. Exptl. Biology and Medicine, Am. Camellia Soc., Am. Hort. Soc., Sigma Xi. Office: Goucher Coll Dept Biol Scis Dulaney Valley Rd Baltimore MD 21204-5109

HABERMEHL, GERHARD GEORG, chemist, educator; b. Seligenstadt, Germany, Feb. 19, 1931; s. Georg Heinrich and Emmy Elisabeth (Fischer) H.; m. Irmentrud Anne Hefner, July 25, 1961; children: Georg, Karin. Diploma in Chemistry, Tech. Hochschule, Darmstadt, Germany, 1957, D of Chemistry, 1960, habilitation, 1964. Rsch. fellow NIH, Bethesda, Md., 1968; prof. organic chemistry Tech. Hochschule, Darmstadt, 1970-80, dean, 1970-72; faculty mem., head chemistry dept. Vet. U., Hannover, Germany, 1980—; vis. prof. sch. pharmacy W.Va. U., Morgantown, 1977, Kyushu U., Fukuoka, Japan, 1985. Author: Rontgenstrukturanalyse Organischer verbindungen, 1973, Gifttiere und ihre Waffen, 1987, Venomous Animals and Their Toxins, 1981, Mitteleurp Giftpflanzen u irhe Wirkstoffe, 1985, Naturstoffe, 1992; editor: Toxicon, Magnetic Resonance in Chemistry; mem. adv. bd. Pesquisa Veterinaria Brasileira; contbr. articles to profl. jours. Fellow Royal Soc. Chemistry; mem. Gesellschaft Deutscher Chemiker, Internat. Soc. Toxinology (pres. 1991-94). Home: Eichhornchensteg 18, D-3000 Hannover-Bothfeld Germany Office: Veterinary U, Bischofscholer Damm 15, D-30173 Hannover Germany

HACKEL, EMANUEL, science educator; b. Bklyn., June 17, 1925; s. Henry N. and Esther (Herbstman) H.; m. Elisabeth Mackie, June 24, 1950 (dec. Apr. 1978); children: Lisa M., Meredith Anne, Janet M.; m. Rachel A. Fisher, Oct. 18, 1981; stepchildren: Daniel E., Tabitha A., and Jessica K. Harrison. Student, N.Y. U., 1941-42; B.S., U. Mich., 1948, M.S., 1949; Ph.D., Mich. State U., 1953. Fisheries biologist Mich. Dept. Conservation, 1949; mem. faculty Mich. State U., East Lansing, 1949—; prof. natural sci. Mich. State U., 1962-74, chmn. dept. natural sci., 1963-74, prof. medicine, 1974—, prof. zoology, 1974—; asst. dean coll., 1958-63; research fellow Galton Lab., Univ. Coll., London, Eng., 1970-71, 77-78; vis. investigator blood group research unit Lister Inst., London, Eng., 1956-57; cons. Mpls. War Meml. Blood Bank, 1983—. Author: Guide to Laboratory Studies in Biological Science, 1951, Studies in Natural Science, 1953, Natural Science, 1955, Vols. 1, 2, 3, 1952-63. Editor: The Search for Explanation-Studies in Natural Science, Vols. 1, 2, 3, 1967-68, Laboratory Manual for Natural Science, Vol. 1, 2, 3, 1967-68, Human Genetics, 1974, Theoretical Aspects of HLA, 1982, Bone Marrow Transplantation, 1983, HLA Techniques for Blood Bankers, 1984, Human Genetics 1984: A Look at the Last Ten Years and the Next Ten, Transfusion Management of Some Common Heritable Blood Disorders, 1992, Advances in Transplantation, 1993. Contbr. articles on genetics, human blood group immunology and chem. nature of blood group antigens, human biochem. genetics, tissue typing, human histocompatability antigens to sci. jours. Served to lt. (j.g.) USNR, 1943-47; now lt. comdr. USNR Ret. Recipient Cooley Meml. award Am. Assn. Blood Banks, 1969, Elliott Meml. award Am. Assn. Blood Banks, 1987. Mem. Assn. Gen. and Liberal Studies (sec.-treas. 1962-65), AAUP, AAAS, Genetics Soc. Am., Am. Soc. Human Genetics, Am. Assn. Blood Banks (dir. 1983-84, chmn. sci. sect. 1983-84), Mich. Assn. Blood Banks (v.p. 1970, pres. 1975-77), Am. Inst. Biol. Sci., Biometric Soc., Transplantation Soc. Mich. (dir. 1975-84), Am. Assn. for Clin. Histocompatability Testing, N.Y. Acad. Scis., Sigma Xi, Phi Kappa Phi. Home: 244 Oakland Dr East Lansing MI 48823-4715 Office: Mich State U Dept Medicine East Lansing MI 48824

HACKEMESSER, LARRY GENE, mechanical engineer; b. Taylor, Tex., June 16, 1950; s. Martin Oscar and Alice Marie (Blom) H.; m. Brenda Carol Engdahl, Mar. 23, 1985; children: Madeline Leigh, Emily Kathleen. AA, Temple Jr. Coll., 1970; BS in Mech. Engring., U. Houston, 1973. Licensed profl. engr., Tex. Assoc. engr. M.W. Kellogg Co., Houston, 1974-79, sr. heat transfer engr., 1979-80, furnace heat transfer engr., 1981-90, sr. prin. furnace engr., 1990-92; chief tech. engr. heat transfer M.W. Kellogg Co., 1992—. Contbr. articles to Environ. Progress. Mem. ASME, Sigma Xi (assoc.). Achievements include patents for reforming exchanger reactor, flexible feed pyrolysis process, flexible pyrolysis furnace with a split flue convection. Office: MW Kellogg Co 601 Jefferson Ave Houston TX 77210-4557

HACKER, DAVID SOLOMON, chemical engineer, researcher; b. Bklyn., June 9, 1925; s. Joseph and Dorothy (Krugman) H.; m. Elaine Samson, Feb. 9, 1949; children: Karen Anne, Julie Ruth. BSChemE, U. Ill., 1949; MSChemE, MIT, 1950; PhDChemE, Northwestern U., 1954. Rsch. engr. GE Co., Evendale, Ohio, 1954-57; sr. rsch. scientist Chgo. Midway Lab., 1957; sr. scientist Ill. Inst. Tech. Rsch. Inst., Chgo., 1957-61; rsch. supr. Gas Rsch. Inst., Chgo., 1961-65; assoc. prof. U. Ill., Chgo., 1965-81; sr. rsch. engr. Amoco Chem. Co., Naperville, Ill., 1981-91, Amoco Corp., Naperville, 1991—; cons. Symatronics, Grayslake, Ill., 1973-74, Universal Oil Products, Mt. Prospect, Ill., 1976-78. Author: Manned Re-entry Flight, 1960, Supercritical Fluid, 1981; contbr. 15 articles to profl. publs., including AICE Jour., Chem. Engring. Sci., Indsl. and Engring. Chemistry, AIAA Jour. Commr. Ridgeville Pk Dist., Evansville, Ill., 1960-64; bd. dirs. ACLU, Chgo., 1966-93. Recipient N.Y. State Regent's scholarship, U. Ill., 1943, Shell Oil Fellowship, Northwestern U., 1952, Amoco Corp. Teaching award, U. Ill., Chgo., 1972. Fellow AICE (pres. Chgo. 1982-83); mem. Combustion Inst. (founding), Am. Phys. Soc., Sigma Xi. Achievements include patents in prodn. of disilane from silane, removal of methyl acetate from offgas in IA manufacture, swirl combustion chamber to improve steel prodn, recovery of diethyl carbonate by extractive distillation, novel cooling for combustion chambers in gas turbines. Home: 343 Beech St Highland Park IL 60035 Office: Amoco Corp PO Box 400 Naperville IL 60566

HACKERMAN, NORMAN, retired university president, chemist; b. Balt., Mar. 2, 1912; s. Jacob and Anne (Raffel) H.; m. Gene Allison Coulbourn, Aug. 25, 1940; children: Patricia Gale, Stephen Miles, Sally Griffith, Katherine Elizabeth. AB, Johns Hopkins U., 1932, PhD, 1935. Asst. prof. Loyola Coll., Balt., 1935-39; research chemist Colloid Corp., 1936-40; asst. chemist USCG, S.I., 1939-41; asst. prof. Va. Poly. Inst., Blacksburg, 1941-43; research chemist Kellex Corp., 1944-45; asst. prof. chemistry U. Tex., 1945-46, assoc. prof., 1946-50, prof., 1950-70, chmn. dept., 1952-61, dir. corrosion research lab., 1948-61, dean research and sponsored programs, 1960-61, v.p., provost, 1961-63, vice chancellor acad. affairs, 1963-67, pres., 1967-70, prof. emeritus chem., 1985—; prof. chemistry Rice U., Houston, 1970-85, Disting. prof. emeritus, 1985—, pres., 1970-85, pres. emeritus, 1985—; chmn. Gordon Corrosion Research Conf., 1950; cons. in corrosion, 1946—; chmn. Inter Soc. Corrosion Com., 1956-58, Gordon Research Conf. on Surface Chemistry, 1959; mem. nat. sci. bd. NSF, 1968-80, chmn., 1974-80; mem. Def. Sci. Bd., 1978-85; chmn. sci. adv. bd. Welch Found., 1982—; mem. Nat. Bd. Grad. Edn., 1971-75; chmn. bd. energy studies Nat. Acad. Scis./NRC Commn. Natural Resources, 1974-77; mem. Energy Research Adv. Bd., 1980-82; mem. Tex. Gov.'s Task Force on Higher Edn., 1981-82; trustee MITRE Corp., 1980-85. Recipient Whitney award Nat. Assn. Corrosion Engrs., 1956, Joseph J. Mattiello Meml. lectr. Fedn. Socs. Paint Tech., 1964, Gold medal Am. Inst. Chemists, 1978, Mirabeau B. Lamar award Assn. Tex. Colls. and Univs., 1981, Disting. Alumnus award Johns Hopkins U., 1982, Alumni Gold medal for disting. service to Rice U., 1984, Vannevar Bush award NSF, 1993, Nat. Medal of Sci., Nat. Sci. Found., 1993. Fellow AAAS (Phillip Hauge Abelson prize 1987), Am. Acad. Arts and Scis., N.Y. Acad. Scis.; mem. Am. Chem. Soc. (bd. editors 1956-62, exec. com. colloid div. 1955-58, chmn. chemistry and public affairs com. 1982-88, S.W. Regional award 1965, Charles Lathrop Parsons award 1987), Electrochem. Soc. (pres. 1957-58, Palladium medal 1965, Edward Goodrich Acheson award 1984), Faraday Soc., Nat. Corrosion Engrs. (bd. dir. 1952-55, chmn. com. edn. Corrosion Research Council 1957-60), Argonne Univs. Assn. (chmn. bd. trustees 1969-73), Nat. Acad. Scis., Am. Philos. Soc., Sigma Xi, Phi Lambda Upsilon, Alpha Chi Sigma, Phi Kappa Phi. Editor Jour. Electrochem. Soc., 1969-89; mem. editorial bd., mem. adv. edn. bd. Corrosion Sci., 1965-70; mem. editorial bd. Catalysis Reviews, 1968-73. Home: 2001 Pecos St Austin TX 78703-2119 Office: The Robert A Welch Fedn 4605 Post Oak Place Dr Ste 200 Houston TX 77027-9759

HACKETT, CAROL ANN HEDDEN, physician; b. Valdese, N.C., Dec. 18, 1939; d. Thomas Barnett and Zada Loray (Pope) Hedden; B.A., Duke, 1961; M.D., U. N.C., 1966; m. John Peter Hackett, July 27, 1968; children: John Hedden, Elizabeth Bentley, Susanne Rochet. Intern. Georgetown U. Hosp., Washington, 1966-67, resident, 1967-69; clinic physician DePaul Hosp., Norfolk, Va., 1969-71; chief spl. health services Arlington County Dept. Human Resources, Arlington, Va., 1971-72; gen. med. officer USPHS Hosp., Balt., 1974-75; pvt. practice family medicine, Seattle, 1975—; mem. staff, chmn. dept. family practice Overlake Hosp. Med. Ctr., 1985-86; clin. instr. U. Wash. Bd. dirs. Mercer Island (Wash.) Preschool Assn., 1977-78; coordinator 13th and 20th Ann. Inter-profl. Women's Dinner, 1978, 86; trustee Northwest Chamber Orch., 1984-85, King County Acad. Family Practice, 1993. Mem. Am. Acad. Family Practice, King County Family Practice, King County Med. Soc. (chmn. com. TV violence), Wash. Med. Soc., DAR, Bellevue C. of C., NW Women Physicians (v.p. 1978), Seattle Symphony League, Eastside Women Physicians (founder, pres.), Sigma Kappa, Wash. Athletic Club, Lakes Club, Seattle Yacht Club. Episcopalian. Home: 4304 E Mercer Way Mercer Island WA 98040-3826 Office: 1414 116th Ave NE Bellevue WA 98004

HACKETT, EARL RANDOLPH, neurologist; b. Moulmein, Burma, Feb. 16, 1932; s. Paul Richmond and Martha Jane (Lewis) H.; m. Shirley Jane Kanehl, May 25, 1953; children—Nancy, Raymond, Susan, Lynn, Laurie, Richard, Alicia. B.S., Drury Coll., Springfield, Mo., 1953; M.D., Western Res. U., 1957. Diplomate Am. Bd. Psychiatry and Neurology, Am. Bd. Electrodiagnostic Medicine. Intern, then resident in neurology Charity Hosp., New Orleans, 1957-62; resident in internal medicine VA Hosp., New Orleans, 1958-59; mem. faculty La. State U. Med. Sch., New Orleans, 1962—, prof. neurology, 1973-88, head dept., 1977-88; clin. prof. neurology U. Mo., Columbia, 1988—; mem. med. adv. bd. Myasthenia Gravis Found. Fellow Am. Acad. Neurology; mem. Am. Assn. Electrodiagnostic Medicine, Soc. Clin. Neurologists, Mo. Med. Assn., Greene County Med. Soc. Methodist. Home: 2517 S Brentwood Blvd Springfield MO 65804-3201 Office: 1965 S Fremont Ave Ste 2800 Springfield MO 65804-2243

HACKETT, JOSEPH LEO, microbiologist, clinical pathologist; b. Springfield, Ohio, Jan. 11, 1937; s. John Roger and Alice Pearl (Parker) H.; m. Phyllis Ann Boice, Apr. 27, 1963; children: Amy, Ron, Beth, Susan. MS, Ohio State U., 1963, PhD, 1968. Rsch. asst. Ohio State U., Columbus, 1962-67; quality control mgr. Courtland Abbott Labs., Chgo., 1967-69; micro sect. head Reference Lab. Abbott Labs., North Hollywood, Calif., 1969-72; quality control supr. Pfizer Diagnostics, Maywood, N.J., 1972-74; staff microbiologist FDA, Rockville, Md., 1974-80, br. chief, 1980-91, assoc. div. dir., 1991—; mem. microbiological area com. Nat. Com. Clin. Lab. Stds., Villanova, Pa., 1991—, susceptibility subcom., 1986-90. Contbr. articles to profl. jours. Mem. Am. Soc. for Microbiology. Office: FDA 1390 Piccard Dr Rockville MD 20850

HACKETT, KEVIN JAMES, insect pathologist; b. Phila., Apr. 24, 1947; s. James Patrick and Betty Corrine (Hulsey) H.; m. Kathleen Ruth Schmitt, Mar. 22, 1969; children: Ryan Hale, Aislinn Elizabeth. BS, Rutgers U., 1969, MS, 1971; PhD, U. Calif., Berkeley, 1980. Ea. coord. John Muir Inst. Environ. Studies, Washington, 1979-83; rsch. entomologist insect biocontrol lab. USDA Agrl. Rsch. Svc., Beltsville, Md., 1983—. Author: The Mycoplasmas, 1990; contbr. articles to jour. Sci. Founder, coord. Ind. Peace Studies U. Calif., Berkeley, 1972-74. Mem. AAAS, Internat. Orgn. Mycoplasmology (team leader spiroplasma working team 1990—, Derrick Edward award 1988), Soc. Invertebrate Pathology, Am. Soc. Microbiology, Entomology Soc. Am. Unitarian. Achievements include development of insect cell spiroplasma coculture, hypothesis for spiroplasma evolution. Office: Insect Biocontrol Lab Rm 214 Bldg 011A BARC-W Beltsville MD 20705

HACKETT, ROBERT MOORE, engineering educator; b. Carthage, Tenn., Feb. 10, 1936; s. William Arlis and Bartie Hart (Moore) H.; m. Barbara Jean Highers, July 26, 1957 (div. Mar. 1973); children: Kimberly Elizabeth, William David, Leigh-Ann; m. Patricia Ann Keen, Apr. 10, 1978; 1 child, Jennifer Moore. BS in Civil Engring., Tenn. Tech. U., 1960; MS in Civil Engring., Carnegie Mellon U., 1966, PhD in Civil Engring., 1968. Registered profl. engr., Pa. Civil engr. Tennessee Valley Authority, Knoxville, Tenn., 1960-62; engr. Brown Engring. Co., Huntsville, Ala., 1962-64; asst. to prof. civil engring. Vanderbilt U., Nashville, Tenn., 1967-79; res. aerospace engr. U.S. Army Missile Lab., Huntsville, Ala., 1979-82; prof. civil engring. U. Ala., Huntsville, Ala., 1982-85; prof., chmn. civil engring. dept. Ohio U., Athens, Ohio, 1985-86, Russ Prof. Civil Engring., 1985; prof., chmn. civil engring. dept. U. Miss., University, Miss., 1986—; cons. Los Alamos (N.Mex.) Nat. Lab., 1983—. Co-author: Computer Methods of Structural Analysis, 1969; contbr. articles on mechanics of composite materials in numerous tech. jours., 1968—. Fellow ASCE (mem. editorial bd. Jour. Materials in Civil Engineering 1988—), AIAA (aasoc.); mem. Am. Soc. Engring. Edn., Am. Acad. Mechanics, Am. Soc. Testing and Materials. Democrat. Episcopalian. Achievements include research in computational modeling of the thermomechanical response of advanced composite material systems. Home: 2235 Lee Loop Oxford MS 38655 Office: U Miss Dept Civil Engring University MS 38677

HACKNEY, ANTHONY C., nutrition and physiology educator, researcher; b. Cin., Jan. 29, 1956; s. Frank and Jackie (Beach) H.; m. Grace R. Griffith, Aug. 13, 1978; children: Sarah E., Zachary C. BA, Berea Coll., 1979; MA, PhD, Kent State U., 1986. Asst. prof. Iowa State U., Ames, 1986-88; postdoctoral fellow Naval Health Rsch. Ctr., San Diego, 1988; assoc. prof. nutrition and exercise physiology U. N.C., Chapel Hill, 1988—. Mem. edit. bd. Internat. Jour. Sports Nutrition, 1992—, Biology of Sport Jour., 1991—; contbr. articles to profl. jours. Mem. Am. Coll. Sports Medicine, Am. Physiol. Soc., Sigma Xi. Democrat. Methodist. Office: U NC CB # 8700 Chapel Hill NC 27599

HACKWOOD, SUSAN, electrical and computer engineering educator; b. Liverpool, Eng., May 23, 1955; came to U.S., 1980; d. Alan and Margaret Hackwood. BS with honors, DeMontfort U., Eng., 1976; PhD in Solid State Ionics, DeMontfort U., Eng., 1979; PhD (hon.), Worcester Poly. Inst., 1993; DSc (hon.), DeMontfort U., 1993. Rsch. fellow DeMonfort U., Leicester, Eng., 1976-79; postdoctoral rsch. fellow AT&T Bell Labs., Homdel, N.J., 1980-81; mem. tech. staff AT&T Bell Labs., Homdel, 1981-83, supr. robotics tech., 1983-84, dept. head robotics tech., 1984-85; prof. elec. and computer engring. U. Calif., Santa Barbara, 1985-89, dir. Ctr. Robotic Systems in Microelectronics, 1985-89; dean Coll. Engring. U. Calif., Riverside, 1990—. Editor: Jour. of Robotic Systems, 1983, Recent Advances in Robotics, 1985; contbr. over 82 articles to tech. jours.; 7 patents in field. Mem. IEEE. Office: U Calif Coll Engring Riverside CA 92521

HADANI, ITZHAK, experimental psychologist, human factors engineer; b. Jerusalem, Sept. 24, 1943; came to U.S., 1989; s. Isaschar and Helen Dahan; m. Rachel Hadani (Ebeschitz), Mar. 9, 1971; children: Yael, Yoav J., Jonathan I. BA in Psychology, Econs., Hebrew U., Jerusalem, 1971, MA in Psychology, 1975; DSc in Biomed. Engring., Technion, Haifa, Israel, 1980. Instr. Technion-IIT, Haifa, 1975-78, rsch. fellow, 1978-87, sr. rsch. fellow, 1978-88; vis. rsch. assoc. Calif. Inst. Tech., Pasadena, 1978, Eindhoven (The Netherlands) U., 1986-87; sr. rsch. fellow Haifa U., 1988-89; assoc. prof. Rutgers U., New Brunswick, N.J., 1989—; cons. Rsch. Inst. Dayton U., Williams AFB, Ariz., Phoenix, 1992. Contbr. articles to profl. jours. Recipient Freund Found. prize for popular sci. writing (Israel), 1987. Achievements include discovery of correlation between perceived depth and interocular distance; development of optical improvement in night vision goggles, of solutions of the indeterminate scale and correspondence problems in visual perception, of navigation theory in space perception, of visimove LED display utilizing visual motion smear effect (Liberty Sci. Ctr., N.J.). Home: 3 Cliffwood Pl Metuchen NJ 08840 Office: Rutgers U Lab Vision Rsch Busch Campus Psychology Bldg New Brunswick NJ 08903

HADAWAY, JAMES BENJAMIN, physicist; b. Atlanta, Oct. 11, 1962; s. Ira Evans and Mary Jean (Geiger) H. BS in Physics, Ga. Inst. Tech., Atlanta, 1984, MS in Applied Physics, 1987. Sr. rsch. assoc. Ctr. Applied Optics, U. Ala., Huntsville, 1987—; mem. Optics Modil Info. Systems Adminstrv. Rev. Com., Oak Ridge, Tenn., 1978—;cons., Huntsville, 1987—. Vice chmn. Madison County Young Reps., Huntsville, 1990. 2d lt. USAF, 1985-86. Mem. Optical Soc. Am., Internat. Soc. Optical Engring., Huntsville Electro-Optic Soc. Achievements include design of first super-high resolution x-ray telescopes for solar astronomy, of first in-space optical properties measurement system; design and fabrication of first practical all-reflective zoom telescope. Home: 357 Oakland Rd Madison AL 35758-1979 Office: U Ala Ctr Applied Optics Huntsville AL 35899

HADDAD, EDWARD RAOUF, civil engineer, consultant; b. Mosul, Iraq, July 1, 1926; came to U.S., 1990; s. Raouf Sulaiman Haddad and Fadhila (Sulaiman) Shaya; m. Balquis Yousef, July 19, 1961; children: Reem, Raid. BSc, U. Baghdad, Iraq, 1949; postgrad., Colo. State U., 1966-67. Project engr., cons. Min. Pub. Works, Baghdad, 1949-63; arbitrator Engring. Soc. & Ct., Kuwait City, Kuwait, 1963-90; tech. advisor Royal Family, Kuwait, 1987-90; cons. pvt. practice Haddad Engring., Albuquerque, 1990—. Organizer Reps. Abroad, Kuwait, 1990. Recipient Hon. medal Pope Paul VI of Rome, 1973. Mem. ASCE, NSPE, KC (chancellor 1992), Am. Arbitration Assn., Sierra Internat. (trustee), Lions (bd. dirs. 1992), Inventors Club (bd. dirs. 1992). Address: 143(A) General Arnold NE Albuquerque NM 87123

HADDAD, GEORGE ILYAS, engineering educator, research scientist; b. Aindara, Lebanon, Apr. 7, 1935; came to U.S., 1952, naturalized, 1961; s. Elias Ferris and Fahima (Haddad) H.; m. Mary Louella Nixon, June 28, 1958; children—Theodore N., Susan Anne. B.S. in Elec. Engring, U. Mich., 1956, M.S., 1958, Ph.D., 1963. Mem. faculty U. Mich., Ann Arbor, 1963—, assoc. prof., 1965-69, prof. elec. engring., 1968—, Robert J. Hiller prof., 1991—, dir. electron physics lab., 1968-75, chmn. dept. elec. engring. and computer sci., 1975-87, 91—, dir. ctr. for high-frequency microelectronics, 1987—; cons. to industry. Contbr. articles to profl. jours. Recipient Curtis W. McGraw research award Am. Soc. Engring. Edn., 1970, Excellence in Research award Coll. Engring., U. Mich., 1985, Disting. Faculty Achievement award U. Mich., 1985-86, S.S. Attwood award, 1991. Fellow IEEE (editor proc. and trans.); mem. Am. Soc. Engring. Edn., Am. Phys. Soc., Sigma Xi, Phi Kappa Phi, Eta Kappa Nu, Tau Beta Pi. Office: U Mich Dept Elec Engr & Computer Sci Ann Arbor MI 48109-2122

HADDAD, HESKEL MARSHALL, ophthalmologist; b. Baghdad, Iraq, Sept. 26, 1930; came to U.S., 1953, naturalized, 1962; s. Moshe M. and Masuda (Cohen) H.; m. Doris I. Fatzer, July 4, 1963; children: Ava Masuda, Andreas Moshe, Michael Albert. Student, Royal Coll. Medicine, Baghdad, 1945-50; M.D., Hebrew U., Jerusalem, 1953. Diplomate: Am. Bd. Pediatrics, Am. Bd. Ophthalmology. Intern Donolo Hosp., Jaffo-Tel Aviv, Israel, 1950-51; rotating intern Hadassah U. Hosp., Jerusalem, 1951-53; pediatric resident Children's Med. Center, Boston, 1953-56; fellow in pediatric endocrinology Johns Hopkins Hosp., Balt., 1956-58; fellow in clin. endocrine br. Nat. Inst. Arthritis and Metabolic Diseases, NIH, Bethesda, Md., 1958-59; pediatrician sect. clin. endocrinology NIH, 1959-60; asst. prof. pediatrics sch. medicine Howard U., Washington, 1959-60; resident, asst. dept. ophthalmology sch. medicine Washington U., St. Louis, 1960-64; leave of absence, 1962-63; fellow pediatric ophthalmology Inst. Visual Sci., San Francisco, 1962; research fellow Hôpital des Quinze-Vingts, Laboratoire de Physiologie de Vision, Ecole des Hautes Etudes, Paris, 1962-63; ophthalmologist Hôpital Beni Messous, Algiers, Algeria, 1964; asst. attending ophthalmic surgeon, also asst. prof. ophthalmology Mt. Sinai Hosp. and Sch. Medicine, N.Y.C., 1964-67; dir. dept. ophthalmology Beth Israel Med. Center, N.Y.C.; also assoc. prof. ophthalmology Mt. Sinai Sch. Medicine, 1967-71; clin. prof. ophthalmology N.Y. Med. Coll., 1971—. Author: Endocrine Exophthalmos, 1973, Metabolic Eye Diseases, 1974, Metabolic-Pediatric Eye Diseases, 1979, Metabolic Ophthalmology: Diagnostic Techniques Vols. I and II, 1985, Jews of Arab and Islamic Countries: History, Problems and Solutions, 1984, (autobiography) Flight from Babylon, 1986; editor-in-chief: Metabolic Ophthalmology, 1976-79, Metabolic and Pediatric Ophthalmology, 1979-82, Metabolic, Pediatric and Systemic Ophthalmology, 1982—; contbr. numerous articles and revs. to profl. jours. Pres. Am. Com. for Rescue and Resettlement of Iraqui Jews, World Orgn. Jews from Arab Countries, Parents' Assn. of Sch. of Performing Arts, 1980-83. Fellow ACS, Am. Inst. Chemists; mem. Am. Endocrine Soc., Am. Fedn. Clin. Research, Assn. Research Ophthalmology and Vision, AMA, New York County Med. Soc., AAAS, Am. Acad. Ophthalmology, N.Y. Acad. Medicine, N.Y. Acad. Scis., N.Y. Soc. Clin. Ophthalmology, Soc. Eye Surgeons, Société Française d' Ophthalmologie, German Ophthal. Soc., Internat. Soc. Metabolic Eye Disease (founder, sec.-treas. 1973—), World Soc. on Systemic Ophthalmology (founder, sec.-treas. 1982, chmn.), N.Y. County Med. Soc. (chmn. com. fgn. med. grads. 1985-90, del. N.Y. State Med. Soc. 1985-86). Office: 1125 Park Ave New York NY 10128-1243

HADDAD, INAD, physician; b. Beirut, Lebanon, June 2, 1953; came to U.S., 1978; s. Andraos Y. and Juliette H. Student, St. Joseph U., 1971-74, U. Paris, 1975-78. Physician Emma L. Bixby Hosp., Adrian, Mich., 1981—, chmn. dept. medicine, 1986, dir. critical care unit, 1985-87, chief of med. and dental staff; chief med. dental staff Bixby Med. Ctr., 1988; bd. dirs., founder Cardiac Rehab. Clinic Adrian, 1986—. Chmn. Lanawee County Dem. Party, 1991—. Named Dem. of Yr., Lenawee County Dem. Party, 1990. Mem. AMA, Mich. State Med. Soc., Lenawee County Med. Soc. (pres. 1991), Lenawee C. of C. (bd. dirs.), Mich. Doctors Polit. Action Com. (bd. dirs.), Am. Soc. Internal Medicine, Mich. soc. Internal Medicine. Roman Catholic. Avocations: chess, tennis, history, internat. affairs. Home: 415 Meadowbrook Dr Adrian MI 49221-1319 Office: 4204 W Maple Ave Adrian MI 49221-1382

HADDAD, JAMES HENRY, chemical engineering consultant; b. Willimantic, Conn., Jan. 30, 1923; s. William Addy and Nellie (Birbarie) H.; m. Isabel Serrano, Feb. 3, 1962; children: Frederick William, Francis Xavier. BS in Engring., Yale U., 1944. Chem. engr. Conn. Hard Rubber Co., New Haven, 1943-44; engr. rsch. dept. Mobil Rsch. Devel. Corp., Paulsboro, N.J., 1944-52; engr. engring. dept. Mobil Rsch. Devel. Corp., N.Y.C., 1952-70; engring. cons. Mobil Rsch. Devel. Corp., Princeton, N.J., 1971-89; ind. cons. worldwide chem. processing/solids systems Chem. Processing/Solids Systems, Princeton Junction, N.J., 1989—. Contbr. articles to profl. publs.; patentee in field. Mem. budget com., trustee Princeton Area Communities United Way, 1977-90. Mem. Am. Chem. Soc., Am. Inst. Chem. Engrs., Alpha Chi Sigma. Republican. Roman Catholic. Avocation: swimming. Home and Office: 45 Van Wyck Dr Princeton Junction NJ 08550

HADDOCK, FRED T., astronomer, educator; b. Independence, Mo., May 31, 1919; s. Fred Theodore Sr. and Helen (Sea) H.; m. Margaret Pratt, June 24, 1941 (div. Sept. 1976); children: Thomas Frederick, Richard Marshall. SB, MIT, 1941; MS, U. Md., 1950; DSc (hon.), Rhodes Coll., 1965, Ripon Coll., 1966. Physicist U.S. Naval Rsch. Lab., Washington, 1941-56; assoc. prof. elec. engring. and astronomy U. Mich., Ann Arbor, 1956-59, prof. elec. engring., 1959-67, prof. astronomy, 1959-88, emeritus prof., 1988—; lectr. radio astronomy Jodrell Bank U. Manchester, Eng., 1962; vis. assoc. radio astronomy Calif. Inst. Tech., 1966; vis. lectr. Raman Inst., Bangalore, India, 1978; sr. cons. Nat. Radio Astron. Obs., W.Va., 1960-61; founder, dir. U. Mich. Radio Astron. Obs., 1961-84. Author: (chpts. in books) Space Age Astronomy, 1962, Radio Astronomy of the Solar System, 1966; contbr. articles to prof. jours. and publs. Mem. Union Radio Sci. Internat., nat. chmn. commn. on radio astronomy, 1954-57; trustee Associated Univs., Inc., 1964-68; prin. investigator, five Orbiting Geophys. Observatories, 1960-74, and Interplanetary Probe 9, 1964-77; co-investigator on Voyager planetary probes, 1970-86, NASA, Washington; mem. astronomy adv. panel NSF, Washington, 1957-60, 63-66. With USN, 1944-45. Fellow IEEE (life), Am. Astron. Soc. (v.p. 1961-63); mem. Internat. Astron. Union (commn. on radio astronomy 1948—), NAS (adv. panel astronomy facilities 1962-64), AIA (hon. mem. Huron Valley chpt. 1980—), Sigma Xi (past pres. U. Mich. chpt. 1956—). Achievements include design and development of first submarine periscope radar antenna, 1943-44; early discoveries in microwave astronomy, gaseous nebulae in 1953 and early space detection of kilometer waves from galaxy and the sun, 1962. Home: 3935 Holden Dr Ann Arbor MI 48103-9415 Office: U Mich Astronomy Dept Ann Arbor MI 48109-1090

HADDOCK, ROBERT LYNN, information services entrepreneur, writer; b. Vallejo, Calif., May 12, 1945; s. Orville Walter and Lee Ellen (Alexander) H. BA, Union Coll., 1967; postgrad., NYU, 1977-81. Editor So. Pub. Assn., Nashville, 1969-74, controller, 1974-75; mktg. analyst Bus. Publs. div. Prentice-Hall, Englewood Cliffs, N.J., 1975-78, bus. mgr., 1978-81; bus. mgr. Ziff-Davis Pub. Co., N.Y.C., 1981-82, dir. bus. devel., 1982-83; pres. Personal Access, Inc., N.Y.C., 1983-84; v.p., dir. product devel. Citicorp Global Report, N.Y.C., 1984-86, v.p., dir. mktg., 1986-88; v.p., dir. product devel. Citibank, N.A., N.Y.C., 1989-90; v.p., dir. product devel. and mktg. Enhanced Telephone Svcs., Inc., N.Y.C., 1990-91; pres. M-Power Corp., N.Y.C., 1991—. Author: The Broken Web, 1973, How to Stop Smoking, 1974; inventor database accessing system, 1983, enhanced telephone, 1989. Mem. IEEE, Am. Assn. Artificial Intelligence, Info. Industry Assn., Mensa. Home: 105 W 13th St Apt 15F New York NY 10011-7848

HADDOX, MARI KRISTINE, biomedical scientist; b. Mpls., Nov. 25, 1950. BSc, U. Minn., 1976; PhD, U. Ariz., 1980. Rsch. fellow U. Minn. Med. Sch., Mpls., 1973-76; rsch. assoc. U. Ariz. Med. Sch., Tucson, 1977-80, adj. asst. prof., 1980-81; asst. prof. U. Tex. Med. Sch./Grad. Sch. Biomed. Scis., Houston, 1981-86, assoc. prof., 1986-92, prof. pharmacology, 1992—; permanent mem. study sect. biochemistry NIH, 1985-89, sr. reviewers res., 1989—; reviewer U.S.-Israel Binational Sci. Found., 1985—. Contbr. articles to profl. publs. Grantee Am. Diabetes Assn., 1982-84, NIH, 1982—, Juvenile Diabetes Found., 1984-86, Am. Cancer Soc., 1985-90. Mem. AAAS, Assn. Women in Sci. Office: U Tex Med Sch PO Box 20708 6431 Fannin St 5.036 Houston TX 77225

HADDOX, MARK, electronic engineer. Student The Sorbonne, U. Paris, 1974; BS in Biophysics, Ohio State U., 1975; BSEE, U. Mich., 1977. Electronic engr. Jodon, Inc., Ann Arbor, 1978-80, chief engr., 1980, v.p. engring., 1980—. Mem. IEEE. Achievements include patents for Method and System for Monitoring Position of a Fluid Actuator Employing Microwave Cavity Resonant Principles; for Method and Apparatus for Measuring Engine Compression Ratio, Clearance Volume, and Related Cylinder Parameters; for Method and Apparatus for Measuring Engine Compression Ratio; (with J. Gillespie) for Ignition Timing Control for Internal Combustion Engines; research on diesel engine emissions, on measurement of engine cylinder minimum volume and compression ratio, and on train safety instrumentation. Home: 1243 King George Blvd Ann Arbor MI 48108-1759 Office: Jodon Engring Inc 62 Enterprise Dr Ann Arbor MI 48103-9562

HADFIELD, JAMES IRVINE HAVELOCK, surgeon; b. Rickmansworth, Herts, England, Dec. 7, 1930; s. Geoffrey and Sarah Victoria Eileen (Irvine) H.; m. Ann Pickernell Milner, May 1, 1957; children: Esme Victoria, Helen Sarah, Geoffrey Irvine. Student, Brasenose Coll., Oxford, England, 1948-52; student, St. Thomas' Hosp. Med. Sch., 1952-55; MA BM BCH, Oxford U., 1955. Intern St. Thomas's Hosp.; resident St. Thomas's & Leicester Royal Infirmany; house surgeon St. Thomas' Hosp., London, 1955; lectr. dept. anatomy St. Thomas' Hosp. Med. Sch., London, 1957; RSO sr. registrar Leicester Royal Infirmary, Leicester, England, 1960-62; surg. tutor Oxford U., Oxford, England, 1962-66; hon. cons. surgeon Radcliffe Infirmary, Oxford; bd. dirs., vice chmn. Seltzer PLC; chmn. med. exec. com. Bedford Gen. Hosp., 1980-85; hon. cons. surgeon Radcliffe Infirmary, 1963-66. Contbr. articles to profl. jours. Trustee Bedford Charity, Bedford, England, 1970; gov. Harpur Trust, 1985. Hon. fellow Assn. Surgeons Pakistan, 1985, Assn. Surgeons India, 1989. Fellow Royal Coll. Surgeons England (cert.), Royal Coll. Surgeons Edinburgh (cert.), Internat. Coll. Surgeons, Assn. Surgeons Gt. Britain, Brit. Assn. Urological Surgeons, Anatomical Soc. Gt. Britain-Ireland, Leander Club (Heney on Thames), Vincents Club (Oxford). Conservative. Mem. Ch. of England. Avocations: fishing for trout and salmon, shooting game. Home: Bakers Barn, Stagsden West End, Bedford England MK43852 Office: Bedford Gen Hosp, Kempston Rd, Bedford England MK429DJ

HADFIELD, M. GARY, neuropathologist, educator; b. Ogden, Utah, Aug. 4, 1935; s. Milton Albert and Opal Ortell (Davis) H.; m. Kathleen Halverson, Dec. 2, 1966; children: Laura Hadfield Rackam, Mark, Rosalyn, Alan. BA, Brigham Young U., 1960; MD, U. Utah, 1964. Intern in pathology Cornell Med. Ctr./NYU Hosp., N.Y.C., 1964-65; resident in pathology —, —, 1965-66; trainer in neuropathology Montefiore Hosp. and Med. Ctr., Bronx, N.Y., 1966-68; fellow in neurochemistry NYU/Bellevue Hosp., N.Y.C., 1968-69, asst. prof., 1969-70; assoc. prof. Med. Coll. Va., Richmond, 1970-80, prof., 1980—. Office: Pathology Dept Box 17 MCV Sta Richmond VA 23293

HADJICOSTIS, ANDREAS NICHOLAS, physicist; b. Paphos, Cyprus, Nov. 5, 1948; s. Nicholas and Angelica (Papados) H. BA magna cum laude, U. Minn., 1971; MSc, Mich. State U., 1974; PhD, U. Denver, 1979. Sr. scientist Spl. Rsch. Group, J&J, Somerset, N.J., 1979-82; transducer R&D mgr. Technicare Ultrasound, Englewood, Colo., 1982-85; v.p. engring. Nutran, Englewood, 1985-90; tech. staff SBU Advanced Tech. Labs., Bothell, Wash., 1990—. Contbr. articles to profl. jours. Grantee Fulbright Found., 1967. Mem. IEEE, Am. Soc. of Elecrocardiography, Acoustical Soc. Am., N.Y. Acad. Scis. Achievements include patents for short ringdown ultrasonic transducer, passive ultrasonic needle probe locator; rsch. includes medical ultrasonic transducer design. Office: ATL PO Box 3003 Bothell WA 98041

HADJIKOV, LYUBEN MANOLOV, civil engineer; b. Skrebatno, Bulgaria, Aug. 13, 1939; s. Manol Bojov and Slavka Mavrodieva (Tscholeva) H.; m. Dimitrijka Ivanova Ignatova; children: Manol Ljubenov, Ivan Ljubenov. Degree civil engring., Higher Inst. Architecture, Sofia, Bulgaria, 1965, PhD in Civil Engring., 1975. Dir. constrn. plant Sofia-West, Goze Deltschev, 1965-69; rsch. fellow Inst. of Tech. Mechanics, Sofia, 1969-79; sr. fellow Inst. of Mechanics and Biomechanics, Sofia, 1979-89, dep. dir., 1989—; dept. head South-West U., Blagoevgrad, 1992—; expert EQE Internat., San Francisco, 1991, Ministry of Edn. and Sci., Sofia, 1989. Contbr. articles to profl. jours. Sgt. Bulgarian Artillery, 1958-60. Recipient Kiril and Method Prize State Coun., Bulgaria, 1987. Mem. Internat. Soc. for Computational Methods (U.K.), Interant. Assn. for Boundary Element Methods, Internat. Inst. Engring. Safety (dep. chmn. sci. coun. 1993), Optical Soc. Am. Achievements include patent for compound of ureaformaldeyhe resin (Bulgarian), Method of Specklegram Processing (Bulgarian). Home: Kalina Veskova 11, 1111 Sofia Bulgaria Office: Bulgarian Acad Sci, Acad Georgy Bontchev Bl 4, 1113 Sofia Bulgaria

HADJISTAMOV, DIMITER, chemical engineer; b. Sofia, Bulgaria, Dec. 18, 1940; arrived in Switzerland, 1972; s. Bojan Hadjistamov and Elena Zacharieva; m. Emilia Petkova, Dec. 10, 1991. Diploma in chemical engring., Tech. U., Dresden, Fed. Republic Germany, 1967, Dr.Ing.Chem., 1971. Rsch. engr. Micafil AG, Zürich, Switzerland, 1972-73, Sarlab AG, Zürich, 1974; chem. process devel. engr. Ciba-Geigy AG, Basel, Switzerland, 1975—; sci. specialist Ciba-Geigy AG, Basel, 1991—; plant mgr. Ciba-Geigy AB, Schweizerhalle, Switzerland, 1983, head rheology group, Basel, 1986—; mem. exec. com. Swiss Rheology Group, Zürich, 1990—. Mem. AAAS, Schweizerischer-Chemische Gesselschaft, Verein Deutscher Ingenieure, Deutsche Rheologische Gesellschaft, Soc. Rheology U.S.A. Avocations: bridge, judo, sailing. Home: Helvetierstr 15, 4125 Riehen Switzerland Office: Ciba Geigy AG, WSH 2093.215, 4133 Schweizerhalle Switzerland

HADLEY, GLEN L., electrical engineer; b. Fairfield, Iowa, June 21, 1952; s. D. Harold and Wilda I. (Deweese) H.; m. Arian J. Zimmerman, May 30, 1978; children: Benjamin E., Ephram I. BA cum laude, Ctrl. Coll., 1973; BSEE, Iowa State U., 1980. Registered profl. engr., Iowa. Instrument and control engr. Ariz. Pub. Svc. Co., Phoenix, 1980-91, Iowa Electric Light & Power, Cedar Rapids, Iowa, 1991—; cons. Instrument Soc. Am. Setpoint 67 XX Coms., 1992. Author and instr. tng. course. With USNR, 1974-76. Mem. Instrument Soc. Am. (sr., sect. v.p. 1990-91), Tau Beta Pi, Eta Kappa Nu. Libertarian. Unitarian. Achievements include research in MCSA as a predictive maintenance tool; on-line weighing systems.

HADLEY, MARY, nutritionist; b. Milton, Ontario, Can., Jan. 28, 1949; came to U.S., 1990; d. Osmund and Gwen (McMinn) H. BSc, U. Guelph, Guelph, Ontario, Can., 1972, MSc, 1974, PhD, 1989. Biochem. technologist Chem. Inst. of Can., U. Guelph (Ont.), Can., 1974-90; asst. prof. N.D. State U., Fargo, 1990—. Contbr. articles to profl. jours. Mem. Am. Coll. Nutrition, Sigma Xi. Office: North Dakota State U Dept Food and Nutrition Fargo ND 58105

HAEBERLIN, HEINRICH RUDOLF, electrical engineering educator; b. Basel, Switzerland, Feb. 9, 1947; s. Rudolf and Bethly (Buergin) H.; m. Ruth Huerzeler, Mar. 31, 1982; children: Andreas, Kathrin. MS in Elec. Engring., Swiss Fed. Inst. Tech., Zurich, 1971, PhD in Elec. Engring., 1978. Asst. Swiss Fed. Inst. Tech., Zurich, 1971-77, asst. in chief, 1978-79; R&D engr. Zellweger Uster AG, Hombrechtikon, Switzerland, 1979-80; prof. elec. engring. Ingenieurschule Polytech., Burgdorf, Switzerland, 1980—. Author: Photovoltaik-Strom aus Sonnenlicht fuer Inselanlagen und Verbundnetz, 1991; contbr. articles to profl. jours. Mem. Schweiz Elektrotechnischer Verein, Swiss Commn. for PV-Stds., Sonnenenergie-Fachverband. Achievements include contbns. to increase reliability and reduce electromagnetic interference problems of grid-connected inverters for photovoltaic installations; creation of a test ctr. for PV-inverters and PV-systems. Office: Ingenieurschule Polytechnic, 1 Ilcoweg, CH-3400 Burgdorf Switzerland

HAEMMERLIE, FRANCES MONTGOMERY, psychology educator, consultant; b. Gainsville, Fla., Feb. 2, 1948; d. Henry John and Ruth Elizabeth (Collins) H.; Robert L. Montgomery, June 16, 1979. BA, U. Fla., 1972; MS, Fla. State U., 1976, PhD, 1978. Prof. U. Mo., Rolla, 1978—; rsch. fellow Ctr. for Applied Engring., U. Mo., Rolla, 1984-87. Contbr. articles to profl. jours, chpts. to books. Sec. Rolla Jr. High Parent-Student-Tchr. Assn., 1989-90. Recipient Teaching awards U. Mo., 1980-85, 87-91, Amoco Teaching award, 1981-82, Faculty Excellence award, 1986-89, Reade Beard Faculty Excellence award, 1989-90, John Stafford Brown, 1991-92, 92-93. Mem. Am. Psychol. Assn. (membership chmn. div. 12 sect. IV 1990—), Southwestern Psychol. Assn. (placement chmn. 1986, program chmn. 1988-89), Psi Chi (outstanding adv. award 1980, 86, profl. svc. award 1983). Achievements include research in promoting technology development in rural settings and human adaptation to technological environments. Home: RR 4 Box 322 Rolla MO 65401-9320 Office: U Mo Dept Psychology 110 Hss Rolla MO 65401

HAENGGI, DIETER CHRISTOPH, psychologist, researcher; b. Breitenbach, Solothurn, Switzerland, Mar. 22, 1960; came to U.S., 1989; s. Eugene Oskar and Helene Margrit (Kuebler) H. MS, U. Fribourg, Switzerland, 1986, PhD, 1988. Cert. in cognitive, clin. and applied psychology. Rsch. asst., rsch. assoc. U. Basel, Switzerland, 1986-89; rsch. assoc. U. Pitts., 1989-90, U. Basel, 1990-91, U. Oreg., Eugene, 1991-92, U. Colo., Boulder, 1992—. Author monograph; contbr. articles to profl. jours. Swiss Nat. Sci. Found. rsch. grantee, 1989, 91; U. Basel publ. grant, 1989. Mem. Am. Psychol. Soc., European Soc. for Cognitive Psychology, Psychonomic Soc., Soc. for Text and Discourse. Office: U Colo Inst Cognitive Sci Box 344 Boulder CO 80309-0344

HAENSEL, VLADIMIR, chemical engineering educator; b. Freiburg, Germany, Sept. 1, 1914; came to U.S., 1930; s. Paul and Nina (Tugenhold) H.; m. Mary Magraw, Aug. 28, 1939 (dec. 1979); children: Mary Ann (Mrs. Michael J. Ahlen, dec. 1982), Katherine (Mrs. C.K. Webster); m. Hertha Skala, Sept. 14, 1986. BS, Northwestern U., 1935, PhD, 1941, DSc, 1957; MS, MIT, 1939; DSc (hon.), U. Wis., Milw., 1979. Rsch. chemist Universal Oil Products Co. (name changed to UOP Inc.), Des Plaines, Ill., 1937-64, v.p., dir. rsch., 1964-72, v.p. sci. and tech., 1972-79, cons., 1979—; prof. U. Mass., Amherst, 1979—. Contbg. author several sci. books; contbr. more than 120 articles to profl. jours.; patentee in field. Recipient award Chgo. Jr. C. of C., 1944, award Precision Sci. Co., 1952, Profl. Progress award Am. Rsch. & Engring. Co., 1965, Modern Pioneers in Creative Industry award NAM, 1965, Perkin medal, 1967, Nat. medal Sci., 1973. Mem. NAS (award for chemistry in svc. to society 1991), NAE, Am. Chem. Soc., Indsl. Rsch. Inst., Catalysis Soc., Sigma Xi, Phi Lambda Upsilon, Tau Beta Pi. Home: 83 Larkspur Dr Amherst MA 01002 Office: U Mass Dept Chem Engring Amherst MA 01003

HAEUSSERMANN, WALTER, systems engineer; b. Kuenzelsau, Germany, Mar. 2, 1914; came to U.S., 1948; s. Otto and Margarethe (Henn) H.; m. Ruth Franziska Knos, Mar. 24, 1940. ME, Inst. Tech., Darmstadt, Germany, 1938, PhD, 1944. Dir. guidance & control lab. Army Ballistic Missile Agy., Huntsville, Ala., 1954-60, dir. astrionics lab. Marshall Space Flight Ctr., Huntsville, 1960-69, dir. cen. systems engring., 1969-72, assoc. dir., tech. asst., 1972-78; cons. various aerospace cos., Huntsville, 1978—. Contbr. over 20 articles to profl. jours. Recipient Superior Achievement award Inst. Navigation, 1970, Medal of Merit State of Baden-Wuerttemberg, Germany, 1985, Wernher von Braun distinction German Aerospace Soc., 1988, Exceptional Civilian Svc. award Dept. of Army, 1959, NASA Medal for Outstanding Leadership, 1963, NASA Exceptional Svc. medal, 1969. Fellow AIAA, Am. Astro. Soc.; mem. Inst. Navigation, Internat. Fedn. Automatic Control in Aerospace, German Soc. for Air and Space Travel. Achievements include patents for Direct Current Motor, Attitude Control Systems for Space Vehicles, Electric Motor, Space Vehicle Attitude Control Mechanism, Satellite Motion Simulator, Velocity Measurement System, and Magnetic Field Control. Home: 1607 Sandlin Ave Huntsville AL 35801-2067

HÅFORS, AINA BIRGITTA, wood conservation chemist; b. Kalmar, Sweden, Oct. 25, 1934; d. Fridolv and Brita Laurentia (Theorin) Bobeck; m. Arne Eigil Håfors, May 27, 1956. MS, U. Stockholm, Sweden, 1961, student, 1961-92; student, U. Umeå, Sweden, 1992—. Conservation analyst, rsch. chemist Wasa Bd., Stockholm, 1961-64; conservation analyst, rsch. chemist Swedish Nat. Maritime Mus., Stockholm, 1964-78, conservation mgr., 1978-92; conservation mgr. Vasa Mus., Stockholm, 1992—; mem. ref. group Scandinavian Archaeometry Ctr., Gothenburg, Sweden, 1989—. Author: (with others) Archaeological Wood, 1990; contbr. articles to profl. jours. Mem. Am. Chem. Soc. (Cellulose Paper and Textiles div.), Swedish Chem. Soc., Internat. Counsel Mus. Achievements include research in treatment of waterlogged archaeological wood with polyethyleneglycols. Office: Vasamuseet, Box 27131, S 102 52 Stockholm Sweden

HAGAR, WILLIAM GARDNER, III, photobiology educator; b. Chester, Pa., Aug. 14, 1940; s. William Gardner and Florence (Allcutt) H.; m. Dorothy Marie Sollinger, June 8, 1963; children: Doreen Marie, Cheryl Ann, William Robert, Jennifer Lynn. BS in Chemistry, Widener U., 1962; PhD in Biochemistry, Temple U., 1972. Teaching asst. chemistry Temple U., Phila., 1963-68; Carnegie fellow Carnegie Inst. Washington, Stanford, Calif., 1971-74; asst. prof. biology U. Mass., Boston, 1974-79, assoc. prof. biology, 1980—; sci. coord. Upward Bound Math. Sci., U. Mas., 1991, 92; mem. design team Nat. TRIO Workshop Conv., Washington, 1992. Contbr. articles to Plant Physiology, other sci. publs. Com. mem. Concerned Area Residents Preservation Tinicum Marsh, 1972; organizer, coach numerous youth sports teams, Newton, Mass., 1978-84; bd. dirs. Newton Conservators, 1992-93; judge high sch. sci. fairs, Boston area, 1990-93. Capt. U.S. Army, 1969-71, Vietnam. Grantee NSF, 1983, 88-91, 92—; recipient Healey Pub.

Svc. grant for acid rain rsch. U. Mass., 1989. Mem. AAAS, Am. Soc. Photobiology (charter), Am. Chem. Soc., Nat. Sci. Tchrs. Assn. Achievements include patent for cordless telephone data logger. Home: 248 Winchester St Newton Highlands MA 02161 Office: U Mass Dept Biology Morrisey Bldg Boston MA 02125

HAGBERG, DANIEL SCOTT, ceramic engineer; b. Rockford, Ill., July 14, 1965; s. John A. and Karin S. Hagberg; m. Jody L. Hagberg, July 11, 1992. BS in Ceramic Engring., U. Ill., 1988, MS, 1991. Cooperative edn. rsch. asst. Argonne (Ill.) Nat. Lab., 1985-87; vis. scientist Leeds (United Kingdom) U., 1988; rsch. asst. U. Ill., Urbana, 1988—. Fellow Office Naval Rsch., 1988-91. Mem. Am. Ceramic Soc., Materials Rsch. Soc., Internat. Soc. fro Hybrid Microelectronics. Office: Beckman Inst 405 N Mathews Urbana IL 61801

HAGEDORN, HENRY HOWARD, entomology educator; b. Milw., Apr. 4, 1940; s. Henry John and Virginia Colette (Hanson) H.; m. Magdalene Offen, June 20, 1964; children: Katrina Louise, Michael Alfred. BS, U. Wis., 1965, MS, 1966; PhD, U. Calif., Davis, 1970. Asst. prof. U. Mass., Amherst, 1973-77; assoc. prof. Cornell U., Ithaca, N.Y., 1977-87, prof., 1987-88; prof. U. Ariz., Tucson, 1988—, dir. ctr. for insect sci., 1989-93, acting head Dept. Entomology, 1993—. Mem. editorial bd. Archives Insect Biochemistry and Physiology, 1986—, Jour. Insect Physiology, 1993—. Named Von Humboldt Sr. Scientist Alexander von Humboldt-Stiftung, 1981. Fellow Am. Assn. for the Advancement Sci. Office: U Ariz Dept Entomology 410 Forbes Bldg Tucson AZ 85721

HAGELIN, JOHN SAMUEL, theoretical physicist; b. Pitts., June 9, 1954; s. Carl William and Mary Lee (Stephenson) H.; m. Margaret Cowhig, Nov. 22, 1985. AB summa cum laude, Dartmouth Coll., 1975; MA, Harvard U., 1976, PhD, 1981. Scientific assoc. European Orgn. Nuclear Rsch., CERN, Geneva, 1981-82; research assoc. Stanford (Calif.) Linear Accelerator Ctr., 1982-83; assoc. prof. physics Maharishi Internat. Univ., Fairfield, Iowa, 1983-84, prof. physics, dir. doctoral physics program, 1984—; pres. Enlightened Audio Designs Corp., 1991—; dir. Inst. Sci., Tech. and Pub. Policy, 1992—. Contbr. numerous articles to scientific jours. Presdl. candidate Natural Law party, 1992. Tyndall fellow Harvard U., 1979-80, Kilby Young Innovator award 1992. Mem. Iowa Acad. Scis. Avocation: research. Home and Office: Maharishi Internat U Fairfield IA 52556

HAGEMAN, BRIAN CHARLES, researcher; b. Burbank, Calif., June 10, 1952; s. Thomas John and Shirlee Aleene (Hamner) H. Student, Ariz. State U., 1988—. Constrn. engr. Fluor Constrns., Indonesia, 1975-77, Saudi Arabia, 1978-80; constrn. engr. Palo Verde Nuclear Plant Bechtel Power Corp., Ariz., 1980-86; owner, pres. Hageman Resources, Tempe, Ariz., 1986—. Mem. Assn. of Energy Engrs., Scottsdale C. of C. Achievements include devel. of a working model of a hydraulic engine, based on physics expts. in hydraulic expansion. Home: 1232 E Henry #2 Tempe AZ 85281 Office: Hageman Resources 1232 E Henry # 2 Tempe AZ 85281

HAGEMANN, DOLORES ANN, water company official; b. Parkston, S.D., June 5, 1935; d. Jacob George and Margaret Marie (Mayer) Schumacher; m. Norbert Bernard Hagemann, June 8, 1954; children: Douglas, Pamela, Susan. AS, Des Moines Community Coll., 1984. Cert. notary pub., Iowa. Sales rep. Stanley Home Products, Westfield, Mass., 1970-76; owner, mgr. Hagemann Gen. Store, Lidderdale, Iowa, 1974-77; motor rt. carrier Des Moines Register, 1977-82; accounts receivable clk. City Water Dept., Lidderdale, 1981—; owner, designer Dolores' Silk Flower Shop, Lidderdale, 1986—; bd. dirs. Lidderdale Apts., Inc., sec., 1974-91. Author: (with other) The Official Carroll County Democrat Cookbook, 1984. Com. person Carroll County Dems., 1970—, sec., 1985-86, 2d vice chair, 1989-90, 1st vice chair, 1990-92, chair, 1992—, chmn. chairs and vice chairs assn. 5th Congl. Dist. Iowa, 1990-92; mem. Iowa Dem. Party Election Rev. Com., 1991—; hospice mem. Community Hospice of Stewart Meml. Community Hosp., 1988—; counselor Carroll Help Line, 1982-87; mem. adv. bd. We. Iowa Transit, 1990—; mem. Carroll County steering com. Child Support Pub. Awareness Project, 1992—. Mem. Am. Assn. Ret. Persons, Holy Family Parish Guild (chair person 1976), Des Moines Community Coll. Alumni, Stewart Meml. Community Hosp. Aux. Democrat. Roman Catholic. Avocations: reading, gardening, canning, traveling, sewing. Home: PO Box 68 Lidderdale IA 51452-0068

HAGEN, DANIEL RUSSELL, physiologist; b. Springfield, Ill., Sept. 29, 1952; s. Robert William and Russella Mae (Lane) H.; m. Rosemary Ellen Simonetta, Mar. 25, 1978; children: Matthew, Mark, Lane, Elise. BS, U. Ill., 1974, PhD, 1978. Rsch. assoc. Cornell U., Ithaca, N.Y., 1978; asst. prof. State U., University Park, Pa., 1978-84, assoc. prof., 1984-93, prof., 1993—; vis. assoc. prof. U. Wis., Madison, 1988-89. Mem. editorial bd. Jour. of Animal Sci., 1983-86, 1993-96; contbr. over 27 articles to profl. jours. Mem. AAAS, Am. Soc. for Animal Sci., Soc. for Study of Reprodn., Soc. for Study of Fertility, Sigma Xi. Office: Pa State U 324 Henning Bldg University Park PA 16802

HAGEN, LAWRENCE JACOB, agricultural engineer; b. Rugby, N.D., Mar. 6, 1940; s. Lars and Alice (Hannem) H. BS, N.D. State U., 1962, MS, 1967; PhD, Kans. State U., 1980. Agrl. engr. USDA, Manhattan, Kans., 1967—. Contbr. tech. articles to profl. pubs. Capt. USAF, 1963-69. Mem. Am. Soc. Agrl. Engrs., Soil & Water Conservation Soc. Am. Office: USDA ARS Kans State U Rm 105 Waters Hall Manhattan KS 66506

HAGEN, STEPHEN JAMES, physicist; b. Harare, Zimbabwe, Apr. 7, 1962; came to U.S., 1969; s. David and Elisabeth Hagen. BA, Wesleyan U., 1984; PhD, Princeton U., 1989. Faculty rsch. assoc. U. Md., College Park, 1989-92; staff fellow NIH, Bethesda, Md., 1992—. Contbr. articles to Phys. Rev., Physica C. Garden State fellow State of N.J., 1984-88, fellow GE Found., 1984-85. Mem. AAAS, Am. Phys. Soc., Phi Beta Kappa, Sigma Xi. Democrat. Achievements include rsch. in field of transport properties of single-crystal high-temperature superconductors, measure of fundamental properties of these superconductors to establish 2-dimensional nature; 1st detailed study of Hall effect anomaly in high-temperature superconductors; physics of biological molecules, especially heme proteins. Office: NIH Bldg 5 Rm 106 Bethesda MD 20892

HAGENIERS, OMER LEON, mechanical engineer; b. Stekene, Belgium, Nov. 12, 1944; arrived in Can., 1948; s. Omer Alphonse and Maria (DeClerc) H.; m. Marilyn Laeona Miner, Sept. 2, 1967; children: Michelle Liane, Amy Louise. MA in Sci., U. Windsor, Ont., Can., 1969, PhD in Mech. Engrin., 1972. Registered prof. engr., Ont. Exec. v.p. Diffracto Ltd., Windsor, 1969-90, pres., 1990—. Contbr. articles to profl. jours. Mem. ASME, Soc. Mfg. Engrs., Soc. Photographic and Instrumentation Engrs. Achievements include numerous patents in field. Office: Diffracto Ltd, 2835 Kew, Windsor, ON Canada N8T 3B7

HAGENSTEIN, WILLIAM DAVID, consulting forester; b. Seattle, Mar. 8, 1915; s. Charles William and Janet (Finigan) H.; m. Ruth Helen Johnson, Sept. 2, 1940 (dec. 1979); m. Jean Kraemer Edson, June 16, 1980. BS in Forestry, U. Wash., 1938; MForestry, Duke, 1941. Registered profl. engr., Wash., Oreg. Field aid in entomology U.S. Dept. Agr., Hat Creek, Calif., 1938; logging supt. and engr. Eagle Logging Co., Sedro-Woolley, Wash., 1939; tech. foreman U.S. Forest Svc., North Bend, Wash., 1940; forester West Coast Lumbermen's Assn., Seattle and Portland, Oreg., 1941-43, 45-49; sr. forester FEA, South and Central Pacific Theaters of War and Costa Rica, 1943-45; mgr. Indsl. Forestry Assn., Portland, 1949-80; exec. v.p. Indsl. Forestry Assn., 1956-80, hon. dir., 1980-87; pres. W.D. Hagenstein and Assocs., Inc., Portland, 1980—; H.R. MacMillan lectr. forestry U. B.C., 1952, 77; Benson Meml. lectr. U. Mo., 1966; S.J. Hall lectr. indsl. forestry U. Calif. at Berkeley, 1973; cons. forest engr. USN, Philippines, 1952, Coop. Housing Found., Belize, 1986; mem. U.S. Forest Products Trade Mission, Japan, 1968; del. VII World Forestry Congress, Argentina, 1972, VIII Congress, Indonesia, 1978; mem. U.S. Forestry Study Team, West Germany, 1974; mem. sec. Interior's Oreg.-and Calif. Multiple Use Adv. Bd., 1975-76; trustee Wash. State Forestry Conf., 1948-92, Keep Oreg. Green Assn., 1957—, v.p., 1970-71, pres., 1972-73; adv. trustee Keep Wash. Green Assn., 1957—; co-founder, dir. World Forestry Ctr., 1965-89, v.p., 1965-79; hon. Dir. for Life, 1990. Author: (with Wackerman and Michell) Harvesting Timber Crops, 1966; Assoc. editor: Jour. Forestry, 1946-53; columnist Wood Rev., 1978-82; contbr. numerous articles to profl. jours. Trustee Oreg. Mus. Sci. and Industry, 1968-73. Served with USNR, 1933-37. Recipient Hon. Alumnus award U. Wash. Foresters Alumni Assn., 1965, Forest Mgmt. award Nat. Forest Products Assn., 1968, Western Forestry award Western Forestry and Conservation Assn., 1972, 79, Gifford Pinchot medal for 50 yrs. Outstanding Svc. Soc. Am. Foresters, 1987, Charles W. Ralston award Duke Sch. Forestry, 1988. Fellow Soc. Am. Foresters (mem. coun. 1958-63, pres. 1966-69, Golden Membership award 1989); mem. Am. Forestry Assn. (life, hon. v.p. 1966-69, 74-92, William B. Greeley Forestry award 1990), Commonwealth Forestry Assn. (life), Internat. Soc. Tropical Foresters, Portland C. of C. (forestry com. 1949-79, chmn. 1960-62), Nat. Forest Products Assn. (forestry adv. com. 1949-80, chmn. 1972-74, 78-80), West Coast Lumbermen's Assn. (v.p. 1969-79), David Douglas Soc. Western N. Am., Lang Syne Soc., Hoo Hoo Club, Xi Sigma Pi (outstanding alumnus Alpha chpt. 1973). Republican. Home: 3062 SW Fairmount Blvd Portland OR 97201-1439 Office: Ste 803 921 SW Washington St Portland OR 97205

HAGER, LOWELL PAUL, biochemistry educator; b. Girard, Kans., Aug. 30, 1926; s. Paul William and Christine (Selle) H.; m. Frances Erea, Jan. 22, 1949; children: Paul, Steven, JoAnn. AB, Valparaiso U., 1947; MA, U. Kans., 1950; PhD, U. Ill., 1953. Postdoctoral fellow Mass. Gen. Hosp., Boston, 1953-55; asst. prof. biochemistry Harvard U., Cambridge, Mass., 1955-60; mem. faculty U. Ill., Urbana, 1960—, prof. biochemistry, 1965—, head biochem. div., 1967-89, dir. Biotech. Ctr., 1987—; chmn. physiol. chemistry study sect. NIH, 1965—; vis. scientist Imperial Cancer Rsch. Fund, 1964; cons. NSF, 1976. Editor life scis. Archives Biochemistry and Biophysics, 1966—; assoc. editor Biochemistry, 1973—; mem. editorial bd. Jour. Biol. Chemistry, 1874—. With USAAF, 1945. Guggenheim fellow U. Oxford, Eng., 1959-60, Max Planck Inst. Zellchemie, 1959-60. Mem. Am. Chem. Soc., Am. Soc. Biol. Chemists, Am. Soc. Microbiology (chmn. physiology div. 1967). Rsch. in enzyme mechanisms, intermediary metabolism, tumor virus. Home: 801 W Delaware Ave Urbana IL 61801-4808

HAGFORS, TOR, national astronomy center director; b. Oslo, Dec. 18, 1930; came to U.S., 1963; s. Vidar Johan and Hanna Viktoria (Edmundson) H.; m. Gillian Patricia Hart, Jan. 3, 1953; children: John, Toril, Martin, Vivien. M.Engring., U. Trondheim, Norway, 1955; Ph.D., U. Oslo, 1959. Scientist Norwegian Def. Research Labs., Kjeller, 1955-59, 61-63; research asst. Stanford U., Calif., 1961-63; staff mem. Lincoln Labs. MIT, Lexington, Mass., 1963-69, 71-73; dir. Jicamarca Obs., Lima, Peru, 1969-71; prof. elec. engring. U. Trondheim, 1973-82; dir. Nat. Astronomy and Ionosphere Ctr. Cornell U., Ithaca, N.Y., 1982-92; dir. Max Planck Institut fü Aeronomie, Katlenburg-Lindau, Ger.; mem. Fachbeirat, Max-Planck-Inst. für Aeronomie, Lindau-Harz, Germany, 1977-82, Swedish Space Rsch. Bd., Stockholm, 1978-82; bd. dirs. N.E. Radio Observatory Corp., Max Planck Institut fur Aeronomie. Author, editor: Radar Astronomy, 1967, High Latitude Space Plasma Physics, 1983. Fellow IEEE; mem. Am. Astron. Soc., Am. Geophys. Union, Internat. Union Radio Sci. (Van der Pol Gold medal 1987, von Humboldt fellow 1989-90), Max Planck Soc. Office: Max Planck Inst Aeronomie, Postfach 20, D-3411 Katlenburg-Lindau Germany

HAGGARD, WILLIAM HENRY, meteorologist; b. Woodbridge, Conn., Nov. 20, 1920; s. Howard Wilcox and Josephine Cecelia (Foley) H.; m. Blanche Woolard, Mar. 21, 1944 (div. May 1967); children: William Henry Jr., Robert H.; m. Martina Wadewitz, Oct. 1, 1967. BS in Physics, Yale U., 1942; cert. in profl. meteorology, MIT, 1942; MS in Meteorology, U. Chgo., 1946; postgrad., Fla. State U., 1958-59. Instr. meteorology N.C. State U., Raleigh, 1946-47; rsch. meteorologist U.S. Weather Bur., 1947-48; forecaster USWB Nat. Airport, 1949-50; instr. U.S. AID, Washington, 1950-51; staff weather rsch. project U.S. Navy, Norfolk, Va., 1951-54; forecaster nat. airport, chief adv. svcs. br. Office of Climatology U.S. Weather Bur., Washington, 1954-59, asst. chief Office of Plans, 1960-61; dep. dir. Nat. Weather Records Ctr., Asheville, N.C., 1961; dir. Nat. Climatic Ctr., Asheville, 1963-75; pres. Climatol. Cons. Corp., Asheville, 1976—; mem. weather com. U.S. Power Squadron, Raleigh, N.C., 1988—. Contbr. articles to tech. jours., 1947-90. Bd. dirs. ARC, Asheville, 1965-70, United Way, Asheville, 1964-70. Capt. USN, 1942-45, with Res. 1951-54. Recipient Tech. Adminstr. award NOAA, Washington, 1970. Fellow Am. Meteorol. Soc. (cert. cons. meteorologist, bd. dirs. pvt. sector meteorology sect. 1989—, mem. cert. cons. meteorologist bd. 1983-88), Nat. Coun. Indsl. Meteorologists (pres. 1988-89, bd. dirs. 1987-90). Republican. Presbyterian. Avocations: sailing, photography. Office: Climatological Cons Corp 150 Shope Creek Rd Asheville NC 28805-9718

HAGGERTY, JAMES JOSEPH, writer; b. Orange, N.J., Feb. 1, 1920; s. James Joseph and Anna (Morahan) H.; student pub. schs.; m. Marian Smith Mitten, Nov. 20, 1962 (dec. Jan. 1989); children: Karin, James Joseph, Brian. Reporter Orange (N.J.) Daily Courier, 1938-40; mil. editor Am. Aviation Publs., 1948-53; aviation editor Collier's, 1953-56; free-lance writer on sci. and aerospace subjects, 1956—; editor Aerospace Year Book, 1957-70; aerospace editorial cons. Aerospace Industries Assn., 1974—, NASA, 1975—, Pres. Commn. on Space Shuttle Challenger Accident, 1986. Served with USAAF, 1942-48. Decorated D.F.C., Air medal with clusters. Mem. Aviation Space Writers Assn. (past pres.), AAAS, Air Force Assn, Bethesda Country Club, Touchdown Club (Washington), Culpepper (Va.) Country Club. Author: First of the Spacemen, 1960; Spacecraft, 1961; Flight, 1964; The U.S. Air Force: A Pictorial History in Art, 1965; Man's Conquest of Space, 1965; Food and Nutrition, 1966; Apollo Lunar Landing, 1969; Hail To The Redskins, 1973; Aviation's Mr. Sam, 1973. Address: 502 H St SW Washington DC 20024

HAGHIGHAT, ALIREZA, nuclear engineering educator; b. Shiraz, Fars, Iran, Aug. 2, 1956; came to U.S., 1978; s. Hassan Haghighat and Pari (Amini) Amini; m. Mastaneh Javadi, June 21, 1978; 1 child, Aarash. BS, Pahlavi U., Shiraz, 1978; MS, U. Wash., 1981, PhD, 1986. Postdoctoral rsch. faculty Pa. State U., University Park, 1986-89, asst. prof., 1989-93, assoc. prof., 1993—, faculty advisor Am. Nuclear Soc. student chpt.; faculty advisor student chpt. Pa. State U., 1990—, faculty advisor for the Nuclear Engring. Honor Soc., 1990—; cons. Battelle, Columbus, Ohio, 1988. Reviewer Nuclear Sci. and Engring. Jour., 1991, Am. Nuclear Soc. Ann. Meeting and Topical Meetings, 1990—; contbr. articles to profl. jours. including Nuclear Sic. and Engring., Nuclear Tech., Procs. of the Internat. Confs. and Advances in Maths. and Computation and Reactor Physics. Scholar grad. sch. U. Wash., 1985. Mem. Am. Nuclear Soc. (sec. math. and computation divsn. 1990, newsletter editor math. & computation divsn., mem. tech. com. several confs.), Alpha Nu Sigma, Sigma Xi. Achievements include development of thermal hydraulics and neutronics codes; codevelopment of new vector/parallel algorithms for the SN transport theory method; research in developing new methodologies for the neutron fluence calculations. Office: Pa State U 231 Sackett Bldg University Park PA 16802

HAGLUND, MICHAEL MARTIN, neurosurgeon; b. Mpls., Sept. 27, 1958; s. William Arthur and Jean Christine (Rallis) H.; m. Christine Lynn Rosen, June 6, 1980; 1 child, Sean Michael. BS in Chemistry, Pacific Luth. U., 1980; MD, U. Wash., 1987, PhD in Physiology and Biophysics, 1988. Resident in neurological surgery Sch. Medicine U. Wash, Seattle, 1987-88, 90-92, chief resident neurological surgery, 1993—; rsch. fellow neurobiology Harvard Med. Sch., Boston, 1989-90; sr. registrar Atkinson Manley's Hosp., Wimbledon, London, Eng., 1992-93; pres., mem. sci. adv. bd. Optimedix, Inc., Seattle, 1993—. Contbr. articles to sci. and med. jours., chpts. to book. Active S.W. London Vineyard, 1992-93. Lt. commdr. USNR. Klingenstein Found. fellow, 1990-92; Initiative grantee Wash. Tech. Ctr., 1992-93. Fellow Am. Epilepsy Soc., Am. Assn. Neurological Scis.; mem. Soc. for Neuroscience. Mem. Westgate Chapel Ch. Achievements include two patents pending on use of optical imaging with contrast-enhancing dyes to visualize tumor tissue; use of optical imaging to visualize human cognitive function. Office: U Wash Dept Neurological Surgery RI-20 Seattle WA 98195

HAGMANN, ROBERT BRIAN, computer scientist; b. Waltham, Mass., Dec. 2, 1948; s. Otto and Katherine (Coffey) H.; m. Joanne Barbano, Feb. 5, 1977; children: Annelise, Ian, Patrick. BS, MIT, 1970; MS, Purdue U., 1972; PhD, U. Calif., Berkeley, 1983. Mem. rsch. staff Xerox Corp., Palo Alto, Calif., 1983-91; sr. staff engr. Sun Soft Co., Mountain View, Calif., 1991—; cons. assoc. prof. Stanford (Calif.) U., 1986—; presenter at profl. confs. Contbr. articles to profl. jours. Mem. IEEE, Assn. Computing Machinery. Office: Sun Soft Co Mail Stop MTV1-208 2550 Garcia Ave Mountain View CA 94043

HAGSON, CARL ALLAN, utilities executive; b. Haggenas, Jamtland, Sweden, Nov. 4, 1921; s. Olof and Elin (Eriksson) H.; m. Marianne Lallerstedt, Nov. 24, 1967. MS in Elec. Engring., Royal Tech. Inst., Stockholm, 1947. With Swedish State Power Bd., 1947-50; with Swedish Assn. Electricity Supply Undertakings, Stockholm, 1950-86, gen. mgr., 1967-86; pub. electrotech. jour. ERA, 1975-86; owner cons. bus. Stockholm, 1989—; mem. govt. coms. on energy consumption forecasting, energy conservation, energy tariffs, 1972-82; mem. tech. council Nat. Bd. Phys. Planning and Bldg., 1970-85; mem. com. action Internat. Electrotech. Commn., 1978-84. Served with Swedish Army, 1942-43, 45. Decorated knight Royal Order of Vasa, 1974, knight Finnish Lions Order 1987. Mem. IEEE, Union Internat. Producteurs Distributeurs d'Energie Electrique (dir. com. 1967-86), Swedish Electrotech. Commn. (bd. dirs. 1967-88, pres. 1973-88), Swedish Inst. Testing and Approval Elec. Equipment (bd. dirs. 1967-88), Conv. Nat. Socs. Elec. Engrs. Western Europe (exec. com. 1972-75, pres. 1973-74), Swedish Assn. Elec. Engrs. (bd. dirs. 1952-60, 69-76, pres. 1972-76, hon.), Swedish Union Grad. Engrs. (bd. dirs. 1954-66, pres. 1961-66, hon.), Swedish Taxpayers Assn. (bd. dirs. 1964-87), Assn. Electroheat Promotion (chmn. 1987—), SEF Tariffs Commn. (chmn. 1986-91). Home: Torstenssonsgatan 10, 114 56 Stockholm Sweden Office: Box 3192, 103 63 Stockholm Sweden

HAGSTEN, IB, animal scientist, educator; b. Assens, Denmark, May 18, 1943; came to U.S., 1971, naturalized, 1980; s. Kresten and Marie (Jakobsen) H.; m. Patricia Ellen Dettman, July 13, 1968; children: Ellen Marie, Scot (dec.), Lisa R. BA in Agr., Bygholm Landbrugskole, Horsens, Denmark, 1965; BS, Royal Danish Agr. U., Copenhagen, 1971; MS, Purdue U., 1973, PhD, 1975. Cert. animal scientist. Farm laborer, foreman various livestock farms, Denmark, Eng., Germany, Can., 1958-65; teaching asst. Royal Danish Agr. U., 1969-70; rsch. assoc. Nat. Danish Rsch. Found., Copenhagen, 1971; cons. nutritionist M.D. King Milling Co., Pittsfield, Ill., 1976-77; acting product mgr. Am. Hoechst Corp., Somerville, N.J., 1978, tech. specialist, 1977-83; profl. sales rep. Hoechst-Roussel Agri-Vet. Co., Gladstone, Mo., 1983-89, tech. svc. specialist, 1989-90, sr. profl. svc. specialist, 1990—; cons. Shell Farm, Inc., Ørum, Denmark, 1970-71, Agri-Bus. Tng. & Devel., Inc., Roswell, Ga., 1979—, Nat. Renderer's Assn., Hong Kong, 1989; adj. prof. Rutgers U., New Brunswick, N.J., 1981-84, U. Mo., Columbia, 1990—; pres. Personal Growth Alternatives, 1982—. Author: Energy Metabolism Evaluations, 1971; contbr. articles to profl. jours. and popular publs. Bd. dirs. MACOS handicapped support group, Macomb, Ill., 1976-77; co-chair Community Hunger Walks (CROP), Western N.J., 1978-82; mem. family curriculum bd. Lopatcong Twp. Sch., Phillipsburg, N.J., 1982; vice moderator Pilgrim Presbyn. Ch., Phillipsburg, 1980-83, elder, 1979-83; elder Gashland Presbyn. Ch., Gladstone, 1990—; regional exec. bd. mem. United Marriage Encounter (Mo., Kans.), 1983—; mem. Core of Advocates, Coll. Vet. Medicine Kans. State U. Sgt. Danish King's Royal Guard, 1959-61. Mem. Danish Soc. Animal Sci., Am. Soc. Agrl. Cons. (bd. dirs. 1978-81, 92-94, chmn. ethics com. 1992-94, Disting. Svc. award 1980), Am. Soc. Agrl. Cons. Internat. (charter), Nat. Feed Ingredient Assn., Am. Registry Profl. Animal Scientists (chmn. ethics com. 1982-85). Republican. Avocations: people, travel, reading. Home: 7212 N Woodland Ave Kansas City MO 64118-2263 Office: Hoechst-Roussel Agri-Vet Co 7212 N Woodland Ave Kansas City MO 64118-2263

HAGSTROM, STIG BERNT, materials science and engineering educator; b. Barkeryd, Sweden, Sept. 21, 1932; came to U.S., 1976; s. Johan A. and Nanny (Svanberg) H.; m. Brita-Stina Felldin Hagstrom, June 23, 1957; children: Anders, Mats, Karin, Elisabet. BS, Uppsala (Sweden) U., 1957, PhD, 1964; Dr.Techn. (hon.), Linkoping (Sweden) U., 1987. Asst. prof. Uppsala U., 1962-64; assoc. prof. Chalmers U. Tech., Goteborg, Sweden, 1966-69; prof. Linkoping U., 1969-78, Stanford U., Palo Alto, Calif., 1987—; researcher U. Calif., Berkeley, 1965-66; mgr. Xerox Palo Alto Rsch. Ctr., 1976-87; dir. Stanford (Calif.) Ctr. for Materials Rsch., 1989—; chmn. materials sci. and engring. dept. Stanford U., 1987-91; vice rector Linkoping U., 1970-76. Contbr. articles to profl. jours. Fellow Am. Phys. Soc.; mem. Royal Swedish Acad. Engring. Scis., Royal Norwegian Acad. Sci. and Letters, Royal Swedish Acad Sci. Achievements include 4 patents for information storage; research on photoelectron spectroscopy for chemical analysis. Home: 1365 Bay Laurel Dr Menlo Park CA 94025-5804 Office: Stanford U Ctr Materials Rsch 105 McCullogh Bldg Stanford CA 94305-4045

HAHIN, CHRISTOPHER, metallurgical engineer, corrosion engineer; b. Buffalo, Dec. 26, 1945; s. Leo Paul and Nancy (Morabito) H.; children: Bonnie L., Terence J. BS, Mich. State U., 1968; MS, U. Ky., 1974. Cert. profl. engr., Ill., Calif. Missile facilities engr. USAF, Strategic Air Command, Minot AFB, N.D., 1968-72; rsch. metallurgist U.S. Army Corps of Engrs. Constrn. Engr. Rsch. Lab., Champaign, Ill., 1974-81; prin. engr. Container Corp. Am., Carol Stream, Ill., 1981-84; chief metallurgist Avondale Ind., Danly Machine Div., Chgo., 1984-86; engr. bridge investigations Bur. Materials and Phys. Rsch., Ill. Dept. Transp., Springfield, 1987—; prin. assoc. Materials Protection Assn., Springfield, Ill., 1984—. Author: Book Science Baseball, 1983; contbr. Advanced Casting Technology Conf., Kalamazoo, 1986, handbook ASM Metals Handbook vol. 13, 1987; patentee in field. With USAF 1968-72 Minot AFB, N.D. Fellow Ashland Oil U. Ky., 1971; decorated A.F.C.M. 1st Oak Leaf Cluster; recipient U.S. Army Spl. Act Svc., 1981. Mem. Am. Soc. for Metals, Nat. Corrosion Engrs. Assn., Am. Welding Soc., Am. Soc. for Testing and Materials. Democrat. Achievements include development of corrosion cost prediction models for utilities and structures using air, water, soil and air pollution data; determined lifting, earth moving and towing requirements for combat engineer vehicles using actual battle scenarios; introduced leaded free machining steels for use in mass production of improved carburized die set bushings for stamping industry; development of improved pin and link eyebar designs and materials for bridges; developed accurate and rapid method for determination of fatigue damage in bridges using stress frequency histograms and linear damage rule; developed as-welded notch toughness test for steel weldments using natural notches and other similar code authority qualification tests. Office: Ill Dept Transp 126 E Ash St Springfield IL 62704-4792

HAHN, BENJAMIN DANIEL, research executive; b. Embden, N.D., Oct. 24, 1932; s. Benjamin D. and Laura E. (Martin) H.; m. Eleanor B. Anseth, June 8, 1957; children—Lezlie, Deann, Bobette, Lara, Amy. BBA, Concordia Coll., 1960, MA in Acctg. U. N.D., 1961. CPA, N.D. Staff acct. Broeker Hendrickson, Fargo, N.D., 1961-63; asst. adminstrn., contr. N.D. State Hosp., Jamestown, 1963-74; pres. the Neuropsychiat. Inst., Fargo, 1974—; faculty Jamestown Coll., part-time, 1963-74. Bd. dirs. Bethany Home, Fargo, 1975-81, S.E. Mental Health Ctr., 1975-81. With U.S. Army, 1954-56. Am. Coll. Hosp. Adminstrs. fellow, 1976. Mem. Am. Inst. CPA's, Med. Group Mgmt. Assn. Lutheran. Club: Fargo Country. Lodge: Elks. Home: 21 35th Ave NE Fargo ND 58102-1204 also: 700 1st Ave Fargo ND 58102*

HAHN, ERWIN LOUIS, physicist, educator; b. Sharon, Pa., June 9, 1921; s. Israel and Mary (Weiss) H.; m. Marian Ethel Failing, Apr. 8, 1944 (dec. Sept. 1978); children: David L., Deborah A., Katherine L.; m. Natalie Woodford Hodgson, Apr. 12, 1980. B.S., Juniata Coll., 1943, D.Sc., 1966; M.S., U. Ill., 1947, Ph.D., 1949; D.Sc., Purdue U., 1975. Asst. Purdue U., 1943-44; research assoc. U. Ill., 1950; NRC fellow Stanford, 1950-51, instr., 1951-52; research physicist Watson IBM Lab., N.Y.C., 1952-55; assoc. Columbia U., 1952-55; faculty U. Calif., Berkeley, 1955—, faculty rsch. lectr., 1959, prof. physics, 1961—, assoc. prof., then prof. Miller Inst. for Basic Research, 1958-59, 66-67, 85-86; vis. fellow Brasenose Coll., Oxford (Eng.) U., 1981-82, Eastman vis. prof., 1988-89; cons. Office Naval Rsch., Stanford, 1950-52, AEC, 1955—; spl. cons. USN, 1959; adv. panel mem. Nat. Bur. Standards, Radio Standards div., 1961-64; mem. NAS/NRC com. on basic rsch.; advisor to U.S. Army Rsch. Office, 1967-69; faculty rsch. lectr. U. Calif., Berkeley, 1979. Author: (with T.P. Das) Nuclear Quadrupole Resonance Spectroscopy, 1958. Served with USNR, 1944-46. Recipient prize Internat. Soc. Magnetic Resonance, 1971, award Humboldt Found., Germany, 1977, Alumni Achievement award Juniata Coll., 1986, citation U. Calif., Berkeley, 1991; co-winner prize in physics Wolf Found., 1984; named to Calif. Inventor Hall of Fame, 1984; Guggenheim fellow, 1961-62, 69-70, fellow NSF, 1961-62; vis. fellow Brasenose Coll., Oxford U., 1969-70, life hon. fellow, 1984—. Fellow AAAS, Am. Phys. Soc. (past mem. exec. com. div. solid state physics, Oliver E. Buckley prize 1971); mem. NAS (co-recipient Comstock prize in electricity, magnetism and radiation 1993), Slovenian Acad. Scis. and Arts (fgn.), French Acad. Scis. (fgn. assoc.). Home: 69 Stevenson Ave Berkeley CA 94708-1732 Office: U Calif Dept Physics 367 Birge Hall Berkeley CA 94720

HAHN, GEORGE LEROY, agricultural engineer, biometeorologist; b. Muncie, Kans., Nov. 12, 1934; s. Vernon Leslie and Marguerite Alberta (Breeden) H.; m. Clovice Elaine Christensen, Dec. 3, 1955; children—Valerie, Cecile, Steven, Melanie. B.S., U. Mo., Columbia, 1957, Ph.D., 1971; M.S., U. Calif., Davis, 1961. Agrl. engr., project leader and tech. advisor Agrl. Research Service, U.S. Dept. Agr., Columbia, Mo., 1957, Davis, Calif., 1958-61, Columbia, 1961-78, Clay Center, Nebr., 1978—. Contbr. articles to tech. jours. and books on impact of climatic and other environ. factors on livestock prodn., efficiency, and well-being, evaluation of methods of reducing impact and techniques for measuring dynamic responses and stress in meat animals. Recipient award Am. Soc. Agrl. Engrs.-Metal Bldgs. Mfrs. Assn., 1976. Fellow Am. Soc. Agrl. Engrs. (dir. prof. coun. 1991-93); mem. Am. Meteorol. Soc. (award for outstanding achievement in bioclimatology 1976), Internat. Soc. Biometeorology, Am. Soc. Animal Sci. Office: US Meat Animal Rsch Ctr PO Box 166 Clay Center NE 68933-0166

HAHN, HONG THOMAS, mechanical engineering educator; b. Dolma-Myun, Kyungki-Do, Republic of Korea, Feb. 5, 1942; came to U.S., 1966; s. Baek Hyo and Sang Soon (Lee) H.; m. Hoon Pat Paek, Sept. 16, 1967; children: Heryun, Hejin, Jeanie. BS, Seoul Nat. U., Republic of Korea, 1964; MS, Pa. State U., 1968, PhD, 1971. Rsch. engr. U. Dayton (Ohio) Rsch. Inst., 1974-77, Air Force Materials Lab., Wright-Patterson AFB, Ohio, 1972-74, 77-78; mech. engr. Lawrence Livermore (Calif.) Nat. Lab., 1978-79; prof. Washington U., St. Louis, 1979-86, Pa. State U., University Park, 1986-91, UCLA, 1992—; cons. Lawrence Livermore Nat. Lab., Livermore, 1979—, UN Devel. Programme, 1987-88, Dow Chem., 1990—; disting. lectr. NASA-Va. Poly. Inst. Composites Program, Blacksburg, 1988; lectr. SW mechanics Series, 1989; Hughes Aircraft Co. chair in mfg. engring., UCLA, 1992. Co-author: Introduction to Composite Materials, 1980; editor: Composite Materials-Fatigue & Fracture, 1986, Jour. Composite Material, 1981—; contbr. over 100 articles to profl. jours. Mem. sci. adv. bd. Swedish Inst. Composites, Pitea, Sweden, 1989-91. Lt. Korean Army, 1964-66. Recipient Outstanding Rsch. award Pa. State Engring. Soc., 1991; Harry and Arlene professorship in engring. Pa. State U., 1991. Mem. ASME, AIAA, Am. Soc. Composites, Materials Rsch. Soc., Soc. for the Advancement of Material and Process Engring., Soc. Mfg. Engrs., Am. Ceramic Soc., Sigma Xi. Office: UCLA MANE Dept Engring IV Los Angeles CA 90024-1597

HAHN, KLAUS-UWE, aerospace engineer; b. Braunschweig, Germany, July 15, 1954; s. Guenter Karl Georg and Waltraut Wilma (Krueger) H.; m. Karin Christine Heintsch, Feb. 26, 1979; children: Anika Gesche, Timo Alexander. Engr., Coll. Tech., Wolfenbuettel, 1975; diploma, U. Braunschweig, 1981, D, 1988. Rsch. asst. Inst. for Flight Guidance and Control, Braunschweig, 1981-83; researcher Spl. Rsch. Group on Flight Safety U. Braunschweig, 1983-89; scientist German Aerospace Rsch. Establishment, Braunschweig, 1989—; German rep. Group for Aero. Rsch. and Tech. in Europe, 1991—. Mem. AIAA, Deutsche Gesellschaft fur Luft-und Raumfahrt (vice-chmn. working group flight performance 1991—). Achievements include research in aerospace scis. Home: Grauer Hof 7C, 38176 Wendeburg Federal Republic of Germany Office: German Aerospace Rsch Establishment, Flughafen, 38108 Braunschweig Germany

HAHN, RICHARD RAY, academic administrator; b. Rapid City, S.D., July 12, 1930; m. Joan Fager, May 24, 1953; children—David H., Carol L., Donald R., Kathleen J. B.S., Bethany Coll., 1952; M.S., Kans. State U., 1954, Ph.D., 1957. With Harvest Queen Mills, 1956-67; with A.E. Staley Mfg. Co., Decatur, Ill., 1967-87; v.p. research and devel. A.E. Staley Mfg. Co.; asst. to dean agrl. U. Ill., Urbana, 1987-89; dir. Kans. Agr. Value Added Processing Ctr., Manhattan, 1989-92; head dept. grain sci. Kans. State U., Manhattan, 1992—; cons. Agrl. Processing and Tech. Mgmt. Mem. Inst. Food Technologists, Am. Assn. Cereal Chemists (pres. 1988-89). Office: Kans State Univ 202 Shellenberger Hall Manhattan KS 66506

HAHN, WALTER GEORGE, naval architect; b. South Bend, Ind., May 1, 1948; s. Walter W. and Mildred (Siefert) H.; m. Diane A. Hull, 1974 (div. 1979); m. June E. Egner, Dec. 5, 1981; children: R. James De Rosa, Michele De Rosa. Student, U. Miami (Fla.), 1966-67, No. Mich. U., 1967-68, U. Mich., 1968-71. Loftsman Lydia Yachts, Stuart, Fla., 1976-77; purchasing mgr. Monterey Marine, Stuart, 1977-79; boat builder Stuart Yacht Builders, 1979-80; naval architect Lydia Yachts, Stuart, 1980-82; pvt. practice naval architecture Stuart, 1982-86; pres. W.G. Hahn, Inc., Stuart, 1986—; mgr. Design Group, Stuart, 1989—. Treas. E.C. "Buck" Blackman campaign, Stuart, 1984. Mem. AIAA, Soc. Naval Architects and Marine Engrs. (chmn. elect S.E. sect.). Achievements include design of fastest tournament sportfishing vessel "Renegade." Office: W G Hahn Inc PO Box 7018 Stuart FL 34996

HAHNE, C. E. (GENE HAHNE), computer services executive; b. Savannah, Ga., Sept. 21, 1940; s. Charles Eugene and Hortense (Kavanaugh) h.; m. Brenda Wike, Nov. 25, 1983; children: Gregory, Christopher, David, Stephanie. BS in Indsl. Mgmt., Ga. Inst. Tech., 1963. Sales mgr. Shell Oil Co., Cleve., 1969-72; head office rep. Shell Oil Co., Houston, 1973-75, mgr. tng., human resources and products, 1979-93; mgr. tng. and recruiting Shell Chem., Houston, 1976-78; CEO, chmn. bd. dirs. Intercom, The Woodlands, Tex., 1993—; mem. adv. coun. U. Houston, 1984-85, curriculum com., 1990—; chair mktg. com. Houston C.C., 1987—; speaker in field Europe, Can., S.Am., U.S. Author; Mgmt. Handbook, 1981, Training Handbook, 1986, Sales Trng. Handbook, 1989; contbr. articles to profl. jours. Mem. bus. and industry coun. Houston Community Coll., 1986—, mem. adv. coun. Tex. A&M U., College Station, 1985-88; chair fundraising com. Brookwood Community, Houston, 1986-88; bd. dirs. Interact, Houston, 1983, U. Tex., Austin, 1984-85. Recipient Speaker's award United Way, 1988, Pres. award Houston Community Coll., 1987, Tex. Vocat. Excellence award, 1990. Mem. ASTD (nat. bd. treas. 1981-86, Gordon M. Bliss award 1989, Torch award 1987, Lifetime Recognition award 1989, Disting. Contbn. to Community/Nation award 1984, James Ball award 1981), World Future Soc., Sales and Mktg. Execs. of Houston, Soc. for Human Resource Mgmt. Republican. Home: 20210 Atascocita Lake Dr Humble TX 77346-1659

HAILE, BENJAMIN CARROLL, JR., chemical engineer, mechanical engineer; b. Shanghai, China, Apr. 6, 1918; came to U.S., 1925; s. Benjamin Carroll and Ruth Temple (Shreve) H.; m. Lola Pauline Lease, Dec. 28, 1957; children: Thomas Benjamin, Ronald Frederick. BS, U. Calif., Berkeley, 1941; cert., Harvard-MIT, 1945; postgrad., U. So. Calif., 1950-51. Registered profl. chem. and mech. engr., Calif. Chem. engr. Std. Oil of Calif. (Chevron), San Francisco, El Segundo, Calif., 1941-43, 46-48; sr. project chem. engr. C.F. Braun & Co., Alhambra, Calif., 1948-50, 54-56, 67-71, 72; contract chem. and mech. engr. Dow, Stearns-Roger, Fluor, et al, Tex., Colo., Ill., 1951-54, 56-57; sr. process engr. Aerojet-Gen. Corp., Sacramento and Covina, Calif., 1957-67; mech. engr. So. Calif. Edison Co., Rosemead, 1972-84; pvt. practice chem. engr. Fontana and Montclair, Calif., 1986, 88, 92; sr. mem. tech. staff Ralph M. Parsons Co., Pasadena, Calif., 1971, 88-91. 2d lt. USAAF, 1943-46. Mem. NSPE (life, Sacramento chpt. pres. 1960-62), Am. Inst. Chem. Engrs. (mem. emeritus), Toastmasters Internat. (chpt. v.p. 1979, outstanding Toastmaster 1984), Psi Upsilon. Republican. Achievements include project, process and mech. engring. design of oil refineries, chem. plants, others, with estimated cumulative present value of one billion dollars during lifetime; project leader and process development of new fluid bed adsorption process for air separation; economic optimization studies of complete aerospace programs; static electricity protection study for Polaris propellant manufacturing facility; project leader and designer of one of world's largest boring machines, 1952-53. Home: 159 N Country Club Rd Glendora CA 91741

HAILE, JAMES MITCHELL, engineering educator; b. Atlanta, Dec. 7, 1946; s. Charles Mitchell and Gertrude (Jones) H.; m. Patricia Ann West, Mar. 6, 1971; 1 child, James Mitchell II. BS, Vanderbilt U., 1968; PhD, U. Fla., 1976. Asst. prof. to prof. chem. engring. Clemson (S.C.) U., 1976-88, prof., 1989—; vis. assoc. prof. chem. engring. Cornell U., Ithaca, N.Y., 1982-83; dept. head chem. engring. U. Tulsa, 1988-89; vis. scientist Chalk River (Ont.) Nat. Lab., 1982, Inst. Thermal and Fluid Dynamics, Ruhr U., Germany, 1986. Author: Molecular-Dynamics Simulation, 1992; editor: Molecular-Based Study of Fluids, 1983; editor Jour. of Molecular Simulation, 1986-90; contbr. articles to Jour. of Chem. Physics, Molecular Physics, Fluid Phase Equilibria and others. Lt. USNR, 1968-72. Recipient Presdl. Young Investigator award NSF, 1984-89. Mem. Am. Phys. Soc., Am. Chem. Soc., History of Sci. Soc., Sigma Xi. Office: Clemson U Chem Engring Dept Clemson SC 29634

HAILE GIORGIS, WORKNEH, civil engineer; b. Gedamgue, Tegulete, Ethiopia, Mar. 16, 1930; parents: Workneh Aschenaki and Atsede Syoum; m. Jember Teferra, Nov. 16, 1969; children: Workneh, Teferra, Memmenasha, Lelo. BCE, Carnegie Inst. Tech., Pittsburgh, Pa., 1952; MCE, Carnegie Inst. Tech., 1954, PhD, 1956. Design engr. Tippetts, Abbett, McCarthy & Stratton, N.Y.C., 1956-58; engr. constrn. supervision War Reparation Commn., Addis Ababa, Ethiopia, 1958; dir. water resource devel. Ministry Pub. Works and Communications, Addis Ababa, 1959-60, vice minister, 1960, vice minister public works, 1960-62, gen mgr. Ethiopian Hwys. Authority, 1962-65, minister pub. works, 1965-69; Lord Mayor Addis Ababa, Addis Ababa, 1969-74; polit. prisoner Marxist Mil. Govt., 1974-82; chmn. constrn. housing and urban devel. research council Ethiopian Sci. and Tech. Commn., Addis Ababa, 1982-85; cons. road and road transport UN Econ. Commn. for Africa, Addis Ababa, 1985-86; pvt. cons. civil engring. Addis Ababa, 1987-90; advisor Brit. Water Aid Br., Ethiopia, 1990—. Named Man of the Year Internat. Road Fedn., 1970. Mem. ASCE. Mem. Ethiopian Orthodox Ch. Avocation: swimming. Home: PO Box 1296, Addis Ababa Ethiopia

HAILMAN, JACK PARKER, zoology educator; b. St. Louis, May 6, 1936; s. David E. and Katharine Lillard (Butts) H.; m. Elizabeth Bailey Davis, Aug. 26, 1958; children—Karl Andrew, Peter Eric. A.B., Harvard U., 1958; Ph.D., Duke U., 1964. NIH postdoctoral fellow U. Tubingen, Germany, 1964, Rutgers U., 1964-66; asst. prof. zoology U. Md., 1966-69; assoc. prof. U. Wis., Madison, 1969-72; prof. U. Wis., 1972—; hon. research assoc. Smithsonian Instn., Washington, 1966-69. Author: Ontogeny of an Instinct, 1967, Optical Signals, 1977; co-author: Introduction to Animal Behavior, 1967; co-editor: Fascinating World of Animals, 1971. Bd. dirs. County chpt. ACLU, 1968. Served with USN, 1958-61. James P. Duke fellow, 1961; NIH fellow, 1962, 64; NIH research grantee, 1966-69; NSF research grantee, 1970-85; Fulbright research scholar U. Trondheim, Norway, 1987. Fellow AAAS, Am. Ornithologists Union, Animal Behavior Soc. (pres. 1981-82), Norwegian Acad. Scis. Office: U Wis Dept Zoology Birge Hall 430 Lincoln Dr Madison WI 53706

HAILS, ROBERT EMMET, aerospace consultant, business executive, former air force officer; b. Miami, Fla., Jan. 20, 1923; s. Daniel Troy and Jean (Burke) H.; m. Ethel Fitzgerald Gayle, Mar. 2, 1957; children: Robert Emmet Jr., Merrily G., Florence T. Hails Patton, Laura Hails Smith. BS in Aero. Engring., Auburn U., 1947; MS in Indsl. Engring., Columbia U., 1950; postgrad., C&CS Air U., 1955; postgrad. AMP, Harvard U. Sch. Bus., 1965. Enlisted USAAF, 1942, commd. 2d lt., 1944, advanced through grades to lt. gen., 1974, combat pilot Pacific Theater, 1944-45; assigned to SAC, 1947-48; inspector gen. Hdqrs. USAF, 1950-53; program devel. officer Marcel Dassault Mystere IV Jet Aircraft, French Air Force Am. embassy, Paris, 1953-55; air staff project officer F-104/F-105 aircraft HQ USAF, 1956-60; comdr. procurement dist. USAF, San Francisco, 1960-62; mil. asst. for weapons systems acquisition Office Sec. AF, 1962-66; systems program mgr. A-7D aircraft Air Force Systems Command, 1966-68; dep. chief staff maintenance engring. Air Force Logistics Command, 1968-71; comdr. Def. Pers. Support Ctr. Def. Log. Agy., Phila., 1971-72; comdr. Air Logistics Ctr. USAF, Warner Robins AFB, Ga., 1972-74; vice comdr. Tactical Air Command Langley AFB, Va., 1974-75; dep. chief staff systems and logistics Hdqrs. USAF, Washington, 1975-77; ret. USAF, Washington, 1977; mgmt. cons. Atlanta, 1978-80; sr. v.p. internat. ops. LTV Corp., Dallas, 1980-84; pres. Hails Assocs. Inc., Macon, Ga., 1984—. Regional exec. Boy Scouts Am.; mem. Auburn U. Alumni Engring. Coun.; bd. advisors Wesleyan Coll. Decorated DSM with 2 oak leaf clusters, legion of Merit with 2 oak leaf clusters, Air medal with 2 oak leaf clusters; Order of Nat. Security (Korea). Mem. AIAA, Air Force Assn., Daedalians, Auburn U. SPADES, Army-Navy Country Club (Arlington, Va.), Idle Hour Golf and Country Club, Omicron Delta Kappa, Sigma Alpha Epsion. Roman Catholic. Office: PO Box 5290 Macon GA 31208-5290

HAILSTONE, RICHARD KENNETH, chemist, educator; b. Aurora, Ill., Dec. 17, 1946; s. Kenneth William and Margaret (Jennings) H.; m. Woneta J. Reed, Aug. 17, 1964; children: Brenda Annette, Stephen Richard. BS in Chemistry, No. Ill. U., 1970; MS in Phys. Chemistry, Ind. U., 1972. Scientist Eastman Kodak Rsch. Labs., Rochester, N.Y., 1972-81, sr. scientist, 1982-90; assoc. prof. Rochester Inst. Tech., 1990—; scientist Kodak Ltd. Rsch. Labs, London, 1981-82. Contbr. articles to Jour. Imaging Sci. NDEA Title IV fellow, 1970-72. Mem. AAAS, Am. Phys. Soc., Am. Chem. Soc., Soc. for Imaging Sci. and Tech. Achievements include research in image recording and detection in silver halide photographic materials, computer simulation techniques to eludicate important mechanisms in image recording and detection. Office: Ctr for Imaging Sci Rochester Inst Tech Rochester NY 14623-0887

HAIMO, DEBORAH TEPPER, mathematics educator; b. Odessa, Ukraine, July 1, 1921; d. Joseph Meir and Esther (Vodovoz) Tepper; m. Franklin Tepper Haimo, Feb. 27, 1944 (dec. June 3, 1982); children: Zara Tepper, Ethan Tepper, Leah Tepper, Nina Tepper, Varda Tepper. AB magna cum laude, Radcliffe Coll., 1943, AM, 1943; PhD, Harvard, 1964; DSc (hon.), Franklin and Marshall Coll., 1991. Acting head math. dept. Lake Erie Coll., 1943-44; instr. Northeastern U., 1944-45; lectr. Washington U., St. Louis, 1948-61; lectr., asst. prof., asso. prof. So. Ill. U., Edwardsville, 1961-68; prof. math. U. Mo., St. Louis, 1968—; chmn. dept. math. U. Mo., 1969-76; mem. Inst. Advanced Study, 1972-73; mem. coll. level exam. program Ednl. Testing Service, 1976-85; mem. team for feasibility study grad. programs Seoul Nat. U., Korea, 1974; mem. coll. bd. devel. com. in advanced placement calculus, U. Mo., 1986-89; mem. math. nat. adv. com. Common Core of Learning, Conn. Dept. Edn. Editor: Orthogonal Expansions and their Continous Analogues, 1968; mem. editorial bd. Am. Math. Monthly, 1978-81, Soc. Indsl. and Applied Math. Jour. on Math. Analysis, 1971—; contbr. articles to profl. jours. Trustee Radcliffe Coll., 1975-81; mem. bd. overseers Harvard U., 1990—. Recipient Radcliffe Alumnae Recognition award, 1993; grantee: NASA rsch., 1966-69, NSF, 1969-71, 88-91, USAF sci. rsch., 1971-74, 88-91; named NSF sci. faculty fellow, 1964-65. Mem. AAAS, AAUP, Am. Math. Soc., Math. Assn. Am. (bd. govs.-at-large 1974-77, 1st v.p. 1986-88, pres.-elect 1990, pres. 1991-93), Coun. Sci. Soc. Pres., Coordinating Bd. Math. Scis. (exec. com. at-large), Soc. Indsl. and Applied Math., Assn. for Women in Math., Assn. Mems. Inst. Advanced Study (trustee), Phi Beta Kappa, Sigma Xi, Phi Kappa Phi. Home: 7201 Cornell Ave Saint Louis MO 63130-3025

HAINES, ANDREW LYON, engineering company executive; b. Potsdam, N.Y., Oct. 29, 1942; s. Howard B. and Grace S. Haines; m. Elizabeth Elmblade, June 21, 1969; children: Karen, Heather. AB, Princeton U., 1964; MA in Teaching, Harvard U., 1965; MS, Stanford U., 1968; DSc, George Washington U., 1973; MS, Am. U., 1985. Mem. tech. staff Mitre Corp., McLean, Va., 1969-76, group leader, 1976-82, assoc. dept. head, 1983-87, dept. head, 1987—; adj. prof. George Washington U., Washington, 1973-78. Contbr. articles to profl. jours. 1st lt. U.S. Army, 1965-67. Mem. AIAA, Air Traffic Control Assn., Ops. Rsch. Soc. Am. Home: 9226 Vernon Dr Great Falls VA 22066

HAINES, MICHAEL JAMES, asphalt company official; b. Elmer, N.J., Apr. 20, 1957; s. James William Haines and Edna Mae (Thysens) Edmonds; m. Sandra Jean Rozell, May 21, 1977; children: Michael John, Matthew James. Grad. high sch., Mullica Hill, N.J. Chem. operator Major Chem. and Latex Co., Inc., Boston, 1976-78, Master Latex Co., Boston, 1978-79, Marsons Fastener Corp., Chelsea, Mass., 1979-81, Amicon Corp., Woburn, Mass., 1981-83; chem. technician Avco Spl. Materials, Lowell, Mass., 1983-84; mgr. quality control Upaco Adhesives Inc., Richmond, Va., 1984-86; plant supt. Asphalt Emulsion Inc., Richmond, 1986—. Republican. Roman Catholic. Achievements include development of high-strenght, lightweight composite materials for aircraft components, polymer modified asphalt emulsion for roadway overlay microseal, assisting in development of ablative panels for space shuttle elevons. Office: Asphalt Emulsion Inc 1524 Valley Rd Richmond VA 23222

HAINES, MICHAEL ROBERT, economist, educator; b. Chgo., Nov. 19, 1944; s. James Joshua and Anne Marie (Welch) H.; m. Patricia Caroline Foster, Aug. 19, 1967 (div. Dec. 1986); children: James, Margaret. BA, Amherst Coll., 1967; MA, U. Pa., 1968, PhD, 1971. Asst. prof. econs. Cornell U., Ithaca, N.Y., 1972-79; vis. lectr. econs. U. Pa., Phila., 1979, rsch. assoc. prof. Sch. Pub. and Urban Policy, 1979-80; assoc. prof. econs. Wayne State U., Detroit, 1980-86, prof., 1986-90; Banfi Vintners Disting. prof. econs. Colgate U., Hamilton, N.Y., 1990—; cons. NIH, Bethesda, Md., 1980-84, 90, 91, 93. The World Bank, Washington, 1983; rsch. affiliate Population Studies Ctr. U. Mich., 1990—; rsch. assoc. Nat. Bur. of Econ. Rsch., 1987—. Author: Economic-Demographic Interrels. in Developing Agrl. Regions, 1977, Fertility and Occupation, 1979, Fatal Years, 1991; contbr. articles to profl. jours. NIH grantee, 1974-77, 78-82, 89—. Mem. Internat. Union for Sci. Study Population, Econ. History Assn. (bd. editors 1987-91), Social Sci. History Assn. (bd. dirs. 1983-85, treas. 1985—), Am. Econ. Assn., The Cliometrics Soc. (bd. editor 1988—), Population Assn. Am., Am. Statis. Assn. Episcopalian. Avocations: numismatics, wine, book collecting. Office: Colgate U Dept Econs 13 Oak Dr Hamilton NY 13346-1338

HAINLINE, ADRIAN, JR., biochemist; b. Blandinsville, Ill., Mar. 16, 1921; s. Adrian and Blanche Elizabeth (Anderson) H.; m. Velma Jeanette Taylor, Dec. 21, 1942; children: Bryan E., Mark A., David C., Joy E., Carol S. BEd, Western Ill. U., 1942; MS, Denver U., 1948; PhD, U. Mch., 1952. Chemist E.I. DuPont de Nemours, Wilmington, Del., 1942-45; clin. chemist Cleve. Clinic, 1951-64, St. Luke's Hosp., Kansas City, Mo., 1965-67; clin. chemist stds. sect. Internat. Cholesterol Standardization, Coronary Drug Lab., 1967-87; chief clin. chemistry CDC, Atlanta, 1967-87; cons. FDA Clin. Chemistry and Hem. Devices Panel, Rockville, 1974-78; chmn. NHLB(NIH) Lipid Rsch. Clinics quality control com., Bethesda, Md., 1980-87; mem. CDC Licensure Rev. Bd., Atlanta, 1972-85; adv. bd. Chem. Rubber Handbook, Clin. Lab. Data, Cleve., 1964-67. Contbr. articles to profl. jours.; editor: Lipid Research Clinics Lab Method Manual, 1982. With Chem. Corp., U.S. Army, 1945-46. Recipient Superior Performance awards USPHS, 1977, 79, 85,. 87, Disting. Alumni award Western Ill. U., 1977, Merit Svc. award S.E. sect. Am. Assn. Clin. Chemistry, 1991. Mem. Am. Assn. Clin. Chemistry (chair nat. meeting 1959), Am. Bd. Clin. Chemistry (dir. 1972-78), Am. Chem. Soc. Presbyterian. Achievements include 1 patent. Home: 6 Westgate Cir Jacksonville IL 62650

HAIR, JAY DEE, association executive; b. Miami, Fla., Nov. 30, 1945; s. Wilbur B. and Ruth A. Johnson; m. Rebecca McDaniel, May 17, 1970; children: Whitney, Lindsay. B.S., Clemson U., 1967, M.S., 1969; Ph.D., U. Alta., 1975; postgrad., Govt. Execs. Inst., Sch. Bus. Adminstrn., U. N.C., 1980. Asst. prof. wildlife biology Clemson (S.C.) U., 1968-69; assoc. prof. zoology/forestry, adminstr. fisheries and wildlife scis. N.C. State U., Raleigh, 1977-81, adj. prof. zoology and forestry, 1982—; pres. Nat. Wildlife Fedn., Washington, 1981—; spl. asst. to asst. sec. for fish, wildlife and parks Dept. Interior, 1978-80; mem. Nat. Petroleum Coun., Dept. Energy, 1981-84; rep. Natural Resources Coun. Am., 1981; bd. dirs. Sol Feinstone Environ. Awards Program, 1982—, Rails-to-Trails, 1986—, Global Tomorrow Coalition, 1987—, Riggs Nat. Bank, 1989—; mem. Globescope Adv. Coun., 1984—, Nat. Pub. Lands Adv. Coun., 1984-86; mem. Nat. Wetland Policy Forum, 1987-88; vice chmn. Nat. Groundwater Forum, 1984-86; Nat. Nonpoint Source Inst., 1985—, Keystone Ctr., 1985—, Clean Sites, 1986—; mem. biotech. sci. adv. com. EPA, 1987-89; mem. Acad. Mgmt., 1988—; mem. nat. adv. bd. Sci. Journalism Ctr., U. Mo.-Columbia, 1987-88, bd. cons., 1987-88 ; bd. govs. Nat. Safety Coun. Environ. Health and safety Inst., 1986—; U.S. bd. dirs. Internat. Union for Conservation of Nature and Natural Resources, 1989—. Editor: Ecological Perspectives of Wildlife Management, 1977; contbr. articles to profl. jours. Bd. govs. Nat. Shooting Sports Found., 1981—; mem. conservation com. Boy Scouts Am., 1981; pres. N.C. Cued Speech Assn., 1978-81; chmn. bd. dirs. Cued Speech Ctr., Inc., 1979-81; bd. dirs. Parents and Profls. for Handicapped Children, N.C., 1979-81 ; mem. Wake County Hearing-Impaired Assn., (N.C.), 1978—; pres. Alexander Graham Bell Nat. Assn. Deaf, 1979. Served to 1st lt. U.S. Army, 1970-71. Named S.C. Wildlife Conservationist of Yr. Gov.'s Annual Conservation Awards Program, 1977, N.C. Conservationist of Yr., 1980; recipient N.C. Gov.'s award for pub. svc.; Alumni Disting. Svc. award Clemson U., 1986, Centennial Disting. Alumni award, 1989; 4-H Alumni award Purdue U., 1988; Edward J. Cleary award Nat. Acad. Environ. Engrs., 1989. Mem. Wildlife Soc. (Disting. Service award Southeastern sect. 1980), Soc. Am. Foresters, AAAS, Internat. Assn. Fish and Wildlife Agys., Ducks Unltd., Wildlife Soc., N.C. Acad. Scis., N.C. Wildlife Fedn., S.C. Wildlife Fedn. (F. Bartow Culp disting. svc. to conservation award 1978, Disting. Svc. award Southeast sect. 1980, Nat. Wildlife Fedn. Outstanding Affiliate award 1978), Am. Fisheries Soc., Am. Forestry Assn., Scabbard and Blade, Tiger Brotherhood, Blue Key, Phi Kappa Delta. Methodist. Office: Nat Wildlife Fedn 1400 16th St NW Washington DC 20036-2217

HAIRSTON, NELSON GEORGE, animal ecologist; b. Davie County, N.C., Oct. 16, 1917; s. Peter Wilson and Margaret Elmer (George) H.; m. Martha Turner Patton, Aug. 19, 1942; children: Martha Patton, Nelson George, Margaret Elmer. A.B., U. N.C., 1937, M.A., 1939; Ph.D., Northwestern U., 1948. Instr. U. Mich., 1948-52, asst. prof., 1952-57, asso. prof., 1957-61, prof., 1961-75; dir. Mus. Zoology, 1967-75; William R. Kenan, Jr. prof. biology and ecology U. N.C., Chapel Hill, 1975-88; adviser, cons.; mem. expert adv. com. on parasitic diseases WHO, Philippines, Iraq, Switzerland, Egypt, Sudan, Kenya, Tanzania, Rhodesia, South Africa, Ghana, Western Samoa, 1964-81; mem. tropical medicine and parasitology study sect. NIH, Bethesda, Md., 1965-70; chmn. adv. com. on biol. and med. scis. NSF, Washington, 1972. Contbr. articles on ecology to profl. jours.; (2) books on ecology. Served with AUS, 1941-46. NSF rsch. grantee, 1960-68. Mem. AAAS, Am. Soc. Naturalists, Ecol. Soc. Am. (Eminent Ecologist award 1991), Soc. for Study Evolution, Brit. Ecol. Soc., Am. Soc. Ichthyologists and Herpetologists, Sigma Xi. Home: 2-206 Carolina Meadows Chapel Hill NC 27514 Office: U NC Dept Biology Coker Hall Chapel Hill NC 27599-3280

HAJARE, RAJU PADMAKAR, chemical engineer; b. Bombay, June 2, 1960; came to U.S., 1981; s. Padmakar Bhavanishankar and Vidya Padmakar (Soste) H.; m. Lucy Ann Passafiume, July 1, 1989. BS, U. Bombay, 1981; MS, U. Louisville, 1983, PhD, 1985; MBA, La. State U., 1991. Teaching asst. U. Louisville, 1981-86, postdoctoral fellow, 1986-87; plant engr. Exxon Chem. Ams., Baton Rouge, 1988-90, sr. engr., 1990-91, supr. process control sect., 1991—; presenter at profl. confs. Contbr. articles to rpfol. pbuls. Cons. applied econs. program Jr. Achievement, Baton Rouge, 1989; amb. Sci. Edn. in High Sch., Baton Rouge, 1992. Mem. AICE, Instrument Soc. Am. Hindu. Home: 406 W Woodgate Ct Baton Rouge LA 70808 Office: Exxon Chem Ams PO Box 241 Baton Rouge LA 70821-0241

HAJEK, ANN ELIZABETH, insect pathologist; b. San Francisco, Apr. 27, 1952; d. Ernest Emil and Dorothy Fern (Moller) H.; m. James K. Liebherr, July 1, 1984; 1 child, Lisa. MS, U. Calif., Berkeley, 1980, PhD, 1984. Rsch. affiliate, rsch. entomologist USDA Agrl. Rsch. Svc., Ithaca, N.Y., 1985-90; rsch. assoc., sr. rsch. assoc. Boyce Thompson Inst., Ithaca, N.Y., 1990—; vis. fellow Cornell U., Ithaca, N.Y., 1984—. Contbr. articles to profl. jours. Grantee USDA Competitive grants CSRS, 1987, 90, 93; recipient Cert. of Merit USDA, ARS, Ithaca, N.Y., 1988. Mem. Soc. Invert Pathology (sec., treas. microbial control div. 1990—), Entomological Soc. Am., N.Y. Entomological Soc., Sigma Xi. Democrat. Achievements include research in the epizootiology of a newly discovered fungal pathogen of gypsy moth in N. Am. Home: 205 Salem Dr Ithaca NY 14850 Office: Boyce Thompson Inst Tower Rd Ithaca NY 14853-1801

HAJEK, THOMAS J., aerospace engineer. BS in Aero. and Mech. Engring., Ill. Inst. Tech., 1975, MS, 1980. Rschr. compressor aerodynamics Borg Warner Rsch. Ctr., 1977-81; from turbine aerodynamics engr. to dept. head turbines CAE, 1981-85; mgr. turbine tech. comml. engine bus. unit Pratt & Whitney, Hartford, Conn., 1985-89, area dir. internat. mktg., program mgr.

various NASA sponsored projects, 1989—. Recipient Gas Turbine award ASME, 1991. Achievements include research in the effects of rotation on heat transfer in rotating coolant passages, structural analysis, thermal mechanical fatigue, stress analysis, ceramic coatings, experimental heat transfer studies, advanced cooling concepts and cooling systems design. Office: Comm Engine Business Pratt & Whitney Hartford CT 06108*

HAJIAN, GERALD, biostatistician, engineer; b. Newark, Jan. 15, 1940; s. Zakar and Rose (Bakalian) H.; m. Christina Langadinos, June 5, 1966 (dec. July 1983); 1 child, Eleanore Joyce; m. Patricia Jane Foster, Jan. 18, 1986. BSEE, Newark Coll. Engring., 1961; MS, Rutgers U., 1965; PhD, Columbia U., 1972. Assoc. programmer IBM Corp., Poughkeepsie, N.Y., 1961-63; asst. prof. math. King's Coll., Wilkes-Barre, Pa., 1971-73; biostatis. Am. Cyanamid, Princeton, N.J., 1973-76; sr. biostatistician Burroughs Wellcome Co., Research Triangle Park, N.C., 1977—. Mem. Am. Statis. Assn., Soc. of Toxicology, Biometric Soc., Soc. for Risk Mgrs., Sigma Xi. Office: Burroughs Wellcome Co Research Triangle Park NC 27709

HAJJAR, DAVID P., biochemist, educator. BA in Biochemistry, Am. Internat. Coll., 1974; MS, U. N.H., 1977, PhD, 1978; postdoctoral, Cornell U., 1981. Rsch. fellow in biochemistry U. N.H., Durham, 1975-78; rsch. asst. in pathology med. coll. Cornell U., Ithaca, N.Y., 1978-81, asst. prof. biochemistry and pathology med. coll., 1981-84, assoc. prof. med. coll., 1984-89, prof. med. coll., 1989—; sr. investigator N.Y. Heart Assn., 1981-84; mem. pathology A study sect. NIH, 1991—; mem. review com. coun. on arteriosclerosis Am. Heart Assn., 1992—. Contbr. over 85 articles to profl. and sci. jours. Nat. Merit scholar, 1969; Predoctoral fellow N.H. chtp. Am. Heart Assn., 1976-78; PHS postdoctoral rsch. fellow NIH, 1978-81, New Investigator Rsch. award, 1981-84; recipient Tchr.-Scientist award Andrew Mellon Found., 1981-83, FASEB (AAP) Warner-Lambert/Parke Davis award, 1991. Office: Cornell U-Dept of Biochemistry Univ Medical College 1300 York Ave New York NY 10021*

HAJJAR, KATHERINE AMBERSON, physician, pediatrician; b. Rochester, N.Y., Oct. 29, 1952; d. James Burns Amberson and Shirley Elizabeth (Huber) Kuntz; m. David Phillip Hajjar, May 26, 1984; children: Esther Katherine, Amanda Elizabeth. AB, Smith Coll., 1974; MD, Johns Hopkins U., 1978. Diplomate Am. Bd. Pediatrics, Am. Bd. Pediatric Hematology-Oncology. Resident pediatrics Children's Hosp. Pitts., 1978-81, chief resident pediatrics, 1981-82; fellow pediatric hematology-oncology Johns Hopkins U., Balt., 1982-84; asst. prof. Cornell U. Med. Coll., N.Y.C., 1984-89, assoc. prof., 1989—; chief Pediatric Hematology-Oncology Cornell U. Med. Ctr., 1992—; mem. thrombosis study sect. Am. Heart Assn., Dallas, 1990-92; mem. rev. com. NIH, Bethesda, Md., 1992—. Contbr. articles to profl. jours. Recipient Rsch. Career Devel. award NIH, 1989, Irvine S. Page award Am. Heart Assn., 1991; named Established Investigator Am. Heart Assn., 1989-94; scholar Syntex Corp., 1989-92. Fellow Am. Acad. Pediatrics; mem. Soc. for Pediatric Rsch., Am. Soc. Hematology, Am. Soc. Biochemistry and Molecular Biology. Achievements include research on endothelial cell biology, thrombosis and fibrinolysis. Office: Cornell U Med Coll S-600 1300 York Ave New York NY 10021

HAJOS, ZOLTAN GEORGE, chemist; b. Budapest, Mar. 3, 1926; came to U.S., 1957; s. Imre Henrik and Elizabeth Maria (Teichner) H.; m. Irene Edith Pal, Feb. 19, 1955. MS, Tech. U., Budapest, 1947; DSc, U. Budapest, Budapest, 1949. Asst. prof. Tech. U., 1948-57; rsch. assoc. Princeton (N.J.) U., 1957-60; rsch. fellow Hoffmann-La Roche, Inc., Nutley, N.J., 1960-70; rsch. assoc. U. Vt., Burlington, 1972-73, U. Toronto, Ont., Can., 1973-74; prin. scientist R.W. Johnson Pharm. Rsch. Inst., Raritan, N.J., 1975-90. Contbg. author: Carbon-Carbon Bond Formation, 1977; contbr. articles to profl. jours. Mem. Am. Chem. Soc., Sigma Xi. Roman Catholic. Achievements include patents in field of total and asymmetric synthesis of medicinal-organic compounds; glycosides, hydrophenanthrenes, steroidal hormones, heterocyclics, (i.e. furanes, dioxanes and purines); asymmetric synthesis of chiral synthons. Home: Pauler u 2 II 21, H-1013 Budapest Hungary

HAKIM-ELAHI, ENAYAT, obstetrician/gynecologist, educator; b. Teheran, Iran, Nov. 23, 1934; came to U.S., 1959, naturalized, 1973; s. Mohamed-Ali and Masoomeh Rahimi; M.D., Med. Sch., Teheran, 1959; lic. physician, Maine, Conn., Vt., N.Y., N.H., Calif.; diplomate Am. Bd. Ob-Gyn. m. Renate Emsters, Nov. 15, 1967; 1 child, Cristina. Intern, Queens Hosp. Ctr., N.Y.C., 1960, resident in internal medicine, 1961, resident in ob-gyn, 1961-64, resident in radiotherapy of gynecologic cancer, Am. Cancer Soc. fellow Queens div., 1965; resident in gynecology Cancer Rsch. Inst., Columbia-Presbyn. Med. Ctr., N.Y.C., 1964-65; practice medicine specializing in ob-gyn, N.Y.C., 1968—; mem. staff N.Y. Hosp., N.Y.C.; med. dir. Margaret Sanger Ctr., N.Y.C., 1973—, Planned Parenthood of N.Y.C., 1977—, LaGuardia Hosp., Forest Mills, 1993—; asst. prof. ob-gyn Cornell U. Med. Coll., N.Y.C., 1973—; dir. dept. of ob-gyn LaGuardia Hosp., 1990—, med. dir., 1993—. Served with U.S. Army, 1965-67. Fellow ACS, Am. Coll. Obstetricians and Gynecologists, Internat. Coll. Surgeons, Am. Fertility Soc.; mem. Am. Soc. Gynecol. Laparoscopists, Am. Soc. Colposcopy and Cervical Neoplasia, Am. Pub. Health Assn., Am. Coll. Physician Execs., Assn. of Reproductive Health Profls., Royal Soc. Medicine (London), World Med. Assn., N.Y. State Med. Soc., Queens Gynecol. Soc. Contbr. articles to profl. jours.

HAKKILA, EERO ARNOLD, nuclear safeguards technology chemist; b. Canterbury, Conn., Aug. 4, 1931; s. Jack and Ida Maria (Lillquist) H.; m. Margaret W. Hakkila; children: Jon Eric, Mark Douglas, Gregg Arnold. BS in Chemistry, Cen. Conn. State U., 1953; PhD in Analytical Chemistry, Ohio State U., 1957. Staff mem. Los Alamos (N.Mex.) Nat. Lab., 1957-78, assoc. group leader safeguard systems, 1978-80, dep. group leader, 1980-82, group leader, 1982-83, project mgr. internat. safeguards, 1983-87, program coord., 1987—. Editor: Nuclear Safeguards Analysis, 1978; contbr. numerous articles to profl. jours. Fellow Am. Inst. Chemists; mem. N.Mex. Inst. Chemists (pres. 1971-73), Am. Chem. Soc., Am. Nuclear Soc. (exec. com. fuel cycle and waste mgmt. div. 1984-86), Inst. Nuclear Materials Mgmt. Avocations: skiing, fishing, rockhounding. Office: Los Alamos Nat Lab PO Box 1663 Los Alamos NM 87545-0001

HAKKINEN, RAIMO JAAKKO, aeronautical engineer, scientist; b. Helsinki, Feb. 26, 1926; came to U.S., 1949, naturalized, 1960; s. Jalmari and Lyyli (Mattila) H.; m. Pirkko Loyttyniemi, July 16, 1949; children—Bert, Mark. Diploma in aero. engring., Helsinki U. Tech., 1948; M.S., Calif. Inst. Tech., 1950, Ph.D. cum laude, 1954. Head tech. office Finnish Aero. Assn., Helsinki, 1948; instr. engring. Tampere Tech. Coll., Finland, 1949; design engr., aircraft div. Valmet Corp., Tampere, 1949; research asst. Calif. Inst. Tech., 1950-53; mem. research staff MIT, 1953-56; aerodynamics engr. Western div. McDonnell Douglas Astronautics Co., Santa Monica, Calif., 1956-64, chief scientist phys. scis. dept., 1964-70; chief scientist flight scis. McDonnell Douglas Research Labs., St. Louis, 1970-82, dir. research, flight scis., 1982-90; prof. mech. engring., dir. fluid mechanics lab. Washington U., St. Louis, 1991—; lectr. engring. UCLA, 1957-59; vis. assoc. prof. aeros. and astronautics MIT, 1963-64; cons., 1990—. Contbr. articles to profl. jours. Served with Finnish Air Force, 1944. Fellow AIAA (mem. fluid dynamics com. 1969-71, honors and awards com. 1975-83, tech. activities com. 1975-78, dir. at large 1977-70); mem. Am. Phys. Soc., Engring. Soc. in Finland, Calif. Inst. Tech. Alumni Assn., Sigma Xi. Home: 5 Old Colony Ln Saint Louis MO 63131-1509 Office: Washington U Campus Box 1185 1 Brookings Dr Saint Louis MO 63130

HAKOLA, HANNU PANU AUKUSTI, psychiatry educator; b. Lapua, Finland, Feb. 22, 1932; s. Aukusti Jalmari and Toini Kyllikki (Tikkanen) H.; m. Maija-Leena Salo, Apr. 19, 1954; children: Jouni, Marja, Jorma, Jaakko. MD, U. Turku, Finland, 1956, MA, 1960, MD, PhD, 1972. Diploma in health adminstrn. Nat. Bd. Health, Finland, 1979. Asst. physician Neuropsychiat. Clinic, Turku, 1956-60; chief psychiatrist Harjamäki Hosp., Siilinjärvi, Finland, 1960-69, Niuvanniemi Hosp., Kuopio, Finland, 1969-83; prof. forensic psychiatry U. Kuopio, Finland, 1983—; med. dir. Harjamäki Hosp., Siilinjärvi, 1960-69, Niuvanniemi Hosp., Kuopio, 1969—. Author: On Environmental Conditions of Criminal Psychopaths, 1959, Clinical Aspect of a New Hereditary Disease, 1972, Polycystic Lipomembranous Osteodysplasia with Sclerosing Leukoencephalopathy, 1990; editor: Symposium on Forensic Psychiatry, 1988; inventor Carbamazepine in violent schizophrenics, 1982, Duraljan superfamily, 1989; editorial bd. Med. Jour. Duodecim, 1975-81; contbr. over 200 articles to profl. jours. and chpts. to books. Bd. dirs. Kuopio U. Ctrl. Hosp., 1985-89, 93—. Decorated knight Finnish Order of White Rose; comdr. Finnish Order of Lion; Paulo Found. grantee, Helsinki, Finland, 1971, Aaltonen Found. grantee, Tampere, Finland, 1973; recipient Prize, Acta Psychiat. Scandinavia, 1972. Mem. Finnish Med. Assn. (del. com. 1964—, exec. bd. 1980-84), Med. Assn. North-Savo (hon.), Rotary (pres. Puijo club 1974-75, Paul Harris fellow 1982). Lutheran. Avocations: music, hunting, Joutenlahti farm. Home: Satamakatu 3 D 49, SF 70100 Kuopio Finland Office: Niuvanniemi Hosp, SF 70240 Kuopio Finland

HAKOLA, JOHN WAYNE, mechanical engineer, consultant, small business owner; b. Bklyn., Feb. 18, 1932; s. Wayne E. and Esther (Lorentzen) H.; B.A.E., Poly. Inst. Bklyn., 1961, M.S., 1965; postgrad. bus. adminstrn. Adelphi U., 1977-78; m. Patricia Anne Torrington, Mar. 1, 1956 (div. Sept. 1981); children—John W., Wayne Edward, Karen Elizabeth; m. Monica Bragg Spinelli, Oct. 9, 1982. Registered profl. engr., N.Y.; cert. memorialist, 1986. Engr., sr. engr. Sperry Gyroscope Corp., Lake Success, N.Y., 1956-69; cons. engr., Bklyn., 1968—; pres., owner Bklyn. Monument Co. Inc., J.R. Pitbladdo Inc., Bklyn., 1969—; asst. prof. engring. Hofstra U., Hempstead, N.Y., 1981-86, dir. engring. graphics, 1986-91, dir. freshman engring., 1991—. Mem. Nat., N.Y. State (past v.p. Suffolk County chpt.) socs. profl. engrs., AIAA, IEEE, ASEE, Am. Def. Preparedness Assn., Monument Builders N.Am. (v.p.), N.Y. Monument Builders Assn. (past pres.), Asso. Granite Craftsmen's Guild N.Y. (past pres.), Newcomen Soc. N.Am. Office: Brooklyn Monument Co Inc 242 25th St Brooklyn NY 11232-1397

HALBERG, CHARLES JOHN AUGUST, JR., mathematics educator; b. Pasadena, Calif., Sept. 24, 1921; s. Charles John August and Anne Louise (Hansen) H.; m. Ariel Arfon Oliver, Nov. 1, 1941 (div. July 1969); children—Ariel (Mrs. William Walters), Charles Thomas, Niels Frederick; m. Barbro Linnea Samuelsson, Aug. 18, 1970 (dec. Jan. 1978); 1 stepchild, Ulf Erik Hjelm; m. Betty Reese Zimprich, July 27, 1985. B.A. summa cum laude, Pomona Coll., 1949; M.A. (William Lincoln Honnold fellow), UCLA, 1953, Ph.D., 1955. Instr. math. Pomona Coll., Claremont, Calif., 1949-50; assoc. math. UCLA, 1954-55; instr. math. U. Calif.-Riverside, 1955-56, asst. prof. math., 1956-61, assoc. prof. math., 1961-68, prof. math., 1968—, vice chancellor student affairs, 1964-65; dir. Scandinavian Study Center at Lund (Sweden) U., 1976-78; docent U. Goteborg, Sweden, 1969-70; bd. dirs. Fulbright Commn. for Ednl. Exchange between U.S. and Sweden, 1976-79. Author: (with John F. Devlin) Elementary Functions, 1967, (with Angus E. Taylor) Calculus with Analytic Geometry, 1969. Served with USAAF, 1945-46. NSF fellow U. Copenhagen, 1961-62. Mem. Math. Assn. Am. (chmn. So. Calif. sect. 1964-65, gov. 1968), Am. Math. Soc., Swedish Math. Soc., Sigma Xi, Phi Beta Kappa. Home: PO Box 2724 Carlsbad CA 92018-2724

HALBERSTAM, ISIDORE MEIR, meteorologist; b. Bklyn., Apr. 29, 1945; s. Boruch and Hana (Spira) H.; m. Livia Steiner, June 15, 1971; children: Tobi, Batsheva. BA, Yeshiva U., 1966, MHL, 1969; MS, NYU, 1971, PhD, 1973. Rsch. asst., postdoctoral scholar Goddard Inst. of Space Studies, N.Y.C., 1969-74; prin. rsch. scientist Jet Propulsion Lab., Pasadena, Calif., 1974-80; chief scientist Hughes STX, Lexington, Mass., 1980—; Seasat scatterometer team NASA, Pasadena, 1975-79. Contbr. articles to profl. jours. Mem. AAAS, Am. Meteorol. Soc. Democrat. Jewish. Achievements include discovery of revised maximum of temperature profile above snow surface; research investigating various methods of data assimilation, relationship between surface winds and scatterometer measurements. Office: Hughes STX 109 Massachusetts Ave Lexington MA 02173

HALDANE, FREDERICK DUNCAN MICHAEL, physics educator; b. London, Sept. 14, 1951; came to U.S., 1981; BA, Cambridge U., Eng., 1973, Ph.D. in Physics, 1978. Physicist Inst. Laue-Langevin, Grenoble, France, 1977-81; asst. prof. physics U. So. Calif., L.A., 1981-85; mem. tech. staff AT&T Bell Labs., Murray Hill, N.J., 1985-87; prof. physics U. Calif., San Diego, 1987-90, Princeton (N.J.) U., 1990—; trustee Aspen (Colo.) Ctr. for Physics, 1985-90, mem. adv. bd., 1990—. Contbr. articles to profl. jours. Alfred P. Sloan Found. fellow, 1984. Fellow Am. Phys. Soc. (Oliver E. Buckley prize 1993), Am. Acad. Arts and Scis. Achievements includer research in theoretical condensed matter physics; contributions to the understanding of quatum magntism and the frachoral quantum hall effect. Office: Princeton U Dept of Physics Jadwin Hall Princeton NJ 08544*

HALDANE, GEORGE FRENCH, mechanical engineer, consultant; b. Detroit, Apr. 15, 1940; s. George Ward and Lourie (Roulen) H.; m. Wanda M. Munday, July 8, 1967; children: Michelle, Steven. BSME, Calif. State U., L.A., 1964. Registered profl. engr., Calif. Test engr. Douglas Aircraft, Long Beach, Calif., 1965-75; mech. engr. Naval Weapons Sta. Seal Beach, Corona, Calif., 1980-88, Douglas Aircraft, Long Beach, 1988-90; proprietor George Haldane Cons. Engr., La Palma, Calif., 1990—. Mem. AIAA, La Palma C. of C., North Org County Computer Club, Beta Pi. Achievements include calibration reports for transonic tunnel, expert for dimensional measurements and force measurements, designer for C-17 test fixturing for fatigue airframes. Home and Office: George Haldane Cons Engr 8031 Janeen Cir La Palma CA 90623

HALDEMAN, CHARLES WALDO, III, aeronautical engineer; b. Phila., June 9, 1936; s. Charles Waldo Jr. and Anna Freemont (Douglass) H.; m. Louise Stephenson, June 27, 1959; children: Charles Waldo, George Stephenson. BS, MIT, 1959, MS, 1959, ScD, 1964. Rsch. asst. MIT Naval Supersonic Lab., Cambridge, Mass., 1959-64; staff engr. MIT Aerophysics Lab., Cambridge, Mass., 1964-79, assoc. dir., 1979-82; staff engr. MIT Lincoln Lab., Lexington, Mass., 1982-87, sr. staff engr., 1987—; v.p. engring., cons. Megatech Corp., Billerica, Mass., 1971-82. Contbr. numerous articles and rsch. papers to scientific jours.; patentee in field including Wavy Tube Heat Pumping, Vacuum Cleaning. Mem. Am. Inst. Aero. and Astronautics. Avocations: farming, hunting. Office: MIT Lincoln Lab 244 Wood St Lexington MA 02173

HALE, PAUL NOLEN, JR., engineering administrator, educator; b. Galveston, Tex., Dec. 5, 1941; s. Paul Nolen Hale and Margaret (Wentzel) Carroll; m. Frances Anne Andrews, Jan. 26, 1968; children: Tammy Lynn, Eric Timothy. BS in Indsl. Engring., Lamar Tech., Beaumont, Tex., 1965; MS in Indsl. Engring., U. Ark., 1966; PhD in Indsl. Engring., Tex. A&M U., 1970. Registered profl. enrg., La., Tex. Asst. prof. indsl. engring. La. Poly. Inst., Ruston, 1966-67, Tex. A&M U., College Station, 1970-71; from asst. prof. to prof. indsl. engring. La. Tech. U., Ruston, 1971-84, head dept. indsl. engring., 1982-84, head dept. biomed. engring., 1984-85, 87—, dir. rehab. engring. ctr., 1984—; spl. tech. asst. Western Electric, Shreveport, La., 1966; safety cons. Continental Can Co., Hodge, La., 1972; v.p. C.H.& D. Tech. Cons., Ruston, 1967—; pres. Mgmt. Support Corp., Ruston, 1973—; keynote speaker Conf. of Assn. Driver Educators for Disabled, Ky., 1988; standards com. adaptive driving devices, Soc. Auto. Engrs., 1984—; adv. com. bioengring. div. NSF, 1989-92; program evaluator bioengring. Accreditation Bd. Engring. and Tech., 1990—. Author: (with others) Hazard Control Information Handbook, 1983, Ergonomics in Rehabilitation, 1988; editor: Rehabilitation Technical Services, 1989; co-editor: (proceedings) Technology for the Next Decade, 1989; mem. editorial bd. Assistive Tech., 1988—. Pres. Wesley Chapel Water System, Ruston, 1972-82; mem. Mayor's Com. for Disabled, Ruston, 1985-88; adminstrv. bd. Trinity Meth. Ch., Ruston, 1984-87; bd. dirs. Safety Coun. Greater Baton Rouge, 1984-87. Recipient Meritorious Svc. award Gov.'s Conf. for Handicapped, La., 1984. Mem. Soc. Biomed. Engring. (steering com. 1987—), Biomed. Engring. Soc. (mem. student affairs com. 1989-91, mem. soc. affairs com. 1991-93), RESNA (bd. dirs. 1987—), Coun. of Chairs of Biomed. Engring. (chair 1992-93). Home: RR 4 Box 103 Ruston LA 71270-9804 Office: La Tech U PO Box 3185 Ruston LA 71272

HALEVI, EMIL AMITAI, chemistry educator; b. Bklyn., May 22, 1922; s. Mordecai and Rose (Taran) H.; m. Ada Rauch, Jan. 29, 1947; children: Jonathan Gad, Dalia. AB in Chemistry with honors, U. Cin., 1943; MSc, Hebrew U., Jerusalem, 1949; PhD, Univ. Coll., London, 1952. Instr. Hebrew U., 1952-54, lectr., 1954-55; sr. lectr. Technion-Israel Inst. Tech., Haifa, 1955-60, assoc. prof. chemistry, 1960-64, prof. chemistry, 1964-91, prof. emeritus, 1991—, chmn. dept. chemistry, 1983-85, dean grad. sch., 1969-71; mem. organizing com. for several internat. confs. and sumposia, 1972, 88, 89, 90, 92; mem. Commn. on Phys. Organic Chemistry, Internat. Union, 1981—. Author: Orbital Symmetry and Mechanism - the Ocams View, 1992; co-editor spl. issue Israel Jour. Chemistry, 1980; contbr. over 60 articles to profl. publs. Fellow Royal Soc. Chemistry; mem. Am. Chem. Soc., Israel Chem. Soc., Swiss Chem Soc., N.Y. Acad. Scis. Jewish. Home: 36 Danya St, Haifa 34980, Israel Office: Technion-Israel Inst Tech, Dept Chemistry, Technion City Haifa 32000, Israel

HALEVY, ABRAHAM HAYIM, horticulturist, plant physiologist; b. Tel-Aviv, Israel, July 17, 1927; s. Naftali and Henia (Ginzburg) H.; m. Zilla Horngrad, Aug. 20, 1952 (dec. 1981); children: Avishag, Noa, Itai; m. Esther Passal, Apr. 15, 1991. MS., Hebrew U., Jerusalem, 1955, PhD, 1958. Research asso. Dept. Agr. Plant Ind. Sta., Beltsville, Md., 1958-59; lectr. in horticulture and plant physiology, Hebrew U., Rehovot, Israel, 1960-64, sr. lectr., 1964-67, asso. prof., 1967-71, prof., 1971—; head dept. ornamental horticulture, Hebrew U. of Jerusalem, 1967-83, 86—; vis. prof., Mich. State U., 1964-65, U. Calif., Davis, 1970-71, 75-76, 82-84, 86, 87, 89, 91; sci. adv. to Israeli Ministry of Edn. Served with Israel Def. Forces, 1948-50. Fellow Am. Soc. Horticultural Sci. (A. Laurie award 1960-63); mem. Internat. Soc. Horticulture (council), Am., Scandinavian, Socs. Plant Physiologists. Jewish. Editorial bd. Scientia Horticulturae, Plant Growth Regulators; editor-in-chief Israel Jour. Botany, editor Handbook of Flowering, Vols. 1-6, 1985-89, Flowering Newsletter, 1986-89; contbr. over 250 sci. papers in field to publs. Home: 1 Barazani, Ramat Aviv Gimel, Tel Aviv Israel Office: Hebrew Univ Agricult Faculty, PO Box 12, Rehovot 76 100, Israel

HALEY-OLIPHANT, ANN ELIZABETH, science educator; b. Centerville, Ind., Jan. 29, 1957; d. William Howard and Shirley Anne (Wilson) Haley; m. Robert Chalres Oilphant, Apr. 14, 1979; children: Kristen Rae, Matthew Adler. MEd, U. Cin., 1987, EdD, 1989. Sci. tchr. Hazelwood (Mo.) West Jr.-Sr. High Sch., 1979-81, Kings Mills (Ohio) local schs., 1987-92; instnl. coord. Mo. Botanical Garden, St. Louis, 1981-82; grad. rsch. asst. U. Cin., 1983-87, adj. instr., 1986; adj. instr. Miami U., Oxford, Ohio, 1986, vis. asst. prof., 1992—; project dir. Miami U., Ohio's NSF State Systemic Inst. in Math. and Sci., 1993—; chair secondary sci. Nat. Bd. Profl. Teaching Standards, 1991—; evaluator, cons. GE Aircraft Engines, Evendale, Ohio, 1987-91; cons. Biol. Scis. Curriculum Study, Colorado Springs, Colo., 1990-91. Recipient Ohio Tchr. of Yr. award Chief State Bd. Supts., 1990, Presdl. award for excellence in math. and sci. teaching NSF/NSTA, 1991. Mem. AAAS (evaluator, author), Nat. Assn. Rsch. in Sci. Teaching, Am. Edn. Rsch. Assn., Nat. Sci. Tchrs. Assn. Home: 1323 Chaucer Pl Maineville OH 45039 Office: Miami U 421 McGuffey Hall Oxford OH 45056

HALFORD, GARY ROSS, aerospace engineer, researcher; b. Fayette County, Ill., Dec. 7, 1937; s. Herbert Chaffin and Faye Samantha (Meyerholz) H.; m. Patricia Ann McKee, Feb. 8, 1959; children: Kirk Lind, Gwen, Shawn Lynn. BS, U. Ill., 1960, MS, 1961, PhD, 1966. Apprentice Caterpillar Tractor Co. Rsch. Ctr., East Peoria, Ill., 1955, engr., 1956-57; engr. Deere & Co., Moline, Ill., summer 1961; rsch. assoc. T&AM Dept., U. Ill., Urbana, 1960-66; rsch. engr. Lewis Rsch. Ctr., NASA, Cleve., 1966-82, sr. rsch. engr., 1982-91, sci. technologist, 1991—. Co-editor, author: Low Cycle Fatigue - Directions for the Future, 1988; editor, author: Lewis Structures Technology, LST '88, 3 vols., 1988; contbr. articles to profl. jours. Recipient Silver Snoopy award NASA Astronaut Corp., 1991; NASA Dir.'s fellow, 1989. Mem. AIAA, ASTM, SAE, ASME, Am. Acad. Mechanics, Am. Soc. Metals, Metal Properties Coun. Achievements include developing over two dozen high-temperature fatigue life prediction models for aerospace structural durability assessment; co-developer of Strainrange Partitioning as a creep-fatigue life prediction tool. Home: 8439 Celianna Dr Strongsville OH 44136 Office: NASA Lewis Rsch Ctr Struc Div M/S 49-7 21000 Brookpark Rd Cleveland OH 44135-3191

HALKIAS, CHRISTOS CONSTANTINE, electronics educator, consultant; b. Monastiraki, Doridos, Greece, Aug. 23, 1933; s. Constantine C. and Alexandra V. (Papapostolou) H.; m. Demetra Saras, Jan. 22, 1961; children: Alexandra, Helen-Joanna. B.S. in Elec. Engring., CCNY, 1957; M.Sc. in Elec. Engring., Columbia U., 1958, Ph.D., 1962. Prof. elec. engring. Columbia U., N.Y.C., 1962-73; prof. electronics Nat. Tech. U. Athens, Greece, 1973—, Fulbright vis. prof., 1969, dir. infomatics div., 1983-86; dir. Nat. Research Found., Athens, 1983-87; cons. Nat. Bank of Greece, Athens, 1980-89, Ergo Bank, Athens, 1975-91. Author: Electronic Devices and Circuits, 1967; Integrated Electronics, 1972; Electronic Fundamentals and Applications, 1976, Design of Electronic Filters, 1988; contbr. articles to profl. jours. Recipient D.B. Steinman award CCNY, 1956; Higgins fellow Columbia U., 1958. Mem. IEEE (sr., Centennial medal 1984, chmn. Greek sect. 1982-86), Sigma Xi. Home: 4 Kosti Palama St, Paleo Psyhico, 15452 Athens Greece Office: Nat Tech U Athens, Zographou, 15773 Athens Greece

HALL, BRAD BAILEY, orthopaedic surgeon, educator; b. Lubbock, Tex., Nov. 16, 1951; s. John Robert and Anna Ruth (Marks) H.; m. Carol Lynn Martin, Dec. 20, 1975; children: Clint Berkeley, Kathryn Lynn. Student, Ariz. State U., 1970-72; MD, Tex. Tech U., 1977. Diplomate Am. Bd. Orthopaedic Surgery; lic. Am. Acad. Disability Evaluating Physicians. Resident in orthopaedic surgery Mayo Clinic, Rochester, Minn., 1977-82, cons. in orthopaedic surgery, 1982-83; spine fellow U. Toronto, Ont., Can., 1982; pvt. practice, San Antonio, 1983—; clin. assoc. prof. U. Tex. Health Sci. Ctr., San Antonio, 1984—, program dir. lumbar spine segmental internal fixation, 1991; orthopaedic staff physician Audie L. Murphy Meml. VA Hosp., San Antonio, 1985—; chmn. dept. orthopaedics St. Luke's Luth. Hosp., San Antonio, 1986-88, dir. residency, 1987—, chief staff, bd. dirs., mem. fin. com., chmn. exec. staff com., 1992—; mem. adv. panel physicians AcroMed Corp., Cleve., 1990; presenter in field. Contbr. articles and abstracts to med. jours., chpts. to books. Grantee Orthopaedic Rsch. and Edn. Found., 1979; scholar Mayo Found., 1982. Fellow Am. Acad. Disability Evaluating Physicians; mem. AMA (physicians recognition award 1984-87, 88-94), Orthopaedic Rsch. Soc., Am. Acad. Orthopaedic Surgeons, Mid-Am. Orthopaedic Assn., N.Am. Spine Soc., Nat. Assn. Disability Evaluating Profls., So. Med. Assn., Clin. Orthopaedic Soc., Western Orthopaedic Assn., So. Orthopaedic Soc., Tex. Spine Soc. (bd. dirs. 1989—), Tex. Med. Assn., Tex. Orthopaedic Assn., Bexar County Med. Soc., Tex. Med. Found., Tex. Orthopaedic Assn., Physicians Who Care. Office: South Tex Orthopaedic and Spinal Surgery Assocs 9150 Huebner Rd Ste 350 San Antonio TX 78240

HALL, BRONWYN HUGHES, economics educator; b. West Point, N.Y., Mar. 1, 1945; d. Richard Roberts and Elizabeth (Flandreau) Hughes; m. Robert Ernest Hall, June 25, 1966 (div. Apr. 1983); children: Christopher Ernest, Anne Elizabeth. BA, Wellesley Coll., 1966; PhD, Stanford U., 1988. Programming analyst Lawrence Berkeley (Calif.) Lab., 1963-70; sr. programmer econometric programming Harvard U., Cambridge, Mass., 1971-77; owner, opr. Time Series Processor, Palo Alto, Calif., 1976—; from rsch. economist to faculty rsch. fellow Nat. Bur. Econ. Rsch., Stanford, Calif., 1977—; from asst. prof. to assoc. prof. U. Calif., Berkeley, 1987—; mem. data base rev. com. U.S. SBA, Washington, 1983-84; ind. econometric programming cons. ednl. instns., Cambridge, 1970-77. Mem. editorial bd. Econ. of Innovation and New Tech., Uxbridge, Eng., 1989—; contbr. articles on econs. to profl. publs. Sloan Found. dissertation fellow, 1986-87, Nat. fellow Hoover Inst. on War, Revolution, and Peace, Stanford U., 1992-93; NSF rsch. grantee, 1989—. Mem. Am. Econ. Assn. (mem. census adv. com. 1990—), Am. Fin. Assn., Am. Statis. Assn., Econometric Soc., Assn. for Computing Machinery. Avocations: tennis, opera, skiing. Home: 123 Tamalpais Rd Berkeley CA 94708-1948 Office: U Calif Dept Econs 611 Evans Hall Berkeley CA 94720

HALL, CARL WILLIAM, agricultural and mechanical engineer; b. Tiffin, Ohio, Nov. 16, 1924; s. Lester and Irene (Routzahn) H.; m. Mildred Evelyn Wagner, Sept. 5, 1949; 1 dau., Claudia Elizabeth. B.S., B. in Agricultural Engring. summa cum laude, Ohio State U., 1948; M.M.E., U. Del., 1950; Ph.D., Mich. State U., 1952. Registered profl. engr., Mich., Ohio. Instr. U. Del., 1948-50, asst. prof., 1950-51; asst. prof. Mich. State U., 1951-53, assoc. prof., 1953-55, prof., 1955-70, chmn. dept. agrl. engring., 1964-70; dean, dir. research (Coll. Engring.); prof. mech. engring. Wash. State U., Pullman, 1970-82, pres. WSU Rsch. Found., 1973-82; dep. asst. dir. Directorate for Engring. NSF, 1982-90; ret., 1990; with ESCOE, Inc., Washington, 1979; dist. vis. prof. Ohio State U., 1991; rsch. cons. U. P.R., 1957, 63, del. to USSR, 1958, 87; cons. U. Nacional de Colombia, 1960; cons. dairy engring., India, 1961, cons. food engring., Taiwan, 1961; Mission to Ecuador, 1966; U.

Nigeria, 1967; cons. UNDP/SF Project 80 (higher edn. Latin Am.), 1964-70, world food and nutrition study Nat. Acad. Sci., 1976-77, mem. engrng. edn. del. to People's Republic of China, 1978, Indonesia, 1978, 1993; co-chmn. NRC-India Nat. Sci. Acad. Workshop, New Delhi, 1979; with ACA, Inc. (cons. engrng.), 1956-70, pres., 1962-70; chmn. Nat. Dairy Engrng. Conf., 1953-66; mem. U.S. sci. exchange del. to USSR, 1958, 87; mem. postgrad. edn. select com. U.S. Navy, Monterey, Calif., 1975; rsch. fellow Jap. Soc. Promotion Sci., 1991. Author: Drying Farm Crops, 1957, Agricultural Engineering Index 1907-60, 1961-70, 71-80, 81-90, (with others) Drying of Milk and Milk Products, 1966, 71, Agricultural Mechanization for Developing Countries, 1973; co-editor: Agricultural Engineers Handbook, 1960, Processing Equipment for Agricultural Products, 1963, 2d edit., 1979, Spanish edit., 1968, Milk Pasteurization, 1968, Ency. of Food Engineering, 1971, 86, Drying Cereal Grains, 1974, 2d edit., 1991, Dairy Technology and Engineering, 1976, Errors in Experimentation, 1977, Dictionary of Drying, 1979, Drying and Storage of Agricultural Products, 1980, Biomass as an Alternative Fuel, 1981, Dictionary of Energy, 1983, Food and Energy, 1984, Food and Natural Resources, 1988, Biomass Handbook, 1989, (with others) Drying and Storage of Grains, 1992, Literature of Agricultural Engineering, 1992; editor, emeritus: Drying Technology: Marcel Dekker, Inc.; contbr. yearbooks, encys., handbooks, over 400 articles to profl. jours. Served with AUS, 1943-46, ETO. Decorated Bronze Star; recipient Disting. Faculty award Mich. State U., 1963, Centennial Achievement award Ohio State U., 1970; Massey-Ferguson Edn. medal, 1976, Max Eyth medal, Germany, 1979; medal du Merite, France, 1979, Silver medal, Paris, 1980, Cyrus Hall McCormick medal, 1984, Disting. Svc. award and medal NSF, 1988, Excellence in Drying award IDS, 1990, Food Engrng. award and medal, 1993. Fellow AAAS (life), ASME (life, v.p. rsch. 1993—), NSPE (NSF engr. of yr. 1986), Accreditation Bd. Engrng. and Tech., Am. Soc. Agrl. Engrs. (life, pres. 1974-75), Am. Soc. Engrng. Edn. (life), Am. Inst. MEd. and Biol. Engrng.; mem. Am. Inst. Biol. Scis., Wash. Soc. Profl. Engrs. (nat. dir. 1975-79), Va. Soc. Profl. Engrs. (pres. No. Va. chpt. 1987-88), Nat. Acad. Engrng., Internat. Commn. Agrl. Engrs. (v.p. 1965-74), Engrs. Council for Profl. Devel. (exec. com., dir., sec. 1973-74, chmn. engrng. accreditation commn. 1979-80), 99th Inf. Div. Assn., Inst. Food Tech., Sigma Xi, Alpha Zeta, Tau Beta Pi, Phi Kappa Phi, Gamma Sigma Delta, Phi Lambda Tau. Achievements include rsch. in energy, drying, food engrng., properties of materials and biomass. Office: Engring Info Svcs 2454 N Rockingham St Arlington VA 22207-1033

HALL, CHARLES ADAMS, infosystems specialist; b. Damoh, India, Aug. 6, 1949; s. Keith Burckle and Virginia (Bevan) H.; m. Nancy Louise Dahl, June 7, 1980; 1 child, Loren Jarrett. BA, Hiram (Ohio) Coll., 1972; AA, Ind. Vocat. Tech. Coll., 1983. Programmer Superior Supply, Inc., Marion, Ind., 1983-85, data processing mgr., 1985-89, dir. mgmt. info. systems, 1990-92; systems adminstr. Hi-Way Dispatch, Marion, 1992—; programmer Freel & Mason, Marion, 1985; programmer, chief programmer Bruce, Hall & Assocs., Marion, 1986-88. Developer computer game. Sec., treas. bd. dirs. Health Environ. for All Life, Marion, 1988—. With U.S. Army, 1973-76. Mem. Christian Ch. (Disciples of Christ). Office: PO Box 896 Marion IN 46952-0896

HALL, CHARLES ALLAN, numerical analyst, educator; b. Pitts., Mar. 19, 1941; s. George Orbin and Minnie (Carter) H.; m. Mary Katherine Harris, Aug. 11, 1962; children—Charles, Eric, Katherine. B.S., U. Pitts., 1962, M.S., 1963, Ph.D., 1964. Sr. mathematician Bettis Atomic Power Lab., West Mifflin, Pa., 1966-70; assoc. prof. math. stats. U. Pitts., 1970-78, prof., 1978—; exec. dir. Inst. Computational Maths. and Applications, 1978-89; cons. GM Rsch., 1971—, Westinghouse Electric Corp., Pitts., 1974-81, Pitts. Corning Co., 1980-86, Contraves, 1986-88, NASA Lewis, 1989-91, Idaho Nat. Engrng. Lab., 1990-92. Author: (with T. Porsching) Numerical Analysis of Partial Differential Equations, 1990; contbr. articles to prof. jours. Served to 1st lt. U.S. Army, 1964-66. Home: PO Box 83 Murrysville PA 15668-0083 Office: 603 Thackeray Hall U Pitts Pittsburgh PA 15260

HALL, CHRISTOPHER, materials scientist, researcher; b. Henley-on-Thames, England, Dec. 31, 1944; s. Victor and Doris (Gregory) H.; m. Sheila McKelvey, Sept. 1, 1966; children: Liza, Ben. BA, Oxford U., 1966, MA, PhD, 1970. Rsch. assoc. Case Western Res. U., Cleve., 1970-71; sr. rsch. assoc. U. East Anglia, England, 1971-72; lectr. bldg. engring. U. Manchester, England, 1972-83; head rock and fluid physics Schlumberger Cambridge (United Kingdom) Rsch., 1983-88, scientific advisor, 1990—; head chem. tech. Dowell Schlumberger, St. Etienne, France, 1988-90. Author: Polymer Materials, 1981, 89; contbr.: Civil Engineering Materials, 1983, 88; contbr. articles and revs. to profl. jours. Fellow Royal Soc. Chemistry; mem. Am. Chem. Soc., Am. Ceramic Soc. Achievements include 1 U.S. patent; research in inorganic materials, oil field chemistry. Home: 9A Church St Stapleford, Cambridge CB2 5DS, England Office: Schlumberger Cambridge Rsch, PO Box 153, Cambridge CB3 0EL, England

HALL, CLIFFORD CHARLES, pharmaceutical biochemist; b. Madison, Wis., Feb. 2, 1953; s. Charles Robert and Thelma Esther (Hoyt) H.; m. Linda Elaine Wills, Feb. 27, 1982 (div. Dec. 1991); 1 child, Grant Remington. BS in Chemistry, U. Wis., 1975, PhD in Pharmacology, 1981. Cert. biochem. pharmacologist. Rsch. asst. U. Wis., Madison, 1975-80; postdoctoral fellow U. Pa., Phila., 1981-86; pharm. biochemist Hoffmann-La Roche, Nutley, N.J., 1986—. Contbr. articles to profl. jours. Mem. N.J. Coun. For Children's Rights, Butler, N.J., 1992. Predoctoral fellow Pharm. Mfrs. Assn., 1978. Mem. AAAS, Sigma Xi. Democrat. Achievements include research in photoaffinity labelling of the peptidyltransferase center of ribosomal RNA, 1st purification of EF TU from staphylococcus aureus, a human phathogen. Home: 104 E Passaic Ave Bloomfield NJ 07003 Office: Hoffmann La Roche 340 Kingsland St Nutley NJ 07110

HALL, DALE EDWARD, electrochemist; b. Herkimer, N.Y., July 1, 1947; s. Donald Elton and Margaret (Bell) H.; m. Nancy Jane Ely, Aug. 29, 1970; 1 child, Pamela Kristin. BS cum laude, Rensselaer Poly. Inst., 1969, PhD in Physical Chemistry, 1973. Rsch. scientist G.A.F. Corp., Rensselaer, N.Y., 1973-74, Exxon Rsch. & Engring. Co., Linden, N.J., 1975-76; prin. scientist Internat. Nickel Co., Sterling Forest, N.Y., 1977-84; group leader Am. Cyanamid Co., Stamford, Conn., 1984-86; exec. dir. vis. com. advt. tech. Nat. Inst. of Standards and Tech., Gaithersburg, Md., 1987—. Contbr. articles to jours. Electrochemistry, Materials Sci. Mem. Electrochem. Soc. (bd. dirs. 1990-92, chmn. indsl. electric and electrochem. engring. div. 1990-92). Achievements include patents in electrochemical technology; research in hydrogen production and storage, electrolytic industries. Office: Nat Inst Standards and Tech Adminstrn Bldg A 813 Gaithersburg MD 20899

HALL, DAVID LEE, engineering executive; b. Blockton, Iowa, June 15, 1946; s. Clarence Beryl and Dorothy (Madeline) H.; m. Mary Jane Moyer; children: Sonya, Cristin. BA, U. Iowa, 1967; MS, Pa. State U., 1968, PhD, 1976. Staff scientist Lincoln lab. MIT, Boston, 1976-77; mgr. analysis sect. Computer Scis. Corp., Silver Spring, Md., 1977-79; dir. rsch. and devel. HRB Systems, Inc., State College, Pa., 1979-88, dir. rsch. and ops., 1988—; cons. Tech. Tng. Corp., Torence, Calif., 1985-92; tech. reviewer Ben Franklin Tech. Ctr. Pa., State College, 1988-92. Author: Mathematical Techniques for Multi-Sensor Data Fusion, 1992; co-editor AIAA Software Engineering Progress Series, 1992. With USAF, 1968-72. Fellow AIAA (assoc., chmn. software systems tech. com. 1987-90); mem. IEEE, Am. Astron. Soc., Elks, Sigma Xi. Republican. Lutheran. Achievements include research in codifying terminology and algorithms for multi-sensor data fusion. Office: HRB Systems Inc Science Park Rd State College PA 16804

HALL, DONALD NORMAN BLAKE, astronomer; b. Sydney, New South Wales, Australia, June 26, 1944; came to U.S., 1967; s. Norman F.B. and Joan B. Hall. B.Sc. with honors, U. Sydney, 1966; Ph.D. in Astronomy, Harvard U., 1970. Research assoc. Kitt Peak Nat. Obs., Tucson, 1970-72, assoc. astronomer, 1972-76, astronomer, 1976-81; dep. dir. Space Telescope Sci. Inst., Balt., 1982-84; dir. Inst. Astronomy U. Hawaii, Honolulu, 1984—; mem. space sci. bd. Nat. Acad. Sci., 1984-88, mem. astronomy adv. com. NSF, 1984-87; mem. astrophysics council NASA, 1984-88. Mem. Am. Astron. Soc. (Newton Lacey Pierce prize 1978), Internat. Astron. Union. Office: U Hawaii Inst Astronomy 2680 Woodlawn Dr Honolulu HI 96822-1897

HALL, FREDERICK KEITH, chemist; b. Leeds, Eng., Jan. 3, 1930; naturalized, 1976; s. Frederick Stanley and Mary Elizabeth (Stocks) H.; m. Patricia Ellison, Aug. 25, 1956; children: Simon Keith, Stephanie Jane, Andrew Nicholas. B.S. with 1st class honors, U. Manchester, 1951; Ph.D., U. Leeds, 1954; grad., Advanced Mgmt. Program, Harvard U., 1979. Research chemist Courtaulds (Can.) Ltd., 1956-58, asst. tech. mgr., 1958-60, tech. mgr., 1960-63, plant mgr., 1963-66; dir. tech. service Internat. Paper Co., 1966-70, asst. dir. research center, 1970-72, dir. primary process, 1972-75, corp. dir. research, 1975-77; dir. S & ED labs., 1977-79; chief scientist S & T labs., 1979—. Served with Brit. Army, 1953-55. Fellow TAPPI (pres. 1991-93), Royal Soc. Chemistry, Textile Inst., Am. Inst. Chemists; mem. Chem. Inst. Can., Can. Pulp and Paper Assn., Tuxedo Club. Office: Internat Paper TechCorp Rsch Ctr Long Meadow Rd Tuxedo Park NY 10987

HALL, GEORGE JOSEPH, JR., geologist, educator, consultant, geotechnical engineer; b. Little Rock, Ark., Nov. 20, 1952; s. George Joseph Sr. and Jeanette Patsy (Fowler) H.; m. Anita Jo Carter, July 30, 1977; children: Miranda, Chase, Mallory. BS in Geology, Ark. Tech. U., 1975; MSCE, Okla. State u., 1983. Registered profl. engr., Okla.; registered profl. geologist, Ark. Staff geologist U.S. Army C.E., Little Rock, 1975-77; staff geologist U.S. Army C.E., Tulsa, 1977-82, geotechnical engr., 1982-88, dist. geologist, 1988—; adj. prof. hydrogeology U. Tulsa, 1992; mem. field adv. group for ground water modeling C.E., 1992; mem. ctrl. Okla. aquifer liaison com. U.S. Geol. Survey, 1989-92. Bd. dirs. Okla. Airmen Flying Club (pvt. pilot lic., instrument rating), Tulsa, 1979-80; scuba diver (cert.0. Mem. ASCE, Assn. Engring. Geologists, Assn. Ground Water Scientists and Engrs., Internat. Assn. Hydrogeologists (peer reviewer 1992), Tulsa Geologic Soc., Ark. Ground Water Assn., Okla. Ground Water Assn. Republican. Methodist. Office: US Army CE PO Box 61 Tulsa OK 74121-0061

HALL, GRACE ROSALIE, physicist, educator, literary scholar; b. Meriden, Conn., July 15, 1921; d. George John and Grace Cleora (Gleason) White; m. Eldon Conrad Hall, July 2, 1948; children: Brent Channing, Pamela Rosalie, Craig Gleason, Gordon Timothy. BS in Chemistry, Eastern Nazarene Coll., 1946; MA in Physics, Boston U., 1946, doctoral studies in physics, 1946-53; MA in English, Simmons Coll., 1975. Bookkeeper Cherry & Webb Co., Providence, 1939-42; sec. to registrar Eastern Nazarene Coll., Quincy, Mass., 1942-44, instr. physics, chemistry, 1945-46; teaching fellow physics Boston U., 1946-49; instr. physics lab. Northeastern U., Boston, 1956-57; asst. prof. physics Eastern Nazarene Coll., Quincy, 1957-61, asst. prof. chemistry, 1969, asst. prof. phys. sci., 1974; asst. prof. phys. sci., 1974; instr. Shakespeare Barrington (R.I.) Coll., 1984; tchr. Westwood (Mass.) Seminar, 1975; ch. sch. dir. First Parish, Westwood, 1977-81; chair seminar U. Ky., 1988. Author: (chpt. in book) Webs & Wardrobes, 1987; contbr. articles to profl. jours. Dir. Norfolk Assn. for Retarded Citizens, 1978-79; judge High Sch. Sci. Fairs, North Quincy, Mass., 1960-64, 69-76, Regional Sci. Fairs, Bridgewater, mass., 1960-62. Recipient Faculty scholarship Eastern Nazarene Coll., 1943-45, Libr. Family of Yr. award City of Quincy, 1960; named to R.I. Honor Soc. Mem. MLA (session participant 1978, 84), Shakespeare Assn. of Am. (seminar participant 1988-93), Christianity and Lit. Assn., Children's Lit. Assn. (scholarly rsch. and writing 1978—), MIT Women's League (editor activities guide and newsletter 1989—, research and writing), New Eng. Hist. Geneal. Soc., Mus. Fine Arts. Avocations: photography, painting, recycling, grandparenting, snorkeling.

HALL, HOUGHTON ALEXANDER, engineering professional; b. Kingston, Jamaica, W.I., Aug. 17, 1936; came to U.S., 1985; s. James Alexander and Clarice Viola Hall; m. Grace Yvonne Anglin, Feb. 22, 1964; children: Andrew Geoffery, Christine Elizabeth. BS, U. W.I., Kingston, 1958, diploma in chem. tech., 1959, diploma in mgmt., 1977. Registered profl. engr., Fla.; chartered engr. Great Britain. Elec. engr. Jamaica Pub. Svc. Co., Kingston, 1960-84; dir. R&D Ministry of Sci., Tech. and the Environ., Kingston, 1984-85; elec. engr. Electric Dept. City of Tallahassee, 1985-90; supr., substation engring. Electric Dept., City of Tallahassee, 1990—. Mem. IEEE (sr.), NSPE, Inst. of Elec. Engrs., Tallahassee Sci. Soc. (charter pres. 1989—). Baptist. Avocations: electronics, scientific pursuits. Home: 4335 Sherborne Rd Tallahassee FL 32303-7607 Office: City of Tallahassee 2602 Jackson Bluff Rd Tallahassee FL 32304-4498

HALL, HOWARD LEWIS, nuclear chemist; b. Winchester, Va., Sept. 12, 1963; s. Harvey Eugene and Marie Wavilee (Price) H.; m. Mary Lee Davis, Dec. 28, 1986. BS in Chemistry, Coll. Charleston, 1985; PhD in Nuclear Chemistry, U. Calif., Berkeley, 1989. Postdoctoral Lawrence Livermore (Calif.) Nat. Lab., 1989-91, chemist, sect. leader, 1991—; vis. lectr. nuclear chemistry U. Calif., Berkeley, 1992—. Contbr. articles to profl. jours. Recipient Grad. fellowship NSF, 1985-88, Bishop Robert Smith award Coll. Charleston, 1985, Undergrad. fellowship Dept. of Energy, 1984. Mem. Am. Chem. Soc., Am. Nuclear Soc., Am. Inst. Chemistry, Alpha Chi Sigma (dist. counselor 1985-90), Sigma Xi. Achievements include development of automated rapid radiochemistry apparatus for study of short radionuclides; first to show direct proof of delayed fission process; to show use of delayed fission to study fission properties far from stability; to participate in solution-phase chemical studies of hahnium (element 105). Home: 231 Pistachio Pl Brentwood CA 94513 Office: Lawrence Livermore Nat Lab L-233 Radiation Analytical Scis Livermore CA 94551

HALL, HOWARD PICKERING, mathematics educator; b. Boston, July 8, 1915; s. George Henry and Elizabeth Isabel (McCallum) H.; m. Ellen Marguerite Ide, June 25, 1945 (dec. 1984); children: Charlotte McCallum, Stephanie Wilson, Lindsey Louise, Gretchen Elizabeth. AB, Harvard U., 1936, MS, 1937, DSc, 1951. Registered structural engr., Ill., 1953. Instr., civil engring. Brown U., Providence, 1937-38; structural analyst Mark Linenthal, Engr., Boston, 1938-39; instr., asst. prof., assoc. prof. civil engring. Northwestern U., Evanston, Ill., 1939-56; design engr, field engr. Porter, Urquart, Skidmore, Owings, Merrill, Casablanca, Fr. Morocco, 1951-53; dean, sch. engring., acad. v.p. Robert Coll., Istanbul, Turkey, 1956-68; dir. of studies, acting headmaster St. Stephen's Sch., Rome, 1968-72; prof. math. Iranzamin Internat., Tehran, Iran, 1973-80; math. tchr. Vienna Internat. Sch., 1980-83, Copenhagen Internat. Sch., 1983-86; cons. S.J. Buchanan, Bryan, Tex., Eng., 1955. Contbr. articles to profl. jours. Served to Capt. U.S. Army, 1942-46, ETO. Recipient Clemens Herschel award Boston Soc. Civil Engrs., 1954. Mem. AAAS, Sigma Xi. Home: 301 SW Lincoln St Apt 1101 Portland OR 97201-5031

HALL, JOHN LEWIS, physicist, researcher; b. Denver, Aug. 21, 1934; s. John Ernest and Elizabeth Rae (Long) H.; m. Marilyn Charlene Robinson, Mar. 1, 1958; children—Thomas Charles, Carolyn Gay, Jonathan Lawrence. BS in Physics, Carnegie Mellon U., 1956, MS in Physics, 1958, PhD in Physics, 1961; PhD (hon.), U. Paris XIII, 1989. Postdoctoral research assoc. Nat. Bur. Standards, Washington, 1961-62; physicist Nat. Bur. Standards, Boulder, Colo., 1962-75, sr. scientist, 1975—; cons. Los Alamos Sci. Labs, N.Mex., 1963-65; lectr. U. Colo., Boulder, 1977—; cons. numerous firms in laser industry, 1974—. Contbr. numerous tech. papers to profl. jours.; patentee in laser tech.; editor: Laser Spectroscopy 3, 1977. Recipient IR-100 award IR Mag., 1975, 77; Gold Medal Nat. Bur. Standards, Gaithersburg, Md., 1974, Stratton award, 1971, E.U. Condon award, 1979; Gold medal Dept. Commerce, Washington, 1969, Presdl. Meritorious Exec. award, 1980; Meritorious Alumnus award Carnegie Mellon U., Pitts., 1985. Fellow Optical Soc. Am. (bd. dirs. 1980-82, Charles H. Townes award 1984, Fredrick Ives medal 1991), Am. Phys. Soc. (Davisson-Germer award 1988); mem. NAS, Comite Consultatif pour la Definition du Metre. Office: JILA-Nat Bur of Standards 325 Broadway St Boulder CO 80303-3328

HALL, MICHAEL L., nuclear engineer, computational physicist; b. Biloxi, Miss., Jan. 25, 1962; s. Alfred C. and Janice E. (Graham) H.; m. Mary E. Waller, July 29, 1989. BS in Nuclear Engring., N.C. State U., 1983, PhD in Nuclear Engring., 1988. Mem. tech. staff Los Alamos (N.Mex.) Nat. Lab., 1988—; mem. Manuel Lujan award com. Space Nuclear Power Systems Symposium, Albuquerque, 1992—. Editor transactions in field; contbr. articles to profl. jours. Dept. Energy fellow, 1984-87. Mem. Am. Nuclear Soc. (First Place award Sr. Design Competition 1983), Mensa, Intertel, Sigma Xi, Phi Kappa Phi, Tau Beta Pi. Office: Los Alamos Nat Lab PO Box 1663 MS-K551 Los Alamos NM 87545

HALL, PAMELA ELIZABETH, psychologist; b. Jacksonville, Fla., Sept. 10, 1957; d. Gary Curtiss and Ollie (Banko) H. BA, Rutgers U., 1979; MS in Edn., Pace U., 1981, D in Psychology, 1984. Lic. psychologist, N.Y., N.J., Calif. Psychology extern St. Vincent's Med. Ctr., N.Y.C., 1981-82; intern in clin. psychology Elizabeth (N.J.) Gen. Med. Ctr., 1982-83, staff psychologist, 1983-85; staff psychologist J.F.K. Med. Ctr., Edison, N.J., 1985-87; pvt. practice Summit and Perth Amboy, N.J., 1985—; sr. supervising psychologist Muhlenberg Med. Ctr., Summit, N.J., 1987-90; founder, pres. N.J. Soc. for Study of Multiple Personality & Dissociation, Perth Amboy, 1988—. Mem. Mayor's Com. on Substance Abuse, Perth Amboy, 1987. Named Henry Rutgers scholar, 1979. Mem. Am. Soc. Clin. Hypnosis, Internat. Soc. for Study of Multiple Personality Dissociation (pres. N.J. chpt. 1988—), Pace U. Alumni Assn., Rutgers U. Alumni Assn., Psi Chi. Avocations: crew, swimming, fine arts. Home: PO Box 1820 Perth Amboy NJ 08862-1820

HALL, PAMELA S., environmental consulting firm executive; b. Hartford, Conn., Sept. 4, 1944; d. LeRoy Warren and Frances May (Murray) Sheely; m. Stuart R. Hall, July 21, 1967. BA in Zoology, U. Conn., 1966; MS in Zoology, U. N.H., 1969, BS summa cum laude, Whittemore Sch. Bus. and Econs., U. N.H., 1982; student spl. grad. studies program, Tufts U., 1986-90. Curatorial asst. U. Conn., Storrs, 1966; rsch. asst. Field Mus. Natural History, Chgo., 1966-67; teaching asst. U. N.H., Durham, 1967-70; program mgr. Normandeau Assocs. Inc., Portsmouth, N.H., 1971-79, marine lab. dir., 1979-81, programs and ops. mgr., Bedford, N.H., 1981-83, v.p., 1983-85, sr. v.p., 1986-87, pres., 1987—. Mem. Conservation Commn., Portsmouth, 1977-90, Wells, Estuarine Rsch. Res. Review Commn., 1986-88, Great Bay (N.H.) Estuarine Rsch. Res. Tech. Working Group, 1987-89; trustee Trust for N.H. Lands, 1990—; trustee N.H. chpt. Nature Conservancy, 1991—; incorporator N.H. Charitable Fund, 1991—; bd. advisors Vivamos Mejor, USA, 1990—. Graham Found. fellow, 1966; NDEA fellow, 1970-71. Mem. ASTM, Am. Mgmt. Assn., Water Pollution Control Fedn., Am. Fisheries Soc., Estuarine Rsch. Fedn., Nat. Assn. Environ. Profls., Sigma Xi. Home: 4 Pleasant Point Dr Portsmouth NH 03801-5275 Office: Normandeau Assocs Inc 25 Nashua Rd Bedford NH 03110-5500

HALL, RICHARD CLAYTON, psychologist, consultant, researcher; b. Pitts., Apr. 29, 1931; s. Clayton LeClaire and Genevieve (Gorman) H.; m. Doris Margaret Bjorkland, Aug. 26, 1963; children: Karen, Janice, Dorothy. BS in Psychology with honors, Trinity Coll., 1952; MS, U. Pitts., 1959, PhD, 1963. Lic. psychologist, Pa. Rsch. psychologist Polk (Pa.) Ctr., 1963-68, dir. behavior modification programs, 1968-75, chmn. subcom. human rights for behavior mgmt. procedures, 1987-89, staff psychologist, 1989-91; ind. researcher Key West, Fla., 1975-84, Polk, Pa., 1985—. Contbr. articles to profl. jours. With U.S. Army, 1953-55. NSF Coop. Grad. fellow, 1959. Mem. Am. Psychol. Assn., Sigma Xi, Pi Gamma Mu. Democrat. Episcopalian. Avocation: soloist at ch., civic operetta groups. Home: 101 Elm St Polk PA 16342

HALL, STEPHEN GROW, research neuroscientist; b. Danbury, Conn., Oct. 2, 1955; s. Peter Chamberlaine and Patricia (Grow) H.; m. Pamela Sherwood, Dec. 9, 1989. BS in Chemistry, Grand Canyon U., Phoenix, 1990; postgrad., Purdue U., 1990—. Neurosurg. rsch. asst. Barrow Neurol. Inst., Phoenix, 1987-90; neuroscientist, rsch. asst. in biol. scis. Purdue U., West Lafayette, Ind., 1990—, mem. exec. com. neurosci. program, 1990—. Spl. adv. Tippecanoe County Superior Ct., Lafayette, Ind., 1992-94 With Army N.G., 1973-80. Biomed. U.S. Dept. Energy rsch. fellow Argonne Nat. Lab., 1987. Mem. Soc. for Neurosci., Am. Assn. for Cancer Rsch., Am. Chem. Soc., Ind. Acad. Sci. (rsch. grantee 1992-94). Achievements include research on cellular, molecular and genetics of neural development, cancer genetics, neuro-oncology. Office: 1392 Lilly Hall of Life Sci Purdue U Biol Scis Dept West Lafayette IN 47907-1392

HALL, TIMOTHY C., biology educator, consultant; b. Darlington, Durham, Eng., Aug. 29, 1937; came to U.S., 1965; s. Gilbert Leslie and Dorothea Olive (Lindemann) H.; m. Sandra Severn, Aug. 20, 1960; children: Alexandra Vikki Anna, Liza Bryony, Peter Marcus Jeremy. BSc with honors, U. Nottingham, Eng., 1962, PhD in Plant Physiology, 1965. Louis W. and Maud Hill postdoctoral fellow dept. hort. sci. U. Minn., St. Paul, 1965-66; asst. prof. horticulture U. Wis., Madison, 1966-70, assoc. prof., 1970-75, prof., 1975-82, adj. prof. biophysics and genetics, 1982-84; dir. Agrigenetics Advanced Rsch. Div., Madison, 1980-84, Agrigenetics Rsch. Corp., Boulder, Colo., 1981-84; Disting. prof., head dept. biology Tex. A&M U., College Station, 1984-92, dir. Inst. Devel. and Molecular Biology, 1992—; sr. biotech. cons. Rhône-Poulenc Agrochimie, Lyon, France, 1985—; chair, organizer Gorden Conf. on Plant Molecular Biology, 1987; cons. plant biotech. Internat. Paper Co., Tuxedo Park, N.Y., 1987. Editor: (with J.W. Davies) Nucleic Acids in Plants, 2 vols., 1979, (with L. van Vloten-Doting and G.S.P. Groot) Molecular Form and Function of the Plant Genome, 1985; mem. editorial bd. Oxford Surveys Plant Molecular and Cell Biology, 1983—, Transgenic Rsch., 1991—, Plant Jour., 1991—; contbr. numerous articles to profl. jours., book chpts.; patentee in field. Pilot Royal Air Force, 1956-58. Grantee NIH, NSF, USDA, NATO, Rhône-Poulenc Agrochimie, Internat. Paper Co., Tex. Advanced Tech. Program, Rockefeller Found. Fellow Indian Virological Soc.; mem. AAAS, Am. Assn. Cereal Chemists, Am. Soc. Biol. Chemists, Am. Soc. for Biochemistry and Molecular Biology, Am. Soc. for Microbiology, Am. Soc. Plant Physiologists (chair session on nucleic acids, genetics 1975, chair com. on recombinant DNA 1978-79), Am. Soc. for Virology, Biochem. Soc., Internat. Soc. for Molecular-Plant Microbe Interactions, Internat. Soc. for Plant Molecular Biology, Squash Club Tex. A&M U., Sigma Xi. Avocations: squash, racquetball, bridge, travel. Office: Inst Devel-Molecular Biol Tex A&M U College Station TX 77843-3155

HALL, TREVOR JAMES, physicist, educator, consultant; b. Eng., Sept. 22, 1956; s. Harry Samuel and Pauline Ada (Stanley) H.; m. Isia Brecciaroli, Feb. 22, 1989. MA, Cambridge (Eng.) U., 1977; PhD, Univ. Coll., London, 1980. Chartered engr. Optical physicist Cambridge Cons., 1979-80; lectr. physics Queen Elizabeth Coll., London, 1980-84, King's Coll. London, 1984-90; reader in physics King's Coll., London, 1990—; cons. Optoelectric Assocs., London, 1990—; com. mem. advanced devices and materials Sci. and Engring. Rsch. Coun., Dept. Trade and Industry, U.K., 1991-93, com. mem. atomic and molecular physics, 1992—. Exec. editor Optical Computing and Processing jour., 1990—; contbr. chpts. to books, and sci. papers to jours. Chair governing body Waverly Sch., London, 1988. Rsch. grantee Sci. and Engring. Rsch. Coun., Dept. Trade and Industry, 1980—. Mem. Instn. Elec. Engrs. U.K. (mem. profl. group com.), Inst. Physics U.K. (com. mem.). Avocations: opera, classical music, travel, good food and wine. Office: Kings Coll London, Physics Dept Strand, London WC2R 2LS, England

HALLAK, JOSEPH, optometrist; b. Beirut, Lebanon, July 25, 1939; came to U.S., 1970; s. Ovadia and Olga (Tarrab) H.; m. Amy L. Fleissler, Aug. 8, 1971; children: Deborah Gail, Elliot Aaron, Daniel Scott. Optical Engr., U. Paris, Orsay, France, 1969; PhD, Sorbonne U., Paris, 1972; OD, SUNY, N.Y.C., 1975. Staff mem. Nassau County Med. Ctr., L.I., N.Y.; pvt. practice Hicksville, N.Y. Contbr. articles to profl. jours. Mem. Internat. Acad. Sports Vision, N.Y. Acad. Scis., Lions. Republican. Office: 183 Broadway # 308 Hicksville NY 11801

HALLAUER, ARNEL ROY, geneticist; b. Netawaka, Kans., May 4, 1932; s. Roy Virgil and Mabel Fern (Bohnenkemper) H.; m. Janet Yvonne Goodmanson, Aug. 29, 1964; children: Elizabeth, Paul. B.S., Kans. State U., 1954; M.S., Iowa State U., 1958, Ph.D., 1960. Rsch. agronomist UDSA, Ames, Iowa, 1958-60; geneticist UDSA, Raleigh, N.C., 1961-62; rsch. geneticist UDSA, Ames, 1963-89; prof. Iowa State U., 1989—, C.F. Curtiss Disting. prof. agr., 1991—. Author: (with J.B. Miranda) Quantitative Genetics in Maize Breeding, 1981, 2d edit., 1988. 1st lt. U.S. Army, 1954-56. Recipient Applied Rsch. and Ext. award Iowa State U., 1981, Genetics and Plant Breeding award Nat. Coun. Plant Breeding, 1984, Disting. Svc. to Agriculture award Gamma Sigma Delta, 1990, Gov.'s Sci. medal State of Iowa, 1990, Burlington No. Career Rsch. Achievement award Iowa State Found., 1991, Faculty citation Iowa State Alumni Assn., 1987, Henry A. Wallace award, 1992; USDA grantee, 1982, 85, 87, 90. Fellow Am. Soc. Agronomy (Agronomic Achievement award for crops 1989, Agronomic Rsch. award 1992), Crop Sci. Soc. (Dekalb Pfizer Crip Sci. award 1981), Iowa Acad. Sci.; mem. NAS, Nat. Agri-Mktg. Assn. (nat. award for excellence in rsch. 1993), Am. Genetic Assn., Am. Statis. Assn., Iowa State U.

Alumni Assn., Phi Kappa Phi, Sigma Xi, Gamma Sigma Delta. Republican. Lutheran. Home: 516 Luther Dr Ames IA 50010-4735 Office: Iowa State U 1505 Dept Agronomy Ames IA 50010

HALLBAUER, ROBERT EDWARD, mining company executive; b. Nakusp, B.C., Can., May 19, 1930; s. Edward F. and Lillian Anna (Kendrick) H.; m. Mary Joan Hunter, Sept. 6, 1952; children: Russell, Catherine, Thomas. BS in Mining Engring., U. B.C., 1954. Registered profl. engr., B.C. Various engring. and supervisory positions Placer Devel., Salmo, B.C., 1954-60; mine supr. Craigmont Mines Ltd., Merritt, B.C., 1960-64, mine mgr., 1964-68; v.p. mining Teck Corp., Vancouver, B.C., 1968-79, sr. v.p., 1979—; pres., chief exec. officer Cominco Ltd., Vancouver, 1986—, also bd. dirs. Recipient Edgar A. Scholz medal B.C. and Yukon Chamber of Mines, 1984. Mem. Assn. Profl. Engrs., Can. Inst. Mining, Metallurgy and Petroleum (Inco medal 1992). Home: 6026 Glenwynd Pl, West Vancouver, BC Canada V7W 2W5 Office: Cominco Ltd, 200 Burrard St #500, Vancouver, BC Canada V6C 3L7

HALLBERG, BENGT OLOF, fiber optic company executive and specialist; b. Stockholm, Dec. 31, 1943; s. Olle E.S. and Anne-Marie K. H.; m. Lena M. Tengelin, June 13, 1975; children: Niklas O., Mattias A., Andreas E. MS in Physics, Royal Inst. Tech., Stockholm, 1978. Constrnl. engr. AB Svenska Bostäder, Stockholm, 1965-76; scientist Inst. of Optical Rsch., Stockholm, 1976-81; pres. BOH Optical AB, Stockholm, 1981—. Inventor airborne multispectral radiometer, fiber optic communication system based on WDM; patentee frequency and output regulation in laser diodes. Mem. Optical Soc. Am., Am. Phys. Soc., Internat. Soc. Optical Engring., European Optical Soc., Swedish Optical Soc. Avocations: sailing, downhill skiing, shooting. Office: Boh Optical AB, Kärrgränd 108, S-162 46 Vällingby Sweden

HALLER, ARCHIBALD ORBEN, sociologist, educator; b. San Diego, Jan. 15, 1926; s. Archie O. and Eleanor (Brizzee) H.; m. Hazel Laura Zimmermann, Feb. 15, 1947 (dec. 1985); children: Elizabeth Ann, Stephanie Lynn, William John; m. Maria Camila Omegna Rocha, Apr. 12, 1986 (div. 1987); m. Maria Cristina Del Peloso, Sept. 16, 1989; stepchildren: Graziella, Camila. B.A., Hamline U., 1950; M.A., U. Minn., 1951; Ph.D. (Univ. fellow), U. Wis., 1954. Project assoc. research U. Wis.-Madison, 1954-56; prof. rural sociology, also sociology, Indsl. Relations Research Inst. Indsl. Relations Rsch. Inst., Inst. of Enviorn. Studies, 1965—; chmn. dept. rural sociology U. Wis.-Madison, 1970-72; from assoc. prof. to prof. sociology Mich. State U., East Lansing, 1956-65; Fulbright prof. rural U. Brazil, 1962, U. Sao Paulo, 1987-90; Fulbright travel grantee Univs. Sao Paulo Brasilia, Pernambuco, Ceara, 1974, U. Sao Paulo, Pernambuco, Paraiba and Ceara, Brazil, 1979; vis. prof. U. Wis., Madison, 1964, Brigham Young U., Provo, Utah, 1973; disting. vis. prof. rural sociology Ohio State U., 1982-83; vis. fellow Australian Nat. U., 1981; cons. UNESCO, 1989, Govt. of Brazil, Amazonian Rsch., 1991—, others. Author: The Occupational Aspiration Scale: Theory, Structure and Correlates, 1963, 71, the Socioeconomic Macroregions of Brazil--1970, 1983; co-editor: (with R.M. Hauser et al) Social Structure and Behavior: Essays in Honor of William Hamilton Sewell, 1982; editor spl. issues Luso-Brazilian Rev.; author rsch. monographs and tech. articles; contbr. articles to profl. jours. Mem. Mich. Com. Mental Health Policies, 1961-62; sociology fellowship panel Council on Internat. Exchange Scholars, 1977-81. Served with USNR, 1943-46. Decorated Grande Oficial Ordem do Merito do Trabalho, Brazil. Fellow AAAS; mem. Am. Sociol. Assn., Internat. Rural Sociol. Assn.,Internat. Sociol. Assn., Latin Am. Sociol. Assn., Midwest Sociol. Assn., Sociol. Rsch. Assn., Latin Am. Studies Assn., N.Y. Acad. Sci., Rural Sociol. Soc. (pres. 1969-70, rep. AAAS 1973-86, Disting. Rural Sociologist 1990), Univ. Club, Sigma Xi, Gamma Sigma Delta. Home: 529 Edward St Madison WI 53711-1207

HALLER, EDWIN WOLFGANG, physiologist, educator; b. Stuttgart, Germany, May 19, 1936; 2 children. BA, Park Coll., 1959; PhD in Physiology, Western Res. U., 1967. R&D chemist Lucidol div. Wallace and Tiernan, Inc., 1959-61; rsch. assoc. physiologist Sch. Medicine U. Md., Balt., 1967-69, asst. prof., 1969-71, asst. prof. physiology, 1971-72; assoc. prof. physiology, biology Sch. Medicine U. Minn., Duluth, 1972—; vis. scientist Inst. Animal Physiology, Agrl. Rsch. Coun., Cambridge, Eng., 1977-78; cons. Am. Assn. Accreditation Lab. Animal Care, 1978-88, coun. accreditation, 1988—; vis. prof. Health Sci. Ctr., SUNY-Syracuse, 1990. Mem. AAAS, Endocrine Soc., Soc. Neurosci., Sigma Xi (pres. U. Minn. chpt. 1992-93). Achievements include research on central nervous system regulation of gonadotropin secretion, neurophysiological aspects of neurosecretion, regulation of secretion of vasopressin and oxytocin from the posterior pituitary. Office: U Minn Med-Molecular Physiology 10 University Dr Duluth MN 55812

HALLER, EUGENE ERNEST, materials scientist, educator; b. Basel, Switzerland, Jan. 5, 1943; s. Eugen and Maria Anne Haller; m. Marianne Elisabeth Schlittler, May 27, 1973; children: Nicole Marianne, Isabelle Cathrine. Diploma in Physics, U. Basel, 1967, PhD in Physics, 1970. Postdoctoral asst. Lawrence Berkeley (Calif.) Lab., 1971-73, staff scientist, then sr. staff scientist, 1973-80, faculty sr. scientist, 1980—; assoc. prof. U. Calif., Berkeley, 1980-82, prof. materials sci., 1982—; co-chmn. Materials Rsch. Soc. Symposia, Boston, 1982, 89, Internat. Conf. on Shallow Levels in Semiconductors, Berkeley, 1984; rev. com. instrument div. Brookhaven Nat. Lab., Upton, N.Y., 1987-93; mem. Japanese tech. panel on sensors NSF-Nat. Acad. Sci., Washington, 1988; vis. prof. Imperial Coll. Sci., Tech. and Medicine, London, 1991. Editorial adv. bd. Jour. Phys. and Chem. Solids, 1993—; contbr. to numerous profl. publs. Sr. scientist award Alexander von Humboldt Soc., Germany, 1986; rsch. fellow Miller Inst. Basic Rsch., Berkeley, 1990. Fellow Am. Phys. Soc.; mem. AAAS, Materials Rsch. Soc., Swiss Phys. Soc., Sigma Xi. Achievements include patents in surface passivation of semiconductors, and synthesis of crystalline carbon nitride potentially a superhard material. Office: U Calif Berkeley 286 Hearst Mining Bldg Berkeley CA 94720

HALLER, IVAN, chemist; b. Budapest, June 8, 1934; came to the U.S., 1957; s. Tibor and Magdolna (Neubauer) H.; m. Flora E. Woolf, May 22, 1965; children: Paul J., Drew A. BSChE, U. Tech. Scis., 1956; PhD in Phys. Chemistry, U. Calif., Berkeley, 1961. Rsch. staff mem. IBM T.J. Watson Rsch. Ctr., Yorktown Heights, N.Y., 1960—. Contbr. articles to profl. jours. Organizer, coach Am. Youth Soccer Orgn. Region 75, Chappaqua, N.Y., 1975-82; chmn., treas. troop 57, Boy Scouts Am., Chappaqua, 1985-86, 86—. Mem. Am. Chem. Soc., Am. Phys. Soc., Am. Soc. for Mass Spectroscopy. Achievements include 8 U.S. patents for semiconductor processing, microlithographic resists, and thin-film transistor/liquid crystal displays; invented resist system for electron-beam microlithography. Home: 901 Hardscrabble Rd Chappaqua NY 10514 Office: IBM TJ Watson Rsch Ctr PO Box 218 Yorktown Heights NY 10598

HALLER, WILLIAM PAUL, analytical chemist, robotics specialist; b. Orange, N.J., Nov. 23, 1957; s. William Charles and Patricia Marie (Scavone) H.; m. Christina Marie Mangion, Apr. 30, 1983; children: Robert William, Alicia Ann. BS in Biochemistry, Fairleigh Dickinson U., 1980. Rsch. chemist Internat. Paint Co., Union, N.J., 1980-83; analytical quality assurance chemist Ortho Pharm. Corp., Raritan, N.J., 1983—, quality assurance robotics specialist, 1985—; steering com. Johnson & Johnson Intercompany Robotics Interest Group, Raritan, 1985-87. Co-author: Advances in Laboratory Automation-Robotics, 1986, 90. Recipient Johnson & Johnson Achievement award, 1991. Mem. AAAS, Am. Chem. Soc. (exec. com.), Lab. Robotics Interest Group N.J. Democrat. Roman Catholic. Achievements include development of protocols and criteria for the validation of robotic systems in the analytical lab., automate analytical methods to robotic systems, customized apparatus to help in automating analytical methods to robotic systems. Office: Ortho Pharmaceutical Corp PO Box 300 Rt 202 Raritan NJ 08869-0602

HALLETT, MARK, physician, neurologist, health research institute administrator; b. Phila., Oct. 22, 1943; s. Joseph Woodrow and Estelle (Barg) H.; m. Judith E. Peller, June 26, 1966; children—Nicholas L., Victoria C. B.A. magna cum laude, Harvard U., 1965, M.D. cum laude, 1969. Diplomate Am. Bd. Psychiatry and Neurology. Resident in neurology Mass. Gen. Hosp., Boston, 1972-75; Moseley fellow Harvard U., London, 1975-76; lectr., assoc. prof. neurology Harvard U., Boston, 1976-84; head clin. neurophysiology lab. Brigham and Women's Hosp., Boston, 1976-84; clin. dir. Nat. Inst. Neurol. Disorders and Stroke, NIH, Bethesda, Md., 1984—. Author: (with others) Entrapment Neuropathies, 1990; contbr. numerous articles to profl. jours. Bd. dirs. Easter Seal Rsch. Found., Chgo., 1985-87; mem. med. adv. bd. Nat. Parkinson Found., Miami, 1985—, Dystonia Med. Rsch. Found., Chgo., 1989—. Mem. Am. Assn. Electrodiagnostic Medicine, Am. Acad. Neurology, Am. Neurol. Assn., Am. EEG Soc., Soc. for Neurosci., Phi Beta Kappa, Alpha Omega Alpha. Democrat. Jewish. Home: 5147 Westbard Ave Bethesda MD 20816-1413 Office: Nat Inst Neurol Disorders & Stroke NIH Bldg 10 Rm 5N226 Bethesda MD 20892

HALLEY, JAMES ALFRED, physician; b. Marietta, Ohio, Oct. 15, 1948; s. John Crawford Jr. and Shirley June (Davey) H.; m. Mary Elaine Bennett, July 2, 1972; children: Aaron Bennett, Jason Allen, Joel Adam. BA, W.Va. U., 1970; DO, W.Va. Sch. Osteo. Medicine, 1978. Diplomate Am. Bd. Family Practice. Commd. ensign USN, 1971, advanced through grades to commdr., 1988; intern Bethesda (Md.) Naval Hosp., 1978-79; dept. head, sr. med. officer USS Austin, Norfolk, Va., 1979-80; family practice resident Naval Hosp., Jacksonville, Fla., 1980-82; emergency room physician Naval Region Med. Clinic, Quantico, Va., 1982-83; lab. dept. head, family practice physician Mayport (Fla.) Br. Clinic USN, 1983-85; family physician St. Vincent's Med. Ctr., Jacksonville, Fla., 1984-86, Navy Br. Med. Clinic, Jacksonville, 1985; emergency room physician, acting head Naval Hosp., Jacksonville, 1986; family physician, sr. med. officer Naval Br. Med. Clinic, Atsugi, Japan, 1986-89; family physician, residency instr. Naval Hosp., Jacksonville, 1989-92; mem. back-up team to Pres. Carter, U.S. Govt./USN, 1980; residence faculty family practice dept. Naval Hosp., Jacksonville, 1989-92, behavior sci. coord., 1990, quality assurance coord., 1990, family practice liaison for internal medicine and urology dept., 1990—; mem. med. support team Op. Desert Storm USS Guam, 1991; asst. prof. medicine Uniform Svc. U. Health Scis., Bethesda, 1991—. Treas., v.p. Ridgecrest Civic Assn., Orange Park, Fla., 1981-84; asst. cub master, asst. scoutmaster, mem. troop com. Boy Scouts Am., Atsugi, Orange Park, 1987—; asst. coach soccer, basketball Welfare and Recreation Dept., Atsugi, 1986-89; head coach boys soccer Orange Park Soccer Assn., 1991; com. chmn. pastor parish rels. com. Asbury United Meth. Ch., 1993—, mem. adminstrv. bd., trustee. Decorated Nat. Def. medal (2), Navy Sea Svcs. award, Navy Overseas Svc. award (3), Army Commendation award, South West Asea medal (2), Navy Unit Citation (2), Kuate Liberation medal, Navy Commendation award. Fellow Am. Acad. Family Physicians; mem. Am. Osteo. Assn., Assn. Mil. Surgeons of U.S., Assn. Mil. Osteo. Physicians, Fla. Osteo. Med. Assn., Westside Jacksonville Bus. Leaders Assn. Republican. Methodist. Avocations: gardening, skiing, golf, camping, tennis. Office: Family Practice Dept 7628-6 103d St Jacksonville FL 32210

HALLGREN, RICHARD EDWIN, meteorologist; b. Kersey, Pa., Mar. 15, 1932; s. Edwin Leonard and Edith Marie H.; m. Maxine Hope Anderson, Apr. 17, 1954; children—Scott, Douglas, Lynette. BS, Pa. State U., 1953, PhD, 1960; DSc (hon.), SUNY, 1989. Systems engr. IBM Corp., 1960-64; sci. adv. to asst. sec. of commerce, 1964-66; dir. world weather systems ESSA, Rockville, Md., 1966-69; asst. adminstrn. ESSA, 1969-70; asst. adminstr. NOAA, Rockville, 1970-71; asso. adminstr. environ. monitoring and prediction NOAA, 1971-73, asst. adminstr. for ocean and atmospheric scis., 1977-79; dep. dir. Nat. Weather Service, Silver Spring, Md., 1973-77; dir. Nat. Weather Service, 1979-88; exec. dir. Am. Meteorol. Soc., 1988—; permanent U.S. rep. World Meteorol. Orgn. Contbr. articles to sci. jours. Served with USAF, 1954-56. Recipient Arthur S. Flemming award U.S. C. of C., 1968, Gold medal Dept. Commerce, 1969, Internat. Meteorol. Orgn. prize World Meteorol. Orgn., 1990; named Meritorious Sr. Exec., 1980, Disting. Sr. Exec., 1986; alumni fellow Pa. State U., 1987. Fellow AAAS, Am. Meteorol. Soc. (pres., C.F. Brooks award); mem. Oceanographic Soc., Am. Geogphys. Union, Sigma Xi. Lutheran. Home: 6121 Wayside Dr Rockville MD 20852-3546 Office: Am Meteorol Svc 1701 K St Ste 300 Washington DC 20006

HALLIDAY, WILLIAM ROSS, retired physician, speleologist, writer; b. Atlanta, May 9, 1926; s. William Ross and Jane (Wakefield) H.; m. Eleanore Hartvedt, July 2, 1951 (dec. 1983); children: Marcia Lynn, Patricia Anne, William Ross III; m. Louise Baird Kinnard, May 7, 1988. BA, Swarthmore Coll., 1946; MD, George Washington U., 1948. Diplomate Am. Bd. Vocat. Experts. Intern Huntington Meml. Hosp., Pasadena, Calif., 1948-49; resident King County Hosp., Seattle, Denver Children's Hosp., L.D.S. Hosp., Salt Lake City, 1950-57; pvt. practice Seattle, 1957-65; with Wash. State Dept. Labor and Industries, Olympia, 1965-76; med. dir. Wash. State Div. Vocat. Rehab., 1976-82; staff physican N.W. Occupational Health Ctr., Seattle, 1983-84; med. dir. N.W. Vocat. Rehab. Group, Seattle, 1984, Comprehensive Med. Rehab. Ctr., Brentwood, Tenn., 1984-87; dep. coroner, King County, Wash., 1964-66. Author: Adventure Is Underground, 1959, Depths of the Earth, 1966, 76, American Caves and Caving, 1974, 82; co-author: (with Robert Nymeyer) Carlsbad Cavern: The Early Years, 1991; editor Jour. Spelean History, 1968-73; contbr. articles to profl. jours. Mem. Gov.'s North Cascades Study Com., 1967-76; mem. North Cascades Conservation Coun., v.p., 1962-63; pres. Internat. Speleological Found., 1981-87; asst. dir. Internat. Glaciospeleological Survey, 1972-76. Served to lt. comdr. USNR, 1949-50, 55-57. Fellow Am. Coll. Chest Physicians, Am. Acad. Compensation Medicine, Nat. Speleological Soc. (bd. govs. 1988—, chmn. Hawaii Speleological Survey 1989—, chmn. 1st, 3d, 6th Internat. Symposia on Vulcanospeleology, chmn. Internat. Union Speleology Working Group on Volcanic Caves 1990—), Western Speleological Survey (dir. 1957-83, dir. rsch. 1983—), Explorers Club; mem. Soc. Thoracic Surgeons, AMA, Am. Congress Rehab. Medicine, Am. Coll. Legal Medicine, Wash. State Med. Assn., Tenn. State Med. Assn, King County Med. Soc., Am. Fedn. Clin. Rsch., Am. Spelean History Assn. (pres. 1968), Brit. Cave Rsch. Assn., Nat. Trust (Scotland). Clubs: Mountaineers (past trustee), Seattle Tennis.

HALLILA, BRUCE ALLAN, welding engineer; b. Washington, D.C., Nov. 2, 1950; s. Esko Ensio and Gertrude Naomi (Tilley) H.; m. Pamela Joan Guerin, Dec. 18, 1982; children: Gregory Michael Decedue, April Patrice, Andrew Allan, Joshua Scott. BSME, BS in Welding Engring., LeTourneau U., 1974. Welding engr. Chgo. Bridge & Iron Co., Houston, 1975-77, Avondale Shipyards, Inc., New Orleans, 1977-80; asst. shipbuilding supt. Avondale Shipyards, Inc., 1980-82; steel supt. Halter Marine, Inc., New Orleans, 1982; welding supt. Bell Halter, Inc., New Orleans, 1982-84; sr. welding engr. Avondale Industries, Inc., New Orleans, 1984-86; chief welding engr. Avondale Industries, Inc., 1986—; vice chmn. welding com. Ogden Corp., N.Y.C., 1984-86; welding cons. Gas Tech. Cons., Inc., Metairie, La., 1990—; CWI test proctor Am. Welding Soc., Miami, 1979—; welding industry cons. State of La VoTech Welding Coun., Metairie, 1982—; panel mem. welding R & D, Maritime Adminstrn. Mem. com. troop 33 Boy Scouts Am., 1991—. recipient Gov.'s award State of La., Baton Rouge, 1982. Mem. Am. Welding Soc. (mem. D3 com., Proposer award 1982, Dist. Meritorious award 1987, 92) named Disting. Mem. 1989), Am. Bur. Shipping (spl. com. on materials and welding), Delta Sigma Psi. Republican. Avocations: woodworking, welding, photography, mutual fund/stock market. Home: 8725 Carriage Rd New Orleans LA 70123-3605 Office: Avondale Industries Inc PO Box 50280 New Orleans LA 70150

HALLINAN, JOHN CORNELIUS, mechanical engineering consultant; b. Phila., Feb. 12, 1919; s. John Joseph and Ellen Bridget (Sullivan) H.; m. Eleanor Ruth Denny, July 7, 1945; children: Ann, Mary, Kathleen, Claire, Joan, John, Patricia, Mark, Michael, Joseph, William, Theresa. BSME, Villanova U., 1940. Design and lab. engr. Am. Bosch, Springfield, Mass., 1946-47; lab. mgr. Baldwin Lima Hamilton, Eddystone, Pa., 1947-54; rsch. engr. Caterpillar Inc., Peoria, Ill., 1954-62, lab. mgr., 1962-72, engring. mgr., 1972-85; engring. cons., Washington, Ill., 1985—. Contbr. articles to profl. jours. Trustee St. Patrick Parish, Washington, 1960—, lector, 1978—. With USN, 1943-46. Named Engr. of Yr., Peoria Engring. Coun., 1975. Mem. ASME (chmn. ctrl. Ill. sect. 1962-63, other sectional and regional offices, chmn. Soichiro Honda medal com., Honors award Internal Combustion Engine divsn. for disting. tech. svc. to diesel engine industry 1992), Soc. Automotive Engrs. Achievements include direction and management of the design and development of large engines, turbocharging of engines, conversion of diesel to spark ignited engines. Home and Office: 700 Crestview Dr Washington IL 61571-1605

HALPERIN, JOHN JACOB, neurology educator, researcher; b. Montreal, Que., Can., Jan. 25, 1950; came to U.S., 1967.; s. David M. and Maizie (Pottel) H.; m. Toula Jaravinos, June 15, 1975; 1 child, Daniel Mark. SB in Physics, MIT, 1971; MD, Harvard U., 1975. Diplomate Am. Bd. Internal Medicine, Am. Bd. Psychiatry and Neurology, Am. Bd. Electrodiagnostic Medicine. Intern, resident in medicine U. Chgo., 1975-77; resident in neurology Mass. Gen. Hosp., Boston, 1977-80, fellow, 1980-83; asst. prof. SUNY, Stony Brook, 1983-89, assoc. prof., vice chmn. dept., 1989-91, acting chmn. dept., 1990-91; chmn. dept. North Shore U. Hosp., Manhasset, NY, 1992—; assoc. prof. Cornell U. Med. Coll., 1992-93, prof. 1993—. Contbr. numerous articles to med. jours., chpts. to books. Fellow Am. Acad. Neurology, Am. Assn. for Electrodiagnostic Medicine (edn. com. 1989—, examiner 1991); mem. Soc. for Neurosci., Am. Acad. Clin. Neurophysiology, Am. Neurol. Assn. Achievements include research on electrodiagnosis, nervous system Lyme disease. Office: North Shore U Hosp Dept Neurology Manhasset NY 11030

HALPERIN, JOSEPH, entomologist; b. Lodz, Poland, July 13, 1922; came to Israel, 1946; s. Haim and Helene (Goldlust) H.; m. Sara Resnik, May 18, 1951 (div. Nov. 1957); 1 child: Efraim. MA, Hebrew U., 1957, PhD, 1971. Tchr., schoolmaster Belorussian Ministry Edn., Dubrovna, USSR, 1940-41; various schs. Israel, 1949-58; head Lab. for Forest Entomology and Protection, Ilanot, Israel, 1958-87; cons. extension service Israel Ministry of Agriculture, Tel Aviv, 1987-89, Standards Instn. Israel, Tel Aviv, 1987—. Contbr. numerous articles to profl. jours. Founder, mem. anti-Nazi clandestine unit, White Russia, 1941-43; comdr. guerilla group, Poland, 1944; founder, head youth group for Aliya to Israel, Poland, 1945. Served with Israel Army, 1947-49. Recipient Efficiency award Israeli Ministry of Agriculture, 1970. Mem. Entomology Soc. Israel. Home: Hapalmah 5, Nes Ziyona 70 400, Israel Office: The Volcani Ctr, Dept Entomology, Bet Dagan 50250, Israel

HALPERIN, JOSEPH, nuclear chemist; b. Gomel, Byelorussia, Apr. 1, 1923; came to U.S., 1923.; s. Aaron and Sarah (Markovitz) H.; m. Sita Hamilton, Apr. 13, 1951. BS, U. Chgo., 1943, DPhil, 1951. Chemist Oak Ridge (Tenn.) Nat. Lab., 1951-87; cons., 1987—. Contbr. articles to profl. jours. With Corps Eng., 1943-46. Fellow AAAS; mem. Am. Chem. Soc., Am. Phys. Soc., Am. Nuclear Soc. Achievements include contributions in measurement and evaluation of neutron cross sections significant to the operation of nuclear reactors and to the production of important isotopic products; research in isotopic oxygen-transfer in certain oxidation-reduction reactions. Home and Office: 8904 SW 42d Pl Gainesville FL 32608

HALPERN, BRUCE PETER, physiologist, consultant; b. Newark, Aug. 18, 1933; s. Leo and Thelma (Rubin) H.; m. Pauline Touber Anklowitz, June 9, 1956; children: Michael Touber, Stacey Rachael. A.B., Rutgers U., 1955; M.Sc., Brown U., 1957, Ph.D., 1959. Asst. prof. physiology SUNY Health Sci. Ctr., Syracuse, N.Y., 1961-66; assoc. prof. psychology, neurobiology and behavior Cornell U., Ithaca, N.Y., 1966-73, prof., 1973—, chmn. dept. psychology, 1974-80, 91—; mem. Adv. Panel Sensory Physiology and Perception NSF, 1976-79; mem. adv. com. Nat. Inst. Neurol. and Communicative Disorders and Stroke, NIH, 1978-79, 85-87, Internat. Commn. on Olfaction and Taste, Union of Physiol. Scis., 1986—; Fogarty sr. internat. fellow, vis. prof. oral physiology Osaka U., 1982-83; chmn. Gordon Conf. on Chem. Senses: Taste and Smell, 1987-90; PHS-NIMH postdoctoral fellow physiology, rsch. assoc., lectr. psychology cornell U., Ithaca. N.Y., 1959-61. Exec. editor Chem. Senses, 1984-88; contbr. articles to profl. jours. NIMH grantee, 1958-62; NIH grantee, 1963-72; NSF grantee, 1972-90. Mem. Am. Physiol. Soc., Assn. Chemoreception Scis. (exec. chair 1982-83). Office: Cornell U Uris Hall Ithaca NY 14853-7601

HALPERN, JACK, chemist, educator; b. Poland, Jan. 19, 1925; came to U.S., 1962, naturalized; s. Philip and Anna (Sass) H.; m. Helen Peritz, June 30, 1949; children: Janice Henry, Nina Phyllis. BS, McGill U., 1946, PhD, 1949; DSc (hon.), U. B.C., 1986. Postdoctorate overseas fellow NRC, U. Manchester, Eng., 1949-50; instr. chemistry U. B.C., 1950, prof., 1961-62; Nuffield Found. traveling fellow Cambridge (Eng.) U., 1959-60; prof. chemistry U. Chgo., 1962-71, Louis Block prof. chemistry, 1971-83, Louis Block Disting. Service prof., 1983—; vis. prof. U. Minn., 1962, Harvard, 1966-67, Calif. Inst. Tech., 1968-69, Princeton U., 1970-71, Max. Planck Institut, Mulheim, Fed. Republic Germany, 1983—, U. Copenhagen, 1978; Sherman Fairchild Disting. scholar Calif. Inst. Tech., 1979; guest scholar Kyoto U., 1981; Firth vis. prof. U. Sheffield, 1982, Phi Beta Kappa vis. scholar, 1990; R.B. Woodward vis. prof. Harvard U., 1991; numerous guest lectureships; cons. editor Macmillan Co., 1963-65, Oxford U. Press; cons. Am. Oil Co., Monsanto Co., Argonne Nat. Lab., IBM, Air Products Co., Enimont, Rohm and Haas; mem. adv. panel on chemistry NSF, 1967-70; mem. adv. bd. Am. Chem. Soc. Petroleum Research Fund, 1972-74; mem. medicinal chemistry sect. NIH, 1975-78, chmn., 1976-78; mem. chemistry adv. council Princeton U., 1982—; mem. univ. adv. com. Ency. Brit., 1985—; mem. chemistry vis. com. Calif. Inst. Tech., 1991—; mem. German-Am. Acad. Coun., 1993—. Assoc. editor: Inorganica Chimica Acta, Jour. Am. Chem. Soc.; co-editor: Collected Accounts of Transition Metal Chemistry, vol. 1, 1973, vol. 2, 1977; mem. editorial bd. Jour. Organometallic Chemistry, Accounts Chem. Research, Catalysis Revs., Jour. Catalysis, Jour. Molecular Catalysis, Jour. Coordination Chemistry, Gazzetta Chimica Italiana, Organometallics, Catalysis Letters, Kinetics and Catalysis Letters; co-editor Monographs on Chemistry; consulting editor Proceedings NAS, Oxford Univ. Press, Internat. Series Monographs on Chemistry; contbr. articles to rsch. jours. Trustee Gordon Rsch. Confs., 1968-70; bd. govs. David and Arthur Smart Mus., U. Chgo., 1988—; bd. dirs. Ct. Theatre, 1989—. Recipient Young Author's prize Electrochem. Soc., 1953, award in inorganic chemistry Am. Chem. Soc., 1968, award in catalysis Noble Metals Chem. Soc., London, 1976, Wilhelm von Hoffman medal German Chem. Soc., 1988, Humboldt award, 1977, Richard Kokes award Johns Hopkins U., 1978, Willard Gibbs medal, 1986, Bailar medal U. Ill., 1986, Chemical Pioneer's award Am. Inst. Chemists, 1991, Paracelsus prize Swiss Chem. Soc., 1992; Alfred P. Sloan rsch. fellow, 1959-63. Fellow AAAS, Royal Soc. London, Am. Acad. Arts and Scis., Chem. Inst. Can., Royal Soc. Chemistry London (hon.), N.Y. Acad. Scis., Japan Soc. for Promotion Sci.; mem. NAS (fgn. assoc. 1984-85, mem. coun. 1990—, chmn. chemistry sect. 1991—, v.p. 1993—), Am. Chem. Soc. (editorial bd. Advances in Chemistry series 1963-65, 78-81, chmn. inorganic chemistry div. 1971, award for disting. svc. in advancement inorganic chemistry 1985), Max Planck Soc. (sci. mem. 1983—), Art Inst. Chgo., Renaissance Soc. (bd. dirs. 1985—), Sigma Xi. Home: 5630 S Dorchester Ave Chicago IL 60637-1722 Office: U Chgo Dept Chemistry Chicago IL 60637

HALPIN, DANIEL WILLIAM, civil engineering educator, consultant; b. Covington, Ky., Sept. 29, 1938; s. Jordan W. and Gladys E. (Moore) H.; m. Maria Kirchner, Feb. 8, 1963; 1 son, Rainer. B.S., U.S. Mil. Acad., 1961; M.S.C.E., U. Ill., 1969, Ph.D., 1973. Research analyst Constrn. Engring. Research Lab., Champaign, Ill., 1970-72; faculty U. Ill., Urbana, 1972-73; mem. faculty Ga. Inst. Tech., Atlanta, 1973-85, prof., 1981-85; A.J. Clark prof., dir. Constrn. Engring. and Mgmt. U. Md., 1985-87; dir. div. Constrn. Engring. and Mgmt. Purdue U., 1987—; cons. constrn. mgmt.; vis. assoc. prof. U. Sydney, Australia, 1981; vis. prof. Swiss Fed. Inst. Tech., 1985; vis. scholar Tech. U., Munich, 1979; vis. lectr. Ctr. Cybernetics in Constrn., Bucharest, Romania, 1973; cons. office tech. assessment U.S. Congress, 1986-87; mem. JTEC Team to evaluate constrn. tech., Japan, 1990. Author: Design of Construction and Process Operations, 1976, Construction Management, 1980, Planung und Kontrolle von Bauproduktionsprozessen, 1979, Constructo - A Heuristic Game for Construction Management, 1973, Financial and Cost Control Concepts for Construction Management, 1985, Planning and Analysis of Construction Operations, 1992. Served with C.E., U.S. Army, 1961-67. Decorated Bronze Star; recipient Walter L. Huber prize ASCE, 1979, Peurifoy Constrn. Rsch. award, 1992; grantee NSF, Dept. Energy. Mem. ASCE (past sect. pres. 1981-82, chmn. constrn. rsch. coun. 1985-86, Peurifoy Constrn. Rsch. award 1992), NSF, Am. Soc. Engring. Edn., Sigma Xi. Methodist.

HALSTEAD, BRUCE WALTER, biotoxicologist; b. San Francisco, Mar. 28, 1920; s. Walter and Ethel Muriel (Shanks) H.; m. Joy Arloa Mallory, Aug. 3, 1941 (div.); m. Terri Lee Holcomb, June 25, 1988; children by previous marriage: Linda, Sandra, David, Larry, Claudia, Shari. A.A., San Francisco City Coll., 1941; B.A., U. Calif.-Berkeley, 1943; M.D., Loma Linda U., 1948. Research asst. in ichthyology Calif. Acad. Scis., 1935-43; instr. Pacific Union Coll., 1943-44; mem. faculty Loma Linda (Calif.) U.,

1948- 58; research assoc. Lab. Neurol. Research, Sch. Medicine, 1964—; dir. World Life Research Inst., Colton, Calif., 1959—, Internat. Biotoxicol. Center; research assoc. in ichthyology Los Angeles County Mus., 1964-66; instr. Walla Walla Coll., summers 1964-65; cons. to govt. agys., pvt. corps; mem. editorial staff Exerpta Medica, 1959-63, Toxicon, 1962-67; mem. joint group experts on sci. aspects marine pollution UN; Dir. Nat. Assn. Underwater Instrs., Internat. Underwater Enterprises, Internat. Bots., Inc. Author: Poisonous and Venomous Marine Animals of the World, 5 vols., 1966; others.; contbr. 280 articles to profl. jours. Fellow AAAS, Internat. Soc. Toxicology (a founder), N.Y. Acad. Scis., Royal Soc. Tropical Medicine and Hygiene; mem. Am. Inst. Biol. Scis., Am. Micros. Soc., Am. Soc. Ichthyologists and Herpetologists, Am. Soc. Limnology and Oceanography, Inst. Radiation Medicine (hon. cons.), Chinese Acad. Mil. Med. Sci., numerous others. Republican. Adventist. Office: World Life Rsch Inst 23000 Grand Terrace Rd Colton CA 92324-4999

HALSTEAD, JOHN IRVIN, II, analytical chemist; b. Oakharbor, Apr. 26, 1965; s. John Irvin Sr. and Melody Ann (Siffring) H. BS in Engring., Ariz. State U., 1988. Lab. technician Allied Signal Aerospace Co., Phoenix, Ariz., 1984-87; hazardous waste disposal technician S.W. Hazard Control Inc., Tucson, Ariz., 1989; quality control chemist Cyprus Miami Mining Co., Claypool, Ariz., 1989-90, analytical chemist, 1990--. Voting mem. City of Globe Recysling Com., 1992--. Mem. Sons Am. Legion, The Soc. for Creative Anachronism, Ariz. State U. Alumni Assn. Democrat. Lutheran. Home: PO Box 2421 Globe AZ 85502 Office: Cyprus Miami Mining Co PO Box 4444 Claypool AZ 85532

HALSTEAD, PHILIP HUBERT, geologist, consultant; b. N.Y.C., July 30, 1933; s. Walter Wright and Lucille (Joliff) H.; m. Anita Joy Vickerstaff, 1963 (div. 1971); m. Sheila Casey, June 17, 1972; children: Duncan, E. Kaitrin, Megan. Degree in Geol. Engring., Colo. Sch. Mines, 1954. Registered profl. geologist, Calif; cert. petroleum geologist. Dist. geologist Am. Overseas Petroleum, N.Y.C., 1957-69; staff geologist Chevron Overseas Petroleum, San Francisco, 1969-73; exploration mgr. Norwegian State Oil Co., Stavanger, Norway, 1973-79; geol., geophys. cons. Halstead Exploration, Denver, 1979-80; v.p. exploration Energetics, Denver, 1980-82; dist. v.p. Primary Fuels, Denver, 1982-84; v.p. Scientific Software Intercomp, Denver, 1984—; bd. dirs. Statex, Norwegian Geophys. Co., Stavenger, 1974-78. Contbr. articles to profl. jours. Mem. Am. Assn. Petroleum Geologists, Geol. Soc. Am., Soc. Petroleum Engrs., Soc. Exploration Geophysicists, Rotary. Achievements include research in North Sea faulting, optimizing oil field production with seismic, oil field development optimization with seismic, oil and gas reserves of Papna New Guinea. Home: 23675 Currant Dr Golden CO 80401

HALSTEAD, RONALD LAWRENCE, soil scientist; b. Alta., Can., Sept. 18, 1923. BSA, U. Man., Can., 1950; PhD in Soil Sci. and Microbiology, U. Wis., 1954. Rsch. scientist chem. divsn. Can. Dept. Agriculture, 1954-59, rsch. scientist Soil Rsch. Inst., 1959-75, rsch. coord. Plannin and Evaluation Directorate, Rsch. Br. and Ctrl. Experimental Farm, 1975-82, dir. gen. program coord. directorate, 1982-85, dir. gen. insts., 1985-87; ret., 1987. Fellow Can. Soc. Soil Sci.; mem. Agr. Inst. Can. (Fellowship award 1990), Internat. Soc. Soil Sci. Achievements include research in soil organic and inorganic phosphorous, sewage sludge disposal in soils. Home: 1094 Bedbrook St, Ottawa, ON Canada K2C 2R7*

HALTER, EDMUND JOHN, mechanical engineer; b. Bedford, Ohio, May 10, 1928; s. Edmund Herbert and Martha (Demske) H.; student Akron U., 1946-48; B.S. in Mech. Engring., Case Inst. Tech., Cleve., 1952; M.S. in Mech. Engring., So. Meth. U., 1965; m. Carolyn Amelia Luecke, June 29, 1955; children—John Alan, Amelia Katherine, Dianne Louise, Janet Elaine. Flight test engr., analyst Chance Vought Aircraft, Dallas, 1952-59; chief research and devel. engr. Burgess-Manning Co., Dallas, 1959-68; engring. specialist acoustics Vought div. LTV Aerospace Corp., Dallas, 1968-69; mgr. continuing engring. Maxim Silencer div. AMF Beaird, Inc., Shreveport, La., 1969-72; chief research and devel. engr. Burgess-Manning div. Burgess Industries, Dallas, 1972-79; chief engr. Vibration & Noise Engring. Corp., Dallas, 1979-92, v.p. engring., 1993—; cons. Organizer Citizen Noise Awareness Seminar, Irving, 1977. Mission ptnrs., coord. Nothern Tex., Nothern La. Synod of Evang. Luth. Ch. Am. Served with USNR, 1946-49. Registered profl. engr., Tex., Ohio; cert. fallout shelter analyst. Mem. Inst. Environ. Scis. (pres. S.W. chpt. 1977-78), Indsl. Silencer Mfrs. Assn. (chmn. 1975-77), Acoustical Soc. Am., Nat., Tex. socs. profl. engrs., ASME. Republican. Lutheran. Contbr. articles to profl. jours. Patentee in field. Home: 200 Hillcrest Ct Irving TX 75062-6900

HALULA, MADELON CLAIR, microbiologist; b. Kansas City, Kans., Jan. 18, 1955; d. Frank Raymond and Patricia Joy (Griffin) H.; m. Norman A. Palmer, Aug. 11, 1979. BA with honors, Immaculate Heart Coll., 1976; PhD, Stanford U., 1986. Rsch. asst. div. infectious disease Stanford (Calif.) U., 1980-81; postdoctoral fellow dept. microbiology Va. Commonwealth U., Richmond, 1986-88, asst. prof. dept. microbiology, 1988-91; microbiologist Nat. Inst. Allergy and Infectious Disease, NIH, Bethesda, Md., 1991—; reviewer Plasmid Jour., Proc. NAS. Contbr. articles to profl. publs. NSF honor mention, 1980. Mem. AAAS, Am. Soc. Microbiology, Sigma Xi. Office: Nat Inst Allergy and Infectious Disease NIH Solar Bldg 4C-10 Bethesda MD 20892

HALUSHYNSKY, GEORGE DOBROSLAV, systems engineer; b. Lviv, Ukraine, Jan. 10, 1935; came to U.S., 1949; s. Bohdan and Irene (Mryc) H.; m. Mary Stephany Onuferko, Aug. 6, 1960; children: Helene Irene, Martha Christine. BSEE, Case Western Reserve U., 1956; MEA, George Washington U., 1970. Engr. RCA, Camden, N.J., 1958-63, Bunker-Ramo Corp., Silver Spring, Md., 1963-66; sr. engr. Page Comms., Washington, 1966-68, Vitro Corp., Silver Spring, 1968-76; asst. prof. lectr. George Washington U., Washington, 1976-77; sr. prof. staff Johns Hopkins U. Applied Physics Lab., Laurel, Md., 1977—. Author tech. publs. Recipient commendation awards USN, 1990-91. Mem. Ops. Rsch. Soc. Am., Mil. Ops. Rsch. Soc. Office: Johns Hopkins U Applied Physics Lab Johns Hopkins Rd Laurel MD 20723

HALUSKA, GEORGE JOSEPH, biomedical scientist; b. N.Y.C., Apr. 12, 1947. BS, Cornell U., 1978, MS, 1981, PhD, 1985. Rsch. asst. Oreg. Regional Primate Rsch. Ctr., Beaverton, 1985-88, asst. scientist, 1988—. Contbr. articles to profl. jours. Mentor Washington County Roundtable for Youth, Beaverton, 1990—; Oreg. Biomed. Rsch. Network, Beaverton, 1986-88. Rsch. grantee Med. Rsch. Found. of Oreg., 1987, NIH, 1992. Mem. Soc. for Study of Reproduction (chmn. animal care com. 1986-87). Office: Oreg Regional Primate Rsch Ctr 505 NW 185th Ave Beaverton OR 97006

HALVORSEN, STANLEY WARREN, neuropharmacologist, researcher; b. Niobrara, Nebr., Feb. 21, 1953; s. Harold Albert and June (Pease) H.; m. Dawn Jan Ferguson, June 23, 1976; children: Aubrey Diane, Christopher Warren. B Pharmacy cum laude, Wash. State U., 1976; PhD Pharmacology, U. Wash., 1984. Registered pharmacist, Wash. Pharmacist USPHS Hosp., San Francisco, 1975, S.I., N.Y., 1976-77; pharmacist USPHS Clinic, Jacksonville, Fla., 1977-79, Seattle Pub. Health Hosp., 1983-84; postdoctoral researcher U. Calif., San Diego, 1984-89, asst. rsch. biologist, 1989-90; asst. prof. SUNY, Buffalo, 1990—. Ad hoc reviewer Jour. Molecular Pharmacology, 1985—, Brain Rsch. jour., 1990—, NSF, 1992; contbr. articles to profl. publs. Mem. Williamsville (N.Y.) PTA, 1990. Ltr. USPHS, 1976-79. Fellow Muscular Dystrophy Soc., N.Y.C., 1984-86; grantee Am. Heart Assn., 1992—, NSF, 1992—, Am. Assn. Coll. Pharmacy, 1991. Mem. AAAS, Soc. Neurosci. Achievements include research on regulation of ion channel function and expression in nerve cells; role of trophic factors in development and maintenance of synaptic components. Office: SUNY Buffalo Dept Biochem Pharmacology 448 Hochstetter Hall Amherst NY 14260

HALVORSEN, THOMAS GLEN, mechanical engineer; b. Mpls., Feb. 8, 1954; s. Howard Thomas and Jean (Sandberg) H.; m. Linnae Marie Johnson, Apr. 8, 1978; children: Bethany, Tara, Erik. BA, Luther Coll., Decorah, Iowa, 1975; BSME, U. Minn., 1991. Devel. engr. TSI, Inc., St. Paul, 1991—. Mem. ASME. Lutheran. Achievements include patent on reconfigurable hardswitch display. Office: TSI Inc 500 Cardigan Rd Saint Paul MN 55126-3996

HALVORSON, ARDELL DAVID, research leader, soil scientist; b. Rugby, N.D., May 31, 1945; s. Albert F. and Karen (Mygland) H.; m. Linda Kay Johnston, Apr. 11, 1966; children: Renae, Rhonda. BS, N.D. State U., 1967; MS, Colo. State U., 1969, PhD, 1971. Soil scientist Agr. Rsch. Svc., USDA, Sidney, Mont., 1971-83; soil scientist Agr. Rsch. Svc., USDA, Akron, Colo., 1983-88, rsch. leader, 1988—. Contbr. over 100 articles to profl. jours., chpts. to books. Fellow Am. Soc. Agronomy (assoc. editor 1983-87), Soil Sci. Soc. Am. (chmn. div. S-8, 1989), Soil and Water Conservation Soc. (chpt. pres. 1991); mem. Crop Sci. Soc. Am., Lions (treas. Akron club 1987-92), Masons, Elks. Office: USDA Agrl Rsch Svc PO Box 400 Akron CO 80720

HALVORSON, GARY ALFRED, soil scientist; b. Klamath Falls, Oreg., July 18, 1949; s. Alfred Rueben and Dorothy Edna (Boxrud) H.; m. Bonita Rose Hill, July 17, 1976; children: Janet, Daniel, Anne, Clifford. BA, St. Olaf Coll., Northfield, Minn., 1971; MS, Oreg. State U., 1975, PhD, 1979. Cert. profl. soil scientist. Rsch. asst. Oreg. State U., Corvallis, 1972-78; assoc. soil scientist N.D. State U., Mandan, 1979-88, soil scientist, 1989—; interim supt. Land Reclamation Rsch. Ctr., 1991—; Timeryazov Agrl. Acad. rsch. fellow, Moscow, 1978. Contbr. articles to profl. jours. Pres. Heart River Luth. Ch., Mandan, 1991-93. Mem. Am. Soc. Agronomy, Soil Sci. Soc. Am., Am. Soc. Surface Mining and Reclamation, N.D. Acad. Sci., Lions (sec. 1988-93). Democrat. Lutheran. Office: ND State U PO Box 459 Mandan ND 58554-0459

HALVORSON, HARLYN ODELL, marine life administrator, biological sciences educator; b. Mpls., May 17, 1925; s. Halvor Orin and Selma (Halvorson) H.; m. Jean Erickson, Aug. 26, 1954; children: Lisa, Philip. BS cum laude, U. Minn., 1948, MS in Biochemistry, 1950; PhD in Bacteriology, U. Ill., 1952. Instr. U. Mich. Med. Sch., 1952-54, asst. prof. dept. bacteriology, 1954-56, assoc. prof., 1956-62, prof., 1962-71; chmn. Lab. Molecular Biology, U. Wis., 1966-70; prof. biology, dir. Rosenstiel Basic Med. Scis. Rsch. Ctr. Brandeis U., Waltham, Mass., 1971-87; instr. physiology Marine Biol. Lab., Woods Hole, Mass., 1962-65, 67, investigator, 1964—, mem. exec. com., 1970-76, instr. microbial ecology, 1979, 81-84, trustee, 1978—, pres., dir., 1987—; mem. adv. bd. Q.M. rsch. NAS-NRC, 1958—; vis. prof. bacteriology, U. Wash., 1959; instr. molecular basis of regulation, Bergen, Norway, 1968, differentiation Hebrew U., Jerusalem, 1965, RNA-DNA hybridization, Naples, Italy, 1966; career prof. NIH, 1963-71; vis. investigator Lab. Enzymology Central Nat. de Recherche Sci., Gif-sur-Yvette, France, 1965-66; mem. com. mariner sterilization, cons. NASA; mem. space sci. bd., planetary biol. and chem. evolution NRC; rsch. adv. com. USDA; mem. Fogarty Internat. Fellowship Rev. Panel; bd. dirs. Biosci. Information Svc., 1982-85, chmn. bd., 1986-87; adv. bd. Nat. Downs Syndrome Soc., 1980-84; adv. bd. Stazione Zoologica, Naples, 1988—. Author: Microbial Dormancy, 1964; assoc. editor, editor series on molecular biology Harper Row; assoc. editor Jour. Bacteriology, Accounts of Chem. Rsch., Analytical Biochemistry, Archives Biochemistry and Biophysics, Bacteriol. Revs., Archives Chemistry and Biophysics. Mem. bd. higher edn. Luth. Coll., 1964-67; trustee Found. Microbiology, 1972, treas., 1983-87; bd. visitors Boston U., 1979-86; trustee Selman Waksman Found. With USN, 1943-46. USPHS fellow, 1951-52, Merck sr. postdoctoral fellow Pasteur Inst., Paris, 1955-56. Mem. NAS, NAS Inst. Medicine, Am. Inst. Biol. Scis. (cons., lectr. 1970, editor 25th anniversary vol.), Am. Acad. Microbiology (bd. govs.), Am. Soc. Microbiology (v.p. 1975-76, pres. 1976-77, chmn. pub. sci. affairs bd. 1979-89, coun. policy com., councillor-at-large, com. on monographs, lectr. 1969, chmn. physiology div. 1961), Assn. Biomed. Rsch. (founding mem., exec. com. 1980-84), Internat. Rsch. Orgn., Coun. Sci. Soc. Pres.' (chmn.), Am. Acad. Sci., Am. Acad. Arts and Sci., Assn. Am. Med. Colls. (planning com.), Am. Chem. Soc., Am. Cancer Soc., (etiology adv. panel), Sigma Xi, Alpha Chi Sigma.

HALVORSON, WILLIAM ARTHUR, economic research consultant; b. Menomonie, Wis., June 26, 1928; s. George Henry and Katherine Eileen (Dietsche) H.; m. Patricia Janet von Trebra, Dec. 27, 1951; children: Robert, James, Janet, Audrey, Katherine. Student Stout Inst., 1945-46, U. Mich., 1948; B.B.A., U. Wis., 1950, M.B.A., 1951. Registered investment advisor. Asst. group actuary N.Y. Life Ins. Co., N.Y.C., 1951-56; cons. actuary Milliman & Robertson, Inc., San Francisco, 1956-61, Milw., 1961-83, exec. v.p., 1972-81; founder Halvorson Research Assocs., econ. and investment research cons., 1983—; gen. ptnr. HRA Partnership Ltd., 1992-96. Contbg. author: Group Insurance Handbook, 1965; pub. of investment letter. Served with AUS, 1946-47. Recipient Alumni award Menomonie High Sch., 1945. Fellow Soc. Actuaries (v.p. 1973-75, pres. 1977-78); mem. Wis. Actuarial Club (pres. 1964-65), Am. Acad. Actuaries (sec. 1971-73, pres. 1981-82), Beta Gamma Sigma, Phi Kappa Phi, Chi Phi. Roman Catholic. Clubs: Watertown Golf; Cherokee Golf (Madison); Wyndemere Country (Naples, Fla.). Home: 2550 Windward Way Naples FL 33940-4068 Office: 2900 14th St N Naples FL 33940-4501

HAM, GEORGE ELDON, soil microbiologist, educator; b. Ft. Dodge, Iowa, May 22, 1939; s. Eldon Henry and Thelma (Ham) H.; m. Alice Susan Bormann, Jan. 11, 1964; children: Philip, David, Steven. BS, Iowa State U., 1961, MS, 1963, PhD, 1967. Asst. prof. dept soil sci. U. Minn., St. Paul, 1967-71, assoc. prof., 1971-77, prof., 1977-80; prof., head dept. agronomy Kans. State U., Manhattan, 1980-89; assoc. dean Coll. Agr., assoc. dir. Kans. Agr. Expt. Sta., 1989—; bd. dir. Kans. Crop Improvement Assn., Manhattan, Kans. Fertilizer and Chem. Inst., Topeka, Kans. Crops and Soils Industry Coun., Manhattan; cons. Internat. Atomic Energy Agy., Vienna, Austria, 1973-79. Assoc. editor Agronomy Jour., 1979-84. Contbr. articles to profl. jours. and biol. nitrogen fixation rsch. Asst. scoutmaster Indianhead coun. Boy Scouts Am., St. Paul, 1977-80; pres. North Star Little League, St. Paul, 1979-80. Sgt. U.S. Army, 1963-69. Fellow AAAS, Am. Soc. Agronomy, Soil Sci. Soc. Am.; mem. Crop Sci. Soc. Am. Sigma Xi, Gamma Sigma Delta, Phi Kappa Phi. Home: 2957 Nevada St Manhattan KS 66502-2355 Office: Kans State U Agr Expt Sta 113 Waters Hall Manhattan KS 66506

HAM, INYONG, industrial engineering educator; b. Hwangzu, Korea, Dec. 22, 1925; came to U.S., 1954, naturalized, 1975; s. Dukjung and Kwangdo (Kim) H.; m. Hyunduk Kim, Nov. 14, 1949; children: Taewoo, Taewuk. B.Engring., Seoul Nat. U., Korea, 1948; M.Sc., U. Nebr., 1956; Ph.D., U. Wis., 1958, hon. doctorate, Nanjing Aero. Inst., People's Republic China, 1988. Prof. indsl. engring. Pa. State U., University Park, 1958—, FANUC prof., 1989-92, Disting. prof., 1991—; dir. Mfg. Rsch. Ctr., 1990-92; dir. industry and asst. min. of industry Republic of Korea, Seoul, 1960-62; cons. Asian Productivity Orgn., Tokyo, UN Indsl. Devel. Orgn., World Bank, others; Fulbright prof. USSR, 1981; cons. prof. Xian Jiatong U., Beijing Inst. Tech., Harvin Inst. Tech.; hon. prof. Jiling U. Tech., Yinbin U., People's Republic China; chair, vis. prof. U. Tokyo, 1989, Russel Severance Springer vis. prof. U. Calif., Berkeley, 1991. Author: Design of Cutting Tools, 1968; Group Technology, 1985. Recipient CAM-I award Computer Aided Mfg. Internat., 1978, Disting. Svc. citation, U. Wis., 1985. Fellow ASME, Inst. Indsl. Engring. (mfg. systems div. award 1981); mem. N.Am. Mfg. Rsch. Inst. (pres. 1985-86), Internat. Inst. Prodn. Engring. Rsch. (coun. 1983-85, pres.-elect 1994), Soc. of Mfg. Engrs. (Internat. Edn. award 1985, Albert M. Sargent Progress award 1990), Korea Scientist and Engrs. in Am. (pres. 1973-74). Home: 980 Mccormick Ave State College PA 16801-6529 Office: Pa State U Dept Indsl Engring University Park PA 16802

HAM, JAMES MILTON, engineering educator; b. Coboconk, Ont., Can., Sept. 21, 1920; s. James Arthur and Harriet Boomer (Gandier) H.; m. Mary Caroline Augustine, June 4, 1955; chidren: Peter Stace, Mary Martha, Jane Elizabeth (dec.). B.A.Sc., U. Toronto, 1943; S.M., MIT, 1947, Sc.D., 1952; D.ès Sc.A., U. Montreal, 1973; D.Sc., Queen's U., 1974, U. N.B., 1979, McGill U., 1979, McMaster U., 1980; LL.D., U. Man., 1980, Hanyang U., Seoul, Korea, 1981, Concordia Coll., 1983; D.Eng., N.S. Tech. U., 1980, Meml. U., 1981, U. Toronto, 1991; D Sacred Letters, Wycliffe Coll., U. Toronto, 1983; DSc, U. Guelph, 1992. Lectr., housemaster Ajax div. U. Toronto, 1945-46; rsch. assoc. MIT, 1949-51, asst. prof. elec. engring., 1951-52; mem. faculty U. Toronto, 1952-88, head elec. engring., 1964-66, dean faculty applied sci. and engring., 1966-73, chmn. research bd., 1974-76, dean grad. studies, 1976-78, pres., 1978-83, prof. sci., tech. and pub. policy, 1983-88, prof. and pres. emeritus, 1990—; adv. to pres. Can. Inst. for Advanced Research, Toronto, 1988-90; v.p. Can. Acad. Engring., 1988-89, pres., 1990-91; fellow New Coll., 1962; vis. scientist U. Cambridge (Eng.) and USSR, 1960-61; dir. Shell Can. Ltd.; fellow Brookings Instn., 1983-84; chmn. Indsl. Disease Standards Panel, 1985-87. Author: (with G.R. Slemon) Scientific Basis of Electrical Engineering, 1961, Royal Commission on Health and Safety of Workers in Mines, 1976. Bd. govs. Ont. Res. Fedn. Served with Royal Canadian Navy, 1944-45. Decorated Officer Order of Can., 1981; recipient Sci. medal Brit. Assn. Advancement Sci., 1943; Centennial medal Can., 1967; Engring. Alumni medal, 1973; Engring. medal Assoc. Profl. Engrs. Ont., 1974, Gold medal, 1984; Queen's Jubilee medal, 1977; Order of Ont., 1989; confederation medal, Can. 1992. Fellow Engring. Inst. Can. (Sir John Kennedy medal 1983), IEEE (McNaughton medal 1977), Can. Acad. Engring.; mem. Sigma Xi. Home: 135 Glencairn Ave, Toronto, ON Canada M4R 1N1

HAMADA, HIROKI, science educator; b. Tsushima-cho, Ehime-ken, Japan, Aug. 3, 1952; s. Kikutarou and Konami Hamada; m. Hideko Itazaki, Nov. 30, 1981; children: Manabu, Megumi. MS, Hiroshima (Japan) U., 1981, PhD in Sci., 1987. Asst. prof. Okayama (Japan) U. of Sci., 1983-88, assoc. prof., 1988—. Office: Okayama U Sci, 1-1 Ridai-cho, Okayama 700, Japan

HAMADA, JIN, biologist; b. Kyoto, Japan, Nov. 5, 1945; s. Minoru and Chikako (Honda) H.; m. Kaoru Yasutomi, Jan. 13, 1980; children: Miho, Haruna, Ran, Momoyo. BAgr, Kyoto U., 1968, MAgr, 1970, DAgr, 1979. Postdoctoral fellow dept. biology U. Pa., Phila., 1975-77; rsch. assoc. Toyama (Japan) Med. and Pharm. U., 1980—; lectr. Bukkyo U., Kyoto, 1978-80, Toyama U., 1981—, Toyama Gen. Nurse Sch., 1987-89. Author: Biology of Conjugales, 1990, 2d edit., 1991, Effects of Chemicals on Algae, 1990, Imported tRNA, 1976. Advisor Soc. for the Protection of Natural Environ. of Imizu, Toyama, 1987—. Mem. Internat. Phycological Soc., Phycological Soc. Am., Japanese Soc. Phycology, Botanical Soc. Japan, Genetics Soc. Japan. Buddhism. Avocations: classical music, looking arts, taking photographs of flowers. Home: 9-44 Minamitaikoyama, Kosugi-machi, Imizu-gun Toyama 939-03, Japan Office: Toyama Med and Pharm Univ, 2630 Sugitani, Toyama 930-01, Japan

HAMADA, KEINOSUKE, chemistry educator emeritus; b. Fukuyama, Hiroshima, Japan, Nov. 22, 1924; s. Jitsuo and Fumiko Hamada; m. Kazu Iguchi, Jan. 17, 1942; children: Kaori, Satoshi. BS, Hiroshima U., 1951, DSc, 1961. From asst. to lectr. chemistry Hiroshima U., 1957-65; asst. prof. chemistry Kinki U., Osaka, 1965-68; postdoctoral fellow Toronto (Can.) U., 1968-70; from asst. to prof. chemistry Nagasaki U., 1968-90, prof. emeritus, 1990—; researcher in Raman and IR spectroscopy. Author: Chemical Reaction Based on New Thermodynamics, 1986, Q&A on Basic Chemistry, 1989, Thermodynamic Consideration of Electrical Power Generated by Temperature Difference of Ocean Water, 1989, Absolute Potential of Standard Hydrogen Electrode-Water Cell, 1989, Molecular Structures of Ethylene, Allene and Ethylene Oxide and Bridged Pi-Bond, 1989. Grantee sci. rsch., Japanese Govt., Tokyo, 1972. Mem. Chem. Soc. Japan. Avocation: golf. Home: Keyamyo 20-134, Tarami-cho Nagasaki 859-04, Japan

HAMADA, NOBUHIRO A., electrical engineer, researcher; b. Osaka, Japan, Feb. 3, 1944; s. Atsumi and Kiku (Akao) H.; m. Tamiko Iio, Nov. 3, 1971; children: Yuki, Kazuaki. BS in Elect. Engring., Osaka U., 1966, MS in Elec. Engring., 1968, PhD, 1977. Control systems researcher Hitachi Rsch. Lab., Hitachi, Ibaraki, Japan, 1968-70; software engring. researcher Hitachi, Ltd., Hitachi, Ibaraki, 1971-77, researcher distributed control systems, 1978-81, sr. researcher on local area network and various systems, 1981-90, chief researcher Urban Systems, 1990-92; chief engr. systems engring. div. Hitachi, Ltd., Tokyo, 1992—; vis. scholar Carnegie-Mellon U., 1977-78; lectr. micro-computers Engring. Dept. Ibaraki U., 1984-91. Inventor: Automated Warehouse, 1974, Distributed Control System, 1985. Mem. IEEE (sr. mem.), Assn. for Computing Machinery, Am. Assn. for Artificial Intelligence, Japan Inst. Electrical Engring, Info. Processing Soc. Japan, Soc. of Instrument and Control Engring., Inst. Systems, Controls & Info. Engring. Buddhist. Avocations: personal computing, classical music, history. Home: 810 Isobe, Ibaraki, Hitachi-Ota 313, Japan Japan Office: Hitachi Ltd, Systems Engring Div, 4-6 Kanda-Surugadai, Chiyodaku, Tokyo 101, Japan

HAMAGUCHI, SATOSHI, mathematician; b. Yokohama, Japan, Nov. 22, 1959; s. Takeshi and Chiyono (Hashimoto) H.; m. Angelina G.H. Yap, Sept. 28, 1991. BS in Physics, U. Tokyo, 1982, MS in Physics, 1984; MS in Math., NYU, 1986, PhD in Math., 1988. Rsch. fellow Inst. for Fusion Studies U. Tex., Austin, 1988-90; rsch. staff mem. IBM Thomas J. Watson Rsch. Ctr., Yorktown Heights, N.Y., 1990—; adj. asst. prof. Columbia U., N.Y.C., 1991—. Contbr. articles to profl. jours. Recipient Fulbright scholarship U.S.-Japan Ednl. Commn., 1984. Mem. Am. Phys. Soc. Achievements include development of theory on plasma turbulence, transport and plasma-material interactions. Office: IBM TJ Watson Rsch Ctr PO Box 218 Yorktown Heights NY 10598

HAMAKER, RICHARD FRANKLIN, engineer; b. Lynchburg, Va., Jan. 10, 1924; s. John Irvin and Ray (Parker) H.; m. Marjorie Wrigley, Dec. 23, 1944 (div. 1974); children: Laurel Elisa, Lawrence Walter. BS in Mechanical Engring., MIT, 1946, MS in Mechanical Engring., 1953. Architect, engr. Lynchburg, Va., 1947-50; exec. officer MIT Dynamic Analysis Lab., Cambridge, Mass., 1951-53; magr. data processing Bendix Guided Missle div., Mishawaka, Ind., 1953-59; mgr. computer ctr. Mobil Oil, N.Y.C., 1960-61; systems engr. ITT, Huntsville, Ala., 1962-65; adv. engr. IBM, Research Triangle Park, N.C., 1965-74; owner Hamaker Woodcrafts, Durham, N.C., 1975—; speaker Am. Mgmt. Assn. seminars, N.Y., 1958-62, IBM customer exec. seminars, N.Y., Calif., 1959. Tech. dir. Durham Savovards, 1974-76; mill wright Friends West Point, Durham, 1975-77. Lt. (j.g.) USNR, 1943-46. Mem. Ind. Scholars Assn., Sigma Xi, Tau Beta Pi, Torch Club. Home: 3500 Chapel Hill Durham NC 27707

HAMBLIN, DANIEL MORGAN, economist; b. Kansas City, Kans., Oct. 7, 1942; s. Enright Morgan and Helen Ruth (Cain) H.; m. Rebecca Menn, Dec. 6, 1969 (div. Aug. 1988); children: Caroline Helen, David Thorpe; m. Judith Killen, Nov. 21, 1990 (div. Mar. 1992). BA in Maths., U. Kans., 1972, BA in Econ. with hon., 1972; PhD in Applied Econs., SUNY, Buffalo, 1979. Systems engr. Smoot Co., Kans. City, 1968-70; asst. prof. econs. U. Wis., Parkside Kenosha, 1977-80; rsch. assoc. Oak Ridge (Tenn.) Nat. Lab., 1980-82, leader demand analysis group, 1982-86; project mgr. Battelle Columbus (Ohio) Div., 1986-89; pres Dan Hamblin & Assocs., Inc., Conway, Ark., 1989—. Editor, co-author natural gas utility integrated resource planning rev. Gas Rsch. Inst., 1992—; contbr. (book chpts.) Forecasting U.S. Electricity Demand, 1985, Energy Sources: Conservation and Renewables, 1985. With USN, 1961-65. Mem. AAAS, Am. Econ. Assn., Am. Statis. Assn., Ops. Rsch. Soc. Am., Internat. Assn. Energy Econs., Iron and Steel Soc. Democrat. Achievements include developer of stock market game, stock transactions tape processor, engineering/econ. end-use-based load forecasting and conservation policy models, new technology market forecasting tools, commercial HVAC equipment, choice forecasting tools, others. Home: 16 Ironwood Conway AR 72032-3626 Office: 915 Oak St Ste 106 Conway AR 72032-4371

HAMBLING, MILTON HERBERT, medical virologist consultant, lecturer; b. Southampton, Hampshire, Eng., Mar. 15, 1926; s. Herbert William Stanley and Kathleen Anna (Milton) H.; m. Diane Alderton, Mar. 29, 1952; children: Susan Gillian, Peter Timothy William. MB BS, St. Bartholomews Hosp. Med. Sch., London, 1951. M.R.C.S., England; L.R.C.P., London; Diplomate (obstetrics) R.C.O.G.; M.D., London; Diploma in Bacteriology, London. House surgeon, house physician Redhill County Hosp., Redhill, Surrey, Eng., 1951-52; sr. house officer in pathology Queen Elizabeth Hosp., Birmingham, Eng., 1955-56; registrar in pathology Radcliffe Infirmary, Oxford, Eng., 1956-57; grad. asst. in bacteriology Oxford (Eng.) U., 1957-58; asst. bacteriologist Pub. Health Lab. Svc., Colindale London, 1959-62, sr. bacteriologist, 1962-64; cons. virologist, dept. head Pub. Health Lab. Svc., Leeds, Eng., 1964—; ret., 1991; honorary cons. virologist The Yorkshire Regional Health Authority, West Yorkshire Police, The Yorkshire Clinic; sr. clin. lectr. Microbiology Dept., U. Leeds; mem. The Regional Working Party on AIDS, PHLS Working Group on AIDS, Leeds Standing AIDS Action Group. Contbr. numerous articles on virology to medical and profl. jours. Served to flight lt. with RAF, 1953-55. Recipient Grade A Merit award Distinction Award Com. Dept. of Health, London, 1987. Fellow The Royal Coll. of Pathologists; mem. British Med. Assn., Med. Defence Union, Yorkshire Bone Marrow Transplant Group. Mem. Ch. of England. Home: 2 Elmete Close, Leeds LS8 2LD, England

HAMBURG, DAVID A., psychiatrist, foundation executive; b. Evansville, Ind., 1925. M.D., Ind. U., 1947, D.Sc. (hon.), 1976; D.Sc. (hon.), Rush U., 1977, Mt. Sinai Sch. Medicine, 1980, U. Rochester, 1981, U. Ill., Chgo., 1984, Albert Einstein Sch. Medicine, 1985, U. Pitts., U. So. Calif., Hahnemann U., 1986; LHD (hon.), Ramapo Coll., 1991, Duke U., 1993. Diplomate Am. Bd. Psychiatry and Neurology. Intern Michael Reese Hosp., Chgo., 1947-48, resident in psychiatry, 1949-50; resident in psychiatry Yale U.-New Haven Hosp., 1948-49; staff psychiatrist Brooke Army Hosp., San Antonio, 1950-52; practice medicine specializing in psychiatry, 1950-75; research psychiatrist Walter Reed Army Inst. Research, Washington, 1952-53; assoc. dir. Psychosomatic and Psychiat. Inst., Michael Reese Hosp., Chgo., 1954-56; fellow Center for Advanced Study in Behavioral Scis., Palo Alto, Calif., 1957-58, 67-68; chief Adult Psychiat. Br. NIMH, Bethesda, Md., 1958-61; prof., chmn. dept. psychiatry Stanford U. Med. Sch., 1961-72, Reed-Hodgson prof. human biology, 1972-76; Sherman Fairchild Disting. scholar Calif. Inst. Tech., Pasadena, 1974-75; pres. Inst. Medicine Nat. Acad. Scis., Washington, 1975-80; dir. div. health policy research and edn., John D. MacArthur prof. health policy and mgmt. Harvard U., Cambridge, Mass., 1980-82; pres. Carnegie Corp., N.Y.C., 1983—; adv. com. med. research WHO, 1975-86; mem. exec. panel adv. com. Chief of Naval Ops., 1984-92; chmn. sci. adv. bd. NIMH, 1986-87; sec. Energy Adv. Bd., 1990—; Ctr. for Naval Analysis, 1984—. Bd. dirs. Rockefeller U., Mt. Med. Ctr., N.Y.C.; trustee Stanford U., 1988—, Internat. Devel. Rsch. Centre, Ottawa, Can., 1990—, Am. Mus. Natural History, N.Y.C., 1990—. Served as capt. M.C. U.S. Army, 1950-53. Recipient numerous awards including: Pres.'s medal Michael Reese Med. Ctr., 1974; A.C.P. award, 1977; MIT Bicentennial medal, 1977. Mem. Am. Psychiat. Assn. (Vestermark award 1977, Disting. Svc. award 1991), Nat. Acad. Scis. (com. on internat. security and arms control 1981-86), AAAS, (pres. 1984-85, chmn. bd. 1985-86), Am. Psychosomatic Soc., Assn. Research Nervous and Mental Disease (pres. 1967-68), Am. Acad. Arts and Scis., Phi Beta Kappa, Alpha Omega Alpha. Address: Carnegie Corp NY 437 Madison Ave New York NY 10022

HAMBY, PETER NORMAN, nuclear engineer; b. Milw., Nov. 19, 1959; s. Norman Ray and Lucille Carolyn (Albrecht) H. BS, U. N.Y.; BA, Gov.'s State U., University Park, Ill.; MA, U. Chgo.; cert. in nuclear ops., Memphis State U., 1985. Radiol. engr. Nuclear Ops. div. Commonwealth Edison Co., Chgo., 1985-87, gen. nuclear health physicist, 1987-90, prin. nuclear health physicist, 1990-92; prin. radiol. engr. Commonwealth Edison Co., Chgo., 1992—. Contbr. articles to profl. jours. Mem. Am. Nuclear Soc., Health Physics Soc., Robotics/Mfrs. Users Group, AMS, Robotic Industries Assn. Home: 17550 Roy St Lansing IL 60438-2019 Office: Commonwealth Edison HPSD 1237-Edison PO box 767 Chicago IL 60690

HAMEKA, HENDRIK FREDERIK, chemist, educator; b. Rotterdam, Holland, May 25, 1931; came to U.S., 1960, naturalized, 1963; s. Dirk C. and Johanna (Mannebeck) H.; m. Charlotte C. Procacci, Aug. 2, 1972. Drs., U. Leiden, The Netherlands, 1953, D.Sc., 1956; M.A. (hon.), U. Pa., 1971. Rsch. assoc. U. Rome, Italy, 1956-57; fellow Carnegie Inst. Tech., 1957-58; rsch. physicist N. V. Philips Lamps, Eindhoven, The Netherlands, 1958-60; asst. prof. chemistry Johns Hopkins, 1960-62; assoc. prof. chemistry U. Pa., 1962-67, prof. chemistry, 1967—; disting. vis. rsch. prof. USAF Acad., 1986-87. Author: Advanced Quantum Chemistry, 1965, Introductory Quantum Theory, 1967, Physical Chemistry, 1977; Contbr. numerous articles to sci. jours. Recipient Alexander von Humboldt prize, 1981; Alfred P. Sloan Research fellow, 1963-67. Achievements include research on theory of molecular structure and optical and magnetic properties of molecules; calculations of spin-orbit and spin-spin coupling; theory of resonance optical rotation, spectral predictions. Home: 1503 Argyle Rd Berwyn PA 19312-1905 Office: U Pa Dept Chemistry Philadelphia PA 19104

HAMEL, DAVID CHARLES, health and safety engineer; b. Plattsburgh, N.Y., Feb. 19, 1953; s. Charles Joseph and Helena Mary (Tormey) H. B.S., Fla. Inst. Tech., 1977. Cert. paramedic, Fla. Sr. safety engr. United Space Boosters, Kennedy Space Ctr., Fla., 1977-81; safety and health supr. STC Documation, Inc., Palm Bay, Fla., 1981-82, Martin Marietta Corp., Ocala, Fla., 1983; adminstr. safety, health and environ. affairs Hughes Aircraft Co., Titusville, Fla., 1983-86; pres. Pi Assocs. Inc., Orlando, Fla. and Cary, N.C., 1986—; tchr. Nat. Safety Council, Brevard Safety Council, Cocoa, Fla., 1978-84, U. Fla., 1987-91; cons. Continental Shelf, Inc., Tequesta, Fla., 1976-77. Emergency med. technician Harbor City Vol. Ambulance Squad, Melbourne, Fla., 1974-81, paramedic, 1981-91. Recipient Group Achievement award NASA, 1979, Cert. of appreciation, NASA, 1979, 81. Mem. Am. Soc. Safety Engrs. (treas. 1980-81), Am. Indsl. Hygiene Assn., World Safety Orgn. Avocations: skiing, horseback riding, hiking, camping.

HAMER, WALTER JAY, chemical consultant, science writer; b. Altoona, Pa., Nov. 5, 1907; s. Jessie James and Naomi Gertrude (Roland) H.; m. Alma Robinson, Mar. 19, 1941; 1 child, Margaret. BS, Juniata Coll., Huntingdon, Pa., 1929, DSc (hon.), 1966; PhD, Yale U., 1932. Asst. instr. Juniata Coll., 1926-29; postdoctoral fellow Yale U., New Haven, 1932-34; rsch. assoc. MIT, Cambridge, 1934-35; rsch. chemist Nat. Bur. Standards, Washington, 1935-50, chief electrochemistry, 1950-70, dir. Electrolyte Ctr., 1968-72; chem. cons., Washington, 1972—; adj. prof. Georgetown U., Cath. U., govt. agys. commerce and agr., 1940-50; rsch. chemist Manhattan Project, Washington, 1943-45; adj. examiner Civil Svc. Commn., 1948-50; cons. U.S. Dept. Def., 1951-53; vis. panel mem. Electrochemistry Lab., U. Pa., Phila., 1962-63; lectr. univs. and govt. agys. Author: (monographs) Standard Cells, 1965, Theoretical Activity Coefficients, 1968; co-author: (monograph) Halogen Acids Electrolytic Conductances, 1970; editor: Electrochemical Constants, 1953, The Structure of Electrolytic Solutions, 1959. Recipient cert. of merit Manhattan Project, 1945, OSRD, 1945; Superior Accomplishment award U.S. Dept. Commerce, 1954, 62, 65, Disting. Svc. gold medal, 1966; 1st prize for paper IEEE, 1955. Mem. The Electrochem. Soc., Inc. (hon., v.p. 1960-63, pres. 1963-64, Robert T. Foley award Nat. Capital sect. 1991), Yale Chemists Assn. (pres. 1958-61), Am. Camellia Soc. (accredited judge), Cosmos Club. Republican. Episcopalian. Achievements include discovery of the electromotive series of the elements in Molten Systems, of the primary pH Standard for Aqueous Systems from 0 to 60 degrees Celsius, of the ionization constant of water from 0 to 60 degrees Celsius; research in determining the Faraday Constant, method to set standards for electrolytic conductance, maintenance of U.S. national standard of voltage. Home and Office: Apt 305 407 Russell Ave Gaithersburg MD 20877-2829

HAMERMESH, MORTON, physicist, educator; b. N.Y.C., Dec. 27, 1915; s. Isador J. and Rose (Kornhauser) H.; m. Madeline Goldberg, 1941; children—Daniel S., Deborah R., Lawrence A. B.S., Coll. City N.Y., 1936; Ph.D., N.Y.U., 1940. Instr. physics Coll. City N.Y., 1941, Stanford, 1941-43; research asso. Radio Research Lab., Harvard, 1943-46; asst. prof. physics N.Y.U., 1946-47, asso. prof., 1947-48; sr. physicist Argonne Nat. Lab., 1948-50, asso. dir. physics div., 1950-59, dir. physics div., 1959-63, assoc. lab. dir. basic research, 1963-65; prof. U. Minn., Mpls., 1965-69, 70-86, prof. emeritus, 1986—; head Sch. Physics and Astronomy, 1965-69, 70-73; prof. physics, chmn. dept. physics State U. N.Y., Stony Brook, 1969-70. Author: Group Theory, 1962 (with B.F. Bayman), Review of Undergraduate Physics, 1986; translator: Classical Theory of Fields (by Landau and Lifshitz), 1951; numerous papers in field. Fellow Am. Phys. Soc.; mem. Research Soc. Am. Office: Univ Minn Physics Dept Minneapolis MN 55455

HAMERTON, JOHN LAURENCE, geneticist, educator; b. Brighton, Eng., Sept. 23, 1929; arrived in Can., 1969; s. Bernard John and Nora (Casey) H.; m. Irene Tuck; children: Katherine, Sarah. BS in Zoology, London U., 1951, DSc in Human Genetics, 1968. Sr. sci. officer Brit. Mus. (Natural History), London, 1956-60; lectr. Guys Hosp. Med. Sch., London, 1960-62, sr. lectr., 1962-69; assoc. prof. pediatrics dept. U. Man., Winnipeg, Can., 1969-72, prof. pediatrics dept., 1972, assoc. dean. med. sch., 1978-81, prof. dept. human genetics, 1985—, head. dept., 1985—, Disting. prof., 1987—, mem. univ. bd. govs., 1976-82, 87-90, 91-93; vis. prof. Hebrew U., Jerusalem, 1975; rsch. prof. Med. Rsch. Coun. Can., 1981-82; v.p. Man. Health Rsch. Coun., Winnipeg, 1988—, chmn., 1990—; v.p. XIVth Internat. Congress of Genetics, Toronto, 1988. Author: Chromosomes in Medicine, 1962, Human Cytogenetics, Vols. I and II, 1971; contbr. over 150 articles to profl. jours. Chmn. Can. Sheep Coun., 1988, 89, 90, Man. Livestock Performance Testing Bd., 1988-90. Recipient Robert Roessler de Villiers award Leukemia Soc. Am., 1956, Huxley Meml. medal Imperial Coll. Sci. and Tech., 1958, Teddy rsch. award Children's Hosp. Winnipeg Rsch. Found., Inc., 1987. Fellow Inst. Biology (U.K.), Can. Coll. Med. Geneticists; mem. Royal Coll. Sci. (assoc.), Man. Sheep Assn. (pres. 1986-89). Avocation: sheep farming. Home: RR 2 Box 111, Dugald, MB Canada R0E 0K0 Office: U Man Dept Genetics, 250-770 Bannatyne Ave, Winnipeg, MB Canada R3E 0W3*

HAMID, SYED HALIM, chemical engineer; b. Bhopal, India, Feb. 24, 1953; arrived in Saudi Arabia, 1978; s. Wasim Hamid Rizvi and Safia Naim Redhwi, Feb. 4, 1978; children: Faisal, Ahmad, Salma, Ali. MS, King Fahd U. Petroleum & Minerals, Dhahran, Saudi Arabia, 1980; PhD, City U., London, 1988. Rsch. asst. Rsch. Inst., King Fahd U. of Petroleum and Minerals, 1980-85, sr. rsch. asst., 1985-89, assoc. prof. and project mgr., 1989—; mem. UN Environ. Program, Com. on Effects of Ozone Depletion, N.Y.C., 1989—. Editor: Handbook of Polymer Degradation, 1992; contbr. articles to profl. jours. Recipient Cert. of Appreciation World Oil and Gas Conf. Orgn., 1981; vis. scholar U. Conn., 1991, 92. Mem. ASTM, N.Y. Acad. Scis., Am. Chem. Soc., Soc. of Plastic Engrs. Achievements include research of effect of ozone depletion on material lifetime, increased durability of polymers in near equatorial regions, role of degradable plastics in harsh climate. Home and Office: Rsch Inst King Fahd, U of Petroleum/Minerals, Dhahran 31261, Saudi Arabia

HAMILL, BRUCE W., psychologist; b. Boston, June 12, 1941; s. Ernest Clark and Mary Elizabeth (Kinsman) H.; m. Martha Elizabeth Tudbury, Feb. 26, 1966; 1 child, Jonathan Morrison. BA in Econs., Tufts U., 1964; MA in Psychology, Johns Hopkins U., 1972, PhD in Psychology, 1974. Asst. to dean Johns Hopkins U., Balt., 1967-70, rsch. assoc. in neurology Sch. Medicine, 1974-75, sr. staff psychologist Applied Physics Lab., 1979—; asst. prof. psychology Towson (Md.) State U., 1974-77; rsch. psychologist U.S. Army Rsch. Inst., Alexandria, Va., 1977-79; sci. officer Intergovtl. Per. Mobilization Act, Office Naval Rsch., Arlington, Va., 1987-88. Editor: The Role of Language in Problem Solving, Vol. 1, 1985, Vol. 2, 1987; contbr. articles to profl. pubis. 1st lt. U.S. Army, 1964-67, Vietnam. NIH postdoctoral rsch. fellow, 1974-75, Parsons Rsch. fellow, 1988-89. Mem. Am. Psychol. Soc., Cognitive Sci. Soc., Sigma Xi. Achievements include research on perceptual and cognitive issues in design and use of computer-based displays and automated systems. Office: Johns Hopkins U Applied Physics Lab Johns Hopkins Rd Laurel MD 20723-6099

HAMILL, CAROL, biologist, writing instructor; b. San Diego, July 15, 1953; d. William David Sr. and Katharine Louise (Garlock) H.; m. Apr. 1, 1978 (div. Feb. 1980); 1 child, Jason John Voutas. Student, San Diego State U., 1972-73, Worcester State Coll., 1975-78; BS in Natural Scis., Worcester State Coll., Riverside, 1980; postgrad., Calif. State U., San Bernardino, 1986-92. Lab. asst. Worcester State Coll., 1976-78; ind. biologist Calif., 1983-86; lab. asst. Calif. State U., San Bernardino, 1987-88; mem. adj. faculty Riverside C.C., 1990-93; lab. asst. dept. biochemistry U. Calif., Riverside, 1979-80, instr., 1992—; freelance writer, 1988—; speaker in field. Contbr. articles to Off Duty, Natural Food and Farming, Country Rev., Sr. Highlights, Our Town, Palm and Pine, and others, over 50 articles to nat. and local mags. Mem. publicity com. Orange Empire Rwy. Mus., Perris, Calif., 1989—. Mem. AAAS, N.Y. Acad. Scis., Ecol. Soc. Am. Democrat. Episcopalian. Achievements include research on desert plant Mohave Yucca (Yucca schidigera). Home: 11681 Dalehurst Rd Moreno Valley CA 92555 Office: PO Box 7960 Moreno Valley CA 92552-7960

HAMILTON, JOHN CARL, astronomer, telescope operator; b. Norfolk, Va., Dec. 31, 1955; s. Martin Ridley and Helen Ines (Larsen) H.; m. Ginger Leialoha Wright, Mar. 24, 1984; 1 child, Scott Kauluwela. BS in Physics with honors, U. Tex.-Austin, 1977, BA in Astronomy with honors, 1977; MS in Astronomy, U. Hawaii-Manoa, 1980. Lectr. U. Hawaii-Manoa, Honolulu, 1977-80; observer Mees Solar Obs., Haleakala, Hawaii, 1980-81; operator Laser-Ranging Experiment U. Hawaii L.U.R.E. Obs., Haleakala, 1981-82; telescope operator NASA Infr-Red Telescope Facility I.R.T.F., Mauna Kea, Hawaii, 1982-84, Can.-France Hawaii Telescope, Mauna Kea, Hawaii, 1984—. Mem. Astron. Telescope Operators Rsch. Astronomy, Soc. Photo-Optical Instrumentation Engrs., Astron. Soc. Pacific, Am. Astron. Soc. Home: 27-985 Old Mamalahoa Hwy Pepeekeo HI 96783 Office: Can-France Hawaii Telescope PO Box 1597 Kamuela HI 96743

HAMILTON, LEONARD DERWENT, physician, molecular biologist; b. Manchester, Eng., May 7, 1921; came to U.S., 1949, naturalized, 1964; s. Jacob and Sara (Sandelson) H.; m. Ann Twynam Blake, July 20, 1945; children—Jane Derwent, Stephen David, Robin Michael. B.A., Balliol Coll., Oxford U., Eng., 1943, B.M., 1945, M.A., 1946, D.M., 1951; M.A., Trinity Coll., Cambridge U., Eng., 1948, Ph.D., 1952. Diplomate Am. Bd. Pathology. USPHS research fellow U. Utah, 1949-50; mem. staff Sloan-Kettering Inst., N.Y.C., 1950-79, head isotope studies sect., 1957-64, assoc. scientist, 1965-79; mem. staff Meml. Hosp., N.Y.C., 1950-65; mem. faculty Sloan-Kettering div. Grad. Sch. Med. Scis. Cornell U., 1956-64; sr. scientist, head div. microbiology Med. Research Ctr. Brookhaven Nat. Lab., Upton, N.Y., 1964-76; head biomed. and environ. assessment div. Office. Environ. Policy Analysis, 1973—; attending physician Hosp. Med. Research Ctr., 1964-85; dir. WHO Collaborating Ctr. for Assessment of Health and Environ. Effects of Energy Systems, 1983—, WHO focal point on health and environ. effects of energy systems and mem. expert adv. panel on environ. hazards, 1983—; prof. medicine Health Sci. Ctr., SUNY, Stony Brook, 1968—; cons. HEW, Ctr. Disease Control, Nat. Inst. Occupational Safety and Health, epidemiology study of Portsmouth Naval Shipyard, 1978-88; vis. fellow St. Catherine's Coll., Oxford U., 1972-73; mem. internat. panel experts on fossil fuel UN Environment Programme, 1978, panel on nuclear energy, 1978-79, panel on renewable sources and comparative assessment of different sources, 1980; mem. various coms. Nat. Acad. Sci.-NRC, Washington, 1975-80; mem. N.Y.C. Mayor's Tech. Adv. Com. on Radiation, 1963-77, N.Y.C. Commr. of Health Tech. Adv. Com. on Radiation, 1978—; mem. energy panel WHO Commn. on Health & Environment, 1990-91; mem. Interant. Expert Group 3, Comparative Environ. and Health Effects of Different Energy Systems for Electricity Generation, 1990-91; sr. expert Symposium on Electricity and the Environ., Helsinki, Finland, 1991. Editor: Gerrard Winstanley, Selections from His Works, 1944; Physical Factors and Modification of Radiation Injury, 1964; The Health and Environmental Effects of Electricity Generation—a Preliminary Report, 1974. Recipient Fed. Lab. Consortium award, 1990; Am. Cancer Soc. scholar, 1953-58; Commonwealth Fund grantee, 1955-62. Mem. Am. Assn. Cancer Rsch., Am. Soc. Clin. Investigation, Am. Soc. for Investigative Pathology, Soc. for Risk Analysis, Harvey Soc., Internat. Soc. Environ. Epidemiology, Cosmos Club (Washington). Club: Cosmos (Washington). Home: Childs Ln Old Field Setauket NY 11733 Office: Brookhaven Nat Lab Upton NY 11973

HAMILTON, LINDA HELEN, clinical psychologist; b. N.Y.C., Dec. 2, 1952; d. Peter and Helen (Casey) Homek; m. Terrence White, Aug. 10, 1974 (div. 1983); m. William Garnett Hamilton, Dec. 29, 1984. BA summa cum laude, Fordham U., 1984; MA, Adelphi U., 1986, PhD, 1989. Lic. psychologist, N.Y. Dancer N.Y.C. Ballet, 1969-88; clin. psychologist Fair Oaks Hosp., Summit, N.J., 1989-90, Miller Inst. for Performing Artists, N.Y.C., 1989—; pvt. practice N.Y.C., 1991—; rsch. assoc. Miller Inst. Performing Artists, N.Y.C., 1987—; chair dance com. MedArt U.S.A., N.Y.C., 1990-92; cons. psychologist Sch. Am. Ballet, N.Y.C., 1991—; advice columnist Dance Mag., 1992—; co-leader Performing Arts Medicine Delegation to Russia and Ea. Europe, 1992. Contbr. articles to profl. jours. Miller Inst. Performing Artists grantee, 1987. Mem. APA (Daniel E. Berlyne award 1993), Soc. for Exploration of Psychotherapy Integration, Phi Kappa Phi. Avocations: travel, reading, swimming, opera, ballet. Office: 30 W 60th St New York NY 10023

HAMILTON, MICHAEL BRUCE, engineer; b. Portsmouth, Ohio, Nov. 21, 1939; s. Lewis Robert and Frances Louise (Wente) H.; m. Marilee Thomas,May 27, 1962; children: Christopher Ian, Suzanne Elizabeth. BSME, U. Cin., 1962; MBA, Gannon U., 1974. Engr. Pratt & Whitney Aircraft, West Palm Beach, Fla., 1962-65; engr., cons. Bellan Engring., Cin., 1965-71; engr., diesel design engr. GE Electric Locomotives, Erie, Pa., 1971-77; engr. turbine design GE Electric Aircraft Engines, Cin., 1977-85, mgr. mech. systems, 1985-88, program mgr., 1988-91, mgr. devel., 1991—. Author reports. Mem. ASME (treas. 1975-77). Republican. Presbyterian. Home: 7964 Jolain Dr Cincinnati OH 45242 Office: GE Aircraft Engines G132 1 Neumann Way Cincinnati OH 45215

HAMILTON, ROBERT BURNS, civil engineer; b. Elmhurst, Ill., Feb. 19, 1949; s. Donald Bain and Ellen Mary (Burns) H.; m. Diane Dolores Miller, Aug. 7, 1971; children: Mark M., Rita L., Luke R., Ann Marie. BSCE, U. Detroit, 1972; MSCE, U. Ill., 1973; MBA, Lake Forest Coll., 1984. Registered profl. engr., Ill., Wis., Ind., Mich. Project engr. Keifer & Assocs., Inc., Chgo., 1973-78; mgr. divsn. engring. Harland Bartholomew & Assocs., Inc., Northbrook, Ill., 1978-81; pres. Gewalt-Hamilton Assocs., Inc., Northbrook, Ill., 1981—. Commr. Park Ridge (Ill.) Recreation and Pk. Dist., 1981—, pres., 1984-85, 89-90. Capt. U.S. Army, 1974-75. Mem. ASCE, NSPE, Am. Pub. Works Assn. (membership chmn. 1983-85, pres.'s com. 1991—), Inst. Transp. Engrs. Republican. Roman Catholic. Home: 625 S Fairview Park Ridge IL 60068 Office: Gewalt-Hamilton Assocs Inc 3100 Dundee Rd Ste 404 Northbrook IL 60062

HAMILTON, STEPHEN STEWART, mechanical engineer, consultant; b. Jackson, Miss., Sept. 18, 1962; s. Robert Buck and Virginia Leigh (Rehfeldt) H.; m. Sharon Danielle McNair, Apr. 16, 1988; 1 child, Samantha Nicole. BSME, Miss. State U., 1985. Registered profl. engr., Ga. Asst. mgr. Katsushiro Rome Corp., Rome, Ga., 1990-91; design engr. Lockheed Aero. Svcs. Co., Marietta, Ga., 1985-90; sr. design engr. Lockheed Aero. Svcs. Co., Marietta, 1991—; pres. and chief engr., S.E. Design Cons., Kennesaw, Ga., 1989—. Mem. NSPE, ASME. Home: 4857 Wilkie Way Acworth GA 30102 Office: Lockheed Aero Svcs Co Dept 73-05 Zone 0199 86 S Cobb Dr Marietta GA 30083

HAMILTON, VIRGINIA MAE, mathematics educator, consultant; b. Winchester, Ind., Apr. 15, 1946; d. Charles and Mildred Alene (Horseman) Campbell; m. William Earl Hamilton, Dec. 27, 1974; 1 child, Michelle Annette. BS in Math., Ball State U., 1968, MA in Math., 1974. Math. tchr. Osborn High Sch., Manassas, Va., 1968-71; grad. asst., math. Ball State U., Muncie, Ind., 1971-74, math. tchr., 1977-84, dir. testing and placement, dir. Math. Learning Ctr., 1984-87; math. tchr. WesDel High Sch., Gaston, Ind., 1974-76; math. prof. Shawnee State U., Portsmouth, Ohio, 1987—; cons. placement testing several univs., Calif., Ind., Ohio, 1986—; cons. in-svc. Scioto County Schs., Portsmouth, 1989—; mentor-tchr. Minority Edn. Advocates, Muncie, 1985-87 (Outstanding Teaching award 1987); presenter Ohio Acad. Sci., Portsmouth, 1988; assessment chair nat. project to reform Devel. Math., 1992—; speaker at various confs. on math. and assessment. Author: Testbank for Fundamentals of Mathematics, 1989, Testbank for Elementary Algebra, 1989, Testbank for Intermediate Algebra, 1990, Prepared Tests for Elementary Algebra, 1990, (computer software) Dose Calc, 1984, Arithmetic Skill Builder, 1987; editor: (testbanks) Keedy-Bittinger Worktext Trilogy, 1986, Intermediate Algebra, 1986, Introductory Algebra, 1986. Mem. NEA, Nat. Coun. Tchrs. Math., Nat. Assn. Devel. Educators (chmn. com. on math. placement 1990—), Math. Assn. Am., Ohio Coun. Tchrs. Math., Ohio Assn. Devel. Educators (chmn. spl. interest group 1989—, svc. award 1992, treas. 1992—), Ohio Edn. Assn., Am. Math. Assn. 2-Yr. Colls., Am. Assn. Higher Edn. Avocations: crochet, plaster craft. Office: Shawnee State U 940 2nd St Portsmouth OH 45662-4347

HAMILTON, WALLIS SYLVESTER, hydraulic engineer, consultant; b. Palmyra, N.J., Feb. 14, 1911; s. Wallis Henry and Hazel Fanny (Corle) H.; m. Eva Mary Blichfeldt, June 13, 1937; 1 child, Mary Susan Hamilton Waxler. BSCE, Carnegie Inst. Tech., 1935, MSCE, 1939; PhD in Mechanics and Hydraulics, U. Iowa, 1943. Registered profl. engr., Ill. Engring. aide Tenn. Valley Authority, Knoxville, 1935-36; hydrographer U.S. Geol. Survey, Chattanooga, 1936-37; asst. engr. Hydraulic Rsch. Lab., Carnegie Inst. Tech., Pitts., 1939-42; instr. Carnegie Inst. Tech., Pitts., 1939-42; prof. Northwestern U., Evanston, Ill., 1943-76; sr. hydraulic specialist Harza Engring. Co., Chgo., 1976-83; cons. Wilmette, Ill., 1983—; prof. Emeritus, Northwestern U., Evanston, 1976; commr. Wilmette Sewer Commn., 1988—. Contbr. articles to profl. jours. Mem. ASCE (life, Hydraulics Structures medal 1990). Achievements include research in method of broken characteristics applied to St. Venant equations for flood routing in natural channels; in physical model tests used to predict the behavior of many types of prototype-aquifers, rivers, beaches, shore protection structures, dams, stilling basins. Home and Office: WS Hamilton Cons 4058 Fairway Dr Wilmette IL 60091

HAMILTON, WILLIAM DONALD, zoologist, educator; b. Cairo, Aug. 1, 1936; s. Archibald M. and Bettina M. (Collier) H.; m. Christine A. Friess, 1967; 3 children. Student, U. Cambridge, U. London. Lectr. genetics Imperial Coll., London, 1964-77; prof. evolutionary biology U. Mich., Ann Arbor, 1978-84; Royal Soc. Rsch. prof. zoology, fellow New Coll., Oxford, Eng., 1984—; prof. mus. zoology U. Mich., 1978-84. Contbr. articles to profl. jours. Recipient Frink medal Zool. Soc., 1991, Crafoord Prize, Royal Swedish Acad. Sciences, 1993. Mem. Am. Acad. Arts and Scis. (fgn.), Royal Soc. Scis. Uppsala (Darwin medal 1988, Linnean medal for Zoology 1989). Office: Oxford Univ, Dept of Biological Sciences, Oxford England OX1 2Jd*

HAMLETT, JAMES GORDON, electronics engineer, management consultant and educator; b. Utica, N.Y.. BSEE, Syracuse U., 1947-49; BSBA, SUNY, Syracuse, 1985; MBA, City U., Seattle, 1991. Cert. vocat. edn. tchr., N.Y.; 1st class radiotelephone lic. with ship radar endorsement, FCC. Engr.-writer Warner, N.Y., Inc., Syracuse, 1952-54; vocation edn. tchr. evenings adult edn. Syracuse Cen. Tech. High Sch., 1956-62; project leader GE, Syracuse, 1955-90; mgmt. cons. Syracuse, 1990—; vol. job interview cons. to laid off employees GE, 1991—; lectr. Syracuse U., 1980-81, Queens U., Kingston, Ont., Can., 1976; presenter in field. Author: Your Television Set, 1953, Engineering-Related Abbreviations, 1980-84 (VIP award 1980); editor Syracuse SCANNER, IEEE, 1959-68. Prin. Flood Control Com., Town of Onondaga, N.Y., 1962; tennis coach U.S. Jaycees, North Syracuse, N.Y., 1968; mem. steering com., sec., exec. com. L.C. Smith Coll. Engring. and Computer Sci, Syracuse U., 1991—. With U.S. Army, 1942-45, ETO. Fellow Soc. for Tech. Communications (internat. stem mgr. 1980, exec. com.); mem. IEEE (life sr., exec. com., Cert. 1981), Small Bus. United, N.Y. Acad. Scis. (cert. 1985), Am. Mgmt. Assn., Internat. Platform Assn., Syracuse GE Engrs. Assn., Syracuse U. Alumni Assn., Empire State Coll. Alumni Assn. (pres. Syracuse area alumni/student assn.),City U. Alumni (life), Vets. Battle of the Bulge (life, historian, treas.). Avocations: tournament tennis (Wimbledon, Eng. 1969), reading management practice, music. Home: 330 Everingham Rd Syracuse NY 13205-3258

HAMLIN, KURT WESLEY, manufacturing engineer; b. Lansing, Mich., Sept. 27, 1950; s. Lloyd P. And Gladys M. (Andrews) H.; children: William L., Daniel L. AS, Lansing Community Coll., 1976; BA in Mgmt., Sangamon State U., 1985. Chief engr. Holiday Corp., Bloomington, Ill.; program supr. Ind. Vocat. Tech. Coll., Lafayette, Ind.; team leader Subaru-Isuzu Automotive, Inc., Lafayette, Ind., sr. staff assoc. engr., group leader; indsl. maintenance adv. bd. Ind. Vocat. Tech. Coll., Lafayette. Mem. Soc. Fire Protection Engrs., Masons. Office: Subaru-Isuzu Automotive Inc 5500 SR 38 Box 5689 Lafayette IN 47903

HAMLIN, SCOTT JEFFREY, physicist; b. Rochester, N.Y., May 16, 1961; s. Thomas Henry and Virginia (Comfort) H. AS in Physics, Rochester Inst. Tech., 1984, BS in Physics, 1986. Chem. mfg. oper. Eastman Kodak Co., Rochester, 1980-81; design specialist GE, Binghamton, N.Y., 1985-86; rsch. physicist Kigre, Inc., Hilton Head, S.C., 1986-87, chief scientist, 1990—; rsch. laser physicist Laser Tech. Assocs., Inc., Johnson City, N.Y., 1987-90. Author: Efficient Diode Pumped Er: Glass Laser, 1989; contbr. to profl. publs. Bd. dirs. Cotton Hope Plantation Regime, Hilton Head, 1992—. Mem. IEEE, Internat. Soc. Optical Engrs. (co-chair solid state lasers 1992, 93), Optical Soc. Am., Biomed. Optics Soc., Lasers and Electro-Optics Soc. Home: 1924 Cotton Hope Plantation Hilton Head Island SC 29926 Office: Kigre Inc 100 Marshland Rd Hilton Head Island SC 29926

HAMM, ROBERT MACGOWAN, psychologist; b. Phila., Apr. 30, 1950; s. James Robert and Marian Todd (Miller) H.; m. Ingrid Ellen Young, Aug. 25, 1984; children: James Eric, Laura Elizabeth; 1 stepchild, Ursula Blue Moore. AB, Princeton U., 1972; PhD, Harvard U., 1979. Sr. rsch. assoc. Harvard Bus. Sch., Allston, Mass., 1979-80; researcher Higher Order Software, Inc., Cambridge, Mass., 1980-81; rsch. assoc. Ctr. for Rsch. on Judgment and Policy/U. Colo., Boulder, 1981-85, Inst. of Cognitive Sci./U. Colo., Boulder, 1986-92; NRC sr. assoc. Army Rsch. Inst., Ft. Leavenworth,

Kans., 1988-91; asst. prof. Coll. of Medicine/U. Okla., Oklahoma City, 1992—; vis. asst. prof. Sch. of Bus., U. Iowa, Iowa City, 1985-86; owner Sixth Day Cons., Boulder, 1981-88, Decision Analysis and Judgment Rsch., Boulder, 1988-92. Co-author: Medical Choices, Medical Chances, 1981, Surgical Intuition, 1993, Surgical Scripts, 1993; contbr. articles to profl. jours. and publs. Grantee Milton Fund, Harvard Med. Sch., 1992; recipient Dissertation Assistance award NSF, Harvard, 1978-79, rsch. contracts Army Rsch. Inst., U. Colo., 1986-89, U.S. Geol. Survey, 1989-90. Mem. Am. Psychol. Assn., Am. Psychol. Soc., Psychonomic Soc., Soc. for Math. Psychology, Judgment/Decision Making Soc., Soc. for Med. Decision Making. Office: U Okla Family Medicine/Sch Medicine PO Box 26901 Oklahoma City OK 73190

HAMMACK, WILLIAM S., chemical engineering educator; b. Greencastle, Ind., Nov. 2, 1961. BSChemE, Mich. Techol. U., 1984; MSChemE, U. Ill., Urbana, 1986, PhD in Chem. Engring., 1988. Asst. prof. dept. chem. engring. Carnegie Mellon U., Pitts., 1988—; mem. ChemCARS com. Advanced Photon Source, Argonne; mem. adv. bd. nat. high pressure facility CHESS; reviewer Ben Franklin Tech. Ctr.; presenter numerous seminars. Reviewer Science Mag., Jour. Physics and Chemsitry of Solids; contbr. articles to profl. jours. Recipient Exxon Solid-State Chemistry Fellowship award, 1993. Mem. AAAS, AICE (reviewer jour.), Am. Chem. Soc., Am. Phys. Soc., Am. Crystallographic Assn., Am. Geophys. Union, Materials Rsch. Soc. Achievements include research in pressure-induced amorphization using in situ x-ray diffraction and Raman spectroscopy, schemes in halide glasses, ionic glasses and amorphous molecular solids, solid state amorphizations, production of amorphous metallic alloys at ambient temperatures and the fundamental ordering in quasicrystalline solids and amorphous alloys. Office: Carnegie-Mellon U Dept of Chemical Engineering 5000 Forbes Ave Pittsburgh PA 15213-3890*

HAMMEL, HAROLD THEODORE, physiology and biophysics educator, researcher; b. Huntington, Ind., May 8, 1921; s. Audry Harold and Ferne Jane (Wiles) H.; m. Dorothy King, Dec. 29, 1948; children: Nannette, Heidi. BS in Physics, Purdue U., 1943; MS in Physics, Cornell U., 1950, PhD in Zoology, 1953. Jr. physicist Los Alamos (N.Mex.) Lab., 1944-46, staff physicist, 1948-49; from instr. to asst. prof. U. Pa., Phila., 1953-61; assoc. prof., fellow John B. Pierce Lab. Yale U., New Haven, 1961-68; prof. Scripps Instn. of Oceanography U. Calif., San Diego, 1968-88, emeritus prof., 1988—; adj. prof. physiology and biophysics Ind. U., Bloomington, 1989—; fgn. sci. mem. Max Planck Inst. for Physiol. & Clin. Rsch., 1978—; U.S. sr. scientist Alexander von Humboldt Found., 1981. Author: (with Scholander) Osmosis and Tensile Solvent, 1976; contbr. over 200 articles to profl. jours. Fellow AAAS; mem. Am. Phys. Soc., Am. Physiol. Soc., Am. Soc. mammology, Norwegian Acad. Sci. & Letters. Democrat. Achievements include first measurement of phloem sap pressure in higher plants, and of xylem sap pressure in higher plants; explanation of freezing without cavitation in evergreen plants; extension and application of kinetic theory to Hulett's theory of solvent tension; research in theory of adjustable set point and gain for regulation of body temperature in vertebrates, in temperature regulation, in osmoregulation, and in osmosis and fluid transport in plants. Home: 1605 Ridgeway Dr Ellettsville IN 47429-9474 Office: Ind U Biomedical Science Program Bloomington IN 47405

HAMMER, DONALD ARTHUR, ecologist; b. Minot, N.D., Sept. 17, 1942; s. Archie C. and Ila C. (Alg) H.; m. Joan M. Nelson, June 5, 1965. BS, U. N.D., 1965; MS, S.D. State U., 1968; PhD, Utah State U., 1972. With U.S. Fish and Wildlife Svc., N.D., S.D, 1964-70; asst. prof. U.Maine, Orono, 1971-72; sr. wetlands ecology mgr. TVA, Knoxville, 1972--; pres. The Trumpeter Swan Soc., 1985-87; coord. N.Am. IAWQ Macrophytes in Waste Treatment, 1989—. Author: Creating Freshwater Wetlands, 1991; editor: Constructed Wetlands for Wastewater Treatment, 1989; contbr. articles to profl. jours. Bd. dirs. Norris Watershed Bd., Norris, 1984—. Named Outstanding Conservationist of Yr. Tenn. Cons. League, 1980, Outstanding Conservationist Nat. Inst. for Urban Wildlife, 1988. Mem. Soc. Wetlands Scientist, Ecology Soc. Am., The Wildlife Soc. (pres. Tenn. chpt. 1978-80), Water Environ. Fedn. Office: Regional Waste Mgmt TVA 400 W Summit Hill Dr Knoxville TN 37902

HAMMER, JAMES HENRY, physicist; b. Phoenix, Sept. 13, 1951; s. Marvin Hutchinson and Mary Alice (Gaissert) H.; m. Heather Leslie, June 23, 1979; children: Eric, Joseph, Leslie. BS, Ariz. State U., 1973; PhD, U. Calif., Berkeley, 1979. Staff scientist Lawrence Livermore (Calif.) Nat. Lab., 1979-81, program leader compact torus accelerator, 1991—. Contbr. articles to profl. jours. V.p. Livermore Valley Edn. Found., Livermore, 1991—. Recipient Roy Lester Frank Meml. award U. Calif., Berkeley, 1978. Mem. AAAS, Am. Phys. Soc. Democrat. Methodist. Achievements include co-invention of compact torus accelerator; patent pending on concept to tap electric power from the solar wind. Home: 4128 Guilford Ave Livermore CA 94550 Office: Lawrence Livermore Nat Lab L-630 PO Box 808 Livermore CA 94551-0808

HAMMER, JOHN MORGAN, surgeon; b. Oak Park, Ill., Sept. 5, 1913; s. Julius Morgan and Mayta Olivet (Hobba) H.; widowed; children: Sonja, John, Theodore, William, Harriett. BS, Mich. State U., 1935, MS, 1948; MD, U. Chgo., 1939. Diplomate Am. Bd. Surgery. Intern Harper Hosp., Detroit, 1940-41, surgical resident, 1945-48; active staff Bronson and Borgess Hosps., Detroit, 1948-75, Spohn, Meml., Humana Hosp., Corpus Christi, Tex., 1975—. Contbr. more than 40 articles to profl. jours. Maj. U.S. Army (Med. Corps), 1942-45, ETO. Fellow ACS, Ctrl. Surg. Assn., Internat. Cardiovascular Surgery, Assn. for Surgery of the Alimentary Tract. Home and Office: 224 Oleander Corpus Christi TX 78401

HAMMERSCHMIDT, BEN L., environmental scientist; b. Hays, Kans., June 11, 1947; s. Ben L. and Cletis G. (Gilbert) H.; m. Marjorie J. Neerman, Aug. 16, 1969; children: Eric, Ann. BS, Ft. Hays Coll., 1969; MEd, U. Nebr., 1977; postgrad., Kennedy Western U. Cert. tchr. sci. and math., Nebr. Tchr. environ. edn. Nebraska City (Nebr.) Pub. Schs., 1971-90; instr. sci. and math. Peru State Coll., Nebraska City, 1984-87; environ. specialist Am. Meter, 1990—; environ. cons. and trainer Am.-Midwest Environ. and Safety, Nebraska City, 1992—. Author, editor (manual) Studying and Observing Interations of Life, 1980. Recipient J. L. Higgins award Nebr. Dept. Environ. Quality, 1980; named Nat. Sci. Tchr. Nat. Sci. Tchr. Assn., 1986; grantee NSF, 1973-74, Dept. Edn., 1977-82. mem. AAAS, Nat. Assn. Environ. Mgrs., Nebr. Acad. Sci., Environ. Trainer Assn. Home: 102 Arborview Dr Nebraska City NE 68410

HAMMERSCHMIDT, RONALD FRANCIS, environmental director; b. Hays, Kans., Jan. 13, 1951; s. Joseph F. and Louise C. (Miller) H.; m. Mary A. Rziha, Aug. 11, 1973; children: Lisa M., Amy L., Becky A. BA in Chemistry, St. Mary of the Plains Coll., 1973; PhD in Analytical Chemistry, U. Nebr., 1978. Rsch. scientist Vets. Adminstrn., Omaha, Nebr., 1978; dir., agrl. labs. Harris Labs., Lincoln, Nebr., 1978-80; sr. lab. scientist Kans. Health and Environ. Labs., Topeka, 1980-87; dir. bur. environ. remediation Kans. Dept. Health and Environ., Topeka, 1987-90, dep. dir. environ., 1990—; dir. region VII Assn. State and Terr. Solid Waste Mgmt. Officials, Washington, 1992. Office: Kans Dept Health and Enviro Bldg 740 Forbes Field Topeka KS 66620

HAMMES, GORDON G., chemistry educator; b. Fond du Lac, Wis., Aug. 10, 1934; s. Jacob and Betty (Sadoff) H.; m. Judith Ellen Frank, June 14, 1959; children: Laura Anne, Stephen R., Sharon Lyn. A.B., Princeton, 1956; Ph.D., U. Wis., 1959. NSF postdoctoral fellow Max Planck Inst. fur physikalische Chemie, Göttingen, Germany, 1959-60; from instr. to assoc. prof. Mass. Inst. Tech., Cambridge, 1960-65; prof. Cornell U., Ithaca, N.Y., 1965-88; chmn. dept. chemistry Cornell U., 1970-75, Horace White prof. chemistry and biochemistry, 1975-88, dir. biotech. program, 1983-88; prof. U. Calif., Santa Barbara, 1988-91, vice chancellor, 1988-91; prof. Duke U., Durham, N.C., 1991—; vice chancellor Duke U. Med. Ctr., Durham, N.C., 1991—; mem. physiol. chemistry sect., physical biochemistry study sect., tng. grant com. NIH; bd. counselors Nat. Cancer Inst., 1976-80; mem. adv. coun. chemistry dept. Princeton, 1970-75, Poly. Inst. N.Y., 1977-78, Boston U., 1977-88; mem. NRC, U.S. nat. com. for biochemistry, 1989—. Author: Principles of Chemical Kinetics, Enzyme Catalysis and Regulation, (with I. Amdur) Chemical Kinetics: Principles and Selected Topics; editor: Biochemistry, 1992—; also articles. NSF sr. postdoctoral fellow, 1968-69; NIH Fogarty scholar, 1975-76. Mem. NAS, Am. Acad. Arts and Scis., Am. Chem. Soc. (award biol. chemistry 1967, editorial bd. jours., exec. com. div. phys. chemistry 1976-79, exec. com. div. biol. chemistry 1977-78, com. profl. tng. 1985-92, task force on biotech. 1989-90), Am. Soc. Biochemistry and Molecular Biology (coun., pres., editorial bd. jour.), Phi Beta Kappa, Sigma Xi, Phi Lambda Upsilon. Home: 11 Staley Pl Durham NC 27705-2421

HAMMOND, CHARLES BESSELLIEU, obstetrician-gynecologist, educator; b. Ft. Leavenworth, Kans., July 24, 1936; s. Claude G. and Alice (Sims) H.; m. Peggy R. Hammond, June 21, 1958; children: Sharon L., Charles B. Student, The Citadel, 1957; B.S., M.D., Duke U., 1961. Diplomate Am. Bd. Ob-Gyn. Intern in surgery Duke U., 1961-62, resident in ob-gyn, 1962-63, 66-69, fellow in reproductive endocrinology, 1963-64, asst. prof. dept. ob-gyn, 1969-73, assoc. prof., 1973-78, prof., 1978-81, E.C. Hamblen prof., 1981—, chmn., 1980—. Contbr. in field. Served with USPHS, 1964-66. Mem. Am. Fertility Soc. (pres. 1985), Am. Coll. Obstetricians and Gynecologists, Assn. Profs. Obstetrics and Gynecology, Am. Gynecol. and Obstet. Soc. (pres. 1993—), Soc. Gynecologic Investigation, N.C. Med. Soc., N.C. Soc. Obstetricians and Gynecologists (pres. 1985). Presbyterian. Home: 2827 Mcdowell Rd Durham NC 27705-5604 Office: Duke U Med Ctr Box 3244 Durham NC 27710

HAMMOND, CHARLES EARL, chemist, researcher; b. Indpls., Apr. 16, 1960; s. Dickey Earl and Ruth Marie (Ortt) H. BS in Chemistry, Ind. U., South Bend, 1985; PhD in Inorganic Chemistry, Ind. U., Bloomington, 1989. Chemistry lab. supr. Ind. U., South Bend, 1983, physics lab. instr., 1983-85; assoc. instr. Ind. U., Bloomington, 1985-88, grad. rsch. asst., 1986-89; rsch. chemist Vista Chem. Co., Austin, Tex., 1989—, product steward surfactants, 1993—. Contbr. numerous articles to profl. jours. Recipient Chemistry Merit award Ind. U., Bloomington, 1988, William H. Nebergall Meml. award Ind. U., Bloomington, 1988, Competent Toastmaster award, Toastmasters Internat., 1992. Mem. Am. Chem. Soc. (sec. So. Ind. Sect. 1988-89, Fin. Chmn. Southwest Region 1993, chmn. Ctrl. Tex. Sect. 1990-92, Gilbert H. Ayres award, 1992), Am. Oil Chemists Soc. Office: Vista Chem Co 12024 Vista Park Dr Austin TX 78726

HAMMOND, C(LARKE) RANDOLPH, healthcare executive; b. Anniston, Ala., July 3, 1945; s. Clarke MacAlpin and Edna Odell (Webb) H.; m. Carolyn Jane Milam, Oct. 26, 1974; children: Chadwick, Kyle, Amanda. BS in BA, U. So. Miss., 1968; M.Health Adminstrn., U. Ala., Birmingham, 1978. CPA, Ark. Asst. adminstr., controller Helena (Ark.) Hosp., 1971-74; asst. adminstr. S. Highlands Hosp., Birmingham, 1974-77; adminstr. Brookwood Health Svcs., Rocky Mt., N.C., 1977-79; v.p. ops. Brookwood Health Svcs., Birmingham, 1979-81; pres., chief exec. officer Medlab, Birmingham, 1981-88; asst. v.p. Hoffman-LaRoche, Birmingham, 1988-90; sr. v.p.. Roche Biomed. Lab., Birmingham, 1988-90; v.p. Jemison Investment Co., Birmingham, 1990-92; chief exec. officer, pres. Textile Resource & Mktg., Dalton, Ga., 1990-92; sr. v.p. ops. Diagnostic Health Corp., Birmingham, 1992—; bd. dirs. Uromed Techs. Inc., Clearwater, Fla., Express Oil, Pinnacle Oil Inc., Birmingham, Ala., Puckett Labs., Hattiesburg, Miss., Touch & Know, Altanta. Pres. Diabetes Trust Fund, Birmingham, 1989, bd. dirs., 1984-89; chmn. deacons Valleydale Bapt. Ch., Birmingham, 1983, strategic planning com., 1983; chmn. bd. Liberty Recreation Ctr., Inc., Pensacola, 1989—. With USN, 1968-70. Mem. Am. Assn. Health Care Execs., Health Fin. Mgmt. Assn., U. Ala. Health Adminstrs. Alumni Soc. (pres. 1989), U. Ala. Alumni Assn. (bd. dirs. 1988-90), The Club, The Summit Club, N. River Yacht Club, Inverness Country Club, Greystone Country Club. Republican. Baptist. Avocations: flying, scuba diving. Home: 2125 Cameron Cir Birmingham AL 35242-6412 Office: Diagnostic Health Corp 22 Inverness Ctr Pky Ste 400 Birmingham AL 35242

HAMMOND, EARL GULLETTE, food science educator; b. Terrell, Tex., Nov. 21, 1926; s. Joseph Carrol and Kate (Gullette) H.; m. Johnie Gray Wright, Sept. 17, 1951; children: Bruce, Linda, Pamela, Christopher. BS in Chemistry, Tex. U., 1948, MA in Chemistry, 1950; PhD in Agrl. Biochemistry, U. Minn., 1953; Honoris Causa Doctoris, U. Agr., Olztyn, Poland, 1990. Prof. food sci. Iowa State U., Ames, 1953—. Pres. Ames Ecumminical Housing-HUD, 1986, Stonehaven, Ames, 1992. Pfc. U.S. Army, 1944-45. Mem. Am. Chem. Soc., Am. Oil Chemists Soc. (Chang award 1992, Bailey award 1993), Am. Dairy Sci. Assn. (Pfizer award 1980), Iowa Acad. Sci. (Disting. fellow 1992), Inst. Food Technologists. Democrat. Baptist. Achievements include research in identification of flavors in oxidized fats, mechanisms of flavor formation in cheese, glyceride structure of fats and oils, production of mutant soybeans with altered fat composition. Home: 3431 Ross Rd Ames IA 50010 Office: Iowa State U Food Sci & Human Nutrition Ames IA 50011

HAMMOND, GEORGE SIMMS, chemist; b. Auburn, Maine, May 22, 1921; s. Oswald Kenric and Marjorie (Thomas) H.; m. Marian Reese, June 8, 1945 (div. 1977); children: Kenric, Janet, Steven, Barbara, Jeremy; m. Eva L. Menger, May 22, 1977; stepchildren—Kirsten Menger-Anderson, Lenore Menger-Anderson. BS, Bates Coll., 1963; MS, PhD, Harvard U., 1947; DSc (hon.), Wittenberg U., 1972, Bates Coll., 1973; Dr. honoris causa, U. Ghent, 1973, Georgetown U., 1985, Bowling Green State U., 1990, Weizman Inst. Sci., 1993. Postdoctoral fellow UCLA, 1947-48; mem. faculty Iowa State Coll., 1948-58, prof. chemistry, 1956-58; vis. asso. prof. U. Ill., summer 1953; prof. organic chemistry Calif. Inst. Tech., Pasadena, 1958-72; div. chemistry and chem. engring. Calif. Inst. Tech., 1968-72; Arthur Amos Noyes prof. chemistry; vice chancellor natural scis. U. Calif.-Santa Cruz, 1972-74, prof. chemistry, 1972-78; exec. dir. for biosci., metals and ceramics Allied Corp., Morristown, N.J., 1978-88; cons., 1988—; mem. chem. adv. panel NSF, 1962-65; fgn. sec. Nat. Acad. Scis., 1974-78. Author: (with J. S. Fritz) Quantitative Organic Analysis, 1956, (with D.J. Cram) Organic Chemistry, 1958, (with J. Osteryoung, T. Crawford and H. Gray) Models in Chemical Science, 1971; co-editor: Advances in Photochemistry, 1961; mem. editorial bd. Jour. Am. Chem. Soc, 1967—. Recipient James Flack Norris award in phys. organic chemistry, 1968. Mem. NAS, Am. Chem. Soc. (award in petroleum chemistry 1960, Priestly medal 1976), Am. Acad. Arts and Scis., Materials Rsch. Soc., Inter-Am. Photochem. Soc., European Photochem. Soc., Phi Beta Kappa, Sigma Xi. Home: 27 Timber Ln Painted Post NY 14870

HAMMOND, HAROLD LOGAN, pathology educator, oral pathologist; b. Hillsboro, Ill., Mar. 18, 1934; s. Harold Thomas and Lillian (Carlson) H.; m. Sharon Bunton, Aug. 1, 1954 (dec. 1974); 1 child, Connie; m. Pat J. Palmer, June 3, 1986. Student Millikin U., 1953-57, Roosevelt U., Chgo., 1957-58; DDS, Loyola U., Chgo., 1962; MS, U. Chgo., 1967. Diplomate Am. Bd. Oral Pathology. Intern, U. Chgo. Hosps., Chgo., 1962-63, resident, 1963-66, chief resident in oral pathology, 1966-67; asst. prof. oral pathology U. Iowa, Iowa City, 1967-72, assoc. prof., 1972-80, assoc. prof., dir. surg. oral pathology, 1980-83, prof., dir., 1983—; cons. pathologist Hosp. Gen. de Managua, Nicaragua, 1970-90, VA Hosp., Iowa City, 1977—. Cons. editor: Revista de la Asociation de Nicaragua, 1970-71, Revista de la Federacion Odontologica de Centroamerica y Panama, 1971-77. Contbr. articles to sci. jours. Recipient Mosby Pub. Co. Scholarship award, 1962. Fellow AAAS, Am. Acad. Oral Pathology; mem. Am. Men and Women of Sci., N.Y. Acad. Scis., AAUP, Internat. Assn. Oral Pathologists, Internat. Assn. Dental Rsch., Am. Dental Assn., Am. Assn. for Dental Rsch. Avocations: collecting antique clocks, collecting gambling paraphernalia, collecting toys. Home: 1732 Brown Deer Rd Coralville IA 52241-1157 Office: U Iowa Dental Sci Bldg Iowa City IA 52242-1001

HAMMOND, HOWARD DAVID, retired editor; b. Phila., Feb. 10, 1924; s. Clarence Elwood Jr. and Myrtle Iva (Sprowles) H. BS, Rutgers U., 1945, MS, 1947; PhD, U. Pa., 1952. Asst. prof. U. Del., Newark, 1957-58, Howard U., Washington, 1958-68; from asst. prof. to assoc. prof. SUNY, Brockport, 1968-83; assoc. editor N.Y. Bot. Garden, Bronx, 1984-92. Co-editor: Floristic Inventory Tropical Countries, 1989. Mem. Am. Inst. Biol. Scis., Bot. Soc. Am., Torrey Bot. Club (editor 1976-82, 88-92, pres. 1992), Sigma Xi. Home: Apt 33 4025 Lake Mary Rd Flagstaff AZ 86001-8608

HAMMOND, RICHARD HORACE, geologist; b. Sioux City, Iowa, Sept. 17, 1950; s. Harry Horace and Marie (Hughes) H.; m. Sarah Ann Chadima, Mar. 16, 1985; children: Gabriel, Zachary, Paul. Student, U. S.D., 1973, 92. Cert. profl. geologist Am. Inst. Profl. Geologists. Geologist Midwest Test Labs., Sioux City, 1973-76, S.D. Geol. Survey, Vermillion, 1977—; cons. geologist Hammond Geol. Svcs., 1977—. Co-editor: Geological Soc. of America Special Publ., 1992; co-author: Standard Guides for Monitoring Well Installation at Environmental Work Sites, 1990-92. Chmn. Union County Dem. Party, Elk Point, S.D., 1978-80; treas. United Ministry U. S.C., Vermillion, 1991—. Farber Fund Travel grantee U. S.D. Found., 1990. Mem. ASTM (standards devel. com., com. on groundwater 1985—), Geol. Soc. Am., Am. Soc. Groundwater Scientists and Engrs., Am. Geophys. Union, Am. Soc. for Pub. Adminstrn. (treas. Siouxland chpt. 1991—). Democrat. Methodist. Home: 114 Willow St Vermillion SD 57069 Office: SD Geol Survey Sci Ctr U SD Vermillion SD 57069

HAMMONDS, ELIZABETH ANN, environmental engineer; b. West Chester, Pa., Sept. 27, 1968; d. J.A. and Susan A. (Earl) H. BS in Commerce and Engring., Drexel U., 1991. Rsch. asst. coop. Wyeth-Ayerst Pharms., Paoli, Pa., 1988; safety coop. Ethicon, Inc., Newtown, Pa., 1989, environ. coop., 1990, assoc. engr., 1991—. Named Acad. All-Am., Nat. Secondary Edn. Coun., 1986. Mem. Am. Prodn. and Inventory Control Soc., Am. Indsl. Hygiene Assn., Air & Waste Mgmt. Assn., Internat. Indsl. Engrs. Assn., Key and Triangle Soc., Gamma Sigma Sigma. Office: Ethicon Inc 110 Terry Dr Newtown PA 18940

HAMNER, HOMER HOWELL, economist, educator; b. Lamont, Okla., Oct. 22, 1915; s. Homer Hill and Myrtle Susan (Edwards) H.; m. Winnie Elvyn Heafner, May 8, 1943 (dec. Aug. 23, 1946); 1 dau., Jean Lee (Mrs. Richard L. Nicholson); m. Marjorie Lucille Dittus, Nov. 24, 1947; 1 dau., Elaine (Mrs. Alan M. Yard). A.A., Glendale Coll., 1936; A.B., U. So. Calif.; A.B. (Gen. Achievement scholarship 1936-37), 1938, J.D., 1941, A.M., 1947, Ph.D., 1949. Fellow and teaching asst., dept. econs. U. So. Calif., 1945-49; prof. and chmn. dept. econs. Baylor U., 1949-55; editor research com. Baylor Bus. Studies, 1949-55, lectr. summer workshop, 1954; prof., chmn. dept. bus. adminstrn. and cons. U. Puget Sound, Tacoma, Wash., 1955-58; dir. sch. bus. adminstr. and econs. U. Puget Sound, 1959-63, Edward L. Blaine chair econ. history, 1963—; also occasional lectr. Roman Forums, Ltd., Los Angeles, 1936-40; lectr. Am. Inst. Banking, 1949-50; lectr. Southwest Wholesale Credit Assn., 1949, James Connally AFB, 1950; cons. State of Wash. tax adv. council, 1957-58, State of Wash. Expenditures Adv. Council, 1960. Author: Population Change in Metropolitan Waco, 1950; reviewer, contbr. to Jour. of Finance. Served with U.S. Army, 1941-44. Fellow Fund. Econ. Edn., Chgo., Inst. on Freedom, Claremont Men's Coll.; mem. AAUP, Am. Econ. Assn., Southwest Social Sci. Assn. (Tex. chmn. membership com.), Nat. Tax Assn., Am. Finance Assn., Am. Acad. Polit. and Soc. Sci., Order of Artus, Waco McLennan County Bar Assn. (hon.), Phi Beta Kappa, Phi Kappa Phi, Omicron Delta Gamma, Delta Theta Phi, Phi Rho Pi (degree highest achievement 1936). Methodist. Home: 4404 N 44th St Tacoma WA 98407-6604

HAMPER, BRUCE CAMERON, organic chemist, agricultural chemist; b. Evergreen Park, Ill., July 17, 1955; s. Sidney Charles and Grace (Foster) H.; m. Mary Beth Wilczak, Oct. 15, 1955; children: Laura Jane, Henry William. BA in Chemistry, Kalamazoo Coll., 1977; PhD in Organic Chemistry, U. Ill., 1984. Sr. rsch. chemist Monsanto Co., St. Louis, 1984-87, rsch. specialist, 1987-89, rsch. group leader, 1990—; bd. dirs. chem. adv. bd. U. Mo., Columbia. Contbr. articles to profl. jours. Mem. AAAS, Am. Chem. Soc., Internat. Soc. Heterocyclic Chemistry. Achievements include patents in preparation of herbicidal halo-alkyl pyrazoles; in preparation of phenyl (methylsulfonyl) pyrazoles as herbicides. Home: 132 Wildwood Ln Kirkwood MO 63122 Office: Monsanto Co 800 N Lindbergh Blvd Saint Louis MO 63167

HAMPTON, JAMES WILBURN, hematologist, medical oncologist; b. Durant, Okla., Sept. 15, 1931; s. Hollis Eugene and Ouida (Mackey) H.; m. Carol McDonald, Feb. 22, 1958; children: Jaime, Clay, Diana, Neal. B.A., U. Okla., 1952, M.D., 1956. Intern U. Okla. Hosps., 1956-57; also resident; instr. to prof. U. Okla., Oklahoma City, 1959-77, clin. prof. medicine, 1977—, mem. admissions bd., 1965—, subcom., 1985—, head hematology/oncology, 1972-77; head hematology rsch. Okla. Med. Rsch. Found., Oklahoma City, 1972-77; dir. cancer program and med. oncology Baptist Med. Ctr., 1977-85, med. dir. Cancer Ctr. S.W., 1985—; cons. NIH, Biomed. and Nat. Cancer Inst.; vis. prof. Karolinska Inst., Stockholm; vis. scientist U. N.C., 1966-67; pres. Stewart Wolf Soc., 1990-92; founder Robert Montgomery Bird Soc., 1973—. Contbr. over 100 articles to profl. jours. Chmn. network com. Cancer Prevention and Control for Am. Indians/Alaska Natives Nat. Cancer Inst., 1990—; chmn. Fine Home Awards Com., 1987-90; bd. dirs. Heritage Hills, Oklahoma City, 1972-90, Am. Cancer Soc., mem. at large coms., nat. bd. dirs., Cancer in the Socio-economically Disadvantaged, 1990—, chmn. div. svc. and rehab. com., 1988-91; co-chmn. Save St. Paul's Episcopal Cathedral com., 1983, chmn. bishop's Okla. com. on Indian work, mem. privince VII Indian com., alt. del. Diocesan conv. for Okla., 1991, others. NIH Career Devel. Award., 1966-76. Fellow ACP; mem. Am. Fedn. Clin. Research (pres. midwest sect. 1970-71), Central Soc. Clin. Research (asso. editor Jour. Lab. and Clin. Medicine 1975-76), Okla. County Med. Soc. (editor bull. 1981—, bd. dirs. 1989-91, chmn. 1991—), Internat. Soc. Thrombosis and Hemostasis, Assn. Am. Indian Physicians (pres. 1978-79, 88-89); Am. Physiol. Soc., Assn. Am. Pathologists, Am. Soc. Hematology, Am. Soc. Clin. Oncology, So. Soc. Clin. Investigation, Am. Psychosomatic Soc., Clubs: Oklahoma City Golf and Country, Blue Cord, Chaine des Rotisseurs. Home: 1414 N Hudson Ave Oklahoma City OK 73103-3721 Office: Cancer Ctr of the Southwest at Bapt Med Ctr 3300 NW Expressway Oklahoma City OK 73112

HAMPTON, MATTHEW JOSEPH, engineering technician; b. Denver, Apr. 6, 1958; s. Gordon Marion and Rita Mae Hampton. Engring. technician P.S.C. Inc., Fairfax, Va., 1985—. Contbr. articles to profl. jours. With U.S. Army, 1977-85. Office: PSC Inc 10560 Arrowhead Dr Fairfax VA 22030

HAMSTRA, STEPHEN ARTHUR, mechanical engineer; b. Grand Haven, Mich., Feb. 27, 1959; s. Calvin Arthur and Jo Ann (Sternberg) H.; m. Melinda Ruth Sytsma, Aug. 24, 1978; children: Chris, Alison, Sean. BSME, Mich. State U., 1981. Registered profl. engr. Mich. Pres. Hamstra Assocs., Inc., Holland, Mich., 1984-86; dir. engring. Van Dyken Mechs., Grand RApids, Mich., 1986-88; sr. engr., dir. GMB Arch. Engring., Holland, 1988—. Councilman Zeeland (Mich.) City Coun., 1990-93; commr. Planning Commn., Zeeland, 1991-93. Mem. NSPE, Am. Soc. Heating, Refridgeration and Air Conditioning Engrs., Internat. Soc. Pharm. Engrs., Mich. Soc. Profl. Engrs. Republican. Office: GMB A E PO Box 2159 Holland MI 49422-2159

HAMZEHEE, HOSSEIN GHOLI, nuclear engineer; b. Jan. 4, 1956; s. Ali G. and Nosrat (Sabery) H.; m. Maryam Ghaemi, June 22, 1985; 1 child, Sean B. BSME, U. Calif., Irvine, 1981; MSME, Calif. State U., Long Beach, 1986. Engring. cons. Pickard, Lowe & Garrick, Inc., Irvine, 1981-85, cons., 1985-87; reliability engr. N.E. Utilities, Hartford, Conn., 1985-87; supr. systems analysis Tex. Utilities Electric Co., Dallas, 1987—. Mem. Am. Nuclear Soc. Home: 2042 Robin Hill Ln Carrollton TX 75007-1650 Office: TU Electric 400 Olive St Ste 81 Dallas TX 75201-4007

HAN, CHINGPING JIM, industrial engineer, educator; b. Shanghai, People's Republic China, Aug. 24, 1957; came to U.S., 1983; s. Bao-San Zhang and Xiao-xian Han; m. Man-xia Maria Zhang, Feb. 22, 1982; 1 child, George Zong-qi Han. PRC, BSME, Dalian Inst. Tech., Dalian, 1982; MS in Indsl. Engring., Pa. State U., 1985, PhD, 1988. Rsch. asst. Pa. State U., University Park, 1984-87; project coord. Nat. Forge Co., Irvine, Pa., 1988; asst. prof. mfg. systems engring. Fla. Atlantic U., Boca Raton, 1988-93; assoc. prof. mfg. systems engring. Fla. Atlantic U., Boca Rabon, 1993—; cons. Nat. Forge Pa., 1989—, KDS, Deerfield Beach, Fla., 1991, Motorola, Inc., Boynton Beach, Fla., 1991, Ford Motor Co., Livonia, Mich., Xerox Co., Webster, N.Y. Contbr. articles to profl. jours., procs. Grantee NSF, 1990—, Material Handling Rsch. Ctr., 1990—. Mem. Soc. Mfg. Engrs., N.Am. Mfg. Rsch. Inst., Inst. Indsl. Engrs., Inst. Mgmt. Scis. Avocations: classical music, travel. Home: 7946 Villa Nova Dr Boca Raton FL 33433-1029 Office: Fla Atlantic Univ 500 NW 20th St Boca Raton FL 33431-6498

HAN, JE-CHIN, mechanical engineering educator; b. Kaoshiung, Taiwan, Sept. 15, 1946; came to U.S., 1972; s. Swin and Yeh-Yen (Cheng) H.; m. Su-Huei, Aug. 3, 1973; children: George, Jean. BS, Nat. Taiwan U., 1970; MS,

Lehigh U., 1973; ScD, MIT, 1976. Registered profl. engr., Tex. Rsch. asst. Lehigh U., Bethlehem, Pa., 1972-73, MIT, Cambridge, Mass., 1973-76; process devel. engr. Ex-Cell-O Corp., Walled Lake, Mich., 1976-80; asst. prof. Tex. A&M U., College Station, Tex., 1980-84, assoc. prof., 1984-89, prof., 1989-93, endowed HTRI prof., 1993—; adv. bd. mem. Pacific Ctr. Themal Fluids Engring., Honolulu, 1981—; consulting rsch. GE-Aircraft Engines, Cin., 1989—. Contbr. articles to profl. jours. Mem. North Am. Taiwanese Profs. Assn., Chgo., 1985—; chmn. Taiwanese Assn. of Am., College Station, Tex., 1986-87; advisor Internat. Student Programs-Taiwanese Student Assn., College Station, 1986-91. TEES Young fellow Tex. engring. Experiment Sta., 1984, TEES fellow, 1987, TEES Sr. fellow, 1988, Halliburton prof., 1991. Fellow ASME; mem. Am. Soc. Engring. Edn., Am. Inst. Aero. and Astronautics. Achievements include devel. of a new in-mold steam heating process to improve the surface quality of structural foam parts; a new angled rib type of turbulence promoters for turbine blade internal coolant passages. Home: 1503 Brittany Dr College Station TX 77845 Office: Tex A&M U Mech Engring Dept College Station TX 77843

HAN, JIAWEI, computer scientist, educator; b. Shanghai, China, Aug. 10, 1949; came to U.S., 1979; arrived in Can., 1987; s. Yu-chang Han and Jia-zhi Wang; m. Yandong Cai, July 3, 1979; 1 child, Lawrence. BSc, USTC, Beijing, China, 1979; MSc, U. Wis., 1981, PhD, 1985. Asst. prof. Northwestern U., Evanston, Ill., 1986-87, Simon Fraser U., Burnaby, B.C., Can., 1987-91; assoc. prof. Simon Fraser U., Burnaby, 1991—. Contbr. articles to profl. jours. Grantee NSF, 1986-88, NSERC, 1988—. Mem. IEEE (conf. program com. 1990-92), ACM, Assn. Logic Programming. Office: Simon Fraser Univ, Sch Computing Sci, Burnaby, BC Canada V5A 1S6

HAN, OKSOO, biochemistry educator; b. Okgu-gun, Republic of Korea, Sept. 20, 1960; s. Jinsuk and Okju (Kim) H.; m. Jaehwa Byun Han, May 12, 1989; 1 child, Joohyung. BS, Seoul Nat. U., 1982, MS, 1984; PhD, U. Minn., 1989. Postdoctoral rschr. U. Wis., Madison, 1989-90; prof. Chonnam Nat. U., Kwangju, Republic of Korea, 1990—. Author: Enzymes Dependent on Pyridoxal, 1991. Mem. Am. Chem. Soc. Office: Dept Genetic Engring, Chonnam Nat U, Yongbong-dong Buk-gu 500-757, Republic of Korea

HAN, RUIJING, chemical engineer; b. Tian Jin, China, Aug. 3, 1963; s. Chien and Minfang (Hsu) H. MS, U. Md., 1989, PhD, 1992. Rsch. asst. dept. chem. engring. U. Md., College Park, 1986—. Contbr. articles to Jour. of Aerosol Sci. and Tech. and Jour. of Aerosol Sci. Mem. Sigma Xi (assoc.). Office: Univ Md Dept Chem Engring College Park MD 20742

HAN, SANG HYUN, metallurgist; b. Seoul, June 26, 1959; came to U.S., 1987; s. Kyu-Ho Han and Myung-Ahk Jang; m. Moon-Sun Huh, Feb. 18, 1987; 1 child, Hye-Jung. MS, Seoul Nat. U., 1985; PhD, Iowa State U., 1992. Metallurgist Ames (Iowa) Lab., Iowa State U. Mem. Sigma Xi. Office: Iowa State Univ Ames Lab 258 Spedding Hall Ames IA 50011

HANAFUSA, HIDESABURO, virologist; b. Nishinomiya, Japan, Dec. 1, 1929; came to U.S., 1961; s. Kamehachi and Tomi H.; m. Teruko Inoue, May 11, 1958; 1 dau., Kei. B.S., Osaka (Japan) U., 1953, Ph.D., 1960. Research asso. Research Inst. for Microbial Diseases, Osaka U., 1958-61; postdoctoral fellow virus lab. U. Calif., Berkeley, 1961-64; vis. scientist College de France, Paris, 1964-66; assoc. mem., chief dept. viral oncology Public Health Research Inst. of City N.Y. Inc., 1966-68, mem., 1968-73; prof. Rockefeller U., 1973—, Leon Hess prof., 1986—. Mem. editorial bd. Jour. Virology, 1975—, Molecular Cell Biology, 1984—; contbr. articles to profl. jours. Recipient Howard Taylor Ricketts award, 1981, Albert Lasker Basic Med. Rsch. award, 1982, Asahi Press prize, 1984, Clowes Meml. award, 1986, Culture Merit award, Japan, 1991, Alfred Sloan prize, 1993; Nat. Cancer Inst. grantee, 1966—. Mem. Am. Soc. Microbiology, Am. Soc. Virology, AAAS, N.Y. Acad. Sci., Am. Soc. Biol. Chemistry, Am. Assn. Cancer Research, Nat. Acad. Sci. Achievements include research on retroviruses and oncogenes. Home: 500 E 63rd St New York NY 10021-7946 Office: Rockefeller U 1230 York Ave New York NY 10021-6399

HANAI, TOSHIHIKO, chemist; b. Osaka, Japan, Mar. 19, 1941; s. Syunjiro and Kimi Hanai. BS in Chemistry, Ritsumeikan U., 1965; PhD, Kyoto U., 1974. Asst. Kyoto (Japan) U., 1965-74; postdoctoral rsch. assoc. Northeastern U., Boston, 1975-76, Colo. U., Boulder, 1976-77; postdoctoral rsch. assoc. U. Montreal, 1977-79, rsch. asst., 1979-84; mgr. R&D Gasukurokogyo, Tokyo, 1984-89; sr. rsch. scientist Internat. Inst. Tech. Analysis, Kyoto, 1989—; mgr. R&D Towa Sci. Ltd., Hiroshima, Japan, 1990—, Irica Kiki Ltd., Kyoto, 1990—. Author: Experimental HPLC, 1977, Chromatography Separation System, 1981, Phenols and Organic Acids, 1982, New Experimental HPLC, 1988, Liquid Chromatography in Biomedical Analysis, 1991. Mem. Am. Chem. Soc., Chem. Inst. Canada, Chem. Soc. Japan, Japan Soc. Analytical Chemistry, Pharm. Soc. Japan. Achievements include research in molecular recognition by computer and chromatography. Home: 3-492 Matsumi Eclairer 2-913, Kanagawa-ku Yokohama 221, Japan Office: Internat Inst Tech Analysis, Inst Pasteur de Kyoto 5F, Kyoto 606, Japan

HANAMIRIAN, VARUJAN, mechanical engineer, educator, journalist, publisher; b. Istanbul, Turkey, June 23, 1952; s. Kurgin and Etil Sona (Azat) H. Dip. in Mech. Engring., U. Stuttgart, Fed. Republic Germany, 1983. Postal mgr. Foto Annemie, Stuttgart, 1977; tchr. Berlitz Sch., Stuttgart, 1980; creative dir. Unver Werbeagentur, Stuttgart, 1986; scientific asst. Fraunhofer-Gesellschaft, Stuttgart, 1987; course leader Volkshochschule, Stuttgart, 1987; educator, cons.; translator various govtl. and pvt. instns. and offices. Mem. adv. bd. Produktion weekly, 1981-85. Organizer Orgn. Com. EM 1986, Stuttgart; interviewer various market rsch. assn., 1979—; adminstr. various offices, 1980-87. Mem. Verein Deutscher Ingenieure. Club: Allgemeiner Deutscher Automobil, München. Avocation: chess.

HANCOCK, DON RAY, researcher; b. Muncie, Ind., Apr. 9, 1948; s. Charles David and June Lamoine (Krey) H. B.A., DePauw U., 1970. Community worker Fla. Meth. Spanish Ministry, Miami, 1970-73; seminar designer United Meth. Seminars, Washington, 1973-75; info. coord. SW Rsch. and Info. Ctr., Albuquerque, 1975—; cons. State Planning Coun. on Radioactive Waste Mgmt., Washington, 1980-81; task force mem. Gov.'s Socioecon. Com., Santa Fe, 1983; pub. adv. bd. WIPP Socioecon. Study, Albuquerque, 1979-81. Writer mag. articles. Bd. chmn. Roadrunner Food Bank, Albuquerque, 1981-92, N.Mex. Coalition Against Hunger, 1978-85; bd. dirs. Univ. Heights Assn., Albuquerque, 1977-82, 85, 88-89, 90-93, United Meth. Bd. of Ch. and Society, Washington, 1976-80. Democrat. Office: SW Rsch and Info Ctr PO Box 4524 Albuquerque NM 87196-4524

HANCOCK, JAMES BEATY, interior designer; b. Hartford, Ky.; s. James Winfield Scott and Hettie Frances (Meadows) H.; BA, Hardin-Simmons U., 1948, MA, 1952. Head interior design dept. Thornton's, Abilene, Tex., 1945-54; interior designer The Halle Bros. Co., Cleve., 1954-55; v.p. Olympic Products, Cleve., 1955-56; mgr. interior designer Bell Drapery Shops of Ohio, Inc., Shaker Heights, 1957-78, v.p., 1979—; lectr. interior design, Abilene and Cleve.; works include 6 original murals Broadway Theater, Abilene, 1940, mural Skyline Outdoor Theatre, Abilene, 1950, cover designs for Isotopics mag., 1958-60. Mem. Western Reserve Hist. Soc., Cleve. Mus. Art, Decorative Arts Trust. Served with AUS, 1942-46. Recipient 2d place award for oil painting West Tex. Expn., 1940, hon. mention, 1940. Mem. Abilene Mus. Fine Arts (charter), Cleveland Circle of the Decorative Arts Trust (charter mem.). Home and Office: 530 Sycamore Dr Cleveland OH 44132-2150

HANCOCK, JOHN C., pharmacologist; b. Lockwood, Mo., Aug. 20, 1938; s. Daniel L. and Cordelia O. (Chandler) H. BS, U. Mo.; MS, U. Tex., Galveston, PhD. Instr. U. Conn., Storrs, 1968-69; asst. prof. U. Conn., Farmington, 1969-71, La. State U., New Orleans, 1971-73; assoc. prof. La. State U., Farmington, 1973-77; prof. East Tenn. State U. Coll. Medicine, Johnson City, 1977—, dep. chair, 1985—; peer rev. panel Am. Heart Assn., Tenn., 1991—; presenter in field. Author (software) Autonomic Pharmacology; contbr. articles to profl. jours. Grantee NIH. Mem. Am. Soc. Pharmacol. Exptl. Therapeutics, Neurosci./Am. Heart Assn., Soc. Exptl. Biology and Medicine, Soc. Neuroscience (Applachian chpt.), Sigma Xi. Achievements include research on the role of sensory peptides in the regulation of blood pressure, on physiopathology of ganglion transmission in hypertension characteristics of ganglion transmission in the rat.

HANCOCK, THOMAS EMERSON, educational psychologist, educator; b. Hollywood, Calif., July 7, 1949; s. Waldo Emerson and Lillian Rose (Wilkinson) H.; m. Lisa Kathleen Day, June 21, 1980. MEd, Seattle Pacific U., 1986; PhD, Ariz. State U., 1990. Co-founder, co-dir. Heritage Sch., Seattle, 1980-86; asst. prof. ednl. psychology Grand Canyon U., Phoenix, 1988—; faculty rsch. assoc. Air Force Office Sci. Rsch., Armstrong Labs., 1991, 93; rsch. cons. USAF, Higley, Ariz., 1992. Author Allyn and Bacon Pubs., 1990-91; contbr. articles to profl. publs. Grantee Grand Canyon U., 1989, 90. Mem. APA, Am. Psychol. Soc., Am. Ednl. Rsch. Assn., Control Systems Group. Republican. Achievements include empirical tests of higher level human learning with hierarchical perceptual control theory, matching instructional feedback with levels of learning goal or perceptual control, scale for process approach of language acquistion. Office: Grand Canyon Univ 3300 W Camelback St Phoenix AZ 85017

HANCOCK, WILLIAM MARVIN, engineering executive; b. Portsmouth, Va., Feb. 10, 1957; s. William H. and Marjorie E. (Davis) H. BA in Computer Sci., Thomas A. Edison Sr. Coll., 1992; MS in Computer Sci., Greenwich U., 1993, postgrad., 1993—. Programmer Tex. Instruments, Dallas, 1972-74; cons. Digital Equipment Corp., Dallas, 1979-82; div. analyst Standard Oil of Ohio, Dallas, 1982-84; v.p. engring. New Leaf Techs., Arlington, Tex., 1984-90, Network 1 Inc., Arlington, Tex., 1990—; U.S. network expert Am. Nat. Stas. Inst., N.Y.C., 1985-87; stas. editor Internat. Orgn. for Stas., Geneva, Switzerland, 1986-88. Author: Designing and Implementing Ethernet Networks, 1988, Network Concepts and Architectures, 1989, Issues and Problems in Computer Networks, 1990, Advanced Ethernet/802.3 Management and Performance, 1992, Computer Consulting is a Very Funny Business, 1993. With USN, 1974-79. Recipient Arnold Fletcher award, 1992, Tech. Excellence award Decus, 1992. Mem. IEEE, DECUS, NSPE. Achievements include design of over 3900 computer networks. Six-time world aikido champion. Office: Network -1 Inc PO Box 8370 Long Island City NY 11101

HAND, CADET HAMMOND, JR., marine biologist, educator; b. Patchogue, N.Y., Apr. 23, 1920; s. Cadet Hammond and Myra (Wells) H.; m. Winifred Werdelin, June 6, 1942; children—Cadet Hammond III, Gary Alan. B.S., U. Conn., 1946; M.A., U. Calif. at Berkeley, 1948, Ph.D., 1951. Instr. Mills Coll., 1948-50, asst. prof., 1950-51; research biologist Scripps Inst. Oceanography, 1951-53; mem. faculty U. Calif. at Berkeley, 1953—, prof. zoology, 1963-85, prof. emeritus, 1985—; dir. Bodega Marine Lab., 1961-85; Cons. NIH, 1964-66, NSF, 1964-69; mem. atomic safety and licensing bd. panel Nuclear Regulatory Commn., 1971-92, adminstrv. judge atomic safety and licensing bd. panel, 1980-92; NSF sr. postdoctoral fellow, 1959-60; Guggenheim fellow, 1967-68. Contbr. articles to profl. jours. Fellow Calif., Wash. acads. scis.; mem. No. Calif. Malacozool. Soc. (pres. 1963-87), Soc. Systematic Zoology, Ecol. Soc. Am., Ray Soc. (Gt. Britain), Am. Soc. Zoologists (chmn. div. invertebrate zoology 1977-78), Am. Soc. Limnology and Oceanography. Home: 305 McChristian Ave Bodega Bay CA 94923-9723 Office: Bodega Marine Lab Bodega Bay CA 94923

HAND, PETER JAMES, neurobiologist, educator; b. Oak Park, Ill., Jan. 5, 1937; s. James Harold and Edna Mae (Watson) H.; m. Mary Minnis, Sept. 16, 1958; children—Katherine Patricia, Carol Jane, Margaret Anne, Robin Lynn, Stephen Douglas, Peter James; m. Carol Louise Corson, Oct. 23, 1976; m. Christine L. Arnold, Sept. 19, 1986. V.M.D., U. Pa., 1961, Ph.D., 1964. Mem. faculty U. Pa., Phila., 1964—, prof. anatomy, 1979—, head dept. anatomy, 1980-87, 91—; mem. NIH rev. com. Regional Primate Ctrs., 1985-89. Contbr. articles on neurobiology to profl. publs. Pres. USO Council, Cape May, N.J., 1972-73, nat. del.; trustee Mid-Atlantic Ctr. for Arts, Cape May, 1973-74; bd. dirs. Cape May Taxpayers Assn., 1972-74, Univ. City Hist. Soc., Phila., 1978-80. NIH grantee, 1970-82, 86—. Mem. Am. Assn. Anatomists, Am. Assn. Vet. Anatomists, Soc. Neurosci. (pres. Phila. chpt. 1984-85), Internat. Brain Rsch. Orgn., World Assn. Vet. Anatomists, Internat. Assn. for Study of Pain, Am. Assn. Acupuncture, Internat. Coll. Acupuncture and Electro-Therapeutics, Sigma Xi, Alpha Psi (trustee 1965-87). Republican. Home: PO Box 144 Wycombe PA 18980-0144 Office: U Pa Sch Vet Medicine Philadelphia PA 19104

HANDEL, YITZCHAK S., psychologist, educator; b. Bklyn., Aug. 7, 1940; s. Sol and Pauline (Kreisel) H.; m. Noemi Lowinger, Nov. 18, 1967; children: Eliezer Tzvi, Sara Leah, Devorah Yehudit, Raphael Alexander, Ben-Tzion Yaacov. BA, Yeshiva U., 1962, MS, 1965; MS, City Coll. N.Y., 1970; PhD, Yeshiva U., 1977. Cert. profl. psychologist, sch. psychologist. Tchr. Talmud, bible, law Yeshiva of Hartford (Conn.), 1965-67, Yeshiva U. High Sch., N.Y.C., 1967—; assoc. prof. psychology and Jewish edn. Yeshiva U., N.Y.C., 1988—; pvt. practice psychologist N.Y.C., 1980—; asst. dir. grad. Jewish edn. Ferkauf Grad. Sch., N.Y.C., 1977-80; dir. grad. sch. Azrieh Grad. Inst., N.Y.C., 1980—; ednl. advisor, cons. Yeshiva and U. Students for the Spiritual Survival of Soc. Jewery, 1992—. Author: (with others) Educational Principles of the Haggada, 1984; contbr. articles to profl. jours. Recipient Bernard Revel Meml. award Yeshiva Coll. Alumni Assn., N.Y.C., 1990. Mem. Am. Psychol. Assn., Assn. Orthodox Jewish Scientists. Jewish. Office: Azrieli Grad Inst Yeshiva U 245 Lexington Ave New York NY 10016

HANDELSMAN, MITCHELL M., psychologist, educator; b. Phila., Nov. 21, 1954; s. David and Eleanore (Welsh) H.; m. Margie Krest, May 26, 1986. BA, Haverford Coll., 1976; MA, U. Kans., 1979, PhD, 1981. Lic. psychologist, Colo. Asst. prof. Cen. Mo. State U., Warrensburg, 1981-82; asst. prof. U. Colo., Denver, 1982-87, assoc. prof. psychology, 1987-93, prof., 1993—. Contbr. articles to profl. jours., chpts. to books. Named Colo. Prof. of the Yr. Coun. for Advancement and Support of Edn., 1992, Tchr. of the Yr. U. Colo., Denver, 1992, recipient Award for Rsch. Excellence, 1986. Mem. APA (congl. fellow, Washington 1989-90, chmn. task force on ethics in the undergrad. psychology curriculum 1992—), Am. Psychol. Soc., Phi Beta Kappa. Democrat. Home: 1280 Glencoe St Denver CO 80220 Office: Univ of Colo Dept Psychology Box 173 PO Box 173364 Denver CO 80217-3364

HANDLER, ENID IRENE, health care administrator, consultant; b. N.Y.C., Oct. 17, 1932; d. Solomon and Fran S. (Bernstein) Ostrov; m. Murry Raymond Handler, Nov. 22, 1952; children: Lowell S., Lillian Handler Koch, Evan Elliott. BS, Queens Coll., 1968; MS in Adminstrv. Medicine, Columbia U., 1973. Adminstrv. asst. Ctr. Preventive Psychiatry, White Plains, N.Y., 1962-64; adminstrv. dir. Phelps Mental Health Ctr., North Tarrytown, N.Y., 1973-85; ind. cons. E. Handler Assoc., Cortlandt Manor, N.Y., 1986-92; prsenter to profl. orgns. Contbr. articles and book revs. to profl. jours. Mem. adv. bd. Marymount Coll., North Tarrytown, 1983, Iona Coll., New Rochelle, N.Y., 1983; mem. adv. bd. Search for Change, Inc., White Plains, 1987-90; founding mem. Assn. Cortlandt Townsmen Town Planning Group, Cortlandt Manor, 1970-73; bd. dirs. Keon Sch., Montrose, N.Y., 1986-88; chairperson North Westchester County Mental Health Coun., 1974-80; pres. Westchester Assn. Vol. Agys., 1981-82; mem. Westchester County Community Svcs. Bd., 1980-86. NIH fellow Columbia U., N.Y.C., 1971-72. Fellow Am. Orthopsychiat. Assn.; mem. NAFE, Columbia U. Alumni Assn., Adult Continuing Edn. Alumni Queens Coll. (founder). Avocations: music, travel, tennis. Home and Office: Enid Handler Assoc 9318 Laurel Springs Dr Chapel Hill NC 27516

HANDWERKER, A. M., retired transportation executive; b. Chgo., Mar. 15, 1928; s. Fred and Celia H.; m. Betty Jean Ellingson, Nov. 28, 1948; children: Michael L., Sharon J. Behtash, Nancy G. Karam, James A. B.S. in Indsl. Engring., Ill. Inst. Tech., 1950. With Chgo. & North Western Transp. Co., Chgo., 1950-88, v.p. rates and divs., 1980-85, v.p. corp. analysis, 1985-88. Author: (with others) A Guide to Railroad Cost Analysis, 1964. Chgo. Jaycees scholar Ill. Inst. Tech., 1946. Mem. Ops. Research Soc. Am., Inst. Mgmt. Sci., Soc. for Advancement Mgmt., Am. Ry. Engring. Assn., Cost Analysis Orgn., Am. Statis. Assn., Soc. Indsl. and Applied Maths., Transp. Rsch. Forum, Am. Math. Soc., Math. Assn. Am., Alpha Pi Mu. Repubican. Methodist. Club: Union League of Chgo. (bd. dirs. Civic and Arts Found. 1987). Avocations: tennis, hunting, travel. Office: Chgo & North Western Transp Co 1 North Western Ctr 165 N Canal St Chicago IL 60606-1512

HANISCH, KATHY ANN, psychologist; b. Spencer, Iowa; d. Gordon Lee and Carol Jean (Richards) H. BA, U. No. Iowa, 1985; MA, U. Ill., 1988, PhD, 1990. Grad. asst. Dept. Psychology, U. Ill., Champaign, 1985-90; psychology prof. Iowa State U., Ames, 1990—; cons. in field, 1988—. Contbr. articles to Jour. Applied Psychology, The Indsl.-Orgnl. Psychologist, Ergonomics, Jour. Vocat. Behavior. Literacy tutor Ames Pub. Libr., 1991—. Mem. APA, Am. Psychol. Soc., Soc. for Indsl. and Orgnl. Psychology (Ghiselli award 1989), Phi Kappa Phi. Achievements include discovery that constructs of work and job withdrawal identified are empirically supported, that experts' mental models of systems can be used to design tng. programs and/or systems to aid novice users, that retirement attitudes are related to withdrawal. Home: 1420 Illinois Ave Ames IA 50014 Office: Iowa State U W212 Lagomarcino Hall Ames IA 50011

HANISKO, JOHN-CYRIL PATRICK, electronics engineer, physicist; b. Detroit, Mar. 17, 1937; s. John Joseph and Pauline Victoria (Vrabel) H. BEE, U. Detroit, 1963, MSEE, 1965; MA, Wayne State U., 1972, PhD in Physics, 1988. Engr. Burroughs, Detroit, 1962-65; rsch. engr. Boeing, Seattle, 1965-67; sr. engr. Eastman Kodak, Rochester, N.Y., 1967-68; staff engr. Kent-Moore Corp., Warren, Mich., 1971-73; rsch. engr. Udylite, Warren, 1973-75; cons. Southfield, Mich., 1975-76; project engr. Bendix, Troy, Mich., 1976-80; staff engr. TRW, Farmington Hills, Mich., 1980—. Contbr. articles to profl. jours. Mem. Cath. League for Civil and Religious Rights, Phila., Nat. Tax Limitation Com., Washington, Nat. Right to Life Com., Washington. Named Design of Yr. EDN Mag., 1977. Mem. IEEE (sr.). Roman Catholic. Achievements include 5 patents for Electrical Control Apparatus for Internal Combustion Engines, for Sequential Injection Timing Apparatus, for Voltage Controlled Oscillator Having Ratiometric and Temperature Compensation, For Automotive Anti-theft Device. Home: 21888 Murray Crescent Dr Southfield MI 48076-1619 Office: TRW 24175 Research Dr Farmington MI 48335-2642

HANKINS, HESTERLY G., III, computer systems analyst; b. Sallisaw, Okla., Sept. 5, 1950; s. Hesterly G. and Ruth Faye (Jackson) H. BA in Sociology, U. Calif., Santa Barbara, 1972; MBA in Info. Systems, UCLA, 1974; postgrad., Golden Gate U., 1985-86, Ventura Coll., 1970, Antelope Valley Coll., 1977, La Verne U., 1987. Cert. community coll. tchr., Calif. Applications programmer Xerox Corp., Marina Del Rey, Calif., 1979-80; computer programmer Naval Ship Weapon Systems Engring. Sta. of Port Hueneme, Oxnard, Calif., 1980-84; spl. asst. to chief exec. officer Naval Air Sta. of Moffett Field, Mountain View, Calif., 1984-85; mgr. computer systems project Pacific Missile Test Ctr., Oxnard, 1985-88, MIS Def. Contract Adminstrn. Svcs. Region, Oxnard, 1988—; instr. bus. West Coast U., Camarillo, Calif., 1985; core adj. faculty Nat. U., L.A., 1988—; lectr. bus. Golden Gate U., Los Altos, Calif., 1984; instr. computer sci. Chapman Coll., Sunnyvale, 1984, Ventura (Calif.) Coll., 1983-84; cons. City of L.A. Police Dept., Allison Mortgage Trust Investment Co.; minority small bus. assn. cons. UCLA. Author: Campus Computing's Accounting I.S. As a Measurement of Computer Performance, 1973, Campus Computer, 1986, Network Planning, 1986, Satellites and Teleconferencing, 1986, Quotations, 1992, Quotable Expressions and Memorable Quotations of Notables, 1993. Mem. St. Paul United Meth. Ch., Oxnard, Calif., 1986-87; fundraiser YMCA Jr. Rodeo, Lake Casitos, Calif.; key person to combine fed. campaign United Way. Named one of Outstanding Young Men in Am. U.S. Jaycees, 1980, Internat. Leader of Achievement and Man of Achievement, Internat. Biog. Centre, Cambridge, Eng., 1988. Mem. Nat. Assn. Accts., Calif. Assn. Accts., Intergovtl. Council on Tech. Info. Processing, Assn. Computing Machinery (recipient Smart Beneficial Suggestion award 1984), IEEE, Fed. Mgrs. Assn., Alpha Kappa Psi (sec. 1972-73). Office: National Univ 9920 S La Cienega Blvd Inglewood CA 90301-4423

HANKINS, RALEIGH WALTER, microbiologist; b. Tokyo, July 18, 1958; s. Roland Raleigh and Maruko (Okawa) H.; m. Kazue Suzuki, Sept. 26, 1991. BA, Johns Hopkins U., 1980; MPhil, Yale U., 1983, PhD, 1985. Med. tchnician Health Scis. Rsch. Inst., Yokohama, Japan, 1980-81; postdoctoral assoc. Dept. Epidemiology and Pub. Health Yale U., New Haven, 1985-86; rsch. scientist Health Scis. Rsch. Inst. Ctr. Technol. Devel., Yokohama, 1986-89, chief scientist, 1989—. Mem. AAAS, Am. Soc. Microbiology, N.Y. Acad. Scis., Japan Molecular Biology Soc. Avocations: fishing, gardening. Office: Health Scis Rsch Inst, 106 God0-Cho Hodogaya-ku, Yokohama Kanagawa 240, Japan

HANKINSON, RISDON WILLIAM, chemical engineer; b. St. Joseph, Mo., Dec. 11, 1938; s. William Augusta and Rose Mary (Thompson) H.; BS, U. Mo., Rolla, 1960, MS, 1962; PhD (Am. Oil fellow), Iowa State U., 1972; Hon. degree, Chem. Engr., U. Mo., Rolla, 1982; Registered engr., Okla.; m. Lyla Pollard, June 4, 1960; children: Kenneth, Michelle, Michael, Mark, Douglass. Instr. chem. engring. U. Mo., Rolla, 1960-62; instr. chem. engring. Iowa State U., 1964-67; engr. Phillips Petroleum Co., Bartlesville, Okla., 1967-69, group leader, 1969-70, cons., 1970-78, prin., thermodynamics, 1978-80, prin. process engr., 1980-82, sr. staff assoc., 1982-85, mgr. engring. scis. br. tech. systems devel., 1985, mgr. tech. systems br. engring. and svcs., 1985-87; mgr. comml. systems, Phillips 66 Natural Gas Co., 1987-91; div. mgr. client applications and tech. svcs. Phillips Petroleum Co., 1991-92, mgr. architecture and new tech. CIT, 1992-93, sr. scientist rsch. and devel., 1993—; adj. prof. math. Okla. State U., 1967-75, Bartlesville Wesleyan Coll., 1969-71. V.p. Tech. Careers Adv. Com., 1972-73, pres., 1973-74; v.p. Vol. Okla. Overseas Mission Bd., 1970-71; cub scout leader; tchr. religious edn., minister of Eucharist, lector, Roman Cath. Ch., 1976—; chmn. bd. dirs. Alcohol and Drug Center Inc., 1984-87; bd. dirs., 1987-91. Served from 2d lt. to 1st lt. AUS, 1962-63. Recipient Outstanding Alumnus Achievement award Iowa State U., 1971; named Outstanding Young Engr. in Okla., 1970, Outstanding Engr. in Okla., 1984. Fellow Am. Inst. Chem. Engrs. (dir., past pres. Bartlesville sect., Achievement award 1990); mem. Okla. Soc. Profl. Engrs. (v.p. membership 1988-89, exec. v.p. 1989-90), Am. Petroleum Inst. (chmn. phys. properties com. static measurement 1979-82, founder, co-chmn electronic flow measurement com. 1989-91), Hilcrest Country Club, Elks, KC (grand knight, coun. 1987-89, state ch. activities dir. 1989-91, faithful navigator 4th degree 1991-93, dir. state program 1993—), Kiwanis. Contbr. articles to profl., sci. jours. Home: 701 Sooner Park Dr Bartlesville OK 74006-8954 Office: Phillips Petroleum Co 415 Information Center Bartlesville OK 74004

HANKOUR, RACHID, civil engineer; b. Ahfir, Morocco, Jan. 31, 1960; s. Mohamed and Fatiha (Gharram) H.; m. Caroline Fish, Oct. 26, 1991. BS, U. Sci. and Tech. of Oran, Algeria, 1984; Masters, Tufts U., 1988, PhD, 1991. Faculty Tufts U., Medford, Mass., 1990—; geotech. engr. G.E.I. Cons., Winchester, Mass., 1991-92; geotech. engr. HSA, Cambridge, Mass., 1990. Vol. Ability Based on Long Experience, Boston, 1991-92. Recipient Algerian scholarship, Earle F. Littleton scholarship, Medford, 1991. Mem. Am. Soc. Civil Engring., Boston Soc. Civil Engrs., Tau Beta Pi, Sigma Xi. Office: Civil Engring Dept Tufts U Medford MA 02155

HANKS, ALAN R., chemistry educator; b. Balt., Nov. 30, 1939; s. Raymond Hanks and Lillian (Simon) Miller; m. Beverly Jean Hinson, Jan. 17, 1961; children: Craig, Denise, Leta. BS in Physics, West Tex. State U., 1962; MS in Biophys. Chemistry, N. Mex. Highlands U., 1964; PhD in Biophysics, Pa. State U., 1967. Nuclear med. sci. officer Armed Forces Inst. Pathology, Washington, 1967-69; asst. to full prof. biochemistry/biophysics Tex. A&M U., Coll. Sta., Tex., 1969-82; state chemist, seed commnr., prof. Purdue U., West Lafayette, Ind., 1982—. Contbr. articles to profl. jours. Fellow Assn. Ofcl. Analytical Chemists (chmn. methods bd. 1986-89, bd. dirs. 1990—, sec.-treas. 1992-93); mem. Assn. Am. Feed Control Ofcls. (chmn. minerals com. 1985—, lab. methods and svcs. com. 1988—), Assn. Am. Plant Food Control Ofcls. (chmn. Magruder check sample com. 1988-90, bd. dirs. 1989—, chmn. environ. policy com. 1990—, pres.-elect 1991-92, pres. 1992-93), Optimists. Avocations: fishing, gardening, sports, travel. Home: PO Box 2627 West Lafayette IN 47906-0627 Office: Purdue U 1154 Biochemistry Bldg West Lafayette IN 47907-1154

HANLEY, JOSEPH ANDREW, civil engineer; b. Burlington, Vt., Sept. 9, 1966; s. William Richard and Joan Isabelle (Marshall) H.; m. Michelle Boucher, Oct. 3, 1992. AS in Civil Engring. Tech., Vt. Tech. Coll., 1986; BS in Civil Engring., U. Vt., 1989. Engr. in tng. Staff civil engr. Vt. Agy. Transp., Montpelier, 1989-91; project mgr. Mass. Hwy. Dept., Boston, 1991-

92; staff civil engr. Gannett Fleming, Braintree, Mass., 1992—. Mem. ASCE (assoc.). Home: 126 Warren Ave Milton MA 02186

HANLEY, MICHELLE BOUCHER, civil engineer; b. Lowell, Mass., Mar. 15, 1968; d. Roland Joseph and Theresa Marie (Boudreau) B.; m. Joseph Andrew Hanley, Oct. 3, 1992. BS in Civil Engring., U. Vt., 1990. Engr. in tng. Civil engr. I Maguire Group Inc., Hartford, Conn., 1990-91; civil engr. I, transp. program planner II Mass. Hwy Dept., Boston, 1991-93, transp. program planner III, 1993—. Mem. ASCE. Roman Catholic. Home: 126 Warren Ave Milton MA 02186 Office: Mass Hwy Dept Rm 4150 10 Park Plz Boston MA 02116

HANN, JAMES, science administrator; b. Jan. 18, 1933; s. Harry Frank and Bessie Gladys H.; m. Jill Margaret Howe, 1958; 2 children. Student, IMEDE, Lausanne, Switzerland. With James Hann & Sons, 1950-52, Royal Artillery, 1952-54, United Dairies, 1954-65, IMEDE, 1965-66; mng. dir. Hanson Dairies, Liverpool, 1966-72, Seaforth Maritime, Aberdeen, 1972-86; chmn. Bauteil Engring., Glasgow, 1986-88, Exacta Holdings, Selkirk, 1986—, Associated Fresh Foods, Leeds, 1987-89, Strathclyde Inst., Glasgow, 1987-90, Scottish Nuclear, Glasgow, 1990—; mem. Offshore Energy Tech. Bd., 1982-85, Offshore Tech. Adv. Group, 1983-86, Nationalised Industries Chmn.'s Group, 1990—; chmn. dept. Scottish Transport Group, 1987—; commr. Northern Lighthouse Bd., 1990—; dir. William Baird, 1991—. Fellow Inst. Petroleum, Inst. Dirs. Mem. Beaver Club, Burgess of Guild. Office: Scottish Nuclear Ltd, Scottish Nuclear Peel Park, East Kilbride G7Y 5PR, Scotland*

HANN, WILLIAM DONALD, microbologist, biology educator; b. Balt., July 11, 1928; s. William Levi and Helene Charlotte (Krebs) H.; m. Emma Osgood Jones, Apr. 11, 1953; children: Alexander Dwight, Susan Elizabeth, Randolph Warren, Christine Laura. BS in Biology, Wilson Tchrs. Coll., Washington, 1952; MS in Bacteriology, George Washington U., 1956, PhD in Microbiology, 1964. Med. bacteriologist U.S. Army Chem. Corps Biol. Labs., Ft. Detrick, Md., 1952-53; assoc. in microbiology George Washington U., Washington, 1956-65, part-time instr., 1965-66, lectr. in microbiology, 1966-67; resident rsch. assoc. Nat. Naval Med. Ctr., Bethesda, Md., 1964-65; staff fellow NIH, Bethesda, 1965-67; asst. prof. biology Bowling Green (Ohio) State U., 1967-73, assoc. prof., 1973-92, assoc. prof. emeritus, 1992—; vis. asst. prof. biology U. Toledo, 1970; part-time instr. U.S. Army Acad. Health Scis., San Antonio, 1970-80; part-time instr. Nat. Def. U., Washington, 1977-85; vis. assoc. prof. microbiology Mich. State U., East Lansing, 1984-85; bd. dirs., v.p. Wood County Tb and Respiratory Disease Assn., 1971-72; mem. environ. com. Health Planning Assn., Wood County, 1971-75; bd. dirs., pres., N.W. Ohio Lung Assn., 1972-85; various positions including bd. dirs., treas., Am. Lung Assn., Ohio, 1975-85. Contbg. author: Magills Survey of Science, 1991; contbr. articles to profl. jours. Various positions Boy Scouts Am., 1964—, including scoutmaster, dist. chmn., coun. exec. bd., dist. commr.; area rep. Youth for Understanding N.W. Ohio, 1977-84; dir. Liturgical Art Show, St. Mark Luth. Ch., Bowling Green, 1991. With U.S. Army, 1950-52, col. Res., to 1982. Recipient Lamb award Luth. Coun. in U.S.A., 1987, Meritorious Svc. medal U.S. Army, Silver Beaver award Boy Scouts Am., 1980; grantee NIH, 1972-80, Bowling Green State U., 1968-91. Mem. AAAS, AAUP, Am. Soc. Microbiology (edn. chmn. Ohio 1975-84), Assn. Mil. Surgeons U.S. (life), N.W. Ohio Electron Microscopy Soc., Res. Officers Assn. (life), Soc. Armed Forces Med. Lab. Scientists, Soc. for Basic Irreducible Rsch., Sigma Xi, Phi Sigma Pi, Beta Beta Beta, Alpha Eta. Democrat. Achievements include development of a graduate program in blood banking; research in effects of pathogenicity of microorganisms, including viruses, chlamydia and spirochetes. Home: 216 Western Ave Bowling Green OH 43402-2627 Office: Bowling Green State U Dept Biol Scis Bowling Green OH 43403

HANNA, DAVID, optics scientist, educator. Prof. dept. physics U. Southampton, Eng. Recipient Max Born medal and prize, Optical Soc. Am., 1993. Office: Univ of Southhampton, Dept of Physics, Highfield Southhampton S09 5NH, England*

HANNA, GEORGE PARKER, environmental engineer; b. Manhattan, Kans., Mar. 25, 1918; s. George Parker and Agnes (McNamara) H.; m. Jayne Schindler, June 22, 1944; children: Janet, Judith. BSCE, Ill. Inst. Tech., 1940; MSCE, NYU, 1942; PhD in Environ. Engring., U. Cin., 1968. Registered profl. engr., N.Y., Ohio, Nebr., Calif. Design engr. Nussbaumer, Clark & Velzy Engrs., Buffalo, 1953; sr. civil engr. I Creole PEtroleum Corp., Maracaibo, Venezuela, 1954-59; prof. civil engring. Ohio State U., Columbus, 1959-69; chmn. civil engring. U. Nebr., Lincoln, 1969-72, dean coll. engring. & tech., 1971-79; prof. civil engring. Calif. State U., Fresno, 1979-89, dir. engring. rsch. inst., 1986—; ptnr., cons. engr. Hanna, Longley & Assocs., Fresno, 1983-85; environ. lectr. Peoples Rep. of China, 1985. Chmn. Citizens Adv. Com. on Solid Waste, Fresno, 1990-92, Rsch. Sub-com. Valley Drainage Program, Fresno County, 1991-92; adj. mem. tech. adv. com. Met. Water Plan, Fresno, 1991-92. Sci. faculty fellow NSF, 1966. Mem. ASCE (pres. Nebr. sect. 1974-75), NSPE (v.p. 1978-79), Am. Acad. Environ. Engrs. (pres. 1981-82), Calif. Soc. Profl. Engrs. (pres. 1989-90). Achievements include built and directed water resources ctr. at Ohio State U., consolidated engring. & tech. programs at U. Nebr. Lincoln and Omaha into a single Coll. of Engring. & Tech. for U. Nebr., developed Engring. Rsch. Inst. Calif. State U. Office: Calif State U Fresno CA 93740-0094

HANNA, M(AXCY) G(ROVER), JR., civil engineer; b. Greenwood, S.C., June 3, 1940; s. Maxcy Grover Sr. and Nina Dukes (Scott) H.; m. Linda Chandler, Sept. 4, 1969; children: Elizabeth Ruth, Max G. Hanna III. BS in Civil Engring., Clemson U., 1963. Registered profl. engr., Va., Wis., Ga. Field engr. Tidewater Constrn. Corp., Norfolk, Va., 1965-70; sr. field engr. Stone & Webster, Boston, Mass., 1970-74; constrn. engr., contract adminstr. Blount Bros. Corp., Montgomery, Ala., 1974-78; resident engr. Stanley Cons., Muscatine, Iowa, 1978-81; constrn. mgr. Simons-Eastern Co., Decatur, Ga., 1981-83, C.E. Maguire, Inc., New Britain, Conn., 1983-84; constrn. engr., constrn. mgr. Oglethorpe Power Corp., Tucker, Ga., 1984-93; resident engr. Sandwell, Inc., Atlanta, 1993—. Mem. ASCE, NSPE, Ga. Soc. Profl. Engrs. (chmn. profl. engr. in constrn. com., chmn. Energy Action Com., pres., v.p., sec. chpt.). Lutheran. Home: PO Box 2434 Silsbee TX 77656 Office: Ste 300 Sandwell 2690 Cumberland Pk Atlanta GA 30339

HANNA, MILFORD A., agricultural engineering educator; b. West Middlesex, Pa., Feb. 26, 1947; s. Clayton S. and Clara (Burrows) H.; m. Lenora J. Uhrmacher, May 13, 1978; children: Michelle L., Charles C., Susan R., Andrew A. BS, Pa. State U., 1969, MS, 1971, PhD, 1973. Registered engr.-in-tng., Nebr. Asst. prof. Calif. Poly. State U., San Luis Obispo, Calif., 1973-75; asst. prof. U. Nebr., Lincoln, 1975-79, assoc. prof., 1979-85, prof., 1985—, disting. prof. food engring., 1990—, dir. Indsl. Agrl. Products Ctr., 1991—. Contbr. more than 70 articles to profl. jours. Recipient Rsch. award of Merit, Gamma Sigma Delta, 1991. Mem. Am. Soc. Agrl. Engrs. (chmn. Food Processing Engring. Inst. 1991-92, Engr. of Yr. Nebr. sect. 1991), Am. Assn. Cereal Chemists, Coun. for Agrl. Sci. and Tech., Kiwanis (gov. Nebr.-Iowa dist. 1992-93). Achievements include patent pending on starch-based plastic foams. Office: U Nebr 211 L W Chase Hall Lincoln NE 68583-0726

HANNA, WILLIAM FRANCIS, geophysicist; b. Chgo., July 27, 1938; s. Francis Owen and Doris Edith (Cranford) H.; m. Naomi Vivian Purdy, July 23, 1977; stepchildren: James Edward Spedden, Kenneth Alan Spedden. BS, Ind. U., 1960, AM, 1962, PhD, 1965. Lic. geophysicist, geologist. Vis. asst. prof. geophysics Stanford (Calif.) U., 1964-65; geophysicist U.S. Geol. Survey, Menlo Park, Calif., 1965-74, Denver, 1977-83, Reston, Va., 1983—; dep. chief Office Geochemistry and Geophysics, 1974-77; chief Br. Regional Geophysics, 1978-83; instr. Archeol. Soc. Va., Manassas, 1992; thesis com. Coll. William and Mary, Williamsburg, 1990. Recipient Meritorious Svc. award Dept. Interior, 1980. Fellow Geol. Soc. Am.; mem. Am. Geophys. Union, Soc. Exploration Geophysicists, European Assn. Exploration Geophysicists. Achievements include research in solid-earth geophysics. Office: US Geol Survey 927 Nat Ctr Reston VA 22092

HANNAH, ROBERT WESLEY, infrared spectroscopist; b. Niagara Falls, N.Y., May 10, 1931; s. John Wesley and Alice Minerva (Tooker) H.; m. Barbara M. Richter, Sept. 28, 1957; children: Matthew, Rebecca, John, Stephen, Clare, Robert. BS in Chemistry, Niagara U., 1953; PhD in Phys. Chemistry, Purdue U., 1957. Infrared spectroscopist Alcoa Rsch., New Kensington, Pa., 1957-62; applications specialist Perkin-Elmer Corp., Norwalk, Conn., 1962-70; applications mgr. Perkia-Elmer Corp., Norwalk, Conn., 1970-80, sr. scientist, 1980—, mgr. applied rsch., 1985—; chair elect Ctr. for Proceeding Analyt. Chemistry, U. Washington, Seattle, 1991, 92, chair, 1992-93. Author: (with others) Fundamentals of IR Sampling, 1988, Introduction to IR Data Handling, 1989; contbr. articles to profl. jours. Mem. Coblentz Soc. (sec. 1969-92, Williams-Wright award 1985). Roman Catholic. Achievements include patents for Software System for Automatic Interpretation of IR Spectra. Home: 5 Wilderness Rd Newtown CT 06470 Office: Perkin Elmer Corp Main Ave Norwalk CT 06859-0284

HANNAMAN, GEORGE WILLIAM, JR., nuclear/electrical engineer; b. Johnson County, Ind., June 12, 1943; s. George William and Mary Louise (Early) H.; m. Elizabeth Ann Rodda, Aug. 16, 1969; children: Andrew William, Lisa Beth, Jill Kathleen. BSEE, Iowa State U., 1965, MS in Nuclear Engring., 1971, PhD in Nuclear Engring., 1974. Registered profl. engr., Calif. Supr. apparatus repair divsn. Westinghouse Elec. Corp., Milw./Cleve., 1966-70; sr. reactor operator Iowa State U., Ames, 1971-74; staff engr. Gen. Atomic, San Diego, 1974-81; sr. exec. engr. NUS Corp., Rancho Benardo, Calif., 1981-88; sr. staff mem. Sci. Applications Internat. Corp., San Diego, 1988—; mem. com. for U.S./USSR Coop. on Nuclear Safety, U.S. Nat. Acad. Sci., Washington, 1987-88; sec. com. for containment selection Modular High Temperature Gas Cooled Reactor, U.S. Dept. Energy, Washington, 1990; sec. com. for reliability data CEGA Corp., San Diego, 1990. Contbr. articles to profl. jours. Coach Little League, Youth Soccer, San Diego, 1980-85. Elec. Power Rsch. Inst. grantee, 1983-88, 91—, Nuclear Regulatory Rsch./Elec. Power Rsch. Inst. grantee, 1987-88. Mem. IEEE, Am. Nuclear Soc. (exec. com. 1986-89), Sigma Xi. Republican. Episcopalian. Achievements include development of framework and models for human reliability assessments; application of assessments to define products and activities to reduce risk in complex technical systems. Office: Sci Applications Internat 10210 Campus Point Dr San Diego CA 92121

HANNEMA, DIRK, information technology executive; b. Hilversum, The Netherlands, Sept. 19, 1953; arrived in Switzerland, 1979; s. Dirk P. and Maria E. (van Lint) H.; m. Bernadette S. Oeuvray, Dec. 27, 1984; children: Thierry Alexandre, Gwenael Loïc, Robin Emmanuel. Degree in mech. engring., U. de Córdoba, Argentina, 1976; MSME, Ga. Inst. Tech., 1979; degree in engring., Ministry Edn., The Netherlands, 1980; grad., Internat. Inst. for Mgmt Devel, Lausanne, Switzerland, 1987. Project mgr. ABB, Baden, Switzerland, 1980-82; from project mgr. to sr. project mgr. Ascom, Solothurn, Switzerland, 1983-86; asst. to mng. dir. Ascom Banking Systems, Solothurn, 1987-88, mng. dir., 1989-92; mng. dir. Ascom Corp. Devel., Solothurn, 1992-93; mgr. strategic projects Landis & Gyr, Zug, Switzerland, 1993—. Avocations: family, tennis, hiking, sailing, reading. Home: Keltenstrasse 35, Solothurn CH-4500, Switzerland Office: Landis & Gyr Bldg Control, Grafenauweg 10, CH-6301 Zug Switzerland

HANNEMAN, RODNEY ELTON, metallurgical engineer; b. Spokane, Wash., Mar. 14, 1936; s. Christie Luther and Viva Helen (Sugrue) H.; married; 3 children. BS in Phys. Metallurgy, Wash. State U., 1959; MS in Metallurgy, MIT, 1961, PhD, 1964; grad., GE Mgmt. Devel. Inst., 1979. With GE Co., Schenectady, 1963-81; mgr. materials characterization lab. GE Co., 1977-80, mgr. materials programs, 1980-81; v.p. research, devel. and energy resources Reynolds Metals Co., Richmond, Va., 1981-85; v.p. quality assurance and tech. op. Reynolds Metals Co., 1985—; mem. vis. com. dept. materials sci. and engring. MIT, 1975-80, mem. adv. bd. Materials Processing Ctr., 1980—; mem. adv. bd. U. Va., 1982-87, chmn. indsl. adv. bd. grad. engring. program, 1983-86; chmn. rsch. coordinating coun. Gas Rsch. Inst., 1985-87, adv. coun. 1988—; bd. dirs. Materials Properties Coun., 1982-90; mem. ind. adv. bd. ASME, 1987—. Mem. found. bd. Sci. Mus. Va., 1989—; v.p. Civic Assn., 1990-92. Recipient Alumni Achievement award Wash. State U., 1978; Joint Engring. Council award, 1984. Mem. AIME, MAPI, SAE, Am. Soc. Metals (Geisler award 1971, Engring. Materials Achievement award 1973), Am. Chem. Soc. (Chem. Innovator award 1970, Edison medallion 1979), Indsl. Rsch. Inst., Aluminum Assn. (chmn. tech. com. 1989—), Sigma Xi. Achievements include patents in field. Office: Reynolds Metals Co 6601 W Broad St Richmond VA 23230-1701

HANNEMAN, TIMOTHY JOHN, aerospace engineer; b. Jefferson, Iowa, Mar. 18, 1950; s. John Herman and Anna Marie (Walker) H. BA in Mgmt., U. Redlands, 1985, MBA, 1989. Cert. profl. mgr., 1990. Electromech. drafter various cos., 1970-76; instr., coord. drafting program Iowa Ctrl. C.C., Ft. Dodge, 1976-79; revision control adminstr. Rockwell-Collins Avionics, Cedar Rapids, Iowa, 1979-81; engring. design analyst engring. ops. Rocketdyne, Canoga Park, Calif., 1981-82, adminstr. design svcs. Peacekeeper program, 1982-87, structural devel. engr., 1987, project office coord., 1987-88, project mgr. engring. ops. Nat. Aerospace Plane Program, 1988-89, project mgr. engring. ops engring. and test, 1989—; adv. bd. mem. mech. drafting program Iowa Ctrl. C.C., Ft. Dodge, 1974-75. Exec. editor, pubs. rep. Modern Engineering for Design of Liquid Propellant Rocket Engines, 1992. Recipient Spl. Oscar award San Fernando Valley Engrs. Coun., Woodland Hills, Calif., 1991. Fellow Inst. Advancement Engring.; mem. AIAA (membership committeeman 1992—). Mem. Ch. of Christ. Home: 151 N Via Colinas Westlake Village CA 91362 Office: Rockwell Internat Corp Rocketdyne Div PO Box 7922 6633 Canoga Ave MS AB49 Canoga Park CA 91309-7922

HANNING, GARY WILLIAM, utility executive, consultant; b. Sherman, Tex., Aug. 30, 1942; s. William Homer and Mary Maxine (Harshbarger) H.; m. Robin Dale Smith, June 8, 1974; children: TJ, Lorissa Diane. BS, Rollins Coll., 1974; MBA, Stetson U., 1976. Mgr., co-owner Hanning Water Systems, Denison, Tex., 1963-66; engring. technician Gen. Dynamics, Ft. Worth, 1966-67; engr. supr. Bendix Field, Pasadena, Calif., 1967-70; engr. Philco-Ford Corp., Cape Kennedy, Fla., 1970-73, Jet Propulsion Lab., Pasadena, Calif., 1973-74; sect. mgr. Planning Rsch. Corp., Kennedy Space Ctr., 1974-77; pres. S.S.S. Water Systems, Inc., Denison, 1978-83, Texoma Svcs. Corp., Pottsboro, Tex., 1980—; bd. dirs. Ind. Water and Sewer Co. Tex. Inc., Austin, Texoma Valley Coun. Boy Scouts Am., Sherman, Tex.; entrepreneur Bells Discount Supply, Tex., 1983-87. Contbr. articles to profl. jours. Mem. City Coun., Pottsboro, Tex., 1992-93. With USN, 1960-63. Mem. Tanglewood Golf Assn. (sec.-treas. 1992), Am. Legion, C. of C. Mem. Ch. of Christ. Avocations: inventing, camping, reading, golfing, boating, hunting. Office: Texoma Svcs Corp Hwy 120 # 561 Pottsboro TX 75076

HANRATTY, CARIN GALE, pediatric nurse practitioner; b. Dec. 31, 1953; d. Burton and Lillian Aleskowitz; m. Michael Patrick Hanratty, May 22, 1983; 1 child, Tyler James. BSN, Russell Sage Coll., 1975; postgrad., U. Calif., San Diego, 1980. Cert. CPR instr.; cert. NALS; cert. specialist ANA. PNP day surgery unit Children's Med. Ctr., Dallas, 1981-85; clin. mgr. pediatrics Trinity Med. Ctr., Carrollton, Tex., 1985-86; pediatric drug coord. perinatal intervention team for substance abusing women and babies Parkland Meml. Hosp., Dallas, 1990—. Guest talk show Morning Coffee, Sta. KPLX-FM, various TV programs. Rep. United Way, 1988; blood donor chair Parkland Hosp., 1990—, chair March of Dimes, 1992; bd. mem., med. cons. Kidnet Found. Mem. ARC (profl., life), Nat. Assn. PNPs (v.p. Dallas chpt. 1982-83), Tex. Nurses Assn. Avocations: sewing, swimming. Office: Parkland Meml Hosp care Pediatric Nurse Practitioners 5201 Harry Hines Blvd Dallas TX 75235-7793

HANSBROUGH, JOHN FENWICK, surgery educator; b. Front Royal, Va., Sept. 4, 1945; s. Lyle J. and Helen (Trimble) H.; m. Wendy Butler, Dec. 24, 1986; children: John Fenwick Hansbrough Jr., Elizabeth Butler Hansbrough. BS in Chemistry, U. Wis., 1967; MD magna cum laude, Harvard U., 1972. Intern, then resident in surgery U. Colo. Med. Sch., 1972-77; from asst. to assoc. prof. U. Colo. Med. Ctr., Denver, 1977-84; prof. surgery U. Calif., San Diego, 1984—, dir. Regional Burn Ctr., 1984—; Scientific Adv. Bd. DepoTech, La Jolla, Calif., 1992—. Author: Wound Coverage with Biologic Dressings and Cultured Skin Substitutes, 1992; contbr. articles, book chpts. to prof. jours., publs. Recipient 3 rsch. grants NIH, 1984—. Mem. ACS, Am. Burn Assn., Wound Healing Soc., Surg. Infection Soc. Achievements include devel. of dermal and composite skin substitutes using cultured cells and matrix supports for use on burn patients. Office: U Calif San Diego Med Ctr 200 W Arbor Dr San Diego CA 92103

HANSEL, JAMES GORDON, chemical engineer; b. N.Y.C., Oct. 17, 1937; s. Gordon Franklin and Edith (Bradshaw) H.; m. Sarah Craig Martin, Dec. 27, 1964; 1 child, Claire E. BS in Engring., Stevens Inst. Tech., 1959, MSME, 1960, ScD, 1964. Mem. rsch. faculty Princeton (N.J.) U., Guggenheim Labs., 1964-69; rsch. engr. Exxon Rsch., Linden, N.J., 1969-72; mgr. new catalyst devel. Engelhard Corp., Menlo Park, N.J., 1972-81; sr. engring. assoc. Air Products and Chems., Inc., Allentown, Pa., 1981—; adj. assoc. prof. Columbia U., N.Y.C., 1976-80; vis. lectr. Stevens Inst. Tech., Hoboken, N.J., 1970-76; cons. on safety to major corps., 1987—; adj. prof. chem. engring. Pa. State U., State College, 1992—. Author: Theory of Experiments, 1967; contbr. chpt. to Book of Knowledge encyc., 1979; contbr. articles to profl. jours. Mem. Am. Inst. Chem. Engrs. (tech. com. on reactive chems.), Internat. Standards Orgn. (tech. com. on hydrogen vehicles), Nat. Fire Prevention Assn. (tech. coms. on explosion prevention systems, oxygen enrichment hazards), N.Y. Acad. Sci., Sigma Xi, Tau Beta Pi. Achievements include development of Flamchek method of flammability prediction, four patents on applications of oxygen, key participation in development of Three Way Conversion catalyst and automotive engine control system used in over 100 million automobiles worldwide. Home: 829 Frank Dr Emmaus PA 18049 Office: Air Products & Chems Inc 7001 Hamilton Blvd Allentown PA 18105

HANSEL, WILLIAM, biology educator; b. Vale Summit, Md., Sept. 16, 1918; s. John W. and Helen M. (Sperlein) H.; m. Milbrey Downey, Aug. 16, 1942; children: Barbara, Kay. MS, Cornell U., 1947, PhD, 1949. Asst. prof. Cornell U., Ithaca, N.Y., 1949-52, assoc. prof., 1952-61, prof., 1961-90, Liberty Hyde Bailey prof., 1983-90, chmn. physiology dept., 1978-83; Gordon D. Cain prof. La. State U., Baton Rouge, 1990—; scientific adv. Merck, Sharp and Dohme, Rahway, 1980-85, Smith, Kline, Beecham, Westchester, Pa., 1986-91. Author: Genetic Engineering of Animals, 1990; contbr. more than 280 articles to profl. jours. Maj. U.S. Army, 1941-46, ETO. Fellow AAAS; mem. Soc. Study Reprodn. (pres. 1976), Am. Physicol. Soc., Endocrine Soc., Soc. Exptl. Biology and Medicine (treas. 1975), Gamma Sigma Delta, Sigma Xi, Phi Kappa Phi. Achievements include isolation and identification of causative agent of bovine x-disease; developed a successful technique for estrous cycle regulation in catlle, pioneered in development of assays for hormones in blood of animals; discovered control mechanisms for corpus luteum function in cattle; demonstrated the relationships between nutrition and reproduction in cattle. Office: Veterinary Sci/La State U 251 Dalrymple Hall Baton Rouge LA 70803

HANSEN, ALAN LEE, architect; b. Inglewood, Calif., Dec. 2, 1951; s. Grant Lewis and Iris Rose (Heyden) H.; m. Karon Leslie Hargrove, Oct. 16, 1981; children: Jonathan David, Christopher Robert, Garrett Michael. B.Arch. with honors, U. Md., 1974. Registered architect, Va., Md., D.C. Archtl. intern BRW Inc., McLean, Va., 1972, 73; staff architect R.E. Deslauriers AIA, San Diego, 1974-75; project architect Swaney Kerns Architects, Washington, 1975-78; prin. Kerns Group Architects, P.C., Washington, 1978-87; prin. Hansen Architects and Assocs., P.C., Sterling, Va., 1988—; design critic U. Md., College Park, 1978—, Cath. U., Washington, 1981—, Va. Poly. Inst. and State U., 1990—. Mem. Site Plan Rev. Com. Arlington, Va., 1983-87; mem. Planning Commn., Arlington, 1984-87; tchr. youth Potomac Chapel Bible Ch., McLean, Va., 1984-86; mem. Reston Bible Ch., 1987—. Recipient 1st award Housing Mag., 1982, Honor award Am. Plywood Assn., 1977, 2 Honor awards No. Va. chpt., 1977, 85, Va. state, 1983, 84. Mem. AIA (bd. dirs. 1988-91, sec., 1992, treas. 1993), Nat. Council Archtl. Registration Bds., Nat. Trust for Hist. Preservation, Com. for Dulles (bd. dirs. 1989-92). Club: Antique Auto of Am. (Hershey, Pa.). Office: Hansen Architects & Assoc PC 2 Pidgeon Hill Dr Ste 340 Sterling VA 20165-6104

HANSEN, CHRISTIAN GREGORY, architectural engineer; b. St. Joseph, Mo., July 1, 1957; s. Milton Gregory and Dorothy Lee (Deistelkamp) H.; m. Teresa Lynn Spear, 1977 (div. Nov. 1980); m. Linda Marie Shadduck, Feb. 5, 1982; children: Andrew Christian, Benjamin William. BS in Archtl. Engring., Kans. State U., 1984. Registered profl. engr., Mich., Mo., Ohio, Pa., Wis.; cert. energy mgr. Grant program adminstr. State of Kans., Topeka, 1984-86, State of Iowa, Des Moines, 1986-87; project engr. Viron Corp., Kansas City, Mo., 1987-89, project mgr., 1989-91; project mgr. Energy Masters Corp., Overland Park, Kans., 1991—; mem. energy com. City of Topeka, 1984-86. Author: Technical Assistance Report Guidelines, Kansas, 1984, Iowa, 1986. With U.S. Army NG, USNR, 1982-88. Mem. Am. Soc. Heating, Refrigeration and Air Conditioning Engrs. (energy award judge Iowa chpt. 1987), Soc. Energy Engrs. (sr.), Assn. Energy Engrs. Republican. Office: Energy Masters Corp 7301 College Blvd Ste 270 Overland Park KS 66210-1895

HANSEN, DALE J., science administrator, plant biochemist; b. Idaho Falls, Idaho, Sept. 3, 1939; s. Afton Elwin and Ruth (Grover) H.; m. Ruth L. Wheeler, Oct. 21, 1965; children: Evan, Eric, Ryan, Kristi. BS in Agronomy, U. Idaho, 1963, MS in Botany, 1966; PhD in Plant Physiology, Ohio State U., 1969. Postdoctoral fellow U. Ill., Urbana, 1969-72; scientist Monsanto Corp., St. Louis, 1972-81; rsch. dir. Agrigenetics Rsch. Corp., Boulder, Colo., 1981-87, pres., 1982-86; dir. R&D AgriDyne Techs. Inc., Salt Lake City, 1987—. Mem. AAAS, Am. Soc. Agronomy, Plant Growth Regulator Soc. Am. (pres. 1984), Sigma Xi, Alpha Zeta, Phi Sigma. Home: 1537 Spring Ln Salt Lake City UT 84117 Office: AgriDyne Techs Inc 417 Wakara Way Salt Lake City UT 84108-1255*

HANSEN, FINN, electrical engineer; b. Copenhagen, May 21, 1945; s. Hans Henning and Nina (Rasmussen) H.; m. Margit Hansen, Aug. 1, 1987; children by previous marriage: Tine, Jakob, Thomas Duus. BEE, Copenhagen Engring coll., 1971. Electrician Danish Cen. Com., 1967; project engr. Det Danske Stallvalsevaerk, Fr. Vaerk, 1973-75; project engr. Nea-Lindberg A/s, Copenhagen, 1975-77, engring. mgr., 1977-80; service mgr. ISS Securities A/S, Copenhagen, 1980-87; regional mgr. Dansk Erhvers Rengoring A/S, Copenhagen, 1987-88, engring. mgr. KONE Elevator A/S, Copenhagen, 1988—. Avocations: jogging, modern lit. Home: Lyngborghave 20, Birkerod, Copenhagen DK 3460, Denmark Office: KONE Elevator A/S, Lygten 37, Copenhagen DK 2400, Denmark

HANSEN, JAMES E., physicist, meteorologist, federal agencey administrator; b. Mar. 29, 1941. BA in Physics and Math. with highest distinction, U. Iowa, 1963, MS in Astronomy, 1965; postgrad., U. Kyoto and Tokyo U., 1965-66; PhD in Physics, U. Iowa, 1967. NAS-NRC resident rsch. assoc. Goddard Inst. for Space Studies, N.Y.C., 1967-69; NSF postdoctoral fellow Leiden Observatory, Netherlands, 1969; rsch. assoc. Columbia U., 1969-72; mem. staff, space scientist, mgr. planetary and climate programs Goddard Inst. for Space Studies, 1972-81, head, 1981—; adj. assoc. prof. dept. geol. scis. Columbia U., 1978-85, adj. prof., 1985—; co-prin. investigator AEROPOL Project, 1971-74; co-investigator Voyager Photopolarimeter Experiment, 1972-85; prin. investigator Pioneer Venus Orbiter Cloud-Photopolarimeter Experiment, 1974-78, co-investigator, 1978—; prin. investigator Galileo Photopolarimeter Radiometer Experiment, 1977—, Earth Observing System Interdisciplinary Investigation, 1989—. Author: (with others) Radiation in the Atmosphere, 1978, Carbon Dioxide Review, 1982. Am. Geophys. Union fellow, 1992; recipient Goddard Spl. Achievement award, 1977, Group Achievement award NASA, 1982, 93, Exceptional Svc. medal NASA, 1984, Presdl. Rank award NASA, 1990. Achievements include research in radiative transfer in planetary atmospheres, interpretation of remote sounding of planetary atmospheres, development of simplified climate models and 3-D global climate models, climate mechanisms such as the role of clouds in climate, current climate trends from observational data and projections of man's impact on climate. Office: Goddard Inst Space Studies 2880 Broadway New York NY 10025

HANSEN, MICHAEL ROY, chemist; b. Bremerton, Wash., July 27, 1953; s. Roy Vernon and Bonnie Jean (St. Cyr) H.; m. Valerie Jean Paulson, Feb. 14, 1984 (div. Aug. 14, 1987); 1 child, Ryan Ernest. BS in Chemistry, BS in Molecular Biology, U. Wash., 1978. Rsch. scientist Weyerhaeuser, Federal Way, Wash., 1987—; chemistry instr. Highline Community Coll., Des Moines, Wash., 1991—. Patentee in field. With U.S. Army, 1972-75. Mem. AAAS, Am. Chem. Soc., Soc. Plastics Engrs., N.Y. Acad. Soc. Achievements include co-author 5 patents and 16 other patents pending. Office: Weyerhaeuser Technology Ctr 32901 Weyerhaeuser Way S Federal Way WA 98003

HANSEN, PER BRINCH, computer scientist; b. Copenhagen, Nov. 13, 1938; came to U.S., 1970, naturalized, 1992; s. Jorgen Brinch and Elsebeth (Ring) H.; m. Milena Marija Hrastar, Mar. 27, 1965; children: Mette, Thomas. MS, Tech. U. Denmark, Copenhagen, 1963, Dr. Tech., 1978. Systems programmer Regnecentralen, Copenhagen, 1963-70, mgr. software devel., 1967-70; rsch. assoc. Carnegie-Mellon U., Pitts., 1970-72; assoc. prof. Calif. Inst. Tech., Pasadena, 1972-76; chmn. dept. computer sci. U. So. Calif., L.A., 1976-77, prof., 1976-84, Henry Salvatori prof., 1982-84; prof. U. Copenhagen, 1984-87; disting. prof. Syracuse (N.Y.) U., 1987—; cons. Burroughs, Honeywell, IBM, JPL, Mostek, TRW, others. Author: Operating System Principles, 1973, The Architecture of Concurrent Programs, 1977, Programming a Personal Computer, 1982, On Pascal Compilers, 1985; contbr. articles to profl. jours.; mem. editorial bd. Acta Informatica, Concurrency, Software, Lecture Notes in Computer Sci.; inventor Concurrent Pascal, Edison, Joyce programming langs. Recipient Chancellor's medal Syracuse U., 1989; grantee NSF, Army Rsch. Office, Office Naval Rsch., Rome Air Devel. Ctr. Fellow IEEE; mem. Assn. for Computing Machinery. Avocations: history, photography, jazz. Home: 5070 Pine Valley Dr Fayetteville NY 13066-9723 Office: Syracuse U 4-116 CST Syracuse NY 13244

HANSEN, PETER JAMES, reproductive physiologist, researcher; b. Oak Park, Ill., Nov. 23, 1956; s. Peter Aloysious and Cathleen Ann (Forristal) H.; m. Nancy Ann Donovan, Mar. 15, 1980; 1 child, Meghan. BS, U. Ill., 1978; MS, U. Wis., 1980, PhD, 1983. Postdoctoral rsch. assoc. U. Fla. Dept. Biochemistry and Molecular Biology, Gainesville, 1983-84; asst. prof. U. Fla. Coll. Vet. Medicine, Gainesville, 1984-86; asst. prof. U. Fla. Dairy Sci. Dept., Gainesville, 1986-89, assoc. prof., 1989-93; prof. U. Fla. Dairy Sci Dept., Gainesville, 1993—; panel mem. USDA CSRS Competitive Grants Animal Sci. Rev. Panel, Washington, 1990, 93. Contbr. articles to profl. jours. Named Individual Postdoctoral fellow NIH, 1984. Mem. Am. Dairy Sci. Assn. (Young Scientist award 1991), Soc. for Study of Reproduction, Am. Soc. Animal Sci., Am. Soc. Reproductive Immunology. Democrat. Roman Catholic. Achievements include patents in use of interferons to enhance pregnancy rate of farm animals. Home: 3519 NW 27th Terr Gainesville FL 32605 Office: Univ Fla PO Box 110920 Gainesville FL 32611-0920

HANSEN, ROBERT J., mechanical engineer; b. Houston, July 14, 1940; s. Robert John and Opal Ester (Bloomer) H.; m. Patricia A. Poster, Sept. 10, 1966; 1 child, Krista L. BS, Stanford U., 1962; ScD, MIT, 1969. Nat. Rsch. Coun. postdoctoral fellow Naval Rsch. Lab., Washington, 1969-71, rsch. engr., 1971-80, sect. leader, 1980-87; divsn. dir. Office of Naval Rsch., Arlington, Va., 1987-92; chief scientist applied rsch. lab. Pa. State U., State College, 1992—. Contbr. over 70 articles on fluid dynamics, acoustics and signal processing to profl. jours. Achievements include 4 patents. Office: ARL Pa State U PO Box 30 State College PA 16804

HANSEN, ROBERT JOSEPH, civil engineer; b. Tacoma, May 27, 1918; s. Joseph and Olaug (Axness) H.; m. Eleanor Swaim Welch, Dec. 26, 1948; children: Eric Charles, Karen Welch. B.S., U. Wash., 1940; Sc.D., MIT, 1948. Research engr. NRC, 1940-43; Princeton U., 1943-45, Arthur D. Little Co., Cambridge, Mass., 1945; NRC predoctoral fellow, 1946-47; research asso. MIT, 1947-48, mem. faculty, 1948—, prof. civil engring., 1957—, dep. dir. Project Transp., 1964-67; ptnr. Hansen, Holley & Biggs, Inc. (cons. engrs.), Cambridge, 1955-88, prin., 1975-88; ptnr. Newmark, Hansen & Assos., Cambridge and Urbana, Ill., 1958-68; cons. biomechanics Mass. Gen. Hosp., 1956-60; mem. security resources panel Exec. Office of Pres., 1957; mem. sr. adv. panel Air Force Ballistic Div., USAF, 1958-60; mem. exec. com. Adv. Com. CD, Nat. Acad. Scis., 1959—. Author: (with others) Structural Design for Dynamic Loads, 1959; also articles, chpts. in books.; editor: Seismic Design for Nuclear Power Plants, 1970. Recipient Army-Navy cert. of appreciation, 1948; Disting. Service citation Dept. Def., 1969. Fellow ASCE (Moisseiff award 1974, Raymond C. Reese research prize 1975, Innovation Civil Engring award 1989); mem. Boston Soc. Civil Engrs., Sigma Xi, Tau Beta Pi. Home: 25 Cambridge St Winchester MA 01890-3703 Office: MIT Cambridge MA 02139

HANSEN, ROSS N., electrical engineer; b. LaChappel, France, June 20, 1956; s. Robert C. and Maria (Ogrodnik) H.; m. Karlyn J. Bliss, May 25, 1984; children: Breanne L., Brittney R., Brigett H. BSAE, U. Colo., 1979. Registered profl. engr., Colo. Engr.-in-tng. Ellerbe, Mpls., 1980-84; project engr. Dunham Assocs., Reno, Nev., 1984-88, Cator Ruma & Assocs., Lakewood, Colo., 1989-91; sr. elec. engr. U. Colo. Facilities Mgmt., Boulder, 1991—; vis. lectr. U. Minn., St. Paul, 1982-84. Co-author elec. portion: U. of Colo. Electrical Standards for Construction. Mem. Nat. Soc. Archtl. Engrs. (founding mem.), Soc. Fire Protection Engrs., Illuminating Engring. Soc. (program chair 1986-88). Office: Univ of Colo Facilities Mgt Stadium 255 Campus Box 53 Boulder CO 80309

HANSEN, SHIRLEY JEAN, energy consulting executive, professional association administrator; b. Anacortes, Wash., Apr. 18, 1928; d. Elmond and Lazetta Ruth (Poyns) Bushaw; m. James Christian Hansen, Dec. 27, 1947; children: James Douglas, Stephen Clarke, Christian Mark, Margaret Jean Hansen Bon. MA, Cen. Mich. U., 1968; PhD, Mich. State U., 1975. Tchr. Northshore Sch. Dist., Bothell, Wash., 1961-64; tchr. Midland (Mich.) Pub. Schs., 1964-68, adminstr., 1968-73; assoc. exec. dir. Am. Assn. Sch. Adminstrs., Arlington, Va., 1975-80; dir. schs. and hosp. conservation div. U.S. Dept. Energy, Washington, D.C., 1980; pres. Hansen Assocs., Inc., Annapolis, Md., 1980—; chmn. affiliate group Nat. Assn. State Energy Officials, Washington, 1989—; mem. adv. bd. Assn. Energy Engrs., Atlanta, 1991—; chmn. bd. dirs. indoor air quality certification, Environ. Engrs. and Mgrs. Inst., Atlanta, 1992. Author: Managing Indoor Air Quality, 1990, Performance Contracting for Energy and Environ. Systems, 1992; contbr. more than 100 articles and papers to profl. and popular pubs. Mem. Coun. Human Concerns Rep. Nat. Com., Washington, 1979-80. Recipient Spl. Recognition U.S. Congress House Com. Edn. & Labor, 1976; named Disting. Alumnae, Mich. State U., East Lansing, 1989, Environ. Profl. Yr. Assn. Energy Engrs., 1993. Methodist. Achievements include recognition as leading authority on energy efficiency financing, indoor air quality; condition of pub. sch. facilities in Am. and the status of energy concerns and efficiency related to pub. sch. insts.; credited with passage of law to help pub. and non-profit institutions operate more efficiently at a savings of over $15 billion for institutions and energy resources for our nation by 1991 according to U.S. Department of Energy.

HANSEN, WILL, civil engineer, educator, consultant; b. Tonder, Denmark, Jan. 16, 1953; s. Hans-Christian and Jenny Dorothea H.; m. Ruth Deborah Shaffer, Aug. 20, 1983. MSCE, Tech. U. Denmark, 1977; PhD in Civil Engring., U. Ill., 1983. Lectr. civil engring. U. Mich., Ann Arbor, 1982-83, asst. prof. civil engring., 1983-89, assoc. prof. civil engring., 1989-90, 92—; prof. concrete materials Aalborg (Denmark) U., 1990-92. Editorial bd. Nordic Countries Concrete Rsch. Jour., 1990—; contbr. articles to profl. jours. Masuda Internat. fellow Kobe U., Japan, 1989. Mem. ASCE, Am. Ceramic Soc. (chmn. program com. 1989-90), Am. Concrete Inst. Office: Univ of Mich Civil Engring Dept 2330 G G Brown Bldg Ann Arbor MI 48109

HANSOHM, DIRK CHRISTIAN, economics educator, editor; b. Kiel, Germany, June 3, 1956; s. Gerhard Otto Fritz and Lena Ernestine (Finck) H. Vordiplom, Christian-Albrechts U, Kiel, 1978; diploma in econs., U. Bremen, Fed. Republic of Germany, 1983, PhD in Polit. Sci., 1991. Researcher Sudan Econ. Rsch. Group U. Bremen, 1984-92, lectr., 1985—, economist dept. devel. studies, 1991—; editor peripherie, Munster, Fed. Republic of Germany, 1985—; dir. Info. Ctr. Africa, Bremen, 1987-91; editor Afrika-Hefte, Bremen, 1988—, African Devel. Perspectives Yearbook, Bremen, 1989-92, Bremer Afrika Studien, 1991—; cons. Sudan Economy Rsch. Group, Bremen, 1986—, Internat. Labour Office, Geneva, 1992—, African Regional Labour Adminstrn. Ctr., Harare; cons. for internat. orgns., 1986—. Author: The Agriculture of the Sudan, 1991, Small Industry Development in Africa, Lessons from Sudan,1 992; author various pubs. on devel. econs. Director Information Zentrum Afrika, Bremen, 1987-91. Mem. Sudan Studies Soc. of U.K., Sudan Studies Assn. U.S., Wissenschaftliche Vereinigung Entwicklungstheorie und Entwicklungspolitik, European Assn. Devel. Rsch. and Tng. Insts. (working group industrialisation). Home: Inselstrasse 57, 2800 Bremen Germany Office: U Bremen Dept Devel Studies, Bibliothekstrasse, 1800 Bremen 33, Germany

HANSON, BRETT ALLEN, nuclear engineer; b. Kittery, Maine, Mar. 12, 1963; s. Dennis Wilfred and Joan Alberta (Patton) H.; m. Karen Anne Leahy, Aug. 24, 1989. MSEE, Fla. Inst. Tech., 1988, MS in Physics, 1990; postgrad., U. Va., 1990—. Rsch. asst. Fla. Inst. Tech., Melbourne, 1989-90; grad. rsch. asst. U. Va., Charlottesville, 1990—. Mem. IEEE, Soc. Physics Students, Am. Nuclear Soc., Dielectrics and Elec. Insulation Soc., Sigma Pi Sigma. Home: 1224 Smith St Apt D Charlottesville VA 22901-4164

HANSON, CURTIS JAY, structural project engineer; b. Milw., June 17, 1954; s. Donald Edwin and Mary Jean (Hoerig) H.; m. Cheri Lee Froemming, Sept. 12, 1981. BS in Engring., U. Wis., Milw., 1977, MS in Engring., 1987. Registered profl. engr., Wis. Sales engr. Riopelle Engring. Sales, Inc., Milw., 1977-78; design engr. Graef, Anhalt, Schloemer & Assocs. Inc., Milw., 1978-84; engring. supr. Newport News Shipbldg., Brookfield, Wis., 1984-92; project engr. Harnischfeger Corp., Milw., 1992—. Mem. ASCE, NSPE, Wis. Soc. Profl. Engrs. (membership chmn. Milw. North chpt. 1990-92, v.p. 1990-92, program chmn. 1992-93, pres. 1993—, Engr. of Yr. in Industry 1992). Office: Harnischfeger Corp 4400 W National Ave Milwaukee WI 53201

HANSON, DAVID ALAN, software engineer; b. Great Lakes, Ill., Nov. 23, 1956; s. John Berry and Roberta (Sargent) Sizemore; m. Dianna Lynn Hixson, May 20, 1989; stepchildren: Stanley Bob Capps, Jon Allen Capps. BS, U. So. Miss., 1978, MS, 1980. Programmer USDA/Miss. State U., Starkville, Miss., 1979-80; engr. Automated Warehouse, Tex. Ins., Dallas, 1980-83; evening instr. Dallas Community Coll. Dist., 1981-90; software engr. FSAS Systems E-Systems, Dallas, 1983-85, sr. software engr., instr. tng. dept., 1985-90; sr. software engr., lead engr. Spl. Systems E-Systems, Greenville, Tex., 1990—. Author: (tech. manuals) Introduction to C Language, 1985, Introduction to UNIX O.S., 1986, The Ada Language, 1986, Advance C Language, 1987, Advance Ada, 1988. Mem. Tex. Rep. Party, Dallas, 1986-89, PTA, Sulphur Springs, Tex., 1990-91, Dallas, 1989-90, Civil Air Patrol, Miss., 1972-83; cadet leader Civil Air Patrol, Dallas, 1988-90; asst. leader Boys Scouts Am., 1988-90. Recipient scholarship Air Force ROTC, U. So. Miss., 1976, assistantship dept. computer sci., 1978, Billy Mitchell award Civil Air Patrol, Gulfport, Miss., 1973, Amelia Earhart award, 1975. Mem. Toastmasters (sec. 1990-91, pres. 1991—), Sulphur Springs Country Club, Ultralight Assn. (Dallas chpt.). Lutheran. Avocations: flying, water skiing, golf, piano, travel abroad. Home: 1317 Carter St Sulphur Springs TX 75482-4422 Office: E-Systems Inc PO Box 6056 Greenville TX 75403-6056

HANSON, FLOYD BLISS, applied mathematician, computational scientist, mathematical biologist; b. Bklyn., Mar. 9, 1939; s. Charles Keld and Violet Ellen (Bliss) H.; m. Ethel Louisa Hutchins, July 27, 1962; 1 child, Lisa Kirsten. BS, Antioch Coll., 1962; MS, Brown U., 1964, PhD, 1968. Space technician Convair Astronautics, San Diego, 1961; applied mathematician Arthur D. Little, Inc., Cambridge, Mass., 1961; physicist Wright-Patterson AFB, Dayton, Ohio, 1962; assoc. research scientist Courant Inst., N.Y.C., 1967-68; asst. prof. U. Ill.-Chgo., 1969-75, assoc. prof., 1975-83, prof., 1983—, assoc. dir. Lab. for Advanced Computing, 1990—; faculty rsch. participant Argonne Nat. Lab., 1985-87, faculty rsch. leave, 1987-88. Contbr. articles in field to profl. jours. NSF research grantee, 1970-83, 88—; NSF equipment grantee, 1973; Nat. Ctr. for Supercomputer Applications Computer grantee, 1986—. Mem. Soc. Indsl. and Applied Math., Assn. for Computing Machinery, Computer Soc. of IEEE, Control Soc. of IEEE, Resource Modeling Assn. Home: 5435 S East View Park Chicago IL 60615-5915 Office: U Ill Dept Math Stats & Computer Sci M/C 249 PO Box 4348 Chicago IL 60680-4348

HANSON, HAROLD PALMER, physicist, government official, editor, academic administrator; b. Virginia, Minn., Dec. 27, 1921; s. Martin Bernhard and Elvida Elaine (Paulsen) H.; m. Mary Jean Stevenson, June 22, 1944; children: Steven Bernard, Barbara Jean. B.S., Superior (Wis.) State Coll., 1942; M.S., U. Wis., 1944, Ph.D., 1948. Mem. faculty U. Fla., 1948-54, dean grad. sch., 1969-71, v.p. acad. affairs, 1971-74, exec. v.p., 1974-78, exec. v.p. emeritus, 1990—; mem. faculty U. Tex., Austin, 1954-69, prof. physics, 1961-69, chmn. dept., 1962-69; provost Boston U., 1978-79; exec. dir. Com. on Sci. and Tech., U.S. Ho. of Reps., Washington, 1979-82, 84-90; provost Wayne State U., Detroit, 1982-84; summer rsch. physicist Lincoln Labs., MIT, 1953, Gen. Atomic Co., San Diego, 1964; summer vis. lectr. U. Wis., 1957; Fulbright rsch. scholar, Norway, 1960-61. Editor DELOS, 1991—. Bd. dirs. N. Central Fla. Health Planning Coun.; mem. steering com. Fla. Ednl. Computer Network. With USNR, 1944-46. Decorated St. Olav's medal Norway, Order of North Star 1st class Sweden; U. Fla. presdl. scholar, 1976. Fellow Am. Phys. Soc.; mem. Sigma Xi, Sigma Pi Sigma, Omicron Delta Kappa. Clubs: Town and Gown (Austin); Rotary. Office: Univ Florida 215 Williamson Hall Gainesville FL 32605

HANSON, HUGH, ecologist, educator; b. Lewis, Kans., Dec. 16, 1915; s. Charles Cleveland and Blanche (Malin) H.; m. Thelma Klotz, Aug. 6, 1941. BS in Edn., Kans. State Tchrs. Coll., 1939; MS in Zoology, U. Ill., 1941, PhD in Zoology, 1948. Prof. zoology Ariz. State U., Tempe, 1948-78, chmn. dept. zoology, 1957-62, acting head div. life scis., 1960-62, prof. emeritus, 1978—; pres. Maricopa Audubon Soc., Greater Phoenix area, 1953-55; mentor Sigma Xi., 1948—. Bd. dirs. Sertoma, Emporia, Kans., 1991—; mem. pres.'s club, founders' soc. Emporia State U. Found., 1991—; bd. dirs. Plumb Place, Emporia, 1991—. With USN, 1942-45. Mem. Nature Conservancy, Friends of the Great Plains Studies, Current Club.

HANSON, JOHN M., civil engineering and construction educator; b. Brookings, S.D., Nov. 16, 1932; m. Mary Josephson, Jan. 16, 1960. B.S.C.E., S.D. State U., 1949; M.S. in Structural Engring., Iowa State U., 1957; Ph.D. in Civil Engring., Lehigh U., 1964. Profl. engr. Ill., N.C., Colo., Oreg. Structural engr. J.T. Banner & Assoc., Laramie, Wyo., 1957-58, Phillips, Carter, Osborn, Denver, 1958-60; research inst. prof. Lehigh U., Bethlehem, Pa., 1960-65; engr., asst. mgr. structural devel. Portland Cement Assn., Skokie, Ill., 1965-72; rsch. dir., v.p., pres. Wiss, Janney, Elstner, Northbrook, Ill., 1972-92; disting. prof. civil engring. and constrn. N.C. State U., Raleigh, 1992—. Contbr. articles to profl. jours. Served to lt. USAF, 1953-55, Korea. Recipient Disting. Engring. award S.D. State U., 1979; Profl. Achievement citation Iowa State U., 1980. Fellow ASCE (State of the Art award 1974, Reese award 1976, 88, Y.T. Lin award 1979), Am. Concrete Inst. (bd. dirs. 1981-84, v.p. 1988-89, pres. 1990, Bloem award 1976); mem. Prestressed Concrete Inst. (bd. dirs., Korn award 1978), Transp. Rsch. Bd., Internat. Assn. Bridge and Structural Engrs. (pres. 19931—), NAE (elected 1992). Lutheran. Office: NC State U Dept Civil Engring Box 7908 Raleigh NC 27695-7908

HANSON, JOHN MARK, ecologist, researcher; b. Ottawa, Ont., Can., Apr. 14, 1955; s. Albert John and Mary Margaret (Pender) H.; m. Catherine Mary Merlin, May 31, 1980; children: Margaret Anne, Jennifer Theresa, Brian Joseph. MSc, U. Ottawa, 1980; PhD, McGill U., 1985. Postdoctoral fellow U. Alta., Edmonton, Can., 1985-89; vis. fellow Halifax (N.S.) Fisheries Rsch. Lab., 1989-90; rsch. scientist Dept. Fisheries and Oceans, Moncton, N.B., Can., 1990—, editor secondary publs. div. marine and anadromous fish, 1991-93; referee Can. Jour. Fisheries and Aquatic Scis., Ottawa, 1986—, Trans. Am. Fish. Soc., Bethesda, Md., 1990—; mem. groundfish subcom. Can. Atlantic Fisheries Sci. Adv. Com., Dartmouth, N. S., 1990—. Co-author: Atlas of Alberta Lakes, 1990; contbr. articles to Jour. Wildlife Mgmt., Can. Jour. Fisheries and Aquatic Scis., Jour. Animal Ecology, Jour. Freshwater Biology, Can. Jour Zoology. Postgrad. fellow Natural Scis. and Engring. Rsch. Coun. Can., 1981-83, postdoctoral fellow, 1986-88, vis. fellow, Halifax, 1989-90. Mem. Can. Soc. Zoologists, Ecol. Soc. Am., Am. Soc. Naturalists, Am. Fisheries Soc. (cert. fisheries scientist). Achievements include research in aquatic ecology, biology of freshwater clams, ecology of crayfish in lakes, codfish ecology in Gulf of St. Lawrence. Office: Gulf Fisheries Ctr, PO Box 5030, Moncton NB Canada E1C 9B6

HANSON, KENNETH MERRILL, physicist; b. Mt. Vernon, N.Y., Apr. 17, 1940; s. Orville Glen and Marion (Chamberlain) H.; m. Earle Marie Low, June 1964 (div. July 1989); children: Jennifer Anne, Keith Merrill. BE in Physics, Cornell U., 1963; MS in Physics, Harvard U., 1967, PhD in Physics, 1970. Rsch. assoc. Lab. of Nuclear Studies, Ithaca, N.Y., 1970-75; mem. staff Los Alamos (N.Mex.) Nat. Lab., 1975—. Author: (with others) Radiology of Skull and Brain, 1979, Image Recovery, 1987; contbr. articles to profl. jours. Recipient Award Excellence, Dept. Energy, 1986. Mem IEEE (sr.), Am. Phys. Soc., Soc. Photo Optical Instrumentation Engrs. (program com. imaging conf. 1984—). Achievements include research in image analysis and elementary particle physics. Office: Los Alamos Nat Lab MSP 940 Los Alamos NM 87545

HANSON, LOWELL KNUTE, seminar developer and leader, information systems consultant; b. Langford, S.D., Sept. 28, 1935; s. Hans Jacob and Katherine Sofie (Hoines) H.; m. Mary Lou Heeney, Oct. 24, 1964; children: Victoria Lynn Hanson Wheeler, Thomas Lowell, Ronald Richard. BSEE, S.D. Sch. Mines and Tech., 1961. Field engr. supr. Control Data, Sunnyvale, Calif., 1961-62; mgr. systems test Control Data, Mpls., 1962-69; mgr. product mgmt. Control Data, France, 1969-71; mgr. test and integration Control Data, Mpls., 1971-74; communications cons. Control Data, Republic of South Africa, 1974-76; prin. applications engr. Control Data, Mpls., 1976-85; prin. systems engr. Martin Marrietta, Washington, 1986; pres. Viking Svcs., Centreville, Va., 1986—; LAN seminar dir. The Am. Inst., N.Y., 1987, 89, dir. and cons. Bus. Communication Rev., Hinsdale, Ill., 1990—. Author, presenter Trouble Shooting LANs seminar, 1988, Hands on LAN, 1990, Maintaining and Trouble Shooting Novell LANs, 1990. Pres. Homeowner Assn., Maple Grove, Minn., 1983-84. Fellow IEEE; mem. Toastmasters (pres. Maple Grove chpt. 1981-82, Capital area gov. 1986, Toastmaster of Yr. Maple Grove chpt. 1983, past pres. worldwide 84, Club of Yr. 84 Toastmasters). Republican. Avocations: reading, personal development, hiking, plays, movies. Home and Office: Viking Svcs 6418 Overcoat Ln Centreville VA 22020-2314

HANSON, RICHARD EDWIN, civil engineer; b. Sioux City, Iowa, July 22, 1931; s. Gustav Edwin and Delia Thelma (Horton) H.; m. Joann Gager Terhune, Nov. 6, 1954 (div. Jan. 1971); children: Richard Edwin Jr., John William, Tamara Terhune; m. Lillie Gwenette Capitanio, Feb. 21, 1987. BSCE, Iowa State U., 1953; postgrad., U.S. Army Gen. Staff Coll. Registered profl. engr., Iowa. Engr.-in-tng. U.S. Army Corps Engrs., Washington, 1957-58; sr. co. engr. Dickinson Constrn. Co., Chgo., 1959-62; asst. chief constrn. mgmt. Goddard Space Flight Ctr., NASA, Greenbelt, Md., 1963-69; chief project mgmt. U.S. Postal Svc., Washington, 1969-70; chief Air Force project mgmt. U.S. Army Corps Engrs., Washington, 1971-77; chief of constrn. U.S. Army Corps Engrs., Balt., 1977-81; chief of constrn. South Atlantic div. U.S. Army Corps Engrs., Atlanta, 1982-85; chief of constrn. U.S. Army Corps Engrs., Washington, 1986-91; dir. constrn. ops. Pacific Ocean div. U.S. Army Corps Engrs., Honolulu, 1991—. Editor: Corps Engrs. Constrn. Newsletter, 1988-91. Pres. Walbrooke Manor Citizens Assn., Lanham, Md., 1963-64. 1st lt. C.E., U.S. Army, 1953-57. Mem. ASCE, NSPE, Soc. Am. Mil. Engrs. (bd. dirs. Washington chpt. 1988-91), Beta Theta Pi. Republican. Avocations: flying, golf, skiing, fishing, sailing. Home: 44-125 Kahinani Way Kaneohe HI 96744

HANSON, ROBERT JAMES, electrical test engineer, consultant; b. Baudette, Minn., Aug. 17, 1936; s. Severt and Doris Emma (Flatner) H.; m. Beverly Joanne Johnson, July 8, 1961; children: James, Carey, Gregory. BSBA, BS in Indsl. Engring., U. N.D., 1959; BSEE, U. Wash., 1963; MSEE, U. So. Calif., 1966. With dept. aerospace Boeing, Seattle, 1959-63, 66-77, with dept. B-1B, 1983-85, with dept. comml. avionics systems, 1987-91; with dept. electronics radar Rockwell, Anaheim, Calif., 1963-66; with dept. marine systmes Honeywell, Seattle, 1977-81; with dept. power supply products Eldec, Seattle, 1981-83; test engr. Goodyear, Akron, Ohio, 1985-87; pres. Americom Svcs., Bainbridge Island, Wash., 1988—; lectr. on test, concurrent engring. and mfg. at convs. nationwide. Contbr. articles to profl. jours. Mem. Surface Mount Tech. Assn., Am. Soc. Test Engrs. Home and Office: 883 Park Ave NE Bainbridge Is WA 98110

HANSON, ROLAND STUART, industrial engineer, educator; b. Kearny, N.J., July 2, 1932; s. Warren Robert and Adolpha (Koyen) H.; m. Susan Elva Berg, Feb. 3, 1955; children: Gary Alan, Lisa Susan. BS in Engring. and Mgmt., Fairleigh Dickinson U., 1957, MBA, 1960. Registered profl. engr., Oreg., Calif. Sr. indsl. engr. Beckman Instruments, Inc., Palo Alto, Calif., 1963-65; mgmt. cons. A.T. Kearney & Co., Inc., San Francisco, 1965-70; dir. indsl. engring. St. Vincent Hosp. and Med. Ctr., Portland, Oreg., 1970-73; sr. systems cons. Oreg. Assn. Hosps., Portland, 1973-78; sr. indsl. engr. Tektronix, Inc., Beaverton, Oreg., 1978-80; pres. R.S. Hanson & Assocs., Portland, 1980-81; chair engring. tech. Ga. So. U., Statesboro, 1981—; mem. tech. accreditation com. Accreditation Bd. for Engring. and Tech., N.Y.C., 1987—. Contbr. articles to profl. jours. Fellow AAAS; mem. Inst. Indsl. Engrs. (sr., trustee 1975-79, Indsl. Engr. of Yr. 1970, 72), Am. Soc. for Engring. Edn. Democrat. Unitarian. Office: Ga So U Landrum Box 8045 Statesboro GA 30460-8045

HANSON, RONALD WINDELL, cardiologist; b. Jeffersonville, Ind., Apr. 30, 1947; s. Erwin D. and Bernice (Windell) H. B.S. summa cum laude, Ariz. State U., 1968, M.S., 1969, Ph.D. in Physics, 1972; M.D., U. Ala., 1977. Diplomate Am. Bd. Internal Medicine, Am. Bd. Cardiovascular Diseases. Asst. prof. physics U. Ala., 1972-74; resident in internal medicine and cardiology Good Samaritan Hosp., Phoenix, 1977-82; practice medicine specializing in cardiology, Gadsden, Ala., 1982—; entrepreneur in comml. real estate and gas and oil exploration. Served to lt. col. CAP. Fellow Am. Coll. Cardiology; mem. Phi Kappa Phi. Office: 300 Bapt Med Dr Ste 500 Gadsden AL 35903

HANSON, WILLIAM BERT, physics educator, science administrator; b. Warroad, Minn., Dec. 30, 1923; s. Bert Hanson and Viola Mae Carlquist; m. Wenonah Ann Dahlquist, Mar. 14, 1946 (dec. Sept. 1989); children: Bryan, Craig, David, Karen; m. Annelies Ruth Hanson, Jan. 5, 1990. BAChemE, U. Minn., 1944, MS in Physics, 1949; PhD in Physics, George Washington U., 1954. Rsch. physicist Nat. Bur. Standards, Washington, 1949-54, Boulder, Colo., 1954-56; rsch. scientist Lockheed Missiles and Space Co., Palo Alto, Calif., 1956-62; prof. S.W. Ctr. for Advanced Studies, Dallas, 1962-69, U. Tex. at Dallas, Richardson, 1969—. Contbr. over 150 articles to sci. jours. Mem. nat. bd. Planned Parenthood, 1970. Lt. USN, 1944-46. Fellow Am. Geophys. Union (John Adam Fleming medal 1985); mem. European Geophys. Union, Internat. Acad. Astronautics. Achievements include performance of only in-situ measurements of the Mars ionosphere; research for NASA missions Viking to Mars, Atmosphere Explorers, Dynamics Explorer, others. Home: 7831 La Sobrina Dr Dallas TX 75248-3138 Office: U Tex at Dallas 2601 N Floyd Rd Richardson TX 75080-1407

HANSON-PAINTON, OLIVIA LOU, biochemist, educator; b. Tulsa, June 11, 1949; d. Louis Benton and Dorothy Marie (Sewell) Hanson; m. Ronald Phillip Painton, Feb. 26, 1977; 1 child, Marie Elizabeth. BS, U. Okla., 1972; PhD in Biochemistry, U. Okla. Health Scis. Ctr., 1982. Asst. prof. dept. biochemistry U. Okla. Health Scis. Ctr., Oklahoma City, 1988-89, asst. prof. dept. pathology, 1989-91; asst. prof. U. Cen. Okla., Edmond, 1991—. Contbr. Internat. Jour. Devel. Biology, Modern Pathology, Jour. Cell Physiology, Analytical Biochemistry. Fellow Okla. Med. Rsch. Found., 1978, Am. Heart Assn., 1986, 87. Mem. Am. Soc. for Biochemistry and Molecular Biology, Am. Chem. Soc., Am. Soc. for Cell Biology, Sigma Xi. Achievements include research in characterization of calmodulin gene from Drosophila. Home: 717 NW 39 Oklahoma City OK 73118 Office: U Cen Okla 100 N University Dr Edmond OK 73034

HANSSON, GUNNAR CLAES, medical biochemist; b. Lessebo, Sweden, Jan. 30, 1951; s. Henry V. and Asta G. (Claesson) H.; m. Ulrica I. Sterky; children: Fredrik C., Patrik C., Elisabeth S. MD, U. Gothenburg, 1976, PhD, 1981. Postdoctoral scholar U. Gothenburg, 1982, rsch. assoc., 1985-87, asst. prof. molecular biology, 1987—; postdoctoral scholar NIH, Bethesda, Md., 1983-84; prof. analytical biochemistry Karblinska Inst., Stockholm, 1992. Contbr. over 80 articles to profl. jours. Mem. Gothenburg Med. Assn., Swedish Med. Assn. Achievements include devel. of the biosynthetic pathway for a cancer associated antigen, technique for determination of sugar specific attachment of bacteria and virus, techniques for combined separation characterization of complex saccharides, characterization of cancer-associated mucins, cloning of intestinal mucin. Office: U Gothenburg, Dept Med Biochemistry Medicinareg 9, 41390 Gothenburg Sweden

HANTON, SCOTT DOUGLAS, physical chemist; b. Saginaw, Mich., Sept. 25, 1963; s. Douglas Oliver and Ann Ruth (Curnow) H.; m. Helen Jean

Murphy, July 13, 1985. BS in Chemistry, Mich. State U., 1985; PhD in Physical Chemistry, U. Wis., 1990. Sr. rsch. chemist Air Products and Chems., Allentown, Pa., 1990—; Contbr. articles to profl. jours. Mem. Am. Chem. Soc. Achievements include first reaction cross sections for J state specific metal cations and total gas management system for excimer laser users.

HAPPER, WILLIAM, JR., physicist, educator; b. Vellore, India, July 27, 1939; came to U.S., 1941, naturalized, 1961; s. William and Gladys (Morgan) H.; m. Barbara Jean Baker, June 10, 1967; children—James William, Gladys Anne. B.S., U. N.C., 1960; Ph.D., Princeton U., 1964. Rsch. assoc. Radiation Lab., Columbia U., N.Y.C., 1964; asst. prof. physics Radiation Lab., Columbia U., 1967-70, assoc. prof., 1970-74, prof., 1974-80; dir. Radiation Lab., Columbia U. (Radiation Lab.), 1976-79; prof. Princeton (N.J.) U., 1980-91, 93—; dir. office energy rsch. of Energy Dept. Energy, Washington, D.C., 1991—; chmn. Jason/Mitre, 1987-90; trustee Mitre Corp.; cons. in field. Alfred P. Sloan fellow, 1967; Recipient Alexander von Humboldt award Fed. Republic of Germany, 1976. Fellow Am. Phys. Soc. Home: 559 Riverside Dr Princeton NJ 08540-4007 Office: Princeton U Dept Physics Princeton NJ 08544

HAQ, BILAL UL, national science foundation program director, researcher; b. Gorakhpur, India, Oct. 8, 1943; came to U.S., 1968; s. Fazli and Sorraya (Rabbani) H.; m. Nazli Azam, June 11, 1975. MSc, U. Panjab, Pakistan, 1963; PhD, U. Stockholm, 1967, DSc, 1972. UNESCO scholar U. Vienna, Austria, 1964-65; Swedish Internat. Devel. Authority rsch. scholar U. Stockholm, 1965-68; rsch. scientist Woods Hole (Mass.) Oceanographic Inst., 1968-82, Exxon Prodn. Rsch. Co., Houston, 1982-88; program dir. NSF, Washington, 1988—; rsch. assoc. Smithsonian Inst., Washington, 1988—; vis. prof. U. Copenhagen, Geol. Survey of Denmark, 1991; keynote speaker Bur. Mineral Resources, Canberra, Australia, 1988, Internat. Geol. Congress, Washington, 1989, Kyoto, Japan, 1992; vis. com. mem. for Brit. Geol. Survey, U.K. Natural Environment Rsch. Coun., Swindon, 1990-91; mem. panel U.S. Japan Coop. Program Natural Resources, Tokyo, 1991-92; on assignment White House Office Mgmt. and Budget, 1992, World Bank environment dept., 1993. Author, editor: Introduction to Marine Micropaleontology, 1978, Marine Geology and Oceanography of Arabian Sea, 1984, Ocean Drilling on Exmouth Plateau, 1990, Calcareous Nannoplankton, 1984, Nannofossil Biostratigraphy, 1984. Fellow Geol. Soc. Am., Geol. Soc. London; mem. Am. Assn. Petroleum Geologists (disting. lectr. 1989), Am. Geophys. Union. Achievements include research in global sea level and environmental change. Office: NSF 1800 G St NW Washington DC 20550-0002

HAQ, IFTIKHAR UL, mechanical engineer, consultant; b. Distt Faisalabad, Punjab, Pakistan, Apr. 1, 1939; s. Muhammad and Maryam Ali; m. Razia Iftikhar, Oct. 10, 1963; children: Maimoona, Saad Ul, Asad Ul. BS in Mech. Engring., U. Engring. and Tech., Lahore, Pakistan, 1958. Jr. engr. Machinery Pool Orgn. Wapda, Lahore, 1960-62, sr. engr., workshop mgr., 1962-71; chief mech. engr. Ministry of Works, Sokoto, Nigeria, 1971-75; gen. mgr. Mechanized Constrn. Pakistan, Lahore, 1976-79; chief exec. Engring. Svcs. Internat., Dubai, United Arab Emirates, 1979-85, Regent Enterprises (Pvt.) Ltd., Lahore, 1985-87; mng. ptnr. Engring. Gen. Cons., Lahore, 1987—. Active Human Rights Commn. Pakistan, Lahore, 1988—, Azad Pakistan Group, Lahore, 1989-90. Mem. ASME, Instn. Engrs. (exec.), Pakistan Engring. Congress (sec.), All Pakistan Music Conf. Muslim. Home: 209 Ahmad Block, New Garden Town, Lahore 54600, Pakistan Office: Engring Gen Cons, Apt 10 2d Fl Auriga, Gulberg Lahore 54666, Pakistan

HARA, MASANORI, chemist; b. Takasago, Japan, Nov. 8, 1952; came to U.S., 1984; s. Hisashi and Fumiko (Nigaki) H.; m. Masami Koyama, Feb. 26, 1992. MS, Kyoto U., 1977, PhD, 1981. Postdoctoral fellow McGill U., Montreal, 1981-84; asst. prof. Rutgers U., New Brunswick, N.J., 1984-89, assoc. prof., 1989—. Editor: Polyelectrolytes: Science and Technology, 1992; contbr. articles to profl. jours. Office: Rutgers U MMS Piscataway NJ 08855-0909

HARA, YASUO, physics educator and researcher; b. Kamakura, Japan, Mar. 30, 1934; s. Ichiro and Sugako (Ishimaru) H.; m. Saeko Watanabe, Mar. 24, 1963; children: Michiyo Kawano, Natsuyo. BS, U. Tokyo, 1957, MS, 1959, PhD, 1962. Rsch. assoc. Tokyo U. Edn., 1962-63; rsch. fellow Calif. Inst. Tech., Pasadena, 1963-64; rsch. assoc. U. Chgo., 1964-65; mem. Inst. for Advanced Study, Princeton, N.J., 1965-66; assoc. prof. Tokyo U. Edn., 1966-75; dean U. Tsukuba, Japan, 1981-85, prof., 1975—, v.p. for acad. affairs, 1992—; mem. Univ. Accreditation Com., Tokyo, 1991—. Recipient Nishina Meml. prize Nishina Meml. Found., Tokyo, 1977. Home: 3-15-22 Yakumo Meguro-ku, Tokyo 152, Japan Office: U Tsukuba, 1-1-1 Tennoudai, Tsukuba 305, Japan

HARADA, YOSHIYA, chemistry educator; b. Hofu, Yamaguchi, Japan, Jan. 5, 1934; s. Tomosuke and Setsuko Hayashi; m. Yoshiko Ozawa, Dec. 12, 1964; children: Mariko, Fumiyo, Yuko. BSc, Tokyo U., 1957, MSc, 1959, DSc, 1965. Rsch. assoc. Tokyo U., 1961-69, assoc. prof., 1969-83, prof. chemistry, 1983—, dean Coll. Arts and Scis., 1991-93. Author: Quantum Chemistry, 1978, Chemical Thermodynamics, 1984; contbr. articles to profl. jours. Mem. Chem. Soc. Japan (award 1984), Phys. Soc. Japan. Avocations: travel, gardening. Home: 3-4-10 Kamitakada, Nakano, Tokyo 164, Japan Office: Coll Arts and Scis U Tokyo, Komaba, Meguro, Tokyo 153, Japan

HARALSON, JOHN OLEN, utility company executive; b. Lepanto, Ark., July 16, 1954; s. Leonard Bell and Aquila (Kendrick) H.; m. Pamela Kay Michael, Oct. 7, 1972; children: Jennifer Diann, Jonathan David. Grad., Perryville (Ark.) High Sch. I & C tech. Ark. Power & Light Co., Russellville, 1976-79; I&E head tech. Ark. Kraft Paper, Morrilton, 1979; test engr. Westinghouse Inst. Svc. Co., Perry, Ohio, 1980 81; start up engr. elec. Bechtel Power Corp., Grand Gulf, Miss., 1981-87; sr. engr. I&C/elec. Bechtel Power Corp., Chattanooga, 1986-87; specialist Houston Lighting & Power, Wadsworth, Tex., 1987; tech. cons. Power Mgmt. Corp., Northglen, Colo., 1982-83. With USN, 1972-76. Mem. Am. Nuclear Soc., Nat. Rifle Assn., Masons (Scottish and York Rites). Home: 16 Murex St Bay City TX 77414 Office: Houston Lighting & Power Hwy 521 Wadsworth TX

HARAMUNDANIS, KATHERINE LEONORA, computer and data processing company executive; b. Boston, Jan. 25, 1937; d. Sergei Illarionovich and Cecilia Helena (Payne) Gaposchkin; m. John Haramundanis, Mar. 6, 1958; children: George John, Sergei Edward. BA, Swarthmore Coll., 1958; postgrad., Boston U. Rsch. assoc. Smithsonian Astrophys. Obs., Cambridge, Mass., 1958-74; tech. writer Wang Labs., Lowell, Mass., 1974-77; advanced devel. mgr. Digital Equipment Corp., Marlboro, Mass., 1977—; judge Soc. for Tech. Communication, 1989, 92. Contbr. articles to profl. jours. Recipient Spl. Svc. award Smithsonian Instn., 1966, Merit award Smithsonian Astrophys. Obs., 1972. Mem. AAAS, IEEE Computer Soc., Assn. for Computing Machinery (treas. Spl. Interest Group for System Documentation 1993), Soc. for Tech. Comm. (exec. coun. 1993), Assn. Computational Linguistics, Linguistic Soc. Am., Am. Astron. Soc., Am. Archeol. Soc., Am. Soc. Oriental Rsch. Home: PO Box 1365 Westford MA 01886-4865

HARARY, FRANK, mathematician, computer science educator; b. N.Y.C., Mar. 11, 1921; s. Joseph and Mary (Laby) H.; children: Miriam, Natalie, Judith, Thomas, Joel, Chaya. B.A., Bklyn. Coll., 1941, M.A., 1945, D.A. (hon.), 1962; Ph.D., U. Calif., Berkeley, 1948; M.A. status, U. Oxford, Eng., 1973; D.Sc., M.Sc. (hon.), U. Aberdeen, Scotland, 1975; Fil.Dr. in Social Scis. (hon.), U. Lund, Sweden, 1978; M.A. status, U. Cambridge, Eng., 1981; D.Sc. (hon.) in Computer Sci., U. Exeter, Eng., 1992. Mem. faculty dept. math. U. Mich., Ann Arbor, 1948-86, prof., 1964-86, prof. emeritus, 1987—; faculty assoc. Research Center for Group Dynamics, Inst. Social Research, 1950-82; Disting. vis. prof. math. and computer sci. N. Mex. State U., Las Cruces, 1986, Disting. prof. computer sci., 1987—; mem. Inst. Advanced Study, Princeton, N.J., 1957-59; vis. prof. math. Univ. Coll. London, 1962-63; vis. prof. stats. London Sch. Econs., 1966-67; vis. prof. psychology U. Melbourne, 1969; vis. prof. combinatorics U. Waterloo, Ont., Can., fall 1970; vis. prof. elec. engring. U. Chile, Santiago, 1970; vis. prof. math. U. Copenhagen, 1970, Technion, Haifa, Israel, 1973, U. Niamey, Niger, 1975, U. Newcastle, Australia, 1976, Simon Bolivar U., Caracas, Venezuela, 1977, Ain Shams U., Cairo, 1992, U. Rome, 1992, U. Jyvaskyla, Finland, 1992; fellow Wolfson Coll., U. Oxford, 1973-74, Churchill Coll., Cambridge U., 1980-81; vis. prof. chemistry U. Paris, 1971, Texas A&M U., Galveston, 1993; vis. prof. geography U. Lagos, Nigeria, 1975; vis. prof. Tallinn Bot. Gardens, Estonian Acad. Scis., 1989; colloquium lectr. Edinburgh Math. Soc., St. Andrews, 1972; inaugural lectr. S.E. Asian Math. Soc., Singapore, 1972; lectr. Malaysian Math. Soc., Penang, Malaysia, 1974; lectr. Inst. Mgmt. Scis., Phila., 1958, IRE, N.Y.C., 1958, AAAS, N.Y.C., 1960, Linguistic Soc. Am., N.Y.C., 1960, Am. Math. Soc., N.Y.C., 1960, 1st Caribbean Combinatorial Conf., Kingston, Jamaica, 1970, 2d Caribbean Combinatorial Conf., Barbados, 1977, 6th Caribbean Combinatorial Conf., Trinidad, 1991, Soc. Indsl. and Applied Math., Santa Barbara, Calif., 1968, 5th Brit. Combinatorial Conf., Aberdeen, Scotland, 1975, 7th S.E. Conf. on Graph Theory and Computing, Baton Rouge, 1977, Ont. Math. Meetings, St. Catherine, 1977, Bremer Konferenz zur Chemie, Bremen, Germany, 1978, European Assn. for Theoretical Computer Sci., Udine, Italy, 1978, Math. Assn. Am., Bradenton, Fla., 1974, Valparaiso, Ind., 1980, Holland, Mich., 1980, Brookings, S.D., 1983, Serbian Chem. Soc., Kragujevac, Yugoslavia, 1980, Greek Math. Soc., Athens, 1983, Assn. for Math. Applied to Econ. and Social Scis., Catania, Sicily, 1983, Conf. on Graph Theory, Steiner Systems and Their Applications, Santa Tecla, Sicily, 1986, 89, Calcutta Math Soc., 1986, Allahabad Math Soc. 1986, Hong Kong Math Soc., 1986, 1st Japan Conf. on Graph Theory, Hakone, 1986, 1st China-U.S. conf. on Graph Theory, Jinan, 1986, Chem. Soc. Japan, Shizuoka, 1987, 30th Anniversary Conf. of Thessaloniki Grad. Sch. Bus. Adminstrn., 1987, 1st Internat. Conf. on Artificial Intelligence, Hong Kong, 1988, 7th Sunbelt Conf. on Social Networks, Tampa, Fla., 1989, 4th Internat. Conf. On Mathematical Chemistry, Galveston, Tex., 1989, 9th Conf. on Discrete Math., Clemson U., 1989; 7th Internat. Conf. on Graph Theory, Western Mich. U., Kalamazoo, 1992; disting. vis. lectr. U. Cen. Fla., Orlando, 1984, 90; invited Reunion of Nobel Laureates, Lindau, Germany, 1989, disting. scientist in residence N.Y. Acad. Scis., 1977; Humboldt Found. fellow, Munich, Germany, summers 1978, 79; dir. summer schs. NATO, Frascati, Italy, 1962, Varenna, Italy, 1964; participant numerous internat. symposia in honor of 70th birthday including Estonia Math. Soc., Tartu and Kaariku, 1991, Graphs and Hypergraphs, Varenna, 1991, Graph Theory and Applications, Durban and Itala, South Africa, 1991, Graph Theory and Theoretical Chemistry, Saskatoon, Sask., Can., 1991, Graph Theory and Computer Sci., Guadalajara, Mex., 1991, Grh Theory and Mech. Engring. Taiwan, 1991. Editor, founder: Jour. Combinatorial Theory, 1966—; editor: Discrete Mathematics, 1970—, Jour. Math. Sociology, 1970-78, Bull. Calcutta Math. Soc, 1976—, Jour. Combinatorics, Info. and Systems Scis, 1976—; hon. editor in chief, founder: Jour. Graph Theory, 1977—, Social Networks, 1978-81, Networks, 1979—, Discrete Applied Mathematics, 1979—, Math. and Computer Modelling, 1980—, Caribbean Jour. Math. and Computing Scis., 1980—, Math. Social Scis., 1980-87, Jour. Information and Optimization Scis., 1987-89, Applied Math. Letters, 1988—, Computers and Math. with Application, 1988—, Jour. Mathematical Chemistry, 1991—, Proyecciones, 1992—; author: (with R. Norman and D. Cartwright) Structural Models, 1965, Graph Theory, 1969, (with E. Palmer) Graphical Enumeration, 1973; (with Per Hage) Structural Models in Anthropology, 1983, (with F. Buckley) Distance in Graphs, 1990, (with P. Hage) Exchange in Oceania, 1991; Editor: A Seminar on Graph Theory, 1967, Graph Theory and Theoretical Physics, 1967, Proof Techniques in Graph Theory, 1969, New Directions in the Theory of Graphs, 1973, (with R. Bari) Graphs and Combinatorics, 1974, Topics in Graph Theory, 1979, (with J. S. Maybee) Graphs and Applications, 1985. Mem. Am. Math. Soc., London Math Soc., Glasgow Math Soc., Edinburgh Math Soc., Can. Math Soc., S.E. Asian Math Soc., Malaysian Math Soc., Calcutta Math Soc. (v.p. 1978—), Math. Assn. Am., Soc. for Indsl. and Applied Math, Allahabad Math. Soc. Office: NMex State U Dept Computer Sci Box 30001 Dept 3CU Las Cruces NM 88003

HARARY, KEITH, psychologist; b. N.Y.C., Feb. 9, 1953; s. Victor and Lillian (Mazur) H.; m. Darlene Moore, Oct. 22, 1985. BA in Psychology, Duke U., 1975; PhD, Union Inst., 1986. Crisis counselor Durham (N.C.) Mental Health Ctr., 1972-76; rsch. assoc. Psychical Rsch. Found., Durham, 1972-76; rsch. assoc. dept. psychiatry Maimonides Med. Ctr., Bklyn., 1976-79; dir. counseling Human Freedom Ctr., Berkeley, Calif., 1979; rsch. cons. SRI Internat., Menlo Park, Calif., 1980-82; design cons. Atari Corp., Sunnyvale, Calif., 1983-85; pres., rsch. dir. Inst. for Advanced Psychology, San Francisco, 1986—; freelance sci. journalist, 1988—; lectr. in field; adj. prof. Antioch U., San Francisco, 1985, 86; guest lectr. Lyceum Sch. for Gifted Children, 1985-89; vis. researcher USSR Acad. Scis., 1983; rsch. cons. Am. Soc. for Psychical Rsch., 1971-72, Found. for Rsch. on Nature of Man, 1972. Co-author: Berkeley Personality Profile, 1994, 30-Day Advanced Psychology Series, 1989-91, The Mind Race, 1984, 85; contbr. articles to Omni, Jour. Am. Soc. Psychical Rsch., Psychology Today, Exceptional Human Experience, Magical Blend, ASPR Newsletter, Jour. Near Death Studies, others. Mem. Am. Psychol. Soc., Am. Psychol. Assn. Achievements include first to develop reflective approach to personality profiling; development of advanced human potential research, including original training methodologies in extended perception and communication. Home and Office: 2269 Chestnut St # 875 San Francisco CA 94123

HARBAUGH, LOIS JENSEN, secondary science educator; b. Elmhurst, Ill., Sept. 16, 1942; d. G. E. and Dorothy G. (Madsen) Jensen; m. Lou L. W. Harbaugh Jr., Aug. 8, 1964; children: Michelle, Bill. BA, Wheaton Coll., 1964; MAT in Sci. Edn., U. Tex., Dallas, 1978. Cert. composite secondary sci. tchr., Tex. Tchr. Troy Mills (Iowa) Sch., 1965-66, Richardson (Tex.) Jr. High Sch., 1969-71; tchr., chair sci. First Bapt. Acad., Dallas, 1975-81, Lake Highlands Jr. High Sch., Richardson, 1981—; mem. Tex. state textbook com. Tex. Edn. Agy. 1990. Bd. dirs. Crisis Pregnancy Ctr., Dallas, 1983-88; bd. educators Found. for Thought and Ethics, Richardson, 1988—. Christa McAuliffe fellow Dept. Edn., 1988; grantee Recognizing Innovation for Student Edn. (RISE) Found., 1989; recipient Nat. Radio Astronomy Observatory (NRAO) Inst. award NSF, 1988, Newmast award NASA, 1989, Tchr. Cons. award Tex. Instruments, 1991. Mem. Nat. Sci. Tchrs. Assn., Nat. Sci. Suprs. Assn. (sec. 1988-92), Sci. Tchrs. Assn. Tex., Richardson Assn. Tex. Profl. Educators (pres. elect 1993), Tex. Earth Sci. Tchrs. Assn. Office: Lake Highlands Jr High Sch 10301 Kingsley Rd Dallas TX 75238

HARBER, M(ICHAEL) ERIC, industrial engineer; b. Tulsa, Okla., Nov. 14, 1965; s. Charles C. and Joyce F. (Allen) H.; m. Alyson Kelley, Sept. 1, 1990. Student, Oxford U., 1987; BS in Indsl. Engring. and Mgmt., Stanford U., 1988. Cert. in systems integration. Rsch. asst. Amoco Rsch. Ctr., Tulsa, 1985; quality control and design asst. Nutter engring. divsn. Patterson-Kelly, Tulsa, 1986; indsl. engring. coop. intern space systems divsn. Gen. Dynamics, San Diego, 1987; indsl. engring. cons. Hewlett-Packard, Cupertino, Calif., 1988; mgmt. assoc., asst. mgr. backcards divsn. Citicorp, San Mateo, Calif., 1988-90; sr. mgmt. analyst El Camino Healthcare System, Mountain View, Calif., 1990—; co-chairperson, mentor program adv. bd. Silicon Valley Indsl. Engring., Sunnyvale, Calif., 1993—; mem. adv. task force Ergonomic-Repetive Strain Injury, Mountain View, 1993—; speaker in field. Recipient Engr.'s ring Order of Engr., 1993. Sr. mem. Inst. Indsl. Engrs. (chpt. pres. 1992-93); mem. NSPE, Am. Soc. Quality Control, Soc. Health Systems, Healthcare Info. Mgmt. Systems Soc., Soc. Engring. Mgmt. Systems, Calif. Soc. Profl. Engrs. Presbyterian. Avocations: scuba diving, volleyball, writing poetry, golf, Judo. Home: 20800 Homestead Rd Cupertino CA 95014 Office: El Camino Healthcare System 2500 Grant Rd Mountain View CA 94039

HARBORT, ROBERT ADOLPH, JR., computer science educator; b. Emory, Ga., Jan. 29, 1947; s. Robert A. and Mary Alice (Mitchell) H.; m. Lucy Patrick, Mar. 16, 1968, (div. Aug. 1976). BS, Emory U., 1968; MS, Ga. Inst. Tech., 1975; PhD, Emory U., 1987. Registered profl. engr. Staff physicist Emory U. Med. Sch. Radiology, Atlanta, 1970-72; with med. computing Emory U., Atlanta, 1972-75; with div. biomed. engring. Crawford Long Hosp., Atlanta, 1975-79; with anesthesia rsch. Emory U. Med. Sch., Atlanta, 1979-83; asst. prof. computer sci. So. Coll. Tech., Marietta, Ga., 1983-87, dept. chmn. computer sci., 1983-90, assoc. prof. computer sci., 1987-93, prof., 1993—. Contbr. articles to profl. jours. mem. IEEE (Atlanta chpt. exec. com. 1990-92), Assn. for Computing Machinery (Atlanta chpt. chmn. 1977-84, sec. 1972-77), Am. Soc. for Engring. Edn., Sigma Xi. Democrat. Quaker. Home: Box 21510 Atlanta GA 30322-1001 Office: So Coll Tech Computer Sci Dept 1100 S Marietta Pkwy Marietta GA 30060-2896

HARDAGE, PAGE TAYLOR, health care administrator; b. Richmond, Va., June 27, 1944; d. George Peterson and Gladys Odell (Gordon) Taylor; m. Thomas Brantley, July 6, 1968; 1 child, Taylor Brantley. AA, Va. Intermont Coll., Bristol, 1964; BS, Richmond Profl. Inst., 1966; MPA, Va. Commonwealth U., Richmond, 1982. Cert. tchr. Competent toastmaster, dir. play therapy svcs. Med. Coll. Va. Hosps., Va. Commonwealth U., Richmond, 1970-90; dir. Inst. Women's Issues, Va. Commonwealth U., U. Va., Richmond, 1986-91; adminstr. Childhood Lang. Ctr. at Richmond, Inc., 1991—; bd. dirs. Math. and Sci. Ctr. Found., Richmond, Emergency Med. Svcs. Adv. Bd., Richmond. Treas. Richmond Black Student Found., 1989-90, Leadership Metro Richmond Alumni Assn.; bd. dirs. Richmond YWCA, 1989-91; group chmn. United Way Greater Richmond, 1987; bd. dirs. Capital Area Health Adv. Coun. Mem. NAFE, ASPA, Adminstrv. Mgmt. Soc., Internat. Mgmt. Coun. (exec. com.), Va. Recreation and Park Soc. (bd. dirs.), Va. Assn. Fund Raising Execs., Rotary Club of Richmond. Unitarian. Avocations: bridge, target shooting, aerobics, pub. speaking. Office: Childhood Lang Ctr at Richmond Inc 4202 Hermitage Rd Richmond VA 23227-0136

HARDENBURG, ROBERT EARLE, horticulturist; b. Ithaca, N.Y., July 27, 1919; s. Earle Volcart and Aline (Crandall) H.; m. Jean Marie Swett, Oct. 3, 1943; children: Kathryn, Mary Ann. BS, Cornell U., 1941, MS, 1947, PhD, 1949. Assoc. horticulturist USDA, Beltsville, Md., 1949-53, horticulturist, 1953-58, sr. horticulturist, 1958-61, prin. horticulturist, 1961-67, lab. chief, 1967-81; pvt. practice cons. Venice, Fla., 1981-90; ret., 1990. Author: Commercial Storage of Fruit, Vegetable, and Nursery Stock, 1986. Maj. U.S. Army, 1941-46. Fellow Am. Soc. Hort. Scis. (sect. chmn. 1972-73); mem. Produce Mktg. Assn. (life, bd. dirs. 1960-62, Disting. Svc. award 1963), Refrigeration Rsch. Found. (mem. sci. adv. coun. 1974-91, Cert. Appreciation 1991), Rotary (dir., editor Venice, Fla. 1981-93), Lions (treas. College Park, Md. chpt. 1955-62). Republican. Methodist. Achievements include research on handling, transportation, storage and packaging of fruits, vegetables and nursery stocks. Home: 648 Bird Bay Dr W Venice FL 34292-4026

HARDER, RUEL TAN, mechanical engineer; b. Iloilo, Philippines, Jan. 2, 1946; came to U.S. 1968; s. Rodrigo Jardenico and Jesusa (Tan) H.; m. Margo Patience Guib, Aug. 1, 1970; children: Reed, Ruel, Rainee. BSME, Mapua Inst. Tech., 1966. Registered profl. engr., Wash. Jr. mech. engr. LaPaz Engring. Co., Iloilo, 1966-67; assoc. engr. The Boeing Co., Seattle, 1968-70; chief engr. Seattle Steam Co., 1971—. Mem. Internat. Dist. Heating and Cooling Assn. Republican. Roman Catholic. Achievements include design and development of steam flowmeter testing using steam as the testing medium. Home: 6807 51st Ave S Seattle WA 98118 Office: Seattle Steam Co 1319 Western Ave Seattle WA 98101

HARDGROVE, GEORGE LIND, JR., chemistry educator; b. Barberton, Ohio, Dec. 2, 1933; s. George Lind and Chesley Janice (Peck) H.; m. Gretchen Marie Grosenick, June 18, 1961; children: George William, Anne Elizabeth. AB, Oberlin Coll., 1956; PhD, U. Calif., Berkeley, 1959. Chemistry educator St. Olaf Coll., Northfield, Minn., 1959—. Contbr. articles to Crystallographica. Fellow USPHS, 1965-66. Mem. Am. Chem. Soc., Am. Crystallographic Assn., Midwest Assn. of Chemistry Tchrs. in Liberal Arts Colls. Episcopalian. Home: 117 S Orchard Northfield MN 55057 Office: St Olaf Coll Dept Chemistry Northfield MN 55057

HARDIN, CLIFFORD MORRIS, retired executive; b. Knightstown, Ind., Oct. 9, 1915; s. James Alvin and Mabel (Macy) H.; m. Martha Love Wood, June 28, 1939; children: Susan Carol (Mrs. L.W. Wood), Clifford Wood, Cynthia (Mrs. Robert Milligan), Nancy Ann (Mrs. Douglas L. Rogers), James. BS, Purdue U., 1937, MS, 1939, PhD, 1941, DSc (hon.), 1952; Farm Found. scholar, U. Chgo., 1939-40; LLD, Creighton U., 1956, Ill. State U., 1973; Dr. honoris causa, Nat. U. Colombia, 1968; DSc, Mich. State U., 1969, N.D. State U., 1969, U. Nebr., 1978, Okla. Christian Coll., 1979. Instr. U. Wis., 1941-42, asst. prof. agrl. econs., 1942-44; assoc. prof. agrl. econs. Mich. State Coll., 1944-46, prof., chmn. agrl. econs. dept., 1946-48, dir. expt. sta., 1949-53, dean agr., 1953-54; chancellor U. Nebr., 1954-69; sec. U.S. Dept. Agr., Washington, 1969-71; vice chmn. bd., dir. Ralston Purina Co., St. Louis, 1971-80; dir. Center for Study of Am. Bus., Washington U., St. Louis, 1981-83, scholar-in-residence, 1983-85; cons., dir. Stifel, Nicolaus & Co., St. Louis, 1980-87; bd. dirs. Gallup, Inc., Lincoln, Nebr., Halifax Corp., Alexandria, Va.; bd. dirs. Omaha br. Fed. Res. Bank of Kansas City, 1961-67, chmn., 1962-67. Editor: Overcoming World Hunger, 1969. Trustee Rockefeller Found., 1961-69, 72-81, Winrock, Internat., Morrilton, Ark., 1984—, Am. Assembly, 1975—, U. Nebr. Found., 1975—; mem. Pres.'s Com. to Strengthen Security Free World, 1963. Mem. Assn. State Univs. and Land-Grant Colls. (pres. 1960, chmn. exec. com. 1961).

HARDIN, JAMES W., botanist, herbarium curator, educator; b. Mar. 31, 1929. BS, Fla. Southern Coll., 1950; MS, U. Tenn., 1951; PhD, U. Mich., 1957. Instr. U. Mich., 1956-57; from asst. prof. and curator of herbarium to assoc. prof. and curator of herbarium N.C. State U., Raleigh, 1957-68, prof. and curator of herbarium, 1968—; vis. prof. Mountain Lake Biological Sta. U.Va., summers 1962, 64, 83, U. Okla. Biological Sta., summers 1967, 70; mem. exec. com. Flora Southeastern U.S., 1966—; endangered species com. N.C. Dept. Natural & Econ. Rsch., 1973-74, natural areas adv. com., 1973-79; mem. plant conservation sci. com. N.C. Dept. Agriculture, 1980—, chmn. 1987—; mem. endangered species com. N.C. Wildlife Resources Commn., 1976-78, N.C. State Mus. Natural Hist., 1975-78; pres. Highlands Biological Station, Inc., 1963-69, trustee, 1958-69, sec., 1960-63; invited symposium speaker. Author: Human Poisoning, 1974, Textbook of Dendrology, 1991, editor ASB Bulletin, 1980-86, editorial com. Am. Jour. Botany, 1964-66; editorial bd. Brittonia, 1964-67, Brimleyana, 1975—; reviewer jours. in field. Trustee Highlands Biological Found., 1976—. Mem. Am. Soc. Plant Taxonomists (pub. policy com. 1976-78, editorial bd. 1964-67, editor-in-chief Systematic Botany 1985-91, pres. elect 1991-92, pres. 1992-93, past pres. 1993—, Cooley award 1958), Southern Appalachian Botanical Club (v.p. 1959-60, pres. 1964-65), Botanical Soc. Am. (editorial com. 1964-66, chair southeastern sect. 1968-69), Assn. Southeastern Biologists (Meritorious Teaching award 1991, chmn. local arrangements 1966, 77, v.p. 1968-69, pres. 1970-80, editor 1980-86), Internat. Assn. Plant Taxonomy, Soc. Economic Botany (chmn. local arrangements 1979), Torrey Botanical Club, Coun. Biology Editors (reference style com. 1986-89), Gamma Sigma Delta (sec.-treas. N.C. chpt. 1972-73), Phi Kappa Phi, Sigma Xi (exec. com. N.C. chpt. 1962-63, sec. 1965-66, treas. 1966-67, v.p. 1967-68, program chmn. 1968-69, pres. 1969-70). Office: N C State Univ Dept Botany Raleigh NC 27695-7612

HARDIN, JAMES WEBB, molecular biologist, medical center administrator; b. Baton Rouge, Nov. 16, 1946; s. James William and Ora (Felps) H.; m. Gwendolyn Elsa Stafford, June 3, 1972; children; Margaret Elizabeth, James William. BS, La. State U., 1968; PhD, Purdue U., 1972. Postdoctoral fellow Harvard U., Cambridge, Mass., 1972-74; postdoctoral fellow Baylor Coll. Medicine, Houston, 1974-75, asst. prof., 1975-80; assoc. prof. U. Ark. Med. Scis., Little Rock, 1980—; assoc. dir. Ark. Cancer Rsch. Ctr., Little Rock, 1986—; mem. study sect. NIH, Bethesda, Md., 1982-87. Recipient Rsch. Career Devel. award NIH, 1982, Rsch. grant, 1985-88, 93—, Rsch. grant ACS, 1991—. Mem. Am. Soc. Cell Biology, Am. Assn. Cancer Rsch., Sigma Xi. Democrat. Methodist. Office: Univ Ark Med Sci 4301 W Markham Little Rock AR 72205

HARDIN, WILLIAM BEAMON, JR., electrical engineer; b. Lumberton, N.C., Jan. 28, 1953; s. William Beamon and Virginia Ruth (Conner) H.; m. Mary Wanda Livingston, Jan. 22, 1971; children: William David, Christopher Wayne, David Wayne. BEE, Thomas A. Edison State Coll., 1990. Cert. plant engr., worldwide energy mgr., lighting efficiency profl.; lic. elec. contractor, N.C. Elec. engr. Barnhill Electric Engr. Co., Fayetteville, N.C., 1973-76; indsl. contractor Red Springs, N.C., 1976-78; plant engr. Waverly Mills, Laurinburg, N.C., 1978-81; chief elec. engr. Sharon-Harris Nuclear Power Plant, Raleigh, N.C., 1981-82; group engr. J.P. Stevens & Co., Inc., Great Falls, S.C., 1982-86; mgr. plant engring. and energy mgr. Springs Industries, Laurel Hill, N.C., 1986—; advisor indsl. adv. com. Alternative Energy Corp., Raleigh, N.C., 1988—, state edn. adv. com., 1990—, Indsl. Energy Textile Focus Group N.C., Raleigh, 1991, indsl. maintenance and elec. installation adv. com. Richmond Community Coll. Engring. Dept., Rockingham, N.C., 1990—. Chmn. Laurinburg (N.C.) C. of C. Edn. Com.,

1991—. Named Vol. of Yr. Laurel Hill Sch., 1992. Mem. Cogeneration Inst. Environ. Engrs. and Mgrs., Am. Inst. Plant Engrs., Environ. Engrs. and Mgrs., Environ. Engrs. and Mgrs. Inst. Assn. Energy Engrs. (Regional Energy Engr. of Yr. 1992, Regional Energy Mgr. of Yr. 1990, Energy Engr. of Yr. 1989, 90), Elec. Industry Evaluation Panel, Scotch Meadows Country Club. Republican. Baptist. Achievements include development of Energy Patrol program involving students in efficient school energy usage by monitoring lights, leaks and thermostat settings. Home: 8141 Braemar Circle Laurinburg NC 28352

HARDING, CHARLTON MATTHEW, electrical engineer; b. Hartford, Conn., Dec. 1, 1957; s. Charles Harry and Nancy Louise (Day) H.; m. Elizabeth Giacoumis, May 29, 1983; 1 child, Christopher Andreas. BSEE, Cornell U., 1985. Mem. tech. staff Sch. Elec. Engring. Cornell U., Ithaca, N.Y., 1981-85; mgr. device fabrication Opto-Electronics Ctr. McDonnell Douglas Corp., Elmsford, N.Y., 1986-93; engr. Westchester Indsl. Specialty Optical Co., Inc., Peekskill, N.Y., 1993—; referee Electronics Letters, Stevenage, Hertshire, U.K., 1992-93. Contbr. articles to profl. publs. Presbyterian. Achievements include co-designer of high output etched mirror laser diodes; invention of low divergence laser diode, high power laser diode array; demonstrated that ion beam etching can be used to produce reliable laser diodes. Home: 149 Nardin Rd Lake Peekskill NY 10537 Office: Westchester Indsl Specialty Optical Co Inc 8 John Walsh Blvd Ste 311 Peekskill NY 10566

HARDING, KAREN ELAINE, chemistry educator and department chair; b. Atlanta, Sept. 5, 1949; d. Howard Everett and Ruth Evangeline (Lund) H.; m. Bruce Roy McDowell, Aug. 30, 1975. BS in Chemistry, U. Puget Sound, Tacoma, 1971; MS in Environ. Chemistry, U. Mich., 1972; postgrad., Evergreen State Coll., 1972, 84, Yale U., 1986, Columbia U., 1991. Chemist Environ. Health Lab., Inc., Farmington, Mich., 1972-73, U. Mich. Med. Sch., Ann Arbor, 1973-75; instr. chemistry Schoolcraft Coll., Livonia, Mich., 1975-77; chair chemistry dept. Pierce Coll., Tacoma, 1977—; adj. prof. U. Mich., Dearborn, 1974-77; instr. S.H. Alternative Learning Ctr., Tacoma, 1980-83, Elderhostel, Tacoma, 1985-89. Mem. County Solid Waste Adv. Com., Tacoma, 1989—, Superfund Adv. Com., Tacoma, 1985-89, Sierra Club, Wash., 1989—; mem., past pres. Adv. Com. Nature Ctr., Tacoma, 1981-87. Faculty Enhancement grantee Pierce Coll., 1990; recipient Nat. Teaching Excellence award, 1991. Mem. NW Assn. for Environ. Studies (treas. 1985—), Am. Chem. Soc., Ft. Steilacoom Running Club (race dir. 1986—). Avocations: running, skiing, backpacking, bicycling, reading. Office: Pierce Coll 9401 Farwest Dr SW Tacoma WA 98498-1999

HARDT, ROBERT MILLER, mathematics educator; b. Pitts., June 24, 1945; s. Otto Christ and Charlotte Louis (Miller) H.; m. Lois Mary Ryder, July 6, 1968; children: Susanna Louise, Stephen Lincoln, Heidi Frances. BS, MIT, 1967; PhD, Brown U., 1971. Rsch. assoc. Brown U., Providence, R.I., 1971; instr. then asst. prof. U. Minn., Mpls., 1971-72, assoc. prof. then prof., 1977, 81-89; William Moody prof. math. Rice U., Houston, 1988—; vis. mem. Inst. for Advanced Study, Princeton, N.J., 1976, Institut des Hautes Etudes Scientifiques, Bures-sur-Yvette, France, 1978, 81; vis. prof. U. Melbourne, Australia, 1979, U. Wuppertal, Federal Republic of Germany, 1984, Stanford U., Calif., 1987. Contbr. articles to jours. in math. field. Recipient Cochran Leadership award MIT, 1967; grantee NSF, Mpls., Houston, 1971—. Mem. Am. Math. Soc. Democrat. Mem. Unitarian Ch. Avocations: basketball, horseback riding. Office: Rice U Math Dept Houston TX 77251

HARDY, GREGG EDMUND, aerospace engineer; b. Middletown, Conn., Mar. 14, 1966; s. Edmund Ellwood and Marsha Claire (McGeorge) H. BS in Engring., Princeton U., 1988. Assoc. mem. tech. staff GE Astro, East Windsor, N.J., 1988-89; project officer spacecraft ops. Phillips Lab., Albuquerque, 1989-91, lead engr., 1991—; tchr. Sylvan Learning Ctr., Maui, Hawaii, 1990. Pres. Princeton U. Jazz Ensemble, 1985-86. Capt. USAF, 1988—. Recipient Rotary Club award, 1984; Cum and Forster scholar, 1983. Mem. AIAA, Sigma Xi (assoc.). Home: 20 Joy St Boston MA 02114 Office: Phillips Lab Litt Kirtland AFB Albuquerque NM 87117-6008

HARDY, HENRY REGINALD, JR., geophysicist, educator; b. Ottawa, Ont., Can., Aug. 19, 1931; came to U.S., 1966; s. Henry Reginald Sr. and Lois Irene (Moreland) H.; m. Margaret Mary Lytle, June 5, 1954; children: William Reginald, David Alexander. BS, McGill U., 1953; MS, Ottawa U., 1962; PhD, Va. Polytechnic Inst., 1965. Scientific officer Canadian Dept. of Energy, Mines and Resources, Ottawa, 1953-60, rsch. scientist, 1960-66; assoc. prof. Pa. State U., University Park, 1966-70, prof., 1970—, chmn. Geomechanics Section, 1976-90; dir. Mining and Mineral Resources Rsch. Inst., 1990—; cons. UN, Ankara, Turkey, 1984. Editor proceedings 5 internat. confs. on acoustic emission, 2 internat. conf. on salt mechanics, 1975-91. Vis. professorship U. Aachen, Fed. Republic of Germany, 1980; sr. vis. fellowship Tohoku U., Japan, 1986. Mem. ASTM, Am. Geophys. Union, Can. Assn. Physicists, Internat. Soc. Rock Mechanics, Am. Soc. Nondestructive Testing, Acoustic Emission Working Group. Avocations: antiques, sport cars, travel. Office: Pa State U Rm 110 MS Bldg University Park PA 16802

HARDY, JOHN W., optics scientist. Recipient Albert A. Michelson medal Franklin Inst., 1992. Office: Litton Itek Optical Sys 10 Maguire Rd Lexington MA 02173*

HARDY, JOYCE MARGARET PHILLIPS, plant physiologist, educator; b. Mullen, Nebr., Dec. 25, 1958; d. Lynn Eugene and Joellen Ann (Loudon) Phillips; m. Robert E. Hardy, Aug. 17, 1980; children: Johannah, James. BA, Chadron State Coll., 1981, MA, 1986; PhD, Brigham Young U., 1992. Instr. Chadron (Nebr.) State Coll., 1991-92, asst. prof. sci., 1992—. Contbr. articles to sci. jours. Mem. Am. Soc. Plant Taxonomists, Am. Soc. Plant Physiologists, Bot. Soc. Am., Soc. Range Mgmt., Phi Kappa Phi, Sigma Xi. Office: Chadron State Coll 10th St and Main St Chadron NE 69337

HARDY, MAJOR PRESTON, JR., analytical chemist; b. Bklyn., Nov. 18, 1937; s. Major Preston Hardy and Gladys Nancy (Howard) Powell; m. Sarah Ann Farmer, Jan. 12, 1957; children: Victor, Victoria, Leila. BS, SUNY, 1979. Ordained in Jehovah's Witnesses Ch., 1955. Prin. chemist Walker & Whyte, Inc., N.Y.C., 1956-79; lab. mgr. Rodman & Yaruss Refining Co., Inc., N.Y.C., 1979-82; v.p., tech. dir. Jewelers Assay, Inc., N.Y.C., 1982—. Mem. Am. Inst. Chemists. Achievements include discovery of innovative methodology in the analysis of precious metals; design of analytical laboratories. Home: GPO Box 020636 Brooklyn NY 11202-0014

HARDY, MAURICE G., medical and industrial equipment manufacturing company executive; b. 1931. Dir. European ops. Pall Corp., 1962-72, v.p. European ops., 1972-75; exec. v.p. Pall Corp., Glen Cove, N.Y., 1975-85, pres., COO, 1985-89, pres., CEO, 1989—, also bd. dirs. Office: Pall Corp 30 Sea Cliff Ave Glen Cove NY 11542-3634

HARDY, RALPH W. F., biochemist, biotechnology executive; b. Lindsay, Ont., Can., July 27, 1934; s. Wilbur and Elsie H.; m. Jacqueline M. Thayer, Dec. 26, 1954; children: Steven, Chris, Barbara, Ralph (dec.), Jon. B.S.A., U. Toronto, 1956; M.S., U. Wis.-Madison, 1958, Ph.D., 1959. Asst. prof. U. Guelph, Ont., Can., 1960-63; research biochemist DuPont deNemours & Co., Wilmington, Del., 1963-67, research supr., 1967-74, assoc. dir., 1974-79, dir. life scis., 1979-84; pres. Bio Technica Internat., Inc., Kansas City, Kans., 1984-86; pres., chief exec. officer Boyce Thompson Inst., Inc. at Cornell, Ithaca, N.Y., 1986—; dep. chmn. Bio Technica Internat., Inc., 1986-90, cons., bd. dirs., 1990—; mem. exec. com. bd. agr. NRC, 1982-88, mem. commn. life scis., 1984-90, bd. biology, 1984-90, com. on biotech., 1988—, bd. sci. technol. internat. devel., 1990-93, chmn. com. on biol. control, 1992—, chmn. com. on biol. nitrogen fixation, 1992—; mem. com. genetic experimentation Internat. Coun. Sci. Union, 1981—; mem. sci.adv. com. U. S. Dept. Energy, 1991—; mem. alternative agr. res. commercialization bd. USDA, 1992—. Author: Nitrogen Fixation, 1975, A Treatise on Dinitrogen Fixation, 3 vols, 1977-79; mem. editorial bd. sci. jours.; contbr. over 150 articles to sci. jours. Mem. biotech. exec. bd. Cornell U., 1986—, adv. coun. Vet. Coll., 1989—, mem. governing bd. Cornell Ctr. for Environment, 1991—; chmn. Nat. Agr. Biotech. Coun., 1988-93. Recipient Gov. Gen.'s Silver medal, 1956, Sterling Henricks award 1986; WARF fellow, 1956-58; DuPont fellow, 1958-59. Mem. Indsl. Biotech Assn. (bd. dirs. 1986-89), Agr. Rsch. Inst. (bd. govs. 1988-91), Am. Chem. Soc. (exec. com. biol. chemistry div., Del. award 1969), Am. Soc. Biol. Chemists and Molecular Biologists, Am. Soc. Plant Physiology (exec. com. 1974-77), Am. Soc. Agronomy, Am. Soc. Microbiology. Episcopalian. Home: 330 The Parkway Ithaca NY 14850-2249 Office: Boyce Thompson Inst Tower Rd Cornell Univ Ithaca NY 14853-1801

HARDY, RICHARD ALLEN, mechanical engineer, diesel fuel engine specialist; b. Cleveland, Ohio, Sept. 16, 1928; s. Harry and Mae Hardy; m. Lois L. Fawcett, May 16, 1953 (dec. Dec. 1990); children: Pamela, Richard, James, Thomas; m. Irene F. Habic, Feb. 14, 1992. BSME, Case Inst. Tech., 1952. Founder, CEO Fluid Mechanics Inc., Cleve., 1957—. Designed and built largest dynamic fuel-injection pump test stand in Western hemisphere. Cpl. U.S. Army, 1946-48. Recipient Weatherhead 100 award Cleve., 1989, Aircraft Design award AIAA, 1992. Mem. Assn. of Diesel Specialists (various coms. 1960—). Roman Catholic. Avocations: racquetball, scuba. Home: 26875 Hilliard Blvd Cleveland OH 44145

HARDY, SALLY MARIA, retired biological sciences educator; b. San Juan, P.R., June 12, 1932; d. Obdulio Roberto Cordero and Maria Teresa (Judice) Perez; m. Anthony Michael Hardy, Apr. 22, 1962; children: Ricardo Antonio, Maria Isabel. BS, Midland Coll., Fremont, Nebr., 1952; MS, Fordham U., 1956, PhD, 1958. Tech. asst. Postgrad. Hosp., N.Y.C., 1952-54; tchr. Ursuline Acad., N.Y.C., 1957-58; asst. prof. Marymount Manhattan Coll., N.Y.C., 1957-62, 67-68, Queensborough Community Coll., CUNY, 1967-69; rsch. assoc. Am. Mus. Natural History, N.Y.C., 1967-73; assoc. prof. Bergen Community Coll., Bergen, N.J., 1969-70, Rutgers U., Piscataway, N.J., 1970-79; prof. biol. scis. La. State U., Shreveport, 1979-88; ret., 1988; mem. minority com. Grad. Record Exam. Bd., 1976-79; cons. Office Tech. Assessment, U.S. Ho. of Reps., 1985-87. Mem. editorial bd. Jour. Allied Health Professions, 1983-86; contbr. articles to profl. publs. Pres. Hispanic Youth Civic Assn., N.Y.C., 1954-57, PTA, Mendham, N.J., 1972-74; mem. Mendham Borough Bd. Edn., 1973-77, Mendham Borough Bd. Health, 1979. Recipient Faculty Merit award Rutgers U., 1978, Nuestro award Hispanic mag., 1978, award Assn. for Advancement Chicanos and Native Ams. in Sci., 1982. Mem. AAAS (chmn. Office Opportunities in Sci. 1979-85), N.Y. Acad. Scis., Assn. for Puerto Ricans in Sci. and Engring. (pres. 1981-84, bd. dirs. 1984-87), Sigma Xi, Pi Epsilon, Lambda Tau. Roman Catholic. Home: PO Box 4173 Emerald Isle NC 28594

HARDY, WALTER NEWBOLD, physics educator, researcher; b. Vancouver, B.C., Mar. 25, 1940; s. Walter Thomas and Julia Marguerite (Mulroy) H.; m. Sheila Lorraine Hughes, July 10, 1959; children: Kevin James, Steven Wayne. BSc in Math and Physics with honors, U. B.C., 1961; PhD in Physics, Univ. B.C., 1965. Postdoctral fellow Centre d'Etudes Nucleaires de Saclay, France, 1964-66; mem. tech. staff N.Am. Rockwell, Thousand Oaks, Calif., 1966-71; assoc. prof. physics U. B.C., 1971-76, prof., 1976—; vis. scientist Ecole Normale Superieure, Paris, 1980-81, 85. Contbr. articles to sci. jours.; patentee precision microwave instrumentation. Recipient Stacie prize NRC of Can., 1978, Gold medal B.C. Sci. Coun., 1989; Rutherford Meml. scholar, 1964; Alfred P. Sloan fellow, 1972-74; Can. Coun. Rsch. fellow, 1984-86. Mem. Can. Assn. Physicists (Herzberg medal 1978, gold medal for achievement in physics, 1993), Am. Phys. Soc. Office: UBC, Dept Physics, Vancouver, BC Canada V6T 1Z1

HARE, DAPHNE KEAN, medical association director, educator; b. Palmerton, Pa., Jan. 19, 1937; d. Clare Hibberd and Lucile (Lawrence) Kean; m. Peter Hewitt Hare, May 30, 1959; children: Clare Kean, Gwendolyn Meigs. BA in Physics with honors, Barnard Coll., 1958; MD, Cornell U., 1962. Intern and resident Buffalo Gen. Hosp., 1963-65; NIH fellow SUNY, Buffalo, 1965-68, asst. prof. medicine and biophysics, 1968-76, assoc. prof., 1976-82; assoc. chief of staff for edn. Buffalo VA Med. Ctr., 1975-79, VA clin. investigator, 1972-75; assoc. chief for Grad. Med. Edn. VA Cen. Office, Washington, 1979-82, chief for Med. Edn., 1982-89, dir. Affiliated Residencey Programs Svc., 1989—; mem. NIH tng. grant com., Washington, 1971-75, rsch.-edn. com. Buffalo United Way, 1974-75; cons. Johann Gutenberg U., Mainz, Fed. Republic of Germany, 1969-70, Free U. West Berlin, 1969-70, biol. faculty, Moscow State U., 1989—, Englehardt Inst. Molecular Biology, Moscow, 1990—; mgr. Internat. Exchange Programs, Urals State Med. Inst., 1990—; trustee Biomedical Scis. Exchange Program, Inc., Salisbury Cove, Maine, 1991—; founder Daphne K. Hare Fund for Russian-Am. Med. Edn. Cooperation, 1991—; mem. editorial bd. Physiological Revs., 1978-82; co-dir. Internat. Biomed. Agency, Ekaterinburg, Russia, 1993—. Contbr. articles to profl. jours. Founder, pres. 1970-71, Buffalo chpt. NOW; bd. dirs. Western N.Y. chpt. ACLU, 1968-76 (chairperson 1974-78); mem. editorial bd. Signs, N.Y.C., 1974-79; mem. profl. resources com., Am. Med. Women's Assn., 1976-78; mem. exec. coun. Fed. Orgns. for Profl. Women, Wash., 1976-78; v.p. bd. dirs. 1661 Crescent Pl. NW, Inc., Washington, 1981-86; mem. legis. com. Am. Women in Sci., Washington, 1986-88. Recipient Rice fellowship Barnard Coll., 1958, Borden prize Cornell U. Med. Coll., 1962, David Worthen Med. Edn. award VA and Assn. Am. Med. Colls., 1991, Meritorious Svc. award, VA, 1993; NIH grants, 1970-73, VA merit rev. grants, 1975-83. Mem. Biophys. Soc. (founder com. on profl. opportunities for women, chair 1973-74), Biophysics Soc. (mem. coun. 1974-77, chair grievance com. 1974-82, exec. com. 1975-77), Assn. Am. Med. Colls. (adv. bd. for women in mgmt. 1979-80). Home: 219 Depew Ave Buffalo NY 14214-1621

HARE, KEITH WILLIAM, computer consultant; b. Salem, Ohio, Jan. 14, 1958. BS, Muskingum Coll., 1980; MS, Ohio State U., 1985. Programmer, analyst Denison U., Granville, Ohio, 1980-84; sr. cons. JCC Consulting, Inc., Granville, 1985—; com. mem. ANSI X3H2 com. database langs., Washington, 1988—; active DECUS, Shrewsbury, Mass., 1983—. Office: JCC Consulting Inc 600 Newark Rd PO Box 381 Granville OH 43023

HARELL, GEORGE S., radiologist; b. Vienna, Austria, Apr. 27, 1937; came to U.S., 1940; s. Isidore and Zinaida (Hilferding) Silbermann; m. Carol Deane Wright, Mar. 20, 1970; children: Mark, Ben. AB, Oberlin (Ohio) Coll., 1959; MD, Columbia U., 1963. Resident in radiology Med. Sch., Stanford (Calif.) U., 1967-71, asst. prof., 1971-78, assoc. prof., 1978-82; radiologist dept. radiology East Jefferson Gen. Hosp., Metairie, La., 1982-84, chmn. dept. radiology, 1984—; project officer NIH, Washington, 1965-67. Author: (chpt.) The Oesophagus, 1986, 92. Lt. comdr. USAF/USPHS, 1963-66. James C. Picker Found. grantee, 1972-74, NIH grantee, 1977-80, 82-85, Am. Heart Assn. grantee, 1981-83. Mem. Soc. Computed Body Tomography, Radiol. Soc. N.Am., Soc. Gastrointestinal Radiologists, Phi Beta Kappa. Office: East Jefferson Gen Hosp 4200 Houma Blvd Metairie LA 70006-2907

HARIGEL, GERT GÜNTER, physicist; b. Kassel, Hesse, Germany, Mar. 29, 1930; came to Switzerland, 1965; s. Walter Eduard Wilhelm Hermann and Regina Martha (Schmidt) H.; m. Natalia Michailovna Kolb-Seletski, June 20, 1970 (div. 1986). PhD magna cum laude, U. Würzburg, 1961, Diplom-Physicist with honors, 1958. Physicist Max-Planck-Inst. for Physics, Munich, 1961-63, German Electron Synchrotron DESY, Hamburg, 1963-65; sr. physicist European Lab. for Particle Physics, CERN, Geneva, Switzerland, 1965—; vis. prof. physics Columbia U., N.Y.C., 1983, U. Hawaii, Honolulu, 1987, 88. Contbr. over 125 articles to sci. publs. Mem. coun. Geneva Internat. Peace Rsch. Inst., 1991. Mem. N.Y. Acad. Scis., Am. Phys. Soc., European Phys. Soc., German Phys. Soc., Nat. Geog. Soc. Lutheran. Office: European Lab Particle Phys, PPE-Division, CH-1211 Geneva 23, Switzerland

HARIJAN, RAM, technology transfer researcher; b. Keecheri, Kerala, India, June 3, 1938; s. Narayanan and Devaki Nambiar; m. Lakshmi VP, Aug. 19, 1977; 1 child, Devi. BA with honors, Madras U., India; MA, Southampton U., Eng.; PhD, Reading U., Eng. Lectr. Kerala (India) U.; mining officer Singareni Collieries, India; sch. tchr. Barnstaple Grammar Sch., Eng.; lectr. Bosworth Coll., Eng.; tutor cons. Open U., Eng.; researcher Centre for Studies in Tech. Transfer, Eng.; involved in formulating computerisation policies of Indian Govt., 1985-89. Chmn. North Devon Dist. Labour Party, 1972-77, North Devon Assn. Racial Equality, 1978-80; vol. social worker Helping the Disabled and Disadvantaged. Avocations: bridge, chess.

HARJU, WAYNE, electrical engineer; b. Milw., Mar. 8, 1945; s. Wayne Arvid and Agnes Anastasia (Lewandowski) H.; m. Maureen Phyllis Sarro, July 7, 1973; children: Susan Agnes, Stephanie Lynn, William Wayne. BEE, Milw. Sch. Engring., 1966; MS in Power Engring., Rensselaer Poly. Inst., 1972. Registered profl. engr. N.Y.; cert. cogeneration profl. Power transformer design engr. Gen. Electric, Pittsfield, Mass., 1968-77; staff mem. R&D Ctr. Gen. Electric, Niskayuna, N.Y., 1977-78; application engr. utility systems engring. dept. Gen. Electric, Schenectady, N.Y., 1978-79, sr. facilities engr., 1979-87; chief elec. engr./mgr. dept. elec. engring. Galson Engrs./Architects P.C., East Syracuse, N.Y., 1987-93; head elec. engr. Tighe & Boud, Inc., Westfield, Mass., 1993—. V.p. parish coun. St. Joseph's Ch., Liverpool, N.Y., 1971-72. Staff sgt. USAF, 1966-68. Mem. IEEE (sr.), Assn. Energy Engrs. (sr.), Nat. Soc. Profl. Engrs. (pres. Berkshire chpt. 1974-75, 75-76), Internat. Elec. Testing Assn. (affiliate), Elfun Soc., Am. Legion, Eta Kappa Nu, Tau Omega Mu. Republican. Roman Catholic. Achievements include patent dockets for power transformer design. Home: 7349 Eastgate Circle Liverpool NY 13090 Office: Galson Engrs Architects PC 6601 Kirkville Rd East Syracuse NY 13057

HARKINS, HERBERT PERRIN, otolaryngologist, educator; b. Scranton, Pa., Aug. 13, 1912; s. Percy Stoner and Myra (Perrin) H.; B.S., Lafayette Coll., 1934; M.D., Hahnemann Med. Coll., 1937; M.Sc., U. Pa., 1942; m. Anna Catherine Shepler, July 16, 1938; children—Herbert P., Sally Anne, Nancy Shepler. Lectr. otolaryngology Hahnemann Med. Coll., 1939-44, asso. prof., 1944-51, prof. head dept. otolaryngology, 1951; asst. prof. otolaryngology Grad. Sch. Medicine, U. Pa., 1951—; sr. staff otolaryngology Lankenau Hosp. Bd. Studies in Higher Edn. Trustee, Lafayette Coll. Served as comdr. U.S. Navy, 1945-48; Res. Diplomate Am. Bd. Otolaryngology. Fellow A.C.S., Am. Otorhinol. Soc. Plastic Surgery; mem. Am. Soc. Ophthalmic and Otolaryngologic Allergy (pres.), Am., Pa. acads. ophthalmology and otolaryngology, Coll. Physicians Phila., Phila. Laryngol. Soc., Phila County Med. Soc., AMA, Am. Laryngol., Rhinol. and Otol. Soc. Clubs: Union League, Phila. Country, Bachelors Barge. Contbr. numerous articles on ear, nose and throat to med. jours. Home: 701 Woodleave Rd Bryn Mawr PA 19010-1708 Office: Lankenau Med Bldg Wynnewood PA 19096

HARKINS, WILLIAM DOUGLAS, mechanical engineer; b. Phila., July 26, 1926; s. William Dickey and Edna Stewart (Gallagher) H.; m. Grace Coryell, Aug. 14, 1948; children: William, Edward, Janet, Scott. BS, U.S. Naval Acad., 1947; BS in Aero. Engring., U.S. Naval Postgrad. Sch., 1953; Engr., Calif. Inst. Tech., 1954. Registered profl. engr., Va. Sr. prin. engr. Syscon Corp., Arlington, Va., 1975-92; cons. WDH Assocs., McLean, Va., 1992—. Contbr. articles to profl. jours. Mem. McLean CBD Planning Com., 1977-90, McLean Citizens Assn., 1970—; pres. Lewinsville Citizens Assn., McLean, 1974-78. Capt. USN, 1947-75. Recipient Mgmt. Improvement award Office of the Pres. of U.S., 1973. Fellow AIAA;mem. NSPE, Va. Soc. Profl. Engrs., Sigma Xi. Episcopalian. Home: 1630 Warner Ave Mc Lean VA 22101

HARKONEN, WESLEY SCOTT, physician; b. Mpls., Dec. 17, 1951; s. Wesley Sulo and Frances (Fedor) H.; m. Barbara Jean Harkonen, Feb. 14, 1986; children: Kirsten, Alan. BA summa cum laude, U. Minn., 1973, MD, 1977. Resident internal medicine U. Minn., Mpls., 1977-81; fellow allergy and immunology U. Calif., San Francisco, 1983-85, fellow clin. pharmacology, 1984-85; project dir. Xoma Corp., Berkeley, Calif., 1983-87; rsch. assoc. Stanford U., 1987-88; assoc. med. dir. Becton Dickinson, Mountain View, Calif., 1988-89; v.p. med. affairs Calif. Biotechnology, Mountain View, Calif., 1989-91; v.p. med. and regulatory affairs Univax Biologics, Rockville, Md., 1991—. Author: Traveling Well, 1984; contbr. articles to profl. jours. J. Thomas Livermore Rsch. award, U. Minn., 1977. Mem. Am. Fedn. for Clin. Rsch. Office: Univax Biologics 12280 Wilkins Ave Rockville MD 20852-1843

HARLAN, JAMES PHILLIP, environmental engineer; b. New Orleans, Oct. 14, 1958; s. John Carter and Patricia (Sloop) H.; m. Sheliah Celeste Baker, Sept. 5, 1987 (div. Dec. 1990); 1 child, Benjamin Cade; m. Jayme Suzanne Roberts, Dec. 31, 1991; 1 stepchild, Michael Byron Edward Akers. BSME, U. Mo., Rolla, 1981; postgrad., Mid. Tenn. State U., 1992-93. Registered profl. engr., Tex., W.Va., Ohio, Ky. Systems engr., plant engr. Tex. Oil and Gas, Canton, Okla., 1982-84; pvt. practive real estate develop. Austin, Tex., 1985; project mgr. McClelland Engrs., St. Louis, 1986; project mgr., design engr. ERM Southwest, Houston, 1987; sr. engr. Entrix, Houston, 1988; project mgr. Chester Environ., Nashville, 1988—. Instr. Cabell County Literacy, Huntington, W.Va., 1990. Mem. NSPE, Pi Tau Sigma. Republican. Methodist. Office: Chester Environ 2501 Hillsboro Rd Ste 4 Nashville TN 37212

HARLAP, SUSAN, epidemiologist, educator; b. Hull, Eng., June 4, 1939; came to U.S., 1988; d. Thornton Jack and Lydia Margaret (Brentall) Legg; m. Jacob Haim Harlap, Dec. 30, 1970 (div. 1989); children: Michal Margaret, Gital Miriam. MBBS, Royal Free Hosp.-London U., 1963. House officer in internal medicine, hematology Royal Free Hosp., London, 1963; house officer in gen. and thoracic surgery Whittington Hosp., London, 1964; house officer in internal medicine, sarcoidosis Royal No. Hosp., London, 1964-65; family physician Brit. Nat. Health Svc., London, 1965-67; mem. faculty Hebrew U.-Hadassah Med. Sch., Jerusalem, 1967-90, prof. epidemiology, 1985-90; prof. epidemiology Cornell U., N.Y.C., 1991—; chief epidemiology svc. Meml. Sloan Kettering Cancer Ctr., N.Y.C., 1990—; cons., temporary adviser WHO, Geneva, 1979-88; vis. epidemiologist Kaiser Permanente Med. Ctr., Walnut Creek, Calif., 1978-79; vis. scientist epidemiology and biometry bf. Nat. Insts. Child Health and Human Devel., NIH, Bethesda, Md., 1979; vis. scientist Family Health Internat., Durham, N.C., 1988-89; vis. epidemiologist Alan Guttmacher Inst., N.Y.C.,1989-90. Author: Gender of Infants Conceived on Different Days of the Menstrual Cycle, 1979, Preventing Pregnancy, Protecting Health, 1991; contbr. chpts. to books, articles to peer-reviewed publs.. Fellow Am. Coll. Epidemiology; mem. Soc. Epidemiologic Rsch., Am. Soc. Preventive Oncology, Internat. Epidemiol. Assn. Jewish. Home: 504 E 63d St Apt 32M New York NY 10021 Office: Meml Sloan Kettering Cancer Ctr 1275 York Ave New York NY 10021

HARLESS, JAMES MALCOLM, corporate executive, environmental consultant; b. Beaumont, Tex., June 9, 1948; s. Charles Malcolm and Wilma Elena (Jordan) H.; m. Margaret Ruth Millenbach, Oct. 1, 1977; children: Jared Mathew, Benjamin Wesley. BA in Chemistry, Rice U., 1970; PhD in Organic Chemistry, U. Tex., 1975. Cert. hazardous materials mgr.; registered environ. profl. Scientist Radian Corp., Austin, Tex., 1976-79, group leader, 1979-81, dept. head, 1981-82, program mgr., 1982-84; pres. Environ. Rsch. Group, Inc., Ann Arbor, Mich., 1984-85; cons. Quantum Cons., Ann Arbor, Mich., 1985-86; pres. Techna Corp., Plymouth, Mich., 1986—; active U.S. Intergovtl. Sci., Engring., and Tech. Advs. Panel Task Force for Sci. and Tech. Assessments Hazardous Waste Mgmt., Washington, 1979; mem. Small Quantity Generator Edn. Adv. Panel, Waste Systems Inst. Mich./Mich. DNR, Lansing, Mich., 1986—, co-sponsor, com. chair Environ. Tech. 90 Conf., Dearborn, Mich., 1990; lab. design cons., Austin, 1980-83; presenter in field. Contbr. 48 articles to profl. jours. Treas., trustee Wildlife Rescue, Inc., Austin, 1981-83; trustee The Emerson Schs., Ann Arbor, 1988-89, chmn. bd. trustees, 1989-92. Capt. USAR, 1971-79. Mem. AAAS, ASTM, Am. Chem. Soc. (mem. task force on lab. waste mgmt. 1982—, symposium organizer and chair 1982, 94, environ. directory advs. com. 1989—), Sci. Apparatus Mfrs. Assn. (exposition planning and coordinating com. 1983-84), Nat. Water Well Assn., Inst. Hazardous Materials Mgmt., Liquid Indsl. Control Assn. Mich. Achievements include patent for permeation testing apparatus. Office: Techna Corp 44808 Helm St Plymouth MI 48170-6026

HARLING, OTTO KARL, nuclear engineering educator, researcher; b. N.Y.C., Oct. 1, 1931; s. John G. and Hedwig (Kahlhorn) H.; m. Elizabeth T. Trafford, Aug. 3, 1959; children: Elizabeth, Maura, Ottilie, Kurt. BS, Ill. Inst. Tech., 1953; MS, U. Heidelberg, Fed. Republic Germany, 1955; PhD, Pa. State U., 1962. Physicist Curtis-Wright Corp., Quehanna, Pa., 1956-59; project engr. HRB-Singer, Inc., State College, Pa., 1959-62; sr. physicist GE, Hanford, Wash., 1962-65; rsch. assoc. Battelle Pacific N.W. Lab., Hanford, 1965-72, staff scientist, 1973-76; sr. rsch. assoc. MIT, Cambridge, 1972-73, dir. MIT nuclear reactor lab. and prof. nuclear engring., 1976—; prin.

Harling Cons. Svcs., Hingham, Mass., 1989—. Editor: Use and Development of Low and Medium Flux Research Reactors, 1984, Neutron Beam Design, Development and Performance for Neutron Capture Therapy, 1989; contbr. more than 100 articles to profl. jours. Office: MIT 138 Albany St Cambridge MA 02139-4296

HARMAN, WILLARD NELSON, malacologist, educator; b. Geneva, N.Y., Apr. 20, 1937; s. Samuel Willard and Mary Nelson (Covert) H.; m. Susan Beth Mead, June 12, 1968 (div. 1980); children—Rebecca Mary, Willard Wade; m. Barbara Ann Stong, June 8, 1981; children—Jessica Mary, Samuel Willard. Student, Hobart Coll., 1954-55; B.S., Coll. Environ. Sci. and Forestry, SUNY, 1965; Ph.D., Cornell U., 1968; postgrad., Marine Biol. Lab., Woods Hole, Mass., 1968. Asst. prof. SUNY, Oneonta, 1968-69, assoc. prof., 1969-76, prof. biology, 1976—, chmn. dept. biology, 1981-89, dir. Biol. Field Sta., 1989—; resource advisor N.Y. State Dept. Environ. Conservation, Albany, 1980—. Contbr. articles to profl. jours. Rep. Otsego County Republican Com., N.Y., 1973-76; chmn. planning bd., Springfield, N.Y., 1984—. Served with USN, 1956-61. Recipient Chancellor's award SUNY, 1974-75, Quality award EPA, 1989, Excellence award SUNY, 1990. Mem. Soc. Limnology and Oceanography, N.Am. Benthological Soc., Soc. for Exptl. and Descriptive Malacology, Am. Malocological Union, Otsego County Conservation Assn. (bd. dirs. 1970—, pres. 1974-78, 80-81, chmn. lake com. 1981—). Episcopalian. Avocations: sailing; fishing; scuba diving; skiing. Home: RD 2 Box 829 Cooperstown NY 13326 Office: Biol Field Sta RR 2 Box 1066 Cooperstown NY 13326-9330

HARMON, CHARLES WINSTON, energy engineer; b. McGaheysville, Va., Aug. 8, 1948; s. Eston Bayne and Nancy Louise (Burner) H.; m. Linda Sue Leet, Mar. 11, 1972; children: Thomas Leet, Jakob Burner. BS in Indsl. Engring., Va. Poly. Inst. and State U., 1971. Coord. material handling, mech. equipment and energy Best Products Co., Inc., Richmond, Va., 1974-80, mgr. energy mgmt. dept., 1980-90; energy cons. Dacobe, Richmond, 1990; demand side mgmt. engr. Old Dominion Electric Coop., Glen Allen, Va., 1990-93; mgr. systems J. W. Fergusson & Co., Inc., Richmond, 1993—. Webelos leader Boy Scouts Am., Mechanicsville, Va., 1990-92; soccer coach Atlee Youth Sports, Mechanicsville, 1986-91. Sgt. U.S. Army, 1971-74. Mem. Assn. Energy Engrs. (sr., cert. energy mgr., cert. lighing efficiency prof.), Demand Side Mgmt. Soc. Assn. Energy Engrs. (charter), Va. Assn. Energy Engrs. (charter). Republican. Presbyterian. Home: 2107 Chartwell Ct Mechanicsville VA 23111 Office: JW Fergusson & Co Inc 4107 Castlewood Rd Richmond VA 23234

HARMONY, MARLIN DALE, chemistry educator, researcher; b. Lincoln, Nebr., Mar. 2, 1936; s. Philip and Helen Irene (Michal) H. A.A., Kansas City (Mo.) Jr. Coll., 1956; B.S. in Chem. Engring., U. Kans., 1958; Ph.D. in Chemistry, U. Calif.-Berkeley, 1961. Asst. prof. U. Kans., Lawrence, 1962-67, assoc. prof., 1967-71, prof., 1971—; chmn. U. Kans., 1980-88; panel mem. NRC-Nat. Bur. Standards., 1969-78; mem. review panel NSF, 1977, 92. Author: Introduction to Molecular Energies and Spectra, 1972; contbg. editor: Physics Vade Mecum, 1981; mem. editorial bd. Structural Chemistry; contbr. articles to profl. jours. Postdoctoral fellow NSF Harvard U., 1961-62. Fellow AAAS; mem. Am. Chem. Soc., Am. Phys. Soc., Sigma Xi, Alpha Chi Sigma, Phi Lambda Upsilon, Tau Beta Pi. Democrat. Home: 1033 Avalon Rd Lawrence KS 66044-2505 Office: U Kans Dept Chemistry Lawrence KS 66045

HARMS, DAVID JACOB, agricultural consultant; b. Springfield, Ill., Jan. 10, 1943; s. Jacob Dietrich and Onita Ruth (Schnapp) H.; m. Bobette Sue Speckhart, Nov. 18, 1967; children: Jacob, Johanna, Anika. BS in Agr., U. Ill., 1967. Field rep. Monsanto Corp., St. Louis and Manhattan, Kans., 1967-70; product mgr. Masonite Corp., Chgo., 1970-73; biochem. rep. P.P.G. Industries, Chgo., 1973-76; pres. Crop Pro-Tech, Inc., Naperville, Ill., 1976—; v.p., evaluation of rsch. for United Soybean Bd./Global Harvest Enterprises, Inc., 1993; pres. Profl. Crop Cons. Ill., 1984-86, treas., 1993—. Contbg. author tng. manual for privatization of agriculture in Egypt. Named to Cons. Hall of Fame, Ag Cons. and Fieldman Mag., 1984. Mem. Nat. Alliance Ind. Crop Cons. (pres. 1991, chair steering com., Communicator of Yr. award 1992, mem. editorial bd. AgriFin. mag. 1992-93). Achievements include patents in field; research in herbicide by hybrid sensitivity, herbicide/hybrid/insecticide sensitivity. Office: Crop Pro-Tech Inc 33 W Bailey Rd Naperville IL 60565-2376

HARMS, JOHN MARTIN, electrical engineer; b. Tokyo, June 25, 1961; s. Walter William and Ellen Jeanne (Brugge) H.; m. Shawn Atkins, Feb. 29, 1992. BS, U. Tex., 1985; postgrad., St. Mary's U., 1990—. Elec. engr. Automatic Test Systems Kelly AFB, San Antonio, 1985-90; Elec. engr. DET 3 Space Command Kelly AFB, San Antonio, 1990—. Mem. IEEE, Soc. Automotive Engrs., Assn. Old Crows. Lutheran. Office: DET 3 PO Box 680098 San Antonio TX 78268-0098

HARNAD, STEVAN ROBERT, cognitive scientist; b. Budapest, Hungary, June 2, 1945; grew up in Montreal; came to U.S., 1969; s. Istvàn Hesslein and Susan (Simoni-Suss) H. BA, McGill U., Montreal, 1967; MA, McGill U., 1969, Princeton U., 1971; PhD, Princeton U., 1991. Rsch. psychologist N.J. Bur. Rsch. Neurology and Psychiatry, Princeton, 1971-73, Psychiatry Dept., Rutgers Med. Sch., Piscataway, 1973-76; founder, editor Behavioral and Brain Scis. Jour., N.Y.C., 1977—; vis. rsch. fellow Princeton U., 1989-91, Anthropology Dept., Rutgers U., 1992-93; assoc. rschr. Laboratoire Cognition et Movement URA CNRS, Marseille, France, 1993—; co-prin. investigator Pew Sci. Program, Faculty Rsch. Collaboration Grant, Princeton, 1989-91. Editor: Categorical Perception, 1987, Peer Commentary on Peer Review, 1987; co-editor: The Selection of Behavior, 1988, Lateralization in the Nervous System, 1977, Origins and Evolution of Language and Speech, 1976; editor, founder Psycoloquoy Electronic Jour., 1989—. NIH grantee, 1980-81, Alfred P. Sloan Found. grantee, 1979-80. Mem. Soc. for Philosophy and Psychology (pres. 1986-87), Am. Psychol. Soc., Princeton Music Club (pres. 1989-90), Broadmead Swim Club (v.p. 1987-88). Home: 45 Billie Ellis Ln Princeton NJ 08540 Office: Princeton Univ Cognitive Sci Lab 221 Nassau St Princeton NJ 08540

HARNEY, PATRICK JOSEPH DWYER, meteorologist, consultant; b. Fulton Chain, N.Y., June 30, 1908; s. Patrick Joseph and Laura Frances (Brooks) H. BSEE, Clarkson U., 1931; MS in Meterology, Calif. Tech. U., 1935. Cert. consulting meterologist, N.Y.; registered profl. engr., N.Y. Weather caster, radio operator Western Air Express, Burbank, Cheyenne, 1936-38; air mass meterologist U.S. Weather Bur., Mt. Washington, N.H., 1938-41; instrument devel. user U.S. Weather Bur., Chgo., 1945-49; bomber radar and war zone user U.S. Signal Corps., Panama, 1941-42; airplane lightning rsch. engr. Cook Rsch. Lab., Skokie, Ill., 1951-52; atmospheric physicist USAF, Cambridge Rsch. Directorate, Bedford, Mass., 1952-70; owner, mgr. Energy Users Cons. Svcs., Utica, N.Y., 1973—; asst. to flight dir. United Airlines-Western Air Express, Cheyenne, 1932-34; radio weather map aide Calif. Tech. Weather Lab., PAsadena, 1935-36; airplane icing rsch. investigator Mt. Washington Office U.S. Weather Bur., 1940-41. Author: Sea Breeze Study, 1945, The Thunderstorm Project Report, 1950; editor: USAF Project Reports, 1955, 70. Mem. Citizens Ecol. Com., Utica, 1970—; participant Ch. Men's Club. Mem. NSPE (registered profl. engr. N.Y.), , IEEE (sec. atenna and propagation group 1960s), ASHRAE, Am. Meterological Soc. (cert. meteorological cons.), Assn. Am. Weather Observers, Inst. Aeronautical Scis., Rsch. Soc. Am., Mohawk Valley Environ. Info. Exch. (engrs. exec. com., amateur radio lic.). Roman Catholic. Achievements include patents for kithen tool, airplane true air temperature indicator, corner reflector. Home and Office: Energy Users Cons Svcs 21 Sunnyside Dr Utica NY 13501

HARNISH, RICHARD JOHN, psychologist; b. Arnold, Pa., Apr. 1, 1963; s. Richard C. and Alice J. (Tirdil) H. BA in Psychology, Pa. State U., 1985; MA in Social Psychology, Mich. State U., 1987, PhD in Social Psychology, 1992. Undergrad. rsch. asst. Pa. State U., State College, 1983-85; teaching asst. Mich. State U., East Lansing, 1985-89, instr., 1988, 89, rsch. asst., 1989-90; project dir. ARBOR, Inc., Media, Pa., 1990—; presenter in field. Contbr. articles to Jour. Applied Social Psychology, Jour. Rsch. in Personality, Jour. Personality and Social Psychology, Personality and Individual Differences. Mem. APA, Am. Psychol. Soc., Soc. for Personality and Social Psychology, Soc. for Consumer Psychology. Office: ARBOR Inc Arbor Corp Ctr One West Third St Media PA 19063

HAROCHE, SERGE, optics scientist. Recipient Albert A. Michelson medal Franklin Inst., 1993. Home: 45 rue d'Ulm F 75230, Paris CEDEX 05, France

HARPEN, MICHAEL DENNIS, physicist, educator; b. Toledo, July 6, 1949; s. Walter John and Jeanne (Beaber) H.; m. Lydi Hong, July 24, 1981; children: Dennis, Mary. BS, Xavier U., 1971, MS, 1971; PhD, U. Cin., 1979. Postdoctoral scholar UCLA, 1979-81; prof. radiology U. South Ala., Mobile, 1981—; cons. NIH, Washington, 1989—, Radiol. Physics, 1981—. Contbr. articles to profl. publs. Mem. Am. Assn. Physicists in Medicine (Outstanding Publ. award), Am. Coll. Radiology, Soc. Magnetic Resonance in Medicine, Assn. Univ. Radiologists. Republican. Roman Catholic. Achievements include patent for improved method for assessment of free thyroxine and thyroxine binding globulin. Home: 9255 Lakewoods Semmes AL 36575 Office: U South Ala Dept Radiology 2451 Fillingim St Mobile AL 36617

HARPER, DOYAL ALEXANDER, JR., astronomer, educator; b. Atlanta, Oct. 9, 1944; s. Doyal Alexander and Emliy (Brown) H.; m. Carolyn James, Mar. 11, 1967; children: Scott Alexander, Nathan Todd, Amy Claire, Evan James. BA in Elec. Engring., Rice U., 1966, PhD in Space Sci., 1971. Asst. prof. astronomy and astrophysics U. Chgo., 1971-76, assoc. prof., 1976-80, prof., 1980—; dir. Yerkes Obs., from 1982. Mem. Am. Astron. Soc. (Newton Lacy Pierce prize 1979), Astron. Soc. Pacific. Rsch. includes infared observations of galaxies, stars, planets, star formation regions, far infared detectors, cryogenics, optical systems. Office: Yerkes Obs Williams Bay WI 53191-0258

HARPER, JUDSON MORSE, university administrator, consultant, educator; b. Lincoln, Nebr., Aug. 25, 1936; s. Floyd Sprague and Eda Elizabeth (Kelley) H.; m. Patricia Ann Kennedy, June 15, 1958; children: Jayson K., Stuart H., Neal K. B.S., Iowa State U., 1958, M.S., 1960, Ph.D., 1963. Registered profl. engr., Minn. Instr. Iowa State U., Ames, 1958-63; dept. head Gen. Mills, Inc., Mpls., 1964-69, venture mgr., 1969-70; prof., dept. head agrl. and chem. engring. Colo. State U., Ft. Collins, 1970-82, v.p. for rsch., 1982—, interim pres., 1989-90; cons. USAID, Washington, 1972-74, various comml. firms., 1975—. Author: Extrusion of Foods, 1982, Extrusion Cooking, 1989; editor newsletter Food, Pharm. & Bioengring. News, 1979-83, LEC Newsletter, 1976-89; contbr. articles to profl. publs.; patentee. Mem. sch. bd. St. Louis Park, Minn., 1968-70. Recipient Food Engring. award Dairy and Food Industry Supply Assn. and Am. Soc. Agrl. Engrs., 1983, Cert. of Merit, U.S. Dept. Agr.-Office Internat. Coop. and Devel., 1983, Svc. award CIATECH, Chihuahua, Mex., 1980, Disting. Svc. award Colo. State U., 1977, Charles Lory Pub. Svc. award, 1993, Prof. Achievement citation Iowa State U., 1986. Fellow Inst. Food Technologists (Internat. award 1990); mem. AAAS, Am. Inst. Chem. Engring. (dir. 1981-84), Am. Soc. Agrl. Engrs. (com. chmn. 1973-78, hon. engr. Rocky Mountain region), Am. Chem. Soc., Am. Soc. Engring. Edn. (com. chmn. 1976-77). Mem. Ind. United Methodist Ch. Home: 1818 Westview Rd Fort Collins CO 80524-1891 Office: Colo State U Office VP Rsch Office VP Rsch Fort Collins CO 80523

HARPER, ROBERT JOHN, JR., chemist, researcher; b. Savannah, Ga., Aug. 16, 1930; married; 6 children. BS, Fordham U., 1952; PhD in Organic Chemistry, Ohio State U., 1957. Rsch. chemist organic/metallic chemistry Ethyl Corp., Baton Rouge, 1958-62; rsch. chemist fluoroaromatic chemistry Materials Lab., Wright-Patterson AFB, Ohio, 1962-63; rsch. chemist So. Regional Rsch. Ctr. USDA, New Orleans, La., 1963-73, rsch. leader, cellulose chemistry So. Regional Rsch. Ctr., 1973-85, lead scientist So. Regional Rsch. Ctr., Agrl. Rsch. Svc., 1985—. Mem. Am. Chem. Soc., Am. Assn. Textile Chemists & Colorists (Olney medal 1991). Achievements include research in cellulose and textile chemistry with emphasis on durable press, flame retardant, smolder resistance, speciality dyeing and weather resistance, organo metallics, organo synthesis, flouroaromatic chemistry. Office: SRRC ARS USDA PO Box 19687 New Orleans LA 70179*

HARPSTER, ROBERT EUGENE, engineering geologist; b. Olney, Ill., Sept. 25, 1930; s. Christian Edward and Margaret (Tatum) H.; m. Carol Ann Dewald, Nov. 25, 1977; step-children: Larry Britt, Charla Britt. BS, Beloit Coll., 1952; MA, U. Tex., 1957. Registered geologist Calif.; cert. engring. geologist Calif., cert. quality assurance lead auditor. Petroleum geologist Geo Svc. Co., Abilene, Tex., 1952; engring. soil instr. Corp Engrs., Ft. Belvoir, Va., 1952-54; project geologist Bechtel Corp., Vernon, Calif., 1956-57; sr. project engr. geologist dept. water resources State Calif., Sacramento, 1957-73; sr. project engr. geologist Woodward-Clyde Cons., San Francisco, 1973-80, v.p. quality assurance applied sci., 1980-88; consulting quality assurance mgr. MACTEC, Las Vegas, Nev., 1988--; mem. tech. rev. bds. U.S. Gov. and pvt. industries, San Francisco, 1972--; instr. Antelope Community Coll., Lancaster, Calif., 1969-72; del. People to People, USSR, 1991. Author: Selected Clays used for Dam/Fills Construction, 1979, Methods of Investigating Faults, 1979. coach swimming Antelope Valley YMCA, Lancaster, 1969-72. Sgt. U.S. Army, 1952-54. Fellow Geological Soc. Am., Assn. Engring. Geologist (mem. chmn., v.p. 1961-63, vice chmn. 1959-62), Am. Soc. Civil Engrs., Am. Soc. for Quality Control, Earthquake Engr. Rsch. Inst., Interanl Clay Mineral Soc. Achievements include field methods for investigating faulting, and development of x-ray diffraction studies for relative age dating of paleosoils. Home: 5735 Buena Vista Ave Oakland CA 94618 Office: CER Ste 664 101 Convention Center Dr Las Vegas NV 89109

HARRELL, BARBARA WILLIAMS, public health administrator; b. Montgomery, Ala., Feb. 5, 1949; d. Orrin Ted and Edith Marvlyn (Harris) Williams; m. James Lester Harrell, Mar. 30, 1971; children: James L. Jr., Jayne Leslie, Kerry Orrin-Ted, Barry Earl. BS, Ala. State U., 1971; MPA, Auburn U., 1987. Pub. health rep. Ala. Dept. Pub. Health, Montgomery, 1972-88, health svcs. adminstr., 1988—. Active NOW, Montgomery, 1974, LWV, Montgomery, 1988. Named Outstanding Young Women Am., 1978. Mem. Am. Soc. Pub. Adminstrn., Pi Sigma Alpha, Alpha Kappa Alpha (sec. 1979-80), Civitans. Democrat. Methodist. Office: Ala Dept Pub Health 434 Monroe St Montgomery AL 36130-3017

HARRELL, DOUGLAS GAINES, chemical engineer; b. Louisville, Nov. 16, 1958; m. Carolyn Dorothy Youse, May 24, 1980. BChemE, U. Del., 1980; PhD in Chem. Engring., U. Pa., 1987. Devel. engr. Arco Chem. Co., Newtown Square, Pa., 1986; rsch. engr. Himont, Inc., Wilmington, Del., 1987-90; scientist Himont, Inc., Ferrara, Italy, 1990-92; project leader Himont, Inc., Wilmington, 1992—. Mem. AICE, Am. Chem. Soc., Tau Beta Pi. Achievements include expertise on transient operation of sheripol loop reactors for polypropylene production. Office: Himont Inc 912 Appleton Rd Elkton MD 21921

HARRELL, THOMAS HICKS, psychologist; b. Winston-Salem, N.C., Jan. 5, 1951; s. Thomas G. and Ivis (Hicks) H. BA, East Carolina U., 1973, MA, 1975; PhD, U. Ga., 1979. Lic. psychologist. Asst. prof. Fla. Inst. Tech., Melbourne, 1979-82, assoc. prof., assoc. dean, 1983-90, prof. and assoc. dean, 1991—; pres. Psychologistics, Inc., Melbourne, 1983—; cons. Holmes Regional Med. Ctr., Melbourne, 1979—, Nat. Computer Systems, Mpls., 1989—, Heritage Health Corp., Melbourne, 1989—, others. Contbr. articles to profl. jours.; co-author 12 computer-based programs in psychol. assessment. Leadership mem. Greater Brevard Area C. of C., Melbourne, 1988—; bd. dirs. Ensemble Theatre of Fla., Melbourne, 1985-90, Brevard County Mental Health Assn., Melbourne, 1980-85. Recipient Pub. Svc. award City of Palm Bay, Fla., 1987, Fla. Inst. Tech., Melbourne, 1987. Mem. Am. Psychol. Assn., Assn. for Advancement of Behavior Therapy, Sigma Xi, Chi Beta Phi, Psi Chi. Achievements include co-devel. of teaching, rsch. and comml. applications; recognized for leadership in field of computerized psychol. assessment. Office: Fla Inst Tech Sch of Psychology 150 W University Blvd Melbourne FL 32901

HARRIETT, JUDY ANNE, medical equipment company executive; b. Walterboro, S.C., July 22, 1960; d. Billy Lee and Loretta (Rahn) H. BS in Agrl. Bus./Econs., Clemson U., 1982. Sales rep. III Monsanto Corp., Atlanta, 1982-85; surg. stapling rep. Ethicon, Inc., Johnson & Johnson Corp., Somerville, N.J., 1985-87; dist. sales mgr. Imed Corp., San Diego, 1987—, regional tng. coord., 1992-93; mem. pres. adv. panel, 1991, 92. Author: Time and Territory Management, 1984. Com. mem. Multiple Sclerosis Fund Raising Benefit, Knoxville, Tenn., 1988, 89, Women's Ctr. Benefit, Knoxville, 1990. Mem. NAFE. Republican. Avocations: golf, reading, snow skiing, water skiing, travel. Home: 21620 Mayhew Rd Mooresville NC 28115 Office: Imed Corp 9775 Business Park Ave San Diego CA 92131

HARRIMAN, PHILIP DARLING, geneticist, science foundation executive; b. San Rafael, Calif., Nov. 24, 1937; s. Theodore Darling and Luciel Harriet (Muller) H.; m. Jenny Elizabeth Flack, June 12, 1959; 1 child, Marc Stuart. BS in Physics, Calif. Inst. Tech., 1959; PhD in Biophysics, U. Calif., Berkeley, 1964. Postdoctoral fellow U. Cologne (Germany), 1964-65, Pasteur Inst., Paris, 1965-66, Cold Spring Harbor (N.Y.) Lab., 1966-68; asst. prof. biochemistry Duke U. Med. Ctr., Durham, N.C., 1968-75; assoc. prof. biology U. Mo., Kansas City, 1975-77; program dir. genetic biology NSF, Washington, 1977—; sr. scientist office asst. dir. biology, behavioral scis., 1981-82; vis. prof. Johns Hopkins U. Med. Sch., Balt., 1987-88. Contbr. articles to profl. jours. Legis. asst. Congressman Dave McCurdy of Okla., 1980-81. 1st lt. USAFR, 1962—. Congl. fellow U.S. Office Personnel Mgmt., 1980-81. Fellow AAAS; mem. Genetics Soc. Am., Am. Soc. Microbiology. Unitarian. Achievements include research in genetics, genetic engineering, and molecular biology. Office: NSF Rm 325 Washington DC 20550

HARRINGTON, WILLIAM FIELDS, biochemist, educator; b. Seattle, Sept. 25, 1920; s. Ira Francis and Jessie Blanche (Fields) H.; m. Ingeborg Leuschner, Feb. 24, 1947; children: Susan, Eric, Peter, Robert, David. B.S., U. Calif. at Berkeley, 1948, Ph.D., 1952. Research chemist virus lab. U. Calif. at Berkeley, 1952-53; Nat. Found. Infantile Paralysis postdoctoral fellow Cambridge (Eng.) U., 1953-54; Nat. Cancer Inst. postdoctoral fellow Carlsberg Lab., Copenhagen, Denmark, 1954-55; asst. prof. chemistry Iowa State U., 1955-56; biochemist Nat. Heart Inst., 1956-60; prof. biology Johns Hopkins, Balt., 1960—; chmn. dept. biology Johns Hopkins, 1973-83, Henry Walters prof. biology, 1975—; dir. McCollum Pratt Inst., 1973-83, Inst. Biophys. Rsch. on Macromolecular Assemblies, 1989-90; vis. scientist Wiezmann Inst., Rehovot, Israel, 1959, vis. prof., 1970; vis. prof. Oxford U., 1970; Mem. adv. panel physiol. chemistry NIH, 1962-66, adv. panel biophys. chemistry study sect., 1968-72; bd. sci. councillors Nat. Inst. Arthritis and Metabolic Diseases, 1968-72; mem. vis. com. for biology Brookhaven Nat. Lab., 1969-73; adv. bd. Fedn. Advanced Edn. in the Scis., 1975—; adv. coun. Nat. Inst. Arthritis, Muskuloskeletal and Skin Diseases, NIH, 1987-90. Co-editor: Monographs on Physical Biochemistry, 1970—; Bd. editors: Jour. Biol. Chemistry, 1963-68, Mechanochemistry and Motility, 1970-72, Biochemistry, 1971-77, 83—, Jour. Phys. Biochemistry, 1973—, Analytical Biochemistry, 1978—. Fellow Am. Acad. Arts and Sci.; mem. Biophysics Soc., Nat. Acad. Scis., Soc. Biol. Chemists, Phi Beta Kappa, Sigma Xi. Home: 10182 Tracey Beth Ct Ellicott City MD 21042-1633 Office: Johns Hopkins University Dept Biology 3400 N Charles St Baltimore MD 21218

HARRINGTON, WILLIAM PALMER, retired civil engineer; b. Alpine, Tex., Mar. 29, 1925; s. William Lee and Sadie Lou (Crawford) H.; m. Willetta Rose McKenzie, July 27, 1950; children: Patricia Louise, Lee Ann. BSCE, Tex. Tech U., 1956. Registered profl. engr., Tex. Sr. engr. asst. Tex. Hwy. Dept., San Angelo, 1956-60, resident engr., 1960, lab. engr., 1960-66, dist. constrn. engr., 1966, asst. dist. engr., 1966-87. Sgt. USAAF, 1943-45, PTO. Mem. NSPE, Tex. Soc. Profl. Engrs., San Angelo Rotary Club. Home: PO Box 61051 San Angelo TX 76906

HARRIS, ARTHUR HORNE, biology educator; b. Middleborough, Mass., May 18, 1931; s. Frank Arthur and Winifred Stevens (Deane) H.; div.; children—Tina Melissa, Rebecca Ann, Megan Aneen. B.A., U. N.Mex., 1958, M.S., 1959, Ph.D., 1965; postgrad. U. Ariz., 1959-60. Asst. prof. Ft. Hays Kans. State Coll., 1962-65; from asst. prof. to prof. U. Tex.-El Paso, 1965—, curator vertebrate paleobiology, 1967—, co-dir. resource collections lab. environ. biology, 1980-93, , curator higher vertebrates, 1983—, dir. lab environ. biology, 1993—. Co-author: The Mammals of New Mexico, 1975, The Faunal Remains from Arroyo Hondo Pueblo, 1984; author: Late Pleistocene Vertebrate Paleocology of the West, 1985. Served with U.S. Army, 1951-53. NSF grantee, 1967-70; Nat. Geog. Soc. grantee, 1971-73, 84-86; recipient Faculty Research award U. Tex.-El Paso, 1976. Mem. Am. Soc. Mammalogists, Soc. Vertebrate Paleontology, Southwestern Assn. Naturalists (editor 1978-82), Soc. Systematic Biology, Am. Quaternary Assn. Democrat. Home: 665 Stedham Cir El Paso TX 79927-4202 Office: University of Texas Laboratory for Environmental BIO Dept of Biology El Paso TX 79968

HARRIS, BENJAMIN, psychologist; b. Dec. 12, 1949. BA, Hampshire Coll., 1971; PhD, Vanderbilt U., 1975. Asst. prof. Radford (Va.) Coll., 1976-78; asst. prof. Vassar Coll., Poughkeepsie, N.Y., 1978-85; vis. asst. prof. Bowdoin Coll., Brunswick, Maine, 1985-86; assoc. prof. U. Wis. Parkside, Kenosha, 1986-93; fellow Inst. for Rsch. in the Humanities, 1989-90, U. Wis. Teaching Improvement Coun., 1987-88; co-founder Forum for History of Human Sci., 1988-89. Adv. editor Contemporary Psychology, 1991—; editor: (hist. issue) Journal of Social Issues; contbr. articles to profl. jours., chpts. to several books. State exec. com. Assn. U. Wis. Profls., AFL-CIO, 1992—. Fellow APS (charter); mem. History of Sci. Soc., Historians of Am. Communism, Sigma Xi. Office: U Wis Parkside Kenosha WI 53141

HARRIS, CHAUNCY DENNISON, geographer, educator; b. Logan, Utah, Jan. 31, 1914; s. Franklin Stewart and Estella (Spilsbury) H.; m. Edith Young, Sept. 5, 1940; 1 child, Margaret. AB, Brigham Young U., 1933; BA, Oxford U., 1936, MA, 1943, DLitt, 1973; postgrad., London Sch. Econs., 1936-37; PhD, U. Chgo., 1940; DEcon (honoris causa), Catholic U., Chile, 1956; LLD (honoris causa), Ind. U., 1979; DSc (honoris causa), Bonn U., 1991, U. Wis., Milw., 1991. Instr. in geography Ind. U., 1939-41; asst. prof. geography U. Nebr., 1941-43; asst. prof. geography U. Chgo., 1943-46, assoc. prof., 1946-47, prof., 1947-84, prof. emeritus, 1984—, dean social scis., 1955-60, chmn. non western area programs and internat. studies, 1960-66, dir. ctr. for internat. studies, 1966-84, chmn. dept. geography, 1967-69, Samuel N. Harper Disting. Svc. prof., 1969-84, spl. asst. to pres., 1973-75, v.p. acad. resources, 1975-78; del. Internat. Geog. Congress, Lisbon, 1949, Washington, 1952, Rio de Janeiro, 1956, Stockholm, 1960, London, 1964, New Delhi, 1968, Montreal, 1972, Moscow, 1976, Tokyo, 1980, Paris, 1984, Sydney, Australia, 1988, Washington, 1992; v.p. Internat. Geog. Union, 1956-64, sec.-treas., 1968-76; mem. adv. com. for internat. orgns. and programs Nat. Acad. Scis., 1969-73, mem. bd. internat. orgns. and programs, 1973-76; U.S. del. 17th Gen. Conf. UNESCO, Paris, 1972; exec. com. div. behavioral scis. NRC, 1967-70; mem. coun. of scholars Libr.of Congress, 1980-83, Conseil de la Bibliographie Géographique Internationale, 1986—. Author: Cities of the Soviet Union, 1970; editor: Economic Geography of the U.S.S.R, 1949, International List of Geographical Serials, 1960, 71, 80, Annotated World List of Selected Current Geographical Serials, 1960, 64, 71, 80, Soviet Geography: Accomplishments and Tasks, 1962, Guide to Geographical Bibliographies and Reference Works in Russian or on the Soviet Union, 1975, Bibliography of Geography, Part I, Introduction to General Aids, 1976, Part 2, Regional, vol. 1, U.S., 1984, A Geographical Bibliography for American Libraries, 1985, Directory of Soviet Geographers 1946-87, 1988; contbr. Sources of Information in the Social Sciences, 1973, 86, Encyclopedia Britannica, 1989; contbg. editor: The Geog. Rev., 1960-73, Soviet Geography, 1987-91, Post-Soviet Geography, 1992—; contbr. articles to profl. jours. Recipient Alexander Csoma de Körösi Meml. medal Hungarian Geog. Soc., 1971, Lauréat d'Honneur Internat. Geog. Union, 1976; Alexander von Humboldt Gold Medal Gesellschaft für Erdkunde zu Berlin, 1978; spl. award Utah Geog. Soc., 1985; Rhodes scholar, 1934-37. Fellow Japan Soc. Promotion of Sci.; mem. Assn. Am. Geographers (sec. 1946-48, v.p. 1956, pres. 1957, Honors award 1976), Am. Geog. Soc. (coun. 1962-74, v.p. 1969-74; Cullum Geog. medal 1985), Am. Assn. Advancement Slavic Studies (pres. 1962, award for disting. contbns. 1978), Social Sci. Rsch. Coun. (bd. dir. 1959-70, vice-chmn. 1963-65, exec. com. 1967-70), Internat. Coun. Sci. Unions (exec. com. 1969-76), Internat. Rsch. and Exchs. Bd. (exec. com. 1968-71), Nat. Coun. Soviet and East European Rsch. (bd. dir. 1977-83), Nat. Coun. for Geog. Edn. (Master Tchr. award 1986); hon. mem. Royal Geog. Soc. (Victoria medal 1987), Geog. Socs. Berlin, Frankfurt, Rome, Florence, Paris, Warsaw, Belgrade, Japan, Chgo. (Disting. Svc. award 1965, bd. dir. 1954-69, 82-90), Polish Acad. Scis. (fgn. mem.). Home: 5649 S Blackstone Ave Chicago IL 60637-1871 Office: U Chgo Dept Geography 5828 S University Ave Chicago IL 60637-1583

HARRIS, ERIC WILLIAM, neuroscientist, pharmaceutical executive; b. Maietta, Ga., Sept. 5, 1951; s. Lloyd Webb and Harriet (Hanson) H.; m. Laurie Elizabeth Braman, June 18, 1988 (div. Mar., 1992); 1 child, Cooper William. BA, U. Va., 1973, PhD, 1979. Rsch. asst. dept. neurosurgery U. Va. Sch. Medicine, Charlottesville, 1976-79; postdoctoral fellow dept. psychobiology U. Calif., Irvine, 1979-82, sr. rsch. assoc., 1982-86; sr. rsch. scientist Pennwalt Pharm., Rochester, N.Y., 1986-88; prin. investigator Fisons Pharms., Rochester, 1988—; adj. prof. pharmacology U. Rochester; reviewer, referee European Jour. Pharmacology, Brain Research, Jour. of Neurochemistry, Pharmcology, Biochemistry and Behavior, 1986—. Mem. steering com. Rochester Alliance Promoting Sci. (Edn.), 1990—, Fathers Rights Assn. of N.Y., 1991—. Grantee: pre and post doctoral awards NIH, 1976-82. Mem. AAAS, Soc. for Neurosci., N.Y. Acad. Scis., Internat. Brain Rsch. Orgn. Achievements include first to propose that the drug Remacemide was a pro drug for an NMDA antagonist; did seminal work in the involvement of NMDA receptors in the process of neuronal long term potentiation. Office: Fisons Pharms PO Box 1710 Rochester NY 14603

HARRIS, GUY HENDRICKSON, chemical research engineer; b. San Bernardino, Calif., Oct. 2, 1914; s. Edwin James and Nellie Mae (Hendrickson) H.; m. Elsie Mary Dietsch, Mar. 15, 1940; children: Alice, Robert, Mary, Sara. BS, U. Calif., Berkeley, 1937; AM, Stanford U., 1939, PhD, 1941. Analytical chemist Shell Devel. Co., Emeryville, Calif., 1937-38; organic chemist William S. Merrell Co., Cin., 1941-45; rsch. chemist Fibe Board, Emeryville, 1945-46; from organic chemist to assoc. scientist The Dow Chem. Co., Pittsburg, Calif., 1946-62; assoc. scientist The Dow Chem. Co., Walnut Creek, Calif., 1964-82; sr. lectr. U. Ghana, Legon Accra, 1962-64; pvt. practice cons. Concord, Calif., 1982-88; rsch. engr. U. Calif., Berkeley, 1988—. Contbr. K & O Encyclopedia Chem. Tech., 1964, 74, 84, Reagents in Mineral Tech., 1990. Fellow AAAS, Royal Soc. Chemistry; mem. Am. Inst. Mining Engrs. Roman Catholic. Achievements include 49 patents in field of mineral processing reagents in particular Z200 (R) agricultural chemicals and process for manufacture. Home: 1673 Georgia Dr Concord CA 94519 Office: Univ California 386 Hearst Mining Bldg Berkeley CA 94720

HARRIS, HOWARD ALAN, laboratory director; b. Cleve., Aug. 22, 1939; s. Benjamin Richard and Mary Alice (Bauman) H.; m. Carolyn Sher Harris, Apr. 9, 1976; 1 child, Lauren Beth. AB, Western Res. U., 1961; MS, Yale U., 1963, PhD, 1966; JD, St. Louis U., 1972. Sr. rsch. chemist Shell Oil Co., Wood River, Ill., 1966-74; dir. police lab. N.Y. C. Police Dept., 1974-85; dir. Pub. Safety Lab. Monroe County, Rochester, N.Y., 1985—; adj. prof. John Jay Coll., N.Y.C., 1975-85. Fellow Am. Acad. Forensic Scis. (chair crime sect. 1989); mem. Am. Soc. Crime Lab. Dirs. (pres. 1985), Am. Chem. Soc., Northeastern Soc. Forensic Scientists (founder 1977). Achievements include patents in field; research in forensic science and a method for identifying LSD. Office: Monroe County Pub Safety Lab Pub Safety Bldg Rm 524 Rochester NY 14614

HARRIS, IRA STEPHEN, medical and health sciences educator; b. Bklyn., July 13, 1945; s. Simon and Vera (Vichness) H.; m. Arlene Cramer, Dec. 25, 1971; children: Elliot, David, Sara. BS, Fairleigh Dickinson U., 1968; MS, L.I. U., 1970, Profl. Diploma magna cum laude, 1978. Sci. educator 158Q Marie Curie High Sch., Bayside, N.Y., 1968-76; tchr. math., sci. and social studies, media specialist Campbell Jr. High Sch. 218Q, Flushing, N.Y., 1976-79, Beard Jr. High Sch. 189Q, Flushing, 1979-86; tech. specialist Carson Intermediate Sch. 237Q, Flushing, 1986—; judge sci. fair N.Y. Acad. Scis., N.Y.C., 1985—. Commodore Newbridge Boat Club, Bellmore, N.Y.; v.p., edn. chmn. Bellmore Jewish Ctr.; pres. East Bay Civic Assn., Bellmore. Mem. N.Y. Acad. Scis. Republican. Home: 2729 Claudia Ct Bellmore NY 11710-4740

HARRIS, JAMES C II, aerospace engineer; b. Louisville, Feb. 11, 1967; s. James C. and Linda M. (Fehribach) H. BS, St. Louis U., 1990. Performance engr. Trans World Airlines, St. Louis, 1990—. Lutheran. Achievements include research in NTSB investigation with the primary source of reading the flight data recorder and data analysis. Office: Trans World Airlines 11495 Natural Bridge Rd Saint Louis MO 63141

HARRIS, JAMES HERMAN, pathologist, neuropathologist, consultant, educator; b. Fayetteville, Ga., Oct. 19, 1942; s. Frank J. and Gladys N. (White) H.; m. Judy K. Hutchinson, Jan. 30, 1965; children: Jeffrey William, John Michael, James Herman. BS, Carson-Newman Coll., 1964; PhD, U. Tenn.-Memphis, 1969, MD, 1972. Diplomate Am. Bd. Pathology; sub.-cert. in anatomic pathology and neuropathology. Resident and fellow N.Y.U.-Bellevue Med. Ctr., N.Y.C., 1973-75; adj. asst. prof. pathology N.Y. U., N.Y.C., 1975—; asst. prof. pathology and neuroscis. Med. Coll. Ohio, Toledo, 1975-78, assoc. prof., 1978-82, dir. neuropathology and electron microscopy lab., 1975-82; cons. Toledo Hosp., 1979-82, assoc. pathologist/neuropathologist, dir. electron microscopy pathology lab., 1983—, mem. overview com., credentials com., appropriations subcom. medisgroup, interqual task force; chmn. clin. support services com., vice chmn. med. staff quality rev. com.; cons. neuropathologist Mercy Hosp., 1976—, U. Mich. dept. pathology, 1984—; cons. med. malpractice in pathology and neuropathology; mem. AMA Physician Rsch. and Evaluation Panel; mem. ednl. and profl. affairs commn., exec. council Acad. Medicine; mem. children's cancer study group Ohio State U. satellite; chmn. tech. and issues subcom. of adv. com. Blue Cross, mem. task force on Cost Effectiveness N.W. Ohio; chmn. med. necessity appeals com. Blue Cross/Blue Shield; adv. bd. PIE Mut. Ins. Co. Chmn. steering com. Pack 198, Boy Scouts Am.; chmn. fin. com., dir. bldg. fund campaign First Baptist Ch., Perrysburg, Ohio; faculty chmn. Med. Coll. Ohio United Way Campaign; mem. adv. com. Multiple Sclerosis Soc. NW Ohio. Recipient Outstanding Tchr. award Med. Coll. Ohio, 1980; named to Outstanding Young Men Am., U.S. Jaycees, 1973; USPHS trainee, 1964-69, postdoctoral trainee, 1973-75; grantee Am. Cancer Soc., 1977-78, Warner Lambert Pharm. Co., 1978-79, Miniger Found., 1980, Toledo Hosp. Found., 1985, Promedica Health Care Found., 1986. Mem. Am. Profl. Practice Assn., Am. Pathology Found., Am. Soc. Law and Medicine, Coll. Am. Pathologists, Lucas County Acad. Medicine (bar acad. liaison com.), Ohio State Med. Assn. (fed. key contact), Am. Assn. Neuropathologists (profl. affairs com., awards com., program com., constitution com.), Internat. Acad. Pathologists, Ohio Soc. Pathologists, Coll. Am. Pathologists, EM Soc. Am., Sigma Xi. Author med., sci. papers; reviewer Jour. Neuropathology and Exptl. Neurology. Republican. Avocations: tennis, real estate rehabilitation, gardening, white water rafting, hang-gliding. Home: 9105 Nesbit Lakes Dr Alpharetta GA 30202 Office: Toledo Hosp Dept Pathology 2142 N Cove Blvd Toledo OH 43606-3895

HARRIS, JOSEPH MCALLISTER, chemist; b. Pontiac, Ill., July 27, 1929; s. Fred Gilbert and Catherine Marguerite (McAllister) H.; m. Margot Jeanette L'Hommedieu, Feb. 17, 1952; children: Timothy, Kaye, Paula, Bruce, Anne, Martha, Rebecca. BA, Blackburn Coll., Carlinville, Ill., 1952; postgrad., So. Ill. U., 1953-54, U. Ill., 1956-61. Technician Olin Ind., Inc., Energy, Ill., 1953-54; quality control staff Union Starch and Refining Co., Granite City, Ill., 1954; rsch. asst. Ill. State Geol. Survey, Urbana, 1954-61; chemist II Water Pollution Control Bd., Annapolis, Md., 1961-63; phys. chemist Ball Bros. Rsch., Inc., Muncie, Ind., 1963-66; engr. Radio Corp. Am., Marion, Ind., 1966-70; chemist OA Labs., Inc., Indpls., 1973-86; chemist OA Labs. & Rsch., Inc., Indpls., 1986-93, cons., 1993—. Bd. dirs. Tri-County Hearing Assn. for Children, Muncie, 1967-70. Mem. Am. Chem. Soc., AAAS, Soc. Applied Spectroscopy. Republican. Presbyterian. Avocations: gardening, camping. Home: 800 E Washington St Muncie IN 47305-2533

HARRIS, JULES ELI, medical educator, physician, clinical scientist, administrator; b. Toronto, Ont., Can., Oct. 12, 1934; came to U.S., 1978; s. George Joseph and Ida (Teska) H.; m. Josephine Leikin; children: Leah, Daniel, Adam, Sheira, Robin, Naomi. MD, U. Toronto, 1959. Intern, then resident Toronto (Can.) Gen. Hsp., 1959-65; asst. prof. medicine M.D. Anderson Hosp. Med. Ctr., Houston, 1966-69; prof. medicine U. Ottawa (Ont.), 1969-78; prof. medicine, prof. immunology Rush Med. Coll., Rush U., Chgo., 1978—; mem. gov.'s adv. bd. for cancer control, State of Ill., 1988—; chmn. bd. trustees Ill. Cancer Coun., Chgo., 1987-88. Author: Immunology of Malignant Disease, 1975; editor: New Perspectives in Large Bowel Cancer, 1989. Mem. internat. bd. govs. Ben Gurion U. of Negev, Beer-Sheva, Israel, 1986—. Fellow ACP, Royal Coll. Physicians Can. (cert. in internal medicine); mem. Am. Soc. Clin. Oncology (chmn. pub. rels. com. 1987-89), Univ. Club, Alpha Omega Alpha. Jewish. Office: Rush-Presbyn-St Luke's Med Ctr 1725 W Harrison St Chicago IL 60612-3828

HARRIS, KAREN L., psychologist; b. Peoria, Ill., Feb. 15, 1963; d. Richard L. and Jacqueline R. (Hand) Sears; m. Howard P. Harris, May 18, 1985; children: Jessica, Kenneth. PhD, U. Ill., 1990. Asst. prof. psychology Western Ill. U., Macomb, 1990—. U. Rsch. Coun. rsch. grantee Western Ill. U., 1991. Office: Western Ill U Dept Psychology Macomb IL 61455

HARRIS, KENNETH KELLEY, chemical engineer; b. Huntingburg, Ind., Feb. 19, 1960; s. Kenneth Dwayne Kelley and Mary Charlene (Harris) Putnam; m. Bonnie Dorothy Gennuso, Oct. 23, 1983 (div. Aug. 1987); 1 child, Kenneth Kelley II; m. Amy Lynn Tucker, Dec. 26, 1992. BS in Forestry, Purdue U., 1982, BSChemE, 1992. Commd. 2d lt. U.S. Army, 1985, advanced through grades to 1st lt., 1987; tank platoon leader 1-8 cavalry 2d brigade U.S. Army, Ft. Hood, Tex., 1985-88, asst. logistics officer 1st cavalry div. 2d brigade, 1987-88, brigade adjutant, 1988; battery comdr. 1-138 air def. artillery, battery exec. officer, battalion chem. officer Ind. Army Nat. Guard, 1988-92; coop. engr. GE Plastics, Mt. Vernon, Ind., 1989-92; project engr. Internat. Paper, Moss Point, Miss., 1992-93, sr. project engr., 1993—. Editor newspaper Owl Sheet, 1981; newspaper reporter Purdue Exponent, 1992. Cubmaster Boy Scouts Am., Mobile, Ala., 1992—; Mem. Purdue Young Reps., W. Lafayette, 1980. Hoosier scholar. Mem. AIChE (assoc.), Toastmasters Internat., Ind. Red Leg Assn. Baptist. Home: 3815 Cabana Blvd N Apt 108 Mobile AL 36609 Office: Internat Paper 6901 Grierson St Moss Point AL 39563

HARRIS, LEONARD ANDREW, civil engineer; b. San Francisco, Apr. 11, 1928; 3 children. BS, Stanford U., 1950; MS, U. Ill., 1953, PhD, 1954. Rsch. assoc. civil engring. U. Ill., 1953-65; from engr. to sr. engring. specialist applied mech. Space and Info. Systems Divsn., N.Am. Aviation Corp., 1954-58, supv., 1958-59, prin. sci. structural sci., 1959-61, dir., 1961-67, asst. mgr. sci. and tech., 1967-68; mgr. structure and design N.Am. Rockwell Corp., 1968-70, mgr. shuttle tech., 1970-72; mgr. material and structure Office Aero. and Space Adminstrn., NASA, 1972—; lectr. aero. engring. U. So. Calif., 1956-60; guest lectr. UCLA, 1959-65; mem. indsl. adv. bd. com. mil. handbook on sandwich construction, 1959-65; mem. ad hoc com. design with brittle material Mat Adv. Bd. NAS, 1964; mem. rsch. adv. com. space vehicle structure NASA, 1965-67. Fellow Internat. Aero. & Astronautical (assoc.); mem. ASCE. Achievements include rsch. in experimental and theoretical studies of buckling of plates and shells; static and fatigue strength of welds in structural materials; brittle behaviour of materials. Office: NASA Aeronautics & Space Tech 300 E St SW Washington DC 20546 also: 5515 Trent Chevy Chase MD 20815*

HARRIS, LOUIS SELIG, pharmacologist, researcher; b. Boston, Mar. 27, 1927; s. Max Selig and Pearl (Oppochinski) H.; m. Ruth Irma Schaufus, Aug. 22, 1952; 1 child, Charles Allan. BA, Harvard U., 1954, MA, 1956, PhD, 1958. Sect. head, sr. rsch. biologist Sterling-Winthrop Rsch. Inst., Rensselaer, N.Y., 1958-66; lectr. in pharmacology Albany (N.Y.) Med. Coll., 1959-66; from assoc. prof. to prof. U. N.C., Chapel Hill, 1966-73; Harvey Haag prof., chmn. Med. Coll. Va./Va. Commonwealth U., Richmond, 1972—; acting assoc. dir. Nat. Inst. on Drug Abuse, Rockville, Md., 1987-88; Sterling Drug vis. prof., 1983; mem. com. on problems of drug dependence Nat. Acad. Scis., NRC, 1973-77; mem. Com. on Problems of Drug Dependence, Inc., 1977—, chmn., 1990-92; hon. prof. Beijing Med. U., People's Republic of China, 1990. Editor: (monograph) NIDA Monographs, Proceedings, Committee on Problems of Drug Dependence, 1979—; author chpts. in books. Recipient Hartung Meml. award U. N.C., 1981, Univ. Excellence award Med. Coll. Va./Va. Commonwealth U., 1984, Outstanding Faculty award, 1984, Nathan B. Eddy award Com. on Problems of Drug Dependence, 1985, Abe Wikler award Nat. Inst. on Drug Abuse, 1991, Gov.'s award on Drug Abuse Rsch., 1992, Presdl. medallion Va. Commonwealth U., 1993. Fellow Am. Coll. Neuropsychopharmacology; mem. AAAS, AAUP, Am. Soc. for Pharmacology and Exptl. Therapeutics, Am. Chem. Soc., Am. Assn. for Med. Sch. Pharmacology, Am. Pain Soc. (charter 1977), Am. Pharm. Assn., Am. Soc. for Clin. Pharmacology and Therapeutics, Med. Electronics Soc., Assn. Harvard Chemists, Internat. Anesthesia Rsch. Soc., Elisha Mitchell Sci. Soc., Internat. Narcotic Enforcement Officers Assn., Soc. for Neurosci., Internat. Soc. Biochem. Pharmacology, Acad. of Scis., Internat. Soc. for the Study of Pain, Collegium Internationale Neuro-Psychopharmacologicum, Va. Acad. Sci., Harvard Club Boston, Cosmos Club Washington. Achievements include research in field. Home: 7830 Rockfalls Dr Richmond VA 23225-1049 Office: Va Commonwealth U MCV Sta Box 27 Richmond VA 23298

HARRIS, MARTIN SEBASTIAN, JR., architect; b. Sea Cliff, N.Y., Dec. 13, 1932; s. Martin Sebastian and Anne Charlotte (Liffmann) H.; m. Carolyn Jonker, July 29, 1958; 1 child, Glynis Anne. AB in Architecture cum laude, Princeton U., 1954, MFA in Architecture and Urban Planning, 1959; postgrad., SUNY, Albany, 1965-67. Registered architect, Vt., N.H., N.Y. Draftsman Svenska Riksbyggen, Stockholm, 1958; designer, draftsman Perkins and Will, architects, White Plains, N.Y., 1959-62; planning and econ. devel. cons. Ebasco Svcs., Inc., N.Y.C., 1962-64; assoc. Engelhardt and Leggett, ednl. cons., Purdys Station, N.Y., 1964-66; pvt. practice, Vergennes, Vt., 1962—; chmn. Ednl. Cons. Svcs., 1966—; instr. constrn. mgmt. and code compliance C.C. Vt.; vis. lectr. Middlebury Coll., Dartmouth Coll.; instr. Castleton State Coll.; adj. prof. Rensselaer Poly. Inst., Troy, N.Y.; appt.'d constrn. arbitration panel Am. Arbitration Assn.; constrn. cons. (Lithuania) Vols. in Overseas Co-op. Assistance, Washington. Co-founder, assoc. editor New Eng. Builder; columnist Burlington Free Press, New Eng. Farmer, Vt. Bus. World, Valley Voice, Rutland Herald, RFD Vt. Farm News, Green Mountain Dispatch, Vt. Sch. Bds. Assn. Newsletter, Rutland Voice. Chmn. local sch. bd., Union High Sch. Bd., local planning bd.; mem. sch. supervisory union exec. com.; local dir. CD, vice chmn. Vt. Housing Assn.; pres. Addison County Housing Assn., Vt. Housing Investment Fund; mem. Vt. 1122 Rev. Com. for Health Facilities; town energy coord.; mem. legis. com. on self-help housing, State Farmland Preservation Com. 1st lt. U.S. Army, 1955-57, Korea; mem. USAR and Vt. Army NG, 1958-92. Recipient award for Sudbury (Vt.) Sch., 1981, Major Samuel Howard House, Benson, Vt., 1985, sweater factory, Shelburne, Vt., 1986; IAESTE fellow, 1958, D'Amato fellow Princeton U., 1958-59. Republican. Methodist. Home and Office: RD 2 Box 2716 Vergennes VT 05491

HARRIS, MARTIN STEPHEN, aerospace engineering executive; b. Greenville, S.C., Nov. 23, 1939; s. Vitruvius Aiken and Clara Margaret (Thackston) H.; m. Helen C. Dean, Sept. 7, 1963 (div. May 1980); children: Dean, Susan, James; m. Prudence Cooper Bolstad, Jan. 20, 1990 (dec. Mar. 10, 1993). BS in Physics, Furman U., 1962; MS in Physics, Fla. State U., 1967; ret., USAF, 1982. Commd. 2d lt. USAF, 1962, advanced through grades to maj., 1973, ret., 1982; sr. project engr. Hughes Aircraft Co., El Segundo, Calif., 1982-84, section head, 1984-86, space vehicle mgr., 1986-89, asst. program mgr., 1989—. Mem. Sigma Alpha Epsilon.

HARRIS, MILES FITZGERALD, meteorologist; b. Brunswick, Ga., Feb. 2, 1913; s. James Madison and Louise (Fitzgerald) H.; m. Marguerite Bertice Leonard, May 13, 1938; children: Ann Louise, Theresa Geraldine, Emily Leland. BSc in Meteorology, NYU, 1944, MSc in Meteorology, 1957. Weather observer U.S. Weather Bur., Macon, Ga., 1932-35, Savannah, Chattanooga,, Macon,, Washington, 1937-42; cadet/clk. South Atlantic Steamship Line, Savannah, 1935-37; meteorologist U.S. Weather Bur., Washington, 1944-45; hurricane forecaster U.S. Weather Bur., San Juan, P.R., 1945-48; spl. projects meteorologist U.S. Weather Bur., Washington, 1948-51, rsch. meteorologist, 1951-61, head editing and pub. br., 1961-66; phys. scientist, chief Sci. Info. Br. Environ. Sci. Svcs. Adminstrn., Washington, 1966-70; editor Mon. Weather Review, 1968-70; editor Am. Meteorol. Soc., Boston, 1970-83, ret., 1983; Editor, writer, cons. Earth Sci. Curriculum Project, Boulder, Colo., 1964-67. Author: Man Against Storm, 1962, Getting to Know the World Meteorological Organization, 1966, Opportunities in Meteorology, 1972, Investigating the Earth, 1967-84; contbg. author Ency. of Earth Scis., 1976; contbg. author and editor John Hale, A Man Beset by Witches, 1992. Mem. Am. Meteorol. Soc. Democrat. Congregationalist. Avocations: local history, writing and research on John Hale, The Salem Witchcraft Trials. Home: 40 Lothrop St Beverly MA 01915-5150

HARRIS, PATRICK DONALD, physiology educator; b. Nebraska City, Nebr., Mar. 30, 1940; s. Donald Wilson and Theresia Marie (Bierl) H.; m. Doris Jean, July 18, 1959; children: Donna Beth, Wesley Mark, Kennet Fulton. BSEE with honors, U. Mo., 1962, MSEE with honors, 1963; PhD in Physiology, Northwestern U., Evanston, Ill., 1967. Nat. Heart & Lung Inst. postdoctoral fellow dept physiology Sch. Medicine, Ind. U., Indpls., 1967-68; asst. prof. physiology Sch. Medicine U. Mo., Columbia, 1968-71, assoc. prof. Sch. Medicine, 1971-77, assoc. prof. Grad. Sch., 1974-77, prof. Sch. Medicine and Grad. Sch., 1977-81, investigator Dalton Rsch. Ctr., 1974-80, assoc. dir. Dalton Rsch. Ctr., 1980-81; vis. assoc. biomed. engring. div. engring. and applied scis. Calif. Inst. Tech., Pasadena, 1977-78; prof., chmn. dept. physiology and biophysics Sch. Medicine, U. Louisville, 1981—, dir. Ctr. Applied Microcirculatory Rsch., Health Scis. Ctr., 1986—; pres. Micro-Med Inc., 1989—; spl. rsch. fellow Nat. Heart and Lung Inst., 1970-72; mem. nat. com. Commn. on Life Scis., NRC, 1984-90; sci. program cons. Nat. Heart, Lung and Blood Inst., 1972-93; mem. advanced tech. coun. Louisville Urban Area and Commonwealth of Ky., 1987—; bd. dirs. Jewish Hosp. Heart & Lung Inst., 1990—, chmn. sci. affairs com., 1990—. Referee Jour. Microvascular Rsch., 1969-86, bd. editors, 1979-86; referee Anesthesia and Analgesia-Current Researches, 1970-74; bd. editors Proc. Soc. Exptl. Biol. Medicine, 1975-78; referee Am. Jour. Physiology, 1974—, bd. editors, 1986-93; referee Sci., 1979-84; bd. editors Microcirculation, 1980-85; referee Circulation Rsch., 1981-85, Jour. AMA, 1982-84, Hypertension, 1982-88; bd. editors Circulatory Shock, 1986-93; contbr. articles to profl. jours. Adult leader Boy Scouts Am., 1967-68, 75-84; vol. parole officer State of Mo., Columbia, 1974-81; bd. dirs. Ky. affiliate Am. Heart Assn., 1987-93; active MGC Cath. Ch., Louisville, 1981—, Cath. Archdiocese, Louisville, 1981—; coach, v.p. Daniel Boone Little League Baseball, 1973-77; mem. hosp. worship svc. program Interfaith Coun. Columbia, 1974-81; mem. athletic bd. Mother of Good Counsel ath. Ch., Louisville, 1982-86, chmn., 1982-84, mem. adult formation program, 1981-93. Grantee Lilly Rsch. Labs., 1969-70, Nat. Inst. Gen. Med. Scis., 1969-71, Nat. Heart and Lung Inst., 1972-77, NIH, 1970-90, NSF, 1980-81, VA, 1984—, Merck Rsch. Inst., 1985, Am. Heart Assn., 1975-89, Humana Corp., 1986-90, Commonwealth Ctr. Excellence, 1987—; recipient Rsch. Career Devel. award Nat. Heart and Lung Inst. Fellow Am. Heart Assn.; mem. IEEE (mem. exec. com. Columbia chpt. engring. in medicine and biology group 1970-75), Microcirculatory Soc. (pres. V World Congress for Microcirculation 1987-91), European Soc. Microcirculation, Am. Physiol. Soc. (mem. program exec. com. 1984-90), Soc. Exptl. Biology and Medicine, Shock Soc., Assn. Chairpersons Depts. Physiology, Ky. Heart Assn. (mem. rsch. peer rev. com. 1981-91, bd. dirs. 1987-93), Jefferson County Med. Soc., Tau Beta Pi. Roman Catholic. Avocation: golf. Home: 9014 Billingsgate Pl Louisville KY 40242-2440 Office: Micro-Med Inc 707 Lyndon Ln # 2 Louisville KY 40222-4680*

HARRIS, ROBERT BERNARD, biochemist; b. N.Y.C., Apr. 8, 1952; s. Melville and Frances (Bernard) H.; m. Lesley K. Harris, Jan. 21, 1978; children: Lindsay M., David A. BA, U. Rochester, 1974; PhD, NYU, 1979. Postdoctoral assoc. U. Colo., Boulder, 1980-82, sr. rsch. assoc., 1982-84, asst. prof., 1984-86; asst. prof. Va. Commonwealth U., Richmond, 1987-91, assoc. prof., 1991—; founder prin. Commonwealth Biotechs., Inc., Richmond, 1992; pres. CBI, 1992—; cons. Glycomed, Inc., Alameda, Calif., 1991—, Glaxo Rsch. Labs., Research Triangle Park, N.C., 1990—. Contbr. more than 70 articles to profl. jours. Mem. Brandermill Community Assn.; coach Midlothian Youth Soccer League. Established investigatorship award Am. Heart Assn., 1987-92; rsch. grantee USPHS NIH, 1986—, Am. Heart Assn., 1989-92. Achievements include patents in field. Office: Va Commonwealth Univ Dept Biochemistry PO Box 614 Richmond VA 23298

HARRIS, ROBERT JAMES, engineer; b. Sioux Falls, S.D., Nov. 28, 1940; s. Lloyd James and Margaret Loretta (Monner) H.; m. Judy Ellen Barnes, June 9, 1962; children: Taryl, Kimmi, Todd. BS in Civil Engring., S.D. State U., 1963; MS in Applied Mechanics, Denver U., 1968. Registered profl. engr., Colo., Idaho. Design engr. Boeing, Seattle, 1963, N. Am. Aviation, Tulsa, 1964-65; rsch. engr. Martin Marietta, Denver, 1966-71, Aerojet Nuclear Co., Idaho Falls, 1972-75; gen. mgr. Energy Inc., Idaho Falls, 1976-88; dept. mgr. EG&G, Morgantown, W.Va., 1988-92; unit mgr. EG&G, Idaho Falls, 1992—. Contbr. articles to profl. jours. Pres., bd. dirs. PTA, Idaho Falls, 1972-85; mem. bldg. com. Sch. Dist., Idaho Falls, 1976-78; coach Little League, Idaho Falls, 1976-82, Grid Kid Football, Idaho Falls, 1976-81. Mem. Civitan Club (bd. dirs., pres. Idaho Falls chpt. 1972-88, Club key 1977). Avocations: skiing, fishing, hunting. Home: 2805 S Blvd Idaho Falls ID 83404-9999 Office: EG&G PO Box 1625 Idaho Falls ID 83415-9999

HARRIS, ROGER SCOTT, energy consultant; b. Greenfield, Mass., Nov. 12, 1959; s. Robert A. and Norma L. (Day) H.; m. Linda P. Harris, Apr. 5, 1986. BA in Govt., Western New Eng. Coll., 1982. Cert. energy conservation svc. auditor and inspector, Mass. Energy auditor Mass-Save, Inc., Springfield, Mass., 1982-83; tech. cons. Hampden County Energy Office, Springfield, 1982-88, tech. dir., 1988—; mem. adv. com. Energy Conservation Svc., State of Mass., 1987-89. Contbg. editor: Residential Heating Systems, 1987, 93, Air-to-Air Heat Exchangers, 1987; contbr. articles to profl. jours. Chmn. Ashfield (Mass.) Energy Resources Commn., 1977-83; town rep. Franklin County Energy Task Force, Greenfield, Mass., 1978-83; commr. Springfield Hist. Commn., 1991—. Home: 65 Buckingham St Springfield MA 01109-3926 Office: Hampden County Energy Office Hall of Justice 50 State St Springfield MA 01103

HARRIS, SHARI LEA, mathematics educator; b. Macon, Mo., May 17, 1964; d. Walter Edward and Darlene (Tipton) H. BSE in Math. Edn. cum laude, Northeast Mo. State U., 1986; MS in Applied Math., U. Mo., 1991. Cert. secondary math. tchr., Mo. Cons. micromputer lab. Northeast Mo. State U., Kirksville, 1982-84, tutor coll. algebra, 1984-85, instr. math. lab., 1985; teaching asst. U. Mo., Columbia, 1986-87, 89-91; math. tchr. Highland High Sch., Ewing, Mo., 1987-88; substitute tchr. Quincy (Ill.) Sr. High Sch., 1988-89; math. tchr. Kemper Mil. Acad., Boonville, Mo., 1990; instr. math. S.E. Mo. State U., Cape Girardeau, 1991—. John J. Pershing scholar Northeast Mo. State U., Kirksville, 1982-86. Mem. Am. Math. Soc., Grad. Profl. Coun., Nat. Coun. Tchrs. of Math., Kappa Mu Epsilon, Alpha Phi Sigma. Avocations: singing, concerts, playing piano and organ.

HARRIS, WESLEY L., federal agency administrator; b. Richmond, Va., Oct. 29, 1941; s. William M. and Rosa P. (Minor) H.; m. Myrtle Ann Satterwhite, June 14, 1960 (div. Mar. 1985); children: Wesley Jr., Zelda, Kamau, Kalomo; m. Sandra Maria Butler, Sept. 2, 1985; 1 child, Tosha. B in Aeronautical Engring. with honors, U. Va., 1964; MA in Aeronautical Scis., Princeton U., 1966, PhD in Aeronautical Scis., 1968. Asst. prof. aerospace engring. U. Va., 1968-70; assoc. prof. physics Southern U., 1970-71; assoc. prof. aerospace engring. U. Va., 1971-72, dir. Office Minority Edn., 1975-78; assoc. prof. aeronautics & astronautics/ocean engring. MIT, 1973-79, assoc. prof. aeronautics and astronautics, 1980-81, prof. aeronautics & astronautics, 1981-85; mgr. computational methods Office Aeronautics & Space Tech. NASA HQs, 1979-80; dean sch. engring. U. Conn., Storrs, 1985-90; v.p. U. Tenn. Space Inst., Tullahoma, 1990-93; assoc. adminstr. Office of Aeronautics NASA HQ, Washington, 1993—; mem. adv. groups Nat. Rsch. Coun. Commn. Engring. and Tech. Systems, Bd. Engring. Edn., Bd. Army Sci. and Tech., Air Force Studies Bd., Com. Aeronautical Techs.; mem. adv. com. NSF, U.S. Army Sci. Bd.; advisor univs.; nat. adv. com. dept. engring. Hampton U., 1989—. Author more than 100 tech. papers. Trustee Sci. Mus. Conn., 1985-90; bd. dirs. Conn. Pre-Engring. Program, Inc., 1986-90; adv. bd. dirs. Am. City Bank, Tullahoma, 1990—; bd. vis. sch. engring. Duke U., 1991—; vis. com. dept. aeronautics and astronautics MIT, 1988—, past vis. com. dept. mech. engring., 1985-88. Recipient Herbert S. and Jane Gregory Disting. Lectr. award Coll. Engring. U. Fla., 1992. Fellow AIAA; mem. AAAS, Am. Helicopter Soc., Am. Physical Soc., Math. Assn. Am., Nat. Mgmt. Assn., Nat. Tech. Assn. Democrat. Avocation: squash. Office: NASA HQs/Code R 300 E St SW Washington DC 20546

HARRIS, WILLIAM JOSEPH, genetics educator; b. Dundee, Angus, Scotland, Nov. 17, 1944; s. James Robertson and Sarah Jane (Orr) H.; m. Linda McPherson, July 3, 1969; children: Neill, Gordon, Kathleen. BSc Biochemistry with 1st class honors, St. Andrews (Fife, Scotland) U., 1966; PhD, Dundee U., 1969. Head in vitro toxicology Inveresk Rsch. Inst., Edinburgh, Scotland, 1978-82; head biotech. Inveres Rsch. Internat., Edinburgh, Scotland, 1982-87; lectr. biochemistry U. Aberdeen, Scotland,

1969-78, prof. genetics, 1987-92; pres., chief scientific officer Scotgen Biopharm., Inc., Scotland, 1992—; chmn., mng. dir. Scotgen Ltd., Aberdeen, 1987—; tech. dir. Cogent Biotech. Investments, London, 1982-88; rsch. dir. Bioscot Ltd., Edinburgh, 1986-88; bd. dirs. NCIMB, Ltd., Aberdeen. Contbr. chpt. to Strategies in Industrial Biotechnologies, 1987; contbr. articles to Molecular and Gen. Genetics, Biotech., others. Recipient Endeavour prize ICI Ltd., 1969; rsch. grantee U.K. Med. Rsch. Coun., 1989-90, U.K. Ministry of Agr., 1990—, U.K. Serc Biotech. Directorate, 1991—. Mem. U.K. Biochem. Soc., U.K. Soc. for Gen. Microbiology, U.K. Genetical Soc., Am. Soc. Microbiology. Achievements include U.S. and foreign patent for method of detecting presence of HCMV-specific IGM, DNA probe dipsticks in urinary tract infections; foreign patent for use of human monoclonals in treating respiratory viral infections. Home: 18 Queen St, Carnoustie Angus, Scotland DD7 6DR

HARRIS, YVETTE RENEE, psychologist, educator; b. Japan, Sept. 4, 1957; came to U.S., 1961; d. Joe Curtis and Lucille (Fletcher) H. BA, Troy State U., 1979; MA in Clin. Psychology, Fisk U., 1981; PhD in Psychology, U. Fla., 1988. Psychologist Counseling and Testing Ctr. U. Cen. Fla., Orlando, 1981-84; assist. prof. psychology Miami U., Oxford, Ohio, 1988—. Contbr. articles to profl. jours. Adv. bd. mem. Visions Day Care, Cin., 1989—; mem. Dominican Community Svcs., Cin., 1989—. Mem. NAACP (edn. chair 1983-84), Am. Psychol. Soc., Soc. for Rsch. Child Devel., Sigma Xi. Roman Catholic. Office: Dept Psychology Benton Hall Miami Univ Oxford OH 45056

HARRISON, AIDAN TIMOTHY, chemist; b. York, England, July 26, 1960; came to U.S., 1988; s. Bernard and Mary Winifred (Brown) H. BSc, Liverpool Polytech., England, 1982; PhD, Warwick U., Coventry, England, 1986. Rsch. intern Imperial Chem. Industries, Runcorn, Eng., 1979-80, Shell Rsch. Ltd., Sittingbourne, Eng., 1981; postgrad. rsch. assoc. Warwick U., Coventry, 1982-85, postdoctoral rsch. fellow, 1986-88; dir. NMR facility Cornell U., Ithaca, N.Y., 1988—. Contbr. articles to profl. jours. Bd. dirs. Commonland Community Residents Assn., Ithaca. Grantee NATO Advanced Sci. Inst. 1988. Mem. AAAS, Am. Chem. Soc., Royal Soc. Chemistry. Office: Cornell U Chemistry Dept Baker Lab Ithaca NY 14853

HARRISON, ANNA JANE, chemist, educator; b. Benton City, Mo., Dec. 23, 1912; d. Albert S.J. and Mary (Jones) H. Student, Lindenwood Coll., 1929-31, L.H.D. (hon.), 1977; A.B., U. Mo., 1933, B.S., 1935, M.A., 1937, Ph.D., 1940, D.Sc. (hon.), 1983; D.Sc. (hon.), Tulane U., 1975, Smith Coll., 1975, Williams Coll., 1978, Am. Internat. Coll., 1978, Vincennes U., 1978, Lehigh U., 1979, Hood Coll., 1979, Hartford U., 1979, Worcester Poly. Inst., 1979, Suffolk U., 1979, Eastern Mich. U., 1983, Russell Sage Coll., 1984, Mt. Holyoke Coll., 1984, Mills Coll., 1985; L.H.D. (hon.), Emmanuel Coll., 1983; D.H.L., St. Joseph Coll., 1985, Elms Coll., 1985, R.I. Coll., 1990. Instr. chemistry Newcomb Coll., 1940-42, asst. prof., 1942-45; asst. prof. chemistry Mt. Holyoke Coll., 1945-47, asso. prof., 1947-50, prof., 1950-76, prof. emeritus, 1979—, chmn. dept., 1960-66, William R. Kenan, Jr. prof., 1976-79; Mem. Nat. Sci. Bd., 1972-78; Disting. Vis. prof. U.S. Naval Acad., 1980. Author: (textbook with Edwin S. Weaver) Chemistry: A Search to Understand, 1989; contbr. articles to profl. jours. Recipient Frank Forrest award Am. Ceramic Soc., 1949; James Flack Norris award in chem. edn. Northeastern sect. Am. Chem. Soc., 1977; AAUW Sarah Berliner fellow Cambridge U., Eng., 1952-53; Am. Chem. Soc. Petroleum Research Fund Internat. fellow NRC Can., 1959-60; recipient Coll. Chemistry Tchr. award Mfg. Chemists Assn., 1969. Mem. AAAS (dir. 1979-85, pres.1983, chmn. bd. 1984-85), Am. Chem. Soc. (chmn. div. chem. edn. 1971, pres. 1978, dir. 1976-79, award in chem. edn. 1982), Internat. Union Pure and Applied Chemistry (U.S. nat. com. 1978-81), Vols. in Tech. Assistance (bd. dirs. 1990—), Sigma Xi (bd. dirs. 1988-91). Address: Mt Holyoke Coll Dept Chemistry South Hadley MA 01075

HARRISON, BARRY, economics educator, researcher, author; b. South Shields, Eng., Dec. 18, 1951; s. Richard and Hazel Mary (Aire) H.; m. Lea Lääts, Aug. 16, 1975; children: Paul Barry, Matthew Richard, Simon Joseph. BA, U. Nottingham, Eng., 1974; MA, U. Sheffield, Eng., 1987. Lectr. econs. Wyggeston and Queen Elizabeth I Coll., Leicester, Eng., 1976-80, head dept., 1980-89; sr. lectr. Nottingham Trent U., 1989—; lectr. U. Sheffield, 1989-91, Nimbas, Utrecht, The Netherlands, 1991—. Author: Economics, 1986, Economics Revision Guide, 1989, Introductory Economics, 1992. Mem. U.K. Econs. Assn. Avocations: chess, running, squash.

HARRISON, CHARLES WAGNER, JR., applied physicist; b. Farmville, Va., Sept. 15, 1913; s. Charles Wagner and Etta Earl (Smith) H.; m. Fern F. Perry, Dec. 28, 1940; children—Martha R., Charlotte J. Student, U.S. Naval Acad. Prep. Sch., 1933-34, U.S. Coast Guard Acad., 1934-36; BS in Engring., U. Va., 1939, EE, 1940; SM, Harvard U., 1942, M in Engring., 1952, PhD in Applied Physics, 1954; postgrad., MIT, 1942, 52. Registered profl. engr., Va., Mass. Engr. Sta. WCHV, Charlottesville, Va., 1937-40; commd. ensign U.S. Navy, 1939, advanced through grades to comdr., 1948; research staff Bur. Ships, 1939-41, asst. dir. electronics design and devel. div., 1948-50; research staff U.S. Naval Research Lab., 1944-45, dir.'s staff, 1950-51; liaison officer Evans Signal Lab., 1945-46; electronics officer Phila. Naval Shipyard, 1946-48; mem. USN Operational Devel. Force Staff, 1953-55; staff Comdg. Gen. Armed Forces Spl. Weapons project, 1955-57; ret. U.S. Navy, 1957; cons. electromagnetics Sandia Nat. Labs., Albuquerque, 1957-73; instr. U. Va., 1939-40; lectr. Harvard U., 1942-43, Princeton U., 1943-44; vis. prof. Christian Heritage Coll., El Cajon, Calif., 1976. Author: (with R.W.P. King) Antennas and Waves: A Modern Approach, 1969; contbr. numerous articles to profl. jours. Fellow IEEE (Electronics Achievement award 1966, best paper award electromagnetic compatibility group 1972); mem. Internat. Union Radio Sci. (commn. B. and H), Electromagnetics Acad., Famous Families Va., Sigma Xi. Home: 2808 Alcazar St NE Albuquerque NM 87110-3516

HARRISON, EDWARD THOMAS, JR., chemist; b. Norfolk, Va., Mar. 4, 1929; s. Edward Thomas and Mabel (Weaver) H.; B.S., Va. State U., 1951, M.S., 1958; Ph.D., George Washington U., 1981; m. Bertha Mae Neal, Dec. 30, 1962; children—April, Edward. With NIH, 1959-92, biologist Nat. Inst. Dental Research, 1962-64, chemist, 1964—. Served with AUS, 1951-53. USPHS fellow, 1960-61. Mem. AAAS, Am. Inst. Biol. Scis., Tissue Culture Assn., Va. Acad. Scis., N.Y. Acad. Scis., Fedn. Am. Scientists, Internat. Platform Assn., Brazilian-Am. Cultural Inst., Alpha Phi Alpha. Democrat. Episcopalian. Author articles in field. Home: 438 Quackenbos St NW Washington DC 20011-1308

HARRISON, JAMES OSTELLE, ecologist; b. Harrison, Ga., June 17, 1920; s. James Drew and Marie (Mills) H.; m. Katherine Deal, Jan. 12, 1942 (div. 1970); m. Joyce Rape, Mar. 21, 1971; children: Michael James, Juliet. BA, Mercer U., 1949; MS, U. Ga., 1953; PhD, Cornell U., 1962. Assoc. entomologist United Fruit Co., Palmar Sur, Costa Rica, 1956-62; asst. prof. biology Mercer U., Macon, Ga., 1962-64, assoc. prof., 1964-67, prof., 1967-85, prof. emeritus, 1986—; columnist Macon Telegraph, 1976-80; editorial assoc. Seabreeze Mag., St. Simons Island, Ga., 1985—; ecol. cons. Avland Devel. Co., Macon, 1970-71. Contbr. articles to profl. jours. Chmn. Area Water Quality Adv. Com., Macon, 1979-82, City Energy Adv. Com., Macon, 1979-80; pres. Ocmulgee Monument Assn., Macon, 1981-83. Capt. USAAF, 1942-47, North Africa, ETO. Mem. Nat. Audubon Soc., Nat. Wildlife Fedn. Republican. Baptist. Avocations: photography, hiking, camping, writing. Home: 1179 Matthews Pl Macon GA 31210-3425 Office: Mercer U 1400 Coleman Ave Macon GA 31207-0003

HARRISON, JAMES WILBURN, gynecologist; b. Martin, Tenn., Mar. 23, 1918; s. Woodie and Georgia Harrison; m. Babs Wise Dudley, Jan. 29, 1948; children: James Wilburn, James Michael, Babs Suzanne, Linda Denise. Student, U. Tenn., Martin, 1936-37, U. Tenn., Knoxville, 1937-38; MD, U. Tenn., Memphis, 1941. Diplomate Am. Bd. Ob-gyn. Asst. resident Brooke Gen. Hosp., Ft. Sam Houston, Tex., 1947; resident, sr. resident Letterman Gen. Hosp., San Francisco, 1949-51; commd. officer U.S. Army, 1943, advanced through grades to col., ret., 1954; chief staff St. Michael Hosp., Texarkana, Ark., Wadley Regional Med. Ctr., Texarkana, Tex., So. Clinic, Texarkana, Ark. Chmn. Bowie County Child Welfare Bd.; mem. N.E. Tex. Mental Health Bd. Decorated Army Commendation medal, Legion of Merit. Fellow ACS (life), ICS (life), Am. Coll. Ob-gyn. (life), Assn. Mil. Surgeons U.S. (life), Tex. Soc. Ob-gyn. (life); mem. AMA (life), Tex. Med. Assn. Methodist. Avocations: collecting, travel, military history. Home: 4009 Pecos St Texarkana TX 75503-2857 Office: So Clinic 300 E 6th St Texarkana AR 75502-5296

HARRISON, MARCIA ANN, biology educator; b. Rutland, Vt., July 26, 1955; d. Richard James and Catherine (Foley) H.; m. Gregory M. Pitaniello, July 8, 1978; children: Richard, Michael. BS, U. Vt., 1977; MS, U. Mich., 1978, PhD, 1983. Lectr. Washington U., St. Louis, 1983, NASA rsch. assoc., 1983-86; asst. prof. biology Marshall U., Huntington, W.Va., 1986-92, assoc. prof., 1992—; cons. editor McGraw/Hill Ency. of Sci. and Tech., 1987-92; cons. scientist C.E., Huntington, 1990. Contbr. articles to profl. jours. Mem. Am. Soc. Plant Physiologists, Am. Soc. for Space and Gravitational Biology, W.Va. Acad. Sci. Office: Marshall U 3d Ave Huntington WV 25755

HARRISON, ROBERT HUNTER, psychology educator; b. Manapay, Sri Lanka, Oct. 21, 1929; came to U.S., 1944; s. Max Hunter and Minnie (Hastings) H.; m. Judy Feldman, June 28, 1959 (dec. May, 1968)l children: David, Amy; m. Virginia Ann Johnson, Jan. 21, 1984; 1 child, Anna. BA, Oberlin Coll., 1951; PhD, Pa. Steate U., 1957. Lic. psychologist, Mass. Rsch. assoc. MIT, Cambridge, 1955-57; post doctoral fellow Mass. Mental Health Ctr., Boston, 1957-59; rsch. assoc. Harvard Med. Sch., Boston, 1959-87; asst. prof. U. Mass., 1962-66; assoc. prof. Boston U., 1966—; cons. Rsch. Inst. Ednl. Problems, Cambridge, 1965—, Headache Found., Boston, 1973-85; consulting editor Jour. Dreaming, Boston, 1988—. Contbr. over 30 articles to profl. jours. Grantee NIMH, 1959-62, 1967, 1992. Democrat. Mem. Soc. Friends. Office: Boston U Psychology Dept 64 Cummington St Boston MA 02215

HARRISON, ROBERT VICTOR, auditory physiologist; b. Bristol, Somerset, Eng., Oct. 11, 1951; s. Leonard Stanley and Marion Eugenie (Oram) H.; m. Debra Jean-Marie Bertollo, Oct. 25, 1986; children by previous marriage: Victoria May, Danielle Zoe; children by present marriage: Brittany Jade, Adrienne Lee. BSc with honors, U. Birmingham, U.K., 1973; PhD, U. Keele, 1978; DSc, U. Birmingham, 1991. Attache de recherche INSERM, France, 1981-83, charge de recherche, 1983-84; assoc prof. U. Toronto, 1984-89; sr. scientist dept. otolaryngology Hosp. for Sick Children, Toronto, Ont., 1984—; prof. dept. otolaryngology and physiology U. Toronto, 1989—; dir. cochlear implant program Hosp. for Sick Chidlren, Toronto, 1989—; pres. Ont. Derfress Rsch. Found., 1990-92. Author: The Biology of Hearing and Deafness, 1988; contbr. over 150 articles to profl. jurs. Z.W.O. travelling fellow, Holland, 1982, Dale Fund, London Physiol. Soc., 1978. Mem. Acoustical Soc. Am., Soc. for Neurosci., Assn. for Rsch. in Otolaryngology, Collegium Oto-Rhino-Laryngologicum Amicitae Sacrum, Sigma Xi. Achievements include research on animal models of hearing loss. Office: Hosp for Sick Children, 555 University Ave, Toronto, ON Canada M5G 1X8

HARRISON, WALTER ASHLEY, physicist, educator; b. Flushing, N.Y., Apr. 26, 1930; s. Charles Allison and Gertrude (Ashley) H.; m. Lucille Prince Carley, July 17, 1954; children: Richard Knight, John Carley, William Ashley, Robert Walter. B. Engring. Physics, Cornell U., 1953; M.S., U. Ill., 1954, Ph.D., 1956. Physicist Gen. Elec. Research Labs., Schenectady, 1956-65; prof. applied physics Stanford(Calif.) U., 1965—, chmn. applied physics dept., 1989-93. Author: Pseudopotentials in the Theory of Metals, 1966, Solid State Theory, 1970, Electronic Structure and the Properties of Solids, 1980; editor: The Fermi Surface, 1960. Guggenheim fellow, 1970-71; recipient von Humboldt Sr. U.S. Scientist award, 1981; vis. fellow Clare Hall, Cambridge U., 1970-71. Fellow Am. Phys. Soc. Home: 817 San Francisco Ct Stanford CA 94305 Office: Stanford U Dept Applied Physics Stanford CA 94305

HARRISON, WILLARD W., chemist, educator; b. McLeansboro, Ill., July 28, 1937. BA, So. Ill. U., 1958, MA, 1960; PhD, U. Ill., 1964. Asst. prof. chemistry U. Va., 1964-69, assoc. prof. chemistry, 1969-74; vis. sci. Max-Planck Inst. for Plasmaphysics, Munich, West Germany, 1975-76; chmn. dept. chemistry U. Va., 1978-81; sr. rsch scholar Fulbright-Hayes U. Paris, 1981; vis. scholar dept. chemistry Stanford U., 1982; assoc. provost foe acad. support U. Va., 1982-86; vis. scholar dept. geology Stanford U., 1987-88; prof. chemistry U. Va., 1974-88; dean, Coll. Liberal Arts and Scis. U. Fla., 1988—. Recipient Lester W. Strock award Soc. for Applied Spectrscopy, Frederick, Md., 1993. Office: Univ of Florida Dept of Chemistry Gainesville FL 32611*

HARRUFF, LEWIS GREGORY, industrial chemist; b. Marion, Ohio, Apr. 18, 1947; s. Ralph Lewis and Opal Marie (Hobson) H.; m. Maria De Los Angeles Vazquez, July 1, 1978. BS, Wayne State U., 1972; PhD, U. Oreg., 1977. Lab. scientist ITT Corp., Shelton, Wash., 1977-82; prodn. mgr. Enzyme Systems Products, Pleasanton, Calif., 1982-85; sr. scientist Saudi Aramco, Dhahran, Saudi Arabia, 1985—. Contbr. articles to profl. jours. Jour. Am. Chem. Soc., Prehydrolysis Soda AQ Pulping, Tappi. Mem. Am. Chem. Soc., N.Y. Acad. Scis. Achievements include development of new method of environmentally sound wood pulping, of process improvements in pulping medical diagnostics and petroleum refining; patent for One Pot Conversion of Vanillin to 5-Hydroxyvanillin. Home: PO Box 9053, Dhahran Saudi Arabia Office: Saudi Aramco, PO Box 62, Dhahran Saudi Arabia

HARSHBARGER, JOHN CARL, pathobiologist; b. Weyers Cave, Va., May 9, 1936; s. John Carl and Alma (Baker) H. BA, Bridgewater Coll., 1957; MS, Va. Poly. Inst., 1959; PhD, Rutgers U., 1962. Postdoctorate in insect pathology USDA, Beltsville, Md., 1962-64; researcher in pathobiology Univ. Calif., Irvine, 1964-67; dir. registry of tumors in lower animals Smithsonian Instn., Washington, 1967—. Co-editor: National Cancer Institute Monograph 31, 1969, Annals of the New York Academy of Science 298, 1977; contbr. more than 100 articles on diseases of invertebrate and cold blooded vertebrate animals to profl. jours. Fellow AAAS; mem. Am. Assn. Cancer Rsch., Soc. for Invertebrate Pathology (pres. 1987-89), Cosmos Club. Home: 4501 8th St S Arlington VA 22204 Office: Smithsonian Instn NHB W216A MRC 163 Washington DC 20560

HART, BARRY THOMAS, environmental chemist; b. Rochester, Australia, Aug. 5, 1940; s. Richard J. and Lillias A. (Whalebone) H.; m. Margaret Frances Dixon, Aug. 24, 1962; children: Trudy, Peter, David, Michael, Ashley. BS with honors, Monash U., 1968, PhD, 1972. Dir. Water Studies Ctr. Monash U., Melbourne, Australia, 1980—. Fellow Royal Australian Chem. Inst. Office: Water Study Univ, PO Box 197, Caulfield East 3145, Australia

HART, DEAN EVAN, research optometrist; b. Westbury, N.Y., Nov. 4, 1957; s. Ronald Warren and Beatrice (Rosenblitt) H. AAS in Ophthalmic Dispensing, Erie Community Coll., 1980; BS in Gen. Studies, N.Y. Inst. Tech., 1981; MA in Biology, Hofstra U., 1983; postgrad. researcher, U. Calif., Berkeley, 1985; OD, SUNY Coll. Optometry, N.Y.C., 1987. Lic. in ophthalmic dispensing and contact lens fitting, N.Y.; lic. optician, Fla.; cert. Am. Bd. Opticianry. Dir. Low Vision Clinic, instr. refraction and optics Columbia U. at Harlem Hosp. Med. Ctr., N.Y.C., 1988—; assoc. rsch. scientist Harkness Eye Inst., Columbia U. Coll. Physicians and Surgeons, 1989—; dir., asst. med. advisor Tristate Consumer Ins. Co.; numerous presentations in field. Contbr. articles and abstracts to sci. jours., chpts. to books. Grantee Optometric Ctr. N.Y., 1984, Sola-Barnes-Hind, 1984, 85; travel fellow Nat. Eye Inst., 1985. Fellow Am. Acad. Optometry; mem. Am. Optometric Assn., Contact Lens Assn. Ophthalmologists, Internat. Soc. for Contact Lens Rsch., Assn. for Rsch. in Vision and Ophthalmology, N.Y. Acad. Sci., Contact Lens Soc. N.Y. State. Home and Office: Woodbury Optical Group 185 Woodbury Rd Hicksville NY 11801-3029

HART, HELEN MAVIS, planetary astronomer; b. Billings, Mont., Nov. 1, 1954; d. Howard Malcolm and Ardis Pearl (Reiner) H. BA in Physics, U. Mont., 1981; PhD in Astrophysical Planetary and Atmospheric Scis., U. Colo., 1989. Driver Perkle Refridgerated Freight Lines, Madison, Wis., 1976-77, Internat. Transport Co., Rochester, Minn., 1977-78; tutor physics U. Mont., Missoula, 1980-82; ops. astronomer Space Telescope Sci. Inst., Balt., 1989—; dir. local arrangements Case for Mars II Conf., Boulder, 1984; chair com. Implication of SDI for Nat. Security debate, Boulder, 1987. Mentor Boulder Valley Schs. Talented and Gifted Program, 1985-86. Recipient Mars prize Planetary Soc., 1984, 85; Fox scholar U. Mont., 1979, 80; fellow U. Colo., 1984, 85. Mem. Am. Astronomical Soc., Am. Geophys. Soc. (June Bacon-Bercy scholar 1982), Pi Mu Epsilon. Office: Space Telescope Sci Inst 3700 San Martin Dr Baltimore MD 21218

HART, JEAN HARDY, information systems specialist, consultant; b. Cleve., Jan. 19, 1942; d. Gilbert Elliott and Jessie (Peterson) Brown; m. Richard Pierpont Thomas, June 16, 1962 (div. Sept. 1974); children: Perry Glenn, Geb Weller, Hans Richard; m. Howard Phillips Hart, Jan. 19, 1988; stepsons: Colin, Guy. BA, Goddard Coll., 1973. Tech. communicatiaons specialist The Mitre Corp., Boston, 1973-75; site adminstr., tech. communications expert The Mitre Corp., Madrid, 1975-77; mgr. internat. programs Honeywell Info. Systems, Newton, Mass., 1977-78; account mgr. sales and contracts Honeywell Info. Systems, 1978-79, tech. analyst, 1979-81; dist. mgr. Europe Honeywell Info Systems, 1981-84; mgr. third party svc. Honeywell Info. Systems, Boston, 1984-85, 85-86; resident mgr. Honeywell Info. Systems, Beijing, 1985; dir. fed. accounts Honeywell Info. Systems, McLean, Va., 1986-88; bus. advisor CIA, McLean, Va., 1989-90; dir. Hartwell Mgmt., Ltd., London and Keezletown, Va., 1991—; internat. bus. cons. U.S. Govt., Washington, 1991—; info. profl. various cos., Washington, 1991—; lectr. numerous worldwide colls. and govt. orgns., 1968—; strategy cons. to provost Coll. Integrated Sci. and Tech., James Madison U., Harrisonburg, Va., 1992. Author, producer: (videotape) Chief Justice Warren Burger, 1987; contbr. articles to profl. jours. Mem. AAUW, Am. Assn. Info. Profls. Episcopalian. Avocations: woodworking, writing, sewing. Home and Office: Hartwell Mgmt Ltd PO Box 80 Dyke VA 22935

HART, JOHN AMASA, wildlife biologist, conservationist; b. Amery, Wis., May 3, 1950; s. Nathaniel I. and Joanne (Velz) H.; m. Terese Kay Butler, June 12, 1977; 3 children. BA, Carleton Coll., 1972; PhD, Mich. State U., 1986. Assoc. zoologist Wildlife Conservation Internat.-N.Y. Zool. Soc., Bronx, 1986—. Achievements include long-term study and conservation of rain forests of Zaire, first studies of Okadi, African rain forest giraffe, in wild. Office: Wildlife Conservation Inst NY Zool Soc 185th St at Southern Blvd Bronx NY 10460

HART, JOHN FINCHER, construction management company executive; b. Amarillo, Tex., Feb. 27, 1937; s. Phil Charles and Ethel (Fincher) H.; m. Marilyn Kegg, 1962 (div. 1972); children: Christopher H., Karen A. Wendy L.; m. Patricia Eck, June 1978. BS in Indsl. Mgmt., Ga. Tech., 1961; MBA in Corp. Fin. Mgmt., Ga. State U., 1978. Registered profl. engr., Calif. Jr. civil engr. Western Pacific R.R., San Francisco, 1961-64; assoc. transp. engr. Calif. Pub. Utility Commn., San Francisco, 1964-66; project engr. RMK-BRJ Constructors, South Vietnam, 1966-69; sr. field engr. Parsons Brinckerhoff-Tudor-Bechtel, San Francisco, 1969-72; mgr. project controls Parsons Brickerhoff-Tudor-Bechtel, San Francisco, 1975-81; chief adminstrn. engr. Christenson & Foster Contractors, Santa Rosa, Calif., 1972-74; project mgr. Bechtel Civil, Las Vegas, Nev., 1981-87; sr. project mgr. Parsons Brinckerhoff Constrn. Svc., Herndon, Va., 1987—; guest lectr. Ga. State U., Atlanta, 1978-81, U. Nev., Las Vegas, 1983-87, Am. U., Washington, 1988; bus. feasiblity cons., Las Vegas, 1986-87; vocat. speaker various high schs., N.Y., 1990. Contbr. articles to profl. jours. Co. coord. United Way, Atlanta, 1980. With U.S. Army Res., 1955-62. Mem. ASCE, Am. Soc. Mil. Engrs., Mensa. Achievements include devel. of major claims program for Atlanta Rapid Transit System; devel. of CPM Systems Procedures for Christensen & Foster Contractors; design of standard rock-crushing system for use throughout South Vietnam; devel. of contractor quality control/owner quality assurance program for NE region U.S. postal service contracts; devel. of innovative railroad construction survey method enabling one survey crew to take the place of four planned crews; design and construction management of 4 major airports. Home: 4805 Fincher Rd Del Valle TX 78617 Office: New Airport Project Team Bldg 4218 Ave F Austin TX 78719

HART, ROBERT GERALD, physiology educator; b. Cumberland, Md., Feb. 20, 1937; s. Harry and Margaret (Welsh) H.; m. Claudia Dawn Heichel, Jan. 11, 1986; children: Robert, Kevin, Scott, Steven. BS, Duquesne U., 1963, MS, 1965; PhD, U. Ill., 1967. Assoc. prof. Slippery Rock (Pa.) U., 1967-72, prof., 1973—; rsch. assoc. U. Ill., Champaign-Urbana, summer 1969, vis. scientist, summers 1974, 75, 76, 78; tchr. nursing program Pa. State U., Sharon, 1982, 87, 88; cons. St. Francis Coll., Loretto, Pa., 1989. Reviewer Biology of Reproduction, 1972—; contbr. articles to Biology of Reproduction, BioScience, Am. Jour. Obstetrics and Gynecology. Recipient Travel grants NIH, Paris, 1969, Munich, 1972. Mem. Sigma Xi (grant in aid 1968-69). Office: Dept Biology Slippery Rock Univ Slippery Rock PA 16057

HART, STANLEY ROBERT, geochemist, educator; b. Swampscott, Mass., June 20, 1935; s. Robert Winfield and Ruth Mildred (Standley) H.; m. Joanna Smith, Sept. 1, 1956 (div. Dec. 1978); 1 dau., Jolene Kaweah; m. Pamela Coulouras Shepherd, Nov. 4, 1980; children—Elizabeth Ann, Nathaniel Charles. B.S., Mass. Inst. Tech., 1956, Ph.D., 1960; M.S., Calif. Inst. Tech., 1957. Staff mem. Carnegie Instn., Washington, 1960-75; prof. dept. earth and planetary sci. Mass. Inst. Tech., Cambridge, 1975-89; sr. scientist Woods Hole (Mass.) Oceanographic Instn., 1989—; mem. U.S. Nat. Com. for Geochemistry, 1973-76, chmn., 1975; mem. ocean crust panel Internat. Phase of Ocean Drilling, 1974-76; mem. U.S. nat. com. Internat. Geol. Correlations Program, 1974-76. Assoc. editor: Jour. Geophys. Rsch., 1966-68, Revs. of Geophysics, 1970-72, Geochimica et Cosmochimica Acta, 1970-76; editorial bd.: Physics of the Earth and Planetary Interiors, 1977-92, Earth and Planetary Sci. Letters, 1977-87, Chem. Geology, 1985—; contbr. articles in field to profl. jours. Fellow Geol. Soc. Am., Am. Geophys. Union; mem. NAS, Geochem. Soc. (councillor 1981-83, v.p. 1983-85, pres. 1985-87, V.M. Goldschmidt award 1992). Home: 53 Quonset Rd Falmouth MA 02540-1656 Office: Woods Hole Ocean Instn Dept Geology & Geophysics Woods Hole MA 02543

HART, TERRY JONATHAN, communications executive; b. Pitts., Oct. 27, 1946; s. Jonathan Smith Hart and Lillian Dorothy (Zugates) Hart Pierson; m. Wendy Marie Eberhardt, Dec. 20, 1975; children: Amy, Lori. B of Mech. Engring., Lehigh U., 1968, DEng (hon.), 1988; MS, MIT, 1969; MEE, Rutgers U., 1978. Mem. tech. staff AT&T Bell Labs., Whippany, N.J., 1968-69, 73-78, supr., 1984—, head cellular systems strategic planning, 1989—; astronaut NASA Johnson Space Ctr., Houston, 1978-84; captured solar maximum satellite NASA Johnson Space Ctr., 1984, div. mgr. Telstar 4 Satellite Program. Patentee in field. Served to lt. col. USAF Air N.G., 1969—. Recipient N.J. Disting. Service medal, NASA Space Flight medal, Pride of Pa. medal. Mem. IEEE, Sigma Xi, Tau Beta Pi. Avocations: skiing; golf. Office: AT&T Rte 202/206N Rm 2A127 Bedminster NJ 07921

HARTE, REBECCA ELIZABETH, computer scientist, consultant; b. Camp LeJeune, N.C., Jan. 28, 1967; d. Franklin James and Rebecca Irene (Adams) H. BA in Maths., Hollins Coll., 1989. Math. asst. Hollins Coll., Roanoke, Va., 1987-89; chief information officer BRTRC, Vienna, Va., 1989—. Mem. AAAS, Math. Assn. Am., Am. Math. Soc., Assn. for Computing Machinery, Phi Beta Kappa, Sigma Xi. Home: 5200 Olley Ln Burke VA 22015-1747 Office: BRTRC 8260 Willow Oaks Court Fairfax VA 22031

HÄRTEL, CHARMINE EMMA JEAN, industrial/organizational psychology educator, consultant; b. Anchorage, Oct. 9, 1959; d. Charles Marion and Addleen Betty Jean (Bosket) Crane; m. Günter Franz Härtel, June 28, 1980; 1 child, Jameson Brice Günter. BA in Psychology summa cum laude, U. Colo., 1986; MA, PhD, Colo. State U., 1991. Asst. prof. U. Tulsa, 1991—; cons. Team Performance Lab., U. Ctrl. Fla., Orlando, 1992, Nova Engring, Tulsa, 1992, DeLine Box Co., Windsor, Colo., 1990, City of Ft. Collins, Colo., 1987-90; contractor Naval Tng. Systems Ctr. Human Factors Div., Orlando, 1990-91, Civil Aeromed. Inst., Tng. and Orgnl. Rsch. Lab., Oklahoma City, 1992. Mem. editorial bd. The Indsl.-Orgnl. Psychologist, 1992—; contbr. articles to profl. jours. Recipient Richard M. Suinn Commendation award for excellence in rsch. and the advancement of psychology Colo. State U., 1990, Martin E.P. Seligman Applied Rsch. award Colo. State U., 1990; postgrad. rsch. appointee Naval Tng. Systems Ctr., 1990. Mem. APA, Acad. Mgmt. (personnel rsch. div., orgnl. behavior div.), Am. Psychology Soc., Soc. for Indsl. Orgnl. Psychology, Psi Chi. Achievements include development of Problem Identification Validation decision making strategy used in Navy aircrew training programs. Office: Univ Tulsa Dept Psychology Tulsa OK 74104-3189

HARTER, DAVID JOHN, radiation oncologist; b. Milw., Apr. 12, 1942; s. Herbert George and Marion Bertha (Kahl) H.; m. Diane Leigh Kuebler; children: Renée, Andrew, Susannah Lee. BA, U. Wis., Milw., 1964; MD, U. Wis., Madison, 1968. Diplomate Am. Bd. Radiology. Dir. radiation oncology Immanuel Radiation Treatment Ctr., Omaha, 1979—, chmn. dept. radiology, 1989—; CEO Midwest Cancer Care, P.C., 1991—; asst. clin. prof. radiology U. Nebr. Coll. Medicine, Omaha, 1978—; pres. Harter Land and Lumber Co., Green County, Va., 1986—; dir. Panasiatic Corp., Seattle, 1988—. Pres. Nebr. div. Am. Cancer Soc., 1983-84, bd. dirs., 1980—; pres. Omaha Pub. Art Commn.; mem. adv. bd. U. Nebr. Sheldon Meml. Art Gallery, 1989—. Mem. AMA, Am. Coll. Radiology, Am. Soc. Therapeutic Radiologists, Am. Radium Soc., Am. Soc. Clin. Oncology, Gilbert H. Fletcher Soc., Masons. Club: Omaha Country, Omaha; Doctors of Houston. Home: 9927 Essex Rd Omaha NE 68114-3873 Office: Immanuel Med Ctr 6901 N 72d St Omaha NE 68122

HARTER, DONALD HARRY, research administrator, medical educator; b. Breslau, Germany, May 16, 1933; came to U.S., 1940; naturalized, 1945; s. Harry Morton and Leonor Evelyne (Goldmann) H.; m. Lee Grossman, Dec. 18, 1960 (div. 1976); children: Kathryne, Jennifer, Amy, David; m. Rikki Horne, May 18, 1985 (div. 1986); m. Marjorie Brandt Dahlin, Oct. 12, 1990. A.B., U. Pa., 1953; M.D., Columbia U., 1957. Diplomate: Am. Bd. Psychiatry and Neurology. Intern in medicine Yale-New Haven Med. Center, 1957-58; asst. resident, then resident neurology N.Y. Neurol. Inst., 1958-61; guest investigator Rockefeller U., 1963-66; mem. faculty Columbia Coll. Physicians and Surgeons, 1960-75, prof. neurology and microbiology, 1973-75; vis. fellow Clare Hall, Cambridge, Eng., 1973-74; attending neurologist N.Y. Neurol. Inst., Presbyn. Hosp., 1973-75; Charles L. Mix prof. Northwestern U., 1975-85, Benjamin and Virginia T. Boshes prof. neurology, 1985-88, chmn. dept. neurology, 1975-87; chmn. dept. neurology Northwestern Meml. Hosp., Chgo., 1975-87; dir. rsch. scholars program Howard Hughes Med. Inst./NIH, Bethesda, 1989—; vis. sci. officer Howard Hughes Med. Inst., 1986-87, sr. sci. officer, 1987—; clin. prof. neurology George Washington U. Sch. Medicine and Health Scis., 1987—; mem. adv. com. on fellowships Nat. Multiple Sclerosis Soc., 1976-79, chmn., 1977-79, rsch. programs adv. com., 1989—; mem. Nat. Commn. on Venereal Disease, HEW, 1970-72; mem. med. adv. bd. Am. Parkinson Disease Assn., 1976-90, Myasthenia Gravis Found., 1980-87; mem. sci. adv. coun. Nat. Amyotrphic Lateral Sclerosis Found., 1978-85; mem. sci. rev. com. Amyotrophic Lateral Sclerosis Assn., 1987-91, chmn. 1989-91; mem. bd. sci. counselors Nat. Inst. Dental Rsch. NIH, 1990—; sr. sci. advisor Amyotrophic Lateral Sclerosis Assn., 1992—. Mem. editorial bd. Neurology, 1976-82, Anns. of Neurology, 1983-89; mem. adv. bd. Archives of Virology, 1975-81. Recipient Joseph Mather Smith prize Columbia U., 1970, Lucy G. Moses award, 1970, 72; Am. Cancer Soc. scholar, 1973-74; USPHS spl. fellow, 1963-66, Guggenheim fellow, 1973. Mem. Am. Soc. Clin. Investigation, Am. Neurol. Assn., Assn. Univ. Profs. Neurology, Infectious Disease Soc. Am., Am. Acad. Neurology, Am. Soc. Microbiology, Am. Assn. for History of Medicine, Am. Soc. Virology, Royal Soc. Medicine (U.K.), Cosmos Club, Yale Club N.Y.C., Univ. Club Chgo., Univ. Club Washington, Phi Beta Kappa, Sigma Xi. Home: 2475 Virginia Ave NW Apt 503 Washington DC 20037-2639 Office: Howard Hughes Med Inst 4000 Jones Bridge Rd Chevy Chase MD 20815-6789 also: Howard Hughes Med Inst 1 Cloister Ct Bethesda MD 20814

HARTL, DANIEL LEE, genetics educator; b. Marshfield, Wis., Jan. 1, 1943; s. James W. and Catherine E. (Stieber) H.; m. Carolyn Teske, Sept. 5, 1964 (div. Apr. 1978); children: Dana Margaret, Theodore James; m. Christine Blazynski, July 23, 1980; 1 child, Christopher Lee. BS, U. Wis., 1965, PhD, 1968. Postdoctoral fellow in genetics U. Calif., Berkeley, 1968-69; asst. prof. genetics and cell biology U. Minn., St. Paul, 1969-73, assoc. prof., 1973-74; assoc. prof. biol. scis. Purdue U., West Lafayette, Ind., 1974-78, prof., 1978-81; prof. genetics Washington U. Sch. Medicine, St. Louis, 1981-92, James S. McDonnell prof. genetics, head genetics dept., 1984-91, dir. div. biology and biomed. scis., 1986-89; prof. biology Harvard U., Cambridge, Mass., 1993—; mem. genetics study sect. NIH, Washington, 1976-80, genetic basis of disease rev. com. NIH, 1983-87. Author: Principles of Population Genetics, 1980, Human Genetics, 1983, General Genetics, 1985, Primer of Population Genetics, 1988, Basic Genetics, 1988; assoc. editor Ann. Revs. Inc., 1984-89, Molecular Biology and Evolution, 1983—, Molecular Phylogenetics and Evolution, 1993—, Molecular Ecology, 1993—, Genetics, 1977-85, BioSci., 1974-80, Theoretical Population Biology, 1975-81, Molecular Ecology, 1992—, Molecular Phylogeretics and Evolution, 1992—. Recipient Career Devel. award NIH, 1974-79. Mem. Genetics Soc. Am. (pres. 1989), Phi Beta Kappa. Office: 219 Biological Labs Harvard U 16 Divinity Ave Cambridge MA 02138

HARTLEY, JAMES MICHAELIS, aerospace systems, printing and hardwood products manufacturing executive; b. Indpls., Nov. 25, 1916; s. James Worth and Bertha S. (Beuke) H.; m. E. Lea Cosby, July 30, 1944; children: Michael D., Brent S. Student Jordan Conservatory of Music, 1934-35, Ind. U., Purdue U., Franklin Coll. With Arvin Industries, Inc., 1934-36; founder, pres. J. Hartley Co., Inc., Columbus, Ind., 1937—; founder, pres. Hartley Group, 1989—. Pres. Columbus Little Theatre, 1947-48; founding dir. Columbus Arts Guild, 1960-64, v.p., 1965-66, dir., 1971-74; musical dir., cellist Guild String Quartet, 1963-73; active Indpls. Mus. of Art; founding dir. Columbus Pro Musica, 1969-74; dir. Regional Arts Study Commn., 1971-74; v.p. Ind. Coun. Rep. Workshops, 1965-69, pres., 1975-77; pres. Bartholomew County Rep. Workshop, 1966-67. Served with USAAF, 1942-46. Mem. AAAS, NAM, Nat. Fedn. Ind. Bus., U.S. C. of C., Phi Eta Sigma (honoris causa). Office: J Hartley Co Inc 101 N National Rd Columbus IN 47201-7848

HARTLEY, JOE DAVID, communications specialist; b. Hickory, N.C., Jan. 6, 1959; s. Ralph Ronald and Evelyn Ines (Reinyhardt) H.; m. Janice Theresa Bailey, June 1, 1979; children: Nicholas, Nathaniel, Christina, Jonathan. AAS, Catawba Valley Tech. Inst., Hickory, 1979; student, U. N.C. Cert. electronics technician. Technician Dixie Radio and TV, Hickory, 1975-79; comm. specialist Duke Power Co., Charlotte, N.C., 1979—. Baptist. Home: 210 Dutchman Dr Mount Holly NC 28120 Office: Duke Power Co 6733 Craig St Charlotte NC 28214

HARTLINE, BEVERLY KARPLUS, physicist, science educator; b. Princeton, N.J., June 13, 1950; d. Robert and Elizabeth Jane (Frazier) Karplus; m. Frederick Flanders Hartline, Apr. 15, 1972; children: Jason, Jeffrey. BA in Chemistry/Physics, Reed Coll., 1971; PhD in Geophysics, U. Wash., 1978. Vis. asst. prof. Hampshire Coll., Amherst, Mass., 1977-78; rsch. news writer AAAS, Washington, 1978-80; phys. scientist NASA-Goddard Space Flight Ctr., Greenbelt, Md., 1980-82; staff scientist Lawrence Berkeley Lab., Berkeley, Calif., 1983-85; asst. dir. Continuous Electron Beam Acceleration Facility, Newport News, Va., 1985-89, assoc. dir., project mgr., 1989—; mem. math. dept. adv. com. Christopher Newport U., Newport News, 1992—; coord. mem. NASA working group, 1980-82, 81-82. Contbr. articles to profl. jours. Mem. Va. Adv. Com. on Girls Edn., 1992—. Recipient Woodrow Wilson award, 1971, Quality award NASA, 1982. Mem. AAAS, Am. Phys. Soc., Project Mgmt. Inst., Am. Women in Sci. Achievements include serving as project mgr. of CEBAF accelerator and physics equipment constrn. project; serving as dir. of precoll. edn. programs at CEBAF. Office: SURA/CEBAF 12000 Jefferson Ave Newport News VA 23606

HARTMAN, CHARLES EUGENE, nuclear engineer; b. Spangler, Pa., Sept. 2, 1945; s. Kenneth E. and Betty L. (Parry) H.; m. Betty L. Craver, Mar. 23, 1968; children: Gregory, Gary. AA, Pa. State U., DuBois, 1965; B Engring. Tech., Pa. State U., Middletown, 1970. Registered profl. engr., Pa. Elec. engr. GPU Nuclear, Middletown, 1970-78, sr. engr., 1978-85, mgr. engring., 1985—; sr. reactor oper. U.S. Nuclear Regulatory Commn., Middletown, 1978-79; tech. adv. com. Pa. State U.-Harrisburg, Middletown, 1990-92. With U.S. Army, 1965-67. Mem. IEEE. Office: GPU Nuclear PO Box 480 Middletown PA 17057

HARTMAN, DAVID ROBERT, chemistry educator; b. Streator, Ill., Sept. 17, 1940; s. Robert William Sr. and Georgeia Gertrude (Hopper) H.; m. Sandra Ann Davis, July 1, 1967; children: Timothy James, Stephen Vincent. BS, North Ctrl. Coll., Naperville, Ill., 1962; MS, Va. Poly. Inst. and State U., 1964, PhD, 1967. Asst. prof. dept. chemistry Western Ky. U., Bowling Green, 1966-73, assoc. prof., 1973-90, prof., 1990—; with health careers opportunity program U. Ky., Lexington, summers 1984, 85; grant rev. com. Am. Heart Assn., Ky. affiliate, Louisville, 1983-92. Scoutmaster Boy Scouts Am., Bowling Green, 1980—. Mem. Am. Chem. Soc., Ky. Acad. Sci. (treas. 1991-93), Sigma Xi, Phi Lambda Upsilon, Phi Kappa Phi. Methodist. Office: Western Ky U Dept Chemistry Bowling Green KY 42101

HARTMAN, HOWARD LEVI, mining engineering educator, consultant; b. Indpls., Aug. 7, 1924; s. Howard Levi and Catherine Gladys (Miller) H.; m. Bonnie Lee Sherrill, June 8, 1947; children: Sherilyn Hartman Knoll, Greg Alan. Student, Colo. Sch. Mines, 1942-44; BS, Pa. State U., 1946, MS, 1947; PhD, U. Minn., 1953. Registered profl. engr., Colo., Pa. Instr. Pa. State U., University Park, 1947-48, prof., dept. head and assoc. dean, 1957-67; mining engr. Phelps Dodge Corp., Bisbee, Ariz., 1948-49; state mine dust engr. Mine Inspector's Office, Phoenix, 1949-50; instr. U. Minn., Mpls., 1950-54; from asst. prof. to assoc. prof. Colo. Sch. Mines, Golden, 1954-57; prof., dean Calif. State U., Sacramento, 1967-71, Vanderbilt U., Nashville, 1971-80; Drummond endowed chair mining engring. U. Ala., Tuscaloosa, 1980-89; adj. prof. engring. Calif. State U., Sacramento, 1989—; chmn. Fed. Metal and Nonmetallic Mine Safety Bd. Rev., Washington, 1971-75; Warren lectr. U. Minn., 1965; Disting. lectr. Can. Inst. Mining and Metallurgy, 1966; cons. engr. Standard Oil N.J., Tulsa, 1961-64, Ingersoll-Rand Co., Bedminster, N.J., 1964-66, Inst. for Technol. Research, São Paulo, Brazil, 1977, 85, Bechtel Corp., San Francisco, 1982—. Author, editor various books including: Mine Ventilation and Air Conditioning, 1971, 82, Introductory Mining Engineering, 1987; contbr. articles to profl. jours. Mem. Human Rights Commn., Nashville, 1975-79. Served to lt. (j.g.) USNR, 1942-44. Recipient Faculty Service award Nat. Univ. Continuing Edn. Assn., 1985, Howard N. Everson award Soc. Mining, Metallurgy and Exploration, 1992. Mem. Soc. Mining Engrs. (disting. mem., chmn. com., Mineral Industries Edn. award 1965, Book Pub. award 1982, Hartman Mine Ventilation award 1989); mem. Am. Soc. for Engring. Edn., Met. Opera Guild, Sigma Xi, Kappa Sigma. Democrat. Presbyterian (elder). Club: Yosemite Assn. (El Portal, Calif.). Avocations: hiking, opera, color photography. Home: 4052 Alex Ln Carmichael CA 95608-6728 Office: Calif State Univ Sch Engring & Computer Sci 6000 J St Sacramento CA 95819-2605

HARTMAN, JAMES AUSTIN, geologist; b. Lanark, Ill., Jan. 29, 1928; s. Llewelyn John and Gladys Mae (Doyle) H.; m. Zoe Marie Wiley, June 16, 1951; children: Victoria Lynn, Lester James. BS, Beloit (Wis.) Coll., 1951; MS, U. Wis., 1955, PhD, 1957. Registered profl. geol. scientist; cert. petroleum geologist. Geologist Reynolds Jamaica (W.I.) Mines, Jamaica, W.I., 1951-53, Union Carbide Ore Co., Parimaribo, Surinam, 1956-57; various positions Shell Oil Co., New Orleans, 1957-86; cons. New Orleans, 1986—. Bd. mgmt. YMCA, Metairie, 1972-74; pres. Jefferson Com. for Better Schs., Metairie, 1961-63, pres. Westgate PTA, Kenner, La., 1964-65. With U.S. Army, 1946-47. Union Carbide Rsch. fellowship U. Wis., 1954-56. Fellow Geol. Soc. Am.; mem. Am. Assn. Petroleum Geologists (hon., sec. 1981-83, Disting. Svc. award 1985), New Orleans Geol. Soc. (hon., 2d v.p. 1975-76, pres. elect 1984-85, pres. 1985-86, Outstanding Mem. 1977), Gulf Coast Assn. Geol. Socs. (hon., v.p. 1987, pres. 1988), Soc. Petroleum Engrs., Sigma Xi. Republican. Episcopalian. Achievements include research in heavy minerals in Jamaican Bauxite, titanium mineralogy of Bauxites, petroleum geology. Home: 4512 Newlands St Metairie LA 70006-4138 Office: 4300 I-10 Service Rd # 103V Metairie LA 70001-7405

HARTMAN, KEITH WALTER, physicist; b. Flint, Mich., Dec. 27, 1954; s. Raymond A. and Esther Irene (Pieper) H.; m. Robin A. McCloskey, Sept. 10. 1988. BA in Physics, U. Mich., Flint, 1981; PhD in Physics, Pa. State U., 1990. Rsch. assoc. Pa. State U., University Park, Pa., 1990-92, Stanford (Calif.) U., 1992—. Achievements include research in hadronic production of direct photons and neutral pions; research in departures from standard model physics in very rare neutral kaon decays. Office: Stanford University Physics Dept SLAC Bin 63 Palo Alto CA 94309

HARTMAN, KENNETH OWEN, chemist; b. Phila., Oct. 20, 1939; s. Owen Wister and Marie (Mason) H.;; m. Patricia Mamet, Sept. 25, 1965; children: Megan C., Pamela M. BS in Chemistry, Lehigh U., 1961; PhD in Chemistry, Pa. State U., 1965. Rsch. fellow Mellon Inst., Pitts., 1965-67; sr. rsch. chemist Hercules, Inc., Rocket Center, W.Va., 1967-73, supr. propellant chemistry, 1973-82, sr. scientist, 1983-85, tech. mktg. specialist, 1986-89, tech. dir., 1989—. Contbr. articles to profl. jours., confs. Mem. AIAA, Am. Chem. Soc., Am. Def. Preparedness Assn., Combustion Inst. Home: 609 N 2d St Lavale MD 21502 Office: Hercules Inc ABL Box 210 Rocket Center WV 26726

HARTMAN, MARK LEOPOLD, internist, endocrinologist; b. Downey, Calif., May 21, 1956; s. Raymond William and Marlene (Keske) H.; m. Laurie Ann Komornik, Dec. 22, 1979; children: David Raymond, Marian Elizabeth, Lisa Greta. AB, Dartmouth Coll., 1978; MD, U. Conn., 1983. Diplomate Am. Bd. Internal Medicine. Resident in internal medicine Dartmouth-Hitchcock Med. Ctr., Hanover, N.H., 1983-86; fellow in endocrinology U. Va. Health Scis. Ctr., Charlottesville, 1986-88; rsch. asst. prof. medicine U. Va. Sch. Medicine, Charlottesville, 1988-92, asst. prof. medicine, assoc. dir. clin. lab. dept. medicine, 1992—. Contbr. articles to Jour. Clin. Endocrinology and Metabolism, Am. Jour. Physiology, Jour. Clin. Investigation, others. Mem. Christ Community Ch., Charlottesville, 1987—. Grantee NIH, 1988-92, 92—. Mem. ACP, Endocrine Soc., Am. Diabetes Assn.,Am. Fedn. Clin. Rsch., Christian Med. and Dental Soc. Achievements include research in physiology of growth hormone secretion and its metabolic actions, particularly the role of declining growth hormone secretion in the aging process. Office: Univ Va Health Scis Ctr Dept Medicine Box 511 Charlottesville VA 22908

HARTMAN, PHILIP EMIL, biology educator; b. Balt., Nov. 23, 1926; married, 1955; 3 children. BS, U. Ill., 1949; PhD in Med. Microbiology, U. Pa., 1953. Instr. med. microbiology U. Pa., Phila., 1953-54; NIH fellow Carnegie Inst., 1954-55; mem. faculty dept. bacteriology and immunology Harvard U. Med. Sch., Boston, 1955-56; Am. Cancer Soc. fellow animal morphology lab. U. Brussels, 1956-57; from asst. prof. to assoc. prof. biology Johns Hopkins U., Balt., 1957-65, prof., 1965—, now William D. Gill prof. biology. Home: 1604 Ralworth Rd Baltimore MD 21218-2232 Office: Johns Hopkins U 7 Biology E Baltimore MD 21218

HARTMANN, DAVID PETER, electrical engineer, educator, consultant; b. Fond du Lac, Wis., Sept. 7, 1934; s. Arnold Joseph and Stella (Desimowich) H.; m. Margaret Ellen Meier, Dec. 26, 1959; children: Jennifer Susan, Laure Ann, James Arnold, Elizabeth Marie. MSEE, U. Wis., 1956, PhDEE, 1974. Registered profl. engr., Wis. Elec. engr. Allis Chalmers Mfg. Co., West Allis, Wis., 1957-58; systems engr. AC Electronics Div. GMC, Milw., 1958-60; applications engr. Gen. Electric Co., Phoenix, 1960-63, Collins Radio, Cedar Rapids, Iowa, 1963; computer engr. AC Electronics Div. GMC, Milw., 1963-67; program dir. U. Wis., Madison, 1967-74; elec. engr. Bonneville Power Adminstrn., Portland, 1974—; adj. prof. U. Portland, 1981-82; Harmonics instr., U. Wis., Madison, 1984—, U. Tex., Austin, 1990—. Mem. IEEE (sr.), Congress of Rsch. on Large High Voltage Power Systems. Home: 2320 SW Kanan St Portland OR 97201-2040 Office: Bonneville Power Adminstrn PO Box 491 Vancouver WA 98660-0491

HARTMANN, GREGOR LOUIS, technical translator, writer; b. Louisville, Apr. 8, 1951; s. Eugene Edward and Gwendolyn Patricia (Heuser) H.; m. Sally Jean Constable, Dec. 2, 1989. BA in Journalism, U. Ky., 1973. Reporter Fremont (Calif.) Argus, 1974-76; English tchr. Kokusai Kanko Coll., Nagoya, Japan, 1978-80; pres. REM Video, Inc., San Francisco, 1980-83; word processor San Francisco, 1983-89, tech. translator, 1989; tech. translator Chgo., 1990, Bklyn., 1990—. Author short stories pub. in popular mags. Mem. Am. Translators Assn., Nat. Space Soc. (pres. San Francisco chpt. 1988-89), Phi Beta Kappa. Achievements include development of "post-atomic" literary genre which uses science and technology for fresh language and developed an original scientific mythology. Home and Office: 137 Oak St Ridgewood NJ 07450

HARTMANN, JÜRGEN HEINRICH, physicist; b. Karlstadt, Bavaria, Germany, May 16, 1945; s. Heinrich and Angelina (Schmitt) H. Student, Technische Hochschule, Munich, 1966-67; diploma in physics, U. Würzburg, Fed. Republic Germany, 1973; D Natural Scis., U. Kaiserslautern, Fed. Republic Germany, 1978. Sci. asst. U. Kaiserslautern, 1978-79, hybrid circuit devel. staff, 1979-91; continuous improvement process staff Robert Bosch GmbH, Reutlingen, Fed. Republic Germany, 1991—. Mem. Internat. Soc. for Hybrid Microelectronics, Deutscher Verband für Schweisstechnik. Home: Kaiserstrasse 27, 72764 Reutlingen Germany Office: Robert Bosch GmbH, Tübingerstrasse 123, 72762 Reutlingen Germany

HARTMANN, LUIS FELIPE, health science association administrator, endocrinologist; b. La Paz, Bolivia, Nov. 25, 1923; s. José A. and Victoria (Lavadenz) H.; m. Beatriz de Grandchant Luzio; children: Beatriz, Felipe, Carlos, Isabel. MD, U. San Simon, Cochabamba, Bolivia, 1949; Degree in Endocrinology, Inst. Exptl. Biology and Medicine, Buenos Aires, Argentina, 1954; MS in Endocrinology, London, 1961. Medico cirujano H. Researcher Inst. Exptl. Biology and Medicine, 1951-54; prof. Biology Sch. Pharmacy U. San Simon, 1954-55; prof. med. diagnosis Sch. Medicine, La Paz, Bolivia, 1956-71; rector U. San Andrés, La Paz, 1972-74; prof. medicine UMSA Endocrinology, La Paz, 1975-84; pres. Bolivian Univs. Bur., La Paz, 1982-83; chief med. dept. Univ. Hosp., La Paz, 1984-86; pres. Nat. Acad. Scis., La Paz, 1986—; bd. dirs. Inst. Human Genetics, La Paz, Univ. Hosp., La Paz, med. chief women's ward, 1965-85; hon. prof. U. Gral Ballivian, Trinidad, Beni, 1972, U. Autonoma Guadalajara, Mex., 1973; med. adv. Polyclinic Nat. Social Security, La Paz, 1963-64. Author: Fisiologia Médica, 1965, Citogenética Médica, 1970 (award). Pres. Com. Mal. Coordi. Desarro. C. Biol., Bolivia, 1975-85; v.p. Fund. Bol. Capacit. Dem., Bolivia, 1985. Named Huesped Illustre, Ciudad de Sucre, 1973, Ciudad de Tarija, 1974; Eisenhower Exchange fellow, Phila., 1962. Fellow ACP, Third World Acad. Scis., N.Y. Acad. Scis.; mem. Acad. Mex. Cirujia, Acad. Nat. Scis. (pres. 1966), Univ. Council Edn. (v.p. 1971-72). Roman Catholic. Clubs: Golf, La Paz, Tennis, La Paz. Home: Av Julio C Patino 755, La Paz Murillo Bolivia Office: Bolivian Nat Acad Scis, Avda 16 de Julio 1732, La Paz Bolivia

HARTMANN, RUDOLF, electro-optical engineer; b. Hannover-Münden, Germany, Sept. 19, 1929; came to U.S., 1956; s. Carl Gottlieb and Mathilde (Reinhardt) H.; m. Ingrid-Kunze, Dec. 21, 1957; children: Martin, Linda. BS in Indsl. Optics, U. Göttingen, Fed. Republic Germany, 1952. Cert. precision instrument optician. Precision instrument optician ISCO (Joseph Schneider Co.), Göttingen, 1947-52, Kern & Cie., Aarau, Switzerland, 1952-56, C.P. Goerz, Am. Opt. Co., Inwood, N.Y., 1956-57; optical engr. Bell & Howell Co., Chgo., 1957-60, optical lab. mgr., 1960-70, dir. optical engring., 1970-78; sr. mem. profl. staff mgmt. Martin Marietta Electronics & Missile Group, Orlando, Fla., 1978—. Editor 3 books, proceedings, critical revs.; contbr. articles to profl. jours.; inventor, 11 patents in field. Judge 42d Internat. Sci. & Engring. Fair, Orlando, 1991. Fellow Internat. Soc. Optical Engring. (edn. and standards com.); mem. ASME, Optical Soc. Am. (standards com.), Am. Nat. Standards Inst. (vice chair PH3), Internat. Standards Orgn. (convenor ISO/TC172/SC3/WG2). Avocations: photography, foreign travel, cycling, swimming. Home: 2179 Turkey Run Winter Park FL 32789-6145

HARTMANN, WERNER, physicist; b. Ansbach, Fed. Republic Germany, May 17, 1955; s. Johann and Maria Frieda (Fischer) H.; m. Angelika Maria Luise Emmert, Oct. 11, 1982; children: Florian, Carolin. MS in Physics, U. Erlangen, Fed. Republic Germany, 1981, PhD in Physics, 1986. Asst. prof. U. Erlangen, 1986-91; rsch. physicist Siemens AG Corp. Rsch., Erlangen, 1991—. Mem. IEEE, Am. Phys. Soc., Deutsche Physikalische Gesellschaft. Office: Siemens AG Corp Rsch, Paul Gossen Str 100, 91052 Erlangen Germany

HARTMANN-JOHNSEN, OLAF JOHAN, internist; b. Aalesund, Norway, Aug. 22, 1924; s. Odd and Helga Elisabeth (Hartmann) Johnsen; M.B., B.S., U. Queensland (Australia), 1956; M.D., Oslo U., 1974; m. Mary Essil Archibald, 1956 (dissolved 1968); children: Sally, Helga Elisabeth; m. Mary Eldbjørg Hestad, May 23, 1969; children: Olaf Johan, Else Margrete. Physician, Royal Brisbane (Australia) Hosp., 1956-63, Oslo Univ. Hosp., 1964-65, Bundaberg Gen. Hosp., 1966, Hornsby Dist. Hosp., 1967-70, Upton Hosp., Slough, Eng., 1970, Blacktown Dist. Hosp., 1971-73, Ullevål Hosp., 1974-77, Vefsn Hosp., 1977-78, Kragerø Hosp., 1978-79; physician-in-chief St. Joseph's Hosp., Porsgrunn, 1979-82; cons. physician, chief med. officer Nesset County, 1982-91; govt. med. officer, 1982-91; tutor in medicine U. Oslo Med. Sch., 1975-77; communal med. officer Rauma County Aandalsnes, 1991-92; cons. in gen. practice and community med., 1991—; staff physician Psychiatric Hosp., Hjelset, 1993—. Served with Royal Norwegian Air Force, 1942-47. Decorated King Haakon VII medal, Norwegian War Svc. medal, several British campaign medals. Mem. Norwegian Med. Assn., Coll. Norwegian Internists, N.Y. Acad. Scis. Conservative. Lutheran. Contbr. articles to med. jours. Home: Ranvik, 6460 Eidsvaag Norway Office: Psykiatrisk Storaudeling, Fylkessjukehusset 1 Molde, 6450 Hjelset Norway

HARTONG, MARK WORTHINGTON, military officer, engineer; b. Rochester, Pa., Nov. 2, 1958; s. Robert Mark and Elizabeth Agnes (Norwood) H. BSME with distinction, Iowa State U., 1980; MSc in Computer Sci., Naval Postgrad. Sch., Monterey, Calif., 1985. Profl. engr. Commd. ensign USN, 1980, advanced through grades to lt. comdr.; damage control asst. U.S.S. Sculpin, 1980-82; student Naval Postgrad. Sch., 1983-85; ship supt., docking officer Mare Island Naval Shipyard, Vallejo, Calif., 1985-88; officer in charge Ship Repair Unit, Manamma, Bahrain, 1988-89; 7th fleet type desk officer U.S. Naval Ship Repair Facility, Subic Bay, 1989-90; systems engr. Def. Info. Systems Agy.-Ctr. for Engring., Reston, Va., 1990—. Mem. U.S. Naval Inst., Pi Tau Sigma, Phi Kappa Phi, Tau Beta Pi. Lutheran. Office: Def Info Systems Agy 1860 Wiehle Ave Reston VA 22090

HARTQUIST, E(DWIN) EUGENE, electrical engineer; b. Cortland, N.Y., Mar. 15, 1941; s. Edwin A. and Polly (Loveless) H.; m. Eleanor Gressel, June 30, 1962; children: Kimberly Ann, Christine Beth. BSEE, U. Rochester, 1969; MSEE, Ohio State U., 1972. Technician GE, Syracuse, N.Y., 1961-63; technician U. Rochester, N.Y., 1963-69, rsch. engr., 1974-86; chief engr. radio obs. Ohio State U., Columbus, 1969-72; engr. Coaxial Comm., Columbus, 1973; systems engr. Gen. Ionex Corp., Rochester, 1973-74; rsch. engr. Cornell U., Ithaca, N.Y., 1986—. Mem. IEEE (sr.), Sigma Xi. Office: Cornell U 192 Enging and Theory Ctr Ithaca NY 14853

HARTSAW, WILLIAM O., mechanical engineering educator; b. Tell City, Ind., Oct. 17, 1921; s. William A. and Hazel (Barr) H.; m. Delma Stuckey, June 30, 1946; 1 son, Mark Alan. BS in Mech. Engring., Purdue U., 1946, MS in Engring., 1953; PhD, U. Ill., 1966. Instr. engring. U. Evansville, Ind., 1946-52, asst. prof., 1952-54, assoc. prof., 1954-63, head engring. dept., 1958-61, dir. engring., 1958-68, dir. Sch. Engring., 1961-68, prof. engring., 1963-85; Disting. prof. mech. engring. emeritus, 1985-92; dean engring. U. Evansville, Ind., 1968-76, chmn. dept. mech. engring., 1977-85, disting. prof. mech. engring., 1985-92; disting. prof. emeritus mech. engring. U. Evansville, 1992—; vice chmn. Evansville Environ. Protection Agy., 1980—. Author: The Peltier Effect, 1958, Low Cycle Fatigue Strength Investigation of a High Strength Steel, 1966, Increased Productivity without Added Costs, 1989, Teamwork Enhances Productivity, 1990. Mem. exec. bd. Buffalo Trace coun. Boy Scouts Am., 1969—, chmn. svc. com., 1975-76; mem. Evansville Urban Transp. Advisory, 1975—, Nat. Agenda del., 1990-92. Served with USAAF, 1942-43. Recipient Alumnus Certificate of Excellence U. Evansville, 1972, Tech. Achievement award Tri-State Council for Sci. and Engring., 1979; Lilly Found. fellow, 1960; NSF fellow, 1961-62. Mem. ASME (faculty advisor student chpt. 1968-92, nat. com. div. solar energy com. on components 1975—, vice chmn. faculty advisors region VI 1975-76, chmn. faculty advisors region VI 1976-78, vice chmn. Evansville sect. 1980-81, chmn. Evansville sect. 1981-82, v.p. elect region VI 1982-83, v.p. region VI 1983-85, advisor to regional v.p. 1985-87, mem. nat. bd. on issues mgmt. 1986-89, nat. bd. on profl. devel. 1988—, nat. nominating com. 1989-91, Centennial Svc. award 1980, Centennial medal 1980, Disting. Svc. award 1989, Faculty Advisor of the Yr. region IV 1992), ASTM, AAUP, AAAS, Am. Soc. Engring. Edn., ASHRAE (life, pres. Evansville chpt. 1966-67), Am. Soc. Metals, Phi Kappa Phi, Phi Beta Chi, Phi Tau Sigma. Methodist. Home: 1407 Green Meadow Rd Evansville IN 47715-6055

HARTSFIELD, BRENT DAVID, environmental engineer; b. Tallahassee, Fla., Aug. 21, 1956; s. Albert Clinton and Thelma Muriel (Jenkins) H. BS in Environ. Engring. with honors, U. Fla., 1978; grad., Dale Carnegie Course. Cert. profl. engr., Fla. Surveyor Thomas E. Jenkins & Assocs., Bonifay, Fla., 1976; engring. technician Dept. Transp. State of Fla., Tallahassee, 1977; environ. engr. Dept. Environ. Protection, Tallahassee, 1981—; engring. technician U. Fla., Gainesville, 1978; environ. engr. Camp, Dresser and McKee, Ft. Lauderdale, Fla., 1979-81. Sunday Sch. tchr. First Bapt. Ch., Tallahassee, 1992; tutor Gilchrist Elem. Sch., Tallahassee, 1992; mentor Cobb Mid. Sch., Tallahassee, 1992. Mem. NSPE, ASCE, Fla. Engring. Soc., Hazardous Materials Control Resources Inst. Republican. Baptist. Office: Fla Dept Environ Protection 2600 Blair Stone Rd Tallahassee FL 32399

HARTSFIELD, HENRY WARREN, JR., astronaut; b. Birmingham, Ala., Nov. 21, 1933; s. Henry Warren and Alice Norma (Sorrell) H.; m. Judy Frances Massey, June 30, 1957; children: Judy Lynn, Keely Warren. BS, Auburn U., 1954; postgrad., Duke U., 1954-55, Air Force Inst. Tech., 1960-61; MS, U. Tenn., 1970; DSc (hon.), Auburn U., 1986. Commd. 2d lt. USAF, 1955, advanced through grades to col., 1974; assigned to tour with 53d Tactical Fighter Squadron USAF, Bitburg, Fed. Republic Germany, 1961-64; instr. USAF Test Pilot Sch., Edwards AFB, Calif., 1965-66; assigned to Manned Orbiting Lab. USAF, 1966-69; astronaut, NASA Lyndon B. Johnson Space Ctr., 1969—, mem. support crew Apollo 16, Skylabs 2, 3, 4 missions, pilot STS-4; comdr. STS-41D, STS-61A, ret., 1977; civilian astronaut NASA; dep. dir. Flight Crew Ops. Directorate, 1987-89; dir. tech. integration and analysis Office Space Flight, NASA Hqrs., 1989-90; dep. dir. ops. space sta. projects Marshall Space Flight Ctr. NASA, 1990-91; mgr. man-tended capability phase Space Sta. Freedom Program, 1991—. In space: 483 hours. Decorated Meritorious Service medal, D.S.M. NASA, 1982, 88, Space Flight medal NASA, 1982, 84, 85; recipient Nat. Geog. White Space Trophy, 1973. Mem. Soc. Exptl. Test Pilots, Air Force Assn., Sigma Pi Sigma. Office: MTC Phase Mgr SSFP Code MS Lyndon B Johnson Space Ctr Houston TX 77058

HARTSTEIN, ALLAN MARK, physicist; b. N.Y.C., Oct. 5, 1947; s. Arnold Hartstein and Freda Anne (Jaffe) Bernstein; m. Phyllis Mirman, June 27, 1971; children: Marc, Jonathan. BS, Calif. Inst. Tech., Pasadena, 1969; PhD, U. Pa., 1973. Rsch. staff mem. IBM/TJ Watson Rsch. Ctr., Yorktown Heights, N.Y., 1976—; vis. scientist IBM-TJ Watson Rsch. Ctr., 1974-76. Contbr. articles to profl. jours., publs. Fellow Am. Phys. Soc.; mem. IEEE (sr.). Achievements include 10 patents, 14 patent publs. in field; pioneering exptl. work in semiconductor physics including the discovery and study of 2-dimensional impurity bands, fabrication of first 1-dimensional MOSFET and shortest channel length MOSFET. Office: IBM-TJ Watson Rsch Ctr PO Box 218 Yorktown Heights NY 10598

HARTWELL, LELAND HARRISON, geneticist, educator; b. Los Angeles, Oct. 30, 1939; s. Majorie (Taylor) H. BS, Calif. Inst. Tech., 1961; PhD, MIT, 1964. Postdoctoral fellow Salk Inst., 1964-65; asst. prof. U. Calif., Irvine, 1965-67, assoc. prof., 1967-68; assoc. prof. U. Washington, Seattle, 1968-73, prof., 1973—; rsch. prof. Am. Cancer Soc., 1990—. Recipient Eli Lilly award, 1973, GM Sloan award, 1991, Hoffman LaRoche Mattia award, 1991, Gairdner Found. Internat. award, 1992, Brandeis U. Rosenstiel award, 1993; Guggenheim fellow, 1983-84; Am. Bus. Cancer Rsch. grantee, 1983—; Am. Cancer Soc. scholar. Mem. NAS, AAAS, Am. Soc. Microbiology, Am. Soc. Cell Biology, Genetics Soc. Am. (pres. 1990). Office: Dept of Genetics SK-50 U Wash Seattle WA 98195

HARTWELL, ROBERT CARL, chemical engineer; b. Chgo., July 30, 1961; s. Robert C. and Corinne M. Hartwell; m. Kristi L. Eye, Feb. 4, 1989. BSChE, Purdue U., 1983. Registered profl. engr., Ill., Ind. Constrn. supr. Urschel Devel. Corp., Valparaiso, Ind., 1983-85; project engr. Eichleay Engrs. Inc., Chgo., 1985-90; sr. project engr. Eichleay Engrs., Inc., Chgo., 1990—. Mem. NSPE, Am. Inst. Chem. Engrs., Am. Chem. Soc., Ill. Soc. Profl. Engrs. (v.p.1992-93, pres. 1993—). Achievements include development and implementation of ultrafiltration/membrane separation processes for treatment of industrial waste machine coolants and lubricants and toxicity evaluation and reduction for industrial wastewater discharges. Office: Eichleay Engrs Inc 600 S Federal St Chicago IL 60605

HARTZELL, CHARLES R., research administrator, biochemist, cell biologist; b. Butler, Pa., Aug. 12, 1941; s. Charles R. and Ada Grace (Giles) H.; m. Marguerite K. Getty; children: Scott David, Amy Lynette. BS, Geneva Coll., 1963; PhD, Indiana U., 1967. Post-doctoral fellow Ind. U., Bloomington, 1967; rsch. fellow Commonwealth Sci. and Industry Rsch. Orgn., Melbourne, Australia, 1967-68; rsch. fellow, asst. rsch. prof. U. Wis., Madison, 1968-71; asst. prof. Pa. State U., University Park, 1971-75, assoc. prof., 1975-78; sr. rsch. scientist Alfred I. DuPont Inst., Wilmington, Del., 1978-80, dir. rsch., 1981—; dir. rsch. Nemours Children's Clinic, Jacksonville, Fla., 1987—; rsch. mgr. The Nemours Found., Jacksonville, 1987—. Contbr. over 70 articles to profl. jours. NIH fellow, 1968-70; established investigator Am. Heart Assn., 1970-75. Mem. Am. Chem. Soc., Am. Soc. Biochemistry and Molecular Biology, Biophys. Soc., Am. Soc. Cell Biology, AAAS. Republican. Presbyterian. Avocations: ballroom dancing, music, carpentry, exercise. Office: Alfred I DuPont Inst PO Box 269 Wilmington DE 19899-0269 also: Nemours Children's Clinic PO Box 5720 Jacksonville FL 32207

HARTZELL, HARRISON CRISS, JR., biomedical research educator; b. Phila., Dec. 4, 1946; s. Harrison Criss and Jane Ann (Price) H.; m. Martha Ann Kroon, Feb. 1, 1969; children: Laura Brook, Robyn Elizabeth, Catherine Anne. BA magna cum laude, Lawrence U., 1968; PhD with distinction, Johns Hopkins U., 1973. Rsch. fellow in neurobiology Harvard Med. Sch., Boston, 1973-76; asst. prof. dept. anatomy Emory U. Sch. Medicine, Atlanta, 1976-81, assoc. prof. dept. anatomy and cell biology, 1981-85, prof. dept. anatomy and cell biology, 1985—, prof. dept. physiology, 1990—; vis. prof. U. Paris, Orsay, France, 1985-86, 87, 89; mem. physiology study sect. NIH, 1988-91; presenter in field. Editorial bd.: Am. Jour. Physiology, 1990—, Jour. Cardiac Electrophysiology, 1990-94; contbr. chpts. to books and articles to profl. jours. including Jour. Gen. Physiology, Sci., Jour. Physiology, Nature and others. Named Rsch. fellow Muscular Dystrophy Assn., 1973-75; recipient NIH Rsch. Career Devel. award NIH, 1978-83, Albert E. Levy Faculty Rsch. award Sigma Xi, 1982, NIH Merit award Nat. Heart Lung & Blood Inst., 1988, James Clark Foye scholarship in chemistry Lawrence U. Mem. AAAS, Internat. Soc. for Heart Rsch., Am. Heart Assn., Biophysical Soc., Phi Sigma. Achievements include research in signal transduction; mechanisms of regulation of ion channel and contractile protein function by hormones, neurotransmitters and intracellular components. Office: Emory Univ Sch Medicine Dept Anatomy & Cell Biology Atlanta GA 30322

HARVEY, AUBREY EATON, III, industrial engineer; b. Charlottesville, Va., Oct. 20, 1944; s. Aubrey Eaton Jr. and Jaquelin Ambler (Nicholas) H.; m. Elizabeth Dillard Pettit, June 6, 1964; children: Eleanor Taylor, Philip Ambler. BS, U. Ark., 1966; MA, U. Va., 1970; PhD, U. Ark., 1974. Asst. prof. Tex. A&M U. Indsl. Engring. Dept., College Station, 1973-74, Miami U. Dept. Systems Analysis, Oxford, Ohio, 1974-78; analyst computer svc. Norfolk and Western Railway, Roanoke, Va., 1978-80, systems analyst computer svc., 1980-83; ops. rsch. analyst Norfolk (Va.) Southern Corp., 1983-90, sr. ops. rsch. analyst, 1991; rsch. assoc. Va. Polytech Inst. and State U., Blacksburg, 1991-93, rsch. scientist, 1993—; cons. Ark. Dept. Labor, Little Rock, 1971-72, Ark. Health Systems Found., Little Rock, 1972-73; adj. faculty Va. Polytech Inst. and State U., 1980-85. Contbr. articles to profl. jours. Pres. U. Va. Law and Grad. Young Reps., Charlottesville, 1969; treas. Va. Young Reps., Richmond, 1970, 71. Mem. Ops. Rsch. Soc. Am., Inst. Indsl. Engrs. (divsn. dir. 1983-84, Disting. Svc. 1985), Sigma Xi, Alpha Pi Mu, Omega Rho. Episcopalian. Achievements include development of a track quality index; developed a consensus measure. Home: 7107 Deerwood Rd Roanoke VA 24019-2113 Office: Mgmt Systems Labs 1900 Kraft Dr Blacksburg VA 24060

HARVEY, BRYAN LAURENCE, crop science educator; b. Newport, Gwent, Wales, U.K., Nov. 1, 1937; came to Can., 1948; s. Laurence W.J. and Irene E.D. (Stoneman) H.; m. Eileen Bernice Pfeifer, Sept. 24, 1961; children—Donald, James. B.S.A., U. Sask., 1960, M.Sc., 1961; Ph.D. U. Calif.-Davis, 1964. Asst. prof. crop sci. U. Guelph, Ont., Can., 1964-66; from asst. prof. to prof. crop sci. U. Sask., Saskatoon, 1966—, head dept. crop sci. and plant ecology, 1983, dir. Crop Devel. Ctr., 1983—, asst. dean Coll. Agr., 1980-83; vis. prof. U. Nairobi, Kenya, 1975; chmn. Can. Expert Com. Grain Breeding, 1984-89; chmn. Can. Expert Com. on Plant gene resources, 1986-93; chmn. Can. Adv. Com. on Variety Registration, 1987—; chmn. Can. Prairie Registration Recommending Com. for Grain, 1989—; chmn. Can. Adv. Com. on Plant Breeders Rights. Developed 16 varieties of barley; contbr. articles to sci. jours. on plant genetics. Fellow Agrl. Inst. Can. (pres. 1984-85, Fellowship award 1990), Am. Soc. Agronomy, Crop Sci. Soc. Am.; mem. Assn. Faculties of Agr. Can. (pres. 1982-83), Barley Genetics Congress (pres. 1986-91), Am. Barley Workers (pres. 1970-74), Sask. Seed Growers Assn. (hon. life), Rotary (pres. 1987-89). Office: U Sask, Dept Crop Sci & Plant Ecology, Saskatoon, SK Canada S7N 0W0

HARVEY, JOHN ASHMORE, civil engineer; b. Plainfield, N.J., Apr. 12, 1964; s. Ernest Albert and Joan Elizabeth-Rose (Ashmore) H.; 1 child, Kierstyn Nicole. BSCE, Mich. Tech. U., 1986, MSCE, 1988. Registered profl. engr., Mich. Transp. engr. Conn. Dept. of Transp., Newington, Conn., 1988-90; project engr. JCK & Assocs., Inc., Novi, Mich., 1990-92, Envirodyne Engrs., Inc., Lansing, Mich., 1992—. Mem. NSPE, ASCE, Assn. State Flood Plain Mgrs., Am. Water Works Assn. Methodist. Achievements include original research in establishing flood frequencies for Frazil related riverine freeze-up events leading to further understanding and exploration of the unusual phenomena. Home: 4278 Marr Ave Warren MI 48091 Office: Envirodyne Engrs Inc Ste 200 6500 Centurion Dr Lansing MI 48917

HARVEY, JOHN MARSHALL, agricultural scientist; b. Fresno, Calif., Jan. 13, 1921; s. Fred Marshall and Louise Marie (Roth) H.; m. Cora Gertrude Verboon, Aug. 17, 1968. BA, Calif. State U., Fresno, 1942; MA, Stanford U., 1948; PhD, U. Calif., Berkeley, 1950. With USDA, Fresno, 1950—, sr. plant pathologist, 1960-64, supervisory plant pathologist, 1964-87, cons., 1987—; cons. various orgns., 1987—. Contbr. chpts. to books, articles and papers to profl. publs. Tech. sgt. M.C. U.S. Army, 1942-45, ETO, NATOUSA. Recipient Disting. Svc. award Produce Packaging Assn., Washington, 1966, Mentors Ann. award Calif. Grape and Tree Fruit League, 1989. Mem. AAAS, ASHRAE, Am. Phytopathological Soc., Internat. Inst. Refrigeration. Home: 5461 E Heaton Ave Fresno CA 93727

HARVEY, LOUIS JAMES, architect; b. Corvallis, Oreg., June 8, 1948; s. Robert Gordon and Grace Elaine (Harrington) H.; m. Wendy Howard Baker, Oct. 11, 1980 (div. Jan. 1990); children: Veronica Lynn, Sarah Louise, Susan Katherain. BS, Portland State U., 1979. Pres., CEO Intercoastal Resources Ltd., Vancouver, B.C., Can., 1980-81; process systems engr. Foxboro Can. Ltd., Vancouver, 1981-85; process systems cons. IMS Cons., Inc., Vancouver, 1986-90; pres., CEO computer integrated systems architect Harvey, House & Assoc., Inc., Cary, N.C., 1990—. With USAF, 1969-73. Grantee Oregon Dept. Energy Alternative Energy Resource Mgmt., 1978. Mem. Instrument Soc. Am., Indsl. Computing Soc., Soc. Mfg. Engrs., Sigma Nu. Republican. Achievements include developing and implementing fast track methodology for applications design, integration and use of plant-wide computer integrated manufacturing systems for industrial manufacturing sites based on UNIX, OSF.OPEN standards and user requirements. Office: Harvey House & Assoc Inc 8255 Chapel Hill Rd Cary NC 27513

HARWIT, MARTIN OTTO, astrophysicist, educator, museum director; b. Prague, Czechoslovakia, Mar. 9, 1931; came to U.S., 1946, naturalized, 1953; s. Felix Michael and Regina Hedwig (Perutz) Haurowitz; m. Marianne Mark, Feb. 1, 1957; children: Alexander, Eric, Emily. B.A. in Physics, Oberlin Coll., 1951; M.A. in Physics, U. Mich., Ann Arbor, 1953, postgrad., 1952-54; Ph.D. in Physics, Mass. Inst. Tech., 1960. NATO postdoctoral fellow Cambridge (Eng.) U., 1960-61; NSF fellow Cornell U., Ithaca, N.Y., 1961-62; asst. prof. astronomy Cornell U., 1962-64, assoc. prof., 1964-68, prof., 1968-87, prof. emeritus, 1988—, chmn. dept. astronomy, 1971-76, co-dir. program for history and philosophy of sci. and tech., 1985-87; dir. Nat. Air and Space Mus., Smithsonian Inst., Washington, 1987—; E.O. Hulburt fellow Naval Research Lab., Washington, 1963-64; Nat. Acad. Sci. exchange visitor Czechoslovak Acad. Sci., Prague, 1969-70; v.p., dir. Spectral Imaging Inc., Concord, Mass., 1971-77; external mem. Max Planck Inst. Radioastronomy, Bonn., W. Ger., 1979—; cons. NASA.; chair for space history Nat. Air and Space Mus., Smithsonian Instn., 1983; chmn. astrophysics mgmt. ops. working group, NASA, 1985-87;. Author: Astrophysical Concepts, 1973, 2d edit. 1988 (transl. into Chinese 1981), (with N.J.A. Sloan) Hadamard Transform Optics, 1979, Cosmic Discovery-The Search, Scope and Heritage of Astronomy, 1981 (transl. into German and French 1982). With U.S. Army, 1955-57. Recipient Alexander von Humboldt Found. sr. U.S. scientist award Max Planck Inst. Radioastronomy, 1976-77; NSF grantee, 1963-68; Research Corp. grantee, 1970-75; NASA grantee, 1965—; Air Force Cambridge (Mass.) Research Labs. grantee, 1969-74. Fellow AAAS, Am. Phys. Soc. (chmn. div. history of physics 1986-87, chmn. astrophysics div. 1988-89), Royal Astron. Soc.; mem. Soc. for History of Tech., Am. Astron. Soc. Avocations: Research on infrared astron. observations from spacecraft rockets and aircraft; theoretical astrophysics; cryogenic optics; Hadamard transform optics; history of astron. discoveries. Home: 511 H St SW Washington DC 20024-2725 Office: Nat Air & Space Mus 6th Street Ave Washington DC 20560-0001

HARWOOD, IVAN RICHMOND, pediatric pulmonologist; b. Huntington, W.Va., July 3, 1939. BA, Dartmouth Coll., 1961; MD, U. W.Va., 1965. Diplomate Nat. Bd. Med. Examiners; lic. physician, Calif., Can.; cert. Am. Bd. Pediatrics. Intern in pediatrics U. W.Va. Hosp., Morgantown, 1965-66; resident in pediatrics Yale-New Haven (Conn.) Hosp., 1966-68, sr. resident outpatient dept., 1968-69; chief pediatrics USAF Hosp. 3646, Del Rio, Tex., 1968-70; asst. prof. pediatrics U. Calif. Med. Ctr., San Diego, 1971-78, chief pediatric pulmonary div., 1972—, dir. pediatric intensive care unit, 1972-78, assoc. adj. prof. pediatrics, 1978-86, prof., 1987—; mem. patient care rev. and numerous other coms. U. Calif. Med. Ctr., 1976—; co-dir. Cystic Fibrosis Ctr., San Diego, 1972-73, dir., 1973; mem. Cystic Fibrosis Young Adult Com., Atlanta, 1974-80, chmn., 1976-80, Cystic Fibrosis Ctr. Com., 1986-89, vice-chmn., 1990—; mem. San Diego County Tuburculosis Control Bd., 1974-78; presenter, lectr. in field. Producer: (videos) Issues in Cystic Fibrosis Series; mem. rev. bd., CF Film, 1980; contbr. chpts. to books, and numerous articles to profl. jours. Mem. Air Quality Adv. Com., State of Calif., 1974-80; mem. Genetically Handicapped Persons Program Adv. Com., Calif., 1977-87; mem. adv. bd. Grossmont Coll. Inhalation Therapy Sch., San Diego, 1975-76; mem. inpatient adolscent adv. com., Mercy Hosp., 1982-85. U. Calif. fellow in pediatric cardiology, 1970-71; recipient 1st Prize Internat. Rehab. Film Library Competition, 1980. Mem. Calif. Med. Assn. (patient care audit com. 1975-78), Nat. Cystic Fibrosis Found. (mem. adv. com. 1976-80, planning ad hoc com. 1976-77, patient registry subcom. 1986—), San Diego Found. for Med. Care (major med. rev. com. 1978-84), San Diego Lung Assn. (pediatric com. 1976-80, chmn. Project Breath-Easy 1976-78). Home: PO Box 431 Jamul CA 91935-0431 Office: U Calif Pediatric Ctr 4130 Front St San Diego CA 92103-2016

HARWOOD, JOHN SIMON, chemist, university official, consultant; b. York, Eng., Oct. 6, 1960; came to U.S., 1968; s. Ernest Stewart and Joan Mavis (Dudding) H. BS in Chemistry, U. Tex., Arlington, 1983, DSc in Chemistry, 1988. Postdoctoral fellow U. Tex. Med. Sch., Dallas, 1988-89, U. Calif., Berkeley, 1989-90; dir. nuclear magnetic resonance facility U. Ga., Athens, 1991-93, U. Ill., Chgo., 1993—. Mem. AAAS, Am. Chem. Soc., Assn. Mgrs. Magnetic Resonance Labs. Democrat. Achievements include research on solution-state structure determinations of proteins and natural products. Office: U Ill Dept Chemistry M/C 111 Box 4348 Chicago IL 60680-4348

HARWOOD, KIRK EDWARD, aeronautical engineer; b. Wurtzburg, Germand, Dec. 21, 1964; came to the U.S., 1965; s. Frederick Crosby and Michelle Lidya (Linder) H. BS in Aero. Engring., Embry-Riddle Aero. U., 1988. Aircraft analyst Aerospace Sci. and Tech. Ctr., Wright-Patterson AFB, Ohio, 1989-93, B-2 flying qualities flt. test engr., 1993—. Editor, author: Foxhound Weapon System Study, 1992. Capt. USAF, 1988-93. Mem. AIAA. Home: Apt 105 1850 West Ave J-12 Lancaster CA 93534

HARWOOD, ROBIN LOUISE, psychologist; b. Glendale, Calif., Sept. 13, 1958; d. Baxter and Ramona Kathleen (Koch) H. PhD, Yale U., 1991. Postdoctoral assoc. NIH, Washington, 1991; asst. prof. U. New Orleans, 1992—; ad hoc reviewer Psychol. Bull., 1990—; cons. W.T. Grant Consortium on Social Competence, 1989; guest reviewer Smith Richardson Grant Found., 1988. Contbr. articles to profl. jours. Hon. custodian Nature Conservancy Fund, State of Va., 1992—. Recipient Rsch. Tng. award NIH, 1991, Nat. Rsch. Svc. award NIH, 1990-91, Dissertation Rsch. grant NSF, 1989-91, pre-doctoral fellowship Bush Ctr. in Child Devel. and Social Policy, 1984-91, Yale U. Dissertation award, 1990. Mem. APA, Am. Anthrop. Assn., Soc. for Rsch. in Child Devel., La. Infant Mental Health Assn. (co-founder), Smithsonian Inst. Democrat. Achievements include innovative rsch. on models of normative devel. used to understand mother-infant relationships and child devel. in a minority population (Puerto Rico). Office: Psychology Dept U New Orleans New Orleans LA 70148

HARWOOD, THOMAS RIEGEL, pathologist; b. Knoxville, Tenn., Dec. 9, 1926; s. Thomas Everett and Grace Deborah (Thomas) H.; m. Violet Rose Jaeggi, July 9, 1949 (div. 1975); m. Phyllis Gail Bredthauer, Jan. 3, 1976; children: Joseph David, Thomas Mark, Shannon Donnelly. BS, Georgetown U., 1949; MD, Vanderbilt U., 1953. Diplomate Am. Bd. Pathology. Intern Vanderbilt U. Hosp., Nashville, Tenn., 1953-54, resident, 1954-55; resident Med. Coll. of Va. Hosp., Richmond, 1955-56; asst. Vanderbilt Dept. Pathology, Nashville, 1954-55; instr. pathology Med. Coll. Va., Richmond, 1955-56; instr. to assoc. prof. pathology Northwestern U., Chgo., 1956-90, assoc. prof. emeritus, 1990—; pathologist Reference Pathology Labs., Nashville, 1990—. Contbr. articles to profl. jours., chpts. to books. Ltd. comdr. USNR, 1943-65: PTO. Recipient Borden award for undergrad., rsch. Vanderbilt U., 1953. Fellow Am. Soc. Clin. Pathology (emeritus); mem. AMA, Am. Soc. for Investigative Pathology, Am. Assn. Blood Banks, Ill. Assn. Blood Banks (pres. 1979-80), Chgo. Med. Soc. (trustee 1983-84, pres. North Side br. 1982-83). Episcopalian. Home: 916 Williamsburg Village Dr Jackson TN 38305

HASBROOK, A. HOWARD, aviation safety engineer, consultant; b. Trenton, N.J., July 15, 1913; s. Albert Howard and Mabel (Naar) H.; m. Christel Anna Schneider, 1938 (div. 1955); children: Barbara Elaine, Howard Richard Jay; m. Virginia Randolph Whiting, 1955. Grad. high sch., DuBois, Pa. Safety engr., Calif. Flight instr., engring. test pilot USAAF, 1942-45; agrl., charter, airline & test pilot, comml. flight examiner, 1945-50; assoc. dir. Av-CIR Cornell U., 1950-55; chief crash safety, sr. rsch. scientist FAA Civil Aeromed. Inst., 1960-67, chief fligh performance, sr. rsch. scientist, 1968-75; aviation safety cons., profl. engr., accident investigator, analyst & reconstructionist, rsch. pilot, engring. test pilot, flight instr., 1975—, also accident prevention counselor FAA, 1975—, assoc. prof. Embry-Riddle Aerospace U., 1982, expert witness in accident litigation. Presenter lectures/tech. papers before numerous orgns.; contbr. more than 100 tech. papers & reports to profl. jours. Served with U.S. Army, 1933-34. A. Howard Hasbrook sect. established in his honor by Wright State U. Med. Sch. Libr., 1990; recipient Flight Safety Found. award, 1958, Gen. Spruance award, 1970, Harry G. Mosely award, 1972; Fellow in Aerospace Medicine, 1972. Fellow Aerospace Med. Assn. (Hasbrook award named in his honor); mem. Internat. Soc. Air Safety Investigators, Quiet Birdmen, OX-5 Aviation Pioneers, Nat. Forensic Ctr. Home and Office: Safety Engring & Rsch HC 30 Box 813 Prescott AZ 86301

HASCHKE, PAUL CHARLES, analytical chemist; b. Rockford, Ill., May 18, 1954; s. Gilbert Alfred and Pauline Victoria (Castiglioni) H.; m. Floy Angela Sparks, May 26, 1979; children: Steven Harrison, David William. BS in Chemistry, No. Ill. U., 1977; MS in Chemistry, Ill. Inst. Tech., 1982. Chemistry technician Sundstrand Aviation, Rockford, Ill., 1973; technician Peabody Testing, Byron, Ill., 1976; rsch. assoc. Stepan Chem. Co., Northfield, Ill., 1977; analytical chemist Commonwealth Edison, Maywood, Ill., 1978—. Mem. PTA, Wheaton, Ill., 1989-92; presenter of programs of chem. demonstrations Pleasant Hill Elem. Sch., 1990, 91, 92. Recipient Award for Vol. Svc., Pleasant Hill Elem. Sch., 1991, 92. Mem. Am. Chem. Soc. (pub. outreach program 1991-92), Edison Electric Inst., Ion Chromatography Task Force (vice chmn. 1991, 92), Commonwealth Edison Garden Club. Roman Catholic. Achievements include two patents for devising a novel method of treating non-union bone fractures with accompanying soft tissue injury, determining fluoride and chloride in component cooling water by gradient ion chromatography. Office: Commonwealth Edison 1319 S First Ave Maywood IL 60153-2496

HASEGAWA, AKINORI, chemistry educator; b. Matsuzaka, Mie, Japan, Jan. 3, 1941; s. Michio and Fumiko (Nakamura) H.; m. Tomoko Kuru, Nov. 6, 1966; children: Yuki, Teruaki. BEd, Hiroshima (Japan) U., 1963, MSc, 1965, DSc, 1970. Rsch. fellow Hiroshima U., 1966-77, lectr., 1977-81, assoc. prof., 1981-83; assoc. prof. Kogakkan U., Ise, Mie, Japan, 1983-85, prof. chemistry, 1985—, head univ. press, 1989-91, chmn. liberal arts dept., 1990-91; vis. rsch. fellow Hokkaido (Japan) U., 1971-72; postdoctoral fellow Tenn. U., 1975-76; vis. fellow Leicester (Eng.) U., 1981, hon. vis. prof., 1988. Author: Radical Ionic Systems, 1991, Handbook of Radiation Chemistry, 1991; contbr. articles to profl. jours. Grantee Matsunaga Found., Tokyo, 1972, Found. for Sci. Promotion, Tokyo, 1978; grantee-in-aid Ministry Edn., Japan, Tokyo, 1980-81, 86-89. Mem. Chem. Soc. Japan, Japanese Assn. Crystal Growth, Am. Chem. Soc. Office: Kogakkan U, 1704 Kodakujimoto-cho, Ise Mie 516, Japan

HASEGAWA, HIDEKI, electrical engineering educator; b. Tokyo, June 22, 1941; s. Zen-ichi and Umeko Hasegawa; m. Moriko Hasegawa, Mar. 11, 1967; children: Ken-ichi, Yukiko, Eiji. BE, U. Tokyo, 1964, ME, 1966, PhD in Electronics, 1970. Lectr. Hokkaido U., Sapporo, Japan, 1970-71, asst. prof., 1971-80, prof., 1980—; dir. Interface Quantum Electronics Rsch. Ctr., Hokkaido U., 1991—; vis. rsch. fellow U. Newcastle-upon-Tyne, U.K., 1973-74. Editor: Passivation of Metals and Semiconductors, 1990; contbr. over 150 papers to sci. publs. Recipient Max Planck Rsch. Award, 1990. Mem. IEEE, Am. Phys. Soc., Inst. Electronics, Info. and Communication Engrs., Japan Soc. Applied Physics. Achievements include research in crystal growth, characterization, processing and device applications of III-V compound semiconductors with emphasis on surfaces and interfaces. Home: 3-6 Toko-cho, Ebetsu-shi Hokkaido 067, Japan Office: Hokkaido U Dept Elec Engring, N-13 W-8, Sapporo 060, Japan

HASEGAWA, RYUSUKE, materials scientist; b. Nagoya, Japan, Feb. 7, 1940; came to U.S., 1964; s. Tokusaburo and Teru (Hanai) H.; m. Alice Pamela Quayle, Dec. 17, 1967; children: Sergei Pol, Linnea Marie. MS, Calif. Inst. Tech., 1968, PhD, 1969. Rsch. fellow Calif. Inst. Tech., Pasadena, 1969-72; postdoctoral fellow IBM, Yorktown Heights, N.Y., 1973-75; sr. staff physicist, group leader Allied-Signal, Morristown, N.J., 1975-80, rsch. assoc., then sr. rsch. assoc., 1980-85; v.p. Nippon Amorphous Metals div. Allied-Signal, Tokyo and Morristown, 1985-89; dir. magnetics rsch., Far East bus. devel. amorphous metals Allied-Signal, Parsippany, N.J., 1989—; adj. prof. U. Tokyo, 1988-89; bd. dirs. Nipon Amorphous Metals Co., Ltd., Tokyo. Editor: Glassy Metals: Magnetic, Chemical and Structural Properties, 1983; co-editor: Amorphous Magnetism II, 1977; contbr. tech. papers to refereed sci. jours. Fellow IEEE; mem. Am. Phys. Soc., Materials Rsch. Soc., Sigma Xi. Presbyterian. Achievements include 19 patents in amorphous magnetic materials and their applications. Office: Allied Signal Amorphous Metals 6 Eastmans Rd Parsippany NJ 07054

HASEGAWA, TADASHI, chemical educator; b. Chuouku, Tokyo, Japan, Oct. 25, 1948; s. Shuji and Rei (Yoshino) H.; m. Yoshie Kobayashi, Oct. 3, 1975; 1 child, Masashi. DSc, Tokyo Kyoiku U., 1976. Rsch. assoc. Mich. State U., East Lansing, 1978-80; asst. prof. Tokyo Gakugei U., Koganie, Japan, 1980-82, assoc. prof., 1982—. Contbr. articles to profl. jours. including Bull. Chem. Soc. Japan, Jour. Chem. Soc. Chem. Comm., Jour. Chem. Soc. Perkin Trans. 1, Jour. Photochemistry and Photobiology, Tetrahedron Letters, Jour. Am. Chem. Soc. Grantee Japanese Ministry of Edn., 1981. Mem. Chem. Soc. of Japan (editorial bd.), Am. Chem. Soc., Japanese Photochemistry Assn., Soc. of Synthetic Organic Chemistry Japan. Achievements include photocyclization of Alpha and Beta Oxoamides, photocyclization via remote hydrogen transfer in dicarbonyl systems, Stern-Volmer quenching kinetics in photoreaction systems with strong internal filter. Office: Tokyo Gakugei U, 4-1-1 Nukui-kitamachi, Koganei Tokyo 184, Japan

HASELTINE, FLORENCE PAT, research administrator, obstetrician, gynecologist; b. Phila., Aug. 17, 1942; d. William R. and Jean Adele Haseltine; m. Frederick Cahn, Mar. 12, 1964 (div. 1969); m. Alan Chodos, Apr. 18, 1970; children: Anna, Elizabeth. BA in Biophysics, U. Calif., Berkeley, 1964; PhD in Biophysics, MIT, 1964-69; MD, Albert Einstein Coll. of

Medicine, 1972. Diplomate Am. Bd. Ob-Gyn. Asst. prof. dept. ob-gyn. and pediatrics Yale U., New Haven, 1976-82, assoc. prof. dept. ob-gyn. and pediatrics, 1982-85; dir. Ctr. for Population Research, Nat. Inst. Child Health and Human Devel. NIH, Bethesda, Md., 1985—; Chmn. Woman's Council of the Am. Fertility Soc. Co-author: Woman Doctor, 1976, Magnetic Resonance of the Reproductive System, 1987; co-editor 11 books on reproductive scis. Fellow Am. Coll. Ob-Gyn; mem. Inst. of Medicine, AAAS (bd. dirs.), Am. Fertility Soc., Soc. Gynecol. Investigation, Soc. Study of Reproduction, Endocrine Soc., Soc. Reproductive Endocrinologists, Soc. for Advancement Women's Health Rsch. (co-founder, pres. bd. dirs.), Soc. Cell Biology. Office: NIH 900 Rockville Pike 6100 Executive Blvd Rm 8B07D Bethesda MD 20892

HASENBERG, THOMAS CHARLES, physicist; b. Kenosha, Wis., Feb. 4, 1958; s. Mark Thomas and Gloria Ann (Stefani) H.; m. Elizabeth Anne Tompkins, Sept. 8, 1990. MS in Material Sci., U. So. Calif., 1981, PhD, 1986. Undergrad. student U. Notre Dame, Notre Dame, Ind., 1976-80; mem. tech. staff IBM, Los Gatos, Calif., 1980, Gen. Dynamics, Pomona, Calif., 1981; rsch. fellow U. So. Calif., L.A., 1980-86; mem. tech. staff Aerospace Corp., El Segundo, Calif., 1985-86; mem. tech. staff Hughes Rsch. Labs., Malibu, Calif., 1986-91, sr. staff physicist, 1991—. Co-author: Ultra-Fast Optical Phenomena, 1992, 91. Recipient Grad. Rsch. award U. So. Calif., 1986, Rsch. Award for Coll. Scientists, U. So. Calif., 1983-84. Mem. Optical Soc. of Am., Am. Phys. Soc., Sigma Xi. Roman Catholic. Achievements include patents in Bistable Lasers, Q-Switched Lasers, In-GaASP Lasers. Office: Hughes Rsch RL65 3011 Malibu Canyon Rd Malibu CA 90265

HASHA, DENNIS LLOYD, research chemist, spectroscopist; b. San Diego, Oct. 5, 1953; s. Cecil Raleigh and Betty Mae (Tussay) H.; m. Susan Amy Toth, Sept. 29, 1992; children: Dennis F. Kundinger Jr., Sarah Suzanne Priem. BA, U. Calif., San Diego, 1976; PhD, U. Ill., 1981. Rsch. leader Dow Chem. USA/Analytical Scis., Midland, Mich., 1981—; faculty cons. Saginaw (Mich.) Valley State U., 1989—. Achievements include research in unimolecular reactions in dense liquids: observed transition from inertial regime to diffusive region for collisional activated process thereby providing experimental proof of the prediction of stochastic models of unimolecular reactions in dense liquids.

HASHEMI-YEGANEH, SHAHROKH, electrical engineering educator, researcher; b. Tehran, Iran, May 6, 1958; came to U.S., 1977; s. Gol and Gohar (Manooteh) H.-Y. BS, UCLA, 1981, MS, 1983, PhD, 1988. Rsch. cons. UCLA, 1984-85, rsch. asst. elec. engring. dept., 1985-87, staff rsch. assoc. II, 1987-88; asst. prof. Ariz. State U., Tempe, 1988—. Contbr. articles to profl. jours. Grantee Air Force Engring. Found. and IEEE, 1991-92. Mem. IEEE (sr., exec. com. waves and devices Phoenix sect. 1992—, R.W.P. King prize paper award 1991), Sigma Xi, Eta Kappa Nu. Office: Elec Engring Dept Ariz State U Tempe AZ 85287-5706

HASHIM, GEORGE A., immunologist, biomedical researcher, educator; b. Damour, Lebanon, May 28, 1931; came to U.S., 1957; m. Audrey E. Mailhiot, Aug. 5, 1962; children: Laura E., Charles E., Sami G. MS, Columbia U., 1963, PhD, 1967. Post-doctoral fellow Salk Inst., La Jolla, Calif., 1967-69; dir. biology rsch. Continental Rsch. Inst., N.Y.C., 1969-72; assoc. prof. microbiology Columbia U., N.Y.C., 1976-85; dir. exptl. immunology St. Luke's-Roosevelt Hosp. Ctr., N.Y.C., 1972-93; sr. rsch. scientist/scholar Columbia U., N.Y.C., 1985-93; assoc. rsch. dir. Coun. Tobacco Rsch., USA, Inc., N.Y.C., 1993—; mem. study sect. NIH, Bethesda, Md., 1984-89. Author 7 book chpts.; editor 6 books; mem. editorial bd. Jour. Neurochemistry, Jour. Neuroscience Rsch., Jour. Immunology; contbr. over 150 articles to profl. jours. Recipient Andres Bello Orden, Govt. Venezuela, 1980, grants NIH, NSF, 1967—. Recipient Andres Bello Orden, Govt. Venezuela, 1980, grant NIH, NSF, 1967—. Mem. Am. Soc. Neurochemistry (treas. 1986-92, pres. 1993-95), Am. Assn. Immunologists, Am. Soc. Biology Chemists and Molecular Biologists, Internat. Soc. Neurochemistry, Sigma Xi. Achievements include several patents and notable reseacrh in peptides therapy and multiple sclerosis. Office: Coun for Tobacco Rsch USA Inc 900 Third Ave New York NY 10022

HASHIMOTO, ANDREW GINJI, bioresource engineer; b. Wailuku, Maui, Hawaii, Aug. 13, 1944; s. Jack and Marion (Kitagawa) H.; m. Merle Euguchi, Sept. 2, 1967; children: Meri, Riki, Noelle, Joel. MS, Purdue U., 1968; PhD, Cornell U., 1972. Agrl. engr. Ag Rsch. Svc., USDA, Ithaca, N.Y., 1969-76; rsch. leader U.S. Meat Animal Rsch. Ctr., USDA, Clay Center, Nebr., 1976-86; asst. prof. Cornell U., Ithaca, 1972-76; assoc. prof. U. Nebr., Lincoln, 1976-82, prof., 1982-86; prof. and head bioresource engineering Oreg. State U., Corvallis, 1986—; sci. adv. bd. Unisyn, Inc., Seattle, 1990—. Editor Bioresource Tech. Jour., 1986—; contbr. articles to profl. jours. USPHS fellow, 1966-67; recipient Arthur Fleming award Jaycees, 1983. Mem. AAAS, Am. Soc. Agrl. Engrs., Am. Soc. Engring. Edn., Internat. Assn. on Water Pollution Rsch. and Control, Sigma Xi, Gamma Sigma Delta. Office: Oreg State Univ Bioresource Engring Dept Gilmore Hall Rm 116 Corvallis OR 97331-3906

HASHIMOTO, KUNIO, architect, educator; b. Tokyo, Jan. 6, 1929; s. Shinichi and Tomiko (Fuse) H.; m. Masako Takahashi, Dec. 24, 1969; 1 child, Mami. BArch, Tokyo Nat. U. Fine Art and Music, 1948. Registered architect, Japan. Participate assoc. Yokokawa Komusho Architects & Engrs., Tokyo, 1948-50; assoc. Takashi Matsumoto Architect & Assocs., Tokyo, 1950-54, Baker, Butler & Tripllet Architects & Engrs., Tokyo, 1954-56; chief architect R. Kitadai Architect & Assocs., Tokyo, 1956-62; prin. architect K. Hashimoto Architect & Assocs., Tokyo, 1962—; lectr. Shibaura Inst. Tech., Tokyo, 1958-67, assoc. prof. architecture, 1967-79, prof. architecture, 1979—. Former mem. Nomination Com. of Annon-Grand Prix, Tokyo, 1960-65, Com. on Constrn. of City Mus., 1980-81; judge Jury Competition for Refreshing Ctr. of Health Orgn., Atami, Japan, 1986-87. Recipient Honorable Mention award Jury Competition for City Hall, 1951, 2d prize for Model House Chiba Prefectural Gov., 1956, 1st prize for Model House Asahi Newspaper Office, 1957. Mem. Japan Architect Assn. (chmn. coms. 1963-66, 68-71, 74-78, bd. dirs. 1970-74, 81-85), Kawaski Architects Club (bd. dirs. 1980—), Archtl. Inst. Japan, Japan Inst. Architect. Clubs: Tokyo Shinagawa. Lodge: Rotary. Avocations: hunting, travelling. Home: 903 Obata Kanramachi, Kanragun Gumma-ken 370-22, Japan also: 1305 4-18-30 Shibaura, Minato-ku Tokyo 108, Japan Office: K Hashimoto Architect & Assocs, 3-15-18 Meguro, Meguro-ku Tokyo 153, Japan

HASHIMOTO, NOBUYUKI, electro-optical engineer; b. Kohenji, Suginami, Japan, Aug. 16, 1958; s. Kiyoshi and Toshiko Hashimoto. B. in Engring., Waseda U., Tokyo, 1981, MS, 1983. Engr. Citizen Watch Co., Ltd., Tokyo, 1983—; co-founder Electro-Holography Com. of Japan, 1991—. Contbr. articles to profl. jours. Mem. Optical Soc. Am., Internat. Soc. for Optical Engring., Japan Soc. Applied Physics. Achievements include patents for Circuit for Driving the Laser Diode; patent pending for System for Correcting the Optical Abberation; development of color liquid crystal television, of real-time optical effect generator for computer graphics, and of holographic 3 dimensional television system with LCTV-SLM. Home: 5-16-23 Higashifujisawa, Iruma Saitama 358, Japan Office: Citizen Watch Co Ltd Tech, Lab 840 Shimotomi, Tokorozawa Saitama 359, Japan

HASHIMOTO, SHIORI, biomedical scientist; b. Kanagawa, Japan, Oct. 23, 1952; came to U.S., 1988; d. Seisyo and Haru Hashimoto. MD, Tokyo Women's Med. Coll., 1980; Diploma in Neurology (hon.), Japanese Bd. Neurology, 1986. Resident Tokyo Women's Med. Coll., 1980-83, asst., 1984—; postdoctoral fellow North Shore U. Hosp., Manhasset, N.Y., 1988—. Achievements include research in cloning and determining the human Ig-B gene sequence. Home: 145 E 48th St # 30G New York NY 10017 Office: North Shore U Hosp 350 Community Dr Manhasset NY 11030

HASHIMOTO, TSUNEYUKI, materials scientist; b. Zentsuji, Kagawa, Japan, July 31, 1950; s. Saburo and Fujie (Takagi) H.; m. Ayako Gomi, Mar. 20, 1980; 1 child, Takanori. BSc, U. Tokyo, 1974, MSc, 1976, DSc, 1979. Researcher Hitachi (Ibaraki, Japan) Ltd., 1979-84, 85-89, sr. researcher, 1989—; vis. scientist Argonne (Ill.) Nat. Lab., 1984-85. contbr. articles to profl. jours. Mem. Phys. Soc. Japan, Japan Inst. Metals, Atomic Energy Soc. Japan. Avocations: music, skiing. Office: Hitachi Ltd Energy Rsch Lab, 7-2-1 Omika-cho, Hitachi Ibaraki 319-12, Japan

HASKELL, BARRY GEOFFRY, communications company research administrator; b. Lewiston, Maine, Sept. 1, 1941; s. George Raymond and Dorothy Mae (Libbey) H.; m. Ann Kantrow, Sept. 13, 1964; children: Paul Eric, Andrew. AA, Pasadena City Coll., 1962; BSEE, U. Calif., Berkeley, 1964, MSEE, 1965, PhD, 1968. Electronics engr. Lawrence Livermore (Calif.) Lab., 1965; rsch. asst. Electronics Rsch. Lab. U. Calif., Berkeley, 1965-68; mem. tech. staff AT&T Bell Labs., Holmdel, N.J., 1968-76, head radio communication rsch. dept., 1976-83, visual communications cons., 1984-86, head visual communications rsch. dept., 1987—; adj. prof. Rutgers U., New Brunswick, N.J., 1976, 79, CCNY, 1983-84, Columbia U., N.Y.C., 1987, 93; negotiator Internat. Standards Orgn., Am. Nat. Standards Inst.; negotiator videotelephone com. Consultative Com. on Internat. Tel. & Tel., U.S. core mem., 1983—. Co-author: Image Transmission Tech., 1979, Digital Pictures, 1988; assoc. editor Circuits and Systems for Video Tech., 1991—; contbr. over 50 papers to tech. jours.; patentee in field. Fellow IEEE (orgn. com. Picture Coding Symposium 1988—, assoc. editor Image Comm. 1989—); mem. Phi Beta Kappa, Sigma Xi. Avocations: sailing, skiing, guitar playing. Office: AT&T Bell Labs HO 4C-538 Holmdel NJ 07733

HASKELL, CHARLES MORTIMER, medical oncologist, educator; b. L.A., June 7, 1939; s. Maurice Mortimer and Beryl (Rentfro) H.; m. Christine Wursten, Sept. 4, 1959; 1 child, Candace M. BA, U. Calif., Santa Barbara, 1961; MD, U. Calif., San Francisco, 1965. Diplomate Am. Bd. Internal Medicine. Resident in medicine U. Calif., San Francisco, 1965-67, instr. medicine, 1970-71; clin. assoc. Nat. Cancer Inst., Bethesda, Md., 1967-70; asst. prof. medicine UCLA Sch. Medicine, 1971-76, assoc. prof. medicine, 1976-82, prof. medicine, 1982—; dir. Wadsworth Cancer Ctr., VA West L.A., 1978—; mem. adv. com. oncologic drugs FDA, Rockville, Md., 1979-82; mem. oncology com. Am. Bd. Internal Medicine, Phila., 1980-85; editorial bd. Jour. Clin. Oncology, Boston, 1992—. Co-author: Cancer Chemotherapy, 3d edit., 1980; editor: Cancer Treatment, 1980, 3d edit., 1990. Mem. Physicians for Social Responsibility. Asst. surgeon USPHS, 1967-70. Fellow ACP; mem. Am. Soc. Clin. Oncology (chair constitution and by-laws com. 1976-81, 84-90), Alpha Omega Alpha. Unitarian-Universalist. Achievements include pharmacology and clinical use of L-asparaginase, intra-arterial therapy with doxorubicin immunotherapy studies in breast cancer, monclonal antibodies against CEA for immunoscintigraphy. Office: West LA VA Med Ctr 111 N Cancer Ctr Los Angeles CA 90073

HASKIN, LARRY ALLEN, earth and planetary scientist, educator; b. Olathe, Kans., Aug. 17, 1934; s. Harvard Glenn and Mary Virginia (Callaway) H.; m. Mary Anita Gehl, Dec. 21, 1963; children: Dierk Allen, Rachel Lee, Jean Marie. B.A., Baker U., 1955; Ph.D., U. Kans., 1960. Asst. prof. Ga. Inst. Tech., 1959-60; instr. U. Wis., Madison, 1960-61; asst. prof. U. Wis., 1961-65, asso. prof., 1965-68, prof. chemistry, 1968-73; cons. NASA, 1970-73, Argonne Nat. Lab., 1960-68; chief planetary and earth scis. div. NASA-JSC, 1973-76; prof. earth and planetary scis., chemistry Washington U., St. Louis, 1976-90, R.E. Morrow Disting. prof. earth and planetary scis., prof., 1986—, chmn. dept., 1976-90; mem. mercury rev. panel Nat. Acad. Scis., 1970-71; mem. U.S. Nat. Com. on Geochemistry, 1975-78; mem. NASA Solar System Exploration Com., 1983-87, mem. mgmt. council, 1984-86, adv. com. Space and Earth scis. 1985-88; mem. NRC Com. Planetary and Lunar Exploration, 1985-88; mem. NASA Adv. Coun., 1988-90. Recipient Exceptional Sci. Achievement award NASA, 1971; Guggenheim fellow Max Planck Inst. for Nuclear Physics, Heidelberg, Germany, 1966-67. Fellow The Meteoritical Soc.; mem. Am. Chem. Soc., Geochem. Soc. (v.p. 1985-87, pres. 1987-89), Am. Geophys. Union, AAAS, Phi Beta Kappa, Sigma Xi. Research on trace inorganic elements in meteoritic, lunar and terrestrial matter.

HASKINS, CARYL PARKER, scientist, author; b. Schenectady, Aug. 12, 1908; s. Caryl Davis and Frances Julia (Parker) H.; m. Edna Ferrell, July 12, 1940. Ph.B., Yale U., 1930; Ph.D., Harvard U., 1935; D.Sc., Tufts Coll., 1951, Union Coll., 1955, Northeastern U., 1955, Yale U., 1958, Hamilton Coll., 1959, George Washington U., 1963; LL.D., Carnegie Inst. Tech., 1960, U. Cin., 1960, Boston Coll., 1960, Washington and Jefferson Coll., 1961, U. Del., 1965, Pace U., 1974. Staff mem. rsch. lab. Gen. Electric Co., Schenectady, 1931-35; rsch. assoc. MIT, 1935-45; pres., rsch. dir. Haskins Labs., Inc., 1935-55, dir., 1935—, chmn. bd., 1969-87; dir. E.I. du Pont de Nemours, 1971-81; research prof. Union Coll., 1937-55; pres. Carnegie Instn. of Washington, 1956-71, also trustee, 1949—; Asst. liaison officer OSRD, 1941-42, sr. liaison officer, 1942-43; exec. asst. to chmn. NDRC, 1943-44, dep. exec. officer, 1944-45; sci. adv. bd. Policy Council, Research and Devel. Bd. of Army and Navy, 1947-48; cons. Research and Develop. Bd., 1947-51, to sec. def., 1950-60, to sec. state, 1950-60; mem. Pres.'s Sci. Adv. Com., 1955-58, cons., 1959-70; mem. Pres.'s Nat. Adv. Commn. on Libraries, 1966-67, Joint U.S.-Japan Com. on Sci. Coop., 1961-67, Internat. Conf. Insect Physiology and Ecology, 1971-73; panel advisers Bur. East Asian and Pacific Affairs, Dept. State, 1966-68; mem. Sec. Navy Adv. Com. on Naval History, 1971-83, vice chmn., 1975-83. Author: Of Ants and Men, 1939, The Amazon, 1943, Of Societies and Men, 1950, The Scientific Revolution and World Politics, 1964; contbr. to anthologies and tech. papers.; Editor: The Search for Understanding, 1967; Chmn. bd. editors: Am. Scientist, 1971-83 ; chmn. publs. com., 1971-83 . Trustee Carnegie Corp. N.Y., 1955-80, hon. trustee, 1980—, chmn. bd., 1975-80; trustee Rand Corp., 1955-65, 66-75, adv. trustee 1988—; fellow Yale Corp., 1962-77; regent Smithsonian Instn., 1956-80, regent emeritus, 1980—, mem. exec. com., 1958-80; bd. dirs. Council Fgn. Relations, 1961-75, Population Council, 1955-80; bd. dirs. Ednl. Testing Service, 1958-61, 67-71, chmn. bd., 1969-71; trustee Center for Advanced Study in Behavioral Scis., 1960-75, Thomas Jefferson Meml. Found., 1972-78, Council on Library Resources, 1965—, Pacific Sci. Center Found., 1962-72, Asia Found., 1960—, Marlboro Coll., 1962-77, Wildlife Preservation Trust Internat., Inc., 1976—, Nat. Humanities Center, 1977—; trustee Woods Hole Oceanographic Instn., 1964-73, mem. council, 1973—; bd. dirs. Franklin Book Programs, 1953-58; mem. Save-The-Redwoods League, 1943—, mem. council, 1955—; mem. vis. coms. Harvard, Johns Hopkins; bd. visitors Tulane U. Yale Corp. fellow, 1962-77; recipient Presdl. Cert. Merit U.S., 1948, King's medal for Service in Cause of Freedom Gt. Britain, 1948, Joseph Henry medal Smithsonian Inst., Centennial medal Harvard U., 1991. Fellow AAAS (bd. dirs. 1971-75), Am. Phys. Soc., Am. Acad. Arts and Scis., Royal Entomol. Soc., Entomol. Soc. Am., Pierpont Morgan Library; mem. NAS, Washington Acad. Scis., Nat. Geog. Soc. (trustee 1964-84 , honorary trustee, 1984—, fin. com. 1972-85, com. on rsch. and exploration 1972—, exec. com. 1972-84), Royal Soc. Arts (Benjamin Franklin fellow), Faraday Soc., Met. Mus. Art, Am. Mus. Natural History (trustee 1973-89, bd. mgmt. 1973-89), Am. Philos. Soc. (councillor 1976-78, 81-83), Brit. Assn. Advancement Sci., Linnean Soc. London, Internat. Inst. Strategic Studies, Asia Soc., Japan Soc., Biophys. Soc., N.Y. Zool. Soc., N.Y. Acad. Scis., Audubon Soc., N.Y. Bot. Garden, P.E.N., Pilgrims, Phi Beta Kappa, Sigma Xi (nat. pres. 1966-68, dir. 1966-83), Delta Sigma Rho, Omicron Delta Kappa. Episcopalian. Clubs: Somerset (Boston), St. Botolph (Boston); Century (N.Y.C.), Yale (N.Y.C.); Mohawk (Schenectady); Metropolitan, Cosmos (Centennial award 1978, bd. mgmt. 1973-76); Lawn (New Haven). Home: 22 Green Acre Ln Westport CT 06880-5027 Office: Haskins Labs Inc 1545 18th St NW Washington DC 20036-1345

HASLANGER, MARTIN FREDERICK, pharmaceutical industry professional, researcher; b. Dayton, Ohio, Mar. 27, 1947; s. John Frederick and M. Isabelle (McEowen) H.; m. Martha Louise Anderson, June 29, 1969; children: Andrea Louise, Jonathan Frederick. BS in Chemistry, Denison U., 1969; PhD in Chemistry, U. Mich., 1974. Postdoctoral fellow (with E.J. Corey) chemistry dept. Harvard U., Cambridge, Mass., 1974-76; rsch. investigator Squibb Inst. Med. Rsch., Princeton, N.J., 1976-80, sr. rsch. investigator, 1980-81, group leader, 1981-85; assoc. dir. Schering-Plough Rsch., Bloomfield, N.J., 1985-88, dir. chem. rsch., 1988-92; dir. chemistry, biochemistry pharm. Lilly Rsch. Labs., Indpls., 1992—. Referee Jour. Medicinal Chemistry, 1981—, Jour. Organic Chemistry; contbr. articles to profl. jours., including Jour. Am. Chem. Soc., Jour. Organic Chemistry, Pharacologist, European Jour. Pharmacology, others. Mem. AAAS, Am. Chem. Soc., Am. Heart Assn., N.Y. Acad. Scis., Phi Lambda Upsilon. Achievements include discovery of new drug; design and synthesis of enzyme inhibitors and receptor antagonists, peptide memetics, use of data and computational chemistry to design and optimize small molecule-large molecule interactions; 45 patents. Home: 12597 Chyverton Cir Carmel IN 46032

HASPEL, ARTHUR CARL, podiatrist, surgeon; b. Bklyn., May 18, 1945; s. Ephriam and Sophie (Rabinowitz) H.; m. Anna Kiperman, Feb. 2, 1969; children: Mark Steven, Alan Charles. BS, L.I. U., 1967; D of Podiatric Medicine, Ohio Coll. Podiatric Medicine, 1972. Cert. Am. Bd. Ambulatory Foot Surgery, Am. Bd. Quality Assurance and Utilization Rev. in Podiatric Medicine, Am. Inst. of Foot Medicine in Surgery & Podiatric Medicine. Practice medicine specializing in podiatry Chgo., 1972-77, Hallandale, Fla., 1977—, Miami Shores, Fla., 1990—; lectr. in field. Author HMOs- Long Term Effects, 1986, (with others) Lions and Retinitis Pigmentosa, 1982, Procedural Podiatric Service Codes, 1986. Bd. govs. Hillel Community Day Sch., 1980-82; mem. land acquisition and devel. com. Beth Torah Congregation, 1986; active Am. Red Magen David for Israel (charter 1st v.p. South Broward profl. chpt. 1980, 1st v.p. 1980—, fin. sec. 1980—, southeastern U.S. steering com. 1980-82), Highland Lakes Homeowners Assn. Fellow Acad. Ambulatory Foot Surgery (trustee, 1984-89, 91—, mem. profl. standards com., research and devel. com., seminars com., ins. com., pres. Region IX, 1981-85, sec.-treas. 1979-81, mem.-at-large 1978-79, mem. adv. bd. 1985—, seminar coordinator 1979-85, chmn. statewide referral service, coordinator continuing med. edn. State of Fla. 1979-85), Am. Assn. Hosp. Podiatrists, Am. Podiatric Circulatory Assn., Am. Soc. Podiatric Medicine; mem. Am. Coll. Foot Surgeons, Fla. Podiatric Med. Assn. (pres. 1990-91, 1st v.p. 1989-90, v.p. 1988-89, treas. 1987-88, sec. 1986-87, exec. com. sec. 1986-87, exec. bd. 1986, chmn. banner com. 1981, 86, chmn. pub. health com., 1982, co-chmn. membership com., 1983, 84, 85, mem. sci. and surg. program com. 1984—, chmn. sunshine com. 1986, chmn. profl. standards rev. orgn. 1982), Broward County Podiatric Med. Assn. (co-chmn. legis. appreciation night 1979-81, sec. 1981-83, sec. May Day com., sec. continuing med. edn. cert., March of Dimes Walk-a-Thon, health fairs, ethical advt. com.), Am. Podiatric Circulatory Assn., Am. Podiatric Med. Assn., Ill. Podiatry Assn., Fla. Podiatric Med. Assn. (del. to Am. Podiatric Med. Assn. 1989-91), Am. Pub. Health Assn., Ohio Coll. Podiatric Medicine Alumni Soc., Hallandale C. of C., Lions (charter treas. Fla. community hearing bank 1979, treas. 1979-82, 83-85, pres. 1982-83, v.p. 1985—, exec. dir. 1983-84, mem. exec. com. 1979-84, TV and radio liaison 1982-84, bd. dirs. Hallandale club 1978-80, treas. 1978-80, charter treas. dist. 35A for Retinitis Pigmentosa 1980, treas. 1980-82, mem. exec. com. 1980-82, co-chmn. state com. on deaf and hearing impaired, multiple dist. 35, sec. Fla. Lions deaf profject 1980-82), Optimists (soccer coach, tee ball coach), Kappa Tau Epsilon. Office: 1105 E Hallandale Beach Blvd Hallandale FL 33009-4431

HASSAN, AFTAB SYED, scientific research director; b. Lahore, Punjab, Pakistan, Apr. 20, 1952; came to U.S., 1976; s. Maqsud Syed and Saliha Aktar Hassan. BSCE with distinction, U. Engring. and Tech., Lahore, 1973; postgrad. in aerodyns., Colo. State U., 1976; MS, George Washington U., 1977; PhD, Columbia Pacific U., 1985. Scientist in ocean, coastal and environ. engring. George Washington U., 1977-84; tech. asst. George Washington U., Washington, 1979-84, asst. prof., 1980-85; chmn. math. and sci. Emerson Prep. Inst., Washington, 1979-89; acad. coord. Ctr. for Minority Student Affairs, Georgetown U. Med. Sch., Washington, 1983-87; v.p. Met. Acctg. Assocs., Washington, 1987-88; acctg. mgr. Washington Info. Group, 1988-91; owner Met. Acctg. and Rsch., Washington, 1988-91; sr. tech. editor and author Betz Pub. Co., Rockville, Md., 1991—, designer new products, dir. sci. rsch., 1991—; bd. dirs. French Pastries, Inc., Washington, McPherson News and Gift, Washington; rsch. assoc. Chesapeake Bay Tidewater Adminstrn. in Fish Physiology, part-time 1979-88; prof. reasoning and problem solving Sch. for Summer and Continuing Edn., Georgetown U., 1983—; acad. specialist Drew/UCLA Med. Sch. Program, 1985-87; acad. cons. for Grad. Record Exam/Med. Coll. Admission Test, Morgan State U., 1989—; workshop leader Tulane MedRep Program, Charles R. Drew Ctr. Ednl. Achievement, others. Major author, dir. sci. rsch.: A Complete Preparation for the MCAT, 6th edit., 1992, 93, Preparing for the D.A.T., 1992, Dental Admission Test-The Betz Guide, 1993, Optometry Admission Test-The Betz Guide, 1993, Problem Solving Software for the MCAT-Biological Sciences, 1993. Recipient Merit award Nat. Assn. Chiefs of Police, Leaders in Community Svc. award Am. Biog. Inst., 1990. Mem. ASCE, NSPE, Am. Soc. Engring. Edn., Am. Inst. Profl. Bookkeepers, Soc. Am. Mil. Engrs., Nat. Soc. Tax Profls., Nat. Law Enforcement Acad. (hon.), Nat. Assn. Advisors for Health Professions, Nat. Assn. Fgn. Student Advisors, Nat. Sci. Tchrs. Assn., Nat. Assn. Profl. Educators. Avocations: exotic cooking, swimming, collecting currency. Home: Ste 291 4401-A Connecticut Ave NW Washington DC 20008

HASSAN, HOSNI MOUSTAFA, biochemistry, toxicology and microbiology educator, biologist; b. Alexandria, Egypt, Sept. 3, 1937; came to U.S., 1961; s. Moustafa Hosni and Sania M. (El-Harir) H.; m. Awatif El-Domiaty, July 12, 1961 (div. May 1983); children: Jehan, Suzanne; m. Linda C. McDonald, Dec. 16, 1992. BSc, Ain Shams U., Cairo, 1959; PhD, U. Calif., Davis, 1967. Asst. prof. Cairo High Polytech. Inst., 1968-70; assoc. prof. U. Alexandria, 1970-72; vis. prof. McGill U., Montreal, Can., 1972-74; assoc. prof. McGill U. Med. Sch., Montreal, Can., 1979-80; rsch. assoc. biochemistry Duke U. Med. Ctr., Durham, N.C., 1975-79; assoc. prof. N.C. State U., Raleigh, 1980-84, prof., 1984—. Mem. editorial bd. Free Radicals in Biology and Medicine, 1984—; author: (chpts.) Enzymatic Basis of Toxicology, 1980, Biological Role of Copper, 1980, Advances in Genetics, 1989, Stress Responses in Plants, 1990, FEMS Microbiol. Reviews, 1993; author, co-author over 100 rsch. publs. NIH fellow, 1967, Fulbright sr. fellow Paris, 1987-88; NIH/NSF grantee N.C. State U., 1982, 83—. Fellow Am. Inst. Chemists, Sigma Xi; mem. Am. Soc. Biol. Chemists and Molecular Biology, Am. Soc. for Microbiology. Democrat. Achievements include discovery of the toxicity and mutagenicity of oxygen free radicals and the protective role of superoxide dismutases and hydroperoxidases; the mechanism of regulation of the synthesis of the enzyme superoxide dismutase in bacteria. Home: 1309 Swallow Dr Raleigh NC 27606-2414 Office: NC State U Biochem Dept 344 Polk Hall Box 7622 Raleigh NC 27695-7622

HASSAN, JAWAD EBRAHIM, power engineer, consultant; b. Manama, Bahrain, Dec. 25, 1949; s. Ebrahim Ali Rasrumani and Fatima Ahmed Baqer; m. Shams Sadiq Albaharna, May 2, 1979; children: Muhannad, Majid, Lama. BS, Engring. Coll., Alleppo, Syria, 1974; MS, Trinity Coll., Dublin, Ireland, 1978. Registered power engr. Projects engr. Bahrain Supply Electricity Directorate, Manama, 1975-77, head of constrn., 1979-83, mgr. planning and devel., 1984-91; dir. Water Transmission Directorate, Manama, 1992—; mem. corp. planning com. Ministry of Works, Power and Water, Bahrain, 1992—, elec. and electronics specifications com. Ministry Commerce, Bahrain, 1988—, energy conservation com. Bahrain Supply Electricity Directorate, 1981-87. Contbr. articles to profl. jours. Mem. Planetory Soc., Pasadena, Calif., 1989. Mem. IEEE (sr.), ASTM, Electricity Rsch. Assn., Bahrain Society Engineers. Home: PO Box 20101, Manama Bahrain Office: Ministry Works Power Water, PO Box 2, Manama Bahrain

HASSAN, MOHAMMAD HASSAN, electrical engineering educator; b. Elkazimiya, Iraq, June 1, 1956; came to U.S., 1984; BSEE, U. Baghdad, 1977; MS in Electronics, U. of Tech., 1979; MSEE, Wayne State U., 1986, PhD in Elec. and Computer Engring., 1988. Registered profl. engr., Mich. Instr. Kuwait U., Kuwait City, 1981-84; instr. part-time Wayne State U., Detroit, 1988; assoc. prof. Lawrence Tech. U., Southfield, Mich., 1988—; adj. prof. Wayne State U., Detroit, 1992—; pres. New Innovative Technologies, Farmington Hills, 1992—. Contbr. articles to profl. publs. Mem. IEEE, Optical Engring. Soc., IEEE Computer Soc., Sigma Xi, Tau Beta Pi, Eta Kappa Nu. Achievements include rsch. advances in uncertain reasoning and its applications in decision making and object recognition, research in real-time image processing, research advances in automated vehichles, in smart vehichles, electric hybrid vehichles and pollusion control. Office: Lawrence Tech U 21000 W Ten Mile Rd Southfield MI 48075

HASSELBACHER, PETER, rheumatologist, educator; b. N.Y.C., Mar. 25, 1946; s. Frank X. and Gloria V. (Insel) H.; m. Martha J. Young, Dec. 29, 1969; children: David A., Matthew J. BA with high honors, Wesleyan U., 1968; MD, Columbia U., 1972. Diplomate Am. Bd. Internal Medicine, Am. Bd. Rheumatology, Am. Bd. Geriatrics, Am. Bd. Diagnostic Lab. Immunology. Resident in medicine Columbia-Presbyn., N.Y.C., 1972-74; fellow in rheumatology U. Pa., Phila., 1974-76, from assoc. to asst. prof. medicine, 1976-78; from asst. prof. to assoc. prof. medicine Dartmouth Coll., Hanover,

N.H., 1978-84; chief divsn. rheumatology U. Louisville, 1984-92, prof. medicine, 1984—; mem. Ctr. for Health Svcs. and Policy Rsch., U. Louisville, 1990—. Mem. editorial bd. Jour. Rheumatology; contbr. 44 articles to profl. jours. Bd. dirs. Arthritis Found. Ky., Louisville, 1984—, U. Louisville Athletic Assns., 1990—; exec. com. Frazier Rehab. Hosp., Louisville, 1989-92. Fellow ACP (health and pub. policy com. 1991—), Am. Coll. Rheumatology (councilor ctrl. region 1986-90); mem. Ky Med. Soc. (com. on care of elderly), Jefferson County Med. Soc. (legis. com., indigent care com. 1991—). Office: U Louisville Dept Medicine Louisville KY 40292

HASSELL, PETER ALBERT, electrical and metallurgical engineer; b. Springfield, Mass., Aug. 8, 1916; s. Cornelius and Erica (Farkasch) H.; m. Elizabeth Heffner, July 11, 1941; children: Cornelius, Kenneth Joseph. Student, Case Sch. Applied Sci., 1935-38, Fenn Coll., 1940-41, U. Wis., 1946-48. Lab. technician Tocco Div., Ohio Crankshaft, Cleve., 1940-42; radar officer USN, USS Mindanao and Puget Sound Navy Yard, 1942-45; application engr. Allis Chalmers Mfg. Co., Milw., 1945-48, Lindberg Engring. Co., Chgo., 1948-51, Westinghouse Electric Co., Chgo., 1951-54; high freqency div. mgr., then asst. to pres. Ajax Magnethermic Corp, Warren, Ohio, 1954-83; retired, 1983; cons. Hassell Assocs., 1983—. Contbr. articles to profl. jours. and tech. papers. Recycle coord. Breckenridge Village, Willoughby, Ohio, 1989-91; mgmt. counselor SCORE/SBA, Willoughby, 1987—, Ctr. Indsl. Heat Treating Processes, U. Cin., 1990; active with Ret. Srs. Vol. Program; mem. Willoughby Keep America Beautiful Team. Lt. USN, 1941-45. Mem. ASM Internat., Am. Foundrymens Soc., Soc. Automotive Engrs., Am. Assn. Iron & Steel Engring., Am. Ordnance Assn., Great Lakes Hist. Soc., Nat. Assn. Ptnrs. in Edn. Unitarian. Avocations: sailing, iceboating, bicycling, cross country skiing., boomeranging. Home: 194 N Ridge Dr Willoughby OH 44094-5639

HASSELRIIS, FLOYD NORBERT, mechanical engineer; b. N.Y.C., Nov. 6, 1922; s. Malthe and Ruth (Borch) H.; m. Helen Mihalyfi, Sept. 23, 1950; children: Lauren, Norbert. BS in Mech. Engring., Columbia U., 1943; MS, U. Del., 1951. Diplomate Am. Acad. of Environ. Engrs.; registered profl. engr., N.Y., N.J., Mass., Vt. Fla. Instr. in mech. engring. Cooper Union, N.Y.C., 1948-53; project engr. Am. Hydrotherm Corp., 1953-55, chief engr., 1955-71; chief engr. Combustion Equipment Assocs., 1971-80; sr. engr. Gershman, Brickner & Bratton, 1985-89, Doucet & Mainka, P.C., 1989-92; cons. engr. Forest Hills, N.Y., 1992—. Contbr. over 30 articles to profl. jours. Mem. ASME (exec. com. solid waste processing divsn., chmn. rsch. com. on indsl. and mcpl. waste 1976—), Am. Inst. Chem. Engrs., Nat. Soc. Profl. Engrs., Air & Waste Mgmt. Assn. Achievements include research in combustion of wastes, particularly in regard to dioxins and heavy metals, impact of waste composition and combustion process on emissions and environmental impact of emissions and ash residues. Home and Office: 52 Seasongood Rd Forest Hills NY 11375-6033

HASSIALIS, MENELAOS DIMITIOU, mineral engineer; b. N.Y.C., Dec. 25, 1909; s. Dimitri Athanaslou and Maria (Mantsalk) H.; m. Ruth Elizabeth Arnowitz, June 10, 1931 (dec.); children: Joan I. Buchor, Peter John. BA cum laude, Columbia Coll., 1931; MA, Columbia U., 1933; DS (hon.), Bard Coll., 1954. Assoc. mineral engr. Columbia U., N.Y.C., 1944, asst. prof., 1945, assoc. prof., 1947, prof., 1951, Krumb prof., 1954, exec. officer, 1951-87; dir. USAEC Lab. Columbia U., 1951-58, vice chmn. Inst. Study Scis. in Human Affairs, 1966-70; pres. Pacific Uranium Mines Co., 1959-61; v.p. Tech. Investors Corp., 1961-68; dir. Ambrosia Lakes Uranium Corp., Kerr McGee Nuclear Fules Corp.; chmn. bd. Sandvik Steel Inc., 1973-87; chmn. Disston Corp., 1976; mem. Am. delegation to Geneva Confs., 1955, 58; head UN mission, Turkey, 1964. Co-author: Microscopy, 1945, Handbook of Mineral Engineering, 1945; contbr. articles to profl. jours. Dir., v.p. Valley Hosp., Ridgewood, N.J. Recipient citation govt. Pakistan, 1958, citation Tech. U. W. Berlin, 1968, medal Freiberg Acad., Germany, 1969, Reud Econ. award Nat. Inst. for Minerals and Petroleum, 1992, various citations from govts. Egypt, USSR, Sweden. Fellow Exploeres Club; mem. Am. Inst. Mineral, Metallurgical Petroleum Engrs., Am. Chem. Soc., Mineral & Metallurgical Soc. Am., Soc. Mineral Engrs. Independent. Greek Orthodox. Achievements include design of nine and mill of Gooseberry mine in Nev., Oil-shale mine for Mobil Oil; cons. of Manhattan project, chemical warforce svc. colved fire bomb problem and mortor bas problem. Home and office: 122 Phelps Rd Ridgewood NJ 07450

HASSNER, ALFRED, chemistry educator; b. Czernowitz, Romania, Nov. 11, 1930; s. Sigmund and Mina Hassner; m. Cyd Schachter; children: Lillian, Lawrence. BSc, U. Nebr., 1952, PhD, 1956. Rsch. fellow Harvard U., Cambridge, Mass., 1956-57; from asst. prof. to prof. U. Colo., Boulder, 1957-75; leading prof. SUNY, Binghamton, 1975-84; prof. Bar-Ilan U., Ramat Gan, Israel, 1984—; vis. prof. U. Wurzburg, Fed. Republic Germany, 1971, Weizmann Inst., Rehovot, Israel, 1982-83, U. Calif., Berkeley, 1989; cons. 3M Co., St. Paul, TAMI, Haifa. Mem. editorial bd. Jour. Organic Chemistry; editor: Small Ring Heterocycles, 1983-84, Advances in Asymetric Synthesis, 1993-94; contbr. over 200 articles to profl. jours. Fellow Von Humboldt Found., 1971, Nat. Cancer Inst., 1972-73, Lady Davis Found., 1979, Meyerhoff fellow Weizmann INst., 1982; recipient Fulbright Sr. award NRC, 1984. Mem. Am. Chem. Soc., Israel Chem. Soc. (pres. 1991—), Royal Soc. Chemistry. Achievements include research in stereochemistry, organic synthesis, heterocyclic chemistry, anticancer agents. Office: Bar-Ilan U, Chemistry Dept, Ramat Gan 52900, Israel

HAST, ROBERT, hematologist; b. Stockholm, Apr. 2, 1945; s. Nils and Brita Hast; m. Beatrice Krafft, June 14, 1969; children: Nils, Cecilia, Gustav. MD, Karolinska Inst., Stockholm, 1971, PhD, 1979. Asst. prof. medicine Karolinska Hosp., Stockholm, 1981-85, assoc. prof. medicine, 1985-87; assoc. prof. medicine Danderyd (Sweden) U. Hosp., 1987—; head div. hematology Danderyd Hosp., Stockholm, 1987—. Contbr. articles to profl. jours. Mem. Swedish Soc. Med. Sci., Swedish Soc. Hematology, Internat. Soc. Hematology. Avocations: tennis, golf. Office: Danderyd Hosp, S 182 88 Danderyd Sweden

HASTERLO, JOHN S., civil engineer, researcher; b. Omaha, Apr. 15, 1963; s. Stanley and Kathryn M. H.; m. Constance S. Anthone, Aug. 7, 1987; children: Samuel J., Ian P. BSCE, U. Nebr., 1992. Environ. rsch. engr. City of Omaha Pub. Works, 1992—. Mem. ASCE, Inst. Transp. Engrs. Home: 13514 Spring St Omaha NE 68144

HASTINGS, HAROLD MORRIS, mathematics educator, researcher, author; b. Dayton, Ohio, Nov. 21, 1946; s. Julius M. and Celia A. (Morse) H.; m. Gretchen E. Saalbach, June 2, 1968; children: Curtis, Matthew. BS, Yale U., 1967; MA, Princeton U., 1969, PhD, 1972. From instr. to assoc. prof. math. Hofstra U., N.Y., 1968-81, prof., 1981—, dept. chmn., 1985-90, 93—, assoc. dean., 1990-93; vis. assoc. prof. SUNY, Binghamton, 1974-75, U. Ga., Athens, 1978-79; prin. Hastings, Saalbach Assocs., Inc., Garden City, N.Y., 1983—; mem. working group on supercomputers NASA, Greenbelt, Md., 1985-90. Author: (with D. Edwards) Cech and Steenrod Homotopy Theory, 1974; editor: (with M. Kochen) Advances in Cognitive Science, 1988; contbr. articles to profl. jours. Patentee in field for computerized acoustic fetal monitor; research in strong pro homotopy theory in algebraic topology, fractal models in ecology. Pres., v.p. Garden City Lay Ecumenical Com., N.Y., 1983—. Grantee NSF, 1977, 80, Woodrow Wilson Found, NAS. Mem. Am. Math. Soc., Assn. Computing Machinery, Soc. Math. Biology. Avocations: running, photography, music. Office: 109 Adams Hall 103 Hofstra U Hempstead NY 11550

HASTINGS, S. ROBERT, architect, building researcher; b. Plainfield, N.J., Nov. 16, 1945; s. Stanley Robert and Elaine (Mansfield) H.; m. Dagmar Tuzilova, Dec. 9, 1972; 1 child, Robert Alex. BArch, Cornell U., 1968, MS, 1972. Registered architect Mass. Architect Bedar and Alpers, Boston, 1972-73, Fuller and Sadao, CAmbridge, Mass., 1973-74, Abrash and Eddy, Reston, Va., 1974-75; researcher Nat. Inst. Standards and Tech., Gaithersburg, Md., 1975-80; rsch. mgr. EMPA Dübendorf (Switzerland), 1980-88; program dir., lectr. Solarachitektur ETH, Zürich, Switzerland, 1989—; operating agt. Internat. Energy Agy., Paris, 1986—; bd. dirs. AG Jolar Noldrhine Westfallen, Düsseldorf, Germany; sec. bd. officers PLEA Internat. Solar Energy Soc., Caulfield, Austria, 1988-91. Author, editor: Passive Solar Commercial and Industrial Buildings, 1993, Sonnenfiber, 1991; author: Window Design Strategies, 1977. With U.S. Army, 1969-71. Recipient Superior Accomplishment award Nat. Bur. Standards, 1976, Outstanding Performace award, 1977, Communication award, 1978. Mem. Interant. Solar Energy Soc. (Swiss Nat. div.), Interant. Assn. for Solar Energy Edn. Episcopalian. Home: Erikastrasse 18, CH-8304 Wallisellen Switzerland Office: Solararchitektur, ETH Honggerberg, CH-8093 Zurich Switzerland

HATA, KOICHI, heat transfer researcher; b. Hiroshima, Japan, July 12, 1949; s. Fujito and Etsuko Hata; m. Michie Harada, Oct. 22, 1978; children: Akiko, Ryoko. BS, Doshisha U., Kyoto, Japan, 1972; PhD, Kyoto U., 1991. Tech. officer Inst. Atomic Energy, Kyoto U., 1972-80, instr., 1980—. Mem. ASME (Best Paper award heat transfer divsn. 1990, Melville medal 1991). Home: 9-10 Seifu-cho, Otsu Shiga 520-02, Japan Office: Inst of Atomic Energy, Kyoto Univ, Uji Kyoto 611, Japan

HATADA, KAZUYUKI, mathematician, educator; b. Maebashi, Gunma, Japan, Dec. 23, 1951; s. Kiyoshi and Tokiko H.; m. Kumiko Yoshikawa, Dec. 15, 1985; 1 child, Hidehiko. BS, U. Tokyo, 1974, MS, 1976, DSc, 1979. Rsch. fellow faculty sci. U. Tokyo, 1979-80; assoc. prof. math. faculty edn. Gifu U., Gifu City, Japan, 1981—; vis. prof. U. Paris XI, autumn 1993. Contbr. articles to profl. mathematical jours. Recipient Insignia of Dedications, Cambridge, 1988, Silver medal, 1989, Gold medal for 1st 500, 1990, The Internat. Order of Merit, 1990, 20th Century Award for Achievement, 1993. Mem. World Inst. of Achievement (life), Math. Soc. of Japan, Am. Math. Soc. (reviewer), Astro. Soc. of Japan. Achievements include proofs of the 10 new congruences enjoyed by the Hecke operators on SL (2,Z); discovery that the Hecke rings act naturally on the integral homology groups and l-adic cohomology groups of suitable smooth projective toroidal compactifications of the higher dimensional modular varieties and the investigation of this properties; obtained the new expressions of the local zeta functions of the compactified Hilbert Modular schemes in terms of the action of the Hecke rings; study of the parabolic cohomology, others. Home: 6-2 Chiyoda 2 chome, Maebashi Gunma 371, Japan Office: Gifu U Dept Math Faculty Edn, 1-1 Yanagido, Gifu 501-11, Japan

HATANAKA, HIROSHI, neurosurgeon; b. Toyama Prefecture, Japan, Apr. 20, 1932; s. Taichi and Hana (Takahashi) H.; m. Anita Louisa Beck, Oct. 15, 1973; children: Elsa, Clara. MD, U. Tokyo, 1957, D of Med. Scis., 1963. Resident in surgery and neurosurgery U. Tokyo Hosp., 1958-62, asst. in surgery, 1963-64; Fulbright scholar, surg. rsch. fellow Harvard U., 1964-67; clin. and rsch. fellow Mass. Gen. Hosp., Boston, 1964-67, vis. fellow neurosurgery, 1971-72; asst. in neurosurgery U. Tokyo Hosp., 1967-73; prof. neurosurgery Teikyo U., Tokyo, 1971—; vis. resident in neurosurgery Montreal Neurol. Inst. and Hosp. at McGill U., summer 1966; disting. vis. prof. Ohio State U., Columbus, 1991-92. Mem. Japanese Cancer Assn., Japanese Surg. Soc., Japanese Congress Neurosurgeons, World Fedn. Neurol. Surgeons, Japan Neurosurg. Soc., Japan Soc. Practicing Surgeons, Internat. Coll. Surgeons, Japan Assn. Cancer Rsch., Asian and Australasian Soc. Neurol. Surgeons, Japan Soc. Cen. Nervous System Computed Tomography, Japan Radiol. Soc., Internat. Soc. Radiology, Japan EEG Soc., Japan Neuropathol. Soc., Japan Neurochemistry Soc., Internat. Soc. for Neutron Capture Therapy (founder, 1st pres. and sec.), Japan Soc. Clin. Imaging (exec. bd.), J.E. Purkyne Czech. Med. Assn. (hon.), Czech Neurosurg. Soc. (hon.). Mem. Society of Friends. Home: 3-27-E1005 Nakadai, Itabashi-ku, Tokyo 174 Japan Office: Teikyo U Hosp, 2-11-1 Kaga, Itabashi-ku, Tokyo Japan

HATANO, SADASHI, molecular biology educator; b. Kobe, Japan, Apr. 12, 1929; s. Yoriaki and Kimiko Hatano; m. Kimie Hatano, May 27, 1958; children: Kazuko, Fumiko. BS, Osaka (Japan) U., 1954, MS, 1956, DSc, 1959. Rsch. assoc. Osaka U., 1959-60; rsch. assoc. Nagoya (Japan) U., 1961-71, assoc. prof., 1971-75, prof., 1975-93, prof. emeritus, 1993—. Editor: Cell Motility, 1979, 86, Molecular Biology of Physarum, 1986. Avocations: travel, photography, music.

HATCH, JEFFREY SCOTT, chemist; b. Hackensack, N.J., Aug. 13, 1964; s. Gerald Alfred Hatch and Kathleen Gay (Kehoe) Cassimore. BS in Chemistry, Trenton State Coll., 1986. Lab. technician Mobile Oil Corp., Pennington, N.J., 1983, 84; tchr. chemistry, physics Notre Dame High Sch., Lawrenceville, N.J., 1986-87; chemist Solkatronic Chemicals, Inc., Morrisville, Pa., 1987-91; site mgr. Solkatronic Chemicals, Inc., Chandler, Ariz., 1991--. Mem. Bordentown Environ. Com., N.J., 1991. Office: Solkatronic Chemicals Inc 7007 Sundust Rd Chandler AZ 85226

HATCH, MARK BRUCE, software engineer; b. Lynn, Mass., July 20, 1959; s. Carroll Bruce and Claire Adelle (Sherys) H. BS, U. Mass., 1981; MS, U. Mass., Lowell, 1990. Cons. Bassook & Brisk, Wayland, Mass., 1988-90; dir. R&D Toltran Ltd., Lake Zurich, Ill., 1990—; instr. computer sci. North Shore C.C., Lynn, Mass., 1988—. Mem. Internat. Asns. Machine Trans., Assn. Machine Trans. in Americas. Achievements include U.S. and fgn. patents for improved trans. system in the field of machine trans. based on a multi-lingual approach to computer analysis; design of a complete lang.-ind. multi-lingual machine trans. system. Home: 43 Archer St Lynn MA 01902

HATCH, MARY WENDELL VANDER POEL, laboratory executive, interior decorator; b. N.Y.C., Feb. 6, 1919; d. William Halsted and Blanche Pauline (Billings) Vander Poel; m. George Montagu Miller, Apr. 5, 1940 (div. 1974); children: Wendell Miller Steavenson, Gretchen Miller Elkus; m. Sinclair Hatch, May 14, 1977 (dec. July 1989). Pres. Miller Richard, Inc., Interior Decorators, Glen Head, N.Y., 1972—; bd. dirs. Eye Bank Sight Restoration, N.Y.C., 1975—, pres., 1980-88, hon. chair, 1988—; bd. dirs. Manhattan Eye Ear and Throat Hosp., N.Y.C., 1966—, v.p., 1978-90; sec. Cold Spring Harbor Lab., N.Y., 1985-89, 92—, bd. dirs., 1985-90; chair DNA Learning Ctr., 1991—; bd. dirs. Cold Spring Harbor Lab, 1991—, sec., 1992—. V.p. North Country Garden Club, Nassau County, N.Y., 1979-81, 1983-85; dir. Planned Parenthood Nassau County, Mineola, N.Y., 1982-84, Hutton House C.W. Post Coll., Greenvale, N.Y., 1982—; chair Hutton House, 1992—. Recipient Disting. Trustee award United Hosp. Fund, 1992. Mem. Colony Club (N.Y.C.), Church Club (N.Y.C.), Piping Rock Club (Long Island), Order St. John Jerusalem (N.Y.C.). Republican. Episcopalian. Home: Mill River Rd # 330 Oyster Bay NY 11771-2712

HATFIELD, (DAVID) BROOKE, analytical chemist, spectroscopist; b. Denver, Aug. 3, 1958; s. James Carl and Anne Margret (Chetterbock) H.; m. Carina Marie Cirrincione, Aug. 10, 1986. BA in Chemistry, U. Colo., Denver, 1983; PhD in Chemistry, U. Ariz., 1993. Chem. technician U.S. Geol. Survey, Denver, 1979-84, chemist, 1984-85, rsch. chemist, 1985-87; grad. teaching asst. dept. chemistry U. Ariz., Tucson, 1987-90; rsch. assoc. Lunar and Planetary Labs., U. Ariz., 1990—; mem. lab. design com. U.S. Geol. Survey, Denver, 1985-87. Contbr. articles to profl. jours. Achievements include design and construction of an automated instrument for selenium, arsenic and antimony determinations in geologic materials; a new approach to data storage/retrieval for spectroscopic modeling of non-equilibrium chemical systems. Office: U Ariz Lunar and Planetary Labs Tucson AZ 85721

HATHAWAY, ALDEN MOINET, II, electrical engineer; b. Honolulu, Dec. 8, 1958; s. Alden Moinet and Anna Harrison (Cox) H.; m. Carol Ann Smith, Mar. 26, 1983; children: Alden Moinet III, Mary Pamelia, Anna Margaret. BSEE, U. Va., 1982; postgrad., Washington U., 1986-88. Cert. energy mgr., lighting efficiency profl. Asst. mktg. engr. Ga. Power Co., Columbus, 1982-84; lighting engr., Sylvania Lighting Divsn. GTE, St. Louis, 1984-89; sr. lighting engr., Sylvania Lighting Divsn. GTE, Washington, 1989-91; mgr. energy programs, Sylvania Lighting Divsn. GTE, Danvers, Mass., 1991—; elect. engr. Lighting Assocs., St. Louis, 1989; green lights retail adv. group EPA, Washington, 1992—; Ashrae/ IES Energy Com., N.Y., 1992—. Contbr. articles to profl. jours. including Pub. Utilities Fortnightly, Intelligent Bldgs. Quarterly, Mainlighter. Mem. IEEE, Assn. Energy Engrs., Illuminating Engring. Soc. Republican. Episcopalian. Achievements include patent for master/satellite fluorescent lighting system. Office: 100 Endicott St Danvers MA 01923

HATHAWAY, DAVID ROGER, physician, medical educator; b. Lafayette, Ind., Jan. 8, 1948; s. Ralph Roger Hathaway and Marjorie Alice Friend; m. Elaine Mary Green, Aug. 3, 1974; children: Julia E., Alison S. AB, Ind. U., 1970, MD, 1975. Diplomate Am. Bd. Internal Medicine, Subsplty. Cardiovascular Diseases. Clin. asst. NHLBI/NIH, Bethesda, Md., 1976-77; intern Ind. U. Med. Ctr., Indpls., 1975-76, resident, 1976-77, chief resident, 1979-80, from asst. prof. to assoc. prof., 1980-86, prof., 1986—, chief cardiovascular divsn., dir. Krannert Inst. Cardiology, 1990—. Lt. comdr. USPHS, 1977-79. Fellow Am. Coll. Cardiology; mem. Am. Fedn. for Clin. Rsch. (pres. 1987-88), Am. Soc. for Clin. Investigation, Assn. Am. Physicians (sec. 1991—), Assn. Univ. Cardiologists, Phi Beta Kappa, Alpha Omega Alpha. Achievements include patents for composition and method for delivery of drugs, method for preventing restenosis following reconfiguration of body vessels. Home: Krannert Cardiology Inst 7966 N Illinois St Indianapolis IN 46202-4800 Office: Krannert Inst Cardiology 1111 W 10th St Indianapolis IN 46202-4800

HATHAWAY, RUTH ANN, chemist; b. Sidney, Ohio, Dec. 6, 1956; d. Earl Eugene and Mary Helen (Smith) Schmidt; m. Bruce Alan Hathaway, May 16, 1981. BS in Sci., Huntington Coll., 1979; postgrad., Purdue U., 1979-80. Instr. Harvey Mudd Coll., Claremont, Calif., 1981-82; head chemist So. Indsl. Products, Cape Girardeau, Mo., 1983-86; cons. Cape Girardeau, 1987-89; alterationist Patricks Cleaner, Cape Girardeau, 1988-89; lab. dir. Delta-Y Electric Co., Sedgewickville, Mo., 1989-91; chemist Environ. Analysis South, Cape Girardeau, 1991—. Editor: Safety Considerations in Microscale Lab, 1991; contbr. articles to profl. jours. Mem. exec. bd. dirs. NAACP, Cape Girardeau, 1986—; chmn. disaster com. ARC, Cape Girardeau, 1986-91; dir. S.E. Mo. Regional Sci. Fair, 1992—. Mem. Am. Chem. Soc. (div. chem. health and safety 1989—, nat. chemistry week coord., 1988—), Am. Inst. Chemists (local emergency planning com., chmn. S.E. Mo. 1989—). Republican. Home: 1810 Georgia St Cape Girardeau MO 63701-3816 Office: Environ Analysis South 1810 E Plaza Way Cape Girardeau MO 63701-5842

HATHERILL, JOHN ROBERT, toxicologist, educator; b. Waterford, Mich., Aug. 20, 1953; s. John William and Anna Marie (Morin) H. MS, Ea. Mich. U., 1978; PhD, U. Mich., 1985. Med. technologist U. Mich., Ann Arbor, 1976-78, sr. clin. chemist, 1979-86, rsch. assoc., 1980-85; sr. scientist Ciba-Geigy Pharms., Summit, N.J., 1986-87; rsch. dir. Stanford (Calif.) U., 1987-89; prof. U. Calif., Santa Barbara, 1990—; adj. prof. UCLA, 1990—. Contbr. articles to profl. jours., chpts. to books. Judge, mem. adv. bd. Calif. State Sci. Fair, L.A., 1992. Fellow World Ctr. for Exploration (bd. dirs. 1985); mem. Soc. Toxicology, Am. Soc. Clin. Pathologists, N.Y. Acad. Sci., Sigma Xi, Gamma Alpha. Achievements include research on oxygen free radicals and lipid peroxidation. Home: 3055 Paseo del Descanso Santa Barbara CA 93105

HATHEWAY, ALLEN WAYNE, geological engineer, educator; b. L.A., Sept. 30, 1937; s. Clarence Wilman and Marie Elizabeth (Sisto) H.; m. E. Anne Sellars, Apr. 4, 1959 (div. Jan. 1990); children: Shannon, Brian, Steven. AB in Geology, UCLA, 1961; MS in Geol. Engring., U. Ariz., 1966, PhD in Geol. Engring., 1971, profl. degree in geol. engring., 1982. Registered profl. engr., Ariz., Calif., Mass., profl. geologist, Calif., Maine; cert. engr. geology, Calif. Rsch. assoc. Lunar & Planetary Lab., Tucson, 1967-69; staff engr. Law Engring. Co., L.A., 1969-71; project mgr. geotech. br. U.S. Forest Svc., Arcadia, Calif., 1971-72; project mgr. Woodward-Clyde Cons., L.A., 1972-74; project geologist Shannon & Wilson Cons. Engrs., Burlingame, Calif., 1974-76; v.p., ch. geologist Haley & Aldrich Cons. Engrs., Cambridge, Mass., 1976-81; prof. geol. engring. U. Mo., Rolla, 1981—. Author: (tech. manual U.S. Army: Geotechnical Field Handbook, 1991; co-author: (textbook) Geology and Engineering, 3d edit., 1988, (tech. manual) Geophysical Methods for Hazardous Waste Site Characterization, 1992; editor, author: (tech. manual) AASHTO Manual on Subsurface Investment, 1988; editor jour. series Cities of the World, 1982—. Col. USAR, 1961-91. Recipient Cert. Appreciation Gov. Mo., 1989. Fellow ASCE (Calif. Outstanding Young Civil Engr. 1973, Mead prize 1975), Geol. Soc. Am. (chmn. engring. geol. divsn. 1980, Burwell award 1981), Geol. Soc. London; mem. Assn. Engring. Geologists (pres. 1985), Theta Xi, Sigma Gamma Epsilon, Sigma Xi,, Soc. Scabbard and Blade. Republican. Home: RFD 4 Box 66 Rolla MO 65401 Office: U Mo Dept Geol Engring 129 McNutt Hall Rolla MO 65401-0249

HATHEWAY, ALSON EARLE, mechanical engineer; b. Long Beach, Calif., Nov. 15, 1935; s. Earle Miller and Carla (Barnhart) H.; m. Robin Lewis, Aug. 24, 1968; children: Jason Teale, Teale. BSME, U. Calif., Berkeley, 1959. Registered profl. engr., Calif. Engr. Boeing Aerospace Co., Seattle, 1959-60, Ford Aerospace Co., Newport Beach, Calif., 1960-66; mgr. Xerox Corp., Pasadena, Calif., 1966-72, Hughes Aircraft Co., Culver City, Calif., 1972-76, Gould Inc., El Monte, Calif., 1976-79; pres. Alson E. Hatheway Inc., Pasadena, 1979—; instr. U. La Verne, Calif., 1989. Editor: Procs. Structural Mechanics of Optical Systems II, 1987, Procs. Precision Instrument Design, 1989; contbr. articles to profl. jours. Mem. AIAA (sr., chmn. San Gabriel Valley sect. 1992-93), ASME, NSPE, Soc. Photo-Optical Instrumentation Engrs. (fellow, instr. 1987—, conf chmn. 1987, 89, 91, program chmn. 1990, 91), Am. Soc. Precision Engrs., Calif. Soc. Profl. Engrs. (treas. 1965-66), Optical Soc. So. Calif. (pres. 1986-87), Opto-Mech. Engring. and Precision Instrument Design Working Group (chmn. 1992-93), Assn. Old Crows. Achievements include patents on optical scanner, precision transducer and optical calibration target; findings on Optical Analog and calibrated elasticity. Home: 419 S Meridith Ave Pasadena CA 91106 Office: 595 E Colorado Blvd Ste 400 Pasadena CA 91101

HATLEY, LARRY J., plant manager; b. Tahlequah, Okla., Aug. 4, 1946; s. Jerome Edward and Josephine (Hubbard) H.; m. Kathy Anne Ogden, Apr. 10, 1971. BS in Physics, Math., Northeastern State U., 1969. Registered profl. engr., Okla. Results engr. Okla. Gas and Electric Co., Harrah, 1969-73; supt. tech. svcs. Okla. Gas and Electric Co., Muskogee, 1973-85; plant mgr. Okla. Gas and Electric Co., Harrah, 1985-87, Muskogee, 1987—. Mem. Okla. Soc. Profl. Engrs. (pres. 1989). Home: 2909 Robin Ln Muskogee OK 74403 Office: Okla Gas and Electric PO Box 1270 Muskogee OK 74402

HATSCHEK, RUDOLF ALEXANDER, electronics company executive; b. Grafenberg, Austria, May 10, 1918; s. Rudolf Bernhard and Maria (Zischka) H.; m. Erika Lucia Satory, Jan. 10, 1946. Student, U. Prague, Czechoslovakia, 1936-40, U. Graz, Austria, 1945-46; Doctorate, U. Graz, 1946. Biochemist Interpharma AG, Prague, 1940-45; head of lab. Fux, Vienna, Austria, 1946-54; v.p. engring. BCF, Vienna, 1954-59; gen. mgr. instrumentation dept. AVL Engine Inst., Graz, 1959-65; v.p. research and devel. Vibro-Meter S.A., Fribourg, Switzerland, 1965-77; v.p. engring. div. ASULAB S.A., Neuchatel, Switzerland, 1978-83; cons. advanced piezo-electric applications in medicine, chronometry, automation and telecommunication; introduced piezo-electric aircraft-engine vibration monitoring. Mem. ASME, IEEE, Internat. Soc. Optical Engring., Swiss Phys. Soc., N.Y. Acad. Scis. Home: 3 Rue Vogt, CH1700 Fribourg Switzerland

HATTEM, ALBERT WORTH, physician; b. High Point, N.C., May 20, 1951; s. Henry Albert and Stella Jane (Penfield) H.; m. Deborah Elaine Bellew, Nov. 9, 1974. BA, U. S.C., 1985, MD, 1989. Officer McColl (S.C.) Police Dept., 1970-71; mgr. Norris Ambulance Svc., Spartanburg, S.C., 1971-72; spl. events mgr. Coca-Cola Co., Spartanburg, 1972-73; tng. officer, paramedic Emergency Med. Svcs., Spartanburg, 1973-76; tng. coord. emergency med. svcs. divsn. S.C. Dept. Health, Columbia, 1976-83; intern U. Tenn. Med. Ctr., Knoxville, 1989-90, resident in ob-gyn., 1990-93; obstetrician-gynecologist Paradise Valley Women's Care, Las Vegas, Nev., 1993—. Mem. faculty S.C. Heart Assn., Columbia, 1981-89; adv. coun. S.C. Emergency Med. Svcs., Columbia, 1983-89. Fellow ACOG (jr.), ACS (assoc.); mem. AMA (v.p. student sect. 1985-86), Am. Med. Student Assn. (chpt. v.p. 1985-86). Avocations: home computers, record collecting, boating, sailing. Office: Paradise Valley Women's Care 2080 E Flamingo Rd Ste 100 Las Vegas NV 89109

HATTIS, DAVID BEN-AMI, architect; b. Haifa, Palestine, Aug. 27, 1934; came to U.S., 1951; s. Ben William and Gertrude (Ben-Ami) H.; m. Rina Irene Frank, Aug. 10, 1965; 1 child, Eleanne. AB summa cum laude, Swarthmore Coll., 1955; M of Architecture, Harvard U., 1960. Registered architect Israel. Architect Gelfman & Hattis, Haifa, Israel, 1961-67; rsch. architect IRNES, Inc., Montreal, Ont., Can., 1967-69, NAt. Bur. Stas., Washington, 1969-71; pres. Bldg. Tech. Inc., Silver Spring, Md., 1972—. Contbr. articles to profl. jours. Mem. ASCE, NAt. Inst. Bldg. Sci., NAt. Trust Hist. PReservation, Cosmos Club, Phi Beta Kappa. Office: Bldg Tech Inc 1109 Spring St Silver Spring MD 20910

HATTON, DANIEL KELLY, computer scientist; b. Hayward, Calif., Jan. 9, 1945; s. Thomas Kelly Hatton and Armetta Pauline (Munkers) Hatton O'Connor; m. Gerda Monika Gotter, June 3, 1966; children: Heidi Monika Hatton Nealey, Anna Maria. BA, Calif. State U., Hayward, 1972; MS, Am. U., 1985; ScD, Nova U., 1991. Enlisted pvt. U.S. Army, 1964, advanced through grades to lt. col., 1993, chief info. officer med. dept., 1974-92, chief info. officer med. dept. Europe, 1993—; adj. prof. computer sci. U. Md., 1986-87, Troy (Ala.) State U., 1991—; adj. faculty Schiller Internat. U., 1993—. Contbr. articles to profl. jours. Vol. tutor, Fla., Ala., Ga., 1990—. Decorated Meritorious Svc. medal, Joint Svc. Commendation medal; Dept. of Def. scholar, 1983-85, 88-91. Fellow Svc. Corps. Silver Cadeusus Soc. (founder); mem. Officers Benefit Assn. Achievements include research in re-engineering computer science based projects, design of modern fiber-optic network. Home: 75 Hermann Loensweg, 6902 Sandhausen Germany Office: Hq 7th MEDCOM Med/Dental Activities Unit 29218 Box 319 APO AE 09102

HATTON, THURMAN TIMBROOK, JR., retired horticulturist, consultant; b. Bartow, Fla., Feb. 4, 1922; s. Thurman Timbrook Sr. and Pearl Catherine (Holliday) H.; m. Eileen Marie Snowber, Jan. 25, 1947 (dec. Jan. 1976); children: Mary, Nina, Alexa Michele; m. Marilyn Mae Memory, July 12, 1979. BS, U. Fla., 1943, MS, 1949; PhD, Wash. State U., 1953. Teaching fellow U. Fla., Gainesville, 1948-49; asst. prof. U. P.R., Mayaguez, 1949-50; instr. Wash. State U., Pullman, 1950-53; extension specialist N.C. State U., Raleigh, 1953-55; leader investigations Market Quality Rsch. Lab. USDA, Miami, Fla., 1955-68; leader rsch. Export and Quality Improvement Unit USDA, Orlando, Fla., 1968-89; cons. Chuluota, Fla., 1989—; adj. prof. U. Fla., 1972-89; cons. Agrl. Rsch. Svc. USDA, Japan, 1974-86, Pub. Law 480 Projects, India and Pakistan, 1977, 81, 85, U.S. AID, India, 1982. Author 3 book chpts.; contbr. numerous articles to profl. jours. Bd. dirs. Am. Youth Exchange Program, Miami, 1962-68. Col. U.S. Army, 1943-48, ETO. Recipient rsch. award Fla. Fruit and Vegetable Assn., 1975, Exceptional Svc. award Fla. Citrus Packers, 1989, Award of Merit Fla. Citrus Commn., 1989. Fellow Am. Soc. for Hort. Sci.; mem. Fla. State Hort. Soc. (pres. 1988), Fla. Mango Forum (pres. 1963), InterAm. Soc. for Tropical Horticulture, Internat. Hort. Soc., Internat. Soc. for Citriculture, Sigma Xi. Achievements include development of maturity standards for Florida avacados, of a cold treatment for quarantine purposes in the export of grapefruit. Home and Office: PO Box 660068 350 Willingham Rd Chuluota FL 32766-0068

HATTORI, AKIRA, economics educator; b. Kasaoka, Japan, July 8, 1949; s. Setsuo and Chiho Hattori; m. Mizue Tsurufuji, Mar. 21, 1981; children: Yasuyuki, Tomoko. BA, Kagawa U., Takamatsu, Japan, 1972; MA, Kyushu U., Fukuoka, Japan, 1974; postgrad., Kyushu U., 1977. Rsch. fellow dept. econs. Kyushu U., Fukuoka, 1977-79; vis. scholar Stanford (Calif.) U., 1987, 92; prof. internat. macroecons. Fukuoka (Japan) U., 1987—. Co-author: World Economy and International Trade, 1985, Disarmament, Economic Conversion and Management of Peace, 1991, Economic Issues of Disarmament, 1993; editor: International Finance, 1986; co-editor: International Economic Policy Coordination in the 1990's, 1991;. Bd. dirs. Economists Allied for Arms Reduction, N.Y.C., 1993—, gen. sec., Japan, 1989—. Mem. Am. Econ. Assn., Royal Econ. Soc., Am. Polit. Sci. Assn., Internat. Studies Assn., Western Econ. Assn. Internat., Japan Soc. Money and Banking. Home: 1-6-31-103 Jonan Ku, Higashiaburayama, Fukuoka 814-01, Japan Office: Faculty Commerce Fukuoka U, 8-19-1 Nanakuma Jonan ku, Fukuoka 814-01, Japan

HATZAKIS, MICHAEL, electrical engineer, research executive; b. Chania, Crete, Greece, Jan. 1, 1928; came to U.S., 1956; s. John and Poly (Lionakis) H.; m. Mary Giannickos, Sept. 29, 1955; children: Michael Jr., Helene. BSEE, NYU, 1964, MSEE, 1967. Technician Radio Engring. Labs., Long Island City, N.Y., 1958-61; technician IBM T.J. Watson Rsch. Ctr., Yorktown Heights, N.Y., 1961-67, staff mem., 1967-76, mgr., 1976-88, IBM fellow, 1988-91; ret., 1991; dir. Microelectronics Inst. at Democritos, Athens, Greece, 1988—. Mem. NAE, IEEE (sr., Cledo Brunetti award 1987), Am. Vacuum Soc., Electrochem. Soc., Materials Rsch. Soc. Democrat. Greek Orthodox. Office: Microelectronics Inst at NCSR Demokritos, PO Box 60228s, 153-10 Aghia Paraskevi Athens Greece

HATZFELD, JACQUES ALEXANDRE, biologist, researcher; b. Chameyrat, France, July 4, 1940; s. Henri Hatzfeld and Roselene Leenhardt; m. Antoinette Klein, July 5, 1966; children: Laure, Vincent. Degree in agronomic engring., Montpellier and genetics, Inst. Nat. Agronomique, Paris, 1966; degree in microbiology, Sci. U. Orsay, France, 1967; PhD, Sci. U. Paris, 1972. Fellow Anticancer Nat. League, Paris, 1968-69; researcher Nat. Ctr. Sci. Rsch., Paris, 1969—, rsch. dir., 1986—; project leader European Concerted Action on Human Bone Marrow Stem Cell, Paris, 1990—. Contbr. articles to profl. jours. Avocations: skiing, rock climbing, wind surfing, music. Office: CNRS, 7 rue Guy Môquet, 94801 Villejuif Cedex 94801, France

HATZIADONIU, CONSTANTINE IOANNIS, electrical engineer, educator; b. Syros, Greece, May 11, 1960; s. John C. and Mary N. (Strates) H.; m. Donna G. Bowman, Jan. 23, 1990. Diploma in elec. engring., U. Patras, Greece, 1983; PhD in Elec. Engring., W.Va. U., 1987. Registered profl. engr. Cons. Greece, 1983-84; rsch. asst. W.Va. U., Morgantown, 1984-87; vis. prof. So. Ill. U., Carbondale, 1987-89, asst. prof., 1989—. Contbr. articles to profl. jours. Mem. IEEE, Greek Engrs. Assn., Sigma Chi. Achievements include development of algorithms for simulation and analysis of high voltage DC systems; design of devices for flexible AC transmission systems. Office: So Ill U Dept Elec Engring Carbondale IL 62901

HATZIKIRIAKOS, SAVVAS GEORGIOS, chemical engineer; b. Volos, Magnesia, Greece, Feb. 3, 1961; s. Georgios S. and Stavroula (Ekizoglou) H.; m. Christina Diles, Jan. 28, 1989; 1 child, George. Diploma in chem. engring., Aristotlean U. Salonica, Thessaloniki, 1983; MA, U. Toronto, Ont., Can., 1988; PhD, McGill U., Montreal, Que., Can., 1991. Rschr. U. Toronto, 1987-88, McGill U., 1988-91; prof. chem. engring. U. B.C., Vancouver, Can., 1991—. Lt. Greek Army, 1983-86. Office: Univ British Columbia, Dept Chem Engring, Vancouver, BC Canada V6T 1Z4

HATZILABROU, LABROS, mechanical engineer, metallurgical engineer; b. Thessaloniki, Greece, July 17, 1958; came to U.S., 1984; s. Thomas Hatzilabrou; m. Mina-Georgia Drakou, 1984; children: Thomas, Rhothie, Anthony. BSME, U. Thessaloniki, 1981; MS in Metallurgical Engring., U. Pitts., 1988. Teaching fellow U. Thessaloniki, 1982-84; grad. researcher U. Pitts., 1984-89; engring. mgr. Elec. Melting Svcs., Massillon, Ohio, 1989—; grad. student rep. U. Pitts., 1985-86, student com., 1988-89, recitation instr. freshmen program, 1988-89. Author: Optimization Study of a Foundry, 1981. Scout master Boy Scouts Greece, Thessaloniki, 1974-84; steering com. St. George Ch., Massillon, 1991-93. Recipient scholarship Greek Nat. Scholarship Found., 1979, Fulbright Found., 1983-84, Bodossaki Found., 1984-86. Mem. ASM Internat., AIME Iron & Steel Soc., Am. Welding Soc., Am. Foundrymen Soc. Greek Orthodox. Achievements include development of a process map for the FOSBEL welding process, new product lines for special furnaces, precast shapes, push-out systems, slag skimmers, etc. Home: 2945 Colony Woods Cir SW Canton OH 44706 Office: Elec Melting Svcs Co 1000 Nave Rd SE Massillon OH 44648

HAUBEN, MANFRED, physician; b. N.Y.C., Apr. 9, 1959; s. Richard and Zora (Soumerai) H. BA in Chemistry, NYU, 1980; MD, N.Y. Med. Coll., 1984, MPH, 1990, DTMH, 1989. Diplomate Nat. Bd. Med. Examiners. Resident dept. pathology Columbia Presbyn. Med. Ctr., N.Y.C., 1985-86; resident in preventive medicine and pub. health Our Lady Med. Ctr., N.Y.C., 1987-89, fellow clin. preventive medicine and chief resident, 1989-90; assoc. med. dir. Sterling-Winthrop, Inc., N.Y.C., 1990—; clin. asst. prof. community and preventive medicine N.Y. Med. Coll., 1993—. Contbr. articles to profl. jours. Mem. Internat. Soc. Pharmacoepidemiology, Drug Info. Assn., N.Y. Acad. Scis., Alpha Omega Alpha.

HAUBER, FREDERICK AUGUST, ophthalmologist; b. Pitts., July 3, 1948; s. Michael H. and Cecilia (Azinger) H.; m. Cathy Lu Rosellini, Aug. 3, 1981; 1 child, Elizabeth Alexandra. BS in Microbiology cum laude, U. Pitts., 1970; MD, U. Tenn., 1974. Intern U. South Fla., Tampa, 1975, resident in ophthalmology, 1982; pvt. practice Pasco Eye Inst., New Port Richey, Fla., 1983—; asst. clin. prof. U. South Fla., Tampa, 1984—; researcher, speaker in field, 1990—. Advisor health care cost containment com., Tarpon Springs, Fla., 1988; founder Pasco County Diabetes Assn. Fellow ACS. Achievements include patent for achromatic intraocular lens; first to insert glaucoma pressure regulator; development of binary optical intraocular lens, color vision eye chart system. Office: Pasco Eye Inst 5347 Main St New Port Richey FL 34652

HAUBER, JANET ELAINE, mechanical engineer; b. Milw., July 21, 1937; d. Ralph Joseph and Ethel Esther (Forsyth) H. BME, Marquette U., 1965; MS, Stanford U., 1967, PhD, 1970. Rsch. metallurgist dept. chemistry Lawrence Livermore (Calif.) Nat. Lab., 1970-73, project leader, 1973-74, sect. leader, facility mgr., 1974-76, dep. div. leader, 1976-78, dep. div. leader mech. engring. dept., 1978-86, dep. assoc. dept. head, 1986-87, sect. leader, 1987—. Contbr. articles to profl. jours. Ford Foudn. fellow Stanford U., 1965, ASTM fellow, 1967. Mem. AAWU (v.p. 1992-93), Soc. Women Engrs., Sigma Xi. Office: Lawrence Livermore Nat Lab PO Box 808 Livermore CA 94550

HAUBERT, ROY A., former engineer, educator; married. BA, U. Okla., M in Physical Sci.; postgrad., U. Calif., Berkeley. Systems engr. Gen. Dynamics, San Diego, 1955-70; rsch. assoc. N.Mex. Rsch. Inst., Alamogordo, 1975-78; engr. Comarco Engring. Co., China Lake, Calif., 1978-80, WESTEC Svcs., Inc., San Diego, 1980-82, Planning Rsch., Inc., China Lake, 1982-83; staff vol. Scripps Institution of Oceanography-Aquarium U. Calif. at San Diego, La Jolla, 1983—; chief maintenance Computer, Simulation, Data Acquisition and Foreign Tech. Analysis Lab., Air Rsch. and Devel. Command; br. chief comm. surveillance and electronic warfare U.S. Army Intelligence Security Agy. Contbr. pubs. and tech. manuals in field. Chmn. Immigration and Naturalization Svc. New Citizen Ceremony, 1989, 90; active Calif. Capitol Caucus Club, 1989-90. Recipient Disting. Svc. award Calif. State Assembly, 1986, Outstanding Svcs. award, 1987, Cert. for Svc., 1988, Svc. award Atomic Energy Commn.,. Mem. AIAA, Assn. Former Intelligence Officers, Air Force Assn., Assn. Old Crows, Confederate Air Force (Air Group One). Achievements include developments in infra-red research for thermal imaging analysis, research program for information storage and retrieval, extended position location tracking circuitry for MIRAN computer, brain pulse monitoring research program, AEC Plowshare advisory program, instrumentation. Home: 645 Farview St El Cajon CA 92021

HAUCK, FREDERICK HAMILTON, retired naval officer, astronaut, business executive; b. Long Beach, Calif., Apr. 11, 1941; s. Philip and Virginia (Hustvedt) H.; m. Dolly Bowman, Aug. 27, 1962 (div.); children: Whitney Irene, Stephen Christopher; m. Susan Cameron Bruce, June 27, 1993. B.S., Tufts U., 1962; M.S., MIT, 1966. Commd. ensign USN, 1962, advanced through grades to capt., 1983; pilot Attack Squadron 35, USS Coral Sea, 1968-70; instr. pilot Attack Squadron 42, Oceana, Va., 1970-71; test pilot Naval Air Test Ctr., Patuxent River, Md., 1971-74; ops. officer Carrier Air Wing 14, Miramar, Calif., USS Enterprise, 1974-76; exec. officer Attack Squadron 145, Wash., 1976-78; astronaut NASA, Houston, 1978-89; space shuttle pilot shuttle transp. system mission 7, 1983; space shuttle comdr. STS-51A, 1984; assoc. adminstr. for external rels. NASA, 1986; space shuttle comdr. STS-26, 1988; dir. Navy Space Systems (OP-943), Washington, 1989-90, ret., 1990; pres., CEO Internat. Tech. Underwriters, Bethesda, Md., 1990—; mem. comml. space transp. adv. com. Dept. Transp.; mem. comml. programs adv. com. NASA, 1991-92; chair COMSTAC task group on Soviet entry into world space markets Dept. Transp. Trustee Tufts U.; bd. govs. St. Albans Sch. Decorated Def. D.S.M. (2), Def. Superior Svc. medal, Legion of Merit, DFC, Air medal (9), Navy Commendation Medal with Gold Star and Combat V, NASA D.S.M., NASA medal for Outstanding Leadership, NASA Space Flight medal (3), Presdl. Cost Saving Commendation, AIAA Haley Space Flight award, Lloyd's of London Silver medal for meritorious svc., Am. Astronautical Soc. Flight Achievement award (2), Federation Aeronautique Internationale Yuri Gagarin Gold medal, Federation Aeronautique Internationale Komarov Diploma (2), Tufts U. Presdl. medal, Delta Upsilon Disting. Alumnus award; named Navy's Outstanding Test Pilot for 1972. Fellow Soc. Exptl. Test Pilots, AIAA (assoc.); mem. Assn. Space Explorers (v.p. 1991—), Army and Navy Club, Winter Harbor (Maine) Yacht Club. Office: INTEC 4800 Montgomery Ln 11th Fl Bethesda MD 20814

HAUFLER, CHRISTOPHER HARDIN, botany educator; b. Niskayuna, N.Y., Apr. 20, 1950; s. J. Hervie and Patricia (DeLearie) H. BA, Hiram Coll., 1972; MA, Ind. U., 1974, PhD, 1977. Assoc. instr. Indiana U., 1972-76., asst. prof, 1977; postdoctoral fellow Gray Herbarium, Harvard U., 1977-78, Mo. Bot. Garden, St. Louis, 1978-79; asst. prof. U. Kans., Lawrence, 1979-84, assoc. prof., 1984-90, chmn., dept. botany, 1985—, prof., 1990—; faculty sponsor undergrad. biology club, 1980—, search coms., 1980—, field facilities com., 1980—, cirriculum com., 1980—, greenhouse com. 1980—, chair, 1984-88, honors and awards com. 1980-87, space com., 1980-86, chair departmental admissions and awards com., 1981-84, biology core review com. 1983-85, biol. scis. resource ctr., 1984-87, biol. scis. exec. com., 1985—, steward Evolutionists, 1985, sec., 1986; mem. panel systematic biology NSF, 1987—. Reviewer Index to Plant Chromosome Numbers, 1979-89; presenter papers in field. William R. Ogg Departmental fellow, 1976-77; Rsch. fellow Gray Herbarium, 1977-78; NEA Postdoctoral fellow Mo. Botanical Garden, 1978-79. Mem. Bot. Soc. Am. (nominating com.for officers pteridological sect. 1981, 82, sec./treas. 1983-89, program organizer 1985, 87, 89, chair 1991-93, symposium organizer 1984, 85, 87, editor Annual Bibliography Am. Pteridology 1978-82, sec. 1991—, Best Paper award 1979, 80, 82, 83, 84), Am. Fern Soc. (nominating com. for officers 1980, 83, assoc. editor Am. Fern. Jour. 1986—), Am. Inst. Biol. Scis., Am. Soc. Plant Taxonomists (rsch. awards com. 1989-91, program dir. 1990-93, editorial bd. Systematic Botany 1985-87), Nat. Geog. Soc., Soc. Systematic Biologists (editorial bd. 1992—), Internat. Assn. Pteridologists (sec. 1987—, compiler Internat. Report Pteridological Rsch., 1984-88), Brit. Pteridological Soc., New England Botanical Club, Soc. for Study Evolution, Sigma Xi. Office: Univ Kans Dept of Botany Lawrence KS 66045*

HAUG, EDWARD JOSEPH, JR., mechanical engineering educator, simulation research engineer; b. Bonne Terre, Mo., Sept. 15, 1940; s. Edward Joseph and Thelma (Harrison) H.; m. Carol Jean Todd, July 1, 1979; 1 child, Kirk Anthony. BSME, U. Mo., Rolla, 1962; MS in Applied Mechanics, Kans. State U., 1964, PhD in Applied Mechanics, 1966. Rsch. engr. Army Weapons Command, Rock Island, Ill., 1969, chief system analysis, 1970, chief systems rsch., 1971-72, chief concepts & tech., 1973-76; prof. U. Iowa, Iowa City, 1976—, Carver Disting. prof., 1990—, dir. Ctr. for Computer Aided Design, 1983—. Author 11 books on computer aided design and dynamics; editor 5 books; contbr. more than 170 papers to profl. jours. Served to Capt. U.S. Army, 1966-68. Recipient Innovative Info. Tech. award Computerworld/Smithsonian Instn., 1989, Colwell Merit award Soc. Automotive Engrs., 1989. Mem. ASME (Design Automation award 1991, Machine Design award 1992), Am. Acad. Mechanics. Achievements include patents for Constant Recoil Automatic Cannon, and for Real-Time Simulation System. Home: 1407 Eastview Dr Coralville IA 52241-1011 Office: U Iowa Ctr Computer Aided Design 208 ERF Iowa City IA 52242

HAUG, ROAR BRANDT, architect; b. Oslo, Oct. 26, 1928; s. Elias and Dudu Synøve (Brandt) H.; m. Kari Synnøve Kristoffersen, Sept. 19, 1959; children: Vibeke Synnøve, Hilde Merete, Hege Cecilie. Degree in Exam Artium, Vestheim Coll., Oslo, 1947; degree in Econ., Oslo Comml. Coll., 1948; MS in Architecture, Tech. U. Norway, 1953; MArch, U. Mich., 1958. Architecture, Norway. Architect Architect Guttorm Bruskeland A/S, Oslo, 1954-57, Platou Architects A/S, Oslo, 1957, Alfred Easton Poor & Assoc., N.Y.C., 1958-59; chief architect Norconsult A/S, Addis Ababa, Ethiopia, 1959-62; architect, mng. dir. Roar Haug-Byggplan, Oslo, 1962—; chmn. bd. dirs. Norplan A/S, Oslo, I/S Railway Engring., Oslo, Byggplan A/S, Oslo, I/S Defcon, Oslo; bd. dirs. I/S Petroconsult, Drammen, Norway. Prin. works include Pub. Sch., Nannestad, Norway, 1968, 24 Old-Age Dwellings, Eidsvoll, 1969, Adminstrv. Bldg. for Dist. Gov., Svalbard, 1976, Main Post Office Bldg., Førde, Norway, U.S. MAB Support Facilities Norway, Stegler's Gorge Power Devel. Project, Tanzania, C. of C., Addis Abeba, Ethiopia, various pub. schs., plants and adminstrv. bldgs. 2d lt. Corps of Engrs., 1953-54, Norway. Fulbright scholar Inst. Internat. Edn., U. Mich., 1957; recipient various prizes from archtl. competitions. Mem. Royal Norwegian Yachtclub, Norwegian Architect's League, Norwegian Town Planning and Housing Assn., Norwegian Poly. Assn., Norwegian Soc. Preservation Hist. Bldgs., Norwegian Soc. Conservation Nature. Avocations: theatre, literature, skiing, tennis, sailing. Office: Roar Haug-Byggplan, Bogstadveien 27B, 0355 Oslo Norway

HAUGEN, ROBERT KENNETH, product developer; b. Detroit, July 12, 1947; s. Olaf Kenneth and Grace Elizabeth (O'Connor) H.; m. Tally Glossbrenner, June 29, 1985; 1 child, Robert F. BS, U. Ill., 1969, MS, 1973, PhD, 1977. Product testing specialist Montgomery Ward, Chgo., 1966-67; researcher Lawrence Livermore (Calif.) Labs., 1968-69; instr. U. Ill., Urbana, 1969-74; researcher Tri-Met Sanitary Dist., Urbana, 1971-74; writer Environ. & Tech. Project, La Salle, Ill., 1974-83, project coord., 1983-86; institutional product mgr. St. Charles (Ill.) Mfg., 1987—. Author: Laboratory Fume Hood Design and Use, 1989. Recipient Community Achievement award Kishwaukee Community Coll., Malta, Ill., 1986. Mem. Am. Soc. testing and Materials (subcom. chmn. 1989—), Am. Chem. Soc., Phi Beta Kappa. Achievements include development of series of electrical alarms for fume hoods; conducted research regarding fume hood containment testing. Office: St Charles Mfg 1611 E Main Saint Charles IL 60174

HAUGH, CLARENCE GENE, agricultural engineering educator; b. Spring Mills, Pa., Oct. 11, 1936; s. Clarence Glenn and Estella Jane (Baney) H.; m. Patricia Anne Breon, June 16, 1962; children: Amy Elizabeth, Jennifer Lea, Mitchell Breon. BS in Agrl. Engring., Pa. State U., 1958; MS in Agrl. Engring., U. Ill., 1959; PhD, Purdue U., 1964. Registered profl. engr., Fla. Asst. prof. U. Fla., Gainesville, 1964-65; asst. prof. Purdue U., Lafayette, Ind., 1965-68, assoc. prof., 1968-72, prof., 1972-79; prof. Va. Poly. Inst. & State U., Blacksburg, 1979—; dept. head Va. Poly. Inst. & State U., 1979-86; mem. Nat. Engring. Accreditation Commn., 1985-90; trustee Chippokes Plantation, Surry, Va., 1980-91; cons. King Faisall U., Hofhuf, Saudi Arabia, 1984-86. Patentee in field; contbr. over 50 articles to profl. jours. Asst. scoutmaster Boy Scouts Am., Blacksburg, 1985-91; deacon, elder, trustee Covenant Presbyn. Ch., Lafayette, 1968-76; adminstrv. bd. Blacksburg Meth. Ch., 1979-82. Served as 1st lt. USAF, 1958-64. Republican. Methodist. Fellow Am. Soc. Agrl. Engrs. (bd. dirs. 1989-91, chmn. 12 tech. coms., Young Rschr. award 1976); mem. Am. Soc. Engring. Edn. (chmn. agrl. engring. sect. 1982-83), Inst. Food Technologists, Soc. Rheology, Internat. Soc. Agromaterials Sci. and Engring. (sci. bd. 1992—), Arnold Air Soc., Rotary, Masons, Shriners, Sigma Xi, Tau Beta Pi, Alpha Epsilon, Gamma Sigma Delta, Phi Tau Sigma, Alpha Zeta. Republican. Methodist. Lodges: Rotary, Masons, Shriners. Avocations: sailing, backpacking, collecting antiques. Home: 406 Murphy St Blacksburg VA 24060-2539 Office: Va Poly Inst & State U Dept of Agrl Engring 314 Seitz Hall Blacksburg VA 24061

HAUGH, DAN ANTHONY, mechanical engineer; b. Lawrence, Kans., Feb. 16, 1953; s. Oscar Martin and Rita (Rosso) H.; m. Jay McLaughlin, Mar. 19, 1983; children: Alden Elizabeth, Emily Marston. BSME, U. Kans., 1978. Engr. R & D Boeing, Wichita, Kans., 1978-80, specialist engr. propulsion dept., 1980-83; midwest engring. mgr. PSI Bearings, Inc., Wichita, Kans., 1983-85; midwest sales mgr. Jamaica Bearings Co., Inc., Lawrence, 1983-91; dir. engring. Jamaica Bearings Co., Inc., New Hyde Park, N.Y., 1991—; bd. dirs. Aeros, Inc., Lawrence, 1986—. Bd. dirs. West Hills Home Assn., Lawrence, 1989-91. Mem. ASME, Internat. Gas Turbine Inst., Profl. Aviation Maintenance Assn., Robot Inst. Am., Sch. Engring. Soc., U. Kans. Alumni Assn. Home: 1512 University Dr Lawrence KS 66044-3148

HAULBROOK, ROBERT WAYNE, computer systems engineer; b. Atlanta, Feb. 23, 1967; s. Roy Wayne and Hattie Ruth (Powell) H. BS, U. S.C., 1989; MS, N.C. State U., 1991. Registered engr. in tng., Ga. Systems engr. Parents' Resource Inst. for Drug Edn., Atlanta, 1991—. Mem. Nat. Soc. Profl. Engrs. Republican. Achievements include producing statistics used in Newsweek mag. article on increased hallucinogen usage among Am. high sch. seniors. Home: 410 Chase Lake Dr Jonesboro GA 30236 Office: Pride Inc 50 Hurt Pla Ste 210 Atlanta GA 30303

HAUN, JAMES WILLIAM, chemical engineer, retired food company executive, consultant; b. Birmingham, Ala., Sept. 8, 1924; s. James Cecil and Eva (Walker) H.; m. Lucia Land, Sept. 6, 1946; children: James William, Lucy Margaret, Daniel Victor, Robert Paul. BSChemE, U. Tex., 1946, MS, 1948, PhD, 1950; grad. Advanced Mgmt. Program, Harvard U., 1961. Registered profl. engr. Instr. chem. engring. U. Tex., 1948-49; successively research engr., sr. research engr. and research group leader plastics div. Monsanto Chem. Co., 1950-56; with Gen. Mills, Inc., Mpls., from 1956; dir. corp. engring. Gen. Mills, Inc., from 1960, v.p., 1963-75, v.p. engring. policy, 1975-85, now ret.; mem. environ. engring. com. Sci. Adv. Bd. EPA, 1982-90; mem. food industry adv. com. U.S. Dept. Energy, 1978-81; mem. indsl. adv. council U. Minn. Inst. Tech.; dept. chem. engring. U. Calif., Berkeley, 1971-77; chmn. Internat. Centre for Industry and Environ., Paris and Nairobi, 1977-80; bd. dirs. World Environ. Center, N.Y.C., 1977-85. Author: Guide to the Management of Hazardous Waste, 1991. Bd. dirs. Center for Parish Devel., Chgo., chmn., 1980-85; bd. advisors U. Minn. Grad. Sch. Served with USMCR, 1942-46. Humble Oil & Refining Co. fellow U. Tex., 1949-51; named Engr. of Year Minn. Soc. Profl. Engrs., 1974. Fellow Am. Inst. Chem. Engrs. (emeritis); mem. NAM (dir., chmn. environ. quality com. 1973-79), C. of C. U.S. (com. on environ.), Sigma Xi, Omega Chi Epsilon, Phi Lambda Upsilon. Home: 6912 E Fish Lake Rd Osseo MN 55369-5447

HAUPT, CARL P., retail drugs executive; b. Chgo., Aug. 14, 1940; s. Carl and Clara (Brandt) H.; m. Donna A. Peltzer, Apr. 23, 1971; children: Corinne M., Christine J. BS in Edn., No. Ill. U., 1966. Mgr. spl. svcs. Jewel Cos. Inc., Chgo., 1970-86; dir. risk adminstrn. Am. Drug Stores, Oak Brook, Ill., 1986—. Alderman City of Des Plaines, Ill., 1987—; treas. Good Shepherd Luth. Ch., 1983-88, elder, 1988-93, pres., 1993—; commr. Waycinden Baseball League, 1989-92. With USAF, 1959-63. Mem. Internat. Assn. Indsl. Accidents Bds. and Commns., Nat. Coun. Self Insurers (bd. dirs. 1984-86), Ill. Self Insurers Assn. (pres. 1984-86), Calif. Self Insurers Assn. Republican. Avocations: baseball, golf, bowling. Home: 96 Jeffery Ln Des Plaines IL 60018-1219 Office: Am Drug Stores Inc 2100 Swift Dr Oak Brook IL 60521-1559

HAUPT, H. JAMES, mechanical design engineer; b. Palmerton, Pa., Jan. 3, 1940; s. Harry C. and Mary L. (Patrick) H.; m. Betty S. Niemi, Sept. 5, 1970; children: Nadine R., Heather J. AAS, Broome Tech. Coll., 1964; BS, Ill. Inst. Tech., Chgo., 1971. Lic. profl. engr. Mech. engr. Argonne (Ill.) Nat. Lab., 1964—. Co-contbr. articles to profl. publs. AEC scholar, Washington, 1969. Mem. Am. Nuclear Soc., Ill. Profl. Engrs., Phi Theta Kappa, Tau Beta Pi. Republican. Roman Catholic. Avocation: golf. Home: 3215 Saddle Dr Joliet IL 60435-1142 Office: Argonne Nat Lab 9700 Cass Ave Apt 310 Lemont IL 60439-4807

HAUPT, RANDY LARRY, electrical engineering educator; b. Johnstown, Pa., Aug. 11, 1956; s. Howard and Anna Mae Haupt; m. Sue E. Slagle, Feb. 17, 1979; children: Bonny Ann, Amy Jean. BSEE, USAF Acad., 1978; MS in Engring. Mgmt., Western New Eng. Coll., 1981; MSEE, Northeastern U., 1983; PhD in Electrical Engring., U. Mich., 1987. Registered profl. engr., Colo. Commd. 2d lt. USAF, 1974, advanced through grades to maj., 1990—; project engr. OTH-B Radar electronic systems divsn. USAF, Hanscom AFB, Mass., 1978-80, rsch. engr. in microwave antennas Rome Air Devel. Ctr., 1980-84; from instr. to asst. prof. USAF Acad., Colo., 1987-91; dir. rsch. dept. electrical engring. USAF, 1990-91, chief comms. divsn. dept. electrical engring., 1991—; assoc. prof. USAF Acad., 1991—; vis. rsch. engr. Los Alamos Nat. Lab., 1992; presenter numerous papers and tech. reports in field. Contbr. numerous articles to engring. jours. Nordic ski team coach USAF Acad., 1992-93. Recipient USAF Rsch. and Devel. award, 1983, 87, Frank J. Seiler award for rsch. excellence, 1990, 92, Founder's Gold medal, 1993, 6 Rome Air Devel. Ctr. Sci. Achievement awards; named Outstanding Mil. Educator in Electrical Engring., 1992, USAF Mil. Engr. of Yr., 1993, Fed. Engr. of Yr., Nat. Soc. Profl. Engrs., 1993; rsch. grantee Rome Air Devel. Ctr., 1988-90, Frank J. Seiler Rsch. Lab., 1990—, Cray Rsch., Inc., 1991-92, Phillips Lab., 1992—. Mem. IEEE (sr., student br. counselor 1988-90, reviewer Transactions on Antennas and Propagation 1984—), Am. Soc. for Engring. Edn. (reviewer), Applied Computational Electromagnetics Soc., Tau Beta Pi. Achievements include research in electromagnetics, scattering, antennas, electro-optics, numerical methods, chaos theory, radar, systems engineering, communications systems. Office: DFEE Ste 2F6 2354 Fairchild Dr U S A F Academy CO 80840

HAUPTMAN, HERBERT AARON, mathematician, educator, researcher; b. N.Y.C., Feb. 14, 1917; s. Israel and Leah (Rosenfeld) H.; m. Edith Citrynell, Nov. 10, 1940; children: Barbara, Carol Hauptman Fullerton. BS in Math., CCNY, 1937; MA, Columbia U., 1939; PhD, U. Md., 1955, PhD (hon.), 1985; PhD (hon.), CCNY, 1986, U. Parma, Italy, 1989, D'Youville Coll., 1989, Bar-Ilan U., Israel, 1990, Columbia U., 1990, Tech. U., Lodz, Poland, 1992. Statistician U.S. Census Bur., Washington, 1940-42; civilian instr. electronics and radar U.S. Army Air Force, Boca Raton, Fla., 1942-43, 46-47; physicist, mathematician Naval Rsch. Lab., Washington, 1947-70; head math. biophysics lab. Med. Found., 1970-72, exec. v.p., rsch. dir., 1972-85; pres., res. dir. Med. Fedn. Buffalo, 1985-87, pres., 1988—; also bd. dirs., 1972—; prof. biophys. sci. SUNY, Buffalo, 1970—, prof. computer scis., 1992—, chmn. bd. dirs.; with N.Y. State Inst. on Superconductivity, 1988—; mem. sci. adv. bd. BioCryst, 1989—; math. instr. U. Md., 1958-70; chmn. Inter Congress Symposium Direct Methods in Crystallography, Buffalo, 1976; pres. Assn. Ind. Rsch. Insts., 1979, 80, mem. nominating com., 1982; mem. U.S.A. Nat. Com. for Crystallography, 1979-81, 82-85, 88, 89; mem. nat. adv. com. Comprehensive Regional Ctr. for Minorities CCNY, 1989—. Author: (with J. Karle) Solution of the Phase Problem., 1953, Crystal Structure Determination: The Role of the Cosine Seminvariants, 1972; editor: Dir. Methods in Crystallography, Proceeding of the 1976 Intercongress Symposium, 1978; contbr. chpts. to books, articles to profl. jours. Trustee Buffalo Gen. Hosp., 1990—; chmn. communications com. Philos. Soc. Washington, 1966-67, corr. sec., 1967-69, pres., 1969-70. Served to lt. (jg.) USNR, 1943-46. Sr. fellow for Travel, Lectures and Rsch. in Italy NATO, 1973; grantee NSF, 1972-92; recipient Belden prize (gold medal) in Math., 1935, RESA award in Pure Scis., 1959, Citizen of Yr. award Buffalo Evening News, 1986, Schoelkopf award Am. Chem. Soc., 1986, Gold Plate award Am.Acad. Achievement, 1986, Nat. Libr. Medicine medal, 1986, Law Sch. award Maimomides Chabad House, 1986, others, (with J. Karle) Patterson award, 1984, Nobel Prize in Chemistry, 1985; honoree Western New York Man of Yr. Buffalo C. of C., 1986, YMCA Dinner, 1986, 90th Nobel Ann. Dinner, 1991; inductee Nobel Hall Sci. Mus. Sci. and Industry, 1986, Townsend Harris Hall Fame, 1989; guest of honor Roswell Park Meml. Inst., 1985, YMCA Luncheon, 1986, others; invited guest Am. Nobel Convocation, 1987, 88, Weizmann Nat. Dinner, 1988, others. Fellow Washington Acad. Scis., Jewish Acad. Arts and Scis. (medal 1986); mem. AAAS, Am. Math. Soc., Am. Phys. Soc., Am. Crystallographic Assn. (mem. Fankuchen award com. 1988), Math. Assn. Am., U.S. Nat. Acad. Scis., Cosmos Club, Saturn Club (guest of honor 1985), Sigma Xi (sec. Buffalo chpt. 1971-72), Phi Beta Kappa. Avocations: stained glass art, swimming, hiking. Office: Med Found Buffalo 73 High St Buffalo NY 14203-1196

HAUPTMANN, RANDAL MARK, molecular biologist; b. Hot Springs, S.D., July 6, 1956; s. Ivan Joy and Phyllis Maxine (Pierce) H.; m. Beverly Kay Suko, May 22, 1975; 1 child, Erich William. BS, S.D. State U., 1979; MS, U. Ill., 1982, PhD, 1984. Postdoctoral researcher Monsanto Corp. Rsch., St. Louis, 1984-86; asst. prof. No. Ill. U., DeKalb, 1988—, dir. plant molecular biology ctr., 1989—; vis. rsch. scientist U. Fla., Gainesville, 1986-88; cons. Amoco Life Sci. Techs., Naperville, Ill., 1990—. Author: (with others) Methods in Molecular Biology, 1990; contbr. articles to profl. jours. Mem. Internat. Assn. Plant Tissue Culture, Internat. Soc. Plant Molecular Biology, AAAS, Am. Soc. Plant Physiologists, Tissue Culture Assn. (Virginia Evans award 1982), Sigma Xi, Gamma Sigma Delta. Republican. Office: Amoco Tech Co Amoco Rsch Ctr 150 W Warrenville Rd Naperville IL 60563-8473

HAUSER, GEORGE, biochemist, educator; b. Vienna, Austria, Dec. 13, 1922; came to U.S., 1939.; s. Hans Joseph and Juliane Therese (Gleissner) H.; m. Louise Jean Russo, July 2, 1955. BS, Ohio State U., 1949; PhD, Harvard U., 1955. Mem. faculty Harvard Med. Sch., Boston, 1952-55, from rsch. assoc. to prof., 1955-93, prof emeritus, 1993—; from asst. biochemist to biochemist McLean Hosp., Belmont, Mass., 1993, sr. biochemist, 1993—; mem. adv. & editorial bds. Jour. Neurochemistry, 1977-86, dep. chief editor, 1986-92; interim dir. Ralph Lowell Labs., McLean Hosp., Belmont, 1983-93; reviewer many sci. jours.; spl. cons. NIH, NSF. Co-editor: Inositol & Phosphoinositides: metabolism & metabolic regulation. Mem., treas. Dem. Ward Com., Newton, Mass., 1976—. With U.S. Army, 1943-48. Grantee Nat. Insts. Health, 1965-92, Nat. Sci. Found., 1980-82; fellow Japan Soc. for the Promotion of Sci., 1988. Mem. Biochem. Soc., Am. Soc. Biochemistry and Molecular Biology, Internat. Soc. Neurochemistry, Am. Soc. Neurochemistry (coun. 1983-87), Soc. Neuroscience, Soc. Complex Carbohydrates. Democrat. Jewish. Achievements include research in normal and drug-modified lipid metabolism, receptor-mediated transmembrane signaling, phosphoinositide metabolism and function, second messenger generation and function, cell membranes, protein kinase C. Home: 47 Windermere Rd Auburndale MA 02166 Office: McLean Hosp 115 Mill St Belmont MA 02178

HAUSER, HELMUT OTMAR, biochemistry educator; b. Graz, Austria, Nov. 18, 1936; s. Othmar Otto and Juliana Theresia (Steifer) H.; m. Ingrid Maria Schoeller, Jan. 16, 1958; children: Christina, Helen. PhD in Chemistry, U. Graz (Austria), 1963. Brit. Coun. rsch. fellow Agrl. Rsch. Coun., Cambridge, Eng., 1966-68; mgr. biophys. rsch. Unilever Rsch. Lab., Colworth/Welwyn, U.K., 1968-74; lectr. dept. biochemistry Swiss Fed. Inst. Tech., Zürich, Switzerland, 1975-82, prof. biochemistry, 1983—; vis. prof. U. Boston, 1979; disting. vis. rsch. scientist Nat. Rsch. Coun. Can., Ottawa, 1987; cons. Ciba-Geigy, 1980-87. Mem. adv. bd. Chem. Phys. Lipids, 1983—, Biochim. Biophys. Acta, 1987—, Biochim. Biophys. Acta. Revs. on Biomembranes, 1988—; contbr. articles to profl. jours. including Biochemistry, Biochim. Biophys. Acta, Jour. Biol. Chemistry, Jour. Molecular Biology, Chem. Phys. Lipids, European Jour. Biochemistry, PNAS, Nature, Biochemistry Jour.; author, co-author 41 rev. papers to sci. jours. and books. Chmn. Chem. Soc., Zürich, 1982. Mem. Swiss Biochem. Soc. (chmn. 1991—), IUPAC (steering com. on biophys. chemistry 1991—). Achievements include contribution to understanding of the structure and motion of lipids in biological membranes; discovery that fat absorption in the small intestines is catalyzed by proteins of the brush border membrane. Office: Swiss Fed Inst. Tech Zürich, Universitaetstrasse 16, 8092 Zurich Switzerland

HAUSER, JULIUS, retired drug regulatory official; b. Chgo., Aug. 17, 1914; m. Sylvia Ann Gross (dec. 1972); children: Michael George, Robert Mason; m. Irene Shirley Berde, Dec. 1, 1974. SB in Chemistry, U. Chgo., 1934, postgrad., 1941. Food and drug inspector FDA, Chgo., 1939-49; food and drug officer FDA, Washington, 1949-55, asst. to med. dir., 1955-66, asst. for regulations, 1966-69; mgr. regulatory compliance Parke Davis and Co., Washington, 1969-70; dir. regulatory affairs Warner Lambert Co., Washington, 1970-79. Contbr. articles to profl. jours. Pres. Washington Area Group for the Hard Hearing, Bethesda, Md., 1990. Mem. AAAS, Am. Chem. Soc.

HAUSER, KURT FRANCIS, anatomy and neurobiology educator; b. Pitts., May 28, 1955; s. Richard and Esther Henrietta (Murdoch) H.; m. Barbara Gail Conrad, May 5, 1984; children: Alan Conrad, Steven Conrad. BA, Rutgers Coll., 1977; PhD, U. of Medicine and Dentistry, Newark, 1983. Postdoctoral fellow Columbia U., N.Y.C., 1983-86; instr. Penn State U., Med. Ctr., Hershey, Pa., 1986-87, asst. prof., 1987; asst. prof. U. Ky. Med. Ctr., Lexington, 1987-92, assoc. prof., 1992—; ad hoc reviewer NSF Grants, Washington, 1990-91 and various jours. in field. Author: Opiates and Neural Development, 1992, Development of the Nervous System: Effect, 1992; contbr. articles to profl. jours. Grantee Nat. Inst. Drug Abuse, 1990-93, Am. Heart Assn., 1990-93, U. Ky., 1990; recipient fellowship Whitehall Found., 1983-86, scholarship U. Medicine and Dentistry N.J., 1980. Mem. Soc. for Neurosci., Am. Soc. Cell Biology, Am. Assn. Anatomists, Endocrine Soc. Achievements include discovery that glia, the support cells in brain, are a primary target for opiate drugs of abuse during development; discovery of implications for maternal drug use during pregnancy and lactation. Office: U Ky Sch of Medicine 800 Rose St Lexington KY 40536

HAUSER, SIMON PETRUS, hematologist; b. Morschwil, St. Gallen, Switzerland, Nov. 13, 1954; came to U.S., 1990; s. Anton Karl and Maria Frida (Klaus) H.; m. Christina Elisabeth Gressbach, Dec. 22, 1989. Diploma, U. Zurich, Switzerland, 1980, MD, 1981, postgrad., 1982-84. Intern internal medicine U. Zurich Med. Clinic, 1984-87; chief resident Zurich Hochgebirgsklinsk, Clavadel Davos, Switzerland, 1987-88; resident ctrl. lab. hematology Inselspital & Univ., Bern, Switzerland, 1988-89; postdoctoral fellow dept. medicine U. Ark. Med. Scis., Little Rock, 1990-92, asst. prof., 1992—; advisor on unproven methods in oncology Swiss Cancer League, Bern, 1989—; rep. Study Group on Unproven Methods in Oncology, Swiss Cancer League, Swiss Soc. Oncology, Bern, 1982—. Contbr. articles to profl. jours. Mem. Swiss Soc. Hematology, Swiss Soc. Oncology, Swiss Soc. Magnesium Rsch., Multinat. Assn. Supportive Care in Cancer. Achievements include research in cancer patients and unproven methods, stromal regulation of hematopoiesis by cytokines. Office: U Ark Med Scis VA Med Ctr 4300 W 7th St Slot 182 Little Rock AR 72205

HAUSHEER, FREDERICK HERMAN, internist, cancer researcher, pharmaceutical company officer; b. Mount Ayr, Iowa, Oct. 29, 1955; s. Herman Joseph and Margaret Jean (Ford) H.; m. Ann Benage, 1984; children: Derek Louis, Stefanie Ann, Kristin Nicole. BS in Biology, Graceland Coll., 1977; MS in Biophysics and Physiology, U. Ill., 1979; MD with honors, U. Columbia, 1982. Diplomate Nat. Bd. Med. Examiners, Am. Bd. Internal Medicine. Med. oncology sr. scientist Frederick (Md.) Cancer Rsch. Inst., 1987-88; asst. prof. Johns Hopkins Oncology Ctr., Balt., 1988—, U. Tex. Health Ctr., San Antonio, 1988-92; chief drug discovery Cancer Therapy and Rsch. Ctr., San Antonio, 1988-92; asst. dir. Inst. Drug Devel., San Antonio, 1991-92; assoc. prof. U. Tex. Health Ctr., San Antonio, 1992—; CEO BioNumerik Pharms., Inc., San Antonio, 1992—; chmn. BioMed. Rsch. Supercomputing, Mpls., 1992—; mem. sci. review bd. NSF, 1989-92; cons. Nat. Cancer Inst., 1988-90; mem. pharmaceutical exec. bd. IBM, 1992. Author: over 75 rsch. pubs.; reviewer Cancer Rsch. Inst. jour., 1989—, Jour. Am. Chem. Soc., 1991—, Investigational New Drugs jour., 1988—, Biopolymers jour., 1991—; contbr. articles to profl. jours. Recipient Cray Rsch. award Cray Rsch. Inc., 1989-91, Clin. fellowship Am. Cancer Soc., 1985-86. Fellow ACP; mem. IEEE (Forefronts in Large Scale Computation award 1992), AMA (Physician's Recognition award 1992-95), AAAS, Drug Info. Assn., Am. Chem. Soc., Am. Assn. Cancer Rsch. Achievements include 3 drug related patents. Office: BioNumerik Pharms Inc 8122 Datapoint Dr Ste 1250 San Antonio TX 78229

HAUSMAN, BOGUMIL, computer scientist; b. Szczecin, Poland, Mar. 20, 1956; arrived in Sweden, 1981; s. Stanisław and Władysława (Zieniewicz) H.; m. Marzena Nowak, Oct. 18, 1980; 1 child, Alexander. MS in Electronics, Tech. U. Gdańsk, Poland, 1979; Tech. Lic. in Computer Sci., Royal Inst. Tech., Stockholm, 1986, PhD in Computer Sci., 1990. Researcher dept. physics and biophysics Med. Acad. Gdańsk, 1979-81; researcher dept. telecommunication and computer systems Royal Inst. Tech., 1982-85; researcher logic programming and parallel systems lab. Swedish Inst. Computer Sci., 1985-91; sr. rschr. computer sci. lab. Ellemtel, Alvsjo, Sweden, 1992—; referee Internat. Jour Parallel Programming, 1989-90, also various internat. confs. Contbr. articles to profl. publs. Mem. info. office Solidarity Trade Union, Stockholm, 1982-88. Mem. Assn. for Logic Programming. Avocations: martial arts, antiques, science fiction, parapsychology. Office: Ellemtel Telecom Systems Labs, Box 1505, S 125 25 Älvsjö Sweden

HAUSMAN, STEVEN JACK, health science administrator; b. Phila., May 20, 1945; s. Leo and Bella Hausman. BA, U. Pa., 1967, MS, 1968, PhD, 1972. Postdoctoral fellow Inst. for Cancer Rsch., Phila., 1972-75; staff fellow Nat. Inst. on Aging, Balt., 1975-77; spl. asst. to assoc. dir. Nat. Inst. Arthritis, Metabolism and Digestive Diseases, Bethesda, Md., 1977-78, dir. ctrs. program, 1978-86; dep. dir. extramural program Nat. Inst. Arthritis and Musculoskeletal and Skin Diseases, Bethesda, 1986-90, dep. dir., 1990—. Mem. Am. Assn. Immunologists, Tissue Culture Assn., Am. Chem. Soc., Am. Soc. for Cell Biology. Office: NIAMS-NIH Bldg 31 Rm 4C-32 Bethesda MD 20892

HÄUSSINGER, DIETER LOTHAR, medical educator; b. Nördlingen, Bavaria, Germany, June 22, 1951; s. Konrad and Margot (Wöhler) H. MD, U. Munich, 1976. Asst. dept. biochemistry U. Munich, 1979; asst. dept. gastroenterology U. Freiburg, Fed. Republic of Germany, 1979-84, pvt. lectr. dept. gastroenterology, 1984-88; prof. U. Hosp. Freiburg, 1988—; Heisenberg prof. German Sci. Found., Freiburg, 1985-90, Schilling prof., 1992—; cons. Inst. for Medizin Prufungen, Mainz, Fed. Republic of Germany, 1989—. Editor: Glutamine Metabolism, 1984, pH Homeostasis, 1988, Mammalian Amino Acid Transport, 1992; contbr. numerous articles to profl. jours. Recipient Wewalka price Internat. Soc. Ammonia Metabolism, 1984, Heisenberg award German Sci. Found., 1984, Leibniz-Laureate, 1990, Thannhauser prize Soc. Digestive Diseases, 1989. Mem. Biochem. Soc. London (editor 1991—), Gesellschaft Biolog. Chemie. Lutheran. Office: U Hosp Freiburg, Hugstetterstr 55, D-7800 Freiburg Germany

HAUXWELL, GERALD DEAN, chemical engineer; b. Indianola, Nebr., Sept. 24, 1935; s. Lawrence F. and Mildred E. (Wing) H.; m. Ingrid M.O. Postner, Dec. 20, 1964. BS in Chem. Engring., U. Colo., 1958; PhD in Chem. Engring., Oreg. State U., 1971. Registered profl. engr., Oreg., Va., Md. Program engr. GE Atomic Power, San Jose, Calif., 1962-64; rsch. engr. E.I. DuPont, Wilmington, Del., 1965-68; rsch. assoc. E.I. DuPont, Richmond, Va., 1971-80, competitor analyst, 1980-88, mgr. intellectural property, 1988-90, mgr. tech. assessment, 1990—; prin. cons. Adler Tech. Cons., Richmond, 1976—; adj. prof. Va. Commonwealth U., Va. State U., U. Va., Richmond, 1975—; rsch. fellow Dow Chem., 1968, Shell Co., 1970. Elder Wilmington Christian Ch., Wilmington, 1968; pres. Neberry Homeowners Assn., Richmond, 1975. Lt. (j.g.) USN, 1959-62. Mem. NSPE, Soc. Competitor Intelligence Profs. (bd. dirs. 1992—), Am. Inst. Chem. Engrs. (bd. dirs. 1973-75), Am. Chem. Soc., Va. Soc. Profl. Engrs., Phi Kappa Phi, Alpha Chi Sigma. Achievements include patent for Yarn Guide. Home: 11400 Edenberry Dr Richmond VA 23236 Office: E I DuPont de Nemours & Co 13807 Village Mill Dr Midlothian VA 23113

HAVASY, CHARLES KUKENIS, microelectronics engineer; b. Schenectady, N.Y., Aug. 15, 1970; s. Gerard Francis and Geraldine (Kukenis) H. BS in Elec. Engring., Rensselaer Poly. Inst., 1991, MS, 1992. Rsch. asst. Ctr. for Mfg. Productivity and Tech. Transfer Rensselaer Poly. Inst., Troy, N.Y., 1991-92; process devel. engr. Solid States Electronics Directorate Wright Labs., Wright-Patterson AFB, Ohio, 1992—. 2d lt. USAF, 1991—. Recipient Bronze award Soc. Am. Mil. Engrs., 1991; Theodore Von Karman scholar Air Force Assn., 1991. Mem. IEEE. Republican. Roman Catholic. Home: 2608 Brown Bark Dr Beavercreek OH 45324

HAVEL, JOHN EDWARD, biology educator; b. Red Bank, N.J., Oct. 23, 1950; s. Jerome Mathew and Doris Louise (Fuess) H.; m. Susan Jane Robords, May 25, 1985. BA, Drake U., 1973, MA, 1980; PhD, U. Wis., 1984. Teaching asst. U. Wis., Madison, 1980-82, rsch. asst., 1982-84, rsch. assoc., 1984-85; vis. asst. prof. U. West Fla., Pensacola, 1984-85, Cen. Mich. U., Mt. Pleasant, 1985-87; postdoctoral fellow U. Windsor, Ont., Can., 1987-89; asst. prof. biology S.W. Mo. State U., Springfield, 1989—; instr. Cen. Mich. U. Biol. Sta., Beaver Island, 1989, 92; vis. scientist U. Guelph, Ont., 1993. Contbr. articles to profl. publs. Grantee NSF, 1990, Mo. Water Resources, 1991, 92. Mem. Am. Soc. Limnology and Oceanography, Ecol. Soc. Am., Internat. Soc. Theoretical and Applied Limnology, N.Am. Berthological Soc., Soc. Environ. Toxicology and Chemistry, Sigma Xi. Office: SW Mo State U 901 S National Ave Springfield MO 65804-0095

HAVEL, RICHARD JOSEPH, physician, educator; b. Seattle, Feb. 20, 1925; s. Joseph and Anna (Fritz) H.; m. Virginia Johnson, June 28, 1947; children: Christopher, Timothy, Peter, Julianne. BA, Reed Coll., 1946; MS, MD, U. Oreg., 1949. Intern Cornell U. Med. Coll., N.Y.C., 1949-50; resident in medicine Cornell U. Med. Coll., 1950-53; clin. assoc. Nat. Heart Inst., NIH, 1953-54, research assoc., 1954-56; faculty Sch. Medicine, U. Calif., San Francisco, 1956—; prof. medicine Sch. Medicine, U. Calif., 1964—; assoc. dir. Cardiovascular Research Inst., 1961-73, dir., 1973-92; chief metabolism sect., dept. medicine, 1967—; dir. Arteriosclerosis Specialized Center of Research, 1970—; mem. bd. sci. counselors Nat. Heart, Lung and Blood Inst., 1976-80; chmn. food and nutrition bd. NRC, 1987-90. Contbr. chpts. to books, numerous articles to profl. jours.; editor: Jour. Lipid Research, 1972-75; co-editor: Adv. Lipid Res., 1991—; mem. editorial bd.: Jour. Biol. Chemistry, 1981-85, Jour. Arteriosclerosis, 1980—; mem. bd. cons. editors: Am. Jour. Medicine, 1981-86. Served with USPHS, 1951-53. Recipient Theobald Smith award AAAS, 1960, Bristol-Myers award for nutrition rsch., 1989. Mem. NAS, Inst. Medicine NAS, Am. Acad. Arts and Scis., Am. Fedn. for Clin. Rsch. (pres. 1965-66), Am. Soc. Clin. Nutrition (McCollum award 1993), Am. Heart Assn. (established investigator 1956-61, chmn. coun. on arteriosclerosis 1977-79, Disting. Achievement award 1993), Assn. Am. Physicians, Western Soc. for Clin. Rsch. (v.p. 1964), Am. Physiol. Soc., Am. soc. for Physicians, Phi Beta Kappa, Alpha Omega Alpha. Office: U Calif San Francisco Cardiovascular Rsch Inst San Francisco CA 94143-0130

HAVEWALA, NOSHIR BEHRAM, chemical engineer; b. Bombay, India, Dec. 19, 1938; came to U.S. 1961; s. Behram Dadabhai and Piroja Fakirji (Todiwala) H.; m. Carol Jean Ames, Dec. 19, 1963; children: Zarine N. Andolino, Tonia N. Fletcher. BS in Chem. Engring., Nagpur U., India, 1961; MS in Chem. Engring., U. Maine, 1962; PhD in Chem. Engring., N.C. State U., Raleigh, 1969. Product devel. engr. Kimberly Clark Corp., Munising, Mich., 1962-65; process engr. Corning (N.Y.) Inc., 1965-67, project mgr. bioengring., 1969-73, mgr. chem. engring., 1973-76, mgr. engring. svcs., 1976-80, mgr. chem. process tech., 1980-84, mgr. engring. rsch., 1984—; mem. ind. adv. group N.C. A&T State Univ., Greensboro, 1987—; chaired sessions on ceramic materials and fiber optics and microelectronic processing at Am. Inst. Chem. Engrs. meeting. Contbr. articles to profl. jours., chpts. to books. Mem. Am. Inst. Chem. Engrs., Am. Chem. Soc., Am. Ceramic Soc., Soc. for Info. Display, Phi Kappa Phi. Achievements include patents for immobilized enzymes and for annealing of liquid crystal display glass. Home: 15 Highland Dr Corning NY 14830-2425 Office: Corning Inc SP-PR-11 Corning NY 14831

HAVNER, KERRY SHUFORD, civil engineering and materials science educator; b. Huntington, W.Va., Feb. 20, 1934; s. Alfred Sidney and Jessie May (Fowler) H.; m. Roberta Lee Rider, Aug. 28, 1954; children: Karen Elese Smith, Clark Alan, Kris Sidney. BSCE, Okla. State U., 1955, MS, 1956, PhD, 1959. Registered profl. engr., Okla. Stress analyst Douglas Aircraft Co., Tulsa, 1956; from instr. to asst. prof. civil engring. Okla. State U., Stillwater, 1957-62; sr. stress and vibration engr. Garrett Corp., Phoenix, 1962-63; sect. chief solid mechs. rsch. missile/space systems divsn. McDonnell-Douglas Corp., Santa Monica, Calif., 1963-68; lectr. civil engring. U. So. Calif., L.A., 1965-68; from assoc. prof. to prof. civil engring. N.C. State U., Raleigh, 1968-82, prof. civil engring. and materials sci., 1982—; sr. vis. dept. applied math. and theoretical physics U. Cambridge, 1981, 89. Author: Finite Plastic Deformation of Crystalline Solids, 1992; contbg. author: Mechanics of Solids, The Rodney Hill 60th Anniversary Volume, 1982; contbr. articles to Jour. Applied Math. and Physics, Jour. of Mechs. and Physics of Solids, Acta Mechanica, Procs. Royal Soc., others; bd. editors Mechs. of Materials, Internat. Jour. Plasticity. 1st lt. U.S. Army, 1961. Rsch. grantee NSF, 1971, 74, 76, 78, 81, 83, 87, 91; vis. fellow Clare Hall, 1981. Fellow ASCE (sec. engring. mechs. divsn. 1983-85, chmn. 1987-88, chmn. engring. mechs. adv. bd. 1990-91, chmn. TAC-CERF awards com. 1991—; assoc. editor Jour. Engring. Mechs. 1981-83), Am. Acad. Mechanics (assoc. editor Mechanics. 1991—); mem. ASME, Soc. Engring. Sci., Soc. Indsl. and Applied Math., Sigma Xi. Democrat. Methodist. Achievements include development of widely recognized theories and analyses of anisotropic hardening and finite deformation in crystalline materials, particularly metals. Home: 3331 Thomas Rd Raleigh NC 27607 Office: NC State U Dept Civil Engring Box 7908 Raleigh NC 27695

HAVRILEK, CHRISTOPHER MOORE, technical specialist; b. Hopkinsville, Ky., Feb. 6, 1959; s. Frank and Judith Ann (Moore) H. BS, Murray State U., 1984. Chemist Halliburton Indsl. Svcs., Duncan, Okla., 1985-89, ops. supr., 1989-90; tech. specialist Brown and Root, Inc., Houston, 1990—. Democrat. Baptist. Home: 211 Ottawa Bend Dr #112 Morris IL 60450 Office: Brown and Root Inc PO Box 3 Houston TX 77001-0003

HAWES, NANCY ELIZABETH, secondary educator; b. Phila., Oct. 28, 1944; d. Charles E. and Margaret M. (Cassel) H. BS in Edn., Millersville (Pa.) State Coll., 1966; MAT, Purdue U., 1970; M.Div., Ea. Bapt. Theol. Sem., Phila., 1979. Ordained deacon A.M.E. Zion Ch., 1978, elder, 1980. Tchr. math. Penncrest High Sch./Rose Tree Media (Pa.) Sch. Dist., 1966-68; asst. pastor Wesley A.M.E. Zion Ch., Phila., 1975-82; pastor St. John A.M.E. Zion Ch., Bethlehem, Pa., 1982-88, Mt. Tabor A.M.E. Zion Ch., Avondale, Pa., 1988-90; assoc. pastor Wesley A.M.E. Zion Ch., Phila., 1990—; tchr. math. Upper Merion Area Sch. Dist., King of Prussia, Pa., 1968—; sponsor, Upper Merion Area High Sch. Math. Team, 1987—. Mem. Nat. Coun. Tchrs. Math., Math. Assn. Am., Pa. Coun. Tchrs. Math., Assn. Tchrs. Math. of Phila. and Vicinity. A.M.E. Zion Ch. Office: Upper Merion Area High Sch 435 Crossfield Rd King Of Prussia PA 19406

HAWK, CHARLES SILAS, computer programmer, analyst; b. Tulsa, Mar. 8, 1953; s. Samuel Shumway and Patricia Lee (Livingston) H.; m. Emelia Elizabeth Presley. BS in Zoology, U. Tex., 1980. Cons. Med. Systems, Houston, Dallas, 1982-87, Mfg. Systems, Dallas, 1987-89, The Hawk Group, Dallas, 1987—. Del. Tex. Rep. Conv., Houston, 1984. Commd. USNR, 1984. Grantee Merit scholarship U. Tex., 1974. Mem. AAAS, IEEE, N.Y. Acad. Scis., Tau Kappa Epsilon, Mason. Mem. Christian Ch. Home and Office: 1655 Randolph Pl Memphis TN 38120

HAWK, CLARK WILIAMS, mechanical engineering educator; b. Berea, Ohio, Sept. 16, 1936; s. Harry Lyle and Catherine (Williams) H.; m. Julia Ann Milthaler, Nov. 7, 1959; children: Sandra Lynn Smith, Brian Clark. BSME, Pa. State U., 1958; MSME, Purdue U., 1967, PhD, 1970. Registered mech. engr., Calif., Ala. Project engr. Propulsion Lab., Wright-Patterson AFB, Ohio, 1958-59; project engr. Rocket Propulsion Lab., Edwards AFB, Calif., 1959-67, sect. chief, researcher, 1969-71, br., sect. chief, 1971-81; div. dir. Astronautics Lab., Edwards AFB, Calif., 1983-91; lectr. U. So. Calif., L.A., 1971; prof. mech. engring., dir. Propulsion Rsch. Ctr. U. Ala., Huntsville, 1991—; mem. NRC Earth To Orbit Propulsion Options, Washington, 1991-92; mem. NRC Com. on Advanced Space Techs., Washington, 1992—. Contbr. articles to profl. jours. Recipient Gilbert award Antelope Valley YMCA, 1982, Meritorious Svc. award L.A. County Bd. Edn., 1985, Cert. of Merit, L.A. Sch. Trustees Assn., 1985, Excellence award Air Force Space Div., 1986. Assoc. fellow AIAA (assoc. editor Jour. Propulsion and Power 1986-90); mem. NSPE, Pi Tau Sigma. Achievements include conceiving and creating LPIAG and SPIAG which brought together U.S. liquid rock engine manufacturers to solve problems of mutual concern; creation of JANNAF Rocket Nozzle Tech. Com. to establish a community of interest in carbon-carbon nozzle technology area. Home: 179 Stoneway Trail Madison AL 35758-8543

HAWKING, STEPHEN W., astrophysicist, mathematician; b. Oxford, England, Jan. 8, 1942; s. F. and E.I. Hawking; m. Jane Wilde, 1965; two sons, one daughter. BA, Oxford U., DSc (hon.), 1978; PhD, Cambridge U.; DSc (hon.), U. Chgo., 1981, Notre Dame U., 1982, NYU, 1982, Leicester U., 1982. Research asst. Inst. Astronomy, Cambridge, Eng., 1972-73; research asst. dept. applied maths. and theoretical physics Inst. Astronomy, Cambridge, 1973-75, reader in gravitational physics, 1975-77, prof., 1977-79, Lucasian prof. math., 1979—. Author: (with G.F.R. Ellis) The Large Scale Structure of Space-Time, 1973; A Brief History of Time: From the Big Bang to Black Holes, 1988, Black Holes and Baby Universes: And Other Essays, 1993; co-editor: 300 Years of Gravitation, 1987. Decorated comdr. Brit. Empire, 1981; recipient Eddington medal Royal Acad. Sci., 1975, Pius XI Gold medal Pontifical Acad. Sci., 1975, Danne Heinemann prize for math. and physics Am. Phys. Soc.-Am. Inst. Physics, 1976, William Hopkins prize Cambridge Philos. Soc., 1976, Maxwell medal Inst. Physics, 1976, Einstein award Strauss Found., 1978, Albert Einstein medal Albert Einstein Soc. of Berne, 1979, Wolf Prize in physics, 1988, Britannica award, 1989. Fgn. mem. Am. Philos. Soc., AAAS; fellow Royal Soc. (Hughes medal 1976). Address: DAMTP U Cambridge, Silver St, Cambridge England CB3 9EW

HAWKINS, DARROLL LEE, civil engineer; b. Louisville, May 1, 1948; s. Robert Lee Hawkins and Jonnie Charlotte (Snider) Hair; m. Patricia Nannette Hachter, Apr. 20, 1968; children: Leah Nannette, James Robert. BLS in Environ., U. Louisville, 1980, MS in Community Devel., 1983, PhD in Urban Infrastructure, 1993. Registered profl. engr., Ky., Ind., Pa., Va., W.Va.; cert. wetland-delineator. Engring. technician estimating and specifications U.S. Army Engrs. Dist. Louisville, 1968-71, engring. technician basin hydrology, 1971-78, project mgr. engring. tech., 1978-80, project mgr. sci. specifications, 1980-82, project mgr., sci. specialist, 1982-85, chief no. sect.

regulatory br., 1985-91; mgr. Louisville office environ. cons. divsn. Commonwealth Tech., Inc., 1991—; sec., treas. Nat. Wetland Specialist Regulation Bd., Louisville, pres., 1990-92. Editor: Storm Water Monitoring, 1991. Mem. Louisville Forum, 1988-92. Mem. ASCE, Soc. Am. Mil. Engrs. (1st v.p. Kentuckiana post 1991-92), Ky. Soc. Profl. Engrs. (scholarship com. 1992), Nat. Coun. Examiners for Engring. and Surveying, Inland Rivers Ports and Terminals Inc., Propeller Club Am., Louisville C. of C. (chmn. water com. 1991-92). Republican. Baptist. Achievements include research on passive benefits of urban wetlands, analytical model for urban policy devel. relative to deindustrialization. Office: Commonwealth Tech Inc Environ Cons Divsn 11001 Bluegrass Pkwy #330 Louisville KY 40299-2368

HAWKINS, DAVID ROLLO, SR., psychiatrist, educator; b. Springfield, Mass., Sept. 22, 1923; s. James Alexander and Janet (Rollo) H.; m. Elizabeth G. Wilson, June 8, 1946; children: David Rollo Jr., Robert Wilson, John Bruce, William Alexander. B.A., Amherst Coll., 1945; M.D., U. Rochester, N.Y., 1946. Intern Strong Meml. Hosp., Rochester, 1946-48; Commonwealth Fund fellow in psychiatry and medicine U. Rochester, 1950-52; instr. psychiatry U. N.C. Sch. Medicine, 1952-53, asst. prof., 1953-57, asso. prof. psychiatry, 1957-62, prof., 1962-67; prof., chmn. dept. psychiatry U. Va. Sch. Medicine, 1967-77, Alumni prof. psychiatry, 1967-79, asso. dean, 1969-70; psychiatrist-in-chief U. Va. Hosp., 1967-77; prof. psychiatry Pritzker Sch. Medicine, U. Chgo., 1979-90, U. Ill., 1990—; clin. prof. psychiatry U. N.C., Chapel Hill, 1992—; dir. liaison and consultation svcs. dept. psychiatry Michael Reese Hosp., Chgo., 1979-87, chmn., 1987-92; assoc. attending physician N.C. Meml. Hosp., Chapel Hill, 1952-62, attending physician, 1962-67; cons. Watts Hosp., Durham, 1952-67, VA Hosp., Fayetteville, N.C., 1956-67, Eastern State Hosp., Williamsburg, Va., 1971—, VA Hosp., Salem. Va., 1969-79, mem. deans com., 1971-77; spl. rsch. fellow Inst. Psychiatry, U. London, 1963-64, Fogarty internat. rsch. fellow, 1976-77, U.S.-USSR and Romania health exch. fellow, 1978. Rev. editor Psychosomatic Medicine, 1958-70; assoc. editor Psychiatry, 1970-92. Mem. small grants com. NIMH, 1958-62; mem. nursing research study sect. NIH, 1965-67; mem. Gov.'s Commn. Mental, Indigent and Geriatric Patients, 1968-72; mem. research evaluation com. Va. Dept. Mental Hygiene and Hosps., 1970-73 ; mem. behavioral sci. test com. Nat. Bd. Med. Examiners, 1970-73. Served as capt. MC Aus, 1948-50. Fellow Am. Coll. Psychoanalysts (charter bd. regents 1979-81, treas. 1989-91, pres.-elect 1992, pres. 1993), Am. Psychiat. Assn.; mem. Am. Psychosomatic Soc. (mem. coun. 1959), AMA, Group for Advancement Psychiatry (bd. dirs. 1987-89), Assn. Am. Med. Colls. (coun. acad. socs. 1973-78), Am. Psychoanalytic Assn., Am. Coll. Psychiatrists, AAAS, Va. Psychoanalytic Soc., Washington Psychoanalytic Soc., Chgo. Psychoanalytic Soc., Ill. Psychiat. Soc. (coun. 1981-82, pres.-elect 1987, pres. 1988-90), AAUP, Soc. Neurosci., Am. Assn. Chmn. Depts. Psychiatry (sec.-treas. 1971-73, pres. 1974-75), Sleep Rsch. Soc., Nat. Bd. Med. Examiners (exam. com. 1983-87), Phi Beta Kappa, Sigma Xi, Alpha Omega Alpha. Address: 405 Deming Rd Chapel Hill NC 27514

HAWKINS, JAMES DOUGLAS, JR., structural engineer, architect; b. Dallas, Jan. 29, 1959; s. James Douglas Sr. and Evelyn Carolyn (Roos) H. BArch, Tex. Tech. U., 1986, BSCE, 1986. Registered profl. engr. Landscape architect asst. New Leaf Environ. Systems, Ft. Worth, 1980-82; clk. Lone Star Gas Co., Dallas, 1983; architect asst. Al Cox, Architect, Dallas, 1984; technician Geotech Lab. Tex. Tech U., Lubbock, 1985; constrn. labor George Chadick, Dallas, 1987; structural engr. and architect U.S. Army Corps of Engrs., Jacksonville, Fla., 1988—. Co-author: Overton Revitalization, 1985. Vol. Habitat for Humanity, Jacksonville, 1988—. Mem. AIA, ASCE, Jacksonville Jaycees. Achievements include rehab. buildings for public housing and design development of residents following Hurricane Hugo; evaluation and design repairs following Hurricane Andrew. Home: 3000 Coronet Ln # 162 Jacksonville FL 32207-5125 Office: US Army Corps Engrs 400 W Bay St Jacksonville FL 32202

HAWKINS, JANET LYNN, school psychologist; b. July 3, 1956; d. James Crawford Jr. and OBeria (Aiken) H. BS in Spanish, Psychology, Furman U., 1978; MEd in Secondary Edn., Converse Coll., 1981; EdS in Counseling, Sch. Psychology, Wichita State U., 1986. Cert. sch. psychologist, S.C. Tchr. Spartanburg Sch. Dist. 5, Duncan, S.C., 1979-80, Wichita (Kans.) Sch. Dist., 1982-85; psychology intern Mulvane (Kans.) Sch. Dist., 1985-86; sch. psychologist Sch. Dist. greenville (S.C.) County, 1986—; presenter at profl. confs. Mem. NOW, Coun. Exceptional Children, Nat. Assn. Sch. Psychologists, Piedmont Assn. Sch. Psychologists, S.C. Assn. Sch. Psychologists, Sierra Club, Nat. Wildlife Fedn., Phi Kappa Phi. Avocations: music, art, nature hiking. Office: Sch Dist Greenville County 206 E Church St Greer SC 29651-2534

HAWKINS, JOSEPH ELMER, JR., acoustic physiologist; b. Waco, Tex., Mar. 4, 1914; s. Joseph Elmer and Maude Burke (Schlenker) H.; m. Jane Elizabeth Daddow, Aug. 24, 1939; children: Richard Spencer Daddow, Peter Douglas Huntington, James Marion Davis, William Alexander Parmley, Priscilla Ann (Mrs. Philip A. Leach). Student, Altes Realgymnasium, Munich, 1929-30; AB, Baylor U., 1933; postgrad., Brown U., 1933-34; BA in Physiology, U. Oxford, 1937, MA, 1966, DSc, 1979; PhD Med. Sci., Harvard U., 1941. Teaching fellow physiology Harvard Med. Sch., 1937-41, instr., 1941-45; asst. investigator NDRC-OSRD, 1941-43; spl. research asso. Harvard Psycho-Acoustic Lab., 1943-45; asst. prof. physiology Bowman Gray Sch. Medicine, Wake Forest Coll., 1945-46; research asso., head neurophysiology Merck Inst. for Therapeutic Research, Rahway, N.J., 1946-56; asso. prof. otolaryngology N.Y.U. Sch. Medicine, 1956-63; prof. otorhinolaryngology U. Mich., Ann Arbor, 1963-84, prof. emeritus, 1984—; chmn. grad. program in physiol. acoustics U. Mich., 1969-81; asso. dir. Kresge Hearing Research Inst., 1979-82; disting. vis. prof. biology Baylor U., Waco, Tex., 1985—; mem. NIH sensory diseases study sect., 1958-61, communicative disorders research tng. com., 1965-69, communicative scis. study sect., 1975-79; mem. Nat. Library Medicine Communicative Disorders Task Force, 1977-79; lectr. Armed Forces Inst. Pathology, 1969-74; cons. various pharm. cos. Contbr. to: Ency. Brit, 1974; Editor: (with M. Lawrence and W.P. Work) Otophysiology, 1973, (with S.A. Lerner and G.T. Matz) Aminoglycoside Ototoxicity, 1981; contbr. sci. articles to profl. jours. Pres. Fleming Creek Neighborhood Assn., Washtenaw County, Mich., 1973-74; mem. Bd. Edn., Cranford, N.J., 1958-61. Rhodes scholar Tex. and Worcester Coll., U. Oxford, 1934-37; USPHS spl. fellow Öronkliniken, Sahlgrenska sjukhuset U. Göteborg, Sweden, 1961-63; NAS exch. lectr. to Yugoslavia and Bulgaria, 1977; Chercheur étranger de l'INSERM, Lab. d'Audiologie Expérimentale, U. Bordeaux II, 1978; recipient Disting. Achievement award Baylor U., 1982, City of Pleven, Bulgaria, medal, 1982, U. Bordeaux medal, 1983, Humboldt Rsch. award for sr. U.S. scientists U. Würzburg, 1991; named Hon. Citizen award Bordeaux, 1991. Fellow AAAS, Acoustical Soc. Am.; mem. Am. Physiol. Soc., Assn. for Rsch. in Otolaryngology (award of merit 1985), Collegium Oto-rhino-laryngologicum Amicitiae Sacrum, Bárány Soc., European Workshop for Inner Ear Biology, Am. Assn. for History Medicine, History of Sci. Soc., Am. Otol. Soc. (assoc.), Pacific Coast Oto-ophthalmol. Soc. (hon.), Prosper Menière Soc. (hon.), Connétalie de Guyenne (Bordeaux) (assoc.), Phi Beta Kappa, Sigma Xi. Anglican. Home: 4004 E Joy Rd Ann Arbor MI 48105-9609 Office: U Mich Med Sch Kresge Hearing Rsch Inst Ann Arbor MI 48109-0506

HAWKINS, PAMELA LEIGH HUFFMAN, biochemist; b. Washington, Oct. 7, 1950; d. Lauria Carl and Maryalice (Flinner) Huffman; m. James Lee Hawkins, Mar. 7, 1981 (div. Aug. 1993). BS in Biochemistry, Va. Polytech. Inst. & State U., 1972; MS in Biochemistry, Pa. State U., 1975. Sci. info. specialist Informatics, Inc., Rockville, Md., 1972; asst. rsch. scientist Union Carbide Corp., Tarrytown, N.Y., 1975; assoc. rsch. scientist Am. Hosp. Supply Corp., Gibbstown, N.J., 1976-78; rsch. scientist Am. Hosp. Supply Corp., Miami, Fla., 1978-85; R & D scientist Baxter Healthcare Corp., Miami, 1985-93, sr. rsch. scientist, 1993—. Contbr. articles to profl. jours. Bd. dirs. Kairos Prison Ministry, Miami, 1988-91. Recipient Baxter Diagnostics Tech. award for Thromboplastin-IS, 1990, Baxter Internat. Tech. award, 1991. Fellow Am. Inst. Chemists; mem. Am. Chem. Soc., Mortar Bd., Phi Sigma, Gamma Sigma Delta, Phi Lambda Upsilon. Lutheran. Achievements include U.S. and European patent for fresh blood (unfixed) hematology control, patent pending for improved extraction methods for preparing thromboplastin reagents, production of thromboplastin IS, Innovin. Office: Baxter Diagnostics Inc PO Box 520672 Miami FL 33152-0672

HAWKINS, SHANE V., electronics and computer engineer; b. Canton, Ohio, Oct. 16, 1957; s. Howard Edward and Martha Jean (Van Horn) H. BSEE, W.Va. U., 1979. Dir. tech. ops. CADSA/SpaceLink, Webster, Tex., 1982-86; dir. devel. Metrocast, San Diego, 1985-89; mng. ptnr. Shadetree Software, San Diego, 1987—; v.p. TI-IN Network, San Antonio, 1989-92; pres. SR Comm., Atlanta, 1993—; mem. SALT, 1989—; cons. Brit. Telecom, U.K., 1988-89. Contbr. Tech. Updates newsletter, 1988—; contbr. articles to profl. jours.; reviewer in field. Mem. Soc. Broadcast Engrs., Am. Mgmt. Assn., Assn. Computing Machinery, Soc. Cable TV Engrs., IEEE Computer Soc., Soc. Applied Learning, Assn. Curriculum Devel. Achievements include patent on scanning pager system; research in voice over frame relay technology, satellite based frame relay routing methods. Office: 7510 Sunset Blvd Ste 1063 Los Angeles CA 90046-3418

HAWLEY, SANDRA SUE, electrical engineer; b. Spirit Lake, Iowa, May 7, 1948; d. Byrnard Leroy and Dorothy Virginia (Fischbeck) Smith; m. Michael John Hawley, June 7, 1970; 1 child, Alexander Tristin. BS in Elec. Engring., U. Dayton, 1981; BS in Math. and Stats., Iowa State U., 1970; MS in Stats., U. Del., 1975. Rsch. analyst State of Wis., Madison, 1970-71; rsch. asst. Del. State Coll., Dover, 1972-73; asst. prof. math. and statis. Wesley Coll., Dover, 1974-81, chmn. dept. math. and computer sci., 1978-80; elec. engr. Control Data Corp., Bloomington, Minn., 1982-85; sr. elec. engr. Custom Integrated Circuits, 1985-89; sr. lead engr. Cardiac Pacemakers, Inc., 1989-90; mgr. Corp. Tech. Rosemount Inc., 1990—. Contbr. articles to profl. jours. Elder Presbyn. Ch. U.S.A., 1975—, mem. session Oak Grove Presbyn. Ch., Bloomington, 1985-88; chair Presbtery of Twin Cities Area Coun. on United Action, 1989-92, adminstrv. comm., 1989-91, commr. to Synod of Lakes & Prairies, 1990, Gen. Assembly, 1991, com. on coun., 1992; Gen. Assembly Coun., 1992—; NSF scholar U. Dayton, 1981. Mem. IEEE, Soc. Women Engrs. Home: 7724 W 85th Street Cir Minneapolis MN 55438-1311 Office: Rosemount Inc 12001 Technology Dr Eden Prairie MN 55344-3695

HAWTHORNE, MARION FREDERICK, chemistry educator; b. Ft. Scott, Kans., Aug. 24, 1928; s. Fred Elmer and Colleen (Webb) H.; m. Beverly Dawn Rempe, Oct. 30, 1951 (div. 1976); children: Cynthia Lee, Candace Lee; m. Diana Baker Razzaia, Aug. 14, 1977. B.A., Pomona Coll., 1949; Ph.D. (AEC fellow), U. Calif. at Los Angeles, 1953; D.Sc. (hon.), Pomona Coll., 1974; PhD (hon.), Uppsala U., 1992. Research asso. Iowa State Coll., 1953-54; research chemist Rohm & Haas Co., Huntsville, Ala., 1954-56; group leader Rohm & Haas Co., 1956-60; lab. head Rohm & Haas Co., Phila., 1961; vis. lectr. Harvard, 1960, Queen Mary Coll., U. London, 1963; vis. prof. Harvard U., 1968; prof. chemistry U. Calif. at Riverside, 1962-68, U. Calif. at Los Angeles, 1968—; vis. prof. U. Tex., Austin, 1974; mem. sci. adv. bd. USAF, 1980-86, NRC Bd. Army Sci. and Tech., 1986-90; disting. vis. prof. Ohio State U., 1990; mem. dir.'s external adv. bd. div. M, Los Alamos (N.Mex.) Nat. Lab., 1991—. Editor: Inorganic Chemistry, 1969—; Editorial bd.: Progress in Solid State Chemistry, 1971—, Inorganic Syntheses, 1966—, Organometallics in Chemical Synthesis, 1969—, Synthesis in Inorganic and Metalorganic Chemistry, 1970—. Recipient Chancellors Research award, 1968, Herbert Newby McCoy award, 1972, Am. Chem. Soc. award in Inorganic Chemistry, 1973, Tolman Medal award, 1986, Nebr. sect. Am. Chem. Soc. award, 1979, Disting. Service in the Advancement of Inorganic Chemistry award Am. Chem. Soc., 1988, Disting. Achievements in Boron Sci. award, 1988, Bailar medal, 1991, Polyhedron Medal and prize, 1993; sr. scientist Alexander von Humboldt Found., Inst. Inorganic Chemistry U. Munich, 1990—; Sloan Found. fellow, 1963-65, Japan Soc. Promotion Sci. fellow, 1986; named Col. Confederate Air Force, 1984. Fellow AAAS; mem. U.S. Nat. Acad. Scis., Am. Acad. Arts and Scis., Aircraft Owners and Pilots Assn., Sigma Xi, Alpha Chi Sigma, Sigma Nu. Club: Cosmos. Home: 3415 Green Vista Dr Encino CA 91436-4011

HAWTHORNE, SIR WILLIAM (REDE), aerospace and mechanical engineer, educator; b. May 22, 1913; s. William and Elizabeth H.; ed. Trinity Coll., Cambridge, M.I.T.; m. Barbara Runkle, 1939; 1 son, 2 daus. Devel. engr. Babcock & Wilcox Ltd., 1937-39; sci. officer Royal Aircraft Establishment, 1940-44; with Brit. Air Commn., Washington, 1944; dep. dir. engine research Ministry of Supply, 1945; assoc. prof. mech. engring. MIT, 1946, George Westinghouse prof. mech. engring., 1948-51, Jerome C. Hunsaker prof. aero. engring., 1955-56; master Churchill Coll., Cambridge, 1968-83, now faculty; Hopkinson and ICI prof. applied thermodynamics U. Cambridge, 1951-80, head dept. engring., 1968-73; chmn. Home Office Sci. Adv. Council, 1967-76, Adv. Council Energy Conservation, 1974-79; dir. Dracone Devels. Ltd. Bd. govs. Westminster Sch., 1956-76. Recipient R. Tom Sawyer award ASME, 1992. Fellow Royal Soc.; mem. NAE (fgn. assoc.), NAS (fgn. assoc.). Office: Churchill Coll, Cambridge England also: 19 Chauncey St Cambridge MA 02138

HAXTON, DONOVAN MERLE, JR., astronomer; b. Mason City, Iowa, June 27, 1941; s. Donovan M. and Edna O. (Peterson) H.; m. Judith Lee Morton, Sept. 8, 1963; children: Lilah Jane, Ruth Joanne. AA in Physics, North Iowa C.C., Mason City, 1961; BA in Astronomy, U. Iowa, 1963. Sr. programmer, analyst AEGIUS, naval tactical data systems USN Fleet Computer Programming Ctr., Virginia Beach, Va., 1963-71; flight dynamic specialist Flight Dynamics Group Code 552 Goddard Space Flight Ctr., Greenbelt, Md., 1971-84; rsch. specialist, astronomer space telescope project Goddard Space Flight Ctr., Greenbelt, 1984—. Assoc. adviser Explorer Post 1275, Boy Scouts Am., Goddard Space Flight Ctr./NASA, 1976—. Achievements include computing of launch and reentry analysis for the Viking Mars Lander/Orbiter, computing of launch maneuver analysis for the Voyager I and II missions to Jupiter, Saturn, Uranus and Neptune; designed and developed engineering support systems for Hubble Space Telescope project; was launch mission coordinator (flight dynamics group) for first 13 space shuttle flights. Home: 9722 Whiskey Botton Rd Laurel MD 20723 Office: Goddard Space Flight Ctr Code 440.8 Code 440.8 Greenbelt MD 20771

HAY, ELIZABETH DEXTER, embryology researcher, educator; b. St. Augustine, Fla., Apr. 2, 1927; d. Isaac Morris and Lucille (Lynn) H. AB, Smith Coll., 1948; MA (hon.), Harvard U., 1964; ScD (hon.), Smith Coll., 1973, Trinity Coll., 1989; MD, Johns Hopkins U., 1952, LHD (hon.), 1990. Intern in internal medicine Johns Hopkins Hosp., Balt., 1952-53; instr. anatomy Johns Hopkins U. Med. Sch., Balt., 1953-56, asst. prof., 1956-57; asst. prof. Cornell U. Med. Sch., N.Y.C., 1957-60; asst. prof. Harvard Med. Sch., Boston, 1960-64, Louise Foote Pfeiffer assoc. prof., 1964-69, Louise Foote Pfeiffer prof. embryology, 1969—, chmn. dept. anatomy and cellular biology, 1975-93; prof. dept. cell biology, 1993—; cons. cell biology sect. NIH, 1965-69; mem. adv. coun. Nat. Inst. Gen. Med. Sci., NIH, 1978-81; mem. sci. adv. bd. Whitney Marine Lab., U. Fla., 1982-86; mem. adv. coun. Johns Hopkins Sch. Medicine, 1982—; chairperson bd. sci. counselors Nat. Inst. Dental Rsch., NIH, 1984-86; mem. bd. sci. counselors Nat. Inst. Environ. Health Sci., NIH, 1990-93. Author: Regeneration, 1966; (with J.P. Revel) Fine Structure of the Developing Avian Cornea, 1969; editor: Cell Biology of Extracellular Matrix, 1981, 2d edit., 1991; editor-in-chief Developmental Biology Jour., 1971-75; contbr. articles to profl. jours. Mem. Scientists Task Force of Congressman Barney Frank, Massach, 1982-92. Recipient Disting. Achievement award N.Y. Hosp.-Cornell Med. Ctr. Alumni Council, 1985, award for vision research Alcon, 1988. Mem. Soc. Devel. Biology (pres. 1973-74), Am. Soc. Cell Biology (pres. 1976-77, legis. alert com. 1982—, E.B. Wilson award 1989), Am. Assn. Anatomists (pres. 1981-82, legis. alert com. 1982—, Centennial award 1987, Henry Gray award 1992), Am. Acad. Arts and Scis., Johns Hopkins Soc. Scholars, Nat. Acad. Sci., Inst. Medicine, Internat. Soc. Devel. Biologists (exec. bd. 1977), Boston Mycol. Club. Home: 14 Aberdeen Rd Weston MA 02193-1733 Office: Harvard Med Sch Dept Cell Biology 220 Longwood Ave Boston MA 02115-6092

HAY, RICHARD CARMAN, anesthesiologist; b. Queens, N.Y., June 9, 1921; s. Richard Carman and Frances Pauline (Woodbury) H.; B.S., U. Vt., 1944, M.D., 1946; m. Martha Fambrough, Mar. 2, 1957; children—Richard C., William W., Anne H., Sandra L., Bradford T., Holly K. Practice medicine, specializing in anesthesiology, Houston. Served with M.C., U.S. Army, 1948-50. Mem. AMA, Tex. Med. Soc., Harris County Med. Soc., Am. Soc. Anesthesiologists, Tex. Soc. Anesthesiologists. Republican. Baptist. Office: 1102 Deerfield Rd Richmond TX 77469-6574

HAYAISHI, OSAMU, director science institute; b. Stockton, Calif., Jan. 8, 1920; s. Jitsuzo and Mitsu (Uchida) H.; m. Takiko Satani, Nov. 3, 1947; 1 child, Mariko. MD, Osaka (Japan) U., 1942, PhD, 1949; DSc (hon.), U. Mich., 1980; DMed, Karolinska Inst., 1985. Asst. prof. Wash. U., Seattle, 1952-54; chief. sect. toxicology NIH, Bethesda, Md., 1954-58; prof. Kyoto (Japan) U., 1958-83; pres. Osaka Med. Coll., 1983-89; dir. Hayaishi bioinfo. transfer project Rsch. Devel. Corp. Japan, 1983-89; dir. Osaka Biosci. Inst., 1987—; pres. Internat. Union Biochemistry, 1973-76, Found. Asian Oceanian Biochemistry, 1981-83; chmn. Japan Nat. Com. Biochemistry. Author: Oxygenases, 1962, Molecular Mechanism of Oxygen Activities, 1974, Biochemical and Medical Aspects of Active Oxygen, 1977, Biochemical Imaging, 1986, Clinical and Nutritional Aspects of Vitamin, 1987. Capt. medicine, 1942-45. Recipient Order of Culture award Japan, 1972, Louis and Bert Freedman Found. award N.Y. Acad. Sci., 1976, Wolf Medicine prize Wolf Found., Israel, 1986, 1st Order of Merit Grand Cordon of Sacred Treasure, 1993. Mem. Japan Acad. (Asahi award sci. and culture 1965), Nat. Acad. Scis., Am. soc. Biol. Chemist, Deutsche Acad. der Naturfors cher Leopoldia. Achievements include discovery of oxygenases; discovery of molecular mechanisms of sleep-wake regulation by prostaglandins D2 and E2. Home: 1-29-205 Izumigawacho, Kyoto 606, Japan Office: Osaka Bioscience Inst, 6-2-4 Furuedai, Suita 565, Japan

HAYAKAWA, KAN-ICHI, food science educator; b. Shibukawa, Gumma, Japan, Aug. 12, 1931; came to U.S., 1961, naturalized, 1974; s. Chygoro and Kin (Hayakawa) H.; m. Setsuko Maekawa, Feb. 18, 1967. BS, Tokyo U. Fisheries, 1955; PhD, Rutgers U., 1964. Rsch. fellow Canners' Assn. Japan, 1955-60; asst. prof. food sci. Rutgers U., New Brunswick, N.J., 1964-70; assoc. prof. food sci. Rutgers U., 1970-77, prof. food engring., 1977-82, Disting. prof. food engring., 1982—; OAS vis. prof. U. Campinas, Brazil, summers 1972, 73; vis. fgn. rsch. fellow Tokyo U., 1992; cons. to food processing cos. Organizer, chmn., participant NSF sponsored U.S.-Japan Coop. Conf., Tokyo, 1979; lectr. Industry R & D Inst. and Nat. Taiwan U., both Taiwan, June 1982, Wuxi Inst. of Light Industry, China, 1986, Tokyo U. of Fisheries, May 1992; vis. investigator, Tokyo U., June 1992. Co-editor: Heat Sterilization of Food, 1983. Contbr. articles to books, profl. jours. and encys.; developer new math methods for predicting safety of food processes; found theoretical and exptl. theorems on heat and mass transfer in biol. material with or without strain--stress formation. Rsch. grantee USPHS, 1966-73, Nabisco Found., 1975-76, NSF, 1981-82, travel grantee NSF, 1972, Rutgers Rsch. Found., 1977, rsch. grantee Advanced Food Tech. Ctr., 1985—, John von Neumann Nat. Supercomputer Ctr., 1989-90, Pitts. Nat. Supercomputer Ctr., 1990—, U.S. Army Natick R&D Ctr., 1992—. Fellow Inst. Food Technologists; mem. AAAS, Am. Inst. Chem. Engrs., ASHRAE (chmn. tech. com. on thermophys. property values of food 1981-85, mem. com. 1981—), Am. Soc. Agrl. Engrs., Can. Inst. Food Sci. and Tech., Sigma Xi. Home: 631 Lake Dr Princeton NJ 08540-5634 Office: Rutgers U Dept Food Sci PO Box 231 New Brunswick NJ 08903-0231

HAYASHI, GEORGE YOICHI, airline executive; b. Yokohama, Japan, July 7, 1920; s. Daisaku and Ginko (Yamada) H.; m. June 6, 1955; children: Fuyeko, Kazuhiko. LLB, U. Tokyo, 1942; BS in Langs. cum laude, Georgetown U., Washington, 1951; postgrad., Georgetown U., 1951-52. Lectr. Sch. Fgn. Svc. Georgetown U., 1951-52; 1st sec. Embassy of Japan, London, 1955-59; dir. gen. Kanto (Tokyo-Yokohama) Maritime Bur., Japan, 1963-65, Lighthouse Bur., Japan, 1965-67; dep. dir. gen. Civil Aviation Bur., Japan, 1967-69; dep. commandant Japanese Coast Guard/Maritime Safety Agy., Japan, 1969-70; sr. v.p. All Nippon Airways, Tokyo, 1971-81; chmn. Internat. Airport Svc. Co. Ltd., Tokyo, 1981-87; sr. adviser All Nippon Airways Trading Co., Ltd., Tokyo, 1987—; bd. dirs., adviser Nippon Rent-A-Car Svc. Co. Ltd.; bd. dirs. Transp. Bus. Consultants, Tokyo. Author: Introduction to Norway and Iceland, 1954; translator Scandinavian lit. Chmn. parish coun. Kikuna Roman Cath. Ch., Yokohama, 1984-90. Lt. Japanese naval air wing, 1942-45. Knight Order St. Olav, 1963; decorated Order of Sacred Treasure, Emperor of Japan, 1990. Mem. Nordic Cultural Soc. Japan (chair 1974—), Norway Japan Soc. (v.p. 1979—), Georgetown U. Club. Japan (pres. 1980-85), Tokyo Club, Internat. Bar Assn. London, Hodogaya Country Club, Pacific Asia Travel Assn. (past exec. dir.). Avocations: literature, travel, photography, classical music, golf. Home: 1-3-13 Fujizuka Kohoku-ku, Yokohama 222, Japan Office: All Nippon Trading Co Ltd, 3-3-2 Kasumigaseki Chiyoda-ku, Tokyo 100, Japan

HAYASHI, MITSUHIKO, physics educator; b. Okazaki, Aichi Pref, Japan, Sept. 3, 1930; s. Katsuzo and Rakuko (Morita) H.; m. Etsuko Ito, Oct. 18, 1964; children: Mayura, Nao. BS, Nagoya (Japan) U., 1958, MS, 1960; PhD, Tokyo Inst. Tech., 1971. Rsch. assoc. Nagoya U., 1960-70, asst. prof., 1970-75, assoc. prof., 1975-76; prof. Toyama (Japan) Med. and Pharm. U., 1976—; lectr. Kinjo Women's Coll., Nagoya, 1962-73, Toyama U., 1985-90; cons. Noritake China Co., Nagoya, 1973-76; vis. scientist U. Wash., Seattle, 1980-81, 82, Tech. U. Denmark, Lyngby, 1986; guest prof. Delft U. Tech., The Netherlands, 1987. Author: Introductory Physics, 1966, Ultrafine Particles 1984; contbr. articles to profl. jours. Mem. Phys. Soc. Japan, Crystallographic Soc. Japan, Biophys. Soc. Japan, Surface Sci. Soc. Japan. Presbyterian. Avocation: concerts. Home: 3-103 2556-4 Gofuku, Suehirocho Toyama 930, Japan Office: Toyama Med & Pharm U, 2630 Sugitani, Toyama 930-01, Japan

HAYASHI, SHIZUO, physics educator; b. Takasaki, Japan, Mar. 23, 1922; s. Kaisuke and Hana (Kasuya) H.; m. Ihori Kiyono, Dec. 28, 1931; children: Hironao, Yurika. BS, Tokyo Bunrika U., 1945, ScD, 1961. Asst. prof. Gunma U., Maebashi, Japan, 1949-50; assoc. prof. Gunma U., 1950-64, prof. physics, 1964-87, dean, 1978-82, prof. emeritus, 1987—; prof. Takasaki Art Ctr. Coll., Yoshii, Japan, 1988—. Author: Basic Physics, 1980, Rheology, 1973; contbr. articles to sci. jours. Mem. Phys. Soc. Japan, Chem. Soc. Japan, Biophys. Soc. Japan, Soc. Polymer Sci. Japan. Achievements include research in linear and nonlinear rheology of concentration polymer system, growth habit of polymer single crystle, motion of protein molecules in gel. Home: 63 Tokiwa cho, Takasaki 370, Japan Office: Takasaki Art Ctr Coll, 2229 Iwasaki, Yoshii 370-21, Japan

HAYASHI, TADAO, engineering educator; b. Toyohashi, Aichi-ken, Japan; s. Yoshio and Matsue Hayashi; m. Fumi Imoto, May 15, 1950; children: Hidetaka, Fumihiko; m. Sumiko Sano, Oct. 25, 1981. BS, Tokyo U. Lit. and Sci., 1948, MS, 1950; D Engring., Tokyo Inst. Tech., 1963. Rsch. asst. Naniwa U., Osaka, Japan, 1950-52; rsch. assoc. Ohio State U., Columbus, 1952-54; rsch. assoc. U. Osaka Prefecture, 1955-63, asst. prof., 1964-68, prof. engring., 1969-86, prof. emeritus, 1986—. Co-author: Electroplating, 1980, Properties of Electrodeposits, 1986, Testbook of Electroplating, 1987. Recipient Gold medal Electrochem. Soc. Japan, 1977; Best Paper award Metal Finishing Soc. Japan, 1966, 84, Silver medal award Am. Electroplaters and Surface Finishers Soc., 1986, Sci. Achievement award, 1990. Mem. Electroplaters Tech. Assn., Electroplaters and Surface Finishers Soc. (internat. liaison rsch. bd. 1982—), Internat. Soc. Electrochemistry, Surface Finishing Soc. Japan. Avocations: tennis, golf. Home: 4-17-17 Tezukayama-Minami, Nara 631, Japan

HAYASHI, TAIZO, hydraulics researcher, educator; b. Nagoya, Aichi, Japan, June 28, 1920; s. Masaharu and Michiko (Kawakami) H.; m. Kyoko Abe, May 14, 1953. B of Engring., Tokyo Imperial U., 1942; DEng, U. Tokyo, 1953; D Honoris Causa, Inst. Nat. Poly., Toulouse, France, 1979. Asst. prof. Tokyo Imperial U., 1942-46; rsch. mem. U. Tokyo, 1946-50; from assoc. prof. to prof. Chuo U., Tokyo, 1950-91, emeritus prof., 1991—; vis. prof. MIT, Cambridge, 1967, Iowa Inst. Hydraulic Rsch., Iowa City, 1967-68, Calif. Inst. Tech., Pasadena, 1968, Ecole Nat. Poly. Lausanne, Switzerland, 1987; guest prof. Saitama U., Urawa, Japan, 1978-86; head Hayashi Applied Hydraulic Rsch. Office, Tokyo, 1991—; spl. adviser, civil engring. cons. INA Corp., 1991—. Author: Water Waves, 1957, Water Hammer and Surge Tanks, 1958; co-author: Water Hammer, 1962; co-editor: Advanced Hydraulics, 1980. Pres. Flow Visualization Soc. Japan, Tokyo, 1984-85. Recipient Pierre Fermat medal Acad. Scis. Toulouse, 1963, Silver medal Soc. for Encouragement Rsch. and Invention, Paris, 1968, Purple Ribbon medal Cabinet Japan, 1983. Fellow ASCE, Internat. Water Resources Assn.; mem. Acad. Scis. Toulouse (corr. mem. 1963—), Internat. Assn. Hydraulic Rsch. (hon., pres. 1972-75), Engring. Acad. Japan, Japan Soc. Civil Engrs. (hon., dir. 1960-62, Thesis prize 1966). Avocation: music. Home: 1-17-2 Tamagawa-Denenchofu, Seyagaya-ku Tokyo Japan Office: Hayashi Applied

Hydraulic Rsch, Yasuke Bldg Sekigushi 1-19-6, Bunkyo-ku Tokyo 112, Japan

HAYASHI, TAKAO, mathematics educator, historian, Indologist; b. Shiozawa, Niigata, Japan, Sept. 1, 1949; s. Saburo Sakurai and Reiko (Hayashi) H.; m. Kinue Odaira, June 29, 1980; 1 child, Makoto. BSc, Tohuku U., Sendai, Japan, 1974, MA, 1976; postgrad., Kyoto (Japan) U., 1976-79; PhD, Brown U., 1985. Researcher Mehta Rsch. Inst. for Math. and Math. Physics, Allahabad, India, 1982-83; lectr. Kyoto Women's Coll., 1985-87; lectr. history of sci. Doshisha U., Kyoto, 1988-89, assoc. prof., 1989—. Contbr. articles on Indian math. to profl. jours. Fellow Japan Soc. for Promotion Sci., 1979; jr. rsch. fellow Am. Inst. Indian Studies, NSF, 1982. Mem. Am. Math. Soc., Indian Soc. for History Math., History of Sci. Soc. Japan, Japanese Assn. Indian and Buddhist Studies. Office: Doshisha U, Kami-Kyo-Ku, Kyoto 602, Japan

HAYASHI, TAKEMI, physics educator; b. Nagoya, Aichi, Japan, Oct. 8, 1938; s. Katsuzo and Rakuko (Morita) H.; m. Mariko Tsurumi, Sept. 24, 1978; children: Akiko, Kazutaka. BS, Nagoya (Japan) U., 1961, MS, 1963, DSc, 1966. Rsch. fellow Hiroshima (Japan) U., 1966-83, lectr., 1983; assoc. prof. Kure Nat. Coll. Tech., Kure-Hiroshima, Japan, 1983-85, prof., 1985-91; prof. Kogakkan U., Ise, Japan, 1991—. Contbr. articles to profl. jours. Mem. AAAS, Phys. Soc. Japan, Am. Phys. Soc., N.Y. Acad. Scis. Avocation: tennis. Office: Kogakkan U, 1704 Kodakujimoto-cho, Ise Mie 516, Japan

HAYASHI, YOSHIHIRO, architect; b. Hikone, Japan, Sept. 29, 1940; s. Genzo Ōdachi and Koshie (Yoshida) H. B of Engring., Kogakuin U., Tokyo, 1965; MA, U. Mich., 1975. Supt. Sato Hide Komuten Co. Ltd., Tokyo, 1965-72; dir. Yoshihiro Hayashi Architects and Assoc., Tokyo, 1975—. Author: Zosaku, 1981. Recipient Nika-Ten award Nika Orgn., Japan, 1981, 83. Fellow Archtl. Inst. Japan, Japan Inst. Architects, Tokyo Soc. Architects and Bldg. Engrs. Roman Catholic. Avocations: sculpture, oil painting. Home: 3-35 2-Chome Shibaura Minato, Tokyo 108, Japan Office: 1-2-19 Hamamatu-Cho Minato, Tokyo 105, Japan

HAYASI, NISIKI, physicist; b. Niigata City, Japan, Mar. 12, 1929; came to U.S., 1963; s. Matsuki and Fuku (Fukushima) H.; m. Chikako Nomura, Nov. 21, 1952; children: Fujio, Kay Keiko Makishi. AB, First Coll., Tokyo, 1949; SB, SM, U. Tokyo, 1952, PhD, 1962. Tech. ofcl. Japanese Ministry Transp., Tokyo, 1952-6l; tech. ofcl. Japanese Premier's Office, Tokyo, 196l-62, br. mgr., 1962-64; postdoctoral assoc. NASA Ames Rsch. Ctr., Moffett Field, Calif., 1964-65; program mgr., staff scientist Lockheed Corp., Marietta, Ga., 1965-70; mgr. advanced tech. Langston div. Harris Corp., Cherry Hill, N.J., 197l-74; sr. mng. dir. Fgn. Ops. div. SPS Techs., Inc., Jenkintown, Pa., 1975; exec. dir. Tech. Transfer Cons., Cherry Hill, N.J., 1975—; pres. Culti Corp., Cherry Hill, 1977—; vis. asst. prof. U. Cin., 1963-64; exec. cons. Tomoku Co., Ltd., Tokyo, 1974-78; NAS-NRC postdoctoral rsch. assoc. 1964, 65. Contbr. numerous articles to profl. jours.; patentee in U.S., France, Netherlands, Germany, Japan. Pres. Japanese Sch. Greater Phila. PTA, 1973-74; dir. Japanese Assn. of Greater Phila., 1990-91. Japanese Govt. overseas fellow, 1963; U. Pa. scholar, 1971-73. Fellow AIAA (assoc.); mem. Ambs. Club, Sigma Gamma Tau. Office: Culti Corp 16 Locust Grove Rd Cherry Hill NJ 08003

HAYDEN, W(ALTER) JOHN, botanist, educator; b. Putnam, Conn., July 12, 1951; s. Walter Burrill and Ethel Harriet (Pitkin) H.; m. Sheila Mae Doble, Sept. 12, 1970; children: Sean Walter, Joseph William. BA, U. Conn., 1973; MS, U. Md., 1974, PhD, 1980. Instr. botany U. Md., College Park, 1978-80; prof. botany U. Richmond, Va., 1980—; mem. adv. bd. Lewis Ginter Bot. Garden, Richmond, 1982-90. Contbr. articles to Jour. of Arnold Arboretum, Systematic Botany, Am. Jour. Botany, others. Grantee Jeffress Meml. Trust, 1982, U.S. Dept. Interior, 1985, 86, NSF, 1984, 86, others. Mem. AAAS, Bot. Soc. Am., Am. Soc. Plant Taxonomists, Va. Acad. Sci. (officer botany sect. 1986-89, Horsely rsch. award 1991), Internat. Assn. Wood Anatomists. Achievements include confirmation of systematic placement of genus Picrodendron in family Euphorbiaceae; showing Hawaiian genus Neowawraea to be a species of Flueggea; naming 4 new species of Amanoa from South America. Office: U Richmond Dept Biology Richmond VA 23173

HAYES, ALBERTA PHYLLIS WILDRICK, retired health service executive; b. Blakeslee, Pa., May 31, 1918; d. William and Maude (Robbins) Wildrick; diploma Wilkes Barre Gen. Hosp. Sch. Nursing, 1938-41; student Wilkes Coll., 1953-54, Pa. State U., 1969—; m. Glenmore Burton Hayes, Oct. 9, 1942; children—Glenmore Rolland, William Bruce. Nurse, Monroe County Gen. Hosp., East Stroudsburg, Pa., 1941-44; pvt. duty nurse, 1944-56; with White Haven (Pa.) Center, 1956-82, dir. residential services, 1966-82, ret., 1982. Pres. Tobyhanna Twp. Sch. PTA, 1948-49, Top-o-Pocono Women of Rotary, 1975-76; nurse ARC, 1955; adv. council Luzerne County Foster Grandparent Program, 1977—, Health Services Keystone Job Corps, Drums, Pa., 1977—; active Tobyhana Twp. Zoning Hearing Bd., Pocono Pines, Pa. Mem. Am. Assn. Mental Deficiency, Am. Legion Aux. (unit pres. 1946-47). Club: Pocono Mountains Women's (Blakeslee). Home: PO Box 11 Blakeslee PA 18610-0011

HAYES, A(NDREW) WALLACE, toxicologist; b. Corning, Ark., Aug. 29, 1939; s. Andrew Wallace and Helen (Latimer) H.; m. Sandra June Smith, Dec. 28, 1963; children: Andrew Wallace III, Helen Cathleen, Benjamin Bailey. AB, Emory U., 1961; MS, Auburn U., 1964, PhD, 1967. Diplomate Am. Bd. Toxicology. Postdoctoral fellow Vanderbilt U., Nashville, 1966-68; prof. U. Ala., Tuscaloosa, 1968-75, U. Miss. Med. Ctr., Jackson, 1975-80; dir. Rohm & Haas, Spring House, Pa., 1980-84; v.p. RJR Nabisco, Winston-Salem, 1984-92; prof. toxicology Bowman Gray Sch. Medicine, Winston-Salem, 1993; v.p. The Gillette Co., Boston, 1993—; adv. coun. Auburn U., 1987—. Editor: Mycotoxin Teratogencity, 1981, Toxicology of Eye, Ear and Other Senses, 1985, Extrapolation of Doscimelric Relationships for Inhaled Gas, 1989, Principles and Methods in Toxicology, 3d edit., 1993; contbr. articles to profl. jours. Mem. nat. coun. Fla. Coll., Temple Terrace, 1980—; chmn. Math & Sci. Alliance, Winston-Salem, 1992—. NASA fellow, 1964-66; NIH rsch. career devel. awardee, 1971-76; NATO fellow, 1977; recipient Cert. of Merit EPA, 1985. Fellow Acad. Toxicol. Scis.; mem. AAALAC (trustee 1987-89), Soc. Toxicology (coun. 1982-84), Am. Inst. Nutrition, Am. Soc. Pharmacology and Exptl. Therapeutics, Am. Chem. Soc. Office: The Gillette Co Corp Product Integrity Prudential Tower Bldg Boston MA 02199

HAYES, BRIAN PAUL, editor, writer; b. Somers Point, N.J., Dec. 10, 1949. Book rev. editor Balt. Sun, 1970-73; staff editor Scientific Am., N.Y.C., 1973-80, assoc. editor, 1980-84; freelance sci. writer, 1984-89; editor Am. Scientist, Research Triangle Park, N.C., 1989-93. Columnist Sci. Am., 1983-85, Scis., 1990—; editor: Made in America, 1989. Home: 211 Dacian Ave Durham NC 27701

HAYES, CHARLES FRANKLIN, III, museum research director; b. Boston, Mass., Mar. 6, 1932; m. Nannette J. Rhodes; children: Marna Brewster, Tavia Frances. AB in Anthropology, Archaeology, and Ethnography, Harvard U., 1954; MA in Anthropology, U. Colo., 1958. Rsch. asst. Glen Canyon Archeol. Survey U. Utah., 1957; rsch. asst. ShoShone Indian Land Claims U. Colo., 1957; jr. anthropologist Rochester (N.Y.) Mus. and Sci. Ctr., 1959-61, assoc. curator anthropology, 1961-66, curator anthropology, 1966-79, coord. curator, mus. dir., 1970-79, dir. rsch., 1979—, also instr. Sch. Sci. and Man; asst. lectr. U. Rochester, 1961-69, assoc. lectr., 1970-73; lectr. anthropology St. John Fisher Coll., 1986—. Contbr. 70 publs. on museology and archeology. Trustee Seneca Iroquois Nat. Mus., Salamanca, N.Y., 1977—; mem. restoration com. New City Hall, Rochester, 1977. 2nd lt. USAF, 1954-56, USAFR, 1956-67. Fellow N.Y. State Archeol. Assn. (sec. 2 yrs., v.p. 2 yrs., pres. 1967-69, chair publs. 1965-67, editor rschs. and transactions 1966-67, co-editor 1976-77, chmn. awards and fellowships com. 1975-77, sec. Lewis H. Morgan chpt. 2 yrs., pres. 4 yrs., exec. com. 20 yrs.); mem. Am. Am. Anthrop. Assn., Am. Assn. Mus. (U.S. rep. to Internat. Coun. Mus. 1974-76), Soc. for Am. Archeology (N.Y. state rep. com. for pub. understanding archeology 1971), Soc. for Hist. Archeology, N.Y. State Assn. Mus., N.Y. Archeol. Coun. (steering com. 1972, treas. 1973-74), Soc. for Profl. Archeologists, Northeast Mus. Conf. Office: Rochester Museum & Science Ctr Research Division 657 East Ave PO Box 1480 Rochester NY 14603-1480

HAYES, CHARLES VICTOR, nuclear engineer; b. Johnson City, N.Y., July 26, 1955; s. Lawrence Patrick and Ellen Jane (Calvert) H.; m. Rhonda Lombard, Oct. 20, 1989; 1 child, Jason Alexander. BS in Physics, SUNY, Binghamton, 1978. Lic. sr. reactor operator. Ops. engr. Consol. Edison, Buchanan, N.Y., 1982-86, sr. tng. instr., 1986-88, event analysis engr., 1988—. Contbr. articles to Am. Jour. Physics, Power 132, Nuclear News. Group comdr. CAP, N.Y. Wing, 1971—. With USN, 1978-82. Mem. MENSA, Am. Nuclear Soc. Achievements include design of PC based computer program to perform nuclear plant trip/event analysis within 35 minutes of event. Home: 38 Sharon Dr Fishkill NY 12524-1317 Office: Consol Edison Indian Point 2 Broadway Ave Buchanan NY 10511-1006

HAYES, DAVID KIRK, nuclear engineer; b. Lubbock, Tex., June 20, 1963. Student, Rensselaer Poly. Inst., 1989—. Licensed sr. reactor operator. With USN, 1982-88. Fellow U.S. Nuclear Regulatory Commn., 1992. Mem. Am. Nuclear Soc., Health Physics Soc., Tau Beta Pi, Alpha Nu Sigma. Home: Box 9 Troy NY 12180

HAYES, GORDON GLENN, civil engineer; b. Galveston, Tex., Jan. 2, 1936; s. Jack Lewis and Eunice Karen (Victery) H. BS in Physics, Tex. A&M U., 1969. Registered profl. engr., Alaska, Tex. Rsch. technician Shell Devel. Co., Houston, 1962-68; rsch. assoc. Tex. Trans. Inst., College Station, 1969-71, asst. rsch. physicist, 1971-74, assoc. rsch. physicist, 1974-80; traffic safety specialist Alaska Dept. Transp. & Pub. Facilities, Juneau, 1981-83, state traffic engr., 1983-85, traffic safety standards engr., 1985-90; owner Alaska Roadsafe Cons., Juneau, 1990-92, Hayes Highway Consulting, Carson City, Nev., 1992—. Author of numerous pubs. in the hwy. safety field; producer of numerous documentary films in the hwy. safety field. Petty officer USN, 1953-57. Mem. ASCE, Nat. Com. on Uniform Traffic Control Devices (signs tech. com.) Inst. Transp. Engrs., Mensa. Avocations: nordic skiing, fishing, kayaking, hunting, mountaineering. Home: 3817 S Carson St # 804 Carson City NV 89701

HAYES, JOHN MARION, civil engineer; b. Wingate, Ind., May 18, 1909; s. William Lucas and Margaret Eliza (Gallaher) H.; m. Coye Matilda Cunningham, June 22, 1935; children: Marian Sue Hayes Jernigan, Julia Kethleen Hayes Casey. BSCE, Purdue U., 1931; MS, U. Tenn., 1944, D in Civil Engring., 1946. Structural engr. TVA, Knoxville, Tenn. and Chattanooga, 1935-46; dist. bridge engr. U.S. Bur. Pub. Rds., Little Rock, 1946-48; assoc. prof. Purdue U., West Lafayette, Ind., 1948-58; prof. structural engr. Purdue U., West Lafayette, 1958-75, prof. emeritus, 1975—. Fellow ASCE (hon., dir. dist. 9 1964-66, v.p. zone III, 1969-70), NSPE, Am. Concrete Inst., AAAS, Am. Welding Soc., ASTM, Am. Soc. Engring. Edn., Am. Inst. Steel Constrn. Inc., Ind. Sci. Engring. Found. Inc., SAR (pres. Ind. Soc. 1990-92, nat. trustee 1992-94), Lions, Chi Epsilon Sigma Xi. Republican. Methodist. Avocations: traveling, languages, reading, photography. Home: 312 Highland Dr West Lafayette IN 47906-2406

HAYES, ROBERT DEMING, electrical engineer, consultant; b. Lexington, Ky., Mar. 11, 1925; s. Roy Bagley and Essie May (Brigman) H.; m. Nancy Ellen Taylor, Nov. 12, 1924 (dec. Mar. 1972); children: William, Katherine Hayes Rottersman, Carol E., Jennifer Hayes Whitehead; m. Jean Copeland, Apr. 29, 1977. BEE, U. Ky., 1948, MS in Physics, 1950; MEE, Ga. Inst. Tech., 1957, PhD, 1964. Field engr. Western Electric Co., Winston-Salem, N.C., 1950-54; rsch. engr. Ga. Inst. Tech., Atlanta, 1954-66, prof. elec. engring., 1958-66; head advanced engring. Harric Corp., Melbourne, Fla., 1966-68; prof. elect. engring. Ga. Inst. Tech., Atlanta, 1968-76; also prin. rsch. engr., 1976—; pres. RDH, Inc., Marietta, Ga., 1980—; lectr. Chung Shan Inst., Taiwan, 1975; adj. prof. Fla. Inst. Tech., Melbourne, 1966-68, Naval Postgrad. Sch., Monterey, Calif., 1988; cons. U.S. Army, USAF, USN, Def. Advanced Rsch. Programs Agy., 1965—. Author: (chpt.) Radiometeric Measurements, 1987; co-author: Millimeter-Wave Radar Clutter, 1992; contbr. articles to profl. jours. Chmn. Cobb County Planning and Zoning, Marietta, 1969-73, Cobb Bd. Tax Assessors, Marietta, 1986, 87, 90, 91. Cpl. U.S. Army Air Corps, 1943-46. Mem. IEEE (life, v.p. Atlanta sect. 1960), Kiwanis Internat. (gov. 1987-88, Tablet of Honor award 1988, charter Hixson fellow), Golden Key, Sigma Pi Sigma, Sigma Xi, Eta Kappa Nu (charter pres. 1947), Phi Kappa Tau (Outstanding Adviser 1982). Home: 605 Chestnut Hill Rd Marietta GA 30064 Office: RDH Inc PO Box 1392 Marietta GA 30061

HAYES, ROBERT GREEN, chemical educator, researcher; b. Phila., Oct. 23, 1936; s. James Edward and Marie Jeanette (Green) H.; m. Linda Joyce Shumaker, Aug. 20, 1960; children: Kevin, Brian, Amy, Derek, Sarah. BS, U. Pitts., 1958; PhD, U. Calif., Berkeley, 1962. Instr. dept. chemistry U. Notre Dame, Ind., 1961-62, asst. prof. dept. chemistry, 1962-68, assoc. prof. dept. chemistry, 1968-74, prof. dept. chemistry, 1974—. Contbr. articles to profl. jours. Recipient Postdoctoral fellowship NSF, 1962, Sr. Postdoctoral fellowship NATO, 1972, 76, Travel award Fulbright Commn., 1975-76. Mem. AAAS, Am. Chem. Soc. Democrat. Mem. United Ch. of Christ. Achievements include first to show that porphyrin free base has two kinds of nitrogen atoms; research in application of electron paramagnetic resonance to transition metal complexes; in lanthanide organametallic complexes; in application of X-ray photoelectron spectroscopy to valence electronic structure of compounds. Home: 1909 Peachtree Ln South Bend IN 46617 Office: Univ Notre Dame Dept Chemistry Notre Dame IN 46556

HAYES, WILBUR FRANK, biology educator; b. Rhinelander, Wis., Nov. 10, 1936; s. Wilbur Mead and Evelyn (Stritesky) H.; m. Dawn Olivia Waldorf, July 21, 1979 (div. Feb. 1991); stepchildren: Lynn, Robert, Dana, Richard, Gary, Kevin. BA, Colby Coll., 1959; MS, Lehigh U., 1961, PhD, 1965. Postdoctoral fellow Yale U., New Haven, 1965-67; asst. prof. biology Wilkes Coll., Wilkes-Barre, Pa., 1967-71, assoc. prof., 1971—; vis. prof. Northeastern U., Boston, 1987-88. Contbr. articles to profl. jours. Chmn. bd. dirs. Northeastern Pa. chpt. A. Heart Assn., Wilkes-Barre, 1986-87. Mem. Am. Soc. Zoologists, Pa. Acad. Sci., Electron Microscope Soc. Am., Sigma Xi (pres. Wilkes Coll. club chpt. 1976-77, sec., treas. 1984-87, 88-91). Republican. Congregationalist. Achievements include co-discovery of tendon receptor organ in Limulus; discovery of peripheral synaptic endings in Limulus chemoreceptors. Avocations: downhill skiing, photography, travel. Home: 47 Stanley St Wilkes Barre PA 18702-2308 Office: Wilkes U Dept Biology Wilkes Barre PA 18766

HAYMAKER, CARLTON LUTHER, JR. (BUD HAYMAKER), metrology engineer; b. Roanoke, Va., Dec. 14, 1944; s. Carlton Luther and Rebecca Frances (Kidd) H.; m. Linda Gaynell Williamson, Feb. 28, 1966; children: Scott Allen, Mark William. Student, U. Va., 1963-65; BBA, Bellevue Coll., 1976; cert. program mgmt., West Coast U., 1986; postgrad., Pepperdine U., 1991. Cert. mgr. Surveyor U.S. Forest Svc., Roanoke, Va., 1963-66; asst. mgr. avionics and avionics repair Sky Harbor Air Svc., Omaha, Nebr., 1974-79; maintenance electrician John Deere Tractor, Waterloo, Iowa, 1979-82; metrology engr. Rocketdyne divsn. Rockwell, Internat., Canoga Park, Calif., 1982—; divsn. rep. Govt.-Industry Data Exch. Program, Corona, Calif., 1986—, Nat. Conf. Standards Labs., Boulder, Colo., 1986—. Editor (newsletter) Rockwell Valley Communique, 1987-88. With USAF, 1966-74. Mem. Nat. Mgmt. Assn. (bd. dirs. Rockwell Valley chpt., sec. 1988-89, v.p. fin. 1992-93, exec. v.p. 1993—), Nat. Space Soc., Planetary Soc. Republican. Home: 2548 Fallon Cir Simi Valley CA 93065 Office: Rockwell Internat Rocketdyne Divsn 6633 Canoga Ave BA68 Canoga Park CA 91309-7922

HAYNES, DEBORAH GENE, physician; b. York, Neb., Feb. 18, 1954; d. Gene Eldridge and Margaret Lucille (Manchester) Haynes; m. Russell Larry Beamer, Mar. 3, 1979; children: Staci E. Beamer, Lindsay M. Beamer, Stephanie L. Beamer. BA in Biology cum laude, Wichita State U., 1976; MD, U. Kans., Wichita, 1979. Diplomate Am. Bd. Family Practice. Resident St. Joseph Hosp., Wichita, 1979-82; instr. dept. family and community medicine St. Joseph Family Practice Residency, U. Kans., Wichita, 1982-84, asst. prof. dept. family and community medicine, 1984-85; pvt. family practice Northeast Family Physicians, Wichita, 1985—; clin. asst. prof. U. Kans. Sch. Medicine, Wichita, 1985—. Med. adv. com. Blue Cross/Blue Shield of Kans., Topeka, 1990—; vis. com. Wichita State U., Coll. Health Profls., Wichita, 1988-92. Recipient P.G. Czarlinsky award for Disting. Clin. Svc., U. Kans., 1979. Fellow Am. Acad. Family Physicians (del. 1991—, commn. on edn. 1991—, Mead Johnson award 1981); mem. AMA, Kans. Acad. Family Physicians Found. (trustee 1991-92, pres. 1990-91), Kans. Acad. Family Physicians (pres. elect 1988-89, pres. 1989-90), Kans. Med. Soc., Med. Soc. Sedgwick County (del. 1982), Alpha Omega Alpha. Presbyterian. Avocation: reading. Home: 1015 N Linden Cir Wichita KS 67206 Office: 8100 E 22nd St N Bldg 2200 Wichita KS 67226

HAYNES, DUNCAN HAROLD, pharmacology educator; b. Owosso, Mich., June 27, 1945; s. Alfred Cleveland and Mary Frances (MacDonald) H.; m. Celeste Ann Howard, Oct. 10, 1965 (div. 1969); 1 child, Norman Douglas; m. Gisela Busche, June 28, 1974; children: Karl Harold, Ellen Ursula. BS, Butler U., 1966; PhD, U. Pa., 1970. Postdoctoral fellow Max-Planck Inst. für biophysikalische Chemie, Göttingen, Fed. Republic Germany, 1970-73; asst. prof. pharmacology U. Miami (Fla.) Sch. Medicine, 1973-77, assoc. prof. pharmacology, 1977-82, prof. pharmacology, 1982—. Active various pub. sch.-related and community activities. Rsch. grantee USPHS, 1973—, Fla. affiliate Am. Heart Assn., 1974, 78, 88; recipient Gov.'s Award for Outstanding Contbn. to Sci. and Tech., Fla., 1989. Fellow Am. Coll. Clin. Pharmacology; mem. Biophys. Soc., Am. Soc. Pharmacology & Exptl. Therapeutics, Soc. Clin. Investigation, Am. Physiol. Soc., Soc. Gen. Physiologists, Assn. Mil. Surgeons U.S., Soc. Armed Forces Med. Scientists, Controlled Release Soc., Ctr. for Health Tech. Republican. Episcopalian. Achievements include 17 U.S. and foreign patents for protecting invention of phospholipid-coated microdroplets and microcrystals as an injectable delivery system for water-insoluble drugs; invention of simple fluorimetric test for calcium handling abnormality in human blood platelets; pioneering use of calcium channel blocking drugs as anti-platelet drugs in thrombosis. Home: 4051 Barbarossa Ave Miami FL 33133-6628 Office: U Miami Sch Medicine Dept Pharmacology PO Box 016-189 Miami FL 33101

HAYNES, HAROLD EUGENE, JR., medical physicist; b. Newnan, Ga., Aug. 12, 1956; s. Harold Eugene, Sr. and Margaret Frances (Morgan) H. BEE, Auburn U., 1979; MS, Ga. Tech., 1986. Cert. therapeutic radiol. physics, Am. Bd. Radiology. Staff physicist Baylor Coll. Medicine, Houston, 1986-89; med. physicist Archbold Med. Ctr., Thomasville, Ga., 1989-92; mem. radiotherapy adv. com. Thomas Tech., Thomasville, Ga., 1990—, nuclear protection delegation to Russia and Ukraine, People to People, 1992. Vol. Wakulla County Disaster Relief Team, Homestead, Fla., 1992. Lt. USNR, 1979-83. Mem. IEEE, Am. Assn. Physicists in Medicine, Health Physics Soc., Am. Coll. Radiology. Office: Singletary Oncology Ctr 116 Mimosa Dr Thomasville GA 31792

HAYNES, JOEL ROBERT, molecular biologist; b. Williamson, W.Va., July 3, 1954; s. Robert Newton and Pauline (Charles) H.; m. Gail Eleanor Sears, Aug. 1, 1981; children: Elise, Kaitlin, Courtney. BS, Western Ky. U., 1976; PhD, U. Cin., 1981. Rsch. scientist Connaught Labs., Ltd., Willowdale, Ont., Can., 1982-86, sr. scientist, dir., 1988-91; asst. mem. St. Jude's Children's Rsch. Hosp., Memphis, 1986-88; sr. scientist Agracetus, Inc., Middleton, Wis., 1991—. Contbr. articles to profl. jours. Bd. dirs. Middleton Outreach Ministry, 1992—; firefighter Middleton Fire Dist., 1992—. Episcopalian. Home: 1502 Parmenter St Middleton WI 53562 Office: Agracetus Inc 8520 University Green Middleton WI 53562

HAYS, HERSCHEL MARTIN, electrical engineer; b. Neillsville, Wis., Mar. 2, 1920; s. Myron E. and Esther (Marquardt) H.; E.E., U. Minn., 1942; grad. student U. So. Calif., 1947; children—Howard Martin, Holly Mary, Diane Esther, Willet Martin Hays II. Elec. engr. City of Los Angeles, 1947-60; pres. Li-Bonn Corp. Served as radio officer, 810th Signal Service Bn., U.S. Army, 1942-43; asst. signal constrn. officer, E.T.O., 1943-45, tech. supr. Japanese radio systems, U.S. Army of Occupation, 1945-46; mem. tech. staff, Signal Corps Engring. Labs., U.S. Army, 1946; col. U.S. Army, ret. Signal Officer Calif. N.G. 1947-50. Registered profl. engr. Calif. Mem. Eta Kappa Nu, Pi Tau Pi Sigma, Kappa Eta Kappa. Republican. Episcopalian. Home: 603 Alhambra Rd Venice FL 34285-2502

HAYS, PAUL B., science educator, researcher; b. Battle Creek, Mich., May 25, 1935; s. Paul Beckman and Doritha (Green) H.; m. Ruth Ann Hays, Aug. 20, 1978; children: Jeffrey Scott, Michael Alan, Cynthia Lynn Tomlinson. BS in Aerospace, U. Mich., 1958, MS in Aerospace, 1960, PhD in Aerospace, 1964. Assoc. prof., dept. aerospace engring. U. Mich, Ann Arbor, 1969-75, assoc. prof., dept. atmospheric, oceanic and space scis., 1972-75, dir. space physics rsch. lab., 1984-89, prof. aerospace engring., 1975—, prof., dept. atmospheric, oceanic and space scis., 1975—, chair., dept. atmospheric, oceanic and space scis., 1991-92; sabbatical leave Nat. Ctr. Atmospheric Rsch., Boulder, Colo., 1989-90; mem. subcommittee NASA adv. coun., Washington, 1988; pres. PBH Cons., Inc., Ann Arbor, 1991—. Regional editor: Planetary and Space Sci., 1986-89. Recipient Orin B. Scott award, U. Mich., 1963; named Dwight F. Benton Prof. of Advanced Technology. Fellow Am Geophys. Union; mem. Commn. on Solar-Terrestrial Physics, Phi Kappa Phi, Tau Beta Pi, Sigma Xi. Achievements include patent in Circle-to-Line Interferometer Optical System. Office: U Mich 2455 Hayward St Ann Arbor MI 48109-2143

HAYS, RONALD JACKSON, naval officer; b. Urania, La., Aug. 19, 1928; s. George Henry and Fannie Elizabeth (McCartney) H.; m. Jane M. Hughes, Jan. 29, 1951; children: Dennis, Michael, Jacquelyn. Student, Northwestern U., 1945-46; B.S., U.S. Naval Acad., 1950. Commd. ensign U.S. Navy, 1950, advanced through grades to adm., 1983; destroyer officer Atlantic Fleet, 1950-51; attack pilot Pacific Fleet, 1953-56; exptl. test pilot Patuxent River, Md., 1956-59; exec. officer Attack Squadron 106, 1961-63; tng. officer Carrier Air Wing 4, 1963-65; comdr. All Weather Attack Squadron, Atlantic Fleet, 1965-67; air warfare officer 7th Fleet Staff, 1967-68, tactical aircraft plans officer Office Chief Naval Ops., 1969-71; comdg. officer Naval Sta., Roosevelt Roads, P.R., 1971-72; dir. Navy Planning and Programming, 1973-74; comdr. Carrier Group 4, Norfolk, Va., 1974-75; dir. Office of Program Appraisal, Sec. of Navy, Washington, 1975-78; dep. and chief staff, comdr. in chief U.S. Atlantic Fleet, Norfolk, Va., 1978-80; comdr. in chief U.S. Naval Force Europe, London, 1980-83; vice chief naval ops. Dept. Navy, Washington, 1983-85; comdr. in chief U.S. Pacific Command, Camp H.M. Smith, Hawaii, 1985-88; pres., chief exec. officer Pacific Internat. Ctr. for High Tech. Rsch., Honolulu, Hawaii, 1988-92; tech. cons., 1992—. Decorated D.S.M. with 3 gold stars, Silver Star with 2 gold stars, D.F.C. with silver star and gold star, Legion of Merit, Bronze Star with combat V, Air Medal with numeral 14 and gold numeral 3, Navy Commendation medal with gold star and combat V. Baptist. Home and Office: 869 Kamoi Pl Honolulu HI 96825-1318

HAYSLETT, PAUL JOSEPH, computer programmer, consultant; b. Newmarket, Suffolk, U.K., Mar. 7, 1963; came to U.S., 1964; s. John Paul Hayslett and Roseanne (Borchetta) Brandon. BS in Physics, Yale U., 1985. Sr. tech. assoc. AT&T Bell Labs., Holmdel, N.J., 1985-86; cons. Anistics, Inc., N.Y.C., 1986-87; systems analyst Random House, N.Y.C., 1987-90; programmer CD Plus, N.Y.C., 1990-91; ind. cons. Port Washington, N.Y., 1991—. Mem. IEEE, Assn. Computing Machinery. Democrat. Roman Catholic. Achievements include design of database for RH sales force, database for RH reference works, others. Home and Office: 68 Graywood Rd Port Washington NY 11050

HAYSSEN, VIRGINIA, mammalogist, educator; b. Milw., Feb. 28, 1951. BA, Pomona Coll., 1973; PhD, Cornell U., 1985. Researcher Boston Biomed. Rsch. Inst., 1974-76, E.K. Shriver Ctr., Waltham, Mass., 1976-78; asst. prof. biol. scis. Smith Coll., Northampton, Mass., 1985—. Author: Asdell's Patterns of Mammalian Reproduction, 1993; contbr. articles to profl. publs. Office: Smith Coll Dept Biology Northampton MA 01063

HAYTON, WILLIAM LEROY, pharmacology educator; b. Mt. Vernon, Wash., June 16, 1944; s. Richard Leroy and Irene Cecelia (Dickson) H.; m. Jane G. Axelson, Aug. 19, 1967 (div. 1993); children: Michael Alan, Brian Jeffrey. BS in Pharmacy, U. Wash., 1967; PhD in Pharms., SUNY, Buffalo, 1971. Asst. prof., assoc. prof., then prof. Wash. State U., Pullman, 1971-90; prof. pharms., chmn. dept. Ohio State U. Coll. Pharmacy, Columbus, 1990—. Contbr. over 75 articles to profl. jours. Bd. dirs. Pullman United Way, 1975-78; asst. scoutmaster Boy Scouts Am., Pullman, 1980-87. Grantee Nat. Inst. Environ. Health Scis., NIH, 1982-85, ERA, 1986-88, USAF Office Sci. Rsch., 1988-89, FDA, 1991-94. Fellow Am. Assn. Pharm.

Scientists; mem. Soc. Toxicology, Am. Soc. Pharmacology and Exptl. Therapeutics, Soc. Environ. Toxicology and Chemistry. Achievements include development of a mathematical model of chemical exchange at the fish gill, pharmacokinetic models for fish; determination of model parameters for a variety of environmental conditions. Office: Ohio State U Coll Pharmacy 500 W 12th Ave Columbus OH 43210-1291

HAYWOOD, H(ERBERT) CARL(TON), psychologist; b. Taylor County, Ga., July 2, 1931; s. Howard Chapman and Rosebud (Smith) H.; m. Nancy Patricia Roberts, Oct. 5, 1951 (div. Mar. 1971); children: Carlton, Terence, Elizabeth, Kristin. A.B., San Diego State Coll., 1956, M.A., 1957; Ph.D., U. Ill., 1961. Mem. faculty George Peabody Coll. (merged with Vanderbilt U. 1979), Nashville, 1962-93, prof. psychology, 1969-93, prof. spl. edn., 1975-79, dir. mental retardation research tng. program, 1968-70; dir. Inst. Mental Retardation and Intellectual Devel., 1970-73, Office Research Adminstrn., 1974-76, John F. Kennedy Center Research Edn. and Human Devel., 1971-83; prof. neurology Vanderbilt U. Sch. Medicine, 1971-93; prof. psychology, dean grad. sch. edn. and psychology Touro Coll., N.Y.C., 1993—; vis. prof. U. Toronto, 1965-66; sr. fellow Vanderbilt Inst. Pub. Policy Studies, 1983-88; chmn. Nat Mental Retardation Research Center Dirs., 1979-82; adv. bd. Ill. Inst. Developmental Disabilities, Chgo., 1970-78, Eunice Kennedy Shriver Center Mental Retardation, Waltham, Mass., 1973-80, Tenn. Dept. Mental Health, 1964-92 ; mem. nat. child health and human devel. council NIH, 1983-88; cons. President's Com. on Mental Retardation, 1968-73; mem. sci. rev. com., health research facilities br., div. edn. and research facilities NIH, 1967-71. Author: (with Brooks and Burns) Bright Start: Cognitive Curriculum for Young Children, 1992; editor: Brain Damage in School Age Children, 1968, Social Cultural Aspects of Mental Retardation, 1970, (with Begab and Garber) Prevention of Retarded Development in Psychosocially Disadvantaged Children, 1981, (with J.R. Newbrough) Living Environments for Developmentally Retarded Persons, 1981, (with D. Tzuriel) Interactive Assessment, 1992; editorial bd.: Jour. Abnormal Child Psychology, 1973-89, Contemporary Psychology, 1982-85, Acta Paedologica, 1983-87, Jour. Mental Deficiency Research, 1984—, Internat. Rev. Rsch. in Mental Retardation, 1982—; contbr. articles on child devel., motivation and mental retardation to profl. jours. Served with USN, 1950-54. Fellow Am. Assn. Mental Retardation (v.p. psychology 1975-77, 1st v.p. 1978-79, pres. 1980-81), Am. Psychol. Assn. (pres. Div. 33 1978-79, mem. Council of Reps. 1980-82); mem. Internat. Assn. Cognitive Edn. (pres. 1988-92), Soc. Research Child Devel., Inst. Medicine, Psychonomic Soc. Democrat. Episcopalian. Office: Touro Coll 350 Fifth Ave Ste 5122 New York NY 10118

HAZEL, JOANIE BEVERLY, elementary educator; b. Medford, Oreg., Jan. 20, 1946; d. Ralph Ray Lenderman and Vivian Thelma (Holtane) Spencer; m. Larry Aydon Hazel, Dec. 28, 1969. BS in Edn., So. Oreg. Coll., Ashland, 1969; MS in Edn., Portland State U., 1972; postgrad., U. Va., 1985. Elem. tchr. Beaverton (Oreg.) Schs., 1972-76, Internat. Sch. Svcs., Isfahan, Iran, 1976-78; ESL instr. Lang. Svcs., Tucker, Ga., 1983-84; tchr. Fairfax (Va.) Schs., 1985-86; elem. tchr. Beaverton (Oreg.) Schs., 1990—. Mem. U.S. Hist. Soc., Platform Soc., Smithsonian Instn., Am. Mus. Natural Hist., Nat. Mus. Women in Arts, U.S. Hist. Soc., The United Nations, The Colonial Williamsburg Found., Wilson Ctr., N.Y. Acad. Scis., Beta Sigma Phi. Home: 9247 SW Martha St Portland OR 97224-5577

HAZELRIGG, GEORGE ARTHUR, JR., engineer; b. Summit, N.J., Oct. 28, 1939; s. George Arthur Hazelrigg and Dorothy Hetty (Howell) Orr; m. Lauretta Blanche Powell, Aug. 31, 1968; children: George A. III, Geoffrey A. BS, N.J. Inst. Tech., 1961, MS, 1963; MA, Princeton U., 1966, MSE, 1968, PhD, 1969. Engr. Curtiss-Wright, Wood Ridge, N.J., 1961-63, Jet Propulsion Lab, Pasadena, Calif., 1966-67; staff sci. Gen. Dynamics, San Diego, 1968-71; rsch. staff Princeton U., 1971-75; dir., systems engr. Econ, Inc., Princeton, 1976-82; dep. div. dir. NSF, Washington, 1982—; prof. of systems engring. (sabbatical) Inst. for Advanced Engring., Seoul, 1993; dir. ECON, Inc., Princeton, 1974-84; cons. Princeton Synergetics, Inc., 1986—. Editor: Opportunities for Academic Research in a Low Gravity Environment, 1986; assoc. editor: Jour. Spacecraft and Rockets, 1977-82. Named Disting. Alumnus, N.J. Inst. Tech., Newark, 1989. Mem. AIAA, ASME, AAAS, Am. Soc. for Engring. Edn., Tau Beta Pi. Avocation: commercial pilot. Home: 161 Autumn Hill Rd Princeton NJ 08540-2911 Office: NSF 1800 G St NW Washington DC 20550-0002

HAZELTINE, BARRETT, electrical engineer, educator; b. Paris, Nov. 7, 1931; came to U.S., 1932; s. L. Alan and Elizabeth (Barrett) H.; m. Mary Frances Fenn, Aug. 25, 1956; children—Michael B., Alice W., Patricia F. BSE, Princeton U., 1953, MSE, 1956; PhD, U. Mich., 1962; ScD (hon.), SUNY, Stony Brook, 1988. Registered profl. engr., R.I. Asst. prof. engring. Brown U., 1959-66, assoc. prof., 1966-72, prof., 1972—; asst. to dean Brown U. (The Coll.), 1962-63, asst. dean, 1968-74, assoc. dean, 1974-93; Robert Foster Cherry chair for disting. teaching Baylor U., 1991-92; prof. U. Botswana, 1993; lectr., vis. prof. U. Zambia, Lusaka, 1970-71; -76-77; vis. prof. U. Malawi-Poly., Blantyre, 1980-81, 83-84, 88-89; asst. to mgr. rsch. labs., space and info. systems div. Raytheon Co., 1964-65, cons., 1965-67; cons. R.I. Utilities Commn., 1977-80, others. Author: Introduction to Electronic Circuits and Applications, 1980. Editor: The Weaver. Trustee Stevens Inst. Tech. Recipient award for excellence in instrn. Western Electric, 1968; grantee NSF, Dept. Edn.; grantee Met. Life Ins. Ednl. Found.; Fulbright fellow 1988-89, 93. Mem. IEEE (sr., chmn. Providence sect. 1971-72), Providence Engring. Soc. (pres. 1977-78), Am. Soc. Engring. Edn., Sigma Xi, Tau Beta Pi. Congregationalist (deacon). Clubs: Providence Art, Providence Review. Achievements include patents for color recognition system. Home: 60 Barnes St Providence RI 02906-1502 Office: Brown U Div Engring Providence RI 02912

HAZELTINE, RICHARD DEIMEL, physics educator, university institute director; b. Jersey City, June 12, 1942; s. L. Alan and Elizabeth (Barrett) H.; m. Cheryl Pickett, June 27, 1964; children: Richard Eliot, Susannah Elizabeth. AB, Harvard Coll., 1964; MS, U. Mich., 1966, PhD, 1968. Lectr. physics U. Mich., 1968; rsch. scientist U.S. Naval Rsch. Lab, 1969; with Inst. Advanced Study, Princeton, 1969-71; rsch. scientist assoc. V U. Tex. Austin, 1971-75, rsch. scientist fusion rsch. ctr., 1975-83, asst. dir. inst. fusion studies, 1982-86, prof. physics, 1986—; vis. scientist Aspen Ctr. Physics, 1970; acting dir. inst. fusion studies U. Tex. Austin, 1987-88, 91, dir., 1991—; mem. rev. panel Magnetic Fusion Sci. Fellowship Program, Edge physics working group U.S. Dept. Energy; chmn. and member numerous profl. coms. and boards; cons. in field. Author: Plasma Confinement, 1992; assoc. editor Revs. Modern Physics; mem. editorial bd. Phys. Rev. A, 1978-79, The Physics of Fluids, 1978-80; contbr. over 100 articles to profl. jours. Scholar Harvard Coll., Horace H. Rackman predoctoral fellow. Fellow Am. Phys. Soc.; mem. Sigma Xi, Phi Kappa Phi. Office: U of Texas at Austin Moore Hall Rm 11.222-61500 26th & Speedway Austin TX 78712

HAZEWINKEL, MICHIEL, mathematician, educator; b. Amsterdam, Netherlands, June 22, 1943; s. Jan Hazewinkel and Geertrude Hendrika Werner; m. M.T. de Jong, Sept. 10, 1969 (div. 1990); children: Maarten M., Annette A.; m. Dausa Zvirenaite, Mar. 8, 1991; 1 child: Jan-Algimantas. Degree in math., U. Amsterdam, Netherlands, 1965; PhD, U. Amsterdam, 1969. Rsch. fellow Steklov Inst. Math., Moscow, 1969-70; lectr. Netherlands Sch. Econs., Rotterdam, 1970-72; prof. math. Erasmus U. Rotterdam, 1972-82; head dept. pure math. Ctr. Math. and Computer Sci., Amsterdam, 1982-88; head dept. algebra, analysis and geometry Ctr. Math. and Computer Sci., 1988—; prof. extraordinarius in math., Erasmus U. Rotterdam, 1982-85, U. Utrecht, Netherlands, 1985—; lectr. at profl. confs.; co-founder European Consortium for Mathematics in Industry. Author: Formal Groups and Applications, 1978; co-author: Stochastic Systems: the Mathematics of Filtering and Identifications and Applications, 1981, Current Developments in the Interface: Economics, Econometrics, Mathematics, 1982, Mathematics of Biology, 1985, Mathematics and Computer Science I, II, 1986, Deformation Theory Algebras and Structures and Applications, 1988, others; editor: Encyclopedia of Mathematics, 10 vols., 1987—, Handbook of Algebra, 7 vols.; mng. editor (jours.) Acta Aplicandae Mathematical, Nieuw Archief Voor Wiskunde, (book series) Mathematics and its Applications; assoc. editor Chaos, Solutions and Fractals; contbr. to numerous profl. publs. Fulbright-Hays grantee, Harvard U., 1973. Mem. IEEE, London Math. Soc., Italian Math. Union, Am. Math. Soc., Math. Assn. Am., Japan Math. Soc., Soc. Math. France, German Math. Soc., N.Y. Acad. Scis., Am. Phys. Soc., European Phys. Soc., European Math. Soc., numerous other internat. math. and physics orgns. Home: 18 Burg Jacoblaan, 1401 BR Bussum The Netherlands Office: Math Ctr, PO Box 4079, 1009 AB Amsterdam The Netherlands

HE, CHENGJIAN, aerospace engineer; b. Zhouning, Fujian, China, Sept. 15, 1957; came to U.S., 1986; s. Kailin and Shiying (Ruan) H.; m. Lie Ling, July 14, 1983; 1 child, Lingmin. MS, Nanjing (China) Aero. Inst., 1984; PhD, Ga. Inst. Tech., 1989. Instr. Nanjing Aero. Inst., 1984-86; rsch. asst. Ga. Inst. Tech., Atlanta, 1986-89, postdoctoral fellow, 1989-92, rsch. engr., 1992; sr. engr. Lockheed Engring. and Scis. Co., Hampton, Va., 1992—; PhD adv. com. Ga. Inst. Tech., 1992—. Contbr. articles to Jour. Aircraft, Jour. Am. Helicopter Soc., others. Recipient scholarship award Am. Vertical Flight Found., 1987, 88. Mem. AIAA, Am. Helicopter Soc. (Lichten award chmn. Atlanta chpt. 1992—). Achievements include development of generalized dynamic wake model to predict complicated dynamic behavior of rotor downwash. Office: Lockheed Engring Scis Co 144 Research Dr Hampton VA 23666

HE, DUO-MIN, physics educator; b. Shanghai, China, Feb. 5, 1941; s. Liang He and Wen-Fan Gu; m. Yi-Ding Feng, Apr. 17, 1970; 1 child, Xin. Grad., Nanjing U. Dept Info. Physics, 1964. Researcher Architecture Acad. China, Beijing, 1964-74; vice-dir. U. Wuhan Indsl. Tech., Inst. Laser Application, China, 1974-81, 84-85; rsch. prof. Nanjing (China) Aeronautic Inst., 1985-88; prof., lab. head Nanjing U. Sci. & Tech., Modern Optics Tech. Ctr., Nanjing, 1988—; vis. scholar SUNY, Stony Brook, 1981-84, Florence U., Italy, 1992; guest prof. Ocean U., Ocean Optics Info. Lab., Quingdao City, China, 1987—. Contbr. articles to profl. jours. Mem. Optical Soc. China, Acoustic Soc. China, Optical Soc. Am., Am. Soc. Experimental Mechanics. Avocations: stamp collecting, swimming, photography. Home: Drum Tower, Da Zhong Xin Cun 21, Nanjing City 210008, China Office: Nanjing U Sci & Tech, Modern Optics Tech Cntr, Nanjing 210014, China

HE, XING-FEI, physicist; b. Chenghai, Guangdong, People's Republic of China, Mar. 1, 1958; arrived in Can., 1991; s. Pei-Yao He and Shan-Shan Du; m. Chun Peng, Dec. 17, 1986. BS, Zhongshan U., Guangzhou, 1982, MS, 1985; PhD, Australian Nat. U., Canberra, 1992. Rsch. fellow Inst. Microelectronics Zhongshan U., 1985-88; PhD scholar Australian Nat. U., Canberra, 1988-91; NSERC Internat. fellow dept. physics U. B.C., Vancouver, Can., 1992—. Contbr. 35 rsch. articles to profl. jours. Izaak W. Killam Meml. fellow (hon.), 1992—. Mem. Am. Phys. Soc. Achievements include foundation of fractional derivative spectrum techniques; development of fractional-dimensionality theories for studying anisotropic interactions. Avocations: poetry, painting, photography, classical music.

HEACOCK, E(ARL) LARRY, electrical engineer; b. Tuscola, Ill., Jan. 27, 1935; s. Earl Rice and Helen Irene (Kaga) H.; m. Nancy Louise Voelkel, Sept. 2, 1956; children: Gregory Laurence, Kent Alan, Christopher Charles. BSEE, U. Ill., 1957, MSEE, 1966. Engr. Ill. Bell Telephone Co., Springfield, Ill., 1957-62; rsch. engr. Environ. Sci. Svcs. Adminstrn., Washington, 1962-69; systems engr. European Space Agy., Noordwijk, Holland, 1970-72; meteosat project mgr. European Space Agy., Toulouse, France, 1972-76; dir. system devel. NOAA, Washington, 1976-85, dir. satellite ops., 1985—. Editor: Space Applicatons at the Crossroads, 1983, Weather Satellites: Systems, Data, and Environmental Applications, 1990. Capt. USAF, 1957-62. Fellow Brit. Interplanetary Soc.;mem. AIAA, Am. Astron. Soc. (dir., pres. 1989-90). Office: NOAA Washington DC 20023

HEACOX, RUSSEL LOUIS, mechanical engineer; b. Big Timber, Mont., Feb. 7, 1922; s. Charles Lewis and Gladys Ellen (Gibson) H.; m. Jacqueline J. Jewett, Sept. 22, 1944 (dec. 1974); children: William J., Teri Bertoli; m. Ketty Hansine Jorgenson, Dec. 22, 1976. BSME, U. Wash., 1950. Registered profl. engr., Calif. Equipment design engr. P & Z Co. Inc., South San Francisco, Calif., 1966-74; owner Heacox Engring. Designs, Tiburon, Calif., 1974—. Capt. USMC, 1943-46, PTO. Mem. NSPE, Crane Certification Assn., Scottish Rite Shrine Assn., Masons. Achievements include patent for Slurry Trench Excavation Bucket and Spotter Assembly, for Diesel Pile Driver Muffler System. Home and Office: 131 Esperanza St Tiburon CA 94920

HEAD, ELIZABETH SPOOR, retired medical technologist; b. Galveston, Tex., July 10, 1928; d. Robert Newcomb and Bernice Lillian (Lumley) Spoor; m. Foy Paul Head, Feb. 23, 1952; children: Robert Paul, Phillip Lee, Elisabeth Anne. Student, North Tex. State U., 1945-47, U. Tex. Med. Br., Galveston, 1947-48; BS in Health Care Scis. with high honors, U. Tex., Galveston, 1984. Cert. med. technologist. Med. technologist, lab dir. dept. dermatology U. Tex. Med. Br., Galveston, 1948-91; ret., 1991. Editor newsletter Nerium News, 1986—; contbr. chpts. to books, articles to profl. jours. Rep precinct election judge, Galveston; mem. Altar Guild, Trinity Episcopal Ch., Galveston. Mem. Am. Soc. Med. Technologists, Galveston Dist. Soc. Med. Technologists (pres. 1976-77), Internat. Oleander Soc. (pres. 1978-82, corr. sec. 1985—, editor newsletter 1986—), Galveston Hist. Found., Friends Moody Gardens (pres., chmen. bd. 1990-92), Wednesday Lit. Club (pres. 1975-77, 1992-93). Avocations: gardening, bridge, writing, art, flower photography. Home: 4610 R 1/2 Galveston TX 77551 Office: U Tex Med Br Dept Dermatology Galveston TX 77551

HEAD, GREGORY ALAN, mechanical engineer, consultant; b. Dallas, Mar. 2, 1955; s. A. Lee and Georgia M. Head. BSME, Brigham Young U., 1981; MS in Engring. Mgmt., U. Alaska, Anchorage, 1988; postgrad., U. Tex., Arlington, 1990—. Registered profl. engr., Tex. Engr. tech. Hercules Aerospace, Inc., Salt Lake City, 1978-79, LTV Aerospace Inc., Dallas, 1979-80; engr. Hercules Aerospace Inc., Sale Lake City, 1980-82, CMH-Vitro, Anchorage, 1982-84; petroleum engr. Arco Alaska, Anchorage, 1984-87; cons. FAH World Wide Photographers, Anchorage and Dallas, 1987—; sr. v.p. systems divsn. FAH Corp., Denton, Tex., 1988—; cons. Alaska Mountaineering Assn., Anchorage, 1990—; capt. Arctic Adventurers, 1988—. Author: Arctic Lands and Uses, 1989. Missionary, Ch. of Jesus Christ of Latter Day Saints, Washington, 1975-77; mem. Mountain Rescue Team, Anchorage, 1989—; emergency med. technician State of Alaska, 1987—. Named Photographer of the Yr., FAH Worldwide Photo, Inc., 1990. Mem. ASME, Nat. Geog. Soc., Nat. Assn. Pvt. Enterprise, Am. Soc. Profl. Photographers, Brigham Young U. Football Alumni Assn., Suzuki Moto Cross Team. Republican. Avocations: expeditionary hiking and mountain climbing, civic activities, motor cross racing, writing, international travel. Office: FAH Corp Systems Divsn 1800 N Carroll Blvd # C Denton TX 76201-3098 also: PO Box 91107 Anchorage AK 99509

HEAD, JONATHAN FREDERICK, cell biologist; b. Syracuse, N.Y., Nov. 23, 1949; s. Arthur Everard and Lillian Myrtle (Hendra) H.; m. Priscilla Catherine Tambone, July 28, 1984; 1 child, Catherine Elizabeth. BS in Zoology, Syracuse U., 1971; MA in Biology, Bklyn. Coll., 1977; PhD in Biology, Fordham U., 1985. Rsch. asst. Naylor Dana Inst. Disease Prevention/Am. Health Found., Valhalla, N.Y., 1974-78, Cornell U. Med. Coll., N.Y.C., 1978; rsch. asst. Mt. Sinai Sch. Medicine, N.Y.C., 1978-84, rsch. assoc., 1984-86, rsch. asst. prof., 1986-87; dir. tumor cell biology Ctr. Clin. Scis./Internat. Clin. Labs., Nashville, 1986-89; pres. Mastology Rsch. Inst., Baton Rouge, 1989—; adj. asst. prof. Tulane U. Sch. Medicine, New Orleans, 1989—; adj. prof. Delta State U., Cleveland, Miss., 1992—; researcher and lectr. in field of cancer. Contbr. articles, abstracts and chpts. to sci. publs. Mem. State of La. Adoption Community Adv. Bd. Mem. AAAS, Am. Assn. Cancer Rsch., Am. Soc. Clin. Oncology. Methodist. Home: 6144 Hagerstown Dr Baton Rouge LA 70817-3917 Office: Mastology Rsch Inst 1770 Physicians Park Dr Baton Rouge LA 70816-3225

HEAD, JULIE ETTA, computer engineer; b. Louisville, July 13, 1966; d. Henry Everett and Charlene (Ritsert) H. BS, U. Louisville, 1988, M of Engring., 1989. Draftsman Paramount Foods, Louisville, 1985-89; computer engr. NASA, JFK Space Ctr., Fla., 1989—; guide, speaker NASA V.I.P. Launch Support, 1992—; recruiter NASA, 1992—. Achievements include project leader for the development of custom circuit card maintenance for a shuttle payload's checkout system. Office: NASA CS-GSD-22 Kennedy Space Center FL 32899

HEAD, WILLIAM IVERSON, SR., retired chemical company executive; b. Tallapposa, Ga., Apr. 4, 1925; s. Iverson and Ruth Britain (Hubbard) H.; m. Mary Helen Ware, June 12, 1947; children: William Iverson, Connie Suzanne Head Toohey, Alan David. BS in Textile Engring., Ga. Inst. Tech., 1949; D of Textile Engring. (hon.), World U., 1983; PhD in Indsl. Mgmt., Columbia Pacific U., 1988. Research and devel. engr. Tenn. Eastman Co., Kingsport, 1949-56, quality control-mfg. sr. engr., 1957-67, dept. supt., 1968-74; supt. acetate yarn dept., mem. of bus. team Chems. div. Eastman Kodak Co., Kingsport, 1975-85; info. officer U.S. Naval Acad., 1983—; mem. adv. bd., rsch. assoc. Point One Adv. Group, Inc., 1988—. Patentee textured yarns tech. in U.S., Great Britain, Fed. Republic of Germany, Japan and France. Capt. USNR, 1943-83. Decorated Navy Commendation medal, Selective Svc. System Meritorious Svc. medal, 1980. Mem. Internat. Soc. Philos. Enquiry (pers. cons. 1978-79, v.p. 1979-80, sr. rsch. fellow and internat. pres. 1980-85, diplomate and trustee 1986—, chmn. bd. trustees 1987—), Prometheus Soc., Internat. Platform Assn., Naval Res. Assn., Assn. Naval Aviation, Mil. Order World Wars, Res. Officers Assn. (pres. Tenn. dept. 1981-82, nat. councilman 1991—, nat. coun. steering com. 1993—), Ret. Officers Assn., VFW, Mensa (pres. Upper East Tenn. 1976-79), Sons of Revolution, Internat. Legion of Intelligence. Unitarian. Home and Office: 4035 Lakewood Dr Kingsport TN 37663-3374

HEAD-GORDON, TERESA LYN, chemist; b. Akron, Ohio, Sept. 28, 1960; d. Richard Donald and Betty Jane (Hill) G.; m. Martin Paul Head-Gordon, Aug. 24, 1985; children: Genevieve Eva, Nadine Jane. BS in Chemistry, Case Western Res. U., 1983; PhD, Carnegie-Mellon U., 1989. Postdoctoral researcher AT&T Bell Labs., Murray Hill, N.J., 1990-92; prin. investigator Lawrence Berkeley Labs., Berkeley, Calif., 1992—. Contbr. articles to Jour. Am. Chem. Soc., Biopolymers, Proceedings Nat. Acad. Sci. Mem. Am. Phys. Soc., AAAS, Am. Chem. Soc. Office: Lawrence Berkeley Labs Donner 314 Cyclotron Rd Berkeley CA 94720

HEADLEY, OLIVER ST. CLAIR, chemistry educator; b. St. Peter, Barbados, July 5, 1942; s. Reginald Beresford and Ivy Daphne (Walker) H.; m. Hortense Beatrice Sergeant, Aug. 15, 1965; children: Eric J.W., Keith A. BSc in Spl. Chemistry, U. of the W.I., 1964; PhD in Inorganic Chemistry, U. Coll., London, 1967. Asst. lectr. in chemistry U. of the W.I., St. Augustine, Trinidad, 1967-68, lectr. in chemistry, 1968-77, sr. lectr. in chemistry, 1977-89, dean. faculty of nat. sci., 1985-89, reader in solar energy, 1989-91; prof. chemistry U. of the W.I., Cave Hill, Barbados, 1991—; univ. dean, faculty of nat. scis. U. of the W.I., 1987-89; advisor in solar energy Coll. of Arts, Scis. and Tech., Kingston, Jamaica, 1979-80; summer sch. lectr. Internat. Ctr. for Heat and mass Transfer, Belgrade, Yugoslavia, 1982; vis. scholar Commn. of the European Communities, Brussels, 1986, Disting. scholar in residence Pa. State U., Harrisburg, 1991. Contbr. articles to profl. jours. Barbados scholarship Ministry of Edn., 1961, Commonwealth scholarship Commonwealth Scholarship Commn., 1966; recipient Guinnes award for scientific achievement Arthur Guinness Ltd., 1982, Spl Commendation award Orgn. of Am. States, 1985. Fellow Caribbean Acad. Scis.; mem. Internat. Solar Energy Soc. (bd. dirs. 1985-88, 91—), Am. Chem. Soc., Caribbean Solar Energy Soc. (chairperson). Adventist. Office: U WI, PO Box 64, Bridgetown Barbados

HEAL, GEOFFREY MARTIN, economics educator; b. Bangor, Wales, Apr. 9, 1944; s. Thomas John and Gwen Margaret (Owen) H.; children: Bridget, Natasha Chichilnisky-Heal. BA, Cambridge U., 1966, PhD, 1969. Dir. studies Christs Coll., Cambridge U., 1967-73; prof. econs. Sussex U., Brighton, Eng., 1973-81, head dept. econs., 1976-81; mng. editor Rev. Econ. Studies, London, 1973-78; dir. Economists Adv. Group, London, 1975-80; prof. Essex U., Colchester, Eng., 1981-83; exec. dir. Fin. Telecommunications, London, 1984-89; prof. grad. sch. bus. Columbia U., N.Y.C., 1983—; sr. vice dean grad. sch. bus. Columbia U., 1991—; cons. U.K. Dept. Energy, London, 1973-76, U.S. Dept. Energy, Washington, 1976-78, OPEC Sec. Gen., Vienna, Austria, 1979-81. Author: The Theory of Economic Planning, 1973, Public Policy and the Tax System, 1979, Economics Theory and Exhaustible Resources, 1979, Linear Algebra and Linear Economics, 1980, The Evolving International Economy, 1987, Oil In The International Economy, 1991. Fellow Econometric Soc., 1973, Royal Soc. Arts, 1984; NSF grantee. Home: 335 Riverside Dr New York NY 10025-3421 Office: Columbia Univ Bus Sch Uris Hall New York NY 10027

HEALD, PAUL FRANCIS, mechanical engineer, consultant; b. Rock Ferry, Cheshire, Eng., July 21, 1944; s. Raymond and Veronica (Butcher) H.; m. Norma Cotterall, Nov. 19, 1966; children: Karen Rachel, Matthew James. Degree in Marine Engring., Riversdale Coll., Liverpool, Eng., 1962; degree in Mech. Engring., Liverpool Coll., 1965. Registered profl. engr., N.H., Maine, Mass. Marine engr. Elder Dempster Lines, Ltd., Liverpool, 1960-63; HVAC (heating, ventilating and air conditioning) engr. Littlewoods Mail Order Stores, Ltd., Liverpool, 1963-65, group energy conservation engr., 1970-76; product application engr. Woods Fans, Inc., Colchester, Eng., 1965-68; sr. heating engr. Revo Domestic Appliances, Liverpool, 1968-70; dir. engring. Foss Mfg. Co., Inc., Hampton, N.H., 1978-89; cons. engr. Heald Engring. Assocs., Rye, N.H., 1989—. Mem. ASHRAE, Chartered Inst. Bldg. Svc. Engrs. Home and Office: Heald Engring Assocs 44 Birchwood Dr Rye NH 03879

HEALEY, CHRIS M., earth scientist. Recipient Julian Boldy Meml. award Can. Inst. Mining and Metallurgy, 1990. Office: care Can Inst Mining Metallurg, 3400 de Maisonneuve Blvd W, Montreal, PQ Canada H3Z 3B8*

HEALEY, MICHAEL CHARLES, fishery ecologist, educator; b. Prince Rupert, B.C., Can., Mar. 31, 1942; s. Arthur George and Elizabeth Healey; m. Judy Takahashi, July 2, 1966; children: Christopher Graham, Larissa Anne. BSc, U. B.C., 1964, MSc, 1966; PhD, U. Aberdeen, Scotland, 1969. Postdoctoral fellow Pacific Biol. Sta., Nanaimo, B.C., 1969-70; rsch. scientist Freshwater Inst., Winnipeg, Man., 1970-71, program leader, 1971-74; program leader Pacific Biol. Sta., Nanaimo, 1974-82, 83-88; sr. policy fellow Woods Hole (Mass.) Oceanog. Inst., 1982-83; vis. scientist U. B.C., Vancouver, 1988-89; coord. Fisheries & Ocean Can., Vancouver, 1989-90; prof. & dir. Westwater Rsch. Ctr., Vancouver, 1990—; bd. dirs. Westwater Rsch. Ctr., Vancouver, Rawson Acad. Aquatic Scis., Ottawa, Ont.; chmn. com. Royal Soc. of Can., Ottawa, 1993. Editor: Canadian Aquatic Resources, 1987. Fed. Rsch. grantee Fed. Tri-Coun., 1993—. Mem. Am. Fisheries Soc., Internat. Limnological Assn. Office: University of British Columbia, 1933 W Mall, Vancouver, BC Canada V6T 1Z2

HEALTON, BRUCE CARNEY, data processing executive; b. Montebello, Calif., Oct. 22, 1955; s. Donald Carney and Doris May (Kubler) H.; m. Deborah Louise Stevens, Nov. 26, 1977; children: Alexander Carney, Michaela Shawn. BA of Bus., Western Ill. Univ., 1977; Cert. Brokerage Ops., N.Y. Inst. of Fin., 1986. Programmer Westinghouse Learning Corp., Iowa City, 1977-78; contract programmer Cutler-Williams, Mpls., 1978-79; programmer/analyst Northwest Computer Svcs., Mpls., 1979-81; cons. Cytrol, Edina, Minn., 1981-90; cons./pres. Elegant Tech. Solutions, Brooklyn Park, Minn., 1990—; treas. Minn. Joint Computer Conf., Mpls., 1990-91, asst. treas. 1989-90; sec. Twin Cities ACM, Mpls., 1987-89. Mem. IEEE (cons. software com. 1989-90), Assn. for Computing Machinery, Twin Cities Assn. for Computing Machinery (sec. 1987-89, chair 1993). Office: Elegant Technology Solution 8480 Yates Ave N Minneapolis MN 55443-2186

HEALY, BERNADINE P., physician, scientist; b. N.Y.C., Aug. 2, 1944; d. Michael J. and Violet (McGrath) Healy; m. Floyd Loop, Aug. 17, 1985; children: Bartlett Anne Bulkley, Marie McGrath Loop. AB summa cum laude, Vassar Coll., 1965; MD cum laude Harvard Med. Sch., 1970. Diplomate Am. Bd. Med. Examiners, Am. Bd. Cardiology, Am. Bd. Internal Medicine (bd. dirs. 1983-87); lic. physician, Md., Ohio. Intern in medicine Johns Hopkins Hosp., Balt., 1970-71, asst. resident, 1971-72; staff fellow sect. pathology Nat. Heart, Blood & Lung Inst., NIH, Bethesda, Md., 1972-74, —; fellow cardiovascular div. dept. medicine Johns Hopkins U. Sch. Medicine, Balt., 1974-76, fellow dept. pathology, 1975-76, asst. prof. medicine and pathology, 1976-81, assoc. prof. medicine, 1977-82, asst. dean for postdoctoral programs and faculty devel., 1979-84, assoc. prof. pathology, 1981-84, prof. medicine, 1982-84; active staff medicine and pathology Johns Hopkins Hosp., from 1976, dir. CCU, 1977-84; dep. dir. Office Sci. and Tech. Policy, Exec. Office of Pres. White House, Washington, 1984-85; chmn. Rsch. Inst. The Cleve. Clinic Found., 1985-91; dir., NIH,

Bethesda, Md., 1991-93; vice chmn. Pres.' Coun. Advisers on Sci. and Tech., 1990-91; mem. Spl. Med. Adv. Group, Dept. Veterans Affairs, 1990-91, chmn. adv. panel for Basic Rsch. for 1990's, Office Tech. Assessment, 1990-91, mem. NHLBI Task Force on Atherosclerosis, 1990; mem. Vis. Com. Bd. Overseers Harvard Med. Sch. and Sch. of Dental Medicine, Boston, 1986-91; councillor Harvard Med. Alumni Assn., 1987-90; mem. Nat. Adv. Bd. Johns Hopkins Ctr. for Hosp. Fin. and Mgmt., 1987-91; mem. Bd. Overseers Harvard Coll., 1989—; chmn. Office of Tech. Assessment Panel New Devels. in Biotech., U.S. Congress, 1986-87; mem. U.S.-Brazil Panel on Sci. and Tech., 1987; mem. White House Sci. Council, 1988-89; cons. Nat. Heart, Lung and Blood Inst., NIH, 1976-91, mem. Adv. Com. to Dir., NIH, 1986-91; chmn. steering com. Post-CABG Clin. Trial, 1987-91; bd. dirs. Medtronic, Inc., Mpls., Nat. City Corp., Cleve., Nova Pharms., Balt.; mem. adv. bd. Bayer Fund for Cardiovascular Rsch., N.Y.C., 1987-89; trustee Edison BioTech. Ctr., Cleve., 1990—; chmn. Ohio Coun. on Rsch. and Econ. Devel., 1989-91. Editorial cons. numerous jours.; abstract reviewer; editorial bd. Jour. Cardiovascular Medicine, 1980-91, Am. Jour. Medicine, 1986-91, Am. Jour. Cardiology, 1981-82, Circulation, 1981—, Jour. Am. Coll. Cardiology, 1982-84. Contbr. articles to profl. jours. Matthew Vassar scholar, 1962-65, Harvard Nat. scholar, 1965-70; Eloise Ellery fellow, 1965-66, Stetler Research fellow, 1976-77; recipient Nat. Bd. Ann. award for Medicine, Med. Coll. Pa., 1983. Mem. Am. Fedn. Clin. Research (pres. 1983-84), Am. Heart Assn. (award 1983-84, 90, pres. 1988-89, fellow Coun. on Clin. Cardiology, Coun. on Circulation, dir. 1983-84), Am. Coll. Cardiology (bd. govs. 1979-82), ACP, Assn. Am. Med. Colls., Internat. Acad. Pathology, Am. Med. Women's Assn., Assn. for Women in Sci., Am. Soc. Clin. Investigation, Am. Bd. Internal Medicine (bd. govs. 1986—), Inst. Medicine, NAS, Johns Hopkins U. Soc. Scholars, Inst. Medicine NAS, Phi Beta Kappa, Alpha Omega Alpha. Office: Cleve Clinic Found 1 Clin Ctr S1-134 9500 Euclid Ave Cleveland OH 44195-5210

HEALY, KEVIN E., biomedical engineering educator. BSChemE, U. Rochester, 1983; MSE, U. Pa., 1985, PhD in Bioengineering, 1990. Rsch. assistant U. Pa., Phila., 1984-86, rsch. fellow, 1986-89; assist prof. biomaterials Northwestern U., Chgo., Ill., 1989—; assist prof. biomed. engring. Northwestern U., Evanston, Ill., 1990—; founder BioEngineered Materials, Inc., Evanston, Ill., 1991—; prin. investigator, program dir. Northwestern U.; lectr. in field. Editorial participation Jour. Biomedical Materials Rsch., Biotechnology and Bioengineering, Jour. Bone and Joint Surgery, Internat. Jour. Oral and Maxillofacial Implants; pub. numerous papers. Teaching fellow, U. Pa., 1984; Bayne scholar Northwestern U. Dental Sch., 1992. Mem. AAAS, Biomed. Engring. Soc., Soc. Biomaterials, Am. Soc. Artificial Internal Organs, Materials Rsch. Soc. Achievements include invention of surface modification of titanium to control cell response, processing technique to produce porous absorbable scaffolds useful for tissue and organ regeneration, method to co-dissolve absorbable polymers and water-soluble proteins; research in biomimetic materials and tissue engring., osteogenic cell attachment to degradable polymers, interaction of titanium with biological environments, biological (peptide) modification of implant materials. Office: Northwestern U Dept Biol Materials and Biomed Engring 311 E Chgo Ave Chicago IL 60611-3008

HEAP, JAMES CLARENCE, retired mechanical engineer; b. Trinidad, Colo.; s. James and Elsie Mae (Brobst) H.; m. Alma Mae Swartzendruber. Registered profl. engr., Wis. Sr. mech. engr. Cook Electric Research Lab, Morton Grove, Ill., 1955-56; assoc. mech. engr. Argonne (Ill.) Nat. Lab., 1956-66; sr. project engr. Union Tank Car Co., East Chicago, Ind., 1966-71; sr. engr. Thrall Car Mfg. Co., Chicago Heights, 1971-77; research design engr. Graver Energy Systems, Inc., East Chicago, Ind., 1977-79; mech. cons. design engr. Pollak & Skan, Inc., Chgo., 1979-83, ret., 1983; cons. mech. design and stress analysis, 1965-83. Author: Formulas for Circular Plates Subjected to Symmetrical Loads and Temperatures, 1966; contbr. tech. papers to profl. jour.; patentee in field. Served with USAF, 1946-47. Mem. ASME, Christian Businessmen's Com. U.S., The Gideon's Internat. Home: 1406 Ashton Ct Goshen IN 46526-4679

HEARN, MILTON THOMAS, biomedical scientist; b. Adelaide, Australia, Feb. 17, 1943; s. Maurice and Mary (Bewley) H.; m. Tanice Bojana Stoyanoff, Dec. 20, 1968; children: Sean Stephen, Cyrma Maria. PhD, Adelaide U., 1970, DSc, 1983. NRC fellow U. B.C., Vancouver, 1969-71; ICI fellow Oxford (Eng.) U., 1971-75; sr. rsch. fellow U. Otago, Dunedin, New Zealand, 1975-81; prin. rsch. fellow St. Vincent's Inst. Med. Rsch., Melbourne, Australia, 1981-86; prof. biochemistry Monash U., Clayton, Australia, 1986—, prof., dir. ctr. for bioprocess tech., 1987—; mem. editorial adv. bd. 6 scientific jours.; cons. to pharm. and biotech. cos. Author: HPLC of Proteins and Peptides, 1991, Peptide and Protein Reviews, 1982-88, Protein Purification, 1990; contbr. chpt.: Physical Biochemistry, 1989; contbr. over 280 articles to profl. jours.; editor 14 books. Fellow Royal Australian Chem. Inst., Australian Acad. Tech. Scis.; mem. Am. Chem. Soc., Endocrine Soc., Acad. Tech. Scis. (nat. sci. and tech. adv. com.). Achievements include 30 patents; research in biotechnology, bioseparation science, biopharmaceuticals, protein chemistry, and protein surface interactions. Office: Monash U, Wellington Rd, 3168 Clayton Australia

HEATH, GREGORY ERNEST, electrical engineer; b. Mineola, N.Y., Apr. 15, 1941; s. Ernest Alfred and Audrey (Johnson) H.; m. Myrna Eileen Plousha, Dec. 17, 1966; children: Gregory, Michael. AB, ScB, Brown U., 1963, ScM, 1965; PhD, Stanford U., 1979. Lectr. Stanford (Calif.) U., 1969-70; prof. elec. engring. Brown U., Providence, 1970-75; sr. engr. Raytheon Co., Portsmouth, R.I., 1976; prin. investigator MIT Lincoln Lab., Lexington, 1976—; referee NATO Advanced Rsch. Workshop, Bad Windsheim, Fed. Republic Germany, 1983, IEEE, 1985-92. Contbr. to sci. publs. Instr. Opportunities Industrialization Ctr., Providence, 1971-75; bd. dirs. Newton Athletic Assn., Newton, Mass., 1978—; tutor, mentor Mass. Pre.-Engring. Program, Boston, 1993. Named MIT Black Achiever, Greater Boston YMCA, 1993. Mem. IEEE, AAAS, Internat. Neural Network Soc., Sigma Xi, Omega Psi Phi. Achievements include development and application of signal processing, parameter estimation, feature extraction, pattern recognition and neural network learning to automatic target recognition. Home: 19 Judith Rd Newton Centre MA 02159-1715 Office: MIT Lincoln Lab KB 309 Lexington MA 02173-0073

HEATH, TED HARRIS, environmental engineer; b. Durango, Colo., May 24, 1951; s. Parrish Richard Sr. and Dixie (Harris) H.; m. 1974 (div. 1991); children: Geoffrey Edwin, Sally Jane, Anne Elizabeth. BS in Engring., Calif. State U., Long Beach, 1979. Registered profl. engr., Calif. Mech. engr. USN, Mare Island Naval Shipyard, Vallejo, Calif., 1980-81; rsch. engr. So. Calif. Edison Co., Rosemead, 1981-85, sr. environ./regulatory licensing engr., 1985—. Contbr. articles to profl. publs. With USN, 1970-74. Mem. ASME, Nature Conservancy, Zool. Soc. San Diego, Pacific Coast Elec. Assn., Tau Beta Pi. Office: So Calif Edison Co 2131 Walnut Grove Ave Rosemead CA 91770

HEATH, TIMOTHY GORDON, bioanalytical chemist; b. Newark, N.J., Sept. 19, 1962; s. Gordon Neil and Ellen Ida (Rattasep) H.; m.Melissa Marie Gallaher, Oct 1, 1988. BS, Calvin Coll., 1984; PhD, Mich. State U., 1990. Post doctoral fellow Parke-Davis Pharm., Ann Arbor, Mich., 1990-92; bioanalytical chemist Marion Merrell Dow, Inc., Kansas City, Mo., 1992--. Contbr. articles to profl. jours. Yates Meml. Found. fellow, 1989, Coll. Nat. Sci. fellow, 1989-90. Mem. Am. Chem. Soc., Am. Soc. for Mass Spectrometry, Am. Assn. for Advancement Sci. Republican. Baptist. Achievements include development of gas phase ion-molecule reactions using triple quadrupole mass spectrometry for providing structural information of gas phase ions. Information provided by this technique not always accessible by conventional triple quadrup mass spectrometry techniques. Office: Marion Merrell Dow Inc PO Box 9627 Kansas City MO 64134-0627

HEATHCOCK, CLAYTON HOWELL, chemistry educator, researcher; b. San Antonio, Tex., July 21, 1936; s. Clayton H. and Frances E. (Lay) H.; m. Mabel Ruth Sims, Sept. 6, 1957 (div. 1972); children: Cheryl Lynn, Barbara Sue, Steven Wayne, Rebecca Ann; m. Cheri R. Hadley, Nov. 28, 1980. BSc, Abilene Christian Coll., Tex., 1958; PhD, U. Colo., 1963. Supr. chem. analysis group Champion Paper and Fiber Co., Pasadena, Tex., 1958-60; asst. prof. chemistry U. Calif.-Berkeley, 1964-70, assoc. prof., 1970-75, prof., 1975—, chmn., 1986-89; chmn. Medicinal Chemistry Study Sect., NIH, Washington, 1981-83; mem. sci. adv. coun. Abbott Labs., 1986—. Author: Introduction to Organic Chemistry, 1976; editor-in chief Organic Syntheses, 1985-86, Jour. Organic Chemistry, 1989—; contbr. numerous articles to profl. jours. Recipient Alexander von Humboldt U.S. Scientist, 1978, Allan R. Day award, 1989, Prelog medal, 1991. Mem. AAAS, Am. Acad. Arts and Scis., Am. Chem. Soc. (chmn. div. organic chemistry 1985, Ernest Guenther award 1986, award for creative work in synthetic organic chemistry 1990, A.C. Cope scholar 1990), Chem. Soc. London, Am. Soc. Pharmacognosy. Home: 20 Highgate Ct Kensington CA 94707-1115 Office: U Calif Dept Chemistry Berkeley CA 94720

HEAUSLER, THOMAS FOLSE, structural engineer; b. Buffalo, Dec. 16, 1959; s. William Alfred and Aline (Folse) H.; m. Mary Aton, Aug. 14, 1982; children: Emily Kinloch, Abigail Folse. BS in civil Engring., Tulane U., 1981, M of Engring., 1983. Registered profl. civil engr. Calif., Wash., La., profl. structural engr., Calif., Wash. Structrual engr. Waldemar A. Nelson and Co., New Orleans, 1981-83; civil, structural engr. Burk and Assocs., New Orleans, 1983-85; structural engr. Dura-Cast Bldg. Systems, Inc., Brentwood, Calif., 1986-88; project engr. Dasse Design, Inc., San Francisco, 1988-91; sr. structural engr. Burns and McDonnell, Kansas City, Mo., 1991-- ; seismology com. mem. Structural Engrs. No. Calif., San Francisco, 1991--. Registered disaster svc. worker Calif. Office Emergency Svcs., 1990--; mem. Bacchus Found. Kansas City, 1991. Mem. Structural Engrs. No. Calif., Am. Soc. Civil Engrs., Am. Inst. Steel Construction, Am. Concrete Inst., Tau Beta Pi, Alpha Tau Omega, Loch Lloyd Country Club, Kansas City, So. Yacht Club, New Orleans. Republican. Episcopalian. Achievements include research in guidelines for seismic rehabilitation of buildings. Home: 4301 W 126th Ter Leawood KS 66209 Office: Burns and McDonnell 4800 E 63d St Kansas City MO 64130

HEAVIN, MYRON GENE, aeronautical engineer; b. Greencastle, Ind., Mar. 1, 1940; s. Halbert and Jean Mildred (Eastham) H.; m. Sharyl Lynn, Aug. 21, 1964; children: Bruce Alan, Holley Jean, Kristie Lynn. BS in Aero. Engring., Purdue U., 1963. Assoc. engr. Rocketdyne, Canoga Park, Calif., 1963-65; engr. Douglas Aircraft, Long Beach, Calif., 1965-82, sect. chief, 1982-91, sr. prin. engr., 1991—. Programmer comml. aircraft performance computer programs, 1965-87, PC computer program operational performance for airlines. Presbyterian. Achievements include working with airlines to transmit aircraft performance using electronic media. Home: 7712 Duquesne Pl Westminster CA 92683 Office: McDonnell Douglas Corp 3855 Lakewood Blvd Long Beach CA 90846

HEBARD, FREDERICK V., plant pathologist; b. Phila., Mar. 24, 1948; s. Frederick Vanuxem Hebard and Elizabeth (Fales) Vaughan; m. Chieko Eda, Jan. 20, 1977 (div. Sept. 1985); m. Dayle Zanzinger, June 21, 1986; children: Kyla Tyne, Paige Tierney. BS, Columbia U., 1973; MS, U. Mich., 1976; PhD, Va. Poly. Inst., 1982. Rsch. plant pathologist USDA-ARS, Prosser, Wash., 1982-83; rsch. specialist U. Ky., Lexington, 1986-89; supt. Am. Chestnut Found., Meadowview, Va., 1989—. Contbr. articles to profl. publs. Mem. Am. Phytopathol. Soc., Soc. Am. Foresters, Meadowview Civic Club, Sigma Xi. Achievements include discovery of existence of low levels of blight resistance in some American chestnut trees, abiltiy of chestnut blight fungus to form mycelia fan is central to blight resistance and hypovirulence, the small diameter of American chestnut sprouts in the forest is the cause of the low incidence of blight in them and thus of their continued surivial, breeding of many chestnut trees. Office: Am Chestnut Found Rt a Box 17 Meadowview VA 24361

HEBEL, GAIL SUZETTE, obstetrician/gynecologist; b. St. Augustine, Fla., Nov. 6, 1949; d. Lawrence George and Jean (Faulk) H. BS, La. State U., 1971; postgrad., Fla. State U., 1975-76; MD, La. State Med. U., New Orleans, 1981. Rsch. chemist Exxon Corp., Baton Rouge, 1971-75; intern ob.-gyn. Univ. Hosp., Jacksonville, Fla., 1982; regional staff physician Health & Human Resources, Lafayette, La., 1982-86; intern in internal and emergency medicine Tulane U., New Orleans, 1985-86, resident ob-gyn., 1987; obstetrician/gynecologist Douglas Women Ctr., Lithia Springs, Ga., 1991—; asst. med. dir. Dept. Health and Human Resources, Lafayette, 1984-85; cons. Douglas Women's Ctr., Lithia Springs, 1991—. Mem. United Way, New Orleans, 1988-89, Habitat, New Orleans, 1989. Mem. AMA, Acad. Ob.-Gyn., Am. Med. Women's Assn., Conrad Collins Soc. Office: Douglas Womens Ctr 880 Crestmark Dr Ste 200 Lithia Springs GA 30057

HEBERLEIN, DAVID CRAIG, physicist; b. San Antonio, Jan. 8, 1942; s. Frank Albert and Marcia Elizabeth (Bassett) H.; m. Martha Lois Walkden, Dec. 23, 1967; 1 child, Elizabeth Edith. BS in Physics, U. Va., 1963; PhD in Physics, U. Fla., 1969. NSF postdoctoral fellow Wesleyan U., Middletown, Conn., 1969-71; rsch. physicist USN Sci. Tech. Ctr., Suitland, Md., 1971, U.S. Army Mobility Equipment Command, Ft. Belvoir, Va., 1971-82; tech. dir. mobile elec. power U.S. Army Troop Support Command, St. Louis, 1984-86; dep. dir., dir. combat engring. U.S. Army Belvoir RD&E Ctr., 1982-84, assoc. tech. dir. R&D, 1986-88, dir. countermine systems, 1988—; mem. panel Naval Rsch. Adv. Com., Arlington, Va., 1990-91; fellow-by-courtesy GWC Whiting Sch. Engring. Johns Hopkins U., Balt., 1992—. Contbr. articles to Internat. Symposia. Mem. Am. Phys. Soc., Sigma Xi (pres. 1972—). Achievements include development of theoretical codes coupled to field measurements that give exceptionally high pressures and impulses from dispersed powders in air; development of electromagnetic signature duplication and projection equipment that causes premature activiation of enemey ordance; development of solid state measurements on composite materials that defined the shock compression equations of state; development of high resolution capacitive strain gauge to measure phase changes in solid methane and nitrogen; research includes enhanced output from dispersed exp. powder, electromagnetic signature duplication and production, equations of state for composite materials, pulse temperature measurement techniques, and solid phase charges in methane and nitrogen. Home: 6504 Elmdale Rd Alexandria VA 22312 Office: US Army Belvoir RD&E Ctr SATBE-N Fort Belvoir VA 22060-5606

HEBERT, BENJAMIN FRANCIS, mechanical engineer; b. Portland, Maine, Nov. 14, 1966; s. Jon Berry and Priscilla Val (Roma) H.; m. Jennifer Becker, June 6, 1992. BS, Tufts U., 1989, MS, 1992. Cert. engr.-in-tng., Mass. Scientist Cambridge (Mass.) Collaborative, Inc., 1991—. Mem. ASME, Acoustical Soc. of Am., Tau Beta Pi. Home: 15 Governors Ave Apt 21 Medford MA 02155-3029 Office: Cambridge Collaborative Inc 689 Concord Ave Cambridge MA 02138-1002

HEBERT, LEONARD BERNARD, JR., contractor; b. Jeanerette, La., Aug. 8, 1924; s. Leonard B. Sr. and Katherine R. (Rader) H.; m. Hilda Girard, Nov. 28, 1946 (div. 1980); children: Suzanne Lynne, Andre Lane, Yvette Ann; m. Catherine Kempa, June 14, 1984. BE, Tulane U., 1944. Registered civil engr., La., Miss. Field engr. Humble Oil & Refining Co., Houston, 1946-48; exec. v.p. Gurtler-Hebert & Co., Inc., New Orleans, 1948-72; chief exec. officer, pres. Leonard B. Hebert Jr., & Co., Inc., New Orleans, 1972—; chief exec. officer Profl. Constrn. Svcs. Inc., New Orleans, 1974—. Lt. USNR, 1944-46. Republican. Roman Catholic. Office: 7933 Downman Rd New Orleans LA 70126-1296

HEBSON, CHARLES STEPHAN, hydrogeologist, civil engineer; b. Jersey City, N.J., July 20, 1957; s. James Donald and Catherine Margaret (Haubert) H.; m. Margaret Mary Rinderle, Oct. 25, 1986; children: Stephen, Anne. ScB, Brown Univ., 1979; MA, Princeton Univ., 1981, MSE, 1983, PhD, 1985. Registered profl. engr., Maine; cert. geologist, Maine. Hydraulic engr. U.S. Geological Survey, West Trenton, N.J., 1983-84; rsch. hydrologist Univ. Coll., Galway, Ireland, 1984-86; rsch. hydraulic engr. USDA Agri. Rsch. Svc., Fort Collins, Colo., 1986-88; sr. hydrologic engr. Robert G. Gerber, Inc., Freeport, Maine, 1988—. Contbr. articles to profl. jours. Trustee Falmouth Conservation Trust, Falmouth, Maine, 1989—; instr. Holy Martyns Ch. Religion Edn., Falmouth, 1989—. Fellow Garden State, N.J. Dept. Higher Edn., 1979-83. Mem. ASCE, Am. Geophysical Union, Am. Inst. Hydrology (prof.). Home: 19 Carmichael Ave Falmouth ME 04105 Office: Robert G Gerber Inc 17 West Street Freeport ME 04032-1133

HECHT, HARRY GEORGE, chemistry educator; b. Powell, Wyo., May 6, 1936; s. George W. and Ruth (Neilson) H.; m. Glenda Burr, Sept. 18, 1959; children: David H., James B., Daniel R. BS, Brigham Young U., 1958, MS, 1959; PhD, U. Utah, 1962. Asst. prof. chemistry Tex. Tech. U., Lubbock, 1962-66; mem. staff Los Alamos (N.Mex.) Nat. Lab., 1966-73; head chemistry S.D. State U., Brookings, 1973-80, prof. chemistry, 1980—; exch. prof. People's Republic of China, Kunming, Yunnan, 1988. Author: Magnetic Resonance Spectroscopy, 1966, Mathematics in Chemistry, 1990; co-author: Reflectance Spectroscopy, 1965; contbr. articles to jours. Pvt. U.S. Army, 1959. Recipient Outstanding Alumni award for Profl. Achievement Northwest Coll., 1990; Fulbright scholar Commonwealth Sci. and Indsl. Rsch. Orgn., 1965-66; rsch. fellow Alexander von Humboldt Stiftung, 1971-72. Mem. Am. Chem. Soc. (chmn. local sect. 1987-88), S.D. Acad. Scis. (pres. 1981-82), Phi Kappa Phi, Sigma Xi. Office: SD State U Dept Chemistry PO Box 2202 Brookings SD 57007

HECK, ANDRÉ, astronomer; b. Jalhay, Belgium, Sept. 20, 1946; M in Math. and Edn., Liège State U., 1969, PhD, 1975, DSc, 1985; cert. mktg., mgmt., comm. technics, Strasbourg 3 U.; habilitation Strasbourg 1 U., 1986. Asst. prof. Liège State U., 1970-78; visitor Paris Obs., 1971; astronomer Strasbourg Obs., 1976; resident astronomer European Space Agy. IUE Ground Obs., Villafranca del Castillo, Spain, 1977-80, acting IUE obs. controller, 1980-81, dep. obs. controller, 1980-83; astronomer Strasbourg Obs., 1983—, dir., 1988-90. Mem. Internat. Astron. Union (working groups on astron. data, standard stars, modern astron. methodology, vice chmn. 1988-91, chmn. 1991—), Am. Astron. Soc. (IAU rep. to CODATA), European Astron. Soc., Société Française des Spécialistes d'Astronomie, Am. Mgmt. Assn., Soc. Indsl. Applied Math., Assn. Computational Machinery, Vistas in Astronomy (editorial bd.). Contbr. articles to profl. jours.; author books in field. Discoverer Comet 1972 VIII. Office: Observatoire Astronomique, 11 rue de l'Universite, F-67000 Strasbourg France

HECK, DAVID ALAN, orthopaedic surgery educator, mechanical engineering educator; b. Syracuse, N.Y., Nov. 20, 1952; s. William C. and Shirley W. (Wolthausen) H.; m. Kimberly Kay North, Sept. 27, 1980; children: William Donald, Andrew David, Daniel Robert. BS in Elect. and Computer Engring. cum laude, Clarkson Coll. Tech., 1973; MD, SUNY, Syracuse, 1977. Cert. AM. Bd. Orthopaedic Surgery. Intern in gen. surgery U. Minn., Mpls., 1977-78; resident in orthpaedic surgery SUNY, Syracuse, 1978-82; resident in orthopaedic biomechanics Mayo Clinic, Rochester, Minn., 1982-83; asst. prof. Ind. U. Sch. Medicine, Indpls., 1983-87, assoc. prof., 1987—; attending physician Ind. U. Med. Ctr., Indpls., 1983—, VA Med. Ctr., Indpls., 1983—, Riley Hosp., 1983—, Wishard Meml. Hosp., 1983—; adj. asst. prof. Sch. Mech. Engring. Purdue U., West Lafayette, Ind., 1984-87, assoc. prof., 1987—; chief, orthopaedic surgery sect., VA Med. Ctr., 1983—, medipro advisor, 1986; bd. dirs. Indian Creek Hills, Inc., The Orthopaedic Rev. Course; lectr. various profl. orgns.; dir. Orthopaedic Biomechanics Lab. Ind. U. Med. Ctr., 1984—; mem. residency applicants rev. com., Ind. U., 1983—, orthopaedic chief's of services com., 1983—, search and screen com., 1984-86, adult ambulatory care com., 1986-90, orthopaedic basic sci. com. 1986-87, chmn. orthopaedic edn. com., 1987—, quality assurance com., 1988-91, med. admissions com., 1989-91, total quality mgmt. chmn., 1993—. Mem. editorial bd. Jour. Anthroplasty; editorial reviewer Clin. Orthopaedics; contbr. numerous articles to profl. jours. Sports medicine advisor White River Park, 1984-86; bd. dirs. Hand Surgery Rsch. & Edn. Found. Mem. AMA, Am. Acad. Orthopedic Surgery (outcome com. 1990—, com. on comps. 1988-90), Am. Coll. Sports Medicine, Am. Soc. Metals, Am. Soc. Biomechanics, Ind. Med. Assn., Marion County Med. Soc., Orthopaedic Rsch. Soc., Knee Soc. (ex-officio, com. on evaluation, outcome com. chmn. 1991—), 7th Dist. Med. Soc., Eta Kappa Nu, Tau Beta Pi. Avocations: camping, canoeing, shooting, swimming. Home: 11440 Valley Meadow Dr Zionsville IN 46077-9342

HECK, JONATHAN DANIEL, toxicologist; b. Vienna, Austria, July 28, 1952; came to U.S., 1959; s. Harry Harlan and Jean (Jenkins) H.; m. Anne Carroll Wohlford; 1 child, Samuel Louis. BA, U. N.C., Greensboro, 1975; PhD, U. Tex., Houston, 1983. Diplomate Am. Bd. Toxicology. Toxicologist Lorillard Tobacco Co., Greensboro, 1983-85, mgr. life scis., 1985—. Author: Advances in Inorganic Biochemistry, 1984; co-author: Metal Ions in Biological Systems, 1986. Mem. AAAS, Am. Coll. Toxicology, Soc. Toxicology, Environ. Mutagen Soc. Achievements include rsch. in metals carcinogenesis, tobacco smoke toxicology, food ingredients toxicology. Office: Lorillard Tobacco Co Rsch Ctr 420 English St PO Box 21688 Greensboro NC 27420

HECKADON, ROBERT GORDON, plastic surgeon; b. Brantford, Ont., Can., Jan. 30, 1933; s. Frederick Gordon and Laura (Penrose) H.; B.A., U. Western Ont., 1954, M.D., 1960; postgrad. U. Toronto, 1960-66, U. Vienna, 1966; m. Camilla Joyce Russell, July 11, 1959; children—David, Louise, Peter, William, Barbara. Intern, Toronto Gen. Hosp., 1960-61; asst. resident Toronto Western Hosp., 1961, Toronto Wellesley Hosp., 1962, Toronto Gen. Hosp., 1962-63; resident in plastic surgery St. Michael's Hosp., Toronto, 1963, Toronto Western Hosp., 1964, Toronto Gen. Hosp., 1964, Toronto Hosp. for Sick Children, 1965; asst. resident orthopedics Toronto East Gen. Hosp., 1965-66; practice medicine specializing in plastic surgery, Windsor, Ont., Can., 1966—; chief med. staff Hotel Dieu; mem. staff Grace Hosp., Met. Hosp. (all Windsor); surveyor Can. Coun. on Health Facilities Accreditation. Served with RCAF, 1951-56. Fellow A.C.S.; mem. Canadian Med. Assn., Ont. Med. Assn., Essex County Med. Assn., Windsor Acad. Surgery, Royal Coll. Physicians and Surgeons, Can. Soc. Plastic Surgeons.

HECKER, RICHARD JACOB, research geneticist; b. Miles City, Mont., Mar. 26, 1928; s. John T. and Elizabeth R. H.; m. Diane M. Lauer, Aug. 12, 1958; children: Carol, John, Ann, Douglas. BS, Mont. State U., 1958; PhD, Colo. State U., 1964. Geneticist U.S. Dept. Agr., Ft. Collins, Colo., 1959-64; rsch. geneticist U.S. Dept. Agr., Salinas, Calif., 1964-65; rsch. geneticist U.S. Dept. Agr., Ft. Collins, 1965-73, rsch. leader, 1973—; affiliate prof. Colo. State U., 1965—. Contbr. numerous articles to profl. jours. Fellow Am. Soc. Agronomy, Crop Sci. Soc. Am.; mem. Western Soc. Crop Sci., Am. Soc. Sugar Beet Technologists (Meritorious Svc. award 1981), KC. Roman Catholic. Achievements include development of genetic resistance in sugarbeet to Rhizoctonia solani. Home: 1100 Morgan St Fort Collins CO 80524-3835 Office: Colo State U Crops Rsch Lab Fort Collins CO 80523

HECKER, SIEGFRIED STEPHEN, metallurgist; b. Tomaszów, Poland, Oct. 2, 1943; came to U.S., 1956; s. Robert and Maria (Schaller) Mayerhofer; m. Janina Kabacinski, June 19, 1965; children: Lisa, Linda, Lori, Leslie. BS, Case Inst. Tech., 1965, MS, 1967; PhD, Case Western Res. U., 1968. Postdoctoral assoc. Los Alamos Sci. Lab., 1968-70, mem. staff, 1973-80, assoc. div. leader, 1980-81, dep. div. leader, 1981-83, div. leader, 1983-85, chmn. Ctr. for Materials Sci., 1985-86; dir. Los Alamos Nat. Lab., 1986—; sr. rsch. metallurgist Gen. Motors Rsch. Labs., Warren, Mich., 1970-73. Author, editor: Formability, 1977. Bd. dirs. Carrie Tingley Hosp.; bd. regents U. N.Mex., 1987—. Recipient E. O. Lawrence award Dept. Energy, 1984; named One of 100 Top Innovators, Sci. Digest, 1985. Fellow Am. Soc. Metals (mem. nat. commn. superconductivity 1989-91); mem. NAE, Metall. Soc. (bd. dirs. 1983-84), Los Alamos Ski Club (pres. 1980-81). Republican. Roman Catholic. Avocation: skiing. Home: 117 Rim Rd Los Alamos NM 87544-2967 Office: Los Alamos Nat Lab PO Box 1663 Los Alamos NM 87545*

HECKSCHER, AUGUST, journalist, author, foundation executive; b. Huntington, N.Y., Sept. 16, 1913; s. Gustav Maurice and Louise (Vanderhoef) H.; m. Claude Chevreux, Mar. 19, 1941; children: Stephen August, Philip Hofer, Charles Chevreux. BA, Yale U., 1936; MA, Harvard U., 1939; LLD, Fairleigh Dickinson U., 1962; LHD, NYU, 1962, Temple U., 1964, Brandeis U., 1964; LittD, C.W. Post Coll., 1963, Adelphi Coll., 1963; DHL, Parsons Sch. Design, 1987. Govt. instr. Yale U., 1939-41; editor Auburn (N.Y.) Citizen-Advisor, 1946-48; editorial writer, chief editorial writer N.Y. Herald Tribune, 1948-56; dir. Twentieth Century Fund, 1956-67; mem. editorial bd. The Am. Scholar. Author: These Are The Days, 1936, A Pattern of Politics, 1947, The Politics of Woodrow Wilson, 1956, Diversity of Worlds (with Raymond Aron), 1957, The Public Happiness, 1962, When La Guardia Was Mayor (with Phyllis Robinson), 1979, Alive in the City—Memories of an Ex-commissioner, 1974, Open Spaces: The Life of American Cities, 1977, St. Paul's: The Life of A New England School, 1981, Woodrow Wilson, 1991. Spl. cons. on arts for Pres. Kennedy, 1962-63; art commr. City of N.Y., 1957-62, pks. commr., 1967-71; chmn. bd. dirs. Internat. Coun. Mus. of Modern Art, 1958-63, Nat. Repertory Theatre Found.; trustee Internat. House, St. Paul's Sch.; past trustee Lavanburg Found.; mem. governing bd. Yale U. Press.; v.p. Mcpl. Arts Soc.; mem. N.Y. State Coun. on the Arts; vice-chmn. bd. dirs. Urban Am.; past pres. Woodrow

Wilson Found.; chmn. bd. trustees New Sch. Social Rsch., 1966-68. Recipient Joseph Henry medal Smithsonian Instn., 1990; decorated officer French Legion of Honor, Moroccan Order of Quissam Alouit; Jonathan Edwards Coll. fellow. Fellow AAAS; mem. AIA (hon.), Century Assn., Coun. Fgn. Rels., ACLU (past bd. dirs.), Phi Beta Kappa. Address: 333 E 68th St New York NY 10021

HEDGE, JEANNE COLLEEN, health physicist; b. Scottsburg, Ind., May 30, 1960; d. Paul Russell and Barbara Jean (Belshaw) H. BS in Environ. Health, Purdue U., 1983. Chemistry and health physics technician Marble Hill Nuclear Generating Sta., Pub. Svc. Ind., Madison, 1983-84; radiation protection asst. Pub. Svc. Electric and Gas Co., Hancock's Bridge, N.J., 1984-85, radiation protection technician, 1985-89, engr., 1989-90, lead engr., 1990-91, sr. staff engr., 1991—; mem. People to People Internat. Citizen Ambassador Exch., People's Republic China, 1988. Mem. NOW, Am. Nuclear Soc., Health Physics Soc.

HEDGES, RICHARD H., epidemiologist; b. Louisville, July 16, 1952; s. Houston and Frances Ruth (Zemo) H.; m. Donna Jean Hough. BA, U. Ky., 1974; MA, Ea. Ky. U., 1975, MPA, 1983; PhD, U. Ky., 1986; postgrad., Capital U. Law. Rehab. specialist Commonwealth of Ky., Somerset, 1976-81; chief health planner Commonwealth of Ky., Frankfort, 1981-82; asst. prof. U. Ky., Lexington, 1985-87; rsch. assoc. dept. med. behavioral sci. U. Ky. Coll. Medicine, Lexington, 1982-85; program adminstr. Rollman Psychiat. Inst., Cin., 1987-88; asst. prof. Ohio U., 1988-92, assoc. prof., 1992—; dir. div. on aging Ohio U. Health Promotion and Rsch. Contbr. articles to profl. jours. Fellow NIMH, 1984-86. Mem. ABA, APHA, Am. Soc. Law and Medicine, Soc. for Epidemiol. Rsch., Am. Coll. Rehab. Medicine, Health Svcs. Rsch., Am. Coll. Health Care Execs., Nat. Health Lawyers Assn., Assn. Behavioral Medicine, Pi Sigma Alpha, Phi Delta Phi. Democrat. Episcopalian. Avocations: backpacking, volleyball, bicycling, sailing. Home: 7 Fairview Ave Athens OH 45701-1706 Office: Ohio U Health Sci Health Svcs Peden Tower Athens OH 45701

HEEGER, ALAN JAY, physicist; b. Sioux City, Iowa, Jan. 22, 1936; s. Peter J. and Alice (Minkin) H.; m. Ruthann Chudacoff, Aug. 11, 1957; children: Peter S., David J. B.A., U. Nebr., 1957; Ph.D., U. Calif., Berkeley, 1961. Asst. prof. U. Pa., Phila., 1962-64; asso. prof. U. Pa., 1964-66, prof. physics, 1966-82; prof. physics U. Calif., Santa Barbara, 1982—, dir. Inst. for Polymers and Organic Solids, 1983—; dir. Lab. for Rsch. on Structure of Matter, 1974-81, acting vice provost for rsch., 1981-82; Morris Loeb lectr. Harvard U., 1973; pres., founder UNIAX Corp., Santa Barbara. Editor-in-chief Synthetic Metals jour.; contbr. sci. articles to profl. jours. Recipient John Scott medal City of Phila., 1989; Alfred P. Sloan fellow, Guggenheim fellow; Govt. grantee. Fellow Am. Phys. Soc. (Oliver E. Buckley prize in solid state physics 1983). Patentee in field. Office: U Calif Dept Physics Santa Barbara CA 93103 also: UNIAX Corp 5375 Overpass Rd Santa Barbara CA 93111

HEFFELFINGER, DAVID MARK, optical engineer; b. Ft. Worth, Jan. 10, 1951; s. Hugo Wagner and Betty Lu (Graf) H.; m. Barbara Lynne Putnam, May 1, 1980; children: Jakob, Leon, Stacy. MS in Physics, Wayne State U., 1984. Project scientist GM Rsch. Lab., Warren, Mich., 1978-90; grad. rsch. asst. Wayne State U., Detroit, 1982-84; optical group leader Bio-Rad Labs., Richmond, Calif., 1990—. Contbr. articles to Jour. Applied Physics, Bull. Am. Phys. Soc. Recipient Vaden Miles award Wayne State U., 1982. Mem. AAAS, Internat. Soc. Optical Engring., Optical Soc. Am. Achievements include research in laser ablation and photoacoustic interactions, laser scanning and imaging. Office: Bio Rad Labs 2000 Alfred Nobel Dr Hercules CA 94547

HEFFRON, W(ALTER) GORDON, computer consultant; b. New Orleans, July 25, 1925; s. Walter Gordon Heffron and Adrienne Inez (Parker) Roy; m. Mary Frances Gilmore (div. May 1992); children: Kathleen M., Elizabeth G. Heffron Mackall, Laura W., D. Parker. BE in Elec. Engring., Tulane U., 1947; MSEE, Purdue U., 1950; DSc, George Washington U., 1969. Registered profl. engr., N.Y., Calif. Instr. elec. engring. Tulane U., New Orleans, 1947-48; analytical engr. GE, Schenectady, 1949-55; sr. dynamics engr. Convair, San Diego, 1955-56; dept. head Melpar, Inc., Falls Church, Va., 1956-64, Bellcomm, Inc., Washington, 1964-72, AT&T Bell Labs., Holmdel, N.J., 1972-85; pres. Microbase Systems, St. Thomas, V.I., 1986—; study dir. manned space flight NASA, Washington, 1967, USAF Studies Bd., Washington, 1976. Contbr. articles to sci. jours. Mem. IEEE (sr.), AIAA. Republican. Episcopalian. Achievements include patent for method for simulating helicopter flight; research on Apollo descent guidance, guidance system for Mars probe, polar perturbation and lunar orbits.

HEFLICH, ROBERT HENRY, microbiologist; b. Cairo, N.Y., Nov. 10, 1946; s. Henry George and Mary Ann (Ljutich) H.; m. Mary C. Garvey, Mar. 27, 1971. MS, Rutgers U., 1970, PhD, 1976. Rsch. assoc. Mich. State U., East Lansing, 1976-79; microbiologist VA Hosp., Little Rock, 1979-83, Nat. Ctr. for Toxicological Rsch., Jefferson, Ark., 1983—. Co-editor: Genetic Toxicology, 1991; contbr. over 100 articles to profl. publs. With U.S. Army, 1970-72. Bush predoctoral fellow Rutgers U., 1975. Mem. Am. Soc. Microbiology, Environ. Mutagen Soc., Sigma Xi (treas. Cen. Ark. chpt. 1991—). Democrat. Office: Nat Ctr Toxicol Rsch Div Genetic Toxicology Jefferson AR 72079

HEFNER, JERRY NED, aerospace engineer; b. Norfolk, Va., July 3, 1944; s. Ira Eugene and Martha (Birch) H.; m. Betty Meredith, June 12, 1965; children: Terri Lynne Hefner Thames, Jerry Ned Jr. BSME, Old Dominion U., 1966; MSME, N.C. State U., 1968. Aerospace engr. NASA-Langley Rsch. Ctr., Hampton, Va., 1966-79, asst. br. head viscous flow br., 1979-83, mgr. viscous drag reduction program, 1983-89, mgr. gen. aviation and transport tech., 1986-89, br. head civil aircraft br., 1986-89, asst. div. chief fluid mechanics div., 1989—. Editor tech. books/proceedings; contbr. articles to profl. jours. Mem. coun. St. Mark Luth. Ch., Yorktown, Va., 1992—, evangelism bd., 1992—. Recipient Lampe award Hampton Roads Engring. Club, 1966. Fellow AIAA (assoc., chmn. aircraft design tech. com. 1992—), Phi Kappa Phi. Achievements include patent for turbulent drag reduction concept. Home: 108 Kittywake Dr Newport News VA 23602 Office: NASA Langley Rsch Ctr Fluid Mechanics div Mail Stop 197 Hampton VA 23681

HEFTMANN, ERICH, biochemist; b. Vienna, Austria, Mar. 9, 1918; came to U.S., 1939; s. Salomon and Rosa (Seifert) H.; m. Lily Rubin (div. 1966); children: Rex, Lisa, Erica; m. Brigitte Hedwig Sander, Mar. 14, 1968; children: Karen, David. BS, NYU, 1942; PhD, U. Rochester, 1947. Cert. Clin. Chemist. Biochemist USPHS, Boston, 1947-48, NIH, Bethesda, Md., 1948-63; biochemist USDA, Pasadena, Calif., 1963-70, Berkeley, Calif., 19790-83; editor Jour. of Chromatography, Amsterdam, The Netherlands, 1983—. Author (books) Biochemistry Steroids, 1960, Steroid Biochemistry, 1970, Chromatography of Steroids, 1976; editor (books) Chromatography, 1961, 67, 75, 83, 92, Modern Methods of Steroid Analysis, 1973. Recipient Humboldt Prize German Govt., 1975. Fellow AAAS; mem. Am. Chem. Soc., Am. Soc. of Biol. Chemists. Home: PO Box 928 Orinda CA 94563

HEGARTY, WILLIAM PATRICK, chemical engineering educator; b. Phila., Apr. 13, 1929; s. William Francis and Mary Zella (Haggart) H.; m. Elizabeth Giacobbe, June 4, 1957 (dec. May 1989); children: Patrick S., Molly K. BS in Chem. Engring., U. Mich., 1953. Design engr. Std. Oil Devel. Co., Linden, N.J., 1953-58; devel. engr. Std. Oil Co., Whiting, Ind., 1958-60, Hercules, Wilmington, Del., 1960-62; process engr. Olin Matheson Co., New Haven, 1962-67; devel. engr. Technichem Co., Wallingford, Conn., 1967-70; prin. engring. assoc. Air Products, Allentown, Pa., 1970-92; assoc. prof. Pa. State U., State College, 1992—. 2nd lt. U.S. Army, 1946-48. Recipient Kirkpatrick Chem. Engring. Achievement award McGraw Hill Chem. Engring. Mag., 1987. Fellow Am. Inst. Chem. Engrs. Roman Catholic. Achievements include 17 patents in synthetic fuels, acid gas processing, sulfur recovery and oxygen prodn. Home: 140 E Doris Ave State College PA 16801 Office: Pa State U Dept Chem Engring 112A Fenske Lab University Park PA 16802-4400

HEGDE, ASHOK NARAYAN, neuroscientist; b. Karnataka, India, Apr. 12, 1959; came to the U.S., 1990; s. Narayan V. and Susheela Hegde; m. Lalita Bhat, Mar. 20, 1982; 1 child, Monica. BS, UAS Bangalore, 1080, MS, 1983; PhD, Ctr. Cellular Mol. Biology, India, 1990. Rsch. scientist Ctr. for Cellular and Molecular Biology, Hyderabad, India, 1988-90; rsch. assoc. Columbia U. Ctr. for Neurobiology and Behavior, N.Y.C., 1990—. Contbr. articles to profl. jours. Fellow Grass Found., 1992; recipient Travel award Coun. Scientific and Indsl. Rsch., 1989, 90; grantee Lady Tata Meml. Trust, 1991-92. Mem. Soc. Neurosci., N.Y. Acad. Scis. Office: Columbia U Ctr Neurosci and Behavior 722 W 168 St New York NY 10032

HEGEDUS, L. LOUIS, chemical engineer, research and development executive; b. Budapest, Hungary, Apr. 13, 1941; came to U.S., 1968; s. Lajos and Anna (Kellessy) H.; m. Eva Judith Brem, Mar. 28, 1968; children: Caroline Nora, Monica Michelle. MS in Chem. Engring. Tech. U. Budapest, 1964, D honoris causa, 1991; PhD, U. Calif., Berkeley, 1972. Rsch. engr. Rsch. Inst. Organic Chem. Industry, Budapest, 1964-65; group leader Daimler-Benz AG, Mannheim, Fed. Republic Germany, 1965-68; supr. catalysis rsch. Gen. Motors Research Labs., Warren, Mich., 1972-80; v.p. Rsch. div. W.R. Grace Co., Columbia, Md., 1980—; Allan P. Colburn lectr. U. Del., 1976; Union Carbide lectr. SUNY, Buffalo, 1983; B.F. Dodge lectr. Yale U., 1988; J.A. Gerster lectr. U. Del., 1988, Regents lectr. UCLA, 1991, Mason lectr. Stanford U., 1991, Cary Lectr. Ga. Inst. Tech., 1993, Hulburt Lectr., Northwestern Univ., 1993; mem. adv. bd. chem. thermal bioengring. div. NSF, 1985; mem. adv. bd. dept. chem. engring. Princeton U., 1980-92, U. Calif., Berkeley, 1985—, U. Wis., Madison, 1987—, Lawrence Berkeley Lab. Ctr. for Advanced Materials Surface Sci. Program, 1990—; mem. governing bd. Council Chem. Rsch., 1987-89, 91—, chmn., 1993-94; mem. bd. on chem. sci. and tech., NRC, 1991—, chmn. com. critical techs., 1992. Author: Catalyst Poisoning, 1984; editor 2 books on catalysis; mem. editorial bd. Indsl. and Engring. Chem. Rsch., 1992—, Catalysis Letters, 1993—; contbr. articles to profl. jours. Fellow Am. Inst. Chem. Engrs. (editorial bd. jour. 1978-83, 85-88, R.H. Wilhelm award 1988, Profl. Progress award 1980, Chem. Engr. of Yr. award Detroit 1978); mem. NAE, Am. Chem. Soc. (Chemtech Leo Friend award 1981, editorial bd. Industrial and Engineering Chemistry Research, 1992—), Md. Acad. Scis. (sci. council 1987-91). Avocation: flying. Home: 6625 Paxton Rd Rockville MD 20852-3659 Office: WR Grace & Co Rsch Div 7379 Rt 32 Columbia MD 21044

HEGGERS, JOHN PAUL, surgery and microbiology educator, microbiologist, retired army officer; b. Bklyn., Feb. 8, 1933; s. John and May (Hass) H.; m. Rosemarie Niklas, July 30, 1977; children: Arn M., Ronald R., Laurel M., Gary R., Renee L., Annette M. BA in Bacteriology, Mont. State U., 1958; MS in Microbiology, U. Md., 1965; PhD in Bacteriology and Pub. Health, Wash. State U., 1972. Diplomate Am. Bd. Bioanalysis. Med. technologist U.S. Naval Hosp., St. Albans, N.Y., 1951-53; bacteriologist Hahnemann Hosp., Worcester, Mass., 1958-59; commd. 2d lt. U.S. Army, 1959, advanced through grades to lt. col., 1975; mem. staff dept. bacteriology 1st U.S. Army Med. Lab., N.Y.C., 1959-60; chief clin. lab. U.S. Army Hosp., Verdun, France, 1960-63; chief virology and rickettsiology div. dept. microbiology 3d U.S. Army Med. Lab., Ft. McPherson, Ga., 1965-66; instr. bacteriology Basic Lab. Sch., Ft. McPherson, 1965-66; chief diagnostic bacteriology 9th Med. Lab., Saigon, Vietnam, 1966-67; chief microbiology div. dept. pathology Brooke Gen. Hosp., Ft. Sam Houston, Tex., 1967-69; chmn. dept. microbiology U.S. Army Sch. Med. Tech., Ft. Sam Houston, 1967-69; instr. bacteriology evening div. San Antonio Jr. Coll., 1969; lab. scis. officer Office Surgeon Gen., Washington, 1972-74; microbiologist spl. mycobacterial disease br. div. geog. pathology Armed Forces Inst. Pathology, Washington, 1973; spl. asst. to dir. Armed Forces Inst. Pathology, 1973-74; chief clin. rsch. lab. clin. rsch. svc. Madigan Army Med. Ctr., Tacoma, 1974-76, asst. chief clin. investigation svc., 1976-77; instr. immunology, parasitology and mycology Clover Park Vocat. Tech. Inst., 1976-77; ret., 1977; assoc. prof. dept. surgery U. Chgo., 1977-80, prof., 1980-83; prof. surgery Wayne State U., Detroit, 1983-88; prof. surgery and microbiology U. Tex. Med. Br., 1988—; dir. clin. microbiology Shriners Burn Inst., Galveston, Tex., 1988—. Author: Current Problems in Surgery, 1973, Quantitative Bacteriology, 1991; contbr. articles to profl. jours.; contbg. editor: Jour. Am. Med. Tech., 1972—. Pres. Aloe Rsch. Found., 1989-92. Decorated Bronze Star; Legion of Merit; recipient certificate of appreciation A.C.S., 1969, certificate of appreciation Armed Forces Inst. Pathology, 1974; Valley Forge Honor certificate Freedoms Found., 1974; Fisher award in med. tech. Am. Med. Technologists, 1968, 82; Gerard B. Lambert award, 1973; Ednl. Found. Research award Am. Soc. Plastic and Reconstructive Surgery, 1978, Alumni Achievement award Wash. State Univ., 1993. Fellow N.Y. Acad. Sci., Am. Acad. Microbiology, Royal Soc. Tropical Medicine and Hygiene, Am. Geriatrics Soc.; mem. Nat. Registry Microbiologists (chmn. exec. council 1976-79), Am. Soc. Microbiology (chmn. com. tellers 1974-75), Wash. Soc. Am. Med. Technologists (pres. 1975-77), Wash. Soc. Med. Tech. (chmn. sect. microbiology sci. assembly, dir. 1975-77), Assn. Mil. Surgeons U.S., Am. Soc. Clin. Pathologists (asso.), Am. Med. Technologists (disting. service award 1975, exceptional merit award 1976, nat. dir. 1979-80, nat. sec. 1980-82, nat. v.p. 1982-84, Technologist of Yr. 1983), Am. Burn Assn. (President's continuing edn. award 1981, chmn. rsch. com., At Large award 1986, Robert B. Lindberg award 1991, 92), Plastic Surgery Research Council, Surg. Infection Soc. (charter), Ill. State Soc., Med. Technologists (v.p. 1979), AVMA, Internat. Soc. for Burn Injuries, Mason (32 degree), Shriner, Sigma Xi. Office: Shriners Burns Inst 815 Market St 610 Texas Ave Galveston TX 77550-2788

HEIDORN, DOUGLAS BRUCE, medical physicist; b. Rochester, N.Y., June 8, 1957; s. Herman Hugh and Diane Evans (Nicely) H.; m. Iris C. Casas, Nov. 16, 1985; children: Kimberly, Deanna. AB, U. Chgo., 1979; MS, Rice U., 1983, PhD, 1985. Postdoctoral fellow Los Alamos (N.Mex.) Nat. Lab., 1985-89; rsch. fellow U. Mich., Ann Arbor, 1989-91; med. physicist St. Josephs Regional Med. Ctr., Lewiston, Idaho, 1991—. Contbr. articles to profl. jours. Mem. Am. Assn. Physicists in Medicine. Office: Saint Joseph's Med Ctr 504 6th St Lewiston ID 83501

HEIKKINEN, RAIMO ALLAN, electronics executive; b. Espoo, Finland, Apr. 7, 1955; s. Aulis and Lahja (Kallionpää) H.; m. Tuula Martta Sinikka Tossavainen, Sept. 7, 1984; children: Katja, Minna, Sanna. Electronics Engr., Tech. Coll., Helsinki, 1975. Designer Finnish Broadcasting Corp., Helsinki, 1976-79; mgr. Lohja (Finland) Electronics, 1979-81; product mgr. Kaukomarkkinat Oy, Espoo, 1981-84, sales mgr., 1984-87, gen. mgr., 1987-90; mng. dir. Finn Metric Oy, Espoo, 1990—; bd. dirs. Elkomit Ry., Helsinki, Metric As, Norway. Avocations: tennis, jogging, skiing, gardening, cooking. Office: Finn Metric Oy, Riihitontuntie 2, 02200 Espoo Finland

HEILICSER, BERNARD JAY, emergency physician; b. Bklyn., Jan. 19, 1947; s. Murray and Esther (Dubrow) H.; m. Marcia Cherry, June 2, 1976; children: Micah, Seth, Jacob. BA, SUNY, Binghamton, 1968; MS, Hahnemann Med. Coll., Phila., 1971; DO, Coll. Osteo. Medicine/Surgery, Des Moines, 1976. Diplomate Am. Bd. Emergency Medicine. Instr. anatomy and physiology U. Pa. and Hahnemann Med. Coll., Phila., 1971-73; staff physician Va. Inst. Tech., Blacksburg, 1977-78; asst. prof. emergency medicine Chgo. Coll. Osteo. Medicine, 1979; emergency physician St. Margaret Hosp., Hammond, Ind., 1979-83, Michael Reese Med. Ctr., Chgo., 1989-91, Ingalls Hosp., Harvey, Ill., 1983—; project med. dir. South Cook County Emergency Med. Svc., Harvey, 1984—; mem. faculty Chgo. Osteo. Med. Ctr., 1987—; faculty trauma nurse specialist St. James Hosp., Chicago Heights, Ill., 1980—; preceptor nurse practitioners Purdue U., Hammond, 1981—; fellow The Ctr. for Clin. Med. Ethics, U. Chgo., 1993—. Vol. fireman Flossmoor (Ill.) Fire Dept., 1985—, Matteson (Ill.) Fire Dept., 1980—. Fellow Ctr. Clin. Med. Ethics, U. Chgo., 1993—. Fellow Am. Coll. Emergency Physicians; mem. Am. Osteo. Assn., Nat. Assn. Emergency Med. Svcs. Physicians, Nat. Assn. Emergency Med. Technicians, Prehosp. Care Providers Ill., Sigma Sigma Phi. Jewish. Avocations: running, basketball. Office: Ingalls Hosp One Ingalls Dr Harvey IL 60426

HEILMEIER, GEORGE HARRY, electrical engineer, researcher; b. Phila., May 22, 1936; s. George C. and Anna I. (Heineman) H.; m. Janet S. Faunce, June 24, 1961; 1 dau., Elizabeth. BEE, U. Pa., 1958; MS in Engring., Princeton U., 1960, MA, 1961, PhD, 1962. With RCA Labs., Princeton, N.J., 1958-70; dir. solid state device rsch. RCA Labs., 1965-68, dir. device concepts, 1968-70; White House fellow, spl. asst. to sec. def. Washington, 1970-71; asst. dir. def. rsch. and engring. Office Sec. Def., 1971-75; dir. Def. Advanced Projects Agy., 1975-77; v.p. rsch., devel. and engring. Tex. Instruments Inc., 1978-83, sr. v.p., chief tech. officer, 1983-91; pres., chief exec. officer Bell Communications Rsch., Livingston, N.J., 1991—; mem. adv. group on electron devices; bd. dirs. TRW. Patentee in field. Mem. vis. com. MIT, Sch. Engring. and Applied Sci. of U. Pa. Recipient IR-100 New Product award Indsl. Rsch. Assn., 1968, 69; Sec. Def. Disting. Civilian Svc. award, 1975, 77; Arthur Flemming award U.S. Jaycees, 1974; Nat. medal Sci. NSF, 1991, Indsl. Rsch. Inst. medal, 1993. Fellow IEEE (David Sarnoff award 1976, Outstanding Achievement award Dallas chpt. 1984, Philips award 1985, Founder's award 1986, Japan Computers and Comm. prize 1990, Pres. Nat. medal of Sci. 1991); mem. NAE (Founders award 1992), U. Pa. Alumni Assn., Princeton U. Grad. Alumni Assn., Sigma Xi, Tau Beta Pi, Eta Kappa Nu (Outstanding Young Engr. in U.S. award 1969, Vladimir Karapetoff Eminent Mem. award 1993). Methodist. Office: Bell Communications Rsch 290 W Mt Pleasant Ave Livingston NJ 07039-2747

HEIMER, WALTER IRWIN, psychologist, educator; b. N.Y.C., June 24, 1927; s. Joseph and Dorothy (David) H.; m. Eva Bettina Schoenberger; children: Jessica Heimer Wagener, Annette S. BA, Allegheny Coll., 1949; MA, New Sch. Social Rsch., N.Y.C., 1953, PhD, 1962. Lic. psychologist, N.Y. Rsch. asst. New Sch. Social Rsch., 1954-57; asst. psychology dept. Swarthmore (Pa.) Coll., 1957-58; rsch. psychologist USN Tng. Device Ctr., Port Washington, N.Y., 1958-59; instr. Fairleigh Dickinson U., Madison, N.J., 1959-62; asst. prof. Hofstra u., Hempstead, N.Y., 1963-67; assoc. prof. L.I. U., Brookville, N.Y., 1967—; vis. prof., sr. engr. Grumman Aerospace Corp., Bethpage, N.Y., summers, 1984-86, vis. faculty U.S. Naval Pers. and Tng. Ctr., San Diego, summer, 1979; profl. appointment U.S. Naval Tng. Device Ctr., Port Washington, N.Y., summers, 1963-69. Contbr. articles to profl. jours. Human Resources, Inc. grantee, 1967-69. Mem. APA (pub. manual task force com. 1973-83), AAAS, Am. Psychol. Soc., ea. Psychol. Assn. Jewish. Achievements include research in perception, memory and learning. Home: 102 Oakland Ave Port Washington NY 11050 Office: L I U Psychology Dept C W Post Campus Brookville NY 11548

HEIMLICH, HENRY JAY, physician, surgeon; b. Wilmington, Del., Feb. 3, 1920; s. Philip and Mary (Epstein) H.; m. Jane Murray, June 3, 1951; children: Philip, Peter, Janet and Elizabeth (twins). B.A., Cornell U., 1941, M.D., 1943; D.Sc. (hon.), Wilmington Coll., 1981, Adelphi U., 1982, Rider Coll., 1983. Diplomate: Am. Bd. Surgery, Am. Bd. Thoracic Surgery. Intern Boston City Hosp., 1944; resident VA Hosp., Bronx, 1946-47, Mt. Sinai Hosp., N.Y.C., 1947-48, Bellevue Hosp., N.Y.C., 1948-49, Triboro Hosp., Jamaica, N.Y., 1949-50; attending surgeon div. surgery Montefiore Hosp., N.Y.C., 1950-69; dir. surgery Jewish Hosp., Cin., 1969-77; prof. advanced clin. scis. Xavier U., Cin., 1977-89; assoc. clin. prof. surgery U. Cin. Coll. Medicine, 1969—; pres. Heimlich Inst.; mem. Pres. Commn. on Heart Disease, Cancer and Stroke, 1965; Pres. Nat. Cancer Found., 1963-68, bd. dirs., 1960-70; founder, pres. Dysphagia Found. Author: Postoperative Care in Thoracic Surgery, 1962, (with M.O. Cantor, C.H. Lupton) Surgery of the Stomach, Duodenum and Diaphragm, Questions and Answers, 1965; also; contbr. chpts. to books; numerous articles to med. jours.; Producer: films Esophageal Replacement with a Reversed Gastric Tube (awarded Medaglione Di Bronzo Minerva 1961), Reversed Gastric Tube Esophagoplasty Using Stapling Technique, How to Save a Choking Victim: The Heimlich Maneuver, 1976, 2d edit., 1982, How To Save a Drowning Victim: The Heimlich Maneuver, 1981, Stress Relief: The Heimlich Method, 1983; video: Dr. Heimlich's Home First Aid Video, 1989 (Vira award 1989); mem. editorial bd.: films Reporte's Medicos. Bd. dirs. Community Devel. Found., 1967-70; bd. dirs. Save the Children Fedn., 1967-68, United Cancer Council, 1967-70. Served to lt. (s.g.) USNR, 1944-46. Recipient Lasker Award for Pub. Service, Lasker Found., 1984, China-Burma-India Vets. Assn. Americanism award, 1988. Fellow ACS (chpt. pres. 1964), Am. Coll. Chest Physicians, Am. Coll. Gastroenterology; mem. Soc. Thoracic Surgeons (founding mem.), AMA (cons. to jour.), Cin. Soc. Thoracic Surgery, N.Y. Soc. Thoracic Surgery, Soc. Surgery Alimentary Tract, Am. Gastroent. Assn., Pan Am. Med. Assn., Collegium Internat. Chirurgiae Digestive, Central Surg. Assn. Developer Heimlich Operation (reversed gastric tube esophagoplasty) for replacement of esophagus; inventor Heimlich chest drain valve, Heimlich Micro-Trach (HMT for COPD, emphysema and cystic fibrosis); developer Heimlich Maneuver to save lives of victims of food choking and drowning (listed in Random House, Oxford Am. and Webster dictionaries); developer Computers for Peace, a program to maintain peace throughout world and a caring world. Office: Ste 410 2368 Victory Pky Cincinnati OH 45206

HEIN, FRITZ EUGEN, engineer, consultant, architect; b. Bruex, Czechoslovakia, Mar. 25, 1926; came to U.S., 1956; s. Friedrich and Maria (Lehner) H.; m. Gertraud Marie Conrad, Dec. 28, 1954; children: Carmen, Wolfgang. Student, O.v. Miller, Munich, 1953. Registered architect, Bavaria, Wis.; registered profl. engr., Wis. Architect, engr. U. Wuerzburg, 1953-56; design engr. USN, Great Lakes, Ill., 1956-60; dir. of design USN, Washington and Madrid, 1971-79; facilities engr. U.S. Army, Karlsruhe, Fed. Republic of Germany, 1979-82; chief engr. C.E., 1984-88; project mgr. C.E., Frankfurt, Fed. Republic of Germany, 1982-84; pvt. practice cons. Naples, Fla., 1988—. Author: Drydock Launch Facilities, 1961, Guide to Airborne Sound Control, 1971, Design Manual DM 1 & 18, 1974, Navy Civil Engineer, 1977, The Military Engineer, 1980. Bd. dirs. Bldg. Rsch. Adv. Bd., Washington, 1976-79; advisor Camp David constrn. modifications White House, Washington, 1962. Decorated D.S.M.; recipient Appreciation award U.S. Asst. Sec. of Def., 1976. Mem. ASTM (com. Washington chpt. 1976-79), Am. Nat. Metric Coun., Acoustial Soc. Am., Joint U.S./Spain Mil. Group, Contamination Control Assn. Achievements include development of air to ground missiles, 1945; design of the Geodesic Dome for the New South Pole Station; introduction of cogeneration power plants in U.S.; conception of a launch facility at equatorial sites. Home: 6805-H Carnation Rd Richmond VA 23225-5256 Office: Wettersteinplatz 2, 8000 Munich 90, Germany

HEIN, ILMAR ARTHUR, electrical engineer; b. Woodstock, Ill., Aug. 15, 1959; s. Arthur and Vera (Sulluste) H. MSEE, U. Ill., 1983, PhD in Elec. Engring., 1990. Microwave engr. Hughes Aircraft Co., El Segundo, Calif., 1983-85; postdoctoral rsch. assoc. dept. elec. engring. U. Ill., Urbana, 1990—. Contbr. articles to profl. jours. Mem. Am. Inst. Ultrasound in Medicine (Terrence Matzuk Meml. award 1988), IEEE, Acoustical Soc. Am. Office: Univ Illinois Dept Elec Engring 1406 W Green St Urbana IL 61801

HEIN, JOHN WILLIAM, dentist, educator; b. Chester, Mass., Sept. 29, 1920; s. Rudolf Jacob and Mercedes Viola H.; m. Jeannette Marie BeVier, Dec. 16, 1944. B.S., Am. Internat. Coll., 1941; D.M.D., Tufts U., 1944; Ph.D., U. Rochester, 1952; A.M. (hon.), Harvard, 1962; D.Sc. (hon.), Am. Internat. Coll., 1979, Tufts U., 1993. Student instr. oral pathology Tufts Coll. Dental Sch., 1943-44; head div. dental research U. Rochester, 1948-52, sr. fellow dental research, 1949-52, instr. pharmacology, 1951-53, asst. prof. dental research, 1952-55, asst. prof. pharmacology, 1954-55, chmn. dept. dentistry and dental research, 1952-55; instr. anatomy and physiology Eastman Sch. Dental Hygiene, 1950-55, lectr. dental research, 1953-55; research specialist Bur. Biol. Research, Rutgers U., 1955-59; dental dir. Colgate Palmolive Co., 1955-59; prof. preventive dentistry, dean Sch. Dental Medicine, Tufts U., 1959-62; dir. Forsyth Dental Center, 1962-91; prof. dentistry Harvard Dental Sch., 1962-67. Trustee Am. Internat. Coll., 1960-76. Served to capt. AUS, 1942-47. Fellow AAAS, Internat. Coll. Dentists (regent 1967-72, pres. U.S. 1975-76, internat. pres. 1983-84); mem. ADA, Mass. Dental Soc. (pres. 1964-65), Internat. Assn. Dental Research (treas. 1978-82), Am. Assn. Dental Research (treas. 1985-88), Am. Acad. Dental Sci., New Eng. Dental Soc. (hon. pres. 1978), Am. Soc. Dentistry for Children, Assn. Ind. Research Insts. (1st v.p. 1980, pres. 1981-83), Royal Soc. Medicine (hon.), Sigma Xi, Omicron Kappa Upsilon, Delta Sigma Delta. Club: Harvard (Boston). Home: 3 Bridge St PO Box 156 Medfield MA 02052-1539 Office: Forsyth Dental Ct 140 The Fenway Boston MA 02115

HEINE, JAMES ARTHUR, utilities plant manager; b. Chgo., Dec. 1, 1939; s. Arthur Bernard and Lucille Agnes (Doran) H.; m. Carol Agnes Escherich, Apr. 11, 1964; children: James A. Jr., Janet M., Jeffrey R. BSME, Chgo. Tech., 1961; MS, Nat. Louis U., Evanston, Ill., 1990; student, Olivet U., 1991—. Lic. EPA, Ill. Supt. combustion engr. and environ. control Republic Steel Corp., Chgo., 1961-86; chief engr. cast metals & facilities mgmt. L.B. Knight Engrs, Architects, and PLanners, Chgo., 1988-91; mgr. utilities Argonne (Ill.) Nat. Lab., 1991—; adj. faculty Coll. Bus. and Mgmt. Nat. Louis U., 1990—, Joliet Jr. Coll., 1993—; cons. L.B. Knight Engrs.

Architects and Planners, Chgo., 1991—. Pres., v.p. Oak Forest (Ill.) Baseball Asn., 1982-83, Oak Forest Flag Football, 1980-81; asst. baseball coach Moraine Valley C.C., 1983-84; mem. environ. com. Calumet (Ill.) Area Indsl.Commn., 1992. Mem. Am. Foundrymen's Soc. (sub com. energy issues 1983—), Assn. Iron and Steel Engrs. (past officer). Republican. Roman Catholic. Achievements include research in steel mill energy, iron and steel engring. Home: 7661 Wheeler Dr Orland Park IL 60462 Office: Agronne Nat Lab 9700 S Cass Ave Argonne IL 60439

HEINEKEN, FREDERICK G., biochemical engineer; b. Chgo., Oct. 22, 1939; s. Frederick W.G. Heineken and Marie Helene Faber Heineken; divorced; 1 child, Christopher P. BS, Northwestern U., 1962; PhD, U. Minn., 1966. Sr. biochem. engr. Monsanto, St. Louis, 1966-71; postdoctoral fellow U. Colo., Denver, 1972-74, rsch. assoc., instr., 1974-76; sr. project engr. Cobe Labs., Lakewood, Colo., 1977-79, dept. head, 1979-81, therapy scientist, 1981-84; cons. Heineken & Assocs., Potomac, Md., 1985—; program dir. NSF, Washington, 1985—. Trustee 1st Universalist Ch., Denver, 1980-83, vice-moderator, 1984. Recipient Young Investigator award, NIH, 1974. Mem. AICE, AAAS, Am. Chem. Soc. (councilor 1990—), Assn. for Advancement of Med. Instrumentation, Am. Soc. for Artificial Organs, St. Louis Ski Club (pres. 1971). Home: National Science FOundation Engineering 7908 Turncrest Dr Potomac MD 20854 Office: NSF 1800 G St NW Washington DC 20550

HEINEMAN, WILLIAM RICHARD, chemistry educator; b. Lubbock, Tex., Oct. 15, 1942; s. Ellis Richard and Edna (Anderson) H.; m. Linda Margaret Harkins, Oct. 25, 1969; children: David William, John Richard. BS, Tex. Tech. U., 1964; PhD, U. N.C., 1968. Rsch. chemist Hercules, Inc., Wilmington, Del., 1968-70; rsch. assoc. Case Western Res. U., Cleve., 1970-71, The Ohio State U., Columbus, 1971-72; asst. prof. U. Cin., 1972-76, assoc. prof., 1976-80, prof., 1980-88, dist. rsch. prof., 1988--; mem. adv. bd. Analytical Chemistry, Washington, 1984-86, The Analyst, Eng., 1987-- Selective Electrode Revs., 1987-92, Fresenius Jour. Analytical Chemistry, 1991-94, Analytical Chimica Acta, 1991-93, Applied Biochemistry and Biotechnology, 1991—; U.S. editor Biosensors and Bioelectronics, Eng., 1987--;m coun. Gordon Rsch. Confs. Author: Experiments in Instrumental Methods, 1984, Chemical Instrumentrumentation, 1989; editor: Laboratory Techniques in Electroanalytical Chemistry, 1984, Chemical Sensors and Microinstrumentation, 1989. Recipient Humboldt prize Humboldt Soc., 1989, Rieveschl award U. Cin., 1988, Japan Rsch. award Japan, 1987; named Disting. Scientist Tech. Socs. Coun., 1984; fellow Japan Soc. for Promotion of Sci., 1981. Mem. Am. Chem. Soc. (treas. analytical chem. divsn. 1983-86, councilor 1984—, named Chemist of Yr. 1983), Soc. for Electroanalytical Chemistry (pres. 1984-85, bd. dirs. 1984-90). Office: U Cin Dept Chemistry Mail Location 172 Cincinnati OH 45221-0172

HEINEMEYER, PAUL HUGH, quality assurance specialist; b. Valejo, Calif., Nov. 14, 1954; s. Harley Albert and Anne (Glenewinkle) H.; Sharon Diane Lange, Dec. 30, 1988; 1 child, Conrad. BS, S.W. Tex. State U., 1980. Quality analyst U.S. Gypsum Co., New Braunfels, Tex., 1978-82; quality supt. USG Corp., New Braunfels, 1982-87; quality assurance supt. APG Lime Corp., New Braunfels, 1987—; in-house cons. APG/USG, New Braunfels, 1982—. Achievements include research for development of new products.

HEINEN, CHARLES M., retired chemical and materials engineer. Student, Chrysler Inst.; BSChemE, U. Mich., MSChemE, 1942. Various positions Chrysler Corp., 1934-42, lab. supr. Manhattan project, 1942-45, from materials engr. to dir. dir. emissions/fuel economy and materials engring., 1945-78, former dir. rsch. and materials engring., cons., dir. automotive rsch. group. Recipient Soichiro Honda medal ASME, 1990. Fellow Soc. Automotive Engrs. (fuels and lubricants activities com.), Am. Soc. Metals (com. govt. and pub. affairs, chmn. numerous coms.); mem. Soc. Testing and Materials, U.S. C. of C. (former mem. com. on environment), Engring. Materia Coun., Motor Vehicle Mfg. Assn. U.S. (former mem. quality com.), Coord. Rsch. Coun. (past chmn. group combustion of exhaust gas), Air Pollution Control Assn. Achievements include research in combustion, control devices, fuel economy efforts in light materials and power rating devices, and alternate engines such as turbines, electric motors and vehicles powered with hydrogen and alcohol. Home: 4595 Burnley Dr Bloomfield Hills MI 48304*

HEINEN, JAMES ALBIN, electrical engineering educator; b. Milw., June 23, 1943; s. Albin Jacob and Viola (DeBuhr) H. BEE, Marquette U., 1964, MS, 1967, PhD, 1969. Registered profl. engr., Wis. Data analyst Med. Sch. Marquette U., Milw., 1963, teaching asst. elec. engring. dept., 1964-65, 65-66, research asst., 1966, NASA trainee, 1966-69, research assoc. Provost's Office, 1970, asst. prof. and grad. adminstr., 1971-73, assoc. prof., chmn. elec. engring. dept., 1973-76, assoc. prof., 1976-80, prof. elec. engring. and computer sci., 1980-87, prof., dir. grad. studies elec. and computer engring., 1987—; cons. in field. Contbr. numerous articles and revs. on elec. engring. and computer sci. to profl. jours. Recipient Outstanding Engring. Tchr. award Marquette U., 1979, Teaching Excellence award Marquette U., 1985; Arthur J. Schmitt fellow, 1971, 77, 80. Mem. IEEE (various coms., tech. reviewer Trans. Automatic Control 1969—, Trans. Circuits and Systems Soc. 1980—, Signal Processing Soc. 1980—, sr. mem., Meml. award Milw. sect. 1981, assoc. editor Trans. Circuits and Systems 1983-85), Am. Soc. Engring. Edn. (local arrangements chmn. ann. meeting 1978), Sigma Xi (local exec. com. 1979-80, 88-89), Tau Beta Pi, Eta Kappa Nu (Most Oustanding Elec. Engring. Tchr. in U.S. award 1974), Pi Mu Epsilon, Alpha Sigma Nu. Home: 8200 W Menomonee River Pky Wauwatosa WI 53213-2537 Office: Marquette U 1515 W Wisconsin Ave Milwaukee WI 53233-2286

HEINER, LEE FRANCIS, physicist; b. Balt., Nov. 9, 1941; s. Frank Joseph and Constance Lee (Kelley) H. BA, Johns Hopkins U., 1963; MS, George Washington U., 1973. With Carderock Div. Naval Surface Warfare Ctr., Annapolis, Md., 1963—. Author over 50 tech. reports in field. Chmn. Muscular Dystrophy Telethon, Annapolis, Md., 1979. 1st lt. U.S. Army, 1963-65. Recipient Letter of Commendation, Pres. Gerald R. Ford, 1976. Mem. Acoustical Soc. Am., U.S. Naval Inst. Achievements include patent applications related to ship silencing. Home: 1181 Bayview Vista Rt 6 Annapolis MD 21401

HEINKE, GERHARD WILLIAM (GARY HEINKE), environmental engineering educator; b. Korneuburg, Austria, Dec. 11, 1932; arrived in Can., 1953; s. Erich and Maria (Schwabe) H.; m. Judy Donath, Aug. 1, 1952 (div. 1989); children: Elizabeth, Richard; m. Karlene Heinke, Sept. 30, 1989. BASc in Civil Engring., U. Toronto, 1956, MASc in Civil and Environ. Engring., 1961; PhD in Chem. Engring., McMaster U., 1969. With U. Toronto, 1968—, prof. civil engring., 1974—, chair dept. civil engring., 1974-84, dean faculty applied sci. and engring., 1986-93; dir. inst. environ. studies, prof. civil and structural engring. Hong Kong U. Sci. & Tech., 1993—; mcpl., environ. cons. 1956-65; cons. fed., provincial, mcpl. Can. govts., industry and consulting engring. cos., 1968—. Author: (with J.G. Henry) Environmental Science and Engineering, 1989; contbr. over 100 articles to profl. jours. Fellow Can. Soc. Civil Engring. (Albert E. Berry medal 1992); mem. Am. Water Works Assn., Am. Soc. Engring. Edn., Water Environment Fedn., Assn. Profl. Engrs. Ontario, Assn. Profl. Engrs. NWT. Office: U Toronto Dept Civil Engineering, 35 St George St, Toronto, ON Canada M5S 1A4*

HEINMILLER, ROBERT H., JR., communications company executive; b. Cleve., Dec. 17, 1940; s. Robert Harry and Venita Laynell (Law) H.; m. Nancy Daniels, 1961 (div. 1963); 1 child, Patricia Minsk; m. Susan Kathleen Kubany, Dec. 23, 1978. BS in Physics, MIT, 1962. Rsch. specialist Woods Hole (Mass.) Oceanographic Inst., 1962-76; rsch. asst. MIT, Cambridge, 1976-82; v.p. Omnet Inc., Newton, Mass., 1982—. Mem. AAAS, Am. Geophys. Union, Oceanography Soc., Am. Soc. Limnology and Oceanography. Achievements include developing deepsea oceanographic mooring technology; created and developed SCIENCEnet, a network for earth scientists. Home: 14 Newton St Weston MA 02193 Office: Omnet Inc 154 Wells Ave Pembroke MA 02359

HEINTZ, ROGER LEWIS, biochemist, educator, researcher; b. Jackson Center, Ohio, Mar. 15, 1937; s. Claude O. and Ruth A. (Thompson) H.; m. Judith A. Fisher, Aug. 11, 1962; children—Claude R., Robert A., James S., Steven G. B.S., Ohio No. U., 1959; M.S., Ohio State U., 1961; Ph.D., U. Wis.-Madison, 1964. NIH postdoctoral fellow U. Ky., Lexington, 1964-66, Am. Heart Assn. research fellow, 1966-68; asst. prof. Iowa State U., Ames, 1968-75; prof. biochemistry, biophysics SUNY-Plattsburgh, 1975—, chmn. biochemistry, biophysics sect., 1978—, chmn. dept. biol. sci., 1986—. Bd. dirs. Plattsburgh Little Theatre, 1978-83; coach Babe Ruth baseball. Grantee NIH, 1967-82, SUNY Research Found., 1976-82. Mem. AAAS, Am. Soc. for Biochemistry and Molecular Biology, Am. Chem. Soc., Sigma Xi, Phi Lambda Upsilon. Democrat. Avocation: amateur theatre. Home: 133 Broad St Plattsburgh NY 12901-2602 Office: SUNY Dept Chemistry Plattsburgh NY 12901

HEINY, RICHARD LLOYD, chemical engineer; b. Gunnison, Colo., Nov. 25, 1929; s. Frantz Lloyd and Dorothy Louise (Dye) H.; m. Suzanne Heiny, Mar. 8, 1958; children: Christopher L., Michael A., Katherine S. BSChemE, U. Kans., 1950; MSChemE, Pa. State U., 1951, PhD in Chem. Engring., 1954. Instr. Pa. State U., Dept. Chem. Engring., State College, 1952-54; chem. engr. group leader Dow Chem. Co., Midland, Mich., 1954-62, sr. systems engr., 1962-64, bus. mgr., 1964-84, dir. discovery devel. dept., 1984-86; vice chmn. Omni Tech Internat. Ltd., Midland, 1986—; past rep. for Dow Chem. Co. at rsch. confs. on statis. methods; founder Aspirin Found. Am., Washington, bd. dirs., 1980-83; mem. bd. govs. Synthetic Organic Chem. Mfrs. Assn., Washington, 1980-82. Contbr. articles to profl. jours. Mem. Midland City Pks. and Recreation Commn., 1990—; cubmaster Boy Scouts Am., Midland, 1968-71; fin. chmn. Rep. State Rep., Midland, 1970-80. Summerfield scholar, U. Kans., 1946-50; article named outstanding contbn. to chem. engring. thermodynamics for yr., 1953. Mem. Am. Inst. Chem. Engrs. (Chem. Engr. of Yr. Mid-Mich. sect. 1987), Kiwanis, Sigma Xi, Alpha Chi Sigma, Phi Lambda Upsilon, Tau Beta Pi, Sigma Tau. Republican. Methodist. Achievements include supervision of development of numerous processes for manufacturing organic chemicals. Home: 17 Brown Ct Midland MI 48640-4317 Office: Omni Tech Internat Ltd 2715 Ashman St Midland MI 48640-4460

HEINZ, DON J., agronomist; b. Rexburg, Idaho, Oct. 29, 1931; s. William and Berniece (Steiner) H.; m. Marsha B. Hegsted, Apr. 19, 1956; children: Jacqueline, Grant, Stephanie, Karen, Ramona, Amy. BS, Utah State U., 1958, MS, 1959; PhD, Mich State U., 1961; grad., Stanford U. Exec. Program, 1982. Assoc. plant breeder Experiment Sta. Hawaiian Sugar Planters' Assn., Aiea, 1961-66, head dept. genetics and pathology, 1966-78, asst. dir., 1977-78, v.p. and dir., 1979-85, pres., dir. experiment sta., 1986—; cons. Phillippines, Egypt, Colombia, Reunion; mem. adv. com. plants Hawaii Dept. Agr., 1970—, Pres. Nat. Commn. Agriculture and Rural Devel. Policy, 1988. Contbr. articles to sci. jours. on sugarcane breeding, cytogenetics, cell and tissue culture techniques. Served with USAF, 1951-54. Mem. AAAS, Internat. Soc. Sugar Cane Technologists (chmn. com. germplasm and breeding 1975-86), Sigma Xi. Mem. LDS Ch. Home: 224 Ilihau St Kailua HI 96734-1654 Office: Hawaiian Sugar Planters PO Box 1057 99-193 Aiea Heights Dr Aiea HI 96701

HEINZ, RONEY ALLEN, civil engineering consultant; b. Shawano, Wis., Dec. 29, 1946; s. Orville Willard and Elva Ida (Allen) H.; m. Judy Evonne Olney, Oct. 30, 1965. BSCE, Mont. State U., 1973. Surveyor U.S. Army Corps Engrs., Seattle, 1966-73; civil engr. Hoffman, Fiske, & Wyatt, Lewiston, Idaho, 1973-74, Tippetts-Abbott-McCarthy-Stratton, Seattle, 1977-79; asst. editor Civil Engring. Mag. ASCE, N.Y.C., 1974-77; constrn. mgr. Boeing Co., Seattle, 1979-83; owner, gen. mgr. Armwavers Ltd., South Bend, Wash., 1983—; pres. Great Wall Inc., Elma, Wash., 1993—; mem. dams and tunnels del. to China, People to People Internat., Spokane, 1987. Asst. editor Commemorative Book Internat. Congress on Large Dams, 1987; contbr. articles to profl. publs., including Civil Engring. Mag., Excavator Mag., Internat. Assn. for Bridge and Structiral Engring., Japan Concrete Inst., others. Dir. Canaan Christians Fund, Aberdeen, 1993—; bd. dirs. Seaman's Ctr., Aberdeen, Wash., 1990—. Recipient First Quality award Asphalt Paving Assn. Wash., 1991. Mem. ASCE (sec. met. sect. 1975-76, assoc. mem. forum), ASTM (Student award 1973). Republican. Lutheran. Achievements include management of first commercial installation worldwide of sediment control by water jets. Office: Armwavers Ltd PO Box 782 South Bend WA 98586

HEINZ, TONY FREDERICK, physicist; b. Palo Alto, Calif., Apr. 30, 1956. BS in Physics, Stanford U., 1978; PhD in Physics, U. Calif., Berkeley, 1982. Fellow NSF, Berkeley, 1978-81, IBM, Berkeley, 1982; rsch. staff mem. IBM Rsch. Divsn., Yorktown Heights, N.Y., 1983—, rsch. mgr., 1987—; co-chair program Quantum Electronics and Laser Sci. Conf., 1992. Fellow Am. Phys. Soc.; mem. Am. Vacuum Soc., Optical Soc. Am. (topical editor jour.). Achievements include research in surface dynamics, ultrafast laser spectroscopy, nonlinear optics. Office: IBM Rsch Divsn TJ Waston Rsch Ctr PO Box 218 Yorktown Heights NY 10598-0218

HEINZ, WALTER RICHARD, sociology educator; b. Munich, Germany, Nov. 21, 1939; s. Walter and Dorothea (Schwarz) H.; m. Eva C. Drust, 1965; children: Oliver, Nina. MA in Psychology, Munich U., 1964; PhD in Sociology, Regensburg U., Fed. Republic Germany, 1969. Asst. prof. Dept. Sociology, U. Munich, 1964-65; with Dept. Sociology, U. Calif., Berkeley, 1965-66; rsch. assoc., Harkness Fellow Dept. Social Rels., Harvard U., Cambridge, Mass., 1967; asst. prof. Dept. Sociology, U. Regensburg, 1967-72; vis. prof. Dept. Sociology, U. Minn., 1971; prof. sociology Dept. Social Scis., U. Bremen, Fed. Republic Germany, 1972—; vis. prof. Dept. Anthropology, Sociology, Simon Fraser U., Vancouver, B.C., Can., 1982-83; fellow Netherlands Inst. Advanced Study, Wassenaar, The Netherlands, 1989—; v.p. for rsch. Bremen U., 1984-86; chmn. Collaborative Rsch. Programme "Status Passages and Social Risks in the Life Course", Bremen, 1988—; vis. prof. Life Course Ctr., U. Minn., Mpls., 1992; noted scholar U. B.C., Vancouver, 1992. Editor: Work, Personality, Socialization, 1978, Youth and Labor Markets, 1985, Theoretical Advances in Life Course Research, 1991, Life Course and Social Change: Comparative Perspectives, 1991, Institutions and Gatekeeping in the Life Course, 1993. Mem. R.C. Alienation, Internat. Sociol. Assn. (bd. dirs. 1986—), Am. Sociol. Assn., Deutsche Gesellschaft fur Soziologie. Home: Humboldtstr 91, 2800 Bremen Federal Republic of Germany Office: Dept Social Sci Univ Bremen, Wienerstrasse, 2800 Bremen Germany

HEIRD, WILLIAM CARROLL, pediatrician, educator; b. Decatur, Tenn., Jan. 27, 1936; s. C.T. and Mary Edna (Ward) H.; m. Jane Ray, Aug. 21, 1960. BS, Maryville Coll., 1958; MS, Vanderbilt U., 1963, MD, 1964. Intern Vanderbilt U. Med. Ctr., Nashville, 1964-65; resident Babies Hosp. Columbia-Presbyn. Med. Ctr., N.Y.C., 1965-67; asst. prof. pediatrics Coll. Physicians and Surgeons Columbia U., N.Y.C., 1971-77, assoc. prof. pediatrics Coll. Physicians and Surgeons, 1977-89; prof. pediatrics Baylor Coll. Medicine, Houston, 1990—; pediatrician Children's Nutrition Rsch. Ctr., Houston. Co-editor: Protein and Energy Needs During Infancy, 1987; editor: Nutritional Needs of the 6-to-12 Month Old, 1991; contbr. numerous articles to profl. publs., chpts. to books. Capt. USAF, 1967-69. Mem. Am. Pediatric Soc., Soc. for Pediatric Rsch., Am. Soc. Clin. Nutrition, Am. Acad. Pediatrics. Home: 2001 Holcombe Blvd Houston TX 77030 Office: Children's Nutrition Rsch 1100 Bates St Houston TX 77030

HEISEL, RALPH ARTHUR, architect; b. St. Louis, Sept. 17, 1935; s. Ralph Alonzo and Marie Lucille (Hadfield) H.; m. Janet Clevenger Scott, Aug. 4, 1962; children: Jean Marie, Arthur Scott. BS, Ga. Inst. Tech., 1957, BArch, 1958; MArch, U. Pa., 1961. Registered architect, N.Y., Fla., Ga., Conn., N.J. Designer Bodin and Lamberson, Architects, Atlanta, 1961, Fry Drew & Ptnrs., Architects, London, 1962-64; sr. assoc. I.M. Pei and Ptnrs., Architects, N.Y.C., 1964-86; pres. Heisel Assocs., Architects P.C., N.Y.C., 1986—; vis. critic various univs. Prin. works (for I.M. Pei & Ptnrs.) include Paul Mellon Ctr. for the Arts, The Choate Sch., Wallingford, Conn., Johnson & Johnson Baby Products Co. Hdqrs., N.J., Sunning Plaza Office and Apt. Complex, Hong Kong, Raffles City Hotel, Office and Shopping Complex, Singapore, The Morton H. Meyerson Symphony Ctr., Dallas, (for Heisel Assocs.) Barell Residence, Kingspoint, N.Y., St. David's Episc. Ch., N.Y., Wildwood Pla., Atlanta, Winsland House, Singapore. Mem. bd. dirs. Palmer House Group Home for the Handicapped, Larchmont, N.Y., 1980—, bd. of archtl. rev., Larchmont. Served to 1st lt. USAF, 1958-60. Recipient Design award, N.J. Bus. and Industry Assn., 1982. Mem. AIA (design awards com., Nat. Design awards 1974, 91), N.Y. State Assn. Architects, Nat. Trust for Hist. Preservation, Nat. Coun. Archtl. Regis. Bds., Univ. Club (Larchmont). Home: 2 Acorn Ln Larchmont NY 10538-1901 Office: Heisel Assocs Architects PC 611 Broadway New York NY 10012-2608

HEISTEIN, ROBERT KENNETH, obstetrician/gynecologist; b. Newark, Oct. 14, 1940; s. Samuel M. and Elizabeth M. (Jellinek) H.; B.A., U. Vt., 1962, M.D., 1966; m. Vallery Gubner, Aug. 26, 1967; children—Jonathan, Erica, Michael. Intern, Newark Beth Israel Med. Center, 1966-67, resident in ob-gyn 1967-70, attending staff, 1972—; asst. chief, dept. ob-gyn Patuxent River Naval Hosp., Md., 1970-72; pvt. practice medicine, specializing in ob-gyn Millburn, N.J., 1972—; mem. staffs St. Barnabas Med. Ctr., Livingston, N.J., Newark Beth Israel Med. Center, Overlook Hosp., Summit, N.J.; clin. instr. N.J. Med. Sch., 1974—. Served with USNR, 1970-72. Diplomate Am. Bd. Ob-Gyn. Fellow Am. Coll. Obstetricians and Gynecologists, ACS, Internat. Coll. Surgeons, Am. Fertility Soc., N.J. Acad. Medicine, Am. Soc. Abdominal Surgeons; mem. AMA, Pan Am. Med. Assn., N.J. Med. Soc., Essex County Med. Soc., Am. Assn. Gynecol. Laparoscopists, Royal Soc. Medicine. Office: 68 Essex St Millburn NJ 07041-1611 also: 23 Green Village Rd Madison NJ 07940

HEISTER, STEPHEN DOUGLAS, aerospace propulsion educator, researcher; b. Ashland, Ohio, Dec. 17, 1958; s. James Elvin and Patricia Ann (Latimer) H.; m. Edith Katherine Schmidt, Aug. 13, 1988. MS in Aero. Engring., U. Mich., 1983; PhD in Aero. Engring., UCLA, 1988. Engr. Lockheed Aircraft, Burbank, Calif., 1981-82; mem. tech. staff Aerospace Corp., El Segundo, Calif., 1983-90; asst. prof. Purdue U., West Lafayette, Ind., 1990—. Author: (chpt.) Space Propulsion Analysis and Design, 1992; assoc. editor Jour. Propulsion and Power, 1991—; contbr. articles to profl. jours. Recipient grant Air Force Office Sci. Rsch., 1992, rsch. grant, 1992. Mem. AIAA, Aerospace Industries Assn., Solid Rocket Tech. Com. Office: Purdue Univ Grissom Hall West Lafayette IN 47907

HEIT, RAYMOND ANTHONY, civil engineer; b. Norfolk, Va., Sept. 12, 1936; s. Lawrence H. and Cecelia H. (Klauke) H.; m. H. Carlee Langford, Oct. 25, 1969; children: Christopher C., Amy C. CE, U. Cin., 1959. Assoc. prof. Tex. A&M U., Galveston, 1967-68; project mgr. Union Carbide Corp., Texas City, Tex., 1967-68; prin. Raymond A. Heit, Cons. Engrs., Houston, 1968-74; dir. engring. stds. Brown & Root, Inc., Houston, 1974—. Fellow ASCE, mem. ASTM, Am. Nat. Standards Inst., Constrn. Specifications Inst. Home: 1609 Cranway Dr Houston TX 77055-3116 Office: Brown & Root Inc PO Box 3 Houston TX 77001-0003

HEITMEYER, MICKEY E., conservationist; b. Moberly, Mo., Sept. 9, 1956; m. Mary Sue; children: Jessica, Jacob. AA in Wildlife Conservation, Moberly Area Jr. Coll., 1976; BS in Fisheries/Wildlife magna cum laude, U. Mo., 1978, PhD in Wildlife Biology, 1985; MS in Wildlife Ecology, Okla. State U., 1980. Area employee Mo. Dept. Conservation, Hunnewell, Mo., 1976; field rsch. asst. Mo. Cooperative Wildlife Rsch. Unit/Mo. Dept. Conservation, Hermann, 1977; rsch. technician Mo. Dept. Conservation, Ashland, Mo., 1977; rsch. asst. Okla. Cooperative Wildlife Rsch. Unit/Okla. State U., Stillwater, 1978-80; rsch. asst. Gaylord Meml. Lab. Sch. Forestry, Fisheries and Wildlife, U. Mo., Puxico, 1981-84; postgrad. rsch. biologist Dept. Wildlife/Fisheries Biology U. Calif., Davis, 1984-88; dir. rsch. and outreach Calif. Waterfowl Assn., Sacramento, 1988-90; dir. ops. Pacific Flyway, Ducks Unltd./Western Regional Office, Sacramento, 1990-92; group mgr. conservation programs Ducks Unltd./Nat. Hdqtrs., Memphis, Tenn., 1992—; teaching asst. ornithology U. Mo., Columbia, 1978; guest lectr. Okla. State U., 1979-80, U. Mo. 1981-82, U. Calif. Davis, 1986, 88, 91, others; presenter in field; mem. numerous profl. coms. including Rice Straw Task Force, U. Calif., Davis, Dept. Agrl. Engring., 1991-92, Calif. Wetlands Policy Forum, Calif. State C. of C., 1990-92, Habitat Protection and Land Resources Com., Internat. Assn. Fish and Wildlife Agys., 1990—, others. Contbr. articles to profl. jours.; assoc. editor California Waterfowl publ., 1988-90; mng. editor Pacific Whistler newsletter, 1990-92; editorial bd. Ducks Unlimited, 1992; reviewer numerous jours. Recipient Spl. Recognition award Cen. Valley Habitat Joint Venture, North Am. Waterfowl Mgmt. Plan, 1992, Best Paper award 45th Midwest Fish and Wildlife Conf., St. Louis, 1983, Nat. Wildlife Fedn. scholarship, 1979, Stamper scholarship, U. Mo., Columbia, 1976, scholarship Moberly Area Jr. Coll., 1976. Mem. The Wildlife Soc. (Publ. award 1991), AAAS, Am. Ornithologist's Union, Colonial Waterbird Group, Cooper Ornithol. Soc., Mo. Acad. Scis., Soil and Water Conservation Soc., Soc. Wetland Scientists, Wilson Ornithol. Soc., Gamma Sigma Delta, Sigma Xi. Office: Ducks Unlimited Inc One Waterfowl Way Memphis TN 38120

HEITSCHMIDT, RODNEY KEITH, rangeland ecologist; b. Hays, Kans., Oct. 28, 1944; s. Harold W. and Wilma I. Heitschmidt; m. Judy S. Heitschmidt, Nov. 27, 1944; children: Jason K., Dustin L. BS, Ft. Hays State U., 1967, MS, 1968; PhD, Colo. State U., 1977. Prof. Tex. Agrl. Experiment Sta., Vernon, 1977-90; rsch. leader USDA Agrl. Rsch. Svc., Miles City, Mont., 1990—. Past bd. dirs. Vernon Youth Soccer League, Kid League Baseball, Boys Club; mem. Vernon Ind. Sch. Dist. Adv. Bd.; past pres. Wilbarger County United Way, Booster Club, Wilbarger County Ag Workers; pres. bd. dirs. Hillcrest Country Club, past pres. HCCC Mens Golf Assn.; active First United Meth. Ch. Capt. USAF, 1969-74. Mem. Soc. for Range Mgmt. (dir. Tex. sect. 1988-91, activities com. 1980, Outstanding Rangeman com. 1981-83, R & D com. 1984-86, grazing mgmt. session chmn. 1983, 85, 90, 91, assoc. editor Jour. Range Mgmt. 1987—), Am. Inst. Biol. Scis., Ecol. Soc. Am., Coun. for Agrl. Sci. & Tech., Phi Kappa Phi, Beta Beta Beta. Methodist. Home: 1116 S Merriam Miles City MT 59301 Office: USDA Ft. Keogh Livestock & Range Montana State Univ Rte 1 Box 2021 Miles City MT 59301

HEJAZI, SHAHRAM, biomedical engineer; b. Apr. 9, 1963. BS, SUNY, Buffalo, 1984, MS, 1986, PhD, 1990. Rsch. asst. SUNY, 1984-87, 1988-90, rsch. assoc., 1990-92; rsch. engr. Watson Rsch. Ctr. IBM, York Town Heights, N.Y., 1987-88; sr. engr. Eastman Kodak Co., Rochester, N.Y., 1992—. Contbr. articles to publs. Mem. IEEE, SPIE. Achievements include research in multiwave length infrared imaging, neural network. Office: Eastman Kodak Co 100 Carlson Rd Rochester NY 14653-9015

HEJTMANCIK, MILTON RUDOLPH, medical educator; b. Caldwell, Tex., Sept. 27, 1919; s. Rudolph Joseph and Millie (Jurcak) H.; B.A., U. Tex., 1939, M.D., 1943; m. Myrtle Lou Erwin, Aug. 21, 1943; children: Kelly Erwin, Milton Rudolph, Peggy Lou; m. 2d, Myrtle M. McCormick, Nov. 27, 1976. Resident in internal medicine U. Tex., 1946-49, instr. internal medicine, 1949-51, asst. prof., 1951-54, assoc. prof., 1954-65, prof. internal medicine, 1965-80, dir. heart clinic, 1949-80, dir. heart sta., 1965-80; chief of staff John Sealy Hosp., 1957-58; chief staff U. Tex. Hosps., 1977-79; prof. medicine Tex. A&M Coll. Medicine, 1981-82; cardiologist Olin E. Teague VA Hosp., Temple, Tex., 1981-82, VA Clinic, Beaumont, Tex., 1982-86. Served from 1st lt. to capt., M.C., AUS, 1944-46; ETO. Recipient Ashbel Smith Outstanding Alumnus award U.Tex. Med.Br., 1991, Titus Harris Disting. Svc award, 1992. Diplomate in cardiovascular diseases Am. Bd. Internal Medicine. Fellow ACP, Am. Coll. Chest Physicians, Am. Coll. Cardiology; mem. Am. (fellow council clin. cardiology), Tex. (pres. 1979-80), Galveston Dist. (pres. 1956) heart assns., AMA (Billing's Gold medal 1973), Am. Fedn. Clin. Research, AAAS, Tex. Acad. Internal Medicine (gov. 1971-73, v.p. 1973-74, pres. 1976-77), N.Y. Acad. Scis., Tex. Club Cardiology (pres. 1972), Galveston County (pres. 1971), Tex. (del. 1972-80) med. assns., Am. Heart Assn. (pres. Tex. affiliate 1979-80), Phi Beta Kappa, Sigma Xi, Alpha Omega Alpha, Phi Eta Sigma, Mu Delta, Phi Rho Sigma. Contbr. articles to profl. jours. Home: 500 N Spruce St Hammond LA 70401-2549

HELANDER, CLIFFORD JOHN, state agency administrator; b. Lake Forest, Ill., July 7, 1948; s. Orvo Axel and Thelma Viola (Kaski) H.; m. Lesley Kelton Fairman, Aug. 3, 1988. B in Sociology, Occidental Coll., 1970; MPA, U. So. Calif., 1979. Rsch. analyst dept. motor vehicles State of Calif., Sacramento, 1981-87, rsch. mgr., 1987—; com. rep. Gov.'s Policy Coun. on Drug and Alcohol Abuse, Sacramento, 1990-92; DDP adv. com. Dept. Alcohol and Drug Programs, Sacramento, 1988-92; com. rep. Gov.'s Adv. Coun. on Alcohol, Drugs and Traffic Safety, Sacramento, 1982-83. Contbr. articles to Jour. Safety Rsch.; author: Development of a California DUI Management Information System, 1989, An Evaluation of the California Habitual Traffic Offender Law, 1986, The California DUI Countermeasure System Evaluation, 1986, (with others) Annual Report of the California DUI Management Information System, 1992. Fellow U. So. Calif., 1977-79. Mem. Sacramento Statis. Assn. Democrat. Achievements include research in drunk driving and traffic safety. Home: 1451 Joby Ln Sacramento CA 95864 Office: California Dept Motor Vehic 2415 1st Ave Sacramento CA 95818

HELD, COLBERT COLGATE, retired diplomat; b. Stamford, Tex., Sept. 3, 1917; s. John Adolf and Annie (Hardie) H.; m. Mildred McDonald, Nov. 23, 1940; children: Melinda Ann Brunger, Joanne Dee Cummings. BA, Baylor U., 1938; MA, Northwestern U., 1940; PhD, Clark U., 1949. Asst. prof. Miss. Coll., Clinton, 1939-40, Tarkio (Mo.) Coll., 1940-42; prof. West Tex. State U., Canyon, 1949-50, U. Nebr., Lincoln, 1950-56, 67-69; fgn. svc. officer U.S. Dept. State, 1957-67, 69-75; diplomat-in-residence Baylor U., Waco, Tex., 1978—. Co-author: Europe, 2d edit., 1952; assoc. author: World Political Geography, 2d edit., 1957; author: Middle East Patterns, 1989; contbr. articles to profl. publs. Lt. col. USAF, 1942-46. Mem. Assn. Am. Geographers. Democrat. Baptist. Home: 4800 Crestwood Dr Waco TX 76710-1702 Office: Baylor U Polit Sci Waco TX 76798-7276

HELD, GEORGE ANTHONY, architect; b. Paterson, N.J., Sept. 4, 1949; s. George William and Carmella (De Negri) H.; m. Patricia Anne Corrado, Sept. 5, 1976; children: Nicole, Ryan. BS in Archtl. Tech., N.Y. Inst. Tech., 1972, BArch, 1977. Registered profl. architect, N.J., N.Y. Architect intern Gerard J. Oakley, AIA, Teaneck, N.J., 1972-76; ptnr. Aybar Partership, Ridgefield, N.J., 1976-88; owner George A. Held, AIA & Assocs., Clifton, N.J., 1988—. Mem. AIA, N.J. Soc. Architects, Architects League N.J. (sec. 1986-88, Dirs. award 1985, 88, Vegliante award 1986). Roman Catholic. Home: 47 Westview Rd Wayne NJ 07470 Office: George A Held AIA & Assocs 457 Crooks Ave Clifton NJ 07011

HELD, PAUL G., molecular biologist; b. Hornell, N.Y., Apr. 7, 1960; s. John Edward and Nadine Nancy (Woodard) H.; m. Susan Lynn Madziarz, Aug. 20, 1983; children: Jason Edward, Nora Elizabeth, Sara Evelyn. BS, Albany Coll. of Pharmacy, 1983; MS, Albany Med. Ctr., 1990, PhD, 1990. Registered pharmacist, Utah. Postdoctoral fellow in molecular biology U. Vt., Burlington, 1989—. Mem. AAAS, Sigma Xi. Home: 130 Cross Pkwy Burlington VT 05401 Office: Univ of Burling Med Alumni Bldg Burlington VT 05405

HELD, WILLIAM JAMES, civil engineer; b. Pitts., Mar. 18, 1944; s. John Matthew and Gladys Ann (Carlisle) H.; m. Rosemary Talak, May 27, 1967; children: Eric John, Adam Paul. BCE, U. Dayton, 1967; MCE, U. Pitts., 1979. Registered profl. engr., Pa.; registered land surveyor, Pa. Supervising engr. and environ. awareness instr. Duquesne Light Co., Pitts., 1967—. Pres. Seneca Valley Quarterback Club, Harmony, Pa., 1991-93; folk group dir. St. Matthias Ch., Evans City, Pa., 1979—. Recipient Appreciation award Elec. Power Rsch. Inst., 1990. Democrat. Roman Catholic. Office: Duquesne Light Co One Oxford Ctr 301 Grant St Pittsburgh PA 15279

HELDENBRAND, DAVID WILLIAM, civil engineer; b. Denver, Aug. 8, 1950; s. Willard Bruce and V. Leone (Hollertz) H.; m. Andrée Elizabeth Norman, Nov. 1, 1975 (dec. Oct. 1989); 1 child, Katie; m. Judith Ann Montgomery, June 8, 1991; children: Jacque, Katie, Rhett. BS in Environ. Biology, U. Colo., 1972, BS in Civil Engring., 1978. Registered profl. engr., Tex. Engr. Zorich-Erker Engring., Denver, 1977-79, Charles R. Haile, Houston, 1979; dist. engr. United Tex. Transmission Co., Houston, 1980-85; prin. engr. CH & A Corp., Kingwood, Tex., 1985-93; pres. Peak Engring., 1993—. Mem. vestry Ch. of Good Shepherd-Episcopal, Kingwood, 1985-90. Mem. AAAS, ASTM, Assn. Groundwater Scientists and Engrs., Nat. Soc. Profl. Engrs., Am. Gas Assn., Nat. Fire Protection Assn. Home: 2214 Lakeville Dr Kingwood TX 77339

HELDMAN, DENNIS RAY, engineering educator; b. Findlay, Ohio, June 12, 1938; s. Merritt L. and Lavonne (Smith) H.; m. Joyce M. Anspach, Dec. 21, 1956 (div. 1989); children: Cynthia Ann, Candace Lee, Craig Stanton; m. Louise A. Campbell, July 1, 1990. BS in Dairy Tech., Ohio State U., 1960, MS in Dairy Tech., 1962; PhDAE, Mich. State U., 1965. Rsch. assoc., instr. Mich. State U., Ea. Lansing, 1962-66, asst. prof., 1966-69, assoc. prof., 1969-71, prof., 1971-84, ACE fellow in acad. adminstrn., 1974-75, chmn. dept., 1975-79; v.p. process R & D Campbell Inst. for Rsch. and Tech., Camden, N.J., 1984-86; exec. v.p. sci. affairs Nat. Food Processors Assn., Washington, 1986-91; prin. Weinberg Consulting Group Inc, Washington, 1991-92; prof. food processing engring. U. Mo., Columbia, 1992—; mem. agrl. sci. adv. bd. Ohio State U., Columbus, 1984—. Author: Food Process Engineering, 1975 (co-author) 2d edit., 1981; co-author: Introduction to Food Engineering, 1984; co-editor Jour. of Food Process Engineering, 1975—, 2d edit. 1993; Handbook of Food Engineering, 1992; cons. editor McGraw-Hill Ency. Sci. and Tech., 1981-91. Recipient Disting. Alumni award Ohio State U. Coll. Agr., 1978. Fellow Am. Soc. Agrl. Engrs. (bd. dirs. 1974-78, Paper award 1966, 68, Food Engring. award 1981), Inst. Food Technologists (mem. exec. com. 1990); mem. Am. Assn. Cereal Chemists, Am. Inst. Chem. Engrs. Office: U Mo 253 Agri Engring Bldg Columbia MO 65211

HELFRICK, ALBERT DARLINGTON, electronics engineering educator, consultant; b. Camden, N.J., June 10, 1945; s. Eugene G. and Irma (Darlington) H.; m. Toni Venezia, May 6, 1989; children: A. Karl, Rachel. BS, Upsala Coll., East Orange, N.J., 1969; MS, N.J. Inst. Tech., 1973; PhD, Clayton (Mo.) U., 1988. Registered profl. engr., N.J. Sr. rsch. engr. Singer-Kearfott Div., Little Falls, N.J., 1969-72; sr. engr. Kay Elemetrics, Pine Brook, N.J., 1972-77; sr. project engr. Cessna Aircraft, Boonton, N.J., 1977-84; prin. engr. RFL Industries, Boonton, 1984-89; cons. engr. Boonton, 1989-92; prof. electronics engring. Embry-Riddle Aero. U., Daytona Beach, Fla., 1992—; com. mem. Radio Tech. Commn. for Aeros., Washington, 1980-85; mem. adj. faculty Upsala Coll., 1972-73, Kean Coll. N.J., 1979-81, Fairleigh Dickinson U., 1986-87. Author: Practical Repair and Maintenance of Communications Equipment, 1983, Modern Aviation Electronics, 1984, Electronic Instrumentation and Measurement Techniques, 1985, Modern Electronic Instrumentation and Measurement Techniques, 1990, Electrical Spectrum and Network Analyzers, 1991; also other 40 articles. Sgt. U.S. Army, 1969-71, Vietnam. Recipient award RF Design mag., 1988. Fellow Radio Club Am. (bd. dirs. 1989-90, sec. 1990-91); mem. IEEE (sr.). Achievements include patents in magnetic recording tape erasure, a method of frequency synthesis, antenna coupling device. Home: 2925 Betty Dr De Land FL 32720 Office: Embry-Riddle Aero U 600 Clyde Morris Blvd Daytona Beach FL 32114

HELGASON, SIGURDUR, mathematician, educator; b. Akureyri, Iceland, Sept. 30, 1927; came to U.S., 1952; d. Helgi and Kara (Briem) Skulason; m. Artie Gianopulos, June 9, 1957; children: Thor Helgi, Anna Loa. Student, U. Iceland, 1946, D honoris causa, 1986; MS, U. Copenhagen, 1952, D honoris causa, 1988; PhD, Princeton U., 1954. C.L.E. Moore instr. MIT, Cambridge, 1954-56, asst. prof. math., 1960-61, assoc. prof. math., 1961-65, prof. math., 1965—; lectr. Princeton (N.J.) U., 1956-57; Louis Block asst. prof. math. U. Chgo., 1957-59; vis. mem. Inst. Advanced Study, Princeton, 1964-66, 74-75, 83-84. Author: (books) Differential Geometry, Lie Groups and Symmetric Spaces, 1978, Groups and Geometric Analysis, 1984, others; editor Progress in Math., 1980-86, Perspectives in Math. Academic Press, Cambridge, 1985—; contbr. articles to profl. jours. Decorated Major Knight's Cross of Icelandic Falcon, 1991; recipient Jessen diploma Danish Math. Soc., 1982, Gold medal U. Copenhagen, 1951; Guggenheim fellow, 1964-65. Mem. Am. Acad. Arts and Scis., Royal Danish Acad. Scis. and Letters, Icelandic Acad. Scis., Am. Math. Soc. (Steele prize 1988). Avocations: music, photography. Office: MIT Dept Math 77 Massachusetts Ave Cambridge MA 02139-4307

HELGESON, JOHN PAUL, physiologist, researcher; b. Barberton, Ohio, July 25, 1935; s. Earl Adrian and Marguerite (Dutcher) H.; m. Sarah Frances Slater, June 10, 1957; children: Daniel, Susan, James. AB, Oberlin Coll., 1957; PhD, U. Wis., 1964. NSF postdoctoral fellow Dept. of Chemistry, U. Ill., Urbana, 1964-66; from asst. to prof. botany and plant pathology U. Wis., Madison, 1966—; plant physiologist USDA Agrl. Rsch. Svc. plant disease resistance unit, Madison, 1966-90, rsch. leader, 1990—; program dir. USDA, Washington, 1982-83; vis. scientist Lab. of Cell. Biologists, Versailles, France, 1985-86. Lt. USAF, 1957-60. Mem. AAAS, Internat. Soc. Plant Molecular Biologists, Am. Soc. Plant Pysiologists. Achievements include development of tissue culture procedures for studying interactions of plants and fungi, of somatic hybridizations to obtain new disease resistances in plants. Office: USDA Plant Disease Res Rsch Ctr Univ of Wisconsin Madison Dept of Plant Pathology Madison WI 53706

HELINGER, MICHAEL GREEN, mathematics educator; b. Syracuse, N.Y., Feb. 5, 1947; s. Harley George and Marion Irene (Green) H.; m. Susan Jessie McRae, Apr. 13, 1974 (div. Feb. 1987). BS with distinction, Clarkson U., Potsdam, N.Y., 1968; MS, Rensselaer Poly. Inst., Troy, N.Y., 1969. Instr. Clinton Community Coll., Plattsburgh, N.Y., 1969-73, asst. prof., 1974-86, assoc. prof. math., 1987—; life ins. agt. William LaCount Assocs., Franklin United Life, Plattsburgh, 1980-84; owner, mgr. Sue's Beauty Salon, Plattsburgh, 1984; pres., treas. Helinger Rentals, Inc., Plattsburgh, 1976—; owner B&M Firewood Co., 1990-92; mem. acad. affairs com. Clinton C.C., 1970-72, 75—, chmn. 1971-72, 86-90, 92—, gen. edn. com., 1990-92, chmn., 1990-92. Vol. Clinton Correctional Facility, Dannemora, N.Y., 1971-75, 88—; apptd. ministerial servant Jehovah's Witnesses, 1991—. Recipient Cert. of Appreciation, Clinton Correctional Facility, Dannemora, 1989, 92. Mem. Math. Assn. Am., N.Y. State Math. Assn. Two-Yr. Colls. (cert. appreciation 1993), Math. League (team founder, coach 1984—, mem. state com. for Math. League Exam 1992—), Ski Club (advisor 1985—), Pi Mu Epsilon. Avocations: collecting art, antiques, coins, skiing, tennis. Home: 20 Riley Rd Peru NY 12972-9419 Office: Clinton C C Plattsburgh NY 12901

HELINSKI, DONALD RAYMOND, biologist, educator; b. Balt., July 7, 1933; s. George L. and Marie M. (Naparstek) H.; m. Patricia G. Doherty, Mar. 4, 1962; children—Matthew T., Maureen G. B.S., U. Md., 1954; Ph.D. in Biochemistry, Western Res. U., 1960; postdoctoral fellow, Stanford U., 1960-62. Asst. prof. Princeton (N.J.) U., 1962-65; mem. faculty U. Calif., San Diego, 1965—; prof. biology U. Calif., 1970—, chmn. dept., 1979-81, dir. Ctr. for Molecular Genetics, 1984—; mem. com. guidelines for recombinant DNA research NIH, 1975-78. Author papers in field. Mem. Am. Soc. Biol. Chemists, Am. Soc. Microbiology, AAAS, Am. Acad. of Arts and Scis., Nat. Acad. Scis., Genetics Soc. Office: U Calif Ctr for Molecular Gen 9500 Gilman Dr San Diego CA 92093

HELLAND, DOUGLAS ROLF, intergovernmental organization computer executive; b. St. Paul, Nov. 14, 1945; arrived in Switzerland, 1971; s. Erling Olaf Johan and Thordis (Tanner) H.; m. Gertrud Margarete Hahnen, July 3, 1980. ScB in Applied Math., Brown U., 1967. Programmer UN Secretariat, N.Y.C., 1967-71; systems programmer UN Internat. Computing Ctr., Geneva, 1971-74, chief tech. support, 1974-76, chief devel., 1976-80, chief network svcs., 1980—, dep. to dir., 1982—. Mem. Assn. for Computing Machinery, IEEE Computer Soc. Presbyterian. Avocation: mountain hiking. Home: Chemin du Lin 9, CH 1292 Chambésy Switzerland Office: UN Internat Computing Ctr, Palais des Nations, CH 1211 Geneva 10, Switzerland

HELLEN, PAUL ERIC, electrical engineer; b. Mpls., Sept. 16, 1955; s. Osmond Joseph and Lillian Jane (Mueske) H.; m. Carol Ann Yoder, July 9, 1983; children: Eric M., Kathryn J., Michael P. BSEE, Old Dominion U., 1986. Quality assurance lead auditor No. States Power, Mpls., 1986-88; elec. engr. No. States Power, Red Wing, Minn., 1988—. With USN, 1977-83. Mem. IEEE (com. mem. for battery standards 1989—). Office: No States Power 1717 Wakonade Dr E Welch MN 55089

HELLENBRECHT, EDWARD PAUL, mechanical engineer; b. N.Y.C., Jan. 3, 1942; s. Edward M. and Monica A. (Murray) H.; m. Shannon L. Jensen, June 17, 1962 (div. 1971); 1 child, Rhona S.; m. Renée Beaulieu, June 9, 1979; children: Michael R., Renée A. BSME, U. New Haven, 1965; MS in Mgmt., Rensselaer Polytech. Inst., 1970. Dep. program mgr. Gen. Dynamics, Groton, Conn., 1962-77; chief marine engr. Seatrain Shipbuilding Corp., Bklyn., 1977-78; process engr. mgr. Chesebrough-Ponds Inc., Clinton, Conn., 1978-87; engring. mgr. Ragu Foods Co., Trumbull, Conn., 1987-92; dir. engring. Lancome Mfg. Div. Cosmair Inc., Piscataway, N.J., 1992—. Rep. Piscataway C. of C., 1992—. Mem. Jr. Chamber Internat. Senate. Office: Cosmair Inc 81 New England Ave Piscataway NJ 08854

HELLER, AUSTIN NORMAN, chemical and environmental engineer; b. Elizabeth, N.J., Aug. 18, 1914; s. Samuel Sidney and Bessie (Rosenfield) H.; m. Frances Sandler, Mar. 21, 1943; children: Richard David, Susan Starr. AB in Chemistry, Johns Hopkins U., Balt., 1938; MS, Iowa State U., 1941. Diplomate Am. Acad. Environ. Engrs.; registered profl. engr., N.Y. Rsch. asst. environ. sci. Rutgers U., New Brunswick, N.J., 1935-38, 39; chemist, bacteriologist Wallace and Tiernan Co., Belleville, N.J., 1942; rsch. assoc. dept. civil engring. N.Y.U. Coll. Engring., N.Y.C., 1946-48; supr. indsl. waste devel. sect., coord. long range planning Allied Chem. Corp., N.Y.C., 1948-61; dep. chief tech. assistance br. Air Pollution div. USPHS, Cin., 1961-66; cons. E.F. Drew and Co., Boonton, N.J., 1946-48; U.S. del. OECD, Sci. Div., Air Pollution Rsch. Survey Techniques Group, Paris, 1962-66, Surgeon Gen., Belgian Govt., 1965, Royal Commn. for Air Purification, Govt. Sweden, 1965; pres. Austin N. Heller, Inc., Annapolis, Md., 1977-88, cons. 1991-92; environ. adv. com. Fed. Energy Adminstrn., Washington, 1973-75; adj. prof. environ. engring. Cooper Union Coll., 1966-67; mem. adv. coun. dept. chem. engring. Princeton (N.J.) U., 1967-70; adj. assoc. prof. environ. Columbia U.,. Contbr. articles to profl. jours. Trustee Engirng. Index, Inc., N.Y.C., 1969-72; expert testimony Pres. Nixon's Adv. Bd. on Water Pollution and Ocean Dumping, Washington, 1974; commr. Dept. Air Resources, N.Y.C., 1966-70; sec. Dept. Natural Resources and Environ. Control, Dover, Del., 1970-73; exec. dir. N.Y. State Coun. Environ. Advisers, N.Y.C., 1973-75; asst. adminstr. conservation U.S. Energy R&D Adminstrn., Washington, 1975-76. Lt. USN, 1942-46, PTO; lt. comdr. USN Rsch. Res., 1948-61. Recipient Cert. of award corp. planning seminar Am. Mgmt. Assn., N.Y.C., 1960, Engring. award ASME, N.Y.C., 1967, 15th Ann. Honor award N.Y. State Soc. Profl. Engrs., N.Y.C., 1968, Humanitarian award Children's Asthma Rsch. Inst. N.Y.C., 1969; Wallace and Tiernan rsch. fellow Iowa State U., Ames, 1940-41. Fellow APHA, Am. Chem. Soc., Am. Inst. Chemists; mem. AAAS, Am. Inst. Chem. Engrs., Am. Water Works Assn. (life), Air Pollution Control Assn. (bd. dirs. 1960-63, 67-70), Fedn. Water Pollution Control Assn. (life), N.Y. Acad. Sci., Masons. Republican. Jewish. Achievements include patents for Cyclic Method for Removal of Impurities from Coke Over Tar by Water Washing, Recovery of Phenolics from Industrial Wastes, Process for Production of High Grade Naphthalene and Preparation of B-Naphthol from Acidic Waters Therefrom, Solvent Dephenolization of Aqueous Solutions; development of the use of process research and development as a primary method to solve industrial waste problems in chemical and allied industries at a profit; first use of a telemetry/computer system to measure, on a continuous basis, the air quality of urban atmospheres. Home and Office: 2675 Claibourne Rd Annapolis MD 21403-4250

HELLER, DONALD FRANKLIN, chemical and laser physicist; b. N.Y.C., Mar. 29, 1947; s. George M. and Florence (Gelb) H.; m. Mary Rose Noberini, May 28, 1972 (div. Sept. 1987); children: Katherine Ann, Elizabeth Rose. BS in Chemistry, U. Calif., Berkeley, 1969; PhD in Chem. Physics, U. Chgo., 1972. Postdoctoral rschr. U. Chgo., 1972-73, U. Calif., Berkeley, 1973-75; project assoc. U. Wis., Madison, 1975-76; vis. prof. McGill U., Montreal, Que., Can., 1976-77; group leader Allied-Signal, Inc., Morristown, N.J., 1977-88; CEO Light Age, Inc., Somerset, N.J., 1989—; conf. chair Interdisciplinary Laser Sci. Conf., Monterey Calif., 1991; mem. spl. study sect. NIH, Washington, 1991-92. Co-editor: Advances in Laser Science-IV, 1989; contbr. over 100 articles to profl. jours. Swift fellow U. Chgo., 1972; NSF rsch. grantee, 1990; recipient award for tunable laser system instrument NASA, 1984-90. Mem. Am. Phys. Soc. (steering com. laser sci. topical group), Am. Chem. Soc., Optical Soc. Am., N.Y. Acad. Scis., Soc. Photoinstrumentation Engring. Achievements include patents for lasers and their applications; invention of diode laser injection seeding; first to demonstrate diode laser pumping of tunable solid-state lasers, to determine femtosecond vibrational relaxation in polyatomic molecules, to demonstrate nonlinear frequency conversion of tunable solid-state lasers. Home: 740 Watchung Rd Bound Brook NJ 08805 Office: Light Age Inc 2 Riverview Dr Somerset NJ 08873

HELLER, JOHN PHILLIP, petroleum scientist, educator; b. N.Y.C., Apr. 15, 1923; s. Edward and Anna (Lang) H.; m. Janet Sterling, Sept. 13, 1946; children: William Edward, Richard Vincent, Ruth Ellen. BS, Queens Coll., 1944; PhD, Iowa State U., 1953. Sr. rsch. physicist Magnolia Petroleum Co. (Mobil), Dallas, 1953-60; rsch. assoc. Mobil Oil Co., Dallas, 1960-79; sr. scientist Petroleum Recovery Rsch. Ctr. N.Mex. Tech., Socorro, 1979—. Co-author: (chpt.) Reservoir Characterization II, 1990, Foams in Petroleum Industry, 1992; contbr. articles to profl. jours. Sgt. U.S. Army, 1944-46, ETO. Recipient EOR Pioneer award Soc. Petroleum Engrs., 1988. Mem. AIChE, Soc. Petroleum Engrs., Am. Phys. Soc., Am. Chem. Soc., Sigma Xi (pres. N.Mex. tech. chpt. 1992-93). Home: 509 Mesa Loop Socorro NM 87801 Office: PRRC-NMex Tech Socorro NM 87801

HELLER, KENNETH JEFFREY, physicist; b. Port of Spain, Trinidad, Nov. 7, 1943; s. George M. and Florence (Gelb) H.; m. Patricia Margaret Autry, Sept. 29, 1972. BA, U. Calif., Berkeley, 1965; PhD, U. Wash., 1973. Physicist Naval Rsch. Lab., Corona, Calif., 1965; tchr. U.S. Peace Corps, Nigeria and Kenya, 1966-68; rsch. asst., teaching asst. U. Wash., Seattle, 1968-73; rsch. assoc., lectr. U. Mich., Ann Arbor, 1973-78; asst. prof. U. Minn., Mpls., 1978-82, assoc. prof., 1982-86, prof., 1987—; mem. users exec. com. Fermilab, Batavia, Ill., 1984-86, bd. dirs., 1988-92; trustee Univs. Rsch. Assn., 1985-88. Editor: High Energy Spin Physics, 1988; contbr. articles to profl. jours. Mem. Am. Phys. Soc., Symposium of High Energy Spin Physics (internat. adv. com. 1988—). Achievements include discovery of large polarization in high energy particle production technique for the precise measurement of hyperon magnetic moments; application of the quark model to understand the mechanism for polarized particle production technique of spin transfer to high energy hyperons. Office: U Minn Sch Physics & Astronomy Minneapolis MN 55455

HELLERMAN, LEO, retired computer scientist and mathematician; b. Bklyn., Feb. 8, 1924; s. Azriel and Rebecca (Hellerman) H.; children: David Seth, Lisa Beatrice Hellerman Kopchik, Daniel Asa. BEE, CCNY, 1946; PhD, Yale U., 1958. Patent examiner U.S. Patent Office, Washington, 1948-50; engr. IBM, Poughkeepsie, N.Y., 1956-73, 75-82, Böblingen, Fed. Republic Germany, 1974, Kingston, N.Y., 1983-87; ret., 1988; lectr. Fachtagung Struktur und Betrieb von Rechensystemen, Braunschweig, Fed. Republic Germany, 1974. Contbr. articles to profl. jours. Mem. Am. Math. Soc., Math. Assn. Am., Sigma Xi. Achievements include patent for logic performing device; development of first algebraic symbol manipulation program, of first statistical design of electronic circuits; a theory of computational work, of moment free methods for processing discrete distributions. Home: 1 Feller Rd Rhinebeck NY 12572-2307

HELLMAN, HARRIET LOUISE, pediatric nurse practitioner; b. Rochester, N.Y., Feb. 11, 1950; d. W. Frank Jr. and Louise B. Fowler; m. Claude Bourgoin, Aug. 30, 1971 (div. Dec. 1981); m. Samuel V. Hellman, Feb. 20, 1983; children: Zachary, Alyssa. BSN, U. Maine, 1972, postgrad., 1973-74. Cert. pediatric nurse practitioner 1974. Pub. health nurse Portland (Maine) City Health Dept., 1972-73, pediatric nurse practitioner, 1973-78; maternal child health unit tchr. Maine Med. Ctr., Portland, 1978; dir. adolscent medicine, pediatric nurse practitioner USPHS Nena Health Ctr., N.Y.C., 1978-83; dir. new family program Maternity Ctr. Assn., N.Y.C., 1990-91; pediatric nurse practitioner East End Pediatrics, East Hampton, N.Y., 1991—; Allied health profl. Southampton (N.Y.) Hosp., 1993—; supr. City Coll. Med. Sch., N.Y.C., 1979-83, Pace U., N.Y.C., 1981-83; chmn. health com. Mayor's Task Force on Adolscent Pregnancy, N.Y.C., 1982-83; cons. Head Start Spl. Svcs. for Children, N.Y.C., 1983; treas. adv. bd. Ctrl. Park West Nursery, N.Y.C., 1990. Mem. NAt. Assn. Pediatric Nurse Practitioners & Assocs., N.Y. State Coalition Nurse Practitioners, N.Y. Acad. Scis. Achievements include certified expert witness child abuse diagnosis. Office: East End Pediatrics 94 Pantigo Rd East Hampton NY 11937

HELLMUTH, GEORGE FRANCIS, architect; b. St. Louis, Oct. 5, 1907; s. George W. and Harriet M. (Fowler) H.; m. Mildred Lee Henning, May 24, 1941; children: George William, Nicholas Matthew, Mary Cleveland, Theodore Henning, Daniel Fox. B.Arch., Washington U., 1928, M.Arch., 1930; Steedman traveling fellow, 1931; diploma, Ecole des Beaux Arts, Fontainebleau, France. Founder Hellmuth, Yamasaki & Leinweber, 1949-55, Hellmuth, Obata & Kassabaum, 1955-78, HOK Internat., Inc., 1977-86; numerous offices including, St. Louis, N.Y.C., San Francisco, Dallas, Washington, Tokyo, Japan, Kuwait City, Kuwait, Kansas City, Mo., Tampa, Fla., Los Angeles, Hong Kong and London; pres. Bald Eagle Co., Gladden, Mo.; Chmn. St. Louis Landmarks and Urban Design Commn., 1950-70. Prin. archtl. works include: King Saud U., Riyadh, Saudi Arabia; King Khaled Internat. Airport, Riyadh; (outside U.S.) Nile Tower, Cairo, Egypt, U. West Indies, Trinidad, Spanish Honduras secondary sch. system, Am. Embassy, El Salvador, Am. embassy housing, Cairo, Canadian medium and maximum prisons, Taipei World Trade Ctr., Taiwan, Housing for Royal Saudi Naval Forces, Saudi Arabia, Military Secondary Schools, Saudi Arabia, Air Def. Command Hdqtrs. Complex, Saudi Arabia, Burgan Bank Hdtrs., Kuwait, Asoka Dev., Kuala Lumpur, Chesterton Retail Mall, U.K.; prin. archtl. works include: (U.S.) Nat. Air and Space Mus., Washington, Marion Fed. Maximum Security Prison, (Ill.), IBM Advanced systems Lab, Los Gatos, Calif., Dallas/Ft. Worth Regional Airport, U. Wis. Med. Center, Madison; Internat. Rivercenter, New Orleans; SUNY Health Scis. Complex, Buffalo, The Galleria/Post Oak Center, Houston, E.R. Squibb Co, Lawrenceville, N.J., McDonnell Planetarium, St. Louis, Dow Research and Devel. Facility, Indpls.; Commonwealth P.R. Penal System; Duke U. Med. Center, Durham, N.C.; Lubbock Regional Airport, (Tex.), Lambert-St. Louis Internat. Airport, St. Louis, D.C. Courthouse, St. Louis U. Sch. Nursing, No. Ill. U. Library, Mobil Oil Hdqtrs., Fairfax, Va., Cities Service Research Ctr., Tulsa, Marriott Corp. Hdqtrs., Bethesda, Md., McDonnell Douglas Automation Ctr., St. Louis, Moscone Conv. Ctr., San Francisco, Piers 1, 2, 3., Boston, Clark County Dentention Ctr., Las Vegas, Nev., Pillsbury Research and Devel. Facility, Mpls., Saturn Automotive Facility, Tenn., Burger King World Hdqrs., Miami, Exxon Research and Egrning. Ctr., Clinton, N.J., Incarnate Word Hosp., St. Louis, Kellogg Co. Hdqrs., Battle Creek, Mich., Fleet Ctr., Providence, Phillips Point, West Palm Beach, Fla., Sohio Corp. Hdqrs., Cleve., Lincoln Tower, Miami, 2000 Pennsylvania Ave., Washington, Providence Park, Fairfax, Va., Tower One, Houston, Griffin Tower, Dallas, ARCO Tower, Denver, Levi's Plaza, San Francisco, Southwestern Bell Telephone Hdqrs., St. Louis, Met. Life Bldg., St. Louis, Burger King Hdqrs., Miami, Fla., Saturn Automotive Facility, Tenn., Living World Edn. Ctr., St. Louis Zoo, BP Hdqrs, London, U. Ala.-Birmingham Hosp., Univ. Hosp. at St. Louis U., Mo., Saint Louis Galleria Expansion, 801 Grand Office Bldg., Des Moines, Moore Bus. Forms Hdqrs., Lake Forest, Ill., many other indsl. and bus. corporate hdqrs., research centers. Recipient First Honor award AIA, 1956; knight Sovereign Mil. Order Malta in U.S.A. Fellow AIA. Home: 5 Conway Ln Saint Louis MO 63124-1279 Office: 1831 Chestnut St Saint Louis MO 63103-2220

HELLWIG, HELMUT WILHELM, air force research director; b. Berlin, May 7, 1938; came to U.S., 1966; s. Adolf H.W.M. and Walburga (Hieber) H.; m. Thekla Maria Polzin, Dec. 28, 1960; children: Frank G., Peter M. MS, Tech. U. Berlin, 1963, PhD, 1966, hon. doctorate, U. Besancon, 1988. Research physicist Heinrich Hertz Inst., Berlin, 1963-66, U.S. Army Electronics Command, Ft. Monmouth, N.J., 1966-69; research physicist Nat. Bur. Standards, Boulder, Colo., 1969-74, sect. chief, 1974-79; pres. Frequency and Time Systems, Inc., Beverly, Mass., 1979-86; assoc. dir. Nat. Bur. Standards (now Nat. Inst. Standards and Tech.), Gaithersburg, Md., 1986-90; dir. Nat. Measurement Lab. Nat. Bur. Standards, Gaithersburd, 1987-88; dir. Air Force Office of Sci. Rsch., Bolling AFB, Wash., 1990—. Patentee in field. Contbr. articles to profl. jours. Recipient Sci. award Dept. Def., 1968, IR-100 award, 1976, Condon award Dept. Commerce, 1979. Fellow IEEE; mem. Am. Phys. Soc., Internat. Union Radio Sci. (chmn. Com. A, 1984), Sigma Xi. Republican. Office: Nat Inst Standards & Tech Adminstrn Bldg A526 Gaithersburg MD 20899

HELM, CHARLES GEORGE, entomologist, researcher; b. East St. Louis, Ill., Sept. 11, 1949; s. Walter Kermit and Dolores Bertha (Perschbacher) H.; m. Patricia Ann Fendley, Mar. 7, 1970. BS magna cum laude, Ea. Ill. U., 1971; MS, U. Ill., 1973. Tech. asst. Ill. Natural History Survey, Champaign, 1973-74, jr. profl. scientist, 1974-78, asst. supportive scientist, 1978-83, assoc. supportive scientist, 1983-89, rsch. biologist, 1989—; sr. rsch. specialist in agriculture U. Ill., Champaign, 1991—; state survey coord. Coop. Agrl. Pest Survey, Champaign, 1991—; chmn. Regional Rsch. Project S-219, Champaign, 1992. Author: (chpt.) Sampling Leafhoppers in Soybeans, 1980, Natural Enemies and Host Plants of Heliothis, 1989; co-editor: World Bibliography of Soybean Entomology, 1988; contbr. articles to profl. jours.

Mem. So. Poverty Law Ctr. Grantee USDA, 1987, 90, 91, 92, Am. Soybean Assn., 1987; U. Ill. Grad. fellow, 1971-72; recipient Oberly award for Excellence in Agrl. and Related Scis., 1989. Mem. Entomological Soc. Am., Nature of Ill. Found., Nature Conservancy. Achievements include development of maturity group III soybean lines with resistance to insect feeding, evaluation of soybean germplasm for insect resistance, effects of weed competitive and insect defoliation on soybean yield. Office: Ill Natural History Survey 172 NRB 607 E Peabody Champaign IL 61820

HELM, JOHN LESLIE, mechanical engineer, company executive; b. Red Wing, C Apr. 10, 1921; s. Leslie Cornell and Dora (McGuigan) H.; B.S. in Mech. Engring., Columbia U., 1943, M.S., 1944; postgrad. in Nuclear Engring., U. Conn., 1956-57; m. Nancy Ellen Molle, May 15, 1954; children—John Leslie, Juli-Ann, Catherine Marie. Asst. in mech. engring. Columbia U., 1943-44; process engr. Metals Disintegrating Co., Elizabeth, N.J., 1944-45; project engr. Aero Manuscripts Inc., 1945-46; staff engr., central engring. dept. Gen. Foods Corp., White Plains, N.Y., 1946-52; with Gen. Dynamics Corp., Groton, Conn., 1952-74, spl. tech. asst. Office of Pres., Electric Boat div., 1965-72, gen. mgr. Gen. Dynamics Energy Systems, 1972-74; founder, pres., chief exec. officer Proto-Power Mgmt. Corp., Groton, 1974-82, also dir.; pres. chief exec. officer Proto-Power Corp., subs. of Kollmorgen Corp., 1982-89; dir. Electronic Assocs., Inc., West Long Branch, N.J., 1983-89; founder, pres., CEO Transplex Inc., 1990—. Mem. Groton Bd. Edn., 1967-77, chmn., 1971-72. Recipient citation for work on Manhattan Project, War Dept., 1945; registered profl. engr., N.Y. Mem. ASME. Republican. Roman Catholic. Clubs: Shonnecosset Yacht, Off Soundings; Princeton, Columbia (N.Y.C.), N.Y. Yacht. Home: 116 Tyler Ave Groton CT 06340-5923 Office: 591 Poquonnock Rd Groton CT 06340

HELMES, LESLIE SCOTT, architect; b. Fort Snelling, Minn., Oct. 27, 1945; s. Leslie Charles and Marilyn (Tomlinson) H.; m. Julie Williams, Sept. 15, 1967 (div. Dec. 4, 1974). BArch, U. Minn., 1968. Registered architect. Instr. U. Minn., Mpls., 1969-73; program coord. Mpls. Inst. Arts, 1970-74; dir. edn. programs Minn. Hist. Soc., St. Paul, 1974-77; dir. phys. planning Freerks Sperl Flynn Architects, St. Paul, 1977-80, Smiley Glotter Assocs., Mpls., 1980-82, Park Nicollet Med. Ctr., Mpls., 1982-85; sole proprietor Helmes Architects, Mpls., 1985-86; v.p. Skaaden Helmes Architects, Mpls., 1986-91; bd. dirs. Friends of Gillette Children's Hosp., St. Paul; bus. adv. coun. CAD/Mpls. Rehab. Ctr., 1990-92. Author: Metapoems, 1989, (poem) Kaldron, 1990; contbr. articles to profl. jours. Bd. dirs., founder Frosty Sch. for Mentally Retarded/Autistic/Down Syndrome, Mpls., 1973—; master coach Internat. Spl. Olympics, Washington, 1988-91, decorations commn., 1990-91. Recipient 1st Pl. award Gamut Mag., 1982. Mem. Am. Inst. Architects, Inland Lake Yachting Assn., Minn. Soc. Am. Inst. Architects, Profl. Ski Instrs. Am. Avocations: antique rubber stamp collector, sailing, reading. Home: 862 Tuscarora Ave Saint Paul MN 55102-3706 Office: Skaaden Helmes Architects 401 N 3d St Ste 100 Minneapolis MN 55401-1334

HELMHOLZ, AUGUST CARL, physicist, educator emeritus; b. Evanston, Ill., May 24, 1915; s. Henry F. and Isabel G. (Lindsay) H.; m. Elizabeth J. Little, July 30, 1938; children: Charlotte C.K. Colby, George L., Frederic V., Edith H. Roth. A.B., Harvard Coll., 1936; student, Cambridge U., 1936-37; Ph.D., U. Calif., Berkeley, 1940; Sc.D. (hon.), U. Strathclyde, 1979. Instr. physics U. Calif.-Berkeley, 1940-43, asst. prof., 1943-48, assoc. prof., 1948-51, prof., 1951-80, emeritus, 1980—, chmn. dept., 1955-62; rsch. physicist Lawrence Berkeley Lab., 1940—; mem. Vis. Scientist Program, 1966-71; governing bd. Am. Inst. Physics, 1964-67. Recipient Citation U. Calif., Berkeley, 1980; Berkeley fellow, 1988—, Guggenheim fellow, 1962-63. Fellow Am. Phys. Soc.; mem. AAAS, Am. Assn. Physics Tchrs., AAUP, Phi Beta Kappa, Sigma Xi. Home: 28 Crest Rd Lafayette CA 94549-3349 Office: U Calif Dept Physics Berkeley CA 94720

HELMLE, RALPH PETER, computer systems developer, manager; b. Detroit, Sept. 12, 1962; s. Ronald and Ingeborg (Kalb) H. BSME, Lawrence Tech. U., 1987; MS in Systems Engring., Oakland U., Rochester, Mich., 1991. Registered profl. engr., Mich. Student designer Kent-Moore Stamping & Fabrication, Detroit, 1978-79; hardware tool and process engr. Fisher Body div. GM Corp., Warren, Mich., 1979-82; indsl. systems engr. Fisher Guide div. GM Corp., Troy, Mich., 1982-85; sr. indsl. systems engr. Inland Fisher Guide div., GMC, Troy, Mich., 1985—. Mem. Lambda Iota Tau. Home: 11584 Adams Dr Warren MI 48093 Office: Inland Fisher Guide div GMC 1401 Crooks Rd Troy MI 48084

HELMS, MARY ANN, critical care nurse, consultant; b. Compton, Calif., Jan. 7, 1935; d. Raymond Whitfield and Amanda Zelpha (Hancock) Spencer; m. Willard Ford Helms, Mar. 15, 1958; children: Michael Steven, Steven Allen. AA in Nursing, El Camino Coll., 1971; BSN, Calif. State U., L.A., 1976; MA in Mgmt., St. Mary's Coll., 1978; MSN, Ariz. State U., 1985; PhD, Columbia Pacific U., 1993. RN; cert. clin. specialist, critical care nurse, quality assurance, CCRN. Med. sec., bookkeeper Palm Springs (Calif.) Med. Clinic, 1956-61; office mgr. William R. Stevens Ins. Agy., Santa Ana, Calif., 1961-63, I.J. Weinrot & Son Ins. Agy., L.A., 1963-67; staff nurse Veteran's Adminstrn. Hosp., 1971; staff nurse Kaiser Found. Hosp., Harbor City, Calif., 1971-76; supr., coord. pediatrics Maricopa County Gen. Hosp., Phoenix, 1976-80; critical care nurse Phoenix Baptist Hosp., 1980-81, critical care mgr., 1981-89, clin. nurse specialist, 1989—, critical care cons., 1986—. Mem. ANA, AAAS, Am. Cons's. League, N.Y. Acad. Sci., Am. Soc. Women Accts., Natural History Mus., Met. Mus. Art, Smithsonian Instn., Phoenix Zoo, Phoenix Art Mus., Cousteau Soc., Calif. State U. Alumni Assn., KAET Public Broadcasting System, Am. Assn. Quality Assurance Profl. Ariz. Nurses Assn., Am. Assn. Legal Nurse Cons., Ariz. State U. Alumni Assn., Hastings Ctr., Phi Kappa Phi, Alpha Gamma Sigma, Sigma Theta Tau. Mem. LDS Ch. Research on effects of noise pollution on physical and mental health of citizenry, phenylketonuria testing in Los Angeles, measurement of attitudes toward children in pediatric nurses, nursing practice, physiological changes with back massage, incidence of prolonged Q-T interval in critically ill patients, assessment of artierial circulation in vascular surgery patients; use of autotransfusion in hip and knee surgery patients; use of pulse oximetry in pre and post operative patients; side effects of patient-controlled analgesia; conf. medication histories in hospitalized patients; correlation of patient medication histories to nurses and physicians. Home: 1007 E Michelle Dr Phoenix AZ 85022-6048 Office: 6025 N 20th Ave Phoenix AZ 85015

HELSER, TERRY LEE, biochemistry educator; b. Indpls., Dec. 23, 1944; s. Lester Freeman and Frances Elizabeth (Arnold) H.; m. Wanda Joan Ralston, Aug. 11, 1968; children: Aron Trent, Janelle Lynn. BA, Manchester Coll., 1967; PhD, U. Wis., 1972. Postdoctoral fellow U. Calif., Irvine, 1972-75; rsch. fellow Brown U., Providence, 1975-77; vis. prof. Hershey (Pa.) Med. Ctr., Penn State U., 1980; asst. prof. Millersville (Pa.) State Coll., 1977-80; NSF fellow SUNY, Albany, 1984; vis. prof. U. N.C., Chapel Hill, 1986-87; assoc. prof. SUNY, Oneonta, 1980—; chmn. Com. on Rsch. of Faculty Senate, Oneonta, 1988-91. Contbr. articles to profl. jours. Head judge Odyssey of the Mind Problems, 1988-93; judge CASSC Sci. Fair, Oneonta, 1988, Sci. Olympiad, Oneonta, 1990. Named Disting. Biologist, Hartwick Co., 1990; grantee W.B. Ford Found., 1989, NSF, 1984, 80, Rsch. Corp., 1982-84. Mem. AAAS, Am. Soc. for Microbiology, Sigma Xi. Democrat. Achievements include devel. of two ednl. card games "The Gene Expression Game" for the lactose operon and "The Elemental Chemistry Game" for introductory chemistry courses, numerous word puzzles. Office: Chemistry Dept SUNY Coll Oneonta NY 13820-4015

HELSLEY, CHARLES EVERETT, geologist, geophysicist; b. Oceanside, Calif., June 24, 1934. BS, Calif. Inst. Tech., 1956, MS, 1957; PhD in Geology, Princeton U., 1960. Asst. prof. geology Calif. Inst. Tech., 1960-62; asst. prof. Case Western Res. U., 1962-63; from asst. to assoc. prof. S.W. ctr. advancement studies U. Tex., Dallas, 1963-69, prof. geoscience, 1969-76, assoc. head geoscience divsn., 1971-72, head geoscience program, dir. inst. geoscience, 1972-75; prof. geology and geophysics U. Hawaii, Manoa, dir. Hawaii inst. geophysics; adj. prof. So. Meth. U., 1963-76, marine sci. inst. U. Tex., Galveston, 1973-76. Mem. Geol. Soc. Am., Am. Geophys. Union. Achievements include research in rock magnetism and paleomagnetism and their implications regarding continental drift, marine geophysics, magnetostratigraphy, geothermal resource exploration. Office: U Hawaii Hawaii Inst Geophysics 2525 Correa Rd Honolulu HI 96822-2285*

HEM, JOHN DAVID, research chemist; b. Starkweather, N.D., May 14, 1916; s. Hans Neilius and Josephine Augusta (Larsen) H.; m. Ruth Evans, Mar. 11, 1945; children: John David Jr., Michael Edward. Student, Minot State Coll., 1932-36, N.D. State U., 1937-38, Iowa State U., 1938; BS, George Washington U., 1940. Analytical chemist U.S. Geol. Survey, Safford, Ariz., 1940-42, 43-45, Roswell, N.Mex., 1942-43; dist. chemist U.S. Geol. Survey, Albuquerque, 1945-53; rsch. chemist U.S. Geol. Survey, Denver, 1953-63, Menlo Park, Calif., 1963—; rsch. advisor U.S. Geol. Survey, 1974-79, mem. water rsch. adv. com., 1984—. Author: Study and Interpretation Chemistry of Natural Water, 3d rev. edit., 1985; contbr. articles to profl. publs., chpts. to books. Recipient Meritorious Svc. award U.S. Dept. Interior, 1976, Disting. Svc. award U.S. Dept. Interior, 1980, Sci. award Nat. Water Well Assn., 1986, O.E. Meinzer award Geol. Soc. Am., 1990, Special award Internat. Assn. Geochemistry and Cosmochemistry, 1992. Mem. Am. Chem. Soc., Am. Geophys. Union, Am. Water Works Assn., Geochem. Soc., Soc. Geochemistry and Health. Democrat. Lutheran. Achievements include wide usage of the results of research in aqueous chemistry of aluminum, manganese and iron. Home: 3349 St Michael Ct Palo Alto CA 94306-3056 Office: US Geol Survey MS 427 345 Middlefield Rd Menlo Park CA 94025-3591

HEMANN, RAYMOND GLENN, aerospace company executive; b. Cleve., Jan 24, 1933; s. Walter Harold Marsha Mae (Colbert) H.; BS, Fla. State U., 1957; postgrad. U.S. Naval Postgrad. Sch., 1963-64, U. Calif. at Los Angeles, 1960-62; MS in Systems Engring., Calif. State U., Fullerton, 1970, MA in Econs., 1972, cert. in tech. mgmt. Calif. Inst. Tech., 1990; m. Lucile Tinnin Turnage, Feb. 1, 1958; children: James Edward, Carolyn Frances; m. Pamela Lehr, Dec. 18, 1987. Aero. engring. aide U.S. Navy, David Taylor Model Basin, Carderock, Md., 1956; analyst Fairchild Aerial Surveys, Tallahassee, 1957; research analyst Fla. Rd. Dept., Tallahassee, 1957-59; chief Autonetics div. N.Am. Rockwell Corp., Anaheim, Calif., 1959-69; v.p., dir. R. E. Manns Co., Wilmington, Calif., 1969-70; mgr. Avionics Design and Analysis Dept. Lockheed-Calif. Co., Burbank, 1970-72, mgr. Advanced Concepts div., 1976-82; gen. mgr. Western div. Arinc Research Corp., Santa Ana, 1972-76; dir. Future Requirements Rockwell Internat., 1982-85; dir. Threat Analysis, Corp. Offices, Rockwell Internat., 1985-89; pres., chief exec. officer Advanced Systems Rsch., Inc., 1989—; adj. sr. fellow Ctr. Strategic and Internat. Studies, Washington, 1987—; cons. various corps. U.S. govt. agys.; sec., bd. dirs. Calif. State U., Fullerton, Econs. Found., 1989—; mem. naval studies bd. panels Nat. Acad. Scis., 1985—, Arms Control Working Group; asst. prof. ops. analysis dept. U.S. Naval Postgrad. Sch., Monterey, Calif., 1963-64, Monterey Peninsula Coll., 1963; instr. ops. analysis Calif. State U., Fullerton, 1963, instr. quantitative methods, 1969-72; program developer, instr. systems engring. indsl. rels. ctr. Calif. Inst. Tech., 1992—; lectr. Brazilian Navy, 1980, U. Calif., Santa Barbara, 1980, Yale U., 1985, Princeton U., 1986, U.S. Naval Postgrad. Sch., 1986, Ministry of Def., Taiwan, Republic of China, 1990; Calif. Inst. Tech., 1992; mem. exec. forum Calif. Inst. Tech., 1991—. Chmn. comdr.'s adv. bd. CAP, Calif. Wing, 1990—; reader Recording for the Blind, 1989—. With AUS, 1950-53. Syde P. Deeb scholar, 1956; recipient honor awards Nat. Assn. Remotely Piloted Vehicles, 1975, 76; named to Hon. Order Ky. Cols., 1985. Comml., glider and pvt. pilot. Fellow AAAS; assoc. fellow AIAA; mem. IEEE, Ops. Rsch. Soc. Am., Air Force Assn., Nat. Coalition for Advanced Mfg. (adv. bd. 1990—), N.Y. Acad. Scis., Assn. Old Crows, Phi Kappa Tau (past pres.). Episcopalian. Contbr. articles to profl. jours. and news media.

HEMMING, BRUCE CLARK, plant pathologist; b. Pocatello, Idaho; s. Parley Lynn and Vernetta (Clark) H.; m. Caroline McDaniel, May 20, 1973; children: Eric M., Heidi, Heather, Crystal Lynn, Keri Lynn. BS in Microbiology, Brigham Young U., Provo, Utah, 1974, MS in Biochemistry, 1977; PhD in Plant Pathology, Mont. State U., Bozeman, 1982. Staff rsch. assoc. dept. chemistry Brigham Young U., 1977-78; sr. rsch. biologist molecular biology Monsanto Co., St. Louis, 1982-84, rsch. specialist plant molecular biology, 1984-89, project leader biocontrol crop protection, 1989, sr. rsch. specialist crop protection, 1989-91; pres. Microbe Inotech Labs., Inc., 1991—; chmn. regional com. USDA, Washington, 1986-89, mem. tech. subcom. on biocontrol expt. sta. com. on policy, 1987-89; panel mem. Nat. Rsch. Coun. Briefing, Washington, 1987; disting. guest lectr. Coll. Sci. Utah State U., Logan, 1989. Author: Methods in Enzymology, 1979; co-editor: Iron Chelation in Plants and Soil Microorganisms, 1993; mem. editorial bd. Biology of Metals, Springer-Verlag, 1988-91. Troop committeeman Boy Scouts Am., Manchester, Mo., 1985. Mem. AAAS, Am. Phytopath. Soc., Am. Chem. Soc., Am. Soc. Microbiology, Nat. Registry of Environ. Profls. Achievements include development of first microbial recombinant marker system tested in U.S. environment. Office: Microbe Inotech Labs Inc 1840 Craig Rd Saint Louis MO 63146

HEMPFLING, GREGORY JAY, mechanical engineer; b. Terre Haute, Ind., Sept. 7, 1961; s. John G. and Sandra (Sutton) H. BSME, Rose-Hulman Inst. Tech., Terre Haute, Ind., 1983; M in Engring. Mgmt., George Washington U., Washington, 1991. Asst. engr. Cherne Contracting Corp., New Washington, Ind., 1983-84; engr. I Newport News Shipbuilding, 1984-86, engr. II, 1986-89, engring. supr., 1989—; chmn. Industry Rels. Com. (ea. Va. sect.), 1990-92. Mem. ASME (chmn. industry rels. com. 1990-92, rep. Peninsula Engring. Coun. 1992—), SAE (Marine Vehicle Systems panel 1989—), Soc. Naval Architects and Marine Engrs., Am. Mgmt. Assn., Naval Submarine League. Avocations: skiing, golf. Home: 559 Kristy Ct Newport News VA 23602-9025 Office: Newport News Shipbuilding 4101 Washington Ave Newport News VA 23607-2734

HENAGER, CHARLES HENRY, civil engineer; b. Spokane, Wash., July 11, 1927; s. William Franklin and Mary Agnes (Henry) H.; m. Dorothy Ruth Parker, May 6, 1950; children: Charles Henry, Jr., Donald E., Roberta R. BS in Civil Engring., Wash. State U., 1950. Registered profl. engr., Wash. Instrumentman Wash. State Dept. Hwys., Yakima, 1950-52; engr. Gen. Electric Co., Richland, Wash., 1952-62; shift supr., reactor GE, Richland, Wash., 1962-63, sr. engr., 1963-65; sr. devel. engr. Battelle Pacific N.W. Labs., Richland, 1965-68, sr. rsch. engr., 1968—. Contbr. articles to profl. jours.; patentee in field. With USN, 1945-46. Fellow Am. Concrete Inst. (tech. activities com. 1987-89, Del Bloem award 1986), ASTM (subcom. 1980-92), ASCE (pres. Columbia sect. 1961-62); mem. Kennewick Swim Club (pres. 1962-63), Sigma Tau, Tau Beta Pi, Phi Kappa Phi. Republican. Methodist. Avocations: stamp and coin collecting, calligraphy. Home: 1306 N Arthur Pl Kennewick WA 99336-1545 Office: Battelle Pacific NW Labs Battelle Blvd Richland WA 99352

HENCHAL, ERIK ALEXANDER, microbiologist; b. Natick, Mass., Apr. 3, 1953; s. Harold Francis and Evelyn Lucille (Raven) H.; m. Laraine Carol Stab, Nov. 5, 1976. BS in Microbiology, U. Maine, Orono, 1975; PhD in Microbiology, Pa. State U., 1980. Commd. 2d lt. Med. Svc. Corps. U.S. Army, 1980; advanced through grades to major U.S. Army, 1988; prin. investigator Walter Reed Army Inst. Rsch., Washington, 1980-84, 87-90; asst. dept. chief Armed Forces Rsch. Inst. Med. Sci., Bangkok, Thailand, 1984-87; staff officer HQ, U.S. Army Med. R&D Command, Ft. Detrick, Frederick, Md., 1990—; adj. instr. clin. virology U. Md., College Park, 1988. Co-author: Proceedings of the International Conference on Dengue and Dengue Hemorrhagic Fever, 1983, Rapid Detection of Dengue and Japanese Encephalitis Viruses Using Nucleic Acid Hybridization: Proceedings of the Four Australian Arbovirus Symposium, 1987; contbr. articles to Jour. Wildlife Diseases, Am. Jour. Tropical Medicine and Hygiene, Jour. Gen. Virology, SE Jour. Tropical Medicine and Pub. Health, Jour. Virological Methods, Molecular and Cellular Probes, Virology. Decorated Overseas Svc. Ribbon, Meritorious Svc. medal, Nat. Def. Svc. ribbon. Mem. AAAS, Amk. Soc. for Tropical Medicine and Hygiene, Am. Soc. for Virology. Republican. Achievements include discovery and application of the first complete set of Dengue virus specific monoclonal antibody reagents; development of three different rapid diagnostic assays for Dengue viruses. Office: US Army Med Rsch Inst Infectious Diseases Virology Div Ft Detrick Frederick MD 21702

HENDEE, JOHN CLARE, college dean, natural resources educator; b. Duluth, Minn., Nov. 12, 1938; s. Clare Worden and Mary Myrtle (Parker) H.; m. Frances Katherine Gasperson; children: John Jr., James, Landon, Joy, Joni, Jared. BS in Forestry, Mich. State U., 1960; MF in Forestry Mgmt., Oreg. State U., 1962; PhD in Forestry, Econs. and Sociology, U. Wash., 1967. With USDA Forest Svc., 1961-85; with timber mgmt. dept. Waldport and Corvallis, Oreg., 1961-64; fire rsch. forester Pacific S.W. Forest Experiment Sta., Berkeley, Calif., 1964; recreation rsch. unit leader Pacific N.W. Forest Expt. Sta., Seattle, 1967-76; legis. affairs staff Washington, 1977-78; asst. sta. dir. Southeastern Forest Experiment Sta., Asheville, N.C., 1978-85; dean Coll. Forestry, Wildlife, and Range Sci. U. Idaho, Moscow, 1985—, prof. forest resources and resource recreation and tourism, 1985—; dir. Idaho Forest, Wildlife and Range Experiment Sta., 1985—; mem. affiliate faculty in forestry U. Wash., Seattle, 1968-76; vice chmn. for sci. 4th World Wilderness Congress; owner, mgr. family tree farm, Dreary, Idaho. Co-author: Wildlife Management in Wilderness, 1978, Introduction to Forestry, rev. 6th edit., 1991; sr. co-author: Wilderness Management, 1978, rev. edit., 1990; contbr. numerous articles to profl. jours. Bd. dirs. Internat. Wilderness Leadership Found.; active Boy Scouts Am. Recipient Spl. Merit award Keep Am. Beautiful, 1972, Nat. Conservation Achievement award Am. Motors, 1974, Spl. award for Wilderness Rsch. and Edn. Nat. Outdoor Leadership Sch., 1985, Merit award USDA-Forest Service, 1979, 80, 85, Lifetime Achievement award Am. Soc. Pub. Adminstrn., 1988; Fed. Congl. fellow, Washington, 1976-77. Mem. Am. Forestry Assn., Nat. Assn. Profl. Forestry Schs. and Colls. (chmn. western div. 1987-89), Wildlife Soc., Soc. Am. Foresters (edn. and communication working group chmn. 1986-89). Mem. Unitarian Ch. Avocations: family, tree farming, hiking, skiing, fishing. Office: Univ of Idaho Coll Forestry Wildlife & Range Scis Moscow ID 83843

HENDEL, FRANK J(OSEPH), chemical and aerospace engineer, educator, technical consultant; b. Sambor, Poland, Dec. 2, 1918; came to U.S., 1950, naturalized, 1955; s. Emil and Henrietta (Sprecher) H.; children: Anna H. (Mrs. Gary Carrillo), Emily E. (Mrs. Edward Winfield), Erica F. Ph.D., P.E. Tech. U. Lwow, 1941. Chief chem. engr. Wigton-Abbott Corp. (engrs. and constructors), Plainfield, N.J., 1950-56; head chem. engring. Aerojet Gen. Corp., Azusa, Calif., 1956-61; staff scientist, space div. N.Am. Aviation, Downey, Calif., 1961-64; mem. tech. staff Jet Propulsion Lab., Calif. Inst. Tech., Pasadena, 1964-67; prof. aerospace sci. and aero. engring., dept. aero. and mech. engring. Calif. Poly. State U., San Luis Obispo, 1967-85, prof. emeritus, 1985-89; with Space Systems div. Gen. Dynamics Corp., San Diego, 1986-89, tech. cons., 1989-90; ret.; cons. Nat. Acad. Sci., 1961-63, Space and Missile Test Ctr., Vandenberg AFB, 1968-75; mem. faculty Inst. Aerospace Safety and Mgmt., U. So. Calif., 1970-73; lectr. aerospace, propulsion and ordnance systems UCLA Extension, 1964-67, researcher Laramie (Wyo.) Energy Tech. Ctr., U.S. Dept. Energy, 1977-78; aero. engr. Flight Test Ctr., Edwards AFB, Summer 1980; research specialist Lockheed Missile and Space Co., Sunnyvale, Calif., summers 1981, 82; gen. engr. U.S. Navy Pacific Missile & Test Ctr., Point Mugu, summer 1983; project engr. United Techs., San Jose, Calif., summer 1984. Abstractor: Chem. Abstracts, 1952-82; contbr. articles to profl. jours. and books; designer presentations on Liquid Rocket Boosters on Space Shuttles. Polyglot San Diego Aerospace Mus., 1989-92. Fellow AIAA (assoc., chmn. symposium on alt. fuel resources 1976, mem. tech. com. on elec. propulsion, subcom. on space missions 1988-91); mem. NSPE, Am. Soc. Engring. Edn. Achievements include design and engring. of several chem., indsl. pyrotechnics and ordnance for space vehicles and missiles. Home: 5440 Ralston St Apt 310 Ventura CA 93003-6002

HENDERSON, ARVIS BURL, data processing executive, biochemist; b. Abilene, Tex., Oct. 24, 1943; s. Arvis Vernon and Aubra Lee (Patton) H.; m. Mary Ann Pickett, Mar. 17, 1966 (div. Sept. 1983); 1 child, Michelle Rene; m. Jo Nell Hartsell, July 2, 1985. AA, San Angelo Coll., 1964; BA, U. Tex., 1966; MAS, So. Meth. U., 1969; PhD, U. Tex. Health Sci. Ctr., 1976. Postdoctoral fellow U. Tex., Austin, 1976-80; dir. rsch. lab. Instrumentation Specialities Co., Lincoln, Nebr., 1980-81; asst. prof. pediatrics U. Tex. Health Sci. Ctr., Houston, 1981-84; dir. sci. computing S.W. Found. for Biomed. Rsch., San Antonio, 1984-91; assoc. v.p. info. tech. U. Tex., San Antonio, 1991—. Contbr. articles on biomed. research to profl. jours., chpts. to books. Chmn. Alamo Area Quality Workforce Planning Com., 1990-92. Served to capt. USAF, 1966-72. Recipient Research Service award NIH, 1976-79; fellow U. Tex., 1976-80, Clayton Found. Biochemistry Inst., 1980. Mem. NIH spl. study sect. 9, Data Processing Mgmt. Assn., Assn. Systems Mgmt., Assn. for Computing Machinery. Republican. Baptist. Avocation: photography. Home: 8707 Timber Ledge St San Antonio TX 78250-4174 Office: U Tex 6900 NW Loop 1604 San Antonio TX 78249-0677

HENDERSON, BRIAN EDMOND, physician, educator; b. San Francisco, June 27, 1937; s. Edward O'Brien and Antoinette (Amstutz) H.; m. Judith Anne McDermott, Sept. 3, 1960; children—Sean, Maire, Sarah, Brian John, Michael. B.A., U. Calif.-Berkeley, 1958; M.D., U. Chgo., 1962. Resident Mass. Gen. Hosp., Boston, 1962-64; chief arbovirology Ctr. Disease Control, Atlanta, 1969-70; assoc. prof. pathology U. So. Calif., Los Angeles, 1970-74, prof. pathology, 1974-78, prof. preventive medicine, dept. chmn., 1978-88, dir. Comprehensive Cancer Ctr., 1983—; cons. WHO, South Pacific Commn., U.S.-Japan-Hawaii Cancer Program; mem. Charles S. Mott selection com. Gen. Motors Cancer Research Found., 1982-88; bd. councillors Nat. Cancer Inst., 1979-82; mem. sci. council IARC, 1982-86. Contbr. articles to profl. jours., chpts. to books; mem. editorial bd. Jour. Clin. Oncology; assoc. editor: Cancer Research. Served to lt. col. USPHS, 1964-69. Nat. Acad. Sci. disting. scholar to China, 1982; recipient Richard & Hinda Rosenthal Found. award Am. Assn. Cancer Research, 1987. Fellow Los Angeles Acad. Medicine; mem. AAAS, Infectious Disease Soc. Am., Am. Epidemiol. Soc., Alpha Omega Alpha. Democrat. Roman Catholic. Office: U So Calif School of Medicine Dept of Preventive Medicine 1975 Zonal Ave Los Angeles CA 90033-1039

HENDERSON, DANIEL GARDNER, electrical engineer; b. Norfolk, Va., Oct. 25, 1941; s. Mac Daniel and Edith (Bosemberg) H.; m. Lois Barr, June 24, 1966; children: Sondra Kaye, Janine Michelle. AA in Sci., Chipola Jr. Coll., Marianna,Fla., 1963; BSEE, U. Tenn., 1965. With assoc. staff Applied Physics Lab. Johns Hopkins U., Laurel, Md., 1965-71; with sr. staff Applied Physics Lab. Johns Hopkins U., Laurel, 1971-81, prin. staff Applied Physics Lab., 1981—. Mem. Assn. Old Crows (Gold Cert. of Merit 1983, Cert. of Appreciation 1986), U.S. Naval Inst., Tailhook Assn., Tau Beta Pi. Democrat. Baptist. Avocations: old movies, short wave radio, fishing. Home: 7921 Anfred Dr Laurel MD 20723-1136 Office: Johns Hopkins U Applied Physics Lab Johns Hopkins Rd Laurel MD 20723

HENDERSON, JOHN GOODCHILDE NORIE, electrical engineer; b. Phila., Dec. 11, 1945; s. Nancy Margaret Fisher, May 22, 1971. BSEE, U. Pa., 1967; MSE, Princeton U., 1969. Mem. tech. staff RCA Labs., Princeton, N.J., 1967-81, group head, 1981-91; chief rschr. Hitachi Am., Ltd., Princeton, N.J., 1991—; chmn. testing task force FCC Adv. Com. on Advt. TV, Washington, 1992-93; lectr. in field. Contbr. articles to profl. jours. and confs. Bd. dirs. Cape May Pt. (N.J.) Taxpayer's Assn., 1991-92; tchr. Minorities in Engring. Program, Princeton; organist Beadle Meml. Ch., Cape May Pt., 1970—. Mem. IEEE (chmn. consumer electronics conf. 1988), Corinthian Yacht Club Cape May (rear com. 1977), Sigma Xi., Tau Beta Pi. Achievements include patents for Consumer Electronics, TV, Advanced TV. Office: Hitachi Am Ltd 307 College Rd E Princeton NJ 08540

HENDERSON, KAYE NEIL, civil engineer, business executive; b. Birmingham, Ala., June 10, 1933; s. Ernest Martin and Mary (Head) H.; B.S., Va. Mil. Inst., 1954; B.A. with honors, U. South Fla., 1967; m. Betty Jane Belanus, June 26, 1954; children: David Scott, Alan Douglas, Helen Kaye. Registered profl. engr., Fla. Mgmt. trainee Gen. Electric Co., Schenectady, 1954; sales engr. Fla. Prestressed Concrete, Tampa, 1956-57; field engr. Portland Cement Assn., Tampa, 1957-63; gen. mgr. residential and comml. sales Tampa Electric Co., 1963-66; v.p. Watson & Co., architects and engrs., Tampa, 1966-69; v.p. Reynolds, Smith & Hills, architects, engrs. and planners, Jacksonville, Fla., 1969-78, sr. v.p., 1978-86, dir., 1976-86; sec. transp. State of Fla., 1987-89; pres., chief exec. officer, Old World Svc., Inc., Tallahassee, 1989—; bd. dirs. Stauros Ctr. Econ. Edn. Vice chmn. Temple Terrace Planning and Zoning Bd., 1962-67; pres. Guidance Center Hillsborough County, 1969; mem. adv. bd. Multi-State Transp. System, 1976-82; mem. Duval County Republican Exec. Com., 1970-72; bd. dirs. Salvation Army Home and Hosp. Coun., 1964-69; mem. found U. South Fla., Fla. Taxwatch; mem. exec. com. Floridians for Better Transp. 1st lt. USAF, 1954-56. Recipient Service awards Greater Tampa C. of C., 1964-66; named Outstanding Young Man of Tampa Jr. C. of C., 1965; Outstanding Young Man of Am., U.S. Jr. C. of C., 1967; Boss of Year award Am. Bus. Women's Assn., 1978. Mem. Fla. Engring. Soc., Am. Rd. and Transp. Builders Assn.,

Gov's Club, Killearn Golf and Country Club, Timuquana Country Club, Univ. Club, Rotary (dir. 1984-85, Paul Harris fellow 1987), Phi Kappa Phi (life), Tau Beta Pi. Republican. Episcopalian. Home: 1743 Armistead Pl Tallahassee FL 32312-3453 Office: 1680 Metropolitan Cir Tallahassee FL 32308-3755

HENDERSON, RICHARD MARTIN, chemical engineer; b. Winston-Salem, Dec. 12, 1934; s. Billy Martin and Marion Lucille (Dunn) H.; m. Patricia Lucille Green, Dec. 27, 1958 (div. 1978); children: Marian Patricia, Richard Martin; m. Janice Lee Ferris, Apr. 3, 1981. BBA, Wake Forest U., 1957. Cert. quality engr. Jr. chem. engr. R.J. Reynolds Tobacco Co., Winston-Salem, 1957-58, asst. chem. engr., 1958-65, product devel. group leader, 1965-66, devel. sect. head, 1966-80, div. mgr., 1981—. Pres. Y Men's Club of Winston-Salem, 1989, Moravian Music Found., Winston-Salem, 1983; active United Way of Forsyth County, 1958—, Winston-Salem Arts Coun., 1958—. With Signal Corps, U.S. Army, 1960-61. Recipient Merit award, Moravian Music Found., 1984; Ky. Col. Mem. Am. Soc. for Quality Control (founding dir./sr. mem.), Forsyth Country Club. Republican. Moravian. Achievements include patents for the method of and apparatus for automatically analyzing the degradation of processed leaf tobacco, for the method and apparatus for automatically determining the stem content of baled tobacco, for the method and apparatus for automatically determining the basis weight and moisture content of paper and paper-like substance, for the method and apparatus for automatically sampling a material and transporting it from one location to another remote location. Home: 717 Mitch Dr Winston Salem NC 27104-5127 Office: RJ Reynolds Tobacco Co 401 N Main St Winston Salem NC 27101-3818

HENDERSON, WILLIAM BOYD, engineering consulting company executive; b. Elkton, Tenn., Apr. 8, 1928; s. William Warren and Berneece Bramlett (Boyd) H.; m. Virginia D. Ferguson, Aug. 22, 1953; children: William Boyd II, Kimberly Ann, Karen Beth. BCE, Vanderbilt U., 1950; M in Engring., U. Pitts., 1955-59. Engring. mgr. Westinghouse Electric Corp., numerous locations, 1954-74; pres. Buell div. Envirotech Corp., Lebanon, Pa., 1974-80; pres., chmn. bd. Rettew Assocs., Inc., Newmanstown, Pa., 1980-83; pres., chief exec. officer, chief oper. officer Trident Engring. Assocs., Annapolis, Md., 1983—; bd. dirs. Sowers Printing Co., Lebanon, 1979—, Dauphin Bank, Lebanon, 1976-82; cons. Self-Air Pollution Equipment, Lebanon, 1979-83; mem. Md. Gov.'s TMI Commn., Annapolis, 1984—. Contbr. articles to various nuclear power confs., 1968-74. Bd. dirs. United Way, Lebanon, 1975-81; trustee United Meth. Ch, 1965—. Served to lt. USN, 1950-54. Republican. Lodge: Rotary. Avocations: hunting, fishing, gardening. Office: Trident Engring Assocs Inc 2010 Industrial Dr Annapolis MD 21401-1673

HENDON, MARVIN KEITH, psychologist; b. Miami, Fla., Oct. 18, 1960; s. James William and Esther (Holts) H.; m. Deborah Faye Moore, Mar. 17, 1990. BA, U. Fla., 1980, MS, 1982, PhD, 1985. Lic. psychologist Fla. Psychologist Psylab Psychol. Svc., Sarasota, Fla., 1986-87; pvt. practice psychology Sarasota, Fla., 1987—; psychol. cons. Child Protection Team, Sarasota, 1988—; outpatient therapist Mental Health Resource Ctr., Jacksonville, 1984-86; therapist U. Counseling Ctr, U. Fla., Gainesville, 1979-83; adj. prof. Manatee Community Coll., 1985. Columnist Insights into Human Behavior, 1987. Adv. bd. One Ch. One Child, 1990—; mem. sch. bd. Westcoast Sch. for Human Devel., 1987—. Republican. Office: 240 N Washington Blvd Sarasota FL 34236-5922

HENDRICK, JOHN MORTON, science and engineering editor; b. Trinidad, Colo., July 16, 1917; s. Jack Morton and Alice (Kirkpatrick) H.; m. Margaret Beekman Hambley, Jan. 26, 1958. Student, Vanderbilt U., 1934-35; AB in Econs., U. Mich., 1938. Shop-order writer, schedule-status reporter Lockheed Aircraft Corp., Burbank, Calif., 1939-55; owner, mgr. D&C Airlines, Detroit, 1955-56; change-board scheduler, drafting procedures writer The Martin Co., Denver, 1956-61; operating stats. exhibits planner Lockheed Missiles and Space Co., Sunnyvale, Calif., 1961-62; engring. scheduler Hiller Aircraft Co., Palo Alto, Calif., 1962-64; contbg. writer, editor Air Transport World mag., Washington, 1965; editor Aries Corp., McLean, Va., 1966-67, Volt Info. Scis., Inc., Lanham, Md., 1967-69; cons. Franklin Engine Co., Inc., Syracuse, N.Y., 1970-73; tech. editor Applied Sci. Assocs., Inc., McLean, 1973-74, CIA, Langley, Va., 1974-90, Joint Publs. Rsch. Svc., Arlington, Va., 1990—. Contbr. revisions to The Pentagon and the Art of War, 2nd. edit., 1985. Republican. Protestant. Home: 523 Epping Forest Rd Annapolis MD 21401-6537

HENDRICKS, LEONARD D., emergency medicine physician, consultant; b. Chgo., Feb. 29, 1952; s. Leonard D. and Edith V. (Elliott) H.; m. Gail Williams, Aug. 26, 1989. BS in Engring., U. Ill., 1974; MD, U. Wis., 1979. Diplomate Am. Bd. Emergency Medicine. Med. dir. Cuyahoga County Corrections Facility; emergency physician Meridia Huron Hosp., East Cleveland, Ohio; asst. dir. emergency medicine Kaiser Permanente Hosp., Parma, Ohio; emergency physician Western Res. Care System, Youngstown, Ohio; dir. emergency medicine St. Joseph Riverside Hosp., Warren, Ohio; cons. Friedman, Domiano and Smith Law Firm, Cleve.; instr. emergency medicine Case Western Res. U., Cleve., Northeastern Ohio U., Rootstown; instr. ACLS, Am. Heart Assn.; instr. advanced trauma life support ACS; instr. pediatric ALS, neonatal resuscitation, Am. Acad. Pediatrics. Fellow Am. Coll. Medicine, Am. Coll. Emergency Physicians; mem. Am. Coll. Physician Execs., Soc. Acad. Emergency Medicine.

HENDRICKS, TERRY JOSEPH, mechanical engineer; b. Culver City, Calif., Oct. 19, 1954; s. Donald Barnabas and Lorna Maria (Fritz) H.; m. Deborah Lynne Colip, Mar. 19, 1983; 1 child, Matthew Ryan. BS in Physics summa cum laude, U. Lowell, 1976; MS in Engring., U. Tex., 1979, postgrad., 1990—. Registered profl. engr., Tex., Calif. Sr. engr. Martin Marietta Aerospace Co., Denver, 1980-82; sr. devel. engr. Varo, Inc., Garland, Tex., 1982-85; rsch. specialist Lockheed Missiles and Space Co., Sunnyvale, Calif., 1985-89; sr. scientist W.J. Schafer Assocs., Pleasanton, Calif., 1989-90; tech. session chmn. 6th Internat. Conf. on Thermoelectric Energy Conversion, Arlington, Tex., 1986. Contbr. articles to AIAA Jour. Thermophysics and Heat Transfer, conf. proceedings. R.C. Baker fellow U. Tex., 1976, U. Tex. fellow, 1992-93. Mem. ASME (assoc.), AIAA, IEEE, Sigma Pi Sigma, Phi Kappa Phi. Roman Catholic. Achievements include patent for method and apparatus for fabricating a thermoelectric array. Office: U Tex Ctr Energy Studies 10100 Burnet Rd Austin TX 78758

HENDRICKSON, WAYNE A(RTHUR), biochemist, educator; b. Spring Valley, Wis., Apr. 25, 1941; m. Gerry L., 1969; children: Helen Margaret, Inga Marie. BA, U. Wis., River Falls, 1963; PhD in Biophysics, Johns Hopkins U., 1968. Rsch. assoc. Johns Hopkins U., Balt., 1968-69; postdoctoral rsch. assoc. Naval Rsch. Lab., 1969-71, rsch. biophysicist, 1971-84; prof. biochemistry and molecular biophysics Columbia U. Coll. Physicians and Surgeons, N.Y.C., 1984—, 1984—; investigator Howard Hughes Med. Inst., $D, $D, 1986—; sci. adv. bd. mem. Progenics Pharms., 1987—; sci. policy bd. Stanford Synchrotron Radiation Lab., 1991-92; Stanford Linear Accelerator Ctr., sci. policy com. 1992—; proposal evaluation bd. Advanced Photon Source, 1989—; biomed. adv. com. for Pitts. Supercomputing Ctr., 1987-92; DOE Synchrotron Rev. Com., 1987-88; proposal rev. panel Cornell High Energy Synchrotron Source, 1987—; mem. NSF Molecular Adv. Panel, 1980-83, NIH Biophys. Chemistry Study Sect., 1986-89. Mem. editorial bd. Jour. Biomolecular Structure and Dynamics, 1986-91; assoc. editor Jour. of Molecular Biology, 1987-93; editor Current Opinion in Structural Biology, 1989—, Macromolecular Structures, 1991—, Structure, 1993—; contbr. numerous articles to profl. jours. Mem. NSF Molecular Adv. Bd., 1980-83, NSF Biophys. Chemistry Study Sect., 1986-89. Recipient Biol. Scis. award Washington Acad. Scis., 1976, Meritorious Civilian Svc. award U.S. Navy, 1978, Arthur S. Flemming award Outstanding Young Fed. Employees, 1979. Fellow AAAS; mem. Am. Crystallographic Assn. (chmn. biol. macromolecules group 1980, A.L. Patterson award 1981, Fankuchen award com. 1982), Am. Soc. Biochemistry and Molecular Biology (Fritz Lipmann award 1991), Biophys. Soc. (coun. mem. 1987-90), mem. publs. com. 1989—), Internat. Union Crystallography (commn. on biol. macromolecules 1981-87, commn. on crystallographic computing 1984-87, commn. on synchrotron radiation 1990—). Achievements include rsch. in macromolecular structure and function, in principles of protein structure, dynamics and assembly, in properties of specific proteins, in fraction methods, in crystallographic computing, and in synchrotron radiation. Office: Columbia U Dept Biochemistry & Biophys 630 W 168th St New York NY 10032-3702

HENDRICKSON, WILLIAM GEORGE, business executive; b. Plainview, Minn., May 31, 1918; s. Clarence and Hildegarde (Heaser) H.; m. Virginia M. Price, Sept. 1, 1942; children: Robert, Thomas, Donald, Julie Ann. BS, St. Mary's Coll., Winona, Minn., 1939; MS, U. Detroit, 1941; PhD, U. Wis., 1946; D Humanities, St. Mary's Coll., Winona, Minn., 1991. Scientist Wis. Alumni Research Found., Madison, 1946-54, dir. devel., 1954-61; v.p. Ayerst Labs. div. Am. Home Products Corp., N.Y.C., 1961-67, exec. v.p., 1967-69; group v.p. Am. Home Products Corp., N.Y.C., 1969-80; chmn. emeritus bd. St. Jude Med., Inc., St. Paul; bd. dirs. Rsch. Corp. Techs., Tucson, IntelliNet, Naples, Fla. Mem. Am. Chem. Soc., N.Y. Acad. Scis., Country Club N.C., Quail Creek Country Club, Sigma Xi. Republican. Roman Catholic.

HENDRICKX, ANDREW GEORGE, research physiologist; b. Butler, Minn., July 14, 1933; B.S. in Biology, Concordia Coll., Minn., 1959; M.S., Kans. State U., 1961, Ph.D. in Zoology, 1963. Head sect. embryology Southwest Found., San Antonio, 1964-68, assoc. scientist, chmn., 1969; assoc. research physiologist Calif. Primate Research Ctr., Davis, 1969-73, research physiologist, 1973—, assoc. dir., 1978-97, dir., 1987—; prof. Sch. Medicine, U. Calif.-Davis, 1971—; adviser WHO, 1977—. Author: Embryology of the Baboon, 1971; numerous articles. Served with U.S. Army, 1953-55. NDEA scholar, 1959-62; recipient Disting. Alumni award Concordia Coll., 1977. Mem. Teratology Soc. (sec. 1979-83, pres. 1987), Am. Soc. Primatologists (pres. 1982-84), AAUP, Am. Assn. Anatomists, Internat. Soc. Primatologists, others. Office: U Calif Davis Calif Primate Rsch Ctr Davis CA 95616

HENDRIE, JOSEPH MALLAM, physicist, nuclear engineer, government official; b. Janesville, Wis., Mar. 18, 1925; s. Joseph Munier and Margaret Prudence (Hocking) H.; m. Elaine Kostell, July 9, 1949; children: Susan Debra, Barbara Ellen. BS, Case Inst. Tech., 1950; PhD, Columbia U., 1957. Registered profl. engr., N.Y., Calif. Asst. physicist Brookhaven Nat. Lab., Upton, N.Y., 1955-57, assoc. physicist, 1957-60, physicist, 1960-71, sr. physicist, 1971—, chmn. steering com., project chief engr. high flux beam reactor design and constrn., 1958-65, acting head exptl. reactor physics div., 1965-66, project mgr. pulsed fast reactor project, 1967-70, assoc. head engring. div., dept. applied sci., 1967-71, head, 1971-72, chmn. dept. applied sci., 1975-77, spl. asst. to dir., 1981—; dir. Tenera Corp., 1984-85, Houston Industries, Inc., Houston Lighting and Power Co., Entergy Ops. Inc.; dep. dir. licensing for tech. rev. U.S. AEC, 1972-74; chmn. U.S. Nuclear Regulatory Commn., Washington, 1977-79, 81, commr., 1980, mem. adv. com. on enforcement policy, 1984-85; lectr. nuclear power plant safety MIT, Ga. Inst. Tech., Northwestern U., summers 1970-77; cons. radiation safety com. Columbia U., 1964-72; mem. adv. com. reactor safeguards AEC, 1966-72, chmn., 1970; U.S. mem. sr. adv. group on reactor safety standards IAEA, 1974-78; mem. nat. rsch. coun. com. Internat. Cooperation in Magnetic Fusion, 1983-85; cons. AEC, Nuclear Regulatory Commn., 1974-75, GAO, 1975-77, Electric Power Rsch. Inst., 1982, various nuclear utilities, 1981—. Mem. editorial adv. bd. Nuclear Tech, 1967-77. Served with AUS, 1943-45. Recipient E.O. Lawrence award, 1970; decorated comdr. Order of Leopold II (Belgium), 1982. Fellow Am. Nuclear Soc. (dir. 1976-77, v.p. 1983-84, pres. 1984-85), ASME; mem. IEEE, Nat. Acad. Engring., Am. Phys. Soc., ASTM (com. on rsch. and tech. planning 1985-90), Am. Concrete Inst., Inst. Nuclear Power Operation (adv. coun. 1984-90), Nat. Soc. Profl. Engrs., Sigma Xi, Tau Beta Pi. Achievements include research and publications on physics nuclear reactors, nuclear power plant safety, engineering design reactors, electrical power transmission, chem. physics nitrogen dissociation process, structure oxygen molecule. Office: Brookhaven Nat Lab Upton NY 11973

HENDRIX, KENNETH ALLEN, systems analyst; b. Marietta, Ga., Sept. 22, 1959; s. Thomas Cal and Violet (Hendricks) H.; m. Helen Sarah Sirett, Jan. 3, 1987. BS in Mech. Engring., U. Tenn., 1982. Registered profl. engr. mech. engring. Mech. engr. Carolina Eastman Co., Columbia, S.C., 1982-87; systems analyst Eastman Chem. Co., Kingsport, Tenn., 1987—. Mem. ASME (sec.-treas. Holston sect. 1991-92), Kingsport Sertoma Club (sec., v.p. 1990-93). Achievements include development of environmental management information system for Tenn. Eastman div. Eastman Chem. Co. Office: Eastman Chem Co PO Box 1973 Eastman Rd Kingsport TN 37662

HENDRIX, RONALD WAYNE, physician, radiologist; b. St. Louis, June 4, 1943; s. Arthur W. and Lida (Martin) H.; m. Miriam Jensen, June 14, 1969. AB, Wash. U., St. Louis, 1965, MD, 1969. Diplomate Am. Bd. Nuclear Medicine, Am. Bd. Radiology. Intern Wash. U., Barnes Hosp., St. Louis, 1969-70; resident U. Chgo., 1970-73, fellow in nuclear medicine, 1973-74; staff radiologist Symmes Hosp., Arlington, Mass., 1976-77; asst. prof. radiology Northwestern U. Med. Sch., Chgo., 1977-84, assoc. prof. radiology, 1984—; attending physician Northwestern Meml. Hosp., Chgo., 1977—, chief, musculoskeletal radiology, 1977—; dir. radiology Rehab. Inst. of Chgo., 1986—. Contbr. articles to profl. jours.; contbg. author to several books. Pres. LaSalle St. Ch., 1982-84, treas., 1984-86, chmn. fin. com., 1986-92. Lt. comdr. USN, 1974-76. Mem. Radiol. Soc. of N.Am., Am. Roentgen Ray Soc., Assn. of U. Radiologists, Am. Coll. Radiology, Internat. Skeletal Soc. Office: Northwestern Meml Hosp 710 N Fairbanks Ct Chicago IL 60611-3013

HENEBRY, MICHAEL STEVENS, toxicologist; b. Decatur, Ill., Jan. 19, 1946; s. Bernard Stevens and Lucille (Dolores) H.; m. Virginia Godelsoson Azuela, Jan. 5, 1984; children: Jeffrey Adams, James Stevens. BA in Biology, Millikin U., Decatur, Ill., 1968; MS in Zoology, Ea. Ill. U., 1978; PhD in Aquatic Toxicology, Va. Poly. Inst. and State U., 1981. Teaching asst. zoology Ea. Ill. U., Charleston, 1974-76; teaching asst. biology U. Mich. Biol. Sta., Pellston, 1977-78; rsch. asst. zoology U. Mich. Biol. Sta., 1976-81; rsch. asst. biology Va. Poly. Inst. and State U., Blacksburg, 1967-80; teaching asst. biology Va. Poly. Inst. and State U., 1979-80; asst. prof. biology Ottawa U., Kans., 1981-82, Clarke Coll., Dubuque, Iowa, 1982-83; aquatic toxicologist/ecologist Ill. Natural History Survey, Champaign, Ill., 1983-88; acquatic toxicologist, lab. supr. ecotoxicology lab. Ill. EPA, Springfield, 1988—; cons. in field; lectr. in field; conductor seminars in field. Contbr. articles to profl. jours. Organizer Citizens Utility Bd., Chgo., 1976—, Am. Fedn. State, County and Mcpl. Workers, Springfield, 1983—. With USAF, 1969-73. Mem. ASTM, Soc. Protozoologists, Am. Microscopical Soc., Soc. Environ. Contamination and Toxicology, Ecol. Soc. Am., Midwest Pollution Control Biologists, Sigma Xi. Roman Catholic. Avocations: photography, travel, Eastern philosophy, Western religion. Home: 3345 S 3d St Springfield IL 62703 Office: Ill Environ Protection Agy 2200 Churchill Rd Springfield IL 62794-9276

HENEGHAN, SHAWN PATRICK, research scientist, educator; b. New Orleans, Dec. 3, 1951; s. Thomas F. and Ruth D. (Unk) H. BS, San Diego State U., 1977; PhD, U. So. Calif., 1982. Postdoctoral assoc. U. Iowa, 1982-85; analyst Ctr. for Naval Analysis, Alexandria, Va., 1985-87; scientist U. Dayton, Ohio, 1987—. Contbr. numerous articles to profl. jours. With U.S. Navy, 1972-78. NSF undergrad. fellow, 1976, Quad Chems. fellow, 1977-80. Mem. AIAA, Am. Phys. Soc., The Combustion Inst. Home and Office: U Dayton 1614 Meriline Ave Dayton OH 45410

HENEHAN, JOAN, chemical engineer, consultant; b. Louisville, Mar. 16, 1959; d. Robert William and Cornelia Catherine (Eckerle) Owens; m. Paul Vincent Henehan, Apr. 25, 1987. BS, U. Louisville, 1983. Registered profl. engr., Colo. Rsch. engr. Dow Chem. Co., Freeport, Tex., 1983-87; v.p. Urie Environ. Health, Inc., Wheat Ridge, Colo., 1988—. Mem. AICE, NSPE, Am. Indsl. Hygiene Assn. Republican. Roman Catholic. Office: Urie Environ Health Inc 11407 W 170 Frontage Rd N Wheat Ridge CO 80033

HENGGAO, DING, federal official; b. Nanjing City, Jiangsu, China, 1931. Alt. mem. CCP 12th Ctrl. Com., 1985-87; Minister Commn. Sci., Tech. and Industry for Nat. Def., 1985—; mem. CCP 13th Ctrl. Com., 1987—. Mem. Chinese Communist Party. Office: St Science & Tech Comm, 54 Sanlihe Fusingmenwai, Beijing 100862, China also: State Commn Sci Tech & Industry, Nat Def, Bejing China*

HENGLEIN, FRIEDRICH ARNIM, chemistry educator; b. Cologne, Germany, May 23, 1926; s. Friedrich August and Gertaute (Christ) H.; m. Gudrun Fröhlich-Henglein, June 20, 1961 (div.); children: Frank Arwed, Friederike, Franziska; m. Maritza Gutiérrez, May 18, 1990. Diploma in chemistry, U. Karlsruhe, Germany, 1949; D of Natural Scis., U. Mainz, Germany, 1951. Rsch. asst. Max Planck Inst. Chemistry, Mainz, 1949-53; physicist Farbenfabriken Bayer, Elberfeld, Germany, 1953-55; lectr. U. Cologne, 1955-58; scientist Mellon Inst., Pitts., 1958-60; prof. Tech. U., Berlin, 1960—; dir. Hahn-Meitner-Inst., Berlin, 1960—; advisor Nat. Ctr. Sci. Rsch., France, 1980-82, Max Planck Soc., Germany, 1985—, Solar Rsch. Inst., Hannover, Germany, 1988, Jour. Phys. Chemistry, 1991; vis. prof. Mellon Inst., 1961-63, U. Fla., Gainesville, 1972, U. Kyoto, Japan, 1972, U. Paris, 1975, U. Notre Dame, Ind., 1980. Author: Introduction to Radiation Chemistry, 1961, translated Japanese, 1972; contbr. numerous articles to profl. publs. Sgt. signal corps German mil., 1943-44. Recipient Golden Heyrovsky medal Acad. Sci. Prague, 1988, J.J. Weiss medal Assn. Radiation Rsch., 1978. Mem. Gesellschaft Deutscher Chemiker, Bunsengesellscheft for Physikalische Chemie. Achievements include rsch. on stripping mechanism in chem. dynamics, polarography of short-lived radicals, photocatalysis by colloidal semiconductors and metals, size quantization effect in semiconductor and metal particles, surface chemistry of colloidal particles. Home: Gardeschutzenweg 90A, 12203 Berlin Germany Office: Hahn Meitner Inst Berlin, Glienickerstr 100, 14109 Berlin Germany

HENINGER, GEORGE R., psychiatry educator, researcher; b. L.A., Nov. 15, 1934; s. Owen P. and Rachel (Cannon) H.; m. Julie Hawkes, June 27, 1957; children: Steven, Catharine, Karen, Brian. BS, U. Utah, 1957, MD, 1960. Diplomate Am. Bd. Psychiatry and Neurology. Intern Boston City Hosp., 1960-61; resident in psychiatry Mass. Mental Health Ctr., 1961-63, chief resident, 1963-64; clin. assoc., clin. neuropharmacology rsch. ctr. St. Elizabeth's Hosp. NIMH, Washington, 1964-65; program specialist, office of dir. NIMH, Bethesda, Md., 1965-66; asst. prof. Psychiatry, assoc. chief rsch. ward Yale U., New Haven, Conn., 1966-71, assoc. prof., 1971-76, chief rsch. ward, 1971-78, prof. Clin. Psychiatry, 1976-78, prof. Psychiatry, dir. Abraham Ribicoff Rsch. Facilities, 1978—, assoc. chmn. rsch. dept. Psychiatry, 1988—; cons. NIMH, 1975-86, 88-93, NIH, 1987, McGill U., 1989, VA, 1990-93, Nat. Rsch. Coun. Can., 1991-93, Nat. Inst. Aging, 1992-93, Wellcome Trust, 1992-93, Pfizer Inc., Merck, Sharp & Dohme, Inc., The Upjohn Co., Hoffman La Roche, Inc., Burroughs Wellcome Co., Bristol-Meyers Co., Squibb Corp., Kali DuPhar, Inc.; bd. sci. advisors, Neurogen Corp. Reviewer manuscripts Archives Gen. Psychiatry, Am. Jour. Psychiatry, Psychiatry Rsch., Biological Psychiatry, Jour. Affective Disorders, Jour. Clin. Psychopharmacology, Life Scis., Neurochemistry Internat., Psychiatry, Schizophrenia Bulletin, Psychoneuroendocrinology, Jour. Am. Med. Assn. Sr. asst. surgeon USPHS, 1964-66. Recipient Rsch. Sci. Devel. award Type II, 1971; grantee NIMH 1971, 74, 77, 82, 85, 89, 91. Fellow Am. Coll. Neuropsychopharmacology, Am. Psychiatric Assn.; mem. AAAS, Am. Psychopathological Assn., Am. Coll. Neuropsychopharmacology, Soc. Neurosci., Soc. Biol. Psychiatry, Psychiat. Rsch. Soc., N.Y. Acad. Sci., Conn. Psychiatric Soc., Phi Kappa Phi, Alpha Omega Alpha, Sigma Xi. Avocation: running. Office: Yale U 34 Park St New Haven CT 06511

HENIS, JAY MYLS STUART, research scientist, research director; b. N.Y.C., July 9, 1938; s. Samuel Henry and Grace (Haber) H.; m. Evelyn Jean Rosenfelder, June 16, 1963; children: Nancy Gail, David Michael Alan. BA, Alfred U., 1959; PhD in Phys. Chemistry, Syracuse U., 1964. Rsch. assoc. Brown U., Providence, 1964-66; rsch. chemist Monsanto Co., St. Louis, 1964-91; pres., cons. Henis Techs. Inc., St. Louis, 1991—. Contbr. more than 35 rsch. articles to sci. and profl. jours. Recipient Soc. Plastics Engrs. Award for Best New Product, 1983, Thomas & Hochwald Award in Sci., Monsanto Co., St. Louis, 1989. Mem. Am. Chem. Soc. (St. Louis chpt. award 1986, Nat. award for separation sci., Dallas 1988), Am. Soc. Chem. Engrs. (Kirkpatric award for chem. engring., Chgo. 1982 , Am. Membrane Soc., European Membrane Soc., Internat. Soc. for Molecular Recognition. Achievements include over 9 U.S. and foreign patents; invention of first commercial gas separation membranes; development of new methods for spinning selective high strength hollow fibers, modification techniques for improved gas, liquid and biological separation properties and improved biocompatibility of membranes, microencapsulation technologies for living cells and organisms, new controlled delivery systems for proteins and peptides, advanced liquid membranes, and separation systems for environmental control. Office: Henis Techs Inc 501 Marford Dr Saint Louis MO 63141

HENIZE, KARL GORDON, astronaut, astronomy educator; b. Cin., Oct. 17, 1926; s. Fred R. and Mabel (Redmon) H.; m. Caroline Rose Weber, June 27, 1953; children: Kurt Gordon, Marcia Lynn, Skye Karen, Vance Karl. Student, Denison U., 1944-45; B.A., U. Va., 1947, M.A., 1948; Ph.D., U. Mich., 1954. Observer U. Mich. Lamont-Hussey Obs., Bloemfontein, Union South Africa, 1948-51; Carnegie postdoctoral fellow Mt. Wilson Obs., Pasadena, Calif., 1954-56; sr. astronomer in charge photog. satellite tracking stas. Smithsonian Astrophys. Obs., Cambridge, Mass., 1956-59; assoc. prof. dept. astronomy Northwestern U., Evanston, Ill., 1959-64, prof., 1964-72; scientist-astronaut NASA Johnson Space Ctr., Houston, 1967-86, sr. scientist, 1986—; guest observer Mt. Stromlo Obs., Canberra, Australia, 1961-62; prin. investigator astronomy expts. for Gemini 10, 11, 12 and Skylab 1, 2, 3, 1964-78; mem. astronomy subcom. NASA Space Sci. Steering Com., 1965-68; adj. prof. dept. astronomy U. Tex., Austin, 1972—; team leader NASA Facility Definition Team for Starlab Telescope, 1974-78; chmn. NASA working group for Spacelab Wide-Angle Telescope, 1978-79; jet pilot tng. Vance AFB, Enid, Okla., 1968-69; mem. support crew Apollo 15 and Skylab 1, 2, 3, 1970-73, mission specialist ASSESS 2 Spacelab simulation, 1976-77, mission specialist Shuttle flight 51F (Spacelab 2), 1985. Served with USNR, 1944-46; lt. comdr. Res., ret. Recipient Robert Gordon Meml. award, 1968; NASA medal for exceptional sci. achievement, 1974, Space Flight medal, 1985, Flight Achievement award Am. Astronautical Soc., 1985. Mem. Am., Royal, Pacific astron. socs., Internat. Astron. Union, Phi Beta Kappa. Research on planetary nebulae, emission-line stars, ultraviolet stellar spectra, space debris. Home: 18630 Point Lookout Dr Houston TX 77058-4037 Office: NASA Space Physics Br Johnson Space Ctr Houston TX 77058

HENKEL, CHRISTIAN JOHANN, astrophysicist; b. Dusseldorf, Fed. Republic Germany, Jan. 8, 1950; s. Christian Wilhelm and Renate (Haeberle) H.; m. Ping Yen, Aug. 22, 1985; 1 child, Melanie. PhD, U. Bonn, Germany, 1980. Postdoctoral fellow Max Planck Inst. for Radioastronomy, Bonn, 1980, Univ. Calif., Berkeley, 1982, Bell Labs., Holmdel, N.J., 1982-83; staff scientist Max Planck Inst. for Radioastronomy, 1984—. Contbr. 100 articles to Astronomy and Astrophysics, Astrophysical Jour. Mem. Astronomische Gesellschaft, Am. Astron. Soc. Achievements include detection of extragalactic CS, CH_3OH, HNC, C_2H, CN, HC_3N, HNCO, N_2H^+, SiO, CH_3CN, CH_3C_2H, of most luminous H_2O maser known to date; co-explanation of OH megamaser phenomenon; first detection of a maser from a carbon star; determination of interstellar oxygen and carbon isotope ratios inside and outside the galaxy establishing a 12C/13C abundance gradient across the galactic plane; one of first molecular surveys in lenticular and elliptical galaxies; research in studies of the cool dense gas in active galactic nuclei, studies of "hot cores" (sites where massive stars are forming) with respect to density, temperature and chemistry. Office: Max Planck Inst Radioastronomy, Auf dem Hugel 69, 53121 Bonn Germany

HENKIN, ROBERT IRWIN, neurobiologist, internal medicine, nutrition and neurology educator, scientific products company executive; b. L.A., Oct. 5, 1930; s. William and Ida Mildred (Scher) H.; m. Marsha Lynn Jacobs, May 15, 1964 (div. Jan. 1982); children: Amanda Joan, Michael Jonathan, David Gorman, Joshua Adam, Elizabeth Madeline, Hannah Deborah. AB cum laude, U. So. Calif., 1951; MA, UCLA, 1953, PhD, 1956, MD, 1959. Intern in medicine U. Calif. Hosp., L.A., 1959-60; resident in medicine Jackson Meml. Hosp., U. Miami (Fla.), 1960-61; commd. officer USPHS, 1961, advanced through grades to sr. surgeon, resigned, 1975; rsch. assoc. Nat. Inst. Mental Health, NIH, Bethesda, Md., 1961-63, sr. investigator, 1963-69; chief sect. on neuroendocrinology Nat. Heart and Lung Inst., NIH, Bethesda, Md., 1969-75; dir. Ctr. Molecular Nutrition and Sensory Disorders Georgetown U. Med. Ctr., Washington, 1975-85, assoc. prof. pediatrics and neurology, 1975-82, dir. Taste and Smell Clinic, 1985—, prof., 1982—; pres., chief exec. officer Sialon Corp., Washington, 1987—; cons. USDA/NIH, 1975—, Hooker Chemical Co., Buffalo, 1976-77, Washington Conf. for Zinc, 1985—, Florasynth, N.Y.C., 1986—, Squibb Pharm. Co., N.Y.C., 1986-87. Author: Zinc, 1975; contbr. articles to profl. jours. Patentee saliva and taste diagnostics. Composer music for solo speaker, chorus and orch. The Hollow Men, Ode for Strings performed L.A. Chamber Symphony, 1959. Recipient Vicennial medal Georgetown U., 1984; Atwater Kent fellow UCLA, 1957; grantee Dept. Def., USDA, NIH and various NIH insts., 1969—. Fellow Am. Coll. Nutrition; mem. Biophys. Soc. (charter), Am. Physiol. Soc., Inst. of Nutrition, Am. Soc. Clin. Investigation, Composers Guild Am., Cosmos Club, Phi Beta Kappa, Sigma Xi (nat. lectr. 1984-87). Avocations: tennis, running, skiing. Home: 6601 Broxburn Dr Bethesda MD 20817-4709 Office: Ctr Mol Nutrn/Sensory Disorders Taste and Smell Clin 5125 Macarthur Blvd NW # 20 Washington DC 20016-3300

HENLE, ROBERT ATHANASIUS, engineer; b. Virginia, Minn., Apr. 27, 1924; s. Robert Alois and Ethel (O'Donnel) H.; m. Eleanor Bonnel, Sept. 9, 1950 (dec. 1972); children—Robert, David, Barbara, John. B.S.E.E., U. Minn., 1949, M.S.E.E., 1951. Fellow components div. IBM, Fishkill, N.Y., 1964-70; with corp. tech. com. IBM, Armonk, N.Y., 1970-73; fellow IBM, Fishkill, N.Y., 1973-75, mgr. advanced tech. components div., 1975-80; dir. silicon tech. lab. IBM, Yorktown Heights, N.Y., 1980—; mem. adv. bd. Nat. Security Agy., 1968-73. Patentee in field. Contbr. articles to profl. jours. Served to lt. (j.g.) USN, 1944-46. Fellow IBM, 1964. Fellow IEEE (Edison medal 1987); mem. NAE, N.Y. Acad. Sci., AAAS, Sigma Xi. Republican. Roman Catholic. Avocations: sailing; skiing; tennis. Home: RR2-255 Sunset Trail Clinton Corners NY 12514-9629 Office: IBM T J Watson Rsch Ctr Research Div PO Box 218 Yorktown Heights NY 10598-0218

HENLEY, ERNEST MARK, physics educator, university dean emeritus; b. Frankfurt, Germany, June 10, 1924; came to U.S., 1939, naturalized, 1944; s. Fred S. and Josy (Dreyfuss) H.; m. Elaine Dimitman, Aug. 21, 1948; children: M. Bradford, Karen M. B.E.E., CCNY, 1944; Ph.D., U. Calif. at Berkeley, 1952. Physicist Lawrence Radiation Lab., 1950-51; research assoc. physics dept. Stanford U., 1951-52; lectr. physics Columbia U., 1952-54; mem. faculty U. Wash., Seattle, 1954—; prof. physics U. Wash., 1961—, chmn. dept., 1973-76, dean Coll. Arts and Scis., 1979-87, dir. Inst. for Nuclear Theory, 1990-91; researcher and author numerous publs. on symmetries, nuclear reactions, weak interactions and high energy particle interactions; chmn. Nuclear Sci. Adv. Com., 1986-89. Author: (with W. Thirring) Elementary Quantum Field Theory, 1962, (with H. Frauenfelder) Subatomic Physics, 1974, 2nd edit. 1991, Nuclear and Particle Physics, 1975. Bd. dirs. Pacific Sci. Ctr., 1984-87, Wash. Tech. Ctr., 1983-87; trustee Associated Univs., Inc., 1989—. Recipient sr. Alexander von Humboldt award, 1984, T.W. Bonner prize Am. Physics Soc., 1989, Townsend Harris medal CCNY, 1989; F.B. Jewett fellow, 1952-53, NSF sr. fellow, 1958-59, Guggenheim fellow, 1967-68, NATO sr. fellow, 1976-77. Fellow Am. Phys. Soc. (chmn. div. nuclear physics 1979-80, pres. elect. 1991, pres. 1992), AAAS (chmn. physics sect. 1989-90); mem. Nat. Acad. Scis., Sigma Xi. Office: U Wash Physics Dept FM 15 Seattle WA 98195

HENLEY, TERRY LEW, computer company executive; b. Seymour, Ind., Nov. 10, 1940; s. Ray C. and Barbara Marie (Cockerham) H.; children: Barron Keith, Troy Grayson; m. Jennifer L. Baldwin, Sept., 1991. BS, Tri-State U., 1961; MBA, Loyola U., 1980, D in Psychology, 1982. Rsch. and devel. engr. Halogens Rsch. Lab., Dow Chem. Co., Midland, Mich., 1961-63, lead process engr., polymer plant, Bay City, Mich., 1964, supt. bromide-bromine plants, Midland, 1964-68; nat. sales mgr. Ryan Industries, Louisville, 1968-70; internat. sales mgr. Chemineer, Inc., Dayton, Ohio, 1970-77; cons. mktg., Xenia, Ohio, 1977-78; pres. Med-Systems Mgmt., Inc., Dayton, Ohio, 1978—; pres. United Telemanagement, Inc., Dayton, 1991—; bd. dirs. Grandcor Hosp. Author: Chemical Engineering, 1976; contbr. articles in field to profl. jours.; patentee in field. Recipient C. of C. Civic award, 1990. Mem. ASTM (com.), Am. Hosp. Assn., Computer Based Patient Record Inst., Internat. Graphoanalysis Soc., Radiology Bus. Mgmt. Assn., Am. Med. Peer Rev. Assn., Am. Inst. Chem. Engrs., Am. Mgmt. Assn., Med. Group Mgmt. Assn., Ohio Handwriting Analysts Assn., Soc. for Ambulatory Care Profls., Healthcare Fin. Mgmt. Assn. Home: 278 N Childrens Home Rd Troy OH 45373 Office: Med-Systems Mgmt Inc 8290 N Dixie Dr Dayton OH 45414-2775

HENN, FRITZ ALBERT, psychiatrist; b. Alden, Pa., Mar. 26, 1941; s. Fredrich and Luise (Kimm) H.; m. Suella Henn, Aug. 1, 1964; children: Sarah, Stephen. BA, Wesleyan U., Middleton, Conn., 1963; PhD, Johns Hospkins U., 1967; MD, U. Va., 1971. Dir. rsch. tng. U. Iowa Hosps. and Clinics, Iowa City, 1975; asst. prof. U. Iowa, Coll. of Medicine, Iowa City, 1974-78, assoc. prof., 1978-81, prof. dept. psychiat., 1981; prof., chmn. SUNY, Stony Brook, 1982—; dir. L.I. Rsch. Inst., Stony Brook, 1982-83, Inst. of Mental Health Rsch., Stony Brook, 1983—; pres. Winter Conf. on Brain Rsch., 1990-92. Mem. editorial bd. Jour. Neurochemistry, 1980-90, Archives Gen. Psychiatry, 1983—. Cons. Project Dawn Justice Dept., 1973-74. Fellow Life Ins. Medicine Rsch. Fund, 1968-71, Falk fellow Am. Psychiat. Assn., 1972-74. Mem. AMA, Am. Coll. Psychiatrists, Am. Coll. Neuropsychopharmacology, Soc. for Neurol. Sci., Psychiat. Rsch. Soc. (pres. 1992), Am. Soc. Neurochemistry, Sigma Xi, Alpha Omega Alpha. Office: SUNY Stony Brook Dept Psychiatry HSC-T10 Stony Brook NY 11794

HENNESSEY, AUDREY KATHLEEN, information systems educator; b. Fairbanks, Apr. 4, 1936; d. Lawrence Christopher and Olga Virginia (Strandberg) Doheny; m. Gerard Hennessey, Mar. 10, 1963; children: Brian, Kate. BA, Stanford U., 1957; HSA, U. Toronto, Ont., Can., 1968; PhD, U. Lancaster, Eng., 1982. Asst. dir. European sales Univ. Soc., Heidelberg, Fed. Republic Germany, 1959-61; landman's asst. Union Oil Co. Calif., Anchorage, 1962-63; adminstr. group pension Mfgs. Life Ins., Toronto, 1963-65; instr. office systems Adult Edn. Ctr., Toronto, 1965-68; lectr. office systems Salford Coll. Tech., Lancashire, Eng., 1968-70; st. lectr. data processing Manchester Polytechnic, Eng., 1970-79; lectr. computation U. Manchester, Eng., 1979-82; assoc. prof. computer sci. Tex. Tech. U., Lubbock, 1982-86, assoc. prof. info. systems, 1987—, dir. ISOA bus. adminstrn., 1987—; vis. instr. Fed. Law Enforcement Tng. Ctr., Glynco, Ga., 1984-88. Author: Computer Applications Project, 1982; editor procs.: Office Document Architecture Internat. Symposium, English version, 1991; contbr. articles to profl. jours.; inventor in field; 5 patents pending. Organizer Explorer Scouts Computer Applications, Lubbock, 1983-85. Recipient various awards Tex. Instruments, 1982-86, Xerox Corp., 1985, Halliburton, 1986, Systems Exploration, 1987, State of Tex., 1988—, Knowledge-Based Image Analysis award USN Space Systems, 1991—, Immunization Tracking System award Robert Wood Johnson Found., 1993. Mem. Data Processing Mgmt. Assn. (pres. chpt. 1989, Disting. Info. Sci. award 1992), SME, ACM. Office: ISOA-Tex Tech U MS 2101 Lubbock TX 79409

HENNESSY, JOHN FRANCIS, III, engineering executive, mechanical engineer; b. N.Y.C., Nov. 27, 1955; s. John Francis Jr. and Barbara (McDonnell) H. AB, Kenyon Coll., 1977; BSME, Rensselaer Poly Inst., 1978; MS, MIT, 1988. Registered profl. engr., N.Y., N.J., Mass., Va., Del., Calif. Project engr. Syska & Hennessy, N.Y.C., 1978-83; project mgr. Syska & Hennessy, San Francisco, 1983-86; v.p. Syska & Hennessy, L.A., 1986-87; v.p. Syska & Hennessy, Cambridge, Mass., 1987-88, sr. v.p., 1988-89; CEO Syska & Hennessy, N.Y.C., 1989—, chmn., 1992—; chmn., bd. dirs. N.Y. Bldg. Congress, 1989—; chmn. Times Square Subway Sta. Improvement Corp., N.Y.C., 1989—. Mem. USO of Met. N.Y.; mem. adv. bd. Salvation Army of N.Y.; mem. Bldg. Futures Coun. Sloan fellowship, 1987. Mem. ASHRAE, NSPE, ASME, Coun. on Tall Bldgs. and Urban Habitat, Univ. Club, Olympic Club, Union League Club (N.Y.C.), Met. Club (Washington), Lyford Cay Club (Nassau), Winged Foot Golf Club (Mamaroneck, N.Y.), Nat. Golf Links of Am., Princeton Club (N.Y.C.). Roman Catholic. Avocations: golf, tennis, skiing, squash. Office: Syska & Hennessy ll W 42d St New York NY 10036

HENNET, REMY JEAN-CLAUDE, geochemist, environmental consultant; b. Courtetelle, Jura, Switzerland, July 11, 1955; came to U.S., 1981; s. Fernand and Georgette (Chèvre) H.; m. Christel Gertrud Boettcher, 1989; children: Margo, Louis. Diplôme de Géologie, Université de Neuchâtel, Switzerland, 1980; Certificat déHydrogéologie, Université de Neuchatel, Switzerland, 1981; PhD, Princeton Univ., 1987. Researcher Woods Hole (Mass.) Oceangraphic Inst., 1987-89; sr. geochemist S.S.Papadopulos & Assocs., Bethesda, Md., 1989—; mem. working group 91 Sci. Com. on Oceanic Rsch., 1989-92; mem. Expert Panel on PCB fate and behavior in environ., 1990—. Author: Predictions, Petroleum Generation, 1992; contbr. articles to profl. jours. Recipient scholarship Woods Hole Oceanographic Inst., 1987. Mem. Internat. Soc. Study Orgins of Life, Internat. Humic Substances Soc., Am. Chem. Soc., Am. Geophysical Union. Achievements include synthesis of amino acids under hydrothermal conditions. Office: SSP&A 7944 Wisconsin Ave Bethesda MD 20814

HENNIG, CHARLES WILLIAM, psychology educator; b. Queens, N.Y., May 7, 1949; s. Charles Joseph and Evelyn Mary (Gerstel) H.; m. Mary Christina Shamrock, Jan. 9, 1982; 1 child, Brian Steve. BA, SUNY, Buffalo, 1971; MS, Tulane U., 1976, PhD, 1978. Grad. teaching asst. Tulane U., New Orleans, 1974-78; vis. asst. prof. psychology U. Okla., Norman, 1978-79, Centre Coll. Ky., Danville, 1979-80; asst. prof. Salem (W.Va.) Coll., 1980-83, assoc. prof., 1983-88, prof., 1988-89, chair psychology, 1983-89; prof., chair psychology McMurry U., Abilene, Tex., 1989—; bd. dirs. African Elephant Rsch. and Survival Ranch, Abilene, 1990—. Contbr. articles to profl. jours. Vol. United Way of Abilene, 1990—; mem. Abilene Zool. Soc., 1990—. With U.S. Army, 1972-74. Mem. APA, Am. Psychol. Soc., Animal Behavior Soc., Psychonomic Soc., Midwestern Psychol. Assn., Southeastern Psychol. Assn., Abilene Psychol. Assn. (sec.-treas. 1990-91, pres. 1992-93), Psi Chi. Republican. Roman Catholic. Avocations: travel, reading. Home: 4701 Stonehedge Rd Abilene TX 79606-3429 Office: McMurry U Psychology Dept PO Box 86 Abilene TX 79697-0086

HENNING, KATHLEEN ANN, manufacturing systems engineer; b. Suffern, N.Y., Oct. 30, 1963; d. Frederick and Ann Elizabeth (O'Malley) H. BS in Indsl. Engring., Lehigh U., 1985; MS in Indsl. Engring./CIMS, Ga. Inst. Tech., 1986. Indsl. engring. asst. Lehigh U., Bethlehem, Pa., 1982-85; indsl. engr. GM, Warren, Ohio, 1983-85; computer integrated mfg. rsch. asst. Ga. Inst. Tech., Atlanta, 1985-86; computer integrated mfg. cons. Coopers & Lybrand, N.Y.C., 1986-87; sr. integrated mfg. engr. CRS Sirrine Engrs., Inc., Greenville, S.C., 1987—. Contbr. articles to profl. jours. Recipient Best Paper on Quality Mgmt. award Am. Soc. Quality Control, 1992; GM scholar, 1983-85. Mem. NAFE, S.C. Jr. C. of C. (Newsletter of Yr. award 1991), Cons. Engrs. S.C. (edn. com. 1990-93), Greenville Jr. C. of C. (Presdl. award of honor 1991), Soc. Mfg. Engrs. (instr. simulation 1989), Inst. Indsl. Engrs., Soc. Computer Simulation, Tau Beta Pi, Alpha Pi Mu. Office: CRS Sirrine Engrs PO Box 5456 1041 E Butler Rd Greenville SC 29606-5456

HENNINGER, POLLY, neuropsychologist, researcher; b. Pasadena, Calif., Apr. 1, 1946; d. Paul Bennett and Mary (MacNair) Johnson; m. Richard Henninger Jr., 1966 (div. 1983); children: Marguerite, Nathan; m. Clyde Pechstedt, 1985 (div. 1992). BA, Ind. U., 1967, Pomona Coll., 1977; MA, U. Toronto, 1969, PhD, 1982. Registered psychologist, Ont., Can. Postdoctoral fellow Calif. Inst. Tech., Pasadena, 1982-84; asst. prof. Pitzer Coll., Claremont, Calif., 1984-87, Brock U., St. Catharines, Ont., 1987-91; vis. assoc. div. biology Calif. Inst. Tech., Pasadena, 1984—; asst. dir. neuropsychol. svcs. Ctr. for Aging Resources, Fuller Theol. Sem., Pasadena, 1991-92; cons. Ch. of Our Savior, San Gabriel, Calif., 1992. Contbr. chpts. to books, articles to profl. jours. Recipient fellowships and grants. Mem. APA (div. 40 chair rsch. selection 1986-89), Am. Psychol. Soc., Can. Psychol. Assn., Calif. Psychol. Assn., Internat. Neuropsychol. Soc. Democrat. Episcopalian. Home: 126 S San Marino Ave Pasadena CA 91107 Office: Calif Inst Tech Div Biology 156-29 Pasadena CA 91125

HENRICKSON, EILER LEONARD, geologist, educator; b. Crosby, Minn., Apr. 23, 1920; s. Eiler Clarence and Mabel (Bacon) H.; m. Kristine L. Kuntzman; children: Eiler Warren, Kristin, Kurt Eric, Ann Elizabeth. BA, Carleton Coll., 1943; PhD, U. Minn., 1956. Geologist U.S. Geol. Survey, Calif., 1943-44; instr. Carleton Coll., 1946-47, 48-51, asst. prof., 1951-53, 54-56, assoc. prof., 1956-62, prof., 1962-70, Charles L. Denison prof. geology, 1970-87, chmn. dept., 1970-78, wrestling coach, 1946-58, 83-87; prof. geology, chmn. dept. Colo. Coll., 1987—; instr. U. Minn., 1947-48, 53-54; vis. lectr. numerous univs., Europe, 1962; cons. Jones & Laughlin Steel Corp., 1946-58, Fremont Mining Co., Alaska, 1958-61, G.T. Schieldahl Co., Minn., 1961-62, Bear Creek Mining Co., Mich., 1965-66, U. Minn. Messenia Expdn., 1966-75, Exxon Co., 1977-78, Cargill Corp., Mpls., 1983-84, Leslie Salt Co., San Francisco, 1985-86, various other cos.; research scientist, cons. Oak Ridge (Tenn.) Nat. Lab., 1985-86; cons. Argonne Nat. Lab., 1966-78, research scientist, summers, 1966-67; field studies metamorphic areas, Norway and Scotland; dir. young scholars program NSF, 1988-90. Author: Zones of Regional Metamorphism, 1957. Dir. Northfield Bd. Edn., 1960-63; steering com. Northfield Community Devel. Program, 1966-67. Served as 1st lt. USMCR, 1943, AUS, 1944-46. Fulbright research scholar archeol. geology, Greece, 1966-87. Mem. AAAS, Mineral Soc. Am., Nat. Assn. Geology Tchrs., Minn. Acad. Sci (vis. lectr.), Am. Geol. Inst., Geol. Soc. Am., Soc. Econ. Geologists, Rocky Mountain Assn. Geologists, Nat. Wrestling Coaches and Ofcls. Assn., Archaeol. Inst. Am. (vis. lectr.), Sigma Xi. Rsch. in archael. geology, Greece and North Africa, 1977-78, in mineral potential of Greece and Egypt, 1978-79, on ore deposits and archael. geology, province of copper and tin in artifacts in N.Am. and world. Home: 19560 Four Winds Way Monument CO 80132-9309 Office: Colo Coll Dept Geology Colorado Springs CO 80903

HENRIKSSON, JAN HUGO LENNART, architect, educator; b. Halmstad, Sweden, Feb. 28, 1933; s. John Hugo and Anna Kristina (Göransson) H.; m. Eva Anita Olson, Nov. 28, 1959 (div. June 1981); children: Lars Jonas, Mattias, Jan Andreas. Diploma in Bldg. Engring., Tekniska Gymnasiet, Gothenburg, Sweden, 1954; MArch, Royal Inst. Tech., Stockholm, 1960, PhD in Architecture, 1982. Cert. architect. Architect Peter Celsing Arkitektkontor AB, Stockholm, 1962-74; ptnr. AFHJ Arkitektkontor AB, Stockholm, 1974-78; owner Jan Henriksson Arkitektkontor AB, Stockholm, 1978—; prof. architecture U. Lund, Sweden, 1984-87, Royal Inst. Tech., Stockholm, 1987—. Author: Lägenheter på verkstadsgolvet; prin. works include Sveriges Riksbank, Stockholm, rebuilding and new bldg., Volvo Olofström new bldg. Bd. dirs. Malmstens Verkstadsskola. Sgt. Swedish Armed Forces, 1954. Recipient Stadsbyggnadspris, Malmö kommun, 1986, commemorative medal of honor ABI USA, Disting. Leadership award ABI USA, decorated FIDA Internat. Order Merit, 1990. Mem. Svenska Arkitekters Riksförbund, Sveriges Praktiserande Arkitekter, Arkitektförbundet, Stockholms Byggnadsförening. Mem. Social Dem. Party. Avocations: classical guitar, sports. Home: Riksrådsvägen 45, 121 60 Johanneshov Stockholm Sweden Office: Götgatan 48, 118 26 Stockholm Sweden

HENRY, CHARLES JAY, library director; b. Washington, June 17, 1950; s. Charles J. and June (Statz) H.; m. Nancy C. Todd, Oct. 4, 1986. BA, Northwest Mo., 1972; MA, Columbia U., 1977, MPhil, 1980, PhD, 1987. Instr. Columbia U., N.Y.C., 1981-82; asst. to dean Columbia Coll., N.Y.C., 1982-85; asst. dir., divsn. humanities, hist. Columbia Librs., N.Y.C., 1985-91; dir. libr. Vassar Coll., Poughkeepsie, N.Y., 1991—. Co-author: Computing and Humanities: New Dir., 1990; contbr. articles to profl. jours; panel mem., speaker in field. Lectrs., symposia peace edn. UN, Peace Edn. Columbia U. Fulbright scholar Vienna, 1980-81; Lilian Becker scholar Middlebury Coll., 1977; MacArthur Found. grantee, 1984-87; Presidents fellow Columbia U. 1978-79, 79-80; recipient Best Paper award humanities architecture divsn. Conf. Cybernetics and Systems Rsch., Vienna, 1992. Mem. AAAS, Am. Libr. Assn., Assn. Computers & Humanities, Am. Soc. for Information Sci., N.Y. Acad. Sci., Coalition for Networked Info. (project leader, 1991—). Democrat. Achievements include rsch. in cybernetics and systems rsch. Office: Vassar College PO Box 452 Poughkeepsie NY 12601

HENRY, GARY NORMAN, astronautical engineer, educator; b. Fort Wayne, Ind., Nov. 3, 1961; s. Norman Thomas and Elaine Cathrine (Schabb) H. BS in Astronautical Engring., USAF Acad., 1984; MS in Aero./Astronautical Engring., Stanford U., 1988. Commd. 2d lt. USAF, 1984, advanced through grades to capt., 1988; project engr. USAF Weapons Lab., Kirtland AFB, N.Mex., 1984-87; asst. prof. astronautics USAF Acad., Colorado Springs, 1989-93; flight test engr. USAF Test Pilot Sch., Edwards AFB, Calif., 1993—. Contbr. articles to profl. jours. Mem. AIAA (sr. mem.), Am. Soc. for Engring. Edn., Internat. Bd. Cert. Fin. Planners. Achievements include research director of 1st successful DOD land based hybrid rocket flight. Office: USAF Test Pilot Sch USAF TPS/EDB Edwards CA 93523-5000

HENRY, HOWELL GEORGE, electrical engineer; b. Athens, Ga., June 1, 1948; s. Hugh Fort and Emmaline (Rust) H.; m. Stephanie Anne Strine, June 17, 1978; children: Leigh-Anne, Howell Joseph, Caitlin Elizabeth. BS in Physics, Duke U., 1970; MSEE, U. Notre Dame, 1972, PhDEE, 1979. Engr. Cin. Electronics Corp., 1972-74; project engr. Turnbull Control Systems Ltd., Worthing Sussex, U.K., 1974-76; mem. tech. staff Tex. Instruments, Dallas, 1979-82; fellow engr. Westinghouse Electric Corp., Balt., 1982—. Contbr. articles to profl. jours. Pres., treas. Woodbridge Valley Improvement and Civic Assn., Balt., 1985—; den leader Cub Scouts Boy Scouts Am., Balt., 1989—. Mem. IEEE. Home: 1418 Lincoln Woods Dr Baltimore MD 21228 Office: Westinghouse Electric Corp MS3K11 Box 1521 Baltimore MD 21203

HENRY, JOSEPH PATRICK, chemical company executive; b. Mansfield, Ohio, Mar. 3, 1925; s. Harold H. and Louise A. (Droxler) H.; student Bowling Green State U., 1943-44; B.S., Ohio State U., 1949; m. Jeanette E. Russell, Oct. 26, 1957; 1 dau., Jeanette Louise. Ohio sales mgr. NaChurs Plant Food Co., Marion, Ohio, 1949-55; organizer, pres. Growers Chem. Corp., Milan, 1955—, Sandusky Imported Motors, Inc. (Ohio), 1958-78; pres. Homestead Motors, Inc., 1978-83; co-owner Homestead Inn Restaurant, Homestead Farms; v.p. Homestead Inn, Inc. Motels, South Avery Corp. Motels; dir. Erie County Bank, Vermilion, Ohio., Soc. Bank of Firelands. Served with USMCR, 1943-46; PTO. Mem. Nat. Fedn. Ind. Bus. (nat. adv. council), AAAS,Ohio Farm Bur. Fedn., Milan C. of C., Aircraft Owners and Pilots Assn., Internat. Flying Farmers, Ohio Restaurant Assn., Ohio Motel-Hotel Assn., Ohio Licensed Beverage Assn., Am. Horse Show Assn., Nat. Trust for Historic Preservation, N.A.M., Internat. Platform Assn., Huron County Hist. Soc., Ohio Farm Bur., (pres.), Ohio, Internat. (dir. 1978-84) Arabian horse assns. Clubs: Antique Automobile Am., Sports Car Am., N. Am. Yacht Racing Union, Sandusky Yacht, Sandusky Sailing, Catawba Island. Developer (with V.A. Tiedjens) foliage fertilization and direct to seed fertilization of comml. field crops. Home: 128 Center St Milan OH 44846-9757 Office: Growers Chem Corp PO Box 1750 Milan OH 44846-1750 Home: Homestead Farms RR 1 Milan OH 44846-9801

HENRY, KEITH EDWARD, chemical engineer; b. Nashville, May 17, 1953; s. William Calloway and Maxine Marilyn (Kirchman) H.; m. Kimberly Elizabeth Wallace, Aug. 30, 1975; children: Sarah Elizabeth, David Edward, Patrick Wallace. BS in Chemistry, U. Del., 1975; MS in Chemistry, U. Pitts., 1980, MBA, 1990. Power and fuel engr. ctrl. engring. dept. U.S. Steel Chems. divsn. USX Corp., Pitts., 1975-80, rsch. engr., 1980-82, application engr. process engring. dept., 1982-83, mktg. mgr. coal chems., 1983-84, systems coord., 1984-85; mktg. mgr. Aristech Chem. Corp., Pitts., 1985-88, mgr. tech. svcs., 1988-92, divsn. mgr. quality and tech. svc., 1992—. Elder Meml. Pk. Presbyn. Ch., Allison Park, Pa., 1987-89, deacon, 1981-85; fund raiser local charities, 1992-93. Mem. ASTM (chmn. task group 1988—), Chem. Mfrs. Assn. (phenol safety panel 1988—). Republican. Office: Aristech 600 Grant St Rm 1066 Pittsburgh PA 15219-2704

HENRY, KENT DOUGLAS, analytical chemist, hardware engineer; b. Lancaster, Pa., Sept. 15, 1964; s. Glenn S. Henry and Ella Mae (Nolt) Murray; m. Patricia Anne Thissell, Jan. 1, 1993. BS in Chemistry, Lebanon Valley Coll., 1986; PhD, Cornell U., 1991. R&D chemist Hewlett-Packard Sci. Instruments Div., Palo Alto, Calif., 1991—. Contbr. articles to Jour. Am. Chem. Soc., Organic Mass Spectrometry, other profl. publs. Pres., sec. Friends of Palo Alto Jr. Mus., 1992—. Mem. Am. Chem. Soc., Am. Soc. Mass Spectrometry, Bay Area Mass Spectrometry Soc. Brethren. Achievements include instrument coupling electrospray ionization with FTMS; unit mass resolution of 10 KD biomolecules. Office: Hewlett Packard Co 1601 California Ave Palo Alto CA 94304

HENRY, LEANNE JOAN, physicist; b. Pitts., Jan. 31, 1960; d. Albert Eugene and Joan (McIntyre) H. BS in Chemistry, Carnegie-Mellon U., 1982; MS in Physics, U. Pitts., 1984, PhD, 1989. Physicist Air Force Fgn. Aerospace Sci. and Tech. Ctr., Wright-Patterson AFB, Ohio, 1989—. With USAF, 1989—. Home: 3798-B Catalina Dr Beavercreek OH 45431 Office: FASTC TATE Wright Patterson AFB OH 45433-6508

HENRY, MATTHEW JAMES, biochemist, plant pathologist; b. North Platte, Nebr., Apr. 2, 1954; s. LeRoy Jerome and Lois Ophelia (Butts) H.; m. Margaretha Emma Schalk, May 8, 1982; children: Matthew Alan, Colin Anderson, Garrett Jeffery, Keith Elliot. BA, U. Denver, 1976; PhD, U. Md., 1982. Rsch. biologist E.I. DuPont de Nenours Co., Wilmington, Del., 1982-89; rsch. scientist Eli Lilly and Co., Greenfield, Ind., 1989-90; rsch. group leader crop disease mgmt. DowElanco, Indpls., 1990—; rev. bd. USDA Biocontrol Lab., Beltsville, Md., 1990. Editorial bd. Pesticide Sci., 1988—; contbr. articles to sci. jours. Bd. dirs. Nurse Midwifery Ctr. Del., Wilmington, 1986-88. Mem. Am. Phytopathological Soc. Office: DowElanco PO Box 68955 9410 Zionsville Rd Indianapolis IN 46140

HENRY, NORMAN WHITFIELD, III, research chemist; b. Phila., May 8, 1943; s. Norman Whitfield and Ethel (Black) H.; m. Joy Lessner, Apr. 15, 1967; 1 child, Heather. BA in Chemistry, Lafayette Coll., Newa, 1966; MS in Chemistry, U. Del., 1977. Cert. prof. chemist; cert. clin. chemist and indsl. hygienist. Chemist DuPont, Newark, Del., 1967-70, rsch. chemist, 1970-80, sr. rsch. chemist, 1980—; chemistry instr. Del. Tech., Stanton, 1981—, indsl. hygiene instr., 1981—. Author: Chemical Protective Clothing, 1990; contbr. articles to Am. Indsl. Hygiene. Mem. Del. State Radiation Authority, 1979—; bd. dirs. Appalachian Compact of Users of Radio Isotopes, State College, Pa., 1988—. Mem. ASTM (award of merit), Am. Chem. Soc. (treas. local sect. 1985), Am. Indsl. Hygiene Assn., Nat. Registry in Clin. Chemistry, Am. Bd. Indsl. Hygiene. Republican. Episcopalian. Achievements include rsch. on protective clothing, analytical and environmental chemistry. Home: 129 Ballantrne Dr Elkton MD 21921 Office: DuPont Stine-Haskell Rsch Ctr PO Box 50 Elkton Rd Newark DE 19714

HENRY, THEODORE LYNN, nuclear scientist, educator; b. Springfield, Ill., Jan. 4, 1949; s. Douglas Alan and Melba Mourine (McClain) H.; m. Barbara Jo Burns, June 21, 1971 (div. Sept. 1987); children: Michael, Kristopher; m. Kernelle Marie McFarland, May 28, 1988; 1 child, Alison. BA, Sangamon State U., 1973, MA, 1989. Lab. mgr. Sangamon State U., Springfield, Ill., 1971-73; exec. Bank of Springfield, 1973-74; chief budget analyst State of Ill. Dept. Rehab. Svcs., Springfield, 1974-90; nuclear scientist State of Ill. Dept. Nuclear Safety, Springfield, 1990—; mem. adj. faculty Sangamon State U., Springfield, 1987-91. Mem. Sangamon County Rep. Found., Springfield, 1991—. Staff sgt. USAF, 1969-71, CBI. Mem. AAAS, Astron. Soc. of the Pacific, Lions. Lutheran. Achievements include design of computer interface linking PC based system to microdensitometer. Home: 4013 Hazelcrest Rd Springfield IL 62703-5237 Office: State Ill Dept Nuclear Safety 1035 Outer Park Dr Springfield IL 62704-4462

HENSGEN, HERBERT THOMAS, medical technologist; b. Cin., May 28, 1947; s. Herbert and Carolyn Elizabeth (Stites) H. BS, U. Cin., 1973, MS, 1978; AAS, Cin. Tech. Coll., 1981. Reg. med. technologist. Grad. teaching asst. U. Cin., 1976-77; instr. Edgecliff Coll., Cin., 1977-78; tech. Our Lady of Mercy Hosp., Cin., 1979-81, med. lab. tech., 1981-84, med. technologist, 1984-86; rsch. asst. Children's Hosp. Med. Ctr., Cin., 1986—; instr. Cin. Tech. Coll, 1984-85. Contbr. article to Gen. and Comparative Endocrinology; co-author abstracts for Soc. for Pediatric Rsch., Endocrine Soc. Deacon Madisonville Bapt. Ch., Cin., 1977. Mem. Am. Soc. Clin. Pathologists, Am. Soc. Zoologists, Am. Mensa, Ltd., N.Y. Acad. Scis. Achievements include production of data suggesting lack of insulin-like growth factor-1 (IGF-I) may mediate growth retardation in the neonatal rat; discovery of evidence that IGF-I may be one of several growth factors regulating differentiation of the fetal brain; demonstration that the antigonadal effect of prolactin in the lizard Anolis carolinensis is directed toward the smaller ovarian follicles; research on effects of IGF-I and its binding proteins on fetal and neonatal development. Home: 7420 Drake Rd Cincinnati OH 45243-1422 Office: Children's Hosp Med Ctr Elland & Bethesda Aves Cincinnati OH 45229

HENSHILWOOD, JAMES ANDREW, chemist, researcher; b. Epsom, Surrey, Eng., Mar. 13, 1964; s. Conrad Peter and Gwendolen Joy (Mills) H.; m. Maria Jane Bell, June 6, 1992. BS with honors, Thames Poly. U., 1986; PhD, Southampton U., 1989. Rsch. student Beecham Pharms., Epsom, 1984-85; rsch. asso. Wayne State U., Detroit, 1989-91; rsch. assoc. U. Sheffield, United Kingdom, 1992—. Contbr. articles to profl. jours. Mem. Royal Soc. Chemistry, Am. Chem. Soc. Mem. Ch. of England. Achievements include co-discovery of versatile and high yielding method for obtaining 6TI and 2TI cycloadditions. Office: U Sheffield, Dept Chemistry, Sheffield S3 7HF, England

HENSLEY, JARVIS ALAN, petroleum engineer; b. Odessa, Tex., Sept. 12, 1954; s. John Newton and Syble Maxine (Bilbrey) H.; m. Jana Lynece Bass, Mar. 2, 1974; 1 child, Heather Nicole. BS in Petroleum Engring., Tex. Tech. U., 1981. Registered profl. engr., Okla. Ops. engr. Arco Oil & Gas, 1981-85, Pacific Enterprises Oil Co., 1985-89, Diamond Energy, Tulsa, 1989—. Mem. NSPE, Soc. Petroleum Engrs., Okla. Soc. Profl. Engrs. Republican. Baptist. Office: Diamond Energy 8908 S Yale Ave Ste 340 Tulsa OK 74137-3561

HENSON, ANNA MIRIAM, otolaryngology researcher, medical educator; b. Springfield, Mo., Nov. 7, 1935; d. Bert Emerson and Esther Miriam (Crank) Morgan; m. O'Dell Williams Henson, Aug. 1, 1964; children: Phillip, William. BA, Park Coll., Parkville, Mo., 1957; MA, Smith Coll., 1959; PhD, Yale U., 1967. Instr. Smith Coll., Northampton, Mass., 1960-61; rsch. assoc. Yale U., New Haven, 1967-74; instr. U. N.C., Chapel Hill, 1975-78, rsch. asst. prof., 1978-83, rsch. assoc. prof., 1983-86, rsch. prof. dept. surgery Sch. Medicine, 1986—; mem. study sect. on hearing rsch. NIH, Bethesda, Md., 1990-93. Contbr. articles to profl. jours. Fulbright scholar, Australia, 1959-60; NIH grantee, 1975—. Mem. Assn. for Rsch. in Otolaryngology, Sigma Xi. Office: U NC CB # 7090 Taylor Hall Chapel Hill NC 27599

HENSON, C. WARD, mathematician, educator; b. Worcester, Mass., Sept. 25, 1940; s. Charles W. and Daryl May (Hoyt) H.; m. Faith deMena Travis, August 31, 1963; children: Julia Rebecca, Suzanne Amy, Claire Victoria. AB, Harvard U., 1962; PhD, MIT, 1967. Asst. prof. Duke U., Durham, N.C., 1967-74, N.Mex. State U., Las Cruces, 1974-75; asst. prof. U. Ill., Urbana, 1975-77, assoc. prof., 1977-81, prof., 1981—, chmn. dept. math., 1988-92; vis. assoc. prof. U. Wis., Madison, 1979-80; vis. prof. RWTH Aachen, Fed. Republic Germany, 1985-86, Univ. Tübingen, Fed. Republic Germany, 1992-93. Mem. Assn. for Symbolic Logic (sec.-treas. 1982—), Am. Math. Soc., Assn. for Computing Machinery, Math. Assn. Am., London Math. Soc., European Assn. Theoretical Computer Sci. Office: U Ill Dept Math 1409 W Green St Urbana IL 61801-2917

HENSON, EARL BENNETTE, biologist; b. Charleston, W.Va., June 13, 1925; s. Earl Bennette and Lillian (Davison) H.; m. Ruth Marie Mayer, June 26, 1956; children: Katherine Anne, Karl Edward. BS, Marshall U., 1949; MS, U. W.Va., 1950; PhD, Cornell U., 1954. Tutor St. John's Coll., Annapolis, Md., 1958-62; aquatic biologist USPHS/EPA, Cin., 1962-65; prof. U. Vt., Burlington, 1970-91, emeritus prof., 1991—; sr. biologist Aquatic Inc., 1992—. Author: Trophic Status and Lake Champlain, 1977. Author: Trophic Status of Lake Champlain, 1977. Fellow Ohio Acad. Sci.; mem. Am. Soc. Limnology and Oceanography, Internat. Assn. Great Lakes Rsch., Internat. Assn. Theoretical and Applied Limnology, N.A. Benthological Soc. Home: 102 Adams St Burlington VT 05401

HENSON, PETER MITCHELL, physician, immunology and respiratory medicine executive; b. Moreton in Marsh, England, Aug. 25, 1940; came to U.S., 1967; S. Edwin Roland Witham and Irene May (Whitmee) H.; m. Janet Elizabeth Neilan, Dec. 22, 1962; children: Neil, Ruth. BVM&S with William Dick medal for best graduate, Edinburg (Scotland) U., 1963, BS in Bacteriology with first class honors, 1964; PhD, U. Cambridge, England, 1967. Rsch. fellow dept. exptl. pathology Scripps Clinic and Rsch. Found., La Jolla, Calif., 1967-69, assoc. dept. exptl. pathology, 1969-72, assoc. mem. dept. immunopathology, 1972-77; dir. rsch. dept. pediatrics Nat. Jewish Hosp. and Rsch. Ctr., Nat. Asthma Ctr., Denver, 1977-80, vice-chmn. dept. pediatrics, 1980-82, asst. v.p. profl. svcs., 1981-82, v.p. biomedical svcs., 1982-84; prof. pathology sch. medicine U. Colo., Denver, 1977—, prof. dept. medicine, sch. medicine, 1980—, co-head pulmonary dvsn. dept. medicine, health scis. ctr., 1986-87, assoc. dean hosp. affairs health scis. ctr., 1985—; exec. v.p. biomedical affairs Nat. Jewish Ctr. Immunology and Respiratory Medicine, Denver, 1985-92; exec. v.p. acad. affairs U. Colo., Nat. Jewish Ctr. Immunology and Respiratory Medicine, Denver, 1992—. Mem. editorial bd. Am. Jour. Pathology, 1982-88, Am. Review for Respiratory Disease, Pulmonary Pharmacology, 1988-92 and other profl. jours. Recipient Am. Heart Assn. Established investigatorship, 1969-70, Research Career Devel. award NIH, 1970-75, Parke Davis award Am. Assn. Pathologists, 1980; named Margaret A. Regan prof. Pulmonary Inflammation, 1991. Mem. Royal Coll. Vet. Surgeons, Brit. Soc. Immunology, Am. Assn. Pathology, Am. Assn. Immunologists, Am. Thoracic Soc., Am. Soc. Cell Biology, Am. Acad. Allergy and Immunology, Reticuloendothelial Soc. (Marie T. Bonazinga award 1991). Achievements include research in inflammation. Office: Nat Jewish Ctr Immunology and Respiratory Medicine 1400 Jackson St Denver CO 80206

HENTGES, DAVID JOHN, microbiology educator; b. LeMars, Iowa, Sept. 18, 1928; s. Romaine Francis and Geneva Mae (Kruger) H.; m. Kathleen Edwina Mullan, Dec. 28, 1957; children: Stephen Edward, Kathleen Marie, Margaret Ann. BS, U. Notre Dame, 1953; MS, Loyola U., Chgo., 1958, PhD, 1961. Asst. prof. Creighton U. Sch. Medicine, Omaha, 1964-67, assoc. prof., 1967-68; assoc. prof. U. of Mo. Sch. of Medicine, Columbia, 1968-72, prof., 1972-81, interim chmn., 1976-79; prof., chmn. Tex. Tech U. Sch. Medicine, Lubbock, 1981—. Editor: Human Intestinal Microflora, 1983, Medical Microbiology, 1986; regional editor Microbial Ecology in Health and Disease, 1987—; editorial bd. Infection and Immunity, 1983-92; contbr. chpts. to books and articles to profl. jours. Lay gen. chmn. Diocesan Cath. Appeal, Lubbock, 1989, steering com., 1985—. Named Knight Order of the Holy Sepulchre, 1990. Fellow Am. Acad. of Microbiology; mem. Am. Soc. Microbiology, Assn. for Gnotobiotics, Soc. for Microbial Ecology and Dis. (pres. 1987-89), Serra Internat. (club pres. 1983-84, dist. gov. 1987-88, Serran of Yr. 1988), Sigma Xi. Republican. Roman Catholic. Avocations: gardening, fly fishing. Home: 4601 88th St Lubbock TX 79424-4107 Office: Tex Tech U Health Sci Ctr Dept Microbiology Lubbock TX 79430

HEPGULER, YASAR METIN, architectural engineering educator, consultant; b. Istanbul, Turkey, Oct. 25, 1931; arrived in Switzerland, 1962; s. Hasan Sevki and Hatice Inayet Hepguler; m. M. Guner Berktan, Apr. 12, 1953 (div. 1966); m. Rahmiye Alev Arz, Apr. 24, 1968. MS and diploma in archtl. engring., Tech. U. of Istanbul, 1953. Mng. dir., co-owner SITE Collective Design Team, Istanbul, 1954-66; gen. sec. TMMOB Inst. of Architects, Istanbul, 1955-56; chmn., owner MHM Internat., Istanbul and Ankara, Turkey, 1966-70; chief advisor Pub. Security, Riyadh, Saudi Arabia, 1969-75; chmn., owner Architects Assn. Ltd., Zurich, Geneva, Istanbul, 1969—; invited prof. and lectr. ETH, Zurich, 1985, Colo. Poly., 1988-89, Calif. State Poly. U., 1990-91; cons. GE, 1966, Chrysler, 1969, Internat. Sheraton, 1985, Bayer, Indsl. Devel. Bank, Turkey, 1969-74; mng. dir. Architects Assn. Ltd., Zurich, Geneva, Istanbul and London. Contbr. numerous articles to profl. jours. Named Hon. Representant of Nation by Pres. of Turkey, 1981; winner 158 design awards, 44 1st prizes in archl. design competitions (world record). Fellow TMMOB Inst. Architects; mem. Inst. Inst. Architects, Swissair Travel Club, Am. Club Switzerland (Geneva chpt.). Avocations: foreign exchange market, politics, classical music, walking. Office: Architects Assn Ltd, PO Box 5218, 8022-CH Zurich Switzerland

HEPPA, DOUGLAS VAN, computer specialist; b. Bklyn., May 26, 1945; s. Joseph Charles and Antoinette Palmer (Vanasco) H.; m. Barbara Zanlunghi. BS in Social Sci., Poly. Inst. N.Y., 1968, BS in Math., 1971, MS Insl. & Applied Math., 1973, postgrad., 1983—. Assoc. engr. Raytheon Co., Portsmouth, R.I., 1968-70; systems engr. PRD Electronics, Syosset, N.Y., 1970-71; mathematician USN, New London, Conn., 1971; asst. computer engr. George Sharp, N.Y.C., 1972-73; programmer N.Y.C. Dept. Social Svcs., 1975; quantitative analyst N.Y.C. Fire Dept., 1976-80, assoc. staff analyst, 1980-81, computer specialist, 1981—; pres. Algorithim Devel. Co., Queens, N.Y., 1985—. Mem. Math. Assn. Am., Am. Mgmt. Assn., Soc. for Indsl. & Applied Math., Assn. for Computing Machinery, IEEE, Am. Math. Soc. Avocations: fishing, swimming, boating, amateur radio. Home: 64 08 60 Rd Maspeth NY 11378 Office: NYC Fire Dept 250 Livingston St Brooklyn NY 11201-5812

HEPPNER, DONALD GRAY, JR., immunology research physician; b. Lynchburg, Va., Jan. 17, 1956; s. Donald Gray Sr. and Nathalie (Ward) H.; m. Mary Virginia Leach, June 12, 1983; children: Charlotte Nathalie, Virginia Dearing. BA in Biochemistry/German Lit., U. Va., 1978, MD, 1983. Diplomate Am. Bd. Internal Medicine, Am. Bd. Infectious Diseases. Intern in internal medicine U. Minn. Hosps. and Clinics, Mpls., 1983-84, resident in internal medicine, 1984-86; rsch. assoc. Dight Lab., U. Minn., Mpls., 1987; with emergency medicine dept. Abbot North Western Hosp., Mpls., 1986-88; fellow infectious diseases U. Md., Balt., 1988-90; infectious disease officer Dept. Immunology, Walter Reed Army Inst. of Rsch., Washington, 1990-93; attending physician Walter Reed Army Med. Ctr., Washington, 1991-93. Mem. Com. of Fgn. Rels., Charlottesville, Va., 1983—. Maj. U.S. Army, 1990—. Fellow ACP; mem. Am. Soc. Tropical Medicine and Hygiene, Assn. of Mil. Surgeons of the U.S. Achievements include development and testing of human malaria vaccines. Office: Dept Immunology US Army Med Component AFRIMS APO AP 96546-5000 also: US Army Med Component AFRIMS, 315/6 Rajavithi Rd, Bangkok 10400, Thailand

HERB, CYNTHIA JOHNSON, programmer; b. Newark, N.Y., June 17, 1957; d. Daniel Edward and Dolores Margaret (Himes) Johnson; m. Mark Steven Herb, Sept. 27, 1980; children: Kathryn Elizabeth, Michael Robert. B in Math., Siena Coll., 1978. Jr. programmer IBM Corp., Kingston, N.Y., 1978-79; assoc. programmer IBM Corp., Kingston, 1979-80, sr. assoc. programmer, 1980-83, staff programmer, 1983-85, devel. mgr., 1985-89, advisory programmer, 1989-92, sr. programmer, 1992—. League coord. Am. Youth Soccer Orgn., Saugerties, N.Y., 1991, 92. Republican. Episcopalian. Home: 2924 Winchester Dr Round Rock TX 78664 Office: IBM Corp Burnett Rd MS 2900 Austin TX 78756

HERB, RAYMOND G., physicist, manufacturing company executive; b. Navarino, Wis., Jan. 22, 1908; s. Joseph and Annie (Stadler) H.; m. Anne Williamson, Dec. 26, 1945; children—Stephen, Rebecca, Sara, Emily, William. PhD in Physics, U. Wis., Madison, 1935, DSc (hon.), 1988; PhD (hon.), Lund U., 1993. Assoc. prof. U. Wis., Madison, 1941-45, prof., 1945-61, Charles Mendenhall prof., 1961-72; chmn. bd., pres. Nat. Electrostatics, Middleton, Wis., 1965—. Contbr. articles to profl. jours. Recipient Tom W. Bonner award Am. Physical Soc., 1968. Fellow Am. Phys. Soc.; mem. NAS. Office: Nat Electrostatics Corp PO Box 310 7540 Graber Rd Middleton WI 53562

HERBEL, LEROY ALEC, JR., telecommunications engineer; b. Ft. Carson, Colo., July 24, 1954; s. LeRoy Alec and Mabel Bertha (Huffman) H. BS, S.W. Mo. State U., 1976; MEd, Ga. So. U., 1978; MS in Telecommunications, Golden Gate U., 1987, MBA, 1990. Asst. mgr. toy dept. Dillard's Dept. Store, Springfield, Mo., 1971-76; material controller GTE of the South, Durham, N.C., 1979-80; asst. prof. mil. sci. Army ROTC, U. N.H., Durham, 1982-85; tech. instr. Northern Telecom Inc., Raleigh, N.C., 1988-91; sr. engr. No. Telecom Inc., Raleigh, N.C., 1991—; adj. prof. DeKalb (Ga.) C.C., 1978-79, N.C. Wesleyan Coll., Rocky Mount, 1991. Scoutmaster Troop 213 Boy Scouts Am., Cary, N.C., 1990-93, asst. dist. commr. Dan Beard dist., 1992—, mem. merit badge staff Nat. Jamboree, 1993. Capt. U.S. Army, 1980-88; maj., USAR, 1988—. Recipient Scoutmaster award of merit Boy Scouts Am., 1991, Disting. Leadership citation Boy Scouts Am., 1991, Scoutmaster Key award Boy Scouts Am., 1992. Mem. Telephone Pioneers of Am., Phi Delta Kappa. Avocations: golf, running, trains, camping, music. Office: Northern Telecom Inc PO Box 13010 Durham NC 27709-3010

HERBERT, VICTOR DANIEL, medical educator; b. N.Y.C., Feb. 22, 1927; s. Allan Charles and Rosaline (Margolis) H.; children from previous marriages: Robert, Steven, Kathy, Alissa, Laura. BS in Chemistry, Columbia U., 1948, MD, 1952, JD, 1974. Intern Walter Reed Army Med. Ctr., N.Y.C., 1952-53; resident Montefiore Hosp., Bronx, N.Y., 1954-55; asst. instr., rsch. fellow Albert Einstein Coll. of Medicine, Bronx, 1955-57; rsch. asst. in hematology Mt. Sinai Hosp., N.Y.C., 1958-59; from instr. to asst. prof. Med. Sch.'s Thorndike Lab. Harvard U., Boston, 1959-64; assoc. prof., prof. pathology and medicine Columbia U., N.Y.C., 1964-72; clin. prof. pathology and medicine Mt. Sinai Sch. of Medicine, N.Y.C., 1964—; prof. SUNY Downstate Med. Ctr., Bklyn., 1976-84. Author: Nutrition Cultism: Facts & Fictions, 1981; co-author: Vitamins & Health Food: The Great American Hustle, 1981, Genetic Nutrition: Designing a Diet Based on Your Family Medical History, 1993; editor, author: The Mount Sinai School of Medicine Complete Book of Nutrition, 1990; contbr. over 650 articles to profl. jours. Lt. col. U.S. Army, ETO, 1944-46, Korea, 1952-54, Vietnam, 1965, 66, 73, Mid. East. 1990-91. Recipient Middleton award U.S. Dept. Vets. Affairs, 1978, Commr.'s Citation, FDA, 1984. Fellow ACP; mem. Am. Soc. Hematology (Parliamentarian 1975—), Am. Fedn. Clin. Research, Am. Soc. Clin. Investigation, Assn. Am. Physicians, Am. Inst. Nutrition (Fellow award 1993), Am. Soc. Clin. Nutrition (pres. 1980-81, Herman award 1986, McCollum award 1972, Van Slyke award 1990). Avocations: theatre, Judo. Office: Mt Sinai Sch of Medicine 130 W Kingsbridge Rd Bronx NY 10468-3992

HERCULES, DAVID MICHAEL, chemistry educator, consultant; b. Somerset, Pa., Aug. 10, 1932; s. Michael George and Kathryn (Saylor) H.; m. Nancy Catherine Miller, Sept. 23, 1957 (div. 1968); 1 dau., Kimberly Ann; m. Shirley Ann Hoover, Dec. 14, 1970; children: Sherri Kathryn, Kevin Michael. B.S., Juniata Coll., 1954; Ph.D., MIT, 1957. Asst. prof. Lehigh U., 1957-60; assoc. prof. Juniata Coll., Huntington, Pa., 1960-63; asst. prof. MIT, 1963-68, assoc. prof., 1968-69; assoc. prof. U. Ga., Athens, 1969-74, prof., 1974-76; prof. dept. chemistry U. Pitts., 1976—, chmn., 1980-89, Miles Prof., 1990—; mem. vis. com. for chemistry Lehigh U., 1980-84 ; vis. prof. Mich. State U., 1972; chmn. Gordon Research Conf. on Electron Spectroscophy, 1974, Gordon Research Conf. on Analytical Chemistry, 1966; co-chmn. Internat. Conf. Chemiluminescence, 1972; univ. rep. Council on Chem. Research, 1980-88 ; mem. program com. Pitts. Conf. on Analytical Chemistry and Applied Spectroscopy, 1977—; mem. vis. scientist program NSF, 1964-76. Mem. editorial bds.: Applied Spectroscopy, 1963-65, Analytical Chemistry, 1964-67, Jour. Electron Spectroscopy, 1971-77, Environ. Analytical Chemistry, 1973—, Spectrochimica Acta, 1973-83, Talanta, 1974-80, Spectroscopy Letters, 1975—, The Scis., 1979-84, Trends in Analytical Chemistry, 1980-88, Jour. Trace and Microprobe Techniques, 1980—; patentee (in field). Recipient Benedetti-Pichler award Am. Microchem. Soc., 1987, Achievement in Analytical Chemistry award Ea. Analytical Symposium, 1988, prize Alexander von Humboldt Found., 1984, Disting. Alumnus award Juniata Coll., 1989, Pres.'s Disting. Rsch. award U. Pitts., 1990; John Simon Guggenheim Meml. fellow, 1973. Mem. Am. Chem. Soc. (Petroleum Research Fund adv. bd. 1978-80, chmn. div. analytical chemistry 1977-78, analytical chemistry award 1986, Arthur W. Adamson award disting. svc. in advancement of surface chemistry 1993), Soc. Applied Spectroscopy (Lester W. Strock medal New Eng. sect. 1981), Am. Vacuum Soc., Photoelectric Spectrometry Group, Pa. Acad. Scis., Spectroscopy Soc. Pitts., Soc. Analytical Chemists Pitts., Sigma Xi. Home: 1134 Fox Chapel Rd Pittsburgh PA 15238-2016 Office: U Pitts Dept Chemistry Pittsburgh PA 15260

HERCZYNSKI, ANDRZEJ, physicist, educator; b. Warsaw, Poland, Apr. 29, 1956; came to U.S., 1980; s. Ryszard and Elvira (Calman) H. MS in Math., Warsaw U., 1980; MS in Physics, Lehigh U., 1983, PhD in Physics, 1987. Lectr. math. Lehigh U., Bethlehem, Pa., 1987; postdoctoral rsch. assoc. U. Colo., Boulder, 1987-90; asst. to exec. dir. Am. Inst. Physics, Woodbury, N.Y., 1990—. Contbr. articles to profl. jours. Mem. adv. bd. Internat. Ednl. Orgn., Patchogue, N.Y., 1992—. Recipient fellowship Lehigh U., 1985-86. Mem. Am. Phys. Soc., Soc. Indsl. and Applied Math. Home: 65 N Woodhull Rd Huntington NY 11743 Office: Am Inst Physics 500 Sunnyside Blvd Woodbury NY 11797

HEREMANS, JOSEPH PIERRE, physicist; b. Leuven, Belgium, Jan. 8, 1953; came to U.S., 1984; s. Joseph Felix and Marie Therese (Bracke) H.; m. Claire Pierre Mali, July 1, 1978; children: Hilde Anne, Joseph Paul. Elec. Engr., U. Louvain, Belgium, 1975, PhD in Applied Physics, 1978. Aspirant Belgium Nat. Sci. Found., Louvain, 1978-80, charge de recherche, 1980-83; rsch. scientist GM Rsch., Warren, Mich., 1984-85, group leader, 1985-87, sect. mgr., 1987–; invited prof. U. Louvain, 1989; vis. scientist U. Tokyo, 1982, MIT, Cambridge, 1980-81. Editor: Growth, Characterization and Properties of Ultrathin Magnetic Films and Multilayers, 1989; contbr. articles to profl. jours. Fellow Am. Phys. Soc.; mem. AAAS, Materials Rsch. Soc., Sigma Xi. Achievements include patents in field. Office: GM Rsch Physics Dept 30500 Mound Rd Warren MI 48090-9055

HERGERT, DAVID JOSEPH, engineering educator; b. Cin., Mar. 19, 1952; s. Walter and Mary (Kreiner) H.; m. Susan Joan Keith, May 30, 1980; children: Lucas, Benjamin, Marita. BSME, U. Cin., 1982; PhD, Pacific Western, 1992. Test engr. Cin. Gas and Electric, 1974-78; project engr. Square D, Oxford, Ohio, 1979-83; systems engr. GM, Cin., 1985; assoc. prof. Miami U., Hamilton, Ohio, 1985—; cons. Tri Mfg., Lebanon, Ohio, 1986—, Miller Brewery, Trenton, Ohio, 1992—, Rixan Assocs., Dayton, Ohio, 1987—. Contbr. articles to profl. publs. Mem. adv. bd. Hamilton High Sch., 1986—, D. Russel Lee Vocat. Sch., 1986, U. Cin., 1987. Recipient competition for excellence award IBM, 1989. Mem. Am. Soc. Engring. Edn., DEC Users (Decus). Achievements include development of a fuzzy logic control system for the expandable styrene foam industry, of a robotic simulator presently being used in over 100 universities. Office: Miami U 1601 Peck Blvd Hamilton OH 45011-3399

HÉRITIER, CHARLES ANDRÉ, physicist, computer systems consultant, educator; b. St. Quentin, France, Dec. 25, 1931; s. Charles T. and Helene (Böröcs) H.; married May 15, 1954 (widowed Apr. 1982); children: Isabelle, Dominique, Diane; m. Anne-Marie Emery, Aug. 9, 1991. Grad. in physics, U. Neuchatel (Switzerland), 1954; PhD, U. Fribourg (Switzerland), 1962. Rsch. assoc. U. Fribourg (Switzerland), 1958-63; computer architect IBM, Yorktown Heights, N.Y., 1965-67; mem. faculty IBM European Systems Rsch. Inst., Geneva, 1969-74; systems engr. IBM Switzerland, Geneva, 1963-65, systems cons., 1974-93; assoc. prof. computer sci. U. Geneva, 1973—; mem. working group European Share Orgn., 1970-74. Contbr. articles to profl. jours. Mem. IEEE, European Physics Soc., Swiss Physics Soc., Swiss Informatics Soc. Home: Chemin du Rucher, CH-1261 Genolier Switzerland Office: U Geneva Ctr Univ Info, Rue General-Dufour 24, CH-1204 Geneva Switzerland

HERMAN, BARBARA HELEN, pediatric psychiatrist, educator; b. N.Y.C., June 4, 1950. BA, SUNY, Binghamton, 1972; MA, Bowling Green State U., 1974, PhD in Biopsychology, 1979. Fellow pharmacology Addiction Rsch. Found., Palo Alto, Calif., 1978-82; assoc. dept. pharmacology sch. medicine Emory U., 1982-83; chief immunocytochemistry brain rsch. ctr. Children's Hosp. Nat. Med. Ctr., Washington, 1983-86, chief lab. ops., 1986-88, chief, 1988—; asst. rsch. prof. psychiatry and behavioral sci. sch. medicine George Washington U., 1983-89, asst. rsch. prof. child health and devel., 1984-89, assoc. rsch. prof. pediatrics, psychiatry and behavioral sci., 1989—; mem. strategic planning steering com. Children's Hosp. Nat. Med. Ctr., 1986, chair dept. psychiatry children's rsch. inst. com., 1990—, rsch. tng. dir. dept. psychiatry, 1990—, dir. grand rounds, 1990—; mem. senate com. rsch. George Washington U., 1988-89; mem. com. autism office spl. svc. and state affairs divsn. spl. edn. and pupil pers. svc. D.C. Pub. Schs., 1988—; cons. Calif. State Devel. Rsch. Insts., 1988—, Disneyland Project Frontiers Med. Neuroscience, 1989—. Mem. AAAS, N.Y. Acad. Sci., Soc. Neuroscience, Women Neuroscience, Sigma Xi. Achievements include research on the role of neuropeptides, particularly opioid peptides, in the development of the central nervous system and the expression of the functions of the central nervous system. Office: Brain Research Ctr Chld Hospital Natl Medical Ctr 111 Michigan Ave NW Washington DC 20010*

HERMAN, LARRY MARVIN, psychotherapist; b. Pitts., Feb. 15, 1951; s. Albert Sanford and Miriam (Pearl) H.; m. Sandy Lee Checkler, Apr. 28, 1988; 1 child, Barry Craig, 1 stepchild, Mark. BA, W.Va. U., 1973, MS, 1974. Lic. profl. counselor; lic. mental health counselor, Fla. Career counselor Community Coll. Allegheny County, Monroeville, Pa., 1975; job devel. specialist Ohio State Rehab. Svcs. Commn., Steubenville, Ohio, 1975-77; mental health therapist St. John Med. Ctr., Steubenville, 1977-88; pvt. practitioner Stream, Inc., Steubenville, 1985-88; sr. therapist The Cloisters of Pine Island, Pineland, Fla., 1988-90; pvt. practice Ft. Myers, Fla., 1990—; program dir. adult psychiat. svcs. Deering Hosp., Miami, Fla., 1992—; dir. adult svcs. Grant Ctr., 1992—; pvt. practice San Jose and Costa Rica, 1992—; cons. Stream, Inc., Steubenville, 1985-88. Author relaxation program: Power From Within, 1990. Mem. AACD, ACA, Fla. Alcohol and Drug Abuse Assn. Avocations: reading, golf, target shooting. Office: 9333 SW 152d St Miami FL 33157

HERMAN, MICHAEL HARRY, physicist, researcher; b. Hartford, Conn., June 8, 1954; s. Richard Allen and Barbara Jane (Weinstein) H.; m. Susan Barbara Blum, Apr. 5, 1981; children: Edward, Beth, Andrea. BA in Physics, Grinnell Coll., 1976; PhD in Physics, Pa. State U., 1982. Sr. engr., materials tech. Intel Corp., Santa Clara, Calif., 1982-84; sr. device physicist, tech. devel. Intel Corp., Santa Clara, 1984-87; staff scientist Charles Evans & Assocs., Redwood City, Calif., 1987-89; sr. scientist Charles Evans & Assocs., Redwood City, 1989-91; head of characterization Power Spectra, Inc., Sunnyvale, Calif., 1991—. Contbr. chpt. to Analysis of Microelectronic Materials and Devices, 1991; contbr. articles to Jour. Electronic Materials, Jour. Applied Physics. Mem. IEEE, Am. Phys. Soc., Materials Rsch. Soc., Soc. Photometric Instrumentation Engrs., Phi Beta Kappa, Phi Kappa Phi. Home: 10401 N Blaney Ave Cupertino CA 95014-2333 Office: Power Spectra 919 Hermosa Ct Sunnyvale CA 94086-4103

HERMAN, ROBERT, physics educator; b. N.Y.C., Aug. 29, 1914; s. Louis and Marie (Lozinsky) H.; m. Helen Pearl Keller, Nov. 24, 1939; children: Jane Barbara, Lois Ellen, Roberta Marie. BS cum laude, CCNY, 1935; MA, Princeton U., 1940, PhD, 1940; hon. degree in engring., U. Karlsruhe, 1984. Fellow physics dept. CCNY, 1935-36, instr. physics, 1941-42; research asst. Moore Sch. Elec. Engring., U. Pa., 1940-41; supr. chem. physics group, physicist, asst. to dir. Applied Physics Lab., John Hopkins U., 1942-55; cons. physicist GM Research Labs., Warren, Mich., 1956, asst. chmn. basic sci. group, 1956-59, dept. head theoretical physics dept., 1959-72, traffic sci. dept., 1972-79; prof. physics Ctr. for Studies in Statis. Mechanics U. Tex., Austin, 1979-84, prof. civil engring., 1979—, L.P. Gilvin Centennial prof. in civil engring., 1982-84, L. P. Gilvin Centennial prof. emeritus, 1986—; vis. prof. U. Md., 1955-56; Regents lectr. U. Calif., Santa Barbara, 1975; Smeed Meml. lectr. U. Coll. London, 1983; mem. Assembly Math. and Phys. Scis. NRC, 1977-80, mem. Commn. on Engring. and Tech. Systems, 1980-83, mem. com. on sociotech. systems NRC, com. on resources for math. scis., 1981-84, infrastructure innovation com., 1987—; cons. in field. Assoc. editor: Revs. of Modern Physics, 1953-55, Ops. Research Soc. Am, 1960-74; founding editor: Transp. Sci, 1967-73; Author: (with Robert Hofstadter) High Energy Electron Scattering Tables, (with Ilya Prigogine) Kinetic Theory of Vehicular Traffic; contbr. articles to profl. jours.; patentee in field. Recipient numerous awards including Naval Ordnance Devel. award, 1945, Lanchester prize Johns Hopkins U. and Ops. Rsch. Soc. Am., 1959, medal Université Libre de Bruxelles, 1963, Townsend Harris medal CCNY, 1963, Magellanic Premium Am. Philos. Soc., 1975, Prix Georges Vanderlinden Belgian Royal Acad., 1975, Award in Phys. & Math. Scis., N.Y. Acad. Scis., 1981, Transp. Sci. Lifetime Achievement award ORSA, 1990, Henry Draper medal NAS, 1993, ORSA/TIMS John Von Neumann OR Theory prize, 1993; named Republic of China lectr. Nat. Sci. Coun., 1990. Fellow Am. Phys. Soc., Washington Acad. Sci., Franklin Inst. (John Price Wetherill gold medal 1980); mem. NAE, Am. Acad. Arts and Scis., Ops. Rsch. Soc. Am. (pres. 1980-81, Philip McCord Morse Meml. lectr. 1989-91, George E. Kimball medal 1976), Washington Philos. Soc., Phi Beta Kappa, Sigma Xi. Achievements include prediction of present temperature of the residual cosmic black-body radiation, a vestige of the initial explosion of the Big Bang Universe; use of high energy electron scattering resulting in theories on proton and neutron charge structure; research on the existence of definite electron trapping states in solids; development of theory of influence of vibration-rotation interaction on the intensity of infrared molecular spectra, of a theory of the stability and flow of single lane traffic; of a kinetic theory of multi-lane traffic flow and a two-fluid model of urban traffic. Office: U Tex at Austin Austin TX 78712-1076

HERMAN, WILLIAM ELSWORTH, psychology educator; b. Chgo., Dec. 3, 1948; s. Elsworth William John and Alice Loretta (Rousseau) H.; m. Judith Cheryl Rosenthal, Oct. 5, 1974; children: Bryan Keith Herman, Jennifer Marie Herman. BS in Geography, Mich. State U., 1970; MA in Ednl. Psychology, Ea. Mich. U., 1977, MA in Guidance and Counseling, 1980; PhD in Ednl. Psychology, U. Mich., 1987. Substitute tchr. Mt. Morris (Mich.) Pub. Schs., 1970-71; admission counselor Suomi Coll., Hancock, Mich., 1973-76; adult edn. tchr. Dundee (Mich.) Pub. Schs., 1976-78, Huron Pub. Schs., New Boston, Mich., 1976-78, Airport Community Schs., Carleton, Mich., 1977-78; prof. Edn. and Psychology Madonna U., Livonia,

Mich., 1978-93; prof. psychology dept. Potsdam Coll./SUNY, 1993—; tchr. bi-lingual ednl. leadership graduate program Madonna U., Taiwan, summer, 1989-93; editorial rev. bd., consortium editor Issues & Inquiry in Coll. Learning and Teaching, 1987—. Contbr. articles to profl. jours. Fulbright lectureship Moscow State Pedagogical U., Russia, 1993; recipient Point of Excellence award Kappa Delta Pi, 1992. Mem. Am. Psychol. Soc., Assn. Tchr. Educators (nat. del. 1992-93, Profl. Jour. com. 1993-96), Am. Ednl. Rsch. Assn., Mich. Assn. Tchr. Educators (exec. bd. dirs. 1991-93). Office: SUNY Potsdam Coll Dept Psychology Potsdam NY 13676-2294

HERMANN, JOHN ARTHUR, physicist; b. Melbourne, Victoria, Australia, Mar. 29, 1943; s. Arthur Frederick and Esme Grace (Ennis) H.; m. Leona Marian Sarah Kindlan, Sept. 28, 1976; children: Diana Catherine, Gareth James. BSc, U. Melbourne, 1968; MSc, Monash U., 1972; PhD in Theoretical Physics, Queen's U., No. Ireland, 1976. Spl. rsch. asst. U. Manchester (Eng.) Inst. Sci. and Tech., 1977-79; SRC postdoctoral researcher and tutor physics Imperial Coll. Optics Sec., U. London, 1979-81; rsch. scientist Materials Rsch. Lab. Def. Sci. and Tech. Orgn., Melbourne, 1982-85; sr. rsch. scientist Def. Sci. and Tech. Orgn., Adelaide, 1986-90, prin. rsch. scientist, 1990—; com. mem. 1st, 2d Internat. Nonlinear Optics Conf., Hawaii, 1990, 92. Spl. issue editor Jour. Modern Optics, 1990; mem. editorial adv. bd. Internat. Jour. Nonlinear Optical Physics, 1991—, spl. issue editor, 1993; mem. conf. coms., reviewer sci. jours.; contbr. articles to profl. jours. Sec., treas. Econs. Reform Australia, S. Australia, 1993—. Recipient coop. rsch. award Dept. Industry, Technology and Commerce, 1991, 92. Mem. Optical Soc. Am., Australian Optical Soc. Avocations: musical appreciation, gardening, jogging. Office: DSTO Optoelectronics div, Box 1500, Salisbury SA 5108, Australia

HERMANS, HUBERT JOHN, psychologist, researcher; b. Maastricht, Limburg, The Netherlands, Oct. 9, 1937; s. Mathias John and Jeannette Maria (Spronck) H.; m. Petronella Cornelia Jansen, Nov. 28, 1961; children: Matthieu, Désirée. B, Cath. U. Nijmegen, The Netherlands, 1962, MA, 1965, D in Psychology, 1967. Asst. psychologist Asthma Ctr., Groesbeek, The Netherlands, 1963-65; staff mem. Psychol. Lab., Nijmegen, 1965-72, lectr., 1972-80, prof. psychology, 1980—; vis. prof. Louvain (Belgium) U., 1975, Duquesne U., Pitts., 1979; chmn. adv. com. Han Fortmann Ctr. Human Growth, Nijmegen, 1982—. Author: (test-constrn.) Achievement Motivation Test for Adults, 1967, Achievement Motivation Test for Children, 1971, (assessment procedure) Self-Confrontation Method; contbr. personality theory and valuation theory to profl. jours. With The Netherlands mil., 1958-60. U.S. travel grantee Dutch Orgn. Advancement Pure Rsch., The Hague, The Netherlands, 1968; fellow Netherlands Inst. Advanced Studies in the Humanities and Social Sci., Wassenaar, The Netherlands, 1976-77. Mem. Soc. for Personology (1st internat. mem.), Dutch Orgn. Psychologists. Avocations: playing several musical instruments, folkloristic customs. Home: Bosweg 18, 6571 CD Berg en Dal The Netherlands Office: Cath U Nijmegen, Montessorilaan 3, 6525 HR Nijmegen The Netherlands

HERMANSON, THEODORE HARRY, software engineer; b. Hershey, Pa., July 24, 1965; s. Theodore Folke and Winifred Emerick (Kreider) H. BS, Lebanon Valley Coll., 1987. Ops. rsch. anaylst U.S. Army, Aberdeen, Md., 1987-89, New Cumberland, Pa., 1989-90; software engr. Turtle Beach Systems, Inc., York, Pa., 1990—. Co-author: (software packages) Sample Vision, ver 2.0, 1991, Recording Studio Professional, 1991, Wave for Windows, 1992.

HERMSEN, ROBERT WILLIAM, engineering consultant; b. Baker, Oreg., Apr. 25, 1934; s. William Henry and Alphia (Busick) H.; m. Janet Marie Grexton, May 25, 1957; children: Carol Anne, Jeanne Marie, Susan Elise, Kathleen Margaret. BS, Oreg. State U., 1956; PhD, U. Calif., Berkeley, 1962. Rsch. engr. Titanium Metals Corp., Henderson, Nev., 1956-57; staff scientist United Techs., San Jose, Calif., 1961-84, mgr. combustion R & D, 1984-92; pvt. practice Palo Alto, Calif., 1993—. Contbr. 20 articles to profl. jours. NSF fellow, 1959-61. Fellow AIAA (assoc.); mem. Am. Inst. Chem Engrs., Combustion Inst. Achievements include research in combustion of solid and liquid propellants, atmospheric pollution and hazards of energetic materials. Home: 3563 Evergreen Dr Palo Alto CA 94303

HERNADY, BERTALAN FRED, thermonuclear engineer; b. Budapest, Hungary, July 6, 1927; came to U.S., 1957; s. Bertalan and Zsofia (Feher) H.; m. Antonia Ilona Kozari, Jan. 5, 1957 (div. Apr. 1984); children: Beatrice, Eric. Diploma, Tech. U. Budapest, 1950; MS, Wayne State U., 1963, postgrad., 1964. Design engr. Thermotech. Design Bur., Budapest, 1950-56; sr. safety analysis engr. Atomic Power Devel. Assoc., Detroit, 1961-67; with nuclear power dept. Combustion Engring. Co., Windsor, Conn., 1967-71; with United Nuclear/Gulf Nuclear Fuels Corp., Elmsford, N.Y., 1971-74, Met. Edison Co., Reading, Pa., 1974-80, N.Y. Power Authority, N.Y.C., 1980-90, Energetics, Inc., Washington, 1990-91, Westinghouse Savannah River Co., Aiken, S.C., 1991—; mem. faculty Lawrence Inst. Tech., Detroit, 1965-67; adj. mem. faculty fusion program com. Electric Power Rsch. Inst., Palo Alto, Calif., 1976-80. Contbr. Fast Reactor Handbook, 1964; contbr. numerous articles to profl. jours. Mem. Am. Nuclear Soc. Achievements include research in nuclear reactor systems safety analysis and experimental validation, nuclear fuel performance analysis. Office: Westinghouse SRC Centennial Corporate Ctr 992W-1 Aiken SC 29803

HERNANDEZ, JOHN PETER, physicist, educator; b. Madrid, Sept. 6, 1940; came to U.S., 1958; s. Juan and Carmela (Garcia) H.; m. Maria Luisa Garcia. BSEE, Manhattan Coll., 1962; MSc, Stanford U., 1963; PhD, U. Rochester, 1967. Mem. faculty U. N.C., Chapel Hill, 1966—, prof. physics, 1977—, sr. Fulbright lectr. U. Autonoma de Madrid, 1972-73, W.R. Kenan Jr. rsch. prof., 1991-92; Kenan rsch. prof. Oxford (Eng.) U., 1982-83. Contbr. rsch. papers to refereed jours. Mem. Am. Phys. Soc., Sigma Xi, Eta Kappa Nu. Achievements include research in theoretical condensed matter physics, chemical physics, statistical physics; electronic properties of disordered materials, light particle localization in fluids, metal-nonmetal and liquid vapor transitions. Office: Univ NC Dept Physics/Astronomy Phillips Hall Chapel Hill NC 27599-3255

HERNANDEZ, MEDARDO CONCEPCION, chemist; b. Manila, June 8, 1947; came to U.S., 1969; s. Sancho Enriques and Trinidad (Concepcion) H.; m. Lowella Alino Hernandez, May 8, 1972; children: Medardo Jr., Wesley Adams. BS in Chem. Engring., Mapua Inst. Tech., Manila, 1968. Registered chem. engr., Philippines. Quality control chemist West Chem. Products, Inc., Long Island City, N.Y., 1970-72, R&D chemist, 1972-79; lab. mgr. West Chem. Products, Inc., Tenafly, N.J., 1979-81; tech. affairs mgr. West Agro, Inc., Des Plaines, Ill., 1981-82, R&D mgr., 1982—; cons. Resource Internat., Inc., Bedford Park, Ill., 1989-92. Roman Catholic. Achievements include patent for D-limonene based aqueous cleaning compositions; research in sodium polyacrylate in phosphate free chlorinated alkaline cleaners, stable high foaming wetting agents in chlorinated alkaline liquid products. Home: 2244 Mayfair Ave Westchester IL 60154

HERNANDEZ, WILBERT EDUARDO, physician; b. Progreso, Mex., Mar. 17, 1916; s. Alonso C. and Adolfina (Camara) H.; came to U.S., 1947, naturalized, 1949; B.S., U. Yucatan, 1937; M.D., Hahnemann Med. Coll. Phila., 1941; m. Jayne Rhodes, Oct. 4, 1941; children: Mary Jayne (Mrs. Clarence Deldrige), Patricia (Mrs. James Wheeler). Intern, Wyoming Valley Hosp., Wilkes-Barre, Pa., 1941-42; gen. practice medicine, Merida, Mex., 1942-46, Allentown (Pa.) State Hosp., 1947-48, Wilkes-Barre, 1948-51; specialized in anesthesiology St. Catherine's Hosp., Bklyn., 1955-57; chief dept. anesthesia Wyoming Valley Hosp., 1957-82; assoc. in anesthesiology Geisinger Med. Group, Geisinger Wyo. Valley Med. Ctr., 1982-83; med. dir. Blue Cross of Northeastern Pa., 1970-87; ret., 1987. Served as capt. M.C., AUS, 1951-53. Fellow Am. Coll. Anesthesiologists; mem. AMA, Pa. Med. Soc., Luzerne County Med. Soc., Am. Soc. Anesthesiologists, Pa. Soc. Anesthesiologists. Author: The Blood of the Conquistador, 1967. Contbr. articles to profl. jours. Home: 1172 Scott St Wilkes Barre PA 18705-3722

HERNANDEZ-MARTICH, JOSE DAVID, population and conservation biologist, educator; b. San Cristobal, Dominican Republic, June 26, 1955; came to U.S., 1983; s. Ernesto Hernandez-Dume and Luz Celeste Martich-Boissard; m. Nancy Rosibel Santana, Aug. 4, 1984; children: Nicole Rosibel, David Alexander. BS in Agronomy with high honors, Instituto Politecnico Loyola, San Cristobal, 1974; BS in Biology magna cum laude, Universidad Autonoma, Santo Domingo, Dominican Republic, 1982; MS in Zoology, U. Ga., 1988, PhD in Ecology, 1993. Tchr. biology Fray Pedro de Cordoba High Sch., San Cristobal, 1975-77; sci. illustrator Domican-German Project for Crop Protection, others, Santo Domingo, 1977-85; provisional dir. Wildlife Dept., Santo Domingo, 1981; asst. prof. genetics and gen. biology Universidad Autonoma, Santo Domingo, 1979-93; rsch. coord. Wildlife Dept., Santo Domingo, 1978-83; population and conservation biologist Savannah River Ecology Lab., Aiken, S.C., 1985-93; cons. Environ. Edn. Dept., Santo Domingo, 1978-83; mem. adv. bd. Grupo Jaragua for the Conservation of Protected Areas, Santo Domingo, 1992-93. Contbr. chpt. to book, articles to profl. jours.; sci. illustrator books and articles. Dir. Theatrical Group Los Peregrinos, San Cristobal, 1975-80. U.S. AID scholar, 1983-86; Am. Mus. Natural History collection study grantee, 1990. Mem. Soc. Study Evolution, Am. Soc. Ichthyologists and Herpetologists, Sigma Xi, Phi Beta Delta. Achievements include founding of a bank of Drosophila to supply material for teaching and research on ecology and genetics of this genus in La Hispaniola; discovery of striking genetic differences among populations of eastern mosquitofish in different locales. Office: Savannah River Ecology Lab Drawer E Aiken SC 29801

HERNANDEZ-SANCHEZ, JUAN LONGINO, electrical engineering educator; b. Los Andes, Chile, Mar. 15, 1928; s. Juan D. and Elba C. (Sanchez) Hernandez; m. Corina Poblete-Espinoza, May 8, 1954; children: Glenn W., Nestor M., Veronica C. Civil./Elec. Engr., U. Tecnica F. Santa Maria, Valparaiso, Chile, 1952; PhD in Elec. Engring., U. Pitts., 1962. Engr. ENDESA, Santiago, Chile, 1952-53; test engr. GE Co., Schenectady, N.Y., 1953-54; engr. C. Chilena de Electricidad, Santiago, 1954-56; prof. U. T.F. Santa Maria, Valparaiso, 1956-66, prof. automatic control, 1968—; prof. U. Tecnica del Estado, Santiago, 1966-68. Contbr. articles to profl. jours.; translator 3 books. GE Co. scholar, 1953; U.S. Internat. Agy. grantee, 1960-62. Mem. IEEE (sr. mem.), Asociacion Chilena de Control Automatico (pres. 1987-88), Colegio de Ingenieros de Child, Internat. Neural Network Soc. Home: Bulnes 1270, Quilpue Chile Office: UTFSM, PO Box 110-V, Valparaiso Chile

HERNDON, DAVID N., surgeon. Chief staff Shriners Burn Inst., Galveston, Tex., 1981—. Office: Shriners Burns Inst 610 Texas Ave Galveston TX 77550-2788*

HERNDON, JAMES HENRY, orthopaedic surgeon, educator; b. Los Angeles, Oct. 31, 1938; s. James Greene and Kathleen Theresa (Murphy) H.; m. Geraldine Grace Armiger, Feb. 26, 1971; children: Jennifer, Jonathan. BS, Loyola U., Los Angeles, 1961; MD, UCLA, 1965; MA, Brown U., 1979; MBA, Boston U., 1990. Diplomate Am. Bd. Orthopaedic Surgery. Intern Hosp. of U. Pa., 1965-66, resident in surgery, 1966-67; resident in orthopedics Children's Hosp.-Mass. Gen. Hosp. Boston, 1967-70; chief resident in orthopedics Mass. Gen. Hosp., 1970; asst. clin. prof. orthopaedic surgery Mich. State U., Grand Rapids, 1974-77, assoc. clin. prof., 1977-78; prof., chmn. dept. orthopaedics Brown U., Providence, 1979-88 ; surgeon-in-chief dept. orthopaedic surgery R.I. Hosp., Providence, 1979-88 ; silver prof., chmn. dept. Orthopaedic surgery U. Pitts., chief dept. orthopaedics and rehab. Presbyn. U. Hosp., Pitts., Montifiore U. Hosp.; site visitor Residency Rev. Commn., 1981—; examiner Am. Bd. Orthopaedic Surgery, Chgo., 1977—, pres., 1990-91. Reviewer Jour. Bone and Joint Surgery, 1975—; contbr. articles to profl. jours., chpts. to books; also author books. Trustee Meeting Street Sch., Providence, 1984-88, Harmarville Rehab. Hosp., Pitts., 1989—; mem. bd. govs. Arthritis Found., Providence, 1984-88, Pitts., 1989—. Served to maj. U.S. Army, 1971-73. Recipient Edith and Carl Lasky Meml. award UCLA Med. Sch., 1965, Bronze award Am. Congress Rehab. Medicine, 1972, clin. research award N.Y. Med. Soc., 1974. Fellow ACS, Am. Acad. Orthopaedic Surgeons; mem. Am. Orthopaedic Assn., Orthopaedic Research Soc., Am. Bd. Orthopedic Surgery (bd. dirs., pres. 1991-92), Residency Rev. Com. Orthopedic Surgery, Am. Soc. Surgery of Hand (chmn.). Clubs: Agawam Hunt (Providence), Hope (Providence), Longue Vue (Pitts.). Office: Sch Medicine U Pitts Ste 1010 Kaufmann Bldg 3471 5th Ave Pittsburgh PA 15213

HERNDON, ROY CLIFFORD, physicist; b. Washington, Sept. 25, 1934. BS, Washington and Lee U., 1955; PhD, Fla. State U., 1962. Staff physicist Lawrence Livermore (Calif.) Lab., 1962-67; prof. Nova U., Ft. Lauderdale, 1967-75; dir. CBTR Ctr. for Biomed. & Toxicological Rsch., Fla. State U., Tallahassee, 1983—; exec. dir. Fla. Hazardous Waste Adv. Coun., Tallahassee, 1980-82; mem. adv. bd. Fla. State U. System, Tallahassee, 1988—. Author: (with others) Methods of Computational Physics, 1966, Land Use: A Spatial Approach, 1980, Theories of Electrons in Disordered Systems, 1982; contbr. over 100 articles to profl. jours. Mem. Phi Beta Kappa. Office: CBTR Fla State U 2035 E Paul Dirac Dr Tallahassee FL 32310-3760

HERNES, GUDMUND, federal official; b. Mar. 25, 1941; s. Asbjorn and Tronesvold (Brynhild) H.; 1 child, Stein. PhD in Sociology, 1971. Prof. U. Bergen, 1971, U. Oslo, 1989—; doctor U. Umea, 1982; fellow Ctr. Advanced Study in Behavioural Scis., Palo Alto, Calif., 1974-75; state sec. Planning Secretariat, 1980-81; head rsch. FAFO; Minister of Ch., Edn. and Rsch. Norway. Author: Lituauia. Living Conditions, 1991; contbr. articles to profl. jours. Mem. Norwegian Acad. Scis. Mem. Labour Party. Office: Ministry of Educ Rsch & Ch Affrs, Akersgt 42 POB 8119 DEP, 0032 Oslo 1, Norway*

HERNON, RICHARD FRANCIS, engineer; b. N.Y.C., July 14, 1940; s. Francis Augusta and Mary Columba (Francis) H.; m. Susan Teresa Hartnett (dec. Jan. 1975); m. Jane Margaret Murphy, June 12, 1977; children: Robert Elizabeth, Richard Jr., Patrick. BCE, Rutgers U., 1962; MCE, N.J. Inst. Tech. (formerly Newark Coll. Engring.), 1969. Registered profl. engr., N.J. Constrn. engr. N.J. Dept. Transp., Trenton, 1964-74, fleet mgr., 1974-76; county engr. Hudson County, Jersey City, 1976-80; dir. engring. and planning N.J. Transit Waterfront Transp., Jersey City, 1980—. Pres. Rutgers Engring. Soc., New Brunswick, 1977, Island Heights (N.J.) Voters and Taxpayers, 1973-75; chmn. Island Heights Planning Bd., 1973-75. Served to 1st Lt. U.S. Army, 1962-64. Mem. ASCE. Roman Catholic. Home: 69 Hadley Ave Toms River NJ 08753-7769 Office: NJ Transit 2 Journal Sq Fl 8 Jersey City NJ 07306-4006

HERR, JOHN CHRISTIAN, cell biologist, educator; b. Dubuque, Iowa, June 29, 1948; s. King George and Julia Marie (Hansen) H.; m. Mary Jo Haberman, Sept. 1, 1978; children: Christian Craig, Austin King. BA, Grinnell Coll., 1970; PhD, U. Iowa, 1978; postgrad., U. Wash., 1978-81. Asst. prof. U. Va., Charlottesville, 1981-87, assoc. prof., 1987-92, prof. cell biology, 1992—; mem. population rsch. study sect. Nat. Inst. Child Health and Human Devel., 1990-91; mem. small bus. study sect. Nat. Inst. Diabetes, Digestive & Kidney Diseases, 1988; mem. adv. com. U.S. AID, 1990, 91. Contbr. articles to profl. jours. Grantee NIH, 1990—, Mellon Found., 1991—, Conrad. Mem. Soc. Cell Biology, Am. Soc. Immunol. Reproduction, Am. Soc. Andrology, Soc. Study Reproduction. Achievements include invention novel probe for sexual assault analysis, probe for identification of human origin of blood or tissue, human acrosomal sperm antigen for use in a contraceptive vaccine. Office: U Va Dept Anatomy & Cell Biology Box 439 Med Ctr Charlottesville VA 22908

HERRICK, ELBERT CHARLES, chemist, consultant; b. Joliet, Mont., Oct. 16, 1919; s. Charles Albert and Marie (Johnson) H.; m. Doris Christine Brock, June 1, 1962; children: David, Dennis, Douglas, Donna. BSChemE, Mont. State U., 1941; degree of ChemE, Princeton U., 1942; PhD in Organic Chemistry, MIT, 1949. Rsch. chemist Cen. Rsch. Dept. E.I. duPont de Nemours, 1949-54; assoc. rsch. chemist Houdry Process Corp., 1955-58; supr. chem. rsch. Climax Molybdenum Co., Mich., 1958-59; sr. rsch. chemist R&D div. Sun Oil Co., 1959-61; sr. rsch. chemist Textile Fibers div. Dow Chem. Co., 1962-64; cons. chemist and chem. engrs. pvt. practice, 1964-65; organic sect. head Great Lakes Rsch. Corp., 1965-67; dir. chem. rsch. Escambia Chem. Corp., 1967-69; sr. rsch. chemist Air Products and Chems., Inc., 1969-77; sr. chem. engr., scientist Tracor Jitco, Inc., 1977; environ. systems scientist The MITRE Corp., 1977-88; sr. staff specialist Dynamac Corp., 1988-89; pvt. practice Monkton, Md., 1989—. Patentee in field; contbr. articles to profl. jours. Lt. USAF, 1942-45, ETO. Fellow Am. Inst. Chemists; mem. Am. Inst. Chem. Engrs., Am. Chem. Soc., N.Y. Acad. Scis., Sigma Xi, Tau Beta Pi, Phi Kappa Phi. Republican. Adventist. Avocations: gardening, reading. Home and Office: 2740 Florence Rd Woodbine MD 21797-7841

HERRICK, PAUL E., aerospace technology educator, researcher; b. Mt. Pleasant, Mich., May 31, 1963; s. Myron T. and Doris L. (Burt) H. BS, Utah State U., 1985; MS, Ariz. State U., 1989. Design engr. Bede Jet Corp., Chesterfield, Mo., 1990-91; asst. prof. aerospace tech. Parks Coll. of St. Louis U., Cahokia, Ill., 1991—. Mem. AIAA, Exptl. Aircraft Assn. (chpt. pres. 1992—), U.S. Parachute Assn., Tau Alpha Pi. Achievements include measurements of turbulence decay and pressure loss between wind tunnel turbulence smoothing screens. Office: Parks Coll of St Louis U 500 Falling Springs Rd Cahokia IL 62206

HERRICK, THOMAS EDWARD, aeronautical/astronautical engineer; b. Steubenville, Ohio, Feb. 26, 1958; s. Edward John and Elizabeth Mildred (O'Brien) H.; m. Susan Rose Moore, Mar. 17, 1984; children: James, Timothy, Heather. BS in Aero. and Astro. Engring., Ohio State U., 1980. Mfg. engr. Tex. Instruments, Abilene, 1985-89; sr. process engr. Wilcox Elec., Inc., Kansas City, Mo., 1989—. Big bro. Big Bros./Big Sisters, Columbis, 1979. 1st lt. USMC, 1980-85. Navy ROTC scholar, 1976-80. Methodist. Home: 5301 N Brighton Ave Kansas City MO 64119 Office: Wilcox Elec Inc 2001 NE 46th St Kansas City MO 64116

HERRICK, TODD W., manufacturing company executive; b. Tecumseh, Mich., 1942. Grad., U. Notre Dame, 1967. Pres., chief exec. officer Tecumseh (Mich.) Products Co. Office: Tecumseh Products Co 100 E Patterson St Tecumseh MI 49286-2041

HERRIES, EDWARD MATTHEW, civil engineer; b. Newburgh, N.Y., May 21, 1969; s. Alexander North and Marilyn (McAvoy) H. BS in Civil Engring., Bucknell U., 1991. Tech. asst. Weis Performing Arts Ctr., Bucknell U., Lewisburg, Pa., 1989-91; staff engr. Wehran Envirotech, Middletown, N.Y., 1991—. Mem. ASCE. Republican. Roman Catholic. Home: 300 Concord Ln Middletown NY 10940 Office: Wehran Envirotech 666 E Main St Middletown NY 10940

HERRING, BRUCE E., engineering educator; b. Fremont, Ohio, Sept. 6, 1934; s. Harold W. and Eloise E. (Hanson) H.; m. Nancy Jane Kelly, June 9, 1955; children: Melanie Jane Herring Smith, Rylan Bruce. B Indsl. Engring., Ohio State U., 1958; MSME, N.Mex. State U., 1963; PhD, Okla. State U., 1972. Asst. prof. indsl. engring. Auburn (Ala.) U., 1965-73, assoc. prof., 1973-87, prof. indsl. and mfg. engring., 1987—; prof. computer sci. Tuskegee (Ala.) U., 1980-84; cons. various computer applications, 1972—. Author: Production Management Simulator, 1972. Mem. City Coun., Auburn, 1972-76; asst. scoutmaster Boy Scouts Am., Auburn, 1972—; bd. dirs. Auburn Mus. Soc., 1976-92. NSF sci. faculty fellow, 1970; named Outstanding Instr. Indsl. Engring. Student Engring. Coun., Auburn U., 1969-70, 88-89, 92-93. Mem. Am. Inst. Indsl. Engrs., Soc. Computer Simulation. Achievements include development of Ala. Resources Info. System and several other info. systems for state of Ala. and industry. Home: 442 Hare Ave Auburn AL 36830 Office: Auburn U Indsl Engring Dept 207 Dunstan Hall Auburn AL 36849

HERRING, KENNETH LEE, editor scientific society publications; b. West Palm Beach, Fla., Jan. 12, 1954; s. Grover Cleveland and Dorothy Lorene H.; m. Michele Nathan, Dec. 26, 1976; children: Julian Keir, Evan Ross, Damian Joel. BA, Tulane U., 1975; postgrad., U. Ga., 1977-78. Technical editor Internat. ComputaPrint Corp., Annandale, Va., 1979; sci. editor Am. Psychol. Assn., Washington, 1979-85, sci. officer, 1986-90; dir. communications Am. Psychol. Soc., Washington, 1990—; sci. editor Moshman Assocs., Inc., Bethesda, Md., 1985-86; info. specialist Pub. Tech., Inc., Washington, 1986; co-editor, Nat. Found. for Brain Rsch., Washington, 1992—, Nat. Coalition for Rsch. in Neurol. Disorders, Washington, 1992—; freelance editor Internat. Neural Network Soc., Washington, 1992—. Editor: Thesaurus of Psychological Index Terms, 3rd and 4th edits., 1982, '85, American Psychological Association Guide to Research Support, 1987. Active PTA, McLean, Va., 1986—. Mem. Am. Psychol. Soc. (charter), Nat. Sci. Writers Assn., D.C. Sci. Writers Assn., Phi Beta Kappa, Psi Chi. Office: Am Psychol Soc Ste 1100 1010 Vermont Ave NW Washington DC 20005-4907

HERRING, PAUL GEORGE COLIN, aeronautical engineer; b. Norwich, Norfolk, U.K., July 25, 1943; s. Paul and Joyce (Smith) H.; m. Mavis Ann Bailey, May 17, 1969; children: Lynne Paula, Anne Marie. BSc in Aero. Engring., U. Bath, U.K., 1968. Chartered engr., U.K. Engring apprentice Brit. Aircraft Corp., Bristol, 1962-68, aerodynamicist, 1968-74; section leader aero. rsch. Brit. Aerospace, Bristol, 1974-80, group leader aero. rsch., 1980-88; head aerodynamics and vulnerability rsch. Brit. Aerospace PLC, Bristol, 1988—; mem. aerodynamics com. Engring. Scis. Data Unit, London, 1986-93; panelist AGARD Fluid Dynamics, 1993—. Contbr. articles to profl. jours. Mem. AIAA, Royal Aero. Soc. (mem. aerodynamics group com. 1991—). Anglican Ch. Achievements include research in aerodynamics; main areas of research have been in weapon aerodynamics, development of semi-empirical prediction methods, high speed aerodynamics test programs. Office: Brit Aerospace, Sowerby Rsch Ctr PO Box 5, Filton Bristol BS127QW, England B5127QW

HERRINGTON, DANIEL ROBERT, chemist; b. Erie, Pa., Sept. 22, 1946; s. Bailey David and Eva Sophia (Miller) H.; m. Linda Ann Brooks, Aug. 22, 1970 (div. 1978); m. Peggy Ann Winkle, May 24, 1980; children: Daniel Garland Sidlo, John David Sidlo. BA in Chemistry, Thiel Coll., Greenville, Pa., 1968; MS in Inorganic Chemistry, Carnegie-Mellon U., 1971, PhD in Inorganic Chemistry, 1972. Rsch. chemist Standard Oil (Ohio), Cleve., 1973-79, rsch. assoc., 1979-81, rsch. mgr., 1981-82, rsch. supr., 1982-83; tech. dir. Standard Oil (Ohio)/BP Am., Cleve., 1983-89; asst. rsch. mgr. BP Chems., London, 1989-91; rsch. mgr. BP Am., Cleve., 1991—; mem. NSF Workshop-Rsch. Needs in Catalysis, U. Md., 1978. Mem. AAAS, Am. Chem. Soc. (convassing com. grad. edn. award 1982-85), N.Y. Acad. Sci. Achievements include 12 U.S. patents. Office: BP Rsch 4440 Warrensville Center Rd Cleveland OH 44128

HERRLINGER, STEPHEN PAUL, flight test engineer, air force officer, educator; b. Louisville, Ky., Nov. 23, 1959; s. John Howard and Josephine Doris (Martin) H.; m. Julie Louise Nelson, Feb. 4, 1989. BS in Chemistry, U. Akron, 1981; BS in Aero. Engring., USAF Inst. Tech., 1985; MS in Engring. Mgmt., Golden Gate U., 1989; M in Aero. Sci., Embry Riddle Aero. U., Tyndall AFB, 1992. Registered Engr. in Tng., Ohio. Commd. 2d lt. USAF, 1981, advanced through grades to capt., 1985; rsch. chemist USAF Rocket Propulsion Lab., Edwards AFB, Calif., 1981-83; aerodynamic engr. advanced cruise missile 4200 Test and Evaluation Squadron USAF, Edwards AFB, 1985-86; chief advanced cruise missile aerodynamics sect. 31st Test and Evaluation Squadron USAF, Edwards AFB, 1986-87, chief advanced cruise missile performance, environ. sect., 1987-89; projct mgr E-9A surveillance aircraft program 4484th Test Squadron, Tyndall AFB, Fla., 1989-91; dir. missile scoring flight test 4484th Test Squadron, Tyndall AFB, 1991-92; dir. C-27A operational flight test 84th Test Squadron USAF, Tyndall AFB, 1992—; mem. adv. bd. USAF-USN Noncoop. Missile Scoring, Eglin AFB, Fla., 1990-92; adj. instr. Gulf Coast C.C., U. West Fla. Contbr. articles to Jour. Organic Chemistry, Soc. Flight Test Engrs. Jour., Jour. Aircraft. Leader youth group Calif. Luth. U. Chapel, Thousand Oaks, 1986-89; guitarist Messiah Luth. Ch., Panama City, Fla., 1990—; vol. leader N.W. Fla. Spl. Olympics, Tyndall AFB, 1992. Decorated USAF Commendation medal, Achievement medal with 1 oak leaf cluster. Mem. AIAA, Air Force Assn., Soc. Flight Test Engrs. Achievements include patent for aerodynamic fairing/nosecone for M-130 CHAFF/Flare dispenser design. Home: 305 S Star Ave Panama City FL 32404 Office: 84th Test Squadron Test Ops USAF Tyndall AFB FL 32403

HERRMANN, GEORGE, mechanical engineering educator; b. USSR, Apr. 19, 1921. Diploma in Civil Engring., Swiss Fed. Inst. Tech., 1945, PhD in Mechanics, 1949. Asst., then asso. prof. civil engring. Columbia, 1950-62; prof. civil engring. Northwestern U., 1962-69; prof. applied mechanics Stanford, 1969—; cons. SRI Internat., 1970-80. Contbr. 230 articles to profl. jours; editorial bd. numerous jours. Fellow ASME (hon. mem. 1990, Centennial medal 1980); mem. ASCE (Th. v. Karman medal 1981), Nat.

Acad. Engring., AIAA (emeritus). Office: Stanford U Div Applied Mechanics Durand Bldg 281 Stanford CA 94305-4040

HERRMANN, JUDITH ANN, microbiologist; b. Peoria, Ill., Dec. 25, 1943; d. Howard Joseph and Florence Regina (McCurdy) D.; m. Ernest Carl Herrmann, Jr., July 3, 1970. BS, Iowa State U., 1966. Asst. lab. dir. Peoria City/County Health Dept., 1966-67; tech. specialist Meth. Med. Ctr., Peoria, 1970-73; rsch. tech. Mayo Clinic, Rochester, Minn., 1968-70; pres. Mobilab, Inc., Peoria, 1974—, Daily Labs., Peoria, 1976—. Author: Encyclopedia of Pharmacology and Therapeutics, 1984, (book chpt.) Viral Infections: A Clinical Approach, 1976; contbr. to Jour. Food Protection, 1978, Antiviral Rsch. jour., 1984. Recipient Robert M. Scott award, Ill. Dept. of Pub. Health, Springfield, 1989. Mem. Internat. Assn. for Milk, Food and Environ. Sanitarians, Am. Soc. Microbiology, Ill. Assn. for Milk, Food, and Environ. Sanitarians. Home: 5116 N Big Hollow Rd Peoria IL 61615-3512 Office: Daily Labs 5120 N Big Hollow Rd Peoria IL 61615-3597

HERRMANN, WALTER, retired laboratory administrator; b. Johannesburg, Republic of South Africa, May 2, 1930; came to U.S., 1953; s. Gottlob Friedrich and Gertrud Louise (Retzlaff) H.; m. Betty Allard (div.); children: Peter Friedrich, Inga Louise; m. Ednarae B. Gross. BSc in Engring. cum laude, U. Witwatersrand, Republic South Africa, 1950; PhD in Mech. Engring., U. Witwatersrand, 1955. Rsch. engr. MIT, Boston, 1953-55, sr. rsch. engr., 1957-64; lectr. U. Cape Town, Rep. South Africa, 1955-57; div. supr. Sandia Nat. Labs., Albuquerque, 1964-67, dept. mgr., 1967-82, dir. engring. scis., 1982-90, dir. shock physics rsch., 1990-93; retired Sandia Nat. Labs., 1993; W.W. Clyde prof. U. Utah, Salt Lake City, 1971-72. Contbr. articles to profl. jours. Mem. ASME, Am. Phys. Soc., Nat. Acad. Engring.

HERRMANNSFELDT, WILLIAM BERNARD, physicist; b. Chgo., Apr. 22, 1931; s. Bernard Ernst and Carolyne (Mueller) H.; m. Marcia Esther Bowman, June 12, 1954; children: Glen A., Paul W. AB, Miami U., 1953; PhD, U. Ill., 1958. Physicist Los Alamos (N.Mex.) Nat. Lab., 1958-62, Stanford (Calif.) Linear Accelerator, 1962-74, 76—; acting sect. leader US AEC, Washington, 1974-76; ptnr. Electron Optics Simulations, Los Alamos, 1987—; mem. fusion policy adv. com. DOE, Washington, 1990, mem. FEAC panel in inertial fusion energy, 1992. Contbr. articles to profl. jours. Fellow Am. Phys. Soc. (exec. com. div. physics beams 1991-94). Achievements include experimental determination of the nature of beta decay; laser alignment of the 2-mile linear accelerator; electron optics simulation program. Office: Stanford Linear Accelerator MS 26 PO Box 4349 Stanford CA 94309

HERRUP, KARL, neurobiologist; b. Pitts., July 16, 1948; s. J. Lester and Florence Bernice (Hersh) H.; m. Claire Morse, Aug. 20, 1972 (div. Jan. 1989); children: Rachael, Adam, Alex; m. Leslie Reinherz, Mar. 1, 1992. BA in Biology magna cum laude, Brandeis U., 1970; PhD in Neuro- and Behavioral Sci., Stanford U., 1974. Postdoctoral fellow in neurogenetics Harvard Med. Sch./Children's Hosp., Boston, 1974-77; postdoctoral fellow in pharmacology Biozentrum, Basel, Switzerland, 1978; asst. prof., then assoc. prof. human genetics Sch. Medicine Yale U., New Haven, 1978-84, assoc. prof. biology, 1986-88; assoc. prof. neurology Mass. Gen. Hosp., Boston, 1988-92; assoc. prof. neurosci. Harvard Med. Sch., Boston, 1988-92; dir. div. devel. neurobiology Eunice Kennedy Shriver Ctr. for Mental Retardation, Waltham, Mass., 1988-92; prof. Alzheimer Rsch. Ctr. Case Western Sch. Medicine, Cleve., 1992—; mem. staff Yale Comprehensive Cancer Ctr., New Haven, 1987-88. Contbr. articles to profl. publs.; mem. editorial bd. Internat. Jour. Devel. Neurobiology, Neurobiology of Aging. Fellow NSF, 1978, Med. Found., 1976, Jane Coffin Childs Meml. Rsch. Fund, 1974; recipient faculty award Andrew W. Mellon Found., 1982. Mem. Soc. for Neurosci. (mem. social issues com. 1987-90, program com. 1989-92, edn. com. 1992—, sec. Conn. chpt. 1982-84, v.p. 1987-88), Soc. for Devel. Biology, Sigma Xi. Office: Case Western Res Med Sch Alzheimer Rsch Lab 10900 Euclid Ave Cleveland OH 44106

HERSCHBACH, DUDLEY ROBERT, chemistry educator; b. San Jose, Calif., June 18, 1932; s. Robert Dudley and Dorothy Edith (Beer) H.; m. Georgene Lee Botyos, Dec. 26, 1964; children: Lisa Marie, Brenda Michele. BS in Math., Stanford U., 1954, MS in Chemistry, 1955; AM in Physics, Harvard U., 1956, PhD in Chem. Physics, 1958; DSc (hon.), U. Toronto, 1977, Cornell Coll., 1988, Framingham State Coll., 1989, Adelphi U., 1990, Dartmouth Coll., 1992, Charles U., Prague, 1993. Jr. fellow Harvard U., Cambridge, Mass., 1957-59, prof. chemistry, 1963-76, Frank B. Baird prof. sci., 1976—, mem. faculty council, 1980-83, master Currier House, 1981-86; asst. prof. U. Calif., Berkeley, 1959-61, assoc. prof., 1961-63; cons. editor W.H. Freeman lectr. Haverford Coll., 1962; Falk-Plaut lectr. Columbia, 1963; vis. prof. Gottingen (Germany) U., summer, 1963, U. Calif., Santa Cruz, 1972; Harvard lectr. Yale U., 1964; Debye lectr. Cornell U., 1966, Rollefson lectr. U. Calif., Berkeley, 1969, Reilly lectr. U. Notre Dame, 1969, Phillips lectr. U. Pitts., 1971; disting. vis. prof. U. Ariz., 1971, U. Tex., 1977, U. Utah, 1978, Gordon lectr. U. Toronto, 1971, Clark lectr. San Jose State U., 1979, Hill lectr., Duke U., 1988, Priestly lectr. Pa. State U., 1990, Kaufman lectr. U. Pa., 1990, Polanyi lectr. U. N.C., 1991, Dreyfus lectr. Dartmouth Coll., 1992, Paulins lectr., Caltech., 1993. Assoc. editor: Jour Phys. Chemistry, 1980-88.. Guggenheim fellow U. Freiburg, Germany, 1968; vis. fellow Joint Inst. for Lab. Astrophysics U. Colo., 1969; Fairchild Disting. scholar Calif. Inst. Tech., 1976; Sloan fellow, 1959-63, Exxon Faculty fellow, 1980—; recipient pure chemistry award Am. Chem. Soc., 1965, Centenary medal, 1977, Pauling medal, 1978; Spiers medal Faraday Soc., 1976, Polanyi medal, 1981, Langmuir prize, 1983, Nobel Prize in Chemistry, 1986, Nat. Medal of Sci. NSF, 1991, Heyrovsky medal 1992; named to Calif. Pub. Edn. Hall of Fame, 1987. Fellow Am. Phys Soc. (chmn. chem. physics div. 1971-72), Am. Acad. Arts and Scis.; mem. AAAS, Am. Chem. Soc., Nat. Acad. Scis., Royal Soc. Chemistry (fgn. hon. mem.), Am. Philos. Soc., Phi Beta Kappa (orator Harvard U. 1992), Sigma Xi. Office: Harvard U Dept Chemistry 12 Oxford St Cambridge MA 02138-2900

HERSH, ROBERT TWEED, biology educator; b. Cleve., Oct. 22, 1927; s. Amos Henry and Roselle (Karrer) H.; m. Sally Newton Six, Dec. 18, 1959; children: Jennifer, Christopher. AB, Columbia U., 1947, MA, 1951; PhD, U. Calif., 1956. Physicist DuPont, Wilmington, Del., 1957-58; asst. prof. U. Kans., Lawrence, 1958-62, assoc. prof., 1962-67, prof., 1967—; chmn. biochemistry dept., U. Kans., 1971-78, dir. human biology, 1986—. Contbr. articles to profl. jours. Grantee NIH, U. Kans., 1961-70; recipient Career Devel. award NIH. Mem. AAAS, N.Y. Acad. Sic., Biophys. Soc. Home: 3030 W 9th St Lawrence KS 66049

HERSHEY, ALFRED DAY, geneticist; b. Owosso, Mich., Dec. 4, 1908; s. Robert Day and Alma (Wilbur) H.; m. Harriet Davidson, Nov. 15, 1945; 1 son, Peter. B.S., Mich. State U., 1930, Ph.D. in Chemistry, 1934, D.M.S., 1970; D.Sc. (hon.), U. Chgo., 1967. Asst. bacteriologist Washington U. Sch. Medicine, St. Louis, 1934-36; instr. Washington U. Sch. Medicine, 1936-38, asst. prof., 1938-42, asso. prof., 1942-50; mem. staff, genetics research unit Carnegie Inst. of Washington, Cold Spring Harbor, N.Y., 1950-62; dir. Carnegie Inst. of Washington, 1962-74; ret., 1974. Contbr. articles to profl. jours. Recipient Nobel prize in Medicine (joint), 1969; Albert Lasker award Am. Pub. Health Assn., 1958; Kimber Genetics award Nat. Acad. Scis., 1965. Mem. Nat. Acad. Scis. Address: RD 1640 Moores Hill Rd Syosset NY 11791

HERSHEY, ROBERT LEWIS, mechanical engineer, management consultant; b. Chgo., Dec. 18, 1941; s. Maurice and Rose Beverly (Barrish) H. BSME summa cum laude, Tufts U., 1963; MSME, MIT, 1964; PhD in Engring., Cath. U. Am., 1973. Registered profl. engr., D.C., N.Y.; cert. mfg. engr. Engr. Bell Telephone Labs., Whippany, N.J., 1963-67; acoustics mgr. Weston Instruments, Inc., Poughkeepsie, N.Y., 1967-68; sr. scientist Bolt Beranek & Newman, Washington, 1968-71; acoustics program mgr. Booz Allen & Hamilton, Bethesda, Md., 1971-79; program v.p. Sci. Mgmt. Corp., Washington, 1979-80, divsn. v.p., 1980-88; exec. engr. AEA O'Donnell, Inc., Washington, 1988—; sec. Engring. Registration Bd., D.C., 1987—, D.C. Profl. Coun., Washington, 1974; mem. coordinating com. on productivity Am. Assn. Engring. Socs., Washington, 1984-88. Author: How to Think With Numbers, 1982; contbr. articles in energy and environment to profl. jours.; patentee tempo enhancement device. Sci. policy analyst George Bush Presdl. Campaign, Washington, 1988, 92; pres. Hamilton House Assn. Resident Tenants, Washington, 1988, 90—; mem. Joint Bd. on Sci. Engring. Edn., Washington, 1972-78. Recipient of the Design award Machinery Mag., 1963. Mem. ASME (chmn. Washington chpt. 1978-79), Nat. Energy Resources Orgn., Mensa, Capital PC Users Group, Acoustical Soc. Am. (chmn. Washington chpt. 1982-83), D.C. Soc. Profl. Engrs. (pres. 1975-76, nat. dir. 1980-86, Young Engr. of Yr. 1974), D.C. Coun. Engring. and Archtl. Socs. (del. 1969—, pres. 1978-79, Pres.'s award 1989, Nat. Capital award 1974), Soc. Mfg. Engrs. (chmn. Washington Robotics Internat. chpt. 1986-87), Washington Coal Club, MIT Club of Washington (pres. 1979-80), Washington Tufts U. Alumni Club (v.p. 1970-71), Tau Beta Pi (pres. Tufts student chpt. 1962-63, v.p. Washington alumni chpt. 1988-89), Sigma Xi. Republican. Avocations: chess, tennis, sports cars, golf. Home: Apt 1033 1255 New Hampshire Ave NW Washington DC 20036-2328

HERSZTAJN MOLDAU, JUAN, economist; b. Cochabamba, Bolivia, Dec. 28, 1945; arrived in Brazil, 1948; s. Leon and Liselotte (Moldau) Hersztajn. B, U. São Paulo (Brazil), 1967; M., Vanderbilt U., 1973, PhD, 1976. Researcher Inst. de Pesquisas Economicas, São Paulo, 1973—; asst. prof. econs. U. São Paulo, 1974-86, assoc. prof. econs., 1986-92, prof. econs., 1992—. Author: Avaliačao de Projetos, 1981, A Teoria da Escolha, 1988; contbr. articles to profl. jours., including Econs. and Philosophy, Jour. Econ. Theory. Recipient Haralambos Simeonidis award Brazilian Nat. Assn. Grad. Schs., 1986, Haralambos Simeomidis Hon. Mention award, 1992. Mem. Soc. for Promotion of Econ. Theory, Conselho Regional de Economia, Am. Econ. Assn. Avocations: swimming, cycling. Home: Av Moaci 973, 04083 São Paulo Brazil Office: U Sao Paulo Economia, Av Luciano Gualberto 908, 05508 São Paulo Brazil

HERTZ, DANIEL LEROY, JR., entrepreneur; b. Montclair, N.J., Feb. 27, 1930; s. Daniel Leroy and Elizabeth Nielsen (Beet) H.; m. Valerie A. Smith, Mar. 15, 1956 (div. 1962); m. Isabel Waud Hurd, Apr. 18, 1970; children: Valerie H. Boyle, Suzanne E., Daniel L. III, Seana L. Burdge. Degree in mech. engring., Stevens Inst. Tech., 1952, MS in Mech. Engring. (hon.), 1982. Sales engr. C.E. Conover & Co., Fairfield, N.J., 1953-58; pres. Seals Eastern, Red Bank, N.J., 1958—; mem. adv. bd. polymer tech. cons. Tex. A&M U., College Station, 1990—, CHEMTECH, Washington, 1983-91, Elastomerics, Atlanta, 1984-92. Contbr. chpts. to Intermediate Rubber Technology, 1983, Handbook of Elastomers, 1988, Vanderbilt Handbook, 1990, Engineering with Rubber, 1992, Rubber Products Manufacturing Technology, 1993. Mem. vis. com. mech. engring. dept. Stevens Inst. Tech., 1992; sec. Riverside Drive Assn., Red Bank, 1980-85. Cpl. U.S. Army, 1950-51, Korea. Mem. Am. Chem. Soc. (treas. rubber divsn. 1988-90), N.Y. Rubber Group (chmn. 1983), Rumson Country Club, Seabright Tennis Club. Republican. Episcopalian. Achievements include 5 U.S. patents. Home: 734 Navesink River Rd Red Bank NJ 07701 Office: 134 Pearl St Red Bank NJ 07701

HERTZ, PATRICIA RITENOUR, analytical chemist; b. Atlanta, Aug. 13, 1966; d. Donald V. and Laura H. Ritenour; m. William J. Hertz, July 14, 1990. BS in Chemistry with high honors, Coll. William and Mary, 1988; PhD, Duke U., 1992. Recitation instr. Duke U., Durham, N.C., 1990; exploratory analytical chemist Procter & Gamble, Cin., 1992—. Contbr. articles to Applied Spectroscopy, Analytical Chemistry. Am. Chem. Soc./Dow Chem. summer fellow, 1991; Duke U. fellow, 1988-89, 89, 90, 92. Mem. Am. Chem. Soc., Soc. Applied Spectroscopy, Phi Lambda Upsilon, Kappa Kappa Gamma. Methodist. Achievements include research in spectral characterization of coal liquids in organized media via synchronous flourescence; demonstrated first application of phase-resolved flourescence spectroscopy to characterize complex, multicomponent system.

HERTZBERG, ABRAHAM, aeronautical engineering educator, university research scientist; b. N.Y.C., July 8, 1922; s. Rubin and Paulien (Kalif) H.; m. Ruth Cohen, Sept. 3, 1950; children: Eleanor Ruth, Paul Elliot, Jean R. BS in Aero. Engring., Va. Poly. Inst., 1943; MS in Aero. Engring., Cornell U., 1949; postgrad., U. Buffalo, 1949-53. Engr. Cornell Aero. Lab., 1949-57, asst. head aerodynamics research, 1957-59, head aerodynamics research, 1959-65; dir. aerospace & energetics rsch. program U. Wash., 1966-93, prof. astronautics emeritus, 1966—; prin. investigator numerous fed. rsch. grants; cons. Aerospace Corp., past mem. sci. adv. bd. USAF, Olin-Rocket Rsch., STI Optronics; past mem. electro-optics panel SAB, mem. various ad hoc coms.; mem. space systems & tech. adv. com., research and tech. subcom., past mem. research and tech. adv. council NASA; mem. plasma dynamics rev. panel NSF, U.S. Army.; honored speaker Laser Inst. Am., 1975, Citizens of Sendai, 1991; mem. theory adv. com. Los Alamos Nat. Lab.; vis. lectr. Chinese Acad. Scis., Beijing, 1983, 88; Paul Vieille lectr. 7th Internat. Shock Tube Symposium, 1969, 89, 17th Internat. Symposium on Shock Waves and Shock Tubes, 1989. Editor Physics of Fluids, 1968-70; contbr. numerous articles on modern gas dynamics, high powered lasers, controlled thermonuclear fusions processes, space laser solar energy concepts, space energy concepts and new ultra velocity propulsion concepts to profl. jours. Served with AUS, 1944-46. Honored speaker Laser Inst. Am. Fellow AIAA (Dryden lectr. 1977, Agard lectr. 1978, Plasmadynamics and Lasers award 1992); mem. NAE, Am. Phys. Soc., Internat. Acad. Astronautics, Sigma Xi. Achievements include patents in field. Office: U Washington Aerospace & Engring Rsch Bldg FL-10 Seattle WA 98195

HERTZOG, ROBERT WILLIAM, pathologist, consultant, educator; b. Danville, Pa., Oct. 31, 1939; s. Robert Lee and Thelma Isabelle (McKean) H.; m. Florence Rebecca Smoot, May 26, 1962; children: Brian, Sheryl, Brent. BS, Morgan State U., Balt., 1963; MD, U. Md., Balt., 1967. Diplomate in forensic, anatomic and clin. pathology Am. Bd. Pathology. Staff pathologist Armed Forces Inst. Pathology, Washington, 1973-76; chief accident pathology sect., registrar Am. Registry Accident Pathology, 1974-75, chief missile trauma pathology, 1975-76, chief forensic pathology, 1976; attending pathologist Millard Fillmore Hosp., Buffalo, 1976; clin. asst. prof. dept. pathology Sch. Medicine SUNY, Buffalo, 1977—; dir. Sch. Med. Tech., Millard Fillmore Hosp., 1982-89. Contbr. articles to profl. jours. Lt. col. U.S. Army, 1967-76. Fellow Am. Coll. Pathologists, Am. Soc. Clin. Pathologists, Am. Acad. Forensic Scis. Home: 34 Ruskin Ct East Aurora NY 14052 Office: Ctr for Lab Medicine 115 Flint Rd Williamsville NY 14221

HERZ, GEORGE PETER, industrial consultant, chemical engineer; b. Vienna, Austria, Oct. 27, 1928; s. Armin J. and Barbara Maria (Trucker) H. PhB in Liberal Arts, U. Chgo., 1947, BSc in Chemistry, 1948; LLB, Temple U., 1957, JD, 1972. Bar: D.C. 1957. Rsch. chemist Sherwin-Williams Co., Chgo., 1948-50; rsch. chemist Rohm and Haas Co., Phila., 1954-57, mktg. engr., 1957; sales engr., mgr. Rohm and Haas Co., Paris, 1958-66; gen. mgr. Rohm and Haas Co., Stockholm, 1966-72, Vienna, 1973-78; cons. Rohm and Haas Co., Tokyo, 1978-86; internat. cons. Newark, Notts, Eng., 1986—. Editor: Temple Law Quarterly, Phila., 1956-57; contbr. articles to profl. publs. Capt. S.C., U.S. Army, 1950-53. Mem. Inst. Chem. Engrs., Am. Chem. Engrs., Inst. Chem. Industry (Eng.), Phi Delta Phi, Sigma Chi.

HERZ, JOSEF EDWARD, chemist; b. Lucerne, Switzerland, Oct. 14, 1924; came to U.S., 1947; s. Albert Adolf and May Therese (Priggen) H.; m. Julieta Abdala, May 5, 1960; children: May Isabel, Barbara Therese. BSc, ETH, Zurich, Switzerland, 1947; AM, Harvard U., 1950, PhD, 1952. Rsch. assoc. E.R. Squibb & Sons, New Brunswick, N.J., 1951-58; dir. devel. and rsch. adminstrn. Syntex S.A., Mexico City, 1958-65; prof., head chemistry dept. CIEA-IPN, Mexico City, 1965-81; prof., head dept. biotech. rsch. inst. UNAM, Mexico City, 1980-88; vis. prof. dept. nuclear medicine U. Conn. Health Ctr., Farmington, 1983-84; sr. rsch. scientist div. nuclear medicine George Washington U., Washington, 1987-89; sr. rsch. scientist dept. biochemistry Rice U., Houston, 1989-90; vis. scientist dept. human biol. chemistry and genetics U. Tex. Med. Br., Galveston, 1991-92; cons., 1993—; royal soc. vis. scientist Imperial Coll. Sci., London, 1969; prin. scientist task force on contracept rsch. WHO, Geneva, 1975-82; scientific advisor Internat. Found. Sci., Stockholm, Sweden, 1979—. Contbr. 70 articles to profl. publs. Grantee Royal Soc. London, 1969, Ford Found., 1973, Ministry of Sci. Fed. Republic Germany, 1979. Mem. NRA, Am. Chem. Soc., Mex. Acad. Scientific Rsch., Sigma Xi. Achievements include 33 patents in field. Home: 8421 Hearth Dr Apt 24 Houston TX 77054

HERZBERG, GERHARD, physicist; b. Hamburg, Germany, Dec. 25, 1904; emigrated to Can., 1935, naturalized, 1945; s. Albin and Ella (Biber) H.; m. Luise H. Oettinger, Dec. 29, 1929 (dec.); children: Paul Albin, Agnes Margaret; m. Monika Tenthoff, Mar. 21, 1972. Dr. Ing., Darmstadt Inst. Tech., 1928; postgrad., U. Goettingen, U. Bristol, 1928-30; D.Sc. hon causa, Oxford U., 1960; D.Sc., U. Chgo., 1967, Drexel U., 1972, U. Montreal, 1972, U. Sherbrooke, 1972, McGill U., 1972, Cambridge U., 1972, U. Man., 1973, Andhra U., 1975, Osmania U., 1976, U. Delhi, 1976, U. Bristol, 1975, U. Western Ont., 1976; Fil. Hed. Dr., U. Stockholm, 1966; Ph.D. (hon.), Weizmann Inst. Sci., 1976, U. Toledo, 1984; LL.D., St. Francis Xavier U., 1972, Simon Fraser U., 1972; Dr. phil. nat., U. Frankfurt, 1983, others. Lectr., chief asst. physics Darmstadt Inst. Tech., 1930-35; research prof. physics U. Sask., Saskatoon, 1935-45; prof. spectroscopy Yerkes Obs., U. Chgo., 1945-48; prin. research officer NRC Can., Ottawa, 1948; dir. div. pure physics NRC Can., 1949-69, disting. research scientist, 1969—; Bakerian lectr. Royal Soc. London, 1960; holder Francqui chair U. Liege, 1960. Author books including: Spectra of Diatomic Molecules, 1950; Electronic Spectra and Electronic Structure of Polyatomic Molecules, 1966, The Spectra and Structures of Simple Free Radicals, 1971, (with K.P. Huber) Constants of Diatomic Molecules, 1979. Appt. to Queen's Privy Coun. for Can., 1992. Recipient Faraday medal Chem. Soc. London, 1970, Nobel prize in Chemistry, 1971; named companion Order of Can., 1968, academician Pontifical Acad. Scis., 1964. Fellow Royal Soc. London (Royal medal 1971), Royal Soc. Can. (pres. 1966, Henry Marshall Tory medal 1953), Hungarian Acad. Sci. (hon.), Indian Acad. Scis. (hon.), Am. Phys. Soc. (Earle K. Plyler prize 1985), Chem. Inst. Can.; mem. Internat. Union Pure and Applied Physics (past v.p.), Am. Acad. Arts and Scis. (hon. fgn. mem.), Am. Chem. Soc. (Willard Gibbs medal 1969, Centennial fgn. fellow 1976), Nat. Acad. Sci. India, Indian Phys. Soc. (hon.), Japan Acad. (hon.), Chem. Soc. Japan (hon.), Royal Swedish Acad. Scis. (fgn., physics sect.), Nat. Acad. Sci. (fgn. assoc.), Faraday Soc., Am. Astron. Soc., Can. Assn. Physicists (past pres., Achievement award 1957), Optical Soc. Am. (hon., Frederic Ives medal 1964). Home: 190 Lakeway Dr, Rockcliffe Pk, Ottawa, ON Canada K1L 5B3 Office: Nat Rsch Coun, Ottawa, ON Canada K1A 0R6

HERZFELD, CHARLES MARIA, physicist; b. Vienna, Austria, June 29, 1925; came to U.S., 1942, naturalized, 1949; s. August Alfred and Frieda Auguste (Poehlman) H.; children: Charles Christopher, Thomas Augustine, Paul Vincent; m. Shannon Stock Shuman, June 9, 1990. BS in Chem. Engring. cum laude, Cath. U. Am., 1945; PhD (Carnegie Found. fellow), U. Chgo., 1951. Lectr. chemistry Cath. U. Am., 1946; lectr. gen. sci. Coll. U. Chgo., 1946-47; lectr. physics DePaul U., Chgo., 1948-50; physicist Ballistic Research Lab., Aberdeen, Md., 1951-53, Naval Research Lab., Washington, 1953-55; lectr. physics U. Md., 1953-57, prof. physics, 1957-61; cons. chief heat and power div. Nat. Bur. Standards, 1955-56, acting asst. chief, 1956-57, chief heat div., 1957-61, asso. dir. bur., 1961; asst. dir. Advanced Research Project Agy., Dept. Def., 1961-63, dir. ballistic missile def., 1963; dep. dir. Advanced Research Projects Agy., 1963-65, dir., 1965-67; tech. dir. def. space group ITT, Nutley, N.J., 1967-74; tech. dir. aerospace-electronics-components-energy group ITT, 1974-76, tech. dir. telecommunications and electronics group N.Am., 1978-79; v.p., dir. research ITT Corp., 1979-83, v.p., dir. research and tech., 1983-85; vice chmn. Aetna, Jacobs and Ramo, N.Y.C., 1985-90; dir. def. rsch. and engring. Dept. Def., Washington, 1990-91; cons. to Office Sci. and Tech. Policy, Exec. Office Pres. of U.S., Washington, 1991; chmn. bd. Westronix Co., Midvale, Utah, 1985-88; mem. Def. Sci. bd., 1968-83, Def. Policy Bd., 1985-90, Nat. Commn. on Space, 1985-86; cons. in field; fellow Hudson Inst., 1970-90; mem. Brookings Inst. 5th Conf. for Career Execs. in Fed. Govt., 1958, mem. chief of Naval Ops. exec. panel, 1970—; mem. Tech. Review Bd. Hong Kong, Nat. Security Advisory Bd., Los Alamos Nat. Lab. Editor: Temperature, Its Control in Science and Industry, vol. III, 1962; contbr. articles to profl. jours. Recipient Flemming award, 1963; Meritorious Civilian Service medal Dept. Def., 1967. Fellow AAAS, Am. Phys. Soc., Conf. on Sci., Philosophy and Religion, Coun. Fgn. Rels.; mem. Explorers Club, Inst. for Strategic Studies (London), Cath. Assn. Internat. Peace (pres. 1959-61), Cosmos Club (Washington), Sigma Xi.

HERZOG, CATHERINE ANITA, process development engineer; b. Dearborn, Mich., Feb. 6, 1960; d. John and Marian Helen (Hardy) H. BSEE, U. Mich., Dearborn, 1982. Project engr. Hughes Aircraft, Tucson, 1982-83, project test engr., 1983-85, responsible engring. activity, 1985-87; process engr. Litton Data Systems, Colorado Springs, Colo., 1987-89; process devel. engr. Cray Computer Corp., Colorado Springs, 1989—. Big sister Big Bros./Big Sisters, Tucson, 1982-87, Colorado Springs, 1988—. Named Big Sister of Yr., Big Bros./Big Sisters, Tucson, 1987. Mem. Internat. Hybrid Micro-Electronic Soc. (membership chairperson 1989-91, pres. 1991—). Republican. Home: 10420 Teachout Rd Colorado Springs CO 80908 Office: Cray Computer Corp 1110 Bayfield Dr Colorado Springs CO 80906

HESELTON, KENNETH EMERY, energy engineer; b. Corning N.Y., Nov. 17, 1943; s. Richard Linsmore and Dorothy Bertha (Schoonover) H.; m. Susan B. Benkert, July 4, 1965. BS in Marine Engring., USMMA, 1965. Registered profl. engr., Md.; cert. energy mgr. Engr. Hercules Inc., Wilmington, Del., 1968-72; engr. Power and Combustion, Inc., Balt., 1972-89, v.p., 1989—; pres. Md. Nat. Cert. Pipe Welding Bur., 1989—. Office: Power and Combustion Inc 7909 Philadelphia Rd Baltimore MD 21237

HESP, B., pharmaceutical executive; b. Buxworth, Derbyshire, UK. BA in Chemistry, Oxford U., 1963, MA, PhD, 1968. Rsch. asst. British Cotton Industry Rsch. Assn., 1958-59; rsch. chemist Natural Products Rsch. Group, 1968-72, Anthelmintics Project Team, 1973; mgr. chemistry section CNS Disorders, 1973-78; head Natural Products Rsch., 1978-79, Imperial Chemical Industries PLC., 1968-79; dir. medicinal chemistry Zeneca Pharmaceuticals Group, 1979-91, v.p. biomedical rsch. group, 1991—; com.mem. Drug Discovery Subsection, PMA, 1992—; mem. sci. adv. bd. Keystone Symposia, 1992—; chmn., organizer Leukotrienes session Nat. Medicinal Chemistry Symposium, 1988, The Linear Pathway of Archidonic Acid Metabolism; chmn. Gordon Rsch. Conf. on Medicinal Chemistry, 1987; vice chmn., 1986. Mem. AAAS, Am. Chemical Soc., N.Y. Acad. Sci., Soc. for Drug Rsch., The Chemical Soc. Office: ICI Americas Inc Biomedical Rsch Dept Concord Pike & New Murphy Rd Wilmington DE 19897*

HESS, ANN MARIE, systems engineer, electronic data processing specialist; b. Grants Pass, Oreg., Mar. 29, 1944; d. Wilbur Lill and Elaine Esther Groner; m. William Charles Hess, July 25, 1969; children: David William, William Albert. BSEE, BS in Math., Oregon State U., 1968. Engr. Lawrence Livermore Lab., Livermore, Calif., 1968-69; mgr., owner RBR Scales, Inc., Anaheim, Calif., 1969-84; lead engr. Rockwell Internat., Seal Beach, Calif., 1984-86, '87-88; software engr. Hughes Aircraft Co., Fullerton, Calif., 1986-87; sr. engr. Logican Eagle Tech., Inc., Eatontown, N.J., 1988-91; owner Holistic Eclectic Software Svc., Inc., Orange, Calif., 1991-93; engr. Jacobs Engring Group, 1993—. Active Calif. Master Chorale, Santa Ana, 1990-92. Mem. IEEE, Am. Soc. Quality Control, Phi Kappa Phi, Eta Kappa Nu, Tau Beta Pi. Lutheran. Avocations: singing, art, gardening. Office: JEGIPEP 251 S Lake Ave Pasadena CA 91101

HESS, CECIL F., engineering executive; b. Sosúa, Dominican Republic, Apr. 29, 1949; came to U.S., 1975; s. Kurt Luis and Ana Julia (Silva) H.; m. Josefina E. Acosta, June 30, 1968; children: Rachel, Karen. BS, U. Catolica Madre y Maestra, Dominican Republic; MS, U. Calif., Berkeley, 1973, PhD, 1979. Sr. scientist Spectron Devel. Labs., Inc., Costa Mesa, Calif., 1979-88; dir. engring. Metrolaser, Irvine, Calif., 1988—. Patentee laser-based particle sizing system; contbr. articles to sci. jours. Laspau scholar, 1972-73, Orgn. Am. States scholar, 1975-78. Achievements include patents in field. Office: Metrolaser 18006 Skypark Cir Irvine CA 92714-5328

HESS, CHARLES EDWARD, environmental horticulture educator; b. Paterson, N.J., Dec. 20, 1931; s. Cornelius W. M. and Alice (Debruyn) H.; children: Mary, Carol, Nancy, John, Peter; m. Eva G. Carroad, Feb. 14, 1981. BS, Rutgers U., 1953; MS, Cornell U., 1954, PhD, 1957; DAgr (hon.), Purdue U., 1983; DSc (hon.), Delaware Valley Coll., Doylestown, Pa., 1992. Asst. prof. Purdue U., West Lafayette, Ind., 1958-61, assoc. prof., 1962-64, prof., 1965; research prof., dept. chmn. Rutgers U., New Brunswick, N.J., 1966, assoc. dean, dir. N.J. Agrl. Exptl. Sta., 1970, acting dean Coll. Agrl. and Environ. Sci., 1971, dean Cook Coll., 1972-75; dean Coll. Agr. and Environ. Scis., U. Calif.-Davis, 1975-89; assoc. dir. Calif. Agrl. Exptl. Sta., 1975-89; asst. sec. sci. and edn. USDA, Washington, 1989-91; prof. dept. environ. horticulture U. Calif., Davis, 1991—, dir. internat. programs Coll. Agrl. and Environ. Scis., 1992—; cons. AID, 1965, Office

Tech. Assessment, U.S. Congress, 1976-77; chmn. study team world food and nutrition study Nat. Acad. Scis., 1976; mem. Calif. State Bd. Food and Agr., 1984-89; mem. Nat. Sci. Bd., 1982-88, 92—, vice chmn., 1984-88; co-chmn. Joint Coun. USDA, 1987-91. Mem. West Lafayette Sch. Bd., Ind., 1963-65, sec., 1963, pres., 1964; mem. Gov.'s Commn. Blueprint for Agr., 1971-73; bd. dirs. Davis Sci. Ctr., 1992—; trustee Internat. Svc. for Nat. Agrl. Rsch., The Hague, Netherlands, 1992—. Served with AUS, 1956-58. Mem. U.S. EPA (biotechnology sci. adv. com. 1992—), AAAS (chmn. agriculture sect. 1989-90), Am. Soc. Hort. Sci. (pres. 1973), Internat. Plant Propagators Soc. (pres. 1973), Agrl. Research Inst., Phi Beta Kappa, Sigma Xi, Alpha Zeta, Phi Kappa Phi. Office: U Calif Coll Agrl & Environ Scis Dept Environ Horticulture Davis CA 95616

HESS, EARL HOLLINGER, laboratory executive, chemist; b. Lancaster, Pa., June 16, 1928; s. Abram Myer and Ruth Stoner (Hollinger) H.; m. Anita F. Swords, Sept. 2, 1951; children—Kenneth Earl, Bonita Sue, Carol Denise. BS cum laude, Franklin and Marshall Coll., 1952; PhD, U. Ill., 1955. Teaching asst. U. Ill., Urbana, 1952-54; asst. prof. chemistry Franklin and Marshall Coll., Lancaster, Pa., 1955-57; group leader chemistry research Gen. Cigar Co., Lancaster, 1957-61; pres., chmn., chief exec. officer Lancaster Labs., Inc., 1961—; chmn. bd. Mountain States Analytical, Salt Lake City, 1990—; mem. Pa. Gov.'s Small Bus. adv. coun., 1988-92, Pa. Gov's. Commn. on Families and Children, 1991-92; testifier before congl. coms. and govt. agencies; trustee Franklin and Marshall Coll., 1990-92. Contbr. articles to profl. jours., chpts. to books. Patentee in field. Bd. dirs. Bethany Theol. Sem., Oak Brook, Ill., 1981-92; chmn. Pa. Del./White House Conf. Small Bus. (commr. gov.'s conf. on small bus. 1988). Recipient Disting. Pennsylvanian award Wm. Penn Com., 1982, Alumni citation Franklin and Marshall Coll., 1988; Socony-Mobil Research fellow, 1954-55. Fellow Am. Coun. Ind. Labs. (spl. svc. award 1979, Roger W. Truesdail award 1983, pres. 1985-86, Fellow award 1992); mem. ASTM, Am. Chem. Soc., Am. Pub. Health Assn., AAAS, N.Y. Acad. Scis., Am. Assn. Lab. Accreditation (dir. 1982-92, chmn. 1988-90), Am. Coun. Ind. Labs. Edn. Inst. (chmn. bd. dirs. 1992—, slaes and mktg. exec. of yr. award 1989), Lancaster C. of C. and Industry (chmn. bd. dirs. 1985, Exemplar award 1988), Pa. C. of C. (vice chmn. 1984, bd. dirs. 1984—, Bus. Leader of Yr. award 1988), U.S. C. of C. (bd. dirs., mem. small bus. adv. coun. 1980—, chmn. legis. policy com. 1985-89, econ. policy com. 1987—, chmn. environ. com. 1988—, ea. regional vice chair 1991—), Commonwealth Found. (bd. dirs.), Bus. Execs. Nat. Security, Conestoga Valley Found. (bd. dirs. 1991—), Phi Beta Kappa, Sigma Xi, Phi Lambda Upsilon. Home: 2435 New Holland Pike Lancaster PA 17601-5935 Office: Lancaster Labs Inc 2425 New Holland Pike Lancaster PA 17601-5994

HESS, IDA IRENE, statistician; b. Central City, Ky., Aug. 27, 1910; d. Bartley Eugene and Mary Anna (Surring) H. AB, Ind. U., 1931. Tchr. pub. schs., Central City, 1933-42; geodetic computer U.S. Geol. Survey, Washington, 1942-44; math. Ordnance Devel. Div./Bur. Standards, Washington, 1945-46; assoc. in stats. George Washington U., Washington, 1947-48; statistician Bur. of Census, Washington, 1946-47, 48-54; asst. head, head sampling sect. Survey Rsch. Ctr./U. Mich., Ann Arbor, 1954-63, 63-81, head of sampling, emeritus, 1981—; cons. in sampling Nat. Coun. Applied Econs. Rsch., New Delhi, India, 1962, Mich.-Peru Project, Lima, 1970. Contbr. articles to profl. jours./publs.; author (monograph) Sampling for Social Research Surveys, 1947-80, 1984. Grantee Div. Community Health Svcs./USPHS, 1962-65. Fellow Am. Statis. Assn. (chmn. various coms.); mem. AAAS, Internat. Assn. Survey Statisticians, Women's Rsch. Club/U. Mich. (pres. 1980-81, various coms.). Methodist. Achievements include development and direction of probability sampling procedures for social rsch. surveys conducted by the Survey Rsch. Ctr., Inst. for Social Rsch., U. Mich. Home: 715 S Forest Ave Apt 302 Ann Arbor MI 48104-3151 Office: U Mich Survey Rsch Ctr 426 Thompson St Ann Arbor MI 48104-2380

HESS, LAVERNE DERRYL, research laboratory scientist; b. Stockton, Ill., Oct. 28, 1933; s. James and Gertrude (Posey) H.; m. Mary Daune McDermott, Jan. 20, 1956; children: Donald, Patti, Susan, Daniel, Bart, Jennifer. BA, U. Calif., Riverside, 1961, PhD, 1965. Mem. tech. staff Hughes Research Labs., Malibu, Calif., 1965-78, head tech. staff, 1978-85; asst. mgr. Chem.-Physics dept. Hughes Research Labs., Malibu, Calif., 1985, acting mgr., 1986-87, sr. scientist, 1987-89; program dir. div. materials rsch. NSF, Washington, 1990—. Contbr. numerous articles to sci. jours. Mgr. Little League Baseball, Little League Basketball, Thousand Oaks, Calif., 1968-78. Served with U.S. Army, 1954-56. Mem. Materials Research Soc. Avocations: bridge, jogging, biking, sailing, guitar. Home: 2049 Golf Course Dr Reston VA 22091 Office: NSF Div Materials Rsch 1800 G St NW Washington DC 20550

HESS, LEONARD WAYNE, obstetrician/gynecologist, perinatologist; b. Richlands, Va., Nov. 23, 1949; s. Ralph Eugene and Lucille Cindy (Kennedy) H.; m. Sarah Mahala Leedy, Nov. 27, 1969 (div. July 1988); children: Gregory Scott, Lauren Ashley; m. Darla Irma Bakersmith, July 20, 1988; 1 child, Ever Marie. BSChemE, Va. Poly. Inst., 1973; MD, Va. Commonwealth U., 1977. Diplomate Nat. Bd. Med. Examiners, Am. Bd. Ob-Gyn., also sub.-bd. Maternal-Fetal Medicine. Intern U.S. Naval Hosp., Portsmouth, Va., 1977-78; resident in ob-gyn. U.S. Naval Hosp., Portsmouth, 1978-81; fellow in maternal-fetal medicine Naval Med. Command, Walter Reed Army Med. Ctr., Washington and Bethesda, 1981-83; staff dept. ob-gyn. U. Health Scis., Bethesda, 1981-85; dept. ob-gyn. U.S. Naval Hosp., Portsmouth, 1985-87; comdr. USNR, 1987-88; asst. prof. dept. ob-gyn. U. Miss. Med. Ctr., Jackson, 1987-91; assoc. prof. ob-gyn. U. Mo. Med. Ctr., Columbia, 1991—, head obstetrics and maternal-fetal medicine, 1991—; mem. Med. Ethics Com., U.S. Naval Hosp., Portsmouth, 1985-87, chmn. Obstetrical Spl. Care Unit Com., 1985-87; mem. Patient Care Com., U. Miss. Med. Ctr., Jackson, 1988-91, Infection Control Com., 1988-91; head obstetrics and maternal fetal medicine U. Mo. Med. Ctr., Columbia, 1991—. Cons. editor Obstetrics and Gynecology, 1988, Am. Jour. Obstetrics and Gynecology, 1988, Am. Jour. Med. Genetics, 1989; contbr. numerous articles to profl. jours. Mem. Am. Coll. Obstetricians and Gynecologists, Soc. Perinatal Obstetricians, Am. Inst. Ultrasound in Medicine, Assn. Profs. Gynecology and Obstetrics, Cen. Assn. Obstetricians and Gynecologists, Am. Soc. Human Genetics, So. Med. Assn., Winifred L. Wiser Soc., AMA, Miss. State Obstet. and Gynecol. Soc., Cen. Med. Soc., Gynecic Soc., Med. Soc. Va., Portsmouth Acad. Medicine, Med. and Surgical Soc. of Md., Miss. State Med. Assn., Assn. Mil. Surgeons, Miss. Perinatal Assn., So. Perinatal Assn. Republican. Episcopalian. Office: U Mo Med Ctr Health Sci Ctr N613 Columbia MO 65212

HESS, LINDA CANDACE, process control engineer; b. St. Louis, May 21, 1952; d. Arthur Edward and Alice M. (Bonham) H. BSEE, Washington U., St. Louis, 1974; MBA, So. Ill. U., 1988. Registered profl. engr., Mo. From engr. I to gen. engr. Monsanto Chem. Co., St. Louis, 1974-88; group leader, prin. engr. The Chem. Group of Monsanto, St. Louis, 1988—. Contbr. articles to Intech, Advances in Instrumentation 1986. Mem. IEEE, Instrument Soc. Am. (sr.; program chair 1985-86, Outstanding Svc. award 1986). Office: Monsanto 800 N Lindbergh Blvd F4WB Saint Louis MO 63167-0001

HESS, RONALD ANDREW, aerospace engineer, educator; b. Norwalk, Ohio, Mar. 12, 1942; s. Robert Andrew and Catherine Ann (Caruso) H.; m. Connaught Ann McCormack, Sept. 7, 1967; children: Christian Anthony, Catherine Ann. BS in Aerospace Engring., U. Cin., 1965, MS in Aerospace Engring., 1967, PhD in Aerospace Engring., 1970. Registered profl. engr., Calif. Asst. prof. dept. aero. Naval Postgrad. Sch., Monterey, Calif., 1970-76; rsch. scientist NASA Ames Rsch. Ctr., Moffet Field, Calif., 1976-82; assoc. prof. dept. mech., aero. and materials engring. U. Calif., Davis, 1982-84, prof., 1984—. Assoc. editor Jour. Aircraft, 1977—; contbr. over 70 tech. papers to profl. jours. Mem. AIAA (assoc. fellow, tech. com. guidance navigation and control 1984-86, tech. com. atmospheric flight mechanics 1988-90), IEEE (sr.), Systems, Man and Cybernetics Soc. (v.p. 1989-91, chmn. manual control tech. com. 1986—, assoc. editor IEEE Transactions Systems, Man and Cybernetics 1979—), Sigma Xi, Tau Beta Pi. Achievements include rsch. on developing models of human pilot behavior, aircraft control system design, aircraft handling qualities assessment. Office: U Calif Mech Aero Materials Engring Davis CA 95616

HESS, ULRICH EDWARD, electrical engineer; b. Schenklengsfeld, Germany, Sept. 22, 1940; came to the U.S., 1958; s. Adolf Wilhelm and Margarete Anna (Theobald) H.; m. Robyn Gayl Cook, Feb. 6, 1972 (div. 1987); children: Oliver, Gabriele. BSEE, U. Calif., Berkeley, 1967, MSEE, 1969. Cert. tchr., Calif. Sr. engr. Applied Tech., Sunnyvale, Calif., 1972-75; elec. engr. Hewlett Packard Co., Loveland, Colo., 1975-78; engring. mgr. Hewlett Packard Co., Ft. Collins, Colo., 1978-82, mem. tech. staff, 1982-83; mem. tech. staff Hewlett Packard Co., San Diego, 1983-85, Corvallis, Oreg., 1985—. With USAF, 1959-63. Mem. Tau Beta Pi. Achievements include patents in field.

HESS, WILMOT NORTON, science administrator; b. Oberlin, Ohio, Oct. 16, 1926; s. Walter Norton and Rachel Victoria (Metcalf) H.; m. Winifred Esther Lowdermilk, June 16, 1950; children—Walter Craig, Alison Lee, Carl Ernest. B.S. in Elec. Engring., Columbia, 1946; M.A. in Physics, Oberlin Coll, 1949, D.Sc., 1970; Ph.D., U. Calif., Berkeley, 1954. Staff Lawrence Radiation Lab., U. Calif., Berkeley and Livermore, 1954-59; head plowshare div. Lawrence Radiation Lab., U. Calif., Livermore, 1959-61; dir. theoretical div. Goddard Spaceflight Center (NASA), Greenbelt, Md., 1961-67; dir. sci. and applications Manned Spacecraft Center, Houston, 1967-69; dir. NOAA Research Labs. (Commerce Dept.), Boulder, Colo., 1969-80, Nat. Center for Atmospheric Research, Boulder, 1980-86; dir. high energy and nuclear physics U.S. Dept. Energy Office of Energy Research, Germantown, Md., 1986-93, assoc. dir., 1993—; adj. prof. U. Colo., 1970-78. Contbr. articles to profl. jours.; editor: Introduction to Space Science, 1965; author: Radiation Belt and Magnetosphere, 1968, (with others) Weather and Climate Modification, 1974; assoc. editor: (with others) Jour. Geophys. Research, 1961-67, Jour. Atmospheric Sci, 1961-67, Jour. Am. Inst. Aeros. and Astronautics, 1967-69. Served with USN, 1944-46. Fellow Am. Geophys. Union, Am. Phys. Soc.; mem. NAE. Home: 14508 Pebble Hill Ln Gaithersburg MD 20878-2473 Office: High Energy and Nuclear Physics Mail Stop ER (GTN) Washington DC 20585

HESSE, CHRISTIAN AUGUST, mining industry consultant; b. Chemnitz, Germany, June 20, 1925; s. William Albert and Anna Gunhilda (Baumann) H.; B. Applied Sci. with honors, U. Toronto (Ont., Can.), 1948; m. Brenda Nora Rigby, Nov. 4, 1964; children: Rob Christian, Bruce William. In various mining and constrn. positions, Can., 1944-61; jr. shift boss N.J. Zinc Co., Gilman, Colo., 1949; asst. layout engr. Internat. Nickel Co., Sudbury, Ont., 1949-52; shaft and tunnel engr. Perini-Walsh Joint Venture, Niagara Falls, Ont., 1952-54; constrn. project engr. B. Perini & Sons (Can.) Ltd., Toronto, Ottawa, and New Brunswick, 1954-55; field engr. Aries Copper Mines Ltd., No. Ont., 1955-56; instr. in mining engring. U. Toronto, 1956-57; planning engr. Stanleigh Uranium Mining Corp. Ltd., Elliot Lake, Ont., 1957-58, chief engr., 1959-60; subway field engr. Johnson-Perini-Kiewit Joint Venture, Toronto, 1960-61; del. Commonwealth Mining Congress, Africa, 1961; with U.S. Borax & Chem. Corp., 1961-90; mng. dir. Yorkshire Potash, Ltd., London, 1970-71, gen. mgr., pres. Allan Potash Mines Ltd., Allan, Sask., Can., 1974, chief engr. U.S. Borax & Chem. Corp., L.A., 1974-77, v.p. engring., 1978-81, 87-90, v.p. and project mgr. Quartz Hill molybdenum project, 1981-90; v.p. Pacific Coast Molybdenum Co., 1981-90, v.p. mining devel., 1984-90. Sault Daily Star scholar, Sault Sainte Marie, Ont., Can., 1944. Fellow Inst. Mining and Metallurgy; mem. SME/AIME, Can. Inst. Mining and Metallurgy (life), Assn. Profl. Engrs. Ont., Prospectors and Developers Assn., N.W. Mining Assn., Alaska Miners Assn., L.A. Tennis Club. Lutheran. Office: 2701 Lake Hollywood Dr Los Angeles CA 90068-1629

HESSE, THURMAN DALE, welding metallurgy educator, consultant; b. Plymouth, Wis., Nov. 28, 1938; s. Leonard Ferdinand and Eileen (Thurman) H.; m. Virginia Raynoha, Sept. 5, 1959; children: Daniel Jacob, David Tyler, Laura Alice. BS, Wis. State Coll. & Inst. Tech., 1962; MS, Stout State U., 1965; postgrad., U. Wis., 1974-75. Tchr. welding State Vocat. Tech. & Adult Edn., Madison, 1965—, tchr. machine shop, 1966—, tchr. welding tech., 1968—; welding instr. indsl. div. Madison (Wis.) Area Tech. Coll., 1966—; weld test condr. Wis. Dept. Industry Labor & Human Rels., Madison, 1976—; owner Tech. Welding Svcs., Cottage Grove, Wis., 1978—; lectr. U. Wis. Engring. Extension, Madison, 1978-85. Producer videotape on welding career options; contbr. articles to Welding Jour. Mem. coun. St. Stephens Luth. Ch., Monona, Wis., 1982-85. Mem. ASTM, Am. Welding Soc. (cert. welding insepctor, chmn. bd. dirs. Madison-Beloit chpt., membership com. 1989—, Howard Adkins award 1975, Dist. 12 dir. 1992—), Am. Soc. Metals. Home: 2302 Whiting Rd Cottage Grove WI 53527-9724 Office: Madison Area Tech Coll 3550 Anderson St Madison WI 53704-2520

HESSEN, MARGARET TREXLER, internist, educator; b. Allentown, Pa., Jan. 6, 1956; d. John Peter and Virginia Ruth (Hamilton) Trexler; m. Scott Edward Hessen, Aug. 15, 1981; 1 child, Scott Trexler. AB, Mt. Holyoke Coll., 1978; MD, Jefferson Med. Coll., 1982. Diplomate Am. Bd. Internal Medicine. Resident in internal medicine Laukenau Hosp., 1982-85; fellow in infectious diseases Med. Coll. Pa., Phila., 1985-87, instr. medicine, 1987-88, asst. prof., 1988-89, clin. asst. prof., 1989—; pvt. practice Drexel Hill and Bryn Mawr, Pa., 1989—; dir. med. edn. Jefferson Park Hosp., Phila., 1987-89. Contbr. articles to Am. Jour. Medicine, Jour. Infectious Disease, Antimicrobial Agts. and Chemotherapy, also chpt. to books. Grantee Am. Fedn. for Aging Rsch., 1988, Allegheny Singer Rsch. Inst., 1988; Mary Dewitt Petit fellow Alumni Assn. Med. Coll. Pa., 1988. Fellow ACP; mem. AMA, AAAS, Infectious Disease Soc. Am. (assoc.), Am. Soc. for Microbiology, Alpha Omega Alpha. Achievements include research in experimental antibiotic therapeutics and pharmacodynamics, particularly post antibiotic effect in vitro and in vivo, and in immunology of aging, particularly T-lymphocyte function.

HESSER, JAMES EDWARD, astronomy researcher; b. Wichita, Kans., June 23, 1941; arrived in Can., 1977; s. J. Edward and Ina (Lowe) H.; m. Betty Hinsdale, Aug. 24, 1963; children: Nadja Lynn, Rebecca Ximena, Diana Gillian. BA, U. Kans., 1963; MA, Princeton U., 1965, PhD, 1966. Rsch. assoc. Princeton (N.J.) U. Obs., 1966-68; from jr. astronomer to assoc. astronomer Cerro Tololo Inter-Am. Obs., La Serena, Chile, 1968-77; sr. rsch. officer Dominion Astrophys. Obs., NRC, Victoria, B.C., Can., 1977—; assoc. dir. Cerro Tololo Inter-Am. Obs., 1974-76; assoc. dir. Dominion Astrophys. Obs., NRC, Victoria, B.C., Can., 1984-86, dir., 1986—. Editor: CTIO Facilities Manual, 1973, 2d rev. edit., 1978, Star Clusters, 1980; co-editor: Late Stages of Stellar Evolution, 1974; contbr. more than 175 articles to profl. and sci. jours. Mem. Am. Astron. Soc. (councilor 1985-88, v.p. 1991-94), Astron. Soc. Pacific (bd. dirs. 1981-84, v.p. 1985-86, pres. 1987-88), Can. Astron. Soc., Internat. Astron. Union, Royal Astron. Soc. Can. Avocations: reading, walking, cooking, gardening. Home: 1874 Ventura Way, Victoria, BC Canada V8N 1R3 Office: Dominion Astrophys Obs NRC, 5071 W Saanich Rd, Victoria, BC Canada V8X 4M6

HESSMAN, FREDERICK WILLIAM, aeronautical engineer; b. Bowling Green, Ohio, July 23, 1939; s. Lowell Dell and Alma Grace (Limmer) H.; m. Genese Sandra Van der Putten, June 8, 1963; children: Laurie Genese, Dirk Frederick. BS in Aeronautical Engring., Ohio State U., 1962, MS in Aeronautical Engring., 1964. Fluid dynamics specialist Rockwell Internat., Columbus, Ohio, 1962-67, aeropropulsion analyst, 1968-72, supr. aeromechanics group, 1973-74, missile configuaration rsch. and devel., 1975-84; sr. missile systems engr. Rockwell Internat., Duluth, Ga., 1985—; panel mem. corp. fluid dynamics, Rockwell Internat., L.A., 1966-88. Author tech. papers in field. Mem. Twin Mills Homeowners Assn., Duluth, 1984-88. Mem. AIAA, Sigma Theta Tau. Republican. Home: 3606 Wildwood Farms Dr Duluth GA 30136 Office: Rockwell Tactical Systems Div 1800 Satellite Blvd Duluth GA 30136

HETRICK, DAVID LEROY, nuclear engineering educator; b. Scranton, Pa., Jan. 26, 1927; s. LeRoy and Wilma (Shoemaker) H.; m. Margaret Emma Wetsel, Aug. 28, 1948; children: Carol, Nancy, Amy. BS, Rensselaer Poly. Inst., 1947, MS, 1950; PhD, U. Calif., L.A., 1954. Instr. Rensselaer Poly. Inst, Troy, N.Y., 1947-50; physicist Rockwell Internat., Canoga Park, Calif., 1950-59; assoc. prof. physics Calif. State U. Northridge, L.A., 1960-63; prof. nuclear engring. U. Ariz., Tucson, 1963-92, prof. emeritus, 1992—; adminstrv. judge U.S. Nuclear Regulatory Commn., Washington, 1974—; cons. Los Alamos Nat. Lab, 1989—; cons. numerous orgns. Author: Dynamics of Nuclear Reactors, 1971; contbr. numerous articles to profl. jours. Fellow AAAS, Am. Nuclear Soc. (awards for svc. to minorities 1978, 1986); mem. Am. Phys. Soc. Democrat. Unitarian. Home: 8740 E Dexter Tucson AZ 85715 Office: Univ Ariz Nuclear Engring Dept Tucson AZ 85721

HETTCHE, L. RAYMOND, research director; b. Balt., Mar. 24, 1938; s. Leroy and Dorothy (Curtain) H.; m. Patricia Durkan, July 1965; children: Lisa, Kathleen, Matthew, Craig. BSCE, AB in Math., Bucknell U., 1961; MSCE, Carnegie-Mellon U., 1961, PhD in CE, 1965. Asst. prof. Rutgers U., New Brunswick, N.J., 1964-66; resident rsch. assoc. Nat. Bur. Standards, Washington, 1966-68; structural engr. metallurgy div. Naval Rsch. Lab., Washington, 1968-71, head thermomech. effect sect., 1971-73, head mech. br. metallurgy div., 1973-75, supt. materials sci. div., 1975-81; now, dir. Applied Rsch. Lab. Pa. State U., State College, 1981—; navy rep. Tech. Working Group Export Control, Washington, 1979-81; navy rep. subgroup P materials panel for metals Tech. Cooperation Program, Washington, 1977-81; session chmn. Submarine Tech. Symposium, Columbia, Md., 1990. Contbr. numerous articles to profl. jours. Tau Beta Pi Nat. fellow, 1961-63; NSF fellow, 1963; recipient Outstanding Achievement award Am. Def. Preparedness Assn., 1986. Mem. ASME (com. on mgmt. 1976-80), Acoustical Soc. Am., Am. Soc. Engring. Edn., Sigma Xi. Office: Pa State U Applied Rsch Lab PO Box 30 State College PA 16804-0030

HETZEL, DONALD STANFORD, chemist; b. Phila., July 1, 1941; married, 1964. BA, Wesleyan U., 1963; MS, U. Ill., 1965, PhD in Organic Chemistry, 1968. Rsch. chemist, chem. divsn. Pfizer Inc., 1967-72, dir. licensing health care products, 1972-74; dir. corp. rsch. Howmedia Inc., 1974-76, v.p. corp. rsch., 1976-81; v.p. corp. rsch. Becton, Dickinson & Co., 1981—. Mem. Am. Chem. Soc., Soc. Biomedical Rsch., N.Y. Acad. Sci. Achievements include research in organic synthesis; biomedical polymers; new technology aquisition; research management. Office: Becton Dickinson & Co One Becton Dr Franklin Lakes NJ 07417*

HEUER, MARGARET B., data processing coordinator; b. Juneau, Alaska, Sept. 12, 1935; d. William George and Flora (Rusk) Allen; m. Joseph Louis Heuer; children: Leilani, Joseph, Daniel, Suzanne, Karen, Mark, Jerina. AA, San Bernardino Valley Coll., 1980. Cert. data processing, computer repair and maintenance, microcomputer support specialist. Coord. microcomputers lab. Oakton Community Coll., Skokie, Ill., 1981—. Office: Oakton Community Coll 7701 Lincoln Ave Skokie IL 60077-2800

HEUER, MARVIN ARTHUR, physician, science foundation executive; b. Mankato, Minn., Mar. 11, 1947; s. Marvin Ernst and Elaine Olive (Melahn) H.; m. Kathryn Ann Klejbuk, Nov. 28, 1975; children: David Walter, Michael Arthur. BA, Mankato State U., 1969; BS, U. Minn., Mpls., 1973, MD, 1973. Intership, resident family practice St. John's Hosp., St. Paul, 1973-74; ptnr. Family Med. Group practice Park Rapids (Minn.)/Walker Clin. LTD, 1974-80; assoc. med. dir. Smith Kline & French Corp., Phila., 1980-81, group dir. clin. rsch., 1981-82, acting v.p. world-wide ops., 1982-84; v.p. med. affairs Ayerst Labs., Am. Home Prodn., N.Y.C., 1984-87; v.p. R&D Wallace Labs., Cranbury, N.J., 1987-89; v.p., dir. clin. rsch. Worldwide Smithkline Beecham Corp., London, 1989-91; CEO Heuer Assocs., North Oaks, Minn., 1991—; physician Westview Clinic, West Saint Paul, Minn., 1991—; dir. clin. rsch. Health Span Corp., Mpls., 1993—; pres. KMB Co.; clin. asst. prof. Robert Wood Johnson Med. Sch., Dept. of Family Medicine, New Brunswick, N.J., 1981—; clin. assoc. prof. U. Minn. Med. Sch., Dept. of Family Practice, 1992—; mem. biotech. adv. bd. Mankato State U.; mem. drug utilization rev. panel Dept. Health, Minn., 1992—. Contbr. 12 articles on drugs to profl. jours., tng. manual, Med. Monitors Guide 1983. Dir. youth activities St. Matthews Luth. Ch., Moorestown, N.J., 1981-86, trustee 1983-92, coun. mem., 1984-87, alt. bd. mem. 1986; fin. com. Incarnation Luth. Ch., St. Paul, Minn., 1991—, property com., 1991—. Fellow Am. Bd. Family Practice; mem. AMA, Am. Assoc. Physician Execs, Am. Acad. Family Physicians, Minn. Med. Soc., Pharm. Mfrs. Assn (del. clin. safety 1985—), Minn. Acad. Family Practice (del., rsch network), Am. Coll. Cardiology, Am. Coll. Physicians, Am. Rheumatologic Assn., Med. Alley Assn., Nat. Geographic Soc., Drug Info. Assn., Soc. Clin. Trials. Republican. Avocations: private pilot, scuba diving, running, surfing, tennis. Home: 28 Nord Circle Rd North Oaks MN 55127-6512 Office: Westview Clinic 156 Emerson Ave W Saint Paul MN 55118-2599

HEWITT, JOHN STRINGER, nuclear engineer; b. Kincardine, Ont., Can., Feb. 5, 1939; s. Albert Edwin and Mabel Priscilla (Stringer) H.; m. Alice Marlene Morton, July 7, 1962. BS in Engring. Physics, Queen's U., Kingston, Ont., 1961; MSc, U. Birmingham, U.K., 1962, PhD, 1966. Lic. profl. engr., Ont. Asst. prof. chem. engring. and applied chemistry U. Toronto, 1969-75, assoc. prof. chem. engring., 1975-82, assoc. dean Faculty of Applied Sci. and Engring., 1981-84, prof. applied nuclear studies, 1982-87; v.p. technologies ECS Power Systems, Inc., Ottawa, Ont., 1987-89; pres., cons. Stringer Hewitt Assocs., Inc., Ottawa, Ont., 1990—; tech. devel. mgr. Can. Space Sta., Can. Space Agy., Ottawa, 1990-93; sr. tech. adv. Nat. Res. Can., Ottawa, 1993—; vis. scientist Chalk River Nuclear Labs., Atomic Energy of Can., Chalk River, Ont., 1978-79; cons. Tech. Adv. Panel on Nuclear Safety, Ontario Hydro, Toronto, 1990-93, R&D Adv. Panel to Atomic Energy of Can., Ltd., Ottawa, 1991-93. Contbr. articles to profl. jours. Pres. Carleton Condominium Corp. #250, Ottawa, 1990-93; chmn. Metro Toronto Sci. Fair Com., 1973, judge, 1974, 77, 83, 84; coord. United Appeal Campaign, U. Toronto, 1970, 71, 83, 84. Athlone fellow Brit. Bd. Trade, Birmingham, 1961-63; NRC scholar, 1963-65. Fellow Can. Nuclear Soc. (founding mem., pres. 1983-84); mem. Can. Pugwash Group (treas. 1990—), Sci. for Peace (founding mem., dir. 1981-86). Achievements include patents on measurement of slurry consistencies (Can. and U.S.); patent applied for nuclear reactor plant and nuclear reactor cooling system.

HEWITT, KENNETH, geography educator. BA, Cambridge U., 1961, MA, 1963; PhD, London U., 1967. Instr. Outward Bound Sea Sch., Aberdovey, 1957-58; lectr. geography City of London Coll., 1963-66, Sidney Webb Coll. Edn., 1963-66; postdoctoral fellow, lectr. geography U. Toronto, 1967-69, asst. prof. geography, 1969-71, assoc. prof. dept. geography and Inst. Environ. Scis. and Engring., 1971-73; chmn., prof. dept. human ecology and social scis. Rutgers U., 1973-76; prof. dept. geography Wilfrid Laurier U., Waterloo, Ont., Can., 1976—; assoc. natural hazards rsch. project Toronto-Chgo.-Clark U., 1967-72; cons. disaster divsn. UNESCO, Paris, 1972, UN Environ. Program Conf., 1977, B.C. Hydro Internat., 1991; del., ICSU/SCOPE rep. planning meeting UNESCO, Norway, 1974, U.N.U./Internat. Mountain Soc. Workshop, Berne and Ruderalp, Switzerland, 1981; advisor Pakistan Sci. Found., 1975; dir. field survey of Swat Kohistan Earthquake, Ministry of Info. and Comm., Pakistan; mem. adv. delegation NAS, Peshawar, Pakistan, 1976, appraisals com. Ont. Coun. Grad. Studies, 1978-80, chmn., 1980-91; mem. adjudications com. Social Scis. and Human Resources Coun., 1981-82, chmn., 1982-83; mem. program com. Assn. Am. Geographers, 1984-85, sub-com. on glaciers Nat. Rsch. Coun., 1985; dir. Ctr. Hazards and Devel. Rsch., Wilfrid Laurier U., 1985-87; prin. investigator snow and ice hydrology project Upper Indus Basin, Karakoram, Himalaya, 1985-88; cons. disaster mgmt. com. Commonwealth Sci. Coun., 1988-89; disting. vis. prof. U. Brit. Columbia, 1989; guest prof. Geographisches Inst., U. Koln, Germany, 1992-93. Author: (with Ian Burton) The Hazardousness of a Place: A Regional Ecology of Damaging Events in Southern Ontario, 1971, (with F. Kenneth Hare) Man and Environment: Conceptual Framework, 1973, Lifeboat: Man and a Habitable Earth (An Introduction to Human Ecology), 1976, Spartan Walls: Civilians and Civil Ecology at the 'Sharp End' of War, 1985; editor: Proceedings: National Seminar on Ecology, Environment and Afforestation, 1975; editor and contbr.: Interpretations of Calamity, Risks and Hazards Series, vol. 1, 1983; contbr. chpts. to books, numerous articles to profl. jours. With Royal Artillery, 1955-57. Recipient Scholarly Distinction in Geography award Can. Assn. Geographers, 1991; Rsch. fellow Emergency Measures Orgn., 1967-69; rsch. grantee Can. Coun., 1976, SSHRC, 1980, 81, 85, 87-90, 90-91, Internat. Devel. Rsch. Ctr. Contract, 1984, IDRC, 1987, 89-90, WMO Forecasting Project WAPDA, 1987. Office: Wilfrid Laurier U, Dept Geography, Waterloo, ON Canada N2L 3C5*

HEWLETT, WILLIAM (REDINGTON), manufacturing company executive, electrical engineer; b. Ann Arbor, Mich., May 20, 1913; s. Albion Walter and Louise (Redington) H.; m. Flora Lamson, Aug. 10, 1939 (dec. 1977); children: Eleanor Hewlett Gimon, Walter B., James S., William A., Mary Hewlett Jaffe; m. Rosemary Bradford, May 24, 1978. BA, Stanford

U., 1934, EE, 1939; MS, MIT, 1936; LLD (hon.), U. Calif., Berkeley, 1966, Yale U., 1976, Mills Coll., 1983; DSc (hon.), Kenyon Coll., 1978, Poly. Inst. N.Y., 1978; LHD (hon.), Johns Hopkins U., 1985; EngD (hon.), U. Notre Dame, 1980, Utah State U., 1980, Dartmouth Coll., 1983; PhD, Rand Grad. Inst.; D Electronic Sci. (hon.), U. Bologna, Italy, 1989; HHD(hon.), Santa Clara U., 1991. Electromedical researcher, 1936-39; co-founder Hewlett-Packard Co., Palo Alto, Calif., 1939, ptnr., 1939-46, exec. v.p., dir., 1947-64, pres., 1964-77, chief exec. officer, 1969-78, chmn. exec. com., 1977-83, vice chmn. bd. dirs., 1983-87, emeritus dir., 1987—; mem. internat. adv. council Wells Fargo Bank, 1986-92; trustee Rand Corp., 1962-72; trustee Carnegie Inst., Washington, 1971-90, trustee emeritus, 1990—, chmn. bd. 1980-86; dir. Overseas Devel. Council, 1969-77; bd. dirs. Inst. Radio Engrs. (now IEEE), 1950-57, pres. 1954; coord. chpt. on rsch. in industry for 5-Yr. Outlook Report, NAS, 1980-81; mem. adv. coun. on edn. and new techs. The Tech. Ctr. of Silicon Valley, 1987-88; past bd. dirs. Chrysler Corp., FMC Corp., Chase Manhattan Bank, Utah Internat. Inc. Contbr. articles to profl. jours.; patentee in field. Trustee Stanford U., 1963-74, Mills Coll., Oakland, Calif., 1958-68; mem. Pres.'s Gen. Adv. Com. on Fgn. Assistance Programs, Washington, 1965-68, Pres.'s Sci. Adv. Com., 1966-69; San Francisco regional panel Commn. on White House Fellows, 1969-70, chmn., 1970; pres. bd. dirs. Palo Alto Stanford Hosp. Ctr., 1956-58, bd. dirs., 1958-62; dir. Drug Abuse Council, Washington, 1972-74, Kaiser Found. Hosp. & Health Plan Bd., 1972-78; chmn. The William and Flora Hewlett Found., 1966—; bd. dirs. San Francisco Bay Area Council, 1969-81, Inst. Medicine, Washington, 1971-72, The Nat. Acads. Corp., 1986—, Monterey Bay Aquarium Rsch. Inst., 1987—, Univ. Corp. for Atmospheric Rsch. Found., 1986-88. Lt. col. AUS, 1942-45. Recipient Calif. Mfr. of Yr. Calif. Mfrs. Assn., 1969, Bus. Statesman of Yr. Harvard Bus. Sch. No. Calif., 1970, Medal of Achievement Western Electronic Mfrs. Assn., 1971, Industrialist of Yr. (with David Packard) Calif. Mus. Sci. and Industry and Calif. Mus. Found., 1973, Award with David Packard presented by Scientific Apparatus Makers Assn., 1975, Corp. Leadership award MIT, 1976, Medal of Honor City of Boeblingen, Germany, 1977, Herbert Hoover medal for disting. service Stanford U. Alumni Assn., 1977, Henry Heald award Ill. Inst. Tech., 1984, Nat. Medal of Sci. U.S. Nat. Sci. Com., 1985, Laureate award Santa Clara County BUs. Hall of Fame Jr. Achievement, 1987, World Affairs Coun. No. Calif. award, 1987, Degree of Uncommon Man award Stanford U., 1987, Laureate award Nat. Bus. Hall of Fame Jr. Acievement, 1988; Decorated Comdr.'s Cross Order of Merit Fed. Republic Germany, 1987, John M. Fluke Sen. Meml. Pioneer award, Electronics Test Mag., 1990, Silicon Valley Engring. Hall of Fame award Silicon Valley Engring. Coun., 1991, Exemplary Leader award Am. Leadership Forum, 1992, Alexis de Tocqueville Soc. award United Way, Santa Clara County, 1991, Nat. Inventors Hall of Fame award Nat. Inventors Hall of Fame Found. Akron, 1992. Fellow NAE (Founders award 1993), IEEE (life fellow, Founders medal with David Packard 1973), Franklin Inst. (life, Vermilye medal with David Packard 1976), Am. Acad. Arts and Scis.; mem. NAS (panel on advanced tech. competition 1982-83, president's circle 1989—), Instrument Soc. Am. (hon. life), Am. Philos. Soc., Calif. Acad. Sci. (trustee 1963-68), Assn. Quadrato della Radio, Century Assn. N.Y.C. Office: Hewlett-Packard Co 1501 Page Mill Rd Palo Alto CA 94304-1100

HEYDERMAN, ARTHUR JEROME, engineer, civilian military employee; b. Bklyn., N.Y., Jan. 1, 1946; s. Herbert Robert and Sally (Baron) H.; m. Renee Linda Pearlman, July 4, 1967; children: Brian Douglas, Deborah Ann, Cathy Ruth. BS in Applied Math., Polytech. Inst. of Bklyn., 1966; MS in Applied Math., Polytech Inst. of Bklyn., 1973; postgrad., Stevens Inst. Tech., 1982, The Brookings Inst., 1992, The Wharton Sch. Bus., 1993. Nuclear weapons engr. U.S. Army Armaments R&D Ctr., Picatinny Arsenal, N.J., 1971-83; asst. tech. dir. U.S. Army Armaments R&D Ctr., Picatinny Arsenal, 1983-84, chief prodn. program planning, 1984, assoc. tech. dir., 1984-86; armaments rsch. and devel. prog. mgr. U.S. Army Armaments Munitions and Chem. Command, Rock Island, Ill., 1986-93, chief of rsch. devel., test and evaluation integration, 1993—; bd. dirs., sec./treas., pres. Iowa-Ill. chpt. Am. Def. Preparedness Assn., Rock Island; lt. col. nuclear weapons officer U.S. Army Res., Ft. Sheridan, Ill., 1989-93; pres. OPICON, Bettendorf, Iowa, 1989—; nat. coun. Am. Def. Preparedness Assn.; coun. mem. Quad-Cities Engring. and Sci. Coun.; dir. Quad-Cities chpt. ACLU; adj. faculty U.S. Army Command and Gen. Staff Coll., Ft. Leavenworth, Kans., 1981-89. Contbr. tech. papers on weapons and weaponry assessement to profl. meetings. Pres., bd. dirs. Sussex County Jewish Ctr., Newton N.J., 1979-86; fund raiser United Jewish Fedn., Davenport, Iowa, 1986—, mem. RIA Com. for the Diasbled, Rock Island, 1987—; dir. Quad City chpt. ACLU. Capt. U.S. Army, 1968-71, Vietnam. Decorated with Bronze Star, U.S. Army, Vietnam, 1970, Cross of Gallantry, Republic of Vietnam, 1970, Hon. Order St. Barbara, U.S. Army Field Artillery Assn. Mem. U.S. Army Acquisition Corps, Assn. of the U.S. Army (v.p. Ft. Armstrong chpt. 1993—), Soc. Am. Military Engrs. (scholar 1966), Soc. Am. Military Comptrollers, Federally Employed Women, Nat. Soc. Scabbard and Blade, (chpt. v.p. 1965-66), Reserve Officers Assn., Quad. City Chpt. Polytech. Alumni Assn. (pres. 1989—), Mensa, Intertel. Democrat. Jewish. Avocations: horticulture, art, music, bonsai, theater, cooking. Home: 1430 Grapler Ct Bettendorf IA 52722-1847

HEYDORN, KAJ, science laboratory administrator; b. Århus, Denmark, Mar. 10, 1931; s. Jakob Heydorn and Erna (Jensen) Jakobsen; m. Kirstine Svenningsen, Mar. 31, 1965; children: Siri, Arne. MS, Tech. U. Denmark, 1954, Dr. Technices, 1980. Chartered chem. engr. Sci. asst. Niels Bohr Inst., Copenhagen, 1955-56; rsch. scientist Atomic Energy Commn., Copenhagen, 1956-65; visiting scientist Gen. Atomic Corp., San Diego, 1965-66; sect. leader AEC, Copenhagen, 1966-76; div. head Risø Nat. Lab., Roskilde, Denmark, 1976—; mem. adv. com. Nucleotecnica, Chile, 1992—; expert mem. Nuclear Cert. Com., Europe, 1982—; cons. Sci. and Technol. Standards, Europe, 1986—; tech. coop. expert Internat. Atomic Energy Agy., UN, 1989—; coun. mem. Hamdard U., Karachi, Pakistan, 1987—; tech. assessor Danish Accreditation Bd., 1992—. Author: Aspects of Precision and Accuracy in Neutron Activation Analysis, 1978, Neutron Activation Analysis for Clinical Trace Element Research, 1984; editor: (6 vols.) Modern Trends in Activation Analysis, 1987; assoc. editor Radioanalytical Nuclear Chems. jour., 1978—; mem. editorial bd. Trace and Microprobe Techniques jour., 1981—, Isotopenpraxis, 1987-92; contbr. over 100 articles to profl. jours. Nat. rep. Internat. Chemometrics Soc., Graz, Austria, 1977—; founding mem. Internat. Soc. Trace Element Rsch. in Humans, Detroit, 1984—. Lt. Denmark Civil Def., 1954-56. Mem. ASTM (task group mem. 1985—), Acad. Tech. Scis. (life), Internat. Conf. Com. Modern Trends in Activation Analysis. Avocations: travel, photography. Office: Risø Nat Lab Isotope div, PO Box 49, Roskilde DK-4000, Denmark

HEYEN, BEATRICE J., psychotherapist; b. Chgo., June 23, 1925; d. Carl Edwin and Anna W. (Carlson) Lund; m. Robert D. Heyen, June 16, 1950 (dec. Feb. 1981); children: Robin, Jefferson, Neil; m. Robert Christiansen, Nov. 24, 1984. BS, U. Chgo., 1949. Instr. Boone (Iowa) Jr. Coll., 1959-64, Rochester (Minn.) Jr. Coll., 1967-68, Winona (Minn.) State Coll., 1965-68; dir. social svc. State Clinic, Kirksville, Mo., 1968-71; supr., dir. Family Counseling Agy., Joliet, Ill., 1971-85; pvt. practice Muskegon, Mich., 1985—; cons. Homes for Aged, Programs for Aged, Winona, 1965-68, Spl. Programs and Individuals in Psychotherapy, Muskegon, 1984—; dir. Christiansen Fine Art Gallery, North Muskegon. Mem. Gov.'s Com. on Status of Women, Iowa, 1957-62, Gov.'s Com. on Aging, Minn., 1966-68. Grantee for Pilot Projects in Svc. to Women 1971-84. Mem. NASW, Acad. Cert. Social Workers, C.G. Jung Inst. (Chgo.). Methodist. Avocations: ecological interests, day lily gardening, contemporary art. Home: 1610 N Weber Rd Muskegon MI 49445-9629

HEYERDAHL, THOR, anthropologist, explorer, author; b. Larvik, Norway, Oct. 6, 1914; s. Thor and Alison (Lyng) H.; m. Liv Coucheron Torp, Dec. 24, 1936; children—Thor, Bjorn; m. Yvonne Dedekam-Simonsen, Mar. 7, 1949; children—Anette, Marian, Bettina. Realartium, Larvik Coll., 1933; postgrad., field study, U. Oslo, Polynesia, Brit. Columbia, 1937, Ph.D. (hon.), Oslo U., 1961. Ethnol. collection and research primitive man, his habits Polynesia and British Columbia, 1937-40; prod. documentary film Kon-Tiki, 1951; leader, organizer Norwegian archeol. expdn., Galapagos, 1953; research Andes region, 1954; prod. film Galapagos, 1955; leader, organizer Norwegian archaeol. expdn., Easter Island and the East Pacific, 1955-56, Ra expdns., 1969-70, Tigris Expdns., 1977-78; leader, organizer archaeol. expdns., Republic of Maladives, 1982-84, Easter Island, 1986-87; Mem., lectr. Internat. Congress Americanists, Cambridge, 1952, São Paulo, 1954, San Jose, 1958, Vienna, 1960, Barcelona, Madrid, Sevilla, 1964, Internat. Congress Anthropology and Ethnology, Paris, 1960, Moscow, 1964, Internat. Pacific Sci. Congress, Honolulu, 1961, Tokyo, 1965, Vancouver, 1976. Author: Paa Jakt Efter Paradiset, Oslo, 1938, Kon-Tiki (Am. edit.), 1950, American Indians in the Pacific: The Theory Behind the Kon-Tiki Expedition, 1952, Archaeological Evidence of pre-Spanish visits to the Galapagos Island, 1956, Aku-Aku, The Secret of Easter Island (Am. edit.), 1958, Reports of the Norwegian Archaeological Expedition to Easter Island and the East Pacific: Vol. 1, Archaeology of Easter Island, 1961, Vol. 2, Miscellaneous Papers, 1965, Sea Routes to Polynesia, 1968, The Ra Expeditions, 1972, Fatu-Hiva, Back to Nature (Am. edit.), 1975, Zwischen den Kontinenten, 1975, The Art of Easter Island, 1975, Early Man and The Ocean (Am. edit.), 1979, Tigris, 1979; The Maldive Mystery, 1985, Easter Island: The Mystery Solved, 1989, (with C. Rallins) The Kon-Tiki Man, 1990; contbr. articles to sci. and popular mags. Founder, bd. mem. Kon-Tiki Mus., Oslo; internat. patron United World Colls.; trustee World Wildlife Fund Internat. Decorated grand officer Order Al Merito della Repubblica Italiana; Order of Merit First Class Egypt; grand officer Royal Alaouites Order, Morocco; Kirll i Metodi Order of 1st Class Bulgaria; comdr. with star Order St. Olav, Norway; Order Golden Ark Netherlands; recipient Retzius medal Royal Swedish Anthropol. and Geog. Soc., 1950, Vega gold medal, 1962; Mungo Park medal Royal Scottish Geog. Soc., 1951; Oscar for camera achievement Nat. Acad. Motion Picture Arts and Scis., 1951; Prix Bonaparte-Wyse Société de Geographie Paris, 1951; Elish Kane gold medal, Geog. Soc. Phila., 1952; Lomonosov medal Moscow U., 1962; Patron's Gold medal Royal Geog. Soc., 1964; co-recipient Internat. Pahlavi environ. prize UN, 1978; named hon. prof. El Instituto Politécnico Nacional Mexico, others. Fellow N.Y. Acad. Scis.; mem. Belgian, Brazilian, Peruvian, Russian, Swedish (hon.) anthropol. geog. socs., Norwegian Acad. Sci., Norwegian Geog. Soc. (hon.), World Assn. World Federalists (v.p.), Worldview Internat. (v.p.), Explorers Club (hon. dir.). Office: care Mng Editor Simon & Shuster 1230 Ave Of The Americas New York NY 10020-1513 other: Colla Micheri, 17020 Laiguеglia Italy

HEYL, ALLEN VAN, JR., geologist; b. Allentown, Pa., Apr. 10, 1918; s. Allen Van and Emma (Kleppinger) H.; student Muhlenberg Coll., 1936-37; BS in Geology, Pa. State U., 1941; PhD in Geology, Princeton U., 1950; m. Maxine LaVon Hawke, July 12, 1945; children: Nancy Caroline, Allen David Van. Field asst., govt. geologist Nfld. Geol. Survey, summers 1937-40, 42; jr. geologist U.S. Geol. Survey, Wis., 1943-45, asst. geologist, 1945-47, assoc. geologist, 1947-50, geologist, Washington and Beltsville, Md., 1950-67; staff geologist, Denver, 1968-90; cons. geologist 1990—; disting. lectr. grad. coll. Beijing, China and Nat. Acad. Sci., 1988; chmn. Internat. Commn. Tectonics of Ore Deposits. Fellow Instn. Mining and Metallurgy (Gt. Brit.), Geol. Soc. Am., Am. Mineral. Soc., Soc. Econ. Geologists; mem. Inst. Genesis of Ore Deposits, Geol. Soc. Wash., Colo. Sci. Soc., Rocky Mountain Geol. Soc., Friends of Mineralogy (hon. life), Evergreen Naturalist Audubon Soc., Sigma Xi, Alpha Chi Sigma. Lutheran. Contbr. numerous articles to profl. jours., chpts. to books. Home: PO Box 1052 Evergreen CO 80439-1052

HEYMAN, JULIUS SCOTT, neuroscientist; b. Phila., Mar. 26, 1960; s. Louis Sidney and Irma Beverly (Green) H.; m. Michelle Louise Tiger, Aug. 23, 1986; children: Zachary, Jacob. BSc in Food and Nutrition Scis., Drexel U., 1983; PhD in Pharmacology and Toxicology, U. Ariz., 1989; postgrad., Med. Sch., Thomas Jefferson U., 1991—. Vis. rsch. fellow Princeton (N.J.) U., 1988-90, lectr. dept. psychology, 1990; rsch. assoc. dept. anesthesia Hosp. of U. Pa., Phila., 1990-91; lectr. dept. pharmacology Jefferson Med. Coll., Phila., 1992—. Field editor European Jour. Pharmacology, 1988—; contbr. articles to profl. jours. including Brain Rsch., Jour. Pharmacology and Exptl. Therapeutics, Trends in Pharmacological Scis., European Jour. Pharmacology, Neurosci. Letters, Life Scis.; author: (with others) Recent Progress n Chemistry and Biology of Centrally Acting Peptides, 1988; author abstracts; contbr. to procs. Rsch. fellow Nat. Rsch. Svc. Award-NIMH, 1988-90; recipient rsch. prize Am. Gastrointestinal Assn., 1987, travel award Western Pharmacological Soc., 1987. Mem. AAAS, Soc. for Neurosci., Mid-Atlantic Pharmacology Soc., Am. Med. Students Assn. Jewish. Home: 1041 Bolton Ct Bensalem PA 19020 Office: Jefferson Med Coll Dept of Pharmacology Philadelphia PA 19107

HEYMAN, MELVIN BERNARD, pediatric gastroenterologist; b. San Francisco, Mar. 24, 1950; s. Vernon Otto and Eve Elsie Heyman; m. Jody Ellen Switky, May 8, 1988. BA in Econs., U. Calif., Berkeley, 1972; MD in Medicine, UCLA, 1976, MPH in Nutrition, 1981. Diplomate Am. Bd. Pediatrics, Am. Bd. Pediatric Gastroenterology. Intern, resident Los Angeles County-U. So. Calif. Med. Ctr., 1976-79; fellow UCLA, 1979-81; asst. prof. U. Calif., San Francisco, 1981-88, assoc. prof., 1988—, chief pediatric gastroenterology and nutrition, 1990—; assoc. dir. Pediatric Gastroenterology/Nutrition, San Francisco, 1986-89; mem. cons. staff San Francisco Gen. Hosp., Oakland (Calif.) Children's Hosp., Sonoma County Med. Ctr., Santa Rosa, Calif., Natividad Med. Ctr., Salinas, Calif., Scenic Gen. Hosp., Modesto, Calif. Contbr. articles to profl. jours. Chmn. scientific adv. com. San Francisco chpt. Nat. Found. Ileitis and Colitis, 1987—, bd. dirs., 1986—. Research grantee Children's Liver Found., 1984-85, John Tung grantee Am. Cancer Society, 1985-89. Avocations: skiing, swimming, hiking, tennis. Office: U Calif Dept Pediatrics M 680 Box 0136 San Francisco CA 94143-0136

HEYN, ARNO HARRY ALBERT, retired chemistry educator; b. Breslau, Germany, Oct. 6, 1918; s. Myron and Margarete M.E.C. (Cierpinski) H.; m. Helen A. Pielemeier, Mar. 14, 1942; children: Evan A., Margaret L., Robert E. B.S., U. Mich., 1940, M.S., 1941, Ph.D., 1944. Exptl. chemist Sun Oil Co., Norwood, Pa., 1944-47; from instr. to prof. chemistry Boston U., 1947-84, prof. emeritus, 1984; vis. scientist Brookhaven Nat. Lab., summers 1954-56; acad. guest Eidg. Techn. Hochschule, Zurich, 1965, Gesellschaft F. Kernforschung, Karlsruhe, 1973, 80, 81, 82, Landesanst. F. Wasserbiologie, Vienna, 1973; sci. adviser Boston Dist. U.S. FDA, 1967-72. Contbr. articles profl. jours. Fellow AAAS; mem. Am. Chem. Soc. (councilor 1967-94, chmn. coun. com. on constn. and bylaws 1983-85, coun. policy com. 1986-91, vice-chmn. 1987-88, com. on coms. 1992-94, Henry Hill award N.E. sect. 1986, editor Nucleus 1989—), AAUP (treas. Boston U. chpt. 1979-83), Sigma Xi, Phi Lambda Upsilon, Sub Sig Outing Club (Boston). Avocation: locksmithing. Home: 21 Alexander Rd Newton MA 02161

HIBST, HARTMUT, scientist, chemistry educator; b. Duisburg, Germany, Apr. 4, 1950; s. Matthias and Kaethe (Roessler) H. Diploma in chemistry, U. Giessen, Fed. Republic Germany, 1973, grad., 1977. Sci. asst. dept. inorganic chemistry U. Giessen, 1973-77; scientist Cen. Rsch. of BASF AG, Ludwigshafen, Fed. Republic Germany, 1978-81, head thin film rsch. div., 1982—; lectr. inorganic chemistry and tech. U. Mannheim, Fed. Republic Germany, 1989—; cons. expert Commn. of the European Communities, Brussels, 1990-91. Author: Magnetic Recording Materials; contbr. articles to profl. jours.; patentee in field. Mem. Gesellschaft Deutscher Chemiker, Verband Angestellter Akademiker. Avocations: surfing, violoncello, biking, painting, modern art and design. Home: Branichstr 23, 6905 Schriesheim Germany Office: BASF Aktiengesellschaft, ZAA/F-M320, D-6700 Ludwigshafen Germany

HICKEY, FRANCIS ROGER, physicist, educator; b. Troy, N.Y., June 8, 1942; s. Frank R. and Ann M. (O'Malley) H.; m. Paula Williamson, Aug. 29, 1964; children: Sharon Ann, Kevin Derus. BS, Siena Coll., 1964; MS, Clarkson U., 1967, PhD, 1970. From asst. to assoc. prof. Physics Hartwick Coll., Oneonta, N.Y., 1969-83; prof. Physics Hartwick Coll., Oneonta, 1983—; adv. bd. Sci. Discovery Ctr. of Oneonta, 1989—; nat. councillor Soc. Physics Students, 1974-75. Contbr. articles to profl. jours. Mem. Am. Phys. Soc., Am. Assn. Physics Tchrs. Roman Catholic. Achievements include development of Physics Educational Computer Program. Home: RR 3 Box 417 Oneonta NY 13820-9461 Office: Hartwick Coll Physics Dept Oneonta NY 13820

HICKEY, ROSEMARY BECKER, retired podiatrist, lecturer, writer; b. Miami, Ariz., July 6, 1918; d. Morris Louis and Sara (Frankel) Becker; m. Richard Charles Hickey, Dec. 19, 1959 (div. Dec. 1977); children: Morris Richard, David Richard. BA, Cen. YMCA, 1940; D in Podiatric Medicine, Ill. Coll. Podiatric Medicine, 1946; MA, Tex. Women's U., 1974; MSW, U. Houston, 1985. Cert. social worker, Tex.; lic. counselor, Tex. Pvt. practice podiatry Chgo., 1946-68; pvt. practice counseling Dallas, 1974-77; children's protective services investigator DHR Harris County Children's Protective Services, Houston, 1978-83; instr. Chgo. Coll. Chiropody, 1959. Editor/pub. Cognate mag., 1959—; contbr. articles to profl. jours. Asst. coordinator AARP Tex. 5-county area, 1984—; adv. svcs. social worker Travelers Aid of Houston, 1985; tutor Literacy Program Kern County Library, Boys and Girls Club of Bakersfield; mem. Kern County HIV+ AIDS Consortium. Recipient diplome of honor and medaille Internat. Recherches de Podologie, 1952. Mem. Royal Soc. Health, Internat. Assn. Platform Speakers, NASW, Am. Soc. Foot Roentology, Am. Coll. Foot Roentgenology, Am. Coll. Foot Orthopedists (sci. chmn., v.p. midwest div. 1948-66, past nat. sec., sci. award 1957, editor jour. 1964), Fellows Pedic Rsch. Soc. (past holder various offices), No. Tex. Assn. Unitarian-Universalist Socs. (past pres.), Mensa, Valley Writers Network, Spectator Amateur Press Soc., Sierra Club, Kaypro Users Group. Avocations: writing, teaching, camping, music. Home: 1451 N Peach Ave #182 Fresno CA 93727

HICKMAN, HUGH VERNON, physics educator; b. Washington, June 3, 1947; s. Jack Wallis Hickman and Mary Cecelia (Regar) McCoy. BSEE, U. South Fla., 1984, PhD, 1989. Entrepreneur, 1969-80; vis. prof. elec. engring. U. South Fla., Tampa, 1989-90; vis. prof. computer sci. Eckerd Coll., St. Petersburg, Fla., 1990-91; prof. physics Hillsborough Community Coll., Tampa, 1991—. Contbr. articles to Am. Jour. Physics, Applied Physics Letters, Solid State Communications, IEEE Transactions on Magnetics. Mem. AAAS, IEEE, Am. Assn. Physics Tchrs., Ye Mystic Krewe of Gasparilla, Phi Kappa Phi. Republican. Roman Catholic. Achievements include research in temporal dynamics. Home: 5010 Dante Ave Tampa FL 33629 Office: Hillsborough Community Coll PO Box 30030 Tampa FL 33630

HICKOX, GARY RANDOLPH, pulp and paper engineer; b. Waycross, Ga., Oct. 16, 1952; s. Everett Lloyd and Alma (Douglas) H. BCE with high honors, So. Tech. Inst., Marietta, Ga., 1975. Registered profl. engr., Ga., Ala., N.Mex., Tex., Colo., Miss., La. Engr. technologist Internat. Paper Co., Mobile, Ala., 1975-79, project engr., 1979-81, design engr., 1981-85, dept. engr., 1985-89, sr. dept. engr., 1989-93, acting mgr. design-paper group, 1993—. Mem. TAPPI (project officer finishing div.-winding 1989-90, sec. 1990-91, vice chmn. 1991—, tech. cons. to instrn. videos and manuals 1991), Ducks Unltd., Rocky Mountain Elk Found., Safari Club Internat., NRA (life mem.), Colo. Bowhunters Assn. (life mem.), United Bowhunters N.Mex. (life mem.), U.S. Shooter Devel. Fund (com. 1984 mem., inner cir. mem.), Tau Alpha Pi. Avocations: hunting, traveling.

HICKS, DALE R., agronomist, educator; b. Odin, Ill., Oct. 10, 1938; married; 3 children. BS, U. Ill., 1960, MS, 1966, PhD in Agronomics, 1968. Asst. farm advisor U. Ill., 1964-65, teaching asst. stats., 1966-68; from asst. prof. to assoc. prof. U. Minn., St. Paul, 1968-76, prof. agronomics, 1976—, extension agronomist, 1968—. Fellow Am. Soc. Agronomy (Robert E. Wagner award 1991), Crop Sci. Soc. Am. Achievements include research in growth regulator effects on yield, genotypes and yield of both corn and soybeans, production practices and maximum yield. Office: Univ of Minnesota Saint Paul MN 55108*

HICKS, GEORGE WILLIAM, automotive and mechanical engineer; b. Ypsilanti, Mich., Jan. 15, 1948; s. Troy Diamond Sr. and Clara (Sehl) H.; m. Carol Ann Kohorst, Aug. 5, 1967; children: Lorelei Lynn, Dawn Marie, Heather Nicole. BSME, U. Mich., 1977. Registered profl. engr., Mich. Test and devel. engr. Chrysler Corp., Chelsea, Mich., 1976-81; sr. engr. Alexander Proudfoot Co., Chgo., 1981; mech. engr. Polytechnic, Lincolnwood, Ill., 1981-82; mgr. tech. svcs. Shackson Assocs., Ann Arbor, Mich., 1982-84; forensic engr. Joscelyn & Treat, Ann Arbor, 1984-86; staff cons. Packer Engring., Troy, Mich., 1986-90; owner, prin. cons. Ingenium Svcs., Rochester Hills, Mich., 1990—; cons. Backplane Tech., Clinton, 1984-86, Shackson Assocs., Ann Arbor, 1984-86. Author: Safety Standards and The Rehabilitation Vehicle, 1991; editor: Roll Over Protective Structures Manual, 1989; lectr. in field. Mem. ASME, NSPE, Nat. Mobility Equipment Dealers Assn., Engring. Soc. Detroit, Assn. Driver Educators for the Disabled, Mich. Assn. Traffic Accident Investigators, Soc. Automotive Engrs. Office: Ingenium Svcs 3889 Mildred Ave Rochester Hills MI 48309-4269

HICKS, JOCELYN MURIEL, laboratory medicine specialist; b. Leamington Spa, Warwickshire, Eng., Aug. 17, 1937; came to U.S., 1965; d. Harold Archie and Muriel Ellen (Cumberland) Bingley; m. John Geoffrey Hicks, Aug. 15, 1959 (div. Nov. 1965); m. Melvin Blecher, May 1, 1973. BS, U. London, 1959, MSc, 1962; PhD, Georgetown U., 1971. Fellow Georgetown U. Med. Ctr., Washington, 1969-71; dir. clin. chemistry Children's Hosp. Nat. Med. Ctr., Washington, 1971-75, chmn. dept. lab. medicine, 1975-90, chief of lab. medicine and pathology, 1990—; asst. prof. George Washington U. Med. Ctr., Washington, 1972-74, assoc. prof., 1975-81, prof., 1981—; mem. profl. staff The Hosp. for Sick Children, Washington, 1984—; pres. Children's Faculty assocs. Children's Hosp., Washington, 1989-90, chmn. bd. dirs. 1990-93, clin. affiliate Cath. U. Am., Washington, 1982—; cons. Eastman Kodak Co., Rochester, N.Y., Miles Diagnostics, Tarrytown, N.Y. Author: Selected Analytes of Clinical Chemistry, 1984, Textbook of Pediatric Clinical Chemistry, 1984, Directory of Rare Analyses, 1986, 87, 90, 92, The Neonate, 1974; co-author: Biochemical Basis of Pediatric Disease, 1992; contbr. articles to profl. jours. Recipient Kone award Assn. Clin. Biochemists, 1987. Mem. Am. Assn. for Clin. Chemistry (bd. dirs. 1978-81, pres. 1981-82, commr. publs. commn. 1982-87, Joseph H. Roe award 1976, Bernard Gerulat Meml. award 1983, Fisher award 1984, cert. of honor, Van Slyke award 1988, Miriam Reiner award 1991, Outstanding Contbns. to Clin. Chemistry 1993), Acad. Clin. Lab. Physicians and Scientists. Avocations: tennis, bridge, travel, reading, cooking. Home: 4329 Van Ness St NW Washington DC 20016-5625 Office: Childrens Nat Med Ctr 3800 Reservoir Rd NW Bldg 2phc Washington DC 20007-2196*

HICKS, MICHAEL DAVID, nuclear engineer; b. Williamsport, Pa., Nov. 9, 1958; s. Richard Charles and Louise Kay (Werner) H. AS in Engring. Sci., Adirondack C.C., 1983; BSEE, SUNY, Buffalo, 1985. Elec. engr. Raytheon/Sedco Systems, Melville, N.Y., 1985-86, Knolls Atomic Power Lab., Schenectady, N.Y., 1986-89; sr. elec. engr. Newport News Reactor Svcs., Idaho Falls, Idaho, 1989-90; sr. nuclear engr. EBASCO Plant Svcs., Langhorne, Pa., 1990-91, ABB Combustion Engring., Windsor, Conn., 1990-91; program mgr. Idaho field office U.S. Dept. Energy, Idaho Falls, 1991—. Lutheran. Office: US Dept Energy MS 7135 785 Doe Pl Idaho Falls ID 83401

HICKS, THOMAS ERASMO, nuclear engineer; b. York, Pa., May 9, 1957; s. Paul Willet and Evelyn (Pezzetta) H.; m. Claudia Ellen Alt, Apr. 2, 1983; children: Thomas, Tyler. BS in Systems Engring., U.S. Naval Acad., 1979. Commd. ensign USN, 1979, advanced through grades to lt., resigned, 1984, submarine officer, 1979-84; resident inspector U.S. Nuclear Regularory Commn., Wilmington, N.C., 1984-85; ops. engr. Perry (Ohio) Nuclear Power Plant, 1985-88; sr. engr. So. Tech. Svcs. Inc., Bethesda, Md., 1988-90, div. dir., 1990—. Contbr. articles to profl. jours. Bd. dirs. Huntington Terrace/Bethesda Civic Assn., 1991—. Mem. Am. Nuclear Soc., Engrs. for Edn. Home: 5603 Madison St Bethesda MD 20817 Office: So Tech Svcs Inc 3 Metro Ctr Ste # 610 Bethesda MD 20814

HICKS, WALTER JOSEPH, electrical engineer, consultant; b. Lawrence, Mass., Mar. 10, 1935; s. Walter Francis and Ethel Mary (Royds) H.; m. Faith Winifred McCrum, Apr. 4, 1959; children: Janet Lee, Walter David, Pamela Jean. BSEE, MIT, 1957, MSEE, 1957; PhD in Plasma Physics, N.Mex. State U., 1969. Elec. engr. Raytheon Co., Bedford, Mass., 1957-67; radar system engr., dept. mgr. Raytheon Co., 1970-74; tech. advisor Raytheon Co., Lowell, Mass., 1974-84; cons. engr. Raytheon Co., Bedford, 1984—; mem. sci. adv. bd. USAF, Washington, 1983. Patentee in field. Elder United Presbyn. Ch., Newton, Mass., 1978-82. Home: 7 Pinewood Rd Acton MA 01720-4409 Office: Raytheon Co Hartwell Rd Bedford MA 01730-2407

HICKSON, ROBIN JULIAN, mining company executive; b. Irby, Eng., Feb. 27, 1944; s. William Kellett and Doris Matilda (Martin) H.; m. P. Anne Winn, Mar. 28, 1964; children: Richard, Sharon, Nicholas, Steven. BS in Mining Engring. with honors, U. London, 1965; MBA, Tulane U., 1990. Mining engr. N.J. Zinc Co., Austinville, 1965-70; divisional mgr. N.J. Zinc

Co., Jefferson City, Tenn., 1970-71; spl. project engr. Kerr McGee Corp., Grants, N.Mex., 1971-72; gen. mgr. Asarco, Inc., Vanadium, N.Mex., 1972-78; gen. mgr. Gold Fields Mining Corp., Ortiz, N.Mex., 1978-83, Mesquite, Calif., 1982-86; v.p. Freeport Mining Co., New Orleans, 1986-91, Freeport Indonesia, Inc., 1991-92; pres. Freeport Rsch. and Engring. Co., New Orleans, 1992-93; sr. v.p. Cyprus Copper Co., Tempe, Ariz., 1993—. Author: (with others) Interfacing Technologies in Solution Mining, 1981. Mem. Instn. Mining and Metallurgy, Am. Inst. Mining and Metallurgy, Mining and Metall. Soc., N.Mex. Mining Assn. (bd. dirs. Santa Fe, N.Mex. chpt. 1975-83), Calif. Mining Assn. (bd. dirs. Sacramento chpt. 1982-86), Beta Gamma Sigma. Episcopalian. Avocations: ornithology, travel. Home: 12246 S Honah Lee Ct Phoenix AZ 85044 Office: PO Box 22015 1501 W Fountainhead Pkwy Tempe AZ 85285-2015

HIDA, GEORGE TIBERIU, ceramic engineer; b. Cluj, Romania, June 9, 1946; came to U.S. 1987; s. Tiberiu and Ilana (Lazarovics) H.; m. Veronica-Irma Torok, Feb. 17, 1967 (div. 1975); m. Rodica Silvia Jeleapov, Sept. 19, 1975; children: Sven, Sever. MS, Poly. Inst., Bucharest, Romania, 1970; PhD, Technion, Haifa, Israel, 1987. Glass engr. TV Screen Factory, Bucharest, 1970-73; ceramic engr. Ceramic Engring. divisn. Rsch. Inst. Electrotechnic Ind., Bucharest, 1973-75; rsch. engr. Rsch. Inst. Electrotechnic Ind., Bucharest, 1975-77, rsch. scientist, 1977-80, sr. rsch. scientist, 1980-82; rsch. assoc. prof. Technion/U. Buffalo, Haifa/Buffalo, 1983-88; dir. tech. Benchmark Structural Ceramics, Buffalo, 1988—; lectr. Coll. for Bldg. Materials, Bucharest, 1975-79, Rsch. Inst. Electrotechnic Ctr., 1978-81. Contbr. articles to profl. jours. Served with Israeli Army, 1984-86. Mem. ASTM, Am. Ceramic Soc., Am. Chem. Soc., N.Y. Acad. Scis., Nat. Inst. Ceramic Engring., Am. Assn. Combustion Synthesis. Achievements include introduction of mechanochemical activation to combustion-synthesis to achieve full control of combustion occurence predetermined composition and structure of the product; 12 patents issued. Home: 63 Eastwick Dr Amherst NY 14221 Office: Benchmark Structural Ceram 350 Nagel Dr Buffalo NY 14225

HIDA, TAKEYUKI, mathematics researcher, educator; b. Okazaki, Japan, Nov. 12, 1927; s. Koichi and Fumi (Mitsui) Ota; m. Minami Hida, Mar. 30, 1953; children—Misachi Isogawa, Fumikazu Hida. B.S., Nagoya U., 1952; Ph.D., Kyoto U., 1961. Instr. Aichi-Gakugei U., Okazaki, Japan, 1952-59; asst. prof. Kyoto U., Japan, 1959-64; prof. math. Nagoya U., Japan, 1964-91, dean sch. sci., 1976-77, Meijo U., 1991—. Author: Stationary Stochastic Processes, 1970, Brownian Motion, 1980, White Noise, 1993. Recipient Chunichi Culture prize, The Chunichi (newspaper company), 1980. Mem. Math. Soc. Japan, Am. Math. Soc., Internat. Statis. Inst. (chmn. com., conf. stochastic processes) Avocation: painting. Home: 6-2 Hirabari-danchi, Tenpaku-cho, Tenpaku-ku, Nagoya 468, Japan Office: Meijo U Dept Math, Shiogamaguchi, Tenpaku, Nagoya 468, Japan

HIDEAKI, OKADA, information systems specialist; b. Tokyo, Japan, Oct. 27, 1935; s. Toyoji and Matu Okada; m. Tomoko Okada, Sept. 5, 1964; children: Yujin, Fumiko. BS in Chemistry, Tokyo Inst. Tech., 1959. Mgr. Kyowa Hakko Kogyo Inc., Tokyo, 1973-79, head of div., 1980-87; sr. system cons. NIX Systems Rsch., Tokyo, 1988-89; dir. PRIDE Japan Inc., Tokyo, 1990—; asst. chmn. Japan Inst. Office Automation, Tokyo, 1989-91; chmn. Bus. Rsch. Inst. Inc., Tokyo, 1988—. Author: Methodologies for Constructing the Strategic Information Systems, 1990; editor: Handbook for the Intelligent Factories, 1990; contbr. articles to profl. jours. Mem. Am. Mgmt. Assn., Acad. Assn. Organized Sci., Ops. Rsch. Soc. Japan. Avocations: noh (playing), shodo (calligraphy), paintings, tennis. Home: 403 Rumine Fujigaoka, 1-18-16 Fujigaoka, Midoriku Yokohama 227, Japan Office: PRIDE Japan Inc, 1-18-13 Masujima Bldg, Higasigotanda Shinagawa-ku Tokyo 141, Japan

HIEBERT, ERNEST, plant pathologist, educator; b. Rosenfeld, Man., Can., May 28, 1941; married; 1 child. BSA in Plant Sci. with honors, U. Man., 1964; MS in Plant Pathology, Purdue U., 1967, PhD in Plant Virology, 1969. Asst. prof. plant pathology dept. U. Fla., Gainsville, 1969-74, assoc. prof. plant pathology dept., 1974-81, prof. virology dept. plant pathology, 1981—; vis. scientist Agriculture Can., Vancouver, 1977-78. Assoc. editor: Phytopathology. Grantee NSF, 1972-74, 75-78, 78-79, 84-87, USDA, 1982-85, 84-87, BARD, 1985-88. Recipient Ruth Allen award, Am. Phytopathological Soc., 1993. Mem. Am. Phytopath. Soc. (Ruth Allen award 1993), Am. Soc. Virology, Am. Soc. Microbiology, Can. Phytopath. Soc. Achievements include research in molecular biology of plant viruses and their nonstructural proteins, cloning of viral genomes; molecular biology of the potyviruses; He developed procedures for the purification of numerous viruses and for the purification of the nonsturctural proteins (cylindrical, amorphous, and nuclear inclusions) associated with potyvirus infections, an essential step for their subsequent characterization. Office: U Florida Dept Plant Pathology Gainesville FL 32611*

HIEFTJE, GARY MARTIN, analytical chemist, educator; b. Zeeland, Mich., Oct. 1, 1942. AB, Hope Coll., 1964; PhD in Analytical Chemistry, U. Ill., 1969. Rsch. asst. phys. chemistry Ill. State. Geol. Survey, 1964-65; from asst. prof. to prof. analytical chemistry Ind. U., Bloomington, 1969-85, assoc. chmn. dept. chemistry, 1978-80, disting. prof. analytical chemistry, 1985—; cons. Lawrence Livermore Lab., 1970—, Upjohn Co., 1979—, Los Alamos Nat. Lab., 1983—, Am. Cyanamid, 1984—, LECO, 1987—; sr. fellow Sci. and Eegring. Rsch. Coun., Eng., 1983. Recipient Can. Test award Chem. Inst. Can., 1979, Anachem award, 1984, Lester W. Strock medal, 1984, Pitts. Analytical Chemistry award, 1986, Theophilus Redwood award Royal Soc. Chemistry, 1986, Tracy M. Sonneborn award Ind. U., 1987. Fellow AAAS; mem. Am. Chem. Soc. (Chemical Instrumentation award 1985, Analytical Chemistry award 1987), Optical Soc. Am., Soc. Applied Spectros (Meggars award 1984, Lester W. Strock award 1992), Sigma Xi. Achievements include research in basic mechanisms in atomic emission, absorption and fluorescence flame spectrometric analysis, development of flame methods of analysis, computer interfacing and control in analysis, chemical instrumentation, correlation spectroscopy. Office: Ind Univ Dept of Chemistry Bloomington IN 47401*

HIESTAND, NANCY LAURA, biology researcher; b. Phila., June 28, 1956; d. Ed Alexander and Jean Lois (Tyrrell) H. BA, Mount Holyoke Coll., 1978; MS, U. Conn., 1986, PhD, 1989. Postdoctoral rsch. assoc. dept. biology Psychology dept., U. Guelph, Ont., Can., 1990-93; postdoctoral rsch. assoc. Biology dept., Wesleyan U., Middletown, Conn., 1990—; postdoctoral rsch assoc. dept. physiology and neurobiology U Conn., Storrs, 1993—. Author: (with others) The Inevitable Bond: Examining Scientist-Animal Interactions, 1992; contb. articles to profl. jours. Mary Lyon scholar Mount Holyoke Coll., 1978. Mem. Animal Behavior Soc. (membership com. 1991—), Am. Assn. of Zool. Parks and Aquariums, Am. Soc. Zoologists, Am. Behavior Soc., Sigma Xi. Democrat. Achievements include discovery that rats showed numerical sensitivity when foraging on a radial maze (with Hank Davis); that mice selected for different levels of nes building showed differential circadian rhythms (with Abel Bult); that in research on neotenous behavior in the dog and wolf, young wolves and adult dogs were similarly capable at problem solving, but adult wolves out-performed them on more complex tasks. Office: U Conn U-42 Dept Physiol/Neurobiology Storrs Mansfield CT 06269-3042

HIGA, TATSUO, marine science educator; b. Nago, Okinawa, Japan, May 3, 1939; s. Tatsusho and Tsuru (Uehara) H.; m. Tomiko Hokama, Aug. 12, 1965; children: Takenobu, Akiko, Kenji. BSc in Engring., Yokohama (Japan) Nat. U., 1963; MSc in Chemistry, Ohio State U., 1968, PhD in Chemistry, 1971. Asst. engr. Noren Sugar Mill, Gushikawa, Okinawa, 1963-64; chemistry tchr. Okinawa Tech. High Sch., Naha, 1964-66; teaching asst. Ohio State U., Columbus, 1968-71; postdoctoral assoc. U. Hawaii, Honolulu, 1971-75; assoc. prof. U. Ryukyus, Nishihara, Okinawa, 1976-82, prof. marine sci., 1982—; cons. Harbor Br. Oceanographic Inst., Ft. Pierce, Fla., 1985-86; vis. prof. U. Copenhagen. Patentee in field. Mem. Am. Chem. Soc., Chem. Soc. Japan, Pharm. Soc. Japan, Am. Soc. Pharmacognosy. Avocations: travel, table tennis. Office: U Ryukyus Dept Marine Sci, Senbaru 1, Nishihara Okinawa 903-01, Japan

HIGASHIDA, YOSHISUKE, mechanical engineer; b. Hiratsuka, Kanagawa, Japan, Jan. 11, 1942; s. Hideki and Aya (Hirota) H.; m. Ruriko Kubotera, July 10, 1973; children: Joji, Tomoki. BS, Yokohama (Japan) Nat. U., 1967, MS, 1969; PhD, U. Ill., 1976. Rschr. Sumitomo Heavy Industries, Ltd., Hiratsuka, Japan, 1969-83, mgr., sr. engr., 1983-90, sr. chief rschr., 1990-91, sr. chief engr., 1991—; lectr. Yokohama Nat. U., 1991—. Contbr. articles to profl. jours. including Trans. of the Japan Inst. of Metals and Welding Jour. Achievements include 5 patents for shapes of bolts with longer fatigue lives, plastic injection molding machine, others. Home: 5-5-6 Zengyo, Fujisawa-Shi 251, Japan Office: Sumitomo Heavy Industries, 6-1 Asahi-Cho, Ohbu 474, Japan

HIGASHIGUCHI, MINORU, aerospace electronic engineering educator; b. Japan, Mar. 25, 1930; s. Shinji and Fujie (Sagara) H.; m. Yoshiko Kugimura, July 22, 1958; children: Takeshi, Yutaka. B in Engring., U. Tokyo, 1953, D in Engring., 1959. Lectr. faculty engring. U. Tokyo, 1957-58, assoc. prof. Aero. Rsch. Isnt., 1958-64; assoc. prof. Inst. Space and Aero. Scis., 1964-71; prof. engring. U. Tokyo, 1972-81, prof., 1982-90; prof. Tokyo Engring. U., Hachioji, 1990—; mem. Japan Aircraft Accident Inspection Com. Mem. Japan Inst. Elec. Engrs., Japan Inst. Electronics Info. and Communication Engrs., Japan Soc. Instrument and Control Engring. (Devel. of Fiber Optic Gyro prize 1984), Japan Civil Aviation Coun. Home: 7-19-33 Tsukimino, Yamato-shi Kanagawa 242, Japan Office: Tokyo Engring U, 1404-1 Katakura, Hachioji Tokyo 192, Japan

HIGBY, EDWARD JULIAN, safety engineer; b. Milw., June 9, 1939; s. Richard L. Higby and Julie Ann (Bruins) O'Kelly: m. Frances Ann Knoodle, 1959 (div. 1962); 1 child, Melinda Ann Mozader. BS in Criminal Justice, Southwestern U., Tucson, 1984. Tactical officer Miami Police Dept., Fla., 1967-68; intelligence officer Fla. Div. Beverages, 1968-72; licensing coord. Lums Restaurant Corp., Miami, 1972-73; legal asst. Walt Disney World, Lake Buena Vista, Fla., 1973-78; loss control cons. R.P. Hewitt & Assocs., Orlando, Fla., 1978-79; safety coord. City of Lakeland, Fla., 1979—. Author: Safety Guide for Health Care, 1979. Bd. dirs. Tampa Area Safety Coun., 1983—, pres., 1990-91, Imperial Traffic Safety Coun., Lakeland, 1983-89; bd. dirs. Parent Resources & Info. on Drug Edn., Lakeland, 1989—; mem. Bay Lake City Coun., 1974-76, mayor, 1975-76; bd. dirs. Greater Lakeland chpt. ARC, 1980-86, chmn. bd. dirs., 1983-84. 85-86, chmn. health svcs., 1980-86; mem. budget com. United Way Cen. Fla., 1983-85; mem. Fla. League Cities, 1974-76, Tri-County League Cities, 1974-76, Orange County Criminal Justice Coun., 1974-78, Cen. Fla. Safety Coun., 1978-79, Fla. Pub. Health Assn., World Safety Orgn.; mem. Polk County Disaster Coordination Com.; bd. dirs. Employers Health Care Group Polk County, 1987-89, ARC Polk County chpt., 1990—. With U.S. Army, 1963-64. Named Vol. of Yr., Greater Lakeland chpt. ARC, 1983-84. Fellow Am. Biog. Inst. Rsch. Assn., Internat. Biog. Assn.; mem. Fla. Sheriffs Assn. (hon. life), Internat. Assn. Identification (life, Fla. div.), Pub. Risk and Ins. Mgmt. Assn., NRA, Fla. Fedn. Safety, Risk and Ins. Mgmt. Soc., Am. Soc. Safety Engrs. (chpt. pres. 1984-85, Safety Profl. of Yr. 1984-85), Heartland Safety Soc. (pres. 1983), Fla. Citrus Safety Assn. (pres. 1981-83), Nat. Fire Protection Assn., Am. Indsl. Hygiene Assn. Republican. Club: Lakeland Rifle and Pistol. Avocations: hunting, fishing. Office: 520 N Lake Parker Ave Lakeland FL 33801

HIGDON, CHARLES ANTHONY, elementary education educator; b. Owensboro, Ky., Mar. 17, 1947; s. Joseph Charles and Mary Ellen (Fulkerson) H.; m. Freda Jean Danhauer, June 18, 1983; children: Kathryn Reneé, Brittany Leigh, Kimberly Ellen. BS in Edn., Brescia Coll., 1970; M, Murray State U., 1976; postgrad., Western Ky. U., 1986-90. Cert. tchr., Ky. Tchr. math. Lourdes Elem. Sch., Owensboro, 1970-76, Hawesville (Ky.) Elem. Sch., 1976—; cons. Hancock County Bd. Edn., Hawesville, 1986—. Chmn. Hancock County Bd. Adjustments, Hawesville, 1983—; mem. planning and zoning bd. Hancock County, Hawesville, 1986—. Tate C. Page Ednl. Leadership Meml. scholar Western Ky. U., 1989. Mem. NEA, ASCD, Nat. Coun. Tchrs. Maths., Ky. Coun. Tchrs. Maths., Hancock County Tchrs. Assn. (pres. 1988-89, v.p. 1991—), Ky. Edn. Assn. Democrat. Roman Catholic. Avocations: skiing, archery, classic muscle cars, antique math. and reading textbooks.

HIGGINBOTHAM, LLOYD WILLIAM, mechanical engineer; b. Haydentown, Pa., Nov. 24, 1934; s. Clarence John and Nannie Mae (Piper) H.; m. Genevieve Law, Oct. 17, 1953 (div.); 1 child, Mark William; m. Mary Bannaian, July 23, 1966; 1 child, Samuel Lloyd. With rsch. and devel. TRW Inc., Cleve., 1953-57; pres. Higginbotham Rsch., Cleve., 1957-64; pres., chief exec. officer Lloyd Higginbotham Assocs., Woodland Hills, Calif., 1964—; cons. grad. engring. programs UCLA, Calif. State U., L.A., U. So. Calif.; pres. adv. com. Pierce Coll., L.A.; adv. com. So. Calif. Productivity Ctr.; cons. various Calif. legislators. Mem. Town Hall Calif.; pres. San Fernando Valley Joint Com. Engrs., 1992-93. Recipient Community Svc. award City of Downey, Calif, 1974, Archimedes award NSPE, Outstanding Contbr. Recognition, 1986, Outstanding Leadership Recognition, 1987, William B. Johnson Meml. Internat. Interprofl. award, 1992. Fellow Inst. Advancement of Engring. (exec. dir. 1984—); mem. Soc. Carbide and Tool Engrs. (chmn. 1974-76), Soc. Mfg. Engrs. (chmn. San Fernando Valley chpt. 1977-79, numerous awards), San Fernando Valley Joint Coun. Engrs. (advisor, pres. 1981-82, 92-94), Profl. Salesmen's Assn., Am. Soc. Assn. Execs., L.A. Coun. Engrs. and Scientists (exec. mgr. 1984—), L.A. Area C. of C., Toastmasters, Masons. Republican. Avocations: golf, spectator sports. Office: Higginbotham Assocs 24300 Calvert St Woodland Hills CA 91367-1113

HIGGINS, PETER THOMAS, government information management executive; b. Hackensack, N.J., Aug. 17, 1943; s. Joseph Alexander and Rita Barth (Buckley) H.; m. Kathleen Mary Melehan, June 6, 1970; 1 child, Kelton Charles. BS in Math., Marist Coll., 1967; MS in Math., Computer Sci., Stevens Inst. Tech., 1968. Front desk clk. Carlyle Hotel, N.Y.C., 1964-67; sci. programmer CIA, Washington, 1968-74, project engr., 1974-80, ops. engr. mgr., 1981-86; congl. fellow U.S. House of Reps., Washington, 1986-87, U.S. Senate, Washington, 1987; mgr. rsch. and devel. CIA, Washington, 1987-89, chief info. officer, 1989-92; dep. asst. dir. engring. FBI, Washington, 1992—; speaker in field. Local leader Jaycees, McLean, Va., 1970; vol. Dem. NAt. Conv., Atlantic City, N.J., 1964; Sunday sch. tchr. Holy Trinity Ch., Washington, 1983-91. Mem. Am. Polit. Sci. Assn. (Fgn. Affairs Congl. fellow 1986), Congl. Fellows Alumni Steering Com. Roman Catholic. Avocations: reading, walking, automobiles. Office: FBI 9th & Pa Ave Washington DC 20535

HIGGINS, RICHARD J., microelectrical engineer. BS in Metallurgy, MIT, 1960; PhD in Materials Sci., Northwestern U., 1965. Prof. physics U. Oreg., Eugene, founding dir. materials sci. inst.; prof. elec. engring. Ga. Tech., Atlanta, 1987—, dir. microelectronics rsch. ctr., 1988—; vis. rschr. Bell Labs, 1972-73, U. Paris, 1981, Thomson-CSF, Paris, 1982, MIT, 1983-85; cons. Analog Devices, 1983-85, Tektronix, 1985-87, NSF, UNESCO, Ford Found. Author: Electronics with Digital and Analog ICs, 1983, Digital Signal Processing in VLSI, 1989, 2 others; contbr. 80 tech. articles to profl. jours. Fellow Am. Phys. Soc. Achievements include research in GaAs heterostructure devices for next-generation ultra-high speed integrated circuits and optoelectronics; roadmap studies for future semiconductor devices, electronic packaging, optoelectronic packaging. Office: Ga Inst Tech Microelectronics Rsch Ctr Office of Interdisciplinary Prgm Atlanta GA 30332-0269*

HIGGINSON, ROY PATRICK, psychologist; b. Middlewich, Cheshire, Eng., Sept. 4, 1946; came to U.S., 1981; s. Joseph and Gwendoline May (Walton) H.; life ptnr. Allen Gilbert. PhD, Washington State U., 1985. Asst. prof. Iowa State U., Ames, 1985-92; pvt. rschr., 1992-93; adv. bd. mem. Child Lang. Data Exchange System, Pitts., 1987—; occasional reviewer NSF, Washington, 1990. Jour. Child Lang., Cambridge, Mass., 1990; vis. prof. U. Calif., San Diego, 1991, 93. Compiler: (book) CHILDES/BIB Annotated Bibliography, 1991, (book) CHILDES/BIB Supplement, 1993, (electronic database) CHILDES/BIB version 3.0, 1993; contbr. articles to Jour. Child Lang. and Child Lang. Teaching and Therapy. Recipient Linguistics Inst. fellowship Linguistic Soc. Am., Washington, 1983, MacArthur fellowship Ctr. for Rsch. in Lang., U. Calif., San Diego, 1991; grantee Basil and Ella Jerrard Fund, Wash. State U., 1983, Sigma Xi, 1983, Carnegie Mellon U., Pitts., 1985-93. Fellow Am. Psychol. Soc.; mem. Soc. for Rsch., Child Devel., Linguistic Soc. Am., Cognitive Sci. Soc. (assoc.), Internat. Neural Network Soc. Achievements include compilation of most extensive and comprehensive database of research literature in child language and language disorders.

HIGHLAND, HAROLD JOSEPH, computer scientist; b. N.Y.C., Apr. 26, 1917; s. Joseph Francis and Frances (Bernstein) H.; m. Esther Harris, June 16, 1940; 1 child, Joseph Harris. BS, CCNY, 1938, MS, 1939; PhD, NYU, 1942. Editor, pub. several consumer and trade publs., N.Y.; dean Arthur T. Roth Grad. Sch., dir. computer labs. L.I. U., 1958-66; prof. computer sci. SUNY, Farmingdale, 1966-78, Disting. prof., 1978-81, ret., 1981; prof. statistics and rsch. Hoftsra U., N.Y., 1968-70; adj. prof. computer sci. Hofstra U., 1974-75; Fulbright prof. computer sci. and ops. rsch. Helsinki (Finland) U. Tech., 1970-71; prof. computer sci. Nat. Tng. Ctr. U.S. Customs, 1971-73; mng. dir. Compulit Inc.; adviser Computer Security Tech. Com. of Chinese Computer Fedn., Beijing; chmn. internat. com. on info. security edn. and tng., mem. tech. com. on info. security Internat. Fedn. for Info. Prcoessing; mem. NATO Sci. Affairs Adv. Bd.; mem. Info. Security Rsch. Ctr., Queensland U. of Tech., Brisbane, Australia; pres. Virus Security Inst.; chmn. task force Implementation of ACM Internat. Plan; cons. to various govt. agys. in the U.S. and abroad. Author: Modelling and Simulation using GPSSII, 1971, Probability Models, 1972, Protecting Your Microcomputer System, 1984, Computer Virus Handbook, 1990; co-author (with Esther Harris Highland) CBASIC/CB86 With Graphics, and other; founding editor, editor-in-chief jour. Computers & Security, emeritus, 1990—; mem. editorial bd. Info. Systems Security, Virus Bull. (Eng.), Security Letter, Computer Law and Security Report (U.K.), SIGSAC Security, Audit and Control, CVIG News (Australia). Recipient ednl. film award Nat. Audio Visual Assn., 1961, ann. award for profl. writing Sci. Rsch. Soc., 1963, citation medal U. Helsinki, 1970, Sitka award Finnish Ministry Edn., 1971; award Winter Simulation Conf., 1976, spl. recognition, 1983; Commn. of Honor award SUNY, 1978, Disting. Svc. award Inst. Mgmt. Sci., 1989, Kristian Beckman award Internat. Fedn. Info. Processing, 1993; faculty fellow CCNY, 1938, fellow Ford Found., 1964-65; computer grantee IBM, 1962. Fellow Irish Computer Soc.; mem. AAAS, IEEE Computer Soc., Assn. for Computing Machinery (SIGSIM spl. recognition 1973), Info. Systems Security Assn., Computer Profls. for Social Responsibility, Internat. Assn. Cryptographic Rsch., Soc. Basic Irreproducible Rsch., N.Y. Acad. Scis. Achievements include research in computers and statistics. Home: 562 Croydon Rd Elmont NY 11003-2814

HIGHTOWER, JESSE ROBERT, chemical engineer, researcher; b. Cleveland, Tenn., Apr. 9, 1939; s. Jesse Robert and Frances (Heidler) H.; m. Dorothy Turnage, June 24, 1961; 1 child, Alison. BSChemE, U. Miss., Oxford, 1961; PhD, Tulane U., 1964. Registered profl. engr. Devel. staff mem. Oak Ridge (Tenn.) Nat. Lab., 1964-74, program dir., 1974-76, sect. head, 1976-88, assoc. div. dir. chem. tech. div., 1988-92; dir., center for waste mgnt. Martin Marietta Energy Systems, Inc., Oak Ridge, Tenn., 1992—; mem. adv. bd. chem. tech. dept. Pellissippi State Community Coll., Knoxville, Tenn., 1985-89. Bd. dirs. Planned Parenthood Assn. of So. Mountains, Oak Ridge, 1972; pres. bd. dirs. Clinch River Home Health, Inc., Clinton, Tenn., 1992. Named Outstanding Adv. Bd. Mem. Pellissippi State Community Coll., 1990. Fellow AICE (pres. nuclear engring. div. 1993, Engr. of Yr. 1982). Home: 104 Scenic Dr Oak Ridge TN 37830 Office: Martin Marietta Energy Systems K-25 Site PO Box 2003 Oak Ridge TN 37831-7357

HILBURN, JOHN CHARLES, geologist, geophysicist; b. Dallas, Sept. 16, 1946; s. William Grant and Catherine (Thorwald) H.; 1 child, John C. Jr. BS in Geol. Scis., U. Tex., Austin, 1978. Mfg. mgr. Scorpio, Inc., Austin, 1972-74; rsch. engr., scientist U. Tex., Austin, 1974-78; corp. v.p. Reeves, Inc., Houston, 1978-79; mgr. acqusitions S.AM. Western Geophys. Corp., Houston, 1979-80; sr. mktg. geophysicist GECO Geophys. Co., Inc., Houston, 1980-85; pres. John Hilburn & Assocs., Inc., Austin, 1985—; sec.-treas. Hilburn Assocs., Inc., Austin, 1980—; dir. Rusty Pelican Restaurant, Houston. Mem. Soc. Exploration Geophysicists, European Assn. Exploration Geophysicists, Can. Soc. Exploration Geophysicists, Am. Assn. Petroleum Geologists, Geol. Soc. Am. Avocations: gem cutting, stamp collecting, skiing, scuba diving, cooking. Home and Office: 6302 Mountain-climb Dr Austin TX 78731-3908

HILDEBRAND, WILLIAM CLAYTON, chemist; b. Hartford City, Ind., Aug. 14, 1951; s. William Hiram and J. Maxine (Smith) H.; m. Margaretta Louise Urbanc, Oct. 22, 1977; children: Christopher Ian, Andrea Leigh. BS in Chemistry, Rose Hulman Inst., 1973; postgrad., Purdue U., 1982-84. Cert. hazardous material mgr. Chemist Demert and Dougherty, Chgo., 1973-74, Hydrosol, Inc., Chgo., 1974-75, Perma-Line Corp., Chgo., 1975-81; tech. dir. Superior Oil Co., Inc., Indpls., 1982--; cons. Time-Life Books, Montreal, Can., 1987; mem. Hazardous Materials Tech. Expert Adv. com. Mem. Local Emergency Planning Com., Indpls., 1988--, St. Louis, 1988--; coach Fall Creek Little League, Indpls., 1990. Mem. Theta Xi (pres. 1971-72). Home: 5119 Chatham Pl Indianapolis IN 46226 Office: Superior Oil Co Inc 400 W Regent St Indianapolis IN 46225

HILDEBRANSKI, ROBERT JOSEPH, civil engineer; b. Hammond, Ind., Oct. 4, 1966; s. Robert Charles and Elizabeth Ann (Karwatka) H. Student, Mich. State U., 1984-86; BS, U. Ill., 1989. Design engr. Consoer Townsend & Assocs., Chgo., 1989-90, constrn. inspector, 1989, survey crew chief, 1992, resident engr., 1990—; speaker Denver Concrete Corrosion Prevention Seminar. Lector St. Johns Ch., Glenwood, Ill., 1993—. Home: 18020 Commercial Lansing IL 60438 Office: Consoer Townsend & Assocs 303 E Wacker Dr Ste 600 Chicago IL 60601

HILDEBRANT, ANDY MCCLELLAN, electrical engineer; b. Nescopeck, Pa., May 12, 1929; s. Andrew Harmon and Margaret C. (Knorr) H.; m. Rita Mae Yarnold, June 20, 1959; children: James Matthew, David Michael, Andrea Marie. Student, State Tchrs. Coll., Bloomsburg, Pa., 1947-48, Bucknell U., 1952-54, UCLA, 1955-57, Union Coll., 1965-70. Rsch. analyst Douglas Aircraft Co., Santa Monica, Calif., 1954-57; specialist engring. GE, Johnson City, N.Y., 1957-58, Ithaca, N.Y., 1958-64; elec. engr. GE, Utica, N.Y., 1964-70, Sylvania Electro Systems, Mountain View, Calif., 1970-71, Dalmo-Victor Co., Belmont, Calif., 1971-72, Odetics/Infodetics, Anaheim, Calif., 1972-75, Lear Siegler, Inc., Anaheim, 1975-78, Ford Aerospace, Newport Beach, Calif., 1978-79, THUMS Long Beach Co., Long Beach, Calif., 1979—; elec. engring. cons. Perkin-Elmer, Auto Info. Retrieval, Pi-Gem Assn., Pasadena, Calif., Palo Alto, Calif., 1971-73. Patentee AC power modulator for a non-linear load. Juror West Orange County Mpcl. Ct., Westminster, Calif., 1979, U.S. Dist. Ct., L.A., 1991-92. With USN, 1948-52. Recipient Cert. Award in Indsl. Controls Tech., Calif. State U., Fullerton, 1991-92. Mem. Orange County Chpt. Charities (sec. 1988), KC (past grand knight 1987-88). Republican. Roman Catholic. Avocations: woodworking, camping, hiking. Home: 20392 Bluffwater Cir Huntington Beach CA 92646-4723 Office: THUMS Long Beach Co 300 Oceangate Long Beach CA 90802-6801

HILDNER, ERNEST GOTTHOLD, III, solar physicist, science administrator; b. Jacksonville, Ill., Jan. 23, 1940; s. Ernest Gotthold Hildner Jr. and Jean (Johnston) Duffield; m. Sandra Whitney Shellworth, June 29, 1968; children: Cynthia Whitney, Andrew Duffield. BA in Physics and Astronomy, Wesleyan U., 1961; MA in Physics and Astronomy, U. Colo., 1964, PhD in Physics and Astronomy, 1971. Experiment scientist High Altitude Obs., Nat. Ctr. Atmospheric Rsch., Boulder, Colo., 1972-80, vis. scientist, 1985-86; chief solar physics br. NASA Marshall Space Flight Ctr., Huntsville, Ala., 1980-85; dir. space environment lab. NOAA Environ. Rsch. Labs., Boulder, 1986—; mem. com. on solar and space physics NRC, Washington, 1986-90; chmn. Com. on Space Environment Forecasting, fed. coord. for meteorology, Washington, 1988—. Contbr. rsch. papers in solar and interplanetary physics, 1971—; co-inventor spectral slicing X-ray telescope with variable magnification. Mem. AAAS, Am. Geophys. Union (assoc. editor Geophys. Rsch. Letters 1983-85), Am. Astron. Soc. (councillor solar physics div. 1979-80), Internat. Astron. Union, Sigma Xi. Achievements include patent for Spectral Slicing X-Ray Telescope with Variable Magnification. Office: NOAA Space Environment Lab 325 Broadway (R/E/SE) Boulder CO 80303-3328

HILE, MATTHEW GEORGE, psychologist, researcher; b. Cleve., Mar. 21, 1953; s. Robert Henry and Charlotte Laura (Simon) H.; m. Allison McGhee, Aug. 14, 1976; 1 child, Simon Robert. MS, U. Ky., 1978, PhD, 1984. Lic. psychologist, Mo. Unit psychologist St. Louis Devel. Disease Treatment Ctr., St. Louis, 1981-82, chief psychologist, 1982-84, unit dir., 1983; postdoctoral fellow Mo. Inst. Mental Health, St. Louis, 1984-87, rsch. asst.

prof. Psychology, 1987-89, asst. prof. Psychology, 1989—. Contbr. articles to profl. jours. Mem. nat. adv. bd. Foster Child Health Passport Enhancement Project, Menlo Park, Calif., 1991—, St. Louis Children's Hosp. Ethics Com., St. Louis, 1991. Mem. APA, Am. Assn. on Mental Retardation, Soc. for Computers in Psychology. Achievements include development of first decision support expert system in psychology. Office: MIMH 5247 Fyler Ave Saint Louis MO 63139-1494

HILER, EDWARD ALLAN, academic administrator, agricultural engineering educator; b. Hamilton, Ohio, May 14, 1939; s. Earl and Thelma (Kolb) H.; m. Patricia Burke; children: Karen, Richard, Scott. BS, MS in Agrl. Engring., Ohio State U., 1963, PhD in Agrl. Engring., 1966. Registered profl. engr., Tex. Instr. Ohio State U., Columbus, 1963-64, grad. fellow, 1964-65; instr. Ohio Agrl. Research and Devel. Ctr., Wooster, 1965-66; asst. prof. agrl. engring Tex. A&M U., College Station, 1966-69, assoc. prof. agrl. engring., 1969-73, prof. agrl. engring., 1973-74, prof., head agrl. engring. dept., 1974-88, dep. chancellor for acad. program planning and rsch., 1989-92; interim chancellor Tex. A&M U. System, 1991-92, dep. chancellor for acad. and rsch. programs, 1992-93; vice chancellor, dean agr. and life scis., dir. Tex. Agrl. Experiment Sta., 1992—; cons. Office of Tech. Assessment Food Industry Adv. Com., Congress of U.S., 1978-82, Office of Water Research and Tech., Dept. of Interior, 1980; project advisor Gas Research Inst., Chgo., 1984-86. Editor: Biomass Energy, 1985; contbr. articles to profl. jours. Recipient Disting. Alumnus award coll. engring. Ohio State U., 1978. Mem. Am. Soc. Agrl. Engrs. (bd. dirs. 1985-87, 90—, pres.-elect 1990-91, pres. 1991-92, past pres., chmn. bd. trustee ASAE found. 1987-90, 92-93), Am. Soc. Engring. Edn. (chmn. agrl. engring. div. 1976-77), Accrediting Bd. Engring. and Tech. (bd. dirs. 1985-88), Nat. Acad. Engring. Presbyterian. Avocations: golf, photography, reading novels. Office: Tex A&M U Dean & Vice Chancellor Agriculture Program Office College Station TX 77843-2142

HILGARTNER, C(HARLES) ANDREW, theorist; b. Austin, July 10, 1932; s. Henry Louis and Constance (Stark) H.; m. Carol Howe, Aug. 15, 1955 (div. Aug. 1981); children: Stephen Hallett, Elizabeth, James Mitchell, Thomas Kimball, Margaret; m. Martha Ann Taylor Bartter, Jan. 17, 1986. AB cum laude, Amherst Coll., 1954; MD, Johns Hopkins U., 1958. Intern Barnes Hosp., St. Louis, 1958-59; fellow in biochemistry U. Rochester, N.Y., 1959-61, fellow in biology, 1961-62, fellow in brain rsch., 1962-64, instr. biochemistry, 1964-68; scholar Rochester, 1968-85; dir. Hilgartner and Assocs., 1982—. Contbr. articles to sci. and profl. jours. Bd. dirs. Rochester Oratorio Soc., 1968-72, Marion Civic Chorus, 1985-92. Mem. AAAS, Inst. Gen. Semantics (assoc. rsch. dir. 1969-71, mem. edit. bd. 1979—), Internat. Soc. for Gen. Semantics, Kiwanis. Home and Office: Hilgartner and Assocs 2413 NE Street Kirksville MO 63501

HILGENFELD, ROLF, chemist; b. Göttingen, Fed. Republic Germany, Apr. 3, 1954; s. Hans-Adolf and Sigrid (Hardt) H.; m. Anna-Liisa Peltola, Jan. 30, 1981; 1 child, Jaana. Diploma, U. Göttingen, 1981; D. in Natural Scis., Free U., Berlin, 1987. Rsch. stipendiate Max-Planck Inst. for Exptl. Medicine, Göttingen, 1979-81; rsch. asst. Free U., Berlin, 1981-86; scientist protein crystallography Hoechst Cen. Rsch., Frankfurt, Fed. Republic Germany, 1986-88; vis. scientist Biocenter of the Univ., Basel, Switzerland, 1987-88; group leader protein crystallography, protein engring. Hoechst Cen. Rsch., Frankfurt, 1988—; lectr. Free U., Berlin, 1985-91, Friedrich-Schiller U., Jena, Fed. Republic Germany, 1990—. Author: Host-Guest Complex Chemistry, 1982; contbr. articles to profl. jours. Mem. Am. Crystallographic Assn., Brit. Crystallographic Assn., European Assn. for Crystallography of Biol. Macromolecules, Molecular Graphics Soc., Protein Soc. Office: Ctrl Rsch Hoechst AG, PO Box 800320, D 65926 Frankfurt 80, Germany

HILKERT, ROBERT JOSEPH, cardiologist; b. Akron, Ohio, Nov. 12, 1958; s. William Albert and Mary Louise (Ahern) H.; m. Deborah Lynn Toppmeyer, Sept. 12, 1987. BS, Ohio State U., 1981; MD, Med. Coll. Ohio, 1985. Diplomate Nat. Bd. Med. Examiners., Am. Bd. Internal Medicine. Internal medicine intern Univ. Health Ctr. Hosps. of Pitts., 1985-86, resident in internal medicine, 1986-88; fellow dept. physiology-dept. internal medicine U. Pitts., 1988-90; clin. and rsch. fellow cardiac unit Mass. Gen. Hosp., Boston, 1990-92; rsch. fellow Harvard Med. Sch., Boston, 1990-92; fellow sect. cardiology and cardiovascular inst. Boston U. Med. Ctr., 1992—. Contbr. articles to Jour. Biol. Chemistry, Trends in Cardiovascular Medicine, Archives of Biochemistry, others. Recipient Summa award, 1978; grantee NIH, 1981, Am. Heart Assn., 1989-90, 91-92. Mem. ACP, AAAS, Am. Med. Athletic Assn., Mass. Med. Soc., Am. Fedn. for Clin. Rsch., Paul Dudley White Soc., Helix, Alpha Omega Alpha, Phi Kappa Phi. Achievements include research in transcriptional regulation of endothelial cell genes; genetics of essential hypertension; biochemical isolation and biophysical properties of Ca2 release channels from skeletal and cardiac sarcoplasmic reticulum. Office: Boston U Med Ctr 88 E Newton St D8 Boston MA 02118

HILL, BRUCE MARVIN, statistician, scientist, educator; b. Chgo., Mar. 13, 1935; s. Samuel and Leah (Berman) H.; m. Linda Ladd, June 18, 1958; children—Alec Michael, Russell Andrew, Gregory Bruce; m. Anne Edith Gardiner Bruce, Aug. 5, 1972. B.S. in Math., U. Chgo., 1956; M.S. in Stats., Stanford U., 1958, Ph.D. in Stats., 1961. Mem. faculty U. Mich., Ann Arbor, 1960—, assoc. prof. stats. and probability theory, 1964-70, prof., 1970—; vis. prof. bus. Harvard U., 1964-65; vis. prof. systems engring. U. Lancaster, U.K., 1968-69; vis. prof. stats. U. London, 1976; vis. prof. econs. U. Utah, 1979; vis. prof. math. U. Milan, U. Rome, 1989. Editor Jour. Am. Statis. Assn., 1977-83, Jour. Bus. and Econ. Stats, 1982—; contbr. articles to profl. jours., chpts. to books on stats, encys. Grantee NSF, 1962-69, 81-86, 89—, USAF, 1971-73, 87-89, 89—. Fellow Am. Statis. Assn. (pres. Ann Arbor chpt. 1986-91), Inst. Math. Stats.; mem. AAUP, Am. Math Assn., Rsch. Club U. Mich., Psi Upsilon, Sigma Chi. Home: 1657 Glenwood Rd Ann Arbor MI 48104-4133 Office: U Mich Dept Stats Ann Arbor MI 48109-1027

HILL, CRAIG LIVINGSTON, chemistry educator, consultant; b. Pomona, Calif., Feb. 24, 1949; s. Theodore Hill and Isabel Temple (Chemeshaw) Kimble; m. Linda Sue Ross. BA with high honors, U. Calif., San Diego, 1971; PhD in Chemistry, MIT, 1975; postdoctoral, Stanford (Calif.) U., 1977. Asst. prof. U. Calif., Berkeley, 1977-83; assoc. prof. Emory U., Atlanta, 1983-88, prof., 1988—. Cons. editor: JAI Press, 1988—; editor: New Jour. of Chemistry, 1989—, Alkane Functionalization Jour., 1989; editor: book Activation and Functionalization of Alkanes, 1989; mem. editorial adv. bd. in field.; contbr. numerous articles to profl. jours. NSF grantee, 1984—, Army Rsch. Office grantee, 1987—; NIH fellow, 1986—. Mem. Am. Chem. Soc. (Charles H. Stone award 1992), Royal Soc. Chemistry, AAAS, Internat. Union Applied Chemistry, N.Y. Acad. Scis. Avocations: travel, geography, alpine skiing, speed skating. Home: 2941 Cravey Dr NE Atlanta GA 30345-1421 Office: Emory Dept Chemistry 1515 Pierce Dr Atlanta GA 30322

HILL, DAVID LAWRENCE, research corporation executive; b. Boonville, Miss., Nov. 11, 1919; s. David Alexander and Mabel Clair (Brown) H.; B.S., Calif. Inst. Tech., 1942; Ph.D. (Socony Vacuum Co. fellow), Princeton U., 1951; m. Mary M. Shadow, Dec. 31, 1950; children—David A., Mary C., Robert L., John F., Cynthia A., Sandra E., James A. With U. Chgo. Metall. Lab. and Argonne Nat. Lab., 1942-46, assoc. physicist, group leader, 1944-46; asst. prof. physics Vanderbilt U., Nashville, 1949-52, assoc. prof., 1952-54; guest scholar Inst. Theoretical Physics, Copenhagen, summer 1950; cons. theoretical physics U. Calif., Los Alamos (N.Mex.) Sci. Lab., 1952-54, staff mem., 1954-58, group leader theoretical nuclear physics, 1955-58; mgmt. cons., 1958-60; pres. Phys. Sci. Corp., Fairfield, Conn., 1960-62, Nanosecond Systems, Inc., Fairfield, 1963-72, Particle Measurements, Inc., Southport, Conn., 1965-81, Harbor Research Corp., 1978—; chmn. bd. Integrated Total Systems, Inc., Hingham, Mass., 1968-81; pres. Southport Computers, Inc., Conn., 1973-81, Valutron N.V., Netherlands Antilles, 1980—; pres. Patent Enforcement Fund, Inc., Southport, Conn., 1990—; lectr. in field; sci. advisor to Vice Presdl. nominee, Senator Estes Kefauver, 1956; incorporator, exec. v.p. dir. Los Alamos Investment Corp., 1956-58; cons. physicist in field. Adv. com. on sci. and tech. of Adv. Council of Dem. Nat. Com., 1959-61. Fellow Am. Phys. Soc., AAAS; mem. IEEE, Fedn. Am. Scientists (nat. chmn. 1953-54), Sigma Xi. Contbr. articles to profl. jours. Office: Patent Enforcement Fund Inc PO Box L Southport CT 06490-0569

HILL, FREDRIC WILLIAM, nutritionist, poultry scientist; b. Erie, Pa., Sept. 2, 1918; s. Vaino Alexander and Mary Elvira (Holmstrom) H.; m. Charlotte Henrietta Gummoe, Apr. 1, 1944; children: Linda Charlotte, James Fredric, Dana Edwin. B.S., Pa. State U., 1939, M.S., 1940; Ph.D., Cornell U., 1944. Rsch. asst. Pa. State U., 1939-40, Cornell U., 1940-44; head nutrition div. rsch. labs. Western Condensing Co., Appleton, Wis., 1944-48; assoc. prof., then prof. animal nutrition and poultry husbandry Cornell U., 1948-59; prof. poultry husbandry, chmn. dept. U. Calif., Davis, 1959-65, prof. nutrition, 1965—, chmn. dept. nutrition, 1965-73, assoc. dean Coll. Agr., 1965-66, assoc. dean rsch. and internat. programs, 1976-80, coordinator internat. programs, 1976-80, prof. nutrition emeritus, 1989—; Mem. subcom. hormonal relationships and applications com. on Animal Nutrition, NRC, 1953, subcom. poultry nutrition, 1953-74; mem. Food and Nutrition Bd., 1975-78; commr. Calif. Poultry Improvement Commn., 1959-65; participant 8th Easter Sch. Agrl. Scis., U. Nottingham, Eng., 1961, World Conf. Animal Prodn., Rome, Italy, 1963, U.S. AID-Nat. Acad. Sci. Seminar on Protein Foods, Bangkok, 1970, USIA Asia Seminars on Food, Population and Energy, 1974-75; Japan Soc. Promotion Sci. vis. prof. Nagoya U., 1974-75; vis. scientist FDA, 1975, 88, Nutrition Inst., USDA, 1975; cons. Institut National de Recherche Agronomique, France, 1982; plenary speaker 3d Asian-Australian Animal Sci. Congress, Seoul, Republic of Korea, 1985. Contbr. articles profl. jours.; Editorial bd.: Poultry Sci. Jour, 1960-64; editorial bd.: Jour. of Nutrition, 1964-68; editor, 1969-79. Fellow Danforth Found., 1938; recipient Nutrition Rsch. award Am. Feed Mfrs. Assn., 1958, Newman Internat. Rsch. award Brit. Poultry Assn., 1959; Guggenheim Found. fellow Nat. Inst. Med. Rsch., Mill Hill, Eng., Hebrew U., Jerusalem, 1966-67; Alumni fellow Pa. State U., 1983. Fellow AAAS, Poultry Sci. Assn. (Rsch. prize 1957, Borden award 1961), Am. Inst. Nutrition (councillor 1982-85); mem. Soc. Exptl. Biology and Medicine, Nutrition Soc. (Gt. Britain), Coun. Biology Editors, World's Poultry Sci. Assn., Am. Inst. Biol. Scis., Am. Soc. Animal Sci., Am. Chem. Soc., Fedn. Am. Socs. for Exptl. Biology (publs. com. 1987-93, chmn. 1991-93), Sigma Xi, Phi Eta Sigma, Gamma Sigma Delta, Phi Kappa Phi, Delta Theta Sigma, Gamma Alpha. Clubs: Cosmos (Washington); El Macero (Calif.). Home: 643 Miller Dr Davis CA 95616-3618 Office: U Calif Dept Nutrition Davis CA 95616

HILL, HENRY ALLEN, physicist, educator; b. Port Arthur, Tex., Nov. 25, 1933; s. Douglas and Florence (Kilgore) H.; m. Ethel Louise Eplin, Aug. 23, 1954; children—Henry Allen, Pamela Lynne, Kimberly Renee. B.S., U. Houston, 1953; M.S., U. Minn., 1956, Ph.D., 1957; M.A. (hon.), Wesleyan U., 1966. Research asst. U. Houston, 1952-53; teaching asst. U. Minn., 1953-54, research asst., 1954-57; research assoc. Princeton U., 1957-58, instr., then asst. prof., 1958-64; assoc. prof. Wesleyan U., Middletown, Conn., 1964-66; prof. physics Wesleyan U., 1966-74, chmn. dept., 1969-71; prof. physics U. Ariz., 1966—; chmn. bd. Zetetic Inst., 1992—; researcher on nuclear physics, relativity and astrophysics. Editorial advisor Internat. Jour. Pure and Applied Physics, 1993—; contbr. articles to profl. jours. Sloan fellow, 1966-68. Fellow Am. Phys. Soc.; mem. AAAS, Am. Astron. Soc., Royal Astron. Soc., Optical Soc. Am., Am. Geophys. Union. Office: U Ariz Dept Physics Bldg 81 Tucson AZ 85721

HILL, JACK DOUGLAS, electrical engineer; b. Fort Frances, Ont., Can., Nov. 28, 1937; came to U.S. 1961; s. John Lawson and Elizabeth (Stoddart) H.; m. Margaret Ann McCarthy, Sept. 19, 1959; children: Bruce Douglas, Patricia Ann. BSEE, U. Man., 1959, MSEE, 1960; PhD, Purdue U., 1965. Rsch. scientist Def. Rsch. Bd., Ottawa, Ont., 1960-61; instr. Purdue U., West Lafayette, Ind., 1961-64; mem. tech. staff Bellcom, Inc., Washington, 1964-65; sr. program mgr. Battelle-Columbus, Ohio, 1965—. Contbr. articles to profl. jours.; co-author (monographs): A Unified Systems Engineering Concept, 1972, An Assault on Complexity, 1973. U. Man. scholar, 1957, Nat. Rsch. Coun. of Can. grantee, 1959. Mem. IEEE (sr., v.p. systems 1975-77); mem. Nat. Mil. Intelligence Assn., Armed Forces Commn. Electronics Assn., Assn. Old Crows. Achievements include development of unified planning approach for renewal of USAF logistics management systems. Home: 712 Olde Settler Pl Columbus OH 43214 Office: Battelle 505 King Ave Columbus OH 43201

HILL, JOHN CHRISTIAN, physics educator; b. Blacksburg, Va., Apr. 13, 1936; s. Henry Harris and Olivia Tutwiler (Olivia) H.; m. Fay Gish, May 20, 1967; 1 child, Christina. BS, Davidson Coll., 1957; PhD, Prudue U., 1966. Asst. prof. Tex. A&M U., College Station, 1968-75; from asst. prof. to prof. Iowa State U., Ames, 1975—; program dir. Ames Lab. nuclear physics Iowa State U., 1977-91; reviewer proposals U.S. Dept. Energy, Washington, 1980—. Contbr. articles to profl. jours. Mem. AAAS, Am. Physical Soc., Am. Chem. Soc., Sigma Xi. Presbyterian. Achievements include discovery of 10 new short-lived redioavtive nuclei; experiments in relativistic heavy ions at bevalac, AGS, CERN-SPS & SIS accelerators. Home: RR 4 Ames IA 50014 Office: Iowa State U Dept Physics & Astronomy Ames IA 50011

HILL, JOSEPH CALDWELL, microwave engineer, consultant; b. Boston, July 19, 1944; s. Seth Mirch and Catherine Elizabeth (Flahive) H.; m. Alison Mason Chase, Apr. 15, 1983. BS, Boston Coll., Chestnut Hill, Mass., 1967; MSEE, U. So. Calif., 1969; PhD, Tufts U., 1986. Design engr. Varian Assocs., Beverly, Mass., 1969-71, 73-79; rsch. assoc. dept. psychology U. So. Calif., L.A., 1971-73; sr. design engr. SMDO div. Raytheon, Northboro, Mass., 1979-81; engring. mgr. Wincom Corp., Lawrence, Mass., 1981-83; co-founder, v.p. engring. Enon Microwave, Topsfield, Mass., 1983-91, bd. dirs., 1983—; cons. Hill Engring., Boxford, Mass., 1991—. Contbr. articles to profl. jours. Mem. IEEE, Sigma Xi. Achievements include patent on double balanced mixer; research on mmicrowave imaging using correlation techniques, high power solid state microwave switches. Home: 41 High Ridge Rd Boxford MA 01921

HILL, KELVIN ARTHUR WILLOUGHBY, biochemistry educator; b. Blaenau Ffestiniog, Wales, Nov. 14, 1957; came to U.S., 1979; s. Kenneth Charles W. and Hazel May (Sanders) H.; m. Marcia Claudette Bell, Aug. 15, 1982; 1 child, Roland Arthur W. BS, Andrews U., 1980; PhD, U. Notre Dame, 1986. Postdoctoral fellow MIT, Cambridge, Mass., 1986-89; asst. prof. Loma Linda (Calif.) U., 1989—; mem. adv. coun. U. Notre Dame (Ind.) Coll. of Sci., 1989-92. Contbr. articles to profl. jours. Tuition scholar Andrews U., 1980; Searle Rsch. fellow U. Notre Dame, 1982-85, Postdoctoral fellow Loma Linda U., 1986-89; rsch. grantee NSF, 1991-93, equipment grantee Loma Linda U., 1991. Mem. AAAS, Am. Chem. Soc., Am. Soc. Biochem. Molecular Biology, Protein Soc. Seventh-day Adventist. Office: Dept Biochemistry Loma Linda U Loma Linda CA 92350

HILL, MARTIN JUDE, engineer, consultant; b. Limerick, Ireland, Nov. 3, 1966; s. Joseph and Brenda (Sadlier) H.; m. Fiona Margaret Shine, Mar. 21, 1992; 1 child, Aoife Mary. BSEE, Univ. Coll. Cork, Ireland, 1987, M Engring. Sci., 1989. Rsch. engr. Nat. Microelectronics Rsch. Ctr., Cork, 1989-90; project leader Hyperion Energy Systems Ltd., Cork, 1989-91; mng. dir. Environ. Measuring Solutions Ltd., Cork, 1991—. Author: PV Battery Handbook, 1991; contbr. articles to profl. publs. Mem. Instrument Soc. Am., Internat. Solar Energy Soc. Achievements include contributions to WHO, UNESCO and European Community programs for Photovaltaic system application and development; research in implementation of control systems using MMS (MAP3.0) protocol for European Community Computer Networks for Manufacturing Applications project. Office: Environ Measuring Solutions, Unit 18B Enterprise Ctr N Mall, Cork Ireland

HILL, MARY ANN, metallurgical engineer; b. Fresno, Calif., Oct. 25, 1961; d. William Hoyt and Mary Louise (Waters) Kincade; m. Jeffrey Owen Hill, Aug. 20, 1983; 1 child, Zachary Benjamin. BSMetE, N.Mex. Tech. Coll., 1982; MSMetE, U. Wash., 1985. Staff mem. Los Alamos (N.Mex.) Nat. Lab., 1987—, materials liaison Nat. High Magnetic Field Lab., 1991—, supr. Metallography Lab., 1990—. Contbr. articles to profl. jours. and conf. procs. Mem. ASTM (com. mem.), Am. Soc. Metals (sec. local chpt. 1991-92). Office: Los Alamos Nat Lab Mail Stop G770 Los Alamos NM 87545

HILL, PATRICK RAY, power quality consultant; b. Bad Constatt, Germany, Sept. 5, 1950; s. Basil Clayton and Gladys Marie (Burdin) H.; m. Patricia Ann Rath, Dec. 18, 1981; children: Danielle K. Irvin, Michael E. Irvin, Chandra K., John T.C. AAS, North Idaho Coll., 1981, AS in Liberal Arts, 1986. Applications engr. Transtector Systems, Inc., Hayden Lake, Idaho, 1981-84; dir. engring. No. Techs., Inc., Spokane, Wash., 1985-89; co-founder RayAnn Corp., Bothell, Wash., 1989—. Sgt. U.S. Army, 1969-79, Vietnam. Mem. IEEE (assoc.), Nat. Fire Protection Assn. Republican. Baptist. Avocations: PC computers, sailing, travel, reading.

HILL, PAUL W., biochemical engineer; b. Jacksonville, Fla., Oct. 23, 1967; s. Robert Paul and Mary Alice (Mullane) H. AB, Dartmouth Coll., 1990; BE, ME, Thayer Sch. Engring., 1992. Rsch. assoc. R. W. Johnson Pharm. Rsch. Inst., Raritan, N.J., 1992—. Mem. AICE. Democrat. Office: R W Johnson Pharm Rsch Inst 1000 Rte 202 S Raritan NJ 08869

HILL, RICHARD CONRAD, engineering educator, energy consultant; b. Schenectady, N.Y., Oct. 15, 1918; s. Walter Bradford and Elizabeth (Conrad) H.; m. Elizabeth Crowley, Oct. 25, 1944; children: Judith, Martha, Carol, David, Jonathan. BSME, Syracuse U., 1941. Engr. GE, Lynn, Mass., 1941-46; prof., dir. dept. indsl. cooperation, dean of engring. U. Maine, Oorono, 1946-92. Contbr. articles on energy saving to various pubs. Pres. Trustees Water Co., Old Town, Me., 1972-92. Recipient Pub. Svc. award U. Maine, Orono, 1972; named Energy Innovator U.S. Dept. Energy, 1985. Mem. ASME. Achievements include several U.S. patents on wood combustion systems. Home: 501 College Ave Orono ME 04473-1211

HILL, ROBERT JAMES, mechanical engineer; b. San Mateo, Calif., Jan. 14, 1951; s. Jettie Brushwood and Lois Marie (Steiner) H.; m. Judith Ann Meyers, Sept. 5, 1981; children: Ryan, Brittany, Tyler. BSME, San Jose State U., 1975; MSME, Santa Clara U., 1979. Engr. GE, Sunnyvale, Calif., 1979-88; sr. engr. GE, San Jose, Calif., 1988-93, Martin Maretta, San Jose, Calif., 1993—. Mem. ASME. Achievements include research in non-linear evaluation of a reactor spent fuel pool for ACI code compliance with consolidated fuel loadings, Spacepin computer program devel. Office: Martin Marietta 6835 Via Del Oro M/C S-15 San Jose CA 95119

HILL, ROBERT LEE, biochemistry educator, administrator; b. Kansas City, Mo., June 8, 1928; s. William Alfred and Geneva Eunice (Scurlock) H.; m. Helen Amarette Root, Oct. 24, 1948 (div.); children—Sterrette L., Amarette L., Geneva L., Rebecca M.; m. Deborah Anderson, Apr. 10, 1982. A.B., U. Kans., 1949, M.A., 1951, Ph.D., 1954. Research instr. U. Utah, Salt Lake City, 1956-57, asst. research prof., 1957-60, assoc. research prof., 1960-61; assoc. prof. biochemistry Duke U., Durham, N.C., 1961-65, prof., 1965-74, James B. Duke prof., 1974—, chmn. dept. biochemistry, 1969-93. Co-author: Principles of Biochemistry, 6th edit., 1978, 7th edit., 1983; co-editor: The Proteins, 3d edit., Vol. I, 1974, Vol. II, 1976, Vol. III, 1977, Vol. IV, 1979, Vol. V, 1982. NIH fellow, 1953-54, 54-56; recipient William C. Rose Award in Biochemistry Am. Soc. for Biochemistry and Molecular Biology, 1991. Fellow Am. Acad. Arts and Scis.; mem. NAS, Am. Soc. Biol. Chemists (pres. 1976-77), Inst. Medicine, Internat. Union Biochemistry (U.S. Nat. Com. 1982-91, gen. sec. 1985-91), Assn. Med. Depts. Biochemistry (pres. 1982-83), Am. Chem. Soc., Assn. Am. Med. Colls. (adminstrv. bd. council acad. socs. 1979-85, chmn. 1983-84), AAAS, Sigma Xi, Alpha Omega Alpha. Office: Duke U Med Ctr Dept Biochemistry Durham NC 27710

HILL, ROGER EUGENE, physicist; b. San Bernardino, Calif., Feb. 12, 1936; s. George Eugene and Alice Marie (Greek) H.; m. Bette Cerf Ross, Aug. 14, 1955 (div. Dec. 1974); children: Catherine Marie, Teresa Jean, Diana Louise; m. Louise Mary Jackson, May 8, 1993. BS, St. Mary's Coll., Moraga, Calif., 1957; PhD, U. Calif., Berkeley, 1964. Rsch. assoc. U. Chgo., 1963-66; prin. sci. officer Rutherford High Energy Lab., Chilton, Didcot, Berks, U.K., 1966-68; vis. scientist CERN, Geneva, 1968; tech. mgr. Geonuclear Nobel Paso, S.A., Paris, 1969-78; sr. rsch. scientist U. N.Mex., Albuquerque, 1982-86; sect. leader Los Alamos (N.Mex.) Nat. Lab., 1986—; sci. liaison officer Joint Verification Experiment, USSR, 1988; tech. advisor U.S. Delegation Nuclear Testing Talks, Geneva, 1988-89. Contbr. articles and papers to profl. jours. Recipient Excellence award Dept. Energy, 1990. Mem. Am. Phys. Soc., Am. Assn. Physics Tchrs. Office: Los Alamos Nat Lab Ms D406 Los Alamos NM 87545

HILL, RONALD CHARLES, surgeon, educator; b. Parkersburg, W.Va., Sept. 4, 1948; s. Lloyd E. and Margaret (Pepper) H.; m. Lenora Jane Rexrode, June 12, 1971; children: Jeffrey, Mandy. BA, W.Va. U., 1970, D of Medicine, 1974. Lic. physician W.Va., N.C.; diplomate Am. Bd. Surgery, Am. Bd. Thoracic Surgery, Nat. Bd. Med. Examiners. Intern dept. of surgery Duke U. Med. Ctr., Durham, N.C., 1974-75; resident in surgery Duke U., Durham, N.C., 1974-85, rsch. assoc., 1976-79, teaching scholar, 1984-85; asst. prof. surgery W.Va. U., Morgantown, 1985-90, assoc. prof., 1990—; cons. VA Med. Ctr., Clarksburg, W.Va., 1985—; dir. surg. rsch. dept. surgery W.Va. U., 1986-88, student coord. dept. surgery, 1986—; mem. ad hoc com. Merit Rev. Bd. for Cardiovascular Studies, VA, Washington, 1988-90. Contbr., co-contbr. numerous book chpts. and articles to profl. publs. (Lange Med. Book award 1971, 73, 74, Merck Med. Book award 1974). Mem.-at-large adminstrv. bd. Drummond Chapel United Meth. Ch., Morgantown, 1987-89, 92—; mem. Morgantown C. of C., 1991-92. Recipient Roche Award Scholastic Med. award, 1972, Lange Med. Book award, 1971, 73, 74, Merck Med. Book award, 1974; named Outstanding Young Man of Am., 1976. Fellow ACS, Southeastern Surg. Congress, Assn. Acad. Surgery, Sabiston Soc., Am. Coll. Cardiology, Am. Coll. Chest Physicians, Am. Coll. Angiology, So. Thoracic Surg. Assn., Soc. Thoracic Surgeons; mem. Am. Heart Assn. (v.p. W.Va. affiliate), Soc. Univ. Surgeons, Am. Assn. Thoracic Surgery, W.Va. Med. Assn., Mended Hearts, Lakeview Country Club, Pines Country Club, Phi Beta Kappa, Alpha Omega Alpha, Alpha Epsilon Delta. Republican. Avocations: fishing, photography. Home: 10 Flagel St Morgantown WV 26505-2240 Office: WVa U Med Ctr Dept of Surgery Medical Center Dr Morgantown WV 26506

HILL, RUSSELL GIBSON, chemical engineer, consultant; b. N.Y.C., June 25, 1921; s. Frank James and Maybelle Bertha (Wood) H.; m. Elizabeth Starr Cummings, May 28, 1949; children: Andrea van Waldron, Bradford Wray. BSChemE, RensseLaer Polytechnic Inst., 1943. Lic. profl. engr., N.Y. Rsch. engr. Heyden Chem. Corp., Fords, N.J., 1946-47, Stauffer Chem. Corp., Chauncey, N.Y., 1947-51; engr. Arabian Am. Oil Co., N.Y.C., 1951; dir. project engring. Scientific Design Co., N.Y.C., 1951-74; mgr. process safety Oxirane Internat., Princeton, N.J., 1974-80, ARCO Chem. Co., Newtown Square, Pa., 1980-85; cons. Princeton, 1985—. Editor (newsletter) Safety and Health News, 1993; editorial bd. Process Safety Progress, 1993; Contbr. articles to profl. jours., chpts. to books. Vestryman Christ Ch., Bronxville, N.Y. 1st lt. U.S. Army, 1943-46. Mem. AICHE (directory editor 1992, program com. 1983—), Old Guard of Princeton, Nassau Club, U. Club of Winter Park. Office: Russell G Hill Assocs 5 Brook Dr W Princeton NJ 08540

HILL, WALTER EDWARD, JR., geochemist, extractive metallurgist; b. Moberly, Mo., June 4, 1931; s. Walter Edward and Louise Katherine (Sours) H.; m. Beverly Gwendolyn Kinkade, Sept. 8, 1951; children: Walter III, Michele, Janet, Sean, Christopher. BA in Chemistry, U. Kans., 1955, MA in Geology, 1964. Cert. safety instr., Mine Safety and Health Adminstrn. Mgr. standards div. Hazen Rsch. Inc., Golden, Co., 1974-79; lab dir. Earth Scis. Inc., Golden, Colo., 1979-80; tech. svcs. supr. Texasgulf Inc., Cripple Creek, Colo., 1980-82; gen. mgr. Calmet, Fountain, Colo., 1983; tech. svs. mgr. Marathon Gold, Craig, Colo., 1984-85; cons., 1985-86; chief chemist Nev. Gold Mining, Winnemucca, 1987-88; ops. mgr. Apache Energy & Minerals, Golden, Colo., 1989-91; cons., 1991—; speaker in field. Contbr. 28 articles to profl. jours. Sgt. U.S. Army, 1950-52. Fellow Am. Inst. Chemists (life); mem. Assn. Exploration Geochemists, Kans. Geol. Soc., Denver Mining Club. Republican. Roman Catholic. Achievements include analytical quality control of minerals exploration and processing; development and manufacture of rock and mineral analytical standards. Home and Office: 1486 S Wright St Lakewood CO 80228-3857

HILLARY, EDMUND PERCIVAL, diplomat, explorer, bee farmer; b. Auckland, N.Z., July 20, 1919; s. Percy and Gertrude H.; ed. U. Auckland; LL.D. (hon.), Victoria U., B.C., Can., U. Victoria, (New Zealand); m. Louise Mary Rose, 1953 (dec. 1975); 1 son, 2 daus. (1 dec.). Bee farmer; dir. Field Edni. Enterprises of Australasia Pty. Ltd.; went to Himalayas on N.Z. Garwhal expdn., 1951, with Brit. expdn. to Cho Oyu, 1952, Brit. Mt. Everest Expdn. under Sir John Hunst, 1953; leader N.S. Alpine Club Expdn. to Barun Valley, 1954, N.Z. Antarctic Expdn., 1956-58; reached South Pole, 1957; leader Himalayan Expdns., 1961, 63, 64; leader climbing expdn. on Mt. Herschel, Antarctica, 1967, River Ganges Expdn., 1977; built hosp. for

Sherpa tribesmen, Nepal, 1966; high commr. to India, also accredited to Bangladesh, Bhutan and Nepal, 1985—. Served with Royal N.Z. Air Force, 1944-45; PTO. Decorated Gurkha Right Hand 1st Class; Star of Nepal 1st Class; recipient Cullum Geog. medal, 1954; Hubbard medal, 1954; Polar medal, 1958; Founders Gold medal Royal Geog. Soc., 1958; James Wattle Book of Yr. award N.Z., 1975, Centennial award Nat. Geographic Soc., 1988. Author: High Adventure, 1955; (with Sir Vivian Fuchs) The Crossing of Antarctica, 1958, No Latitude for Error, 1961; (with Desmond Doig) High in the Thin Cold Air, 1963, Schoolhouse in the Clouds, 1965, Nothing Venture, Nothing Win, 1975, From the Ocean to the Sky, 1978, (with Peter Hillary) Two Generations, 1983. Office: High Commn New Zealand, 25 Golf Links, New Delhi 110003, India

HILLEBRAND, JULIE ANN, biotechnology executive; b. N.Y., Jan. 27, 1962; d. Raymond and Caroline Lucy (Mulcahy) H. BS, Chatham Coll., 1984; M in Hyg., U. Pitts., 1987. Technician Grad. Sch. Pub. Health U. Pitts., 1984-89; mem. tech. svc. staff Fisher Sci., Pitts., 1989-91, asst. product mgr., 1991—. Contbr. articles to profl. jours. Mem. Chatham Coll. Alumnae Orgn. Office: Fisher Sci 711 Forbes Ave Pittsburgh PA 15219

HILLEL, DANIEL, soil physics and hydrology educator, researcher, consultant; b. L.A., Sept. 13, 1930; s. Morris Jacob and Sarah Frances (Fromberg) Bugeslov; m. Michal Artzy; children: Adi, Ron, Sari, Ori, Shira. BS, U. Ga., 1950; MS, Rutgers U., 1951; PhD, Hebrew U., Jerusalem, 1958; postgrad., U. Calif., Berkeley, 1960-61; DSc honoris causa, U. Guelph, 1992. Founding mem. Kibbutz Sdeh-Boker, Negev Region, Israel, 1952-56; advisor land devel. Govt. of Burma, Rangoon, 1957-58; head soil tech. Agrl. Rsch. Orgn., Bet Dagan, Israel, 1959-65; head soil and water scis. Hebrew U., Rehovot, Israel, 1966-76; prof. soil physics and hydrology U. Mass., Amherst, 1977—; vis. scientist Japan Soc. for Promotion Sci., Tottori and Fukuoka, 1972, 82, 90; nat. lectr. Sigma Xi, 1987-89; cons. agrl. applications Internat. Atomic Energy Agy., Vienna, Austria, 1971-72, environ. dept. The World Bank, Washington, 1987-90; v.p. soil physics com. Internat. Soil Sci. Soc., Hague, The Netherlands, 1964-68, 82-86. Author: Computer Simulation of Soil-Water Dynamics, 1977, Fundamentals and Applications of Soil Physics, 1980, Negev: Land, Water and Life in a Desert Environment, 1982, Out of the Earth: Civilization and the Life of the Soil, 1991; contbr. over 200 articles to profl. jours.; patentee in field. Recipient Chancellor's medal U. Mass., 1982; John Simon Guggenheim Meml. Found. fellow, 1993. Fellow AAAS, Am. Soc. Agronomy, Soil Sci. Soc. Am. (chmn. soil physics div. 1988-89); mem. Am. Geophys. Union. Office: U Mass 11 Stockbridge Hall Amherst MA 01003

HILLEMAN, MAURICE RALPH, virus research scientist; b. Miles City, Mont., Aug. 30, 1919; s. Robert A. and Edith (Matson) H.; m. Lorraine Witmer, Aug. 3, 1963; children—Jeryl Lynn, Kirsten Jeanne. B.S., Mont. State U., 1941, D.Sc. (hon.), 1966; D.Sc. (hon.), U. Md., 1968, Washington and Jefferson Coll., 1992, U. Leuven, 1984; Ph.D., U. Chgo., 1944. Asst. bacteriologist U. Chgo., 1942-44; research asso. virus labs. E.R. Squibb & Sons, 1944-47, chief virus dept., 1947-48; chief research and diagnostic sects. virus and rickettsial diseases Army Med. Service Grad. Sch., Walter Reed Army Med. Center, 1948-56, asst. chief lab. affairs, 1953-56; chief respiratory diseases Walter Reed Army Inst. Research, Washington, 1956-57; dir. virus and cell biol. research Merck Inst. Therapeutic Research, Merck & Co. Inc., 1957-66, exec. dir., 1966-71, v.p., 1971-78, sr. v.p., 1978-84; dir. Merck Inst., 1984—; dir. virus and cell biology research, v.p. Merck, Sharp & Dohme Research Labs., 1970-78, sr. v.p., 1978—; vis. investigator Hosp. of Rockefeller Inst. for Med. Research, 1951; vis. prof. bacteriology U. Md., 1953-57; adj. prof. virology pediatrics Sch. Medicine U. Pa., 1968—; cons. Children's Hosp. of Phila., 1968—; mem. council div. biol. scis. Pritzker Sch. Medicine, 1977—; John Herr Musser lectr. Musser-Burch Soc., Tulane U. Sch. Medicine, 1969, 19th Graugnard lectr., 1978; Mem., spl. cons. panel respiratory and related viruses USPHS, 1960-64; mem. Nat. Cancer Inst. primate study group, 1964-70; mem. council analysis and projection Am. Cancer Soc., 1971-76; mem. expert adv. panel on virus diseases WHO, 1952—; bd. dirs. W. Alton Jones Cell Sci. Center, Lake Placid, N.Y., 1980-82, Am. Liver Found. (hon.), 1986—, Am. Type Culture Collect, 1992, Nat. Fedn. Infectious Diseases, 1987—, Nat. Cancer Inst., 1991—, Sci. Counselors Paul Erlich Found. (Frankfurt, Germany), 1993—, Jos. J. Stokes Rsch. Inst. U. Pa., 1986—; mem. overseas med. research labs. com. Dept. Def., 1980; mem. virology dept. rev. com. Am. Type Culture Collection, 1980; mem. Ad Hoc Vaccine Subcom. AIDS Program NIH, 1991, AIDS R and D Vaccine Working Group, 1992—, Panel Internat. Task Force NIH Strategic Plan, 1992. Editorial bd.: Internat. Jour. Cancer, 1964-71, Inst. Sci. Information, 1968-70, Am. Jour. Epidemiology, 1969-75, Infection and Immunity, 1970-76, Excerpta Medica, 1971—, Proc. Soc. Exptl. Biology and Medicine, 1976—, Jour. Antiviral Research, 1980—, Vaccine, 1983, Virus Genes, 1986, Vaccine Research, 1990; contbr. 460 articles to sci., profl., med. jours. Phi Kappa Phi fellow, 1941-42; Koessler fellow, 1943-44; Recipient Howard Taylor Ricketts prize, 1945, 83, Distinguished Civilian Service award sec. def., 1957; Walter Reed Army Med. Incentive award, 1960; Dean M. McCann award, 1970; Procter award, 1971; Achievement award Indsl. Research Inst., 1975, Joseph E. Smadel award, 1984, Alumni medal, U. Chgo., 1987, Albert B. Sabin medal, 1988, Nat. Medal Sci., U.S., 1988, San Marino award, 1989, Robert Koch Gold medal, 1989. Fellow Am. Acad. Microbiology, Am. Acad. Arts and Scis.; mem. Nat. Acad. Sci., Am. Soc. Microbiology, Soc. Exptl. Biology and Medicine (mem. editorial and publs. com. 1977—), Tissue Culture Assn. (mem. council 1977—), Am. Assn. Immunologists, Am. Assn. Cancer Research, Infectious Diseases Soc., Permanent sect. Microbiol. Standardization Internat. Assn. Microbiol. Socs., Russian Acad. Biotechnology (hon. fgn. mem.), L'Académie Nationale de Pharmacie (fgn. corr.). Office: Merck Rsch Labs UM4-1 West Point PA 19486

HILLER, WILLIAM CLARK, physics educator, engineering educator, consultant; b. San Antonio, Aug. 17, 1933; s. William John and Marjorie Ruth (Clark) H.; m. Pamela Humphrey, June 4, 1960 (div. July 1970), 1 child, Katherine; m. Elizabeth Anne DuChateau, July 27, 1980; 1 child, Patricia Cathleen. BSME, U. Ariz., 1960, MS, 1976. Registerd profl. engr., Ariz. Reactor engr. Nat. Reactor Testing Lab., Idaho Falls, Idaho, 1960-61; assoc. design engr. Boeing Co., New Orleans, 1962-65; sr. engr. Kennecott Copper Corp., Kearny, Ariz., 1966-67; rsch. engr. Boeing Co., Kennedy Space Ctr., Fla., 1967-69; project test engr. Hughes Aircraft Co., Tucson, 1969-71; prof. Cen. Ariz. Coll., Coolidge, 1976—; Mem. adv. bd. Cen. Ariz. Coll., Coolidge, 1976—. With USN, 1955-57. Recipient Recognition award Langley Rsch. Ctr., 1966. Mem. Tribe of Illini, Barth Rangers, Zeta Psi. Republican. Presbyterian. Achievements include development of a method to expedite the tolerence study of the S-1C propellant delivery system component integration. Home: 11102 W Ironwood Hills Dr Casa Grande AZ 85222 Office: Cen Ariz Coll 8470 N Overfield Rd Coolidge AZ 85228

HILLERY, MARK STEPHEN, physicist; b. Sydney, Australia, Apr. 19, 1951; s. Frances Vincent and Jane Stephanie (Lun) H. BS, MIT, 1973; PhD, U. Calif., Berkeley, 1980. Rsch. asst. U. N.Mex., Albuquerque, 1980-84; asst. prof. Hunter Coll., N.Y.C., 1985-88, assoc. prof., 1989-92, prof., 1993—. Contbr. articles to profl.jours. Capt. USAF, 1977. Mem. Am. Phys. Soc., Optical Soc. Am. Achievements include application of path integral techniques to quantum optics; clarification of the foundations of the quantum theory of nonlinear optics, of properties on nonclassical field states and definition of nonclassical distance; definition and exploration of new forms of field squeezing. Office: Hunter Coll Physics Dept 695 Park Ave New York NY 10021

HILLERY, ROBERT CHARLES, naval engineer, management consultant; b. Waltham, Mass., May 3, 1953; s. Robert Parker Hillery; m. Diane Christine Kelly, Aug. 5, 1989; 1 child, Kathleen. BA in Marine Transp., Mass. Maritime Acad., Buzzards Bay, 1975; MA in Internat. Rels., Salve Regina Coll., Newport, R.I., 1988; MA in Strategic Studies, Naval War Coll., Newport, 1991. Commd. ensign U.S. Navy, 1977, advanced through grades to comdr., 1990; exec. officer USS Charles F. Adams, Mayport, Fla., 1988-90; head enlisted engring. assignments U.S. Navy Bur. of Pers., Washington, 1990-92, head sea spl. programs, 1992—; tech. advisor Navy Pers. R&D Ctr., San Diego, 1990—; mem. membership bd. Surface Navy Assn., Arlington, Va., 1990—; Mgmt. Info. Systems cons. Pers. Systems Ctr. Naval Analysis, 1991; prin. rschr. Navy Enlisted Pers. Assignment model, 1990—. Contbr. articles to profl. jours. Recipient George Washington Honor medal Freedoms Found. at Valley Forge, 1979; MIT fellow, 1991. Mem. U.S. Naval Inst. Office: Bur Naval Personnel PERS 409 Washington DC 20370-5409

HILLIARD, KIRK LOVELAND, JR., osteopathic physician, educator; b. Phila., Mar. 9, 1941; s. Kirk Loveland and Lillian Adele (Hinkle) H.; m. Janet Louise Moyer, Sept. 29, 1970; children: Michael Spence, Stephen Matthew, Allison Day. AB, Haverford Coll., 1963; DO, Phila. Coll. Osteo. Medicine, 1967. Diplomate Am. Coll. Osteo. Internists, Internal Medicine and Med. Diseases of Chest. Intern Doctors' Hosp., Columbus, Ohio, 1967-68, resident, 1970-72, sr. attending, 1977—, dir. respiratory svcs., 1978—; fellow Hahnemann Hosp., Phila., 1972-74; pvt. practice Columbus, 1974—; asst. prof. Ohio U., Athens, 1979-88, assoc. prof., 1988—; med. dir. CP Home Care, Columbus, 1985—; acting dir. Med. Edn. Doctors Hosp., Columbus, Ohio, 1991-92. Capt. M.C., U.S. Army, 1968-70, Vietnam. Named resident trainer of yr. Doctors Hosp., 1977, 79, 81. Fellow Am. Coll. Osteo. Internists; mem. Am. Osteo. Assn., Am. Lung Assn., Am. Legion, Masons, Shriners. Avocations: water and snow skiing, hunting, fishing, scuba diving, horseracing. Office: Setnar Nagy & Assocs 94 W 3d Ave Columbus OH 43207-1034

HILLIER, JAMES, communications executive, researcher; b. Brantford, Ont., Can., Aug. 22, 1915; came to U.S., 1940; s. James Sr. and Ethel Anne (Cooke) H.; m. Florence Marjory Bell, Oct. 24, 1936; children: James Robert, William Wynship. BA, U. Toronto, Ont., Can., 1937, MA, 1938; PhD, U. Toronto, 1941, DSc (hon.), 1978; DSc (hon.), N.J. Inst. Tech., 1981. Rsch. asst. Banting Inst. U. Toronto Med. Sch., 1938-40; head electron microscope rsch. RCA Labs., Camden and Princeton, N.J., 1940-53; adminstrv. engr. corp. rsch. and engring. RCA Corp., Princeton, 1954-55; chief engr. comml. electronic products RCA Corp., Camden, 1955-57; gen. mgr. labs. RCA Corp., Princeton, 1957-58, v.p. labs., 1958-68; v.p. corp. rsch. and engring. RCA Corp., N.Y.C., 1968-69, exec. v.p. rsch. and engring., 1969-76, exec. v.p., sr. scientist, 1976-77, ret., 1977; dir. corp. rsch. Westinghouse Air Brake Co., Pitts. and Alexandria, Va., 1953-54; pres. Indsl. Reactor Labs., Princeton, 1964-65; mem. higher edn. study com. Gov.'s Office, State of N.J., 1963-64; mem. commerce tech. adv. bd. U.S. Dept. Commerce, Washington, 1964-70; chmn. adv. coun. dept. elect. engring. Princeton U., 1963-65; mem. adv. coun. Coll. Engring., Cornell U., Ithaca, N.Y., 1966—; mem. joint consultative com. U.S. AID/Egyptian Acad. Sci. Rsch. & Tech., Cairo, 1978-84. Co-author: Electron Optics and the Electron Microscope, 1945; co-contbr.: Medical Physics, 1944, vol. II, 1950, Colloidal Chemistry, vol. VI, 1946; contbr. Ency. Britannica, 1948. Inducted into Nat. Inventors Hall of Fame, 1980, N.J. Inventors Hall of Fame, 1992; recipient James Loudon Gold medal U. Toronto, 1937, Albert Lasker award APHA, 1960, Commonwealth award, 1980, Presdl. award Microbeam Analysis Soc., 1989. Fellow AAAS (chmn. nomination com. sect. M 1965), IEEE (David Sarnoff award 1967, Founders medal 1981), Am. Phys. Soc. (mem. at large, governing bd. 1964-65); mem. Microscope Soc. Am. (pres. 1944, Disting. Scientist award 1977), Indsl. Rsch. Inst. (bd. dirs. 1960-65, pres. 1964, Inst. medal 1975), Nat. Inventors Hall of Fame Found., Inc. (bd. dirs. 1992—), Rotary (bd. dirs. 1988-91), Nassau Club, Sigma Xi. Achievements include 41 patents in field; co-design of first successful electron microscope in North America, of first commercially available electron microscope in North America; discovery of principle of Stigmator for correcting astigmatism of electron microscope objective lenses; invention of electron microprobe microanalyser; first to picture tobacco mosaic virus, bacterial viruses and ultra-thin section of a single bacterium. Home: 22 Arreton Rd Princeton NJ 08540

HILLION, PIERRE THÉODORE MARIE, mathematical physicist; b. Saint-Brieuc, France, Jan. 31, 1926; s. Pierre Auguste Alexandre and Olive Jane (Marion) H.; Licencié es Sciences, Engineer Ecole Supérieure d'Electricité, 1950, Docteur es Sciences, 1957; m. Jane Garde, July 9, 1955; children: Catherine, Pierre, Joëlle, Hervé . Engr., Le Matériel Electrique Schneider-Westinghouse, 1950-55; math. physicist Section Technique de L'Armée, 1955-64; head math. phys. dept. Laboratoire Central de L'Armement, 1964-83; sci. cons. Centre D'Analyse de Défense, 1983-91; maitre de conferences Ecole Nationale Supérieure des Techniques Avancées, 1976-88; mem. Electromagnetic Acad. MIT. Mem. du bureau Assocation de Parents d'Élèves, 1965-76. With French Army, 1950. Recipient Mérite pour la Recherche et l'Invention, 1965; Palmes Académiques, 1970; Ordre National pour le Mérite, 1978, Legion d'Honneur, 1988. Mem. Societé Mathématique de France, Societé Française de Radioprotection, Syndicat de la Presse Scientifique, Internat. Assn. Math. Physics. Roman Catholic. Contbr. articles on high energy physics, math. physics and numerical analysis to profl. jours. Home: 86 bis Rt de Croissy, 78110 Le Vesinet Yvelines France

HILLIS, WILLIAM DANIEL, scientist, engineer; b. Balt., Sept. 25, 1956; s. William and Argye (Briggs) H.; m. Patricia Hillis, Mar. 14, 1987; children: Asa, Noah. MS, MIT, 1980, PhD, 1988. Founder Thinking Machines Corp., Cambridge, Mass., 1983—. Author: The Connection Machine, 1986; editor: Complex Systems, Future Generation Computer Systems, Machine Vision Applications, Advanced in Applied Math., Ops. Rsch. Soc. Am. Jour. on Computing. Recipient Ramanujan award, 1988, Spirit of Am. Creativity award Found. for a Creative Am., 1991. Fellow AAAS; mem. IEEE, Am. Assn. Artificial Intelligence, Assn. Computing Machinery (Thesis award 1985, Grace Murray Hopper award 1989), Santa Fe Inst. Achievements include inventor of the Connection Machine and holder of 28 patents. Office: Thinking Machines Corp 245 First St Cambridge MA 02142

HILLMAN, LEON, electrical engineer; b. N.Y.C., July 31, 1921; s. Harry and Jennie (Gartenberg) H.; m. Rita Katchen, July 18, 1948; children: David, Deborah. BEE, NYU, 1950. Registered profl. engr., N.J. Radio engr. Communication Devel. Co., Newark, 1940-42; head elec. sect. U.S. Army Engring. Lab., Ft. Monmouth, n.J., 1942-45; rsch. assoc. Elec. Engring. Dept., NYU, N.Y.C., 1946-51; v.p., chief engr. Prodn. Rsch. Corp., Thornwood, N.Y., 1951-56; pres. Automation Dynamics Corp., Northvale, N.J., 1957-71, ADCO Aerospace Inc., Closter, N.J., 1971—; electronics cons. Johnson Controls, Milw., 1949-69; lectr. in field. Contbr. articles to profl. jours. Chmn. United Jewish Appeal, Englewood, N.J., 1960, Demarest, N.J., 1978. Sgt. USAF, 1945-46. Named Hon. Citizen, State of Md., 1957. Mem. IEEE, Am. Phys. Soc., Sigma Xi, Eta Kappa Nu. Achievements include patents for meteorological instruments, industrial controls and water sterilization; design of instruments used in space flight and lunar landing; invention of electronic controlled water treatment system. Office: ADCO Aerospace Inc PO Box 748 Closter NJ 07624-0748

HILLMAN, ROBERT EDWARD, biologist; b. Bklyn., N.Y., Nov. 5, 1933; s. Joseph and Louise Marie (Franz) H.; m. Dale Jean Ryan, June 23, 1962 (div. 1972); children: Jennifer Ann, Stephanie Joyce. BA, Hofstra U., 1955, MA, 1958; PhD, U. Del., 1962. Cert. fisheries scientist Am. Fishery Soc. Bd. Rsch. assoc. U. Del., Newark, 1961-62; rsch. asst. prof. Chesapeake Biology Lab. U. Md., Solomons, Md., 1962-65; rsch. scientist Battelle Meml. Inst., Duxbury, Mass., 1962-72, sr. rsch. scientist, 1972-82, rsch. leader, 1982—. Contbr. articles to profl. jours. Pres. Plymouth (Mass.) Philharmonic Orch., 1976. With U.S. Army, 1955-57. Mem. N.Y. Acad. Scis., Nat. Shellfisheriews Assn. (v.p. 1984, pres. 1986), Sigma Chi. Home: 63 Cooke Rd Plymouth MA 02360 Office: Battelle Meml Inst 397 Washington St Duxbury MA 02332

HILLSMAN, REGINA ONIE, orthopedic surgeon; b. N.Y.C., June 2, 1955; d. David Oka and Laettia Louise (Miller) H.; m. Peter Michael Schmitz, Mar. 18, 1982; children: Michelle, Julien Pierre, Christophe-Gerard Philippe. BA cum laude, Bryn Mawr Coll., 1972; MD cum laude, George Washington U., 1977. Diplomate Am. Bd. Orthopaedic Surgery, 1990. Intern Beth Israel Hosp., Boston, 1977-78; resident in surgery Montefiore Hosp., Bronx, N.Y., 1978-79; resident in orthopedics U. Pa., Phila., 1979-81; chief resident Howard U., Washington, 1981-82; practice med. specializing in orthopedics Los Angeles, 1983-87, N.Y.C., 1987—; clin. liaison Martin Luther King, Los Angeles, Calif. Fellow Am. Acad. Orthopaedic Surgeons; mem. ACS, AMA, Am. Med. Women's Assn., Am. Orthopaedics, N.Y. Acad. Scis., N.Y. State Med. Soc., Westchester Polit. Women's Caucus, Westchester Conservatory Music, Phi Delta Epsilon. Roman Catholic. Avocations: violin, piano. Home: 3614 Post Rd W Ste 238 Westport CT 06880-4754 Office: Community Med Specialists 469 Wolcott Rd Wolcott CT 06716-2613

HILLYARD, RICHIE DOAK, wastewater treatment plant operator; b. Williamsport, Pa., Sept. 25, 1961; s. Carl Matthew and Marlene Roseanna (Hill) H.; m. Lori Ann Wagner, Nov. 5, 1988; children: Jeremy Cousteau, Chad Michael. BA in Biology, Lycoming Coll., 1983. Cert. sewage treatment plant and waterworks operator, Pa. Office mgr. Grand Ctrl. Beer Dist., Avis, Pa., 1983-88; operator Pine Creek (Pa.) Mcpl. Authority, 1988—. Recording sec. Pine Creek Mcpl. Authority Bd., 1990—; dir. race course officials Benefits-People That Love Ctr. and Avis Food Pantry, 1982—. Named Outstanding Young Man of Am., 1989. Republican. Home: 21 E Park St PO Box 549 Avis PA 17721 Office: Pine Creek Mcpl Authority PO Box 608 Avis PA 17721

HILMY, SAID IBRAHIM, structural engineer, consultant; b. Sharkia, Egypt, Nov. 8, 1953; came to U.S., 1979; s. Ibrahim Hilmy and Dawlat Youssef; m. Sahar Omar El Ramly, Aug. 16, 1989; children: Omar, Sharif. BS with honor, Cairo U., Egypt, 1976; MS, Cornell U., 1981, PhD, 1984. Licensed profl. and structural engr., Calif. Rsch. engr. Stanford Rsch. Inst., Menlo Park, Calif., 1984-85; sr. engr. ACMA, L.A., 1985-90, Dames & Moore, L.A., 1990—; v.p. Group 2000 Incorporation, L.A., 1987-91; NSF Seismic Retrofit Rsch. Group, Washington, 1991-93. Contbr. more than 20 articles to profl. jours. Mem. ASCE, Am. Concrete Inst., Earthquake Engring. Rsch. Inst., Nat. Geographic Soc., Egyptian Am. Orgn., Cornell Alumni Club of So. Calif., Sigma Xi. Republican. Muslim. Achievements include developing popular computer programs including STRAND, 1987. Home: 12 Lancewood Irvine CA 92715 Office: Dames & Moore 6 Hutton Centre Dr Santa Ana CA 92707

HILPMAN, PAUL LORENZ, geology educator; b. N.Y.C., Feb. 3, 1932; s. John Henry Hilpman and Emma Eich; m. Carol D. Crater, Nov. 2, 1991. AB in Geology, Brown U., 1954; PhD in geology, Kans. U., 1969. Cert. profl. geologist. Petroleum geologist Kans. Geol. Survey, Lawrence, 1956-65, engring. geologist, 1965-68, environ. geologist, 1968-74; prof. geology U. Mo., Kansas City, 1974—, dir. Ctr. for Underground Studies, 1987—. Contbr. more than 50 articles to profl. jours. Mem. Fellow AAAS, Geol. Soc. Am.; mem. Assn. Engring. Geologists (life). Home: 3204 NW Karen Rd Kansas City MO 64151 Office: Ctr for Underground Studies U Mo Kansas City 5100 Rockhill Rd Kansas City MO 64110

HILSENRATH, JOEL ALAN, computer scientist, consultant; b. Bklyn., Apr. 18, 1965; s. Daniel Wallace and Lee Betty (Batch) H. Cert. computer tech., Manhattan Career Inst., N.Y.C., 1988. Asst. dir. registry Grand Lodge Free and Accepted Masons, N.Y.C., 1988-90; database adminstr. Grand Lodge Free and Accepted Masons, 1991; pres. Xoanon Enterprises, Bklyn., 1991—. Mem. N.Y. Zool. Soc., N.Y.C., Midwood Civic Action Coun., Bklyn. Mem. Amateur Computer Club, Masons. Democrat. Jewish. Avocations: Dr. Who and sci. fiction, Shakespeare. Home: 945 E 15th St Brooklyn NY 11230-3703 Office: Xoanon Enterprises PO Box 679 Brooklyn NY 11230

HILTON, THEODORE CRAIG, computer scientist, computer executive; b. Oakland, Calif., June 14, 1949; s. Theodore Caldwell and Maxine (Donnelly) H.; m. Peggy Estes, May 21, 1990; children from a previous marriage: Christopher, Kelly, Clark. BS in Internat. Rels., Occidental Coll., 1972; BS, Calif. Inst. Tech., 1972; MS in Computer Sci., N.Y. Inst. Tech., 1980. Ptnr., founder Cen. Data Corp., L.A., 1971—, chief exec. officer, 1988—; engr. RSK, L.A., 1972-73; prof. Lake City (Fla.) Coll., 1981-85, dept. chair, 1983-85; prin. rsch. invest. U.S. Dept. Def., L.A., 1985-88; bd. dirs. TBS S.A., Versailles, France; U.S. presenter SOLE Internat. Conv., 1991. Creator: (computer systems) Broadcast Management System, 1972, ICSS, 1974, EBook, 1993; contbr. over 50 articles to profl. jours. Mem. IEEE, IEEE Computer Soc., Am. Mgmt. Assn., Logistics Engrs. Soc., Data Processing Mgmt. Assn., Rotary (Paul Harris fellow). Avocations: hiking, painting, music. Office: Cen Data Corp 145 N Church St Spartanburg SC 29301-5163

HILTY, TERRENCE KEITH, analytical sciences manager; b. San Francisco, May 9, 1951; s. John Britton and Betty Lou (Cravener) H.; m. Ann Louise Helberg, Aug. 17, 1974; 1 child, Sean Austin. BS in Chemistry, Miami U., 1973; MS in Chemistry, U. Mich., 1975, PhD in Chemistry, 1979. Chemist ctrl. R & D Dow Corning Corp., Midland, Mich., 1979-83, chemist fluids bus., 1983-87, synthesis sect. mgr., 1987-91, analytical scis. mgr., 1992—, mem. polit. action com., 1992. Contbr. articles to jours. Inorganic Chemistry, Organometallics, Tetrahedron Letters. Coach Midland Soccer Club, 1986-92. Mem. Am. Chem. Soc. (local sect. variety of com. leadership 1972-89), Analytical Lab. Mgrs. Assn., Sigma Xi. Lutheran. Achievements include patents for Thermo Oxidatively Stable Fluids, Process to Prepare Phenyl Disilanes, Synthesis of Organosilicon Monomers and Polymers. Home: 5107 Natalie Ct Midland MI 48640 Office: Dow Corning Corp PO Box 1767 Midland MI 48640

HILTZ, ARNOLD AUBREY, former chemist; b. Sea View, Canada, July 31, 1924; came to the U.S., 1953; s. Aubrey Claremont and Fannie Mae (Bryanton) H.; m. Margery Jane Beer, July 17, 1946; children: Sharon Lynne, Deborah Jane. BS in Chemistry, Acadia U., Wolfville, Nova Scotia, 1947; PhD in Phys. Chemistry, McGill U., Montreal, Quebec, Canada, 1952. Ordained deacon, 1976, ordained priest, 1976. Rsch. scientific officer Def. Rsch. Bd. Canada, Quebec City, Canada, 1951-53; rsch. chemist Am. Viscose Corp., Phila., 1953-59, group leader, 1959-60; group leader Avisun Corp., Phila., 1960-65; rsch. chemist Borden Chem. Co., Phila., 1965-66; sr. scientist Gen. Electric Co., Phila., 1966-79, mgr. materials applications, 1979—; tutor math. and sci. Rose Tree Media (Pa.) Sch. Dist., 1958-74. Contbr. articles to profl. jours.; patentee in field. Sch. dir. Rose Tree Media Sch. Dist., 1969-74; bd. dirs., treas. Middletown (Pa.) Free Libr., 1964-69; bd. dirs. Sheepscot Island Co., MacMahan Island, Maine, 1983-85; docent Phila. Mus. Art, 1988—. Recipient Silver medal Gov.-Gen. Can. 1942, Canadian Overseas medal 1945, Frank J. Sensenbrenner fellow McGill U. 1949-51, inventor's medal GE 1964. Mem. Am. Chem. Soc. (sci. lectr. 1958—, chem. abstractor 1958-79). Republican. Episcopalian. Avocations: art appreciation, music appreciation, reading. Home: 524 Cedar Ln Swarthmore PA 19081-1105

HIMATHONGKAM, THEP, endocrinologist; b. Bangkok, Thailand, Aug. 29, 1942; s. Chan and Payia H.; m. Jitrapa Kantabutra, Dec. 12, 1970; children: Tanya, Tarin, Tinapa. BA, U. Calif., Berkeley, 1963; MD, U. Wis., 1969. Diplomate Am. Bd. Internal Medicine. Fellow in endocrinology Peter Bent Brigham Hosp. Harvard U. Med. Sch., Boston, 1972-74; asst. prof. endocrinology Ramathibodi Med. Sch. Mahidol U., Bangkok, 1974-78, assoc. prof., 1979-82, prof., 1983—; dir. Theptarin Diabetes Ctr., Bangkok, 1985-90; cons. endocrinologist Phyathai Hosp., Bangkok, 1976—, Samitivej Hosp., Bangkok; bd. dirs. Phyathai Hosp., Theptarin Gen. Hosp., Bangkok (chmn. 1991—). Author: (book) Disease of the Thyroid, 1984. Dir. Duangprateep Foundn., Bangkok, 1990—. Fellow Am. Coll. Physicians; mem. Royal Coll. Physicians Thailand. Avocation: photography. Home: 3850 Rama IV, Prakanong, Bangkok 10110, Thailand Office: Theptarin Hosp, 3850 Rama IV, Prakanong, Bangkok 10110, Thailand

HIMES, GEORGE ELLIOTT, pathologist; b. Huntington, W.Va., Jan. 5, 1922; s. Connell Bradley Sr. and Elizabeth (Skeans) H.; m. Rita T. Wasniewski; children: Rita Ann Brust, Susan Ruth Burger, George Elliot Jr., Brent Lee. Student, U. Cin., 1939-42; DO, Chgo. Coll. Osteo. Medicine, 1942-45. Intern Lamb Mem. Hosp., Denver, 1945-46; resident pathology Chgo. Osteo Hosp., 1946-48; asst. prof. pathology Chgo. Coll. Osteo Med., 1948-56; asst. lab. dir. Chgo. Osteo Hosp., 1948-51; dir. of labs Flint (Mich.) Osteo Hosp., 1951-87, dir. of labs and nuclear med., 1957-80; assoc. prof. pathology Coll. Osteo Med., Des Moines, 1968-89; adj. prof. pathology Coll. Osteo Med., East Lansing, Mich., 1974-84; dir. sch. med. tech. Flint (Mich.) Osteo Hosp., 1975-85; mem. radiation, chem. & biol. safety com. Mich. State U., 1978—; mem. Am. Osteo. Bd. Pathology, 1959-68, Am. Osteo. Bd. Nuclear Medicine, 1974-84. Bd. dirs. ARC, Flint, 1963-74, United Way, 1964-72; pres. Flint Civitan Club, 1954. Mem. Am. Coll. Osteo. Pathologists (past pres. & sec.-treas. 1954-72), Am. Osteo. Assn., Am. Assn. Blood Banks, Mich. Assn. Osteo. Physicians and Surgeons, Coll. Am. Pathologists, Am. Soc. Clin. Pathologists, Soc. Nuclear Medicine, Mich. Soc. Pathologists, Genesee County Osteo. Assn., Flint Golf Club. Avocations: golf, stamps, computing sciences. Home: 444 Luce Ave Flushing MI 48433-1411 Office: Flint Osteopathic Hosp 3921 Beecher Rd Flint MI 48532-3699

HIMMELBLAU, DAVID MAUTNER, chemical engineer; b. Chgo., Aug. 29, 1923; s. David and Roda (Mautner) H.; m. Betty H. Hartman, Sept. 1, 1948; children—Andrew, Margaret Ann. B.S., MIT, 1947; M.B.A., Northwestern U., 1950; Ph.D., U. Wash., 1957. Cost engr. Internat. Harvester Co., Chgo., 1946-47; cost analyst Simpson Logging Co., Seattle, 1952-53; mgr. Excel Battery Co., Seattle, 1953-54; teaching asst., instr. U. Wash., Seattle, 1955-57; successively asst. prof., asso. prof., prof. chem. engring. U. Tex., Austin, 1957—; chmn. dept. U. Tex., 1973-77; pres. RAMAD Corp., CACHE Corp. of Mass., Univ. Fed. Credit Union, 1964-68. Author: Basic Principles and Calculations in Chemical Engineering, 1962, 67, 74, 82, 89, Process Analysis and Simulation, 1968, Process Analysis by Statistical Methods, 1970, Applied Nonlinear Programming, 1974; contbr. numerous articles in field to profl. jours. Served with U.S. Army, 1943-46, 51-52. NSF grantee, 1957-93; NATO Sci. Com. grantee, 1969. Mem. Am. Inst. Chem. Engrs. (dir. 1973-76), Am. Chem. Soc., Am. Math. Soc., Ops. Research Soc. Am., Soc. Indsl. and Applied Mathematics, Sigma Xi, Delta Mu Delta. Club: Headliners (Austin). Home: 4609 Ridge Oak Dr Austin TX 78731-5211 Office: Univ Texas Coll Engring Austin TX 78712

HINCKLEY, LYNN SCHELLIG, microbiologist; b. Poughkeepsie, N.Y., June 28, 1944; d. John Alfred and Hilda Louise (Russell) Schellig; m. Henry Pember, Apr. 15, 1967; children: John, Russell, Clark. BA, U. Conn., 1966. Rsch. asst. U. Conn., Storrs, 1966-67, microbiologist, dir. Conn. Mastitis Lab., 1977—; microbiologist Angell Meml. Animal Hosp., Boston, 1967-68, Pharm-House, Inc., Hope Valley, R.I., 1971-73. Contbr. articles to profl. jours. Mem. Nat. Mastitis Coun. (bd. dirs. 1990—), Nat. Conf. on Interstate Milk Shipments (com. chair 1989—), Am. Assn. Vet. Lab. Diagnosticians, Internat. Assn. Milk, Food and Environ. Sanitarians, Inc., U.S. of Am. Nat. Com. of Internat. Dairy Fedn. (expert group somatic cell count A2B chair 1993—), N.E. Am. Soc. Microbiology, N.E. Dairy Practices Coun. (dir. task force 1991—), Conn. Mastitis Coun. (chair 1987—). Office: U Conn Diagnostic Testing 61 N Eagleville Road Ext # U-203 Storrs Mansfield CT 06269-3203

HINDE, JOHN GORDON, lawyer, solicitor; b. Newcastle, NSW, Australia, Apr. 5, 1939; s. George Gordon and Vera Jean (Francis) H.; m. Judith Anne Walmsley, May 13, 1967; children: Kathryn Anne, Andrew John. BS, U. NSW, Sydney, Australia, 1959, Diploma of Edn., 1960; Diploma of Law (SAB), Sydney U., 1970. Solicitor Supreme Ct. of NSW. Tchr. NSW Dept. Edn., Sydney, 1960-65, Inner London Edn. Authority, 1966, Trinity Grammar Sch., Sydney, 1967-70; barrister at law NSW Bar Assn., Sydney, 1970-71; tech. asst. Spruson & Ferguson, Sydney, 1971-76, prin., 1976—; prin. Williams Niblett, Solicitors, Sydney, 1978—; lectr. in law Sydney U. Law Sch., 1974-77. Fellow Inst. Patent Attys. Australia; mem. ABA (com. 1987—), Am. Intellectual Property Law Assn. (com. mem. 1986—), Law Soc. NSW (com. 1978—), Chartered Inst. Patent Agts. U.K. (overseas mem.), Am. Nat. Club Sydney, Law Coun. Australia (com. 1990—), Australian Jockey Club, Killara Golf Club. Avocations: thoroughbred horse breeding and racing, squash, swimming. Office: Spruson & Ferguson, 31 Market St, Sydney NSW 2000, Australia

HINDERLITER, RICHARD GLENN, electrical engineer; b. Tulsa, Apr. 9, 1936; s. Robert Verl and Aileen (Burton) H.; m. Leila Ratzlaff, June 8, 1958; children: Daniel Scott, Susan Paige, Alison Ann, Matthew Glenn. BSEE, U. Kans., 1958; MSEE, NYU, 1960, PhD in Ops. Rsch., 1973. Staff mem. Bell Labs., Murray Hill, N.J., 1958-62; dept. head Bell Labs., Holmdel, N.J., 1962-72, Whippany, N.J., 1972-82; divsn. mgr. AT&T, N.Y.C., 1982-83, Bellcore, Morristown, N.J., 1984-91. Contbr. articles to Internat. Conf. on Communications, Computer Mag., Internat. Symposium on Subscriber Loops, Internat. Teletraffic Conf. Mem. Zoning Bd. of Adjustment, Chatham Twp. N.J., 1992—; scoutmaster Boy Scouts Am., Chatham Twp., Red Bank, N.J., Wichita, Kans., 1958—. Recipient Silver Beaver award Morris-Sussex Coun. Boy Scouts Am., 1988. Fellow AAAS; mem. IEEE (sr.), Ops. Rsch. Soc. Am. Methodist. Achievements include application of ops. rsch. techniques to large software systems. Home: 49 Hall Rd Chatham NJ 07928

HINDERS, MARK KARL, mechanical engineer; b. Sioux Falls, S.D., Sept. 30, 1963; s. Dean Corwin and Denise Ann (Green) H.; m. Theresa June Rice, Feb. 29, 1988; 1 child, Maxwell Thomas. BS in Aero. Engring., Boston U., 1986, MSME, 1987, PhD in Engring., 1990. Registered engr.-in-tng., Mass. Sr. scientist Mass. Technol. Lab., Belmont, 1991-93; asst. prof. mech. engring. Boston U., 1991-93; asst. prof. Applied Sci. Coll. William and Mary, Williamsburg, Va., 1993—. Capt. USAF, 1987-91. Mem. ASME, AAAS, Am. Phys. Soc. Home: 1604 S Glendale Ave Sioux Falls SD 57105 Office: Coll William and Mary Dept Applied Sci Rm H Small Hall Williamsburg VA 23186

HINDO, WALID AFRAM, radiology educator, researcher; b. Baghdad, Iraq, Oct. 4, 1940; came to U.S. 1966, naturalized 1976; s. Afram Paul and Laila Farid (Meshaka) H.; m. Fawzia Hanna Batti, Apr. 20, 1965; children—Happy, Rana, Patricia, Heather, Brian. MB, ChB, Baghdad U., 1964. Diplomate Am. Bd. Radiology. Instr. radiology Rush Med. Coll, Chgo., 1971-72; asst. prof. Northwestern U., Chgo., 1972-75; assoc. prof. medicine and radiology Chgo. Med. Sch., 1975-80, prof., chmn. dept. radiology, 1980-90, prof. dept. radiology, 1990, dir. radiology rsch. program, 1990—; dir. radiology rsch. program VA Med. Ctr., North Chicago, Ill.; cons. Ill. Cancer Coun. Contbr. articles on cancer treatment to profl. jours. Bd. dirs. Lake County div. Am. Cancer Soc., Ill., 1975-80. Served to lt. M.C., Iraq; Army, 1965-66. Named Prof. of Yr., Chgo. Med. Sch., 1981, 82, 83, 85, 86. Mem. Am. Coll. Radiology, Am. Soc. Acad. Radiologists. Republican. Roman Catholic. Office: Univ of Health Scis Chgo Med Sch 3333 Green Bay Rd North Chicago IL 60064-3095

HINDS, FRANK CROSSMAN, animal science educator; b. Sioux Falls, S.D., Nov. 30, 1930; s. William and Carol (Crossman) H.; m. Donna Fogel, Jan. 24, 1953; children: Eric William, Mark Alan, Matthew Robert. BS in Edn., Ill. State Normal U., 1952; PhD in Animal Sci., U. Ill., 1959. Rsch. assoc. Dixon Springs Exptl. Sta. U. Ill., Robbs, 1959-60, asst. prof., 1960-64; asst. prof. dept. animal sci. U. Ill., Urbana, 1964-70, assoc. prof., then prof. dept. animal sci., 1970-80; prof. U. Wyo., Laramie, 1980—, head dept. animal sci., 1980-87. Vice-pres. Wyo. Territorial Prison Corp., Laramie, 1986-89, pres. 1989-92; chmn. Found. Bd. Wyo. Territorial Pk., 1993—. Fellow AAAS; mem. Am. soc. Animal Sci. (editorial bd., program chair), Soc. Range Mgmt. Office: Univ Wyo Dept Animal Sci Box 3354 Univ Sta Laramie WY 82071

HINDS, GLESTER SAMUEL, program specialist, tax consultant; b. N.Y.C., July 4, 1951; s. Glester Samuel and Kathryne Elizabeth (Ellison) H. BBA, Bernard M. Baruch Coll., 1973; MBA in Fin., Columbia U., 1975. Stock Broker, Ins. Broker, Notary. Staff acct. Peat Marwick Mitchell, N.Y.C., 1975-77; fin. analyst Citicorp, N.Y.C., 1977-79; sr. fin. analyst Am. Express, N.Y.C., 1979-80; owner, cons. Hinds Fin. Svcs., Long Island, N.Y., 1980-87; owner, founder, pres. Emerald Advt. Co., 1985—; program specialist Calif. FTB, Manhasset, N.Y., 1987—; cons. Am. Entrepreneur's Assn., L.A., 1980-89, Mildred Burke Prodns., 1982-84, Worldwide Diamonds Assn., 1983-85, Acad. Fin. Aid Matching Svcs., 1983-87; licensee Creative Capital Pubs., Inc., 1983; with Mail Order Assocs., Inc., 1984—. Editor: Financial Newsletter the H-Club, 1978-82; actor: On Camera TV Acting, 1986; contbr. articles to profl. jours. Founder Heritage Found., Washington, 1981, 82, Ronald Reagan Rep. Ctr., 1989; chman's com. U.S. Senatorial Bus. Adv., Bd., 1981, 82. Recipient Edward M. Paster Meml. award, Sigma Alpha award, Beta Gamma Sigma award, Beta Alpha Psi award, Bernard M. Baruch Coll., 1973; named Toronto Sprots Club Athlete of Yr., 1987. Mem. USA Amateur Athletes, Interval Internat., Am. Mus. Natural History (assoc.), Am. Soc. Notaries, U.S. Olympic Soc., N.Y. Pub. Interest Rsch. Group, Rep. Presdl. Task force, 24K Club, USA Wrestling, Franklin Mint Collector's Soc., Pro-Wrestling Hall of Fame (chmn.), U.S. Tennis Assn. Methodist. Avocations: world travel, financier, champion wrestler, political advisor, collector. Home: 31 Thomas St Coram NY 11727 Office: California Franchise Tax Bd 1129 Northern Blvd Manhasset NY 11030

HINDS, JAMES WILLIAM, aeronautical engineer; b. Colchester, Eng., Oct. 25, 1966; s. John Peter and Elsie Irene Hazel (Claydon) H.; m. Priscilla Joy Darch, June 9, 1990. B Engring. in Aeros. and Astronautics, Southampton (Eng.) U., 1989. Grad. design engr. Matra Marconi Space, Portsmouth, Hampshire, Eng., 1989-90; design engr. Matra Marconi Space UK, Portsmouth, 1990-91, sr. design engr., 1991—. Mem. Royal Aero. Soc. Achievements include research on degradation of monopropellant thrusters due to silica ingestion. Office: Matra Marconi Space UK, Anchorage Rd, Portsmouth P04 8BA, England

HINER, THOMAS JOSEPH, tractor manufacturing administrator; b. Garnett, Kans., Dec. 19, 1960; s. William L. and Alfreda M. (Lickteig) H. Student, Johnson County Community Coll., 1979-81, U. Kans., 1982-86. Adminstrv. asst. Royal Tractor, Industrial Airport, Kans., 1986—. Mem. Assn. for Computing Machinery, N.Y. Acad. Scis., IEEE (affiliate). Home: 13820 Pflumm Olathe KS 66062 Office: Royal Tractor 100 Mission Woods Rd Industrial Airport KS 66031

HINES, DWIGHT ALLEN, II, family practice physician; b. Tyler, Tex., Nov. 19, 1960; s. Dwight Allen and Marilyn Jeane (Holland) H. BA, Baylor U., 1983; MS, Tex. A&M U., 1987; MD, U. Tex. Med. Br., Galveston, 1992. Resident John Peter Smith Hosp., Fort Worth, Tex., 1992—. Contbr. articles to profl. jours. Mem. AMA, Sigma Xi (assoc.), Phi Chi. Baptist. Home: 3006 DeCharles Tyler TX 75701 Office: John Peter Smith Hosp. 1500 S Main St Fort Worth TX 76104

HINGE, ADAM WILLIAM, energy efficiency engineer; b. Utica, N.Y., Feb. 6, 1961; s. Edward Steven and Jacqueline (Baque) H.; m. Kathleen Conlon, Aug. 11, 1984. BS in Mech. Engring., Rensseleaer Polytech. Inst., 1982, MS in Mech. Engring., 1984. Design cons. Solar Systems Design, Voorheesville, N.Y., 1980-82; energy cons. Northeast Utilities, Madison, Conn., 1982-83; engr. Northeast Utilities, Hartford, Conn., 1983-85; program mgr. N.Y. State Energy Office, Albany, 1985-87, mgr. tech. svcs., 1987-89; pres. The Hinge Group, Troy, N.Y., 1989-91; energy utilization mgr. Niagara Mohawk Power Corp., Albany, 1991—; chmn. steering com. Capital Dist. Engrs. Week., Albany, 1988-91; chmn. bd. dirs. N.E. Sustainable Energy Assn., Greenfield, Mass., mem. energy mgmt. coun., 1991-92. Author, editor: Commercial Lighting Efficiency Resource Book, 1991; contbg. editor ASHRAE Handbook, 1991; contbr. articles to profl. jours., 1985—. Mem. adv. bd. Leadership Rensselaer, Troy, N.Y., 1989-92; bd. dirs. RCCA: The Arts Ctr. Named Energy Engr. of Yr., Assn. Energy Engrs., Conn. chpt., 1985. Mem. ASHRAE (scholar 1982, various coms. 1985-92, pres. N.E. N.Y. chpt. 1990-91, Presdl. award of excellence 1991). Achievements include rsch. to quantify and implement energy efficiency, particularly with respect to comml. bldg. energy systems and utility energy efficiency problems. Home: 130 W Sand Lake Rd Wynantskill NY 12198 Office: Niagara Mohawk Power Corp 1125 Broadway Albany NY 12201

HINKELMANN, KLAUS HEINRICH, statistician, educator; b. Bad Segeberg, Germany, June 6, 1932; came to U.S. 1966; s. Emil F. and Elly (Meincke) H.; m. Christa G., July 29, 1966; 1 child, Christoph. PhD, Iowa State U., 1963. Rsch. assoc. Inst. for Forest Genetics, Schmalenbeck, Germany, 1960; sci. asst. U. Freiburg, Germany, 1964-66; assoc. prof. Va. Poly. Inst. and State U., Blacksburg, 1966-72, prof. statistics, 1972—, head dept. statistics, 1982-93. Editor: Design of Experiments, Statistical Models and Genetic Statistics, 1984; co-author: Design and Analysis of Experiments, 1993; editor Biometrics, 1990-93. Fellow AAAS, Am. Statis. Assn. Office: Va Poly Inst and State U Dept Statistics Blacksburg VA 24061-0439

HINKLE, MURIEL RUTH NELSON, naval warfare analysis company executive; b. Bayonne, N.J., Mar. 17, 1929; d. Andrew and Florence Martha Ida (Nuber) Nelson; student Md. Coll. for Women, 1947-49; B.A., U. Md., 1951; m. David Randall Hinkle, June 5, 1954; children: Valerie Nelson, Janet Lee, Sally Ann. Mgr., Wildacres Thoroughbred Horse Farm, Waterford, Conn., 1960-70; illustrator Naval Warfare Predictions/Computer Simulated Naval Engagements, Analysis & Tech., Inc., North Stonington, Conn., 1970-73; pres. Sonalysts, Inc., Waterford, 1973-88, chmn., CEO, 1973—, also founder, past dir. Command Engring. & Tech. Services Co.; pres., CEO, chmn. Stonington Farms Inc. (now Mystic Valley Hunt Club), 1983—, Conn. Nat. Bank Adv. Bd., 1988—; chmn., CEO Angiers Assocs., 1989—, S.I. Devel. Corp., 1989—; cons. anti-submarine warfare cruise missile weapon systems Gen. Electric Co., 1974-76; cons. Def. Nuclear Agy. for Tactical Nuclear Effects in anti-submarine warfare, 1974-75; spl. edn. substitute tchr. Waterford Pub. Schs., 1968-74. Bd. trustees Thames Sci. Center, 1979-82. Recipient commendation for services to submarine force Comdr. Submarine Squadron Ten, 1973, SBA New Eng. Contractor of Yr. award, 1986, SBA Adminstr.'s award for excellence, 1985, 86. Mem. Am. Horse Shows Assn., Nat. Audubon Soc., Submarine Devel. Group Two Wives Club (pres. 1968), Sigma Kappa (pres. Senesk chpt. 1987-89), Navy Wives Club. Republican. Baptist. Co-author: Scope of Acoustic Communications Systems in Naval Tactical Warfare, 1974; Non-Acoustic Anti Submarine Warfare, 1974; Nuclear Weapons Effects in Anti Submarine Warfare, 1974; Measures of Effectiveness, Naval Tactical Communications, 1975; co-author: Destroyer ASW Barrier, 1977. Home: 9 Cove Rd Stonington CT 06378-2304 Office: Sonalysts Inc PO Box 280 215 Parkway N Waterford CT 06385

HINKLE, NANCY C., veterinary entomologist; b. Opelika, Ala., Apr. 2, 1955; d. Charles Grady and Norma Claire (Henderson) H. MS, Auburn (Ala.) U., 1983; PhD, U. Fla., 1992. Entomological rschr. U. Ga., Tifton, 1984-88; entomologist U.S. Dept. Agriculture, ARS, Gainesville, Fla., 1991-92; extension entomologist dept. entomology U. Calif., Riverside, 1992—. Author: (with others) Flea Rearing in Vivo and in Vitro for Basic and Applied Research, 1992, Methods of Laboratory Rearing of Fleas, 1991; subject editor Jour. of Agricultural Entomology, 1993—; contbr. 24 articles to profl. jours. Coord. Bringing Sci. to the Scientists of Tomorrow, Gainesville, 1989-92. Pi Chi Omega scholarship Nat. Pest Control Honorary, 1990. Mem. Entomol. Soc. Am. (John Henry Comstock award 1991, Robert T. Gast award S.E. br. 1992), Southeastern Biol. Control Working Group (sect. rep. 1986-87), Pi Chi Omega (cons. 1990-91), Sigma Xi (grant). Office: Dept Entomology U Calif Riverside CA 92521

HINOJOSA, RAUL, physician, ear pathology researcher; b. Tampico, Tamulipas, Mexico, June 18, 1928; came to U.S., 1962, naturalized, 1968; s. Raul Hinojosa-Flores and Melida (Prieto) Hinojosa; m. Berta Ojeda, Sept. 25, 1953; children—Berta Elena, Raul Andres, Jorge Alberto, María de Lourdes. B.S. in Biology, Inst. Sci. and Tech., Tampico, 1946; M.D., Nat. Autonomous U. Mexico, Mexico City, 1954. Asst. prof. U. Chgo, 1962-68, assoc. prof., 1968—, dir. temporal bone program for ear rsch., 1962—, rsch. assoc., 1968-88; rsch. fellow biophysics Harvard U., Boston, 1963; rsch. assoc. in neuropathology, Harvard U., 1964, rsch. fellow in anatomy, 1965. Editor temporal bone histopathology update Am. Jour. of Otolaryngology, 1989—; contbr. articles to profl. jours. Recipient Rsch. Career Devel. award NIH, 1962-65, rsch. grantee, 1962—, hearing rsch. study sect. grantee, 1988-92. Mem. Electron Microscope Soc. Am., AAAS, Midwest Soc. Electron Microscopists, Assn. Research in Otolaryngology, Am. Otological Soc., N.Y. Acad. Scis. Home: 5316 S Hyde Park Blvd Chicago IL 60615-5714 Office: U Chgo 5841 S Maryland Ave Chicago IL 60637-1470

HINSHAW, HORTON CORWIN, physician; b. Iowa Falls, Iowa, 1902; s. Milas Clark and Ida (Bushong) H.; m. Dorothy Youmans, Aug. 6, 1924; children—Horton Corwin, Barbara (Mrs. Barbara Baird), Dorothy (Mrs. Gregory Patent). A.B., Coll. Idaho, 1923, D.Sc., 1947; A.M., U. Calif., 1926, Ph.D., 1927; M.D., U. Pa., 1933. Diplomate Am. Bd. Internal Medicine, Nat. Bd. Med. Examiners. Asst. prof. zoology U. Calif., 1927-28; adj. prof. parasitology and bacteriology Am. U., Beirut, Lebanon, 1928-31; instr. bacteriology U. Pa. Sch. Medicine, 1931-33; fellow, 1st asst. medicine. Mayo Found., U. Minn., 1933-35, asst. prof., 1937-46, assoc. prof., 1946-49; cons. medicine Mayo clinic, 1935- 49, head sec. medicine, 1947-49; clin. prof. medicine, head div. chest diseases Stanford Med. Sch., 1949-59; clin. prof. medicine U. Calif. Med. Sch., 1959-79, emeritus prof., 1979—; chief thoracic disease svc. So. Pacific Meml. Hosp., 1958-69; dir. med. svcs.and chief staff Harkness Community Hosp. and Med. Ctr., San Francisco, 1968-75; Dir. med. ops. Health Maintenance No. Calif., Inc.; mem. Calif. Com. Regional Med. Programs, 1969-75. Author: Diseases of the Chest, rev. edit., 1980; co-author: Streptomycin in Tuberculosis, 1949; contbr. over 215 articles to med. publs.; co-discoverer antiTB chemotherapy, exptl. and clin., with several drugs. Del. various internat. confs., 1928-59. Recipient Disting. Alumnus award Mayo Found., 1990. Fellow A.C.P., Am. Coll. Chest Physicians; hon. mem. Miss. Valley Med. Assn.; mem. AMA, Nat. Tb Assn. (bd. dirs., chmn. com. therapy, v.p. 1946- 47, 67-68, rsch. com.), Am. Thoracic Soc. (pres. 1948-49, hon. life 1979), Am. Clin. and Climatol. Soc., Minn. Med. Assn., Am. Bronchoesophalogical Assn., Am. Soc. Clin. Investigation, Cen. Soc. Clin. Rsch., Soc. Exptl. Biology and Medicine, Aero-Med. Assn., Am. Lung Assn. (hon., Hall of Fame 1980), Minn. Soc. Internal Medicine, Sigma Xi, Phi Sigma, Gamma Alpha. Mem. Soc. of Friends. Home: Box 546 512 San Rafael Ave Belvedere Tiburon CA 94920

HINSON, JACK ALLSBROOK, research toxicologist, educator; b. Mullins, S.C., Aug. 18, 1944; s. Layton Liston and Will (Allsbrook) H.; m. Joanne Edwards Kidd; children: Edward Thomas, Richard William. BS, Coll. of Charleston, 1966; MS, U. S.C., 1968; PhD, Vanderbilt U., 1972. Postdoctoral fellow Nat. Inst. of Health, Bethesda, Md., 1972-75, sr. staff fellow, 1975-80; rsch. toxicologist Nat. Ctr. Toxicological Rsch., Jefferson, Ark., 1980-90, chief biochem. mechanisms br., 1989-90; adj. prof. U. Ark. Med. Sci., Little Rock, 1980-90; prof., dir. div. toxicology U. Ark. Med. Sci., Little Rock, 1990—; dir. interdisciplinary toxicology program, occupational and environ. health program U. Ark. Med. Sci., 1990—; adj. assoc. prof. U. Tenn. Ctr. for Health Scis., Memphis, 1982-90; vis. fellow Middlesex Hosp. Med. Sch., London, 1982; vis. prof. U. Leiden, The Netherlands, 1986. Contbr. chpts. to books and articles to profl. jours. Mem. Soc. Toxicology (pres. South Ctrl. chpt. 1990-92), Am. Soc. Pharmacology and Exptl. Therapeutics. Episcopalian. Home: 22 Evergreen Ct Little Rock AR 72207-5971 Office: U Ark Med Sci Div Toxicology 4301 W Markham Slot 638 Little Rock AR 72205

HINTON, DAVID OWEN, electrical engineer; b. Guilford County, N.C., May 12, 1938; s. George Owen Hinton and Barbara Elizabeth (Greeson) Wilder; m. Thelma Marie Arrington, Jan. 26, 1963; 1 child, David Scott. BSEE, N.C. State U., 1965. Avionics tech. student USN, 1960-65; electronics officer USN Destroyer, Norfolk, Va., 1965-67; naval flight officer Patrol Squadron, Brunswick, Maine, 1967-70; aircraft maintenance officer USN Rsch. Lab., Patuxent River, Md., 1970-72; project officer U.S. EPA Health Effects Rsch. Lab., Research Triangle Park, N.C., 1972-79; engr. dir. U.S. EPA Atmospheric Rsch. Exposure Assessment Lab., Research Triangle Park, 1991; dep. div. dir. U.S. EPA Atmospheric Rsch. Exposure Assessment Lab., Rsch. Triangle Park, 1992—; mem. Air Sampling Instruments Com., Cin., 1976-84; chmn. Electronics Tech. Adv. Com., Durham, N.C., 1981-85; mem. N.C. State U. Engr. Adv. Program, Raleigh, 1986-88. Author: (with others) Air Sampling Instruments for Evaluation of Atmospheric Contaminants, 1983; contbr. papers, articles to profl. jours. Capt. USPHS, 1977-91. Recipient Nat. Def. medal USN, 1972, Commendation medal USPHS, 1986, Bronze medal U.S. EPA, 1988. Mem. IEEE, Internat. Soc. Exposure Analysis, Am. Conf. Govt. Indsl. Hygienists, Soc. Am. Inventors, Commd. Officers Assn. (sec., treas. 1988), Am. Mil. Engrs., Navy Res. Assn. (life), Res. Officers Assn. (life). Achievements include patents in field.

HINTON, NORMAN WAYNE, information services executive; b. Maysville, Ky., Mar. 8, 1944; s. Eugene Fay and Julia Lafelle (Dalton) H.; m. Juanita Ann Smith Hinton, Nov. 16, 1968; children: Janis Renee Hinton, Brian Wayne Hinton. BA in Bus. Adminstrn., Centre Coll. Ky., 1966. Programmer, systems analyst Union Cen. Life Ins., Cin., 1966-70, Electronic Data Systems Corp., Dallas, 1970—. Bd. mem. S.E. La. Chpt. Am. Red Cross, New Orleans, 1991-92, Orleans Svc. Ctr. Am. Red Cross, 1990-92, Tulane U. Coll. Bus. Adv. Bd., New Orleans, 1989-92; chmn. emergrncy svcs. commn., Orleans Svc. Ctr. Am. Red Cross. Fellow Life Office Mgmt. Assn., Am. Mgmt. Assn.; mem. Beau Chene Country Club, Am. Quarter Horse Assn., Petroleum Club, Assoc. Clubs, Assn. Ky. Cols., Sigma Alpha Epsilon. Mem. Ch. of God Internat. Avocations: quarterhorses, golf, travel. Office: care EDS 5400 Legacy Plano TX 75024

HINTON, TROY DEAN, civil engineer; b. Floydada, Tex., Jan. 5, 1959; s. James Henry and Olive Nan (Gross) H.; m. Kim Lanette Bertrand, May 30, 1981; children: Kyle, Bradley. BS in Agr. Engring., Tex. Tech U., 1981. Registered profl. engr., Tex. Engring. assoc. West Tex. Consultants, Inc., Odessa, Tex., 1981-84, Amarillo, Tex., 1984-85; engr. West Tex. Consultants, Inc., Odessa, 1985-88, Corlett, Probst & Boyd, Inc., Wichita Falls, Tex., 1989—. Mem. Faith Bapt. Ch., Wichita Falls. Mem. NSPE, ASCE, Tex. Soc. Profl. Engrs. (sec.-treas, Mathcounts coord. North Ctrl. Tex. chpt. 1992-93, v.p. 1993—). Achievements include design engineering of water transmission pipeline to West Texas secondary oil recovery operations, of expansion to River Road Wastewater Treatment Plant, Wichita Falls. Office: Corlett Probst & Boyd Inc 1912 Kemp Blvd Wichita Falls TX 76309-3960

HINTON, WILBURT HARTLEY, II, construction engineering executive; b. Denver, July 24, 1942; s. Wilburt Hartley and Helen Catherine (Panak) H.; m. Kathleen Marie Meehan, Dec. 30, 1967; children: Wilburt H. III, Scott A., Stephen B., James N. BS in Arch. Engring., U. Colo., 1964; MBA, U. Colo., Denver, 1974. Registered profl. engr., Colo., Calif., Tenn. Bridge engr. City of L.A., 1964-66; chief bridge engr. URS/Ken White Co., Denver, 1970-74; field constrn. mgr. Auraria Higher Edn., Denver, 1974-75; v.p. Centric Corp., Denver, 1975-85; owner, mgr. Centric-Jones Co., Denver, 1985—. Lt. USN, 1966-70, Vietnam; capt. USNR, 1971-92. Mem. ASCE, Associated Builders and Contractors (chpt. pres., nat. bd. dirs., mem. various coms. 1982—), Associated Prevailing Wage Contractors (nat. bd. dirs. 1988—), Beta Gamma Sigma. Roman Catholic.

HIOE, FOEK TJIONG, physics educator, researcher; b. Jakarta, Indonesia, Apr. 20, 1941; came to U.S., 1967; s. Njan Yoeng and Tai Jin (Woen) H.; m. Siew Hoon Toh, July 24, 1967; children: Wayne, Elliott, Nelson. BSc in Physics, Imperial Coll. Sci. and Tech., London, 1963; PhD in Physics, U. London, 1967. Rsch. chemist U. Calif.-San Diego, La Jolla, 1967-71; rsch. assoc. U. Rochester, N.Y., 1971-75, sr. rsch. assoc., 1975-81; assoc. prof. physics St. John Fisher Coll., Rochester, 1981-84, prof., 1984—. Contbr. over 75 articles to Phys. Revs., Phys. Rev. Letters, Jour. Physics, Physics Letters, Jour. Math. Physics, Jour. Chem. Physics, Jour. Optical Soc. Am. Rsch. grantee U.S. Dept. Energy, 1982-91. Achievements include contribution to understanding of excluded volume effect in polymers, nonlinear dynamics in 2-dimensional Hamiltonian systems, and multiphoton and multilevel coherent effects in laser-atoms interaction. Office: St John Fisher Coll 3690 East Ave Rochester NY 14618

HIRAI, DENITSU, surgeon; b. Yokkaichi, Mie, Japan, July 27, 1943; came to U.S. 1969; s. Denyomu and Shizuo (Tanaka) H.; m. Fumiko Hada, June 14, 1969; 1 child, R. Lisa. MD, U. Tokyo, 1968. Diplomate Am. Bd. Surgery, Am. Bd. Quality Assurance and Utilization Review Physicians. Intern and residency Waterbury (Conn.) Hosp., 1969-74; fellow Mt. Sinai Hosp., 1974-75; asst. chief surgery VA Med. Ctr., Lincoln, Nebr., 1975-80; chief surgery VA Med. Ctr., Lincoln, 1981—; asst. clin. prof. surgery Creighton U., Omaha, 1982-84, asst. prof. surgery, 1984—; clin. instr. U. Nebr., Omaha, 1986-88, clin. asst. prof. surgery, 1988—. Author: Brain Ticklers (Japanese), 1983. Mem. AAAS, AMA, ACS, Am. Soc. Parenteral and Enteral Nutrition, Soc. Am. Gastrointestinal Endoscopic Surgeons, Southwestern Surg. Congress, Soc. Critical Care Medicine, Assn. VA Surgeons. Avocations: photography, Braille transcription. Office: VA Med Ctr 600 S 70th St Lincoln NE 68510-2493

HIRANO, ARLENE AKIKO, neurobiologist, research scientist; b. L.A., Oct. 24, 1962; d. Yasuo and Toyoko (Fujimori) H. BS, U. Calif., Irvine, 1984; PhD, Rockefeller U., 1991. Grad. fellow Rockefeller U., N.Y.C., 1984-91, postdoctoral fellow, 1991; rsch. assoc., postdoctoral fellow Cornell U. Med. Coll., N.Y.C., 1991—. Recipient Nat. Rsch. Svc. award Pub. Health Svc., 1984-90, Excellence in Rsch. award U. Calif., Irvine, 1984, Regents scholar U. Calif., Irvine, 1980-84; Lucille P. Markey fellow Charitable Trust, 1984-90, fellow Rockefeller U., 1984-91, David Warfield fellow in Opthalmology N.Y. Community Trust/N.Y. Acad. Medicine. Mem. AAAS, Soc. for Neuroscience. Office: Cornell Univ Med Coll 1300 York Ave New York NY 10021

HIRANO, KEN-ICHI, metallurgist, educator; b. Hamamatsu-City, Japan, May 18, 1927; s. Etsuhei and Toki (Takabayashi) H.; B.Eng., Tokyo Inst. Tech., 1952; Sc.D., Hokkaido U., 1959; m. Tetsuko Kondo, Dec. 21, 1963; children: Mitsuko, Hiromi. Rsch. assoc. M.I.T., Cambridge, Mass., 1957-62; assoc. prof. Tohoku U., Sendai, Japan, 1962-69, prof., 1969-91, head dept.

metallurgy and materials sci., 1971-91, prof. emeritus, 1991—; prof. Sci. U., Tokyo, Noda, Japan, 1992—; vis. prof. univs. including U. Tokyo, Tokyo Inst. Tech., Nat. Cheng-Kung U., Taiwan; also cons. Recipient Achievement award Japan Inst. Metals, 1970. Fellow Am. Soc. Metals; mem. Phys. Soc. Japan, Metal Soc. of AIME, Japan Inst. Light Metals (pres. 1993—), Japan Inst. Metals (trustee), N.Y. Acad. Sci., Sigma Xi. Club: Japan Polar Philatelist (pres.). Author: Metal Physics, 1975, Properties and Structure of Matter, 1993; editor Jour. Japan Inst Metals, 1971-72, 79-80, 82-83, 87—. Home: A1406 Sunlight Pastoral-1, 3-296 Shin-Matsudo, Matsudo 270, Japan Office: Tokuyama Soda Co Ltd Rsch Lab, 40 Wadai, 300-42 Tsukuba Japan

HIRAYAMA, CHISATO, healthcare facility administrator, physician, educator; b. Kajiki, Kagoshima, Japan, Nov. 6, 1923; s. Shigeki and Hide (Kokusho) H.; m. Utako Nishimura June 10, 1989; children from previous marriage: Toko, Tomoo. M.D., Kyushu U., Fukuoka, Japan, 1947, Ph.D., 1956. Intern Kyushu U. Hosp.; asst. Kyushu U., 1953-62, asst. prof., 1962-66, assoc. prof., 1966-76; prof. medicine Tottori U., Yonago, Tottori, Japan, 1976-89 , dean Sch. Medicine, 1986-88 ; dir. Tottori U. Hosp., 1984-86; dir. Saiseikai Gotsu Hosp., Gotsu, Shimane, Japan, 1989—. Author: Diseases of the Liver, 1977; editor: Plasma Proteins, 1979; Treatment of Hepatobiliary Diseases, 1980; Pathobiology of Hepatic Fibrosis, 1985. Recipient prize Japan Soc. Electrophoresis, 1968. Fellow Japan Soc. Internal Medicine (bd. dirs. 1985-87), Japan Soc. Gastroenterology; mem. Japan Soc. Hepatology (dir. 1980-88), Internat. Assn. Study Liver, N.Y. Acad. Scis. Office: Saiseikai Gotsu Hosp, 1551 Gotsu, Gotsu Shimane 695, Japan

HIRNER, JOHANN JOSEF, mechanical engineer; b. Kleinreifling, Austria, Nov. 8, 1956; s. Karl and Berta (Pichler) H. Degree in mech. engring., Higher Tech. Sch., Waidhofen, Austria, 1976. Cert. engr. in mech. engring. With constrn. Steyr (Austria) Daimler Puch AG, 1977-78; with costing Georg Grabner & Sons, Krems, Austria, 1978-84; technician Grabner GesmbH, Krems, 1984-86; indsl. engr. Semperit/Wimpassing, 1987; technician Penn GmbH, Krems, 1987-90; technician controlling ÖGUSSA, Vienna, Austria, 1990-93; distbr. BBE, Vienna, Austria, 1993—. Mem. Nat. Geog. Soc. Roman Catholic. Avocations: politics, photography, science. Home and Office: Mitteraustrasse 3/6/26, A 3500 Krems Austria

HIRNING, HARVEY JAMES, agricultural engineer, educator; b. Mott, N.D., Nov. 9, 1940; s. Albert Jacob and Sophia (Bader) H.; m. Joyce Lila LaMotte, June 23, 1961; children: Suzann Kaye, James Lawrence. BS, N.D. State U., 1962, MS, 1964; PhD, Iowa State U., 1970. Registered profl. engr., Iowa, N.D. Grad. asst. N.D. State U., Fargo, 1962-63, instr., 1963-64; extension assoc. Iowa State U., Ames, 1964-70, asst. prof., 1970-72; asst. prof. extension U. Ill., Urbana, 1972-75; assoc. prof. agrl. engring., extension N.D. State U., Fargo, 1975-81, prof. agrl. engring., extension, 1981-93; bd. dir. N.D. Power Use Coun., Fargo. Editor extension circulars and bulls. Mem. bd. suprs. Reed Twp., N.D., 1987-93; chair adminstrv. bd. Edgewood Meth. Ch., 1986-90. Named Hon. State Farmer, N.D. Future Farmers Am., 1987; recipient Merit award N.D. Power Use Coun., 1987. Mem. Am. Soc. Agrl. Engrs., Aircraft Owners and Pilots Assn., Cessna Pilots Assn., Sigma Xi, Epsilon Sigma Phi, Alpha Epsilon, Gamma Sigma Delta, Phi Kappa Phi. Avocations: aviation, golf, fishing, hunting.

HIRNLE, PETER, gynecologist, obstetrician, radiation oncologist, cancer researcher; b. Wroclaw, Poland, Aug. 29, 1953; came to Germany, 1982; s. Zbigniew and Ludmila (Skaja) H.; m. Elisabeth, Feb. 10, 1976; 1 child, Christoph. MD, Bonn (Germany) U., 1983; PhD, Tübingen (Germany) U., 1991. Cert. Med. Sci. Project leader U. Bonn Konrad-Adenauer Grant, 1982-83, U. Tübingen Mildred Scheel Grant, 1983-88; head lymphol. lab. U. Tübingen; asst. prof. gynecology U. Tübingen, 1991—. Author: (with others) Lymph Stasis, 1991; chief editor: Our Opinion, Wroclaw, 1981; contbr. articles to profl. jours. German Rsch. Soc. grantee Tübingen, 1985, Erwin Riesch grantee Tübingen, 1988. Mem. AAAS, Am. Assn. Cancer Rsch., European Soc. Therapeutic Radiology and Oncology, Internat. Soc. Lymphology, N.Y. Acad. Scis. Roman Catholic. Achievements include determination of basic conditions for local treatment of lymph node metastases and lymphomas; use of liposomes as drug carriers for endolymphatic diagnosis and therapy. Home: Ursrainer Ring 104, 72076 Tübingen Germany Office: Dept Gynecology, U Schleichstrasse 4, 72076 Tübingen Germany

HIROKAWA, SHOJI, chemistry educator; b. Asahikawa, Hokkaido, Japan, Feb. 18, 1942; s. Masao and Misako (Yamamoto) H.; m. Yoshiko Konno, June 6, 1971; children: Mio, Mahito. BS, Kyoto U., 1965, MS, 1967, ScD, 1976. Rsch. fellow Kyoto (Japan) U., 1970-82; assoc. prof. Kyushu Inst. Design, Fukuoka, 1982-88, prof., 1988—. Contbr. articles to profl. jours. Mem. AAAS, Am. Phys. Soc., N.Y. Acad. Scis, Chem. Soc. Japan, Phys. Soc. Japan. Avocation: reading. Office: Kyushu Inst Design, 4-9-1 Shiobaru Minami-ku, Fukuoka 815, Japan

HIROOKA, MASAAKI, economics educator; b. Tokyo, Japan, Jan. 14, 1931; s. Kanesuke and Hisako Hirooka; m. Iku Takeuchi, Apr. 29, 1958; children: Tamiko, Kuniko, Keiko. B, Kyoto U., Japan, 1954, DEng, 1971. Researcher Sumitomo Chem. Co., Niihama, Ehime, Japan, 1954-65; researcher Sumitomo Chem. Co., Takatsuki, Osaka, Japan, 1965-83, sr. rsch. assoc., 1975-83, sr. rsch. fellow, 1984-89; vis. prof. dept. I.P.I chemistry U. Liverpool, Eng., 1978-80; prof. econs. Kobe (Japan) U., 1989—. Author: Chemical Industry, 1990; adv. editor internat. jour. Rsch. Policy, 1989—; editorial bd. jour. Polymer, 1981—. Mem. Engring. Acad. Japan, Japan Materials Forum (bd. dirs.), Internat. Joseph A. Schumpeter Soc., Soc. Polymer Sci. Japan (auditor, former v.p., Sci. award 1973, Disting. Svc. award 1993), Chem. Soc. Japan (adv. bd.). Avocation: swimming. Home: 82 Kamifusa-cho Koyama, Kita-ku Kyoto 603, Japan Office: Kobe U Econs Dept, 2-1 Rokkodai-cho, Nada-ku Kobe 657, Japan

HIROSE, TERUO TERRY, surgeon; b. Tokyo, Jan. 20, 1926; s. Yohei and Seiko (Ogushi) H.; m. Tomiko Kodama, June 1, 1976; 1 son, George Philamore. B.S., Tokyo Coll., 1944; M.D., Chiba U., Japan, 1948, Ph.D., 1958. Diplomate: Am. Bd. Surgery, Am. Bd. Thoracic Surgery. Intern Chiba U. Hosp., 1948-49, resident in surgery, 1949-52; resident in surgery Am. Hosp., Chgo., 1954; resident in thoracic surgery Hahnemann Med. Coll., Phila., 1955-56, N.Y. Med. Coll., N.Y.C., 1961-62; practice medicine specializing in surgery Chiba, Japan, 1952-53; chief of surgery Tsushimi Hosp., Hagi, Japan, 1958-59; asst. prof. surgery Chiba U., 1959; research fellow advanced cardiovascular surgery Hahnemann Hosp., Phila., 1959; teaching fellow surgery N.Y. Med. Coll., 1959-60, instr., 1961-62; practice medicine specializing in surgery N.Y.C., 1965-89, N.J., 1965-89; dir. cardiovascular lab. St. Barnabas Hosp., N.Y.C., 1975-84; sr. attending surgeon St. Barnabas Hosp., 1965-81; chief vascular surgery Union Hosp., Bronx, N.Y., 1966-67; attending surgeon Flower and Fifth Ave Hosp., N.Y.C., 1973-80, Jewish Hosp. Med. Center, Bklyn., 1976-80, St. Vincent Hosp., N.Y.C., 1976-88, Mamonides Hosp., Bklyn., 1976-78, Passaic Gen. Hosp., 1977-88, Westchester (N.Y.) County Hosp., 1977-78, Yonkers (N.Y.) Profl. Hosp., 1978-79, Westchester Sq. Hosp., 1978-84, Yonkers Gen. Hosp., 1980-89, St. Joseph Hosp., Yonkers, 1980-89; clin. prof. surgery N.Y. Med. Coll., 1974-89. Author: (in Japanese) A Chaos of American Medicine, 1987, Japanese Doctor, 1987, Where American Medicine Is Going, 1988, Major Surgery Without Blood Transfusion, 1990, Problems and Solutions of American Medicine, 1991, Warning for Modern Medical Science (New Medical Ethics), 1992, Comparative Studies of Medical System in the World, 1992; spl. editor Japanese Med. Planner Ltd.; contbr. 400 articles in field of cardiovascular surgery and gen. medicine to Am. and Japanese med. jours., 10 Am. med. monograms, 1968-80. Recipient Hektoen Bronze medal AMA, 1965, Gold medal, 1971. Fellow Am. Coll. Angiology, Am. Coll. Chest Physicians, Am. Coll. Cardiology, Internat. Coll. Surgeons, N.Y. Acad. Medicine; mem. Am. Assn. Thoracic Surgery, N.Y. Soc. Thoracic Surgery, Pan-Pacific Surg. Assn., Internat. Cardiovascular Soc., Am. Geriatric Soc., Am. Fedn. Clin. Rsch., Am. Writers Assn. Achievements include invention of single pass low prime oxygenator; pioneer aortocoronary direct bypass surgery, open heart surgery without blood transfusion.

HIROTA, JITSUYA, reactor physicist; b. Hyogo-ken, Japan, Oct. 31, 1924; s. Jitsuki and Isoko (Adachi) H.; m. Hiroko Terasaka, Apr. 17, 1954; 2 children. Grad. Kyoto U., 1948, Sc.D., 1961. Research scientist Japan Atomic Energy Research Inst., Tokai Research Establishment, 1956-64, prin. scientist, 1965-83, chmn. com. reactor physics, 1967-81; cons. Mitsubishi Atomic Power Industries, Inc., 1984—; mem. com. on exam. reactor safety Nuclear Safety Commn., 1964-83 (award 1987); Japanese rep. nuclear energy agy. com. reactor physics OECD, 1966-81. Author papers in field. Mem. Atomic Energy Soc. Japan (spl. prize 1973), Am. Nuclear Soc. Home: 11-5 Okusawa 4-chome, Setagaya-Ku, Tokyo 158, Japan Office: Mitsubishi Atomic Power Ind Inc, 4-1 Shibakouen 2-chome Minato-Ku, Tokyo 105, Japan

HIROTA, MINORU, chemistry educator; b. Katsuura, Chiba-ken, Japan, Apr. 9, 1933; s. Kenji and Kimi (Unno) H.; m. Reiko Shiomi, Nov. 28, 1962; children: Koichi, Junko. BSc, U. Tokyo, MSc, DSc. Asst. dept. chemistry U. Tokyo, 1961-64; assoc. prof. dept. applied chemistry Yokohama (Japan) Nat. U., 1964-76, prof. dept. synthetic chemistry, 1976—. Author: Molecular Orbital Theory (in Japanese), 1969, Organic Chemistry (in Japanese), 1973; (J.W. Akitt) NMR and Chemistry, 1988. Mem. Chem. Soc. Japan (award 1966), Am. Chem. Soc. Avocations: hiking, skiing, go. Office: Yokohama Nat U, 156 Tokiwadai Hodogaya-ku, Yokohama 240, Japan

HIROYASU, HIROYUKI, mechanical engineering educator; b. Hiroshima, Japan, Jan. 13, 1935; s. Genzaburo and Yoshiko Hiroyasu; m. Yoshiko Ikeda, Jan. 5, 1964; children: Tomoyuki, Shoko. BA in Mech. Engring., Tohoku U., 1957, MS in Mech. Engring., 1959, PhD in Mech. Engring., 1962. Rsch. engr. Toyota Ctrl. Rsch. and Devel. Co. Ltd., Nagoya, Japan, 1962-64; rsch. assoc. dept. mech. engring. U. Wis., Madison, 1964-66; assoc. prof. dept. mech. engring. U. Hiroshima, Japan, 1966-67, prof. dept. mech. engring., 1968—; vis. prof. dept. mech. engring. Wayne State U., Detroit, 1979-80; adv. prof. Shanghai Jiao Tong U., China. Author: Internal Combustion Engines, 1973, 86; editor: (jour.) Atomization, 1992—; reviewer for jours. including Combustion and Flame, Atomization and Spray. Recipient Soichiro Honda medal ASME, 1992, Outstanding Paper award 20th CIMAC, London, 1993. Fellow Soc. Automotive Engrs. (advanced power plant com. 1974-82, diesel engine com. 1988-93, small power plant com. 1989-93, reviewer transaction); mem. Japan Soc. Mech. Engrs. (bd. dirs. 1986, sect. bd. chmn. 1987, councilor 1977, 79, 80, 82, 83, 85, 86, 88, 89, chmn. engine systems 1993, reviewer transaction), Japan Soc. Automotive Engrs. (councilor 1984, 85, 86, 87, 88, 89, 90, chmn. diesel engine com. 1991-93), Combustion Inst., Inst. Liquid Atomization and Spray Systems (pres. Japan 1993), Gas Turbine Soc. Japan, Japan Soc. Energy and Resources, Visualization Soc. Japan, Fuel Soc. Japan, Marin Engring Soc. in Japan (councilor 1980, 81, Best Tech. Paper award 1989), Japan Soc. for Aero. and Space Scis. (councilor 1980,81), Heat Transfer Soc. Japan, Japanese Soc. Engring. Edn. Home: Dept of Mech Engring, 5-9-21 Asaminamiku, Hiroshima 73101, Japan Office: U Hiroshima Dept Mech Engr, 1-4-1 Kagamiyama, Hiroshima 724, Japan

HIROYUKI, HASHIMOTO, engineering educator; b. Yubari, Hokkaido, Japan, Aug. 27, 1938; s. Masaji and Teruko Hashimoto; m. Noriko Hashimoto, Apr. 29, 1966. BSc, Tohoku U., Japan, 1961, MC, 1963; PhD, Tohoku U., 1966. Lectr. Tohoku U., 1968-70, assoc. prof., 1970-78, prof., 1978—. Mem. ASME, AIAA, Japan Soc. Mech. Engring., Electrochem. Soc. Buddhist. Achievements include patents for vibration pumps (Japan). Office: Tohoku U, Katahira 2-1, Sendai 290, Japan

HIRSCH, GARY MARK, energy policy specialist, cogeneration manager; b. Bklyn., June 6, 1950; s. Herman and Bella Rachel (Pollack) H.; children: Melyssa, Benjamin. BA, Evergreen State Coll., 1984. Electrician, elec. supr. various elec. contractors, 1972-84; conservation mgr. Wash. State Energy Office, Olympia, 1989-91, cogeneration mgr., 1991—. Pres. S.E. Olympia Neighborhood Assn. Mem. Assn. Energy Engrs., Cogeneration Inst., Evergreen State Coll. Alumni Assn. (treas.). Office: Wash State Energy Office PO Box 43165 Olympia WA 98504-3165

HIRSCH, IRA J., otolaryngologist, educator. PhD, Harvard U., 1948. Rsch. prof. otolaryngology Washington U., St. Louis. Recipient Gold medal Acoustical Soc. Am., 1992. Office: Washington Univ Sch of Medicine Dept of Otolaryngology 1 Brooking Dr Saint Louis MO 63130*

HIRSCH, JERRY, psychology and biology educator; b. N.Y.C., Sept. 20, 1922; s. Samuel M. and Mollie (Barnett) H.; m. Jean Carol Kronsky, Mar, 1944 (div. 1950); m. Marjorie Jean Barrie, July 29, 1950; 1 child, Wesley Michel. BA (with hon.), U. Calif., 1952, PhD in Psychology, 1955; doctorat honoris causa, U. Rene Descartes Paris V, 1987. Asst. prof. psychology Columbia U., N.Y.C., 1956-60; assoc. prof. psychology U. Ill., Urbana, 1960-63, prof. psychology, 1963—, prof. zoology, 1966-76, prof. ecology, ethology, & evolution, 1976—; lectr. Howard U., Washington, 1984, No. Mich. U., 1991, Oberlin (Ohio) Coll., 1991; Robert Choate Tryon lectr. U. Calif., Berkeley, 1987. Editor: Behavior Genetic Analysis, 1967, 82; coauthor: Defroquer les Charlatans, 1987; editorial adv. bd. Behavior Genetics, 1971-92; contbr. articles to profl. jours. Mem. YMCA, U. Ill., 1967—, YMCA Friday Forum, U. Ill., 1978-81, 87-90; panelist Howard U., Washington, 1989; organizer of racism conf., U. Ill., Urbana, 1993. Recipient postdoctorate NSF, U. Calif., 1955-57, Soc. Sci. Rsch. Coun. Aux. Rsch. award U. Ill., 1962. Mem. Cosmos Club, Am. Psychol. Assn. (rep. to exec. com. div. 6, 1983-86), Animal Behavior Soc. (exec. com., editor, 2d pres. elect, 1st pres. elect, past pres., chmn.), Soc. for Study of Social Biology, Internat. Soc. Comparative Psychology, Internat. Soc. Human Ethology, Am. Psychol. Soc., Behavior Genetic Assn. Achievements include development of Objective Behavioral Population Screening Apparatus. Office: Psychol Dept Univ Illinois 603 E Daniel St Champaign IL 61820

HIRSCH, JOSEPH ALLEN, psychology and pharmacology educator; b. N.Y.C., Aug. 30, 1950; s. Robert Theodore and Gertrude (Bernstein) H.; 1 child, Jason Mathew; m. Karen Weinberg, June, 1989. BS in Chemistry, Bklyn. Coll., 1972; BS in Pharmacy summa cum laude, Columbia U., 1975; PhD in Pharmacology, Downstate Med. Ctr., Bklyn., 1979; cert. in counseling, Postgrad. Ctr. Mental Health, 1989; postgrad., Psychotherapy Study Ctr., 1989-91; MS in Edn. in Psychology, Pace U., 1993, postgrad., 1993—. Lic. pharmacist; cert. counselor. Postdoctoral fellow MIT, Cambridge, Mass., 1979-81; rsch. assoc. Med. Coll. Cornell U., N.Y.C., 1981-83; asst. prof. St. John's U., Jamaica, N.Y., 1984-88; sr. editor McGraw Hill, N.Y.C., 1988-89; dir. profl. rels. Park Row Pubs., N.Y.C., 1989-91; assoc. prof. N.Y. Coll. Podiatric Medicine, N.Y.C., 1991-92; adj. asst. prof. then assoc. prof. L.I. U., Bklyn., 1984—; adj. asst. prof. Coll. of Dentistry, NYU, N.Y.C., 1989—; adj. instr. psychology Pace U., N.Y.C., 1991—, adj. assoc. prof. biology, 1993—. Freelance med. writer; contbr. articles to profl. jours. Recipient award for excellence in pharmacology Merch Sharp and Dohme, 1975. Mem. APA, AAAS, N.Y. Acad. Scis., Am. Orthopsychiat. Assn., Am. Psychol. Soc., N.Y. Neuropsychology Group, Internat. Platform Assn., Psi Chi. Achievements include discernment of neuropharmacol. behavioral effects of various enantiomers of fenfluramine and analogs; neurochemical effects of lithium; effects of hypoxia and thiamine on brain function/behavior; neurochemical and /or consequences of hypoxia and thiamine; rsch. contributed to the patent approval (European) of d-fenfluramine (Isomeride) for the treatment of obesity. Office: Pace U Dept Psychology 41 Park Row New York NY 10038-1502

HIRSCH, JULES, physician, biochemistry educator; b. N.Y.C., Apr. 6, 1927. Student, Rutgers U., 1943-45; MD, U. Tex., 1948. Intern pathology and medicine Duke Hosp., N.C., 1948-50; from asst. resident to resident coll. medicine SUNY, Syracuse, 1950-52; asst. prof. biochemistry, assoc. physician Rockefeller U., N.Y.C., 1954-60, assoc. prof., physician, 1960-67, prof., sr. physician, 1967—. Mem. AAAS, Am. Soc. Clin. Investigation, Assn. Am. Physicians, Am. Fedn. Clin. Rsch., Harvey Soc. Achievements include research in obesity, human behavior, internal medicine, biochemistry and physiology of lipids, lipid metabolism and nutrition. Office: Rockefeller University 1230 York Ave New York NY 10021*

HIRSCH, MARK J., computer engineer; b. Trenton, N.J., June 24, 1953; s. Morris William and Lorraine (Demner) H.; m. Judith Leslie Stone, June 7, 1981; children: Danielle, Rachel. BSEE, Stevens Inst. Tech., 1975; MSEE, Carnegie Mellon U., 1982, PhD, 1989. Engr. Def. Dept., Fort Meade, Md., 1975-81; mem. tech. staff Johns Hopkins Applied Physics Lab., Columbia, Md., 1982-83; sr. mem. tech. staff Mentor Graphics, Warren, N.J., 1989—. Contbr. articles to profl. jours. Mem. IEEE. Home: 61 Garfield St Berkeley Heights NJ 07922

HIRSCH, PETER BERNHARD, metallurgist; b. Berlin, Jan. 16, 1925; immigrated to Eng., 1939, naturalized, 1946; s. Ismar and Regina (Less) H.; BA, Cambridge (Eng.) U., 1946, MA, 1950, PhD, 1951, hon. DSc, Newcastle, 1979, City U., 1979, Northwestern, 1982, ScD, East Anglia, 1983, D Eng. Liverpool, 1991, D Eng. Birmingham, 1993; m. Mabel Anne Kellar Stephens, July 22, 1959; stepchildren: Janet Susan Caldwell, Paul Roderick Noel Kellar. Research on structure of coal Cavendish Lab., Cambridge U., 1950-53, ICI fellow, 1953-55, research on plastic deformation of metals, 1955-58; asst. dir. research in physics Cambridge U., 1957-58, univ. lectr. in physics, 1958-64, univ. reader in physics, 1964-66, fellow Christ's Coll., 1960-66, hon. fellow Imperial Coll., London, 1988, hon. fellow St. Catharine's Coll., 1982; co-editor: Progress in Materials Science, vol. 36, 1992. mem. U.K. Atomic Energy Authority, 1982—, chmn., 1982-84; Isaac Wolfson prof. metallurgy, head dept. metallurgy and sci. of materials Oxford U., 1966-92, fellow St. Edmund Hall, 1966—; hon. prof. Beijing U. of Sci. and Tech., 1986—; chmn. metallurgy and materials com., mem. engring. bd. Sci. Research Council, 1970-73; mem. Council Sci. Policy, 1970-72, Electricity Supply Research Council, 1969-82, Adv. Com. Safety of Nuclear Installations, 1977-82; mem. equipment subcom. Univ. Grants Com., 1977-83; mem. tech. adv. bd. Monsanto Electronic Materials Co., 1985-88; mem tech. adv. com. Advent, 1982-91; dir. Cogent Ltd., 1985-89; chmn. Isis Innovation Ltd., 1988—. Created knight bachelor, 1975; recipient C.V. Boyes prize Inst. Physics and Phys. Soc., 1962; Wihuri (Finland) Internat. prize, 1971; Wolf prize in physics Wolf Found., 1983-84; Arthur von Hippel award Materials Research Soc., 1983, Disting. Scientist award Electron Microscopy Soc. Am., 1986, Holweck prize Inst. Physics and French. Phys. Soc., 1988, Gold medal Japan Inst. Metals, 1989. Fellow Royal Soc. (Hughes medal 1973, Royal medal 1977, council 1977-79), Inst. Physics (council 1968-72), Franklin Inst. (life; Clamer medal 1970), Royal Microscop. Soc. (hon.); mem. Inst. Metals (council 1968-73; Rosenhain medal 1961), Metals Soc. (council 1976-82; Platinum medal 1976), Materials Sci. Club (A.A. Griffith medal 1979), Academia Europaea, Materials Rsch. Soc. India (hon.), Japanese Soc. ELectron Microscopy (hon.), Chinese Electron Microscopy Soc. (hon.). Jewish. Co-author: Electron Microscopy of Thin Crystals, rev. edit., 1977; editor: The Physics of Metals II-Defects, 1975; contbr. articles to profl. jours. Home: 104A Lonsdale Rd, Oxford OX2 7ET, England Office: U Oxford Dept Materials, Parks Rd, Oxford OX1 3PH, England

HIRSCH, RICHARD ARTHUR, mechanical engineer; b. N.Y.C., Jan. 2, 1925; s. Melvin Mordecai and Gertrude Matilda (Schwarz) H.; m. Carol Walter Sampson, June 18; children: Andrew Sampson, Patricia Ann. BAE, Rensselaer Poly. Inst., Troy, N.Y., 1945; MS in Applied Math., Brown U., Providence, 1950. Structural engr. Republic Aviation Corp., Farmingdale, N.Y., 1946-47; devel. engr. Swank, Inc., Attleboro, Mass., 1947-48; vibrations engr. Boeing Vertol, Morton, Pa., 1950-52; chief structures engr. AAI Corp., Balt., 1952-60; asst. tech. dir. Martin Marietta Corp., Balt., 1960-68; assoc. prof. U.S. Naval Acad., Annapolis, Md., 1968-82; prog. mgr. AAI Corp., Balt., 1982—. Contbr. articles to profl. jours. With USN, 1943-46. Fellow ASME (v.p. 1984-88, bd. govs. 1990-92); mem. NSPE, Soc. Engring. Edn. Home: 8220 Marcie Dr Baltimore MD 21208-1944 Office: AAI Corp PO Box 126 Cockeysville Hunt Valley MD 21030-0126

HIRSCH, WALTER, economist, researcher; b. Phila., Apr. 21, 1917; s. Arnold Harry and Ann Belle (Feldstein) H.; m. Leanore Brod, Feb. 12, 1939 (dec. 1985); stepchild, Stephen M. Gold; children: Jeffrey A., Robert A.; m. June Freedman Gold Clark, Dec. 16, 1986. BS in Econs., U. Pa., 1938; LLD (hons.), Chapman Coll., 1968. Economist U.S. Bur. Stats., Washington and N.Y.C, 1946-50, Dept. USAF, Washington, 1950-51, Nat. Prodn. Auth., Washington, 1952-53; dir. indsl. mobilization Bur. Ordnance Dept. USN, Mechanicsburg, Pa., 1954-56; ops. rsch. analyst Bur. Supplies and Accts. Dept. USN, Arlington, Va., 1956-58; economist, ops. rsch. analyst Internat. Security Affairs Office Sec. of Def., Arlington, 1958-61; chief ops., rsch. analyst Gen. Svcs. Adminstrn., Washington, 1961-63; ops. rsch. analyst Spl. Projects Office Sec. of Def., Arlington, 1963-67; dir. ednl. rsch. U.S. Office Edn., San Francisco, 1967-72; cons. on loan to Office of Dean Acad. Planning San Jose (Calif.) State U., 1972-74. Author: Unit Man-Hour Dynamics for Peace or War, 1957, Internal Study for Office Secretary of Defense: Sharing the Cost of International Security, 1961. Vol. De Young Mus., San Francisco, 1981-84, Calif. Palce of Legion of Honor, Phila. Mus. Art, 1984-86; pres. Met. Area Reform Temples, Washington, Nat. Fedn. Temple Brotherhoods; supporter Phila. Orch., San Francisco Symphony, San Francisco Conservatory Music, Curtis Inst. With USAAF, 1942-46. Recipient Meritorious Civilian Svc. award Navy Dept., 1956. Mem. Pa. Athletic Club, Commonwealth Club of Calif., World Affairs Council, Press Club of San Francisco, Phi Delta Kappa. Avocations: collecting art, music, chess, poetry.

HIRSCH, WARREN MITCHELL, chemistry educator; b. Bklyn., Sept. 15, 1945; s. Leon and Anne (Bocian) H.; m. Ellen Berman, July 16, 1978. MA in Chemistry, CUNY, 1973; PhD in Chemistry, Poly. Inst., Bklyn., 1984. With N.Y.C. Bd. Edn., Bklyn., 1967—, chemistry tchr. Edward R. Murrow High Sch., 1976—; adj. asst. prof. chemistry Bklyn. Coll., CUNY, 1985-91. Contbr. articles to profl. jours. Mem. AAAS, Am. Chem. Soc. (bd. dirs. Bklyn. subsect. 1984-92, Nichols Found. Teaching award 1987), N.Y. Acad. Scis., Chemistry Tchrs. Club of N.Y.C. Achievements include discovery of binding constants of cyclodextrins with various ions and molecules. Home: 2509 Avenue K Brooklyn NY 11210-3717 Office: CUNY Bklyn Coll Dept Chem Bedford Ave # H Brooklyn NY 11222-3102

HIRSCHHORN, ELIZABETH ANN, environmental engineer, consultant; b. Phila., June 12, 1962; d. Arthur J. and Rita (Fein) H. BSCE, Johns Hopkins U., 1984. Registered profl. engr., Md. Project mgr. Buchart-Horn, Inc., Balt., 1984-91, Metcalf & Eddy, Inc., Redwood City, Calif., 1991—. Mem. ASCE, NSPE, Am. Water Works Assn., Md. Soc. Profl. Engrs. Office: Metcalf & Eddy Inc 555 Twin Dolphin Dr Ste 400 Redwood City CA 94065

HIRSCHHORN, JOEL STEPHEN, engineer; b. N.Y.C., Sept. 8, 1939; s. Leon and Blanche H.; m. Jacqueline M. Rams; children: Terri, Lesa. B of Engring., Poly Inst. Bklyn., 1961; MS, Poly. Inst. Bklyn., 1962; PhD, Rensselaer Poly Inst., 1965. Rsch. metallurgist Pratt & Whitley Aircraft, North HAven, Conn., 1962-63; prof. U. Wis., Madison, 1965-78; sr. assoc. Congl. Office of Tech. Assessment, Washington, 1978-90; pres. Hirschhorn & Assocs. Inc., Lanham, Md., 1990—; cons. in field. Author: Introduction to Powder Metallurgy, 1969, Technology and Steel Industry Competitiveness, 1980, Serious Reduction of Hazardous Waste, 1986, (with others) Prosperity without Pollution, 1990, (1 chpt.) The Greening of American Business, 1992; contbr. articles to profl. jours. Recipient Engring. News-Rec. award McGraw Hill, 1990, Environ. award Clean Water Fund, 1988, Citizens Clearinghouse for HAzardous Waste, 1989. Mem. NSPE, Am. Soc. Testing & Materials, Air and Waste Mgmt. Assn. Achievements include 2 patents on powder metallurgy alloys; pioneer in pollution prevention and waste reduction. Home: 3231 Coquelin Terr Chevy Chase MD 20815 Office: Hirschhorn & Assocs Inc 4221 Forbes Blvd Lanham MD 20706

HIRSCHHORN, KURT, pediatrics educator; b. Vienna, Austria, May 18, 1926; came to U.S., 1940, naturalized, 1945; s. Emanuel and Helen (Mayberger) H.; m. Rochelle Reibman, Dec. 20, 1952; children—Melanie D., Lisa R., Joel N. Student, U. Pitts., 1944; B.A., N.Y. U., 1950, M.D., 1954, M.S. (Bergquist fellow), 1958. Intern Bellevue Hosp., N.Y.C., 1954-55; resident Bellevue Hosp., 1955-56; fellow N.Y. U., 1956-57, U. Upsala, Sweden, 1957-58; instr. N.Y. U. Sch. Medicine, 1956-58, asst. prof., 1958-63, asso. prof., 1963-66; Arthur J. and Nellie Z. Cohen prof. genetics and pediatrics Mt. Sinai Sch. Medicine, City U. N.Y., 1966-76, Herbert H. Lehman prof., chmn. pediatrics, 1977—; adj. prof. biology N.Y. U., 1966-74; Established investigator Am. Heart Assn., 1960-65; career scientist N.Y.C. Health Research Council, 1965-75. Author numerous sci. publs.; Editor: (with Harry Harris) Advances in Human Genetics, 1969—; editorial bd. 16 sci. jours. Mem. council Village Community Sch., 1968-73, chmn., 1972-73. Served with AUS, 1944-47. Recipient Rudolph Virchow medal, 1974, Alumni Achievement award NYU Sch. Medicine, 1982. Fellow AAAS, Am. Acad. Pediatrics, N.Y. Acad. Medicine; mem. Inst. Medicineof the Nat. Acad. Scis., Am. Coll. Med. Genetics, Am. Soc. Clin. Investigation, Am. Assn. Physicians, Am. Pediatric Soc., Am. Soc. Human Genetics (pres. 1969, dir.), Pediatric Travel Club, Am. Assn. Immunologists, Harvey Soc. (v.p. 1979-80, pres. 1980-81, council 1981-84), Genetics Soc. Am., Environmental

Mutagen Soc. (council 1969-76), Inst. for Soc. Ethics and Life Scis. (dir. 1969-72), Am. Soc. Pediatric Chmn. (council 1983-86), Am. Cancer Soc. (coun. 1989-92), Alpha Omega Alpha. Home: 29 Washington Sq W New York NY 10011-9180 Office: Mt Sinai Sch Medicine 1 Gustave L Levy Pl New York NY 10029-6504

HIRSCHHORN, ROCHELLE, medical educator; b. Bklyn., Mar. 19, 1932; d. Hyman and Anna Reibman; m. Kurt Hirschhorn; children: Melanie D., Lisa R., Joel N. BA, Barnard Coll., 1953; MD, NYU, 1957. Intern NYU-Bellevue Med. Divsn., N.Y.C., 1958-59; rsch. fellow, teaching asst. NYU Sch. Medicine, N.Y.C., 1963-65, assoc. rsch. scientist, 1965-66, instr. in medicine, 1966-69, asst. prof. medicine, 1969-74, assoc. prof. medicine, 1974-79, prof. medicine, 1979—, head divsn. med. genetics, 1984—; hon. fellow Galton Lab. Human Genetics & Biometry Univ. Coll., London, 1971-72; assoc. attending physician in medicine Bellevue Hosp., N.Y.C., 1969-80, Univ. Hosp., NYU Sch. Medicine, 1974-81; attending physician Bellevue Hosp., 1980—, Univ. Hosp., 1981—; mem. numerous NIH coms. & study sects., 1973—. Senator NYU Senate, mem. pediatrics search com., 1987-89, human subjects instl. rev. bd., 1989—, co-dir. second year med. genetics course, 1989-93; trustee AIDS Med. Found./AMFAR; judge Westinghouse Nat. Sci. Talent Search; founding mem. Village Community Sch. Fellow AAAS, Am. Coll. Rheumatology; mem. Am. Soc. for Clin. Investigation, Assn. Am. Physicians, Am. Assn. Immunologists, Am. Soc. Human Genetics (cert. 1987), Am. Fedn. for Clin. Rsch., Interurban Clin. Club (pres. 1987-88), Peripatetic Soc., Soc. for Inherited Metabolic Diseases, Harvey Soc. (coun. 1989-92), Alpha Omega Alpha (councillor Delta of N.Y. 1982—). Achievements include elucidation of pathophysiologic mechanisms, delineation of molecular and biochemical defects of genetic disorders including adenosine de. Office: NYU Medical Center 550 1st Ave New York NY 10016

HIRSCHL, SIMON, pathologist; b. Zagreb, Yugoslavia, Nov. 22, 1935; s. Ludwig and Hilda Hirschl; m. Mirna Kriznic, Apr. 27, 1961; children: Cynthia, Sandra, Melissa, Diane. MD, U. Zagreb, 1961; student, A. Einstein Coll. Medicine, 1963-67. Pathologist Hosp. for Joint Diseases, N.Y.C., 1967-69, Flower and Fifth Ave Hosp., N.Y.C., 1969-77; assoc. prof. pathology N.Y. Med. Coll., Valhalla, N.Y., 1969-78; dir. pathology lab. Lourdes Hosp., Binghamton, N.Y., 1977-90; clin. prof. pathology SUNY, Syracuse, N.Y., 1978-90; dir. pathology Mather Meml. Hosp., Port. Jefferson, N.Y., 1990—, St. Charles Hosp., Port. Jefferson, 1990—; cons. dir. MDS, Inc., Toronto, Ont., Can., 1979-90. Fellow Coll. Am. Pathologists. Office: St Charles Hosp 200 Belle Terre Rd Port Jefferson NY 11777

HIRSCHOWITZ, BASIL ISAAC, physician; b. Bethal, South Africa, May 29, 1925; came to U.S., 1953, naturalized, 1961; s. Morris and Dorothy (Drieband) H.; m. Barbara L. Burns, July 6, 1958; children: David E., Karen, Edward A., Vanessa. BSc, Witwatersrand U., Johannesburg, 1943, MB, B of Surgery, 1947, MD, 1954. Intern, resident Johannesburg Gen. Hosp., 1948-50; house physician Postgrad. Med. Sch., London, Eng., 1950; registrar Central Middlesex Hosp., London, 1951-53; instr., asst. prof. U. Mich., 1953-56; asst. prof. Temple U., 1957-59; assoc. prof. medicine U. Ala. Med. Center, Birmingham, 1959-64; prof. medicine U. Ala. Med. Center, 1964—, prof. physiology, 1970—; Disting. faculty lectr. U. Ala., 1988; chmn. faculty coun. U. Ala. Sch. Medicine, 1989-90; dir. div. gastroenterology medicine U. Ala. Hosp. and Clinics, 1959-87; chmn. exec. com. U. Ala. Hosp., 1986-88. Recipient Charles F. Kettering Prize, Gen. Motors Cancer Found., 1987, Seale Harris award So. Med. Assn., 1992; named to Ala. Acad. Honor, 1991. Fellow AAAS, ACP (Laureate award 1989), Royal Coll. Physicians (Edinburgh), Royal Coll. Physicians (London), Royal Soc. Medicine (hon.), Royal Philatelic Soc., London; mem. AMA, South African, Brit., Ala. Med. Assns., Med. Rsch. Soc. Gt. Britain, Am. Fedn. Clin. Rsch., So. Soc. Clin. Investigation, Am. Physiol. Soc., Biophys. Soc., Am. Gastroent. Assn. (Friedenwald medal 1992), Am. Soc. Gastro-Intestinal Endoscopy (Schindler medal 1974), Am. Coll. Gastroenterology (Disting. Sci. Achievement award 1982), Brit. Soc. Gastro-Intestinal Endoscopy (hon.), Brit. Soc. Gastroenterology (Founders lectr. 1988), Italian Soc. Gastroenterology (corr.), William Beaumont Soc. (Eddy Palmer award for contbns. to endoscopy 1976), Soc. Exptl. Biology and Medicine, Sigma Xi, Alpha Omega Alpha. Office: U Ala Med Ctr Birmingham AL 35294

HIRSCHY, JAMES CONRAD, radiologist; b. Kalaupapa, Hawaii, July 6, 1938; s. Ira Dwight and Florence (Moeller) H.; m. Jill Spiller, Oct. 5, 1965; children: Philip, Julia, Thomas. AB, Princeton U., 1960; MD, Jefferson Med. Coll., 1964. Diplomate Am. Bd. Radiology. Intern Pa. Hosp., Phila., 1965; resident N.Y. Hosp., N.Y.C., 1965-68, asst. radiologist, 1968—; radiologist, out-patient dept. Hosp. for Spl. Surgery, N.Y.C., 1968—; ptnr., pvt. practice N.Y.C., 1968—; cons. Squibb Corp. & Union Carbide, N.Y.C., 1968-88, Exxon Corp., N.Y.C., 1968-90, N.Y. Telephone Co., N.Y.C., 1970—, Life Extension Inst., 1992—. Author: (with others) Computed Tomography of Spine, 1983; contbr. articles to profl. jours. Capt. USAR, 1965-72. Fellow N.Y. Acad. Medicine, Am. Coll. Chest Physicians; mem. N.Y. State Radiol. Soc. (alternate del.), Met. Opera Club (pres. 1990-92). Republican. Roman Catholic. Achievements include research on imaging lumbrosacral spine. Office: Hamilton-Hirschy 61 E 66th St New York NY 10021-6114

HIRSH, NORMAN BARRY, management consultant; b. N.Y.C., Apr. 20, 1935; s. Samuel Albert and Lillian Rose (Minkow) H.; m. Christina M. Poole, Sept. 21, 1957 (div. 1967); children: Richard Scott, Lisa Robin; m. Sharon Kay Girot, Dec. 29, 1973; 1 child, Sharon Margaret. BSME, Purdue U., 1956; cert. in mgmt., UCLA, 1980. Mech. engr. Ford Motor Co., Dearborn, Mich., 1956-58; design engr. Gen. Dynamics, San Diego, 1958-62; mech. engr. aircraft div. Hughes Tool Co., Culver City, Calif., 1962-65, project engr. aircraft div., 1965-69, engr. mgr. aircraft div., 1969-72; dep. program dir. Hughes Helicopters, Culver City, 1972-79, v.p., 1979-84; v.p., gen. mgr. Hughes Helicopters, Mesa, Ariz., 1984-85; exec. v.p. McDonnell Douglas Helicopter Co., Mesa, 1986-90; pres. Rogerson Hiller Corp., Port Angeles, Wash., 1990-93, Rogerson Aircraft Corp. Flight Structures Group, Port Angeles, 1990-93; cons. in field. Served with U.S. Army. Recipient Disting. Engring. Alumnus award Purdue U., 1990, Outstanding Mech. Engring. Alumnus award, 1991. Hon. fellow Am. Helicopter Soc. (chmn. 1986-87); mem. Assn. U.S. Army, Army Aviation Assn. Am., Am. Def. Preparedness Assn., Nat. Aeronautic Assn., Helicopter Assn. Internat. Achievements include development and FAA certification of the NOTAR (No Tail Rotor) concept for the Md52on helicopter; initiated design, development and FAA Certification of the MD Explorer light twin engine helicoptor.

HIRUKI, CHUJI, plant virologist, science educator; b. Fukue, Nagasaki, Japan, June 16, 1931; arrived in Can., 1966; s. Chuichi and Mitsu (Kawamuko) H.; m. Yasuko Hijikata, Dec. 26, 1961; children: Tadaaki, Lisa. BSc, Kyushu U., Fukuoka, 1954, PhD, 1963. Plant pathologist Hatano Tobacco Expt. Sta., Hatano, Japan, 1954-65; asst. prof. U. Alberta, Edmonton, Can., 1966-70, assoc. prof., 1970-76, prof., 1976-91, univ. prof., 1991—; vis. plant pathologist, U. Calif., Berkeley, 1963-64; vis. scientist INRA, Versailles, France, 1972; vis. prof. Agrl. U., Wageningen, The Netherlands, 1973; CSFP vis. prof. U. Queensland, Brisbane, Australia, 1984, vis. prof., 1984-85; chmn. Internat. Working Group Plant Viruses with Fungal Vectors, 1988-93, IUFRO Working Party on Virus and Mycoplasma Diseases, 1990—. Editor: Tree Mycoplasmas and Mycoplasma Diseases, 1988; over 150 scientific rsch. papers, 300 rsch. paper presentations. Fellow U. Wis., 1964-66, The Netherlands Internat. Ctr., 1973; recipient rsch. award Disting. Fgn. Specialist Govt. Japan, 1991, J. Gordin Kaplan award U. Alberta, 1993; named Nat. Sci. Coun. lectr. Republic China, 1989. Fellow Royal Soc. Can., Am. Phytopathological Soc. (Pacific divsn., Lifetime Achievement award 1993), Can. Phytopathological Soc. (pres. 1990-91); mem. N.Y. Acad. Scis., Phytopathological Soc. Japan (award excellence in rsch. 1990). Avocations: reading, classical music, swimming. Home: 152 Windermere Cres, Edmonton, AB Canada T6R 2H6 Office: U Alta, Dept Plant Sci, Edmonton, AB Canada T6G 2P5

HISCOCK, RICHARD CARSON, marine safety investigator; b. Washington, Dec. 18, 1944; s. Earle Francis and Alice Morgan (Carson) H.; m. Nancy Lynn Schafer, Oct. 12, 1968 (div. Jan. 1986). Student, Am. U., 1964-66. Fisherman F/V Benjo, Chatham, Mass., 1977-78; asst. harbormaster Town of Chatham (Mass.), 1977-87; exec. dir. U.S. Lifesaving Mfrs. Assn., North Chatham, Mass., 1984-86; investigator Marine Safety Cons., Fairhaven, Mass., 1987-91; pres. ERE Assocs. Ltd., North Chatham, 1991—; instr. hypothermia, cold water survival, emergency rescue equipment and fishing vessel safety, 1979-87; mem. Comml. Fishing Industry Vessel Adv. Com., 1991; mem. Cape Cod Coastal Zone Mgmt. Adv. Com., 1977-92, chmn., 1986-91; mem. Barnstable County Coastal Resources Com., 1992-93; mem., chmn. Chatham Waterways Adv. Com., 1983-87. Contbr. articles to profl. jours. Recipient Pub. Svc. Commendation, USCG, 1984. Mem. Soc. Naval Architects and Marine Engrs., U.S. Marines Safety Assn. (lic. Merchant Mariner). Achievements include drafting a bill to establish crew licensing, inspection and additional safety requirements of certain fishing industry vessels; rsch. on comml. fishing, uninspected vessel safety, fishing vessel safety and hypothermia. Home and Office: ERE Assocs Ltd 545 Old Harbor Rd North Chatham MA 02650-1134

HITCHCOCK, CHRISTOPHER BRIAN, computer and physical sciences research, development and systems executive; b. Albany, N.Y., Aug. 1, 1947; s. John Dayton and Patricia (Blake) H.; m. Kathryn Anne Fufte, Dec. 27, 1970; 1 child, Jonathan David. BS in Econs. and Acctg., St. John's U., Collegeville, Minn., 1969; MS with honors in Logistics Mgmt., USAF Inst. Tech., 1975. Cert. profl. contract mgr., logistician, profl. cost estimator/analyst. Commd. 2d lt. USAF, 1969, advanced through grades to capt., 1973; div. chief, dep. contracting, Hanscom AFB, Mass., 1977-79; sr. tech. rep. Analytical Systems Engring. Corp., Burlington, Mass., 1979-82, bus. devel. mgr., 1982-83, dir. contracts, 1983-84; sr. contracts mgr., Bolt Beranek and Newman, Inc., Cambridge, Mass., 1984-86, dir. contracts, 1986-92; dir. contracts and subcontracts The Cadmus Group, Inc., Waltham, Mass., 1992—; asst. to dep. base comdr., Hanscom, AFB, 1989—; advanced through grade to lt. col. USAFR, 1991; speaker, moderator seminars in field. Decorated Air Force Commendation medal (2). Mem. Nat. Contract Mgmt. Assn. (chpt. pres.), Soc. Logistics Engrs., Order Daedalians, Armed Forces Communications and Electronics Assn., Soc. Cost Estimating and Analysis, Assn. Old Crows. Roman Catholic. Home: 49 Thomas St Belmont MA 02178-2438 Office: 135 Beaver St Waltham MA 02154-8424

HITCHINGS, GEORGE HERBERT, retired pharmaceutical company executive, educator; b. Hoquiam, WA, Apr. 18, 1905; s. George Herbert and Lillian (Belle) H.; m. Beverly Reimer, May 30, 1933 (dec. 1985); children: Laramie Ruth (Mrs. Robert C. Brown), Thomas Eldridge; m. Joyce Shaver, Feb. 9, 1989. BS, U. Wash., 1927, MS, 1928; PhD, Harvard U., 1933; DSc, U. Mich., 1971, U. Strathcylde, 1977; DSc (hon.), N.Y. Med. Coll., Valhalla, 1981, Emory U., Atlanta, 1981, Duke U., Durham, N.C., 1982, U. N.C., Chapel Hill, 1982, Mt. Sinai Sch. Medicine, CUNY, N.Y.C., 1983, Harvard U., 1987, Med. U. of S.C., 1988; DMS (hon.), U. N.C., 1988; DSc (hon.), L.I. U., 1989; DHL (hon.), Hahnemann U., 1990; DSc (hon.), East Carolina U., 1993. Teaching fellow U. Wash., 1926-28; from teaching fellow to assoc. Harvard U., 1928-39; sr. instr. Western Res. U., 1939-42; with Burroughs Wellcome Co., Research Triangle Park, N.C., 1942-75, rsch. dir., 1955-67, v.p., 1967-75, dir., 1968-84, scientist emeritus, 1975—; prof. pharmacology Brown U., 1968-80; adj. prof. pharmacology and exptl. medicine Duke U., 1970—; adj. prof. U. N.C., 1972—; Hartung lectr., 1972; Dohme lectr., Johns Hopkins U., 1969; Michael Cross lectr. Cambridge U., Eng., 1974; George Hitchings and Gertrude Elion lectr. N.Y. Acad. Scis., 1992; Castle lectr. U. South Fla., 1992; cons. NRC, 1952-53, USPHS, 1955-60, 74-78, Am. Cancer Soc., 1963-66, Leukemia Soc. Am., 1969-73; vis. lectr. Pakistan, Iran, 1976, Japan, India, 1980, Republic South Africa, 1978, 81, Czechoslovakia, France, Pakistan, 1989, Taiwan, Korea, 1990; cons., participant in company roundtables, 1988—. Patentee in fields of chemotherapy, antimetabolites, organic chemistry of heteroycles, nucleic acids, anti tumor, antimalarial, antiviral and anti bacterial drugs. Mem. Am. Cancer Soc., ARC (bd. dirs. Durham County Chpt., 1972-77, 78-84, 85—); bd. dirs. Durham United Fund; founder Greater Triangle Community Found., pres., 1983-85; bd. dirs. Burroughs Wellcome Fund, 1968-93, pres., 1972-91; mem. rsch. and evaluation adv. com. N.C. State Dept. Corrections, 1974-76, drug devel. com. Nat. Cancer Inst., 1975-78, external adv. com. Duke Comprehensive Cancer Ctr., 1978-84; bd. dirs. Med. Found. N.C. Inc., 1984-87; friend Duke U. Libr. (life), 1987—; mem. Hitchings day and symposium Harvard U., 1992. Recipient Gairdner award, 1968; Passano award, 1969; de Villier award, 1969; Cameron prize practical therapeutics, 1972, Bertner Found. award, 1974, Royal Soc. Mullard medal 1976, Papanicolaou Cancer Soc. award, 1979, C. Chester Stock medal, 1981, Disting. Svc. award U. N.C., 1982, Oscar B. Hunter award, 1984, Alfred Burger award, 1984; Disting. Achievement award Modern Medicine, 1973, Gregor Mendel medal Czechoslovakia Acad., 1968, Purkinje medal, 1971; Medicinal Chemistry award, 1972, Ministry of Health medal, Warsaw, Poland, 1988, Inst. Lekow medal, Warsaw, 1988, Nobel prize in Medicine or Physiology, 1988; Albert Schweitzer Internat. prize for Medicine, 1989, Golden Plate award Am. Acad. Achievement, 1989, City of Medicine award Durham, N.C., 1990, Man of the Century award Hoquiam, 1990, Sagamore of the Wabash State of Ind. Gov.'s award, 1990, N.C. Disting. Chemist award, 1991, medal of Honor, Tech. U. Gdansk, 1993. Fellow AAAS, Royal Soc. Chemistry (hon., fgn.); mem. NAS, NRC (com. on growth 1952-53), Am. Acad. Arts and Scis., Am. Soc. Biol. Chemistry, Internat. Transplantation Soc., Am. Assn. for Cancer Rsch. (hon., award 1989), Soc. Exptl. Biology and Medicine, Royal Soc. Medicine (fgn. mem., dir. Found. 1987—), Sci. Rsch. Soc. (pres. 1991), Hope Valley Country Club, Phi Beta Kappa, Sigma Xi (pres. Rsch. Triangle Park Club 1991), Phi Lambda Upsilon. Achievements include patents in chemotherapy, anti-metabolites, organic chemistry of hetercycles, nucleic acids, antitumor, antimalarial, antibacterial drugs. Home: 1 Carolina Meadows Apt 102 Chapel Hill NC 27514 Office: Burroughs Wellcome Co 3030 W Cornwallis Rd Research Triangle Park NC 27709-2700

HITLIN, DAVID GEORGE, physicist, educator; b. Bklyn., Apr. 15, 1942; s. Maxwell and Martha (Lipetz) H.; m. Joan R. Abramowitz, 1966 (div. 1981); m. Abigail R. Gumbiner, Jan. 2, 1982. BA, Columbia U., 1963, MA, 1965, PhD, 1968. Instr. Columbia U., N.Y.C., 1967-69; research assoc. Stanford (Calif.) Linear Accelerator Ctr., 1969-72, asst. prof., 1975-79, mem. program com., 1980-82; asst. prof. Stanford U., 1972-75; assoc. prof. Calif. Inst. Tech., Pasadena, 1979-85, prof. physics, 1985—; mem. adv. panel U.S. Dept. Energy Univ. Programs, 1983; mem. program com. Fermi Nat. Accelerator Lab., Batavia, Ill., 1983-87, Newman Lab., Cornell U., Ithaca, N.Y., 1986-88; mem. rev. com. U. Chgo. Argonne Nat. Lab., 1985-87; chmn. Stanford Linear Accelerator Ctr. Users Orgn., 1990—; mem. program com. Brookhaven Nat. Lab., Upton, N.Y., 1992—. Contbr. numerous articles to profl. jours. Fellow Am. Phys. Soc. Rsch. in elementary particle physics. Home: 1704 Skyview Dr Altadena CA 91001-2143 Office: Calif Inst of Tech Dept Physics 356-48 Lauritsen Pasadena CA 91125

HITOMI, GEORGIA KAY, mechanical engineer; b. Fullerton, Calif., Sept. 30, 1952; d. Wing Wai and Eleanor (Fukami) Jung; m. Bruce Isao Hitomi, Feb. 10, 1974 (div. 1990). AA, L.A. Community Coll., 1976; BA, Calif. State U., Carson, 1989; MA in Speech Communication, Calif. State U., L.A., 1991. Mfg. planner Lockheed Aero. Systems Co., Burbank, Calif., 1980-90; mech. engr. Boeing Co., Everett, Wash., 1990—. Recipient Zonta scholarship, 1993. Mem. NAFE, LAPA. Republican. Avocations: walking, reading, travel. Office: Boeing Co 3003 W Casino Rd Everett WA 98203

HITTI, YOUSSEF SAMIR, biologist; b. Beirut, Feb. 28, 1954; came to U.S., 1982; s. Samir Youssef and May Khalil (Hitti) H.; m. Francoise Caroline Knaack, Nov. 29, 1986; 1 child, Sarah Elise May. BSc, Am. U. Beirut, 1978, MSc, 1981; PhD, Marquette U., 1990. Tchr. sci. Coll. Famille Libanaise, Beirut, 1976-78; teaching asst. Am. U. Beirut, 1979-81, instr., 1981-82; teaching asst. Syracuse (N.Y.) U., 1984-85, Marquette U., Milw., 1985-86; rsch. asst. U. Cin., 1989-90; rsch. assoc. biology and medicine Brown U., Providence, 1990—; adj. instr. Arabic lang. R.I. Coll., Providence, 1992—; instr. Arabic lang. Brown U. Learning Community, 1992. Contbr. sci. articles to profl. publs. including Jour. Biol. Chemistry. Media exec. dir. Coun. Lebanese Am. Orgns., Washington, 1991-92. Recipient fellowship Lebanon Coun. Sci. Rsch., 1982-84, Dr. William Scholl Found., 1988-89. Mem. AAAS, Am. Mus. Natural History, Am. Assn. Tchrs. Arabic, Sigma Xi. Office: Brown Univ Dept Bio Med Box G 69 Brown St Providence RI 02912

HITZ, C. BRECK, optics and laser scientist; b. Palm Springs, Calif., Jan. 1, 1944; s. Chester Wood and Edith (Brechbill) H.; m. Anne Evers, Nov. 16, 1985; children: Emily Anne, Chester William. BS, Pa. State U., 1966; MS, U. Mass., 1969. Rsch. engr. GTE, Mountain View, Calif., 1969-75; owner Photonetics Assn., Pacifica, Calif., 1975—; editorial dir. Laser & Applications Mag., Torrance, Calif., 1982-85; exec. dir. Laser & E-O Mfrs. Assn., Pacifica, 1987—; del. COCOM, Paris, 1988-90; secretariat OPTCON conf., 1988—. Author: (textbook) Understanding Laser Technology, 1985, rev. edit., 1991; contbr. articles to Jour. of Quantum Electronics, Optics Letter, Jour. Applied Physics. NDEA fellow U.S. Govt., 1966-69. Mem. IEEE, Internat. Standards Orgn. (chair laser com. 1990—), Optical Soc. Am., Laser Inst. Am. Achievements include first report and explanation of conflict between intracavity nonlinear optics and modelocking; creation of 24-hour short course on laser tech. Office: LEOMA 123 Kent Rd Pacifica CA 94044

HIWATASHI, KOICHI, biologist, educator; b. Sendai, Japan, Feb. 11, 1921; s. Uzaemon and Kiyoshi Hiwatashi; m. Hisako Hiwatashi, Jan. 6, 1950; 1 child, Fumiko. B of Agr., Morioka (Japan) Agrl. Coll., 1946; BS, Tohoku U., Sendai, Japan, 1949, PhD, 1955. Asst. prof. Yamagata (Japan) U., 1949-53; asst. prof. Tohoku U., Sendai, Japan, 1953-61, prof., 1962-64; prof. Miyagi Coll. of Edn., Sendai, Japan, 1965-69, Tohoku U. and Grad. Sch., Sendai, Japan, 1969-84; prof., chmn. Senshu U. of Ishinomaki, Japan, 1989—; prof. emeritus Tohoku U., Sendai, 1984—, vis. scholar Indiana U., Bloomington, 1961-62, vis. prof. East China Normal U., Shanghai, China, 1984, U. Münster, Germany, 1985, Harbin Normal U., China, 1985. Author: Fertilization, 1969, Genetics of Paramecium, 1982, Origin of Sexuality, 1986, Physiology Reproduction, 1989. Named hon. prof. East China Normal U., Shanghai, 1988. Mem. Internat. Commn. Protozool (pres. 1987-91), Soc. of Protozool (v.p. 1987-88), Internat. Congr. Protozool (pres. 1989), Japanese Soc. Protozool (exec. mem. 1968-91), Genetics Soc. Japan (hon.), Genetics Soc. Am., Zool. Soc. Japan (zool. soc. prize 1976), Japan Soc. Biomed. Gerontology. Avocations: skiing, travel, butterfly collecting, reading. Home: Dainohara 2-6-8, Sendai 981, Japan Office: Senshu U of Ishinomaki, Minami-sakai, Ishinomaki 986, Japan

HIYAMA, TETSUO, biochemistry educator; b. Tokyo, Jan. 28, 1939; s. Yoshio and Toshiko (Sugimura) H.; m. Aiko Narui, Dec. 28, 1980; children: Noriko, Junko. BS, U. Tokyo, Japan, 1962, MA, 1964, PhD, 1967. Asst. prof. U. Tokyo, Japan, 1967-68; postdoctoral fellow U. Pa., Phila., 1967-69; rsch. fellow C. F. Kettering Rsch. Lab., Yellow Springs, Ohio, 1969-71, Carnegie Instn., Stanford, Calif., 1971-74; NASA guest worker Moffet Field, NASA, Calif., 1972-74; rsch. scientist U. Calif., Berkeley, 1974-79; assoc. prof. Saitama U., Urawa, Japan, 1979-87, prof., 1987—; prof. U. Tokyo, 1991—. Mem. AAAS, Am. Soc. Photobiology, Am. Soc. Plant Physiology. Avocations: hiking, skiing, photography. Home: 2-14-6 Hongkomagome, Bunkyo-ku Tokyo, Japan Office: Saitama U, Dept Biochemistry, Urawa 338, Japan

HLUBUČEK, VRATISLAV, chemical engineer; b. Turnov, Czechoslovakia, Jan. 20, 1946; s. Vratislav Hlubuček and Anastázie (Kocourová) Hlubučková; m. Vlasta Režná, Apr. 20, 1975; children: Lucie, Petr. MSc, Tech. U., Prague, Czechoslovakia, 1968; PhD, Tech. Chem. U., Prague, Czechoslovakia, 1973. System engr. Spolana Chemopetrol Corp., Neratovice, Czechoslovakia, 1972-76; head chem. engring. dept. Chemopetrol TIÚ, Neratovice, Czechoslovakia, 1976-91; tech. dir. TIÚ, Neratovice, Czechoslovakia, 1992—; Mem. PhD examining bd. Tech. Chem. U., Prague, 1989—. Author R&D reports; contbr. papers to chem. confs. and articles to profl. jours. Mem. Town Coun. of Neratovice, 1990—. Mem. Am. Chem. Soc., Czechoslovak Soc. Chem. Engring. Achievements include development of energy retrofit and revamping in chemical industry. Home: Na Vysluni 1056, 27711 Neratovice Czechoslovakia Office: TIÚ a s, 27711 Neratovice Czech Republic

HMURCIK, LAWRENCE VINCENT, electrical engineering educator; b. Bridgeport, Conn., Aug. 9, 1952; s. Joseph Peter and Helen Barbara (Klosiewicz) H.; m. Catherine Mary Marczak, Aug. 17, 1990. BS, Fairfield U., 1974; MS, Clarkson U., 1976, PhD, 1980. Registered profl. engr., Conn. Rsch. physicist Diamond Shamrock Corp., Painesville, Ohio, 1980-83; prof. physics and engring. U. Bridgeport, 1983—; cons. in field, 1983—. Co-author: Physics for Scientists and Engineers, 1982, Instructor's Manual for Engineers, 1986; contbr. articles to profl. publs.; book reviewer for jours. in field. Grantee Cottrell Corp., 1985, State of Conn., 1986; recipient Teaching Excellence award Sears Roebuck Co., 1990. Mem. IEEE, Am. Phys. Soc., Am. Soc. Engring. Edn. Roman Catholic. Office: U Bridgeport Dept Physics Bridgeport CT 06601

HO, CHI-TANG, food chemistry educator; b. Fuzhou, Fujian, China, Dec. 26, 1944; came to U.S., 1969; s. Chia-jue and Siu (Lin) H.; m. Mary Shieh, June 29, 1974; children: Gregory, Joseph. MS, Washington U., 1971, PhD, 1974. Postdoctoral assoc. sch. chemistry Rutgers U., New Brunswick, N.J., 1975-76; postdoctoral assoc. dept. food sci. Rutgers U., New Brunswick, 1976-78, asst. prof. dept. food sci., 1978-83, assoc. prof. dept. food sci., 1983-87, prof. dept. food sci., 1987—. Co-editor: Thermal Generation of Aromas, 1989, Food Extension Science and Technology, 1991, Phenolic Compounds and Their Effects on Health I and II, 1992; contbr. over 180 articles to profl. jours. Fellow Am. Chem. Soc. (chmn. flavor subdivision 1990-91); mem. Inst. Food Technologists. Achievements include patents for rosemary anti-oxidant and butter flavor. Home: 32 Jernee Dr East Brunswick NJ 08816-5308 Office: Rutgers U Dept Food Sci New Brunswick NJ 08903

HO, CHUEN-HWEI NELSON, environmental engineer; b. Taichung, Taiwan, Republic of China, Oct. 28, 1949; came to the U.S., 1974; s. Shi-Chen and Chang-Chung (Chow) H.; m. Yulan Ellan Hse, Aug. 25, 1975; 1 child, Jennifer Chin-Lueng. MS, U. Mich., 1978, MS in Environ. Engring., 1981; PhD, Columbia Pacific U., 1992. Asst. sr. engr. Mobil R&D Corp., Princeton, N.J., 1981-82; sr. engr. Mobil Oil Corp., Paulsboro, N.J., 1982-84; environ. coord. Mobil Tech. Ctr., Pennington, N.J., 1984-85; sr. environ. engr. Mobil Chem. Co., Edison, N.J., 1985-86; mgr. safety and environ. Hatco Corp., Fords, N.J., 1986-87; sr. prin. environ. engr. Air Products & Chems. Inc., Allentown, Pa., 1987-89; dir. environ. affairs Graphic Arts Tech. Found., Pitts., 1989—; adv. bd. 3R Corp., Pitts., 1990—; pres. NEJC Consulting Corp., Wexford, Pa., 1991—. Contbr. articles to profl. jours. 2d lt. Chinese Army, 1972-74. Mem. ASCE, Am. Chem. Soc. Achievements include invention of waste minimization and pollution prevention expert systems, zero discharge waste water management system, and spill/emission and emergency reporting logic chart. Home: 2609 Fountain Hills Dr Wexford PA 15090-7815 Office: Graphic Arts Tech Found 4615 Forbes Ave Pittsburgh PA 15213-3796

HO, HIEN VAN, pediatrician; b. Hue, Thuathien, Vietnam, Aug. 1, 1947; came to U.S., 1982; s. Vinh Van Ho and Te (Thi) Truong; m. Huong Xuan Diep, Aug. 2, 1972; children: Hoa, Hieu, Hiep, Stephen Huy. MD, Saigon U., 1972. Diplomate Am. Bd. Pediatrics. Resident in pediatrics Georgetown U. Hosp., Washington, 1983-86; chief resident Arlington Hosp., Va., 1985-86; pediatrician Seven Corners Pediatrics, Falls Church, Va., 1986—; instr. Georgetown U. Hosp., 1986. First lt. Republic of Vietnam Army, 1972-75. Fellow Am. Acad. Pediatrics; mem. N.Y. Acad. Scis., Vietnamese Med. Assn. Home: 10001 Robindale Ct Great Falls VA 22066-1848 Office: Seven Corners Pediatrics 6079 Arlington Blvd Falls Church VA 22044-2707

HO, JIN-MENG, acoustician; b. Quanzhou, Fujian, China, May 10, 1963; s. Ke-Dan and Ke (Zhang) H.; m. Shelley Xu, Dec. 17, 1990. MEE, Tsinghua U., Beijing, China, 1985; PhD, Poly. U., 1990. Rsch. assoc. Poly. U., Farmingdale, N.Y., 1991-92; rsch. scientist SFA, Landover, Md., 1992—; referee The Jour. of the Acoustical Soc. of Am., 1991—. Contbr. articles to profl. jours. Mem. The Acoustical Soc. of Am. Achievements include contbn. to high-frequency techniques for modeling of the acoustic response of submerged elastic structures. Home: 5821 Cherrywood Ln #302 Greenbelt MD 20770 Office: SFA 1401 McCormick Dr Landover MD 20785

HO, JOHN WING-SHING, chemistry educator, researcher; b. Hong Kong, Sept. 10, 1954; came to U.S., 1979; s. Tak-Kam and Sam-Mui (Tong) H. BS in Biochemistry, U. Alberta, Can., 1979; MA in Chemistry, SUNY, Buffalo, 1982, PhD in Chemistry, 1985. Teaching asst. dept. chemistry SUNY, Buffalo, 1979-82, rsch. asst. dept. chemistry, 1982-85; chemistry lectr. Millard Fillmore Coll., SUNY, Buffalo, 1985; postdoctoral fellow SUNY, Buffalo, 1985-87; rsch. assoc. dept. chemistry U. Utah, Salt Lake City, 1987-88, rsch. faculty Ctr. for Human Toxicology, 1988—; vis. prof. dept. applied biology and chem. tech., Hong Kong Poly., 1992; speaker seminars and confs., 1983—. Author: Porphyrins: Separation and Analysis, 1991, Toxicology of Porphyrins, 1991; referee/reviewer Jour. Chromatography, 1990—; reviewer Jour. Chromatogratography (Biomedical Applications), 1990—; contbr. articles to profl. jours. IBR fellow Inst. Basic Rsch., 1986-88; recipient traineeship Health and Human Svcs., 1985-86, grant Dept. of Energy, 1988-91, grant NSF Small Bus., 1989. Fellow Am. Inst. Chemists; mem. AAAS, Am. Chem. Soc., N.Y. Acad. Scis., U.S. Tennis Assn.

HO, LOUIS TING, mechanical engineer; b. Shanghai, China, July 22, 1930; s. Zwei and Joan-sen (Lee) H.; m. Claudine Lee Jiang, Sept. 9, 1961; children: Charlton, Denise. MS in Physics, Cath. U., Washington, 1961, PhD in Acoustics, 1972. Rsch. assoc. U. Md., College Park, 1955-61; physicist Gen. Kinetics, Inc., Arlington, Va., 1961-63; gen. physicist Marine Engring. Lab., Annapolis, 1963-68; sr. project engr. Naval Ship Rsch. and Devel. Ctr., Annapolis, 1968-85; program mgr. David Taylor Rsch. Ctr., Annapolis, 1985-89, br. head, 1989—. Contbr. articles to profl. jours. Mem. Acoustical Soc. Am. Achievements include patents on acoustic measurements, 1972, and noise reduction device, 1980; rsch. interests include noise reduction techniques, sound propagation in fluid and structure, structural vibration, sound radiation of vibrating structure. Office: David Taylor Rsch Ctr Annapolis MD 21402

HO, YIFONG, environmental engineer; b. Taipei, Taiwan, Jan. 12, 1962; came to U.S. 1988; s. PinChi and ChiYun (Lin) H.; m. Mayling Lin, July 19, 1987. BS, Nat. Cheng-Kung U., Taiwan, 1984, MS, 1986. Cons. engr. Taiwan Elec. Power Co., Taipei, 1984-86; tchr.math. ChiZen Jr. High Sch., Taipei, 1986-88; sci. rschr. in environ. engring. Rensselaer Poly. Inst., Troy, N.Y., 1989—; cons. Niagara Mohawk Power Corp., Syracuse, N.Y., 1990-92, N.J. Natural Gas Co., Wall, 1992—. Contbr. articles to profl. jours. Lt. Taiwan Army, 1987-88. Grantee Niagara Mohawk Power Corp., 1990—, N.J. Natural Gas Co., 1992. Mem. World Tae Kwon Do Assn.., Rensselaer Poly. Inst. Badminton Club (pres. 1992-93), Sigma Xi. Achievements include research in combined biodegradation of cyanide and polycyclic aromatic hydrocarbons in manufactured gas plant sites. Office: Rensselaer Poly Inst MRC 236 110 8th St Troy NY 12180

HOAG, DAVID H., steel company executive; b. 1939; married. BA, Allegheny Coll., 1960. With LTV Steel Co., Cleve., 1960—, sales trainee, 1960-61, salesman Cin. and Chgo. dist. sales offices, 1961-68, asst. prodn. mgr. standard pipe, 1968-69, prodn. mgr. standard pipe, then asst. dist. sales mgr. Pitts. dist. sales offices, 1968-75, then prodn. mgr. hot rolled sheet, then mgr. tubular prodn. sales, then gen. sales specialty steels, 1968-75, gen. mgr. mktg., 1975-77, v.p. mktg. services, 1977-79, pres. basic steel Eastern div., 1979-82, exec. v.p., from 1982, now pres., chief operating officer, also bd. dirs.; former group v.p. parent co. LTV Corp., Dallas, now exec. v.p., bd. dirs. Office: LTV Steel Mining Co LTD 25 Prospect Ave NW Cleveland OH 44115-1018 also: LTV Corp LTV Aircraft Product Corp 9314 W Jefferson PO Box 655907 Dallas TX 75265

HOAK, JOHN CHARLES, physician, educator; b. Harrisburg, Pa., Dec. 12, 1928; s. John Andrew and Anna Bell (Holley) H.; m. Dorothy Elizabeth Witmer, Dec. 21, 1952; children: Greta Elizabeth, Laurinda Elaine, Thomas Emory. B.S., Lebanon Valley Coll., 1951; M.D., Hahnemann Med. Coll., 1955. Diplomate: Am. Bd. Internal Medicine. Intern Harrisburg Polyclinic Hosp., 1955-56; resident internal medicine VA Hosp., Iowa City, 1958; resident internal medicine U. Iowa Hosps., 1958-61, research fellow blood coagulation, 1958-59; mem. faculty U. Iowa Med. Sch., 1961-62, 63-84, assoc. prof. internal medicine, 1967-70, prof., 1970-84, dir. div. hematology, 1970-72, dir. div. hematology-oncology, 1973-84; prof. medicine, chmn. dept. U. Vt., Burlington, 1984-87, assoc. dir. Specialized Ctr. Research in Thrombosis, 1986-87; prof. dept. medicine U. Iowa, Iowa City, 1987-89; dir. div. blood diseases and resources Nat. Heart, Lung and Blood Inst., NIH, Bethesda, Md., 1989—; clin. prof. medicine Uniformed Svcs. U. of Health Scis., 1989—; cons. Walter Reed Army Hosp., 1989—; vis. research staff Sir William Dunn Sch. Pathology, Oxford (Eng.) U., 1962-63; prin. research and tng. grants Nat. Heart and Lung Inst., assoc. dir. Specialized Ctr. for Research in Atherosclerosis U. Iowa, 1970-84; research fellow, then advanced research fellow Am. Heart Assn., 1961-63. Mem. editorial bd.: Jour. Lab. and Clin. Medicine, 1972-78, Jour. Arteriosclerosis, 1986—; contbr. articles to profl. jours. Fellow ACP; mem. Am. Soc. Hematology, Internat. Soc. Thrombosis and Haemostasis, Am. Heart Assn., Am. Fedn. Clin. Research, Assn. Am. Physicians, Cen. Soc. Clin. Research, Phi Alpha Epsilon, Alpha Omega Alpha. Office: NHLBI NIH Div Blood Diseases and Resources Federal Bldg Rm 516 Bethesda MD 20892

HOBBIE, RUSSELL KLYVER, physicist; b. Albany, N.Y., Nov. 3, 1934; s. John Remington and Eulin Pomeroy (Klyver) H.; m. Cynthia Ann Borcherding, Dec. 28, 1957; children: Lynn Katherine, Erik Klyver, Sarah Elizabeth, Ann Stacey. B.S. in Physics, Mass. Inst. Tech., 1956; A.M., Harvard U., Ph.D., 1960. Research asso. U. Minn., 1960-62, mem. faculty, 1962—, prof. physics, 1972—, assoc. dean, 1983—, dir. Space Sci. Ctr., 1979—84. Author: Intermediate Physics for Medicine and Biology, 1978, 2d edit. 1987. Mem. Am. Assn. Physics Tchrs. (exec. bd. 1980-83), Am. Phys. Soc., Am. Assn. Physicists in Medicine, AAAS., IEEE. Home: 2151 Folwell Ave Saint Paul MN 55108-1306 Office: 106 Lind Hall 207 Church St SE Minneapolis MN 55455-0134

HOBBS, DAVID E., mechanical engineer. BA in Engring. Sci., Dartmouth Coll., 1963; BS in Mech. Engring., Case Inst. Tech., 1964; MS in Mech. Engring., Rensselaer Poly. Inst., 1967, PhD in Mech. Engring., 1983. With turbine component design group Pratt & Whitney, East Hartford, Conn., 1964-67, with turbine analysis and tech. devel. group, 1967-77, with compressor analysis and tech. devel. group, 1977—. Contbr. articles to profl. jours. Mem. ASME (mem. gas turbine turbomachinery com., axial compressor panel, Gas Turbine award 1990), AIAA. Office: Pratt & Whitney Mail Stop 162 07 400 Main St East Hartford CT 06108*

HOBBS, HORTON HOLCOMBE, III, biology educator; b. Gainesville, Fla., Dec. 17, 1944; s. Horton Holcombe Jr. and Georgia Cates (Blount) H.; m. Susan Claire Krantz, Oct. 12, 1967; children: Heather Renee, Horton Holcombe IV. BA, U. Richmond, 1967; MS, Miss. State U., 1969; PhD, Ind. U., 1973. Instr. Christopher Newport Coll., Newport News, Va., 1973-75; asst. prof. George Mason U., Fairfax, Va., 1975-76; prof. biology Wittenberg U., Springfield, Ohio, 1976—; com. mem. Nongame Wildlife Tech. Adv. Com., Columbus, Ohio, 1989—; trustee Island Cave Rsch. Ctr., 1987—. Author: The Crayfishes and Shrimp of Wisconsin, 1988; life scis. editor: Nat. Speleological Soc. Bull., Huntsville, Ala., 1985—; contbr. to profl. publs. Campaign co-chair County Park Dist., Springfield, 1980. Fellow Nat. Speleological Soc. (bd. govs. 1985-88), The Explorers Club; mem. Crustacean Soc. (coun. mem. 1980-83), Biol. Soc. Wash. (exec. coun. 1976-77), Cave Conservancy of the Virginias (bd. dirs. 1988—). Achievements include development of Ohio's Cave Protection Law; participation in International Speleological Expedition to Costa Rica. Office: Wittenberg U Dept Biology Springfield OH 45501

HOBBS, KEVIN DAVID, software engineer; b. Dallas, Jan. 27, 1957; s. Edward C. and Violet V. Hobbs. AB, U. Calif., Berkeley, 1980. Software engr. Tinsley Labs, Inc., Richmond, Calif., 1979—. Usher Bill Graham Presents, San Francisco, 1980—. Democrat. Office: Tinsley Labs Inc 3900 Lakeside Dr Richmond CA 94806

HOBBS, LEWIS MANKIN, astronomer; b. Upper Darby, Pa., May 16, 1937; s. Lewis Samuel and Evangeline Elizabeth (Goss) H.; m. Jo Ann Faith Hagele, June 16, 1962; children: John, Michael, Dara. B.Engring. Physics, Cornell U., 1960; M.S., U. Wis., 1962, Ph.D. in Physics, 1966. Jr. astronomer Lick Obs., U. Calif., Santa Cruz, 1965-66; mem. faculty U. Chgo., 1966—, prof. astronomy and astrophysics, 1976—; dir. Yerkes Obs. Williams Bay, Wis., 1974-82; bd. dirs. Assn. Univs. Rsch. in Astronomy, Inc., Washington, 1974-85; mem. Space Telescope Inst. Coun., 1982-87; astronomy com. of bd. trustees Univs. Rsch. Assn., Inc., Washington, 1979-83, chmn., 1979-81; bd. govs. Astrophys. Rsch. Consortium, Inc., Seattle, 1984-91; mem. Users Com. for Hubble Space Telescope, NASA, 1990—. Contbr. to profl. jours. Bd. dirs. Mil. Symphony Assn. of Walworth County, 1972-88. Alfred P. Sloan scholar, 1955-60. Mem. Am. Astron. Soc., Am. Phys. Soc., Internat. Astron. Union, Wis. Acad. Scis. Arts and Letters. Office: U Chgo Yerkes Observatory Williams Bay WI 53191-0258

HOBBS, MARVIN, engineering executive; b. Jasper, Ind., Nov. 30, 1912; s. Charles and Madge (Ott) H.; B.S. in Elec. Engring., Tri-State Coll., Angola, Ind., 1930; postgrad. U. Chgo., 1932; PhD in Elec. Engring. Greenwich U., 1990; m. Bernadine E. Weeks, July 4, 1936. Chief engr. Scott Radio Labs., Chgo., 1939-46; cons. engr. RCA, Camden, N.J., 1946-49; v.p. Harvey-Wells Electronics, Southbridge, Mass., 1952-54; asst. to exec. v.p. Gen. Instrument Corp., Newark, N.J., 1958-62; mgr., cons. engr. Design Service Co., N.Y.C., 1963-68; v.p. Gladding Corp., Syracuse, N.Y., 1968-71, cons. corporate devel., 1971-79; mem. adminstrv. group Bell Telephone Labs., Naperville, Ill., 1979-82. Mem. Electronics Prodn. Bd., ODM, Washington, 1951-52; operations analyst Far East Air Force, 1945. Recipient Certificate of Appreciation War Dept., 1945, Certificate of Commendation, Navy Dept., 1947. Registered profl. engr., Ill. Mem. IEEE (life). Author: Basics of Missile Guidance and Space Techniques, 1959, Fundamentals of Rockets, Missiles and Spacecraft, 1962; Modern Communications Switching Systems, 1974, 2d edit., 1981; Modern CB Radio Servicing, 1979; Servicing Home Video Cassette Recorders, 1982; Techniche Moderne Di Riparazione Delle Radio CB, 1982; E.H. Scott—The Dean of DX-A History of Classic Radios, 1985; Video Cameras and Camcorders, 1989; RISC/CISC Development and Test Support, 1992; Servicing Facsimile Machines, 1992. Inventor low radiation radio receiver. Home and Office: 5415 N Sheridan Rd Chicago IL 60640-1949

HOBBS, MICHAEL LANE, chemical engineer, researcher; b. Idaho Falls, Idaho, Oct. 13, 1959; s. Billie Lee and Joy Adrene (Felsted/Cotterell) H.; m. Celestine Kupski Pitcher, Mar. 28, 1981; children: Jared, Jessica, Jacob, Samuel. AAS, Ricks Coll., Rexburg, Idaho, 1978; BS, Brigham Young U., 1984, MS, 1985, PhD, 1990. Missionary LDS Ch., Sapporo, Japan, 1978-80; rsch. asst. Thermochem. Inst., Brigham Young U., Provo, Utah, 1982-84, rsch. asst. Combustion Lab., 1984-85, rsch. asst. Advanced Combustion Engring. Rsch. Ctr., 1987-90; facilities engr. Shell Oil Co., Houston, 1985-87; sr. mem. tech. staff Sandia Nat. Labs., Albuquerque, 1990—. Contbr. articles to AIChE Jour., Fuel, Shock Waves, Progress in Energy and Combustion Science. Spori scholar Ricks Coll., 1979. Mem. AIAA, Sigma Xi, Phi Kappa Phi. Achievements include successfully applied coal structure models to countercurrent fixed-bed coal gasifiers; created large product species data base to predict detonation performance of energetic materials. Office: Sandia Nat Labs PO Box 5800 Albuquerque NM 87185

HOBEL, CALVIN JOHN, obstetrician/gynecologist; b. Leigh, Nebr., Feb. 19, 1937; s. Adolph and Clara (Johns) H.; m. Marsha Lynn Van Campen, Aug. 26, 1967; children: Hillary, Marshall, Cameron. BA, U. Minn., 1959; MD, U. Nebr., 1963. Intern Harbor/UCLA Med. Ctr., Torrance, Calif., 1963, resident, 1964-68; chief obstetrics Harbor Med. Ctr. UCLA, Torrance, 1968-83; from asst. to assoc. prof. ob-gyn. UCLA, 1968-78, prof. ob-gyn., 1978—; dir. maternal fetal medicine Cedars Sinai Med. Ctr., L.A., 1983—; vis. prof. U. Western Autralia, Perth, 1980; vis. scientist U. Paris, 1980; mem. perinatal brain damage com. NIH, 1983. Contbr. articles to profl. jours. Chmn. profl. adv. com. March of Dimes, L.A., 1980-86. Giannini Found. fellow Bank of Am., 1966; NIH grantee. Fellow Am. Coll. Ob-Gyn.; mem. Am. Gynecology and Obstet. Soc., Soc. for Gynecologic Investigation, Perinatal Rsch. Soc. Republican. Achievements include development of the POPRAS perinatal record system. Home: 105 Via Alameda Palos Verdes Estates CA 90274 Office: Cedars Sinai Med Ctr 8700 Beverly Blvd Los Angeles CA 90048

HOBERG, ERIC PAUL, parasitologist; b. San Francisco, Oct. 18, 1953; s. Max Martin and Edna Marie (Bohle) H.; m. Margaret Isabel Dykes, Feb. 23, 1979. BS, U. Alaska, 1975; MSc, U. Saskatchewan, 1979; PhD, U. Wash., 1984. Instr. parasitology dept. vet. microbiology U. Sask., Saskatoon, 1978; rsch. asst. dept. pathobiology U. Wash., Seattle, 1979-84, post doctoral sr. fellow dept. pathobiology, 1984-85; rsch. assoc. vet. parasitology Coll. Vet. Medicine, Oreg. State U., Corvallis, 1985-89; asst. prof. vet. parasitology Atlantic Vet. Coll., U. Prince Edward Island, Charlottetown, 1989-90; zoologist USDA, Agrl. Rsch. Svc., Beltsville, Md., 1990—; vis. scientist Inst. Biol. Problems of the North, Acad. Scis. Russia, Labs. of Ornithology/Parasitology, 1981, 88; mem. Interacademy Exchange program Nat. Acad. Scis., Magadan, Russia, 1988. Assoc. editor for systematics: Jour. of Parasitology, 1993—; editorial bd.: Jour. of the Helminthological Soc. of Wash., 1992—; contbr. articles to Can. Jour. Zoology, Jour. Parasitology. Bd. dirs. Patuxent-Potomac chpt. Trout Unlimited, Md., 1991-92. Recipient Rsch. grant div. polar programs NSF, Antarctica, 1982, Operating grant population biology Nat. Scis. and Engring. Rsch. Coun., Can., 1990, Robert and Virginia Rausch Vis. Professorship, U. Sask., Saskatoon, 1992. Mem. AAAS, Am. Soc. Parasitologists (Henry Baldwin Ward medal 1992, coun. mem. 1993—), Soc. Systematic Biologists, Helminthological Soc. Wash. (exec. coun. 1991—). Achievements include elucidation of higher-level phylogenetic relationships of tapeworms, tapeworm systematics; application of phylogenetic inference to studies of nematodes of vet. importance; discovery of Nematodirus battus in N.Am.; formulation of the Arctic Refugium Hypothesis outlining determinants of speciation and historical biogeography among host parasite systems in the Holarctic. Office: USDA Agrl Rsch Svc BARC East # 1180 10300 Baltimore Ave Beltsville MD 20705

HOBERMAN, SHIRLEY E., speech pathologist, audiologist; b. N.Y.C., Jan. 24, 1917; d. Julius and Eva (Fleischl) Gall; m. Morton Hoberman, June 12, 1938 (dec. May 1979); children: Peter, Judith. BA, Hunter Coll., 1937; MA, Columbia U., 1958. Asst. speech pathologist St. Barnabas Hosp., N.Y.C., 1957-59; rsch. asst. Ittelson Ctr. for Child Rsch., N.Y.C., 1959-63; supr. speech pathology N.Y. State Rehab. Hosp., West Haverstraw, N.Y., 1963-71; speech pathologist Donald Reed Speech Ctr., North Tarrytown, N.Y., 1970-75; dir. speech pathology Yonkers (N.Y.) Gen. Hosp., 1973-78; supr. speech dept. Adelphi U., Garden City, N.Y., 1977-79; speech pathologist Ctrl. Suffolk Hosp., Riverhead, N.Y., 1979-80; pvt. practice speech pathology Hampton Bays, N.Y., 1975—. Contbr. articles to Jour. Mich. State Med. Jour., Jour. Am. Speech and Hearing Assn. Mem. Am. Speech and Hearing Assn. (life mem.), N.Y. State Speech and Hearing Assn. (life mem.). Home: PO Box 241 Hampton Bays NY 11946

HOBSON, GEORGE DONALD, retired geophysicist; b. Hamilton, Ont., Can., Jan. 8, 1923; s. Robert Charles and Agnes Hamilton (Mathieson) H.; m. Arletta Louise Russell, May 21, 1948; children: Robert, Linda, Douglas, Donna. BA, McMaster U., 1946, DSc (hon.), 1991; MA, Toronto U., 1948. Registered profl. geophysicist, Can. Party chief, ptnr. Heiland Exploration Can. Ltd., Calgary, Alta., 1948-55; geophysicist Can. Fina Oil Co., Calgary, 1955-56; chief geophysicist Merrill Petroleums Ltd., Calgary, 1956-57; geophysicist Pacific Petroleums Ltd., Calgary, 1957-58; chief seismic sect. Geol. Survey Can., Ottawa, Ont., 1958-69, chief geophysics div., 1969-71; dir. Polar Shelf Project, Ottawa, 1972-88, sr. advisor, 1988-90. Author or co-author over 200 articles in field. Recipient No. Sci. award and Centennial medal Dept. Indian and No. Affairs, Can., 1991, Ind. Achievement award Am. Soc. Mech. Engrs. Fellow Exploration Geophysicists India, Royal Can. Geog. Soc. (bd. govs. 1987—, Massey medal 1991), Arctic Inst. N.Am. (bd. govs. 1984-91); mem. Sci. Inst. Northwest Territory (bd. govs. 1990—), Soc. Exploration Geophysicists (v.p. 1968), Assn. Profl. Engrs., Geologists, Geophysicists Alta., Can. Soc. Exploration Geophysicists. Mem. United Ch. Can. Avocations: genealogy, barbershop singing. Address: PO Box 161, 5428 Long Island Rd, Manotick, ON Canada K4M 1A3

HOBSON, KEITH LEE, civil engineer, consultant; b. Nevada, Iowa, Feb. 16, 1958; s. Bobby Lee and Marjorie Pearl (Breeden) H.; m. Brenda Hunter, June 6, 1981; children: Leah Christine, Rebecca Marie. BSCE, Iowa State U., 1980; MSCE, U. Mo., 1983. Registered profl. engr., Kans., Mo. Design engr. Black & Veatch, Kansas City, Mo., 1980-85, project engr., 1985—. Youth counselor, program adminstr. Red Bridge United Meth. Ch., Kansas City, 1980—. Mem. ASCE (chpt. program chmn. 1985-86), Water Environ. Fedn., Kans. Water Pollution Control Assn. (editor newsletter 1991-92), Engrs. Club Kansas City (chmn. attendance com. 1986-88, mem. student assistance com.), Toastmasters (Toastmaster of Yr. Club award 1986), Chi Epsilon, Tau Beta Pi. Republican. Home: 3316 W 92d St Leawood KS 66206 Office: Black & Veatch 8400 Ward Pky Kansas City MO 64114-2031

HOBURG, JAMES FREDERICK, electrical engineering educator; b. Pitts., Dec. 30, 1946; s. William Lawrence and Virginia (Stewart) H.; m. Margaret Jean Ryan, Mar. 4, 1978. BS, Drexel U., 1969; SM, MIT, 1971, PhD in Elec. Engring., 1975. Instr. MIT, Cambridge, Mass., 1973-75; asst. prof. elec. engring. Carnegie-Mellon U., Pitts., 1975-80, assoc. prof. elec. engring., 1980-84, prof. elec. and computer engring., 1984—, assoc. head, dept. elec. and computer engring., 1985-91; cons. to rsch. and devel. orgns. Contbr. articles to profl. jours. Recipient teaching award MIT, 1972, Ryan award for Excellence in Undergrad. Edn., Carnegie-Mellon U., 1980; named Outstanding Prof. in Elec. Engring. Dept., Carnegie-Mellon U., 1977, 80, 84, 90. Mem. IEEE, Electrostatics Soc. Am., Am. Soc. Engr. Edn., Sigma Xi, Tau Beta Pi, Eta Kappa Nu. Avocations: Long distance running; walking; mountaineering. Home: 1000 Oak Creek Ln Baden PA 15005-2856 Office: Carnegie-Mellon U Dept Elec and Computer Engring Schenley Park Pittsburgh PA 15213-3830

HOCH, PEGGY MARIE, computer scientist; b. Balt., Dec. 2, 1959; d. Stanley Elijah Hoch, Jr. and Nancy Irene (Bishop) Austin; 1 child, Kiana Mariah Shurkin. AA, Catonsville (Md.) Community, Coll., 1982; BS, Towson State U., 1987; MS, Johns Hopkins U., 1989. Lab. technician McCormick & Co., Hunt Valley, Md., 1980-84; computer scientist U.S. Army Concepts Analysis, Bethesda, Md., 1985-88; sr. assoc. programmer IBM Corp., Rockville, Md., 1989-91; computer programmer Nat. Oceanic and Atmospheric Adminstrn., Silver Spring, Md., 1991—. Author: (software) Design CDRLs for IBM/FAA, 1991. Recipient Nat. Computer Sci. award U.S. Achievement Acad., 1987, Computer Sci. award Towson (Md.) State U., Chemistry award Catonsville Community Coll., 1980. Mem. AIAA, Am. Assn. Artificial Intelligence, Johns Hopkins U. Alumni Assn. Avocations: gourmet cooking, chess, reading, movies, walking. Home: 10551 Twin Rivers Rd Apt D2 Columbia MD 21044 Office: National Weather Svc 1325 E West Hwy Silver Spring MD 20910-3233

HOCHACHKA, PETER WILLIAM, biology educator; b. Therien, Alta., Can., Mar. 9, 1937; s. William and Pearl (Krainek) H.; m. Brenda Clayton, Dec. 12, 1970; children: Claire, Gail, Gareth William. B.Sc. with honors, U. Alta., Edmonton, Can., 1959; M.Sc., Dalhousie U., Halifax, N.S., Can., 1961; Ph.D., Duke U., 1965. Research asst. U. Alta., 1958-69; vis. investigator Woods Hole Oceanographic Inst., Mass., 1962; asst. prof. biology U. Toronto, Ont., Can., 1964-65; postdoctoral fellow Duke U., Durham, N.C., 1964-66; asst. prof. U. B.C., Vancouver, Can., 1966-70; assoc. prof. U. B.C., 1970-75, prof., 1975—; research scientist R-V Alpha Helix of NSF (U.S.) Amazon Expdn. and Bering Sea Expdn., 1967-68, R-V Alpha Helix Guade Lupe Expdn., 1970, Eklund Biol. Sta., McMurdo, Antarctica, 1976-77, 82-83, 93, Palmer Peninsula, 1986; sr. scientist Oceanic Inst., Hawaii, 1970-71, R-V Alpha Helix, Galapagos Expdn., 1970-71, Amazon Expdn., 1976; vis. investigator Inst. Arctic Biology, U. Alaska, 1971, Pacific Biomed. Research Ctr., U. Hawaii, 1975, Nat. Marine Fisheries, Honolulu, 1976, 81, 82, 84, 89, Plymouth Marine Lab., Eng., 1978, Concord Field Sta., Harvard U., 1984; sr. research scientist R-V Alpha Helix Hawaii (Kona Coast) Expdn., 1973; vis. investigator dept. physiology U. Hawaii, 1973, vis. investigator dept. biochemistry, 1976; vis. prof. Friday Harbor Marine Lab., U. Wash., 1975, Harvard U. Med. Sch., 1976-77; mem. R-V Alpha Helix Expdn. to Philippines, 1979; mem. Kenya lungfish program, dept. physiology and biochemistry U. Nairobi, 1979-80; vis. sr. scientist Heron Island Biol. Research Sta., 1983; vis. Q.E. sr. fellow at 27 Australian sci. instns., 1983; mem. U.S. Antarctic Rsch. Program, 1982-83, R/V Polar Duke rsch. expdn., Palmer Peninsula, Antarctica, 1986, high-altitude biochem. adaptaion program, La Raya, Peru, 1982, 87. Author: Strategies of Biochemical Adaptation, 1973, Living Without Oxygen, 1980, Biochemical Adaptation, 1984, Metabolic Arrest and the Control of Biological Time, 1987; editor: The Mollusca, Vol. 1: Metabolic Biochemistry and Molecular Biomechanics, Vol. 2: Environmnental Biochemistry and Physiology, 1983; mem. editorial bd. Molecular Physiology, Am. Jour. Physiology, Biochem. Systematics & Ecology; contbr. articles to profl. jours. Guggenheim fellow, 1977-78, Queen Elizabeth II sr. fellow, Australia, 1983; grantee NRC, 1976; recipient Gold medal for Natural Scis. Council of B.C., 1987, Killam Research prize U. B.C., Vancouver, 1987-88, 88-89, Killiam Meml. Sci. award, 1993. Fellow AAAS, Royal Soc. Can. (Flavelle medal 1990); mem. Soc. Exptl. Biology, Can. Soc. Zoologists, Am. Soc. Biol. Chemists, N.Y. Acad. Scis., Am. Physiol. Soc., Am. Soc. Zoologists, Sigma Xi. Home: 4211 Doncaster Way, Vancouver, BC Canada V6S 1W1 Office: Univ BC, Dept Zoology, Vancouver, BC Canada V6T IZ4

HOCHBERG, IRVING, audiologist, educator; b. Bklyn., Apr. 17, 1934. BA, Bklyn. Coll., 1955; MA, Tchrs. Coll., Columbia, 1957; PhD in Audiology, Pa. State U., 1962. Prof. audiology NYU, 1962-70; prof. Bklyn. coll. CUNY, 1970—, exec. officer speech and hearing sci. grad. sch., 1974—, dir. ctr. rsch. speech and hearing sci. grad. ctr., 1979—; with Danforth Assoc., 1968-71; audiology cons. Goldwater Meml. Hosp., Mt. Sinai Hosp., 1970—, VA Hosp., East Orange, N.J., Bronx, N.Y., 1977—; tng. project dir. office spl. edn. U.S. Dept. Edn., 1974—. Fellow Am. Speech and Hearing Assn., Acoustical Soc. Am., Inst. Soc. Audiology, Acad. Rehab. Audiology. Achievements include research in clinical audiology, psychoacoustic behavior of children and the aged, adaptive testing in audition. Office: City University of New York 33 W 42nd St New York NY 10036*

HOCHBERG, MARK S., cardiac surgeon; b. Providence, R.I., Nov. 26, 1947; s. Robert and Gertrude (Meth) H.; m. Faith Shapiro, June 6, 1976; children: Alyssa T., Asher R. BA, Brown U., 1969; MD, Harvard U., 1973; MD (Honoris Causa), Chongqing Sch. Med. Sci., China, 1987. Diplomate Am. Bd. Thoracic Surgery, Am. Bd. Surgery. Chief resident cardiothoracic surgery Mass. Gen. Hosp., Boston, 1980; clin. fellow in surgery Harvard Med. Sch., Boston, 1980; attending cardiac surgeon Newark Beth Israel Med. Ctr., 1981—, dir. cardiac surgery, 1988—; cons. cardiac surgeon Overlook Hosp., Summit, N.J., 1983—; asst. prof. surgery U. Medicine and Dentistry of N.J., Newark, 1981-87, assoc. prof. surgery, 1987—; chmn. grant rev. com. Am. Heart Assn., N.J. Affiliate, New Brunswick, 1986-88, bd. dirs., 1986—; bd. dirs., com. on med. affairs Corp. of Brown U., Providence, 1987—. Vice pres. Temple B'nai Jeshurun, Short Hills, 1988—. Lt. comdr. USPHS, 1975-77. Fellow ACS; mem. Soc. Thoracic Surgery, Am. Assn. Thoracic Surgery, Racquets Club, Alpha Omega Alpha. Home: 4465 Salem Ln NW Washington DC 20007 Office: Newark Beth Israel Med Ctr 201 Lyons Ave Newark NJ 07112-2094

HOCHENEGG, LEONHARD, physician; b. Innsbruck, Austria, Jan. 24, 1942; s. Hans and Annemarie (Grass) H.; m. Fatima Rendon, Nov. 4, 1976; children: Fatima, Anni, Hans, Franz Dominic, Clara Theres, Eugen. Grad., U. Innsbruck, 1967, MD, 1972. Intern U. Hosp., Innsbruck, Austria, 1967-75; asst. in Pharmacological Inst. U. Innsbruck, 1967-69; practice medicine mental hosp. U. Hosp., Hall, Austria, 1969-72; practice in psychiatric clinic St. Gallen, Switzerland, 1972; psychiatristin clin. U. Innsbruck, 1972; pvt. practician neurology and psychiatry Hall, 1980—. Editor: Heiltees, 1987, Die Kunst Nicht Krank zu Werden, 1988; contbr. articles on psychiatry to profl. jours.

HOCHGRAF, NORMAN NICOLAI, retired chemical company executive; b. N.Y.C., Jan. 9, 1931; s. Robert I. and Elsie A. (Nicolai) H.; m. Gale M. Rudine, Aug. 22, 1953; children: Scott R., Lee H. BS in Chem. Engring., Princeton U., 1952; MS in Chem. Engring., U. Del., 1954, PhD, 1956. Sect. head rsch. Exxon Rsch. & Engring., Linden, N.J., 1960-66; mgr. engring. Exxon Rsch. & Engring., Florham Park, N.J., 1966-72; v.p. mfg. Essochem. Europe, Brussels, 1972-74, v.p. chem. raw materials, 1974-78; v.p. corp. planning Exxon Chem. Co., Darien, Conn., 1978-83, v.p. polypropylene and fabricated products, 1983-86, v.p. tech. and corp. devel., 1986-90, exec. com., 1986-90; mem. adv. com. corp. rsch. Exxon Rsch. & Engring., Clinton, N.J., 1983-90, Exxon Biomed. Scis., Somerville, N.J., 1986-90. Fellow AAAS; mem. Coun. Chem. Rsch. (com. chair. 1988-92, governing bd. 1991). Achievements include use of GLC for determining non-ideal hydrocarbon equilibria; system for developing and teaching bus. strategy; application of systems dynamics analysis to rsch. and devel.; strategy and program for bus. support for K-12 sci. edn. Home: PO Box 189 Bristol ME 04539

HOCHSTER, MELVIN, mathematician, educator; b. Bklyn., Aug. 2, 1943; s. Lothar and Rose (Gruber) H.; m. Anita Klitzner, Aug. 29, 1965 (div. Feb. 1983); 1 child, Michael Adam; m. Margie Ruth Morris, Dec. 20, 1987; 1 child, Hallie Margaret Hochster Morris. B.A., Harvard U., 1964; M.A., Princeton U., 1966, Ph.D., 1967. Asst. prof. math. U. Minn., Mpls., 1967-70, assoc. prof., 1970-73; prof. math. Purdue U., West Lafayette, Ind., 1973-77; prof. math. U. Mich., Ann Arbor, 1977-84, Raymond L. Wilder prof. math., 1984—; guest prof. Math. Inst. Aarhus, Denmark, 1973-74; trustee

Math. Sci. Research Inst., Berkeley, Calif., 1985-87, mem. sci. adv. coun., 1989—; bd. govs. Inst. for Math. and its Application, Mpls., 1985-87. Chmn. editorial com. Math. Revs., 1984-89. Guggenheim fellow, 1982. Fellow Am. Acad. Arts and Scis.; mem. Am. Math. Soc. (Frank Nelson Cole prize 1980), Math. Assn. Am., Nat. Acad. Sci. Office: U Mich Math Dept Angell Hall Ann Arbor MI 48109

HOCHSTRASSER, JOHN MICHAEL, environmental engineer, industrial hygienist; b. Cin., July 19, 1938; s. Alvin Louis and Helen Augusta (Furst) H.; m. Wilma Ruth Reckman, Feb. 27, 1960; children: Ronald, Jennifer, Caroline. BSME, U. Cin., 1963, MS in Environ. Engring., 1972, PhD in Environ. Health, 1976. Registered profl. engr., Ohio, Ill., N.J.; diplomate Am. Acad. Environ. Engrs.; cert. indsl. hygienist Am. Bd. Indsl. Hygiene; registered occupational hygienist, Can.; lic. asbestos safety technician, N.J. Reliability and safety engr. GE Co., Evendale, Ohio, 1963-72; dir. environ. affairs G.D. Searle & Co., Skokie, Ill., 1975-78; dir. indsl. hygiene Tenneco Chems., Inc., Piscataway, N.J., 1978-83; project dir. Roy F. Weston, Inc., West Chester, Pa., 1983-85; dir. health and safety CH2M Hill, Inc., Parsippany, N.J., 1985-89; tech. dir. First Environ., Inc., Riverdale, N.J., 1989-92; dir. environ. health and safety Tastemaker, Cin., 1992—. Mem. ASCE, NSPE, Am. Indsl. Hygiene Assn. (bd. dirs. 1987-90, chair ethics com. 1992-93), Am. Soc. Safety Engrs., Air and Waste Mgmt. Assn., Water Pollution Control Fedn., System Safety Soc., Inst. Environ. Scis., N.Y. Acad. Scis., Soc. for Risk Analysis. Achievements include research on the use of fault tree analysis to solve environmental problems; research of short circuit flow in cyclone dust collectors; established occupational exposure limits for estrogen dusts and availability requirements for incinerators. Home: 11317 Longden Way Union KY 41091 Office: Tastemaker 110 E 70th St Cincinnati OH 45216

HOCOTT, JOE BILL, chemical engineer, educator; b. nr. Big Flat, Ark., Sept. 19, 1921; s. Jeiks Edmonds and Frances Clara (Berry) H.; B.S., U. Ark., 1945; M.S., Okla. State U., 1951. Insp. Maumelle Ordnance Works, U.S. Army Ordnance Dept., Little Rock, 1942-43; head sci. dept. Joe T. Robinson High Sch., Little Rock, 1945-46; instr. chemistry U. Tulsa, 1946-47; teaching fellow Okla. A. and M. Coll., Stillwater, 1947-49; research chem. engr. Deep Rock Petroleum Corp., Cushing, Okla., 1950, Kerr-McGee Oil Corp., Stillwater, 1951; chem. engr. cons. Joe Bill Hocott, Little Rock, 1952-55, 63—; med. technician U. Ark. Med. Center, Little Rock, 1955-56, research asso., 1956-57, instr. internal medicine, 1957-62; head chemistry dept. Little Rock Central High Sch., 1963-66; head sci. dept. Met. Vocat.-Tech. High Sch., Little Rock, 1967-73. Asst. scoutmaster Boy Scouts Am., 1945-46, troop committeeman, 1945-46, 57-58, neighborhood commr., 1969-70. Bd. dirs. Ark. Jr. Sci. and Humanities Symposium, 1965-75, asst. dir., 1972. Mem. Am. Inst. Chem. Engrs., Nat. Soc. Profl. Engrs., Ark., Ark. Jr. (dist. dir. 1966-70) acads. sci., Sigma Xi, Phi Lambda Upsilon, Unitarian. Home: 1010 Rice St Little Rock AR 72202-4536

HODES, JONATHAN EZRA, neurosurgeon, educator; b. N.Y.C., Mar. 21, 1956; s. Marion Edward and Halina Zora (Markowicz) H.; m. Janet M. Wingiman, June 18, 1989; children: Isaac Alexander, Tahlia Louise. MD, Ind. U., 1980, MS, 1982. Cert. Bd. Internal Medicine. Clin. instr. U. Calif., San Francisco, 1988-89; neurointerventional rads fellowship U. Paris VII, France, 1989-91; asst. prof. Wayne State U., Detroit, 1991-92, U. Ky., Lexington, 1992—. Recipient Chateubriand fellowship French Govt., Paris, 1989. Mem. ACS, Am. Assn. Neurol. Surgeons, Am. Coll. Internal Medicine, World Fedn. Interventional and Therapeutic Neuroradiology, Congress Neurol. Surgeons. Achievements include melding together of neurosurgical neurovascular practice with new innovative techniques in neurointerventional radiology, and gamma knife radiosurgery. Office: Univ Ky 800 Rose St Lexington KY 40536-0084

HODES, RICHARD MICHAEL, internist, educator; b. Rockville Center, N.Y., May 30, 1953; s. Elliot Jerome Hodes and Bessie Ida Flynn. BA, Middlebury Coll., 1975; MD, U. Rochester, 1982. Diplomate Am. Bd. Internal Medicine. Intern in internal medicine Balt. City Hosp., 1982-83, resident, 1983-85; resident Black Lion Hosp., Addis Ababa, Ethiopia, 1985—; med. lectr. Addis Ababa U. Contbr. numerous articles to profl. jours. Fulbright fellow, 1985. Avocations: tennis, running, swimming. Office: Dept State Addis Ababa Washington DC 20520*

HODESS, ARTHUR BART, cardiologist; b. N.Y.C., Jan. 15, 1950; s. Samuel and Dora (Rosenkrantz) H.; m. Carol Yasona, Aug. 31, 1969 (div. May 1985); children: Joshua David, Jeremy Scott; m. S. Christina Ellsworth, Dec. 23, 1987; children: Jonathan Ellsworth, Jason Dorian. BA, Boston U., 1970; MD, Columbia U., 1974. Intern Hosp. of U. Pa., Phila., 1974-75, resident in medicine, 1975-77, fellow in cardiology, 1977-79; asst. instr. dept. medicine Hosp. U. of Pa., Phila., 1974-79; instr. physiology, dept. animal biology U. Pa., Sch. Veterinary Medicine, Phila., 1977-78; clin. assoc. dept. medicine U. Pa., Phila., 1979-81; attending cardiologist Brandywine Hosp., Coatesville, Pa., 1979—; dir. intensive care Brandywine Hosp., Coatesville, 1989—, chief of cardiology, 1990—, chmn. dept. medicine, 1991—; pres. Brandywine Valley Cardiovascular Assocs., Thorndale, Pa., 1991—. Contbr. articles to profl. jours. V.p. Chestnut Hollow Homeowners Assn., West Chester, Pa., 1990—; bd. dirs. Beth Israel Congregation, Coatesville, 1991—. Fellow Clin. Coun. Cardiology Am. Heart Assn. Fellow ACP, Am. Coll. Cardiology, Am. Coll. Chest Physicians; mem. Am. Soc. Echocardiography, Phila. Acad. Cardiology, Drinker Soc. for Critical Care in Phila., Cardiac Electrophysiology Group, Soc. Critical Care Medicine. Office: Brandywine Valley Cardio 3456 Lincoln Hwy Thorndale PA 19372-1006

HODGE, BOBBY LYNN, mechanical engineering director; b. Yadkinville, N.C., Oct. 14, 1956; s. Robert Henry and Betty Jean (Martin) H.; m. Robin Mayhue Renegar, June 8, 1979; children: Robert, Adam. AAS with honors, Forsyth Tech. Inst., Winston-Salem, N.C., 1976; B. Engring. Tech., U. N.C., Charlotte, 1978. Design engr. Clark/Gravely Corp., Clemmons, N.C., 1978-79; project engr. Clark/Gravely Corp., 1979-80; design engr. Ingersoll-Rand, Davidson, N.C., 1980-83; devel. engr. Ingersoll-Rand, 1983-85; sr. applications engr. INA Bearing Co., Ft. Mill, S.C., 1985-87, mgr. automotive driveline engring. group, 1987-88, mgr. automotive applications engring., 1988-89, dir. automotive applications engring., 1989—; internat. speaker on design and application of anti-friction bearings. Contbr. articles to profl. jours.; inventor, 8 patents in field. Mem. ASME (assoc.), Soc. Automotive Engrs. (mem. manual transmission com., automatic transmission com., clutch standards com.), Soc. Tribologists and Lubrication Engrs., Raintree Country Club, Deerpark Homeowners Assn. (bd. dirs.). Republican. Baptist. Avocations: golf, hunting, woodworking. Home: 10032 Whitethorn Dr Charlotte NC 28277-9030 Office: INA Bearing Co 308 Spring Hill Farm Rd Fort Mill SC 29715-7700

HODGE, DONALD RAY, systems engineer; b. Springfield, Mo., Aug. 22, 1939; s. William Orin Jr. and Ruth Mildred (Jones) H. BS, Drury Coll., 1961; MS, U. Wis., 1963, PhD, 1968. Analyst Ctr. for Naval Analysis U. Rochester, Arlington, Va., 1968-71; staff mem. NAS, Washington, 1971; ops. rsch. analyst Dept. Army, Washington, 1971-73; sr. scientist The BDM Corp., Vienna, Va., 1973-77; sr. project engr. TRW, Fairfax, Va., 1977-88; sr. scientist Jaycor, Vienna, 1988-89; exec. staff Computer Scis. Corp., Falls Church, Va., 1989—; panel moderator Internat. Command Ctr. Facilities Interoperability, Honolulu, 1987. Contbr. to Van Nostrand's Scientific Encyclopedia, 1976. Recipient Disting. Alumni award Drury Coll., Springfield, Mo., 1980. Mem. IEEE Computer Soc., Ops. Rsch. Soc. Am., Am. Phys. Soc., Mil. Ops. Rsch. Soc. (chmn. tactical command and control workshop). Achievements include development of measure 6B8 of NATO Long Term Defense Program, Defense Transportation System Information System Technical Reference Model. Home: 2907 Farm Rd Alexandria VA 22302-2411 Office: Computer Scis Corp 3160 Fairview Park Dr Falls Church VA 22042-4501

HODGE, PHILIP GIBSON, JR., mechanical and aerospace engineering educator; b. New Haven, Nov. 9, 1920; s. Philip Gibson and Muriel (Miller) H.; m. Thea Drell, Jan. 3, 1943; children: Susan E., Philip T., Elizabeth M. A.B., Antioch Coll., 1943; Ph.D., Brown U., 1949. Research asst. Brown U., 1947-49, asso., 1949; asst. prof. math. UCLA, 1949-53; assoc. prof. applied mechanics Poly. Inst. Bklyn., 1953-56, prof., 1956-57; prof. mechanics Ill. Inst. Tech., 1953-71; prof. mechanics U. Minn., Mpls., 1971-91, prof. emeritus, 1991—; Russell Severance Springer vis. prof. U. Calif., 1976; vis. prof. Stanford U., 1993—; sec. U.S. nat. com. Theoretical and Applied Mechanics, 1982—. Author: 5 books, the most recent being Limit Analysis of Rotationally Symmetric Plates and Shells, 1963, Continuum Mechanics, 1971; also numerous rsch. articles in profl. jours.; tech. editor Jour. Applied Mechanics, 1971-76. Recipient Disting. Service award Am. Acad. Mechanics, 1984; Karman medal ASCE, 1985. NSF sr. postdoctoral fellow, 1963. Mem. NAE, ASME (hon., Worcester Reed Warner medal 1975, ASME medal 1987), Internat. Union Theoretical and Applied Mechanics (asst. treas. 1984-92). Home: 350 Sharon Park Dr # C-21 Menlo Park CA 94025 Office: Stanford U Applied Mech Div Durland Bldg Palo Alto CA 94309

HODGE, PHILIP TULLY, structural engineer; b. L.A., Mar. 2, 1950; s. Philip G. and Thea D. (Drell) H. BS, Ill. Inst. Tech., 1973; MS, U. Minn., 1986. Registered profl. engr., Ind., Ohio, Pa., Md., Va., Fla., N.Y. Engr. trainee Chgo. Bridge & Iron, 1969-72; engr. Schererville (Ind.) Steel, 1972-76; dist. engr. Engineered Structural Products, Highland, Ind., 1976-78; sr. engr. Ceco Corp., Oak Brook, Ill., 1978-84; prin. Habco, Beaver, Pa., 1984—; mem. rsch. com. Steel Joint Inst., Myrtle Beach, S.C., 1978-84. v.p. Ill. Prairie Path, Wheaton, 1981-84. Mem. NSPE, ASCE. Home: 114 McKenney Dr Beaver PA 15009 Office: Habco 114 McKenney Dr Beaver PA 15009-9354

HODGE, WINIFRED, environmental scientist, researcher; b. Washington, Dec. 26, 1950; d. Max Elwyn and Virginia (Davis) H.; m. Yannis Sakellarakis, Dec. 21, 1969 (div. 1984); children: Irene, George. BS, Cornell U., 1983; MS, Ariz. State U., 1986; postgrad., U. Ill., 1988—. Pvt. practice lang. tchr. Athens, Greece, 1976-81; soils lab. technician Cornell U., Ithaca, N.Y., 1983; hydrology technician U.S. Forest Svc., Prescott, Ariz., 1984; prin. investigator U.S. Army Constrn. Engring. Rsch. Lab., Champaign, Ill., 1985—; team leader Ecol. Modeling and Risk Assessment Team. Organizer sci. fair U.S. Army Dependent Sch., Germany, 1987; judge sci. fair Consol. Local Schs., Champaign, 1989, 91. Andrew Mellon Found. grantee, 1982. Mem. AAAS, Am. Polit. Sci. Assn., Agronomy Soc. Am., Internat. Studies Assn. (environ. sect.). Episcopalian. Achievements include development of linear programming model used by USDA Forest Service research station as decision making frame work for optimizing water yield from an experimental watershed in Arizona; co-development of resource management program for Department of Defense which covered integrated natural and cultural resources; co-establishment of integrated training area management program for Dept. of Defense installations; research in environmental management at large-scale construction projects for NATO. Office: US Army Constrn Engring Rsch Lab PO Box 9005 Champaign IL 61826-9005

HODGES, JAMES CLARK, engineer; b. Pueblo, Colo., Feb. 6, 1935; s. Marion Clark and Violet Isobel (Heaslet) H.; m. Eunice Rose Brotherton, June 23, 1961; children: Teresa Ann, Linda Jean. BS in Geology and Engring., Stanford U., 1959. Rsch. engr. SRI Internat., Menlo Park, Calif., 1958, sr. rsch. engr., 1958—. Contbr. articles to profl. publs. Achievements include ionospheric research, underground sonar and radar applied to archaeology. Office: SRI Internat Mail Code BN240 333 Ravenswood Ave Menlo Park CA 94025

HODGES, ROBERT EDGAR, physician, educator; b. Marshalltown, Iowa, July 30, 1922; s. Wayne Harold and Blanche Emma (McDowell) H.; m. Norma Lee Stempel, June 8, 1946; children: Jeannette Louise, Robert William, Karl Wayne, James Wolter. B.A., State U. Iowa, 1944, M.D., 1947, M.S. in Physiology, 1949. Diplomate: Nat. Bd. Med. Examiners, Am. Bd. Internal Medicine. Intern Meml. Hosp., Johnstown, Pa., 1947-48; fellow physiology, also obstetrics and gynecology, then resident in internal medicine State U. Iowa Hosp., 1948-52, dir. metabolic ward, 1952-71; mem. faculty State U. Iowa Med. Sch., 1952-71, prof. internal medicine, 1964-71, chmn. com. nutritional edn., adminstrn. grad. ednl. program nutrition, 1968-71; mem. liaison com. Maximum Security Hosp., Iowa City, 1966-71; prof. internal medicine, chief sect. nutrition U. Calif. Med. Sch., Davis, 1971-80, U. Nebr. Coll. Medicine, Omaha, 1980-82; prof. and dir. nutrition program, dept. family medicine, prof. dept. internal medicine U. Calif. Irvine Sch. Medicine, 1982—; mem. nutrition study sect. NIH, 1964-68; chmn. subcom. ascorbic acid and pantothenic acid ARC, 1966-68; mem. com. nutrition overview and adjustment of food on demand Nat. Acad. Scis.-NRC, 1976; cons. to hosps., other govt. agencies. Author: Nutrition in Medical Practice, 1980, also articles.; Editor: Human Nutrition, A Comprehensive Treatise, 1980; Mem. editorial bds. med. jours. Served to capt. M.C. AUS, 1943-46, 54-56. Fellow ACP; mem. AMA, Am. Heart Assn. (fellow councils atherosclerosis, epidemiology; chmn. com. nutrition 1966-68), Am. Bd. Nutrition (pres. 1973-74), Internat. Soc. Parenteral Nutrition, Am. Soc. Parenteral and Enteral Nutrition, Soc. Exptl. Biology and Medicine, Am. Fedn. Clin. Research, Am. Inst. Nutrition, Am. Soc. Clin. Nutrition (pres. 1966-67), Nutrition Soc. (London). Office: U Calif Irvine Sch Medicine Dept Internal Medicine 101 City Blvd W Orange CA 92668-2901

HODGES, VERNON WRAY, mechanical engineer; b. Roanoke, Va., Dec. 26, 1929; s. Charlie Wayne and Kathleen Mae (Williams) H.; m. Lorraine Patricia Smart, Apr. 1, 1955 (div. 1966); children: Vernon Wray Jr., Gregory Elmer, Michelle Lynn; m. Linda Lou Wall, Feb. 3, 1967; children: Kenneth Wray, Kelly Dianne. BS in Mech. Engring., Va. Poly. Inst. and State U., 1951; MS in Systems Mgmt., U. So. Calif., 1979. Registered profl. engr., Kans., Wash., Calif. Commd. 2d lt. USAF, 1951, advanced through grades to major, 1964, ret., 1965; flight test engr. Boeing Co., Wichita, Kans., 1966-71; sr. engr. Boeing Co., Seattle, 1971-76; systems test engr. Rockwell Internat., Edwards AFB, Calif., 1976-77; sr. engr. Rockwell Internat., Palmdale, Calif., 1981-90, Hughes Helicopters, Inc., Culver City, Calif., 1977-81, Computer Scis. Corp., Edwards AFB, 1990—; pvt. comml. pilot, Lancaster, Calif., 1953—; asst. prof. air sci. Boston U., 1958-61. Elder, deacon Presbyn. Ch. USA, Lancaster, 1966—; active Calif. Rep. Cen. Com., Sacramento, 1977—, Rep. Cen. Com., Washington, 1977—. Recipient Letter of Commendation, USAF. Mem. ASME, NSPE (sec. 1972-75), Air Force Assn., Masons, Shriners. Home: 2731 West Ave J-8 Lancaster CA 93536 Office: Computer Scis Corp Edwards AFB CA 93523

HODGE-SPENCER, CHERYL ANN, orthodontist; b. Dorchester, Mass., Apr. 1, 1952; d. Herbert Thomas and Edwina Catherine (Morey) Hodge; m. John Lawrence Spencer, June 10, 1978; children: Devin Thomas, Ian Nicholas. BS in Biology cum laude, Boston Coll., 1974; DMD, Tufts Sch. Dental Medicine, 1977; MPH, Harvard U. Sch. Pub. Health, 1981; Cert. in Orthodontics, Harvard Dental Sch., 1983. Orthodontist Brockton/Bridgewater, Mass., 1984—; orthodontic cons. Mass. Hosp. Sch., Canton, Mass., 1990—; vice chmn. Bd. of Investment, Bridgewater Savs. Bank, 1989-92. Lt. Dental Corps USN, 1977-80. Recipient Johnson & Johnson Dentistry award, 1977. Mem. Am. Assn. Orthodontists, Mass. Dental Soc., South Shore Dist. Dental Soc. (sec. 1990-92, peer rev. bd. 1990-92), Northeastern Soc. Orthodontists, Harvard Club Boston, Harvard Soc. Advancement Orthodontics, Metro South C. of C., Rotary (bd. dirs. charitable and ednl. fund 1989-92), Pierre Fouchard Acad. Roman Catholic. Avocations: acoustic guitar, singing, cross stitch. Office: 572 Pleasant St Brockton MA 02401-2594

HODGKIN, SIR ALAN LLOYD, biophysicist; b. Feb. 5, 1914; s. G.L. and M.F. (Wilson) H.; m. Marion de Kay Rous, 1944; 1 son, 3 daus. Student, Trinity Coll.; (fellow), Cambridge U., 1936, M.A., Sc.D.; M.D. (hon.), univs. of Berne, Louvain; D.Sc. (hon.), univs. of Sheffield, Newcastle-upon-Tyne, East Anglia, Manchester, Leicester, London, Nfld., Wales, Rockefeller U., Bristol, Oxford; LL.D., U. Aberdeen, Cambridge. Sci. officer radar Air Ministry, also Ministry Aircraft Prodn., 1939-45; lectr., then asst. dir. research Cambridge U., 1945-52; Foulerton research prof. Royal Soc., 1952-69; John Humphrey Plummer prof. biophysics U. Cambridge, 1970-81; master Trinity Coll., Cambridge, 1978-84; mem. Med. Research Council, 1959-63; chancellor U. Leicester, 1971-84. Author: Conduction of the Nervous Impulse, 1963; also sci. papers on nature of nervous conduction, muscle, and vision. Devised (with Andrew Huxley) system of math. equations describing nerve impulse; worked with giant nerve fibers of squid, proving that electricity was direct causal agt. of impulse propagation. Decorated knight Order Brit. Empire, 1972, Order of Merit, 1973; recipient Baly medal, 1955, Nobel prize for medicine or physiology (with A.F. Huxley, J.C. Eccles), 1963, Lord Crook Medal, 1983. Fellow Royal Soc. (Royal medal 1958, Copley medal 1965, pres. 1970-75), Imperial Coll. Sci., Indian Nat. Sci. Acad. (hon.) Girton Coll., Cambridge (hon.); mem. Physiol. Soc. (fgn. sec. 1960-67), Nat. Acad. Scis., Am. Acad. Arts and Scis. (fgn. hon.), Royal Danish Acad. Scis. (fgn.), Leopoldina Acad., Royal Swedish Acad. Scis. (fgn.), Am. Philos. Soc. (fgn.), Royal Irish Acad. (hon.), USSR Acad. Scis. (fgn.), Marine Biol. Assn. U.K. (pres. 1966-76). Office: Cambridge U Physiol Lab, Downing St, Cambridge CB2 3EG, England

HODGKIN, DOROTHY CROWFOOT, chemist; b. Cairo, 1910; m. Thomas L. Hodgkin, 1937 (dec. 1982). Student, Somerville Coll., Oxford, Eng., 1928-32, Cambridge U., 1932-34; ScD (hon.), U. Leeds, U. Manchester, Cambridge U., others, 1932-34; MD (hon.), Modena. Mem. faculty Oxford U., 1934-77, prof. emeritus, 1977—; chancellor Bristol U., 1988—. Determined structure of vitamin B12, cholesterol iodide, and penicillin using x-ray crystallographic analysis. Decorated Order of Merit; First Freedom of Beccles; recipient Nobel Prize in chemistry, 1964; Mikhail Lomonosov gold medal, 1982. Fellow Royal Soc. (Royal medal 1956, Copley medal 1976), Australian Acad. Sci., Akad. Leopoldina; mem. Nat. Acad. Scis., Brit. Assn. Advancement of Sci. (pres. 1977-78); fgn. mem. Royal Netherlands Acad. of Sci. and Letters, Am. Acad. Arts and Scis.; hon. fgn. mem. USSR Acad. Scis., Austrian Acad Scis., Norwegian Acad. Scis. Home: Crab Mill, Ilmington, Shipston-on-Stour Warwickshire, England Office: U Oxford, Oxford OX1 3PS, England

HODGSON, KENNETH P., mining executive, real estate investor; b. Canon City, Colo., Sept. 20, 1945; s. Cecil L. and Jaunita J. (Murrie) H.; m. Rebecca K. Thompson, Feb. 15, 1967; 1 child, Amber K.; m. 2d, Rita J. Lewis, Apr. 22, 1979. Student Metro Coll., 1966-68. With Golden Mining Corp., Utah, 1973-79, Windfall Group Inc., Utah, 1976-77; pres. Houston Mining, Ariz., 1979-82; v.p. Silver Ridge Mining, Inc., Gold Ridge Mining Inc., Ariz., 1979-82; pres. Ken Hodgson & Co., Inc., Canon City, 1983-91; Riken Resources Ltd., 1985—. Recipient numerous safety awards. Mem. AIME. Republican. Presbyterian. Lodges: Moose, Elks.

HOEBEL, BARTLEY GORE, psychology educator; b. N.Y.C., May 29, 1935; s. Edward Adamson and Frances (Gore) H.; m. Cynthia A. Eney, June 22, 1962; children—Valerie, Carolyn, Brett. AB, Harvard, 1957; PhD, U. Pa., 1962; PhD (hon.), Cath. U., Louvain, France, 1991. Mem. faculty psychology dept. Princeton, 1962—, prof, 1970—. Contbr. articles to tech. jours. and books. Fellow AAAS, Am. Psychol. Assn., Am. Psychol. Soc.; mem. Soc. Neurosci., Soc. Study Ingestive Behavior. Unitarian. Home: 207 Hartley Ave Princeton NJ 08540-5615

HOEGGER, ERHARD FRITZ, chemist, consultant; b. Baden, Switzerland, July 3, 1924; came to U.S., 1954; s. Erhard August and Emmy (Streuli) H.; m. Martha Anne Baker, Jan. 4, 1956; 1 child, Heidi Elizabeth. PhD magna cum laude, U. Basle, Switzerland, 1952; fellow, U. Basle, 1953, U. Colo., 1954-55. Asst. to rsch. dir. Givaudan Co., Geneva, Switzerland, 1955-56; from rsch. chemist to staff scientist DuPont Co., Wilmington, Del., 1957-85; v.p., dir. Cecon Group, Inc., Wilmington, 1985—; pres. Paracelsus Co., Wilmington, 1985—. Chromatography Forum Del. Valley (pres. 1990-91), Sigma Xi (chpt. pres. 1983). Achievements include patents in field. Office: Cecon Group Inc 242 N James St Wilmington DE 19804

HOEL, DAVID GERHARD, federal administrator, statistician, scientist; b. Los Angeles, Nov. 18, 1939; s. Paul Gerhard and Hazel Bessie (Helvig) H.; m. Nancy Carolyn Keller, Sept. 3, 1961; children—Erik Gerhard, Brian David, Christian Paul. A.B., U. Calif.-Berkeley, 1961; Ph.D., U. N.C. at Chapel Hill, 1966. Postdoctoral fellow Stanford U., Calif., 1966-67; sr. mathematician Westinghouse Rsch. Labs., Pitts., 1967-68; statistician Oak Ridge Nat. Lab., 1968-70; adj. prof. dept. biostats. U. N.C., Chapel Hill, 1970—; math. statistician Nat. Inst. Environ. Health Scis., Research Triangle Park, N.C., 1970-73, chief biometry br., 1973-81, acting sci. dir., 1977-79, dir. div. biometry and risk assessment, 1981-93, prof., chair Med. U. S.C., Dept. Biostat. & Epi., 1993—; mem. coun. fellows Collegium Ramazzini, 1987; vis. scientist Radiation Effects Rsch. Found., Hiroshima, Japan, 1979-80, dir., 1984-86; mem. NAS sci. bd. on toxicity and environ. health hazards, Washington, 1982-85, NAS com. on biol. effects of ionizing radiation, 1986-89, NAS com. to provide interim oversight of Dept. Energy nuclear weapons complex, 1988-90, NAS com. on environ. epidemiology, 1990, NAS com. on epidemiology and vets. affairs, 1990—, NAS com. on applied and theoretical stats., 1991—, NAS com. on the health effect of mustard gas, 1991—, NAS com. on radiol. safety in lie Marshall Islands, 1992—; chmn. NCRP sci. com. on extrapolation of risks from non-human exptl. systems to man; mem. sci. adv. bd. Nat. Ctr. for Toxicological Rsch., 1977-80. Contbr., co-contbr. articles to profl. publs. Co-editor workshop, conf. proceedings. Recipient NIH Dir. award, 1977; Mortimer Spiegelman Gold medal award Am. Pub. Health Assn., 1977; Pub. Health Service Supr. Service award USPHS, 1980; sr. Exec. Svc. Bonus Nat. Inst. Environ. Health Scis., 1983, 87-91. Fellow Am. Statis. Assn. (sec. biometrics sect. 1979); mem. NAS Inst. Medicine, Internat. Statis. Inst., Royal Statis. Soc., Biometric Soc., Soc. for Risk Analysis coun. mem. 1982-85). Home: 36 S Battery Charleston SC 29401 Office: Med Univ S.C. 171 Ashley Ave DBESS Charleston SC 29425-2503

HOELZER, GUY ANDREW, biologist; b. N.Y.C., Apr. 16, 1956; s. Hiram Howell Hoelzer and Janet Margaret Forbes; m. Michelle Boullianne, May 19, 1981 (div.); m. Mary Ann Steele, Oct. 10, 1986 (div.); 1 child, Adam Mathew. AB in Biology, Williams Coll., 1978, AB in Psychology, 1978; MS in Biology, San Jose State U., 1982; PhD in Ecology and Evolutionary Biology, U. Ariz., 1989. Rsch. asst. N.Y. ocean Sci. Lab., Montauk, N.Y., 1978-79; scuba instr. Moss Landing (Calif.) Marine Lab., 1980-82; grad. teaching asst. U. Ariz., Tucson, 1982-85, grad. teaching assoc., 1985-89, grad. rsch. assoc., 1989; postdoctoral rsch. scientist Columbia U., N.Y.C., 1989-91; asst. prof. U. Nev., Reno, 1991—. Contbr. articles to profl. jours. Recipient Outstanding Paper award Western Soc. Naturalists, 1986, 87, Best Student Paper award Am. Soc. Zoologists, 1988, Hatch Grant, Nev. Agrl. Experiment Sta., 1992, Jr. Faculty Rsch. award U. Nev., 1992. Mem. Animal Behaviour Soc. (runner-up Allee award 1988), AAAS, Molecular Biology and Evolution Soc., Internat. Soc. for Study of Behavioral Ecology. Office: Dept Biology Univ Nev Reno NV 89557

HOENIGSWALD, HENRY MAX, linguist, educator; b. Breslau, Germany, Apr. 17, 1915; s. Richard and Gertrud (Grunwald) H.; m. Gabriele Schoepflich, Dec. 26, 1944; children: Frances Gertrude, Susan Ann. Student, U. Munich, 1932-33, U. Zurich, 1933-34, U. Padua, 1934-36; D.Litt., U. Florence, 1936, Perfezionamento, 1937; L.H.D. (hon.), Swarthmore Coll., 1981, U. Pa., 1988; M.A. (hon.), U. Pa., 1971. Staff mem. Istituto Studi Etruschi, Florence, 1936-38; lectr., research asst., instr. Yale U., 1939-42, 44-45; lectr., instr. Hartford Sem. Found., 1942-43, 45-46; lectr. Hunter Coll., 1942-43, 46; lectr. charge Army specialized tng. U. Pa., Phila., 1943-44, assoc. prof., 1948-59, prof. linguistics, 1959-85, prof. emeritus, 1985—, chmn. dept. linguistics, 1963-70, co-chmn., 1978-79, mem. Ctr. for Cultural Studies, 1987—, chmn. Caldwell Prize com., 1989-91; P-4 Fgn. Service Inst., Dept. State, 1946-47; asso. prof. U. Tex., 1947-48; sr. linguist Deccan Coll., India, 1955; Fulbright lectr., Kiel, summer 1968, Oxford U., 1976-77; corp. vis. com. fgn. lits. and linguistics MIT, 1968-74; chmn. overseers com. to visit dept. linguistics Harvard U., 1978-84; vis. assoc. prof. U. Mich., 1946, 52, Princeton U., 1959-60; vis. assoc. prof. Georgetown U., 1952-53, 54, Collitz prof., 1955; vis. prof. Yale U., 1961-62, U. Mich., 1968; mem. Seminar, Columbia U., 1965—; vis. staff mem., Leuven, 1986; fellow St. John's Coll., Oxford U., 1976-77; del. Comparative Linguistics Internat. Rsch. and Exchs. Bd., 1986—; cons. Etymological Dictionary of Old High German, 1980—; Poultney lectr. Johns Hopkins U., 1991; co-promotor, Leuven, 1992. Author: Spoken Hindustani, 1946-47, Language Change and Linguistic Reconstruction, 1960, Studies in Formal Historical Linguistics, 1973; Editor: Am. Oriental Series, 1954-58, The European Background of American Linguistics, 1979, (with L. Wiener) Biological Metaphor and Cladistic Classification, 1987, (with M.R. Key) General and American Ethnolinguistics, 1989; assoc. editor Indian Jour. Linguistics, 1977—; cons. editor Jour. History of Ideas, 1978—; adv. bd. Lang. and Style, 1968—, Jour. Indo-European Studies, 1973—, Diachronica, 1984—, Lynx, 1988—, Bryn Mawr Classical Rev., 1990—; mem. editorial bd. Oxford Internat. Ency. Linguistics, 1986-91. Am. Council Learned Socs. fellow, 1942-43, 44, Guggenheim fellow, 1950-51, Newberry Library fellow, 1956, NSF and Center Advanced Study Behavioral Scis. fellow, 1962-63, Faculty fellow Modern Langs. Coll. House, 1990-91; Festschrift in his honor, 1987. Corr. fellow British Acad.; mem. AAAS, NAS, Am. Philos. Soc. (rsch. com. 1972-84, libr. com. 1984—, chmn. 1988-94, membership com. class IV 1984-90, chmn. 1987-90, exec. com. 1988—, Henry Allen Moe prize 1991), Am. Acad. Arts and Sci., N.Y. Acad. Scis., Linguistic Soc. Am. (pres. 1958), Am. Oriental Soc. (editor 1954-58, pres. 1966-67), Philol. Soc. (London), Linguistic Soc. India, Societas Linguistica Europaea, Linguistics Assn. Gt. Britain, Internat. Soc. Hist. Linguistics, Indogermanische Gesellschaft, Am. Philol. Assn., Classical Assn. U.S., Società di linguistica italiana, Henry Sweet Soc., Studienkreis Geschichte der Sprachwissenschaft, N.Am. Assn. History of Lang. Scis. Home: 908 Westdale Ave Swarthmore PA 19081-1804 Office: U Pa 618 Williams Hall Philadelphia PA 19104-6305

HOEY, DAVID JOSEPH, aerospace design engineer; b. Vallejo, Calif., Feb. 16, 1967; s. John Edward and Catherine (Douse) H.; m. Heather Ann Guthrie, Apr. 4, 1992. BS in Aero. Engring., U. Calif., Davis, 1990, MS in Aero. Engring., 1991. Student pilot USAF, Enid, Ohio, 1991-92; aero. design engr. USAF, Dayton, Ohio, 1992—. Lt. USAF, 1990—. Mem. AIAA, Aircraft Owners and Pilots Assn.

HOEY, MICHAEL DENNIS, organic chemist; b. Dewitt, N.Y., Nov. 27, 1960; s. Michael Paul and Monica Dolores (Segrue) H.; m. Kimberly Ann Johnston, Mar. 27, 1993. BA, Niagara U., 1983; PhD, Syracuse U., 1989. Sr. chemist Nalco Chem. Co., Naperville, Ill., 1989-91; rsch. chemist Akzo Chem. Co., Dobbs Ferry, N.Y., 1991—. With U.S. Army, 1978-79. Mem. Am. Chem. Soc., The Planning Forum. Republican. Roman Catholic. Achievements include patents pending. Home: 26 Oakwood Village Apt 10 Flanders NJ 07836 Office: Akzo Chems Inc 1 Livingston Ave Dobbs Ferry NY 10522-3401

HOF, PATRICK RAYMOND, neurobiologist; b. Geneva, Switzerland, Aug. 25, 1960; s. Robert and Ariane (Fluckiger) H. BA, Geneva Coll., 1979; MD, U. Geneva, 1985. Asst. Dept. Pharmacology, U. Geneva, 1985-87; postdoctoral fellow Rsch. Inst. of Scripps Clinic, San Diego, 1989; sr. rsch. assoc. Dept. Neurobiology, Mt. Sinai Sch. Medicine, N.Y.C., 1989-90, asst. prof. neurobiology, 1990—, asst. prof. geriatrics, 1991—; doctoral faculty grad. sch. CUNY, 1990—. Contbr. articles to profl. jours. Named Young Scientist award Swiss Soc. for Biol. Psychiatry, Zurich, 1991. Mem. Am. Assn. Neuropathologists, Internat. Soc. Neuropathology, Soc. for Neursci., Internat. Brain Rsch. Orgn. Office: Mt Sinai Sch Medicine Box 1065 Neurobiology One Gustave C Levy Pl New York NY 10029

HOFF, EDWIN FRANK, JR., research chemist; b. Bellville, Tex., Aug. 2, 1938; s. Edwin Frank and Eliza Otto (Bader) H.; m. Jean Estelle Collum, Apr. 14, 1956 (div. Aug. 1982); children: Edwin Frank III, Lisa Louise; m. Madolyn Earline Richardson, Mar. 12, 1988; children: Shawnee, Patrick, Scott. BS in Math. & Chemistry, Cen. State U., Edmond, Okla., 1960; PhD in Chemistry, Univ. North Tex., 1970. Rsch. chemist Black, Syvalls & Bryson Steel Co., Okla. City, 1960-61, Frito-Lay Co., Dallas, 1961-66; rsch. group head Petro-Tex. (now Syntex), Houston, 1970-81; sr. formulation chemist Kocide Chem. (now Griffin Corp.), Valdosta, Ga., 1981—. Contbr. articles to Jour. Am. Chem. Soc., Jour. Organic Chemistry, Tetrahedron Letters. Mem. Inst. Food. Technologists, Am. Chem. Soc. Achievements include 5 U.S., 1 British and 1 South African patents; U.S. and European patents pending. Office: Griffin Corp Rocky Ford Rd Valdosta GA 31601-9476

HOFF, GERALD CHARLES, transportation engineer, planner; b. Waukegan, Ill., Aug. 31, 1938; s. Carl John and Lillian Evelyn (Anderson) H.; m. Yvonne Jean Burnjas, July 29, 1961; children: Michael John, Kristen Anne. BS in Civil Engring., U. Ill., 1960; MS in Urban Traffic and Transp., Northwestern U., 1969. Registered profl. engr., Ill. Traffic rsch. engr. Chgo. Area Expressway Surveillance Project, Oak Park, Ill., 1963-71; mgr. program devel. mgmt. and sceduling unit Ill. Dept. Transp., Chgo., 1971-72, chief operational rsch. unit office rsch. & devel., 1973-74; mem. staff Gov.'s Task Force on Pub. Transp., Chgo., 1972-73; dir. program devel. capital devel. dept. Chgo. Transit Authority, 1974-81, mgr. grant programming and adminstrn., 1981-82, dir. plans and programs capital devel. dept., 1982-84; divsn. mgr. grant devel. and programming gen. devel. dept. Metra, Chgo., 1984-89, dept. head gen. devel., 1989—. Active Bd. Local Improvements, Arlington Heights, Ill., 1979-83, pres., 1983-86; mem. Planning Commn., Arlington Heights, 1986-89. 1st lt. U.S. Army, 1960-63. Fellow Inst. Transp. Engrs. (sect. pres. 1989); mem. Chi Epsilon, Sigma Xi. Home: 1619 Surrey Ridge Arlington Heights IL 60005 Office: Metra 547 W Jackson Chicago IL 60661

HOFF, JULIAN THEODORE, physician, educator; b. Boise, Idaho, Sept. 22, 1936; s. Harvey Orval and Helen Marie (Boraas) H.; m. Diane Shanks, June 3, 1962; children—Paul, Allison, Julia. B.A., Stanford U., Calif., 1958; M.D., Cornell U., N.Y.C., 1962. Diplomate Am. Bd. Neurol. Surgery (sec. 1987-91, chmn. 1991-92). Intern N.Y. Hosp., N.Y.C., 1962-63; resident in surgery N.Y. Hosp., 1963-64, resident in neurosurgery, 1966-70; Asst. prof. neurosurgery U. Calif., San Francisco, assoc. prof. neurosurgery, 1974-78, prof. neurosurgery, 1978-81; prof. neurosurgery U. Mich., Ann Arbor, 1981—; head sect. neurosurgery U. Mich., 1981—; mem. Am. Bd. Neurol-Surgery, 1986-92, chmn., 1991-92. Editor: Practice of Neurosurgery, 1979-85; Current Surgical Management of Neurological Diseases, 1980; Neurosurgery: Diagnostic and Management Principles, 1992, Mild to Moderate Head Injury, 1989; contbr. articles to profl. jours. Served to capt. US Army, 1964-66. Recipient NIH Tchr.-Investigator award, 1972-77, Javits neurosci. investigator award NIH, 1985-99; Macy Faculty scholar, London, 1979. Fellow ACS; mem. Am. Assn. Neurol. Surgeons (v.p. 1991-93, pres.-elect 1992-93, pres. 1993—), Am. Surg. Assn., Congress Neurol. Surgeons (v.p. 1982-83), Am. Acad. Neurosurgeons (treas. 1989-92, sec. 1992—), Cen. Neurosurg. Soc. (pres. 1985-86). Republican. Presbyterian. Home: 2120 Wallingford Rd Ann Arbor MI 48104-4563 Office: U Mich Hosps Kresge Med Rsch Bldg Ann Arbor MI 48109

HOFFBERG, STEVEN MARK, lawyer; b. Bklyn., Nov. 21, 1960; s. Theodore Justin and Miriam Mary (Herman) H.; m. Roberta Jacquelin Sadin, June 19, 1983; children: Sandra Lauren, Elizabeth Shari, Kathryn Victoria. BS in Applied Biology, MIT, 1980, MS in Human Nutrition and Metabolism, 1981; MEE, Rensselaer Poly. Inst., 1985; JD magna cum laude, NYU, 1989. Bar: Conn. 1989, N.Y. 1990, U.S. Ct. Appeals (fed. cir.) 1992. Engring. mgr. Biosystems, Inc., Rockfall, Conn., 1983-85; project mgr. cons. AIW, Inc., Windsor, Conn., 1986-89; atty. Fitzpatrick, Cella, Harper & Scinto, N.Y.C., 1989-91, Cohen, Pontani, Lieberman & Pavane, N.Y.C., 1991—. Contbr. articles to law rev. Mem. N.Y. Acad. Scis., ABA, N.Y. State Bar Assn., N.Y. Patent, Copyright, Trademark Law Assn., Alpha Chi Sigma. Jewish. Achievements include co-invention of advanced computer interface having adaptive pattern recognition. Office: Cohen Pontani Lieberman 551 5th Ave New York NY 10176

HOFFER, JAMES BRIAN, physicist, consultant; b. Madera, Calif., Aug. 2, 1956; s. Robert C. and Jane A. (Rylander) H.; m. Florina Bojeri, Aug. 20, 1983. BS in Physics and Math., Pacific Union Coll., 1977; MS in Physics, Mich. State. U., 1979, PhD in Physics, 1983. Vis. scientist Los Alamos (N.Mex.) Nat. Lab., 1981; instr. Mich. State U., East Lansing, 1983, rsch. assoc., nat. superconducting cyclotron lab., 1983; rsch. assoc., lab. for atmospheric and space physics U. Colo., Boulder, 1983-85; asst. scientist Applied Rsch. Corp., Landover, Md., 1985-86; pres. Hoffer and Assocs., Gaithersburg, Md., 1986—; mem. tech. adv. com. Aviation Week, 1990-91. Author: Utilizing VAX/UMS Utilities and DCL, 1989; contbr. articles to sci. jours. Appointee Consumer Affairs Adv. Com., Montgomery County, Md., 1988—; sci. fair judge, Fairfax, Va., 1989. Mem. Am. Astron. Soc., Sigma Pi Sigma. Achievements include development of technique to reduce computer time required for planetary ring model, of technique to reduce computer time required for modeling of gravitational interactions between pairs of binary stars. Home and Office: 208 Leafcup Rd Gaithersburg MD 20878-2653

HOFFLEIT, ELLEN DORRIT, astronomer; b. Florence, Ala., Mar. 12, 1907; d. Fred and Kate (Sanio) H. A.B., Radcliffe Coll., 1928, M.A., 1932, Ph.D., 1938; D.Sci. (hon.), Smith Coll., 1984. From research asst. to astronomer Harvard Coll. Obs., 1929-56; mathematician Ballistic Research Labs., Aberdeen Proving Ground, Md., 1943-48; tech. expert, 1948-62; lectr. Wellesley Coll., 1955-56; mem. faculty Yale U., 1956—, sr. research astronomer, 1974—; dir. Maria Mitchell Obs., Nantucket, Mass., 1957-78; mem. Hayden Planetarium Com., N.Y.C., 1975-90; editor Meteoritical Soc., 1958-68. Author: Some Firsts in Astronomical Photography, 1950, Yale Bright Star Catalogue, 4th edit., 1982, Astronomy at Yale, 1701-1968, 1992; also rsch. papers. Recipient Caroline Wilby prize Radcliffe Coll., 1938, Grad. Soc. medal, 1964, certificate appreciation War Dept., 1946, alumnae recognition award Radcliffe Coll., 1983, George van Biesbroeck award U. Ariz., 1988; asteroid Dorrit named in her honor, 1987. Fellow AAAS, Meteoritical Soc.; mem. Internat. Astron. Union, Am. Astron. soc. (Annenberg award 1993), Am. Geophys. Union, Astron. Soc. New Haven (hon.), Am. Assn. Variable Star Observers (hon.), Am. Def. Preparadness Assn., N.Y. Acad. Scis., Conn. Acad. Arts and Scis., Nantucket Maria Mitchell Assn. (hon.), Nantucket Hist. Soc., Yale Peabody Mus. Assocs., Astron. Soc. Pacific, Phi Beta Kappa, Sigma Xi, Harvard Club of So. Conn. Home: 255 Whitney Ave Apt 17 New Haven CT 06511-3728 Office: Yale U Obs PO Box 6666 New Haven CT 06511-8101

HOFFMAN, ALAN JEROME, mathematician, educator; b. N.Y.C., May 30, 1924; s. Jesse and Muriel (Schrager) H.; m. Esther Atkins Walker, May 30, 1947 (dec. July 1988); children: Eleanor, Elizabeth Hoffman Perry; m. Elinor Klausner Hershaft, Sept. 2, 1990. AB, Columbia U., 1947, PhD, 1950; DSc (hon.), Technion U., 1986. Mem. Inst. Advanced Study, Princeton, N.J., 1950-51; mathematician Nat. Bur. Standards, Washington, 1951-56; sci. liaison officer Office Naval Research, London, 1956-57; cons. Gen. Electric Co., N.Y.C., 1957-61; research staff mem. IBM Research Ctr., Yorktown Heights, N.Y., 1961—; fellow IBM Research Ctr., 1978—; vis. prof. Technion, Haifa, Israel, 1965, Stanford U., 1980-91, Rutgers U., 1990—; adj. prof. CUNY, 1965-76, Yale U., 1976-85; Phi Beta Kappa lectr., 1989-90. Served with U.S. Army, 1943-46, ETO, PTO. Recipient von Neumann prize Ops. Rsch. Soc. and Inst. Mgmt. Sci., 1992. Fellow N.Y. Acad. Sci., Am. Acad. Arts and Scis.; mem. NAS, Am. Math. Soc. (coun. 1982-84). Office: IBM TJ Watson Rsch Ctr PO Box 218 Yorktown Heights NY 10598-0218

HOFFMAN, ALLAN AUGUSTUS, retired urologist, consultant; b. N.Y.C., Nov. 26, 1934; s. Allan A. Hoffman and Katherine (Winifred) Mackenzie; m. Susan Wilburn, Apr. 22, 1978 (div.); children: Alexandra-Leigh Abbott, Ian Mackenzie, Allan Augustus III. BA, Princeton (N.J.) U., 1956; MD, Harvard U., 1960. Diplomate Am. Bd. of Urology. Intern Univ. Hosps., Cleve., 1960-61, asst. resident in surgery, 1961-64, resident in urology, 1964-67; physician Danville (Va.) Urologic Clinic, 1967-92; urology cons. Annie Penn Meml. Hosp., Reidsville, N.C., 1967-92, Southern VA Mental Health Inst., Danville, 1967-92; pres. med. staff Meml. Hosp. of Danville, 1985-86, bd. dirs., 1987-90; Author: (monograph) History of Dialysis, Water Resource Alternatives, 1978; contbr. articles to profl. jours. Mem. Coun. on the Environment, Commonwealth of Va., 1970-83, Commn. of Game and Inland Fisheries, 1970-83, chmn., 1973; pres. Roanoke River Basin Assn., 1984—, Friends of Roanoke River Basin, 1986—; chmn. adv. com. Sch. of Forestry and Wildlife Resources, Va. Polytech. Inst. & State U., Blacksburg, 1972-85, 87-88, 90-91; dir. Va. Mus. & Natural History Found., 1989—. Capt. U.S. Army 1962-63. Named Water Conservator of the Yr., Va. Wildlife Fedn., 1976; recipient Cert. of Appreciation, Va. Polytech. Inst. & State U., 1974, 83, 85, Silver Trout award Va. Trout Unltd., 1978, Disting. Virginian award Exch. Clubs of Va., 1982, 84. Fellow ACS; mem. AMa, Am. Urological Assn., Am. Soc. Nephrology, Internat. Soc. Nephrology, Danville-Pittsylvania (Va.) Acad. of Medicine (pres. 1982). Republican. Avocations: fishing, photography, video production, sailing. Home: 862 Main St Danville VA 24541-1808

HOFFMAN, BERNARD DOUGLAS, JR., petroleum engineer, company executive; b. Kansas City, Mo., Feb. 27, 1957; s. Bernard D. and Mary A. (Lowe) H.; m. Dianne Lee Murphy, Aug. 12, 1978; children: Emily B., Joshua D. BS in Petroleum Engring., U. Kans., 1980; postgrad., So. Meth. U., 1984-86. Engr.-in-tng. Cities Svc. Co., Wichita, Kans., 1978-79; rsch. asst. Tertiary Oil Recovery Project, Lawrence, Kans., 1978-80; dist. engr. Tex. Oil and Gas Corp., Houston, 1980-82; asst. v.p. InterFirst Bank Houston N.A., 1982-83; v.p. Bank IV Wichita, N.A., 1983-86, WMA Corp., Adama Tech Inc., Wichita, 1986-91; dir. of ops. Home Oil Co., Inc., Wichita, 1991—; bd. dirs. Prairie Homestead, Wichita. Deacon fin. Newport Bapt. Ch., Houston, 1981-83; deacon adv. cabinet 1st Bapt. Ch., Wichita, 1984—; coach Beech Aircraft Youth Baseball, Wichita, 1984-85. Mem. Soc. Petroleum Engrs. (chpt. pres. 1980), Inst. Packaging Profls., Am. Inst. Banking, Kans. Ind. Oil and Gas Assn. Republican. Office: Home Oil Co Inc 3511 N Ohio St Wichita KS 67219

HOFFMAN, CHARLES STUART, molecular geneticist; b. Saranac Lake, N.Y., Oct. 3, 1958; s. Howard and Bertha (Brown) H.; m. Linda Lepnis, June 8, 1980; children: Richard Lepnis, Alexander Louis, Catherine Anna. BS in Life Scis., MIT, 1980; PhD, Tufts U., 1986. Grad. researcher Sackler Sch. Tufts U., Boston, 1980-86; postdoctoral fellow Harvard Med. Sch., Boston, 1986-90; asst. prof. Boston Coll., Chestnut Hill, Mass., 1990—. Contbr. chpt. to book Phosphate Metabolism and Regulation in Microorganisms, articles to Gene, Genetics, Genes and Devel. Recipient First award NIH, 1991-96. Mem. AAAS, Am. Soc. for Microbiology, Genetics Soc. Am. Achievements include patent for export of intracellular substances; identification of genes encoding the adenylate cyclase activation pathway in the fission yeast Schizosaccharomyces pombe. Office: Boston Coll Dept Biology Higgins Hall 315 Chestnut Hill MA 02167

HOFFMAN, DARLEANE CHRISTIAN, chemistry educator; b. Terril, Iowa, Nov. 8, 1926; d. Carl Benjamin and Elverna (Kuhlman) Christian; m. Marvin Morrison Hoffman, Dec. 26, 1951; children: Maureane R., Daryl K. BS in Chemistry, Iowa State U., 1948, PhD in Nuclear Chemistry, 1951. Chemist Oak Ridge (Tenn.) Nat. Lab., 1952-53; mem. staff radiochemistry group Los Alamos (N.Mex.) Sci. Lab., 1953-71, assoc. leader chemistry-nuclear group, 1971-79, div. leader chem.-nuclear chem. div., 1979-82, div. leader isotope and nuclear chem. div., 1982-84; prof. chemistry U. Calif., Berkeley, 1984—; faculty sr. scientist Lawrence Berkeley (Calif.) Lab., 1984—; dir. G.T. Seaborg Inst. for Transactinium Sci., 1991—; panel leader, speaker Los Alamos Women in Sci., 1975, 79, 82; mem. subcom. on nuclear and radiochemistry NAS-NRC, 1978-81, chmn. subcom. on nuclear and radiochemistry, 1982-84; (hon.) mem. commn. on radiochem. and nuclear techniques Internat. Union of Pure and Applied Chem., 1983-87, chmn., 1987-91, assoc. mem. 1992—; mem. com. 2d Internat. Symposium on Nuclear and Radiochemistry, 1988; planning panel Workshop on Tng. Requirements for Chemists in Nuclear Medicine, Nuclear Industry, and Related Fields, 1988, radionuclide migration peer rev. com., Las Vegas, 1986-87, steering com. Advanced Steady State Neutron Source, 1986-90, steering com., panelist Workshop on Opportunities and Challenges in Research with Transplutonium Elements, Washington, 1983; mem. energy rsch. adv. bd. cold fusion panel, Dept. Edn., 1989-90; mem. NAS separations subpanel of separations tech. and transmutation systems panel, 1992-93. Contbr. numerous articles in field to profl. jours. Recipient Alumni Citation of Merit Coll. Scis. and Humanities, Iowa State U., 1978, Disting. Achievement award Iowa State U., 1986; fellow NSF, 1964-65, Guggenheim Found., 1978-79. Fellow Am. Inst. Chemists (pres. N.Mex. chpt. 1976-78), Am. Phys. Soc., AAAS; mem: Am. Chem. Soc. (chmn. nuclear chemistry and technology div. 1978-79, com. in sci. 1986-88, exec. com. div. nuclear chem. and tech. 1987-90, John Dustin Clark award Cen. N.Mex. sect. 1976, Nuclear Chemistry award 1983, Francis P. Garvan-John M. Olin medal 1991), Am. Nuclear Soc. (co-chmn. internat. conf. Methods and Applications of Radioanalytical Chemistry 1987), Norwegian Acad. Arts and Scis, Sigma Xi, Phi Kappa Phi, Iota Sigma Pi, Pi Mu Epsilon, Sigma Delta Epsilon, Alpha Chi Sigma. Methodist. Home: 2277 Manzanita Dr Oakland CA 94611 Office: Lawrence Berkeley Lab MS70A-3307 NSD Berkeley CA 94720

HOFFMAN, DAVID JOHN, physiologist; b. New London, Conn., Sept. 22, 1944; s. John Leslie and Margaret Amy (Stokes) H.; m. Suzanne Elizabeth O'Clair, Aug. 20, 1966; children: Michael David, James Stephen. BS, McGill U., 1966; PhD, U. Md., 1971. Instr. in genetics, embryology U. Md., College Park, 1968-71; postdoctoral fellow/NIH Oak Ridge Nat. Lab., Oak Ridge, Tenn., 1971-73; faculty, biology dept. Boston Coll., Newton, Mass., 1973-74; sr. staff physiologist Health Effects Rsch. Lab/U.S. EPA, Cin., 1974-76; rsch. physiologist Patuxent Wildlife Rsch. Ctr./USDI, Laurel, Md., 1976—. Mem. editorial bd. Archives of Environ. Contamination and Toxicology Jour., 1986—, Jour. Toxicology and Environ. Health, 1989—, Environ. Toxicology and Chemistry, 1990—; editor: Handbook of Ecotoxicology, 1994; contbr. chpts. to books, articles to profl. jours. Recipient dissertation fellowship U. Md., College Park, 1970, spl. achievement award USDI, 1990. Mem. Teratology Soc., Soc. Environ. Chemistry and Toxicology (editoral bd. 1990—), Soc. Exptl. Biology and Medicine, Soc. Toxicology, AAAS, Phi Sigma Soc. Avocations: distance swimming, adult fitness swimming, fishing, boating, birdwatching. Home: 1679 Justin Dr Gambrills MD 21054 Office: Patuxent Wildlife Rsch Ctr USDI Laurel MD 20708

HOFFMAN, DONALD DAVID, cognitive and computer science educator; b. San Antonio, Dec. 29, 1955; s. David Pollock and Loretta Virginia (Shoemaker) H.; m. Geralyn Mary Souza, Dec. 13, 1986; 1 child from previous marriage, Melissa Louise. BA, UCLA, 1978; PhD, MIT, 1983. MTS and project engr. Hughes Aircraft Co., El Segundo, Calif., 1978-83; rsch. scientist MIT Artificial Intelligence Lab, Cambridge, Mass., 1983; asst. prof. U. Calif., Irvine, 1983-86, assoc. prof., 1986-90, full prof., 1990—; cons. Fairchild Lab. for Artificial Intelligence, Palo Alto, Calif., 1984; panelist MIT Corp. vis. com., Cambridge, 1985, NSF, Washington, 1988; conf. host IEEE Conf. on Visual Motion, Irvine, 1989; conf. host Office of Naval Rsch. Conf. on Vision, Laguna Beach, Calif., 1992. Co-author: Observer Mechanics, 1989; contbr. articles to profl. jours. Vol. tchr. Turtle Rock Elem. Sch., Irvine, 1988-90. Recipient Distinguished Scientific award, Am. Psychol. Assn., 1989; grantee NSF, 1984, 87. Mem. AAAS. Avocations: running, swimming, racket sports, ice skating. Office: U Calif Dept Cognitive Sci Irvine CA 92717

HOFFMAN, JERRY CARL, civil engineer; b. Madisonville, Ky., Apr. 15, 1943; s. Carl J. and Lily Pearl (Niswonger) H.; m. Ann Marie Curren, Feb. 12, 1966; children: Mark A., Laura L., Christopher C., Allison N. BS, Western Ky. U., 1972. Registered profl. engr., Ky., Mo., Iowa, Calif., Utah, Okla., Ala. Engr. Burns & McDonnell Engring. Co., Kansas City, Mo., 1972-75, dept. mgr., 1975-78, project mgr., 1978-88; project mgr., dir. Burns & McDonnell Waste Cons., Overland Park, Kans., 1988—; lectr. in field; expert witness on hazardous materials mgmt. issues. Contbr. articles to profl. jours. Mem. air quality forum Mid-Am. Regional Coun., Kansas City, Mo., 1988—; mem. Bd. of Aldermen, Lee's Summit, Mo., 1986-88. With USAF, 1963-67. Mem. ASTM, ASCE, NSPE, Air and Waste Mgmt. Assn., Environ. Audit Roundtable, C. of C. of Kansas City (chmn. environ. com. 1988-90), KC. Roman Catholic. Office: Burns & McDonnell Waste Co 10881 Lowell Overland Park KS 66210

HOFFMAN, KARLA LEIGH, mathematician; b. Paterson, N.J., Feb. 14, 1948; d. Abe and Bertha (Guthaim) Rakoff; BA, Rutgers, U., 1969; MBA, George Washington U., 1971, DSc in Ops. Research, 1975; m. Allan Stuart Hoffman, Dec. 26, 1971; 1 son, Matthew Douglas. Ops. research analyst IRS, Washington, 1970-72; research asst. George Washington U., 1972-75; asso. professorial lectr., 1978-85; NSF postdoctoral research fellow Nat. Acad. Sci., Washington, 1975-76; mathematician Nat. Bur. Standards, Washington, 1976-84; vis. assoc. prof. ops. research U. Md., spring 1982; assoc. prof. systems engring. dept. George Mason U., 1985-86, assoc. prof. ops. research and applied stats., 1986-89, prof. ops. research, 1990—, disting. prof. 1989; mng. ptnr. Optimization Software Assocs.; cons. to govt. agys. Recipient Applied Research award Nat. Bur. Standards, 1984, Silver medal U.S. Dept. Commerce, 1984. Mem. Ops. Research Soc. Am. (sec.-treas. computer sci. tech. sect. 1979-80, vice chmn. sect. 1981, chmn. sect. 1982, vis. professorial lectr. 1980—, chmn. tech. sect. com. 1983-86, council 1985-88, chmn. Lanchster Prize com. 1989, treas. 1993—), Math. Programming Soc. (editor newsletter 1979-82, chmn. com. algorithms 1982-85, council 1985-88, exec. com. 1986-88, chmn. mem. com. 1988-89). Contbr. articles to profl. jours.; assoc. editor Internat. Abstracts of Ops. Research, The Math. Programming Jour., Series B, The Ops. Research Soc. Jour. on Computing, Jour. Computational Optimization and Applications. Home: 6921 Clifton Rd Clifton VA 22024-1525

HOFFMAN, KEVIN WILLIAM, aerospace engineer; b. Edmonton, Alberta, Can., Mar. 16, 1961; came to U.S., 1984; s. John and Dolores Janet Edith (Ziegler) H.; m. Kelly Sharon McKinney, Dec. 12, 1987; children: Ashle Alise, Lindse Leana. Degree in aero./mech. engring. tech., So. Alberta Inst. Tech., 1984; BS in Aero. Engring., U. Ala., 1987. Grad. rsch. asst. Univ. Ala. Flight Dynamics, Tuscaloosa, 1986-87; flight test engr. Gulfstream Aerospace Corp., Savannah, Ga., 1987-88; aircraft designer-product devel. engr. Gulfstream Aerospace Corp., Savannah, 1988-90; group head aircraft design Canadair Aerospace- Bombardier, Inc, Montreal, Que., 1990—; guest lectr. aircraft design, Stanford (Calif.) U., Ga. Tech. U., U. Ala., others, 1987—; prof. aircraft design Ecole Poly., Montreal, 1991—; adv. bd. Aero. Engring. Edn., So. Alberta Inst. Tech., Calgary, Alta., Can., 1992—. Author (course notes) Applied Aerodynamic Techniques in Aircraft Design, 1987, Conceptual Aircraft Design Methods, 1990. Mem. AIAA (sr., dir. programs, Membership Booster award 1989, Outstanding Achievemnt award 1990), Nat. Estimating Soc. (bd. dirs. Savannah, Ga. 1989-90), Ala. Capstone Engring. Soc., U. Ala. Alumni Assn. Republican. Achievements include coordination and conceptual aircraft design on Canadair Global Express Project and next model regional jet; contbrs. to initial designs of supersonic bus. jet at Gulfstream Aerospace. Home: 4854 Perron St, Pierrefonds, PQ Canada H9A 3E5 Office: Canadair Aerospace, 1800 Laurentian Blvd, Montreal, PQ Canada H3C 3G9

HOFFMAN, PAUL FELIX, geologist, educator; b. Toronto, Ont., Can., Mar. 21, 1941; s. Samuel and Dorothy Grace (Medhurst) H.; m. Erica Jean Westbrook, Dec. 4, 1976; 1 child, Guy Samson. BS, McMaster U., 1964; MA, Johns Hopkins U., 1965, PhD, 1970. Lectr. Franklin & Marshall Coll., Lancaster, Pa., 1968-69; rsch. scientist Geol. Survey Can., Ottawa, Ont., 1969-92; lectr. U. Calif., Santa Barbara, 1971-72; prof. U. Victoria, B.C., Can., 1992—; lectr. U. Calif., Santa Barbara, 1971-72; dist. lectr. Am. Assn. Petroleum Geologists, 1979-80; vis. prof. U. Tex., Dallas, 1978, Columbia U., 1990; adj. prof. Carleton U., 1989-92; mem. Internat. Union Geol. Scis. Commn. on Precambrian Stratigraphy, 1976, Internat. Commn. Lithosphere Working Group on Mobile Belts, 1986-90. Fairchild Found. vis. scholar Calif. Inst. Tech., 1974-75; recipient Bownocker medal Ohio State U., 1989. Fellow Royal Soc. Can., Geol. Assn. Can. (past pres.' medal 1976, Logan medal 1992), Geol. Soc. Am.; mem. Am. Geophys. Union, Can. Soc. Petroleum Geologists (R.J.W. Douglas Meml. medal 1991), Nat. Acad. Sci. U.S. (fgn. assoc.). Home: 3018 Blackwood St, Victoria, BC Canada V8T 3X4 Office: U Victoria, Sch Earth and Ocean Scis, PO Box 1700, Victoria, BC Canada V8W 2Y2

HOFFMAN, PAUL JEROME, psychologist; b. San Francisco, June 25, 1923; s. Louie and Bessie (Brodofsky) H.; m. Elaine Stroll, Mar. 18, 1944; children: Valerie, Jonathan, Elisabeth. BA in Exptl. Psychology, Stanford U., 1949, PhD in Psychology and Statistics, 1954. Lic. pscyhologist, Oreg., Calif. Asst. prof. Wash. State U., Pullman, 1953-57; asst. prof. U. Oreg., Eugene, 1957-60, adj. prof., 1967-76; pres. Paul J. Hoffman Psychometrics, inc., Los Altos, Calif., 1978-83, Magic 7 Software Co., Los Altos, Calif., 1984—; pres. and founder Oreg. Rsch. Inst., Eugene, 1960-77; cons. Nat. Bd. Med. Examiners, Phila., 1971, Am. Assn. State Psychol. Bds. nat. examination com., N.Y.C., 1972-78, NIH, Washington, 1980, Internat. Bus. Machines Corp., Armonk, N.J., 1982, Hewlett Packard, Palo Alto, Calif., 1983. Author: (with others) Decision Processes, 1954, Formal Representation of Human Judgement, 1968, Computer Aided Decision Analysis, 1993; contbr. 53 articles to profl. jours. Chair fgn. policy Dem. Ctrl. Com., Oreg., 1960-72; advisor Sen. Wayne Morse, Oreg., 1964-70; chmn. Bob Straub for Gov. Com., Oreg., 1974. Lt. USAF, 1942-46. Grantee NIH, 1958-72, NSF, 1961-63. Fellow AAAS, Am. Psychol. Assn., Psychonomic Soc., Psychometric Soc., Human Factors Soc.; mem. Oreg. Psychol. Assn. (pres. 1962-63), Oreg. Inventor's Coun.. Achievements include copyrights for expert systems software, consensus building software. Home: 1120 Royal Ln San Carlos CA 94070

HOFFMAN, RICHARD GEORGE, psychologist; b. Benton Harbor, Mich., Oct. 6, 1949; s. Robert Fredrick and Kathleen Elyce (Watts) H.; m. Julia Ann May, Dec. 18, 1970; children: Leslie Margaret, Michael Charles, Angela Lynn, Jennifer Elizabeth. BS with honors, Mich. State U., 1971; MA in Psychology, Long Island U., 1974, PhD in Clin. Psychology, 1980. Lic. con. psychologst. Instr. pediatrics U. Va., Charlottesville, 1977-80; asst. prof. pediatrics and family med. U. Kans., Wichita, 1980-84; asst. prof. behavioral sci. U. Minn., Duluth, 1984-90, assoc. prof. behavioral sci., 1990—, dir.

neuropsychology lab., 1986—, co-dir. hypothermia and water safety lab., 1987—, co-dir. neurobehavioral toxicology lab., 1990—; assoc. dir. Child Evaluation Ctr., Wichita, 1981-82; dir. adminstrn. Comprehensive Epilepsy Clinic, Wichita, 1983-84; cons. psychologist U. Assocs., P.A., Duluth, 1984—. contbr. articles to profl. jour. Pres. Home and Sch. Assn., St. Michael's Sch., Duluth, 1986. Rsch. grantee NIH, 1985, USCG, 1986, Sch. Medicine U. Kans., 1984, U. Minn., 1986, U.S. Army Med. Rsch. Command, 1988—, U.S. Naval Med. Rsch. Command, 1988, Great Lakes Protection Fund, 1991—. Fellow Am. Psychol. Soc.; mem. APA, Nat. Acad. Neuropsychologists, Soc. Behavioral Medicine, N.Y. Acad. Scis., KC. Democrat. Roman Catholic. Avocations: bicycling, hiking. Home: 219 Occidental Blvd Duluth MN 55804-1365 Office: U Minn Dept Behavioral Scis Duluth MN 55812

HOFFMAN, ROBERT S., federal agency administrator; b. Evanston, Ill., Mar. 2, 1929; m. Sally Monson Hoffman; 4 children. Student, U. Ill., Moline, 1946-47, U. Mont., 1947-48; BS in Zoology, Utah State U., 1950, DSc (hon.), 1988; MA, U. Calif., Berkeley, 1954, PhD, 1955. Instr. dept. zoology U. Mont., Missoula, 1955-57, from asst. prof. to prof., 1957-1968, curator zoological mus., 1965-68; curator mammals mus. nat. history U. Kans., Lawrence, 1968-86, prof. dept. zoology, 1968-69, prof., chmn. dept. systematics and ecology, 1969-72, prof., 1972-86, acting chmn. divsn. biological scis., 1976-77, assoc. dean coll. liberal arts and scis., 1978-80, 81-82, acting dean, 1980-81; dir. nat. mus. natural history Smithsonian Inst., Washington, 1986-87, asst. sec. rsch., 1988—; rsch. biologist Mont. Fish and Game Dept., 1956; vis. asst. prof. U. B.C., Vancouver, Can., 1960; mem. area com. Soviet and East European Studies, U. Kans., 1983-86; cons. Pakistan Agrl. U., Lyallpur, 1971—, Inst. Artic and Alpine Rsch. U. Colo., Bolder, 1980—, U.S. Atomic Energy Commn., 1966, NIMH, 1965, Boeing Co. Seattle, 1961-62, Quaternary Rsch. Inst., U. Wash., 1981-86, Nat. Adv. Com., Ctr. Internat. Studies, U. Kans., 1986—; mem. various com. U.S. NAS; invited speaker at various symposiums. Bd. editors Acta Zoologica Sinica, 1990—; contbr. over 200 articles to profl. jours. Fellow NSF 1952-53, 54-55, NAS 1963-64. Fellow AAAS; mem. Am. Assn. Mus., Am. Assn. for Quaternary Rsch., Am. Soc. Mamalogists (bd. dirs. 1965—, first v.p. 1973-78, pres. 1978-80, chmn. internat. rels. 1964-68, 72-78), Ecol. Soc. Am., Soc. Study Evolution, Soc. Systematic Zoology, Internat. Coun. Mus., Internat. Union Quaternary Rsch., Internat. Assn. Ecology, Internat. Mountain Soc., Brit. Mammal Soc., All-Union Theriological Soc. (hon.), Sigma Xi, Phi Sigma, Phi Kappa Phi. Office: Smithsonian Institution Sciences 1000 Jefferson Dr SW Washington DC 20560*

HOFFMANN, GÜNTER GEORG, chemist; b. Oberhausen, Germany, July 21, 1954; s. Adolf and Maria (Hitschfel) H.; m. Heike Hoffmann, Mar. 23, 1978; children: Marcel Oliver, Vanessa Ina, David Gerrit. Diplom Chemie, Ruhruniversitaet, 1978, Dr.rer.nat., 1983. Wissenschaftl. hilfskraft Ruhruniversitaet, Bochum, 1978-82, wissenschaftl. mitarbeiter, 1982-83; postdoctoral fellow Universitaet - GHS, Essen, 1983-85, Max-Planck-Institut fuer Strahlenchemie, Muelheim, 1985-86; wissenschaftlicher mitarbeiter Ruhruniversitaet, Bochum, 1986-87; researcher in chemistry Universitaet - GHS, Essen, 1988—; Exerptor Beilstein-Institut, Frankfurt, 1978-89. Contbr. numerous articles to profl. jours. Recipient Bennigsen-Foerder award Ministerin fuer Wissenschaft und Forschung des Landes NRW, 1989. Mem. Am. Chem. Soc., Soc. for Applied Spectroscopy, Gesellschaft Deutscher Chemiker, DASp, Coblentz Soc. Avocations: biology, physics, astronomy, medicine, music, electronics. Home: Wachstrasse 29, D-46045 Oberhausen Germany Office: Universitaet GHS, FB 8 Postfach 10 37 64, D-45117 Essen 1, Germany

HOFFMANN, GUNTHER F., forest products executive; b. Dortmund, Germany, Aug. 8, 1938; came to U.S., 1959; s. Heinrich Karl and Martha A. (Hustert) H.; m. Margaret A. Hoffmann, July 10, 1990; children: Lisa J., Laura M. BS in Nuclear Engring., Calif. State U., 1966. Mktg. mgr. GE, San Francisco, 1966-78; dir. mktg. FMC Corp., San Jose, Calif., 1978-80; v.p., gen. mgr. Canron Corp. of Can., Tacoma, Wash., 1980-85; sr. program mgr. R&D Weyerhauser Co., Tacoma, 1985—. Pres., founder World Affairs Coun., Tacoma, 1988. Mem. ASME, Tech. Assn. Pulp and Paper Industry, Zellcheming, Licensing Execs. Soc. Office: Weyerhauser Co WTC-1H39 Tacoma WA 98477

HOFFMANN, ROALD, chemist, educator; b. Zloczow, Poland, July 18, 1937; came to U.S., 1949, naturalized, 1955; s. Hillel and Clara (Rosen) Safran (stepson Paul Hoffmann); m. Eva Börjesson, Apr. 30, 1960; children: Hillel Jan, Ingrid Helena. BA, Columbia U., 1958; MA, Harvard U., 1960, PhD, 1962; D Tech. (hon.), Royal Inst. Tech., Stockholm, 1977; D.Sc. (hon.), Yale U., 1980, Columbia U., 1982, Hartford U., 1982, CUNY, 1983, U. P.R., 1983, U. Uruguay, 1984, U. La Plata, SUNY, Binghamton, 1985, Colgate U., Lehigh U., 1989, Carleton Coll., 1989, Ben Gurion Coll., 1989, U. of the Negev, 1989, U. Md., 1990, U. Athens, 1991, U. Thessaloniki, Greece, 1991, U. Ariz., 1991, U. Cen. Fla., 1991, Bar Ilan U., 1991; DSc (hon.), U. St. Petersburg, Russia, 1991, U. Barcelona, 1992. Jr. fellow Soc. Fellows Harvard, 1962-65; assoc. prof. Cornell U., Ithaca, N.Y., 1965-68; prof. Cornell U., 1968-74, John A. Newman prof. phys. sci., 1974—; U. Md., 1990. Author: (with R.B. Woodward) Conservation of Orbital Symmetry, 1970, Solids and Surfaces, 1988, (with V. Torrence) Chemistry Imagined; author: (poetry) The Metamict State, 1987, Gaps and Verges, 1990; Chemistry Imagined, 1993. Recipient award in pure chemistry Am. Chem. Soc., 1969, Arthur C. Cope award, 1973, Fresenius award Phi Lambda Upsilon, 1969, Harrison Howe award Rochester sect. Am. Chem. Soc., 1970; ann. award Internat. Acad. Quantum Molecular Scis., 1970, Pauling award, 1974, Nobel prize in chemistry, 1981, inorganic chemistry award; Am. Chem. Soc., 1982, Nat. Medal of sci., 1983, Chem. Scis. award Nat. Acad. Sci., 1986, Priestley medal, 1990, N.N. Semenov Gold medal Acad. Scis. USSR. Mem. NAS (award in chem. scis. 1986), Am. Acad. Arts and Scis., Russian Acad. Scis. (Semenov Gold medal), Internat. Acad. Quantum Molecular Scis., Royal Soc. (fgn.), Indian Nat. Sci. Acad., Royal Swedish Acad. Scis., Finnish Acad. Arts and Letters, Acad. Scis. USSR.

HOFFMEISTER, JANA MARIE, cardiologist. MD, SUNY Upstate Med. Ctr., Syracuse, 1976. Diplomate Am. Bd. Internal Medicine. Intern Albany (N.Y.) Med. Ctr., 1976-78, asst. resident, 1978-80, fellow div. cardiology, 1981-83; fellow div. cardiology Emory U., Atlanta, 1984; fellow coronary angioplasty and interventional cardiology Emory U. Hosp., 1985-86; presenter numerous cardiology confs. Contbr. numerous articles to profl. jours. Mem. AMA, Cardiac Soc. Upstate N.Y., N.Y. State Soc. Internal Medicine, Am. Soc. Cardiovascular Intervention, Am. Coll. Physicians. Home: 7 Reddy Ln Albany NY 12211-1697

HOFMANN, ALBERT JOSEF, physicist; b. Uznach, Switzerland, Feb. 27, 1933; s. Rudolf and Agnes (Mächler) H.; m. Elisabeth I. Bloechlinger, July 20, 1960; children: Bettina, Sabine, Regula. Diploma in physics, Swiss Fed. Inst. Tech., Zürich, Switzerland, 1957, D. in Natural Scis., 1964. Sci. coworker Swiss Fed. Inst. Tech., Inst. for Nuclear Physics, Zürich, 1957-66; rsch. fellow Cambridge (Mass.) Electron Accelerator, Harvard U.-MIT, 1966-72; sr. physicist European Orgn. Nuclear Rsch., Geneva, 1973-83, sr. physicist superproton synchrotron-large electron position ring div., 1987—; prof. applied rsch. synchrotron radiation lab., Stanford linear accelerator ctr. Stanford (Calif.) U., 1983-86. Fellow Am. Phys. Soc. Home: Chemin de l'Erse 20, CH 1218 Grand Saconnex Switzerland Office: CERN SL-div European Orgn, Nuclear Rsch, CH 1211 Geneva 23, Switzerland

HOFMANN, FRIEDER KARL, biotechnologist, consultant; b. Eppstein, Hessen, Fed. Republic of Germany, June 15, 1949; came to U.S., 1984; s. Friedrich Karl and Anna Johannette (Heist) H.; m. Sigrid Marianne Thomae, Sept. 5, 1975. MS, J.W. Goethe U., Frankfurt, Fed. Republic of Germany, 1977, PhD, 1981. Staff scientist, asst. prof. J.W. Goethe U., Frankfurt, 1977-81; sci. mgr. Brunswick Corp., Eschborn, Fed. Republic of Germany, 1982-84; tech. dir. Biotechnetics, San Diego, 1984-90; pres. Hofmann & Co., Oceanside, Calif., 1990—. Author: (with others) Scale-Up and Downstream Processing of rDNA Products, 1991, GMP Production of Monoclonal Antibodies, 1991; contbr. over 40 articles to profl. jours. Recipient Senckenberg prize Senckenberg Rsch. Soc., Frankfurt, Fed. Republic of Germany, 1977; Kirkpatrick Chem. Engring. Achievement Honor award, Chem. Engring., 1989, Parenteral Drug Assn. Jour. award, Parenteral Drug Assn., Pa., 1985. Mem. Am. Chem. Soc., Am. Inst. Chem. Engrs., Tissue Culture Assn., European Soc. for Animal Cell Tech. Achievements include 6 patents for bioreactor and membrane technology; invention and development of tester for membrane filters, of first scalable membrane based animal cell reactor; first integration of upstream and downstream processes in bioreactor system; invention of formulation and procedure to grow animal cells in protein-free nutrient. Office: Hofmann & Co 2360 Autumn Dr Ste C Oceanside CA 92056-3528

HOFSTADTER, DANIEL SAMUEL, aerospace engineer; b. L.A., May 2, 1958; s. Sol and Charlotte (Marcus) H.; m. Shari Ann Linn, Apr. 21, 1991. BS in Physics, U. Calif., Irvine, 1980; MS in Aerospace Engring., U. Ariz., 1983. Mem. tech. staff Hughes Aircraft Co., Tucson, 1983-90; optomech. engr. Kaman Aerospace Corp., Tucson, 1990—. Past bd. dirs. Tucson Clean and Beautiful; del. Ariz. Dem. Conv., Phoenix. Mem. AIAA (Tucson sect. chmn. 1985-86), Sierra Club (exec. com. Rincon Group). Achievements include research in observatory dome heat generation. Office: Kaman Aerospace 3480 E Britannia Dr Tucson AZ 85706

HOFSTADTER, DOUGLAS RICHARD, cognitive, computer scientist, educator; b. N.Y.C., Feb. 15, 1945; s. Robert and Nancy (Givan) H.; m. Carol Ann Brush, 1985; children: Daniel Frederic, Monica Marie. B.S. in Math. with distinction, Stanford U., 1965; M.S., U. Oreg., 1972, Ph.D. in Physics, 1975. Asst. prof. computer sci. Ind. U., Bloomington, 1977-80; assoc. prof. Ind. U., 1980-84; Walgreen prof. Cognitive Sci. U. Mich., Ann Arbor, 1984-88; prof. cognitive sci., computer sci. Ind. U., Bloomington, 1988—; adj. prof. psychology, philosophy, history, philosophy of sci. and comparative lit., dir. Ctr. for Rsch. on Concepts and Cognition, Ind. U. Author: Gödel, Escher, Bach: an Eternal Golden Braid, 1979, Metamagical Themas, 1985, Ambigrammi, 1987; editor: (with Daniel C. Dennett) The Mind's I, 1981; columnist: Metamagical Themas in Sci. Am., 1981-83. Recipient Pulitzer prize for gen. nonfiction, 1980; Am. Book award, 1980; Guggenheim fellow, 1980-81. Mem. Cognitive Sci. Soc., Am. Assn. Artificial Intelligence, Am. Lit. Translators Assn. Office: Ctr Rsch Concepts & Cognition 510 N Fess Ave Bloomington IN 47408-3822

HOGAN, CHARLES CARLTON, psychiatrist; b. Quincy, Ill., Oct. 5, 1921; s. Carlton Monta and Maryanne (Henry) H.; m. Nina Harriet Redman; children: Matthew P., Carlton H., Noelle N. Student, Bradley U., 1939-41, Ul Ill., 1941-42; MD, Columbia U., 1945, D Med. Sci., 1952. Diplomate Am. Bd. Psychiatry and Neurology. Intern Phila. Gen. Hosp., 1945-46; rsch. asst. neurology dept. Columbia U., N.Y.C., 1948-50, lectr. Ctr. for Psychoanalytic Tng., 1979—, psychoanalytic clinician Ctr. for Psychoanalytical Tng., 1948-52; resident N.Y. State Psychiatric Inst., N.Y.C., 1949-50; asst. physician Presbyn. Hosp., N.Y.C., 1951-54, asst. attending psychiatrist, 1954-60; asst. vis. psychiatrist Bronx (N.Y.) Mcpl. Hosp., 1960—; asst. clin. prof. psychiatry Albert Einstein Coll. Medicine, Bronx, 1960—; chmn. profl. adv. com. Riverdale Mental Health Clinic, Bronx, 1969-77; cons. Wiltwyck Sch. for Boys, Yorktown Heights, N.Y., 1971-77. Author: Psychosomatics and Inflammatory Disease of the Colon; editor Fear of Being Fat, 1985, Psychodynamic Technique in the Treatment of the Eating Disorders, 1992; contbr. articles to profl. publs., 12 chpts. to books on psychomatic disorders. Treas. Physicians for Social Responsibility, N.Y.C., 1968. Capt. U.S. Army, 1946-48. Fellow Am. Psychiat. Assn. (life); mem. AMA, Am. Psychoanalytic Assn. (life), N.Y. State Med. Assn., Am. Psychosomatic Soc., N.Y. Acad. Scis., Pan-Am. Med. Assn., Assn. for Psychoanalytic Medicine, World Mental Health Assn., Riverdale Yacht Club (bd. govs.), Undersea and Hyperbaric Med. Assn. Avocations: sailing, diving. Home: 6 Ploughmans Bush Bronx NY 10471-3541 Office: 8 E 96th St New York NY 10128-0820

HOGAN, CLARENCE LESTER, retired electronics executive; b. Great Falls, Mont., Feb. 8, 1920; s. Clarence Lester and Bessie (Young) H.; m. Audrey Biery Peters, Oct. 13, 1946; 1 child, Cheryl Lea. BSChemE, Mont. State U., 1942, Dr. Engring. (hon.), 1967; MS in Physics, Lehigh U., 1947, PhD in Physics, 1950, D in Engring. (hon.), 1971; AM (hon.), Harvard U., 1954; D in Sci. (hon.), Worcester Poly. U., 1969. Rsch. chem. engr. Anaconda Copper Mining Co., 1942-43; instr. physics Lehigh U., 1946-50; mem. tech. staff Bell Labs., Murray Hill, N.J., 1950-51, sub-dept. head, 1951-53; assoc. prof. Harvard U., Cambridge, Mass., 1953-57, Gordon McKay prof., 1957-58; gen. mgr. semi-conductor products div. Motorola, Inc., Phoenix, 1958-60, v.p., 1960-66, exec. v.p., dir., 1966-68; pres., chief exec. officer Fairchild Inst., Mt. View, Calif., 1968-74, vice chmn. bd. dirs., 1974-85; bd. dirs. MEMC HUELS; gen. chmn. Internat. Conf. on Magnetism and Magnetic Materials, 1959, 60; mem. materials adv. bd. Dept. Def., 1957-59; mem. adv. coun. dept. electrical engring. Princeton U.; mem. adv. bd. sch. engring. U. Calif., Berkeley, 1974—, adv. bd. dept. chem. engring. Mont. State U., 1988—; mem. nat. adv. bd. Desert Rsch. Inst., 1976-80; mem. vis. com. dept. electric engring. and computer sci. MIT, 1975-85; mem. adv. coun. div. electrical engring. Stanford U., 1976-86; mem. sci. and ednl. adv. com. Lawrence Berkeley Lab., 1978-84; mem. Pres.'s Export Coun., 1976-80; mem. adv. panel to tech. adv. bd. U.S. Congress, 1976-80. Patentee in field; inventor microwave gyrator, circulator, isolator. Chmn. Commn. Found. Santa Clara County, Calif., 1983-85; mem. vis. com. Lehigh U., 1966-71, trustee, 1971-80; trustee Western Electronic Edn. Fund; mem. governing bd. Maricopa County Jr. Coll.; bd. regents U. Santa Clara. Lt. (j.g.) USNR, 1942-46. Recipient Community Svc. award NCCJ, 1978, Medal of Merit Am. Electronics Assn., 1978, Berkeley Citation U. Calif., 1980; named Bay Area Bus. Man of Yr. San Jose State U., 1978, One of 10 Greatest Innovators in Past 50 Yrs. Electronics Mag., 1980. Fellow AAAS, IEEE (Frederick Philips gold medal 1976, Edison silver medal Cleve. Soc. 19078, Pioneering medal for microwave theory and tech. 1993), Inst. Elec. Engrs. (hon.); mem. NAE, Am. Phys. Soc., Menlo Country Club, Masons, Sigma Xi, Tau Beta Pi, Phi Kappa Phi, Kappa Sigma. Democrat. Baptist. Avocations: woodworking, computer programming. Home: 36 Barry Ln Menlo Park CA 94027-4023

HOGAN, DANIEL BOLTEN, management consultant; b. Lawrence, Mass., Sept. 20, 1943; s. Daniel Edward and Gisela (Bolten) H.; m. Jean Elizabeth Haley, Jan. 25, 1979; children: Matthew Pollard, Sarah Elizabeth, Haley Elizabeth. BA, Yale U., 1965; EdM, Boston U., 1971; JD, Harvard U., 1972, PhD, 1983. Lic. psychologist; bar: Mass., U.S. Dist. Ct. Mass. Rural community devel. worker U.S. Peace Corps, Butha-Buthe, Lesotho, 1967-69; mgmt. cons. Wayland, Mass., 1972-76; rsch. and teaching fellow Dept. Psychology and Social Rels., Harvard U., Cambridge, Mass., 1976-81; clin. fellow and instr. psychology Harvard Med. Sch., Boston, 1981-85; asst. in psychology McLean Hosp., Belmont, Mass., 1982-83, asst. psychologist, 1983, asst. attending psychologist, 1984-86; mgmt. cons. Weston, Mass., 1986-90; v.p. and dir. R & D McBer & Co., Boston, 1990; pres. The Apollo Group, Concord, Mass., 1991—; dir. Standex Internat. Corp., Salem, N.H.; sr. rsch. assoc. Nat. Ctr. for Study of Professions, Washington, 1976-83; mem. com. on use of human subjects in rsch. Harvard U. Faculty of Arts and Scis., 1980-81. Author: The Regulation of Psychotherapists: A Study in the Philosophy and Practice of Professional Regulation, vol. I, 1979, The Regulation of Psychotherapists: A Handbook of State Licensure Laws, vol. II, 1979, The Regulation of Psychotherapists: A Review of Malpractice Suits in the United States, vol. III, 1979, The Regulation of Psychotherapists: A Resource Bibliography, vol. IV, 1979, 83; guest editor (spl. double issue) Law and Human Behavior, 1983, cons. editor, 1985-91; contbr. articles to profl. jours., chpts. books. Trustee Fenn Sch., Concord, Cambridge Ctr. for Behavioral Studies, 1985-91; bd. dirs. New Eng. Tng. Inst., Stonington, Conn., 1976-78, Phillips Exeter Acad. Alumni Coun., Exeter, N.H., 1986-92, Community Change, Inc., Boston, 1972-79, adv. bd., 1979—. Maurice Falk Med. Fund. grantee, 1976-78. Fellow APA, Am. Psychol. Soc., Am. Orthopsychiatric Assn.; mem. Am. Psychology Law Soc. (dir. 1982-84), Internat. Orgn. Devel. Assn., Mass. Psychol. Assn. (bd. profl. affairs 1980-83), Mass. Bar Assn. Avocations: squash, tennis, music, computers, biographies. Office: The Apollo Group 229 Westford Rd Concord MA 01742-5232

HOGAN, JOSEPH CHARLES, university dean; b. St. Louis, May 26, 1922; s. Joseph D. and Anna (Lange) H.; m. Mary Elizabeth Carrere, June 21, 1944; children: Joseph Charles (dec.), Mary E., Susan L., Thomas C., Stephen J., Michael C., Martha A., William G., Daniel C. BSEE, Washington U., St. Louis, 1943; MS, U. Mo., 1949; PhD, U. Wis., 1953. Registered profl. engr., Mo., Ind. Instr. U. N.D., 1947; mem. faculty U. Mo., Columbia, 1947-67, prof., 1958-67, dean engring., 1961-67; prof., dean engring. U. Notre Dame, 1967-81, dean emeritus, 1981—; dir. engring. R&D Ga. Inst. Tech., 1985-87; cons. Columbia Water & Light Dept., 1954-56, North Cen. Assn., 1967-87, Whirlpool Corp., 1971-81, State U. System Fla.; bd. dirs. TII, Inc., Copiague, N.Y., Am. Biogenetics Inc., Notre Dame, Ind.; pres. Mac Engring. & Equipment, Inc., Benton Harbor, Mich., 1976-84, MacWilliams Corp., South Bend, Ind., 1983-84; chmn. water and light adv. bd. City of Columbia, 1957-61. Contbr. research articles profl. jours. Pres. Nat. Consortium for Grad. Degrees for Minorities, 1976-81, pres. emeritus, 1981—. Served as ensign USNR, 1943-45. Recipient Engring. Honor award U. Mo., 1973, Alumni Achievement award Washington U., 1979; U. Wis. fellow, 1951-53. Fellow IEEE (life), Am. Soc. Engring. Edn. (life, v.p. 1974-75, bd. dirs. 1973-75, pres.-elect 1981-82, pres. 1982-83, Centennial medal 1993); mem. NSPE, Mo. Soc. Profl. Engrs., Ind. Soc. Profl. Engrs. (Engr. of Yr. award 1981), Engrs. Coun. for Profl. Devel. (bd. dirs. 1974-79), Sigma Xi, Tau Beta Pi, Eta Kappa Nu, Theta Xi. Home: 12 Sandhill Crane Rd Hilton Head Island SC 29928-5705

HOGAN, JOSEPH THOMAS, podiatrist; b. Cooperstown, N.Y., Nov. 25, 1943; s. James H. and Ann M. (Keery) H. DPM, Ill. Coll. Podiatric Medicine, 1975; BA, U. Notre Dame, 1966. Diplomate Am. Bd. Podiatric Surgery (credentials com. 1987-88), Am. Bd. Podiatric Orthopedics, Am. Acad. Pain Mgmt., Am. Bd. Quality Assurance Utilization Rev. Physicians; m. Kathleen Mary Sullivan, July 22, 1978; children: Kathleen, Margaret, Joseph II, Thomas, John, Mary Claire. Resident, St. Mary Hosp., Phila.; staff podiatrist diabetic foot clinic Wilson Meml. Hosp., Binghamton, N.Y., 1979—; clin. instr. SUNY Health Sci. Ctr., Syracuse, 1979—; ptnr., pvt. practice podiatric medicine and surgery, Binghamton, 1976—; surg. staff Lourdes Meml. Hosp., Wilson Meml. Hosp., Binghamton Gen. Hosp.; mem. faculty Penn. Podiatric Med. Assn. ann. surg. seminar, 1985-88, instr. bd. rev. course, 1986-88; clin. instr. family practice residency program United Health Svcs., 1980—. Bd. dirs. Our Lady of Lourdes Meml. Hosp. Found.; chmn. United Way, 1981-82; v.p. N.Y. So. Tier chpt. Am. Diabetes Assn., 1978-83, bd. dirs., 1984—. Mem. parish coun. St. Vincent de Paul Ch.; mem. Utilization Rev. Com. United Health Svcs., Lourdes Meml. Hosp. Found. Bd.; trustee N.Y. State Podiatric Med. Assn., 1990-92. With U.S. Army, 1967-70. Fellow Am. Assn. Hosp. Podiatrists, Am. Coll. Foot Surgeons, Am. Coll. Foot Orthopedists, Nat. Soc. Conscious Sedation, Am. Coll. Podiatric Med. Review; mem. Am. Coll. Podopediatrics, Am. Soc. Podiatric Angiology, Broome County Hist. Soc., Am. Podiatry Assn., N.Y. State Podiatry Soc. (chpt. pres. 1980-82), Am. Diabetes Assn., Am. Pub. Health Assn., N.Y. Acad. Scis., Am. Assn. Colls. Podiatric Medicine, Am. Legion, Am. Bd. Quality Assurance and Utilization Rev. (co-founder, diplomate), Am. Acad. of Pain Mgmt. (diplomate), NYSPMA (co-chmn. So. tier div. ann. seminar. 1981-88, mem. sci. com. ann. clin. conf. 1989—, podiatric med. div. 1987—, legis. action com. 1990-92, bd. dirs. 1990-92), Edward Frederick Sorin Soc., Pa. Podiatric Med. Assn Surgeons (seminar 1985-87, credentials com. 1987—), Sorin Soc. Roman Catholic. Clubs: Elks, Ancient Order Hibernians, Notre Dame of Triple Cities (pres. 1982-84, editor newsletter, Winner of Yr. award 1991). Home: 608 Dartmouth Dr Vestal NY 13850-2926 Office: 41 Oak St Binghamton NY 13905-4627

HOGAN, MERVIN BOOTH, mechanical engineer, educator; b. Bountiful, Utah, July 21, 1906; s. Charles Ira and Sarah Ann (Booth) H.; m. Helen Emily Reese, Dec. 27, 1928; 1 son, Edward Reese. B.S., U. Utah, 1927, M.E., 1930; M.S., U. Pitts., 1929; Ph.D., U. Mich., 1936, postgrad.; Sterling fellow, Yale U., 1937-38. Registered profl. engr., Conn., Mich., N.Y., Utah, Va. chartered engr., U.K. Design engr. Westinghouse Electric Corp., East Pittsburgh, Pa., 1927-31; asst. prof. mech. engring. U. Utah, Salt Lake City, 1931-36, asso. prof., 1936-39, prof., 1939-56, chmn. dept. mech. engring., 1951-56, prof., 1971-76, prof. emeritus, 1976—; mgr. product design engring. GE, Syracuse, N.Y., 1956-65; mgr. design assurance engring. GE, Phoenix, 1965-70; cons. engr. GE, Waynesboro, Va., 1970-71; cons. Chgo. Bridge & Iron, 1950-56. Author: Mormonism and Freemasonry: The Illinois Episode, 1977, The Origin and Growth of Utah Masonry and Its Conflict with Mormonism, 1978, Mormonism and Freemasonry under Covert Masonic Influences, 1979, Freemasonry and the Lynching at Carthage Jail, 1981, Freemasonry and Civil Confrontation on the Illinois Frontier, 1981, The Involvement of Freemasonry with Mormonism on the American Midwestern Frontier, 1982; contbr. articles to engr. jours., numerous articles to Masonic publs. Recipient Merit of Honor award U. Utah, 1981. Fellow ASME, Inst. Mech. Engrs. (London), Yale Sci., Engring. Assn.; mem. IEEE (sr.), Nat. Eagle Scout Assn., DeMolay Legion of Honor, S.R. in State N.Y., Utah Soc. SAR (pres. 1983-84), Aztec Club, Timpanogos Club, Elfun Soc., Rotary, Masons (33 deg.), Shriners, Prophets, KT, DeMolay, Quatuor Coronati Lodge 2076, Sigma Xi, Phi Kappa Phi, Tau Beta Pi, Pi Tau Sigma, Sigma Nu, Theta Tau, Alpha Phi Omega, Phi Lambda Epsilon. Home: 921 Greenwood Terr Douglas Park Salt Lake City UT 84105 Office: U Utah 3008 Merrill Engring Bldg Salt Lake City UT 84112

HOGAN, STEPHEN JOHN, electrical engineer; b. Columbia, Mo., July 10, 1951; s. Joseph C. and Mary E. (Carrere) H.; m. Ann M. Blicher, Apr. 5, 1975; children: Emily, Meghan, Courtney. BSEE, U. Notre Dame, 1973; MSEE, U. Colo., 1978. Registered profl. engr., Colo., Mass. Rsch. engr. Whirlpool Corp., Benton Harbor, Mich., 1974-77; scientist Solar Energy Rsch. Inst., Golden, Colo., 1978-84; engring. mgr. Spire Corp., Bedford, Mass., 1984—. Contbr. articles to profl. publs. Bd. dirs. Colo.-Sierra Fire Protection Dist., Gilpin County, 1980-84; chief Colo.-Sierra Vol. Fire Dept., 1982-84. Mem. IEEE (sr.), ASTM (chair subcom. 1980-84), Sigma Xi, Eta Kappa Nu. Achievements include patents for novel control circuitry and advance materials developments; establishment of solar cell production facility in China. Home: 21 Natalie Rd Chelmsford MA 01824 Office: Spire Corp 1 Patriots Park Bedford MA 01730

HOGANSON, CURTIS WENDELL, physical chemist; b. Mpls., Sept. 15, 1955; s. Raynold Alroy and Lila Jean (Shirk) H.; m. Nancy Nicole Artus, Apr. 30, 1988; 1 child, Hannah Lynn. BS, N.D. State U., 1977; PhD, Washington State U., 1985. Rsch. assoc. dept. chemistry Mich. State U., East Lansing, 1985-88, 90—; rsch. assoc. Chalmers Inst. Tech., Gothenburg, Sweden, 1988-90. Office: Mich State U Dept Chemistry East Lansing MI 48824

HOHL, MARTIN D., electrical engineer; b. Grinnell, Iowa, May 22, 1958; s. Melvin L. and Doris (Jeffrey) H.; m. Kathy Ann Noonan, May 30, 1984; children: JEffrey M., Jenna K. AS, Marshalltown C.C., 1978; BSEE, Iowa State U., 1981; MBA, St. Ambrose U., 1989. Registered profl. engr. Iowa, Ill. Elec. engr., field engr. Schulmberger, Wharton, Tex., 1981; distbn. engr., engr.-in.tng. EIT, Iowa City, 1982; substa, engr., profl. engr. GE, Rock Island, Ill., 1984; customer svc. engr. Iowa-Ill. Gas & Electric Co., Davenport, Iowa, 1988; program mgr. in power quality Iowa-Ill. Gas & Electric Co., Rock Island, 1989, indsl. engr., 1992—. Mem. IEEE, NSPE (pres. 1990-92), Assn. Energy Engrs. (cert.). Office: Iowa-Ill Gas and Elect Co 1830 2nd Ave Ste 100 Rock Island IL 61201

HOHLOV, YURI EUGENIEVICH, mathematician; b. Donetsk, USSR, Nov. 16, 1954; s. Eugene D. and Nina G. (Popova) H.; m. Tatyana N. Nebotova, July 30, 1981; 1 child, Nina. MS, Kazan (USSR) U., 1976, PhD, 1980. Asst. prof. Kazan U., 1976-78; assoc. prof. Donetsk U., 1980-83, prof. math., 1984-88; researcher, specialist Steklov Math. Inst., Moscow, 1988—; dir. Volga's Sch. for Talents in Math., Kazan, 1976-80; head group in applications of math. in gas industry, Donetsk, 1982-88; exec. dir. Found. Math., Moscow, 1992—; dir. dept. informatics and analysis Russian Basic Rsch. Found., Moscow, 1993—. Author: Mathematical Methods in Informatics, 1986. Recipient Gold medal Ministry of High Edn. of USSR, 1978; Swedish Inst. fellow, 1984; Vis. fellow London Math. Soc., 1991, Rsch. fellow Sci. and Engring. Rsch. Coun., 1992—. Mem. Am. Math. Soc. Office: Steklov Math Inst, Vavilova Str 42, 117966 Moscow USSR also: Russian Basic Rsch Found, Leninskij prosp 32 a, Moscow 117334, Russia

HOHMEYER, OLAV HANS, economist, researcher; b. Minden, Germany, Nov. 23, 1953; s. Karl-Heinz and Ingeborg (Artus) H.; m. Ulrike M. Wehking, Mar. 22, 1985. Grad., U. Bremen, Fed. Republic Germany, 1980, PhD, 1989. Researcher Hochschule für Wirtschaft, Bremen, 1980, U. Oldenburg, Fed. Republic Germany, 1980-82; sr. researcher Frannhofer Inst. für Systemtechnik und Innovationsforschung, Karlsruhe, Fed. Republic Germany, 1982-89, dept. head on tech. change, 1989—; cons. Commn. of the European Communities, Luxemburg, 1990-91. Author: Employment Effects of Energy Conservation Investments, 1985, Technometrie, 1987, Social Costs of Energy Consumption, 1988, Soziale Kosten des Energieverbrauchs, 1988, 89, External Environmental Costs of Electric Power, 1991. Recipient Fraunhoferpreis Fraunhofer-Gesellschaft, Munich, 1988, 91, Deutscher Enegiepreis, Deutsche Energie Gesellschaft, Munich, 1989. Mem. Internat. Input Output Assn., Internat. Soc. Ecol. Econs., European Assn. Environ. and Resource Economists. Lutheran. Achievements include development of new methodology for measuring the comparative levels of technological achievements of nations; of new model integrating branch specific emission coefficients into official input-output tables for Germany; first to analyze the difference in external costs of electricity generation between conventional electric power production and renewables. Office: Fraunhofer Inst für, Systemtechnik, D-7500 Karlsruhe Germany

HOIT, MARC I., civil engineer, educator; b. Gary, Ind., May 7, 1957; s. Leonard and Maida (Siegal) H.; m. Fay Prost; 1 child, Sarah. BSE, Purdue U., 1978; MS, U. Calif., Berkeley, 1980, PhD, 1983. Assoc. prof. dept. civil engring. U. Fla., Gainesville, 1984—; vis. faculty Tokyo U., 1991; sabbatical chair Sony Corp., Tokyo, 1991; cons. various engring. firms. Contbr. to profl. publs. Mem. ASCE (assoc., br. pres. 1988, Edmund Friedman Young Engr. award 1988), Am. Soc. Engring. Edn. (Dow Outstanding Young Engr. award 1987), Fla. Engring. Soc., Tau Beta Pi. Office: Univ Fla Dept Civil Engring 345 Weil Hall Gainesville FL 32611

HOJNACKI, JEROME LOUIS, university dean; b. Stamford, Conn., Mar. 9, 1947; s. Louis L. and Jennie L. (Faski) H.; m. Mary Elizabeth Riley, July 8, 1978. BS, So. Conn. State U., 1969; MS, U. Bridgeport, 1971; PhD, U. N.H., 1975; M Health Adminstrn., Clark U., Worcester, Mass., 1982. Rsch. fellow Harvard U. Sch. Pub. Health, Boston, 1975-78; rsch. cons. Harvard Med. Sch., Boston, 1977-79; asst. prof. to prof. U. Mass., Lowell, 1978-85, dean grad. sch., 1987—; rsch. fellow Mass. Gen. Hosp., Boston, 1992—; grant reviewer NIH Pathology Study Sect., 1993. Contbr. articles to profl. jours., chpts. to books. Mem. Bd. Health, Groton, Mass., 1981. Grantee Coun. Tobacco Rsch., Alcoholic Beverage Med. Rsch. Found., Am. Heart Assn., Nat. Inst. Alcohol Abuse and Alcoholism, pvt. founds., 1979-92. Fellow Am. Assn. Pathologists, Rsch. Soc. Alcoholism, Am. Fedn. Clin. Rsch., Coun. Arteriosclerosis, Am. Heart Assn., Am. Inst. Nutrition, Am. Coll. Nutrition, Am. Soc. Clin. Nutrition. Democrat. Roman Catholic. Achievements include research on metabolic mechanisms by which tobacco products and alcohol influence blood cholesterol (plasma lipoproteins) and coronary heart disease. Office: U Mass 1 University Ave Lowell MA 01854

HOJO, MASASHI, chemistry educator; b. Uchiumi-mura, Ehime-ken, Japan, Feb. 17, 1952; s. Tsugio and Toshiko (Kuroda) H.; m. Mari Hashimoto, Dec. 1, 1987; children: Ken-ichi, Shigefumi. BS, Kobe (Japan) U., 1974; MS, Kyoto (Japan) U., 1976, PhD, 1981. Instr. Kochi U., 1979-87, lectr., 1987-89, assoc. prof., 1989—; rsch. assoc. U. Calgary, Alta., Can., 1982-84, Tex. A&M U., College Station, 1987-88. Contbr. articles to profl. jours. Mem. Am. Chem. Soc., Chem. Soc. Japan, Japan Soc. for Analytical Chemistry, Polarographic Soc. Japan, Internat. Soc. Electrochemistry. Avocation: classical music. Home: 399-13 Mama, Kochi, Kochi 780, Japan Office: Kochi U Dept Chemistry, 2-5-1 Akebono-cho, Kochi 780, Japan

HOKANA, GREGORY HOWARD, engineering executive; b. Burbank, Calif., Nov. 24, 1944; s. Howard Leslie and Helen Lorraine (Walker) H.; m. Eileen Marie Youell, Apr. 29, 1967; children: Kristen Marie, Kenneth Gregory. BS in Physics, UCLA, 1966. Design engr. Raytheon Co., Oxnard, Calif., 1967-74; staff engr. Bunker Ramo Corp., Westlake Village, Calif., 1974-84; mgr. analog engring. AIL Systems, Inc., Westlake Village, 1984-91; mgr. product devel. Am. Nucleonics Corp., Westlake Village, 1991—. Mem. Assn. Old Crows. Democrat. Methodist. Avocations: golf, swimming, photography. Home: 3485 Farrell Cir Newbury Park CA 91320-4333 Office: Am Nucleonics Corp 696 Hampshire Rd Westlake Village CA 91361-2512

HOKIN, LOWELL EDWARD, biochemist, educator; b. Chgo., Sept. 20, 1924; s. Oscar E. and Helen (Manfield) H.; m. Mabel Neaverson, Dec. 1, 1952 (div. Dec. 1973); children: Linda Ann, Catherine Esther (dec.), Samuel Arthur; m. Barbara M. Gallagher, Mar. 23, 1978; 1 child, Ian Oscar. Student, U. Chgo., 1942-43, Dartmouth Coll., 1943-44, U. Louisville Sch. Medicine, 1944-46, U. Ill. Sch. Medicine, 1946-47; M.D., U. Louisville, 1948; Ph.D., U. Sheffield, Eng., 1952. Postdoctoral fellow dept. biochemistry McGill U., 1952-54, faculty, 1954-57, asst. prof., 1955-57; mem. faculty U. Wis.-Madison, 1957—, prof. physiol. chemistry, 1961-68, prof., chmn. dept. pharmacology, 1968—. Contbr. numerous articles to tech. jours., chpts. to numerous books on phosphoinositides, biol. transport, the pancreas and the brain. With USNR, 1943-45. Mem. AAAS, Am. Soc. Biochemistry and Molecular Biology, Biochem. Soc. (U.K.), Am. Soc. Pharmacology and Exptl. Therapeutics, N.Y. Acad. Scis. Home: 5 Nokomis Ct Madison WI 53711-2710 Office: U Wis Med Sch Dept Pharmacol 1300 University Ave Madison WI 53706-1532

HOL, WIM GERARDUS JOZEF, biophysical chemist; b. Wamel, Netherlands, July 6, 1945; s. Wilhelmus W.J. and Gertruda (Van Gelder) H.; m. Carol M. Runner, Apr. 18, 1970; children: Helen P., Lysbeth K., Willem A. MSc, U. Eindhoven, 1966; PhD, U. Groningen, 1971. Assoc. expert in sci. policy UNESCO, Nairobi, Kenya, 1972-74; assoc. prof. U. Groningen, 1974-85, prof. biophys. chemistry, 1985-92; prof. biol. structure, head biomolecular structure program U. Wash., Seattle, 1992—; mem. sci. adv. bd. Netherlands Cancer Inst., Amsterdam, 1986-92. Contbr. numerous articles to profl. jours. Recipient Keilin medal Biochem. Soc. U.K., 1991. Mem. Royal Dutch Chem. Soc. (Gold Medal 1979), European Molecular Biology Orgn., Am. Crystallographic Assn., N.Y. Acad. Sci., Royal Dutch Acad. Sci. Achievements include research on role of alpha-helix dipole in protein structure and function; three dimensional x-ray structure determinations of subtilisin, hemocyanin, trypanosomal enzymes, choleratoxin-like heat-labile enterotoxin from E. Coli, pyruvate dehydrogenase multienzyme complex, quinoproteins. Office: Univ Wash Sch Medicine Health Scis Bldg SM-20 Seattle WA 98195

HOLBROOK, ANTHONY, manufacturing company executive; b. 1940; married. With Advanced Micro Devices Inc., Sunnyvale, Calif., 1973—, former exec. v.p., chief operating officer, pres., chief operating officer, 1986-90, vice chmn., chief tech. officer, 1990—. Office: Advanced Micro Devices Inc PO Box 3453 901 Thompson Pl Sunnyvale CA 94088*

HOLCOMBE, CRESSIE EARL, JR., ceramic engineer; b. Anderson, S.C., Dec. 18, 1945; s. Cressie Earl Sr. and Blanche Elizabeth (Keaton) H.; m. Catherine Joselyn Brockman, Dec. 27, 1966; children: Justin Kent, Eric Benjamin. BS, Clemson U., 1966, MS, 1967; postgrad. U. Mo.-Rolla, 1973. Assoc. devel. engr. Union Carbide Corp., Oak Ridge, 1967-72, devel. engr., 1972-76, mem. devel. staff, 1977-80, mem. advanced devel. staff, 1980-84; advanced devel. staff Martin Marietta Energy Systems, Inc., Oak Ridge, 1984-88, sr. devel. staff, 1988-90, advanced sr. devel. staffman, 1990-92, corp. fellow, 1992—; founder ZYP Coatings Inc., Oak Ridge, 1982—, Orpac, Inc., Oak Ridge, 1984—. Author: (with others) Metallurgical Coatings, Vol. 1, 1976. Contbr. articles to profl. jours., unclassified spl. tech. reports for AEC/Dept. Energy. Patentee on refractory materials, metals and ceramics. Recipient Top Twenty award Materials Engring. mag., 1984, Excellence-Inventor award Martin Marietta Energy Systems, 1985, 90, Indsl. Applications award Excellence Titanium Devel. Assn., 1990; Weapons Complex award Excellence, Dept. Energy, 1986, 89; IR-100 award for innovation Rsch. and Devel. mag., 1987; scholar Vol. Cement Co., 1964, 3-M Co., 1965; indsl. fellow Cabot Corp., 1966. Fellow Am. Ceramic Soc.; mem. Nat. Inst. Ceramic Engrs. (cert.), Keramos, Inventors Forum (v.p. 1985-86), Sigma Xi, Tau Beta Pi, Phi Kappa Phi. Republican. Presbyterian. Achievements include patents and patents pending in field ceramics/metals/refractory materials; studies on low-expansion systems, hydrogen fluoride chemical laser methods, noncarbon furnace, refractory coatings, diamond syntheses, microwave sintering. Office: Martin Marietta Energy Systems Inc PO Box 2009 bldg 9202 Oak Ridge TN 37831-2009

HOLCOMBE, HOMER WAYNE, nuclear quality assurance professional; b. Winston-Salem, N.C., Oct. 7, 1949; s. Calvin Littleberry Holcombe and Mary Elizabeth (Fisher) Portwood; m. Kathleen Lorraine Sheldon; children: Matthew Michael Dickson, Meghan Elizabeth Holcombe, Adam Wayne Holcombe. AAS in Quality Assurance, Metropolitan State Coll., Denver, 1980; BS in Bus. Adminstrn., City U., Seattle, 1982. Cert. nuclear insp. Engring. assoc. Stone & Webster Engring. Corp., Boston, 1974-76; authorized nuclear insp. Hartford Steam Boiler Inspection & Ins. Co., Denver, 1976-80; project quality dir. Morrison-Knudsen Co., Elma, Wash., 1980-84; asst. project mgr. Pullman/Kenith-Fortson Co., Waynesboro, Ga., 1984-87; project quality mgr. MK-Ferguson Co., U.S. DOE Savannah River Site, Aiken, S.C., 1987-90; quality dir. MK-Ferguson of Oak Ridge Co., U.S. DOE Oak Ridge Site, Oak Ridge, Tenn., 1990-92; dep. gen. mgr. MK-Ferguson of Idaho Co., U.S. DOE Idaho Nat. Engring. Lab., Idaho Falls, 1992—; lectr. Aiken (S.C.) Tech. Coll., 1987. With U.S. Navy, 1968-74. Mem. Am. Nuclear Soc., Am. Soc. for Quality Control, Am. Welding Soc. Republican. Office: MK-Ferguson of Idaho Co PO Box 1745 Idaho Falls ID 83403-1745

HOLDAR, ROBERT MARTIN, chemist; b. Ozark, Ark., Feb. 10, 1949; s. Luther and Francess Ethyl (Briscoe) H.; m. Barbara Jean Sobczak, Jan. 5, 1985; children: Luther Edward, William Thomas, Frank King, Samuel Robert. BS in Chemistry, U. Ark., 1976; MS in Chemistry, Tex. A&M U., 1979; postgrad., U. Dallas, 1990—. Chemist Parkem Indsl. Svcs., LaPorte, Tex., 1979-80, Mohawk Labs div. NCH Corp., Irving, Tex., 1980—. Patentee in field. Chmn. Zoning Bd. Adjustments, Irving, 1991—;/ mem. Local Emergency Planning Commn., Dallas, 1991—. With USAF, 1968-73. W.K. Noyce scholar U. Ark., 1975. Mem. Am. Chem. Soc., Nat. Assn. Corrosion Engrs., Am. Soc. Lubrication Engrs. (chmn. North Tex. sect. 1982-83), Irving Rep. Club (editor 1989-92, treas. 1992—), Irving Noon Toastmasters (pres. 1983, Accomplished Toastmaster award 1986). Avocations: snow skiing, scuba, gardening. Home: 2816 Brockbank Dr Irving TX 75062-4523 Office: Mohawk Labs NCH Corp 2730 Carl Rd Irving TX 75062-6453

HOLDEN, ERIC GEORGE, microbiologist; b. Burlington, Vt., Sept. 6, 1963; s. Robert Adams and Madeline Norma (Jackson) H. BA, Hampshire Coll., 1958; MS, U. Vt., 1991. Microbiologist Applied Immune Scis., Menlo Park, Calif., 1991—. Mem. Sigma Xi.

HOLDEN, NORMAN EDWARD, nuclear physicist, philosopher, historian; b. N.Y.C., Feb. 1, 1936; s. Edward Francis and Marie Anne (O'Donnell) H.; m. Gail Rafferty, Aug. 25, 1962; children: Sean, Kristen, Maurisa, Gregory, Megan, Victoria, Keith, Kathleen, Kevin. BS in Physics, Fordham U., 1957; MS in Physics, Cath. U. Am., 1959, PhD in Physics, 1964. Physicist GE-Knolls Atomic Power Lab., Schenectady, N.Y., 1965-74, Brookhaven Nat. Lab. Nat. Nuclear Data Ctr., Upton, N.Y., 1974-90; rsch. coord. high flux beam reactor Brookhaven Nat. Lab., Upton, N.Y., 1990—; mem. cross sect. evaluation working group U.S. Dept. of Energy; advisor UN/IAEA, Vienna, Austria, 1976-77; U.S rep. Internat. Commn. RAdio-Chemistry and Nuclear Techniques, 1984—. Author: Neutron Cross Sections, 1981; contbr. Handbook Chemistry and Physics, 1990—. Mem. Shoreham (N.Y.) Wading River Bd. Edn., 1980-92; pres. spl. edn. PTA, 1978-80. 1st lt. USAF, 1962-65. Mem. Internat. Commn. on Atomic Weights (chmn., sec. Oxford, Eng. 1971-87, subcom. isotopic abundance measurements, working party on natural isotopic fractionation), Internat. Union Pure Applied Chemistry (i-norganic chemistry div., Oxford, 1979-85, 91—), Internat. Com. Radionuclide Metrology. Home: 10 James St Shoreham NY 11786-1827 Office: Brookhaven Nat Lab High Flux Beam Reactor Upton NY 11973

HOLDER, HAROLD D., public health administrator, communications specialist, educator; b. Raleigh, N.C., Aug. 9, 1939. AB in History and Journalism, Samford U., 1961; MA in Comm., Syracuse U., 1962, PhD, 1965. Dir. Comm. Rsch. Ctr., asst. prof. Dept. Journalism and Oral Comm. Baylor U., Tex., 1965-67; rsch. scientist divsn. rsch., dir. systems analysis and program evaluation N.C. Dept. Mental Health, Raleigh, 1967-69; assoc. prof., mem. grad. faculty dept. Sociology N.C. State U., Raleigh, 1967-75; site mgr. The Human Ecology Inst., Wellesley, Mass., 1973-75; sr. scientist The Human Ecology Inst., Chapel Hill, N.C., 1977-87, dir., 1980-87; mgr. devel. City of Portsmouth, Va., 1975-76; lectr. dept. Health Adminstrn. Sch. Pub. Health U. N.C., Chapel Hill, 1980-86; dir. prevention rsch. ctr. Pacific Inst. Rsch. and Evaluation, Berkeley, Calif., 1987—; lectr. health edn. program Sch. Pub. Health U. Calif., Berkeley, 1987—; vis. prof. Newhouse Comm. Ctr., 1966, U. N.C. Chapel Hill, 1979-80; guest lectr. Sch. Pub. Health U. Carolina, Chapel Hill, 1967-68, dept. Psychiatry, 1967-68; instr. advanced exec. systems N.C. Dept. Mental Health, 1967-69; program coord. John Ulmstead Disting. Lecture Series, 1970; ednl. evaluator Syracuse (N.Y.) Sch. Dist., 1971-73; trainer systems approaches and model bldg., dynamo computer simulation, United Meth. Ch., N.Y., 1972-73; sr. analyst Caseway, Inc., Brockton, Mass., 1977-78; mem. psycho-social internal review group Office Sci. Affairs Nat. Inst. Alcohol Abuse and Alcoholism, 1983-87, chmn. 1986-87; cons. Nat. Inst. Alcohol Abuse and Alcoholism, Washington, Tex. Ednl. Agy., Austin, Ctrl. Tex. Regional Edn. Ctr., James Connally Tech. Inst., Waco, Tex., Tech. Edn. Rsch. Ctr., Cambridge, Mass., U. N.C., Chapel Hill, N.C. Dept. Health, Child Advocacy Ctr., Durham, United Meth. Ch., N.Y.C., NIMH, Svcs. Integration Tech. Inst., Boston, Learning Inst. N.C; presenter, chairperson numerous confs. Editor: Advances in Substance Abuse: Behavioral and Biological Rsch., 1987; co-editor: Monitoring Child Service Agencies: A Guide for Child Advocacy Groups, 1974, Control Issues in Alcohol Prevention: State and Local Designs for the 80s, 1984, OSAP Prevention Monograph 4: Research, Action and the Community: Experiences in the Prevention of Alcohol and Other Drug Problems, 1990, Community Prevention Trials for Alcohol Problems: Methodological Issues, 1992; mem. editorial bd. Alcohol Health and Research World, 1990—; contbr. chpts. to books, articles to profl. jours. Bd. dirs. Ctrl. Tex. Rsch. Coun., 1965-67. Newhouse Rsch. fellow Syracuse U., 1961-65. Mem. APHA (alcohol and drug sect.), Am. Sociol. Assn., Am. Assn. Pub. Opinion Rsch.Nat. Soc. Study Comm., Nat. Assn. Pub. Health Policy (mem. alcohol policy com., sec./treas. 1985-86), Soc. Gen. Systems Rsch. (nat. coms. social systems, health systems), Rsch. Soc. Alcoholism (com. nat. advocacy), Soc. Health Svcs. Rsch. Office: Prevention Research Ctr Pacific Inst. Rsch. & Evaluation 2150 Shattuck Ave Ste 1900 Berkeley CA 94704

HOLDSWORTH, JANET NOTT, women's health nurse; b. Evanston, Ill., Dec. 25, 1941; d. William Alfred and Elizabeth Inez (Kelly) Nott; children: James William, Kelly Elizabeth, John David. BSN with high distinction, U. Iowa, 1963; M of Nursing, U. Wash., 1966. RN, Colo. Staff nurse U. Colo. Hosp., Denver, 1963-64, Presbyn. Hosp., Denver, 1964-65, Grand Canyon Hosp., Ariz., 1965; asst. prof. U. Colo. Sch. Nursing, Denver, 1966-71; counseling nurse Boulder PolyDrug Treatment Ctr., Boulder, 1971-77; pvt. duty nurse Nurses' Official Registry, Denver, 1973-82; cons. nurse, tchr. parenting and child devel. Teenage Parent Program, Boulder Valley Schs., Boulder, 1980-88; bd. dirs., treas. Nott's Travel, Aurora, Colo., 1980—; instr., nursing coord. ARC, Boulder, 1979-90, instr., nursing tng. specialist, 1980-82. Mem. adv. bd. Boulder County Lamaze Inc., 1980-88 ; mem. adv. com. Child Find and Parent-Family, Boulder, 1981-89; del. Rep. County State Congl. Convs., 1972-92, sec. 17th Dist. Senatorial Com., Boulder, 1982-92; vol. Mile High ARC, 1980; vol. chmn. Mesa Sch. PTO, Boulder, 1982-92, bd. dirs., 1982—, v.p., 1983—; elder Presbyn. ch. Mem. ANA, Colo. Nurses Assn. (bd. dirs. 1975-76, human rights com. 1981-83, dist. pres. 1974-76), Coun. Intracultural Nurses, Sigma Theta Tau, Alpha Lambda Delta. Republican. Home: 1550 Findlay Way Boulder CO 80303-6922 Office: Teenage Parent Program 3740 Martin Dr Boulder CO 80303-5499

HOLDSWORTH, ROBERT LEO, JR., emergency medical services consultant; b. Boston, Sept. 30, 1959; s. Robert Leo and Anne Marie (Walsh) H.; m. Deborah Ann Hendrickson, Mar. 20, 1993. Student, U. Hartford, West Hartford, Conn., 1977-80; cert. paramedic, Combined Hosps. Program, Hartford, Conn., 1986. Emergency room technician St. Francis Hosp., Hartford, 1981-82; EMT, L&M Ambulance Corp., Hartford, 1981-85; correctional paramedic State of Conn., Hartford, 1981-84; paramedic L&M Ambulance Corp., Hartford, 1983-85, gen. mgr., 1985-87; dir. emergency med. svcs. Lawrence and Meml. Hosp., New London, Conn., 1987-89; pres. Holdsworth & Assocs., East Berlin, Conn., 1988—; chmn. Ea. Regional Mobile Intensive Care Com., Norwich, Conn., 1987-89; pub. rels. dir. Conn. C.I.S.D. team, 1989-92. Author: Multiple Casualty Exercises, 1990, Guidebook on Occupational Exposure to Bloodborne Pathogens, 1992; writer, producer: (video) The Alternative. Mem. Nat Assn. EMT's, Nat. Soc. Emergency Med. Svc. Adminstrs., Nat. Soc. Emergency Med. Svc. Instrs., Nat. Soc. Paramedics, Conn. Soc. Emergency Med. Svc. Instrs., Kiwanis (chmn. pediatric tng. seminar Norwich 1990). Avocations: music, reading, model building, photography. Office: 1224 Mill St East Berlin CT 06023-1140

HO-LE, KEN KHOA, software engineer; b. Ho Chi Minh, Vietnam, Aug. 15, 1955; m. Cuc T.T. Nguyen, 1980; children: Nolan, Clifford, Lily. BE in Mech. Engring., Monash U., 1978, PhD in Mech. Engring., 1988; BSc in Computer Sci., U. Melbourne, 1981. Design engr. Telecom Australia, Melbourne, 1978-83; rsch. scientist Commonwealth Sci. and Indsl. Rsch. Orgn., Melbourne, 1984-88; dir. Abak Computing Pty., Ltd., Melbourne, 1988-90; system mgr. Leading Edge Techs., Melbourne, 1990-92; sr. software engr. Moldflow Pty., Ltd., Melbourne, 1992—; bd. dirs. Atty. med. Pty. Ltd., Melbourne, Abak Computing Pty. Ltd., Melbourne. Tech. referee for and contbr. articles to profl. jours. Colombo Plan scholar, 1973; recipient Sir Alexander Stewart Meml. award, 1991. Mem. Computer Soc. IEEE (assoc.), N.Y. Acad. Scis.

HOLGUIN, LIBRADO MALACARA (LEE HOLGUIN), civil engineer; b. Upland, Calif., May 9, 1957; s. Librado Banuelos and Lupe Malacara (Gomez) H. BSCE, Calif. State Poly. U., 1981; MSCE, Loyola Marymount U., L.A., 1986. Registered profl. engr., Calif. Design engr. L.A. County Sanitation Dist., Whittier, Calif., 1980-89; sr. design engr. Psomas and Assocs., Santa Monica, Calif., 1989-90; sr. project engr. Roy F. Weston, Inc., Woodland Hills, Calif., 1991—; speaker in field. Mem. ASCE (assoc.), Water Pollution Control Fedn. (assoc.). Home: PO Box 7000-420 Redondo Beach CA 90277 Office: Roy F Weston Inc 6400 Canoga Ave Ste 100 Woodland Hills CA 91367-2434

HOLLA, KADAMBAR SEETHARAM, chemist, educator; b. Kasargod, Kerala, India, May 16, 1934; s. Kadambar Parameshwar and Kadambar Bhagirathi Holla; m. Kadambar Sharada, Apr. 25, 1966; children: Kadambar Shabala, Kadambar Vishak. BSc in Tech., Bombay (India) U., 1957, MSc in Tech., 1959; PhD, Ohio State U., 1964. Scientist Tata Oil Mills Co. Ltd., Bombay, 1959-61, 65-72, plant-in-charge, 1972-78; mgr. Tata Oil Mills Co. Ltd., Madras, India, 1978-85, Delhi, India, 1985-88; head, gen. mgr. R & D Tata Oil Mills Co. Ltd., Bombay, 1988—; PhD examiner Bombay U., 1992, Kanpur (India) U., 1990; advisor Tamilnadu Govt. Agr. Dept., Madras, 1984; mem. selection-promotion com. C.F.T.R.I., Mysore, India, 1992; mem. com. B.A.R.C., Bombay, 1991. Contbr. rsch. papers to profl. jours. Fellow R.S.C.; mem. F.I.E., M.T.T. ChemE, Oil Technologists Assn. of India (pres. Western Zone), Sigma Xi. Achievements include patent for plant growth nutriant, neem extract process for industrial scale. Home: Tata Colony, 123 Falcon Crest, Parel, Bombay 400 012, India Office: Tata Oil Mills Co Ltd, Hay Bunder Rd Sewri, Bombay 400 033, India

HOLLAND, CHARLES EDWARD, corporate executive; b. Pottstown, Pa., Aug. 31, 1940; s. Charles Edward and Ethel Viola (Ludwig) H.; m. Linda Beth VandeBerg, Nov. 20, 1982. Student, Messiah Coll., 1962-63; BS in Biology and Chemistry, Albright Coll., 1966; PhD in Zoology, Rutgers U., 1974. Clin. lab. technician Reading Hosp., West Reading, Pa., 1962-66; rsch. assoc. dept. biochemistry St. Louis U. Sch. Medicine, 1972-75; rsch. assoc. dept. pharmacology and surgery U. Ill.Med. Ctr., Chgo., 1975-77; clin. project mgr. Am. Critical Care (Am. Hosp. Supply), McGraw Park, Ill., 1977-81; asst./assoc. dir. clin. rsch. Glaxo Inc., Research Triangle Park, N.C., 1981-84; dir. planning and project mgmt. Glaxo Inc., Research Triangle Park, 1984-86, dir. human resources, 1986-87, dir. strategic planning, 1987-88, dir. dermatology bus. expansion, 1988-89, dir. dermatology bus. and product devel., 1989-91, group dir. dermatology bus. and product devel., 1991—; mem. indsl. adv. bd. Pharmacotherapy jour., 1986—, Geriatric Medicine jour., 1986-89; bd. trustees Glaxo Bus. Sch., 1986-89, chmn. 1986-88. Contbr. articles to profl. jours. USPH fellow Rutgers U., 1966-71; grad. teaching fellow Rutgers U., 1971-72; NIH fellow St. Louis U. Sch. Medicine, 1973-75, U. Ill., 1975-77. Mem. Project Mgmt. Inst. (editor newsletter 1986-87), AAAS, Nat. Psoriasis Found., Soc. Investigative Dermatology, Am. Acad. Dermatology, Lic. Exec. Soc., Nat. Psoriasis Found., N.Y. Acad. Sci., Sigma Xi. Avocations: travel, nature, hiking, fishing, hot air ballooning. Office: Glaxo Inc 5 Moore Dr Research Triangle Park NC 27709

HOLLAND, DAVID LEE, environmental scientist, consultant; b. Plainfield, N.J., May 29, 1947; s. Leo Gerber and Myra Aubiene (Butler) H.; m. Patricia Ann Applegate, June 29, 1968; children: Katherine, Bradley, Christopher. BS, Rutgers U., 1971. Cert. plant A operator, S.C. Technician Hydronics Corp., Princeton, N.J., 1969-72; ops. mgr. Hyon Waste Mgmt. Svcs., Chgo., 1972-74; engr. Ebasco Svcs., Inc., Atlanta, 1974-76; sr. engr. Piedmont Engrs., Greenville, S.C., 1976-79; project mgr. Daniel Engring., Greenville, S.C., 1979-80; sr. environ. cons. Piedmont, Olsen, Hensley, Inc., Greenville, S.C., 1980—. Com. chmn. troop 776 Boy Scouts Am., Simpsonville, S.C., 1987—. Mem. S.C. Water Pollution Control, Water Environment Fedn. Republican. Office: Piedmont Olsen Hensley Inc PO Box 1717 420 E Park Ave Greenville SC 29602-1717

HOLLAND, JOHN LEWIS, psychologist; b. Omaha, Oct. 21, 1919; s. Edward Lewis and Ellen (Deane) H.; m. Elsie Margaret Prenzlow, Aug. 28, 1947 (dec. July 1988); children: Kay, Joan, Robert. BA, U. Omaha, 1942; PhD, U. Minn., 1952; DSc (hon.), Doane Coll., 1981; D of Letters (hon.), U. Nebr., 1988. Instr., dir. counseling ctr. Case We. Res. U., Cleve., 1950-53; chief, vocat. counseling svc. VA Hosp., Perry Point, Md., 1953-56; dir. rsch. Nat. Merit Scholarship Corp., Evanston, Ill., 1957-63; v.p. R&D Am. Coll. Testing, Iowa City, 1963-69; prof. social rels. Johns Hopkins U., Balt., 1969-80; pvt. rschr. Balt., 1980-92. Author: The Psychology of Vocational Choice, 1966, Making Vocational Choices, 1973, 85, (vocat. test) Self Directed Search, 1971, 77, 85. Sgt. USAF, 1942-46. Fellow APA (pres. div. 17 counseling psychology 1969), Am. Psychol. Soc., Ctr. for Advanced Studies in Behavorial Scis. Democrat. Achievements include research of theory of careers which led to the Self-Directed Search -- now the most widely used inventory in the world -- and to an occupational classification that is used by the U.S. Department of Labor and the governments of Canada and Australia. Home and Office: 111 St Albans Way Baltimore MD 21212

HOLLAND, KOREN ALAYNE, chemistry educator; b. South Weymouth, Mass., June 7, 1963; d. Herbert Gerry and Valerie Alayne (Shaw) Lipsett; m. Mitchell Mark Holland, May 14, 1988; 1 child, Alayne Ruth. BA, Skidmore Coll., 1985; PhD, U. Md., 1990. NIH postdoctoral fellow Johns Hopkins U., Balt., 1990-92; asst. prof. chemistry Gettysburg (Pa.) Coll., 1992—. Contbr. articles to profl. jours. Mem. AAAS, Am. Chem. Soc., Nat. Sci. Tchrs. Assn. Unitarian. Office: Gettysburg Coll Dept Chemistry Gettysburg PA 17325

HOLLAND, MICHAEL JAMES, computer services administrator; b. N.Y.C., Nov. 20, 1950; s. Robert Frederick and Virginia June (Wilcox) H.; Anita Garay, Jan. 5, 1981 (Aug. 1989); 1 child, Melanie. BA in Comparative Lit., Bklyn. Coll., 1972. Field med. technician 3rd Marine Div., Okinawa, Japan, 1976-77, 1st Marine Div., Camp Pendleton, Calif., 1978-79; clin. supr. Naval Hosp. Subic Day, Philippines, 1979-81; dept. head Tng. Ctr. USMCR, Johnson City, Tenn., 1981-84; clin. supr. No. Tng. Area, Okinawa, 1984-85, 3rd Marine Air Wing, Camp Pendleton, 1985-88; cons. Naval Regional Med. Command, San Diego, 1988-90; system analyst Naval Med. Info. Mgmt. Ctr. Detachment, San Diego, 1990-92; computer svcs. adminstr. U.S. Naval Hosp., Guam, 1993—. Mem. Fleet Res. Assn., Nat. City C. of C. (com. 1989-91), Assn. for Computing Machinery.

HOLLDOBLER, BERTHOLD KARL, zoologist, educator; b. Erling-Andechs, Germany, June 25, 1936; came to U.S., 1973; s. Karl and Maria (Russmann) H.; m. Friederike Probst, Feb. 9, 1980; children: Jakob, Stefan, Sebastian. Dr. rer. nat., U. Wurzburg, 1965; Dr. habil., U. Frankfurt a.M., 1969. Prof. zoology U. Frankfurt a.M., 1971-72; prof. biology Harvard U., Cambridge, Mass., 1973-90, Alexander Agassiz prof. zoology, 1982-90; prof. U. Wurzburg, Federal Republic of Germany, 1989—; adj. prof. U. Ariz., Tucson; rsch. assoc. Harvard U. Author: (with Edward O. Wilson) The Ants, 1990 (Pulitzer Prize for General Nonfiction). John Simon Guggenheim fellow, 1980; recipient Sr. Scientist award Alexander von Humboldt Found., 1986-87, Gottfried Wilhelm Leibniz prize, 1989. Mem. Am. Acad. Sci., Deutsche Akademie der Naturforscher Leopoldina, Bayerische Akademie der Wissenschaften.

HOLLEMWEGUER, ENOC JUAN, immunologist; b. Lima, Peru, May 11, 1947; came to U.S., 1966; s. Juan Enrique and Clotilde (Loayza) H.; m. Sandra Emma Eriksen, Aug. 10, 1974 (dec. Dec. 1974); m. Karla Rae Nicklas, Dec. 17, 1977; 1 child, Robert Juan. BS, Morningside Coll., 1970; PhD, U. S.D., 1978. Diplomate Am. Bd. Med. Lab. Immunologists. Rsch. fellow Mayo Cinic, Rochester, Minn., 1978-80, Nat. Jewish Hosp., Denver, 1980-81; dir. immunology Orlando (Fla.) Regional Med. Ctr., 1981-89; dir. immunology Nat. Ref. Lab., Nashville, 1989-91, dir. flow cytometry, 1989—; faculty pathology Orlando Regional Med. Ctr., 1981-89, faculty med. tech., 1981-89; faculty immunology Ctrl. Fla. Blook Bank, Orlando, 1984-89. Cub scout leader Boy Scouts of Am., Pack 210, Nashville, 1991—. Recipient Sister City Exch. Student scholarship, 1966-70, Lucille B. Wendt scholarship U. S.D., 1972, Rsch. fellowship Mayo Found., 1978-80, Nat. Jewish Ctr. for Immunology, 1980. Mem. Am. Soc. for Clin. Pathologists, Assn. Med. Lab. Immunologists. Republican. Methodist. Achievements include devel., organization and mgmt. immunology dept. at Orlando Regional Med. Ctr. and Immunology and Flow Cytometry Depts. at Nat. Ref. Lab. Office: Nat Reference Lab 1400 Donelson Pk Ste B-10 Nashville TN 37217

HOLLER, JOHN RAYMOND, electronics engineer; b. Vanderbilt, Pa., Sept. 22, 1918; s. Blaine and Mary Myrtle (Johnson) H.; m. Mavis Arvella Alexander, Sept. 1, 1945; children: Karen, Ronald, Kimberly, Roberta. BSEE with distinction, Purdue U., 1949, MSEE, 1951. Registered profl. engr., Ohio. Instr. Purdue U., W. LaFayette, Ind., 1947-51; engring. specialist Goodyear Aero Space Corp., Akron, Ohio, 1951-62; sr. project engr. Rockwell Internat., Columbus, Ohio, 1962-83; cons. systems engr. Rockwell Internat., Duluth, Ga., 1984-85. V.p., treas. Citizens for Pub. Schs., Gahanna, Ohio, 1967-68; chmn. east side Crop Walk, Columbus. Recipient Electrical Engring. scholarship Eta Kappa Nu, 1949, Engring. scholarship Tau Beta Pi, 1949. Mem. IEEE (sr., life), Sigma Xi. Democrat. Methodist. Achievements include original rsch. on cross-correlation techniques that eventually led to guidance systems like those on Tomahawk Missile; work leading to devel. smart bombs; contbr. to devel. of unique underwater swimmer detection systems. Home: 10850 Leitner Creek Dr # 143 Bonita Springs FL 33923

HOLLEY, EDWARD R., civil engineering educator. BSCE, MSCE, Ga. Inst. Tech., 1960; ScD, MIT, 1965. Joined faculty U. Ill. Urbana-Champaign, 1964; mem. faculty U. Tex. Austin, 1979—, now Stanley P. Finch Centennial prof. engring.; one yr. leave of absence at Bur. Reclamation, Denver, Delft Hydraulics Lab, The Netherlands, U. Queensland, Brisbane, Australia; part-time hydrologist Geologic Survey, 1974—. Mem. ASCE (chmn. fluids com. 1973-74, 75-76, Hilgard Hydraulic prize 1971), Am. Geol. Union, Internat. Assn. Hydraulic Research. Primary profl. and research interests currently environ. fluid mechanics including transport of pollutants in surface & groundwaters, outfalls and difussers, thermal discharges & deepwell disposal. Office: The Univ of Tex at Austin Dept of Civil Engring Austin TX 78712

HOLLIFIELD, CHRISTOPHER STANFORD, engineer, consultant; b. Cleve., Aug. 19, 1959; s. Thomas and Mary (Mollinets) H.; m. Debra A. Skvarenina, Jan. 3, 1980; 1 child, Crystal. Student, Broward (Fla.) County C.C., 1980-81, Emery Riddle U., 1982-84. FAA Airman Cert. Gen. foreman Cleve., Miami, Fla., 1981-86; FAA airman, v.p. ops. AAR Corp., Miami, 1987-92; v.p. Arrow Svc. Internat., Plantation, Fla., 1992—; cons. Borward Motors, Fort Lauderdale, Fla., 1990-92; adv. bd. Fla. Market Realty, West Palm Beach, 1991-92; bd. mem. Miami Racing Team Ltd., 1990—. Dir. Miami Disaster Assn., Homestead, Fla., 1992. Home: 4810 SW 188 Ave Fort Lauderdale FL 33332 Office: Arrow Svcs Internat 1520 NW 66 Ave Plantation FL 33313

HOLLINGER, CHARLOTTE ELIZABETH, medical technologist, tree farmer; b. Meadville, Miss., June 29, 1951; d. John Fielding and Irene Elizabeth (Mullins) H. BS in Biology, U. So. Miss., 1973. Cert. med. Technologist ASCP. Staff med. technologist U. Miss. Med. Ctr., Jackson, 1974-76, Grady Hosp., Atlanta, 1976, Atlanta ARC, 1976-78; staff med. technologist I Emory U. Hosp., Atlanta, 1978-85, staff med. technologist II, 1985-88, asst. chief technologist, 1988—; del. Blood Bank Del. to People's Republic China, People-to-People, Seattle, 1988. Supporter Alliance Theater, Atlanta Symphony, Am. Cancer Soc, Am. Heart Assn., United Way; mem. ASPCA, Humane Soc., Atlanta Zool. Soc., Atlanta Bot. Gardens, Atlanta Humane Soc., Audubon Soc., World Wildlife Fund, Nat. Geog. Soc., Smithsonian Instn., Puppetry Arts Ctr., Nat. Wildlife Fedn., Nature Conservancy, Ga. Conservancy, Sierra Club. Mem. PETA, NOW, Am Assn. Blood Banks, Am. Soc. Clin. Pathologists, People-to-People, Miss. Forestry Assn., Cousteau Soc., Ga. Pub. TV, Handgun Control, U. So. Miss. Alumni Assn., Delta Zeta, Pi Tau Chi. Roman Catholic. Avocations: reading, needlework, traveling, swimming, camping. Home: 2490 Silver King Dr Grayson GA 30221-1470 Office: Emory U Hosp Blood Bank 1364 Clifton Rd NE Ste C190 Atlanta GA 30322-1104

HOLLINGER, MANNFRED ALAN, pharmacologist, educator, toxicologist; b. Chgo., June 28, 1939. BS, North Park Coll., Chgo., 1961; PhD, Loyola U., Chgo., 1967. Postdoctoral fellow Stanford U., Palo Alto, Calif., 1967-69; prof., chmn. dept. med. pharmacology and toxicology U. Calif., Davis, 1969—. Author: Respiratory Pharmacology and Toxicology, 1985, Yearbook of Pharmacology, 1990, 91, 92; asst. editor, field editor Jour. Pharm. Exptl. Therapy, 1978—; cons. editor CRC Press, Boca Raton, Fla., 1989—. Mem. Yolo County Grand Jury, Woodland, Calif.; bd. dirs. Davis Little League. Burroughs-Wellcome fellow Southampton U., U.K., 1986; Fogarty sr. fellow NIH, Heidelberg (Germany) U., 1988. Office: U Calif Dept Pharmacology and Toxicology MSI A Rm 4453 Davis CA 95616-5224

HOLLINGSWORTH, CORNELIA ANN, food scientist; b. Carrollton, Ga., Mar. 6, 1957; d. Robert Allen Jr. and Peggy (Carroll) H. BS, Auburn U., 1979; MS, U. Nebr., 1981, PhD, 1984. Rsch. scientist Armour Food Co., Scottsdale, Ariz., 1984-88; mgr. product devel. Bil Mar Foods, Inc., Zeeland, Mich., 1988-92, dir. R & D, 1992—. Co-author book chpt.; contbr. abstracts to profl. jours. Mem. Inst. Food Technologists (profl.), Am. Meat Sic. Assn. (profl., bd. dirs. 1991-93), Gamma Sigma Delta, Delta Gamma, Phi Tau Sigma (profl., nat. bd. dirs. 1988-91, nat. pres. elect 1993—). Baptist. Avocations: sports, counted cross-stitch, flying, reading. Home: 636 Appletree Dr Holland MI 49423-5461 Office: Bil Mar Foods Inc 8300 96th Ave Zeeland MI 49464-9701

HOLLINGSWORTH, DAVID SOUTHERLAND, chemical company executive; b. Wilmington, Del.. BS in Chem. Engring., Lehigh U., 1948. Chem. engr. research ctr. Hercules Inc., Wilmington; chem. engr. lab. Hercules Inc., Kalamazoo, Mich.; tech. rep. sales office Hercules Inc., New Orleans and Wilmington, 1953-61; asst. sales mgr. paper chems. Hercules Inc., Wilmington, 1961-63, sales mgr. specialty chems., 1963-65, mgr. specialty paper chems. pine and paper chem. dept., 1965-67, dir. sales paper chems., 1967-72, from dir. mktg. to asst. gen. mgr., 1972-74, gen. mgr. new enterprise dept., 1974-75, gen. mgr. food and fragrance devel. dept., 1975-78, dir. worldwide bus. ctr. organics, 1978-79, v.p. planning, 1979-82, group v.p. water-soluable polymers, 1982-83, also bd. dirs., divisional v.p. mktg., 1983-84; pres. Hercules Specialty Chems. Co., Wilmington, 1984-86, vice chmn., 1986; chmn., chief exec. officer Hercules Inc., Wilmington, 1987—; bd. dirs. Del. Trust Co. Trustee Grand Opera House Inc.; bd. dirs. Med. Ctr. Del., Del. Symphony. Recipient Nat. medal Tech. U.S. Dept. Commerce Tech. Adminstrn., 1991. Mem. Chem. Mfrs. Assn. (bd. dirs.), Conf. Bd. Club: Hercules Country. Office: Hercules Inc 1313 N Market St Wilmington DE 19894-0001

HOLLIS, WILLIAM FREDERICK, information scientist; b. Cleve., May 25, 1954; s. Raymond Frederick and Elizabeth (Meyer) H.; m. Jo Anne Kohlenberg, June 25, 1977; children: George Anthony, Dawn Elizabeth. BS, Bowling Green State U., 1976; MLS, Kent State U., 1979, EdD, 1992. Cert. chemistry/physics educator Ohio. Info. specialist B.F. Goodrich Rsch. & Devel. Ctr., Breckville, Ohio, 1979-82; instr. libr. & info. sci. Coll. Wooster (Ohio), 1982-84; sr. info. specialist GenCorp Rsch., Akron, 1984, acting head tech. info., 1985, head tech. info ctr., 1986—; instr. sci. & tech. Stark Tech. Coll., Canton, Ohio, 1983-84. Elder United Ch. of Christ, Suffield, Ohio, 1986-89. Mem. Am. Chem. Soc., Am. Inst. Physics, Am. Soc. Info. Sci., Assn. Ednl. Communications & Tech. Office: GenCorp Rsch 2990 Gilchrist Rd Akron OH 44305

HOLLISTER, CHARLES DAVIS, oceanographer; b. Santa Barbara, Calif., Mar. 18, 1936; s. Clinton B. and Amelia Phipps (Davis) (Danelius) H.; m. Jalien Green, Feb. 8, 1958; children: Robin Jalien Hall, David Hellyer. B.S., Oreg. State U., 1960; Ph.D., Columbia U., 1967. Asst. scientist Woods Hole (Mass.) Oceanographic Inst., 1967-72, assoc. scientist, 1972-79, sr. scientist, 1979—, dean, 1979-89, v.p., 1989—; project dir. High Energy Benthic Boundary Layer Expt., Woods Hole, Mass., 1979-89. Author: (with Bruce C. Heezen) Face of Deep, 1971; contbr. 100 articles to profl. jours. and 3 books on sediment transport and waste disposal. Served with U.S. Army, 1955-57. Recipient Oliver La Gorce Gold medal Nat. Geog. Soc., 1967. Fellow AAAS, Geol. Soc. Am.; mem. Am. Geophys. Union. Home: 108 Woods Hole Rd Falmouth MA 02540-1646

HOLLISTER, FLOYD HILL, electrical engineer, engineering executive; b. Cortland, N.Y., Apr. 28, 1937; s. Floyd Howard and Marion Elizabeth (Hill) H.; m. Nancy Ellen Stoddard, June 4, 1958 (div. 1987); children: James K., William H., Robert E.; m. Lilliana Pejovich, Oct. 22, 1989. AB, Columbia U., 1958; PhD in Elec. Engring., Naval Postgrad. Sch., 1965. Registered control engr., Calif. Commd. USN, 1958, advanced through grades to comdr., ret., 1978; control systems engr. Naval Ocean Systems Ctr., San Diego, 1965-69; spl. asst. to commdg. officer Naval Shore Electronics Engring. Activity-Pacific, Honolulu, 1969-71; staff asst. info. and comm. Directorate Def. Rsch. and Engring., Washington, 1971-73; dir. info. systems divsn. Naval Electronic Systems Command, Washington, 1973-75; program mgr. info. systems Def. Advanced Rsch. Projects Agy., Arlington, Va., 1975-78; v.p. corp. rsch., devel., and engring., dir. computer sci. ctr. Texas Instruments, Dallas, 1978-91; dir. sci. tech. divsn. Software Engring. Inst. Carnegie Mellon U., Pitts., 1991—; mem. engring. sch. adv. bd. U. Ill., Urbana, 1987-92, U. Tex., Arlington, 1987-91; mem. adv. bd. automation and robotics inst. U. Tex., Arlington, 1987-91. Contbr. articles to sci. jours. Mem. IEEE (sr., mem. proceedings editorial bd. 1987-92), Soc. Mfg. Engrs. Achievements include several patents in field, including Method for Passively Determining the Relative Position of a Moving Observer with Respect to a Stationary Object. Office: Carnegie Mellon U Software Engring Inst Pittsburgh PA 15213-3890

HOLLMANN, RUDOLF WERNER, geologist, palaeontology researcher; b. Riga, Latvija, Aug. 25, 1931; s. Werner Franz and Sigrid Elly (Kalep) H.; divorced. Diploma in Geology, U. Tübingen, Fed. Republic of Germany, 1963, D of Natural Scis., 1964. Asst. U. Münster, Fed. Republic of Germany, 1964-69; employee Adminstrn. Pub. Property, Munich, 1969-71, Marine Documentation, Hannover, Fed. Republic of Germany, 1972-76, Natural History Museum, Hannover, Fed. Republic of Germany, 1977-79; cons. Marine Rsch. & Scis., Braunschweig, Fed. Republic of Germany, 1980-86; palaeontological sci. rschr. U. Braunschweig, 1986—; sci. diver World Underwater Fedn., Fed. Republic of Germany, Belgium, 1966—. Contbr. articles to profl. jours. Mem. Palaeontol. Soc. Germany. Lutheran. Avocations: languages (5), literature, zoology, botany. Home: P Fach 2530 Freya Str 82, D-38015 Braunschweig Germany Office: Tech Univ Braunschweig, Pockels Str 4, D-38106 Braunschweig Germany

HOLLOWAY, HARRY, aerospace medical doctor. Diploma Sch. Medicine, U. Okla., 1958. Chief neuropsychiatry SEATO Med. Rsch. Lab., Bangkok; dir. neuropsychiatry Walter Reed Army Inst. Rsch.; chmn. dept. psychiatry Uniformed Svcs. U. Health Scis., acting dean Sch. Medicine, 1990-92, dep. dean, 1992—; chmn. aerospace medicine advisory com. NASA, 1988; cons. World Health Orgn. Recipient Dist. Pub. Svc. award NASA, 1992. Achievements include research of impact of extreme environments on human adaption. Office: Natl Aeronautics & Space Admin 300 E St NW Washington DC 20546*

HOLLOWAY, PAUL FAYETTE, aerospace executive; b. Hampton, Va., June 7, 1938; s. Eldridge Manning and Minnie Powell H.; m. Barbara Jane Menetch, June 23, 1956; children—Paul Manning (dec.), Eric Scott. B.S., Va. Poly. Inst. and State U., 1960; postgrad., U. Va., 1961, Coll. William and Mary, 1962-63; grad. advanced mgmt. program, Harvard U., 1988. With NASA Langley Research Center, Hampton, Va., 1960—; aerospace technologist NASA Langley Research Center, 1960-69, mem. space shuttle task group, 1969—, chief space systems div., 1972-75; acting dep. assoc. adminstr. Office Aeronautics and Space Tech., 1977, dir. for space, 1975-85, dep. dir., 1985-91, dir., 1991—, acting dep. adminstr., 1992-93. Assoc. editor Jour Spacecraft and Rockets, 1972-77; editor in chief, 1978-80; contbr. articles in field to profl. jours. Mem. Poquoson Planning Commn.; v.p. local PTA; active Boy Scouts Am. Recipient Outstanding Leadership medal NASA, 1980, Exceptional Service medal, 1981; Presdl. Rank award for meritorious exec., 1981, Presdl. Rank award for disting. exec., 1987, Equal Opportunity medal, 1992, Disting. Svc. medal, 1992. Fellow AIAA (v.p. publs. 1991—), Am. Astronautical Soc.; mem. Fed. Exec. Inst. Alumni Assn., Internat. Acad. Astronautics, Sigma Gamma Tau, Phi Kappa Phi. Methodist. Home: 16 N Westover Dr Hampton VA 23662-1424 Office: Langley Rsch Ctr Mailstop 106 Hampton VA 23665

HOLLOWAY, PAUL HOWARD, materials science educator; b. Marion, Ind., Oct. 31, 1943; s. Charles D. and Pauline (Poe) H.; m. Bette Lorraine Zubrod, Jan. 10, 1943; children: Michael, Brian, Kimberly. BS, Fla. State U., 1965, MS, 1966; PhD, Rensselaer Poly. Inst., Troy, N.Y., 1972. Metallurgist Gen. Electric Co., Schenectady, 1966-69; staff mem. Sandia Nat. Lab., Albuquerque, 1972-78; assoc. prof. dept. materials sci. and engring. U. Fla., Gainesville, 1978-81, prof. dept. materials sci. and engring., 1981—; dir. MICROFABRITECH, Gainesville; statewide coord. Advanced Microelectronics and Materials Program, Tampa, 1988-91. Editor: Compound Semiconductor Growth, Processing Devices, 1988, Characterization of Metals and Alloys, 1993, Critical Revs. in Solid State and Material Scis.; contbr. chpts. to books, articles to profl. jours. Mem., chmn. Alachua County 4-H Adv. Coun., Gainesville, 1986-88; mem Alachua County Extension Office Adv. Coun., Gainesville, 1986-91. Recipient Muller award U. Wis.-Milw., 1988; named Tchr. of Yr. Coll. Engring. U. Fla., Gainesville, 1988. Fellow Am. Vacuum Soc. (pres. 1987); mem. ASTM (vice-chmn. 1983-86), Am. Soc. Metals, The Mining, Metallurgy, Materials Soc., Alpha Sigma Mu. Office: U Fla Dept Materials Sci Gainesville FL 32611

HOLLOWELL, MONTE J., engineer, operations research analyst; b. Helena, Ark., Dec. 30, 1949; s. Jerry B. and Imogene (Hartsfield) H.; m. Jan Bennett, Nov. 19, 1972; children: J. Brett, Matt J. BS in Math., BA in Physics, Ouachita Baptist Univ., 1972; MS in Indsl. Engring., U. Tex., El Paso, 1978. Air def. officer U.S. Army, 1972-80; industrial engr. PPG Industries, Wichita Falls, Tex., 1980-82; industrial engr. U.S. Army Missile Command, Redstone Arsenal, Ala., 1982-85, gen. engr., 1985—. Mem. Redstone Arsenal Military Ops. Rsch. Soc. Home: 12038 Chicamauga Trail Huntsville AL 35803-1546 Office: Advanced Systems Concepts Redstone Arsenal AL 35898-5242

HOLLYFIELD, JOE G., ophthalmology educator; b. El Dorado, Ark., Aug. 6, 1938; s. Nelwyn (Gilbert) H.; m. Mary Edmonson Rayborn. MS, La. State U., 1963; PhD, U. Tex., 1966. Asst. prof. biology and zoology U. Tex., Austin, 1966-67; postdoctoral fellow Hubrecht Lab. U. Utrecht, The Netherlands, 1967-69; from asst. to assoc. prof. dept. ophthalmology Coll. Physicians & Surgeons/Columbia U., N.Y.C., 1969-77; assoc. prof. ophthalmology Baylor Coll. Medicine, Houston, 1977-79, prof. ophthalmology, 1980—, prof. neurosci., 1989—; sr. scientific investigator Rsch. to Prevent Blindness, 1986. Editor: Experimental Eye Research, 1991—, The Structure of the Eye, 1982, Degenerative Retinal Disorders, 1987; contbr. articles to Nature, Exptl. Eye Rsch. Pres. Internat. Soc. for Eye Rsch., 1988-92, sec., 1984-87; vol. scientist Sci. by Mail Children's Mus. Houston, 1991. Recipient Morris J. and Betty Kaplan Found. award, 1983, Sam and Bertha Brochstein award Retina Rsch. Found., 1985, Marjorie Bellingham Prize, 1981, 93, Honor award for rsch. Alcon Rsch. Found., 1987. Fellow Am. Acad. Ophthalmology; mem. AAAS, Assn. for Rsch. in Vision and Ophthalmology (pres. 1993-94, trustee 1989-93), Internat. Soc. for Eye Rsch., Am. Soc. Cell Biology. Office: Cullen Eye Inst Baylor Coll of Medicine 1 Baylor Plz Houston TX 77030

HOLM, RICHARD HADLEY, chemist, educator; b. Boston, Mass., Sept. 24, 1933; m. Florence L. Jacintho, June 8, 1958; children—Sharon, Eric, Christian, Marg. B.S., U. Mass., 1955; Ph.D., Mass. Inst. Tech., 1959. Instr., then asst. prof. chemistry Harvard U., 1959-65, prof., 1980—; asso. prof. U. Wis., 1965-67; prof. chemistry Mass. Inst. Tech., 1967-75, Stanford U., 1975-80. Sloan Found. fellow, 1964-67. Mem. Am. Acad. Arts and Scis., NAS (Chem. Scis. award 1993), Am. Chem. Soc., Chem. Soc. London. Home: 40 Temple St Cambridge MA 02139-2404 Office: Harvard U Dept Chemistry Cambridge MA 02138

HOLMAN, HALSTED REID, medical educator; b. Cleve., Jan. 17, 1925; s. Emile Frederic and Ann Peril (Purdy) H.; m. Barbara Marie Lucas, June 26, 1949 (div. July 9, 1982); children: Michael, Andrea, Alison; m. Diana Barbara Dutton, Aug. 10, 1985; 1 child, Geoffrey. Student, Stanford U., 1942-43, UCLA, 1943-44; MD, Yale U., 1949. Med. resident Montefiore Hosp., N.Y.C., 1952-55; staff physician Rockefeller Inst., N.Y.C., 1955-60; prof. medicine Stanford (Calif.) U., 1960—, chmn. dept. medicine, 1960-71, co-chief div. family and community medicine, 1987—, dir. clin. scholar program, 1969—, dir. Multipurpose Arthritis Ctr., 1977—; pres. Midpeninsula Health Svc., Palo Alto, Calif., 1975-80; mem. adv. bd. Calif Health Facilities Commn., Sacramento, 1978-81, Office Tech. Assessment, U.S. Congress, 1979-81, Inst. Advancement of Health, N.Y.C., 1982—; Guggenhime prof. medicine, 1960—. Author 1 book; contbr. numerous articles to profl. jours. Recipient Bauer Meml. award Arthritis and Rheumatism Found., N.Y., 1964. Master Am. Coll. Rheumatology; fellow ACP; mem. Assn. Am. Physicians, Am. Soc. Clin. Investigation (pres. 1970), Western Assn. Physicians (pres. 1966),. Democrat. Home: 747 Dolores St Stanford CA 94305-8427 Office: Stanford University Stanford Arthritis Ctr 701 Welch Rd Ste 3301 Palo Alto CA 94304-2203

HOLMBERG, EVA BIRGITTA, research scientist; b. Stockholm, Sweden; came to U.S., 1982; m. Joseph Shaile Perkell, May 2, 1982; 1 child by pervious marriage, Cecilia Lundström. PhD, Stockholm U., 1993. Rsch. scientist MIT, Cambridge, 1982--, Mass. Eye and Ear Infirmary, Boston, 1992--. Author: Elementär Fonetik, 1979; contbr. articles to profl. jours. Home: 15 Iroquois Rd Arlington MA 02174 Office: MIT RLE 50 Vassar St Cambridge MA 02139

HOLMEN, REYNOLD EMANUEL, chemist; b. Essex, Iowa, Oct. 23, 1916; s. John Algot and Clara Amelia (Christensen) H.; m. Betty Jane Heginbottom, June 20, 1942 (dec. 1990); children: Karen C. Maass, John R., Robert C. AB, Augustana Coll., Ill., 1936; MS, U. Mich., 1937, PhD, 1949. Rsch. chemist DuPont Co., Phila. and Flint, Mich., 1937-46; sr. rsch. chemist 3M Co., St. Paul, 1948-55, sect. mgr. tech. info., 1955-57, inorganic sect. mgr., 1957-62, organic scouting mgr., 1959-62, mgr. R&D Lab., Reflective Product divsn., 1962-71, mgr. spl. enterprises dept., 1971-82; v.p. R&D KEMSERCH, Inc., Onamia, Minn., 1984—. Author: Kasimir Fajans: The Man and His Work, 1990. Rackham scholar U. Mich., 1936-37. Mem. Am. Chem. Soc., AAAS, Phi Lambda Upsilon, Sigma Gamma Epsilon. Lutheran. Achievements include 20 U.S. patents; development of first catalytic dehydration of lactic acid to acrylic acid, first catalytic dehydrochlorination of alpha-chloropronic acid to acrylic acid, (with other) first sealed polycellular cube-corner retroreflective sheet. Home: 2225 Lilac Ln White Bear Lake MN 55100-3824

HOLMES, CARL KENNETH, mechanical engineer; b. Dover, N.H., Mar. 19, 1960; s. Carleton and Margaret Laidlaw (Page) H. BS, U. N.H., 1983; MS, MIT, 1991. Edison engr. GE, Louisville, 1983-84, applications/devel. engr., 1984-87; sr. mfg. engr. GE, Somersworth, N.H., 1987-88; rsch. asst. MIT, Cambridge, 1988-90; staff technologist Mobil Solar Energy Corp., Billerica, Mass., 1990—. Recipient McConnell Scholarship U. N.H., Durham, 1983. Assoc. mem. ASME, Sigma Xi. Achievements include research in theoretical and exptl. stress analysis of high field bitter-type magnets, fracture behavior of silicon cut with a high power laser. Home: 3 Bartlett Rd Kensington NH 03827 Office: Mobil Solar Energy Corp 4 Suburban Park Dr Billerica MA 01821

HOLMES, DONNA JEAN, biologist; b. Decatur, Ill., July 26, 1954; d. Roy A. Holmes and Cynthia (Brock) Higgins; m. Ronald E. Meisner, Oct. 2, 1976 (div. 1985). AB, Miami U., Oxford, Ohio, 1976; MS, Bowling Green State U., 1980, PhD, 1987. Rsch. assoc. N.J. Med. Sch., Newark, 1986-89; sci. writing cons. Immunology Rsch. Inst., Annandale, N.J., 1990; asst. ecologist Savannah River Ecology Labs., U. Ga., Aiken, 1990-92; postdoctoral fellow in organismic and evolutionary biology Harvard U., Cambridge, Mass., 1990-93; rsch. assoc., Dept. Biological Scis. Univ. Idaho, 1993—; referee jours. Animal Behavior, 1990—, Jour. of Mammalogy, 1990—. Contbr. articles to profl. jours., chpts. to books. Fellow Nat. Inst. Aging, 1990, Assn. for Women in Sci., 1983; grantee NSF, 1986, Sigma Xi, 1983, 84. Mem. Assn. for Women in Sci., Animal Behavior Soc. (facilitator women's roundtables 1990-92, grants com. 1990-93), Am. Ornithologists' Union, Am. Soc. Mammalogists. Democrat. Unitarian-Universalist. Achievements include description of sternal scent gland in Virginia opossum, seasonal changes in social behavior and scent marking in Virginia opossum; discovery of evidence of role of olfaction in social behavior of Virginia opossum. Home: Concord Field Sta Old Causeway Rd Bedford MA 01730 Office: Harvard U Dept Evolutionary Biology 16 Divinity Ave Cambridge MA 02138

HOLMES, EDWARD W., physician, educator; b. Winona, Miss., Jan. 25, 1941; s. Edward and Mary (Hart) H.; m. Judith L. Swain, Jan. 25, 1980. BS, Washington & Lee U., 1963; MD, U. Pa., 1967. Prof. medicine and biochemistry Duke U., Durham, N.C., 1974-91; investigator Howard Hughes Med. Inst., 1974-87; prof., chmn. dept. medicine U. Pa., Phila., 1991—. Reviewer in molecular medicine. With USPHS, 1968-70. Grantee NIH. Mem. Am. Soc. Clin. Investigation, Assn. Am. Physicians. Office: Hosp U of Pa 34th & Spruce St Philadelphia PA 19104

HOLMES, GEORGE EDWARD, molecular biologist, researcher, educator; b. Chgo., May 8, 1937; m. Norreen Ruth Petersen, Mar. 12, 1967; children: George Petersen, Norreen Eliza. BA in Biology and Chemistry, Wiley Coll., 1960; MS in Natural Sci., Chgo. State U., 1967; postgrad., U. Calif., Davis, 1967-68; PhD in Molecular Biology, U. Ariz., 1973. Med. technologist DePaul Hosp., St. Louis, 1961, Chgo. Hosp., 1961-67; tchr. Chgo. Bd. of Edn., 1965-67; rsch. assoc. Rockefeller U., N.Y.C., 1973-74; asst. prof. dept. microbiology Coll. of Medicine Howard U., Washington, 1974-82, assoc. prof., 1982—. Contbr. articles to Nature, Jour. Virology, Virology, Molecular and Gen. Genetics, Jour. Gerontology, Jour. Mutation Rsch. NIH fellow in Molecular Biology, 1968-73; Nat. Inst. on Aging grantee, 1982-87. Mem. Am. Soc. for Biochemistry and Molecular Biology, Am. Soc. for Virology, Am. Inst. Chemists, Am. Men and Women in Sci. and Medicine, Gerontol. Soc. Am. Lutheran. Achievements include research in molecular and general genetics, nature, mutation, virology and gerontology. Office: Howard U Coll of Medicine Dept Microbiology Washington DC 20059

HOLMES, MARGARET E., health science administrator, researcher; b. Evansville, Ind., Dec. 11, 1942; d. Melvin Archie and Mary Ellis (Sexton) Raney. BS, Pa. State U., 1964; PhD, SUNY, Buffalo, 1973. Postdoctoral scholar U. Mich., Ann Arbor, 1973-74; sr. cancer rsch. scientist Roswell Park Meml. Inst., Buffalo, 1974-78; spl. asst. Nat. Cancer Inst., NIH, Bethesda, Md., 1978-82, health sci. adminstr., exec. sec., 1982-84, program dir., 1984-89, chief cancer ctrs. br., 1989—. Contbr. chpt. to book, articles to Jour. Gen. Microbiology, Environ. Sci. Rsch., Jour. Bacteriology, others. Mem. Sigma Xi. Office: Nat Cancer Inst Exec Pla N Rm 502 Bethesda MD 20892

HOLMES, RICHARD BROOKS, mathematical physicist; b. Milw., Jan. 7, 1959; s. Emerson Brooks Holmes and Nancy Anne (Schaffter) Winship. BS, Calif. Inst. Tech., 1981; MS, Stanford (Calif.) U., 1983. Sr. systems analyst Comptek Rsch., Vallejo, Calif., 1982-83; staff scientist Western Rsch., Arlington, Va., 1983-85; sr. scientist AVCO Everett (Mass.) Rsch. Lab., 1985-88; prin. rsch. scientist North East Rsch. Assocs., Woburn, Mass., 1988-90; sr. mem. tech. staff Rocketdyne div. Rockwell Internat., Canoga Park, Calif., 1990—; cons. North East Rsch. Assocs., 1990. Contbr. Matched Asymptotic Expansions, 1988; contbr. articles to Phys. Rev. Letters, Phys. Rev., Jour. of the Optical Soc. Am. and IEEE Jour. of Quantum Electronics. Mem. No. Calif. Scholarship Founds., Oakland, 1977; mem. Wilderness Soc., Washington, 1989. Stanford fellow Stanford U., 1982; fellow MIT, 1990; recipient Presdl. Medal of Merit, 1992. Mem. AAAS, Am. Phys. Soc., Optical Soc. Am. Achievements include patents for means for photonic communication, computation, and distortion compensation. Office: Rockwell Internat Rocketdyne Div 6633 Canoga Ave # FA40 Canoga Park CA 91309-2703

HOLMES, ROBERT WAYNE, service executive, consultant, biological historian; b. Brush, Colo., July 16, 1950; s. George William Jr. and Reba Mary (Sandel) H. BA, Western State Coll., 1972. Exec. Rose Exterminator Co., San Francisco, 1986-92; founder, owner BFE Cons., 1992—. Author: The Killing River. Mem. Smithsonian Instn., Washington, 1986, Sta. KRMA-TV-PBS, Denver, 1987, Ft. Morgan (Colo.) Heritage Found., 1988, Ctr. for Study of Presidency, Wilson Ctr., Nat. Mus. Am. Indian. Mem. AAAS, N.Y. Acad. Scis., Acad. Polit. Sci., Wilson Ctr. Assoc. Ctr. for Study of the Presidency, Am. Mus. Natural History, Denver Mus. Natural History, Nat. Mus. Am. Indian, Nature Conservancy, FPCN, SoAm. Explorers Club.

HOLMES, THOMAS JOSEPH, aerospace engineering consultant; b. Paterson, N.J., Oct. 22, 1953; s. George Murray and Juliette (Schutz) H. BSEE, BS in Aerospace Engring., Pa. State U., 1975; MS in Aerospace and Astron. Engring., Stanford U., 1976; MS in Indsl. Adminstrn., Carnegie-Mellon U., 1989; Hypnotherapist Cert., Palo Alto Sch. Hypnotherapy, 1991. Assoc. engr. Lockheed Missiles & Space Orgn., Sunnyvale, Calif., 1976-78; engr. Jet Propulsion Lab., Pasadena, Calif., 1978-79; sr. engr., project mgr. Systems Control Technology, Palo Alto, Calif., 1979—. Author: Advances in Control and Dynamic Systems, Vol. 23, 1986; contbr. articles to profl. jours. Counselor Parental Stress Hotline, Palo Alto, 1986. Mem. AIAA, IEEE, U.S. Cycling Fedn., Tau Beta Pi. Avocation: bicycle racing. Home: 2253 Harvard St Palo Alto CA 94306-1359 Office: Systems Control Tech 2300 Geng Rd Palo Alto CA 94303-3317

HOLMQUEST, DONALD LEE, physician, astronaut, lawyer; b. Dallas, Apr. 7, 1939; s. Sidney Browder and Lillie Mae (Waite) H.; m. Ann Nixon James, Oct. 24, 1972. B.S. in Elec. Engring., So. Meth. U., 1962; M.D., Baylor U., 1967, Ph.D. in Physiology, 1968; J.D., U. Houston, 1980. Student engr. Ling-Temco-Vought, Dallas, 1958-61; electronics engr. Tex. Instruments, Inc., Dallas, 1962; intern Meth. Hosp., Houston, 1967-68; pilot tng. USAF, Williams AFB, Ariz., 1968-69; scientist-astronaut NASA, Houston, 1967-73; research assoc. MIT, 1968-70; asst. prof. radiology and physiology Baylor Coll. Medicine, 1970-73; dir. nuclear medicine Eisenhower Med. Ctr., Palm Desert, Calif., 1973-74; assoc. dean medicine, assoc. prof. Tex. A&M U., College Station, 1974-76; dir. nuclear medicine Navasota (Tex.) Med. Ctr., 1976-84, Med. Arts Hosp., Houston, 1977-85; ptnr. Wood Lucksinger & Epstein, Houston, 1980-91, Holmquest & Assocs., Houston, 1991—. Contbr. articles to med. jours. Mem. Soc. Nuclear Medicine, Am. Coll. Nuclear Physicians, Tex. Bar Assn., Am. Fighter Pilots Assn., Sigma Xi, Alpha Omega Alpha, Sigma Tau. Home: 3721 Tangley St Houston TX 77005-2031

HOLONYAK, NICK, JR., electrical engineering educator; b. Zeigler, Ill., Nov. 3, 1928; s. Nick and Anna (Rosoha) H.; m. Katherine R.A. Jerger, Oct. 8, 1955. BS, U. Ill., 1950, MS, 1951, PhD (Tex. Instruments fellow), 1954; DSc (hon.), Northwestern U., 1992. Mem. tech. staff Bell Telephone Labs., Murray Hill, N.J., 1954-55; physicist, unit mgr., mgr. advanced semiconductor lab. Gen. Electric Co., Syracuse, N.Y., 1957-63; prof. elec. engring. and materials research lab. U. Ill., Urbana, 1963—, John Bardenn chair prof. elec. & computer enging. & physics, 1993—; mem. Center Advanced Study, 1977—; series editor Prentice-Hall, Inc., 1962—; cons. Monsanto Co., 1964-89, Nat. Electronics Co., 1963-70, Skil Corp., 1967, GTE Labs. Tech. Adv. Council, 1973, Xerox, 1983-87, Ameritech, 1985-86. Author: (with others) Semiconductor Controlled Rectifiers, 1964, Physical Properties of Semiconductors, 1989. Served with U.S. Army, 1955-57. Recipient Cordiner award GE, 1962, John Scott medal City of Phila., 1975, GaAs Conf. award with Welker medal 1976, Monie A. Ferst award Sigma Xi, 1988, Nat. Medal Sci. NSF, 1990, NAS award Indsl. Application Sci., 1993, ASEE Centennial medal, 1993, 50th Ann. award Am. Elec. Assn., 1993. Fellow Am. Phys. Soc., IEEE (Morris Liebmann award 1973, Jack A. Morton award 1981, Edison medal 1989), Am. Acad. Arts and Scis., Am. Optical Soc. (Charles H. Townes award 1992); mem. AAAS, NAE, NAS, (Indsl. Application of Sci. award 1993), Electrochem. Soc. (Solid State Sci. and Tech. award 1983), Math. Assn. Am., Ioffe Inst. (hon. 1992). Home: 2212 Fletcher St Urbana IL 61801-6915 Office: U Ill Dept Elec/Computer Engring 1406 W Green St Urbana IL 61801-2991

HOLT, DONALD A., university administrator, agronomist, consultant, researcher; b. Minooka, Ill., Jan. 29, 1932; s. Cecil Bell and Helen (Eickoff) H.; m. Marilyn Louise Jones, Sept. 6, 1953; children: Kathryn A. Holt Stichnoth, Steven Paul, Jeffrey David, William Edwin. BS in Agrl. Sci., MS in Agronomy, U. Ill.; PhD in Agronomy, Purdue U. Farmer Minooka, Ill., 1956-63; instr., asst. prof., assoc. prof. then prof. agronomy Purdue U., West Lafayette, Ind., 1964-82; prof., head dept. agronomy U. Ill., Urbana-Champaign, Ill., 1982-83, dir. Ill. Agr. Expt. Sta., assoc. dean Coll. Agr., 1983—; cons. Deere and Co., Ottumwa, Iowa, 1978, NASA, Houston, 1979, Control Data Corp., Mpls., 1978-79, EPA, Corvallis, Oreg., 1981-92. Town Bd. commr., Otterbein, Ind., 1972-76. Fellow AAAS, Am. Soc. Agronomy (pres. 1988), Crop Sci. Soc. Am.; mem. Agrl. Rsch. Inst. (pres. 1991), Am. Forage and Grassland Coun., Ill. Forage and Grassland Coun., Gamma Sigma Delta (internat. pres. 1974-76). Republican. United Methodist. Home: 1801 Moraine Dr Champaign IL 61821-5261 Office: U Ill Agr Expt Sta 211 Mumford Hall 1301 W Gregory Dr Urbana IL 61801-3608

HOLT, ELIZABETH MANNERS, chemistry educator; b. Montclair, N.J., Aug. 2, 1939; d. Theodore Roland and Helen (Whitenack) Manners; m. Smith Lewis Holt, Aug. 24, 1963; children: Alexandra Doepel, Smith Lewis III. BA, Smith Coll., 1961; PhD, Brown U., 1966. Rsch. assoc. Poly. Laereanstalt, Copenhagen, 1965-66; instr. Poly. Inst. Bklyn., N.Y., 1968-69; postdoctoral fellow U. Wyo., Laramie, 1969-74; temporary asst. prof. Colo. State U., Fort Collins, 1974, U. Wyo., Laramie, 1975-78; rsch. assoc. U. Ga., Athens, 1978-80; prof. Okla. State U., Stillwater, 1981—. Recipient Amoco award for teaching Okla. State U., 1984, Fulbright rsch. award Fulbright Commn., Morocco, 1988, 1992—. Mem. Am. Chem. Soc., Am. Crystallographic Assn. Office: Chemistry Dept Okla State Univ Stillwater OK 74074

HOLT, FRANK ROSS, retired aerospace engineer; b. New Haven, Ind., Aug. 3, 1920; s. Lewis Wesley and Emma (Martin) Hollopeter; m. Treva Mae Souder, Sept. 2, 1946; children: Lou Cinda, Erik Lee, Brian Ross, Kevin Wayne. Student, Evansville U., 1946-48; BS in Engring., Purdue U., 1950. Engr. Ford Tractor and Implement Div., Birmingham, Mich., 1954-56; test engr. Northrop Corp., Hawthorne, Calif., 1957-65; tech. supr. NASA test facilities Johnson Space Ctr. Northrop Svcs., Inc., Houston, 1965-83; ret., 1983. Lt. col. USAF, 1942-45, ETO., Korea; Res. ret. Mem. Am. Vacuum Soc. (emeritus), Conway (Ark.) Noon Lions Club (sec.-treas. 1988-91), Res. Officers Assn., Air Force Assn., Mil. Order World Wars (commdr. Searcy-Fairfield Bay chpt. 1993—). Methodist. Home: 256 Hwy 286 E Conway AR 72032-8705

HOLT, WILLIAM HENRY, physicist, researcher; b. San Antonio, Aug. 5, 1939; s. Joseph Marion and Mildred Louise (Ragsdale) H.; m. Margaret Ann Harrell, June 21, 1963; children: Benjamin, Andrew. BS cum laude, St. Mary's U., San Antonio, 1960; MA, U. Tex., 1962, PhD, 1967. Postdoctoral fellow, lectr. U. Man., Winnipeg, Can., 1966-69; rsch. physicist Naval Surface Warfare Ctr., Dahlgren, Va., 1969—. Patentee; contbr. articles and papers to numerous sci. jours. and revs. Tchr. Sunday sch. St. Matthias United Meth. Ch., Fredericksburg, Va., 1973—, past co-chmn. edn., past lay leader, past mem. pastor-parish rels. com., past chmn. coun. on ministries. Mem. Am. Phys. Soc., Can. Assn. Physicists, Sigma Xi, Sigma Pi Sigma, Lions. Office: Naval Surface Warfare Ctr Dahlgren VA 22448-5000

HOLTMAN, MARK STEVEN, chemist; b. Dayton, Ohio, Apr. 3, 1949; s. Ray G. and Anne E. (Nau) H.; m. Patricia R. Blum, Jan. 13, 1980; 1 child, Robert. BS, U. Dayton, 1971; MS, Wright State U., 1973; PhD, Ohio State U., 1979. Sr. rsch. chemist PPG Industries, Barberton, Ohio, 1979-84; engring. specialist BP America, Cleve., 1985-92; corp. quality mgr. Am. Analytical Labs., Inc., Akron, Ohio, 1993—. Mem. Am. Chem. Soc., Am. Soc. for Quality Control. Achievements include 1 U.S. patent and 1 European patent, polymerizates of (allyl carbonate) and aliphatic polyurethanes having acrylic unsaturation; reduction of discoloration of aromatic peroxide initiated polyol (allyl carbonate) polymer. Home: 970 Brookpoint Dr Macedonia OH 44056 Office: Am Analytical Labs Inc 840 S Main St Akron OH 44311-1516

HOLT-OCHSNER, LIANA KAY, psychology educator, researcher; b. Redding, Calif., Dec. 15, 1958; d. James Cravens and Angela Emily (Atkins) Holt; m. Daniel Thomas Ochsner, Aug. 25, 1990; 1 child, Alexandra; stepchildren: Laurie, Joanie. BA, U. So. Calif., 1984, MA, 1986, PhD, 1988. Rsch. asst. U. So. Calif., L.A., 1982-88, asst. lectr., 1987-88; statis. cons., rsch. assoc. Inst. for Prevention Rsch., Pasadena, Calif., 1987-88; asst. prof. psychology Southeastern La. U., Hammond, 1989—; vice chair bd. dirs. Youth Svc. Bur., Hammond, 1990—. Contbr. chpt. to book, articles to Annals of Dyslexia, Jour. of Exptl. Child Psychology, others. Judge coord. Regional Sci. Fair, Hammond, 1991, 92; vol. phone counselor Tangipahoa Crisis Phone, Hammond, 1990-91. Sigma Xi rsch. grantee, 1985, Southeastern La. U. grantee, 1989-90, 90-91, 92-93. Mem. Am. Psychol. Soc., Orton Dyslexia Soc., Soc. for Rsch. in Child Devel., Sigma Xi. Achievements include fostering automatic word recognition in dyslexic children; research on drug use; the effects of gender-related personality traits on drug use among college students. Office: Southeastern La U 563 Hammond LA 70402

HOLTZ, NOEL, neurologist; b. N.Y.C., Sept. 13, 1943; s. Irving and Lillian H.; m. Carol Sue Smith, June 9, 1968; children: Pamela Wendy, Aaron David, Daniel Judah. BA, NYU, 1965, MD, U. Cin., 1969. Diplomate Am. Bd. Psychiatry and Neurology. Intern Cin. Gen. Hosp., 1969-70; resident in internal medicine and neurology Emory U., Atlanta, 1970-71, 73-76; pvt. practice medicine specializing in neurology, Marietta, Ga., 1977—; mem. faculty Emory U. Coll. Medicine, Atlanta, 1977—, asst. prof. dept. neurology, 1977—, assoc. prof., 1987, vice-chmn. dept. medicine, 1988-90, chmn., 1991—; mem. staffs Kennestone Hosp.; dir. neurodiagnostics unit; mem. staff Grady Meml. Hosp.; cons. Ga. Med. Care Found. Neurology. Co-author: Conceptual Human Physiology, 1985. With USN, 1971-73. Mem. Am. Acad. Neurology, Ga. Neurol. Soc. (sec.-treas., pres. 1990-92), Alpha Omega Alpha. Office: 522 North Ave Marietta GA 30060-1147

HOLTZ, TOBENETTE, aerospace engineer; b. Rochester, N.Y., June 20, 1930; d. Marcus and Leah (Cohen) H.; m. Joseph Laurinovics, Dec. 25, 1964. BS in Aeronautical Engring., Wayne State U., 1958; MS in Aero/Astro Engring., Ohio State U., 1964; PhD, U. So. Calif., L.A., 1974. Sr. engr. North Am. Aviation, Columbus, Ohio, 1954-59; rsch. assoc. Ohio State U., Columbus, 1959-60; sr. engr. U. So. Calif. Rsch. Found., Pt. Mugu, 1960-62, Northrop Corp., Hawthorne, Calif., 1962-67; engring. specialist McDonnell Douglas Corp., Huntington Beach, Calif., 1967-75; staff engr. Acurex Corp., Mountain View, Calif., 1975-76; project mgr. Aerospace Corp., El Segundo, Calif., 1976-82; tech. mgr. TRW, Inc., San Bernardino, Calif., 1982—. Contbr. articles to profl. jours. Fellow AIAA (assoc., sect. vice chair 1980-82, 91-92, nat. tech. com. 1991—, organizer nat. tech. confs. 1979, 86, 88, Disting. Svc. award 1983). Office: TRW Inc PO Box 1310 San Bernardino CA 92402

HOLTZAPPLE, MARK THOMAS, biochemical engineer, educator; b. Enid, Okla., Nov. 16, 1956; s. Arthur Robert and Joan Carol (Persson) H.; m. Carol Ann Kamps, Jan. 11, 1992. BS, Cornell U., 1978; PhD, U. Pa., 1981. Capt. U.S. Army, Natick, Mass., 1982-85; asst. prof. Tex. A&M U., College Station, 1986-91, assoc. prof., 1991—. Author papers on models for describing enzymatic hydrolysis of cellulose, pretreatments to enhance enzymatic digestion of cellulose; contbr. articles to profl. jours. Recipient Teaching awards Coll. Engring, 1990, Gen. Dynamics, 1990, Tenneco, 1991, Dow, 1991. Mem. Am. Inst. Chem. Engring. (exec. com. South Tex. sect. 1991-92). Achievements includes patents for torque monitor; orientation insensitive, high-efficiency evaporator; hermetic compressor; biomass process and high-efficiency refrigeration. Home: 1805 Southwood Dr College Station TX 77840 Office: Dept Chem Engring Tex A&M U College Station TX 77843

HOLTZBERG, FREDERIC, chemist, solid state researcher; b. N.Y.C., Apr. 12, 1922; s. Abraham and Edith (Schild) H.; m. Sylvia Nesin, June 25, 1950; 1 child, Margaret Holtzberg-Call. BS, Bklyn. Coll., 1947; PhD, Polytech. U., 1952; Docteur Honoris Causa, U. Grenoble, France, 1987. Mgr. IBM T.J. Watson Sci. and Computing Lab. Columbia U., N.Y.C., 1952-59; tech. asst. to spl. asst. for sci. and tech. Pres. U.S., Washington, 1959-60; cons. Office of Sci. and Tech. Exec. Office of Pres., Washington, 1959-60; staff technician Dir. of Rsch. IBM Rsch. Ctr., Yorktown Heights, N.Y., 1960-61, mem. rsch. staff, 1961-63, group mgr. material sci., 1963-74, dept. mgr. materials sci., 1974-82; adj. prof. U. Grenoble, 1975-76. Mem. editorial adv. bd. Jour. Solid State Chemistry, 1972-89, Rev. de Chimie Minerale, 1976-86; mem. internat. coun. Alloy Phase Diagrams, 1980-86. Recipient Outstanding Contbn. award IBM, 1966, Outstanding Tech. Achievement award IBM, 1989. Fellow Am. Phys. Soc. (Internat. New Materials prize 1991), Am. Chem. Soc., Am. Crystallographic Soc., Sigma Xi. Office: IBM T J Watson Rsch Ctr PO Box 218 Yorktown Heights NY 10598-0218

HOLTZMAN, RICHARD BEVES, health physicist, chemist; b. Chgo., Sept. 24, 1927; s. Samuel and Fannie (Greenspon) H.; m. Marilyn Wasserman, Sept. 20, 1953; children: Faye M. Brislawn, Alan B., Jill R. Larson. BS, U. Chgo., 1946, MS in Chemistry, 1952, PhD in Chemistry, 1953. Rsch. physicist Armour Rsch. Found., Chgo., 1954-59; sr. radiation specialist U.S. Nuclear Regulatory Commn., Glen Ellyn, Ill., 1984-91; chemist Argonne (Ill.) Nat. Lab., 1959-84, 1991-92, health physicist/group leader in internal dosimetry, 1992—. Contbr. more than 150 articles to profl. jours., 1951—. With U.S. Army, 1954-56. Fellow Health Physics Soc. (pres. Midwest chpt. 1981-82); mem. Am. Chem. Soc., Rsch. Soc. Am., Sigma Xi. Jewish. Achievements include discovery of distbn. and doses of naturally occurring radioactivity in man and the environment; development of inspection programs and quality assurance programs for nuclear power plant chemistry labs. and environ. programs. Home: 6108 Carpenter St Downers Grove IL 60516 Office: Argonne Nat Lab 9700 S Cass Ave Argonne IL 60439

HOLTZMAN, WAYNE HAROLD, psychologist, educator; b. Chgo., Jan. 16, 1923; s. Harold Hoover and Lillian (Manny) H.; m. Joan King, Aug. 23, 1947; children: Wayne Harold, James K., Scott E., Karl H. B.S., Northwestern U., 1944, M.S., 1947; Ph.D., Stanford, 1950; L.H.D. (hon.), Southwestern U., 1980. Asst. prof. psychology U. Tex., Austin, 1949-53; assoc. prof. U. Tex., 1953-59, prof., 1959—; dean Coll. Edn., 1964-70, Hogg prof. psychology and edn., 1964—; assoc. dir. Hogg Found. Mental Health, 1955-64, pres., 1970—; Dir. Social Sci. Research Council, 1957-63, Centro de Investigationes Sociales, Mex., 1960-70; cons. USAF, also mem. sci. adv. bd., 1969-71; mem. com. basic research com. NRC, 1968-71; mem. behavioral sci. study sect. USPHS, 1957-59, mental health study sect., 1960, chmn. personality and cognition research rev. com., 1968-72; research adv. panel Social Security Adminstrn., 1961-62; mem. Latin Am. adv. bd. IBM, 1985-89. Author: (with B.M. Moore) Tomorrow's Parents, 1964, Computer Assisted Instruction Testing and Guidance, 1971, (with R. Diaz-Guerrero and J. Swartz) Personality Development in Two Cultures, 1975, Introduction to Psychology, 1978; (with K.A. Heller and S. Messick) Placing Children in Special Education, 1982, (with T. Bornemann) Mental Health of Immigrants and Refugees, 1990, School of the Future, 1992; editor: Jour. Ednl. Psychology, 1966-72. Trustee Ednl. Testing Service, Princeton, 1972-74, 77-80, 83-86, J.W. and Cornelia Scarborough Found., 1977-82, Ctr. for Applied Linguistics, 1978-80, Salado Inst. Humanities, 1980-85, Population Inst., 1979-85, Menninger Found., 1982—, Population Resource Ctr., 1980—; dir. Sci. Research Assocs., 1975-88; pres., bd. dirs. S.W. Ednl. Devel. Lab., 1974-75; mem. adv. com. computing activities NSF, 1970-73; mem. computer sci. and engring. bd. Nat. Acad. Scis., 1971-73, chmn. panel on selection and placement of mentally retarded students, 1979-82; chmn. interdisciplinary cluster on social and behavioral devel. Pres.'s Biomed. Research Panel, 1975-76; bd. dirs. Found.'s Fund for Research in Psychiatry, 1973-77, chmn., 1976-77; dir. Conf. of S.W. Found., 1976-84, pres., 1978-79; mem. nat. adv. mental health council Alcohol, Drug Abuse, and Mental Health Adminstrn., 1978-81; mem. acad. info. systems adv. council IBM, 1982-85. Served to lt. (j.g.) USNR, 1944-46. Faculty Research fellow Social Sci. Research Council, 1953-54; Faculty Research fellow Center Advanced Study Behavioral Scis., 1962-63. Fellow APA; mem. AAAS, Tex. Psychol. Assn. (pres. 1957), S.W. Psychol. Assn. (pres. 1958), Am. Statis. Assn., Interam. Soc. Psychology (pres. 1966-67), Am. Ednl. Rsch. Assn., Internat. Union Psychol. Scis. (sec.-gen. 1972-84, pres. 1984-88, exec. com. 1988-92), Philos. Soc. Tex. (pres. 1982-83), Sigma Xi. Methodist. Home: 3300 Foothill Dr Austin TX 78731-5823

HOLUB, ROBERT FRANTISEK, nuclear chemist, physicist; b. Prague, Czechoslovakia, Sept. 19, 1937; came to U.S., 1966; s. Stanislav and Marie (Prochazkova) H.; m. Johnna S. Thames, Dec. 27, 1977; children: Robert M., John F., Elisabeth J. BS, Charles U., Prague, 1958, MS, 1960; PhD, McGill U., 1970. Research assoc. Fla. State U., Tallahassee, 1970-73; teaching intern U. Ky., Lexington, 1973-74; rsch. physicist Bur. Mines, U.S. Dept. Interior, Denver, 1974—; cons. Internat. Atomic Energy Agy., Vienna, Austria, 1984-89, key participant radon intercalibration program, 1990—; faculty affiliate Colo. State U., Ft. Collins, 1982—. Patentee continuous working level exposure apparatus. Contbr. articles to sci. jours. NRC Can. scholar, 1967-70. Mem. Am. Phys. Soc., Health Physics Soc., Am. Assn. for Aerosol Rsch.

HOLVECK, DAVID P., pharmaceutical company executive; b. 1945. With Corning Glass Works, 1964-79; mgr. digital x-ray Gen. Electric Corp., 1979-83; with Centocor, Inc., Malvern, Pa., 1983—, exec. v.p. N.Am., 1986-87, exec. v.p. in vitro diagnostic products, 1987-88, exec. v.p., pres. diagnostics divsn., 1988—. Office: Centocor Inc 200 Great Valley Pkwy Malvern PA 19355*

HOLZBACH, JAMES FRANCIS, civil engineer; b. Elizabeth, N.J., July 26, 1936; s. Norman Bernard and Mary Elizabeth (Devine) H.; m. Juliette Horwitz, May 21, 1977. BSCE, U. Notre Dame, 1960. Registered profl. engr., N.Y. Commd. ensign C.E. Corps USN, 1960, advanced through grades to lt. comdr., 1967, resigned, 1970; contract adminstr. Teetor-Dobbins Cons. Engrs., Rochester, N.Y., 1970-72; assoc. engr. Monroe County Dept. Engring., Rochester, 1972-90, acting chief constrn., 1990-92, mng. engr., 1993—. Contbr. articles to profl. jours. Mem. ASCE. Home: 50 Westminster Rd Rochester NY 14607-2224 Office: Monroe County Dept Engring 350 E Henrietta Rd Rochester NY 14620-2728

HOLZBACH, RAYMOND THOMAS, gastroenterologist, author, educator; b. Salem, Ohio, Aug. 19, 1929; s. Raymond T. and Nelle A. (Conroy) H.; m. Lorraine E. Cozza, May 26, 1956; children—Ellen, Mark, James. B.S., Georgetown U., 1951; M.D., Case Western Res. U., 1955. Diplomate Nat. Bd. Med. Examiners, Am. Bd. Internal Medicine. Intern, asst. resident U. Ill. Research and Edn. Hosps., Chgo., 1955-56; sr. asst. resident medicine Cleve. Met. Gen. Hosp., 1959-60; fellow in gastroenterology U. Calif., San Francisco, 1960-61; asst. chief gastroenterology Case Western Res U., 1961-63; physician Gastroenterology Unit U. Hosps. of Cleve., 1961-63; instr. medicine Case Western Res. U. Sch. Medicine, Cleve., 1961-64; clin. instr. medicine Case Western Res. U. Sch. Medicine, 1964-71; head gastrointestinal research unit, assoc. physician div. medicine St. Luke's Hosp., Cleve., 1967-73; dir. div. gastroenterology St. Luke's Hosp., 1970-73; head gastrointestinal research unit dept. medicine Cleve. Clinic Found., 1973—; vis. prof. numerous instns. including: Mayo Med. Sch., 1974, U. Calif.-San Diego, 1977, U. Heidelberg, 1978, U. Pa., 1979, U. Zurich, 1980, U. Munich, 1982, U. Minn. Med. Ctr., 1985, med. ctrs. numerous Japanese univs., 1986, 1992, Karolinska Inst., 1986, Royal Soc. London, 1987, Pa. State U. Sch. Med., U. Helsinki, RWTH-Aachen, Düsseldorf, Fed. Republic of Germany, U. Groningen, Utrecht, U. Amsterdam, The Netherlands, 1989, U. Perugia, Italy, Va. Commonwealth U.-Med. Coll. Va., Richmond, Christ Ch. Sch. Medicine, U. Otago, New Zealand, SUNY, Buffalo Sch. Medicine, 1990, Pontifical/Cath. U. of Chile Sch. Medicine, 1991, Hiroshima U. Sch. Medicine, 1992, Kyoto U. Sch. Medicine, 1992, Jikei U. Sch. Medicine, 1992, U. Bologna, Italy, 1992, Sch. Medicine U. Hiroshima, Tokyo, 1992, Sch. Medicine U. Jikei, Tokyo, 1992; lectr. in field. Editorial bd. Gastroenterology jour., 1984-89; contbr. revs. and articles to med. jours. Served to capt. USAF, 1957-59. Recipient Alexander von Humboldt Found. Spl. Program award, 1978, 82. Fellow ACP; mem. Am. Gastroent. Assn. (research com. 1976-79), Central Soc. Clin. Research, Am. Assn. Study Liver Diseases, AAAS, Am. Soc. Biol. Chemists, Am. Physiol. Assn., Biophys. Soc., Internat. Assn. Study of Liver, Am. Fedn. Clin. Research, Midwest Gut Club, Am. Soc. Clin. Nutrition, Cleve. Acad. Medicine (profl. edn. com. 1972—), Ohio State Med. Assn., AMA, Sigma Xi, Alpha Omega Alpha. Unitarian. Home: 39251 Lander Rd Chagrin Falls OH 44022-2146 Office: Cleve Clin Found 9500 Euclid Ave Cleveland OH 44195-0002

HOLZMAN, DAVID CARL, journalist; b. Cambridge, Mass., June 29, 1953; s. Franklyn Dunn and Mathilda Sarah (Wiesman) H. BA, U. Calif., Berkeley, 1975. Editor People & Energy, Washington, 1978-80, Washington editor, 1993—; freelance writer Washington, 1980-86, 91-93; writer Insight Mag., Washington, 1986-91. Block capt. Families United, Washington, 1992; trustee Group Health Assn., 1993—. Recipient 1st prize for feature writing Am. Coll. Radiology, 1989. Democrat. Jewish. Achievements include bicycling across U.S.

HOLZMAN, PHILIP SEIDMAN, psychologist, educator; b. N.Y.C., May 2, 1922; s. Barnet and Natalie (Seidman) H.; m. Hannah Abarbanell, Sept. 18, 1946; children: Natalie Kay, Carl David, Paul Benjamin. B.A., CCNY, 1943; Ph.D., U. Kans., 1952. Diplomate: Am. Bd. Examiners Profl. Psychology. Psychology intern Topeka VA Hosp., 1946-49; psychologist Topeka State Hosp., 1949-51, cons., 1951-58; psychologist Menninger Found., Topeka, 1949-68; dir. research tng. Menninger Found., 1963-68; prof. psychiatry and psychology U. Chgo., 1968-77; prof. psychology dept. psychology Harvard U., 1977-92; prof. dept. psychiatry Med. Sch., 1977-92; Esther and Sidney R. Rabb prof. psychology Harvard U., 1984-92, prof. emeritus, 1992; chief sect. psychology Labs. for Psychiat. Research, McLean Hosp., Belmont, Mass., 1977—, tng. and supervising psychoanalyst Boston Psychoanalytic Soc. and Inst., 1977—; vis. prof. U. Minn., 1965, U. Kans., 1966, Boston U., 1973, Jefferson Med. Coll., 1981, U. Pa., 1987; Mem. small grants com. NIMH, 1960-64, clin. projects research rev. com., 1964-68, clin. program projects research rev. com., 1970-74, treatment devel. and assessment rev. com., 1982-86; cons. Ill. State Psychiat. Inst., 1970-77, mem. adv. coms. classification of mental disorders WHO. Author: (with others) Cognitive Control, 1959, Psychoanalysis and Psychopathology, 1970, (with Karl Menninger) The Theory of Psychoanalytic Technique, rev. edit, 1973; Editor: (with Merton M. Gill) Psychology Versus Metapsychology, 1975, (with Mary Hollis Johnston) Assessing Schizophrenic Thinking, 1979; bd. editors: Psychol. Issues, 1968—, Contemporary Psychology, 1969-76, Bull. of Menninger Clinic, 1961—, also Psychoanalysis and Contemporary Thought, Jour. Psychiat. Rsch.; assoc. editor Schizophrenia Bulletin, Schizophrenia Rsch.; contbr. articles to profl. jours. Mem. Topeka Mayor's Com. on Human Rels., 1963-68; chmn. bd. dirs. Founds.' Fund for Rsch. in Psychiatry; mem. program adv. com. MacArthur Found., sci. adv. bd. NIMH, 1986-92. With AUS, 1943-46. Recipient Career Scientist award NIMH, 1974-77, 92—; Stanley Dean award Am. Coll. Psychiatrists, Lieber prize of Nat. Alliance for Research in Schizophrenia and Depression; Townsend Harris medal CCNY. Fellow APA, AAAS, Am. Acad. Arts and Scis., Am. Coll. Neuropsychopharmacology; mem. Am. Psychoanalytic Assn., Boston Psychoanalytic Soc., Am. Psychopath. Assn., Inst. Medicine of NAS, Sigma Xi. Office: Harvard U William James Hall Cambridge MA 02138 also: McLean Hosp Lab Belmont MA 02178

HOM, THERESA MARIA, osteopathic physician; b. Detroit, Oct. 25, 1957; d. Richard Gay and Elizabeth Marie (Moye) H.; m. Rick L. Anderson, June 30, 1990. BS in Biology, U. Mich., 1979; DO, Mich. State U., 1984. Diplomate Am. Osteo. Bd. Gen. Practice. Intern Oakland Gen. Hosp., Madison, Mich., 1984-85; resident, gen. practice Doctors Hosp., Columbus, Ohio, 1985-86; family physician Madison Clinic, 1986-87, Community Family Health Ctr., Columbus, 1987—; clin. prof. Ohio U. Coll. Osteo. Medicine, Columbus, 1989—; apptd. osteo. mem. Ohio State Med. Bd., 1990-93, supervisory mem., 1993, v.p., 1993. Physician Columbus Free Clinic, 1990. Featured poet Larry's Poetry Forum, Columbus, 1991. Mem. Am. Osteo. Assn., Am. Med. Women's Assn., Ohio Osteo. Assn. (del. 1989, 93), Pax Christi Columbus (coord. program 1989-91), Am. Coll. Osteo. Gen. Practice, Am. Acad. Med. Acupuncturists. Roman Catholic. Avocations: peace activist, poet. Office: Community Family Health Ctr 1043 E Weber Rd Columbus OH 43211-1293

HOM, WAYNE CHIU, civil engineer; b. Manhattan, N.Y., Feb. 10, 1967; s. Ling You and Sau Ying (Tse) H. BCE, Pratt Inst., 1990. Cert. intern engr.,

N.Y. Motor transport operator USMCR, Bklyn., 1986-92; asst. bldg. constrn. engr. N.Y. State Office Gen. Svcs., Bronx, 1990—. Office: NY State Office Gen Svcs 1 Fordham Plz Rm 246E Bronx NY 10458

HOMBURG, JEFFREY ALLAN, geoarchaeologist; b. Oklahoma City, Jan. 30, 1957; s. Leo Paul and Linda Jane (Fisher) St. Onge. BA, U. Okla., 1979; MA, La. State U., 1991. Supr. Archaeol. Rsch. and Mgmt. Ctr., Norman, Okla., 1978-79, Great Plains Mus., Lawton, Okla., 1979-80; crew chief/project dir. New World Rsch., Inc., Pollock, La., 1980-84; project dir. New World Rsch., Inc., Fort Walton Beach, Fla., 1986-87; rsch. asst. La. State U., Baton Rouge, 1984-86; project dir. Statis. Rsch., Inc., Tucson, 1988—; reviewer Ariz. Archaeol. and Hist. Soc., Tucson, 1990-93. Author: Cultural Resources Survey and Overview for the Rillito River Drainage Area, 1989, Playa Vista Archaeological Project Research Design, 1990, Intermontane Settlement Trends in the Eastern Papagueria, 1993, Archaeological Investigations at the LSU Campus Mounds, 1992, Life in the Ballona: Archaeological Investigations at the Admiralty Site (CA-LAN-47) and the Channel Gateway Site (CA-LAN-1596-H), 1992, Archaeological Investigations at Lee Canyon: Kayenta Farmsteads in the Upper Basin, Coconino County, Arizona, 1992, Late Prehistoric Change in the Ballona Wetland, 1993, The Centinela Site: Data Recovery at a Middle Period Creek Edge Site in the Ballona Wetland (CA-LAN-60), Los Angeles County, California, 1993, Comments on the Age of the LSU Mounds: A Reply to Jones, 1993. Chmn. tech. standards com. Profl. Disc Golf Assn., Oklahoma City, 1989-93. Sigma Xi rsch. grantee, La. State U., 1986. Mem. Soc. Am. Archaeology, Soc. Archaeol. Sci., Soil Sci. Soc., Soc. Profl. Archaeologists (cert.). Democrat. Lutheran. Achievements include development of techniques for studying physical and chemical properties of soils that were cultivated during prehistoric times, remote sensing and archaeological recovery methods for forensic investigations; reconstruction of the depositional history of La. State U. Campus Mounds, one of the earliest known Indian mounds in North America. Home: 5330 E Bellevue St Apt 3 Tucson AZ 85712 Office: Statis Rsch Inc PO Box 31865 Tucson AZ 85751-1865

HOMMA, MORIO, microbiology educator; b. Fukushima, Japan, Oct. 9, 1930; s. Toma and Shige (Kakudate) H.; m. Mitsuko Endo, Mar. 29, 1955; children: Naoki, Yukari, Akihiko. MD, Tohoku U., Sendai, Miyagi Prefecture, Japan, 1955, DMS, 1961. Med. diplomate. Rsch. assoc. Tohoku U. Sch. Medicine, Sendai, 1956-68, asst. prof., 1968-73; prof., chmn. dept. bacteriology Yamagata (Japan) U. Sch. Medicine, 1973-83; prof., chmn. dept. microbiology Kobe (Hyogo Prefecture, Japan) U. Sch. Medicine, 1983—, dean, 1989-93. Contbr. articles to rsch. jours. Recipient Kojima Meml. Prize, Kojima-Fukumi Commemoration Inst., Tokyo, 1984. Mem. Soc. Japanese Virologists (pres. 1989-90). Avocations: fishing, Bonsai culture, mushroom hunt. Home: Kita-ku Naruko 2 chome 6-10, Kobe 651-11, Japan Office: Kobe U Sch Med Dept Microbiol, Chuo-ku Kusunoki-cho 7 chome 5-1, Kobe Hyogo 650, Japan

HOMMES, FRITS AUKUSTINUS, biology educator; b. Bellingwolde, Netherlands, May 28, 1934; came to U.S., 1979; s. Aukustinus and Anje (Wester) H.; m. Grietje Renes, June 14, 1958; children: Peter, Anneliek. M.Sc. in Chemistry, U. Groningen, Netherlands, 1958; Ph.D., U. Nijmegen, Netherlands, 1961. Diplomate Am. Bd. Med. Genetics. Research asst. dept. biochemistry U. Nijmegen, 1959-61; instr. U. NiJmegen, 1963-66; postdoctroal fellow dept. biochemistry and biophysics U. Pa., Phila., 1961-63; head lab. dept. pediatrics U. Groningen, 1966-72, assoc. prof., 1972-79; prof. dept. biochemistry and molecular biology Med. Coll. Ga., Augusta, 1979-93, dir. biochem. genetics lab., 1980-93; dir. biochem. genetics lab. NYU, 1993—, prof. dept. pediatrics, 1993—; cons. genetic diseases Dutch Health Council, 1974-79, FDA, 1992—; chmn. Dutch Bioenergetics Study Group, 1975-77,. Author: Inborn Errors of Metabolism, 1973, Normal and Pathological Development of Energy Metabolism, 1975, Models for the Study of Inborn Errors of Metabolism, 1979, Techniques in Human Diagnostic Biochemical Genetics, A Laboratory Manual, 1990; mem. editorial bd. Nutrition and Metabolism, 1975—; contbr. articles to profl. jours.; patentee in field. Chmn. Groningen chpt. Round Table, 1970-71; chmn. No. Dist., Netherlands, 1973-75, mem. nat. bd., 1974-75. Fulbright fellow, 1961-63; recipient medal City of Milan, 1987, Disting. Svc. award SERGG, 1993. Mem. AAAS, European Soc. Pediatric Rsch., Soc. Study of Inborn Errors of Metabolism, Soc. Inherited Metabolic Disease (bd. dirs. 1993—), Am. Soc. Human Genetics (chmn. com. biochem. genetics 1990-91), N.Y. Acad. Sci., Soc. Pediatric Rsch., Am. Soc. Biol. Chemists. Roman Catholic. Lodge: Rotary. Office: Univ Med Ctr Dept Pediatrics Human Genetics Program 550 First Ave New York NY 10016

HOMZIAK, JURIJ, environment, aquaculture and marine resources specialist, consultant; b. Charleroi, Hainaut, Belgium, Dec. 6, 1948; came to U.S., 1951; s. Jakiw and Halyna (Slusarchuk) H.; m. Mary Margaret Texler, July 18, 1981; 1 child, Nicholas. MA, San Diego State U., 1977; PhD, U. N.C., 1985. Rsch. and teaching asst. San Diego State U., 1975-77; NSF and Sea Grant fellow Inst. of Marine Sci. U. N.C., Morehead City, 1977-82; marine biologist Oil Pollution Control Ctr., Jubail, Saudi Arabia, 1983; biologist Comml. Shrimp Culture Internat., Los Fresnos, Tex., 1983-84; dir. of aquaculture AGRO 21, Ltd., Kingston, Jamaica, 1984-85; marine ecologist U.S. Army C.E., Vicksburg, Miss., 1982-83, 86-88; marine resources specialist, assoc. prof. Miss. State U. Sea Grant Adv., Biloxi, 1988-92; environ. officer Bur. Latin Am. and Caribbean, U.S. AID, Washington, 1992—; cons. MariQuest, Inc., Newport Beach, Calif., 1986, Sedgewick James Co., Houston, 1990—; assoc. Barry Vittor and Assoc., Mobile, Ala., 1990—. Contbr. articles to Aquaculture mag. and profl. jours. Mem. Sierra, 1977, Audubon Soc., 1983, Amnesty Internat., 1986; 1st dir. Aquaculture of Jamaica; dir. containment area aquaculture program U.S. Army C.E. With USN, 1967-71, Vietnam and Korea. Van Dyke fellow ARCO, Inc., 1972-75, NSF fellow, 1979-80, N.C. Sea Grant fellow, 1980-81, Farm Found. fellow, 1990-91; USDA grantee Hungary, Bulgaria, 1990. Mem. Am. Fisheries Soc. (various coms.), Ecol. Soc. Am., World Aquaculture Soc., Gulf and Caribbean Fisheries Inst. Achievements include coordination of the development of the Mississippi soft-shell crawfish industry; co-development of commercial fish culture in Jamaica. Home: 13309 Jasper Rd Fairfax VA 22033-1445 Office: US AID LAC/DR/E 2201 C St NW Washington DC 20523-0010

HONAMI, SHINJI, mechanical engineer educator; b. Fukushima, Japan, Mar. 21, 1945; s. Takeo and Takeko H.; m. Utako Nishimura, Oct. 5, 1972; children: Toshiharu, Yasuharu. BS, Keio U., Tokyo, 1967, MS, 1969, PhD, 1974. Lectr. Sci. U. Tokyo Shinjuku, 1974-75, assoc. prof., 1975-87, prof., 1987—; vis. scholar Stanford (Calif.) U., 1979-80. Author: American Society of Mechanical Engineer, 1989. Mem. Am. Soc. Mech. Engrs., Am. Inst. Aero. Astronauts. Office: Science U Tokyo, 1-3 Kagurazaka, Shinjyuku-ku 162, Japan

HONAVAR, VASANT GAJANAN, computer scientist, educator; b. Poona, India, Apr. 28, 1960; came to the U.S., 1982; s. Gajanan N. and Bhavani (Melinkeri) H. PhD in Computer Sci. and Cognitive Sci., U. Wis., 1990. Asst. prof. Iowa State U., Ames, 1990—. Contbr. articles to profl. jours., chpts. to books. Nat. Sci. Talent scholar, 1977; Gold medal Bangalore U., 1982. Mem. AAAI, ACM, N.Y. Acad. Sci., IEEE, Sigma Xi. Office: Iowa State U 226 Atanasoff Hall Ames IA 50011

HONDA, HIROSHI, engineer, energy economist; b. Kanazawa, Ishikawa, Japan, Mar. 22, 1950; s. Eiichi and Toyoko (Tsuchimuro) H.; m. Reiko Nakamura, Mar. 4, 1979; children: Ryo, Naoko. B in Engring., Kyoto (Japan) U., 1972, D in Engring., 1986; MS, Pa. State U., 1976. Registered profl. engr., Tex., Minn. R&D engr. Mitsui Engring. and Shipbuilding Co., Tamano, Okayama, Japan, 1972-74, rsch. engr., 1976-83; mem. staff machinery hdqrs. Mitsui Engring. and Shipbuilding Co., Tokyo, 1983-85; chief researcher Chiba (Japan) Lab. Mitsui Engring. and Shipbuilding Co., Ichihara, 1985-88; assoc. mgr. corp. planning Mitsui Engring. and Shipbuilding Co., Tokyo, 1988-89, assoc. mgr. bus. planning, 1989-91, mgr. bus. devel., 1991; sr. economist, project mgr. internat. cooperation dept. Inst. Energy Econs. Japan, Tokyo, 1991—; dir. energy master plans and confs. Japanese Com. for Pacific Coal Flow, IEE, Tokyo, 1992—. Author, editor: Working in Japan, 1991, International People Tell How to Successfully Deal with Foreign Employees in Japanese Companies, 1993; author: Book for Managers Mono-O-Mirume, 1990; editor: JAPAC News, ASME Japan News. Mem. ASME, Japan Soc. Mech. Engrs., Kyoto U. Mech. Engr.'s Assn. Avocations: tennis, contract bridge, reading. Home: 2-710 Yatsu 6-chome 7-ban, Narashino-shi Chiba 275, Japan Office: Inst Energy Econs Japan Sawa Bldg, 2-2-2 Nishi Shimbashi, Minato-ku Tokyo 105, Japan

HONDA, NATSUO, anesthesiologist, educator; b. Nagasaki, Japan, July 9, 1930; s. Quanji and Masako Honda; m. Nobuko Sagayama, Feb. 11, 1975; 1 child, Naoko. MD, Nagasaki U., 1955, PhD in Physiology, 1960. Intern Nagasaki U. Hosp., 1955-56; asst. dept. physiology Nagasaki U., 1960-62, instr., 1962-64, instr. dept. anesthesiology, 1964-69; rsch. assoc. Presbyn. St. Lukes Hosp., Chgo., 1969-71; asst. prof. Nagasaki U., 1971-80; prof. Oita Med. U., Japan, 1980—, chmn. anesthesia, 1980—, chmn. operating room, 1980—, chmn. intensive care unit, 1984—, chmn. critical care unit, 1989—. Mem. Japan Soc. Anesthesiology (mem. elder com. internat. affiliate), Japan Soc. Clin. Anesthesiology (elder), Japan Soc. Critical Care Medicine, Japan Soc. Intensive Care Medicine (elder), Japan Soc. Pain Clinic (elder), Soc. Cardiovascular Anesthesiologists, Internat. Asns. Study Pain, Japan Alpine Club. Avocations: mountaineering, skiing. Home: 2-203 2-1 Idaigaoka, Hazama-cho, 879-56 Oita-gun Japan Office: Oita Med U, 1-1 Idaigaoka Hazama-cho, Oita 879-55, Japan

HONDA, TOSHIO, cardiologist; b. Matsuyama, Ehime, Japan, Dec. 23, 1956; s. Masao and Sachiko (Ninomiya) H.; m. Akiko Matsuzawa, June 16, 1985; 1 child, Rie. MD, Hamamatsu U., 1981. Tng. physician 2d dept. internal medicine Ehime U., Shizukawa, 1981-82; physician Ehime Rosai Hosp., Niihama, 1982-84, South Ehime Sanatorium, Izume, 1984-85, Ehime Prefectural Cen. Hosp., Matsuyama, 1985-90; physician, cardiologist Takanoko Hosp., Matsuyama, 1990-92, Ehime Prefectural Iyomishima Hosp., Iyomishima, Japan, 1992—. Contbr. articles to profl. jours. Mem. Japanese Circulation Soc. (cert.), Japanese Soc. Internal Medicine (cert.). Avocation: golf. Home: 4-7-3 Sanbancho, Matsuyama, Ehime 790, Japan Office: Ehime Prefectural Iyomishima Hosp, 1684-2 Nakanoshocho, Iyomishima Ehime 799-04, Japan

HONDEGHEM, LUC M., cardiovascular and pharmacology educator; b. Jabbeke, Belgium, Sept. 22, 1944; married; 3 children. MD, U. Louvain, Belgium, 1970, MS in Physiology, 1971; PhD in Pharmacology, U. Calif., San Francisco, 1973. From asst. prof. to assoc. prof. pharmacology U. Calif., San Francisco, 1973-85; prof. medicine and pharmacology, Stahlman chmn. cardiovascular research program Vanderbilt U., 1985—. Mem. Am. Heart Assn., Med. Electronics and Data Soc., Am. Soc. Pharmacology and Exptl. Therapeutics, Sigma Xi. Fields of research include ultrastructural and electrophysiol. aspects of impulse transmission in cardiac tissue; mechanisms of cardiac arrhythmias; effects of antiarrhythmic drugs on normal and abnormal impulse transmission in the heart. Office: Vanderbilt U Sch of Medicine Nashville TN 37132

HONEA, FRANKLIN IVAN, chemical engineer; b. Hope, Ark., Aug. 9, 1931; s. Tholbert Milton and Irene Anna (Hall) H.; m. Marvine Lakedon, Apr. 25, 1959; children: Mary, Rose, John, Paul, Jane, Luci, Joan, Luke, Rita, Love. BSME, U. Calif., Berkeley, 1955; MSME, U. So. Calif., L.A., 1962; PhDChemE, U. Denver, 1969. Registered profl. engr., Tex., Colo. Rsch. specialist space divsn. North Am. Rockwell, Downey, Calif., 1961-66; sr. staff engr. space divsn. Martin-Marietta Corp., Denver, 1966-68; prin. devel. engr. Mason & Hanger-Silas Mason, Inc. (Pantex AEC Plant), Amarillo, Tex., 1970-74; sr. environ. engr. Midwest Rsch. Inst., Kansas City, Mo., 1974-76; project engr. clean coal program U.S. Dept. Energy, Grand Forks, N.D., 1976-87; project mgr. Ill. Clean Coal Inst., Carterville, 1990—; cons. Energy and Environ. Systems, Fertile, Minn., 1988-90; adj. prof. chem. engring. dept. U. Denver, 1968-70. Co-author tech. report U.S. Dept. Energy, 1985; contbr. articles on coal processes, energy, environ. systems control and spacecraft temperature control to profl. jours. NIH predoctoral grantee U.S. HEW Environ. Control Divsn., 1969, Metallic Particle Ignition Rsch. grantee USAF, 1968; recipient inventors award Northrop Corp., 1959. Mem. ASME, ASTM (sub-chair peat energy 1979-82), AIChE, Sigma Xi. Achievements include patents and disclosures on miniature coolant control valve, high explosives incinerator, evaporative heat exchanger. Home: 109 Edgewood Park Marion IL 62959 Office: Ill Clean Coal Inst Ste 200 Coal Development Park Carterville IL 62918-0008

HONEYCUTT, MICHAEL ALLEN, computer consultant; b. Asheville, N.C., July 21, 1962; s. Scovell Elbert and Areatha (Roberts) H. AAS, A-B Tech., 1982. Computer programmer U. N.C., Asheville, 1983-87, computer cons., 1987—; cons. various non-profit orgns., Asheville, 1985—. Vol. ARC, Asheville, 1991—. Recipient Svc. award ARC, 1989. Mem. Western N.C. Word Perfect Users Group (v.p. 1991-91, pres. 1992—). Home: 2 Pelzer St Asheville NC 28804 Office: UNC Asheville One University Hts Asheville NC 28804-3299

HONG, RICHARD, pediatric immunologist, educator; b. Danville, Ill., Jan. 10, 1929. BS, U. Ill., 1949; MD, U. Ill., Chgo., 1953. Sr. rsch. assoc. immunology U. Cin., 1960-65; from asst. to prof. pediatrics and immunology U. Minn., Mpls., 1965-69; prof. pediatrics and microbiology med. ctr. U. Wis., Madison, 1969—, assoc. dean clin. affairs, 1971-74; mem. Transplant Registry, 1968-69. Rsch. Career Devel. award USPHS, 1963-69, fellow, 1960-62. Mem. Am. Soc. Clin. Investigation, Am. Assn. Immunology, Am. Pediatric Soc., Soc. Pediatric Rsch. Achievements include research in immunoglobulin structure and function, immunochemistry of hypogammaglobulinemia, treatment of immunological deficiency states, physiology of immunity. Office: U Wisconsin Asthma & Allergic Disease Ctr 600 Highland Ave Madision WI 53792*

HONG, WAUN KI, medical oncologist, clinical investigator; b. Kyung gi Do, South Korea, Aug. 13, 1942; naturalized Sept. 17, 1976; s. Sung Ku and Bok Young; m. Mi Hwa Yoo, Sept. 9, 1969; children: Edward, Burton James. Student, Yon-Sei U., 1963, MD, 1967. Diplomate Am. Bd. Internal Medicine. Rotating intern Bronx-Lebanon Hosp., N.Y.C., 1970-71; jr. med. resident Boston Vets. Affairs Med. Ctr., 1971-72, sr. med. resident, 1972-73, chief of medical oncology, 1975-84, program dir. hematology/oncology tng. program, 1982-84; teaching assoc. Sch. Medicine Boston U., 1971-73, asst. prof. medicine, 1975-79, assoc. prof. medicine, 1980-84; clin. instr. medicine Cornell U., 1973-75; attending physician in medicine Boston City Hosp., 1978-84; clin. assoc. prof. pharmacology Northeastern U., Boston, 1980-84; internist, prof. medicine M.D. Anderson Cancer Ctr., U. Tex., Houston, 1984-88, chief sect. thoracic med. oncology, 1987-88; dep. head divsn. medicine M.D. Anderson Cancer Ctr., U. Tex., 1992—, chmn. dept. thoracic/head and neck med. oncology, 1993—; mem. scientific adv. bd. The Cancer Inst. of N.J., 1993-98, The San Antonio Cancer Inst., 1993—; cons. Immunogen, Inc., 1993—, Houston Vet. Affairs Med. Ctr., 1992—, Genelabs Tech., Inc., 1992—, Sparta Pharmaceuticals, Inc., 1992—, The Lifescience Corp., 1989-91; adj. prof. medicine Baylor Coll. Medicine, Houston, 1991—;vis. prof. Tufts U. Sch. Medicine, 1993, Dana-Farber Cancer Ctr., 1993, Johns Hopkins Oncology Ctr., 1993, Tex. Tech. U. Sch. Medicine, 1992; lectr. in medicine Tufts U., 1975-84. Author: (with others) Chemoimmuno Prevention of Cancer, 1991, The Biology and Prevention of Aerodigestive Tract Cancer, 1992; editor: (with others) Advances in the Diagnosis and Therapy of Lung Cancer, 1993, Retinoids in Oncology, 1993, Internat. Jour. Radiation Oncology, Biology and Physics, 1992—; assoc. editor Cancer Rsch., 1993—; mem editorial bd. Cancer Prevention Internat., 1993—, PDQ Screening and Prevention, NCI, 1993—, Annals of Surg. Oncology, 1993—, Cancer Rsch. Therapy and Ctrl., 1993—, Jour. Clin. Oncology, 1992—, Cancer Prevention, 1990-93; mem. editorial adv. bd. Jour. Nat. Cancer Inst., 1990—; reviewer Annals of Internal Medicine, Archives of Otolaryngology Head and Neck Surgery, Cancer, Chest, Head and Neck Surgery, Jour. AMA, New England Jour. Medicine. Served as flight surgeon Korean Air Force, 1967-70. Recipient AACR 17th Annual Richard and Hinda Rosenthal Found. award, 1993, pres. citation Am. Soc. for Head and Neck Surgery, 1991; Jr. Med. Oncology fellow Meml. Sloan Kettering Cancer Ctr., 1973-74, Sr. Med. Oncology fellow Cornell U., 1974-75; also numerous federal, industry, and found. grants. Mem. AMA, AAAS, FDA, Am. Radium Soc., Am. Fedn. Clin. Rsch., Am. Assn. Physicians, Am. Assn. Cancer Rsch. (mem. program com., subcom. on clin. investigations, 1993-94, cancer prevention, 1993-94, mem. task force clin. investigations, 1990—), Am. Cancer Soc. (mem. nat. conf. clin. trials, 1993, profl. edn. subcom. on profs. clin. oncology, 1990—), Am. Soc. Clin. Oncology (chmn. cancer prevention and ctrl. com., 1994), Nat. Cancer Inst., Tex. Med. Assn., Radiation Therapy Oncology Group (mem. med. oncology com. 1989—, head and neck com. 1989—), Harris County Med. Soc., Soc. Head and Neck Surgeons. Home: 7603 Moondance Ln Houston TX 77071 Office: U Tex MD Anderson Cancer Ctr 1515 Holcombe Blvd Box 80 Houston TX 77030-4095

HONG, YONG SHIK, aerospace engineer; b. Seoul, Republic of Korea, Nov. 1, 1932; s. Suhn Hung and Kap Shun (Bae) H.; m. Byung Hee Min, 1962; children: Joon-suh, Soo-jin, Won-suh. BSME, Seoul Nat. U., 1955; MSME, U. Ill., 1959; PhDME, U. Wash., 1968. Rsch. specialist Boeing Co., Seattle, 1958-68; rsch. assoc. U. Wash., Seattle, 1966-68; mem. tech. staff Aerospace Corp., El Segundo, Calif., 1968-74; v.p. Agy. for Def. Devel., Seoul, 1974-76; tech. advisor Korean Air Lines, Seoul, 1976-78; dir. rsch. Korean Air, Seoul, 1978-92; advisor Korea Inst. Aero. Tech., Seoul, 1992—; adj. prof. Korea Advanced Inst. for Sci., Seoul, 1976; prof. aerospace engring. INHA U., Inchon, Republic of Korea, 1976—; chmn. sci. and aerospace R&D rev. com. Ministry Sci. and Tech., Seoul, 1987-92, mem. sci. and technol. coun., 1988-92. Author: Gas Turbine Engines, 1983, Satellites and Space Launchers, 1985, Space Propulsion Engineering, 1990, Man's Dream Toward Space, 1991; editor Def. and Tech. Jour., 1986-91. Bd. trustees Hankook Aviation U., Seoul, 1979—. Mem. AIAA, ASME, Korean Soc. Aero. and Space Scis. (pres. 1981-83), Assn. for Unmanned Vehicle Systems. Home: 11-1208 Samho Apt, Bangbae-dong, Seocho-ku, Seoul 137-069, Republic of Korea Office: Korea Inst Aero Tech, CPO Box 864, Seoul Republic of Korea

HONG, ZUU-CHANG, engineering educator; b. Bei-Kun, Yuan-Lin, Taiwan, Apr. 25, 1942; s. San-Lin and Fei-Rien (Jih) H.; m. Sue-Jane Chen, Jan. 15, 1974 (div. Apr. 1979); children: Grace Shau-Wei Hong, Chao-I Hong; m. Hsiu-Ching Chen, Apr. 14, 1982; children: Chao-Tien Hong, Chao-Hun Hong, Chao-Min Hong. BS, Nat. Taiwan U., 1968; MS, U. Calif., Davis, 1971; PhD, U. Ill., 1975. Assoc. to full prof. dept. mech. engring. Nat. Taiwan U., Taipei, 1975-80, prof. and head dept. mech. engring., 1980-82, prof., 1980-85; prof., dean Coll. Engring., Nat. Cen. U., Chungli, Taiwan, 1985-88, prof., 1988—; cons. Chun-San Inst. Sci. and Tech., Lung-Tan, Tao-Yuan, 1981—, Taiwan Inst. Econ. Rsch., Taipei, 1989—. Editor-in-chief: Jour. of the Chinese Soc. of Mech. Engr., 1979-89; prof.-in-charge Automatic Control Lab., Taiwan U., 1976, Computational Fluid Dynamic Lab., Cen. U., 1985; editor-in-chief: (book) Experiments for Mechanical Engineering, 1983; contbr. articles to profl. jours., publs. With Chinese Navy, 1968-69. Recipient Disting. Rsch. award Nat. Sci. Coun., Taipei, 1991-92; Disting. Prof., Nat. Cen. U., Chung-Li, 1989, Disting. Paper of Yr. Soc. of Theoretical and Applied Mechanics, Tainan, 1982. Mem. Chinese Inst. Engrs. (Disting. Paper of Yr. 1979), Chinese Soc. Mech. Engrs. (bd. dirs. 1980-85, Disting. Paper of Yr. 1984, 86), Chinese Soc. Aeronautics and Astronautics (bd. dirs. 1980—), Welding Soc. of Rep. of China (bd. dirs. 1986—), Chinese Assn. of Indsl. Automation (bd. dirs. 1992), others. Achievements include patent in field. Home: # 11 5th Flr, Lane-16 Wen-Chou St, Taiwan, 10616 Taipei Republic of China Office: Dept Mech Engring, Nat Cen Univ, 32054 Chung-Li Taiwan

HONIG, LAWRENCE STERLING, neuroscientist, neurologist; b. Berkeley, Calif., Oct. 26, 1953; s. Arnold and Alice H. AB, Cornell U., 1973; PhD, U. Calif., Berkeley, 1978; MD, U. Miami (Fla.), 1986. Diplomate Am. Bd. Med. Examiners, Am. Bd. Psychiatry and Neurology. Staff scientist MRC Nat. Inst. Med. Rsch., London, 1980; rsch. asst. prof. U. So. Calif., L.A., 1981-83; med. intern Stanford (Calif.) U. Hosp., 1986-87, neurology resident, 1987-90; clin. instr. Stanford U. Med. Sch., 1990-92, clin. asst. prof., 1992—. Author: (with others) Embryonic Mechanisms, 1986, Multiple Sclerosis, 1989; contbr. articles to profl. jours. Anna Fuller Fund fellow Middlesex Hosp. Med. Sch., London, 1978-79, Dana fellow Stanford U., 1990, Walter and Idun Berry fellow, Stanford U., 1991. Fellow Zool. Soc. London; mem. AMA, AAAS, Am. Acad Neurology, Soc. for Neurosci., Sigma Xi. Achievements include research in mechanisms of development, maintenance and regeneration of nervous systems connections. Office: Dept Neurology & Neurol Sci (H3160) Stanford U Med Ctr Stanford CA 94305-5235

HONIKMAN, LARRY HOWARD, pediatrician; b. Washington, Feb. 16, 1936; s. Zuse and Frances (Deckelbaum) H.; m. Elaine Honikman; children: Sheryl, Julie, Amy. BA, U. Va., Charlottesville, 1957, MD, 1961. Intern Royal Victoria Hosp., McGill U., Montreal, Que., Can., 1961-62; resident in pediatrics Children's Hosp. Med. Ctr., Boston, 1964-66, fellow in medicine, 1966-68, asst. prof. in medicine, 1968-73; fellow in medicine dept. pediatrics House of Good Samaritan, Harvard Med. Sch., Boston, 1966-68; rsch. fellow in medicine Harvard Med. Sch., Boston, 1969-70; pvt. practice pediatrics, 1973—; tchr. Boston U. Sch. Medicine, 1977—, Harvard U. Sch. Medicine, 1977—, Tufts U. Sch. Medicine, 1982—; mem. staff pediatrics dept. Children's Hosp. Med. Ctr., 1973—, Brigham and Women's Hosp., Boston, 1976—, Cardinal Cushing Hosp., Brockton, Mass., 1975—, Carney Hosp., Dorchester, Mass., 1977—; lectr. in field.; cardiac cons. dept. pub. health, div. maternal and child health svcs. Commonwealth of Mass., 1968-75. Contbr. articles to profl. jours. including Jour. of Am. Med. Assn., Pediatrics, Am. Jour. Disease Children, Circulation. Capt. U.S. Army, 1962-64. Mem. AMA, AAAS, APHA, Am. Heart Assn., Mass. Heart Assn. (chmn. rheumatic fever adv. com. S.E. chpt. 1973-74, mem. state rheumatic fever svcs. com. 1968-71, co-chmn. rheumatic fever workshop 1966), Mass. Med. Soc., Nutrition Today Soc., N.Y. Acad. Scis., Collegium Internationale Angiologae, Phi Beta Kappa. Home: 6 Old Bridge Ln Sharon MA 02067

HONTS, CHARLES ROBERT, psychology educator; b. Clifton Forge, Va., Sept. 17, 1953; s. George Edward Jr. and Emily (Gordon) H.; m. Rosanna Conlon, Dec. 8, 1972; 1 child, Eric Alan. BS in Psychology, Va. Poly. Inst. and State U., 1974, MS in Exptl. Psychology, 1982; PhD in Exptl. Psychology, U. Utah, 1984. Cert. detection of deception examiner, N.D. Polygraph examiner, 1976-80; teaching asst. dept. psychology Va. Poly. Inst. and State U., Blacksburg, 1981-82; rsch. asst. dept. psychology U. Utah, Salt Lake City, 1982-86, rsch. assoc., 1986-88, adj. asst. prof., 1987—; rsch. psychology, leader rsch. team Def. Polygraph Inst., Ft. McClellan, Ala., 1988-90; assoc. prof. U. N.D., Grand Forks, 1990—; adj. prof. Jacksonville (Ala.) State U. Coll. Grad. Studies and Continuing Edn., 1989-90; numerous presentations in field, 1981—; organizer workshops and confs.; ad hoc editorial cons. Jour. Personality and Social Psychology, 1991-92, Jour. Applied Psychology, 1992; speaker in field, 1983—; instr. Ariz. Sch. Polygraph Sci., Phoenix, 1985-91; cons. to pub. agys.; expert witness, 1983—. Mem. editorial bd. Forensic Reports, 1991—; contbr. articles and abstracts to Forensic Reports, Psychophysiology, Law and Human Behavior, Polygraph, Jour. Police Sci. and Adminstrn., Jour. Applied Psychology, Jour. Rsch. in Personality. Grantee Supply and Svcs. Can., 1991, U. N.D., 1991, Office Naval Rsch., 1992. Mem. Soc. for Psychophysiol. Rsch., Am. Psychology and Law Soc., Am. Psychol. Soc. (charter), Rocky Mountain Psychol. Assn. Achievements include first scientist to demonstrate that polygraph tests could be defeated through the use of physical and mental countermeasures; research on credibility assessment of verbatim statements. Office: U ND Psychology Dept PO Box 8380 Grand Forks ND 58202

HOOD, DOUGLAS CRARY, electronics educator; b. Toledo, Nov. 26, 1932; s. Douglas Crary and Pauline Edna (Thurston) H.; m. Marlene Carole Ashenfelter, Sept. 14, 1984. BSE, Electronics Inst. Technology, Detroit, 1962; MSD, Brotherhood of White Temple, Sedalia, Colo., 1975. Avionics technician USAF, 1952-56; systems design engr. AT&T, Toledo, 1956-59; customer engr. IBM, Detroit, 1962-66; territory mgr. Mgmt. Assistance Inc., Detroit, 1966-71; owner, mgr. Sonny's Rainbow, Toledo, 1969-84; owner, operator Rainbow Lapidary Art Sch., Toledo, 1971-74; maintenance mechanic Lucas Met. Housing Authority, Toledo, 1984-86; electronics educator, dean dept. engring. Stautzenberger Coll., Toledo, 1987-92; dir. RETS Inst. Tech., Toledo, 1992—; master gem cutter and carver, originator of the rainbow cut. Author: various elec. lab. manuals. Mem. IEEE, BOAZ, Toledo Gem Club (pres. 1970-74), Masons, Shriners. Democrat. Avocations: computer programming, photography, astronomy, genealogy, painting. Home: PO Box 207 Tecumseh MI 49286-0207 Office: Rets Inst Tech 1606 Laskey Rd Toledo OH 43612

HOOD, LAMARTINE FRAIN, college dean; b. Johnstown, Pa., Feb. 25, 1937; s. Lamartine and Marion Camm (Frain) H.; m. Emeline Rose Harpster, June 18, 1960; children: Thomas Gregory, Christopher Michael, Sandra Beth. BS, Pa. State U., 1959, PhD, 1968; MS, U. Minn., 1963. Asst. prof. Cornell U., Ithaca, N.Y., 1968-74, assoc. prof., 1974-80, prof. food sci., 1980-86, assoc. dir. Agr. Experiment Station, 1980-83, assoc. dir. office of rsch., 1980-86; dir. N.Y. State Agr. Experiment Sta. Cornell U., Geneva, N.Y.,

1983-86; dean Coll. Agr. Sci., dir. Agr. Expt. Sta., dir. Coop. Ext. Pa. State U., University Park, 1986—; adv. bd. Chase Lincoln Bank, Geneva, 1984-86. Author/editor: Carbohydrates & Health, 1977; contbr. articles to profl. jours. and chpts. for books. Fellow Inst. Food Technologists; mem. Am. Assn. Cereal Chemists (William F. Geddes Meml. Lectureship award 1984, pres. 1987-88, chmn. bd. 1988-89), Geneva State Coll. Club, Rotary, Gamma Sigma Delta, Phi Lambda Upsilon. Home: 1694 Princeton Dr State College PA 16803-3257 Office: Pa State U 201 Agrl Adminstrn Bldg University Park PA 16802-2600

HOOD, LEROY EDWARD, biologist; b. Missoula, Mont., Oct. 10, 1938; s. Thomas Edward and Myrtle Evylan (Wadsworth) H.; m. Valerie Anne Logan, Dec. 14, 1963; children: Eran William, Marqui Leigh Jennifer. B.S., Calif. Inst. Tech., 1960, Ph.D. in Biochemistry, 1968; M.D., Johns Hopkins U., 1964. Med. officer USPHS, 1967-70; staff scientist Pub. Health Svc., Bethesda, Md., 1967-70; sr. investigator Nat. Cancer Inst., 1967-70; asst. prof. biology Calif. Inst. Tech., Pasadena, 1970-73, assoc. prof., 1973-75, prof., 1975-92, Bowles prof. biology, 1977-92, chmn. div. biology, 1980-89; Gates prof. molecular biotech., chmn. bd. U. Wash. Sch. Medicine, Seattle, 1992—; dir. NSF Sci. and Tech. Ctr. for Molecular Biotech., 1989—. Author: (with others) Biochemistry, a Problems Approach, 1974, Molecular Biology of Eukaryotic Cells, 1975, Immunology, 1978, Essential Concepts of Immunology, 1978, The Code of Codes, 1992. Co-recipient, Albert Lasker Basic Medical Research Award, 1987. Mem. NAS, Am. Assn. Immunologists, Am. Assn. Sci., Am. Acad. Arts and Scis., Sigma Xi. Avocations: mountaineering, rockclimbing, photography. Office: Dept Molecular Biotech U Wash Sch Medicine FJ 20 Seattle WA 98195

HOOK, VIVIAN YUAN-WEN HO, biochemist, neuroscientist; b. Oakland, Calif., Mar. 21, 1953; d. Timothy T. and Cheng-Ping (Wang) Ho; m. Gregory R. Hook, July 9, 1976; childrn: Lisa, Michelle. AB, U. Calif., Berkeley, 1974; PhD, U. Calif., San Francisco, 1980. From postdoctoral fellow to sr. scientist NIMH, NIH, Bethesda, Md., 1980-85; asst. prof. Uniformed Svcs. U., Bethesda, 1986-90, assoc. prof., 1991—; biochemistry and neuroscience study sect. Nat. Inst. Drug Abuse, Bethesda, 1989-92. Contbr. articles to profl. jours. NIH grantee, 1987—; Wellcome Sr. Scientist fellow NIH, 1983-86, Pharmacology Rsch. Assoc. fellow, 1980-82. Mem. Soc. for Neurosci., Am. Soc. Biochemistry and Molecular Biology, Endocrinology Soc. Achievements include research in proteases required for synthesis of peptide neurotransmitters and hormones. Office: Uniformed Svcs U Health Sci Dept Biochemistry 4301 Jones Bridge Rd Bethesda MD 20814

HOOP, BERNARD, physicist, researcher, educator; b. San Francisco, Feb. 17, 1939; s. Bernard and Annette (Barbata) H.; m. Nancy Clark Hulbert, June 13, 1965; children: Heidi Ann, Katrina Clark. BS with distinction, Stanford U., 1960; MS, U. Wis., 1962, PhD, 1966. Physicist Mass. Gen. Hosp., Boston, 1967—; assoc. prof. medicine (physics) Harvard Med. Sch., Boston, 1986—; Fulbright lectr., India, 1993. Editor: Physical Principles of Physiological Phenomena, 1987; also over 60 articles. Mem. Am. Phys. Soc., Am. Physiol. Soc. Achievements include rsearch on physics of living systems, control of breathing. Home: 266 Main St Wakefield MA 01880 Office: Mass Gen Hosp Pulmonary Unit Fruit St Boston MA 02114

HOOPER, JOHN EDWARD, retired physicist, researcher; b. Edmonton, Alberta, Can., Dec. 25, 1926; arrived in Denmark, 1952; s. Percival Ralph and Mary Michelina Grant (Ferguson) H.; m. Lizzie Trolle, Sept. 24, 1955 (dec. Apr. 1990); children: Alasdair, Angus. BSc with honors, St. Andrews U., Scotland, 1949; PhD, Bristol U., Eng., 1953. Rsch. visitor Cern Theory Study Group, Copenhagen, 1952-53; Churchill scholar Niels Bohr Inst., Copenhagen, 1954-55, Rask-Oersted scholar, 1955-56, asst., 1957-71; lectr. Neils Bohr Inst., Univ. Copenhagen, Copenhagen, 1972-92; project leader Danish-Swedish Spiral Reader, Stockholm, 1968-71; sci. assoc. Cern, Geneva, 1985-86; vis. Univ. Tenn., Knoxville, 1961-62. Co-author: The Cosmic Radiation, 1958; designer: (software, specifications) A Spiral Reader for Bubble Chamber Film; designer: Development Plant for Nuclear Emulsions; contbr. 80 articles to profl. jours. Dep. Dansk Magisterforening, Copenhagen, 1972-84. Fellow Soc. Antiquaries Scotland; mem. Am. Phys. Soc., N.Y. Acad. Sci. Avocations: archaeology, early medieval history, geneaology. Home: 38 George St, Peebles EH45-8DL, Scotland

HOOTMAN, HARRY EDWARD, nuclear engineer, consultant; b. Oak Park, Ill., June 5, 1933; s. Merle Albert and Rachel Edith (Atkinson) H.; m. Linda P. Smith, Nov. 23, 1963; children: David, Holly, John. B.S. in Chemistry, Mich. Technol. U., 1959, M.S. in Nuclear Engring., 1962; LLB , LaSalle Extension U., 1971. Registered profl. engr., S.C. Rsch. assoc. Argonne (Ill.) Nat. Lab., 1959-62; process engr. Savannah River Plant, Aiken, S.C., 1962-65; rsch. assoc. reactor physics group, nuclear engring. div. Savannah River Lab., Aiken, 1965-87; with New Reactor Devel. Group, 1987-92, adv. engr. Planning, Studies and Analysis, 1992—; cons. transuranic waste disposal and incineration, radioisotope prodn., separation and shielding; instr. Math. and Engring. Dept. U. S.C., Aiken, 1979-80, 90—. Inventor alpha waste incinerator. Bd. dirs. Central Savannah River Area Sci. and Engring. Fair, Inc., Augusta, Ga., 1972-91. Served to sgt. USAF, 1953-57. Mem. Am. Acad. Environ. Engrs., Nat. Soc. Profl. Engrs. (local chmn. 1978-79), Am. Nuclear Soc. (local chmn. 1979-80), Am. Phys. Soc., Sigma Xi. Baptist. Home: 820 Brandy Rd Aiken SC 29801-7281 Office: Savannah River Lab Aiken SC 29808

HOOVER, HERBERT WILLIAM, SR., foundry engineer; b. Allentown, Pa., July 23, 1928; s. Walter Russel and Erma May (Mattern) H.; m. Joan Eleanor Kresge, Oct. 15, 1951; children: Linda Kay Weise, Herbert William Jr. BS in Indsl. Engring., Penn State Coll., 1952. Jr. metall. engr. Ingersoll Rand, Phillipsburg, N.J., 1952-55; asst. rsch. engr. Lebanon (Pa.) Steel Foundry, 1955-57; quality assurance mgr. Strong Steel Foundry, Buffalo, 1957-59; foundry supt. Buffalo (N.Y.) Forge Co., 1958-68; plant mgr. Strong Steel Foundry, Buffalo, 1962-71; foundry supt. Abex Foundry, Medina, N.Y., 1971-82; mgr. foundry engring. Birdsboro (Pa.) Steel Foundry, 1982; supr. patterns and purchased castings Dresser-Rand, Wellsville, N.Y., 1982—. Author: Solidification of Gray Iron, 1959. Chmn. Kenilworth United Ch. of Christ, Town of Tonawanda, N.Y., 1974; treas. Am. Foundrymen's Soc., Buffalo, 1968. With USN, 1946-48. Mem. Masons (Wellsville Lodge #230, master 1990). United Church of Christ. Home: RR 1 Box 54 Wellsville NY 14895-9721 Office: Dresser-Rand Steam Turbine and Motor Generation Div Coates St Wellsville NY 14895

HOPE, AMMIE DELORIS, computer programmer, systems analyst; b. Washington, Nov. 28, 1946; d. Amos Alexander and Amanda Irene (Moore) H. BA cum laude, Howard U., 1976; postgrad., Am. U., 1976-84. Police officer Met. Police Dept., Washington, 1972-73, officer, 1972; tchr. St. Benedict the Moor Cath. Sch., Washington, 1979; adminstrv. asst. Coun. of D.C., Washington, 1979-81; computer programmer, systems analyst IRS, Washington, 1984—. Honoree Civic Assn.; Trustees scholar; Pub. Svc. fellow. Mem. Alpha Kappa Delta. Achievements include research on adminstration of justice, computer applications and systems development. Home: 1904 D St NE Washington DC 20002-6720

HOPE, GEORGE MARION, vision scientist; b. Waycross, Ga., Jan. 24, 1938; s. George Marion and Jessie Candler (Norman) H.; m. Dorothy Marie Hendrix, Aug. 4, 1956; 1 child, Steven Richard. AB, Mercer U., 1965; MA, U. Fla., 1967, PhD, 1971. Asst. prof., rsch. assoc. U. Louisville, 1972-80; assoc. rsch. scientist U. Fla., Gainesville, 1980—; dir. low vision svc. U. Fla. Eye Ctr., U. Fla. Coll. Medicine, 1980—; co-dir. low vision clinic Dept. Ophthalmology U. Louisville, 1972-79. Contbr. numerous articles to profl. jours. Nat. Eye Inst. NIH grantee, 1975-78, 83-87. Mem. AAAS, Assn. Rshc., Vision and Ophthalmology (placement svc. 1972-84), Sigma Xi. Avocations: photography, camping, nature study. Office: U Fla Ctr Low Vision Study Care & Rehab JHMHC Box 100284 Gainesville FL 32610*

HOPE, JAMES DENNIS, SR., components engineer; b. Youngstown, Ohio, Sept. 20, 1948; s. Betty Jane (Morrison) Hope; m. Rosa Martinez, Feb. 7, 1973; 1 child, James Dennis II. BA in Pre-law, Youngstown (Ohio) State U., 1970; BSEE, Am. U., Washington, 1974. Cert. quality engr. Program mgr. F-18 Flair Tex. Instruments Inc., Dallas, 1978-84; sr. liaison engr. Boeing Electronics, Irving, Tex., 1984; sr. quality engr. Varo Inc., Garland, Tex., 1984; sr. components engr. Electrospace Systems Inc., Richardson, Tex., 1984—. Asst. dist. commr. Boy Scouts Am., Garland, 1984-88, Explorer post advisor, 1989—. Lt. col. U.S. Army, 1970-93. Mem. Am. Soc. Quality Control (cert. quality engr. rev. instr. 1984-93), Internat. Inst. Connector and Interconnection Tech. (pres. Dallas chpt., nat. dir. chpt. rels. 1990—). Methodist. Office: Electrospace Systems Inc 1301 E Collins Blvd PO Box 831359 MS 2167-31 Richardson TX 75083-1359

HOPEN, HERBERT JOHN, horticulture educator; b. Madison, Wis., Jan. 7, 1934; s. Alfred and Amelia (Sveum) H.; m. Joanne C. Emmel, Sept. 12, 1959; children: Timothy, Rachel. BS, U. Wis., 1956, MS, 1959; PhD, Mich. State U., 1962. Asst. prof. U. Minn., Duluth, 1962-64; prof. U. Ill., Urbana, 1965-85, prof., acting head, 1983-85; prof. horticulture U. Wis., Madison, 1985—, chmn. dept. horticulture, 1985-91. Mem. AAAS, Am. Soc. for Hort. Sci., Weed Sci. Soc. Am., North Ctrl. Weed Sci. Soc., CAST, Ygdrasil, Sigma Xi. Avocations: reading, gardening. Office: U Wis Dept Hort 1575 Linden Dr Madison WI 53706-1590

HOPFIELD, JESSICA F., neuroscientist, researcher, administrator; b. Princeton, N.J., Dec. 11, 1964; d. John Joseph and Cornelia (Fuller) H. BS in Biology summa cum laude, Yale U., 1986; PhD in Neuroscience, Rockefeller U., 1990; MBA, Harvard U., 1993. Postdoctoral assoc. Rockefeller U., N.Y.C., 1990-91; biomed. cons. Boston, 1991-93; assoc. dir. clin. rsch. Merck & Co., Rahway, N.J., 1993—. Fellow NSF, 1986; Baker scholar Harvard U., 1993. Mem. AAAS, Soc. Neuroscience, Sigma Xi.

HOPFIELD, JOHN JOSEPH, biophysicist, educator; b. Chgo., July 15, 1933; s. John Joseph and Helen (Staff) H.; m. Cornelia Fuller, June 30, 1954; children—Alison (Mrs. Charles C. Lifland), Jessica, Natalie. A.B., Swarthmore Coll., 1954; Ph.D., Cornell U., 1958; DSc (hon.), Swarthmore Coll., 1992. Mem. tech. staff ATT Bell Labs., 1958-60, 73-89; vis. research physicist Ecole Normale Superieure, Paris, France, 1960-61; asst. prof., then asso. prof. physics U. Calif. at Berkeley, 1961-64; prof. physics Princeton U., 1964-80, Eugene Higgins prof. physics, 1978-80; Dickinson prof. chemistry and biology Calif. Inst. Tech., Pasadena, 1980—. Trustee Battelle Meml. Inst., Harvey Mudd Coll., Huntington Med. Rsch. Inst. Guggenheim fellow, 1969, MacArthur Prize fellow, 1983; recipient Golden Plate award Am. Acad. Achievement, 1985, Michelson-Morley prize, 1988, Wright prize, 1989; named Calif. Scientist of Yr., 1991. Fellow Am. Phys. Soc. (Oliver E. Buckley prize 1968, Biol. Physics prize 1985); mem. Nat. Acad. Scis., Am. Acad. Arts and Scis., Am. Philos. Soc., Phi Beta Kappa, Sigma Xi. Home: 931 Canon Dr Pasadena CA 91106-4428 Office: Calif Inst Tech 139-74 Pasadena CA 91125

HOPKINS, DOUGLAS CHARLES, electrical engineer; b. Rochester, N.Y., Oct. 29, 1950; s. Russell William and Mildred Louise (Iocco) H.; m. Linda J. VanIngen, Aug. 31, 1974; children: Christopher, Erin. BSEE, SUNY, Buffalo, 1975, MSEE, 1977; PhD, Va. Polytech Inst., 1989. Rsch. engr. GE Corp., Schenectady, N.Y., 1977-82; sr. engr. Carrier Corp., Syracuse, N.Y., 1982-83; instr. Va. Polytech Inst., Blacksburg, 1983-88; asst. prof. Auburn (Ala.) U., 1988-92; rsch. asst. prof. dept. elec. engring. SUNY, Binghamton, 1993—; cons. PowerTech South, Raleigh, N.C., 1988-92, Brush Wellman, Cleve., 1990-92; summer fellow NASA, Lerc, Cleve., 1991-93. Contbr. more than 35 articles to profl. jours. Cubmaster Boy Scouts of Am., Auburn, 1990-92; legis. rep. PTA, Auburn, 1990-92; co-founder, v.p. Cloverleaf Homeowners Assn., Auburn, 1991-92. Named Most Valuable Prof. student br. IEEE/Internat. Soc. Hybrid Microelectronics, Va. Polytech Inst. and SU, 1986. Mem. IEEE (sr., 1st pl. Tech. Paper award Ala. sect. 1989), Internat. Soc. Hybrid Microelectronics (Best Paper of Session award internat. symposiums 1988, 89, 92), Sigma Xi, Eta Kappa Nu. Achievements include invention of Opto Coupled Power IC in GE Comp.; discovered energy recirculation tech. to apply to elec. power converters. Home: 408 Denal Way Vestal NY 13850 Office: SUNY Dept Elec Engring PO Box 6000 Binghamton NY 13902-6000

HOPKINS, MARK WILLARD, mechanical engineer; b. Wilmington, Del., Apr. 21, 1958; s. Walter Elmond and Jane Rebecca (Klair) H.; m. Mary Caroline Button, Dec. 30, 1982; children: William Jefferson, Anna MacIntyre. BSME, U. Del., 1980; MSME, Princeton U., 1982. Engr., Fibers E.I. DuPont, Wilmington, 1982-84; sr. engr. Photo Products Div. E.I. DuPont, Parlin, N.J., 1984-86; sr. rsch. engr. Composites Div. E.I. DuPont, Newark, 1986—. Contbr. articles to profl. jours.; patentee in field. Mem. ASME, Del. Soc. Profl. Engrs. Office: EI DuPont Fibers Div 400 Bellevue Rd Newark DE 19713

HOPKINS, ROBERT ARTHUR, retired industrial engineer; b. Youngstown, Ohio, Dec. 14, 1920; s. Arthur George and Margaret Viola (Brush) H.; m. Mary Madelaine Bailey, Apr. 6, 1946; 1 child, Marlaine Hopkins Kaiser. BBA, Case Western Reserve U., 1949; cert. loss control engr., U. Calif., Berkeley, 1969. Ins. agt. Nat. Life and Accident Ins. Co., Lorain, Akron, Ohio, 1951-56, San Mateo, Calif., 1951-56; ins. agt., engr. Am. Hardware Mt. Ins. Co., San Jose, Fresno, Calif., 1956-60; loss control engr. Manhattan Guarantee-Continental Ins. Co., Calif., 1967-77. Organizer Operation Alert CD, Lorain, 1951-52; prin. speaker CD, Fresno, 1957; active Pleasant Hill (Calif.) Civic Action Com., 1981-83; civilian coord. Office Emergency Svcs., Pleasant Hill, 1983-85; advisor, coord. airshows and warbird aircraft, 1980—; chmn. bd. Western Aerospace Mus., Oakland, Calif., 1988—; ops. asst. for tower and ops. 50th Anniversary Golden Gate Bridge, San Francisco, 1987; advisor, coord. Travis AFB Air Expo '90, 1990, Naval Air Sta. Alameda (Calif.) 50th Anniversary, 1990; advisor Naval Air Sta. Moffett Field Air Show, 1990, 92; warbird coord. Port of Oakland Airshow, 1987; mem. Smithsonian Mus., Smithsonian Air & Space Mus. With U.S. Army Air Corps, 1942. Recipient Letter of Appreciation Fresno CD, 1957, cert. of appreciation City of Pleasant Hill, 1986. Mem. No. Calif. Safety Engrs. Assn. (v.p., pres., chmn. 1974-77), Confederate Air Force (mem. staff, leader Pacific wing 1980—), Nat. Aero. Assn., Aero. Club No. Calif., Hamilton Field Assn. (dir. ops. Wings of Victory Air Show 1987, coord. 1988, 89—, asst. to pres. 1989—, advisor contr. 1990—), VFW (life, state civil disaster chmn. Area 5 Calif. 1991), Air Force Assn., Kiwanis (chpt. sec.-treas.). Republican. Roman Catholic. Avocations: fishing, reading, writing, aircraft restoration. Home: 48 Mazie Dr Pleasant Hill CA 94523-3310

HOPKINS, THOMAS DUVALL, economics educator; b. Spring Valley, Ill., Mar. 10, 1942; s. Joel Willis and Mildred (Duvall) H.; m. Jane Cole Eveleth, Apr. 20, 1968; children: Edward Eveleth, Catherine Chapin Hopkins. BA, Oberlin (Ohio) Coll., 1964; MA, Yale U., 1965, M of Philosophy, 1967, PhD, 1971. Asst. prof. econs. Bowdoin Coll., Brunswick, Maine, 1968-73; cons. Irwin Mgmt. Co., Inc., Columbus, Ind., 1973-75; asst. dir. Coun. on Wage and Price Stability, Washington, 1975-81, acting dir., 1981; dep. adminstr. Office of Mgmt. and Budget, Washington, 1981-84; assoc. prof. U. Md., College Park, 1984-87; assoc. prof. econs. Am. U., Washington, 1987-88; prof. econs., Arthur J. Gosnell prof. Rochester (N.Y.) Inst. Tech., 1988—; cons. Adminstrv. Conf. U.S., Washington, 1986-88, Office Tech. Assessment, U.S. Congress, Washington, 1987-89, Inst. Liberty and Democracy, Lima, Peru, 1986-91, U.S. Regulatory Info. Svc. Ctr., 1990-92, Congl. Budget Office, 1991; seminar leader Inst. Internat. Edn., Washington, 1987-88; mem. com. on tank vessel design marine bd. NRC, Washington, 1989-91; mem. com. on taxation, fin. and pricing Transp. Rsch. Bd., NRC, 1990-93, com. on public policy for surface freight transp., 1993—; lectr. U.S. Bus. Sch. in Prague, Czech Republic, 1992-93. Co-author: Tanker Spills: Prevention by Design, 1991. Elder Presbyn. Ch., 1986—; mem. coun. Eastman House, Rochester, 1991—. Woodrow Wilson Found. fellow, 1964. Fellow NSF; mem. Am. Econs. Assn., Nat. Economists Club, Assn. for Pub. Policy Analysis and Mgmt. Home: 215 Dorchester Rd Rochester NY 14610-1322 Office: Rochester Inst Tech 92 Lomb Meml Dr Rochester NY 14623-5604

HOPMEIER, MICHAEL JONATHON, mechanical engineer, troubleshooter; b. Detroit, Feb. 17, 1965; s. George Howard and Maxine Ethel (Liebowitz) H. BSME, U. Fla., 1985, MSME, 1988. Engring. asst. dept. physics U. Fla., Gainesville, 1983-85, engring. asst. combustion lab., 1985-87; chief project engr. Envireco, Inc., Gainesville, 1987-88; cons. Gainesville, 1988; engr. Sverdrup Tech., Inc., Eglin AFB, Fla., 1988—; with demolitions HAZMAT staff U. Fla., 1985-88; crisis mgr. Emergency Mgmt., Okaloosa County, Fla., 1988-91. Contbr. articles to profl. jours. Mem. AIAA, ASME (Sgt.-at-Arms 1984-87), Am. Def. Preparedness Assn., Soc. Automobile Engrs., Jaycees (bd. dirs. 1990—, Fla. Jaycee of Yr. 1990-91). Achievements include design and development of novel method of soil decontamination; development widely-used method of technical research at Eglin AFB; initiated and operated concepts technical liason office for special operations forces and conventional defense technologies. Office: Sverdrup Tech Inc PO Box 1935 Eglin A F B FL 32542-0935

HOPPE, RUDOLF REINHOLD OTTO, chemist, educator; b. Wittenberge, Germany, Oct. 29, 1922; s. Rudolf and Meta Hoppe; Diploma in Chemistry, U. Kiel, later Dr. rer. nat. h.c.; Dr. rer. nat., U. Muenster, Dr.h.c. Univ. Ljubljana/Slovenia, 1990; m. Karin Saborowski, 1951; children: Klaus-Dieter, Jens Reimar. Mem. faculty U. Giessen, 1965—, prof. and dir. Inst. Inorganic and Analytical Chemistry, 1965-91. Bd. dirs. Max Planck Inst.; mem. Kuratorium-Gmelin Inst., 1967-86. Recipient Akademie-Preis U. Goettingen, 1962, Henry Moissan Medal Chem. Soc. France, 1986, Josef-Stefan Medal U. Ljubljana, 1986. Mem. German Chem. Soc. (Alfred Stock prize 1974), OTTO-HAHN-Preis für Chemie und Physik, 1989, Austrian Chem. Soc., Royal Netherlands Chem. Soc., Chem. Soc. London, Am. Chem. Soc., Union Internat. Amies di cirque, German Soc. Ornithology, German Soc. Herpetology, German Soc. Natural Scientists and Physicians, German Mineral. Soc., Acad. Deutscher Naturforscher und Ärzte Leopoldina (senator, adj.), Mitglied der Osterreichischen, der Bayerischen Akademie der Wissenschaften. Author: The Chemistry of Noble Gases, 1961; over 600 articles; adv. bd. several chem. jours. Address: Heinrich Buff Ring 58, D-63 Giessen Germany

HOPPEL, THOMAS O'MARAH, electrical engineer; b. Nicktown, Pa., Dec. 14, 1937; s. Martin Michael and Helen Irene (Marr) H.; m. Lois Jean Weiland, Oct. 21, 1961; children: Eric, Christopher, Matthew. BSEE, Pa. State U., 1960. Registered profl. engr., Pa. Cert. Energy Mgr. Materials engr. GTE Sylvania, Emporium, Pa., 1960-65; product engr. GTE Sylvania, Altoona, Pa., 1965-66; sr. engr. GTE Sylvania, Altoona, 1966-81; project engr. Philips ECG Inc., Altoona, 1981-85; indsl. engr. Pa. Electric Co., Altoona, 1985—. Merit badge counselor Boy Scouts Am., cubmaster 1979-80; pres. St. Mary's Choir, 1988-91; advisor Jr. Achievement, 1967-68. Mem. IEEE (program chair 1963-64), Assn. Energy Engrs., Demand Side Mgmt. Soc., Altoona Engring. Soc. (pres. 1976-77, 88-89). Democrat. Roman Catholic. Achievements include patent for a vacuum tube part to reduce high pitch squeal from televisions. Home: 507 Garber St Hollidaysburg PA 16648 Office: Pa Electric Co 405 W Plank Rd Altoona PA 16603

HOPPENSTEADT, FRANK CHARLES, mathematician, university dean; b. Oak Park, Ill., Apr. 29, 1938; s. Frank Carl and Margaret (Goltermann) H.; m. Leslie Thomas, Dec. 27, 1986; children: Charles, Matthew, Sarah. BA, Butler U., 1960; MS, U. Wis., 1962, PhD, 1965. Instr. math. U. Wis., Madison, 1965; asst. prof. math. Mich. State U., East Lansing, 1965-68, dean Coll. Natural Sci., 1986—; assoc. prof. NYU, N.Y.C., 1968-76, prof., 1976-79; prof. U. Utah, Salt Lake City, 1977-86, chmn. dept. math., 1982-85. Author: Mathematical Methods in Population Biology, 1982, An Introduction to Mathematics of Neurons, 1986, Mathematics in Medicine and the Life Sciences, 1991, Analysis and Simulation of Chaotic Systems, 1993. Mem. Am. Math. Soc. (chmn. applied math. com. 1976-80), Soc. Indsl. and Applied Maths., Sigma Xi. Office: Mich State Univ Dean Coll Natural Sci 103 Natural Sci East Lansing MI 48824

HOPPER, ARTHUR FREDERICK, biological science educator; b. Plainfield, N.J., Sept. 7, 1917; s. Arthur Frederick and Catherine (Hoenig) H.; m. Amy Patricia Hull, Dec. 28, 1940 (dec. Nov. 1982); children: Arthur Frederick, Geoffrey Victor, Christopher James, Gregory Lorton; m. Patricia Ann Vennett, Sept. 6, 1986. AB, Princeton U., 1938; MS, Yale U., 1942; PhD, Northwestern U., 1948. Instr. Northwestern U., Evanston, Ill., summer 1948; asst. prof. Wayne U., Detroit, 1948-49; asst. prof. to prof. Rutgers U., New Brunswick, N.J., 1949-80, dir. biol. scis. grad. program, 1973-75; rsch. assoc. Brookhaven (N.Y.) Nat. Lab., 1961-68; visiting prof. U. Liège Med. Sch., Belgium, 1967-68; prof. emeritus Rutgers U., New Brunswick, 1980—; rsch. assoc. Detroit Cancer Inst., 1948-49; scientist aboard Columbia U. R/V "Vema", summer, 1955, 58; vis. investigator Battelle N.W., Richland, Wash., summer 1970, Jackson Meml. Lab., Bar Harbor, Maine, summers, 1971, 73. Author: Foundations of Animal Development, 1st ed. 1979, 2nd ed. 1985; contbr. articles to profl. jours. Chmn. troop 53 Boy Scouts Am., Bedminster, N.J., 1953-58; v.p., pres. Bedminster Bd. Edn., 1957-63; coach, mgr. Far Hills Little League Baseball, 1954-56; pres. Somerset County (N.J.) Bd. Edn., Somerville, N.J., 1960-63; radiology def. coord. Somerset County, 1959-63; bd. dirs. Palm Beach Country Kidney Assn., Lake Worth, Fla., 1988-90. 1st lt. USAF, 1943-46. Recipient rsch. grants NSF, USPHS, Am. Cancer Soc., Lalor Found., Rutgers U. Rsch. Coun., 1950-80. Mem. AAAS, Am. Soc. Zoologists, Soc. Devel. Biologists, Sigma Xi. Home: 231 Cocoanut Row Palm Beach FL 33480-4132

HOPPER, JACK RUDD, chemical engineering educator; b. Highlands, Tex., May 12, 1937; s. Bonnie Preston and Rosa Mae (Simmons) H.; m. Marilyn Joyce Spears, May 30, 1958; children: Connie, Bradley. Student, Lee Coll., 1957; BSChemE, Tex. A&M U., 1959; MChemE, U. Del., 1964; PhD, La. State U., 1969. Rsch. engr. Esso Rsch. and Engring., Baytown, Tex., 1959-67; asst. prof. chem. engring. Lamar U., Beaumont, Tex., 1969-72, assoc. prof. chem. engring., 1972-75, prof. chem. engring., 1975—, chair chem. engring. dept., 1974—, dir. engring. grad. studies, 1989—, liaison hazardous waste alternatives ctr., 1987-88; interim dir Gulf Coast Rsch. Cntr., 1993—; cons. J. M. Montgomery, New Orleans, 1991-92, Texaco Chem., Port Arthur, Tex., 1989-90, Star Enterprise, Port Arthur, 1990-91, Tex. Internat. Ednl. Consortium, Austin, 1991-93. Mem. editorial bd. Waste Mgmt., 1992—; contbr. articles to profl. publs. Officer Rotary, South Park, Tex., 1973-80. Recipient Dow Outstanding Faculty award Am. Soc. for Engring. Edn., 1971, Outstanding Alumni award Lee Coll., 1981. Fellow AICE. Lutheran. Achievements include inventions in field. Office: Lamar U 4400 MLK Pkwy Beaumont TX 77705

HOPPER, KEVIN ANDREW, electrical engineer, educator; b. Springfield, Mo., Feb. 21, 1962; s. Gary Don and Donna Kathleen (Wilkinson) H.; m. Andrea Lynn Bishop, Oct. 20, 1990. BSEE, U. Mo., Rolla, 1985; MBA, Drury Coll., 1988. Mgr. ops. and engring. Sho-Me Power Electric. Coop., Marshfield, Mo., 1985—; adj. faculty member Breech Sch. Bus. Drury Coll., Springfield, 1989—; lectr. numerous seminars. Mem. IEEE, NSPE, Mo. Soc. Profl. Engrs. Republican. Methodist. Office: Sho-Me Power Electric Coop PO Box D Marshfield MO 65706

HOPPERT, GLORIA JEAN, food products executive; b. LaFollette, Tenn., Apr. 2, 1949; d. Fred and Mona Ruth Cawood; m. Herschel M. Hoppert, Aug. 22, 1970; children: Hadden, Freya. BSChemE, U. Mich., 1971; MBA, Xavier U., 1975. Registered profl. engr., Wis., Pa., Nev., Ill. Staff engr., supr. div. shortening and oil Procter & Gamble, Cin., 1972-75, product tech. engr. div. paper, 1975-77; sr. project engr. Joseph Schlitz Brewing Co., Milw., 1977-78; sr. project engr. M&M/Mars Inc., Elizabethtown, Pa., 1978-81, mgr. maintenance sect., 1982-84, mgr. ops. Kudos, 1984-87; plant mgr. Sheba Kal Kan, Inc., Columbus, Ohio, 1987-88; mgr. mfg. Ethel M Chocolates, Las Vegas, Nev., 1988-91; mgr. ops. M&M/Mars, Inc., Chgo., 1991—; keynote speaker U. Dayton, 1980; cons. Transition to Mgmt. Conf., 1983, Mid-Atlantic Change Workshop Conf., N.Y.C. Contbr. articles to profl. jours. Merit scholar Mich. Higher Edn., 1967-71, Coll. Engring. scholar U. Mich., 1971. Fellow Soc. Women Engrs. (bd. dirs. 1978-80, sec. 1979); mem. NSPE. Lutheran. Avocations: reading, riding, travel, adventures. Home: 343 N Elmwood Ln Palatine IL 60067-7711 Office: M&M/Mars Inc 2019 N Oak Park Ave Chicago IL 60635

HOPPES, HARRISON NEIL, corporate executive, chemical engineer; b. Lehighton, Pa., Aug. 11, 1935; s. Charles Harold and Margaret Lois (Troxell) H.; m. Friederike Witte, June 20, 1959; children: Anne Marie, Charles Victor, Michael David, Margaret Louise, John Christian, Daniel James. BS in Chem. Engring., Pa. State U., 1957; MS in Indsl. Mgmt., MIT, 1959; PhD in Bus. Adminstrn., Am. U., 1968. Ops. analyst, project dir. Rsch. Analysis Corp., Bethesda, Md., 1961-69; dir. European field office Rsch. Analysis Corp., Heidelberg, Fed. Republic Germany, 1969-73; dir. div. Gen. Rsch. Corp., McLean, Va., 1973-76; v.p. Gen. Rsch. Corp., McLean, 1976-80; pres. Am. Tech. Assistance Corp., McLean, 1977-80; v.p. Flow Labs, McLean, 1980-85; v.p., pres. facilities group ERC Internat., Fairfax, Va., 1985-89; pres. Ogden Biosvcs. Corp., Gaithersburg, Md., 1989—; chmn. bd. dirs.

Logistics Ops., Fairfax, ERC Internat. Facilities Svc. Corp., Fairfax; bd. dirs. D-K Assocs., Inc., Fairfax. Author: Happes Family to 1800, 1985. Mem. Am. Mgmt. Assn., Pa. German Soc., Wash. Ops. Rsch. and Mgmt. Sci. Council. Republican. Lutheran. Avocations: art, music, genealogical history, local history, Egyptology. Home: 15716 Jones Ln Gaithersburg MD 20878-3563 Office: Ogden Bioservices Corp Montgomery Exec Ctr 6 Montgomery Village Ave Gaithersburg MD 20879

HOPPMEYER, CALVIN CARL, JR., civil engineer; b. New Orleans, Feb. 24, 1960; s. Calvin Carl and Reggie (Rigby) H.; m. Stephanie M. Omes, June 12, 1981; children: Brenna Marie, Kathryn Elaine. BS in Civil Engring., Tulane U., 1982. Specification writer Corps of Engrs., New Orleans, 1981-82; project engr. Burk & Assocs. Inc., New Orleans, 1982-88, Camp Dresser & McKee, Inc., Ft. Lauderdale, Fla., 1988-89; project mgr. Camp Dresser & McKee, Inc., Ft. Myers, Fla., 1989—. Mem. ASCE, NSPE, Am. Pub. Works Assn., Am. Water Works Assn. Republican. Roman Catholic. Office: Camp Dresser & McKee Inc 8191 College Pky # 206 Fort Myers FL 33919

HOPSON, JANET LOUISE, writer, educator; b. St. Louis, Oct. 22, 1950; d. David Warren and Ruth H.; m. Michael Rogers, Oct. 1976 (div. Nov. 1990). BS, So. Ill. U., 1972; MS, U. Mo., 1975. Biology editor Sci. News Mag., Washington, 1974-76; nature columnist Outside Mag., San Francisco and Chgo., 1977-82; lectr. dept. natural scis. U. Calif., Santa Cruz, 1983—; reviewer NSF Grant Rev. Panel for Undergrad. Sci. Edn., Washington, 1991. Author: Scent Signals, 1979; co-author: (coll. textbooks) Biology, 1988, Nature of Life, 1989, 2d edit., 1992, Essentials of Biology, 1990, Biology! Bringing Science to Life, 1991; contbr. articles to profl. jours. Recipient Russell L. Cecil award Arthritis Found., 1980; NIH/Coun. for Advancement of Sci. Writing fellow, 1976; Frank Luther Mott fellow U. Mo., 1973; grad. fellow So. Ill. U., 1972. Mem. Am. Inst. Biol. Scis., Nat. Assn. Sci. Writers, Textbook Authors Assn., No. Calif. Sci. Writers Assn. (pres. 1983-86).

HOPSON, KEVIN MATHEW, chemist; b. Roanoke, Va., May 11, 1959; s. James Mathew and Lois Dorene (Spurlock) H. BS in Microbiology, Va. Poly. Inst. and State U., 1983, BS in Biochemistry, 1983. Process engr. Hercules Inc., Radford, Va., 1983-87, analytical chemist, 1987-90; chemist DHHS, FDA, Phila., 1990—, internat. inspector, 1993—. Recipient City Key award Roanoke Award Com., 1976, 87, Leadership award Nat. Soc. Black Engrs., 1982. Mem. Am. Chem. Soc. (risk coord. 1983—), Am. Inst. Chemists, Am. Inst. Chem. Engrs., Alpha Phi Alpha (life, sec. 1989-90, v.p. 1988), Toastmasters Internat. (sgt.-at-arms 1988-89). Achievements include 2 engring. patents. Home: 1600 Garrett Rd Apt E209 Upper Darby PA 19082 Office: DHHS FDA Sci Br HFR MA160 2nd and Chestnut Sts Philadelphia PA 19106

HOPWOOD, DAVID ALAN, biotechnologist, geneticist, educator; b. Kinver, Staffs, Eng., Aug. 19, 1933; s. Herbert Hopwood and Dora Grant; m. Joyce Lillian Bloom, 1962; 3 children. Student, St. John's Coll., Cambridge, Eng.; PhD (hon.), ETH, Zurich, Switzerland. John Stothert Bye fellow Magdalene Coll. U. Cambridge, 1956-58; rsch. fellow St. John's Coll., 1958-61; univ. demonstrator, 1957-61; lectr. genetics U. Glasgow, 1961-68; John Innes prof. genetics U. East Anglia, 1968—; rschr. genetics dept. John Innes Inst., 1968—; hon. prof. Chinese Acad. Med. Sci., Inst. Microbiology and Plant Physiology, Chinese Acad. Scis., Huazhong Agrl. U., Wuhan, China. Author: (with K.F. Chater) Genetics of Bacterial Diversity, 1989; contbr. numerous articles to profl. jours., chpts. to books. Recipient Chiron Corp. Biotech. Rsch. award Am. Soc. Microbiology, 1993. Fellow UMIST (hon.), Inst. Biology, Indian Nat. Sci. Acad. (fgn.); mem. Genetical Soc. of Great Britain (former pres.), Hungarian Acad. Scis. (hon.), Soc. Genetic Microbiology (hon.). Office: John Innes Inst Dept Genetics, Colney Ln, Norwich NR4 7UH, England*

HORBATUCK, SUZANNE MARIE, optical engineer; b. Rego Park, N.Y., Aug. 5, 1964; d. Wesley Joseph and Shirley (Kurucz) H.; m. Douglas Scott Kindred, Sept. 23, 1989. BS, U. Rochester, 1986, MS, 1987, postgrad. Teaching asst. Inst. Optics U. Rochester, N.Y., 1987; rsch. asst. Inst. Optics U. Rochester, 1986—. Mem. Optical Soc. Am., Internat. Soc. Optical Engrs., Tau Beta Phi. Office: Inst of Optics Univ Rochester Rochester NY 14627

HORGAN, PETER JAMES, energy engineer; b. Neptune, N.J., Dec. 27, 1949; s. Andrew Bothwell and Marion (Lieth) H.; m. Nancy Kaye Crudup, Sept. 18, 1971; children: Shayna Beth, Shaun Michael. AS, Brookdale C.C., Lincroft, N.J., 1975; BA, U. R.I., 1980. Shift supr. R.I. Port Authority, North Kingstown, R.I., 1976-88; energy systems mgr. Conn. Coll., New London, 1988—. Mem. vestry St. Elizabeth's Chapel, Canochet, R.I., 1985, 92, Sunday Sch. tchr., 1986—; asst. cub master pack 1 Cub Scouts of Richmond, R.I., 1992. With USN, 1968-71. Mem. Assn. Energy Engrs., Conn. Assn. Energy Engrs. Episcopalian. Achievements include economizing and modernizing a 1960's vintage central heating plant and upgrading energy management systems; formulating a power/heat generation package for the college. Office: Conn Coll 270 Mohegan Ave New London CT 06320

HORI, YUKIO, scientific association administrator, emeritus engineering educator; b. Tokyo, Aug. 22, 1927; s. Kojiro and Yoshi (Saito) H.; m. Noriko Sunabori, May 15, 1965; children: Gen, Jun, Dan. B.Eng., U. Tokyo, 1951, Dr.Eng., 1960. Instr. U. Tokyo, 1953-55, assoc. prof., 1955-65, prof., 1965-88, emeritus prof., 1988—; exec. dir. Japan Soc. for Promotion of Sci., Tokyo, 1988—. Contbr. articles to profl. jours. Recipient Tokyo Metropolis award, 1984. Mem. ASME, Japan Soc. Mech. Engrs. (pres. 1988-89, awards 1960, 74, 89), Japan Soc. Tribologists (pres. 1990-92, award 1982), Japan Fedn. Engring. Soc. (v.p. 1989-93), Engring. Acad. Japan (v.p. 1993). Avocations: music. Home: Kugayama 3-19-19, Suginami-ku, Tokyo Japan 168 Office: Japan Soc Promotion Sci, Kojimachi 5-3-1 Chiyoda-Ku, Tokyo Japan 102

HORIUCHI, ATSUSHI, physician, educator; b. Tsuru, Yamanashi, Japan, Nov. 12, 1929; s. Masashige and Suzuyo Horiuchi; m. Dec. 1, 1961 (dec. June 1977); 1 child, Tadashi; m. Mar. 30, 1981. MD, Nihon U., Tokyo, 1956, D in Med. Sch., 1961. Clin. fellow Sch. Medicine Nihon U., 1957-62; rsch. fellow Sch. Medicine Yale U., New Haven, 1962-64; instr. Sch. Medicine Nihon U., 1964-73, assoc. prof., 1973-74; prof. Sch. Medicine Kinki U., Osaka, Japan, 1974—. Office: Kinki U Sch Medicine, 377-2 Ohnohigashi, Osakasayama, Osaka 589, Japan

HORKOWITZ, SYLVESTER PETER, chemist; b. Lansford, Pa., Sept. 7, 1921; s. Simeon and Mary (Leshefka) H.; m. Olga Assaf, Sept. 12, 1964. Student, Kans. State Coll., Pittsburg, 1948-51. Chemist Spencer Chem. Co., Pittsburg, 1946-51; chief chemist Spencer Chem. Co., Vicksburg, Miss., 1951-56; rsch. mgr. Spencer Chem. Co., Orange, Tex., 1956-61; v.p. Spencer Chem. Far East, Tokyo, 1961-65; chem. mgr. Far East Gulf Oil Corp., Tokyo, Singapore, Bangkok, 1965-72; cons. chemist New Orleans, 1972—; cons. chemist New Orleans, 1972—; bd. dirs.; chmn. A-Jin Chem. Co., Pusan, Republic of Korea, 1965-68; adv. bd. Pertamina Gulf, Djakarta, Indonesia, 1969-71; bd. dirs. chmn. Gulf Plastics-Singapore. Contbr. articles to profl. jours. With U.S. Army, 1942-46. Mem. ASTM, Am. Oil Chemists Soc., Soc. Plastics Engrs., Am. Chem. Soc. Republican. Byzantine Catholic Ch. Achievements include patents for ethylene/acrylate co-polymers, deconyl peroxide-free radical polymerization initiator, ammonium nitrate prilling tower process. Home: 5700 Ruth St Metairie LA 70003-2330

HORN, JANET, physician; b. Oak Ridge, Aug. 10, 1950; d. Harry and Molly (Rich) Horn; m. Alan R. Yuspeh, June 8, 1975. BA magna cum laude, Vanderbilt U., 1972; MS in Physiology and Biophysics, Georgetown U., 1973; MD, George Washington U., 1978. Diplomate Am. Bd. Internal Medicine, also sub-bd. Infectious Diseases; diplomate Am. Bd. Med. Examiners. Intern George Washington U. Hosp., Washington, 1978-79, resident in obstetrics and gynecology, 1979-81; resident in internal medicine Georgetown U., Washington, 1981-83; fellow in infectious diseases Johns Hopkins Hosp., Balt., 1983-85; mem. med. staff Georgetown U. Hosp., also Sibley Meml. Hosp., Washington, 1985-86, Johns Hopkins Hosp., 1986—, Sinai Hosp. of Balt., 1989—; asst. prof. medicine, div. infectious diseases Johns Hopkins U. Sch. Medicine, 1986—. Mem. editorial bd. Johns Hopkins Med. Grand Rounds, Am. Jour. Gynecologic Health; contbr. articles to profl. jours., chpts. to books. Bd. dirs. Chesapeake AIDS Found., 1989—; chair AIDS Coordinating and Adv. Coun. to Mayor, Balt., 1988—. Recipient Pearl M. Stetler Found. rsch. award Johns Hopkins U., 1987, Merck Found. clinician scientist rsch. award Johns Hopkins U., 1988. Mem. AAAS, ACP, Am. Soc. for Microbiology, Infectious Diseases Soc. Am., Johns Hopkins Med. and Surg. Assn., Phi Beta Kappa, Alpha Omega Alpha. Office: 5500 Newbury St Baltimore MD 21209-3652

HORNBEIN, THOMAS FREDERIC, anesthesiologist; b. St. Louis, Nov. 6, 1930; s. Leonard and Rosalie (Bernstein) H.; m. Gene Schwartz (div. 1968); children: Lia, Lynn, Cari, Andrea, Robert; m. Kathryn Mikesell, Dec. 24, 1971; 1 child, Melissa. BA, U. Colo.; MD, Wash. U. Diplomate Am. Bd. Anesthesiology. Intern King County Hosp., Seattle; resident in anesthesiology Wash. U., St. Louis, USPHS postdoctoral residency; instr. anesthesiology div. Wash. U., 1960-61; asst. prof. U. Wash., Seattle, 1963-67, assoc. prof., 1967-70, prof., 1970—; vice chmn. Dept. Anesthesiology, U. Wash., Seattle, 1972-74, asst. chmn. research 1974-77, chmn. 1978-93, research affiliate Primate Ctr., 1980. Author: Everest the West Ridge, 1966. Mem. bd. trustees Little Sch., Bellevue, Wash., 1982-89. Served to lt. comdr. USN, 1961-63. Recipient George Norlin award U. Colo., Denver, 1970, Alumni Centennial Symposium award 1975, Disting. Teaching award U. Wash., 1982. Fellow AAAS; mem. Am. Physiol. Soc. (editor 1967-73), Am. Soc. Anesthesiologists (Rovenstine lectr. 1989), Assn. Univ. Anesthetists (treas. 1969-72, pres. 1974-75), Soc. Acad. Anesthesia Chairmen, Inst. of Medicine, Phi Beta Kappa, Alpha Omega Alpha. Avocation: mountaineering. Office: U Wash Sch of Medicine Dept Anesthesiology RN-10 1959 NE Pacific St Seattle WA 98195-0001

HORNBY, ROBERT RAY, mechanical engineer; b. La Crosse, Wis., Dec. 2, 1958; s. William James and Nancy Kay Boettcher H.; m. Michal Rae Berrey, Aug. 2, 1980; children: Tabitha Kay, Maria Rae, Felicia Anne, Belinda Jo. BS in Mech. Engring., U. of Wis., Platteville, 1981. Registered profl. engr., Wis. Engring. cons. Geoscan Svcs. Co., Tulsa, Okla., 1983-84; sr. project engr. Howard Rotavator Co., Inc., Muscoda, Wis., 1984; mech. design engr. Rayovac, Portage, Wis., 1984-85; designer Gilman Engring. Co., Janesville, Wis., 1985-86; assoc. mech. design engr. Gilman Engring. Co., Janesville, 1986-87; sr. mech. design engr. Giddings Lewis, Janesville, 1989-92, project mgr., 1992—. Edn. chmn. Good Shepherd Luth. Ch., Janesville, 1985-88; com. chmn. Explorer Post 400 Boy Scouts Am., Janesville, 1985-91; scoutmaster Troop 516, Janesville, 1985—. Recipient Scoutmaster award of merit, Boy Scouts Am., Janesville, 1990, Dist. Award of Merit, 1992; named Outstanding Leader Exploring, Koshkonong Dist. Boy Scouts Am., Janesville, 1991. Mem. ASME (assoc.), Nat. Soc. Profl. Engrs., Soc. Mfg. Engrs. Achievements include development of math. model to predict lateral movement of oil well drill bit while drilling. Home: 3811 Colorado Trail Janesville WI 53546-9548 Office: Giddings Lewis 305 W Delevan Dr Janesville WI 53547-1357

HORNBY-ANDERSON, SARA ANN, metallurgical engineer, marketing professional; b. Plymouth, Devon, Eng., Apr. 17, 1952; came to U.S., 1986; d. Foster John and Joanna May (Duncan) Hornby; m. John Victor Anderson, Sept. 2, 1978 (div. May 1987). BSc in Metallurgy with honors, Sheffield (Eng.) City Poly., 1973, PhD in Indsl. Metallurgy, 1980. Chartered engr. Metallurgist Joseph Lucas Rsch., Solihull, Eng., 1970, William Lee Maleable, Dronfield, Eng., 1972; tech. sales specialist Applied Rsch. Labs, Luton, Beds, Eng., 1973-74; quality assurance metallurgist Firth Brown Tools, Sheffield, 1974-75, rsch. metallurgist high speed steel, 1975; lectr. Sheffield City Poly., 1975-78; grad. metallurgist, strip devel. metallurgist British Steel Corp., Rotherham, Eng., 1978-80; program mgr. Can. Liquid Air, Montreal, 1980-85; group mktg. mgr. Liquid Air Corp., Countryside, Ill., 1986-90; tech. mgr. Liquid Air Corp., Walnut Creek, Calif., 1990-93; mktg. mgr.-metals Can. Liquid Air, Toronto, Ont., 1993—; bd. dirs., chmn. R & D com., mem. publs. com., chmn. promotions and mktg. com. INvestment Casting Inst., Dallas; presenter to confs. in field. Contbr. articles to profl. jours.; patentee in field of metallurgy. Mem. AIME, Inst. Metals (young metallurgists com. 1974-80), Sheffield Metall. Soc. Inst. Metals (sec. 1978-80), Am. Soc. Metals, Am. Foundry Soc., Powder Metals Soc., Am. Iron & Steel Inst. (steering com. 1987—, chmn. topics com. 1988—, sec. 1992, vice chair 1993), Inst. Chartered Engrs. Eng. Mem. Ch. of Eng. Avocations: scuba diving, horseback riding, swimming. Office: Can Liquid Air, 1700 Steeles Ave E, Bramalea, ON Canada L6T 1A6

HORNE, GREGORY STUART, geologist, educator; b. Mpls., June 11, 1935; married; 2 children. A.B., Dartmouth Coll., 1957; Ph.D. in Geology, Columbia U., 1968. Geologist Tidewater Oil Co., 1957-64; geologist Pan Am. Petrol Corp., 1965; assoc. prof. earth sci. Wesleyan U., Middletown, Conn., 1967-80, prof. earth and environ. sci., 1980—, George I. Seney prof. geology and prof. earth and environ. scis.; pres. Essex Marine Lab., 1975—. Mem. Geol. Soc. Am., Am. Assn. Petroleum Geologists, Soc. Econ. Paleontology and Minerals. Office: Wesleyan U Dept Earth and Environ Sci Middletown CT 06457

HORNE, JEREMY, fiber optics, computer research executive; b. Palo Alto, Calif., Dec. 5, 1944; s. Frank Wescott and Mildred Cooley (Wright) H.; m. Deborah Elizabeth Hepburn, June, 1976 (div. Oct., 1979). AB, Johns Hopkins U., 1967; MS, So. Conn. State U., 1969; PhD, U. Fla., 1988. Instr. La. State U., Baton Rouge, 1984-85; adj. instr. Pima C.C., Tucson, 1985-87; tchr. correctional edn. program Ariz. State Prison, Douglas, 1988-89; curriculum coord. summer youth program Cochise Pvt. Industry Coun., Bisbee, Ariz., 1989; adj. faculty Cochise Coll., Douglas and Sierra Vista, Ariz., 1988-91; tech. writer, editor, task leader Sci. Applications Internat. Corp., Sierra Vista, 1990-93; v.p. Griffin Group Internat., Phoenix, 1992—; adj. prof. Cen. Ariz. Coll., Coolidge, Ariz.; writer Info. Gate Keepers, Boston, 1992—. Author: (textbook) Logic: The Theory of Order, 1989; co-author: (mil. handbook) Design Handbook for Fiber Optic Communications Systems, 1993. Active Nat. Sch. Bds. Assn., 1992—. Mem. AAAS, IEEE (mem. fiber optics tech. adv. group 1992—), Soc. Tech. Comm., Computer Soc. of IEEE, Lasers and Electro-Optics Soc. of IEEE, Telecomms. Industry Assn. (mem. standards com. fiber optic systems 1992—), Phi Kappa Phi. Achievements include rsch. in developing computational architecture based on orders of logical operators in a parenthesis-free expression, indications that Peano-based orderings portend significant consequences for current Von-Neuman architectures. Home: 32717 W Santa Cruz Ave Maricopa AZ 85239 Office: Griffin Group Internat Inc 2651 W Guadalupe Rd Ste A-225 Mesa AZ 85202

HORNER, ELAINE EVELYN, elementary education educator; b. Portales, N.M., Feb. 26, 1941; d. Carlton James and Clara C. (Roberson) Carmichael; m. Bill G. Horner, Feb. 2, 1959; children: Billy G. Sr., Frances E. Moreau, Aaron J. BA, Ea. N.Mex. U., 1973, MEd, 1978. Tchr. Artesia (N.Mex.) Jr. High Sch., 1973—. Recipient Cert. Appreciation Office Edn. Assoc., Artesia, 1978, 79, Cert. Support, 1986. Mem. NEA, Nat. Coun. Tchrs. Math., N.Mex. Coun. Tchrs. Math., Artesia Edn. Assn. (v.p. 1987-88), Delta Kappa Gamma (treas. 1988—). Democrat. Baptist. Avocations: reading, golf. Home: 2406 N Haldeman Rd Artesia NM 88210-9435 Office: Artesia Jr High Sch 15th and Cannon Artesia NM 88210

HORNER, JOHN ROBERT, paleontologist, researcher; b. Shelby, Mont., June 15, 1946; s. John Henry and Miriam Whitted (Stith) H.; m. Virginia Lee Seacotte, Mar. 30, 1972 (div. 1982); 1 child, Jason James; m. Joann Katherine Raffelson, Oct. 3, 1986. DSc (hon.), U. Mont., 1986. Rsch. asst. dept. geology Princeton (N.J.) U., 1975-82; curator paleontology Mus. of the Rockies, Mont. State U., Bozeman, 1982—; adj. asst. prof. dept. geology Mont. State U., 1982—; rsch. scientist Am. Mus. Nat. History, N.Y.C., 1980-82. Co-author: Maia: A Dinosaur Grows up, 1985, Digging Dinosaurs, 1988 (N.Y. Acad. Sci. award 1989), Digging Up Tyrannosaurus Rex, 1993, The Complete T-Rex, 1993; contbr. articles to profl. jours. With USMC, 1966-68; Vietnam. MacArthur fellow, 1986. Home: 9304 Cougar Dr Bozeman MT 59715-9515 Office: Mont State U Mus of the Rockies Bozeman MT 59717

HORNICEK, FRANCIS JOHN, JR., cell biophysicist, orthopedic surgeon; b. Pitts., Oct. 4, 1957; s. Francis John and Mary Ann (Zehel) H.; m. Kaye Wagar, July 21, 1984. BA, Washington and Jefferson Coll., 1979; MS, Georgetown U., 1982, PhD, 1983; MD, U. Pitts., 1991. Intern in gen. surgery Jackson Meml. Hosp.- U. Miami, Fla., 1991-92; resident in orthopedic surgery Jackson Meml. Hosp., U. Miami, Fla., 1992—; guest scientist NIH, Nat. Ctr. Drugs and Biologics, Bethesda, Md., 1982-83; postdoctoral fellow U. Miami (Fla.) Sch. Medicine, 1983-85, asst. prof., 1985-87, rsch. assoc., 1991—; postdoctoral fellow U. Pitts. Sch. Medicine, 1988-91, Allegheny Gen. Hosp., Pitts., 1989-91; sr. biochemist U.S. EPA, Washington, 1989. Contbr. to Ency. Immunology; contbr. articles to profl. jours. Mem. AAAS, AMA, Am. Soc. Cell Biology, Soc. Leukocyte Biology, Fla. Med. Soc., Dade County Med. Assn., Sigma Xi. Office: Univ Miami Sch Medicine PO Box 016960 R 12 Miami FL 33101

HORNIG, DONALD FREDERICK, scientist; b. Milw., Mar. 17, 1920; s. Chester Arthur and Emma (Knuth) H.; m. Lilli Schwenk, July 17, 1943; children: Joanna, Ellen, Christopher, Leslie. B.S., Harvard U., 1940, Ph.D., 1943; LL.D., Temple U., 1964, Boston Coll., 1966, Dartmouth Coll., 1974; D.H.L., Yeshiva U., 1965; D.Sc., U. Notre Dame, 1965, U. Md., 1965, Rensselaer Poly. Inst., 1965, Ripon Coll., 1966, Widener Coll., 1967, U. Wis., 1967, U. Puget Sound, 1968, Syracuse U., 1968, Princeton U., 1969, Seoul Nat. U., Korea, 1973, U. Pa., 1975, Lycoming Coll., 1980; D.Eng., Worcester Poly. Inst., 1967. Research asso. Woods Hole (Mass.) Oceanographic Instn., 1943-44; scientist, group leader Los Alamos Lab., 1944-46; asst. prof. chemistry Brown U., 1946-49, asso. prof., 1949-51, prof., 1951-57; dir. Metcalf Research Lab., 1949-57, asso. dean grad. sch., 1952-53, acting dean, 1953-54; vis. prof. Princeton U., 1957, prof. chemistry, 1957-64, chmn. dept., 1958-64, Donner prof. sci., 1959-66; spl. asst. sci. and tech. to Pres. U.S., 1964-69; dir. Office Sci. and Tech., 1964-69; chmn. Fed. Council Sci. and Tech., 1964-69; v.p., dir. Eastman Kodak Co., 1969-70; prof. chemistry U. Rochester, 1969-70; pres. Brown U., Providence, 1970-76; hon. research asso. in applied physics Harvard U., 1976-77, prof. chemistry in pub. health, dir. Interdisciplinary Programs in Health, 1977-90, Alfred North Whitehead prof. chemistry (public health), 1981-86, chmn. dept. environ. sci. and physiology, 1988-90; pres. Water Bd., Cambridge, Mass., 1985—; pres. Radiation Instruments Co., 1946-48; mem. Presdl. sci. advisory com., 1960-69, chmn. 1964-89; chmn. Project Metcalf Office of Naval Rsch., 1951-52; bd. dirs. Chem. Industry Inst. Author articles sci. jours. Bd. overseers Harvard U., 1964-70; bd. dirs. Overseas Devel. Coun., 1969-75; trustee George Eastman House, 1969-71, Manpower Inst., 1969-76; bd. dirs., treas. Overseas Devel. Network, 1984-92. Decorated Disting. Civilian Service medal Korea; Guggenheim fellow, 1954-55; Fulbright fellow, 1954-55; recipient Charles Lathrop Parsons award Am. Chem. Soc., 1967, Engring. Centennial award, 1967, Mellon Inst. award, 1968. Fellow Am. Phys. Soc., Am. Acad. Arts and Scis.; mem. Nat. Acad. Scis., Am. Chem. Soc., AAAS, Am. Philos. Soc., Romanian Acad. (fgn.), Sigma Xi. Home: 16 Longfellow Park Cambridge MA 02138-4831 Office: Harvard U Sch Pub Health 665 Huntington Ave Boston MA 02115-6021

HORNIK, JOSEPH WILLIAM, civil engineer; b. N.Y.C., May 7, 1929; s. Joseph and Josephine (Nemecek) H.; B.C.E., Cooper Union, 1952; grad. studies Columbia U., 1955-61; m. Barbara Joan Simko, Nov. 16, 1957; children: Heidi Josepha, Joseph Jared, Jason William, Heather Justine. Field engr. Stone & Webster Engring. Corp., Roanoke Rapids, N.C. and Portsmouth, Va., 1952-54; sr. engr. Howard, Needles, Tammen & Bergendoff, Jersey City, 1954-56; resident engr. Edwards & Kelcey, Bridgeport, Conn., 1956-59; project engr., project supt. The Austin Co., Bklyn. and San Juan, P.R., 1959-62; resident engr. Seelye, Stevenson, Value & Knecht, Whitehall, N.Y., 1962-65; county engr., county supt. hwys. County of Rockland, New City, N.Y., 1965-90; cons. engr., West Nyack, N.Y., 1967—; village engr. Village of Sloatsburg, N.Y., 1972-81, 85-88, Village of Haverstraw, N.Y., 1982-83, Village of Monroe, N.Y., 1984-92, Village of New Hempstead, N.Y., 1988-90, Village of Nyack, N.Y., 1988—. Mem. Rockland County Planning Bd., 1972-90, Rockland County Drainage Agy., 1972-90, Rockland County Soil and Water Conservation Agy., 1972-90, Rockland County Traffic Safety Bd., 1979-90, Nat. Com. on Uniform Traffic Control Devices, 1985—. Registered profl. engr., N.Y., Conn., Fla., P.R.; registered land surveyor, N.Y. Fellow ASCE; mem. NSPE, N.Y. State County Hwy. Supts. Assn. (dir. 1975-87, v.p. 1980-81, pres. 1982), N.Y. State Soc. Profl. Engrs., Nat. Assn. County Engrs. (dir. 1984-90), Nat. Assn. Counties, Am. Rd. and Transp. Builders Assn., Rockland County Assn. Hwy. Supts. (pres. 1979), Inst. Engrs., Architects and Surveyors of P.R., Soil and Water Conservation Soc. Am., Omega Delta Phi. Clubs: West Nyack Swim and Tennis, West Rock Tennis. Home and Office: 2 Dearborn Rd West Nyack NY 10994

HORNSBY, ANDREW PRESTON, JR., human services administrator; b. Tuskegee, Ala., Nov. 14, 1943; s. Andrew Preston Sr. and Margaret (Moore) H.; m. Wanda Weldon, Dec. 22, 1966; children: Tammy Leigh, Preston. BBA, Auburn U., 1968. Officer-in-charge Food and Nutrition Svcs., USDA, Birmingham, Ala., 1968-69, Montgomery, Ala., 1969-71; supr. child nutrition programs Food and Nutrition Svcs., USDA, Atlanta, 1971-74, sect. chief family nutrition program, 1974-76; br. chief Food and Nutrition Svcs., USDA, Washington, 1976-77, div. dir., 1977-79, dep. adminstr., 1979-82; regional adminstr. Food and Nutrition Svcs., USDA, Trenton, N.J., 1982-87; commr. Ala. Dept. Human Resources, Montgomery, 1987—. Mem. Cabinet Ala. Govs. Office, Montgomery, 1987—; bd. dirs. Ala. Dept. Youth Svcs., 1987—, Montgomery Area United Way, 1989-91. Mem. Am. Pub. Welfare Assn. (bd. dirs. 1989—). Office: State Dept Human Resources 50 N Ripley St Montgomery AL 36130-3901

HORNSTEIN, LOUIS SIDNEY, retired emergency room physician; b. Phila., Feb. 3, 1926; s. John Nathan De Toledano and Maria (De Leucena) Seixas; m. Erika G. Scaaale, Apr. 17, 1946; children: Louis, John, Michaela, Robert. MD, U. Heidelberg, 1953; PhD, U. Vienna, 1954. Intern St. Joseph Hosp., Phila., 1955-56, resident in anesthesiology, 1956-58; pvt. practice Skowhegan, Maine, 1958-81; emergency rm. physician 97th Gen. Hosp., Frankfurt, Germany, 1981-87; ret., 1988; physician Mother Theresa's Sisters of Charity, Calcutta, 1988. Contbr. articles to profl. jours. Sgt. U.S. Army, 1944-46, ETO, col., 1981-87. Decorated Croix de Guerre, Silver Star, Bronze Star, Purple Hearts. Mem. Masons, Shriners, Knights of Malta. Jewish. Home: 220 Water St Skowhegan ME 04976-1707

HOROSZEWICZ, JULIUSZ STANISLAW, oncologist, cancer researcher, laboratory administrator; b. Warsaw, Poland, Jan. 4, 1931; came to U.S., 1961; s. Tytus Michal and Stefania (Domanska) H.; m. Hanna Urszula Kubik, Jan. 12, 1969; children: Nike Joanna, Peter Juliusz. D of Medicine summa cum laude, Acad. of Medicine, Lodz, Poland, 1954, DMSc, 1960. Teaching asst. dept. bacteriology Acad. of Medicine, Lodz, 1950-55, asst. prof., 1955-59, assoc. prof., 1959-61; cancer rsch. scientist Roswell Park Meml. Inst., Buffalo, 1962-64, sr. cancer rsch. scientist, 1964-67, assoc. cancer rsch. scientist, 1967-76, prin. cancer rsch. scientist, 1976-86; assoc. chief oncological urology rsch. N.Y. State Dept. Health, Roswell Park Meml. Inst. Div., Buffalo, 1986-88; dir. exptl. cancer ctr. Millard Fillmore Hosp., Buffalo, 1988—; dir. electron microscopy lab. viral oncology, 1963-66, dir. human fibroblast interferon program Roswell Park Meml. Inst., 1976-82; chmn. Pleuro-Pneumonia Like Organisms subcom. human cancer virus task force Nat. Cancer Inst., Bethesda, Md., 1963-64, mem. Nat. Prostatic Cancer Project working cadre, 1972-74; assoc. rsch. prof. microbiology SUNY, Buffalo, 1966—; rsch. prof. biology Canisius Coll., Buffalo, 1968—, Niagara U., Niagara Falls, N.Y., 1968—. Contbr. over 90 articles to profl. jours.; patentee on specific monoclonal antibody for diagnosis and treatment of human prostate cancer. Rockefeller Found. fellow, 1961-62; rsch. grantee Nat. Cancer Inst., 1979-82; named Citizen of Yr., Am.-Polish Eagle, Buffalo, 1967. Mem. AAAS, Am. Assn. Cancer Rsch., Am. Soc. Microbiology, Polish Soc. for Bacteriology, Am. Cancer Soc., Am. Assn. for Clin. Rsch., N.Y. Acad. Scis. Roman Catholic. Avocations: fishing, bridge, classical music, chess. Home: 2210 N Forest Rd Buffalo NY 14221-1346 Office: Millard Fillmore Hosp 3 Gates Cir Buffalo NY 14209-1194

HOROWITZ, BARRY MARTIN, systems research and engineering company executive; b. Bklyn., Apr. 20, 1943; s. Isaac Harry and Clara Fireda (Weintraub) H.; m. Sheryl Robin Lang, Jan. 24, 1965; children: Hillary, Charles. BSEE, CCNY, 1965; MSEE, NYU, 1967, PhDEE, 1969. Asst. project engr. Bendix Corp., 1965-66, sr. project engr., 1967-69; project engr. Gen. Precision, 1966-67; tech. staff MITRE Corp., McLean, Va., 1969-71, group leader, 1971-74, dept. head, 1974-79; dir. spl. studies MITRE Corp., Bedford, Mass., 1979-80, tech. dir., 1980-84, v.p. strategic programs, 1984-85, v.p. programs, 1985-86, sr. v.p., gen. mgr., 1986, group v.p., gen. mgr., 1986-87, exec. v.p., chief oper. officer, 1987—, also dir.; cons. sci. adv. bd.

USAF, Pentagon, Washington, 1982—, Def. Sc. Bd., Pentagon, 1988—. Contbr. articles to profl. jours. Mem. IEEE, AIAA, Armed Forces Communications and Electronics Assn. (pres. 1987-88, pres.-elect 1990, Gold medal for Engring. 1990), Ctr. Sci. and Internat. Affairs., Eta Kappa Nu, Tau Beta Pi. Avocation: musician. Home: 97 Belcher Dr Sudbury MA 01776-1246 Office: MITRE Corp 202 Burlington Rd Bedford MA 01730-1306

HOROWITZ, BEN, medical center executive; b. Bklyn., Mar. 19, 1914; s. Saul and Sonia (Meringoff) H.; m. Beverly Lichtman, Feb. 14, 1952; children: Zachary, Jody. BA, Bklyn. Coll., 1940; LLB, St. Lawrence U., 1935; postgrad. New Sch. Social Rsch., 1942. Bar: N.Y. 1941. Dir. N.Y. Fedn. Jewish Philanthropies, 1940-45; assoc., ea. regional dir. City of Hope, 1945-50, nat. exec. sec., 1950-53, exec. dir., 1953-85, gen. v.p., bd. dirs., 1985—, bd. dirs. nat. med. ctr., 1980—; bd. dirs. Beckman Rsch. Inst., 1980—. Mem. Gov.'s Task Force on Flood Relief, 1969-74. Bd. dirs., v.p. Hope for Hearing Found., UCLA, 1972—; bd. dirs. Forte Found., 1987-92, Ch. Temple Housing Corp., 1988—, Leo Baeck Temple, 1964-67, 86-89, Westwood Property Owners Assn., 1991—. Recipient Spirit of Life award, 1970, Gallery of Achievement award, 1974, Profl. of Yr. award So. Calif. chpt. Nat. Soc. Fundraisers, 1977; Ben Horowitz chair in rsch. established at City of Hope, 1981. City street named in his honor, 1986. Jewish. Formulated the role of City of Hope as pilot ctr. in medicine, sci., and humanitarianism, 1959. Home: 221 Conway Ave Los Angeles CA 90024-2601 Office: City of Hope 208 W 8th St Los Angeles CA 90014-3208

HOROWITZ, EMANUEL, materials science and engineering consultant; b. N.Y.C., Mar. 29, 1923; s. Barney and Florence (Stein) H.; m. Diane Silverman, Aug. 6, 1950; children: Amy, Andrew, Alice, Alan. BS, CCNY, 1948; MS, George Washington U., Washington, 1956, PhD, 1963. Chemist Smithsonian Instn., Washington, 1949-51; rsch. chemist Nat. Bur. Stds., Washington, 1951-68; dep. dir. Inst. Materials Rsch. Nat. Bur. Stds., Gaithersburg, Md., 1969-78; dep. dir. Nat. Measurement Lab. Nat. Bur. Stds., Gaithersburg, 1978-80; prof. materials sci. and engring., biomaterials Johns Hopkins U., Balt., 1980-89; prof. (part time) Balt., 1989—; U.S. del. Internat. Stds. Orgn., T.C. 150, Surg. Implants, 1990—. Editor: Resource Recovery and Utilization, 1979, Materials and National Policy, 1979. 1st lt. USAAF, 1943-46, PTO. Recipient Profl. Svc. award Alpha Chi Sigma, 1990. Mem. ASTM (trustee 1990—, resources materials subcom. chmn. F-4 surg. implants 1972—, meritorious award 1972, hon. award 1976), Inst. Stds. Rsch., N.Y. Acad. Sci., Washington Acad. Sci., Cosmos Club, Fedn. Materials Socs. (pres. 1983), Nat. Bur. Stds. Alumni Assn. (pres. 1990-92, Gold medal, Rosa award). Achievements include first to report relationship between thermal stability and periodicity of transition metals in metallo-organic polymers. Home: 14100 North Gate Dr Silver Spring MD 20906 Office: John Hopkins U 34th and Charles Sts Baltimore MD 21218

HOROWITZ, ISAAC M., control research consultant, writer; b. Safed, Galilee, Israel, Dec. 15, 1920; Came to U.S., 1951; s. Yeshayahu Y. Horowitz and Feige Loberboim; m. Chana Shankman, Sept. 15, 1945 (div. Dec. 1984); children: Sharon, Ruth, David, Dafna; m. Gloria T. August, Dec. 24, 1984; children: Matanya, Benyakir. BSc with honors, U. Man., Can., 1944; BEE, MIT, 1948; MEE, Poly. Inst. N.Y., 1953, DEE, 1956. Design engr. Taller & Cooper, Bklyn., 1948; electronic engr. Israel Def. Army, Haifa, 1948-50, Halross Instruments, Winnipeg, Can., 1950-51; from instr. to asst. prof. Poly. Inst. Bklyn., 1951-58; sr. staff Hughes Res. Labs., Malibu, Calif., 1958-62; sr. scientist Hughes Aircraft Co., Culver City, Calif., 1962-67; prof. U. Colo., Boulder, 1967-85, U. Calif., Davis, 1985-91; cons. and writer Boulder, 1991—; cons. Israel Aircraft Co., Lod, 1969-72, Flight Dynamics Lab., WPAFB, Dayton, Ohio, 1983-92, Sandia Nat. Labs., Livermore, Calif., 1991—; Prof. Cohen Chair, Weizmann Inst. Sci., Rehovot, Israel, 1969-85; dist. vis. prof. Air Force Inst. Tech., WPAFB, 1983-92. Author: Synthesis of Feedback Systems, 1963, Quantitative Feedback Design Theory, 1993. Recipient Best Paper award Nat. Electronics Conf., 1956, Rufus Oldenberger medal ASME, 1992; rsch. grantee NASA, NSF, AFOSR, 1967-91. Fellow Internat. Elec. and Electronics Engrs. Republican. Hebrew. Achievements include patents in magnetic amplifiers, phase-locked loop; founder and developer of quantitative feedback theory (QFT). Home: 4470 Grinnell Ave Boulder CO 80303 Office: U Calif Davis Dept EECS Davis CA 95616

HOROWITZ, JOSEPH, civil engineer; b. N.Y.C., June 14, 1931; s. Morris and Pearl (Ingwer) H.; m. Carol Barbara Hefler, Mar. 22, 1964; children: Susan Miriam, Gail Sharon. BCE, The Cooper Union, 1953; MBA, CUNY, 1958; MS, Poly. Inst. N.Y., 1970. Lic. profl. engr. N.Y., N.J. Field engr. Andrews and Clark, N.Y.C., 1957-58; chief engr. The Tumpane Co., Ankara, Turkey, 1958-61; engr. assessor N.Y.C. Dept. Assessment, 1961-62; adminstrv. engr. Facil. Engring. Dept. CBS Inc., N.Y.C., 1962-66, mgr. planning and design, 1968-72; dir. facilities engring. Facil. Engring. Dept. CBS, Inc., N.Y.C., 1974-86, dir. environ. engring., 1987--; cons., lectr. in field, 1962--; adj. faculty Rutgers U., CUNY; vis. assoc. prof. Pratt Inst. Sch. Architecture, Bklyn., 1987--. Author: Critical Path Scheduling: Mangement Control Through CPM and PERT, 1967; author: (chpt.) Time Saver Standards, 1973; contbr. articles to profl. jours. Lt. commdr. USNR, 1953-65. Mem. Am. Soc. Civil Engrs., Nat. Soc. Profl. Engrs., Internat. Facilities Mgmt. Assn., Hazardous Materials Control Inst., Water Environ. Fedn., Nat. Assn. Environ. Profls., Environ. Auditing Roundtable, Tau Beta Pi, Chi Epsilon.

HORSBURGH, ROBERT LAURIE, entomologist; b. Coronach, Sask., Can., June 23, 1931; came to U.S., 1974; s. Minno Spencer and Helen (Louise) H.; m. Norma Jean Hodge; children: Karn Irene, Robert Scott, Sandra Lynn. BS, McGill U., Montreal, Can., 1956; MS, Pa. State U., State Coll., 1968, PhD, 1969. Entomologist Nova Scotia Dept. Agr., Kentville, N.S., Can., 1956-66, 1970-74; entomologist Va. Polytech. Inst., Steeles Tavern, Va., 1974-81; lab. supt. entomologist Winchester (Va.) Fruit Research Lab., 1981—. Named Outstanding Extension Entomologist, Entomol. Soc. Am., 1983. Mem. Entomol. Soc. Can. Presbyterian. Home: 786 Mulberry Cir Stephens City VA 22655-2321 Office: Winchester Fruit Research Lab 2500 Valley Ave Winchester VA 22601-2762

HORTON, DONNA ALBERG, technical writer; b. Newport, Ky., Sept. 2, 1935; d. Donald Hyatt and Virginia Margaret (Bauer) Mincey; m. Frederick Albert Alberg, July 16, 1962 (div. 1980); children: Stephanie Alberg OprinFger, Jenee, Nicole Alberg Armstrong; m. William Michael Horton, Sept. 29, 1985. Student, Cin. Conservatory of Music, 1953; BA, Ea. Ky. U., 1957, 60, MA, 1961; postgrad., U. South Fla., 1981. Cert. tchr. Fla., Ky. Tchr. Flint (Mich.) Pub. Sch. System, 1957-59; head dept. art Titusville High Sch., Brevard County, Fla., 1961-63; tchr. Toledo (Ohio) Pub. Schs., 1964-68; feature writer St. Petersburg (Fla.) Times, 1978-79; instr. spl. edn. Pinellas County (Fla.) Pub. Schs., 1979-82; tech. writer, computer specialist MAV/Strategic div. Honeywell Inc., Clearwater, Fla., 1982—; lectr. in field; profl. tech. writer. Contbg. writer Cin. Times Star, 1949-53, Richmond Register, 1953-61, Titusville Star Advocate, 1961-63, Toledo Blade, 1964-68; author children's plays: Holiday Presentation, 1957-59; contbr. articles to profl. jours. and internat. symposia. Mem. Flint Community Chorus, 1957-59; head vols. Largo (Fla.) Pub. Libr., 1975-79; mem. Messiah Choir, Clearwater, 1988-89. Mem. Soc. Logistics Engrs. (treas. 1983-85, workshop organizer), Profl. Tech. Writers Assn., Sigma Xi Mu (chpt. pres. 1954-57), Alpha Chi Epsilon. Democrat. Roman Catholic. Avocations: antiques, computers, music, biking. Home: 3717 Mckay Creek Dr Largo FL 34640-4515 Office: Honeywell Inc 13350 US Hwy 19 N Clearwater FL 34624-7226

HORTON, GRANVILLE EUGENE, nuclear medicine physician, retired air force officer; b. Jean, Tex., July 2, 1927; s. James Granville and Etna (Boyle) H.; B.A., Tex. Technol. Coll., 1950; M.D., U. Tex., 1954; m. Mildred Helen Veale, June 13, 1953; children: Linda Kay, Kevin Bruce, Carson Scott. Intern, Detroit Receiving Hosp., 1954-55; tng. in radioactive isotope techniques Oak Ridge Inst. Nuclear Studies, 1958; practice medicine, Weslaco, Tex., 1955-56, Outlar-Blair Clinic, Wharton, 1956-72; dir. dept. nuclear medicine Nightingale Hosp., El Campo, Tex., 1973-75; mem. staff Horton Med. Clinic, El Campo, 1972-75; part-time research assoc. radioisotope dept. Meth. Hosp., Houston, 1961-66; mem. med. adv. com. and sec. med. staff Caney Valley Meml. Hosp., Wharton, 1956-72; clin. dir. Wharton County TB Assn., 1957-67; commd. lt. col. U.S. Air Force, 1975; postgrad. U.S. Air Force Sch. Aerospace Medicine, 1975; chief aeromed. services Brooks AFB, Tex., 1976-82. Bd. dirs. Wharton County div. Am. Cancer Soc., pres., 1960-61; dir. 8th dist. Tex., Citizens Com. for Hoover Report, 1957-58. Served with USN, 1946-47. Fellow Am. Coll. Angiology (state gov. 1979), Am. Coll. Nuclear Medicine; mem. Am. Coll. Emergency Physicians, Wharton C. of C. (dir., v.p. 1960-61), Am., Tex. (ho. of dels. 1959-61) med. assns., Soc. Nuclear Medicine, Tex. Assn. Physicians Nuclear Medicine, AAAS, Law Enforcement Officers Tex. (asso.), Am. Nuclear Soc., Tex. Med. Found., El Campo C. of C., Phi Chi. Republican. Episcopalian. Lodge: Elks. Contbr. articles to med. publs. Home: 15102 Oakmere St San Antonio TX 78232-4623 Office: Occupational Med Clinic 2200 Mccullough Ave San Antonio TX 78212-3751

HORTON, WILFRED HENRY, mathematics educator; b. Newark, Nottingham, Eng., May 27, 1918; s. Henry and Alice M. (Spence) H.; m. Margaret E. Haskard; children: Richard, Sheila, David, Jennifer. BSc in Math. with honors, U. Coll., Nottingham, 1940; Engr., Stanford U., 1959. With De Havilland Aircraft Co., Hatfield, Eng., 1940-45, Percival Aircraft, Luton, Eng., 1945-50; sr. sci. officer Royal Aircraft Establishment, Farnborough, Eng., 1950-54, prin. sci. officer, 1954-57; assoc. prof. Stanford (Calif.) U., 1959-67; prof. Ga. Inst. Tech., Atlanta, 1967-84; prof. emeritus U. System Ga., 1985—; cons. various orgns. Contbr. articles to profl. jours. and encys. Achievements include design of test facilities for supersonic aircraft, hypersonic wind tunnel.

HORTON, WILLIAM ALAN, structural designer; b. Ft. Benning, Ga., Dec. 29, 1955; s. William Howard and Edith Marjorie (Amick) H.; m. Laurie Lynn Sewell, Sept. 6, 1986. AAS, Midlands Tech. Coll., 1975; BS, David Lipscomb U., 1980. Sr. drafter Nat. Convenience Stores, Inc., Houston, 1981-87, Morris and Assocs., Inc., Houston, 1987-88; sr. designer Fluor Daniel, Inc., Sugar Land, Tex., 1988—. Judge skill olympics Vocat. Indsl. Clubs Am., Houston, 1990, San Antonio, 1991. Mem. Church of Christ. Avocations: computer programming, bicycling.

HORTON, WILLIAM ARNOLD, medical geneticist, educator; b. Kans., Oct. 2, 1945; s. Kenneth Guy and Virginia Ella (Arnold) H.; m. Jean E. Foss, Nov. 20, 1971; 1 child, Joshua. BA, U. Kans., 1967, MD, 1971. Diplomate Nat. Bd. Med. Examiners, Am. Bd. Internal Medicine, Am. Bd. Med. Genetics. Asst. prof. medicine and pediatrics U. Kans. Med. Sch., Kansas City, 1977-81, assoc. prof., 1981-83; assoc. prof. pediatrics and medicine U. Tex. Health Sci. Ctr., Houston, 1983-88, prof., 1988—; bd. dirs. Little People of Am., Houston, Human Growth Found.; ad hoc reviewer NIH, Bethesda, Md., 1988—. Contbr. chpt. to books, articles to profl. jours. Recipient Walter Sutton award, 1969; grantee NIH, 1992—, Shriners, 1991—, 92—. Fellow Am. Acad. Physicians; mem. Am. Soc. Human Genetics, Bone Dysplasia Soc. (founding mem.), Am. Soc. Pediatrics. Achievements include development of novel methods for analyzing growing bone in humans. Office: U Tex Med Scis Ctr 6431 Fannin MSB3-244 Houston TX 77030

HORTON, WILLIAM DAVID, JR., army officer; b. Memphis, Oct. 25, 1953; s. William David and Maynell (Holland) H.; m. Rebecca Jean Griffin, Oct. 11, 1975; children: Elizabeth Anne, Wendy Leigh, William Robert. BS in Physics, U. Miss., 1975; MS in Physics, MIT, 1979. Commd. 2d lt. U.S. Army, 1975, advanced through grades to lt. col., 1992; rsch. coord. directed energy directorate U.S. Army, Redstone, Ala., 1983-85, asst. project mgr., 1985-86; br. chief dir. combat devels. U.S. Army, Ft. Knox, Ky., 1987-91, div. chief dir. combat devels., 1990-91; asst. project mgr. PM survivability systems U.S. Army, Warren, Mich., 1991—. Author reports, manuals. Mem. Armor Assn., 2d Armored Cavalry Assn. (life). Achievements include development of analytical methodology for applying electronic warfare components to ground vehicles, systems engineering methodologies for directed energy weapons on combat vehicles and defeat of smart, top attack munitions, development of electronic warfare design methodologies for ground systems. Home: 441 Skeel Ave Selfridge Air Base MI 48045 Office: US Army PM Survivability SFAE ASM SS E Warren MI 48397

HORVÁTH, CSABA, chemical engineering educator, researcher; b. Szolnok, Hungary, Jan. 25, 1930; came to U.S., 1963; s. Gyula and Roza (Lanyi) H.; children: Donatella, Katalin. Diploma in Chem. Engring., U. Tech. Scis., Budapest, Hungary, 1952, Dr. (hon.), 1986; PhD, J.W. Goethe U., Frankfurt-am-Main, Germany, 1963; MA (hon.), Yale U., 1979. Asst. in chem. tech. U. Tech. Scis., Budapest, 1952-56; chem. engr. Hoechst AG, Frankfurt am Main, 1956-61; research fellow Harvard U., Cambridge, Mass., 1963-64; research assoc. Yale U. Sch. Medicine, New Haven, 1964-69, assoc. prof., 1970-79, prof. chem. engring., 1979—, chmn. dept. chem. engring., 1987—; cons. various govt. and indsl. orgns. Co-author: Introduction to Separation Science, 1973; editor: Series High Performance Liquid Chromatography, 1981—; mem. editorial bds. 9 sci. periodicals; contbr. over 240 rsch. papers and articles to sci. publs. Recipient S. Dal Nogare award Delaware Valley Chromatography Forum, 1978, Tswett medal 15th Internat. Symposium on Advances in Chromatography, 1979, Humboldt sr. U.S. scientist award Humboldt Found., Fed. Republic of Germany, 1982, EAS Chromatography award, 1986, Van Slyke award N.Y. Metro Section Am. Assn. Clin. Chemists, 1992. Fellow Inst. Med. and Biomed. Engrs.; mem. AAAS, AICE, Deutsche Gesellschaft fuer Chemiches Apparatewesen Chemische Technik und Biotechnologie e.v, Am. Chem. Soc. (nat. chromatography award 1983), Am. Ceramic Soc., Hungarian Chem. Soc. (hon.), Hungarian Acad. Scis. (external), Conn. Acad. Sci. and Engring., Conn. Acad. Arts and Scis., Inst. Food Technologists, Sigma Xi. Home: PO Box 2041 41 Temple Ct New Haven CT 06521-2041 Office: Yale U 9 Hillhouse Ave New Haven CT 06520-2159

HORVATH, DIANA MEREDITH, plant molecular biologist; b. N.Y.C., June 2, 1963; d. Bruce L. and Shirley Jane (Wadhams) Ralston; m. Curt Michael Horvath, Sept. 12, 1992. BS, Tufts U., 1985; MS, Northwestern U., 1991, PhD, 1991. Scientist USDA, Gaborone, Botswana, 1985-86, Agrl. Rsch., Govt. of Botswana, Gaborone, 1985-86, postdoctoral fellow The Rockefeller U., N.Y.C., 1991—. Contbr. chpts. to books, articles to Cell, Procs. of Nat. Acad. Sci. ARCS Found. scholar, 1987; NIH tng. grantee, 1988; NSF fellow, 1991; recipient 1st place prize Sigma Xi, 1989. Mem. Am. Soc. Plant Physiologists. Office: Rockefeller U Lab Plant Molecular Biology 1230 York Ave New York NY 10021

HORVATH, JOAN CATHERINE, aeronautical engineer; b. N.Y.C., Feb. 19, 1959; d. David Robert and Martha Horvath; m. Stephen Charles Unwin, June 10, 1988. BS, MIT, 1981; MS, UCLA, 1983. Master's fellow Aerospace Corp., El Segundo, Calif., 1981-83; mem. tech. staff propulsion sect. Jet Propulsion Lab., Pasadena, Calif., 1984-85, mem. tech. staff tracking systems sect., 1985-88, sequence design engr. mission profile sect., 1988-90, cognizant design engr. mission profile sect., 1990-91, asst. sequence team chief mission profile sect., 1991—, mgr. instant sequencing mission profile sect., 1988—; session chair Supercomputing 89 Conf., Reno, 1989, 6th Internat. Conf. on math modeling, St. Louis, 1987. Contbr. articles to profl. jours. Ednl. counselor MIT Ednl. Coun., Cambridge, 1981—; bd. dirs. Pasadena Figure Skaing Club, 1992. Recipient Group Achievement award NASA Hypercube Project, 1989, NASA Magellan Project, 1992. Mem. AIAA, Soc. for Indsl. and Applied Math, Nat. Space Soc. Democrat. Achievements include development of approach for using parallel computers to automate mission operations, approach for combining different mission control functions into one software system in order to reduce costs. Office: Jet Propulsion Lab 4800 Oak Grove Dr Pasadena CA 91109

HORVATH, VINCENT VICTOR, electrical engineer; b. Bethlehem, Pa., Oct. 21, 1942; s. Geza J. and Mary E. (Sik) H.; m. Kathleen McCaughey, Jan. 29, 1966; children: Jeffrey T., Kevin J., Matthew J. BSEE, Lehigh U., 1964, MSEE, 1966, PhD in Elec. Engring., 1970. Engr. info. physics br. R&D ctr. GE Corp., Schenectady, N.Y., 1966-68; supr. and engr. instrument devel. rsch. dept. Bethlehem Steel Corp., Bethlehem, Pa., 1970-84, mgr. advanced mfg. systems div., 1984-87; pres. Bitronics, Inc., Bethlehem, 1987—; sessionchmn. Nat. Forum on the Future of Automated Materials Processing in U.S., Nat. Bur. Stds., 1986; U.S. del. Internat. Energy Agy. Conf., Stockholm, 1979. Contbr. articles to profl. jours. Recipient D.J. Blickwede Rsch. Recognition award, 1983; Ben Franklin Seed grantee, 1989. Achievements include U.S. and fgn. patents covering temperature measurement sensors and systems. Office: Bitronics Inc 261 Brodhead Rd Bethlehem PA 18017

HORWITZ, ALAN FREDRICK, cell and molecular biology educator; b. Mpls., Oct. 26, 1944; s. Burt and Helen (Bolnick) H.; m. Carole Joanne Rosen, Nov. 26, 1972; children: Jeremy J., Rachel T. BA in Chemistry and Math. with hons., U. Wis., 1966; PhD in Biophysics, Stanford U., 1969; MA (hon.), U. Pa., 1978. NIH postdoctoral fellow Lab. Chem. Biodynamics, U. Calif., Berkeley, 1970-72, chemist P-5, 1972-73; scientist Biozentrum der Universitat Basel, Switzerland, 1973-74; asst. prof. dept. biochemistry and biophysics Sch. Medicine, U. Pa., Phila., 1974-78, assoc. prof., 1978-84, prof., 1984-87; prof., head dept. cell and structural biology U. Ill., Urbana, 1987—, chmn. biophysics program, 1978-85, assoc. dir. med. scientist tng. program, 1986-87, dir. cell and molecular biology tng. program, 1988-92; mem. commn. on cell and membrane biophysics Internat. Union Pure and Applied Biophysics, 1975-82; mem. sci. adv. com. biochemistry and chem. carcinogens Am. Cancer Soc., 1977-81; mem. spl. study sects. Nat. Inst. Gen. Med. Scis. NIH, 1980—, mem. rev. com. cellular and molecular basis of disease Nat. Inst. Gen. Med. Sci., 1984-88; mem. spl. study sect. Nat. Inst. Aging, Nat. Cancer Inst. NIH, 1980—, mem. biotech. rev. panel, 1988; mem. innovative aging rsch. com. VA, 1983; lectr. various nat. and internat. symposia; steering com. Howard Hughes Inst. Med. Rsch., U. Pa., adv. com. for H.M. Watts, Jr. Neuromuscular Disease Rsch. Inst., U. Pa., 1986. Mem. editorial bd.: Jour. Cell Biology, 1989—, Jour. Cell Sci., 1990—, Cell Adhesion and Communication, 1992—; assoc. editor: Devel. Biology, 1989; contbr. articles to profl. jours. Recipient Dr. William Daniel Stroud Established Investigator award Am. Heart Assn., 1975-80, NIH MERIT award; prin. investigator, grantee in field, 1974. Mem. Biophys. Soc., Am. Soc. Biol. Chemists, Am. Soc. Cell Biology (pub. policy com. 1986—, nominating com. 1988), Am. Assn. of Anatomy. Avocations: music, outdoor activities, hiking. Home: 3410 S Persimmon Circle Urbana IL 61801 Office: U Ill Dept Cell & Structural Biology Urbana IL 61801

HORWITZ, DAVID A., medicine and microbiology educator. BA, U. Mich., 1958; MD, U. Chgo., 1962. Intern, resident Michael Reese Hosp., Chgo., 1966; rheumatology fellow Southwestern Med. Sch. U. Tex., 1969; instr. internal medicine Southwestern Med. Sch. U. Tex., Dallas, 1968-69; from asst. prof. to assoc. prof. medicine Sch. Medicine U. Va., Charlottesville, 1969-79, prof. medicine, 1979-80; prof. medicine and microbiology, chief divsn. rheumatology and immunology sect. Sch. Medicine U. So. Calif., L.A., 1980—; vis. prof. Clin. Rsch. Ctr., Harrow, Eng., 1976-77; vis. investigator Inperial Cancer Rsch. Fund, London, 1988-89. Contbr. articles to profl. jours. Achievements include research in human blood lymphocytes with iC3b (type 3) complement receptors, characterization of immunologic function in homosexual males with PGL and AIDS, abnormalities in patients with active systemic lupus erythematosus, human CD8 lymphocytes stimulated in the absence of CD-4 cells enhance IgG production by antibody-secreting B cells. Office: U Soc Calif Divsn Rheumatology & Immunology 2011 Zonal Ave # 711 Los Angeles CA 90033-1054

HORWITZ, DAVID LARRY, pharmaceuticals company executive, researcher, educator; b. Chgo., July 13, 1942; s. Milton Woodrow and Dorothy (Glass) H.; m. Gloria Jean Madian, June 20, 1965; children: Karen, Laura. AB, Harvard U., 1963; MD, U. Chgo., 1967, PhD, 1968; MBA, Lake Forest Grad. Sch. Mgmt., 1991. Diplomate Am. Bd. Internal Medicine. Resident in internal medicine U. Chgo. Hosp., 1971-72; fellow in endocrinology U. Chgo., 1972-74, asst. prof., 1974-79; assoc. prof. U. Ill., Chgo., 1979-90, clin. prof. medicine medicine, 1990-92; med. dir. Baxter Healthcare Corp., Deerfield, Ill., 1982-91, v.p. med. and profl. affairs, 1991-92; v.p. med. and regulatory affairs SciClone Pharms., San Mateo, Calif., 1992—. Contbr. articles to profl. jours. Bd. dirs. No. Ill. affiliate Am. Diabetes Assn., 1976-92, pres. 1987-89. Comdr. USNR, 1969-71. Recipient Research and Devel. award Am. Diabetes Assn., 1974-76; Outstanding Young Citizen of Chgo. award Chgo. Jr. C. of C., 1976; Outstanding Young Citizen Ill. award Ill. Jaycees, 1977. Fellow ACP; mem. Endocrine Soc., Am. Diabetes Assn. (chmn. com. on planning and orgn. 1986-88), Am. Assn. Clin. Nutrition. Research in clinical diabetes and insulin physiology. Office: SciClone Pharms 901 Mariners Island Blvd Ste 315 San Mateo CA 94404

HORWITZ, SUSAN BAND, molecular pharmacologist. BA, Bryn Mawr Coll., 1958; PhD in Biochemistry, Brandeis U., 1963. Postdoctoral fellow dept. pharmacology, sch. medicine Tufts U., 1963-65, Emory U., 1965-67; rsch. assoc. dept. medicine Albert Einstein Coll. Medicine, N.Y.C., 1967-68, instr. dept. pharmacology, 1968-70, asst. prof. dept. medicine, 1970-75, asst. prof. dept. cell biology, 1973-75, assoc. prof. depts. molecular pharmacology and cell biology, 1980—, co-chair dept. molecular pharmacology, 1985—, Rose C. Falkenstein prof. cancer rsch., 1986—, assoc. dir. cancer rsch. ctr., 1991—; mem. pharmacology-toxicology rsch. team Nat. Inst. Gen. Med. Sci., 1975-80; adv. com. Irma T. Hirschl Scientist award, 1979-85; bd. scientific counselors divsn. cancer treatment NCI, 1981-86, 87-90, mem. review com. Outstanding Investigators Grant award, 1984—, ad hoc review com. in vitro and in vivo disease-oriented screening project, 1986—; guest reviewer sci. adv. com. Damon Runyon/Walter Winchell Rsch. Fund, 1983, 88; vice chair Gordon Conf. Chemotherapy of Exptl. and Clin. Cancer, 1986, chair, 1987, mem. coun., 1990-93; Sterling Drug vis. prof. dept. pharmacology Boston U., 1987; mem. Charles F. Kettering selection com. Gen. Motors Cancer Rsch. Found., 1988-89, awards assembly, 1991—. Contbr. articles to profl. jours., chpts. to books. Recipient Rsch. Career Devel. award 1970-75, award Pharm. Mfrs. Assn., 1972, Irma T. Hirschl Career Scientists award, 1975-80; grantee Merck, 1970, Nat. Cancer Inst., 1985-92, 92—, Bristol-Myers, 1988-93; named Outstanding Woman Scientist metro N.Y.C. chpt. Assn. Women in Science. Mem. Am. Soc. Pharmacology and Exptl. Therapeutics (com. edn. and profl. affairs 1973-77), Am. Soc. Microbiology (vice chair anitmicrobial chemotherapy), Am. Chem. Soc., Am. Assn. Cancer Rsch. (biochem. program com. 1983-84, Clowes award selection com. 1986-87, bd. dirs. 1987-90, spl. confs. com. 1989-92, chmn. Rhoads award selection com. 1990-91, co-chair conf. in cancer rsch. membrane transport in multidrug resistance, devel. and disease, 1991, Cain Meml. award 1992), Am. Soc. Cell Biology, Harvey Soc. (mem. coun. 1991—). Office: A Einstein Coll of Medicine Dept of Molecular Pharmacology 1300 Morris Park Ave Bronx NY 10461*

HOSENEY, RUSSELL CARL, chemistry educator; b. Coffeyville, Kans., Dec. 3, 1934. BS, Kans. State U., 1957, MS, 1960, PhD in Cereal Chemistry, 1968. Rsch. chemist crops rsch. divsn. Hard Winter Wheat Quality Lab., Agrl. Rsch. Svc. USDA, 1956-70; rsch. assoc. cereal chemistry Kans. State U., 1967-71, assoc. prof. grain sci., 1971-75, prof. grain sci., 1975—. Contbr. 250 articles to profl. jours. Mem. Am. Chem. Soc. (hon.), Am. Assn. Cereal Chemists (nat. program chmn. 1980, chmn. carbohydrate divsn. 1987, nat. pres. 1989-90, bd. dirs. 1988-91, mem. editorial bd., Thomas Burn Osborne medal 1991), Inst. Food Technologists (hon.). Achievements include research in wheat and flour quality including the role of proteins, carbohydrates and lipids; 7 patents in field. Office: Kansas St Univ Dept of Grain Science & Industry Manhattan KS 66506*

HOSKEN, RICHARD BRUCE, systems engineer; b. Toronto, Ont., Can., Jan. 26, 1942; came to U.S. 1947; s. Walter James and Annie Matilda (Edwards) H.; m. Mary Louise Frey, June 25, 1966; children: Brian Kent, Ryan Heath. BSET, Fla. Tech. U., Orlando, 1976; MS in Computer Sci., Fla. Inst. Tech., Melbourne, 1984; ScD, Nova U., Ft. Lauderdale, Fla., 1992. Sys. designer/programmer Shuttle & Apollo Sys. IBM, Kennedy Space Ctr., Fla., 1966-78; sr. sys. engr. Shuttle Launch Processing Sys. Computer Sci. Corp., Kennedy Space Ctr., 1978-83; staff sys. engr. artificial intelligence sys. tech. svcs. div. Grumman Corp., Kennedy Space Ctr., 1983—. NASA Achievement award, Shuttle Mission STS-51A, 1984, Grumman Achievement award, 1987, Outstanding Performance award, 1987, Project Sterling awards, 1988. Mem. IEEE Computer Soc., Nat. Mgmt. Assn., Fla. Space Coast Coun., Missile, Space and Range Pioneers, Nat. Space Soc., Assn. Computing Machinery. Republican. Avocations: astronomy, microcomputers, metaphysical science, model rocketry, tennis. Home: 7406 Windover Way Titusville FL 32780 Office: Grumman Corp Mail Code GTS-647 Kennedy Space Center FL 32815

HOSKINS, L. CLARON, chemistry educator; b. Logan, Utah, July 27, 1940; s. Leo Cooper and Luella (Hall) H.; m. Flora Jensen, June 18, 1960; children: Duane, Daryl, Sandra. BS, Utah State U., 1962; PhD, MIT, 1965. Asst. prof. chemistry U. Alaska, Fairbanks, 1965-68, assoc. prof., 1968-75, prof., 1975—, head dept., 1984—. Mem. Am. Phys. Soc., Sigma Xi, Phi

Kappa Phi, Pi Mu Epsilon. Home: 3291 Bluebird Ave Fairbanks AK 99709-2369

HOSKINS, WILLIAM JOHN, obstetrician/gynecologist, educator; b. Harlan, Ky., May 10, 1940; s. Lonnie S. and Joanne (Huff) H.; m. Betty Jean Gay, Sept. 10, 1960 (div. 1985); children: Tonya J., William John Jr.; m. Iffath Abbasi Ahson, Nov. 9, 1985; children: Ahad A., Mariya A. BA, U. Tenn., Knoxville, 1962; MD, U. Tenn., Memphis, 1965. Diplomate Am. Bd. Ob-Gyn., Am. Bd. Gynecol. Oncology. Commd. lt. USN, 1966, advanced through grades to capt.; intern Jacksonville (Fla.) Naval Hosp., 1966-67; med. officer Destroyer Squadron 8 USN, Mayport, Fla., 1967-68; resident in ob-gyn Oakland (Calif.) Naval Hosp., 1968-71; mem. staff, dept. ob -gyn Pensacola (Fla.) Naval Hosp., 1971-74; fellow in gynecol. oncology U. Miami, Fla., 1974-76; dir. Gynecol. Oncology Nat. Naval Med. Ctr., Bethesda, Md., 1976-86; assoc. prof. ob-gyn Uniformed Svcs. U., Bethesda, 1976-86; ret. USN, 1986; assoc. chief gynecology svc. Meml. Sloan-Kettering Cancer Ctr., N.Y.C., 1988-90, chief gynecology svc., 1990—; assoc. prof. ob-gyn Cornell U. Med. Ctr., N.Y.C., 1986-90; prof. ob-gyn. Cornell U. Med. Coll., N.Y.C., 1990—, vice chmn. gynecol. oncology group, 1993—, vice chmn. gynecologic oncology group, 1993—; chmn. ovarian com. Gynecol. Oncology Group, Phila., 1984-89, vice chmn., 1993—. Editor: Principles and Practice of Gynecology & Oncology, 1992, Cancer of the Ovary, 1993; contbr. 191 articles to profl. jours., also chpts. to books. Fellow Am. Coll. Obstetricians and Gynecologists (v.p. Navy sect. 1982-83), ACS; mem. Soc. Gynecol. Oncologists (sec.-treas. elect 1992, coun. mem. 1988—), Soc. Gynecol. Surgeons, Internat. Gynecol. Cancer Soc., Am. Radium Soc. Republican. Muslim. Office: Meml Sloan-Kettering Cancer Ctr Gyn Svc 1275 York Ave New York NY 10021-6094

HOSMANE, NARAYAN SADASHIV, chemistry educator; b. Gokarn, Karnatak, India, June 30, 1948; came to U.S. 1976; s. Sadashiv Ganapati and Lalita (Kurse) H.; m. Sumathy Rao, May 6, 1976; children: Suneil Narayan, Nina Narayan. BS, Karnatak U., Dharwar, India, 1968, MS, 1970; PhD, Edinburgh (Scotland) U., 1974. Rsch. asst. Queen's U. of Belfast (No. Ireland), 1974-75; rsch. scientist Lambeg (No. Ireland) Indsl. Rsch. Inst., 1975-76; rsch. assoc. Auburn (Ala.) U., 1976-77, U. Va., Charlottesville, 1977-79; asst. prof. Va. Poly. Inst. and State U., Blacksburg, 1979-82; asst. prof. So. Meth. U., Dallas, 1982-86, assoc. prof., 1986-89, full prof., 1989—; chemist cons. Vertically Integrated Tech., Inc., Dallas, 1990—; invited speaker in field; chmn. 1st Boron-USA (BUSA-I) Workshop, 1988. Author: (with others) Boron Chemistry, 1980, Advances in Boron and The Boranes, 1988, Advances in Organometallic Chemistry, 1990, Electron Deficient Boron and Carbon Clusters, 1991, Pure and Applied Chemistry, 1991, Chemical Reviews, 1993, Trends in Inorganic Chemistry, 1993; contbr. over 100 peer rev. sci. articles to profl. jours. including Jours. of Am. Chem. Soc., Inorg. Nuclear Chem. Recipient Cert. Merit, Bronze Plaque award Internat. Biog. Ctr., Cambridge, Eng., 1987, Cert. of Achievement in Sci., Am. Men & Women of Sci., 1986; So. Meth. U. Seed grantee, 1983-84, Rsch. Corp. grantee, 1983-85, Petroleum Rsch. Fund Type G grantee, 1984-86, NSF grantee, 1985—, Robert A. Welch Found. grantee, 1985—, Petroleum Rsch Fund Type B, 1987-89, 90—. Fellow Royal Soc. Chemistry, Am. Inst. Chemists; mem. Am. Chem. Soc. Soc., Soc. of Sigma Xi (Outstanding Rsch. award 1987). Achievements include synthesis of over 70 main group metal-lacarboranes of pentagonal bipyramidal geometries and their coordinated complexes; discovery of zirconium-, hafnium-, and yttrium-carborane sandwich compounds which are envisioned as potential precursors to Ziegler-Natta catalysts, and of main group of metal sandwich compounds of the C2B4-carborane ligands. Home: 2802 Pear Tree Ln Garland TX 75042-5635 Office: Southern Meth U Chemistry Dept Airline Rd Dallas TX 75275

HOSNI, MOHAMMAD HOSEIN, mechanical engineering educator; b. Yazd, Iran, June 25, 1955; came to U.S., 1978; s. Bemanali and Hajer Hosni; m. Fakhry Hosni, Jan. 8, 1983; children: Mehrdad, Mina. MS, La. State U., 1984; PhD, Miss. State U., 1989. Asst. prof. So. U., Baton Rouge, 1984-86; postdoctoral researcher Miss. State U., Starkville, 1989-90; asst. prof. Kans. State U., Manhattan, 1990—; cons. Coast Machinery, Baton Rouge, 1984-86. Contbr. articles to Internat. Jour. Heat and Mass Transfer, Jour. Thermophysics and Heat Transfer, Jour. Heat Transfer, Exptl. Heat Transfer, Exptl. Thermal and Fluid Sci. Rsch. project grantee ASHRAE, 1992. Mem. ASME, AIAA, Am. Soc. Engring. Edn. Home: 3409 Stonehenge Dr Manhattan KS 66502

HOSPODOR, ANDREW THOMAS, electronics executive; b. Endicott, N.Y., Jan. 30, 1937; s. Andrew and Verna (Yurick) H.; m. Rose Marie Pitarra, June 28, 1958; children: Andrew D., Sarah E., Peter J. B.S.M.E., Cornell U., 1960; M.S.M.E., Lehigh U., 1963, M.B.A., 1967. Product specialist Air Products Inc., Emmaus, Pa., 1960-66; mgr. RCA, Camden, N.J., 1966-77; dir. mktg. RCA, Burlington, Mass., 1977-79, program mgr. command and control, 1979-81, div. v.p., gen. mgr., 1981-85; pres., chief exec. officer RCA Am. Communications, Inc., Princeton, N.J., 1985-87; chmn., chief exec. officer ARINC, Annapolis, Md., 1987—. Home: 361 Berkshire Dr Riva MD 21140-1433 Office: ARINC 2551 Riva Rd Annapolis MD 21401-7435

HOSTETLER, KARL YODER, internist, endocrinologist, educator; b. Goshen, Ind., Nov. 17, 1939; s. Carl Milton and Etta LaVerne (Yoder) H.; m. Margaretha Steur, Dec. 17, 1971; children: Saskia Emma, Kirsten Cornelia, Carl Martijn. BS in Chemistry, DePauw U., 1961; MD, Western Res. U., 1965. Diplomate Am. Bd. Internal Medicine, Am. Bd. Endocrinology and Metabolism. Intern, resident in medicine Univ. Hosp. Cleve., 1965-69; fellow endocrinology Cleve. Clinic Found., 1969-70; postdoctoral fellow, lipid chemistry U. Utrecht, The Netherlands, 1970-73; asst. prof. medicine U. Calif., San Diego, 1973-79, assoc. prof. medicine, 1979-82, prof. medicine, 1982—; founder, sr. v.p. Vical Inc., San Diego, 1987-92. Assoc. editor: Jour. of Clin. Investigation, 1993—; contbr. over 100 articles to scholarly and profl. jours. Pres. San Diego County chpt. Am. Diabetes Assn., 1982-83. Recipient fellowship John Simon Guggenheim Found., 1980-81, Japan Soc. for Promotion of Sci., Tokyo, 1986. Mem. Am. Soc. Clin. Investigation, Am. Soc. Biochemistry and Molecular Biology, Western Assn. Physicians, Am. Soc. Microbiology. Achievements include research in phospholipid chemistry and biochemistry, in mechanism of cardiolipin biosynthesis, in the biological basis of drug induced lipid storage, and in lipid prodrug antivirals. Office: U Calif San Diego Dept Medicine 0676 La Jolla CA 92093

HOTES, ROBERT WILLIAM, cognitive scientist; b. Cleve., Jan. 29, 1942; s. Norbert William and Florence Lew (White) H.; m. Lynn Edith Waibel, Dec. 21, 1969. BA, St. Paul Coll.; MA, U. Dallas, Irving, Tex., 1970; PhD, U. North Tex., 1982. Cert. sr. profl. in human resources; registered orgn. devel. cons. Standards and evaluation specialist Bell Helicopter Textron, Ft. Worth, 1978-82; mgr. human resource devel. Portion-Trol Foods, Inc., Mansfield, Tex., 1982-84; dir. tng. Advanced Telemktg., Irving, 1984-85; sr. tng. analyst, scientist The Allen Corp. Am., Hurst, Tex., 1985-86; prin. cons. The Evans Group, Dallas, 1986-87; mgr. tng. devel. Data Comm., Inc., Richardson, Tex., 1987-88; sr. instructional technologist Wilcox Electric, Kansas City, Mo., 1988—; pres., bd. dirs. Metroplex Orgn. Devel. Inst., Dallas, 1985-88, Orgn. Devel. Inst. Kansas City, 1989—; mem. adv. bd. computers and cognitive systems U. North Tex., Denton, 1989-90; presenter in field; adj. prof. So. Ill. Univ., 1991. Author, editor: Managing Learning Resources Centers, 1976, Leadership in Higher Education, 1982, Foundations of the Community College M., 1985; also articles. Recipient chpt. devel. award ASTD, 1985. Fellow Am. Psychol. Soc.; mem. Assn. Mgmt. (v.p. logistics 1989-90, Orgn. Devel. award 1991), Nat. Soc. for Performance and Instrn. (chmn. memberships svcs.). Achievements include development of an algorithm for evaluating human resource programs using a data envelopment analysis approach, applications of cognitive psychology to aviation training.

HOTZ, HENRY PALMER, physicist; b. Fayetteville, Ark., Oct. 17, 1925; s. Henry Gustav and Stella (Palmer) H.; m. Marie Brase, Aug. 22, 1952; children: Henry Brase, Mary Palmer, Martha Marie. B.S., U. Ark., 1948; Ph.D., Washington U., St. Louis, 1953. Asst. prof. physics Auburn U., Ala., 1953-58, Okla. State U., Stillwater, 1958-64; assoc. prof. Marietta Coll., Ohio, 1964-66; physicist, scientist-in-residence U.S. Naval Radiol. Def. Lab., San Francisco, 1966-67; assoc. prof. U. Mo., Rolla, 1967-71; physicist Qanta Metrix div. Finnigan Corp., Sunnyvale, Calif., 1971-74; sr. scientist Nuclear Equipment Corp., San Carlos, Calif., 1974-79, Envirotech Measurement Systems, Palo Alto, Calif., 1979-82, Dohrmann div. Xertex Corp., Santa Clara, Calif., 1982-86; sr. scientist Rosemount Analytical Div. Dohrmann, 1983-91; cons. Burlingame, Calif., 1991—; cons. USAF, 1958-62; mem. lectr. selection com. for Hartman Hotz Lectrs. in law, liberal arts U. Ark. Served with USNR, 1944-46. Mem. Am. Phys. Soc., Am. Assn. Physics Tchrs., AAAS, Phi Beta Kappa, Sigma Xi, Sigma Pi Sigma, Pi Mu Epsilon, Sigma Nu. Methodist. Lodge: Masons. Home: 290 Stilt Ct Foster City CA 94404-1323 Office: Hotz Assocs 525 Almer Rd Ste 201 Burlingame CA 94010

HOU, GUANG KUN, computer science educator; b. Mei Zhou, Guangdong, People's Republic China, Feb. 15, 1941; m. Xing Hua Xiong, Mar. 8, 1968; children: Yu Lun, Yu Li. MA in Maths., Moscow U., 1965. With Phys. Inst. Academia Sinica, Beijing, 1966-70; with 207th Inst. Ministry of Spacef-light & Aviation, Beijing, 1970-77; prof. Zhongshan U., Guangzhou, People's Republic China, 1977—; v.p. Thinking Sci. Inst. of Guangdong; dir. Artificial Intelligence Inst., Knowledge Engring. Inst. Mem. Am. Math. Soc. Avocations: classical poems, classical music. Office: Dept Computer Sci, Zhongshan U, Guangzhou 510275, China

HOU, JIASHI, mathematician, educator; b. Shanghai, China, Mar. 9, 1955; s. Jisou Hou and Huifeng Shen; m. Xiaobo Lu, Nov. 20, 1982; children: Helen, Katherine. MS, Rensselaer Poly. Inst., 1984, PhD, 1989. Instr. Shanghai U. Sci. and Tech., 1982; teaching and rsch. asst. Tensselaer Polytech. Inst., Troy, N.Y., 1983-86; mem. rsch. staff Columbia U., N.Y.C., 1986-90; asst. prof. math. Old Dominion U., Norfolk, Va., 1990—. Contbr. articles to profl. jours. Mem. ASME (Best Jour. Paper award Bioengring. Div.). Office: Old Dominion U Dept Math & Statistics Norfolk VA 23529

HOU, JIN CHUAN, mathematics educator; b. Sichuan, People's Republic China, Oct. 29, 1954; s. Zhi Lang and Zhi Xian (Cao) H.; m. Xiu Ling Zhang, Sept. 18, 1979; children: Wen Tao, Wen Jun. Grad., Shanxi Tchrs. Coll., Linfen, People's Republic China, 1975, Shaanxi Normal U., Xian, People's Republic China, 1979; SM, East China Normal U., Shanghai, People's Republic China, 1982; PhD, Fudan U., Shanghai, 1986. Tchr. Qinxiang Mid. Sch., Fenxi county, People's Republic China, 1973-75, Shanxi Tchrs. Coll., Linfen, 1978-79, 81-84; prof. Shanxi Tchrs. U., Linfen, 1986—; vis. scholar Ind. U., Bloomington, 1988-89. Editor-in-chief Jour. of Shanxi Tchrs. U. (Natural Sci.), 1988—; reviewer math. Revs., 1988—, Zentralblatt für Mathematik, 1987—; contbr. articles to profl. jours. . Recipient prize for achievements in sci. rsch. Shanxi Commn. Sci. and Tech., 1983, 84, prize for excellent rsch. papers written by young scientists Sci. and Tech. Soc. Shanghai, 1985, prize Fund for Young Scientists, Academia Sinica, 1986, Outstanding Teachers In Shanxi Province, Shanxi Provincial Govt., 1987, prize for young scientists Chinese Assn. Sci. and Tech., 1991, Outstanding Returned Students and Scholars award State Edn. Commn., Ministry Pers. Adminstrn., 1991, Outstanding Expert award Shanxi Provincial Govt., 1991. Mem. Math. Soc. Shanxi Province (coun. 1990—), Math. Soc. Am. Avocations: reading novels, music, shuttlecock. Office: Shanxi Tchrs U, Dept Math, Linfen 041004, China

HOUGH, JACK VAN DOREN, otologist; b. Lone Wolf, Okla., Sept. 12, 1920; s. Chapman Ernest and Hazel (Van Doren) H.; m. Joan Ingle, Dec. 29, 1943; children: Ted Chapman, Jack Van Doren Jr., Timothy Ingle, David Alliston. BS, Southeastern State U., 1939; MD, U. Okla., 1943. Diplomate, Am. Bd. Otorhinolaryngology. Intern USN Hosp., Farragut, Idaho, 1944; resident, then fellow in otolaryngology U. Okla. Hosps., Oklahoma City, 1946-50; clin. instr. otorhinolaryngology U. Okla. Health Scis. Ctr., Oklahoma City, 1950-51; now clin. prof. otorhinolaryngology, head and neck surgery U. Okla. Health Scis. Ctr.; pvt. practice Oklahoma City, 1951—; developer surg. techniques and instruments for hearing restoration and middle ear reconstrn., electromagnetic hearing devices, cochlear implants. Contbr. sci. articles and textbook chpts. to med. publs. Past ruling elder, Cen. Presbyn. Ch., Oklahoma City; founder, Covenant Community Ch. Oklahoma City, 1980, now session moderator. Decorated Bronze Star medal; recipient Harris P. Mosher award, Triologic Soc., numerous awards from profl. orgns.; inducted into Okla. Hall of Fame, 1991. Mem. Am. Bd. Otolaryngology, Am. acad. Otolaryngology-Head and Neck Surgery, Am. Otological Soc., Am. Triological Soc., Ostosclerosis Study Group, AMA, Oklahoma County Med. Assn., Okla. Med. Assn., Okla. Acad. Medicine, Osler Soc., Am. Acad. Ophthalmologic and Otolaryngologic Allergy, Christian Med. Soc., Am. Audiology Soc., Pan-Am. assn. Otorhinolaryngolgoy and Bronchoesophagology, Politzer Soc., Am. Sci. Affiliation, Von Bekesy Soc., numerous other profl. orgns. Office: Hough Ear Inst 3400 NW 56th St Oklahoma City OK 73112-4452

HOUGHTON, JAMES RICHARDSON, glass manufacturing company executive; b. Corning, N.Y., Apr. 6, 1936; s. Amory and Laura (Richardson) H.; m. May Tuckerman Kinnicutt, June 30, 1962; children: James DeKay, Nina Bayard. AB, Harvard U., 1958, MBA, 1962. With Goldman, Sachs & Co., N.Y.C., 1959-61; v.p., European area mgr. Corning Glass Internat., Zurich, Switzerland, Brussels, Belgium, 1964-68; with Corning Glass Works (name changed to Corning Inc. 1989), 1962—, v.p., gen. mgr. consumer products div., 1968-71, vice chmn. bd., dir., chmn. exec. com., 1971-83, chmn. bd., chief exec. officer, 1983—; bd. dirs. Met. Life Ins. Co., J.P. Morgan Co., Inc., Dow Corning Corp. Trustee Corning Inc. Found., Corning Mus. Glass, Pierpont Morgan Libr., N.Y.C., Bus. Coun. of N.Y. State, Met. Mus. Art;l mem. Bus. Com. for Arts, N.Y.C.; dir. Coun. on Fgn. Rels.; mem. Trilateral Commn. Bus. Coun. With U.S. Army, 1959-60. Mem. Bus. Roundtable. Episcopalian. Clubs: Corning Country; River, Harvard, Univ., Links (N.Y.C.); Brookline (Mass.) Country; Tarratine (Dark Harbor, Maine); Augusta (Ga.) Nat. Golf; Rolling Rock, Laurel Valley Golf (Ligonier, Pa.). Home: The Field 36 Spencer Hill Rd Corning NY 14830-9417 Office: Corning Inc Houghton Park CB-09-8 Corning NY 14831

HOUGHTON, RAYMOND CARL, JR., computer science educator; b. Greenfield, Mass., May 26, 1947; s. Raymond Carl and Phyllis Irene (Richason) H.; m. Jan Marie Laws, Sept. 22, 1973; children: Raymond James, April Monica, Amy Rose. BS in Math., Norwich U., 1969; MS in Computer Sci., George Washington U., 1975; MSEE, Johns Hopkins U., 1980; PhD in Computer Sci., Duke U., 1991. Computer operator Norwich U., Northfield, Vt., 1967-69; specialist programmer power transformer dept. GE Co., Pittsfield, Mass., 1969-70; mathematician armament dept. GE Co., Burlington, Vt., 1972-73; mem. tech. staff Computer Scis. Corp., Silver Spring, Md., 1974-75; data systems analyst computer security applications div. Nat. Security Agy., Ft. Meade, Md., 1975-78; computer scientist Inst. Computer Scis. and Tech./Nat. Bur. Standards, Gaithersburg, Md., 1978-83; instrnl. rsch. asst. dept. computer sci. Duke U., Durham, N.C., 1984-91; assoc. prof. dept. math. and computer sci. Augusta (Ga.) Coll., 1987-93; lectr. Skidmore (N.Y.) Coll., 1993—; bd. advisers, columnist Software Engring: Tools, Techniques, Practice, 1990—; speaker at profl. confs. Contbr. to profl. publs. 1st lt. U.S. Army, 1971-72, Vietnam. Recipient Certs. of Recognition, U.S. Dept. Commerce, 1981, 83, Letter of Appreciation, Defence Comms. Agy., 1976. Mem. IEEE, N.Y. Acad. Scis., Assn. Computing Machinery. Democrat. Lutheran. Achievements include description of software development tools. Office: Skidmore Coll Dept Math & Computer Sci Saratoga Springs NY 12866

HOUGHTON, WOODS EDWARD, agricultural science educator; b. Albuquerque, Aug. 22, 1956; s. Paul Wayne and Elizabeth (Culver) H.; m. Cindy Anderson, Jan. 15, 1984; children: Jannetta Lynn, Joshua Woods, Jeremiah Wayne. BS in Animal Sci., N.Mex. State U., 1978, MS in Entomology, 1982. Cert. farrier, N.Mex. Mgmt. operator Sandia Stables, 1972-75; forest aid U.S. Forest Svc., Ariz., 1975-76; student rsch. asst. entomology dept. N.Mex. State U., Las Cruces, 1977-79, rsch. asst. I, 1979-80, rsch. asst. II, 1980-85; program dir. DeBaca County Coop. Ext. Svc., Ft. Sumner, N.Mex., 1985-90; agr. ext. agt. Coop. Ext. Svc., Carlsbad, N.Mex., 1990—; textbook rev. com. Carlsbad Pub. Schs.; ex officio mem. Eddy County Fair Bd. Recipient Svc. Recognition award Revitalization of Rural Am., 1990, Outstanding Young Citizen award DeBaca C. of C., 1990, Cert. of Appreciation DeBaca County Commrs., 1990. Mem. N.Mex.-Tex. Profl. Entomologists (bd. dirs. 1985-88), Nat. Assn. County Agr. Agts., N.Mex. County Agts. Assn. (profl. tng. com.), Entomol. Soc. Am., Nt. HOrse Judging Coaches Assn. Home: 4306 Old Cavern Hwy Carlsbad NM 88220 Office: NMex State U Coop Ext Svc 1304 W Stevens Carlsbad NM 88220

HOUK, ROBERT SAMUEL, chemistry educator; b. New Castle, Pa., Nov. 23, 1952; s. Robert H. and Rose B. Houk; m. Linda Lembke, Oct. 3, 1981; children: Andrew, Mary. BS, Slippery Rock State Coll., 1974; PhD, Iowa State U., 1980. Asst. prof. chemistry Iowa State U., Ames, 1981-87, assoc. prof., 1987-91, prof., 1991—; cons. Perkin Elmer Sciex, Norwalk, Conn., 1982—. Author: Handbook of ICP-MS, 1992; contbr. articles to profl. jours. Recipient M.F. Hasler award Spectroscopy Soc. Pitts., 1993, Wilkinson Teaching award Iowa State U., 1993. Mem. Am. Chem. Soc. (award in chem. instrumentation 1993), Soc. for Applied Spectroscopy (L.W. Strock award 1986, Maurice F. Hasler award 1993), Am. Soc. for Mass Spectrometry. Achievements include construction of first inductively coupled plasma-mass spectrometer for trace elemental analysis. Office: Iowa St Univ Ames Lab Ames IA 50011

HOUNSFIELD, GODFREY NEWBOLD, scientist; b. Aug. 28, 1919; s. Thomas H. Ed., City and Guilds Coll., London; diploma, Faraday House Elec. Engring. Coll., London; M.D. (hon.), U. Basel, 1975; D.Sc. (hon.), City U., 1976, U. London, 1976; D. of Tech. (hon.), U. Loughborough, 1976; DHC, Cambridge U., 1992. Joined EMI Ltd., Hayes, Middlesex, Eng., 1951, head med. systems sect., cen. research labs., 1972-76, sr. staff scientist, 1977—; professorial fellow in imaging scis. Manchester U., 1978—. Contbr. articles to sci. jours. Recipient Nobel prize in Physiology or Medicine, 1979; MacRobert award, 1972; Wilhelm-Exner medal Austrian Indsl. Assn., 1974; Ziedses des Plantes medal Physikalishe Medizinische Gesellschaft, Würzburg, 1974; Prince Philip Medal award CGLI, 1975; ANS Radiation Industry award Ga. Inst. Tech., 1975; Lasker award Lasker Found., 1975; Duddell Bronze medal Inst. Physics, 1976; Golden Plate award Am. Acad. Achievement, 1976; Reginald Mitchell Gold medal Stoke-on-Trent Assn. Engrs., 1976; Churchill Gold medal, 1976; Gairdner Found. award, 1976; decorated comdr. Order Brit. Empire, 1976, knight, 1981. Fellow Royal Soc. Led design team for 1st large all-transistor computer to be built in Gt. Britain; invented EMI-scanner computerized transverse axial tomography system for X-ray exam.; developed new X-ray technique (EMI-scanner system). Office: Thorn EMI Research Labs, Dawley Rd, Hayes Middlesex UB3 1HH, England

HOURAN, JAMES PATRICK, counselor, psychiatric technician; b. Decatur, Ill., Aug. 21, 1969; s. Michael Grant and Alice Ruth (Turner) H. BA in Psychology, Ill. Benedictine Coll., Lisle, 1991. Instr. Decatur Recreation Dept., 1985; office asst. Sec. of State State of Ill., Springfield, summer 1990; rsch. asst. Psychology/Sociology Dept. Ill. Benedictine Coll., 1988-91; biofeedback therapist, counselor St. Mary's Hosp., Decatur, 1991—. Co-contbr. articles to profl. jours. Recipient Eileen J. Garrett Rsch. scholarship in parapsychology, 1992. Mem. AAAS, Psi Chi, Pi Gamma Mu. Roman Catholic. Home: 2423 Wakefield Dr Decatur IL 62521 Office: St Marys Hosp Dept Mental Health 1800 E Lake Shore Dr Decatur IL 62521

HOUSE, JOHN W., otologist; b. Los Angeles, July 12, 1941; s. Howard and Helen House; m. Patricia Roberts, June 10, 1967; children: Hans, Chris, Kurt. BS, U. So. Calif., Los Angeles, 1964; MD, U. So. Cal., Los Angeles, 1967. Resident Glendale (Calif.) Adventist Hosp., 1971-72, L.A. County Med. Ctr., 1972-74; fellow Otologic Med. Group, L.A., 1974, pvt. practice, 1975—; pres. House Ear Inst., L.A., 1987—. Mem. editorial bd. Am. J. Otology, 1986—; contbr. articles to jours. in field. Admissions com. interviewer, U. So. Calif. Sch. Medicine, Los Angeles, 1976—; mem. Los Angeles County Sheriff's Res. Med. Co. Capt. U.S. Army, 1969-71. Recipient Hocks Meml. award Am. Tinnitus Assn., 1988; named Tchr. of Yr., U. So. Calif. Family Practice Dept., 1987. Fellow Am. Acad. Otolaryngology/Head and Neck Surgery; mem. Am. Neurotology Soc. (program chmn. 1976—), AMA, Am. Soc. Mil. Otolaryngologists, Pan-Am. Assn. Otorhinolaryngology Broncho Esophagology, Jonathan Club (Los Angeles). Avocations: skiing, computers, running, swimming. Office: House Ear Inst 2100 W 3rd St Los Angeles CA 90057-1902*

HOUSE, LON WILLIAM, engineer, energy economist; b. Travis, Calif., June 14, 1952; s. Ottis Lee Dilworth and Ann Ewing (Porter) House; m. Cheryl Ann Breslin, July 5, 1980; children: Jordan David, Ariel Dawn, Samara Diane. BS, U. N.Mex., 1974, MA, 1976; MS, Portland State U., 1978; PhD, U. Calif., Davis, 1982. Cert. energy mgr. Elec. generation specialist Calif. Energy Commn., Sacramento, 1980-85; chief resource planner Calif. Pub. Utilites Commn., San Francisco, 1985-90; prin. cons. Henwood Energy Svcs., Sacramento, 1990-93; adj. lectr. Coll. Engring., U. Calif., Davis, 1983-90. Contbr. articles to profl. jours. Tuition com. mem. Placerville (Calif.) Christian Sch., 1992. Mem. Assn. Energy Engrs. (sr.), Sigma Xi. Republican. Achievements include research in the integration of alternative technologies and conservation measures into utility resource planning. Home: 4901 Flying C Rd Cameron Park CA 95682

HOUSER, DONALD RUSSELL, mechanical engineering educator, consultant; b. River Falls, Wis., Sept. 2, 1941; s. Elmont Ellsworth and Helen (Doris) H.; m. Colleen Marie Collins, Dec. 30, 1967; children: Kelle, Kerri, Joshua. BS, U. Wis., 1964, MS, 1965, PhD, 1969. Registered profl. engr., Ohio. Instr. U. Wis., Madison, 1967-68; from asst. prof. to prof. Ohio State U., Columbus, 1968—; dir. Gear Dynamics and Gear Noise Rsch. Lab., Columbus, 1979—; v.p. Gear Rsch. Inst., Lisle, Ill., 1990—. Author: (with others) Gear Noise, 1991; contbg. editor Sound and Vibration Mag.; contbr. articles to profl. jours. Elder St. Andrews Presbyn. Ch., Columbus, 1972-75. Mem. ASME (legis. liaison Ohio coun. 1976-80, Century II medallion 1980), Am. Gear Mfrs. Assn. (acad.), Soc. Automotive Engrs. Roman Catholic. Achievements include development of technology for measuring gear transmission error under load. Office: Ohio State U 206 W 18th Ave Columbus OH 43210-1154

HOUSER, VINCENT PAUL, medical research director; b. Springfield, Mass., Mar. 25, 1943; s. Joseph Mahlen and Margarete (Donahue) H.; m. Frances L. Roberts, Aug. 28, 1965 (div. 1983); 1 child, Jennifer Rae; m. Sandra B. Lowe, Dec. 4, 1991. BA with honors, Boston Coll., 1965; MS, U. Mass., 1967, PhD, 1969. Lic. psychologist, Pa. Psychology rsch. assoc. VA Hosp., Perry Point, Md., 1970-72, rsch. psychopharmacologist, dir. psychotropic drug lab., 1972-75; sr. pharmacologist and acting head psychopharmacology lab. Lederle Labs., Pearl River, N.Y., 1975-77; prin. scientist and group leader CNS sect. Schering Corp., Bloomfield, N.J., 1977-78; assoc. dir. clin. rsch. dept. internal medicine Hoechst-Roussel Pharms. Inc., Somerville, N.J., 1978-83, assoc. dir. clin. rsch. dept. psychopharmacology, 1983-86; clin. rsch. dir. psychotropics Beecham Labs., Bristol, Tenn., 1987-88; dir. cen. nervous system therapeutic area Solvay Pharms., Marietta, Ga., 1989—. Cons. to 5 sci. jours.; contbr. 40 sci. articles to profl. jours. Fellow APA; mem. AAAS, N.Y. Acad. Sci., N.J. Acad. Sci., N.J. Psychol. Assn., Ea. Psychol. Assn., Soc. Clin. Trials. Achievements include research in fixated behavior and its alteration by psychotropic agents, evaluation of the potential for interactions of paroxetine with diazepam, potential benzazepine antipsychotic with unique interations on dopaminergic systems. Home: 4436 Dobbs Crossing Marietta GA 30068

HOUSHOLDER, GLENN THOLEN, pharmacology educator; b. Granbury, Tex., Dec. 29, 1925; d. Herman and Jewell (Glenn) T.; m. Dwight Eugene Housholder, June 12, 1948. BS, U. Houston, 1947; PhD, U. Tex., 1970. Asst. prof. U. Tex. Dental Br., Houston, 1970-74, assoc. prof., 1974-81, prof., 1981—; grad. faculty U. Tex. Grad. Sch. Biomed. Scis., Houston, 1972—; cons. Lamar U. Sch. of Dental Hygeine, Beaumont, Tex., 1977-78. Contbg. author: (textbooks) Pharmacology and Therapeutics in Dentistry, 1980, 85, 89, Oral Surgery, Oral Medicine, Oral Pathology, 1993; contbr. articles to profl. jours. Sec. CBB Civic Assn., Houston, 1981-87. Mem. AAAS, N.Y. Acad. Sci., Am. Chem. Soc., Am. Assn. Dental Schs., Assn. Women in Sci., Sigma Xi, Iota Sigma Pi, Omicron Kappa Upsilon. Office: Univ Tex Health Scis Dental Branch 6515 John Freeman Ave Houston TX 77225

HOUSNER, GEORGE WILLIAM, civil engineering educator, consultant; b. Saginaw, Mich., Dec. 9, 1910; s. Charles and Sophie Ida (Schust) H. B.S. in Civil Engring., U. Mich., 1933; Ph.D., Calif. Inst. Tech., 1941. Registered profl. engr., Calif. Engr. U.S. Corps Engrs., Los Angeles, 1941-42; ops. analyst 15th Air Force, Libya and Italy, 1943-45; prof. engring. Calif. Inst. Tech., Pasadena, 1945—, now prof.emeritus; earthquake engring. cons. Pasadena, 1945—; mem. Gov.'s Earthquake Council, 1971-76, Los Angeles County Earthquake Commn., 1971-72, adv. panel on Earthquake Hazard

Nat. Acad. Scis., 1981-83; chmn. Com. on Earthquake Engring., NRC, 1983-92, Com. on Internat. Decade Natural Hazard Reduction, NRC, 1986-88. Author 3 textbooks, numerous tech. papers. Recipient Disting. Civilian Service award U.S. War Dept., 1945, Bendix Research award Am. Soc. Engring. Edn., 1967, Nat. Medal Sci., 1988. Mem. NAE (Founders award 1991), NAS, Seismol. Soc. Am. (pres. 1977-78, medal 1981), ASCE (von Karman medal 1972, Newmark medal 1981), Internat. Assn. Earthquake Engring. (pres. 1969-73), Earthquake Engring. Research Inst. (pres. 1954-65). Office: Calif Inst Tech 211 Thomas Lab Pasadena CA 91125

HOUSTON, DEVIN BURL, biomedical scientist, educator; b. Roswell, N.Mex., Jan. 13, 1957; s. Louis Burl and Leah June (Hall) H.; m. Patricia Wynette Carter, Sept. 9, 1978; 1 child, Dustin Burl. BA, Hendrix Coll., 1979; BS, U. South Ala., 1981, PhD, 1987. Postdoctoral fellow U. Va. Health Sci. Ctr., Charlottesville, 1986-90; postdoctoral fellow St. Louis U. Med. Ctr., 1990-93, asst. rsch. prof., 1993—. Author: Marijuana Cannabinoids: Neurobiology and Neurophysiology, 1992. Recipient fellowship Am. Heart Assn., 1988-90. Mem. Soc. for Neuroscience, Internat. Cannabis Rsch. Soc. Office: St Louis Univ Med Ctr Pharmacol and Physiol Sci 1402 S Grand Blvd Saint Louis MO 63104

HOUSTON, JOHN F., chemist; b. Newport, Vt., Feb. 3, 1952; s. John F. and Arvilla R. (Smith) H.; m. Elizabeth H. McMullen, Aug. 31, 1991. BS, Johnson State Coll., 1975. Wastewater treatment plant operator Village of Orleans (Vt.), 1980-82; chemist Vt. State Dept. of Health, Burlington, 1982-83, 84—; aquatic compliance monitoring chief Vt. Agy. Natural Resources, Montpelier, 1983-84. Office: State Dept of Health 195 Colchester Ave Burlington VT 05401

HOUSTON, THOMAS DEWEY, environmental engineer; b. Ludowici, Ga., July 28, 1930; s. Thomas Dewey and Eva (Bland) H.; m. Margaret Ann Adamson, June 11, 1955; children: T. Dewey, Roger Weyman. Student, South Ga. Coll., 1947-49, Ga. Inst. Tech., 1956-57. Registered Environ. Profl.; cert. mgr. hazardous materials. Jr. engr. Atlantic Coastline R.R., Wilmington, N.C., 1953-56; elec. engr. Savannah (Ga.) Dist. Corps. of Engrs., 1958-69; gen. engr. Ft. Stewart, Ga., 1961-63; chief engring. divsn. Post Engr., Ft. Stewart, 1963-70; chief bldgs. and grounds divsn. Directorate of Engring., Ft. Stewart, 1970-82, chief facilities engring. divsn., 1983-85, chief environ. and natural resources divsn., 1985—; mem. rail maintenance task force and policy com. Dept. Army, Washington, 1970-83; mem. solid waste task force CG RDC, Brunswick, Ga., 1983—; environ. adv. bd. Long County, Lucowici, 1990—. Contbr. articles to environ. jours. Chmn. Long County Forestry Bd., Ludowici, 1967—; v.p. Long County Jaycees, Ludowici, 1963. With U.S. Army, 1950-52. Mem. Acad. Hazardous Materials Mgrs. (diplomate). Baptist. Home: Rt 1 Box 85 Ludowici GA 31316 Office: Environ Natural Resources Directorate Engring Fort Stewart GA 31314

HOUZE, GERALD LUCIAN, JR., metallurgist; b. Pitts., Sept. 21, 1936; s. Gerald Lucian and Margaret Gladys (Gillard) H.; m. Margaretta Mary Dascalos, Nov. 8, 1958 (div. Oct. 1979); children: Eric Charles, David Wayne, Jonathan Brice; m. Patricia Marie Wells, May 5, 1980. BSMetE, Purdue U., 1958; MS, Case Inst., 1961, PhD, 1965. Jr. rsch. metallurgist United Engring. & Foundry, Canton, Ohio, 1958-59; sr. rsch. metallurgist Allegheny Ludlum Corp., Brackeridge, Pa., 1964-71, chief rsch. metallurgist, 1971-73, mgr. phys. metallurgy, 1973-78, dir. rsch., 1978-86, dir. rsch. and analytical labs., 1986-88, asst. v.p. R&D, 1988—; mem. ASM Internat. Govt. and Pub. Affairs Com., Materials Park, Ohio, 1985—, sec. 1986-87, chmn. 1988-89. Contbr. articles to profl. jours. Bd. dirs. Allegheny Valley Hosp. Found., Natrona Heights, Pa., 1986—. Recipient Disting. Engring. Alumnus Purdue U., West Lafayette, Ind. 1984. Felllow Am. Soc. Metals Internat.; mem. Metall. Soc. of AIME, Sigma Xi, Tau Beta Pi. Achievements include pioneer of high speed cinematography of 60Hz magnetic domain, motion in ferromagnetic materials; elucidated the mechanism of anomalous losses in grain oriented silicon steel; directing the efforts to develop a direct strip casting process which would eliminate ingot or slab casting and hot rolling ops. in the prodn. of flat rolled steel products. Office: Allegheny Ludlum Corp Tech Ctr Brackenridge PA 15014

HOVANEC, B. MICHAEL, chemist, researcher; b. Riverside, Calif., May 26, 1952; s. Bernard Michael and Nancy Marianne (McHale) H.; m. Deborah Kay Morrow, June 24, 1972 (div. 1976); 1 child, Clayton Michael; m. Patricia Marie Gutierrez, July27, 1979; children: Christopher Michael, Michelle Marie. BA, U. Calif., Riverside, 1975; Cert. in Hazardous Materials Mgmt., U. Calif., Irvine, 1990. Chemist Indsl. Polymers, Cucumonga, Calif., 1976-77, Lever Bros., City of Commerce, Calif., 1977-78; sr. chemist West Coast Tech. (IT Corp.), Cerritos, Calif., 1978-80; sr. analytical chemist Rockwell Internat., Newport Beach, Calif., 1980-86; sr. staff chemist West Coast Analytical Svc., Santa Fe Springs, Calif., 1986—; instr. Calif. State Univ., Long Beach, 1992—. Contbr. articles to profl. jours. Mem. Am. Chem. Soc., Am. Welding Soc., Am. Soc. for Mass Spectroscopy. Achievements include methods development and research in Inductively Coupled Plasma Mass Spectrometry for last 6 years bringing this technology on-line in a commercial lab. Office: West Coast Analytical Svc 9840 Alburtis Ave Santa Fe Springs CA 90670-3208

HOVANEC, TIMOTHY ARTHUR, aquatic researcher; b. Riverside, Calif., Aug. 17, 1956; s. Bernard M. and Nancy M. Hovanec; m. Linda S. Kelley. BS in Biology, San Diego State U., 1978, MS in Biology, 1986; postgrad., U. Calif., Santa Barbara, 1993—. Mgr. Aquatic Systems Inc., San Diego, 1980-90, dir. Aquaria Inc., Moorpark, Calif., 1990—. Co-contbr.: Aquaculture of Striped Bass, 1984. Bd. dirs. Pet Industry Joint Action Coun., 1993—. Mem. Am. Water Works Assn., Am. Soc. Limnology and Oceanography. Office: Aquaria Inc 6100 Condor Dr Moorpark CA 93021

HOVNANIAN, H. PHILIP, biomedical engineer; b. Aleppo, Syria, Dec. 17, 1920; s. Philip and Rosa (Jebejian) H.; m. Siran Norian, June 10, 1948; children: Rosemary Janice, Joan Anita, John Philip. B.S., Am. U., Beirut, 1942, postgrad., 1945-47; postgrad., Brown U., 1947-49; M.S., State Coll., Boston, 1951; Ph.D., U. Beverly Hills, Calif. Registered profl. engr., N.Y., Mass.; chartered physicist (U.K.). Prin. investigator, rsch. grant Nat. Heart Inst., NIH; faculty dept. physics Am. U., Beirut, 1942-47, Brown U., 1947-49; sr. engr. Western Electric Co., Haverhill, Mass., 1951-52; asst. chief engr. Calidyne Co., Winchester, Mass., 1952-54; sr. physicist, project head, asst. rsch. dir. Boston Electronics div. Norden-Ketay Corp., 1954-56; ptnr., R & D dir. physics Neutronics Rsch. Co., Waltham, Mass., 1956-58; sr. staff scientist Avco Corp., 1958, dir. med. sci. dept., 1959-66; mgr. lunar biosci. NASA, Washington, 1966-67; mgr. biomed. engring. and biophysics Kollsman Instrument Corp., Syosset, N.Y., 1967-68; v.p., dir. biomed. products Cavitron Corp., Cooper Med. Corp., Cooper Vision, Inc., 1969-85; dir. scientific devel. Vital Signs Inc., Totowa, N.J., 1990—; corp. dir. Vital Signs Corp.; v.p. rsch. Sonokinetics Inc., Hoboken, N.J.; guest lectr. biomed. engring. Northeastern U., MIT-Harvard Study Group on Biomed. Engring.; rsch. assoc. in surg. rsch. Lahey Clinic Found.; mem. workshop interaction between industry and biomed. engring. NAE; former mem. ob-gyn. devices panel and panel on ophthalmic, ear, nose and throat devices and dental devices FDA; cons. BEI Med. Systems, Hackensack, N.J., 1993—. Contbr. tech. papers to profl. jours.; patentee in field. Trustee Haigazian Coll. of Beirut, Lebanon. Fellow Inst. Physics (chartered, Brit.), Am. Soc. Laser Med. and Surgery, Am. Acad. Dental Electrosurgery (hon.); mem. Optical Soc. Am., Am. Inst. Physics, IEEE (sr. mem., profl. group biomed. electronics), Internat. Fedn. Med. Electronics, Biomed. Engring. Soc., Rsch. Soc. Am., Internat. Microscopy Assn., Am. Inst. Ultrasound in Medicine, Am. Soc. Microbiology, N.Y. Acad. Scis., AAAS, Assn. for Advancement Med. Instrumentation, Am. Dental Trade Assn. (stds. & materials) ADA (standards coms. dental UV curing lights and electrosurgery devices), Am. Inst. Biol. Scis., Armenian Missionary Assn. of Am. (pres. and chmn. bd. dirs.), Sigma Xi. Congregationalist (United Ch. of Christ) (chmn. bd. trustees, deacon, moderator). Lodge: Masons. Home and Office: Unit 1B 3902 Manhattan College Pkwy Bronx NY 10471

HOWARD, AUGHTUM LUCIEL SMITH, retired mathematics educator; b. Almo, Ky., Nov. 10, 1906; d. Leander E. and Anna (Wright) Smith; m. Noel Judson Howard, Jan. 6, 1929; children: Carl Eugene, Robert Alvin. BA, Georgeown Coll., 1926; postgrad., U. Mich., 1927; MS, U. Ky., 1938, PhD, 1942. Lab technician Parke Davis Drug Co., Detroit, 1926-27; tchr. Marshall County High Sch., 1927-29; grad. asst. math. dept. U. Ky., 1936-41, fellow, 1941-42; assoc. prof. math. Ky. Wesleyan Coll., 1942-46, prof., 1946-58; assoc prof. Eastern Ky. State Coll., Richmond, 1958-62, prof., 1962-73; mem. curriculum study com.; commn. on pub. edn., State of Ky., 1961. Tchr. adult Sunday sch. class Richmond 1st Christian Ch., 1971-78, deacon, 1986-87, elder, 1989. Mem. AAUP, Math Assn. Am. (chmn. Ky. sect. 1944-46, lectr. 1953-55, sec.-treas. 1949-51, 69-71, cert. of meritorious svc. 1988), Richmond Women's Club, Sigma Xi. Home: 206 Pembroke Dr Richmond KY 40475-2240

HOWARD, BERNARD EUFINGER, mathematics and computer science educator; b. Ludlow, Vt., Sept. 22, 1920; s. Charles Rawson and Ethel (Kearney) H.; m. Ruth Belknap, Mar. 29, 1942. Student Middlebury Coll., 1938-40; B.S., MIT, 1944; M.S., U. Ill., 1947, Ph.D., 1951. Staff mem. Radiation Lab, MIT, Cambridge, 1942-45; asst. math. U. Ill., Champaign-Urbana, 1945-49; sr. mathematician Inst. Air Weapons Rsch., U. Chgo., 1951, asst. to dir. Inst. for Systems Rsch., 1952-56, assoc. dir., 1956-60, assoc. dir. Labs. for Applied Sci., 1958-60; dir. Sci. Computing Ctr. U. Miami, Coral Gables, Fla., 1960-64, prof. math. and computer sci., 1960-91, prof. emeritus, 1991—; chmn. bd. dirs. Sociocybernetics, Inc.; exec. sec. Air Force Adv. Bd. Simulation, 1951-54; cons. Systems Rsch. Labs, Inc., Dayton, Ohio, 1963-67, acting dir. math. scis. div., 1965; cons. Variety Children's Rsch. Found., Miami, 1964-66, Fla. Power & Light Co., Miami, 1968, Shaw & Assocs., 1964-75; vis. fellow Dartmouth Coll., Hanover, N.H., 1976; co-investigator Positron Emission Tomography Ctr., U. Miami Dept. Neurology/Mt. Sinai Med. Ctr., 1981-84. Creator Parabolic-Earth Radar Coverage Chart, 1944; co-creator: (with Henry W. Kunce) Sociocybernetics, 1971, Optimum Curvature, 1964, Optimum Torsion, 1974, (with J.F.B. Shaw) Principles in Highway Routing, (with James M. Syck) Twisted Splines, 1992. Chmn. bd. dirs. Blue Lake Assn., Inc., Miami, 1969—. Am. Soc. Engring. Edn.-Office of Naval Research fellow Naval Underwater Systems Ctr., 1981, 82. Mem. Am. Math. Soc., Soc. Indsl. and Applied Math. (treas. S.E. sect. 1964), Am. Phys. Soc., Assn. Computing Machinery (chpt. chmn. 1969-70), IEEE, AAUP (chpt. sec. 1974—), Sigma Xi, Phi Kappa Phi, Pi Mu Epsilon, Alpha Sigma Phi. Home: 7320 Miller Dr Miami FL 33155-5504 Office: U Miami Coral Gables FL 33124

HOWARD, DONIVAN R(ICHARD), engineering executive; b. Linton, Ind., Oct. 26, 1937; s. Joseph Randolph H. and Eva Yvonne (Hunter) Stalcup; m. Clarisa Faye Harper, Aug. 1, 1987; children: Sandra L., Linda A., Tracy G., James L., Suzanne J., Clint C. BSEE, Purdue U., 1958; MSEE, Calif. Tech. Inst., 1959; PhD in Electrical Engring., Purdue U., 1964. Mem. tech. staff Hughes Aircraft Co., Culver City, Calif., 1958-59, Firestone Tire & Rubber Co., Southgate, Calif., 1959-60, TRW Systems, Redondo Beach, Calif., 1960-63; various positions The Aerospace Corp., El Segundo, Calif., 1963-79, div. gen. mgr., 1979-81; pres. bd Systems Inc., Torrance, Calif., 1981-87; tech. dir. BD Systems Inc., Torrance, Calif., 1987—. Mem. AIAA, Soc. Photographic Instrumentation Engrs., Sigma Xi. Achievements include patent for remote attitude determination technique. Office: BD Systems Inc 385 Van Ness Ave Ste 200 Torrance CA 90501

HOWARD, KENNETH IRWIN, psychology educator; b. Chgo., Oct. 19, 1932; s. Simon and Florence (Bergman) H.; m. April Rose Zweig, Dec. 15, 1979; children: Deborah, Peter, Lisa, David, Rebecca, Matthew. BA, U. Calif., Berkeley, 1954; PhD, U. Chgo., 1959. Prof. psychology Northwestern U., Evanston, Ill., 1967—. Contbr. over 150 articles to profl. publs.; author 3 books in field. 1st lt. U.S. Army, 1954-56. NIMH rsch. scientist, 1991—; recipient Disting. Rsch. Career award Soc. for Psychotherapy Rsch., 1991. Democrat. Jewish. Office: Northwestern U Dept Psychology Evanston IL 60208

HOWARD, RICHARD RALSTON, II, medical health advisor, researcher, financier; b. Winnfield, Kans., May 26, 1948; s. Richard Ralston and Ione (Mayer) H. BBA, Loyola U., New Orleans, 1970; MPH, Tulane U., 1977, MS, 1984, DrPH, 1988. Researcher Loyola U., 1973; educator Dominican Coll., New Orleans, 1977; educator Sch. Pub. Health Tulane U., New Orleans, 1978-82, researcher Sch. Medicine, 1979-88; med. health advisor Howard Med. Clinic, Slidell, La., 1982-91; founder, CEO The Howard Found., New Orleans, 1993—. NIH grantee, 1979; VA grantee, 1984. Mem. Internat. Platform Assn., Am. Assn. Individual Investors, New Orleans Spl. Interest Group of Computerized Investing (dir.), Beta Beta Beta. Achievements include research on the impact of the health food industry on nutrition awareness, cocaine testing through quantitative tear analysis, vitamin C and ophthalmic wound healing. Home: 3531 Nashville Ave New Orleans LA 70125-4339

HOWARD, ROBERT FRANKLIN, observatory administrator, astronomer; b. Delaware, Ohio, Dec. 30, 1932; s. David Dale and Clarine Edna (Morehouse) H.; m. Margaret Teresa Farnon, Oct. 4, 1958; children—Thomas Colin, Alan Robert, Moira Catharine. B.A., Ohio Wesleyan U., 1954; Ph.D., Princeton U., 1957. Carnegie fellow Mt. Wilson and Palomar Obs., Pasadena, Calif., 1957-59; staff mem. Mt. Wilson and Palomar Obs., 1961-81; asst. prof. U. Mass., Amherst, 1959-61; asst. dir. for Mt. Wilson Mt. Wilson & Las Campanas Obs., Pasadena, 1981-84; dir. Nat. Solar Obs., Tucson, 1984-88, astronomer, 1988—. Editor: Solar Magnetic Fields, 1971; editor: (jour.) Solar Physics, 1987—; contbr. articles to profl. jours. Mem. Am. Astron. Soc., Internat. Astron. Union. Office: Nat Solar Obs PO Box 26732 950 N Cherry Ave Tucson AZ 85726

HOWARD, ROBERT P., propulsion engineering scientist; b. Jamestown, Tenn., Feb. 22, 1955; s. Curtis T. and Hattie B. (Young) H.; m. Martha J. Choate, May 19, 1979; children: Bradley R., Scott C. BSEE, Tenn. Tech. U., 1978, MSEE, 1982, PhD in Engring., 1987. Instr. elec. engring. dept. Tenn. Tech. u., Cookeville, 1980-85, instr., 1987; cons. Benjamine Industries div. Thomas Industries, Sparta, Tenn., 1985; rsch. cons. Arnold Air Force Base, Tenn., 1985-86; sr. engr., scientist Sverdrup Tech., Inc., Arnold AFB, 1987—. Contbr. to AIAA jours., JAANF confs., AGARD, France, Air Waste and Mgmt. Confs. Cubmaster pack 342 Elk River dist. Boy Scouts Am., Manchester, Tenn., 1990—; asst. coach Nat. Little League Program, Tullahoma, Tenn., 1990-92; pres. Hickerson Elem. Student Improvement Assn., Manchester, 1991—. Recipient Arnold AFB Tech. Achievement award. Mem. AIAA Inc. (sr.), Sigma Xi. Achievements include research in NO and OH measurement for liquid propellant rockets, arc heater facilities, impulse facility flows, and automobile engine exhausts, molecular electronic resonance techniques, laser measurement and gaseous diagnostic techniques, measurement applications to solid and liquid rocket engines. Home: Rt 7 Box 7657 Manchester TN 37355

HOWARD, RUSSELL DUANE, aerospace engineer; b. Phoenix, Aug. 19, 1956; s. Lynn Edward and Doris Irene (Morrissette) H.; m. Janice Elaine Voss, Feb. 20, 1983 (div. 1988). BS in Math., Calif. Inst. Tech., 1977, MS in Aero. Engring., 1981; MA in Astronomy, U. Calif., Berkeley, 1979; PhD in Aerospace Systems, MIT, 1990. Applications engr. Sperry Flight Systems, Glendale, Ariz., 1979-80; assoc. dir. Space Systems Lab., U. Md., College Park, 1990—; cons. Am. Rocket Co., Camarillo, Calif., 1989-90. Mem. Sigma Xi. Office: U Md Space Systems Lab 382-2100E College Park MD 20742

HOWARD, SUSAN CAROL PEARCY, biochemist; b. Enid, Okla., Jan. 7, 1954; d. William Silas and Bettie Josephine (Tribble) Pearcy; m. Jeffrey Lane Howard, Apr. 16, 1977; children: Heather Carol, Brandon Lane. BS, Okla. State U., 1975, PhD, 1980. Sr. rsch. biochemist Monsanto Corp. Rsch., St. Louis, 1986-90, rsch. specialist, 1990—. Contbr. articles to profl. publs. Jane Coffin Childs fellow, 1980-81. Mem. Am. Soc. Cell Biology, Sigma Xi, Phi Kappa Phi, Phi Lambda Upsilon. Achievements include 2 patents in field. Home: 35 Worthy Ct Saint Louis MO 63026 Office: Monsanto Co 800 N Lindbergh Blvd Saint Louis MO 63167

HOWDEN, DAVID GORDON, engineering educator; b. Scarborough, Yorkshire, Eng., Aug. 22, 1937; came to U.S., 1967; s. Leslie Irving and Alice (Smith) H.; m. Elsie Ragnhild Elisabet Svensson, Oct. 6, 1962; 1 child, Nils Roger. BSc, U. Birmingham, Eng., 1959, PhD, 1962. Asst. prof., researcher Centro Tecnico de Aeronavica, Sao Paulo, Brazil, 1963-65; sr. researcher Dept. Mines Tech. Svcs., Ottawa, Ont., Can., 1965-67; assoc. mgr. Battelle-Columbus (Ohio) Labs., 1967-77; assoc. prof. engring. Ohio State U., 1977—; chmn. project NRC, Nat. Materials Adv. Bd., Washington, 1978, com. mem., 1980; chmn. com. Welding Rsch. Coun., N.Y.C., 1977-88. Author: American Welding Society - Welding Safety and Health, 1980, American Welding Society - Welding Structural Design, 1988; editor: Niobium, 1982, American Welding Society - Welding Qualification of HSLA Steel Weldments, 1992. Mem. Internat. Inst. Welding (del. 1978-92), Am. Welding Soc. (nat. v.p. 1990-92, Clarence E. Jackson hon. lectr. 1977, Adams Meml. membership award 1979, James Lincoln Gold medal 1983), Welding Rsch. Coun. (chmn. 1974-92), Am. Soc. Metals Internat., Brazilian Metals Soc. Achievements include research on welding/joining of steels, gas/metal chem. reactions in arc welding, welding safety and health, joining/welding of materials. Home: 3285 Wilson Rd Sunbury OH 43074 Office: Ohio State U Dept Welding 190 W 19th Ave Columbus OH 43210

HOWE, BURTON BROWER, pharmacologist; b. East Orange, N.J., Sept. 20, 1936; s. Burton Charles and Mildred Kathryn (Brower) H.; A.B., Upsala Coll., 1963; M.A., Montclair State Coll., 1973; m. Jane Elizabeth Rathslag; children: Diane, Kathleen. Sr. scientist, dept. pharmacology Warner-Lambert Research Inst., Morris Plains, N.J., 1962-77; sr. research pharmacologist ICI Americas Inc., Wilmington, Del., 1977—. Mem. AAAS, Am. Heart Assn., N.Y. Acad. Scis., Physiol. Soc. Phila., Am. Soc. Pharmacology and Exptl. Therapeutics, Mid-Atlantic Pharmacology Soc., Sigma Xi. Contbr. articles to profl. jours.; chpts. to sci. books. Achievements include research in cardiovascular and autonomic pharmacology, physiology and pharmacology of the coronary circulation. Home: 811 Summerset Dr Hockessin DE 19707-9336 Office: Biomed Research Div ICI Ams Inc Wilmington DE 19897

HOWE, DANIEL BO, aviation executive, retired naval officer; b. New Orleans, June 23, 1944; s. Kenneth Edwin and Wilma Mayme (Mayo) H.; m. Sandra Nell Langford, Feb. 3, 1968; children: Daniel Bo II, Emily Nelwyn. BA in Sociology, U. N.C., 1966. Commd. ensign USN, 1966, advanced through grades to comdr., 1975, pilot in various locations, 1966-72, ret., 1987; pilot Delta Airlines, Atlanta, 1972-73; v.p. flight ops. Express Airlines I, Memphis, 1987-90; mgr. system tng. Fed. Express Corp., Memphis, 1990-91, sr. mgr. feeder ops., 1991—. Mem. NRA, Tailhook Assn., Ret. Officers Assn., Ducks Unlimited (com. mem. Memphis br. 1987—), Delta Upsilon. Methodist. Home: 632 Brinsley Dr Collierville TN 38017 Office: Fed Express Corp PO Box 727 Memphis TN 38194

HOWE, JOHN PRENTICE, III, health science center executive, physician; b. Jackson, Tenn., Mar. 7, 1943; s. John Prentice and Phyllis (MacDonald) H.; m. Jill Olmsted, Aug. 19, 1967; children: Lindsey Warren, Brooke Olmsted, John Prentice IV. BA, Amherst Coll., 1965; MD, Boston U., 1969. Diplomate Am. Bd. Internal Medicine, internal medicine and cardiovascular disease. Research assoc. cellular physiology Amherst Coll., 1963-64; research assoc. cardiovascular physiology Boston U. Sch. of Medicine, 1966-67, lectr. medicine, 1972-73; intern Boston City Hosp., 1969-70, asst. resident, 1970-71; research fellow in medicine Harvard U., 1971-73, Peter Bent Brigham Hosp., 1971-73; survey physician Framingham Cardiovascular Disease Study, Nat. Heart and Lung Inst., 1971; asst. clin. prof. medicine U. Hawaii, 1973-75; asst. prof. medicine U. Mass., 1975-77, assoc. prof., 1977-85, vice chmn. dept. medicine, 1975-78, asst. dean continuing edn. for physicians, 1976-78, assoc. dean profl. affairs and continuing edn., 1978-80, acad. dean, 1980-85, vice chancellor, 1980-85, acting chmn. dept. anatomy, 1982-85; pres., prof. medicine U. Tex. Health Scis. Ctr., San Antonio, 1985—; assoc. chief div. medicine U. Mass. Hosp., 1975-78, dir. patient care studies dept., 1975-80, chief of staff, 1978-80. Mem. editorial bd. Archives Internal Medicine, 1991—; contbr. numerous articles to profl. jours., chpts. to books. Mem. med. adv. com. Hospice San Antonio; trustee S.W. Found. for Biomed. Rsch., San Antonio Med. Found., S.W. Rsch. Inst. Maj. M.C., U.S. Army, 1973-75. Alfred P. Sloan scholar Amherst Coll., 1962-65; recipient Ruth Hunter Johnson award Boston U. Sch. of Medicine, 1969. Fellow ACP, Am. Coll. Cardiology, Am. Coll. Chest Physicians; mem. AMA (alt. del. ho. of dels. 1989—), Am. Heart Assn. (fellow coun. clin. cardiology), Tex. Med. Soc. (coun. med. edn. 1986—, ho. of dels. 1989—), Tex. Soc. Biomed. Rsch. (pres.), Bexar County Med. Soc. (exec. com. 1985—), Japan-Am. Soc. San Antonio (bd. dirs.), Assn. Acad. Health Ctrs. (bd. dirs. 1991—), Alpha Omega Alpha, Omicron Kappa Epsilon. Avocations: tennis; skiing.

HOWE, JULIETTE COUPAIN, chemist; b. Woonsocket, R.I., Aug. 16, 1944; d. Camille Jacques and Octavie Henriette (Beaujean) Coupain; m. John Thomas Howe, June 2, 1979; children: Laura Samantha, Kathryn Elizabeth. BS, U. R.I., 1965; MS, U. Md., 1978, PhD, 1987. Biologist USDA, Beltsville, Md., 1965-77, rsch. biologist, 1977-79, rsch. nutritionist, 1979-80, rsch. chemist, 1980--. Co-author: Nutritional Bioavailability of Ca., 1985; contbr. articles to profl. jours. XII Internat. Congress Nutrition Travel grantee, 1981. Mem. Am. Inst. Nutrition, Am. Chem. Soc., Sigma Xi. Office: USDA Bldg 308 Rm 213 BARC-East Beltsville MD 20705

HOWE, MARTHA MORGAN, microbiologist, educator; b. N.Y.C., Sept. 29, 1945; d. Charles Hermann and Miriam Hudson (Wagner) M. A.B., Bryn Mawr Coll., 1966; Ph.D., MIT, 1972. Postdoctoral fellow Cold Spring Harbor Lab, N.Y., 1972-74; asst. prof. bacteriology U. Wis., Madison, 1975-77, assoc. prof., 1977-81, prof., 1981-84, Vilas prof., 1984-86; Van Vleet prof. virology U. Tenn., Memphis, 1986—; mem. genetic biology rev. panel NSF, 1980-82; mem. gen. research support rev. com. NIH, Bethesda, 1982-86. Assoc. editor Virology, 1983—; mem. editorial bd. Jour. Bacteriology, 1985-90; contbr. articles to profl. jours. and books. Recipient Rsch. Career Devel. award NIH, 1978; H.I. Romnes Faculty fellow U. Wis., 1981; Amoco Teaching award U. Wis., 1981; Eli Lilly award Am. Soc. Microbiology, 1985. Fellow Am. Acad. Microbiology (bd. govs. 1991-93); mem. Am. Soc. Microbiology (chmn. div. H, 1983, councillor div. H, 1989-91, chmn. com. on awards 1990—), Am. Soc. Biochemistry and Molecular Biology, Genetics Soc. Am. (bd. dirs. 1989-91, program com. 1989-90). Office: U Tenn Dept Microbiology and Immunology 858 Madison Ave Memphis TN 38163-0001

HOWELL, CHARLES MAITLAND, dermatologist; b. Thomasville, N.C., Apr. 14, 1914; s. Cyrus Maitl and Lilly Mae (Ammons) H.; m. Betty Jane Myers, Feb. 12, 1949; children—Elizabeth Myers, Pamela Jane. B.S., Wake Forest U., Winston-Salem, N.C., 1935; M.D., U. Pa., 1937. Intern Charity Hosp., New Orleans, 1937-38; resident in medicine Burlington County Hosp., Mt. Holley, N.J., 1938-39; sch. physician Lawrenceville (N.J.) Sch., 1939-42; resident in pathology N.C. Baptist Hosp., Winston-Salem, 1947-48; resident in dermatology Columbia-Presbyn. Med. Ctr., N.Y.C., 1948-50; resident in allergy Roosevelt Hosp., N.Y.C., 1950-51; practice medicine specializing in dermatology Winston-Salem, 1951—; mem. staff N.C. Bapt., Forsyth Meml. hosps.; mem. faculty Bowman Gray Sch. Medicine, Wake Forest U., 1951-86, head. sect., 1984-86, prof. dermatology, 1967-84, prof. emeritus, 1984, head sect., 1961-86, acting head sect., 1984-86. Served as officer M.C. AUS, 1942-46. Fellow Am. Acad. Dermatology, Am. Acad. Allergy; mem. N. Am. Clin. Dermatol. Soc., N.Y. Acad. Scis. Democrat. Baptist. Clubs: Old Town (Winston-Salem); Bermuda Run Country (Clemmons, N.C.). Home: 1100 E Kent Rd Winston Salem NC 27104-1116 Office: 340 Pershing Ave Winston Salem NC 27103-2501

HOWELL, JOHN REID, mechanical engineer, educator; b. Columbus, Ohio, June 13, 1936; s. Frederick Edward and Hilma Lavilla (Kief) H.; m. Arlene Elizabeth Pollitt, June 20, 1959 (div. 1974); m. Susan Gooch Conway, May 20, 1979; children: John Reid Jr., Keli Dianne, David Lee. BSChemE, Case Inst. Tech., 1958, MSChemE, 1960, PhD, 1962. Registered profl. engr. Aerospace engr. NASA Lewis Research Ctr., Cleve., 1961-68; assoc. prof. U. Houston, 1969-73, prof., 1973-78; dir. Energy Inst. U. Houston, 1975-78; vis. prof. mechanical engring. U. Tex., Austin, 1978-79, prof., 1979-82, E.C.H. Bantel prof., 1982-90, Baker-Hughes prof., 1990—, chmn. mech. engring. dept., 1986-90, dir. Ctr. for Energy Studies, 1988-91. Co-author: Thermal Radiation Heat Transfer, 1981, 3d edit., 1992, Design of Solar Thermal Systems, 1984, Fundamentals of Engineering Thermodynamics, 1987, 2d edit., 1992; also numerous articles. Commr. Renewable Energy Resources Commn., Austin, 1980-81. Served to 1st lt. USAF, 1962-65. Recipient Spl. Service award NASA, 1965; named to Hon. Order Ky. Cols., 1980. Fellow ASME (Heat Transfer Meml. award 1991), AIAA (Thermophysics award 1990); mem. Am. Soc. Engring. Edn. (Ralph Coats Roe award 1987). Office: U Tex Dept Mech Engring ETC 7 142D Austin TX 78712

HOWELL, RALPH RODNEY, pediatrician, educator; b. Concord, N.C., June 10, 1931; s. Fred Lee and Grace Mary (Blackwelder) H.; m. Sarah Vosburg Esselstyn, Nov. 19, 1960; children: Grace Meyer, Elizabeth Eriksson, John Esselstyn. BS, Davidson Coll., 1953; MD, Duke U., 1957. Intern Duke U., 1957-58, resident in pediatrics, 1958-59, research fellow in pediatrics and medicine, 1959-60; clin. assoc. and staff NIH, Bethesda, Md., 1960-64; assoc. prof. pediatrics Johns Hopkins U., Balt., 1964-72; pediatrician-in-chief Univ. Children's Hosp. at Hermann, Houston, 1972-87; chmn. med. bd. Univ. Children's Hosp. at Hermann, 1972-87; David Park prof. U. Tex. Med. Sch., Houston, 1972-89, chmn. dept. pediatrics, 1972-87; prof., chmn. dept. pediatrics U. Miami Sch. Medicine, 1989—; sec. med. staff Jackson Meml. Hosp., Miami, 1992-93, v.p. med. staff, 1993-94; cons. pediatrics M.D. Anderson Hosp. and Tumor Inst., 1972-89; mem. metabolism study sect. NIH, 1973-77; mem. nat. clin. adv. com. Nat. Found. March of Dimes, 1973-79; nat. med. adv. bd., v.p. Muscular Dystrophy Assn., chmn. sci. adv. bd., 1989—; chmn. maternal and child health adv. com. NIH, 1983-86; vis. prof. Baylor Coll. Medicine Inst. Molecular Genetics, Houston, 1988; chief pediatrics Childrens Hosp. Ctr., U. Miami and Jackson Meml. Med. ctr., Miami, 1989—. Author: (with G.H. Thomas) Selected Screening Tests for Genetic Metabolic Diseases, 1973, (with F.H. Morriss, L.K. Pickering) Role of Human Milk in Infant Nutrition, 1986; contbr. articles to profl. jours. Trustee Jackson Lab., Bar Harbor, Maine; dir. Caldwell B. Esselstyn Found., Troy, N.Y., 1987-92, pres. 1992—. Served to sr. surgeon USPHS, 1960-64. Fellow Am. Acad. Pediatrics (com. on genetics), mem. AMA, Am. Pediatric Soc., Soc. Pediatric Rsch., Houston Pediatric Soc. (pres. 1978-79), Tex. Med. Assn., Soc. Inborn Errors of Metabolism (pres. 1981), Miami Pediatric Soc., Fla. Med. Assn., Am. Coll. Med. Genetics (bd. dirs. 1991—), Pi Kappa Alpha. Home: L'Hermitage Villa 66 2000 S Bayshore Dr Miami FL 33133-3250 Office: U Miami Sch Medicine Dept Pediatrics D-820 PO Box 016820 Miami FL 33101

HOWELL, RICHARD PAUL, SR., transportation engineer; b. Sarasota, Fla., Nov. 20, 1927; s. Paul Augustus and Mary Amanda (Snead) H.; m. Judith Kay Eshelman, Sept. 6, 1958; children: Richard Paul, Thomas Bradford, Robert Greggson, Mary Amanda. BS in Civil Engring., Mich. State U., 1949. Registered profl. engr., Ohio, Mass., R.I., Conn., N.Y., N.J., Pa., Del., D.C., Md. Track supr. to div. engr. Pa. R.R. and successor co. Penn Cen. R.R., 1949-71; chief R.R. engr. to v.p. Parsons DeLeuw, Inc., Washington, 1971—; dep. gen. mgr. Northeast Corridor Rail Improvement Program, 1978-82; tech. advisor to financing banks Eurotunnel, London, 1989-90; mem. Mich. State U. Alumni Engring. Coun., 1968-72. Contbr. articles on transp. to profl. publs. Dist. chmn. Md. gubernatorial campaign, 1967. Lt. USNR, 1945-46, Civil Engr. Corps Res. USNR. Recipient Toulmin medal Am. Soc. Mil. Engrs., 1979; named Railroader of Mo., Progressive Railroads, 1979. Mem. Am. Ry. Engring. Assn., Transp. Rsch. Bd., Camp Hill Jr. C. of C. (pres. 1961-62), Masons, Phi Delta Theta. Republican. Presbyterian. Avocations: golf, racquetball, sailing, skiing, travel. Home: 15205 Hannans Way Rockville MD 20853-1815 also: 27 South Terr Chautauqua NY 14722 Office: Parsons DeLeuw Inc 1133 15th St NW Washington DC 20005-2710

HOWELL, TERRY ALLEN, agricultural engineer; b. Dallas, Sept. 7, 1947; s. Levi Lowe III and Lila Lee (Allen) H.; m. Mary Sue Parkerson, Feb. 22, 1969; children: Terry A. Jr., Lisa K. Dreibrodt, Michael S. BS, Tex. A&M U., 1969, MS, 1970, PhD, 1974. Rsch. asst. Tex. A&M U., College Station, 1969-70, rsch. assoc., 1971-74; asst. prof. N.Mex. State U., Las Cruces, 1975, Tex. A&M U., College Station, 1976-79; agr. engr. USDA ARS, Fresno, Calif., 1979-83, Bushland, Tex., 1983--. Co-author: Irrigation of Agricultural Crops, 1991, Design and Operation of Farm Irrigation Systems, 1980, Limitations to Effective Water Use in Crop Production, 1983, Modification of the Aerial Environment Crops, 1979; co-editor, co-author: Management of Farm Irrigation Systems, 1991. Tchr. Paramount Bapt. Ch., Amarillo, 1985—, deacon, 1987—; troop com. chmn. Boy Scouts Am., Amarillo, 1991-93. Fellow ASAE (chmn. soil and water divsn. 1987-88, Paper award 1972, 74, 80); mem. Am. Soc. Civil Engrs. (chmn. irrigation water requirements com. 1990-93, soil and water divsn. editor 1993—), Am. Soc. Agronomy, Soil Sci. Soc. Am., Coun. for Agr. Sci. and Tech. Office: USDA ARS PO Drawer 10 Bushland TX 79012

HOWELL, WILLIAM EVERETT, computer professional, consultant; b. Summit, N.J., Aug. 11, 1956; s. Everett Frederick and Mabel (Christensen) H. BS in Computer Sci., Rensselaer Poly. Inst., 1983. Mgr. data systems Mech. Tech., Inc., Latham, N.Y., 1977-81; mgr. maj. account software Digital Equipment Corp., Big Flats, N.Y., 1981-83; dir. computer svcs. U. N.C., Chapel Hill, 1983—; adv. bd. Claris Corp., Santa Clara, Calif., 1991—, Am. Heart Assn., Chapel Hill, N.C., 1992—; cons. Duke U., Durham, N.C., 1992, U. N.C. Sch. Medicine, Chapel Hill, 1992. Contbr. papers, articles to tech. publs. Mem. Help Desk Inst. (pres.). Office: U NC Computer Sci Chapel Hill NC 27599-3175

HOWER, PHILIP LELAND, semiconductor device engineer; b. Reading, Pa., Apr. 9, 1934; s. Frank B. and Gladys (Fox) H.; m. Suzanne Mulvey, Apr. 28, 1962; children: Benjamin L., Suzanne E. BSEE, Lehigh U., 1956; MSEE, U. So. Calif., 1958; PhDEE, Stanford U., 1967. Mem. tech. staff Fairchild R&D, Palo Alto, Calif., 1966-71; adv. engr. Westinghouse R&D, Pitts., 1971-81; prin. scientist Unitrode Corp., Watertown, Mass., 1981-92; prin. engr. Unitrode Integrated Cirs., Merrimack, N.H., 1992—. Contbr. 35 articles to profl. jours. Fellow IEEE; mem. IEEE Power Electronics Soc. (William E. Newell award 1986). Achievements include 6 patents in the field of semiconductor device design. Home: 315 Border Rd Concord MA 01742 Office: Unitrode Integrated Cirs 7 Continental Blvd Merrimack NH 03054-0399

HOWERTON, ROBERT MELTON, telecommunications engineer; b. Greensboro, N.C., Aug. 8, 1957; s. Zachariah Hampton and Ruth Gandy (Melton) W. BA in History, Wake Forest U., 1979; postgrad., Campbell U., 1979-82; BSEE, N.C. State U., 1990. Devel. engr. IBM, Research Triangle Park, N.C., 1985-90; telecommunications planner Wachovia Corp., Winston-Salem, 1990-92; systems engr. Microdyne Corp., Alexandria, Va., 1992—. Mem. Tau Beta Pi, Eta Kappa Nu, Phi Alpha Theta. Episcopalian. Home: 2208 Farrington Ave Alexandria VA 22303 Office: Microdyne Corp 3601 Eisenhower Ave Alexandria VA 22304

HOWES, TREVOR DENIS, metallurgical engineering educator, researcher. BSME, Univ. Coll., Cardiff, Wales, 1960; PhD in Mech. Engring., U. Bristol, Eng., 1963; diploma, Manchester Bus. Sch., U. Manchester, Eng., 1968. Aeronautical engring. apprentice Bristol Aero-Engines, Ltd., U.K., 1952-57; tech. officer Imperial Chem. Industries Ltd., Severside, U.K., 1963-65; project engr., then mgr. engring. divsn. Strachan and Henshaw Ltd.-Dickinson Robinson Group, U.K., 1965-71; rsch. fellow U. Bristol, 1971-85, mng. dir. Inst. Grinding Tech., 1985-88, dir. mfg. group, 1985-88; prof. metallurgy U. Conn., Storrs, 1988—, dir. Ctr. for Grinding Rsch. & Devel., 1988—, dir. Advanced Tech. Ctr. for Precision Mfg., 1990—. Author (with C. Andrew & T.R.A. Pearce) Creep-Feed Grinding, 1985; contbr. articles to profl. jours. Fellow Royal Soc. Arts; mem. ASME, Japan Soc. Precision Engring., Conn. Acad. Sci. & Engring., Am. Soc. for Precision Engring., Abrasive Engring. Soc. (bd. dirs.), Soc. Mfg. Engrs. (sr. mem.), CIRP-Internat. Instn. for Prodn. Engring. Rsch. (corr. mem.), Romania Societatea Inginerilor Mechanici, Inst. Material Sci.-U. Conn., Instn. Mech. Engrs. Office: University of Connecticut Longley Bldg U-119 Middle Turnpike Rt 44 Storrs Mansfield CT 06269

HOWITT, ANDREW WILSON, scientist; b. Boston, June 20, 1960; s. Andrew James and Jane (Walsh) H. BSEE, Notre Dame, 1983; MS, MIT, 1987. Engr. Digital Equipment Corp., Maynard, Mass., 1983-85; scientist Sensimetrics Corp., Cambridge, Mass., 1991—. Instr. Taoist Tai Chi Soc. of USA. Office: Sensimetrics 64 Sidney St Ste 100 Cambridge MA 02139

HOWLAND, PAUL, industrial engineer; b. Detroit, July 1, 1948; s. Oren E. and Virginia (Chamberlain) H.; m. Sheryl Ann Henson, Feb. 1, 1976; children: Brian Scott, Jonathan Robert, Laura Marie. AA in Electronics, Cerritos Coll., 1978; BS in Indsl. Tech., Calif. State U., Long Beach, 1983. Steam engr., instrument mechanics in power plant div. L.A. County, L.A., 1969-77; instrument shop supr. Lever Brothers Co., L.A., 1977-81; engring. supr. Varec div. Emerson Electric, Cypress, Calif., 1981-84; project mgr. Hewlett Packard, Fullerton, Calif., 1984-87; energy systems mgr. Kendall McGaw Labs., Irvine, Calif., 1987-89; ctrl. plant mgr. Univ. Calif., Irvine, 1989—. With USN, 1965-69. Mem. Assn. Energy Engrs. Home: 5859 Eberle St Lakewood CA 90713 Office: U Calif Central Plant Irvine CA 92717

HOWLING, ALAN ARTHUR, physicist, researcher; b. Harrogate, Yorkshire, Eng., Oct. 26, 1958; came to Switzerland, 1986.; s. Arthur Howling and Shirley Dickinson Grunnill. BA with honors, Pembroke Coll., Oxford U., 1981; MSc, Wolfson Coll., Oxford U., 1982, PhD, 1985. Rsch. physicist Centre de Recherches en Physique des Plasmas Ecole Polytechnique Federale, Lausanne, Switzerland, 1986—. Contbr. articles to profl. jours. Pembroke open scholar, 1979. Home: Ch de Pierrefleur 36, Lausanne Switzerland 1004 Office: Centre de Recherches, en Physique des Plasmas, Ave des Bains 21, Lausanne Switzerland CH1007

HOY, MARJORIE ANN, entomology educator, researcher; b. Kansas City, Kans., May 19, 1941; d. Dayton J. and Marjorie Jean (Acker) Wolf; m. James B. Hoy; 1 child, Benjamin Lee. A.B., U. Kans., 1963; M.S., U. Calif.-Berkeley, 1966, Ph.D., 1972. Asst. entomologist Conn. Agrl. Expt. Sta., New Haven, 1973-75; research entomologist U.S. Forest Service, Hamden, Conn., 1975-76; asst. prof. entomology U. Calif., Berkeley, 1976-80, assoc. prof. entomology, 1980-82, prof. entomology, 1982-92, prof. emeritus, 1992—; Fischer, Davies and Eckes prof., dept. entomology and nematology U. Fla., Gainesville, 1992—; chairperson Calif. Gypsy Moth Sci. Adv. Panel, 1982—; mem. genetics resources adv. com. USDA, 1992—. Editor or co-editor: Genetics in Relation to Insect Management, 1979, Recent Advances in Knowledge of the Phytoseiidae, 1982, Biological Control of Pests by Mites, 1983, Biological Control in Agricultural IPM Systems, 1985; contbr. numerous articles to profl. jours. NSF fellow U. Calif.-Berkeley, 1963-64. Fellow AAAS, Royal Entomol. Soc. London; mem. Entomol. Soc. Am. (mem. Pacific br. governing bd. 1985, Bussart award 1986, Founder's Meml. award 1992), Am. Genetic Assn., Internat. Orgn. Biol. Control (v.p. 1984-85), Am. Inst. Biol. Scis., Acarological Soc. Am. (governing bd. 1980-84, pres. 1992), Soc. for Study of Evolution, Phi Beta Kappa, Sigma Xi (chpt. sec. 1979-81). Avocations: hiking, gardening, snorkeling. Home: 4320 SW 83d Way Gainesville FL 32608 Office: U Fla Dept Entomology/Nematology Bldg 970 Hull Rd Gainesville FL 32611

HOYE, ROBERT EARL, urban policy educator, health care management consultant; b. Warwick, R.I., Jan. 12, 1931; s. S. Earl and Alice (Landry) H.; m. Patricia Buswell, Aug. 20, 1955; children: Robert Earl Jr., Joanne D., Peter M., Kathleen B. BA, Providence Coll., 1953; MS, St. John's U., N.Y.C., 1955; PhD, U. Wis., Madison, 1973. Instr. St. John's U., 1953-55; dir. guidance Middleboro (Mass.) Pub. Schs., 1955-56, Rutland (Vt.) Pub. Schs., 1956-57; dean Champlain (Vt.) Coll., 1957-58; supt. Frontier Regional Sch. Dist., Deerfield, Mass., 1958-60; New Eng. dir. Sci. Rsch. Assocs. subs. IBM, Chgo., 1960-65; nat. dir. Learning Systems div. Xerox Corp., N.Y.C., 1965-66; dir. Instrnl. Media Lab. U. Wis., Milw., 1966-73; asst. v.p. U. Louisville, 1974-81, prof. urban policy, coord. grad. program in health systems, 1981—, prof. edn., 1992—; cons. to mgmt., Louisville, 1966—. Author: Index to Computer Based Learning, 1973; editor Edn. Jour., 1968-73; also articles. Recipient cert. of merit San Diego State U., 1983, Grad. Teaching Excellence award U. Louisville, 1984, gold medal Project Innovation, 1984. Fellow Am. Acad. Med. Adminstrs. (diplomate, chmn. editorial bd. 1986—), Royal Soc. Health (Statesman in Healthcare Adminstrn. award 1992). Democrat. Roman Catholic. Home: 2238 Wynnewood Cir Louisville KY 40222-6342 Office: U Louisville Louisville KY 40292

HOYT, STANLEY CHARLES, retired research administrator, entomologist; b. Oakland, Calif., Oct. 4, 1929; s. Fred C. and Elphie V. (Stohl) H.; m. Beverley J. Carpenter, Aug. 6, 1955; children: Kathleen, Kristine, David. BS, U. Calif., Berkeley, 1951, PhD, 1958. Asst. entomologist Wash. State U., Wenatchee, 1957-63, assoc. entomologist, 1963-70, entomologist, 1970—, supt., 1983-93. Editor: IPM of Insect Pests of Pome and Stone Fruits, 1983; contbr. 131 publs. to profl. jours. Bd. dirs. Wenatchee Valley Col. found., 1989—, Community Resource Ctr., Wenatchee, 1979-90, Sta. KFAE Pub. Radio, Tri Cities, 1990—, United Way, Wenatchee, 1984-88. Recipient Fulbright award, New Zealand, 1970. Mem. Entomol. Soc. Am. (pres. Pacific br. 1982-83, CIBA-GEIGY Recognition award 1973, C.W. Woodworth award Pacific br. 1989), Rotary. Avocations: photography, flute playing, gardening, fishing. Office: Washington State U Tree Fruit Rsch Ctr 1100 Western Ave N Wenatchee WA 98801-1299

HRANITZKY, E. BURNELL, medical physicist; b. Yoakum, Tex., Dec. 11, 1941; s. E. Leo and Lucy Leona (Berkofsky) H. MS, U. Tex., 1969. Diplomate Am. Bd. Radiology, Am. Bd. Med. Physics. Physicist M.D. Anderson Hosp./U. Tex., Houston, 1963-76; pres. Radiation Assocs., Inc., Houston, 1976—; cons. physicist Med. & Radiation Physics, Inc., San Antonio, 1988-92. Contbr. articles to profl. publs. Office: Radiation Assocs Box 2 Barker TX 77073

HRIBAR, LAWRENCE JOSEPH, entomologist, researcher; b. Rochester, Pa., Nov. 22, 1960; s. Larry C. and Frances E. Hribar. BS, Pa. State U., 1982; MS, U. Tenn., 1984; PhD, Auburn U., 1989. Rsch. asst. Pa. State U., State College, 1982; grad. rsch. asst. U. Tenn., Knoxville, 1982-84; grad. rsch. asst. Auburn (Ala.) U., 1985-89, lab. technician, 1989-90; postdoctoral assoc. La. State U., Baton Rouge, 1990-91, Fla. Med. Entomology Lab., Vero Beach, 1991—. Mem. Entomol. Soc. Am., Am. Arachnol. Soc., N.Am. Benthol. Soc., Soc. Vector Ecology, Am. Mosquito Control Assn. Office: Fla Med Entomology Lab 200 9th St SE Vero Beach FL 32962

HRUDEY, STEVE E., civil engineer, educator. Prof. civil engring. U. Alta., Edmonton, Can. Recipient Albert E. Berry medal Canadian Soc. Civil Engring., 1990. Office: U Alta, Dept Civil Engring, Edmonton, AB Canada T6G 2M7*

HRYCAK, PETER, mechanical engineer, educator; b. Przemysl, Poland, July 8, 1923; came to U.S., 1949, naturalized, 1956; s. Eugene and Ludmyla (Dobrzanska) H.; m. Rea Meta Limberg, June 13, 1949; children: Maria (dec.), Michael Paul, Orest W.T., Alexandra Martha. Student, U. Tubingen, Germany, 1946-48; B.S. with high distinction, U. Minn., 1954, M.S., 1955, Ph.D., 1960. Registered profl. engr., N.J. Adminstrv. asst. French Mil. Govt. in Germany, 1947-49; instr. mech. engring. U. Minn., Mpls., 1955-60; mem. tech. staff Bell Telephone Labs., Murray Hill, N.J., 1960-65; sr. project engr. Curtiss-Wright Corp., Woodridge, N.J., 1965; assoc. prof. mech. engring. N.J. Inst. Tech., 1965-68, prof., 1968-93, prof. emeritus, 1993—, dir. jet rsch. lab., 1966-93; Participant in Internat. Conf. on Engring. and Applied Sci. Contbr. articles to profl. jours.; one of original Telstar designers. Bd. dirs. Ukrainian Congress Com. Am., Mpls., 1956-60, Plast Camp, East Chatham, N.Y., 1963-68; v.p. Ukrainian Music Found., 1977-93. NASA grantee, 1967-68; NSF grantee, 1982-84. Mem. ASME, AIAA, AAUP, Inst. Environ. Scis. (sr.), Am. Soc. Engring. Edn., Ukrainian Engrs. Soc. Am. (pres. 1966-67), Am. Geophys. Union, Shevchenko Sci. Soc., Ukrainian Acad. Arts and Scis. in U.S.A., Am. Chem. Soc., Sigma Xi, Pi Tau Sigma, Tau Beta Pi. Home: 19 Roselle Ave Cranford NJ 07016-2532 Office: NJ Inst Tech 323 Martin Luther King Blvd Newark NJ 07102

HRYCIW, ROMAN D., civil engineering educator; b. Phila., Sept. 23, 1958; s. Theodosij and Lucia (Stojkewycz) H.; m. Olena M. Prasicky, Dec. 26, 1981; children: Dmytri, Demyan. BS, Drexel U., 1981; MS, Northwestern U., 1984, PhD, 1986. Engr. in tng. Environ. technician Roy F. Weston Cons., West Chester, Pa., 1977-78; geotech. technician U.S. Army Engrs., Phila., 1978-79; geotech. lab. technician Woodward-Clyde Cons., Plymouth Meeting, Pa., 1980; rsch. asst. Northwestern U., Evanston, Ill., 1981-86; asst. prof. U. Mich., Ann Arbor, 1986-92, assoc. prof., 1992—; bd. dirs. U.S. Univs. Coun. Geotech. Engring.; cons. Egyptian Antiquities Orgn., Cairo, 1992, Dow Chem., U.S. Bur. Reclamation, Pitts. & Midway Coal Co., Martin Marietta, others. Contbr. over 30 articles to jours. Scoutmaster, bd. dirs. Plast Scouting Orgn., 1973-92, internat. bd. dirs., 1988-90. Fellow Woodward-Clyde, 1981, Water P. Murphy fellow Northwestern U., 1981. Mem. ASCE (Arthur Casagrande Profl. Devel. award 1993), ASTM, Internat. Soc. Soil Mechs. and Found. Engrs., Chi Epsilon (James M. Robbins Excellence in Teaching award 1990). Achievements include rsch. in soil dynamics and earthquake engring., soil improvement, soil testing and in-site investigation, soil mechs. and slope stabilization. Home: 1118 Ferdon Rd Ann Arbor MI 48104 Office: U Mich Dept Civil Environ Engring 2366 GG Brown Bldg Ann Arbor MI 48109-2125

HSIAO, FEI-BIN, aerospace engineering educator; b. Taichung, Taiwan, Republic of China, Feb. 10, 1953; s. For-Ru and Sho-Yuan (Tu) H.; m. Huey-Mei Tseng, June 10, 1980; children: Edwin Yih-Wei, Ellen Yih-Lynn, Ernie Yih-Hao. MS, Nat. Tsing-Hua U., Hsinchu, Republic of China, 1979; PhD, U. So. Calif., L.A., 1985. Assoc. prof. Nat. Cheng Kung U., Tainan, Republic of China, 1985-90, prof., 1990—; editorial cons. Radio Control World Mag., Taipei, Republic of China, 1988—. Contbr. articles to profl. jours. including AIAA, Jour. of Aircraft, Physics of Fluids, Experiments in Fluids, Jour. Fluids and Structures. Mem. AIAA (M. Barry Carlton award 1991), Assn. Unmanned Vehicle Systems, Chinese Soc. Mech. Engrs., Assn. Chinese Engrs., Aero. and Astronautical Soc. (Republic of China, membership com. 1989-90, conf. organizer 1991-92). Achievements include research in low Reynolds number aerodynamics, control of flow separation and mixing enhancement, windtunnel design and testing, remotely piloted vehicle research, satellite communication/design, flight vehicle stability/design. Home: 8F-1 88 Chang Rong Rd Sec 3, Tainan 70101, Taiwan Office: Nat Cheng Kung U, 1 Ta Hsieh Rd, Tainan 70101, Taiwan

HSIAO, MING-YUAN, nuclear engineer, researcher; b. Kaohsiung, Taiwan, Feb. 23, 1954; came to U.S., 1978; s. Fei and Hwang-Fang H.; m. Shwu Chuen Lee. MS, U. Ill., 1980, PhD, 1983. Postdoctoral fellow Los Alamos (N.Mex.) Nat. Lab., 1983-84; asst. prof. Pa. State U., University Park, 1984-90; sr. engr. Commonwealth Edison Co., Chgo., 1990—; Author, reviewer jour. publs. Univ. fellow U. Ill., 1981-83, Profl. Devel. fellow U.S. Dept. Energy, 1987; recipient Faculty Summer Rsch. award Argonne (Ill.) Nat Lab., 1986-89. Mem. Am. Nuclear Soc. (Mark Mills award 1984), Am. Phys. Soc., Sigma Xi, Phi Tau Phi. Office: Commonwealth Edison Co 125 S Clark St Ste 900 Chicago IL 60603-5104

HSIAO, WILLIAM C., economist, actuary educator; b. Beijing, China, Jan. 17, 1936; came to U.S., 1948; children: Roderick, Douglas. BA Math, Physics, Ohio Wesleyan U., 1959; MPA, Harvard U., 1972, MA Econs., 1974, PhD, 1982. Fin. dir., actuary Conn. Gen. Life Ins., 1959-68; dep. chief actuary U.S. Social Secuity Adminstrn., Washington, 1968-71; mem. faculty Harvard Bus. Sch., Cambridge, Mass., 1974-77; assoc. prof. Harvard U., Cambridge, Mass., 1975-84, prof., 1985—; cons. in field to Com. Fin., Com. Aging, U.S. Congress, Washington, 1974—, World Bank, 1983—, Internat. Labor Orgn., 1991—, World Health Orgn., 1989—, State of N.Y., 1984-86, Gov. State of Vt., 1974-77, State of N.H., 1976-79, United Meth. Nat. Bd. Pensions, 1991—, Commonwealth Sch., Boston, 1983-91, Home Savs. Bank, Boston, 1981-89; hon. prof. Beijing Med. U., People's Republic China, 1985, Shanghai Med. U., People's Republic China, 1985; referee Am. Econ. Rev., Inquiry, New Eng. Jour. Medicine, Jour. Human Res. Contbr. articles to profl. jours. Named Man of Yr. McGraw Hill Publs., 1989. Fellow Actuarial Sci.; mem. NAS (Inst. Medicine), Soc. Actuaries (bd. dirs. 1991—), Nat. Acad. Social Ins. (founding). Office: Harvard Univ Holyoke 726 1350 Massachusetts Ave Cambridge MA 02138

HSU, BERTRAND DAHUNG, mechanical engineer; b. Anwhei, People's Republic of China, Feb. 25, 1933; came to U.S., 1979; s. Edward Yuan Hsu and Pearl (Soong) Wei; m. Jane Juifang Hsiang, Jan. 15, 1955; children: Ning, Ann. BSME, Tsinghua U., Peking, People's Republic of China, 1953, PhD (equiv.), 1966. Asst. and lectr. Tsinghua U., 1953-71, prof. and deputy dir. dept. automobile engring., 1971-78; project engr. GE Transportation Systems, Erie, Pa., 1979-88, tech. leader, 1988-90, mgr. coal fueled diesel program, 1990—. Contbr. articles to Jour. ASME, Internat. Congress on Combustion Engines, Soc. Automotive Engrs. Jour., Jour. Mech. Engring. Inst. Recipient speakers award Internal Combustion Engine div. ASME, 1988, Outstanding Tech. Contbn. Engring. award GE Power Sector, 1984. Mem. Soc. Automotive Engrs., The Combustion Inst. Achievements include 1 Canadian and 2 U.S. patents for Fuel Injector System; development of first 1000 RPM coal water slurry fueled diesel engine combustion systems. Office: Transp Systems div GE 2901 East Ave Bldg 143 Erie PA 16531-0001

HSU, CHENG, decision sciences and engineering systems educator; b. Taipei, Republic of China, May 11, 1951; came to U.S., 1976; s. Chung-Yu and Te-Zeng (Yeh) H.; m. Ihsin Lydia Wu, Oct. 24, 1979; 1 child, Diana. BS in Indsl. Engring., Tunghai U., Taichung, Republic of China, 1973; MS, Ohio State U., 1978, PhD, 1983. Info. engr. China Tech. Cons., Inc., Taipei, 1975-76; grad. rsch. asst. Ohio State U., Columbus, 1977-80, grad. teaching assoc., 1980-82; asst. prof. decision scis. and engring. systems Rensselaer Poly. Inst., Troy, N.Y., 1982-88, assoc. prof., 1988—, dir. undergrad. programs, 1989-91; cons. Coopers and Lybrand, Albany, N.Y., 1988, Digital Equipment Corp., Nashua, N.H., 1991. Grantee GM, DEC, Johnson & Johnson, 1986-89, Aluminum Co. of Am., Digital Equipment Corp., 1992—, GE, GM and IBM, 1986-92, 92—, AT&T, 1987, NATO, 1988, State of N.Y., 1988, NSF, 1991—. Mem. IEEE, ACM, Inst. Mgmt. Sci., Ops. Rsch. Soc. Am., Soc. Mfg. Engrs., Prodn. and Ops. Mgmt. Soc., N. Am. Chinese Bus. Educators Assn. (bd. dirs. 1988-90). Republican. Home: 5 Christine Ct Latham NY 12110-3734 Office: Rensselaer Poly Inst 5219 CII Troy NY 12180-3590

HSU, CHENG-TZU THOMAS, civil engineering educator; b. I-Lan, Taiwan, Republic of China, Dec. 18, 1941; came to U.S., 1969; s. Moun-Chun Chung and A-Mei Hsu; m. Ursula Cruz Trivino; children: Anthony Trivino, Jennifer M.T. BSE, Cheng-Kung U., Tainan, Taiwan, 1964; MSc, Coll. Chinese Culture, Taipei, Taiwan, 1967, Yale U., 1972; MEngring., PhD, McGill U., Montreal, Que., Can., 1969, 74. Registered profl. engr., Taiwan, Que., Ont. Bridge design engr. Beauchemin-Beaton-Lapointe, Inc., Montreal, 1974; civil engring. specialist Bechtel and Co., Montreal and Toronto, 1974-78; asst. prof. N.J. Inst. Tech., Newark, 1978-83, assoc. prof., 1983-86, prof. civil engring., 1986—, assoc. chmn. grad. studies, 1988—; cons. engr. Paulus, Sokolowski & Sartor, Inc., Warren, N.J., 1980-81; cons. Shih Engring. Co., Fairfield, N.J., 1982, Samuel M. Ruth & Assocs., Newark, 1984; hon. prof. Wuhan (China) U. Tech., 1992. Editor: Advances in Structural Concrete Design, 1983; contbr. over 80 articles to profl. jours. 2nd lt. U.S. Army, 1964-65. Yale U. fellow, 1969-71, miron fellow McGill U., Montreal, 1971-73. Mem. ASCE (Raymond C. Reese rsch. prize 1987), Am. Concrete Inst. (com. mem. 1979—), Sigma Xi, Chi Epsilon, Tau Beta Pi. Roman Catholic. Achievements include research in limit analysis-RC beams and frames, moment-curvature for reinforced concrete, load-deformation of RC columns, load-deformation of steel fiber RC beams. Office: NJ Inst Tech Dept Civil & Environ Engring 323 King Blvd Newark NJ 07102

HSU, CHIEH SU, applied mechanics engineering educator, researcher; b. Soochow, Kiangsu, China, May 27, 1922; came to U.S., 1947.; s. Chung yu and Yong Feng (Wu) H.; m. Helen Yung-Feng Tse, Mar. 28, 1953; children—Raymond Hwa-Chi, Katherine Hwa-Ling. BS, Nat. Inst. Tech., Chungking, China, 1945; MS, Stanford U., 1948, Ph.D., 1950. Project engr. IBM Corp., Poughkeepsie, N.Y., 1951-55; assoc. prof. U. Toledo, 1955-58; assoc. prof. Univ Calif.-Berkeley, 1958-64, prof., 1964—, chmn. div. applied mechanics, 1969-70; mem. sci. adv. bd. Alexander von Humboldt Found. of Fed. Republic Germany, Bonn, 1985—. Author: 98 tech. papers; contbg. author: Thin-Shell Structures, 1974, Advances in Applied Mechanics, vol. 17, 1977; author: Cell-to-Cell Mapping, 1987; tech. editor Jour. Applied Mechanics, N.Y.C., 1976-82; assoc. editor profl. jours. Recipient Alexander von Humboldt award Fed. Republic Germany, 1986; Guggenheim Found. fellow, 1964-65; Miller research prof., U. Calif.-Berkeley, 1973-74. Fellow ASME (Centennial award 1980) Am. Acad. Mechanics; mem. Acoustical Soc. Am., Soc. Indsl. and Applied Math., U. S. Nat. Acad. Engring., Acad. Sinica, Sigma Xi. Office: U Calif Dept Mech Engring Berkeley CA 94720

HSU, CHIEN-YEH, electrical engineer, speech and hearing scientist; b. Nan-Tou, Taiwan, China, Mar. 10, 1963; came to U.S. 1987; s. Hsi-Hsueh and Li-Yu (Wu) H.; m. Chen-Jui Chao, Jan. 23, 1991. MS in Elec. Engring., Ohio State U., 1989, PhD in Speech and Hearing Sci., 1993. Grad. rsch. assoc. Nisonger Ctr., Ohio State U., Columbus, 1988-89; grad. rsch. assoc. dept. elec. engring. Ohio State U., Columbus, 1989, grad. rsch. assoc. div. speech and hearing sci., 1989-93; rsch. assoc. dept. Speech and Hearing Sci. U. Ill., Urbana-Champaign, 1993—. Lt. Chinese Army, 1985-87. Recipient Excellent Acad. Performance Nat. Cheng-kung U., 1982, 84; Mrs. Hu, yamchou scholar, Ministry of Interior, 1984; recipient Medal of Brilliant Reputation Ministry of Nat. Def., Taiwan, 1987, Graduate Student Alumni Rsch. award Ohio State U., 1992. Mem. IEEE (student mem.), Acoustical

Soc. Am. (student mem.). Achievements include development of research laboratory for testing the cognitive and motor performance of children with learning and behavior problems; development of a computer controlled signal generating and processing system for psychoacoustics experiments. Home: Apt Z4 2403 W Springfield Ave Champaign IL 61821 Office: U Ill Urbana Champaign 901 S 6th St Champaign IL 61820

HSU, CHUNG YI, neurologist; b. Taipei, Taiwan, China, Oct. 14, 1944; s. Huo and Jane (Wu) H.; m. Amy Yang, Sept. 27, 1974; children: Alice L., Virginia, Charles Y. MD, Nat. Taiwan U., Taipei, Republic of China, 1970; PhD, U. Va., 1975. Diplomate Am. Bd. Psychiatry and Neurology. NIH fellow Diabetes Rsch. Ctr., U. Va., Charlottesville, 1975-77; fellow dept. pharmacology Med. U. S.C., Charleston, 1977, intern dept. medicine, 1977-78, resident dept. neurology, 1978-80, chief resident dept. neurology, 1980-81, fellow clin. neuropharmacology, 1981, dir. neuropharmacology dept. neurology, 1981-89; dir. neuropharmacology div. restorative neurology Baylor Coll. Medicine, Houston, 1989-93; head cerebrovascular disease sect., dept. neurology Washington U. Sch. Medicine, St. Louis, 1993—; mem. adv. panel on drug info. U.S. Pharmacopeial Conv., Rockville, Md., 1985-90; mem. CNS adv. panel Eastman-Kodak/Sterling, Malvern, Pa., 1988—; mem. study sect. NIH Ninds, 1988—. Editor N.Am. Taiwanese Profs.' Assn., 1992—; contbr. articles to profl. jours. Pres. Chinese Assn., Charleston, 1984-85. 2d lt. Republic of China Navy, 1970-71. Grad. fellow U. Va. Sch. Medicine, Charlottesville, 1971-75; recipient Nat. Rsch. Svc. award USPHS, 1977, 81, NIH Tchr. Investigator Devel. award, 1983-88, NIH Javits Neurosci. Investigator award, 1991—. Fellow Am. Acad. Neurology, Stroke Coun., Am. Heart Assn.; mem. Am. Neurol. Assn., World Congress Neurology (stroke rsch. group), Am. Fedn. Clin. Rsch. (sr.), Internat. Soc. Cerebral Blood Flow and Metabolism, So. Med. Assn., Neurotrauma Soc. (pres-elect 1991-92, pres. 1992—). Avocation: literature. Home: 538 Conway Village Saint Louis MO 63141 Office: Washington U Sch Medicine Dept Neurology Box 8111 660 S Euclid Ave Saint Louis MO 63110

HSU, JULIE MAN-CHING, pediatric pulmonologist; b. Shanghai, China, May 16, 1933; came to U.S., 1959; MD, Nat. Taiwan U. Coll. Medicine, 1958. Diplomate Am. Acad. Pediatrics. Med. staff pulmonary medicine Children's Hosp. of Mich., Detroit, 1982—, dir. Cystic Fibrosis Ctr., 1989—, dir. pulmonary medicine, 1990—. Recipient Cystic Fibrosis Care award United Way S.E. Mich., 1989, 90, 91, 92, 93. Fellow Am. Acad. Pediatrics; mem. Am. Thoracic Soc. Office: Childrens Hosp of Mich Childs Hosp if Michigan 3901 Beaubien Detroit MI 48201

HSU, MING-YU, engineer, educator; b. Kweiyang, Kweichow, China, Dec. 4, 1925; s. Pei-Kung and Wan-Ju (Hsiao) H.; m. Chih-Ju Yao, Jan. 1, 1952; children: Chi-Hsing, Chi-Yun, Chi-En, Chi-Che, Chi-Cheng. BE, Nat. Kweichow U., 1948; Dipl.Engr., Delft Tech. U., The Netherlands, 1959. Registered profl. engr., Ill., Ga., Fla., S.C. Prof. Cheng-Kung U., Tainan, Taiwan, 1960-68; dir. Land Devel. Commn., Taipei, 1960-68; engring. cons. Ministry of Housing & Utilities, Sehba, Libya, 1968-71; sr. engr. Philipp Holzmann Ag., Hamburg, Fed. Republic of Germany, 1971-74, Weber, Griffith & Mellican, Galesburg, Ill., 1974-80; chief engr. Chatham Engring. Co., Savannah, Ga., 1980-82; sr. cons. Hussey, Gay, Bell & DeYoung, Inc., Savannah, 1982—; prof. Savannah Coll. of Art and Design, 1986—; designed and constructed numerous indsl. office, apt. and comml. bldgs., marine structures including docks, loading platforms, marinas, shipyards and water and waste water treatment structures. Contbr. articles on structural engring. to profl. jours. Mem. Nat. Soc. Profl. Engrs., ASCE. Home: 1115 Wilmington Island Rd Savannah GA 31410-4508 Office: Hussey Gay Bell & DeYoung 329 Commercial Dr Savannah GA 31406-3617

HSU, PETER CHEAZONE, chemical engineer; b. Changhua, Taiwan, China, June 3, 1951; came to U.S. 1981; s. Chinann and Doh (Yeh) H.; m. Carol S. Lin, Jan. 2, 1979; children: Victoria, Eleanor. BS, Nat. Cheng Kung U., 1973, MS, 1976; PhD, Mich. State U., 1988. Rsch. assoc. Mich. State U., East Lansing, 1988; rsch. engr. Dept. Energy, U.S. Govt., Morgantown, W.Va., 1989-90; process engr. The Pritchard Corp., Overland Park, Kans., 1990—; cons. Amoco Oil Co., Naperville, Ill., 1989. Contbr. articles to profl. jours. Mem. Am. Inst. Chem. Engrs. Achievements include 1 patent. Office: The Pritchard Corp 10950 Grand View Dr Overland Park KS 66210

HSU, ZUEY-SHIN, physiology educator; b. Shining, Taiwan, Republic of China, Dec. 13, 1930; s. Kua and Mun Mei (Kuo) H.; m. Pan Tsu Wu, Feb. 1, 1964; 1 child, Sheng Chin. M.D., Nat. Taiwan U., 1956. Intern in internal medicine Nat. Taiwan U., 1956-57; asst. Kaohsiung Med. Coll., Kaohsiung City, Republic of China, 1957-59, instr., 1959-62, assoc. prof. legal medicine, 1962-68, assoc. prof. physiology, 1968-72, prof. physiology, 1972—, acting dir. dept. pharmacology, 1972-73, dir. dept. pharmacology, 1973-74, dir. dept. physiology, 1972-85. Inventor method of detoxicating heterologous blood for transfusion, 1978; a new Immunological method for desensitizing allergic individuals, 1981, preparation of tumor vaccine 1987. Nat. Sci. Council of Taipei grantee, 1967. Fellow Inst. Med. Sci., Tokyo U., Internat. Biographical Assn. (life); mem. Internat. Parliament for Safety and Peace, Maison Internationale des Intellectuels and Academi Midi, Formosan Med. Assn., Chinese Physiol. Soc., Chinese Soc. Immunology, Endocrine Soc. of Republic of China, Chinese Pharmacological Soc. Home: 24 157 Ln Fu-Herng 1st Rd, Kaohsiung Taiwan Office: Kaohsiung Med Coll, 100 Shih-Chuan 1st Rd, Kaohsiung 80708, Taiwan

HU, SENQI, psychologist, educator; b. Shanghai, People's Republic of China, Sept. 4, 1952; came to U.S., 1986; s. Suiyu Yao and Xinyang Hu; m. Jin-Di Wang, Dec. 27, 1979; 1 child, Zhong-Min. MD, Shanghai Coll. Chinese Med., 1977; PhD, Pa. State U., 1990. Resident surgeon Shanghai First Hosp., 1977-78; asst. rsch. fellow Shanghai Inst. Chinese Medicine, 1980-85; adj. asst. prof. Pa. State U., State College, 1986; asst. prof. psychology Humboldt State U., Arcata, Calif., 1990—. Contbr. articles to profl. publs. Mem. Am. Psychol. Soc., Soc. for Psychophysiol. Rsch. Office: Humboldt State U Dept Psychology Arcata CA 95521

HU, STEVE SENG-CHIU, scientific research company executive, academic administrator; b. Yangchou City, Kiangsu Province, Peoples Republic of China, Mar. 16, 1922; s. Yubin and Shuchang (Lee) H.; m. Lily Li-Wan Liu, Oct. 2, 1977; children: April, Yendo, Victor. MS, Rensselaer Poly. Inst., 1940; PhD, MIT, 1942; postgrad., UCLA, 1964-66. Mng. tech. dir. China Aircraft/China Motor Programs, Douglas Aircraft Co., Calif. and N.J., 1943-48, Kelly Engring Co., N.Y. and Ariz., 1949-54; systems engr., meteorol. sci. dir. R.C.A., Ariz., 1955-58; rsch. specialist Aerojet Gen., Calif., 1958-60; rsch. scientist Jet Propulsion Lab., Calif., 1960-61; mng. tech. dir. Huntsville div. Northrop Corp., Calif. and Ala., 1961-72; pres. Century Rsch., Inc.; bd. dirs. Am. Tech. Coll., pres., U. Am. United Rsch. Inst., Gardena, San Bernardino, Calif., 1973—; pres. U. Am. Found. and U. Am. Rsch. Found., Calif. and Taiwan, Republic of China, 1981—; bd. dirs., exec. v.p. Am. Astronautical Soc., Wash., 1963-70; cons. Hsin-Hwa Nuclear Reactor Program, Taiwan, 1954-58; prof. Auburn (Ala.) U., U. Ala., U. Ariz., U. So. Calif., L.A., 1957-73. Fellow Calif. Inst. Tech., 1943-44. Recipient Cert. of Merit and Cash award Commn. Aeronautical Affairs, Republic of China, 1945. Mem. Am. Astronautical Soc., AIAA, Nat. Assn. Tech. Schs. Office: Century Rsch Bldg Office Sect 16935 S Vermont Ave Gardena CA 90247

HU, TSAY-HSIN GILBERT, aerospace engineer; b. Taipei, Taiwan, Republic of China, Mar. 5, 1956; came to U.S., 1980; s. Chi-Hor and Han (Lin) H.; m. Mann-Shya Grace Lee, July 16, 1982; 1 child, Daniel Deng-Yuan. BSME, Nat. Taiwan U., Taipei, 1978; MEME, Stevens Inst. Tech., 1982; PhD, Rensselaer Poly. Inst., 1985. Sr. engr. AeroStructures, Inc., Arlington, Va., 1985-88; prin. engr. Lockheed Engring. & Sci. Corp., Houston, 1988-90; engring. specialist Grumman Space Sta. Integration Div., Reston, Va., 1990—; speaker various confs. Contbr. articles to profl. jours. Mem. Citizens Against Waste. Scholar Taiwan Metal & Mining Corp. Mem. AIAA, Rensselaer Poly. Inst. Alumni Assn., Stevens Inst. Tech. Alumni Assn., Chinese Profl. Club. Avocations: table tennis, swimming, reading. Home: 1042 Ware St SW Vienna VA 22180-6476 Office: Grumman Space Sta Integration Div 1760 Business Center Dr Reston VA 22090-5318

HU, XIMING, engineering educator; b. Shanghai, China, Dec. 25, 1946; s. Yaochen and Yejing (Cho) Y.; m. Cuiping Bai, Apr. 1, 1976; children: Zheng Hu. MS, Shanghai Maritime U., China, 1981; PhD, Iowa State U., 1992. Mechanical engr. Shanghai Factory, 1976-82; lectr. Shanghai Maritime U., 1982-88; rsch. and teaching asst. Iowa State U., Ames, 1988-92; vis. scientist Gallent Intelligent Tech., El Monte, Calif., 1992--. Contbr. articles to profl. jours. Iowa Power Co. grantee, 1988, Outdoor Techs. Group Co. grantee, 1992. Mem. Soc. for Experimental Mechanics, Sigam Xi, Phi Kappa Phi.

HUA, LULIN, technological company executive, research scientist. BS in Chemistry, U. Sci. and Tech. China, Beijing, MS in Chemistry, 1966. Asst. rsch. chemist Inst. Metal Chinese Acad. Sci., 1966-71, Astronomy Observatory Chinese Acad. Sci., 1971-76; rsch. chemist Hong-Xing Chem. Inst. China, 1976-82; vis. scholar Dept. Chemistry U. Md., 1982-84; vis. assoc. prof. Dept. Chemistry U. Ariz., 1984-85; vis. rsch. assoc. Dept. Chemistry Rensselaer Poly. Inst., 1985-86; postdoctoral fellow Dept. Chemistry Georgetown U., 1986-87; rsch. assoc. Dept. Chemistry U. Tenn., 1988-90; v.p. Micro Imaging Systems, Inc., Md., 1990—. Contbr. articles to profl. jours. Recipient Third award Hong-Xing Chem. Inst. 1980, Third award Hong-Xing Inst. 1979, Third award Shaanxi Astronomu Observatory 1972, Second award Nat. Sci. and Tech. Com. China. mem. Am. Chem. Soc., Analytical Chem. and Applied Spectroscopy Soc., Am. Assn. for the Advancement Sci., Interant. Soc. for the Origin Life. Office: PO Box 175 Silver Spring MD 20905

HUA, TONG-WEN, chemistry educator, researcher; b. Shanghai, China, Sept. 27, 1929; s. Bing-yuan and Yishan (Wu) H.; m. Yizhong Dong, May 7, 1955; children: Dong Mouqun, Dong Liqun. BS, Yanjing U., Beijing, 1951; MS, Peking U., Beijing, 1954. Teaching asst. chemistry dept. Yanjing U., Beijing, 1951-52; teaching asst. dept. chemistry Peking U., Beijing, 1954-56, lectr., 1956-60, assoc. prof., 1961-85, prof., 1986—; vis. scholar U. Conn., Storrs, 1982-83; dir. Div. Inorganic Chemistry Peking Univ., Beijing, 1956-85, Rsch. Ctr. Higher Chem. Edn., Beijing, 1986—. Author: (textbook) Principles of General Chemistry, 1989, 2d edit., 1993; translator: (textbook) Masterson's Chemical Principles, 1980, (reference book) Pimentel's Opportunities in Chemistry Today and Tomorrow, 1990; chief editor (jour.) Univ. Chemistry (Daxue Huaxue); contbr. articles to profl. jours. Recipient scholarship State Edn. Commn., 1989, Sci. and Tech. award State Petroleum Co., 1981. Mem. Internat. Union Pure and Applied Chemistry, Com. on Teaching of Chemistry (nat. rep.), Chinese Chem. Soc. (bd. dirs. 1986-90, standing com. 1991—). Home: Peking U Zhong Guan Yuan 41-208, Beijing 100871, China Office: Peking U, Dept Chemistry, Beijing 100871, China

HUANG, DENIS K., chemical engineer, consultant; b. Canton, China, May 14, 1925; came to U.S., 1948; s. Shui Fu and Wai Men Wong; married; 1 child, Lloyd K. BS in Math., St. John's U., 1944; BSChemE, U. Calif., Berkeley, 1950; MSChemE, U. Maine, Orono, 1951; DChemE, Poly. Inst. Bklyn., 1958. Head chemist Internat. Paper Co., Phila., 1958-62; sr. rsch. chemist Simoniz Co., Chgo., 1962-65; sr. rsch. engr. Westvaco Corp., Laurel, Md., 1965-78; process engring. cons. Fed. Paper Bd., Augusta, Ga., 1978-90; cons. Tech. Cons. Internat., Augusta, Ga., 1990—; cons. UNDP to China, 1983, OAS, 1972, Argentina UNIDO, 1970; tech. expert to India. Patentee in field; contbr. articles to encyclopaedia, profl. jours. Mem. TAPPI, Sigma Xi, Phi Lambda Upsilon. Avocations: body building, tennis. Home: 3641 Nassau Dr Augusta GA 30909-2645

HUANG, EDWIN I-CHUEN, physician, environmental researcher; b. Lin-Lin, Hunan, China, Sept. 10, 1933; came to U.S., 1970; s. Chu-Ouh and Wan-Lan (Chaing) H.; m. Hwei-Mei Lai, Apr. 4, 1963; children: David, Sherman, Jennifer. Student, Nat. Def. Med. Ctr., China, 1951-53; MD, Nat. Def. Med. Coll., Taipei, China, 1960; MPH, U. Tex., 1971. Resident in gen. practice, pediatrics Taiwan, China, 1960-66; staff physician Chaiyi (China) Christian Hosp., 1964-66; staff physician occupational medicine Air Am./Air Asia Inc., Southeastern Asia, 1966-69; intern Kenmore Mercy Hosp., Buffalo, 1973-74; staff physician USPHS Hosp., Clinton, Okla., 1974-76; staff physician dept. correction and rehab. Ga. Diagnostic Ctr., Jackson, 1977; attending physician Walton County Hosp., DeFuniak Springs, Fla., 1978-79; pres. Countryside Med. Clin., Inc., DeFuniak Springs, 1978-79; staff physician VA Med. Ctr., Bath, N.Y., 1979—; coord. Agt. Orange project VA Med. Ctr., Bath., 1979—, Ionizing Radiation program, 1979—. Author: Emotional Stress and Psycho-somatic Disorders, 1971, A Collection of 48 Selected Papers Published from 1980 to 1983, 1985; co-author: The Death of Chaing Kai-Shek, 1975, English-Chinese Aerospace Science/Technology Dictionary, 1977; contbr. articles to profl. jours. Chmn. bd. trustees Chaing Tzu Ming Edn. Foun., Bath, 1987; sustaining mem. Republic Nat. Com., Washington, 1986—; hon. fellow Truman Libr. Inst., Mo., 1976; mem. Am. Mus. Natural History, 1980—, The Smithsonian Assn., 1985—, U.S.-China People's Friendship Assn., 1975—, Soc. Asian Am. Culture Affairs. Served with Chinese Air Force, 1962-63. Named Hon. Prof. Heng Yang Med. Coll. (China), 1992. Mem. AMA, AIAA, APHA, Am. Coll. Physicians (assoc.), Am. Def. Preparedness Assn., Am. Chinese Med. Soc., Am. Acad. Family Physicians, Am. Naval Inst., Chinese Med. Assn., Aero. and Astronautical Soc. Republic of China, Fla. Med. Assn., Walton County Med. Soc., Aerospace Med. Assn., Nat. Geog. Soc., Acad. Polit. Sci., Air Force Assn., So. Tier Chinese Assn., Assn. for Asian Studies, VA Physician Assn., Planetary Soc., Nat. Space Soc., Air & Space Soc., Western Returned Student Assn. (China, hon.). Republican. Mem. Christian Ch. Home: 8660 State Rt 415 Avoca NY 14809 Office: VA Adminstrn Med Ctr Bath NY 14810

HUANG, ENG-SHANG, virology educator, biomedical engineer; b. Chia-Yi, Taiwan, Republic of China, Mar. 17, 1940; came to U.S., 1968; s. Juong-Sun and King-fa (Ong) H.; m. Shu-Mei Huong, Dec. 26, 1965; children: David Y., Benjamin Y. BS, Nat. Taiwan U., Taipei, Taiwan, 1962, MS, 1964; PhD, U. N.C., 1971. Asst. prof. U. N.C., Chapel Hill, 1973-78, assoc. prof., 1978-86, prof., 1986—; virology program leader Cancer Rsch. Ctr., Chapel Hill, 1979-91; mem. virology study sect. DRG/NIH, Bethesda, Md., 1978-82; mem. AIDS basic rsch. rev. com. Nat. Inst. Allergy & Infectious Diseases/NIH, 1988-90; chmn. Internat. Sci. Promotion Com., U.S. chpt., 1988—. Contbr. articles to Molecular Biology of Human Cytomegalovirus, Devel. Abnormality Induced by Cytomegalovirus Infection, Interaction between Cytomegalovirus and Human Immunodeficiency Virus. Chmn. membership com. Soc. Chinese Bioscientists in Am., Washington, 1988-89. Lt. ROTC, 1964-65. NIH fellow, 1971-73, Rsch. Career Devel. award NIAID, NIH, 1978-83; grantee in field. Mem. AAAS, Am. Soc. Microbiology, N.Y. Acad. Sci., Am. Cancer Rsch. Democrat. Achievements include development of mouse model to study the developmental abnormality induced by cytomegalovirus infection in humans; research in inhibition of human cytomegalovirus DNA replication by DHDG. Office: U NC Lineberger Cancer Ctr Chapel Hill NC 27599-7295

HUANG, EUGENE YUCHING, civil engineer, educator; b. Changsha, China, Nov. 28, 1917; came to U.S., 1948, naturalized, 1962; s. Sam and Yi Yun (Chao) H.; m. Helen W. Woo, Aug. 20, 1955; children—Martha, Pearl, William, Mary, Priscilla, Stephen. M.S., U. Utah, 1950; D.Sc., U. Mich., 1954. Registered profl. engr., Ill., Mich. Asst. engr. Chinese Nat. Hwy. Adminstrn., 1941-45, asso. engr., 1945-48; research asst. Engring. Research Inst., U. Mich., 1953-54; research asst. prof. civil engring. U. Ill., Urbana, 1954-58; asso. prof. U. Ill., 1958-63; prof. transp. engring. Mich. Tech. U., Houghton, 1963-84, prof. emeritus transp. engring., 1984—; cons. transp. systems design, soil mechanics, 1954—. Author: Overview of the American Transportation System, 1976; contbr. numerous articles on transp. design systems and research on materials for pavement to profl. jours. Recipient Faculty Research award Mich. Tech. U., 1967. Fellow ASCE; mem. Am. Soc. Engring. Edn., AAAS, Assn. Asphalt Paving Technologists, Inst. Mgmt. Sci., Am. Ry. Engring. Assn., ASTM, NRC (transp. research bd. 1954), Sigma Xi, Chi Epsilon, Tau Beta Pi, Phi Tau Phi. Episcopalian. Home: 400 Garnet St Houghton MI 49931-1420

HUANG, JASON JIANZHONG, ceramic engineer; b. Nantong, Jiangsu, People's Republic of China, Oct. 12, 1965; came to U.S., 1987; s. Dechang and Zhifeng (Zhang) H.; m. Rulong Chen, Aug. 17, 1990. BS, Hunan U., People's Republic of China, 1985; MS, UCLA, 1988; PhD, Ohio State U., 1992. Grad. rsch. asst. UCLA, 1987-88; grad. rsch. assoc. Ohio State U., Columbus, 1988-92, postdoctoral fellow, 1993—; cons. Superconductive Components, Inc., Columbus, 1992-93, Owens-Corning Fiberglas Corp. Tech. Ctr., Granville, Ohio, 1993. Contbr. articles to profl. publs. State Edn. Commn. of China fellow, 1985; recipient Hon. Mention Grad. Student award N.Am. Thermal Analsis Soc., 1991. Mem. Am. Ceramic Soc. (Honorable mention of Kreidl award from Glass and Optical Materials Div. 1993), Sigma Xi, Phi Beta Delta. Achievements include development of ceramic powder coating via melt precipitation technique, melt-precipitation for processing ceramic superconducting wires, non-Newtonian effects in structural relaxation of systems far from equilibrium. Office: Ohio State U Dept Mat Sci 378 Watts Hall 2041 College Rd Columbus OH 43210

HUANG, JIM JAY, chemist; b. Taipei, Taiwan, July 26, 1946; s. Chi Yuan and Chao (Chen) H.; m. Elona Theresa Tombrello, Aug. 13, 1977; children: Eric Jason, Calvin Edwin, Cheryl Theresa, Rachel Carolyn. BS, Chung-Hsing U., Taichung, Taiwan, 1969; PhD, U. Ala., 1985. Rsch. scientist Burroughs Wellcome Co., Rsch. Triangle Pk., N.C., 1980—; reviewer Jour. of Organic Chemistry, 1985—. Contbr. articles to Jour. of Organic Chemistry and Jour. of Heterocyclic Chemistry. Mem. Am. Chem. Soc. Achievements include research in peptide and nucleoside and hereocyclic chemistry. Home: 116 Mariposa Dr Cary NC 27513-5330 Office: Burroughs Wellcome Co 3030 Cornwallis Rd Research Triangle Park NC 27709

HUANG, JOSEPH CHEN-HUAN, civil engineer; b. Nanking, China, Oct. 18, 1933; came to U.S., 1962, naturalized, 1972; M.S. in Structural Engring., Va. Poly. Inst. and State U., 1964; m. Elizabeth C. Huang, Sept. 3, 1966; children: Edith, Eleanor, Evelyn, Edna. Registered profl. engr. N.Y., N.J., Pa., Del., Md., Va., W.Va., N.C., Fla., D.C. Project engr. Green Assos., Inc., Balt., 1964-68; pres. Gen. Engring. Consultants, Inc., Balt., 1968-76; chmn., chief exec. officer Highlights Corp., Towson, Md., 1976—. Mem. ASCE, Am. Concrete Inst., Nat. Soc. Profl. Engrs. Author; Prestressed Steel Structures; also tech. papers. Home: 3506 Templar Rd Randallstown MD 21133-2428 Office: 1248 E Joppa Rd Towson MD 21204 also: 1045 Taylor Ave Towson MD 21204 also: 825 N Hammonds Ferry Rd Ste B Linthicum Heights MD 21090-1350

HUANG, JU-CHANG, civil engineering educator; b. Kaohsiung, Taiwan, Jan. 3, 1941; came to U.S., 1964; s. Ti Huang and Bih Lin; m. Amy H. Hung, Feb. 5, 1965 (dec. 1983); children: Alina E., Nancy E.; m. Jolynn C. Chen, Nov. 26, 1986. BS in Civil Engring., Nat. Taiwan U., 1963; MS in Environ. Engring., U. Tex., 1966, PhD in Environ. Engring., 1967. Diplomate Am. Acad. Engrs.; profl. engr., Mo., Tex. Asst. prof. U. Mo., Rolla, 1967-70, assoc. prof., 1970-75, prof., dir. Environ. Rsch. Ctr., 1975-92; chief environ. engr. Austin, Smith & Assocs., Honolulu, 1972-73; prof. Hong Kong U. Sci. and Tech., 1993—; environ. engring. cons. U.S. EPA, WHO, UN, various industries & consulting firms, U.S. and overseas, 1970—; environ. cons. & fellow Indsl. Tech. Rsch. Inst., Hsin-Chu, Taiwan, 1989-90; sr. environ. cons. Super Max Engring. Co., Taipei, Taiwan, 1990—, Formosa Plastics Groups, Inc., Taipei, 1991—. Contbr. over 100 tech. articles to profl. jours. ASEE-Ford Found. fellow, 1972; recipient Young Engr. of Yr. award Mo. Soc. Profl. Engrs., 1976. Fellow Am. Soc. Civil Engrs. (Walter Huber Rsch. award 1979), Water Environment Fedn. Home: 1405 Highland Dr Rolla MO 65401 Office: Hong Kong U Sci&Tech, Civil Engring Dept, Kowloon Hong Kong

HUANG, MEI QING, physics educator, researcher; b. Wuhan, Hubei, People's Republic China, Jan. 20, 1942; came to U.S., 1988; parents Gong Li and Hui Qin Xia Huang; m. Jin Song Chen, Jan. 6, 1938; children: Qun Chen, Li Chen. Grad. dept. physics, U. Sci. and Tech. China, Beijing, 1964. Asst. prof. dept. physics U. Sci. and Tech. China, 1964-70; asst. prof. dept. physics U. Sci. and Tech. China, Hefei, 1970-78, instr., 1978-87, assoc. prof., 1987—, head div. magnetism, 1986-88; rsch. assoc. dept. MEMS Carnegie-Mellon U., Pitts., 1983-85, 88-91, rsch. scientist in advanced materials, 1991—; participant Chinese-Am. coop. program in atomic, molecular and condensed matter physics Chinese Acad. Sci. and Am. Physics Soc., 1988; presenter at nat. and internat. confs., including 30th, 34th, 35th, 36th, 37th, 38th Ann. Conf. on Magnetism and Magnetic Materials, Internat. Conf. on Rare Earth Applications and Devels., 11th Internat. Workshop on Rare Earth Magnets and Their Applications. Contbr. articles to Physica, Jour. Appleid Physics, Jour. Magnetism and Magnetic Materials, Jour. Less Common Metals. Recipient 3d prize of sci. and tech. Acad. Sci. China, 1988. Mem. Chinese Phys. Soc., Am. Phys. Soc. Achievements include patent pending for Cerium-free Mischmetal Fe-B-o Permanent Magnets; research on magnetic properties and structure of magnetic recording powder using magnetic measurements, electron microscopic investigation and Mossbauer spectrum analysis, magnetic properties and structure of rare earth intermetallic properties using X-ray diffraction and magnetic measurement, influence of hydrogen on the magnetic characteristics of $R_2Fe_{14}B$ system, magnetic and structural properties of $R_2Fe_{17}N_x$, $R_2(Fe, Co)_{17}N_x$ and $(Sm, R)_2Fe_{17}N_x$ nitrides, sintering studies of permanent magnet materials, and metal bonded $Sm_2Fe_{17}N_x$ type magnets. Home: 4531 Forbes Ave Apt 605 4531 Forbes Ave Apt 210 Pittsburgh PA 15213-3535 Office: Carnegie Mellon U Mellon Inst Box 60 Pittsburgh PA 15213

HUANG, PAN MING, soil science educator; b. Pu-tse, Taiwan, Sept. 2, 1934; arrived in Can., 1965; s. Rong Yi and Koh (Chiu) H.; m. Yun Yin Lin, Dec. 26, 1964; children: Daniel Chian Yuan, Crystal Ling Hui. BSA, Nat. Chung Hsing U., Taichung, Taiwan, 1957; MSc, U. Man., Winnipeg, Can., 1962; PhD, U. Wis., Madison, 1966. Cert. prof. agrologist. Asst. prof. soil sci. U. Sask., Saskatoon, Can., 1965-71, assoc. prof., 1971-78, prof., 1978—; nat. vis. prof., head dept. soil sci. Nat. Chung Hsing U., 1975-76; councilor Clay Minerals Soc., 1985-88. Author: Soil Chemistry, 1991; contbr. over 180 articles to profl. jours. Bd. dirs. Saskatoon Chinese Mandarin Sch., 1977-79, Saskatoon Soc. for Study Chinese Culture, 1983—. 2d lt. Taiwan Mil. Tng. Corps, 1957-59. Grantee The UN Environment Programme, Nat. Scis. and Engring. Rsch. Coun. Can. and numerous other agys., 1965—. Fellow Can. Soc. Soil Sci., Soil Sci. Soc. Am. (rep. clay minerals soc. 1979-83, chmn. dir. S-9 1982-83, bd. dirs. 1983-84, assoc. editor 1987-92, editor spl. publ. 1986, rep. to Internat. Union Pure and Applied Chemistry 1990—, award com. 1986, Marion L. and Christie M. Jackson Soil Sci. award com. 1990—, fellow com. 1992—, chair elect div. S-2 1993), Am. Soc. Agronomy; mem. Internat. Soc. Soil Sci. (chmn. working group 1990—), Am. Chem. Soc., N.Y. Acad. Sci., Sigma Xi. Avocations: music, reading. Home: 130 Mount Allison Cres, Saskatoon, SK Canada S7H 4A5 Office: U Sask, Dept Soil Sci, Saskatoon, SK Canada S7N 0W0

HUANG, PEISEN SIMON, mechanical engineer; b. Shanghai, Peoples Rep. China, Aug. 21, 1962; came to U.S., 1990; s. Shizhong and Zhenfang (Ji) H. BS, Shanghai Jiao Tong U., 1984; M of Engring., Tohoku U., 1988, postgrad., 1988—; PhD, U. Mich., 1993. Rsch. asst. U. Mich., Ann Arbor, 1990-93; asst. prof. SUNY Stony Brook, 1993—. Contbr. articles to profl. jours. Mem. Optical Soc. Am., Soc. Mfg. Engrs. Achievements include research on method and apparatus for angle measurement based on the internal retlection effect, multi degree-of-freedom geometric error measurement system. Office: SUNY Stony Brook Dept Mech Engring Stony Brook NY 11794-2300

HUANG, SUNG-CHENG, electrical engineering educator; b. Canton, China, Oct. 26, 1944; came to U.S., 1967; s. Hip-chung Wong and Chung Huang; m. Caroline S. Soong, Sept. 4, 1971; children: Michael, Dennis. BSEE, Nat. Taiwan U., Taipei, 1966; DSc, Wash. U., 1973. Postdoctral rsch. assoc. Biomed. Computer Lab. Wash. U., St. Louis, 1973-74; project engr. Picker Corp., Cleve., 1974-77; asst. prof. Sch. Medicine UCLA, 1977-82, assoc. prof. Sch. Medicine, 1982-86, prof. Sch. Medicine, 1986—; Edward Farber lectr. U. Chgo., 1986. Mem. editorial bd. Jour. Cerebral Blood Flow, 1989-92; dep. chief editor Jour. Cerebral Blood Flow and Metabolism, 1993—; contbr. over 150 articles to scholarly and profl. jours. Recipient George Von Hevesy Prize World Congress of Nuclear Medicine and Biology, 1982; grantee U.S. Dept. Energy, 1977—, NIH, 1977—. Mem. AAAS, IEEE, Soc. Nuclear Medicine, Soc. Cerebral Blood Flow. Achievements include patent for spread beam overlap method; development of various tracer techniques used for positron emission tomographic (PET) studies in nuclear medicine; research in computer tomographic image construction technique. Office: UCLA Sch Medicine Div Nuclear Medicine and Biophysics 405 Hilgard Ave Los Angeles CA 90024

HUANG, YANG-TUNG, electronics educator, consultant; b. Taitung, Republic of China, Aug. 15, 1955; m. Yuan-Guang Thea, May 30, 1987; children: Yu-An, Yu-Ping. BSc in Electrophysics, Nat. Chiao Tung U., Hsin Chu, Taiwan, Rep. of China, 1978, MSc in Electronics, 1982; PhD in Elec. Engring., U. Ariz., 1990. Registered profl. engr., Republic of China. Lectr.

China Jr. Coll. of Tech., Taipei, Republic of China, 1982-83; lectr. Nat. Chiao Tung U., 1983-85, dep. dir., 1984-85, assoc. prof., 1990—, dep. dir., 1991-92; cons. Hopax Industries Co., Ltd., Kaohsiung, Taiwan, 1990-91. 2d lt. Chinese Army, 1978-80. Mem. IEEE, Optical Soc. Am., Soc. Photo-Optical Instrumentation Engrs. Achievements include patents in field. Office: Nat Chiao Tung U, 1001 Ta-Hsueh Rd, Hsinchu 300, Taiwan

HUANG, ZHI-YONG, honey bee biologist; b. Yin-tian, China, Jan. 1, 1962; s. Zhou-liang and Feng-ying (Shen) H.; m. Gui-jie Wang, Dec. 19, 1986; children: David Yang-long, Melissa Shan-mei. BS, Hunan Agr. Coll., Changsha, China, 1982; PhD, U. Guelph, Ont., Can., 1988. Postdoctoral fellow U. Mo., Columbia, 1988-90; postdoctoral fellow U. Ill., Urbana, 1990-91, sr. rsch. scientist, 1991—; mem. computer drawing com. U. Mo., Columbia, 1989. Editor: Advances and Trends in Modern Ecology, 1992; co-author: Advances in Honey Bee Ecology; contbr. articles to profl. jours. Mem. Nature Conservancy, 1991; editor Chinese Illini, Urbana, 1991. China-Cornell fellow, 1993—; Soden Meml. fellow, 1986, Beatty-Munro Family Meml. fellow, 1984, 86, Gordon F. Townsend fellow, 1986. Mem. Am. Assn. Advancement Sci., SINO Ecologists Club Overseas (v.p. 1990-91), Entomological Soc. Am., Apicultural Club (pres. 1986-87), Sigma Xi. Achievements include discovery that honey bee workers similar to cells in higher organisms interact with each other to obtain information about their colony, that worker bees visit brood cells non-randomly; research in mechanism of food provisioning in nursing larval bees, mechanism of hypopharyngeal gland activation. Office: U Ill 320 Morrill Hall Urbana IL 61801

HUBBARD, ARTHUR THORNTON, chemistry educator, electro-surface chemist; b. Alameda, Calif., Sept. 17, 1941; s. John White and Ruth Frances (Gapen) H.; children: David A., Lynne F. BA, Westmont Coll., 1963; PhD, Calif. Inst. Tech., 1967. Prof. chemistry U. Hawaii, Honolulu, 1967-76, U. Calif., Santa Barbara, 1976-86; Ohio eminent scholar and prof. chemistry U. Cin., 1986—, dir. Surface Ctr., 1986—; chmn. Ohio Sci. and Engring. Roundtable. Assoc. editor Jour. Colloid and Interface Sci., 1993—. Grantee NSF, NIH, Air Force Office Sci. Research, U.S. Dept. Energy. Mem. Am. Chem. Soc. (assoc. editor jour. Langmuir 1984-90), Electrochem. Soc. (David C. Grahame award 1993). Office: U Cin Dept Chemistry Surface Ctr Cincinnati OH 45221-0172

HUBBARD, BESSIE RENEE, mechanical engineer, mathematician; b. Fayetteville, N.C., Sept. 23, 1961; d. Kenneth Brigman and Ellen Merle H. BSME, N.C. State U., 1983, MME, 1985, BS in Applied Math., 1989. Registered profl. engr., N.C. Mech. engr. N.C. State Univ., Raleigh, 1985—; spl. engr. cons. United Daughters of Confederacy, Raleigh, 1989—; mem. faculty indsl. ventilation conf. N.C. State U. Author: (with others) NCSU Guidelines for Construction, 1988, 91. Sec. Nat. Soc. Daughters of Revolution, Garner, N.C., 1991—; editor Cumberland County Geneal. Soc., Fayetteville, 1991—. Mem. ASHRAE, ASME (chpt. historian 1987-88), NSPE, N.C. Soc. Engrs. (Order of Engr. 1987), Order of Crown of Charlemagne, Jamestowne Soc., Nat. Soc. Dau. Colonial Wars, Tau Beta Pi, Pi Alpha Alpha. Republican. Home: 116 E Ransom St Fuquay-Varina NC 27526 Office: NC State U Phys Plant Campus Box 7219 Raleigh NC 27695

HUBBARD, CHARLES RONALD, engineering executive; b. Weaver, Ala., Feb. 4, 1933; s. John Duncan Hubbard and Athy Pauline (Lusk) Thorpe; m. Betty Lou McKleroy, Dec. 29, 1951; 1 son, Charles Ronald Hubbard II. BSEE, U. Ala., 1960. Mktg. mgr. Sperry Corp., Huntsville, Ala., 1969-71, head engring. sect., 1971-74; sr. staff engr. Honeywell Inc., Clearwater, Fla., 1974-76, mgr., 1976-79, chief engr., West Covina, Calif., 1979-83, assoc. dir. engring., 1983-84, assoc. dir. advanced systems, 1984-87, assoc. dir. programs, 1987-88; v.p. govt. systems div. Integrated Inference Machines, Anaheim, Calif., 1988-91; pres. Synergy Computer Systems, Anaheim, 1991—. Served as staff sgt. USAF, 1953-57. Recipient Outstanding Fellow award U. Ala., 1991. Mem. IEEE (sect. chmn. 1972-73). Methodist. Home: 5460 E Willowick Cir Anaheim CA 92807-4642 Office: Synergy Computer Systems 5460 Willowick Cir Anaheim CA 92807

HUBBARD, HAROLD MEAD, research institute executive; b. Beloit, Kans., Apr. 16, 1924; s. Clarence Richard and Elizabeth (Mead) H.; m. Doreen J. Wallace, Aug. 13, 1948 (div. 1975); children—Stuart W., David D. B.S., U. Kans., 1948, Ph.D., 1951; DSc (hon.), Regis Coll. Instr. chemistry U. Kans., Lawrence, 1949-51; rsch. chemist, rsch. mgr., lab. mgr. E. I. DuPont de Nemours & Co., Inc., Wilmington, Del., 1951-69; dir. phys. sci. Midwest Rsch. Inst., Kansas City, Mo., 1970-75, v.p. rsch., 1976-78, sr. v.p. ops., 1979-82, exec. v.p., 1983-90; dir. Solar Energy Rsch. Inst., 1982-90; vis. sr. fellow Resources for the Future, 1990-91; bd. dirs. Guaranty State Bank. With U.S. Army, 1942-45. Mem. Mo. Acad. Sci. (councillor at large 1977-80), Tech. Transfer Soc. (v.p. 1978-79), Am. Chem. Soc., AAAS, N.Y. Acad. Scis., Am. Solar Energy Soc., Sigma Xi, Delta Upsilon. Unitarian. Clubs: Rockhill Tennis, Rolling Hills Country. Home: 2938 Newark St NW Washington DC 20008-3338 Office: Solar Energy Rsch Inst 1617 Cole Blvd Golden CO 80401-3305 also: Solar Energy Rsch Inst Washington Office Portal Bldg Ste 710 Washington DC 20024-2188

HUBBARD, KENNETH GENE, climatologist; b. Bridgeport, Nebr., Mar. 21, 1949; s. Harold D. and Bessie B. (Arrants) H.; m. Susan Elizabeth Alcorn, Aug. 15, 1971; children: Carter M., Benjamin W. MS, S.D. Sch. Mines and Tech., 1973; PhD, Utah State U., 1981. Meteorologist Geophys. Fluid Dynamics Lab., Princeton, N.J., 1973-74, Utah Water Rsch. Lab., Logan, 1974-77; climatologist Utah Dept. Agriculture, Logan, 1977-81, U. Nebr., Lincoln, 1981—; dir. High Plains Climate Ctr., Lincoln, 1987—. Co-author: Automated Weather Station Networks, 1993; contbr. chpts. to books, articles to publs. Fellow Ctr. for Great Plains Studies; mem. AAAS, Am. Soc. Agronomy, Am. Meteorol. Soc. (com. agrl. and forest meteorology 1986-89), Am. Assn. State Climatologists (pres. 1985-86), World Meteorol. Orgn. (rapporteur 1992—). Achievements include rsch. in automated weather data network devel. Office: U Nebr 242 Chase Hall Lincoln NE 68583-0728

HUBBELL, DOUGLAS OSBORNE, chemical engineering consultant; b. Norfolk, Va., Jan. 6, 1952; s. Harold Arthur and Mary Virginia (Osborne) H.; m. Marilyn Patricia Morrison, Sept. 10, 1966; children: Erin Marie, Mary Patricia. BSChemE, Va. Poly. Inst. and State U., 1964, MSChemE, 1965; PhDChemE, Princeton U., 1969; postgrad., Swiss Fed. Inst. Tech., Zurich, 1970. Devel. engr. Hoechst Fibers, Spartanburg, S.C., 1970-87; quality cons. Qualpro Inc., Knoxville, 1987-93; dir. process engring. Mayfair Mills, Inc., Spartanburg, S.C., 1993—. Mem. Am. Soc. Quality Control, Fiber Soc. (pres. 1987), Sigma Xi. Republican. Episcopalian. Office: Mayfair Mills Inc Arcadia SC 29320

HUBBS, CLARK, zoologist, researcher; b. Ann Arbor, Mich., Mar. 15, 1921; s. Carl Leavitt and Laura Cornelia (Clark) H.; m. Catherine Vickery Symons; children: Laura Ellen Hubbs Tait, John Clark, Ann Frances Hubbs Weissman. BA, U. Mich., 1942; PhD, Stanford U., 1951. Instr. zoology U. Tex., Austin, 1949-52, asst. prof., 1952-57, assoc. prof., 1957-63, prof., 1963-88, Regents prof., 1988-91, Regents prof. emeritus, 1991—, chmn. biology dept., 1974-76, chmn. zoology dept., 1978-86, with grad. faculty dept. marine sci., 1987-91; curator ichthyology Tex. Meml. Mus., 1978—; vis. prof. U. Okla., Kingston, 1970-84; bd. dirs. Hubbs/Sea World Rsch. Inst., San Diego; faculty advisor U. de Nuevo Leon, Monterey, Mex., 1985-87; biology advisor Bd. Higher Edn., Little Rock, 1987, Jackson, Miss., 1983; leader Rio Grande Fishes Recovery Team, U.S. Interior Dept., Albuquerque, 1978—; mem. adv. com. Fish, Wildlife and Parks, U.S. Interior Dept., Washington, 1975-77; mem. sci. adv. com. Bass Anglers Sportsmans Soc., Montgomery, Ala., 1974—; mem. sci. adv. bd. Tex. Utilities, Dallas, 1971—; chmn. inland task force, power plant sitting com., Office of Gov., Austin, Tex., 1971-72; mem. nuclear power adv. com. Tex. Energy Adv. Council, Austin, 1978-80; U.S. rep. European Ichthyological Congress, 1985-88; bd. dirs. Tex. Nature Conservancy, 1988—. Mng. editor Copeia, 1971-84; author over 200 sci. articles on fish biology. Mem. Nat. Rsch. Coun. Com. on Glen Canyon releases into the Colo. River, 1991-94. Served with U.S. Army, 1942-46, PTO. Named Educator and Researcher of Yr., Tex. chpt. Am. Fisheries Soc., 1978; recipient Excellence award Am. Fisheries Soc., 1988; Clark Hubbs Endowed Professorship in Zoology established in his honor, Dept. Zoology. Mem. Am. Soc. Ichthyologists and Herpetologists (pres. 1987, Lifetime Achievement award 1992), Tex. Acad. Scis. (pres. 1972-73), S.W. Assn. Naturalists (pres. 1966-67, W.F. Blair Eminent Naturalist 1990). Office: U Tex Dept Zoology Austin TX 78712

HUBEL, DAVID HUNTER, physiologist, educator; b. Windsor, Ont., Can., Feb. 27, 1926; s. Jesse Hervey and Elsie (Hunter) H.; m. Shirley Ruth Izzard, June 20, 1953; children: Carl Andrew, Eric David, Paul Matthew. BSc, McGill U., 1947, MD, 1951, DSc (hon.), 1978; AM (hon.), Harvard U., 1962; DSc (hon.), U. Man., 1983; DHL (hon.), Johns Hopkins U., 1990. Intern Montreal Gen. Hosp., 1951-52; asst. resident neurology Montreal Neurol. Inst., 1952-53, fellow clin. neurophysiology, 1953-54; asst. resident neurology Johns Hopkins Hosp., 1954-55; sr. fellow neurol. scis. group Johns Hopkins U., 1958-59; faculty Harvard U. Med. Sch., 1959—, George Packer Berry prof. physiology, chmn. dept., 1967-68, George Packer Berry prof. neurobiology, 1968-82, John Franklin Enders univ. prof., 1982—; George H. Bishop lectr. exptl. neurology Washington U., St. Louis, 1964; Jessup lectr. biol. scis. Columbia, 1970; James Arthur lectr. Am. Mus. Natural History, 1972; Ferrier lectr. Royal Soc. London, 1972; Harvey lectr. Rockefeller U., 1976; Weizmann meml. lectr. Weizmann Inst. Sci., Rehovot, Israel, 1979; George Eastman prof. Oxford, Eng., 1991-92; Fenn lectr. 30th internat. congress Internat. Union Psychol. Sci., Vancouver, B.C., Can., 1986; researcher brain mechanisms in vision.; bd. syndics Harvard U. Press, 1979-83. Served with AUS, 1955-58. Recipient Trustees award Rsch. to Prevent Blindness, 1971, Lewis S. Rosenstiel award for disting. work in basic med. rsch., 1972, Karl Lashley prize Am. Philos. Soc., 1977, Louisa Gross Horwitz prize Columbia U., 1978, Dickson prize in medicine U. Pitts., 1979, Ledlie prize Harvard U., 1980, Nobel prize, 1981, Outstanding Sci. Leadership award Nat. Assn. for Biomed. Rsch., 1990, City of Medicine award, 1990; fellow Harvard Soc. Fellows, 1971—. Fellow Am. Acad. Arts and Scis.; mem. Nat. Acad. Sci., Am. Physiol. Soc. (Bowditch lectr. 1966), Deutsche Akademie der Naturforscher leopoldina, Soc. for Neurosci. (Grass lecture 1976), Assn. for Rsch. in Vision and Ophthalmology (Friedenwald award 1975), Johns Hopkins U. Soc. Scholars, Am. Philos. Soc. (Karl Spencer Lashley prize 1977), Royal Soc. London. Home: 98 Collins Rd Newton MA 02168-2235 Office: Harvard Med Sch Dept Neurobiology 220 Longwood Ave Boston MA 02115-5717

HUBER, BRIAN EDWARD, molecular biologist; b. Huntington, N.Y., June 5, 1954; s. Valentine F. Huber and Cathrine T. Schiavoni; m. Anne S. Schmitt, Aug. 21, 1976; children: Jason F., Matthew A. BS, Villanova U., 1976, MS, 1979; PhD, George Washington U., 1983. Diplomate Am. Bd. Clin. Pharmacology. Sr. staff fellow NIH, Bethesda, Md., 1983-86; sr. scientist Wellcome Rsch. Labs., Research Triangle Park, N.C., 1986-91, asst. dir. cell biology, 1991—; professorial lectr. dept. pharmacology George Washington U., Washington, 1985—; adj. prof. Duke U. Med. Ctr., Durham, N.C., 1991—, U. N.C., Chapel Hill, 1992—; presenter at profl. confs. Editorial bd. Cambridge Cancer Studies, 1992—; contbr. chpts. to books, articles to refereed jours. including Hepatology, Biochem. Pharmacology, others. Fellow Pharmacology Rsch. Assoc. Tng. program, NIH, 1983-85; recipient Pritsella award Advances in Pharmacology, 1985. Mem. AAAS, Am. Assn. Cancer Rsch., Am. Soc. Pharmacology and Exptl. Therapeutics (chmn. molecular therapeutics com. gene therapy sect.), N.Y. Acad. Sci., Soc. Exptl. Biology and Medicine, Sigma Xi. Roman Catholic. Achievements include development of entities for treatment of cancer and HIV therapy, research in gene therapy, eucaryotic molecular biology, multidrug resistance, biotechnology, AIDS, molecular mechanisms in drug design and action. Office: Wellcome Rsch Labs Div Cell Biology 3030 Cornwallis Rd Research Triangle Park NC 27709

HUBER, DAVID LAWRENCE, physicist, educator; b. Highland Park, N.J., July 31, 1937; s. Howard Frederick and Katherine Teresa (Smith) H.; m. Virginia Hullinger, Sept. 8, 1962; children: Laura Theresa, Johanna Jean, Amy Louise, William Hullinger. BA, Princeton U., 1959; MA, Harvard U., 1960, PhD, 1964. Instr. U. Wis., Madison, 1964-65, asst. prof., 1965-67, assoc. prof., 1967-69, prof., 1969—; bd. dirs. Synchrotron Radiation Ctr., Stoughton, Wis., Phys. Sci. Lab., Stoughton; disting. vis. prof. U. Mo., Kansas City, 1988. A.P. Sloan fellow, 1972-73, Nat. Assn. State Univs. and Land Grant Colls. fellow Office of Sci. and Tech. Policy, Washington, 1990-91. Mem. Sigma Xi, Phi Beta Kappa. Office: U Wis Synchrotron Radiation Ctr 3731 Schneider Rd Stoughton WI 53589-3097

HUBER, DOUGLAS CRAWFORD, pathologist; b. S. Charleston, W.Va., June 11, 1939; s. Abram Paul and Mary Ashley (Grow) H.; m. Deena Rae Freedman, Aug. 8, 1969; children: Adam Crawford, Laura Kristen; m. Angelika Madelon Pohl, June 3, 1961 (div. 1965); 1 child, Heidemarie Jutta. Student, Harvard U., 1958-59; AB, Emory U., Atlanta, 1960; MD, Emory U. Sch. of Med., Atlanta, 1964. Assoc. pathologist Baldwin County Hosp., Milledgeville, Ga., 1971-72, Leary Lab., Boston, 1972-73; lab. dir. Homer D. Cobb Mem. Hosp., Phenix City, Ala., 1973-79; gen. practitioner Leonard Morse Hosp., Natick, Mass., 1979-80; lab. dir. Douglas Gen. Hosp., Douglasville, Ga., 1980—; med. dir. Roche Biomedical Lab., Atlanta Div., Tucker, Ga., 1989—; deputy state commr. Coll. Am. Pathologists Lab. Inspection Program, Skokie, Ill., 1976-79; v.p. Ala. Assn. Pathologists, Birmingham, 1979. Pres. Nam Vets of Ga., 1982-85; capt. with U.S Army, 1965-67. Fellow Coll. Am. Pathologists, Am. Soc. Clinical Pathologists. Home: 795 Tanglewood Trl NW Atlanta GA 30327-4570

HUBER, GARY ARTHUR, aerospace engineer; b. Reichenberg, Germany, Dec. 15, 1944; came to U.S., 1950; s. Emil and Johanna (Elstner) H.; m. Kathleen Marie Kleinschnitz, May 29, 1972; children: Gary E., Ann M., Michael P., Stephen K., Katherine E. BS in Indsl. Engring., U. Fla., 1967. Support scheduler tech. support directorate Kennedy Space Ctr., Fla., 1967-73, project engr. test requirements and scheduling, 1973-77, sr. systems engr. launch processing systems, 1978-80, sr. engr. ground systems div., 1980-87, sr. engr. spacelab and experiments div., 1980-87, tech. asst. to chief payload processing divsn., 1987-92, tech. asst. to dir. payload ops., 1993—; mem. many Apollo, Space Shuttle launch teams; mem. Space Shuttle Payload Mgmt. Team. Author: Apollo Aircraft Support Plans, 1968-72. Pres. Brevard Cath. Singles Club, Brevard County, Fla., 1970-72. Recipient Silver Snoopy award Astronaut Corps, 1975; Apollo Soyuz honoree Atmospheric Electicity Team, 1975, Group Achievement award First Spacelab Mission. Mem. Am. Inst. Indsl. Engrs. (charter). Republican. Office: JFK Space Ctr CS Orlando FL 32899

HUBER, JOHN HENRY, III, economic scientist, researcher; b. New Orleans, June 5, 1946; s. John Henry Jr. and Eldoris Margarite (Rockvoan) H.; m. Terry Sue Rayno, May 7, 1974; children: Suesan, Lilith, Michael. BS, U. New Orleans, 1975, MS, 1979; PhD, La. State U., 1986. Dept. head fin. instns. Office of Econ. Devel. Securities Divsn., New Orleans, 1987—; econ. advisor to La. Legislature, Gov. of La. Author: Econometrics: An Introduction to Maximum Likelihood Methods, 1975, A Textbook of Economics, 1990. Quartermaster VFW, Metairie, La., 1970. Lt. U.S. Army, 1964-69, Vietnam. Decorated Purple Heart. Democrat. Roman Catholic. Achievements include development of mathematical methods of applying theories of econometrics to computer models. Home: 3112 Ridgeway Dr Metairie LA 70002-5056

HUBER, PAUL WILLIAM, biochemistry educator, researcher; b. Medford, Mass., July 23, 1951; s. William Francis and Catherine (Sheridan) H. BS, Boston Coll., 1973; PhD, Purdue U., 1978. NIH postdoctoral fellow U. Chgo., 1979-81, rsch. assoc., 1982-85; asst. prof. U. Notre Dame, Ind., 1985-92, assoc. prof., 1992—. Contbr. articles to profl. jours. Mem. AAAS, Am. Soc. Biochemistry and Molecular Biology. Home: 1215 E Irvington South Bend IN 46614-1417 Office: U Notre Dame Dept Chemistry/Biochemistry Notre Dame IN 46556

HUBER, ROBERT, biochemist; b. Munich, Feb. 20, 1937; s. Sebastian and Helene (Kebinger) H.; m. Christa Huber, 1960; children: Ulrike, Martin, Robert, Julia. Diploma, Tech. Universität Munich, 1960, PhD, 1963, habil., 1968; D (hon.), Louvain, Belgium, 1987, U. Ljubljana, Slovenia, 1989; D for Medicine and Surgery (hon.), U. 'Tor Vergata', Rome, 1991. Former prof. Tech. U. Munich; prof., dir. Max-Planck-Inst. für Biochemie, Martinsried, Fed. Republic Germany, 1971—. Editor Jour. Molecular Biology. Recipient E. K. Frey medal Gesellschaft für Chirurgie, 1972, Otto Warburg medal Gesellschaft für Biologische Chemie, 1977, Emil van Behring medal U. Marburg, 1982, Keilin medal Biochem. Soc. London, Richard Kuhn medal Soc. German Chemists, 1987, E. K. Frey-E. Werle meml. medal, 1989, Kone award Assn. Clin. Biochemists, 1990, Sir Hans Krebs medal, 1992; co-recipient Nobel prize for chemistry, 1988. Mem. Am. Soc. Biol. Chem. (hon.), Ges. fü Biol. Chem., Deutsche Chem. Ges., Swedish Soc. Biophys. (hon.), Japanese Biochem. Soc. (hon.), Deutsche Akad. Naturforscher Leopoldina, Croation Acad. Sci., European Molecular Biology Orgn. (mem. coms.), Bavarian Acad. Scis. Office: Max Planck Inst Biochemistry, Am Klopferspitz 18A, 82152 Martinsried Germany

HUBER, ROBERT JOHN, electrical engineering educator; b. Payson, Utah, July 10, 1935; s. Robert Earl and Pamella (Lewis) H.; m. Virginia Goldberg, Sept. 20, 1957; children—Robert William, John Lee, Scott Edward. B.S., U. Utah, 1956, Ph.D. in Physics, 1961. Assoc. physicist Argonne Nat. Lab., Idaho Falls, Idaho, 1960-66; physicist microelectronics research and devel. ctr. Gen. Instrument Corp., Salt Lake City, 1966-68, gen. mgr., 1968-71; dir. microcircuit lab. Inst. Biomed. Engring., U. Utah, Salt Lake City, 1971-76, research prof., 1977-78, prof. elec. engring., 1978—; cons., 1971—. Co-author: Chemically Sensitive Field Effect Transistors, 1980. Co-editor: Solid State Chemical Sensors, 1985. Patentee in field (4). Mem. IEEE (state chair 1976; Tech. Achievement award 1980). Home: 1145 E Millbrook Way Bountiful UT 84010-2025 Office: U Utah Salt Lake Salt Lake City UT 84112

HUBERT, WALTER, psychologist; b. Velen, Germany, Jan. 4, 1956; s. Paul and Anna (Albersmann) H.; m. Mathilde Möller, July 15, 1988. Diploma, U. Münster, 1984, PhD, 1988. Psychologist U. Münster, 1984—. Home: Masurenweg 8, D-48147 Münster Germany Office: U Münster Inst Psychology, Rosenstr 9, D-48143 Münster Germany

HUBLER, H. CLARK, writer, retired educator; b. Portland, Oreg., July 26, 1910; s. W.H. and Elsie Rowena (Clark) H.; m. Reta Allinson, June 12, 1934; children: Keith, Bonnie, Thomas, Rowena, Craig. BA in Sci. Edn., Western Wash. U., 1937; EdD in Sci. Edn., Columbia U., 1949. Life cert. pub. sch. tchr., Wash. Tchr. Wash. State Pub. Schs., Aberdeen and Seattle, 1934-45; cons., sci. Columbia U., N.Y.C., 1945-47; asst. prof. Tchrs. Coll. Conn., New Britain, 1947-49; prof. Wheelock Coll., Boston, 1949-63; Fulbright lectr. Philippines, 1963-64; ednl. advisor in sci. Ohio U., South Vietnam, 1966-69; prof. Ohio U., Athens, 1969-76; lumber grader Long-Bell Lumber Co., Longview, Wash., 1927-31. Author: Working With Children in Science, 1957, Science for Children, 1974, Overpopulation..., 1985, Nuclear Energy..., 1988 and others. Pres. Coun. Elem. Sci., 1955-56, PTA, Scioto County, Ohio, 1982. Named Disting. Svc. Nat. Sci. Tchrs. Assn., 1983, Sci. Edn. Recognition, Nat. Assn. Rsch. Sci. Teaching, 1958; recipient Honor medal, South Vietnam, 1969. Fellow AAAS. Home and Office: 1610 28th St Portsmouth OH 45662

HUCK, MATTHEW L., process development engineer; b. Syracuse, N.Y., Apr. 19, 1961; s. Ludwig A. and Margaret (Brooks) H.; m. Wanda Leigh Englert, July 6, 1985 (div. May 1992); 1 child, Brian. BS, Rochester (N.Y.) Inst. Tech., 1983. Photolithographic engr. Am Microsystems Inc, Santa Clara, Calif., 1983-85, sect. mgr., 1985-87; R & D engr. Am Microsystems Inc, Pocatello, Idaho, 1987-90, R & D photolithographic engring. staff, 1990-93; R & D photolithographic engring. sr. staff Am. Microsystems Inc., Pocatello, Idaho, 1993—; sec. Bacus, Inc., San Jose, Calif., 1985-87. Contbr. articles to profl. jours. Mem. Soc. Photo-Optical Instrumentation Engrs. (active BACUS and Microlithography groups). Republican. Avocations: camping, bicycling, music. Home: 2750 Castle Peak Way Pocatello ID 83201 Office: Am Microsystems Inc 2300 Buckskin Rd Pocatello ID 83201

HUCKABY, JAMES L., chemical engineer; b. Louisville, Ky., Jan. 28, 1958; s. Bill Herbert and Ellen (Litwin) H.; m. Alisa Dunbar, June 21, 1986; children: Samuel Stephen, James Reps. BS in Chemistry, U. Ky., 1981, MS in Chem. Engring., 1986, PhD in Chem. Engring., 1991. Sr. engr. Westinghouse Hanford Co., Richland, Wash., 1992—. Mem. Am. Inst. Chem. Engrs., Tau Beta Pi, Phi Beta Kappa. Home: 1524 Ridge View Ct Richland WA 99352 Office: Westinghouse Hanford Co PO Box 1970 R2-11 Richland WA 99352

HUCKINS, HAROLD AARON, chemical engineer; b. Cambridge, Mass., Nov. 28, 1924; s. Harold Aaron and Julia E. (Nugent) H.; m. Elizabeth L. Kearns, Nov. 15, 1952; children: Richard W., Robert M., Christopher N., Patricia A., Leslie K. BSChemE, Northeastern U., 1945; ASME, Lowell Inst., 1946; postgrad., Boston U., 1947-49, U. Pitts., 1950-52. Chem. process engr., asst. project mgr. Monsanto Chem. Co., Boston-Everett, Mass., 1945-49; sr. process engr., group leader Koppers Co. Chem. Div., Pitts., 1949-53; mgr. pilot plants, project mgr. Sci. Design Co., Inc., N.Y.C., 1953-66; v.p. tech. ops. Oxirane Chem. Co., Princeton, N.J., 1966-73; v.p. tech. assessment Halcon SD Group, N.Y.C., 1973-85; pres. Princeton Advanced Tech., Inc., 1985—; dir., program chmn. Assn. Cons. Chemists & Chem. Engrs. div., N.Y.C., 1990-93, program chair 1992-93; dir. Materials Tech. Inst., St. Louis, 1976-85. Co-author: The Chemical Plant, 1966; contbr. articles to profl. jours. Fellow Am. Inst. Chem. Engrs. (dir. mgmt. div. 1981-82, dir. materials engring. sci. div. 1976-80, chmn. materials engring. and sci. div. 1992-93, chmn. chem. tech. materials com. 1983-84, chmn. John Fritz medal commn. 1989); mem. Am. Soc. Materials, Am. Chem. Soc., Am. Ceramic Soc., Nat. Assn. Corrosion Engrs. (chmn. 1984), Comml. Devel. Assn., Mensa Internat. Achievements include 8 patents for chemical process technology. Office: Princeton Advanced Tech Inc 4 Bertram Pl Hilton Head Island SC 29928

HUCKSTEP, APRIL YVETTE, chemist; b. Aliquippa, Pa., May 12, 1961; d. Charles Jr. and Geraldine (Wilson) H. Cert., Parkway Tech. Coll., Oakdale, Pa., 1979; BS, Pa. State U., 1984; MS, U. Akron, 1991. Technician Arco Chem. Inc., Newton Square, Pa., 1979-81; technician DiversiTech, Akron, 1986; chemist Goodyear Tire and Rubber, Akron, 1986; intern Lord Corp., Erie, Pa., 1989; chemist, grad. asst. U. Akron, 1988—. U. Akron fellow. Mem. NAACP, Am. Chem. Soc., Soc. Women Engrs., Am. Inst. Chem. Engrs., Nat. Soc. Black Engineers. Democrat. Baptist. Avocations: jogging, aerobics, music. Home: 868 Monaca Rd Monaca PA 15061-2831 Office: U Akron Polymer Sci Polymer Sci Bldg Akron OH 44304

HUDDLESTON, PHILIP LEE, physicist; b. St. Louis, Jan. 12, 1947; s. Joseph Berl and Myrtle (Craig) H.; m. Angela Jeaninne Gryting, Aug. 10, 1973. BS, Washington U., 1967; MA, Boston U., 1969, PhD, 1974. Asst. prof. physics and math. Edward Waters Coll., Jacksonville, Fla., 1975-76; asst. prof. physics Parks Coll., St. Louis U., Cahokia, Ill., 1976-79; sci. programmer McDonnell Douglas Automation Co., St. Louis, 1979-81; scientist McDonnell Douglas Rsch. Labs., St. Louis, 1981-89; sr. scientist McDonnell Douglas Techs. Inc., San Diego, 1989-92; sr. tech. specialist McDonnell Douglas Aerospace, St. Louis, 1992—. Referee IEEE Jours., IEE Jours., AEU Jour., Radio Sci. Jour.; reviewer Math. Revs.; contbr. articles to profl. jours. Mem. IEEE (sr., pres. St. Louis combined group), AAAS, Soc. Indsl. and Applied Math, Optical Soc. Am., Sigma Xi. Achievements include research in computational and applied physics with an emphasis on modeling of electromagnetic and optical phenomena. Home: 5471 Kenrick Parke Dr Saint Louis MO 63119-5027 Office: McDonnell Douglas Aerospace PO Box 516 MC 0642263 Saint Louis MO 63166

HUDGENS, KIMBERLYN NAN, industrial engineer; b. Hartwell, Ga., June 18, 1964; d. Kenneth Howard and Nan (Skelton) H.; m. Douglas Howard Abrams, June 30, 1990. BS in Indsl. Engring., Auburn U., 1988; BA in Math., Ogelthorpe U., 1988. Registered engr. in tng. Field rep. Law Assocs., Inc., Atlanta, 1988-89; assoc. quality engr. ABB Power T&D Co., Athens, Ga., 1989-91; rep. mem. svcs. Atlanta Gas Light Co., 1992—. Tutor Adult Literacy Program Gwinett Tech., Lawrenceville, Ga., 1992—. Mem. Inst. Indsl. Engrs., Nat. Soc. Profl. Engrs. Democrat. Methodist. Home: 3310 Hart Way Snellville GA 30278

HUDIK, MARTIN FRANCIS, hospital administrator, educator; b. Chgo., Mar. 27, 1949; s. Joseph and Rose (Ricker) H; 1 child, Theresa Abraham. BS in Mech. and Aerospace Engring., Ill. Inst. Tech., 1971; BPA, Jackson State U., 1974; MBA, Loyola U., Chgo., 1975; postgrad. U. Sarasota, 1975-76. Cert. health care safety mgr., hazard control mgr., hazardous materials mgr., OSHA hazardous materials response instr., hazardous materials incident comdr., disaster coord., police instr., Ill., security certification instr., Ill. With Ill. Masonic Med. Ctr., Chgo., 1969—, dir. risk mgmt., 1974-79, asst. adminstr., 1979—; part-time sr. lt. tng. div. Cicero (Ill.) Police Dept., sr. lt. Tng. and Internal Affairs Div., 1971—; instr. Nat. Safety Council Safety Tng. Inst., Chgo., 1977-85; cons. mem. Council Tech. Users Consumer Products, Underwriters Labs., Chgo., 1977—; instr., lt. U.S. Def. Civil Preparedness Agy. Staff Coll., Battle Creek, Mich., 1977-85; liaison officer to Cook County, asst. dir. Emergency Svcs. and Disaster Agy., Town of Cicero, 1988—, asst. dir.; pres. bd. dir. Cook County Emergency Mgmt. Coun., 1991-92; bd. dirs. Northside Community Fed. Credit Union, 1992-93. Pres. sch. bd. Mary Queen of Heaven Sch., Cicero, 1977-79, 84-86; pres. Mary Queen of Heaven Ch. Council, 1979-81, 83-86; pres. I.M.M.C. Employee Club, 1983-86. Ill. State scholar, 1969-71; recipient Meritorious Svc. award Town of Cicero, 1990, Spl. Svc. award Underwriters Labs. , 1992, Outstanding Svc. award People of Cook County, Ill. Emergency Health Svcs., 1993, Emergency Mgmt. award Cook County Sheriffs Dept., 1993. Mem. Am. Coll. Healthcare Execs., Am. Soc. Hosp. Risk Mgmt., Nat. Fire Protection Assn., Am. Soc. Safety Engrs. (profl.), Am. Soc. Law and Medicine, Ill. Hosp. Security and Safety Assn. (co-founder 1976, founding pres. 1976-77, hon. dir. 1977-82), Cath. Alumni Club Chgo. (bd. dirs. 1983-84, 86), Mensa, Masons (Berwyn, Ill. chpt.), Pi Tau Sigma, Tau Beta Pi, Alpha Sigma Nu. Republican. Roman Catholic. Lodges: KC (Cardinal council), Masons. Home: 2116 S 51st Ct Cicero IL 60650-2345 Office: Ill Masonic Med Ctr 836 W Wellington Ave Chicago IL 60657-5192

HUDSON, COURTNEY MORLEY, interior landscape designer; b. Shreveport, La., Nov. 19, 1955; d. Morley Alvin and Lucy (North) H. Student, So. Meth. U., 1973-74; BS in Horticulture Tech., La. State U., 1978. Owner Interiorscapes, Shreveport, 1978-83; dir. interior landscaping Lambert's, Dallas, 1983; owner Coco Hudson Co., Dallas, 1983—. Del. La. Rep. Conv., 1975. Mem. Am. Soc. Landscape Architects, North Tex. Interiorscape Assn. (pres. 1984-85), Assn. Landscape Contrs. Am. (exec. bd. 1984-86), Habitat for Humanity (vol.), Network Exec. Women in Hospitality, Comml. Real Estate Women. Avocations: sailboat racing, underwater photography, snow skiing, traveling. Office: 5501 LBJ Freeway Ste 304 Dallas TX 75240

HUDSON, HALBERT AUSTIN, JR., retired manufacturing engineer, consultant; b. Orange, Va., Dec. 20, 1923; s. Halbert Austin and Lillian Naomi (Cook) H.; m. Dorothy Alma Keilholz, Aug. 25, 1945; children: Janis Lee Hudson Bamberger, Paul Frederick. M in Engring., U. Cin., 1949. Registered profl. engr., Wis. Apprentice Cin. Milicron Co., 1942-49; signal engr. So. Ry. System, Knoxville, Tenn., 1950-53; indsl. engr., mech. mgr., project engring. mgr., tech. mgr. Procter & Gamble Co., Cin. and Green Bay, Wis., 1953-83; ret., 1983; ind. cons. engr Green Bay, 1983—; v.p. Green Bay Water Commn., 1967-77. Scoutmaster Cin. area Boy Scouts Am., 1953-59; vol. driver elderly transport ARC, Green Bay, 1983-90, mem., sec. bd. dirs., 1986-88. Capt. U.S. Army, 1943-46, ETO. Decorated Bronze Star medal, Purple Heart medal. Mem. NSPE, Assn. Energy Engrs. (chpt. pres. 1986-87), Green Bay Engring. Soc. (v.p. 1972), U. Cin. Alumni Assn., Infantry Officer Candidate Sch. Alumni Assn., Kiwanis Golden K., Pi Tau Sigma. Methodist. Achievements include patent for a milling machine arbor vibration dampener. Home and Office: 847 Spence St Green Bay WI 54304

HUDSON, JACK ALAN, electrical engineer; b. Indpls., Oct. 25, 1947; s. Marion Bradbury Hudson and Wilma Josephine (Jones) Black; m. Susan Deborah Kraft, May 24, 1975; children: Scott Andrew, Keith Alan. BSEE, Purdue U., 1969, MS in Computer Sci., 1973; MS in Engring., U. Pa., 1971; PhD in Elec. Engring., U. N.Mex., 1985. Registered profl. engr., N.Mex. Engr. GE, Phila., 1969-71; grad. researcher Purdue U., West Lafayette, Ind., 1971-74; mem. tech. staff Sandia Nat. Labs., Albuquerque, 1974-89, sr. mem. tech. staff, 1989—. Contbr. to profl. publs. Exec. bd., treas. Albuquerque Childbirth Edn. Assn., 1984-86; exec. com. N.Mex. chpt. Muscular Dystrophy Assn., Albuquerque, 1985-87; elder 1st Presbyn. Ch., Albuquerque, 1987-89; Cub pack com. chmn. Albuquerque area Boy Scouts Am., 1988-93, troop asst. scoutmaster, 1993—. Mem. IEEE, Assn. Computing Machinery, Sigma Xi. Achievements include technology transfer of research on computer aided design tools algorithms and software to public corporation for commercialization. Office: Sandia Nat Labs Albuquerque NM 87185

HUDSON, KENNETH SHANE, electrical engineer; b. Charleston, W.Va., Aug. 25, 1967; s. Kenneth H. and Elaine E. (Smith) H. BSEE magna cum laude, W.Va. Inst. Tech., 1989; postgrad., Marshall U., 1990—. Survey asst. Appalachian Power Co., Charleston, summer 1986, drafting asst., summer 1987; coop. engr. U.S. Dept. Def., Balt., summer 1988; elec. engr. Rhone-Poulenc AG Corp., Balt., 1989—. Mem. W.Va. Inst. Tech. Alumni Assn. (v.p. Charleston chpt. 1991—), Tau Beta Pi, Eta Kappa Nu, Alpha Chi. Home: 114 Stone Dr Elkview WV 25071-9326

HUDSON, THOMAS, robotics company executive; b. Karlsruhe, Fed. Republic Germany, May 22, 1954; came to U.S., 1971; s. Marlin Ridgeway Hudson and Brigitte Jutta (Imhof) Hunter. Diploma, Cornell U., 1980, Granton Inst. Tech., Toronto, Ont., Can., 1983; AA, Fla. Community Coll., 1987; student, U. North Fla., 1989. Asst. mgr. Publix, Jacksonville, Fla., 1980-83; tutor Fla. Community Coll., Jacksonville, 1983-86, VA, Jacksonville, 1987; pres., chief exec. officer Atlantis Co., Jacksonville, 1987-89; founder, pres., chief exec. officer Robotronix, Jacksonville, 1989—; cons. in field. Author: Atlantis, 1988; inventor, Volksrobot, 1989. Mem. Am. Soc. Student Actuaries, Am. Math. Soc., Math. Assn. Am., Soc. Indsl. and Applied Math., Nat. Svc. Robot Assn., Math. and Stat. Club U. North Fla. Avocations: mathematics, electronics, robotics, science fiction. Home: 2404 Looking Glass Ln Jacksonville FL 32210-3625 Office: Robotronix 5115 Jammes Rd Jacksonville FL 32210 7718

HUDSON, WILLIAM RONALD, transportation engineering educator; b. Temple, Tex., May 17, 1933; s. Clarence W. and Nan S. Hudson; m. Martha Ann Collins, May 6, 1936; children: Stuart William, Alan David, Paul Collin. BS, Tex. A&M U., 1954, MS, 1955; PhD, U. Tex., Austin, 1965. Registered profl. engr., Tex., Ariz., Kans., Ind., Ill.; registered pub. surveyor, Tex. Civil engr. S.J. Buchanan and Assocs., Cons. Engrs., Bryan, Tex., 1957-58; asst. chief rigid pavement research br. AASHO Road Test, Nat. Acad. Scis., Ottawa, Ill., 1955-61; asst. project engr. Nat. Coop. Hwy. Research Program Project I-1 Nat. Acad. Scis., 1964-65; supervising designing research engr. Hwy. Design Div. Tex. Hwy. Dept., Austin, 1961-63; instr. civil engring., research engr. U. Tex., Austin, 1963-65; asst. prof., research engr. Ctr. for Hwy. Research, U. Tex., Austin, 1965-68, acting dir., 1969; asst. dean engring. U. Tex., 1969-70, assoc. dean engring., 1970-72, assoc. prof. civil engring., dir. Pavement Systems Research Lab., 1968-73, prof. civil engring., 1973—, dir. Council for Advanced Transp. Studies, 1972-75, internat. tech. dir. Brazilian Hwy Project, 1975-80, coordinator of chair of free enterprise, 1977-78, area coordinator for transp., dir. Pavement Systems Research Lab., 1977-81, Dewitt C. Greer Centennial prof. engring., area coordinator transp., 1981—; mem. student affairs dean's council, 1969-70, U. Tex., Austin, chmn. univ. com. on fin. aid to students, 1970-71; program mgr. long-term pavement performance study Strategic Hwy. Rsch. Program, 1987-90. Author: Pavement Management Systems, 1978; over 250 tech. reports and papers; over 300 oral presentations at nat. and internat. meetings. Maj. USAF, 1955-57. Recipient Engring. Found. Adv. Council award, 1977-78; also various awards for research papers; named keynote speaker several confs. on pavement and transp. topics. Mem. ASCE (chmn. exec. com. hwy. div. 1983-84, chmn. econ. affairs Tex. sect. 1983-85; numerous other com. activities; J. James R. Croes medal 1968), Am. Soc. Testing and Materials, Am. Soc. Engring. Edn., Assn. Asphalt Paving Technologists, Am. Concrete Inst., Transp. Research Bd., Nat. Acad. Scis. (numerous com. activities on pavement mgmt. and design), Internat. Soc. Soil Mechanics and Found. Engring., NSPE, Tex. Soc. Profl. Engrs., Tex. Assn. Coll. Tchrs., Sigma Xi, Chi Epsilon, Tau Beta Pi. Republican. Office: Univ Tex Dept of Civil Engring ECJ 6.10 Austin TX 78712-1076

HUDSPETH, WILLIAM JEAN, neuroscientist; b. Burbank, Calif., July 24, 1935; s. Weldon Benton and Corinne Lenore (Murray) H.; divorced; children: William Benton, Roger Ashley, Ian Holmes. BA, San Jose State U., 1960, MA, 1962; PhD, Claremont Grad. Sch., 1966. Postdoctoral scholar Brain Rsch. Inst., UCLA, 1966-68; assoc. prof. biopsychology U. Waterloo, Ont., Can., 1968-72; rsch. asst. prof. psychiatry N.Y. Med. Coll., N.Y.C., 1972-74; assoc. prof. behavioral biology U. Nev. Sch. Medicine, Reno, 1974-81; assoc. prof. psychology U. No. Colo., Greeley, 1984-89; assoc. prof. cognitive neurosci. Radford (Va.) U., 1989—; proprietor Neuropsychometric Labs., Radford, 1981—; prin. investigator NRC Can., U. Waterloo, 1968-71, Ont. Mental Health Found., U. Waterloo, 1971-72; predoctoral rsch. fellow NIMH, Claremont Grad. Sch., 1962-65. With USN, 1952-56, Korea. Mem. AAAS, Am. Psychol. Soc., N.Y. Acad. Scis., Sigma Xi. Achievements include patent for neurocybernetic devices. Home: 415-B Sanford St Radford VA 24141 Office: Radford U Ctr Brain Rsch Russell Hall Rm 433 Radford VA 24142

HUDY, JOHN JOSEPH, psychologist; b. Milw., June 10, 1956; s. Jack Donald and Florence Rosemary (Harlow) H. BA in Psychology, U. Wis., Milw., 1979; MA in Exptl. Psychology, U. South Fla., 1981, PhD in Indsl. Psychology, 1985. Chief of manpower City of Clearwater (Fla.), 1982-87; project mgr. CORE Corp., Pleasant Hill, Calif., 1987-91; rsch. psychologist Acumen Internat., San Rafael, Calif., 1991—; pres. Mgmt. Systems Inc., Alameda, Calif., 1991—; bd. dirs. U.S.F. Psychology Found., Tampa, Fla., 1982-87. Reviewer Acad. Mgmt. Rev. jour.; contbr. articles to profl. jours. Speaker, writer Zero Population Growth, Berkeley, Calif., 1987-91. Recipient Kemper-Knapp Found. scholarships U. Wis. Milw., 1975, 76. Mem. APA, Acad. of Mgmt. Achievements include rsch. on computerized expert-system mgmt. assessment and feedback systems, data from U.S., France, Japan, Great Britain showing trait distributions relevant to effective performance. Home: 1353 Park Ave Alameda CA 94501 Office: Acumen Internat 3950 Civic Ctr Dr Ste 310N San Rafael CA 94903

HUEBNER, JOHN STEPHEN, geologist; b. Bryn Mawr, Pa., Sept. 9, 1940; s. John Mudie and Elizabeth (Converse) H.; m. Emily Mayer Zug, June 16, 1962; children: Christopher Converse, Jeffrey Worrell. A.B. magna cum laude, Princeton U., 1962; Ph.D., Johns Hopkins U., 1967. Research geologist U.S. Geol. Survey, 1967—; cons. NASA, 1976-78; lectr. George Washington U., 1971; sec.-treas. Am. Geol. Inst., 1974-75. Assoc. editor: Jour. Geophys. Research, 1977-79; Contbr. articles profl. jours. Pres. Wood Acres Citizens Assn., 1977-78. Fellow Mineral. Soc. Am. (bd. dirs. 1985-88, recipient MSA award 1978); mem. AAAS, Geochem. Soc. (treas. 1972-75), Am. Geophys. Union, Geol. Soc. Washington (v.p. 1991, pres. 1992), Sigma Xi. Club: Cosmos (Washington). Home: 6102 Cromwell Dr Bethesda MD 20816-3410 Office: 959 National Center Reston VA 22092

HUELKE, DONALD FRED, anatomy and cell biology educator, research scientist; b. Ill., Aug. 20, 1930; s. Arthur August and Laura Sophia (Malon) H.; m. Jean Louise Kilbert (dec.); children: Donna, David; m. Linda Sue Westhoven, Sept. 12, 1943. B.S. in Zoology, U. Ill., 1952, M.S. in Physiology, 1954; Ph.D. in Anatomy, U. Mich., 1957. Teaching fellow U. Ill., Urbana, 1952-54; Teaching fellow U. Mich., Ann Arbor, 1954-57, instr. in anatomy, 1957-60, asst. prof. anatomy, 1960-63, assoc. prof. anatomy, 1963-68, prof., 1968—; cons., Gen. Motors, Warren, Mich., 1965—, Ford Motor Co., Dearborn, Mich., 1966-82, Chrysler, Highland Pk., Ill., 1980—. Recipient Disting. Achievement award Mich. Driver & Traffic Safety Edn. Assn., 1978, 89. Mem. Assn. Advancement Automotive Medicine (award of merit 1982), Soc. Automotive Engrs. (Ralph Isbrandt award 1980, award for excellence in oral presentation 1977, 79, 80, 82, 89, 91), Internat. Assn. Auto and Traffic Medicine, Am. Trauma Soc. Republican. Lutheran. Home: 510 Heritage Dr Ann Arbor MI 48105-2556 Office: U Mich Dept Anatomy Ann Arbor MI 48109 also: 2901 Baxter Rd Ann Arbor MI 48109

HUENING, WALTER CARL, JR., retired consulting application engineer; b. Boston, Feb. 10, 1923; s. Walter Carl and Gladys (Whittemore) H.; m. Margaret Laurence McGeary, Aug. 5, 1944 (dec. 1986); children: Peter Carl, Susan Laurence Huening Locke; m. Elizabeth Ann Young Wright, Apr. 9, 1988. BSEE magna cum laude, Tufts U., 1944. Registered profl. engr., N.Y., Ohio. Instr. elec. engring. Tufts U., Medford, Mass., 1946-48; distbn. engr. plant engring. dept. GE, Lynn, Mass., 1948-50; application engr. indsl. power engring. GE, Schenectady, N.Y., 1952-56; product planner protective devices dept. GE, Plainville, Conn., 1956-58; design engr. vacuum cleaner dept. GE, Cleve., 1958-59; application engr. comml. and mcpl. dept. GE, Schenectady, 1960-62, application engr. steel mill, 1962-68, cons. application engr. indsl. power engring., 1968-89; mem. U.S. nat. com. Internat. Electrotech. Commn., tech. advisor on Tech. Com. 73 matters, 1972-89. Contbr. tech. papers to jours. and chpts. to books; patentee vacuum cleaner latch. Lt. comdr. USNR, 1944-46, 50-52, ret. Fellow IEEE (life, R. H. Kaufmann award 1988, Indsl. and Comml. Power Systems Dept. Achievement award 1989, prizes for papers 1970, 82); mem. Tau Beta Pi. Republican. Avocations: photography, collecting recorded traditional jazz music. Address: 1229 Godfrey Ln Schenectady NY 12309

HUESTIS, MARILYN ANN, toxicologist, clinical chemist; b. Agana, Guam, Nov. 10, 1948; d. John Paca and Margaret Ann (Bolgiano) Thomas; m. Michael Milchanowski, Dec. 21, 1969 (div. Oct. 1979); children: Michael Thomas Milchanowski, Allison Barnes Milchanowski; m. Walter Wess Huestis, May 22, 1982. AB, Mt. Holyoke Coll., 1970; MS, U. N.Mex., 1979; PhD, U. Md., 1992. Chemist, toxicologist N.Mex. Reference Lab., Albuquerque, 1979-80; tech. dir. Meridian Pathology Lab., Meridian, Miss., 1980-82; asst. toxicologist Office of Chief Med. Examiner, Balt., 1988-89; chief toxicologist Nichols Inst., San Diego, 1983-88; lab. insp., instr. Nat. Lab. Cert. Program Substance Abuse and Mental Health Svcs. Adminstrn., Bethesda, Md., 1989—; assoc. Duo Rsch., Inc., Annapolis, Md., 1989-92; rsch. staff fellow Addiction Rsch. Ctr., Nat. Inst. Drug Abuse, Balt., 1989-92; toxicology cons. Westlake Village, Calif., 1992—. Manuscript reviewer Jour. Analytical Toxicology, Forensic Sci. Review, Am. Hosp. Formulary Svc.; contbr. articles to profl. jours. Recipient Rsch. fellowship U. Md., 1988-92, Mary Lyon award Mt. Holyoke Coll., 1969, 70. Fellow Am. Acad. Forensic Scis. (Irving Sunshine Rsch. award 1992); mem. Soc. Forensic Toxicologists, Am. Assn. Clin. Chemistry, Internat. Assn. Forensic Toxicologists, Calif. Assn. Toxicologists. Episcopalian. Achievements include research in mechanisms of actions of drugs of abuse, analytical methods of drugs, performance effects of drugs, predictive models for interpreting drug levels, quality control of analytical methods. Office: 1030 Brookview Ave Westlake Village CA 91361

HUETING, JUERGEN, internist; b. Bad Oeynhausen, Fed. Republic Germany, May 21, 1956; s. Karl and Lotte H. Grad. magna cum laude, Groton Sch., Mass., 1974; grad. summa cum laude, Immanuel Kant Gymnasium, Bad Oeynhausen, 1975; MD, U. Giessen, Fed. Republic Germany, 1985. Intern, then resident in internal medicine U. Giessen, 1983-87; mem. staff clin. cardiology Max Planck Soc., Bad Nauheim, Fed. Republic Germany, 1987—; mem. sci. coun. Internat. Coll. Angiology; mem. coun. clin. cardiology Am. Heart Assn. Contbr. articles to prof. jours. Recipient Assisi Medicine award in Cardionephrology, 1991. Avocations: organ recitalist, art, photography, tennis. Home: Hintergasse 9, 6250 Limburg 4 Federal Republic of Germany Office: Kerckhoff-Klinik, Benekestrasse 2-8, 61231 Bad Nauheim Germany

HUETTNER, RICHARD ALFRED, lawyer; b. N.Y.C., Mar. 25, 1927; s. Alfred F. and Mary (Reilly) H.; children—Jennifer Mary, Barbara Bryan; m. 2d, Eunice Bizzell Dowd, Aug. 22, 1971. Marine Engrs. License, N.Y. State Maritime Acad., 1947; B.S., Yale U. Sch. Engring., 1949; J.D., U. Pa., 1952. Bar: D.C. 1952, N.Y. 1954, U.S. Ct. Mil. Appeals 1953, U.S. Ct. Claims 1961, U.S. Supreme Ct. 1969, U.S. Ct. Appeals (fed. cir.) 1982, also other fed. cts, registered to practice U.S. Patent and Trademark Office 1957, Canadian Patent Office 1968. Engr. Jones & Laughlin Steel Corp., 1954-55; assoc. atty. firm Kenyon & Kenyon, N.Y.C., 1955-61; mem. firm Kenyon & Kenyon, 1961—; specialist patent, trademark and copyright law. Trustee N.J. Shakespeare Festival, 1972-79, sec., 1977-79; trustee Overlook Hosp., Summit, N.J., 1978-84, 86-89, vice chmn. bd. trustees, 1980-82, chmn. bd. trustees, 1982-84; trustee Overlook Found., 1981-89 , chmn. bd. trustees, 1986-89, emeritus trustee, 1991; trustee Colonial Symphony Orch., Madison, N.J., 1972-82, v.p. bd. trustees 1974-76. pres. 1976-79; chmn. bd. overseers N.J. Consortium for Performing Arts, 1972-74; mem. Yale U. Council, 1978-81; bd. dirs. Yale Communications Bd., 1978-80; chmn. bd. trustees Center for Addictive Illnesses, Morristown, N.J., 1979-82; rep. Assn. Yale Alumni, 1975-80, chmn. com. undergrad. admissions, 1976-78, bd. govs., 1976-80, chmn. bd. govs., 1978-80; chmn. Yale Alumni Schs. Com. N.Y., 1972-78; assoc. fellow Silliman Coll., Yale U., 1976—; bd. dirs., exec. com. Yale U. Alumni Fund, 1978-81; mem. Yale Class of 1949 Council, 1980—; bd. dirs. Overlook Health Systems, 1984—. Served from midshipman to lt. USNR, 1945-47, 52-54; cert. JAGC 1953; Res. ret. Recipient Yale medal, 1983, Disting. Svc. to Yale Class of 1949 award, 1989, Yale Sci. and Engring. Meritorious Svc. award, 1992. Fellow N.Y. Bar Found.; mem. ABA, N.Y. State Bar Assn., Assn. Bar City N.Y., N.Y. Patent Trademark Copyright Law Assn. (chmn. com. mtgs. 1961-64, chmn. com. econ. matters 1966-69, 72-74), AAAS, N.Y. Acad. Scis., N.Y. County Lawyers Assn., Am. Intellectual Property Law Assn., Internat. Patent and Trademark Assn., Am. Judicature Soc., Yale Sci. and Engring. Assn. (v.p. 1973-75, pres. 1975-78, exec. bd. 1972-79), Fed. Bar Coun. Clubs: Yale (N.Y.C.); Yale of Central N.J. (Summit) (trustee 1973-88, pres. 1975-77), Morris County Golf (Convent, N.J.); The Graduates (New Haven). Home: 150 Green Ave Madison NJ 07940-2513 Office: Kenyon & Kenyon One Broadway New York NY 10004-1007

HUEY, BEVERLY MESSICK, psychologist; b. Havre de Grace, Md., Nov. 14, 1961; d. Merrill Anthony Jr. and Betty Jane (Ehrman) Messick; m. Richard Wesley Huey Jr., Dec. 7, 1985; 1 child, Lauren Nicole. BS in Psychology and Sociology, Va. Poly. Inst. and State U., Blacksburg, 1983; MA, George Mason U., 1985, PhD, 1990. Sr. staff officer Com. on Human Factors Nat. Rsch. Coun., Washington, 1986—; study dir. U.S. Nat. Com. Internat. Union Psychol. Sci., Washington, 1986—; student contractor/psychologist U.S. Army Human Engring. Lab., Md., 1985-89; cons. and data analyst Counseling Ctr. for Reproductive Psychology, Reston, Va., 1985-86. Co-editor: Human Performance Models for Computer-Aided Engineering, 1990, Workload Transition, 1993; author reports in field. Mem. mut. ministry com. and long-range planning com. Luth. Ch. of Good Shepherd, Bel Air, Md., 1991—. Mem. Am. Psychol. Soc., Human Factors and Ergonomics Soc., Psi Chi, Alpha Kappa Delta. Democrat. Home: 1403 Gunston Rd Bel Air MD 21015 Office: Nat Rsch Coun 2101 Constitution Ave NW Washington DC 20418

HUFF, CHARLES WILLIAM, psychologist, educator; b. Tampa, Fla., Oct. 9, 1955; s. Charles William Huff and Fanny Lou (Burke) Sand; m. Sheryl Joy Goetzinger, Aug. 15, 1982. BS, Ga. So. U., 1978; MA, Coll. of William and Mary, 1982, Princeton U., 1984; PhD, Princeton U., 1987. Teaching asst. Coll. William and Mary, Williamsburg, Va., 1981-82, Princeton (N.J.) U., 1982-84; tech. rsch. staff Bell Communications Rsch., Murray Hill, N.J., summer 1984; computer cons. dept. psychology Princeton U., 1984-86; consulting staff Rsch. Svcs., Princeton Computer Ctr., 1984-86; postdoctoral fellow com. for social sci. rsch. in computing Carnegie Mellon U., Pitts., 1986-88; asst. prof. dept. psychology St. Olaf Coll., Northfield, Minn., 1988-92, assoc. prf., 1993—. Editor: Social Issues in Computing: Putting Computers in their Place, 1993; assoc. editor Social Sci. Computer Review, 1992—; contbr. articles to profl. jours. Recipient Prin. Investigator award NSF, 1991, Joyce Found., 1991. Democrat. Methodist. Home: 409 W Woodley St Northfield MN 55057 Office: St Olaf Coll Dept of Psychology 1520 St Olaf Ave Northfield MN 55057-1098

HUFF, RONALD GARLAND, mechanical engineer; b. Toledo, Ohio, Dec. 29, 1930; s. Blenn Chalmer and Helen Ester (Schling) H.; m. Nancy Carroll Warns, June 29, 1957; children: Dennis Lee, Deborah Lynn. BSME, U. Toledo, 1953. Aero. engr. Nat. Adv. Com. for Aeronautics, Cleve., 1955-58; aerospace tech. NASA, Cleve., 1958-87; cons./proprietor Ronald G. Huff & Assocs., Cleve., 1986—. Contbr. articles to profl. jours. Photographer North Olmsted Band Boosters, Ohio, 1974-80; active PTA, North Olmsted, 1969-72. 1st lt. U.S. Army, 1953-55. Mem. ASME (chmn. winter ann.meeting 1986), AIAA. Congregationalist. Achievements include patent on supersonic jet noise suppressor; method for measuring internal hot gas side wall temperatures in thin wall generatively cooled rocket engines. Home and Office: Huff & Assocs 3741 Cinnamon Way Westlake OH 44145-5717

HUFF, WELCOME REX ANTHONY, chemical researcher; b. Indpls., Mar. 26, 1967; s. Welcome Charles and Judith Kathleen (Payton) H. BS in Chemistry with honors, Ind. U., 1989. Undergrad. researcher G.E. Ewing Group, Ind. U., Bloomington, 1988-89; grad. researcher D.A. Shirley Group, U. Calif., Berkeley, 1989—. Fundraiser Multiple Sclerosis Soc., Monterey, Calif., 1990. H.G. Day Summer Rsch. scholar Ind. U. Chemistry Dept., 1988, H.G. Day Acad. Yr. Rsch. scholar Ind. U. Chemistry Dept., 1988. Mem. AAAS, NRA, Am. Chem. Soc., Am. Phys. Soc., Internat. Platform Assn., Alpha Chi Sigma. Home: 25 Neva Ct Oakland CA 94611 Office: Lawrence Berkeley Lab MS 2-300 1 Cyclotron Rd Berkeley CA 94720

HUFFMAN, D. C., JR, pharmacology educator, health science association administrator. BS in Pharmacy, U. Ark., 1966; PhD in Pharmacy Adminstrn., U. Miss., 1971. Prof. Coll. Pharmacy, chmn. dept., adminstrn. v.p. pharmacy and mgmt. svcs. U. Tenn.; presenter numerous seminars. Contbr. articles to profl. jours. Fellow NDEA, Am. Found. Pharm. Edn.; mem. Am. coll. Apothecaries (exec. v.p.), Nat. Assn. Retail Druggists (sr. v.p. pharmacy practice and mgmt. svcs.), Rho Chi. Office: American College of Apothecaries 205 Daingerfield Rd Alexandria VA 22314

HUFFMAN, HENRY SAMUEL, building mechanical engineer; b. Springfield, Ill., Aug. 18, 1926; s. Henry Samuel Sr. and Ruth (Kincaid) H.; m. Dorothy Fayrene Progrelis, June 28, 1945 (dec. Feb. 1978); children: Michael Blaine, David Berkley, Reid Barton, Cheryl Ann, Richard Samuel; m. Judith Ann Brown, Nov. 22, 1980. BS in Engring., U. Mo., 1949. Registered profl. engr., Mo., Minn., La., Calif., Ariz., Ill., Nev., Ind., Miss., D.C. Mgmt. trainee Brown Shoe Co., St. Louis, 1949-51; repair depot mgr. GE Appliances, St. Louis, 1951-53; distbn. agt. Sinclair Oil Co., Wentzville, Mo., 1953-55; project engr. Honeywell Regulator Co., St. Louis, 1955-57, Bank Bldg. and Equipment Co., St. Louis, 1957-67; sr. staff engr. Fru-Con Engrs. Inc., St. Louis, 1967-92, cons., 1992—; cons. Bristol-Meyers Squibb, Co., Evansville, 1992—, Proctor & Gamble, New Orleans, 1991—, Fuin Colnon Constrn., St. Louis, 1991—. Cubmaster Boy Scouts Am., 1955. With AAF, 1944-45. Mem. ASHRAE, Nat. Fire Protection Assn., Assn. of Energy Engrs. (cert. energy mgr.), Engrs. Club of St. Louis. Baptist. Achievements include design of unique air distribution/variable volume air conditioning systems.

HUFFMAN, KENNETH ALAN, operations researcher, mathematician; b. Richmond, Ind., July 12, 1941; s. William Raymond and Sarah Elenor (Miller) H.; m. Melena Robinson, Jan. 26, 1965; children: Michael Alan, Gregory Clark, Graeleigh. BA in Math., Miami U., Oxford, Ohio, 1963; MS in Ops. Rsch., Naval Postgrad. Sch., Monterey, Calif., 1969. Commd. officer U.S. Navy, various locations, 1963-83; assoc. Booz-Allen & Hamilton, Inc., Bethesda, Md., 1983-87; def. analysis officer, dir. def. ops. divsn. U.S. Mission to NATO, Brussels, 1988—. Contbr. articles to profl. jours. Bd. dirs. Village Preservation and Improvement Soc., Falls Church, Va., 1984-87. Mem. Armed Forces Communs.-Electronics Assn. (programs officer Brussels 1989-91), Ops. Rsch. Soc. Am., Sigma Xi. Office: US Mission to NATO PSC 81 Box 55 APO AE 09724

HUFTON, JEFFREY RAYMOND, chemical engineer; b. Pitts., Feb. 11, 1964; s. Harry Lee and Kathleen Julia (Skvarla) H.; m. Bonnie Lyn McIntosh, July 19, 1986; children: Travis Seldon, Katlyn Jean. BS in Chem. Engring., Pa. State U., 1986, PhD in Chem. Engring., 1992; M.S. in Chem. Engring., U. Tex., 1988. Postdoctoral fellow U. New Brunswick, Fredericton, N.B., Can., 1992—; presenter Pa. State U. NASA Space Grant Program, 1991. Contbr. articles to profl. jours. GE/Pa. State U. grantee, 1990; Lawrence J. Ostermayer Meml. scholar, 1984-85. Mem. Am. Inst. Chem. Engrs., Am. Chem. Soc., Pa. State U. Grad. Student Assn. (del. 1988-91), Sigma Xi, Tau Beta Pi. Democrat. Achievements include research in field of chemical separations; experimental studies of gas adsorption systems, design of adsorption-based separation processes, computer simulation of gas/solid systems. Home: 273 McWilliams Rd Trafford PA 15085

HUG, RUDOLF PETER, chemist; b. Winterthur, Switzerland, June 5, 1944; s. Roger Ernest and Charlotte Lydia (Peter) H.; m. Barbara Maria F. Hubatka, Sept. 17, 1983; children: Anna-Maria, Floriana, Rudolf Bruno, Patrik Roger. PhD, U. Zurich, 1972. Postdoctoral fellow Eid Inst Reaktorforsch., Wührenlingen, Switzerland, 1973, U. East Anglia, Norwich, Great Britain, 1974, U. Munich, 1975; dept. head prodn. Roche Basel, Switzerland, 1976-83; dept. head rsch. Givaudan-Roure, Dubendorf, Switzerland, 1984—. Contbr. articles to profl. jours. mem. Swiss Chem. Soc., Am. Chem. Soc., Royal Soc. Chemistry, German Chem. Soc., Lions. Achievements include patent for development of industrial syntheses of pharmaceuticals. Home: Saint Gallerstr 130, 8404 Winterthur Switzerland Office: Givaudan-Roure Rsch, 8600 Dubendorf Switzerland

HUGGARD, RICHARD JAMES, federal agency administrator; b. Norton, N.B., Can., June 18, 1935; s. Russell Clyde and Lillian Erna (Gillies) H.; m. Marjorie MacRae, Apr. 16, 1960; children: Richard James, Lesley Anne. Diploma, N.S. Agrl. Coll., Truro, 1956; BS in Agriculture, McGill U., 1958; MS, U. Ill., 1965. Beef fieldman N.S. Dept. Agriculture & Mktg., Truro, 1958-65, livestock supt., 1965-73, dir. livestock br., 1973-75, dir. ext. br., 1975-86; chief dir. ops. N.S. Dept. Agriculture & Mktg., Halifax, 1986-90, exec. dir. adminstrn., 1990-91, dep. minister, 1991—; dir. Royal Agrl. Winter Fair, Toronto, Ont., 1991—. Sec.-treas. St. John's Laymen's Assn., Truro, 1987—. Recipient Erland Lee award Federated Women's Insts. Can., 1992. Agrl. Inst. Can. Fellowship award, 1992; mem. Can. Soc. Animal Sci. (Cert of merit 1981), Can. Soc. of Ext. (hon., life), Can. Agrl. Hall of Fame Assn. (v.p. 1992—), N.S. Inst. Agrologists. Home: NSDAM, 59 Shannon Dr, Truro, NS Canada B2N 3V7 Office: NS Dept Agriculture & Mktg, PO Box 190, 1690 Hollis St, Halifax, NS Canada B3J 2M4

HUGGINS, CHARLES BRENTON, surgical educator; b. Halifax, N.S., Can., Sept. 22, 1901; s. Charles Edward and Bessie (Spencer) H.; m. Margaret Wellman, July 29, 1927; children: Charles Edward, Emily Wellman Huggins Fine. BA, Acadia U., 1920, DSc (hon.), 1946; MD, Harvard U., 1924; MSc, Yale U., 1947; DSc (hon.), Washington U., St. Louis, 1950, Leeds U., 1953, Turin U., 1957, Trinity Coll., 1965, U. Wales, 1967, U. Mich., 1968, Med. Coll. Ohio, 1973, Gustavus Adolphus Coll., 1975, Wilmington (Ohio) Coll., 1980, U. Louisville, 1980; LLD (hon.), U. Aberdeen, 1966, York U., Toronto, 1968, U. Calif., Berkeley, 1968; D of Pub. Service (hon.), George Washington U., 1967; D of Pub. Service (hon.) sigillum magnum, Bologna U., 1964. Intern in surgery U. Mich., 1924-26, instr. surgery, 1926-27; with U. Chgo., 1927—, instr. surgery, 1927-29, asst. prof., 1929-33, assoc. prof., 1933-36, prof. surgery, 1936—, dir. Ben May Lab. for Cancer Research, 1951-69, William B. Ogden Disting. Service prof., 1962—; chancellor Acadia U., Wolfville, N.S., 1972-79; Macewen lectr. U. Glasgow, 1958, Ravdin lectr., 1974, Powell lectr., Lucy Wortham James lectr., 1975, Robert V. Day lectr., 1975, Cartwright lectr., 1975. Trustee Worcester Found. Exptl. Biology; bd. govs. Weizmann Inst. Sci., Rehovot, Israel, 1973—. Decorated Order Pour le Mérite Germany; Order of The Sun Peru; recipient Nobel prize for medicine, 1966, Am. Urol. Assn. award, 1948, Francis Amory award, 1948, AMA Gold medals, 1936, 40, Société Internationale d'Urologie award, 1948, Am. Cancer Soc. award, 1953, Bertner award M.D. Anderson Hosp., 1953, Am. Pharm. Mfrs. Assn. award, 1953, Gold medal Am. Assn. Genito-Urinary Surgeons, 1955, Borden award Assn. Am. Med. Colls., 1955, Comfort Crookshank award Middlesex Hosp., London, 1957, Cameron prize Edinburg U., 1958, Valentine prize N.Y. Acad. Medicine, 1962, Hunter award Am. Therapeutic Soc., 1962, Lasker award for med. research, 1963, Gold medal Virchow Soc., 1964, Laurea award Am. Urol. Assn., 1966, Gold medal Worshipful Soc. Apothecaries of London, 1966, Gairdner award Toronto, 1966, Chgo. Med. Soc. award, 1967, Centennial medal Acadia U., 1967, Hamilton award Ill. Med. Soc., 1967, Bigelow medal Boston Surg. Soc., 1967, Disting. Service award Am. Soc. Abdominal Surgeons, 1972, Sheen award AMA, 1970, Sesquicentennial Commemorative award Nat. Library of Medicine, 1986; Charles Mickle fellow, 1958. Fellow ACS (hon.), Royal Coll. Surgeons Can. (hon.), Royal Coll. Surgeons Scotland (hon.), Royal Coll. Surgeons England (hon.), Royal Soc. Edinburgh (hon.), La Academia Nacional de Medicina (Mexico, hon.); mem. NAS (Charles L. Meyer award for cancer research 1943), Am. Philos Soc. (Franklin medal 1985), Am. Assn. Cancer Rsch., Can. Med. Assn. (hon.), Alpha Omega Alpha. Home: 5807 S Dorchester Ave Chicago IL 60637-1729 Office: U Chgo Ben May Lab Cancer Rsch 5841 S Maryland Ave Chicago IL 60637-1470

HUGGINS, JAMES ANTHONY, biology educator; b. Flint, Mich., Oct. 1, 1953; s. James Polk and Wava Jean (Phillips) H.; m. Cathy Sue Hester, June 2, 1973; 1 child, Kyle Anthony. BS in Agr., Ark. State U., Jonesboro, 1975, MS in Biology, 1977; PhD in Biology, Memphis State U., 1985, postgrad., 1991—. Asst. prof. Miss. Indsl. Coll., Holly Springs, 1980; instr. Memphis State U., 1981-82; asst. prof. Shelby State Community Coll., Memphis, 1982-87; prof., chair div. sci. Union U., Memphis, 1987—; frequent speaker creation/evolution confs. Author: Intraspecific Variation Within the Prairie Vole, 1984, Morphologic Variation in Shrews, 1989, Chromosome Morphology in Shrews, 1991. Pres. Bible Sci. Assn., Memphis, 1989—; deacon chair Berclair Bapt. Ch., Memphis, 1989-91, 93-94. Named one of Outstanding Young Men of Am. 1988-89. Mem. Soc. Systematic Zoologists, Tenn. Med. Tech. Soc., Sigma Xi. Baptist. Achievements include research in morphologic variation in mammals as affected by environment and interspecific interaction chromosome morphology of shrews, sympatric speciation. Home: 5115 Battle Creek Dr Memphis TN 38134-4307 Office: Union U Bapt Meml Hosp Campus 999 Monroe Ave Memphis TN 38104-3110

HUGHES, ARLEIGH BRUCE, microbiologist, educator; b. San Marcos, Tex., Mar. 15, 1930; s. Arlie P. and Ada B. (Hughes) H.; m. Charlene J. Rubin. BA, Tex. U., 1951, MA, 1955; PhD, Tex. A&M U., 1968. Asst. prof. So. Colo. State Coll., Pueblo, 1968-70; prof. microbiology Maui C.C., Kahului, Hawaii, 1970—. Mem. Am. Soc. Microbiology. Home: 158 Haulani St Pukalani HI 96768 Office: Maui C C Divsn Sci and Math 310 Kaahumanu Ave Kahului HI 96732

HUGHES, IAN FRANK, steel company executive; b. Douglas, Eng., Apr. 14, 1940; came to U.S. 1967; m. Doreen Hughes, July 12, 1962; children: Robert Ian, Vanessa Marian. BS in Metallurgy, London U., 1962, PhD in Metallurgy, 1966. Research scientist Richard Thomas & Baldwins Steel Co., Eng., 1958-62; research fellow Nat. Phys. Lab., Eng., 1965-67; research engr. Inland Steel Co., East Chicago, Ind., 1967-69, sr. research engr., 1969-72, asst. dir., 1972-76, assoc. dir., 1976-80, dir., 1980-84, gen. mgr., 1984-86, v.p., 1986—, v.p. tech., 1987—. Office: Inland Steel Industries Inc 3210 Watling St MC 8-100 East Chicago IN 46312-1716*

HUGHES, JAMES ARTHUR, electrical engineer; b. Wayne, Nebr., Feb. 15, 1939; s. James Wallace and Ruth Genevieve H.; m. Judy Lorraine Gaskins, July 18, 1967; children: Robert Linn, Benjamin Reed, Barnaby James. BSEE, U. Nebr., 1967. Electronic technician, space tech. labs. TRW, Redondo Beach, Calif., 1963-67; mem. tech. staff systems group TRW, 1967-80, sect. mgr. electronics and def. div., 1980-82, systems engr. space and tech. group, 1982--. Designer solid state thermostat, pn generator. Deacon First Bapt. Ch. Lakewood, Long Beach, Calif., 1975-76, 78-80, 87-89; mem. exec. bd. parent-tchr. fellowship, Grace Sch., Rossmoor, Calif., 1981-87. With USN, 1959-63. Mem. AAAS, IEEE, Nat. Soc. Profl. Engrs. Republican. Avocations: sailing, youth sports, photography, personal computing. Office: TRW Space and Def One Space Pk S/1869 Redondo Beach CA 90278

HUGHES, JOHN RUSSELL, physician, educator; b. DuBois, Pa., Dec. 19, 1928; s. John Henry and Alice (Cooper) H.; m. Mary Ann Dick, June 14, 1958; children: John Russell Jr. (dec.), Christopher Alan, Thomas Gregory, Cheryl Ann. A.B. summa cum laude, Franklin and Marshall Coll., 1950; B.A. with honors, Oxford (Eng.) U., 1952, M.A. with honors, 1955, D.M. hon., 1976; Ph.D., Harvard U., 1954; M.D., Northwestern U., 1975. Neurophysiologist, NIH, 1954-56; dir. electroencephalography dept. Meyer Hosp., SUNY, 1956-63; dir. div. lab. scis., including electroencephalography Northwestern U. Med. Center, 1963-77, prof. neurology, 1968—; dir. EEG and Epilepsy Clinic, U. Ill. Med. Center, 1977—; staff U. Ill. Hosp., Community Hosp., Geneva, Delnor Hosp., St. Charles; dir.neurophysiology Humana-Michael-Reese Med. Ctr., 1992—; cons. Chgo. VA Westside Hosp., Mercyville and Copley Meml. Hosp., Aurora, Ill., others; participant debate on brain death BBC-TV; bd. dirs. Am. Bd. EEG and Neurophysiology; participant Am. Med. EEG Assn.; rep. Internat. Fedn. EEG and Clin. Neurophysiology lectr. tour to Africa, 1989; keynote speaker Internat. Course of Neurophysiology, Oxford Univ., 1993. Author: Functional Organization of the Diencephalon, 1957, Atlas on Cerebral Death and Coma, 1976, EEG in Clinical Practice, 1982, EEG Evoked Potentials in Psychiatry and Behavioral Neurology, 1983; contbr. articles to profl. jours. Command Surgeon, USAR, 1986-90, with Army Med. R & D Command, 1990—, mobilization replacement for maj. gen., comdr. Recipient Alumni award Franklin and Marshall Coll., 1978. Mem. Am. Electroencephalography Soc. (treas. 1965-68), Eastern Electroencephalography Soc. (sec.-treas. 1961-64), Central Electroencephalography Soc., Am. Med. EEG Assn. (bd. dirs.), Am. Bd. EEG and Neurophysiology (bd. dirs.), Internat. EEG and Clin. Neurophysiology (bd. dirs.), Am. Acad. EEG (bd. dirs.), Chgo. Acad. Medicine, Am. Epilepsy Soc., Am. Physiol. Soc., Soc. Neuroscis., Am. Acad. Neurology, Phi Beta Kappa, Sigma Xi (lectr. 1960—). Rsch. on coding in central nervous system, new theory on neural mechanisms in olfaction, electro-clin. correlations in different types of epilepsy, organic aspects in juvenile delinquency. Home: 720 Roslyn Ter Evanston IL 60201-1722 Office: U Ill Consultation Clinic Epilepsy 912 S Wood St Chicago IL 60612-4303

HUGHES, LAUREL ELLEN, psychologist, educator, writer; b. Seattle, Oct. 30, 1952; d. Morrell Spencer and Eleanore Claire (Strong) Chamberlain; m. William Henry Hughes Jr., Jan. 27, 1973; children: Frank, Ben, Bridie. BA in Psychology, Portland State U., 1980, MS in Psychology, 1986; D in Clin. Psychology, Pacific U., 1988. Lic. psychologist, Oreg. Counselor Beaverton (Oreg.) Free Meth. Ch., 1982-85; psychotherapist Psychol. Svc. Ctr., Portland, Oreg., 1986, Psychol. Svc. Ctr. West, Hillsboro, Oreg., 1987-89; pvt. practice Beaverton, 1990—; adj. mem. faculty Portland C.C., 1990-91, U. Portland,1992—, CU/Seattle. 1993—; vis. asst. prof. U. Portland, 1991-92; psychol. cons. children's weight control group St. Vincent's Hosp., Portland, 1991. Author: How To Raise Good Children, 1988, How To Raise a Healthy Achiever, 1991; contbr. articles to profl. jours. Tchr. Sunday sch. Beaverton Free Meth. Ch., 1983-88; mother helper Walker Elem. Sch., Beaverton, 1988-90, 92—; foster parent Washington County, Oreg. 1976-77, 79-80; vol. disaster mental health svcs. ARC, 1993—. Mem. APA, Oreg. Psychol. Assn. (bd. dirs. 1990-91, editor jour. 1990-91). Avocations: knitting, gardening, football, reading. Office: 4320 SW 110th Beaverton OR 97005

HUGHES, MARK LEE, pharmacist; b. Guthrie Center, Iowa, Sept. 24, 1960; s. Darwin M. and Sheryl Ann (Jorgensen) H.; m. Linda Sue McDonald, Apr. 28, 1984. BS, Drake U., 1983. Registered pharmacist, Ill., Iowa. Retail pharmacist Peoples Drug Store, Bloomington, Ill., 1984-87; hosp. pharmacist North Iowa Med. Ctr., Mason City, 1987-89, asst. dir. pharmacy, 1989-90, dir. pharmacy, 1991—; cons. pharmacist Hospice of North Iowa, Mason City, 1987—. Mem. Am. Soc. Hosp. Pharmacists, Iowa Pharmacists Assn., Iowa Soc. Hosp. Pharmacists, Kappa Psi. Republican. Methodist. Avocations: tennis, golf, running, skiing, reading. Home: 2 Boulder Rd Mason City IA 50401-2503 Office: North Iowa Med Ctr 910 N Eisenhower Ave Mason City IA 50401-1597

HUGHES, MICHAEL WAYNE, systems engineer; b. St. Joseph, Mo., Nov. 1, 1957; s. Deever Wayne and Mary Margaret (Magin) H.; children: Elizabeth Ann, Matthew Michael, Rachel Louise. BS in Civil Engring., Cornell U., 1992. Systems engr. Burlington No. R.R., Ft. Worth, 1991—. Mem. ASCE, NSPE, Chi Epsilon, Tau Beta, Phi Kappa Phi. Home: PO Box 792 Saint Joseph MO 64502 Office: Burlington No R R 777 Main St Fort Worth TX 76102

HUGHES, THOMAS JOSEPH, mechanical engineering educator, consultant; b. Bklyn., Aug. 3, 1943; s. Joseph Anthony and Mae (Bland) H.; m. Susan Elizabeth Weh, July 1, 1972; children: Emily Susan, Ian Thomas, Elizabeth Claire. B.M.E., Pratt Inst., Bklyn., 1965; M.M.E., Pratt Inst., 1967; M.A. in Math., U. Calif.-Berkeley, 1974, Ph.D. in Engring. Sci., 1974. Mech. design engr. Grumman Aerospace, Bethpage, N.Y., 1965-66; research and devel. engr. Gen. Dynamics, Groton, Conn., 1967-69; lectr., asst. research engr. U. Calif.-Berkeley, 1975-76; assoc. prof. structural mechanics Calif. Inst. Tech., Pasadena, 1976-80; assoc. prof. mech. engring. Stanford U., Calif., 1980-82, prof., 1983—; chmn. div. applied mechanics Stanford U., 1984-88, chmn. dept. mech. engring., 1988-89; founder, chmn. CENTRIC Engring. Systems, Inc., 1990—; cons. in field. Author: A Short Course in Fluid Mechanics, 1976, Mathematical Foundations of Elasticity, 1983, The Finite Element Method: Linear Static and Dynamic Finite Element Analysis, 1987; editor: Nonlinear Finite Element Analysis of Plate and Shells, 1981, Computational Methods in Transient Analysis, 1983; editor Jour. of Computer Methods in Applied Mechanics and Engring., 1980—; contbr. numerous articles to profl. jours. Fellow Am. Acad. Mechanics, ASME (Melville medal 1979); mem. ASCE (Huber prize 1978), AIAA, Soc. Engring. Sci., U.S. Assn. for Computational Mechanics (pres. 1990-92), Sigma Xi, Phi Beta Kappa. Office: Stanford U Dept Mech Engring Durand Bldg Stanford CA 94305

HUGHES, THOMAS WILLIAM, technical company executive; b. Cin., Jan. 11, 1950; s. Thomas Paul and Elaine N. (Schmidt) H.; m. Shirley Carter, Feb. 6, 1971 (div. Sept. 1985); children: Eric Lee, Jill Elaine; m. Wendy Lenore Mitchell, May 25, 1985 (div.); 1 child, Heather Marlene; m. Nora Lee London, Aug. 5, 1990; 1 child, Devin Abram. BS in Chem. Engring., U. Cin., 1973; MS in Engring. Mgmt., U. Dayton, 1981. Rsch. engr. Monsanto Rsch. Corp., Dayton, Ohio, 1973-76, rsch. group leader, 1976-78, contract mgr., 1978-81; product mgr. Monsanto Rsch. Corp., Miamisburg, Ohio, 1982-85, prodn. control mgr., 1985-88; mgr. mfg. support EG&G Mound Applied Techs., Miamisburg, 1988-91, mgr. reconfiguration, 1991—; prof. Coll. Engring. U. Dayton, 1984—. Author: Treatability Manual, 5 vols., 1980, Engineering Handbook for Hazardous Waste Incineration, 1981; contbr. 50 articles and reports to profl. jours. Leader Boy Scouts Am., Kettering, Ohio, 1978-81. Mem. Am. Inst. Chem. Engrs. (treas., chmn. 1980-84), Soc. Creative Anachronism (pres. 1987-88, awards 1987, 91). Republican. Achievements include first U.S. commercialization of hazardous waste disposal (of chlorinated wastes) in cement kilns; research in designing and permitting hazardous waste incinerators under RCRA and wastewater permits under NPDES. Home: 220 Brookside Dr Springboro OH 45066 Office: EG&G Mound Applied Techs 1 Mound Rd Miamisburg OH 45066

HUGHSON, MARY HELEN, electrical engineer; b. Cheyenne, Wyo., Sept. 21, 1965; d. Lester E. and Helen (Neal) H. BS in Elec. and Computer Engring., Clarkson U., 1987; postgrad., U. Rochester, 1989-93. Engr. Rochester (N.Y.) Tel. Corp., 1987—. Soccer commr. Town of Ontario, N.Y., 1990-92, soccer coach, 1983-92.

HUGUENEL, EDWARD DAVID, microbiologist; b. N.Y.C., Oct. 1, 1954; s. Edward G. and Katherine P. (Fleming) H.; m. Anne M. Higgins, Nov. 11, 1978; children: Colin, Brynn. BA, Rutgers Coll., 1976; MS, Boston Coll., 1978; PhD, Princeton U., 1984. Rsch. fellow Harvard Med. Sch., Boston, 1984-86; sr. rsch. scientist Molecular Diagnostics, Inc., West Haven, Conn., 1986-91; sr. staff scientist Miles Pharms. Rsch. Ctr., West Haven, 1992—; adj. asst. prof. Emmanuel Coll., Boston, 1984-85. Contbr. articles to profl. jours. Asst. Cub Scouts, Guilford, Conn., 1992; mentor Bridges Program for Inner City Children, Guilford and New Haven, 1992. Mem. AAAS, Am. Soc. Microbiology, N.Y. Acad. Scis. Achievements include 2 patents, and 4 patents pending. Office: Miles Rsch Ctr 400 Morgan Ln West Haven CT 06516

HUIE, CARMEN WAH-KIT, chemistry educator; b. Hong Kong, Aug. 7, 1959; came to U.S. 1975; s. Shui and May-Kuen Lee. BS in Chem. summa cum laude, U. Wis., Superior, 1980, BS in Math. magna cum laude, 1980; PhD in Chemistry, Iowa State U., 1986. Rsch. asst. EPA, Duluth, Minn., summer 1978; teaching asst. U. Wis., Superior, 1977-80, Iowa State U., Ames, 1980-81; rsch. asst. dept. energy Ames (Iowa) Lab., 1981-85; vis. asst. prof. SUNY, Binghamton, 1986-88, asst. prof. chemistry, 1988—; rsch. collaborator IBM, Endicott, N.Y., 1988-90, Lourdes Hosp., Binghamton, 1990—; lectr. XXVII Colloquium Spectroscopium Internationale, Lofthus, Norway, 1991. Recipient Citations for Lester Strock award Soc. Applied Spectroscopy, 1990. Mem. Am. Chem. Soc., Sigma Xi. Achievements include development of novel laser-based imaging systems for the study of fundamental processes occuring within the laser microprobe and graphite furnace atomizer; development and applications of new and improved analytical techniques for the extraction , separation and detection of important biomolecules, e.g., porphyrins, bilirubins and lipids. Home: 201 Evergreen St Apt 3-3c Vestal NY 13850 Office: SUNY-Binghamton Dept Chemistry Binghamton NY 13902-6000

HUISMAN, TITUS HENDRIK JAN, biochemist, educator; b. Leeuwarden, The Netherlands, Sept. 1, 1923; came to U.S., 1959; s. Pieter and Catherine Poppolina (Drifhout) H.; m. Geertruida A.E. Tjemmes, Dec. 27, 1950; children: Jeanette Huisman Reed, Theodore Schleider. MS in Chemistry, U. Groningen, The Netherlands, 1946, PhD in Chemistry, 1948; DS in Biochemistry, U. Utrecht, The Netherlands, 1950. Head pediatric biochemistry rsch. lab. U. Groningen, 1951-59; assoc. prof. biochemistry and pathology Med. Coll. Ga., Augusta, 1959-61, prof. biochemistry and pathology, 1961-63, Regents' prof. biochemistry, 1964—, prof. medicine, dir. Comprehensive Sickle Cell Ctr., 1972—, chmn. dept. cell and molecular biology, 1977-90; prin. investigator Hemoglobin Rsch. Lab. VA Med. Ctr. Augusta, 1964-81; mem. ad hoc rev. com. Sickle Cell Anemia program NIH, 1972-74, mem. hematology study sect. on genetics Nat. Heart, Lung and Blood Inst., Bethesda, Md., 1975-79; mem. Human Gene Nomenclature Com. Author: The Hemoglobinopathies: Techniques of Identification, 1977, The Chromatography of Hemoglobin, 1980; editor in chief Hemoglobin Internat. Jour. Hemoglobin Rsch., 1976—; mem. editorial bd. various jours. including Am. Jour. Hematology, 1976—. With Dutch Underground Forces, 1940-45. Named Hon. Citizen City of Brugge, Belgium, 1968; recipient Van Slyke award in clin. chemistry, 1971; Silver medal of Colloquium St. Jan's Hosp., Brugge, 1972; named hon. prof. Chinese Acad. Med. Scis., Beijing, Peking Union Med. Coll., Beijing, Shanghai Children's Hosp. Fellow Nat. Acad. Clin. Biochemistry; mem. AAAS, Am. Soc. Biol. Chemists, Am. Soc. Clin. Chemists, Am. Soc. Human Genetics, Am. Soc. Hematology, Am. Chemistry Soc., Am. Soc. Exptl. Biology and Medicine, Biochem. Soc. Eng., Biochem. Soc. Holland, Dutch Soc. Clin. Chemists, European Soc. Hematology (hon.), Royal Dutch Acad. Sci. (corr.), Soc. Clin. Rsch., N.Y. Acad. Scis., Turkish Soc. Hematology (hon.), Macedonian Acad. Scis. and Arts (fgn. mem., Skopje), Sigma Xi. Mem. Dutch Reformed Ch. Home: 2318 Walton Way Augusta GA 30904-6124 Office: Med Coll Ga Dept Biochem Cell and Molecular Biology 1435 Laney-Walker Blvd Augusta GA 30912-2100

HUIZENGA, JOHN ROBERT, nuclear chemist, educator; b. Fulton, Ill., Apr. 21, 1921; s. Harry M. and Josie B. (Brands) H.; m. Dorothy J. Koeze, Feb. 1, 1946; children—Linda J., Jann H., Robert J., Joel T. A.B., Calvin Coll., 1944; Ph.D., U. Ill., 1949. Lab. supr. Manhattan Wartime Project, Oak Ridge, 1944-46; instr. Calvin Coll., Grand Rapids, Mich., 1946-47; assoc. scientist Argonne Nat. Lab., Chgo., 1949-57; sr. scientist Argonne Nat. Lab., 1958-67; professorial lectr. chemistry U. Chgo., 1963-67; prof. chemistry and physics U. Rochester, 1967-78, Tracy H. Harris prof. chemistry and physics, 1978-91, Tracy H. Harris prof. emeritus chemistry and physics, 1991—, chmn. dept. chemistry, 1983-88; vis. prof. Joliot-Curie Lab., U. Paris, 1964-65, Japan Soc. for Promotion of Sci., 1968; chmn. Nat. Acad. Sci.-NRC Com. on Nuclear Sci., 1974-77; mem. energy rsch. adv. bd. Dept. Energy, 1984-90; numerous adv., vis. coms. to univs., govt. and nat. labs. Author: (with R. Vandenbosch) Nuclear Fission, 1973; (with W.U. Schröder) Damped Nuclear Reactions, 1984; Cold Fusion: The Scientific Fiasco of the Century, 1992; contbr. articles to profl. jours. Fulbright fellow Netherlands, 1954-55; Guggenheim fellow Paris, 1964-65; Guggenheim fellow Berkeley, Calif., 1973; Guggenheim fellow Munich, W.Ger., 1974; Guggenheim fellow Copenhagen, 1974; recipient E.O. Lawrence award AEC, 1966; named Disting. Alumnus Calvin Coll., 1975. Fellow AAAS, Am. Phys. Soc., Am. Acad. Arts and Scis.; mem. NAS (chmn. NAS-NRC com. on nuclear and radiochemistry 1988-91), Am. Chem. Soc. (award for nuclear applications in chemistry 1975), Phi Beta Kappa, Sigma Xi, Phi Kappa Phi. Home: 51 Huntington Meadow Rochester NY 14625-1810 Office: U Rochester Dept Chemistry Rochester NY 14627

HUJAR, RANDAL JOSEPH, software company executive, consultant; b. L.A., May 28, 1959; s. Theodore Casmir and Josephine (Camin) H.; m. Kerrie Biancalana, Oct. 17, 1981; children: Tabitha, Allison, Colleen. BS in Fin. and Mktg., U. Santa Clara, 1981. Cons., programmer I.P. Sharp Assocs., Vienna, Austria, 1979-80; mktg. rep. Hewlett-Packard, Cupertino, Calif., 1981-83, mgr. internat. mktg., 1983-84; mgr. product line Hewlett-Packard, Roseville, Calif., 1985-86; dir. mktg. Ashton-Tate, East Hartford, Conn., 1986-88; mgr. product mktg. IBM, Milford, Conn., 1989-90; dir. 1-2-3 product line mktg. Lotus Devel. Corp., Cambridge, Mass., 1990-91; mng. ptnr., founder, pres. Lyriq Internat. Corp., Milford, Conn., 1991—. Contbr. articles to profl. jours. Com. mem. Boston Computer Mus., 1991—. Avocations: golf, tennis, hiking, strategic games. Home: 64 Brookhaven Dr Glastonbury CT 06033

HULL, DAVID GEORGE, aerospace engineering educator, researcher; b. Oak Park, Ill., Mar. 27, 1937; s. John Lawrence Hull and Elizabeth Christine (Carstensen) Meyer; m. Meredith Lynn Kiesel, June 2, 1962 (div. July 1980); children: David, Andrew, Matthew; m. Vicki Jan Poole, June 30, 1983; children: Katherine, Emily. BS, Purdue U., 1959; MS, U. Wash., 1962; PhD, Rice U., 1967. Staff assoc. Boeing Sci. Research Labs., Seattle, 1959-64; research assoc. Rice U., Houston, 1964-66; asst. prof. U. Tex., Austin, 1966-71, assoc. prof., 1971-77, prof., 1977-85, M.J. Thompson Regents prof., 1985—; cons. several aerospace cos. Reviewer several engring. jours.; contbr. over 50 articles to profl. jours. Recipient, co-recipient over 30 grants and contracts. Assoc. fellow AIAA (atmospheric flight mechanics tech. com. 1974-77, guidance and control tech. com. 1984-87); mem. Delta Tau Delta (treas. Purdue U. 1958-59). Avocations: handball, softball, weekend farming. Office: U Tex ASE/EM 60600 Austin TX 78712

HULL, EDWARD WHALEY SEABROOK, freelance writer, consultant; b. Washington, Mar. 10, 1923; s. Edward Seabrook III and Marie Hortense (Marshall) H.; m. Nellie Phinizy Fortson, June 25, 1944; children: Edward, John, Thomas, Nellie Phinizy Hull Price. Student, Union Coll., Schenectady, 1939-42; M Marine Affairs, U. R.I., 1970; PhD in Marine Sci., U. S.C., 1987. Chief copy boy Times-Herald, Washington, 1947; corr. Washington Bur., McGraw-Hill, 1948-51; editor Whaley-Eaton Fgn. Letter, Washington, 1951-53; bur. chief McGraw-Hill World News, London, 1954-56; assoc. editor Missiles & Rockets mag., Washington, 1956-58; ea. rep. Diversey Engring., Inc., Washington, 1958-60; editor, pres., dir. Nautilus Press, Inc., Washington, 1960-73; cons., Washington and, S.C., 1969-85; freelance writer, Yonges Island, S.C., 1987—; dir. S.C. Writers Workshop, Columbia and Charleston, 1991—. Author: Rocket to the Moon, 1958, The Bountiful Sea, 1964; editor: Peenemunde to Canaveral (D.K. Kuzel), 1962; contbr. numerous articles on rocketry, space flight, oceanography, also others, to profl. publs. 1st lt. USMCR, 1942-45, PTO. Decorated Air medal; fellow Woodrow Wilson Internat. Ctr. for Scholars, 1970-71. Mem. Am. Soc. Internat. Law, Nature Conservancy, Appalachian Trail Conf., Wilson Ctr. Assocs. (assoc.), Nat. Press Club (Washington). Democrat. Episcopalian. Avocations: photography, canoeing, woodworking, fishing, orchids. Home: 7717 White Point Rd Hollywood SC 29449-6203

HULL, GORDON FERRIE, physicist; b. Hanover, N.H., May 23, 1912; s. Gordon Ferrie and Wilhelmine (Brandt) H.; m. Mona Jerusha Cutler, June 24, 1937; children: Gordon, Mona, David, Jonathan, Berney. AB, Dartmouth Coll., 1933, MA, 1934; PhD, Yale U., 1937. Rsch. physicist Bell Telephone Labs., Murray Hill, N.J., 1937-44; prof. physics Dartmouth Coll., Hanover, N.H., 1944-55; sci. officer Office Naval Rsch., London and Washington, 1949-51; sci. attaché Am. Embassy, Bern, Switzerland, 1951-52; pres. Hull Assocs., Sci. Cons. to INdustry and Ednl. Insts., Concord and Rockport, Mass., 1955—. Contbr. over 50 articles to profl. publs. Chmn. Rockport Planning Bd., 1975-90. Fellow AAAS, Am. Phys. Soc. Achievements include microwave and optical analogies. Home: 2 Jewett St Rockport MA 01966 Office: Hull Assocs 2 Jewett St Rockport MA 01966

HULL, JEROME, JR., horticultural extension specialist; b. Canfield, Ohio, Dec. 15, 1930; s. Jerome and Doris (Humes) H.; m. Suzanne Lotze, Mar. 23, 1957; children: Amy, Meredith, Molly, Andrew. BS, Mich. State U., 1952; MS, Va. Poly. Tech. Inst., 1955; PhD, Mich. State U., 1958. Orchard mgr. Whitehouse Fruit Farm, Canfield, 1955-56; grad. asst. Mich. State U., East Lansing, 1957-58, extension horticulturalist, 1964—; extension horticulturalist Purdue U., West Lafayette, Ind., 1959-64. Clk., elder East Lansing Trinity Ch. With U.S. Army, 1953-55. Named Fruit Man of Yr. Mich. State Pomesters Assn., 1983; recipient Outstanding Specialist award Mich. State U. Specialists Assn., 1988. Fellow Am. Soc. for Hort. Sci. (Bittner award com. 1973-75, nominating com. 1979-81, v.p. to exec. v.p. 1983-85, adv. com. 1985-87); mem. Mich. State Hort. Soc. (exec. coord. 1968-70, sec. 1971—, Stewart Grant fellow 1975, Disting. Svs. award. 1983). Achievements include discovery of ability of gibberellin to inhibit flower initiation in tart cherry and peach; established ability of gibberellin to reduce inhibitory effect of virus on sour cherry tree vegetative growth. Home: 2638 Raphael Ave East Lansing MI 48823 Office: Mich State U Dept Horticulture East Lansing MI 48824

HULSE, RUSSELL ALAN, astrophysicist, plasma physicist; b. N.Y.C., Nov. 28, 1950. BS, Cooper Union, 1970; MS, U. Mass., 1972, PhD in Physics, 1975. Rsch. assoc. Nat. Radio Astron. Obs., 1975-77, mem. tech. staff, 1977-80, staff rsch. physicst II, 1980-84; rsch. physicts plasma physics lab Princeton (N.J.) U., 1984—. Recipient Nobel Prize in Physics Nobel Found., 1993. Mem. Am. Physical Soc., Am. Astron. Soc., Soc. Indsl. and Applied Math. Achievements include research in computer modeling of transport and atomic processes in tokamak controlled thermonuclear fusion plasmas; advanced computational environments. Office: Princeton U Plasma Physics Lab James Forrestal Research Campus P Box 451 Princeton NJ 08543*

HULTMARK, GORDON ALAN, civil engineer; b. Chgo., Apr. 14, 1944; s. John Harold and Carolyn Bernice (Nelson) H.; m. Sarah Delle Carsey, Mar. 29, 1969; chidlren: Rifka, Menachem. BS in Civil Engring., U. Ill., 1967. Structural engr. Wilson, Andros, Roberts & Noll, Chgo., 1967-68; sales trainee Ceco Corp., Chgo., 1968-69; design engr. Mobile Homes Mfg. Assn., Chgo., 1969-70; plant engr. Union Tank Car Co., Chgo., 1970-77; project engr. Menasha Corp., Otsego, Mich., 1977-78; corp. project engr. Plainwell (Mich.) Paper Co., 1978—; speaker in field. Chair Plainwell Constrn. Rev. Bd., 1985—; treas. Right to Life of Van Buren County, Bangor, Mich., 1987-89; del. Mich. Rep. Conv., 1988. Mem. ASCE, Audubon Soc. Republican. Jewish. Home: 49392 CR 384 Grand Junction MI 49056 Office: Simpson Plainwell Paper Co 200 Allegan St Plainwell MI 49080

HULTS, SCOTT SAMUEL, engineer; b. Clear Lake, S.D., June 26, 1964; s. Eugene Samuel and Pamela Rogene (Hovey) H.; m. Penny Jean Odland, June 7, 1986; children: Karl Eugene, Madison Kristine. BA, Augustana Coll., 1986; BSME, S.D. State U., 1988. Engr. I, indsl. gas mktg. No. States Power Co./Gas Utility, St. Paul, Minn., 1988-89, engr. II, indsl. gas mktg., 1989-92, gas supply cons., 1992—; v.p. Blue Flame Comml. and Indsl. Com., Mpls., 1991-92. Editor: Commercial and Industrial Gas Reference Manual, 1991. Coord., campaign worker United Way, St. Paul, 1990-91; treas., coun. mem. Lord of Life Luth. Ch., Oakdale, Minn., 1989—. Mem. ASME, Assn. of Energy Engrs. (cogeneration inst.-environ. engrs. and mgrs. inst.), Twin Cities Energy Engrs. Mem. Evangelical Luth. Ch. of Am. Office: NSP Gas 825 Rice St Saint Paul MN 55117

HUMBER, WILBUR JAMES, psychologist; b. Winnepeg, Manitoba, Can., June 21, 1911; came to U.S., 1922; s. Arthur W. and Annie Humber; m. Jean Adriansen, May 25, 1945; children: Philip, Scott, Michael. BA, Macalester Coll., 1930; MA, U. Chgo., 1937; PhD, U. Minn., 1942. Diplomate clin. psychology. Dean Kalamazoo (Mich.) Coll., 1941-43; prof. Lawrence U., Appleton, Wis., 1943-46; psychologist Rohrer, Hibler & Replogle, Chgo., 1947-52; sr. ptnr. Humber, Mundie & McClary, Milw., 1952—; bd. dirs. Hopkins Savs. & Loan, Milw., Pope Sci. Corp., Menomonee Falls, Wis. Author: Development of Human Behavior, 1951, Introduction to Social Psychology, 1968; editorial bd. Jour. of Consultation. Bd. dirs. Lakeside Children Ctr., Milw., 1966-86; pres. Wis. Mental Health Assn., Milw., 1962-63. Fellow APA, N.Y. Acad. Sci., AAAS; mem. Wis. Psychol. Assn. (founder, 1st pres., Disting. Contbn. award 1984), Milw. Club, Milw. Country Club. Congregationalist. Avocations: tennis, writing, travel. Home: 10260 N Range Line Ct Mequon WI 53092-5346 Office: Humber Mundie & McClary 111 E Wisconsin Ave Ste 1950 Milwaukee WI 53202-4889

HUMES, H(ARVEY) DAVID, nephrologist, educator; b. Honolulu, Nov. 20, 1947; s. William and Nancy Humes; m. Dolores Humes; 1 child, Michael David. BA, U. Calif., Berkeley, 1969; MD, U. Calif., San Francisco, 1973. Diplomate Am. Bd. Internal Medicine. Intern Moffit Hosp. and U. Calif. Hosps., San Francisco, Calif., 1973-74; resident U. Calif. Hosps., San Francisco, 1974-75; clin. fellow nephrology U. Pa. Hosp., Phila., 1975-76; rsch. fellow lab. kidney & electrolyte physiology Peter Bent Brigham Hosp., Boston, Mass., 1976-77; instr. Harvard U., Boston, Mass., 1977-78; asst. prof. medicine Harvard U., Boston, 1978-79; asst. prof. internal medicine U. Mich., Ann Arbor, 1979-82, assoc. prof. internal medicine, 1982-86, prof. internal medicine, 1986—; founder EpiGenesis, Inc.; cons. Sandoz Pharm., Bristol-Meyers-Squibb, Sterling-Winthrop, AmGen.; instr., asst. prof. Peter Bent Brigham Hosp., Boston, 1977-79; dir., chief Nephrology Rsch. Labs., U. Mich., Ann Arbor, 1980-81; chief med. svc. VA Med. Ctr., Ann Arbor, 1983—. Contbr. articles to profl. jours. Grantee Nat. Kidney Found., 1981-83, 84-85, 87-88; PHS, 1982-93, 87-91; VA, 1982-83, 83-87, 87-90, 90-93. Fellow, ACP; mem. AAAS, Am. Physiol. Soc., Am. Soc. Biol. Chemists, Am. Soc. Renal Chemistry & Metabolism, Am. Soc. Clin. Investigation, Am. Heart Assn., Am. Soc. Nephrology, Am. Fedn. Clin. Rsch., Internat. Soc. Nephrology, Nat. Kidney Found., Nat. Kidney Found. Mich., Cen. Soc. Clin. Rsch., Midwest Salt & Water Club, Alpha Omega Alpha, Phi Beta Kappa. Achievements include research in cellular basis of acute renal failure, molecular basis of renal repair in acute renal failure, and molecular basis of kidney tubulogenesis, and development of an implantable, bioartificial kidney. Office: VA Med Ctr 2215 Fuller Rd Ann Arbor MI 48105-2300

HUMIEC, FRANK STANLEY, chemist; b. Niagara Falls, N.Y., June 11, 1933; s. Frank and Mary (Taborska) H.; children: Jerry, Mary Beth. BS, Niagara U., 1959; MS, Fairleigh-Dickinson U., 1968. Chemist Hooker Chem. Co., Niagara Falls, N.Y., 1951-58, Mellon Inst., Pitts., 1958-63, Hoffman LaRoche, Nutley, N.J., 1968-85; cons. chemist AT&T Bell Labs., Murray Hill, N.J., 1986—. Contbr. articles to profl. jours. Achievements include patents in field. Home: 6-14 Essex Pl Fair Lawn NJ 07410

HUMPHREY, LEONARD CLAUDE, electrical engineer; b. Jewett, N.Y., Mar. 20, 1928; s. Leonard Claude and Susan Naomi (Rambo) H.; m. Rosemary Busic, Aug. 23, 1952; children: Mark D., Jonathan B., Elizabeth L. BS in Physic, Union Coll., 1949; PhD, Cornell U., 1968. Physicist GE, Schenectady, N.Y., 1949-52; engr. Gen. Dynamics, Rochester, N.Y., 1957-59; sr. engr. GE, Syracuse, N.Y., 1959-78, Norden Systems, Norwalk, Conn., 1978-84; prin. engr. GE, Valley Forge, Pa., 1984—; adj. prof. U. Bridgeport, Conn., 1978-84. Contbr. articles to profl. jours. Asst. dir., col. Fed. Emergency Mgmt. Assn., Fairfield County, Conn., 1978-84; elder Great Valley Presbyn. Ch., Malvern, Pa. Lt. USAF, 1952-78; ret. col. USAFR. Mem. Valley Forge Ret. Officers (pres. 1990—), Sigma Xi, Sigma Pi Sigma. Achievements include patents for Improved Direction Finder, Electromagnetic Sensor. Home: 4 Carriage Ln Chester Springs PA 19425

HUMPHREY, PHILIP STRONG, university museum director; b. Hibbing, Minn., Feb. 26, 1926; s. Watts Sherman Humphrey and Katharine (Strong) Osborne; m. Mary Louise Countryman, Jan. 1, 1946; children: Margaret Hubbard, Stephen Strong. BA, Amherst Coll., 1949; MS, U. Mich., 1952, PhD, 1955. Asst. curator Peabody Mus. Natural History, New Haven, 1957-62; asst. prof. zoology Yale U., New Haven, 1957-62; chief curator ornithology Nat. Mus. Natural History, Washington, 1962-65, chmn. vertebrate zoology, 1965-67; prof. zoology, chmn. dept. U. Kans., Lawrence, 1967-69; dir. Mus. Nat. History-U. Kans., Lawrence, 1967—, prof. systematics and ecology, 1969—. Co-author: The Darwin Reader, 1956, The Birds of Isla Grande, 1970. Adv. bd. Lawrence Conv. and Visitors' Bur., 1988. Cpl. USAAF, 1944-47. Guggenheim fellow, Argentina, 1960-61. Fellow AAAS, Am. Ornithologists' Union; mem. Soc. Systematic Zoology, Wilson Ornithol. Soc. Office: U Kans Mus Natural History Dyche Hall University Of Kansas KS 66045

HUMPHRIES, JOAN ROPES, psychologist, educator; b. Bklyn., Oct. 17, 1928; d. Lawrence Gardner and Adele Lydia (Zimmermann) Ropes; m. Charles C. Humphries, Apr. 4, 1957; children: Peggy Ann, Charlene Adele. BA, U. Miami, 1950; MS, Fla. State U., 1955; PhD, La. State U., 1963. Part-time instr. psychology dept. U. Miami, Coral Gables, Fla., 1964-66; prof. dept. psychology Miami-Dade Community Coll., 1966—. Pres. Inst. Evaluation, Diagnosis and Treatment, Miami, 1987—, bd. dirs. 1975—, v.p. 1975-87. Registered lobbyist State of Fla., 1986—. Recipient cert. Appreciation Miami-Dade Community Coll. Mem. AAUP (pres. Miami-Dade Community Coll. chpt. 1988—, v.p., sec., mem. exec. bd., Fla. Conf. v.p. 1986-89, mem. exec. bd., 1986-88), AAUW (former v.p. Tumiami branch) Biofeedback Soc. of Am. (pres. 1990—), Biofeedback Assn. Fla. (pres. 1990—), Internat. Platform Assn. (gov. 1979—, Silver Bowl award 1993), Am. Psychol. Assn., AAUW (life, appreciation award 1979), Am. Psychol. Soc. (charter 1989), Mexico Beach C. of C. (bus. 1991—), North Campus Speaker's Bur. (award for community lecture series), Internat. Soc. for Study Subtle Energies and Energy Medicine (charter 1990), Dade County Psychol. Assn. (Fla. chpt.), Colonial Dames 17th Century, N.Y. Acad. Scis. (life), Regines in Miami, Soc. Mayflower Descs. (elder William Brewster colony), Phi Lambda (founder's plaque 1976, appreciation award 1987), Phi Lambda Pi. Democrat. Clubs: Country of Coral Gables (life), Jockey (life). Editorial staff, maj. author: The Application of Scientific Behaviorism to Humanistic Phenomena, 1975, rev. edit., 1979; researcher in biofeedback and human consciousness. Home: 1311 Alhambra Cir Coral Gables FL 33134-3521 Office: Miami Dade Community Coll North Campus 11380 NW 27th Ave Miami FL 33167

HUNDHAUSEN, ROBERT JOHN, mining engineer; b. Cape Girardeau, Mo., Sept. 28, 1916; s. Herman Henry and Charlotte Virginia (Heekman) H.; married, 1937 (div. 1965); children: Robert J. Jr., Thomas G., William H., Phyllis Hundhausen Taylor; m. Adele Johannah Eck, May 2, 1968. Grad., Colo. Sch. Mines, 1934-38. Mining engr. N.J. Zinc Co., Gilman, Colo., 1938-40; with U.S. Bur. Mines, 1942-55; gen. mgr. Dawn Mining Co., Ford, Wash., 1955-57; prin. Real Estate & Mining, Hayden, Idaho, 1958-81. Contbr. articles to profl. jours.; inventor in field. Mem. AAAS, Am. Inst. Mining Engrs., N.W. Mining Assn., N.Y. Acad. Scis. (life), Colo. Sch. Mines Alumni Assn., Spokane Club, Hayden Lake Country Club, Sigma Xi, Kappa Sigma. Presbyterian. Avocations: golf, swimming, painting. Home: RR 2 Box 164 Hayden ID 83835 *Died July 7, 1993.*

HUNG, GEORGE KIT, biomedical engineering educator; b. Shanghai, China, Nov. 1, 1947; s. Ernest K. and Dorothy Hung. MS, U. Calif., Berkeley, 1971, PhD, 1977. Engr. Standard Oil Co. of Calif., Taft, 1968-69; asst. dir. U. Calif., Berkeley, 1976-77; postdoctoral rsch. assoc. NASA Ames Rsch. Ctr., Moffett Field, Calif., 1977-78; asst. prof. elec. engring. Rutgers U., Piscataway, N.J., 1978-83, assoc. prof. elec. engring., 1983-86, assoc. prof. biomed. engring., 1986-91, prof. biomed. engring., 1991—. Cons. editor Med. Electronics Jour., Pitts., 1988—; contbr. articles to profl. jours., chpts. to books. Grantee NSF, 1980-81, NIH, 1981-84, 90-93, N.J. Commn. on Sci. and Tech., 1989-90, Essilor Internat., 1989-90. Mem. IEEE (sr.), Assn. Rsch. in Vision and Ophthalmology, Sigma Xi. Office: Rutgers U Dept Biomed Engring Piscataway NJ 08855-0909

HUNG, KUEN-SHAN, anatomy educator; b. Chia-Yi, Taiwan, Jan. 13. 1938; s. Tien-Won and Guam (Tsai) H.; Shirley Hwang, Nov. 30, 1968; children: Irene, Melissa. BS, Nat. Taiwan U., 1960; PhD, U. Kans., 1969. Asst. prof. U. S.D., Vermillion, 1969-71, U. So. Calif., L.A., 1971-74; asst. prof., then assoc. prof. dept. anatomy U. Kans. Med. Ctr., Kansas City, 1974-83, prof. dept. anatomy, 1983—. Contbr. articles to profl. pubis. Pres. Taiwanese Assn. of Greater Kansas City, 1976. 2d lt. Taiwanese infantry, 1960-61. NIH grantee, 1976-83, Am. Lung Assn., 1974-76, Kans. affiliate Am. Heart Assn., 1981-83, 84-87, 90-93. Mem. Am. Assn. Anatomists, Microscopy Soc. Am., Am. Soc. Cell Biology, Sigma Xi. Achievements include characterization and experimental models of lung endocrine cell hyperplasia. Office: Kans U Med Ctr Dept Anatomy and Cell Biol 3901 Rainbow Blvd Kansas City KS 66160

HUNG, PAUL PORWEN, biotechnologist, educator, consultant; b. Taipei, Taiwan, Sept. 30, 1933; s. Yao-Hsun and Shiu-Chin (Wu) H.; m. Nancy Kay Clark, May 5, 1956; children: Pauline E., Eileen K., Clark D. BS in Arts and Sci., Millikin U., 1956; PhD in Biochemistry, Purdue U., 1960. Head molecular virology and biology Abbott Labs., North Chicago, 1960-81; gen. mgr. Bethesda Rsch. Lab., Gaithersburg, Md., 1981-82; asst. v.p. Wyeth Ayerst Labs., Radnor, Pa., 1982—; adj. prof. Northwestern U. Med. Sch., Chgo., 1975-86; mem. Nat. Vaccine Adv. Com., Washington, 1990—; cons. Am. Inst. Biol. Sci., Washington, 1992, UN Indsl. Devel. Orgn., Vienna, Austria, 1981; adj. prof. Author: (chpt.) Recombinant DNA, 1991, Hepatitis Vaccine, 1991; contbr. over 250 articles to profl. jours. Mem. Am. Soc. Biochemistry and Molecular Biology, Am. Assn. Cancer Rsch., Internat. Assn. Biol. Standardization, Am. Inst. Chemists. Achievements include patents in field. Home: 506 Ramblewood Dr Byrn Mawr PA 19010 Office: Wyeth Ayerst Labs PO Box 8299 Philadelphia PA 19101

HUNG, RU J., engineering educator; b. Hsin Chu, Taiwan, June 17, 1934; came to U.S., 1966; s. Hsio F. and Lan (Chang) H.; m. Nancy L. Hung, Sept. 24, 1966; children: Elmer, Christine. BS, Nat. U. Taiwan, 1957; MS, U. Osaka, 1966; PhD, U. Mich., 1970. Registered profl. engr. Rsch. engr. Hsin Chu Advanced Tech Inst., 1957-64; vis. scholar U. Osaka, Japan, 1964-66; rsch. asst. U. Mich., Ann Arbor, 1966-69, postdoctoral fellow, 1970; rsch. assoc. NASA/Ames Rsch. Ctr., Moffett Field, Calif., 1971-72; prof. U. Ala., Huntsville, 1972—; cons. U.S. Army, NASA, Integraph, Northrop, Firestone, Chinese Acad. Sci. Contbr. over 200 articles to profl. jours. Recipient Rsch. Achievement award NASA, new tech. award NASA; Japan Ministry of Edn. fellow, 1964; Nat. Acad. Scis. rsch. fellow, 1972, Sigma Xi rsch awards. Fellow Royal Meteorol. Soc. (Gt. Britain); assoc. fellow AIAA (award of achievement 1989); mem. Nat. Soc. Profl. Engrs., Am. Geophys. Union, Am. Meteorol. Soc., Pi Tau Sigma, Sigma Xi (rsch. awards). Home: 2610 Guenevere Ave SE Huntsville AL 35803-1936 Office: U Ala Huntsville AL 35899

HUNG, WILLIAM MO-WEI, chemist; b. Chekiang, China, Sept. 17, 1940; came to U.S., 1966; s. Jordon T. and I-Hsing (Chang) H.; m. Julia Tsui, July 20, 1968; children: Berwyn, Calvin. BS, Nat. Chung-Hsing U., 1963; PhD, U. Mass., 1970. Rsch. and teaching asst. U. Mass., Amherst, 1967-70; rsch. assoc. The Ohio State U., Columbus, 1970-74; sr. rsch. chemist Hilton-Davis Chem. Co., Cin., 1974-80, dir. chem. rsch., 1980-84, sr. rsch. fellow, 1984-87; staff rsch. scientist CIBA Vision Corp., Atlanta, 1987-92, rsch. fellow, 1992—. Contbr. articles to profl. jours. 2d lt. Air Force of Republic of China, 1963-64, Taiwan. Fellow The Am. Inst. Chemists, Am. Chem. Soc. Achievements include patents for ultraviolet radiation absorbing agent for bonding to an ocular lens, tinted contact lenses with reactive dyes, novel color formers for transfer imaging dye and imaging systems. Home: 4062 Dover Ave Alpharetta GA 30201-1282 Office: CIBA Vision Corp 5000 Mcginnis Ferry Rd Alpharetta GA 30202-3919

HUNGERFORD, HERBERT EUGENE, nuclear engineering educator; b. Hartford, Conn., Oct. 3, 1918; s. Herbert Eugene and Doris (Emmons) H.; m. Edythe Lugene Green, Nov. 4, 1949. B.S. in Physics, Trinity Coll., Hartford, 1941; M.S. in Physics, U. Ala., 1949; Ph.D. in Nuclear Engring., Purdue U., 1964; part-time grad. student, U. Tenn., 1951-55, Wayne State U., 1956-61. Tchr. sci. Brent Sch., Baguio, Philippines, 1941; tchr. math. Choate Sch., 1945-46; head physics dept. Marion Mil. Inst., 1946-48; grad. instr. U. Ala., 1948-49; physicist Oak Ridge Nat. Lab., 1950-55; shielding specialist, head shielding and health physics sect. Atomic Power Devel. Assocs., 1955-62; research assoc. Purdue U., 1963-64, assoc. prof., 1964-68, prof. nuclear engring., 1968-83; prof. emeritus, 1983—; on leave Argonne Nat. Lab., 1977-78; adj. prof. mech. engring. Fla. Inst. Tech., 1984—; cons. in field; v.p., sec., bd. dirs. Hungerford Nuclear, Inc., Vero Beach, Fla., 1984—; sci. columnist Vero Beach Sun newspaper, 1993—. Author chpts. in books, articles. Prisoner of War, 1941-45. Mem. Am. Nuclear Soc. (sec. shielding and dosimetry div. 1960-62, div. vice chmn. 1969-70, div. chmn. 1970-71, mem. standards com. 1959-82, Presdl. citation for meritorious svc. 1993), Am. Phys. Soc., Lafayette Organ Soc. (pres. 1971-72), Amateur Organists Assn. Internat., Health Physics Soc., Am. Assn. Physics Tchrs., Kiwanis (bd. dirs. 1966-68, Presdl. citation 1970), Sigma Xi, Sigma Pi Sigma. Episcopalian (vestryman 1965-68). Club: Kiwanis (dir.) (1966-68). Achievements include invention of lattice model stochastic radiation transp.; pioneering use of serpentine and calcium borate as high temperature shield materials. Home: 2104 4th Ct SE Vero Beach FL 32962-7315

HUNNICUTT, RICHARD PEARCE, metallurgical engineer; b. Asheville, N.C., June 15, 1926; s. James Ballard and Ida (Black) H.; B.S. in Metall. Engring., Stanford, 1951, M.S., 1952; m. Susan Haight, Apr. 9, 1954; children—Barbara, Beverly, Geoffrey, Anne. Research metallurgist Gen. Motors Research Labs., 1952-55; sr. metallurgist Aerojet-Gen. Corp., 1955-57; head materials and processes Firestone Engring. Lab., 1957-58; head phys. scis. group Dalmo Victor Co., Monterey, 1958-61, head materials lab., 1961-62; v.p. Anamet Labs., Inc., 1962-82, exec. v.p., 1982—; partner Pyrco Co. Author: Pershing, A History of the American Medium Tank T20 Series, 1971, Sherman, A History of the American Medium Tank, 1978, Patton, A History of the American Main Battle Tank, vol. 1, 1984, Firepower, A History of the American Heavy Tank, 1988, Abrams, A History of the American Main Battle Tank, vol. 2, 1990, Stuart, A History of the American Light Tank, vol. 1, 1992. Served with AUS, 1943-46. Mem. Electrochem. Soc., AIME, Am. Soc. Metals, ASTM, Am. Welding Soc., Am. Soc. Lubrication Engrs. Research on frictional behavior of materials, development of armored fighting vehicles. Home: 2805 Benson Way Belmont CA 94002-2938 Office: 3400 Investment Blvd Hayward CA 94545-3811

HUNSICKER, HAROLD YUNDT, metallurgical engineer; b. Frankfort, Ind., Dec. 22, 1914; s. Samuel Madison and Carrie May (Yundt) H.; m. Margaret Alice Kerns, Aug. 20, 1939; children: Patricia Tiitto, James, Susan Dryburgh. BS in Chem. Engring., Purdue U., 1936; MS in Metall. Engring., Case Inst. Tech., 1939. Rsch. engr. Aluminum Co. Am. (Aluminum Rsch. Labs.), Cleve., 1936-48, asst. div. chief, 1948-58; chief phys. metall. div. Alcoa-Aluminum Rsch. Labs., New Kensington, Pa., 1958-70; mgr. phys. metall. div. Alcoa Rsch. Labs., Alcoa Center, Pa., 1970-77, tech. advisor, 1977-79; cons. Pitman-Dunn Labs., Frankford Arsenal, U.S. Army, Phila., 1975-79; mem. metall. adv. bd. Pa. State U., University Park, 1970-79. Author: Aluminum, Vol. I, 1967, Metals Handbook, Vol. II, 1972, Metals Handbook, Desk Vol., 1983. Named Disting. Engring. Alumnus Purdue U., West Lafayette, Ind., 1967. Fellow Am. Soc. Metals Internat.; mem. AIME (legion of honor Metals Soc.), Sigma Xi. Presbyterian. Achievements include pioneer development of alumnimum alloys for bearings of automotive, marine and stationary engines, compressors and machinery; development of higher performance aluminum alloys for aircraft/aerospace castings and wrought products (sheet, plate, forgings, extrusions), used today; directed programs in alloy devel. for aluminum cans, architectural and automotive products, welding and brazing alloys and specialty electrical conductors; headed research programs to apply advanced techniques for analysis of metallurgical microstructures, effects on properties, improvement and control; attained 18 patents on aluminum alloys. Home: 508 Chester Dr New Kensington PA 15068-3304

HUNT, CHARLES BUTLER, geologist; b. West Point, N.Y., Aug. 9, 1906; s. Irvin Leland and Annie (Butler) H.; m. Alice Parker, Oct. 20, 1930; children: Eugene Parker, Anne Butler Hunt Kathan. AB, Colgate U., 1928; postgrad., Yale U., 1928-30. With U.S. Geol. Survey, 1930-53, 55-61; prof. geography Johns Hopkins U., Balt., 1961-73; cons. geologist, 1973—; exec. dir. Am. Geol. Inst., 1953-55; mem. adv. panel earth sci. div. NSF, 1960-63, chmn., 1962-63, mem. divisional com. math. and phys. sci., 1964-65; disting. vis. prof. N.Mex. State U., 1973-74; vis. scholar U. Utah. Author: (descriptive geology texts) Mt. Taylor New Mexico, 1938, Pike Co., Kentucky, 1938, Henry Mountains, Utah, 1953, Colorado Plateau, 1956, La Sal Mountains, Utah, 1958, Colorado River, 1968, Death Valley, California, 1968; (textbooks) Physiography of the United States, 1967, Natural Regions of the United States and Canada, 1974, Surficial deposits of the United States, 1986; editor G.K. Gilbert field notes about Lake Bonneville, 1982, about the Henry Mtns., Utah, 1988; also articles on storage nuclear wastes, historical archeology of tin cans and bottles, military geology in WW II. Mem. Phi Beta Kappa, Sigma Xi. Home and Office: 2131 Condie Dr Salt Lake City UT 84119-5307

HUNT, CHARLES EDWARD, electrical engineer, educator; b. Oakland, Calif., Apr. 21, 1953; s. Edward Howard and Lila Richardson (Wicker) H.; m. Carla M. Davis, Mar. 27, 1977; children: Nathan, Edward, Katherine, Alicia, Deborah, Lila. BSEE, U. Utah, 1980, MSEE, 1983; PhD, Cornell U., 1986. Staff engr. dept. computer sci. U. Utah, Salt Lake City, 1979-83; rsch. asst. dept. elec. engring. Cornell U., Ithaca, N.Y., 1983-86; asst. prof. dept. elec. and computer engring. U. Calif., Davis, 1986-92, assoc. prof. dept. elec. and computer engring., 1992—; staff cons. engring. rsch. div. Lawrence Livermore (Calif.) Nat. Lab., 1986—. Contbr. articles on silicon-on-insulator and semiconductor wafer bonding to profl. publs. Grantee IBM, 1991—. Mem. IEEE (Svc. award 1980), Electrochem. Soc., Materials Rsch. Soc., Tau Beta Pi. Achievements include co-development of storage-logic array, fabrication of world's sharpest silicon needles, development of silicon self-aligned field-emission vacuum micro-triode. Home: 1224 Fordham Dr Davis CA 95616 Office: Univ Calif Dept Elec/Computer Engring Davis CA 95616

HUNT, DARWIN PAUL, psychology educator; b. Lima, Ohio, July 16, 1926; s. I. Paul and Helen R. (Drees) H.; m. Sallie Sue Brubaker, June 20, 1953. BA, Miami U., 1950; PhD, Ohio State U., 1960. Rsch. psychologist USAF, Wright-Patterson AFB, Ohio, 1951-62; assoc. prof. U. Dayton, 1962-67; prof. N.Mex. State U., Las Cruces, 1967-92, head psychol. dept., 1967-72, prof. emeritus, 1992—; pres. Human Performance Enhancement, Inc., Las Cruces, 1990—; disting. vis. prof. Inst. Higher Edn., Limerick, Ireland, 1986; lectr. in field, 1972-92. Contbr. chpts. to books Perception of Exertion, 1986, Psychophysics in Action, 1989, Human Self Assessment, 1992. With USN, 1944-46. Rsch. grantee NIH, 1968, U.S. Army Rsch. Inst., 1978, 81. Fellow APA, Am. Psychol. Soc., Human Factors Soc.; mem. Ergonomics Soc. Achievements include 7 copyrights for software/hardware for self-assessment testing, user's guides. Home: 2005 Huntinton Dr Las Cruces NM 88001 Office: Human Performance Enhance 345 N Water St Las Cruces NM 88011

HUNT, DONNELL RAY, agricultural engineering educator; b. Danville, Ind., Aug. 11, 1926; s. Ray Hadley and Sarah Leona (Booty) H.; m. Dorothea Marie May, Sept. 2, 1951; children: David Carter, DeAnne Elizabeth. B.S., Purdue U., 1951; M.S., Iowa State U., 1954; Ph.D., Iowa State, 1958. Registered profl. engr., Ill. Instr. to assoc. prof. agrl. engring. Iowa State U., Ames, 1951-60; assoc. prof. U. Ill., Urbana, 1960-68, prof., 1968—; asst. dean, dir. coop. edn. program Coll. Engring. U. Ill., 1986—; cons. in field. Author: Farm Power and Machinery Management, 8th edit., 1983, Farm Machinery Mechanisms, 1972, Engineering Models for Agricultural Production, 1986. Served with U.S. Army, 1945-46. Fulbright awardee Ireland, 1968-69. Fellow Am. Soc. Agrl. Engrs.; mem. Am. Soc. Engring. Edn. Republican. Presbyterian. Office: U Ill Dept Engring 1308 W Green St Urbana IL 61801-2936

HUNT, HAROLD RAY, chemical engineer; b. Riverside, Calif., Oct. 12, 1945; s. Noel Rex and Virginia Elaine (Peterson) H.; m. Patricia Oris Proffitt, June 12, 1971; 1 child, Alan Auberon. BSChE with distinction, U. N.Mex., 1968. Registered profl. engr., Okla. Summer employee Phillips Petroleum Co., Bartlesville, Okla., 1966, rsch. engr., 1967-80, sr. rsch. engr., 1980-88, sr. process engr., 1988—. Contbr. articles to profl. jours. Treas. tech. career adv. com. Bartlesville Dist. Sci. Fair, 1990—. Mem. NSPE, Okla. Soc. Profl. Engrs. (Outstanding Engr. 1992). Achievements include 22 U.S. patents. Office: Phillips Petroleum Co 8 Phillips Bldg Bartlesville OK 74004

HUNT, JAMES H., JR., safety and environmental executive; b. Miami, Fla., June 29, 1948; s. James H. and Melinda (Suelter) H.; Kathryn Gale, Apr. 21, 1972; children: Arianna, James, Bryan. BA in Psychology, R.I. Coll., 1971. Sales mgr. Gorham Co., Smithfield, R.I., 1972-84, engr. prodn. control, 1984-86, risk mgmt. dir., 1986—. Mem. Am. Soc. Safety Engrs., R.I. Safety Assn., Grace Found. Home: 158 Douglas Rd Warwick RI 02886

HUNT, JANET ROSS, research nutritionist; b. Glendale, Ariz., Sept. 23, 1952; d. Halver Vincent and Shirley (Little) Ross; m. Eugene Daniel Mahalko, Apr. 18, 1973 (div. Feb. 1986); m. Curtiss Dean Hunt, Oct. 11, 1986; children: Vanessa, Carilla, Renatta, Brian. BS, Brigham Young U., 1973, MS, 1975; PhD, U. Minn., St. Paul, 1987. Registered dietitian. Instr. Brigham Young U., Provo, Utah, 1975-76, U. Tex. Health Sci. Ctr., Dallas, 1976-78; chief rsch. dietitian Human Nutrition Rsch. Ctr. USDA Agrl. Rsch. Svc., Grand Forks, N.D., 1978-87, rsch. nutritionist, 1988—. Contbr. over 40 articles to nutrition jours. Mead Johnson scholar Am. Dietetic Assn. Found., 1973-74. Mem. Am. Dietetic Assn. (bd. dirs. 1991-93), Am. Soc. Clin. Nutrition, Am. Inst. Nutrition. Office: USDA Agrl Rsch Svc Human Nutrition Rsch Ctr PO Box 7166 University Sta Grand Forks ND 58202

HUNT, PATRICIA JACQUELINE, mathematician, system manager, graphics programmer; b. Pasadena, Calif., Feb. 20, 1961; d. Daniel Joseph and Jacqueline (Vautrain) Collins; m. Daniel Phillip Hunt, Oct. 10, 1987. BS in Applied Math., U. Calif., Santa Barbara, 1983; MS in Applied Math., Naval Postgrad. Sch., 1988. Mathematician Computer Ctr. Naval Postgrad. Sch., Monterey, Calif., 1983-87; computer analyst Metro Info. Svcs., Virginia Beach, Va., 1988-90; system mgr. Lockheed Engring. and Sci.

Corp., Hampton, Va., 1990—; instr. NASA Engr.'s Week, 1992. Mem. Math. Assn. Am., Assn. Computing Machinery (graphics spl. interest group). Avocations: amateur radio, ballet, volleyball, piano. Home: 317 Willow Bend Ct Chesapeake VA 23323 Office: Lockheed Engring/Sci Corp 155 Research Dr Hampton VA 23666

HUNT, PHILIP GEORGE, computer consultant; b. Bolton, Eng., Apr. 2, 1957; came to U.S., 1981; s. John Joseph and Lilian (Boyer) H.; m. Carol Brown, Mar. 24, 1979 (div. 1990); children: Hazel R., Elizabeth. BSc, U. Manchester, 1978. Lic. electronic engr. Engr. Ferranti Computer Systems, Wythenshawe, Eng., 1978-81, Apollo Computer, Chelmsford, Mass., 1983-85; sect. mgr. Gould Modicon, Andover, Mass., 1981-83; ind. cons. Nixdorf, Digital Equipment, Adi Systems, Cayman Systems, So. N.H., 1986-92; sr. prin. engr. Alacron Inc., Nashua, N.H., 1992-93; systems engring. mgr. Xedia Corp., Wilmington, Mass. Mem. IEEE, Assn. Computing Machinery. Achievements include patents for memory architecture of modvue a graphical man machine interface, V.M.E. bridge via yabus; co-develop Cayman Systems Gatorbox. Home and Office: 80 Emerson Ave Hampstead NH 03841

HUNT, RONALD DUNCAN, veterinarian, educator, pathologist; b. Los Angeles, Oct. 9, 1935; s. Charles H. and Margaret (Duncan) H. B.S., U. Calif.-Davis, 1957, D.V.M. with highest honors, 1959; student, UCLA, 1954-55. Research fellow pathology Harvard Med. Sch., 1963-64, research assoc. pathology, 1964-69, prin. assoc. pathology, 1969-72, assoc. prof. comparative pathology, 1972-77, prof. comparative pathology, 1977—; dir. Animal Resources Center Harvard Med. Sch., 1979-89; dir. New Eng. Regional Primate Research Center, Southborough, Mass., 1976—. Author: (with T.C. Jones) Veterinary Pathology, 1972, 5th edit., 1983, (with T.C. Jones and U. Mohr) Endocrine System, Respiratory System, Digestive System, Urinary System, Genital System; contbr. numerous articles on research on vet. pathology to profl. jours.; editorial bd.: Lab Animal Medicine, 1969—, Jour. Med. Primatology, 1977—, Internat. Life Scis. Inst, 1981—, Am. Jour. Vet. Research, 1978-80. Trustee Charles Louis Davis DVM Found., 1979—; exec. com. Tufts U. Sch. Vet. Medicine, 1979-82. Served with Vet. Corps U.S. Army, 1959-63. Mem. Am. Coll. Vet. Pathologists, AVMA, Internat. Acad. Pathology, Am. Soc. Exptl. Pathology, Am. Soc. Clin. Pathologists, Mass. Soc. Pathologists, Am. Assn. Lab. Animal Sci., New Eng. Soc. Pathologists, Internat. Primatological Soc., N.Y. Acad. Scis., Am. Assn. Accreditation of Lab. Animal Care (mem. exec. com. 1989).

HUNT, THEODORE WILLIAM, civil engineer; b. Troy, N.Y., Feb. 4, 1909; s. William Theodore and Ida May (Bradshaw) H.; m. Louise Johnson Hunt, June 22, 1934 (dec.); 1 child, William Theodore. BS in Civil Engring., Mo. Sch. of Mines, 1933. Civil engr. Corps. Engrs. U.S. Army, numerous locations, 1934-54; architectural concrete cons. Portland Cement Assn., Chgo., 1954-72; concrete cons. Boston, 1972-84; concrete cons. semi-retired Rolla, Mo., 1985--. Fellow Am. Soc. Civil Engrs. (life), Am. Concrete Inst. Office: Box 1566 Rolla MO 65401

HUNTEN, DONALD MOUNT, planetary scientist, educator; b. Montreal, Que., Can., Mar. 1, 1925; came to U.S., 1963, naturalized, 1979; s. Kenneth William and Winnifred Binnmore (Mount) H.; m. Isobel Ann Rubenstein, Dec. 28, 1949; children: Keith Atherton, Mark Ross. B.Sc., U. Western Ont., 1946; Ph.D., McGill U., 1950. From research asso. to prof. physics U. Sask. (Can.), Saskatoon, 1950-63; physicist Kitt Peak Nat. Obs., Tucson, 1963-77; sci. adv. to asso. adminstr. for space sci. NASA, Washington, 1976-77; prof. planetary scis. U. Ariz., Tucson, 1977-88, Regents prof., 1988—; cons. NASA, 1964—. Author: Introduction to Electronics, 1964; (with J.W. Chamberlain) Theory of Planetary Atmospheres, 1987; contbr. articles to profl. jours. Recipient Public Service medal NASA, 1977, 85, medal for exceptional sci. achievement, 1980. Mem. Am. Phys. Soc., Can. Assn. Physicists (editor 1961-63), Am. Geophys. Union, Am. Astron. Soc. (chmn. div. planetary scis. 1977), Internat. Astron. Union, Internat. Union Geodesy and Geophysics, Internat. Assn. Geomagnetism and Aeronomy, AAAS, Nat. Acad. Scis., Explorers Club. Club: Cosmos (Washington). Home: 10 Calle Corta Tucson AZ 85716-4910 Office: U Ariz Dept Planetary Scis Tucson AZ 85721

HUNTER, ALAN GRAHAM, reproductive physiologist, educator; b. Pawtucket, R.I., Apr. 1, 1934; s. Lester Sisson and Margaret (Fitzpatrick) H.; m. Carolyn Whittingham, Aug. 27, 1955; children: Beth Allyn, David. BS, U. R.I., 1955, MS, 1958; PhD, Mich. State U., 1963. Instr. U. R.I., Kingston, 1955-58; rsch. assoc. Mich. State U., Lansing, 1958-63; prof. reproductive physiology U. Minn., St. Paul, 1963—; vis. prof. Cornell U., Ithaca, 1970, U. Cambridge, Eng, 1985. Contbr. 208 articles to profl. jours. Coach youth hockey, Roseville, Minn., 1975-85, youth baseball, Roseville, 1975-85. Recipient award of merit Gamma Sigma Delta, St. Paul, 1991; named Outstanding Tchr. of 1990, Disting. Tchr., 1992. Mem. Am. Dairy Sci. Assn., Soc. Study of Reprodn., Soc. Study of Fertility, Sigma Xi (pres. 1990). Methodist. Achievements include patent on early pregnancy diagnosis in cattle. Home: 2104 S Rosewood Ln Saint Paul MN 55108 Office: U Minn 495 Am Sci/Vet Medicine Saint Paul MN 55108

HUNTER, LARRY LEE, electrical engineer; b. Versailles, Mo., Mar. 5, 1938; s. Donnan Kleber and Molly Opal (Roe) H.; m. Marcella Ann Avey, Feb. 1, 1959; children: Cynthia Lynn Hunter Chapman, Stuart Roe. BSEE, U. Mo., 1963; MBA, Fla. Inst. Tech., 1984. System test engr. McDonnell Aircraft Corp., St. Louis, 1963-65; design engr. Magnavox Co., Urbana, Ill., 1965-66, research and devel. engr., 1966-68; project engr. LTV Electrosystems, Garland, Tex., 1968-69, systems engr., 1969-70; program mgr. Dorsett Electronics, Tulsa, 1970-73; program mgr. Harris Corp., Melbourne, Fla., 1973-75, bus. area mgr., 1975-85; v.p. mktg., engring., program mgmt. Teledyne Lewisburg, Tenn., 1985-88; gen. mgr. Precision Cable div. Precision Cable div. AMP Inc., Harrisburg, Pa., 1989—. Inventor thermometer; contbr. articles to profl. jours. Mem. IEEE, Eta Kappa Nu. Republican. Methodist. Avocations: hunting, fishing, golf. Home: 6303 Auburn Rd Mechanicsburg PA 17055-1982

HUNTER, NANCY QUINTERO, obstetrician/gynecologist; b. Maracaibo, Venezuela, May 18, 1961; came to U.S., 1976; d. George Albert and Jean Louise (Gimbert) Quintero; 1 child, Kathleen Jean. BS, Tulane U., 1982, MD, 1986. Intern Baylor Coll. Medicine, Houston, 1986-87, resident in ob-gyn., 1987-90; pvt. practice ob-gyn. Florence and Darlington, S.C., 1990-91, La Grange, Ga., 1992, Florence, 1993—. Fellow Am. Coll. Obstetricians and Gynecologists (bd. cert.); mem. AMA, S.C. Med. Soc. Office: 506 E Cheves St Florence SC 29501

HUNTER, RICHARD GRANT, JR., neurologist, executive; b. San Antonio, Aug. 22, 1938; s. Richard Grant Hunter Sr. and Elton (Hall) Samouce; m. Margaret Jenkins (div. Dec. 1984); m. Sandra K. Meisel-Conrad; children: Richard Grant III, Brandon Scott. BS, Ga. Inst. Tech., 1960; MBA, U. Pa., 1961; MD, U. Va., 1968. Diplomate Am. Coll. of Physicians, Am. Soc. Neuroimaging, Am. Bd. Internal Medicine, Am. Bd. of Psychiatry and Neurology. Systems analyst N.C. Nat. Bank, Charlotte, 1962-63; ops. rsch. analyst U.S. Geol. Survey, Washington, 1963-64; med. intern Grady Meml. Hosp., Atlanta, 1968-69; resident medicine Emory U. Hosp., Atlanta, 1969-71, resident psychiatry, 1971-72; resident neurology U. Va. Hosp., Charlottesville, 1972-75; neurologist U. Va. Hosp., Martha Jefferson Hosp., Charlottesville, 1975-83; chief of neurology Gwinett Med. Ctr., Atlanta, 1983-89; pres. Dixon Biotherapeutics Inc., Atlanta, 1985-86, neuropsychiatrist, behavioral neurologist, 1989—; pres. Charlottesville Neurology Inc., 1975—; bd. dirs. Dixon Biotherapeutics Inc., Atlanta, Charlottesville Neurology Inc. With U.S. Army, 1961-62. Fellow Am. Acad. Neurology, ACP; mem. AMA, Behavioral Neurology Soc., Med. Soc. Va., Med. Assn. Ga., Va. Neurol. Soc., Ga. Neurol. Soc., Am. Psychiat. Assn., Transcultural Psychology, Farmington Country Club, Farmington Hunt Club, River Bend Gun Club, Milrock Gun Club, Alpha Omega Alpha, Med. Honor Soc., Engring. Honor Soc. Episcopalian. Avocations: polo, combat pistol, skeet, trap.

HUNTER, ROBERT FABIO, civil engineer; b. Bklyn., July 6, 1960; s. Fabio and Dolores (Hayes) H. BCE, U. Detroit, 1984, MBA, 1990; post-grad., Wayne State U., 1991—. Registered profl. engr., Mich. City engr. City of Mt. Clemens, Mich., 1984-92; environ. engr. Camp, Dresser & McKee, Inc., Detroit, 1992—. Chmn. City of Mt. Clemens Affirmative Action Adv. Com., 1990. Mem. ASCE, NSPE, Am. Pub. Works Assn. Home: 21A Dickinson Mount Clemens MI 48043 Office: Camp Dresser & McKee Inc 1 Woodward Ave Detroit MI 48226

HUNTER, WILLIAM SCHMIDT, engineering executive, environmental engineer; b. Bellwood, Pa., Mar. 24, 1931; =. William Franklin and Mary Mildred (Schmidt) H.; m. Barbara Ruth Crosland (dec. Mar. 1959); m. Sandra Lee Showalter, Aug. 26, 1961 (dec. Sept. 1991); children: Felicia Fawn, Clarissa Cay, Patricia Schmidt. BSME, Lehigh U., 1953. Registered profl. engr., Pa. Project engr. W.Va. Pulp & Paper Co., Tyrone, Pa., 1953-56; plant engr. W.Va. Pulp & Paper Co., Williamsburg, Pa., 1956-61; project engr. St. Regis Paper Co., DeFeriet, N.Y., 1961-62; plant engr. Ga. Pacific Corp., Lyons Falls, N.Y., 1962-67; sr. project engr. Allied Paper Co., Kalamazoo, Mich., 1967-69, Hammermill Paper Co., Erie, Pa., 1969-80; chief engr. Stora Papyrus Newton Falls, N.Y., 1980-81, v.p. engring., 1981-90, dir. environ. affairs, 1990—. Bd. dirs. Altoona (Pa.) Engring. Soc., 1955-58, Am. Cancer Soc., Lewis County, N.Y., 1965; vice chmn. Solid Waste Disposal Authority, St. Lawrence County, N.Y., 1992-93; com. chmn. Explorer Scouts, Williamsburg, Pa., 1959; councilman Williamsburg Borough Coun., 1961; pres., bd. trustees Edinboro (Pa.) Presbyn. Ch., 1978-79. Mem. ASME, TAPPI. Avocation: amateur radio. Home: Rte 1 Box 63 Colby Ln Star Lake NY 13690 Office: Stora Papyrus Newton Falls Main St Newton Falls NY 13666

HUNTINGTON, HILLARD GRISWOLD, economist; b. Boston, Apr. 10, 1944; s. Hillard Bell and Ruth Smedley (Wheeler) H.; m. Honor Mary Griffin, Sept. 30, 1972; children: Honora Redmond, Emma Anne Hillard. BS, Cornell U., 1967; MA, SUNY, Binghamton, 1972, PhD, 1974. Staff economist Fed. Energy Adminstrn., Washington, 1974-77; dir., sr. economist Data Resources, Inc., Washington, 1977-80; exec. dir. Energy Modeling Forum Stanford (Calif.) U., 1980—; vol. U.S. Peace Corps., Pub. Utilities Authority, Monrovia, Liberia, 1967-69; vis. rsch. assoc. Inst. Devel. Studies, U. Nairobi, Kenya, 1972-73; mem. joint U.S.-U.S.S.R. Nat. Acad. Sci. Panel on Energy Conservation, 1986-90; mem. peer rev. panel Nat. Acid Precipitation Assessment Program Task Force, Ctrs. for Excellence Govt. Can.; consultant to Argonne Nat. Lab., Electric Power Rsch. Inst., numerous others. Editor Macroeconomic Impacts of Energy Shocks, 1987, N. Am. Natural Gas Markets: selected tech. studies, 1989. Mem. Internat. Assn. Energy Econs. (v.p. publs. 1990—, program chmn. ann. conf.), Am. Statis. Assn. (com. on energy stats. 1992—), Am. Econ. Assn. Home: 305 Hermosa Way Menlo Park CA 94025-5821 Office: Stanford U 406 Terman Ctr Stanford CA 94305

HUNTLEY, MARK EDWARD, biological oceanographer; b. Seattle, May 7, 1950; s. James Robert Huntley and Patricia Mary (Barricklow) Kissel; m. Patricia Darlene McFarlane, June 21, 1973 (div. 1980); children: Seth, Timothy; m. Kimberly Batcheller Brown, Sept. 19, 1981 (div. 1992); children: Swan Fairchild, Flannery Elizabeth, Zara Edith, Fletcher Wells. BSC with honors, U. Victoria, B.C., Can., 1976; PhD, Dalhousie U., Halifax, N.S., Can., 1980. Postdoctoral fellow Inst. Marine Resources, Scripps Instn. Oceanography, U. Calif. San Diego, La Jolla, Calif., 1980-82, asst. rsch. biologist, 1982-84; adj. lecturer Scripps Instn. Oceanography, U. Calif. San Diego, La Jolla, 1984—; asst. rsch. biologist marine biology rsch. div. Scripps Instn. Oceanography, La Jolla, 1984-87, assoc. rsch. biologist, 1987—; pres. Aquasearch, Inc., San Diego, 1984-88, chief oper. officer, 1988-93, CEO, 1993—, also chmn. bd. dirs., 1988—; deputy coord. water rsch. project U. Calif. San Diego, La Jolla, 1988-90; chmn. bd. dirs. Aquasearch, Inc., San Diego, 1984—; chief scientist Rsch. Antarctic Coastal Ecosystem Rates, La Jolla, 1986-87, 89, 91-92; exec. and steering com. mem. Global Ocean Ecosystem Dynamics, Washington, 1989—. Editor: Biological Treatment of Agricultural Wastewater, 1989; inventor Aquasearch Growth Module, 1989. Grantee Nat. Sci. Found. Office Naval Rsch., 1980—. Mem. Am. Soc. Limnology and Oceanography, Oceangraphy Soc. Avocations: tennis, gardening, music, body surfing. Office: Scripps Instn Oceangraphy 0202 La Jolla CA 92093

HUNTOON, ROBERT BRIAN, chemist, food industry consultant; b. Braintree, Mass., Mar. 1, 1927; s. Benjamin Harrison and Helen Edna (Worden) H.; m. Joan Fairman Graham, Mar. 1, 1952; children: Brian Graham, Benjamin Robert, Elisabeth Ellen, Janet Lynne, Joelle. BS in Chemistry, Northeastern U., 1949, MS, 1961. Analytical chemist Mass. Dept. Public Health; microbiologist Met. Dist. Commn. Mass. Dept. Public Health, Boston, 1950-53; rsch. and devel. chemist Heveatex Corp., Melrose, Mass., 1953-56; with Gen. Foods Corp., 1956-70; acting quality control mgr. Gen. Foods Corp., Woburn, Mass., 1965-67; head group rsch. and devel. Gen. Foods Corp., Tarrytown, N.Y., 1967-70; dir. quality control U.S. Flavor div. Internat. Flavors & Fragrances, Teterboro, N.J., 1970-83, mgr. tech. svcs., 1983-87, mgr. product devel., 1987-89, cons., 1989—; ind. cons. product devel., 1989—. Contbr. articles on flavor and food quality control to profl. and co. publs.; patentee gelatin compositions and mfg. processes. Served with USCG, 1945-46. Mem. Essential Oils Assn. (com. mem.), Flavor and Extracts Mfg. Assn. (com. mem.), Am. Chem. Soc., Inst. Food Technologists, Internat. Platform Assn., Indsl. Mgmt. Club (v.p. 1967) (Woburn), Croton Yacht Club, Saugus River Yacht Club (treas. 1967-68). Republican. Presbyterian. Office: 7 Scotland Hill Park Chestnut Ridge NY 10977-5908

HUNTRESS, WESLEY THEODORE, JR., government official; b. Washington, Apr. 11, 1942; s. Wesley Theodore and Elizabeth Agnes (Moran) H.; m. Roseann Albano, June 22, 1973; 1 child, Garret. BS, Brown U., 1964; PhD, Stanford U., 1968. Scientist Jet Propulsion Lab., Pasadena, Calif., 1968-88; dep. dir earth sci. NASA, Washington, 1988-90, dir. solar system exploration, 1990-93, assoc. adminstr. space sci., 1993—. Home: NASA Code SL 600 Independence Ave Washington DC 20546

HUOT, RACHEL IRENE, cell biologist; b. Manchester, N.H., Oct. 16, 1950; d. Omer Joseph and Irene Alice (Girard) H. BA in Biology cum laude, Rivier Coll., 1972; MS in Biology, Cath. U. Am., 1976, PhD in Biology, 1980. Sr. technician Microbiol. Assocs., Bethesda, Md., 1974-77; chemist Uniformed Svcs. Univ. of Health Scis., Bethesda, 1977-79; biologist Nat. Cancer Inst., Bethesda, 1979-82; postdoctoral fellow S.W. Found. for Biomed. Rsch., San Antonio, 1982-85, asst. scientist, 1985-87, staff scientist, 1987-88; instr. U. Tex. Health Sci. Ctr., San Antonio, 1988-89; asst. prof., dir. basic urologic rsch. La. State U., New Orleans, 1990—; judge sr. div. Alamo Regional Sci. Fair, San Antonio, 1989-90. Contbr. articles to profl. jours. Vol. ARC; active Stephen Ministry. NSF grantee, 1972-74; NIH Rsch. Svc. award, 1983-86. Mem. AAS, LWV, AAUW, Soc. for Microbiology, Fedn. Am. Scientists, Sci. Club (pres. 1971-72), Tissue Culture Assn., N.Y. Acad. Scis., St. Vincent De Paul Soc., Sierra Club, Sigma Xi, Iota Sigma Pi, Delta Epsilon Sigma. Democrat. Roman Catholic. Avocations: drawing, painting, reading, Volksmarching. Home: 3440 Edenborn Ave Apt 5 Metairie LA 70002-3334 Office: La State U Sch Medicine Dept Urology 1542 Tulane Ave New Orleans LA 70112-2865

HURD, SUZANNE SHELDON, federal agency health science director; b. Elmira, N.Y., Dec. 17, 1939; d. Victor Sheldon H. BS, Bates Coll., 1961; MS, U. Wash., 1963, PhD, 1967. Post-doctoral fellow U. Calif., Berkeley, 1967-69; grants assoc. NIH, Bethesda, Md., 1969-70; health sci. adminstr. Nat. Heart, Lung and Blood Inst., Bethesda, 1970-78, dep. dir. div. lung diseases, 1979-84; dir. div. lung diseases Nat. Heart, Lung and Blood Inst., Bethesda, 1984—. Mem. Am. Thoracic Soc. Office: Nat Heart Lung and Blood Inst Div Lung Diseases 5333 Westbard Ave # 6A16 Bethesda MD 20892-0001

HURDIS, EVERETT CUSHING, chemistry educator; b. Providence, R.I., Feb. 5, 1918; s. Charles Everett and Elsie Wild (Cushing) H.; m. Marion Elsie Smith, Mar. 16, 1946; children: Helen, Cindy, Holly, Sara. BS in Chemistry, Brown U., 1939; MA, Princeton (N.J.) U., 1941, PhD, 1943. Rsch. chemist U.S. Rubber Co., Passaic, N.J., 1943-58; group mgr. Koppers Co., Verona, Pa., 1958-62; chemistry tchr. Tex. Woman's U., Denton, 1962—. Republican. Episcopalian. Home: 621 E University Dr Denton TX 76201-2034 Office: Tex Woman's U Chemistry Dept Box 23973 TWU Sta Denton TX 76204

HURLBURT, HARLEY ERNEST, oceanographer; b. Bennington, Vt., Apr. 12, 1943; s. Paul Rhodes and Evelyn Arlene (Lockhart) H. BS in Physics (scholar), Union Coll., Schenectady, 1965; MS, Fla. State U., 1971, PhD in Meteorology, 1974. NASA trainee Fla. State U., 1970-72; postdoctoral fellow advanced studies program Nat. Center Atmospheric Research, Boulder, Colo., 1974-75; staff scientist JAYCOR, Alexandria, Va., 1975-77; oceanographer Naval Oceanographic and Atmospheric Rsch. Lab. (merged with Naval Rsch. Lab., 1992), Stennis Space Center, Miss., 1977—, br. head, 1983-85; adj. prof. of Marine Sci. U. So. Miss., Stennis Space Ctr., 1993—; mem. nat. adv. panels NASA satellite surface stress working group, 1981-84, minerals mgmt. service interagy. adv. group, 1982-89, world ocean circulation experiment working group on numerical modeling, 1984—, USN space oceanography working group, 1986-89; co-chmn. working group on global prediction systems, ocean prediction workshop, 1986; working group on acoustic monitoring of world ocean Sci. Com. Oceanic Rsch., 1991—; project leader to develop the world's first eddy-resolving global ocean prediction model for the USN, 1987—. Contbr. numerous articles to profl. jours. V.p. Burgundy Citizens Assn., 1976-77. Weather officer USAF, 1965-69. Recipient Disting. Scientist medal 13th Internat. Colloquium, Liege, Belgium, 1981, Publ. award for best basic research paper Naval Ocean Research and Devel. Activity, 1980, 90; Office Naval Research grantee, 1975-77, 84—, Dept. Energy grantee, 1975-78, Tex. A&M U. grantee, 1976, Office of Naval Technology grantee, 1987-93, Space Warfare Systems grantee, 1989—, Defense Advanced Rsch. Projects Agy. grantee, 1993—. Mem. Am. Meteorol. Soc., Am. Geophysical Union, Phi Sigma Kappa, Sigma Xi (Kaminski Publ. award 1991), Sigma Tau, Chi Epsilon Pi. Methodist. Home: 507 Hermitage Ct Pearl River LA 70452-3903 Office: Naval Rsch Lab Code 7320 Bldg 1103 Stennis Space Center MS 39529

HURLEY, LAURENCE HAROLD, medicinal chemistry educator; b. Birmingham, U.K., Jan. 29, 1944; s. Harold Harcourt and Mary (Cottrell) H.; children: Bridget, Nicole. BPharm, U. Bath, U.K., 1967; PhD, Purdue U., 1970. Apprentice pharmacist Boots the Chemist, Birmingham, 1963-64; hosp. pharmacist Birmingham Gen. Hosp., 1967; postdoctoral fellow U. B.C., Vancouver, 1970-71; asst. prof. U. Md., Balt., 1971-73; from asst. to assoc. to full prof. U. Ky., Lexington, 1973-80; prof. U. Tex., Austin, 1981—, Henry Burlage prof., 1983-86, James Bauerle prof., 1986-88, George Hitchings prof. drug design, 1988-91, George Hitchings regents chmn. drug design, 1992—; cons. Upjohn Co., Kalamazoo, 1979—, Smith Kline French, Phila., 1984-87, Abbott Labs., 1992—; mem. sci. adv. bd. Sun Pharm. Corp.; chmn. bioorganic and natural products study sect. NIH, 1986-88. Sr. editor Jour. Medicinal Chemistry; contbr. numerous articles to profl. jours. Recipient George Hitching award in inovative drug design, 1988, Volweiler Rsch. Achievement award, 1989, Rsch. Achievement award in medicinal chemistry Am. Pharm. Assn., 1992; named Outstanding Investigator Nat. Cancer Inst., 1989. Fellow AAAS; mem. Am. Chem. Soc. Democrat. Home: 5915 Northwest Pl Austin TX 78731 Office: U Tex Coll Pharmacy Austin TX 78712

HURLEY, RICHARD KEITH, chemist; b. Greensboro, N.C., July 2, 1956; s. Vernon Castle and Carolyn Nell (Costner) H.; m. Susan Elizabeth Lishness, June 30, 1990. BA in Chemistry, U. N.C., 1989. Analytical technician Ciba-Geigy, Greensboro, N.C., 1984-87; rsch. chemist ENCAS Analytical Labs., Winston-Salem, N.C., 1989-91; analytical program mgr., chemical hygiene officer Wildlife Internat. Ltd., Easton, Md., 1991—. Mem. Am. Chem. Soc. (agrochems. sect.). Home: 15 Apple Ln Preston MD 21655 Office: Wildlife Internat Ltd 8651 Brooks Dr Easton MD 21601

HURLEY, THOMAS DANIEL, biochemist; b. New Haven, Oct. 28, 1961; s. Thomas Joseph and Charlotte L. (Schweitzer) H.; m. Joyce Ann Harts, Jan. 25, 1992. BS, Penn State U., 1983; MS, Creighton U., 1985; PhD, Ind. U., 1990. Grad. student Creighton U., Omaha, 1983-85; rsch. technician J. T. Baker Chem. Co., Phillipsburg, N.J., 1985; grad. student Ind. U. Med. Sch., Indpls., 1985-90; postdoctoral fellow Johns Hopkins Med. Sch., Balt., 1990-92; asst. prof. Ind. U. Med. Sch., Indpls., 1992—. Contbr. articles to profl. jours. Recipient Kinsley award Ind. U. Grad. Sch., 1991, Rsch. Scientist Devel. award, 1992. Mem. Sigma Xi. Democrat. Achievements include determination of the three-dimensional structure of human alcohol dehydrogenase; determination of the key enzyme involved in ethanol metabolism. Office: Ind Univ Med Sch 635 Barnhill Dr Indianapolis IN 46202

HURST, CHRISTON JAMES, microbiologist; b. Cin., Sept. 11, 1954; s. Hubert Leon and Joy Ethel (Strack) H.; m. Pei-Fung Liew, Mar. 16, 1980; children: Rachel, Allen. BS, U. Cin., 1976; PhD, Baylor Coll. Medicine, 1979. Rsch. assoc. Baylor Coll. Medicine, Houston, 1980, U. Cin., 1980; microbiologist U.S. EPA, Cin., 1980—. Editor: Modeling the Environmental Fate of Microorganisms, 1991, Modeling the Metabolic and Physiologic Activities of Microorganisms, 1992; contbr. articles to profl. jours. Coach Odyssey of the Mind, Cin., 1991—. Mem. AAAS, Am. Soc. Microbiology (divsn. chmn. 1991-92, mem. editorial bd. 1989—, found. lectr. 1991, 92, 93), Am. Soc. Virology, Internat. Assn. Water Quality. Achievements include research in application of mathematical modeling in microbiology. Office: US EPA 26 Martin L King Dr W Cincinnati OH 45268

HURST, HARRELL EMERSON, pharmacology and toxicology educator; b. Somerset, Ky., Mar. 4, 1949; s. Hugh Thomas and Emmer Fugate (Davidson) H.; m. Carol Jo Fathergill, May 11, 1974. BS in Chemistry, U. Ky., Lexington, 1972, MS in Toxicology, 1976, PhD in Toxicology, 1978. Instr. U. Louisville Sch. Medicine, 1978-80, asst. prof., 1980-84, assoc. prof., 1984-91, prof., 1991—; dir. U. Louisville Therpeutics and Toxicology Lab., 1982—; cons. Mass. Spectrometry and Drug Disposition. Author: Drug Metabolism and Disposition, 1990-91. Pres. Ohio Valley Regional Soc. of Toxicology, 1986. Mem. Am. Chem. Soc., Am. Soc. for Mass Spectrometry, Soc. of Toxicology, Am. Soc. Pharmacology and Exptl. Therapeutics, Internat. Soc. for Study of Xenobiotics, Sigma Xi. Office: Univ Louisville Dept Pharmacology Louisville KY 40292

HURST, MICHAEL OWEN, biochemistry educator; b. Kansas City, Mo., Aug. 31, 1955; s. George Owen and Betty Lee (Fain) H. BS, U. Mo., Kansas City, 1977; PhD, Iowa State U., 1983. Rsch. assoc. U. Colo., Boulder, 1983-84; vis. asst. prof. Ball State U., Muncie, Ind., 1984-85, U. Wis., River Falls, 1985-88; asst. prof. biochemistry Ga. So. U., Statesboro, 1988—. Contbr. articles to profl. jours. Grantee Ga. So. U., 1989, 91, U. Wis.-River Falls, 1986. Mem. Am. Chem. Soc., Sigma Xi (local v.p. 1992—), Phi kappa Phi (local publicity officer 1990-91, local sec. 1989-90). Home: 564 E Main Apt 128 Statesboro GA 30458 Office: Ga So U Dept Chemistry LB8064 Statesboro GA 30460

HURST, MICHAEL WILLIAM, psychologist; b. Medford, Oreg., Dec. 9, 1947; s. William George and Betty Muriel (Stevens) H.; m. Patricia C. Scully, Aug. 22, 1970 (div. 1981); 1 child, Michelle P.; m. Renee Catherine Sancoff, Aug. 20, 1988; children: Rachel C., James M. BS, MIT, 1970; MEd, Boston U., 1972, EdD, 1974. Lic. psychologist, Mass., N.H.; cert. employee assistance profl. Rsch. assoc. Powell Assocs., Inc., Cambridge, Mass., 1970-72; rsch. and teaching fellow dept. counselor edn. Boston U., 1972-73, post-doctoral fellow and instr. dept. psychosomatic medicine, 1973-74, asst. prof. psychiatry, 1974-80, assoc. prof. psychiatry, 1980—; attending psychologist Univ. Hosp., Boston, 1978—; pres. Hurst Assocs., Inc., Boston, 1978-91; cons. to numerous orgns., 1978—; v.p. Am. PsychMgmt./Hurst Assocs., Inc., 1991-92; managed behavioral health care cons., 1992-93; pres. Med. Communications Tech., Inc., 1993—. Cons. editor: Behavioral Medicine, 1982-91; patentee in field; contbr. articles to profl. jours. Bd. dirs., treas. Mass. Spl. Needs Assn., Inc., Groton, 1973-76; bd. dirs., steering com. Mass. Psychol. Health Plan Inc., Groton, 1979-82; pres., v.p. Pine Acre Park Assn., Hampstead, N.H., 1982-85. Fellow Mass. Psychol. Assn.; mem. Am. Psychol. Assn., Am. Psychopathological Assn. (Morton Prince prize 1978), N.Y. Acad. Scis., Phi Delta Kappa, Delta Upsilon Alumni Assn. (bd. dirs. 1979-82). Republican. Episcopalian. Avocations: flying, shooting, boating. Home: 3 Eastwood Dr Windham NH 03087-1638

HURT, GEORGE RICHARD, electrical engineer; b. Memphis, Dec. 3, 1946; s. Aubrey Mack and Eileen Cecelia (Davidson) H.; m. Ruth Ann Rattie, June 27, 1970; children: Richard Joseph, Valerie Renee. BSEE, Christian Bros. Coll., 1968. Project engr. NASA/Apollo Checkout Team, Kennedy Space Ctr., Fla., 1968-70, NASA/Digital Fly-By-Wire, Dryden FLight Rsch., Calif., 1970-71; design engr. NASA/KSC Launch Proc. System, Kennedy Space Ctr., 1971-84; engring. mgr. NASA/Core System

Project, Kennedy Space Ctr., 1984—. Mem. Kennedy Mgmt. Assn. Democrat. Roman Catholic. Achievements include design of large computerized launch processing system for space shuttle, large real time checkout incorporating standards. Office: NASA DL-DSD Kennedy Space Ctr Cape Canaveral FL 32899

HURWICZ, LEONID, economist, educator; b. Moscow, 1917; came to U.S., 1940; LLM, U. Warsaw, Poland, 1938; DSc (hon.), Northwestern U., 1980; Dr honoris causa, U. Autónoma de Barcelona, Spain, 1989; D Econs honoris causa, Keio U., Tokyo, 1993. Rsch. assoc. Cowles Commn. U. Chgo., 1944-46; from assoc. prof. to prof. Iowa State U., Ames, 1946-49; prof. econ., math. and stats. U. Ill., 1949-51; prof. econ., math. and stats. U. Minn., Mpls., 1951—, Regents' prof., 1969-88, Regent's prof. emeritus, 1988—, Carlson prof. econs., 1989-92, prof. econs., 1992—; vis. prof. econs. Stanford (Calif.) U., 1955-56, 58-59, Harvard U., Cambridge, Mass., 1969-71, U. Calif., Berkeley, 1976-77, Northwestern U., Evanston, Ill., 1988-89; Fisher lectr. U. Copenhagen, 1963; hon. prof. Cen. China U. Sci. and Tech., Wuhan, 1984; vis. lectr. People's Univ., Beijing, People's Republic of China, 1986, Tokyo U., 1982; vis. Fulbright lectr. Bangalore U., India, 1965-66. Author and editor: (with K.J. Arrow); Studies in Resource Allocation Processes, 1977; (with K.J. Arrow and H. Uzawa) Studies in Linear and Non-Linear Programming, 1958; (with J.S. Chipman et al) Prefences, Utility and Demand, 1971, (with D. Schmeidler and H. Sonnenschein) Social Goals and Social Organization, 1985; contbr. articles to profl. jours. Sherman Fairchild Disting. scholar Calif. Inst. Tech., 1984-85; recipient Nat. medal Sci. NSF, 1990; fellow Ctr. Adv. Studies in the Behavioral Scis., 1955-56. Fellow Econometric Soc. (pres. 1969), Am. Econ. Assn. (disting.; Ely lectr. 1972); mem. NAS, Am. Acad. Arts and Scis. Office: Univ Minn Dept Econs 271 19th Ave S Minneapolis MN 55455-0430

HURWITCH, JONATHAN WILLIAM, energy consultant; b. Flushing, N.Y., May 24, 1955; s. Stanley and Ann (Silverstein) H.; m. Susan Frances Michelson, Aug. 21, 1977; children: Beryl Ayn, Sara Michelle. BS in Chemistry, U. Miami, 1977; MS in Chemistry, Georgetown U., 1980; MBA in Mgmt., Va. Polytech Inst., 1985. Rsch. scientist Battelle Columbus Lab., Washington, 1979-85; project mgr. Battelle Pacific Northwest Labs., Washington, 1985-87; asst. v.p. Energetics Inc., Columbia, Md., 1987—; coord. Utility Battery Group, 1989—; cons. Electric Power Rsch. Inst., Palo Alto, Calif., 1988—; tech. conf. chmn. Intersociety Energy Conversion Engring. Conf., San Diego, 1984. Contbr. articles to profl. jours. Fundraiser Children's Hosp. Nat. Med. Ctr., Washington, 1987-91; featured family Arthritis Found. Nat. Telethon, Nashville, 1988; conf. chmn. Am. Juvenile Arthritis Orgn., Orlando, Fla., 1989. Mem. IEEE. Jewish. Achievements include development of analytical methodologies and conceptual designs to evaluate and deploy electric energy storage technologies. Office: Energetics Inc 7164 Gateway Dr Columbia MD 21046

HURWITZ, CHARLES EDWIN, oil company executive; b. Kilgore, Tex., May 3, 1940; s. Hyman and Eva (Engler) H.; m. Barbara Raye Gollub, Feb. 24, 1963; children: Shawn Michael, David Alan. B.A., U. Okla., Norman, 1962. Chmn. bd., pres. Investam. Group, Inc., Houston, 1965-67, Summitt Mgmt. & Research Corp., Houston, 1967-70; chmn. bd. Summit Ins. Co. of N.Y., Houston, 1970-75; with MCO Holdings, Inc. (and predecessor), Los Angeles, from 1978, chmn. bd., chief exec. officer, from 1980, dir., from 1978; chmn. bd., pres., CEO, dir. Maxxam Inc., Houston; chmn. bd., pres. Federated Devel. Co.; dir. MCO Resources, Inc. Jewish. Office: Maxxam Inc 5847 San Felipe Houston TX 77057

HUSAR, EMILE, civil engineer, consultant; b. N.Y.C., Aug. 21, 1915; s. Elias and Tekla (Melech) H.; m. Lillian Semko, 1960; 1 child; 2 stepchildren. BCE, CCNY, 1938, MCE, 1940. Lic. profl. engr. and land surveyor, N.Y., N.J. Jr. engr. N.Y.C. Park Dept., 1940; jr. engr. def. constrn. Panama Canal, 1941; constrn. supt., designer, estimator gen. contractor, 1946-49; resident bldg. insp. N.Y.C. Bd. Edn., 1950; estimator bldgs. N.J. Turnpike, 1951; dir. pub. works, borough engr. sec. planning bd. bldg. insp. Leonia, N.J., 1952-64; chmn. bd. Stuyvesant Catering Corp., 1960-64; dir. pub. works, twp. engr. Twp. of Berkeley Heights, N.J., 1965-67; asst. city engr. East Orange, N.J., 1967-70; office and project engr. Clinton Bogert Assocs., Cons. Water & Sewer Engrs., 1970-76; borough engr., dir. pub. works Roselle, N.J., 1977; registered rep. First Investors Corp., 1967-69, Sage, N.J., 1969-76; pvt. practice as bldg. constrn. estimator and home inspection cons., 1976—; resident engr. Lawler, Matusky & Skelly, cons. engrs., N.Y.C., 1979-83, Singstad, Hurks & Assocs., 1984-85, RBA Group Engrs. & Planners, 1986-87, Shah Assocs. Cons. Engrs., 1987-88, Goodkind & O'Dea, Cons. Engrs., 1984, Massand Cons. Engrs., 1989-90. Pres. Ukranian Nat. Home of N.Y.C., 1956-62; conciliator N.J. Pub. Adv., 1981-84. 1st lt. C.E., U.S. Army, 1942-46. Recipient Presdl. citation for S.W. Pacific and Papuan Campaigns. Fellow ASCE; mem. Nat. Soc. Profl. Engrs., N.J. Soc. Profl. Engrs., 401 Investors Club (pres. 1965-75). Home and Office: 411 Charles Pl Leonia NJ 07605-1309

HUSAR, RUDOLF BERTALAN, mechanical engineering educator; b. Martonos, Yugoslavia, Oct. 29, 1941; came to U.S., 1966; s. Ga'bor and Ilona Barna Huszar; m. Janja Djukic, Oct. 8, 1967; children: Maja, Attila. Ing. Mech. Engring., Univ. if Zagreb, Croatia, 1962; Dipl. Ing. Mech. Engring., Tech. Univ., Germany, 1966; PhD in Mech. Engring., Univ. Minn., 1971. Design technician W. Hofer, Krefeld, W. Germany, 1962-63; rsch. asst. Tech. U., Berlin, Germany, 1963-66; from rsch. asst. to assoc. Univ. Minn., Mpls., 1966-71; rsch. fellow Calif. Inst. Tech., Pasadena, Calif., 1971-73; prof. Wash. U., St. louis, 1973—; vis. prof. U. Stockholm, Sweden, 1976; co-chmn. Interagency Com. Health & Environ. Effects Advanced Energy Tech., 1970; coop. program mem., Devel. and Appln. Space Tech. Air Pollution, EPA/NASA, 1978; dir. Ctr. for Air Pollution Impact and Trend Analysis (CAPITA), St. Louis, 1979—; com. mem. Atmospheric-Biosperic Interactions, Nat. Acad. Sci., 1979-81. Editor: Atmospheric Environment, 1980, Iclojaras, 1980; adv. bd. Environmental Science Technology, 1980. Rsch. fellow Univ. Glasgow, Scotland, 1965, Rsch. fellow Univ. Minn., 1966-71; EPA grantee, 1973—, NOAA grantee, 1991—, U.S. Dept. Defense grantee, 1989-92. Mem. Air & Waste Mgmt. Assoc., Ges. Aerosolforschung. Office: Wash U CAPITA Campus Box 1124 Saint Louis MO 63130

HUSE, DIANE MARIE, dietitian; b. Mpls., June 21, 1944; d. Gordon Simmons and Mildred L. (Johnson) H. BS in Dietetics, U. Minn., 1966, MS in Clin. Nutrition and Biochemistry, 1972. Registered dietitian. Dietetic intern Henry Ford Hosp., Detroit, 1966-67; therapeutic dietitian St. Barnabas Hosp., Mpls., 1967-68; clin. dietitian dept. pediatrics, eating disorders program, and cardiovascular health clinics for the young Mayo Clinic, Rochester, Minn., 1971—; instr. nutrition Mayo Med. Sch., 1976-85, asst. prof. nutrition, 1985—. Contbr. numerous articles and abstracts to profl. jours. Sec. Hist. Preservation Com., Rochester, 1987—. Mem. Am. Dietetic Assn., Am. Coll. Sports Medicine, Sports and Cardiovascular Nutrition Practice Group, Pediatric Nutrition Practice Group, Ctr. for Adolescent Obesity (bd. advisors 1984—), Minn. State Dietetic Assn. (sec. 1977-79), Minn. Heart Assn. (Sch. Lunch Task Force 1988-90), Rochester Dist. Dietetic Assn. (exec. bd. mem. 1972, 78, 80, program chmn. 1974-75, state meeting chmn. 1975, pres. 1975-76). Republican. Lutheran. Office: Mayo Clinic 200 1st St SW Rochester MN 55905-0001

HUSEMANN, ANTHONY JAMES, science educator; b. Balt., Feb. 16, 1955; s. Bernard William and Loretta Mary (Winkler) H.; m. Lynn F. Hauter, Sept. 14, 1980; children: Rebecca L., Aaron J., Joel A. BSc in Biology with honors, St. Mary's Coll. Md., 1976; MEd summa cum laude, U. S.C., 1991. Cert. sci. tchr., S.C. Grad. teaching asst., rsch. asst. Bowling Green (Ohio) State U., 1977-78; tchr. Sonoma and Mendocino County Schs., Willits, Calif., 1980-83, Rincon Valley Jr. High Sch., Santa Rosa, Calif., 1983-84; missionary tchr., sci. chair Triple C Sch., Grand Cayman, 1984-87; instr. biology Columbia (S.C.) Bible Coll., 1987; tchr. adult basic sci. Richland Schs., Columbia, 1988-91; tchr. biology, chemistry, phys. sci. South Carolina High Schs., 1988-91; tchr., sci. chair Triple C Sch., 1991-92; prin. Truth for Youth Sch., 1992—; asst. prof. Edn., Adult Edn. Internat. Coll. Cayman Islands, 1992—. Author: Willits News, 1980, 81, Self Guided Nature Tour, Westminster Woods, 1982, Nature Guide-Alliance Redwoods Camp, 1983; co-author: Field Studies-Lake Sonoma Project, 1983; contbr. to environ. impact statements. Coach Cayman Islands Nat. Swim Team, 1986-87; dir. day camp, tchr. Sunday sch. 1st Presbyn. Ch., Columbia, 1991; mem. Royal Life Saving Soc., Grand Cayman, Brit. Red Cross Soc., Grand Cayman, 1984-87. Nat. Merit scholar. Mem. Pvt. Schs. Prins. Assn., Nat. Assn. Biology Tchrs., Gideons Internat. (sec. Grand Cayman chpt.), Ednl. Community C. of C., Cayman Tchrs. Assn. Republican. Home: Box 370 G, Georgetown BWI Grand Cayman Island, Cayman Islands

HUSTED, RUSSELL FOREST, research scientist; b. Lafayette, Ind., Apr. 4, 1950; s. Robert Forest and Miriam Ruth (Jackson) H.; m. Nancy Lee Driscoll, Oct. 25, 1969 (div. Feb. 1986); children: Jacqueline Marie, Randall Forest; m. Ruth Elaine Hurlburt, Nov. 12, 1988. BS in Chemistry with highest distinction, Colo. State U., 1972; PhD in Pharmacology, U. Utah, 1976. Post-doctoral fellow dept. medicine U. Iowa, Iowa City, 1976-79, rsch. scientist dept. medicine, 1979-81, 1982—; asst. prof. U. Conn. Sch. Medicine, Farmington, 1981-82. Contbr. articles to profl. jours. Mallinckrodt scholar Colo. State U., 1968. Mem. AAAS, Am. Soc. Nephrology, Am. Physiol. Soc., Soc. Gen. Physiology, N.Y. Acad. Sci., Sigma Xi. Democrat. Methodist. Office: Univ Iowa 317 Medical Laboratories Iowa City IA 52242

HUSTON, FRED JOHN, automotive engineer; b. Muskegon, Mich., June 12, 1929; s. Fred and Sadie (Borgman) Huston; m. Jacqueline Terry, Apr. 28, 1957; children: Sandra, William. BSME, Mich. Tech. U., 1952. Engr. trainee IHC (Navistar), Ft. Wayne, Ind., 1952, test engr., 1953; mech. engr. asst. Aberdeen (Md.) Proving Ground, 1953-55; test engr. IHC (Navistar), Ft. Wayne, 1956, project engr., 1956-82; supr., chassis engr. M.A.N. Truck & Bus Corp., Cleveland, N.C., 1983-86; design engr. Thomas Built Buses, Inc., High Point, N.C., 1987-88, supr. body design, 1988-89, supr. chassis design, 1989-91, sr. staff engr., 1992—. With U.S. Army, 1953-55. Mem. Soc. Automotive Engrs. Methodist. Avocations: automobile restoration, swimming. Home: 603 Westchester Dr High Point NC 27262-7426

HUSTON, JEFFREY CHARLES, mechanical engineer, educator; b. Johnstown, Pa., Jan. 30, 1951; s. Charles Virgil and Pauline (Brubaker) H.; m. Patricia Ann Lemmon, June 1, 1974; children: Tiffany, Roger. BS, Ill. Inst. Tech., 1972; MS, W.Va. U., 1973, PhD, 1975. Registered profl. engr., Iowa. Mech. engr. Morgantown (W.Va.) Energy Rsch. Ctr., 1975-76; asst. prof. W.Va. U., Morgantown, 1975-76; asst. prof. Iowa State U., Ames, 1976-80, assoc. prof., 1980-87, prof. aerospace engring., engring. mechs., biomed. engring., 1987—; cons. in field. Contbr. articles to profl. jours.; inventor hip pod protective clothing, 1990. Witness testimony pub. hearing on safety of ATVs U.S. Consumer Product Safety Commn., Milw., 1985. Fellow W.Va. Found., 1972-75. Mem. ASME, NSPE, Am. Soc. Engring Edn. (awards chair 1987-89, Mickol award 1984), Iowa Profl. Engring. Soc. (Order of Engr. 1987), Soc. Automotive Engrs. (recreational vehicle com. 1980—, Ralph Teetor award 1980), Sigma Xi. Democrat. Methodist. Avocations: golf, hiking, puzzles. Home: 535 Valley West Ct West Des Moines IA 50265-4047 Office: Iowa State U AE & EM 3022 Black Engineering Ames IA 50010

HUSTON, RIMA, chemist; b. Kermanshah, Iran, Jan. 14, 1941; came to U.S. 1959; d. Avak and Shoushan (Golnazarian) Hacopian; m. Joseph P. Huston, Oct. 16, 1965; children: Alexander, Leila. BS in Phys. Sci., U. Md., 1965; PhD in Chemistry, U. Zurich, Switzerland, 1982. Rsch. scientist Arthur D. Little, Inc., Cambridge, Mass., 1965-66, Tyco Labs., Waltham, Mass., 1966-69, ETH-EAWAG, Zurich, 1972-76; rsch. scientist Redevcor SA, Zurich, 1985--, dept. head., 1985--. Contbr. articles to profl. jours. Mem. Am. Chem. Soc. Home: Tramstrasse 91, CH-8050 Zurich Switzerland Office: Redevcor SA, Ueberlandstrasse 241, CH-8050 Zurich Switzerland

HUT, PIET, astrophysics educator; b. Utrecht, Holland, Sept. 26, 1952; came to U.S., 1981.; s. Jan Lambertus Hut and Jenneke Johanna Hut-Broekroelofs; m. Eiko Ikegami, July 26, 1991. MS, U. Utrecht, 1977; PhD, U. Amsterdam, Holland, 1981. Asst. prof. astronomy dept. U. Calif., Berkeley, 1984-85; mem. Inst. for Advanced Study, Princeton, N.J., 1981-84, prof., 1985—; Contbr. articles to prof. jours. Mem. Am. Astron. Soc., Dutch Astron. Club. Office: Inst for Advanced Study Olden Ln Princeton NJ 08540

HUTA, HENRY NICHOLAUS, manufacturing and service company executive; b. Traunstein, Bavaria, Fed. Republic of Germany, Nov. 16, 1947; came to U.S., 1963; s. Mykola and Berta (Hoffmann) H.; m. Kay W. Crouch (div. 1985); stepchildren: David, Scott; m. Sharon L. Huta, Jan. 2, 1986; children: Nicholaus Henry, Garrett Thomas. AS with honors, Suffolk County Community Coll., 1976; BS magna cum lude, L.I. U., 1976; MS with honors, West Coast U., 1980; postgrad., Peter F. Drucker Grad. Sch. Mgmt., 1986. CPA, Calif. Sr. cons. Arthur Young & Co., N.Y.C., 1976-78; pres. R&B Info. Systems, Inc., Los Angeles, 1978-80; mgr. cons. Price Waterhouse & Co., Los Angeles, 1980-81; v.p. fin. and ops. Bay Distbrs., Los Angeles, 1981-83; v.p., gen. mgr. Cal Fruit Inc., Los Angeles, 1983-84; v.p., asst. to chmn. Ducommun Inc., Los Angeles, 1984-86; pres., chief exec. officer Pacific Diversified Capital Co. subs. San Diego Gas & Electric, San Diego, 1986—; pres., chief exec. officer Phase One Devel., Inc., San Diego, 1989—, also bd. dirs.; pres., chief exec. officer Wahlco Environ. Systems, Inc., L.A., 1989—, also bd. dirs.; bd. dirs. Wahlco, Inc., L.A., Creative Nail Design Inc., GTI Corp. Lance cpl. USMC, 1964-67, Vietnam. Exec. fellow Chapman U., Orange, Calif. Mem. AICPA, Calif. State Soc. CPAs, Presidents Assn., Am. Mgmt. Assn. (mem. exec. com.), Assn. Energy Engrs., Turnaround Mgmt. Exec. Republican. Home: 30001 Hillside Ter San Juan Capistrano CA 92675-1536

HUTCHEON, CIFFORD ROBERT, engineer; b. N.Y.C., June 10, 1913; s. Forbes Gerard and Bertha Johanna (Von Biela) H. m. Mary D. Kearny, June 1, 1939; children: Pamela M., David F. BME, Polytechnic Inst. Bklyn., 1948. Asst. regional mgr. Latin Am. Carrier Corp., Internat. Div., N.Y.C., 1947-49; product mgr. Carrier Corp., Internat. Div., San Juan, P.R., 1950-51, Havana, Cuba, 1952-53; asst. gen. mgr. Stewart Mfg. Co., Cedar Grove, N.J., 1953-54; owner, chmn. C.R. Hutcheon, Inc., Bloomfield, N.J., 1954—. Mem. ASHRAE (life), Rotary (dist. com. chmn. 1979-85). Republican. Episcopalian. Office: CR Hutcheon Inc 225 Belleville Ave Bloomfield NJ 07003-3584

HUTCHERSON, KAREN FULGHUM, nursing administrator; b. Winston-Salem, N.C., Oct. 1, 1951; d. John Fulghum and Viola Sprinkle Shaw; m. Victor J. Hutcherson, Dec. 18, 1970; children: Shannon Renae, Ashley Michelle. Diploma, N.C. Bapt. Hosp. Sch. Nursing, 1972; BSN, N.C. A&T State U., 1981; MBA, Wake Forest U., 1990. RN. Staff nurse N.C. Bapt. Hosp., Winston-Salem, 1972; oncology nurse clinician Cancer Ctr., Wake Forest U., Winston-Salem, 1972-81; oncology nurse educator Bowman Gray Sch. of Med., Wake Forest U., Winston-Salem, 1981-87, dir. nursing cancer ctr., 1982-87; clin. coord. nursing svcs. Bowman Gray Sch. Medicine, Wake Forest U., Winston-Salem, 1987—; curriculum coord., primary instr. Cancer Ctr., Bowman Gray Sch. Medicine, Wake Forest U., 1980-87, mem. numerous coms.; mem. speakers bur. A.H. Robbins Pharms. Co., 1983-88; cons. S.E. Cancer Control Consortium, Winston-Salem, 1987-90. Author: Patient Education in Understanding Cancer: An Introductory Handbook, 1986; co-author: Understanding Cancer Treatment: A Guide for You and Your Family, 1988, Cancer Chemotherapy Guidelines, 5th edit., 1985. Chmn. western div. nursing com. N.C. Am. Cancer Soc., 1981-82, speakers bur., 1982, bd. dirs., 1988-92; mem. spl. rev. com. clin. commun. oncology program Nat. Cancer Inst., Bethesda, Md., 1987. Recipient Leadership award Babcock Grad. Sch. Mgmt., Wake Forest U., 1992. Mem. ANA, Am. Acad. Ambulatory Nursing Adminstrn., Med. Ctr. Nursing Assn., Piedmont Oncology Assn. (numerous coms.), Oncology Nursing Soc. (mem. com.), Nat. League for Nursing, Am. Orgn. Nursing Execs., Med. Group Mgmt. Assn., S.E. Cancer Control Consortium, N.C. Nurses Assn. (legis. com. 1989, vice chmn. commun. health coun., del. conv. 1987, 88, 89, 90, 91), Sigma Theta Tau. Home: 754 Lacock Ave Rural Hall NC 27045-9742 Office: Bowman Gray Sch of Med Medical Center Blvd Winston Salem NC 27157-0001

HUTCHESON, PHILIP CHARLES, computer programmer, analyst; b. Birmingham, Ala., Feb. 3, 1948; s. Charles Emerson H. BA, U. S.Fla., 1973. Computer programmer analyst U.S. Gen. Acctg. Office, Atlanta, 1977—; mem. adv. com. on persons with disabilities U.S. GAO. Named Outstanding Handicapped Fed. Employee of Atlanta, Fed. Exec. Bd., 1980; recipient A. P. Jarrell award for excellence Ga. Rehab. Assn., 1977. Mem. IEEE, ACM, Persons with Disabilities (adv. coun.). Libertarian. Methodist. Home: 1814 Oriole Trl Lithia Springs GA 30057-2725 Office: US Gen Acctg Office 101 Marietta St NW Ste 2000 Atlanta GA 30323-0201

HUTCHINS, CARLEEN MALEY, acoustical engineer, violin maker, consultant; b. Springfield, Mass., May 24, 1911; d. Thomas W. and Grace (Fletcher) Maley; m. Morton A. Hutchins, June 6, 1943; children: William Aldrich, Caroline. A.B., Cornell U., 1933; M.A., NYU, 1942; D. Engring. (hon.), Stevens Inst. Tech., 1977; D.F.A. (hon.), Hamilton Coll., 1984; DSc (hon.), St. Andrews Presbyn. Coll., 1988; LLD (hon.), Concordia U., Montreal, Que., Can., 1992. Tchr. sci. Woodward Sch., Bklyn., 1934-38, Brearley Sch., N.Y.C., 1938-49; asst. dir., asst. prin. All Day Neighborhood Schs., N.Y.C., 1943-45; sci. cons. Coward McCann, Inc., 1956-65, Girl Scouts, Nat. Recreation Assn., 1957-65; permanent sec. Catgut Acoustical Soc., Montclair, N.J., 1962—. Author: Life's Key, DNA, 1961; Moon Moth, 1965; Who Will Drown the Sound, 1972; (with others) Science Through Recreation, 1964. Editor: (2 vols.) Musical Acoustics, Part I, Violin Family Components, 1975; Musical Acoustics, Part II, Violin Family Functions, 1976; The Physics of Music, 1978; contbr. articles to profl. jours. in Sci. Am., Jour. of the Acoustical Soc. Am., Am. Viola Soc., Catgut Acoustical Soc. Martha Baird Rockefeller Fund for Music grantee, 1966, 68, 74; Guggenheim fellow, 1959, 61; recipient several spl. citations in music. Fellow AAAS (electorate nominating com. 1974-76), Audio Engring. Soc. (life), Acoustical Soc. Am. (membership com. 1980-86, exec. council 1984-87, medal and awards com. 1987-89, nominating com. 1987-88, Silver Acoustics Medal 1981, tech. com. music. acoustics 1964—, chmn. pres.'s ad hoc com. 1987-88, archives com. 1988—); mem. So. Calif. Violin Makers Assn. (hon.), Viola da Gambda Soc. Am. (hon.), Scandinavian Violin Makers Assn. (hon.), Audio Engring. Soc., Guild Am. Luthiers, Am. Viola Soc., Violoncello Soc., Amateur Chamber Music Players Assn., Am. Philos. Soc. (award violin acoustics 1968, 81), Sigma Xi, Pi Lambda Theta, Alpha Xi Delta. Clubs: Three O'Clock, Dot & Circle, others. Home: 112 Essex Ave Montclair NJ 07042-4121 Office: Catgut Acoustical Soc Inc 112 Essex Ave Montclair NJ 07042

HUTCHINS, JAMES BLAIR, neuroscientist; b. Odessa, Tex., Oct. 21, 1958; s. John Sayles Hutchins and Patricia Ruth (Dyer) Haefele; m. Rosemary Theresa Hoffman, Feb. 22, 1985. BA, U. Colo., 1980; MA, U. Calif., 1982; PhD, Baylor Coll. Medicine, 1985. Postdoctoral fellow Vanderbilt U. Sch. Medicine, Nashville, 1985-87, rsch. asst. prof., 1987-89; asst. prof. U. Miss. Med. Ctr., Jackson, 1989--. Contbr. articles to profl. jours. Participant Sci. Edn. Programs, Jackson, 1989--; vol. Vanderbilt U. Children's Hosp., Nashville, 1987-88. NIH rsch. grantee 1989--, Pfeiffer Found rsch. grantee 1987-89. Mem. Soc. for Neuroscience (pres. Miss. chpt. 1992-93), Am. Soc. for Cell Biology, Assn. for Rsch. in Vision and Ophthalmology. Democrat. Episcopalian. Achievements include studies of nervous system development and characterization of the presence and role of the neurotransmitter acetylcholine in the human retina. Office: U Miss Med Ctr 2500 N State St Jackson MS 39216-4505

HUTCHINS, PAUL FRANCIS, JR., energy engineer; b. Jacksonville, Fla., Nov. 5, 1949; s. Paul Francis and Mary Elizabeth (Simpson) H.; m. Leslie Jane Zerwer, Nov. 14, 1972; children: Christopher Brett, Kathryn Anne. BS, U. Tenn., 1972, PhD, 1978; MS, U. Cen. Fla., 1976. Registered profl. engr., Fla.; cert. energy mgr. Staff engr. Martin Marietta Aerospace, Orlando, Fla., 1972-76; grad. rsch. asst. U. Tenn., Knoxville, 1976-77, Oak Ridge (Tenn.) Nat. Lab., 1977-78; dir. energy analysis Reynolds, Smith & Hills, Inc., Jacksonville, 1978—; mem. environ. control com. Gen. Devel. Corp., Jacksonville and Miami, 1991. Mem. ASHRAE (mem. tech. com. 9.5 1991—), Assn. Energy Engrs., Tau Beta Pi, Pi Kappa Phi, Sigma Pi Sigma. Republican. Methodist. Avocations: golf, fishing, computers. Office: Reynolds Smith & Hills Inc 4651 Salisbury Rd Jacksonville FL 32256

HUTCHINSON, JOHN WOODSIDE, applied mechanics educator, consultant; b. Hartford, Conn., Apr. 10, 1939; s. John Woodside and Evelyn (Eastburn) H.; m. Helle Vilsen, Aug. 28, 1964; children: Leif, David, Robert. B.S., Lehigh U., 1960; M.S., Harvard U., 1961, Ph.D., 1963; D.Sc. (hon.), Royal Inst. Tech., Stockholm, 1985, Tech. U. Denmark, Lyngby, 1992. Asst. prof. Harvard U., Cambridge, Mass., 1964-69, Gordon McKay prof. applied mechanics, 1969—; cons. to various industries; cons. Mobil Solar, Arthur D. Little, IBM, Polaroid. Contbr. articles to profl. jours. Guggenheim Found. fellow, 1974. Fellow ASME (Arpard L. Nadai award 1991); mem. AAAS, ASTM (Irwin medal 1982), NAE,, NAS.

HUTCHINSON, MARTHA LUCLARE, pathologist; b. Alton, Ill., Oct. 26, 1941; d. Elmer Frank and LuClare (Hall) H.; m. Marshall Edward Kadin, June 15, 1980. BS, Iowa State U., 1963; PhD, Purdue U., 1970; MD, Case Western Res. U., 1974. Intern Cleve. Met. Gen. Hosp., 1974-75; resident U. Calif., San Francisco, 1975-77, U. Wash., Seattle, 1977-79; asst. prof. Purdue U., West Lafayette, Ind., 1970, U. Wash., Seattle, 1979-84; assoc. prof. pathology Tufts U., Boston, 1984—; dir. cytopathology New Eng. Med. Ctr. Hosps., Boston, 1984—. Co-investigator devel. and testing of automated devices to facilitate cytology (pathology) diagnosis, 1986—. NIH grantee, 1986, 89, 92. Mem. AMA, Am. Soc. Clin. Pathology, Coll. Am. Pathologists, Internat. Soc. Analytical Cytology, Internat. Acad. Pathology, Am. Soc. Cytology, Internat. Soc. Analytical Cytology. Home: 103 Clinton Rd Brookline MA 02146-5812 Office: New Eng Med Ctr 750 Washington St Boston MA 02111-1533

HUTCHINSON, THOMAS EUGENE, biomedical engineering educator; b. York, S.C., Aug. 1, 1936; m. Colleen Ray, 1959; 2 children. BS, Clemson U., 1958, MS, 1959; PhD in Physics, U. Va., 1963. Teaching asst. U. Va., 1960-61; research fellow AEC, 1962; sr. scientist 3M Co., 1963-66, research specialist, 1966-67; assoc. prof. chem. engring. and material sci. U. Minn., 1967-74, prof., 1974-76; prof. bioengring. and chem. engring. U. Wash., from 1976, assoc. dean engring. research, from 1982; William Stanfield Calcott prof. biomed. engring. U. Va., 1982—; cons. various orgns. including 3M Co., RCA, North Star Research, 1967-73, on metallurgy, fed. agys., 1988; chmn. Gordon Research Conf., 1970; vis. prof. Cavendish Lab. Cambridge, Eng., 1971; sr. research fellow U. Glasgow, Scotland, 1974—; chmn. Bettelle Conf. Microprobe Analysis, 1980, Acad. Conf. on the Future of Sci. in the Southeast, 1987, Southeastern Univs. Reasearch Assn. Materials Sci. com., 1984—, SURA Conf. on Synchrotron Radiation, 1986, 87, SURA select com. on the Future of Materials Service; mem. Nat. Sci. Found. select com. for Materials Service Ctrs., 1985-88; dir. SURA-COM Satellite Video Network, 1988. Patentee Eyegaze Response Computer Aid (ERICA), 1987. 2nd lt. Civil Air Patrol, 1987. Recipient Disting. Service award Va. Track Assn., 1988. Elected sr. fellow Biomedical Engring. Soc.; mem. Electron Microscopy Soc. Am., Am. Vaccuum Soc., Va. Advanced Tech. Assn. (bd. dirs. 1985—), Cosmos Club. Fields of research include microprobe analysis of biol. tissue-an application of physics tools to solution of questions of ion transport in excitable cells, eye gaze computer interface for the handicapped, psychological and drug impairment testing using eye gaze response analysis. Home: Hardendale Box 168 Ivy VA 22945 Office: Univ Va Dept Bio-Engineering Cobb Chemical Bldg Charlottesville VA 22903

HUTCHINSON, WILLIAM BURKE, surgeon, research center director; b. Seattle, Sept. 6, 1909; s. Joseph Lambert and Nona Bernice (Burke) H.; m. Charlotte Rigdon, Mar. 25, 1939; children: Charlotte J. Hutchinson Reed, William B., John L., Stuart R., Mary Hutchinson Wiese. BS, U. Wash., Seattle, 1931; MD, McGill U., 1936; HHD (hon.), U. Seattle, 1982. Diplomate: Am. Bd. Surgery. Intern Balt. City Hosp., 1936-37; resident Union Meml. Hosp., Balt., 1937-39, James Walker Meml. Hosp., Wilmington, N.C., 1939-40; surgeon Swedish Hosp. and Med. Ctr., Seattle, 1941—, Providence Hosp., Seattle, 1941—; pres., founding dir. Pacific Northwest Research Found., Seattle, 1955—; founding dir. Fred Hutchinson Cancer Research Ctr., Seattle, 1972-85; dir. Surg. Cancer Cons. Service, 1982—; clin. prof. surgery emeritus U. Wash.; pres. 13th Internat. Cancer Congress, 1978-82; mem. Yarborough com. for writing Nat. Cancer Act, 1970. Contbg. editor, 13th Internat. Cancer Congress. Recipient 1st Citizen of Seattle award, 1976, Alumnus Summa Laude Dignatus award U. Wash., 1983, Wash. State award of Merit, 1988, Disting. Achievement award U. Wash., 1993. Fellow ACS; mem. AMA, King County Med. Soc., Seattle Surg. Assn., North Pacific Surg. Assn., Pacific Coast Surg. Assn., Western Surg. Assn., Soc. Surg. Oncologists, NRC, Alpha Sigma Phi. Clubs: Men's University (Seattle); Seattle Golf and Country. Home: 7126-55th Ave So Seattle WA 98118 Office: Pacific NW Rsch Found 720 Broadway Seattle WA 98122-4327

HUTCHISON, CLYDE ALLEN, III, microbiology educator; b. N.Y.C., Nov. 26, 1938; divorced; 3 children. BS, Yale U., 1960; PhD in Biophysics, Calif. Inst. Tech., 1969. From asst. prof. to assoc. prof. U. N.C., Chapel Hill, 1968-78, prof. bacteriology Sch. Medicine, 1978—, now Kenan prof. microbiology and immunology. Recipient Rsch. Career Devel. award NIH, 1973-78, MERIT award NIH, 1987—; NIH grantee, 1969—. Fields of research include genetics of viruses; mammalian genome structure ; restriction enzymes; DNA sequencing, site-directed mutagenesis. Office: U NC Dept Microbiology CB 7290 Chapel Hill NC 27599

HUTCHISON, GEORGE BARKLEY, epidemiologist, educator; b. Lexington, Ky., Oct. 18, 1922; s. George Barkley and Aliena Hale (Hunter) H. AB magna cum laude, Harvard Coll., 1943; MD, Harvard U., 1951, MPH, 1960. Intern Mass. Meml. Hosp., Boston, 1951-52; resident Lahey Clinic, Boston, 1952-55; asst. med. dir. Health Ins. Plan, N.Y.C., 1956-59, N.Y.C. Health Dept., 1957-59; from asst. to assoc. prof. Harvard Sch. Pub. Health, Boston, 1960-66; epidemiologist Michael Reese Hosp., Chgo., 1966-71; assoc. prof. U. Chgo., 1969-71; prof. Harvard Sch. Pub. Health, Boston, 1972-88, emeritus prof., 1988—; mem. study sect. NIH, Bethesda, Md., 1984-87, rsch. com. NAS, Washington, 1980-85. Contbr. over 50 articles to profl. jours. Capt. U.S. Army 1943-46. Mem. AAAS, APHA, Am. Statis. Assn., N.Y. Acad. Medicine, Phi Beta Kappa. Democrat. Home: 115 Saint Francis Ct Louisville KY 40205

HUTCHISON, JAMES ARTHUR, JR., engineering company executive; b. Gainesville, Mo., Oct. 25, 1917; s. James Arthur and Dora Ethel (James) H.; m. Imogene Cox, Dec. 5, 1946; children—Judith Lynn, Janet Gayle, James Arthur III. B.S. in Mech. Engring., Okla. State U., 1940; B.S. in Aero. Engring., Spartan Sch. Aeronautics, 1942; B.S. in Acctg., Okla. Sch. Accountancy, 1963. Registered profl. engr. Del., N.J., Md., V.I. Asst. chief engr. Spartan Aircraft Co., Tulsa, 1943-51; ptnr., owner H & H Engring. & Constrn. Co., Tulsa, 1951-68; sr. liaison engr. ILC Industries Inc. Apollo Astronaut Program, Dover, Del., 1968-72; v.p. Diamond State Engring. Inc., Dover, 1972-78; founder, chmn. bd. dirs. The JAED Corp., Smyrna, Del., 1978—. Chmn. bd. trustees 1st Baptist Ch., Dover, 1979-85. Recipient Gold Seal Del. Bd. Edn., 1974. Mem. Am. Inst. Steel Constrn., Am. Concrete Inst., Am. Soc. Heating Engrs., Central Del. Pilots Assn. (pres. 1977-78). Republican. Avocations: aircraft flight instruction. Office: The JAED Corp 19 Village Sq Smyrna DE 19977-1836

HUTCHISON, ROBERT B., chemist; b. Freeport, Ill., June 14, 1935. BS, Kent State U., 1957; PhD in Organic Chemistry, U. Calif., Berkeley, 1960. Rsch. chemist Miami Valley labs. Procter & Gamble Co., 1960-63; asst. prof. chemistry Bowling Green State U., 1963-67; rsch. mgr. Henkel Corp. Emery Group, Cin., 1967-74, dir. ctrl. rsch., 1974-78, dir. corp. rsch. devel., 1978-79; v.p. rsch. and devel. Emery Group, 1979—. Grantee USPHS, 1965-69. Mem. Am. Chem. Soc., Soc. Cosmetic Chemistry. Achievements include structure studies of naturally occurring compounds; reaction study of compounds leading to electron deficient nitrogen species. Office: Emery Tech Ctr 4900 Este Ave Cincinnati OH 45232-1419*

HUTCHISON, VICTOR HOBBS, biologist, educator; b. Blakely, Ga., June 15, 1931; s. Joseph Victor and Veva (Hobbs) H.; m. Theresa Dokos, Dec. 14, 1952; children—Victoria Ann, John Christopher, David Michael, Kenneth Hobbs. B.S., N. Ga. Coll., 1952; M.A., Duke, 1956, Ph.D., 1959. Instr. Duke, 1957-58, faculty fellow, So. Fellowship Fund fellow, 1958-59; mem. faculty U. R.I., 1959-70, prof. biology, 1968-70; dir. Inst. Environ. Biology; 1966-70; prof., chmn. dept. zoology U. Okla., Norman, 1970-80; George Lynn Cross research prof. zoology U. Okla., 1979—; research prof. Universidad de Los Andes, Bogotá, Colombia, 1965-66; prin. investigator Nat. Geog. Soc.-U. R.I. herpetological expdn. to Colombia, 1964-65, Nat. Geog. Soc.-U. Okla. expdns. to Lake Titicaca, 1975, to Cameroon, 1981. Research and articles on heat tolerances of lower vertebrates, effects of day-length on metabolism and temperature tolerance of lower vertebrates, physiology of lower vertebrates, physiol. ecology of amphibians and reptiles, respiration in amphibians, behavioral thermoregulation. Guggenheim fellow, 1965-66. Fellow AAAS; mem. Am. Inst. Biol. Sci., Am. Soc. Ichthyologists and Herpetologists (pres. 1988), Am. Physiol. Soc., Ecol. Soc. Am., Herpetologists League (exe. com. 1968-71), Soc. Study Amphibians and Reptiles (bd. govs. 1986-88), Explorers Club, Sigma Xi, Phi Sigma, Phi Kappa Phi. Achievements include demonstration of facultative endothermy in brooding pythons; research on role of skin in amphibian respiration; development of standardized method for determination of critical thermal maximum in animals. Home: 2010 Crestmont Ave Norman OK 73069-6414 Office: U Okla Dept Zoology Norman OK 73019

HUTH, PAUL CURTIS, ecosystem scientist, botanist; b. Kingston, N.Y., Feb. 18, 1947; s. Berthold Carl and Ruth Doris (Persons) H.; m. Ann Louise Friess, May 22, 1983. BS, SUNY, New Paltz, 1972, MA, 1979. Lab. technician N.Y. State Agrl. Experiment Sta., Highland, 1967-76; quality control inspector N.Y. State Dept. Agr. and Markets, Albany, 1980; regional enumerator N.Y. State Crop Reporting Svc., Albany, 1980-81; field supr. Fed. Crop Ins. Corp. USDA, Harrisburg, Pa., 1981-82; ecosystem rsch. scientist Mohonk Preserve, Inc., New Paltz, 1983-88, dir. rsch., 1989—; cons. Hudsonia, Ltd., Bard Coll., Annendale, N.Y., 1984—; chair rsch. and records com. John Burroughs Natural History Soc., High Falls, N.Y., 1986—; coop. weather observer Nat. Oceanic and Atmospheric Adminstrn., Mohonk Lake, N.Y., 1990—. Editorial bd. Up River/Down River mag., 1990-92; contbr. articles to Environ. Entomology, Northeastern Geology. Bd. dirs. Eastern N.Y. chpt. Nature Conservancy, Albany, 1988—, Klyne Esopus Hist. Soc. Mus., Ulster Park, N.Y., 1989-91; mem.-at-large Ulster County Environ. Adv. Bd., Kingston, 1990—, John Burroughs Assn., N.Y.C., 1992—, 1st v.p. Mem. Am. Soc. Ichthyologists and Herpetologists, Ecol. Soc. Am., Am. Assn. Applied Sci., Am. Soc. Environ. History, Am. Ornithologists' Union, Am. Soc. Mammalogists, Am. Inst. Biol. Sci., Linnaean Soc. N.Y., Torrey Botan. Club, N.Y. Flora Assn. (adv. coun.), Royal Horticultural Soc. Achievements include specialized, long-term research on Shawangunk Mountain ecosystem in southeastern N.Y. Home: Esopus Ave Box 45 Esopus NY 12429-0045 Office: Mohonk Preserve Inc Mountain Rest Rd New Paltz NY 12561-2917

HUTTEMAN, ROBERT WILLIAM, civil engineer; b. Webster, N.Y., Sept. 22, 1965; s. Charles William and Mary Jane (Hurzsy) H. BS in Physics, Buffalo State U., 1987; BSCE, U. Buffalo, 1989. Structural engr. Bergmann Assocs., Rochester, N.Y., 1989-90; engr. Joseph C. Lu, P.E., P.C., Rochester, N.Y., 1990—. Mem. ASCE, Sigma Phi Epsilon (chaplain 1986-87). Home: 1361 Schlegel Rd Webster NY 14580 Office: Joseph C Lu PE PC 2230 Penfield Rd Penfield NY 14526

HUTTER, GARY MICHAEL, environmental engineer; b. Harvey, Ill., May 10, 1948; s. Samuel and Ann H.; m. Shelley Hamilton, June 16, 1973; 1 child, Emily. BSME, U. Ill., 1970, MS in Environ. Engring., 1977; PhD in Environ. and Occupational Health, U. Ill., Chgo., 1991. Registered profl. engr., Ill.; cert. safety profl. R & D engr. Universal Oil Products Co., Des Plaines, Ill., 1970-73; test engr. Ford Motor Co., Dearborn, Mich., 1973-75; compliance engr. Ford Motor Co., Dearborn, 1977-81; environ. engr. Ill. EPA, Ottawa, 1975-76; sr. engr. Euclid (Ohio) Inc., 1981-84, Triodyne Inc., Niles, Ill., 1984—; pres. Triodyne Environ. Engring. Inc., Niles, 1990—. Contbr. articles to profl. jours. Recipient Grad. Student award, Am. Chem. Soc., 1989; U.S. EPA fellow, Washingt, 1976. Mem. Am. Nat. Stds. Inst., Nat. Safety Coun., Soc. Automotive Engrs., Water Pollution Control Assn., Air and Waste Mgmt. Assn., Phi Kappa Phi. Achievements include research into industrial hygiene problems in machine tool industry. Home: 3802 Michael Ln Glenview IL 60025

HUTTON, DAVID GLENN, environmental consultant, chemical engineer; b. Tarentum, Pa., Jan. 23, 1936; s. D. Ray and Z. Alberta (Rieger) H.; m. Judith Ann Gaumer, Dec. 27, 1965; children: Steven L, Michael W. BSChemE, Pa. State U., 1957, MSChemE, 1960. Rsch. engr. DuPont Co., Deepwater, N.J., 1960-68, sr. rsch. engr., 1968-71, sr. process engr., 1971-78, process assoc., 1978-81; tech. specialist DuPont Co., Wilmington, Del., 1981-82; environ. cons. DuPont Co., Newark, Del., 1982-92, D.G. Hutton, Inc., Newark, 1992—. Author chpt.: Carbon Adsorption Handbook, 1978; contbr. articles to profl. jours.; patentee in field. Chmn. Citizens Adv. Coun., Newark High, 1985-89, Secondary Citizens Adv. Coun., Christina Sch. Dist., 1985-90; treas. Fairfield Civic Assn., Newark, 1990-92. Mem. Water Environ. Fedn., AICE, Soc. Environ. Toxicology and Chemistry (assoc. mem.). Methodist. Home and Office: 107 Locust St Newark DE 19711

HUTTON, LARRIE VAN, cognitive scientist, neural networker; b. Potsdam, N.Y., Jan. 16, 1946; s. Robert Byron and Hope E. (Holliday) H.; m. Jean Ellen Trevethan; m. Robin Lynn Eichenbaum, May 30, 1980. BS, Mich. State U., 1969; PhD, Ohio U., 1979. Asst. prof. dept. psychology and dept. computer sci. Marietta (Ohio) Coll., 1980-88; instr. continuing profl. program Johns Hopkins U., Laurel, Md., 1988-91, sr. scientist Applied Physics Lab., 1988-92; adj. asst. prof. dept. psychology Johns Hopkins U., Balt., 1989—; CEO RobiNets Homuncular Solutions, Balt., 1993—. Contbr. chpts. to books, articles to APL Tech. Digest and others. Vol. math. tutor Job Corps, Laurel, 1990-91. With U.S. Army, 1970-72. Mem. APA, Internat. Neural Network Soc., Sigma Xi, Phi Kappa Phi. Democrat. Unitarian/Universalist. Achievements include demonstration of equivalence of psychophysical models of choice responding and thermodynamic laws; development of a set of microcomputer-based neural network modules; neural network replacement for cubic spine algorithm. Home and Office: RobiNets Homuncular Solu 2114 Bank St Baltimore MD 21231

HUTZ, REINHOLD JOSEF, physiologist; b. Salzburg, Austria, Mar. 18, 1956; came to the U.S., 1959; s. Josef and Eva (Strauch) H.; m. Irene Maria O'Shaughnessy, May 21, 1983; children: Erika, Michael. BS, Loyola U., 1978, MS, 1980; PhD, Mich. State U., 1983. Rsch. assoc. Wis. Regional Primate Rsch. Ctr., Madison, 1983-86, affiliate scientist, 1987—; asst. prof. U. Wis., Milw., 1986-92, assoc. prof., 1992—; assoc. adj. prof. Med. Coll. Wis., Milw., 1991—. Cons. editor: Am. Jour. Primatology, 1990—; referee Endocrinology, Biology of Reproduction, 1983—; contbr. articles to Jour. Med. Primatology, Jour. Endocrinology, Biology of Reproduction. Mem. exec. com. Milw. Donauschwaben German Soc., Milw., 1991—, music dir., 1991—. Grantee NIH, 1992. Mem. Internat. Soc. Primatologists (treas.), Endocrine Soc., Soc. Study Reproduction, AAAS, Am. Soc. Primatologists (program chair 1990-92). Democrat. Roman Catholic. Achievements include research in effects of estrogen on ovarian tissue, contraceptive development. Home: 4830 N Bartlett Ave Whitefish Bay WI 53217 Office: U Wis 3209 N Maryland Milwaukee WI 53201-0413

HUVOS, PIROSKA EVA, molecular biology researcher, chemistry educator; b. Budapest, Hungary; came to U.S., 1978; d. Paul Tibor and Piroska Dawn (Schwartz) H.; m. Peter Hardwicke; children: Catherine, Pamela. BS in Biology, Biochemistry, Eotvos Sci. U.; PhD in Biochemistry, Semmelweis Med. Sch., 1969. Rsch. assoc. Duke U., Durham, N.C., 1978-79; sr. rsch. assoc. Brandeis U., Waltham, Mass., 1980-81, 84-85; adj. asst. prof. Southern Ill. U., Carbondale, 1988—. Contbr. articles to profl. jours. Rsch. fellow Nat. Inst. Med. Rsch., London, 1972-73, King's Coll., London, 1973-76, St. George's Med. Sch., London, 1976-78. Mem. Biochem. Soc. London, Sigma Xi.

HUXLEY, SIR ANDREW (FIELDING), physiologist, educator; b. London, England, Nov. 22, 1917; s. Leonard and Rosalind (Bruce) H.; m. Jocelyn Richenda Gammell Pease, July 5, 1947; children: Janet Rachel, Stewart Leonard, Camilla Rosalind, Eleanor Bruce, Henrietta Catherine, Clare Marjory Pease. B.A., Cambridge (Eng.) U., 1938, M.A., 1941, Sc.D. (hon.), 1978; M.D. (hon.), U. Saar, 1964; D.Sc. (hon.), U. Sheffield, Eng., 1964, U. Leicester, Eng., 1967, London U., 1973, U. St. Andrews, Scotland, 1974, U. Aston, Birmingham, Eng., 1977; LL.D. (hon.), U. Birmingham, 1979, Marseille U., 1979, York U., 1981, U. Western Australia, 1982, NYU, 1982, Oxford U., 1983, U. Pa., 1984, Dundee U., 1984, Harvard U., 1984, U. Keele, 1985, East Anglia U., 1985, Humboldt U., East Berlin, 1985, Md. U., 1987, Brunel U., 1988, U. Hyderabad, 1991, Glasgow U., 1993, Ulm U., 1993. Mem. research staff Anti-Aircraft Command, 1940-42, Admiralty, 1942-45; fellow Trinity Coll. Cambridge, Cambridge, 1941-60, '90—; hon. fellow Trinity Coll., Cambridge, 1967-90, master, 1984-90, dir. studies, 1952-60; demonstrator Cambridge U., 1946-50, asst. dir. research, 1951-59, reader exptl. biophysics, 1959-60; Jodrell prof. physiology U. Coll. London, 1960-69, Royal Soc. research prof., 1969-83, hon. fellow, 1980; emeritus prof. physiology U. London, 1983—; Herter lectr. Johns Hopkins U., 1959; Jesup lectr. Columbia U., 1964; Forbes lectr., 1966; Croonian lectr. Royal Soc., 1967, Florey lectr., 1982, Blackett Meml. lectr., 1984; Fullerian prof. Royal Inst., London, 1967-73; Hans Hecht lectr., Chgo., 1975; Sherrington lectr. Liverpool U., 1976-77; Centenary Colloquium lectr. Berlin Inst. Physiology, 1977; Cecil H. and Ida Green vis. prof. U. B.C., 1980; 6th annual Darwin Lecture, 1982, Romanes Lecture, Oxford U., 1983; Tarner lectrs. Trinity Coll., Cambridge, 1988; Maulana Abul Kalam Azad Meml. Lecture, New Delhi, 1991; C.G. Bernhard lecture, Stockholm, 1993. Author: Reflections on Muscle, 1980; editor: Jour. Physiology, 1950-57, chmn. bd. Publs. on analysis of nerve conduction (with Hodgkin), physiology of striated muscle, devel. of interference microscope and ultramicrotome. Trustee Brit. Mus. (Natural History), 1981-90, Sci. Mus., 1984-88. Created knight bachelor, 1974; decorated Order of Merit, 1983; recipient (with A.L. Hodgkin and J.C. Eccles) Nobel prize for physiology or medicine, 1963; Imperial Coll. Sci. and Tech. hon. fellow, 1980; Queen Mary and Westfield Coll. fellow, 1987. Fellow Royal Soc. (Copley medal 1973, council 1960-62, 77-79, 80-85, pres. 1980-85), Royal Acad. Engring. (hon.), Inst. Biology (hon.), Royal Soc. Can. (hon.), Royal Soc. Edinburgh (hon.), Indian Nat. Sci. Acad. (fgn.); mem. Physiol. Soc. (hon., rev. lectr. on muscular contraction 1973), Internat. Union Physiol. Scis. (pres. 1986-93), Brit. Biophys. Soc., Royal Acad. Scis., Letters and Fine Arts Belgium (assoc.), Muscular Dystrophy Group Gt. Britain and No. Ireland (chmn. med. research com. 1974-81, v.p., 1981—), Royal Instn. Gt. Britain (hon.), Anat. Soc. Gt. Britain and Ireland (hon.), Am. Philos. Soc., Brit. Assn. Advancement Sci. (pres. 1976-77), NAS (U.S.) (fgn. assoc.), Royal Acad. Medicine Belgium (assoc.), Dutch Soc. Scis. (fgn.), Am. Soc. Zoologists (hon.), Royal Irish Acad. (hon.), Japan Acad. (hon.), Nature Conservancy (coun. 1985-88). Home and Office: Manor Field, 1 Vicarage Dr Grantchester, Cambridge CB3 9NG, England

HUXLEY, HUGH ESMOR, molecular biologist, educator; b. Birkenhead, Eng., Feb. 25, 1924; s. Thomas Hugh and Olwen (Roberts) H.; B.A., Christ's Coll., Cambridge (Eng.) U., 1948, M.A., 1950, Ph.D., 1952, Sc.D., 1964; D.Sc. (hon.), Harvard U., 1969, U. Chgo., 1974, U. Pa., 1975, U. Leicester, 1989. Research student molecular biology unit Med. Research Council, Cavendish Lab., Cambridge, 1948-52, sci. staff, 1954-55; external staff Med. Research Council, dept. biophysics U. Coll., London, 1956-61, Med. Research Council Lab. Molecular Biology, Cambridge, 1962-87, dep. dir., 1977-87; Commonwealth fund fellow dept. biology Mass. Inst. Tech., Boston, 1952-54; fellow Christ's Coll., Cambridge U., 1954-56, hon. fellow, 1981, fellow King's Coll., 1961-67, fellow Churchill Coll., 1967-87; prof. biology, dir. Rosenstiel Basic Med. Scis. Research Ctr., Brandeis U., Waltham, Mass., 1987—. Radar rsch. officer RAF, 1943-47. Decorated mem. Order Brit. Empire, 1948; recipient Feldberg prize, 1963, Hardy prize, 1965, Louisa Gross Hurwitz prize, 1971, Internat. Feltrinelli prize, 1974, Gairdner award, 1975, Baly medal Royal Coll. Physicians, 1975, Royal medal Royal Soc. London, 1977, E.B. Wilson medal Am. Soc. Cell Biology, 1983, Albert Einstein award World Cultural Council, 1987, Franklin medal, 1990, Disting. Scientist award Electron Microscopy Soc. Am., 1991. Fellow Royal Soc., 1960; mem. Physiol. Soc., Brit. Biophys. Soc., European Molecular Biology Orgn.; hon. fgn. mem. Am. Acad. Arts and Scis., Danish Acad. Scis., Leopoldina Acad., NAS (hon. fgn. assoc.). Editor: Progress in Biophysics and Molecular Biology, 1960-66; editorial bd. Jour. Cell Biology, 1959-63, Jour. Molecular Biology, 1962-70, 79-86, 90—, Jour. Cell Sci., 1966-70. Research, publs. on ultrastructures of straited muscles, especially by electron-microscopy and X-ray diffraction leading to sliding filament theory of contraction (with Jean Hanson; simultaneously proposed by A.F. Huxley and R. Niedergerke): studies on electron microscopy of viruses, ribosomes and other nucleic-acid containing structures. Home: 349 Nashawtuc Rd Concord MA 01742-1616 Office: Brandeis U Rosenstiel Ctr 415 S St Waltham MA 02254

HUYNH, ALEX VU, electrical engineer; b. Saigon, Vietnam, Sept. 29, 1968; came to U.S., 1975; s. Ba and Bich-Lien Thi (Le) H. BS, Tex. Tech U., 1990, MS, 1991. Rsch. asst. Tex. Tech U., Lubbock, 1990—. Named NASA/Tex. Space grantee, 1990-93; recipient Achievement Rewards for Coll. Scientists scholarship, 1992. Mem. IEEE, NSPE, Optical Soc. Am., Internat. Soc. for Optical Engring., Eta Kappa Nu (treas. 1989-90), Sigma Xi. Republican. Roman Catholic. Achievements include implementation of an optical quadratic neural network in barium titanate; development of an entropy measure for characterization of information flow through multi-layer linear neural networks. Home: 408 N Meredith Ave Dumas TX 79029 Office: Tex Tech U Dept Elec Engring Optical Systems Lab Lubbock TX 79409

HUYNH, NAM HOANG, physics educator; b. Cantho, An Binh, Vietnam, Mar. 7, 1949; came to U.S., 1975; s. Tu Van and Ngoc Thi (Phan) H.; m. Thuy Bich Nguyen, May 22, 1969; children: Khanh Hoang, Nguyen Hoang, Khoa Hoang. BS, U. Ctrl. Okla., 1985, MS, 1990. Registered profl. engr., Okla. Layout drafter Constrn. Machinery Inc. Corp., Oklahoma City, 1979-80, designer, 1980-82, engr., 1982-84, project engr., 1984—; adj. instr. Physics Okla. Jr. Coll., Oklahoma City, 1991-92. News editor Vietnamese Cath. Community, Oklahoma City, 1986-88, edn. dir., 1988-90. Mem. Royal Vietnamese Naval Mutual Assn. (chmn. 1988-92). Republican. Roman Catholic. Achievements include participation in designing storage silo to reduce segregation problem of materials in asphalt mix. Home: 941 Cooper Ln Yukon OK 73099-2805

HUYNH NGOC, PHIEN, computer science educator, consultant; b. Quangngai, Vietnam, May 2, 1944; s. Huynh Vien and Trinh Thi La; m. Tran Thi Huong; children: Phiem, Tue Anh. BS in Edn., Hue (Vietnam) U., 1968, BA in Math., 1970; MS, Asian Inst. Tech., Bangkok, 1976, D of Tech. Sci., 1978. High sch. tchr. Hue, 1968-71; instr. Hue U., 1971-74; asst. prof. Asian Inst. Tech., Bangkok, 1978-81, assoc. prof., 1982-90, prof., 1990—; cons. Mekong Com., ESCAP, UNDP, Statis. Inst. Asia and Pacific, Bangpakong Indsl. Park 2, Co. Ltd.; project dir. UNDP Project, Bangkok, 1986-91; chmn. computer sci. div. Asian Inst. Tech., Bangkok, 1986-90, chmn. software devel., 1988—;. Mem. editorial bd. Jour. Hydrology, Holland; contbr. over 150 papers and tech. reports to profl. jours. Mem. Assn. for Computing Machinery, Assn. for Advancement Modelling and Simulation (rep.), N.Y. Acad. Scis. Office: Asian Inst Tech, PO Box 2754, 2754 Bangkok 10501, Thailand

HUYSMAN, ARLENE WEISS, psychologist, educator; b. Phila.; d. Max and Anna (Pearlene) Weiss; B.A., Shaw U., 1973; M.A., Goddard Coll., 1974; Ph.D., Union Inst. Grad., 1980; m. Pedro Camacho; children: Pamela Claire, James David. Actress, dir. Dramatic Workshop, N.Y.C., 1956-68; music and drama critic and columnist Orlando (Fla.) Sentinel Star, 1966-68; psychodramatist Volusia County Guidance Center, Daytona Beach, Fla., 1966-68; free-lance journalist, 1968-70; psychodramatist Psychiat. Inst., Jackson Meml. Hosp., Miami, 1972-77, dir. Adult Day Treatment Center, 1974-77, dir. Lithium Clinic, 1976-77; psychodramatist South Fla. State Hosp., Hollywood, 1971-72; psychotherapy supr., neurosci. program coord. Miami Beach Community Hosp. (formerly St. Francis Hosp.), 1984—, clin. dir. Family Workshop, 1985—, clin. dir. Adult Day Treatment Ctrs., 1987—; adj. asst. prof. Med. Sch., U. Miami, 1976—; adj. prof., Union Inst., 1992—; mem. adv. panel Fine Arts Council Fla., 1976-77; mem. Fla. Gov.'s Task Force on Marriage and the Family Unit, 1976, 89-90; vol. Rec. for Blind, 1974—. Recipient Best Dirs. award and Best Actress award Fla. Theatre Festival, 1967. Mem. Am. Psychol. Assn., Fla. Psychol. Assn., Dade County Psychol. Assn. (bd. dirs.), Mental Health Assn. Dade County, Internat. Assn. Group Psychotherapy, Union Inst. Grad. Alumni Assn. (bd. dirs., southeastern rep.), Am. Soc. Aging, Am. Assn. Group Psychotherapy and Psychodrama, Moreno Acad., Fedn. Partial Hospitalization Study Groups, World Fedn. Mental Health, Fla. Assn. Practicing Psychologists (bd. dirs., pres. 1987-88, treas. 1990—). Office: Ctr Psychol Growth 3050 Biscayne Blvd Miami FL 33137-4143

HUZEL, DIETER KARL, retired aerospace engineer; b. Essen, Germany, June 3, 1912; came to U.S., 1946; s. Alfred Emil and Frieda (Garbe) H.; m. Irmgard Klebba, Nov. 3, 1945. Diploma engring., T.U., Stuttgart, Germany, 1937. Project engr. Siemens-Schuckertwerke, Berlin, 1937-42; mgr. V-2 launch facility Electro-Mech. Werke, Peenemuende, Germany, 1943-44, tech. asst. to Dr. W. v.Braun, 1944-45; project engr. U.S. Army Res. and Devel. Div., Fort Bliss, 1946-50; supr. Rocketdyne div. Redstone Engine Devel., N.Am. Aviation, Canoga Park, Calif., 1950-61; asst. chief engr. Sturn S-II, Space div. Rockwell Internat., 1961-75; project engr. advanced projects Rocketdyne div. Rockwell Internat., Canoga Park, 1975-76, ret., 1976. Author: Peenemuende to Canaveral, 1961, 81; (with David Huang) The Design of Liquid Propellant Rocket Engines, NASA, 1967, 71, updated AIAA, 1992, Methods to Extend Mechanical Component Life, AIAA, 1993; contbr. articles to profl. jours. Vol. Boyscouts. Fellow AIAA (assoc.). Achievements include patents for process of liquified gas pumping and for static pressure investigator.

HWA, TERENCE TAI-LI, physicist; b. Shanghai, China, May 30, 1964; came to U.S., 1979; s. Chia-Xu and Rebecca Tsu-Yung (Zhang) H. BS in Physics, Biology, Elec. Engring., Stanford U., 1986; PhD, MIT, 1990. Post-doctorate fellow Harvard U., 1990-93; mem. Inst. for Advanced Study, Princeton, N.J., 1993—. Recipient Apker award Am. Physics Soc., 1986, Outstanding Young Rschr. award Oversea Chinese Physics Assn., 1993; pre-doctoral fellow IBM, 1989, post-doctoral fellow, 1991. Mem. Tau Beta Pi. Office: Harvard U Dept Physics Cambridge MA 02138

HWANG, CORDELIA JONG, chemist; b. N.Y.C., July 14, 1942; d. Goddard and Lily (Fung) Jong; m. Warren C. Hwang, Mar. 29, 1969; 1 child, Kevin. Student Alfred U., 1960-62; BA, Barnard Coll., 1964; M.S., SUNY-Stony Brook, 1969. Rsch. asst. Columbia U., N.Y.C., 1964-66; analytical chemist Veritron West Inc., Chatsworth, Calif., 1969-70; asst. lab. dir., chief chemist Pomeroy, Johnston & Bailey Environ. Engrs., Pasadena, Calif., 1970-76; chemist Met. Water Dist. So. Calif., Los Angeles, 1976-79, rsch. chemist 1980-91, sr. chemist 1992—; mem. Joint Task Group on Instrumental Identification of Taste and Odor Compounds, 1983-85, instr. Citrus Coll., 1974-76; chair Joint Task Group on Disinfection by-products: chlorine, 1990. Mem. Am. Chem. Soc., Am. Water Works Assn. (cert. water quality analyst level 3, Calif.-Nev.), Am. Soc. for Mass Spectometry. Office: Met Water Dist So Calif 700 Moreno Ave La Verne CA 91750-3399

HWANG, DANNY PANG, aerospace engineer; b. Hsinchu, Taiwan, Oct. 14, 1936; came to U.S., 1962; s. Yin T. and Chin (Wong) H.;m. Li Hwa Chen, June 2, 1962; children: Charles, Vanessa. MS, N.C. State U., 1964; PhD, Ga. Tech., 1971. Instr. Penn Valley C.C., Kansas City, Mo., 1970-77; asst. prof. Ind. Inst. Tech., Ft. Wayne, 1977-78, assoc. prof., 1978-79; aerospace engr. Lewis Rsch. Ctr./NASA, Cleve., 1979—. Contbr. numerous articles to porfl. jours. Pres. Taiwanese Assn. of Kansas City, 1972, Taiwanese Assn. Greater Cleve., 1980. Mem. AIAA. Achievements include rsch. in the devel. and application of computational fluid dynamics computer codes such as panel method, Euler and Navier-Stokes Solver, and kinetic theory. Office: Lewis Rsch Ctr/NASA 21000 Brookpark Rd Cleveland OH 44135-3191

HWANG, JENN-SHIN, civil engineer; b. Ping-Tung, Taiwan, Republic of China, Nov. 21, 1956; came to U.S., 1982; s. Jau-Shane and Jean-Fong (Yang) H.; m. Shu-Chen Chou, Apr. 15, 1989. MS, SUNY, Buffalo, 1984, PhD, 1988. Registered profl. engr. Calif., Wash. Rsch. assoc. Nat. Ctr. for Earthquake Engring. Rsch., Buffalo, 1988-90; bridge engr. Washington State Dept. Transp., Olympia, 1990-91; civil engr. Office of Earthquake Engring. Calif. Dept. Transp., Sacramento, 1991—. Contbr. articles to Jour. Engring. Mechs., Jour. Structural Engring. Mem. ASCE (assoc., joint tech. com 343 on concrete bridges 1992—, tech. com. on repair, rehab. and inspection of bridges 1992—, tech. com. on steel bridges 1992—), Earthquake Engring. Rsch. Inst. Achievements include rsch. in system characteristics and dynamic performance of one of the two most versatile earthquake simulators in U.S., programs conducted using earthquake simulator regarding seismic resistant design of low-rise bldg. structures. Office: Calif Dept Transp Office Earthquake Engring 1801 30th St West Bldg Sacramento CA 95816

HWANG, SHYSHUNG, mechanical engineer; b. Taichung, Taiwan, Oct. 28, 1938; came to U.S., 1964; s. Chingpao and Shang (Chang) H.; m. Cecilia H. Chen, Nov. 25, 1972; children: Katherine Y., Jennifer Y., Victoria Y. BS, Taiwan U., Taipei, Taiwan, 1962; PhD, U. Rochester, 1967. Assoc. scientist Xerox, Webster, N.Y., 1967-70, scientist, 1970-74, sr. scientist, 1974-84, sr. mem. rsch. staff, 1984—, project mgr., 1970—. Recipient scholarships Taiwan Mech. Inc., 1960-62, Spl. Merit award, 1981, 84. Mem. ASME, Imaging Sci. and Tech., Sigma Xi. Achievements include invention of Electrostatic Recording Apparatus. Office: Xerox 800 Phillips Rd 114-21D Webster NY 14580

HWANG, WOEI-YANN PAUCHY, physics educator; b. Miao-Li, Taiwan, Republic of China, Aug. 25, 1948; s. A.S. and T.-M. (Chang) H.; m. Jane Su-Chen, June 4, 1977; children: Justin Han-Che, Irving Hua-Hsuan. BS in Physics, Taiwan U., Taipei, Taiwan, 1971; PhD in Physics, U. Pa., 1977. Rsch. assoc. U. Wash., Seattle, 1978-81; asst. prof. Ind. U., Bloomington, 1981-83, assoc. prof., assoc. scientist, 1983-86; vis. assoc. prof. Carnegie-Mellon U., Pitts., 1986-88; prof. Taiwan U., Taipei, 1987—; adj. prof. Carnegie-Mellon U., Pitts., 1988—; collaborator Los Alamos (N.Mex.) Nat. Lab., 1986—. Co-author: (textbook) Relativistic Quantum Mechanics and Quantum Fields, 1991; editor Chinese Jour. of Physics, 1988—; co-editor of 7 books; series editor Trends in Particle and Nuclear Physics, 1993—. Recipient Humboldt award, 1990-91, Outstanding Rsch. award Nat. Sci. Coun., 1989-91, 91-93. Mem. Am. Phys. Soc., Internat. Astron. Union, Phys. Soc. Republic of China, Sigma Xi. Office: Taiwan U Dept Physics, Taipei 10764, Taiwan

HWANG, WOONBONG, mechanical engineering educator, consultant; b. Seoul, Korea, May 15, 1958; s. Jong Heul and Sook Hee (Kim) H.; m. Boyoung Lee, Aug. 2, 1986; children: Jeehye, Jeeyun. BS, Han Yang U., Seoul, 1982; MS, SUNY, Buffalo, 1985, PhD, 1988. Rsch. assoc. Pohang (Republic of Korea) Inst. Sci. & Tech., 1988-89; asst. prof., 1989—; adj. researcher Rsch. Inst. Indsl. Sci. & Tech., Pohang, 1989—. Mem. ASTM, Am. Soc. for Composites (founder), Korean Soc. Mech. Engrs., Korean Soc. for Composites, Korean Soc. Aero. and Space Engrs., Soc. for Advancement Material and Process Engring., Soc. Plastics Engrs. Home: Jeekok-Dong Kyosoo Apt C-801, Pohang 790-330, Republic of Korea Office: Pohang Inst, Sci/Tech Dept, Mech Engring Hyoja-Dong San, PO Box 125, Pohang 790-600, Republic of Korea

HWU, REUBEN JIH-RU, chemistry educator; b. Taipei, Taiwan, Republic of China, Apr. 17, 1954; came to U.S., 1978; s. Jen Cheng and Cheng Mei (Tseng) Hu; m. Alice Show-Mei Leu, Aug. 24, 1978; children: Grace, Jennifer. BS, Nat. Taiwan U., 1976; PhD, Stanford U., Palo Alto, 1982. Assoc. prof. Johns Hopkins U., Balt., 1988-91, asst. prof., 1982-88; prof. Nat. Tsing Hua U., Hsinchu, Republic of China, 1990—; rsch. fellow Academia Sinica, Taipei, 1990—; rsch. cons. SUNY, Albany, 1987—; assoc. mem. IUPAC Commn. on Nomenclature, 1989—. Contbr. articles to jours. including Jour. Am. Chem. Soc., Jour. Chem. Soc. Perkin Trans., Jour. Organic Chemistry, Tetrahedron, Chemistry Rev. Alfred P. Sloan fellow Sloan Found., 1986-90; recipient Ming-Yu-Wen-Hwa award Ming-Yu-Wen-Hwa Found., 1976, Fu-Luen award Fu-Luen Found., 1975, Disting. Young Chemist award Fedn. Asian Chemical Socs., 1993-93. Mem. Am. Chem. Soc., Royal Soc. Chemistry, Chinese Chem. Soc., Chinese Am. Chem. Soc. (pres. 1991-93). Achievements include 2 patents on new methods for prostanoid synthesis and development of new prostaglandins; first to define the term of counterattack reagent in chemistry and establish the research field; research on concept of bulky proton and demonstrate its applicability. Office: Nat Tsing Hua U, Dept Chemistry, Hsinchu Taiwan 30043

HYATT, DAVID ERIC, environmental scientist; b. Bryson City, N.C., Apr. 27, 1957; s. John Shelton Sr. and E. Frances (Bryson) H.; m. Carol Ann Jones, Oct. 16, 1976; children: Laura, Jessie. BS cum laude, Western Carolina U., 1981; MS, U. Tenn., 1984. Researcher U.S. Forest Svc., Rocky Mountain Forest and Range Sta., Ft. Collins, Colo., 1979; park interpreter U.S. Nat. Park Svc., Cades Cove, Tenn., 1982-83; phys. scientist Def. Mapping Agy., Washington, 1984-87, Engring Topographic Labs.-U.S. Army, Ft. Belvoir, Va., 1987; environ. scientist Office of Waste Programs Enforcement U.S. EPA, Washington, 1987-89, environ. scientist Office of Policy, Planning and Evaluation, 1989-90; acting tech. coord. EMAP Integration and Assessment U.S. EPA, Research Triangle Park, N.C., 1990—; reviewer Ecol. Risk Assessment Framework EPA, Washington, 1991—, steering com. Ecol. Valuation Forum, 1992—; mem. FHM assessment workgroup U.S. Forest Svc., Research Triangle Park, 1992—, EPA Ecol. Risk Assessment Oversight Group, 1993—; organizing com. 1st Internat. Symposium on Ecosystem Health and Medicine, 1993—; co-chair Internat. Symposium Ecol. Indicators, 1990. Editor: Biological Populations as Indicators of Large-Scale Environmental Change, 1992; co-editor: Ecological Indicators, 1992, Environmental Monitoring and Assessment Program, 1992. Mem. AAAS, Ecol. Soc. Am., Internat. Soc. Ecol. Econs., Great Smoky Mountains Natural History Assn., Sigma Gamma Epsilon. Achievements include pursuit of developments in assessment science, incorporation of sound science into a relevant context for environmental policy and decision-makers, new methods and techniques for environmental data assessment and valuation. Office: US EPA EMAP Research and Assessment Ctr MD 75 Research Triangle Park NC 27709

HYATT, WALTER JONES, aerospace engineer; b. Greensboro, N.C., Feb. 2, 1918; s. Frederick Carlyle and Myrtle Ann (Cook) M.; m. Mary A. Malloy, July 18, 1942; children: Mary Valerie, Catherine Maureen. BS, U. St. Louis, 1938; postgrad, Calif. Tech., 1942-43. Engr. Curtiss-Wright Corp., St. Louis, 1938-39; engr. field supr. Lockheed Aircraft Corp., 1939-46, instr. tng. program Burbank High Sch., 1943-45; nat. sales mgr. Jackson Chem. Co. Calif., 1947-50; div. mgr. Pacific Airmotive Corp., Kansas City, 1950-51; sales engr. Adel Precision Prods., 1950-57; est. W.J. Hyatt Co., 1957-66; est. Aqualite Corp., 1958; ptnr. W.J. Hyatt & Assocs., Beverly Hills, 1966—; v.p. Penwal Industries, Chino, 1981-92; pres. Air Hydraulics Control, Canoga Park, 1981-87; bd. dir. Hi Temp Insulation, Camarillo, Insulfab, Camarillo; tech. writer Aero Publs., 1943-45; ptnr. Struc. Engring. Svc. for Architects, 1946; est. Walter J. Hyatt, desk model and exhibits. Author: (with others) War Production Coun. Inspection Handbook; coord. and Lockheed rep. Micro-Inch Finish Standards for aircraft ind. (1943), estab. X-ray and Magnetic Design Stds. for Design Manual and USAF Design Handbook, 1944. Past dir. L.A. C. of C. (mfg. group), Aircraft Distbrs. and Mfrs. Assn. Tech. rep. USAF and USN, 1942-45, World War II. Named Man of Yr., Kansas City C. of C., 1950. Mem. Air Force League, Navy League Assn. of U.S. Army, Magic Castle (Hollywood), Rotary. Achievements include development of procedures and organization of maintenance trouble reports for all aircraft; patents for Synthetic Rubber Compound for Rotatins Seals, 3-Way Auto Rear View Safety Mirror. Home: 630 N Foothill Rd Beverly Hills CA 90210 Office: PO Box 943 Beverly Hills CA 90213

HYDE, JAMES FRANKLIN, industrial chemical consultant; b. Solvay, N.Y., Mar. 11, 1903; s. Burton DeForest and Amelia (Bennett) H.; m. Hildegard Erna Lesche, June 25, 1928; children: Ann Hildegard, James F. Jr., Sylvia Hyde Schuster. AB, Syracuse U., 1923, MA, 1925; PhD, U. Ill., 1928. Postdoctoral fellow Harvard U., 1928-30; chemist Corning Glass Works, 1930-51; sr. scientist, chem. researcher Dow Corning Corp., 1951-75; indsl. chem. rsch. cons. Marco Island, Fla., 1975—; abstractor Glasstechnische Berichte Chem. Abstracts, Ceramic Abstracts. Contbr. articles to profl. jours. Recipient Mich. Patent Law Assn. award, 1963, Perkin medal Am. sect. Soc. Chem. Industry, 1971, Midgley award Detroit sect. Am. Chem. Soc., 1974, Fire of Genious award Saginaw Valley Patent Law Assn., 1982, Midland Matrix Festival award for exellent in sci., 1978; named Whitehead Meml. Lectr. engring. sect. NRC, 1971; elected to Plastics Hall of Fame, 1975. Mem. Am. Chem. Soc., Am. Inst. Chemists, AAAS, N.Y. Acad. Scis., Alpha Chi Sigma, Sigma Xi, Phi Beta Kappa. Achievements include approximately 120 patents and publications for silicone and glass related technology; invention of vapor phase process for forming vitreous silica used in making astronomical telescope mirror blanks and optical fibers; pioneering research in organo-silicon chemistry which initiated the silicone industry. Home: 544 Yellowbird St Marco Island FL 33937

HYDER, GHULAM MUHAMMAD ALI, physicist; b. Vill-Tepri, Bangladesh, Dec. 31, 1953; came to U.S., 1981; s. Muhammad A. Ali and Musammat (Halima) Begum; m. Kamrun Nahar, May 14, 1979; children: Homaira Parveen, Muntasir Hyder. MSc, Jahangirnagar, Savar, Bangladesh, 1978; PhD, Wayne State U., 1987. Med. physicist Bangladesh Atomic Energy Commn., Dhaka, Bangladesh, 1978-79; lectr. Jahangirnagar U., Savar, Bangladesh, 1979-81; grad. teaching asst. and rsch. asst. Wayne State U., Detroit, 1981-87, postdoctoral fellow, 1987-88; R & D physicist Chevrolet - Pontiac - Can. Group, Warren, Mich., 1988-91. Mem. Am. Phys. Soc., Laser Inst. Am., Wayne State U. Bangladesh Assn. (pres. 1983-85), Bangladesh Assn. Mich. (gen. sec. 1991-92, v.p. 1993). Achievements include patent for use of oxygen in galvanized sheet laser welding; development of use of gap to strengthen weld in a laser lap welding process. Home: 27115 Selkirk St Southfield MI 48076-5144 Office: Engring Physics 434 W 8 Mile Rd Ferndale MI 48220

HYDER, MONTE LEE, chemist; b. Maryville, Tenn., June 16, 1936; s. Frank Kimsey and Dorothy Jane (Weaver) H.; m. Penelope Lee Fletcher, June 20, 1964; children: Robert Lee, Edward Jordan, John Benjamin. BS in Chemistry, Rice U., 1958; PhD in Chemistry, U. Calif., Berkeley, 1962. Rsch. chemist DuPont Savannah River Lab., Aiken, S.C., 1962-67, rsch. supr., 1967-76, rsch. mgr., 1976-84, rsch. assoc., 1984-89; sr. adv. scientist Westinghouse Savannah River Co., Aiken, 1989—; mem. subcom. Com. on Nuclear Air and Gas Treatment. Mem. Am. Chem. Soc., U.S. Chess Fedn. (sec. 1975-78). Office: Westinghouse Savannah River PO Box 616 Aiken SC 29802

HYER, CHARLES TERRY, civil and mining engineer; b. Flatwoods, W.Va., Jan. 25, 1946; s. John Delfred and Elva Gay (Wyatt) H.; m. Karen Sue Prouse, Dec. 22, 1968; children: Charles Michael, Jennifer Susan. BSCE, W.Va. Inst. Tech., 1967; MBA, Bristol U., 1992. Registered profl. engr., W.Va., Ky., Va., Tenn. Civil engr. W.Va. Dept. Hwys., Charleston, 1968-73, Hawley Coal Mining Corp., Landgraff, W.Va., 1974-76; chief engr. H & F Mining, Bluefield, W.Va., 1976-78; mgr. engring. Sovereign Pocahontas Co., Bluefield, 1978-81; v.p. Inspiration Coal Inc., Knoxville, Tenn., 1981-92; pres. Charles T. Hyer and Assocs., Bristol, Tenn., 1992-93; sr. project mgr. Ogden Environ. and Energy Svcs. Co., Inc., Oak Ridge, Tenn., 1993—. Deacon 1st Presbyn. Ch., Bristol, 1990-92. Mem. NSPE (chpt. pres. 1978), Soc. for Mining, Metallurgy and Exploration Inc. Home: 64 Fairway Dr Bristol TN 37620

HYLA, JAMES FRANKLIN, rheumatologist, educator; b. Natick, Mass., Dec. 1, 1945; s. Walter John and Helen (Kipec) H.; m. Sharon Ann Frayer, June 21, 1969. BS in Engring. Physics, Cornell U., 1967; postgrad. in aeros. and astronautics, MIT, 1967-68; MD, U. Rochester, 1972. Diplomate Am. Bd. Internat. Medicine, Am. Bd. Rheumatology, Nat. Bd. Examiners. Med. intern Mary Imogene Bassett Hosp., Columbia U., Cooperstown, N.Y., 1972-73, resident in medicine, 1973-75; fellow in rheumatology U. Mich. Hosp., Ann Arbor, 1975-77; pvt. practice Syracuse, N.Y., 1977—; attending physician Community Gen. Hosp., Syracuse, 1977—, St. Joseph's Hosp., Syracuse, 1977—; attending physician Crouse Irving Meml. Hosp., Syracuse, 1977—, sec.-treas. med. starr, 1990-92; clin. assoc. prof. SUNY Health Sci. Ctr., Syracuse, 1978—; dir. Arthritis Ctr., St. Camillus Health and Rehab. Ctr., 1988—; mem. med. and sci. com., bd. dirs. Cen. N.Y. chpt. Arthritis Found., 1978—, v.p. devel. and fundraising, 1981-83. Contbr. articles to med. jours. Regional coord. Cornell Summer Job Network, Ithaca, N.Y., 1985—; mem. Cornell U. Coun., 1987-91; bd. dirs. Fedn. Cornell Clubs, 1984-88; regional dir. bd. Cornell Alumni Fedn., 1991—. Recipient nat. vol. svc. citation Arthritis Found., 1983. Fellow Am. Coll. Rheumatology; mem. ACP, Am. Soc. Internal Medicine, N.Y. State Med. Soc., Onondaga County Med. Sco., Cornell Club Cen. N.Y. (bd. govs. 1979—, corr. sec. 1981-82, v.p. 1982-83, pres. 1983-85, scholarship chmn. 1985—, treas. 1991—), Tau Beta Pi, Phi Eta Sigma. Democrat. Home: 118 Grenfell Rd Dewitt NY 13214-1624 Office: Rheumatology Assocs Cen NY 310 S Crouse Ave Syracuse NY 13210

HYLANDER, WALTER RAYMOND, JR., retired civil engineer; b. Memphis, July 22, 1924; s. Walter Raymond and Mary Howard (Douglass) H.; m. Marjorie Jean Gunter, Mar. 8, 1951; children: Walter Raymond, Joyce Elizabeth. BS, U.S. Mil. Acad., 1945; MS in Civil Engring., MIT, 1950. Registered profl. engr., N.Y., Miss. Commd. 2d lt., U.S. Army, 1945, advanced through grades to col., 1969, ret., 1973; tng. dir. Bechtel Power Corp., Grand Gulf, Miss., 1974-76; tng. and edn. mgr. Saudi-Arabian Bechtel Co., Jubail, 1976-77; tng. dir. St. Regis Paper Co., Montecello, Miss., 1978-79; chief civil engr. Bechtel Power Corp., Grand Gulf, 1979-86; chmn. Panel of Experts on Mine Warfare, NATO, London, 1962-65; sr. advisor on engr. tng., Vietnam, 1967-68; mem. U.S. Army Com. on Mil. History, West Point, N.Y., 1972-73; mem. U.S. ACDA, Washington, 1968-69. Contbr. articles to profl. jours. Fellow ASCE; mem. Soc. Am. Mil. Engrs., Nat. Assn. Model Railroaders, La. Miss. Christmas Tree Assn., Phi Kappa Phi. Methodist. Avocations: growing Christmas trees, model railroading, Civil War history. Home: Rosswood Plantation Lorman MS 39096

HYLKO, JAMES MARK, health physicist; b. Detroit, Sept. 11, 1961; s. James John and Frances Rose (Gorski) H. BS in Biochemistry, Ea. Mich. U., 1984; MPH in Health Physics, U. Mich., 1986. Lab. tech. dept. chemistry Ea. Mich. U., Ypsilanti, 1980-84; environ. radiochemist Argonne (Ill.) Nat. Lab., 1984; radiochemist U. Mich., Ann Arbor, 1984-86; health physics tech. Monticello (Minn.) Nuclear Sta., 1985; rsch. scientist U. Va., Charlottesville, 1986-88; health physicist Fluor Daniel Inc., Chgo., 1988-92, Roy F. Weston, Inc., Albuquerque, 1992—; mem. nuclear/ednl. planning com. Am. Power Conf., Chgo., 1989-91; guest lectr. Purdue U., 1991; invited speaker Inst. Atomic Energy, Swierk-Otwock, Poland, 1991. Contbr. articles to Jour. Radiation Protection Mgmt., Nuclear Tech., and book revs. to profl. jours. Judge N.Mex. Regional and State Sci. and Engring. Fair, 1993. Fellow Inst. Nuclear Power Ops., 1986. Mem. Health Physics Soc., Am. Nuclear Soc., Toastmasters (pres. Fluor Daniel chpt. 1990). Home: 5801 Eubank NE # 153 Albuquerque NM 87111 Office: Roy F Weston Inc Ste 800 6501 Americas Pkwy NE Albuquerque NM 87110

HYMAN, ALBERT LEWIS, cardiologist; b. New Orleans, Nov. 10, 1923; s. David and Mary (Newstadt) H.; m. Neil Steiner, Mar. 27, 1964; 1 son, Albert Arthur. B.S., La. State U., 1943; M.D., 1945; postgrad., U. Cin., U. Paris, U. London, Eng. Diplomate: Am. Bd. Internal Medicine. Intern Charity Hosp., 1945-46, resident, 1947-49, sr. vis. physician, 1959-63; resident Cin. Gen. Hosp., 1946-47; instr. medicine La. State U., 1950-56, asst. prof. medicine, 1956-57; asst. prof. Tulane U., 1957-59, assoc. prof., 1959-63, assoc. prof. surgery, 1963-70, prof. research surgery in cardiology, 1970—, prof. clin. medicine Med. Sch., 1983—, adj. prof. pharmacology Med. Sch., 1974—; dir. Cardiac Catheterization Lab., 1957—; sr. vis. physician Touro Hosp., Touro Infirmary, Hotel Dieu; chief cardiology Sara Mayo Hosp.; cons. in cardiology USPHS, New Orleans Crippled Children's Hosp., St. Tammany Parish Hosp., Covington La. area VA, Hotel Dieu Hosp., Mercy Hosp., East Jefferson Gen. Hosp., St. Charles Gen. Hosp.; electrocardiographer Metairie Hosp., 1959-64, Sara Mayo Hosp., Touro Infirmary, St. Tammany Hosp.; cons. cardiovascular disease New Orleans VA Hosp.; cons. cardiology Baton Rouge Gen. Hosp.; Barlow lectr. in medicine U. So. Calif., 1977; mem. internat. sci. com. IV Internat. Symposium on Pulmonary Circulation, Charles U., Prague. Mem. editorial bd. Jour. Applied Physiology; contbr. over 250 articles to profl. jours. Recipient award for rsch. of the Hadassah, 1980, Vis. Scientist award Wellcome Found., Univ. Coll., London, 1991, Disting. Achievement award Am. Heart Assn., 1992. Fellow ACP, Am. Coll. Chest Physicians, Am. Coll. Cardiology, Am. Fedn. Clin. Research; mem. Am. Heart Assn. (fellow council on circulation, fellow council on clin. cardiology, mem. council on cardiopulmonary medicine, regional rep. council clin. cardiology, chmn. sci. com. of cardiopulmonary council 1981, chmn. cardiopulmonary council, mem. research com. bd. dirs., editorial bd. mem. Circulation Research, edit. bd. mem. Am. Jour. Physiology, Heart Disease and Stroke, Jour. Applied Physiology, Dickinson Richards Meml. Lectr. 1986, Disting. Scientific Achievement award 1992), La. Heart Assn. (v.p. 1974, Albert L. Hyman Ann. Rsch. award, Wellcome Rsch. Found. Vis. Scientist award Univ. Coll., London 1992, Disting. achievement award outstanding sci. contbns. to cardiopulmonary medicine), Am. Soc. Pharmacology and Exptl. Therapeutics, So. Soc. Clin. Investigation (chmn. membership com.), So. Med. Soc. (Seale-Harris award 1988), Am. Physiol. Soc., N.Am. Soc. Pacing and Electrophysiology, N.Y. Acad. Scis. Nat. Am. Heart Assn. (vice chmn. research com.), AAUP, Alpha Omega Alpha. Achievements include research in cardiopulmonary circulation. Home: 5467 Marcia Ave New Orleans LA 70124-1052 Office: 3601 Prytania St New Orleans LA 70115-3641

HYMAN, BRUCE MALCOLM, ophthalmologist; b. N.Y.C., May 22, 1943; s. Malcolm A. and Sylvia S. H.; A.B., Columbia U., 1964; M.D., N.Y. U., 1968. Intern in surgery Albert Einstein Coll. Medicine/Bronx Mcpl. Hosp., 1968-69; resident in ophthalmology Manhattan Eye, Ear and Throat Hosp., N.Y.C., 1971-74; pvt. practice medicine specializing in ophthalmology, N.Y.C., 1974—; tchr. attending surgeon Manhattan Eye, Ear and Throat Hosp., 1974—; med. cons. U.S. Seaplane Pilots Assn., 1975—, Health Ins. Plan Greater N.Y., 1977—; ophthalmologist to Hotel Trades Council, Hotel Assn. N.Y.C., 1974—; attending ophthalmologist Roosevelt Hosp., N.Y.C., 1979—, dir. adult outpatient ophthalmology, 1980—; police surgeon N.Y.C., 1977—, dep. chief police surgeon, 1978—; attending ophthalmologist Doctors Hosp., 1979—, Le Roy Hosp., 1979—, St. Luke's Hosp., 1980—; outpatient ophthalmologist N.Y. Hosp., 1975-77; clin. ophthalmologist Columbia Coll. Physicians and Surgeons, 1981—. Served with USPHS, 1969-71. Diplomate Am. Bd. Ophthalmology. Fellow ACS; mem. N.Y. State, N.Y. County med. socs., Am. Acad. Ophthalmology and Otolaryngology. Contbr. articles to profl. jours. Office: 133 E 64th St New York NY 10021

HYMAN, LEONARD STEPHEN, financial executive, economist, author; b. N.Y.C., June 5, 1940; s. Milton and Elsie (Reiter) H.; m. Judith N. Siegel, July 4, 1965; children: Andrew S., Robert C. BA, NYU, 1961; MA, Cornell U., 1965. Fin. analyst Chase Manhattan Bank, N.Y.C., 1965-72; ptnr. H.C. Wainwright & Co., N.Y.C., 1972-77; v.p. Wainwright Securities, N.Y.C., 1977-78; v.p., head utility rsch. group Merrill Lynch Capital Markets, N.Y.C., 1978—, first v.p., 1987—; bd. advisors, Exnet, 1992—. Author: America's Electric Utilities, 1983, The New Telecommunications Industry, 1987; contbr. Electric Power Strategic Issues, 1983, The Future of Electrical Energy, 1986, Deregulation and Diversification of Utilities, 1988; contbr. article to profl. jours; mem. editorial bd. Forum for Applied Research and Public Policy, 1993—. Mem. Pa. Task Force on Electric Utility Efficiency, Harrisburg, 1982-83; mem. adv. com. U.S. Congress-Office Tech. Assessment, Washington, 1983-84, 86-87, 87-88, 92—. Mem. AAAS, NASA (lunar energy enterprise case study task force 1988-89), N.Y. Soc. Security Analysts, Fin. Analysts Fedn., Inst. Chartered Fin. Analysts, U.S. Energy Assn., Phi Beta Kappa. Democrat. Jewish. Avocations: photography, travel, music, canoeing. Home: 34 Fremont Rd Tarrytown NY 10591-1118 Office: Merrill Lynch World Hdqrs World Fin Ctr North Tower New York NY 10281-1319

HYMAN, LESLIE GAYE, epidemiologist; b. Balt., Sept. 4, 1952; s. Nathan Bernard and Joy Charlotte Rhoda (Goldberg) H. BA in Biology, Brandeis U., 1974; M in Health Sci., Johns Hopkins U., 1979, PhD in Epidemiology, 1981. Rsch. asst. dept. epidemiology Johns Hopkins U., Balt., 1975-78; epidemiologist Nat. Eye Inst., Bethesda, Md., 1978-80; rsch. asst. prof. dept. preventive medicine SUNY, Stony Brook, 1981—, instr. Dept. Preventive Medicine, 1981-91, course dir., 1992—; epidemiologic cons. Retinal Vascular Ctr., The Wilmer Inst., Johns Hopkins Hosp., Balt., 1980-81; participant Soc. for Epidemiologic Rsch., New Haven, Conn., 1979, Symposium on Occupationally Induced Macular Degeneration Nat. Inst. for Occupational Safety and Health, Cin., 1982, Nat. Workshop on Long-Term Visual Health Risks of Optical Radiation FDA, Rockville, Md., 1983, 16th Cambridge (Eng.) Ophthal. Symposium, 1986; manuscript reviewer Arch Ophthal, Am. Jour. Indsl. Medicine, Current Eye Rsch., JAMA. Author: (with others) Possible Role of Optical Radiation in Retinal Degenerations, 1986, The Microenvironment and Vision, 1987, Basic and Clinical Aspects of Malignant Melanoma, 1987, The Susceptible Visual Apparatus, 1991, Age-Related Macular Degeneration; contbr. articles to Invest. Ophthalmol. Vis. Sci. Suppl., Am. Jour. Epidemiology, APHA, Nat. Eye Inst. Symposium on Eye Disease Epidemiology, Internat. Congress of Eye Rsch., among others. Grantee Nat. Eye Inst., 1986-93, Hoffmann-LaRoche, Inc., 1986-91, Tissue Banks Internat., Inc., 1983-91, NIH, 1986—, The Macular Found., Inc., 1991-92. Mem. APHA, Am. Coll. Epidemiology, Assn. for Rsch. in Vision and Ophthalmology (clin. rsch. sect., sect. chairperson 1987-89, program com. 1986-89), Internat. Epidemiol. Assn., Soc. for Epidemiologic Rsch. Achievements include research in risk factors for age related maculopathy, keratoplasty evaluation study, biochemical analysis for risk factors for age related macular degeneration and risk factors for lens opacities, diabetic retinopathy. Office: SUNY Dept Preventive Med Health Scis Ctr 36099 Stony Brook NY 11794

HYMAN, SCOTT DAVID, physicist, educator; b. Stamford, Conn., June 9, 1960; s. Leon Irving and Frances (Skydel) H.; m. Phoebe Dara Swedlow, Nov. 5, 1989. BS in Physics, MIT, 1982; MS, U. Md., 1985, PhD in Physics, 1989. Reliability engr. IBM, Essex Junction, Vt., 1982-83; contracting scientist Sachs/Freeman Assocs., Bowie, Md., 1983; rsch. asst., teaching asst. dept. physics U. Md., College Park, 1983-89; postdoctoral assoc. Nat. Rsch. Coun.-NIH, Bethesda, Md., 1990-91; cons. Environ. Rsch. Inst. Mich., Arlington, Va., 1991; asst. prof. physics Sweet Briar (Va.) Coll., 1992—; vis. asst. prof. physics Randolph Macon Coll., Ashland, Va., 1991. Contbr. articles on nuclear physics, artificial neural networks to profl. publs. Vol. sci. instr. Somerset Elem. Sch., Chevy Chase, Md., 1992. Mem. Am. Phys. Soc., Internat. Neural Network Soc. Democrat. Jewish. Office: Sweet Briar Coll Dept Physics PO Box 28 Sweet Briar VA 24595

HYMES, DELL HATHAWAY, anthropologist; b. Portland, Oreg., June 7, 1927; s. Howard Hathaway and Dorothy (Bowman) H.; m. Virginia Margaret Dosch, Apr. 10, 1954; 1 adopted child, Robert Paul; children: Alison Bowman, Kenneth Dell; 1 stepchild, Vicki (Mrs. David Unruh). BA, Reed Coll., 1950; MA, Ind. U., 1953, PhD, 1955; postgrad., UCLA, 1954-55. Instr., then asst. prof. Harvard U., 1955-60; assoc. prof., then prof. U. Calif., Berkeley, 1960-65; prof. anthropology U. Pa., 1965-72, prof. folklore and linguistics, 1972-88, prof. sociology, 1974-88, prof. edn., 1975-88, dean U. Grad. Sch. Edn., 1975-87; prof. anthropology and English U. Va., 1987—, Commonwealth prof. anthropology, 1990—, Commonwealth prof. English, 1990—; bd. dirs. Social Sci. Rsch. Coun., 1965-67, 69-70, 71-72. Author: Language in Culture and Society, 1964, The Use of Computers in Anthropology, 1965, Studies in Southwestern Ethnolinguistics, 1967, Pidginization and Creolization of Languages, 1971, Reinventing Anthropology, 1972, Foundations in Sociolinguistics, 1974, Studies in the History of Linguistics, 1974, Soziolinguistik, 1980, Language in Education, 1980, In Vain I Tried to Tell You, 1981, (with John Fought) American Structuralism, 1981, Essays in the History of Linguistic Anthropology, 1983, Vers la Competence de Communication, 1984; assoc. editor: Jour. History Behavioral Scis, 1966—, Am. Jour. Sociology, 1977-80, Jour. Pragmatics, 1977—; contbg. editor: Alcheringa, 1973-80, Theory and Society, 1976—; editor: Language in Society, 1972-92. Trustee Ctr. for Applied Linguistics, 1973-78. With AUS, 1945-47. Fellow Ctr. Advanced Study Behavioral Scis., 1957-58, Fellow Clare Hall, Cambridge, Eng., Guggenheim fellow, 1969-70, Nat. Endowment for Humanities sr. fellow, 1972-73. Fellow Am. Folklore Soc. (pres. 1973-74), Brit. Acad.; mem. AAAS (coun. 1979-80), Am. Anthrop. Assn. (exec. bd. 1968-70, pres. 1983), Am. Assn. Applied Linguistics (pres. 1986), Linguistic Soc. Am. (exec. bd. 1967-69, pres. 1982), Coun. on Anthropology and Edn. (pres. 1978), Consortium Social Sci. Assns. (pres. 1984-85). Home: 205 Montvue Dr Charlottesville VA 22901-2022

HYNAN, LINDA SUSAN, psychology educator; b. Ft. Sill, Okla., Nov. 20, 1953; d. Christy J. and Barbara Jean (Camp) Genzel; m. Edward F. Hynan, Feb. 3, 1973; 1 child, Patrick Shane. MS, U. Ill., 1982, PhD, 1993. Teaching asst., rsch. asst. dept. psychology U. Ill., Urbana, 1980-92; rsch. asst. dept. psychology Del. State Coll., Dover, 1983—; asst. prof. dept. psychology Baylor U., Waco, Tex., 1991—; cons. Infosphere Devel. Systems, Waco, Tex., 1986—; reviewer Allyn & Bacon/Simon & Schuster, Needham Heights, Mass., 1992. Contbr. chpt. to book Cognitive Bias, 1990, articles to profl. jours. Fellow U. Ill., 1988-89. Mem. APA, Am. Ednl. Rsch. Assn., Am. Psychol. Soc., Am. Statis. Assn., Ea. Psychol. Assn., McLennon County Psychol. Assn., Midwestern Psychol. Assn., Psychometric Soc., Soc. for Judgement and Decision-Making, Soc. for Applied Multivariants Rsch., Soc. for Math. Psychology, Southwestern Psychol. Assn., Ctrl. Tex. Women's Alliance, Thyroid Found. Am., Phi Kappa Phi. Achievements include devel. of EPL program using light pen to collect data for multiattribute decision-making experiments, other text presentations of info. on computer screen; rsch. in theories of bias in probability judgements. Home: 1312 Western Ridge Dr Waco TX 76712 Office: Baylor U Psychology Dept and Inst Grad Statistics PO Box 97334 Waco TX 76798

HYNES, NANCY ELLEN, reading educator; b. Jersey City, N.J., June 13, 1956; d. Timothy Joseph and Alice Mae (Menig) H. BA, Jersey City State Coll., 1978, MA, 1979. Cert. nursery tchr., N.J., elem. tchr., N.J., prin., supr., N.J., reading tchr., N.J. Tchr. St. Bridget's Sch., North Bergen, N.J., 1978; tchr. reading, writing Bd. Edn. East Orange (N.J.), 1979-86; unit coord. HSPT program Bd. Edn. City of East Orange (N.J.), 1985; dir. Nancy's Sch. Dance, West N.Y., 1981-85; tchr. computer Bd. Edn. City of

Elizabeth (N.J.), 1986—; also reading tchr., 1992—; tchr. dance Westfield (N.J.) Workshop for Arts, 1989, N.J. Workshop for Arts, Westfield, 1990; tchr. Huntington Learning Ctr., Woodbridge, N.J., 1987—; mem. com. Computer Curriculum Guide, 1989, Language Arts Curriculum Guide, 1989. Chmn. Dance for Heart, 1985. Grantee Bd. Edn. City of Elizabeth, 1989. Mem. ASCD, Phi Delta Kappa (historian 1989-91), Secondary Sch. Women's Club (Elizabeth). Roman Catholic. Avocations: skiing, dancing, reading, tennis. Home: 8 Tulip Dr Fords NJ 08863-1173 Office: Elizabeth Bd Edn Broad St Elizabeth NJ 07202-2692

HYNES, RICHARD OLDING, biology educator; b. Nairobi, Kenya, Africa, Nov. 29, 1944; s. Hugh Bernard Noel and Mary Elizabeth (Hinks) H.; m. Fleur Marshall, July 29, 1966; children: Hugh Jonathan, Colin Anthony. BA with honors, U. Cambridge, Eng., 1966, MA, 1970; PhD, MIT, 1971. Asst. prof. biology MIT, Cambridge, 1975-78, assoc. prof., 1978-83, prof. Dept. Biology, 1983—, assoc. head Dept. Biology, 1985-89, head, 1989-91, dir. Ctr. for Cancer Rsch., 1991—; investigator Howard Hughes Med. Inst., Bethesda, Md., 1988—. Author: Fibronectins, 1990; editor Tumor Cell Surfaces and Malignancy, 1979, Surfaces of Normal and Malignant Cells, 1979; contbr. articles to profl. jours. Guggenheim Found. fellow, 1982. Fellow AAAS, Royal Soc. London; mem. Am. Soc. Cell Biology, Soc. for Devel. Biology. Office: MIT EI7-227 Ctr for Cancer Rsch Cambridge MA 02139

IACHELLO, FRANCESCO, physicist educator. Prof., physics Yale U., New Haven, Conn. Office: Yale Univ PO Box 6666 New Haven CT 06511

IACOBUCCI, GUILLERMO ARTURO, chemist; b. Buenos Aires, May 11, 1927; s. Guillermo Cesar and Blanca Nieves (Brana) I.; m. Constantina Maria Gullich, Mar. 28, 1952; children: Eduardo Ernesto, William George. MSc, U. Buenos Aires, 1949, PhD in Organic Chemistry, 1952. Came to U.S., 1962, naturalized, 1972. Research chemist E.R. Squibb Research Labs., Buenos Aires, 1952-57; research fellow in chemistry Harvard U., Cambridge, Mass., 1958-59, prof. phytochemistry U. Buenos Aires, 1960-61; sr. research chemist Squibb Inst. Med. Research, New Brunswick, N.J., 1962-66; head bio-organic chemistry labs. Coca-Cola Co., Atlanta, 1967-74, asst. dir. corp. research and devel., 1974-87, mgr. biochemistry and basic organic chemistry group, 1988—; adj. prof. chemistry Emory U., 1975—. John Simon Guggenheim Meml. Found. fellow, 1958. Fellow Am. Inst. Chemists; mem. AAAS, Assn. Harvard Chemists, Am. Chem. Soc., Internat. Union Pure and Applied Chemistry, N.Y. Acad. Scis., Am. Soc. Pharmacognosy, Phytochemical Soc. N.Am., Smithsonian Instn., Planetary Soc., Sigma Xi. Achievements include structure/activity correlations and molecular design of sweeteners; use of enzymes in asymmetric organic synthesis; natural products chemistry; contbr. articles on organic chemistry to sci. jours. Patentee in field. Home: 160 N Mill Rd NW Atlanta GA 30328-1837 Office: Coca Cola Co PO Box 1734 Atlanta GA 30301-1734

IACONO, JAMES MICHAEL, research center administrator; b. Chgo., Dec. 11, 1925; s. Joseph and Angelina (Cutaia) I.; children: Lynn, Joseph, Michael, Rosemary. BS, Loyola U., Chgo., 1950; MS, U. Ill., 1952, PhD, 1954. Chief Lipid Nutrition Lab. Nutrition Inst. Agrl. Rsch. Svc. USDA, Beltsville, Md., 1970-75; dep. asst. adminstrv. nat. program Agrl. Rsch. Svc. USDA, Washington, 1975-77, assoc. adminstr. office human nutrition, 1978-82; dir. Western Human Nutrition Rsch. Ctr. Agrl. Rsch. Svc. USDA, San Francisco, 1982—; adj. prof. nutrition Sch. Pub. Health UCLA, 1987—. Author over 80 rsch./tech. publs. and chpts. in books relating to nutrition and biochemistry and lipids. With U.S. Army, 1944-46. Recipient Rsch. Career Devel. award NIH, 1964-70. Fellow Am Heart Assn. (coun. on arteriosclerosis and thrombosis), Am. Inst. Chemists; mem. Am. Inst. Nutrition, Am. Soc. Clin. Nutrition, Am. Oil Chemists Soc. Office: USDA ARS Western Human Nutrition Rsch Ctr PO Box 29997 San Francisco CA 94129-0997

IACOPONI, MICHAEL JOSEPH, systems engineer; b. New Bedford, Mass., Mar. 10, 1959; s. Gino Anthony and Mary Louise (Collins) I. BSEE, Tufts U., 1981; MSEE, Va. Poly. Inst. & State U., 1983. Staff engr. Harris Corp., Melbourne, Fla., 1983—; prin. investigator advanced architecture onboard processor Harris Corp., 1987-89, prin. investigator space processing R&D, 1988-92, chief architect advanced fault tolerant data processor, 1988-92, fault tolerant MIPS multiprocessor, 1990-92. Inventor method of preserving integrity of a codeword, hierarchical variable die size gate array. Mem. IEEE, Assn. Computing Machinery. Avocations: skiing, tennis, classical piano, traveling, theater. Home: 111 Cypress Brook Cir Apt 801 Melbourne FL 32901-8730 Office: Harris Corp PO Box 94000 Melbourne FL 32901

IANNICELLI, JOSEPH, chemical company executive, consultant; b. N.Y.C., Aug. 25, 1929; s. Peter and Catherine (Gugliotti) I.; m. Betty Peterson, June 28, 1978; children: Mark, Rex, Gina. SB, MIT, 1951, PhD, 1955. Rsch. chemist Textile Fibers, E.I. DuPont, Wilmington, Del., 1955-60; tech. dir. Clay Div. J.M. Huber, Macon, Ga., 1960-70; founder, chief exec. officer Aquafine Corp., Brunswick, Ga., 1970—, Aero-Instant Corp., Brunswick, Ga., 1988—; co-founder IMPEX Corp., Brunswick, Ga., 1988—; cons. Consol. Goldfields Australia, Sydney, 1976-78, Rio Tinto, Madrid, 1980-82, Hoganes, Malmo, Sweden, 1984. Author: Evaluation and Comparison of Crossfield and Solenoid Field Magnetic Filters, 1981; co-author: A Survey-Benneficiation of Industrial Minerals, 1980; contbr. over 30 articles to profl. jours. Pres. Ga. Tidewater Conservation Assn., Brunswick, 1991-92; bd. dirs. Jekyll Island (Ga.) Citizens Assn., 1992—, pres., 1993—; foreman Glynn County Grand Jury, Brunswick, 1989. Recipient Rsch. grant NSF, 1980, 84, Elec. Power Rsch. Inst., 1980. Fellow Am. Inst. Chemists; mem. Tech. Assn. of Pulp and Paper Industry (chmn. pigments com. 1971-72). Achievements include over 100 patents including paramagnetic separator and process, silane modified organo clays, mercaptan scrubber. Home: 28 Saint Andrews Dr Jekyll Island GA 31527 Office: Aquafine Corp 157 Darien Hwy Brunswick GA 31525

IANZITI, ADELBERT JOHN, industrial designer; b. Napa, Calif., Oct. 10, 1927; s. John and Mary Lucy (Lecair) I.; student Napa Jr. Coll., 1947, 48-49; m. Doris Moore, Aug. 31, 1952; children: Barbara Ann Ream, Susan Therese Shifflett, Joanne Lynn Lely, Jonathan Peter, Janet Carolyn Kroyer. AA, Fullerton Jr. Coll., 1950; student UCLA, 1950, Santa Monica Community Coll., 1950-51. Design draftsman Basalt Rock Co. Inc. div. Dillingham Heavy Constrn., Napa, 1951-66, chief draftsman plant engring., 1966-68, process designer, 1968-82, pres. employees assn., 1967; now self-employed indsl. design cons. V.p., Justin-Siena Parent-Tchr. Group, 1967. Mem. Aggregates and Concrete Assn. No. Calif. (vice-chmn. environ. subcom. 1976-77), Constrn. Specifications Inst., Native Sons of the Golden West, World Affairs Coun. No. Calif., Internat. Platform Assn., Commonwealth of Calif. Club. Republican. Roman Catholic. Home and Office: 2650 Dorset St Napa CA 94558-6110

IAQUINTO, JOSEPH FRANCIS, electrical engineer; b. Phila., Nov. 9, 1946; s. Francis Edward Iaquinto and Maria Carmina (Mancini) Feldman; m. Jo-Carol Maniscalco, Nov. 21, 1977; children: Joseph Michael, Jonathan Franklin. BSEE, Drexel U., 1969; MSEE, Stanford U., 1971. Registered professional engineer, Pa., Va. Teaching asst. Stanford (Calif.) U., 1969-71; sr. project engr. GM Corp., 1971-75; regional system engring. mgr. Memorex Corp., King of Prussia, Pa., 1975-77; sr. prin. engr. Computer Sci. Corp., Falls Church, Va., 1977-80; dir. devel. Tesdata Systems Corp., Tyson's Corner, Va., 1980-82; staff engr. HRB-Singer Co., Lantham, Md., 1982-84; sr. staff engr. Lockheed Electronics Co., Vienna, Va., 1984-86; mem. tech. staff MRJ div. Perkin Elmer, Oakton, Va., 1986-89; system engr. Ford Motor Co., Dearborn, Mich., 1989-93; engring. mgr. A.C. Nielsen, Dunedin, Fla., 1993—. Author: Memorex 1380 Internal and Lesson Plan, 1977, Simulation of Microwave Propagation in the Atmosphere, 1987; co-author: (with H. Brandt) Control Engineering Application to Automobiles, 1973; author: (with others) Secure Internetwork Data Communications, 1979, Mission Planning System Specification, 1985; contbr. articles to tech. publs. Instr. ARC, Mich. and Pa., 1971-77; treas. Macomb County Young Reps., Sterling Heights, Mich., 1975; councilman Longacre PTA, Farmington, Mich., 1990-92. Recipient acad. scholarship Phila. Sch. System. Mem. IEEE, Nat. Soc.Profl. Engrs., Inst. Soc. Am. Roman Catholic. Achievements include development of first microprocessor based direct digital engine fuel control algorithm at GM, of first microcomputer based direct digital wheel lock control algorithm at GM; co-invention of a classified secure network inter network communications protocol; co-conversion of classical signal processing algorithms to massively parallel computer algorithms; modification of system engineering technology to suit automotive electronics applications. Home: 23950 Fairview St Farmington MI 48335-3117 Office: AC Nielsen Co 375 Patricia Ave Dunedin FL 34698

IBATA, KOICHI, chemistry research manager; b. Yokohama, Kanagawa, Japan, Mar. 30, 1947; s. Shiroh and Fusako (Yoshida) I.; m. Sachiko Oikawa, June 29, 1986; children: Harue, Keiichi. B of Tech., Yokohama Nat. U., Japan, 1969; DSc, Tokyo Inst. Tech., Japan, 1975. Chemistry researcher Kuraray Co. Ltd., Kurashiki, Japan, 1975-85; electronics developer KOA Corp., Tokyo, Japan, 1985-87; chemistry rsch. mgr. Graphtec Corp., Yokohama, Japan, 1987—. Patentee in field; contbr. papers to profl. publs. Mem. Chem. Soc. Japan, Crystallographic Soc. Japan, Japan Soc. Applied Physics, Japan Soc. Analytical Chemistry, Am. Phys. Soc., Am. Chem. Soc., IEEE, Soc. for Imaging Sci. & Tech., The Planetary Soc. Buddhist. Avocations: playing piano, playing game of go, astron. observation, badminton. Home: 7 Terakubo Naka-ku, Yokohama Kanagawa 231, Japan Office: Graphtec Corp, 503-10 Shinanocho, Totsuka-ku, Yokohama Kanagawa 244, Japan

IBAYASHI, TSUGUIO, legal educator; b. Shiga Prefecture, Japan, 1931. LLM, Hitotsubashi U., 1960. With Japan Fedn. Econ. Orgns., 1960-85, asst. dir. fin. affairs dept., 1971-75, sep. asst. to chmn., 1975-79, dir. Office Chmn., 1979-82, dir. internat. econ. affairs dept., 1982-85; mng. dir. Japan Inst. for Social and Econ. Affairs, Tokyo, 1985-91, sec.-gen., 1985-91; prof. law Toyama U., Tokyo, 1991—. Author over 20 books on acctg. prins., comml. law, taxation, and internat. econs. Office: 11-4 3 chome Minami, Meguro-ku Tokyo Japan

IBBOTT, GEOFFREY STEPHEN, physicist; b. London, Mar. 23, 1949; s. Frank Alfred and Gladys Josephine (Gilbert) I.; m. Suzan Helen Doro, Feb. 14, 1969 (div. 1971); 1 child, Brian Richard; m. Diane Lorraine McCollum, Dec. 2, 1989. BA, U. Colo., 1979; MS, U. Colo. Health Sci., 1981; PhD, Colo. State U., 1993. Cert. therapeutic radiol. physicist, radiation oncology physicist. Sr. instr., med. physicist U. Colo. Health Scis. Ctr., Denver, 1974-90; lectr., med. physicist Yale-New Haven Hosp., 1990—; mem. examination panel Am. Bd. Med. Physics, 1992—. Assoc. editor Jour. of Med. Physics, 1982—; contbg. author: The Selection and Performance of Radiologic Equipment, 1985; author: (booklet) Performance Evaluation of Hyperthermia Equipment, 1989; contbr. articles to profl. jours. Recipient Meml. award for profl. achievement Rocky Mountain chpt., Health Physics Soc., 1973. Mem. Am. Assn. Physicists in Medicine (bd. dirs. 1982-84), Am. Coll. Radiology, Am. Soc. Therapeutic Radiology and Oncology, Radiation Rsch. Soc. Achievements include research on radiation response of mouse taste organ. Office: Yale-New Haven Hosp 20 York St New Haven CT 06504

IBEGBU, CHRIS CHIDOZIE, immunologist, virologist; b. Lagos, Nigeria, Dec. 24, 1957; came to U.S., 1980; s. Emmanuel C. and Mathilda C. I.; 1 child, Brandon Ikenna. BS, U. Waterloo, Ont., Can., 1980; PhD, Atlanta U., 1986. NRC postdoctoral fellow Ctrs. for Disease Control, Atlanta, 1987-90; immunologist, virologist dept. pediatrics Emory U. Sch. Medicine, Atlanta, 1990—; adv. bd. Pediatric NCCLS Flow Cytometry Guidelines, Washington, 1992—. Author book chpt.: Role of CD4 in HIV Infection, 1991. Publicity sec. Onitsha Ado Club of Atlanta, 1990—. EERC award Children's Rsch. Ctr., Atlanta, 1992. Mem. N.Y. Acad. Scis., Am. Soc. Microbiology, Internat. AIDS Soc., AAAS. Achievements include definition of the structural features of CD4 (the receptor for HIV) required for binding to human immunodeficiency virus that cuases AIDS. Office: Emory Univ Sch of Medicine Dept Pediatrics 69 Butler St SE Atlanta GA 30303

IBEN, ICKO, JR., astrophysicist, educator; b. Champaign, Ill., June 27, 1931; s. Icko and Kathryn (Tomlin) I.; m. Miriam Genevieve Fett, Jan. 28, 1956; children: Christine, Timothy, Benjamin, Thomas. BA, Harvard U., 1953; MS, U. Ill., 1954, PhD, 1958. Asst. prof. physics Williams Coll., 1958-61; sr. research fellow in physics Calif. Inst. Tech., Pasadena, 1961-64; assoc. prof. physics MIT, Cambridge, 1964-68, prof., 1968-72; prof. astronomy and physics, head dept. astronomy U. Ill., Champaign-Urbana, 1972-84, prof. astronomy and physics, 1972-89; holder of Eberly family chair in astronomy Pa. State U., 1989-90; disting. prof. astronomy and physics U. Ill., Urbana, 1989—; vis. prof. astronomy Harvard U., 1966, 68, 70; vis. fellow Joint Inst. for Lab. Astrophysics U. Colo., 1971-72; vis. prof. astronomy and astrophysics U. Calif. at Santa Cruz, 1972; vis. prof. physics and astronomy Inst. for Astronomy U. Hawaii, 1977; mem. adv. panel, astronomy sect. NSF, 1972-75; mem. vis. com. Aura Observatories, 1979-82; vis. scientist astronomical council Union Soviet Socialist Rep. Acad. of Sci., 1985; sr. vis. fellow Australian Nat. U., 1986; vis. prof. U. of Bologna, Italy, 1986; sr. research fellow U. of Sussex, Eng., 1986; George Darwin lectr. Royal Astronomical Soc., London, 1984; McMillin lectr. Ohio State U., 1987; vis. eminent scholar Univ. Ctr. Ga., 1988; guest prof. Christian Albrechts Universität zu Kiel, 1990. Contbr. articles to profl. jours. John Simon Guggenheim Meml. fellow, 1985-86; recipient Eddington medal Royal Astron. Soc., 1990. Fellow Japan Soc. for Promotion of Sci.; mem. Am. Astron. Soc. (councilor 1974-77, Henry Norris Russell lectr. 1989), U.S. Nat. Acad. of Scis., Internat. Astronom. Union. Home: 3910 Clubhouse Dr Champaign IL 61821-5554 Office: U Ill Dept of Astronomy 1002 W Green St Urbana IL 61801-3074

IBERS, JAMES ARTHUR, chemist, educator; b. Los Angeles, June 9, 1930; s. Max Charles and Esther (Imerman) I.; m. Joyce Audrey Henderson, June 10, 1951; children—Jill Tina, Arthur Alan. B.S., Calif. Inst. Tech., 1951, Ph.D., 1954. NSF post-doctoral fellow Melbourne, Australia, 1954-55; chemist Shell Devel. Co., 1955-61, Brookhaven Nat. Lab., 1961-64; mem. faculty Northwestern U., 1964—, prof. chemistry, 1964-85, Charles E. and Emma H. Morrison prof. chemistry, 1986—. Mem. NAS, Am. Acad. Arts and Sci., Am. Chem. Soc. (inorganic chemistry award 1979, disting. svc. in the advancement of inorganic chemistry award, 1992), Am. Crystallographic Assn. Home: 2657 Orrington Ave Evanston IL 60201-1760 Office: Northwestern U Dept Chemistry Evanston IL 60208-3113

IBRAHIM, FAYEZ BARSOUM, chemist; b. Cairo, July 28, 1929; came to U.S., 1957; s. Barsoum and Labiba (Wassef) I.; m. Roswitha E. Runke, Mar. 5, 1960; children: Michael, Robert, Nanette. BS, U. Cairo, 1952; MS, U. Houston, 1959; PhD, Rutgers U., 1978. Rsch. chemist Miles Labs., Elkhart, Ind., 1960-62; group leader Philips Roxane, St. Joseph, Mo., 1962-63; rsch. chemist Chemagro (Mobay) Corp., Kansas City, Mo., 1963-67; sect. leader analytical Carter Wallace Inc, Cranbury, N.J., 1968-79; sect. head analytical Revlon Health care Group, Tuckahoe, N.Y., 1979-83; lab. mgr. G & W and Able Labs., South Plainfield, N.J., 1984-90; sect. head analytical R&D Ganes Chems. Inc., Pennsville, N.J., 1990—; teaching asst. U. Houston, 1957-59. Mem. Am. Chem. Soc. Republican. Mem. Coptic Orthodox Ch. Achievements include research in UV method for fenthion, decomposition of di-syston on fertilizer, determination of barbiturates and analgesics. Home: 6 Nathan Dr North Brunswick NJ 08902 Office: Ganes Chems 33 Industrial Park Rd Pennsville NJ 08070

IBRAHIM, KAMARULAZIZI, physics educator, researcher; b. K.Kangsar, Perak, Malaysia, Mar. 13, 1959; s. Ibrahim and Kamsiah (Md. Taib) I.; m. Lili Hanum Yusof, Aug. 24, 1980; children: Muhammad, Khadijah, Sulaiman, Soleh. BSc in Physics, Ind. U., 1981; MSc in Applied Sci., So. Methodist U., 1984; PhD in Physics, Heriot Watt U., Edinburgh, Scotland, 1989. Lectr., assoc. prof. U. Sci. Malaysia, Minden, Penang, 1984—; referee Jour. Fizik Malaysia, 1992—; federated scheme vis. ICTP, Trieste, Italy, 1990. Recipient Overseas Rsch. award Com. of Vice-Chancellors and Prins. U.K., Eire, 1986-89. Mem. Inst. Energy Malaysia, Malaysian Solid State Sci. and Tech. Soc. (coun. 1991—). Achievements include proposition and testing of a new structure for space charge limited current measurement; research of solar energy studies in Malaysia. Office: Univ Sci Malaysia, Sch Physics Minden N, 11800 Penang Malaysia

IBRAHIM, MOHAMMAD FATHY KAHLIL, aerospace engineer; b. Cairo, Sept. 9, 1946; arrived in Canada, 1970; s. Fathy Khalil and Sakina (Attia) I.; m. Zeinab Abdul-Samad, Sept. 24, 1971; children: Ahmad, Omar, Noha, Yomna. BA in Aero. Engring., Cairo U., 1969; MA, U. Toronto, 1972; PhD in Space Sci., York U., Toronto, 1978. Teaching asst. York U., 1973-78, postdoctoral fellow, 1978-80, rsch. assoc., 1985-86; rsch. scientist Meteorol. and Environ. Planning Ltd., Toronto, 1980-81, Atmospheric Environ. Svc., Toronto, 1981-85; assoc. prof. aero. engring. Al-Fateh U., Tripoli, 1986-90; prin. Trajectory div. Orbit Engring., Toronto, 1990—; cons. scientist Orbit Engring., 1990—. Contbr. articles to profl. jours. Mem. Assn. Profl. Engrs. Ontario, Internat. Energy Found., AIAA. Islam. Home: 256 John Garland Blvd # 127, Toronto, Canada M9V 1N8 Office: Orbit Engring, 127-256 John Garland Blvd, Etobicoke, ON Canada M9V 1N8

IBRAHIM, MOUNIR BOSHRA, mechanical engineering educator; b. Aswan, Egypt, Feb. 18, 1947; came to U.S., 1980; s. Boshra and Samsuma Ibrahim; m. Nagwa William, Feb. 14, 1974; children: Joseph, Victor, Mary, Andrew. BSME, Cairo U., Egypt, 1968; PhD, Bradford U., U.K., 1977. Registered profl. engr., Iowa. Vis. rsch. fellow Tech. U. Denmark, Copenhagen, 1973; asst. prof. of Mech. Engring. U. Petroleum and Minerals, Dhahran, Saudi Arabia, 1978-83; vis. asst. prof. of Mech. Engring. Iowa State U., Ames, 1983-84, Mich. Tech. U., Houghton, 1984-85; assoc. prof. Mech. Engring. Cleve. (Ohio) State U., 1985—. Contbr. articles to profl. jours. Egyptian Govt. scholar, Cairo, 1960-68, Danish Internat. Devel. Agy. scholar, Denmark, 1973, U.K. Atomic Energy Authority scholar, 1974-78. Sr. mem. AIAA (treas. 1991-92, chmn. 1992-93); mem. ASME. Achievements include advancements in thermal energy storage systems; unsteady flows and heat transfer in applications such as gas turbines and Stirling engines. Office: Cleve State U E 24th & Euclid Ave Cleveland OH 44115

ICHIMURA, TOHJU, pediatrician, educator; b. Tokyo, Japan, Oct. 18, 1931; s. Kumesaburo and Hiro Ichimura; m. Eiko Kobayashi, June 8, 1960; children: Satomi, Rie, Akiko. MD, Tokyo Med. and Dental U., 1957; PhD, Sch. Medicine, Tokyo, 1962. Med. diplomate. Asst. dr. Tokyo Med. and Dental U., 1962-64, asst. prof., 1966-72; assoc. prof. Sch. Medicine Dokkyo U., Tochigi, Japan, 1973-75; prof. pediatrics Sch. Medicine Dokkyo U., Tochigi, 1975—. Editor jour. Allergy Practice, 1981—; contbr. articles to profl. jours. Faculty scholar Berlin Open U., 1964-66. Mem. Japanese Soc. Pediatric Otorhino-Laryngology (pres. 1990), Japanese Pediatric Soc., Japanese Soc. Allergology, Japanese Soc. Pediatric Allergology, Japanese Soc. Pediatric Pulmonology. Buddhist. Avocations: golf, igo, collecting books on intercultural contact. Home: 4-14-8 Saiwaicho, Mibu Shimotsuga Tochigi 321-02, Japan Office: Dokkyo U Sch Medicine, 880 Kitakobayashi Mibu, Shimotsuga Tochigi 321-02, Japan

ICKES, WILLIAM, psychologist, educator; b. Salt Lake City, Nov. 10, 1947; s. William Keith and Shirley Doris (Hallman) I.; m. Mary Jo Renard, Sept. 4, 1967; children: Marcus, John, William. BS, Brigham Young U., 1969; PhD, U. Tex., 1973. Asst. prof. psychology U. Wis., Madison, 1973-79; asst. prof. psychology U. Mo., St. Louis, 1979-81, assoc. prof. psychology, 1981-82; assoc. prof. psychology U. Tex., Arlington, 1982-87, prof., 1987—; vis. prof. U. Wash., Seattle, 1992. Action editor: Jour. Personality and Social Psychology, 1978-79, cons. editor, 1984—; mem. editorial bd: Jour. Social and Clin. Psychology, 1982—; co-editor: New Directions in Attribution Research, 1976, 78, 81, Personality Roles and Social Behavior, 1982; editor: Compatible and Incompatible Relationships, 1985. Fellow NDEA Title IV, 1969-72; recipient award Nat. Coun. Tchrs. English, 1965. Mem. Am. Psychol. Soc., Soc. Exptl. Social Psychology, Midwestern Psychol. Assn., Southwestern Psychol. Assn. Achievements include development of the unstructured interaction paradigm and a method for measuring empathic accuracy. Office: U Tex Dept Psychology Arlington TX 76019-0528

IDAN, MOSHE, aerospace engineer; b. Vilnius, Lithuania, Oct. 23, 1957; arrived in Israel, 1971; s. Abraham Alesin and Ita (Gurvitz) Milner. PhD in Aerospace Engring., Stanford (Calif.) U., 1990. Lectr. Technion, IIT, Haifa, Israel, 1991—. Contbr. articles to profl. jours. Postdoctoral fellow Technion, IIT, 1990, Rothschild fellowship Rothschild Found., 1986. Office: Technion IIT, Faculty of Aerospace Engr, Haifa 32000, Israel

IDELSOHN, SERGIO RODOLFO, mechanical engineering educator; b. Parana, E. Rios, Argentina, Nov. 15, 1947; s. Francisco and Edith Isabel (Barg) I.; m. Lelia Eva Zielonka, July 24, 1971; children: Muriel Veronica, Sebastian, Alejandra. Diploma in Tech., Tech. Sch. Parana, Argentina, 1965; Diploma in Engring., U. Nacional Rosario, Argentina, 1970; PhD, U. Liege, Belgium, 1974. Sr. scientist U. Liege, 1971-74; assoc. prof. U. Nacional Rosario, 1975-80, prof. mech. engring., 1980—; chmn. mechanics lab. Instituto de Desarrollo Technológico para la Industria Química, Santa Fe, Argentina, 1981—; prin. rsch. scientist Conicet, Santa Fe, Argentina, 1981—; prof. U. Nacional Litoral, Santa Fe, 1989—; chmn. Computational Mech. Lab., Sante Fe, 1981—; vis. prof. Inst. Advanced Study, Princeton, N.J., 1987-88, U. Paris VI, 1989-90, Universidad Politecnica de Cataluna, 1991-92. Editor: Mecanica Computacional, 1985; contbr. articles to profl. jours. Mem. Argentina Assn. Computer Mechanics (pres. 1985—), Internat. Assn. Computer Mechanics, Internat. Assn. Applied Mechanics, Am. Math. Soc., Internat. Union Theoretical and Applied Mechanics. Avocations: yachting, tennis, paddle. Office: INTEC, Guemes 3450, 3000 Santa Fe Argentina

IENNACO, JOHN JOSEPH, civil engineer, consultant; b. Longmeadow, Mass., July 17, 1964; s. Frank A. and Fay (Fleming) I.; m. Joanne Marie DeSanto, Sept. 26, 1992. BS, U. R.I., 1992. Engr.-In-Tng. Cert., 1992. Civil engr. Engring. Cons., S. Portland, Maine, 1992—; assoc. mem. ASCE, R.I., 1991—. Recipient Francis Connell scholarship, Kingston, R.I., 1991, Ebstein scholarship, Kingston, R.I., 1992. Mem. Chi Epsilon Civil Engring. Hon. Soc., Tau Beta Pi Engring. Hon. Soc. Home: PO Box 298 Raymond ME 04071 Office: DeLuca-Hoffman Assocs Inc 778 Main St Ste 8 South Portland ME 04106

IFFT, EDWARD MILTON, government official; b. Grove City, Pa., July 19, 1937; s. John T. and Edith M. (Patterson) I.; m. A. Jeanne Felts, Aug. 12, 1967; children—John R., Sharon E. B.S., Antioch Coll., Yellow Springs, Ohio, 1960; Ph.D., Ohio State U., 1967. Phys. sci. officer U.S. Arms Control & Disarmament Agy., Washington, 1967-73; dep. dir. Arms Control Office Dept. State, Washington, 1973-78; chief, internat. program policy NASA, Washington, 1978-81; sr. policy adviser U.S. Start Del., Washington and Geneva, 1982-84; State Dept. rep. U.S. Nuclear & Space Talks Del., Washington and Geneva, 1985-91; dep. dir. U.S. On-Site Inspection Agy., 1991—. Contbr. articles to profl. publs. Gen. Motors scholar, 1955-60, US/USSR Exchange scholar Consortium of U.S. Univs., 1964-65; NSF fellow, 1960-63. Mem. Am. Phys. Soc., Internat. Inst. Strategic Studies, Am. Assn. Advancement Slavic Studies. Home: 6825 Wheatley Ct Falls Church VA 22042-4025 Office: US On-Site Inspection Agy Washington DC 20041-0498

IFFY, LESLIE, medical educator; b. Budapest, Hungary, May 17, 1925; came to U.S., 1969; s. Zoltan and Rozsa (Lantos) I.; m. Maureen B. Deeney. MD, U. Budapest, Hungary, 1949. Diplomate Am. Bd. Ob-Gyn. Resident, fellow Orszagos Testnevelesi es Sportegeszsegügy Intezet Hosp. Ministry of Health, Budapest, 1951-56; fellow U. Wash., Seattle, 1964; asst. prof. Temple U., Phila., 1969-70; assoc. prof. U. Ill., Chgo., 1971-72, Jefferson Med. Coll., Phila., 1972-73; prof. U. Medicine and Dentistry of N.J., Newark, 1974—; dir. obstetrics U. Hosp., Newark, 1974—. Contbr. over 160 articles to profl. jours. and chpts. to books; editor: Perinatology Case Studies, 1978, 85, Obstetrics and Perinatology, 1981 (in English and Spanish), Operative Perinatology, 1984 (in English, Spanish and Japanese), Operative Obstetrics, 2d edit., 1992. Recipient Dr. Robert Jardine Rsch. prize U. Glasgow, 1963, Ford Found. rsch. fellowship, Seattle, 1964, hon. fellowship Hungarian Obstet. Soc., 1986. Fellow Am. Coll. Obstetricians and Gynecologists, Royal Coll. Surgeons (Can.); mem. Cen. Assn. of Obstetricians and Gynecologists (life), Chgo. Gynecol. Soc., Am. Coll. Legal Medicine (bd. dirs.), Royal Coll. Physicians (Edinburgh, Scotland; licentiate), Royal Coll. Surgeons (Edinburgh; licentiate), Royal Faculty Physicians and Surgeons (Glasgow, Scotland; licentiate). Avocations: music, chess, literature. Home: 5 Robin Hood Rd Summit NJ 07901-3718 Office: NJ Med Sch U Medicine & Dentistry 150 Bergen St Newark NJ 07103-2406

IGLEWSKI, BARBARA HOTHAM, microbiologist, educator; b. Freeport, Pa., Mar. 23, 1938; married, 1965; 2 children. BS, Allegheny Coll., 1960; MS, Pa. State U., 1962, PhD in Microbiology, 1964. Instr. Oregon Health Sci. U., 1968-69, asst. prof., 1969-73, assoc. prof., 1973-79, prof. microbiology, 1979—; mem. bacterial and mycotic disease study sect., NIH, 1979-83;mem. rsch. and tng. com. Nat. Cystic Fibrosis Found., 1981-84, vaccine related biol. product com., 1981-82. Fellow Pa. State U., 1964-65, U. Colo. Med. Ctr., 1965-66, Pub. Health Rsch. Inst. N.Y., 1966-68; sr. fellow Walter Reed Army Inst. Rsch., 1976-77. Mem. Am. Soc. Microbiology. Achievements include research in bacterial toxins and pathogenesis of gram negative bacteria, including Pseudomonas aeruginosa and Legionella pneumophila. Office: Univ of Rochester Dept of Microbiology & Immunol 601 Elmwood Ave Box 672 Rochester NY 14642*

IGLSEDER, HEINRICH RUDOLF, space engineer, researcher; b. Salzburg, Austria, May 15, 1957; arrived in Germany, 1977; s. Heinrich and Berta (Wiltschiko) I.; m. Christina Scholle, Oct. 20, 1989; 1 child, Christian-Andreas. Engr. diploma, Tech. U., Munich, 1983, D of Engring., 1986; D of Tech., Tech. U., Vienna, 1987. Cert. space technology engr. Sci. asst. Ctr. Space Tech. and Microgravity Zarm, Bremen, Germany, 1987-90; project mgr., head space tech. div. Zarm, Bremen, 1990-92, head predevel. divsn. space tech. and advanced projects, 1992—; project scientist Chair for Astronautics and Space, Munich, 1987—; cons. Space Tech. and Micro Systems, Sittensen, Germany, 1993—. Contbr. articles to profl. jours. Roman Catholic. Home: Amselweg 14, D-27419 Sittensen Germany Office: Zarm, Am Fallturm, D-28359 Bremen Germany

IGUCHI, IENARI, physicist, educator; b. Tokyo, Oct. 8, 1941; s. Motonari and Akiko Iguchi; m. Yurika Hayakawa, Jan. 20, 1973; children: Takahiko, Erika. B of Engring., U. Tokyo, 1965, M of Engring., 1967; PhD, Stanford U., 1971. Rsch. assoc. U. Tokyo, 1973-78; assoc. prof. U. Tsukuba, Japan, 1979-87, prof. physics, 1988—. Contbr. articles to profl. jours. Fellow Japan Soc. Promotion of Sci., 1978, faculty fellow IBM/IBM Japan, 1983; Fulbright grantee, 1965. Mem. Am. Phys. Soc., Phys. Soc. Japan, Japan Soc. Applied Physics, R&D Assn. Future Electron Devices. Avocations: music, skiing, hiking, fishing. Home: 1458-2 Hanamuro, Tsukuba 305, Japan Office: Univ Tsukuba, Inst Materials Sci, 1-1-1 Tennoudai, Tsukuba 305, Japan

IHARA, HIROKZU, systems engineer; b. Nagano, Japan, Jan. 17, 1937; s. Hiroji and Haruno (Takanashi) I.; m. Atsuko Kanda, Oct. 12, 1963; children: Izumi, Takeshi, Mariko. B. Engring., Shinsyu U., Nagano, 1959; D. Engring., Tokyo Inst. Tech., 1987. Engr. Ohmika Works, Hitachi (Japan) Ltd., 1959-69, sr. engr., 1969-76; sr. researcher Systems Devel. Lab. Hitachi, Kawasaki, Japan, 1976-77, chief researcher, 1977-80, dept. mgr., 1980-83, dep. gen. mgr., 1983-86; sr. chief engr. space systems div. Hitachi, Yokohama, Japan, 1986-92; chief engr. corp. tech. Hitachi, Tokyo, 1992-93; bd. dir., gen. mgr. R & D ctr. Hitachi Med. Corp., Kashiwa, Japan, 1993—; vis. prof. Self Def. Acad. of Japan, Yokosuka, 1987-91. Editor: Encyclopedia of Electrical & Electronics, vol. 8, 1982; contbr. articles to IEEE Proceedings, IEEE Computer. Recipient Yr. Achievement award Inst. Electronics, Info., and Communication Engrs., Tokyo, 1980; named for Best Paper Inst. System, Communication, Info. and Automation, Tokyo, Soc. Info. Control Engrs., Tokyo, 1985, Inst. Elec. Engrs., Tokyo, 1987, Compcon Fall '84, 1984. Fellow IEEE; mem. AIAA (sr.). Achievements include development of computer control system for iron and steel making process, automobile assembly, factory automation, medical systems; of train computer control and managing system for Shinkansen and subway; of fault tolerant computing. Home: 1-15-11 Minaminaruse, Machida 194, Tokyo Japan Office: Hitachi Med Corp R & D Ctr, 2-1 Shintoyofuta Kashiwa 227, Chiba Japan

IHRIG, EDWIN CHARLES, JR., mathematics educator; b. Washington, June 26, 1947; s. Edwin Charles and Lenore (Kokas) I.; m. Laurie Heather McColgan, July 6, 1974; 1 child, Karen Ann. BS, U. Md., 1969, MA, 1970; PhD, U. Toronto (Can.), 1974. Postdoctoral fellow math. dept. U. New Brunswick, Fredericton, Can., 1974-75; asst. prof. math. dept. Dalhousie U., Halifax, N.S., Can., 1975-76, McMaster U., Hamilton, Ont., Can., 1976-79; assoc. prof. math. dept. Ariz. State U., Tempe, 1979-85, prof. math. dept., 1985—. Contbr. articles to Gen. Relativity, Nuclear Physics, Combinatorics, Differential Geometry, Group Theory. Home: 1032 E Riviera Dr Tempe AZ 85282-5533 Office: Ariz State U Dept Math Tempe AZ 85287

II, JACK MORITO, aerospace engineer; b. Tokyo, Mar. 20, 1926; s. Iwao and Kiku Ii; came to U.S., 1954, naturalized, 1966; BS, Tohoku U., 1949; MS, U. Washington, 1956; M in Aero. Engring., Cornell U., 1959; PhD in Aero. and Astronautics, U. Wash., 1964; PhD in Engring., U. Tokyo, 1979; children: Keiko, Yoshiko, Mutsuya. Reporter, Asahi Newspaper Press, Tokyo, 1951-54; aircraft designer Fuji Heavy Industries Ltd. Co., Tokyo, Japan, 1956-58; mem. staff structures rsch. Boeing Co., Seattle, 1962—. Mem. AIAA, Japan Shumy and Culture Soc. (pres. 1976—), Sigma Xi. Mem. Congregational Ch. Contbr. numerous articles on aerodyns. to profl. jours. Office: The Boeing Co M/S 67-HC Seattle WA 98124

IIDA, SHUICHI, educator, physicist; b. Kobe, Hyogo-Ken, Japan, Jan. 30, 1926; s. Shunzoh and Sono (Ueda) I.; m. Kyoko Matsuoka, Apr. 29, 1955; children: Mariko Takahara, Junko Kose. BS in Physics, U. Tokyo, 1947, PhD in Physics, 1952. Asst. prof. physics U. Tokyo, 1952-58, assoc. prof., 1958-68, prof., 1968-86, prof. emeritus, 1986; prof. Teikyo U., Sagamiko, Kanagawa, Japan, 1988-89, Utsunomiya, 1989—; vis. prof. AT&T Bell Labs., Murray Hill., N.J., 1961-63. Contbr. numerous articles to profl. jours. Mem. Magnetics Soc. IEEE, Japan Soc. Powder and Powder Metallurgy, N.Y. Acad. Sci., Magnetics Soc. Japan, Japan Inst. Metals, Physics Soc. Japan, Am. Inst. Physics. Achievements include patents for ferrites; founder of new frame in physics. Home: 4-23-11 Funabashi, Setagaya-ku, Tokyo 156, Japan Office: Teikyo U Sch Sci & Engring, 1-1 Toyosatodai, Utsunomiya Tochigi 320, Japan

IIDA, YUKISATO, lawyer; b. Tokyo, Aug. 24, 1918; s. Sueharu and Kazuko I.; m. Turuko Aoki, May 1, 1948; children: Hidesato, Toshiyuki, Fumisato. Author: English-Japanese Dictionary of Patent Terms, 1973, Translation of English Patent Specifications, 1981, Japanese-English Dictionary of Patent Terms, 1982, Drafting of English Patent Specifications, 1983, Manual of Foreign Patent Application, 1985. Recipient Yellow Ribbon medal Japanese Govt., 1978, 5th Order of Merit of the Rising Sun, 1988. Home: 5-18-13 Koenji-Minami, Suginami-Ku, Tokyo 166, Japan Office: Iida & Kuriu World Times Bldg 3F, 10-7 Ichibancho Chiyoda-Ku, Tokyo 102, Japan

IIJIMA, SHIGETAKA, astronomy educator; b. Kôfu, Yamanashi, Japan, Jan. 13, 1919; s. Shigeichi and Seki (Maruyama) I.; m. Ai Isobe, Sept. 25, 1942; children: Takae, Tokie, Kazue. BEng, Tokyo Inst. Tech., 1942; DSc, U. Tokyo, 1962. Engr.-in-chief Tokyo Astron. Obs., 1950-79; lectr. U. Tokyo, 1963-66, asst. prof., 1966-70, prof., 1970-79; prof. U. Electro Communication, Chôfu, Japan, 1981-84; prof. Musashi Inst. Tech., Tokyo, 1984-89, lectr., 1989—. Author: Introduction to Electrical Circuits, 1985. Mem. Astron. Soc. Japan (v.p. 1973-75), Internat. Astron. Union (v.p. commn. 31, 1976-79, pres. 1979-82), Geodetic Soc. Japan (hon.). Avocations: photography, travel, driving, do-it-yourself projects. Home: 4-23-6 Ōsawa, Mitaka Tokyo 181, Japan Office: Musashi Inst Tech, 1-28-1 Tamazutsumi, Setagaya-ku Tokyo 158, Japan

IIZUKA, JUGORO, physics educator, researcher; b. Misato-machi, Gumma-gun, Japan, Apr. 5, 1933; s. Tokimatsu and Kiwa (Aoki) I.; m. Etsuko Ohkawara, Aug. 28, 1958; children: Erito, Yuri, Makito. BS, Tokyo U. Edn., 1956; MS, Tokyo U. Edn., Tokyo, 1958; PhD in Theoretical Physics, U. Pa., 1961. Rsch. assoc. Enrico Fermi Inst., U. Chgo., 1960-62; rsch. assoc. Tokyo U. Edn. (now U. Tsukuba), 1962-65; asst. prof. Kanazawa (Japan) U., 1965-68; asst. prof. physics Nagoya (Japan) U., 1968—. Contbr. theoretical articles on elem. particle physics to sci. jours. Recipient Nishina Meml. prize Nishina Meml. Found., 1976; Yomiuri Yukawa fellow Yomiuri-Shinbun-sha, 1958, fellow Sakkokai Found., 1965-68. Mem. Phys. Soc. Japan, Am. Phys. Soc. Avocations: reading, walking, travel. Home: II-301 G-Haitsu Anjo, Hananoki-cho 7-18, Anjo-shi Aichiken 446, Japan Office: Nagoya U Dept Physics, Furo-cho, Chikusa-ku Nagoya 464-01, Japan

IKA, PRASAD VENKATA, chemist; b. Eluru, India, Apr. 8, 1958; came to U.S., 1981; s. Dalayya and Savitri (Divi) I.; m. Suguna Venkata Adigopula, Aug. 21, 1987; children: Deepika V., Amit V. MS, SUNY, Albany, 1983, PhD, 1985. Rsch. engr. Norton Co., Troy, N.Y., 1985-88; sr. rsch. engr. Norton Co., Worcester, Mass., 1988-90, rsch. supr., 1990-93; new product devel. mgr. Radiac Abrasives, Inc., Aurora, Ill., 1993—. Contbr. articles to profl. jours. Nat. sci. talent search scholar Nat. Coun. Ednl. Rsch. and Tng., New Delhi, 1975. Mem. Am. Chem. Soc. Hindu. Achievements include patent for Bismaleimide-Triazine resin bonded superabrasive wheels. Home: 1517 Westminster Dr Apt 103 Naperville IL 60563 Office: Radiac Abrasives Inc 400 N Highland Ave Aurora IL 60506

IKALAINEN, ALLEN JAMES, environmental engineer; b. Clinton, Mass., Aug. 30, 1945; s. Allen Carl and Anne (McBreen) I.; m. Barbara Joy Henwood, Sept. 4, 1976; children: Thomas, Stephen. BS in Civil Engring., Worcester Poly., 1967; MS in Environ. Engring., Northeastern U., 1978. Registered profl. engr., Maine, Mass. Engr. U.S. Army Engrs. New Eng., Waltham, Mass., 1967-70, U.S. EPA Region I, Boston, 1971-81; sect. chief H. S. EPA Region I, Boston, 1981-84, br. chief, 1984-85; project mgr. E. C. Jordan Co., Portland, Maine, 1985-88; dept. mgr. ABB Environ. Svcs., Portland, 1989-90, divsn. mgr., 1990-91, office mgr., cons. engr., 1991—; com. mem. Vision 2000 Adv. Group, Portland, 1990. Contbr. papers to rsch. assns. Active Vol. Fire Dept., Princeton, Mass., 1960-67. Mem. ASCE, NSPE, Soc. Am. Mil. Engrs., Am. Def. Preparedness Assn., Hazardous Waste Action Coalition. Democrat. Roman Catholic. Home: 36 Farnell Dr Yarmouth ME 04096 Office: 110 Free St Box 7050 Portland ME 04112-7050

IKEDA, KAZUYOSI, physicist, poet; b. Fukuoka, Japan, July 15, 1928; s. Yosikatu and Misao (Misumi) I.; m. Mieko Akiyama, Nov. 20, 1956; children: Hiroko Ikeda Kori, Yosihumi. 1st degree Rigakusi, Kyushu U., Fukuoka, Japan, 1951, DSc, 1957. Asst. dept. physics Kyushu U. Faculty of Sci., Fukuoka, 1956-60, assoc. prof. dept. physics, 1960-65; assoc. prof. dept. applied physics Faculty Engring. Osaka (Japan) U., 1965-68, prof. theoretical physics dept. applied physics, 1968-89, prof. theoretical and math. physics dept. math. scis., 1989—. Author: Statistical Thermodynamics, 1975, Mechanics Without Use of Mathematical Formulae—From a Moving Stone to Halley's Comet, 1980, Invitation to Mechanics—From the Fundamentals of Calculus to the Motion of a Comet, with Appendix on a comet in ancient times, 1985, collection of poems: Bansyoo Hyakusi, 1986, Basic Mechanics, 1987, Basic Thermodynamics - from entropy to osmotic pressure, 1991, The World of God, Creation and Poetry, 1991, Poems on the Hearts of Creation, 1993; contbr. over 100 articles to profl. jours., serialized poems, essays on poetry. Hon. founder Olympoetry Movement, 1992—. Recipient Internat. Man of Yr. award Internat. Biog. Ctr., 1992, 25-Yr. Achievement award, 1992, Most Admired Man of Decade, 1992, Silver Shield of Valor, Am. Biog. Inst., 1992. Fellow United Writers' Assn., World Lit. Acad.; mem. N.Y. Acad. Scis., World Inst. Achievement, Lifetime Achievement Acad. (Golden Acad. award 1991), Phys. Soc. Japan (com. mem. 1970—, chmn. Osaka br. 1976-77, 83-84, editor jour. 1976-78), Internat. Poets Acad. (Internat. Eminent Poet award, 1993), World Acad. Arts and Culture, World Congress of Poets, Confedn. Chivalry (mem. grand coun. 1991—, Internat. Order of Merit award 1990, Chevalier Grand Cross 1991). Home: Nisi 7-7-11 Aomadani, Minoo-si, Osaka 562, Japan Office: Osaka U Faculty Engring, Dept Math Sci 2-1 Yamadaoka, Suita-si, Osaka 565, Japan

IKEUCHI, SATORU, astrophysicist, educator; b. Himeji, Hyogo, Japan, Dec. 14, 1944; s. Taro and Hisa (Todo) I.; m. Yasuko Jindai, Nov. 23, 1972; 1 child, Rie. BS, Kyoto U., Japan, 1968, PhD, 1975. Rsch. assoc. Kyoto (Japan) U., 1972-77; asst. prof. Hokkaido U., Sapporo, Japan, 1977-84; assoc. prof. U. Tokyo, 1984-88, prof., 1988-92; chmn. Nat. Astron. Obs., Tokyo, 1988-92; prof. Osaka U., Japan, 1992—; councilor Astron. Soc. Japan, Tokyo, 1980—; cons. Japan Sci. Com., Tokyo, 1985—. Author: Investigating the Cosmic Structures, 1988, Large Scale Structures in the Universe, 1988, Bubbly Universe, 1989, Nature Seen From Sky, 1991; editor Publs. Astron. Soc Japan, 1989-92. Mem. Astron. Soc. Japan (bd. dirs. 1985-92), Phys. Soc. Japan. Home: 5-11 Kawashima Tamagashira, Nishigyo-ku, Kyoto 615, Japan Office: Dept Physics Osaka Univ, Machikaneyama 1-1, Toyonaka Osaka 560, Japan

IKUMA, YASURO, ceramics educator, researcher; b. Sasayama, Hyogo, Japan, Mar. 17, 1948; s. Kazumasa and Chie (Yamamoto) I.; m. Keiko Kudo, Sept. 6, 1975; children: Ryota, Naoko, Harumi. BS, Tokyo Inst. Tech., 1970, MS, 1972; PhD, U. Utah, 1980. Researcher Calpis Food Industry Co., Shibuyaku, Tokyo, Japan, 1972-73, computer system programmer, 1973-75; postdoctoral assoc. U. Utah, Salt Lake City, 1979-81; rsch. assoc. Ikutoku Tech. U., Atsugi, Kanagawa, Japan, 1981-83, asst. prof., 1983-85; assoc. prof. Kanagawa Inst. Tech., Atsugi, Kanagawa, Japan, 1985—; test program cons. Japan Ctr. Exam. Rsch., Minatoku, Japan, 1981—. Author: (with others) Basic Science of Ceramics, 1989, Chemistry in Industry, 1990, Inorganic Chemistry for Graduate Students, 1992. Rsch. grantee Nippon Sheet Glass Co., 1987, Japanese Ministry Edn. Sci. Culture, 1989-91. Mem. Am. Ceramic Soc., Am. Chem. Soc., Ceramic Soc. Japan (mem. editorial bd. 1988-90), Surface Sci. Soc. Japan (mem. editorial bd. 1985-89). Avocation: backpacking. Home: 3-14-17 Morinosato, Atsugi Kanagawa 243-01, Japan Office: Kanagawa Inst Tech, 1030 Shimoogino, Atsugi Kanagawa 243-02, Japan

ILGEN, DANIEL RICHARD, psychology educator; b. Freeport, Ill., Mar. 16, 1943; s. Paul Maurice and Marjorie V. (Glasser) I.; m. Barbara Geiser, Dec. 26, 1965; children—Elizabeth Ann, Mark Andrew. B.S. in Psychology, Iowa State U., 1965; M.A., U. Ill., 1968, Ph.D. in Indsl.-Organizational Psychology, 1969. Asst. prof. dept. psychology U. Ill., Urbana, 1969-70; instr. Dutchess County Community Coll., Poughkeepsie, NY, 1971-72; asst. prof. to prof. dept. psychol. scis. Purdue U., West Lafayette, Ind., 1972-83; area head indsl.-organizational psychology Purdue U., 1978-83; Hannah prof. organizational behavior depts. mgmt. and psychology Mich. State U., East Lansing, 1983—; vis. assoc. prof. dept. mgmt. and orgn. U. Wash., Seattle, 1978-79. Co-author: (with J.C. Naylor and R.D. Pritchard) A Theory of Behavior in Organizations, 1980; (with E.J. McCormick) Industrial Psychology, 1985; contbr. chpts. to books and articles to profl. jours.; acting editor, mem. editorial bd. Organizational Behavior and Human Performance; assoc. editor Organizational Behavior and Human Decision Processes. Served as capt. M.I., U.S. Army, 1970-72. Purdue U. Found. grantee, 1973-75, 76-77, 81-82, U.S. Army Rsch. Inst. grantee, 1974-82, Office Naval Rsch. grantee, 1982-86, 90—. Fellow Am. Psychol. Assn. (edn. tng. com., coun. reps. 1985-87), Soc. Indsl. and Organizational Psychology of Am. Psychol. Assn. (pres. 1987-88), Am. Psychol. Soc.; mem. Midwest Psychol. Assn., Ind. Psychol. Assns., Acad. Mgmt., Soc. Organizational Behavior, Sigma Xi. Office: Mich State U Depts Mgmt and Psychology East Lansing MI 48824-1117

ILGEN, MARC ROBERT, aerospace engineer; b. New Haven, Oct. 19, 1961; s. William David and Eleanor Muriel (Roberts) I.; m. Babette Ann Kaiser, Sept. 12, 1992. BS in Physics, Haverford Coll., 1983; MS in Nuclear Engring., UCLA, 1986, PhD in Aerospace Engring., 1992. Sr. mem. tech. staff TRW Space and Def., Redondo Beach, Calif., 1986-91, Aerospace Corp., El Segundo, Calif., 1991—. Contbr. chpt. to book; author publs. in field. Fellow TRW Space and Def., 1988-91, Aerospace Corp., 1991-92. Mem. AIAA (sr.). Office: Aerospace Corp M4/957 PO Box 92957 Los Angeles CA 90009

ILIAS, SHAMSUDDIN, chemical engineer, educator; b. Faridpur, Bangladesh, Apr. 1, 1951; s. Abdul Aziz Mollah and Zaeda Khatun; m. Shahnaz Ilias, Oct. 28, 1979; children: Shayerah, Samia. BSc in Chem. Engring., U. Engring. and Tech., Dhaka, Bangladesh, 1974; MSc in Chem. Engring., U. Petroleum and Minerals, Dhahran, Saudi Arabia, 1979; PhD in Chem. Engring., Queen's U., Kingston, Can., 1986. Plant engr. BFFWT, Dhaka, 1974-77; asst. rsch. engr. UPM Rsch. Inst., Dhahran, 1979-82; rsch. assoc. U. Cin., 1986-90; asst. prof. chem. engring. N.C. A&T U., Greensboro, 1990—. Contbr. articles to Jour. Membrane Sci., Jour. Aerosol Sci., others. Grantee U.S. Dept. Energy, 1991-94, 92-93. Mem. AIChE, Am. Chem. Soc., Sigma Xi. Office: NC A&T State U Dept Chem Engring Greensboro NC 27411

ILLMAN, SÖREN ARNOLD, mathematician, educator; b. Helsinki, Finland, May 12, 1943; s. Arne and Hebe Dorotea (Nordström) I.; m. Kerstin Gunhild Anna Johansson, Aug. 25, 1968; children: Erik Jerker, Johanna Kristel. MS, U. Helsinki, 1966; PhD, Princeton U., 1972. Asst. U. Helsinki, 1965-66, prof., 1975—; rsch. asst. Acad. of Finland, 1971-72, jr. rsch. fellow, 1972-75; mem. Inst. for Advanced Study, Princeton, N.J., 1974-75; vis. rsch. fellow Inst. des Hautes Etudes Scientifiques, Bures-sur-Yvette, France, 1977, Math. Inst. Oxford U., Eng., 1978, Forschungs Inst. für Matematik, ETH, Zurich, Switzerland, 1982, 85; vis. scholar U. Mich., Ann Arbor, 1983; vis. prof. Dept. Math., Purdue U., West Lafayette, Ind., 1983; vis. prof. Rsch. Inst. Math. Scis., Kyoto (Japan) U., 1986-87; vis. prof. Max-Planck-Inst. für Math., Bonn, Fed. Republic of Germany, 1989, 92; vis. prof. Dept. Math., Princeton (N.J.) U., 1991-92; sr. mem. Math. Scis. Rsch. Inst., Berkeley, Calif., 1992; vis. prof. Forschungs Institut für Mathematik, ETH, Zürich, 1993. Contbr. articles to math. jours. Fulbright grantee, 1967-68, Acad. of Finland grantee, 1982, 86, 91-92. Fellow Finnish Soc. Scis. and Letters (chmn. math.-phys. sect. 1988-91); mem. Finnish Math. Soc. (sec. 1966, vice chmn. 1991—), Am. Math. Soc. Avocations: music, travel, movies. Home: Johannesbrinken 1B 46-47, Helsinki Finland Office: U Helsinki Dept Math, PO Box 4 Hallituskatu 15, SF-00014 Helsinki Finland

ILLNER-CANIZARO, HANA, physician, oral surgeon, researcher; b. Prague, Czechoslovakia, Nov. 2, 1939; came to U.S., 1968; d. Evzen Pospisil and Emilie (Chrastna) Pospisilova; m. Pavel Illner, June 14, 1963 (div. 1981); children: Martin Illner, Anna Illner; m. Peter Corte Canizaro, Nov. 1, 1982. MD, Charles U., Prague, 1961. Diplomate State Bd. Oral Surgery, 1963. Resident in oral surgery Inst. of Health, Pribram, Czechoslovakia, 1961-63; attending physician Oral Surgery Clinic, Prague, 1963-68; rsch. assoc. dept. surgery U. Tex. Southwestern Med. Sch., Dallas, 1969-72, instr. surgery, 1972-74; instr. surgery U. Wash. Sch. Medicine, Seattle, 1974-77; asst. prof. surgery Cornell U. Med. Coll., N.Y.C., 1977-81, assoc. prof. surgery, 1981-83; assoc. prof. surgery Tex. Tech U. Health Scis. Ctr., Lubbock, 1984-88, prof. surgery, 1988—; site visitor NIGMS Postdoctoral Tng. Grant, Bethesda, Md., 1987. Mem. editorial bd. Circulatory Shock, N.Y.C., 1981—; manuscript reviewer Surgery, Gynecology and Obstetrics, Chgo., 1985—; contbr. chpts. to books, numerous articles to profl. jours. NIH grantee, 1979-83, 87-92; Tex. Tech U Health Scis. Ctr. grantee, 1985, 86; U.S. Dept. of Army grantee, 1988-90; Fogarty Sr. Internat. fellow, 1991-92. Mem. Shock Soc. Avocations: remodeling of historical homes, gardening, skiing, pottery. Home: 4622 8th St Lubbock TX 79416 Office: Tex Tech U Health Scis Ctr 3601 4th St Lubbock TX 79430

ILTEN, DAVID FREDERICK, chemist; b. Marshalltown, Iowa, July 24, 1938; s. Fred H. and Olga Katherine (Keiper) I.; m. Veronika Maria Thamm, May 18, 1968; children: Paul, Stephan, Eric. BA, Yale U., 1960; PhD (NIH fellow 1960-64), U. Calif.-Berkeley, 1964. Teaching and rsch. prof. U. Frankfurt (W. Ger.), 1964-68; staff chemist IBM Corp., E. Fishkill, N.Y., 1968-72; teaching and rsch. prof. phys. chemistry Tech. U. Berlin, 1972-74, U. Regensburg, 1974-78; systems analyst chem. application programs and computer-aided ednl. systems Control Data Corp., Frankfurt, 1978-89; with thermal physical property dept., info. and data bases dept., dir. mktg. Deutsche Gesellschaft für Chemische Apparatewesen, Chemische Technik und Biotechnologie, eV, 1989— lectr. univs. Frankfurt, Heidelberg, Stuttgart, Md. Author papers in field. Nat. Merit scholar, 1956; Gen. Motors Corp. fellow, 1956-60, Humboldt fellow, 1964-66. Mem. AAAS, ASTM, Am. Chem. Soc., Am. Phys. Soc., Am. Inst. Chem. Engrs., Electrochem. Soc., N.Y. Acad. Scis., Assn. Yale Alumni (bd. govs.), Bunsen Gesellschaft, Gesellschaft Deutscher Chemiker, Verein Deutscher Ingenieure, Sigma Xi. Republican. Lutheran. Clubs: Mory's (Yale U.), Yale of Germany (pres.). Achievements include research in computational chemistry and chemical engineering, intermolecular forces. Home: Schifferstr 22, D 60594 Frankfurt 1, Germany Office: DECHEMA, Theodor-Heuss-Allee 25, D 60594 Frankfurt 97, Germany

ILTIS, HUGH HELLMUT, plant taxonomist and evolutionist, educator; b. Brno, Czechoslovakia, Apr. 7, 1925; came to U.S., 1939, naturalized, 1944; s. Hugo and Anne (Liebscher) I.; m. Grace Schaffel, Dec. 20, 1951 (div. Mar. 1958); children: Frank S., Michael George; m. Carolyn Merchant, Aug. 4, 1961 (div. June 1970); children: David Hugh, John Paul. B.A., U. Tenn., 1948; M.A., Washington U., St. Louis and Mo. Bot. Garden, 1950, Ph.D., 1952. Rsch. asst. Mo. Bot. Garden, 1948-52; asst. prof. botany U. Ark., 1952-55; mem. faculty U. Wis.-Madison, 1955—, prof., 1967—, curator univ. herbarium, 1955-68, dir. univ. herbarium, 1969—; vis. prof. U. Va. Biol. Sta., 1959; lectr. in field; expdns. to Costa Rica, 1949, 89, Peru, 1962-63, Mexico, 1960, 71, 77, 78, 79, 80, 81, 82, 84, 87, 88, 90, 92, Guatemala, 1976, Ecuador, 1977, St. Eustatius, Puerto Rico, 1989, USSR, 1975, 79, Nicaragua-Honduras, 1991, Venezuela, 1991, Hawaii, 1967; mem. adv. bd. Flora N.Am., 1970-73, Gov. Wis. Commn. State Forests, 1972-73. Author articles flora of Wis., Capparidaceae, biogeography, evolution of maize, human ecology, especially innate responses to, and needs for, natural beauty and diversity, nature preservation, especially Latin Am. Co-instigator Reserva Biosfera Sierra de Manantlán, Jalisco, Mex. (co-discoverer Zea Diploperennis). Recipient Biology award U. Tenn., 1948, Feinstone Environ. award SUNY Syracuse, 1989, Spl. Conservation award Nat. Wildlife Fedn., 1992. Fellow AAAS, Linnean Soc. (London); mem. Am. Inst. Biol. Scis., Bot. Soc. Am., Am. Soc. Plant Taxonomists, Internat. Assn. Plant Taxonomy, Conservation Internat., Soc. Study Evolution, Ecol. Soc. Am., Wis. Acad. Arts, Sci. and Letters, Forum for Corr.-Internat. Center Integrative Studies, Nature Conservancy (dir. Wis. chpt., Nat. award 1963), Wilderness Soc., Sierra Club, Nat. Parks Assn., Citizens Natural Resources Assn. Wis., Natural Resource Def. Council, Environ. Def. Fund, Friends of Earth, Zero Population Growth, Sigma Xi, Phi Kappa Phi. Home: 2784 Marshall Pky Madison WI 53713-1023

IMADO, FUMIAKI, control engineer; b. Komagome, Tokyo, Mar. 1, 1945; s. Bunzo Ishiwara and Yoshi Imado; m. Mikiko Wakabayashi, Oct. 11, 1976; children: Eisuke, Tomoko, Kenji. BS, U. Tokyo, 1968, PhD, 1975. Registered profl. engr. Engr. Ctrl. Rsch. Lab. Mitsubishi Electric Co., Amagasaki, Hyogo, 1973-77, sr. engr., 1977-84, chief engr., 1984-91, chief engr., mgr., 1991—; lectr. Chun Shan Inst. Sci. and Tech., Lung-tan, Taiwan, 1990—, Ryukoku U., Otsu, Shiga, 1991—; guest lectr. nat. Aerospace Lab., Mitaka, Tokyo, 1993—. Contbr. articles to Jour. GN & C, Jour. Aircraft, Computers Math. with Appl. Mem. AIAA, Japan Soc. Aeronautics and Astronautics (bd. dirs. 1990-92, editor jour. 1983—), Japan Soc. Biomechanism, Soc. for Instrument and Control Engrs. Achievements include patent for artficial satellite, missile guidance and control field; rsch. in finding optimal fighter maneuvers against PNG missiles; design of missile guidance systems; development of softwares in the fields, satellites, bombs. Home: 2-5-25-610 Kohama, Takarazuka Hyogo, Japan 665

IMAI, NORIYOSHI, chemist, researcher; b. Toyama, Japan, Aug. 11, 1933; s. Hideichi and Misao (Yoshino) I.; m. Eiko Mizukoshi, Aug. 18, 1957; children: Yumi, Youichi. B in Chemistry, Kyoto U., Japan, 1956. Ink chemist Toyo Ink Mfg. Co., Ltd., Tokyo, 1956-63, mgr. printing ink rsch. group, 1963-73, tech. dir. printing ink rsch. div., 1973-82; gen. mgr. Imaging Rsch. Lab., Tokyo, 1982-86, Corp. Tech. Planning Div., Tokyo, 1986-88, Basic Rsch. Lab., Tokyo, 1988-91; dir., gen. mgr. tech. devel. adminstrn. Toyo Ink Mfg. Co., Ltd., Tokyo, 1991—; mem. organizing com. RadTech Japan, Tokyo, 1986—. Author: Photochemistry, 1983, 1987; inventor of waterless printing, 1970, UV-curing ink, 1979. Mem. Japanese Soc. Printing Sci. and Tech. (mem. editorial bd. 1980—), Soc. Polymer Sci. (steering com. of info. and rec. mem. 1980—), Soc. Colour Material (low temperature drying ink award 1975). Avocations: classical music, violin, horse back riding. Home: 2-403 540-1 Kamiochiai, Yono 338, Japan Office: Toyo Ink Mfg Co Ltd, 3-13 Kyobashi 2-chome, Chuo-ku Tokyo 104, Japan

IMAM, M. ASHRAF, materials scientist, educator; b. Patna, Bihar, India, Sept. 7, 1945; came to U.S. 1970; s. Naimuddin Ahmad and Zakia (Begum) Ahmad; m. Shamim Akhtar, June 22, 1979; children: Nabil S., Rahil U., Mariam S. BS, Ranchi U., India, 1966; MS, Carnegie-Mellon U., 1972; DSc, George Washington U., 1976. Lectr. E.P. U., Dacca, Pakistan, 1969-70; rsch. assoc. George Washington U., Washington, 1976-78, rsch. scientist, 1978-81, adj. prof., 1981—; guest scientist Nat. Inst. Standard, Gaithersburg, Md., 1974—; sr. rsch. scientist Geo-Centers Inc., Newton, Mass., 1981-84; metallurgist Naval Rsch. Lab., Washington, 1984—. Contbr. articles to profl. jours.; editor: Structure and Deformation of Boundaries, 1986, Advances in Low-Carbon High Strength Ferrous Alloys, 1993. MRL fellow Carnegie Mellon U., 1971-72, CSIR fellow, 1966-68, others. Mem. ASM, The Minerals, Metals, Materials Soc. (titanium com. 1980—, phys. metallurgy com. 1980—, mech. metallurgy com. 1980—), Sigma Xi. Achievements include 3 patents. Home: 8454 Van CT Annandale VA 22003 Office: Naval Rsch Lab 4555 Overlook Ave SW Washington DC 20375-5320

IMAMURA, TORU, molecular cell biologist; b. Kagoshima, Japan, Feb. 25, 1956; s. Hiroshi and Eiko (Ishikawa) I.; m. Reeko Urabe, Mar. 15, 1986;

children: Mayumi, Ryota. BS, Tokyo U., 1979, MS, 1981, PhD, 1984. Researcher Fermentation Rsch. Inst., Tsukuba, 1984-88, sr. scientist, 1988—; vis. scientist ARC, Rockville, Md., 1988-90; biotech. cons. Ministry Internat. Trade and Industry, Chiyoda and Tokyo, 1991—; chmn. radiation safety com. Fermentation Rsch. Inst., 1987-88, chmn. pollution prevention com., 1990-91. Contbr. articles to profl. jours. Mem. AAAS, Japan Biochem. Soc., Japan Cell Biology Soc., Japan soc. Pharma. Sci. Avocations: art, tennis, reading, travel, music. Home: 2-20-12-1301, Senju-Azuma, Adachi 120, Japan Office: Nat Inst Biosci & Human Tech, 1-1 Higashi, Tsukuba 305, Japan

IMAMURA, TSUTOMU, physicist, educator; b. Nishinomiya, Japan, May 28, 1927; s. Arao and Haruko (Ishida) I.; m. Masako Takeda, Nov. 21, 1955; children: Akira, Satoshi, Motoko. BS, Osaka U., 1951, DSc, 1957. Asst. Osaka (Japan) U., 1952-61, asst. prof., 1961; rsch. assoc. U. N.C., Chapel Hill, 1958-60, vis. assoc. prof., 1966-67; lectr., rsch. assoc. Boston U., 1960-61; prof. Kwansei Gakuin U., Nishinomiya, 1961—, dean faculty of sci., 1975-77. Author: Mathematics for Stochastic Fields, 1976, Fourier Transform and Physics, 1976, Green Functions and Physics, 1978, Theory of Functions and Physics, 1981. Trustee Kwansei Gajuin U., Nishinomiya, 1989-92. Mem. Phys. Soc. Japan, Am. Phys. Soc. Avocation: Shogi. Home: 7-31 Matsugaoka-cho, Nishinimoya 662, Japan Office: Kwansei Gakuin U, Uegahara, Nishinomiya 662, Japan

IMBAULT, JAMES JOSEPH, electromechanical engineering executive; b. Muskegon, Mich., Oct. 31, 1944; s. Joseph Lionel and Ruth Pauline (Schutter) I.; m. Vallery Ann Rumisek, Dec. 29, 1967; children—Michelle, Allan. A.S., Muskegon Community Coll., 1965; B.S. Mech. Engring. with honors, Mich. Tech. U., 1967; postgrad. UCLA, 1972-73. Registered profl. engr., Calif. Sr. mem. tech. staff RCA, E.A.S.D., Van Nuys, Calif., 1968-74; mech. engring. mgr. Litton Italia SPA, Rome, Italy, 1974-78; electromech. engring. mgr. Incosym, Inc., Westlake Village, Calif., 1978-82, v.p., 1982-88, also dir.; sr. program mgr. Precision Products div. Northrop, Norwood, Mass., 1988—. Patentee in field. Recipient Meritorious Performance award Muskegon Community Coll., 1965; Mich. Tech. U. scholar, 1965-67. Mem. Nat. Soc. Profl. Engrs., ASME, AIAA, Mich. Tech. U. Alumni Assn., Pi Tau Sigma, Tau Beta Pi, Phi Kappa Phi. Democrat. Home: 12 Hayden Dr Foxboro MA 02035-1127 Office: Northrop Corp 100 Morse St Norwood MA 02062-4681

IMBERSKI, RICHARD BERNARD, zoology educator; b. Amsterdam, N.Y.. BS, U. Rochester, 1959, PhD, 1966. Mem. faculty dept. zoology U. Md., College Park, 1967—. Grantee Dept. Agr., NSF. Achievements include research on genetics and devel. biology. Office: U Md Dept Zoology College Park MD 20742

IMHOFF, DONALD WILBUR, chemist; b. West Salem, Ohio, Dec. 8, 1939; s. Harry E. and Florence I. (Miller) I.; m. Marianne Fry, Dec. 22, 1962 (div. Dec. 1974); 1 child, Kimberly; m. J. Maxine Herring, Aug. 6, 1981. BS, Manchester Coll., 1961; MS in Chemistry, Ohio State U., 1964, PhD in Chemistry, 1966. Sr. analytical chemist Ethyl Corp. R&D, Baton Rouge, 1966-76, rsch. assoc., 1976-83, supr., 1983-88, rsch. advisor, 1988-89, sr. R&D advisor, 1989—. Mem. ASTM (mem. com. on electronic materials 1989—). Republican. Methodist. Achievements include research in analytical application of nuclear magnetic resonance and infrared spectroscopies, in measurement of low-level impurities in polysilicon by infrared spectroscopy. Office: Ethyl Corp PO Box 341 Baton Rouge LA 70821

IMORDE, HENRY K., earth scientist. Recipient Selwyn G. Blaylock medal Can. Inst. Mining and Metallurgy, 1991. Office: care Xerox Tower Ste 1210, 3400 de Maisonneuve Blvd W, Montreal, PQ Canada H3Z 3B8*

IMPERIAL, JOHN VINCE, systems engineer, consultant; b. Honolulu, Feb. 5, 1958; s. Filomeno and Vicenta (Lapinig) I.; m. Lynn Sau-Mee Ching, Aug. 13, 1988. BSEE, U. Hawaii, 1982; MBA with distinction, Hawaii Pacific U., 1993. Rsch. engr. Lockheed Missiles and Space Co., Sunnyvale, Calif., 1982-86; design engr. SECON Inc., Redondo Beach, Calif., 1986; mem. tech. staff TRW Def. Systems Group, Redondo Beach, 1986-88; systems engr. SAIC Comsystems, Hickam AFB, Hawaii, 1988-89; sr. systems engr. Computer Scis. Corp., Honolulu, 1989-92, Camp Smith, Hawaii, 1992-93; sr. systems engr. I-Net, Inc., Wheeler Army Airfield, Hawaii, 1993—; cons. to various orgns., Honolulu, 1990—. Coach Am. Youth Soccer Orgn., Pearl City, Hawaii, 1990—; bd. dirs., coach SAY Soccer, Honolulu, 1991. Mem. IEEE, Soc. Photooptical and Instrumentation Engrs., Computer Measurement Group, Assn. for Computing Machinery. Achievements include design of parallel pipelined polygon processor board. Home: 94-294 Lupua Pl Mililani Town HI 96789-2151 Office: I-Net Inc DISA-PAC Wahiawa HI 96854-5120

IMRE, PAUL DAVID, mental health administrator; b. N.Y.C., May 30, 1925; s. Maximilian and Bluma (Datz) I.; B.S., U. Ill.-Urbana, 1950; M.A., N.Y. U., 1951; M.P.H., Johns Hopkins U., 1963; m. Jo Ellen Varner, Aug. 16, 1956; children—David Maximilian, Robert Bruce. Extern City Hosp., Welfare Island, N.Y., 1951-52; intern Springfield State Hosp., Sykesville, Md., 1952-53; chief psychologist Cherokee Mental Health Inst., Iowa, 1953-54; staff and chief psychologist Spring Grove State Hosp., Catonsville, Md., 1954-62; research assoc. Johns Hopkins U., 1964-72; dir. mental health ctr. Balt. County Dept. Health, Catonsville, 1970-88 ; cons. Md. State Dept. Health and Mental Hygiene, 1954-70, Children's Guild of Md., 1962-70, Jewish Child and Family Service, Balt., 1957-62; pvt. practice psychology, Columbia, Md. and Balt., 1951—. Served with inf., AUS, 1943-46; ETO. Research grantee; cert. psychologist, Md. Mem. Md. Psychol. Assn. (pres. div. II 1982-83, cert. of recognition 1978), Am. Psychol. Assn., Am. Pub. Health Assn. Club: Johns Hopkins (Balt.). Home: 10418 Green Mountain Cir Columbia MD 21044-2456

IMTIAZ, KAUSER SYED, aerospace engineer; b. Karachi, Pakistan, Jan. 26, 1952; came to U.S., 1973; s. Syed Imtiaz Ahmad and Rehana Imtiaz; m. Lubna Kauser, July 25, 1982; children: Hina Kauser, Yusra Kauser. BS in Aerospace Maintenance Engring., St. Louis U., 1977; MS in Aerospace Engring., Wichita State U., 1979. Lic. pvt. pilot and airframe and powerplant mechanic. Sr. engr. Beech Aircraft Corp., Wichita, Kans., 1978-82, 84-89; aircraft engr. Saudia Airlines Royal Fleet, Jeddah, Saudi Arabia, 1982-84; tech. specialist McDonnell Douglas Helicopter Co., Mesa, Ariz., 1989-91; prin. engr. Advance Civil Space Systems, Boeing Aerospace Co., Huntsville, Ala., 1991—. Author: (computer codes) Fatigue Life Calculation & Fracture Control, 1980. Mem. AIAA (sr. mem.), Pi Mu Epsilon, Alpha Sigma Nu. Muslim. Achievements include development of methodology and computer codes for crack propagation and aircraft life prediction; research in manned and unmanned missions to the Moon and and Mars; structural analysis of state of the art aerospace programs such as Space Station Freedom, Beech Starship, V22 Osprey, McDonnell Douglas Apache AH-64, NOTAR, LHX and C-17, Airforce TTTS project. Office: Boeing Def & Space Group 499 Boeing Blvd MS JR-34 Huntsville AL 35824

INADA, TADAHICO, mechanical engineer; b. Gifu, Japan, Feb. 4, 1947; s. Isaku and Sueko (Yoshida) I.; m. Francoise Vautherin, July 8, 1976. BME, Osaka U., Japan, 1969; MME, Osaka U., 1971. Dir. NASDA, Washington, 1978-83, Paris, France, 1983-87; dir. NASDA, Tokyo, 1988-91, dir. office of policy and strategy, 1991—. Office: Nat Space Devel Agy, 2-4-1 Hamamatsu-cho, Tokyo 105, Japan

INAGAMI, TADASHI, biochemist, educator; b. Kobe, Japan, Feb. 20, 1931; m. Masako Araki, Nov. 12, 1961. B.S., Kyoto U., Japan, 1953, D.Sc., 1963; M.S., Yale U., 1955, Ph.D., 1958. Research staff Yale U., New Haven, 1958-59, research assoc., 1962-66; research staff Kyoto U., Japan, 1959-62; instr. biochemistry Nagoya City U., Japan, 1962; asst. prof. biochemistry Vanderbilt U., Nashville, 1966-69, assoc. prof., 1969-74, prof. biochemistry, dir. hypertension ctr., 1975-91, Stanford Moore prof. biochemistry, 1991—, prof. medicine, 1992—. Contbr. numerous articles to profl. jours. Fulbright fellow, 1954-55; recipient Roche Vis. Prof. award, 1980, Humbolt Found. award, 1981, Ciba award Am. Heart Assn., 1985, Spa award Belgium Nat. Funds Scientific Research, 1986, Sutherland prize Vanderbilt U., 1990. Fellow High Blood Pressure Rsch. Coun.; mem. Am. Soc. Biol. Chemists, Am. Physiol. Soc., Endocrine Soc., Am. Chem. Soc., Am. Heart Assn. (mem. coun.), Am. Soc. Cell Biology, Soc. Neurosci., Japan Endocrine Soc. (hon. mem.). Office: Vanderbilt U Sch Medicine Spl Ctr Rsch Hypertention Garland Ave Nashville TN 37232

INAZUMI, HIKOJI, chemical engineering educator; b. Oda, Japan, Feb. 14, 1923; s. Chugoro and Tani Inazumi; m. Yoshiko Inazumi, Oct. 13, 1956; children: Toru, Yasushi. B in Engring., Tokyo Inst. Tech., 1945, D in Engring., 1956. Asst. prof. chem. engring. Kanazawa (Japan) U., 1950-58, prof., 1958-59; prof. Shizuoka U., Hamamatsu, Japan, 1959-67; prof. Tokyo Inst. Tech., 1967-83, prof. emeritus, 1983—; prof. chem. engring. Chiba Inst. Tech., Narashino, Japan, 1983—; chmn. bd. edn. Tokyo Inst. Tech., 1974-76. Author: Chemical Engineering, 1972; contbr. articles to profl. jours. Councilor Sci. Coun. Ministry Edn., Tokyo, 1973-75. Home: 21-7 Miyama 9-chome, Funabashi 274, Japan Office: Chiba Inst Tech, 17-1 Tsudanuma 2-chome, Narashino 274, Japan

INCAPRERA, FRANK PHILIP, internist; b. New Orleans, Aug. 24, 1928; s. Charles and Mamie (Bellipanni) I.; BS, Loyola U. of South, 1946; MD, La. State U., 1950; m. Ruth Mary Duhon, Sept. 13, 1952; children: Charles, Cynthia, James, Christopher, Catherine. Diplomate Am. Bd. Internal Medicine. Intern, Charity Hosp., New Orleans, 1950-51, resident, 1951-52; resident VA Hosp., New Orleans, 1952-54; practice medicine specializing in internal medicine, New Orleans, 1957—; adminstrv. mgr. Internal Medicine Group, New Orleans, 1973—; med. dir. Owens-Ill. Glass Co., New Orleans, 1961-85, Kaiser Aluminum Co., Chalmette, La., 1975-84, Tenneco Oil Co., Chalmette, 1978-84, Lutheran Nursing Home, 1990—; assoc. med. dir. Cigna Health Plan of La., 1991—; co-founder Med. Ctr. E. New Orleans, 1975; clin. assoc. prof. medicine Tulane U. Sch. Medicine, 1971-87, clin. prof. medicine, 1987—; med. dir. Luth. Nursing Home, 1990—; adv. bd. Healthcare New Orleans, 1991—; mem. New Orleans Bd. Health, 1966-70. Bd. dirs. Meth. Hosp., 1971—, v.p., 1992—, Lutheran Home New Orleans, 1976-80, Chateau de Notre Dame, 1977-82, New Orleans Opera Assn., 1975—; mem. New Orleans Human Relation Com., 1968-70; bd. dirs. Emergency Med. Svcs. Coun., 1977-86, pres., La. southeastern region, 1979-81; bd. dirs. New Orleans East Bus. Assn., 1980—, v.p., 1981-83; bd. dirs. Luth. Towers, 1988-89, Peace Lake Towers, 1988-89, La. State U. Med. Ctr. Found. Bd., 1989-91; mem. pastoral care adv. com. So. Bapt. Hosp., 1982-83; mem. pres's. adv. bd. coun. Loyola U. of South, 1989—. Capt. USAF, 1955-57. Fellow ACP, Am. Geriatrics Soc.; mem. AMA, Am. Coll. Physician Execs., Am. Coll. Physicians (Laureate award 1993), La. Med. Soc. (v.p. 1975-76), Orleans Parish Med. Soc. (sec. 1972-74), New Orleans Acad. Internal Medicine (pres. 1969), La. Occupational Medicine Assn. (pres. 1971-72), La. State Med. Soc. (v.p. 1975-76), La. Soc. Internal Medicine (exec. com. 1975—, pres. 1983-85), New Orleans East C. of C. (dir. 1979-85), La. State U. Med. Sch. Alumni Assn. (pres. 1989-90), Order of St. Louis, Blue Key, Delta Epsilon Sigma, Optimists Club (bd. dirs. 1964-69, New Orleans). Home: 2218 Lake Oaks Pky New Orleans LA 70122-4345 Office: 5640 Read Blvd New Orleans LA 70127

ING, DEAN CHARLES, novelist, consultant; b. Austin, Tex., June 17, 1931; s. Dean Emory and Louise (Hardin) I.; m. Rhoda Margaret Barrier, June 8, 1952 (div. 1958); children: Diana Capri, Laura Victoire; m. Gina Baker, Aug. 21, 1959; children: Dina Valerie, Dana Christie. BA in Speech, Fresno State U., 1956; MA in Speech, San Jose State U., 1970; PhD in Communication Theory, U. Oreg., 1974. Tech. editor Gilfillan, L.A., 1956-57; tech. writer Aerojet-Gen. Corp., Folsom, Calif., 1957-62; publ. engr. Lockheed, Sunnyvale, Calif., 1962, sr. rsch. engr., 1966-70; programs engr. United Tech. Corp., Sunnyvale, 1962-66; asst. prof. Mo. State U., Maryville, 1974-77; free-lance novelist, 1977—. Author: Soft Targets, 1979, Pulling Through, 1983, The Ransom of Black Stealth One, 1989; co-author: Mutual Assured Survival, 1984, The Future of Flight, 1985; cons. editor: High Frontier, 1983. With USAF, 1951-55. Achievements include completion of novel presaging Tehran crisis by 6 weeks; development of survival hardware made public domain as civic duty by publishing; design and building of trendsetting car; design of Aerojet's first toroidal solid rocket, 1960.

INGBER, DONALD ELLIOT, pathology and cell biology educator; b. Oceanside, N.Y., May 1, 1956; s. David and Helen Edith (Horowitz) I.; m. Ellen Dolnansky; 1 child, Mark Jacob. BA, MA, Yale U., 1977, MPhil, 1981, MD, PhD, 1984. Postdoctoral fellow Harvard Med. Sch., Children's Hosp., Boston, 1984-86; instr. pathology Harvard Med. Sch., Brigham and Women's Hosp., Boston, 1986-88, asst. prof. pathology, 1988-92; assoc. prof. pathology Harvard Med. Sch., Children's Hosp., Boston, 1992—; cons. in biotechnology Neomorphics, Boston, 1988-92, Biosyn R, Balt., 1987-88, Digene Diagnostics, Balt., 1990, Collaborative Rsch., Bedford, 1991—, Advanced Tissue Sciences, La Jolla, Calif., 1992—. Contbr. articles to profl. jours., chpts. to books; writer TV scripts; artist cartoons for postcards; assoc. producer TV spl. R.C. Bates Travelling fellow, 1976; recipient Johnson & Johnson Rsch. Fund award, 1990, Whitaker Health Scis. Fund award, 1990, Am. Cancer Soc. Faculty Rsch. award, 1991. Mem. Am. Soc. Cell Biology, Tissue Culture Soc., Am. Soc. for Space and Gravitational Biology. Achievements include patents for angiogenesis inhibitors; 10 patents pending; development of concept of a "Tensegrity Model" for cell and tissue architecture; discovery of a new class of anti-cancer drugs (Angioinhibins) which inhibit blood vessel growth, including TNP-470; demonstration of importance of extracellular matrix as a mechanochemical regulator in morphogenesis. Office: Children's Hosp 300 Longwood Ave Boston MA 02115-5737

INGBER, MARC STUART, applied mechanics educator; b. Ann Arbor, Mich., Apr. 20, 1950; s. George and Shirlee (Teitel) I.; m. Jeanine Annette Huckell, Jan. 11, 1979; children: Hillary Ann, Allison Brook. BS in Engring., Tulane U., 1972; PhD, U. Mich., 1984. Rsch. engr. The Boeing Co., Seattle, 1977-78; asst. prof. Iowa State U., Ames, 1984-88; asst. prof. U. N.Mex., Albuquerque, 1988-91, assoc. prof., 1991—; cons. Sandia Nation Labs., Albuquerque, 1987—, Los Alamos (N.Mex.) Nat. Labs., 1989—. Editor: Boundary Element Technology VII, 1992; contbr. articles to profl. jours. Mem. Am. Acad. Mechanics, Am. Soc. Engring. Edn., Internat. Soc. Boundary Elements (internat. scientific com. 1992), Soc. Rheology, Tau Beta Pi, Sigma Xi. Achievements include development of munerical computer codes: analysis of limited and super cavitation on axisymmetric bodies at angle of attack, the hydrodynamic interactions among submerged particles; determination of criterion for selective withdrawal. Office: U NMex Dept Mechanical Engring Albuquerque NM 87131

INGELS, JACK EDWARD, horticulture educator; b. Indpls., Mar. 28, 1942; s. Carl Eugene and Mary Louise (Fultz) I. BS, Purdue U., 1964; MS, Rutgers U., 1966; postgrad., Ball State U., 1968-70. Rsch. asst. Rutgers U., New Brunswick, N.J., 1964-66; prof. SUNY, Cobleskill, 1966-89, disting. teaching prof., 1990—; hort. cons. J.C. Penney Corp., N.Y.C., 1966-69; landscape designer, 1966—; hort. and/or landscape cons. numerous small cos., 1970—; pres. J. Ingels Assoc., 1991—. Author: Landscaping: Principles and Practices, 4th edit., 1991, Ornamental Horticulture: Principles and Practices, 2d edit., 1993. Chmn. Cobleskill Restoration and Devel., Inc., 1991—, bd. dirs., 1988—; mem. Schoharie County Coun. on Arts, Cobleskill, Albany Inst. of History and Art. Mem. Associated Landscape Contractors Am., Northeastern N.Y. Nursery Assn., Genesee-Finger Lakes Nursery Assn., Univ. Club (Albany, N.Y.), Moose, Elks. Avocations: gourmet cooking, landscape garden history, travel. Home: Jay Ridge Apts Cobleskill NY 12043 Office: SUNY Cobleskill NY 12043

INGERMAN, PETER ZILAHY, infosystems consultant; b. N.Y.C., Dec. 9, 1934; s. Charles Stryker and Ernestine (Leigh) I.; AB, U. Pa., 1958, MS in Elec. Engring., 1963; PhD, Greenwich U., 1991; m. Carol Mary Pasquale, Dec. 19, 1970 (div. May 1980). CLU; cert. data processor, computer programmer, systems profl., emergency med. technician. Rsch. investigator U. Pa., Phila., 1958-63; tech. dir. programming research, Westinghouse, Balt., 1963-65; mgr. RCA, Cherry Hill, N.J., 1965-71, staff 1971-72; sr. staff cons., Equitable Life Assurance Soc. of U.S., N.Y.C., 1972-77; ind. cons., 1977—; adj. prof. computer sci. Pratt Inst. Tech., 1968-73; mem. working groups Internat. Fedn. Info. Processing, 1962—; rep. Conf. Data Systems Langs., 1967-71, Am. Nat. Standards Inst., 1960-69. Bd. dirs. Phila. Health Plan, Inc., 1975-77, Crossroads Runaway Program, Inc., 1981-82, Willingboro Emergency Squad, 1986-89, Compliance, Inc., 1989—, Providence House, 1991—, vice chair, 1993—. Fellow Brit. Computer Soc.; mem. IEEE (Sr.), AAAS, Assn. Computing Machinery, N.J. Acad. Scis., Data Processing Mgmt. Assn., Mensa, Am. Cryptogram Assn., Brit. Engring. Coun. (chartered engr.), Am. Guild Organists, Triple Nine Soc., Sigma Xi (life), Upsilon Pi Epsilon. Author: A Syntax-Oriented Translator, 1966, Russian transl., 1969; contbr. papers to publs.; patentee electronic circuits. Office: 40 Needlepoint Ln Willingboro NJ 08046-1997

INGHAM, DAVID R., physicist; b. N.Y.C., May 12, 1942; s. Harrington V. and Ruth E. (Triggs) I.; m. Dara M. St. George, Feb. 6, 1971, divorced; children: Laurence, Anne. BA, UCSB, 1965, PhD, 1971. Staff scientist Inst. for Kernphysik, Jülich, Germany, 1971-74, Phys. Inst. U. Heidelberg, Germany, 1974-76; rsch. asst. U. Minn., Mpls., 1976-77; mem. rsch. staff MIT, Middleton, Mass., 1977-81; sr. scientist Tech. for Communications Internat., Mountain View, Calif., 1981-85; cons. Palo Alto, Calif., 1985-86; rsch. specialist Lockheed Missiles and Space Ctr., Sunnyvale, Calif., 1985-. Contbr. articles to profl. jours. Mem. IEEE, Am. Phys. Soc., Applied Computational Electromagnetics Soc. Democrat. Achievements include development of a method for computation of diffraction by trailing edges. Home: 486 James Rd #9 Palo Alto CA 94306

INGHAM, ELAINE RUTH, ecology educator; b. St. Paul, June 26, 1952; d. Clarence Mortimer and Ruth Rowena (Sweet) Stowe; m. Russell Elliot Ingham, May 31, 1975; children: Jennifer Lynna, Scott Christopher. BA in Biology and Chemistry cum laude, St. Olaf Coll., 1974; MS in Microbiology, Tex. A&M U., 1977; PhD in Microbiology, Colo. State U., 1981. Grad. teaching asst. Tex. A&M U., College Station, 1976-77; grad. rsch. asst. Colo. State U., Ft. Collins, 1977-81, postdoctoral fellow Natural Resource Ecology Lab., 1981-85; rsch. assoc. Inst. Ecology U. Ga., Athens, 1985-86; rsch. assoc. dept. botany and plant pathology Oreg. State U., Corvallis, 1986-89, asst. prof. dept. botany and plant pathology, 1989-93; assoc. prof. dept botany and plant pathology, 1993—; cons. EPA, Corvallis, 1986; bd. dirs. Restoration, Conservation Biology Coop. Project, Corvallis, 1990—, dir. Soil Microbial Biomass Svc., 1992—; assoc. editor Applied Soil Ecology, 1992—; reviewer numerous jours and NSF panels; presenter numerous orgns. and univs. including Towson State U., Swedish U., Colo. State U.,U. Ga., U. Nebr., Oreg. State U., EPA, Ecol. Soc. Am., AAAS. Contbr. articles to Can. Jour. Microbiology, Applied Environ. Microbiology, Jour. Protozoology, Soil Biology and Biochemistry, New Phytologist, Plant and Soil, Microbial Ecology, Ecol. Monographs, Crop Protection, Can. Jour. Soil Sci., Biol. Fertility Soil, Jour. Applied Ecology, Biogeochemistry, Ecology, Proc. Royal Soc., Agrl. Ecosystem and Environ., Pedobiologia, Microorganisms, Plants and Herbivores and Methods in Soil Ecology. Treas. Gymnastics Fantastics, Corvallis, 1990; bd. sec. Heart of Valley Children's Choir, 1992-94. Grantee NSF, 1974, 84-85, 86, 88-91, 89-90, Nat. Agrl. Pesticide Impact Assessment Program, 1989-93, 94-95, EPA, 1989-90, 92-94, U.S. Forest Svc., 1990—. Mem. Ecol. Soc. Am. (chair Western chpt. 1991-93, vice chair soil ecology sect. 1993—), Soil Ecology Soc. (pres. 1989-92, editor newsletter 1992-94), Internat. Symposium on Soil Ecology (pres., host internat. meetings 1991), Sigma Xi, Phi Sigma. Republican. Home: 2864 NW Monterey Pl Corvallis OR 97330-3436 Office: Oreg State Univ Dept Botany/ Plant Pathology Cordley Hall 2082 Corvallis OR 97331-2902

INGLE, JAMES CHESNEY, JR., geology educator; b. Los Angeles, Nov. 6, 1935; s. James Chesney and Florence Adelaide (Geldart) I.; m. Fredricka Ann Bornholdt, June 14, 1958; 1 child, Douglas James. B.S. in Geology, U. So. Calif., 1959, M.S. in Geology, 1962, Ph.D. in Geology, 1966. Registered geologist, Calif. Research assoc. Univ. So. Calif., 1961-65; vis. scholar Tohoku U., Sendai, Japan, 1966-67; asst., assoc. to full prof. Stanford U., Calif., 1968—; W.M. Keck prof. earth scis. Stanford U., 1984—, chmn. dept. geology, 1982-86; co-chief scientist Leg 31 Deep Sea Drilling Project, 1973, co-chief scientist Leg 128 Ocean Drilling Program, 1989; geologist U.S. Geol. Survey W.A.E, 1978-81. Author: Movement of Beach Sand, 1966; contbr. articles to profl. jours. Recipient W.A. Tarr award Sigma Gamma Epsilon, 1958; named Disting. lectr. Am. Assn. Petroleum Geologists, 1986-87, Joint Oceanographic Institutions, 1991; A.I. Leverson award Am. Assn. Petroleum Geologists, 1988. Fellow Geol. Soc. Am., Calif. Acad. of Scis.; mem. Cushman Found. (bd. dirs. 1984-91), Soc. Profl. Paleontol. and Mineralogists (Pacific sect. 1958—, prs. elect. 1993), Am. Geophys. Union.

INGLEHART, LORRETTA JEANNETTE, physicist; b. Cleve., July 14, 1947; d. Trevor Gordon and Lois Jeannetta (Kolterman) Kopp. BS, Wayne State U., 1979, MS, 1982, PhD, 1984. Vis. prof. Lab. d'Optique Physique Ecole Supervieure de Physique ef de Chimie, Paris, 1984, 86, 89; assoc. research scientist Johns Hopskins U., Balt., 1985-87, asst. prof. materials sci., 1987—; professorial lectr. physics Am. Univ., Washington, 1986-87; guest scientist Nat. Bur. Standards, Gaithersburg, Md., 1985—. Contbr. articles on thermal wave NDE to profl. jours., 1980—. Fulbright scholar, France, 1986; rsch. grantee Bendix Corp., 1979-81. Mem. AAAS, Am. Phys. Soc., Optical Soc., Sigma Xi. Office: Johns Hopkins U 34 and Charles St Baltimore MD 21218

INGOLD, KEITH USHERWOOD, chemist, educator; b. Leeds, Eng., May 31, 1929; s. Christopher Kelk and Edith (Usherwood) I.; m. Carmen Cairine Hodgkin, Apr. 7, 1956; children: Christopher Frank (dec.), John Hilary, Diana Hilda. B.Sc. with honors in Chemistry, Univ. Coll., London, 1949; D.Phil., Oxford (Eng.) U., 1951; D.Sc. (hon.), U. Guelph, 1985; LLD (hon.), Mt. Allison U., 1987; DSc (hon.), St. Andrews U., Scotland, 1989, Carleton U., 1992. Postdoctoral fellow NRC Can., 1951-53, rsch. officer, 1955-77, assoc. dir. chemistry, 1977-90; adj. prof. U. Guelph, Ont., Can., 1985-87, Brunel U., U.K., 1983—, Carleton U. Ottawa, Can., 1991—; postdoctoral fellow U. B.C., 1953-55; vis. scientist Chevron Research Co., Richmond, Calif., 1966, Univ. Coll., London, 1969, 72, Ford Motor Co., 1971, Esso Research and Engring. Co., Linden, N.J., 1973, U. Western Ont., 1975, 1993, Iowa State U., 1975, U. Bologna, Italy, 1975 (Mangini prize 1990), U. Adelaide, Australia, 1979, U. Grenoble, France, 1983, Australian Nat. U., 1987, U. Freiburg, Germany, 1990, U. Essen, Germany, 1990, U. Dusseldorf, Germany, 1991, U. Leiden, The Netherlands, 1992. Recipient Can. Silver Jubilee medal, 1977, Humboldt Sr. Rsch. Fellowship award, Germany, 1989, Veris award, 1989, Lansdown Visitor award U. Victoria, B.C., 1990, Izaak Walton Killam Meml. prize Can. Coun., 1992; Carnegie fellow U. St. Andrews, Scotland, 1977; vis. fellow Japan Soc. for Promotion Sci., 1982, Italian Nat. Rsch. Coun., 1983; Nat. Sci. Coun. Republic China lectr., 1992. Fellow Royal Soc. Can. (treas. 1979-81, Centennial medal 1982, Henry Marshall Tory medal 1985), Royal Soc. (London, Davy medal 1990), Chem. Inst. Can. (medal 1981, Syntex award for phys. organic chemistry 1983), Univ. Coll. (London); mem. Am. Chem. Soc. (award petroleum chemistry 1968, Pauling award 1988, Arthur C. Cope scholar 1992, James Flack Norris award phys. organic chem 1993), Chem. Soc. (award kinetics and mechanism 1978) (London), Can. Soc. Chem. (v.p. 1985-87, pres. 1987-88, Alfred Bader award in organic chemistry 1989), Royal Soc. Chemistry (Ingold lectr. 1990). Achievements include research papers in free radical chemistry. Home: 72 Ryeburn Dr, Gloucester, ON Canada K1G 3N3 Office: Nat Rsch Coun of Can, Steacie Inst for Molecular Scis, Ottawa, ON Canada K1A 0R6

INGRAM, WILLIAM THOMAS, III, mathematics educator; b. McKenzie, Tenn., Nov. 26, 1937; s. William Thomas and Virginia (Howell) I.; m. Barbara Lee Gordon, June 6, 1958; children: William Robert, Kathie Ann, Mark Thomas. BA, Bethel Coll., 1959; MS, La. State U., 1961; Ph.D., Auburn U., 1964. Instr. Auburn U., Ala., 1961-63; instr. math. U. Houston, 1964-65, asst. prof., 1965-68, assoc. prof., 1968-75, prof., 1975-89; prof., chmn. U. Mo.-Rolla, 1989—. Contbr. articles to profl. jours. Mem. Am. Math. Soc., Math. Assn. Am. Presbyterian. Avocation: photography. Home: 826 Oak Knoll Rolla MO 65401 Office: U Mo-Rolla Dept Math and Statistics Inst Applied Math Rolla MO 65401

INGRODY, PAMELA THERESA, mechanical engineer; b. Detroit, Mar. 7, 1962; d. Gerald Leroy and Anna Jo (Yann) Hale; m. Ronald Gerard Ingrody, June 23, 1984; 1 child, Eric Hale. BS in Mech. Engring., U. Mich., Dearborn, 1984, MS, 1986. Registered profl. engr. Mich. Rsch. asst. U. Mich., Dearborn, 1983-84; student engr. Detroit Edison Co., 1982-85, asst. engr., 1985-89, assoc. engr., engr., 1989-91, account exec., 1991—, sr. mktg. engr., 1992—, supr. customer info. tech., 1993—. Judge Engring. Soc. Sci. Fair, Detroit, 1992; tchr. Our Lady of Sorrow, Farmington, Mich., 1988, Jr. Achievement, Bloomfield, Mich., 1988. Merit scholar Inst. Nuclear Power Ops., 1981-84. Mem. ASME, Assn. Energy Engrs., Women's Econ. Club Detroit. Roman Catholic. Home: 35101 Oakland Farmington MI 48335 Office: Detroit Edison Co 2000 2d Ave Detroit MI 48226

INMAN, CULLEN LANGDON, telecommunications scientist; b. N.Y.C., June 24, 1933; s. Claude Colbert and Myra Eugenia (Langdon) I.; m. Patricia Anne McDonough, Dec. 23, 1965; children: Cathleen Elaine, Elisabeth Myra. AB in Math., Univ. Calif., 1956; PhD in Physics, N.Y.U., 1965. Rsch. assoc. Goddard Inst. Space Studies, N.Y.C., 1965-66; group leader systems and analysis Am. Inst. Physics, N.Y.C., 1967-72; cons., 1972-74; programmer analyst N.Y. Telephone/Bell Labs., N.Y.C., 1974-89; mem. tech. staff Nynex Corp. Contbr. articles to profl. jours. Mem. Am. Physical Soc., Am. Math. Soc., Assn. Computing Machinery. Achievements include research of first calculation of binding energy of a neutron star, calculation of weak induced magnetic moment of the muon's neutrino.

INMAN, DANIEL JOHN, mechanical engineer, educator; b. Shawano, Wis., May 10, 1947; s. Glen and Wilma (Sidebotham) I.; m. Catherine Little, Sept. 18, 1982; children: Jennifer W., Angela W., Daniel J. BS, Grand Valley State, Allendale, Mich., 1970; MAT, Mich. State U., 1975, PhD, 1980. Instr. physics Grand Rapids (Mich.) Ednl. Park, 1970-76; technical staff Bell Labs., Whippany, N.J., 1978; rsch. asst. Mich. State U., East Lansing, 1976-79, 79-80, instr., 1978-79; prof. SUNY, Buffalo, 1980; chmn. U. Buffalo, 1989-92; Samuel Herrick prof. engring., sci., mechanics Va. Tech., 1992—; dir. Mech. Systems Lab., Buffalo, 1984; adj. prof. Brown U., Providence, 1986; cons. Kistler Instrument Corp., Amherst, N.Y., 1985. Author: Vibration: Control Stability and Measurement, 1988, 90, Eng Vibration, 1994; assoc. editor ASME Vibration, Acoustics, Stress and Reliability in Design, 1984—, ASME Jour. Applied Mechanics, SEM Jour. of Theoretical and Exptl. Modal Analysis, 1986, Mechanics of Structures and Machines, Jour. Intelligent Material Systems and Structures, 1992—; tech. editor ASME Jour. Vibration and Acoustics. Presdl. Young Investigator NSF, 1984-89. Fellow ASME (chair Buffalo sect. 1986-87), AIAA (fellow); mem. IEEE (Control Systems Soc.), Am. Acad. Mechanics, Soc. Indsl. and Applied Math., Am. Helicopter Soc., Soc. Engring. Sci. (bd. dirs. 1992—). Home: 3545 Deer Run Rd Blacksburg VA 24060 Office: Va Tech Inst Dept Engring Sci and Mechanics Blacksburg VA 24061-0219

INMAN, JAMES CARLTON, JR., psychological counselor; b. Bossier City, La., Nov. 10, 1945; s. James Carlton and Laura Evelyn (Jones) I.; m. Linda Leigh Dugas, Jan. 20, 1968; children: Amy Denise, John David. BS, McNeese State U., Lake Charles, La., 1969; BA, McNeese State U., 1981, MA, 1981. Lic. profl. counselor, La., rehab. counselor, La.; cert. rehab. counselor, rehab. therapist; bd. cert. pain practitioner; diplomate Am. Bd. Profl. Disability Cons. Master interviewer La. Dept. Labor, Lake Charles, 1976-83; master counselor La. Dept. Labor, 1983-87; voc. cons. Crawford Health & Rehab., Lake Charles, 1987-89; pvt. practice Lake Charles, 1989—; rehab. counselor La. div. Rehab. Svcs., Lake Charles, 1989-90; voc. expert Social Security Adminstrn., Alexandria, La., 1989—. With USAF, 1969-75. Fellow Am. Psychol. Soc. (charter), Am. Acad. Pain Mgmt., Royal Anthropol Inst. Gt. Brit., Soc. Antiquaries of Scotland, Royal Soc. of Antiquaries of Ireland, Internat. Biographical Assn.; mem. AACD, Am. Congress Rehab. Medicine, Am. Orthopsychiat. Assn., Nat. Rehab. Assn., N.Y. Acad. Scis., Internat. legion of Intelligence, Internat. Soc. for Philos. Enquiry, Mensa. Republican. Roman Catholic. Avocations: reading, stamp collecting, horticulture, painting. Home and Office: 1908 23d St Lake Charles LA 70601-6565

INOUE, MICHAEL SHIGERU, industrial engineer, electrical engineer; b. Tokyo, June 27, 1936; came to U.S., 1956; s. Takajiro and Kazu (Morimoto) I.; m. Mary Louise Shuhart, Sept. 23, 1965; children: Stephen M., Rosanne E., Marcus S., Joanne K., Suzanne T. BSEE magna cum laude, U. Dayton, 1956; MS, Oreg. State U., 1963, PhD, 1967. Registered profl. engr., Oreg., Calif.; cert. data processor. Sr. rsch. engr. Black and Decker Mfg. Co., Towson, Md., 1960-62; prof. Oreg. State U., Corvallis, 1966-82; v.p. Kyocera Internat., Inc., San Diego, 1982—. Co-author: Introduction to Operation Research & Management Science, 1975, Circulo de Qualidad, 1982, Pacific Saury, 1971. Recipient Grad. Rsch. award IBM, 1965. Mem. Inst. Indsl. Engrs. (sr. mem., IE of Yr. award 1976), Am. Cer. Soc., Inst. Mgmt. Scis., Tau Beta Pi, Alpha Pi Mu, Sigma Xi. Republican. Roman Catholic. Home: 5154 Via Playa Los Santos San Diego CA 92124-1555 Office: Kyocera International Inc 8611 Balboa Ave San Diego CA 92123-1580

INOUÉ, SHINYA, microscopy and cell biology scientist, educator; b. London, Eng., Jan. 5, 1921; came to U.S, 1948, naturalized, 1989; s. Kojiro and Hideko (Yano) I.; m. Sylvia McCandless, July 18, 1952; children: Heather C., Jonathan H., Christopher W., Stephen K., Theodore D. Rigakushi, Tokyo U., 1941; MA, Princeton U., 1950, PhD, 1951; MA (hon.), Dartmouth Coll., 1959, U. Pa., 1966. Instr. U. Wash. Med. Sch., Seattle, 1951-53; asst. prof. Tokyo Met. U., 1953-54; rsch. assoc., assoc. prof. U. Rochester, N.Y., 1954-59; instr. Marine Biol. Lab., Woods Hole, Mass., 1961—, NATO Summer Schs., Cannes, Stressa, Szeged, 1967, 70, 75; prof., chmn. Dartmouth Med. Sch., Hanover, N.H., 1959-66; prof. U. Pa., Phila., 1966-89; sr. Disting. Scientist Marine Biol. Lab., Woods Hole, 1980—; cons. Am. Optical Co., 1954-60, NSF, 1962-65, NIH, 1965-70, Hamamatsu Photonics K.K., Hamamatsu City, Japan, 1988—; pres. Universal Imaging Corp., Falmouth, Mass., West Chester, Pa., 1984-87, chmn. bd. dirs., 1987-93. Author: Video Microscopy, 1986; co-editor: Molecules and Cell Movement, 1975; contbr. articles to profl. jours.; mem. editorial bd. several sci. jours., 1964—; ad hoc reviewer, advisor on sci. and tech. NSF, NIH, many univs., founds.; patentee in optics. Trustee Marine Biol. Lab., 1970-77, 81-85, 92. Recipient Rosenstiel award Brandeis U., 1988, Brown-Hazen award State of N.Y., 1988; Guggenheim Found. fellow, 1971-72; cancer rsch. scholar Am. Cancer Soc., N.Y.C., 1955-58. Fellow AAAS, Am. Acad. Arts Scis., Royal Microscopical Soc. (hon.); mem. NAS, Biophys. Soc. (coun. 1968-71), Soc. Gen. Physiologists (coun., pres. 1962-65, 69-70), Am. Soc. Cell Biology (coun. 1970-73, E.B. Wilson award 1992). Avocations: reading, photography. Home: 40 Shore St Falmouth MA 02540-3146 Office: Marine Biol Lab Water St Woods Hole MA 02543-1024

INOUÉ, TAKAO, logician, philosopher; b. Motegi, Tochigi, Japan, May 28, 1957; s. Hideo and Ikuko (Mogaki) I. BSc, Sci. U. Tokyo, 1983, M Tech., 1985; Propaedeuse Math., U. Utrecht, The Netherlands, 1987. Contbr. articles, revs. and abstracts to profl. jours. Scholar Dutch Ministry Edn. and Sci., 1989. Mem. Am. Math. Soc., Assn. for Symbolic Logic, Philosophy of Sci. Soc. Japan, Math. Soc. Japan, Japanese Assn. for Philosophy of Sci., History of Sci. Soc. Japan, Dutch Math. Soc., The Mind Assn. Home: Ina Boudier-Bakkerlaan, 117-II, 3582 XP Utrecht The Netherlands

INOUE, TAKESHI, psychiatrist; b. Kumamoto, Japan, Sept. 3. 1932; s. Tatsuichi and Shigeye (Matsushita) I.; M.D., Kumamoto U., 1957, Ph.D., 1962; m. Michiko Takeshita, Oct. 8, 1961; children: Hiroko, Takao. Lectr. dept. biochemistry Kumamoto U. Med. Sch., 1957-61; lectr. U. Calif. Med. Sch., San Francisco, 1961-73, postdoctoral fellow, 1968; postdoctoral fellow dept. biochemistry Temple U. Med. Sch., Phila., 1968-70; asst. prof. dept. neuropsychiatry Kumamoto U. Med. Sch. (Japan), 1971-83; dir. Minamata Hosp., 1983-85, dir., 1985—. Mem. AAAS, Japanese Neurochem. Soc., N.Y. Acad. Scis., Sigma Xi. Home: 16-15 Musashigaoka 2-Chohome, Kumamoto 862, Japan Office: 4051 Hama, Minamata-shi 867, Japan

INOUE, YOSHIO, engineering educator; b. Tokyo, Feb. 13, 1929; s. Kishiro and Toyo (Watanabe) I.; m. Mitue Nagatani, Apr. 30, 1960; 1 child, Hirotaka. M.Engring., Waseda U., Tokyo, 1957; Dr.Engring., Waseda U., 1978. Instr. engring. Kanto Gakuin U., Yokohama, Japan, 1962-65, asst. prof. engring., 1965-70, prof. engring., 1970—, dir. Inst. Archtl. Environ. Engring., chief prof. Grad. Sch., 1991-93. Author: Dakutokei-no-Sekkei, Vol. I, 1991, Vol. II, 1993. Mem. ASHRAE, Soc. Heating, Air Conditioning and San. Engrs. Japan. Office: Kanto-Gakuin Univ, 4834 Mutsuura-cho Kanazawa-Ku, Yokohama Japan 236

INOUYE, DAVID WILLIAM, zoology educator; b. Phila., Jan. 7, 1950; s. William Yoshio and Eleanor (Ward) I.; m. Bonnie Ann Gregory, May 31, 1969; children—Brian, Kevin. B.A., Swarthmore Coll., 1971; Ph.D., U. N.C., 1976. Asst. prof. Dept. Zoology Univ. Md., College Park, 1976-82, assoc. prof., 1982—; treas. Rocky Mountain Biol. Lab, Crested Butte, Colo., 1984-87. Bd. dirs. ACLU, Prince George's County, Md., 1984-87. NSF grantee, 1976-85. Democrat. Quaker.

INTRILIGATOR, DEVRIE SHAPIRO, physicist; b. N.Y.C.; d. Carl and Lillian Shapiro; m. Michael Intriligator; children: Kenneth, James, William, Robert. BS in Physics, MIT, 1962, MS, 1964; PhD in Planetary and Space Physics, UCLA, 1967. NRC-NASA rsch. assoc. NASA, Ames, Calif., 1967-69; rsch. fellow in physics Calif. Inst. Tech., Pasadena, 1969-72, vis. assoc., 1972-73; asst. prof. U. So. Calif., 1972-80; mem. Space Scis. Ctr., 1978-83; sr. rsch. physicist Carmel Rsch. Ctr., Santa Monica, Calif., 1979—; dir. Space Plasma Lab., 1980—; cons. NASA, NOAA, jet Propulsion Lab; chmn. NAS-NRC com. on solar-terrestrial rsch., 1983-86, exec. com. bd. atmospheric sci. and climate, 1983-86, geophysics rsch. bd., 1983-86, geophysics study com., 1983-86; U.S. nat. rep. Sci. Com. on Solar-Terrestrial Physics, 1979-81; mem. adv. com. NSF Div. Atmospheric Sci. Contbr. articles to profl. jours. Recipient 3 Achievement awards NASA, Calif. Resolution of Commendation, 1982. Mem. NAS-NRC, Am. Phys. Soc., Am. Geophys. Union, Cosmos Club. Achievements include being a participant Pioneer 10/11 missions to outer planets; pioneer Venus Orbiter, 6, 7, 8 and 9 heliocentric missions. Home: 140 Foxtail Dr Santa Monica CA 90402-2048 Office: Carmel Rsch Ctr PO Box 1732 Santa Monica CA 90406-1732

INYANG, HILARY INYANG, geoenvironmental engineer, researcher; b. Uyo, Nigeria, Nov. 8, 1959; came to U.S., 1982; s. Inyang Amos and Abigail (Affiong) I.; m. Robin Ann Tenney, July 2, 1988; 1 child, Imikan Hilary. BS in Geology with honors, U. Calabar, Nigeria, 1981; BS, N.D. State U., 1985, MS, 1986; PhD, Iowa State U., 1989. Soil technician Midwest Soil Testing Lab., Fargo, N.D., 1984; program asst. FHWA Tech. Transfer Ctr., Fargo, N.D., 1985-86; rsch. asst. Spangler Geotechnical Lab., Ames, Iowa, 1986-88; resident researcher Wis. Dept. of Transp., Madison, summer 1989; vis. asst. prof. Purdue U., West Lafayette, Ind., summer 1990; asst. prof. U. Wis., Platteville, 1988—; sr. geoenvironmental engr. U.S. EPA, Washington, 1991—. Mem. editorial bd. Internat. Jour. Surface Mining and Reclamation, Rotterdam, The Netherlands, 1991—, Rsch. Conservation and Recycling, Amsterdam, The Netherlands, 1992—, Jour. Environ. Systems, N.Y.C., 1992—, Internat. Jour. Pub. Works, Washington, 1991—, Internat. Jour. Environ. Issues, Rotterdam, 1992—, CRC Reviews Environ. Control, Fla., 1992—; contbr. articles to profl. jours. Team mem. Dubuque (Iowa) Tennis Club, 1989-91; chmn. engring. minority com. U. Wis., Platteville, 1990-91; affiliate mem. United Meth. Ch., Falls Church, Va., 1992—. Recipient scholarship U.S. Dept. or Energy, 1986-88, scholarship Fed. Govt. of Nigeria, 1983-85, Geology award Shell-British Petroleum Co., 1978. Fellow AAAS (EPA Environ. Sci. and Egnring.), Geol. Soc. London; mem. ASCE (co-editor publ. 1992, Essay 1st prize 1984), ASTM, Transp. Rsch. Bd., Geol. Soc. Am., World Rock Boring Assn., Internat. Soc. Rock Mech., Internat. Assn. Engring. Geol. Achievements include development of techniques for protection of foundation systems in contaminated land, new device for impact strength measurements on rocks. Home: PO Box 2106 Falls Church VA 22042 Office: US EPA OS341 401 M St SW Washington DC 20460

IODICE, ELAINE, software engineer; b. Cambridge, Mass., Oct. 16, 1947; d. Arthur Peter and Mary Elizabeth (Stefanelli) I.; m. Ian Gordon Prittie, July 27, 1980. BFA, Boston Conservatory Music, 1969; BS, U. Victoria, B.C., Can., 1982. Programmer Software Products Internat., San Diego, 1982-83; mem. tech. staff Hughes Aircraft Co., San Diego, 1983-85, 86-87; cons. Reading, Eng., 1985; sr. software engr. Automated Systems Inc. (name now Cadence Design Systems), San Diego, 1987-92, Mentor Graphics, San Jose, Calif., 1992-93; assoc. engr. Cooper & Chayan Tech. Inc., Cupertino, Calif., 1993—. Contbr. articles to profl. publs. Mem. Assn. for Computing Machinery, Am. Assn. Artificial Intelligence. Office: Cooper & Chyan Tech 1601 Sarratoga-Sunnyvale Rd # 255 Cupertino CA 95014

IONESCU TULCEA, CASSIUS, research mathematician, educator; b. Bucharest, Rumania, Oct. 14, 1923; naturalized, 1967; s. Ioan and Ana (Caselli) Ionescu T. M.S., U. Bucarest, 1946; Ph.D., Yale, 1959. Mem. faculty U. Bucarest, 1946-57, assoc. prof., 1952-57; research assoc. Yale U., 1957-59, vis. lectr., 1959-61; assoc. prof. U. Pa., 1961-64; prof. U. Ill., Urbana, 1964-66, Northwestern U., 1966—. Author: Hilbert Spaces (in Rumanian), 1956, A Book on Casino Craps, 1980, A Book on Casino Blackjack, 1982; co-author: Probability Calculus (in Rumanien), 1956, Calculus, 1968, An Introduction to Calculus, 1969, Honors Calculus, 1970, Topics in the Theory of Liftings, 1969, Sets, 1971, Topology, 1971, A Book on Casino Gambling, 1976. Recipient Asachi prize Rumanian Acad., 1957. Office: Northwestern U Lunt Bldg Evanston IL 60208

IOVANNISCI, DAVID MARK, biochemist researcher; b. Syracuse, N.Y.; s. Thomas Anthony and Ellen Louise Iovannisci. BA in Biochemistry, Canisius Coll., 1973; MS in Microbiology, U. Ky., 1980, PhD in Biochemistry, 1984. Clin. microbiologist, emergency lab. technician Deaconess Hosp., Buffalo, 1974-78; lab. technician Cen. Ky. Blood Ctr., Lexington, 1978; postdoctoral fellow Harvard Med. Sch., Boston, 1984-87; assoc. scientist Applied Biosystems, Foster City, Calif., 1987-91, project leader DNA diagnostics, 1991—. Author: (with others) Proceedings Vth International Yeast Symposium, London, Ontario, 20-25 July 1980, Purine Metabolism in Man-IV. Part A., 1984, Biology of Parasitism, 1988; contbr. articles to profl. jours. including Jour. Parasitol., Jour. Biol. Chemistry, Molecular Cell Biology, Mol. Biochem. Parasitol., Mol. and Biochem. Parasitol., Clin. Chemistry; author abstracts. Postdoctoral fellowship Nat. Inst. Allergy and Infectious Diseases, NIH, 1987. Mem. AAAS, Am. Soc. for Microbiology, Am. Soc. for Human Genetics. Office: Applied Biosystems 850 Lincoln Centre Dr Foster City CA 94404

IOVINE, CARMINE P., chemicals executive; b. Bklyn., Jan. 16, 1943; s. Felice and Bellovia (Sivillo) I.; m. Elizabeth Ann McKinney, June 19, 1965; children: Matthew C., Anthony P., Peter Michael. BS in Chemistry, Manhattan Coll., 1964; MS in Chemistry, Rutgers U., 1972. Project supr. exploratory rsch. Nat. Starch & Chem. Co., Bridgewater, N.J., 1965-72, rsch. assoc. corp. rsch., 1972-80, v.p., dir. synthetic polymer rsch., 1980-89, corp. v.p. rsch. and devel., 1989—; bd. dirs. N.J. R. & D. Coun., 1991—, also mem. edn. coun.; bd. dirs. Chem. Industry Inst. Tech., Durham, N.C., 1989—. Bd. dirs. Somerset County Bus. and Edn. Partnership, Somerville, N.J., 1990—. Mem. Am. Chem. Soc., N.Y. Acad. Scis. Office: Nat Starch & Chem Co 10 Finderne Ave Bridgewater NJ 08807

IP, JOHN H., cardiologist; b. Hong Kong, Jan. 25, 1960. BS, U. Ill., 1980; MD, U. Chgo., 1985. Diplomate Am. Bd. Internal Medicine, Nat. Bd. Med. Examiners. Intern NYU, resident; fellow Mt. Sinai Med. Ctr., N.Y.C.; dir. electrophysiology sect. Thoracic Cardiovascular Inst., Lansing, Mich.; cons. AMA, 1991; assoc. dir. cardiac electrophysiology Mt. Sinai Hosp., N.Y.C., 1991-92. Editor Cardiovascular Rev. and Report; contbr. articles to profl. jours. Fellow Am. Coll. Cardiology. Office: Thoracic Cardiovascular Inst 405 W Greenlawn St Lansing MI 48910

IQBAL, ZAFAR, physicist; b. Calcutta, Bengal, India, Aug. 24, 1941; s. Abdul and Fatima (Ahmed) Motalib; m. Lolita Navarro, May 23, 1989. BS, U. Dacca, East Pakistan, 1960, MS, 1961; PhD in Physics, U. Cambridge, Eng., 1967. Rsch. fellow U. Warwick, Coventry, Eng., 1967-68; staff physicist Army Rsch. and Devel. Ctr., Dover, N.J., 1969-79; guest rsch. prof. Swiss Fed. Inst. Tech., Zurich, Switzerland, 1979-83; sr. rsch. fellow U. Zurich, 1979; sr. rsch. physicist Allied Signal Corp., Morristown, N.J., 1983-91; rsch. scientist, project leader Allied Signal Corp., Morristown, 1992--; advisor Nat. Rsch. Coun. Postdoctoral Program, Washington, 1973-77; reviewer Nat. Sci. Found., Washington, 1985--; edit. bd. Condensed Matter and Materials Commun, N.Y., 1992--. Editor: Vibrational Spectroscopy of Phase Transitions, 1984, (jour.) Synthetic Metals, 1989; contbr. articles to profl. jours. Recipient Paul A. Sipple medal Army Sci. Conf., 1971, Individual Achievement award Allied Signal, 1991; Alexander von Humboldt fellow, 1978, Rsch grantee Defense Advanced Rsch. Projects Agy., 1991-92. Mem. Am. Phys. Soc., Materials Rsch. Soc. Achievements include high temperature superconductor tech. patents and discovery of new high temperature copper oxide and carbon superconductor, structure of polymer superconductor bromine poly sulphur nitride, process to fabricate nanocrystalline silicon, mechanisms of structural phase transitions in key materials. Office: Allied Signal Inc 101 Columbia Rd Morristown NJ 07962

IQBAL, ZAFAR MOHD, cancer researcher, biochemist, pharmacologist, toxicologist, consultant; b. Hyderabad, India, Dec. 12, 1938; came to U.S., 1965, naturalized, 1973; s. M.A. and Haleemunissa (Begum) Rahim. BSc, Osmania U., 1958, MSc, 1962; PhD, U. Md., 1970. Asst. prof. pharmacology Case Western Res. U., Cleve., 1974-76; assoc. dir. ERC programs in occupational toxicology U. Ill. Med. Ctr., Chgo., 1980-81, assoc. prof. microbiology, 1977-80, assoc. prof. occupational and environ. health, 1976—, assoc. prof. preventive medicine, 1982—; mem. com. radiation biohazards U. Ill., campus rsch. bd., recombinent DNA, 1978—, chmn. com. on rsch. risks from HIV; chmn. Inst. Biosafety Com., 1986—; cons., book, and grant reviewer, lectr. in continuing edn. Contbr. over 50 articles to profl. jours. Sponsor, trainer India-U.S. Exchange Scientists, NSF, 1981, 85-86; spl. advisor RRL (India) Dirs., 1980—. Fogarty Internat. fellow Nat. Cancer Inst./NIH, 1970-71, staff fellow, 1971-74; Coun. Sci. and Indsl. Rsch. of India fellow, 1963-65. Mem. AAAS, Am. Assn. Cancer Rsch., Am. Pancreatic Assn., N.Y. Acad. Scis., Am. Chem. Soc., Soc. Toxicology, Sigma Xi. Office: U Ill PO Box 6998 M/C 922 Chicago IL 60680

IRAVANCHY, SHAWN, aerospace engineer, astrophysicist, consultant; b. Tehran, July 14, 1965; came to U.S., 1983; s. Hamid and Parvin (Kaveh) I. BS in Astrophysics, U. Okla., 1990; MS in Aerospace Engring., Calif. State U., 1993; postgrad., Calif. State U., Long Beach, 1993—, PhD in Aerospace Engr. Rsch. asst. astro & physics U. Okla., Norman, 1985-86; aerospace engr., systems mgr. I.N. Assocs., Los Alamitos, Calif., 1990—. Mem. AIAA, Planetary Soc. Republican. Avocation: pvt. pilot. Home: 5821 Lemon St Cypress CA 90630 Office: IN Assocs 10601 Calle Lee #187 Los Alamitos CA 90720

IRFAN, MUHAMMAD, pathology educator; b. Lahore, Punjab, Pakistan, Apr. 22, 1928; parents Muhammad and Sadar (Begum) Noman; m. Tahira Irfan, Dec. 10, 1970; children: Soma Sadaf, Ali Khurram, Asad Kaleem, Uzma Naheed. B.V.Sc., Punjab. Vet. Coll., Lahore, 1948; PhD, U. of London, 1958; MA, Trinity Coll., Dublin, Ireland, 1963. Cert. in vet. pub. health, Berlin. Lectr. in pathology U. Khartoum, 1959-60; lectr. in pathology Trinity Coll., Dublin, 1960-63, sr. lectr., 1966-68; prin. rsch. officer Ghana Acad. Scis., Accra, 1963-68; prof. of pathology U. of Agrl., Faisalabad, Pakistan, 1968-72, 77-79, 83-88, acting vice chancelor, 1979, dean, 1970-72, 79, 83-88; prin. Coll. Vet. Scis., Lahore, 1972-76, 79-82; animal husbandry commr. Govt. of Pakistan, Islamabad, 1976-77. Author: Curriculum Development, 1988, also textbooks, 1988-93; editor-in-chief Pakistan Vet. Jour., 1982-88; author, editor monographs, sci. papers in field. Vice pres. Agrics Housing Soc., Lahore, 1980-93. Fellow Pakistan Acad. Med. Scis., 1987; recipient Tamgha-e-Pakistan award Govt. of Pakistan, 1971, Golden Star award Nat. Farm Guide Coun., Pakistan, 1987, Cert. of Honor, Pakistan Poultry Assn., 1991. Fellow Royal Coll. Vet. Surgeons. Avocations: tennis, badminton, chess. Home: 84-H Gulberg-III, Lahore Pakistan Office: U Agr, Faculty Vet Scis, Faisalabad Pakistan

IRGON, JOSEPH, physical chemist; b. Polonne, Ukraine, Russia, Dec. 30, 1919; s. Joseph and Edith (Galperin) Irga; m. Thelma Pugach, Apr. 11, 1948; children: Deborah Amadei, Judith Wolf, Adam Irgon. BS in Chem. Engring., Northeastern U., Boston, 1943; PhD in Phys. Chemistry, MIT, 1948. Chief rsch. analyst Reaction Motors Inc., Rockaway, N.J., 1952-56; v.p. Fulton-Irgon Corp., Bernardsville, N.J., 1956-59; rsch. dir. Fulton-Irgon div. Lithium Corp. Am., Denville, N.J., 1959-62; pres. Proteus Inc., Mountain Lakes, N.J., 1962-68; v.p. Ocean Recovery Systems, Inc., Morristown, N.J., 1968-73; dir. Joseph Irgon & Assocs., E. Brunswick, N.J., 1973—; pres. City Resources, Inc., East Brunswick, N.J., 1990—. Editorial bd. Am. Rocket Soc., Washington, 1952-62; author: First Handbook of Rocket Propellants, 1958, First Handbook of Ocean Materials, 1964; contbr. articles to profl. jours. Mem. ASME, Am. Chem. Soc., Marine Tech. Soc. (charter mem.). Achievements include discovery of new principle in chemical catalysts; co-invention of supersonic aircraft pilot escape system, and submarine emergency buoyancy systems; introduction of principle in light refraction, refractive gradient field; research in resource recovery, historical review of human chemical warfare test program at MIT during World War II. Home: 144 Emmans Rd Flanders NJ 07836-9042 Office: Joseph Irgon & Assocs 23 St Georges Rd East Brunswick NJ 08816-4626

IRI, MASAO, mathematical engineering educator; b. Tokyo, Jan. 7, 1933; s. Jin-ichi and Yasumi (Kuga) I.; m. Yumi Mizoo, Mar. 19, 1960; children: Chika, Masato, Yuka. B of Engring., U. Tokyo, 1955, M of Engring., 1957, Dr. of Engring., 1960. Research asst. Kyushu U., Fukuoka, Japan, 1960, asst. prof., 1960-62; assoc. prof. U. Tokyo, 1962-73, prof. math. engring., 1973-93, univ. senator, 1986-87, dean faculty engring., 1987-89, v.p., 1989-91, prof. emeritus, 1993; prof. Chuo U., 1993—. Author: Network Flow, Transportation and Scheduling, 1969; also 18 books in Japanese, 9 books transl. into Japanese; editor-in-chief Jour. Ops. Rsch. Soc. Japan, 1980-82; Asian regional editor Jour. Circuits, Systems and Computers, 1990—, Optimization Methods and Software, 1992—; adv. editor Networks, 1975—, European Jour. Operational Rsch., 1981—, Zeitschrift für Ops. Rsch.; assoc. editor Math. Programming, 1976-88, Math. Programming Series A, 1989—, Internat. Jour. Computational Geometry and Applications, 1990—, Investigación Operativa, 1992—; editorial bd. Discrete Applied Math., 1979—, European Jour. Combinatorics, 1981—, Advances in Applied Math., 1982—, Annals of Ops. Rsch., 1983—, Japan Jour. Indsl. & Applied Math, 1984—, Investigação Operacional, 1985-88, Discrete and Computational Geometry, 1985-91, Applied Math. Letters, 1989—, Computational Optimization and Applications Informatica, 1991—, Yugoslav Jour. Ops. Rsch., 1992—, Numerical Algorithms 1993—; editor monographs in Japanese: Math. Programming, 1980—, New Applied Math., 1973—, Applied Math. series, 1993-94, Computer Sci., 1981-83, Modern Math., 1970—; Handbook of Information Systems, 1989, Encyclopaedia of Mathematical Sciences, 1989. Recipient Matsunaga prize Matsunaga Found., 1965, New Tech. prize Inst. Chem. Engrs. Japan, 1968. Fellow IEEE (sr. mem. 1984—), Ops. Rsch. Soc. Japan (v.p. 1984-85, pres. 1992-94, Toray Sci. and Tech. prize 1993); mem. Math. Programming Soc., Japanese Soc. for Quality Control, Life Support Tech. Soc. (v.p. 1985-87), Math. Soc. Japan, Linguistic Soc. Japan, Inst. Electronics Info. and Communication Engrs. Japan (Paper prize 1969, 76, Achievement prize, 1989, editor-in-chief trans. ser. A 1993-94), Info. Processing Soc. Japan (Paper prize 1981, 88, 90), Soc. Instruments and Control Engrs. Japan, Japan Soc. for Indsl. and Applied Math. (v.p. 1992-93), Engring. Acad. Japan, N.Y. Acad. Sci., Geog. Info. Systems Assn., Internat. Fedn. Operational Rsch. Socs. (v.p. 1983-85), Assn. Asian-Pacific Operational Rsch. Socs. (pres. 1992-94), Sigma Xi. Home: Higashi-komagata 4-11-1-402, Sumida-ku, Tokyo 130, Japan Office: U Tokyo Dept Math Engring, 7-3-1 Hongo, Bunkyo-ku, Tokyo 113, Japan

IRR, JOSEPH DAVID, geneticist; b. Pitts., Sept. 19, 1934; s. Joseph T. and Margaret C. (McDonald) I.; m. Mary C. Edmunds, May 25, 1963; children: Caren, Hans. BS, U. Pitts., 1962; PhD, U. Calif., 1967. Postdoctoral fellow U. Wash., Seattle, 1967-68; assoc. prof. Marquette U., Milw., 1969-74; rsch. fellow Harvard Med. Sch., Boston, 1974-77; prin. scientist E.I. duPont de Nemours Co., Wilmington, Del., 1978-90; sr. mgr. new product planning DuPont Merck Pharm. Co., Wilmington, 1991—; adj. prof. U. Del., Newark, 1981—; mem. sci. com. Am. Indsl. Health Coun., Washington, 1979-83, chair mutagenicity com., 1980-82. Contbr. rsch. papers on methods for mutagen detection, human gene expression, genetic control mechanisms to sci. publs. Chmn. Wauwatosa (Wis.) Dem. party, 1972-74; chmn. Del. March of Dimes, 1981-84. Mem. AAAS, Am. Soc. Microbiology, Genetic Toxicology Assn. (pres. 1982-83. Achievements include 8 patents on cellular therapy, discovery of positively controlled gene expression system, pioneering in field of quantitative statistical methods for analysis of mutation induction in cultured animal and bacterial cells. Office: DuPont Merck Pharm Co Barley Mill Plz Wilmington DE 19880-0025

IRRTHUM, HENRI EMILE, chemicals executive, engineer; b. Luxemberg, Oct. 23, 1947; s. Camille Mathias and Suzanne Marie (Konter) I.; children: Christophe, Thierry. Cert. D'Etudes secondaires, Lycee De Garcons Luxembourg, 1966; ingenieur civil chimiste diplome, Universite De L'etat A Liege, Belgium, 1971. Registered profl. engr., Switzerland. From engr. to sr. process engr. photo products dept. Dupont Luxembourg, 1971-74, from asst. to area supt., 1976-77; tng. assignment asst. area supt. polymer products dept. E.I. DuPont, Parkersburg, W.Va., 1978-79; with tng. assignment and mktg. depts. E.I. DuPont, Wilmington, Del., 1979-80; with tng. assignment and sales dept. E.I. DuPont, Paris, 1980-82; product mgr. polymer products dept. DuPont de Nemours Internat., S.A., Geneva, 1982-84; mgr. high performance films Dupont de Nemours Internat. S.A., Geneva, 1985-86, 87, dir. info. storage electronics dept., 1987-88, dir. info. storage and component materials electronics dept., 1988-91; dir. titanium dioxide ops. chems. dept. E.I. DuPont, Wilmington, 1991—. Mem. Assn. Mfrs. Polyester Film (pres.

1989-91), European Polyester Film Mfrs. Assn. Roman Catholic. Avocations: tennis, hiking, swimming, golf, bicycling. Office: DuPont Chems Brandywine Bldg Rm 14348-4 Wilmington DE 19898

IRUVANTI, PRAN RAO, endocrine biochemist, researcher; b. Kovvur, India, Oct. 2, 1957; came to U.S., 1985; s. Venkata Sayi and Tulasibai (Angara) Iruvanti; m. Shubhada Gobburu, Dec. 14, 1987; 1 child, Sirisha. MSc, Osmania U., 1978; PhD, U. Hyderabad, India, 1984. Rsch. assoc. U. Hyderabad, India, 1984-85; rsch. fellow Med. Coll. Ga., Augusta, 1985-89, asst. rsch. scientist, 1989—. Reviewer European Jour. Cancer and Clin. Oncology, 1984; contbr. articles to Indian Jour. Med. Rsch., Biology of Reproduction, Anatomical Record, Molecular Cell Endocrinology, Molecular Basis of Reproduction. Sec. student housing U. Hyderabad, India, 1983. Jr. and sr. rsch. fellow Indian Coun. Med. Rsch., 1980-84; Biomed. Rsch. Support grantee Med. Coll. Ga., 1990, 91. Mem. AAAS, N.Y. Acad. Scis., Endocrine Soc. USA, Soc. for the Study of Reproduction. Hindu. Achievements include demonstration of effectiveness of contraceptive steroids on glutamate-GABA metabolism in brain; first demonstration of role of RU38486 as contraceptive agent in mammals; development of transformed ovarian granulosa cell line. Address: 8415 Franklin Ave #29 Clive IA 50325

IRVIN, GEORGE WILLIAM, economics educator; b. N.Y.C., Dec. 12, 1940; s. Warren Edward and Margurite (Hominal) I.; m. Jean Edna, May 1, 1976 (div. 1981); m. Maria-Jose Peralta, July 14, 1986; children: Marc, Leonora. BA with honors, Oxford (Eng.) U., 1963; MSc in Econs., London Sch. Econs., 1965; PhD, U. London, 1975. Rsch. fellow Inst. Devel. Studies, U. Sussex, Brighton, Eng., 1966-71; ptnr. Sussex Devel. Cons., Brighton, Eng., 1971-73; sr. lectr. Inst. Social Studies, The Hague, The Netherlands, 1973-83, assoc. prof., 1983—; former cons. internat. orgns. including World Bank, European Commn., Brussels, ILO, Geneva. Author: Roads and Redistribution, 1975, Modoern Cost-Benefit Methods, 1978; editor: Toward an Alternative for Central America, 1983, The Future of Central American Integration, 1989; also articles. Mem. Assn. European Rsch. Caribbean and C.Am. (founder, pres. 1983-85). Avocations: tennis, flying. Office: Inst Social Studies, Badhuisweg 251, 2509 LS The Hague The Netherlands

IRVINE, JOHN HENRY, science analyst; b. Hamburg, Germany, Feb. 28, 1951; s. Henry John Irvine and Retha Irene (Kusch) Mowat. BSc in Social Sci. and Tech., Loughborough U., 1973; MSc in Sci. Policy, U. Sussex (Eng.), 1974. Lectr. Mgmt. Sch. Imperial Coll. U. London, 1976-78; fellow sci. policy rsch. unit U. Sussex, 1978-83, lectr. sci. and tech. policy studies, 1983-86; dir. policy studies Tech. Change Ctr., London, 1986-87; sr. fellow sci. policy rsch. unit U. Sussex, 1987-90; vis. sr. fellow in PRISM group The Wellcome Trust, London, 1993—; dir. Sci. Policy Rsch. Consultants, Hove, Eng., 1990—; cons. Shell Rsch. Ltd., 1988-90, UNESCO, 1989, Australian Prime Minister's Sci. Coun., 1989, Ciba Found., 1989, Danish Found. for Sci. Film and Journalism, 1989, U.K. Nat. Audit Office, 1989, Brit. Petroleum, 1990, European Commn., 1990, Netherlands Ministry Edn. and Sci., 1990-93, UNDP, 1991, World Bank, 1991-93, Swedish NUTEK, 1991, ARA Cons. Group (Canada) Inc., 1992—, South African Found. Rsch. Devel., 1992, Can. Nat. Rsch. Coun., 1993; vis. lectr. U.K. Civil Svc. Coll., 1986—; mem. editorial bds. Sci. and Pub. Policy, Rsch. Policy and Scientometrics. Author: Evaluating Applied Research: Lessons from Japan, 1988; co-author: (with B.R. Martin) Foresight in Science: Picking The Winners, 1984, 2d edit., 1986; Research Foresight: Priority-Setting in Science, 1989; (with B.R. Martin, P.A. Isard) Investing in the Future, 1990; editor: (with I. Miles) The Poverty of Progress: Changing Ways of Life in Industrial Societies, 1982; (with I. Miles, J. Evans) Demystifying Social Statistics, 1979, 2nd edit., 1981, Japanese edit., 1984, Korean edit., 1990; Equipping Science for the 21st Century, 1993; contbr. articles to profl jours. Mem. Internat. Sci. Policy Found. Com. Mgmt. Avocations: diving, tennis, travel. Home: 62 Carlisle Rd, Hove BN3 4FS, England Office: PRISM, The Wellcome Trust, 183 Euston Rd, London NW1 2BE, England

IRVINE, THOMAS FRANCIS, JR., mechanical engineering educator; b. Northmont, N.J., June 25, 1922; s. Thomas Francis and Marie (Boeggeman) I.; children by previous marriages: Laura, Kay, Phoebe, Jill, Sadi, Thomas, Tanya; m. Sondra Raines, Oct. 31, 1992. B.S., Pa. State U., 1946; M.S., U. Minn., 1951, Ph.D., 1956. Research asso. Pa. State U., 1947-49; research asst. instr. U. Minn., 1950-56, asst. prof. mech. engring., 1956-58, assoc. prof., 1958-59; prof. mech. engring. U. N.C., 1959-61; prof. engring., dean Coll. Engring., SUNY-Stony Brook, 1961-71, prof., 1971-92; prof. emeritus, 1992—; pres. Rumford Pub. Co.; cons. in heat transfer, 1954—; mem. exec. bd. heat transfer div. ASME, 1964-68, chmn. exec. com., 1968-69; mem. orgn. com. Internat. Center Heat and Mass Transfer, 1966-86, chmn., 1972-74, 82-86. Editor: (with J.P. Hartnett) Advances in Heat Transfer, vols. I-XXIV, 1964-83, Pergamon Unified Engineering Series, Heat Transfer Soviet Research, Heat Transfer Japanese Research, Fluid Mechanics-Soviet Research, Previews in Heat and Mass Transfer, McGraw-Hill Hemisphere Series in Thermal and Fluids Engineering, 1976-82; tech. editor: Jour. Heat Transfer, 1960-63; editorial adv. bd. Internat. Jour. Heat and Mass Transfer, 1959—, Letters in Heat and Mass Transfer, 1974—; contbr. articles to profl. jours. Served with AUS, 1942-46. Fellow ASME (Heat Transfer Meml. award 1992), AAAS, Internat. Ctr. Heat and Mass Transfer; mem. AIAA, Am. Soc. Engring. Edn., Sigma Xi, Tau Beta Pi, Eta Kappa Nu, Sigma Tau, Theta Tau. Home: 161 Mills Pond Rd Saint James NY 11780

IRVING, DOUGLAS DORSET, behavioral scientist, consultant; b. El Paso, Tex., Jan. 30, 1944; s. Douglas Dorset and Mary Louise (Lindlof) I.; m. Terry Kathryn Rodgers, June, 1963 (div. 1975). BA, Rice U., 1965, PhD, 1970. Lic. psychologist, Ariz. Programmer analyst pvt. practice, Houston, 1963-70; ast. prof. New Sch. of Behavioral Studies, Grand Forks, N.D., 1970-74; cons. pvt. practice, Mpls., 1974-75; dir. rsch. and evaluation S.W. Community Mental Health, Las Cruces, N.Mex., 1975-78; dir. rsch. and devel. N. Ariz. Comprehensive Guidance, Flagstaff, 1978-91; cons. pvt. practice, Flagstaff, 1991—; cons. S. Rio Grande Health Planning Coun., Las Cruces, 1975-77; chmn., newsletter editor Ariz. Evaluators Network, 1981-85; com. chmn., bd. dirs. Nat. Coun. Commn. Mental Health Ctrs., 1984-87. Contbr. articles to Jour. Theoretical Biology, Jour. Community Mental Health. Challenge del. Tex. Dem. Convention, 1968; del. N.D. Dem. Non-Partisan League Conv., 1972. Fellow AAAS, Am. Anthropol. Assn.; mem. Am. Psychol. Assn., Am. Sociol. Assn. Achievements include development of Fortran compiler, other programming projects; urban anthropology field studies; helped restructure tchr. tng. in N.D.; pub. early holographic/ neural net model; development and management of MISs for Behavioral Health Treatment. Office: Psychotherapy & Evaluation Svcs 405 N Brewer Ste 9 Flagstaff AZ 86001

IRWIN, DONALD BERL, psychology educator; b. Jefferson, Iowa, Aug. 2, 1945; s. George Henry and Lurene (Terrill) I.; m. Marilyn Anne Ulfers, Nov. 18, 1967; 1 child, Benjamin James. BS, Iowa State U., 1968, MS, 1969, PhD, 1975. Asst. prof. psychology U. Wis., Waukesha, 1971-74; instr. psychology Des Moines Area C.C., Ankeny, Iowa, 1974—. Co-author: Psychology-The Search for Understanding, 1987, Developmental Psychology, 1994. Mem. sch. bd. Ctrl. Dallas Community Sch., Minburn, Iowa, 1984-93; leader Dallas County 4-H, 1977—. NDEA grad. fellow Iowa State U., 1968-71; recipient Disting. Tchr. award Des Moines Area C.C. Found., 1988. Mem. APA, Am. Psychol. Soc., Iowa Psychol. Assn., Midwest Psychol. Assn., Sigma Xi, Psi Beta (nat. pres. 1992-93). Methodist. Home: 2046 F Ave Perry IA 50220-8071 Office: Des Moines Area CC 2006 S Ankeny Blvd Ankeny IA 50021

IRWIN, RICHARD DENNIS, electrical engineering educator; b. Albany, Ga., Mar. 27, 1958; s. Vernon Hugh and Martha Lucille (Carson) I.; m. Charlotte Anita Yancey, Mar. 8, 1981; children: Katherine Virginia, Thomas Ralph, Elizabeth Martha. BSEE, Miss. State U., Starkville, 1980, PhD, 1986. Registered profl. engr., Ohio. Instr. Miss. State U., 1983-86; assoc. sr. staff engr. Control Dynamics Co., Huntsville, Ala., 1986-87; asst. prof. Ohio U., Athens, 1987-90, assoc. prof., 1990—; cons. Control Dynamics Co., Huntsville, 1988, Systran, Dayton, Ohio, 1991, Wright State U., Dayton, 1990-92, Nichols Rsch., Huntsville, 1992; mem. steering com. Southeastern Symposium on System Theory, 1988—. Contbr. articles to Jour. Guidance, Control, Dynamics, Jour. Astron. Scis. Recipient Outstanding Achievement award Ohio Soc. Profl. Engrs., 1989; faculty fellow NASA, 1988, 89, 90; NASA grantee, 1988—. Mem. IEEE, AIAA, Am. Astron. Soc., Am. Soc. Engring. Edn., Jaycees, Sigma Xi, Phi Kappa Phi, Tau Beta Pi, Eta Kappa Nu. Democrat. Achievements include development of frequency domain system identificaiton techniques for flexible systems; demonstration of control system design using experimental data models. Office: Ohio U Stocker 349 Athens OH 45701

ISA, SALIMAN ALHAJI, electrical engineering educator; b. Okene, Kwara, Nigeria, Aug. 13, 1955; came to U.S., 1983; s. Isa Onusagba and Mariyamoh (Anawureyi) I. MSEE, Syracuse U., 1984, PhD, 1989. Elec. engr. Radio Oyo (NYSC), Ibadan, Nigeria, 1979-80, Aladja Steel Plant, Warri, Nigeria, 1980-82; grad. teaching asst. Syracuse (N.Y.) U., 1985-89; asst. prof. S.C. State U., Orangeburg, 1989—. Contbr. articles to profl. jours. Bd. dirs. Rev. Ravanel Scholarship Fund, Charleston, S.C., 1990—; pres. Nigerian Student Union, Syracuse U., 1986-89. Mem. IEEE, Material Rsch. Soc., Am. Vacuum Soc., Phi Beta Delta Honor Soc. Office: SC State Univ PO Box 7355 300 College St Orangeburg SC 29117

ISAAC, WALTER LON, psychology educator; b. Seattle, May 31, 1956; s. Walter and Dorothy Jane (Emerson) I.; m. Susan Victoria Wells. BS, U. Ga., 1978; MA, U. Ky., 1983; postgrad., U. Ga., 1988-89; PhD, U. Ky., 1989. Advanced EMT Athens (Ga.) Gen. Hosp., 1977-79; teaching asst., rsch. asst. U. Ky., Lexington, 1979-87, instr. gifted student program, 1985, 87; instr. evening classes U. Ga., Athens, 1988, temp. asst. prof., 1989; asst. prof. psychology, mem. grad. faculty East Tenn. State U., Johnson City, 1989—; reviewer McGraw-Hill Pub. Co., Cambridge, Mass., 1990—. Contbg. author: Aging and Recovery of Function, 1984; contbr. articles to profl. jours. Bd. dirs. Upper East Tenn. Sci. Fair, Inc., 1992—. Mem. AAAS, Am. Psychol. Soc., Am. Assn. Lab. Animal Sci. (southeastern br.), Southeastern Psychol. Assn., Soc. for Neurosci., Sigma Xi (grantee 1987). Avocations: fishing, photography, canoeing. Home: 905 Carroll Creek Rd Johnson City TN 37601-2401 Office: East Tenn State U Dept Psychology PO Box 70649 Johnson City TN 37614

ISAACS, PHILIP KLEIN, retired chemist; b. 1927; m. Sarah Rose Max, June 14, 1953; children: Eli, Michael, Julie. BA, Bard Coll., 1948; MA, Columbia U., 1950; PhD, U. Cin., 1951. Chemist Dewey and Almy div. WR Grace, Cambridge, Mass., 1951-62; rsch. assoc. WR Grace Rsch. Ctr., Clarksville, Md., 1962-66, Machteshim Chem. Co., Beershava, Israel, 1966-68; group leader Israel Fiber Inst., Jerusalem, 1968-74, chemist, 1987-92; chief technologist Office of Chief Scientist, Jerusalem, 1974-86; sr. lectr. Hebrew U., Jerusalem, 1969-74; ret., 1992. Contbr. article to profl. jours. UN fellow UN Indsl. Devel. Orgn., 1971. Mem. Am. Chem. Soc. Achievements include 31 patents in the fields of polymer, synthesis, latex polymerization, polymer stabilization, paper coating, adhesives, polymer crosslinking, battery separators, food tech., textile flameproofing, water base paints, urethane polymers. Home: 2 Dov Kimche St, Jerusalem 92549, Israel

ISAACSON, MICHAEL, civil engineering educator. BA in Engring. with distinction, U. Cambridge, 1971, MA, 1975, PhD, 1975. Profl. engr., B.C., Can., chartered engr., Great Britain. With civil engring. dept. U. B.C., Vancouver, 1976—, prof., head civil engring. dept.; mem. code design offshore prodn. structures tech. com. Can. Stds. Assn.; mem. com. environ. forces Internat. Ship and Offshore Structures Congress; cons. in field. Coauthor: Mechanics of Wave Forces on Offshore Structures, 1981; assoc. editor Can. Jour. Civil Engring., Internat. Jour. Offshore and Polar Engring. Recipient R.A. McLachlan award Assn. Profl. Engrs. and Geoscientists B.C., 1992. Mem. ASCE, Can. Soc. Civil Engring. (chair engring. mechanics divsn., bd. dirs., Camille A. Dagenais award 1992), Engring. Inst. Can., Royal Instn. Naval Architects, Internat. Assn. Hydraulic Rsch., Internat. Soc. Offshore and Polar Engrs. (PACOMS award 1992). Achievements include research in coastal and ocean engineering. Office: U British Columbia, Dept of Civil Engring, Vancouver, BC Canada V6T 1W5*

ISAACSON, MICHAEL SAUL, physics educator, researcher; b. Chgo., July 4, 1942; s. Henry and Lillian (Goldstein) I.; children: Zoe, Serge, Livia. BS in Engring. Physics, U. Ill., 1965; SM, U. Chgo., 1966, PhD in Physics, 1971. Staff scientist biology divsn. Brookhaven Nat. Lab., Upton, N.Y., 1971-72; asst. prof. physics U. Chgo., 1973-78; assoc. prof. Cornell U., Ithaca, N.Y., 1979-87, prof., assoc. dir. applied & engring. physics, 1988—; mem. adv. bd. Harvard Med. Sch., NIH Biomedical Rsch., Cambridge, Mass., 1973-79, N.Y. Dept. Health, NIH-Biol. Microscopy and Image Reconstruction Resource, Albany, N.Y., 1980—. Editor: Electron Beam Spectroscopy at High Spatial Resolution, 1988; jour. editor Nearfield Optics, 1993; mem. editoral bd. Ultramicroscopy, 1975—; contbr. chpts. to books and articles to profl. jours. Recipient Humboldt award Alexander Von Humboldt Found., 1992; Faculty fellow Sloan Found., 1972-75; rsch. grantee NSF, Air Force Office Sci. Rsch., Def. Advanced Rsch. Projects Agy. Mem. Am. Assn. Engring. Educators, (Electron) Microscopy Soc. Am. (dir. 1986-88, pres. 1992-93, Burton award 1976). Achievements include patents for Electron Beam Lithography at Nanometer Resolution, Near Field Super Resolution Optical Microscopy; first motion picture of moving atoms; first demonstration of field emission gun SEM, electron energy loss spectroscopy of biological molecules in an electron microscope. Office: Cornell U Clark Hall 201 Ithaca NY 14853

ISAACSON, RICHARD EVAN, microbiologist; b. Chgo., Oct. 25, 1947; s. Edward Kenneth and June Lorraine (Rosenfeld) I.; m. Barbara Lee Southon, Dec. 26, 1970; children: William Jonathan, Amanda Joan, Daniel Edward. BS, U. Ill., 1969, PhD, 1974. Microbiologist Nat. Animal Disease Ctr., Ames, Iowa, 1974-78; asst. prof. U. Mich., Ann Arbor, 1978-83; mgr. Pfizer, Inc., Groton, Conn., 1983-89; assoc. prof. U. Ill., Urbana, 1989—; mem. editorial bd. Infection and Immunity, Washington, 1983-86, Animal Biotechnology, N.Y.C., 1991—. Editor: Recombinant DNA Vaccines, Rationale and Strategies, 1992; contbr. over 90 articles to profl. jours. Recipient NIH grant, 1979, USDA grant, 1982, 90, 93. Mem. AAAS, Am. Soc. Microbiology, Am. Acad. Microbiology, Sigma Xi. Achievements include discovery and development of first federally licensed recombinant DNA vaccine, EcoBac TM. Office: Univ Ill Vet Pathobiology 2001 S Lincoln Ave Urbana IL 61801-6199

ISBERG, RALPH, molecular biologist, educator. Prof. molecular biology dept. Tufts U., Boston. Recipient Eli Lilly and Co. Rsch. in Microbiology Immunology award Am. Soc. for Microbiology, 1993. Office: Tufts Univ Dept Molecular Biology 136 Harrison Ave Boston MA 02111*

ISBERG, REUBEN ALBERT, radio communications engineer; b. Chugwater, Wyo., Dec. 11, 1913; s. Albert Gust and Laura Carolina (Thun) I.; m. Dorothe Louise Hall, Feb. 23, 1936; children: Jon Lewis, Barbara Louise Isberg Johnson, Edward Russel. AB in Phys. Sci., U. No. Colo., 1935. Registered profl. engr., Calif. Radio and TV engr. W2XBS/WNBT-NBC, N.Y.C., 1939-42; electronic devel. engr. div. war rsch. Columbia U., Mineola, N.Y., 1942-46; chief engr. KRON-TV, San Francisco, 1946-52; ind. cons. TV engr. various locations, 1952-54; sr. engr. Ampex Corp., Redwood City, Calif., 1954-60; statewide communications engr. U. Calif., Berkeley, 1960-67; ind. cons. radio communications engr. Berkley, 1967—; chair subcom. for FM radio stereo standards NSRC, Washington,1 960-61; mem. com. for establishing 2500 MHz instrnl. TV svc., FCC, 1965-67. Contbr. to profl. publs. Named Honored Alumnus, U. No. Colo., 1993. Fellow IEEE (chair awards com. vehicular tech. soc. 1984-90, Avant Garde medal and cert. 1991), Audio Engring. Soc., Soc. Motion Picture and TV Engrs., Radio Club Am.; mem. Acoustical Soc. Am., Soc. Cable TV Engrs., Inst. Radio Engrs. (chair San Francisco sect. 1951). Republican. Congregationalist. Achievements include work on guided radio communications in subways, mines, ships and buildings, U.S. and Can. patents for tunnel distributed antenna system with signal taps coupling approximately the same amount of energy. Home: Apt B 127 32200 SW French Prairie Rd Wilsonville OR 97070

ISBISTER, WILLIAM HUGH, surgeon; b. Manchester, Eng., Apr. 7, 1934; s. William and Eleanor Gwyneth (Pritchard) I.; m. Magdalena Richter, Mar. 29, 1961; children: William, Gwyneth-Ann, Michael. M.B.,Ch.B., Manchester U., 1958, MD, 1964. From house surgeon to clin. asst. in surgery Manchester Royal Infirmary, 1958-64; resident surg. officer Park Hosp., Davyhulme, 1965-67; sr. surg. registrar Manchester Regional Hosp. Bd., 1967-69; rsch. fellow in surgery and John M. Wilson Meml. scholar Cleve. Clinic Found., 1969-70; sr. surg. registrar, prof. surg. unit Manchester Regional Hosp. Bd., 1970-71; sr. lectr. surgery U. Queensland, Brisbane, 1972-75; from Found. prof. surgery to dep. dean Wellington Sch. Medicine, 1975-90; sr. specialist colorectal surgeon Wellington Hosp., 1975-90, chmn. dept. gen. surgery, 1981-88, 89-90; specialist colorectal surgeon and chmn. dept. surgery King Faisal Specialist Hosp. and Rsch. Ctr., Riyadh, Saudi Arabia, 1990—; mem. Nat. Health and Med. Rsch. Coun. (Australia) Assessing Panel, Med. Rsch. Coun. N.Z. Assessing Panel, Cancer Soc. of N.Z. Assessing Panel, Dept. Vets. Affairs Assessing Panel, Australia and N.Z. Jour. Surgery Rev. Panel, N.S.W. Cancer Coun. Assessing Panel, Anti-Cancer Coun. of Victoria Nat. Coun. Assessing Panel; cons. in field; lectr. in field. Contbr. numerous articles to profl. jours. Fellow Am. Soc. Colon and Rectal Surgeons; mem. Brit. Med. Assn., N.Z., Surg. Rsch. Soc. Australasia (life mem.), Internat. soc. Univ. Colon and Rectal Surgeons, Wellington Postgrad. Med. Soc., Saudi Gastroenterology Assn., The Wellington Club, Riyadh Surg. Club, Riyadh Gastroenterol. Club. Avocations: jazz, opera, computing. Home and Office: King Faisal Specialist Hosp, PO Box 3354, Riyadh 11211, Saudi Arabia

ISCHAY, CHRISTOPHER PATRICK, engineer; b. Willoughby, Ohio, Nov. 21, 1964; s. Donald Francis and Catherine Louise (Ellensohn) I. BS in Electronic Engring. Tech., DeVry Inst. of Tech., Phoenix, 1987. Radwaste/ Health physics dept. project mgr. Palo Verde Nuclear Generating Sta., Wintersburg, Ariz., 1987-91; project mgr. EG&G Idaho, Idaho Falls, 1991—. Mem. Idaho Am. Nuclear Soc. Office: EG&G M/S 3950 PO Box 1625 Idaho Falls ID 83415

ISE, NORIO, chemistry educator; b. Kyoto, Japan, Oct. 19, 1928; s. Jiro and Kinu (Haruta) I.; m. Nobuko Otsuki, Nov. 25, 1963; children: Tadashi, Kiyoshi, Naoko. BS, Kyoto U., Japan, 1954, MS, 1956, PhD, 1959. Assoc. prof. Kyoto U., Japan, 1962-70, prof., 1970-92; guest prof. Johannes-Gutenberg U., Mainz, Germany, 1981; Turner Alfrey vis. prof. Mich. Molecular Inst., Midland, Mich., 1984; dir. Fukui Rsch. Lab. Rengo Co. Ltd., 1992—; mem. Macromolecular Div., Internat. Union Pure and Applied Chemistry, Oxford, Eng., 1981-87, titular mem., 1989-93; coun. mem. Internat. Assn. Colloid & Interface Scientists, Wageningen, Holland, 1990-94. Co-author: Introduction to Polymer Chemistry, 1970; editor, author: An Introduction to Speciality Polymer, 1983; translator: Wasan, Japanese Mathematics, 1993. Recipient The Chem. Soc. of Japan award, 1986. Avocation: music. Home: 23 Nakanosaka, Kamigamo Kita-ku, Kyoto 603, Japan Office: Fukui Rsch Lab 10-8-1 Jiyugaoka, Kanazu-Cho, Sakai-Gun Fukui 919-06, Japan

ISELY, DUANE, biology and botany educator; b. Bentonville, Ark., Oct. 24, 1918; s. Dwight and Blessie Elise (Dort) I.; m. Helen Sue Pearson, Apr. 3, 1940 (div. 1965); children—Deanna Sue, Karl; m. Mary Elizabeth Holman, July 17, 1977. B.A., U. Ark., 1938, M.S., 1939; Ph.D., Cornell U., 1942. Sr. seed analyst Ala. Dept. Agr., 1943-44; extension assoc. Iowa State U., Ames, 1944-46, asst. prof., 1946-49, assoc. prof., 1949-56, prof., 1956-81, disting. prof. scis. and humanities, 1981—. Author: Herbaceous Flora of TVA Reservoirs, 1945; Seed Analysis, 1954; Weed Identification and Control, 1961; Legumimosae of United States, 4 vols., 1973-90; editor Iowa State Jour. of Research, 1979-88. Chmn. Ames Conservation Council; bd. dirs. Iowa chpt. Nature Conservancy. NSF grantee, 1972, 80, 83, 86, 91. Mem. Internat. Soc. Plant Taxonomy, Am. Soc. Plant Taxonomists, Soc. Econ. Botany, Assn. Ofcl. Seed Analysts (pres. 1954). Developed new methods for agriculture see analysis and was major source for training seed analysts. Home: 3012 Story St Ames IA 50010-3503 Office: Iowa State U Dept Botany Ames IA 50011

ISEMURA, MAMORU, biochemist, educator; b. Kobe, Hyogo-ken, Japan, Jan. 26, 1941; s. Fujinaga and Fujinaga (Souko) Seihitiro; m. Satoko Isemura, Oct. 18, 1967; children: Masako, Satoshi. BS, Osaka (Japan) U., 1963, MS, 1965, DSc, 1968. Assoc. prof. Niigata (Japan) U. Sch. Medicine, 1971-80, Tohoku U. Sch. Medicine, Sendai, Miyagi, Japan, 1980-86; prof. Shizuoka (Japan) Women's U., 1986-90; prof. Sch. Food and Nutritional Scis. U. Shizuoka, 1987—. Mem. Japanese Biochem. Soc., Japanese Cancer Assn., N.Y. Acad. Scis. Home: 220-60 Kusanagi, Shimizu, Shizuoka 424, Japan Office: U Shizuoka, 52-1 Yada, Shizuoka 422, Japan

ISENBERG, JOHN FREDERICK, optical engineer; b. Appleton, Wis., 1953. BA in Physics, Lawrence U., 1975; MS in Optics, U. Rochester, 1978. Rsch. asst. dept. physics U. Rochester, N.Y., 1975, rsch. asst. dept. biomath., 1976-78; devel. programmer Sinclair Optics, Pittsford, N.Y., 1977; sr. engr. Optical Rsch. Assocs., Pasadena, Calif., 1978-88, staff engr., 1988—. Eastman Kodak grad. scholar U. Rochester, 1975. Mem. Optical Soc. So. Calif. Home: 1672 E Locust St Pasadena CA 91106 Office: Optical Rsch Assocs 550 N Rosemead Blvd Pasadena CA 91107

ISENHOWER, WILLIAM MARTIN, civil engineering educator; b. Baytown, Tex., Oct. 31, 1952; s. Jodie and Mary Douglass (Williams) I.; m. Eleanor Anne Hexamer, June 30, 1990. BSCE, U. Tex., 1974, MS, 1979, PhD, 1986. Registered profl. engr., Tex. Asst. engr. Dames & Moore, Houston, 1974-75; staff engr. Woodward-Clyde Cons., Houston, 1979-81; sr. design engr. Tex. Dept. Transp., Austin, 1986-90; asst. prof. civil engring. U. Ariz., Tucson, 1990—; expert on mission UN Devel. Program, Pune, India, 1984, New Delhi, India, 1984; summer faculty program Waterways Expt. Sta., Vicksburg, Miss., 1992. Contbr. articles to profl. jours. Recipient Taylor award U. Tex., Austin, 1974; NSF rsch. initiation awardee, 1992. Mem. ASCE, Internat. Soc. for Soil Mechanics and Found. Engring. Home: 3220 E Calle de la Punta Apt 11 Tucson AZ 85718-2142 Office: Univ of Ariz Dept Civil Engring Tucson AZ 85721

ISHIDA, ANDREW, neurobiologist; b. L.A., July 1951. PhD, UCLA, 1981. Asst. prof. U. Calif., Davis, 1986-92, assoc. prof., 1992—. Contbr. articles to profl. jours. Rsch. grantee NIH, 1988. Mem. Soc. Neurosci., Internat. Brain Rsch. Orgn.

ISHIDA, OSAMI, microwave engineer; b. Mikkabi-Cho, Japan, Aug. 25, 1948; s. Hiroshi and Teru (Shimizu) I.; m. Chitoe Kitagawa, Apr. 7, 1977; children: Takashi, Kazuaki. BS, Shizuoka U., Hamamatsu, Japan, 1971, MS, 1973; PhD, Shizuoka U., 1992. Engr. Mitsubishi Electric Corp., Kamakura, Japan, 1973-82, sr. engr., 1982-86, engring. mgr., 1986-88, mgr. aperture antenna group, 1988-92, dep. mgr. opto and microwave electronics dept., 1993—; tech. program com. mem. 3d Asia-Pacific Microwave Conf., Tokyo, 1989-90; steering com. 6th Asia-Pacific Microwave Conf., 1993—. Assoc. editor spl. issue IEICE, 1990; inventor, patentee crossed waveguide type polarization separator. Mem. Inst. Electronics, Info. and Communication Engrs. (transactions editorial com. mem. 1988-92), IEEE. Office: Mitsubishi Electric Corp, Eo & Mw Sys Lab 5-1-1 Ofuna, Kamakura 247, Japan

ISHIHARA, OSAMU, electrical engineer, physicist, educator; b. Osaka, Japan, Nov. 15, 1948; came to U.S., 1974; s. Mamoru and Tomoe (Maeda) I.; m. Yohko Miyake, May 5, 1974; children: Reiko, Takeki, Yuko, Sachiko. BS, Yokohama (Japan) Nat. U., 1972, MS, 1974; PhD, U. Tenn., 1977. Postdoctoral fellow U. Saskatchewan, Saskatoon, Can., 1977-80, profl. rsch. assoc., 1980-84; assoc. prof. dept. elec. engring. Tex. Tech. U., Lubbock, 1985-87, assoc. prof. dept. elec. engring. and physics, 1987-89, prof. dept. elec. engring. and physics, 1989—; vis. prof. dept. engring. Yokohama Nat. U., 1992, vis. prof. dept. elec. engring. and comp. sci. Kumamoto (Japan) U., 1993; grad. advisor Dept. Elec. Engring., Tex. Tech. U., Lubbock, 1986—, faculty advisor Eta Kappa Nu, 1987—; faculty advisor Assn. Japanese Students, 1990—; bd. dirs. South Plains Regional Sci. and Engring. Fair, Lubbock, 1990—. Contbr. rsch. articles to profl. jours. Japan Soc. Promotion of Sci. fellow, 1974-77. Mem. IEEE (sr. mem., Nuclear and Plasma Sci. Soc.), Am. Phys. Soc. (Plasma Physics div.), Phys. Soc. Japan, Sigma Xi, Phi Kappa Phi, Phi Beta Delta. Avocations: tennis, astronomy. Office: Texas Tech Univ Dept of Elec Engring Lubbock TX 79409-3102

ISHII, AKIRA, medical parasitologist, malariologist, allergist; b. Kochi, Japan, July 11, 1937; s. Katsuhiko and Fusae Ishii; m. Fuyuko Ishii, Mar. 20, 1968; children: Ken, Shin, Taku. MD, U. Tokyo, 1964, D Med. Sci., 1969; MSc, U. London, 1970. Cert. malaria advanced epidemiology. Rsch. assoc. Inst. Infectious Disease, Tokyo, 1969-74; asst. prof. Toyko Med. and Dental U., 1974-78, Inst. Med. Sci., U. Tokyo, 1978-79; prof. Miyazaki

(Japan) Med. Coll., 1979-84, Okayama (Japan) U. Med. Sch., 1984-90; dir. dept. parasitology NIH, Tokyo, 1990—; assoc. prof. Jichi Med. Coll., 1992—; com. mem. Japanese Internat. Coop. Agy., Tokyo, 1978-89; panel mem. U.S.-Japan Coop. Med. Program Parasitology, China-Japan Parasitology Seminar. Editor Nettai, Japanese Jour. Sanitary Zoology, Japanese Jour. Tropical Med. Hygiene; editor in chief Japanese Jour. Parasitology. Mem. Japanese Soc. Parasitology (councilor, Koizumi prize), Japanese Soc. Tropical Medicine (councilor), Japanese Soc. San. Zoology (Society prize, councilor), Japanese Soc. Allergologists (councilor), Japanese Soc. Infectious Disease (councilor), Japanese Soc. Internat. Health (councilor). Avocations: mountain trips, tennis, golf. Office: NIH, 1-23-1 Toyama, Shinjuku-ku Tokyo 162, Japan

ISHII, YOSHINORI, geophysics educator; b. Tokyo, Mar. 14, 1933; s. Kichijiro and Kei Ishii; m. Hiroko Hisamune, Nov. 24, 1963; children: Yutaka, Makoto, Akira. BS, U. Tokyo, 1955, ED, 1977. Exploration geophysicist Teikoku Oil Co., Tokyo, 1955; rsch. geophysicist Japan Petroleum Exploration Co., Tokyo, 1955-67, sr. geophysicist, 1970-71; sr. geophysicist Japan Nat. Oil Corp., Tokyo, 1967-70; assoc. prof. geophysics U. Tokyo, 1971-78, prof. geophysics, 1978—; mem. Sci. Coun. of Japan, Tokyo, 1988-91. Author: Introduction to Remote Sensing, 1981, Geophysical Engineering, 1988; co-author several books; contbr. numerous articles to profl. jours. Mem. Soc. Exploration Geophysicists of Japan (pres. 1984-85, 1988-89, Best Paper award, Tokyo, 1976), Remote Sensing Soc. Japan, (v.p. 1981-88, pres. 1990-92), Japanese Assn. for Petroleum Tech. (v.p. 1982-86). Avocations: golf, computer. Home: 8-2-14 Hisagi, Zushi, Kanagawa 249, Japan Office: U Tokyo Engring Faculty, 7-3-1 Hongo, Bunkyo-ku, Tokyo 113, Japan

ISHIKAWA, YUICHI, mechanical engineer, researcher; b. Tokyo, Jan. 5, 1942; s. Shoji and Shigeko Ishikawa; m. Junko Enami, Nov. 23, 1973; children: Hirono, Masae, Yukiyo. BS, Chiba (Japan) U., 1964, MS, 1967; PhD, Pa. State U., 1971. Researcher Noritake, Ltd., Nagoya, Japan, 1964-65; vis. rsch. scientist U. Gottingen, Fed. Republic of Germany, 1971-72; researcher Hitachi, Ltd., Tokyo, 1972-78; sr. researcher Hitachi, Ltd., Tsuchiura, Japan, 1978-88, chief researcher, 1988—; vis. rsch. scientist Brookhaven Nat. Lab., L.I., 1982, Nat. Inst. for Materials, Tsukuba, Japan, 1990—. Contbr. articles to profl. jours. Fulbright scholar U.S.-Japan Edn. Com., 1967. Fellow Corrosion Engring. Soc. of Japan; mem. Am. Vacuum Soc., Nat. Assn. Corrosion Engrs., Electrochem. Soc. Avocations: tennis, skiing, reading. Home: 13-8 Senba-cho, Mito-shi 310, Japan Office: Hitachi Ltd, Mech Engring Rsch Lab, 502 Kandatsu-machi, Tsuchiura 300, Japan

ISHIKAWA-FULLMER, JANET SATOMI, psychologist, educator; b. Hilo, Hawaii, Oct. 17, 1925; d. Shinichi and Onao (Kurisu) Saito; m. Calvin Y. Ishikawa, Aug. 15, 1950; 1 child, James A.; m. Daniel W. Fullmer, June 11, 1980. BE, U. Hawaii, 1950, MEd, 1967; MEd, U. Hawaii, 1969, PhD, 1976. Diplomate Am. Acad. Pain Mgmt. Instr. Honolulu Bus. Coll., 1953-59; instr., counselor Kapiolani Community Coll., Honolulu, 1959-73; prof., dir. counseling Honolulu Community Coll., 1973-74, dean of students, 1974-77; psychologist, v.p., treas. Human Resources Devel. Ctr., Inc., Honolulu, 1977—; cons. United Specialties Co., Tokyo, 1979, Grambling (La.) State U., 1980, 81, Filipino Immigrants in Kalihi, Honolulu, 1979-84, Legis. Reference Bur., Honolulu, 1984-85, Honolulu Police Dept., 1985; co-founder Waianae (Hawaii) Child & Family Ctr., 1979—. Co-author: Family Therapy Dictionary, 1991, Manabu: The Diagnosis and Treatment of a Japanese Boy with a Visual Anomaly, 1991; contbr. articles to profl. jours. Commr. Bd. Psychology, Honolulu, 1979-85. Mem. AACD, APA, Hawaii Psychol. Assn., Pi Lambda Theta (pres. 1969-70, v.p. 1968-69, sec. 1967-68), Delta Kappa Gamma (sec., scholarship 1975, Outstanding Educator award, 1975, Thomas Jefferson award, 1993, Francis E. Clark award, 1993). Avocations: jogging, tennis, dancing. Home: 154 Maono Pl Honolulu HI 96821 Office: Human Resources Devel Ctr 1750 Kalakaua Ave Apt 809 Honolulu HI 96826-3725

ISHIMARU, AKIRA, electrical engineering educator; b. Fukuoka, Japan, Mar. 16, 1928; came to U.S., 1952; s. Shigezo and Yumi I.; m. Yuko Kaneda, Nov. 21, 1956; children: John, Jane, James, Joyce. BSEE, U. Tokyo, 1951; PhDEE, U. Wash., 1958. Registered profl. engr., Wash. Engr. Electro-Tech. Lab, Tokyo, 1951-52; tech. staff Bell Telephone Lab, Holmdel, N.J., 1956; asst. prof. U. Wash., Seattle, 1958-61, assoc. prof., 1961-65, prof. elec. engring., 1965—; vis. assoc. prof. U. Calif., Berkeley, 1963-64; cons. Jet Propulsion Lab., Pasadena, Calif., 1964—, The Boeing Co., Seattle, 1984—. Author: Wave Propagation & Scattering in Random Media, 1978, Electromagnetic Wave Propagation, Radiation and Scattering, 1991; editor: Radio Science, 1982; editor-in-chief Waves in Random Media, U.K., 1990. Recipient Faculty Achievement award Burlington Resources, 1990; Boeing Martin professorship, 1993. Fellow IEEE (mem. editorial bd., Region VI Achieveemnt award 1968, Centennial Medal 1984), Optical Soc. Am. (assoc. editor jour. 1983); mem. Internat. Union Radio Sci. (commn. B chmn.). Home: 2913 165th Pl NE Bellevue WA 98008-2137 Office: U Wash Dept Elec Engring FT-10 Seattle WA 98195

ISHIMARU, CAROL ANNE, plant pathologist; b. Detroit, Feb. 22, 1954; d. Thomas Joseph and Audrey Georgina (Brietenbeck) Shea; m. Dan Takao, June 18, 1977; 1 child, Nickolas. BS, Mich. State U., 1979, PhD, 1985. Rsch. asst. Mich. State U., East Lansing, 1979-85; rsch. assoc. U. Nebr., Lincoln, 1985-87, USDA, Corvallis, Oreg., 1987-89; asst. prof. plant pathology Colo. State U., Ft. Collins, 1989—. Contbr. articles to profl. jours. Mem. Am. Soc. Microbiology, Assn. for Women in Sci., Am. Phytopathol. Soc. Achievements include research in gene cloning and chemical analysis; iron uptake systems preserved in bacteria that cause soft rot of vegetables. Office: Colo State Univ Dept Plant Pathology Fort Collins CO 80523

ISHIMARU, HAJIME, physics and engineering educator; b. Sapporo, Hokkaido, Japan, Feb. 21, 1940; s. Osamu and Sumiko (Matsumoto) I.; m. Masako Kodera, Feb. 23, 1969; children: Dan, Goh. BS in Physics, Hokkaido U., Sapporo, Japan, 1963; MS in Physics, Tohoku U., Sendai, Japan, 1965; DS in Physics, Nagoya (Japan) U., 1968, DSc, 1970; D in Engring., Tokyo U., 1980. Cert. vacuum sci. and Tech. Rsch. assoc. Tokyo U., Tokyo, 1969-72; rsch. assoc. Nat. Lab. High Energy Physics, Tsukuba, Ibaraki, Japan, 1972-75, assoc. prof., 1975-84, prof., 1984—; prof. The Grad. U., Tsukuba, Ibaraki, Japan, 1988—; cons. Superconductive Super Collider Lab., Dallas, 1991—, Synchrotron Radiation Rsch. Ctr., Hsinchu, Taiwan, 1986—. Author: Aluminum Vacuum Technol., 1988; editorial bd. Vacuum Journal, 1974. Steering com. Sakusa Village, Sakura, 1982, ACCS cable TV svc. Tsukuba, 1985—; active PTA, Sakura, 1984, Namiki Football Club, Sakura, 1984. Recipient Vacuum Tech. award Vacuum Soc. Japan, Tokyo, 1979, 82, Best Shop Note award Am. Vacuum Soc., N.Y.C., 1978, Remarkable Patent award Ministry of Sci. and Tech., Tokyo, 1983, 85, Short Note award British Vacuum Coun., London, 1985, Takagi award Precision Measurements Tech. Found., 1991. Home: 2079 Uenomuro, Tsukuba Ibaraki 305, Japan Office: Nat Lab for High Energy, Physics, 1-1 Oho, Tsukuba Ibaraki 305, Japan

ISHIWA, SADAO CHIGUSA, geneticist, educator; b. Tsu, Mie, Japan, July 8, 1936; s. Tetsujirou and Yoshiko (Nakao) Chigusa; m. Hiromi Ishiwa, July 6, 1963; children: Akiko, Naoki. BS in Genetics, Kyoto U., Japan, 1961, MS in Genetics, 1964; PhD in Population Genetics, Purdue U., 1968. Rsch. assoc. N.C. State U., Raleigh, 1968-69; asst. prof. biology Ochanomizu U., Tokyo, Japan, 1969-70, assoc. prof., 1970-88, prof., 1988—, head dept. biology, 1991-92; vis. prof. U. of Air, Chiba, Japan, 1987-93. Author: (with others) Population Biology of Genes and Molecules, 1991. Mem. Genetics Soc. Am., Genetics Soc. Japan, Molecular Biology Soc. Japan, Internat. Soc. Molecular Evolution, Soc. Molecular Biology and Evolution. Avocations: mountaineering. Home: 1-14-11 Josuihoncho, Kodaira City, 187 Tokyo Japan Office: Ochanomizu U, 2-1-1 Ohtsuka, 112 Tokyo Japan

ISHIZAKA, KIMISHIGE, immunologist, educator; b. Tokyo, Dec. 3, 1925; married; 1 child. MD, U. Tokyo, 1948, D of Med. Sci., 1954. Resident mem. NIH, Tokyo, 1950-53, chief div. immunoserology, 1953-62; research fellow microbiology Sch. Medicine, Johns Hopkins U., Balt., 1959, prof. medicine and microbiology, 1970-81, prof. immunology and medicine, dir. sub-dept. immunology, 1981—; from asst. prof. to assoc. prof. microbiology U. Colo., Denver, 1962-70, chief div. serology Children's Asthma Research Inst. and Hosp., 1962-63, dir. dept. basic sci., 1963-70; research fellow chemistry Calif. Inst. Tech., 1957-59; mem. adv. com. immunology WHO; fgn. assoc. Nat. Acad. Sci. Recipient Passano Found. award, Paul Ehrlich and Ludwig-Darmstaedter prize Fed. Republic of Germany, Internat. award Gairdner Found. Can., Emperor's award Japan, Borden award Assn. Am. Med. Colls., Achievement award ACP; named to Order Cultural Merit (Japan). Fellow AAAS, Am. Acad. Allergy (hon.); mem. Am. Assn. Immunology, Soc. Exptl. Biology and Medicine. Office: La Jolla Inst for Allergy and Immunology 11149 N Torrey Pines Rd La Jolla CA 92037-1079

ISHIZAKI, TATSUSHI, physician, parasitologist; b. Tochigi, Japan, Mar. 8, 1915; s. Takaji and Satoko I.; M.D., Tokyo (Japan) U., 1939, Ph.D., 1951; diploma pub. health, Singapore U., 1959; m. Yuriko Sase, Nov. 5, 1946; children: Terumi, Michiharu. With U. Tokyo Sch. Medicine, 1939, sr. asst. dept. phys. therapy and medicine, 1946-55; chief 2d div. dept. parasitology Nat. Inst. Health, Japan, 1953-67, chief dept., 1967-74; lectr. clin. allergy U. Tokyo Sch. Medicine, 1956-71; prof. clin. immunology Dokkyo U. Sch. Medicine, 1973-82, prof. emeritus, 1982—, chmn. clin. profs., 1978-80; panelist Japan-U.S.A. Coop. Study Parasitology, 1965-74. Hon. fellow Am. Coll. Allergists; mem. Japanese Soc. Allergy (exec. com., pres., hon. fellow 1981—), Korean Soc. Allergy (hon.), Japanese Soc. Tropical Medicine (hon. fellow, exec. com.), Japanese Soc. Internal Medicine (exec. com.), Japan-German Assn. Protozoan Diseases (pres.), Research, publs. on skin tests for various antigens; standardization of criteria of positive skin test basic phenomena of skin reaction especially mast cell degranulation mechanism analysis of onset of asthma attacks especially related to air pollution, weather; analysis of mechanisms of occupational allergy, others. Home: 8-15 Torimachi, Mibumachi, Tochigiken Japan Office: Dokkyo U Sch Medicine, Mibumachi, Tochigi Japan

ISKANDRIAN, AMI SIMON (EDWARD), cardiologist, educator; b. Baghdad, Iraq, Oct. 21, 1941; came to U.S., 1971; s. Simon S. and Marian (Demerjian) I.; m. Greta P. Parhad, Apr. 2, 1967; children: Basil, Susie, Kristen. MD, U. Baghdad Med. Sch., 1965. Co-dir. Phila. Heart Inst., 1986—; dir. nuclear cardiology Presbyn. Med. Ctr., Phila., 1986—; clin. prof. medicine U. Pa. Med. Sch., Phila., 1986—; cons. nuclear cardiology AMA, to various jours. Author, editor: Nuclear Cardiac Imaging: Principles and Applications, 1986; author/co-author 500 sci. papers, abstracts, chpts.; assoc. editor ACCEL, 1978-83; mem. editorial bd. Catheterization and Cardiovascular Diagnosis, Am. Jour. Cardiology, Am. Heart Jour., Jour. Cardiac Rehab., Am. Jour. Noninvasive Cardiology, Jour. Am. Coll. Cardiology, Jour. Nuclear Cardiology. Lt. Iraqi Army, 1965-67. Fellow ACP, Am. Coll. Chest Physicians, Am. Coll. Cardiologists, Am. Heart Assn. (clin. coun.), Soc. Nuclear Medicine (ednl. com. cardiovascular coun. 1990); mem. AAAS, Am. Fedn. for Clin. Rsch., Am. Soc. Nuclear Cardiology, European Soc. Cardiology (working group on exercise physiology and electrocardiography). Presbyterian. Office: Phila Heart Inst 51 N 39th St Philadelphia PA 19104

ISLAM, JAMAL NAZRUL, mathematics and physics educator, director; b. Jhenidah, Jessore, Bangladesh, Feb. 24, 1939; s. Mohammed Sirajul Islam and Rahat (Ara) Begum; m. Suraiya Solaiman, Nov. 13, 1960; children: Sadaf Saaz, Nargis Naaz. BSc with honors, St. Xavier's Coll., Calcutta, India, 1957; BA with honors and MA, Trinity Coll., Cambridge, Eng., 1960, PhD, 1964, DSc, 1982. Rsch. assoc. dept. physics U. Md., College Park, 1963-65; postdoctoral fellow U. Cambridge, 1965-66, mem. staff Inst. Theoretical Astronomy, 1967-71; vis. assoc. dept. physics Calif. Inst. Tech., Pasadena, 1971-72; sr. rsch. assoc. dept. astronomy U. Wash., Seattle, 1972-73; temp. lectr. dept. maths. King's Coll., London, 1973-74; rsch. fellow dept. applied maths. U. Coll., Cardiff, Eng., 1975-78; lectr., reader dept. maths. The City U., London, 1978-84; prof. U. Chittagong, Bangladesh, 1984-89, dir., prof. Rsch. Ctr. Math. and Phys. Scis., 1989—; vis. mem. Inst. for Advanced Study, Princeton, N.J., 1968, 73, 84; vis. scientist ICTP, Trieste, Italy, 1991, Dept. of Applied Math. and Theoretical Physics, U. Cambridge, Eng., 1991, Inst. des Hautes Etudes Sci., Bures-sur-Yvette, France, 1991. Author: Ultimate Fate of the Universe, 1983, Rotating Fields in General Relativity, 1985, An Introduction to Mathematical Cosmology, 1992; co-editor: Classical General Relativity, 1984; contbr. articles to profl. jours. Fellow Bangladesh Acad. Scis. (Gold medal 1986), Third World Acad. Scis., Islamic Acad. Scis.; mem. Clare Hall (life). Islamic. Avocations: music, reading. Home: Sabza-Zar, 28, Surson Road, Chittagong Bangladesh Office: U Chittagong Rsch Ctr, Math and Phys Scis, Chittagong Bangladesh

ISLAM, M. RAFIQUL, petroleum engineering educator; b. Rajshahi, Bangladesh, Jan. 3, 1959; came to U.S. 1982; s. Muhammad Sulaiman and Hamida Banu. Diplôme d'éngineur, Inst. Algerien du Pétrole, Boumerdes, Algiers, 1982; MS, U. Alberta, Edmonton, Can., 1985, PhD, 1987. Sr. rsch. engr. Petroleum Recovery Inst., Calgary, Alberta, 1987-88; rsch. engr. NOVA HUSKY Rsch. Corp., Calgary, 1988-90; prin. scientist Emertec Devel. Inc., Calgary, 1990; assoc. prof. S.D. Sch. Mines and Tech., Rapid City, 1991--; cons. Rosebud Sioux Water Resources, Texaco Rsch., Houston, 1991; tech. editor SPE Reservoir Engring., Dallas, 1992—; edt. rev. J. Can. Petroleum Tech., Calgary, 1988-92. Editor AIChE Reprint Series, 1990; assoc. editor Jour. Petroleum Sci. and Engring.; contbr. articles to profl. jours. Reviewer Na. Sci. and Engring. Rsch. Coun., Ottawa, Can., 1989, NSF, Washington, 1992--. Recipient Diploma Honor Nat. Petroleum Honor Soc., 1992; Dept. Environ. and Natural Resources rsch. grantee, 1992, 93, NSF rsch. grantee 1992, Energy Mines and Resource rsch. grantee, 1992, S.D. Gov.'s Office for Econ. Devel. grantee, 1992, 93. Mem. Soc. Petroleum Engrs., Am. Inst. Chem. Engrs., Can. Inst. Mining, Can. Soc. Chem. Engrs. Achievements include development of a novel technique for cleaning petroleum contaminated soils, technique for recovering from heavy oil reservoirs with bottomwater, comprehensive numerical simulator for horizontal wells. Office: SD Sch Mines and Tech Dept Geological Engring Rapid City SD 57701

ISLAM, MOHAMMED N., optics scientist; b. Newark, Feb. 8, 1960. BSEE, MIT, 1980, MSEE, 1982, ScD in Elec. Engring., 1985. With systems engring. and antenna design groups IBM, Oswego, N.Y., 1978-79; recitation instr. in electromagnetic fields and energy MIT, Cambridge, Mass., 1981; mem. spl. studies in physics group Lawrence Livermore (Calif.) Nat. Lab., 1982-83, cons., 1985—; with femtosecond laser lab. AT&T Bell Labs., Holmdel, N.J., 1983, cons., 1983-85, rsch. scientist in photonics switching rsch. dept., 1985-92; tenured assoc. prof. elec. engring. and computer sci. dept. U. Mich., Ann Arbor, 1992—. Author: Ultrafast Fiber Switching Devices and Systems, 1992. Fannie and John Hertz fellow 1981-85; recipient Adolph Lomb medal Optical Soc. Am., 1992. Mem. SPIE (program com. integrated photonics rsch. 1991-93, adv. editor to Optics Letters 1991—, program co-chair nonlinear guided-wave optics, elected program co-chair conf. high speed optical switching 1994, program sub-com. chair integrated photonics rsch. 1994, gen. chair nonlinear guided-wave optics 1995). Achievements include U.S. patents 4,700,339, 4,995,690, 4,932,739, 5,020, 050, 5,101,456, 5,078,464, 5,115,488. Address: 49 Cresci Blvd Hazlet NJ 07730 Office: U Mich Dept Elec Engring and Computer Sci 1301 Beal Ave Ann Arbor MI 48109*

ISLER, NORMAN JOHN, aircraft engine company administrator, consultant; b. Passaic, N.J., May 8, 1929; s. John and Irene Agnes (Good) I.; m. Margaret Jane Evans, Feb. 18, 1955; children: David, Barbara, Ann, Beth. BME, Clarkson U., Potsdam, N.Y., 1951; M in Engring. Mgmt., Northeastern U., 1966. Registered profl. engr., N.Y. Mgr. evaluation engring. GE Small Aircraft Engine Dept., Lynn, Mass., 1958-70, mgr. quality control, 1970-72; mgr. Metroliner program GE Transp. Systems Div., Erie, Pa., 1972-77; mgr. locomotive design GE Transportation Systems Div., Erie, Pa., 1977-80; mgr. tech. requirements GE Small Aircraft Engine Dept., Lynn, 1980-87, mgr. product support engring., 1987-89, mgr. engine program, 1989—. Contbr. articles to profl. jours. Pres. Topsfield (Mass.) Hist. Soc., 1989—; mem. Topsfield Pers. Bd., 1991; mem. Topsfield Historic Dist. Commn., 1989—; trustee Wellman Trust. 1st lt. U.S. Army, 1951-53, Korea. Decorated Army Commendation medal. Mem. ASME. Roman Catholic. Avocations: sailing, skiing. Home: 135 Perkins Row Topsfield MA 01983-1909

ISMAN, MURRAY, entomology educator. Prof. U. B.C., Vancouver. C. Gordon Hewitt award Entomological Soc. Can., 1991. Office: Univ of British Columbia, Vancouver, BC Canada V6T 1Z2*

ISONO, KIYOSHI, biochemistry educator; b. Kawasaki, Kanagawa, Japan, Aug. 16, 1931; s. Hajime and Haru (Narukawa) I.; m. Michiko Morinaga, Feb. 21, 1960; children: Jun, Mariko. BS, U. Tokyo, 1953, PhD, 1961. Scientist Riken, Inst. Phys. and Chem. Rsch., Tokyo, 1959-70; sr. scientist Riken, Inst. Phys. and Chem. Rsch., Tokyo and Wako, Japan, 1970—; dept. head Riken, Inst. Phys. and Chem. Rsch., Wako, 1978-92, chmn., dept. head conf., 1989-90; prof. dept. marine sci. Tokai U., Shimizu, Japan, 1992—, dept. head, grad. sch. of marine sci., 1993—; councilor Okochi Meml. Found., Tokyo, 1987—. Mem. editorial bd. Jour. Antibiotics, 1979—; researcher invention polyoxin fungicide, 1968; discover new antibiotics. Recipient Nihon Nogaku Sho, Japan Soc. Agrl. Scis., Tokyo, 1989, Yomiuri Nogaku Sho Yomiuri-Shinbun Press, Tokyo, 1989. Mem. Japan Soc. Biosci., Biotech. and Agrochemistry (dir. 1987-89). Avocations: golf, skiing, photography. Home: 324-1 Orido, Shimizu 424, Japan Office: Tokai U Dept Marine Sci, 3-20-1 Orido, Shimizu 424, Japan

ISRAEL, ALLEN CHARLES, psychology educator; b. N.Y.C., Sept. 27, 1943; s. Karlton and Helen (Holtzer) I.; m. Alice C. Friedlander, Apr. 11, 1970 (dec. 1985); children: Sara M. Isrel, Daniel E. Israel. BA, Harpur Coll., 1964; PhD, SUNY, Stony Brook, 1971. Lic. psychologist. Prof. psychology U. Albany, SUNY, 1971—; Author: Behavior Disorders of Childhood, 1st edit., 1984, 2d edit., 1991; contbr. articles to profl. jours. Rsch. grantee NIH, NIMH, N.Y. State Health Dept. Jewish. Office: Psychology Dept U Albany 1400 Washington Ave Albany NY 12222

ISRAEL, WERNER, physics educator; b. Berlin, Oct. 4, 1931; s. Arthur and Marie (Kappauf) I.; m. Inge Margulies, Jan. 26, 1958; children—Mark Abraham, Pia Lee. B.Sc., U. Cape Town, 1951, M.Sc., 1954; Ph.D., Trinity Coll., Dublin, 1960; D.Sc. (hon.), Queen's U., Kingston, Ont., 1987. Asst. prof. physics U. Alta., 1958-68, prof., 1968-85, Univ. prof., 1985—; Sherman Fairchild Disting. scholar Calif. Inst. Tech., 1974-75; vis. prof. Dublin Inst. Advanced Studies, 1966-68, U. Cambridge, 1975-76, Institut Henri Poincare, 1976-77, U. Berne, 1980, Kyoto U., 1986; vis. fellow Gonville and Caius Coll., Cambridge, 1985; Alta. fellow Can. Inst. for Advanced Research, 1986—. Editor: Relativity, Astrophysics and Cosmology, 1973; co-editor: General Relativity, An Einstein Centenary Survey, 1979, 300 Years of Gravitation, 1987. Recipient Izaak Walton Killam Meml. Prize, 1984. Fellow Royal Soc. Can., Royal Soc. (London); mem. Internat. Astron. Union, Can. Assn. Physicists (medal of Achievement in Physics 1981), Internat. Soc. Gen. Relativity and Gravitation. Jewish. Office: U Alta, Avadh Bhatia Physics Lab, Edmonton, AB Canada T6G 2J1

ISRAELACHVILI, JACOB NISSIM, chemical engineer; b. Tel Aviv, Israel, Aug. 19, 1944; came to U.S., 1986; s. Haim Israelachvili and Hela (Noma) Galili; m. Karina Haglund, Sept. 14, 1971; children: Josefin, Daniela. BA, U. Cambridge, 1968, MA & PhD, 1972. Prof. U. Calif., Santa Barbara, 1986—; v.p. Internat. Assn. Colloid & Interface Scientists, 1986-89. Author: Intermolecular and Surface Forces, 1985, 2d edit., 1991; contbr. rsch. articles to profl. pubs. Fellow Australian NAt. U., Canberra, 1974-86, Rsch. fellow U. Stockholm, Sweden, 1972-74; recipient Matthew Flinders medal, 1986. Fellow Royal Soc. London, Australian Acad. Sci.; mem. Alpha Chi Sigma. Office: Santa Barbara CA 93106

ISSARAGRISIL, SURAPOL, hematologist; b. Rajburi, Thailand, July 18, 1950; s. Kimhong and Somporn (Romkejpikul) I.; m. Ratana Buranayotkul, Nov. 25, 1978; children: Ruchuta, Vitchaya, Pakapon. BS, Mahidol U., 1971, MD, 1974; Diploma, Siriraj Hosp., Bangkok, 1978; Cert. Exptl. Hematology, U. Ulm, Fed. Republic Germany, 1982. Instr. dept. medicine Siriraj Hosp./Mahidol U., 1978-80, asst. prof. dep. medicine, 1980-83, assoc. prof., 1983-89, prof. medicine, 1989—; Alexander von Humboldt Rsch. fellow dept. clin. physiology U. Ulm, 1980-82; vis. fellow Fred Hutchinson Cancer Rsch. Ctr., Seattle, 1985-86; dir. Chulabhorn Bone Marrow Transplantation Ctr., Siraraj Hosp., 1988—, asst. dean for rsch., 1990-92. Editor in chief: Thai J. Hematology Transfusion Medicine, 1990-92; author: Blood Diseases in SEA, 1987; contbr. articles to profl. jours. Grantee China Med. Bd., N.Y., 1983-91, Nat. Heart, Lung and Blood Inst./NIH, Bethesda, Md., 1988—, Volkswagen Found., Hannover, Fed. Republic Germany, 1988—; recipient Nat. Best Rsch. Work award Nat. Rsch. Inst. Thailand, 1993. Mem. AAAS (internat. mem.), Hematology Soc. of Thailand (officer, sec.. 1993—), Internat. Soc. Exptl. Hematology, Am. Soc. Hematology (corr.). Buddhist. Home: 179/49 Bangkoknoi-Talingchan St, Muban Khunnondniwes, Bangkok 10700, Thailand Office: Div Hematology/Dept Med, Siriraj Hospital, 2 Prannok, 10700 Bangkok Thailand

ISSELBACHER, KURT JULIUS, physician, educator; b. Wirges, Germany, Sept. 12, 1925; came to U.S., 1936, naturalized, 1945; s. Albert and Flori (Strauss) I.; m. Rhoda Solin, June 22, 1955; children: Lisa, Karen, Jody, Eric. AB, Harvard U., 1946, MD cum laude, 1950. Intern, then resident Mass. Gen. Hosp., Boston, 1950-53; investigator NIH, 1953-56; chief gastrointestinal unit Mass. Gen. Hosp., 1957-89, chmn. com. rsch., 1967, dir. Cancer Ctr., 1987—; prof. medicine Harvard Med. Sch., 1966—, chmn. exec. com. depts. medicine, 1968—, Mallinckrodt prof. medicine, 1972—, chmn. univ. cancer com., 1972-87; mem. governing bd. NRC, 1987-90; mem. sci. bd. FDA, 1993. Editor-in-chief: (Harrison) Principles of Internal Medicine, 1976, 91—. Recipient Award for Disting. Achievement in Nutrition Bristol-Myers Squibb, 1991, Sci. Bd. FDA, 1993—. Fellow ACP (John Phillips award for Disting. Achievement in Clin. Medicine 1989), Am. Acad. Arts and Scis.; mem. Nat. Acad. Scis. (chmn. food and nutrition bd. 1983-88, mem. exec. com., mem. coun. 1987-90), Assn. am. Physicians (pres. 1977-78). Rsch. in structure and function of intestinal cells, membrane changes in malignant cells, and colon cancer. Home: 20 Nobscot Rd Newton MA 02159-1323 Office: Mass Gen Hosp 32 Fruit St Boston MA 02114-2698

ISTRICO, RICHARD ARTHUR, physician; b. Bklyn., Oct. 18, 1951; s. Arthur Ralph and Gloria Rose (Petrocelli) I.; m. Candace P. Conforti, July 20, 1974; children: Jonathan Richard, Daniel Robert. BS, Pace U., 1973; DO, Phila. Coll. Osteo. Medicine, 1978. Chief intern Interboro Hosp., 1978-79; resident in family practice Baptist Med. Ctr., 1979-80, chmn. utilization review com., 1981-82, mem. pharmacy formulary com., 1981-82; panel physician N.Y. State Athletic Commn., 1981—; team physician USA/ABF Olympics Com., Colorado Springs, Colo., 1981—; med. dir. N.Y. Golden Gloves, 1986—, N.Y. Met. Boxing Assn., 1988, Golden Hoops Basketball Tournament, 1986—; attending physician Deepdale Hosp., Little Neck, N.Y., 1980—, Parkway Hosp., Forest Hills, N.Y., 1989—; active mortality and morbidity com. Deepdale Hosp., Little Neck, 1990—, edn. com. chmn. USA Boxing, 1993. Mem. AMA, Am. Osteo. Assn. (bd. cert.), Am. Coll. Sports Medicine, Am. Osteo. Acad. Sport Medicine, Am. Coll. Gen. Practitioners (cert.), Fla. Osteo. Med. Soc., Beta Beta Beta. Republican. Roman Catholic. Avocations: jogging, weight tng., bass guitar, restoring cars. Office: 158-01 Crossbay Blvd Howard Beach NY 11414

ITAHARA, TOSHIO, chemistry educator; b. Osaka, Japan, Jan. 20, 1949. B Engring., Kyoto U., 1972, M Engring., 1974, D Engring., 1977. Lectr. Kagoshima (Japan) U., 1977-79, assoc. prof., 1979-90, prof. chemistry, 1990—. Fellow Royal Soc. Chemistry; mem. Am. Chem. Soc., Chem. Soc. Japan, Japanese Biochem. Soc., Soc. Synthetic Organic Chemistry Japan. Home: Kamifukumoto-Cho 2549-2, Kagoshima 891-01, Japan Office: Kagoshima U, Coll Liberal Arts, Korimoto 1-21-30, Kagoshima 890, Japan

ITAMI, JINROH, physician; b. Mitsu, Okayama, Japan, Feb. 21, 1937; s. Yoshito and Kinuko (Takahashi) I. MD, Okayama U., 1963. Lic. oncologist. Physician Kobe Clinic, Kobe City, Japan, 1974-83, Kurashiki Meml. Hosp., Kurashiki City, Japan, 1984-86, Shibata Hosp., Kurashiki City, 1986—. Author: Meaningful Life Therapy, 1988; producer Subliminal tape for cancer treatment. Mem. Japanese Med. Assn., Japan Soc. Cancer Therapy, Japanese Congress Morita Therapy, Assn. Meaningful Life Therapy (pres. 1986—). Avocations: mountain climbing, jogging, humor. Office: Shibata Hosp, 6108 Tamashima-Otoshima, Okayama Kurashiki 713, Japan

ITNYRE, JACQUELINE HARRIET, programmer; b. Camden, N.J., May 13, 1941; d. John Harold and Harriet Geraldine (Rankine) Bruynell; m. Thomas James Itnyre, Oct. 13, 1968 (dec. 1978); children: Beth Thierry, John. AS in Engring., Mercer County Coll., 1961; BA in Liberal Studies, San Jose State U., 1980, MLS, 1981. Media ctr. mgr. Milpitas (Calif.) Unified Sch. Dist., 1975-81; tech. libr. Lockheed Missiles and Space Co.,

Sunnyvale, Calif., 1981, programmer, 1982-83; with ground support dept. Challenger-Space Lab 2 Lockheed Missiles and Space Co., Palo Alto, Calif., 1984-85; systems mgr. gen. clin. rsch. ctr. Stanford (Calif.) U. Med. Sch., 1985-87, scientific programmer div. epidemiology, 1988—. Libr. organizer Ravenswood City Sch. Dist., East Palo Alto, Calif., 1989. Edna B. Anthony scholar San Jose State U., 1981. Mem. Assn. for Computing Machinery, ALA, Nature Conservancy, Sierra Club. Avocations: cycling, travel, sewing, drawing. Home: 2463 Louis Rd Palo Alto CA 94303-3608

ITO, KENTARO, electrical engineering educator; b. Nagoya, Aichi, Japan, Nov. 13, 1939; s. Takashi and Mitsu I.; m. Nobuko Nakashima, Mar. 21, 1969; children: Akiko, Shigeru. B in Elec. Engring. with first class honors, Nagoya Inst. Tech., 1962; M in Elec. Engring., Tokyo Inst. Tech., 1964, D in Engring., 1967. Lectr. Shinshu U., Nagano, 1967-68, assoc. prof., 1968-80, prof., 1980—; chmn. electronic engring. dept., 1983-84, 88-89, co-chmn. electrical and electronic engring. dept., 93, chmn. grad. studies on microdevices, 1991-92; vis. fellow So. Meth. U., Dallas, 1971-72; tech. adviser Nagano Prefecture, 1983—; vis. scholar of several institutes of technology in Europe, 1990; mem. Thin Film Solar Cells com. New Energy and Indsl. Devel. Orgn., Tokyo, 1992—. Author: (with others) Handbook of Solar Cells, 1985, (with K. Takahashi) Principles of Sensors, 1990; contbr. articles to profl. jours.; inventor in field. Recipient grants-in-aid Iwatani Found., Tokyo, 1981, Hoso-Bunka Kikin, Tokyo, 1984, Ministry of Edn., Sci. and Culture, Tokyo, 1969, 78, 81, 84-92. Mem. Japan Soc. Applied Physics (thin film and surface physics div., exec. sec. 1979-83, 89-91), Inst. Electronics, Info. and Communication Engrs. (chmn. component parts and materials group 1993—), Inst. Elec. Engrs. Japan. Achievements include research in heterojunction solar cells and hydrogen detectors. Office: Shinshu U Faculty Engring, 500 Wakasato, Nagano 380, Japan

ITO, SHIGEMASA, electronics executive; b. Osaka, Japan, Sept. 27, 1942; came to U.S., 1988; s. Yoshimasa and Chieko Ito; m. Kiyoe Kita, Nov. 5, 1967; children: Hitomi, Waka, Ikuyo. B in Electronics, Osaka U. Engr. Yokohama (Japan) Works Hitachi Ltd., 1965-67, chief quality control, 1967-74, asst. mgr. product engring., 1978-80, asst. mgr. part prodn., 1980-82, mgr. prodn. dept., 1982-83, dir. oversea's co., 1987-88; gen. mgr. Hitachi Home Electronics, Inc., Anaheim, Calif., 1988—; pres. Hitachi Consumer Products Mex., Tijuana, 1988—. Patentee in field. Bd. dirs. Civic Com. Japan, 1982; chief Fire Brigade Gifu Works of Hitachi, 1984. Mem. Odawara Golf Club. Avocations: golf, fishing, sailing, swimming, collecting. Home: 970 Paseo Entrada Chula Vista CA 91910-6721 Office: Hitachi Home Electronics Inc 1855 Dornoch Ct San Ysidro CA 92173-3206

ITSKEVICH, EFIM SOLOMONOVICH, physicist; b. Moscow, Dec. 4, 1922; s. Solomon Haskel and Roda Michel (Shafir) I.; m. Maria Masing, Nov. 26, 1960; 1 child, Igor. Student, Moscow State U., 1946-51, 57, D in Physics, 1971. Engr. Inst. for Measure and Devices, Moscow, 1951-55; sr. sci. rschr. Inst. for Phys.-Tech. and Radio-Tech. Measurements, Moscow, 1955-58, Inst. for High Pressure Physics Acad. Sci., Troitsk, Russia, 1958-75; head of lab. Inst. for High Pressure Physics Acad. Sci., Troitsk, 1975-91, scientist in chief, 1991; prof. Moscow Phys.-Tech. Inst., 1973. Contbr. over 150 articles to profl. jours. Soldier, 1941-45. Mem. Moscow Phys. Soc. Achievements include creation of the hydrostatic pressure chamber and investigation of the single crystals at low temperatures; co-author of the experimental discovery of electron topological transitions in metals. Home: 32 Kerchenskaya Str, 113461 Moscow Russia

ITZHAK, YOSSEF, neuropharmacologist; b. Plovdive, Bulgaria, May 18, 1950; came to U.S., 1988; s. Meir and Esther (Garti) I.; m. Tova Alexi, Sept. 5, 1976; children: Milly, Amira. MS, Hadassah Med. Sch., Jerusalem, 1978; PhD, Sackler Med. Sch., Tel-Aviv, 1982. Teaching fellow Hadassah Med. Sch., Jerusalem, 1976-78; post-doctoral fellow NYU Med. Ctr., N.Y.C., 1981-84; rsch. scientist Meml. Sloan Kettering Ctr., N.Y.C., 1984-85; asst. prof. Hadassah Med. Sch., Jerusalem, 1985-88; assoc. prof. U. Miami (Fla.) Med. Sch., 1988—; adj. prof. Barry U., Miami, 1992—. Editor: The Sigma Receptors, 1993; contbr. chpt. The Opiate Receptors, 1988; contbr. articles to profl. jours. Proceedings Nat. Acad. Sci. USA, Molecular Pharmacology, Jour. Pharmacology Exp. Ther. Recipient Rsch. awards Nat. Parkinson Found., Miami, 1989, Nat. Alliance Mentally Ill, Arlington, Va., 1990, Nat. Inst. Drug Abuse NIH, 1991. Mem. Soc. for Neuroscience, Coll. on Problems Drug Dependence. Achievements include characterization of CNS receptors for psychoactive drugs opiates and PCP; elucidation of Sigma receptors and their relevance in CNS disorders. Office: Univ Miami Med Sch Dept Biochemistry Miami FL 33101

IVANOV, LYUBEN DIMITROV, naval architecture researcher, educator; b. Varna, Bulgaria, Apr. 14, 1941; came to U.S., 1991; s. Dimitar Dimov and Petra Christova (Grozdeva) I.; m. Svetlana Zekova, Aug. 14, 1965 (div. July 1977); children: Ognyan, Iskra; m. Irina Radeva, Aug. 18, 1977; stepchildren: Ivelin, Michaela. Diploma for Naval Architecture, Higher Naval Sch., Varna, Bulgaria, 1964; PhD, Leningrad Shipbuilding Inst., USSR, 1970. Chartered engr., U.K. Designer Inst. for Shipbuilding, Varna, 1964-66; asst. Tech. Univ., Varna, 1966, reader, head of dept., 1970-74, vice-dean for rsch., 1975-76, vice-dean for continuing edn., 1985-86, dean of faculty of shipbuilding, 1987-89, reader on ship structures, 1989-91; sr. engr. Am. Bur. Shipping, N.Y.C., 1991—; vis. researcher Univ. Newcastle upon Tyne, U.K., 1974-75; dep. dirs. Inst. for Shipbuilding, Varna, 1986-87, mng. dir. 1987-89; v.p. Bulgarian Shipbuilding Corp., Varna, 1987-88. Mem. editorial bd. Marine Structures Jour., 1988-93. Founder, sec. Union of Bulgarian Scientists in Shipbuilding, Varna, 1982. Recipient badge of Honor, Presidium of the Union of Bulgarian Scientists, Sofia, 1984. Mem. Royal Instn. Naval Architects/U.K. (mem. internat. standing com. practical design of ships and mobile units symposium 1987-93), Soc. Naval Architects and Marine Engrs. Achievements include research in application of probabilistic methods in ship structures design and analysis. Home: 910 Spring Valley Rd Maywood NJ 07607 Office: Am Bur Shipping 2 World Trade Ctr Fl 106 New York NY 10048-0203

IVANOVITCH, MICHAEL STEVO, economist; b. Cetinje, Yugoslavia, Sept. 9, 1939; m. Elena Maria Balsinde, Apr. 11, 1987; children: Alexandra, Nicholas, Alexander. Diploma in Law, U. Belgrade, Yugoslavia, 1961; MBA, Columbia U., 1972, M of Philosophy, 1976, PhD, 1977. Rsch. assoc. Columbia U. Inst. on Western Europe, N.Y.C., 1977-78; prof. Columbia U. Grad. Sch. Bus., N.Y.C., 1978-88; internat. economist Fed. Res. Bank of N.Y., N.Y.C., 1978-79; prin. adminstr., sr. economist Orgn. for Econ. Cooperation and Devel., Paris, 1979-89; pres. MSI Global, Inc., N.Y.C. and Paris, 1989—; adj. prof. Columbia U.; advisor Credit Lyonnais, Paris, 1989—, Euroforum, Madrid, 1989—, Dai-Ichi Life Ins. Co., Tokyo and London, 1989—, Kuwait Investment Office, London, 1989—, The Yasuda Life Ins. Co., London and Tokyo, 1990—, The Meiji Life Ins. Co, Tokyo, 1991—. Democrat. Russian Orthodox. Avocation: music. Office: MSI Global Inc 340 W 57th St New York NY 10019-3706

IVER, ROBERT DREW, dentist; b. Miami, Fla., Feb. 6, 1947; s. William Henry and Jeanette (Minden) I.; m. Lisa Marie Stettner-Iver, May 5, 1974. Student, Ohio State U., 1965-66, U. Miami, 1966-68; DDS, Georgetown U., 1972. Pvt. practice dentistry Miami Beach, Fla., 1974—. Lt. USNR, 1968-81. Fellow ADA, Fla. Dental Assn., East Coast Dist. Dental Soc., Acad. Gen. Dentistry, Miami Beach Dental Soc. Avocations: sports fishing, ham radio operating. Office: 1205 Lincoln Rd Ste 203 Miami FL 33139-2365

IVERSEN, JAMES DELANO, aerospace engineering educator, consultant; b. Omaha, Apr. 1, 1933; s. Alfred and Asta Marie (Jorgensen) I.; m. Margery Lynn Peters, Aug. 20, 1960; children—David S., Philip W. B.S., Iowa State U., Ames, 1956, M.S., 1958, Ph.D., 1964. Prof. aerospace engring. Iowa State U., Ames, 1958—; chmn. aerospace engring. Iowa State U., 1987-90; aerodynamicist Sandia Corp., Albuquerque, 1966, NASA Ames Research Ctr., Mountain View, Calif., 1973-74, Boeing Comml. Airplane Co., Seattle, 1978, Skibsteknisk Lab., Lyngby, Denmark, 1981; educator Aarhus U., Denmark, 1991-92; cons. to govtl. agys., industry. Co-author: Wind As A Geological Process, 1985; also articles to profl. jours. Bd. dirs. Black Elk-Neihardt Park, Blair, Nebr., 1984-89, Danish Immigrant Mus., Elk Horn, Iowa, 1985-91, 92—, pres. 1989-91. Recipient Faculty citation Iowa State U., 1982, Disting. Alumnus award Dana Coll., 1987. Fellow AIAA (sr. scholar 1956, sect. pres. 1975-77); mem. Am. Soc. Engring. Edn., Engring. Accreditation Commn. of the Accreditation Bd. for Engring. and Tech., Sigma Xi, Phi Kappa Phi, Tau Beta Pi, Sigma Gamma Tau. Republican. Lutheran. Lodge: Rotary (Ames). Avocations: photography; genealogy. Office: Iowa State U 304 Town Engr Ames IA 50010

IVERSON, FRANCIS KENNETH, metals company executive; b. Downers Grove, Ill., Sept. 18, 1925; s. Norris Byron and Pearl Irene (Kelsey) I.; m. Martha Virginia Miller, Oct. 24, 1945; children: Claudia (Mrs. Wesley Watts Sturges), Marc Miller. Student, Northwestern U., 1943-44; B.S., Cornell U., 1946; M.S., Purdue U., 1947, Dr. (hon.); Dr. (hon.), U. Nebr. Research physicist Internat. Harvester, Chgo., 1947-52; tech. dir. Illium Corp., Freeport, Ill., 1952-54; dir. mktg. Cannon-Muskegon Corp., Mich., 1954-61; exec. v.p. Coast Metals, Little Ferry, N.J., 1961-62; v.p. Nucor Corp. (formerly Nuclear Corp. Am.), Charlotte, N.C., 1962-65, pres., chief exec. officer, dir., 1965-85, chmn., chief exec. officer, 1985—, also bd. dirs.; bd. dirs. Wachovia Corp., Wal-Mart Stores Inc., Wikoff Color Corp. Contbr. articles to profl. jours. Served to lt. (j.g.) USNR, 1943-46. Named Best Chief Exec. Officer in Steel Industry, Wall St. Transcript, 1993; recipient Nat. Metal of Tech., 1991. Mem. AIME, NAM, Am. Soc. Metals, Quail Hollow Country Club. Office: Nucor Corp 2100 Rexford Rd Charlotte NC 28211

IVERSON, RICHARD MATTHEW, earth scientist; b. Albert Lea, Minn., Nov. 1, 1954; s. Roger Duane and Ona Lee Elsie (Whitman) I. BS, Iowa State U., 1977; MS, Stanford U., 1980, Stanford U., 1981; PhD, Stanford U., 1984. Hydrologist U.S. Geol. Survey, Vancouver, Wash., 1984—. Mem. Am. Geophys. Union (erosion and sedimentation com. 1986—), Geol. Soc. Am. (E. B. Burwell award 1991), Am. Assn. Advancement of Sci, Mazamas Mountaineers Club, Portland, Oreg. Achievements include discovery of dynamic pore-pressure fluctuations in rapidly shearing granular materials. Office: US Geological Survey 5400 MacArthur Blvd Vancouver WA 98661

IVERSON, ROBERT LOUIS, JR., internist, physician, intensive care administrator, medical educator; b. Borden, Ind., Sept. 3, 1944; s. Robert L. and Agnes Maxine (Knight) I.; m. Elsa Maschmeyer, Sept. 3, 1967 (div. 1982); children: Nathan, Kirsten; m. Deborah A. Budd, June 16, 1984; 1 child, Richard. Student, Wabash Coll., 1962-64; BA, Ind. U., 1970, MD, 1974, Intern, 1974-75. Diplomate Am. Bd. Internal Med., diplomate in critical care medicine, Am. Bd. Internal Med. Resident (internal med.) Methodist Hosp., Indpls, 1975-77; fellow in critical care med. U. So. Calif. Shock Rsch. Unit, Ctr. for Critically Ill, L.A., 1977; visiting lectr. U. So. Calif., L.A., 1977; co-dir. critical care, teaching staff, Dept. of Med. Methodist Hosp., Indpls, 1977-84; asst. prof. Med. Wayne State U., Detroit, 1984—; assoc. dir. Intensive Care, Harper Hosp., Detroit, 1984-86; dir. Intensive Care Unit Hutzel Hosp, Detroit, 1986—, chief Dept. Critical Care Med., 1988—; participant Ind. Malpractice Review Panels, 1981-85. Author: (with others) Respiratory Care of the Neurosurgical Patient, 1983, Septic Shock in Critical Care Clinics, 1988; contbr: abstracts and articles to profl. jours. Med. advisor to Ind. Coun. Emergency Response Teams, 1980-85, mem. (singing) Ind. Symphonic Choir, 1970-84, bd. trustees, 1983-84; hon. dep. sheriff, Marion County Sheriff's Dept., 1982-84. With U.S. Army, 1964-67, Vietnam. Fellow Am. Coll. Physicians, Am. Coll. Chest Physicians; mem. AMA, Soc. Critical Care Med., Soc. for Parenteral and Enteral Nutrition, Mich. Area Radio Enthusiasts, Wayne County Med. Soc. (elected del. 1990-91), Phi Beta Kappa. Avocations: music, shortwave radio communications, sailing. Home: 1955 Wellesley Dr Detroit MI 48203-1428 Office: Dept Critical Care Med Hutzel Hosp 4707 St Antoine St Detroit MI 48201-1498

IVINS, MARSHA S., aerospace engineer, astronaut; b. Balt., Apr. 15, 1951; d. Joseph L. Ivins. BS in Aerospace Engring., U. Colo., 1973. Lic. pilot. Engr. NASA-Lyndon B. Johnson Space Ctr., 1974—, with crew sta. design br., 1974-80, engr. flight simulation, 1980—, astronaut, 1985—, mission specialist shuttle flight STS-32, 1990, mission specialist shuttle Atlantis Flight, 1992. Mem. Exptl. Aircraft Assn., 99's, Internat. Aerobatic Club. Address: NASA Johnson Space Ctr Astronaut Office Houston TX 77058

IVY, EDWARD EVERETT, entomologist, consultant; b. Hollis, Okla., Sept. 24, 1913; s. James Thomas and Betty (Minnear) I.; m. Elizabeth Alberta Slater, Feb. 23, 1935 (dec. Mar. 1981); children: James, Betty. BS, Okla State U., 1934; PhD, Tex. A&M U., 1951. Registered profl. entomologist, all 50 states, Can., Mex. Research entomologist USDA, College Station, Tex., 1940-55; salesman pesticides Mich. Chem. Corp., St. Louis, Mich., 1955-63; research entomologist Pennwalt Corp., Phila., 1963-75; cons. in pesticide devel. various nations, 1975—. Contbr. numerous articles to profl. jours. Recipient AR100 award for invention of Penncap M, 1973. Mem. Entomol. Soc. Am., Am. Registry Profl. Entomologists, Sigma Xi. Presbyterian. Home and Office: 1771 Broadway St Apt 217 Concord CA 94520-2639

IWAKURA, YOSHIO, chemistry educator; b. Tokyo, Mar. 31, 1914; s. Ichisaku and Ishi Iwakura; m. Hama Ohshima, Oct. 15, 1940; children: Tomoyuki, Naoyuki, Shigehuki, Ken. B Engring., Tokyo Inst. Tech., 1939, D Engring., 1947. Asst. prof. Tokyo Inst. Tech., 1948-55, prof., 1955-64; prof. U. Tokyo, 1962-74, prof. emeritus, 1987—; prof. Seikei U., Tokyo, 1974-82, lectr., 1982-89, prof. emeritus, 1982—. Mem. Chem. Soc. Japan (v.p. 1971-73, award 1968), Soc. Synthetic Organic Chemistry Japan (pres. 1973-75), Soc. Polymer Sci. Japan (pres. 1974-76). Home: 2-18-14 Takaido Nishi, Suginamiku, Tokyo 168, Japan

IWAMOTO, MASAKAZU, chemistry educator; b. Shimabara, Nagasaki, Japan, Oct. 26, 1948; s. Masatomo and Hiroko M.; m. Tomoko Kaetsu, Apr. 22, 1978; children: Fumiko, Michiharu, Nobuko, Shouko. B in Engring., Kyushu U., 1971, M in Engring., 1973, DEng, 1976. Asst. prof. Nagasaki U., 1976-81, assoc. prof., 1981-87; prof. chemistry Miyazaki U., 1987-90, Hokkaido U., 1990—. Recipient Nippon Kagakukai award, 1982, Catalysis Soc. Japan award, 1986, Found. for New Tech. award, 1990. Avocations: skiing, tennis. Home: Ainosato 3-1-14-29, Sapporo 002, Japan Office: Hokkaido U, Catalysis Rsch Ctr, N11 W10 Kitaku, Sapporo 060, Japan

IWANSKI, MYRON LEONARD, environmental engineer; b. Milw., June 9, 1950; s. Patrick L. and Eleanor (Literski) I.; m. Karis Brewer, Apr. 3, 1976; children: Renee, Stephanie, Carmen. BS in Civil and Environ. Engring., U. Wis., 1972, MS in Civil and Environ. Engring., 1974, MS in Water Resource Mgmt., 1975. Registered profl. engr., Tenn. Engr. Water Resources Engrs., Springfield, Va., 1974-77; project mgr. TVA, Chattanooga, 1977-80; environ. project mgr. TVA, Knoxville, 1980-88, supr. environ. rsch. and devel., 1984-88, lead environ. auditor, 1988—; mem. interagy. policy com. Nat. Acid Precipitation Assessment Program, Washington, 1986-88; mem. Oak Ridge (Tenn.) Environ. Quality Adv. Bd., 1991—. Contbr. articles to profl. jours. Councilman, vice mayor City of Norris, Tenn., 1987-89; co-founder, co-chair Citizens for Better Schs., Anderson County, 1988-90. Mem. ASCE, Air and Waste Mgmt. Assn. (environ. audit com.), Edison Electric Inst. Audit Task Force, Chi Epsilon (pres. 1972). Achievements include development of fed. agy. position on acid precipitation. Home: 102 Deerfield Ln Oak Ridge TN 37830 Office: TVA 400 W Summit Hill Dr Knoxville TN 37902

IWASAKI, TOSHIO, hydraulics engineer, investigator; b. Fukuoka, Japan, Feb. 5, 1921; s. Mineasaburo and Ito Iwasaki; m. Kimiko Nakashiki, Mar. 20, 1946; children: Kazuko Hosoya, Mitsutaka, Hideyasu. BS, Tokyo Imperial U., 1943, DEng, 1956. From asst. to asst. prof. Tohoku U., Sendai, Japan, 1950-60, prof. emeritus, 1960—; prof. Ashikaga (Japan) Inst. Tech., 1984-93, prof. civil engring. emeritus, 1993—; mem. coun. Internat. Assn. Hydraulic Rsch., 1978-81. Adviser design for def. works. Recipient Assoc. of Japan Harbour Engg Investigation award, 1979, Kahoku-Bunka Shō award Kahoku-Bunka Jigyo-Dan, 1983, Investigation award Tsunami Commn., Victoria, Can., 1985. Mem. Japan Soc. Civil Engrs. (hon.). Home: 1-9-2 Zempukuji Suginami, Tokyo 167, Japan

IWASHIMIZU, YUKIO, mechanical engineering educator; b. Inazawa, Aichi, Japan, Aug. 20, 1943; s. Minoru and Ishi (Satou) I.; m. Midori Nagata, Apr. 16, 1972; 1 child, Ohmi. B. in Engring., Kyoto U., Kyoto, Japan, 1966; M. in Engring., Kyoto U., 1968, D. in Engring., 1974. Rsch. assoc. dept. aeronautical engring. Kyoto U., 1968-74; assoc. prof. dept. mech. engring. Ritsumeikan U., Kyoto, 1975-79; prof. dept. mech. engring. Ritsumeikan U., 1980—; dept. chmn. dept. mech. engring. Ritsumeikan U., Kyoto, 1983, 89, faculty com. faculty sci. and engring., 1985, 87. Contbr. articles to profl. jours. Fellow Kyoto Sci. Club; mem. Japan Soc. Mech. Engrs., Soc. Materials Sci. Japan, Japanese Soc. for Non-destructive Inspection. Office: Ritsumeikan U Faculty Sci, & Engring Tojiinkita machi, Kita ku Kyoto Japan 603

IWASHITA, TAKEKI, physics educator, researcher; b. Tokyo, Aug. 7, 1931; s. Yasuhei and Toshi (Tani) I.; m. Kiyoko Hayashi, Apr. 20, 1967; 1 child, Yoko. MS, Tokyo U., 1962, DSc, 1965. Asst. staff Tokyo Inst. Tech., 1965-68; assoc. prof. Tokyo Gakugei U., 1968-80, prof., 1980—. Contbr. articles to profl. jours. Mem. The Phys. Soc. Japan. Home: 1-11-11 Matsugaoka, Tokorozawa, Saitama 359, Japan Office: Tokyo Gakugei U, 4-1-1 Nukuikitamachi, Koganei, Tokyo 184, Japan

IWATA, BRIAN ANTHONY, psychologist; b. Scotch Plains, N.J., Aug. 20, 1948; s. Harry Masato and Margaret (Nakagawa) I.; m. Margaret Irene Moore, May 30, 1970; children: Christina, Mary. BA, Loyola Coll., Balt., 1970; MA, Fla. State U., 1972, PhD, 1974. Lic. psychologist, Fla., Md. Asst. prof., then assoc. prof. Western Mich. U., Kalamazoo, 1974-78; assoc. prof. Sch. Medicine Johns Hopkins U., Balt., 1978-86; prof. psychology U. Fla., Gainesville, 1986—; cons. NIH, 1980-84, NIMH, 1988—. Editor 5 textbooks; contbr. articles to profl. publs. Fellow APA (div. pres.), Am. Psychol. Soc., Am. Assn. Mental Retardation; mem. Assn. Behavior Analysis (pres.), Assn. Advancement of Behavior Analysis (pres.), Fla. Assn. Behavior Analysis (pres.), Soc. Exptl. Analysis of Behavior (v.p.). Home: 4312 NW 12th Pl Gainesville FL 32605 Office: Univ Fla Dept Psychology Gainesville FL 32611

IWATA, KAZUAKI, manufacturing engineer educator. Prof. mechanical engring. Osaka U., Japan. Recipient Frederick W. Taylor Rsch. medal Soc. Manufacturing Engrs., 1990. Office: Osaka Univ Dept Mech Engring, 18-8 Hatsumachi, Neyagawa-shi Osaka 572, Japan*

IWATA, KAZUO, business executive; b. Feb. 18, 1910; s. Tsuramatsu and Kocho Iwata; m. Emiko Sumiyoshi, 1939; 3 children. Ed. Tokyo U. Vice pres. Japan Machinery Exporters Assn., 1973—; mng. dir. Japan Fedn. Employers' Assn., Fedn. of Econ. Orgns., 1976-80; mgr. Japanese Com. for Econ. Devel., 1976—; pres. Toshiba Corp., 1976-80, chmn. bd., 1980—; vice chmn. Electronic Industries Assn. Japan, 1977—; dir. Japan Electric Machinery Assn., 1978—; chmn. Japan Soc. for Promotion Machine Industry, 1981—; mem. indsl. structure deliberative council Ministry Internat. Trade and Industry, 1977—. Recipient long-standing disting. service in electric industry prize, 1976, prize Japanese Minister Internat. Trade and Industry, 1979; decorated Order Rising Sun, 2d class. Avocations: golf; pottery; decorative fish. Address: Osaka Univ, 18-8 Hatsumachi Neyagawa Shi, Osaka 572, Japan

IWATANI, YOSHINORI, physician, educator, researcher; b. Osaka, Japan, May 16, 1952; s. Nobuyuki and Teruko Iwatani; m. Atsuko Iwatani, Sept. 24, 1982; children: Shuko, Kento. MS, Kyushu U., Fukuoka, Japan, 1979; DMS, Osaka U., 1983. Rsch. fellow Univ. Toronto, Can., 1984-86; rsch. asst. Osaka Univ. Med. Sch., 1983-84, 86-89, asst. prof., 1989—, assoc. prof., 1993—. Mem. Japanese Soc. Clin. Pathology (specialist, councilor 1989—), Japanese Soc. Clin. Immunology (councilor 1991—), Japan Endocrine Soc. (councilor 1989—), Japan Thyroid Assn. (councilor 1988—, Shichijo prize 1989), Am. Endocrine Soc., Am. Thyroid Assn. (century mem. 1991—). Home: 3-13-9 Nageremachi Hirakonu, Osaka 547, Japan Office: Osaka U Med Sch Lab Medicine, 2-2 Yamadaoka, Suita Osaka 565, Japan

IYER, POORNI RAMCHANDRAN, toxicologist; b. Bangalore, India, Nov. 13, 1961; came to U.S. 1985; d. Pannangudi Sivram Ramchandran and Periyakulam Narayanaswamy Visalakshi. DVM, Bombay Vet. Coll., India, 1984; PhD, Tex. A&M U., 1989. Resident vet. officer Dairy Devel. Corp.-Aarey Milk Colony, Maharashtra, India, 1984-85; grad. asst. Coll. Vet. Medicine, Tex. A&M U., College Station, 1985-89; rsch. assoc. Dept. Vet. Anatomy - SVM Inst. Environ. Studies, La. State U., Baton Rouge, 1990—. Contbr. articles to profl. jours. Mem. Soc. of Environ. Toxicology and Chemistry, Assn. for Women in Sci., Am. Assn. Vet. Anatomists. Office: La State Univ Dept Vet Anatomy Sch Vet Medicine Baton Rouge LA 70803

IYODA, MITSUHIKO, economics educator; b. Aichi, Japan, Oct. 1, 1943; m. Masako Chojahara, Jan. 7, 1975; 1 child, Muneyoshi. BA in Econs., Wakayama Nat. U., Japan, 1965; MA in Econs., Osaka City U., Japan, 1968. Assoc. prof. Momoyama Gakuin U., Osaka, Japan, 1972-82; prof., 1982—; vis. fellow Lancaster U., Eng., 1982-83; dir. Rsch. Inst. Momoyama Gakuin U., Osaka, 1985-89, dean faculty econs., 1990-92. Co-author: (textbook) Introduction to Economics, 1989. Mem. Japan Assn. Econs. and Econometrics, Internat. Assn. for Rsch. in Income and Wealth, Japan Soc. Econ. Policy. Avocations: tennis, music. Home: 3-6-14 Momoyama-Dai, Sakai-Shi Osaka 590-01, Japan Office: Momoyama Gakuin U., 237-1 Nishino, Sakai-Shi Osaka 588, Japan

JABAGI, HABIB DAOUD, consulting civil engineer; b. Ramleh, Palestine, Apr. 23, 1922; came to U.S., 1988; s. Daoud Hanna and Fumia (Salim) J.; m. Annie Ibrahim Haddad, June 30, 1950; children: Jumana, Suha, Jinan, Sahar. BSCE, Am. U. Beirut, 1945; PhD in Civil Engring., N.W. London U., 1978. Cert. in engring. and contract mgmt. Land surveyor Govt. of Palestine, Jaffa, 1945; engr. Lydda (Palestine) Mcpl. Coun., 1952-64; co-owner bldg. and contracting co., Amman, Jordan, 1952-64; project mgr. C.C.C., Riyadh, Saudi Arabia, 1964-77; v.p. devel. and constrn. co., Unayza, Saudi Arabia, 1978-81; asst. project mgr. Queen Alia Internat. Airport Amman Dar El Handasa Cons. Engrs., 1981-84; contract mgr. ICICO, Amman, 1984-87; cons. civil engr., Oak Park, Ill., 1987—. Mem. ASCE, Engr. Unions and Socs. of Jordan. Lebanon and Iraq. Home: 7305 Lake St Apt 608 River Forest IL 60305-2226 Office: 427 Lenox St Oak Park IL 60302-1339

JABLOWSKY, ALBERT ISAAC, civil engineer; b. Bklyn., Oct. 6, 1944; s. Louis and Esther (Charloff) J.; m. Diane Ellen Salb, Nov. 5, 1967; children: Lorraine, Erica. BS, Cornell U., 1967; MSCE, Poly. U., 1972. Registered profl. engr. Jr. asst. civil engr. N.Y. State Dept. Transp., N.Y.C., 1967-72, sr. civil engr., 1972-79; project mgr. N.Y. State Dept. Transp., Queens, 1979-87, project supr., 1987—. Bd. dirs. Flatbush Pk. Jewish Ctr., Bklyn., 1990. Mem. ASCE, Cornell Soc. Engrs., N.Y. State Assn. Transp. Engrs., N.Y. Acad. Scis., Mensa. Achievements include supervision, management, inspection, and inventory of local and state bridges in N.Y.C., State design projects including Queensboro bridge rehabilitation contract 4, Pulaski Bridge, Williamsburg bridge cable suspension system, Queens Midtown viaduct rehabilitation, Major Deegan Expressway corridor bridge rehabilitation 40 bridges, Bruckner Expressway viaduct rehabilitation, and Henry Hudson Parkway rehabilitation. Home: 2440 E 66th St Brooklyn NY 11234-6709 Office: NY State Dept Transp 47-40 21st St Long Island City NY 11101

JABRE, EDDY-MARCO, architect; b. Beirut, July 10, 1948; came to U.S., 1986; s. Farid and Lody (Abourizk) J.; m. Ifrandate Alame, Oct. 12, 1974; children: Moune, Joe, Marc. Architect DPLG, Ecole Nationale Superieure Des Beaux Arts, Paris, 1975; urban planner, Universite Paris XII, Vincennes, 1976. Ptnr. Atelier D'Etudes Raspail, Paris, 1975-78; dir. F.S.K., Riyadh, Saudi Arabia, 1978-80; sr. ptrn., pres. Copreco, Riyadh, Saudi Arabia, 1980-86; pres. Nedcorp, Winchester, Mass., 1986—, Maine Nedcorp, Winchester, Mass., 1987—, Nedcorp Custom Homes, Winchester, Mass., 1989—. Avocations: swimming, reading. Office: Nedcorp 16 Grove St Winchester MA 01890-3842

JACK, STEVEN BRUCE, forest ecologist, educator; b. Anderson, Ind., Apr. 11, 1960; s. Ronald Merle and Alta Ruthe (Crose) J.; m. Barbara Elaine Maness, June 18, 1983; 1 child, Joshua David. AB in Physics, Math., Erskine Coll., 1982; MS in Forest Resources, U. Fla., 1986; PhD in Forest Ecology, Utah State U., 1990. Postdoctoral fellow Utah State U., Logan, 1990-91; asst. prof. forest ecology Tex. A&M U., College Station, 1991—; referee various forest and ecol. rsch. jours., 1989—. Contbr. articles to profl. jours. Mem. Ecol. Soc. Am., Soc. Am. Foresters, Internat. Assn. Vegetation Sci., Xi Sigma Pi, Phi Kappa Phi, Omicron Delta Kappa. Office: Dept Forest Sci Tex A&M U College Station TX 77843-2135

JACKS, JEAN-PIERRE GEORGES YVES, chemical engineering sales executive; b. St. Germain En Laye, France, July 6, 1948; came to U.S., 1952; s. Joseph Albert and Edith Marie (Bertrand) Jacks; m. Cynthia Evans Burns, July 31, 1971; children: Eric, David. BSChemE, U. Calif., Berkeley, 1970; MSChemE, MIT, 1971. Registered profl. engr., Tex. Asst. dir. MIT Sch. of Chem. Engring. Practice, Bound Brook, N.J., 1971-72; tech. data engr. M.W. Kellogg Co., Houston, 1972-75; process engr. The M.W. Kellogg Co., Houston, 1975-78, sr. process engr., 1978-80, sales coord., 1980-81, sales rep., 1981-83, comml. rep., 1983-86, sales mgr., 1986-89, comml. v.p., 1989—. Contbr. articles to publs. including Hydrocarbon Processing Mag. and Can. Jour. of Chem. Engring. Treas. Glenshire Community Assn., Houston, 1978. Mem. Am. Inst. Chem. Engrs., Am. Chem. Soc. Republican. Roman Catholic. Office: MW Kellogg Co 601 Jefferson Ave Houston TX 77210-4557

JACKSON, BENNIE, JR., nuclear engineer; b. Edgefield, S.C., Dec. 22, 1947; s. Bennie and Annie Mae (Robinson) J.; m. Andrea Janice Dungy, July 1980 (div. Feb. 1985); m. Georgia May Mitchell, Aug. 17, 1985. BS in Math., Voorhees Coll., Denmark, S.C., 1970; MS in Nuclear Engring., Radiol. Scis., Ga. Inst. Tech., 1972. Health physics engr. Savannah River Plant, Aiken, S.C., 1972-74; environ. effects engr. Westinghouse Electric Corp., Pitts., 1974-76; prin. radiation engr., instr. visual/non-destructive testing Commonwealth Edison Co., Wilmington, Ill., 1984-88, instr. radiation protection, 1988—. With USN, 1977-83. Mem. Health Physics Soc., Am. Soc. for Non-Destructive Testing, Am. Soc. Quality Control, Am. Nuclear Soc. Avocations: reading, drawing. Office: Commonwealth Edison Co PTC 36400 Essex Rd Wilmington IL 60481-9500

JACKSON, CARL ROBERT, obstetrician/gynecologist; b. Mpls., Jan. 8, 1928; s. Carl J. and Mildred J. (Johnson) J.; children: Amy, Carrie, Tom; m. Carol Franklin, 1992. BA, Gustavus Adolphus Coll., 1951; MD, Jefferson Med. Coll., 1956. Diplomate Am. Bd. Ob-Gyn. Intern St. Mary's Hosp., Duluth, Minn., 1956-57; resident and postdoctoral fellow in ob-gyn U. Wis.-Madison, 1957-61; pvt. practice medicine specializing in ob-gyn. Madison, 1961—; mem. active staff Madison Gen. Hosp., 1961—, vice chief staff, 1972-74; mem. attending staff Univ. Hosps., Madison; assoc. clin. prof. ob-gyn. U. Wis., 1971—, mem. high risk obstet. team, 1974-75; chmn. Physicians Alliance, Dane County, Wis.; mem. Madison Ob-Gyn. Ltd., 1961—; chmn. Madison Med. Ctr., 1979-86; bd. dirs. Physicians Plus HMO, 1988—. Chmn., founder Physician Friends of Gustavus Adolphus Coll., St. Peter, Minn., 1984; mem. peer rev. coun. Office of Ins. Commr., State of Wis. Am. Cancer Soc. fellow, 1960-61; recipient Excellence in Teaching award, U. Wis., 1979, 91. Mem. Am. Coll. Ob-Gyn., Cen. Assn. Obstetricians and Gynecologists, AMA, Wis. Ob-Gyn. Soc. (past pres.), State Med. Soc. (Svc. Recognition award 1990), Dane County Med. Soc., Madison Club, Kiwanis. Republican. Lutheran. Home: 3089 Timberlane Verona WI 53593 Office: Madison Med Ctr 20 S Park St Madison WI 53715-1348

JACKSON, CURTIS MAITLAND, metallurgical engineer; b. N.Y.C., Apr. 20, 1933; s. Maitland Shaw and Janet Haughs (Dunbar) J.; m. Cordelia Ann Shupe, July 6, 1957; children: Carol Jackson Adams, David Curtis. B.S. in Metall. Engring., NYU, 1954; M.S., Ohio State U., Columbus, 1959, Ph.D. (Battelle staff fellow), 1966. Registered profl. engr., Ohio. Prin. metall. engr. Columbus div. Battelle Meml. Inst., 1954-61, project leader, 1961-67, asso. chief specialty alloys, 1967-77, asso. mgr. phys. and applied metallurgy, 1977—; researcher in metall. tech. Chmn. bd.: Wire Jour, 1976-77; dir., 1973-78; Contbr. tech. articles profl. jours. Mem. troop com. Boy Scouts Am., 1975-83 , asst. scoutmaster, 1978-83 ; advisor Order of DeMolay, 1954-57, 78-86; mem. ofcl. bd. Methodist Ch., 1957-66. Recipient IR-100 award Indsl. Research Mag., 1976; recipient certificate of appreciation Soc. Mfg. Engrs., 1977, awards Order of DeMolay, 1955, 1978, 83. Mem. Wire Found. (dir. 1974-86), Wire Assn. Internat. (v.p. 1973-76, pres. 1976-77, dir. 1970-78, Mordica Meml. award 1977, J. Edward Donnellan award 1978, Meritorious Tech. Paper award 1981), N.Y. U. Metall. Alumni Assn. (pres. 1966-68), Am. Inst. Mining, Metall. and Petroleum Engrs. (chmn. Ohio Valley sect. 1964-66, chmn. North Central U.S. region 1965-66), Am. Soc. Metals, Am. Vacuum Soc., NYU, Ohio State U. alumni assns., Sigma Xi, Alpha Sigma Mu, Phi Lambda Upsilon. Club: NYU. Home: 5088 Dalmeny Ct Columbus OH 43220-2693 Office: 505 King Ave Columbus OH 43201-2681

JACKSON, D. MICHAEL, research entomologist, educator; b. Bryan, Ohio, July 10, 1949; s. Herbert Louis and Jennie Sofia (Fritz) J.; m. Rebecca Ann Norwood, Oct. 2, 1982; children: Jenny Ann, David. BS, Mich. State U., 1971; MS, Wash. State U., 1975, PhD, 1978. Adj. postdoctoral fellow U. Fla. and USDA Agrl. Rsch. Svc., Gainesville, 1978-79; rsch. entomologist USDA Agrl. Rsch. Svc., Oxford, N.C., 1980—; assoc. prof. N.C. State U., Raleigh, 1980—. Editor, chmn. editorial bd. Tobacco Sci.; contbr. over 100 articles and abstracts to sci. jours. Recipient cert. of merit USDA Agrl. Rsch. Svc., 1990. Mem. Entomol. Soc. Am., N.C. Entomol. Soc. (pres. 1990-91), Oxford Jaycees (pres. 1988-89, Disting. Svc. award 1991), Men's Garden Club Oxford (pres. 1983, 92), Kiwanis (pres. Oxford 1992-93), Sigma Xi. Home: 205 Grace St Oxford NC 27565 Office: USDA Agr Rsch Svc Crops Rsch Lab PO Box 1168 Oxford NC 27565

JACKSON, DOUGLAS WEBSTER, environmental scientist, consultant; b. Houston, Sept. 5, 1949; s. Jame Lee and Ruby Mae (Dowell) J.; m. Catherine Ann Wagner, Apr. 6, 1977; children: William Douglas, Elizabeth Ann. BS in Forestry/Environ. Mgmt., Stephen F. Austin State U., 1975, MS in Biology/Aquatic Ecology, 1977. Cert. environ. profl. Rsch. asst. Stephen F. Austin State U., Nacogdoches, Tex., 1975-77; ecologist S.W. Rsch. Inst., Houston, 1977-79; prin. investigator, contract cons. Sandia Nat. Labs., Albuquerque, 1979-86; project mgr. Rollins Environ. Svcs., Houston, 1986-87, regional mgr., 1987-88; project mgr. Internat. Tech. Corp., Houston, 1988-90; ops. mgr. Separation Systems Cons., Inc., Houston, 1990—; Contbr. chpts to books, articles to profl. jours. Mem. Nat. Assn. Environ. Profls. (cert.). Office: SSCI 16811 El Camino Real # 214 Houston TX 77058

JACKSON, EARL, JR., medical technologist; b. Paris, Ky., Sept. 4, 1938; s. Earl Sr. and Margaret Elizabeth (Cummins) J. BA, Ky. State U., 1960; postgrad., U. Paris, 1978. Clin. rsch. coord. Harvard U., Boston, 1962-64; chem. devel. specialist Electro-Power Pacs, Corp., Cambridge, Mass., 1964-67; sr. rsch. tech. Mass. Gen. Hosp., Boston, 1967-81, med. tech. specialist, 1981—. Contbr. articles to profl. jours. Mem. AAAS, NY. Acad. Scis., Am. Assn. Clin. Chemistry, Am. Soc. Clin. Microbiology, N.E. Assn. for Microbiology and Infectious Disease. Democrat. Home: 89 Oxford St Somerville MA 02143-1617 Office: Mass Gen Hosp Fruit St Boston MA 02114-2620

JACKSON, EDWIN L., electrical engineer; b. Albany, Tex., Jan. 21, 1930; s. Eugene L. and Elsie (Payne) J.; m. Ruby Re Corn, June 7, 1953; children: Elisabeth, Elaine, Gene. BSEE with honors, U. Okla., 1952, postgrad., 1953. Registered profl. engr., Tex. Sr. design engr. Gen. Dynamics, 1956-62; program mgr. E Systems, 1963-68; gen. mgr. div. static power Varo, Inc., Garland, Tex., 1962-63, 68-72, v.p., gen. mgr. div. electronic systems, 1989—; v.p. Forney Engring. Co., 1972-83; sr. v.p., gen. mgr. Dallas Corp., 1983-88. Deacon 1st Bapt. Ch., Garland, 1979—; bd. dirs. Exodus, Dallas, 1990—; del. corp. coun. Econ. Devel. Com., Garland, 1991. Lt. USN, 1953-56. Mem. Navy League U.S. (life), Garland C. of C. (bd. dirs. 1991). Democrat. Home: 3126 Pecan Ln Garland TX 75041-4456 Office: Varo Inc 2800 W Kingsley Rd Garland TX 75041-2499

JACKSON, FRED, oil executive; b. Transylvania, Minor, July 21, 1952; s. Fred Sr. and Marie J.; children: Tawanna, Tanecia. Internat. Degree of Master/Mates Pilots, Md. U., 1959. With Internat. Mktg. Corp., Phila., 1982—; owner Fred Jackson and Co., Fred Jackson Airlines, Fred Jackson Indsl., Fred Jackson Computer, Fred Jackson Investment Firm, others; creator Baxter Internat. Inc. Contbr. articles to jours. Republican. Home and Office: 8407 Williams Ave Philadelphia PA 19150-1920

JACKSON, JIMMY LYNN, engineer, consulting spectroscopist; b. Claremore, Okla., Sept. 14, 1957; s. Michael R. and Frances L. (Harrison) J. Field svc. engr. Labtest Equipment Co., L.A., 1979-84, Spectro A.I., Inc., Fitchburg, Mass., 1984—. Office: Spectro AI Inc 160 Authority Dr Fitchburg MA 01420

JACKSON, KENNETH ARTHUR, physicist, researcher; b. Connaught, Ont., Can., Oct. 23, 1930; s. Arthur and Susanna (Vatcher) J.; m. Jacqueline Della Olyan, June 20, 1952 (div.); children: Stacy Margaret, Meredith Suzanne, Stuart Keith; m. Camilla M. Maruszewski, June 21, 1980 (div.). BS, U. Toronto, 1952, MS, 1953; PhD, Harvard U., 1956. Postdoctoral fellow Harvard U., Cambridge, Mass., 1956-58; asst. prof. metallurgy Harvard U., 1958-62; mem. tech. staff Bell Labs., Murray Hill, N.J., 1962-67; head material physics research dept. Bell Labs., 1967-81, head optical materials research dept., 1981-89; prof. materials sci. and engring. U. Ariz., 1989—; lectr. Welch. Found., 1970, 85; mem. research adv. panel Air Force Office Sci. Research, 1976-82, space application bd. Nat. Acad. Sci., 1974-82. Contbr. articles to profl. jours. Recipient Mathewson Gold medal AIME, 1966. Fellow The Metallurgical Soc.-AIME, The Am. Phys. Soc.; mem. Internat. Orgn. Crystal Growth (treas. 1978-86), Am. Assn. Crystal Growth (pres. 1968-75, council), Materials Research Soc. (v.p. 1975-77, pres. 1977-78, council), Am. Soc. Metals, Am. Phys. Soc., AAAS, Engring. Council for Profl. Devel. (mem. council), Fedn. Materials Socs. (trustee). Patentee in field. Office: U Ariz 4715 E Ft Lowell Rd Tucson AZ 85712-1201

JACKSON, MICHEL TAH-TUNG, phonetician, linguist; b. Phila., May 28, 1961; s. James Spurgeon and Maria May (Shen) J. BA, Yale u., 1983; PhD, UCLA, 1988. Assoc. software engr. Microsoft Co., Bellevue, Wash., 1984; lectr. Yale U., New Haven, Conn., 1988-89; asst. prof. Ohio State U., Columbus, 1989—. Recipient Rsch. grant NSF, 1988, Nat. Rsch. Svc. award NIH, 1990-91. Mem. Internat. Phonetics Assn., Internat. Clin. Linguistic and Phonetics Assn., Linguistic Soc. Am., Acoustical Soc. Am. Office: Ohio State Univ Dept Speech and Hearing Sci 1070 Carmack Rd Columbus OH 43210

JACKSON, PETER EDWARD, chemist; b. Melbourne, Australia, May 21, 1962; s. Peter Lee J. and Shirley June (Marshall) Dunmall. BS with honors, U. New South Wales, 1984, PhD, 1988. Rsch. asst. U. New South Wales, Sydney, 1988; applications chemist Waters Chromatography, Milford, Mass., 1989-90, Sydney, 1991—; mem. Australian Standard Working Group, Sydney, 1987; com. mem. New South Wales RACI Analytical Group, Sydney, 1992—. Co-author: Ion Chromatography: Principles and Applications, 1990; contbr. articles to Jour. Chromatography. Mem. Am. Chem. Soc., Royal Australian Chem. Inst. Achievements include research in expanding application range of ion chromatography. Office: Millipore Waters, Private Bag 18, Lane Cove 2066, Australia

JACKSON, RAYMOND CARL, cytogeneticist; b. Medora, Ind., May 7, 1928; s. Thornton Comadore and Flossie Oliva (Booker) J.; m. T. June Snyder, Oct. 24, 1947; children: Jeffrey Wayne, Rebecca June. AB, Ind. U., 1952, AM, 1953; PhD, Purdue U., 1955. Instr. to asst. prof. U. N.Mex., Albuquerque, 1955-58; asst. prof. of Botany U. Kans., Lawrence, 1958-60, assoc. prof. of Botany, 1961-64, prof. of Botany, 1964-71, prof. and chmn. Botany, 1969-71; prof. and chmn. biol. scis. Tex. Tech U., Lubbock, 1971-78, Horn prof. of Biol. Scis., 1990—; chmn. interdepartmental PhD Program in Genetics, U. Kans., chmn. dept. Botany, U. Kans., 1969-71; speaker and presenter in field. Contbr. numerous articles to profl. jours. Staff sgt. USAF, 1947-49. Mem. Genetics Soc. Am., Genetics Soc. of Can., Soc. for the Study of Evolution, Botanical Soc. of Am. (BSA Merit award 1992), Am. Soc. Plant Taxonomists, Internat. Orgn. of Plant Biosystematists, Delta Phi Alpha, Sigma Xi, Phi Sigma. Republican. Achievements include research on pairing control genes and their comparative effects at the diploid and polyploid levels, in genetics, cytogenetics, and gametic selection in Haplopappus gracilis, and in cytogenetics of diploid Triticum species. Home: 3726 64th Dr Lubbock TX 79413-5312 Office: Dept Biol Scis Tex Tech Univ Lubbock TX 79409

JACKSON, ROBERT BENTON, IV, environmental engineer; b. Atlanta, June 14, 1965; s. Robert Benton and Linda Arlene Jackson. BS, U. Ga., 1988, MS, 1990. Registered engr.-in-tng., Ga. Soil and concrete technician Hill-Fister Cons. Engrs., Clarkston, Ga., 1988; grad. researcher U. Ga., Athens, 1988-90; prodn. engr. Purina Mills, Inc., Tampa, Fla., 1989; environ. engr. U.S. EPA, Athens, Ga., 1990—, mem. sci. adv. bd. Office R%D, 1991—; mem. environ. permitting com. Athens/Clarke County Govt., 1992; speaker in field. Contbr. chpts. to books, article to jour. Hugar F. Wilkes scholar, 1988. Mem. NSPE, Sigma Xi, Gamma Sigma Delta. Achievements include findings of audio-tropic responses of grasses, agriculutres net sink potential for CO2. Home: U Ga PO Box 2497 Athens GA 30612 Office: US EPA 960 College Station Rd Athens GA 30613

JACKSON, ROBERT LORING, science and mathematics educator, academic administrator; b. Mitchell, S.D., June 8, 1926; s. Olin DeBuhr and Edna Anna (Hanson) J.; m. Elizabeth Denise Koteski; children: Charles Olin, Catherine Lynne, Cynthia Helen. BS, Hamline U., 1950; MA, U. Minn., 1959; PhD, 1965. Tchr. math. and sci., pub. schs., Heron Lake, Minn., 1950-52; tchr. math. Lakewood (Colo.) Sr. High Sch., 1952-53, Nouassuer Air Force Sch. Casablanca, Morocco, 1953-54, Baumholder (Germany) Elem. Sch., 1954-55, U. Minn. Univ. Lab. Sch., Mpls., 1955-60; asst. prof. sci. and math. edn. U. Minn., Mpls., 1965-66, assoc. prof., 1966-70, prof., 1970—, head sci. and math. edn., 1980-84, assoc. chmn., dir. undergrad. studies, curriculum and instrn., 1984-88, assoc. chmn., 1989-92; vis. prof. Hamline U., St. Paul, 1958, Mont. State U., Bozeman, 1981, Bethel Coll., St. Paul, 1981, No. Mich. U., Marquette, 1983-84; cons. math. Minn. Dept. Edn., St. Paul, 1960-62. Bd. dirs. Oratorio Soc. Minn., Minn. Chorale, Mpls., 1973-88, pres., 1978-80. With U.S. Army, 1944-46. Decorated Purple Heart; recipient First Alumni award 1988, Disting. Teaching award Coll. Edn., U. Minn., 1984. Mem. Minn. Coun. Tchrs. Math., Nat. Coun. Tchrs. Math., Math. Assn. Am., Internat. Platform Assn. Methodist. Co-author: (book/man series) Laboratory Mathematics, 1975-76. Home: 2710 Dale St N Apt 101 Saint Paul MN 55113-2384 Office: U Minn 159 Pillsbury Dr SE Minneapolis MN 55455-0208

JACKSON, SHERRY DIANE, internist; b. Detroit, June 30, 1943; d. William and Julia (Berkan) Barris; m. Stuart Evan Seigel, Sept. 24, 1989. BS, Wayne State U., 1964; MA, U. Mich., 1970, MD, 1983. Diplomate Am. Bd. Internal Medicine. Intern and resident in internal medicine Henry Ford Hosp., Detroit, 1983-86, fellow in rheumatology, 1986-87; fellow in rheumatology med. sch. George Washington U., 1989-90; med. officer Nat. Heart, Lung & Blood Inst. NIH, Bethesda, Md., 1988-90; asst. prof. clin. medicine Coll. Physicians and Surgeons, Columbia U., N.Y.C., 1990—; founder, dir. Cholesterol Care Ctr., N.Y.C., 1991—; chmn. task force on women and heart disease Am. Heart Assn., N.Y.C. affiliate, 1991—, bd. dirs.; cons. Women in Need, N.Y.C., 1991—. Mem. Women's Med. Assn., Internat. Women's Forum (co-chair health com. 1993—), N.Y. Acad. Scis., N.Y. County Med. Soc., N.Y. Women's Agenda (chair health edn. subcom. 1993—). Home: 1095 Park Ave New York NY 10128 Office: 1150 Park Ave New York NY 10128

JACKSON, TERRANCE SHELDON, editor; b. N.Y.C., Aug. 17, 1964; s. Alexander Wallace Jackson and Lezlie Sylvia (Melville) Linder. BS in Computer Sci., U. Pa., 1987. System analyst IBM, Colorado Springs, Colo., 1987-90; editor Advanced Knowledge and Self-Awareness Press, N.Y.C., 1990—. Author: AIDS/HIV is Not a Death Sentence, 1992, Putting It All Together, 1991. Mem. AAAS, IEEE, N.Y. Acad. Sci. Office: AKASA Press PO Box 313 New York NY 10027

JACO, WILLIAM H., mathematical association executive; b. Grafton, W.Va., July 14, 1940; s. William Howard Sr. and Catherine Virginia (White) J.; children: William, Brent; m. Linda Kanewske, May 6, 1978; children: John, Andrew. BA magna cum laude, Fairmont (W.Va.) State Coll., 1962; MA, Pa. State U., 1964; PhD, U. Wis., 1968. Project mathematician Ordinance Rsch. Lab., University Park, Pa., 1962-64; asst. prof. U. Mich., Ann Arbor, 1968-73; asst. prof. Rice U., Houston, 1970-73, assoc. prof., 1973-78, prof., 1978-82; head dept. math. Okla. State U., Stillwater, 1982-87, prof. math., 1982-88; exec. dir. Am. Math. Soc., Providence, R.I., 1988—; mem. Joint Policy Bd. for Math., Washington, 1988—, Bd. Math. Scis., Washington, 1987-90, Inst. for Advanced Study, 1971-72, 78-79, 86; vice-chmn R.I. Math. Scis. Edn. Coalition, Providence, 1990; professorial rsch. fellow U. Melbourne, Australiam 1987-88. Author: Lectures on Three-Manifolds, 1977; co-author: Seifert Fibered Manifolds, 1979; editor: Contemporary Math., 1985-88; contbr. articles to profl. jours. Active Bd. Edn. Devel. Fund, Providence, 1991—; mem. adv. bd. Roger Williams Coll. Sch. Sci. and Math., Bristol, R.I., 1990—. Graduate fellow NSF, 1964-67, Postdoctoral fellow NSF, 1971-72; Rsch. grantee, NSF, 1968-88. Office: Am Math Soc PO Box 6248 201 Charles St Providence RI 02940-6248

JACOB, FRANÇOIS, biologist; b. Nancy, France, June 17, 1920; s. Simon and Therese (Franck) J.; m. Lysiane Bloch, Nov. 27, 1947 (dec. 1984); children: Pierre, Laurent, Odile, Henri. M.D., Faculty of Medicine, Paris, 1947; D.Sc., Faculty of Scis., Paris, 1954; D.Sc. (hon.), U. Chgo., 1965. Asst. Pasteur Inst., 1950-56, head dept. cellular genetics, 1960—, pres., 1982-88; prof. cellular genetics Coll. of France, 1964—. Author: The Logic of Life, 1970; The Possible and the Actual, 1981, The Statue Within, 1987. Recipient Charles Leopold Mayer prize, 1962; Nobel prize in physiology and medicine (with A. Lwoff and J. Monod), 1965. Mem. Académie des Sciences (Paris); fgn. mem. Royal Danish Acad. Scis. and Letters, Am. Acad. Arts and Scis., Nat. Acad. Scis. (U.S.), Am. Philos. Soc., Royal Soc. (London), Académie Royale de Médecine de Belgique, Acad. Scis. Hungary, Royal Acad. Scis. Madrid. Rsch. on genetics bacterial cells and viruses; contbr. to mechanisms of info. transfer (messenger RNA) and genetic basis of regulatory circuits, early stages of the mouse embryo. Office: Pasteur Inst, 25 Rue du Dr Roux, 75724 Paris Cedex 15, France

JACOB, GEORGE (PRASAD), electronics and telecommunications engineer; b. Tamil, Nadu, India, Dec. 2, 1950; s. Thovalai Krishna Velayya Pillai Jacob and Thankamma Mary (Chacko) J.; m. Mary Eapen, Jan. 22, 1979; children: Sushma, Mariam, George. B in Electronic Engring., Coll. Engring., Madras, India, 1973, M in Electronic Engring., 1975; postgrad., U. Oslo, 1985—. Assoc. lectr. in electronics, communications engring. Coll. Engring., Madras, 1975-81; head electronics sect. William V.S. Tubman Coll. of Tech., Harper, Liberia, West Africa, 1982-84, instr. electronics, 1981-84; rsch. scholar U. Oslo, Blindern, 1985—; cons. Tangerud Miljø/Renhold, Oslo, 1988—. Co-author Digital TV jour., 1980. Mem. IEEE, World Wildlife Fund, Norwegian Physics Assn., Norwegian Rsch. Workers Assn., Norwegian Engrs. Assn. Mem. Congress Party. Indian Christian. Avocations: reading, writing, jogging, tennis, cycling. Home: Krokhaugen 4, Sandvika 1300, Norway

JACOB, NINNI SARAH, health physicist; b. Tiruvalla, Kerala, India, Dec. 2, 1951; came to U.S., 1977; d. Kochyil Abraham and Rachel (John) Philip; m. James N. Jacob, Sept. 2, 1976; children: Sneha Elizabeth, Sushil Chacko. MSc in Physics, Madras (India) Christian Coll., 1973; MS in Health Physics, Purdue U., 1980. Jr. lectr. Madras Christian Coll., 1974-76; lectr. Women's Christian Coll., Madras, 1976; radiation physicist trainee Dartmouth Med. Sch., Hanover, N.H., 1977; teaching/rsch. asst. Purdue U., West Lafayette, Ind., 1979-81; health physicist Harvard Sch. of Pub. Health, Boston, 1981, Harvard U., Cambridge, Mass., 1984; radiation safety officer R.I. Nuclear Sci. Ctr., U. R.I., Narragansett/Kingston, 1984—; keeper of the list Campus Radiation Safety Officers, 1991—. Contbr. articles to profl. confs. Chmn. Christian Action Com. St. Paul's Ch., Wickford, R.I., 1987-89, Stephen min., 1990-92; sr. warden Ch. South India Congregation of Boston, 1991-93. Sci. talent search scholarship Govt. of India, 1968-73. Mem. Health Physics Soc., Eta Sigma Gamma, Pi Sigma Pi. Office: RI Atomic Energy Commn South Ferry Rd Narragansett RI 02882

JACOB, ROBERT ALLEN, surgeon; b. Cleve., July 25, 1941; s. John B. and Elaine Irene (Puleo) J.; m. M. Elaine Sheppard, Aug. 23, 1980; children: Kristen Elizabeth, Alexandra Elaine. BA, Case Western Res. U., 1963; MSc, Ohio State U., 1966, MD, 1969. Diplomate Am. Bd. Orthopaedic Surgeons, Am. Acad. Pain Mgmt. Orthopaedic surgeon Ford, Fadel & Jacob, P.S.C., Louisville, 1976—; pres. med. staff Sts. Mary & Elizabeth Hosp., Louisville, 1991—; med. dir. Ky. Pain Therapy Ctr. Contbr. articles to profl. jours. Local fundraising chmn. Orthopaedic Edn. and Rsch. Found., Sts. Mary & Elizabeth Hosp., 1990-91, bd. dirs. Kentuckiana Hemophilia Found., 1987-89, Hosp. Found., 1991—. Maj. U.S. Army, 1974-76. Recipient Outstanding Svc. award Kentuckiana Hemophilia Found., 1986, Cert. of Recognition, 1987. Fellow Am. Acad. Orthopaedic Surgeons; mem. Ky. Med. Assn., Southern Med. Assn., Jefferson County Med. Assn., Louisville Orthopaedic Assn., Louisville Soc. Physicians and Surgeons (v.p. 1980, pres. 1984). Republican. Roman Catholic. Avocations: photography, horticulture, scientific instrument collector. Home: 8808 Denington Dr Louisville KY 40222-5011 Office: Ford Fadel & Jacob PSC 1900 Bluegrass Ave Ste 203 Louisville KY 40215-1144

JACOB, STANLEY WALLACE, surgeon, educator; b. Phila., 1924; s. Abraham and Belle (Shulman) J.; m. Marilyn Peters; 1 son, Stephen; m. Beverly Swarts; children—Jeffrey, Darren, Robert; m. Gail Brandis; 1 dau., Elyse. M.D. cum laude, Ohio State U., 1948. Diplomate: Am. Bd. Surgery. Intern Beth Israel Hosp., Boston, 1948-49; resident surgery Beth Israel Hosp., 1949-52, 54-56; chief resident surg. service Harvard Med. Sch., 1956-57, instr., 1958-59; asso. vis. surgeon Boston City Hosp., 1958-59; Kemper Found. research scholar A.C.S., 1957-60; asst. prof. surgery U. Oreg. Med. Sch., Portland, 1959-66; asso. prof. U. Oreg. Med. Sch., 1966—; Gerlinger prof. surgery Oreg. Health Scis. U., 1981—. Author: Structure and Function in Man, 5th edit, 1982, Laboratory Guide for Structure and Function in Man, 1982, Dimethyl Sulfoxide Basic Concepts, 1971, Biological Actions of DMSO, 1975, Elements of Anatomy and Physiology, 1989; contbr. to: Ency. Brit. Served to capt. M.C. AUS, 1952-54; col. Res. ret. Recipient Gov.'s award Outstanding N.W. Scientist, 1965; 1st pl. German Sci. award, 1965; Markle scholar med. scis., 1960. Mem. Phi Beta Kappa, Sigma Xi, Alpha Omega Alpha. Achievements include co-discovery of therapeutic usefulness of dimethyl sulfoxide. Home: 1055 SW Westwood Ct Portland OR 97201-2708 Office: Oreg Health Scis U Dept Surgery 3181 SW Sam Jackson Park Rd Portland OR 97201

JACOBOWITZ, ELLEN SUE, museum administrator; b. Detroit, Feb. 21, 1948; d. Theodore Mark and Lois Clairesse (Levy) J. BA, U. Mich., 1969, MA, 1970; student, Courtauld, London, 1970-71, Bryn Mawr, 1976—. Curator Phila. Mus. Art, 1972-90; administr. Cranbrook Inst. Sci., Bloomfield Hills, Mich., 1991—. Author: THe Prints of Lucas Van Leyden, 1983, American Graphics: 1860-1940, 1982. Past bd. dirs. Nat. Coun. Jewish Women, Detroit, 1990—, Netherlands Am. Amity Trust, Washington, 1982—, Print Coun. Am., Balt., 1972—; me. Am. Jewish Com., 1991—. Mem. Detroit Econ. Club, Womens Econ. Club, Am. Assn. Mus., Assn. Sci.-Tech. Ctrs., Graphic Arts Coun., Leadership Oakland. Avocations: tennis, sailing, opera, dance, theater. Office: Cranbrook Inst Sci POB 801 1221 N Woodward Ave Bloomfield Hills MI 48303-0801

JACOBS, DAN, biometrician, ecologist; b. Washington, July 28, 1955; s. Marcus and Ada (Löwenherz) J.; m. Sharon Harriet White, Sept. 8, 1985; children: Jennifer Eva, Sarah Esther. BS, Rutgers U., 1977; MS, Frostberg State Coll., 1980; PhD, U. Md., 1985. Faculty rsch. asst. dept. animal sci. U. Md., College Park, 1980-85, rsch. assoc. faculty Md. Sea Grant, 1985—; statistician USDA, Hyattsville, Md., 1985; cons. NIH, Rockville, Md., 1991. Editor: Chesapeake Bay Environmental Directory, 1987; contbr. articles to sci. publs. Bd. dirs. Beth Tikva Synagogue, Rockville, 1990-92, corr. sec., 1992-93. Mem. Ecol. Soc. Am., Biometrics Soc., Am. Statis. Assn. Democrat. Jewish. Achievements include development of numerical taxonomy software. Home: 4303 Havard St Wheaton MD 20906 Office: Md Sea Grant Univ Md 0102B Skinner Hall College Park MD 20742

JACOBS, DONALD WARREN, dentist; b. Waynesburg, Pa., Apr. 6, 1932; s. Donald Ray and Nellie Fayette (Church) J.; m. Diane Jeanette Marshall, June 28, 1958; children: Donald Marshall, Carol Anne Jacobs Nagle. BS, Waynesburg Coll., 1953; DDS, U. Pitts., 1961. Pvt. practice York, Pa., 1963—. Author: Implant Materials, 1961, Esthetic Dentistry, 1970. Lt. (j.g.) USN, 1953-57; lt. Dental Corps USN, 1958-63. Recipient Presidential Achievement award Pres. of U.S., Washington, 1980. Fellow Acad. Gen. Dentistry, Internat. Coll. Dentists, Am. Coll. Dentists, Royal Soc. Health, Pierre Fauchard Acad.; mem. ADA, Pa. Dental Assn. (del. chmn. coms. 1972-83), York County Dental Soc. (pres.), Fifth Dist. Dental Soc. (pres.), York C. of C., Tall Cedars of Lebanon, Masons, Shriners, Delta Sigma Phi, Psi Omega. Republican. Presbyterian. Avocations: golf, hunting, fishing. Home: 1265 Detwiler Dr York PA 17404-1107 Office: 2801 N George St York PA 17402-1000

JACOBS, JONATHAN LEWIS, physician; b. N.Y.C., Oct. 29, 1954; s. Walter Jerome and Gwendolyn (Liebman) J.; m. Carolyn Douglas, May 21, 1989; 1 child, William Bradford. BA, Yale U., 1976, MD, 1980. Diplomate Am. Bd. Internal Medicine, Am. Bd. Infectious Diseases. Intern N.Y. Hosp.-Cornell Med. Ctr., N.Y.C., 1980-81, resident, 1981-83, fellow in infectious diseases, 1983-86, attending physician, 1986—, med. dir. Ctr. Spl. Studies, 1988—, dir. Office AIDS Clin. Program Mgmt., 1990—; founder multidisciplinary program for people with HIV infection. Contbr. articles to sci. publs.; dir. AIDS films. Recipient Humanitarian award East Manhattan C. of C., 1988. Mem. Am. Fedn. Clin. Rsch., Alpha Omega Alpha. Home: 17 E 95th St Apt 2R New York NY 10128-0731 Office: New York Hosp 525 E 68th St New York NY 10021-4873

JACOBS, MARK ELLIOTT, electronics engineer; b. Atlantic City, N.J., June 11, 1940; s. Irving I. and Rose J.; m. Irene G., June 13, 1964; children: David, Deborah. BS, Carnegie Mellon U., 1962, MS, 1963, PhD, 1966. Mem. tech. staff AT&T Bell Labs., Whippany, N.J., 1965-71, tech. supr., 1971-89; tech. supr. AT&T Bell Labs., Dallas, 1989—; mem. adv. com. Internat. Telecomms. Energy Conf., 1988—. Contbr. articles to profl. jours. Mem. IEEE, AAAS. Office: AT&T Bell Labs 3000 Skyline Dr Mesquite TX 75149

JACOBS, NICHOLAS JOSEPH, microbiology educator; b. Oakland, Calif., Mar. 29, 1933; s. Nicholas Joseph and Mary (Hughes) J.; m. Judith Manniello, Feb. 10, 1958; 1 child, Charles. BS, U. Ill., 1955; PhD, Cornell U., 1960; MA (hon.), Dartmouth Coll., 1985. Microbiologist U. Chgo., 1960-63; instr. Dartmouth Med. Sch., Hanover, N.H., 1964, asst. prof. microbiology, 1964-70, assoc. prof., 1971-80, prof. microbiology, 1980—. Grantee, NIH, NSF, USDA. Achievements include rsch. in microbiology, medicine and plant biology. Office: Dartmouth Med Sch Dept Microbiology/Physiol Hanover NH 03755

JACOBS, RALPH RAYMOND, physicist; b. Niagara Falls, N.Y., Dec. 31, 1942; m. Leedia Gordeev, June 5, 1966; children: Aleda Anne, Liana Lizabeth. BS cum laude in Physics, NYU, 1964; MS in Physics, Yale U., 1965, PhM in Physics, 1967, PhD in Physics, 1969. Summer fellow physics dept. NYU, N.Y.C., 1962-63; teaching asst. Yale U., New Haven, 1967-69; mem. tech. staff GTE Labs., Inc., Bayside, N.Y., 1969-72; sr. physicist, project mgr. laser programs U. Calif., Lawrence Livermore Nat. Lab., 1972-80; corp. mgr. for rsch. Spectra-Physics, Inc., San Jose, Mountain View, Calif., 1980-84, engring. mgr. laser products div., 1985-89, dir. corp. tech. devel., 1989-90; dir. new technology initiatives, laser programs U. Calif., Lawrence Livermore Nat. Lab., 1990—; conf. chair Lasers and Electro-Optics Soc. Ann. Meeting, Orlando, Fla., 1989; v.p. for confs. and mem. bd. govs. IEEE/Lasers and Electro-Optics Soc., 1990—. Mem. editorial bd. Laser Focus World, Lasers & Optronics. Recipient GM and NYU Alumni scholarships, Kappa Sigma Frat. Nat. Man of Yr. award, 1963. Fellow IEEE (Lasers and Electro-Optics Soc. v.p. for confs. and bd. govs.), Optical Soc. Am.; mem. Am Phys. Soc., Sigma Pi Sigma (NYU chpt. pres.), Tau Beta Pi (NYU chpt. pres.), Sigma Xi.

JACOBS, RICHARD DEARBORN, consulting engineering firm executive; b. Detroit, July 6, 1920; s. Richard Dearborn and Mattie Phoebe (Cobleigh) J.; divorced; children: Richard, Margaret, Paul, Linden, Susan. BS, U. Mich., 1944. Engr., Detroit Diesel Engine div. Gen. Motors, 1946-51; mgr. indsl. and marine engine div. Reo Motors, Inc., Lansing, Mich., 1951-54; chief engr. Kennedy Marine Engine Co., Biloxi, Miss., 1955-59; marine sales mgr. Nordberg Mfg. Co., Milw., 1959-69; marine sales mgr. Fairbanks Morse Engine div. Colt Industries, Beloit, Wis., 1969-81; pres. R.D. Jacobs & Assocs., cons. engrs., naval architects and marine engrs., Roscoe, Ill., 1981—. Served with AUS, 1944-46. Registered profl. engr., Ill., Mich., Wis., Miss. Mem. Soc. Naval Architects and Marine Engrs. (chmn. sect. 1979-80), Soc. Automotive Engrs., Am. Soc. Naval Engrs., Soc. Am. Mil. Engrs., Soc. Marine Cons., ASTM, Permanent Internat. Assn. Nav. Congresses, Navy League U.S., U.S. Naval Inst., Propeller Club U.S., Nat. Forensic Ctr. Unitarian. Clubs: Country (Beloit); Rockford Polo, Masons. Office: 11405 Main St Roscoe IL 61073-9569

JACOBSEN, EDWARD HASTINGS, physicist; b. Elizabeth, N.J., Jan. 2, 1926; s. Edward H. and Marie (Thomas) J.; m. Victoria Thomas, June 29, 1952 (dec. Nov. 1991). BS in Physics, MIT, 1950, PhD in Physics, 1954. Rsch. physicist GE Rsch. Lab., Schenectady, 1954-60; prof. physics U. Rochester, N.Y., 1961-91, prof. emeritus, 1991—; cons. MIT Lincoln Labs., Lexington, Mass., 1976—; vis. prof. biophysics biology dept. MIT, Cambridge, 1967-68, Columbia U., N.Y.C., 1968-69. Contbr. articles to profl. jours. With USN, 1944-46. NIH career devel. fellow MIT and Columbia U., 1968-69. Mem. AAAS, Am. Phys. Soc., Sigma Xi (hon.). Episcopalian. Achievements include patent for jaming enemy radar; notable findings that sound waves can be generated by piezoelectric crystals at ultra microwave frequencies and also amplified by electrons in solids; co-founder of a lens which could provide atomic resolution when used in an electron microscope. Office: PO Box 391533 770 Mass Ave Cambridge MA 02139

JACOBSEN, GERALD BERNHARDT, biochemist; b. Spokane, Wash., Nov. 25, 1939; s. Hans Bernhardt and Mabel Grace (Swope) J.; m. Sally-Ann Heimbigner, June 7, 1961 (div. 1976); children: Claire Elise, Hans Edward; m. Jean Eva Robinson, Dec. 5, 1976. BA, Whitman Coll., 1961; MS, Purdue U., 1965, PhD, 1970. Postdoctoral fellow Oreg. State U., Corvallis, 1970-73; rsch. chemist Lamb-Weston, Inc., Portland, Oreg., 1973-85, sr. rsch. chemist, 1985—; presenter at profl. confs. Contbr. articles to profl. jours. Grantee NSF, 1960; NIH grad. fellow, 1965; Herman Frasch postdoctoral fellow Oreg. State U., 1970. Mem. AAAS, Am. Oil Chemists Soc., Am. Chemistry Soc., Assn. Ofcl. Analytical Chemists, Sigma Xi. Achievements include patents for Process for Making A Starch Coated Product, Coated Potato Product Process. Home: 1204 Knollwood Ct Richland WA 99352-9448 Office: Lamb Weston Tech Ctr 2005 Saint St Richland WA 99352-5306

JACOBSEN, GRETE KRAG, pathologist; b. Copenhagen, Sept. 27, 1943; d. Per Krag J. and Marie Louise (Gorrison) Hanssen; m. Joachim Knop, Apr. 27, 1987; children: Nikolaj, Filip. MD, U. Copenhagen, 1977, Dr.med., 1985. Lic. pathologist. Jr. registrar Mcpl. Hosp., Dept. Pathology, Copenhagen, 1971-73, Mcpl. Hosp., Dept. Dermatology, Copenhagen, 1974-75; registrar dept. pathology Gentofte Hosp., 1975-76; sr. registrar Herlev Hosp., Dept. Pathology, Copenhagen, 1976-80; asst. prof. Herlev Hsop., Dept. Pathology, Copenhagen, 1980-82; sr. registrar dept. pathology Hvidovre Hosp., Copenhagen, 1982-85; cons., chief pathologist Gentofte Hosp., Copenhagen, 1985—; lectr. in field. Author: Atlas of Germ Cell Tumours, 1989; contbr. tech. papers and articles to profl. jours.; exhibited art in 12 shows, 1985—. 19 grants in field, 1976—. Fellow Orgn. Danish Pathologists, Med. Soc. Copenhagen; mem. Danish Soc. Pathology (chmn. 1989-92), others. Home: Bispebjerg Parkallé 11, 2400 NV Copenhagen Denmark Office: Kas Gentofte Dept Pathology, Niels Andersensvej, 2900 Hellerup Denmark

JACOBSEN, RICHARD T., mechanical engineering educator; b. Pocatello, Idaho, Nov. 12, 1941; s. Thorleif and Edith Emily (Gladwin) J.; m. Vicki Belle Hopkins, July 16, 1959 (div. Mar. 1973); children: Pamela Sue, Richard T., Eric Ernest; m. Bonnie Lee Stewart, Oct. 19, 1973; 1 child, Jay Michael; stepchild: Erik David Lustig. BSME, U. Idaho, 1963, MSME, 1965; PhD in Engring. Sci., Wash. State U., 1972. Registered profl. engr., Idaho. Instr. U. Idaho, 1964-66, asst. prof. mech. engring., 1966-72, assoc. prof., 1972-77, prof., 1977—, chmn. dept. mech. engring., 1980-85, assoc. dean engring., 1985-90, assoc. dir. Ctr. for Applied Thermodynamic Studies, 1975-86, dir., 1986—, dean engring., 1990—. Author: International Union of Pure and Applied Chemistry, Nitrogen-International Thermodynamic Tables of the Fluid State-6, 1979; Oxygen-International Thermodynamic Tables of the Fluid State-9, 1987, Ethylene-International Thermodynamic Tables of the Fluid State-10, 1988, ASHRAE Thermodynamic Properties of Refrigerants (2 vols.), 1986; numerous reports on thermodynamic properties of fluids, 1971—; contbr. articles to profl. jours. NSF sci. faculty fellow, 1968-69; NSF rsch. and travel grantee, 1976-83; Nat. Inst. Standards and Tech. grantee, 1974-91, Gas Rsch. Inst. grantee, 1986-91, 1992—, Dept. Energy grantee, 1991—. Fellow ASME (faculty advisor 1972-75, 78-84, chmn. region VIII dept. heads com. 1983-85, honors and awards chmn. 1985-91, K-7 tech. com. thermophys. properties 1985—, chmn. 1986-89, 92—, rsch. tech. com. on water and steam in thermal power systems, 1988—, gen. awards com. 1985-91, chmn. 1988-91, com. on honors 1988—, mem. bd. on profl. practice and ethics, 1991—), N.W. Coll. and Univ. Assn. for Sci. (bd. dirs. 1990—), Idaho Rsch. Found. (bd. dirs. 1991—), Soc. Automotive Engrs. (Ralph R. Teetor Edn. award, Detroit 1968), ASHRAE (co-recipient Best Tech. Paper award 1984), Sigma Xi, Tau Beta Pi, Phi Kappa Phi (Disting. Faculty award 1989). Office: U Idaho Coll Engring Office of Dean Janssen Engring Bldg 125 Moscow ID 83844

JACOBSEN, STEPHEN CHARLES, biomedical engineer, educator; b. Salt Lake City; s. Charles Jacob and Evelyn (Madsen) J.; m. Beth Vanderworth; m. Linda Diane Madeira, 1980; children: Peter Stephen, Genevieve. BS, U. Utah, MS; PhD, MIT, 1973. Teaching asst. U. Utah, Salt Lake City, 1966-67, rsch. asst. artificial organs, 1967-68, asst. coord. heart and kidney contracts, 1968-69, dir. ctr. biomed. design (now ctr. engring. design), 1973—, from asst. to full prof. engring., 1973; cons. in field. Contbr. articles to profl. jours.; patentee in field. Served with U.S. Army, 1961-67. Mem. NAE, NAS Inst. Medicine. Office: U Utah Ctr Engring Design 3176 Merrill Engring Bldg Salt Lake City UT 84112*

JACOBSMEYER, JAY MICHAEL, electrical engineer; b. Okaloosa County, Fla., Mar. 13, 1959; s. John Henry and Patricia Ann (McDonough) J.; m. Joyce Ann Deem, June 20, 1981; children: Abigail Ann, Brian James. BS magna cum laude, Va. Poly. Inst. & State U., 1981; MS, Cornell U., 1987. Registered profl. engr., Colo. Commd. 2nd lt. USAF, 1981-90, advanced through grades to capt., 1985; elec. engr. 3397 Tech. Tng. Squadron, Biloxi, Miss., 1981-82; comm. engr. 1st Combat Comm. Group, Wiesbaden, Germany, 1982-85; communications engr. HQ Air Force Space Command, Colorado Springs, 1987-90; resigned USAF, 1990; staff engr. ENSCO, Inc., Colorado Springs, 1990-91, sr. staff engr., 1991-93; co-founder, chief tech. officer Pericle Comm. Co., 1992—. Patent pending wireless data modem; contbr. articles to profl. publs. Maj. USAFR. Decorated Meritorious Svc. medal, Air Force Commendation medal; named Man of Yr., Va. Poly. Inst. and State U., 1981; rsch. grantee, NSF, USN. Mem. IEEE (sr.), Armed Forces Comm. and Electronics Assn. (v.p. 1989-90), Air Force Assn., Omicron Delta Kappa, Eta Kappa Nu. Avocations: road racing, mountain climbing. Home: 2475 Edenderry Dr Colorado Springs CO 80919

JACOBSON, ALAN LEONARD, otolaryngologist; b. N.Y.C., Sept. 2, 1949; s. Harold and Arline (Pass) J.; m. Rona Sheryl Schwarzberg; children: Alyssa Beth, Sara Nina, Elana Joelle. BA, C.W. Post Coll., 1971; MD, Albert Einstein Coll., 1978. Diplomate Nat. Bd. Med. Examiners, Am. Bd. Otolaryngology. Intern in gen. surgery Mt. Sinai Hosp., N.Y.C., 1979, resident in otolaryngology, 1979-82; pvt. practice otolaryngologist Palm Beach Gardens and Jupiter, Fla.; vis. assoc. prof. Diagnostic Scis. U. of Pacific Sch. of Dentistry; asst. clin. prof. Columbia Coll. Physicians and Surgery, 1983-89; Otolaryngology cons. Sleep Disorders Lab., St. Luke's Roosevelt Hosp. Sr. editor: Textbook on Headache and Facial Pain, 1990, Otolaryngology Clinics of North America, Vol. 22-6, 1989; consulting editor Med. Malpractice Mgmt., Nat. Med.-Legal Info. Network; contbr. articles to profl. jours. Mem. AMA, AAAS, Am. Coll. Surgeons, Am. Acad. Otolaryngology Head and Neck Surgery, Am. Soc. Head and Neck Surgeons, Am. Acad. Facial Plastic and Reconstructive Surgeons, Pan Am. Assn. Laryngology and Broncheosophagology, N.Y. Acad. Medicine, N.Y. Acad. Sci., Nat. Migraine Found., Am. Assn. Study of Headache, Fla. State Med. Soc., Palm Beach County Med. Soc., Am. Assn. Clin. Immunology and Allergy, Am. Cancer Soc. (exec. com. Jupiter/Tequesta div.). Office: 3365 Burns Rd Ste 203 Palm Beach Gardens FL 33410 also: 1000 S Old Dixie Hwy Ste 305 Jupiter FL 33458

JACOBSON, ALEXANDER DONALD, technological company executive; b. N.Y.C., Dec. 1, 1933; s. Seymour Irving and Annabelle (Maibach) J.; m. Judith Althea Berman, July 25, 1955 (div. June 1966); children: Juliet Anna, David Clair; m. Rebecca Mary Davies, Oct. 24, 1971. BS, UCLA, 1955, MS, 1958; PhD in Electrical engring., Calif. Inst. Tech., 1964. Microwave engr. Radar div. Hughes Aircraft Co., Culver City, Calif., 1955-59; mem. tech. staff Rsch. Lab. Hughes Aircraft Co., Malibu, Calif., 1959-66, sect. head Rsch. Lab., 1966-70, assoc. dept. mgr. Rsch. Lab., 1970-76; programs mgr. Comml. Product div. Hughes Aircraft Co., Carlsbad, Calif., 1976-77; free lance cons. L.A., 1977-80; pres., CEO Inference Corp., L.A., 1980-90; chmn. Inference Corp., El Segundo, Calif., 1990—. Contbr. articles to profl. jours. Recipient Prize Rank Orgn., 1986. Mem. Tau Beta Pi, Pi Mu Epsilon, Sigma Xi. Achievements include patents for Reflective Liquid Crystal Light Valve, High Resolution Continusly Distributed Silicon Photodiode Substrtate, Three Dimensional Microelectronics, High Speed Hologram. Home: 12256 Canna Rd Los Angeles CA 90049 Office: Inference Corp 550 N Continental Blvd El Segundo CA 90245

JACOBSON, ARTHUR ELI, research chemist; b. N.Y.C., May 2, 1928; s. William and Sue (Sacks) J.; m. Linda Perry Jacobson, Sept. 12, 1964; children: jay Jacobson, Laura Jacobson. BS, Fordham U., 1949; PhD, Rutgers U., 1960. Postdoctoral fellow Albert Einstein Coll. Medicine, N.Y.C., 1959-62; biological coord. Coll. on Problems of Drug Dependence, Washington, 1977—; rsch. chemist Nat. Inst Diabetes, Digestive and Kidney Diseases, NIH, Bethesda, Md., 1962—, deputy chief Lab. of Med. Chemistry, 1991—; exec. com. Comm. on Problems of Drug Dependence, Washington, 1974-81; affiliate prof. Psychol. and Toxicology, Med. Coll. Va., Richmond, 1984—. Author: NIDA Monograph 19, 1992, J. Med. Chem. 35:1323-1329, 1992. Grantee Scientific Rsch. Nat. Inst. Drug Abuse, Bethesda, Md., 1992, Drug Abuse, Nat. Inst. Drug Abuse, 1989—. Mem. AAAS, Coll. on Problems of Drug Dependence (J. Michael Morrison award 1990), Am. Chem. Soc. Office: LMC NIDDK NIH Rm 81-22 9000 Rockville Pike Bldg 8 Bethesda MD 20892

JACOBSON, GARY RONALD, biology educator, researcher; b. Ames, Iowa, Nov. 9, 1947; s. Norman Leonard and Gertrude A. (Neff) J.; m. Paula Lillian Grisafi, Sept. 29, 1984. BS, Iowa State U., 1969; PhD, Stanford U., 1974. Postdoctoral U. Basel, Switzerland, 1974-77, U. Calif. San Diego, La Jolla, 1977-78; asst. prof. dept. biology Boston U., 1979-85, assoc. prof., 1985-91, prof., 1991—. Assoc. editor Jour. Cellular Biochemistry, 1990-92.; contbr. over 40 articles to profl. jours. Alexander von Humboldt fellow, 1986-87; recipient Rsch. grants NIH, 1979—. Fellow AAAS; mem. Am. Soc. Microbiology, Am. Soc. Biochemistry and Molecular Biology. Achievements include first to purify and characterize E.coli mannitol permease. Office: Boston Univ Dept Biology 5 Cummington St Boston MA 02215

JACOBSON, HOWARD NEWMAN, obstetrics/gynecology educator, researcher; b. St. Paul, Aug. 13, 1923; s. Irvin Oliver and Nora Henrietta (Olson) J.; m. Beverly Mott, Aug. 1952 (div. 1960); m. Barbara Jane Dinger, Aug. 20, 1961. BSc in Medicine, Northwestern U., Chgo., 1947, BM, 1950, MD, 1951. Intern Presbyn. Hosp., Chgo., 1950-51, resident in ob-gyn, 1951-52; fellow, rsch. fellow in obstetrics, mem. family clinic Harvard Sch. Pub. Health, Boston, 1952-55; resident Boston Lying-In Hosp. and Free Hosp. for Women, Brookline, Mass., 1955-58; obstetrician, physiologist Lab. Neuroanat. Scis., Nat. Inst. Nervous Disease and Blindness, NIH, Bethesda, Md., 1958-60; instr., asst. prof. Harvard Med. Sch., Boston, 1960-65; assoc. prof. U. Calif., San Francisco, Berkeley, 1965-69; dir. Macy program Med. Sch. Harvard U., 1969-74; prof. dept. community medicine Coll. Medicine and Dentistry N.J., Piscataway, 1974-78; dir. Inst. Nutrition, clin. prof. U. N.C., Chapel Hill, 1978-88; rsch. prof. Coll. of Pub. Health Coll. of Pub. Health U. South Fla., 1988—; prof. dept. ob-gyn U. South Fla. Med. Sch., Tampa, 1990—; cons. Children's Bur., HEW, Washington, 1964-73, GAO, Washington, 1974-83, AMA, 1980-82, 88—; mem. food and nutrition bd. NRC/NAS, Washington, 1971-74; prof. dept. biology and Sch. Home Econs., U. N.C., Greensboro, 1978-88, Ellen Swallow Richards lectr., 1978; cons. pregnancy and nutrition study U. Minn., Mpls., 1979—; adj. prof. dept. food, nutrition and instn. mgmt. East Carolina U. Sch. Home Econs., Greenville, 1981-88; mem. nutrition grad. faculty N.C. State U., Raleigh, 1979-88. Contbr. over 130 articles and abstracts to FMA Today, Jour. Nurse-Midwifery, Clin. Nutrition, Contemporary Internal Medicine, Food and Nutrition News, Nutrition Today, New Eng. Jour. Medicine, chpts. to books. Panel vice chmn. White House Conf. on Food, Nutrition and Health, Washington, 1969; chmn. Quality of Life Conf., Mass. Med. Soc., Boston, 1972; mem. hunger com. Episcopal Ch. S.W. Fla., 1990—; mem. Fla. Healthy Start Initiative Working Group, 1991—. Lt. (j.g.) USNR, 1943-46, PTO. Recipient Agnes Higgins award March of Dimes and APHA, 1987; recipient Career Devel. award NIH, 1963-65. Fellow Am. Coll. Ob-Gyn (assoc.); mem. Am. Soc. Clin. Nutrition, Am. Physiol. Soc., Mass. Med. Soc. (chmn. commn. 1972-74), Fla. Pub. Health Assn. (chmn. sect. 1990-91), Am. Dietetic Assn. (hon.). Democrat. Achievements include co-development of guides for clinical nutrition studies, portable ultrasound for body composition; co-determination of nature of cardiovascular changes at birth; co-introduction of computer assisted methodology in nutrition; co-initiation of modern nutrition standards for healthy pregnancy. Office: U South Fla Coll Pub Health 13201 Bruce B Downs Blvd Tampa FL 33612-3899

JACOBSON, IRA DAVID, aerospace engineer, educator, researcher; b. Bronx, N.Y., May 28, 1942; s. Abraham and Bertha (Badin) J.; m. Judy Angert; children: Donna, Alan. BS in Aeros. & Astronautics, NYU, 1963; MS in Aero. Engring., U. Va., 1967, PhD in Aero. Engring., 1970. Aerospace engr. NASA, Wallops Island, Va., 1963-67; from asst. to assoc. prof. U. Va., Charlottesville, 1967-74, prof., 1974—, dir. Ctr. for Computer-Aided Engring., 1983-91, dir. Ctr. for Innovative Tech., 1985-91, dir. info. tech. and communication, 1991—; cons. various cos. and govt. agys., 1970—. Contbr. over 100 articles to profl. jours. Assoc. fellow AIAA (Outstanding Aerospace Educator award 1981); mem. Am. Soc. Engring. Educators (Outstanding Aerospace Educator award 1981), Nat. Computer Graphics Assn., Sigma Xi, Tau Beta Pi, Sigma Gamma Nu. Achievements include development of aircraft ride quality model; first description of fluid mechanical phenomena associated with boundary layer on spinning bodies at angle of attack to predict Magnus effect. Home: 108 Blueberry Rd Charlottesville VA 22901-8409

JACOBSON, IVAR HJALMAR, data processing executive; b. Ystad, Sweden, Sept. 2, 1939; s. Hjalmar H. and Edith (Persson) J.; m. Kerstin B. Hellman, Oct. 10, 1964; children: Agneta, Katarina, Stefan, Thomas. MSc, Chalmers Inst. Tech., Gothenburg, Sweden, 1962; PhD, Royal Inst. Tech., Stockholm, 1985. Software mgr. Ericsson, Stockholm, 1963-87; pres. Objective Systems, Stockholm, 1987-90, v.p. tech., 1991—; vis. scientist MIT Computer Sci. Lab., Cambridge, Mass., 1983-84. Author: Object-Oriented Software Engineering, A Use Case Driven Approach, 1992, also various conf. proc. Achievements include research on development methods for object-oriented systems. Office: Objective Systems, Box 1128, Kista S-16422, Sweden

JACOBSON, LEONARD I., psychologist, educator; b. Bklyn., Aug. 9, 1940; s. Harry L. and Violet (Natkin) J. A.B. cum laude, CUNY, 1961; Ph.D., SUNY-Buffalo, 1966. Research psychologist Children's Hosp., Buffalo, 1965-66; asst. prof. psychology U. Miami, Coral Gables, Fla., 1966-71, assoc. prof., 1971-76, prof., 1976—; adj. asst. prof. Guidance Ctr.-U. Miami, Coral Gables, 1969-70; prof. pediatrics U. Miami Sch. Medicine, 1980—; cons. Miami Mental Health Ctr., 1968-79, Sunland Tng. Ctr. at Miami, Opa-Locka, 1969-72, Camarillo State Hosp. (Calif.), 1970, Mailman Ctr. for Child Devel.-U. Miami Sch. Medicine, 1972-75, Miami Lighthouse for the Blind, 1975—; mem. outcome study panel Dade-Monroe Mental Health Bd., 1980; cons. Metro-Dade Pub. Safety Dept., 1982; mem. panel of psychologists, State of Fla., 1982—; dir. psychology Psychol. Specialists, P.A., 1987—. Contbr. articles to profl. jours. USPHS clin. fellow, 1962-63; grantee NSF, 1966-68, NIMH, 1967-68, NIH, 1968, Soc. Psychol. Study Social Issues, 1969, NASA, 1969-71. Fellow Am. Assn. Med. Psychotherapists (diplomate); mem. Am. Psychol. Assn., Southeastern Psychol. Assn., Western Psychol. Assn., Fla. Psychol. Assn., AAAS, Assn. Advancement of Behavior Therapy, Am. Assn. Workers for the Blind, Soc. Research in Child Devel., Psychonomic Soc., Soc. Psychotherapy Research, InterAmerican Assn. Psychology, Internat. Assn. Applied Psychology, Sigma Xi, Psi Chi. Republican. Office: U Miami Dept Psychology Coral Gables FL 33124

JACOBSON, LINDA S(UE), astronomy educator; b. Savannah, Ga., Sept. 29, 1962; d. Ray and Doris (Pelt) Richardson; m. James Lawrence Jacobson, Nov. 15, 1980. Grad. high sch., North Ft. Myers, Fla. Ground person planetarium project Spitz Space Systems, Ft. Myers, Fla., 1986; astronomer Nature Ctr. Planetarium, Ft. Myers, 1986-89; instr. astronomy Tropic Isles Elem. Sch., North Ft. Myers, 1991; instr. coll. for kids, lectr. Edison Community Coll., Ft. Myers, 1991, instr. astronomy dept. continuing edn., 1991—; owner, cons. Milky Way Prodns. Interpretive Sky Programs, North Ft. Myers, 1990—; cons. astronomy equipment Boeder's Camera, Ft. Myers, 1990—. Cons. to numerous writers of articles in field. Vol. instr. astrophotography Edison Community Coll., 1979—. Recipient Recognition award S.W. Fla. Navigators Club, 1987, Collier County Pub. Sch. System, 1988, S.W. Fla. chpt. Boy Scouts Am., Camp Miles, 1990. Mem. Am. Astron. Soc. (assoc.), Am. Inst. Physics (assoc.), S.W. Fla. Astron. Soc. Inc. (contbr. column to newsletter 1986—). Home and Office: Milky Way Prodns Interpretive Sky Programs 24 Massachusetts Rd Marco Island FL 33937

JACOBSON, MURRAY M., chemical engineer; b. Boston, Jan. 2, 1915; s. Robert H. and Charlotte (Moses) J.; m. Madelyn Ruth Marder, June 28, 1962; 1 child, Richard. BS, Tufts U., 1935. Engr. GE, Lynn, Mass., 1940; chem. engr. Watertown (Mass.) Arsenal Labs., 1940-54, chief chem. metallurgy lab., 1955-56, chief materials scis. lab., 1957-62; dep. chief materials engring. div. Army Materials Rsch. Agy., Watertown, 1963-66; chief prototype lab. Army Materials and Mechanics Rsch. ctr., Watertown, 1967-69, chief materials test div., 1970-74; tech. mgr. Jacon Industries, Boston, 1975-90, materials engring. cons., 1975—; mem. Com. on Corrosion, U.S. War Adv. Com., 1942-45; invited panelist Nat. Acad. Scis., 1976-79. Recipient Cert. of Commendation U.S. War Dept., 1945. Mem. ASM, Am. Chem. Soc., Nat. Assn. Corrosion Engrs. (Boston chpt. chmn. 1954-55). Home: 285 Clark Rd Brookline MA 02146-5823

JACOBSON, OLOF HILDEBRAND, forensic engineer; b. Kansas City, Mo., Aug. 7, 1955; s. Olof H. and Margaret E. (Detweiler) J.; m. Joanne E. Florance, May 31, 1985; chldren: Joshua, Valerie. BSME, U. Colo., 1977; MS in Applied Mechanics, Colo. Sch. of Mines, 1991. Registered profl. engr., Colo. Mech. engr. Pub. Svc. Co. of Colo., Denver, 1977-80; mgr. of facilities Samsonite Corp., Denver, 1980-88; site mgr. Colo. Power Ptnrs., Brush, Colo., 1989-90; adj. prof. Colo. Sch. of Mines, Golden, 1991—; sr. engr. Knott Labs., Denver, 1991—. Author: (book) Finite Element Analysis of Contact, 1991. Pres. Toastmasters Internat., 1986. Mem. ASME, SAE, NSPE, Profl. Engrs. of Colo., Pi Tau Sigma, Tau Beta Pi. Democrat. Office: Knott Lab 2727 W 2d Ave Denver CO 80219

JACOBSON, RALPH HENRY, laboratory executive, former air force officer; b. Salt Lake City, Dec. 31, 1931; m. Joan Mathews; children: Mary, Matthew, James. Student, U. Utah, 1950-52; B.S., U.S. Naval Acad., 1956; M.S. in Astronautics, Air Force Inst. Tech., 1962; M.S. in Bus. Adminstrn., George Washington U., 1966; Grad., Air Command and Staff Coll., 1966, Indsl. Coll. Armed Forces, 1974, Naval War Coll., 1976. Commd. 2d lt. U.S. Air Force, 1956, advanced through grades to maj. gen., 1979; project officer Ballistics Systems Div., Norton AFB, Calif., 1962-66; action officer Directorate of Plans, Hdqrs. U.S. Air Force, Washington, 1966-69; wing ops. staff officer 14th Spl. Ops. Wing, Nha Trang Air Base, Republic of Vietnam, 1969-70; successively research and devel. project officer, div. chief and dep. dir. research Air Force Spl. Projects, Los Angeles Air Force Sta., 1970-76; comdr. Air Force Satellite Control Facility, Los Angeles Air Force Sta., 1976-79; asst. dep. chief of staff for space shuttle devel. and ops. Office of Dep. Chief of Staff for Research, Devel. and Acquisitions, Hdqrs. U.S. Air Force, Washington, 1979-80, dir. space systems and command, control and communications, 1980-81; vice dir. Office of Sec. of Air Force Spl. Projects, Los Angeles Air Force Sta., 1981-83, dir., 1983-87; pres., CEO The Charles Stark Draper Lab., Inc., Cambridge, Mass., 1987—. Decorated D.S.M. (Defense, Nat. Intelligence Community, Air Force), Legion of Merit with one oak leaf cluster, D.F.C. Fellow AIAA. Office: Charles Stark Draper Lab 555 Technology Sq Cambridge MA 02139-3563

JACOBSON, WILLARD JAMES, science educator; b. Northfield, Wis., May 22, 1922; s. Harold Wilhelm and Julia (Thompson) J.; m. Carol Elizabeth Whitaker, July 21, 1946; children: Susan Jane, Ellen Elise, Thomas Ray. BS, U. Wis., River Falls, 1946; MA, Columbia U., 1948, PhD, 1951. Sci. tchr. Palmyra (Wis.) Pub. Schs., 1946-47; physics tchr. Horace Man Lincoln Sch., N.Y.C., 1948-49; prof. natural sci. Tchrs. Coll., Columbia U.,

N.Y.C., 1951-89, prof. emeritus natural sci., 1989—; Fulbright sr. lectr. U. London.; nat. rsch. coord. Sec. Internat. Sci. Study, N.Y.C., 1981-92; internat. cons., Afghanistan, Eng., Jamaica, Brazil, Can., India, Venezuela; cons. Sci. Curriculum Improvement Study, AAAS, Population Edn. Project, Nutrition Edn. Project. Author: Science for Children, 1980, Population Education, 1979, Science Activities for Children, 1983, (report) Science Achievement in US and 16 Countries, 1988; author, co-author 73 books. Mem. Fedn. Am. Scientists, Washington, Union of Concerned Scientists, Cambridge, Mass., Educators for Social Responsibility, Cambridge. 1st lt. USAAF, 1943-46, ETO. Recipient Robert J. Carlton award Nat. Sci. Tchrs., 1987; named Disting. Sci. Tchr., Nat. Sci. Tchrs., 1985. Fellow AAAS, N.Y. Acad. Sci. (bd. dirs.); mem. Nat. Assn. Rsch. in Sci. Teaching (former pres., Disting. Sci. Edn. Rsch.), Assn. Edn. of Tchrs. in Sci. (former pres.), Coun. Elem. Sci. Internat. (former pres.). Democrat. Achievements include research in nutrition. Home: 106 Morningside Dr New York NY 10027

JACOBSON-KRAM, DAVID, toxicologist. BA in Biology, U. Conn., 1971, PhD in Devel. Biology, 1976; postgrad., MIT, 1979. Teaching asst. biol. scis. group U. Conn., 1971-72; staff fellow Nat. Inst. Aging NIH, 1976-79, sr. staff fellow, 1979; biologist toxic effects br., office toxic substances U.S. EPA, 1979-83, geneticist reproductive effects assessment group, office health and environ. assessment, office rsch. and devel., 1983-88, acting br. chief genetic and molecular toxicology assessment br., 1988; from rsch. asst. prof. to rsch. assoc. prof. dept. radiology, sch. medicine George Washington U., 1979-84; assoc. prof. oncology dept. radiobiology oncology ctr. Johns Hopkins U., 1984-90; v.p. genetic toxicology divsn. Microbiological Assocs. Inc., Rockville, Md., 1988-90, v.p. toxicology group, 1990—; vis. assoc. prof. oncology ctr. Johns Hopkins U., 1990—. Editor Cell Biology and Toxicology, 1986-88. Nat. Inst. Occupational Health and Safety and Small Bus. Innovative Rsch. grantee; Predoctoral trainee fellow NIH, 1973-76. Mem. AAAS, Am. Coll. Toxicology, Radiation Rsch. Soc., Teratology Soc., Genetic Toxicology Assn. (coun. mem. 1984-89, treas. 1984-86, chmn. 1986-88), Environ. Mutagen Soc. (mem. exec. coun. 1988-90), Soc. Toxicology, Phi Kappa Phi. Office: Microbiological Assoc Inc 9900 Blackwell Rd Rockville MD 20850

JACOBY, MARGARET MARY, astronomer, educator; b. Fall River, Mass., Nov. 10, 1930; d. S. Clifford and Cecilia E. (Donohue) J. ScB in Physics, Brown U., 1952, MAT in Astrophysics, 1961; PhD in Astronomy, Georgetown U., 1965. Rsch. asst. Brown Univ., Providence, 1952-56; secondary tchr. Wheeler Sch., Providence, 1956-59; instr. math./astronomy R.I. Coll., Providence, 1960-62; lectr. astronomy dept. Brown U., Providence, 1964-65; prof., chmn. physics dept. Community Coll. R.I., Providence, Warwick, 1965-74; prof., dir. obs. Community Coll. R.I., Warwick, 1974—; cons. astron. edn. in pub./pvt. schs., R.I., 1974—; lectr. NASA Workshops, Warwick, 1978-84; establisher physics dept. Community Coll. R.I.; designer/supr. constrn. Community Coll. R.I. Obs. Author: (textbook) Astronomy: A Guide to the Basics, 1981, 2d edit., 1987, 3d edit., 1991, (12 segment video tape) Astronomical Topics, 1993. Participant educators colloquium Internat. Astron. Union, 1987. Recipient scholarship Brown Univ., Providence, 1948-52, grant NSF, Brown Univ., 1959-60, fellowship Georgetown Univ., Washington, 1962-64. Mem. Am. Astron. Soc. (full), Am. Inst. Phys. Roman Catholic. Achievements include research on total solar eclipses, spectroscopic analysis of Jupiter's atmosphere, hydrogen-alpha solar study, Halley's Comet, variable stars. Office: Community Coll of RI 400 East Ave Warwick RI 02886-1805

JACOBY, ROBERT OTTINGER, comparative medicine educator; b. N.Y.C., June 20, 1939. DVM, Cornell U., 1963; MS, Ohio State U., 1968, PhD in Pathology, 1969. Asst. prof. pathology Ohio State U., 1969; asst., then assoc. prof. Yale U., New Haven, 1971-87, chmn. sect. comparative medicine, dir. divsn. animal care, 1978—, prof. comparative medicine, 1987—. NIH fellow U. Chgo., 1969-71; recipient Rsch. award Am. Assn. Lab. Animal Sci., 1987. Mem. AAAS, Am. Coll. Vet. Pathologists, Am. Vet. Med. Assn., Am. Assn. Pathologists, Internat. Acad. Pathology, Sigma Xi. Achievements include research in pathogenesis of infectious diseases, diseases of laboratory animals, animal models of human disease. Office: Yale U Resource Study Lab Animal Diseases 333 Cedar St New Haven CT 06510*

JACOVICH, STEPHEN WILLIAM, electronics engineer, consultant; b. Waterbury, Conn., Mar. 15, 1946; s. Anthony William and Jane Barbara (Hulik) J.; m. Elaine Crupo, Oct. 11, 1969; 1 child, Marissa Lynn. BSEE, U. Conn., 1968, MSEE, 1974; MBA, U. New Haven, 1987. Registered profl. engr., Conn. Jr. engr. Bristol Co., Waterbury, Conn., 1968, 70-72; engr. Bristol Babcock Inc., Waterbury, Conn., 1972-80, sr. engr., 1980-85, project engr., mgr., 1986—. 1st lt. U.S. Army, 1968-70, Vietnam. Mem. Instrument Soc. Am. (sr.). Roman Catholic. Home: 131 Fair Haven Dr Middlebury CT 06762 Office: Bristol Babcock Inc 1100 Buckingham St Watertown CT 06795

JADVAR, HOSSEIN, biomedical engineer, oncologist; b. Tehran, Iran, Apr. 6, 1961; came to U.S. 1978; s. Ramezan Ali and Fatemeh (Afzal) J. BS, Iowa State U., 1982; MS, U. Wis., 1984, U. Mich., Ann Arbor, 1986; PhD, U. Mich., Ann Arbor, 1988; MD, U. Chgo., 1993. Rsch. asst. dept. human oncology U. Wis., Madison, 1983-84; rsch. asst. dept. elec. engring. U. Mich., Ann Arbor, 1984-88; sr. rsch. engr. Arzco Med. Electronics, Inc., Chgo., 1988-89; sr. rsch. assoc. Pritzker Inst., IIT Ctr., Chgo., 1989-92; resident U. Calif., San Francisco, 1993-94; reviewer study sect. small bus. innovative rsch. program NIH, 1989; session chmn. IEEE/EMBS 11th Ann. Conf., Seattle, 1989. Contbr. articles to profl. jours., chpts. to books. Recipient Med. Student Rsch. award, U. Chgo., 1991. Mem. AMA, IEEE, Assn. Advancement Med. Instrumentation, Biomed. Engring. Soc., Computers in Cardiology (local com. organizing mem. 1990), Tau Beta Pi, Sigma Xi, Eta Kappa Nu. Achievements include patents for esophgeal catheters and method and apparatus for detection of posterior ischemia. Home: 845 E 57th St # 1 Chicago IL 60637-1457 Office: Pritzker Inst IIT Ctr 10 E 32nd St Chicago IL 60616-3813

JAEGER, MARC JULIUS, physiology educator, researcher; b. Berne, Switzerland, Apr. 4, 1929; came to U.S., 1970; s. Francis K. and Jeanne (Perrin) J.; m. Francis Dick, Dec. 1960 (div. 1972); children: Dominic, Olivia; m. Ina Claire Burlingham-Forbes, June 23, 1973. BA, Gymnasium, Berne, 1948; MD, U. Berne, 1954. Diplomate Swiss Bd. Pulmonary Diseases. Resident U. Hosp. of Berne, 1954-63; asst. prof. U. Fribourg, Switzerland, 1963-69; assoc. prof. Coll. of Medicine U. Fla., Gainesville, 1970-76, prof. Coll. of Medicine, 1976—. Contbr. over 50 articles to profl. jours. Democrat. Achievements include five patents for a Method of Separating Solutes, and for a method to Transport Large Amounts of Heat without Coolant; research in mechanics of breathing, separation of gases. Home: 519 NW 19th St Gainesville FL 32603-1509 Office: U Fla Coll of Medicine Gainesville FL 32610

JAEGER, RICHARD CHARLES, electrical engineer, educator, science center director; b. N.Y.C., Sept. 2, 1944; s. O. Fred and Mary Jane (Shatzer) J.; m. Joan Carol Hill, Dec. 28, 1964; children: Peter, Stephanie. BSEE with high honors, U. Fla., 1966, M in Engineering, 1966, PhDEE, 1969. Staff engr. IBM Corp., Boca Raton, Fla., 1969-72, adv. engr., 1972-74, 77-79; rsch. staff mem. IBM Corp., Yorktown Heights, N.Y., 1974-76; assoc. prof. Auburn (Ala.) U., 1979-82, prof. elec. engring. dept., 1982-90, alumni prof., 1983-88, disting. univ. prof., 1990—; dir. Ala. Microelectronics Ctr., Auburn, 1984—; mem. program com. Internat. Solid State Circuits Conf., San Francisco and N.Y.C., 1978—, program vice chmn., 1992, program chmn. 1993; program co-chmn. Internat. VLSI Cirs. Symposium, Kyoto, Japan, 1988-89, conf. co-chmn., Honolulu, 1990; cons. IBM, InSouth, Digital Equipment Corp., Control Data Corp. Author: Introduction to Microelectronic Fabrication, 1988; contbr. more than 150 tech. papers to profl. jours.; patentee in field. Grantee NSF, Semicondr. Rsch. Corp., Dept. Def., Ala. Rsch. Inst., 1979—. Fellow IEEE; mem. Solid State Cirs. Coun. IEEE (pres. 1990-91, v.p. 1988-89, sec. 1984-87), Computer Soc. IEEE (bd. govs. 1985-86, Outstanding Contbn. award 1984). Home: 711 Jennifer Dr Auburn AL 36830-7116 Office: Auburn U Dept Elec Engring 420 Broun Hall Auburn AL 36849-5201

JAENISCH, HOLGER MARCEL, physicist; b. Salt Lake City, Apr. 22, 1963; s. Klaus Peter Reinhardt and Sieglinde Erika (Freimann) J.; m. Theresa Lynn Snyder, May 24, 1985; children: Falco Alexander, Marcel Fabry, Marcus Antone. MS, Columbia Pacific U., 1989, PhD, 1990. Teaching asst. U. Utah Physics Dept., Salt Lake City, 1979-82; laser engr. Com Tel Inc., Salt Lake City, 1982-85; sr. engr. Odetics Inc., Anaheim, Calif., 1985-88; sr. optical engr. Talandic Rsch. Corp., Irwindale, Calif., 1988-89; sr. rsch. assoc. NASA MSFC, Huntsville, Ala., 1989-90, UAH Ctr. Applied Optics, Huntsville, Ala., 1990-91; sr. scientist Nichols Rsch. Corp., Huntsville, Ala., 1991-92, Tec-Masters Inc., Huntsville, 1992—; pres. Light Strahl Cons., Madison, Ala., 1989—. Author: Genesis II: Chaos/Fractals, 1989, Laser Analogy Using Video Feedback, 1990; contbr. tech. papers to publs. Bd. dirs. Huntsville Sr. Citizen Ctr., 1992. Recipient U.S. Army Medal for Excellence in Sci.; recipient Harvey Eckenrode award South Eastern Simulation Conf., 1992. Mem. IEEE, SPIE, OSA, Masons, Shriners, Scottish Rite. Independent. Achievements include development of ROSETA & KABA fractal analysis algorithms; 1 patent. Home: 135 Dexter Cir Madison AL 35758 Office: Tec-Masters Inc 1500 Perimeter Pkwy Huntsville AL 35806

JAFARI, BAHRAM AMIR, petroleum engineer, consultant; b. Tabriz, Iran, May 24, 1940; came to U.S., 1959; s. Aliashraf and Akhtar (Chahardowli) Amirjafari; m. Inger M. Bohlin, Dec. 27, 1966; children: Mitra, Minou. B-SChemE, U. Calif., Berkeley, 1964; MSChemE, U. Okla., 1965, PhD, 1969. Registered profl. engr., Colo. Instr. U. Okla., Norman, 1967-69; rsch. engr. Gulf R&D Co., Harmarville, Pa., 1969-71; chmn. chem. engrs. Tehran (Iran) Polytech. U., 1971-75; mng. dir. Iranian Mgmt. & Engring. Group, Tehran, 1975-79; dir. petroleum dept. Sci. Applications Internat. Corp., Golden, Colo., 1979-86; sr. hydrologist Dames and Moore, Golden, 1986-87; exec. v.p. Techno-Search Internat. Corp., Golden, 1987—. Contbr. articles to profl. jours. Mem. AIChE, Soc. Petroleum Engrs., Profl. Engrs. and Land Surveyors. Achievements include patent for efficient recovery method of oil from tar sands. Home and Office: 2051 Crestvue Cir Golden CO 80401

JAFFE, ARTHUR MICHAEL, physicist, mathematician, educator; b. N.Y.C., N.Y., Dec. 22, 1937; s. Henry and Clarisse Jaffe; m. Nora Frances Crow, July 24, 1971; 1 child, Margaret Collins; m. Sarah Robbins Warren, Sept. 12, 1992. AB, Princeton U., 1959; BA, Cambridge U., 1961; PhD, Princeton U., 1966. Acting asst. prof. math. Stanford U., 1966-67; asst. prof. physics Harvard U., Cambridge, Mass., 1967-69; assoc. prof. Harvard U., 1969-70, prof. physics, 1970-77, prof. math. physics, 1977-85, Landon T. Clay prof. math. and theoretical sci., 1985—, chmn. dept. math., 1987-90; research fellow Princeton U., 1965-66, Stanford Linear Accelerator Center, 1966-67; mem. Inst. for Advanced Study, 1967; vis. prof. Eidgenössische Technische Hochschule, Zurich, 1968; vis. prof. math. physics Princeton U., 1971; vis. prof. Rockefeller U., 1977; Porter lectr. Rice U., 1982; Hahn lectr. Yale U., 1985; Hendrik lectr. Math. Assn. Am., 1985. Author: Vortices and Monopoles, 1980, Quantum Physics, 1981, 87, Quantum Field Theory and Statistical Mechanics, Expositions, 1985, Constructive Quantum Field Theory, 1985; Asso. editor: Jour. Math. Physics, 1970-72; editorial council: Annals of Physics, 1975-77; asst. editor, 1977—; editor: Communications Math. Physics, 1976—; chief editor, 1979—; mem. adv. bd.: Letters in Math. Physics, 1975—; editor: Progress in Physics, 1979-86, Selecta Mathematica Sovetica, 1980—, Reviews in Mathematical Physics, 1990; contbr. articles to profl. jours. Alfred P. Sloan Found. fellow, 1968-70; Guggenheim Found. fellow, 1977-78, 92; award Math. and Phys. Scis., N.Y. Acad. Sci., 1979; Dannie Heineman prize for Math. Physics, 1980. Fellow Am. Phys. Soc., AAAS, Am. Acad. Arts and Scis.; mem. Am. Math. Soc. (exec. com. of coun. 1991—), Internat. Assn. Math. Physics (pres. 1991—). Home: 27 Lancaster St Cambridge MA 02140-2837

JAFFE, DAVID HENRY, computer systems engineer; b. N.Y.C., Oct. 24, 1942; s. Joshua Henry Jaffe and Jean (Fulerton) Muir; m. Laura Murphy, Nov. 24, 1989. BA in Math., San Francisco State U., 1976, MA in Math./Computer Sci., 1979. Prin., owner David Jaffe Enterprises, Oakland, Calif., 1976-86; prin. engr. System Controls Inc., Palo Alto, Calif., 1985-87, Amdahl, Santa Clara, Calif., 19887-88; dir. engring. SF2 Corp., Sunnyvale, Calif., 1988—; adj. prof. San Francisco State U., 1982-92. Mem. IEEE, Assn. Computing Machinery. Achievements include 8 patents on computer communications, fault tolerant computing and RAID. Home: 551 South Rd Belmont CA 94002

JAFFE, ROBERT BENTON, obstetrician-gynecologist, reproductive endocrinologist; b. Detroit, Feb. 18, 1933; s. Jacob and Shirley (Robins) J.; m. Evelyn Grossman, Aug. 29, 1954; children: Glenn, Terri. M.S., U. Colo., 1966; M.D., U. Mich., 1957. Intern U. Colo. Med. Ctr., Denver, 1957-58, resident, 1959-63; asst. prof. Ob-Gyn. U. Mich. Med. Ctr., 1964-68, assoc. prof., 1968-72, prof., 1972-74, dir. steroid rsch. unit, 1964-74; prof. U. Calif., San Francisco, 1974—, also chmn. dept. Ob-Gyn and reproductive scis., dir. Reproductive Endocrinology Ctr., 1974—; mem. nat. adv. council, mem. human embryology and devel. and reproductive biology study sect. Nat. Inst. Child Health and Human Devel.; bd. dirs. Population Resource Center. Author: Reproductive Endocrinology: Physiology, Pathophysiology and Clinical Management, 1978, 2d edit., 1986, 3d edit., 1991, Prolactin, 1981, The Peripartal Period, 1985; contbr. numerous articles to profl. jours.; mem. editorial bd. Jour. Clin. Endocrinology and Metabolism, 1971-75, Fertility and Sterility, 1972-78; editor-in-chief Obstetric and Gynecologic Survey, 1991—. Josiah Macy Found. faculty fellow, 1967-70, 81; USPHS postdoctoral fellow, 1958-59, 63-64; Rockefeller Found. grantee, 1974—; Andrew Mellon Found. grantee, 1978-81. Mem. Endocrine Soc., Soc. Gynecologic Investigation (pres. 1975-76, Pres.'s Disting. Scientist award 1993), Perinatal Research Soc. (pres. 1973-74), Am. Coll. Obstetricians and Gynecologists (awards), Internat. Soc. Neuroendocrinology, Assn. Am. Physicians, Inst. Medicine Nat. Acad. Scis. Democrat. Jewish. Home: 90 Mt Tiburon Rd Belvedere Tiburon CA 94920-1512 Office: U Calif Med Sch Ob-Gyn & Reproductive Sci San Francisco CA 94143

JAFFE, RUSSELL MERRITT, pathologist, research director; b. Albany, N.Y., Jan. 1, 1947. AB cum laude, Boston U., 1972, MD with honors, 1972, PhD in Biochemistry, 1972. Diplomate Am. Bd. Pathology (clin., chem.), Nat. Bd. Med. Examiners. Med. intern Boston U. Med. Ctr., 1972-73; resident in clin. pathology NIH, Bethesda, Md., 1973-75, sr. staff physician clin. pathology dept., 1973-79, chief resident tng. program clin. chemistry sect., 1976-79; fellow health rsch., practice, policy devel. Health Studies Collegium, 1979—; dir. Serammune Physicians Lab., Vienna, Va., 1987—, Princeton BioCenter, 1989-92; prin. faculty Oriental Med. Strategy in Western Med. Practice, HSC, N.Y.C., 1980-85; dir. Great Smokies Med. Lab., Leicester, N.C., 1986-89; rsch. assoc. Commonwealth, Bolinas, Calif., 1976-81; cons. dept. consumer affairs State of Calif., 1980. Assoc. editor The New Physician, 1971-72, sr. assoc. editor, 1972-73. Active Health Policy Coun., N.Y.C., 1980-85; rep. Nat. Nutrition Consortium, 1980-82; bd. govs. Light Found., 1980—. Comdr. USPHS, 1973-79. Recipient Nat. Rsch. award Am. Acad. Med. Preventics, 1979, J.D. Lane award USPHS, 1975, Excellence in Rsch. award Mead Johnson, 1969, Man of Yr. award Hillel Found., 1967. Fellow Am. Coll. Nutrition, Am. In-Vitro Allergy/Immunology Soc., Am. Soc. Clin. Pathologists; mem. APHA, Am. Assn. Clin. Chemists, Am. Fedn. Clin. Rsch., Am. Holistic Med. Assn. (chmn. sci. adv. com. 1978-80), Illuminating Engrs. Soc. N.Am. (rsch. com.), Internat. Coll. Applied Nutrition, Acad. Clin. Lab. Physicians and Scientists. Achievements include patent in field. Home: 1890 Preston White Dr Reston VA 22091-5430 Office: Serammune Physicians Lab AMSA Bldg 2d Fl 1890 Preston White Dr Reston VA 22091

JAFFE, WILLIAM J(ULIAN), industrial engineer, educator; b. Passaic, N.J., Mar. 22, 1910; s. Elias and Ida (Rosensohn) J. BS in Math and Physics, NYU, 1930; MA in Math., Columbia U., 1931, MS in Indsl. Engring., 1941; ScD in Engring., NYU, 1953. Registered profl. engr., Calif. Cons. engring. math. pvt. practice, 1931-41, 45—; naval architect U.S. Navy Phila. Naval Yard, 1941-45; from instr. to disting. prof. N.J. Inst. Tech., Newark Coll. Engring., 1946-75, disting. prof. emeritus, 1975—; mem. bd. standards rev. Am. Nat. Standards Inst., N.Y.C., 1981-89; adj. prof. NYU Grad. Coll. Engring., 1953-54; mem. Clark bd. internat. mgmt. Com. de Orgn. Scientifique, 1957-60; vis. prof. Sangyo Nohritsu Diagaku, Sanno Inst. Mgmt., Tokyo, 1960. Author: L.P. Alford: Evolution of Modern Industrial Management, (with Lillian M. Gilbreth) Management's Past: A Guide to Its Future; editor: Industrial Engineering Terminology; contbr. numerous articles to profl. jours. Fellow AAAS, ASME (bd. standardization, codes and standards ednl. commn., Dedicated Svc. and Centennial awards), Inst. Indsl. Engrs., N.Y. Acad. Medicine (chmn. biomed. engring. sect.), Soc. for Advancement Mgmt.; mem. Am. Math. Soc., Chemists Club N.Y. Home: 1175 York Ave Apt 9E New York NY 10021-7173

JAFFREY, IRA, oncologist, educator; b. N.Y.C., July 28, 1939; s. Mack and Elaine (Schneider) J.; m. Jane Sharon Friedman, Dec. 26, 1964 (div. Mar. 1979); children: Jonathan David, Marc Jason; m. Sandra Read, June 17, 1979; 1 child, Marc Read. AB, Columbia Coll., N.Y.C., 1960; MD, SUNY, Bklyn., 1965. Intern Jewish Hosp., Bklyn., 1965-66; chief resident Elmhurst Gen. Hosp., N.Y.C., 1970; asst. resident Mt. Sinai Hosp., N.Y.C., 1968-69, resident, 1969-70, chief resident, 1970; ednl. fellow Dept. Hematology, Mt. Sinai Hosp., N.Y.C., 1970-71; asst. clin. prof. dept. neoplastic diseases Dept. Neoplastic Dis., Mt. Sinai Hosp., N.Y.C., 1980—; pres. Palisades Oncology Assocs. P.C., Pomona, N.Y., 1972—. Lt. USNR, 1961-65. Oak Ridge (Tenn.) Inst. fellow, 1965. Fellow ACP, Am. Cancer Soc. (pres. Rockland City unit 1973-74), Rockland City Med. Soc. (v.p. 1992, pres. 1993-94). Office: Palisades Oncology Assocs P C Rt 45 Pomona NY 10970

JAGACINSKI, CAROLYN MARY, psychology educator; b. Orange, N.J., Apr. 12, 1949; d. Theodore Edward and Eleanor Constance (Thys) Jagacinski; m. Richard Justus Schweickert, Dec. 27, 1980; children: Patrick, Kenneth. AB with honors in psychology, Bucknell U., 1971; MA in Psychology, U. Mich., 1975, PhD in Psychology and Edn., 1978. Rsch. assoc. U. Mich., Ann Arbor, 1978-79; rsch. assoc. Purdue U., West Lafayette, Ind., 1979-80, vis. asst. prof., 1980-83, rsch. psychologist, 1983-86, vis. lectr., 1986-88, asst. dean, 1988-89, asst. prof. psychology, 1988—. Contbr. articles to profl. jours. U. Mich. predoctoral fellow, 1977-78, dissertation grantee, 1977-78; Exxon Edn. Found. grantee, 1983-84. Mem. APA, Midwestern Psychol. Assn., Soc. for Judgment and Decision Making, Am. Ednl. Rsch. Assn., Psychonomic Soc., Sigma Xi, Psi Chi. Avocations: tennis, reading. Office: Purdue Univ Dept Psychol Scis West Lafayette IN 47907

JAGENDORF, ANDRE TRIDON, plant physiologist; b. N.Y.C., Oct. 21, 1926; s. Moritz Adolph and Sophie Sheba (Sokolsky) J.; m. Jean Elizabeth Whitenack, June 12, 1952; children: Suzanne E., Judith C., Daniel Z.S. B.A., Cornell U., 1948; Ph.D., Yale U., 1951. Merck postdoctoral fellow UCLA, 1951-53; from asst. prof. to prof. Johns Hopkins U., 1953-66; prof. plant physiology Cornell U., Ithaca, N.Y., 1966—, Liberty H. Bailey prof. plant physiology, 1981—. Author papers, revs. in field. Recipient Outstanding Young Scientist award Md. Acad. Sci., 1961, Kettering Rsch. award, 1963; Weizmann fellow, 1962. Fellow Am. Acad. Arts and Scis., AAAS; mem. NAS, Am. Soc. Plant Physiologists (hon., life, pres. 1967, C.F. Kettering award in photosynthesis, 1978, Charles Reid Barnes award 1989), Am. Soc Biol. Chemists, Am. Soc. Photobiology (councilor 1980), Soc. Gen. Physiologists, Am. Soc. Cell Biology, Japanese Soc. Plant Physiologists. Jewish. Office: Cornell U Plant Biology Sect Plant Sci Bldg Ithaca NY 14853

JAGERMAN, DAVID LEWIS, mathematician; b. N.Y.C., Aug. 27, 1923; s. Morris and Helen (Bader) J.; m. Adrienne Israel, Sept. 8, 1951; children: Diane Tharp, Barbara Magic, Laurie Sutter. BEE, Cooper Union, N.Y.C., 1949; MS in Math., NYU, N.Y.C., 1954, PhD in Math., 1962. Jr. engr. Reeves Instrument Corp., N.Y.C., 1951-55; staff scientist Stavid Engring., Plainfield, N.J., 1955-59; design specialist Convair, San Diego, 1957-59; sr. math. staff cons. System Devel. Corp., Santa Monica, Calif., 1959-63; math. cons. disting. mem. technical staff AT&T Bell Labs., Holmdel, N.J., 1963-89; math. cons. NEC USA, Princeton, N.J., 1989—; tchr. indsl. math. St. Peters Coll., 1968-73; prof. math. Stevens Inst. Tech., 1967-75, prof. elec. engring. and computer sci., 1984-90; prof. math. Fairleigh Dickinson U., Rutherford, N.J., 1958-67. Cpl. AC U.S. Army, 1942-45. Mem. IEEE (sr.). Achievements include research in stochastic models, queueing systems and teletraffic analysis. Office: NEC USA 4 Independence Way Princeton NJ 08540

JÄGER-WALDAU, ARNULF ALBERT, physicist; b. Heidelberg, Germany, Jan. 19, 1962; s. Reinhold Friedrich Alois and Mathilde (Erdt) J.-W. Diploma in physics, U. Konstanz, Germany, 1989, Dr. rer. nat., 1993. Sci. asst. U. Konstanz, 1990—, mem. senate, 1991-93. Contbr. articles to sci. jours., chpt. to book. Mem. German Phys. Soc. Achievements include research in thin solid films, thin film solar cells, polycrystalline semiconductors. Office: U Konstanz, Universitätstrasse 10, 78434 Konstanz Germany

JAGGARD, DWIGHT L(INCOLN), electrical engineering educator; b. Oceanside, N.Y., Apr. 14, 1948. BSEE, U. Wis., 1971, MSEE, 1972; PhD, Calif. Inst. Tech., 1976. Rsch. fellow Calif. Inst. Tech., Pasadena, 1976-78; asst. prof. U. Utah, Salt Lake City, 1978-80; asst. prof. U. Pa., Phila., 1980-82, assoc. prof., 1982-88, prof., 1988—, dir. ExMSE program, 1988-90, assoc. dean, grad. edn. and rsch., 1992—; pres. and co-founder Main Line Waves, Inc., Newtown Square, Pa., 1989—; presenter numerous scientific presentations. Co-editor: Recent Advances in Electromagnetic Theory, 1990; contbg. author: Fractal Electrodynamics and Modelling, Chirality in Electrodynamics; editor: Jour. of Electromagnetic Wave Applications, 1991-94; editor spl. section Proceedings of the IEEE on fractals in elec. engring., 1993; patentee in field; contbr. articles to profl. jours. Recipient Lindback award for Disting. Teaching, U. Pa., 1987, S. Reid Warren award for Disting. Teaching, 1985. Fellow IEEE; mem. Optical Soc. Am., Union Radio Scientists Internat., Internat. Neural Netowrk Soc., Sigma Xi. Mem. Christian Reformed Ch. Office: Univ of Pa Moore Sch 6314 Philadelphia PA 19104-6390

JAGGI, NARENDRA K., physics educator, researcher; b. Meerut, India, Feb. 25, 1954; came to U.S., 1982; s. Krishna Lal and Krishnaa (Vanti) J.; m. Hansa Jaggi, Apr. 16, 1978, 1 child, Tonushree. BSc (hons.), Ranchi Univ., 1973; PhD, Univ. Bombay, 1982. Scientist Bhabha Atomic Rsch. Ctr., Bombay, India, 1975-82; fellow Northwestern Univ., Evanston, Ill., 1982-85; asst. physics prof. Northeastern Univ., Boston, Mass., 1985-91; chmn. physics dept. Ill. Wesleyan Univ., Bloomington, 1991—; cons. Sanders Assocs., Nashua, N.H., 1987—, Spire Corp., Bedford, Mass., 1987—, Physical Scis., Inc., Andover, Mass., 1987—, and others. Contbr. articles to profl. jours. Recipient numerous grants U.S. Army, Office Naval Rsch., NASA, SBIR, SDI. Achievements include being co-discoverer of the high temperature superconductor TE Ba2 Ca3 Cu4 Oy with a tC of 122k., contributed to the understanding of electrorheological fluids. Office: Ill Wesleyan Univ PO Box 2900 Bloomington IL 61702

JAGIELLO, GEORGIANA M., geneticist, educator; b. Boston, Aug. 2, 1927; married, 1957. AB, Boston U., 1949; MD, Tufts U., 1955. Intern Rsch. & Edn. Hosps., U. Ill., 1955-56; resident New England Med. Ctr., Boston, 1956-57, rsch. fellow, 1958-60; rsch. fellow in endocrinology Scripps Clinic, La Jolla, Calif., 1957-58; USPHS rsch. fellow in cytogenetics Guy's Hosp., London, 1960-61, sr. lectr. in cytogenetics, 1966-69; asst. prof. U. Ill., Chgo., 1961-66, rsch. prof. pediatrics, 1969-70; prof. obstetrics, gynecology & human genetics Columbia U., N.Y.C., 1970—; mem. inst. advanced study U. Ill., 1964; cons. Guy's Hosp., 1966-69. Exch. fellow in surgery St. Bartholomew's Hosp., London, 1954; recipient Career Devel. award NIH, 1965. Mem. Am. Soc. Cell Biology, Endocrine Soc., Teratology Soc., Environ. Mutagen Soc., Soc. Study Reprodn. Achievements include research in mammalian meiosis, reproductive endocrinology. Office: Columbia U Ctr Reproductive Scis 630 W 168th St New York NY 10032-3702*

JAGODA, JERZY ANTONI, marine engineer; b. Chełmek, Poland, June 9, 1937; s. Antoni Wojciech and Emilia Anna (Kotowska) J.; m. maria Bronislawa Nimkiewicz, July 29, 1961; children: Piotr, Janusz, Marianna, Jerzy. MSc in Naval Architecture, Tech. U., Gdansk, Poland, 1960; postgrad., Tech. U., Gdansk, 1966-69; D in Tech. Sci., Ship Rsch. Inst., Gdansk, 1974. Registered profl. engr., U.K. Designer Ship Design & Rsch. Ctr., Gdansk, 1961-62, sr. designer, 1966-75, head of dept., 1975-76, prin. designer, 1977-81, 82-85; engr. Polish Steamship Co., Szczecin, Poland, 1962-66; rsch. fellow naval and ocean engring. dept. U. Glasgow, Scotland, 1981-82; engr., chief engr. on bd. of several flag seagoing ships, 1986-88; head R&D dept. Polski Rejestr Statkow, Gdansk, 1989—; sworn expert Polish Chamber for Fgn. Trade, Gdynia, Poland, 1975. Author: Methods of Economic Assessment in Shipbuilding Industry, 1978, Methods of Operations Research in the Preliminary Ship Design, 1979, 60 tech. and sci. papers, 1969-93. Mem. NSZZ Solidarnosc, Gdansk, 1980. Tech. U. Berlin fellow, 1983; Sci. Engring. Rsch. Coun. of London grantee U. Glasgow, 1981.

Fellow Royal Inst. Naval Architects U.K.; mem. Polish Soc. Mech. Engrs. (cert. expert in scope computer application in industry Gdansk chpt. 1976), Polish Cybernetic Soc. Roman Catholic. Avocations: gardening, guitar, poetry, travel. Home: ul-Kruczkowskiego 15D m 5, 80-288 Gdańsk Poland Office: Polski Rejestr Statkow, al Gen J Hallera 126, 80-416 Gdańsk Poland

JAHAN, MUHAMMAD SHAH, physicist; b. Rajshahi, Bangladesh, Dec. 21, 1943; came to U.S. 1972; s. Muhammad Jamir and Saratun (Nesa) Uddin; m. Kaniz, Mar. 26, 1967; children: Muhammad Ashif, Ishrat Shampa. MSc, Rajshahi U., 1965; PhD, U. Ala., 1977. Lectr. various colls. and univs., Rajshahi, 1965-72, U. Khartoum, Sudan, 1977-80; rsch. assoc. U. Ala., Tuscaloosa, summer 1980; asst. prof. Memphis State U., 1980-85, assoc. prof., 1985-88, prof. physics, 1988—; vis. staff Los Alamos (N.Mex.) Nat. Lab., 1989-90, cons., 1988-90; cons. Rsch. Triangle Inst., N.C., 1989—, Schering Plough, Memphis, 1982-88. Contbr. articles to profl. jours., chpts. to books. U.S. AID fellow, 1968, Fulbright-Hayes fellow, 1972-77, Dutch-Sudanese Edn. Exchange fellow, summer 1978; recipient Spur award Memphis State U., 1987, 88; Memphis State U. faculty rsch. grantee, 1981, 84, 89, Naval Surface Weapon Md. Ctr. rsch. grantee, 1982, 86. Mem. Am. Phys. Soc., Am. Chem. Soc., Materials Rsch. Soc., Sigma Xi (pres. 1989-92). Achievements include patent for detection of surface impurity phases in high temperature superconductors using thermally stimulated luminescence. Home: 7764 Widgeon Lake Cove Cordova TN 38018 Office: Memphis State U Dept Physics Memphis TN 38152

JAHN, BILLIE JANE, nursing educator, consultant; b. Byers, Tex., Dec. 12, 1921; d. Thomas Oscar and Molly Verona (Kennemer) Downing; student Scott and White Sch. Nursing, 1941-42, U. Mich., 1973-75; BSN, Wayne State U., 1971; MS, East Tex. State U., 1976, PhD, 1982; m. Edward L. Jahn, Dec. 6, 1942; children: Antoinette R., James T., Thomas L., Edward L., Janette E. Staff nurse Warren Meml. Hosp., Centerline, Mich., 1957-61; supr. nursing svc. Mich. Dept. Mental Health, Northville, 1962-71, Franklin County (Tex.) Hosp., 1972-74; instr. nursing Paris (Tex.) Jr. Coll., 1975-80; nurse educator VA, Waco, Tex., 1981-82; exec. v.p., dir., sr. nursing cons. Dos Cabezas, Inc., Mt. Vernon, Waco and Temple, Tex., 1981—; adj. faculty U. Tex.-Arlington, 1985—; mem. dept. phys. medicine and rehab. Scott and White Hosp., Temple, Tex., 1985—; head nurse dept. phys. med. and rehab., nurse researcher biosci., 1990; cons. East Tex. State U., Texarkana, 1978—; adj. faculty U. Tex.-Arlington. Vol., ARC, 1971—; den mother Boy Scouts Am., 1960-62; sec. PTA, Warren, Mich., 1960-62; v.p., Temple, Tex., 1957-58. Mem. AAAS, AAUP, NAFE, Nat. League Nursing, Nat. Assn. Rehab. Nurses (rev. bd. Rehab. Nursing Inst. 1986—, Rsch Grant Panel, 1992—, Rehab. Nursing Found., 1989—), Tex. League Nursing, Am. Assn. Curriculum and Supervision, Phi Delta Kappa, Kappa Delta Pi.

JAIN, HIMANSHU, materials science engineering educator; b. Mainpuri, India, Jan. 20, 1955; came to U.S., 1974; s. Chandra Kumar and Kusuma Devi Jain; m. Sweety Agrawal, Feb. 14, 1990; 1 child, Isha Himani. MS, Banaras U., Varanasi, India, 1972; M of Tech., Indian Inst. Tech., Kanpur, India, 1974; D of Engring. Sci., Columbia U., 1979. Postdoctoral appointee Argonne (Ill.) Nat. Lab., 1980-82; assoc. scientist Brookhaven Nat. Lab., Upton, N.Y., 1982-85; prof. material sci. and engring. Lehigh U., Bethlehem, Pa., 1985—; vis. scientist Indian Inst. Tech., Kanpur, 1985; Humboldt fellow U. Dortmund, Germany, 1991-92. Editor: Current Trends in the Science and Technology of Glass, 1989, Atomic Migration and Defects in Materials, 1991; contbr. over 70 articles to profl. jours. Mem. Am. Ceramic Soc. (chmn. Lehigh Valley chpt. 1990), Ceramic Ednl. Coun. Achievements include discovery of anomalous isotope mass effect in a glass; demonstration of feasiblity of radiation enhanced sintering in a ceramic; 2 patents for electro-optic polymers. Office: Lehigh U 5E Packer Ave Bethlehem PA 18015

JAIN, RAJ KUMAR, electrical engineering researcher; b. Jawad, India, Sept. 9, 1949; came to U.S., 1989; s. Takhat Singh and Sugan Bai (Silot) J.; m. Madhubala Chanodia, Nov. 24, 1970; children: Alok, Rakesh, Kamana. MSc in Physics, Vikram U., Ujjain, India, 1968; MSc Tech. Electronics, Birla Inst. Tech. Sci., Pilani, India, 1970; PhD, Cath. U. Leuven, Belgium, 1976. Sr. sci. asst. Cen. Electronics Engring. Rsch. Inst., Pilani, 1970-71, scientist, 1978-82; rsch. assoc. Cath. U. Leuven, 1971-76, Technische Hochschule Aachen, Germany, 1974; assoc. researcher Venezuelan Inst. Sci. Rsch., Caracas, 1977-78; mgr. Bharat Heavy Electricals Ltd., Bangalore, India, 1982-89; sr. rsch. assoc. NASA Lewis Rsch. Ctr., Cleve., 1989—. Contbr. articles to profl. jours. including Jour. of Applied Physics, Thin Solid Films, Applied Physics Letters, IEEE Transactions on Electron Devices. Mem. IEEE (sr. mem., mem. internat. com. photovoltaic specialists conf. 1988—), IEEE Electron Devices Soc. Achievements include work on use of inalas as passivating layer for indium phosphide solar cells, research on high efficiency solar cell materials and devices for terrestrial and space power applications, and anomalous impurity diffusion in silicon. Home: Apt 727 26101 Country Club Blvd North Olmsted OH 44070 Office: NASA Lewis Rsch Ctr mail stop 302-1 21000 Brookpark Rd Cleveland OH 44135

JAIN, SUNIL, pharmaceutical scientist; b. New Delhi, Oct. 18, 1963; came to the U.S., 1987; s. Roshan Lal and Maya (Gupta) J.; m. Renu Gupta, Jan. 9, 1991. PharmB, U. Delhi, 1985, PharmM, 1987; PhD, U. Conn., 1993. Summer trainee Indian Drugs and Pharms. Ltd., Gurgaon, India, 1983; mfg. chemist Cyper Pharma, New Delhi, 1985; teaching asst. coll. of pharmacy U. Delhi, 1986-87; teaching asst. coll. of pharmacy U. Conn., Storrs, 1987-89, rsch. asst., 1989—; grad. intern Abbott Labs., North Chicago, Ill., 1989; rsch. scientist Burroughs Wellcome Co., Greenville, N.C. Contbr. articles to Pharm. Rsch. Pres. Grad. Residents Assn. U. Conn., Storrs, 1990, resident advisor, 1991-92. Mem. Am. Assn. Pharm. Scientists, AAAS, Rho Chi. Home: 104 Northwood Apt Storrs CT 06268 Office: Burroughs Wellcome Co Pharm R & D Labs PO Box 1887 Greenville NC 27835

JAIN, SURINDER MOHAN, electronics engineering educator; b. Patiala, Punjab, India, Sept. 19, 1945; came to U.S., 1983; s. Chhajju Ram and Kamla Jain; m. Harmit Kaur, June 9, 1974; children: Sumit, Preeti. MSc in Physics, Punjabi U., Patiala, 1967, post M.S. diploma, 1972. Rsch. assoc. Punjabi U., 1967-71, asst. prof., 1972-74, 75-83; vis. prof. Eindhoven (The Netherlands) Tech. U., 1974-75; asst. prof. Sinclair C.C., Dayton, Ohio, 1983-88, prof., head elec. engring., 1985—, tng. coord., 1985—; nat. electronics program evaulator Am. Coun. on Edn., Washington, 1989—; writer spl. programs GE, Ohio Bell Co., Dayton Power and Light Co.; developer tng. ctr. Pace, Inc., Laurel, Md., Tektronix. Author lab. manual, Analog Electronics, 1983. Social sec. Physics Assn., Punjabi U., 1974-83; career expert Explorer program Boy Scouts Am., Dayton, 1988—. Hewlett-Packard grantee, 1988. Mem. IEEE, Am. Soc. Engring. Edn., Tau Alpha Pi (hon.). Jianist. Achievements include development of special computer aided programs for corporate retraining of employees. Home: 1963 Lord Fitzwalter Dr Miamisburg OH 45342-2049 Office: Sinclair CC 444 W 3d St Dayton OH 45402-1421

JAKAB, IRENE, psychiatrist; b. Oradea, Rumania; came to U.S., 1961, naturalized, 1966; d. Odon and Rosa A. (Riedl) J. MD, Ferencz József U., Kolozsvar, Hungary, 1944; lic. in psychology, pedagogy, philosophy cum laude, Hungarian U., Cluj, Rumania, 1947; PhD summa cum laude, Pazmany Peter U., Budapest, 1948; Dr honoris causa, U. Besançon, France, 1982. Diplomate Am. Bd. Psychiatry. Rotating intern Ferencz József U., 1943-44; resident in psychiatry Univ. Hosp., Kolozsvar, 1944-47, resident in neurology, 1947-50; resident internal medicine Univ. Hosp. for Internal Medicine, Pécs, Hungary, 1950-51; chief physician Univ. Hosp. for Neurology and Psychiatry, Pécs, 1951-59; staff neuropathol. rsch. lab. Neurol. Univ. Clinic, Zurich, 1959-61; sect. chief Kans. Neurol. Inst., Topeka, 1961-63; dir. rsch. and edn., 1966; resident psychiatry Topeka State Hosp., 1963-66; asst. psychiatrist McLean Hosp., Belmont, Mass., 1966-67; assoc. psychiatrist McLean Hosp., 1967-74; prof. psychiatry U. Pitts. Med. Sch., 1974-89, prof. emerita, 1989—, co-dir. med. student edn. in psychiatry, 1981-89; dir. John Merck Program, 1974-81; mem. faculty dept. psychiatry Med. Sch., Pecs, 1951-59; asst. Univ. Hosp. Neurology, Zurich, 1959-61; assoc. psychiatry Harvard U., Boston, 1966-69, asst. prof. psychiatry, 1969-74, program dir. grad course mental retardation, 1970-87; lectr. psychiatry, 1974—. Author: Dessins et Peintures des Aliénés, 1956, Zeichnungen und Gemälde der Geisteskranken, 1956; editor: Psychiatry and Art, 1968, Art Interpretation and Art Therapy, 1969, Conscious and Unconscious Expressive Art, 1971, Transcultural Aspects of Psychiatric Art, 1975; co-editor: Dynamische Psychiatrie, 1974; editorial bd.: Confinia Psychiatrica, 1975-81; contbr. articles to profl. jours. Recipient 1st prize Benjamin Rush Gold medal award for sci. exhibit, 1980, Bronze Chris plaque Columbus Film Festival, 1980, Leadership award Am. Assn. on Mental Deficiency, 1980; Menninger Sch. Psychiatry fellow, Topeka, 1963-66. Mem. AMA, Am. Psychol. Assn., Am. Psychiat. Assn., Société Medico Psychologique de Paris, Internat. Rorschach Soc., N.Y. Acad. Scis., Internat. Soc. Psychopathology of Expression (v.p. 1959—), Am. Soc. Psychopathology of Expression (chmn. 1965—, Ernst Kris Gold Medal award 1988), Royal Soc. of Medicine (overseas fellow), Internat. Soc. Child Psychiatry and Allied Professions, Internat. Assn. Knowledge Engrs. (v.p. for medicine), Deutschsprachige Gesellschaft für Psychopathologie des Ausdruckes (hon. Prinzhorn prize 1967). Home and Office: 74 Lawton St Brookline MA 02146-2501

JAKLITSCH, DONALD JOHN, materials engineer, consultant; b. Bklyn., Dec. 25, 1947; s. Ernest John and Mary Ann (Rupp) J.; m. N. Chapin, 1971 (div. 1975); children: Maizy, Zoie; m. Maria Therese Bostic, Apr. 18, 1988. BA, SUNY, New Paltz, 1970; MS, U. Lowell, 1986, DEng, 1990. Chemist U.S. Army Material and Mechanic Rsch. Ctr., Watertown, Mass., 1982-84, chem. engr., 1984-88; chem. engr. U.S. Army Materials Tech. Lab., Watertown, 1988-91, U.S. Army Missile Command, Redstone Arsenal, Ala., 1991—; chmn. civilian welfare coun. U.S. Army Materials Tech. Lab., Watertown, 1988-89; reviewer K-15 com. ASME, 1990. Contbr. articles to profl. jours. Recipient tech. commendation Dept. Army, 1989, 90, 91. Mem. AIAA, Soc. for Advancement Material and Process Engring., Am. Soc. for Composites. Avocations: wilderness backpacking, photography. Home: 153 Cedar Ln New Market AL 35761 Office: US Army Missile Command Attn AMSMI-RD-ST-CM/ D Jaklitsch Redstone Arsenal AL 35898-5247

JAKOBS, KAI, computer scientist; b. Cologne, Germany, May 27, 1957; parents Willy and Leni (Biermann) J. Tech. staff mem. Computer Sci Dept. Info. IV Tech. U., Aachen, Germany, 1985-88, head tech. staff Computer Sci. Dept. Info. IV, 1988—. Contbr. articles to profl. jours. Mem. IEEE, Assn. Computing Machinery. Office: Tech U Computer Sci Dept, Ahornstr 55, 51 Aachen Germany

JAKOBSEN, JAKOB KNUDSEN, mechanical engineer; b. Bording Sogn, Denmark, Aug. 7, 1912; came to U.S., 1952, naturalized, 1958; s. Laust Peder and Inger Marie (Kristensen) J.; m. Eva Koch, Nov. 19, 1941 (dec. 1983); children—Marianne Gyrithe (Mrs. Earl C. Green), Peter Laust (dec. 1969), Claus Michael, Suzanne Elizabeth (Mrs. Paul B. Marsh), Niels-Olaf Sejten, Lars Jakob. M.S. in Mech. Engring, Royal Tech. U. Denmark, 1941. Registered profl. engr., Mich., Calif. Asst. to prof. machine design Royal Tech. U., Denmark, 1941; mech. engr. turbines Brown Boveri et Cie, Switzerland, 1941-43; project engr. co-generation steam power sta. Pub. Power Utilities of Copenhagen, Denmark, 1943-45; mech. engr., asst. to chief engr. turbo-supercharge two-stroke marine Diesel engines Burmeister & Wain, Copenhagen, 1945-52; gas turbine engr. Clark Bros. div. Dresser Industries, Olean, N.Y., 1952-55; staff engr. automotive research Chrysler Corp., Detroit, 1955-60; sr. tech. specialist for R&D of liquid rocket engines for space program Rocketdyne div. Rockwell Internat., Canoga Park, Calif., 1960-77; cons. to industry on rocket engines, aircraft aux. power units, co-generation power plants, aircraft environ. control systems, 1977—. Author: NASA monograph Rocket Engine Turbopump Inducers, 1971; Contbr. articles profl. jours. Mem. AIAA, NSPE, ASTM (com. for erosion by cavitation and impingement 1964—), ASME (recipient Melville Gold medal 1964), Soc. Automotive Engrs., Danish Inst. Civil Engrs. Republican. Lutheran. Home: 10531 Etiwanda Ave Northridge CA 91326-3113

JAKOBSSON, ERIC GUNNAR, biophysicist, educator; b. N.Y.C., Nov. 18, 1938; s. Ejler Gunnar and Susan Nanette (Kane) J.; m. Naomi Fay Dick; children: Beverly, Susan, Eric Jr., Garret, Jonathan, Sarah, Brenda, Linda. BA, Columbia U., 1959, BS, 1960; PhD, Dartmouth Coll., 1969. Postdoctoral fellow Case Western Res. U., Cleve., 1969-71; asst. prof. U. Ill., Urbana, 1972-78, assoc. prof., 1978-91, prof. physiology and biophysics, 1991—; rsch. scientist Nat. Ctr. Supercomputing Applications, Urbana, 1991—; vis. assoc. prof. Duke U., Durham, N.C., 1978; adv. bd. Nat. Ctr. Supercomputing Applications, Urbana, 1987-90. Contbr. articles to sci. jours. Democrat. Presbyterian. Home: 803 W Main St Urbana IL 61801 Office: Nat Ctr Supercomputing Applications Beckman Inst Urbana IL 61801

JAKOVAC, JOHN PAUL, construction executive; b. LaPorte, Ind., May 29, 1948; s. Anthony Jakovac; m. Sondra Jakovac; children: Ashley, Steffanie, Anna. CEO, pres. Commonwealth Corp., Sacramento, 1975—; author, speaker workshop/seminar Non Adviserial Constrn. Projects, 1992. Founder Anthony Jakovac Found. for the Children of AIDS, Sacramento, 1990. Mem. ASTM, Internat. Conf. Bldg. Ofcls., Am. Assn. Cost Engrs., Am. Concrete Inst., Am. Constrn. Inspection Assn., Sacramento Met. C. of C., Rotary of Cameron Park. Office: Commonwealth Corp 1001 6th St Ste 502 Sacramento CA 95814

JAKSCHIK, BARBARA A., science educator, researcher; b. Lipine, Poland, Sept. 12, 1931; came to U.S., 1958; MS, Duquesne U., 1966; PhD, Washington U., 1974. Traniee asst. Duquesne U., Pitts., 1963; resident Mercy Hosp., Pitts., 1964; dir. pharm. St. Mary's Hosp., Huntington, W.Va., 1965-70; rsch. assoc. Washington U., St. Louis, 1974-77, asst. prof., 1977-84, assoc. prof., 1984—; mem. study section Nat. Inst. Health, Washington, 1980, 83, 88, 90; mem. editorial adv. bd. J. Pharm. Exp. THerapy, Washington, 1984-88. Contbr. articles to profl. jours. Grantee Nat. Inst. Health, 1977-92. Mem. Am. Soc. Pharm. and Experimental Therapy, Am. Soc. Advancement of Sci. Achievements include research in leukotrienes and the role of mast cells in inflammation other than immediate hypersensitivity. Office: Washington U Sch Medicine 660 S Euclid Saint Louis MO 63110

JALLER, MICHAEL M., retired orthopaedic surgeon; b. Zurich, Switzerland, Feb. 25, 1924; came to U.S., 1926; s. Arthur and Anna (Eliash) J.; m. Helen F. Cowan, June 24, 1944; children: David, Daniel, Amy. BA, NYU, 1944; MD, U. Lausanne (Switzerland), 1952. Diplomate Am. Bd. Orthopaedic Surgery. V.p. A. Jaller & Co. Inc., N.Y.C., 1944-61; intern Waterbury (Conn.) Gen. Hosp., 1952-53; surg. resident Bridgeport (Conn.) Gen. Hosp., 1953-54; orthopedic resident U. Hosp. of Cin., Cin. Gen. Hosp., 1954-57; instr. orthopaedic surgery U. Hosp., Cin., 1954-57, chief resident in orthosurgery, 1956-57; rsch. fellow in biophysics Weitzmann Inst., Rehovoth, Israel, 1965; chief cons. VA Hosp., Washington, 1966-71; pres. Greater Washington Orthopaedic Group, Silver Spring, Md., 1967-93; ret., 1993; bd. dirs. Rotocast Plastics Inc. Pres. Silver Spring Jewish Ctr., Silver Spring, 1984-91. With U.S. Army, 1942-46. Fellow ACS, Am. Acad. Orthopaedic Surgery, Royal Soc. Health, Washington Art League, Selby Bay Yacht Club. Republican. Home: 4110 NE Joe's Point Rd Stuart FL 34996

JAMES, BRUCE DAVID, chemistry educator; b. Bolton, Lancashire, Eng., Sept. 23, 1942; arrived in Australia, 1969; s. Stephen William and Jennie (Collier) J.; m. Pauline Crocker, Dec. 19, 1964; children: Caroline, Stephen. BSc with honors, U. Sheffield, Eng., 1964, PhD, 1967. Chartered chemist. Tutor U. Sheffield, 1964-67; rsch. fellow Ga. Inst. Tech., Atlanta, 1967-68; asst. prof. Ga. State U., Atlanta, 1968-69; sr. tutor U. Queensland, Brisbane, Australia, 1969-71; prin. tutor U. Queensland, Brisbane, 1971-75; rsch. fellow La Trobe U., Bundoora, Australia, 1975-79; adviser studies La Trobe U., Bundoora, 1975—, sr. lectr., 1979-87, reader chemistry, 1987—; sec. chem. edn. div. Royal Australian Chem. Inst., Melbourne, 1972-74, chair admissions, 1986-90; vis. scientist chem. physics div. Commonwealth Sci. and Indsl. Rsch. Orgn., Melbourne, 1982-83; vis. prof. U. South Fla.,Tampa, 1987-88. Author revs. Metal Tetrahydroborates, 1970, Halogen-Group IIIB Bonds, 1991; book rev. editor Chemistry in Australia, 1991—; contbr. articles to profl. jours. Grantee Australian Rsch. Grants Scheme, 1970-75, 76, 84-85, Australian Rsch. Coun., 1989—. Fellow Royal Australian Chem. Inst., Royal Soc. Chemistry (referee 1987—, corr. mem. 1973-83); mem. Am. Chem. Soc., Fla. Acad. Scis., Australian Football League Park. Avocations: American football, classical music, true crime stories, cricket umpiring. Office: La Trobe U Chemistry Dept, Plenty Rd, Bundoora 3083 Victoria, Australia

JAMES, EARL EUGENE, JR., aerospace engineering executive; b. Oklahoma City, Feb. 8, 1923; s. Earl Eugene and Mary Frances (Godwin) J.; m. Barbara Jane Marshall, Dec. 15, 1945 (dec. Feb. 2, 1982); children: Earl Eugene III, Jeffrey Allan; m. Vanita L. Nix, Apr. 23, 1983. Student Oklahoma City U., 1940-41; BS, U. Okla., 1945; postgrad. Tex. Christian U., 1954-57; MS, So. Meth. U., 1961. Asst. mgr. Rialto Theatre, 1939-42; with Consol. Vultee Aircraft Co., San Diego, 1946-49; with Convair, Ft. Worth, 1949-89, group engr., 1955-57, test group engr., supr. fluid dynamics lab., 1957-81, engring. chief Fluid Dynamics Lab., 1981-89. Asst. dist. commr. Boy Scouts Am., 1958-59; adviser Jr. Achievement, 1962-63; mem. sch. bd. Castleberry Ind. Sch. Dist. (Tex.), 1969-83; chmn. bd. N.W. br. YMCA, 1971. Served to lt. (j.g.) USNR, 1942-46; PTO. Author/editor over 100 engring. reports. Fellow AIAA (assoc.); mem. Air Force Assn. (life), U.S. Naval Inst. (assoc.), Gen. Dynamics Mgmt. Assn., Nat. Mgmt. Assn., Okla. U. Alumni Assn. (life), Tex. Congress Parents and Tchrs. (hon. life), Pi Kappa Alpha, Alpha Chi Sigma, Tau Omega. Methodist. Democrat. Clubs: Squaw Creek Golf, Camera. Lodge: Elks.

JAMES, FRANCIS CREWS, zoology educator; b. Phila., Sept. 29, 1930; divorced; children: Sigrid Bonner, Helen Olsen, Avis James. AB in Zoology, Mount Holyoke Coll., 1952; MS in Zoology, Louisiana State U., 1956; PhD in Zoology, U. Ark., 1970. Summer rsch. asst. Am. Mus. Natural History, NYC, 1950; grad. teaching asst. Louisiana State U., 1952-54; part time instr., botany, zoology, and physical edn. U. Ark., 1960-70; rsch. assoc. U. Ark. Mus., 1971-73; asst. ecology program dir. NSF, Washington, 1973-76, assoc. program dir., 1976; assoc. prof. and curator of birds and mammals Fla. State U., 1977-84, prof. and curator of birds and mammals, 1984—; instr. summer faculty U. Minn., 1978-81; vis. prof. fall semester Cornell U., Ithaca, NY, 1988; rsch. assoc. spring semester, Smithsonian Inst., 1989; adv. coun. of Systematic and Environmental Biology, Smithsonian Foreign Currency Program, 1983-87; mem. nongame wildlife program, Florida Game and Fresh Water Fish Commission, 1985-86; bd. dirs. World Wildlife Fund/ Conservation Found., Cornell Lab. Ornithology, Am. Inst. Biological Sciences; central com. mem. Internat. Ornithological Congress; com. mem. Nat. Rsch. Coun. Editorial Bd.: American Birds, 1978—, Ecology and Ecological Monographs, 1989—, Annual Review of Ecology and Systematics, 1986-90; assoc. editor 1991—; Current Ornithology, assoc. editor, 1982-87, American Midland Naturalist, 1978-84. Contbr. numerous articles to profl. jours. NSF grantee 1979, 80, 83, U.S. Fish and Wildlife Svc. grantee, 1980 (2), FSU Found. grantee, 1982, 91, Nat. Geographic Soc. grantee 1983 (2), 1984, 85, 86, 87, 88, Fla. Game and Fresh Water Fish Comm. grantee, 1986-87, Cayahoga Trust grantee, 1986—, Nat. Fish and Wildlife Found. grantee, 1990-92, R. G. Crews Fund grantee, 1990-91, Conservation and Rsch. Found grantee, 1991. Fellow AAAS, Sigma Xi, Soc. Systematic Zoology; mem. Am. Ornithologist Union (pres. 1984-86, fellow 1976, permanent mem. coun., 1984—, Eliot Coues award 1992), Wilson Ornithological Soc., Am. Inst. Biological Sciences, Cooper Ornithological Soc. Achievements include research of recent projects on the population dynamics of the Red-cockaded Woodpecker and on analyses of population trends in North American landbirds based on data from the Breeding Bird Survey. Home: 2113 Gibbs Dr Tallahassee FL 32303 Office: Florida State Univ Dept Biological Science 106 Conradi Bldg Tallahassee FL 32306

JAMES, GARY DOUGLAS, biological anthropologist, educator, researcher; b. Norwich, Conn., Dec. 6, 1954; s. Godfrey Merchant ans Joan (McIlwaine) J.; m. Kathleen Louise Wilson, July 28, 1979. BA, Wake Forest U., 1976; MA, Pa. State U., 1980, PhD, 1984. Part-time instr. Pa. State U., University Park, 1982-84; postdoctoral assoc. Cornell U. Med. Coll., N.Y.C., 1984-86; asst. prof. physiology and biophysics Med. Coll. Cornell U., N.Y.C., 1986-91, asst. prof. physiology in medicine, 1986-91, assoc. rsch. prof. of physiology in medicine, 1991—, assoc. rsch. prof. of physiology and biophysics, 1991—. Contbr. chpt. to book, articles to profl. jours. Recipient New Investigator Rsch. award NIH, 1986; NIH postdoctoral trainee, 1984. Fellow Human Biol. Coun. (sec.-treas. 1992—), mem. Am. Assn. Phys. Anthropologists, Soc. Study Social Biology, Soc. Behavioral Medicine, Am. Soc. Hypertension, Lambda Alpha. Lutheran. Office: Cornell U Med Coll Cardiovascular Ctr 520 E 70th St New York NY 10021-4896

JAMES, HAROLD LEE, biochemist, researcher; b. Taylorsville, N.C., Oct. 31, 1939; s. Lee Alexander and Shirley Christine (Land) J.; m. Pranee Limdhamarose, May 1, 1965 (dec. Sept. 1986); 1 child, Daniel Lee. BA in Chemistry cum laude, East Tenn. State Coll., 1962; PhD in Biochemistry, U. Tenn., 1968. Rsch. instr. Temple U. Health Scis. Ctr., Phila., 1968-70, rsch. asst. prof., 1976-80, rsch. assoc. prof., 1980-83; rsch. scientist Am. Nat. Red Cross, Bethesda, Md., 1970-72; rsch. assoc. St. Jude Children's Rsch. Hsop., Memphis, 1972-75; assoc. prof. U. Tex. Health Ctr., Tyler, 1983—. Contbr. chpt. to book. Cub scout asst. Boy Scouts Am., Willow Grove, Pa., 1977-79. NIH grantee, 1992. Mem. Am. Soc. Biochemistry and Molecular Biology, Internat. Soc. on Thrombosis and Haemostasis, Am. Heart Assn., Protein Soc., Sigma Xi. Republican. Unitarian. Achievements include discovery of point mutations in human genes for Factor X Friuli, Factor X Wenatchee, Factor VII Padua, Factor VII Detroit and Factor VII Seattle. Office: Univ of Texas Health Ctr Dept Biochem PO Box 2003 Tyler TX 75710

JAMES, LAWRENCE HOY, nuclear physicist; b. Winston-Salem, N.C., Mar. 6, 1956; s. John Clay and Ann Francis (Poston) J.; m. Sandra Kay Smith, May 20, 1989; children: Allison Leigh Clark, Leslie Catherine Clark. BA in Physics, Wake Forest U., 1982, MS in Physics, 1984; PhD in Physics, N.C. State U., 1989. Rsch. asst. Triangle U. Nuclear Lab. Duke U., Durham, N.C., 1986-89; nuclear physicist Troxler Electronic Labs. Inc., Research Triangle Park, N.C., 1990—. Contbr. articles to profl. jours. Mem. ANS. Achievements include definition of a method for determining the effective volume measured by a gamma ray density gauge. Home: 8425 2 Courts Raleigh NC 27613-1255 Office: Troxler Electronic Labs Inc 3008 W Cornwallis Rd Durham NC 27705-5206

JAMES, THOMAS NAUM, cardiologist, educator; b. Amory, Miss., Oct. 24, 1925; s. Naum and Kata J.; m. Gleaves Elizabeth Tynes, June 22, 1948; children: Thomas Mark, Terrence Fenner, Peter Naum. BS, Tulane U., 1946, MD, 1949. Diplomate Am. Bd. Internal Medicine (bd. govs. 1982-88), Bd. Cardiovascular Diseases (bd. dirs. 1972-78). Intern Henry Ford Hosp., Detroit, 1949-50, resident in internal medicine and cardiology, 1950-53, mem. staff, 1959-68; practice medicine specializing in cardiology Birmingham, Ala., 1968-87; mem. staff U. Ala. Hosps., 1968-87; instr. medicine Tulane U., New Orleans, 1955-58, asst. prof., 1959; prof. medicine U. Ala. Med. Ctr., Birmingham, 1968-87, prof. pathology, 1968-73, assoc. prof. physiology and biophysics, 1969-73, dir. Cardiovascular Rsch. and Tng. Ctr., 1970-77, chmn. dept. medicine, dir. divsn. cardiovascular disease, 1973-81, Mary Gertrude Waters prof. cardiology, 1976-87, Disting. prof. of univ., 1981-87; prof. medicine, prof. pathology, pres. U. Tex. Med. Br., Galveston, 1987—, dir. WHO Cardiovascular Ctr., 1988—; mem. adv. coun. Nat. Heart Lung and Blood Inst., 1975-79; pres. 10th World Congress Cardiology, 1986; mem. cardiology del. invited by Chinese Med. Assn. to China, 1978. Author: Anatomy of the Coronary Arteries, 1961, The Etiology of Myocardial Infarction, 1963; Mem. editorial bd.: Circulation, 1966-83; mem. editorial bd.: Am. Jour. Cardiology, 1968-76; assoc. editor, 1976-82; mem. editorial bd.: Am. Heart Jour, 1976-79; Contbr. articles on cardiovascular diseases to med. jours. Served as capt. M.C. U.S. Army, 1953-55. Fellow ACP (gov. Ala. 1975-79, master 1983); mem. AMA, Am. Clin. and Climatological Assn., Assn. Am. Physicians, Am. Soc. Clin. Investigation, Assn. Univ. Cardiologists (pres. 1978-79), Am. Heart Assn. (pres. 1979-80), Am. Coll. Cardiology (v.p. 1970-71, trustee 1970-71, 76-81, First Disting. Scientist award 1982), Am. Soc. Pharmacology and Exptl. Therapeutics, Soc. Exptl. Biology of Medicine, Am. Coll. Chest Physicians, Ctrl. Soc. Clin. Rsch., Internat. Soc. and Fedn. Cardiology (pres. 1983-84), WHO (expert adv. panel on cardiovascular diseases 1988—), So. Soc. Clin. Investigation, Am. Fedn. Clin. Rsch., Phi Beta Kappa, Sigma Xi, Omicron Delta Kappa, Ala. Acad. Honor, Alpha Omega Alpha, Alpha Tau Omega, Phi Chi. Presbyterian. Clubs: Cosmos, Mountain Brook, Galveston Artillery. Office: U Tex Med Br Office of Pres 301 University Blvd Galveston TX 77555-0129

JAMES, WALTER, retired computer information specialist; b. Mpls., June 8, 1915; s. James Edward and Mollie (Gress) Smoleroff; B.Ch.E., U. Minn., 1938, postgrad, 1945-60; m. Jessie Ann Pickens, Dec. 27, 1948; 1 son, Joel Pickens. Process designer Monsanto Chem. Co., St. Louis, 1940-45; instr.

math. U. Minn., Mpls., 1945-60, extension div., 1950—; researcher computer based applied math. 3M Co., St. Paul, 1960-68; info. systems planner State of Minn., St. Paul, 1968-85; ret., 1985. Mem. Am. Math. Assn., AAAS, Sigma Xi. Contbr. articles to profl. jours. Home: 6228 Brooklyn Dr Minneapolis MN 55430-2023

JAMESON, J(AMES) LARRY, cable company executive; b. Elizabethtown, Ky., 1937; s. William Kendrick and Ruth Helen (Krause) J.; m. Mary Louise Wojcik, June 26, 1965; children: Renee, Jennifer, Julie. BA in Math., Bellarmine Coll., 1959; BS in Chem. Engring., U. Detroit, 1963, MBA, 1970. Tech. mgr. automotive products Rinshed Mason et Cie, Paris, 1965-69; ops. mgr. vinyl coated fabrics Inmont Corp., Toledo, 1969-75; v.p., gen. mgr. European ops. Inmont Corp., London, 1975-79; v.p., gen. mgr. automotive finishes products Inmont Corp., Detroit, 1979-83, sr. v.p. worldwide automotive, 1983-86; pres. Coatings & Colorants div. BASF, Clifton, N.J., 1986-93; pres., CEO Pirelli Cable Corp., Florham Park, N.J., 1993—. Mem. Soc. Automotive Engrs., Orchard Lake Country Club. Avocations: golf, tennis, skiing, hunting. Home: 29 Horizon Dr Mendham NJ 07945-2302 Office: Pirelli Cable Corp 325 Columbia Tpk Florham Park NJ 07932

JAMESON, JULIANNE, clinical psychologist; b. L.A., Apr. 22, 1943; d. Walter McClure and Anne Virginia (Walther) J.; m. Russell Larson, Oct. 9, 1970 (div. Mar. 1973). BA, UCLA, 1965; MS, Calif. State U., L.A., 1973; PhD, Calif. Grad. Inst., 1982. Lic. psychologist, Calif. Tchr. L.A. Unified Sch. Dist., 1965-70; sch. psychologist West Covina (Calif.) Unified Schs., 1973-85; ednl. psychologist, Covina, Calif., 1980—; pvt. practice clin. psychology Covina, 1984—; provider Blue Shield, L.A., 1989—, LifeLink, L.A., 1991—, U.S. Hehavioral Health, L.A., 1991—; assessor Nat. Resource Orgn., San Diego, 1992—. Mem. APA. Office: 750 Terrado Pla Ste 121B Covina CA 91724

JAMIESON, JOHN ANTHONY, engineering consulting company executive; b. London, England, Mar. 16, 1929; came to U.S., 1952; s. John Percival and Jean (Kerr) J.; m. Barbara Armstrong, July 6, 1956; children: John Gordon, Sara Felicity, John Douglas. BS summa cum laude, U. London, 1952; PhD, Stanford U., 1957. Scientist, then mgr. electrooptics div. Aerojet-Gen. Corp., Azusa, Calif., 1957-69; asst. dir. U.S. Army Advanced Ballistic Missile Defense Agy., Washington, 1970-73; pres. Jamieson Sci. & Engring., Inc., Washington, 1973—; chmn. Sci. & Engring. Support Group to SDIO, Washington, 1985—; bd. dirs. Optelecom, Inc., Gaithersburg, Md., Space Computer Corp., Los Angeles. Author: Infrared Physics & Engineering, 1963; contbr. chpt. in book and articles to profl. jours. Served with RAF, 1947-50. Recipient Space Systems award AIAA, 1992; Wesix fellow Stanford U., 1954. Republican. Episcopalian. Office: Jamieson Sci & Engring Inc 7315 Wisconsin Ave Bethesda MD 20814-3202

JAMIESON, STUART WILLIAM, cardiologist, educator; b. Bulawayo, Rhodesia, July 30, 1947; came to U.S., 1977; MB, BS, U. London, 1971. Intern St. Mary's Hosp., London, 1971; resident St. Mary's Hosp., Northwick Park Hosp., Brompton Hosp., London, 1972-77; asst. prof. Stanford U., Calif., 1980-83, assoc. prof., 1983-86; prof., head cardiac surgery U. Minn., Mpls., 1986-89, U. Calif., San Diego, 1989—; dir. Minn. Heart and Lung Inst., Mpls., 1986-89; pres. Calif. Heart and Lung Inst., San Diego, 1991—. Co-author: Heart and Heart-Lung Transplantation, 1989; editor: Heart Surgery, 1987; contbr. over 350 papers to med. jours. Recipient Brit. Heart Found. Fellowship award, 1978, Irvine H. Page award Am. Heart Found., 1979, Silver medal Brit. Surg. Soc., 1986. Fellow ACS, Royal Coll. Surgeons, Royal Soc. Medicine, Am. Coll. Chest Physicians, Am. Coll. Cardiology; mem. Royal Coll. Physicians (licentiate), Internat. Soc. for Heart Transplantation (pres. 1986-88), Calif. Heart and Lung Inst. (pres. 1991—). Office: U Calif-San Diego Divsn Cardiothoracic Surgery 200 West Arbor Dr San Diego CA 92103-1910

JAMISON, DAVID W., marine scientist; b. Portland, Oreg., Apr. 23, 1939; s. Edgar W. and Nina (Ray) J.; m. Susan Elizabeth Porter, Dec. 23, 1962 (div. 1974); children: Adam, Elizabeth; m. Nancy Louise Kasper, Apr. 7, 1979; stepchildren: Kevin, Keith, Kelly. BS, Whitman Coll., 1961; postgrad. U. Oreg., 1961-62; MS, U. Wash., 1966, PhD, 1970. Remote sensing scientist Wash. Dept. Natural Resources, Olympia, 1969-70, marine scientist, 1970-74, supr. baseline studies dept. ecology, 1974-78, dir. marine rsch. and devel., 1978-80, mgr. forestry rsch. and devel., 1980-82, chief marine scientist, 1983-88, sr. marine scientist, 1988—; gov.'s rep. U.S. Dept. Interior outer continental shelf rsch. adv. com., 1974-78; cons. NOAA Interagy. Com. on Ocean Pollution Rsch., Devel. and Monitoring, 1981; mem. adv. com. Puget Sound Water Quality Authority, 1985-86, chmn. rsch. com. 1987—; mem. tech. adv. com. Puget Sound Estuary Program, 1985—; mem. tech. work groups, Puget Sound Dredge Disposal Analysis Study, 1985-90; gov's. rep. Pacific Northwest Regional Marine Rsch Bd., 1992—. Bd. dirs. Boston Harbor Assn., 1980-88, chmn. utilities com., 1981—; mem. Thurston County Shorelines adv. com., 1973-74, 82-83. Mem. Am. Soc. Photogrammetry, Marine Tech. Soc., Pacific Estuarine Rsch. Soc., Sigma Xi. Contbr. articles to profl. jours. Office: State of Wash Dept Natural Resources Olympia WA 98504

JAMPEL, ROBERT STEVEN, ophthalmologist, educator; b. N.Y.C., Nov. 3, 1926; s. Carl Edward and Frances (Hirschman) J.; m. Joan I. Myers, Oct. 2, 1952; children—Henry, Delia, James, Emily. A.B., Columbia U., N.Y.C., 1946, M.D., 1950; M.S., U. Mich., Ann Arbor, 1957, Ph.D., 1958. Assoc. in ophthalmology Columbia U., N.Y.C., 1962-69, asst. prof. ophthalmology, 1969-70, prof., chmn. ophthalmology Wayne State U., Detroit, 1970 , dir. Kresge Eye Inst., 1970—. Served to lt. USN, 1952-54. Mem. Am. Acad. Ophthalmology, Assn. Research in Vision, Assn. Univ. Profs. Ophthalmology, Acad. Neurology. Home: 4363 Barchester Dr Bloomfield Hills MI 48302-2116 Office: Hutzel Hosp 4717 St Antoine St Detroit MI 48201-1423*

JAMPOLSKY, ARTHUR, ophthalmologist; b. Bismarck, N.D., Apr. 24, 1919; married, 1957; 3 children. AB, U. Calif., 1940; MD, Stanford U., 1944. Diplomate Am. Bd. Ophthalmology. Chief strabismus clinic Smith-Kettlewell Inst. Visual Sci., Presby Med. Ctr., San Francisco, 1950-60, dir., 1960—; mem. com. on vision Armed Forces-Nat. Rsch. Coun., 1958—, exec. coun., 1960-64; mem. vis. sci. study sect. NIH, 1967-71, chmn., 1970-71; regional cons. ophthalmologist Oak Knoll Naval Hosp., Oakland & Travis AFB; cons. Letterman Gen. Hosp., San Francisco & Calif. State Bd. Health; spl. cons. Nat. Inst. Neurol. Disease & Blindness. Fellow ACS; mem. Am. Optometry Assn., Am. Acad. Ophthalmology and Otolaryngology, Am. Ophthalmology Soc., Am. Assn. Ophthalmology. Achievements include research in binocular vision, strabismus, physiological optics. Office: Smith-Kettlewell Eye Rsch Inst 2232 Webster St San Francisco CA 94115-1897*

JANAH, ARJUN, physicist, educator; b. Calcutta, India, Jan. 11, 1952; came to U.S., 1975; s. Sunil and Sobha (Dutt) J. MSc in Physics, U. Delhi, India, 1975; PhD of Physics, U. Md., 1982. Cert. physics tchr., N.Y. Grad. asst. Dept. Physics U. Md., College Park, 1975-81; vis. scientist Internat. Ctr. for Theoretical Physics, Trieste, Italy, 1981; postdoctoral rschr. Dept. Physics U. Calif., Irvine, 1982-83; vis. asst. prof. Dept. Physics U. Md., College Park, 1983-84; rsch. assoc. Dept. Physics Kansas State U., Manhattan, 1984-85; rschr. World Bank, Washington, 1985-87; tchr. N.Y.C. Bd. Edn., 1987—. Achievements include research on determining the quark charges and distinguishing between unified gauge theories through processes mediated by one and two photons or other neutral gauge bosons. Home: 6924 16 Ave Brooklyn NY 11204

JANCIS, ELMAR HARRY, chemist; b. Daugavpils, Latvia, Mar. 21, 1935; s. Michael and Theophile Olga (Petzholz) J.; m. Maruta Viksnins, Dec. 26, 1964; children: Paul Alexander, Erik Martin, Karl Eduard. BS, Yale U., 1956; PhD, U. Minn., 1968. Jr. rsch. chemsit U.S. Rubber Co., Naugahuck, Conn., 1956-63; rsch. chemist Uniroyal Inc., Naugatuck, 1967-70, rsch. scientist, 1970-79; sr. rsch. scientist Chem. Div. Uniroyal Inc., Middlebury, Conn., 1979—. Mem. Naugatuck Rep. Town Com., 1970, chmn., 1984-86. Mem. Am. Chem. Soc., Soc. Plastics Engrs., Sigma Xi. Lutheran. Achievements include: 14 patents; developed processes for 7 currently commercial products. Office: Uniroyal Chem Co Benson Rd Middlebury CT 06770

JANDINSKI, JOHN JOSEPH, dentist, immunologist; b. Northampton, Mass., July 28, 1946; m. Mary McKenna, Sept. 18, 1972. DMD, Tufts U., 1977; BA, Case Western Reserve U., 1972. Postdoctoral fellow Harvard Med. & Dental Sch., Boston, 1972-76, Duke U. Med. Ctr., Durham, N.C., 1976-79; rsch. assoc. Merck & Co., Rahway, N.J., 1979-81, Johnson & Johnson, New Brunswick, N.J., 1981-82; assoc. prof. NYU Dental Ctr., N.Y.C., 1982-85, U. of Medicine & Dentistry of N.J., Newark, 1985—; dental dir. Children's Hosp. AIDS program. Newark, 1989—; cons. Cistrin Biotech., Pinebrook, N.J., 1991—. Editor: Physical Diagnosis in Dentistry, 1974; contbr. articles to profl. jours. Mem. Light House Keepers, Watch Hill, R.I., 1988—. Recipient Bates Rsch. award Tufts U., 1971. Mem. AAUP (pres. 1992), Am. Assn. Immunology, Clin. Immunology Soc., Am. Acad. Oral Medicine, Sigma Xi, Omicron Kappa Upsilon. Achievements include first demonstration of role of cytokines in periodontal deisease; patent pending for 1-hour diagnostic test for periodontal disease; renowned consultant for oral diseases in pediatric AIDS. Office: NJ Dental Sch 110 Bergen St Newark NJ 07103-2425

JANEVSKI, BLAGOJA KAME, radiologist, educator; b. Gradsko, Macedonia, Feb. 8, 1934; s. Kame Ilija and Mica (Naceva) J.; m. Charlotte Meijers, June 15, 1967; children: Lucienne, Lidia, Dejan. MD, U. Belgrade, Yugoslavia, 1961. Resident radiology U. Hosp., Skopje, Macedonia, 1965-66, St. Annadal Hosp., Maastricht, The Netherlands, 1966-69, U. Hosp., Leiden, The Netherlands, 1969-70; gen. practice Skopje, Macedonia, 1961-65; chief radiologist St. Annadal Hosp., Maastricht, 1970-86; prof. radiology State U. Limburg, Maastricht, 1986—. Author: Angiography of the Upper Extremity, 1982; contbr. articles to profl. jours. and chpts. to medical books. Mem. Netherlands Assn. Radiology, Internat. Coll. of Angiology. Macedonian Orthodox. Home: Chambertinlaan 13, 6213 EV Maastricht The Netherlands

JANEZIC, DARRELL JOHN, environmental engineer; b. Bristol, Pa., Nov. 3, 1963; s. James Joseph and Cecilia (Gallik) J.; m. Lori Teresa Krank, Oct. 20, 1990. BS in Physics, Longwood Coll., Farmville, Va., 1988; BS in Civil Engring., Old Dominion U., Norfolk, Va., 1989. Civil engr. Dewberry & Davis, Fairfax, Va., 1989-91; environ. engr. Edgerton Environ. Svcs., Inc., Cary, N.C., 1991—. Mem. ASCE (assoc.). Republican. Roman Catholic. Office: Edgerton Environ Svcs Inc PO Box 4350 Cary NC 27519

JANI, SUSHMA NIRANJAN, child and adolescent psychiatrist; b. Gwalior, Madhya, Pradesh, India, Sept. 26, 1959; came to U.S., 1983; d. Kirty Ambalal and Purnima Kirty (Bhatt) Dave; m. Niranjan Natwerlal Jani, Mar. 30, 1983; chidlren: Susan Jani, Raja Jani. Inter Sci., Mithibai Coll., Bombay, India; MB;BS, B.J. Med. Coll., Ahmedabad, India; MD in Adult Psychiatry, Ind. U., 1984; MD in Child Psychiatry, Johns Hopkins U., 1987. Diplomate Am. Bd. Psychiatry and Neurology, sub-bd. Child Psychiatry. Child psychiatrist Johns Hopkins Univ. Hosp., Balt.; asst. clin. prof., dir. child & adolescent psychiatry U. Md., Balt.; chief cons. psychiatrist Balt. Detention Ctr., 1988-89, cons. psychiatrist Vets. Hosp., Indpls., 1986-87. Vol. Radha-Krishna Leprosy Camp, Bombay, 1981-83. Mem. AMA, Am. Acad. Child & Adolescent Psychiatry, Am. Psychiatry Assn., Md. Psychiat. Soc., Columbia Assn., India Assn. Hindu. Avocations: reading, knitting, sewing, letter-writing. Home: 10485 Owen Brown Rd Columbia MD 21044 Office: U Md Hosp 630 W Fayette St Baltimore MD 21201

JANICK, JULES, horticultural scientist, educator; b. N.Y.C., Mar. 16, 1931; s. Henry Spinner and Frieda (Tullman) Janick; m. Shirley Reisner, June 15, 1952; children: Peter Aaron, Robin Helen Janick Weinberger. BS, Cornell U., 1951; MS, Purdue U., 1952, PhD, 1954; DS in Agr. (hon.), U. Bologna, Italy, 1990. Instr. Purdue U., West Lafayette, 1954-56, asst. prof., 1956-59, assoc. prof., 1959-63, prof., 1963-88, James Troop Disting. prof. in horticulture, 1988—; dir. Purdue Ctr. for New Crops and Plant Products, 1990—; cons. Food and Agrl. Orgn., Rome, Italy, 1988. Author: Horticultural Science, 4th edit., 1986, Classical Papers in Horticultural Science, 1989; co-author: Plant Science: An Introduction to World Crops, 3d edit., 1981; co-editor: Advances in Fruit Breeding, 1975, Methods in Fruit Breeding, 1983, Advances in New Crops, 1990; editor Hort. Revs., Plant Breeding Revs. Pres. Parlor Club, Lafayette, 1990. Fellow AAAS, Portuguese Hort. Assn., Am. Soc. Hort. Sci. (pres. 1986-87), Sigma Xi, Gamma Sigma Delta. Jewish. Avocation: drawing. Home: 420 Forest Hill Dr West Lafayette IN 47906-2316 Office: Purdue U Dept Hort West Lafayette IN 47907-1165

JANKE, RHONDA RAE, agronomist, educator; b. Junction City, Kans., Apr. 20, 1958; d. Allen Webster and Harriette Eloise (Grove) J. BS in Agronomy, Kans. State U., 1980; MS in Agronomy, Cornell U., 1984, PhD in Agronomy, 1987. Student technician Kans. State U., Manhattan, 1977-79; grad. teaching asst. Cornell U., Ithaca, N.Y., 1981-82, grad. rsch. asst., 1982-87; agronomy coord. Rodale Inst., Kutztown, Pa., 1986-92; rsch. dir. Rodale Inst., 1992—; adj. asst. prof. Pa. State U., University Park, 1988—; sci. adv. bd. Mich. State U., Lansing, 1988—. Editorial bd. Crop Protection Jour., 1989—; author chpts. in books. Regional rep. New World Agr. Group. Recipient Disting. Teaching Award Cornell Grad. Sch., 1982; Lindbergh grantee, 1985. Mem. Am. Soc. Agronomy, Weed Sci. Soc. Am., Brit. Ecol. Soc., Inst. for Alternative Agr., Sigma Xi, Phi Kappa Phi, Gamma Sigma Delta, Alpha Zeta. Avocations: oil painting, gardening, piano, chickens. Office: Rodale Inst Rsch Ctr 611 Siegfriedale Rd Kutztown PA 19530-9749

JANKOVIC, JOSEPH, neurologist, educator, scientist; b. Teplice, Czechoslovakia, Mar. 1, 1948; came to U.S., 1965; s. Jerry and Hana (Meizlik) Vykouk; m. Cathy Sue Inselberg, May 26, 1973; children: Jason, Daniel, Zachary. MD, U. Ariz., 1973. Diplomate Am. Bd. Neurology. Med. intern Baylor Coll. Medicine, Houston, 1973-74, asst. prof. neurology, 1977-84, assoc. prof., 1984-88, prof., 1988—; resident in neurology Columbia U., N.Y.C., 1974-76, chief resident in neurology, 1976-77; dir. Parkinson's Disease Ctr. and Movement Disorder Clinic, Houston, 1977—; attending physician Meth. Hosp., Houston, 1988—. Author numerous articles and book chpts. in field; editor/co-editor 8 med. books; mem. editorial bd. jours. Movement Disorders, Clin. Neuropharmacology. Chmn. sci. adv. bd. Blepharospasm Rsch. Found.; mem. adv. bd. Dystonia Med. Rsch. Found., United Parkinson Found., Internat. Tremor Found., Tourette's Syndrome Med. Com., others. Grantee disease rsch. founds., pharmaceutical cos., NIH. Fellow Am. Acad. Neurology; mem. AMA, Am. Neurologic Assn., Soc. for Neurosci., Movement Disorders Soc. (pres.-elect 1991—). Avocations: tennis, family activities, music. Office: Baylor Coll Medicine 6550 Fannin St Ste 1801 Houston TX 77030-2721

JANKOVIĆ, SLOBODAN, aerodynamics educator, researcher; b. Brussels, Aug. 10, 1932; s. Dušan and Ljubica (Apostolović) J.; m Ksenija, July 3, 1960 (div. 1975); 1 child, Telena; m. Zorica, Sept 18, 1975 (div. 1982);1 child, Ana; m. Melinda Djelmiš, Dec. 30, 1982. BS in Civil Engring., Ecole d' Application, Brussels, 1959; M in Aerospace, U. Beograd, Yugoslavia, 1964, PhD in Aerospace, 1967. Ballistics researcher Military Tech. Inst., Beograd, Yugoslavia, 1960-67; aerodynamic researcher Military Tech. Inst., Beograd, 1979-84; asst. prof. Hight Military Tech. Sch., Zagreb, Yugoslavia, 1967-73; prof. Hight Military Tech. Sch., Zagreb, 1984-91; asst. prof. U. Beograd., 1973-79; aerodynamic researcher Maritime & Defense Inst., Zagreb, 1991—; cons. United Metal Industry, Sarajevo,1975-91; adv. bd. Metorol. Inst., Zagreb, 1985—. Author: (military edition) Exterior Ballistics, 1977, Missile's Aerodynamics, 1979. Col. Yugoslavian army tech. br., 1960-73. Recipient Sci. award, Ministr of Defense, Beograd, 1966, 1971. Mem. AIAA, Croation Soc. Mechanics. Roman Catholic. Home: Av Vukovar 240, Zagreb Croatia

JANKUN, JERZY WITOLD, biochemist; b. Jutrosin, Leszno, Poland, Oct. 29, 1948; came to U.S., 1983; s. Leonard and Izabella (Kaminiarz) J.; m. Ewa Skrzypczak, Mar. 8, 1972; children: Monika, Hanna. MS, A. Mickiewicz U., 1974; PhD, Agrl. U., 1977. Technician Rsch. Lab. Food Ind., Poznan, Poland, 1968-74; asst. prof. Agrl. U., Poznan, Poland, 1977-83; rsch. assoc. Mich. State U., East Lansing, 1983-90; asst. prof. U. Toledo, 1990—. Author: Food and Nourishment, 1981; contbr. articles to profl. jours. Fellow Am. Assn. Cancer Rsch., Internat. Assn. Fibrinolysis and Thrombolysis. Achievements include 5 patents in 23 countries. Office: The U Toledo Chemistry Dept Toledo OH

JANNETT, THOMAS COTTONGIM, electrical engineering educator; b. Birmingham, Ala., Sept. 19, 1958; s. Thomas Cottongim Jr. and Irene Agnes (Davis) J.; m. Karen Suzette Boatwright, Aug. 29, 1981. BS, U. Ala., Birmingham, 1979, MS, 1981; PhD, Auburn U., 1986. Instr. U. Ala., Birimingham, 1981-82, lectr., 1984-86; asst. prof. U. Ala., Birmingham, 1986-91; assoc. prof. U. Ala., 5, 1991--; proposal reviewer Nat. Sci. Found., 1992; articles reviewer, 1992. Contbr. articles to profl. jours. Rsch. grantee Nat. Sci. Found., 1992, Rsch. grantee The Whitaker Found., 1988, Rsch. grantee pvt. industry, 1984-91. Mem. IEEE (articles reviewer 1992). Baptist. Achievements include development of novel med. instrumentation systems for automated drug delivery. Office: U Ala 1150 10th Ave S Birmingham AL 35294

JANOWITZ, GERALD SAUL, geophysicist, educator; b. N.Y.C., Apr. 5, 1943; s. Leo and Yetta (Caress) J.; m. Barbara Susan Kantrowitz, Mar. 23, 1968; 1 child, David. BS, Poly. Inst. Bklyn., 1963; MS, Johns Hopkins U., 1965, PhD, 1967. Asst. prof. Case Western Res. U., Cleve., 1968-75; assoc. prof. N.C. State U., Raleigh, 1975-80, prof., 1980—. Contbr. over 60 articles to profl. jours. Mem. Am. Geophys. Union, Sigma Xi. Home: 116 Buckden Pl Cary NC 27511 Office: NC State U PO Box 8208 Raleigh NC 27695

JANSEN, KATHRYN LYNN, chemist; b. St. Paul, May 18, 1957; d. Bernard Joseph and Sarah Kathryn (Knight) J. BA in Chemistry, Coll. St. Catherine, St. Paul, 1979; PhD in Phys. and Analytical Chemistry, U. Utah, 1985. Chemist Chems. Quality Svcs., Eastman Kodak Co., Rochester, N.Y., 1985-90, lab. supr., 1990—. Contbr. articles to Applied Spectroscopy, Analytical Chemistry, Advances in Lab. Automation Robotics, others. Mem. Am. Chem. Soc. Office: Eastman Kodak Co Bldg 349 Kodak Park Rochester NY 14652-3635

JANSON, RICHARD WILFORD, manufacturing company executive; b. Canton, Ohio, Mar. 4, 1926; s. Wilford Sherwood and Mary Rebecca (Elliott) J.; m. Nancy Louise Davies, Oct. 31, 1955; children: Hollis L., Daniel W., Raymond E., Eric H. BA, Denison U., 1949; MA, Kent State U., 1982, PhD, 1986. Ptnr. Janson Industries, Canton, 1949—; ptnr. Sta. WJAN TV 17 Canton, 1967-77; bd. dirs. Molecular Tech., Canton, 1987-88; treas., bd. dirs. J.C. Tech., Inc., Canton, 1986—; chmn. Edison Bd., Columbus, Ohio, 1983-93; adj. prof. Kent (Ohio) State U., 1987—; charter mem. Thomas Edison Program, 1982—, chmn., 1987—. Author: Model of Spatial Revitalization, 1986; inventor stage hardware, 1980-90; contbr. articles to profl. jours. Trustee The Wilderness Ctr., Wilmot, Ohio, 1967-88; chmn. W. Va. Seating Co., Huntington, 1975-78. With USN, 1945-47. Mem. Am. Geog. Soc. (councilor), Ohio Acad. Sci. (pres.). Democrat. Presbyterian. Achievements include 10 patents for aluminum extrusions, structural applications, and track systems; development of methodology for interregional simultaneous determination of output, real income, and commodity flows for n commodities and k regions. Avocations: teaching, math modeling, writing. Home: 1200 Garfield Ave SW Canton OH 44706-1690 Office: The Janson Industries 1200 Garfield Ave SW Canton OH 44706-1690

JANSSEN, ANDREW GERARD, civil engineer; b. Pasadena, Calif., Oct. 31, 1967; s. Gerard Johann and Gretchen Granger (White) J.; m. Tonya Joan Janssen, Aug. 1, 1992. BS in Civil Engring., Stanford U., 1989, MS in Structural Engring., 1991. Registered profl. engr., Calif. Cost engr. Turner Constrn. Co., San Francisco, 1989; info. systems mgr. dept. civil engring. Stanford (Calif.) U., 1989-90; rsch. asst. geog. info system Stanford U. and USGS, Palo Alto, Calif., 1990-91; structural engr. Rutherford & Chekene, San Francisco, 1991-92; light rail design engr. Kampe Assocs. and Tri-Met, Portland, Oreg., 1992—. Mem. Structural Engring. Assn. No. Calif. Home: 2202 NE 56th Portland OR 97213

JANSSEN, GAIL EDWIN, banking executive; b. Oconto, Wis., Dec. 11, 1930; s. Ernest Janssen and Helen Arlene (Laduron) Janssen Jelinske; m. Janice Faye Detaeje, May 23, 1953; children: Gary Ernest, Joel Thomas. BS in Agr., U. Wis., 1960, BSME, 1962. Registered profl. engr., Wis. Design engr., asst. chief engr. Gehl Co., West Bend, Wis., 1962-69; chief engr., v.p. engring. Badger Northland Inc., Kaukauna, Wis., 1969-71, pres., 1971-78; pres. F & M Bank, Kaukauna, 1978-80; pres., chmn. bd. F & M Bancorp. Inc., Kaukauna, 1980—; bd. dirs. Keller Structure Inc., Kaukauna, White Clover Dairy, Kaukauna. Patentee farm equipment. Treas. High Clift Pk. Assn.; mem., past chmn. Kaukauna Indsl. Pk. Commn.; mem., past chmn. agrl. engring. adv. com. U. Wis., Madison. Recipient Outstanding Young Farmer award Oconto County, 1954; award of distinction U. Wis., 1989, Disting. Svc. award, 1991. Fellow Am. Soc. Agrl. Engrs. (dir. fin., charter mem. Wis. sect., treas., chmn. , Engr. of Yr. award 1973); mem. Wis. Agrl. and Life Scis. Alumni Assn. (life, bd. dirs.), Kiwanis (past pres. Kaukauna club), Rotary (past pres. Kaukauna club), KC. Republican. Roman Catholic. Home: 1612 Oakridge Ave Kaukauna WI 54130

JANSSEN, PAUL ADRIAAN JAN, pharmaceutical company executive; b. Turnhout, Belgium, Sept. 12, 1926; s. Jan Constant and Margriet (Fleerackers) J.; m. Dora Arts, July 1, 1957; children: Graziëlla, Herwig, Yasmine, Pablo, Maroussia. BSc, Sch. Notre Dame de la Paix, 1945; student, U. Louvain, Belgium, 1945-49; MD magna cum laude, State U. Ghent, Belgium, 1951; reaching cert. in chem. pharm., State U. Ghent, 1956. Asst. Inst. Pharmacology and Therapeutics, State U. Ghent, 1950-56, Inst. Pharmacology, U. Cologne, Fed. Republic Germany, 1951-52; pres., dir. rsch. Janssen Pharmaceutica, Beerse, Belgium, 1958-91; vice chmn. Johnson & Johnson Internat., 1979-91; lectr. psychiatry U. Liege, Belgium, 1966; prof. medicinal chemistry Liège Notre Dame de la Paix, Namur, Belgium, 1973; pres. Collegium Internat. Neuro-Psychopharm., 1980-82; vis. prof. med. sci. King's Coll., U. London, 1982-83; councillor Belgian Nat. Coun. Mgmt. Scis., 1984-87, Flemish Coun. Mgmt. Scis., 1987—; lectr., Francqui chair Cath. U. Louvain, 1985-86, vis. prof. Sch. Pharm. Scis., 1987-88, prof. extraordinary, 1988-91; lectr. in field. Recipient hon. doctorate, Johann Wolfgang Goethe U., Frankfurt, Fed. Republic Germany, 1978, U. Lund, Sweden, 1981, Cath. U. Louvain, 1982, Szeged (Hungary) U. Med. Sch., 1984, State U. Ghent, 1984, U. Dublin, Ireland, 1985, U. Düsseldorf, Fed. Republic Germany, Ben Gurion U. of Negev, Israel, 1986, State U. Liege, Belgium, 1988, U. Granada, Spain, 1989, U. Montreal, Can., 1989, U. Antwerp, 1992. U. Pavia, 1992, U. Istanbul, 1992, U. Rome, 1992, China Pharm. U., 1993; recipient numerous awards from profl. orgns. throughout Europe and U.S. Fellow Royal Acad. Medicine Ireland (hon.), Coll. Medicine South Africa (hon.); mem. Royal Soc. Medicine (London), Am. Chem. Soc., N.Y. Acad. Scis., Excerpta Medica Found., European Soc. Study of Drug Toxicity, AAAS, Collegium Internat. Medicinae Psychosomaticae, Am. Coll. Neuropsychopharmacology, Belgian Coll. Neuropsychopharmacology and Biol. Psychiatry, Belgian Soc. Fundamental and Clin. Physiology and Pharmacology, Royal Soc. Sci. Liege, Colombian Soc. Dermatology, numerous others. Home: Antwerpsteenweg, B 2350 Vosselaar Belgium Office: Janssen Pharmaceutica, Turnhoutseweg 30, B 2340 Beerse Belgium

JANSSON, JOHN PHILLIP, architect, consultant; b. Phila., Nov. 27, 1918; s. John A. and Isabelle (Ericson) J.; B.Arch., Pratt Inst., 1947; postgrad. SUNY, 1949; m. Ann C. Winter, Apr. 8, 1944 (div. Oct. 1970); children: Linda Ann, Lora Joan; m. Elizabeth Clow Peer, Jan. 21, 1978 (dec. May 1984). Architect for various firms, 1949-54; pvt. practice architecture, N.Y.C., 1949—; cons. mktg. of products, materials and services to bldg. and constrn. industry, 1949—; exec. v.p. Archtl. Aluminum Mfrs. Assn., N.Y.C., 1954-58; mgr. market devel. Olin-Metals Div., N.Y.C., 1958-62; dir. Pope, Evans & Robbins, cons. engrs., 1970-82; ptnr. Morris Ketchum, Jr. and Assocs., Architects, 1964-68; exec. dir. N.Y. State Council on Architecture, 1968-73, Associated 1973-74; dir. Gruzen & Ptnrs., 1972-74; pres. Bldg. Constrn. Tech., 1975-78; v.p. The Ehrenkrantz Group, 1974-82; dir. U.S. trade mission leader to Nigeria, Dept. Commerce, 1981; cons. N.Y. State Pure Waters Authority, 1968-69; chmn. N.Y. State Architecture-Constrn. Interagy. Com., 1968-74; sec. N.Y. State Gov.'s Adv. Com. for State Constrn. Programs, 1970-71. Mem. N.Y. State Citizens Com. Pub. Schs., 1952-55; v.p. citizens adv. com. Housing Authority, Town Oyster Bay, N.Y., 1966-68; bd. dirs. Bldg. Industry Data Adv. Council, 1976-78, Park-Ten Coop., 1981-82; instr. Outward Bound, Hurricane Island, Rockland, Maine, 1982—; media specialist The Image Ctr. Am's. Cup, 1987. Served to capt. USMCR, 1943-46. Registered architect, N.Y.; lic. Nat. Council Archtl. Registration Bds. operator/navigator passenger-carrying vessels, U.S. Coast Guard. Mem. AIA (architects in govt. com. 1971-77), Am. Arbitration Assn., Constrn. Specification Inst., Nat. Inst. Archtl. Edn., BRAB Bldg. Research Inst., Nat. Inst. Bldg. Scis., Archtl. League N.Y., N.Y. Bldg.

Congress, N.Y. State Assn. Architects (dir.), Soc. Archtl. Historians, Nat. Trust for Historic Preservation, Mus. Modern Art, Victorian Soc. Am., Associated Council Arts, Am. Mgmt. Assn., Soc. Mil. Engrs., Soc. Mktg. Profl. Services, Soc. Value Engrs., Mcpl. Art Soc. N.Y.C., Md. Capital Yacht Club (bd. dirs.). Home: 52 Chesapeake Landing Annapolis MD 21403

JANUARY, DANIEL BRUCE, electronics engineer; b. Tulsa, Okla., Mar. 27, 1953; s. E. Bruce and Virginia (Stines) J.; m. Theresa Ann Carter, June 24, 1973. BS in Elec. Engring., Okla State U., 1975. Electronics engr. McDonnell Douglas Astr., St. Louis, Mo., 1975-78; R&D mgr. Hunter Engring. Co., Bridgeton, Mo., 1978—. Achievements include 5 patents for advanced automotive wheel alignment equipment, including sensors, displays, hardware, software, methodology, and measurement geometry. Office: Hunter Engring 11250 Hunter Dr Bridgeton MO 63044-2391

JANUS, MARK DAVID, priest, psychologist, researcher, consulant; b. Rochester, N.Y., Mar. 31, 1953; s. Casimir Paul and Pearl Joan (Krajnik) J. BA magna cum laude, St. John Fisher Coll., Rochester, 1974; MA, Cath. U., Washington, 1978; PhD, U. Conn., 1992. Ordained priest Roman Cath. Ch., 1979. Clergy Cath. Info. Ctr., Grand Rapids, Mich., 1978-79, Paulist Ctr., Boston, 1979-83; dir. rsch. Covenant House, Boston, 1983-84; chaplain U. Conn., Storrs, 1984-89, lectr. clin. psychology, 1989-90; clinician U. Conn. Health Ctr., Storrs, 1990-91; faculty Ind. U. Med. Ctr., Indpls., 1991-92, Ohio State Coll. of Medicine, Columbus, 1992—; cons. Children's Hosp. Sexual Abuse Team, Boston, 1979-84; rsch. cons. Covenant House, Toronto, Ont., Can., 1985-90; presenter on adolescent and child sexual abuse over 31 juried presentations. Author: (with others) Child Pornography & Sex Rings, 1984, Adolescent Runaways: Causes & Consequences, 1987; contbr. articles to profl. jours.; appearances include Nightline, Can. radio and TV, numerous newspapers. Bd. dirs. Robert F. Kennedy Action Corps, boston, 1982—, Madonna Hall, Marlborough, Mass., 1984-92; active Pastoral Care Com. Diocese of Norwich, Conn., 1988-92. Fellow Am. Orthopsychiat. Assn.; mem. APA, Internat. Soc. Prevention of Abuse and Neglect, Missionary Soc. St. Paul The Apostle. Achievements include research on nature of physical and sexual abuse of homeless and runaway adolescents, on their experiences at home and on the streets, and the pastoral responses to the religious influences present in sexual abuse. Home: 1534 Runaway Bay Dr 2-B Columbus OH 43204 Office: Ohio State Coll Medicine 473 W 12th Ave Columbus OH 43210

JANZEN, NORINE MADELYN QUINLAN, medical technologist; b. Fond du Lac, Wis., Feb. 9, 1943; d. Joseph Wesley and Norma Edith (Gustin) Quinlan. BS, Marian Coll., 1965; med. technologist St. Agnes Sch. Med. Tech., Fond du Lac, 1966; MA, Cen. Mich. U., 1980; m. Douglas Mac Arthur Janzen, July 18, 1970; 1 son, Justin James. Med. technologist Mayfair Med. Lab., Wauwatosa, Wis., 1966-69; supr. med. technologist Dr.'s Mason, Chamberlain, Franke, Klink & Kamper, Milw., 1969-76, Hartford-Parkview Clinic, Ltd., 1976—; coord. health in bus. Hartford Parkview Clinic, 1990-91, drug program coord., 1991—. Substitute poll worker Fond du Lac Dem. Com., 1964-65; mem. Dem. Nat. Com., 1973—. Mem. Am. Soc. Med. Tech. (people to people clin. lab. scientist del. to People's Republic of China 1989), Nat. Soc. Med. Technologists (awards com. 1984-87, 88-91, chmn. 1986-88, nominations com. 1989—), Wis. Assn. Med. Tech. (exec. sec. 1991—), Wis. Assn. Med. Technologists (chmn. awards com. 1976-77, 84-85, 86-87, treas. 1977-81, pres.-elect 1981-82, pres. 1982-83, dir. 1977-84, 85-87, Mem. of Yr. 1982, numerous svc. awards, chair ann. meeting 1987-88), Am. Soc. Med. Technologists, Milw. Soc. Med. Technologists (pres. 1971-72, bd. dir. 1972-73), Communications of Wis. (originator, chmn. 1977-79), Southeastern Suprs. Group (co-chmn. 1976-77), LWV, Alpha Delta Theta (nat. dist. chmn. 1967-69, nat. alumnae dir. 1969-71), Alpha Mu Tau. Methodist. Home: 98 NW Dotty Way # 17298 Germantown WI 53022 Office: Hartford-Parkview Clinic 1004 E Sumner St Hartford WI 53027-1695

JAPIKSE, DAVID, mechanical engineer, manufacturing executive. Student, Technische Hochschule, Aachen, Germany; BS in Engring. Sci., Case Inst. Tech., 1965; MSc in Engring. Sci., Purdue U., 1968, PhD in Engring. Sci., 1969. With Pratt & Whitney, Creare, Inc.; pres., owner Concepts ETI, Inc., Norwich, Vt., 1980—; invited lectr. Poland, Russia, Japan, South Africa, England, Belgium, Germany, France, Venezuela. Co-editor: Sawyer's Gas Turbine Engineering Handbook. Recipient Silver Beaver award for svc. to youth Boy Scouts Am., 1992; Fulbright scholar, 1969; fellow NSF, 1969. Mem. AIAA, ASME (divsn. fluids engring. turbomachinery com. 1973—, internat. gas turbine inst. 1975—, chair several confs., reviewer, James Harry Potter Gold Medal, 1992), Soc. Automotive Engrs., Sigma Xi, Tau Beta Pi. Achievements include research in the fundamental modeling of turbomachinery processes, especially meanline performance codes, the derivation and publishing of two-zone modeling equations suitable for any developing flow, and the introduction of the TEIS model to describe the thermodynamic state change typical of any bladed row. Office: Concepts ETI Inc PO Box 643 Norwich VT 05055*

JAQUES, JAMES ALFRED, III, communications engineer; b. Akron, Ohio, July 6, 1940; s. James Alfred Jr. and Rosemary (McDonald) J.; m. Maryellen McCarthy, June 8, 1963; children: Jacquelyn Dawn, Jody Lynne, James Justin, Jill Danielle. BSEE cum laude, N.C. State U., 1968; MS cum laude, Naval Postgrad. Sch., 1975. Commd. ens. USN, 1961, advanced through grades to lt. comdr.; br. chief Def. Comm. Agy. USN, Alexandria, Va., 1979-83; ret. USN, 1983; asst. program mgr., chief engr. TRW, Inc., Fairfax, Va., 1983-92; chief engr. Applied Quality Comm., Chesapeake, Va., 1992—; system architect Can. Iris Program, Napean, Ont., 1990. Author, editor numerous papers in field. Mem. publs. com. Lake Smith Civic League, Virginia Beach, Va., 1979-86. Mem. IEEE (student chmn. 1973-75), Air Force Comm. Electronics Assn. (local sec.-treas. 1984, standards com. 1991). Republican. Roman Catholic. Home: 1001 Five Forks Rd Virginia Beach VA 23455 Office: Applied Quality Comm Inc 825 Greenbrier Cir Chesapeake VA 23320

JAQUITH, GEORGE OAKES, ophthalmologist; b. Caldwell, Idaho, July 29, 1916; s. Gail Belmont and Myrtle (Burch) J.; BA, Coll. Idaho, 1938; MB, Northwestern U., 1942, MD, 1943; m. Pearl Elizabeth Taylor, Nov. 30, 1939; children: Patricia Ann Jaquith Mueller, George, Michele Eugenie Jaquith Smith. Intern, Wesley Meml. Hosp., Chgo., 1942-43; resident ophthalmology U.S. Naval Hosp., San Diego, 1946-48; pvt. practice medicine, specializing in ophthalmology, Brawley, Calif., 1948—; pres. Pioneers Meml. Hosp. staff, Brawley, 1953; dir., exec. com. Calif. Med. Eye Council, 1960—; v.p. Calif. Med. Eye Found., 1976—. Sponsor Anza council Boy Scouts Am., 1966—. Gold card holder Rep. Assocs., Imperial County, Calif., 1967-68. Served with USMC, USN, 1943-47; PTO. Mem. Imperial County Med. Soc. (pres. 1961), Calif. Med. Assn. (del. 1961—), Nat., So. Calif. (dir. 1966—, chmn. med. adv. com. 1968-69) Soc. Prevention Blindness, Calif. Assn. Ophthalmology (treas. 1976—), San Diego, Los Angeles Ophthal. Socs., Los Angeles Research Study Club, Nathan Smith Davis Soc., Coll. Idaho Assocs., Am. Legion, VFW, Res. Officers Assn., Basenji Assn., Nat. Geneal. Soc., Cuyamaca Club (San Diego), Elks, Phi Beta Pi, Lambda Chi Alpha. Presbyterian (elder). Office: PO Box 511 665 S Western Brawley CA 92227-0511

JARA DIAZ, SERGIO R., transport economics educator; b. Santiago, Chile, Apr. 6, 1951; s. Sergio Jara and Elena Diaz; m. Maria Ofelia Moroni, June 6, 1975; children: Pedro, Francisco. Civil engr., U. Chile, 1974; Magister, U. Catolica, Santiago, 1977; MS, MIT, 1980, PhD, 1981. Instr. U. Chile, Santiago, 1974-77, asst. prof., 1978-83, assoc. prof., 1984-89, prof. transport econs., 1990—; vis. assoc. prof. MIT, Cambridge, 1988-89; advisor to pres. Metro S.A., Santiago, 1990-92; vis. prof. U. Pa., U. London, U. Leeds, U. Montreal, U. P.R., others; cons. in field. Contbr. more than 50 articles to profl. jours. Orgn. Am. States postgrad. fellow, 1980-81; Fulbright scholar, 1988-89; Andes Found. grantee, 1988-89; Norwegian Tech. and Sci. Rsch. Coun. grantee, 1990. Mem. Chilean Soc. Transport Engring. (v.p. 1990-92, pres. 1993-95), World Conf. on Transport Rsch. Soc., Am. Econ. Assn. Office: U Chile Dept Civil Engring, Casilla 228-3, Santiago Chile

JARAIZ, ELADIO MALDONADO, chemical engineering educator; b. Don Benito, Badajoz, Spain, Sept. 21, 1952; s. Eladio J. Jaraiz and Carmen R. Maldonado; m. Rosa Rodriguez, Aug. 15, 1984; children: Jesus, Myriam. Grad. in Chemistry, U. Salamanca, Spain, 1975, PhD in Chemistry, 1979. Tchr. asst. U. Salamanca, 1976-80; rsch. assoc. Oreg. State U., Corvallis, 1980-82; asst. prof. U. Salamanca, 1982-84, assoc. prof., 1984-90; visiting scholar Oreg. State U., Corvallis, 1990-91; prof. U. Salamanca, 1992—. Contbr. articles to profl. jours.; patentee in field. Mem. Real Sociedad Española de Quimica, Am. Inst. Chem. Engrs. Roman Catholic. Avocations: mountain hiking, reading. Home: Avda Campoamor 2, 4C, Salamanca 37003, Spain Office: Univ Salamanca, Dept Chem Engring, Plz Caidos 1-5, Salamanca 37008, Spain

JARDETZKY, OLEG, medical educator, scientist; b. Yugoslavia, Feb. 11, 1929; came to U.S., 1949, naturalized, 1955; s. Wenceslas Sigismund and Tatiana (Taranovsky) J.; m. Erika Albensberg, July 21, 1975; children by previous marriage: Alexander, Theodore, Paul. B.A., Macalester Coll., 1950, D.Sc. (hon.), 1974; M.D., U. Minn., 1954, Ph.D. (Am. Heart Assn. fellow), 1956; postgrad., U. Cambridge, Eng., 1965-66; LL.D. (hon.), Calif. Western U., 1978. Research fellow U. Minn., 1954-56; NRC fellow Calif. Inst. Tech., 1956-57; asso. Harvard U., 1957-59, asst. prof. pharmacology, 1959-66; dir. biophysics and pharmacology Merck & Co., 1966-68, exec. dir., 1968-69; prof. Stanford U., 1969—, dir. Stanford Magnetic Resonance Lab., 1975—, dir. NMR Center, Sch. Medicine, 1983-84; vis. fellow Merton Coll., Oxford (Eng.) U., 1976; cons., vis. prof., lectr. in field; chmn. Internat. Council on Magnetic Resonance in Biology, 1972-74. Contbr. articles to profl. jours.; mem. editorial bd. Jour. Theoretical Biology, 1961-88, Molecular Pharmacology, 1965-75, Jour. Medicinal Chemistry, 1970-78, biochimica biophypica Acta, 1970-86, Revs. on Bioenergetics, 1972-89, Biomembrante Revs., 1972-80, Jour. Magnetic Resonance in Biology and Medicine, 1986—, Jour. Magnetic Resonance, 1993—. Recipient USPHS Career Devel. award, 1959-66, Kaiser award, 1973, Von Humboldt award, 1977, 93; NSF grantee, 1957—; NIH grantee, 1957—; Am. Physiol. Soc. Travelling fellow, 1959. Fellow AAAS; mem. Am. Chem. Soc., Am. Soc. Biol. Chemists, Biophys. Soc., Assn. Advanced Tech. in Biomed. Scis. (pres. 1981-88), Internat. Soc. Magnetic Resonance (chmn. div. of biology and medicine 1986-89), Phi Beta Kappa, Sigma Xi, Alpha Omega Alpha. Home: 950 Casanueva Pl Palo Alto CA 94305-1001 Office: Stanford U Stanford Magnetic Resonance Lab Stanford CA 94305-5055

JARDIN, STEPHEN CHARLES, plasma physicist; b. Oakland, Calif., Aug. 28, 1947; s. William Frank and Eleanor Gertrude) J.; m. Marilyn M. Gee, June 17, 1973; children: Michael, Emily. BS, U. Calif., Berkeley, 1970; PhD, Princeton U., 1976. Deputy head physics dept. Princeton (N.J.) Plasma Physics Lab., 1991—. Contbr. over 80 articles to profl. pubs. Fellow Am. Physical Soc. Achievements include 4 patents. Office: PPPL PO Box 451 Princeton NJ 08543

JARDINE, DOUGLAS JOSEPH, plant pathologist, educator; b. Rantoul, Ill., Nov. 14, 1954; s. Donald G. and Anna M. (Grobbel) J.; m. Anne Marie Mullen, Sept. 15, 1984; children: Leah, Emily, Theresa. BS, Mich. State U., 1976, MS, 1977, PhD, 1985. 4-H youth agt. Mich.-Coop. Extension Svc., Coldwater, 1978-80; agr. agt. Mich.-Coop. Extension Svc., Mt. Clemens, 1980-81, Corunna, 1981-82; extension plant pathologist Kans. State U., Manhattan, 1985—, assoc. prof. plant pathology, 1985—. Mem. Am. Phytopathol. Soc., Manhattan Optimist Club, KC (grand knight), Sigma Xi. Roman Catholic. Office: Kans State U Plant Pathology 454 Throckmorton Hall Manhattan KS 66506

JARGON, JERRY ROBERT, chemical engineer; b. Beckmeyer, Ill., Aug. 2, 1939; s. Henry Paul and Hattie Bertha (Dorries) J.; m. Jeannette Margaret Boam, Oct. 20, 1963; children: Jeffrey Arendt, Jonathan David, Julie Anne. BSChemE, U. Ill., 1963; MSChemE, U. Denver, 1967. Registered prof. engr., Colo. Adv. scientist Marathon Oil Co., Littleton, Colo., 1973-77; sr. reservoir engr. Marathon Oil Co., Littleton, 1977-79, adv. sr. engr., 1979-85, sr. tech. cons., 1985-88, mgr. reservoir, 1988—; engring. adv. bd. U. Denver, 1990—. Editor: Gas Reservoir Engineering, 1993; tech. editor Jour. Petroleum Tech., 1985-86; contbr. articles to profl. jours. Mem. Soc. Petroleum Engrs. (lectr. on well testing, steering com. 1983 Forum, textbook com. 1986—), Sigma Xi. Home: 1695 S Fillmore St Denver CO 80210 Office: Marathon Oil Co PO Box 269 Littleton CO 80160

JARIC, MARKO VUKOBRAT, physicist, educator, researcher; b. Beograd, Serbia, Yugoslavia, Mar. 17, 1952; came to U.S. 1974; s. Vojin and Mileva (Vukobrat) J.; m. Slavica Vukelic, May 30, 1975 (div. 1979); m. Gabriele Weber, Dec. 10, 1982 (div. 1987); m. Tamara Hunt, Nov. 13, 1987 (div. 1990). Dipl., Beogradski U., 1974; PhD, CUNY, 1978. Rsch. assoc. U. Calif., Berkeley, 1978-80, Free U., West Berlin, 1980-82; vis. assoc. prof. Mont. State U., Bozeman, 1982-84; vis. scholar Harvard U., Cambridge, Mass., 1984-86; assoc. prof. Tex. A&M U., College Station, 1986-90, prof. physics, 1990—; vis. mem. Inst. for Theoretical Physics, U. Calif., Santa Barbara, 1987-88; vis. assoc. prof. U. Calif., Santa Cruz, 1989-90; vis. scholar Inst. des Hautes Etudes Scientifiques, Bures-sur-Yvette, France, 1981, 83; rsch. fellow Einstein Ctr. for Theoretical Physics, Rehovot, Israel, 1982, 83; vis. fellow Nonlinear Sci. Inst., Santa Cruz, 1989-90; rsch. assoc. Inst. Physics, Zemun, Yugoslavia, 1990—, Physics Dept. U. Calif., Santa Cruz, 1991—. Editor book series: Aperidicity and Order, 1988—; editor proceedings, Quasicystals, 1990; contbr. articles to profl. jours. Recipient October Prize, City of Belgrade, 1974; Fulbright grantee, 1974-80; Miller Inst. fellow, 1978-80, Alexander von Humboldt fellow, 1981-82. Mem. Am. Phys. Soc. Achievements include research on theory of symmetries and symmetry breaking in condensed matter physics, and in theory of quasicrystal structure, stability, and growth. Office: Tex A&M Univ Physics Dept College Station TX 77843-4242

JARMAN, SCOTT ALLEN, plastics engineer; b. Kinston, N.C., Oct. 27, 1963; s. James Murray and Shirley (Fitzgerald) Stroud; m. Susan Ellen Brenner, June 29, 1985; 1 child, Kristen Leigh. BS in Indsl. Tech., Western Carolina U., 1985. Mfg. engr. Container Systems Inc., Franklinton, N.C., 1985; quality engr. Raychem Corp., Fuquay-Varina, N.C., 1985-87, mfg. engr., 1987-92, sr. plastic engr., 1992—. Mem. Fuquay-Varina Area Recue Squad, 1992—. Mem. Soc. Mfg. Engrs., Soc. Plastics Engrs. Home: 7521 Glen Willow Ct Willow Spring NC 27592 Office: Raychem PO Box 3000 Fuquay-Varina NC 27526

JARON, DOV, biomedical engineer, educator; b. Tel Aviv, Oct. 29, 1935; came to U.S., 1958, naturalized, 1972; s. Meir and Sara (Levit) Yarovsky; m. Brooke E. Boberg, Sept. 16, 1978; children: Shulamit, Tamara. B.S. magna cum laude, U. Denver, 1961; Ph.D., U. Pa., 1967. Sr. research asso. Maimonides Med. Center, Bklyn., 1967-70; dir. surg. research Sinai Hosp. of Detroit, 1970-73; asso. prof. elec. engring. U. R.I., Kingston, 1973-77; prof. U. R.I., 1977-79, coordinator biomed. engring., 1973-79; dir. Biomed. Engring. and Sci. Inst., prof. biomend. engring. and sci. Drexel U., Phila., 1979—; vis. prof. elec. engring. Rutgers U., New Brunswick, N.J., 1968-73; adj. prof. biomed. engring. Wayne State U., 1971-73; adj. prof. physiology Temple U. Sch. Medicine, 1980—; adj. prof. radiology Jefferson Med. Coll., 1983—; dir. Div. Biol. and Critical Systems, NSF, 1991-93. Contbr. articles to sci. jours. NSF, NIH, Office Naval Research, pvt. founds. research grantee. Fellow IEEE, Am. Inst. for Med. and Biol. Engring.; mem. AAAS, AAUP, Biomed. Engring. Soc., Am. Soc. for Engring. Edn., Assn. for Advancement Med. Instrumentation, Internat. Soc. Artificial Organs, Am. Soc. for Artificial Internal Organs, Biophys. Soc., N.Y. Acad. Scis., Engring. in Medicine and Biology of IEEE (pres. 1986-87), Sigma Xi, Tau Beta Pi, Eta Kappa Nu. Achievements include research of cardiac assist devices, cardiovascular dynamics and modeling, biomed. instrumentation. Home: 2957 Tilden St NW Washington DC 20008-1150 Office: NSF Div Biol and Critical Systems 1800 G St NW Washington DC 20550-0002

JAROSZ, BOLESLAW FRANCIS, mechanical engineer; b. Poland, Oct. 3, 1953; s. Mikolaj and Eugenia (Malec) J.; m. Barbara Ann Stepien, July 7, 1990; 1 child, Stephen. BSME, Drexel U., 1977, MS in Engring. Mgmt., 1982. Registered profl. engr., Pa., Md., N.Mex., Calif., N.J., Mass., S.C.; cert. energy mgr. Mech. engr. Naval Facilities Engring. Com., Phila., 1974-76; svc. supr. UTC-Carrier Corp., King of Prussia, Pa., 1977-80; sr. energy analyst Control Data Corp., Phila., 1980-83; sr. facilities engr. Westinghouse Electric, Balt., 1983-88; mgr. Wyeth-Ayerst Internat., Radnor, Pa., 1988-90; chief engr. United Engrs., Phila., 1990—. Pres. Polish Intercollegiate Club, Phila. Recipient Cert. of Appreciation (pvt. pilot), CAP, Unit Citation award CAP. Mem. ASME, Assn. Energy Engrs., Internat. Soc. Pharm. Engring., Am. Soc. Heating, Refrigeration & Air Conditioning Engrs. Office: Raytheon Engrs & Constructors 30 S 17th St Philadelphia PA 19101

JAROSZEWSKI, JERZY W., chemist; b. Lodz, Poland, Nov. 2, 1950; s. Witold R. and Irene (Ozdzinska) J.; m. Margarethe Jaroszewski, Apr. 27, 1973; children: Anna, Julie. MS, U. Copenhagen, 1976, PhD, 1980. Assoc. prof. Royal Danish Sch. Pharmacy, Copenhagen, 1980-83, 1983—; vis. rsch. assoc. NIH, Nat. Cancer Inst., Washington, 1988-89. Editor: NMR Spectroscopy in Drug Research, 1988; contbr. articles to profl. jours., chpts. to books. Home: Ledreborg Alle 16, Gentofte 2820, Denmark

JARRELL, WESLEY MICHAEL, soil and ecosystem science educator, researcher, consultant; b. Forest Grove, Oreg., May 23, 1948; s. Burl Omer and Edith LaVerne (Sahnow) J.; m. Linda Ann Illig, June 24, 1972; children: Benjamin George, Emily Theresa. BA, Stanford U., 1970; MS, Oreg. State U., 1974, PhD, 1976. Grad. rsch. asst. Oreg. State U., 1971-76; asst. prof. soil sci. U. Calif., Riverside, 1976-83, assoc. prof., 1983-88; dir. Dry Lands Res. Inst., 1985-88; assoc. prof. Oreg. Grad. Inst., Portland, 1988-91, prof., 1991—; dept. head, internat. cons. agr., 1992—. Mem. AAAS, Soil Sci. Soc. Am., Am. Soc. Agronomy. Democrat. Lutheran. Contbr. articles to profl. jours. Home: 1920 NW 110th Ct Portland OR 97229-4852 Office: Oreg Grad Inst Environ Sci Engring PO Box 91000 Portland OR 97291

JARROLL, EDWARD LEE, biology educator; b. Huntington, W.Va., Jan. 4, 1948; s. Edward Lee and Janet (Ferzacca) J.; m. Gail Ann Brookshire, Aug. 2, 1975; 1 child, Christopher David. AB, W.Va. U., 1969, PhD, 1977. Postdoctoral fellow Microbiology and Immunology, Oreg. Health Sci. U., Portland, 1977-80; rsch. assoc. Preventive Medicine, Cornell U., Ithaca, N.Y., 1980-81, sr. rsch. assoc., 1981-82; asst. prof. Biology Dept., West Chester (Pa.) U., 1984-85; asst. prof. Biology Dept., Cleve. State U., 1985-88, assoc. prof., 1988-92, prof., chair, 1991—; cons. biologist Life Systems, Inc., Cleve., 1983; adj. staff gen. med. sci. Cleve. Clinic Found., 1991—; vis. prof. infectious diseases W.Va. U., Morgantown, 1991. Author: (with others) Giardiasis, 1990; contbr. articles to profl. jours. Grantee Thrasher Rsch. Fund, 1986-88, U.S. EPA, 1990-92, NIH, Fulbright Scholar, 1993. Mem. Am. Soc. for Microbiology, Soc. of Protozoologists, Alpha Chi. Achievements include research on molecular and biochemical parasitology, applied and environ. microbiology, molecular and biochemistry. Home: 858 Keystone Dr Cleveland Heights OH 44121 Office: Cleve State U Biology Dept 1983 E 24th St Cleveland OH 44115

JARVENKYLA, JYRI JAAKKO, chemical engineer; b. Helsinki, Finland, June 16, 1947; s. Y.T. and M.K. (Vilppula) J.; m. Reija Rantanen, Dec. 22, 1976; children: Joni, Mikko, Marja, Raisa. MS in Chem. Engring., Helsinki Tech. U., 1973. Dep. mgr. Finnish Plastics Industries Fedn., Helsinki, 1973-81; rsch. mgr. Asko Oy, Lahti, Finland, 1981-82; devel. mgr. Uponor Oy, Nastola, Finland, 1982-83; project mgr. Uponor Ab, Fristad, Sweden, 1983-86; sr. researcher Uponor Oy, Nastola, 1986-90; mgr. rsch. Uponor Group, Helsinki, 1991—. bd. dirs. Uponor Techs.; pres. bd. dirs. Fenno Intelligence Bur. FIB Oy, 1988—. Mem. Plastics Assn. (dep. chmn. 1981-82). Avocations: chess, cars. Office: Uponor Group, Corporate Office POB 21, 15561 Nastola Finland

JARVI, GEORGE ALBERT, chemical engineer; b. Seattle, Aug. 31, 1951; s. Albert O. and Mary Lydia (Ryan) J.; m. Carla Irene Tallman, Mar. 19, 1974; children: Eric, David, Lydia, Albert, Mark. BS, Oreg. State U., 1976; MS, Brigham Young U., 1978. Engring. asst. Am. Bechtel, Inc., San Francisco, 1974-75; assoc. chem. engr. Inst. Gas Tech., Chgo., 1978-79, chem. engr., 1979-80; rsch. engr. Engelhard Mineral & Chem. Corp., Newark, 1980-85; project engr. Engelhard Corp., Menlo Park, N.J., 1985-87, sr. project engr., 1987-90; sr. process engr. Fluor Daniel, Inc., Marlton, N.J., 1990—; lectr. in field. Contbr. articles to profl. jours. EMT South River Rescue Squad, 1980-88; unit commr. Boy Scouts Am., Thomas Edison and Bucks County couns., 1980—. Mem. Am. Inst. Chem. Engrs. (chmn. South Jersey sect. 1992-93), Sigma Xi, Tau Beta Pi, Phi Lambda Upsilon. LDS. Achievements includes research on built or operated research apparatus for steam reforming hydrotreating, catalytic reforming, selective hydrogeneration and xylene isomerization; novel electrolytic cell for potassium gold cyanide. Home: 38 Peaceful Dr Morrisville PA 19067

JARVIS, KENT GRAHAM, electrical engineer; b. Provo, Utah, Jan. 25, 1942; s. James Kieth and Marian (Graham) J.; m. Jayanne Christensen, July 12, 1963; children: Ryan Kent, Mark Graham, Gregory K. BSEE, Brigham Young U., 1975. Registered profl. engr., Utah. TV broadcast engr. KUTV, Salt Lake City, 1961-63; engr. self-employed Provo, Utah, 1972-75; head product assurance Signetics Corp., Orem, Utah, 1969-70; engr. power distbn. ops. engr. Utah Power & Light Co., Salt Lake City, 1978-84; sr. engr. Utah Div. of Pacificorp Electric Ops. Group, Salt Lake City, 1984-93; pres. KGJ Enterprises, 1993—; cons. Elec. Cons. Engring. and Entrepreneurs, 1993—; curriculum advisor West Jordan (Utah) High Sch., 1983-85; engr. software devel. Creative Concepts, West Jordan, 1985—; regional test adminstr. Vo-cat.-Indsl. Clubs of Am., West Jordan, 1984. Author/programmer: 1 & 3 Phase Distribution Fault Study Program, 1982, 1 & 3 Phase Motorstart Program, 1983, Lightning Protection Analysis Program, 1988, Excess Capacity Cost Program, 1989, Guidelines for Determining Susceptibility to Ferroresonance, 1990, Winter De-Energization of Three Phase Transformers, 1991. Republican. Mormon. Avocations: chess, classical music, computer programming, reading, sports. Home: 10834 S High Ridge Ln Sandy UT 84092

JARZEMBSKI, WILLIAM BERNARD, biomedical engineer; b. Little Rock, June 25, 1923; s. Thaddeus and Elsie Alec (Sachs) J.; m. Helen Jendry; children: Nancy, Susan, Donna. BSEE, Northwestern U., 1947; PhD, Marquette U., 1971. Registered profl. engr., Ohio, Ky. Ill., Tex. Pres. Jarco Svcs., Tulsa, 1951-63; engr. advanced rsch. Sunbeam Corp., Chgo., 1963-66; direct product engr. Appleton Electric Co., Chgo., 1966-68; prof. biomed. engring. Marquette U., Milw., 1968-74; prof., chair Tex. Tech U. Med. Sch., Lubbock, 1974-85; pvt. practice engring. W.B. Jarzembski & Assocs., Cin., 1985—; mem. neurology panel FDA, Washington, 1975-79; examiner Accreditation Bd. Engrs., Clemson, S.C., 1988-93. Editor: Workshop Computer Lab Resources, 1978; author: (publs.) Current Vector Probe Cerebral Meas., 1971, Pathology of Trans. Cranial Impedance, 1976, Hopsital Eng. Management, 1977; mem. rev. bd. Jour. Clin. Engring. Lt. U.S. Army, 1942-45, ETO. NIH fellow, 1969. Mem. IEEE (sr., ethics panel 1991-92, chair com. 1987, student profl. awareness com. 1991, Profl. Activities award 1985). Achievements include patents for Electromagnetic Control Transducer, Microwave Oscillator for Microwave Oven, Tabletop Electrostatic Copier, Electric Clinical Thermometer; development of device for level measurement liquid oxygen. Home and office: W B Jarzembski & Assocs 454 Hillcrest Wyoming OH 45215

JASIUK, IWONA MARIA, engineering educator; b. Warsaw, Poland, July 18, 1957; came to U.S., 1973; d. Mieczyslaw and Teresa J.; m. Martin Ostoja-Starzewski, Dec. 28, 1991. BS, U. Ill., Chgo., 1980, MS, 1982; PhD, Northwestern U., 1986. Asst. prof. Mich. State U., East Lansing, 1986-92, assoc. prof. engring., 1992—; vis. rsch. scientist Sci. Ctr. at Rockwell Internat., Thousand Oaks, Calif., 1987, Tokyo Inst. Tech., 1990. Contbr. articles to Jour. Mechanics and Physics of Solids, Jour. Applied Mechanics, others; reviewer for several profl. jours. Mem. ASME (assoc.; elasticity com. 1989—), Am. Acad. Mechanics. Achievements include research in micromechanics of heterogeneous materials; study of the effect of matrix-inclusion interface on the local and global response of composite materials. Home: 4183 Indian Glen Okemos MI 48864 Office: Mich State U Materials Sci and Mechanics East Lansing MI 48824

JASKULA, MARIAN JÓZEF, chemist; b. Cracow, Poland, Aug. 26, 1948; s. Aleksander and Zofia (Banczak) J.; m. Teresa Regina Chwistek, June 17, 1973; children: Anna, Maria, Jerzy. MSc, Jagiellonian U., Kraków, Poland, 1971, PhD in Phys. Chemistry, 1979, Dr. habil. in Electrometallurgy, 1992. Asst. Faculty of Chemistry Jagiellonian U., Kraków, Poland, 1971-79, adj. asst. prof., 1979—; privat dozent Reinisch-Westfalische Technische Hochschule, Aachen, Germany, 1992—. Co-author: Physical Chemistry Exercises, 1989; contbr. over 50 papers to sci. publs.; patentee in field; co-author of two student books. Pres. UJ-Tchrs. Trade Union, Kraków, 1976-80, 80-81. Grantee Austrian Govt., 1982-83, Denmark Govt., 1984, 92-93; Humboldt fellow, 1987-90; recipient Polish Ministry for Sci. award, 1979, 85, Friedrich-

Wilhelm Preis award, Aachen, 1992. Mem. Polish Chem. Soc., Polish Chemistry Engrs. Soc. Roman Catholic. Avocations: travel, sport. Home: Obopólna 3/39, 30-039 Kraków Poland Office: Jagiellonian U, 3 Ingardena Str, 30-060 Kraków Poland

JASTREBOFF, PAWEL JERZY, neuroscientist, educator; b. Lebork, Poland, Feb. 25, 1946; came to U.S., 1982; s. Pawel Ludwik and Maria (Czerniawska) J.; m. Malgorzata Maria Nasierowska, June 27, 1970; children: Peter Maciej, Ania Magdalena. MSEE, U. Warsaw (Poland), 1969, MS in Physics, 1971; PhD, Polish Acad. Sci., Warsaw, 1973. Asst. prof. Polish Acad. Scis., Warsaw, 1975-82, assoc. prof. (tenured), 1982-85; rsch. scientist Yale U. Sch. Medicine, New Haven, 1985-90; prof. (tenured) U. Md. Sch. Medicine, Balt., 1990—; vsi. prof. U. Tokyo Sch. Medicine, 1979, Univ. Coll., London, 1990—, Yale U. Sch. Medicine, New Haven, 1990—. Contbr. articles to Neurosci. Rsch., Experientia, Behavioral Neurosci., Jour. Acoustic Soc. Am., Proceeding of NAS USA. Grantee NIH, 1988, 90. Roman Catholic. Achievements include development of an animal model for auditory phantom perception. Office: U Md Dept Surgery 10 S Pine St MSTF Bldg Rm 4-34F Baltimore MD 21201

JASTROW, ROBERT, physicist; b. N.Y.C., Sept. 7, 1925; s. Abraham and Marie (Greenfield) J. A.B., Columbia, 1944, M.A., 1945, Ph.D., 1948; postdoctoral fellow, Leiden U., 1948-49, Princeton Inst. Advanced Study, 1949-50, 53, U. Calif. at Berkeley, 1950-53; D.Sc. (hon.), Manhattan Coll., 1980, N.J. Inst. Tech., 1987. Asst. prof. Yale, 1953-54; cons. nuclear physics U.S. Naval Research Lab., Washington, 1958-62; head theoretical div. Goddard Space Flight Center NASA, 1958-61, chmn. lunar exploration com., 1959-60, mem. com., 1960-62; dir. Goddard Inst. Space Studies, N.Y.C., 1961-81; adj. prof. geology Columbia, 1961-81, dir. Summer Inst. Space Physics, 1962-70; adj. prof. astronomy Columbia (Summer Inst. Space Physics), 1977-82; adj. prof. earth sci. Dartmouth, 1973-92; pres. G.C. Marshall Inst., 1985—; dir., chmn. of bd. Mt. Wilson Inst., 1991—. Author: The Evolution of Stars, Planets and Life, 1967, Astronomy: Fundamentals and Frontiers, 1972, Until the Sun Dies, 1977, God and the Astronomers, 1978, 2d edit., 1992, Red Giants-White Dwarfs, 1991, The Enchanted Loom, 1981, How To Make Nuclear Weapons Obsolete, 1985, Journey to the Stars, 1989; editor: Exploration of Space, 1960; co-editor: Jour. Atmospheric Scis., 1962-74, The Origin of the Solar System, 1963, The Venus Atmosphere, 1969. Recipient Medal of Excellence Columbia, 1962, Grad. Faculties Alumni award, 1967; Arthur S. Flemming award, 1965; medal for exceptional sci. achievement NASA, 1968. Fellow Am. Geophys. Union, A.A.A.S., Am. Phys. Soc.; mem. Internat. Acad. Astronautics, Council Fgn. Relations, Leakey Found., Nat. Space Soc. (bd. govs.). Clubs: Cosmos, Explorers, Century. Home: 10445 Wilshire Blvd # 304 Los Angeles CA 90024-4606 Office: Mt Wilson Observatory Hale Solar Lab 740 Holladay Rd Pasadena CA 91106

JASZCZAK, JOHN ANTHONY, physicist; b. Garfield Heights, Ohio, July 10, 1961; s. Stephen Mathias and Josephine Marie (Iafelice) J.; m. Sherry Lynne Wilkins, Aug. 23, 1986; children: Jacob, Patrick, Benjamin. BS in Physics, Case Western Res. U., 1983; MS in Physics, Ohio State U., 1985, PhD, 1989. Postdoctoral assoc. Argonne (Ill.) Nat. Lab., 1989-91; asst. prof. physics Mich. Technol. U., Houghton, 1991—; adj. curator Seaman Mineral Mus. Mich. Technol. U., Houghton, 1992—; adv. bd. Seaman Mus. Faculty, Houghton, 1991—. Contbr. articles to Phys. Rev. B., Jour. Materials Rsch., Jour. Applied Physics, Mineral. Record. Presdl. fellow Ohio State U., 1988. Mem. Am. Phys. Soc., Materials Rsch. Soc., Copper Country Rock & Mineral Club (v.p. 1991—). Achievements include research in facets and roughening in crystals and quasicrystals, growth twinning in natural graphite crystals, elastic properties of metallic superlattices. Office: Michigan Technological Univ Dept Physics 1400 Townsend Dr Houghton MI 49931

JASZCZAK, RONALD JACK, physicist, researcher, consultant; b. Chicago Heights, Ill., Aug. 23, 1942; s. Jacob and Julia (Gudowicz) J.; m. Nancy Jane Bober, Apr. 15, 1967; children: John, Monica. BS with highest honors, U. Fla., 1964, PhD, 1968. Staff physicist Oak Ridge Nat. 1969-71, AEC postdoctoral fellow, 1968-69; prin. rsch. scientist Searle Diagnostics, Inc., 1971-73, sr. prin. rsch. scientist, 1973, rsch. group leader, 1973-77, chief scientist, 1977-79; assoc. prof. radiology Duke U. Med. Ctr., Durham, N.C., 1979-89, prof., 1989—, assoc. prof. biomedical engring., 1986-91, prof., 1992—; rsch. prof. Inst. of Stats. and Decision Scis., 1992—; founder, chmn. bd. dirs. Data Spectrum Corp., Chapel Hill, N.C.; investigator Nat. Cancer Inst. Grant, 1983—, Dept. Energy Grant, 1989—. Contbr. articles to profl. jours.; patentee in field. Fellow NASA, 1964-67, U. Fla., 1967-68; RCA scholar, 1963-64. Fellow IEEE; mem. Soc. Nuclear Medicine, Am. Phys. Soc., AAAS, Am. Assn. Physicists in Medicine, Soc. Photo-Optical Instrumentation Engrs., Sigma Xi, Phi Beta Kappa, Phi Kappa Phi, Tau Sigma, Sigma Pi Sigma. Home: 2307 Honeysuckle Rd Chapel Hill NC 27514-1716 Office: Duke U Med Ctr PO Box 3949 Durham NC 27710

JATALA, SHAHID MAJID, electronics engineer; b. Sialkot, Punjab, Pakistan, Mar. 24, 1966; s. Abdul Majid and Khurshed (Beghum) J.; m. Sufia Shahid, Sept. 4, 1988; children: Zvaar, Muteeb. BS in Physics with honors, London U., 1987; B of Electronics Engring. with honors, South Bank U., Eng., 1989. Chief design engr. Berk Electronic (Pvt.) Ltd., Sialkot, Pakistan, 1989-91, Silicon Electronic (Pvt.) Ltd., Sialkot, 1991—. Author: Philosophy of Relativity, 1988, Introduction to Quantum Mechanics, 1989, Integrated Electronics, 1991. Fellow Royal Astron. Soc.; mem. IEEE (assoc.), Am. Inst. Physics Tchrs. Islam. Achievements include research on FM communcation using infrared waves. Home: 31 Madingley St Peters Rd, Kingston-Upon-Thames KT1 3JG, England

JAUHIAINEN, PERTTI JUHANI, software engineer; b. Kangasniemi, Finland, May 8, 1960; s. Simo S. and Elvi K. (Halttunen) J. MSc in Computer Sci., U. Linkoping (Sweden), 1986. Software engr. EPITEC AB, Sweden, 1986-88; knowledge engr. Asea Brown Boveri AG, Baden, Switzerland, 1988-89; mgr. product support EPITEC, Inc., Framingham, Mass., 1989-90; reuse engr. Ellemtel Utveckling AB, Stockholm, 1991—; project leader Race Projects (EC), Stockholm, 1991—. Sgt. UN Forces, 1981. Office: Ellemtel Utvecklings AB, Box 1505, 12525 Älvsjö Sweden

JAVAHERI, KATHLEEN DAKIN, construction and environmental professional; b. Jacksonville, Fla., Sept. 4, 1960; d. John Castle and Mary Ellen (Friemuth) Dakin; m. Seyed Mohammad Javaheri, May 15, 1982. BSCET, U. Ala., Tuscaloosa, 1985. Civil engring. tech. Ga. Dept. Transp., Atlanta, 1985-87; constrn. inspector Ebasco Svcs. Inc., Atlanta, 1987-89; traffic engr. Moreland Altobelli Assocs Inc., Atlanta, 1989-90; constrn. supr. Ebasco Svcs. Inc., Atlanta, 1990-92, assoc. environ. engr., 1992—. Office: Ebasco Svcs Inc 145 Technology Park Atlanta Norcross GA 30092

JAVIER, AILEEN RIEGO, pathologist; b. Fabrica, Negros Occidental, Philipines, Apr. 4, 1948; d. Filemon Yanson and Alicia Vazquez (Alteros) R.; m. Mark Anthony Navarro Javier, July 15, 1972; children: Martha Francesca, Nadine Ruth. BS, U. Phillipines, 1967, MD, 1972; MHA, Ateneo de Manila U., 1989. Diplomate Philipine Bd. Pathology. Instr. pathology U. Philipines, Manila, 1972-76, asst. prof., 1976-80, sr. lectr., 1987—; med. specialist, chmn. dept. Lung Ctr. Philipines, Quezon City, 1987-91; lectr. Ateneo de Manila Grad. Sch., Quezon City, 1989—; med. specialist Philipine Children's Ctr., Quezon City, 1981-82, cons. 1986; cons. Polymedic Gen. Hosp., Rizal, Philipines, 1984-86, dept. chmn. 1987—; med. specialist Nat. Kidney Inst., Quezon City, 1984-87, 91—, cons. 1987-91, Bur. Research and Labs, Manila, 1987—; cons. Cardinal Santos Med. Ctr., 1986—; v.p. Philipine Blood Coordinating Council, 1987-88, pres. 1990; bd. dirs. Fetus as a Patient Inst., Philipines, 1990—. Active Goodwill Industries, Quezon City, 1981. Fellow Philippine Soc. Oncologists (sec.-treas. 1990-91); mem. Philippine Med. Assn. (life), Internat. Acad. Pathology, Philippine Soc. Pathologists (treas. 1985-87, pres. 1987-89), Philippine Bible Soc. (life). Baptist. Avocation: playing piano. Office: Nat Kidney Inst, East Ave, Quezon City The Philippines

JAVOR, EDWARD RICHARD, electrical engineer; b. Norwich, Conn., Apr. 18, 1958; s. Edward George and Marie Ida (Paquin) J.; m. Ellen Elizabeth Girotti, Aug. 20, 1983. BSEE, U. Conn., 1981; MS in Computer Sci., Rensselaer Poly. Inst., 1986. Engring. specialist Electric Boat Div. Gen. Dynamics, Groton, Conn., 1981—; mem. TRIDENT Submarine Electromagnetic Compatibility Adv. Bd., Groton, 1988—. Mem. IEEE (electromagnetic compatibility soc.). Home: 39 Occum Ln Uncasville CT 06382

JAWORSKI, DAVID JOSEPH, project engineer; b. Detroit, June 25, 1965; s. Richard Walter and Geraldine Gertrude (Chmielewski) J.; m. Charlotte Ann Zielinski, Sept. 14, 1991. BS in Engring., U. Mich., Dearborn, 1986; MS in Engring., U. Mich., 1988. Cert. engr. in tng., Mich. Project engr. Gen. Motors Corp., Detroit, 1987-88, Chrysler Corp., Auburn Hills, Mich., 1988-90, Ford Motor Co., Dearborn, Mich., 1990—. Recipient 1st place drafting, 1st place photography State of Mich. Affiliation, 1983. Mem. NSPE, IEEE, Eta Kappa Nu. Roman Catholic. Home: 1524 Emmons Birmingham MI 48009-5164 Office: Ford Motor Co 20000 Rotunda Dr Dearborn MI 48121-2053

JAYASUMANA, ANURA PADMANANDA, electrical engineering educator; b. Colombo, Sri Lanka, Dec. 29, 1956; came to U.S., 1980; s. D. Sugathananda and D. Susima (Seneviratne) J.; m. Geetha Gunamalee Polwatte, Sept. 3, 1980; children: Sahan Thusitha, Ruwan Sujith. BSEE, U. Sri Lanka, Moratuwa, Sri Lanka, 1978; MSEE, Mich. State U., 1982, PhD in Elec. Engring., 1984. Electronic engr. Nat. Engring. Rsch. and Devel. Ctr., Jaela, Sri Lanka, 1978-79; asst. lectr. U. Moratuwa, Sri Lanka, 1979-80; asst. prof. elec. engring. Colo. State U., Ft. Collins, 1985-89, assoc. prof. elec. engring., 1989—. Contbr. articles to profl. jours. Named Outstanding Prof., Am. Electronics Assn., 1990, Best Student in Elec. Engring., U. Sri Lanka, 1978; recipient Outstanding Acad. Achievement award Mich. State U., 1983, 82, HSP award for Acad. Excellence, Mich. State U., 1984. Mem. IEEE, Phi Kappa Phi. Achievements include rsch. in digital communication networks and protocols and VLSI testing; findings which include analytical models for timed-token protocols and testable designs for CMOS and BiCMOS. Office: Colorado State U Dept of Elec Engring Fort Collins CO 80523

JAYNE, CYNTHIA ELIZABETH, psychologist; b. Pensacola, Fla., June 5, 1953; d. Gordon Howland and Joan (Rockwood) J. AB, Vassar Coll., 1974; MA, SUNY, Buffalo, 1978, PhD, 1983. Lic. psychologist, Pa. Instr. dept. psychiatry Temple U. Sch. Medicine, Phila., 1982-84, asst. prof., 1984-85, asst. dir. outpatient services, asst. dir. residency tng., 1982-85, clin. asst. prof., 1985—; pvt. practice psychology Phila., 1985—. Contbr. articles to profl. jours. Soc. for Sci. Study Sex scholar, 1981; Sigma Xi grantee, 1981, Kinsey Inst. Dissertation award, 1983. Mem. Am. Psychol. Assn., Ea. Psychol. Assn., Soc. for Sci. Study Sex (bd. dirs. 1984-86).

JEAN, ROGER V., mathematician, educator; b. Montreal, Que., Can., Oct. 20, 1940; s. Paul-Emile and Irène (Mongeau) J. Bachelors degree, U. Montreal, 1968, Masters degree, 1970; Doctorate, U. Paris, 1977, Doctorate d'Etat, 1984. Demonstrator IBM, Que., 1967; prof. Colls. Montreal, 1967-70; prof. math., biomathematician U. Que., Rimouski, 1970—; academician Acad. Creative Endeavors, Russia; critical reviewer Am. Math. Soc., 1987—; lectr. in field. Author: Mesure et Integration, 1975, rev. edit., 1989—, Phytomathematique, 1978—, Morphogenesis (Patterns), 1984—, Phyllotaxis, 1993; editor: Approche Mathematique de la Biologie, 1989; editor-in-chief Jour. Biol. Systems, Singapore, 1991—, Series on Biol. Systems; editorial bd. Symmetry: Culture and Science, 1989—; contbr. over 100 articles to scholarly jours. Recipient Cert. award Systems Rsch. Found., 1987, Herman Weyl prize Acad. Creative Endeavors, 1992; Hon. grantee Gov. of Que., 1969, grantee Can. Coun., 1974, 75, 77. Mem. Can. Soc. Theor. Biol. (v.p.), Internat. Inst. for Interdisciplinary Study Symmetry (founding mem.). Address: Case Postale 375, Rimouski, PQ Canada G5L 7C3

JEANLOZ, RAYMOND, geophysicist, educator; b. Winchester, Mass., Aug. 18, 1952. BA, Amherst Coll, 1975; PhD in Geology and Geophysics, Calif. Inst. Tech., 1979. Asst. prof. Harvard U., 1979-81; from asst. prof. to assoc. prof. U. Calif., Berkeley, 1982-85, prof., 1985—. Recipient Mineral Soc. Am. award, 1988; MacArthur grantee, 1988. Fellow AAAS, Am. Geophysics Union (J.B. Macelwane award 1984); mem. Am. Acad. Arts and Scis. Office: U Calif Dept Geology Berkeley CA 94720-4767

JEANNE, ROBERT LAWRENCE, entomologist, educator, researcher; b. N.Y.C., Jan. 14, 1942; s. Armand Lucien and Ruth (Stuber) J.; m. Louise Grenville Bluhm, Oct. 18, 1976; children—Thomas Lucien, James McClure. B.S. in Biology, Denison U., 1964; postgrad., Justus-Liebig U., Giessen, Fed. Republic Germany, 1964-65; M.A., Harvard U., 1968, Ph.D. in Biology, 1970. Instr. biology U. Va., Charlottesville, 1970-71; asst. prof. biology Boston U., 1971-76; asst. prof. entomology U. Wis., Madison, 1976-79, assoc. prof., 1979-83, prof., 1983—. Research, numerous publs. on social insects. Rotary Found. fellow in internat. understanding, 1964-65; NSF grantee, 1972—; Guggenheim Meml. fellow, 1986-87. Mem. Assn. Tropical Biology, Internat. Union for Study Social Insects, Animal Behavior Soc., Wis. Acad. of Scis., Arts and Letters, Phi Beta Kappa, Sigma Xi. Field rsch. on insects in U.S., Mex., Costa Rica, Panama, Surinam, Brazil, Australia. Office: U Wis Dept Entomology Madison WI 53706

JEANSONNE, GLORIA JANELLE, laboratory administrator, medical technologist; b. Moreauville, La., Aug. 29, 1946; d. Nicholas Joseph and Gloria Martha (Marcotte) J. BA in Edn., U. Southwestern La., 1968, BS in Med. Tech., 1975. Tchr., libr. Avoyelles Parish Sch. Bd., Simmesport, La., 1968-72; med. technologist Lafayette (La.) Gen. Med. Ctr., 1976-82, shift supr., 1982-83, chief med. technologist, 1983-87, lab. dir., 1987—. Mem. Am. Soc. Clin. Pathologists (assoc.), Clin. Lab. Mgmt. Assn., La. Soc. for Med. Tech., Acadiana Area Soc. for Med. Tech. Roman Catholic. Avocations: racquetball, bicycling. Office: Lafayette Gen Med Ctr 1214 Coolidge Blvd Lafayette LA 70503-2696

JEAS, WILLIAM C., electronics and aerospace engineering executive; b. Worcester, Mass., June 9, 1938; m. Irene M. Merkle, June 18, 1961; 1 child, Dean W. BS, U.S. Naval Acad., 1961; MBA, Air Force Inst. Tech., 1970. Commd. 2d lt. USAF, 1961, advanced through grades to col., 1980; various R & D, engring. and prodn. assignments, 1961-85, ret.,, 1986; COO, v.p. mktg. TE Products, Inc., Framingham, Mass., 1986—; bd. dirs., TE Consulting, Inc., Comtec, Inc. Mem. IEEE (sr. mem.), AIAA, SPIE, Armed Forces Communications Electronics Assn., U.S. Naval Acad. Alumni Assn., Air Force Assn. Home: 87 Wesson Ter Northborough MA 01532-1955

JECK, RICHARD KAHR, research meteorologist; b. Iola, Kans., Oct. 6, 1938. BS in Physics, Rockhurst Coll., 1960; MS, St. Louis U., 1963, PhD, 1968. Physicist Naval Rsch. Lab., Washington, 1968-70, 73-90, Smithsonian Radiation Biology Lab., Rockville, Md., 1971-72; rsch. meteorologist FAA, Atlantic City, 1990—. Contbr. articles to profl. jours. Postdoctoral fellow NAS and NRC, 1968-70. Mem. AIAA, Am. Meteorol. Soc. Achievements include patent for Calibration Device for Optical Particle Size Spectrometers, Automatic Directional Control for Wind-Following Rotors.

JEE, MELVIN WIE ON, mechanical engineer, civilian military engineer; b. Houston, Dec. 28, 1959; s. Way Horn nd Yoke Wah (Joe) Jee. BSME, Tex. A&M U., 1982. Engr. in Tng. Mech. engr. Individual Protection Directorate U.S. Army Natick (Mass.) Rsch., Devel. & Engring. Ctr., 1983-87, mech. engr., project mgr. Aero-Mech. Engring. Directorate, 1987—. Mem. ASME, Nat. Wildlife Fedn.

JEFFERIES, JOHN TREVOR, astronomer, astrophysicist, observatory administrator; b. Kellerberrin, Australia, Apr. 2, 1925; came to U.S., 1956, naturalized, 1967; s. John and Vera (Healy) J.; m. Charmian Candy, Sept. 10, 1949; children: Stephen R., Helen C., Trevor R. MA, Cambridge (Eng.) U., 1949; DSc, U. Western Australia, Nedlands, 1962. Sr. research staff High Altitude Obs., Boulder, Colo., 1957-59, Sacramento Peak Obs., Sunspot, N.Mex., 1957-59; prof. adjoint U. Colo., Boulder, 1961-64; prof. physics and astronomy U. Hawaii, Honolulu, 1964-83, dir., Inst. Astronomy, 1967-83; dir. Nat. Optical Astronomy Obs., Tucson, 1983-87; astronomer Nat. Optical Astronomy Obs., 1987-92; cons. Nat. Bur. Standards, Boulder, 1960-62; disting. vis. scientist Jet Propulsion Lab., 1991—. Author: (monograph) Spectral Line Formation, 1968; contbr. articles to profl. jours. Guggenheim fellow, 1970-71. Mem. Internat. Astron. Union, Am. Astron. Soc. Home: 1652 E Camino Cielo Tucson AZ 85718-1105 Office: Nat Optical Astronomy Obs PO Box 26732 Tucson AZ 85726-6732

JEFFERS, DALE WELBORN, computer specialist, systems analyst; b. Duluth, Ga., Feb. 20, 1952; d. Holman C. and Evelyn L. (Waters) Adams; m. Harry Lee Welborn, Aug. 14, 1971 (div. Sept. 1986); children: Thomas Adam Welborn, Myles Brandon Welborn; m. Jackie Carroll Jeffers, Feb. 17, 1990. BS magna cum laude, Piedmont Coll., 1975. Indsl. engr. Chicopee Mfg. Co., Cornelia, Ga., 1976-77, supr. data processing, 1978-79; supr. data processing, wordprocessing Ethicon, Inc., Cornelia, 1980-91, MIS analyst, 1991—. Methodist. Office: Ethicon Inc 70 Clarkesville Hwy Cornelia GA 30531-1054

JEFFERSON, JAMES WALTER, psychiatry educator; b. Mineola, N.Y., Aug. 14, 1937; s. Thomas Hutton and Alice (Withers) J.; m. Susan Mary Cole, June 25, 1965; children: Lara, Shawn, James C. BS, Bucknell U., 1958; MD, U. Wis., 1964. Cert. Am. Bd. Psychiatry and Neurology, Am. Bd. Internal Medicine. Asst. prof. psychiatry U. Wis. Med. Sch., Madison, 1974-78, assoc. prof., 1978-81, prof., 1981-92; Disting. sr. scientist Dean Found. for Health, Rsch. and Edn., Madison, 1992—; clin. prof. psychiatry U. Wis. Med. Sch., Madison, 1992—; co-dir. Lithium Info. Ctr., Madison, 1975—, Obsessive Compulsive Info. Ctr., Madison, 1990—; dir. Ctr. Affective Disorders, Madison, 1983-92. Co-author: Neuropsychiatric Features of Medical Disorders, 1981, Lithium Encyclopedia for Clinical Practice, 1983, 2d edit. 1987, Depression and Its Treatment, 1984, 2d edit., 1992, Anxiety and Its Treatment, 1986. Served to maj. U.S. Army, 1968-71. Fellow ACP, Am. Psychiat. Assn.; mem. Collegium Internat. Neuropsychopharmacologium. Avocations: running, travelling. Office: Dean Found 8000 Excelsior Dr Madison WI 53717-1914

JEFFERSON, MICHAEL L, environmental educator; b. Dec. 1, 1949; children: Michael C. and John A. AA, Crowder Coll., 1993, postgrad., 1993—. Cert. Class A wastewater, Class C water. Pipe layer Tri-Way Constrn., Neosho, Mo., 1965-67; equipment oper., crew foreman Water Wastewater Collection Treatment Constrn., 1974-78, job supt., 1978-83; utilities supt. City of Noel, Mo., 1983-89; instr. environ. resource ctr. Crowder Coll., Neosho, 1989—; adv. bd. U.S. EPA Region VII, Kansas City, Kans., 1990-93, ops. divsn. Water Environ. Fedn., 1990—; adv. coun. Mo. Dept. Natural Resources, Jefferson City, 1990-93; steering com. Regional Planning Commn., West Plains, Mo., 1991-92; com. chmn. ops. divsn. Mo. Water Pollution Control Fedn., 1992—, mem. pub. assistance, oper. assistance coms., Great Plains regional steering com. Water Pollution Control Fedn. Scout master Boy Scouts Am., Noel, 1984-92; tng./camping com. Mokan Area Coun., Joplin. Mo., 1985—; com. mem. United Way, Joplin, 1985—. Recipient Achievement award Mo. Water and Wastewater Conf., 1987, Treatment Plant of Yr. award State of Mo., 1988, EPA Region VII, 1988, Nat. O & M award EPA, 1988, William D. Hatfield award Water Environment Fedn., 1990; named Mo. Operator of Yr. Water Environ. Fedn., 1988, Quarter-Century Operators Club, 1992. Office: Crowder Coll 601 LaClede Neosho MO 64850

JEFFREY, MARCUS FANNIN, control systems engineer; b. Apple Springs, Tex., Dec. 7, 1934; s. Mark and Vada (Richardson) J.; m. Betty Jo Rhodes; children: Mark, Drew (dec.), Scott, Leigh. Grad. high sch., Lufkin, Tex., 1953. Registered profl. engr., Calif. Mgr. control systems Fluor Engrs., Inc., Sugarland, Tex., 1965-88; mgr. combustion sales Micon, Inc., Houston, 1988-89; dir. control systems Fluor Daniel, Sugarland, 1989-90; control systems specialist Bechtel, Houston, 1990-91; sales mgr. Vac System, Inc., Houston, 1991-92; with Tex. ECI, LaPorte, 1993—. With U.S. Army, 1953-56. Achievements include research in compressor control using microprocessors. Home: 12311 Alston Meadows TX 77477

JEFFREY, ROBERT ASAHEL, JR., pathologist; b. Wilsonville, Nebr., Mar. 5, 1922; s. Robert A. and Hazel Irene (Loesch) J.; m. Margaret Cecelia Degnan, Dec. 29, 1945; children: Robert A. III, Mary Francis, Margaret Cecelia, Johanne Marie, John Paul, Helene Angela, Terence Patrick, Joseph Michael, Thomas Matthew, Philip Brian, Caitlin Marie Therese. BS, Creighton U., 1944, MD, 1947. Diplomate Am. Bd. Pathology. Intern Creighton Meml. St. Joseph's Hosp., Omana, 1947-48; asst. resident pathology Stanford U. Sch. Medicine, 1953-55, 55-56, resident, 1956-57; instr. dept. pathology Stanford U. Sch. Medicine, San Francisco, 1957-59; dir. dept. pathology, clin. labs. St. Mary's Hosp. and Med. Ctr., San Francisco, 1959—. Capt. M.C. U.S. Army, 1952-54, Korea. E.B. DeGolia fellow in Pathology Stanford U. Sch. Medicine, 1954-55. Fellow Coll. Am. Pathologists. Republican. Roman Catholic. Office: Saint Marys Hosp Med Ctr Dept Pathology 450 Stanyan St San Francisco CA 94117

JEFFRIES, ROBERT ALAN, physicist; b. Indpls., Nov. 11, 1933; s. Seth Manes and Mary Elizabeth (Christmas) J.; m. Kelly Grisso, June 5, 1954; children: Russell A., D. Craig. B.S., U. Okla., 1954, M.S., 1961, Ph.D., 1965. Project engr. Pontiac Motor div. GM, Mich., 1954-55; mem. staff Los Alamos (N.Mex.) Sci. Lab., 1957-76, group leader, 1976-77, asst. div. leader, 1977-79, program mgr. nat. security programs, 1979-83, leader Arms Control and Verification Office, 1983-86, program dir. Verification and Safeguards, 1986-88, dir. arms control technology, 1988-89; tech. expert Nuclear Testing Talks, Geneva, 1986-90; Dept. Energy mem. U.S.-Russia Bilateral Consultative Commn., 1991—; sci. advisor Joint Staff, 1993—. Chmn. Los Alamos County Econ. Devel. Council, 1967; bd. dirs. Los Alamos Cancer Clinic, 1977-80. Served with USAF, 1955-57. Recipient Excellence award Dept. Energy, 1990. Mem. Am. Phys. Soc., Sigma Xi.

JEFTIC, LJUBOMIR MILE, marine scientist; b. Novi Becej, Yugoslavia, Feb. 25, 1936; s. Mile Jovan and Slavica Miroslav (Elblinger) J.; m. Ivanka Ivan Karacic, June 10, 1961; children: Nikola, Ivan. BSc, U. Zagreb, Yugoslavia, 1960, MSc, 1963, PhD, 1964. Sr. scientist Ctr. for Marine Rsch. R. Boskovic Inst., Zagreb, 1961-81; head dept. phys. planning and environ. protection Ministry Bldg., Housing and Environ. Protection, Zagreb, 1981-85; sr. marine scientist Mediterranean Coord. Unit, UN Environ. Program, Athens, Greece, 1985—; cons. UNEP, UNESCO, WHO, 1976-85. Contbr. articles to profl. publs. Avocations: bridge, reading, music. Office: UNEP, 48 Vas Konstantinou, 11573 Athens Greece

JEGHAM, SAMIR, chemist; b. Sousse, Tunisia, Sept. 12, 1960; arrived in France, 1984; s. Salem and Khadija Jegham; m. Nadine Souchu, June 17, 1989; 1 child, Cerine. MSc, Faculty of Sci., Monastir, Tunisia, 1984; PhD, U. Paris, Orsay, 1988. Lab. head Synthelabo Recherche, Bagneux, France, 1989-90, project leader, 1991—. Contbr. articles to Tetrahedron Letter. Mem. Am. Chem. Soc. Achievements include 3 patents in field (France). Office: Synthelabo Recherche, 31 Ave P Vaillant Couturier, Bagneux 92225, France

JEGLEY, DAWN CATHERINE, aerospace engineer; b. Cuyahoga Falls, Ohio, Aug. 25, 1961; d. Bernard Lawrence and Carole Ann (Martin) J.; m. J. Noel Chiappa, July 26, 1986; 1 child, Catherine. BS, MIT, 1983; MS, George Washington U., 1987. Rsch. engr. NASA Langley Rsch. Ctr., Hampton, Va., 1983—. Contbr. articles to tech. jours. Sr. mem. AIAA. Office: NASA Langley Rsch Ctr Mail Stop 190 Hampton VA 23681

JEITSCHKO, WOLFGANG KARL, chemistry educator; b. Prague, May 27, 1936; came to Germany, 1975; s. Karl Friedrich and Agnes (Maier) J.; m. Marieluise Fichtner, July 19, 1964; children: Andreas Wolfgang, Thomas David, Peter Oliver. Student Techhochschule, Wien 1956-62, Dr. Phil., U. Wien, 1964. With Metallwerke Plansee AG, Reutte/Tyrol, 1962-64; post doctoral fellow U. Pa., Phila., 1964-66; research assoc., lectr. U. Ill., Champaign-Urbana, 1967-69; with cen. research dept. DuPont Co., Wilmington, Del., 1969-75; prof. U. Giessen, Germany, 1975-79; ordinary prof. U. Dortmund, Germany, 1979-82; ordinary prof., dir. U. Muenster, Germany, 1982—. Mem. Am. Chem. Soc., Am. Crystallographic Assn., Acad. Scis. Moscow (Kurnakov medal 1991), Gesellschaft Deutscher Chemiker, Deutsche Gesellsch. Materialkd. Contbr. articles to profl. jours. Office: U Münster Anorganisch-Chem Inst, Wilhelm-Klemm-Str 8, D-48149 Münster Germany

JELALIAN, ALBERT V., electrical engineer; b. Bridgewater, Mass., June 30, 1933; s. Siragan and Zarouhi (Tanelian) J.; m. Mary B. Karoghlanian; children: Alan H., Leslie K. BSEE, Northeastern U., 1957. Reg. profl. engr., Mass. Engr. Raytheon Co., Lexington, Mass., 1957-81; mgr. electro-optics lab Raytheon Co., Sudbury, Mass., 1981-86; asst. dir. Raytheon Co.,

Sudbury, 1986-91, asst. mgr. equipment devel. labs. E/O, 1991-92; pres. Jelalian Sci. & Engring., Bedford, Mass., 1992—. Inventor: holds ten patents relating to aviation safety and military products; contbr. articles to profl. jours; published book on laser radar systems, 1992. Recipient Recognition award NASA, Washington, 1974, Group Achievement award, 1975. Mem. AIAA, IEEE, Infrared Info. Symposium (vice chmn. active systems 1989-91, nat. chmn., 1991—). Republican. Armenian Orthodox. Office: Jelalian Science & Engring 3 Reeves Rd Bedford MA 01730-1334

JELLEY, SCOTT ALLEN, microbiologist; b. Tarrytown, N.Y., July 22, 1960; s. Alfred Paul and Nadine Elaine (Scott) J. BS in Biology, Bucknell U., 1982; MS in Microbiology, Va. Poly. Inst., 1985. Grad. teaching asst. Va. Poly. Inst., Blacksburg, 1983-85, lab. specialist, 1985; scientist Pfizer Cen. Rsch., Groton, Conn., 1986-88; microbiologist Findley Rsch., Inc., Fall River, Mass., 1988-90; microbiologist sterilization scis. group Codman and Shurtleff, Inc., Randolph, Mass., 1990—. Contbr. articles to Applied and Environmental Microbiology. Mem. Assn. for Advancement Med. Instrumentation (sterilization reusable med. devices com.), Am. Soc. for Microbiology, Johnson and Johnson Corp. Microbiol. Com. (scientific liaison focused giving program). Achievements include development of sterilization validation techniques for medical devices. Office: Codman and Shurtleff Inc 41 Pacella Park Dr Randolph MA 02368-1794

JELLIFFE, ROGER WOODHAM, cardiologist, clinical pharmacologist; b. Cleve., Feb. 18, 1929; s. Russell Wesley and Rowena (Woodham) J.; m. Joyce Miller, June 12, 1954; children: Susan, Amy, Elizabeth, Peter. BA, Harvard U., 1950; MD, Columbia U., 1954. Diplomate Am. Bd. Internal Medicine, Am. Bd. Cardiovascular Disease. Intern Univ. Hosps., Cleve., 1954-56; also jr. asst. resident in medicine; Nat. Found. Infantile Paralysis exptl. medicine fellow Case Western Res. U., Cleve., 1956-58; staff physician in medicine VA Hosp., Cleve., 1958-60; resident in medicine VA Hosp., 1960-61; instr. medicine U. So. Calif. Sch. Medicine, L.A., 1961-63; asst. prof. U. So. Calif. Sch. Medicine, 1963-67, assoc. prof., 1967-76, prof. medicine, 1976—; devel. Lab. Applied Pharmacokinetics, 1973—, The USC*PACK Computer Programs, 1973—; cons. Dynamic Scis., Inc., Van Nuys, Calif., 1976—, Simes S.P.A., Milan, 1979—, IVAC Corp., San Diego, 1983-88, Bionica, Sidney, Australia, 1987—. Author: (book) Fundamentals of Electrocardiography, 1990; cons. editor Am. Jour. Medicine, 1972-78, Current Prescribing, 1974-79, Am. Jour. Physiology, 1984-91; contbr. articles to profl. jours.; patentee in field. Advanced Rsch. fellow L.A. County Heart Assn., 1961-64. Fellow ACP, Am. Coll. Med. Informatics, Coun. on Clin. Cardiology, Am. Heart Assn.; mem. Am. Soc. Clin. Pharmacology and Therapeutics, Am. Fed. Clin. Rsch., Assn. Advancement Med. Instrumentation, Am. Med. Informatics Assn. Achievements include research of optimal mgmt. of drug therapy; development of time-shared computer programs and programs for personal computers for optimal mgmt. of drug therapy for hosps.; development of intelligent infusion devices. Office: U So Calif Sch Medicine CSC 134-B 2250 Alcazar St Los Angeles CA 90033

JEN, TIEN-CHIEN, mechanical engineer; b. Ping-Tung, Taiwan, Republic of China, Mar. 23, 1959; s. Shih-yin and Chi-Mon (Chen) J.; m. Mei-Yen Tung, July 26, 1985; 1 child, Liang Sywan Edward. MS, Nat. Tsing-Hua U., Hsinchu, Taiwan, 1987; PhD, UCLA, 1993. Asst. devel. engr. UCLA, 1990-92, postdoctoral fellow, 1992-93, asst. rsch. engr., 1993—. Contbr. articles to Jour. Heat Transfer, Internat. Heat and Mass Transfer. Mem. ASME, Am. Phys. Soc. Office: UCLA MANE Dept Los Angeles CA 90024

JENAB, S. ABE, civil and water resources engineer; b. Isfahan, Persia, Mar. 26, 1936; came to U.S., 1959; s. Mohamad Taghi and Soghra (Beigom) J.; children: Jenia, Jima. MS, Utah State U., 1962, PhD, 1967. Registered profl. engr. Instr. Tehran U., 1962-64; from asst. prof. to assoc. prof. Utah State U., Logan, 1967-69; vis. prof. Utah Stae U., Logan, 1983-85; dir. water resources engring. dept. Agrl. Devel. Bank of Iran, Tehran, 1969-70; pres., mgr. Aabadin Engring. Inc., Tehran, 1973-83; assoc. prof. Tehran, Isfahan and Shiraz Univs., Iran, 1969-83; sr. engr. St. Johns River Water Mgmt. Dist., Palatka, Fla., 1985—; cons. Iranian Plan Orgn., Tehran, 1973-83; adivsor, cons. Ministry of Water and Power, Tehran, 1975-83; cons. Hwy. Dept., Ogden, Utah, 1984-85, Iranian Plan Orgn., 1970-75; numerous cons. positions. Contbr. more than 30 articles to profl. jours. Recipient Hon. medal Internat. Commn. on Irrigation, 1972; grantee U.S. Dept. Agr., 1967, Iranian Plan Orgn., 1970, others. Mem. ASCE, Am. Soc. Agrl. Engrs. Achievements include patents for double mole drains, for hydro-dynamic rocket for cleaning drains and sewer lines and for high technology in slope stability. Home: PO Box 1922 Palatka FL 32178-1922 Office: St Johns River Water Mgmt PO Box 1429 Palatka FL 32178-1429

JENCKS, WILLIAM PLATT, biochemist, educator; b. Bar Harbor, Maine, Aug. 15, 1927; s. Gardner and Elinor (Melcher) J.; m. Miriam Ehrlich, June 3, 1950; children—Helen Esther, David Alan. Grad., St. Paul's Sch., Balt., 1944; student, Harvard, 1944-47, M.D., 1951. Intern Peter Bent Brigham Hosp., Boston, 1951-52; postdoctoral fellow Mass. Gen. Hosp., Boston, 1952-53, 55-56; postdoctoral fellow chemistry Harvard, 1956-57; mem. faculty Brandeis U., 1957—, prof. biochemistry, 1963—. Served as 1st lt., M.C. AUS, 1953-55. Recipient ASBMB-Merck award Am. Soc. Biochem. and Molecular Biology, 1992. Fellow Royal Soc.; mem. NAS, AAAS, Am. Chem. Soc. (award in biol. chemistry 1962), Am. Soc. Biol. Chemists, Am. Acad. Arts and Scis., Alpha Omega Alpha. Home: 11 Revere St Lexington MA 02173-4419 Office: Brandeis Univ Grad Dept Biochemistry Waltham MA 02254

JENDEN, DONALD JAMES, pharmacologist, educator; b. Horsham, Sussex, Eng., Sept. 1, 1926; came to U.S., 1950, naturalized, 1958; s. William Herbert and Kathleen Mary (Harris) J.; m. Jean Ickeringill, Nov. 18, 1950; children: Patricia Mary, Peter D., Beverly J. BSc in Physiology with 1st class honours, Kings Coll. London, 1947; MB, BS with honours, U. London, 1950; PhD in Pharm. Chemistry (hon.), U. Uppsala, Sweden, 1980. Demonstrator pharmacology U. London, 1948-49; lectr. pharmacology U. Calif.-San Francisco, 1950-51, asst. prof. pharmacology, 1952-53; mem. faculty UCLA, 1953, assoc. prof., 1956-60, prof. pharmacology, 1960—, prof. pharmacology and biomath., 1967—, chmn. dept. pharmacology, 1968-89; Wellcome vis. prof. U. Ala., Birmingham, 1984; mem. brain research inst. UCLA, 1961—. Contbr. articles in field. Served to lt. comdr. M.C., USNR, 1954-58. USPHS Postdoctoral fellow, 1951-53, NSF Sr. Postdoctoral fellow; hon. research assoc. Univ. Coll., London, 1961-62; Fulbright Short-Term Sr. Scholar award, Australia, 1983; recipient Univ. Gold medal U. London, 1950. Fellow Am. Coll. Neuropsychopharmacology; mem. AAAS, Am. Soc. Pharmacology and Exptl. Therapeutics, Am. Physiol. Soc., Physiol. Soc. (London), Soc. Neurosci., Am. Chem. Soc. (div. med. chemistry), Western Pharmacology Soc. (pres. 1970), Assn. for Med. Sch. Pharmacology, Am. Soc. Mass Spectrometry, Am. Soc. Neurochemistry, Internat. Union Pharmacology (sect. on toxicology), N.Y. Acad. Sci., West Coast Coll. Biol. Psychology (charter fellow). Home: 3814 Castlerock Rd Malibu CA 90265-5625 Office: UCLA Sch Medicine Dept Pharmacology Ctr Health Scis Los Angeles CA 90024-1735

JENEKHE, SAMSON ALLY, chemical engineering educator, polymer scientist; b. Okpella, Bendel, Nigeria, Mar. 3, 1951; came to U.S., 1974; s. Ally Damisa Jenekhe and Animetu Ejakome (Ajayi) Olowu. BS, Mich. Tech. U., 1977; MS, U. Minn., 1980, MA, 1981, PhD, 1985. Rsch. asst. U. Minn., Mpls., 1977-80; from sr. rsch. scientist to project leader Honeywell Inc. Phys. Scis. Ctr., Bloomington, Minn., 1981-87; asst. prof. U. Rochester, N.Y., 1988-90; assoc. prof. U. Rochester, 1990—; cons. Honeywell Inc., Bloomington, Minn., 1988-90, 92—, McDonnell Douglas Rsch. Laboratories, St. Louis, 1990-92, Gould Electronics, Inc., Eastlake, Ohio, 1992. Editor: Macromolecular Host-Guest Complexes, 1992; contbr. articles to Nature (London), Macromolecules, Chemistry of Materials, Jour. Phys. Chemistry. Rsch. grantee Office Naval Rsch., 1984-87, Air Force Materials Lab., 1986-88, 93—, Naval Air Devel. Ctr., 1987-90. Mem. AAAS, AICE (symposium chmn. 1992), Am. Chem. Soc. (symposium co-chmn. 1989), Materials Rsch. Soc. (symposium chmn. 1992), Sigma Xi, Tau Beta Pi. Achievements include patents on conducting polymers, high temperature polymers and new methods of materials processing; synthesis and processing of polymers for electronics, optoelectronics and photonics; rsch. in. electronic and photonic polymers, polymer photophysics, polymer nanocomposites. Home: 50 S Village Tr Fairport NY 14450 Office: Univ Rochester 206 Gavett Hall Rochester NY 14627-0166

JENG, TZYY-WEN, biochemist; b. Taichung, Taiwan, Nov. 2, 1947; came to U.S., 1974; s. Ching-Po and Yu-Ju (Wong) J.; m. Kwan-Yee Sum; children: Howard L., Way A. BS, Nat. Taiwan U., Taipei, 1970; PhD, U. Calif., Berkeley, 1978. Rsch. assoc. U. Ariz., Tucson, 1979-84, rsch. asst. prof., 1984-86, rsch. specialist and rsch. asst. prof., 1986-88; sr. rsch. biochemist Abbott Labs., Abbott Park, Ill., 1988-90, rsch. investigator, 1991-92, assoc. rsch. fellow, 1992—. Author: Natural Toxins, 1980; contbr. articles to Jour. Molecular Biology. Wilhelm Bernard Fund grantee Internat. Congress on Electron Microscopy, 1982. Mem. N.Y. Acad. Scis. Achievements include patents in field. Office: Abbott Labs 1 Abbott Dr Dept 2RR AP20 Abbott Park IL 60064

JENKINS, ALAN DELOSS, urologic surgeon, educator; b. Carlsbad, N.Mex., Dec. 10, 1949; s. Robert Deloss and Alice Dorothy (Anderson) J.; m. Barbara Jean Sprowls, June 6, 1970; children: Katherine Hamilton, Peter Anders, Andrew Deloss, Matthew Persson. SB in Physics, MIT, 1971; MD, Boston U., 1975. Resident surgery U. Va., Charlottesville, 1975-77, resident urology, 1977-81, asst. prof. urology, 1984-90, assoc. prof., 1990—; rsch. fellow Mayo Clinic, Rochester, Minn., 1981-83; asst. prof. surgery U. Tex. Med. Sch., Houston, 1983-84; mem. exam com. AUA/ABU, Houston, 1988-92. Assoc. sect. editor Jour. Urology, 1985; asst. editor Jour. Endourology, 1987; co-author: Stone Surgery, 1991. Fellow ACS; mem. Am. Urol. Assn. Achievements include application of extracorporeal shock wave lithotripsy to the treatment of lower ureteral calculi. Home: 250 Spring Ln Charlottesville VA 22901 Office: U Va Sch Medicine Dept Urology Box 422 Charlottesville VA 22908

JENKINS, JAMES THOMAS, mechanical engineering researcher; b. Chgo., June 30, 1942; s. Marvin Nicholas and Esther Alice (Nelson) J.; m. Katharine Kelly, Oct. 8, 1983; children: Thomas Nelson, Peter Kelly. BSME, Northwestern U., Evanston, Ill., 1964; PhD in Mechanics, Johns Hopkins U., 1969. Asst. prof. Theoretical and Applied Mechanics, Cornell U., Ithaca, N.Y., 1971-77, assoc. prof., 1977-83, prof., 1983—, chair, 1991—. Contbr. numerous articles to profl. jours. Office: Theoretical & Applied Mechanics Cornell U Ithaca NY 14853

JENKINS, JAMES WILLIAM, osteopath; b. Columbus, Ohio, May 15, 1953; s. William Harvey and Irene Barbara (Kacsor) J.; m. Deborah Susan Dorrance, June 16, 1987. BA in Biology, Calif. State U., Fullerton, 1976; DO, Coll. Osteopathic Med. Pacific, 1984; diploma in emergency medicine, Ohio State U., 1988. Intern Warren (Ohio) Gen. Hosp., 1984-85; resident in emergency medicine Meml. Osteopathic Hosp., York, Pa., 1985-87; rsch. fellow, clin. instr. Coll. Medicine, Ohio State U., Columbus, 1987-88; clin. emergency physician, med. edn. coord. emergency dept. Dr.'s Hosp., Columbus, 1988-89; med. dir. emergency dept. Greenfield (Ohio) Area Med. Ctr., 1989—, clin. emergency/trauma physician, 1991—; emergency med. svc. med. advisor Franklin Twp. Fire Dept., Columbus, 1988-89; clin. asst. prof. Coll. Osteo. Medicine Pacific, Pomona, Calif., 1989. Contbr. articles to profl. publs., chpt. to book. Mem. CPR com. ARC, Santa Ana, Calif., 1978-81; instr. trainer Am. Heart Assn., Santa Ana, 1972-80; instr., course coord. basic trauma life support Am. Coll. Emergency Physicians, Columbus, 1988—. Rsch. grantee Emergency Medicine Found., 1988, Kellogg Found., 1979-80; recipient rsch. fellow award Emergency Medicine Residents Ohio, 1988, Armstrong Lit. award, 1980. Mem. Am. Coll. Emergency Physicians, Am. Osteo. Assn., Beta Beta Beta. Avocations: collecting books, martial arts. Office: Greenfield Area Med Ctr 545 South St Greenfield OH 45123-1400

JENKINS, WILLIAM KENNETH, electrical engineering educator; b. Pitts., Apr. 12, 1947; s. William Kenneth and Edna Mae (Treusch) J.; m. Suzann Heinricher, Aug. 22, 1970. B.S.E.E., Lehigh U., 1969; M.S.E.E., Purdue U., 1971, Ph.D., 1974. Grad. instr., teaching asst. Purdue U., West Lafayette, Ind., 1969-74; research sci. assoc. Lockheed Corp., Palo Alto, Calif., 1974-77, cons., 1983—; asst. prof. elec. engring. U. Ill., Urbana, 1977-80, assoc. prof., 1980-83, prof., 1983—, acting dir. coordinated sci. lab., 1986-87, dir. coordinated sci. lab., 1987—; cons. Ill. State Water Survey, Urbana, 1978, Siliconix, Inc., Santa Clara, Calif., 1979-81, Bell Labs., North Andover, Mass., 1984, AT&T Bell Labs, Lockheed Missiles and Space Co. Fellow IEEE (pres. Circuits and Systems Soc. 1985, editor reprint volume 1986), Acoustics, Speech and Signal Processing Soc. of IEEE (CAS Soc. Disting. Svc. award 1990). Avocations: tennis; swimming; sports cars; amateur musician. Home: 1913 Moraine Dr Champaign IL 61821-5258 Office: U Ill 1308 W Main St Urbana IL 61801-3005

JENKS, GERALD ERWIN, aerospace company executive; b. Kansas City, Kans., Mar. 26, 1945; s. Albert Edwin and Cecil Edith (Johnson) J.; m. Pamela Danette Houghman, Nov. 21, 1964; children: Brendon, Carissa, Bradley. BS, Kans. U., 1968, ME, 1975. Flight test engr. Cessna Aircraft Co., Wichita, Kans., 1968-74; mgr. Flight Rsch. Lab., Lawrence, Kans., 1974-76; staff dir. U.S. Ho. of Reps., Washington, 1976-82; prin. mgr. McDonnell-Douglas, St. Louis, 1982—; mem. Pres. Reagan Transition Team, Washington, 1980; mem. aerospace engring. steering com. Kans. U., Lawrence, 1990-92. 1st lt. USAF, 1970-76. Fellow AIAA; mem. Sigma Gamma Tau. Achievements include implementation of attitude command control system using separate surface stability augmentation on a Beech Model 99, realities of implementing quality/productivity improvement. Home: 1841 Newburyport Rd Chesterfield MO 63005 Office: McDonnell Douglas Mc0924100 PO Box 516 MC 3061360 Saint Louis MO 63106-0516

JENNER, WILLIAM ALEXANDER, meteorologist, educator; b. Indianola, Iowa, Nov. 10, 1915; s. Edwin Alexander and Elizabeth May (Brown) J., m. Jean Norden, Sept. 1, 1946; children: Carol Beth, Paul William, Susan Lynn. AB, Cen. Meth. Coll., Mo., 1938; certificate meteorology U. Chgo., 1943; MEd, U. Mo., 1947; postgrad. Am. U., 1951-58. Instr. U. Mo., 1946-47; rsch. meteorologist U.S. Weather Bur., Chgo., 1947-49; staff Hdqrs. Air Weather Svc., Andrews AFB, Md., 1949-58, Scott AFB, Ill., 1958-84, dir. tng., 1960-84. Mem. O'Fallon (Ill.) Twp. High Sch. Bd. Edn., 1962—, sec., 1964-71, pres., 1971-83, 1985-87, vice pres., 1990—; pres. St. Clair County Regional Vocat. System Bd., 1986-89; vice chmn. southwestern div. Ill. Assn. Sch. Bds., 1987-89, chmn., 1989—; comdr. 507th Fighter Group Assn. Inc., 1987-89; mem. O'Fallon Planning Commn., 1973-84, sec., 1979-81, sub-div. chmn., 1978-84; alderman City of O'Fallon, 1984-93. With AUS, 1942-46. Recipient Disting. Svc. award O'Fallon PTA, 1968, Disting. Svc. award City of O'Fallon, 1985, Community Svc. award O'Fallon Toastmasters Club, 1991, Master Bd. Mem. award Ill. Assn. Sch. Bds., 1991, award of Excellence O'Fallon C. of C., 1991. Merit cert. St. Clair County, 1987, Exceptional Civilian Svc. award Dept. Air Force, 1984, Jenmer Award established by Air Weather Svc., 1984. Fellow Am. Meteorol. Soc.; mem. APA, Am. Psychol. Soc., Wilson Ornithol. Soc., Am. Philatelic Soc., Am. Philatelic Congress, Am. Meteorol. Soc., AAAS, Nat. Soc. Study Edn., Nat. Audubon Soc., Nat. Arbor Day Found., Tree City USA, Nat. Parks and Conservation Assn., Nat. Wildlife Fedn., Nat. Resources Defense Coun., Nature Conservancy, Vt. Inst. Natural Sci., Leadership St. Louis., The World Wildlife Fund, N.Y. Acad. Scis., Internat. Platform Assn., Am. Legion, The Wilderness Soc., The Wildlife Conservation Soc., Wildlife Forever, Rails to Trails Conservancy, Phi Delta Kappa, Psi Chi. Club: O'Fallon Sportsmen's. Lodges: Masons, Shriners, Sierra. Home: 307 Alma St O'Fallon IL 62269-2449

JENNINGS, DAVID THOMAS, III, electronics executive, consultant; b. Denver, Dec. 10, 1949; s. David Thomas Sr. and Frances Adele (Yingling) J.; m. Dee Adele Whiteside, June 1, 1985; 1 child, Adele Elizabeth. Attended, Auburn U., 1967-73. Owner The SoundWorks, Auburn, Ala., 1973-77; sr. engr. tech. The Gyrex Corp., Santa Barbara, Calif., 1977-78; ptnr., product engr. mgr. KDC Electronics, Carpinteria, Calif., 1981-82; chief engr. Browne Med., Carpinteria, 1984-85; owner Design Cons., Santa Barbara, 1978—. Home and Office: 2808 Clinton Terr Santa Barbara CA 93105

JENNINGS, FREDERIC BEACH, JR., economist, consultant; b. Boston, Dec. 29, 1945; s. Frederic Beach III and Ellen (Osgood) J.; m. Lucille Candace Giglio, Aug. 15, 1975; children: Frederic Beach V, Thomas Chapin. BA magna cum laude, Harvard U., 1968; MA in Econs., Stanford U., 1980, PhD in Econs., 1985. Jr. medicare acct. Blue Cross-Blue Shield, Boston, 1968-69; ind. rsch. fellow Inst. Humane Studies, Menlo Park, Calif., 1969-71, 77-78; asst. mgr. Globe Bag Co., South Boston, 1972-73; rsch. asst. Charles River Assocs., Cambridge, Mass., 1973-74; rsch. and teaching fellow Stanford (Calif.) Dept. Econs., 1974-79; instr. econs. Tufts U., Medford, Mass., 1979-83; asst. prof. Bentley Coll., Waltham, Mass., 1985-87; sr. econ. cons. The Mac Rsch. Group, Cambridge, 1987-88, Charles River Assocs., Boston, 1988-91; sr. mgr. Econ. Analysis Group Office of Fed. Tax Svcs. Arthur Andersen & Co., Washington, 1991-92; pres. EconoLogistics, Ipswich, Mass., 1992—; chmn., rep. Stanford Grad. Student Coun., 1974-76; senator Stanford Student Senate, 1975-76; co-pres. Associated Students Stanford U., 1976-77; founder Stanford Grad. Students Assn., 1978-79, The Bentley Participants, Waltham, 1986-87, Full Circle Discussion Group Tufts U., Medford, 1981-84; resident assoc. Residential Edn., Stanford, 1978-79. Author: Democracy in Disarray, 1978, (paper) Value, Exchange and Profit, 1966, (essays) Academy, Society and Personal Growth, 1983, Whither Our Education?, 1983. Mem. Am. Econ. Assn., Cliometrics Soc., Indsl. Orgn. Soc., Western Econ. Assn., Atlantic Econ. Soc., Harvard Travellers Club. Avocations: fly fishing, sailing, skiing, tennis, golf. Home: 261 Argilla Rd Ipswich MA 01938-2615 Office: EconoLogistics 55 Market St Ste # 201 Ipswich MA 01938

JENNINGS, JERRY L., psychologist; b. Binghamton, N.Y., Apr. 27, 1955; s. John J. and Thelma (Hunt) J.; m. Jane A. Peterson, May 27, 1990. MA, U. N.H., 1983, PhD, 1984. Clin. assoc. Dept. Psychiatry, U. Pa., Phila., 1985-88; clin. psychologist Family and Community Svcs., Burlington, N.J., 1985-88; program coord. Wellspring Pain Control Ctr., Moorestown, N.J., 1985-88; lectr. Osteopathic Coll. of Medicine, Phila., 1989—; dir. TAO, Inc./Bustleton Guidance Ctr., Phila., 1988—; lectr. Temple U., 1992—. Contbr. articles to profl. jours. Achievements include studies in field of phenomenological psychology; innovative work in areas of treatment of battering men and dreams.

JENNINGS, MARCELLA GRADY, rancher, investor; b. Springfield, Ill., Mar. 4, 1920; d. William Francis and Magdalene Mary (Spies) Grady; student pub. schs.; m. Leo J. Jennings, Dec. 16, 1950 (dec.). Pub. relations Econolite Corp., Los Angeles, 1958-61; v.p., asst. mgr. LJ Quarter Circle Ranch, Inc., Polson, Mont., 1961-73, pres., gen. mgr., owner, 1973—; dir. Giselle's Travel Inc., Sacramento; fin. advisor to Allentown, Inc., Charlo, Mont.; sales cons. to Amie's Jumpin' Jacks and Jills, Garland, Tex. Investor. Mem. Internat. Charolais Assn., Los Angeles County Apt. Assn. Republican. Roman Catholic. Home and Office: 509 Mt Holyoke Ave Pacific Palisades CA 90272-4328

JENNINGS, ROBERT BURGESS, experimental pathologist, medical educator; b. Balt., Dec. 14, 1926; s. Burgess Hill and Etta (Crout) J.; m. Linda Lee Sheffield, June 28, 1952; children—Carol L., Mary G., John B., Anne E., James R. B.S., Northwestern U., 1947, M.S., B.M., 1949, M.D., 1950. Diplomate Am. Bd. Pathology (trustee 1976-87, pres. 1986-87). Intern Passavant Meml. Hosp., Chgo., 1949-50; resident pathology Passavant Meml. Hosp., 1950-51; mem. faculty Northwestern U. Med. Sch., 1953-75, prof. pathology, 1963-75, Magerstadt prof. and chmn. pathology dept., 1969-75; prof., chmn. dept. pathology Duke U. Med. Sch., Durham, N.C., 1975-89, James B. Duke prof., 1980—; vis. scientist Middlesex Hosp. Med. Sch., London, 1961-62; cons. VA Rsch. Hosp., Chgo.; mem. attending staff Northwestern Meml. Hosp., Chgo., 1969-75; mem. pathology A study sect. USPHS, 1960-65; mem. clin. cardiology adv. com. NIH, 1976-80, mem. cardiovascular and renal study sect., 1992—. Mem. editorial bd. Lab Investigation, 1967—, Archives Pathology, 1970-80, Jour. Molecular and Cellular Cardiology, 1972-89, Exptl. and Molecular Pathology, 1973—, Circulation, 1988-91, 93—, Circulation Rsch., 1976-82, Histopathology, 1977-92, Am. Jour. Pathology, 1983-92, Jour. Applied Cardiology, 1986—, Circulation, 1988-91, Cardiosci., 1990—, Trends in Cardiovascular Medicine, 1991-92, Cardiovascular Pathology, 1991—. Served as lt. (j.g.) USNR, 1951-53. Markle scholar med. scis., 1958-63. Office: Duke U Med Ctr Dept Pathology Durham NC 27710

JENNY, DANIEL P., retired engineer. Recipient Henry C. Turner medal Am. Concrete Inst., 1991. Home: 21 S Donald Ave Arlington Heights IL 60004*

JENSEN, ANDREW ODEN, obstetrician/gynecologist; b. El Paso, Tex., Aug. 30, 1920; s. Andrew Rudolph and Annie Laura (Oden) J.; m. Patricia deMaret Steele, May 10, 1952; children: Elise Ann Jensen Murphy, Nancy Marie Jensen Jett, Andrew Oden Jr. BA, So. Meth. U., 1941; MD, U. Tenn., Memphis, 1949. Diplomate Am. Bd. Ob-Gyn. Intern Episcopal Hosp., Phila., 1949-50; resident Hermann Hosp., Houston, 1950-53; clin. instr. Baylor U. Coll. Medicine, Houston, 1951-53; chief ob-gyn Med. Arts Clinic, Brownwood, Tex., 1954-55; pvt. practice Denison, Tex., 1955-74, Temple, Tex., 1975-86; locum tenens Salt Lake City, 1986—; cons. in ob.-gyn. Lake Cumberland Dist. Health Dept., Somerset, Ky., 1974-75, Perrin AFB, Sherman, Tex., 1956-69; lectr., del. people-to-people program Nat. Congress Ob-Gyn in China, 1984. Bd. dirs. Brownwood Jr. C. of C., Denison C. of C., 1959-62. Fellow Am. Coll. Obstetricians and Gynecologists, Tex. Assn. Ob-Gyn; mem. AMA, Soc. Med. Soc., Grayson County Med. Soc. (pres. 1969-70), Cen. Assn. Obstetricians and Gynecologists, Am. Fertility Assn., West Tex. Ob-Gyn Soc., Rotary (pres. Denison 1966-67, Paul Harris fellow 1990). Democrat. Episcopalian. Avocations: photography, golf. Home: 3330 Wimbledon Dr Cibolo TX 78108-2162 Office: 2161 NW Military Hwy Ste 205 San Antonio TX 78213-1844

JENSEN, ARTHUR SEIGFRIED, consulting engineering physicist; b. Trenton, N.J., Dec. 24, 1917; s. Emil Anthony and Emma Anna (Lund) J.; m. Lillian Elizabeth Reed, Aug. 9, 1941; children: Deane Ellsworth, Alan Forrest, Nancy Lorraine. B.S., U. Pa., 1938, M.S., 1939, Ph.D., 1941; diploma in advanced engring., Westinghouse Sch. Applied Sci., 1972, diploma in computer sci., 1977. Registered profl. engr., Md. Research physicist U.S. Naval Research Labs., Washington, 1941; research physicist RCA Labs., Princeton, N.J., 1945-57; mgr. spl. electron devices Westinghouse Electronic Tube Div., Balt., 1957-65; sr. adv. physicist Electronics Systems Ctr., Balt., 1965-91; cons. physicist Westinghouse Def. and Electronic Systems Center, Balt., 1991—; mem. Md. State Bd. Registration Profl. Engrs., 1979-86, vice chmn., 1983-86; cons. Nat. Acad. Sci., 1970. Contbr. articles to profl. jours.; 25 patents. Served to capt. USN, 1941-46, USNR, 1946-77, ret., 1977—. Recipient Outstanding Service award Engrs. Council of Md., 1986, Govs. Citation, 1986, Westinghouse Special Patent award, 1972. Fellow IEEE (life), Washington Acad. Scis.; mem. AAAS, AIAA, Res. Officers Assn., Ret. Officers Assn., Naval Res. Assn., Am. Phys. Soc., Am. Assn. Physics Tchrs., Soc. Photo-Optical Instrumentation Engrs., Optical Soc. Am., N.Y. Acad. Scis., Md. Acad. Scis. (chmn. awards com.), Nat. Coun. Engring. Examiners (chmn. internat. rels. com.), Infrared Info. Symposium, Am. Legion, Fleet Res. Assn., Sons of Norway, Sigma Xi, Pi Mu Epsilon, Kappa Phi Kappa. Club: U.S. Naval Acad. Officers and Faculty. Achievements include patents in field. Home: 5602 Purlington Way Baltimore MD 21212-2950 Office: Westinghouse Electronic Systems Group Baltimore MD 21203

JENSEN, BJARNE SLOTH, economist; b. Noerre Nebel, Jylland, Denmark, Nov. 16, 1942; s. Carl Peter Jensen and Helga Sveistrup; m. Kathleen Gail Henrikson, May 13, 1978; children: Susanna, Tina. Degree in econ., U. Aarhus, Denmark, 1971; filsosfie doktor, U. Uppsala, Sweden, 1992. Economist Ministry of Housing, Copenhagen, 1971-72; scholar Danish Social Rsch. Coun., Copenhagen, 1972-73; from jr. to sr. scholar Copenhagen Bus. Sch., 1973-79, assoc. prof., 1979—; vis. scholar U. Pa., Phila., 1976-77; fil. dr. U. Uppsala, Sweden, 1992. Contbr. articles to profl. jours. Home: Gylfesvej 7B, DK-3060 Espergaerde Denmark Office: Inst Econ HHK, Nansensgade 19, DK-1366 Copenhagen K, Denmark

JENSEN, BRUCE ALAN, control engineer; b. July 6, 1953; s. Carroll jean and Sara Donna (Macy) J.; m. Rhonda Kaye Meeker, Sept. 1, 1974; children: Shondalea Marie, Eric Alan. BS in Chem. Engring., Iowa State U., 1975, MS in Chem Engring., 1977. Process control engr. Applied Automation Inc., Bartlesville, Okla., 1977-85, sr. applications engr., 1985-89, sr. control cons., 1989-92; product mgr. Johnson Yokogawa Corp., Newnan, Ga., 1992-; panel mem. Chem. Engr. Rsch. Panel, 1985. Co-contbr.: Instrument Engineers Handbook, 3d, 1993; contbr. articles to profl. jours. V.p. Bartlesville Bowling Assn., 1991-92; treas. PMVC, Bartlesville, 1989-92. Mem. Instrument Soc. Am. (batch com. mem. 1990-), Am. Inst. Chem. Engrs., Tau Beta Pi, Omega Chi Epsilon, Phi Lambda Upsilon. Achievements include 8 patents in Fractional Distillation Control. Office: Johnson Yokogawa Corp 4 Dart Rd Newnan GA 30265

JENSEN, CHRISTOPHER DOUGLAS, civil engineer; b. Alliance, Nebr., July 22, 1962; s. Roger August and Carolyn Joyce (Cole) J. BS in Civl Engring., U. Wyo., 1984. Profl. civl engr. Nev. 1990, Wyo. 1991, Nebr. 1991. Civil engr. Bennett-Carter and Assocs., Rock Springs, Wyo., 1985-89, G.C. Wallace, Inc., Las Vegas, Nev., 1989-91, Baker and Assocs., Torrington, Wyo., 1991--. Mem. Am. Soc. Civil Engrs., Nat. Soc. Profl. Engrs. (scholarship com. 1988). Republican. Lutheran. Home: 1000 E 17th Ave Torrington WY 82240 Office: Baker and Assocs 215 E 21st Ave Ste 111 Torrington WY 82240

JENSEN, ELWOOD VERNON, biochemist; b. Fargo, N.D., Jan. 13, 1920; s. Eli A. and Vera (Morris) J.; m. Mary Welmoth Collette, June 17, 1941 (dec. Nov. 1982); children: Karen Collette, Thomas Eli; m. Hiltrud Herborg, Dec. 21, 1983. AB, Wittenberg U., 1940, DSc (hon.), 1963; PhD, U. Chgo., 1944; DSc (hon.), Acadia U., 1976, Med. Coll. Ohio, 1991. Mem. faculty U. Chgo., 1947-90, assoc. prof. biochemistry Ben May Inst. Cancer Rsch., 1954-60, prof., 1960-63, Am. Cancer Soc. research prof. physiology, 1963-69, dir. Ben May Inst., 1969-82, dir. Biomed. Ctr. Population Research, 1972-75, prof. physiology, 1969-73, 77-84, prof. biophysics, 1973-84, prof. biochemistry, 1980-90, Charles B. Huggins disting. svc. prof., 1981-90, emeritus prof., 1990—; research dir. Ludwig Inst. for Cancer Research, 1983-87; scholar-in-residence Fogarty Internat. Ctr. NIH, 1988, Cornell U. Med. Coll., 1990-91; prof. Inst. for Hormone and Fertility Rsch. U. Hamburg, Fed. Republic Germany, 1991—; vis. prof. Max-Planck-Inst. für Biochemie, Munich, Germany, 1958; mem. chemotherapy rev. bd. Nat. Cancer Inst., 1960-62, bd. sci. counselors, 1969-72; mem. Nat. Adv. Coun. Child Health and Human Devel., 1976-80; mem. adv. com. biochemistry and chem. carcinogenesis Am. Cancer Soc., 1968-72, coun. for rsch. and clin. investigation, 1974-77; mem. assembly life scis. NRC, 1975-78; mem. com. on sci., engring. and public policy Nat. Acad. Scis., 1981-82; mem. rsch. adv. bd. Clin. Rsch. Inst. of Montreal, 1987—, Klinik für Tumor Biologie, Freiburg, 1993—; cons. Rockefeller U. Hosp., 1990-92. Editorial-adv. bd. Perspectives in Biology and Medicine, 1966—, Archives of Biochemistry and Biophysics, 1979-84; editorial adv. bd. Biochemistry, 1969-72, Life Scis, 1973-78, Breast Cancer Research and Treatment, 1980—; assoc. editor.: Jour. Steroid Biochemistry, 1974—; contbr. articles to profl. jours. Guggenheim fellow, 1946-47; recipient D.R. Edwards medal, 1970, La Madonnina prize, 1973, G.H.A. Clowes award, 1975, Papanicolaou award, 1975, prix Roussel, 1976, Nat. award Am. Cancer Soc., 1976, Amory prize, 1977, Gregory Pincus Meml. award, 1978, Gairdner Found. award, 1979, Lucy Wortham James award, 1980, Charles F. Kettering prize, 1980, Nat. Acad. Clin. Biochemistry award, 1981, Pharmacia award, 1982, Hubert H. Humphrey award, 1983, Rolf Luft medal, 1983, Renzo Grattarola award, 1984, Fred C. Koch award, 1984, Axel Munthe award, 1985, Humboldt Sr. forschungspreis, 1992. Mem. Nat. Acad. Scis. (council 1981-84), Am. Acad. Arts and Scis., Am. Soc. Biol. Chemists, Am. Chem. Soc., Am. Assn. Cancer Research, Endocrine Soc. (pres. 1980-81), AAAS, Am. Gyn/Ob Soc. (hon.). Clubs: Quadrangle, Chicago Literary, Cosmos. Office: Inst Hormone & Fertility Rsch, Grandweg 64, D-22529 Hamburg Germany

JENSEN, GORDON FRED, university administrator; b. Ogden, Utah, Apr. 18; s. George Fred and Verna (Farr) J.; m. Marian Wilkison, July 20, 1956; children: Susan Jensen Kiser, Rebecca Elisabeth Jensen, Allyson Jensen Egbert. BA in Physics, U. Utah, 1945, MSCE, 1956; MBA, Stanford (Calif.) U., 1959. Engr., sales rep. Western Steel, Salt Lake City, 1946-47; chief engr. Cobusco Steel Corp., Salt Lake City, 1947-52, Jensen Constrn. Co., Salt Lake City, 1952-57; dep. program mgr. Stanford Rsch. Inst., SRI Internat., Menlo Park, Calif., 1959-75; dir. Utah Engring. Expt. Sta., U. Utah, Salt Lake City, 1975—; mobile home outlook expert witness Interstate Commerce Dept., Denver, 1975; mem. indsl. energy study Fed. Energy Adminstrn., Washington, 1974; cons. Signope Corp., Chgo., 1984; rsch. collaberation with Tech. U. Gdansk, Poland, 1992—; speaker in field. Editor: Economic Evaluation of Oil Shade and Tar Sands Located in State of Utah, 1984; co-author: The Case for Industrialized Housing, 1973, Manufactured Housing, 1978. Chmn. Ann. Utah Conf. on Global Econs., Energy, Mining and New Tech., U. Utah, 1976-92. With U.S. Maritime Svc., 1945-46. Named Utah Engring. Educator of Yr. Utah Soc. of Profl. Engrs., 1988. Fellow ASCE (life); mem. NSPE, Salt Lake City Rotary Club (chmn. world peace com. 1993), Theta Tau, Phi Kappa Phi. Republican. Mem. LDS. Home: 1362 Embassy Way Salt Lake City UT 84108 Office: Univ Utah Utah Engring Experiment Sta Rm EMRO 104 Salt Lake City UT 84108

JENSEN, HARLAN ELLSWORTH, veterinarian, educator; b. St. Ansgar, Iowa, Oct. 6, 1915; s. Bert and Mattie (Hansen) J.; m. Naomi Louise Geiger, June 7, 1941; children: Kendra Lee Jensen Belfi, Doris Eileen, Richard Harlan. D.V.M., Iowa State U., 1941; Ph.D., U. Mo., 1971. Diplomate: Charter diplomate Am. Coll. Vet. Ophthalmologists (v.p. 1970-72, pres. 1972-73). Vet. practice Galesburg, Ill., 1941-46; small animal internship New Brunswick, N.J., 1946-47; small animal practice Cleve., 1947-58, San Diego, 1958-62, Houston, 1962-67; faculty U. Mo., Columbia, 1967-80; chief opthalmology, prof. Vet. Sch. U. Mo., 1967-80, prof. emeritus Vet. Sch., 1980—, assoc. prof. ophthalmology Med. Sch., 1972-80; cons. in vet. ophthalmology to pharm. firms; guest lectr., prof. opthalmology U. Utrecht (Netherlands) Vet. Sch., 1973; tchr., lectr. various vet. meetings; condr. seminar World Congress Small Animal Medicine and Surgery, 1973, 77. Author: Stereoscopic Atlas of Clinical Ophthalmology of Domestic Animals, 1971, Stereoscopic Atlas of Ophthalmic Surgery of Domestic Animals, 1974; co-author: Stereoscopic Atlas of Soft Tissue Surgery of Small Animals, 1973, Clinical Dermatology of Small Animals, 1974, contbr. articles to profl. jours. Recipient Gaines award AVMA, 1973. Mem. Am. Vet. Radiology Soc. (pres. 1956-57), Am. Vet. Ophthalmology Soc. (pres. 1960-62), Farm House Frat., Sigma Xi, Phi Kappa Phi, Phi Zeta, Gamma Sigma Delta. Baptist. Club: Rotary (pres. Pacific Beach, Calif. 1960-62, pres. Columbia 1977-78). Achievements include invention of instrument for ear trimming in dogs, 1949, breathing apparatus, 1953, designer sound proof animal hosps.; developer 3-D study program for vet. ophthalmology, 1969. Home: 82 Legend Rd Circle Fort Worth TX 76132-1024

JENSEN, MARVIN ELI, retired agricultural engineer; b. Clay County, Minn., Dec. 23, 1926; s. John M. and Inga C. (Haugness) J.; m. Doris A. Lundberg, Sept. 4, 1947; children: Connie, Jeffrey, Eric. BS in Agr., N.D. State U., 1951, MS in Agrl. Engring., 1952, DSc (hon.), 1988; PhD in Civil Engring., Colo. State U., 1965. Instr., asst. prof. N.D. State U., Fargo, 1952-55; agrl. engr. Soil and Water Rsch. div. USDA, Bushland, Tex., 1955-58; head irrigation and drain sect. Soil and Water Rsch. div. USDA, Ft. Collins, Colo., 1959-61; investigation leader Soil and Water Rsch. div. USDA, Ft. Collins and Kimberly, Idaho, 1961-68; dir. Snake River Conservation Rsch. Ctr. Agrl. Rsch. Service USDA, Kimberly, 1969-78; nat. program leader Agrl. Rsch. Service USDA, Ft. Collins and Beltsville, Md., 1979-87; dir. Colo. Inst. for Irrigation Mgmt. Colo. State U., Ft. Collins, 1987-92; ret.; pres. Internat. Commn. Irrigation and Drainage, New Delhi, 1984-87. Editor: (monograph) Design and Operation of Farm Irrigation Systems, 1980; sr. editor: (manual) Evapotranspiration and Irrigation Water Requirements, 1990. Recipient Disting. Service award USDA, 1983, W.E. Morgan Alumni Achievement award Colo. State U., 1990. Fellow Am. Soc. Agrl. Engrs. (tech. v.p. 1983-86, John Deere Gold medal 1982); mem. NAE, AAAS, ASCE (hon., chmn. irrigation and drainage div. 1976-77, Tipton award 1982, Arid Lands Hydraulic Engring. award 1990), Am. Soc. Agronomy, Soil Sci. Soc. Am. Avocations: golf, photography.

JENSEN, MOGENS REIMER, psychologist; b. Copenhagen, July 11, 1949; came to U.S. 1978; s. Reimer and Inger (Larsen) J.; m. Myltreda L. Palazzo, Apr. 4, 1986; children: Shulamit, Krista Barker. BA, Hebrew U., Jerusalem, 1974; MA, Hebrew U., 1977; PhD, Yale U., 1984. Lic. psychologist, Ga. Sr. rsch. assoc. Hadassah Wizo Can. Rsch. Inst., Jerusalem, 1977-87; postdoctoral fellow Yale Bush Ctr. for Social Policy and Child Devel., New Haven, 1984-87; lectr. dept. psychology Yale U., New Haven, 1984-85; assoc. rsch. scientist dept. psychology Yale U., 1984-87; dir. Nat. Ctr. for Mediated Learning, Atlanta, 1987—, Delphi Health & Sci., Atlanta, 1988-93; with Cognitive Edn. Systems, Roswell, Ga., 1993—. With Israel Def. Forces. Mem. AAAS, Internat. Assn. Cognitive Edn., World Fedn. for Mental Health, Am. Psychol. Soc., Am. Psychol. Assn., N.Y. Acad. Sci., Internat. Neural Network Soc., Nat. Assn. Sch. Psychologists, Soc. for Cognitive Rehab., Calif. Assn. for Mediated Learning. Achievements include research in role of psychobiological factors in health and illness, cognitive and knowledge structure development, change models in school psychology and education. Office: Cognitive Edn Systems 11660 Alpharetta Hwy # 200 Roswell GA 30076

JENSEN, MONA DICKSON, chemist researcher; b. Washington, Apr. 8, 1944; d. William Oscar and Louise (Archer) Dickson; m. Thomas Carl Jensen, June 17, 1967; children: Carla Louise, Aaron Raymond. BS, MIT, 1966; PhD, Cornell U., 1973; MBA, Babson Coll., 1983. Sr. scientist Instrumentation Lab., Lexington, Mass., 1972-79; project mgr. Instrumentation Lab., Lexington, 1979-86, mgr. reagent systems applications, 1986-88, sr. R & D mgr., 1988—; adj. faculty mem. W. Alton Jones Cell Sci. Ctr., Lake Placid, N.Y., 1974-76; spl. reviewer In Vitro, 1980-81; proposal reviewer NSF, 1983-86; subcom. advisor Nat. Com. Clin. Lab. Standards, 1991—. Contbg. author: Cell Culture and Its Application, 1977, Practical Tissue Culture Applications, 1979; contbr. articles to profl. jours. Organist, pianist Island Pond Bapt. Ch., Hampstead, N.H., 1985—. Grantee WHO, 1980-81. Mem. DAR, Am. Chem. Soc., Am. Assn. for Clin. Chemistry, Am. Philatelic Soc., Pilgrim Edward Doty Soc., Beta Gamma Sigma. Republican. Methodist. Achievements include 1 patent, 1 patent pending; development of numerous clinical assay reagents in general chemistry, immunochemistry and coagulation, of instrument/reagent analytical systems; pioneering research on environmental control and cell culture. Office: Instrumentation Lab 113 Hartwell Ave Lexington MA 02173-3190

JENSEN, OLE, energy researcher; b. Hjorring, Jutland, Denmark, Aug. 21, 1932; s. Ole Pedersen and Agnes (Olesen) J.; m. Gerda Christensen, July 4, 1959 (dec. Feb. 1974); children: Jesper, Birgitte, Mette Lise, Hans; m. Inger Brygger, July 4, 1987. MSME, Tech. U. Copenhagen, 1958. R & D engr. Sabroe Refrigeration, Aarhus, Denmark, 1960-63; asst. prof. Tech. U., Copenhagen, 1963-70, assoc. prof., 1970-77; mgr. R & D, Energy Rsch. Program, Copenhagen, 1977; mem. energy R & D coun. Danish Ministry Energy, Copenhagen, 1977—, mem. solar R & D Coun., 1988—, mgr. solar R & D Coun.; mem. exec. com. for energy conservation Internat. Energy Agy., 1980—, mem. solar exec. com., 1988—; mem. steering group Air Infiltration and Ventilation Ctr., 1986—. Editor Kulde (Refrigeration), 1962-68; contbr. articles to profl. jours. Lt. Danish Army, 1960-61. Recipient award Nordic Innovations, 1983. Mem. Danish Assn. Civil Engrs., Danish Heating and Ventilation Engrs., Internat. Bldg. Coun. Achievements include research on energy conservation in buildings, termodynamics of compressors, solar energy in buildings, energy research and development programs and strategies. Home: Haspegaardsveg 81 A, DK2880 Bagsvard Denmark Office: Danish Bldg Rsch Inst, PO Box 119, 2970 Hoersholm Denmark

JENSEN, PAUL ALLEN, electrical engineer; b. Chgo., Aug. 27, 1936. BS, U. Ill., 1959; MS, U. Pitts., 1963; PhD in Elec. Engring., Johns Hopkins U., 1967. Engr. surface div. Westinghouse Electric Corp., 1959-63, from asst. prof. to assoc. prof., 1967-73; prof. elec. engring. ops. research U. Tex., Austin, 1973—, Hughes Tool Co. Centennial prof. mech. engring., 1987—. Mem. Ops. Research Soc. Am., Inst. Mgmt. Sci. Office: The Univ of Tex at Austin Dept of Mech Engring Austin TX 78712

JENSEN, SOREN STISTRUP, mathematics educator; b. Aalborg, Denmark, Jan. 12, 1956; came to U.S., 1981; s. Johannes and Lilly (Christensen) J.; m. Pauline Quek Hwang, Jan. 5, 1985; children: Sine, Elizabeth, Natasha. Cand. Scient. in Math., Physics, Aalborg U., 1980; PhD in Math., U. Md., 1985. Rsch. fellow Aalborg (Denmark) U., 1980-83; rsch. asst., then faculty rsch. asst. Inst. Phys. Sci. and Tech., U. Md., College Park, 1982-85; asst. prof. math. U. Md., Balt., 1985-92, assoc. prof. math., 1992—; vis. asst. prof. Rutgers U., New Brunswick, N.J., 1991. Contbr. articles to profl. jours. Grantee, Fulbright Found., 1981, Office Naval Rsch., 1987—; award recipient Office Sci. Rsch., USAF, 1988, NSF, 1990. Mem. Am. Math. Soc., Soc. for Indsl. and Applied Math. (referee Jour. Numerical Analysis 1988—, NSF 1989, Jour. Computational Physics, Numerische Math. 1990—, comms. on Pure and Applied Math. 1991, Math. and Computer Modelling 1991, Applied Numeric Math 1993). Home: 11818 Snow Patch Way Columbia MD 21044-4414 Office: U Md Dept Math Baltimore MD 21228

JENSEN, UFFE STEINER, nuclear engineer; b. Vinkel, Viborg, Denmark, Sept. 12, 1948; s. Viggo and Jenny Steiner (Sørensen) J.; m. Ellen Jensen; June 9, 1973; children: Marie Steiner Jensen, Anne Steiner Jensen. MS, Tech. U., 1975. Nuclear engr. Elsam, Fredericia, Denmark, 1975; nuclear engr. Elsam, Fredericia, 1978-89, dep. head of planning dept., 1989—; cons. NKI-skolen, København, Denmark, 1975-77; researcher Rsch. Establishment Risø, Roskilde, Denmark, 1977-78; Denmark rep. The Utility Group, Western Europe, 1979—; cons. Nordic Project Export Fund, 1991; mem. NORDEL -devel. project, The Nordic Countries, 1991-92. Chmn. Socialdem. Party, Fredericia, 1982-90; city coun. mem. Fredericia Byråd, 1990—. Mem. Dansk Ingeniør Forening, Dansk Kerneteknisk Selskabb, Am. Chem. Soc. (div. nuclear chemistry and tech.). Home: Lunddalvej 26, DK 7000 Fredericia Denmark Office: Elsam, DK 7000 Fredericia Denmark

JENTZ, JOHN MACDONALD, engineer, travel executive; b. N.Y.C., Dec. 28, 1928; s. John Hellmer and Christine (Macdonald) J. AB cum laude, Harvard U., 1952, postgrad., 1952-53; MS in Bldg. Engring. and Constrn., MIT, 1956. Instr. in civil engring. MIT, Cambridge, Mass., 1956-57; prin. owner, co-founder, pres. Spring Brook Ctr. Inc., Wellfleet, Mass., 1957—; owner Don Jentz Enterprises, Wellfleet, 1961—; prin. owner, co-founder, pres. Kauai Mountain Tours, Lihue, Hawaii, 1980—. Sr. lt. Wellfleet Fire Dept. and Rescue Squad, 1967—; candidate for selectman, Wellfleet. Recipient Brotherhood award NAACP, Atlanta, 1980. Mem. Cape Cod Foresters and Firefighters Assn. Inc., Cen. Mass. Police Assn., MIT Club Cape Cod (co-founder 1976). Avocations: hiking, swimming, weight lifting, classic cars. Home: Dillingsgate Rd PO Box 900 South Wellfleet MA 02663-0900 Office: Spring Brook Ctr Inc Off Rte 6 PO Box 900 Wellfleet MA 02663-0900

JEPPESEN, C. LARRY, lighting company executive; b. Brigham City, Utah, May 5, 1939; s. Charles Blair and Roma Delila (Peterson) J.; m. Londa Lee Morton, Feb. 17, 1962; children: Sean, Heidi, Tamara, Marni, Christian, Nathan, Brenda. Grad., Bonners Ferry High Sch., 1957. Mgr. ops. Olson Farms, North Hollywood, Calif., 1962-70; pres., CEO Dal-Worth/Olson Egg Farms, Keller, Tex., 1970-83, Am. Egg Co., West Chicago, Ill., 1985-86, Chicago-Edison Corp., West Chicago, Ill., 1986—; chmn. bd. Egg Clearinghouse, Inc., Durham, N.H., 1978-80; cons. to egg industry, 1983-85. Mem. Assn. Energy Engrs. Office: Chicago-Edison Corp 29W034 Colford Ave West Chicago IL 60185

JERGE, DALE ROBERT, loss control specialist, industrial hygienist; b. Buffalo, Oct. 15, 1951; s. Herbert L. and Ruth R. (Maxson) J.; m. Susan B. Rinaldo, Jan. 22, 1983; 1 child, Nicholas D. AAS, Erie Community Coll., Amherst, N.Y., 1972; BA in Sociology, SUNY, Buffalo, 1974, MS in Social Scis., 1976, postgrad., 1988. Cert. occupational health and safety technologist; cert. hazardous materials supr.; cert. indsl. pulmonary technologist; cert. safety and security dir.; cert. occupational hearing conservationist. Loss control specialist Twin Fair, Buffalo, 1975-79; indsl. hygienist Continental Ins. Tech. Svcs., Buffalo, 1979—; adj. prof. Niagara County Community Coll., Lockport, N.Y., 1989—. Mem. APHA, Am. Soc. Safety Engrs., Am. Insl. Hygiene Assn., World Safety Orgn. (affiliate mem., cert. safety and security profl., cert. hazardous materials supr.), Coun. for Accreditation in Occupational Hearing Conservation (cert.). Office: CTEK 50 Lakefront Blvd Buffalo NY 14202-4301

JERNAZIAN, LEVON NOUBAR, psychologist; b. Yerevan, Armenia, Jan. 13, 1958; came to U.S. 1990; s. Nubar S. and Luisa (Tatlian) J.; m. Irina Kanayan, Oct. 16, 1982; children: Hayk, David. PhD, Inst. of Psychology, Tbilisi, Ga., 1984. Lic. psychologist, Calif. Prof. Inst. Edn., Yerevan, 1980-82; head dept. psychology Sci. Method's Ctr., Yerevan, 1982-87; head psychol. ctr. Exp-al Ednl. Network, Yerevan, 1987-89; psychologist Canyon Found., Toluca Lake, Calif., 1990—; psychology instr. Glendale (Calif.) Community Coll., 1991—; pvt. practice psychology Glendale, 1992—. Author: Psychological Roots of Compassion, 1987; contbr. numerous articles to profl. jours. Chmn. conflict group City of Glendale Pub. Rels. Coun., 1991—; psychol. cons. Armenian Nat. Dem. Movement, 1988-90. Mem. APA, L.A. County Psychol. Assn. Achievements include founding of a new branch of Sovietological studies - Psychosovietology. Office: 230 N Maryland Ste 311 Glendale CA 91206

JERNDAL, JENS, holistic medicine educator, health center promoter; b. Goteborg, Sweden, Jan. 5, 1934; came to Spain, 1968; s. Ebbe and Ingrid M. (Forsberg) J.; children: C. Patrick, J.O. Mathias, J.T. Christofer. MS, Stockholm U., 1958; BA, Uppsala U., 1959; Diploma, Internat. Coll. Acupuncture, Colombo, Sri Lanka, 1982; MD, 1987; DSc. honoris causa, U. Complementary Medicines, Colombo, Sri Lanka, 1988. Attaché Royal Swedish Ministry Fgn. Affairs, 1960-62; embassy sec. Royal Swedish Embassy, Copenhagen, 1962; 1st sec. Royal Swedish Embassy, Karachi, Pakistan, 1964, Royal Swedish Ministry Fgn. Affairs, 1965-68; investment broker Real Lanzarote SA, Las Palmas, Spain, 1968-79; founder, pres. Dragon's Head Centre of Holistic Medicine, Lanzarote, Spain, 1983—; expert del. UN High Commr. for Refugees, Geneva, 1966-67; pres. Cosmosphical Found., Stockholm, 1977-88; lectr. in astrology and alternative medicine; vis. prof., internat. coord. The Open Internat. U. for Complementary Medicine, Colombo, Sri Lanka, 1988, prof. holistic medicine, 1991. Author: Indonesien, 1958; contbr. articles to profl. jours. Fgn. lang. transmission mgr., broadcaster for Radio Sweden, 1956-57; rep. Assn. Swedish Citizens Residing Abroad, Canary Islands, 1972-75. Decorated Knight of Royal Order of Dannebrog His Majesty the King of Denmark, Knight of Sovereign Order of St. John of Jerusalem (Knights of Malta), Knight Commdr. of Justice of Sovereign Order of St. John of Jerusalem (Knights of Malta), Knight Grand Cross Ordre Souverain et Militaire de la Milice du Saint Sepulcre, Knight Humanity Sovereign World Order of the White Cross, 1991; recipient Albert Schweizer Prize for Medicine, 1990. Mem. Astrol. Assn. Britain, Am. Fedn. Astrologers, Acupuncture Found. Sri Lanka, Medicina Alternativa (life, vis. lectr.), Commonwealth Inst. Acupuncture and Natural Medicines (founding), Sci. and Med. Network, Inst. Dirs. (London). Office: Dragon's Head Centre, apartado 248, E-35500 Arrecife Lanzarote, Spain

JERNE, NIELS KAJ, scientist; b. London, England, Dec. 23, 1911; s. Hans Jessen and Else Marie (Lindberg) J.; m. Ursula Alexandra Kohl, 1964; 2 children. Grad., U. Leiden, U. Copenhagen; hon. degrees, U. Chgo., U. Copenhagen, U. Basel, U. Rotterdam, Columbia U. Research worker Danish State Serum Inst., Copenhagen, 1943-55; chief med. officer WHO, Geneva, 1956-62; prof. biophysics U. Geneva, 1960-62; chmn. dept. microbiology U. Pitts., 1962-66; prof. exptl. therapy J.W. Goethe U., Frankfurt, 1966-69; dir. Basel Inst. for Immunology, 1969-80; prof. Inst. Pasteur, Paris, 1981-82; bd. dirs. Paul Ehrlich Inst. Contbr. articles to profl. jours. Del. Nobel prize for medicine, 1984. Recipient Marcel Benoist prize, Berne, 1979, Paul Ehrlich prize, Frankfurt, 1982. Fellow Royal Soc. London; mem. NAS, Danish Royal Soc. Scis. Copenhagen, Am. Philos. Soc., Acad. des Sciences de l'Inst. de France, Am. Acad. Arts and Scis., Croatian Acad. Scis. and Arts Zagreb. Home: Chateau de Bellevue, F-30210 Castillon-du-Gard France

JERNIGAN, JOHN MILTON, chemist, chemical engineer; b. Troy, Ala., Aug. 27, 1917; s. Joseph Edward and Emma Rosa (Cooper) J.; m. Josephine Miller, Oct. 11, 1941; children: Mary Jo Jernigan Miller, Kay Elaine Jernigan Helton. BSc in Chemistry and Metall. Ceramics, U. Ala., Tuscaloosa, 1939, MSc in Chemistry, 1951. Chemist Ala. Geol. Survey, Tuscaloosa, 1936-39; supr. Engring. Sci. Mgmt. Def. Tng. Program, Tennessee Valley, 1941; supr. engr. Union Carbide Chem. Corp., W.Va., 1946-49; cons. Gorgas underground gasification of coal USBM and Ala. Power Co., 1949-50; tech. svc. dir. Reichold Tuscaloosa, 1950-55; pres., chief exec. officer So. Pine Chems., Inc., Tuscaloosa, 1955-65; asst. state chemist Ga. Dept. Agriculture, Atlanta, 1965-73; chemist Jacksonville (Fla.) Electric Authority, 1973-83; presenter profl. meetings and confs. Co-author: Water Program and Corrosion Control for Fossil Fueled Utilities, 1980. V.p. Ala. Congress Parents and Tchrs., 1963-64; chmn. Interface Excellence, Tampa, Fla., 1982; elder PC U.S.A. Capt. CWS, AUS, 1941-46, ETO, CBI, PTO. Recipient placque for svc. Am. Foundrymans Soc., Birmingham, 1965; scholar Inst. Paper Chemistry, Appleton, Wis., 1939-40. Fellow Am. Inst. Chemists (cert. chemist and chem. engr. Nat. Certification Commn. in Chemistry and Chem. Engring.); mem. Am. Inst. Chem. Engrs., Assn. Food and Drug Ofcls. of So. States (pres. 1972), Masons, Elks (plaque for outstanding svc. as chmn. Ala. Elks Found.). Achievements include patent (with others) for process for separation of pulping black liquors into LI8NIN and recovered pulping chemicals. Home: 4973 Red Pine Ct Jacksonville FL 32210-7913

JERNIGAN, ROBERT WAYNE, statistics educator; b. Jacksonville, fla., Feb. 4, 1951; s. Belton Karl and Ruth (Warren) J.; m. Rose Marie Receveur, Aug. 4, 1973; children: Nicholas, Laura. PhD, U. South Fla., 1978. Asst. prof. stats. Am. U., Washington, 1978-82, assoc. prof. stats., 1982-86, prof. stats., 1986—, chair dept. math. and stats., 1991—; sr. statistician U.S. EPA, Washington, 1984-91. Contbr. monograph, articles to profl. jours. Fellow Wash. Acad. Scis. (sci. achievement award for math. and computer sci. 1986); mem. AAAS, Soc. for Study of Evolution, Am. Statis. Assn., Math. Assn. Am., Inst. Math. Stats., Sigma Xi, Phi Kappa Phi, Pi Mu Epsilon. Office: Am U Dept Math & Stats 4400 Massachusetts Ave NW Washington DC 20016

JEROME, WALTER GRAY, cell biologist; b. Alexandria, Va., Mar. 3, 1949; s. Walter Gray and Nell (Williams) J.; divorced; 1 child, Heather. BA, St. Andrews U., 1971; PhD, U. Va., 1981. Asst. prof. pathology Bowman Gray Sch. Medicine Wake Forest U., Winston Salem, N.C., 1986—; mem. genetics com. Internat. Cat Assn., Harlingen, Tex., 1987—. Contbr. articles to profl. jours. Mem. rsch. subcom. N.C. affiliate Am. Heart Assn., Chapel Hill, 1992; mem. parent adv. coun. High Point (N.C.) U., 1992—; vol. Forsythe County Sch. Community Resource, Winston-Salem, 1990—. Grantee Am. Heart Assn., 1990-93. Mem. Am. Soc. Cell Biology, Microscopy Soc. Am. (edn. com. 1992—), Appalachian Region EM Soc. (pres. 1991-92), Am. Heart Assn. Coun. on Arteriosclerosis, Sigma Xi (chpt. pres. 1990-91). Office: Bowman Gray Sch Medicine Wake Forest U Medical Center Blvd Winston Salem NC 27157

JERRITTS, STEPHEN G., computer company executive; b. New Brunswick, N.J., Sept. 14, 1925; s. Steve and Anna (Kovacs) J.; m. Audrey Virginia Smith, June 1948; children: Marsha Carol, Robert Stephen, Linda Ann; m. 2d, Ewa Elizabet Rydell-Vejlens, Nov. 5, 1966; 1 son, Carl Stephen. Student, Union Coll., 1943-44; B.M.E., Rensselaer Poly. Inst., 1947, M.S. Mgmt., 1948. With IBM, various locations, 1949-58, IBM World Trade, N.Y.C., 1958-67, Bull Gen. Electric div. Gen. Electric, France, 1967-70, merged into Honeywell Bull, 1970-74; v.p., mng. dir. Honeywell Info. Systems Ltd., London, 1974-76; group v.p. Honeywell U.S. Info. Systems, Boston, 1977-80; pres., chief operating officer Honeywell Info. Systems, 1980-82, also bd. dirs.; pres., chief exec. officer Lee Data Corp., 1983-85; with Storage Tech. Corp., 1985-88, pres., chief operating officer, 1985-87, also bd. dirs., vice-chmn. bd. dirs., 1988; pres., chief exec. officer NBI Corp., 1988-92, also bd. dirs.; bd. dirs. Scully Signal Co., Wang Corp.; cons. crisis mgmt. corp. reorganization and internat. bus., 1992—. Bd. dirs. Guthrie Theatre, 1980-83, Charles Babbage Inst., 1980-90, Minn. Orch., 1980-85; trustee Rensselaer Poly. Inst., 1980-85. Served with USNR, 1943-46. Mem. Computer Bus. Equipment Mfrs. (dir. exec. com. 1979-82), Assoc. Industries Mass. (dir. 1978-80). Home and Office: 650 College Ave Boulder CO 80302-7136

JESNESS, BRADLEY L., psychology educator, testing and professional selection consultant; b. Hastings, Nebr., Nov. 18, 1953; s. Robert F. Jesness and Mary Ann (Lindfors) Kjenaas; m. Reneé Cooke, Aug. 25, 1978. BA, Grinnell Coll., 1975; MA, U. Iowa, 1985. Psychology instr. U. N.D., Williston, 1985-86, Minn. Community Coll. System, Mpls., 1986-87, Augsburg Coll., Mpls., 1987-88; mental health counselor Familystyle Homes, St. Paul, 1990-91; psychology instr. Minn. Community Coll. System, Mpls., 1991—; testing cons. U. S.D., Vermillion, 1990—, Mpls., 1990—. Contbr. articles to profl. jours. Mem. Internat. Soc. for Human Ethology, Am. Psychol. Soc. Achievements include foundation of cognitive developmental ethology. Home and Office: 3513 Dupont Ave S #412 Minneapolis MN 55408

JETER, WAYBURN STEWART, retired microbiology educator, microbiologist; b. Cooper, Tex., Feb. 16, 1926; s. Joseph Plato and Beulah (Stewart) J.; m. Margaret Ann McDonald, May 30, 1947; children—Randall Mark, Monette Ann, Marcus Kent. B.S., U. Okla., 1948, M.S., 1949; Ph.D., U. Wis., 1950. Diplomate: Am. Bd. Microbiology. Mem. faculty U. Iowa, 1950-63, assoc. prof., 1958-63; prof. microbiology U. Ariz., Tucson, 1963-89, prof. microbiology emeritus, 1989—, prof. pharmacology and toxicology, 1983-91, prof. pharmacology and toxicology emeritus, 1991—, head dept. microbiology and med. tech., 1967-83, dir. lab. cellular immunology, 1976-91; dir. med. tech. program U. Ariz., 1976-79; vis. prof. immunology and med. microbiology U. Fla., 1980; pres. Scientific Rels. Svcs., Inc., 1988—. Contbr. articles profl. jours. Served with USNR, 1943-46. Fellow AAAS; mem. Am. Acad. Microbiology, Am. Assn. Immunologists, Ariz. Acad. Sci., Am. Soc. Microbiology (mem. council 1975-77), Soc. Exptl. Biology and Medicine, Sigma Xi. Democrat. Presbyterian. Home: 5140 N Via Sempreverde Tucson AZ 85715-5966

JETHWANI, MOHAN, civil engineer; b. Darbello, Sind, India, Mar. 20, 1938; came to U.S., 1962; s. Sitaldas H. Jethwani and Kalawanti (Assudi) Khushalani; m. Joyce Marie Conen, Dec. 19, 1971; 1 child, Monique Marie. BS and MS, Kans. State U., 1964. Lic. profl. engr., N.Y. Constrn. insp. Bombay (India) Santacruz Airport; engr., adminstrv. engr. City of N.Y.; assoc. dep. commr. N.Y.C. Dept. Environ. Protection, 1993—; agy. rep. infrastructure coun. Cooper Union, N.Y.C., 1992—. Contbr. articles to profl. jours. Mem. Community Bd. 11 Q, N.Y.C., 1982—, value engr., 1988-91; dep. commr. N.Y.C. Dept. Parks and Recreation, 1976-77, asst. commr., 1977-88; mem. selection com. Congl. Dist. Mil. Acad. Recruitment, 1988-92. Recipient Dedicated Pub. Svc. award 100 Yr. Assn. Fellow Am. Mgmt. Assn.; mem. Am. Soc. Mcpl. Engrs., Water Environ. Fedn., Am. Water Works Assn., Mcpl. Engrs. (bd. dirs.), Soc. Indo-Am. Engrs. and Architects (bd. dirs.), Bayside Hist. Soc. Hindu. Home: 36-23 216 St Bayside NY 11361 Office: Dept Environ Protection 59-17 Junction Blvd Elmhurst NY 11361

JEUNG, IN-SEUCK, aerospace engineering educator; b. KyungNam, Republic of Korea, Dec. 13, 1952; s. YongHwa and DooNam (Cho) J.; m. Hunjoo Ha, Aug. 21, 1982; children: Audrey Sung-A, Zi-Ung. MS, Seoul Nat. U., Seoul, Republic of Korea, 1977, PhD in Aero. Engring., 1982. Vis. scientist Hosei U., Tokyo, 1977-78; hon. fellow U. Minn., Mpls., 1982-84; asst. prof. Seoul Nat. U., 1984-88, chmn. dept. aerospace engring., 1990-92, assoc. prof. dept. aero. engring., 1988—, dir. machine shop, 1992—. Contbr. articles to profl. jours. Grantee Ministry Fgn. Affairs Japan, Japan Internat. Cooperation Agy., 1977; postdoctoral fellow Korea Sci. and Engring. Found., 1983. Mem. AIAA, Combustion Inst., Korea Soc. Aero. and Space Scis., Korean Soc. Mech. Engrs. Achievements include research in laser diagnostics application on the fluid dynamics of the transiently propagating flames, on numerical computations for the transiently propagating flames. Home: Rex Apt 11-1202 Eachon-Dong, Yongsan Ku, Seoul 140-030, Republic of Korea Office: Dept Aero Engring, Seoul Nat U, Seoul 151-742, Republic of Korea

JEWELEWICZ, RAPHAEL, obstetrician/gynecologist, educator; b. Nowogrodek, Poland, Dec. 26, 1932; came to U.S., 1963; s. Chaim and Chaia (Tawricki) J.; m. Ronnie Oved, July 3, 1955; children: Rachel, Dov, Daniel, Dory. MD, Hebrew U., Jerusalem, 1961. Cert. Am. Bd. Ob-gyn. 1971, 89, reproductive endocrinology, 1973. Intern Hadassah Hebrew U. Hosp., Jerusalem; resident NYU Med. Ctr., Bellevue Hosp., N.Y.C.; assoc. prof. ob-gyn. Columbia U., N.Y.C., 1975—; bd. dirs. div. reproductive endocrinology Columbia U. Coll. Physicians and Surgeons, N.Y.C. Author: Clinical Aspects of Cervical Incompetence, 1989, The Menstrual Cycle: Physiology, Reproductive Disorders and Infertility, 1992; editor ob-gyn. investigation; mem. editorial bd. several sci. jours.; contbr. over 100 articles to profl. jours. Mem. Am. Coll. Ob-gyn., Am. Coll. Surgeons, Am. Fertility Soc., Am. Gynecol. & Obstet. Soc., N.Y. Obstet. Soc., N.Y. Gynecol. Soc. (sec. 1992—), Soc. for Gynecol. Investigation, Soc. for Study Reproduction. Jewish. Avocations: opera, ballet, theater, travel. Home: Church St Alpine NJ 07620 Office: Columbia Presbyn Med Ctr 630 W 168th St New York NY 10032

JEWELL, THOMAS KEITH, civil engineering educator; b. Rochester, N.Y., Feb. 18, 1946; s. Kenneth Saxon Jewell and Cecilia (Snow) Sherman; m. Pamela Bowen, June 8, 1968; children: James, Keith, John. MS in Envrion. Engring., U. Mass., 1975, PhD in Civil Engring., 1980. Registered profl. engr., N.Y. Instr. dept. civil engring. Union Coll., Schenectady, 1978-80, asst. prof., 1980-84, assoc. prof., 1984-90, prof., 1990-91, Carl B. Jansen prof., 1991—, dept. chmn., 1986-90. Author: A Systems Approach to Civil Engineering Planning and Design, 1986, Computer Applications for Engineers, 1991. Capt. U.S. Army, 1968-73. Mem. ASCE (Wesley W. Horner award 1979), Am. Soc. Engring. Edn. (campus activities coord. 1978—), Tau Beta Pi, Chi Epsilon. Presbyterian.

JEWELL, WILLIAM SYLVESTER, engineering educator; b. Detroit, July 2, 1932; s. Loyd Vernon and Marion (Sylvester) J.; m. Elizabeth Gordon Wilson, July 7, 1956; children—Sarah, Thomas, Miriam, William Timothy. B.Engring. Physics, Cornell U., 1954; M.S. in Elec. Engring, MIT, 1955, Sc.D., 1958. Assoc. dir. mgmt. scis. div. Broadview Research Corp., Burlingame, Calif., 1958-60; asst. prof. dept. indsl. engring. and operations research U. Calif.-Berkeley, 1960-63, assoc. prof., 1963-67, prof., 1967—, chmn. dept., 1967-69, 76-80; dir. O.R. Ctr., U. Calif., 1985-87, Engring. Systems Rsch. Ctr. U. Calif., 1987-88, 91-92; dir. Teknekron Industries, Inc., Incline Village, Nev., 1968-86, Creance Capital, Inc., Oakland, Calif., 1993—; cons. ops. rsch. problems, 1960—; guest prof. Eidgenössiches Technische Hochschule, Zurich, 1980-81. Contbr. articles to profl. jours. Trustee New Coll., Berkeley, 1992—. Recipient Halmstead prize, 1982; Fulbright research scholar France, 1965; research scholar Internat. Inst. Applied Systems Analysis, Austria, 1974-75. Mem. Ops. Rsch. Soc. Am., Inst. Mgmt. Scis., Assn. Swiss Actuaries, Internat. Actuarial Assn., Rotary Internat., Mensa, Triangle, Sigma Xi. Home: 67 Loma Vista Dr Orinda CA 94563-2236 Office: U Calif Dept Indsl Engring and Ops Research Berkeley CA 94720

JEWETT, DAVID STUART, federal agency administrator; b. Passaic, N.J., May 2, 1941; s. William Raymond and Jane Elizabeth (Stuart) J.; m. Kathryn Ann Jensen, Mar, 5, 1942; children: Carl, Mark, Eric. BA in Eng., Princeton U., 1963. Nat. security analyst Office Mgmt. and Budget, Washington, 1969-75; sr. program analyst energy research and devel. U.S. Dept. Energy, Washington, 1975-79, budget officer Energy Tech. div., 1977-79, dir. resource mgmt. Fossil Energy div., 1979-85, assoc. dep. asst. sec. Fossil Energy div., 1985—. Chief exec. officer Childbirth Edn. Assn. D.C., 1978-79. Served to lt. cmdr. USN, 1963-69, Vietnam. Decorated Air medal (5). Office: US Dept Energy Coal Technology 1000 Independence Ave SW Washington DC 20585

JEWETT, STEPHEN CARL, marine biologist, researcher, consultant; b. Dexter, Maine, Dec. 31, 1947; s. Stanley Horace Jr. and Freda (Kennedy) J.; m. Shirley Kay Crock, Sept. 6, 1975; children: Stephanie Kristen, Jeffrey Scott. BA, John Brown U., Siloam Springs, Ark., 1971; MS, U. Alaska, 1977, postgrad., 1992—. Fishery biology Alaska Dept. Fish and Game, Kodiak, 1973-74; rsch. asst. U. Alaska, Fairbanks, 1972-73, rsch. assoc. Inst. Marine Sci., 1974—, coord. sci. diving, 1989—; pres. Marine Rsch., cons., Fairbanks, 1982—. Contbg. author: The Gulf of Alaska: Physical Environment, 1987, Environmental Studies in Port Valdez, Alaska, 1988; also articles. Bd. dirs. Totem Park Ch., Fairbanks, 1992—, Young Life, Fairbanks, 1992—. With USN, 1970-71. Mem. Am. Fisheries Soc. (cert.), Am. Acad. Underwater Scis., Nat. Shellfisheries Soc., Western Soc. Naturalists, Sigma Xi. Achievements include research on stable carbon isotopes on the arctic shelf, feeding biology of demersal fishes. Office: Inst Marine Sci U Alaska Fairbanks AK 99775-1080

JEWSBURY, ROGER ALAN, chemistry educator; b. Poole, Dorset, Eng., Feb. 20, 1947; s. Alan and Joan (Duke) J.; m. Maria Eva Leopando, Apr. 14, 1982; children: Mark Oliver, Anna Maria. BSc with 1st class honors, U. Bristol, Eng., 1968, PhD, 1971. Rsch. chemist Courtaulds Ltd., Coventry, Eng., 1971-72; lectr. Science U. of Malaysia, Penang, 1972-77; rsch. assoc. U. Chgo., 1976-79; sr. lectr. U. Huddersfield, Eng., 1979—; sr. tutor/tutor Open U. Eng., 1983—; cons. NCUK, Malaysia, 1988, 91; external project examiner IUT, Lannion, France, 1991—; vis. prof. U. Aix-Marseille, France, 1992—; vis. specialist Brit. Coun. Yugoslavia, 1984. Contbr. articles to profl. jours. Fellow Royal Soc. Chemistry. Office: U Huddersfield, Dept Chemistry, Huddersfield Yorkshire HD1 3DH, England

JEYARAMAN, RAMASUBBU, chemist, educator; b. Madurai, Tamil Nadu, India, Nov. 2, 1944; s. P. Ramasubbu and R. Renganayagi Jeyaraman; m. J. Hemavathy, Feb. 7, 1973; children; Ramkumar, Ravikumar, Latha. BS, Madras U., 1966; MS, Madurai U., 1968; PhD, Annamalai U., 1975. Asst. prof. Annamalai U., Annamalainagar, India, 1973-77, Am. Coll., Madurai, 1977-86; rsch. assoc. Clemson U., U.S.A., 1981-82, U. Mo., St. Louis, 1983-86; prof. chemistry Bharathidasan U., Tiruchirapalli, India, 1986—, head computer ctr., 1987—; mem. acad. coun., R&D com. Am. Coll., 1977-81; mem. ctr. mgmt. com. Sophisticated Instruments Facility, Indian Inst. Sci., Bangalore,, 1990—. Contbr. articles to profl. jours. and chpts. to books. Active Social Svc. League, Madurai, 1963-67. Mem. Am. Chem. Soc. Hindu. Achievements include development of method of isolation of dioxiranes and establishment of role in carcinogenesis mechanisms, synthesis of several novel strained azabicyclic systems; research in their conformations and reactivity. Home: No 2 Sastha St Iyappa Nagar, 620 021 Tiruchirapalli India Office: Bharathidasan U, Palkalaiperur, 620 024 Tiruchirapalli India

JEYENDRAN, RAJASINGAM SIVAPERAGASAM, andrologist, researcher; b. Jaffna, Sri Lanka, Aug. 15, 1948; came to U.S., 1974; child of Maurice M. and Jegathabal (Rajadururi) S.; m. Sivagandhi; 1 child, Krithika. BVSc (DVM) with honors, Veterinary Sch., Sri Lanka, 1972; PhD, U. Minn., 1978. Dir. andrology Inst. Reproductive Medicine, Chgo., 1979-89; assoc. prof. Northwestern U. Med. Sch., Chgo., 1989—, dir. andrology, 1989—. Contbr. to numerous book chpts., publications. Achievements include research emphasis on male infertility, semen analysis, sperm processing and cryopreservation. Home: 1845 Golden Pond Ln Wheaton IL 60187 Office: Northwestern U Med Sch 333 E Superior St Chicago IL 60611

JEZEK, KENNETH CHARLES, geophysicist, educator, researcher; b. Chgo., May 17, 1951; s. Rudolph and June J.; m. Rosanne M. Graziano, Jan. 27, 1984. BSc in Physics with honors, U. Ill., 1973; MSc in Geophysics, U. Wis., 1977, PhD in Geophysics, 1980. Observer Bartol Rsch. Found. Cosmic Ray Lab., McMurdo Sta., Antarctica, 1973-74; postdoctoral fellow Inst. Polar Studies Ohio State U., Columbus, 1980-81; project assoc. Geophysical and Polar Rsch. Ctr. U. Wis., 1981-83; geophysicist U.S. Army Cold Regions Rsch. and Engring. Lab., Hanover, N.H., 1983-85, 87-89; mgr. polar oceans and ice sheets program NASA, Washington, 1985-87; rsch. asst. prof. Thayer Sch. Engring. Dartmouth Coll., Hanover, 1987-89; assoc. prof. geology, dir. Byrd Polar Rsch. Ctr. Ohio State U., 1989—; prin. investigator Greenland, 1982, 85, Greenland Sea, 1988, Greenland Ice Sheet, 1991, 92; geophysicist Ross Ice Shelf, Antarctica, 1974-75, Devon Island Ice Cap, 1975, Camp Century Greenland, 1977, Southern Greenland Ice Sheet, 1981, East Antarctica, 1981-82; field leader Ross Ice Shelf, 1976-77, Dome C East Antarctica, 1978-79; cons. Polar Ice Coring Office, Greenland, 1983; mem. ad hoc com. remote sensing polar regions Nat. Rsch. Coun., 1985-89, geophysical data com., 1987—, glaciology com. polar rsch. bd., 1988—, glaciology rep. SCAR, 1990—, earth studies com. NAS, 1991—, mem. NAS sci. panel review NASA earth obs. system data info. system, 1992—, mem. NAS sci. panel review nat. space sci. data ctr., 1992—; mem. numerous NASA coms.; mem. Environ. Task Force, 1992; lab. coord. Sea Ice Electromagnetic Accelerated Rsch. Initiative Office Naval Rsch., 1992. Assoc. editor Jour. Geophysical Rsch., 1991—; contbr. articles, abstracts to profl. jours. NSF grantee, 1982-83, 83-84, Office Naval Rsch. grantee, 1984-89, CRREL grantee, 1985-86, NASA grantee 1985-87, 87-89, 88-92, 90, 91-93, ONR grantee, 1992, others. Mem. Am. Geophysical Union (chmn. snow, ice and permafrost com. 1992—), Soc. Exploration Geophysicists, Internat. Glaciological Soc. (coun. 1991—), Sigma Xi. Office: Ohio State U Byrd Polar Rsch Ctr 1090 Carmack Rd Columbus OH 43210

JHA, NAND KISHORE, engineering educator, researcher; b. Banerchua, Bihar, India, Aug. 1, 1941; came to U.S., 1981; s. Nageshwar P. and Putli Jha; m. Munna Jha, June 1, 1960; children: Shivdutt, Bhuvdutt, Yagyadutt, Prabhudutt. BSME, B.I.T. Ranchi, Ranchi, India, 1964; MSME, I.I.T., Delhi, India, 1975, PhD, 1977. Lectr. Bhagalpur (India) Coll. Engring., 1964-66; asst. prof. B.I.T. Sindri, India, 1966-77, I.T.T., Delhi, 1977-80; rsch. assoc. Nat. Acad. Scis., Md., 1981; prof. Manhattan Coll., N.Y.C., 1981—; speaker at numerous conferences in field. Author: Computer Aided Design, 1992; editor: Flexible Manufacturing, 1991; contbr. over 60 articles to profl. jours. Recipient Merit award Manhattan Coll., 1991. Mem. ASME (chmn. CIM com. 1988—, organizer, chmn. conf. 1985, 86, 87, 88), Indian Soc. Mech. Engrs., ORSA. Home: 761 East 237 St Bronx NY 10466 Office: Manhattan Coll 4140 Corlear Ave Riverdale NY 10471

JI, XINHUA, physical chemist, educator; b. Tianjin, China, Apr. 14, 1948; came to U.S. 1985; s. Mobo and Shuqing (Liu) Ji; m. Xiurong Dong, Apr. 25, 1975. BS, Tchrs. Coll., Tongliao, China, 1982; PhD, U. Okla., 1990. Tchr. chemistry No. 8 High Sch., Tongliao, 1972-75, Tchrs. Tng. Sch., Tongliao, 1975-78; instr. structural chemistry Tchr.'s Coll., Tongliao, 1982-85; rsch. asst./assoc. U. Okla., Norman, 1985-90; rsch. assoc. biophys. chemistry Ctr. for Advanced Rsch. in Biotech., Rockville, Md., 1991—. Contbr. articles to profl. jours. Mem. Am. Crystallographic Assn., Am. Hon. Chem. Soc., Phi Lambda Upsilon. Achievements include research on Glutathione S-Transferase, a detoxification enzyme containing 434 amino acid residues; research on 21 small and medium sized molecules. Office: Ctr for Adv Rsch in Biotech 9600 Gudelsky Dr Rockville MD 20850

JIA, HONG, programmer, analyst; b. Beijing, China, Dec. 12, 1960; came to U.S., 1988; s. Yin Long and Qing Lian (Wu) J.; m. Jie Cheng, Sept. 5, 1988. BSEE, Tsinghua U., 1983, MSEE, 1986; MS in Med. Physics, U. Chgo., 1991. Software developer Chinese Acad. Med. Sci., Beijing, 1986-88; programmer, analyst U. Chgo., 1991—. Chinese Acad. Med. Sci. Young Scientist Rsch. grantee, 1987. Mem. IEEE, Am. Assn. Physical Medicine. Achievements include investigation and research on computer-aided diagnosis chest radiographs; analyse textures of chest radiographs and use power spectrum method to distinguish normal and abnormal chest radiographs. Home: 5550 S Dorchester Ave #208 Chicago IL 60637

JIA, QUANXI, electrical and material scientist, researcher; b. Mengxian, Henan, China, Sept. 3, 1957; came to U.S., 1988; s. Luzhong and Yuqing (Xi) J.; m. Xuming Wu, Sept. 30, 1985; 1 child, Yixuan. MS, Jiaotong U., Xian, Shaanxi, China, 1985; PhD, SUNY, Buffalo, 1991. Tchr. mid. sch. Mengxian Sch. Dist., 1975-78; rsch. asst. Jiaotong U., 1982-85, rsch. assoc., 1985-87, asst. prof., 1987-88; rsch. assoc. SUNY, 1988, rsch. asst., 1988-91, postdoctoral fellow, 1991—. Author: (with others) Amorphous Si Solar Cells, 1989; contbr. over 60 articles to profl. jours. Link Energy Found. fellow, 1990-91. Mem. Am. Vacuum Soc., Material Rsch. Soc.

JIAN, SONG, government official, science administrator; b. Rongcheng County, Shangdong, People's Republic of China, Dec. 29, 1931; s. Zengjin Song and Yuxian Jiang; m. Yusheng Wang, July 1, 1961; two children. Student, Harbin Tech. U., 1951-53; degree in engring., Moscow Bauman Poly. Inst., 1958, PhD, 1960; DSc, Moscow Nat. Tech. U., 1990. Head lab cybernetics Inst. Math. Acad. Sci., Beijing, 1960-70; head dept. space sci. Acad. Space Tech. Min. Machine Bldg. Industry, 1971-78, v.p. Acad. Space Tech., 1978-81; vice min., chief engr.-scientist Min. Astonautics, Beijing, 1981-84; chmn. State Sci. and Tech. Commn., Beijing, 1984—, State Environ. Protection Com., Beijing, 1986—; councillor State of Peoples Republic China, Beijing, 1986—; prof. Quinhua U., Beijing, Fudan U., Shanghai; disting. vis. prof. Washington U., St. Louis. Author: Engineering Cybernetics, 1980, China's Population: Problem & Prospect, 1981, Recent Developments in Control Theory and its Applications, 1984, Population Control in China, Theory and Application, 1985; contbr. 60 articles on population theory to profl. jours. Mem. China Automation Soc. (pres. 1980—, coun.), China Systems Engring. Soc. (v.p. 1984—), China Acad. Scis., Population Sci. Soc. (v.p. 1984—), Internat. Fedn. Automatic Control (coun.), Nat. Acad. Engring. Mex. (corr.). Achievements include research in population control theory. Office: State Sci and Tech Commn, 15B Fuxing Rd, Beijing 100862, China

JIANG, HONGWEN, physicist; b. Peking, China, July 14, 1960; s. Chen-Lie and Wei-qi (Shen) J.; m. Ping Wang, Jan. 14, 1987; children: Mason, Richard. PhD, Case Western Res. U., 1989; BS, Calif. State U., 1983. Vis. scientist MIT/Nat. Magnet Lab., L.A., 1989-90; rsch. assoc. Princeton U., L.A., 1989-90; asst. prof. of physics UCLA, 1991—. Contbr. articles to profl. jours. including Phys. Rev. Letters and Phys. Rev. B. Named Sloan fellowship Alfred P. Sloan Found., 1992; recipient William L. McMillan award for outstanding contbns. in condensed matter physics, 1993. Mem. Am. Phys. Soc. Achievements include rsch. on the novel properties of two-dimensional electrons in highly correlated systems. Office: UCLA 405 Hilgard Ave Los Angeles CA 90024-1547

JIANG, WENBIN, electrical engineer; b. Shanghai, China, Dec. 4, 1963; came to U.S., 1990; s. Fenggeng and Huimin (Wang) J.; m. Jianwen Jiang, May, 1990; 1 child, Carrie Jiang. BS in Physics, Fudan U., Shanghai, 1985; MS in Optics, Fudan U., 1988; PhD in Engring., U. Calif., Santa Barbara, 1993. Grad. student rschr. Fudan U., 1985-89; invited researcher NTT Basic Rsch. Labs., Tokyo, 1989-90; rsch. asst. U. Calif., Santa Barbara, 1990—; invited speaker IEEE Engring. Found., Banff, Can., 1992. Contbr. articles to profl. jours. Mem. Optical Soc. Am. Achievements include pioneering of mode-locked vertical-cavity surface-emitting lasers, first room temperature operating ML VCSEL, first generated 200 fs laser pulses from a semiconductor laser, Japanese patent on laser apparatus short pulse generation. Office: ECE Dept U Calif Santa Barbara CA 93106

JIAO, JIANZHONG, SR., development engineer, educator; b. Beijing, China, July 11, 1955; came to U.S., 1984; s. Yimen Diao and Jintang Fan; m. Wenqi Wang, Sept. 2, 1980; 1 child, Alex. BS in Mech. Engring., Beijing Poly. U., China, 1980; MS in Applied Physics, Beijing Inst. Posts & Tel., 1983; PhD in Elec. Engring., Northwestern U., Evanston, Ill., 1989. Rsch. & lab. asst. Beijing Poly. U., 1980; lectr. Beijing Inst. Posts & Tel., 1983-84; teaching asst. Renssalaer Poly. Inst., Troy, N.Y., 1985; teaching & rsch. asst. Northwestern U., Evanston, Ill., 1985-89; sr. devel. engr. GM, Anderson, Ind., 1989-92, Troy, Mich., 1992—; faculty mem. Purdue U., Anderson, Ind., 1990-92, Lawrence Technol. U., Southfield, Mich., 1992—. Contbr. numerous articles to profl. jours. Mem. IEEE, Optical Soc. Am., Soc. Photo-Optical Instrumentation Engring. Achievements include inventions Illuminator Device for a display panel, and several other patents pending. Office: GM IFG Divsn 1401 Crooks Rd T-1 Troy MI 48084

JILHEWAR, ASHOK, gastroenterologist; b. Nanded, Maharashtra, India, Jan. 30, 1947; came to U.S., 1977; naturalized 1987; BS, Marathwada U., Aurangabad, India, 1970; MB, Marathwada U., 1970; MD, Govt. Med. Coll., Aurangabad, 1970. Diplomate Am. Bd. Internal Medicine, Am. Bd. Gastroenterology, Am. Bd. Geriatric Medicine, Am. Bd. Quality Assurance and Utilization Rev. Physicians. Rotatory intern Med. Coll. Hosp., Aurangabad, India, 1968-70; resident St. Luke's Hosp. and Royal infirmary, Hudersfeild, Bolton, Eng., 1970-72; med. registrar internal medicine Gen. Hosp., Sligo, Ireland, 1973-77; chief resident PG1 and internal medicine U. Health Scis.-Chgo. Med. Sch. and VA Hosp., 1977-79; clin. instr. U. Heath Scis.-Chgo. Med. Sch., 1978-79; fellow in gastroenterology Michael Reese Hosp., Chgo., 1980-81; mem. exec. com. Meth. Hosp., Chgo, 1985-90, chmn. dept. med., 1988-90; lectr. preventive and social medicine Med. Coll., Aurangabad, 1970; mem. exec. com. Meth. Hosp. Chgo., 1985-90, v.p. med. staff, 1987-88, treas., sec., 1985-87, chmn. dept. medicine, 1988-90; mem. staff dept. medicine Grant Hosp., Chgo., 1986—; med. dir. approved home for intermediate care nursing home, 1986—; med. advisor Office Hearings and Appeals, HHS, 1985—, Cen. Ill. Med. Rev. Orgn., 1993. Fellow Royal Coll. Physicians Can., Am. Coll. Internat. Physicians; mem. AMA, ACP, Am. Gastroenterol. Assn., Royal Coll. Physicians U.K., Royal Coll. Physicians Ireland, Ill. State Med. Assn., Chgo. Med. Soc. (PRO study com., fee mediation subcom. 1992). Home: 5393 N Milwaukee Ave 2d Fl Chicago IL 60630-1251 Office: North Park Stomach Clinic 5393 N Milwaukee Ave Chicago IL 60630-1251

JIM, KAM FOOK, medical writer; b. Canton, China, Nov. 13, 1953; came to U.S., 1972; s. Gum and Lai (Tsui) J.; m. Margaret Liang; May 19, 1981; children: Carol M., Ryan A. BA, NYU, 1976; PhD, SUNY, Buffalo. Postdoctoral fellow Case Western Res. U., Cleve., 1980-81; rsch. assoc. Cornell U. Med. Coll., N.Y.C., 1981-83; postdoctoral scientist Smith Kline French Labs., Swedeland, Pa., 1983-86; cons. Med. Coll. Pa., Phila., 1986-88; sr. scientific writer Wyeth-Ayerst Rsch., Radner, Pa., 1988—. Co-author: Epilepsy and Sudden Death, 1990, Toxic Interactions, 1990; contbr. articles to profl. jours. NYU scholar, 1974-76. Fellow Am. Coll. Clin. Pharm., Am. Soc. for Pharm. Experimental Theorpy, N.Y. Acad. Scis., Am. Assn. for the Advancement Scis.

JIMÉNEZ, BRAULIO DUEÑO, toxicologist; b. Camuy, Puerto Rico, Dec. 21, 1950; s. Braulio and Herminia (Vélez) J.; m. Carmen Cadilla, June 12, 1981; children: Manuel, Thalia, Sean, Beatriz. BS, U. Puerto Rico, 1971, PhD, 1981. Rsch. asst. U. Puerto Rico, Mayaguez, 1976-78; assoc. prof. U. Puerto Rico, 1990—; environ. scientist Environ. Quality Bd., Santurce, Puerto Rico, 1978-80; dir. environ. rsch. Environ. Quality Bd., Santurce, 1980-82; postdoctoral fellow Oak Ridge (Tenn.) Nat. Lab., 1982-84, rsch. assoc., 1986-90; rsch. assoc. U. Tenn., Knoxville, 1985-86; proposal reviewer Nat. Rsch. Coun., Washington, 1990-91; cons. sch. medicine, Ponce, Puerto Rico, 1991-93, Environ. Quality Bd., Puerto Rico, 1993. Contbr. articles to profl. jours. Recipient Puerto Rico Environ. Quality award, 1982, Tech. Communication award Soc. Tech. Communication, 1988, Industry and Acad. award INDUNIV, 1992. Mem. AAAS, Soc. Toxicology. Achievements include research in the use of enzyme biomarkers as indicators of aquatic pollution, effect of environmental variables on induction of detoxification enzymes in fish. Office: U Puerto Rico Sch Pharmacy PO Box 365067 San Juan PR 00936-5067

JIMENEZ, JUAN IGNACIO, electrical engineer, consultant; b. Seville, Spain, Dec. 9, 1958; s. Angel Maria and Encarnacion (Mazario) J.; m. Maria Christina Gomez-Bastero, Feb. 21, 1985. D Elec. Engring., U. Seville, 1983; MBA, U. Madrid, 1986. Cons. Spain, 1983-84; engr. Abengoa, Spain, 1984-85; systems engr. Abengoa, Fed. Republic of Germany, 1985-86; chief engr. Sainco, Spain, 1986-87; project mgr. Trafinsa, Gijon, Spain, 1987-88, Brussels, 1988-89; engring. mgr. Trafinsa, Gijon, 1990—. Contbr. articles to profl. jours. With Spanish mil., 1982-83. Mem. Inst. Transp. Engrs. (assoc.). Avocations: golf, anthropology. Office: Trafinsa, Ctr de la Carbonera St, Gijon 33209, Spain

JINDRICH, ERVIN JAMES, municipal government official; b. Chgo., June 5, 1939; s. Ervin James and Lydia Renata (Ahrens) J.; m. Denise Lobeth Fowler, Mar. 4, 1970; children: Devin Logan, Antonia Elizabeth. Student, U. Ill., 1960; BS, MD, Northwestern U., 1964. Fellow in forensic pathology U. Calif., San Francisco, 1972-73; coroner City and County of San Francisco, 1973-74, County of Marin, Calif., 1975—; pvt. practice as medico-legal cons., Mill Valley, Calif., 1975—. Contbr. articles to Jour. Forensic Scis., Jour. Analytical Toxicology. Bd. dirs. Marin Suicide Prevention Ctr. and Grief Counseling Svcs., Marin, 1977—. Capt. U.S. Army, 1965-67. Decorated Cert. of Achievement. Fellow Am. Acad. Forensic Scis.; mem. Nat. Assn. Med. Examiners. Home: 9 Heuter Ln Mill Valley CA 94941-2701 Office: County of Marin Civic Center Rm 154 San Rafael CA 94903

JING, HUNG-SYING, aerospace engineering educator; b. Pin-Dong, Taiwan, Republic of China, Apr. 19, 1955; s. Min-Yuan and Shiu-Lin (Chen) J.; m. Char-Jane Wang, Oct. 20, 1979; children: Alexander, Catherine, How. MS, Nat. Taiwan U., 1979; PhD, Ohio State U., 1985. Teaching asst. Nat. Taiwan U., Taipei, 1977-78; engr. Pacific Engrs. and Constructors Ltd., Taipei, 1980; instr. Chen Shiu Inst. Tech., Kaohsiung, 1980-81; rsch. asst. Ohio State U., Columbus, 1982-85; assoc. prof. Nat. Cheng Kung U., Taiwan, 1986-92, prof., 1992—; cons. Metal Industries Devel. Ctr., Kaohsiung, 1991-92; spl. researcher Indsl. Tech. Rsch. Inst., Hsin-chu, 1991. Contbr. articles to Internat. Jour. for Numerical Methods in Engring., Earthquake Engring. and Structural Dynamics, Jour. of Sound and Vibration, Jour. AIAA, Internat. Jour. Solids and Structures. Chief editor Nat. Inst. for Compilation and Translation, Taipei, 1989—; design advisor Nat. Sci. and Tech. Mus., Kaohsiung, 1990—; cons. Young Women's Christian Assn., Tainan, 1992. Recipient award for Social Edn. Dept. Edn., 1992. Mem. Soc. for Advancement of Materials and Process Engring., Exptl. Aircraft Assn., Air and Space Smithsonian. Achievements include establishment of fundamental functional with displacement and transverse shear as independent variables, of partial hybrid stress element, of mixed shear deformation theory for laminated shells. Office: Nat Cheng Kung U, Inst Aero and Astro, Tainan Taiwan

JIRSA, JAMES OTIS, civil engineering educator; b. Lincoln, Nebr., July 30, 1938; s. Otis Frank and Anna Marie (Skutchan) J.; m. Marion Ansley Coad, Aug. 7, 1941; children: David, Stephen. BS, U. Nebr., 1960; MS, U. Ill., 1962, PhD, 1963. Registered profl. engr., Tex. Asst. prof. civil engring. U. Nebr., Lincoln, 1964-65; asst. prof. then assoc. prof. Rice U., Houston, 1965-72; assoc. prof. then prof. U. Tex., Austin, 1972-82, Finch prof. engring., 1982-84, Ferguson prof. civil engring., 1984-88, dir. Ferguson Structural Engring. Lab., 1985-88, Janet S. Cockrell Centennial chair in engring., 1988—; research engr. Portland Cement Assocs., 1965; engr. J.J. Degenkolb Assocs., San Francisco, 1980. Contbr. articles to profl. jours. Recipient rsch. award Japanese Soc. for Promotion Sci., 1980; Fulbright scholar, Paris, 1963-64. Fellow Am. Concrete Inst. (TAC chmn. 1985-88, bd. dirs. 1987-90, Alfred Lindau award 1986, Wason medal 1977, Reese award 1977, 79, Bloem award 1990); mem. NAE, ASCE (com. chmn. 1972-81, Reese award 1970, 91, Huber Research prize 1978), Earthquake Engring. Research Inst., Structural Engring Assn. Tex., Internat. Assn. for Bridge and Structural Engrs., Nebr. Czech Orgn. (King Charles award 1983). Office: U Tex Civil Engin Dept 10100 Burnet Rd Austin TX 78712

JOARDAR, KUNTAL, electrical engineer; b. Asansol, India, Dec. 20, 1962; s. Kisori Mohan and Banya (Ghosh) J.; m. Nivedita Guha, May 7, 1989. BTech, Indian Inst. Tech., Madras, India, 1984; PhD, Ariz. State U., 1989. Teaching asst. Ariz. State U., Tempe, 1984-85, rsch. assoc., 1985-89; sr. staff scientist Motorola, Mesa, Ariz., 1989-92, prin. scientist, 1992—; mem. tech. com. profl. conf. Contbr. articles to profl. jours. Recipient Regents' Acad. scholarship Ariz. State U., 1985-88, Best Paper award IEEE, 1987. Mem. IEEE, Sigma Xi, Phi Kappa Phi. Hindu. Achievements include invention of novel test structures and measurement techniques for characterizing semiconductor devices; devel. advanced tools for predicting and improving mfg. yield of integrated circuits. Office: Advanced Custom Tech Ctr Motorola Inc 2200 W Broadway MD M350 Mesa AZ 85202

JOBE, MURIEL IDA, medical technologist; b. St. Louis, Apr. 17, 1931; d. Ernest William and Mable Mary (Hefflinger) Meissner; m. James Joseph Jobe, Sr., May 17, 1952 (dec. 1984); children: James J. Jr., Timothy D. (dec. 1976), Jonathan J., Daniel D. BS, Wash. U., St. Louis, 1971; med. technologist tng., Mo. Bapt. Hosp., St. Louis, 1973-74; postgrad., Webster U., St. Louis, 1981-83. Cytogenetic tech. St. Luke's Hosp., St. Louis, 1963-65; med. technologist Mo. Bapt. Hosp., St. Louis, 1974-76, 82-84, sr. instr., 1976-82, lead technologist, 1985; supr., clin. instr. St. Louis U. Hosp., 1985—; mem. student selection com. Mo. Bapt. Hosp. Med. Technologists, St. Louis, 1975-78; observer Nat. Com. Clin. Lab. Standards, Villanova, Pa., 1989-90, advisor, 1991-92, 93—. Author: (with others) Clinical Hematology: Principles, Procedures, Correlations, 1991, (with others) 8th Revision PER Handbook, a Review Manual for Clinical Laboratory Exams., 1992. Counselor La Leche League; participant Ecology Day; community rels. chmn. The Life Seekers, St. Louis. Mem. Am. Soc. Clin. Pathologists (staff asst. 1984, 86, 88, 89, dir. workshops 1990, 91, bd. dirs. 1990-92, chmn. regional adv. com., adminstrv. bd. of assoc. mem. sect.), Am. Assn. Clin. Chemists, Am. Soc. Med. Tech. (dir. workshop 1984), Mo. Soc. Med. Tech. (pres. 1985-86), Clin. Lab. Mgrs. Assn. Mem. United Ch. of Christ. Avocations: travel, cooking. Office: St Louis U Hosp Hematology Lab 4 FDT 3635 Vista Ave PO Box 15250 Saint Louis MO 63110-0250

JOBES, CHRISTOPHER CHARLES, mechanical engineer; b. Rochester, N.Y., Oct. 11, 1960; s. Charles Oliver and Ruth Elinor (Howell) J.; m. Debra Ann Wineland, June 16, 1984; children: Amber Deanne, Derek Christopher. BS, Geneva Coll., 1982, BS in Mech. Engring., 1983; MS, W.Va. U., 1985, PhD, 1987. Cert. profl. engr., Pa. Computer programmer Monaca, Pa., 1982; specialist engr. Boeing Svcs. Internat., Pitts., 1987-88; mech. engr. U.S. Bur. Mines, Pitts., 1988—. Contbr. articles to ASME Jour. Mech. Design, IEEE Transactions on the IAS, and various conf. proceedings. Mem. NSPE, ASME (vice chair 1992-93, award of merit 1988), SAE, Sigma Xi. Republican. Baptist. Achievements include patents for variable speed rhino robot controller, controllable residential circuit breaker, treating spring mass in kinetostatic analysis, mechanical position and heading system. Office: US Bur Mines PO Box 18070 Cochrans Mill Rd Pittsburgh PA 15236

JOBS, STEVEN PAUL, computer corporation executive; b. 1955; adopted s. Paul J. and Clara J. (Jobs); m. Laurene Powell, Mar. 18, 1991. Student, Reed Coll. With Hewlett-Packard, Palo Alto, Calif.; designer video games Atari Inc., 1974; co-founder Apple Computer Inc., Cupertino, Calif., chmn. bd., 1975-85, former dir.; pres. NeXT, Inc., Redwood City, Calif., 1985—. Co-designer: (with Stephan Wozniak) Apple I Computer, 1976. Office: NeXT Inc 900 Chesapeake Dr Redwood City CA 94063

JOEL, AMOS EDWARD, JR., telecommunications consultant; b. Phila., Mar. 12, 1918; s. Amos Edward and Anna (Potsdamer) J.; m. Rhoda Ethel Fenton; children: Jeffrey, Stephanie, Andrea. BEE, MIT, 1940, MEE, 1942. Registered profl. engr., N.Y. Mem. tech. staff Bell Telephone Labs., N.Y. and N.J., 1940-52; supr. Bell Telephone Labs., Whippany, N.J., 1952-54, dept. head, 1954-61; dir. Bell Telephone Labs., Holmdel, N.J., 1961-67, cons., 1967-83, ret., 1983.; cons. AT&T Bell Communciations Rsch., GTE, IBM, Contel, Pacific Telephone; lectr. in field of switching systems. Author: Electronic Switching Central Office Systems of the World, 1976, Electronic Switching: Digital Central Office Systems of the World, 1982, History of Science and Technology in the Bell System-Switching Technology, 1982; author: (with others) Fundamentals of Digital Switching, 1983, 2d edit., 1990, Electronics, Computers and Telephone Switching, 1990, Future of the Central Office, 1991; contbr. articles to encys. and profl. jours.; holder more than 70 patents. Co-recipient Outstanding Patent awrd N.J. R&D Coun., 1972; recipient Stuart Ballantine medal Franklin Inst., 1981, Centary prize Internat. Telecommunication Union, 1983, Columbian medal City of Genoa, Italy, 1984, Kyoto prize in advanced tech., 1989, Nat. Medal of Tech., Nat. Sci. Found., 1993; named N.J. Inventor of Yr., 1989. Fellow IEEE (life, co-recipient Alexander Graham Bell medal 1976, IEEE medal of honor 1992), Am. Acad. Arts and Scis. (Nat. Medal of Tech. 1993); mem. NAE, Communication Soc. IEEE (pres. 1973-75), Assn. Computing Machinery, AAAS, Sigma Xi. Avocations: organ and keyboard music, railroading. Home: 131 N Wyoming Ave South Orange NJ 07079-1529

JOFFE, BENJAMIN, mechanical engineer; b. Riga, Latvia, Feb. 23, 1931; came to U.S., 1980, naturalized, 1985; s. Alexander and Mery (Levenson) J.; m. Frida Erenshteyn, Aug. 6, 1960; children: Alexander, Helena. ASME, Mech. Tech. Sch., Kransnoyarsk, USSR, 1951; BSME, Polytechnic Inst. Moscow, 1959; MSME, Polytechnic Inst. Riga, 1961; PhD, Acad. Scis., Riga, 1969. Design engr. Electromachine Mfg. Corp., Riga, 1955-59, head engring. dept., 1959-62; sr. design engr. Acad. Scis., Riga, 1962-67; sr. scientist Inst. Physics, Riga, 1967-78; chief design engr. Main Design Bur., Riga, 1978-80; sr. design engr. Elec-Trol, Inc., Saugus, Calif., 1980-81; sr. design engr. VSI Aerospace div. Fairchild, Chatsworth, Calif., 1981-85; mech. engring. mgr. Am. Semiconductor Equipment Tech., Woodland Hills, Calif., 1985-90; mem. tech. staff Jet Propulsion Lab. Calif. Inst. Tech., Pasadena, 1991—. Author: Mechanization and Automatization of Punching Presses at the Plants of the Latvian SSR, 1963, Mechanization and Automatization of Processes of Plastic Parts Production at the Plants of the Latvian SSR, 1964, Mechanization and Automatization of Control and Measuring Operations, 1966, and 5 sci. engring. books; contbr. numerous articles to profl. jours. Recipient Honored Inventor award Latvian Republic, Riga, 1967, 1st prize Latvian Acad. Scis., 1972, Latvian State award in engring. scis., 1974. Mem. ASME (dir. exec. bd.). Republican. Achievements include over 200 patents for discovery of physical and engineering basis for noncontact techniques of orientation, identification and assembly of parts by electromagnetic fields. Home: 22314 James Alan Cir Chatsworth CA 91311-2054 Office: Calif Inst Tech Jet Propulsion Lab 4800 Oak Grove Dr Pasadena CA 91109-8099

JOH, YASUSHI, science administrator, chemist; b. Naruo, Hyogo, Japan, Mar. 19, 1933; s. Kenzo and Kotoko Joh; m. Mariko Omi, Oct. 10, 1979. MS in Chemistry, Osaka (Japan) U., 1958, DSc in Organic Chemistry, 1963, PhD in Organic Chemistry, 1963. Supr. Mitsubishi Rayon Co., Hiroshima, Japan, 1959-70; rsch. assoc. U. Mich., Ann Arbor, 1970-73; sr. rsch. scientist Nippon Zeon Co., Kawasaki, Japan, 1975-85; assoc. dir. med. tech. and mktg. dept. corp. rsch. and devel. Ube Industries Ltd., Tokyo, 1985—. Achievements include 172 patents for Synthetic Polymer and Fibers (acrylic fibers); 278 patents for Artificial Organs (kidneys, hearts, biomaterials and related things); commercialized artificial kidney (dialyzer), artificial heart (ventricular assist device) as the first governmentally approved device, and artificial blood vessels (vascular grafts) in Japan. Home: # 401 998-1 Futo-o-cho, Kohoku-ku, Yokohama 222, Japan Office: Ube Industries Ltd, Ube Bldg 2-3-11, Higashi-Shinagawa, Shinagawa-ku Tokyo 140, Japan

JOHANNES, RICHARD DALE, civil engineering executive; b. Creston, Iowa, Jan. 2, 1952; s. Robert Dale and Teresa Louise (Harvey) J.; m. Jan Lee Mitchell, Oct. 22, 1977; children: Jason Andrew, Katherine Anne. BSCE, Iowa State U., 1974; MSCE, U. Mo., 1981. Registered profl. engr. Mo., Kans., Iowa, Minn., Wyo., Tenn. Chief indsl. wastewater engring. Burns & McDonnell Engrs., Kansas City, Mo., 1974-88; engring. mgr. Groundwater Tech., Inc., Lenexa, Kans., 1989—; presenter at profl. confs. Mem. youth com. Atonement Luth. Ch., Overland Park, Kans., 1986-88; co-den father Boy Scouts Am., Overland Park, 1991-92. Mem. Am. Water Works Assn., Constrn. Specifications Inst. Office: Groundwater Tech Inc 15010 W 106th St Lenexa KS 66215

JOHANNESSON, THOMAS ROLF, engineering educator; b. Helsingborg, Sweden, Aug. 1, 1943; s. Carl A. and Signe B. (Gustafsson) J.; m. Eva-Lisa A. Ohlsson, May 7, 1966; children: Karolina, Per-Ola. MS in Engring. Physics, Chalmers U. Technology, Gothenburg, 1967, PhD in Physics, 1971. Asst. Chalmers U. Tech., 1967-70; dir. lab. Sandvik Coromant, Stockholm, 1970-75; assoc. prof. Linköping (Sweden) U., 1975-82, prof., 1982-93, dept. chmn., 1978-80, dean engring., 1980-81; prof. Lund (Sweden) U., 1993—; bd. dirs. Barracuda Tech., Laholm; v.p. external rels. Linköping U., 1987-93; mem. coun. of sci. Nat. Bd. Tech. Devel., Stockholm, 1985—. Contbr. articles to profl. jours. Bd. dirs. Nat. Patents Authority, Stockholm, 1990—. Recipient Jacob Wallenberg award in Materials Sci., J. Wallenberg Found., Stockholm, 1979. Mem. SAMPE, Swedish Materials Soc., Royal Acad. Engring. Scis. Office: Lund U, Divsn Engring, S-22100 Lund Sweden

JOHANNSEN, CHRIS JAKOB, agronomist, educator, administrator; b. Randolph, Nebr., July 24, 1937; s. Jakob J. and Marie J. (Lorenzsen) J.; m. Joanne B. Rockwell, Aug. 16, 1959; children: Eric C., Peter J. B.S., U. Nebr.-Lincoln, 1959, M.S., 1961; Ph.D., Purdue U., 1969. Program leader Lab. for Applications of Remote Sensing, Purdue U., 1966-69; asst. prof. agronomy Purdue U., 1969-72; assoc. prof. agronomy U. Mo., Columbia, 1972-77, prof., 1977-84; dir. Geographic Resources Ctr., U. Mo., Columbia, 1981-84, Ag Data Network, Purdue U., 1985-87, Lab. for Applications of Remote Sensing, 1985—, Nat. Resources Research Inst., 1987—; vis. prof. U. Calif., Davis, 1980-81; cons. Lockheed Electronics, Houston, 1975-76, NOAA, Columbia, Mo., 1978-80, FAO UN, Nairobi, Kenya, 1983, 87, Rome, 1987, US Agy. Internat. Devel., Eastern Africa, 1983, USDA-Soil Conservation Svc., Washington, 1984, Space Sci. Corp., Washington, 1984-85, IBM, 1991. Pres. coun. St. Andrew's Lutheran Ch., Columbia, 1975-77; asst. scout master Boy Scouts Am., Gt. Rivers coun., Columbia, 1979-84, West Lafayette, 1985-91; pres. Purdue Luth. Ministry, 1989—. Recipient Tech. Innovation Research award NASA, 1979; recipient Disting. Service award Mo. Assn. Soil and Water Conservation Dists., 1982. Fellow Am. Soc. Agronomy, Soil Sci. Soc. Am., Soil Conservation Soc. Am. (pres. 1982-83); mem. World Assn. of Soil and Water, Am. Soc. Photogrammetry and Remote Sensing (Outstanding Svc. award 1992), Internat. Soc. Soil Sci., Geosci. and Remote Sensing Soc. of IEEE, Indian Acad. of Scis., Epsilon Sigma Phi. Home: 209 Cedar Hollow Ct West Lafayette IN 47906-1671 Office: Purdue U Nat Resources Rsch Inst 1158 ENTM West Lafayette IN 47907-1158

JOHANSEN, JACK T., engineering company executive; b. 1943. From various sr. mgmt. positions to pres. Carlbiotech, Copenhagen; sr. v.p. sci. and tech. Millipore Corp., Bedford, Mass., 1987—. Office: Millipore Corp 80 Ashby Rd Bedford MA 01730-2271*

JOHANSEN, ROBERT JOHN, electrical engineer; b. S.I., N.Y., Mar. 30, 1952; s. Odd Ingvold and Theresa Florence (Stanislawiszyn) J. Grad. high sch., Staten Island, N.Y., 1970. Communications technician AAT Electronics Corp., S.I., 1975-78; engr. ITT Mackay Corp., Elizabeth, N.J., 1978-85; product engr. Panasonic Co., Secaucus, N.J., 1985—. Contbr. articles to profl. jours. Mem. People for Perot Campaign, S.I., 1992, United We Stand Am., 1992. Mem. Am. Amateur Radio League, S.I. Amateur Radio Assn. (pres. 1989-91, mem. exec. coun.), IEEE. Home: 61 Burnside Ave Staten Island NY 10302 Office: Panasonic Comm & Systems Co 50 Meadowland Pkwy Secaucus NJ 07094

JOHANSON, DONALD CARL, physical anthropologist; b. Chicago, Ill., June 28, 1943; s. Carl Torsten and Sally Eugenia (Johnson) J. ; m. Lenora Carey, 1988. BA, U. Ill., 1966; MA, U. Chgo., 1970, PhD, 1974; DSc (hon.), John Carroll U., 1979; D.Sc. (hon.), Coll. of Wooster, 1985. Mem. dept. phys. anthropology Cleve. Mus. Natural History, 1972-81, curator, 1974-81; pres. Inst. Human Origins, Berkeley, Calif., 1981—; prof. anthropology Stanford U., 1983-89; adj. prof. Case Western Res. U., 1978-81, Kent State U., 1978-81. Co-author: (with M.A. Edey) Lucy: The Beginnings of Humankind, 1981 (Am. Book award 1982), Blueprints: Solving the Mystery of Evolution, 1989 (with James Shreeve) Lucy's Child: Discovering a Human Ancestor, 1989, (with Kevin O'Farrell) Journey from the Dawn: Life with the World's First Family, 1981; host PBS Nature series; prodr.: (films) Lucy in Disguise, 1982; contbr. numerous articles to profl. jours. Recipient Jared Potter Kirtland award for outstanding sci. achievement Cleve. Mus. Natural History, 1979, Profl. Achievement award, U. Chgo., 1980, Gold Mercury Internat. ad personem award Ethiopia, 1982, Humanist Laureate award Acad. of Humanism, 1983, Disting. Svc. award Am. Humanist Assn., 1983, San Francisco Exploratorium award, 1986, Internat. Premio Fregene award, 1987; grantee Wenner-Gren Found., NSF, Nat. Geog. Soc., L.S.B. Leakey Found., Cleve. Found., George Gund Found., Roush Found. Fellow AAAS, Calif. Acad. Scis., Rochester (N.Y.) Mus., Royal Geog. Soc.; mem. Am. Assn. Phys. Anthropologists, Internat. Assn. Dental Research, Internat. Assn. Human Biologists, Am. Assn. Africanist Archaeologists, Soc. Vertebrate Paleontology, Soc. Study of Human Biology, Societe de l'Anthropologie de Paris, Centro Studi Ricerche Ligabue (Venice), Founders' Coun., Chgo. Field Mus. Natural History (hon.), Assn. Internationale pour l'etude de Paleontologie Humaine, Mus. Nat. d'Histoire Naturelle de Paris (corr.), Explorers Club (hon. dir.), Nat. Ctr. Sci. Edn. (supporting scientist). Office: Inst Human Origins 2453 Ridge Rd Berkeley CA 94709

JOHAR, JOGINDAR SINGH, chemistry educator; b. Rawalpindi, Panjab, India, Jan. 1, 1935; came to U.S., 1962; s. Waryam Singh and Tej (Kaur) J.; m. Manjit Kaur, Sept. 4, 1960; children: Ravi, Jasjot, Vinny. MS with honors, Panjab U., 1958; PhD, U. Fla., 1966. Lectr. Govt. Coll., Ludhiana, Panjab, 1958-62; head math. and sci. divsn. Cleveland (Tenn.) State C.C., 1967-68; prof. chemistry Wayne (Nebr.) State Coll., 1968—, head math. and sci. divsn., 1987—; pres. State Colls. Edn. Assn., Nebr., 1975-83. Mem. Lions (dep. gov. 1985-89). Democrat. Sikh. Home: 207 Maple St Wayne NE 68787 Office: Wayne State Coll Math and Sci Divsn Wayne NE 68787

JOHN, LEONARD KEITH, aerospace and mechanical engineer, consultant; b. Lahore, Pakistan, Apr. 10, 1949; arrived in Can., 1975; s. Edwin Kenneth William and Olive (Khairullah) J.; m. Yvonne Anna Lee-Anan, Dec. 20, 1980; children: Sarah Ashley, Jason William. Full Tech. Cert., Harrow Coll. Tech. and Art, Middlesex, Eng., 1971; B.S. with honors, Hendon Coll., London, 1975; M.Engring., U. Toronto, 1978. Chartered engr., U.K.; registered profl. engr., Ont. Aero. apprentice Westland Helicopters Ltd., Hayes, Middlesex, Eng., 1965-70, rsch. and devel. engr., 1970-71; devel. engr. Westland Helicopters/Hendon, Hayes and Hendon, Eng., 1971-75; sr. devel. engr. non-metallics de Havilland Inc. divsn. of Bombardier Aerospace, Downsview, Ont., 1975-80; group leader composite structure devel. 1980-85, chief advanced composites and nonmetallics, 1985-89; chief advanced composites and chem. tech., 1989—, pres. 620688 Ont. Inc., Toronto, 1985—; cons. to Revenue Can. Taxation-Rsch. and Devel. Investment Tax Credit, 1987-88; lectr. continuing edn. course in advanced materials, Faculty of Applied Sci. and Engring. U. Toronto, 1989—. Contbr. articles to profl. jours.; patentee in field. Inventor Graphite Fibre Violin, violin type mus. instruments; violins exhibited Planete Composite, Bordeaux, France, 1985, Ontario Sci. Centre, Toronto, 1988—, Sec. of State Exhibit, Bravo Can., Toronto, Quebec City and Vancouver, 1988. Recipient Outstanding Svc. award Soc. for Advancement of Material and Process Engring., 1979, F. H. Baldwin award Can. Aeros. and Space Inst., 1982. Mem. Can. Aeronautics and Space Inst., Instn. Mech. Engrs. (Eng., mem. aerospace industries div. 1984—), Assn. Profl. Engrs. Province Ont. Mem. Ch. of England. Avocations: flying, travel, music.

JOHNS, GARY CHRISTOPHER, electronic technologies educator; b. Detroit, Apr. 27, 1958; s. Llewyllyn Waldemare and Bettie Ruth (Baird) Jenks; m. Christine Marie Johns, July 12, 1986; children: Brittany Lynne, Lindsey Leigh. BS in English, Am. lit. and lang., Ea. Mich. U., 1980. Cert. secondary tchr., Mich. Tchr. Notre Dame High Sch., Harper Woods, Mich., 1980-86, St. Claire Shores (Mich.) Adult and Community Edn., 1985—; network supr. Born Ctr., St. Clair Shores, 1988—; technologies trainer Advanced Ctr. Tech. Tng., Bloomfield Hills, Mich., 1989—. Mem. Network Users Internat. Democrat. Presbyterian. Achievements include development of program curriculum for an electronics/industrial technologies course. Office: Born Ctr 23340 Elmira Saint Clair Shores MI 48082

JOHNS, MICHAEL MARIEB EDWARD, otolaryngologist, university dean; b. Detroit, Jan. 27, 1942; s. Trina Lou DelCampo; children: Christina, Michael. BS, Wayne State U., 1964, Grad. Biol. Sci., 1965; MD with distinction, U. Mich., 1969. Diplomate Am. Bd. Otolaryngology. Intern Univ. Hosp., Ann Arbor, Mich., 1969-70, resident in otolaryngology, 1971-75; resident in gen. surgery St. Joseph's Mercy Hosp., Ann Arbor, 1970-71; asst. prof. U. Va. Med. Ctr., Charlottesville, 1977-79, assoc. prof., 1979-82, prof., 1982-84; prof. Johns Hopkins U. Sch. Medicine, Balt., 1984—, dean med. faculty, v.p. medicine, 1990—; mem. Greater Balt. Com., 1991—; co-chmn. Md. Sci. Week Blue Ribbon Panel, Balt., 1992—. Co-author: Head and Neck Cancer, 1990; contbr. articles to profl. jours. Grantee Robert Wood Johnson Found., 1992. Mem. The Center Club. Office: Johns Hopkins U Sch Medicine 720 Rutland Ave Baltimore MD 21205-2196

JOHNSEN, KJELL, accelerator physicist, educator; b. Meland, Norway, June 11, 1921; arrived in Switzerland, 1959; s. Georg Martin and Borghild (Hagen) J.; m. Aase Birgitte Jordal, Dec. 29, 1945; children: Arnlaug, Georg Kjetil, Ottar. Elec. Engr., Tech. U. Norway, Trondheim, 1948, Dr. Techn., 1954. Asst. Tech. Univ. Norway, Trondheim, 1947-48; rsch. asst. Chr. Michelsen Inst., Bergen, Norway, 1948-52; physicist CERN, Geneva, Switzerland, 1952-57; prof. elec. engring. Tech. Univ. Norway, Trondheim, 1957-59; sr. scientist CERN, Geneva, 1959-86, project dir. ISR, 1966-74; part-time prof. physics U. Bergen, Norway, 1972-86; tech. dir. ISA project Brookhaven (N.Y.) Nat. Lab., 1979-82; ret., 1986; mem. adv. bd. AT divsn. Los Alamos (N.Mex.) Nat. Lab., 1987-91; chmn. HERA machine com. evaluation com. Deutsches Elekronen-Synchrotron, Hamburg, Germany, 1984-91; chmn. evaluation com. Norwegian CERN activities Norway Dept. Edn., Oslo, 1991-92. Author numerous publs. in field; co-author: Circular Accelerators and Storage Rings, 1993. Recipient Norsk Data Physics prize Norwegian Phys. Soc., 1981, Robert R. Wilson prize Am. Phys. Soc., 1990. Hon. mem. Norwegian Acad. Tech. Scis. Home: Chemin du Molard, CH-1261 La Rippe Switzerland

JOHNSON, ANTHONY O'LEARY (ANDY JOHNSON), meteorologist, consultant; b. Tampa, Fla., Apr. 19, 1957; s. Paul Bryan and Katie Hobbs (Nunez) J. BS in Meteorology, Fla. State U., 1979. Cert. cons. meteorologist. Paralegal Gregory, Cours, et. al., Tampa, 1977; water resources planner S.W. Fla. Water Mgmt. Dist., Brooksville, 1978; staff meteorologist Sta. WTVT-TV, Tampa, 1979-82, systems mgr., 1982-89, weather office mgr., 1989—; meteorol. cons. Gulf Coast Weather Svc.-Weather Vision, Tampa, 1979—; software devel. mgr. TTI Techs., Inc., Tampa, 1989—; site coord. Space Sci. and Engring. Ctr. U. Wis., Madison, 1989-91. Active Capital Improvements com. Plantation Homeowner's Assn., Tampa, 1991; judge Hillsborough Regional Sci. Fair, Tampa, 1990, 91. Mem. AAAS, Am. Meteorol. Soc. (Seal of Approval for TV weathercasting 1982—, v.p. W. Fla. chpt. 1984-85, pres. 1989-92), Phi Beta Kappa, Pi Mu Epsilon, Chi Epsilon Pi. Republican. Achievements include development of quantitative predictive methods of energy delivery interruption in severe Florida freezes; research on temporal and spatial climatological anomalies on landfalling hurricanes in West Central Florida. Office: Sta WTVT-TV Weather Svc 3213 W Kennedy Blvd Tampa FL 33609-3092

JOHNSON, BARRY LEE, public health research administrator; b. Sanders, Ky., Oct. 24, 1938; s. Otto Lee and Sarah Josephine (Deatherage) J.; m. Billie Reed, Aug. 19, 1960; children—Lee, Clay, Scott, Reed, Sarah. B.S., U. Ky., 1960; M.S., Iowa State U., 1962, Ph.D., 1967. Elec. engr. USPHS, Cin., 1960-70; bioengr. EPA, Cin., 1970-71; bioengr. Nat. Inst. for Occupational Safety and Health, HHS, Cin., 1971-77, rsch. adminstr., 1978-86; asst. adminstr. Agy. for Toxic Substances and Disease Registry, 1986—; also asst. surgeon gen., 1990—. Editor: Behavioral Toxicology, 1974, Neurotoxicology, 1986; editor Neurotoxicology, 1979, Archives Environ. Health jour., 1980, Toxicology and Industrial Health, 1985, Prevention of Neurotoxic Illness, 1987, Advances in Neurobehavioral Toxicology, 1990, Jour. of Clean Tech. and Environ. Scis., 1991. Pres. S.E. Cin. Soccer Assn., 1978-80. Recipient Commendation medal USPHS, 1980, 84, Superior Performance medal USPHS, 1986, Meritorious Svc. award USPHS, 1988, Disting. Svc. medal Asst. Surgeon Gen., 1990; USPHS fellow 1962-65. Mem. Am. Pub. Health Assn., Am. Conf. Govt. Hygienists, Am. Assn. Clin. Toxicology, Internat. Conf. Occupational Health, Am. Coll. Toxicology, Soc. of Occupational and Environ. Health, Nat. Environ. Health Assn., Internat. Assn. for Exposure Analysis. Avocation: amateur sports. Home: 1635 Amber Trl Duluth GA 30136-4941 Office: Agy for Toxic Substances and Disease Registry 1600 Clifton Rd NE Mail StopE-28 Atlanta GA 30333

JOHNSON, BERTRAND H., optical engineer. BSc, mechanical engineering, Rutgers U., New Brunswick, N.J. Dist. mem. tech. staff AT&T Bell Lab. Recipient Engring Excellence award Optical Soc. Am., 1992. Office: AT & T Bell Research Lab 600 Mountain Ave Murray Hill NJ 07974*

JOHNSON, BOBBY JOE, electronic instrumentation technician; b. Ashland, Ky., June 20, 1950; s. Charles F. and Marjorie (Hunt) J.; m. Sandra Francis Bennett, June 28, 1971; children: Carrie Jo, Robert Bennett. Student, Ky. State VoTech, 1970. Process and analytical instrument technician Ashland (Ky.) Oil, Inc., 1970—; instr. U. Ky., Ashland, 1993, Ky. State VoTech, Ashland, 1989—. Cert. tng. officer Ky. State Fire Marshall, Frankfort, 1986; fire fighter Cannonsburg (Ky.) Vol. Fire Dept., 1972. Mem. Internat. Soc. of Cert. Electronic Technicians (cert. electronic technician, Technician of Yr. award 1991), Tri-State Electronic Technicians. Mem. Ch. of God. Achievements include building and designing an automatic petroleum wax tester, designed built and installed several modifications to equipment to improve their accuracy, data recovery, and general performance so as to maintain current certifications from th EPA and ASTM, automatic fire siren control systems for fire departments. Home: 9121 Oak Hill Dr Ashland KY 41102 Office: Ashland Oil Inc PO Box 391 #2 Refinery Ashland KY 41114

JOHNSON, BONNIE JEAN, neuroscientist; b. Waukesha, Wis., Oct. 26, 1961; d. Jack Arthur and Gloria Grace (Roberts) Gross; m. Brian Wesley Johnson, June 28, 1986. BS, Case Western Res. U., 1985; MA, Ohio State U., 1990, PhD, 1993. Rsch. assoc. Ohio State U., Columbus, 1988-93, teaching assoc., 1990-93; postdoctoral rschr. Hahnemann U., Phila., 1993—. Ad hoc reviewer: Psychophysiology Jour., 1991; contbr. articles to Neurosci. Letters, Behavioral Brain Rsch. Recipient Herbert Toops prize Ohio State U., 1992, Grad. Student Alumni award, 1992. Mem. Soc. for Neurosci., Internat. Soc. for Devel. Psychobiology, Am. Psychol. Assn. (Dissertation Rsch. award 1992), Sigma Xi (grant 1992). Office: Hahnemann U Dept Neurology Mail Stop 423 Broad and Vine St Philadelphia PA 19102

JOHNSON, BRET GOSSARD, dietitian, consultant; b. Maryville, Mo., July 21, 1959; s. Thomas Porter and Irene (Gossard) J. BS, U. Nebr., 1981; M in Govtl. Adminstrn., U. Pa., 1985. Asst. dir. Food Svcs. Gracie Square Hosp., N.Y.C., 1988-89, SUNY Downstate Med. Ctr., Bklyn., 1989-90; dir. dietary and Environ. Svcs. Humana Hosp., Greenbrier Valley, Ronceverte, W.Va., 1990—; newsletter editor Food and Nutrition Coun. Greater N.Y., 1989-90; nutrition cons. Greenbrier Valley Hospice, Inc., Lewisburg, W.Va.,

1990—. Recipient Samuel S. Fels scholarship U. Pa. Fels Sch. State and Local Govt., Phila., 1983, 84. Mem. Am. Dietetic Assn., N.Y. Acad. Sci. Home: Rt 5 Box 27 Lewisburg WV 24901 Office: Humana Hosp Greenbrier Valley Davis Stuart Rd Ronceverte WV 24970

JOHNSON, BRIAN DENNIS, engineer; b. Duluth, Minn., Dec. 30, 1950; s. Theodore R. and Eleanore A. (Carlson) J.; m. Mary Beth IRene Elander, July 21, 1973; 1 child, Carl Elander Johnson. BChemE, U. Minn., 1973. Process engr. E.I. DuPont, Chattanooga, 1973-75; quality supr. BASF Wyandotte Corp., Port Edwards, Wis., 1975-78; quality engr. 3M Co., Aberdeen, S.D., 1978-80; quality devel. specialist 3M Co., St. Paul, Minn., 1980-91, product responsibility specialist, 1991—. Contbg. author: Aerosols In the Mining and Industrial Work Environment, Vol. 2, 1983; contbr. articles to tech. publs., jours. Mem. ASTM (chmn. sub-com. 1989-91, Appreciation award 1991), Indsl. Safety Equipment Assn. (chmn. task group 1989-91), Suppliers of Advanced Composite Materials Assn. Achievements include U.S. patent for stretchable laminate constrn., 1990. Office: 3M Aerospace Materials Dept 3M Ctr 209-1W-26 Saint Paul MN 55144-1000

JOHNSON, BRUCE VIRGIL, mechanical engineer, physicist, researcher; b. Nov. 24, 1935; s. Leslie Emanuel and Dora Annabel (Engeseth) J.; m. Peggy Karen Estelle Gaalaas, Nov. 28, 1959; children: Karen Estelle Johnson Ruth, Paul Leslie, Eric Virgil. BS in Mech. Engring., U. Minn., 1958, MS in Mech. Engring., 1960; MS in Physics, U. Conn., 1966, PhD in Fluid Dynamics, 1972. Asst. rsch. engr. United Techs. Rsch. Ctr., East Hartford, Conn., 1960-63, assoc. rsch. engr., 1963-66, rsch. engr., 1966-68, supr. heat & mass transfer tech., 1968—. Author and co-author more than 100 govt. and lab. reports. Mem. bd. dirs., coms. Emanuel Luth. Ch., Manchester, Conn., 1965—, fin. coms., bd. dirs. Manchester Area Conf. Chs., 1980—. Mem. ASME (internat. gas turbine inst., Gas Turbine award 1991), AIAA (sr.). Lutheran. Achievements include experimental and analytical studies in fluid mechanics and heat transfer studies relating to gas turbines and convection in rotating turbine blade coolant passages and rotating disc-cavity configurations. Home: 46 Hamilton Dr Manchester CT 06040 Office: United Techs Rsch Ctr 411 Silver Ln East Hartford CT 06108*

JOHNSON, CAGE SAUL, hematologist, educator; b. New Orleans, Mar. 31, 1941; s. Cage Spooner and Esther Georgianna (Saul) J.; m. Shirley Lee O'Neal, Feb. 22, 1968; children: Stephanie, Michelle. Student, Creighton U., 1958-61, MD, 1965. Intern U. Cin., 1965-66, resident, 1966-67; resident U. So. Calif., 1969-71; instr. U. So. Calif., L.A., 1971-74, asst. prof., 1974-80, assoc. prof., 1980-88, prof., 1988—; chmn. adv. com. Calif. Dept. Health Svcs., Sacramento, 1977—; dir. Hemoglobinopathy Lab., L.A., 1976—; bd. dirs. Sicke Cell Self-Help Assn., L.A., 1982-86. Contbr. numerous articles to profl. jours. Dir. Sickle Cell Disease Rsch. Found., L.A., 1986—; active Nat. Med. Fellowships, Inc., Chgo., 1979—; chmn. rev. com. NIH, Washington, 1986—. Major U.S. Army, 1967-69, Viet Nam. Fellow N.Y. Acad. Scis., Am. Coll. Angiology; mem. Am. Soc. Hematology, Am. Fedn. Clin. Rsch., Western Soc. Clin. Investigation, Internat. Soc. Biorheology, E.E. Just Soc. (sec.-treas. 1985—). Avocation: restoring antique automobiles. Office: U So Calif 2025 Zonal Ave Los Angeles CA 90033-4526

JOHNSON, CALVIN KEITH, research executive, chemist; b. Litchfield, Minn., Dec. 15, 1937; s. Delphin J. and Iva Mae (Watkins) J.; m. Constance S. Hoffman, June 18, 1960; children—Eric O., Judd. F., Malinda K. B.A., Olivet Nazarene Coll., Ill., 1959; Ph.D. in Chemistry, Mich. State U., 1963. Postdoctoral fellow Columbia U., N.Y.C., 1963-64; research chemist 3M Co., St. Paul, 1964-67; group leader CPC Internat., Summit, Ill., 1967-69; mgr. research and devel. Acme Resin Corp. subs. Borden, Inc., Forest Park, Ill., 1969-71, tech. dir., 1971-76, v.p., tech. dir., 1977-85, sr. v.p., tech. dir., 1985—. Patentee in field (10). Contbr. articles to tech. jours. Mem. ch. bd. 1st Ch. of Nazarene, Lemont, Ill., 1969—, Sunday Sch. supt., 1977-83; chmn. bd. Olivet Research Assocs., Kankakee, Ill., 1982—; mem., fundraiser Chickasaw Homeowners Assn., Lockport, Ill., 1979—. NSF fellow, 1961; NIH fellow, 1963. Mem. Am. Chem. Soc., Am. Foundrymen's Soc. (chmn. com.), Soc. Petroleum Engrs., AAAS, Research Dirs. Assn. Chgo., Sigma Xi. Republican. Avocations: gardening; fishing. Home: 1006 E Division St Lockport IL 60441-4507 Office: Acme Resin Corp 1372 Circle Ave Forest Park IL 60130-2419

JOHNSON, CARL RANDOLPH, chemist, educator; b. Charlottesville, Va., Apr. 28, 1937. BS, Med. Coll. Va., 1958; PhD in Chemistry, U. Ill., 1962. NSF rsch. fellow chemistry Harvard U., 1962; from asst. to prof. chemistry Wayne State U., Detroit, 1962-90, Disting. prof., 1990—; Humboldt sr. scientist, 1991; bd. dirs. Organic Synthesis, Inc. Mem. adv. bd. Jour. Organic Chemistry, 1976-81. Alfred P. Sloan fellow, 1965-68. Mem. Am. Chem. Soc. (assoc. editor jour. 1984-89, Harry and Carol Mosher award 1992), Royal Soc. Chemistry. Achievements include research in organic sulfur chemistry, especially sulfoxides and sulfoximines, exploratory synthetic chemistry, synthesis of compounds of potential medicinal activity, organometallic chemistry, synthesis of natural products, enzymes in synthesis. Office: Wayne St Univ Dept Chemistry Detroit MI 48202*

JOHNSON, CECIL KIRK, JR., cement chemist; b. Kansas City, Mo., Apr. 10, 1945; s. Cecil Kirk and Jessie Marie (Larkin) J.; m. Marie Odile Clarke, Aug. 17, 1968; children: Laurence K., Anne M. BS in Biology, N.Mex. Inst. Mining and Tech., Socorro, 1968. Testing engr. Gen. Portland, Inc., Dallas, 1972-75; supervisory chemist Gen. Portland, Inc., Fort Worth, 1975-78; plant chemist Alpha Cement Co., Orange, Tex., 1978-80; quality control mgr. Nat. Cement Co., Ragland, Ala., 1980; chief chemist Mo. Portland Cement Co., Sugar Creek, 1980-91; quality control mgr. Lafarge Corp., Sugar Creek, 1991—. Lt. USN, 1969-72. Mem. ASTM (com. C-1 1986—), Independence Eastview Lions (pres. 1990-91). Roman Catholic. Home: 18611 E 18th Terr N Independence MO 64058-1214 Office: Lafarge Corp PO Box 1017 Independence MO 64051-0517

JOHNSON, CHARLES ERIK, physics researcher; b. L.A., Sept. 6, 1966; s. Charles Richard and Gudrun (Rosmark) J. BS, UCLA, 1988, MS, 1991. Teaching asst. UCLA, 1989-91, rsch. asst. in physics, 1991—. Contbr. articles to profl. jours. Mem. Am. Phys. Soc. Home: 1221 E Oak El Segundo CA 90245 Office: UCLA Dept Physics 405 Hilgard Los Angeles CA 90024

JOHNSON, CHARLES FOREMAN, architect, architectural photographer, planning, architecture and systems engineering consultant; b. Plainfield, N.J., May 28, 1929; s. Charles E. and E. Lucile (Casner) J.; student Union Jr. Coll., 1947-48; B.Arch., U. So. Calif., 1958; postgrad. UCLA, 1959-60; m. Beverly Jean Hinnendale, Feb. 19, 1961 (div. 1970); children: Kevin, David. Draftsman, Wigton-Abbott, P.C., Plainfield, 1945-52; architect, cons., graphic, interior and engring. systems designer, 1953—; designer, draftsman with H.W. Underhill, Architect, Los Angeles, 1953-55; teaching asst. U. So. Calif., Los Angeles, 1954-55; designer with Carrington H. Lewis, Architect, Palos Verdes, Calif., 1955-56; grad. architect Ramo-Wooldridge Corp., Los Angeles, 1956-58; tech. dir. Atlas weapon system Space Tech. Labs., L.A., 1958-60; advanced planner and systems engr. Minuteman Weapon System, TRW, Los Angeles, 1960-64, div. staff ops. dir., 1964-68; cons. N.Mex. Regional Med. Program and N.Mex. State Dept. Hosps., 1968-70; prin. Charles F. Johnson, architect, Los Angeles, 1953-68, Sante Fe, N.Mex., 1968-88, Carefree, Ariz., 1988—; free lance archtl. photographer, Sante Fe, 1971—; tchr. archtl. apprentice program, 1974—; program writer, workshop leader, keynote speaker Mich. Archtl. Design Competition, 1993. Major archtl. works include: residential bldgs. in Calif., 1955-66; Bashein Bldg. at Los Lunas (N.Mex.) Hosp. and Tng. Sch., 1969, various residential bldgs., Santa Fe, 1973—, Kurtz Home, Dillon, Colo., 1981, Whispering Boulders Home, Carefree, 1981, Hedrick House, Santa Fe, 1983, Kole House, Green Valley, Ariz., 1984, Casa Largo, Santa Fe (used for film The Man Who Fell to Earth), 1974, Rubel House, Santa Fe, 1986, Smith House, Carefree, Ariz., 1987, Klopfer House, Sante Fe, 1988, Janssen House, Carefree, 1988, Art Start Gallery, 1988, Dr. Okun's House, 1990, Luterback House, Carefree, 1992, Ballagura House, Santa Fe, 1992, Phillips House, Carefree, 1992, Grayson addition, Carefree, 1992. Pres., Santa Fe Coalition for the Arts, 1977; set designer Santa Fe Fiesta Melodrama, 1969, 71, 74, 77, 78, 81; designed Jay Miller & Friends Fiesta float 1970-88 (winner of 20 awards); presenter design workshop, keynote address, agrl. design competition, Mich., 1993. Mem. Desert Mountain Gulf Club, Delta Sigma Phi. Contbr. articles on facility planning and mgmt. to profl. publs.; contbr. archtl. photographs to mags. in U.S., Eng., France, Japan and Italy, contbr. articles on facility mgmt., planning info. systems, etc. to profl. jours. Internat. Recognized for work in organic architecture and siting buildings to fit the land; named among top 100 Architects, Archtl. Digest, 1991. Avocations: music, photography, collecting architecture books, Frank Lloyd Wright works. Home: PO Box 6070 1598 Quartz Valley Dr Carefree AZ 85377

JOHNSON, CHARLES WAYNE, mining engineer, mining executive; b. Vinita, Okla., Feb. 7, 1921; s. Charles Monroe and Willie Mae (Hudson) J.; m. Cleo Faye Wittee, 1940 (div. 1952); m. Genevieve Hobbs, 1960 (dec. Sept. 1985); m. Susan Gates Johnson, Apr. 19, 1986 (div. 1992); 1 child, Karen Candace Limon. BE, Kensington U., 1974, ME, 1975, PhDE, 1976. Owner El Monte (Calif.) Mfg. Co., 1946-49; co-owner Anjo Pest Control, Pasadena, Calif., 1946-56, Hoover-Johnson Cons. Co., Denver, 1956-59; pres. Vanguard Chem. Co., Denver, 1957-61, Mineral Products Co., Boise, Idaho, 1957-61; owner Crown Hill Meml. Park, Dallas, 1959-61, Johnson Engring., 1961—; pres. Crown Minerals, Victorville, Calif., 1985-92; owner J&D Mining Co., Victorville, 1977—. Contbr. articles to profl. pubs.; patentee in field. Active Rep. VIP Club. Served with USN, 1941-45. Recipient Outstanding Achievement award East Pasadena Bus. Assn., 1948. Mem. Ch. Ancient Christianity. Avocations: prospecting, assaying, environ. pollution research. Office: Johnson Engring PO Box 1423 Thermal CA 92274

JOHNSON, DAN MYRON, biology educator; b. Albion, N.Y., May 4, 1944; s. Joseph Myron and Margaret Louise (Jones) J.; m. Karol Lynn Leasure, June 14, 1967; children: Amber Lynn, Reid Wille. BS, Emory & Henry Coll., 1965; PhD, Mich. State U., 1969. Postdoctoral fellow U. B.C., Vancouver, 1969-70; asst. prof. Rice U., Houston, 1970-76; asst. prof. East Tenn. State U., Johnson City, 1976-82, assoc. prof., 1982-88, prof. biology, 1988—; vis. asst. prof. Kellogg Biol. Sta., Hickory Corners, Mich., 1975; organizing sec. X Internat. Symposium of Odonatology, Johnson City, 1989. Contbr. articles to profl. jours. Pres. Upper East Tenn. Sci. Fair, Johnson City, 1990—. Recipient Found. Rsch. award East Tenn. State U., 1984. Mem. Am. Inst. Biol. Sci., Zool. Soc. Houston (bd. dirs. 1972-76), Ecol. Soc. Am., N.Am. Benthological Soc. (student awards chair 1990—), Societas Internationalis Odonatologica (editor Selysia newsletter 1987—), Sigma Xi, Beta Beta Beta, Sigma Mu, Phi Kappa Phi. Achievements include research on invertebrate predation, larval dragonfly behavior and ecology, and the ecology of freshwater littoral benthic communities. Office: East Tenn State U Dept Biol Scis Johnson City TN 37614-0703

JOHNSON, DANA LEE, biophysical chemist; b. Fleming, Ky., June 8, 1956; s. Roy L. and Margaret C. (Cornett) J.; m. Donna E. Haynes, Aug. 14, 1983. AAS, Rio Grande Coll., 1976, BS magna cum laude, 1983; PhD, Wayne State U., 1987. Lic. med. technologist Am. Soc. Clin. Pathologists. Med. technologist Pleasant Valley Hosp., Point Pleasant, W.Va., 1979-83; grad. rsch. fellow Wayne State U., Detoit, 1983-87; NIH postdoctoral fellow Pa. State U., University Park, 1988-90; rsch. scientist R.W. Johnson Pharm. Rsch. Inst., Raritan, N.J., 1990-91, sr. rsch. scientist, 1991—. Contbr. chpts. to books, papers to profl. jours. NIH postdoctoral fellow, 1988-91. Mem. AAAS, N.Y. Acad. Sci., Protein Soc., Am. Chem. Soc., Am. Soc. Hematology, Phi Lambda Upsilon, Chi Beta Phi. Office: RW Johnson Pharm Rsch Inst Rte 202 Box 300 Raritan NJ 08869

JOHNSON, DANIEL LLOYD, biogeographer; b. Yankton, S.D., Sept. 30, 1953; children: Sam, Eric. BSc in Biology, U. Sask., Can., 1978; MSc in Insect Biology, U. B.C., 1980, PhD in Plant Sci., 1983. Rsch. sci. Agriculture Can. Rsch. Sta., Lethbridge, Alta., Can.; adj. assoc. prof. in biogeography dept. geography U. Lethbridge, Alta., Can. Contbr. articles to profl. jours. U. Sask. Honours scholar, 1978, NSERC Postgrad. scholar, 1981, 82, Izaak Walton Killam Postgrad. scholar, 1982-83; U. B.C. Postgrad. fellow, 1980; recipient C. Gordon Hewitt Outstanding Achievement in Can. Entomology award Entomol. Soc. Can., 1992. Achievements include rsch. in ecology and control of grasshoppers attaching to grassland, cereals and oilseed crops. Office: Agriculture Can Rsch Sta, Lethbridge, AB Canada T1J 4B1 also: U Lethbridge Dept Geography, 4401 University Dr, Lethbridge, AB Canada T1K 3M4*

JOHNSON, DAVID LEE, mechanical building systems engineer; b. Fargo, N.D., Aug. 4, 1958; s. Robert Lee and Barbara Jean (Edwards) J.; m. Jenifer Lynn Bollmeier, June 7, 1980; children: Rachel Lynn, Audrey Leigh. BSME, U. Tex., 1982. Registered profl. engr., Tex., Calif., Conn., W.Va., Nev., Ariz., Okla., Hawaii. Plant engr. J.M. Huber Corp., Borger, Tex., 1982; prodn. mgr. EN Inc., Austin, Tex., 1982-87, sec.-treas., 1987-89, v.p., 1989-93; with Engring. and Constrn. Concepts, Inc., Pflugerville, Tex., 1993—; speaker on ground source heat pump systems. Mem. ASHRAE, NSPE, Tex. Soc. Profl. Engrs. Achievements include research on new technology in building mechanical systems. Home: Rt 2 Box 156DA Pflugerville TX 78660 Office: ECCI PO Box 1155 Pflugerville TX 78660

JOHNSON, DAVID SELLIE, civil engineer; b. Mpls., Apr. 10, 1935; s. Milton Edward and Helen M. (Sellie) J. BS, Mont. Coll. Mineral Sci. Tech., 1958. Registered profl. engr., Mont. Trainee Mont. Dept. Hwys., Helena, 1958-59, designer, 1959-66, asst. preconstrn. engr., 1966-68, regional engr., 1968-72, engring. specialities supr., 1972-89, preconstrn. chief, 1989—, forensic engr., 1965—, traffic accident reconstructionist, 1978—. Contbr. articles on hwy. safety to profl. jours. Adv. bd. mem. Helena Vocat.-Tech. Edn., 1972-73. Fellow Inst. Transp. Engrs.; mem. Nat. Acad. Forensic Engrs. (diplomate), Mont. Soc. Profl. Engrs., NSPE, Transp. Rsch. Bd., Wash. Assn. Tech. Accident Investigators, Corvette Club. Mem. Algeria Shrine Temple. Club: Treasure State (Helena) (pres. 1972-78). Lodges: Elks, Shriners. Avocations: photography, sports car racing. Home: 1921-6 Ave Helena MT 59601 Office: Mont Dept Transp 1921-6 Ave Helena MT 59601

JOHNSON, DAVID SIMONDS, meteorologist; b. Porterville, Calif., June 29, 1924; s. Frank David and Wanda (Simonds) J.; m. Margaret T. McFarland, Nov. 29, 1974 (dec. Dec. 1987). Student, U. Calif.-Berkeley, 1942-43, Reed Coll., 1943-44, Harvard U., 1945; AB, UCLA, 1948, MA, 1949. Meteorol. aide U.S. Weather Bur., Boise, Idaho, 1946-47; rsch. asst. to asst. meteorologist UCLA, 1947-52; assoc. meteorologist Pineapple Rsch. Inst., Honolulu, 1952-56; with U.S. Weather Bur., 1956-65, dir. Nat. Weather Satellite Ctr., 1964-65; dir. Nat. Environ. Satellite Ctr. Environ. Sci. Svcs. Adminstrn., Washington, 1965-70, Nat. Environ. Satellite Svc. NOAA, 1970-80; asst. adminstr. for satellites NOAA, 1980-82; spl. asst. to pres. Univ. Corp. for Atmospheric Rsch., Washington, 1982-83, also cons.; pres. Damar Internat., 1984-86; sr. program officer NAS-NRC, 1986—; mem. working group II com. space rsch. Internat. Coun. Sci. Unions, 1965-69; chmn. panel neutral atmosphere, 1966-69; mem. working group VI com. space rsch., 1965-78; mem. panel edn. and manpower com. atomspheric scis. NAS, 1967-69, com. for study of nation's weather observation system, 1986-87; mem. Gov. Md. Sci. Resources adv. bd., 1963-67; exec. com. panel on satellites World Meterol. Orgn., 1973-82, cons. to sec.-gen., 1982-86. Co-author: Studies of the Structure of the Atmosphere over the Eastern Pacific Ocean in Summer, 1961. With USAAF, 1943-46. Recipient Gold medal Dept. Commerce, 1965, Satellite Silver medal, 1985, Exceptional Svc. award NASA, 1966, award Nat. Civil Svc. League, 1974, William T. Pecora award, 1978, Presdl. Meritorious Svcs. award, 1980, recognition award Space Tech. Hall of Fame, 1992. Fellow AIAA (assoc.), Am. Meterol. Soc. (pres. 1974, councilor 1963-65, 68-70, 73-76, 81-83, exec. com. 1969-70, 73-75, 81-83, planning commn. 1989-93, ad hoc com. on chpts. 1993—, chmn. com. atmospheric measurements 1965-68, Brooks award 1982), Am. Geophys. Union, Am. Astronautical Soc. (bd. dirs. 1988-93, exec. com. 1988-93, nominating & fellows coms., 1990, Achievement award 1982, Lovelace award 1992); mem. AAAS, Internat. Acad. Astronautics, Cosmos Club, Sigma Xi. Mem. United Ch. of Christ. Home: 1133 Lake Heron Dr # 3A Annapolis MD 21403-3566 Office: NAS/NRC 2101 Constitution Ave NW Washington DC 20418-0001

JOHNSON, DAVID W., JR., ceramic scientist, researcher; b. Windber, Pa., Sept. 23, 1942; s. David W. Sr. and Vanessa J. (Shoff) J.; m. Bonnie Kay Respet, June 20, 1964; children: Analee J., Bradley D. BS in Ceramic Sci., Pa. State U., 1964, PhD in Ceramic Sci., 1968. Mem. tech. staff Bell Telephone Labs., Murray Hill, N.J., 1968-83; supr. advanced ceramic processing AT&T Bell Labs., Murray Hill, 1983-88; head metallurgy and ceramics rsch. dept. AT&T Bell Labs., Murray Hill, N.J., 1988—; adj. prof. Stevens Inst. Tech., Hoboken, N.J., 1982—, Nat. Acad. Engring., 1993; Taylor lectr. Pa. State U., University Park, 1989. Contbr. 120 articles to profl. jours. Chmn. Bedminster (N.J.) Twp. Zoning Bd. of Adjustment, 1991-92. Fellow Am. Ceramic Soc. (v.p. 1990-92, treas. 1992, pres. elect 1993, Ross Coffin Purdy award 1978, Fulruth award 1984); mem. AAAS, Am. Soc. Materials, Testing and Materials Soc., Materials Rsch. Soc., NAE. Achievements include patents for Ceramic Processing; research in ceramic powder processing as applied to ferrites, ceramic substrates, sol-gel silica glass and high temperature superconductors. Office: AT&T Bell Labs 600 Mountain Ave Rm 1F-206 Murray Hill NJ 07974-0636

JOHNSON, DAVID WILLIS, food products executive; b. Tumut, New South Wales, Australia, Aug. 7, 1932; came to U.S., 1976; s. Alfred Ernest and Eileen Melba (Burt) J.; m. Sylvia Raymonde Wells, Mar. 12, 1966; children: David Ashley Lawrence, Justin Christopher Kendall, Harley Alistair Kent. B in Econs., U. Sydney, Australia, 1954, diploma in Edn., 1955; MBA, U. Chgo., 1958. Exec. trainee Ford Motor Co., Geelong, Australia; mgmt. trainee Colgate-Palmolive, Sydney, 1959-60, product mgr., 1961, asst. to mng. dir., 1962, brands mgr., 1963, gen. products mgr., 1964-65; asst. gen. mgr., mktg. dir. Colgate-Palmolive, Johannesburg, Republic of South Africa, 1966, chmn., mng. dir., 1967-72; pres. Warner-Lambert/Parke Davis Asia, Hong Kong, 1973-76; pres. personal products div. Warner-Lambert Co., Morris Plains, N.J., 1977, pres. Am. Chicle Div., 1978; exec. v.p., gen mgr. Entenmann's div. Warner-Lambert Co., Bay Shore, N.Y., 1979; pres. specialty foods group Warner-Lambert Co., Morris Plains, 1980-81, v.p., 1980-82; pres., chief exec. officer Entenmann's div. Warner-Lambert Co., Bay Shore, 1982; v.p. Gen. Foods Corp., White Plains, N.Y., 1982-87; pres., chief exec. officer Entenmann's, Inc., Bay Shore, 1982-87; chmn., pres., chief exec. officer Gerber Products Co., Fremont, Mich., 1987-89, chmn., chief exec. officer, 1989-90; pres., chief exec. officer, dir. Campbell Soup Co., Camden, N.J., 1990—; bd. dirs. Colgate-Palmolive Co. Recipient Disting. Alumnus award U. Chgo., 1992. Mem. Am. Bakers Assn. (past bd. dirs.), Grocery Mfrs. Am. (bd. dirs.). Office: Campbell Soup Co World Hdqrs Campbell Pl Camden NJ 08103-1702

JOHNSON, DAVID WOLCOTT, psychologist, educator; b. Muncie, Ind., Feb. 7, 1940; s. Roger Winfield and Frances Elizabeth (Pierce) J.; m. Linda Mulholland, July 7, 1973; children: James, David, Catherine, Margaret, Jeremiah. B.S., Ball State U., 1962; M.A., Columbia U., 1964, Ed.D., 1966. Asst. prof. ednl. psychology U. Minn., Mpls., 1966-69; asso. prof. U. Minn., 1969-73, prof., 1973—; organizational cons., psychotherapist; vis. scholar Western Mich. U., 1989-90. Author: Social Psychology of Education, 1970, (with Goodwin Watson) Social Psychology: Issues and Insights, 1972, Reaching Out, 1972, 5th edit., 1993, Contemporary Social Psychology, 1973, (with F. Johnson) Joining Together, 1975, 4th edit., 1991, (with D. Tjosvold) Productive Conflict Management, 1983, Circles of Learning, 1984, 4th edit., 1993, (with R. Johnson) Learning Together and Alone, 1975, 3d edit., 1991, Human Relations and Your Career, 1978, 3d edit., 1991, Educational Psychology, 1979, Cooperative Learning, 1984, 4th edit., 1991, Structuring Cooperative Learning, 1987, Creative Conflict, 1987, Leading the Cooperative School, 1989, Cooperation and Competition: Theory and Research, 1989, Teaching Students to be Peacemakers, 1991, also film, 1991, Active Learning: Cooperative Learning in the College Classroom, 1991, Learning Mathematics and Cooperative Learning, 1991, Creative Controversy, 1992; (with R. Johnson, E. Holubec) Advanced Cooperative Learning, 1988, 2d edition, 1992, (with R. Johnson, K. Smith) Cooperative Learning: Increasing College Faculty Instructional Productivity, 1991, Cooperation in the Classroom, 6th edit., 1993, Creative Controversy, 1992, Positive Interdependence, 1992, (video) 1992; editor Am. Ednl. Rsch. Jour., 1981-83; contbr. over 250 articles to profl. jours. Bd. dirs. Walk-In Counseling Center, 1971-74. Recipient Gordon Allport award Soc. for Psychol. Study of Social Issues, 1981, Helen Plante award Am. Soc. Engring. Edn., 1984, Outstanding Rsch. award Am. Pers. and Guidance Assn., 1972, Nat. Coun.l for the Social Studies Rsch. award, 1986, Outstanding Rsch. award Am. Assn. Counseling and Devel., 1988, award for Outstanding Contbn. Am. Edn. Minn. Assn. for Supervision and Curriculum Devel., 1990, Outstanding Alumni of Yr. award Ball State U., 1990, Rsch. and Practice award Southwest Ohio Planning Coun. for Insvc. Edn., 1990. Fellow Am. Psychol. Assn.; mem. Am. Sociol. Assn., Am. Ednl. Research Assn., Am. Mgmt. Assn., Am. Assn. for Counseling and Devel. Home: 7208 Cornelia Dr Minneapolis MN 55435-4160 Office: U Minn 330 Burton Hall Minneapolis MN 55455

JOHNSON, DEBORAH CROSLAND WRIGHT, mathematics educator; b. Winston-Salem, N.C., July 17, 1951; d. Clayton Edward and Elizabeth Elliott (Bradley) Crosland; married; children: Jacqueline, Stephanie. BS in Math. Edn. magna cum laude, Appalachian State U., 1973; MEd in Math., U. N.C., Greensboro, 1976, cert., 1984. Cert. tchr., N.C., academically gifted. Tchr. math. Mt. Tabor High Sch., Winston-Salem, 1973-76, McDowell High Sch., Marion, N.C., 1976-78, Cen. Cabarrus High Sch., Concord, N.C., 1978-81; tchr. math. Walter M. Williams High Sch., Burlington, N.C., 1981—, mem. sch. improvement team, 1989—. Active First Presbyn. Ch., Burlington, 1988—. Mem. NEA, N.C. Assn. Educators, Nat. Coun. Tchrs. Math., N.C. Coun. Tchrs. Math., N.C. Assn. Gifted and Talented, Alpha Delta Kappa (hon. tchrs. sorority). Democrat.

JOHNSON, DENISE DOREEN, electrical engineer; b. Balt., Oct. 16, 1964; d. John Edward and Mary Roberta (Robinson) J. BSEE, Gannon U., 1987. Elec. designer Gipe Assocs. Inc., Easton, Md., 1987-88; TD engr. City of Dover (Del.), 1988—. Mem. Assn. of Energy Engrs. Office: City of Dover 860 Buttner Pl Dover DE 19901

JOHNSON, DEWEY E(DWARD), dentist; b. Charleston, S.C., Mar. 19, 1935; s. Dewey Edward and Mabel (Momeier) J.; A.B. in Geology, U. N.C., 1957, D.D.S., 1961. Pvt. practice dentistry, Charleston, 1964-92, assoc. to Stanley H. Karesh, Charleston, D.D.S., 1970-77, tech. market rschr., designer, Charleston, 1970-90, indsl. designer, various orgns., 1965, 75, 77, 88, 91, 92. Served to lt. USNR, 1961-63. Mem. Royal Soc. Health, Charleston C. of C. (cruise ship com. 1969), ADA, Charleston Dental Soc., Hibernian Soc., Charleston Museum, Internat. Platform Assn., Charleston Library Soc., S.C. Hist. Soc., Gibbes Art Gallery, Preservation Soc. of Charleston, Navy League of U.S., Phi Kappa Sigma, Sigma Gamma Epsilon, Psi Omega. Congregationalist. Club: Optimist. Achievements include various scientific and engineering designs; patent in dental matrix device. Home: 142 S Battery Charleston SC 29401-1828

JOHNSON, DONALD LEE, agricultural materials processing company executive; b. Aurora, Ill., Mar. 9, 1935; s. Leonard F. and Fern J. (Johnson) J.; m. Virginia A. Wesoloski, Sept. 3, 1960; children: Joyce E., Janis M., Jolene G., Jay R. AS, Joliet Jr. Coll., 1959; BS, U. Ill., 1962; DSc, Washington U., 1966. Devel. engr. Petrolite Corp., Webster Groves, Mo., 1962-64; sr. devel. engr. A.E. Staley Co., Decatur, Ill., 1965-67, rsch. mgr. chem. div., 1967-75, dept. dir. rsch. div., 1975-87; v.p., dir. rsch. Grain Processing Corp., Muscatine, Iowa, 1987—, Nat. Acad. Engring., Washington, 1993; mem. applied sci. adv. coun. Miami U., Oxford, Ohio, 1987—, departmental vis. com. botany dept. U. Tex., Austin, 1986—, adv. coun. adult vocat. edn. State of Ill., Springfield, 1983-87; chmn. rev. com. Solar Energy Rsch. Inst., Golden, Colo., 1988-89. Contbr. sci. papers to profl. jours.; patentee in field. Staff sgt. USAF, 1953-57. Mem. AAAS, Am. Chem. Soc., Am. Inst. Chem. Engrs., Tech. Assn. Pulp and Paper Industries, Inst. Food Techs., Am. Legion, Rotary. Republican. Avocations: sailboat racing, running. Home: 2684 Samuel Clemens Rd Muscatine IA 52761-9751 Office: Grain Processing Corp 1600 Oregon St Muscatine IA 52761-1404

JOHNSON, DONALD REX, research institute administrator; b. Tacoma, July 19, 1938; s. Richard Carl and Frieda Maria (Dahlstrom) J.; m. Karen Yvonne Neswoog, 1959; children: Eric Richard, Bradley Allen. BS in Physics, U. Puget Sound, 1960; MS in Physics, U. Idaho, 1962; PhD in Physics, U. Okla., 1967. Physicist optial physics div. Nat. Bur. Standards, Gaithersburg, Md., 1967-76, program analyst, 1976-78, dep. dir. for programs Nat. Measurement Lab., 1978-82, dir. Nat. Measurement Lab., 1982-89; dir. tech. svcs. Nat. Inst. Standards & Tech., Gaithersburg, 1989-93; acting dir. Nat. Tech. Info. Svc., Springfield, Va., 1991-92, dir., 1993—; mem. adv. bd. Md. Office Tech., Annapolis, 1990-92. Contbr. articles to profl. jours. Active Bicentennial Coordinating Com., Gaithersburg, 1975-76, Gaithersburg Planning Commn., 1977-78, Montgomery Edn. Connection, Rockville, Md., 1985-92, Montgomery County High Tech. Coun., Rockville, 1985-92; chmn. bd. overseers Ctr. for Advanced Rsch. in Biotech., Rockville,

1990—. Recipient Arthur S. Flemming award Jaycees, 1976, Sr. Exec. Meritorious Svc. award Pres. of U.S., 1981, Disting. Exec. award Pres. of U.S., 1989. Fellow Am. Phys. Soc.; mem. ASTM, ASME (adv. bd. 1991), Internat. Astron. Union, Am. Astron. Union, Sigma Xi. Presbyterian. Avocation: antique car restoration. Home: 15609 Haddonfield Way Gaithersburg MD 20878-3624 Office: Tech Svcs Bldg 221 Rte 270 Gaithersburg MD 20899

JOHNSON, EDWARD A., physician, educator. M.D., U. Sheffield, 1953. James B. Duke prof., chmn. dept. physiology Duke U. Sch. Medicine, Durham, N.C. Office: Duke Univ Sch Medicine Dept Physiology PO Box 3005 Durham NC 27710-0001

JOHNSON, ELMER MARSHALL, toxicologist, teratologist; b. Midlothian, Ill., June 16, 1930; s. Burt and Gertrude Esther (Miller) j.; m. Sharon Ann Coyle, May 9, 1976; children: Mark Dee, Kim Lea, Erik Marshall, Lora Marlys. Student, U. Mex., 1948; diploma, Thornton Jr. Coll., 1950; BS, Tex. A&M U., 1954, MS, 1955; PhD, U. Calif., Berkeley, 1959. Rsch. asst. U.S. Army Surgeon Gen./Tex. A&M U. Rsch. Found., College Station, 1955; instr. anatomy and physiology Contra Costa Coll., San Pablo, Calif., 1958-59; instr. U. Fla. Coll. Medicine, Gainesville, 1960-61, asst. prof., 1961-65, assoc. prof., 1965-68, prof., 1968-70, acting chmn. dept. anatomy, 1969-70; prof.. chmn. dept. anatomy, prof. dept. developmental and cellular biology U. Calif., Irvine, 1970-72; prof., chmn. dept. anatomy and devel. bology, dir. Daniel Baugh Inst. Thomas Jefferson U., Phila., 1972—; founding pres., chmn. bd. dirs. Argus Rsch. Labs., Inc., Perkasie, Pa.; cons. Allied Chem. Co., Dow Chem. Co., Johnson & Johnson, Pub. Utilities, EPA, U.S. Naval Hosp., Phila., Kirkland & Kirkland, Esq., Hoffman-LaRoche, Merck, Inc., Regulatory Liaison Group, NAS, NRC, Columbia Nitrogen Co., Uniroyal, 3M, FMC, J&J, GE, Sumitomo, Chlorine Inst. Assoc. editor Teratology, 1974-81, Jour. Environ. Pathology and Toxicology, 1979—, Reproductive Toxicology, 1989—, Jour Am. Coll. Toxicology. 2d lt. U.S. Army, 1959. USPHS predoctoral fellow, 1953-55; March of Dimes rsch. grantee 1963-68, Growth Soc. rsch. grantee, 1972-74, NIH rsch. grantee, 1963-74, NIH teratology predoctoral tng. grantee, 1955-59, 85-90. Mem. AAAS, Teratology Soc., Am. Assn. Anatomists, Assn. Anatomy Chairmen, Genetic Toxicology Assn., Soc. Toxicology, Am. Coll. Toxicology, So. Soc. Anatomists, Mid Atlantic Reprodn. and Teratology Assn., Sigma Xi. Republican. Unitarian. Office: Thomas Jefferson U 1020 Locust St Philadelphia PA 19107-6799

JOHNSON, ERIC WALTER, neuroscientist; b. Winchester, Conn., May 2, 1958; s. William C. and Marian M. (Miller) J.; m. Jo B. Bascetta, June 8, 1985; children: Ryan William, Ashley Jean, Arin Michael. BA cum laude, Boston U., 1980; PhD, U. Pa., Phila., 1988. Postdoctoral assoc. dept. psychiatry Yale U. Sch. Medicine, New Haven, 1987-90, postdoctoral assoc. sect. neurol. surgery, 1990-92; scientist The Jackson Lab., Bar Harbor, Maine, 1992—; cons. sect. neurol. surgery Yale U. Sch. Medicine, New Haven, Bigerontronix, Inc. Town com. chair A Conn. Party, New Hartford, 1991-92, del. state conv., 1992. U. Pa. grad. fellow, 1986-87, devel. genetics fellow NIH, Jackson Lab. Mem. Soc. Neurosci., Internat. Brain Rsch. Orgn., Am. Epilepsy Soc. Achievements include research into development of several neuroanatomical imaging techniques including quantitative autoradiography, SPECT imaging and genetic analysis of epilepsy. Office: The Jackson Lab 600 Main St Bar Harbor ME 04609

JOHNSON, EUGENE WALTER, mathematician; b. El Paso, Tex., May 25, 1939; s. Walter Albert and Lillian Ann (Martinets) J.; m. Sandra Sue Gilbert, Oct. 16, 1959; 1 dau., Catherine Mary. Student, Riverside City Coll., 1958-60; B.A., U. Calif., Riverside, 1963, M.A., 1964, Ph.D., 1966. Asst. prof. Eastern N.Mex. State U., 1966; asst. prof. math. U. Iowa, Iowa City, 1966-70; asso. prof. U. Iowa, 1970-75, prof., 1975—, chmn. dept., 1976-79. Contbr. articles to profl. jours. Mem. Am. Math. Soc., Math. Assn. Am. Democrat. Home: 3303 Lower West Branch Rd Iowa City IA 52245-4103 Office: Univ Iowa Dept Math Iowa City IA 52242

JOHNSON, F. MICHAEL, electrical and automation systems professional; b. Sacramento, Jan. 14, 1953; s. Carroll Loren and Constance (Latterell) J.; m. Donna Louise Hamilton, June 28, 1975; children: Bryan J., Cassandra L. BSChemE, U. Calif., Davis, 1975. Registered profl. control systems engr., Calif. Field engr., instrumentation Universal Oil Products, 1975-80; project engr. and leader Atkinson System Techs. Co., 1980-85; control systems project leader, system mgr. spl. project Stearns-Roger, Denver, 1985-87; digital systems engr., mgr. spl. software applications CH2M Hill, Denver, 1987-91; dept. mgr., sr. project leader for software and instrumentation Ch2M Hill, Milw., 1991-92, mgr. dept. elec. and I/C, 1992-93, sr. controls cons., joint startup leader, 1993—. Recipient Top Cat award Atkinson Systems Tech. Co., 1985. Mem. Am. Chem. Soc., Instrument Soc. Am., Toastmasters. Avocations: family activities, primitive camping. Office: Ch2M Hill 310 Wisconsin Ave Ste 700 Milwaukee WI 53203-2277

JOHNSON, GARY WAYNE, aerospace engineer; b. Henderson, Tex., July 14, 1950; s. Charles Childers and Allene (Olsen) J.; m. Ellen Duckworth, Jan. 10, 1976; 1 child, James M. BS in Aero. Engring., U. Tex., 1972, MS in Aero. Engring., 1974. Rsch. asst. Wind Tunnel Lab., U. Tex., Austin, 1972-74; engring. analyst LTV Aerospace "Scout" Launch Support, Dallas, 1974; teaching asst. aerospace dept. U. Tex., Austin, 1974-75; engr. Hercules Aerospace, McGregor, Tex., 1975-83; R&D engr. Tracor Aerospace, Austin, 1983-87; tech. program mgr. Hercules Aerospace, McGregor, 1987-89, R&D engr./R&D programs mgr., 1989—; aero. dept. vis. com. U. Tex., Austin, 1988-90; cons. W.J. Schafer Assocs., Chelmsford, Mass., 1991. Contbr. articles to profl. jours. Scoutmaster Boy Scouts Am., Waco, Tex., 1991—. With USN, 1969. Mem. AIAA (sr.), AAAS, NSPE, SAE. Achievements include patent on fuel injector for ducted rocket motor; rapid deployment scheme for towed aircraft decoys; minimum signature ducted rocket fuel propellant, feasibility demonstration of fuel-air powered infrared decoy concept. Home: Rt 2 Box 44A McGregor TX 76657 Office: Hercules Aerospace PO Box 548 McGregor TX 76657

JOHNSON, GARY WILLIAM, environmental scientist, consultant; b. Warwick, R.I., Feb. 23, 1957; s. Donald Milton and Elaine Carin (Soderlund) J.; m. Diane Lynn Farrell, Aug. 1, 1992. BA in Biology, U. R.I., 1979; MS in Environ. Sci., U. New Haven, 1987. Registered environ. profl. Nat. Registry Environ. Profls. Researcher Nat. Marine Fisheries Svc., Narragansett, R.I., 1978-79; asst. scientist N.E. Utilities, Waterford, Conn., 1979-84; assoc. scientist N.E. Utilities, Berlin, Conn., 1984-86; scientist N.E. Utilities, Rocky Hill, Conn., 1986—; prin. scientist Ecologic Risk Mgmt. Svcs., Monroe, Conn., 1989—; guest lectr. U. New Haven, 1990—; lectr. in field. Contbr. numerous articles to profl. jours. Vol. sci. guide East Lyme (Conn.) Jr. High Sch., 1983—; guide, lectr. Audubon Soc., Jamestown, R.I., 1983-85. Mem. AAAS, Soc. for Risk Analysis, Edison Electric Power Industry Biologists. Achievements include development of state of the art computer models to perform quantitative analysis of ecologic and human health risk from exposure to toxic materials; research in condenser biofouling control efforts of the nuclear power industry in the early 1980's. Home: 2 Melanie Dr Waterford CT 06385 Office: NE Utilities Environ Lab Millstone Nuclear Power Sta Rope Ferry Rd Rt 156 Waterford CT 06385

JOHNSON, GEORGE EDWIN, hydraulic engineer; b. Sturgeon Bay, Wis., Feb. 28, 1933; s. George Edwin and Leona Dorothy (Sorenson) J.; m. Thelma Elizabeth Odegaard, Nov. 19, 1955 (div. 1977); children: Marc, Eric, Kathryn; m. Karen Arlene Hartz, Dec. 10, 1977; stepsons: Harry, Quinton. BS in Civil Engrin., U. Wis., 1960; MS in Civil Engring., Iowa State U., 1970. Registered profl. engr., Wis., Iowa, Ill. Hydraulic engr. U.S. Army Engr. Dist., Rock Island, Ill., 1960-67, chief hydrology sect., 1967-69, chief water control sect., 1969-76, chief hydraulics br., 1976-92, chief hydrology and hydraulics br., 1992—; mem. adv. bd. dept. Civil Engring. Iowa State U., Ames, 1980-86, mem. water resource planning com., 1980—. Mem. coun. Village Bd., Milan, Ill., 1972-73; mem., pres. Rock Island Sch. Bd., Rock Island, 1973-76. Sgt. USMC, 1952-55. Named Engr. of Yr., Quad City Engring. and Sci. Coun., 1983. Fellow ASCE; mem. Am. Soc. Mil. Engrs. (Paul Norton award 1991), Rotary Club (pres. 1991-92, Paul Harris fellow 1991), Quad City Engring. and Sci. Coun. (pres. 1990-91). Home: 3 Parkwood Dr Davenport IA 52803 Office: US Army Engr Dist Clock Tower Bldg Rock Island IL 61201

JOHNSON, GEORGE PATRICK, science policy analyst; b. Pine Bluff, Ark., June 16, 1932; s. George Ferman and Anne Lucille (Ferarra) J.; m. Jean Marie, Dec. 27, 1967; children: Heather, Patrick, Margaret. BS, U. Miss., 1954; PhD, Stanford U., 1971. Registered profl. engr., La. Lt. USN, 1954-65; design engr. Port of New Orleans, 1957-58; ops. analyst SRI Internat., Menlo Park, Calif., 1966-71; rsch. engr. U.S. Army Corp of Engrs., Ft. Belvoir, Va., 1971-74; program mgr. NSF, Washington, 1974—; head Europe office NSF, Paris, 1988-90; sci. policy analyst NSF, Washington, 1990—; Contbr. articles and reports to profl. publs. Recipient NSF Special Achievement award, 1971, NSF Outstanding Performance award, 1982, 1985. Mem. AAAS, Sigma Xi. Office: NSF 1800 G St NW Washington DC 20550

JOHNSON, GERALD, III, cardiovascular physiologist, researcher; b. Liberty, Tex., Aug. 16, 1945; s. Gerald Jr. and Jimmie Leah (Hensley) J.; m. Delynda Juanice Wall, Sept. 20, 1985. MS, U. Okla., 1971; PhD, U. Okla., Oklahoma City, 1980. NIH stipendiary U. Okla., Oklahoma City, 1972-76, rsch. assoc., 1979-80; electrophysiologist Childrens Med. Ctr., Tulsa, Okla., 1980-82; post-doct. fellow Oral Roberts U. Sch. Medicine, Tulsa, 1982-84, asst. prof., 1984-88; sr. rsch. fellow Jefferson Med. Coll., Phila., 1988-90; assoc. prof. dept. medicine, health scis. ctr. U. Okla., 1990—; dir. cardiovascular lab. W. K. Warren Med. Rsch. Inst., Tulsa, 1990—; cons. McGee Rehab. Inst., Phila., 1990, Dept. Pediatrics City of Faith Hosp., Tulsa, 1982, Aerobics Ctr. Oral roberts U., Tulsa, 1981; rsch. asst. to assoc. VA Hosp., Oklahoma City, 1970-72, Cen. State Hosp., Norman, Okla, 1969-70, U. Okla. Health Scis. Ctr., Oklahoma City, 1979-80; mem. numerous coms. Oral Roberts U., 1984-88. Contbr. numerous articles to profl. jours.; presenter in field. Grantee The Hearst Found., 1981-82, Am. Heart Assn., 1985-86; recipient Travel award Biofeedback Soc. Am., 1981, Citation Paper awards, 1981, 82. Mem. AAAS, Am. Heart Assn., Am. Physiol. Soc., Am. Soc. Nuclear Cardiology (founding), Fedn. Am. Socs. for Experimental Biology, Soc. Nuclear Medicine, N.Y. Acad. Scis., Soc. of Sigma Xi. Achievements include research on protective effects of exercise training in shock, adrenoceptor relationships in hypertension, morphologic differences in vasculature during hemorrhagic hypotension, protective effects of nitric oxide and sodium nitrite in ischemia/reperfusion role of endothelium in myocardial ischemia reperfusion. Office: W K Warren Med Rsch Inst 6465 S Yale Ave ste 1010 Tulsa OK 74136

JOHNSON, GIFFORD KENNETH, testing laboratory executive; b. Santa Barbara, Calif., June 30, 1918; s. Elvin Morgan and Rosalie Dorothy (Schlagel) J.; m. Betty Jane Crockett, June 10, 1944; children: Craig, Dane, Janet. Student, Santa Monica (Calif.) City Coll., 1938-39, UCLA, 1940, Harvard Bus. Sch., 1944. With N.Am. Aviation, Inc., 1935-41; chief indsl. engr. Consol. Vultee Aircraft Corp., 1941-48; prodn. mgr. to pres. Chance Vought Aircraft Corp., 1950-61; pres., chief operating officer Ling-Temco-Vought, Inc., 1961-64; pres. S.W. Center Advanced Studies, 1965-69; chmn., pres., chief exec. officer Am. Biomed. Corp., 1969-79; exec. v.p. Nat. Health Labs., 1978-81; chmn., chief exec. officer Woodson-Tenent Labs. Inc., Dallas, 1981—. Mem. devel. bd. U. Tex., Dallas; v.p., trustee Excellence in Edn. Found.; pres. C.C. Young Meml. Home; bd. dirs. Tex. A&M Research Found. Mem. Am. Clin. Lab. Assn. (dir.), Navy League (life). Methodist (treas.). Clubs: Northwood Country; Salesmanship (Dallas). Home: 10555 Pagewood Dr Dallas TX 75230-4255 Office: 10300 NC Expy Bldg 4 Suite 220 Dallas TX 75231

JOHNSON, HERMAN LEONALL, research nutritionist; b. Whitehall, Wis., Apr. 1, 1935; s. Frederick E. And Jeanette (Severson) J.; m. Barbara Dale Matthews, July 3, 1960 (dec. May 1971); m. Barbara Ann Badger, Apr. 3, 1976. BA in Chemistry, North Cen. Coll., Naperville, Ill., 1959; MS in Biochemistry & Nutrition, Va. Poly. Inst. and State U., 1961, PhD in Biochemistry and Nutrition, 1963. Rsch. biochemist S.R. Noble Found., Ardmore, Okla., 1963-65; nutrition chemist U.S. Army Med. Rsch., Denver, 1965-74; nutrition physiologist Letterman Army Rsch., Presidio San Francisco, 1974-80, Western Human Nutrition Rsch. Ctr. USDA, Presidio San Francisco, 1980—. Contbr. numerous articles to profl. jours. Trustee 1st Meth. Ch., Rohnert Park, Calif., 1985—. Served with Med. Service Corps, U.S. Army, 1954-56. Named one of Outstanding Young Men of Am., 1975; NIH traineeship Va. Poly. Inst. and State U., Blacksburg, 1961-63. Mem. AAAS, Am. Inst. Nutrition, Am. Soc. Clin. Nutritionists, Am. Coll. Nutritionists, Am. Coll. Sports Medicine, Sebastopol Spinners, Sigma Xi, Phi Lambda, Phi Sigma. Republican. Achievements include research on human nutrition. Home: 256 Alden Ave Rohnert Park CA 94928-3704 Office: USDA Western Human Nutrition Rsch Ctr PO Box 29997 San Francisco CA 94129-0997

JOHNSON, HOWARD PAUL, agricultural engineering educator; b. Odebolt, Iowa, Jan. 27, 1923; s. Gustaf Johan and Ruth Helen (Hanson) J.; m. Patricia Jean Larsen, June 15, 1952; children—Cynthia, Lynette, Malcolm. B.S., Iowa State U., 1949, M.S. in Agrl. Engring., 1950; M.S. in Hydraulic Engring., U. Iowa, 1954; Ph.D., Iowa State U., 1959. Registered profl. engr., Iowa. Engr., Soil Conservation Service, Sioux City, Iowa, 1949; instr. Iowa State U., Ames, 1950-53, 54-59, asst. prof., 1959-60, assoc. prof., 1960-62, prof. agrl. engring., 1962-80, head dept., 1980-88, prof. emeritus; cons., 1960-80. Contbr. numerous articles, papers to profl. lit. Co-editor Hydrologic Modeling, 1981. Patentee flow meter. Pres., Sawyer Sch. PTA, Ames, 1965; precinct rep. Republican party, Ames, 1980. Served with AUS, 1943-46, ETO. Recipient Iowa State U. Gamma Sigma Delta Merit award, 1983; EPA grantee, 1975-80; Anson Marston Disting. Prof. Engring., 1986. Fellow AAAS, Am. Soc. Agrl. Engrs. (div. chmn. 1969-70, tech. council 1974-76; Engr. of Yr. Iowa sect. 1981, Mid-Central sect. 1982). Baptist. Lodge: Rotary. Avocations: reading; photography; fishing. Office: Iowa St U Dept Agri Engring 100 Davidson Hall Ames IA 50010

JOHNSON, JAMES HODGE, engineering company executive; b. Brantley, Ala., May 22, 1932; s. James Hodge Sr. and Julia Grace (McSwean) J.; m. Carol June Farris; 1 child, Mark Farris. BSChemE, U. Ala., Tuscaloosa, 1959; postgrad., Columbia U., 1978. Process engr. to gen. supt. Union Camp Corp., Savannah, Ga., 1959-73; v.p., div. mgr. Branchemco, Inc., Jacksonville, Fla., 1973-75; mill mgr. Champion Internat. Corp., Canton, N.C., 1975-79; resident mgr. Boise Cascade Corp., Rumford, Maine, 1979-81, St. Regis Paper Co., Jacksonville, 1982-83; gen. mfg. mgr. Kraft div. St. Regis Paper Co., West Nyack, N.Y., 1983-84; v.p., div. gen. mgr. Manville Forest Products Co., West Monroe, La., 1984-85; cons. Birmingham, Ala., 1986; dir. project devel. Rust Internat. Corp., Birmingham, 1987—. Rep. precinct leader Ga., Savannah, 1962-65; trustee Ind. Presbyn. Day Sch., Savannah; bd. dirs. Pine Tree Coun. Boy Scouts Am., Maine, 1975-79. With U.S. Army, 1953-56. Disting. Engring. fellow U. Ala., Tuscaloosa, 1988. Mem. Paper Industry Mgmt. Assn. (various positions Dixie div. 1975-79, program chmn. conf. 1979), Tech. Assn. Pulp and Paper Industry, Kiwanis (pres. 1990), Inverness Country Club, Toastmasters (pres. 1960-68), Tau Beta Pi. Republican. Presbyterian. Avocations: golf, fishing, personal computer. Home: 3152 Bradford Pl Birmingham AL 35242-4602 Office: RUST Internat Corp 100 Corporate Pky Birmingham AL 35242-2982

JOHNSON, JAMES LAWRENCE, clinical psychologist, writer; b. Devils Lake, N.D., Sept. 17, 1953; s. Lawrence Tillman and Irene (Fah) J.; m. Paula Lou Sechler, Aug. 28, 1981; children: Daniel, Michael, Alisha. BA, U. N.D., 1975; MA, Azusa Pacific U., 1980; PhD, U.S. Internat. U., San Diego, 1990. Lic. marriage, family, child counselor and psychologist. Writer, editor Campus Crusade for Christ, San Bernardino, Calif., 1975-77; marriage, family, child counselor Foothill Community Mental Health Ctr., Glendora, Calif., 1978-81; social worker Le Roy Boys Home, LaVerne, Calif., 1981-83; counselor Creative Counseling Ctr., Pomona and Claremont, Calif., 1983—, Covina (Calif.) Psycholog. Group, 1981—; freelance writer, 1977-78, 90—; counselor trainer Stephen Ministries, Covina, 1985-86. Contbr. numerous articles to profl. and popular jours. Editor, advisor 1st Bapt. Ch., Covina, 1988, cabinet leader singles ministries, 1980-82. Mem. APA (clin.), Calif. Assn. Marriage and Family Therapists (clin.), Christian Writers' Fellowship, Psi Chi. Avocations: photography, backpacking, basketball, theater, music. Office: Creative Counseling Ctr 250 W 1st St Ste 214 Claremont CA 91711-4743

JOHNSON, JAMES NORMAN, physicist; b. Tacoma, Sept. 6, 1939; s. Ralph Johnson and Bertha Elizabeth (Merry) Mikal; m. Carol Ann Berry, Dec. 19, 1959; children: Kevin Ronald, Kerry Jane, Timothy James. BS, U. Puget Sound, 1961; PhD, Wash. State U., 1966. Staff mem. Sandia Labs., Albuquerque, 1967-73, Los Alamos (N.Mex.) Nat. Lab., 1976—; staff cons. Terra Tek, Inc., Salt Lake City, 1973-76; mem. rev. panel shock and detonation NATO, Brussels, 1990—. Mem. editorial bd. Springer-Verlag, Berlin, 1989-92; editor: Los Alamos Performance Data, 1982, APS Shock Compression, 1989. Recipient Heineman award NATO, 1985. Mem. Am. Phys. Soc., Mu Sigma Delta. Home: 4544 Ridgeway Dr Los Alamos NM 87544 Office: Los Alamos Nat Lab MS-B221 PO Box 1663 Los Alamos NM 87545

JOHNSON, JANET LEANN MOE, statistician, project manager; b. Mpls., July 19, 1941; d. Arnold Olvin and Ruby Victoria (Nelson) Moe; m. Donald Michael Johnson, Sept. 4, 1965; children: Michael John, Jennifer Kay. BA, U. Minn., 1962; MS, Rochester (N.Y.) Inst. Tech., 1983. Teaching asst. dept. maths. U. Minn., Mpls., 1960-62; mathematician Corning (N.Y.) Inc., 1962-63; sci. forecaster Corning (N.Y.) Glass Works, 1963-66, sci. programmer, 1966-70, sr. statis., 1970-79, sr. devel. engr., 1979-85; sr. project engr. Corning Inc., 1985—. Author: An Analysis of the Low-Level Performance Channel Multiplier Arrays, 1969, Effects of Vacuum Space Charge in Channel Multipliers, 1969. Asst. troop leader Webelos, Boy Scouts Am., Painted Post, N.Y., 1979; troop leader Girl Scouts U.S., Painted Post, 1982-83. Mem. Am. Soc. for Quality Control (sr.), Am. Statis. Assn., Soc. Women Engrs. (sr., past sec. rep. and pres. Twin Tiers chpt. 1978—, Engr. of Yr. 1988). Achievements include patent in glass ceramics for dental constructs; research on non-traditional machining and chemical vapor deposition coating. Office: Corning Inc EJ-33 Corning NY 14831

JOHNSON, JEAN ELAINE, nursing educator; b. Wilsey, Kans., Mar. 11, 1925; d. William H. and Rosa L. (Welty) Irwin. B.S., Kans. State U., 1948; M.S. in Nursing, Yale U., 1965; M.S., U. Wis., 1969, Ph.D., 1971. Instr. nursing Iowa, Kans. and Colo., 1948-58; staff nurse Swedish Hosp., Englewood, Colo., 1958-60; in-svc. edn. coord. Gen. Rose Hosp., Denver, 1960-63; rsch. asst. Yale U., New Haven, 1965-67; assoc. prof. nursing Wayne State U., Detroit, 1971-74, prof., 1974-79; dir. Ctr. for Health Rsch., 1974-79; prof. nursing, assoc. dir. oncology nursing Cancer Ctr., U. Rochester, 1979—; Rosenstadt prof. health rsch. Faculty Nursing U. Toronto, 1985. Contbg. author: Handbook of Psychology and Health, vol. 4, 1975; contbr. articles to profl. jours. Recipient Bd. Govs. Faculty Recognition award Wayne State U., 1975, award for disting. contbn. to nursing sci. Am. Nurses Found. and Am. Nurses Assn. Coun. for Nurse Rschrs., 1983, Grad. Teaching award U. Rochester, 1991, Disting. Rschr. award Oncology Nursing Soc., 1992; NIH grantee 1972, 91-95. Fellow AAAS, Acad. for Behavioral Medicine Rsch., Am. Psychol. Assn.; mem. ANA (chmn. coun. for nurse researchers 1976-78, mem. commn. for rsch. 1978-82), Inst. Medicine NAS (com. on patient injury compensation 1976-77, membership com. 1981-86, governing coun. 1987-89), Sigma Xi, Omicron Nu, Phi Kappa Phi. Home: 1412 East Ave Rochester NY 14610-1619 Office: U Rochester Cancer Ctr 601 Elmwood Ave Rochester NY 14642

JOHNSON, JEFFREY ALLAN, industrial engineer; b. West Chester, Pa., Mar. 27, 1967; s. Ronald E. Johnson and Paula M. (Scharek) Neville; m. Susan Lee Booker, Nov. 9, 1991. BSIE, N.C. State U., 1991. Indsl. engr. Am. Express, Greensboro, N.C., 1991—. Mem. Inst. Indsl. Engrs. Republican. Baptist. Avocations: golf, tennis. Office: American Express 6500 Airport Pky PO Box 35029 Greensboro NC 27425

JOHNSON, JEROME LINNÉ, cardiologist; b. Rockford, Ill., June 19, 1929; s. Thomas Arthur and Myrtle Elizabeth (Swanson) J.; m. Molly Ann Rideout, June 27, 1953; children: Susan Johnson Nowels, William Rideout. BA, U. Chgo., 1951; BS, Northwestern U., 1952, MD, 1955. Diplomate Nat. Bd. Med. Examiners. Intern U. Chgo. Clinics, 1955-56; resident Northwestern U., Chgo., 1958-61; chief resident Chgo. Wesley Meml. Hosp., 1960-61; mem., v.p. Hauch Med. Clinic, Pomona, Calif., 1961-88; pvt. practice cardiology and internal medicine Pomona, 1988—; clin. assoc. prof. medicine, U. So. Calif., L.A., 1961—; mem. staff Pomona Valley Hosp. Med. Ctr., chmn. coronary care com. 1967-77; mem. staff L.A. County Hosp. Citizen ambassador, People to People; mem. Town Hall of Calif., L.A. World Affairs Coun. Lt. USNR, 1956-58. Fellow Am. Coll. Cardiology, Am. Geriatrics Soc., Royal Soc. Health; mem. Galileo Soc., Am. Heart Assn. (bd. dirs. L.A. County div. 1967-84, San Gabriel div. 1963-89), Am. Soc. Internal Medicine, Inland Soc. Internal Medicine, Pomona Host Lions. Avocations: photography, swimming, bicycling, medical and surgical antiques, travel. Home: 648 Delaware Dr Claremont CA 91711-3457

JOHNSON, JERRY DOUGLAS, biology educator, researcher; b. Salina, Kans., Sept. 1, 1947; s. Maynard Eugene and Norma Maude (Moss) J.; m. Kathryn Ann Johnson, May 12, 1973; children: George Walker, Brett Arthur. BS in Zoology, Fort Hays State U., 1972; MS in Biology, U. Tex., El Paso, 1975; PhD in Wildlife Sci., Tex. A&M U., 1984. Teaching asst. biology dept. U. Tex., El Paso, 1973-75; instr. biology El Paso C.C., 1975—; adj. asst. prof. U. Tex., El Paso, 1984—; Piper prof. El Paso Community Coll., 1989-90; bd. scientists Chihuahuan Desert Rsch. Inst., Alpine, Tex., 1991—. Co-author: Middle American Herpetology, 1988; contbr. 17 articles to profl. jours. on herpetofauna of the western hemisphere. Bd. dirs. Meml. Park Improvement Assn., 1987—, El Paso Coun. for Internat. Visitors, 1988—, Parks and Recreation Bd., El Paso, 1991—. Grantee Soc. Sigma Xi, 1974, Theodore Roosevelt Found. Am. Mus. Natural History,1979, Exline Corp., 1980. Mem. NSF, Nat. Ctr. for Academic Achievement, Nat. Inst. Gen. Med. Sci., Soc. for Study of Amphibians and Reptiles (elector 1980), Southwestern Assn. of Naturalists (assoc. editor 1977-85, bd. govs. 1985-89), Herpetologists League and 13 other profl. orgns. Home: 3147 Wheeling El Paso TX 79930 Office: El Paso CC Biology Dept PO Box 20500 El Paso TX 79998

JOHNSON, JOHN IRWIN, JR., neuroscientist; b. Salt Lake City, Aug. 18, 1931; s. John Irwin and Ann Josephine (Freeman) J. A.B., U. Notre Dame, 1952; M.S., Purdue U., 1955, Ph.D., 1957. Instr., then asst. prof. Marquette U., Milw., 1957-60; USPHS spl. research fellow U. Wis., Madison, 1960-63; Fulbright-Hays research scholar U. Sydney, Australia, 1964-65; asso. prof. biophysics, psychology and zoology Mich. State U., E. Lansing, 1965-69; prof. Mich. State U., 1969-81, prof. anatomy, 1981—, chmn. dept. biophysics, 1973-78; vis. fellow psychology dept. Yale U., New Haven, 1975-76. Recipient Career Devel. award NIH, 1965-72, research grantee, 1966-79; research grantee NSF, 1969-71, 71-73, 73-76, 78-89, 91—; 3d hon. life mem. Anat. Assn. Australia and N.Z., 1973. Mem. Soc. Neurosci., Am. Assn. Anatomists, Am. Soc. Zoologists, Am. Soc. Mammalogists, Animal Behavior Soc., AAUP, ACLU, Sigma Xi. Home: 2494 W Grand River Ave Okemos MI 48864-1447 Office: Mich State U Dept Anatomy 514A E Fee Hall East Lansing MI 48824-1316

JOHNSON, JOHN ROBERT, petroleum company executive; b. Omaha, Apr. 17, 1936; s. Robert William and Hazel Marguerite (White) J.; BS, Davidson Coll., 1958; m. Margaret Elizabeth Roberts, June 20, 1959; children: Robert Harle, Martha Elizabeth. With Johnson Oil Co. Inc., Morristown, Tenn., 1961—, pres., 1963—; dir. Lakeway Pubs., United So. Bank. Magistrate, Hamblen County Ct., 1968-78, chmn., 1971-72; elder 1st Presbyn. Ch., Morristown; pres. Hamblen County United Fund, 1969; pres. Great Smoky Mountain coun. Boy Scouts Am., 1977-78, 86; mayor Morristown, 1977-87, 91—; mem. Tenn. Adv. Commn. on Intergovtl. Rels., 1987—. Lt. U.S. Army, 1958-61. Recipient Disting. Svc. award Morristown Jr. C. of C., 1966; named Tenn. Mayor of Yr., Tenn. Mcpl. League, 1983. Mem. Morristown C. of C. (pres. 1976), Tenn. Oil Marketers Assn., Rotary. Democrat. Home: 505 Hale Ave Morristown TN 37813-1830 Office: 1206 S Cumberland St Morristown TN 37813

JOHNSON, JOHNNY, research psychologist, consultant; b. Clarksdale, Miss., Jan. 10, 1938; s. Eddie B. and Elizabeth (Ousley) J.; children: Tonya, Anita. Student, Coahoma Jr. Coll., 1957, Hunter Coll., 1964, N.Y.U., 1963; BS, Tenn. State U., 1970, MS, 1974; postgrad., Saybrook Inst., 1987-89. Instr. Dept. of the Navy, Millington, Tenn., 1976-80, edn. specialist, 1980-87, curriculum advisor, 1987-88; prof. human resources mgmt. Pepperdine U., L.A., 1975-77; prof. psychology Shelby State Community Coll., Memphis, Tenn., 1985—. Actor film Elvis, 1989, Memphis, 1990, The Firm, 1993. With USN, 1957-63. Mem. APA (assoc.), Am. Psychol. Soc., Soc. Psychol. Study of Social Issues, Assn. Black Psychologists, Soc. Psychol. Study Gay and Lesbian Issues. Avocations: golf, dog breeding, music, foreign languages, pocket billiards. Home: 773 Margie Dr Memphis TN 38127-2727

JOHNSON, JOSEPH ERLE, mathematician; b. Memphis, Apr. 27, 1951; s. Louis Miller and Harriette Edith (Geiger) J. BS in Applied Math., Ga. Inst. Tech., 1975. Tax examiner IRS, Atlanta, 1975-77; sec., treas. Louis M. Johnson & Co., Memphis, 1977-82; grad. asst. dept. math. scis. Memphis State U., 1983-84; warehouse adminstr. The Julien Co., Memphis, 1986-89; with Venture Constrn. Co., Memphis, 1990-91, Crager Constrn. Co., Memphis, 1991-92; data processing mgr. Finishing Techs., Inc., Chattanooga, Tenn., 1993—; treas. Memphis Astron. Soc., 1980-81. Mem. Soc. for Indsl. and Applied Math. Home: 613 Tremont St Apt 2 Chattanooga TN 37405

JOHNSON, JOYCE MARIE, psychiatrist, epidemiologist; b. Baton Rouge, Jan. 30, 1952; d. Gene Addison and Helen Marie (Kalcik) J.; m. James Albert Calderwood, Mar. 28, 1987; 1 child, James. B.A., Luther Coll., Decorah, Iowa, 1972; M.A., U. Iowa, 1974; D.O., Mich. State U., 1980. Cert. in clin. psychiatry, public health and preventive medicine, and pharmacology. Cooking instr. Kirkwood Community Coll., Iowa City, Iowa, 1974-76; health planner Iowa Regional Med. Program, Iowa City, 1974-76; intern USPHS Hosp., Balt., 1980-81; med. epidemiologist Hepatitis Labs., Ctrs. Disease Control, Phoenix, 1981-83, AIDS, Ctrs. Disease Control, Atlanta, 1983-84; resident in psychiatry NIMH, 1984-87, staff psychiatrist, 1987-88; epidemiologist, acting div. dir. Food and Drug Adminstrn., 1988-93; dir. divsn. nat. treatment demonstrations, Substance Abuse and Mental Health Svcs. Adminstrn., 1993—. Med. Perspectives fellow, New Guinea and Thailand, 1978-79; mem. clin. faculty Mich. State U., 1983—, Georgetown U. Med. Ctr., 1988—; vol. Uniformed Svcs. U. of the Health Scis. Mem. U.S. Public Health Svc., Mensa. Office: 5518 Western Ave Bethesda MD 20815

JOHNSON, KEITH EDWARD, chemist; b. Hampton, Va., Oct. 25, 1968; s. Edward S. and Nellie M. (Horsely) J. BS, Longwood Coll., 1990. Technician's aide Va. Power-Yorktown Power Sta., 1988-89; chemistry technician Va. Power-Surry Power Sta., 1990—. Recipient Joseph Levendusky scholarship Pitts. Engrs. Soc., 1989-90. Office: Va Power Surry Nuclear Power Sta End of Rt 650 Surry VA 23883

JOHNSON, KENNETH LANGSTRETH, engineer, educator; b. Barrow in Furness, Cumbria, U.K., Mar. 19, 1925; s. Frank Herbert and Ellen Howarth (Langstreth) J.; m. Dorothy Rosemary, Sept. 11, 1954; children: Marian Ruth, Hilary Christine, Andrew Roger. BS in Tech., Manchester U., U.K., 1944, MS in Tech., 1949, PhD, 1955. Tech. asst. Rotol Ltd., Gloucester, U.K., 1944-49; asst. lectr. Manchester U., 1949-54; from lectr. to prof. Cambridge U., U.K., 1954-92. Author: Contact Mechanics, 1985; contbr. articles to profl. jours. Recipient Mayo D. Hersey awd., Am. Soc. Mechanical Engineers, 1991. Fellow Instn. Mech. Engrs., Royal Soc. London, Royal Acad. Engring., Am. Soc. Trib., and Lub. Engrs. (hon.). Office: Cambridge U Engring Dept, Trumpington St, Cambridge CB2 1PZ, England

JOHNSON, KENNETH PETER, neurologist, medical researcher; b. Jamestown, N.Y., Mar. 12, 1932; s. Kenneth Peter and Nina (Bengtson) Johnson; m. Jacquelyn Johnson, June 23, 1956; children: Peter, Thomas, Diane, Douglas. B.A., Upsala Coll., East Orange, N.J., 1955; M.D., Jefferson Med. Coll., Phila., 1959. Diplomate: Am. Bd. Psychiatry and Neurology. Intern Buffalo Gen. Hosp., 1959-60; resident Hosp. of Cleve., 1963-65; asst. prof. neurology Case Western Res. U., Cleve., 1968-71, assoc. prof., 1971-74; prof. U. Calif., San Francisco, 1974-81; prof., chmn. U. Md., Balt., 1981—; chief neurology VA Hosp., Balt., 1981-83. Editor: Neurovirology, 1984; contbr. numerous articles in field to profl. jours. Served to lt. U.S. Navy, 1961-63. Recipient Weil award Am. Assn. Neuropathology, 1967; recipient Research Ctr. Devel. award NIH, 1968-73; Zimmerman lectr. Stanford U., 1981. Fellow Am. Neurol. Assn.; mem. Am. Acad. Neurology, Am. Soc. Virology, Am. Clin. and Climatol. Assn., Am. Congress Rehab. Medicine, Am. Soc. Neurorehab., Teratology Soc., Soc. for Exptl. Neuropathology, Internat. Soc. for Neuroimmunology. Lutheran. Home: 49 Seminary Farm Rd Lutherville Timonium MD 21093-4508 Office: U Md Hosp Neurology Dept N4W46 22 S Greene St Baltimore MD 21201-1544

JOHNSON, L. RONALD, geophysicist; b. Broken Bow, Nebr., Oct. 6, 1938; s. Leonard Reuben Johnson and Lela Mae (Mills) Pike; m. Nellie Ann Westover, Feb. 8, 1959; children: Dennis Logan, Laura Lea, Coralie Rondell, Marsha Jo. BS in Math., Kearney State, 1968; MS in Physics, S.D. Sch. Mines, 1970. Grad. rsch. asst. Physics Dept. S.D. Sch. Mines Tech., Rapid City, 1968-70, grad. rsch. assoc., 1970; rsch. sci. Inst. Atmospheric Scis., Rapid City, 1970-77; rsch. scientist II S.D. Sch. Mines and Tech., Rapid City, 1977-81, rsch. scientist III, assoc. prof. physics, 1982—; vis. sci. Lawrence Berkeley (Calif.) Labs., 1981-92; chair Sci. Fair Com., Rapid City, 1988-92; chair Promotion and Tenure Com., Rapid City, 1990. Contbr. articles to profl. jours. Chair Woodland Hills Community Assn., Black Hawk, S.D., 1974-76; sec.-treas. S.D. Massage Assn., 1983-88, Hobie Fleet 198, Western S.D., 1986-88. Mem. Am. Meteorol. Soc., Oahe Yacht Club, Sigma Xi (chpt. pres. 1993-94). Democrat. Avocations: sailing, photography, woodwork, exercise, music. Home: 8004 Woodland Dr Black Hawk SD 57718-9536 Office: SD Sch Mines and Tech 501 E St Joseph St Rapid City SD 57701-3995

JOHNSON, LAWRENCE ALAN, cereal technologist, educator, researcher, administrator; b. Columbus, Ohio, Apr. 30, 1947; s. William and Wyoma (Swift) J.; m. Bernice Ann Miller, June 15, 1969; children: Bradley, David. BS, Ohio State U., 1969; MS, N.C. State U., 1971; PhD, Kans. State U., 1978. Rsch. chemist Durkee Foods div. SCM Corp., Strongsville, Ohio, 1973-75; assoc. rsch. chemist Food Protein R&D Ctr. Tex. A&M U., College Station, 1978-85; prof.-in-charge Ctr. for Crops Utilization Rsch. Iowa State U., Ames, 1985-91; mem. rsch. com. Am. Soybean Assn., St. Louis, 1987-91, Nat. Corn Grower's Assn., St. Louis, 1990-91. Author: (with others) Handbook of Cereals, 1991; editor: (book--procs.) Technologies for Value-Added Products from Proteins and Co-Products, 1989; contbr. over 60 articles to profl. jours. 1st lt. U.S. Army, 1971-73, Vietnam. Mem. Am. Assn. Cereal Chemists (assoc. editor jour. 1982-85), Am. Soc. Agrl. Engrs., Am. Oil Chemists Soc. (assoc. editor jour. 1989—, Archer Daniels Midland Rsch. award 1986), Inst. Food Techs. Republican. Lutheran. Achievements include 7 patents. Home: 2226 Buchanan Dr Ames IA 50010-4368 Office: Iowa State U Ctr Crops Utilization Rsch Ames IA 50011

JOHNSON, LEE FREDERICK, molecular geneticist; b. Phila., Jan. 10, 1946; s. Robert W. and Jeannette (Mollenkof) J.; m. Ann Marie Lester, June 10, 1967; children: Adam, Karl. BS, Muhlenberg Coll., 1967; PhD, Yale U., 1972. Postdoctoral fellow MIT, Cambridge, Mass., 1972-75; asst. prof. Ohio State U., Columbus, 1975-80, assoc. prof., 1980-85, prof., 1985—, chmn., 1990—; panel mem. NSF, Washington, 1980-84, Am. Cancer Soc., Ohio Div., Columbus, 1982-88; bd. dirs. Ohio Cancer Rsch. Assn., Columbus. Author more than 60 scientific papers. Troop com. chmn. Boy Scouts Am., Columbus, 1987-90. Recipient Faculty Rsch. award Am. Cancer Soc., 1980; rsch. grantee NIH, NSF, 1978—. Achievements include rsch. on analysis of structure and expression of the mouse thymidylate synthase gene. Office: Ohio State U Molecular Genetics 484 W 12th Ave Columbus OH 43210

JOHNSON, LENNART INGEMAR, materials engineering consultant; b. Mpls., Dec. 23, 1924; s. Sixten Richard Wilhem and Marie Augusta (Johansson) J.; m. Muriel Grant, Oct. 7, 1961; 1 child, Sandra Lee. BS in Chem. Engring., U. Minn., 1948. Petroleum engr. Northwestern Refining Co., New Brighton, Minn., 1948-49; sr. engr. ordnance div. Honeywell, Hopkins, Minn., 1949-67, prin. materials engr. def. systems div., 1967-69, supr. engring. def. systems div., 1969-87; staff engr. armament systems div Honeywell Inc., Hopkins, Minn., 1987-88; cons. Soc. Automotive Engring., Warrandale, Pa., 1989—; forum leader and presenter, U. Wisc. Engring. Inst., Madison, 1965. Contbr. articles to profl. jours. Mem. credentials com. Hennepin County Rep. Conv., Minn., 1972, alt. del., 1974. Recipient Prize Paper award, Inst. Elec. Engrs. Fellow Am. Inst. Chemists; mem. Soc. Automotive Engrs. (sec. composites com. 1986-87, chmn. 1987-88), Am. Inst. Chem. Engrs. Achievements include development of and research in injection molding technology, urethane and epoxy casting resins, and urethane foaming resins; preparation of numerous Aerospace Material Specifications published by Society of Automotive Engineers. Home and Office: 14109 Mount Terr Minnetonka MN 55345

JOHNSON, LINDA THELMA, computer consultant; b. New Britain, Conn., May 18, 1954; d. Oren and Lois Elizabeth (Armstrong) J.; 1 child, Portia Lauren. BS in Econs., Va. State U., 1978; cert. in computer programming, Morse Sch. Bus., 1978. Programmer analyst Vitro Automation Industries, Silver Spring, Md., 1980-83; sr. analyst Sci. Mgmt. Corp., Lanham, Md., 1984-86; sr. programmer analyst Applied Mgmt. Scis. Inc., Silver Spring, 1986; programmer analyst Computer Data Systems Inc., Rockville, Md., 1986-88; project leader systems cert. dept. Arbitron Co., Laurel, Md., 1988-90; systems analyst Engring. and Econ. Rsch., Inc., Vienna, Va., 1990; computer cons. Comsys Tech. Svcs. Inc., Rockville, 1990, CPU Inc., Fairfax, Va., 1991; quality assurance cons. Cigna Corp., Bloomfield, Conn., 1992; mem. rsch. bd. advisors The Am. Biographical Inst., Inc. Mem. NAFE, NAACP, Am. Bus. Women's Assn. Democrat. Baptist. Avocations: crossword puzzles, horseback riding. Home: 386 Park Ave Bloomfield CT 06002

JOHNSON, MALCOLM PRATT, marketing professional; b. New Haven, Aug. 9, 1941; s. Malcolm and Evelyn (Pratt) J.; m. Patricia J. Johnson, June 27, 1964; children: David, Christopher. BA, Amherst Coll., 1963; PhD, Northwestern U., 1967. Rsch. chemist Union Carbide, Charleston, W.Va., 1966-71; gen. mgr. Humphrey Chem., Baytown, Tex., 1971-77; bus. mgr. S.W. Spl. Chem., Houston, 1977-80; mktg. mgr. Dixie Chem. Co., Houston, 1980—. Contbr. articles to profl. jours. Republican. Methodist. Home: 3 Thunderbird Cir Baytown TX 77521 Office: Dixie Chem Co PO Box 130410 Baytown TX 77219

JOHNSON, MARK ALLAN, electrical engineer; b. Corvallis, Oreg., May 10, 1966; s. William Allen and Lee Ann (Blessing) J. BS, MIT, 1989, MS, 1989, postgrad., 1991—. Coop. edn. student Motorola Rsch. and Devel., Schaumburg, Ill., 1986-89; engr. Fujitsu Labs. Ltd., Kawasaki, Japan, 1989-90; curriculum cons. Kyoto (Japan) Sch. of Computer Sci., 1993. Contbr. articles to profl. jours. Classroom vol. Cambridge (Mass.) Head Start, 1993. Mem. IEEE (fellow 1988-89), Info., Elec. and Commun. Engrs., Acoustical Soc. Am., Phi Beta Kappa, Sigma Xi, Tau Beta Pi, Eta Kappa Nu (pres. 1987-88). Achievements include patents for speech coding and decoding system, speech coding system, also patents pending. Office: MIT Rsch Lab for Electronics Cambridge MA 02139

JOHNSON, MARK DAVID, anesthesiologist, educator; b. Redlands, Calif., July 16, 1953; s. Walter Magnus and Gloria Constance (Mueller) J.; 1 child, Jeremy. BS, Worcester Poly. Inst., 1977; MD, U. Mass., Worcester, 1980. Diplomate Am. Bd. Anesthesiology, Nat. Bd. Med. Examiners; cert. med. technologist, Mass. Intern and instr. surgery U. Mass. Med. Ctr., 1980-81; resident in anesthesia Brigham and Women's Hosp., Boston, 1981-83, fell in obstet. anesthesia, then in cardiac anesthesia, 1983-84, chief resident in anesthesia, 1984; fellow respiratory ICU, Mass. Gen. Hosp., Boston, 1984; clin. fellow in anesthesia Harvard Md. Sch., Boston, 1981-84, instr., 1984-89, asst. prof., 1990-92; assoc. prof. anesthesia and ob-gyn U. Tex. Southwestern Med. Sch., Dallas, 1992—; dir. div. obstetric anesthesia Parkland Hosp., Dallas, 1992—; lectr., vis. prof., 1984—; vis. prof., lectr. U. Rochester, 1991; expert cons. FDA, Washington, 1992; cons. Smith Industries. Contbr. numerous articles, revs. and abstracts to med. jours., chpts. to books. Mem. AMA, Am. Acad. Clin. Anesthesiologists, Am. Med. Technologists Assn., Anesthesia Patient Safety Found., Am. Soc. Anesthesiologists, Am. Soc. Critical Care Medicine, Am. Soc. Regional Anesthesia, Internat. Anesthesia Rsch. Soc., Neuroanesthesia Soc., Ophthol. Anesthesia Soc., Soc. for Ambulatory Anesthesiologists, Soc. for Edn. in Anesthesia, Soc. Obstetrics, Anesthesia and Perinatology, Soc. for Pain Practice Mgmt., Undersea Med. Soc., Am. Trauma Soc., also others. Achievements include research on continuous spinal-cauda equina model systems, isolated working heart prep-electrophysiology of local anesthetic cardiotoxicity, research and development of continuous spinal micro catheter system. Home: 61 Broad Reach T-83 B Weymouth MA 02191

JOHNSON, MARK DEE, pharmacologist, researcher; b. Bryan, Tex., Mar. 10, 1954; s. E. Marshall and Marlys Claire (Van Overstraton) J.; m. Linda Sue Mulcahy, Apr. 5, 1986; 1 child, Ellen Kimberly. BS, Pa. State U., 1977; PhD, Duke U., 1983. Postdoctoral fellow Roche Inst. of Molecular Biology, Nutley, N.J., 1983-86; asst. prof. pharmacology Med. Coll. of Pa., Phila., 1986—. Contbr. articles to profl. jours. Grantee Am. Heart Assn. of S.E. Pa., 1992, Am. Heart Assn. of Del., 1992, Pharm. Mfrs. Assn., 1989, Am. Fedn. for Aging Rsch., 1989. Mem. Am. Soc. for Pharmacology and Exptl. Therapeutics, Soc. for Neurosci., N.Y. Acad. Sci., Sigma Xi. Office: Med Coll of Pa 3200 Henry Ave Philadelphia PA 19129

JOHNSON, MARK HENRY, psychology educator; b. London, June 4, 1960; came to U.S., 1990; BS, Edinburgh U., Scotland, 1982; PhD, Cambridge U., Eng., 1985. Rsch. scientist Med. Rsch. Coun., London, 1985-90, 93—; assoc. prof. Carnegie Mellon U., Pitts., 1991—. Author: Biology and Cognitive Development, 1991; editor: Brain Development and Cognition, 1993. Achievements include research on the relation between brain devel. and cognitive devel., especially with regard to visual attention, perception and memory. Office: Psychology Dept Carnegie Mellon U Pittsburgh PA 15213

JOHNSON, MARLIN DEON, research facility administrator; b. Salt Lake City, June 18, 1938; s. Leonard A. and Ena (Johnson) J.; m. Diane Beverley, July 21, 1957; children: Roger, Brenda, Walter. BS, Utah State U., 1964, MS, 1969, splty. cert., 1984. Design engr. ACF Industries, Albuquerque, 1964-66, Dynalectron corp., Albuquerque, 1969-72; instr. Utah State U., Logan, 1966-69; instl. dir. Uinlah Basin Applied Tech. Ctr., Roosevelt, Utah, 1972-81; regional dir. Uinlah Basin Applied Tech. Ctr., 1988—; dir. secondary sch. Uinlah Sch. Dist., Vernal, Utah, 1981-87, secondary sch. prin., 1987-88. Regional chmn. State-wide Planning Commn., Salt Lake City, 1980, Gov.'s Blue Ribbon Coun., 1981; bd. dirs. Ashley Valley Community Ctr., Vernal, 1989-91. Mem. Assn. Curriculum Devel., Nat. Tech. Edn. Assn., Am. Vocat. Assn., Phi Delta Kappa. Home: 661 W 300 N Vernal UT 84078 Office: Uinlah Basin Applied Tech Ctr Ste 1036 1680 W Hwy 40 Vernal UT 84078

JOHNSON, MAURICE VERNER, JR., agricultural research and development executive; b. Duluth, Minn., Sept. 13, 1925; s. Maurice Verner Sr. and Elvira Marie (Westberg) J.; m. Darlene Ruth Durand, June 23, 1944; children: Susan Kay, Steven Dale. BS, U. Calif., 1953. registered profl. engr. From research engr. to dir. research and devel. Sunkist Growers, Ontario, Calif., 1953-84; v.p. research and devel. Sunkist Growers, Ontario, 1984-90, retired; cons. to pres., 1990—; v.p., dir. Calif. Citrus Quality Council, Claremont. Contbr. articles to profl. pubs.; patentee in field. Sgt. U.S. Army, 1944-46, ETO. Fellow Am. Soc. Agrl. Engrs. (dir. 1969-70); mem. ASME, Am. Inst. Indsl. Engrs., Am. Assn. Advancement Sci., Nat. Soc. Profl. Engrs., Tau Beta Pi. Republican. Avocation: golf. Office: Sunkist Growers 760 E Sunkist St PO Box 3720 Ontario CA 91761

JOHNSON, NED KEITH, ornithologist, educator; b. Reno, Nov. 3, 1932. BS, U. Nev., 1954; PhD in Zoology, U. Calif., 1961. From asst. to assoc. prof. U. Calif., Berkeley, 1962-74, prof. zoology, 1974—, vice chmn. dept. zoology, 1968—, asst. curator birds mus. vertebrate zoology, 1962-63, curator birds, 1963—, acting dir., 1981. Rsch. grantee NSF, 1965—. Mem. Am. Soc. Zoology, Am. Ornithologists' Union (William Brewster Meml. award 1992), Am. Soc. Naturalists, Cooper Ornithol. Soc., Soc. Study Evolution, Soc. Systematic Zoology. Achievements include research in biosystematics, distribution and ecology of New World birds. Office: Univ of California Museum Vertebrate Zoology Berkeley CA 94720*

JOHNSON, NOEL LARS, biomedical engineer; b. Palo Alto, Calif., Nov. 11, 1957; s. LeRoy Franklin and Margaret Louise (Lindsley) J.; m. Elise Lynnette Moore, May 17, 1986; children: Margaret Elizabeth, Kent Daniel. BSEE, U. Calif., Berkeley, 1979; ME, U. Va., 1982, PhD, 1990. Mgr. automated infusion systems R&D hosp. products div. Abbott Labs., Mountain View, Calif., 1986—. Contbr. articles to profl. jours. Fellowship NIH 1980-85; rsch. grantee Abbott Labs. 1989. Mem. IEEE, Biomed. Engring. Soc., Am. Soc. Anesthesiologists, Sigma Xi, Delta Chi (founder, 1st pres. chpt. U. Calif. at Berkeley). Achievements include invention of respiratory monitor, patented automated drug delivery system and critical care disposables. Home: 6649 Canterbury Ct San Jose CA 95129-3871 Office: Abbott Labs Hosp Products Divsn 1212 Terra Bella Ave Mountain View CA 94043-1899

JOHNSON, OMOTUNDE EVAN GEORGE, economist; b. Freetown, Sierra Leone, Mar. 27, 1941; came to U.S., 1961; s. Evan George and Elizabeth O. (Allen) J.; m. Octavia Olayemi John, Oct. 30 1965; children: Olatunde Cheryl, Omoyemi Evan, Olubayo Darryl. BA, UCLA, 1965, MA, 1967, PhD, 1970. Lectr. in econs. Calif. State U., Long Beach, 1967-69; lectr. U. Sierra Leone, Freetown, 1969-73; vis. asst. prof. U. Mich., Ann Arbor, 1973-74; economist IMF, Washington, 1974-79, sr. economist, dep. div. chief, 1979-92, advisor, 1992—; resident rep. IMF, Ghana, 1987-90. Contbr. numerous articles to profl. jours. Mem. AAAS, Am. Econ. Assn., U.S. Chess Fedn., Royal Econ. Soc. of U.K., Nat. Symphony Orch. Assn., N.Y. Acad. Scis., Wash. Performing Arts Soc., Internat. Platform Assn., Metropolitan Opera Guild. Episcopalian. Avocations: chess, piano, reading. Home: 6117 Tammy Dr Alexandria VA 22310-1524 Office: IMF 700 19th St NW Washington DC 20431-0002

JOHNSON, PATRICIA LYN, mathematics educator; b. Upper Sandusky, Ohio, June 28, 1957; d. Chester Ellsworth and Judith Lyn (Adams) Geary; m. James Walter Johnson III, June 12, 1976. BS, Ohio State U., 1981, MS, 1983. Grad. teaching assoc. math. dept. Ohio State U., Columbus, 1980-87, lectr., 1987-88; supr. Math. Learning Ctr., instr. math. Ohio State U., Lima, 1988—; judge math. counts Lima (Ohio) Soc. Profl. Engrs., 1989—; judge Math. Awareness Week, Joint Policy Bd. for Math., Lima, 1990—. Mem. Am. Math. Soc., Math. Assn. Am., Assn. for Women in Math. Methodist. Office: Ohio State U 4240 Campus Dr Lima OH 45804-3597

JOHNSON, PAULINE BENGE, nurse, anesthetist; b. London, Ky., May 10, 1932; d. Chester G. and Bertha M. (Hale) Benge; m. Scottie W. Johnson, Apr. 29, 1950 (dec. 1976); children: Rita Johnson, Nita Johnson Yaw, Gina Johnson Carlson. AA, U. Ky., 1968; diploma, U. Cin. Sch. Nurse Anesthesia, 1971; BS summa cum laude, U. Cin., 1974, M., 1977, D., 1981. RN, Ohio, Ky., Tenn., Ind., W.Va., Fla., Tex., S.C.; cert., lic. RN anesthetist. Staff anesthetist Jewish Hosp., Cin., 1971-72, Mercy North Hosp., Hamilton, Ohio, 1972-86, Ft. Hamilton Hosp., Hamilton, 1972-86, McCullough-Hyde Hosp., Oxford, Ohio, 1986-88; freelance anesthetist multiple hosps. Ohio, Ky., 1982-88; staff anesthetist, ind. contractor Shriner Burn Inst., Cin., 1989; pres., staff anesthetist, ind. contractor multiple hosps. Pauline B. Johnson Co., Inc., Ohio, Ky., Tenn., Ind., W. Va., Fla., Tex., S.C., 1989—. Ch. clk. Lindenwald Bapt. Ch., Hamilton, 1955-72, mem. 1955-85, instr., 1955-76; mem. 1st Bapt. Ch., Hamilton, 1985—, NOW, 1978—, nominating com. major polit. party, Hamilton, 1986-89; mem., med. com. Planned Parenthood, Hamilton, 1987—. Scholar U. Cin., 1969-71, 77-81; recipient Spl. Recognition Higher Edn., Laurel County Homecoming, London, Ky., 1988. Mem. Am. Assn. Nurse Anesthetists (speaker nat. conv. 1982, speaker rsch. forum nat. meeting 1989, mem. nominating com. 1978), Ohio State Assn. Nurse Anesthetists (state bd. dirs. 1989-92, 88-90, 79-80, chair bylaws com. 1991-92, 92-93, nominating com. 1993—, chair edn. com. 1990-91, pres. 1982-84, state editor Highlights 1974-82, co-chair state meeting 1982, pres. dist. 5 Cin. 1978, govt. rels. capt. Greater Cin. chpt. 1976-87, speaker meetings), Kappa Delta Pi (pres. 1974—). Avocations: swimming, picnicking, reading, music, travel. Home: 128 South F St Hamilton OH 45013

JOHNSON, QULAN ADRIAN, software engineer; b. Great Falls, Mont., Sept. 17, 1942; s. Raymond Eugene and Bertha Marie (Nagengast) J.; m. Helen Louise Pocha, July 24, 1965; children—Brenda Marie, Douglas Paul, Scot Paul, Mathew James. B.A. in Psychology, Coll. Gt. Falls, 1964. Lead operator 1st Computer Corp., Helena, Mont., 1966-67; v.p., sec.-treas. Computer Corp. of Mt., Great Falls, 1967-76, dir., 1971-76; sr. systems analyst Mont. Dept. Revenue, Helena, 1976-78; software engr. Mont. Systems Devel. Co., Helena, 1978-80; programmer/analyst III info. systems div. Mont. Dept. Adminstrn., Helena, 1980-82; systems analyst centralized services Dept. Social and Rehab. Services State of Mont., 1982-87, systems and programming mgr. info systems, Blue Cross and Blue Shield of Montana, Helena, 1987—. Mem. Assn. for Systems Mgmt., Mont. Data Processing Assn., Data Processing Mgmt. Assn., Mensa. Club: K.C. (rec. sec. 1975-76). Home: 2231 8th Ave Helena MT 59601-4841 Office: Blue Cross & Blue Shield Info Systems 404 Fuller Ave Helena MT 59601-5006

JOHNSON, RALPH THEODORE, JR., physicist; b. Salina, Kans., Apr. 29, 1935; s. Ralph Theodore and Mary Alice (Wallerius) J.; m. Ruth Elaine Rohrer, Jan. 25, 1958; children: Barbara A., Thomas T., Gregory E., Janet E. MS in Physics, Kans. State U., 1959, PhD, 1964. Staff mem. GE, Cin., 1957-58; rsch. and teaching asst. Kans. State U., 1958-63; from rsch. scientist to rsch. supr. Sandia Nat. Lab., Albuquerque, 1965-85, mgr., 1985—. Contbr. articles to profl. jours. 1st lt. USAF, 1963-65. Mem. Am. Phys. Soc., Sigma Xi. Achievements include patent for radiation detector for measuring neutron fluence; related surface crysallization in amorphous semiconductors to memory phenomenon; identification of importance of moisture on ionic conduction in lithium solid electrolytes. Home: 6601 Arroyo Del Oso NE Albuquerque NM 87109 Office: Sandia National Lab Dept 4307 PO Box 5800 Albuquerque NM 87185

JOHNSON, RICHARD DEAN, pharmaceutical consultant, educator; b. De Kalb, Ill., July 8, 1936; s. Arthur Dean and Evelyn Alice (Telford) J.; B.S., U. Calif., Berkeley, 1960; Pharm. D., U. Calif., San Francisco, 1961, M.S., 1962, Ph.D., 1965; M.B.A., Rockhurst Coll., Kansas City, Mo., 1984; m. Paula Marcellus Jennings, Nov. 3, 1942; children—Janet Telford, Julie Tess, Richard Dean, Jennings Brodie. Sect. head research and devel. Allergan Pharms., Irvine, Calif., 1965-67; dir. regulatory affairs Syntex Labs., Inc., Palo Alto, Calif., 1967-73; mng. dir. licensing Marion Labs., Inc., Kansas City, Mo., 1973-79, v.p. licensing, 1980-82, v.p. corp. devel., 1983-87, v.p. bus. alliances, 1987-89; corp. v.p. Marion Merrell Dow Inc., Kansas City, Mo., 1989-91, ret.; adj. prof. Sch. Pharmacy U. Mo., Kansas City, 1991—, rsch. coun., 1993—; bd. dirs. Dey Labs. Inc., Concord, Calif., Tanabe-Marion Labs., Kans. City, U.S. Biosci. Inc., Blue Bell, Pa., ImmunoPharmaceutics, Inc., San Diego; guest lectr. U. S.C. Coll. Bus. Adminstrn., Columbia, 1975-79. Presdl. exchange exec. U.S. Dept. Commerce, Washington, 1970-71, U.S. Pharmacopeia Com. of Rev., 1990—. Recipient Grad. award Borden Co., 1962; Am. Found. for Pharm. Edn. fellow, 1962-64; Sir Henry S. Wellcome Meml. fellow, 1962-63; Am. Inst. Chemists fellow, 1965-70. Mem. Am. Chem. Soc., Am. Pharm. Assn., Acad. Pharm. Sci., AAAS, Pharm. Mfrs. Assn., Fedn. Internat. Pharmacy, Licensing Exec. Soc., Sigma Xi, Rho Chi. Republican. Clubs: Balboa Bay (Newport Beach, Calif.), Carriage (Kansas City, Mo.), Hallbrook Country Club. Contbr. articles to pharm. jours. Home: 5330 Ward Pky Kansas City MO 64112-2369 Office: Johnson Assocs 222 W Gregory Blvd Kansas City MO 64114-1127

JOHNSON, RICHARD WARREN, chemist; b. Auburn, Calif., Aug. 12, 1947; s. Harold Warren and Ruby L. (Schmoker) J.; m. Theresa Lynn Murray, Aug. 17, 1974; children: Brenna E., Kayla M.L., Savannah R. BA in Chemistry, Park Coll., 1969. Chemist Hazleton Labs. Inc., Madison, Wis., 1969-84; chemist/owner Wis. Analytical and Rsch. Svcs. Ltd., Madison, 1984—. Mem. Am. Assn. of Pharm. Scis. Home: 765 Eddington Sun Prairie WI 53590 Office: WARS Ltd 1202 Ann St Madison WI 53713

JOHNSON, ROBERT ALAN, materials science educator; b. N.Y.C., Jan. 2, 1933; s. George E. and Betty (Durisek) J.; married; children: Sharon L. Duke, Derek A., Todd A. AB, Harvard U., 1954; PhD, Rensselaer U., 1962. Scientist Brookhaven Nat. Lab., Upton, N.Y., 1962-69; prof. U. Va., Charlottesville, 1969—. Editor: Physics of Radiation Effects in Crystals, 1986; contbr. articles to profl. jours. Lt. USN, 1954-57. Fellow Am. Physical Soc.; mem. AAAS, Materials Rsch. Soc., Metallurgical Soc., Sigma Xi. Home: Rte 3 Box 253 Charlottesville VA 22903 Office: U Va Materials Sci Dept Thornton Hall Charlottesville VA 22903

JOHNSON, ROBERT ANDREW, ecologist; b. Peoria, Ill., Sept. 15, 1954; s. Andrew Branting Johnson. MS, U. Ill., 1980; PhD, Ariz. State U., 1989. Contract ecologist Johnson and Assocs. EEI, Inc., Chandler, Ariz., 1979—, Mus. No. Ariz., Flagstaff, 1980-83; teaching asst. Ariz. State U., Tempe, 1983-89, adj. asst. prof. botany, 1991—; vol. U.S. Fish and Wildlife Svc., Phoenix, 1991—. Contbr. articles to profl. publs. Grantee, S.W. Pks. and Monuments Assn., Tucson, 1989, Cactus and Succulent Soc. Am., 1989. Mem. Am. Soc. Naturalists, Ecol. Soc. Am., Soc. Study Evolution, Audubon Soc., Nat. Geographic Soc.

JOHNSON, ROBERT BRITTEN, geology educator; b. Cortland, N.Y., Sept. 24, 1924; s. William and Christine (Hofer) J.; m. Garnet Marion Brown, Aug. 30, 1947; children: Robert Britten, Richard Karl, Elizabeth Anne. Student, Wheaton (Ill.) Coll., 1942-43, 46-47; AB summa cum laude, Syracuse U., 1949, MS, 1950; PhD, U. Ill., 1954. Asst. geologist Ill. Geol. Survey, 1951-54; asst. prof. geology Syracuse U., 1954-55; sr. geologist and geophysicist C.A. Bays & Asso., Urbana, Ill., 1955-56; from asst. prof. to prof. engring. geology Purdue U., 1956-66, head, engring. geology dept., 1964-66; prof. geology DePauw U., 1966-67, head, dept. geology, 1966-67; prof. geology Colo. State U., 1967-88, acting chmn. dept. geology, 1968, chmn. dept., 1969-73, prof. in charge geology programs, dept. earth resources, 1973-77, acting head dept. earth resources, 1979-81, prof. emeritus, 1988—; geologist U.S. Geol. Survey, 1976-88; cons. in field, 1957—; instr. Elderhostel programs, 1991—. Active local Boy Scouts Am., 4-H Club, Sci. Fair, dist. schs. Served with USAAF, 1943-46. Fellow Geol. Soc. Am. (E.B. Burwell Jr. Meml. award 1989); mem. Assn. Engring. Geologists (Claire P. Holdredge Outstanding Publ. award 1990), Internat. Assn. Engring. Geology, Phi Beta Kappa. Republican. Home: 2309 Moffett Dr Fort Collins CO 80526-2122

JOHNSON, ROBERT GAHAGEN, JR., medical researcher, educator; b. Altoona, Pa., Feb. 28, 1952; s. Robert and Ruth (Haverstick) J.; m. Margaret Ann Liu, Sept. 19, 1983; 1 child, Matthew Kuan. BA, U. Pa., 1974, PhD in Biophysics, 1980, MD, 1981. Bd. cert. internal medicine. Resident medicine Mass. Gen. Hosp., 1981-83, clin. and rsch. fellow medicine, 1983-84, asst. in medicine, 1984-87, asst. in neurobiology, 1984-87; instr. medicine Harvard U. Med. Sch., 1984-85, asst. prof. medicine, 1985-87; assoc prof medicine, physiology, biochemistry and biophysics U. Pa. Sch. Medicine, 1987-91; assoc. in medicine Hosp. of the U. Pa., 1987-91; dir. pharmacology Merck Rsch. Lab., Phila., 1991—; mem. neurobiology grad. group Harvard Med. Sch., 1985-87; mem. Inst. Neurol. Sci., U. Pa., 1988-91, mem. grad. group in neurosci., 1988-91, mem. grad. group in biophysics, 1989-91, mem. grad. group in biochemistry, 1989-91, investigator diabetes rsch. ctr., 1989-91, investigator metabolic NMR rsch. resource, 1989-91; assoc. investigator Howard Hughes Med. Inst., 1984-91. Med. Scientist Tng. Program MSTP fellow, 1974-80; grantee Am. Heart Assn., 1977-80, NIH, 1979-84, 81-86. Mem. AAAS, Biophys. Soc., N.Y. Acad. Sci., Neurosci. Soc., Mass. Med. Soc., Soc. Magnetic Resonance in Medicine, Phi Beta Kappa, Alpha Omega Alpha. Home: 4 Cushman Rd Rosemont PA 19010 Office: Merck Rsch Labs Pharmacology WP44 L200 West Point PA 19486

JOHNSON, ROBERT GLENN, physics educator; b. Iowa, Dec. 12, 1922; s. Lyell E. and Lois Ursula (Mills) J.; m. Elizabeth Louise Gulliver, July 17, 1949; children: Burgess, Diane, Janey, Wendy, Miriam. BS, Case Western Res. U., 1947; PhD, Iowa State U., 1952. Project engr. Bendix Aviation Inc., Red Bank, N.J., 1952-55; scientist Honeywell Inc., Mpls., 1955-74, staff scientist, 1974-90; adj. prof. physics U. Minn., Mpls., 1990—. Contbr. articles to profl. publs. Achievements include 23 patents on control technology sensors; pioneering research in silicon microstructure sensor technology; key contributions to understanding of glacial climate change mechanisms. Office: U Minn Dept Geology-Geophys 310 Pillsbury Dr SE Minneapolis MN 55455-0219

JOHNSON, RODNEY WILLIAM, utility executive; b. Detroit, Feb. 5, 1955; s. Robert Underwood and Marjorie Jesse (Gapske) J.; m. Susan Marie Kane, Nov. 24, 1978; children: Hollyn, Kane, Natalie. BSCE, Mich. Tech. U., 1978. Combustion engr. Republic Steel Corp., Cleve., 1978-83; engr. Stone & Webster Engring. Corp., Cherry Hill, N.J., 1984-89; prin. quality engr. Detroit Edison Co., 1989-92, supr. procurement quality assurance, 1992—. Mem. Instrument Soc. Am., Sigma Tau Gamma. Office: Detroit Edison 6400 Dixie Hwy Newport MI 48166

JOHNSON, ROGER WARREN, chemical engineer; b. Huntsville, Ala., Oct. 25, 1960; s. Frederic Allen and Joan (Bickum) J.; m. Margaret Jane Major, June 16, 1984. BChemE, Auburn U., 1984. Process engr. fibers divsn. E.I. DuPont de Nemours & Co., Waynesboro, Va., 1984-86; devel. engr. imaging systems E.I. DuPont de Nemours & Co., Brevard, N.C., 1986-87; R & D engr. Hercules Inc.-A&TP, Oxford, Ga., 1987-92; account mgr. Hercules Inc.-Absorbents and Textile Products, Norcross, Ga., 1992—. Mem. INDA (Assn. of the Nonwoven Fabrics Industry), Auburn Alumni Assn., Phi Kappa Phi. Home: 1410 Mclendon Ave NE Atlanta GA 30307-2129 Office: Hercules Inc Ste 700 3169 Holcomb Bridge Rd Norcross GA 30071

JOHNSON, RONALD SANDERS, physical biochemist; b. Chgo., Mar. 9, 1952; s. Benny Sanders and Lila (Blackwell) J. BA in Chemistry, Northwestern U., 1973, PhD in Biochemistry, 1978. Postdoctoral fellow U. Calif., Berkeley, 1978-81; asst. prof. East CArolina U., Greenville, N.C., 1981-87, assoc. prof., 1987--. Nat. Sci. Found. grantee 1992--; Miller fellow, 1978-80. Mem. Am. Soc. Biochemistry and Molecular Biology, Am. Chemical Soc., Sigma Xi. Office: E CArolina U Biochemisty Dept Moye Blvd Greenville NC 27858

JOHNSON, ROSS JEFFREY, statistician; b. Cleve., July 3, 1955; s. Harold Leroy and Louise Lillian (Huber) J.; m. Lydia Susan Burruel, Apr. 13, 1986; children: Alexander Harold, Adam Ross. BA, U. No. Colo., 1978; MS, Ariz. State U., 1987. Buyer McAuto Systems, Inc., Phoenix, 1982-83; quality statistician Nelco Tech., Inc., Tempe, Ariz., 1988-89; forecast analyst Allied Signal Corp., Phoenix, 1989-91; statis. analyst Carter Hawley Hale, Inc., Tempe, 1991—. 1st lt. USMC, 1978-81. Mem. Am. Statis. Assn. (treas. 1988-89), Alpha Iota Delta. Home: 2007 S Sierra Vista Dr Tempe AZ 85282

JOHNSON, ROY RAGNAR, electrical engineer; b. Chgo., Jan. 23, 1932; s. Ragnar Anders and Ann Viktoria (Lundquist) J.; m. Martha Ann Mattson, June 21, 1963; children: Linnea Marit, Kaisa Ann. B.S. in Elec. Engring, U. Minn., 1954, M.S., 1956, Ph.D., 1959. Research fellow U. Minn., 1957-59; from rsch. engr. to sr. basic rsch. scientist Boeing Sci. Research Labs., Seattle, 1959-72; prin. scientist KMS Fusion, Inc., Ann Arbor, Mich., 1972-74; dir. fusion expts. KMS Fusion, Inc., 1974-78, tech. dir., 1978-91, dept. head for fusion and plasmas, 1985-88; tech. dir. Lawrence Livermore Nat. Lab. Innovation Assocs., Inc., Ann Arbor, 1992—; vis. lectr. U. Wash., Seattle, 1959-60; vis. scientist Royal Inst. Tech., Stockholm, 1963-64. Author: Nonlinear Effects in Plasmas, 1969, Plasma Physics, 1977, Research Trends in Physics, Inertial Confinement Fusion, 1992; contbr. to profl. publs.; patentee in field. Bd. advisors Rose-Hulman Inst. Tech., 1982—. Decorated chevalier Order of St. George; comdr. Order of Holy Cross of Jerusalem. Fellow Am. Phys. Soc.; mem. AAAS, AIAA, IEEE, Nuclear Plasma Scis. Soc. of IEEE (exec. com. 1972-75), N.Y. Acad. Scis., Am. Def. Preparedness Assn., Assn. of Old Crows, Vasa Order Am., Am. Swedish Inst., Torpar Riddar Orden, Swedish Pioneer Hist. Soc., Swedish Coun. Am., Detroit Swedish Coun., Swedish Club of Detroit, Swedish Am. Hist. Soc., Eta Kappa Nu, Gamma Alpha. Lutheran. Home: 1141 Concannon Blvd Livermore CA 94550-0641 Office: Livermore Nat Lab PO Box 808 Livermore MI 48106

JOHNSON, SAMUEL WALTER, II, mechanical engineer; b. Jasper, Ala., Sept. 4, 1948; s. Samuel Walter and Dorothy Francis (Mattison) J.; m. Cecilia Ann Martin, Dec. 14, 1968; children: Samantha Dawn Johnson Romano, Mattison, Joshua. BSME, Auburn U., 1971. Co-op student Ala. Power Co., 1967-70; engr. Fla. Power Corp., Crystal River, 1972-83; team mgr. Inst. of Nuclear Power Ops., Atlanta, 1983-88; div. mgr. maint. United Energy Svcs. Corp., Marietta, Ga., 1988—. Vol. Feed the Hungry Found., Marietta, 1989—, Spl. Olympics, Marietta, 1985—; mem. Eastside Bapt. Caring Ministry, Marietta, 1989—. Mem. ASME (mem. com. on ops. and maint. of nuclear power plants). Republican. Baptist. Home: 1858 Wicks Valley Dr Marietta GA 30062-6770 Office: United Energy Svcs Corp 1110 Northchase Pkwy Marietta GA 30067

JOHNSON, SIDNEY MALCOLM, foreign language educator; b. New Haven, Aug. 17, 1924; s. Everett Caswell and Eleanor (Eckman) J.; m. Lora Louise Dunbar, Sept. 29, 1945; children: Thomas Malcolm, Frederick William, Karl Everett. B.A., Yale U., 1944, M.A., 1948, Ph.D., 1953. Asst. instr. Yale U., 1946-51; instr. U. Kans., 1951-53, asst. prof., 1953-58, asso. prof., 1958-62, prof., 1962-65; prof. German, chmn. dept. Emory U., Atlanta, 1965-72; prof. German, Ind. U., Bloomington, 1972—; chmn. dept. Ind. U., 1972-78, 91-92; dir. Ind.-Purdue U. Study Program, U. Hamburg, W.Ger., 1978-79. Translator (with M. E. Gibbs) Willehalm (Wolfram von Eschenbach), 1984, (with M.E. Gibbs) Titurel and the Songs (Wolfram von Eschenbach), 1988; contbr. articles on medieval German lit. to profl. jours. Served to lt. (j.g.) USNR, 1943-46. Research grantee Am. Philos. Soc., 1963. Mem. MLA, Am. Assn. Tchrs. German, AAUP, Wolfram von Eschenbach Gesellschaft, Gesellschaft fur interkulturelle Germanistik, Internat. Assn. Germanic Studies, Medieval Acad. Home: 2320 E Covenanter Dr Bloomington IN 47401-5402

JOHNSON, STANLEY R., economist, educator; b. Burlington, Iowa, Aug. 26, 1938; 2 children. BA in Agrl. Econs., Western Ill. U., 1961; MS, Tex. Tech. U., 1962; PhD, Tex. A&M U., 1966. Asst. prof. dept. econs. U. Mo., Columbia, 1964-66, assoc. prof. depts. econs. and agrl. econs., 1967-70, prof., 1970-85, chmn. dept. econs., 1972-74; assoc. prof. dept. agrl. econs. U. Conn., Storrs, 1966-67; prof., dir. Ctr. for Agrl. and Rural Devel., dept. econs. Iowa State U., Ames, 1985—; exec. dir. Food and Agrl. Policy Rsch. Inst., 1984—; vis. assoc. prof. agrl. econs. U. Calif.-Davis, 1970, Purdue U., 1971-72; economist Agr. Can., Ottawa, 1975; vis. prof. econs. U. Ga., 1975-76, U. Calif.-Berkeley, 1981; adj. prof. U. Mo., Columbia, 1985—; chmn. bd. Midwest Agribus. Trade Rsch. and Info. Ctr., 1987—; cons. and lectr. in field. Author: Advanced Econometric Methods, 1984; Demand Systems Estimation, 1984; assoc. editor Am. Jour. Agrl. Econs.; mem. internat. editorial bd. Advances in Agrl. Mgmt. and Econs., 1989; mem. editorial bd. Internat. Review Econs. and Finance. contbr. chpts. to books, articles to profl. jours. Adminstr. Iowa State U. USSR All-Union Acad. of Agrl. Scis. Exch. Agreement, 1988—; chmn. bd. dirs. Inst. Policy Reform, 1990—; co-chair World Food Conf., 1988. Recipient Chancellor's award for outstanding research, 1980, Charles F. Curtiss Disting. Professorship Agr. award, 1990, Internat. Svc. award Wilton Park, 1993, numerous grants in econs. Fellow Am. Agrl. Econs. Assn.; mem. V. I. All-Union Acad. Agrl. Scis. (fgn.), Ukranian Acad. Agrl. Scis. (fgn. academician), Mo. Valley Econ. Assn. (bd. dirs. 1977-82, pres.-elect 1979-81). Office: Iowa State U Ctr for Agr and Rural Devel 578 Heady Hall Ames IA 50011-1070

JOHNSON, STEWART WILLARD, civil engineer; b. Mitchell, S.D., Aug. 17, 1933; s. James Elmer Johnson and Grace Mahala (Erwin) Johnson Parsons; m. Mary Anis Giddings, June 24, 1956; children: Janelle Chiemi, Gregory Stewart, Eric Willard. BSCE, S.D. State U., 1956; BA in Bus. Adminstrn. and Polit. Sci., U. Md., 1960; MSCE, PhD, U. Ill., 1964. Registered profl. engr., Ohio. Commd. 2d lt. USAF, 1956, advanced through grades to lt. col.; prof. mechs. and civil engring. Air Force Inst. Tech. USAF, Dayton, Ohio, 1964-75; dir. civil engring. USAF, Seoul, Republic of Korea, 1976-77; chief civil engring. research div. USAF, Kirtland AFB, N.Mex., 1977-80; ret. USAF, 1980; prin. engr. BDM Corp., Albuquerque, 1980—; cons. in space sci., lunar basing NASA, U. N.Mex., N.Mex. State U., Los Alamos Nat. Labs., 1986—; adj. prof. civil engring. U. N.Mex., 1987—; prin. investigator devel. concepts for lunar astron. obs. U. N.Mex., N.Mex. State U., NASA, 1987—; tech. chmn. Space '88 and Space '90 Internat. Confs.; gen. chair Space '94 Internat. Conf., Albuquerque; vis. lectr. Internat. Space U., Japan, 1992; invited lectr., vis. lectr. in field. Editor Engineering, Construction, and Operations in Space, I and II; contbr. articles to profl. jours. Pres. ch. council Ch. of Good Shepherd United Ch. Christ, Albuquerque, 1983-85, chair bd. deacons, 1991-93; trustee Lunar Geotech. Inst., 1990—; mem. adv. bd. Lab. for Extraterrestrial Structures Rsch. Rutgers U., 1990—. Fellow Nat. Acad. Scis. NRC, 1970-71. Mem. AIAA (Engr. of Yr. region IV 1990), AAAS, ASCE (chmn. exec. com. aerospace div. 1979, tech. activities com. 1984, chmn. com. space engring. and constrn. 1987—, mem. nat. space policy com. 1988—, chmn. 1990—, Outstanding News Corr. award 1981, Aerospace Scis. and Tech. Applications award 1985, 90, Edmund Friedman Profl. Recognition award 1989), Am. Geophys. Union, Sigma Xi, Pi Sigma Alpha. Republican. Mem. United Ch. of Christ. Avocations: photography, swimming, walking, gardening, hiking. Office: BDM Internat Inc 1801 Randolph Rd SE Albuquerque NM 87106-4295

JOHNSON, TERRY CHARLES, biologist, researcher; b. St. Paul, Aug. 8, 1936; s. Roy August and Catherine (McKigen) J.; m. Mary Ann Wilhelmy, Nov. 23, 1957; children: James, Gary, Jean. BS, Hamline U., 1958; MS, U. Minn., 1961, PhD, 1964. Postdoctoral fellow U. Calif., Irvine, 1964-66; asst. prof. Med. Sch., Northwestern U., Chgo., 1966-69, assoc. prof., 1969-73, prof., 1973-77; prof. div. biology Kans. State U., Manhattan, 1977—, dir. div. biology, 1977-92, Univ. Disting. prof., 1989; dir. Konza Prairie Rsch. Area, Manhattan, 1977-92, Ctr. for Basic Cancer Rsch., Manhattan, 1980—, Ctr. for Space Life Scis., Manhattan, 1990-92; co-dir. Bioserve Space Techs., Manhattan, 1989—. Recipient Outstanding Tchr. of Yr. award Med. Sch., Northwestern U., 1975, Outstanding Tchr. award Ill. Coll. Pediatric Medicine, 1976, Disting. Grad. Faculty award Kans. State U., 1987, Outstanding Sci. award Sigma Xi Kans. State U. chpt., 1993. Mem. AAAS, Am. Soc. for Gravitational and Space Biology, Am. Soc. for Microbiology, Am. Soc. for Neurochemistry, Am. Soc. for Cell Biology, N.Y. Acad. Sci. Avocation: reading. Home: 205 S Drake Dr Manhattan KS 66502-3029 Office: Kans State U Div Biology Ackert Hall Manhattan KS 66506

JOHNSON, TESLA FRANCIS, data processing executive, educator; b. Altoona, Fla., Sept. 2, 1934; s. Tesla Farris and Ruby Mae (Shockley) J.; m. Eleanor Mary Riggs, Oct. 17, 1975. BSEE, U. S.C., 1958; MS in Ops. Rsch., Fla. Inst. Tech., 1968; PhD in Adminstrv. Mgmt., Walden U., Mpls., 1990. Machinist apprentice Seaboard Airline Ry., 1952-54; asst. computer engr. So. Ry. System, Washington, 1958-61; sr. sci. programmer NCR, Dayton, Ohio, 1961-66; staff programmer IBM, East Fishkill, N.Y., 1966-72; mgr. Jay Turner Co., Grace, Idaho, 1973-74; programmer, analyst Ccybernetics & Systems, Inc., Jacksonville, Fla., 1974-77; systems analyst lst Nat. Bank Md., Balt., 1977-78; sr. systems analyst GM, Detroit, 1978-80; tech. analyst Sunbank Data Corp., Orlando, Fla., 1980-81; mgr. data adminstrn. dept. Martin Marietta Corp., Orlando, 1981—; adj. prof. bus. adminstrn. Valencia Community Coll., Orlando, 1989—, Orlando Coll., 1990—, Fla. Inst. Tech., Melbourne. Mem. choir South Orlando Bapt. Ch. Recipient cert. of appreciation NASA, 1969, Excalibur award. Mem. Acad. Internat. Bus., Tau Beta Pi, Sigma Phi Epsilon. Republican. Avocations: stamp and coin collecting, organ. Home: 11430 Haymarket Ct Orlando FL 32837-9121 Office: PO Box 133850A Orlando FL 32859-0001

JOHNSON, WALTER EARL, geophysicist; b. Denver, Dec. 16, 1942; s. Earl S. and Helen F. (Llewellyn) J.; Geophys. Engr., Colo. Sch. Mines, 1966; m. Ramey Kandice Kayes, Aug. 6, 1967; children—Gretchen, Roger, Aniela. Geophysicist, Pan. Am. Petroleum Corp., 1966-73; seismic processing supr. Amoco Prodn. Co., Denver, 1973-74, marine tech. supr., 1974-76, div. processing cons., 1976-79, geophys. supr. No. Thrust Belt, 1979-80; chief geophysicist Husky Oil Co., 1981-82, exploration mgr. Rocky Mountain and Gulf Coast div., 1982-84; geophys. mgr. ANR Prodn. Co., 1985—; pres. Sch. Lateral Ditch Co.; cons. engr. Bd. dirs. Rocky Mountain Residence, nursing home. Registered profl. engr., cert. geologist, Colo. Mem. Denver Geophys. Soc., Soc. Exploration Geophysicists. Republican. Baptist. Office: 600 17th St Ste 800 Denver CO 80202-5401

JOHNSON, WARREN ELIOT, wildlife ecologist; b. Bangkok, Thailand, Jan. 16, 1961; s. William Lawrence and Nancy Loana (Crane) J.; m. Mariane Bernadette Carrere, July 29, 1989; 1 child, Daniel Carrere. BA, Oberlin Coll., 1983; MS, Utah State U., 1984; PhD, Iowa State U., 1992. Rsch. fellow Utah State U., Logan, 1983-84; administrv., rsch. asst. Patagonia Wildlife Rsch. Ctr. Iowa State U., Ames, 1985-89, temp. instr., 1990-92; postdoctoral scientist Nat. Cancer Inst., Frederick, Md., 1992—. Contbr. articles to profl. jours. Recipient Teaching Excellence award Iowa State U., 1992, Sherry R. Fisher award, 1991, Zaffarano prize Sigma Xi, 1992; IIT Internat. fellow, 1986, Nat. Wildlife Fedn. Internat. fellow, 1985. Mem. Am. Soc. Mammalogy, IUCN Cat Specialist Group, Soc. for Conservation Biology, Wildlife Soc.

JOHNSON, WILLIAM HERBERT, emergency medicine physician, aerospace physician, retired air force officer; b. Elkhart, Ind., Dec. 12, 1928; s. Herbert John and Lorene Wilhemena (Johnson) J.; m. Ann Marie Bacon, Oct. 17, 1964; children: Ernest Michael, Jennifer Lynn. AB, Augustana Coll., 1951; MD, Ind. U., 1958. Intern, Indpls. Gen. Hosp., 1958-59; resident in internal medicine Ind. U. Med. Ctr., Indpls., 1960-61; practice medicine specializing in gen. medicine, East Gary, Ind., 1959-60; asst. surgeon U.S. Steel Co., Gary Works (Ind.), 1959-60; ptnr. Gary Clinic (now Ross Clinic), 1962-69; staff physician student health svcs. Western Mich. U., 1969-74; staff physician Trauma and Emergency Ctr., Bronson Methodist Hosp., Kalamazoo, Mich., 1969—, chmn., 1972-74; asst. clin. prof. medicine Mich. State U. Coll. Human Medicine, East Lansing, 1976-81; past. mem., pres. S.W Mich. Emergency Svcs., P.C., Kalamazoo, 1969-81; past mem. staffs Borgess Med. Ctr., Community Hosp. Assn., Leila Y. Post Montgomery Hosp., Three Rivers Hosp.; med. dir. emergency dept., mem. exec. bd. Elkhart (Ind.) Gen. Hosp., Goshen (Ind.) Gen. Hosp.; med. dir. Emergency Med. System, Elkhart County, Ind.; pres. Elkhart Emergency Physicians, Inc., Emergency Rm. Physicians, Inc. Pres. Corey Lake Improvement Assn., 1978-80. Served with USAF, 1951-53, 61-62; ret. brig. gen. Air N.G., 1961-85; ret. brig. gen. USAF, 1988; part owner Dunlap Urgent Care & Occupational Medicine Clinic, Elkhart, Ind. Decorated Air N.G. Meritorious Service award, twice USAF Legion of Merit; nominee Malcolm C. Grove award USAF Flight Surgeon of Yr., 1971; recipient Minuteman award State of N.J., 1985. Fellow Aerospace Med. Assn. (by-laws com., membership com.); mem. Kalamazoo Acad. Medicine, Mich. Med. Soc., AMA Physician's Recognition award 1983), Am. Coll. Emergency Physicians (pres. Mich. chpt. 1976-78, dir.), Univ. Assn. for Emergency Medicine, Soc. USAF Flight Surgeons (constn., by-laws com.), Am. Coll. of Occupational Medicine, Alliance of Air N.G. Flight Surgeons (dir., mem. membership com., chmn. nominating com., past pres.), Assn. Mil. Surgeons U.S. (life), Mich. Assn. of Professions, Res. Officers Assn., Air Force Assn. (life), Augustana Coll. (Ill.) Alumni Bd. Lutheran. Lodge: Elks, Am. Legion (post Hice-Shutes 170 Three Rivers). Contbg. editor to books, articles to profl. jours.; contbg. editor, mem. editorial bd. Annals of Emergency Medicine, 1972-89 ; mem. editorial bd. Aviation, Space and Environ. Medicine, 1981—, editor book rev. sect., 1984—. Home and Office: 11451 Coon Hollow Rd Three Rivers MI 49093-9225

JOHNSON, WILLIAM HOWARD, agricultural engineer, educator; b. Sidney, Ohio, Sept. 3, 1922; s. Russell Earl and Dollie (Gamble) J.; m. Wyoma Jean Swift, Oct. 2, 1943; children: Lawrence Alan, Cheri Ellen, Dana Sue. B.S., Ohio State U., 1948, M.S., 1953; Ph.D., Mich. State U., 1960. Registered profl. engr. Mem. faculty Ohio Agrl. Expt. Sta., Wooster, 1948-64; mem. faculty Ohio Agrl. Research and Devel. Center, Wooster, 1964-70; prof., asso. chmn. dept. agrl. engring. Ohio Agrl. Research and Devel. Center, 1959-70; part-time prof. Ohio State U., 1964-70; prof., head dept. agrl. engring. Kans. State U., Manhattan, 1970-81; dir. Engring. Experiment Sta. Kans. State U., 1981-87; cons. farm equipment cos. Author: (with B.J. Lamp) Principles, Equipment and Systems for Corn Harvesting, 1966; also articles. Recipient Distinguished Alumnus award Coll. Engring., Ohio State U., 1974; named to Coll. Engring. Kans. State U. Hall of Fame, 1992. Fellow Am. Soc. Agrl. Engrs. (pres. 1986-87), Kans. Engring. Soc. (pres. 1985-86), Sigma Xi, Tau Beta Pi. Achievements include research on soil-plant-machine relationships, harvesting, design for soiltillers, planters, harvesters. Home: 1582 Williamsburg Dr Manhattan KS 66502-4504 Office: Kans State Univ Dept Agrl Engring Seaton Hall Manhattan KS 66506

JOHNSON, WILLIAM SUMMER, chemistry educator; b. New Rochelle, N.Y., Feb. 24, 1913; s. Roy Wilder and Josephine (Summer) J.; m. Barbara Allen, Dec. 27, 1940. Grad., Gov. Dummer Acad., 1932; B.A., Amherst Coll., 1936, Sc.D., 1956; M.A., Harvard U., 1938, Ph.D., 1940; Sc.D, L.I. U., 1968. Instr. Amherst Coll., 1936-37; research chemist Eastman Kodak Co., summer, 1936-39; instr. U. Wis., 1940-42, asst. prof., 1942-44, assoc. prof., 1944-49, Homer Adkins prof. chemistry, 1954-60; prof. dept. chemistry Stanford U., Calif., 1960-78; Jackson-Wood prof. Stanford U., 1974-78, prof. emeritus, 1978—, chmn. dept., 1960-69; vis. prof. Harvard U., 1954-55; mem. exec. bd. Jour. Organic Chemistry, 1954-56; mem. chem. adv. panel NSF, 1952-56; sec. organic sect. Internat. Congress Pure and Applied Chemistry, 1951. Contbr. chpts. to chemistry books, articles to assn. jours.; mem. bd. editors: Organic Syntheses, vol. 34, 1954, Jour. Am. Chem. Soc., 1956-65, Jour. Organic Chemistry, 1954-56, Tetrahedron, 1957-84. Recipient medal Synthetic Organic Chem. Mfrs. Assn., 1963, Nat. Medal of Sci., 1987. Fellow London Chem. Soc.; mem. NAS, Am. Acad. Arts and Scis., Swiss Chem. Soc., Am. Chem. Soc. (chmn. organic div. 1951-52, award in synthetic organic chemistry 1958, Nichols medal 1968, Roussel prize 1970, Roger Adams award 1977, Arthur C. Cope award 1989, Tetrahedron prize 1991), Phi Beta Kappa, Sigma Xi. Home: 191 Meadowood Dr Menlo Park CA 94028-7625 Office: Stanford Univ Dept Chemistry Stanford CA 94305

JOHNSON, WILLIE ROY, industrial psychology educator; b. Arcola, Miss., Apr. 3, 1947; s. Leroy and Magnolia (Thompson) J.; m. Gloria Jones, Aug. 14, 1982; 1 child, Kyle Jamary; m. Erma Elmore, Aug. 28, 1977 (div. Mar. 1982). BA in Psychology, Chgo. State U., 1974, MS, 1976; MA in Psychology, Bowling Green (Ohio) State U., 1980, PhD, 1986. Dir. field staff Ill. Office Edn., Springfield, 1973-77; student intern personnel Inland Steel Co., East Chicago, Ind., 1980; adj. prof. Wayne State U., Detroit, 1984-85; staff mgr. Mich. Bell Telephone Co., Detroit, 1985-86; asst. prof. indsl. psychology Iowa State U., Ames, 1986-92, assoc. prof. mgmt., 1992—; mgr. HRMD Sandman Personnel Systems, Findlay, Ohio, 1982. Contbr. articles to profl. jours. With USAF, 1966-70. Grad. Profl. Opportunities Program minority fellow, Bowling Green, 1980-82. Mem. Am. Psychol. Assn., Acad. Mgmt., Soc. Indsl. and Organizational Psychology, Sigma Xi. Democrat. Methodist. Office: Iowa State U Dept Mgmt 300 Carver Hall Ames IA 50011-2005

JOHNSON, WINSTON CONRAD, mathematics educator; b. Wellborn, Fla., Apr. 27, 1943; s. Charles Winston and Martha Gwendolyn (McLeran) J. BA, U. Fla., 1971, MEd, 1977, EdS, 1979. Cert. maths. tchr., Fla., Ga. Tchr. maths. Columbia High Sch., Lake City, Fla., 1971-76, Lake City (Fla.) Jr. High Sch., 1979-80, Hinesville (Ga.) Middle Sch., 1980-81; prof. maths. Cen. Fla. C.C., Ocala, 1981—. Sgt. USAF, 1966-70. Mem. Nat. Coun. Tchrs. of Maths., Math. Assn. Am., Fla. Devel. Ednl. Assn., Nat. Assn. for Devel. Edn., Fla. Coun. Tchrs. Maths. Democrat. Baptist. Avocations: reading, music, mathematics. Home: PO Box 37 Wellborn FL 32094-0037 Office: Cen Fla Community Coll 3001 SW College Rd 3001 SW College Rd Ocala FL 34474-1388

JOHNSTON, ALLAN HUGH, physicist; b. Seattle, Dec. 1, 1941; s. Harold and Margaret Sayward (King) J.; m. Susan Meland Diamant, June 12, 1959; children: Michael, Jeffrey, Kathryn. BS in Physics, U. Wash., 1963, MS in Physics, 1983. Devel. engr. Ampex Corp., Redwood City, Calif., 1963-65; rsch. specialist Boeing Aero. Co., Seattle, 1965-86; mgr. microelectronics Boeing Def. and Space Group, Seattle, 1986—. Contbr. articles to profl. jours. Recipient Spl. Merit award ASTM, 1986. Mem. IEEE (adcom com. nuclear and plasma scis. soc. 1991—, outstanding poster paper award Nuclear and Space Radiation Effect Conf. 1987). Achievements include identification of super-recovery in ULSI electronics and mechanisms for latchup windows, model devel. for single-particle latchup in space and hardening techniques for GaAs circuits in space. Home: 10021 43d Pl NE Seattle WA 98125

JOHNSTON, ARCHIBALD CURRIE, geophysics educator, research director; b. Charlotte, N.C., May 19, 1945; s. Frontis Withers and Lucy Martin (Currie) J.; m. Jill Diana Stevens, May 23, 1992. BS in Physics with honors, Rhodes Coll., 1967; postgrad., Dartmouth Coll., 1967-68; PhD in Geological Scis./Geophysics, U. Col., 1979. From asst. prof. to assoc. prof. dept. geological scis. Memphis State U., 1979-88, prof. dept. geological scis., 1988—; dir. Ctr. Earthquake Rsch. & Info. Memphis State U., 1979-92, dir. rsch., 1992—; chmn. panel regional seismic networks com. seismology NAS, 1988-90; mem. Nat. Earthquake Prediction Evaluation Coun., 1990—; mem. exec. com. Inc. Rsch. Instns. in Seismology, 1992—; mem. adv. panel branch global siesmology & geomagnetism Nat. Earthquake Info. Ctr., 1988-90; dir. Ctr. Excellence State of Tenn., 1984-92; cons. Lawrence Livermore Nat. Lab., Berkeley, Electric Power Rsch. Inst., Palo Alto, U.S. Army Corps. Engrs., Vicksburg, Miss., Tenn. Tech. Found., Knoxville, Geomatrix Cons., Inc., San Francisco, Law Environ., Atlanta, Battelle Inst., Seattle, SKB-Swedish Nuclear Fuel and Waste Mgmt. Co., Stockholm, Geological Survey Can., Ottawa, Ctr. Earthquake Rsch. in Australia, U. Queensland; Congl. testifier in field. Editor Seismological Research Letters, 1984—; contbr. articles to profl. jours.; pub. over 100 abstracts; invited speaker in field. Capt. USAF, 1968-73. NSF grantee, U.S. Nuclear Regulatory Commn. grantee, TVA grantee, U.S. Geological Survey grantee, Electric Power Rsch. Inst. grantee, Memphis Light, Gas & Water Utility grantee. Mem. AAAS, Am. Geophysical Union, Seismological Soc. Am. (bd. dirs. 1990—, v.p.

1991-92, pres. 1992-93). Office: Memphis State U Ctr Earthquake Rsch Info Memphis TN 38152

JOHNSTON, BRIAN HOWARD, molecular biologist; b. Palo Alto, Calif., June 6, 1949; s. Howard and Virginia (Ward) J.; m. Thomasa Eckert, Mar. 20, 1971 (div. 1974). BA, Pomona Coll., Claremont, Calif., 1970; MS, U. Calif., Berkeley, 1972, PhD, 1980. Postdoctoral fellow U. Calif., San Francisco, 1980-82; postdoctoral fellow MIT, Cambridge, 1983-88, vis. scientist, 1988-89; sr. molecular biologist SRI Internat., Menlo Park, Calif., 1989-93, dir. program in nucleic acid structure and function, 1993—; cons. assoc. prof. Stamford U., 1993—. Contbr. articles to profl. jours., chpts. to books. Recipient postdoctoral fellowships Am. Cancer Soc., U. Calif., San Francisco, 1980-82, NIH, MIT, 1983-85, Med. Found. Boston, MIT, 1985-87. Achievements include discovery of pulsed-laser technique for making psorolen-DNA monoadducts without crosslinks, development of chem. probing techniques for mapping unusual DNA structures; proving triplex structure for H-DNA; co-inventor anti-HIV agent based on on modification of viral "tat" protein, also co-inventor "strand-switching" approach to DNA triplex formation at mixed sequence DNAs. Office: SRI Internat 205-19 333 Ravenswood Ave Menlo Park CA 94025

JOHNSTON, CAROL ARLENE, ecological researcher; b. Syracuse, N.Y., Nov. 18, 1952; d. Philip Gordon and Virginia Dawn J. MS, U. Wis., 1977, PhD, 1982. Rsch. technician Cornell U., Ithaca, N.Y., 1973-75; biol. technician Soil Conservation Svc., Stevens Point, Wis., 1976; rsch. asst. U. Wis., Madison, 1976-78; supr. natural resources Wis. Dept. Natural Resources, Madison, 1979-83; ind. environ. cons. Oak Ridge, Tenn., 1984-85; postdoctoral fellow U. Minn., Duluth, 1985-86, rsch. assoc., 1986-93, sr. rsch. assoc., 1993—; sci. adv. com. SCOPE Ecotones Project, 1989-91; mem. rev. panel wetland rsch. plan U.S. EPA, 1991; mem. Gov.'s Coun. on Geographic Info., St. Paul, 1992—. Guest editor Landscape Ecology jour., 1990; contbr. articles to profl. jours., chpts. to books. Grantee NSF, 1988, 92, 93, USDA, 1992, U.S. EPA, 1987. Mem. Soc. Wetland Scientists (pres. 1992-93), Ecol. Soc. Am., Internat. Assn. Landscape Ecology, Am. Soc. Photogrammetry and Remote Sensing. Quaker. Home: 8806 W Branch Rd Duluth MN 55803 Office: NRRI Univ Minn 5013 Miller Trunk Hwy Duluth MN 55811

JOHNSTON, CLIFFORD THOMAS, soil and environmental chemistry educator; b. Colorado Springs, Colo., Oct. 3, 1955. BSc in Chemistry, U. Calif., Riverside, 1979, PhD in Soil and Environ. Chemistry, 1983. Postdoctoral fellow Los Alamos (N.Mex.) Nat. Lab., 1983-85; asst. prof. dept. soil sci. U. Fla., Gainesville, 1985-90, assoc. prof., 1990—; sabbatical fellow K.U. Leuven, Belgium, 1991-92; cons. for Los Alamos Nat. Lab., Battelle Pacific N.W. Nat. Lab., English China Clay Am., Inc., Chem. Mfrs. Assn. Contbr. chpts. to books, articles to Soil Sci. Soc. Am., Jour., Jour. Phys. Chemistry, Chem. Physics Letters, Clays and Clay Minerals, Environ. Sci. and Tech., others. Recipient grants from USDA, 1992-94, South Fla. Water Mgmt. Dist., 992-93, Dept. of Air Force, 1986-87, 88-91, others. Mem. Am. Chem. Soc., Soil Sci. Soc. Am., Clay Minerals Soc. Achievements include work in colloid and surface chemistry, vibrational spectroscopy of clay minerals, oxides and zeolites. Office: Purdue Univ Agronomy Dept 1150 Lilly Hall West Lafayette IN 47907

JOHNSTON, E. RUSSELL, JR., civil engineer, educator; b. Phila.; s. E. Russell and Ethel (Doherty) J.;m. Ruth Alice Phillips, Dec. 29, 1951; children: E. Russell III, Bruce P. BSCE, U. Del., 1946; ScD, MIT, 1949. Asst. prof., then assoc. prof. civil engring. Lehigh U., Bethlehem, Pa., 1949-57; prof. civil engring. Worcester (Mass.) Poly. Inst., 1957-63; prof. civil engring. U. Conn., Storrs, 1963—, head dept. civil engring., 1972-77. Author: Mechanics for Engineering, 1956, latest ed. 1987, Vector Mechanics for Engineers, 1962, latest ed. 1988, Mechanics of Materials, 1981, latest ed. 1992. Recipient Benjamin Wright award Conn. Soc. Civil Engrs. Home: PO Box 525 Storrs Mansfield CT 06268 Office: Univ Conn Dept Civil Engring Storrs CT 06269

JOHNSTON, FRANK C., psychologist; b. West Hartford, Conn., June 21, 1955; s. Frank C. and Chris (Butler) J.; m. Susan H. Leffert, July 26, 1981. BA, Fairfield U., 1977; MEd, Columbia U., 1979, MA, 1979; PhD, SUNY, Albany, 1984. Sch. psychologist bd. coop. ednl. svcs. Herkimer, N.Y., 1979-80; intern Counseling Ctr., SUNY, Buffalo, 1983-84; psychologist Family Svc. Rochester, N.Y., 1985-87, Child and Youth div. Rochester Mental Health Ctr., 1988; pvt. practice Rochester, 1988—; cons. Brockport (N.Y.) Day Care Ctr., 1989-90, Learning Devel. Ctr., Rochester Inst. Tech., 1989-90. Mem. APA, N.Y. State Psychol. Assn. (managed care task force), Genesee Valley Psychol. Assn. (mem. legal legis. com. 1988-90, mem. ins. com. 1990-92, chmn. ins. com. 1990, 93, pres. elect 1993), Rochester Area Assn. Clin. Psychologists, Nat. Register Health Svc. Providers in Psychology. Office: 480 White Spruce Blvd Rochester NY 14623

JOHNSTON, GERALD ANDREW, aerospace company executive; b. Chgo., July 17, 1931; s. Gerald Ervan and Mary Henrietta (Dowell) J.; m. Jacquelyn Egan, March 6, 1954; children: Jan, Colleen, Jeffrey, Gregory, Steven. Student, San Bernardino Jr. Coll., 1950-51; BS in Engring., U. Calif., L.A., 1956, Cert. of Bus. Mgmt., 1968, MS in Engring., 1972. Jr. engr. Shell Oil Co., 1952-54; test engr. Robinson Aviation, 1955-56; stress analyst N.Am. Aviation, 1955; assoc. engr., tech. dir. Douglas Aircraft Co., Santa Monica, Calif., 1956-68; dir., v.p. gen. mgr. McDonnell Douglas Astronautics, Huntington Beach, Calif., 1968-87; pres. McDonnell Douglas Corp., St. Louis, 1988—, also bd. dirs. Trustee St. Louis U., 1988—. Roman Catholic. Office: McDonnell Douglas Corp PO Box 516 Saint Louis MO 63166-0516*

JOHNSTON, GORDON INNES, aerospace program manager; b. Santa Monica, Calif., Aug. 15, 1953; s. Robert Bethel and Nancy (Cameron) J.; m. Elissa Carol Lane (div. Dec. 1985); 1 child, Sarah Elizabeth; m. Karen Ann Helbrecht, Jan. 10, 1988. BA in Math., Calif. State U., Northridge, 1976, MS in Math., 1980. Cert. Calif. community coll. tchr. Experiment rep. Kett Engring., Pasadena, Calif., 1977-78; math. instr. Pasadena City Coll., 1980-82; experiment rep. Viking and Galileo Jet Propulsion Lab., Pasadena, 1978-80, remote sensing dep. team chief., 1980-87; program mgr. Jet Propulsion Lab., Washington, 1987-89; program mgr. office aeronautics and space tech. NASA, Washington, 1989-92, program mgr. office advanced concepts and tech., 1992—. Contbr. articles to profl. jours. Mem. exec. com. Sierra Club, No. Va., 1991--; mem. Alexandria (Va.) Gypsy Moth Adv. Group, 1991--. Mem. AAAS, AIAA. Office: NASA Office Advanced Concepts and Tech Mail Code CD Washington DC 20546

JOHNSTON, HAROLD S(LEDGE), chemistry educator; b. Woodstock, Ga., Oct. 11, 1920; s. Smith L. and Florine (Dial) J.; m. Mary Ella Stay, Dec. 29, 1948; children: Shirley Louise, Linda Marie, David Finley, Barbara Dial. AB, Emory U., 1941, ScD (hon.), 1965; PhD, Calif. Inst. Tech., 1948. Instr. to assoc. prof. chemistry Stanford (Calif.) U., 1947-56; assoc. prof. Calif. Inst. Tech., Pasadena, 1956-57; prof. U. Calif., Berkeley, 1957—, dean, coll. chemistry, 1966-70; vis. prof. U. Rome, 1964; adv. com. Calif. Statewide Air Pollution Rsch.Ctr., 1969-73, Nat. Ctr. Atmospheric Rsch., 1975-78, FAA High Altitude Pollution Program; vis. adv. com. Brookhaven Nat. Lab., 1970-73; faculty rsch. lectr. U. Calif., Berkeley, 1989. Author: Gas Phase Reaction Rate Theory, 1966, Gas Phase Reaction Kinetics of Neutral Oxygen Species, 1968, Reduction of Stratospheric Ozone by Nitrogen Oxide Catalysts from Supersonic Transport Exhaust, 1971; contbr. articles to profl. jours. Recipient Tyler prize Environ. Achievement, 1983, Disting. Alumni award Calif. Inst. Tech., 1985; grantee Office Naval Research, 1950-56, M.W. Kellogg Co., 1951-53, Standard Oil Calif., 1955-57, Alfred Sloan Found., 1957-59, NSF, 1959-68, 75-78, U.S. Pub. Health Service, 1963-70, EPA, 1970-72, Dept. Transp., 1972-75, Materials and Molecular Research div. Lawrence Berkeley Lab., 1966—. Fellow AAAS, Am. Chem. Soc. (Gold Medal award Calif. sect. 1956, Pollution Control award 1974, award in the Chemistry of Contemporary Technol. Problems 1985), Am. Phys. Soc., Am. Geophys. Union, Nat. Acad. Scis. (adv. panel to Nat. Bur. Standards, 1965-67, com. Motor Vehicle Emissions, 1971-75, Svc. to Soc. award in chemistry 1993), Am. Assn. Arts and Scis., Sigma Xi (nat. lectr. 1973). Home: 132 Highland Blvd Berkeley CA 94708 Office: U Calif Dept Chemistry Berkeley CA 94720

JOHNSTON, JAMES BENNETT, biochemist; b. San Diego, Dec. 31, 1943; s. Thomas Frazier and Mary Hamilton (Meads) J.; m. Margaret Jean Rosenberry, June 7, 1969; children: Mary Elizabeth, Amy Rose. BS, U. Md., 1966; PhD, U. Wis., 1970. NATO-NSF postdoctoral fellow Pasteur Inst., Paris, 1970-71; rsch. fellow U. Kent, Canterbury, Eng., 1972-74; asst. prof. U. Ill., Urbana, 1974-83; sr. scientist Constrn. Engring. Rsch. Lab., Champaign, Ill., 1983-84, Smith Kline and French, Swedeland, Pa., 1984-87; dir. rsch. Enzymatics Inc., Horsham, Pa., 1987—. Achievements include several patents relating to analog to digital color switching in diagnostic devices. Office: Enzymatics Inc 500 Enterprise Dr Horsham PA 19044

JOHNSTON, JOHN THOMAS, engineering executive; b. St. Louis, Jan. 24, 1930; s. Herbert Johnston and Mabel (Farris) Seeley; m. Shirley Wiladean Trulove, Nov. 25, 1950; children: John David, Thomas Daniel. Cert. in med. tech., Washington U., St. Louis, 1960, BS, 1963; MBA, Lindenwood Coll., 1978. Loads engr. McDonnell Aircraft, St. Louis, 1955-66, 67-70, project engr., 1970-83, integrator-engr., 1983-87; chief engr. Lear Jet, Wichita, Kans., 1966-67; mgr. aero tech. E-Systems, Greenville, Tex., 1987-92, mem. tech. staff, 1992-93; pres. Tech. Engring. Cons., St. Louis, 1967-72; designated engring. rep. FAA, Kansas City, Mo., 1967-75; chief exec. officer Midwest Travel Inst., St. Peters, Mo., 1985-90. Contbr. articles to profl. publs.; patentee in field. Fundraiser Lindenwood Coll., St. Charles, Mo., 1987; organizer Jr. Achievement, St. Charles, 1975; juvenile officer Jud. Dist. 11, St. Charles, 1970's. Sgt. U.S. Army, 1950-53. Mem. AIAA. Home: 3 Thornhill Dr Greenville TX 75402-9754

JOHNSTON, LAREA DENNIS, taxonomist; b. Grants Pass, Oreg., July 25, 1935; d. Joel Albert and Laura Belle (Blackman) Dennis; m. Weldon Kearney, May 22, 1964 (dec. 1992). BA, Willamette U., 1957; MA, Oreg. State U., 1959. Asst. curator Oreg. State U., Corvallis, 1959-91; Author: Name Your Poison: A Guide to Cultivated and Native Plants Toxic to Humans, 1972, Gilkey's Weeds of the Pacific Northwest, 1980, (with others) Aquatic Plants of the Pacific Northwest, 1963, Handbook of Northwestern Plants, 1980. Recipient Citation for Advancement of Sci. in Oreg. Oreg. Acad. Sci., 1974. Mem. Sigma Xi. Home: 2260 NW Chinook Dr Corvallis OR 97330-2813

JOHNSTON, RALPH KENNEDY, SR., aerospace engineer; b. San Antonio, Oct. 2, 1942; s. Abraham Russell and Janace Roberta (White) J.; m. Florence Ann Sheehy (div. Mar. 1983); children: Theresa Ann, Ralph Kennedy Jr., Michael Andrew; m. Frances Helen Marshall, Nov. 5, 1984; children: Deanna Lynn, Carolyn Ann. BS in Aerospace Engring., Okla. City U., 1977; DD, Reformed Bapt. Sem., Denver, 1981. Cert. cost engr. Pilot, electronics technician USMC, 1962-66; test engr., test pilot Grumman Aerospace Corp., Houston, 1966-69; tech. writer Northrop Svcs. Inc., Houston, 1969-73; pres. Johnston & Johnston Enterprises, Houston, 1973-75; cost engr. FLUOR Engrs. & Constructors, Houston, 1975-77; systems engr. Rockwell Internat., Houston, 1977-79; sr. engr. I.L.C. Dover, Houston, 1979-80, Martin Marietta Aerospace, Vanden Berg AFB, 1980-83; sr. specialist engr. Boeing Comml. Airplane Co., Seattle, 1984—; sr. instr. customer tng. ops. support, Boeing, Seattle, 1986-88, maintenance tng., 1988-91. Pub. editor Religious News Mag., 1981-82. Reformed Baptist min., 1981. With USMC, 1962-66. Mem. Masons (worshipful master 32 degree), Shriners, Elks, Lions. Republican. Avocations: sci. fiction, Masonic work with youth orgns., fishing, traveling, outdoor activities.

JOHNSTON, T. MILES G., broadcast engineer; b. Lurgan, No. Ireland, Apr. 17, 1957; s. Thomas Carter and Lavinia (Grayson) J. Degree, Campbell Coll., Knock, Belfast, No. Ireland, 1973; BSc in Mech. and Elec. Engring., Ulster Poly., Whiteabbey, 1980. Lab. technician Watson Labs., Toronto, Ont., Can., 1977-78; installation engr. Standard Telephones & Cables, London, 1981-82; vision control engr. BBC TV, London, 1983-88; project engr. Sta. KISS FM, Monaghan, Belfast, 1987-89; chief engr. Belfast Community Radio, 1989; tech. dir. Sky Satelite TV, Isleworth, London, 1989—; cons., assoc. producer SKY News on UFOs, 1992—; bd. dirs. Irish UFO Rsch. Ctr., Lurgan, 1972—; area investigator Bufora, 1974—, Ctr. for Crop Circles Studies, U.K.; producer Irish Era: Ulster Mus. Archive, No. Ireland, 1978—; tech. cons. in field; lectr. in UFO studies; fin. backer, creator various pirate stations, Irish border, 1979—. Dir., researcher Irish UFO News; producer, dir. video Irish Era Pirate Radio, 1978; contbr. articles to profl. jours. Vol. Hosp. Radio Broadcasting, Craigavon, 1981—; active Campaign for Music Radio, No. Ireland, 1981—, chair, 1989. Mem. IEEE, Irish Astron. Assn., Broadcasting Entertainment Comm. and Tech. Union, Radio Acad., Inst. Electrical Engrs. Avocations: rugby, swimming, photography, pirate radio, disc jockey. Home and Office: The Demesne 39 Antrim Rd, Lurgan BT67-9BW, Northern Ireland

JOHNSTONE, JOHN WILLIAM, JR., chemical company executive; b. Bklyn., Nov. 19, 1932; s. John William and Sarah J. (Singleton) J.; m. Claire Lundberg, Apr. 14, 1956; children: Thomas Edward, James Robert, Robert Andrew. BA, Hartwick Coll., Oneonta, N.Y., 1954; DSc (hon.), Hartwick Coll., 1990; grad. advanced mgmt. program, Harvard U., 1970. With Hooker Chem. Corp., 1954-75, group v.p., 1973-75; pres. Airco Alloys div. Airco, Inc., 1976-79; v.p., gen. mgr. indsl. products, then sr. v.p. chems. group Olin Corp., 1979-80; corp. v.p., pres. chems. group Olin Corp., Stamford, Conn., 1980-85; pres. Olin Corp., 1985-87, chief operating officer, 1986-87, pres., chief exec. officer, chmn., 1988—, also chmn. bd.; bd. dirs. Phoenix Home Life Ins. Co., Research Corp., Am. Brands, Inc. Bd. dirs. Am. Productivity and Quality Ctr.; mem. def. policy adv. com. on trade; trustee Hartwick Coll., 1983-91, 92—; The Conf. Bd.; policy com. Bus. Roundtable, 1992—. Mem. Am. Mgmt. Assn., Soc. Chem. Industry, Soap and Detergent Assn. (former chmn. bd. dirs.), Chem. Mfrs. Assn. (chmn. bd. dirs. 1991), Landmark Club, Woodway Country Club, Blind Brook Club, Links Club. Episcopalian. Office: Olin Corp 120 Long Ridge Rd Stamford CT 06902-1839

JOKL, ALOIS LOUIS, electrical engineer; b. Vienna, Austria, Mar. 16, 1924; came to the U.S., 1939; s. Samuel and Ernestine (Fischer) J.; m. Agnes Antoinette Wozniak, Dec. 29, 1951; children: Justine Ann, Martin Louis, James Anthony. B in Engring., U. So. Calif., 1944; PhD, U. Colo., 1973. Registered profl. engr., Va., N.Y. Elec. engr. Westinghouse Electric Corp., Buffalo, 1946-51; chief elec. engr. R&D div. Continental Motors Corp., Detroit, 1955-64; br. chief power tech. div. USA Belvoir RDE Ctr., Ft. Belvoir, Va., 1964-72, chief power generation div., 1972-88, sr. scientist logistics equipment, 1988-89; cons. Alexandria, Va., 1989—; lectr. Cath. U. Am., Washington, 1981—; mem., chief U.S. delegation Quadripartite Working Group Elec. Power Sources, London, Auckland, New Zealand, 1983-89; judge Sch. Sci. Fairs, Alexandria, 1980-91. Contbr. articles to profl. jours. With U.S. Army, 1944-46, ETO. Mem. IEEE (sr. life), Sigma Xi (v.p. Belvoir chpt. 1983-85). Roman Catholic. Achievements include four patents; research in magnetic field calculations, electrical machinery design methods, waveform predictions.

JOLLES, GEORGES EDGAR RENÉ, scientist; b. Vienna, Austria, Apr. 10, 1929; s. Henri and Marguerite (Weinber) J.; m. Bernadette Bergeret, July 4, 1959; children: Charles, Francois, Brigitte. Lic. es Scis., Ecole Superieure de Chimie, Lyons, France, 1950; PhD, U. Paris, 1953; postgrad. U. Louvain, 1953-54, U. Wis., 1954-55. Research assoc. Rhone-Poulenc Group, Paris, 1956-70, dir. pharm. research, 1970-76, research dir. health div., 1976-82, sci. dir., 1982—; mem. French Nat. Research Council, 1970-80. Author: Histochimie Normale et Pathologique, 1969, Drug Design, Fact or Fantasy?, 1984, Immunostimulants, Now and Tomorrow, 1987, In Vitro Methods in Toxicology, 1992; contbr. articles to profl. jours.; patentee in field. Recipient Galien award Medecine Mondiale, 1973; laureate of the French Acad. of Scis., 1993; named Knight of French Nat. Order Merit. Mem. Am. Chem. Soc., Internat. Soc. Chemotherapy, Soc. Chimique de France, Societe de Chimie Biologique, N.Y. Acad. Scis. Roman Catholic. Home: 1 Allée des Pins, 92330 Sceaux France Office: 20 Ave Raymond Aron, 92165 Antony France

JOLLES, PIERRE, biochemist; b. Ruggell, Liechtenstein, Oct. 9, 1927; s. Henri and Erica (Weinberg) J.; m. Jacqueline Thaureaux, Apr. 6, 1957; children: Beatrice, Anne, Marie-Helene. Engr., Ecole Superieure de Chimie Ind, Lyon, France, 1949; Dr. Engring., U. Paris, 1952, Dr. Sci. Physiques, 1964. Researcher Centre National de la Recherche Scientifique (CNRS), Paris, 1952—, dir. rsch., head of rsch. lab and group, 1964—; prof. biochemistry U. Lausanne (Switzerland) Faculty of Sci., 1971-92, U. Florence, Italy, 1992—. Mng. editor or editor: European Jour. Biochemistry, 1979—, Biochimica et Biophysica Acta, 1971—, FEBS Letters, 1989—, Experientia, 1989—; contbr. over 400 articles to internat. jours.; author: Adjuvants, 1973. Recipient Great Prize, French Acad. Scis., 1979. Mem. European Molecular Biology Orgn., Biochem. Soc., Swiss Biochem. Soc., French Soc. of Biochemistry and Molecular Biology, Rheinisch-Westphalische Akademie der Wissenschaften. Office: Protein Lab Univ Paris V, 45 Rue Des Saints-Peres, F-75270 Paris France

JOLLY, DANIEL EHS, dental educator; b. St. Louis, Aug. 25, 1952; s. Melvin Joseph and Betty Ehs (Koehler) J.; m. Paula Kay Haas, Oct. 13, 1972 (div. Mar. 1988); 1 child, Farrell Elisabeth Ehs; m. Barbara Lee Lindahl, May 7, 1988. BA in Biology and Chemistry, U. Mo., Kansas City, 1974, DDS, 1977. Resident in hosp. dentistry VA Med. Ctr., Leavenworth, Kans., 1977-78; pvt. practice Newcastle, Wyo., 1978-79; asst. prof. U. Mo., Kansas City, 1979-87; dir. dental oncology Trinity Luth. Hosp., 1982-87; assoc. prof., dir. gen. practice residency program Ohio State U., Columbus, 1987—; prof., dir. gen. practice residency program, 1993—; bd. dirs. Rinehart Found., U. Mo. Dental Sch., Kansas City, 1985-87; cons. Lee's Summit (Mo.) Care Ctr., 1984-87, Longview Nursing Ctr., Grandview, Mo., 1986-87; sec. Combined Hosp. Dental Staff, Columbus, 1989-90, v.p., 1990-91, pres., 1991-92. Author: (manual) Hospital Dental Hygiene, 1984, Hospital Dentistry, 1985, OSU Manual of Hospital Dentistry, 1989, 90, 91, 92, 93, (booklet) Nursing Home Dentistry, 1986, Dental Oncology, 1986. Mem. regional coun. Easter Seal Soc., Kansas City, 1985-87, mem. profl. adv. coun. Nat. Easter Seal Soc., 1986-92; sec. bd. dirs. Easter Seal Rehab. Ctr., Columbus, 1990—. Fellow Acad. Dentistry Internat., Am. Soc. Dentistry for Children, Am. Assn. Hosp. Dentists (regional v.p. 1993—), Acad. Gen. Dentistry, Am. Soc. Geriatric Dentistry, Acad. Dentistry for Handicapped, Pierre Fauchard Acad.; mem. ADA, Internat. Assn. Dentistry for Handicapped (pres. elect 1992—), Mo. Dental Assn., Internat. Assn. Dental. Handicap, Greater Kansas City Dental Soc., Fedn. Spl. Care Orgns. in Dentistry, Southwest Oncology Group, Internat. Soc. for Oral Oncology, Ohio Dental Assn. Club: Magna Charta Barons. Avocations: photography, skiing, scuba diving, swimming, sailing. Home: 5322 Bay Meadows Ct Columbus OH 43221-5703 Office: Ohio State U Coll Dentistry 305 W 12th Ave Columbus OH 43210-1241

JOLLY, EDWARD LEE, physicist, pulsed power engineer; b. Indpls., Jan. 1, 1939; s. Edward Lee and Edith Frances (Nation) J.; m. Patricia June Stafford, Aug. 30, 1958 (div.); children: Steven, Thomas, Susan; m. Susan Jenifer Woodruff, Feb. 29, 1992; children: David, Kenneth. BS in Physics, N.Mex. Inst. Mining and Tech., 1961; MS in Nuclear Engring., U. N.Mex., 1968. From staff mem. to mgr. major laser facility U. Calif., Los Alamos, N.Mex., 1961-83, mgr. major accelerator devel. and constrn., 1983—. Mem. Am. Phys. Soc. Home: 120 Dos Brazos Los Alamos NM 82544 Office: Los Alamos Nat Lab MS P940 PO Box 1663 Los Alamos NM 87545

JOLLY, WAYNE TRAVIS, geologist, educator; b. Jacksonville, Tex., Aug. 15, 1940; s. Edward B. and Alfreda J. (Sharp) J. B.F.A., U. Tex., 1963; M.A., SUNY, Binghamton, 1967, Ph.D. (univ. fellow), 1970. Postdoctoral fellow U. Sask., Saskatoon, 1970-71; prof. geology Brock U., St. Catharine's, Ont., Can., 1971—; chmn. dept. Brock U., 1980-84; vis. scientist Commonwealth Sci. and Indsl. Research Orgn., Perth, Australia, 1978; vis. prof. U. Western Ont., 1976. Recipient Acad. Excellence award and Tchr. of Yr. award Brock U. Alumni Assn., 1981 NRC Can. grantee, 1971—. Mem. Geol. Soc. Am., Am. Geophys. Union, Geol. Assn. Can. Office: Brock U, Dept Geol Sci, Saint Catharines, ON Canada

JOLY, JEAN-GIL, medical biochemist, internist, administrator, researcher, educator; b. Montreal, Que., Can., Dec. 14, 1940; s. Roland and Simonne (Caron) J.; m. Marjolaine Laurin, Sept. 19, 1964; children: Chantal, Patrick. B.A. summa cum laude, Coll. André Grasset, Montreal, 1960; M.D. magna cum laude, U. Montreal, 1965; M.Sc., McGill U., Montreal, 1968. Cert. med. specialist in internal medicine and med. biochemistry. Med. Rsch. Coun. Can. scholar Hosp. St. Luc, U. Montreal, 1972-77, dir. clin. rsch. ctr., 1976-79, chief med. biochemistry, 1981-85; pres., gen. mgr. Fonds de la Recherche en Santé du Québec, Montreal, 1981-83; v.p. sci. affairs Squibb Can. Inc., Montreal, 1985-90; pres. Strategex-Sante Inc., 1990; prof. medicine U. Montreal, 1982-85; mem. program grants com. Med. Research Council Can., Ottawa, Ont., 1984-85. Contbr. articles to profl. jours., chpts. to books. Rsch. grantee Med. Rsch. Coun. Can., 1972-85. Mem. Am. Fedn. Clin. Rsch., Assn. Study Liver Disease, Soc. Pharmacology and Exptl. Therapeutics, Med. Soc. Alcoholism, Soc. Clin. Nutrition, Internat. Soc. Biol. Rsch. on Alcoholism, Can. Soc. Clin. Investigation (sec.-treas. 1977-79), Club de Recherches Cliniques du Québec (pres. 1980-82). Roman Catholic. Avocations: ice hockey; drawing; painting; outdoor life. Office: Hotel Dieu St Jerome, 290 Montigny, Saint Jerome, PQ Canada J7Z 5T3

JONAS, JIRI, chemistry educator; b. Prague, Czechoslovakia, Apr. 1, 1932; s. Frantisek and Jirina (Vondrak) J.; m. Ana M. Masiulis, June 1, 1968. BSc, Tech. U. Prague, 1956; PhD, Czechoslovak Acad Sci., 1960. Research assoc. Inst. Organic Chemistry, Czechoslovak Acad. Sci., Prague, 1960-63; vis. scientist, dept. chemistry U. Ill., Urbana, 1963-65, asst. prof., 1966-69, assoc. prof., 1969-72, prof., 1972—, dir. sch. chem. scis., 1983-93; dir. Beckman Inst. Advanced Sci. and Tech., 1993—; sr. staff mem. Materials Research Lab. U. Ill., Urbana, 1970—; dir. Beckman Inst. for Advanced Sci. and Tech., 1993—. Mem. editorial bd. Jour. Magnetic Resonance, 1975—; assoc. editor Jour. of Am. Chem. Soc., 1980-83; mem. NSF adv. com. for chemistry, 1985-88, editorial bd. Jour. Chem., 1980-83; mem. NSF adv. com. for chemistry, 1985-88, editorial bd. Jour. Chem. Physics, 1986-89, Ann. Rev. Phys. Chemistry, 1991—, Accts. of Chem. Rsch., 1990-93; contbr. 270 articles in field of chem. phys. to profl. publs. J.S. Guggenheim fellow, 1972-73, Alfred P. Sloan fellow, 1967-69; Univ. Sr. scholar U. Ill., 1985-88; recipient U.S. Sr. Scientist award Alexander von Humboldt Found., 1988. Fellow Am. Acad. Arts and Scis., AAAS, Am. Phys. Soc.; mem. Nat. Acad. Scis., Am. Chem. Soc. (Joel Henry Hildebrand award 1983), Materials Research Soc. Roman Catholic. Clubs: U. Ill. Tennis; NBTC (Naples, Fla.). Office: Univ of Ill Beckman Inst 405 N Mathews Ave Urbana IL 61801

JONAS, RUTH HABER, psychologist; b. Tel-Aviv, Aug. 24, 1935; d. Fred S. and Dorothy Judith (Bernstein) Haber; m. Saran Jonas, Sept. 16, 1956; children: Elizabeth, Frederick. AB, Barnard Coll., 1957; MA, New Sch. for Social Rsch., 1977, PhD, 1987. Lic. psychologist, N.Y. 1st and 2d yr. intern clin. psychology NYU Med. Ctr.-Bellevue Hosp., N.Y.C., 1985-87; postdoctoral rsch. fellow NYU Med. Ctr., N.Y.C., 1987-88; clin. instr. psychiatry NYU Sch. Medicine, N.Y.C., 1987, clin. asst. prof. psychiatry, 1991; sr. psychologist forensic svc. Bellevue Hosp., N.Y.C., 1988—; pvt. practice psychology N.Y.C., 1988—. Fellow Am. Orthopsychiat. Assn. (mem. stroke coun.); mem. APA, N.Y. State Psychol. Soc., Am. Heart Assn. Office: 200 E 33d St # 10B New York NY 10016

JONAS, SARAN, neurologist, educator; b. N.Y.C., June 24, 1931; s. Myron and Margaret (Wurmfeld) J.; m. Ruth Haber, Sept. 16, 1956; children: Elizabeth Ann, Frederick Jonathan. B.S., Yale U., 1952; M.D., Columbia U., 1956. Diplomate Am. Bd. Psychiatry and Neurology, Am. Bd. Internal Medicine. Intern Bellevue Hosp., N.Y.C., 1956-57; resident and fellow in medicine and neurology Bellevue Hosp., 1957-62; practice medicine specializing in neurology N.Y.C., 1964—; from clin. instr. to assoc. prof. clin. neurology NYU Sch. Medicine, 1964-77, prof. clin. neurology, 1977—, acting chmn. dept. neurology, 1987-91; assoc. dir. neurology N.Y.U. Hosp., 1970-86, dir., 1986-91, dir. electroencephalography, 1969—; acting dir. neurology Bellevue Hosp., N.Y.C., 1987-91, assoc. dir. 1991—. Served with USN, 1962-64. N.Y. State fellow in rheumatic diseases, 1962-64. Mem. Am. Acad. Neurology, Am. Med. Electroencephalographic Assn., Am. Electroencephalographic Soc., Assn. for Research in Nervous and Mental Diseases, Am. Heart Assn. (Stroke Council), Am. Epilepsy Soc. Office: 530 1st Ave New York NY 10016-6402

JONCICH, DAVID MICHAEL, energy engineer; b. Owatonna, Minn., Nov. 27, 1946; s. Michael Joseph Joncich and Jane Kathryn (Leach) Hebert; m. Marcy Lee Jones, Aug. 24, 1968; children: Erica Lee, Adam David. BA, DePauw U., 1967; PhD in Physics, U. Ill., 1976. Postdoctoral res. assoc. in physics U. Ill., Urbana, 1976-78; prin. investigator USACERL, Champaign,

Ill., 1978-80, team leader, 1980-91, div. chief, 1991—; co-founder Solar Design Assocs., Champaign, 1976-85; advisor Nat. Renewable Energy Lab., Golden, Colo., 1990-92; reviewer ASME, Boulder, Colo., 1980. Contbr. articles to profl. publs.; dir. revisions for Army Corps Engrs. tech. manual and guide specification. Troop leader Boy Scouts Am., Urbana, 1989-90; violinist Illini Symphony, Urbana, 1990—. HUD grantee, 1976; recipient Meritorious Civilian award Dept. Army, 1991, commendation Dept. Def., 1980. Mem. Assn. Energy Engrs., Sigma Pi Sigma (v.p. 1966-67). Achievements include research in concurrent engineering, storage cooling systems, condensing heat exchangers, photovoltaics for military applications. Home: 705 W Illinois St Urbana IL 61801 Office: USACERL PO Box 9005 Champaign IL 61826-9005

JONES, BARBARA EWER, school psychologist; b. Marion, Ind., Jan. 28, 1942; d. J. Bertrand and Audrey May (Carter) Ewer; m. Jan Alden Fowler. BS, Ind. U., 1965, MS, 1970; MS, Ind. U.-Purdue U., Indpls., 1977; postgrad., U. Indpls., 1986-93. Tchr. Marion Sch. System, 1965-66, Decatur Twp. Sch. System, Indpls., 1967-71; contract substitute tchr. San Bernardino (Calif.) Sch. System, 1967; dir. Univ. Early Childhood Sch., Indpls., 1971-75; psychologist Monroe County Community Sch. Corp., Bloomington, Ind., 1977—, also consultation and insvc. trainer, rschr.; lectr. evening divsn. Butler U., Indpls., 1974-81; mem. com. for study of children of alcoholics in sch. environ., Bloomington, 1990—. Contbr. articles to profl. jours. Recipient awards for photography and watercolors, 1978, 86-89. Mem. Ind. Sch. Psychologists Assn., Ind. Occupational Therapy Assn. Home: RR 2 Box 5 Morgantown IN 46160-9510

JONES, BARCLAY GIBBS, regional economics researcher; b. Camden, N.J., June 3, 1925; s. Barclay Gibbs Jones and Kathryn (Prince) Preston; m. Anne Van Syckel Tompkins, June 8, 1957; children: Barclay Gibbs, Louise Tompkins. BA, U. Pa., 1948, BArch, 1951; MRP, U. N.C., 1955, PhD, 1961. Registered architect, N.C. Community planner Citizens Coun. on City Planning, Phila., 1951; from instr. to asst. prof. Dept. City Regional planning U. Calif. at Berkeley, 1956-61; from assoc. prof. to prof. Dept. City Regional Planning Cornell U., 1961—; program dir. Cornell Inst. for Social and Econ. Rsch., 1983—; exec. com. mem. Nat. Ctr. for Earthquake Engring. Rsch., Buffalo, 1989-91, rsch. com. mem. 1991—. Editor Protecting Historic Architecture and Museum Collections from Natural Disasters, 1986; contbr. articles to profl. jours. Bd. mem. Archtl. Rsch. Ctrs. Consortium, Inc., 1980—; chair Ithaca (City) Landmarks Preservation Commn., N.Y., 1984-91; vice chair, 1992—; pres. Historic Ithaca and Tompkins County, Inc., 1979-81, bd. mem. 1975-81; bd. mem. Nat. Preservation Inst., Inc., 1984—; with U.S. Army, 1943-46. Decorated Purple Heart; recipient Pub. Svc. award Nat. Park Svc., U.S. Dept. Interior, 1988, Disting. Planning Educator award Am. Collegiate Sch. Planning, 1990; fellow U.S. Internat. Coun. on Monuments and Sites, 1986. Mem. AIA, AAAS, AAUP, Am. Inst. Cert. Planners, Am. Planning Assn., Am. Statis. Assn., Earthquake Engring. Rsch. Inst., Nat. Trust Historic Preservation, N.E. Regional Sci. Assn. (pres 1975-76, 87-88), Am. Econ. Assn., Regional Sci. Assn. (coun. 1976-80, pres. 1983, archivist 1984—), Soc. Archtl. Historians, Urban Regional Info. Systems Assn. (pres. 1966-69), Phi Kappa Phi. Republican. Episcopalian. Home: 502 Turner Pl Ithaca NY 14850-5630 Office: Cornell U 106 W Sibley Hall Ithaca NY 14853

JONES, BENJAMIN ANGUS, JR., retired agricultural engineering educator, administrator; b. Mahomet, Ill., Apr. 16, 1926; s. Benjamin Angus and Grace Lucile (Morr) J.; m. Georgeann Hall, Sept. 11, 1949; children: Nancy Kay Jones-Richardson, Ruth Ann Jones-Sommers. BS, U. Ill., 1949, MS, 1950, PhD, 1958. Registered profl. engr., Ill. Asst. prof., asst. ext. engr. U. Vt., Burlington, 1950-52; instr., agrl. engr. U. Ill., Urbana, 1952-54, asst. prof., agrl. engr., 1954-58, assoc. prof., agrl. engr., 1958-64, prof., agrl. engr., 1964-92, prof. emeritus, 1992—, assoc. dir., agrl. exptl. sta., 1973-92; assoc. dir. emeritus, 1992—, U. Ill., Urbana, 1992; cons. various Ill. Drainage Dists., 1958—. Co-author: (textbook) Engineering Application in Agriculture, 1973; contbr. articles to Jour. Soil & Water Conservation, Encyclopedia Britannica, Agrl. Engring., Transactions of ASAE, Proceedings of ASCE, Soil Sci. Soc. Am. Proceedings, Crops and Soils, Jour. Hydrology, Water Resources Bulletin. Merit badge examiner Boy Scouts Am., Burlington, 1950-52; lay mem. Cen. Ill. Con. United Meth. Ch., 1978-81. With USN, 1944-46. NSF fellow. Fellow Am. Soc. Agrl. Engrs. (bd. dirs., trustee); mem. Soil and Water Conservation Soc., Am. Soc. for Engring. Edn., Sigma Xi, Gamma Sigma Delta, Alpha Epsilon. Home: 708 E Sunnycrest Dr Urbana IL 61801-5965 Office: U Ill Agrl Exptl Sta 211 Mumford Hall 1301 W Gregory Dr Urbana IL 61801-3608

JONES, BEVERLY ANN MILLER, nursing administrator, patient services executive; b. Bklyn., July 14, 1927; d. Hayman Edward and Eleanor Virginia (Doyle) Miller. BSN, Adelphi U., 1949; m. Kenneth Lonzo Jones, Sept. 5, 1953; children: Steven Kenneth, Lonnie Cord. Chief nurse regional blood program ARC, N.Y.C., 1951-54; asst. dir., acting dir. nursing M.D. Anderson Hosp. and Tumor Inst., Houston, 1954-55; asst. dir. nursing Sibley Meml. Hosp., Washington, 1959-61; assoc. dir. nursing svc. Anne Arundel Gen. Hosp., Annapolis, Md., 1966-70; asst. adminstr. nursing Alexandria (Va.) Hosp., 1972-73; v.p. patient care svcs., Longmont (Colo.) United Hosp., 1977—; instr. ARC, 1953-57; mem. adv. bd. Boulder Valley Vo.-Tech Health Occupations Program, 1977-80; chmn. nurse enrollment com. D.C. chpt. ARC, 1959-61; del. nursing adminstrs. good will trip to Poland, Hungary, Sweden and Eng., 1980. Contbr. articles to profl. jours. Bd. dirs. Meals on Wheels, Longmont, Colo., 1978-80, Longmont Coalition for Women in Crisis, Applewood Living Ctr., Longmont; mem. Colo. Hosp. Assn. Task Force on Nat. Commn. on Nursing, 1982; mem. utilization com. Boulder (Colo.) Hospice, 1979-83; vol. Longmont Police Bur., Colo.; mem. coun. labor rels. Colo. Hosp. Assn., 1982-87; mem.-at-large exec. com. nursing svc. adminstrs. Sect. Md. Nurses' Assn., 1966-69; mem. U. Colorado Task Force on Nursing, 1990; vol. Champs program St. Vrain Valley Sch. Dist. Mem. Am. Orgn. Nurse Execs. (chmn. com. membership svcs. and promotions, nominee recognition of excellence in nursing adminstrn.), Colo. Soc. Nurse Execs. (dir. 1978-80, 84-86, pres. 1980-81, mem. com. on nominations 1985-86. Home: 853 Wade Rd Longmont CO 80503-7017 Office: PO Box 1659 Longmont CO 80502-1659

JONES, BRUCE HOVEY, physician, researcher; b. St. Paul, Apr. 2, 1947; s. H. Ivor and Jean Elizabeth (Berger) J.; m. Gail Schneider, Dec. 28, 1978 (div. Mar. 1985); m. Tanya Eyre Morgan, Oct. 28, 1989; children: Ian Fisher, Aaron Grayson. BA in History and Sci. cum laude, Harvard U., 1970, MPH, 1986; MA in Biology, Kans. U., 1974; MD, Kans. U., Kansas City, 1977. Bd. Cert. in Preventive Medicine. Intern, Winter Gen. VA Hosp., Stormont Vail Hosp., Topeka, Kans., 1979-80; resident in preventative medicine Walter Reed Army Inst. Rsch., 1986; commd. capt. U.S. Army, 1977, advanced through grades to maj. lt. col., 1989, gen. med. officer, Ft. Jackson, S.C., 1977-79, med. officer, investigator U.S. Army Research Inst. of Environ. Medicine, Natick, Mass., 1980—; DOD rep. to HHS, CDC Adv. Com. on Injury Prevention and Control, chmn, Work Group on Injury Surveillance and Prevention. Author: Exercise and Sports Sci. Revs., vol. 17, 1989; author, contbr. chpts. in books, articles to jours. in field. Hon. freshman scholar Harvard U., 1965; decorated Army Commendation medal with second oak leaf clusters, Army Achievement medal with oak leaf cluster. Fellow Am. Coll. Preventive Medicine, Am. Coll. Sports Medicine; mem. AMA, Mass. Med. Soc., Assn. Mil. Surgeons U.S. (Outstanding Rsch. award 1988). Unitarian. Home: 355 Hudson Rd Stow MA 01775 Office: Chief Occupational Medicine Div Army Rsch Inst Environ Medicine Kansas St Natick MA 01760-5007

JONES, CHARLES LEE, industrial designer; b. Danville, Ill., Dec. 11, 1957; s. Harry Leon and Maye Francis (Green) J. BS in Indsl. Design/Human Factors Engring. with honors, Purdue U., 1981. From designer to sr. designer advanced products Kinball Internat., Jasper, Ind., 1981-84; mgr. venture group Haworth, Inc., Holland, Mich., 1984-87, mgr. corp. indsl. design, 1987-92; gen. mgr. Herman Miller Inc., 1992-93; dir. indsl. design/human interface Xerox Corp., Rochester, N.Y., 1993—; guest lectr. Ohio State U., Columbus, 1982; cons. Ctr. Advanced Aging,Baton Rouge, 1988—. Contbr. articles to profl. jours. Mem. Grand Rapids (Mich.) Arts Museum, 1990; bd. dirs. Holland Area Arts Coun., 1990—. Rsch. grantee Human Factors Dept. Purdue U., 1978; recipient sholarship Citizen Scholar Com., 1976. Mem. Indsl. Design Soc. Am. (honorable mention award 1987, vice chmn. Mich. chpt. 1985—), Design Mgmt. Inst., Human Factors Soc. Achievements include patents for computer integrated desk and acoustical enclosure for office environments; development of interactive ergonomics computer software. Home: 34 Windloft Cir Fairport NY 14450 Office: Xerox Corp 1350 Jefferson Rd Bldg 801-11A Rochester NY 14623

JONES, CHARLES WELDON, biologist, educator, researcher; b. Providence, May 25, 1953; s. Charles Weston and Evelyn Lois (Hall) J. AB, Harvard U., 1975, AM, 1977, PhD, 1980. Helen Hay Whitney postdoctoral fellow Stanford (Calif.) U., 1980, postdoctoral fellow, 1980-82; asst. prof. Bethel Coll., St. Paul, 1982-85, assoc. prof., 1985-90, prof., 1990—; vis. scholar Havard U., Cambridge, Mass., 1988-89; councilor Coun. Undergrad. Rsch., Washington, 1990—. Author: (with others) Levels of Genetic Control in Development, 1981; contbr. articles to Science, Cell, Nature. Recipient grad. fellowship NSF, 1976, grants Rsch. Corp., 1982, NSF, 1984, 85, 90. Mem. AAAS, Nat. Biology Tchrs. Assn., Assn. Biology Lab. Edn., Sigma Zeta (pres. 1987-88). Achievements include first fusion of animal/plant cells. Home: 8585 Yalta Ln NE Circle Pines MN 55014-4075 Office: Bethel Coll 3900 Bethel Dr Saint Paul MN 55112-6902

JONES, COLIN ELLIOTT, physicist, educator; b. Rochester, N.Y., Apr. 6, 1941; s. Frank Lydick and Althea Charlotte (Goeltz) J.; m. Joan Musette Kleppinger, June 14, 1969; children: Devin Gregory, Ian Mathew. BS, Carnegie-Mellon U., 1963; PhD, U. Ill., 1970. Lectr. Princeton (N.J.) U., 1969-70; asst. prof. Lehigh U., Bethlehem, Pa., 1970-76; prin. rsch. scientist Honeywell Corp. Tech. Ctr., Bloomington, Minn., 1976-81; group leader, sr. scientist GM-Hughes Santa Barbar Rsch. ctr., Goleta, Calif., 1981-89; sr. scientist Santa Barbara Focalplane, Goleta, 1989—; mem. conf. com. US Workshop on Phys. Chemistry of Mercury Cadmium Telluride, 1982-89, co-chmn., 1984; presenter in field. Contbr. articles to Jour. Applied Physics, Jour. Vacuum Sci. and Tech. Asst. scoutmaster Boy Scouts Am., Goleta, 1986—. Mem. Am. Phys. Soc., Tau Beta Pi, Phi Kappa Phi, Sigma Xi. Achievements include research in the areas of defects in semiconductors and insulators, material characterization, 1/f noise, infrared detectors and surface passivation of semiconductors. Home: 1035 May Ct Santa Barbara CA 93111 Office: Santa Barbara Focalplane 69 Santa Felicia Dr Goleta CA 93117

JONES, DALE LESLIE, nuclear engineer; b. Boothbay Harbor, Maine, June 2, 1952; s. George A. and Marilyn E. (Petrie) J.; m. Mary Ann Cooper, Sept. 30, 1990; 1 child, Jeremy. BS in Engring. Physics, U. Maine, 1974. Prototype engr. Knolls Atomic Power Lab., Schenectady, N.Y., 1974-77; ISI engr. Combustion Engring., Windsor, Conn., 1977-81; cons. NDE Engring. Cons., Storrs, Conn., 1981-86; engring. mgr. Engring. Svcs. div. Pacific Nuclear, Chgo., 1986—. Office: Pacific Nuclear 1111 Pasquinelli Dr Westmont IL 60559

JONES, DANIEL SILAS, chemistry educator; b. Charlotte, N.C., Nov. 16, 1943. BS, Wake Forest U., 1965; PhD, Harvard U., 1970. Assoc. prof. chemistry U. N.C., Charlotte. Mem. Am. Chem. Soc., Am. Crystallographic Assn. Office: UNCC Dept Chemistry Charlotte NC 28223

JONES, DAVID ALLAN, electronics engineer; b. Akron, Ohio, Oct. 16, 1942; s. Alva Jr. J. and Vera Henrietta (Seevers) Fuchs; m. Elizabeth Ann Cheek, Dec. 28, 1971; 1 child, Lindsey Ashok-Ray. BS in Math., U. Miss., 1965; MEE, Naval Postgrad. Sch., 1972. Commd. ensign USN, 1965, advanced through grades to comdr., 1979, ret., 1986; tech. dir. advanced programs Stanford Telecommunications, Inc., Reston, Va., 1986—. Elder Presbyn. Ch., Annandale, Va., 1987-90. Decorated Commendation medal, Def. Meritorious Svc. medal, Navy Meritorious Svc. medal, Achievement medals. Mem. AIAA, Armed Forces Communications and Electronics Assn., IEEE. Avocations: gardening, walking.

JONES, DAVID ALWYN, geneticist, botany educator; b. Colliers Wood, Surrey, Eng., June 23, 1934; came to U.S., 1989; s. Trefor and Marion Edna Jones; m. Hazel Cordelia Lewis, Aug. 29, 1959; children: Catherine Susan, Edmund Meredith, Hugh Francis. BA, MA in Natural Scis. with honors, U. Cambridge, Eng., 1957; DPhil in Genetics, U. Oxford, Eng., 1963. Chartered biologist, UK. Lectr. in genetics U. Birmingham, Eng., 1961-73; prof. genetics U. Hull, Eng., 1973-89, head dept. plant biology and genetics, 1983-88; prof., chmn. dept. botany U. Fla., Gainesville, 1989—; chmn. membership com. Inst. of Biology, London, 1982-87. Co-author: Variation and Adaptation in Plant Species, 1971, Analysis of Populations, 1976, What is Genetics?, 1976, Zmiennosc i przystosowanie roslin, 1977; contbr. over 100 articles to profl. jours. Chmn. Birmingham-King's Norton Round Table, Birmingham, 1970-71. Grantee Royal Soc. London, Sci. and Engring. Rsch. Coun. of U.K., Natural Environment Rsch. Coun., U.K.; Med. Rsch. Coun. scholar U. Oxford, 1957-61. Fellow Linnean Soc., Inst. Biology; mem. AAAS, Am. Soc. Naturalists, Bot. Soc. Am., Internat. Soc. Chem. Ecology (coun. 1983-84, 89-91, keynote speaker ann. meeting 1984, pres. elect 1986-87, pres. 1987-88, past pres. 1988-89), Brit. Assn. Advancement of Sci. (chmn. coord. com. for cytology and genetics 1974-87), Genetical Soc. Great Britain (convenor ann. meetings profs. of genetics 1983-88), Ecol. Genetics Group, Population Genetics Group, Soc. for the Study of Evolution, Gamma Sigma Delta, Sigma Xi. Achievements include research in practical population biology. Home: 7201 SW 97th Ln Gainesville FL 32608-6302 Office: U Fla Dept Botany 220 Bartram Hall Gainesville FL 32611-2009

JONES, DAVID ROBERT, zoology educator; b. Bristol, Eng., Jan. 28, 1941; came to Can., 1969; s. William Arnold and Gladys Margery (Parker) J.; m. Valerie Iris Gibson, Sept. 15, 1962; children: Melanie Ann, Vivienne Samantha. B.Sc., Southampton U., 1962; Ph.D., U. East Anglia, Norwich, Eng., 1965. Rsch. fellow U. East Anglia, Eng., 1965-66; lectr. zoology U. Bristol, Eng., 1966-69; prof. zoology U. B.C., Vancouver, B.C., Can., 1969—. Contbr. numerous articles to profl. jours. Fellow Killam Found. Can., 1973, 89. Fellow Royal Soc. Can.; mem. Soc. Exptl. Biology, Am. Physiol. Soc., Can. Zool. Soc. (Fry medal 1992), Am. Zool. Soc., Can. Physiol. Soc. Office: U BC, 6270 University Blvd, Vancouver, BC Canada V6T 2A9

JONES, EDWARD DAVID, plant pathologist; b. Rockland, Md., May 8, 1920; s. Eben E. and Rachel Hannah (Williams) J.; m. Barbara Jane Jones, July 19, 1947; children: David R., Kathleen R., Jacalyn A., E. Douglas. BS, U. Wis., 1946, MS, 1947, PhD, 1953. Grad. research asst. U. Wis., Madison, 1946-47, instr. plant pathology, 1948-53; research plant pathologist Red Dot Foods, Inc., Madison, 1953-58; from asst. prof. plant pathology to Henry and Mildred Uihlein prof. Cornell U., Ithaca, N.Y., 1958—; adj. prof. Plattsburg (N.Y.) State U., 1982—. Mgr., coach Youth Baseball and Hockey Assn., Ithaca, 1958-78; coach Am. Legion Baseball, Lake Placid, N.Y., 1966-67; off-ice hockey ofcl. 1980 Winter Olympics, Lake Placid, 1980. Served to 1st lt. USAAF, 1942-45, ETO. Recipient Dedicated Service award N.Y. Seed Potato Growers, 1979, Disting. Service citation N.Y. State Agrl. Soc., 1984; named Man of Yr. New Brunswick Potato Growers, 1982. Mem. Potato Assn. Am. (hon. life, bd. dirs. 1971-74, v.p. 1981-82, pres. 1983-84), Am. Phytopathol. Assn., Alpha Zeta, Sigma Xi. Republican. Presbyterian. Home: PO Box 260 Wild Rose WI 54984-0260 Office: Cornell U Dept Plant Pathology 318 Plant Science Bldg Ithaca NY 14853

JONES, FAY, architect; b. Pine Bluff, Ark., 1922; m. Gus Jones, 1944; 2 children. Student, U. Ark. Apprentice Frank Lloyd Wright, 1953; dean U. Ark. Sch. Architecture, until 1989; pvt. practice Fayetteville, Ark. With USN, World War II. Named winner AIA Gold medal, 1990, winner AIA award Best Am. Bldg. of Decade, 1991. Office: 619 W Dickson Fayetteville AR 72701

JONES, FRANK RALPH, materials scientist engineer; b. Northampton, U.K., Jan. 16, 1944; s. Horace Ralph Henry and Doris Mary (Amos) J.; m. Christine Jones, Oct. 22, 1966; children: Ian Peter, Helen Laura. PhD, U. Keele, 1970. Lab. asst. Scott-Bader and Co. Ltd., Northamptonshire, U.K., 1960-62, chem. asst. then asst. chemist, 1962-66; demonstrator U. Keele, Staffordshire, U.K., 1968-70; Humboldt rsch. fellow Mainz U., Fed. Republic of Germany, 1970-71; rsch. fellow U. York, Yorkshire, U.K., 1971-73; lectr. II Birkenhead Coll. of Tech., Wirral, U.K., 1973-75; lectr. U. Surrey, Guildford, 1975-85; sr. lectr., reader Sheffield U., South Yorkshire, 1985-93, prof., 1993—; vis. scientist DFVLR, Köln, Germany, 1987. Editor: Interfacial Phenomena in Composite Materials, 1989-91; contbr. articles to profl. jours. Fellow Royal Soc. of Chemistry (chartered 1985), Plastics and Rubber Inst. (chmn. London sect. 1984-85, mem. coun. 1984-85, 88-91), Inst. of Materials. Achievements include rsch. on explanation and observation of enhanced residual stresses in composites, mechanisms of environ. stress corrosion of GRP, microporosity/functionality theory of carbon fibre surface chemistry, new results showing incorporation of substrate components into silanes on glass surfaces; development of the application time-of-flight sims to a study of composite fracture processes. Office: U Sheffield Dept Engring Materials, Robert Hadfield Bldg PO Box 600, Sheffield S14DV, England

JONES, GEOFFREY MELVILL, physiology research educator; b. Cambridge, Eng., Jan. 14, 1923; s. Benett and Dorothy Laxton (Jotham) J.; m. Jenny Marigold Burnaby, June 21, 1953; children—Katharine, Francis, Andrew, Dorothy. B.A., Cambridge U., 1944, M.A., 1947, M.D., 1949. House surgeon Middlesex Hosp., London, Eng., 1949-50; sr. house surgeon Addenbrookes Hosp., Cambridge, Eng., 1950-51; sci. med. officer Royal Air Force Inst. Aviation Medicine, Farnborough, Eng., 1951-55; sci. officer Med. Research Council, Eng., 1955-61; assoc. prof. physiology, dir. aviation med. rsch. unit McGill U., Montreal, Que., Can., 1961-68, prof., dir., 1968-88, Hosmer research prof., 1978-91; emeritus prof. physiology McGill U., Montreal, Que., 1991—; adj. prof. clin. neuroscis. U. Calgary, Alta., Can., 1991—. Author: (with another) Mammalian Vestibular Physiology, 1979. Editor (with another) Adaptive Mechanisms in Gaze Control, 1985. Served to squadron leader Royal Air Force, 1951-55. Sr. rsch. assoc. Nat. Acad. Sci., 1971-72; recipient Skylab Achievement award NASA, 1974, 1st recipient Dohlman medal Dohlman Soc. Toronto U., 1987, Quinquennial Gold medal Barany Soc. Internat., 1988, Ashton Graybiel award U.S. Naval Aerospace Labs., 1989, Wilbur Franks Annual award Can. Soc. Aerospace Medicine, Buchanan-Barbour award Royal Aeronautical Soc., 1991. Fellow Canadian Aeronautics and Space Inst., Aerospace Med. Assn. (Harry Armstrong award 1968, Arnold D. Tuttle award 1971), Royal Soc. Can. (McLaughlin medal 1991), Royal Soc. London, Royal Aeronautical Soc. London (Stewart Meml. award 1989). Avocations: tennis, outdoor activities; reading; choral singing. Office: U Calgary Dept Clin Neuroscis, 3330 Hospital Dr NW, Calgary, AB Canada T2N 4N1

JONES, GEORGE HUMPHREY, retired healthcare executive, hospital facilities and communications consultant; b. Kansas City, Mo., July 10, 1923; s. George Humphrey and Mary R. (Marrs) J.; m. Peggy Jean Thompson, Nov. 23, 1943; children: Kenneth L., Daniel D., Kathleen Jones Carrigan, Carol R. Jones Johnson, Janet S. Jones Fitts. Student, U. Mo., Kansas City, 1940-43, Wis. State Coll., 1943. Police officer Kansas City (Mo.) Police Dept., 1947-51; elec. contr. Paramount Elec. Svc., Kansas City, 1947-50; electrician Automatic Temp. Control Co., Kansas City, 1951-57; pres., chief ops. George H. Jones Co., Kansas City, 1957-65; sales mgr. Nycon Inc., Lee's Summit, Mo., 1965; design engr. Midland Wright Corp., Kansas City, 1966; dist. sales mgr. Communications Electronics, Kansas City, 1967; plant ops. supr. Research Med. Ctr., Kansas City, 1968-77; dir. plant ops. and communications Research Med. Ctr., 1977-90; hosp. facilities and communications cons. Overland Park, Kans., 1990—; guest lectr. Nat. U., San Diego, 1987. Mem. Met. Emergency Preparedness Coun.; bd. dirs. Camellot Fine Arts Acad., 1974-76, v.p. bd., 1975, 76; vol. program devel. mid-Am. chpt. Multiple Sclerosis Soc.; vol. emergency ops. and comms. Kansas City Area Hosp. Assn. Fellow Am. Soc. Hosp. Engring., Healthcare Info. and Mgmt. Systems Soc.; mem. Kansas City Area Hosp. Engrs. (pres. 1985, bd. dirs. 1985-89), Am. Legion, Alpha Phi Omega. Presbyterian. Avocations: fishing, photography. Home and Office: 6022 W 86th St Shawnee Mission KS 66207-1521

JONES, GEORGE RICHARD, physicist; b. L.A., Aug. 16, 1930; s. George Michael and Edna Catherine (Hannibal) J.; m. Jeanne Celeste Dougherty, Nov. 22, 1952; children: Michael, George, Ralph, Susan, Karen, Thomas. BS, Western Md. Coll., 1951; MS, Cath. U of Am., 1953, PhD, 1963. Jr. engr. Davies Labs. Inc., Riverdale, Md., 1952-54; physicist Diamond Ordnana Fuel Labs., Washington, 1954-59; rsch. physicist Harry Diamond Labs., Washington, 1959-66, Night Vision Lab., Fort Belvoir, Va., 1966-92, Army Rsch. Lab., Adelphi, Md., 1992—; dir. Twin Pines Savs. and Loan Assn., Greenbelt, Md., 1954-81, bd. pres., 1962-73; cons. in field. Mem. Cath. Interracial Coun., Prince Georges County, 1965; mem. Greenbelt Fair Housing Com., 1963-67. Recipient Md. State Senatorial scholarship, 1947-51. Mem. AAAS, Am. Phys. Soc., Internat. Neural Network Soc., Sigma Xi. Roman Catholic. Achievements include patents in field; being first to observe Magnetic Dipole Transition Optical Spectroscopy.

JONES, GLADYS HURT, retired mathematics educator; b. Selma, Ala., Dec. 13, 1920; d. Naxie Hurt and Osceola Thelma (Martin) Taylor; m. Herbert Whittier Jones, Sept. 14, 1953; 1 child, Brenda Elaine. Student, Talladega (Ala.) Coll., 1937-39; BS, Tenn. A&I U., 1941; MS, Atlanta U., 1947; postgrad., U. Mich., Fla. State U., 1950-51, 72. Tchr. math. and chemistry Ctr. High Sch., Waycross, Ga., 1941-46; instr. math. Morehouse Coll., Atlanta, 1947-53; rsch. asst. U. Mich., Ann Arbor, 1953-57; assoc. prof. math. Fla. A&M U., Tallahassee, 1957-85. Gen. Edn. Bd. fellow U. Mich., 1950-51. Mem. Math. Assn. Am., Nat. Coun. Tchrs. Math., Am. Orchid Soc., Am. Hort. Soc., Plant Life Soc., Am. Iris Soc., beta Kappa Chi, Delta Sigma Theta. Democrat. Avocations: growing orchids, amaryllis and other flowers. Home: 308 Barbourville Dr Tallahassee FL 32301-4215

JONES, HAROLD CHARLES, retired biologist; b. Oberlin, Ohio, May 25, 1903; s. Lynds and Clara Mabelle (Tallmon) J.; m. Alice Alden Drew, Apr. 29, 1932 (dec. June 1988). AB, Oberlin (Ohio) Coll., 1928, MA, 1930; PhD, George Peabody Coll., 1940. Head dept. biology Berry Coll., 1930-40; head dept. biology Ga. Coll., Milledgeville, ret., 1966; naturalist Ga. State Parks, 1967-68. Author: Plant Ecology of the Berry Schools Property, 1940. Home: 180 Pineaway Dr SE Milledgeville GA 31061-9202

JONES, HOBERT W, health physicist; b. Lexington, Ky., Aug. 12, 1957; s. John E., Jr. and Peggy Ann (Pickle) J. BS in physics, U. Ky., 1980; MS in Health Physics, Ga. Inst. Tech., 1985. Cert. health physicist; registered radiation protection technologist. Radiochemical lab. analyst Tenn. Valley Authority, Soddy-Daisy, 1981-84; health physicist Am. Electric Power Svc. Corp., Columbus, Ohio, 1985-91; sr. health physicist EG&G Mound Applied Techs., Miamisburg, Ohio, 1991-92, tech. specialist health physics, 1992-93; health physicist Labyrinth Group, Dayton, Ohio, 1993—. Mem. Am. Chem. Soc., Am. Nuclear Soc. (assoc.), Health Physics Soc. (plenary), Am. Acad. of Health Physicists. Home: 550 E Whipp Rd Centerville OH 45459 Office: Labyrinth Group PO Box 750155 Dayton OH 45475

JONES, HOWARD ST. CLAIRE, JR., electronics engineering executive; b. Richmond, Va., Aug. 18, 1921; s. Howard St. Claire and Martha Lillian (Mason) J.; m. Evelyn Mercer Saunders, Nov. 27, 1946. B.S., Va. Union U., 1943, D.Sc., 1971; certificate engring., Howard U., 1944; M.S.E.E., Bucknell U., 1973. Registered profl. engr., D.C., Va., Md. Indsl. engring. aid Bur. Ships USN, Washington, 1943; electro mech. engring. aide U.S. Bur. Standards, Washington, 1944; electronic physicist U.S. Bur. Standards, 1946-53; electronic scientist, engr., supervisory phys. scientist Harry Diamond Labs., AUS, Washington, 1953-80; tech. cons. microwave electronics, 1980—; Tchr. radio physics Hilltop Radio-Electronics Inst., Washington, 1946-52; assoc. prof. elec. engring. Howard U., 1958-63, adj. prof., 1982; cons. microwave engring., 1965-69; cons. univ. relations, 1983—. Contbr. tech. reports and publs. Served as instr. mech. engring. AUS, 1944-46. Recipient four Sustained Superior Performance or Spl. Act awards Harry Diamond Labs AUS, 1956, 68, 70, 75, Inventor of Year award, 1972; Sec. Army Fellowship award, 1972; Army Research and Devel. award, 1975; Meritorious Civilian Service award, 1976, 80. Fellow IEEE (Harry Diamond Field award 1985), AAAS, Washington Acad. Sci.; mem. ASEE. Achievements include holding 31 U.S. patents microwave field. Home and Office: 3001 Veazey Ter NW Apt 1310 Washington DC 20008-5407

JONES, JACK ALLEN, aerospace engineer; b. Fayetteville, Tenn., Aug. 16, 1935; s. Albin Oscar and Katharine (Pickett) J.; m. Katharine Tubb, Dec. 27, 1961. Student, Fla. State U., 1960; BS, Berry Coll., 1957. Data systems engr. U.S. Army, Redstone Arsenal, Ala., 1957-60; data systems engr. NASA, Marshall Space Flight Ctr, Ala., 1960-74, systems engr., 1974-82, chief engr., 1982-87, 91—, mission mgr., 1987-91. Editor: Vibration Manual, 1962. Mem. Libr. Bldg. Bd., Huntsville, 1978-87, Courthouse Bldg. Bd., Huntsville, 1978-88. Recipient Laurel award Aviation Week, 1991, NASA's

Exceptional Svc. medal, 1991. Mem. AIAA. Achievements include co-invention of statistical analyzer of random processes. Home: 1394 Dug Hill Rd Brownsboro AL 35741 Office: NASA Marshall Space Flight Ctr Huntsville AL 35812

JONES, JACK HUGH, applications engineer, educator; b. Saddle, Ark., Feb. 6, 1944; s. Thurman L. and Lovina B. (Frisbee) J.; m. Beatrice K. Fellows, June 29, 1968; children: Glen R., Gregory S., Kevin M., Parmalee M. Cert. electronics technician, DeVry Inst. Tech., 1973; cert. mgmt., Rockford Coll., 1990. Sr. lab. technician Warner Electric, Marengo, Ill., 1979-80, supr. engring. lab., 1980-85, supr. engring. svcs., 1985-89, application engr., 1989-93; sr. application engr., 1993—. Deacon, bd. dirs. Harvard Assembly of God Ch., 1989-92, past pres. men's ministry. With U.S. Army, 1965-67. Republican. Mem. Assembly of God Ch. Home: 403 S Ayer St Harvard IL 60033-2812 Office: Warner Bernstein Sentel 1300 N State St Marengo IL 60152-2299

JONES, JAMES BEVERLY, mechanical engineering educator; b. Kansas City, Mo., Aug. 21, 1923. BS, Va. Poly. Inst., 1944; MS, Purdue U., 1947, PhD in Mech. Engring., 1951. Asst. mech. engr. engring. bd. U.S. War Dept., Va., 1944-45; from asst. instr. to assoc. prof. mech. engring. Purdue U., 1945-57, prof., 1957-64; svc. engr. Babcock & Wilcox Co., 1948; devel. engr. Gen. Electric Co., 1951-52; sr. project engr. Allison divsn. Gen. Motors Corp., 1953; prof., head dept. mech. engring. Va. Poly. Inst. & State U., Blacksburg, 1964—. NSF faculty fellow Swiss Fed. Inst. Tech., 1961-62. Mem. ASME (James Harry Potter Gold medal 1991), AIAA, Am. Soc. Engring. Edn., Sigma Xi. Achievements include research in fluid mechanics, thermodynamics. Home: 1503 Palmer Dr Blacksburg VA 24060*

JONES, JAMES LAMAR, psychologist; b. Indpls., Dec. 2, 1958; s. James Calvin and Catherine (Robinson) J.; m. Susan Lee Sommerville. BA, Purdue U., 1981; MS, Butler U., 1983; PhD, Clayton U., 1986. Cert. child life specialist, marriage and family therapist, Ind. Substitute tchr. Indpls. Pub. Schs., 1981; adolescent, child life therapist Meth. Hosp., Indpls., 1984-85, supr., adolescent, child life devel., 1985—; pvt. practice Indpls., 1988—; mem. Cochlear Implant Program, Indpls. Mem. AACD, Assn. Care Children's Health. Home: 1625 Park Hurst Dr Indianapolis IN 46229-4130 Office: Meth Hosp 1701 N Senate Ave Indianapolis IN 46202-1299

JONES, JAMES OGDEN, geologist, educator; b. Punkin Center, Tex., July 25, 1935; s. Charles Armond and Onis Velva (Carter) J.; m. Marilyn Felty, Aug. 13, 1961; children: James II, Alan. BS, Midwestern State U., Wichita Falls, Tex., 1962; MS, Baylor U., 1966; PhD, U. Iowa, 1971. Welder Nat. Tank Co., Electra, Tex., 1953-57; geologist Shell Oil Co., Wichita Falls, Tex., 1958-60; grad. teaching asst. Baylor U., Waco, Tex., 1962-64, U. Iowa, Iowa City, 1964-68; geologist Texaco, Inc., Wichita Falls, 1966; geology prof. U. So. Miss., Hattiesburg, 1971; dept. head So. Ark. U., Magnolia, 1971-77; geology program coord. U. Tex., San Antonio, 1978-82, geology prof., 1982—; researcher in field; lectr. in field. Editor, contbr. articles to profl. jours. and field guides in sedimentology and stratigraphy. With U.S. Army, 1968-70, lt. col. USAR. Nat. Teaching fellow, 1971-73. Fellow Geol. Soc. Am. (gen. chmn. ann. meeting 1986, membership com. 1988-90, mgmt. bd. so. cntl. sect.), Tex. Acad. Sci. (chmn. geology sect. 1986-87, vice chmn. 1985-86); mem. South Tex. Geol. Soc. (v.p. 1992-93, pres. elect 1993—), Soc. Ind. Profl. Earth Sci. (cert.), Soc. Sedimentary Geology (chmn. field trip com. ann. meeting 1989), Am. Assn. Petroleum Geologists (field trip chmn. ann. meeting 1984), Nat. Assn. Geology Tchrs. (sec., treas. Tex. section 1992—), Internat. Assn. Sedimentologists, Am. Inst. Profl. Geologists (cert.), Res. Officers Assn., Am. Legion, Sigma Xi, Phi Sigma Kappa, Sigma Gamma Epsilon. Methodist. Avocations: mineral collecting, travel, skiing. Home: 13226 Hunters Lark Dr San Antonio TX 78230-2018 Office: U Tex Dept Geology San Antonio TX 78249-0663

JONES, JAMES RAY, architectural engineering research associate; b. Martin, Ky., June 30, 1958; s. James Loren Jones and Marianna (Hall) Anzaldua; m. Julie Ann Story, Aug. 4, 1979; children: Jared R., Jaci E. BS Arch., U. Mich., 1981, MArch with distinction, 1983, postgrad., 1990—. Adj. lectr. Coll. Arch., U. Mich., Ann Arbor, 1989-91, rsch. assoc., 1983—; cons. Detroit Sci. Ctr., 1989; architect, design Sherman House, Charlotte, Mich., 1991. Coach Ypsilanti Twp. Recreation Dept., Ypsilanti, Mich., 1990-92. Recipeint First Prize paper IEEE/Industry Applications Soc. Lighting, Dearborn, Mich., 1991; W.A. Oberdick fellow U. Mich. Ann Arbor, 1992. Mem. ASHRAE (pres. 1992-93, grad. grant 1992), Illuminating Engrs. Soc., Am. Solar Energy Soc. Achievements include development of prediction methods of electrical demand, building thermal response and whole building energy consumption. Home: 706 N Prospect Ypsilanti MI 48198 Office: U Mich 2000 Bonisteel Blvd Ann Arbor MI 48109

JONES, JANET LEE, psychology educator, cognitive scientist; b. Scottsdale, Ariz., Apr. 4, 1957; d. Gerry L. and Alicia M. (Coppes) J.; m. Alan M. Krajecki, 1986. BA in Psychology magna cum laude, Pomona Coll., 1984; MA in Higher-Order Cognition, UCLA, 1985, PhD in Cognitive Psychology/Psycholinguistics, 1989. Artist, designer Ruth Downs Ltd., 1973-74; text editor Text Craft, Inc., 1975-76; bookkeeper Verde Ind. Newspaper, 1976-77; exec. asst. to v.p. Honeywell, 1978-81; teaching asst. psychology dept. Pomona Coll., Claremont, Calif., 1982-84; teaching fellow and assoc. UCLA, 1985-89; vis. asst. prof. Pitzer Coll., Claremont, 1990; asst. prof. psychology Ft. Lewis Coll., Durango, Colo., 1990—; horse trainer, riding instr., 1973-77; lectr. Calif. State U., Long Beach, 1989; presenter in field. Contbr. articles to profl. jours. Scholar Pomona Coll., 1981-84; grad. fellow UCLA, 1984-85; recipient Gengerelli Disting. Dissertation award, 1989. Mem. Am. Psychol. Soc. (charter), Western Psychol. Assn., Phi Beta Kappa, Sigma Xi. Office: 108A Hesperus Hall Ft Lewis Coll Dept Psych Durango CO 81301

JONES, JEFFERY LYNN, software engineer; b. Aug. 5, 1960; s. Robert Meryl and Ione Dell (Eaves) J. Ptnr. Megabyn Assocs (previously JJ Enterprises), Oklahoma City, 1982—; v.p. Oklahoma Digital Technologies, Inc., Oklahoma City, 1987—; contract cons. Bank Tech Inc., Oklahoma City, 1985-86, Phillips Petroleum Corp., Bartlesville, Okla., 1986-87; cons. Union Oil Co. of Calif., Oklahoma City, 1989—. Co-author: (software) PetroTrak 2000 Lease/Production Petroleum Tracking System, 1991. Pres. Atari Computer Club, Oklahoma City, 1983-84. With USAF, 1978-80. Recipient Paul Harris award Rotary Internat., 1991. Achievements include co-design of PXI 512 Medical Image Processing System and CompuLanx Computerized Forklift Data System. Home and Office: Megabyn Assocs 3120 Chaucer Dr Oklahoma City OK 73120-2228

JONES, JERRY EDWARD, family physician, educator; b. Winchester, Ky., Mar. 15, 1951; s. Henry Edward and Myrna (Ewen) J.; m. Nancy Jo Rosenbaum, May 21, 1971; children: Lisa, Lesley, Ellen. BS, U. Ky., 1973, MD, 1976; MS, U. Iowa, 1981. Resident in family practice U. Iowa, Iowa City, 1976-79, fellow assoc., 1979-81; asst. prof. U. Ky., Lexington, 1981-86, assoc. prof., 1986-89; assoc. prof. U. Ala., Tuscaloosa, 1989—. Editor/author: Primary Care-Parasitic Diseases, 1991; author: (with others) Office Parasitology, 1990, Diarrhea, 1993; reviewer Am. Acad. Family Physicians, Kansas City, Mo., 1986— Primary Care, N.J., 1988—, Archives of Family Medicine, Chgo., 1992—; contbr. articles to profl. jours. Fellow Am. Acad. Family Physicians; mem. APHA, Acad. Family Phyisicans. Achievements include development of a student summer assistantship program in family medicine in the application of clinical parasitology to the office setting. Office: Dept Family Medicine 700 University Blvd East Tuscaloosa AL 35401

JONES, KATHARINE JEAN, research physicist; b. Torrance, Calif., Feb. 18, 1940; d. Harold Thomas and Olive Katharine (Hume) Holtom; m. Noel Duane Jones, June 29, 1963; children: Evan Edward, Leonard L. BA, Pomona Coll., 1961; PhD magna cum laude, U. Berne, Switzerland, 1966; BS in Math. and Physics, Purdue U., 1976, MS in Physics, 1978. Rsch. asst. Labs. Linus Pauling Calif. Inst. Tech., Pasadena, 1963; rsch. assoc. biochemistry dept. Ind. U. Sch. Medicine, Indpls., 1967-70; teaching asst. physics dept. Purdue U., West Lafayette, Ind., 1976-78, rsch. asst. indsl. engring. dept., 1978; rsch. physicist Naval Avionics Ctr., Indpls., 1978—. Contbr. articles to profl. jours.; patentee in field. Fellow NIH, 1964-66, 68-70. Mem. AAAS, IEEE, SPIE, Optical Soc. Am., Soc. for Advancement Material and Process Engring., Sigma Pi Sigma. Avocation: orchids. Home: 9120 Gleannloch Dr Indianapolis IN 46256-2227 Office: Naval Air Warfare Ctr Indianapolis IN 46219

JONES, LAWRENCE RYMAN, retired research scientist; b. Terre Haute, Ind., Jan. 8, 1921; s. Frank Arthur and Mary Naomi (Ryman) J.; m. Mary Jane Proctor, July 9, 1944; children: Trudi Beth, Lawrence R. II. BS, Ind. State U., 1946. Chemist Comml. Solvents Corp., Terre Haute, 1946; chemist, toxicologist St. Anthony Hosp., Terre Haute, 1957-60; mgr. Ryman Farm, Vigo County, Ind., 1957-86; pres. Rymark Labs., Terre Haute, 1970-80; rsch. scientist Internat. Mineral and Chem. Corp., Terre Haute, 1972-86, ret., 1986—; pres. common coun. City of Terre Haute, 1960-72. Patentee in field; contbr. over 200 articles to profl. jours. Pres. Sr. Citizens of Wabash Valley, 1960-64; v.p. Sr. Citizen Housing Devel.; bd. govs. Task Force for Economy; past mem. Am. Chem. Soc., Am. Inst. Chemists. Republican. Presbyterian. Avocations: genealogy, gardening, golf. Home: 1219 E Alamito St Rockport TX 78382-2958

JONES, LEONARD DALE, facilities engineer; b. Cheyenne, Wyo., Feb. 21, 1948; s. Clifford Dale and Mary Mauverine (Hardin) J.; m. Sheila Rae Lansberry, Aug. 14, 1971; children: Shannon Marie, Meghan Anne. BS, Colo. Sch. of Mines, 1971; MBA, Nova U., 1985. Registered profl. engr., Va., Colo. Powerhouse engr. Stauffer Chem. Co., Green River, Wyo., 1978-81, plant svcs. supt., 1981-85; supr. customer svc. engr. Wis. Natural Gas Co., Racine, 1985-87; facilities mgr. Nat. Renewable Energy Lab., Golden, Colo., 1987—; cons. engr. LSJ Assocs., Littleton, Colo., 1991—; adj. instr. Gateway Tech. Inst., Kenosha, 1986. Conty del. Wyo. Rep. Conv., Jackson, 1984; precinct chair Sweetwater County Rep. Party, Green River, 1983-85; mem. Deer Creek Housing task force, Jefferson County Sch. Dist., Littleton, 1988; chmn., mem. Parks and Recreation Bd., Green River, 1980-84; ruling elder Genesis Presbyn. Ch., Littleton, 1990—; mem. ch. extension com. Presytery of Denver, 1989-92. Recipient Pres. award for Exceptional Performance Midwest Rsch. Inst., 1990. Mem. Assn. of Energy Engrs., Masons, Lakewood Foothills Rotary Club, Cheyenne Consistory, AASR. Home: 28 Mesa Oak Littleton CO 80127 Office: Nat Renewable Energy Lab 1617 Cole Blvd Golden CO 80401

JONES, MAITLAND, JR., chemistry educator; b. N.Y.C., Nov. 23, 1937; s. Maitland and Irma (Tillmanns) J.; m. Susan Hockaday; children: Maitland, Hilary, Stephanie. AB, Yale U., 1959, MS, 1960, PhD, 1963. Postdoctoral fellow U. Wis., Madison, 1963-64; instr. chemistry Princeton (N.J.) U., 1964-66, asst. prof., 1966-70, assoc. prof., 1970-73, prof., 1973-83, David B. Jones prof. chemistry, 1983—; vis. asst. prof. Columbia U., 1969-70; vis. prof. Vrije Univ., Amsterdam, 1973-74, 78, Harvard U., 1986. Contbr. 134 rsch. papers to profl. jours. Recipient numerous grants, 1964—. Home: 111 Fitzrandolph Rd Princeton NJ 08540-7203 Office: Princeton U Dept of Chemistry Princeton NJ 08544

JONES, MARK MITCHELL, plastic surgeon; b. Atlanta, Mar. 27, 1951; s. Curtis B. and Julia (Mitchell) J.; m. Regine M.F. Heckel, Jan. 19, 1980; children: Celine Julia Micheline, Cédric André Curtis. Student, Oxford (Ga.) Coll., 1971; BA in Chemistry, Emory U., 1973; DBA, U. Canterbury, Christchurch, New Zealand, 1975; BA, MA, Oxford (Eng.) U., 1977; MD, Med. Coll. Ga., 1979. Surg. intern Med. U. of S.C., Charleston, 1979-80; surg. resident Union Meml. Hosp., Balt., 1980-81; resident in otolaryngology Johns Hopkins Hosp., Balt., 1981-84, mem. staff, instr., 1984-85; resident in plastic surgery Stanford U. Med. Ctr., Palo Alto, Calif., 1985-86; chief resident in plastic surgery, 1987-88; assoc. Calif. Ear Inst., 1988-89; chief surgeon Atlanta Plastic Surgery Specialist, 1989—. Fulbright fellow in plastic surgery, Paris, 1986-87. Home: 985 Foxcroft Rd NW Atlanta GA 30327-2621 Office: Atlanta Plastic Surgery Specialist 2001 Peachtree Rd NE Ste 630 Atlanta GA 30309

JONES, NANCY GALE, retired biology educator; b. Gaffney, S.C., Nov. 12, 1940; d. Louransey Dowell and Sarah Louise (Pettit) J. BA, Winthrop Coll., 1962; MA, Oberlin Coll., 1964; postgrad., Marine Biol. Lab., Woods Hole, Mass., 1964, N.C. State U., 1965, Ohio State U., 1966, Ariz. State U., 1970. Lectr. biology Oberlin (Ohio) Coll., 1964-66; from instr. to asst. prof. zoology Ohio U., Zanesville, 1966-73; media specialist Muskingum Area Vocat. Sch., Zanesville, 1973-74; salesperson Village Bookstore, Worthington, Ohio, 1975. Hortitherapist for retarded adults Habilitation Svcs., Inc., Gaffney, S.C., 1977-80; dir. emergency assistance to needy PEACHcenter Ministries, Inc., Gaffney, 1991—. Mem. Sigma Xi (assoc.). Baptist. Avocations: travel, reading, Siamese cats, yard work. Home: 1643 W Rutledge Ave Gaffney SC 29341-1023

JONES, RANDALL MARVIN, chemist; b. Beaver Falls, Pa., May 13, 1959; s. Edward Henry and Geraldine Barbara (Woodson) J. BS in Chemistry, Carnegie Mellon U., 1980. Technician Johns Hopkins U., Balt., 1980-82; chemist The Upjohn Co., Kalamazoo, 1982-88; teaching asst. Mich. State U., East Lansing, 1988-90; adj. chemistry instr. Duquesne U., Pitts., 1991-92; analytical chemist Carlisle (Pa.) SynTec Systems, 1992—. Recipient National Merit scholarship Armstrong Cork Co., 1976. Mem. Am. Inst. Chemists, Soc. for Analytical Chemists of Pitts., Spectroscopy Soc. Pitts. Office: Carlisle SynTec Systems PO Box 7000 Carlisle PA 17013

JONES, REGI WILSON, data processing executive; b. Paris, Tenn., Sept. 4, 1960; s. Charles W. and Louise J. AS, Jackson State Community Hosp., Tenn., 1981; BS, U. Tenn., 1986; MS, Middle Tenn. State U., 1991. Med. tech. Vol. Gen. Hosp., Martin, Tenn., 1982-86; med. tech. So. Hills Hosp., Nashville, 1986-90, asst. supr. microbiology, 1990-91; systems installation coord. Info. Svcs. Hosp. Corp. Am., Nashville, 1991—; lab cons., Nashville; med. adv. bd. MLO Mag., 1985-86. Asst. to dir. bands U. Tenn., martin, 1985-86, trumpet section leader, lead soloist. Mem. AAAS, N.Y. acad. Sci., Nat. Rifle Assn. (Inst. for Legis. Action), Smithsonian Instn., Sigma Xi, Phi Mu Alpha. Home: 2810 Ellington Cir Nashville TN 37211 Office: Hosp Corp Am Info Svcs 2555 Park Plz Nashville TN 37203-1528

JONES, RENEE KAUERAUF, health care administrator; b. Duncan, Okla., Nov. 3, 1949; d. Delbert Owen and Betty Jean (Marsh) Kauerauf; m. Dan Elkins Jones, Aug. 3, 1972. BS, Okla. State U., 1972, MS, 1975; PhD, Okla. U., 1989. Statis. analyst Okla. State Dept. Mental Health, Okla. City, 1978-80, divisional chief, 1980-83, adminstr., 1983-84; assoc. dir. HCA Presbyn. Hosp., Okla. City, 1984—; adj. instr. Okla. U. Health Sci. Ctr., 1979—; assoc. staff scientist Okla. Ctr. for Alcohol and Drug-Related Studies, Okla. City, 1979—; cons. in field. Assoc. editor Alcohol Tech. Reports jour., 1979-84; contbr. articles to profl. jours. Mem. APHA, Assn. Health Svcs. Rsch., Alcohol and Drug Problems Assn. N.Am., Am. Sleep Disorders Assn., N.Y. Acad. Scis., So. Sleep Soc. (sec.-treas. 1989-91), Phi Kappa Phi. Democrat. Methodist. Avocations: skiing, scuba diving, racewalking, bicycling, painting. Home: 401 NW 19th St Oklahoma City OK 73103-1911 Office: HCA Presbyn Hosp NE 13th at Lincoln Blvd Oklahoma City OK 73104

JONES, RICHARD ERIC, JR., aerospace engineer; b. Poughkeepsie, N.Y., Apr. 14, 1965; s. Richard Eric and Kathleen Mary (Sharrock) J.; m. Lori Lynn Johnson, Oct. 10, 1992. BS in Aero. Engring., Rensselaer Poly. Inst., 1988; MS in Aerospace Engring., San Diego State U., 1990. Aerospace software engr. Loral Command & Control Systems, Colorado Springs, Colo., 1992—. N.Y. State Regents' scholar, 1983-85; RPI scholar Rensselaer Poly. Inst., 1985-87. Mem. AIAA, Nat. Space Soc. Home: 5460 Jennifer Ln Colorado Springs CO 80917

JONES, RICHARD VICTOR, physics and engineering educator; b. Oakland, Calif., June 8, 1929; 3 children. A.B., U. Calif., 1951, Ph.D. in Physics, 1956; M.A. hon., Harvard U., 1961. Sr. engr. Shockley Semiconductor Lab., Beckman Instruments, Inc., 1955-57; from asst. prof. to assoc. prof. applied physics Harvard U., 1957-71, assoc. dean div. engring. and applied physics, 1969-71, dean Grad. Sch. Arts and Sci., 1971-72, prof. applied physics, now Robert L. Wallace prof. applied physics; vis. MacKay prof. U. Calif.-Berkeley, 1967-68. Guggenheim fellow, 1960-61. Office: Harvard U Div Applied Scis 113 Cruft Lab Cambridge MA 02138

JONES, ROBERT EUGENE, physician; b. Bauxite, Ark., Aug. 25, 1926; s. Curtis Whittemore and Rosina (Nelson) J.; m. Frances (Mickey) Harper, June 24, 1949; children: Eric R., Gary B., Gretchen M. Student, Tulane U., 1944-45; BS, U. Ark., Fayetteville, 1947; MD, U. Ark., Little Rock, 1951. Cert. correctional health profl. Intern St. Vincent Infirmary, Little Rock, 1951-52; pvt. practice physician Benton, 1953—; chief of staff Saline Meml. Hosp., Benton, Ark., 1965-66, 68-69; pres. Saline County Med. Soc., Benton, 1965-66, 68-69; med. dir. Ark. Region PHP Health, Pine Bluff, Ark., 1989-90; Benton, 1952—; assoc. dept. community and family medicine U. Ark. for Med. Sci. 1st lt. Med. Corps, U.S. Army, 1953-55. Fellow Royal Soc. Health. Episcopalian. Avocations: collecting Civil War relics, collecting fine art. Home: 723 W Narroway St Benton AR 72015-3653 Office: Mid Delta Rural Health Market St De Valls Bluff AR 72041

JONES, ROBERT LEWIS, soil mineralogy and ecology educator; b. Wellston, Ohio, Jan. 26, 1936; s. Robert Davis and Laura (Lewis) J.; m. Katharine Anne King, July 8, 1958; children: Kevin, Laura, Dylan. B.Sc., Ohio State U., 1958, M.Sc., 1959; Ph.D., U. Ill., 1962. Research assoc. U. Ill., Champaign-Urbana, 1962-64, asst. prof., 1964-67, assoc. prof., 1967-72, prof. soil mineralogy and ecology, 1972—; Fulbright lectr., 1968. Co-author: (with H.C. Hanson) Biogeochemistry of Blue, Snow and Ross Geese, 1978, Mineral Licks, Geophagy and Biogeochemistry of North American Ungulates, 1985. Mem. Soil Sci. Soc. Am., Mineral Soc. Am., Am. Soc. Agronomy. Home: 16 Ashley Ln Champaign IL 61820-7301 Office: N 415 Turner Hall 1102 S Goodwin Ave Urbana IL 61801-4709

JONES, ROBERT MILLARD, engineering educator; b. Mattoon, Ill., Aug. 8, 1939; s. Verner Everett and Geneva (Millard) J.; m. Donna Thomas, Jan. 26, 1963; children: Mark, Karen. BSCE, U. Ill., 1960, MS, 1961, PhD, 1964. Mem. tech. staff The Aerospace Corp., San Bernardino, Calif., 1964, 66-70; prof. So. Meth. U., Dallas, 1970-81, Va. Tech., Blacksburg, Va., 1981—. Author: Mechanics of Composite Materials, 1975. Capt. U.S. Army, 1964-66. Fellow ASME, AIAA (assoc., chmn. structures tech. com. 1975-76); mem. Am. Acad. Mechanics (founder), Am. Soc. Composites (founder, v.p. 1992-93). Home: PO Box 10698 Blacksburg VA 24062 Office: Va Tech ESM Dept Blacksburg VA 24061-0219

JONES, ROBERT THOMAS, aerospace scientist; b. Macon, Mo., May 28, 1910; s. Edward Seward and Harriet Ellen (Johnson) J.; m. Megan Lillian More, Nov. 23, 1964; children: Edward, Patricia, Harriet, David, Gregory, John. Student, U. Mo., 1928; Sc.D. (hon.), U. Colo., 1971. Aero. research scientist NACA, Langley Field, Va., 1934-46; research scientist Ames Research Center NACA-NASA, Moffet Field, Calif., 1946-62; sr. staff scientist Ames Research Center, NASA, 1970-81, research assoc., 1981—; scientist Avco-Everett Research Lab., Everett, Mass., 1962-70; cons. prof. Stanford U., 1981. Author: (with Doris Cohen) High Speed Wing Theory, 1960, Collected Works of Robert T. Jones, 1976, Wing Theory, 1987; contbr. (with Doris Cohen) articles to profl. jours. Recipient Reed award Inst. Aero. Scis., 1946, Inventions and Contbns. award NASA, 1975, Prandtl Ring award Deutsche Gesellschaft für Luft and Raumfahrt, 1978, Pres.'s medal for disting. fed. service, 1980, Langley medal Smithsonian Instn., 1981, Excalibur award U.S. Congress, 1981, Aeronautical Engring. award NAS, 1990. Fellow AIAA (hon.); mem. NAS (award in aero. engring. 1989), NAE, Am. Acad. Arts and Scis. Home: 25005 La Loma Dr Los Altos CA 94022-4507

JONES, ROBERT WILLIAM, physicist; b. Wenatchee, Wash., Dec. 18, 1940; s. Willis M. and Billie (McCleary) J. BS cum laude in Physics, U. Wash., 1964; PhD in Physics, U. Colo., 1969. Rsch. assoc. U. Ariz., Tucson, 1971-72; assoc. prof. U. Petroleum and Minerals, Dhahran, Saudi Arabia, 1974-76; asst. prof., assoc. prof., prof. U. S.D., Vermillion, 1969-80; staff mem., dep. group leader, sect. leader Los Alamos (N.Mex.) Nat. Lab., 1980-87; ptnr. Balcomb Solar Assocs., Santa Fe, N.Mex., 1984—; rsch. prof. Sch. of Architecture, Ariz. State U., 1993—; vis. prof. U. Rome, 1988—. Co-author: The Sunspace Primer, 1984, Passive Solar Heating Analysis, 1984; co-author, editor: Passive Solar Design Handbook, Vol. 3, 1982; editor in chief Passive Solar Jour., 1985-88; editor Sun World Mag., Santa Fe, 1990-91; contbr. 48 articles to profl. jours. Recipient Citation Progressive Architecture, 1982. Mem. Internat. Solar Energy Soc., Am. Solar Energy Soc. (bd. dirs. 1985-88). Achievements include research in physics and solar energy.

JONES, ROGER CLYDE, electrical engineer, educator; b. Lake Andes, S.D., Aug. 17, 1919; s. Robert Clyde and Martha (Albertson) J.; m. Katherine M. Tucker, June 7, 1952; children: Linda Lee, Vonnie Lynette. B.S., U. Nebr., 1949; M.S., U. Md., 1953; Ph.D. U. Md., 1963. With U.S. Naval Research Lab., Washington, 1949-57; staff sr. engr. to chief engr. Melpar, Inc., Falls Church, Va., 1957-58; cons. project engr. Melpar, Inc., 1958-59, sect. head physics, 1959-64, chief scientist for physics, 1964; prof. dept. elec. engring. U. Ariz., Tucson, 1964-89; dir. quantum electronics lab. U. Ariz., 1968-88, adj. prof. radiology, 1978-86, adj. prof. radiation-oncology, 1986-88, prof. of radiation-oncology, 1988-89, prof. emeritus, 1989—; tech. dir. H.S.C. and A., El Paso, 1989—; guest prof. in exptl. oncology Inst. Cancer Research, Aarhus, Denmark, 1982-83. Patentee in field. Served with AUS, 1942-45. Mem. Am. Phys. Soc., Optical Soc. Am., Bioelectromagnetics Soc., IEEE, AAAS, NSPE, Am. Congress on Surveying and Mapping, Eta Kappa Nu, Pi Mu Epsilon, N.Mex. Acad. Sci. Home: 5809 E 3d St Tucson AZ 85711 Office: U Ariz Dept Elec and Computer Engring Tucson AZ 85721

JONES, ROGER WAYNE, electronics executive; b. Riverside, Calif., Nov. 21, 1939; s. Virgil Elsworth and Beulah (Mills) J.; m. Sherill Lee Bottjer, Dec. 28, 1975; children: Jerrod Wayne, Jordan Anthony. BS in Engring., San Diego State U., 1962. Br. sales mgr. Bourns, Inc., Riverside, 1962-68; sales and mktg. mgr. Spectrol Electronics, Industry, Calif., 1968-77, v.p. mktg., 1979-81; mng. dir. Spectrol Reliance, Ltd., Swindon, England, 1977-79; sr. v.p. S.W. group Kierulff Electronics Corp., L.A., 1981-83; v.p. sales and mktg. worldwide electronic techs. div. Beckman Instruments, Fullerton, Calif., 1983-86; pres., ptnr. Jones & McGeoy Sales, Inc., Newport Beach, Calif., 1986—. Author: The History of Villa Rockledge, A National Treasure in Laguna Beach, 1991. Republican. Home: 4 Royal St George Newport Beach CA 92660 Office: 5100 Campus Dr Newport Beach CA 92660

JONES, SAMUEL B., JR., botany educator; b. Roswell, Ga., Dec. 18, 1933; s. Samuel B. and Belle J.; m. Carleen Arrington, June 26, 1955; children—Valerie, Velinda, Douglas. B.S., Auburn U., 1955, M.S., 1961; Ph.D., U. Ga., 1964. Instr., Auburn U., 1957-61; asst. prof. U. So. Miss., Hattiesburg, 1964-67; from asst. prof. to prof. botany U. Ga., Athens, 1967—; owner Piccadilly Farm Nursery, Bishop, Ga., 1975—. Author: Plant Systematics, 1979. Contbr. articles to profl. jours. Served to lt. col. USAR, 1955-81. Mem. Am. Soc. Plant Taxonomists, Internat. Assn. for Plant Taxonomy, Bot. Soc. Am., New Eng. Bot. Club, So. Appalachian Bot. Club, Phi Beta Kappa, Sigma Xi, Gamma Sigma Delta, Phi Kappa Phi. Avocations: gardening; hostas; wildflowers. Office: Univ Ga Dept of Botany Athens GA 30602*

JONES, SHELLEY PRYCE, chemical company executive; b. Cleve., Aug. 19, 1927; s. Shelley Brynt and Ethel (Price) J.; m. Virginia Marie Setlock, Aug. 30, 1967; children: Robert Bruce, Mark Stuart. ME, We. Reserve; MBA, Ohio Christian Coll., 1973; PhD, Ohio Christian U., 1974. Petroleum engr. Nat. Refining Co., Finley, Ohio; gen. mgr. A.E.O. Corp., Ft. Wayne, Ind.; div. mgr. Gen. Tire and Rubber Co., Chgo.; C.E.O. Internat. Sci. Mgmt. Svcs., Wilmington, Del., 1967-76; pres., CEO Setter Chem. Corp., Indiana, Pa., 1976—; ind. bus., mktg., engring. cons., lectr. Mem. AAAS, N.Y. Acad. Scis., Mason, Shriners. Episcopalian. Achievements include design of first scallop shucking machine, emergency and new type low cost housing, pump for pressure washers; developed many new formulations for chemical products. Home: 2783 Melloney Ln Indiana PA 15701

JONES, STEPHEN YATES, pharmacist; b. Raleigh, N.C., Apr. 22, 1958; s. Norwood Godwin and Ruth (Yates) J. BA in Chemistry, U. N.C., 1980, BS in Pharmacy, 1985. Lic. pharmacist, N.C. Pharmacist intern Treasury Drug, Durham, N.C., 1982-85; pharmacist Treasury Drug, Garner, N.C., 1985-90, Phar-Mor, Garner, 1990-93, Kerr Drug, Raleigh, N.C., 1993—. Singer U. N.C. Mens Glee Club, 1982-85. Recipient Upjohn Achievement award Upjohn Pharm., 1985, Pfizer Achievement award, Pfizer Pharm., 1985. Mem. Am. Pharm. Assn., N.C. Pharm. Assn., Wake County Pharm.

Assn. Republican. Baptist. Avocations: piano, organ, travel, singing at weddings, organizing parties. Home: 6133 Wolverhampton Dr Raleigh NC 27603 Office: Kerr Drug # 72 Timber Crossings S Ctr Gorner NC 27529

JONES, THOMAS GORDON, internist; b. N.Y.C., Feb. 1, 1951; s. Edwin Paul and Myra Beatrice (Wilson) J.; m. Leslie P. Beal, Jan. 30, 1977; children: Daniel, Gregory, Carrie. BS in Biology, Tufts U., 1973; MD, Albert Einstein Coll. Medicine, 1977. Diplomate Bd. of Internal Medicine, Bd. Endocrinology, Metabolism. Med. resident Downstate Med. Ctr., Kings County Hosp., Bklyn. VA, N.Y., 1977-80; officer Morris Heights Health Ctr. U.S. Pub. Health Svc., Bronx, N.Y., 1980-83; fellow in endocrinology U. Conn. Health Ctr., Farmington, Conn., 1983-85; pvt. practice internal medicine, endocrinology Farmington, Conn., 1985—. Office: 1035 Farmington Ave Farmington CT 06032

JONES, VERNON QUENTIN, surveyor; b. Sioux City, Iowa, May 6, 1930; s. Vernon Boyd and Winnifred Rhoda (Bremmer) J.; student UCLA, 1948-50; m. Rebeca Buckovecz, Oct. 1981; children: Steven Vernon, Gregory Richard, Stanley Alan, Lynn Sue. Draftsman III Pasadena (Calif.) city engr., 1950-53; sr. civil engring. asst. L.A. County engr., L.A., 1953-55; v.p. Treadwell Engring. Corp., Arcadia, Calif., 1955-61, pres., 1961-64; pres. Hillcrest Engring. Corp., Arcadia, 1961-64; dep. county surveyor, Ventura, Calif., 1964-78; propr. Vernon Jones Land Surveyor, Riviera, Ariz., 1978—; city engr. Needles (Calif.), 1980-87; instr. Mohave Community Coll., 1987—. Chmn. graphic tech. com. Ventura Unified Sch. Dist., 1972-78, mem. career adv. com., 1972-74; mem. engring. adv. com. Pierce Coll., 1973; pres. Mgmt. Employees of Ventura County, 1974. V.p. Young Reps. of Ventura County, 1965. Pres., Marina Pacifica Homeowners Assn., 1973. Mem. League Calif. Surveying Orgns. (pres. 1975), Am. Congress on Surveying and Mapping (chmn. So. Calif. sect. 1976), Am. Soc. Photogrammetry, Am. Pub. Works Assn., County Engr. Assn. Calif. Home: 913E San Juan Ct Riviera AZ 86442-5618

JONES, WALTER HARRISON, chemist; b. Griffin, Sask., Can., Sept. 21, 1922; s. Arthur Frederick and Mildred Tracy (Walter) J.; BS with honors, UCLA, 1944, PhD in Chemistry, 1948; m. Marion Claire Twomey, Oct. 25, 1959 (dec. Jan. 1976). Research chemist Dept. Agr., 1948-51, Los Alamos Sci. Lab., 1951-54; sr. research engr. N.Am. Aviation, 1954-56; mgr. chemistry dept. Ford Motor Co., 1956-60; sr. staff and program mgr., chmn. JANAF-ARPA-NASA Thermochem. panel Inst. Def. Analyses, 1960-63; head propulsion dept. Aerospace Corp., 1963-64; sr. scientist, head advanced tech. Hughes Aircraft Co., 1964-68; prof. aero. systems, dir. Corpus Christi Center, U. W. Fla., Pensacola, 1969-75, prof. chemistry, 1975—; cons. pvt., fed. and state agys.; vis. prof. U. Toronto. Mem. Gov.'s Task Force on Energy, Regional Energy Action Com., Fla. State Energy Office, adv. com. Tampa Bay Regional Planning Council; judge regional and state sci. fairs. Fed. and state grantee; research corp. grantee; Fellow ASEE/ONR, fellow NATO, fellow Am. Inst. Chemists; mem. AIAA, AAUP, AAAS, Am. Astron. Soc. (propulsion com.), Am. Chem. Soc. (chmn. Pensacola sect.), N.Y. Acad. Scis., Am. Phys. Soc., Internat. Solar Energy Soc., Combustion Inst. World Assn. Theoretical Organic Chemists, Am. Ordnance Assn., Air Force Assn., Philos. Soc. Washington, Pensacola C. of C., Phi Beta Kappa, Sigma Xi (pres. local chpt.), Pi Mu Epsilon, Phi Lambda Upsilon (sec. local chpt.), Alpha Mu Gamma, Alpha Chi Sigma (pres. local chpt.). Author: (fiction) Prisms in the Pentagon, 1971; contbr. numerous articles tech. jours., chpts. in books. Patentee in field. Home: 2412 Oak Hills Cir Pensacola FL 32514-5665 Office: U West Fla Dept of Chemistry Pensacola FL 32514

JONES, WALTON LINTON, internist, former government official; b. McCaysville, Ga., Dec. 4, 1918; s. Walton Linton and Pearl Josephine (Gilliam) J.; m. Caroline Wells Schachte, June 5, 1943; children—Walton Linton III, Francis Stephen, Kathleen Caroline. B.S., Emory U., 1939, M.D., 1942. Diplomate Am. Bd. Preventive Medicine. Commd. lt. (j.g.) U.S. Navy, 1942, advanced through grades to capt., 1956; rotating intern U.S. Naval Hosp., Charleston, S.C., 1942-43, aerospace medicine, 1944; flight surgeon USMC Aircraft Squadrons, 1944-47; head aero. med. safety Navy Dept., 1947-53; sr. med. officer U.S.S. Randolph, 1953-55; dir. aero. med. ops. and equipment Bur. Medicine and Surgery, Navy Dept., 1955-64; dir. biotech. and human research div. NASA, 1964-66; ret. U.S. Navy, 1966; civilian dir. biotech and human research div. NASA, Washington, 1966-79, dep. dir., dir. life scis., 1970-75, dir. occupational medicine, 1975-82, dir. occupational health, 1982-85; cons. aerospace medicine, 1985—; mem. exec. com. hearing and bioacoustics Nat. Acad. Scis., 1964-85, chmn., 1970, mem. exec. com. on vision, 1964-85; Kober lectr. Georgetown U., 1968. Leader, mem. com. Nat. Capital Area council Boy Scouts Am., Falls Church, Va., 1956-64. Decorated Legion of Merit; recipient Exceptional Service medal NASA, 1979, Outstanding Leadership medal NASA, 1985. Fellow Aerospace Medicine Assn. (Bauer award 1970, pres. 1980), AIAA (assoc., recipient John Jeffries award 1970), Royal Soc. Health; mem. Internat. Astronautics Acad., Assn. Mil. Surgeons (Founders award 1956), Internat. Acad. Aerospace Medicine.

JONES, WILLIAM KINZY, materials engineering educator; b. Miami, Fla., July 23, 1946; s. Harold Grover and Josephine (Kinzy) Jones; m. Sharon Mattingly, June 6, 1981; children: Kelli, Kinzy, Brent. BS, Fla. State U., 1967, MS, 1968; PhD, MIT, 1972. Mgr. engring. Cordis Corp., Miami, 1977-87; group head C.S. Draper Lab., Cambridge, Mass., 1972-77; dir. biomed. rsch. and innovation ctr. Fla. Internat. U., Miami, 1988; assoc. prof. engring. Fla. Internat. U., Miami, 1987-91; prof. engring. Fla. Internat. U., Miami, 1991—; adv. bd. Nat. Elec. Packaging and Product Conf., Des Plaines, Ill., 1988—; cons. in field. Contbr. articles to profl. jours.; patentee in field. Recipient rsch. award Fla. Internat. U., 1991. Mem. ASME, Internat. Soc. Hybrid Microelectronics (chmn. materials div. 1990, pres. 1992-93, tech. achievement award 1991, fellow, 1992). Republican. Home: 75550 Overseas Hwy # 534 Islamorada FL 33036-4005 Office: Fla Internat U University Park Campus Miami FL 33199

JONES, WILLIAM V(INCENT), health center administrator; b. N.Y.C., May 16, 1952; s. William Vincent and Millie (Lepore) J.; m. Deborah Ann Alunni, Aug. 23, 1975. BS, U. Del., Newark, 1975; MS, Pa. State U., 1978. Grad. asst. health edn. Pa. State U., State College, 1975-76; coord. consumer health edn. St. Vincent Health Ctr., Erie, Pa., 1976-83, supr. laser ctr., 1984—; laser cons.; pres. Resources for Prevention, Erie, 1979-85. Editor Healthscope newsletter, 1977-82, Cardiac Club Quar., 1982—. Chmn. pub. edn. Am. Cancer Soc., Harrisburg, Pa., 1978-81. Recipient awards for photodynamic laser therapy Herr-Roth Urol. Found., 1991, Am. Cyanamid Co., 1991. Mem. Am. Mgmt. Assn., Am. Soc. for Laser Medicine & Surgery, Pa. State U. Alumni Assn. Achievements include FDA controlled research projects for use of lasers in ob/gyn., ENT, urology, general surgery and photodynamic therapy. Home: 3911 Eliot Rd Erie PA 16508 Office: St Vincent Health Ctr 232 W 25th St Erie PA 16544

JONGEWARD, GEORGE RONALD, systems analyst; b. Yakima, Wash., Aug. 9, 1934; s. George Ira and Dorothy Marie (Cronk) J.; m. Janet Jeanne Williams, July 15, 1955; children: Mary Jeanne, Dona Lee, Karen Anne. BA, Whitworth Coll., 1957; postgrad., Utah State U., 1961. Sr. systems analyst Computer Scis. Corp., Honolulu, 1969-71; cons. in field Honolulu, 1972-76; prin. The Hobby Co., Honolulu, 1977-81; sr. systems analyst Computer Systems Internat., Honolulu, 1981—; instr. EDP Hawaii Pacific Coll., Honolulu, 1982—. Mem. car show com. Easter Seal Soc., Honolulu. 1977-82; active Variety Club, Honolulu, 1978-81. Mem. Mensa (local pres. 1967-69). Republican. Presbyterian. Club: Triple-9. Avocations: organ, piano, community theatre, golf, sports-car rallyes. Home: 400 Hobron Ln Apt 2611 Honolulu HI 96815-1206 Office: Computer Systems Internat 841 Bishop St Ste 501 Honolulu HI 96813-3991

JONKE, ERICA ELIZABETH, aerospace engineer; b. Bronx, June 3, 1969; d. Horst Lorenz and Christina Marie (Hanson) J. BS, U. Okla., 1991. Assoc. engr. Lockheed Engring. and Sci. Co., Houston, 1991—. Mem. AIAA. Democrat. Home: 11619 Gullwood Dr Houston TX 77089-6811

JONSSON, LARS OLOV, hospital administrator; b. Karlskoga, Varmland, Sweden, Nov. 14, 1952; s. Olov Mattias and Alice Eugenia (Wallin) J.; m. Eva-Brita Olsson, Aug. 16, 1975; children: Karl Olov Daniel, Brita Karin Charlotta. Candidate Medicine, Uppsala (Sweden) U., 1974, MD, 1978, PhD, 1986. Intern Östersund (Sweden) Hosp., 1978-80, resident anesthesiology, 1980-83, cons. anesthesiology, 1984-88, sr. cons., 1989—, vice dir. central operating room, 1988—, anesthetist-in-chief, 1988—, dir. anesthesiology, 1990-92, chief exec. officer, 1993—; fellow anesthesiology Uppsala (Sweden) U. Hosp., 1983-84; owner EBLO Cons., Östersund, 1985—; vis. prof. U. Western Ont., London, 1988; assoc. prof. dept. anaesthesia and intensive care Uppsala U., Sweden, 1991—; ednl. cons. dept. anesthesiology and internal care, Östersund, 1986-88; intensive med. rsch. cons. Mediplast AB, Solna, Sweden, 1985-91; cons. Siemens-Elema AB, Div. Ventilation, Solna, 1985-86; med. cons. Swedish Ishockey Fedn., Stockholm, 1986-87, ICI-Pharma AB, Gothenburg, Sweden, 1990—, Abbott Scandinavia AB, Kista, Sweden, 1990—. Author: Spontaneous Breathing, 1986; contbr. articles to profl. jours.; editorial fellow Gronkopings Veckoblad, 1978-92. Fellow Swedish Soc. Medicine, Swedish Soc. Anesthesiology and Intensive Care (bd. dirs. 1992—), Scandinavian Soc. Anesthesiology, Ope I.F. (pres. youth sect. 1987-92), Rotary Internat. (fellow). Avocations: sports, sailing. Home: Ugglevägen 12, S-831 62 Östersund Sweden Office: Östersund Hosp, S-831 83 Östersund Sweden

JORDAN, ARTHUR KENT, electronics engineer; b. Phila., Dec. 28, 1932; s. Arthur Harold and Mary Corona (Schoff) J.; m. Mary Frances Baily, July 10, 1965; children: Thomas Baily, Edward Marshall, Elizabeth Anne. BSc in Physics, Pa. State U., 1957; PhD in Elec. Engring., U. Pa., 1972. Rsch. engr. Philco Corp., Phila., 1958-62; aerospace engr. RCA, Princeton, N.J., 1962-64; aerospace physicits GE, Valley Forge, Pa., 1964-69; rsch. fellow U. Pa., Phila., 1969-73; program mgr. Office Naval Rsch., Arlington, Va., 1986—; rsch. physicist Naval Rsch. Lab., Washington, 1973—; vis. scientist MIT, Cambridge, 1989—. Contbr. articles to sci. jours. Fellow IEEE, Optical Soc. Am.; mem. Am. Phys. Soc. Achievements include patents on optical communications system, research on application of electromagnetic inverse scattering theory to design of optical devices and remote sensing. Office: Naval Rsch Lab Code 7227 4555 Overlook Ave Washington DC 20375-5000

JORDAN, GEORGE EUGENE, air force officer; b. Fairfield, Ill., Feb. 5, 1953; s. George William and Lela Violet (Foley) J.; m. Diana Carol Campbell, June 15, 1974; children: Donald Matthew, Jeremy Eugene. BA, Park Coll., Parkville, Mo., 1980; M Forensic Sci., George Washington U., 1987. Commd. USAF, 1980, advanced through grades to maj., 1992; security specialist 81 Security Police Squadron USAF, RAF Bentwaters, Eng., 1975-77; sr. crew supr. 351 Missile Security Squadron USAF, Whiteman AFB, Mo., 1977-80; chief base adminstrn. 305 Combat Support Group USAF, Grissom AFB, Ind., 1980-82, wing exec. officer 305 Air Refueling Wing, 1982-83; analyst hdqs. Air Force Office Spl. Investigations, Bolling AFB, DC, 1984-86; forensic scientist Air Force Office Spl. Investigations, Andrews AFB, Md., 1987-92; comdr. Air Force Office Spl. Investigations, Kunsan AB, Korea, 1992-93; chief advanced tng. Air Force Office Spl. Investigations, Bolling AFB, Washington, 1993—. Mem. Am. Acad. Forensic Scis. Home: 749 University Dr Waldorf MD 20602-3483 Office: HQ AFOSI Acad Bolling AFB Washington DC 20332

JORDAN, GEORGE WASHINGTON, JR., engineering company executive; b. Chattanooga, Mar. 11, 1938; s. George W. and Omega (Davis) J.; m. Fredine Sims, July 20, 1968; 1 child, George W. III. BSEE, Tuskegee U., 1961; postgrad., Emory U. Mgmt. Inst., 1976; MS in Indsl. Mgmt., Ga. Inst. Tech., 1978. Design engr. Boeing Co., Seattle, 1961-64; test engr. Boeing Co., Huntsville, Ala., 1964-65; design engr. GE, Huntsville, 1965-66; engring. mgr. Lockheed Corp., Marietta, Ga., 1966—; mem. flight simulation tech. com. AIAA, Washington 1984-88. Chmn. Atlanta Zoning Rev. Bd.; chmn. fin. com. Christian Fellowship Bapt. Ch., Atlanta. Mem. AIAA, Nat. Mgmt. Assn., Assn. MBA Execs., Am. Def. Preparedness Assn., Alpha Phi Alpha (life). Democrat. Avocations: tennis, bowling, reading. Home: 3609 Rolling Green Rdg SW Atlanta GA 30331-2325 Office: Lockheed 86 S Cobb Dr Marietta GA 30063-0001

JORDAN, GREGORY WAYNE, aeronautical engineer; b. Chgo., Sept. 22, 1937; s. Robert John and Edythe Lydia (Applehans) J.; m. Judith Gay Narland, June 27, 1961; children: Rachel Anne, Catherine Jeanne, Elizabeth Rebecca. BS in Aeronautical Engring., U. Ill., 1960; MBA, U. Chgo., 1967. Internal cons. Lockheed Missile and Space, Sunnyvale, Calif., 1963-65, Hewlwtt-Packard, Palo Alto, Calif., 1967-68; project mgr. Rohr Industries, Chula Vista, Calif., 1968-73; program mgr. Booz, Allen and Hamilton, Cin., 1974; dir. internal bus. United Technologies, Hartford, Conn., 1974-82; dir., bus. analysis and planning Northrop Corp., L.A., 1982--; advisor U.S. Dept. Comml. Spl. Trade Rep., Washington, 1976-84. Capt. USMC 1960-63. Mem. Planning Forum (steering com. 1989-90), Am. Inst. Astronautics and Aeronautic, Am. Defense Preparadness Assn., Sierra Club (sec., treas. 1988--). Office: Northrop 1840 Century Park E Los Angeles CA 90067-1701

JORDAN, MARK HENRY, consulting civil engineer; b. Lawrence, Mass., Apr. 10, 1915; s. Joseph Augustine and Gertrude (O'Connell) J.; m. Louise Sullivan, June 23, 1939; children: Mary Elizabeth (Mrs. Delio Gianturco), Margaret Michaela. B.S., U.S. Naval Acad., 1937; M. Civil Engring., Rensselaer Poly. Inst., 1942, M.S., 1965, Ph.D., 1968. Registered profl. engr., N.J., N.Y. Commd. ensign U.S. Navy, 1937, advanced through grades to capt., 1955; comdr. 6th Seabee Battalion South Pacific, 1943-44; comdr. 103d Seabee Battalion Central Pacific, 1951-52; comdr. Civil Engr. Corps. Sch. Port Hueneme, Calif., 1960-63; ret., 1963; assoc. prof. civil engring. U. Mo., Columbia, 1966-67; prof. civil engring. Rensselaer Poly. Inst., 1968-77, prof. emeritus, 1977—, dean continuing studies, 1967-72, chmn. civil engring., 1972-73; cons. engr. Smith & Mahoney, Albany, N.Y., 1975-78; individual practice as cons. engr., 1978—. Author: (with others) Saga of the Sixth, 1950, Iron Brigade General, 1993. Mem. Rensselaer County Charter Commn., 1969-71; Bd. dirs. United Community Services, Troy, N.Y., 1969-75. Decorated Bronze Star with V, Presdl. Unit citation. Fellow ASCE (life); mem. NSPE, Am. Arbitration Assn., Am. Soc. Engring. Edn., Soc. Am. Mil. Engrs. (local post pres.), Am. Pub. Works Assn., Sigma Xi, Chi Epsilon. Roman Catholic. Club: Fort Orange (Albany, N.Y.). Home: East Rd Brunswick Hills NY 12180-6861 Office: 256 Broadway Troy NY 12180-3237

JORDAN, NEAL FRANCIS, geophysicist, researcher; b. Franklinville, N.Y., July 8, 1932; s. Gerald H. and Nellie B. Roat J.; m. Rose Alice Lagendorfer, June 18, 1955 (dec. Oct. 1992); children: Kirk Gerald, Sarah Elizabeth Jordan Towler. B in Engring. Physics, Cornell U., 1955; MS in Engring., Purdue U., 1959, PhD, 1963. Jr. physicist Cornell Aero Lab., Buffalo, summer 1954; engr. Kodak, Rochester, N.Y., summer 1958; rsch. assoc. Gen. Tech. Corp., West Lafayette, Ind., 1960-63; rsch. engr. Jersey Prodn. Rsch. Co., Tulsa, 1963-68; rsch. assoc. Esso Prodn. Rsch. Co., Houston, 1968-76, supr. geophysics, 1976-84, divsn. mgr., 1984—. Contbr. articles to publs. 1st lt. U.S. Army, 1955-57. Mem. Soc. Exploration Geophysicists, Soc. Engring. Scis. (sec. 1968-69). Achievements include patent for system to use magnetic polymer. Home: 330 Knipp Rd Houston TX 77024 Office: Exxon Prodn Rsch PO Box 2189 Houston TX 77252-2189

JORDAN, ROBERT R., geologist, educator; b. N.Y.C., June 5, 1937; s. Herbert and Irene (Reed) J.; m. Jane H. Jordan, June 28, 1958; children: Richard P., Judith H. AB, Hunter Coll., 1958; MA, Bryn Mawr Coll., 1962, PhD, 1964. Cert. profl. geologist, Del.; lic. geologist, N.C. Geologist Del. Geol. Survey, Newark, 1958-64, asst. state geologist, 1964-69, state geologist, dir., 1969—; instr. U. Del., Newark, 1962-64, asst. prof., 1964-68, assoc. prof., 1968-88, prof., 1988—; commr. Del Water & Air Resouces Commn., Dover, 1966-73; chmn. Del. State Boundary Commn., Newark, 1971—. Contbr. numerous articles to profl. jours. Fellow Geol. Soc. Am.; mem. Am. Inst. Profl. Geologists (editor 1989-90, Galey Mem. Pub. Svc. award 1992), Del. Bd. Registration Geologists, Assn. Am. State Geologists (pres. 1983-84, Achievement award), Am. Assn. Petroleum Geologists (hon., Disting. Svc. award 1988, Cohee Pub. Svc. Ea. award 1990). Office: Del Geol Survey U Delaware Newark DE 19716-7501

JORDEN, JAMES ROY, oil company engineering executive; b. Oklahoma City, Apr. 16, 1934; s. James Roy and Gordon (Peeler) J.; m. Shirley Ann Swan, Nov. 17, 1956; children: Philip Taylor, David Emerson. BS in Petroleum Engring., U. Tulsa, 1957. Engr. Shell Oil Co., various locations, 1957, 1960-81; petrophys. engr. advisor Shell Oil Co., Houston, 1981-85; mgr. petroleum engring. rsch. Shell Devel. Co., Houston, 1985-88, mgr. head office prodn., tech. tng., 1988-93; mgr. CPI tng. Shell Oil Co., Houston, 1993—; mem. industry adv. bd. petroleum engring. dept., U. Tulsa, 1987-92, chmn. 1988; vis. com. petroleum engring. dept. Colo. Sch. Mines, Golden, 1988—. Co-author: Well Logging I., 1984, Well Logging II, 1986; co-inventor in field. lst lt. USAF, 1957-60. Named to Hall of Fame, Petroleum Engring. Dept. U. Tulsa, 1985. Mem. Soc. Petroleum Engrs. (Disting., pres. 1984, Disting. Svc. award 1988, DeGolyer Disting. Svc. medal 1991, bd. dirs. 1975-79, dir. svc. corps 1984-90, life trustee found., treas. found. 1991-92), Soc. Profl. Well Log Analysts. Republican. Presbyterian. Avocations: golf, reading, wine. Home: 10926 Piping Rock Ln Houston TX 77042-2728 Office: Shell Oil Co PO Box 481 PO Box 576 Houston TX 77001

JORGENSEN, ERIC EDWARD, wildlife ecologist, researcher; b. Wausau, Wis., July 21, 1961; s. Lee Arnold and Shirley June (DeByle) J.; m. Naomi Colleen Sisson, Apr. 20, 1985. BS in Metall. Engring., U. Wis., Madison, 1984; MS in Natural Resources, U. Wis., Stevens Point, 1992; postgrad., Tex. Tech. U. Patent examiner U.S. Patent and Trademark Office, Arlington, Va., 1985-89; rsch. asst. U. Wis., Stevens Point, 1990-92; rsch. asst. dept. range and wildlife mgmt. Tex. Tech U., Lubbock, 1992—. Co-author: Handbook of Wildlife Management Techniques for Commercial Cranberry Growers in Wisconsin, 1992. Mem. Sigma Xi, Phi Kappa Phi. Home: 2105 33d St Apt 117 Lubbock TX 79411 Office: Dept Range-Wildlife Mgmt Tex Tech U Lubbock TX 79409

JORGENSEN, JAMES DOUGLAS, research physicist; b. Salina, Utah, Mar. 23, 1948; s. Grant and Nelda (Breinholt) J.; m. Ramona Gurr, June 6, 1970; children: Lynn Nielson, Michael Nielson, Kristeen Nielson, Kathryn Nielson, Karen Nielson, Scott Nielson. BS in Physics, Brigham Young U., 1970, PhD in Physics, 1975. Postdoctoral rsch. asst. Argonne (Ill.) Nat. Lab., 1974-77, asst. physicist solid state div., 1977-80, physicist material sci. div., 1980-89, sr. physicist, 1989—, group leader, 1988—; mem. U.S. Nat. Com. for Crystallography, 1990-92. Mem. editorial adv. bd. Jour. Solid State Chemistry, 1990-91; contbr. over 200 articles to profl. jours. Bishop LDS Ch., Woodridge, Ill., 1984-89. Recipient award for disting. performance at Argonne Nat. Lab., 1983; co-recipient Pacesetter award Argonne Nat. Lab., 1986, Dir.'s award, 1988; materials scis. rsch. competition award for outstanding sci. accomplishments in solid state physics U.S. Dept. Energy, 1987, 91; named honored alumnus Brigham Young U., 1992. Fellow Am. Phys. Soc.; mem. Materials Rsch. Soc., Am. Crystallographic Assn. (B.E. Warren Diffraction Physics award 1991). Office: Argonne Nat Lab Materials Sci Div Bldg 223 Argonne IL 60439

JORGENSEN, JAMES H., pathologist, educator, microbiologist; b. Dallas, July 11, 1946; m. Jane Drummond, Feb. 18, 1978. BA, North Tex. State U., 1969, MS, 1970, PhD, 1973. cert. microbiologist. Rsch. assoc. Shriners Hosp. for Crippled Children, 1970-73; assoc. dir. Bexar County Hosp., 1973-75; instr. dept. pathology and dept. microbiology, Health Sci. Ctr. U. Tex., San Antonio, 1973-75, asst. prof., 1975-78, assoc. prof., 1978-84; cons. microbiologist Audie Murphy V.A. Hosp., San Antonio, 1973—; dir. clin. microbiology labs. Med. Ctr. Hosp., 1975—, prof. dept. pathology, dept. medicine, dept. microbiology, dept. clin. lab. scis., Health Sci. Ctr., 1984—; mem. editorial bd. Antimicrobial Agents and Chemotherapy, 1982-93, Jou. Clin. Microbiology, 1986-94, Diagnostic Microbiology and Infectious Diseases, 1983-87, reviewer 1992-93; mem. adv. bd. Current Perspectives in Infectious Disease, 1993-94, reviewer of numerous sci. jours. Author: Aminoglycoside Susceptibility Testing of Pseudomonas aeruginosa, (with others) In Vitro Detection of Methicillin-Resistant Staphylococci, 1985, A Clinicians Dictionary of Bacteria and Fungi, 1986, Progress and Pitfalls in Staphylococcus Susceptibility Testing, 1987; editor Automation in Clin. Microbiology, 1987, Manual of Clinical Microbiology, 1994. Recipient Becton-Dickenson and Co. award in Clin. Microbiology, 1992; James W. McLaughlin Pre-Doctoral fellow in Infection and Immunity, Med. Branch, U. Tex., 1971-73; Pre-Doctoral scholarship, North Tex. State U., 1969-70. Fellow Infectious Diseases Soc. of Am., Am. Acad. Microbiology; mem. Am. Soc. for Microbiology (Tex. branch chmn. clin. divsn. 1987-88), Southwestern Assn. of Clin. Microbiology, Tex. Infectious Diseases Soc. (pres. 1985-86), South Tex. Assn. of Microbiology Profls. (program dir. 1981-86, pres. 1989-90). Office: Univ of Texas Health Sciences Dept of Pathology 7703 Floyd Curl Dr San Antonio TX 78284*

JORGENSEN, LELAND HOWARD, aerospace research engineer; b. Rexburg, Idaho, Nov. 1, 1924; s. Leland Maeser and Anne Molyneaux (Howard) J.; m. Lynone Watkins, Mar. 24, 1949; children: Leland Ronald Jorgensen, Paul Victor Jorgensen, Jonathan Arthur Jorgensen, Sara Anne Jorgensen. BS in Mech. Engring. with honors, U. Utah, 1948; MS in Mech. Engring. with honors, Stanford U., 1949; PhD in Mech. Engring. with high honors, Calif. Coast U., 1977. Rsrch. engr. NACA-Ames Aero. Lab., Moffett Field, Calif., 1949-59; rsch. scientist NASA-Ames Rsrch. Ctr., Moffett Field, 1959-66, tech. asst. chief thermo and gas dynamics div., 1966-68, tech. asst. chief aeronautics div., 1968-71, aerospace rsch. scientist, 1971-80; aerospace cons. Sandy, Utah, 1980—; mem. NASA aerodynamics panel for space shuttle, 1978-80; cons. U.S. Navy on Agile missile, 1972, USAF on air-SLEW missile, 1973. Author: (publ. for NATO and NASA) Prediction of Static Aerodynamic Characteristics for Slender Bodies Alone and with Lifting Surfaces to High Angles of Attack, 1977, 50 publs. on aerodynamics of aircraft at subsonic, transonic, supersonic and hypersonic speeds. Trustee Saratoga (Calif.) Sch. Dist., 1977-81; v.p. Eagle Scout Assn. Santa Clara (Calif.) Coun. Boy Scouts Am., 1962-72; pres. Neighborhood 5 Granity Community, Sandy, 1990—. Lt. USNR, 1944-46, PTO. Assoc. fellow AIAA; mem. Sons Am. Revolution (pres. Salt Lake City chpt. 1989, pres. Utah Soc. 1992), Sons of the Utah Pioneers (life), Tau Beta Pi, Pi Tau Sigma, Sigma Nu, Theta Tau. Ch. Jesus Christ of Latter-day Saints. Achievements include development of analytical method for computing aerodynamics of missile and airplane-life configurations to very high angles of attack; svc. on com. that approved aerodynamics of the space-shuttle orbiter for the first flight. Office: Dr Leland H Jorgensen Aerospace Cons 3 Ln Montagne Ln Sandy UT 84092

JORGENSON, JAMES WALLACE, chromatographer, educator; b. Kenosha, Wis., Sept. 9, 1952. BS, No. Ill. U., 1974; PhD in Chemistry, Ind. U., 1979. From asst. to assoc. prof. U. N.C., Chapel Hill, 1979-87, prof. chemistry, 1987—. Mem. AAAS, Am. Chem. Soc. (Chromatography award 1993). Achievements include research in chemical separations, fundamental studies of gas chromatography, liquid chromatography, electrophoresis. Office: Univ of N Carolina Dept of Chemistry Chapel Hill NC 27599*

JÖRSÄTER, STEVEN BERTIL, astronomer, educator; b. Stockholm, July 4, 1955; s. Rolf and Linnea Viola (Persson) J. BSc, Stockholm U., 1976, PhD, 1984. Fellow European So. Obs., Munich, 1984-86; postdoctoral Stockholm Obs., 1987, asst. prof., 1988, assoc. prof., 1991—; instrument scientist European Space Agy., Munich, 1989-90. Avocations: long distance skating, yachting, skiing, hiking. Home: Dalbobranten 25, 12868 Skõndal Sweden Office: Stockholm Obs, Saltsjöbaden, 13336 Stockholm Sweden

JORTNER, JOSHUA, physical chemistry scientist, educator; b. Poland, Mar. 14, 1933; s. Arthur and Regina Jortner; m. Ruth Sanger, 1960; 2 children. PhD, Hebrew U. Jerusalem. Instr. dept. phys. chemistry Hebrew U. Jerusalem, 1961-62, sr. lectr., 1963-65; assoc. prof. Tel Aviv U., 1965-66, prof., 1966—, Heinemann prof. chemistry, 1973—, head Inst. Chemistry, 1966-72, dep. rector, 1966-69, v.p., 1970-72; rsch. assoc. U. Chgo., 1962-64; vis. prof. H.C. Orsted Inst., U. Copenhagen, 1974, vis. prof. chemistry, 1978; vis. prof. UCLA, U. Calif., Berkeley, 1975. Author: (with M. Bixon) Intramolecular Radiationless Transitions, 1968, Molecular Crystals, 1969, Electronic Relaxation in Large Molecules, 1969, Long Radiative Lifetimes of Small Molecules, 1969; editor: (with Bernard Pullman) The Jerusalem Symposia on Quantum Chemistry and Biochemistry, Vols. 15-24, 1982-91. Recipient award Internat. Acad. Quantum Sci., 1972, Weizmann prize, 1973, Rothschild prize, 1976, Kolthof prize, 1976, Israel Prize in Chemistry, 1982, Wolf prize, 1988. Fgn. fellow Am. Acad. Arts and Scis; mem. Israel Acad. Scis. and Humanities (v.p. 1980-86, pres. 1986—), Internat. Acad. Quantum Scis., Am. Philos. Soc., Polish Acad. Scis., Romanian Acad. Scis., Royal Danish Acad. Scis. and Letters (fgn. mem.), European Acad. Scis. and Arts (active). Avocations: reading, writing. Office: Tel Aviv U, Dept Phys Chemistry, Ramat-Aviv, Tel Aviv 69978, Israel Also: Israel Acad Scis-Humanities, PO Box 4040, Jerusalem 91 040, Israel

JOSBENO, LARRY JOSEPH, physics educator; b. Elmira, N.Y., Oct. 21, 1938; s. Samuel Joseph and Katherine Louise (Jessup) J.; m. Cecile Ann Quatrano, Sept. 15, 1962; children: Deborah Ann, John Lawrence. BS in

Math., St. Bonaventure U., 1962; MS in Chemistry, U. N.H., 1970. Cert. tchr., N.Y. Tchr. Horseheads (N.Y.) High Sch., 1965-89; prof. Corning (N.Y.) Community Coll., 1989—; vis. scientist Cornell U., Ithaca, N.Y., 1986-87; adj. prof. Elmira Coll.; cons. State Edn. Dept., Albany, N.Y., 1987, Math Matrix, Ithaca, 1987—, Corning Inc., 1989. Author: ARCO Physics Review Book, 1983; contbr. articles to profl. jours. Mem. bd. govs. Notre Dame High Sch., Elmira, 1977-82; mem. bd. trustees Steele Meml. Lab., Elmira, 1985—; obs. presenter Elmira Corning Astron. Soc., Corning, 1968—. Capt. arty. U.S. Army, 1963-65. Mem. The Math. Assn. Am., Am. Physics Soc., Am. Chem. Soc., Am. Physics Tchr. Assn., Sci. Tchrs. Assn. of N.Y. (pres. 1989-90), Alpha Sigma Lambda (tchr. of yr. 1985). Democrat. Roman Catholic. Home: 539 W Franklin St Horseheads NY 14845-2356 Office: Corning Community Coll Corning NY 14830

JOSCELYN, KENT BUCKLEY, criminologist, research scientist, lawyer; b. Binghamton, N.Y., Dec. 18, 1936; s. Raymond Miles and Gwen Buckley (Smith) J.; children: Kathryn Anne, Jennifer Sheldon. BS, Union Coll., 1957; JD, Albany Law Sch., 1960. Bar: N.Y. 1961, U.S. Ct. Mil. Appeals 1962, D.C. 1967, Mich. 1979. Atty., adviser Hdqrs. USAF, Washington, 1965-67; asso. prof. forensic studies Coll. Arts and Scis., Ind. U., Bloomington, 1967-76, dir. Inst. Rsch. in Pub. Safety, 1970-75; head policy analysis div. Hwy. Safety Rsch. Inst., U. Mich., 1976-81, dir. transp. planning and policy, Urban Tech., Environ. Planning Program, 1981-84; prin. firm Joscelyn and Treat, P.C., 1981-93, Joscelyn, McNair and Jeffery, P.C., 1993—; cons. Law Enforcement Assistance Adminstrn., U.S. Dept. Justice, 1969-72; Gov.'s appointee as regional dir. Ind. Criminal Justice Planning Agy., also vice chmn. Ind. Organized Crime Prevention Council, 1969-72; commr. pub. safety City of Bloomington, 1974-76. Served to capt. USAF, 1961-64. Mem. NAS, ABA, D.C. Bar Assn., Mich. Bar Assn., N.Y. State Bar Assn., Internat. Bar Assn., NRC, Transp. Research Bd. (chmn. motor vehicle and traffic law com. 1979-82), Am. Soc. Criminology (life), Assn. for the Advancement Automotive Medicine (life), Soc. Automotive Engrs., Acad. Criminal Justice Scis. (life), Assn. Chiefs Police (assoc.), Nat. Safety Council, Assn. Former Intelligence Officers (life), Product Liability Adv. Coun., Sigma Xi, Theta Delta Chi. Editor Internat. Jour. Criminal Justice. Office: Joscelyn & Treat PC 325 E Eisenhower Pky Ann Arbor MI 48108-3307

JOSEPH, DANIEL DONALD, aeronautical engineer, educator; b. Chgo., Mar. 26, 1929; s. Samuel and Mary (Simon) J.; m. Ellen Broida, Dec. 18, 1949 (div. 1979); children: Karen, Michael, Charles; m. Kay Jaglo, Feb. 9, 1990. M.A. in Sociology, U. Chgo., 1950; B.S. in Mech. Engring, Ill. Inst. Tech., 1959, M.S., 1960, Ph.D., 1963. Asst. prof. mech. engring. Ill. Inst. Tech., 1962-63; mem. faculty U. Minn., 1963—, assoc. prof. fluid mechanics, 1965-69, prof. aerospace engring. and mechanics, 1969-90; Russell J. Penrose prof. U. Minn., Mpls., 1990—. Author 4 books on stability and bifurcation theory and fluid dynamics; editor 3 books; editorial bd. SIAM Jour. Applied Math, Jour. Applied Mechanics, Jour. Non-Newtonian Fluid Mechanics, others; contbr. articles to sci. jours. Guggenheim fellow, 1969-70. Mem. NAS, ASME, NAE, Am. Phys. Soc., Am. Acad. Arts and Scis., Soc. Engring. Sci. (G.I. Taylor medal 1990). Contbns. to math. theory of hydrodynamic stability; rheology of viscoelastic fluids. Home: 1920 S 1st St Apt 2302 Minneapolis MN 55454-1045 Office: U Minn Dept Aerospace Engring Minneapolis MN 55455

JOSEPH, LURA ELLEN, geologist; b. Tulsa, Jan. 24, 1947; d. Don Roscoe and Ruth Elizabeth (Taplin) J. Student, St. Paul Bible Coll., 1965-67, Pan Am. Coll., 1967-68; BA in Anthropology, U. Okla., 1971, MS in Geology, 1981; MA in Psychology with honors, U. Cen. Okla., 1992. Cert. petroleum geologist. Exploration geologist Getty Oil Co., Oklahoma City, 1977-84; geologist Harper Oil Co., Oklahoma City, 1984-86, consulting geologist, 1986-88; sr. geologist Grace Petroleum, Oklahoma City, 1988-93, cons. geologist, 1993—. Author: (with others) Hugo Reservoir I, 1971; contbr. articles to profl. jours. Active adv. coun. New Life Ranch, Inc., Colcord, Okla. Mem. Am. Assn. Petroleum Geologists, Oklahoma City Geol. Soc., Pan Am. Geol. Soc. (pres. 1967-68), Sigma Gamma Epsilon, Psi Chi. Republican. Mem. Independent Evangelical Ch. Avocations: travel, photography, reading, art, ceramics.

JOSEPH, PAUL GERARD, civil engineer, consultant; b. Madras, India, Dec. 10, 1960; came to U.S., 1983; s. Joseph P. Joseph and Margaret Paul; m. Roshini Joseph, Sept. 16, 1989. B of Engring., Coll. Engring., Madras, 1983; MSCE, Purdue U., 1985; MS, MIT, 1987. Profl. engr., Mass. Lab. dir. Civil Test Labs, Needham, Mass., 1989-90; geotech. engr. GEI Cons. Inc., Winchester, Mass., 1990—; com. mem. BSCE Computer Group, Boston, 1992—. Author rsch. reports. Office: GEI Cons Inc 1021 Main St Winchester MA 01890

JOSEPH, PETER MARON, physics educator; b. Ridley Park, Pa., Mar. 26, 1939; s. Joseph Maron and Doris Joseph; m. Susan Leigh Rittenhouse, June 28, 1980. BS in Physics, Lafayette Coll., 1959; PhD in Physics, Harvard U., 1967; M of Philosophy (hon.), U. Pa., 1987. Fellow Sloan Fettering Cancer Ctr., N.Y.C., 1972-73; from instr. to asst. prof. Columbia U., N.Y.C., 1973-80; assoc. prof. U. Md., Balt., 1980-82; assoc. prof. U. Pa., Phila., 1983-91, prof., 1991—. Mem. editorial bd. jours.; contbr. 79 papers, revs. to jours, chpts. to books. Mem. Balt. Symphony Chorus, 1981. Recipient Sylvia Greenfield prize Am. Assn. Physicists In Med., 1985. Fellow IEEE. Achievements include patent for method and apparatus for measuring the applied kilovotage of x-ray series. Office: Hosp U Pa Dept Radiology Philadelphia PA 19104

JOSEPHS, MELVIN JAY, professional society administrator; b. N.Y.C., Apr. 26, 1926; married, 1948. BSc, Rutgers U., 1950, MSc, 1952, PhD in Plant Physiology and Botany, 1954. Plant physiologist Dow Chem. Co., Midland, Mich., 1954-60; assoc. editor Chem. and Engring. News Am. Chem. Soc., 1960-66; mng. editor Environ. Sci. and Tech., 1966-70, Chem. and Engring. News, 1970-73; asst. dir. Product and Program Mgmt. Nat. Tech. Info. Svc., U.S. Dept. Commerce, Washington, 1973-76; chief Toxicology Data Bank, Nat. Libr. Med. HEW, Washington, 1976-78; asst. dir. Nat. Tech. Info. Svc. Office Govt. Agy. Support Dept. Commerce, Washington, 1978-86; exec. dir. Am. Soc. Plant Physiologists, Rockville, Md., 1986—. Mem. Am. Chem. Soc., Cosmos Club, Nat. Press Club, Sigma Xi. Achievements include research on boron nutrition and organic acid content, growth regulators, herbicides, algae, aquatic plants. Office: Am Soc Plant Physiologists 15501 Monona Dr Rockville MD 20855-2768

JOSEPHSON, BRIAN DAVID, physicist; b. Jan. 4, 1940; s. Abraham and Mimi Josephson; m. Carol Anne Olivier, 1976; 1 dau. BA, Cambridge U., 1960, MA, PhD, 1964; DSc (hon.), U. Wales, 1974, Exeter U., 1983. Asst. dir. research in physics Cambridge U., 1967-72, reader, 1972-74, prof. physics, 1974—; vis. faculty Maharishi European Res. U., 1975; vis. prof. dept. computer sci. Wayne State U., 1983; vis. prof. Indian Inst. Sci., Bangalore, 1984, U. Mo.-Rolla, 1987. Author papers on physics and theory of intelligence; co-editor: Consciousness and the Physical World, 1980. Recipient Nobel prize in physics, 1973, New Scientist award, 1969, Guthrie medal, 1972, van der Pol medal, 1972, Elliott Cresson medal, 1972, Hughes medal, 1972, Holweck medal, 1973, Faraday medal, 1982, Sir George Thompson medal, 1984; fellow Trinity Coll., Cambridge, 1962—. Fellow Royal Soc.; mem. IEEE (hon.), Am. Acad. Arts and Scis. (fgn., hon.). Office: U Cambridge Cavendish Lab, Madingley Rd, Cambridge CB3 OHE, England

JOSHI, AMOL PRABHATCHANDRA, chemical engineer; b. Kirkee, India, Feb. 23, 1963; came to U.S., 1984; s. Prabhatchandra Shantaram Joshi. BSChemE, Bombay (India) U., 1984; MSChemE, U. Wyo., 1986; DSc, Washington U., St. Louis, 1991. Rsch. asst. U. Wyo., Laramie, 1984-86; rsch. asst. Washington U., St. Louis, 1986-91, lectr. process thermo dynamics, 1991; sr. devel. engr. Aspen Tech., Cambridge, Mass., 1991—; cons. in field. Co-author: Computer Aided Chemical Engineering, 1993; contbr. articles to profl. publs. Nat. Merit scholar, Govt. of India, 1990. Mem. AICE, Tau Beta Phi. Achievements include application of time-frequency methods to problems in chemical engineering; development of superior algorithm for detecting gross errors in process operational data; development of an algorithm for on-line implementation of time-frequency techniques such as the wavelet transform. Office: Aspen Tech 10 Canal Park Cambridge MA 02141

JOSHI, JAGMOHAN, agronomist, consultant; b. Dhanoa, Panjab, India, Mar. 20, 1933; came to U.S., 1966; s. Gian Chand and Savitri Devi J.; m. Santosh Sharma, Feb. 19, 1961; children: Shallin, Shushen, Shailesh. MS, Panjab U., Chandigarh, India, 1961; PhD, Ohio State U., 1972. Cert. profl. crop scientist. Lectr. Extension Tng. Ctr., Mashobra, India, 1956-61; asst. agrl. officer Ministry Agr., Nairobi, Kenya, 1961-66; rsch. assoc. Ohio State U., Columbus, 1966-73; rsch. assoc. U. Md. Ea. Shore, Princess Anne, 1973-77, rsch. asst. prof., 1977-85, rsch. assoc. prof., 1985—, dir. Soybean Rsch. Inst., 1976—; cons. N.C. Agrl. & Tech. U., Greensboro, 1988, Transkel Washington Bur., 1990; internat. cons. in Zambia, Zimbabwe, Kenya, Nigeria, India, Republic of China, Sri Lanka, and the Caribbean Islands, 1976—; co-team leader China Tech. & Sci. Exchange, USDA, Washington, 1990. Co-author: Soybeans for the Tropics, 1987; contbr. articles to profl. jours. Pres. India Assn. Ea. Shore, Salisbury, 1979; patron Assn. Agrl. Scientists of Indian Origin, Normal, Ala., 1979—. Grantee USDA, NASA, 1973—. Mem. Am. Soc. Agronomy, Crop Sci. Soc. Am., Am. Soybean Assn. Achievements include research on host plant resistance, on cultural control of soybean pests, on winged bean, on agronomy of hydrocarbon producing plants and development of high yielding and promiscuous soybean varieties for Zambia. Office: Univ Md Ea Shore Trigg Hall Princess Anne MD 21853

JOSHI, R. MALATESHA, reading education educator; b. Davangere, India, Jan. 7, 1946; came to U.S., 1970; s. R. Neelakanta and Gowramma J.; m. Rekha M. Joshi, Aug. 17, 1980; children: Neil, Sunil. MA, Ind. State U., 1971; PhD, U. S.C., 1976. Cert. tchr., Ind. Asst. prof. Idaho State U., Pocatello, 1976-78, Oreg. State U., Corvallis, 1978-82; dir. reading ctr., assoc. prof. Fayetteville (N.C.) State U., 1983-90; assoc. prof. reading Okla. State U., Stillwater, 1990-93, prof., 1993—; cons. U. Cin., 1988; bd. dirs. Okla. br. Orton Dyslexia Soc.; keynote speaker Internat. Conf. on Psycholinguistics and Spl. Edn., Xian, China, 1993. Co-author: Reading Problems: Remediation and Consultation, 1992; co-editor: Reading Disorders: Varieties and Treatments, 1982, Neuropsychology and Cognition (vols. I, II), 1982, Dyslexia: A Global Issue, 1984, Reading and Writing Disorders in Different Languages, 1989; editor: Written Language Disorders, 1991, Neuropsychology and Cognition (book series), 1989; founding editor Reading and Writing jour.; contbr. to profl. publs. Grantee NATO, 1980, 82, 87, 91. Mem. Internat. Neuropsychol. Soc., Internat. Reading Assn., Internat. Assn. Rsch. in Learning Disabilities, Orton Dyslexia Soc., Rodin Remediation Soc., Am. Ednl. Rsch. Assn. Office: Okla State Univ 301 Gundersen St Stillwater OK 74078

JOSHI, SATISH DEVDAS, organic chemist; b. Bombay, Maharashtr, India, Sept. 29, 1950; came to U.S., 1982; s. Devdas Ganesh and Premlata (Prabhu) J.; m. Shima Janakimohan Bhadra, May 2, 1974; children: Shruti, Shilpa. BS, Bombay U., 1970, MS, 1972; PhD in Chemistry, Bombay U., Bombay, 1977. Rsch. fellow State U. Gent, Belgium, 1979-81, Louvain Med. Sch., Brussels, Belgium, 1981-82; rsch. assoc. Mt. Sinai Sch. Medicine, N.Y.C., 1982-85; group leader Bachem, Inc., Torrance, Calif., 1985-87; dir. Bachem Biosci. Inc., Phila., 1987-89; pres., chief exec. officer Star Biochems., Torrance, 1989-91; tech. dir. Mallinckrodt Inc., St. Louis, 1991—. Mem. AAAS, ACS, Am. Peptide Soc., Torrance C. of C. Home: 1928 Via Estudillo Palos Verdes Estates CA 90274

JOSKOW, PAUL LEWIS, economist, educator; b. Bklyn., June 30, 1947; s. Jules and Charlotte Joan (Epstein) J.; m. Barbara Zita Chasen, Sept. 10, 1978; 1 child, Suzanne Zoe. B.A., Cornell U., 1968; M.Phil., Yale U., 1971, Ph.D., 1972. Asst. prof. econs. MIT, Cambridge, 1972-75, assoc. prof. econs, 1975-78, prof. econs, 1978—, Mitsui prof., 1989—; vis. prof. J. F. K. Sch. Govt., Harvard U., Cambridge, Mass., 1979-80; Olin vis. scholar Harvard Law Sch., 1988-90; rsch. assoc. Nat. Bur. Econ. Rsch., 1988—; Joel Dean meml. lectr. Oberlin Coll., Ohio, 1983; cons. NERA, White Plains, N.Y., 1972—, The World Bank, 1991-92, Rand Corp., Santa Monica, Calif., 1972-87; pub. mem. Adminstrv. Conf. U.S., Washington, 1980-82; mem. adv. coun. EPRI, Palo Alto, Calif., 1980-84; mem. acid rain adv. com. EPA, 1990—; chmn. rsch. adv. bd. Com. for Econ. Devel., 1991—, sci. adv. bd. Inst. d'Organization Industrielle, Toulouse, France, 1991—; bd. dirs. New Eng. Electric System, Westborough, Mass., State Farm Indemnity Co., Bloomington, Ill., Whitehead Inst. for Biomedical Rsch., Cambridge, Mass. Co-author: Electric Power in the U.S., 1979, Markets For Power, 1983; author: Controlling Hospital Costs, 1981; also numerous articles, chpts.; co-editor, then assoc. editor Bell Jour. Econs., 1976-85; co-editor Jour. of Law, Econs. and Orgn., 1992—; bd. editors Am. Econ. Review, 1993—. Fellow Woodrow Wilson Found., 1968, NSF, 1973, Ctr. for Advanced Studies in Behavioral Scis., Palo Alto, Calif., 1985; recipient Disting. Svc. award Pub. Utility Rsch. Ctr., U. Fla., 1993. Fellow Am. Acad. Arts and Scis., Econometric Soc.; mem. ABA (assoc.), Am. Econ. Assn., Internat. Assn. for Energy Econs. Home: 7 Chilton St Brookline MA 02146-3902 Office: MIT Dept Econs 50 Memorial Dr Cambridge MA 02142-1347

JOSKOW, RENEE W., dentist, educator; b. N.Y.C., Mar. 15, 1960; d. Melvin Lawrence and Eunice Lila (Levine) J. BA, SUNY, Binghamton, 1981; MPH, Columbia U., 1985, DDS, 1985. Gen. practice resident Hackensack (N.J.) Med. Ctr., 1985-86; pvt. faculty practice, gen. practitioner Columbia Sch. of Dental and Oral Surgery, N.Y.C., 1986-90; asst. prof., dentistry Columbia Univ., N.Y.C., 1986—; pvt. practice gen. dentistry N.Y.C., 1990—; cons. alternative delivery systems Columbia Sch. Dental and Oral Surgery, N.Y.C., 1985-86, admission com., 1988-92; chairperson Frederick Birnberg Award Com., 1986-90; clin. program coord. Health of the Pub. grant Columbia Sch. Pub. Health, 1987-88; dir. freshman dental courses Columbia U., N.Y.C., 1987—; workshop leader Columbia U. Sch. Dental and Oral Surgery, 1993, Columbia Presbyn. Med. Ctr., 1993. Guest lectr. Inst. for Child Devel.-Community Outreach Lectr. on Oral Health, Hackensack Med. Ctr., 1986. Recipient L.I. Acad. of Odontology award, N.Y., 1985, Ella Marie Ewell award for Meritorious Svc., Columbia Univ., 1985, Alumni award for Excellence in Preventive Dentistry, Columbia Univ., N.Y., 1985. Fellow N.Y. Acad. Dentistry, Acad. Gen. Dentistry; mem. ADA, Am. Assn. Women Dentists (faculty advisor 1986—), 1st Dist. Dental Soc., Columbia U. Alumni Assn. (com. chair 1989—), Julliard Evening Divsn. Chorale, Columbia Student Honor and Rsch. Soc. (faculty advisor 1986—), Nat. Assn. Women Bus. Owners, Omicron Kappa Upsilon. Avocations: singer/songwriter, co-ed hosp. softball team, co-ed volleyball, tennis, pottery, cooking. Office: 29 W 57th St New York NY 10019

JOURJINE, ALEXANDER N., theoretical physicist; b. Moscow, Russia, July 2, 1953; came to U.S., 1980; s. Nikolai Alexandrovich and Tatiana S. (Russkova) J.; m. Galina I. Vaniasina, Sept. 15, 1975 (div. 1979); m. Ruth Ann DeMidowitz, Oct. 15, 1987; children: Nicholas W. Alexander. Degree in physics, MGU, Moscow, 1979; PhD in Physics, MIT, 1984. Rsch. assoc. dept. physics U. Wis., Madison, 1984-86; pres. Analog Intelligence DA Corp., Winchester, Mass., 1986—; software engring. specialist Wang Labs., Inc., Lowell, Mass., 1988—. Author: Doktorskaya, 1985; contbr. articles to profl. jours. Mem. Am. Phys. Soc., Russian Am. Sci. Student Found. (founder, bd. dirs.). Achievements include 4 U.S. patents. Office: Wang Labs Inc 1 Industrial Ave Lowell MA 01851

JOUZEL, JEAN, researcher; b. Janze, France, Mar. 5, 1947; s. Jean Marie and Marie Ange (Denis) J.; m. Brigitte Yvonne Menand, June 25, 1971; children: Maud, Jean-Noël. Engr. in Chemistry, Ecole Superieure Chimie Indsl., Lyon, France, 1968; D Phys. Scis., Faculte de'Orsay, Paris, 1974. Rsch. scientist French Atomic Energy Lab. Géchimie Isotopique Saclay, Saclay, France, 1968—; head lab. French Atomic Eneryg Lab. Géochimie Isotopique, Saclay, France, 1986—; assoc. dir. Lab. Glaciologie et Géophysique de l'Environment, Grenoble, France, 1989—, Lab. de Modēlisation du climat et de L'environnoment Saclay, 1991—; mem. steering com. PAGES, Internat. Geosphere Biosphere, 1990. Contbr. articles to profl. jours. Co-recipient Philip Morris award for Climatology, 1992. Mem. Academia Europaea, Am. Geophys. Soc., European Geophys. Soc. (coun. mem. 1989). Avocation: soccer. Home: 19 rue St Honoré, 91430 Igny Essonne, France Office: French Atomic Energy, CE de Saclay, 91191 Gif sur Yvette Essonne, France

JOVE, RICHARD, molecular biologist; b. Barcelona, Cataluna, Spain, Feb. 5, 1955; came to U.S., 1960; s. Ricardo and Maria Rosa (Calmet) J.; m. Hua Yu, June 21, 1984. BA, SUNY, Buffalo, 1977, MS, 1978; M in Philosophy, Columbia U., 1981, PhD, 1984. Postdoctoral fellow Rockefeller U., N.Y.C., 1984-88; asst. prof. U. Mich., Ann Arbor, 1988—, dir. molecular oncology program Cancer Ctr., 1992—. Recipient John S. Newberry prize Columbia U., 1984, Jr. Faculty Rsch. award Am. Cancer Soc., 1988-91; Damon Runyon-Walter Winchell Cancer Fund fellow, 1984-87. Mem. The Harvey Soc., Sigma Xi. Office: U Mich Dept Microbiology and Immunology 6643 Med Sci Bldg II Ann Arbor MI 48109

JOWETT, JOHN MARTIN, physicist; b. Edinburgh, Scotland, Dec. 3, 1954; s. Andrew Bogdan and Kathleen (Brown) J.; m. Siobhan Marie Coffey, oct. 2, 1976; children: Andrew Joseph, Aidan John, Michael Stuart. BSc with honors in math. physics, U. Edinburgh, 1976; Math. Tripos Part III, U. Cambridge, Eng., 1977, PhD, 1983. Physicist ISR div. CERN, Geneva, 1980-82, LEP div. CERN, Geneva, 1982-89, SL div. CERN, Geneva, 1990—; vis. scientist Stanford (Calif.) U., 1985-86; sec. large electron-positron ring machine adv. com. CERN, 1987-89; organizer 1st joint U.S.-CERN Sch. on Particle Accelerators, Sta. Margarita di Pula, Italy, 1985; specialist in design and performance of electron-positron colliders, especially high luminosity Large Electron Positron ring and Tau-Charm Factory. Editor, contbg. author: Nonlinear Dynamics Aspects of Particle Accelerators, 1986; also articles. Mem. Am. Phys. Soc. Avocations: skiing, English language. Home: 14 Rue des Digitales, 01710 Thoiry Ain, France Office: SL Div CERN, CH 1211 Geneva 23, Switzerland

JOY, DAVID ANTHONY, computer consultant; b. Houston, Jan. 27, 1957; s. Ralph E. and Lucille (Bailleres) J.; m. Jacquelyn R. Austin, Jan. 28, 1984; children; Patrick, Gwendolyn, Jeramiah. Student, Rice U., 1975-77. Programmer Tex. Electronic Instr., Houston, 1979-81, Cimarron Software, Clear Lake, Tex., 1981-83, Handle Techs., Houston, 1983-86; engr. mgr. Foxboro Co., Houston, 1986-89; cons. Pioneering Controls, Houston, 1989-91, Joy Rsch. & Devel., Spring, Tex., 1991—. Author: (software) Stars Database, 1985; developer computer protocol translation devices for industry. Bus ministry capt. N.W. United Pentecostal Ch., Houston. Recipient award for sci. and engring. fairs U.S. Army, 1974, also numerous pub. speaking and debate awards. Mem. Uniforum. Pentecostal. Avocation: religious teaching. Office: Joy Rsch & Devel 9403 Wallingham Dr Spring TX 77379-4457

JOYCE, EDWARD ROWEN, chemical engineer, educator; b. St. Augustine, Fla., Oct. 20, 1927; s. Edward Rowen and Annie Margaret (Cobb) J.; m. Leland Livingston White, Sept. 11, 1954; children: Leland Ann, Julia, Edward Rowen III, Theo, Adele. BS in Chem. Engring., U. Miss., 1950; M of Engring., U. Fla., 1969; MBA, U. North Fla., 1975. Registered profl. engr., Fla. Petroleum engr. Texaco, Harvey, La., 1953-55; project engr. Freeport Sulphur Co., New Orleans, 1955-59; chem. engr. SCM Corp., Jacksonville, Fla., 1959-81; profl. engr. Jacksonville Electric Authority, 1981-93, ret., 1993; adj. prof. U. North Fla., Jacksonville, 1977—, Jacksonville U., 1989—; newspaper columnist Fla. Times Union, Jacksonville, 1970-87. Co-author: Sulfate Turpentine Recovery, 1971; author booklet; patentee in field. Sci. fair judge Duval County Sch. System, Jacksonville, 1960-92; co-chmn. adv. com. U. North Fla., 1981-85; merit badge advisor Boy Scouts Am., Jacksonville, 1960—; advisor Jr. Achievement, Jacksonville, 1963. Comdr. USN, 1950-53, Korea. Fellow Fla. Engring. Soc. (pres. Jacksonville chpt. 1983); mem. AICE (pres. Peninsular Fla. chpt. 1963-64), Phi Kappa Phi, Alpha Pi Mu, Gamma Sigma Epsilon. Democrat. Episcopalian. Avocations: stamp collecting, coin collecting, water sports, camping. Home: 5552 Riverton Rd Jacksonville FL 32211-1361

JOYCE, TERENCE THOMAS, aerospace engineer; b. Widnes, Cheshire, Eng., Jan. 30, 1946; s. Michael and Marie (Noone) J.; m. Janet Margaret Barrett, Aug. 30, 1980; children: Helen Frances, Katherine Sarah. HNC. in Chemistry, Widnes (Eng.) Tech. Coll., 1969; BSc in Electronics with honors, Manchester (Eng.) U., 1974. Chemist ICI, Widnes, 1966-69; installation mgr. Marconi Communications, Chelmsford, Eng., 1974-80, sr. systems engr., 1980-83; head of ground stas. Brit. Aerospace, Bristol, Eng., 1983-89, prin. engr. ground stas. and RF tech., 1989—; BAC project mgr. for meteorol. coms. SAT. Payload (Eumetsat); K-band subsystem project engr. (polar platform SAT.) for ESA. Author (report) Millimeterwaves, 1985. Teller Liberal Dem. Party, Horsley, 1988; sec. Village Hall; local sch. gov. Mem. Inst. Elec. Engring. (chartered engr.). Roman Catholic. Avocations: classical music, swimming, reading, computers. Home: The Old Sch House The St, Horsley GL6 0PU, England Office: Brit Aerospace, Space Systems, Filton Bristol BS12 7QW, England

JOYCE, WILLIAM H., chemist; b. 1935. BS, Pa. State U., 1957; MBA, NYU, 1971, PhD, 1984. With Union Carbide Corp., Danbury, Conn., 1957—, past exec. v.p. ops., now pres., COO. Recipient Nat. medal of Tech., NSF, 1993. Office: Union Carbide Corp 39 Old Ridgebury Rd Danbury CT 06817*

JOYCE-BRADY, MARTIN FRANCIS, medical educator, physician, researcher; b. Wilmington, Del., Sept. 25, 1953; s. Robert Lawrence and Marjorie Theresa (Martin) Brady; m. Jean Marie Joyce, Sept. 17, 1977; children: Jessica, Erin, Emily. BA in Arts & Scis., U. Del., 1975; MD, U. Md., Balt., 1979. Medicine intern Boston City Hosp., 1979-80, medicine resident, 1980-82, chief med. resident, 1982-83; pulmonary fellow Pulmonary Ctr., Boston U. Sch. Medicine, 1982-87, asst. prof. medicine, 1987—; dir. pulmonary function lab. Boston City Hosp., 1987—; dir. ventilator care unit Jewish Meml. Hosp., Boston, 1988—. Contbr. articles to profl. jours. H. Fletcher Brown scholar Bank of Del., Wilmington, 1975, E.L. Trudeau scholar Am. Lung Assn., 1990-92; program project grantee on lung devel. NIH, 1991—. Mem. AAAS, Am. Soc. for Cell Biology, Mass. Med. Soc., Am. Thoracic Soc., Mass. Thoracic Soc. Achievements include development of novel hypothesis concerning cellular markings and patterns of differentiation in the developing and postnatal lung alveolar epithelium. Office: Pulmonary Ctr 80 E Concord St Boston MA 02118

JOYNER, CLAUDE REUBEN, JR., physician, medical educator; b. Winston-Salem, N.C., Dec. 4, 1925; s. Claude R. and Lytle (Mackie) J.; m. Nina Glenn Michael, Sept. 21, 1950; children: Emily Glenn, Claude Courtney. B.S., U. N.C., 1947; M.D., U. Pa., 1949. Intern Hosp. U. Pa., 1949-50; resident Bowman Grey Med. Sch., 1950; resident U. Pa., 1954-55, fellow in cardiology; Nat. Heart Inst. trainee, 1952-53; asst. instr. medicine Hosp. U. Pa., Phila., 1951-53; instr. Hosp. U. Pa., 1953-56, asso. medicine, 1956-59, asst. prof., 1959-64, assoc. prof., 1964-72; prof. medicine U. Pitts., 1972-87; prof. medicine Med. Coll. Pa., 1987—, vice dean, 1989—; chief medicine Allegheny Gen. Hosp., Pitts., 1972—. Contbr. articles to profl. jours. Served to lt. M.C. USNR, 1950-52. Fellow Am. Coll. Cardiology, ACP, Councils on Circulation, Arteriosclerosis and Cardiovascular Radiology of Am. Heart Assn.; mem. AAAS, Am. Heart Assn., Am. Clin. and Climatol. Soc. Home: Pulpit Rock Little Sewickley Creek Rd Sewickley PA 15143-8340 Office: Allegheny Gen Hosp Pittsburgh PA 15212

JOYNER, WEYLAND THOMAS, physicist, educator, business consultant; b. Suffolk, Va., Aug. 9, 1929; s. Weyland T. and Thelma (Neal) J.; m. Marianne Steele, Dec. 3, 1955; children: Anne, Weyland, Leigh. B.S., Hampden-Sydney Coll., 1951; M.A., Duke, 1952, Ph.D., 1955. Teaching fellow Duke, 1954, rsch. assoc., 1958; physicist Dept. Def., Washington, 1954-57; rsch. physicist U. Md., 1955-57; asst. prof. physics Hampden-Sydney Coll., 1957-59, assoc. prof., 1959-63, prof., 1963—, physics chmn., 1968-82, 85-87; rsch. assoc. Ames Lab. AEC, 1964-65; vis. prof. Pomona Coll., 1965; staff Commn. on Coll. Physics, Ann Arbor, Mich., 1966-67; vis. fellow Dartmouth Coll., 1981; mem. Panel on Preparation Physics Tchrs., 1967-68; nuclear physics cons. Oak Ridge Inst. Nuclear Studies, 1960-67; NASA-Lewis faculty fellow, 1982-84; pres. Piedmont Farms, Inc., 1958-75; ednl. cons. numerous colls. and univs., 1965-75; pres. Windsor Supply Corp., 1966-82, Three Rivers Farms, Inc., 1971—; mgmt. cons., 1966—; pres. Windsor Seed & Livestock Co., 1969-83. Contbr. articles profl. jours. Bd. dirs. Prince Edward Acad., 1971-92, exec. com., 1975-92; trustee Prince Edward Sch. Electoral Bd., 1979-80. NASA prin. investigator, 1985-87. Fellow AAAS; mem. Am. Phys. Soc., Am. Assn. Physics Tchrs., IEEE, Va. Acad. Sci. (past mem. council, sect. pres.), Am. Inst. Physics (regional counselor, past dir. Coll. Program), Phi Beta Kappa, Sigma Xi, Lambda Chi Alpha. Presbyn. (elder). Home: Venable Pl Hampden Sydney VA 23943

JU, JIANN-WEN, mechanics educator, researcher; b. Taipei, Taiwan, Mar. 18, 1958; came to U.S., 1982; s. Jiang and Kwai-Ing (Chen) J.; m. Mali Guo,

Jan. 14, 1985; children: Derek, Tiffany. BS, Nat. Taiwan U., Taipei, 1980; MS, U. Calif., Berkeley, 1983, PhD, 1986. Instr. Chinese Army Engring. Inst., Taipei, 1980-82; teaching asst. U. Calif., Berkeley, 1983-84, rsch. asst., 1984-86, lectr., 1986, postdoctoral rsch. engr., 1986-87; asst. prof. Princeton (N.J.) U., 1987-93; assoc. prof. UCLA, 1993—; cons. Air Force Engring. and Svcs. Ctr., Panama City, Fla., 1990—; mem. rev. panel NSF, Washington, 1991—; chmn., organizer Symposium on Damage Mechanics, 1990, Symposium on Damage Mechanics and Plasticity, Tempe, Ariz., 1992, Symposium on Damage Mechanics and Localization, Anaheim, Calif., 1992, Symposium on Homogenization and Constitutive Modeling of Heterogeneous Materials, Charlottesville, Va., 1993; invited lectr. 40 univs. and profl. socs. Author, editor: Damage Mechanics in Engineering Materials, 1990, Recent Advances in Damage Mechanics and Plasticity, 1992, Damage Mechanics and Localization, 1992, Homogenization and Constitutive Modeling, 1993; mem. editorial bd. Internat. Jour. Damage Mechanics, 1992—; contbr. articles to profl. jours.; author conf. procs. Fed. and indsl. rsch. grantee U.S. Govt., U.S. cos., Japanese cos., 1987—; recipient Presdl. Young Investigator award NSF, 1991. Mem. ASCE (control group 1989-93), ASME (com. mem. 1989—), Am. Acad. Mechanics, N.Y. Acad. Scis., Internat. Assn. for Computational Mechanics. Office: UCLA Dept Civil Engring 3173 I 405 Hilgard Ave Los Angeles CA 90024

JUARBE, CHARLES, otolaryngologist, neck surgeon; b. N.Y.C., Mar. 6, 1954; s. Santiago and Florence (Santos) J.; m. Jeanne Denise B. Laffitte, Oct. 14, 1978; children: Charles, Michael Andrew. Pre-med., Sacred Heart U., 1971-73; MD, U. Ctrl. del Este, Dominican Republic, 1978. Internship San Pablo Hosp., Bayamon, P.R., 1978-79; pub. health svc. Hosp. Ramon Ruiz Arnau, Bayamon, P.R., 1979-81; gen. surg. trainee St. Vincent Hosp., N.Y.C., 1981-85; ENT trainee Manhattan Eye Ear and Throat Hosp., N.Y.C., 1985-88; attending staff San Pablo Hosp., Bayamon, 1988-92, chief of otolaryngology, head and neck surgeon, 1992—; v.p. bd. dirs. Santa Cruz Med. Bldg., Bayamon, 1990-91. Contbr. articles to profl. jours. Vol. Am. Cancer Soc., P.R. Chpt., Hato Rey, 1991. Recipient Letter of Commendation, U.S. Army, Most Valuable Resident award Manhattan Eye, Ear & Throat Hosp., 1988, Physician Recognition award AMA, 1990, Active Tchr. award Am. Acad. Family Physician, 1991. Mem. Am. Acad. Otolaryngology Head and Neck Surgery, Am. Acad. Facial Plastic and Reconstructive Surgery, Am. Rhinologic Soc., Am. Acad. Otolaryngic Allergy, Am. and P.R. Med. Assn., Am. Cancer Soc. (mem. bd. dirs. P.R. chpt.), N.Y. Acad. Sci. Republican. Office: Santa Cruz Med Bldg 73 Santa Cruz Ste 205 Bayamon PR 00959

JUBERG, RICHARD KENT, mathematician, educator; b. Cooperstown, N.D., May 14, 1929; s. Palmer and Hattie Noreen (Nelson) J.; m. Janet Elisabeth Witchell, Mar. 17, 1956 (div.); children: Alison K., Kevin A., Hilary N., Ian C.T.; m. Sandra Jean Vakerics, July 8, 1989. BS, U. Minn., 1952, PhD, 1958. Asst. prof. U. Minn., Mpls., 1958-65; sci. faculty fellow Univerista di Pisa, Italy, 1965-66; assoc. prof. U. Calif., Irvine, 1966-72, U. Sussex, Eng., 1972-73; prof. U. Calif., Irvine, 1974-91, prof. emeritus, 1991—; vis. prof. U. Goteborg, Sweden, 1981; mem. Courant Inst. Math. Scis., NYU, 1957-58. Contbr. articles to profl. jours. With USN, 1946-48, Guam. NSF Faculty fellow, Univ. Pisa, Italy, 1965-66. Mem. Am. Math. Soc., Tau Beta Pi. Democrat. Avocation: bird watching. Office: U Calif Math Dept Irvine CA 92717

JUDD, BURKE HAYCOCK, geneticist; b. Kanab, Utah, Sept. 5, 1927; s. Zadok Ray and Elva (Haycock) J.; m. Barbara Ann Gaddy, Mar. 21, 1953; children: Sean Michael, Evan Patrick, Timothy Burke. BS, U. Utah, 1950, MS, 1951; PhD, Calif. Inst. Tech., 1954. Postdoctoral fellow Am. Cancer Soc. U. Tex., Austin, 1954-56; from instr. to prof. U. Tex., 1956-79, dir. Genetics Inst., 1977-79; geneticist Atomic Energy Commn., Germantown, Md., 1968-69; chief lab. genetics Nat. Inst. Environ. Health Sci., Research Triangle Park, N.C., 1979—; vis. asst. prof. Stanford U., Palo Alto, Calif., 1960; Gosney vis. prof. Calif. Inst. Tech., Pasadena, 1975-76; adj. prof. U. N.C., Chapel Hill, 1979—, Duke U., Durham, 1980—; mem. panel genetic biology NSF, Washington, 1969-73, genetics study sect. NIH, Washington, 1974, 77, 79, 88, com. on germplasm resources NAS, Washington, 1976-77; chmn. human genome initiative rev. panel Dept. of Energy, Washington, 1988. Author: Introduction to Modern Genetics, 1980; editor: Molecular and Gen. Genetics, 1986—; assoc. editor Genetics, 1973-78; contbr. articles to profl. jours. With U.S. Army, 1946-47. Fellow AAAS; mem. Am. Soc. Naturalists (sec. 1968-70), Genetic Soc. Am. (sec. 1974-76, v.p., pres. 1979-80). Avocations: travel, poetry, fiction. Home: 411 Clayton Rd Chapel Hill NC 27514-7613 Office: Nat Inst Eviron Health Sci PO Box 12233 Research Triangle Park NC 27709

JUDD, FRANK WAYNE, population ecologist, physiological ecologist; b. Wichita Falls, Tex., Aug. 23, 1939; married; 2 children. BS, Midwestern State U., 1965; MS, Tex. Tech U., 1968, PhD in Zoology, 1973. Teaching asst. biology Tex. Tech U., 1965-68, rsch. asst., instr., 1969-71; instr. dept. biology Pan Am U., 1968-69, from asst. prof. to prof., 1972-82, prof. biology, dir. Coastal Studies Lab., 1984—; adj. prof. dept. biol. scis. Tex. Tech U., 1983—; vis. prof. dept. wildlife & fisheries science, Tex. A&M U., 1989—. Mem. Am. Soc. Ichthyologists & Herpetologists, Am. Soc. Mammalogists, Ecology Soc. Am., Herpetologists League. Achievements include rsch. in ecology of the coastal zone of southern Tex. and northern Mex., barrier island ecology, black mangrove distribution, oyster reef distribution, tortoise demography. Office: University of Texas Pan American Coastal Studies Laboratory PO Box 2591 South Padre Island TX 78597*

JUDD, WILLIAM REID, computer engineer, graphic artist; b. Salt Lake City, Sept. 11, 1951; s. William R. Jr. and Theda (Whitehead) J.; m. Margarita Pimentel, Mar. 19, 1983. BS in Zoology and Organic Chemistry, Weber State Coll., Odgen, Ut., 1976; M Med. Physics and Computer Sci., U. Calif., Berkeley, 1980; postgrad. research, U. Utah, 1984-87, U. N.Car., 1988—. Computer graphics programmer Lawrence Berkeley Labs., Berkeley, Calif., 1978-81; computer Image-processor Gould-DeAnza Graphics Div., San Jose, Calif., 1981-82; software engr. Robert Bosch Video Corp., Salt Lake City, 1984-88; computer graphics engr. Evans and Sutherland Corp., Data Gen., Research Triangle Park, NC, 1988-90; mem. tech. staff Sun Microsystems, Research Triangle Park, N.C., 1990—. Musician and composer (computer music) Electronic Prokofiev, 1989; artist: (computer graphics) Artificial Reality 1985, Trefoil Knot, 1987, Graphic Library, 1991. Mem. IEEE, Rsch. Triangle Assn. for Computing Machinery Siggraph (sec. 1993-94), Triangle Area Neural Network Soc. (exec. coun. 1993-94). Libertarian. Avocations: math., computer music. Home: 2617 Sweetbriar Rd Durham NC 27704-9547

JUDGE, JOSEPH B., clinical psychologist; b. Bklyn., Sept. 29, 1919; s. Bernard Joseph and Mary Regina (McCabe) J.; m. Beverly Joan Rice, July 9, 1987; stepchildren: Joyce, William, Christopher, Scott, Carrie, Nancy. MS, Iona U., 1969; PhD, Fordham U., 1979. Diplomate Am. Bd. Med. Psychotherapists. Dir. youth program Roman Catholic Diocese of Bklyn., 1945-59, pastor, 1961-71; chief of mental health svcs. Dept. Health, N.Y.C., 1973-80; staff sr. psychologist Maimonides Hosp., Bklyn., 1980-90; pvt. practice College Point, N.Y., 1990—; asst. prof. N.Y. Inst. Tech., N.Y.C., 1972-75, St. Francis Coll., Bklyn., 1956. Contbr. articles to profl. jours. Charter mem. Brownsville Community Coun., Bklyn., 1962-70; pres., chmn. bd. dirs. Riverdale Towers Housing Corp., Bklyn., 1961-74. Mem. APA, N.Y. State Psychol. Assn., Queens Psychol. Assn., Soc. for Advancement of Psychology. Democrat. Achievements include research on relation of human relations skills and personality variables. Home: 20-08 College Point Blvd College Point NY 11356

JUDKINS, WILLIAM SUTTON, environmental engineer; b. McCook, Nebr., Feb. 21, 1948; s. James W. and Margaret S. (Miller) J.; m. Terese L. Lowe, Nov. 23, 1974; 1 child, Valerie. BSCE, U. Va., 1970; MSCE, George Washington U., 1975. Registered profl. engr., Va. Distbn. engr. Duke Power Co., Winston-Salem, N.C., 1970-72; environ. engr. Chesapeake div. Naval Facilities Command, Washington, 1972-78; hydraulic engr. Fed. Emergency Mgmt. Agy., Washington, 1978-88; environ. engr. Naval Facilities Engring. Command, Alexandria, Va., 1988—. Co-author pamphlet: Early Flood Warning System, 1986. Mem. ASCE, Nat. Soc. Profl. Engrs., Md. Soc. Profl. Engrs. (sec. Potomac chpt. 1992—). Office: Naval Facilities Engring 200 Stovall St Alexandria VA 22332-2300

JUHASZ, STEPHEN, editor, consultant; b. Budapest, Hungary, Dec. 26, 1913. Diploma in Mech. Engring., Tech. U., Budapest, 1936; Tekn Lic., Royal Inst. Tech., Stockholm, 1951. Spl. lectr. Royal Inst. Tech., Stockholm, 1949-51; mem. staff fuels rsch. lab. MIT, 1952-53; exec. editor Applied Mechanics Rev. Midwest Rsch. Inst., Kansas City, Mo.; exec. editor Applied Mechanics Rev. S.W. Rsch. Inst., San Antonio, 1953-59, editor, 1960—, dir., 1974—, also cons. Contbr. articles to profl. jours.; patentee in field. Fellow ASME (Edwin F. Church medal 1992), AAAS; mem. AIAA, Balcones Heights Lions Club. Office: SW Rsch Inst PO Box 28510 San Antonio TX 72884

JUHASZ, TIBOR, physicist, researcher; b. Dorog, Komárom, Hungary, Sept. 8, 1958; came to U.S., 1987.; s. Vendel and Ida (Kiss) J.; m. Marta Papp, Aug. 14, 1982; 1 child, Adam. Ms in Physics, JATE U., Szeged, Hungary, 1982, PhD, 1986. Rsch. scientist Tech. U. Budapest, Hungary, 1986-87; postdoctoral rschr. U. Calif., Irvine, 1987-90, rsch. physicist, 1990—; cons. Intelligent Surg. Lasesrs, Inc., 1989—. Contbr. articles to Phys. Review Letters, Phys. Rev. B, Optics Letters, Laserlight in Surgery and Medicine. Achievements include research in ultrashort pulsed lasers and their applications in solid state physics and ophthalmology; investigation of nonequilibrium phenomena in solids; discovery of a new nonequilibrium phonon development state; development of refractive surgery in ophthalmology using ultrashort pulsed lasers. Office: U Calif Dept Physics Irvine CA 92717

JUKES, THOMAS HUGHES, biological chemist, educator; b. Hastings, Eng., Aug. 25, 1906; came to U.S., 1925, naturalized, 1939; s. Edward Hughes and Ann Mary (Barton) J.; m. Marguerite Esposito, July 2, 1942; children—Kenneth Hughes, Caroline Elizabeth (Mrs. Nicholas Knueppel), Dorothy Mavis (Mrs. Robert Hudson). B.S.A., U. Toronto, 1930, Ph.D., 1933; NRC fellow med. scis., U. Calif. at Berkeley, 1933-34; D.Sc. (honoris causa), U. Guelph, 1972. Instr., asst. prof. U. Calif. at Davis, 1934-42; with pharm. div. Lederle Labs., 1942-45; dir. nutrition and physiology research sect. research div. Am. Cyanamid Co., Pearl River, N.Y., 1945-58; dir. research agrl. div. Am. Cyanamid Co., 1958-59, dir. biochemistry, 1960-62; vis. sr. research fellow in biochemistry Princeton, 1962-63; prof. dept. biophysics and med. physics U. Calif., Berkeley, 1963-91, prof. dept. integrative biology, 1991—, prof. emeritus nutritional scis., 1974—, rsch. biochemist Space Scis. Lab., 1963—, assoc. dir., 1968-70; cons. CWS, AUS, 1944-45, NASA, 1969-70; guest lectr. various univs.; Storer lectr. U. Calif. at Davis, 1973; Fred W. Tanner lectr. Inst. Food Technologists, 1979; vis. prof. U. Wis., River Falls, 1985; plenary lectr. Japanese Molecular Biology and Genetics Socs., Nagoya, 1986; cons. Calif. Cancer Adv. Council, 1981—; invited speaker Internat. Symposium on Evolution of Life, Kyoto, Japan, 1990. Author: B Vitamins for Blood Formation, 1952, Antibiotics in Nutrition, 1955, Molecules and Evolution, 1965; mem. editorial bds., Biochem. Genetics, BioSystems; biog. editor: Jour. Nutrition; assoc. editor: Jour. Molecular Evolution; Contbr. articles to profl. jours. Recipient Borden award Poultry Sci. Assn., 1947; Spencer award Am. Chem. Soc., 1976; Agrl. and Food Chemistry award, 1979; Disting. Service award Am. Agrl. Editors Assn., 1978; Cain Meml. award Am. Assn. Cancer Research, 1987; Klaus Schwarz commemorative medal Internat. Assn. Bioinorganic Scientists, 1988. Fellow Am. Soc. Animal Sci., Poultry Sci. Assn., Am. Inst. Nutrition (coun. 1941-45, pub. affairs officer 1978-81, chmn. com. on history 1979-83), Calif. Acad. Scis.; mem. Internat. Coun. Sci. Unions (chmn. biology working group COSPAR 1978-80, chmn. interdisciplinary sci. commn. F 1980-84), Human Genome Orgn., Am. Soc. Biol. Chemists, Soc. for Exptl. Biology and Medicine, Am. Chem. Soc., Trustees for Conservation (San Francisco) (pres. 1970-71), Sigma Xi, Delta Tau Delta. Clubs: Am. Alpine (N.Y.C.), Explorers (N.Y.C.); Chit Chat (San Francisco), Sierra (San Francisco); Faculty (Berkeley). Home: 170 Arlington Ave Kensington CA 94707-1135 Office: U Calif Space Scis Lab 6701 San Pablo Ave Oakland CA 94608

JULESZ, BELA, experimental psychologist, educator, electrical engineer; b. Budapest, Hungary, Feb. 19, 1928; came to U.S., 1956; s. Jeno and Klementin (Fleiner) J.; m. Margit Fasy, Aug. 7, 1953. Dipl. Elec. Engring., Tech. U., Budapest, 1950; Dr. Ing., Hungarian Acad. Sci., Budapest, 1956. Asst. prof. dept. communication Tech U. Budapest, Hungary, 1950-51; mem. tech. staff Telecommunication Research Inst., Budapest, 1951-56; mem. tech. staff Bell Labs., Murray Hill, N.J., 1956-64, head sensory and perceptual processes, 1964-83; rsch. head visual perception rsch. AT&T Bell Labs., Murray Hill, N.J., 1984-89; State of N.J. prof. psychology, dir. lab. of vision rsch. Rutgers U., New Brunswick, N.J., 1989—; continuing vis. prof. biology dept. Calif. Tech. Inst., Pasadena, 1985—. Author: Foundations of Cyclopean Perception, 1971; author over 170 sci. papers on visual perception; discover computer generated random-dot stereogram technique. Fairchild disting. scholar Calif. Inst. Tech., 1977-79, 87, assoc. Neurosci. Research Progam, 1982; MacArthur Found. fellow, 1983-87; Dr. H.P. Heineken prize Royal Netherlands Acad. Arts and Scis., 1985; Karl Spencer Lashley award Am. Philos. Soc., 1989. Fellow AAAS, Am. Acad. Arts and Scis., Optical Soc. Am.; mem. NAS, Goettingen Acad. Scis. (corr.), Hungarian Acad. Scis. (hon.). Home: 30 Valley View Rd Warren NJ 07059 Office: Rutgers U Lab Vision Rsch Busch Campus Piscataway NJ 08854

JULIAN, ELMO CLAYTON, analytical chemist; b. Columbus, Ohio, Mar. 11, 1917; s. Frederick Augustus and Viola (Stalder) J.; m. Pauline Margaret Prescott, Dec. 5, 1989. BA, Ohio State U., 1939. Chemist Ciba Pharm. Products, Summit, N.J., 1940-41; inspector power and explosives U.S. Govt., Kenvil, N.J., 1941-42; chemist Am. Cyanamid, Bound Brook, N.J., 1942-43, Manhattan Project, Oak Ridge, Tenn., 1944-46, S.E. Massengil, Bristol, Tenn., 1946-47, U.S. AEC, New Brunswick, N.J., 1948-53, U.S. Naval Ord. Test Sta., Inyokern, Calif., 1953-61, USPHS, Salt Lake City, 1961-64, U.S. EPA, Cin., 1964-73; ret. Contbr. articles to profl. jours. Mem. AAAS, N.Y. Acad. Sci., Sigma Psi. Achievements include 5 patents in field (with others); development of several methods in analytical chemistry, computer software for laboratory management and accounting, and statistical software for Multivariate Analysis. Home: 600 N Ware Rd Mcallen TX 78501

JULIANO, JOHN JOSEPH, energy systems engineer; b. Winchester, Mass., Aug. 30, 1963; s. Anthony John and Carolyn Ann (Calandrella) J. BS in Nuclear Engring., MIT, 1986; MS in Applied Math., Johns Hopkins U., 1991—. Registered profl. engr., Md. Engring. technician U.S. Army Corps of Engrs., Waltham, Mass., 1985-86; engr. ARINC Rsch. Corp., Annapolis, Md., 1986-91; sr. engr. EA Engring., Sci. and Tech., Arlington, Va., 1991-92; prin. engr. NUS Corp., Gaithersburg, Md., 1992—. Contbr. publs. to profl. jours. Elected to Rep. Ward Com., Boston, 1984; mem. Anne Arundel County Young Reps., Annapolis, Md., 1989—. Recipient Pres.'s award for Outstanding Performance, 1990, Gen. Mgr. award for Outstanding Performance, ARINC Rsch. Corp., 1989, Uniroyal Undergrad. Rsch. award MIT, 1983. Mem. IEEE, Assn. Energy Engrs., Demand Side Mgmt. Soc., Ops. Rsch. Soc. Am. Achievements include devel. of reliability assessment software for electric utility industry; devel. of process storage modeling system for petroleum refineries and flood condition monitoring system for Corps of Engrs. Home: 14103 Gallop Terr Germantown MD 20874 Office: NUS Corp 910 Clopper Rd Gaithersburg MD 20877

JULIANO, PETRONILO OCHOA, chemist; b. Pasay City, Philippines, Aug. 31, 1943; s. Jose Buencamino and Teodora Canicosa (Ochoa) J.; m. Ermelinda Perez Almazan, Nov. 30, 1968; children: Cheryl Lynn, Ronilo Jose, David Fernando. BS in Chemistry, U. Philippines, 1963; PhD, U. Va., 1970. Registered chemist. Scientist II Philippine Atomic Energy Commn., 1965-66; sr. researcher United Labs., Inc., Philippines, 1970-72; sr. rsch. chemist San Miguel Corp., Philippines, 1972-75, mgr. rsch., 1975-83, dir. corp. rsch. and devel., 1983-91; dir. rsch. and devel. San Miguel Foods Inc., Philippines, 1991--. Mem. Integrated Chemist The Philippines (bd. dirs.), Nat. Rsch. Coun. Philippines, Am. Chem. Soc., Inst. Food Technologists. Roman Catholic. Office: San Miguel Foods Inc, PO Box 1755, Makati The Philippines

JULIUS, STEVO, physician, educator, physiologist; b. Kovin, Yugoslavia, Apr. 15, 1929; came to U.S., 1965, naturalized, 1971; s. Dezider and Jelena (Engel) J.; m. Susan P. Durrant, Sept. 18, 1971; children: Nicholas, Natasha. M.D., U. Zagreb, 1953, Sc.D., 1964; M.D. (hon.), U. Goteborg, Sweden, 1979. Intern, then resident in internal medicine Univ. Hosp., Zagreb, 1953-60; sr. instr. internal medicine Univ. Hosp., 1962-64; research asst. U. Mich. Med. Sch., 1961-62, mem. faculty, 1965—, prof. internal medicine, 1974—, assoc. prof. physiology, 1980-83, prof. physiology, 1983—, dir. div. hypertension, 1974—. Co-editor: The Nervous System in Arterial Hypertension, 1976; contbr. articles med. jours. Fellow Am. Coll. Cardiology; mem. Internat. Soc. Hypertension (v.p.), Interam. Soc. Hypertension (treas. 1978-83), Am. Heart Assn. (couns. high blood pressure rsch. and epidemiology), Am. Physiol. Soc. (adv. bd.), Am. Fedn. Clin. Rsch., Soc. Exptl. Biology and Medicine, Coun. for High Blood Pressure Rsch. (adv. bd.). Office: Univ Mich Med Sch Div Hypertension 3918 Taubman Ctr Ann Arbor MI 48109-0356

JUNG, ANDRÉ, internist; b. Geneva, Oct. 9, 1939; s. Charles and Anna (Schifrin) J.; m. Agnes Sideris, Aug. 6, 1973; children: Michel, Anne. BS in Math., U. Geneva, 1962, MD, 1965. Intern, resident, Geneva, Basle, Berne, Switzerland; chargé de recherche Geneva U., 1977-81; head med. dept. City Hosp., Nyon, Switzerland, 1981—; rsch. fellow Mass. Gen. Hosp., Boston and London. Contbr. numerous articles to profl. jours. Mem. Soc. Internal Medicine, European Soc. Intensive Care Medicine, Swiss Soc. Intensive Care Medicine, Swiss Soc. Tropical Medicine, also others. Mem. Orthodox Ch. Home: l Ch de l'Escalade, 1206 Geneva Switzerland Office: Hopital de Zone, Ch Monastier 8, 1260 Nyon Switzerland

JUNG, MANKIL, organic synthetic chemist, educator; b. Ongjin, Korea, Oct. 27, 1950; came to U.S., 1975; s. Hyungsam and Yoosoon (Kim) J.; m. Soonboon Park, May 22, 1983; children: Diana Euncho, Albert Jiwoong. BS in Chemistry, Yonsei U., Seoul, Korea, 1974; MS, MIT, 1978; PhD, U. Oxford, Eng., 1981. Postdoctoral fellow dept. chemistry Harvard U., Cambridge, Mass., 1981-82; rsch. assoc. U. Notre Dame, Ind., 1983-84; asst. prof. Sch. Pharmacy U. Miss., University, 1984-91, assoc. prof. Sch. Pharmacy, 1991—. Contbr. articles to profl. jours. Grantee WHO, 1991. Mem. Am. Chem. Soc. (referee Jour. Organic Chemistry, Jour. Medicinal Chemistry), Am. Soc. Pharmacognosy, N.Y. Acad. Scis., The Oxford Soc. (hon. br. sec. for Miss. 1988—). Achievements include patents for Deoxoartemisinin, a new compound and composition for treatment of malaria; research on synthesis of biologically active natural products like artemisinin, taxol, beta lactams, on chiral photoxidation in organic synthesis. Home: 81 Jeff St Oxford MS 38655-5603 Office: U Miss Sch Pharmacy University MS 38677

JUNG, REINHARD PAUL, computer system company executive; b. Plochingen, Fed. Republic Germany, Sept. 3, 1946; s. Helmut and Ursula (Benz) J.; m. Valerie Ann Houghton, Sept. 12, 1970; children: Kristina Ingrid, Marie-Louise Larissa. Diploma in Electronics, Kerstensteiner Poly., Bad Homburg, Fed. Republic Germany, 1971. Technician Digital Equipment Corp., Munich, 1971-73; software engr. Interdata GmbH, Munich, 1973-75; mgr. support Perkin Elmer Corp., Munich, 1976-79, tech. dir. Germany, 1980-84; founder, mng. ptnr. CAF GmbH, Munich, 1985—, Jung & Jung GmbH, Munich, 1988—. Avocations: Ham radio operator, skiing, tennis, piano playing. Home: Droesslinger Str 10, D-82229 Seefeld/Oberalting Germany Office: CAF Gmbh, Am Bahnhof 4a, 82205 Gilching Germany

JUNGER, MIGUEL CHAPERO, acoustics researcher; b. Dresden, Germany, Jan. 29, 1923; came to U.S., 1941, naturalized, 1946; s. José and Adrienne (Junger) Chapiro; m. Ellen Sinclair, 1960; children: M. Sebastian, A. Carlotta. B.S., MIT, 1944, S.M., 1946; Sc.D. (Gordon McKay scholar), Harvard U., 1951. Postdoctoral rsch. fellow in acoustics Harvard U., 1951-55; partner Cambridge Acoustical Assocs., Inc., 1955-59, pres., 1959-89, chmn. bd. dirs., 1989—; sr. vis. lectr. ocean engring. dept. MIT, Cambridge, 1968-78; vis. prof. U. Technologie de Compiègne, 1975, 77-82. Author: Sound, Structures and Their Interaction, 1972, 2d edit., 1986, rev. edit., 1993, Eléments d'Acoustique Physique, 1978; contbr. articles to profl. jours. Fellow ASME (Rayleigh lectr., Per Bruel Noise Control and Acoustics Gold medal 1992), Acoustical Soc. Am. (Trent-Crede medal). Achievements include patents in field. Home: 90 Fletcher Rd Belmont MA 02178-2017 Office: 200 Boston Ave Medford MA 02155

JUNGREN, JON ERIK, civil engineer; b. Malmo, Sweden, Oct. 19, 1927; came to U.S., 1966; s. Axel Bernhard and Lilly Ottonie (Eliasson) Ljungren; m. Elaine Berry, May 13, 1977. BS, Chalmers Inst. Technology, Gothenburg, Sweden, 1951; MS, Chalmers Inst. Technology, 1952; BA, Royal Inst. Technology, Stockholm, Sweden, 1954; PhD, Columbia Pacific U., 1985. Registered civil engr., Calif., Ariz. Br. mgr. Jacobsen & Widmark, Lund, Sweden, 1954-59; pres., owner Civilingnjoren SVR Jan Ljunggren AB, Lund, 1959-66; supervising engr. Bechtel Corp., San Francisco, 1966-74; engring. mgr. Morrison Knudsen, Holland, U.S.A., 1974-83; chmn. Jungren & Duran, Inc., Santa Ana Heights, Calif., 1983—; cons. engr. Assn. of Calif. Legis. Com., 1987-90. Contbr. articles to profl. jours. Recipient Archtl. award, 1959, 66. Mem. ASCE, Am. Cons. Engrs. Coun., Cons. Engrs. Assn. Calif., Am. Concrete Inst., Internat. Conf. Bldg. Ofcls., Masons, Mensa. Democrat. Home: 2420 Miseno Way Costa Mesa CA 92627 Office: Jungren & Duran Inc 20341 Irvine Ave Ste 5 Santa Ana Heights CA 92707-5628

JUNKINS, JERRY R., electronics company executive; b. Ft. Madison, Iowa, Dec. 9, 1937; s. Ralph Renaud and Selma Jeannie (Kudebeh) J.; m. Marilyn Jo Schevers, June 13, 1959; children: Kirsten Dianne, Karen Leigh. B.E.E., Iowa State U., 1959; M.S. in Engring. Adminstrn., So. Methodist U., 1968. With def. bus. Tex. Instruments, Inc., Dallas, 1959-75, asst. v.p., mgr. equipment group, 1975-77, v.p., mgr. equipment group, 1977-81, exec. v.p., mgr. data systems and indsl. systems, 1981-85, pres., chief exec. officer, 1985-88, chmn, pres., chief exec. officer, bd. dirs., 1988—; bd. dirs. Procter and Gamble Co., Caterpillar Inc. Trustee So. Meth. Univ.; bd. dirs. Dallas Citizens Coun. Mem. Nat. Acad. Engring. Office: Tex Instruments Inc MS 236 13510 N Central Expwy Dallas TX 75243

JURAN, JOSEPH MOSES, engineer; b. Braila, Rumania, Dec. 24, 1904; came to U.S., 1912, naturalized, 1917; s. Jakob and Gitel (Goldenberg) J.; m. Sadie Shapiro, June 5, 1926; children: Robert, Sylvia, Charles, Donald. BS in Elec. Engring., U. Minn., 1924, JD, Loyola U., 1935, DEng (hon.), Stevens Inst. Tech., 1988; DSc (hon.), U. Minn., 1992; LLD (hon.), U. New Haven, 1992. Bar: Ill. 1935; registered profl. engr., N.Y., N.J. With Western Electric Co., Inc., 1924-41; asst. adminstr. Office Lend-Lease Adminstrn., 1941-43, Fgn. Econ. Adminstrn., 1943-45; prof., chmn. dept. adminstrv. engring. N.Y.U., 1945-51; cons. numerous indsl. cos. and govt. agys., 1945—, vis. lectr. numerous Am. and Fgn. univs.; founder, chmn. Juran Inst., Inc., 1979-87, emeritus, 1987—; founder, chmn. Juran Found., Inc., 1986—. Editor: Quality Control Handbook, 4th edit., 1988 (translated into Japanese, Spanish, Russian, Hungarian, Chinese, Portuguese); author numerous books including: (with N.N. Barish) Case Studies in Industrial Management, 1955, Managerial Breakthrough, 1964, (with J.K. Louden) The Corporate Director, 1966, (with F.M. Gryna, Jr.) Quality Planning and Analysis, 1970, 2d edit., 1980, (video cassette series) Juran on Quality Improvement, 1981, Juran on Planning for Quality, 1988, Juran on Leadership for Quality, 1989, Juran on Quality by Design, 1992; lectr., author numerous papers on mgmt. Decorated Order of Sacred Treasure (Japan), 1981; recipient alumni medal U. Minn., 1954, Scroll of Appreciation Japanese Union Scientists and Engrs., 1961, 250th Anniversary medal Czech Higher Inst. Tech., 1965, Wallace Clark medal, 1967, ann. medal Technikhaza Esztergom, Hungary, 1968, medal Fedn. Tech. and Sci. Industries, Hungary, 1968, medal of honor camera Official de la Industria, Madrid, 1970, Plaque Appreciation Republic Korea, 1978, Stevens Medal Stevens Inst. Tech., 1984, Chairman's award Am. Assn. Engring. Socs., 1988, Nat. Medal Tech. U.S. Dept. Commerce Tech. Adminstrn., 1992. Fellow AAAS, Internat. Acad. Mgmt., Am. Soc. for Quality Control (hon., Brumbaugh award 1958, Edwards medal 1962, Eugene L. Grant medal 1967), Am. Inst. Indsl. Engrs. (Gilbreth medal 1981), Am. Mgmt. Assn., ASME (Warner medal 1945); mem. NAE, Sigma Xi, Tau Beta Pi, Alpha Pi Mu; hon. mem. European Orgn. for Quality Control, Romanian Acad. (academician 1992, established Juran award 1992), Australian Orgn. for Quality Control (Juran medal named in his honor 1975), Argentine Orgn. for Quality Control, Philippine Soc. for Quality Control, Spanish Assn. for Quality Control, Brit. Inst. Quality Assurance, Spanish Soc. for Quality Control; mem., sometime officer many profl. assns. Office: Juran Inst Inc 11 River Rd Wilton CT 06897-4057

JURCZYK, JOANNE MONICA, technical analyst; b. Orange, Calif., Dec. 27, 1958; d. Edward Joseph and Helen Imogene (Shelly) J. BSBA in Econs., Chapman U., 1981. Guest rsch. specialist Disneyland-Walt Disney Co.,

Anaheim, Calif., 1985-88, guest rsch. coord., 1988-89, guest rsch. survey ops. supr., 1989-91, indsl. engring. tech. analyst, 1991-92; active Work Exposure Day, Disneyland/U. Disneyland, Anaheim, 1990. Associate Met. Mus. of Art, nationwide, 1991—. Mem. Am. Film Inst. Democrat. Roman Catholic. Avocations: literature, music, theatre.

JURS, PETER CHRISTIAN, chemistry educator; b. Oakland, Calif., Apr. 13, 1943; s. Peter Clyde and Julie (Tanner) J.; m. Elaine Stahlman, July 9, 1983; children: Harold, Christian, Andrew. BS, Stanford U., 1965; PhD, U. Washington, 1969. From asst. to assoc. prof. Pa. State U., University Park, 1969-78, prof., 1978—. Contbr. over 160 articles to profl. jours. Fellow AAAS; mem. Am. Chem. Soc. (Computers in Chemistry award 1990). Office: Pennsylvania State Univ Dept Chemistry 152 Davey University Park PA 16802

JUSTICE, (DAVID) BLAIR, psychology educator, author; b. Dallas, July 2, 1927; s. Sam Hugh and Lou-Reine (Hunter) J.; m. Rita Norwood, July 26, 1972; children: Cynthia, David, Elizabeth. BA, U. Tex., Austin, 1948; MS, Columbia U., 1949; MA, Tex. Christian U., 1963; PhD, Rice U., 1966. Diplomate Am. Bd. Med. Psychotherapists. Reporter Ft. Worth Star-Telegram, 1952-55; sci. writer N.Y. Daily News, 1955-56, Ft. Worth Star-Telegram, 1956-64; sci. editor, columnist Houston Post, 1964-73; exec. asst. to Mayor Houston, 1966-72; prof. psychology Sch. Public Health, U. Tex., Houston, 1968—; sr. psychologist, group therapist, psychiat. residency faculty Tex. Research Inst. Mental Scis., 1973-85; community assoc. Rice U., Lovett Coll.; cons. child abuse Tex. Dept. Human Resources; faculty assoc. Ctr. for Health Promotion, Research and Devel., U. Tex. Health Sci. Ctr., mem. inter-faculty coun., 1991-92; chmn. standing com. on interpersonal violence U. Tex. Health Sci. Ctr. , 1985-89, dir. Ctr. for Prevention of Violence and Injury, 1987-89, chmn. faculty Sch. of Pub. Health, 1990-91, chmn. faculty policy com., 1989-90, faculty marshal, 1990, mem. exec. com., 1991-93, vice chair interfaculty coun., 1992-93; vis. scholar U. Colo., 1990—; founding assoc. Blaffer Gallery U. Houston. Author: Violence in the City, 1969, Detection of Potential Community Violence, 1967, (with Rita Justice) The Abusing Family, 1976, The Broken Taboo: Sex in the Family, 1979, Perspectives in Public Mental Health, 1982, Who Gets Sick: Thinking and Health, 1987, Who Gets Sick: How Beliefs, Moods and Thoughts Affect Your Health, 1988, The Abusing Family, rev. edit., 1990; editor: Your Child's Behavior, 1972; editorial bd.: Internat. Jour Mental Health, 1980—. Gen. chmn. Houston Job Fair, 1967-73; chmn. Houston Manpower Area Planning Council, 1972-74; mem. Tex. Urban Devel. Commn., 1970-72; bd. dirs. Houston Housing Devel. Corp., Tex. Citizens Human Devel., 1979-84, Greater Houston Com. Prevention of Child Abuse, 1982-88; sec. bd. mgrs. Tarrant County Hosp., Dist., 1961-64; pres. Greater Houston Youth Council, 1978-79, Houston Area Council on Sudden Infant Death Syndrome, 1977-78; mem. nat. adv. com. Marine Biomed. Inst., U. Tex. Med. Br., 1971-84; mem. Office of Minority Affairs, Resource Persons Network, HHS, 1988—; mem. community bd. Tex. Youth Council; vestry, chmn. adult edn. St. John The Divine Episc. Ch., 1984-88. Served with USNR, 1945-46. Recipient most outstanding book award Tex. Writers Roundup, 1970, award of recognition City of Houston, 1973, Benjamin Franklin Book award Pubs. Mktg. Assn. Am., 1988, Excellence in Media award Am. Psychol. Assn., 1988, Friends of Fondren Libr. book award Rice U., 1989, 91, Heritage award for child abuse rsch. Child Abuse Prevention Coun., 1989; named One of Five Outstanding Young Men of Tex., 1962; recipient numerous awards for sci. writing. Fellow Am. Coll. Psychology, Am. Inst. Stress; mem. Nat. Assn. Sci. Writers (life; exec. com. 1965-67), Houston Psychol. Assn. (pres. 1975), Am. Public Health Assn. (chmn. mental health sect. 1980-81, governing council 1983-85, action bd. 1985-87, mental health sect. award 1989), Coun. on Behavioral and Social Scis., Am. Assn. Schs. Pub. Health, Phi Beta Kappa (dir. Houston chpt. 1979-89, pres. Houston chpt. 1982-83), Phi Beta Kappa Assocs. Clubs: Dr.'s of Houston, Knights of the Vine. Home: 6331 Brompton Rd Houston TX 77005-3403 Office: 1200 Herman Pressler Dr Houston TX 77030-3900

JUSZCZAK, NICHOLAS MAURO, psychology educator; b. Chorely, Lancashire, Eng., May 19, 1955; came to U.S. 1956; s. Adam and Augusta (Lugnan) J.; 1 child, Amanda; m. Margie Nina Malkin, Oct. 9, 1988; children: Kimberly, Melissa. BA cum laude, Baruch Coll., N.Y.C., 1980; MS, Hunter Coll., N.Y.C., 1984. Researcher Psychophysiology Lab., Baruch Coll., N.Y.C., 1980-88; instr. psychology Baruch Coll. Dept. Psychology, N.Y.C., 1984—; cons. statistics BOE/CUNY Student Mentor Program, 1987—. Contbr. articles to profl. jours. Mem. N.Y. Acad. Sci. Home: 12-22 149th St Whitestone NY 11357

JUTAMULIA, SUGANDA, electro-optic scientist; b. Muara Enim, Indonesia, July 11, 1954; s. Harris Intankusuma and Conny (Julian) J.; m. Xiaoye Sherry Li, June 9, 1990. BS, Bandung Inst. Tech., Indonesia, 1977; PhD, Hokkaido U., Sapporo, Japan, 1985. Rsch. assoc. Pa. State U., University Park, 1985-87, instr., 1988; sr. scientist Quantex Corp., Rockville, Md., 1988-91; mgr. prodn. devel. Kowa Co. Ltd., San Jose, Calif., 1991--. Editor: Selected Papers on Optical Correlators, 1993; co-author: Optical Signal Processing Computing and Neural Networks, 1992. Indonesian Min. Edn. Sci. scholar, 1973-77, Japanese Min. Edn. Monbusho scholar, 1980-85. Mem. IEEE, Soc. Photo-Optical Instrumention Engrs., Optical Soc. Am., Japan Applied Physics Soc. Roman Catholic. Achievements include patents in Spatial Light Modulator and Photonic Device using Electron Trapping Materials, Pseudocolor Electronic System and findings of hybrid computer optical correlator, optical computing algorithm and architecture, stereoscopic display. Home: 38730 Lexington St #274 Fremont CA 94536 Office: Kowa Co Ltd 100 Homeland Ct Ste 302 San Jose CA 95112

JUTRAS, LARRY MARK, engineer; b. Manchester, N.H., Dec. 2, 1965; s. Roger Arther and Denice (Runser) J. BS in Marine Engring., Marine Maritime Acad., 1988. Project engr. Advanced Marine Enterprises, Arlington, Va., 1992—. Lt. USN, 1988-92. Recipient Award of Merit Gen. Dynamics, 1988. Mem. Assn. of Energy Engrs. Office: Advanced Marine Enterprises Ste 1300 1725 Jefferson Davis Hwy Arlington VA 22202

JÜTZ, JAKOB JOHANN, applied optics engineering educator; b. Zug, Switzerland, June 29, 1942; s. Franz Josef and Frieda Maria (Schneider) J.; m. Rita Rosa Maria Weibel, Aug. 2, 1968; children: Martin, Kathrin. MSc in Physics, Swiss Fed. Inst. Tech., Zürich, Switzerland, 1967, PhD, 1970. Rsch. asst. Swiss Fed. Inst. Tech., Zürich, 1970-72; lectr. Coll. Engring. Neu-Technikum Buchs, Buchs, Switzerland, 1972-82; vis. scientist IBM Rsch. Lab., San Jose, Calif., 1983; prof. Coll. Engring. Neu-Technikum Buchs, 1984—; head postdiploma studies in applied optics and optical systems Coll. Engring., Neu-Technikum Buchs, 1989—; head of meeting Precision Engrin. 91, Neu-Technikum Buchs, 1991. Contbr. articles to profl. jours. Pres. Mixed Choir, Buchs, 1975-82; v.p. Orch. Liechtenstein Werdenberg, Vaduz, 1981-82. Mem. Swiss Soc. for Optics and for Electron Microscopy, Swiss Soc. for Precision Engring. (bd. dirs. 1987—, editor procs.), German Soc. for Applied Optics, Optical Soc. Am., Soc. Photo-Optical Instrumentation Engrs. Roman Catholic. Avocations: choir singing, violin, hiking, languages. Home: Steinbergweg 10, CH 9472 Grabs Switzerland Office: Neu Technikum Buchs, Technikumstrasse, CH 9470 Buchs Switzerland

KABACIK, PAWEL, research electrical engineer; b. Wroclaw, Poland, Jan. 1, 1963; s. Tadeusz and Maria (Kozdeba) K. MSEE with distinction, Tech. U. Wroclaw, 1986, postgrad., 1989—. Jr. design engr. Tech. U. Wroclaw, 1987, jr. rsch. asst., 1987-88, rsch. asst., 1988—; vis. fellow Tech. U. Denmark, Lyngby, 1991-92. Co-author: Microstrip Antennas, 1992; contbr. articles to profl. publs. Pres. civic com. Solidarity, Wroclaw, 1990-91. Recipient Inst. Dir.'s award for rsch. Tech. U. Wroclaw, 1989, award for the Young Scientist 7th Nat. URSI Symposium, Gdansk, Poland, 1993; TEMPUS Program grantee European Community, 1991. Mem. IEEE, Planetary Soc. Avocations: economics, management, architecture, touring, sailing. Home: Kilinskiego 32/6, 50-264 Wroclaw Poland Office: Tech U Wroclaw Inst Telecommunications Acoustics, Wybrzeze Wyspianskiego 27, 50-370 Wroclaw Poland

KABAT, ELVIN ABRAHAM, immunochemist, biochemist, educator; b. N.Y.C., Sept. 1, 1914; s. Harris and Doreen (Otis) K.; m. Sally Lennick, Nov. 28, 1942; children: Jonathan, Geoffrey, David. B.S., CCNY, 1932; M.A., Columbia U., 1934, Ph.D., 1937; LL.D. (hon.), U. Glasgow, 1976; Doctoral degree (hon.), U. Orleans (France); Ph.D. (hon.), Weizmann Inst. Sci., Rehovot, Israel; DSc honoris causa, Columbia U., 1987. Lab. asst. immunochemistry Presbyn. Hosp., 1933-37; Rockefeller Found. fellow Inst. Phys. Chemistry, Upsala, Sweden, 1937-38; instr. pathology Cornell U., 1938-41; mem. faculty Columbia U., N.Y.C., 1941—; asst. prof. bacteriology Columbia U., 1946-48, assoc. prof., 1948-52, prof. microbiology, 1952-85, prof. human genetics and devel., 1969-85, Higgins prof. microbiology, 1984-85, Higgins prof. emeritus microbiology, 1985—; mem. adv. panel on immunology WHO, 1965—; lectr. 25th Michael Heidelberger Lecture, Coll. Physicians and Surgeons, Columbia U., 1986; lectr. The Louis Weinstein lecture, Tufts U; 10th anniversary lectr. Metchnikoff Immunology Bldg. Inst. Pasteur, Paris; expert cons. Nat. Cancer Inst., 1975-82, Nat. Inst. Allergy and Infectious Disease, 1983-88, NIH, Office of Dir., 1989-92; Alexander S. Wiener lectr. N.Y. Blood Center, 1979. Author: (with M.M. Mayer) Experimental Immunochemistry, 1948, 2d edition, 1961, Blood Group Substances, Their Chemistry and Immunochemistry, 1956, Structural Concepts in Immunology and Immunochemistry, 1968, 2d edit., 1976, (with T.T. Wu and H. Bilofsky) Variable Regions of Immunoglobulin Chains, 1976, Sequences of Immunoglobulin Chains, (with others) Sequences of Proteins of Immunological Interest, 1983, 4th edit., 1987, 5th edit., 1991 (with T.T. Wu, M. Reid-Miller, H.M. Perry and K.S. Gottesman).; mem. editorial bd.: Jour. Immunology, 1961-76, Transplantation Bull, 1957-60. Recipient numerous awards including: Ann. Research award City of Hope, 1974, award Center for Immunology, State U. N.Y., Buffalo, 1976, Louisa Gross Horwitz award Columbia U., 1977, R.E. Dyer lectr. award NIH, 1979, Townsend Harris medal CCNY, 1980, Philip Levine award Am. Soc. Clin. Pathology, 1982, award for excellence Grad. Faculties Alumni Columbia U., 1982, Disting. Svc. award Columbia U. Coll. Physicians and Surgeons, 1988, Dickson Prize for Medicine U. Pitts, 1986, Academy medal, N.Y. Acad. Medicine, 1989, Nat. Medal of Sci. 1991; named Pierre Grabar Lectr. Societe Francaise d'Immunologie and German Soc. of Immunology; Fogarty scholar NIH, 1974-75. Fellow AAAS, Am. Acad. Allergy (hon.); mem. NAS, Am. Acad. Arts and Scis., Am. Assn. Immunologists (past pres.), Am. Soc. Biol. Chemists, Am. Chem. Soc., Harvey Soc. (pres. 1976-77), Am. Soc. Microbiology, Internat. Assn. Allergists, Soc. Française d'Allergie (hon.), Biochem. Soc. (Eng.), Assn. for Research in Nervous and Mental Diseases, AAUP, Assn. de Microbiologists de Langue Francaise, Société de Biologie, Société de Immunologie (hon.), Japanese Electrophoresis Soc. (hon.), Phi Beta Kappa, Sigma Xi. Home: 70 Haven Ave New York NY 10032-2600 Office: Columbia U Coll Physicians and Surgeons Dept Microbiology 701 W 168th St New York NY 10032-2704

KACHEL, WAYNE M., environmental engineer; b. Lorain, Ohio, Nov. 21, 1946; s. Raymond George and Mary (Semyczyk) K.; m. Sandi West, July 12, 1969; 1 child, Jamie Rae. BS in Math., Waynesburg (Pa.) Coll., 1968; MS in Envrion. Systems Engring., Clemson (S.C.) U., 1971, PhD in Environ. Systems Engring., 1978. Sr. staff engr. Exxon Rsch. and Engring., Florham Park, N.J., 1978-86; environ. advisor Exxon Co. U.S.A., Benicia, Calif., 1986-90; mem. EPA Sci. Adv. Bd., Washington, 1988—; sr. program mgr. Pilko & Assocs., Houston, 1990—; chmn. hazardous waste/groundwater symposium Water Environ. Fedn. Contbr. articles to profl. jours. Capt. U.S. Army, 1971-74. Mem. Am. Inst. of Chem. Engrs., Internat. Assn. of Water Pollution Rsch., Sigma Xi, Chi Epsilon. Achievements include patent on fluid coking with quench elutriation using indsl. sludge. Office: Pilko & Assocs 2707 N Loop W Ste 960 Houston TX 77008

KACZANOWSKI, CARL HENRY, podiatrist, educator; b. Buffalo, Dec. 29, 1948; s. Henry and Mary Theresa (Slowik) K.; A.A., Niagara County Community Coll., 1969; B.S., Ill. Coll. Podiatric Medicine, 1977, D.P.M., 1977; m. Connie J. Padak, Aug. 28, 1971 (div. Apr. 1991). Co-dir. Vt. Foot Clinic, 1977-84, dir., 1984—; sci. cons. dept. surgery Med. Center Hosp. Vt., Burlington, 1978—; chief podiatrist surg. staff Central Vt. Med. Center, Berlin, 1978—; sports medicine cons. Middlebury Coll., 1978—; pvt. practice podiatric medicine and surgery, Burlington, Vt.,1977—; adj. clin. instr. William Scholl Coll. Podiatric Medicine, 1979—; referee students sem. Calif. Coll. Podiatric Med. Diplomate Nat. Bd. Podiatry Examiners (cons. and evaluator). Fellow Fedn. State Med. Bds. (hon.), Fedn. Podiatric Med. Bds. (sec. bd. dirs. 1986); mem. Am. Public Health Assn., Am. Soc. Microbiology, Am. Coll. Sports Medicine, Am. Soc. Podiatric Legal Medicine (founder, charter mem. 1988—), N.Y. Acad. Scis., Pi Omega Delta. Roman Catholic.

KACZYNSKI, DON, metallurgical engineer; b. Fremont, Mich., Apr. 16, 1948; s. Harold Steven and Helena Mary (Fitzner) K.; m. Sandra Lee Braunworth, Apr. 4, 1985. BS in Chemsitry, Mich. Tech. U., 1971, MS in Chem. Metallurgy, 1973; PhD in Metallurgical Engring., Colo. Sch. Mines, 1978. Metallurgist Hanna Mining Co., Nashwauk, Minn., 1977-80, rsch. metallurgist, 1980-81, sr. rsch. metallurgist, 1981-84; supr. R&D Brush Wellman Engineered Materials, Elmore, Ohio, 1984-90, dir. tech., Beryllium mining, 1990—. Contbr. articles to profl. jours. Mem. ASM Internat., Am. Inst. Mining, Electrochemical Soc. Achievements include patents for controlling the morphology of beryllium oxide powders, removing chlorine from copper ores, froth flotation of borate minerals and concentration of iron ores. Home: 6334 N 6th St Oak Harbor OH 43449 Office: Brush Wellman Inc S River Rd Elmore OH 43416

KADANOFF, LEO PHILIP, physicist; b. N.Y.C., Jan. 14, 1937; s. Abraham and Celia (Kibrick) K.; children: Marcia, Felice, Betsy. AB, Harvard U., 1957, MA, 1958, PhD, 1960. Fellow Neils Bohr Inst., Copenhagen, 1960-61; from asst. prof. to prof. physics U. Ill., Urbana, 1961-69; prof. physics and engring., univ. prof. Brown U., Providence, 1969-78; prof. physics U. Chgo., 1978-82, John D. MacArthur Disting. Service prof., 1982—; Mem. tech. com. R.I. Planning Program, 1972-78, mem. human svcs. rev. com., 1977-78; pres. Urban Obs. R.I., 1972-78. Author: Electricity Magnetism and Heat, 1967; co-author: Quantum Statistical Mechanics, 1963; Adv. bd.: Sci. Year, 1975-79; editorial bd.: Statis. Physics, 1972-79, Nuclear Physics, 1980—, Annals of Physics, 1982—; contbr. articles to profl. jours. NSF fellow, 1957-61; Sloan Found. fellow, 1963-67; recipient Wolf Found. prize, 1980, Boltzmann medal Internat. Union Pure and Applied Physics, 1990. Fellow Am. Phys. Soc. (Buckley prize 1977), Am. Acad. Arts and Scis.; mem. Nat. Acad. Scis. Home: 5421 S Cornell Ave Chicago IL 60615-5608 Office: U Chgo James Franck Inst 5801 S Ellis Ave Chicago IL 60637

KADDU, JOHN BAPTIST, parasitologist, consultant; b. Kampala, Uganda, Dec. 26, 1945; s. Polinali Mukasa and Agness (Mwasiti) K.; m. Margaret Dama, 1971; children: Caroline, Ronnie, Brian. BS, U. East Africa, Nairobi, Kenya, 1970; MS, U. Liverpool, Eng., 1972; PhD, U. Nairobi, 1978. Head protozoology dept. East African Trypanosomiasis Rsch. Orgn., Tororo, Uganda, 1970-77; in charge WHO Trypanosome Cryobank, Tororo, Uganda, 1973-77; head protozoology dept. Kenya Trypanosomiasis Rsch. Inst., Kikuyu, Kenya, 1977-79; in charge WHO Trypanosome Cryobank, Kikuyu, Kenya, 1977-79; post doctoral rsch. fellow Internat. Ctr. Insect Physiology and Ecology, Nairobi, 1979-80, rsch. scientist, deputy program leader, 1980-90; sr. lectr. Makerere U., Kampala, Uganda, 1991—; founder, dir. JBK Dairy Farm, Kampala, 1991—; cons. WHO, Indola, Zambia, 1982; asst. sec. gen. 7th Internat. Congress of Protozoology, Nairobi, 1985; founder, dir. Cats Ltd., Kampala, 1991; vis. scientist WHO, 1985. Author: Glossary of Primary Science; pioneering editor East African Soc. of Parasitologists newsletter, 1987; draft designer commemorative postage stamps of Internat. Congress of Protozoology, 1985; author over 40 rsch. papers in the areas of Leishmaniasis, Trypanosomiasis, Chemotherapy, Malaria, Cryobiology. Recipient 2d prize Luganda Lang. Competition, 1960. Mem. East African Soc. Parasitologists (founder), N.Y. Acad. Scis. (cert. 1987), Internat. Ctr. Insect Physiology and Ecology Alumni Assn. (founding mem., cert. 1989), Kageye Farmers Assn. (dir. planning 1992—). Roman Catholic. Avocations: lawn tennis, horse riding, computer software application, wild life. Home: PO Box 3267, Kampala Uganda

KADIN, ALAN MITCHELL, physicist; b. Bklyn., Dec. 7, 1952; s. Harold and Elinor M. (Mendelsohn) K. AB in Physics, Princeton U., 1974; PhD, Harvard U., 1979. Rsch. assoc. SUNY, Stony Brook, 1979-81, U. Minn., Mpls., 1981-83; rsch. physicist Energy Conversion Devices, Inc., Troy, Mich., 1983-87; assoc. prof. U. Rochester, N.Y., 1987—; mem. tech. adv. bd. CVC Products, Inc., Rochester, 1991—. Contbr. more than 60 articles to profl. jours. Grantee NSF, 1989—. Mem. Am. Phys. Soc., IEEE, Am. Vacuum Soc. Achievements include research in low-temperature and high-temperature superconducting devices, thin film deposition, and magnetic materials. Office: U Rochester Dept Electrical Engring Rochester NY 14627

KADISON, RICHARD VINCENT, mathematician, educator; b. N.Y.C., July 25, 1925; married, 1956; 1 child. MS, U. Chgo., 1947, PhD, 1950; hon. doctorate, U. d'Aix-Marseille, 1986, U. Copenhagen, 1987. NRC fellow math. Inst. Advanced Study, 1950-52; from asst. prof. to prof. Columbia U., 1952-64; Kuemmerle prof. math. U. Pa., 1964—. Fulbright research grantee, Denmark, 1954-55; Sloan fellow, 1958-62; Guggenheim fellow, 1969-70. Mem. Am. Math. Soc., Royal Danish Acad. Sci. and Letters (fgn. mem.), Norwegian Acad. Sci. and Letters (fgn. mem.), Sigma Xi. Office: U Pa Dept Math Philadelphia PA 19104-6395

KAELIN, BARNEY JAMES, technological artist; b. L.A., Mar. 25, 1951; s. Al and Marion (Brohman) K.; m. Phyllis Jean Burroughs, Sept. 21, 1981. BA, Loyola U., L.A., 1972. Pres. Merlin Laser Visuals, Beverly Hills, Calif., 1977-80, Laser Magic Prodns., Playa Del Rey, Calif., 1980-92; founder Visual Music Alliance, Hollywood, Calif., 1986-88. Producer video prodn. Laser Viewsic, 1983. Achievements include patent on device for 3-dimensional imaging of laser and/or collimated light, development of methods for imaging laser light in synchronization to music. Office: Laser Magic Prodns 401 Campdell St Playa Del Rey CA 90293

KAERNBACH, CHRISTIAN, psychophysicist; b. Bonn, Germany, Oct. 4, 1960; s. Johannes and Maria (Kondziella) K. Diploma in physics, U. Bonn., 1985, MD, 1988. Rschr. Nat. Inst. Health Rsch., Bordeaux, France, 1988-89; rsch. grantee Nat. Sci. Found. Germany, Bordeaux, 1989-91; rschr. Ruhr-U. Bochum, Germany, 1991—. Contbr. articles to profl. jours. Mem. German Phys. Assn., German Assn. Acoustics, German Assn. Phys. Medicine, Acoutical Soc. Am. Achievements include improvements in adaptive psychophysical methods, psychophysics on auditory grouping and periodicity detection. Office: Ruhr Univ Bochum, Inst Neuroingormabik, 44780 Bochum Germany

KAFKA, TOMAS, physicist; b. Praha, Czechoslovakia, Oct. 15, 1936; came to U.S. 1968; s. Vaclav and Marta K. Promovany Fysik, U. Karlova, Czechoslovakia, 1960; PhD, SUNY, Stony Brook, 1974. Rsch. assoc. SUNY, Stony Brook, 1974-82; rsch. prof. physics Tufts U., Medford, Mass., 1982—. Contbr. articles to profl. jours. Mem. Am. Phys. Soc. Democrat. Achievements include research in experimental high energy particle physics (hadron-hadron interactions, neutrino nucleon interactions, proton decay, cosmic rays). Home: 4 Upland Rd Lexington MA 02173 Office: Tufts Univ High Energy Physics 4 Colby St Medford MA 02155

KAGAN, MARVIN BERNARD, architect; b. N.Y.C., May 17, 1944; s. Samuel L. and Fannie (Cohen) K.; m. Lynne Wolfsont, Sept. 28, 1964; children: Arielle F., David M. BS in Sci., CCNY, 1968, BArch, 1968. Architect U.S. Coast Guard 3rd dist., N.Y.C., 1968-74; sr. architect Amtrak, Washington, 1974-81; prin. Marvin Kagan Architects & Assocs., Reston, Va., 1981-88, Kagan-Sims Architects, P.C., Herndon, Va., 1988—. Bd. dirs. Inlet Ct. Homeowners Assn., Reston, 1982-85 (pres. 1983-84), Golf Course Dr. Homeowners Assn., Reston, 1976. Mem. Assn. Energy Engrs. (pres. D.C. chpt. 1973, sec.-treas. 1972), Am. Inst. Architects (No. Va. chpt.). Office: Kagan Sims Architects PC 209 Elden St Herndon VA 22070

KAGAN, SIOMA, economics educator; b. Riga, Russia, Sept. 29, 1907; came to U.S., 1941, naturalized, 1950; s. Jacques and Berta (Kaplan) K.; m. Jean Batt, Apr. 5, 1947 (div. 1969). Diplom Ingenieur, Technische Hochschule, Berlin, 1931; M.A., Am. U., 1949; Ph.D. in Econs, Columbia U., 1954. Sci. asst. Heinrich Hertz Inst., Berlin, 1931-33; partner Laboratoire Electro-Acoustique, Neuilly-sur-Seine, France, 1933-48; chief French Mission Telecommunications, French Supply Council in N.Am., Washington, 1943-45; mem. telecommunications bd. UN, 1946-47, econ. affairs officer, 1947-48; econs. cons. to govt. and industry; asso. prof. econs. Washington U., St. Louis, 1956-59; staff economist Joint Council Econ. Edn., N.Y.C., 1959-60; prof. internat. bus. U. Oreg., Eugene, 1960- 67; prof. internat. bus. U. Mo., St. Louis, 1967-87, prof. emeritus, 1987—; faculty leader exec. devel. programs Columbia, Northwestern U., NATO Def. Coll., Rome, others. Contbr. numerous articles profl. publs. Served with Free French Army, 1941-43. Decorated Legion of Honor (France). Recipient Thomas Jefferson award U. Mo., 1984. Fellow Latin Am. Studies Assn.; mem. Am. Econ. Assn., Acad. Polit. Sci., Assn. Asian Studies. Clubs: University (St. Louis); Conanicut Yacht (Jamestown, R.I.). Home: 8132 Roxburgh Dr Saint Louis MO 63105-2436 Office: U Mo Saint Louis MO 63121

KAGEYAMA, YOSHIRO, retired mechanical engineering educator; b. Hamamatsu, Japan, Feb. 2, 1922; s. Suezoh and Ritsu (Ikeda) K.; m. Chiyo Ogino, Jan. 8, 1953; children: Yuri, Jun. MS, U. Md., 1961; PhD, U. Nagoya, Japan, 1962. Chief engr. aircraft engine and rocket propulsion divsn. Ishikawajima-Harima Heavy Industries Co., Tokyo, 1948-79; prof. Yamaguchi U., Ube, Yamagushi, Japan, 1979-87, U. East Asia, Shimonoseki, Yamagushi, Japan, 1987-93; ret., 1993; resident rsch. assoc. astronautics labs. George C. Marshall Space Flight Ctr., NASA, Huntsville, Ala., 1968-69. Author tech. papers. Mem. ASME, AIAA, Japan Soc. Aeronautics & Astronautics, Japan Soc. Mech. Engrs. Achievements include development of improvement of axial gas turbines. Home: 1-7-5 Higashiobayama-Cho, Ube Yamaguchi-ken 755, Japan

KAGHAZCHI, TAHEREH, chemical engineering educator; b. Tehran, Iran, Jan. 29, 1947; d. Mehdi and Fakhereh (Khoyloo) K.; m. Morteza Sohrabi, Oct. 25, 1968; children: Maryam, Manijeh. BSChemE, U. Tehran, 1968; PhD in Chem. Engring., U. Bradford, Eng., 1972. Asst. prof. Amirkabir U., Tehran, 1973-77, assoc. prof., 1977-83, prof. chem. engring., 1983—, head food engring. group, 1985—. Author: Transfer Operations, 1989, Mass Transfer, 1992; contbr. articles to profl. publs. Named Disting. Woman Author, Tehran Ministry of Culture and Higher Edn., 1992. Mem. Iranian Acad. Scis., Iranian Petroleum Soc., Am. Chem. Soc. Office: Amrikabir U Tech, Hafez, Tehran 15, Iran

KAHANA, DAVID EWAN, physicist; b. Montreal, Que., Can., Mar. 19, 1959; s. Sidney Henry and Anne Patricia (Stevenson) K. BSc, SUNY, Stony Brook, 1981, PhD, 1987; MSc, Princeton U., 1984. Forschungs mitarbeiter Univ. Regensburg, Fed. Republic Germany, 1987-89; postdoctoral fellow Continuous Election Beath Accelerator Facility, Newport News, Va., 1989-91, Ctr. for Nuclear Rsch.-Kent (Ohio) State U., 1991—. Mem. Am. Phys. Soc. Office: Physics Dept Ctr for Nuclear Rsch Kent OH 44240

KAHLON, JASBIR BRAR, viral epidemiologist, researcher; b. Deolali, India, Jan. 31, 1952; came to U.S., 1978; d. Z. S. and H. K. Brar; m. H. S. Kahlon, Oct. 2, 1977; children: Summer Paul, Navneet, Amandeep. MS, Punjabi U., Punjab, India, 1973; DrPH, U. Ala., Birmingham, 1983. Asst. prof. Khalsa Coll., Punjab; postdoctoral fellow in pediatrics Children's Hosp., Birmingham, 1983-86; rsch. assoc. Sch. Medicine U. Ala., Birmingham, 1986-87; rsch. virologist So. Rsch. Inst., Birmingham, 1987-90, viral epidemiologist, 1991—; cons. Carrington Labs., Dallas, 1987—, Minority Student Sci. Careers Support Program, NIH, Bethesda, Md., 1987—; vis. scientist Sch. Pub. Health U. Ala., Birmingham, 1987—, assoc. scientist Sch. Medicine, 1989—. contbr. articles to Jour. Virology, Jour. Clin. Virology, Jour. Infectious Diseases, Jour. Clin. Investigations, Molecular Pharmacology. Active Bapt. Women's Assn., Birmingham, 1986—. Recipient fellowship Social Health Scis., Venereal Disease sect., Research Triangle Park, N.C., 1984-86; named prin. investigator NIH, 1986-88, Bio-Rad Labs., Richmond, Calif., 1988-89. Mem. AAAS, Internat. Soc. AIDS, Am. Soc. Microbiology, Am. Soc. Virology. Achievements include development of drug for treatment of AIDS. Home: 3344 Dunbrooke Dr Birmingham AL 35243-4820

KAHN, ROBERT E., electrical engineer; b. Dec. 23, 1938. BEE, CCNY; MEE, Princeton U., PhD in Elec. Engring. Mem. tech. staff Bell Telephone Labs.; asst. prof. elec. engring. MIT, Cambridge; sr. scientist Bolt, Beranek & Newman; dir. info. processing techniques DARPA; founder, pres. Corp. Nat. Research Initiatives, Reston, Va., 1986—. Fellow IEEE; mem. Nat. Acad. Engring. Office: Corp for Nat Rsch Initiaves 1895 Preston White Dr Ste 100 Reston VA 22091-5434

KAIFER, ANGEL EMILIO, chemistry educator, researcher; b. Madrid, Sept. 19, 1955; came to U.S., 1979; s. Angel and Emilia (Brasero) K.; m. Carmen Fernandez, June 22, 1979; children: Cristina, Andres. BS in Chemistry, U. Autonoma, Madrid, 1977; PhD in Chemistry, U. P.R., 1984. Chemist R&D dept. Tudor S.A., Guadalajara, Spain, 1978-79; instr. U. P.R., Humacao, 1979-82; post-doctoral assoc. U. Tex., Austin, 1984-85; asst. prof. U. Miami, Fla., 1985-90, assoc. prof., 1990—. Contbr. articles to profl. jours. Recipient Cottrell Corp. Rsch. grant, 1986, Teaching grant NSF, 1990, Rsch. grant NSF, 1990—, grant NATO, 1993. Office: Univ Miami Chemistry Dept Coral Gables FL 33124

KAIFU, NORIO, astronomer; b. Niigata, Japan, Sept. 21, 1943; s. Yasuyoshi and Kiyoko K.; m. Shgemi, May 3, 1968; children: Yohsuke, Kohji, Kenzoh, Takeshi. Grad., U. Tokyo, 1962, PhD, 1972. Rsch. assoc. Dept. Astronomy, U. Tokyo, Japan, 1969-79; assoc. prof. Tokyo Astron. Obs., U. Tokyo, 1979-88; prof., dir. Nobeyama (Japan) Radio Obs., Nat. Astron. Obs., 1988-90; prof., dir. engring. div. Nat. Astron. Obs., Mitaka, Japan, 1990092; vice dir. Nat. Astronomy Obs., Mitaka, 1992—. Co-author: Cosmic Radio Astronomy, 1988, Molecular Processes in Space, 1990; author 10 books; contbr. more than 90 papers to profl. jours. Recipient Nishina Meml. award, 1987. Mem. Nat. Astron. Obs. (bd. dirs. 1988—), Astron. Soc. Japan (bd. dirs. 1980—). Achievements include construction of 45-mm-wave telescope at Nobeyama Radio Observatory; detection of many new interstellar molecules in dark clouds. Office: Nat Astron Observatory, Osawa 2-21-1, Mitaka Tokyo, Japan

KAISER, ARMIN DALE, biochemist, educator; b. Piqua, Ohio, Nov. 10, 1927; s. Armin Jacob and Elsa Catherine (Brunner) K.; m. Mary Eleanor Durrell, Aug. 9, 1953; children: Jennifer Lee, Christopher Alan. B.S., Purdue U., 1950; Ph.D., Calif. Inst. Tech., 1955. Postdoctoral research fellow Inst. Pasteur, Paris, 1954-56; asst. prof. microbiology Washington U., St. Louis, 1956-59; mem. faculty Stanford U., 1959—, prof. biochemistry. Served with AUS, 1945-47. Recipient molecular biology award U.S. Steel Corp., 1971; Lasker award in basic med. sci., 1980. Mem. Nat. Acad. Scis., Am. Acad. Arts and Scis., Am. Soc. Biochemists, Genetic Soc. Am. (Thomas Hunt Morgan medal 1991). Achievements include research on virus multiplication, microbial development. Office: Stanford Univ Biochemistry Dept Stanford CA 94305

KAISER, KURT BOYE, physicist; b. Mpls., Mar. 17, 1942; s. August and Marguerite (Boye) K.; m. Christin Sanders Waters, Jan. 17, 1970; children: Nicole, August. BA in Physics, Dartmouth Coll., 1964; MS in Physics, U. Idaho, 1967. Scientist EG&G Photo Diode, Bedford, Mass., 1967-70; chief engr. Infrared Ind. ENL, Menlo Pk., Calif., 1970-73; engr. mgr. EG&G Optometrics, Salem, Mass., 1973-83; chief scientist EG&G Electro-Optics, Salem, 1983-87; MIS dir. EG&G Salem Ops., Salem, 1984-87; bus. element mgr. EG&G Frequency Products, Salem, 1987—. Active Boxford (Mass.) Planning Bd., 1983-86, 92—, chmn., 1986, Conservation Commn., 1992—. Mem. IEEE. Republican. Episcopalian. Office: EG&G Frequency Products 35 Congress St Salem MA 01970-5567

KAISER, NICHOLAS, physicist, educator; b. Sept. 15, 1954. BSc in Physics, Leeds U., 1978; Pt III maths tripos, Cambridge U., 1981, PhD in Astronomy, 1982. Lindemann fellow U. Calif., Berkeley, 1983; postdoctoral fellow U. Calif., Santa Barbara, 1984, Berkeley, 1984; postdoctoral fellow U. Cambridge, 1985-86, SERC advanced fellowbye, 1986-88; SERC sr. visitor U. Sussex, 1985; assoc. prof. CITA, 1988-90, prof., 1990—. Recipient Helen Warner prize Am. Astron. Soc., 1989, Gerhard Herzberg medal Can. Assn. Physicists, 1993; grantee NSERC, 1988, 91, 93; Ontario fellow CIAR Cosmology Program, 1988—, Steacie fellow, 1991-92. Achievements include research in observational cosmology, galaxy formation, large-scale structure, bulk flows, gravitational lensing. Home: 53 Templeton Ct, Scarborough, ON Canada M1E 2C3 Office: Univ of Toronto, CITA, Toronto, ON Canada M5S 1A7*

KAITAAKE, ANTEREA, minister of education science and technology; b. Tokamauea, Abemama, Kiribati, Nov. 29, 1949; parents Kaitaake Utimawa and Teraennang Rotina; m. June 14, 1974; 5 children. Diploma in Religious Edn., Ataneo U. Querzon City, Manila, Phillipines, 1983. Tchr. Post Tng. Sch., Rep. of Kiribati, 1972-74; staff Govtl. Dept., Rep. of Kiribati, 1974-78; higher exec. officer Outer Island Coun., Rep. of Kiribati, 1979-81; mission tchr. Phillipines, 1982-83; catechist tchr. Rep. of Kiribati, 1983-88; tchr. Cath. Sec. Schs., Rep. of Kiribati, 1989-91; mem. parliament, min. edn. sci. & tech. Rep. Kiribati, 1991—. Mem. Nat. Progressive Party. Roman Catholic. Office: Ministry of Edn, PO Box 263 Bidenibeu, Tarawa Kiribati

KAJI, AKIRA, microbiology scientist, educator; b. Tokyo, Jan. 13, 1930; came to U.S., 1954; s. Kiichi and Chiyo (Hanai) K.; m. Hideko Katayama, July 22, 1955; children: Kenneth, Eugene, Naomi, Amy. BS, Tokyo U., 1953; PhD, Johns Hopkins U., 1958; MS (hon.), U. Pa., 1973. Rsch. fellow Johns Hopkins Hosp., Balt., 1958-59; guest investigator Rockefeller U., N.Y.C., 1959; rsch. assoc. microbiology Vanderbilt Med. Sch., Nashville, 1959-62; vis. scientist Oak Ridge (Tenn.) Nat. Lab., 1962-63; assoc. U. Pa. Med. Sch., Phila., 1963-64, asst. prof. microbiology, 1964-67, assoc. prof., 1967-72, prof., 1972—; permanent mem. bd. sci. councilors Nat. Eye Inst., Bethesda, Md., 1987-92; prof., chair Tokyo U. Faculty Pharm. Scis., 1972-73; vis. prof. Kyoto U. Virus Rsch. Inst., 1985. Contbr. over 180 articles to profl. jours. Recipient Fulbright-Smith-Mundt award, 1954, Helen Hay Whitney award, 1964-69, John Simmon Guggenheim award, 1972-73, Fogarty Internat. Sr. award, 1985-86. Mem. Am. Soc. Biol. Chemistry and Molecular Biology, Am. Soc. Cell Biology, Am. Soc. Microbiology, Am. Soc. Chemistry. Avocations: ice dancing, swimming. Office: U Pa Sch Medicine Dept Microbiology 258 Johnson Pavilion Philadelphia PA 19104-6076

KAKADE, ASHOK MADHAV, civil engineer, consultant; b. Bombay, India, Mar. 27, 1957; came to U.S., 1986; s. Madhav Y. and Puspa Kakade; m. Neha Ashok, May 2, 1986; 1 child, Roheet. BSCE, U. Bombay, India, 1979; MSCE, S.D. Sch. Mines & Tech., 1988. Civil engr. Engring. Constrn. Corp., Bombay, 1979-81; resident engr. McBauchemie Muller GmbH & Co., Bottnop, Germany, 1981-86; rsch. asst. S.D. Sch. Mines & Tech., Rapid City, 1986-88; sr. engr. Haynes & Assoc., Oakland, Calif., 1988—; pres. East Bay Structural Engrs. Soc., Oakland, 1992; examiner Concrete Technician cert. Program, Calif. Mem. ASCE, Am. Concrete Inst., Nat. Assn. Corrosion Engrs. Office: Haynes & Assoc 3803 Randolph Ave Oakland CA 94602

KAKATI, DINESH CHANDRA, physician; b. Gauhati, India, Feb. 1, 1941; arrived in Eng., 1969; s. Nara Kanta and Subhandra (Baishya) K.; m. Bhabani Medhi, Mar. 27, 1974; children: Rita, Rishi. MBBS with distinction, Gauhati Med. Sch., 1967; Diploma in Tropical Medicine and Hygiene, Liverpool (Eng.) U., 1970; Diploma in Thoracic Medicine, London U., 1984, Diploma in Cardiac Medicine, 1984; Diploma in Geriatric Medicine, Royal Coll. Physicians, 1985. Sr. house officer Sunderland (Eng.) Health Authority, 1969-70, med. registrar, 1970-72; med. registrar Addinbrook Hosp., Cambridge, Eng., 1972-74, London Hosp., 1974-77; assoc. specialist N.E. Thames region, Hornchurch, Eng., 1977-84, clin. asst., 1984—. Fellow Royal Soc. Health Eng.; mem. Brit. Med. Assn., Royal Coll. Gen. Practitioners (assoc.). Home: 39 Veny Crescent, Hornchurch, London RM12 6TJ, England Office: St Georges Hosp, Suttons Ln, London E6 S56, England

KAKU, MICHIO, theoretical nuclear physicist; b. San Jose, Calif., Jan. 24, 1947; s. Toshio and Hideko (Maruyama) K. BA, Harvard U., 1968; PhD, U. Calif.-Berkeley, 1972. Rsch. assoc. Princeton U., N.J., 1972-73; assoc. prof. CCNY and Grad. Ctr., 1973-83, prof., 1983—; vis. prof. NYU, 1988, Inst. for Advanced Studies at Princeton U., 1990. Author: Nuclear Power: Both Sides, 1983; Beyond Einstein, the Cosmic Search for the Unified Field Theory, 1986, Introduction to Superstrings, 1988, Strings, Conformal Fields, and Topology, 1991, Quarks, Symmetries, and Strings, 1991, Quantum Field Theory: A Modern Introduction, 1993; contbr. 60 articles to profl. jours. Fellow AAAS, Am. Phys. Soc. Avocations: nuclear arms control, nuclear power. Office: CCNY Physics Dept 138th St at Convent Ave New York NY 10031

KALAIDJIAN, BERJ BOGHOS, civil engineer; b. Jerusalem, Mar. 7, 1936; s. Boghos Hovhanes and Shoghagat Kevork (Sahakian) K.; B.C.E., Am. U., Beirut, Lebanon, 1958; m. Sonia Kouyoumdjian, Aug. 19, 1963; children—Shahe, Vatche. Site engr. Consol. Contractors Co., Beirut, 1958-62, project mgr., 1962-64, asst. area mgr., jr. partner, 1964-69; mng. dir., partner Acmecon, Jeddah Saudi Arabia, 1969—, also dir. Club: Armenian Benevolent Union. Home: 35 Blvd du Larvotto, MC 98000 Monaco Monaco

KALAMOTOUSAKIS, GEORGE JOHN, economist; b. Chios, Greece, July 26, 1936; came to U.S., 1953; s. John S. and Marika (Nikolaides) K.; 1 child, Yannis. B.A., CUNY, 1956, M.A., 1958; Ph.D., NYU, 1966. Instr. Fairleigh Dickinson, U., Teaneck, N.J., 1958-59; asst. prof. Ithaca (N.Y.) Coll., 1959-62; chief economist Brown Engr., N.Y.C., 1963-64; instr. Washington Sq. Coll., NYU, 1963-65; econ. cons. N.Y. State Office Regional Devel., Albany, 1964-66; adv. economist IBM, Armonk, N.Y., 1969-73; internat. economist Am. Standard, Inc., N.Y.C., 1973-76; prof. finance Grad. Sch. Bus., NYU, 1971-77; external dir. Rank-Xerox, Hellas, Greece, Atlantic Union Ins. Co., Athens, Greece; vis. prof. U. Md. European div. USAF, 1960, 67-68; head dept. pub. finance Center of Planning and Econ. Research, Athens, Greece; dir. econ. research Bank of Greece, 1977-79; chief exec. officer, vice-chmn. bd. Bank of Crete, Athens, 1979-84; exec. dir., country head, gen. mgr. Greece, head. Middle Ea. Region, Am. Express Bank Ltd., N.Y.C., 1985—; bd. dirs. Egyptian Am. Bank, Cairo. Contbr. articles to profl. jours.; Author books on internat. fin., Cyprus and self determination, common market and econ. devel. Greece. Bd. dirs., trustee Hellenic Theatre Found., bd. dirs. Aegian U., Greece. Am. Ford Found. Faculty Research fellow, 1962. Mem. Am. Econ. Assn., AAUP (v.p. chpt. 1961), Omicron Delta Epsilon. Home: 124 Lakeview Ave Lynbrook NY 11563-1755 Office: Am Express Co American Express Tower World Fin Ctr New York NY 10285

KALATHAS, JOHN (IOANNIS), airline pilot, physicist, oceanographer; b. Athens, Greece, July 24, 1946; s. Panayotis J. and Calypso (George Violaki) K.; m. Catherine Kalathas, July 2, 1973 (div.); 1 child, Hector. BS in Physics, U. Athens, 1968; Diplome D'Etudes Approfondies in Phys. Oceanography, U. Paris, 1969, Diploma De Docteur En Geophysique, 1970. Lic. profl. pilot Oxford, Eng. Air Tng. Sch., 1974. Researcher Inst. Oceanography, Athens, 1971, 74-75, Environ. Pollution Control Project, Athens, 1974-75; tng. capt., dep. chief pilot, groundschool and simulator instr. Olympic Airways, Athens, 1975—; instr. scuba diving Hellenic Fedn. Underwater Activities, 1968—, dep. chief instr., 1975-77. Served with Hellenic Navy, 1971-73. Mem. Hellenic Speleological Soc. Club: Aero Athens. Home: 2 Petrou Dimaki St, GR-106 72 Athens Greece Office: Athens Airport Hellenikon, Athens Greece

KALAVAPUDI, MURALI, civil engineer; b. Visakhapatam, India, June 27, 1955; came to the U.S., 1981; s. Krishnamacharyulu and Vijayalakshmi (Kandala) K.; m. Manjusha Kalavapudi, Feb. 17, 1979; children: Jayant, Abhijit. BCE, Andhra U., 1979; MCE, U. Md., 1983. Registered environ. mgr. Civil, environ. engr. AEPCO, Inc., Rockville, Md., 1983-88; sr. project engr. Roy F. Weston, Inc., West Chester, Pa., 1988-90; project mgr. Foster Wheeler, Washington, 1990-92; sr. environ. engr., mgr. Energetics, Columbia, Md., 1992—; mem. environ. regulations Coun. for Environ. Quality, Washington, 1990—; advisor Nat. registry for Environ. Profls., Bethesda, Md., 1991—. Author: Hazardous Waste Treatment, 1991; editor: Site Assessment, 1991, Air Toxics in Wastewater Treatment, 1992, Pollution Prevention Handbook, 1993. Mem. Water Environ. Fedn. (co-chair hazardous waste 1988—), NSPE, HMCRI. Hindu. Achievements include research in biophosphorus removal using specific bacterial cultures in Sequencing Batch Reactor, specific bacterial strains applied to innovative wastewater treatment process to reduce soluble heavy metals in industrial wastewater, and hazardous waste management. Home: 8723 Bell Tower Dr Gaithersburg MD 20879-1783 Office: Energetics Inc 7164 Columbia Gateway Dr Columbia MD 21046

KALBFLEISCH, JOHN DAVID, statistics educator, dean; b. Grand Valley, Ont., Can., July 16, 1943. BSc in Math. and Physics, U. Waterloo, 1966, MMath in Stats., 1967, PhD in Stats., 1969. Lectr. dept. stats. U. Waterloo, 1967-68; rsch. assoc. dept. stats. Univ. Coll., London, 1969-70; asst. prof. dept. stats. SUNY, Buffalo, 1970-73; statis. cons. Radiotherapy Oncology Group, 1970-75, Ea. Coop. Oncology Group, 1970-73; assoc. prof. dept. stats. U. Waterloo, 1973-79, prof. dept. stats. and actuarial sci., 1979—, chmn. dept. stats. and actuarial sci., 1984-90, dean faculty of math., 1990—; vis. scientist Institut Jules Bordet, Brussels, 1972, Fred Hutchinson Cancer Rsch. Ctr., Seattle, 1979-80; vis. prof. dept. biostats. U. Wash., 1979-80, dept. stats. N.C. State U., Raleigh, 1982, Ctr. for Stats. U. Lancaster, 1984, dept. biostats. U. Mich., 1987, dept. epidemiology U. Calif., San Francisco, 1988; chmn. stats. scis. grant selection com. Natural Scis. and Engring. Rsch. Coun. Can., 1982-83, mem., 1979-82; acad. advisor statis. methods Soc. Actuaries, 1980-84; chmn. Joint UW/GM Task Force quality of improvement, 1984; referee Annals of Stats., Biometrika, Jour. Roy. Statis. Soc., Comm. in Stats. Am. Jour. Epidemiology, ISI Review, Utilitas Mathematica, others. Author: (with R.L. Prentice) The Statistical Analysis of Failure Time Data, 1980; author: (with others) Encyclopedia of Statistical Sciences, 1985; sr. assoc. editor Can. Jour. Stats., 1981; assoc. editor Can. Jour. Stats., 1984-89, Annals of Stats., 1980-83; contbr. articles to profl. jours. Fellow Am. Statis. Assn. (referee jour.), Inst. Math. Stats.; mem. Statis. Soc. Can. (program coord. meetings 1981), Internat. Statis. Soc., Royal Statis. Soc., Biometrics Soc. Office: University of Waterloo, Faculty of Mathematics, Waterloo, ON Canada N2L 3G1

KALBFLEISCH, JOHN MCDOWELL, cardiologist, educator; b. Lawton, Okla., Nov. 15, 1930; s. George and Etta Lillian (McDowell) K.; m. Jolie Harper, Dec. 30, 1961. AS, Cameron A&M U., Lawton, 1950; BS, U. Okla., 1952, M.D., 1957. Diplomate Am. Bd. Internal Medicine, subsplty bd. cardiovascular disease. Intern U. Va. Hosp., 1957-58; resident and fellow U. Okla. Med. Ctr., 1958-62, instr. medicine, 1964-66, asst. prof., 1966-69, assoc. clin. prof., 1970-78, clin. prof. Tulsa br., 1978—; pvt. practice specializing in cardiology Tulsa, 1969—; pres., chief exec. officer Cardiology of Tulsa, Inc., 1969—; dir. cardiovascular svcs. St. Francis Hosp., Tulsa, 1975—; mem. physician adv. bd. City of Tulsa, 1978-81; bd. dirs. St. Francis Hosp., mem. exec. com., 1987—; treas. Tulsa Med. Edn. Found., 1988-89, v.p., 1990-92, pres., 1992—; med. dir., chmn. bd. Warren Clinics, 1990—. Contbr. articles to profl. jours. Served with USPHS, 1962-64. Fellow ACP (gov.-elect Okla. 1990-91, gov. 1991—), Am. Coll. Cardiology (gov. Okla. 1978-81); mem. AMA, AAAS, Tulsa County Med. Soc., Okla. State Med. Assn., Am. Heart Assn. (teaching scholar 1967-69), Okla. Soc. Internal Medicine (v.p., pres.-elect 1983-84, pres. 1985-86), Am. Soc. Internal Medicine, Am. Fedn. Clin. Rsch., Am. Inst. Nutrition, Delta Upsilon. Republican. Presbyterian. Office: 6585 S Yale Ave Ste 800 Tulsa OK 74136-8374

KALDOR, GEORGE, pathologist, educator; b. Budapest, Feb. 10, 1926; came to the U.S., 1956; s. Julius and Jolanda (Neumayer) K.; m. Gladys Maria Trujillo-Rojas, Apr. 17, 1963; children: Steven Henry, Nickolas John, Eric George. MD, Semmelweiss U., Budapest, 1950. Asst. prof. clin. pathology Semmelweiss Med. U., 1953-56; resident fellow in biochemistry Mass. Gen. Hosp., Boston, 1957-59; resident fellow in oncol. biochemistry McArdle Meml. Lab. U. Wis., Madison, 1959-60; head phys. biochemistry Isaac Albert Rsch. Inst., Bklyn., 1960-65; assoc. prof. physiology Med. Coll. Pa., Phila., 1965-68, prof. physiology and pathology, 1968-75; prof. pathology Wayne State U., Detroit, 1975—; staff physician VA Med. Ctr., Allen Park, Mich., 1991—, Detroit Hosp., 1985—. Author: Physiological Chemistry of Proteins and Nucleic Acids, 1969, Clinical Enzymology, 1983, Aging in Muscle, 1978; contbr. articles to profl jours. Lt. col. USMC, 1984—. Recipient Career Devel. award NIH, 1967-71; grantee Am. Heart Assn., 1967-80. Fellow Am. Soc. Clin. Pathology, Am. Physiol. Soc., Am. Soc. Exptl. Pathology, Am. Biochem. Soc. Republican. Jewish. Achievements include research in the molecular thermodynamic properties of V1 and V3 isomyosins and the application of full analytical and interpretive automation in lipid metabolism. Home: 2451 Beachview Ln West Bloomfield MI 48324 Office: VA Med Ctr Allen Park MI 48101

KALDOR, UZI, chemistry educator; b. Tel Aviv, Jan. 16, 1939. MSc, Hebrew Univ., 1962; DSc, Technion U., 1966. Rsch. assoc. Stanford (Calif.) U., 1966-68; prof. chemistry Tel Aviv U., Israel, 1968—, chmn. chemistry dept., 1976-80; vis. prof. U. So. Calif., L.A., 1975-76; vis. researcher Johns Hopkins U., Balt., 1987-88, U. Fla., Gainesville, 1990-91; cons. U.S.-Israel Binat. Sci. Found., Jerusalem, 1983—; mem. adv. editorial bd. Internat. Jour. Quantum Chemistry, 1991—; head exact scis. Israeli Sci. Found., 1991—. Editor: Many Body Methods in Quantum Chemistry, 1989; contbr. 100 articles to internat. profl. jours. Mem. Am. Phys. Soc., Israeli Chem. Soc. Office: Tel Aviv U, Sch Chemistry, 69978 Tel Aviv Israel

KALEJTA, PAUL EDWARD, chemist; b. Pottstown, Pa., Jan. 27, 1961; s. Thomas Francis and Joan Frances (Deteske) K.; m. Cheryl Marie Yergey, May 21, 1983; children: Jonathan Paul, Stephanie Marie. BS, Pa. State U., 1982. Cert. lab. chemist. Engring. specialist Anchor Hocking Corp., Lancaster, Ohio, 1982-85; chemist Sharpoint, Reading, Pa., 1985-88, Fermtec, Inc., West Chester, Pa., 1988-91, Merck & Co., Inc., West Point, Pa., 1991; staff pharmacist Merck Rsch. Labs., West Point, 1991-93, staff chemist, 1993—. Editor (newsletter) The Bally Beacon, 1990—. Mem. Am. Chem. Soc. Am. Assn. Pharm. Scientists, Pine Forge Sportsman's Club, KC (Grand Knight 1988-90, Svc. 1992, Fr. Tomko 4th degree assembly). Roman Catholic. Home: Douglass Dr Rd #2 Box 958 Boyertown PA 19512-9337 Office: Merck Rsch Labs Mail Stop WP78-110 Sumneytown & W Point Pikes West Point PA 19486-0004

KALEN, JOSEPH DAVID, physicist, researcher; b. N.Y.C., Apr. 9, 1956; s. Richard and Beatrice (Thaw) K. BS, SUNY, Brockport, 1978; PhD, Ohio State U., 1987. Rsch. assoc. N.Y. Hosp., N.Y., 1978; grad. teaching assoc. in physics Ohio State U., Columbus, 1978-82, grad. rsch. assoc. in physics, 1982-87, postdoctoral lectr. in physics, 1987, postdoctoral researcher in material science, 1988-89, postdoctoral researcher in physics, 1990-91; postdoctoral rschr. in nuclear medicine Med. Coll. Va., Richmond, 1992—; cons. in material science Ohio State U., Columbus, 1987-88. Contbr. articles to Physical Review C., Jour. Am. Ceramic Soc., Astrophysical Jour. Letters, 1988—. Recipient Math. award Sigma Xi, Brockport, N.Y., 1977. Mem. Am. Assn. Physicists in Medicine, Soc. Nuclear Medicine. Office: Med Coll Va Radiology Dept Radiation Physics Divsn MCV Station Box 72 Richmond VA 23298

KALENDER, WILLI ALFRED, medical physicist; b. Thorr, Germany, Aug. 1, 1949; s. Heinz W. and Gertraud (Tillig) K.; m. Marlene Schug, Aug. 20, 1975; children: Benjamin, Björn, Christine. MS, U. Wis., 1975, PhD, 1979; Dr. habil., U. Tübingen, Fed. Republic Germany, 1988. Rsch. asst. physics dept. U. Wis., Madison, 1974-75; doktorand Siemens Med. Labs., Erlangen, Fed. Republic Germany, 1976-79, rsch. physicist, 1979-83, head CT applications, 1983-88, head med. physics, 1988—; adj. asst. prof. physics U. Wis., Madison, 1990—; lehrbeauftragter Tech. U. Munich, Germany, 1991—. Contbr. over 200 articles to profl. publs. Achievements include introduction of spiral computed tomography; 11 patents in field. Home: Auf der Höh 5, 8521 Kleinseebach Germany

KALFF, JACOB, biology educator; b. Velsen, The Netherlands, Dec. 20, 1935; s. Albert Willem and Isabella Selly (Bendien) K.; m. Evelyn Mary Rivaz, Dec. 26, 1959; children: Derek, Sarah. BSA, U. Toronto, 1959, MSA, 1961; PhD, Ind. U., 1965. Asst. prof. McGill U., Montreal, Que., 1965-69, assoc. prof., 1969-76, prof., 1977—; dir. McGill U. Limnology Rsch. Ctr., Montreal, 1978; regional dir. Hydrobiologia jour., Ghent, Belgium, 1981. Author over 100 sci. papers. Office: McGill Univ Limnology Rsch Ctr, 1205 Docteur Penfield, Montreal, PQ Canada H3A 1B1

KALIMO, ESKO ANTERO, research institute administrator, educator; b. Helsinki, Finland, July 9, 1937; s. Yrjö Edvard and Hilma (Hiltunen) K.; m. Raija Vuokko Moilanen, Dec. 20, 1964; children: Anna, Antti. D in Social Scis., U. Helsinki, 1969. Lic. psychologist. Asst. tchr. U. Helsinki, 1961-64; sr. sci. Social Ins. Inst., Helsinki, 1964-73; chief sci. WHO, Geneva, 1974-75, sci., 1981-83; dir. rsch. inst. Social Ins. Inst., Helsinki, 1973—; lectr. psychometrics, U. Helsinki, 1964-69; rsch. assoc. Johns Hopkins U., Balt., 1969-70; assoc. prof. U. Helsinki, 1971—; prof. Nordic Sch. Pub. Health, Gothenburg, Sweden, 1989-92; chmn. adv. commn. rsch. Internat. Social Security Assn., Geneva, 1985-93; mem. expert panel WHO, Geneva, 1977-81, 1983—; cons. in social security and health UN agys., 1976—. Contbr. articles to profl. jours. Mem. Med. Rsch. Coun. of Acad. in Finland, Helsinki, 1976-79; chmn. Finnish Coun. for Health Edn., Helsinki, 1981-87; bd. dirs. Finnish Population Assn., Helsinki, 1989; vice chmn. Finland Com. for the Club of Rome, Helsinki, 1985—. 2d lt. Finland Mil., 1963-64. Recipient The Knight's Cross (1st Class) The Order of the Lion of the Govt. of Finland, 1976, The Order of the White Rose of the Govt. of Finland, 1986. Mem. Internat. Sociol. Assn., Internat. Epidemiol. Assn., European Soc. Med. Sociology, Found. for Internat. Studies on Social Security (gov. 1987—), Munkkiniemi Rotary (sec. Helsinki 1985—). Avocation: tennis. Home: Kilonkallionkuja 6, 02610 Espoo Finland Office: Social Ins Inst, PO Box 78, 00381 Helsinki Finland

KALIN, ROBERT, retired mathematics educator; b. Everett, Mass., Dec. 11, 1921; s. Benjamin and Celia (Kraff) K.; m. Shirley Sharney, Oct. 22, 1944; children: Susan Leslie, John Benjamin; m. 2d Madelyn Pildish, Aug. 17, 1962; children: Sandra Kim, Richard Dean. Student, Northeastern U., 1940-43; B.S., U. Chgo., 1947; MA in Teaching, Harvard U., 1948; Ph.D., Fla. State U., 1961. Tchr. math. Holten High Sch., Danvers, Mass., 1948-49, Beaumont High Sch., Hadley Tech. Sch., Soldan-Blewitt High Sch., St. Louis, 1949-52; ednl. statistician Naval Air Tech. Tng. Center, Norman, Okla., 1952-53; test specialist, assoc. in research Ednl. Testing Service, Princeton, N.J., 1953-55; exec. asst. Commn. on Math. of Coll. Entrance Exam. Bd., 1955-56; instr. dept. math. edn. Fla. State U., Tallahassee, 1956-61; asst. prof. Fla. State U., 1961-63, assoc. prof., 1963-65, prof., 1965-90; prof. emeritus Fla. State U., Tallahassee, 1990; assoc. dept. head Fla. State U., 1968-73, program chmn., 1975-78. Co-author: Elementary Mathematics, Patterns and Structure, 11 vols., 1966, (with George Green) Modern Mathematics for the Elementary School Teacher, 1966, (with E.D. Nichols) Analytic Geometry, 1973, Holt School Mathematics, 9 vols., 1974, Holt Mathematics, 9 vols., 1981, rev., 1985, (with M.K. Corbitt) Prentice Hall Geometry, 1990, rev. edit., 1993. Mem., treas. Brownsville-Haywood County Libr. Bd., 1991—; pres. Temple Adas Israel, 1992—. With U.S. Army, 1943-46. Mem. Math. Assn. Am. (sec.-treas. Fla. sect. 1985-91, Svc. award Fla. sect. 1991), Fla. Coun. Tchrs. Math. (pres. 1960-61), Fla. Assn. Math. Educators (pres. 1984-86), Nat. Coun. Tchrs. Math. (chmn. external affairs com. 1972-73), Nat. High Sch. and Jr. Coll. Math. Clubs (gov. 1972-75, pres. 1978-80). Home: 721 Key Corner Brownsville TN 38012-7417

KALINER, MICHAEL ARON, immunologist, allergist, medical association administrator; b. Balt., 1941. MD, 1967. Intern Hosp. Md., 1967-68; resident in medicine U. Calif., San Francisco; asst. in medicine Peter Bent Brigham Hosp., Boston, 1972-73; chief allergy and immunology Keesler AFB (Miss.) Med. Ctr., 1973-75; sr. investigator Nat. Inst. Allergy and Infectious Disease, NIH, Bethesda, Md., 1975—; now also chmn. Am. Bd. Allergy and Immunology Inc. Served to maj. M.C., USAF. Fellow in allergy and immunology Harvard U., 1970-73. Mem. Am. Thoracic Soc., Am. Acad. Allergy and Immunology, Am. Assn. Immunologists., Amn. Assn. Physicians, Am. Soc. Clin. Investigation. Office: NIH Room 10 11C 205 Bethesda MD 20205

KALISKI, MARY, psychologist; b. Bratislava, Czechoslovakia, Dec. 9, 1938; came to U.S., 1950; d. Frank and Margaret (Fleischman) Reichenthal; m. Thomas Kaliski, Sept. 21, 1957; children: Karen, Kenneth. BS summa cum laude, C.W. Post Coll., 1978; MS, profl. diploma, St. John's U., 1980, PhD, 1987. Psychologist North Shore Schs., L.I., 1977-79, Herricks Schs., L.I., 1979—; speaker in field. Chief psychologist Stepfamily Found. L.I., 1987—; bd. dirs. Nassau Psychol. Svcs. Inst., 1989—. Mem. Am. Psychol. Assn., Nassau County Psychol. Assn., Sch. Coun. N.Y. State Psychol. Educators.

KALKHOF, THOMAS CORRIGAN, physician; b. Wellsville, N.Y., Aug. 12, 1919; s. Arthur Albert and Evelyn (Corrigan) K.; m. Mary E. Jones, Mar. 3, 1946 (dec. 1955); children: Thomas E., Susan A., Mark A., Patricia D.; m. 2d Constance N. McCarthy, Apr. 19, 1958; children: Christopher J., Constance M., Craig Alan. B.S., Gannon U., 1943; M.D., Marquette U., 1946. Intern, resident St. Vincent's Hosp., Erie, Pa., 1946-47; pvt. practice nutritional problems, continued breast cancer rehab. and mammography, thermography, gen. geriatrics and psychosomatic medicine Erie, 1947—; med. dir. Twinbrook Med. Ctr., 1960-84; dir. Iroquois Med. Centre, Erie; staff mem. St. Vincent's Health Ctr., Hamot Med. Ctr., Erie; pres., dir. Small

Hosp. Cons., Inc., Erie, 1954—. Past chmn. Pa. Bd. Accreditation Nursing Homes and Related Facilities; past pres. Cath. Social Svcs., Erie; past pres. Erie County Ind. Coun. on Aging.; bd. dirs. Cath. Charities USA Commn. on Aging. With M.C., AUS, 1943-44. Fellow Am. Coll. Health Care Adminstrs., Am. Geriatric Soc., Am. Acad. Family Physicians, Acad. Psychosomatic Medicine; mem. AMA, Pa. Health Care Assn. (past pres.) Acad. Psychomatic Medicine (past pres.), Pa. Acad. Family Physicians (past pres. Erie chpt.), Assn. Physicians in Chronic Disease Facilities (past pres.), Am. Soc. Clin. Hypnosis, Pa., Erie County Med. Socs., Nat. Geriatric Soc. (pres.), Soc. Prospective Medicine, Pa. Thoracic Soc., Ind. Coun. on Aging (past pres.), KC (4 deg.). Republican. Roman Catholic. Home: 3749 E Lake Rd Erie PA 16511-1346 Office: 4401 Iroquois Ave PO Box 7265 Erie PA 16510

KALLFELZ, HANS CARLO, cardiologist, pediatrics and pediatric cardiology educator; b. Frankfurt, W.Ger., July 17, 1933; s. Hans and Diana Westralia (Ratazzi) K.; m. Irmgard Wittkuhn, Sept. 21, 1957; children: Andreas, Arnica Beate, Johannes, Michael. M.D., U. Bonn, (W.Ger.), 1962. Registrar in pediatrics Children's Hosp. of U. Bonn, 1962-67, sr. registrar in pediatrics and pediatric cardiology, 1968-69, head pediatric cardiology, 1969-73; research fellow Hosp. for Sick Children, Great Ormond St., London, 1967-68; chief dept. pediatric cardiology Med. Sch. Hannover (W.Ger.), 1974—, chmn. hosp., 1976-79, 83-85. Author: Pediatrics in Practice and Hospital, 1980; Hypertension, 1984; Pediatric Pneumology, 1984; contbr. articles to profl. jours. Mem. Deutsche Gesellschaft fü r Kinderheilkunde, Deutsche Gesellschaft fur Herz-u-Kreislaufforschung, Deutsche Gesellschaft fur Padiatrische Kardiologie (councillor 1974-76, pres. 1981-83, v.p. 1983-85), Assn. European Pediatric Cardiologists (councillor 1971-78, sci. sec., 1979-83, sec. gen. 1983-88, pres. 1989-92), Deutsch-Oesterreichische Gesellschaft fü r Pediatrische Intensiv Medizin und Neonatologie. Roman Catholic. Club: Akadem Ruder Club Rhenus (sec. gen. 1957-58). Office: Med Sch Hannover Childrens, Hosp, 8 Konstanty-Gutschow St, D 3000 Hannover Germany

KALLIANPUR, GOPINATH, statistician; b. Mangalore, India, Apr. 16, 1925; married, 1953; children: Asha, Kalpana. BA, U. Madras, 1945, MA, 1946; PhD in Math. Stats., U. N.C., 1951. Lectr. stats. U. Calif., 1951-52; with Inst. Advanced Study, 1952-53; reader stats. Indian Statis. Inst., Calcutta, 1953-56; vis. assoc. prof. Mich. State U., 1956-59; assoc. prof. math. Ind. U., 1959-61; prof. stats. Mich. State U., 1961-63; prof. math., stats. U. Minn., Mpls., 1963-79; with dept. stats. U. N.C., Chapel Hill, 1979—, now Alumni Disting Prof. Fellow Inst. Math. Statisticians; mem. AAAS. Office: The Univ of NC Dept Stats Chapel Hill NC 27514

KALMAN, CALVIN SHEA, physicist; b. Montreal, Que., Can., Oct. 29, 1944; s. William and Dorothy Belle Kalman; m. Judith Matilda Mendes Miller, Aug. 28, 1966; children: Samuel Adam de Sola, Benjamin Mordecai Mendes. BS, McGill U., 1965; MA, U. Rochester, 1967, Phd, 1970. Asst. prof. Loyola U. of Montreal, 1968-75; assoc. prof. Concordia U., Montreal, 1975-84, prof., 1984—, chmn. dept. physics, 1983-89; vis. assoc. prof. Ind. U., Bloomington, 1976-77. Author: Preons, Models of Leptons, Quarks and Gauge Bosons as Composite Objects, 1992. Commr., Protestant Sch. Bd. Greater Montreal, 1981-83, 86-87. J. W. McConnell Meml. scholar; grantee Sci. and Engring. Research Council Can., 1968—. Mem. Am. Assn. Physics Tchrs., Am. Physics Soc. Contbr. articles to sci. jours. Office: 1455 De Maisonneuve Blvd W, Montreal, PQ Canada H3G 1M8

KALMAN, RUDOLF EMIL, research mathematician, systems scientist; b. Budapest, Hungary, May 19, 1930; s. Otto and Ursula (Grundmann) K.; m. Constantina Stavrou, Sept. 12, 1959; children: Andrew E.F.C., Elisabeth K. SB, MIT, 1953, SM, 1954; DSc, Columbia U., 1957; DEng (hon.), U. Bologna, 1988; DSc (hon.), U. Kyoto, Japan, 1990; PhD (hon.), Heriot Watt U., Edinburgh, Scotland, 1990, Tech. U. Crete, 1993. Staff engr. IBM Research Lab., Poughkeepsie, N.Y., 1957-58; research mathematician Research Inst. Advanced Studies, Balt., 1958-64; prof. engring. mech. and elec. engring. Stanford U., 1964-67, prof. math. system theory, 1967-71; grad. rsch. prof. Ctr. for Math. System Theory U. Fla., 1971-92; dir. Center for Math. System Theory, U. Fla., 1971—; prof. math. system theory Swiss Fed. Inst. Tech., Zurich, 1973—; sci. adviser Ecole Nationale Superieure des Mines de Paris, 1968—; mem. sci. adv. bd. Laboratorio di Cibernetica, Naples, 1970-73. Author: Topics in Mathematical System Theory, 1969, over 150 sci. and tech. papers.; editorial bd. Internat. Jour. Math. Modelling, Jour. Computer and Systems Scis., Jour. Nonlinear Analysis, Jour. Optimization Theory and Applications, Applied Math. Letters, Math. of Control, Signals and Systems, Jour. Forecasting, Revue Internationale de Systemique, Internat. Jour. Algebra and Computation. Named outstanding young scientist Md. Acad. Sci., 1962; recipient IEEE medal of honor, 1974, Rufus Oldenburger medal ASME, 1976, Centennial medal IEEE, 1984, 1st Kyoto prize Inamori Found., 1985, Steele prize Am. Math. Soc., 1987; Guggenheim fellow IHES Bures-sur-Yvette, 1971. Fellow AAAS; mem. Nat. Acad. Engring. (U.S.), Hungarian Acad. Scis. (fgn.), Académie des Scis., Inst. de France (fgn.). Office: ETH Center, CH-8092 Zurich Switzerland

KALMAZ, EKREM ERROL, environmental scientist; b. Turkey, Jan. 2, 1940; came to U.S., 1962, naturalized, 1979; s. Memet and Ayse K.; student Queens Coll., 1962-63; B.A. in Chemistry, Okla. State U., 1969; M.S. in Environ. Sci. and Engring., U. Okla., 1972, Ph.D. in Engring., 1974; m. Gulgun Durusoy, Oct. 3, 1974; children: Phyllis, Denise. Research asst. Okla. Med. Research Found., Oklahoma City, 1969-72, research asso., 1972-74; postdoctoral fellow, Duke U., 1974-76; asst. prof. dept. engring. sci. and mechanics U. Tenn., Knoxville, 1976-79; sr. environ. scientist Henningson, Durham & Richardson, Inc., engring. cons., Santa Barbara, Calif., 1979-83; sr. rsch. scientist NASA Johnson Space Ctr., Houston, 1983-87, Walls Med. Found. U. Tex., Galveston, 1987-90; assoc. prof. U. South Fla. Coll. Pub. Health, Tampa, Fla., 1990—, adj. assoc. prof. Dept. Civil Engring., Environ. Engring.; sr. research assoc. NRC, Houston; cons. to industry, engrs. and govt. agys; bd. dirs. Advance Environ. Studies. Mem. Am. Chem. Soc., AAAS, Am. Coll. Toxicology, Inst. Environ. Scis. (tech. chmn. water quality impact), Internat. Soc. Ecol. Modeling, N.Y. Acad. Scis., Soc. for Computer Simulation, Am. Inst. Chemists, Sigma Xi. Contbr. articles to profl. jours., chpts. to books.chpts. to books. Home: 8609 Fishermans Point Dr Tampa FL 33637-1859 Office: U South Fla Coll Pub Health MHH-104 1301 Bruce B Downs Blvd Tampa FL 33612

KALPHAT-LOPEZ, HENRIET MICHELLE, electric and electronics engineer; b. Kingston, Jamaica, Sept. 24, 1963; came to U.S., 1967; d. Michael Karl and Maria Beatrice Kalphat; m. Raim Lopez, Apr. 26, 1986; children: Joshua Raim, Danielle Marie. BSEE with honors, Fla. Internat. U., 1986. Quality/reliability engr. Sch. Engring. and Logistics, Texarkana, Tex., 1986-87; product assurance and test engr. Orlando, Fla., 1987-89; project engr. Naval Tng. Systems Ctr., Orlando, 1989—. Sunday sch. tchr., Ch. of the New Covenant, Orlando, 1989-90. Mem. IEEE, Sigma Xi. Republican. Episcopalian. Office: Naval Tng Systems Ctr 12350 Research Hwy Code 232 Orlando FL 32826-3226

KALRA, YASH PAL, soil chemist; b. Gunjial, Punjab, India, Oct. 28, 1940; arrived in Can., 1963; s. Amir Chand and Hemo Devi (Sapra) K.; children: Maneesh, Navita. BSc in Agr., Agra U., Kanpur, Uttar Pradesh, India, 1961, MSc in Agrl. Chemistry, 1963; MSc in Soil Sci., U. Man., Winnipeg, Can., 1967. Head analytical svcs. Forestry Can., Edmonton, Alta., 1967—; mem. safety com. No. Forestry Ctr., 1970-72; chmn. registration com. Environ. Soil Sci. Conf., Can. Land Reclamation Assn./Can. Soc. Soil Sci., 1992; mem. biol. svcs./environ. scis. adv. com. No. Alta. Inst., Tech., Edmonton, 1993—; mem., chemistry exec. Profl. Inst. Pub. Svc. Can., 1984—, exec. Edmonton br., 1991—; panelist in field. Author: (methods manual) Information Report, 1991; referee for several sci. publs.; contbr. articles to profl. jours. Grantee Prime Min. India, 1963, Nat. Rsch. Coun. Can., 1964-66; rsch. fellow Indian Coun. Agrl. Rsch., 1961-63, U. Man., 1963-64. Mem. Can. Soc. Soil Sci. (sec. 1993—), Internat. Soc. Soil Sci., Soil Sci. Soc. Am. (assoc. referee method validation for pH measurement in soil with AOAC, 1991—), Am. Soc. Agronomy, Indian Soc. Soil Sci., Soc. Ind. Foresters, Assn. Ofcl. Analytical Chemists (co-chmn. soil and environ. workshop 1992, 93), Internat Soc. Soil Sci. (treas. workshop working group MO 1992), Western Enviro-Agrl. Lab. Assn. (co-founder 1979, sec./treas. 1981-82, 85-86, vice-chmn. 1982-83, 86-87, chmn. 1983-84, 87-88, first check sample program 1979-81), Group Analytical Labs. (founder 1990, chmn. 1990-92), Profl. Inst. of Pub. Svc. of Can., Coun. on Soil Testing in Plant Analysis (bd. dirs. 1993—), Pacific Regional Soc. Soil Sci., Potash Rsch. Inst. of India. Office: Forestry Can, 5320 122 St, Edmonton, AB Canada T6H 3S5

KALSHER, MICHAEL JOHN, psychology educator; b. Butte, Mont., Oct. 20, 1956; s. Joseph R. and Lavera G. (Peterson) K.; m. Joanne Simmons, Aug. 17, 1991. Student, Mont. Coll. Mineral Sci., Butte, 1974-75; BS in Psychology, Mont. State U., 1979, MS in Psychology, 1986; PhD, Va. Poly. Inst. & State U., 1988. Rsch. asst. behavorial psychology John F. Kennedy Inst. Johns Hopkins U., Balt., 1980-84; asst. prof. psychology Rensselaer Poly. Inst., Troy, N.Y., 1988-93, assoc. prof. psychology, 1993—, faculty rep. alumni assn. admissions com., 1990—, faculty rep. alcohol rev. bd., 1990-92. Contbr. articles to profl. jours. Named Outstanding Young Men in Am.; William J. Hillman scholar, 1983. Mem. Am. Psychol. Soc. (chartered), Ea. Psychol. Assn., Assn. Behavior Analysis, Human Factors Soc. Republican. Roman Catholic. Achievements include reseach in environmental determinants of alcohol consumption, transportation safety, occupational safety, leadership and applied community issues. Office: Rensselaer Poly Inst Psychology Dept 110 8th St Troy NY 12180-3590

KALSHOVEN, JAMES EDWARD, JR., electronics engineer; b. New Orleans, Jan. 1, 1948; s. James Edward and Estell (Leggio) K. BS, U. Tenn., 1969; MS, Rice U., 1971; PhD, U. Tenn., 1974. Researcher IBM, Lexington, Ky., 1968; instr. Rice U., Houston, 1969-71; tchr., researcher U. Tenn., Knoxville, 1971-74; aerospace technologist NASA/Goddard, Greenbelt, Md., 1974—. Contbr. articles to profl. jours. Bd. dirs. Woodstream Homeowner's Recreational Assn., Seabrook, Md., 1989—. Named Inventor of the Yr., 1984. Mem. IEEE, Tau Beta Pi, Sigma Xi. Roman Catholic. Achievements include patents on laser technique to remotely measure atmospheric temperature and pressure from spacecraft, laser technique for measuring physiology of plants and hydrological and geological phenomena using polarized lasers. Office: NASA/Goddard Code 925 Greenbelt MD 20771

KALTENBACH, CARL COLIN, agriculturist, educator; b. Buffalo, Wyo., Mar. 22, 1939; s. Carl H. and Mary Colleen (McKeag) K.; m. Ruth Helene Johnson, Aug. 22, 1964; children: James Earl, John Edward. BS, U. Wyo., 1961; MS, U. Nebr., 1963; PhD, U. Ill., 1967. Prof. animal sci. U. Wyo., Laramie, 1969-89, head dept. animal sci., 1978-80, assoc. dir. coll. agr., 1980-84, assoc. dean and dir., 1984-89; vice dean coll. agriculture, dir. agrl. experiment sta. U. Ariz., Tucson, 1989—; chmn. Exptl. Sta. Com. on Orgn. and Policy, 1986-87. Contbr. more than 200 articles to profl. jours. Recipient Young Scientist award West Sect. Am. Soc. Animal Sci., 1976, Faculty award merit Gamma Sigma Delta, 1980; named Outstanding Alumnus U. Wyo. Coll Agriculture, 1991. Mem. AAAS, Soc. Study Reproduction (treas. 1980-83), Am. Soc. Animal Sci., Soc. for Study of Fertility. Office: U Ariz 314 Forbes Bldg Tucson AZ 85721

KALTHOFF, KLAUS OTTO, zoology educator; b. Iserlohn, Nordrhein-Westfalen, Germany, Feb. 5, 1941; came to U.S., 1978; s. Hugo and Herta (Brenken) K.; m. Karin Dora Losskarn, Dec. 23, 1965; children: Christian, Ulrich, Philipp. B.A., U. Hamburg, 1964; M.A., U. Freiburg, 1967, Ph.D, 1971. Instr. U. Freiburg, W. Germany, 1969-71, asst. prof., 1971-76, assoc. prof., 1976-78; assoc. prof. U. Tex.-Austin, 1978-80, prof. zoology, 1980—; dir. Ctr. for Devel. Biology, 1983—. Recipient prize for outstanding research work Sci. Soc. Freiburg, 1975; hon. guest mem. Arthropodan Embryology Soc. Japan. Mem. AAAS, Soc. for Devel. Biology, European Devel. Biologist Orgn., Gesellschaft fuer Entwicklungsbiologie. Office: U Tex Dept Zoology Patterson Lab Austin TX 78712

KALTSOS, ANGELO JOHN, electronics executive, educator, photographer; b. Boston, Aug. 19, 1930; s. John Angelo and Rita Thomas (Goudas) K.; m. Verna Kay Wilson, June 30, 1952 (dec. Jan. 1973); children: Pamela, Elaine, Gregory, Stephanie, Lenora, Demetra, Dana. Student, Mass. Radio and TV Sch., Boston, 1955-57, Harvard Coll. Extension, 1964, Boston State Coll., 1965-67, U. N.M., 1976, Fitchburg State Coll., 1977. Clk. U.S. Postal Svc., Boston, 1954-57; electronic rsch. technician Crosley div. Avco, Cin., 1957; electronic technician Raytheon Mfg. Co., Waltham, Mass., 1957-63; educator Cambridge (Mass.) Sch. Dept., 1961-81; ind. ethnology rsch. N.Mex., 1969—; mgr. Pampas, Inc., Boston, 1987-90; bd. dirs. Expansion Dance Co., Boston; cons. 5 P.I.E., Albuquerque, 1976—, Indian Tribal Group, N.Mex.; lectr. S.W. Indian Culture in Boston, Cambridge area, 1990—; pres., treas. Spartan Enterprises, Inc., 1965-69. Author: Southwest Indian, 1986; one-man photo exhibits: Christmas Tree Gallery, Manteo, N.C., 1977, The 4th St. Photo Gallery, N.Y.C., 1980, Cambride Rindge and Latin Sch., Mass., 1981, Jay's, Cambridge, Mass., 1983, Here Today Gallery, Boston, 1984, Andover (Maine) Town Hall, 1984, 86, Piedmont Art Assn., Martinsville, Va., 1985-86, Cambalache Gallery, Boston, 1986-87, The 4th St. Gallery, N.Y.C., 1990; contbg. journalist in field. Chmn. No Thank Q Hydro Quebec, Andover, Maine, 1988-91, coord., Dryden, Maine, 1991—; regional and media coord. N.E. Alliance to Protect James Bay, 1990-91, mem. exec. bd., mem. adv. bd., treas., 1991—; mem. senate faculty Cambridge Sch. Dept., 1980-81; sec. New Eng. Model Car Assn. of Raceways, 1966-69; educator Cambridge Adult Ctr., 1990—, Paulist Ctr., Boston, 1991-92. Mem. Appalachian Mountain Club. Greek Orthodox. Avocations: ethnography, entomology, cooking, gardening, hiking. Home: 10 Lesley Ave Somerville MA 02144-2607

KALVIN, DOUGLAS MARK, research chemist; b. N.Y.C., May 19, 1954; s. Harold Roseman and Florence Irene (Cohen) K.; m. Sheryl Rae Shapiro, Oct. 19, 1991. BA in Chemistry cum laude, Duke Univ., 1976; MS in Organic Chemistry, Univ. Mich., 1978; MS in Pharmacology, Baylor Coll., 1984; PhD, Univ. Mich., 1985. Fellow Univ. Wis., Madison, 1985-86; rsch. assoc. Nova Pharmaceutical Corp., Balt., 1987-89; sr. rsch. chemist Abbott Labs., Abbott Park, Ill., 1990—. Contbr. articles to profl. jours. Mem. Am. Chemcial Soc., Div. Organic Chemistry and Medicine Chemistry, Phi Lambda Upsilon. Home: 1201 Lockwood Dr Buffalo Grove IL 60089

KAMAL, MUSA RASIM, chemical engineer, consultant; b. Tulkarm, Jordan, Dec. 8, 1934; arrived in Canada, 1967; s. Rasim Kamal Ismail and Aminah Abu Hadbah; m. Nancy Joan Edgar, Dec. 23, 1961; children: Rammie, Basim. BSc, U. Ill., 1958; M in Engring., Carnegie-Mellon U., 1959, PhD, 1961. Rsch. chem. engr. Cen. Rsch. Labs. Am.-Cyanimd Co., 1961-65, rsch. group leader plastics, 1965-67; assoc. prof. chem. engring. McGill U., Montreal, Que., Can., 1967-73, prof., 1975—, chmn. chem. engring. dept., 1983—, dir. Brace Rsch. Inst., 1986—; dir. microeconomics Devel. Plan, Rabat, Morocco, 1977; pres. Tulkarm Enterprises, Montreal, 1977—; mem. bd. govs. Can. Plastics Inst., Toronto, 1986—. Editor: Weatherability of Plastic Materials, 1967, Advances in Transport Processes, vol. 5, 1990, vol. 6, 1990; contbr. over 300 articles and presentations to profl. jours. Recipient Internat. Edn. award, Best Paper awards Soc. Plastics Engrs., Kuwait prize Allied Sci. Kuwait Found. Advancement Sci. Mem. Soc. Plastics Industry (CANPLAST award, bd. tech. 1986—), Polymer Processing Soc. Internat. (exec. com. 1984—), Can. Inst. Internat. Affairs, Can. Club Montreal, Montreal Amateur Athletics Assn. Achievements include 5 U.S. patents and 2 patents pending; research in processing of plastics and composites, injection and blow molding, polymer characterization and plastics weatherability. Office: McGill U Dept Chem Engring, 3480 University St, Montreal, PQ Canada H3A 2A7

KAMDEM, DONATIEN PASCAL, chemistry educator; b. Bafang, Cameroon, Apr. 10, 1955; s. Donatien and Lucie (Mientchop) K.; m. Melanie L.B. Wandji, Dec. 29, 1989; children: Coretta J.M., Bria Chelsea K. MBA, Laval U., Que., Can., 1987, PhD, 1990. Lab. technician Yaounde U., Cameroon, 1979; asst. mgr. Socapalm, Cameroon, 1980-83; researcher U. Que., Trois-Rivieres, Can., 1985-87; rsch. assoc. Laval U., Que., 1987-91, SUNY, Syracuse, 1991; asst. prof. Mich. State U., East Lansing, 1991—; salesman CISCO, Que., 1989. Contbr. articles to profl jours. Excellence scholar Que. Gov., 1987-90. Mem. Forest Product Rsch. Soc., Tech. Assn. Pulp and Paper Industries, Soc. Wood Scientists and TEch., Am. Chem. Soc. Office: Mich State U Dept Forestry East Lansing MI 48824-1222

KAMEMOTO, HARUYUKI, horticulture educator; b. Honolulu, Jan. 19, 1922; s. Shuichi and Matsu (Murase) K.; m. Ethel Hideko Kono, June 7, 1952; children—David Yukio, Mark Toshio, Claire Naomi. B.S., U. Hawaii, 1944, M.S., 1947; Ph.D., Cornell U., 1950. Asst. in horticulture U. Hawaii, Honolulu, 1944-47, asst. prof. horticulture, 1950-54, assoc. prof., 1954-58, prof., 1958—, chmn. dept., 1969-75; horticulture adviser Kasetsart U., Bangkok, Thailand, U. Hawaii AID contract, 1962-65; UNFAO hort. cons. to, India, 1971, 80. Author: (with R. Sagarik) Beautiful Thai Orchid Species, 1975; contbr. articles to profl. jours. Recipient Gold medal Malayan Orchid Soc., 1964; Norman Jay Coleman award Am. Assn. Nurserymen, 1977, Norman F. Childers award Am. Soc. for Hort. Sci., 1984, Scientist of Yr. award ARCS Found., 1990; Fulbright rsch. fellow Kyoto U., Japan, 1956-57. Fellow AAAS, Am. Soc. Hort. Sci.; hon. mem. Am. Orchid Soc. (Gold medal 1990), Soc. Am. Florists (Alex Laurie award 1982, inducted into Floriculture Hall of Fame 1991), Japan Orchid Soc., Orchid Soc. Thailand (award of honor 1978), Orchid Soc. S.E. Asia; mem. Am. Genetic Assn., Am. Hort. Soc., Bot. Soc. Am., Internat. Soc. Hort., Internat. Assn. Plant Taxonomy, Soc. Advancement Breeding Rsch. in Asia and Oceania, Phi Kappa Phi. Home: 3246 Lower Rd Honolulu HI 96822-1457 Office: U Hawaii 3190 Maile Way Honolulu HI 96822-2279

KAMIJO, FUMIHIKO, computer science and information processing educator; b. Nagano, Japan, Aug. 4, 1934; s. Ryo'ichi and Tsune (Hayash) K.; m. Mariko Okudaira; children: Asuka, Erika, Yoshiki. BS in Engring., Tokyo U., 1957; SM in Mgmt., MIT, 1967. Mgr. IBM Japan Ltd., Tokyo, 1963-73; dir. Info. Tech. Promotion Agy., Tokyo, 1973-84, mng. dir., 1984-91; lectr. U. Tokyo, 1981-88; prof. U. Tokai, Kanagawa, 1990—. Contbr. articles to books, profl. jours. Mem. Indsl. Tech. Coun., Info. Processing Promotion Coun. (adv. com. 1986—), Industry Standard Coun. Home: 4-7-16 Kataseyama, Fujisawa, Kanagawa 251, Japan Office: Tokai U Info Sci Lab, 1117 Kitakaname, Hiratsuka Kanagawa 259-12, Japan

KAMILLI, ROBERT JOSEPH, geologist; b. Phila., June 14, 1947; s. Joseph George and Marie Emma (Clauss) K.; m. Diana Ferguson Chapman, June 28, 1969; children: Ann Chapman, Robert Chapman. BA summa cum laude, Rutgers U., 1969; AM, Harvard U., 1971, PhD, 1976. Geologist Climax Molybdenum Co., Empire, Colo., 1976-79, asst. resident geologist, 1979-80; project geologist Climax Molybdenum Co., Golden, Colo., 1980-83; geologist U.S. Geol. Survey, Saudi Arabian Mission, Jeddah, 1983-87, mission chief geologist, 1987-89; rsch. geologist U.S. Geol. Survey, Tucson, Ariz., 1989—; adj. prof. U. Colo., Boulder, 1981-83. Mem. editorial bd. Econ. Geology; contbr. articles to profl. jours. Henry Rutgers scholar Rutgers U., 1968-69. Fellow Geol. Soc. Am., Soc. Econ. Geologists; mem. Sigma Xi, Phi Beta Kappa. Avocations: travel, swimming, bicycle riding, music, photography. Home: 5050 N Siesta Dr Tucson AZ 85715-9652 Office: US Geol Survey Tucson Field Office U Ariz Gould-Simpson Bldg #77 Tucson AZ 85721

KAMINS, DAVID STONE, programmer, analyst; b. L.A., Aug. 5, 1959; s. Milton and Helen Alma (Stone) K. BS, Calif. Inst. Tech., 1982; MS, Med. U. S.C., Charleston, 1986. Programmer dept. neurology Children Hosp. L.A., 1981; rsch. scientist R&D Labs., Culver City, Calif., 1987—; advisor TOKA Enterprises, North Hollywood, Calif., 1990—. Asst. scoutmaster Boy Scouts Am., Troop 131, Venice, Calif., 1977—. Mem. AAAS, AIAA, Math. Assn. Am. Achievements include discovery of paired echoes of finite impulse response due to timing ripple from clock doubler circuit and restoration of data corrupted in this manner by application of inverse methods. Office: R&D Labs 5800 Uplander Way Culver City CA 90230

KAMIYA, NORIAKI, research mathematician; b. Chigasaki, Japan, Feb. 17, 1948; s. Hiroji and Kazue (Aoki) K. DSc, Rikkyo U., Tokyo, 1990. Lectr. math. Nihon U., Tokyo, 1976-81; rsch. scientist Shimane U., Matsue, Japan, 1981—. Editor: Algebras, Groups and Geometrics. Home: Saiwaicho 23-11, Chigasaki City 253, Japan Office: Shimane U Dept Math, Nishikawatsu, Matsue 690, Japan

KAMIYA, YOSHIO, research chemist, educator; b. Ofuna, Kanagawa, Japan, Mar. 25, 1930; s. Minoru and Kimi (Moro) K.; m. Masuko Nakamura, Mar. 18, 1957; children: Shoshi, Yoshiko, Mari. B in Engring., U. Tokyo, 1953, PhD, 1960. Postdoctoral fellow Nat. Rsch. Coun. Canada, Ottawa, Ontario, 1962-64; assoc. prof. U. Tokyo, 1964-75, prof., 1975-90, prof. emeritus, 1990—; head dept. reaction chemistry U. Tokyo, 1975, 80, 85; chmn. com. coal energy MITI, Tokyo, 1985-93. Author: Autoxidation of Organic Compounds, 1974, Chemistry of Fuel and Combustion, 1987, Chemistry of Petroleum and Coal, 1973, Process Organic Chemistry, 1981, Aspects of Degradation and Stabilization of Polymer, 1978; published 300 scientific papers. Recipient Award for Progressive Rsch. Chem. Soc. Japan, 1961, Award for Disting. Rsch. Fuel Soc. Japan, 1979, Japan Petroleum Inst., 1988. Mem. Inst. Energy Japan (v.p. 1992, pres. 1993—), Engring. Acad. Japan. Achievements include research in catalysis in organic oxidation reaction, effects of solvent and catalyst in coal liquefaction, autoxidation of organic compounds. Home: 590-96 Ozenji Aso-ku, Kawasaki 215, Japan Office: Sci U Tokyo Dept Ind Chem, 1-3 Kagurasaka Shinjuku, Tokyo 162, Japan

KAMMIN, WILLIAM ROBERT, environmental laboratory director; b. Chgo., Mar. 15, 1953; s. Charles Robert and Joan K.; m. Mary Margaret Kottenbach, Aug. 27, 1983; children: William Jay, Mary Margaret. BS in Chemistry, Ft. Lewis Coll., Durango, Colo., 1979; postgrad., U. B.C., Vancouver, 1979. Quality control chemist The Gillette Co., North Chicago, Ill., 1981; lab. technician, operator Durango Waste Water Treatment Plant, 1984-85; chief chemist Gary (Ind.) San. Dist., 1985; chemist No. Labs. and Engring., Valparaiso, Ind., 1985-86; scientist Calif. Analytical Lab., West Sacramento, 1986-89; chief inorganic chemist The Bionetics Corp., Chgo., 1989-90; environ. lab. dir. State of Wash. Dept. Ecology, Manchester, 1990—; Mem. Std. Methods Com., Denver, 1990—; mem. Puget Sound Working Group on Quality Analysis/Quality Control, Seattle, 1991; prin. U.S. EPA Region X Lab. Mgmt. Com., Seattle, 1990—; conferee The Pitts. Conf., Chgo., 1991. Contbr. articles to profl. jours. Mem. Dem. Nat. Com., 1988-90. Recipient Basic Ednl. Opportunity Grant State of Colo., Durango, 1975. Mem. Am. Chem. Soc., Am. Lab. Mgrs. Assn. Achievements include development of environ. methods for determination of mercury and other toxic metals. Home: PO Box 586 Manchester WA 98366 Office: Wash Dept Ecology 7411 Beach Dr Port Orchard WA 98366-8204

KAMO, ROY, engineering company executive; b. Kent, Wash., Mar. 7, 1921; s. Sokichi and Isa Kamo; m. May Satsuki Morinaga, June 22, 1948; children: Kathryn Kumiko, Joanne Tsuyuko, Lloyd Soichi, Richard Taiji. BSME, U. Nebr., 1945; MSME, U. Mich., 1947. Rsch. asst., instr. Ill. Inst. Tech., Chgo., 1946-48; sr. scientist heat power, mgr. fluids and combustion Armour Rsch. Found., Chgo., 1948-64; mgr. prime movers Cummins Engine Co., Columbus, Ind., 1964-65, dir. heat power rsch., 1966-81, exec. dir. advanced engine and systems, 1982-83; pres. Adiabatics, Inc., Columbus, Ind., 1984—, also chmn. bd.; cons. in field. Editor: Adiabatic Engine, 1984, and spl. publs., 1981-90; patentee in field; contbr. numerous articles to profl. jours. Fellow Soc. Automotive Engrs. (chmn. conf. 1984—); mem. ASME (chmn. gas turbine com. 1974—), Sigma Xi, Tau Beta Pi, Pi Tau Sigma. Avocations: fishing, gardening, art collector. Home: 3493 Riverside Dr Columbus IN 47203-1504 Office: Adiabatics Inc 3385 Commerce Dr Columbus IN 47201-2201

KAMPEN, EMERSON, chemical company executive; b. Kalamazoo, Mar. 12, 1928; s. Gerry and Gertrude (Gerlofs) K.; m. Barbara Frances Spitters, Feb. 2, 1951; children—Douglas S., Joanie L. Kampen Dunham, Laura L. Kampen Shiver, Emerson II, Deborah L. Kampen Smith, Cynthia S. Kampen Van Zelst, Pamela E. Kampen Mayes. B.S. in Chem. Engring., U. Mich., 1951; grad. (hon.), Purdue U., 1990. Chem. engr Gt. Lakes Chem. Corp., West Lafayette, Ind., 1951-57, plant mgr., 1957-62, v.p., 1962-67, sr. v.p., 1968, exec. v.p., 1969-71, pres., 1972—, chief exec. officer, 1977—; pres., chief exec. officer, chmn. bd. GLCD, Inc., Ark., 1988, also bd. dirs.; bd. dirs. GLCD, Inc., Bio-Lab, Inc., Decatur, Ga.; pres., dir. GLI, Inc., Newport, Tenn., GHC (Properties) Inc.; bd. dirs. WIL Rsch. Labs., Inc., OSCA, Inc., Lafayette Life Ins. Co., Ind. Nat. Bank, Huntsman Chem. Corp., Salt Lake City, Pub. Svc. Ind., Plainfield, QO Chems., Inc., Chgo., Pentech Chems., Inc., Chgo.; chmn. bd. E/M Corp., Hydrotech Chem. Corp., Great Lakes Chem. (Europe) Ltd.; pres. Ark. Chems.; mem. listed co. adv. com. Am. Stock Exchange. Mem. corp. advising group Huntington's Disease Soc. Am., N.Y.C.; Ind. United Way Centennial Commn.; trustee Ind. U., Bloomington, Purdue U., West Lafayette; bd. dirs. Jr. Achievement Greater

Lafayette Inc., Lafayette Art Assn. Found. Inc., Purdue Rsch. Found., West Lafayette, Lafayette Symphony Found.; dir., v.p. Hoosier Alliance Against Drugs, Indpls. Capt. USAF, 1953. Recipient Bronze medal Wall Street Transcript, 1980, 86, Gold medal, 1983, 85, 88, 90, 92, 93, Silver medal, 1989, 91, Man of Yr. award Nat. Huntington's Disease Assn., 1984, Kavaler award Chem. Mktg. Reporter, 1992, Bronze Medal award Fin. World Mag., 1992, 93, Winthrop-Sears award Chemist's Club, 1993; co-recipient Gold medal Wall Street Transcript, 1984; named 5th Most Involved CEO Chief Exec. Mag., 1986, Sagamore of the Wabash, 1988, Industrialist of Yr., Ind. Bus. Mag., 1991, CEO of Yr., Fin. World Mag., 1992; inductee Ind. Acad., 1992. Fellow Am. Inst. Chemists; mem. Soc. Chem. Industry, Nat. Asn. Mfrs. (bd. dirs.), Chem. Mfrs. Assn., Ind. C. of C. (bd. dirs.), Greater Lafayette C. of C., Ind. Acad., Lafayette Country Club, Skyline Club (Indpls.), Elks, Rotary. Avocations: golf; family events. Home: 168 Creighton Rd West Lafayette IN 47906-2102 Office: Gt Lake Chem Corp PO Box 2200 Us Hwy 52 NW West Lafayette IN 47906-5301

KAMPER, ROBERT ANDREW, physicist; b. Surbiton, Eng., Mar. 14, 1933; 3 children. BA, Oxford (Eng.) U., 1954, MA, DPhil, 1957. Imperial Chem. Industries rsch. fellow Oxford (Eng.) U., 1957-61; physicist Ctrl. Electric Generating Bd., Eng., 1961-63; physicist cryogenics divsn. Nat. Bur. Standards, 1963-74, assoc. chief electromagnetics divsn., 1974-78; chief electromagnetic tech. divsn. Nat. Standards and Tech., 1978—, dir. Boulder labs., 1982—. Fulbright Travel grantee, 1958-59; recipient Arnold O. Beckman award Instrument Soc. Am., 1974, Gold medal U.S. Dept. Commerce, 1975. Fellow IEEE. Achievements include rsch. in cryoelectronics; superconductivity; elec. measurement technique; electron spin resonance. Office: Nat Inst Standards & Tech Boulder Labs Radio Bldg Boulder CO 80303-3328*

KAN, BILL YUET HIM, aeronautical engineer; b. Singapore, Sept. 29, 1941; s. Po Tuen and Shin Ching (Fung) Kan; m. Sau Kwan Young, Nov. 18, 1968. B.Engring., Sydney U., Australia, 1968. Logistic officer Singapore Air Def. Command, 1968-74; tech. svcs. engr. Singapore Airlines, 1974-78, co. planner, mkt. rsch., 1978-74, systems planner, 1984-87; rsch. cons. Pvt. Ent., Singapore, 1987-90; projects mgr. CDIS, Ministry of Edn., Singapore, 1990-92; rschr. Stereocad, Singapore, 1992—. Contbr. articles to profl. jours. Capt. Singapore Air Def. Command, 1968-74. Mem. AIAA, Royal Aero. Soc. Home: 72-C Lorong H Telok Kurau, Singapore 1542, Singapore Office: Stereocad, 47 Beach Rd Ste 07-00, Singapore 0718, Singapore

KAN, YUET WAI, physician, investigator; b. Hong Kong, June 11, 1936; came to U.S., 1960; s. Tong-Po and Lai-Wai (Li) K.; m. Alvera Lorraine Limauro, May 10, 1964; children—Susan Jennifer, Deborah Ann. BS, MB, U. Hong Kong, 1958, DSc, 1980, DSc (hon.), 1987; DSc (hon.), Chinese U., Hong Kong, 1981; MD (hon.), U. Cagliari, Sardinia, Italy, 1981. Investigator Howard Hughes Med. Inst., San Francisco, 1976—; prof. lab. medicine U. Calif., San Francisco, 1977—, Louis K. Diamond prof. hematology, 1983—; mem. basic rsch. adv. com. March of Dimes, White Plains, N.Y., 1985-88; mem. blood diseases and resources adv. com. NIH, 1985-88; dir. Inst. Molecular Biology, U. Hong Kong, 1991—; mem. NIDDK adv. coun. NIH, 1991—; trustee Croucher Found., Hong Kong, 1992—. Contbr. over 210 articles to med. books, jours. Recipient Dameshek award Am. Soc. Hematology, 1979, Stratton Lecture award Internat. Soc. Hematology, 1980, George Thorn award Howard Hughes Med. Inst., 1980, Gairdner Found. Internat. award, 1984, Allan award Am. Soc. Human Genetics, 1984, Lita Annenberg Hazen award for Excellence in Clin. Research, 1984, Waterford award, 1987, ACP's award, 1988, Genetic Rsch. award Sanremo Internat., 1989, Warren Alpert Found. prize, 1989, Albert Lasker Clin. Med. Rsch. award, 1991, Christopher Columbus Discovery award, 1992, City of Medicine award, 1992, Excellence 2000 award, 1993. Fellow Royal Coll. Physicians (London), Royal Soc. (London), Third World Acad. Scis.; mem. Nat. Acad. Scis. U.S.A., Acad. Sinica (Chinese Acad. Scis.), Am. Acad. Arts and Scis., Assn. Am. Physicians, Am. Soc. Hematology (pres. 1990). Avocations: tennis; skiing. Home: 20 Yerba Buena Ave San Francisco CA 94127-1544 Office: U California Rm U-426 500 Parnassus Ave San Francisco CA 94143

KANAL, EMANUEL, radiologist; b. N.Y.C, Apr. 27, 1957; s. Mark and Rachel (Dvorkin) K.; m. Judith Eisenman; children: Eliezer, Aryeh, Gila, Daniella, Avromi, Tzippora. BA cum laude, Yeshiva U., 1977; MD, U. Pitts., 1981. Asst. prof., chief div. magnetic resonance, dept. radiology U. Pitts., 1989—; dir. The Pitts. NMR Inst., 1989—; lectr. and presenter in field. Bd. dirs. Congregation Poale Zedek, Pitts.; mem. exec. com., rec. sec., bd. dirs. Hillel Acad., Pitts.; mem. planning com. Kollel Bais Yitzchok Inst. for Advanced Jewish Edn., Pitts. Recipient numerous articles to profl. jours. Mem. AMA, Am. Coll. Radiology, Bioelectromagnetics Soc., Pa. Med. Soc., Pa. Radiol. Soc., Allegheny County Med. Soc., Pitts. Neuroradiol. Soc., Radiol. Soc. N.Am., Assn. Univ. Radiologists, Soc. for Magnetic Resonance Imaging, Soc. for Magnetic Resonance Imaging, Soc. for Magnetic Resonance in Medicine, Am. Soc. Neuroradiology (sr. mem.), Alpha Omega Alpha. Home: 5534 Forbes Ave Pittsburgh PA 15217 Office: The Pitts NMR Inst 3260 5th Ave Pittsburgh PA 15213

KANAMORI, HIROO, physics and astronomy educator; b. Tokyo, Oct. 17, 1936. Prof. Tokyo U., 1970-72; prof. Calif. Inst. Tech., Pasadena, 1972-89, John E. and Hazel S. prof., 1989-90; dir. Seismological Lab. Calif. Inst. Tech., Pasadena, 1990—. Recipient Arthur L. Day prize and lectureship NAS, 1993. Fellow Am. Geophys. Union; mem. Seismol. Soc. Am., Seimol. Soc. Japan. Office: Calif Inst Tech Dept Geophysics 1201 E Calif Blvd Pasadena CA 91125

KANAROWSKI, STANLEY MARTIN, chemist, chemical engineer, government official; b. Beausejour, Man., Can., Dec. 12, 1912; came to U.S., 1923, naturalized, 1928; s. Joseph and Caroline Kanarowski; m. Pearl Lewus, Aug. 8, 1936 (dec.); children: Stanley Martin Jr., Janice Ellen, Nancy Carol Kanarowski Cioffari. BS, U. Toledo, 1934; postgrad., Ohio State U., 1938-42, U. Akron, 1943-47, NYU, 1954, Xavier U., 1969, U. Wis., U. Mich., U. Ill., U. Mo. Chemist, chief chemist Ohio Dept. Liquor Control, Columbus, 1936-42; sr. cons. chemist Nebr. Ordnance Plant Firestone Tire and Rubber Co., Fremont, 1942-43; asst. dir. corp. gen. lab., chief factory product, chem. engr., rsch. and devel. compounding engr. Firestone Tire and Rubber Co., Akron, Ohio, 1943-49; lab. dir., asst. rsch. and devel. mgr. Fremont (Ohio) Rubber Co., 1949-52; rsch. and devel. chem. engr. Glass Fibers, Inc., Waterville, Ohio, 1952-53; chief rsch. and devel. chemist-engr., mgr. quality control Dairypak Butler, Inc., Toledo, 1953-60; chief chemist No. Ohio Region Lab. Liquor Control Enforcement Div. State of Ohio, Cleve., 1960-62; rsch. and devel. chemist-engr. Consol Paper Co., Monroe, Mich., 1962-63; chemist City of Toledo, 1963-64; project engr., head chemist investigations sect. Ohio River Div. U.S. Army Engr. Div. C.E., 1964-69; project leader, prin. investigator U.S. Army Constrn. Engring. Rsch. Lab., Champaign, Ill., 1969-86; ret. U.S. Army Constrn. Engring. Rsch. Lab., 1986. Mem. U. Ill. Symphony Orch., 1970-86, Montgomery Coll. Symphony Orch., 1987—. Recipient Army-Navy E award, 1943. Mem. Am. Inst. Chem. Engrs., Am. Chem. Soc. (mem. rubber div. 1954—), Am. Def. Preparedness Assn. Achievements include development of paint test kit and test procedures for evaluating quality of paints and coatings before use; materials research in waterproofing underground structures and sealing of joints and cracks in concrete pavements and structures. Address: 1329 Excaliber Ln Sandy Spring MD 20860

KANASEWICH, ERNEST ROMAN, physics educator; b. Eatonia, Sask., Can., Mar. 4, 1931; s. Max and Pauline (Pomeransky) K.; m. Elaine S. Cybak, May 1, 1968l children: Anthony Stephen, Patricia Michelle Anne. B.Sc., U. Alta., 1952, M.Sc., 1960; Ph.D., U. B.C., Vancouver, 1962. Seismologist Geophys. Svc. Internat. Corp., Can., Middle East, 1952-56; postdoctoral fellow U. B.C., Vancouver, 1962-63; asst. prof. dept. physics U. Alta., Edmonton, 1963-67, assoc. prof. dept. physics, 1967-71, asst. chmn. dept. physics, 1967-69, 71-73, 74-75, acting chmn. dept. physics, 1973-74, prof. dept. physics, 1971—, McCalla prof., 1989-90, chmn. dept. physics, 1991—, dir. geophysical obs.; geophys. cons. Fed. Environment Assessment Rev. Office, Hull, Que., Can., 1990—; co-transect leader LITHOPROBE, Edmonton, 1991. Author: Time Sequence Analysis in Geophysics, 1973, 3d edit., 1981, Seismic Noise Attenuation, 1990. Mem. exec. com. St. Andrew's Ukrainian Orthodox Parish, Edmonton, 1988-90. Fellow Roy. Soc. Can., Can. Gepophys. Union (J. Tuzo Wilson medal 1988), Can. Soc. Exploration Geophysicists (hon.), Assn. Profl. Engrs., Geophysicists and Geoloists. Office: U Alta, Dept Physics, Edmonton, AB Canada T6G 2J1*

KANAZIR, DUŠAN, molecular biologist, biochemist, educator; b. Mošorin, Serbia, Yugoslavia, June 28, 1921; s. Todor and Draginja (Stefanovic) K.; m. Mersija Kolakovic; 1 child, Selma. Student, Faculté de Medicine, Paris, 1949; Diploma of Graduation, Faculty of Medicine, Belgrade, Yugoslavia, 1949; PhD in Physiol. Scis., U. Libre, Brussels, 1955. Asst. prof. U. Belgrade, 1957, assoc. prof. Faculty of scis., 1963, prof. Faculty of scis., 1970-87; head lab. biology Inst. Boris Kidrič, Vinča, Yugoslavia, 1950-65, head lab. molecular biology, endocrinology, 1968-74, sci. counselor, 1974; pres. Serbian Acad. Scis. Arts., Belgrade, Yugoslavia; mem. Fed. Commn. on Nuclear Energy, Belgrade, 1956-65, Fed. Coun. for Coordination of Rsch.; expert IAE Agy. Vienna Atomic Ctr., Buenos Aires, 1966-71; counsellor CIBA Sci. Consultation Coun., Belgrade, London, 1970-87. Author chpts. to books. V.p. Yugoslav Pugwash Conf., Belgrade, 1959-61, mem., Ljubljana, 1986—, Yugoslav League for Peace, Equality and Independence of People, Belgrade, 1960-65,. Decorated Merit for Nation with Gold Star Presidium of the SFRY, Belgrade, 1965, Brotherhood and Unity with a Golden Wreath, 1976, Commdr. of Legion of Honour Pres. of Republic of France, 1984. Mem. Belgian Soc. Biochemistry, Intern Soc. Cell Biology (exec. bd. 1961-68), European Soc. Radiology (Presidium 1965-68), European Soc. Biochemistry, Intern Jour. of Photochemistry and Photobiology, Japan Jour. of Radiation Research (editor 1980—). Home: Save Kovačeviča 20/III, 11000 Belgrade Yugoslavia Office: Acad Sci & Arts, Knez-Mihailova 35, 11000 Belgrade Yugoslavia*

KANDASAMY, SATHASIVA BALAKRISHNA, pharmacologist; b. Nagapattinam, Tamilnadu, India, Jan. 16, 1945; came to U.S., 1979; s. Balikrishna and Rethinavalli Kandasamy; m. Anbukarasi Veerappa, Sept. 14, 1973; 1 child, Prathiba. MSc, U. Madras, India, 1968; PhD, Pasteur Inst., Paris, 1974. Lectr. in pharmacology U. Benin Med. Sch., Benin City, Nigeria, 1976-78, Alfateh U. Med. Sch., Tripoli, Libya, 1978-79; vis. prof. Okla. U., Oklahoma City, 1979-80; vis. scientist NASA, Moffett Field, Calif., 1980-84; assoc. prof. in pharmacology Banaras (India) Hindu U., 1984-85; sr. pharmacologist Armed Forces Radiobiology Rsch. Inst., Bethesda, Md., 1986—. Contbr. over 40 articles to profl. jours. Fellow Internat. Union Pharmacology, 1974-75, NRC, 1980, 86. Mem. Am. Soc. Pharmacology and Exptl. Pharmacology, Soc. for Neurosci. Achievements include discovery that radiation-induced fever and stress hormone release are mediated via prostaglandins, antioxidant enzymes play a significant role in protecting the cellular systems against radiation-induced free radicals, nitric oxide is implicated in radiation-induced decrease in hippocampal norepinephrine release. Home: 19561 Ridge Heights Dr Gaithersburg MD 20879 Office: Armed Forces Radiobiology Rsch Inst 8901 Wisconsin Ave Bethesda MD 20889-5603

KANDETZKI, CARL ARTHUR, engineer; b. New Haven, July 21, 1941; s. Arthur Karl and Eleanor Helen (Hecklinger) K.; m. Diane Josephine Yocher, Aug. 18, 1979; 1 child, Charles. BSChemE, Tufts U., 1963. Mgr. mktg. svcs. Dialog Computing, Milford, Conn., 1969-71; dir. market support Trans Com Inc., Bloomfield, Conn., 1971-73; sr. software engr. Gen. Automation, Anaheim, Calif., 1973-77; tech. specialist GE Info. Svcs. Co., Stamford, Conn., 1977-82; programmer Creative Output, Inc., Milford, Conn., 1982-84; cons. L&A Inc., Rye, N.Y., 1985-86; engr. N.E. Nuclear Energy Co., Waterford, Conn., 1986—; chmn. steering com. BWR Computer Software Maintenance, GE, San Jose, Calif., 1990—; presenter in field. Asst. coach Guilford (Conn.) Youth Soccer, 1988-90, 92, coach, 1991; co-chair vegetable dept. Guilford Agrl. Soc., 1975-82. Lt. USNR, 1963-65. Mem. Masons (sec. Halleck chpt. 1973-86, recorder Menunketuck coun. 1983-86). Republican. Lutheran. Office: NNECO-Millstone Unit 1 Rope Ferry Rd PO Box 128 Waterford CT 06385

KANDIL, OSAMA ABD EL MOHSIN, mechanical engineering educator; b. Cairo, Oct. 25, 1944; came to U.S., 1971, naturalized, 1977; s. Abd El Mohsin and Attiat El-Sayed (El-Shazli) K.; BS in Mech. Engring., Cairo U., 1966; MS in Mech. Engring., Villanova U., 1972; PhD in Engring. Mechanics, Va. Poly. Inst., 1974; m. Rawia Ahmed Fouad, Oct. 20, 1968; children: Dalya O., Tarek O. Instr. mech. engring. dept. Cairo U., 1966-70; grad. teaching asst. mech. engring. dept. Villanova (Pa.) U., 1971-72; grad. rsch. asst. engring. sci. and mechanics dept. Va. Poly. Inst., Blacksburg, 1972-74, asst. prof. engring. sci. and mechanics dept., 1975-78; prof., eminent scholar mech. engring. and mechanics dept., chmn. aerospace engring. dept. Old Dominion U., Norfolk, Va., 1978—; vis. prof. King Saud U., Riyadh, Saudi Arabia, 1983-84. NASA grantee, 1975-93, U.S. Airforce Office for Scientific Rsch. grantee, 1992-93, U.S. Army Research Office grantee, 1975-78, Naval Air Devel. Center grantee, 1980-81; NASA-Am. Soc. Engring. Edn. fellow, 1978-79. Fellow AIAA (assoc., tech. com. fluid dynamics, jour. reviewer); mem. AAUP, Am. Acad. Mechanics, Am. Soc. Engring. Edn., Soc. Engring. Scis., Va. Acad. Scis., Soc. Indsl. and Applied Math., Sigma Xi, Phi Kappa Phi. Moslem. Contbr. articles to profl. jours. Home: 7212 Midfield St Norfolk VA 23505-4124 Office: Old Dominion U Aerospace Engring Dept Norfolk VA 23529

KANDT, RONALD KIRK, computer software company executive, consultant; b. Long Beach, Calif., Dec. 1, 1954; s. Ronald Lee and Nancy Elise (Misener) K. BS in Computer Sci., U. Calif., Irvine, 1978, MS in Computer Sci., 1984. Programmer/analyst Jet Propulsion Lab., Pasadena, Calif., 1978; mem. tech. staff Hughes Rsch. Labs., Malibu, Calif., 1979-81, Symbolics, Woodland Hills, Calif., 1982-83, Xerox Vista Lab., Pasadena, 1983-84, USC Info. Scis. Inst., Marina del Rey, Calif., 1984-86; program mgr. Teknowledge, Thousand Oaks, Calif., 1986-88, Perceptronics, Woodland Hills, 1988-89; pres. Integrated Software Environments, Pasadena, Calif., 1990—; sr. mgr. Price Waterhouse, Menlo Park, Calif., 1993—; cons. Rockwell Sci. Ctr., Thousand Oaks, 1990-91, SMS Cos., Orange, Calif., 1990-91, Jet Propulsion Lab., Pasadena, 1990—. Contbr. articles to profl. jours. Mem. IEEE, Assn. Computing Machinery. Republican. Avocations: motorcycles, water sports. Office: Price Waterhouse 68 Willow Rd Menlo Park CA 94025

KANDULA, MAX, mechanical engineer, aerospace engineer; b. Kopperapad, India, Jan. 11, 1948; came to U.S., 1978; s. Raghavaiah and Subbamma (Botla) K.; m. Sue Kandula, May 12, 1977; children: Sushma, Shravan. BSME, U. Kerala, Trivandrum, India, 1969; MSME, Indian Inst. Tech., Madras, India, 1971; PhD in Mech. Engring., U. Ill., Chgo., 1980. Sr. engr. Indian Space Rsch. Orgn., Trivandrum, 1971-77; sr. rsch. assoc. Ill. Inst. Tech., Chgo., 1980-82; asst. prof. NYU, N.Y.C., 1982-83; rsch. scientist U. Houston, 1983-84; advanced systems engring. specialist Lockheed Engring. & Scis. Co., Houston, 1984—. Contbr. articles to Internat. Jour. Heat and Mass Transfer, Jour. Heat Transfer, Internat. Jour. Multiphase Flow, Internat. Jour. Numerical Methods in Engring., AIAA Jour., Jour. Propulsion and Power, Jour. Spacecraft and Rockets. Mem. AIAA (sr.), ASME. Achievements include definition of thermal conductivity function for charring ablators; design of physical model for deposition motion of liquid drops in two-phase flow, mechanisms of burnout in flow boiling of impinging liquid jets; research on fluid dynamics.

KANE, JAMES HARRY, mechanical engineer, educator, researcher; b. Bridgeport, Conn., Mar. 3, 1848. BSME, U. Bridgeport, Conn., 1977, MSME, 1981; PhD in Mech. Engring., U. Conn., 1986. Design engr. Lycoming Div., Textron Corp., Stratford, Conn., 1974-76, structural analyst, 1976-78, supr. computer aided design, 1978-80; instr. U. Bridgeport, 1980-84; v.p. Computer Aided Engring. Assocs., Woodbury, Conn., 1984-86; asst. prof. Worcester (Mass.) Poly. Inst., 1986-88; assoc. prof. Clarkson U., Potsdam, N.Y., 1989—; co-organizer symposia and multiple tech. sessions on various aspects of computational mechanics. Author: (textbook) Boundary Element Analysis in Engineering Continuum Mechanics, 1993; co-editor: Superlarge Problems in Computational Mechanics, 1989, Advances in Boundary Element Analysis, 1992; mem. editorial bd. Internat. Jour. Numerical Methods in Engring., 1990—, Computational Mechanics, An Internat. Jour., 1991—; contbr. articles to profl. jours. Recipient grant NSF, 1987, 91, grant NASA, 1990. Mem. ASME, AIAA, Soc. for Indsl. and Applied Mathematics, Internat. Assn. Boundary Element Methods, Phi Kappa Phi. Home: PO Box 706 Potsdam NY 13676 Office: Clarkson Univ Rm 305 Old Main Bldg Potsdam NY 13699

KANE, JOHN VINCENT, JR., nuclear physicist, researcher; b. Philadelphia, Penn., Feb. 13, 1928; s. John Vincent and Helen (Soden) K.; children: Michael A., Philip M. BS in Physics, Villanova U., 1950; MS, U. Pa., 1952, PhD, 1957. Scientist Brookhaven Nat. Labs., L.I., N.Y., 1957-61, Bell Telephone Labs., Murray Hill, N.J., 1961-66; prof. Mich. St. U., East Lansing, Mich., 1966-68, U. Saskachewan, Saskatoon, 1968-69, U. Munich, Munich, Germany, 1969-70; electronics Extrion Inc., Gloucester, Mass., 1973-74; nuclear physicist MIT, Cambridge, Mass., 1974. Member Am. Radio Relay League, Soaring Soc. Am., Am. Phys. Soc. (life), N.Y. Acad. Scis. Achievements include research in the first use of computer for nuclear physics which is measurements; in nuclear 2D spectroscopy of nuclear reactions with three particles in the final state; in precision electronic fast timing measurements of short lived nuclear states, O17, F17, B10; discovery of the first excited states of the helium 4 nucleus; of principle of low induced drag lens shaped wing planform and hyper-optical QM wave and flow pattern studies using computer graphics.

KANE, KEVIN THOMAS, editor; b. Dubuque, Iowa, Nov. 12, 1952; s. James Michael and Louise Kathryn (Maiers) K.; m. Mary Lee Schneider, July 28, 1985; children: Austin, Sean, Sarah Elizabeth. BA, St. John's U., 1974; MA, U. Iowa, 1980. Devel. editor William C. Brown Pubs., Dubuque, 1981-83, sales rep., 1983-85, project editor, biology, 1985-86, acquisition editor, biology, 1986-91, exec. editor life scis., 1991—; co-chair adv. com. on pollution prevention edn. U.S. EPA, Dubuque, Washington, 1990—. Editor over 200 coll. life sci. textbooks, 1987—. Min. hospitality St. Joseph's Cath. Ch., 1990—. Mem. AAAS, Am. Inst. Biol. Scis., Am. Soc. Zoologists, Dubuque Golf and Country Club. Office: William C Brown Pubs 2460 Kerper Blvd Dubuque IA 52001

KANE, MICHAEL JOEL, physician; b. Erie, Pa., July 2, 1951. BS, U.S. Naval Acad., 1973; MD, N.J. Med. Sch., 1983. Diplomate Am. Bd. Internal Medicine. Med. intern Thomas Jefferson U. Hosp., Phila., 1983-84, resident in medicine, 1984-86; fellow in neoplastic diseases Mt. Sinai Med. Ctr., N.Y.C., 1986-88; attending physician Jefferson Med. Coll., Phila., 1988-91, Med. Ctr. at Princeton, N.J., 1991—. Served to lt. U.S. Navy, 1969-79. Decorated Navy Achievement medal. Fellow ACP, Acad. Medicine of N.J., Am. Soc. Clin. Oncology, Am. Assn. Cancer Rsch., Oncology Soc. N.J., Med. Soc. N.J. Office: Med Ctr at Princeton 253 Witherspoon St Princeton NJ 08540

KANEKO, HISASHI, business executive, electrical engineer; b. Tokyo, Nov. 19, 1933; came to U.S., 1989; s. Shozo and Toshi K.; m. Motoko Washino; children: Satoshi, Makoto, Hajime. BSEE, U. Tokyo, 1956, PhD in Engring., 1967; MSEE, U. Calif., 1962. Rsch. staff NEC Corp., Japan, 1956-60, rsch. mgr., 1962-68, gen. mgr. transmission div., 1970-85, v.p., 1985-89, sr. v.p., 1989—, also bd. dirs.; pres., chief exec. officer NEC Am., N.Y.C., 1989—; rsch. asst. U. Calif., Berkeley, 1960-62; mem. tech. staff Bell Telephone Labs., Holmdel, N.J., 1968-70. Author 4 books in communications; contbr. 100 articles to profl. jours. Holder 70 patents in Japan, 4 in U.S.A. Recipient Kajii Meml. prize Elec. Comm. Assn., Japan, 1979. FEllow IEEE (E.H. Armstrong award 1992); mem. Inst. Electronics, Info. and Communications Engrs. (Achievement award 1985), Engring. Acad. Japan. Office: NEC Am Inc 8 Old Sod Farm Rd Melville NY 11747-3148

KANEKO, MASAO, chemist; b. Yokohama, Kanagawa, Japan, Jan. 28, 1942; s. Yasuji and Yoshie (Yuyama) K.; m. Tsuneko Tsugeyama, Feb. 27, 1970; 1 child, Yuki. BS, Waseda U., 1965, MS, 1967, PhD, 1970. Asst. Waseda U., Tokyo, 1970-71; postdoctoral researcher Free U., Berlin, Federal Republic of Germany, 1971-73; researcher Nissan Petrochemical Co. Ltd., Ichihara, Japan, 1973-75; sr. researcher RIKEN Inst., Wako, Japan, 1975-85; assoc. dir., chem. dynamics lab. RIKEN Inst., Wako, 1985-93; prof. Ibaraki U., Mito-shi, Japan, 1993—. Author: Advanced Polymer Science, 1987; editor-author: Photofunctional Polymers, 1991; contbr. articles to profl. jours. Recipient Alexander von Humboldt Found. award, 1972, Rsch. grant on Solar Energy Sci. Agy. for Sci. and Tech. Japan, 1978—. Mem. Soc. Polymer Sci. Japan (award 1989), Chem. Soc. Japan, Electrochemical Soc. Japan, Catalysis Soc. Japan, Am. Chem. Soc., Japanese Photochemistry Assn. Avocations: studying Japanese history, religion, culture. Office: Ibaraki U, Dept Chemistry, 2-1-1 Bunkyo Mito-shi 310, Japan

KANEKO, YOSHIHIRO, cardiologist, researcher; b. Shizuoka, Japan, Jan. 22, 1922; s. Rokurohei and Yoshino (Momochi) K.; m. Toyo Nozaki, Apr. 8, 1962; children: Kyoko, Eriko, Hiroko. MD, Tokyo U. Med. Sch., Japan, 1945, DMS, 1951. Clin. assoc. dept. internal medicine Tokyo U. Hosp., Japan, 1945-53, instr., 1953-70; rsch. fellow Cleve. Clinic Found., 1958-61, postdoctoral rsch. fellow, 1962-63; asst. prof. 2nd dept. internal medicine Tokyo U. Med. Sch., 1971-73; prof. medicine, chmn. dept internal medicine Yokohama City Univ. Med. Sch., Japan, 1973-87; dir. Yokohama Hypertension Rsch. Ctr., 1987—; hon. dir. Nishi-Yokohama Internat. Hosp., 1987—. Contbr. articles to profl. jours. Com. mem. Pharm. Bur. Japan Ministry Health & Welfare, Tokyo, 1974-87, Med. Affairs Bur., Tokyo, 1976-79. Grantee NIH, 1965-67; recipient award Japanese Kidney Found., 1986, Internat. Soc. Hypertension, 1988. Fellow High Blood Pressure Coun.; mem. Japanese Soc. Hypertension (1st pres. 1978-79, dir. 1978-89), Japanese Soc. Internal Medicine (councilor), Japan Circulation Soc., Japan Soc. Nephrology (dir. 1974-87), Am. Heart Assn. (coun. mem.), Internat. Soc. Hypertension, (coun. 1982-90, chmn. 1988). Avocations: reading, gardening. Home: 2-27-14 Nishishiba, Kanazawa-ku, Yokohama 236, Japan Office: Yokohama Hypertension Rsch Ctr, Deiki 2-8-19-402 Kanazawa-ku, Yokohama 236, Japan

KANES, WILLIAM HENRY, geology educator, research center administrator; b. N.Y.C., Oct. 15, 1934; married. BS in Geol. Engring., CCNY, 1956; MS in Geology, W.Va. U., 1958, PhD in Geology, 1965. Sr. rsch. geologist Esso Prodn. Co., Houston Rsch. Co., 1964-65; sr. exploration geologist, head New Concepts Group Esso Standards, Libya, 1966-67, frontier exploration geologist, administr. Frontier Area Group, 1967-69; asst. prof. geology W.Va. U., Morgantown, 1970-71; assoc. prof. geology U. S.C., Columbia, 1971-74, prof. geology, dir. Earth Scis. and Resources Inst., 1975—, Disting. prof. earth resources, chair Rsch. and Devel. Found., 1984—; NSF Resident Rsch. prof. Acad. Sci. Rsch. and Tech., Cairo, 1976-77; hon. professorial fellow Univ. Coll. Aberystyth U. Wales, 1979-83, Univ. Coll. Swansea U. Wales, 1988-85; vis. professorial fellow Univ. Coll. Swansea, 1977-83; co-dir. Earth Resources Inst. Univ. Coll. Swansea, U. Wales, U.K., 1980-86, Earth Scis. and Resources Inst. U. Bristol, U.K., 1986—; advisor Atomic Energy Establishment, Egypt, 1974-77, Nat. Oil Co., Libya, 1975-78, U.S. Pres., exec. br. Energy Problems and Controls, 1977-78, Nuclear Materials Corp., Egypt, 1977-81; mem. tech. adv. task force Fed. Power Commn. Contbr. numerous articles, papers to profl. publs. 1st lt. C.E., U.S. Army, 1955, 58-59. Recipient Disting. Svc. award U. S.C. Ednl. Found., 1985; grantee NSF, 1971-81, U.S. Dept. Interior, 1972-74, others. Fellow AAAS, Geol. Soc. Am.; mem. Am. Assn. Petroleum Geologists (cert., chmn. rsch. symposium 1976, mem. acad. affairs com. 1973-76, acad. liaison com. 1976—, rsch. com. on pub. affairs 1975—), Am. Geophys. Union, Geol. Soc. Malaysia, Soc. Econ. Paleontologists and Mineralogists, Ptnrs. of Am. U. S.C., Sigma Xi. Office: U SC Earth & Scis Resources Inst Byrnes Internat Ctr 901 Sumter St Columbia SC 29208

KANESHINA, SHOJI, biophysical chemistry educator; b. Tokushima, Japan, Mar. 5, 1942; s. Shigeo and Kimiko (Kishimoto) K.; m. Fumiko Horie, Jan. 4, 1970; children: Katsuhiko, Atsuko, Kimitoshi. M, Tokushima U., 1966; PhD, Kyushu U., 1974. Rsch. assoc. Tokushima U., 1966-74, prof., 1988—; assoc. prof. Kyushu U., Fukuoka, Japan, 1974-85, prof., 1985-88; vis. rsch. prof. U. Utah, Salt Lake City, 1979-81. Contbr. articles to Jour. Colloid Interface Sci., Bull. Chem. Soc. Japan, Biochim. Biophys. Acta. Mem. Am. Chem. Soc., Japan Soc. High Pressure Sci. and Tech., Chem. Soc. Japan, Internat. Assn. Colloid and Interface Scientists. Achievements include research in effect of pressure on the molecular assemblies such as micelles, vesicles and adsorbed monolayers; pressure-anesthetic antagonism. Office: Tokushima U, Minamijosanjima, Tokushima 770, Japan

KANG, BANN C., immunologist; b. Kyungnam, Korea, Mar. 4, 1939; d. Daeryong and Buni (Chung) K.; came to U.S., 1964, naturalized, 1976; A.B., Kyungpook Nat. U., 1959, M.D., 1963; m. U. Yun Ryo, Mar. 30, 1963. Intern, L.I. Jewish Hosp.-Queens Hosp. Center, Jamaica, N.Y., 1964-65, resident in medicine, 1965-67; teaching asso. Kyungpook U. Hosp., Taegu,

Korea, 1967-70; fellow in allergy and chest Creighton U., Omaha, 1970-71; fellow in allergy Henry Ford Hosp., Detroit, 1971-72; clin. instr. medicine U. Mich. Hosp., Ann Arbor, 1972-73; asst. prof. Chgo. Med. Sch., 1973-74; chief allergy-immunology Mt. Sinai Hosp., Chgo., 1975—; asst. prof. Rush Med. Sch., Chgo. 1975-84, assoc. prof., 1984-86; assoc. prof. U. Ky. Coll. Medicine, 1987-92, prof., 1992—; cons. allergy-immunology Edgewater Hosp., Chgo., St. Anthony's Hosp., Chgo., 1976—, Nat. Heart, Lung, Blood Inst., 1979—; mem. Exptl. Transplantation Adv. Bd., Ill., 1985-86, Diagnostic and Therapeutic Tech. Assessment (AMA), 1987—, Gen. Clin. Rsch. Com. (NIH), 1989-93; counselor Chgo. Med. Soc., 1984-86, mem. policy com., adv. com. to health dept. Chgo. and Cook County, 1984-86, adv. com. Ctr. for Biologics and Rsch., FDA, 1993—. Recipient NIH award U. Mich., 1972-73. Diplomate Am. Bd. Internal Medicine, Am. Bd. Allergy-Immunology. Fellow ACP, Am. Acad. Allergy; mem. Am. Fedn. Clin. Research, AMA, Inter-Asthma Assn. Contbr. over 40 articles to profl. jours. Home: 2716 Martinique Ln Lexington KY 40509-9509 Office: U Ky Coll Medicine K528 Albert B Chandler Med Ctr 800 Rose St Lexington KY 40536

KANG, BIN GOO, biologist; b. Yokohama, Japan, Nov. 17, 1936; s. Sung Wook and Chung Soon (Park) K.; m. Soja Kim, Apr. 26, 1969; children: Judith Inja, Evelyn Minja. BS, Yonsei U., Seoul, Republic of Korea, 1959, MS, 1961; MS, Tufts U., Medford, Mass., 1963; PhD, U. Mich., Ann Arbor, 1967. Rsch. assoc. Mich. State U., East Lansing, 1967-69; rsch. scientist Fairchild Garden Rsch. Ctr., Miami, Fla., 1969-74; prof. biology Yonsei U., Seoul, 1974—; vis. prof. Nagoya (Japan) U., 1991-92. Author: Cell Biology, 1981, Symposium Procs., 1989, College Biology, 1990; editor-in-chief Korean Jour. Botany, 1988-90; contbr. articles to profl. jours. Humboldt fellow U. Freiburg, Fed. Republic of Germany, 1977-78, Smithsonian Instn. fellow, Washington, 1984-85. Mem. Bot. Soc. Korea (v.p. 1988-92), Biochem. Soc. Korea (coun. 1984-91), Korean Soc. Molecular Biology (coun. 1989-91), Am. Soc. Plant PHysiologists, Janpanese Soc. Plant Physiologists, Scandinavian Soc. Plant Physiology. Office: Yonsei U, Biology Dept, Seoul Republic of Korea

KANG, JULIANA HAENG-CHA, anesthesiologist; b. Mokpo, Cheonnam, People's Republic of Korea, July 1, 1941; came to U.S., 1965; d. Johan and E-E-Suk (Lee) Kang; married; children: Mee-Kyung, Mee-Ae, Han-Bae. MD, Yonsei U., Seoul, People's Republic of Korea, 1965. Intern Pittsfield (Mass.) Gen. Hosp., 1965-66; asst. prof. biology Yonsei U., 1965; resident in anesthesiology D.C. Gen. Hosp., 1966-67, Yale-New Haven (Conn.) Hosp., 1967-69; asst. prof. anesthesiology U. Conn., Farmington, 1970-75, 82-85; vice chairperson anesthesia dept. Conn. Surgery Ctr., Hartford, Conn., 1985-86; med. dir., chairperson anesthesia dept. Conn. Surgery Ctr., Hartford, 1986—. Mem. Am. Med. Women's Assn., Am. Soc. Ambulatory Surgery Anesthesia, Am. Soc. Anesthesiologists, Conn. Soc. Anesthesiology, Nat. Abortion Rights Action League, Naral Polit. Arm of Pro-Choice. Office: Conn Surgery Ctr 81 Gillett St Hartford CT 06105-2648

KANG, MINHO, engineering executive; b. Kyungnam, Republic of Korea, July 20, 1946; s. Ji Jung and Ok Hee (Lee) K.; m. Ae Soon Choi, July 10, 1971; children: Soo Jin, Soo Young. B, Seoul (Korea) Nat. U., 1969; M, U. Mo., Rolla, 1973; D, U. Tex., 1977. Mem. tech. staff Bell Lab., Holmdel, N.J., 1977-78; v.p., dir. Electronics & Telecomm. Rsch. Inst., Daejon, Republic of Korea, 1978-90; dir. gen. electronics rsch. Ministry of Sci. and Tech., Republic of Korea, 1985-88; CEO Korea Telecomm. S.W. Rsch. Lab, Seoul, 1991-92, Korea Telecomm. Rsch. Ctr., Seoul, 1990-93; tech. advisor Seoul Olympics Orgn. Com., 1987-88; advisor Presdl. Con. on Sci. and Tech., Seoul, 1989-90, CEO Korea Advanced Minicomputer Devel. Consortium Nat. Computerization Bd., Seoul, 1991-93. Co-author: Optical Fiber Communications, 1981, Laser Application, 1983, 2d edit., 1986, Indtroduction to Electrical Communications Technology, 1990, Introduction to ISDN and Broadband Communications Systems, 1991, Broadband Telecommunications Technology, 1993; contbr. articles to profl. jours. Recipient Nat. Medal of Honor, Govt. of Korea, 1982; New Indsl. Tech. Mgmt. grand prize 21st Century Top Mgmt. Club, 1991. Mem. IEEE (Korean sect., sr.), Korean Inst. Telematics and Electronics (bd. dirs. 1978—), Korean Inst. Comm. Scis. (bd. dirs. 1979—), Korea EMI/EMC Soc. (bd. dirs. 1992—), Phi Kappa Phi, Eta Kappa Nu. Achievements include research in optical fiber endface measurement method. Home: 308-1202 Sinbanpo Apt, Seocho-Gu Seoul Republic of Korea Office: Korea Telecomm Quality Assurance Ctr, 1 Wonhyoro-3ga, Yongsan-Gu Seoul Republic of Korea

KANG, SANG JOON, ecologist, educator; b. Pyoson-ri, Chejo-do, Korea, Oct. 3, 1940; s. Dong Choon and Jong Yeo (Ko) K.; m. Yong Ja Woom, Dec. 4, 1964; children: Young Mi, Jin Yong, Se Hyung. MS, Seoul Nat. U., 1968; PhD, Tohoku U., Sendai, Japan, 1982. Assoc. prof. Chuncheon (Korea) Tchrs. Coll., 1968-78; researcher Malaysia Tchrs. Coll., Penang, 1976; vis. scholar Tohoku U., Sendai, Japan, 1978-79, 84-85; vis. scientist UCLA, 1987-88; prof. ecology Chungbuk Nat. U., Cheongju, Republic of Korea, 1978—. Author: Study on Nature of DMZ in Korea, 1987, Plant and Animal of Chungnam, Korea, 1988, Vegetation of Chungbuk, Korea, 1989, History and Culture in Sosan and Taean, 1991. 2nd lt. Korean Army, 1964-68. Mem. Korean Soc. Ecology (editor mem. 1984—), Korean Soc. Botany (editor mem. 1984-86), Korean Soc. Limnology (dir. mem. 1990—), Brit. Ecol. Soc., NRC of Can., Ecol. Soc. Am. Home: Ra-dong 406 Pyung-Wha Apt, Young-dong, Cheongju 360-020, Republic of Korea Office: San 48 Kaesin-dong, Cheongju 360-763, Republic of Korea

KANG, SHIN IL, economist; b. Seoul, Korea, Jan. 7, 1955; s. Min Chang and In Suk (Cha) K.; m. Kyong Ok Chon; children: Young Suk, Kun Suk. BA, Hankuk U. Fgn. Studies, Seoul, 1980; MA, Ohio State U., 1984, PhD, 1986. Rsch. fellow Korea Devel. Inst., Seoul, 1986-89; rsch. coord. Korea Econ. Rsch. Inst., Seoul, 1989-91; asst. prof. Hansung U., Seoul, 1991—; cons. Asian Devel. Bank, Manila, 1988. Author: Privatization in Korea, 1988, Role of R&D in Public and Private Sector in Korea, Analysis of Changes in Korean Firms' Growth and Size, 1990, A Study on the Business Group, 1991; editor: Jour. Korean Econ. Studies, 1990. Fellow Assn. Korean Pub. Enterprises. Home: Yongsanku Subingodong, Shindongah Apt # 16-1103, Seoul Republic of Korea Office: 389-2GA Samsun-Dong Sungbuk gu, Seoul 150-756, Republic of Korea

KANG, SUNG KYEW, medical educator; b. Naju, Chonnam, Republic of Korea, Sept. 5, 1941; parents WounDong and GongRei (Kim) K.; children: Eun-Seok, Hyun-Seok. MD, Chonnam Nat. U., 1967, PhD, 1978. Diplomate Korean Bd. Nephrology, Korean Bd. Internal Medicine. Instr. Chonbuk (Republic of Korea) Nat. U. Med. Sch., 1975-78, assoc. prof., 1978-86, prof., 1986—; vis. prof. Health Sci. Ctr. at Bklyn. SUNY, 1983-84, chmn. internal medicine, 1989—. Served to maj. Korean Army, 1972-75. Recipient Chung-Rham award Korean Assn. Internal Medicine, 1983. Mem. Internat. Soc. Nephrology, Am. Soc. Nephrology, Asian Pacific Congress Nephrology, Honam Soc. Nephrology (chmn.). Office: Chonbuk Nat U Med Sch, 2-20 Kum-Ahm Dong, Chonju Chonbuk 560-182, Republic of Korea

KANICKI, JERZY, chemist, researcher; b. Kalisz, Poland, Aug. 9, 1954; came to U.S., 1983; s. Kazimierz and Genowefa (Podwalna) K.; m. Ariane Michaux, Dec. 3, 1982; children: Nathalie, Eric. MS in Chemistry, Free U. Brussels, 1978, ScD, 1982. Mem. rsch. staff IBM Rsch. Div., Yorktown Heights, N.Y., 1983—; organizer, presenter at nat. and internat. profl. confs. Editor: Amorphous and Microcrystalline Semiconductor Devices, vol. I, 1991, vol. II, 1992; contbr. over 100 articles to profl. publs. Sgt. maj. Belgian army, 1982-83. Mem. Materials Rsch. Soc., Am. Phys. Soc. Roman Catholic. Achievements include research on large-area amorphous, microcrystalline and polycrystalline thin film semiconductors and insulators, thin film transistors liquid crystal displays, solar cells made of amorphous inorganic and organic semiconductors. Office: IBM TJ Watson Rsch Ctr PO Box 218 Yorktown Heights NY 10598

KANN, HERBERT ELLIS, JR., hematologist, oncologist; b. Ft. Worth, Apr. 25, 1939; m. Carol Anne Lamb; children: Susan Blair, Kristen Elizabeth. BA in Chemistry, Duke U., 1960, MD, 1964. Diplomate Am. Bd. Internal Medicine, Am. Bd. Hematology, Am. Bd. Oncology. Sr. investigator Lab. of Molecular Pharmacology Nat. Cancer Inst., Bethesda, Md., 1969-74; co-dir. divsn. hematology-oncology Sch. Medicine Emory U., Atlanta, 1975-77, assoc. prof. medicine, 1974-79; staff physician DeKalb Med. Ctr., Decatur, Ga., 1979—; pvt. practice DeKalb Hematology-Oncology PC, Decatur, 1979—; mem. devel. therapeutics com. Nat. Cancer Inst., Bethesda, 1974-79; chmn. cancer com., chief dept. oncology DeKalb Med. Ctr., Decatur, 1982-84. Fellow ACP; mem. AAAS, Am. Assn. for Cancer Rsch., Am. Soc. Hematology. Home: 2290 Chrysler Ct NE Atlanta GA 30345 Office: DeKalb Hematology-Oncology 2675 N Decatur Rd Ste 701 Decatur GA 30033

KANNAN, RAMANUJA CHARI, civil engineer; b. Thanjavur, Tamil Nadu, India, Jan. 14, 1947; came to U.S., 1984; s. Krishnamachari Ramanujachari and Kalyani (Parthasarathy) Chari; m. Basantha Thyaharaj, Jan. 12, 1976; children: Geoffrey, Michelle. BE, Shivaji U., Kolhapur, India, 1968; M Tech., Indian Inst. Tech., Bombay, 1970; ME, U. Fla., 1972. Registered profl. engr., Fla., Pa., Del., La., N.J. Project mgr. Gherzi Ea. Ltd./Essar Constrn., Bombay, Madras, India, 1973-79; dir. engring. Analabs/Bylander Meihardt Partnership, Singapore, 1979-81; lectr. Nanyang Tech. U., Singapore, 1981-84; sr. assoc. engr. Ardaman & Assocs., Inc., Ft. Myers, Fla., 1984-86; sr. engr. A & E Testing, Inc., St. Petersburg, Fla., 1986-88; pres. R. Kannan & Assocs., Inc., St. Petersburg, 1988-91; v.p. Triegel, Kannan & Assocs., Inc., St. Petersburg, 1991—; charter mem. Singapore Concrete Inst., 1982; grad. asst. U. Fla., Gainesville, 1970-71. Contbr. articles to profl. jours. Vol. Mcpl. Corp. Greater Bombay, 1962-64; counselor Singapore Anti-Narcotics Assn., 1979-81; vol. Pinellas County (Fla.) Schs., 1988—. Recipient Cert. of Appreciation Pinellas County Sci. Fair, St. Petersburg, 1987-92. Fellow ASCE; mem. ASTM (com. mem. 1991—). Achievements include design of the curriculum in soil mechanics and found. engring. at Nanyang Tech. U.; design and setup the soil mechanics lab. facility; design of the industry-inst. coop. program and supervision of student tng. Office: Triegal Kannan & Assocs Inc 3839 4th St N Ste 350 Saint Petersburg FL 33703

KANNAN, RANGARAMANUJAM, polymer physicist, chemical engineer; b. Thirukoilur, India, Sept. 8, 1966; came to U.S., 1987; s. A. Rangaramanujam and Santhalakshmi N. R. MS, Pa. State U., 1989, Calif. Inst. Tech., 1991; PhD, Calif. Inst. Tech., 1993. Rsch. asst. Calif. Inst. Tech., Pasadena, 1989—; vis. collaborator Planck Inst Polymerforschung, Mainz, Germany, 1992. Contbr. articles to sci. publs. Mem. AICE (assoc.), Am. Phys. Soc., Soc. Rheology. Achievements include discovery that oscillatory shear is effective in inducing macroscopic orientation in side-group liquid-crystalline polymers which have great potential in non-linear optics and optical data storage media. Office: Calif Inst Tech Chem Engring 210 41 Pasadena CA 91125

KANNAN, RAVI, mathematician educator. Prof. math. Carnegie Mellon U., Pitts. Recipient Leroy P. Steele prize Am. Math. Soc., 1992. Office: Carnegie-Mellon U Dept Math 5000 Forbes Ave Pittsburgh PA 15213*

KANNANKERIL, CHARLES PAUL, chemical engineer; b. Kumbalanghy, India; came to the U.S. 1967; s. Paul Joseph and Treesa Paul (Pazhamadam) K.; m. Mary Charles Erinjeri, Jan. 4, 1976; children: Charlene, Crystal. BS in Chemistry, U. Kerala, India, 1966; MS in Chemistry, Lowell (Mass.) Tech. U., 1972, MS in Engring., 1974. Project engr. Amerace Ltd., Markham, Ontario, Canada, 1975-80; sr. devel. engr. Bishop Electric Co., Cedar Grove, N.J., 1980-83, Sealed Air Corp., Fair Lawn, N.J., 1983—. Mem. Soc. Plastics Engrs. Democrat. Roman Catholic. Achievements include 6 patents and 3 patents pending. Office: Sealed Air Corp 19-01 State Rt 208 Fair Lawn NJ 07410-2824

KANNEL, JERROLD WILLIAMS, mechanical engineer; b. Quincy, Mass., May 5, 1935; s. Ira Jay and Phyllis Mary (Williams) K.; m. Sharon Ann Grogg, Apr. 25, 1958; children: Melinda Susan, Stephanie Lynne. MS, Mich. State U., 1960; PhD, N.C. State U., 1986. Rsch. engr. Battelle Inst., Columbus, Ohio, 1960-65, prin. engr., 1965-71, assoc. fellow, 1971-77, sr. scientist, 1971-84, rsch. leader, 1985—. Contbr. articles to profl. jours. Vice pres. Civic Assn., Columbus, 1978, pres., 1979. Fellow ASME (chmn. tribiology div. 1990, chmn. honors com. 1992, Gov.'s award 1991); mem. Sigma Xi. Ch. of Christ. Achievements include measurement of contact pressures in a rolling contact bearing; prediction traction between coated rollers; research in causes of ball bearing instability; causes of journal bearing seizure; development of spl. self-lubricating cage for cryogenic ball bearing. Home: 5385 Crawford Dr Columbus OH 43229-4137 Office: Battelle Meml Inst 505 King Ave Columbus OH 43201

KANOFSKY, JACOB DANIEL, psychiatrist, educator; b. Phila., Apr. 16, 1948; s. Philip and Mollie (Edelstein) K. BA in Physics, Temple U., 1965-69; MD, Thomas Jefferson Med. Coll., Phila., 1974; MPH in Epidemiology, Johns Hopkins U., 1978. Diplomate Am. Bd. Psychiatry and Neurology. Intern Met. Hosp., N.Y.C., 1974-75; resident in psychiatry St. Luke's-Roosevelt Hosp. Ctr., Columbia U., N.Y.C., 1978-80, fellow in psychiat. epidemiology, 1980-82; asst. editor-in-chief Med. Tribune, N.Y.C., 1984-85; ward chief rsch. unit Bronx (N.Y.) Psychiat. Ctr., 1986, assoc. clin. dir., 1986-87, acting clin. dir., 1987, pres. med. staff orgn., 1987-89; assoc. dir. schizophrenia rsch. Albert Einstein Coll. Med./Bronx Psychiat. Ctr., 1989-90, sr. rsch. psychiatrist, 1989—, asst. prof. psychiatry, 1986—; asst. prof. epidemiology and social medicine Albert Einstein Coll. Med., 1993—; lectr. in psychiatry Columbia U., N.Y.C., 1980—; attending psychiatrist St. Luke's-Roosevelt Hosp. Ctr., 1980—; contbg. editor Med. Tribune, 1986—. Consulting editor Jour. of the Am. Coll. of Nutrition, 1990—; contbr. over 50 articles to profl. jours. Fellow Am. Coll. Nutrition; mem. Am. Psychiat. Assn. Jewish. Avocations: swimming, hiking, piano. Office: Bronx Psychiat Ctr 1500 Waters Pl Bronx NY 10461-2723

KANOFSKY, JEFFREY RONALD, physician, educator; b. Chgo., Apr. 30, 1946; s. Louis and Shirley (Frank) K.; m. Donna Ann Cohen, May 21, 1972. BS, Ill. Inst. Tech., Chgo., 1968, MS, 1970, PhD in Chemistry, 1972; MD, Rush Med. Coll., Chgo., 1975. Bd. cert. medicine, med. oncology, hematology. Staff physician Hines (Ill.) VA Hosp., 1980—; asst. prof. medicine Loyola U. Stritch Sch. Medicine, Maywood, Ill., 1980-86, assoc. prof., 1986-88, assoc. prof. medicine and biochemistry, 1988-91, prof., 1991—. Rsch. grantee NIH, 1983, 86, Dept. Vets. Affairs, 1982, 84, 87, 90, 93. Fellow ACP; mem. Am. Soc. Cancer Rsch., Am. Soc. Hematology, Am. Soc. Photobiology, Am. Soc. Clin. Oncology, The Oxygen Soc. Achievements include patent in electronic comm; many published biochemical studies of singlet oxygen production. Home: 261 Adelia Elmhurst IL 60126 Office: Hines VA Hosp Box 278 Hines IL 60141

KANOH, MINAMI, neuropsychologist, clinical psychologist; b. Hokkaido, Japan, May 19, 1927; s. Osamu and Tomi (Tanaami) K.; m. Kazuko Senba, Oct. 10, 1968; children: Naoto, Ikuo, Tomoh. BA, Hokkaido U., 1951; postgrad., Tokyo U., 1952-55. Lectr. in social psychology Hokkaido U., Sapporo, 1956-66, assoc. prof., 1966-71, prof. clin. psychology and spl. edn., 1971-91, dean faculty edn., 1977-81, councillor, 1972-87, prof. emeritus, 1991-92; prof. social psychology Sapporo Gakuin U., Ebetsu, Hokkaido, Japan, 1992—. Contby author: On the Universe of Psychology, 1957, Basic Mechanism of Learning, 1967, Dynamics of Activation and Inhibition in Learning Process, 1982, The Mechanism of Psychological Healing, 1990, Toward a Unified Study on Human Mind, 1992; joint editor Japanese Jour. Ednl. Psychology, 1984-86. Mem. Hokkaido Psychol. Assn. (pres.), Japanese Soc. Psychiatry and Neurology, Japanese Soc. Physiol. Psychology and Psychophysiology (pres. 1989). Avocations: hiking, canoeing, sea fishing, reading, old books. Home: Katsuraoka 27-5, Otaru, Hokkaido 047-02, Japan Office: Sapporo Gakuin U Faculty of, Social Info Bunkyodai 11, Ebetsu Hokkaido 069, Japan

KANT, ARTHUR, scientist emeritus, physical chemist; b. N.Y.C., Apr. 28, 1915; m. Charlotte Kaplan, June 13, 1955; children: Laurence, Deborah, Amy. BA, U. Wis., 1938, MA, 1939; DSc, Carnegie Inst. of Tech., 1951. Phys. and radiochemist Nat. Rsch. Def. Comm., Manhattan Pros., Iowa State Coll., Ames, 1942-45; assoc. chemist U.S. Bur. of Mines, Pitts., 1945-49; rsch. chemist Brookhaven Nat. Lab., Upton, N.Y., 1951-53; radio chemist Air Force Cambridge Rsch. ctr., Bedford, Mass., 1953-55; chief nucleonics br. Watertown (Mass.) Arsenal Lab., 1955-83, scientist emeritus, 1983—. Contbr. articles to profl. jours. Mem. Am. Chem. Soc., Sigma Xi, Phi Lambda Upsilon. Achievements include patent for solvent extraction for purification of thoriam from fissionable radioactive elements; discovery of the diatomic molecules of the transition elements and estimation of the thermodynamic properties. Home: 139 Woodridge Rd Wayland MA 01778

KANTHETI, BADARI NARAYANA, aeronautical engineer; b. Andhra Pradesh, India, Oct. 4, 1954; s. Venkata Krishnaiah and Seetaravamma Kantheti; m. Prameela Rani, July 2, 1980; 1 child, Sharath Krishna. BS in Engring. with honors, Regional Engring. Coll., Calcutta, 1979; ME, Indian Inst. Sci., Bangalore, 1983, PhD, 1991. Product engr. Instrumentation Engrs., Hyderabad, India, 1979-80; plant engr. Magnesium Products, Bangalore, 1980; project asst. dept. aerospace engring. Indian Inst. Sci., Bangalore, 1983-85, sci. officer, 1985-89; scientist ISRO-Satellite Ctr., Bangalore, 1989—. Contbr. articles to profl. jours., confs. Recipient Mrs. Sabita Chaudhuri Meml. Gold Medal, 1992; J.N. Tala Meml. fellow Indian Inst. Sci., 1986; scholar Regional Engring. Coll.-Calicut, 1975-78, Indian Inst. Sci., 1981-83. Mem. AIAA, Aero. Soc. India (assoc.), Non-Destructive Testing of India (assoc.), Indian Soc. Tool Engrs., Acoustic Emission Working Group India. Achievements include development of unified approach for evaluation of serr, a useful fracture parameter, fatigue and fracture, composite materials. Home: # 15 MSH Indian Inst Sci, Bangalore 560012, India Office: ISRO-Satellite Ctr, Vimanapura Rd, Bangalore 560017, India

KANTOR, MEL LEWIS, dental educator, researcher; b. N.Y.C., July 13, 1956; s. Irving and Sarah (Schneider) K. BA in Chemistry and Math., CUNY, 1977; DDS, U. N.C., 1981. Diplomate Am. Bd. Oral & Maxillofacial Radiology. Resident Hennepin County Med. Ctr., Mpls., 1981-82, U. Conn. Health Ctr., Farmington, 1982-84; asst. prof. U. N.C. Sch. Dentistry, Chapel Hill, 1984-88, U. Conn. Sch. of Dental Medicine, Farmington, 1988-92; clin. assoc. prof. UMDNJ-N.J. Dental Sch., Newark, 1993—; cons. FDA Dental Selection Criteria Panel, 1985-87; test constructor Nat. Bd. Dental Exam., 1989-93; mem. bd. dirs. Radiology Centennial, Inc., 1992—. Assoc. editor Jour. Dental Edn., 1986—; contbr. articles to Jour. Chem. Physics, Jour. ADA, Jour. Dental Rsch., Oral Surgery, Oral Medicine and Oral Pathology, Jour. Dental Edn. Mem. Internat. Assn. Dental Rsch. (founding mem. diagnostic systems group, group program chmn. 1993-95), Am. Acad. Oral and Maxillofacial Radiology (history commn., long range planning com., media com., radiology centennial com.), Am. Assn. of Dental Schs. (sec. sect. of Oral and Maxillofacial Radiology, 1993-94), Internat. Assn. of Dentomaxillofacial Radiology, Radiological Soc. of N. Am. Office: UMDNJ-NJ Dental Sch 110 Bergen St C827 Newark NJ 07103-2400

KANTOWITZ, BARRY HOWARD, ergonomist, researcher; b. N.Y.C., Aug. 25, 1943; s. Charles and Dinah K.; Susan Rothman, 1968; children: David, Riva. MA, Queens Coll. CUNY, 1967; PhD, U. Wis., 1969. Prof. Purdue U., W. Lafayette, Ind., 1969-87; chief. scientist Battelle Ctr. for Transp. Human Factors, Seattle, 1987—; pres. Puget Sound Human Factors Soc., Seattle, 1989. Editor: Human Information Processing, 1974; author: Human factors, 1983, Research Methods in Psychology 4th ed., 1989, Experimental Psychology 4th ed., 1991. Fellow Am. Psychol Soc., Soc. Engring. Psychologists. Office: Battelle Seattle Rsch Ctr 4000 NE 41st St Seattle WA 98105

KANTROWITZ, ARTHUR, physicist, educator; b. N.Y.C., N.Y., Oct. 20, 1913; s. Bernard A. and Rose (Esserman) K.; m. Rosalind Joseph, Sept. 12, 1943 (div.); children: Barbara, Lore, Andrea; m. Lee Stuart, Dec. 25, 1980. B.S., Columbia U., 1934, M.A., 1936, Ph.D., 1947; DEng (hon.), Mont. Coll. Mineral Sci. and Tech., 1975; D.Sc. (hon.), N.J. Inst. Tech., 1981. Physicist NACA, 1935-46; prof. aero. engring. and engring. physics Cornell U., 1946-56; founder, dir., chmn., chief exec. officer Avco-Everett Research Lab., Everett, Mass., 1955-78; sr. v.p., dir. Avco Corp., 1956-79; prof. Thayer Sch. Engring., Dartmouth Coll., 1978—; vis. lectr. Harvard U., 1952; Fulbright and Guggenheim fellow Cambridge and Manchester univs., 1954; fellow Sch. Advanced Study, MIT, 1957, vis. inst. prof., 1957—; Joseph Wunsch lectr. Technion, Haifa, Israel, 1968; Messenger lectr. Cornell U., 1978; hon. prof. Huazhong Inst. Tech., Wuhan, China, 1980; mem. Presdl. Adv. Group on Anticipated Advances in Sci. and Tech., head task force on sci. ct., 1975-76; mem. tech. adv. bd. U.S. Dept. Commerce, 1974-77; mem. adv. panel NOVA, Sta. WGBH-TV; bd. overseers Center for Naval Analyses, 1973-83; mem. adv. council Israel-U.S. Binational Indsl. Research and Devel. Found., 1978-81; bd. govs. The Technion (hon. life); mem. adv. council NASA, 1979, 80; life trustee U. Rochester; past mem. sci. and engring. adv. com. U. Rochester, Princeton U., Stanford U. and Rensselaer Poly Inst.; vis. prof. U. Calif., Berkeley, 1983. Contbr. articles to profl. jours.; patentee in field. Bd. dirs. Hertz Found, 1972—; mem. bd. advisors Teller Found., 1992—. Recipient award Am. Acad. Achievement, 1966, Theodore Roosevelt medal, 1967, Kayan medal Columbia U., 1973, MHD Faraday Meml. medal UNESCO, 1983. Fellow AAAS, AIAA (Fluid and Plasmadynamics medal 1981, Aerospace Contbn. to Soc. award 1990), Am. Acad. Arts and Scis., Am. Phys. Soc., Am. Astronautical Soc., Am. Inst. for Med. and Biol. Engring.; mem. NAS, NAE, Internat. Acad. Astronautics, Am. Inst. Physics, Sigma Xi. Achievements include high-energy lasers, heart assist devices, MHD, re-entry from space; early work in fusion and molecular beams noteable. Home: 4 Downing Rd Hanover NH 03755-1902

KANUTH, JAMES GORDAN, chemical engineer; b. Lexington, Ohio, June 18, 1953; s. John Gordon and Helena Jane (Castor) K.; m. Darlene Louise Dowell, Oct. 23, 1976; children: Cheri Marlene Tacoronti, Robert Gordon. BSChemE, U. Cin., 1976. Project engr. Joseph E. Seagram and Sons, Inc., Lawrenceburg, Ind., 1976-80; prodn. engr. Monsanto (name changed to Conoco), Alvin, Tex., 1980-81; sr. area engr. utilities Conoco (name changed to Oxy Chem), Alvin, 1981-89; regional mgr. Puckorius and Assocs., Inc. indsl. water treatment cons., League City, Tex., 1989—. Pres. Gulf Coast Energy Conservation Soc., Houston, 1988-89, Galveston County Mcpl. Utility Dist. 3, League City, 1983-88; city councilman City of League City, 1988—; bd. dirs. Houston Galveston Area Coun., 1991; treas. Clear Lake Area Coun. of Cities, Webster, Tex., 1990—. Mem. Nat. Assn. Corrosion Engrs. (com. mem. 1989—), Am. Inst. Chem. Engrs., Cooling Tower Inst. (water treatment com. 1981—). Presbyterian. Avocations: boating, reading, watching my childrens sports. Home: 217 Glen Haven Dr League City TX 77573-4304 Office: Puckorius and Assoc Inc PO Box 678 League City TX 77574-0678

KAO, CHARLES KUEN, electrical engineer, educator; b. Shanghai, China, Nov. 4, 1933; s. Chun-Hsien and Tisung Fong K.; m. May Wan Wong, Sept. 19, 1959; children—Simon M.T., Amanda M.C. B.Sc. in Elec. Engring., U. London, 1957, Ph.D. in Elec. Engring., 1965. Devel. engr. Standard Telephones & Cables Ltd., London, 1957-60; prin. research engr. Standard Telecommunications Lab. Ltd., Harlow, Eng., 1960-70; prof. electronics, chmn. dept. Chinese U. Hong Kong, 1970-74, now vice chancellor; chief scientist Electro Optical Products div./ITT, Roanoke, Va., 1974-81; v.p., dir. engring. Electro Optical Products div./ITT, Roanoke, VA, 1981-83; exec. scientist, dir. research ITT Advanced Tech. Ctr., Shelton, Conn., 1983-87. Author: Optical Fiber Technology II, 1981, Optical Fibers Systems: Technology, Design and Applications, 1982, Optical Fibre, 1988, A Choice Fulfilled--The Business of High Technology, 1991; contbr. articles to profl. jours.; patentee in field. Decorated Commdr. Brit. Empire, 1993; recipient Morey award Am. Ceramic Soc., 1976, Stewart Ballantine medal Franklin Inst., 1977, Rank prize Rank Trust Funds, 1978, LM Ericsson Internat. prize, 1979, gold medal Armed Forces Communications and Electronics Assn., 1980, Internat. New Materials prize Am. Phys. Soc., 1989, Gold medal Internat. Soc. for Optical Engring., 1992; Marconi Internat. fellow, 1985. Fellow IEEE (Morris Liebmann Meml. award 1978, Alexander Graham Bell medal 1985, Faraday medal 1989), Inst. Elec. Engring. (U.K.), Royal Acad. Engring. (U.K.), Royal Swedish Acad. Engring. Scis. (fgn. mem.) Academia Sinica (Taiwan); mem. NAE. Office: Chinese Univ of Hong Kong, Office of Vice Chancellor, Shatin New Ters, Hong Kong

KAO, RICHARD JUICHANG, biostatistician; b. Kaoshung, Taiwan, Sept. 18, 1959; came to U.S., 1984; s. Fan and Ran (Lan) K.; m. Shirley D. Wang, Aug. 11, 1990. MS in Maths., W.Va. U., 1987; MS in Statistics, Temple U., 1990; postgrad. in biostatistics, Columbia U., 1992—. Programmer Tai-Tang Computer Co., Tai Nan, Taiwan, 1983-84; lectr. W.Va. U., Morgantown, 1987; cons. Einstein Med. Ctr., Phila., 1988-89; biostatistician Meml. Sloan-Kettering Cancer Ctr., N.Y.C., 1990—. Contrb. articles to profl. jours. Mem. Am. Statistical Assn. Achievements include research in clinical trials, categorial data analysis, experiment design, mulitvariate analysis. Home: 40 Highview Rd Denville NJ 07834 Office: Meml Sloan-Kettering Cancer Ctr 1275 York Ave Box 60 New York NY 10021

KAPADIA, MEHERNOSH MINOCHEHER, engineering executive; b. Bombay, India, Feb. 14, 1960; came to U.S., 1982; s. Minocheher G. and Baimai (Sarkari) K.; m. Monaz C. Desai, Jan. 7, 1986; 1 child, Sanaya Mehernosh. BE, U. Bombay, Bombay, India, 1981; MS, Pa. State U., 1983; MBA, St. Joseph's U., 1990; cert. in tech. ops., Nat. Tech. U., 1992. Cert. quality engr. Grad. rsch. asst. Pa. State U., University Pk., 1982-83, rsch. engr., 1983; engr. ATT Microelectronic Lightwave, Reading, Pa., 1984-85, planning engr., 1985-88, quality engr., 1988-91, sr. quality engr., 1991-92; sr. statis. process control engr. cellular divsn. ATT Network Systems, Columbus, Ohio, 1993—; lectr. local chpt. Am. Soc. Quality Control Engrs., Reading, 1990; instr. quality courses, Reading, 1990-92. Contbr. articles to profl. jours. Mem. Am. Soc. for Quality Control (sr.). Avocations: model building, exercising, badminton, traveling. Home: 231 Needlewood Ln Reynoldsburg OH 43068 Office: AT&T Network Systems Cellular Engring 6200 E Broad St Columbus OH 43213

KAPANDJI, ADALBERT IBRAHIM, orthopedic surgeon; b. Paris, Apr. 17, 1928; s. Mehmet Ibrahim and Roberte Jeanne (Chevalier) K.; m. Lydie Mauricette Richard, Oct. 12, 1950; children: Martine, Thierry. MD, Faculty of Medicine U. Paris, 1960. Externe hosps. Pub. Assistance, Paris, 1951-56, interne hosps., 1956-59, asst. hosps., 1960-65; prof. anatomy Nurses Sch. Paris Hosp., 1959-60; chef de clinique Faculty Medicine, Paris, 1960-65; pres. Clinique de L'Yvette S.A., Longjumeau, France, 1965—; prof. articular physiology Physiotherapists Sch. Necker Hosp., Paris, 1959-65, Physiotherapists Staff Sch., Bois-Larris, France, 1968-70. Author: The Physiology of the Joints, 3 vols., 1960 (translated into English, Italian, Spanish, German, Dutch, and Japanese), Dessins de Mains, Vol. 1, 1988 (translated in Japanese). Mem. French Soc. Angeiology, French Soc. Orthopaedics and Traumatoloy, French Soc. Hand Surgery (pres. 1987), Rioplatenese Soc. Anatomy (Argentina). Achievements include author of many surgical procedures, among them a technique of fixation of the lower end radius fractures with special intra-focal pins named ARUM. Inventor of joint prosthesis for instance, on the First Carpo-Metacarpal joint, on the Radio-Carpal joint and on the Distal Radio-Ulnar joint. Home: Copernic 7, 91160 Longjumeau France Office: Clinique de L'Yvette, Rte de Corbeil 43, 91160 Longjumeau France

KAPAT, JAYANTA SANKAR, mechanical engineer; b. Calcutta, India, Nov. 12, 1962; came to U.S., 1984; s. Bishnupada and Jharna (Ghosh) K.; m. Mallika Ghosh, Feb. 18, 1984. MS, Ariz. State U., 1988; ScD, MIT, 1991. Vis. asst. prof. Clemson (S.C.) U., 1991—. Nat. Talent Search scholar, Nat. Talent Search Com., India, 1978-84. Mem. AIAA, ASME (assoc.), Soc. Photo-Optical Instrumentation Engrs., Combustion Inst. Home: 43 Daniel Dr Clemson SC 29631 Office: Clemson U 318 Riggs Hall Clemson SC 29634

KAPLAN, ALEXANDER EFIMOVICH, physics educator, engineering educator; b. Kiev, Ukraine, USSR, June 9, 1938; came to U.S., 1979; s. Efim S. and Anna A. (Vilfand) K. MS in Physics, Moscow Phys. Tech. Inst.; postgrad., USSR Acad. Scis., Moscow, 1961; PhD in Physics and Math., Gorkii State U., USSR, 1967. Rsch. scientist Radio R & D Lab., Moscow, 1961-63; PhD in Physics and Math. USSR Acad. Scis., 1963-79; postgrad. MIT, Cambridge, 1979-82; prof. elec. engring. sch. Purdue U., West Lafayette, Ind., 1982-87; prof. elec. and computer engring. dept. Johns Hopkins U., Balt., 1987—; cons. Bell Labs, Homdell, N.J., 1980-81, Los Alamos (N.Mex.) Nat. Lab., 1981, Honeywell Rsch. Ctr., Mpls., 1982; guest scientist Max-Planck-Inst. Quantenoptik, Garching, Fed. Republic Germany, 1981—. Contbr. more than 90 articles to profl. jours. and 3 books. Fellow Optical Soc. Am.; mem. Am. Phys. Soc., Laser & Electro-Optic Soc. Achievements include patent in field. Office: Johns Hopkins U Elec and Comp Engring Dept 34th & Charles Sts Baltimore MD 21218

KAPLAN, ALLEN P., physician, educator, academic administrator; b. Jersey City, N.J., Oct. 27, 1940; m. Lee Kaplan, Aug. 22, 1965; children: Rachel, Seth. AB, Columbia U., 1961; MD, Downstate Med. Coll. Diplomate Am. Bd. Internal Medicine, Am. Bd. Rheumatology, Am. Bd. Allergy and Clin. Immunology; cert. in diagnostic lab. immunology. Head allergic disease sect. NIH, Bethesda, Md., 1972-78; prof. medicine, head divsn. allergy rheumatology & clin. immunology SUNY, Stony Brook, 1978-87, chmn. dept. medicine, 1987—. Editor: Allergy, 1985; contbr. over 200 articles to profl. jours. Lt. comdr. USPHS, 1972-78. Recipient Commendation medal USPHS, 1976. Mem. Am. Acad. Allergy & Immunology (pres. 1989-90), Clin. Immunology Soc. (pres. 1992-93), Internat. Assn. Allergology and Clin. Immunology (sec. gen. 1991—). Office: SUNY Sch Medicine Asthma & Allergic Ctr Stony Brook NY 11794

KAPLAN, JOSEPH, pediatrician; b. Boston, Mar. 7, 1941. Student, Dartmouth U., 1958-60; BA, NYU, 1962; MD, Johns Hopkins U., 1966. Intern, resident in pediatrics Johns Hopkins Hosp., Balt., 1969-72; mem. staff Children's Hosp. Mich., Detroit. Contbr. article to profl. publ. Maj. U.S. Army, 1969-72. Recipient Rsch. Career Devel. award NIH, 1975-80. Office: Children's Hosp Mich 3901 Beaubien Detroit MI 48201

KAPLAN, MARTIN NATHAN, electrical and electronic engineer; b. Beloit, Wis., Nov. 14, 1916; s. Abraham Louis and Eva (Schomer) K.; m. Florence Helen Grumet (div. 1956); 1 child, Kathy Sue; m. Sylvia Greif, Dec. 7, 1963. BSEE, U. Wis., 1942. Sr. electronics engr. Convair, San Diego, 1951-56; rsch. engr. AMF/Sunstrand, Pacoima, Calif., 1956-59; sr. rsch. engr. Ryan Electronics, San Diego, 1959-63; sr. design engr. N.Am. Aviation, Downey, Calif., 1963-66; rsch. specialist Lockheed, Burbank, Calif., 1966-70; mem. tech. staff Aerospace Corp., El Segundo, Calif., 1980-82; rsch. scientist Motorotor, North Hollywood, Calif., 1983—. Lt. (j.g.) USNR, 1943-46. Mem. IEEE (life), Am. Phys. Soc. (life). Achievements include patents for Statorless Homopolar Motor, for Electromagnetic Transmission for Control of Rotary Power in Vehicles; establishment of presence of quantized ether through the use of Lorentz forces; research on force field propulsion for spacecraft. Home and Office: Motorotor 11610 Cantlay St North Hollywood CA 91605-3940

KAPLAN, MARTIN NATHAN, electrical engineer; b. Phila., July 22, 1919; s. Nathan M. and Edith (Zeitlin) K.; m. Doris Chasman, Jan. 1, 1942; children: Judith Bess, Ruth Paula. BSEE, Drexel U., 1943, MSEE, 1953. REgistered profl. engr. Pa., W.Va. Chief engr. U.S. Maritime Svc., 1943-46; prof. Drexel U., Phila., 1947-89, F.C. Powell prof., 1989--; ednl. cons. Detroit Edison Co., 1974, Fidelity Machine Co., Phila., 1959-63. Author: Introduction to Alternative Current Machinery, 1948; contbr. articles to profl. jours. Comdr. U.S. Maritime Svc., 1943-46. Recipient Lindback award Drexel U., 1975, M.N. Kaplan Disting. Faculty award, 1988. Mem. IEEE (life sr.), Am. Soc. Engring. Educators, Am. Ednl. Rsch. Assn., Eta Kappa Nu, Tau Beta Pi, Sigma Xi. Home: 7211 Cresheim Rd Philadelphia PA 19119 Office: Drexel U ECE Dept Philadelphia PA 19104

KAPLAN, MITCHELL ALAN, sociologist, researcher; b. Bklyn., Jan. 26, 1954; s. Murray Robert and Claire (Meshnick) K. BA in Sociology and Psychology cum laude, L.I. U., 1976; MA in Sociology, New Sch. for Social Rsch., 1979; PhD in Sociology, CUNY, 1987. Cert. social rsch. specialist. Rsch. fellow Narcotic and Drug Rsch. Inc., N.Y.C., 1986-89, cons., 1989-90; cons. Am. Found. for AIDS Rsch., N.Y.C., 1989-90; rsch. scientist Rsch. & Tng. Inst. Nat. Ctr. for Disability Svcs., Albertson, N.Y., 1991-92; acad. rsch. cons. Acad. Rsch. Consulting Svcs., Bklyn., 1992—; evaluation cons. office rsch. and ednl. assessment Bklyn. divsn. N.Y.C. Bd. Edn., 1992-93. Co-author: (chpt.) Days with Drug Distribution Which Drugs? How Many Transactions? With What Returns? 1990. Bd. dirs. Greater N.Y. chpt. Dystonia Med. Rsch. Found., Oakland Gardens, N.Y., 1989-91. Nat. Inst. on Drug Abuse fellow, 1986-89. Mem. APHA, Nat. Rehab. Assn., Soc. for Disability Studies, N.Y. Acad. Scis., Am. Sociol. Assn. (cert. med. sociologist, social policy & evaluation rschr. law & social control rschr.), N.Y. State Sociol. Assn., Am. World Health Assn., Am. Assn. Sex Educators, Counselors and Therapists, Am. Assn. for Pub. Opinion Rsch., Nat. Rehab. Counseling Assn., Nat. Rehab. Assn. (job placement div. 1991), Pi Gamma Mu, Psi Chi, Phi Theta Kappa. Democrat. Jewish. Achievements include research in the areas of Aids and intravenous drug use, the relationship between drug use and criminal behavior, drug treatment methods, and vocational rehabilitation and the physically and emotionally disabled. Home and Office: Ste 8K 2560 Batchelder St Brooklyn NY 11235-1555

KAPLAN, OZER BENJAMIN, environmental health specialist, consultant; b. Santiago, Chile, Jan. 3, 1940; naturalized U.S. citizen, 1969; s. David and Raquel (Klorman) K.; m. Adele M. Brandt, Jan. 12, 1974. Student, U. Chile, 1958-59; BS, Calif. Polytech. U., 1964; MS, U. Calif., Davis, 1966, PhD, 1969; MPH, UCLA, 1973. Teaching and rsch. asst. U. Calif., Davis, 1968-69; assoc. prof. soil sci. N.C. A & T State U., Greensboro, 1969-70; assoc. prof. biology Morris Coll., Sumter, S.C., 1970-71; ind. cost/benefit cons. L.A., 1971-72; mem. environ. health task force Inland Counties Health Systems Agy., San Bernardino, Calif., 1974-76; environ. health planning coord. San Bernardino County, Calif., 1974-80; ind. cons. environ. health San Bernardino, 1987—. Author: Septic Systems Handbook, 1986, 2d edit. 1990. V.p. Citizens Against Pass Area Prisons, Riverside County, Calif., 1982-86, Pass Citizens for Sound Planning, Riverside County, 1986-91. Mem. Fedn. Am. Scientists, Am. Soc. Agronomy, Soil Sci. Soc. Am., Calif. Environ. Health Assn. (chmn. land use com., chmn. environ. health sect., Cert. of Appreciation 1976, 77), Nat. Environ. Health Assn., Common Cause, Pub. Citizens, Sigma Xi, Phi Kappa Phi. Achievements include collection and development of data which helped persuade state of California to relocate planned prison from Beaumont to isolated desert location in Blythe; research on solving septic systems problems. Home and Office: PO Box 522 Calimesa CA 92320-0522

KAPLAN, PAUL ELIAS, physiatrist, educator; b. N.Y.C., Oct. 26, 1940; s. Max Victor and Mae (Klein) K.; m. Candia Starling Post, June 18, 1966; children: Steven Post Hitchcock, Heather, Danielle. BA cum laude, Amherst Coll., 1962; MD, UCLA, 1966. Diplomate Am. Bd. Phys. Medicine and Rehab., Am. Bd. Electrodiagnostic Medicine. Intern in internal medicine Ohio State U. Hosp., Columbus, 1966-67; resident in internal medicine Cedars-Sinai Med. Ctr., L.A., 1969-70, UCLA Med. Ctr., 1970-71; NIH fellow, resident in phys. medicine & rehab. U. So. Calif. Med. Ctr., L.A., 1971-73; pvt. practice Beverly Hills, Calif., 1973-74; prof. medicine and internal medicine Inst. of Chgo./Northwestern U., 1974-86; prof., dept. chmn. U. Mo., Columbia, 1986-89; Bert C. Wylie prof., chmn. phys. medicine and rehab. Ohio State U., Columbus, 1989—, dir. residency program, 1992—; ptnr. Assoc. Physiatrists of Ctrl. Ohio, Columbus, 1989—; chmn. sci. adv. com. Arthritis Ctr. at U. Mo., 1992—. Author several textbooks on phys. medicine and rehab.; editor-in-chief jour. Yearbook of Rehab., 1984-89; alt. editor Archives of Phys. Medicine and Rehab., 1988—; contbr. more than 100 articles to profl. jours. Nat. Inst. Disability and Rehab. Rsch. grantee Arthritis Ctr. at U. Mo., 1988-89. Fellow ACP, Am. Acad. Phys. Medicine and Rehab.; mem. Assn. Acad. Physiatrists (pres. 1987-89), Am. Acad. Neurology, Am. Spinal Injury Assn. (sec. 1991—, mem coun. chairpersons). Avocation: bagpipes. Office: Ohio State U Dodd Hall Rehab Ctr 480 W 9th Ave Columbus OH 43201-2346

KAPLAN, SAMUEL, pediatric cardiologist; b. Johannesburg, South Africa, Mar. 28, 1922; came to U.S., 1950, naturalized, 1958; s. Aron Leib and Tema K.; m. Molly Eileen McKenzie, Oct. 17, 1952. MB, BcH., U. Witwatersrand, Johannesburg, 1944, MD, 1949. Diplomate: Am. Bd. Pediatrics. Intern Johannesburg, 1945; registrar in medicine, 1946; lectr. physiology and medicine U. Witwatersrand, 1946-49; registrar in medicine U. London, 1949-50; fellow in cardiology, research assoc. U. Cin., 1950-54, asst. prof. pediatrics, 1954-61, assoc. prof. pediatrics, 1961-66, prof. pediatrics, 1967-87, asst. prof. medicine, 1954-67, assoc. prof. medicine, 1967-82, prof. medicine, 1982-87; prof. pediatrics UCLA, 1987—; cons. NIH; hon. prof. U. Santa Tomas, Manila. Mem. editorial bd. Circulation, 1974-80, Am. Jour. Cardiology, 1976-81, Am. Heart Jour, 1981—, Jour. Electrocardiology, 1977—, Clin. Cardiology, 1979—, Jour. Am. Coll. Cardiology, 1983-87, Progress Pediat. Cardiology, 1990—. Cecil John Adams fellow, 1949-50; grantee Heart, Lung and Blood Inst. of NIH, 1960—. Mem. Am. Pediatric Soc., Am. Soc. Pediatric Research, Am. Heart Assn. (med. adv. bd. sect. circulation), Am. Fedn. Clin. Research, Am. Coll. Cardiology, Internat. Cardiovascular Soc., Am. Acad. Pediatrics, Am. Assn. Artificial Internal Organs, Midwest Soc. Pediatric Research (past pres.), Sigma Xi, Alpha Omega Alpha; hon. mem. Peruvian Soc. Cardiology, Peruvian Soc. Angiology, Chilean Soc. Cardiology, Burma Med. Assn. Achievements include research and publications on cardiovascular physiology, diagnostic methods, cardiovascular complications of pediatric AIDS and heart disease in infants, children and adolescents. Office: UCLA Sch Medicine Dept Pediatric Cardiology Los Angeles CA 90024

KAPOR, MITCHELL DAVID, foundation executive; b. Bklyn., Nov. 1, 1950; s. Jesse and Phoebe L. (Wagner) K.; m. Judith M. Vecchione, June 4, 1972 (div. 1979); m. Ellen M. Poss, Aug. 7, 1983. BA, Yale U., 1971; MA, Beacon Coll., 1978; postgrad., Sloan Sch. Mgmt., MIT, 1979; DHL (hon.), Boston U., 1985, Mass. Sch. Profl. Psychology, 1990; DCS (hon.), Suffolk U., 1988. Freelance cons. Cambridge, Mass., 1978-80; product mgr. Personal Software, Sunnyvale, Calif., 1980; pres. Lotus Devel. Corp., Cambridge, Mass., 1982-84, chmn., 1984-86; chmn. ON Tech. Inc., Cambridge, Mass., 1987-90, Electronic Frontier Found., Inc., Cambridge, 1990—; adj. rsch. fellow Kennedy Sch. Govt., Harvard U., 1992—; chmn. Mass. Commn. on Computer Tech. and Law, 1992. Author: (with others) (software program) Lotus 1-2-3, 1983; columnist Forbes mag. Trustee Kapor Family Found., 1986—, The Computer Mus., Boston, Comml. Internet. Exch., Washington, Computer Sci. and Telecomms. Bd. of NRC. Recipient Disting. Info. Scis. award Data Processing Mgmt. Assn., 1990. Jewish. Office: Electronic Frontier Found 238 Main St Cambridge MA 02142

KAPPAS, ATTALLAH, physician, medical scientist; b. Union City, N.J., Nov. 4, 1926; s. Attie and Sofia (Kozam) K.; m. Oct. 26, 1963; children: Peter, Michael, Nicholas. A.B., Columbia U., 1947; M.D. with honors, U. Chgo., 1950; Sc.D., N.Y. Med. Coll., 1978. Diplomate: Am. Bd. Internal Medicine. Med. intern Univ. Service, Kings County Hosp., N.Y.C., 1950-51; research fellow div. steroid biochemistry and metabolism Sloan Kettering Inst., N.Y.C., 1951-54; asst. resident physician and sr. asst. resident physician Peter Bent Brigham Hosp. Harvard Med. Sch., Boston, 1954-56; assoc. div. steroid biochemistry and metabolism Sloan Kettering Inst., 1956-57; from asst. prof. to assoc. prof. dept. medicine, head div. metabolism and arthritis U. Chgo. Med. Sch., 1957-67; Guggenheim fellow, guest investigator Rockefeller U., N.Y.C., 1966-67; assoc. prof., physician Rockefeller U., 1967-71, sr. physician, 1971-74, prof., 1971-81, Sherman Fairchild prof., 1981—, v.p., 1983-91, physician-in-chief, 1983-91, physician-in-chief emeritus, 1991—; Vincent Astor chair clin. sci. Meml. Sloan-Kettering Cancer Ctr. and Cornell U. Med. Coll., 1979-81, prof. medicine, 1972—, prof. pharmacology, 1972-87; bd. dirs Russell Sage Inst. Pathology Cornell U., 1977-87; vis. com. div. biol. scis. Pritzker Sch. Medicine, U. Chgo., 1977-86; attending physician N.Y. Hosp., 1972—, Meml Hosp. Cancer and Allied Diseases, 1977-91; mem. selection com. John A. Hartford Found. Fellowship program in clin. scis., N.Y.C., 1979-83; co-dir. Rockefeller U.-Cornell U. combined M.D.-Ph.D. program, 1980-85; mem. com. pyrene and selected analogs NRC-Nat. Acad. Sci., Washington, 1981-83, cons. Merck Sharp & Dohme Research Labs., 1974-79, 82-84, Abbot-Ross Labs., 1985-90, Hoffman LaRoche Labs., 1985-87, Glaxo Research Labs., 1988-90; mem. sci. adv. bd. Environ. Scis. Lab. Mt. Sinai Med. Ctr., 1983-87; prof., adj. faculty dept. pediatrics Karolinska Inst., Stockholm, 1987-93; vis. prof. U. Pediatrics U. Vt. Coll. Medicine, Burlington, 1993—. Contbr. articles to profl. jours. Bd. dirs. Vis. Nurse Service N.Y., 1982-86 ; mem. gov.'s com. on rev. sci. studies and devel. pub. policy on problems resulting from hazardous wastes N.Y. State, 1980. Served with U.S. Army, 1945-46. Commonwealth Fund fellow, 1961-62; Guggenheim fellow, 1966-67; recipient Spl. award in clin. pharmacology Burroughs Wellcome Fund, 1973; named Sr. Henry Hallet Dale Meml. lectr. and vis. prof. Johns Hopkins Hosp., 1975; recipient Disting. Service award in med. scis. U. Chgo. Sch. Medicine, 1975; named Pfizer lectr. clin. pharmacology Peter Bent Brigham Hosp., Harvard Med. Sch., 1977; named Pfizer lectr. Pa. State U., 1980; named first Rolf Blomstrand lectr. Karolinska Inst., 1988; first Glaxo lectr. Cornell U. Med. Sch., 1984; Gunner and Lillian Nicholson Found. exchange prof. Karolinska Inst., Stockholm, 1985-86; Barowsky Meml. lectr., N.Y. Med. Coll., 1986; 1st Annual award for excellence in clin. rsch. NIH, 1989. Fellow ACP; mem. Assn. Am. Physicians, Am. Soc. Clin. Investigation, Am. Clin. and Climatol. Assn., Am. Soc. Pharmacology and Exptl. Therapeutics (pub. affairs com., award for exptl. therapeutics 1978), Practitioners Soc. N.Y., Harvey Soc., Endocrine Soc., Interurban Clin. Club, Cosmos Club (Washington), N.Y. Athletic Club, Lotos Club. Home: 1161 York Ave New York NY 10021-7940 Office: Rockefeller U Hosp 1230 York Ave New York NY 10021-6341

KAPPE, DAVID SYME, environmental chemist; b. Phila., Sept. 28, 1935; s. Stanley Edward and Flora (Syme) K.; m. Patricia K. Kappe, Sept. 1957 (dec. June 1992); children: David Jr., Christopher K., Dawn E. Bray. BS in Chemistry, U. Md., 1959; PhD in Phys. Chemistry, Pa. State U., 1965. Phys. sci. aide Nat. Bur. Standards, Washington, 1958-59; cons. Kappe Assocs., Inc., Rockville, 1959-66; chief Hittman Assocs., Inc., Balti., 1966; v.p., rsch. dir. Kappe Assocs Inc., Rockville, 1967-84, pres., 1984-85, chmn., CEO, 1985--; con. USEPA, Washington, 1981-82, project merit reviewer, 1972. Contbr. articles to profl. jours.; patentee wastewater treatment. Troop chmn., scoutmaster Boy Scouts Am., 1968-78; treas. Elkridge Citizens Assn., 1977-78. Lt. USAF res., 1959-64,. Recipient AFROTC Leadership award Md. U., Am. Legion 1959, WPCAP Rsch. award, 1979, AWWA Abel Wolman award CSAWWA, 1989. Mem. Am. Chem. Soc., Am. Inst. Chemists, Am. Water Works Assn. (chmn. Chesapeake sect. 1992-93), Water Environ. Fedn., Scabbard and Blade, Sigma Pi Sigma, Phi Lambda Upsilon. Home: 5200 Massachusetts Ave Bethesda MD 20816 Office: Kappe Assocs Inc 100 Worman's Mill Ct Frederick MD 21701

KAPPMEYER, KEITH K., manufacturing company executive. V.p. tech. USX Corp., Pitts. Recipient Albert Victor Bleininger award Am. Ceramic Soc., 1992. Office: USX Corp 600 Grant St Pittsburgh PA 15219-2701*

KAPRAL, FRANK ALBERT, medical microbiology and immunology educator; b. Phila., Mar. 12, 1928; s. John and Erna Louise (Melching) K.; m. Marina Garay, Nov. 22, 1951; children: Frederick, Gloria, Robert. B.S., Phila. Coll. Pharmacy and Scis., 1952; Ph.D, U. Pa., 1956. With U. Pa., Phila., 1952-66, assoc. in microbiology, 1958-66; assoc. microbiologist Phila Gen. Hosp., 1962-64, chief microbiology research, 1964-66, chief microbiology, 1965-66; asst. chief microbiol. research VA Hosp., Phila, 1962-66; assoc. prof. med. microbiology Ohio State U., Columbus, 1966-69, prof. med. microbiology and immunology, 1969—; cons. Ctr. Disease Control, Atlanta, 1980, Proctor and Gamble Co., 1981-87. Contbr. articles to profl. jours.; patentee implant chamber. Active Ctrl. Ohio Diabetes Assn., 1992-93. With AUS, 1946-47. NIH rsch. grantee, 1959—; Ctrl. Ohio Diabetes Assn. grantee, 1992-93. Fellow Am. Acad. Microbiology, Infectious Diseases Soc. Am.; mem. AAAS, Am. Soc. for Microbiology, Am. Assn. for Immunologists, Soc. for Exptl. Biology and Medicine, Sigma Xi. Democrat. Roman Catholic. Home: 873 Clubview Blvd Columbus OH 43235-1212 Office: Ohio State U Dept Med Micro and Immunol 2166A Graves Hall Columbus OH 43210

KAPUSINSKI, ALBERT THOMAS, economist, educator; b. Greenport, N.Y., Oct. 16, 1937; s. Casimir Thomas and Anne Mary (Olbrys) K.; m. Margaret Catherine Eichler, Sept. 3, 1963 (dec. March, 1982); children: Albert J., George T., Frank P.; m. Theresa Tafuri, Dec. 27, 1987. BBA, St. Johns U., N.Y.C., 1961, MBA, 1966; PhD, NYU, 1981. Economist Lionel D. Edie & Co., 1962-64; mem. faculty Caldwell (N.J.) U., 1964—, from assoc. prof. to prof. econs., 1969—, chmn. bus. dept., 1970-79; mem. Faculty Senate, 1969-75; assoc. sr. economist Hans Klunder Assocs., Hanover, N.H., 1966-69, ENVICO, Windsor, Vt., 1969-73; owner, operator Albert T. Kapusinski & Assoc., 1966—; econ. cons. to various industries in N.Y. and Vt.; mem. faculty Adirondack Coll., 1966, NYU, 1970-71, Merrill-Lynch Tng. Ctr. for Brokers, N.Y.C., 1973-74. Author: The Economy of Greene County, New York, 1972; contbr. articles to profl. jours. Chmn. Pro-Life del. World Population Conf. Forum, Bucharest, Rumania, 1974; bd. advisors U.S. Coalition for Life, Export, Penn., Ednl. Opportunities Fund, 1973-75. Served with USAR, USNG, 1953-61. Recipient K.L. Kiernan award; Gen. Electric Faculty fellow U. Chgo., 1970, Found. for Econ. Edn. fellow, 1971. Mem. Am. Econ. Assn., AAUP, Assn. Social Econs., Inst. Social Rels. Newark (adv. bd. 1972-74), Univ. Devel. Inst. (pres. 1967-68), Omicron Delta Epsilon. Office: Caldwell Coll Caldwell NJ 07006

KARABOTS, JOSEPH WILLIAM, environmental engineering company executive; b. Hartford, Conn., Dec. 5, 1956; s. William and Lucy (Makris) K. BS in Geology and Geophysics, U. Conn., 1981. Rsch. asst. NSF, Aleutian Islands, Alaska, 1981-82; sr. and gen. field engr. Schlumberger Ltd., various cities, 1982-87; sr. engr. Vitro Corp., Newport, North London, R.I., Conn, 1987-89; dir. bus. devel. Briggs Assocs., Rockland, Mass., 1989—. Vol. Spl. Olympics, Providence, 1984—, Providence Waterfront Festival, 1990—, R.I. Dept. Econ. Devel.-Internat. Rels., 1991—. Recipient Citizens award City of Hartford, 1977. Mem. Providence C. of C. (amb. 1989—, fedn. environ. task force 1991—, chmn. water quality issues com.), Trade Club, Toastmasters (sec. 1991—). Greek Orthodox. Avocations: aviator, international affairs. Office: Briggs Assocs Inc 400 Hingham St Rockland MA 02370

KARACAN, ISMET, psychiatrist, educator; b. Istanbul, Turkey, July 23, 1927. BS, U. Istanbul, 1948, MD, 1953; DSC in Medicine, SUNY Downstate Med. Ctr., 1965. Cert. Turkish Bd. Neuropsychiatry, Am. Bd. Psychiatry and Neurology. From assoc. prof. to prof. psychiatry, dir. Sleep Labs. U. Fla., Gainville, 1966-73; prof. psychiatry, dir. Sleep Disorders & Rsch. Ctr. Baylor Coll. Medicine, Tex. Med. Ctr., Houston, 1973—; assoc. chief staff rsch. & devel., dir. Sleep Rsch. Lab. VA Med. Ctr., Houston, 1973—. Recipient Nathaniel Kleitman prize Assn. Sleep Disorders Ctrs., 1981. Fellow Am. Psychiat. Assn., Am. Coll. Physicians; mem. AMA, AAAS, Sleep Rsch. Soc. (pres. 1976-79), N.Y. Acad. Sci., Am. Coll. Neuropsychopharmacol., Brit. Assn. Psychopharmacol. Achievements include rsch. in phychological and physiological mechanisms of male impotence, neurophysiological and biochemical mechanisms responsible for male erectile failure, pharmacology of human sleep. Office: Baylor College of Medicine Sleep Disorders & Rsch Ctr 1 Baylor Plz Houston TX 77030*

KARAKOZOV, SERGEI DMITRIEVICH, mathematics educator; b. Barnaul, Altai, USSR, Apr. 16, 1956; s. Dmitryi I. and Evdokya T. (Dvoryadkina) K.; m. Elena N. Ryashkov; children: Maria, Ksenia. Diploma, U. Math. Faculty, Novosibizsk, USSR, 1978; PhD, Novosibizsk Inst. Math., 1985. Instr. Pedagogical Inst., Barnaul, USSR, 1981-85, asst. prof., 1985—. Contbr. articles to profl. jours. Mem. Am. Math. Soc. Home: Stroiteley 25-22 Altai, 656015 Barnaul Russia Office: Pedagogical Inst, Socyalistichesky 126 Altai, 656015 Barnaul Russia

KARAMOUZ, MOHAMMAD, engineering educator; b. Kerman, Iran, July 16, 1954; came to U.S., 1977; s. Hosein Karamouz and Talat Karbakhsh; m. Mahbubeh Mozaffari, July 21, 1977; children: Mehrdad, Sahar, Saba. BSCE, Shiraz U., 1977; MSCE, George Washington U., 1979; PhD, Purdue U., 1983. Registered profl. engr., N.Y. Field dir. Zarinpay Constrn., Shiraz, Iran, 1975-77; project mgr. K-H Constrn., Arak, Iran, 1977; rsch.-teaching asst. George Washington U., Washington, 1977-79; rsch. assoc., instr. Purdue U., West Lafayette, Ind., 1979-83; asst. prof. Poly. U., Bklyn., 1983-88; assoc. prof., chmn. civil engring. dept. Pratt Inst., Bklyn., 1988-91, prof., interim dean engring., 1991-93, prof. Grad. Ctr. Planning and Environment, 1993—; chmn. bd. dirs. Arch Cons. Engrs., Rego Park, N.Y.; chmn. tech. program WRPMD, Water Forum '92, Balt., 1992. Author water resources bull. Deterministic and Stochastic Operation, 1988, Water Resources Research, A Baysian Stochastic Optimization of Reservoir Operation Using Uncertain Forecast, 1992, ASCE Jour. Water Resources Planning and Mgmt. (WRPM), Optimization and Stilulation of Multiple Reservoir Systems, 1992, Demand Driven Operation of Reservoirs, 1993; contbr. articles to profl. publs. NSF rsch. asst., 1980-83, NSF grantee, 1989-91. Mem. NAS, ASCE (editor publ. 1986, 92, assoc. editor Water Resources and Planning Management divsn., chmn. water resources systems com. 1990-92), Am. Water Resources Assn., Am. Geophys. Union, Ops. Rsch. Soc. Am., Sigma Xi, Chi Epsilon, Tau Beta Pi. Home: 15 Newwoods Rd Glen Cove NY 11542 Office: Pratt inst 200 Willoughby Ave Brooklyn NY 11205

KARAPOSTOLES, DEMETRIOS ARISTIDES, civil engineer; b. Larissa, Greece, Mar. 2, 1936; s. Aristides Demetrios and Zoi-Lili Aristides (Papadimitriou) K.; m. Toula Haralambos Giagtzoglou, Oct. 24, 1976 (dec. Aug. 1989); children: Zoi-Lili, Aristides. Diploma Cathedral Sch. St. Paul, Garden City, N.Y. 1954; student pre-engring. Bethany Coll., W.Va., 1954-56; BS in Civil Engring. U. Mich., 1961; profl. degree in civil engring. Nat. Tech. U., Athens, 1962. Registered profl. engr. Cons. profl. civil engr., Thessaloniki, 1964-75; assoc. prof. civil engring. Larissa's Inst. Tech., 1975; asst. head hydraulics div. Ministry Pub. Works, Larissa, 1976—. Lt. C.E., Greek Army, 1962-64. Fellow ASCE; mem. Tech. Chamber of Greece, Assn. Civil

Engrs. of Greece. Greek Orthodox. Avocations: photography, short-wave radio.

KARATO, SHUN-ICHIRO, geophysicist; b. Fukuoka, Japan, Sept. 4, 1949; came to U.S., 1989; s. Yoshio and Tomiko (Hidaka) K.; m. Yoko Tomoda, Dec. 3, 1978; children: Toshihiko, Yukako. PhD, Univ. Tokyo, 1977. Asst. prof. Univ. Tokyo, 1977-91; rsch. fellow Australian Nat. Univ., Canberra, Australia, 1981-85; assoc. prof. Univ. Minn., Mpls., 1989-92, prof., 1992—; editorial bd. Tectonophysics, 1992. Editor: Rheology of Solids and of the Earth, 1989; contbr. articles to profl. jours. Co-chmn. Internat. Com. Lithosphere, 1985-90. Mem. Internat. Assn. Seismology and Physics of the Earth's Interior, N.Y. Acad. Scis. Office: Univ Minn Dept Geology and Geophysics 310 Pillsbury Dr SE Minneapolis MN 55455

KARATSU, OSAMU, telephone company research and development executive; b. Tokyo, Apr. 25, 1947; s. Hajime and Sumako (Narumi) K. BS, Tokyo U., 1970, MS, 1972, PhD in Physics, 1975. Researcher Musashino Labs. Nippon Telegraph and Telephone Pub. Corp., Tokyo, 1975-79, staff researcher, 1979-83; sr. staff researcher Atsugi (Japan) Labs. Nippon Telegraph and Telephone Corp., 1983-86; rsch. group leader LSI Labs. Nippon Telegraph and Telephone Corp., Atsugi, 1987, sr. rsch. mgr., 1989-90, exec. mgr., 1991—; sr. mgr. Nippon Telegraph and Telephone Corp. Hdqrs., Tokyo, 1988-89; chmn. LSI design lang. standardization com., Tokyo, 1987—. Author: Introduction to Very Large Scale Integration Design, 1983; Microelectronics Series, 1985, Encyclopedia of Information Science, 1990. Mem. IEEE, Japan Soc. Applied Physics, Am. Phys. Soc., Inst. Electronic and Communication Engrs. Japan, Inst. Elec. Engring. Japan. Avocations: playing and listening to classical music. Home: 3-5-16 Nishi-Azabu # 102, Minatoku Tokyo 106, Japan Office: NTT LSI Labs 3-1 Morinosato, Wakamiya, Atsugi-shi Kanagawa 243-01, Japan

KARCZMARZ, KAZIMIERZ, botany educator; b. Wola Sernicka, Lublin, Poland, July 1933; s. Michał and Franciszka (Bodzak) K. Doctor, Maria Curie-Skłodowska U., Lublin, 1975. Asst. dept. systematics and plant geography Maria Curie-Skłodowska U., 1956-62, adj., 1962-72, privatdocent, 1972-78, assoc. prof., 1978-91, full prof. botany, 1991—; curator Cryptogamic Herbarium, U. Lublin, 1957—. Author: A Monograph of Genus Calliergon, 1971 (award 1972); editor: Biological Investigations of Terrestrian and Aquatic Ecosystems of the Roztocze and Eastern Carpathians, 1990. Recipient Gold Cross of Merit, Ministry of Edn., 1977, Bachelor Cross of Order of Renaissance of Poland, Ministry of Edn., 1978, medal of Commn. of Nat. Edn., Ministry of Edn., 1982. Mem. Polish Acad. Scis. (mem. editorial bd. 1972—, cons. in paleobotany 1974—), Polish Botany Soc. (chmn. bryological sect. 1973-89), Polish Hydrobiol. Soc., Internat. Assn. Bryologists, Fedn. Assn. Polish Tchrs. of Higher Schs. and Edn., Nature Conservation Lodge (cons.). Roman Catholic. Avocations: hunting, history of European painting, touring, books. Home: Grazyny 23/4, 20-602 Lublin Poland Office: Inst Biology MCS U, Akademicka 19, 20-033 Lublin Poland

KARDAS, SIGMUND JOSEPH, JR., secondary education educator; b. Phila., Jan. 14, 1940; s. Sigmund Joseph Sr. and Mary Olga (Sambor) K. BSc in Geology, Villanova U., 1962; postgrad., U. Madrid, 1971. Lic. tchr. N.J., Mass. Prof. sci. English U. Madrid, 1964-65; tchr. sci. Am. Sch. Madrid, 1966-69; chmn. dept. sci. Am. Sch. of Las Palmas, Spain, 1969-71; prof. sci. English, phys. anthropology and palaeontology U. La Laguna, Spain, 1971-78; tchr. sci. Mid. Twp. High Sch., Mays Landing, N.J., 1980-81, Trenton (N.J.) Pub. Schs., 1982—; bd. dirs. Lab. Investigaciones Sobre Biorritmos Humanos, Tenerife, Spain; editorial assoc. Metron Publs., Princeton, 1982-84; vis. scientist Senckenberg Inst., Wilhelmshaven, Fed. Republic Germany, 1964. Contbr. articles to profl. jours. Grantee NATO Paleoclimate Conf., 1963, Internat. Biorhythm Rsch. Assn., 1977, NSF Geology Inst., 1985; named Hon. Rsch. Assoc. Japanese Biorhythm Lab., Tokyo, 1976. Mem. Fedn. Study of Environ. Factors (adv. bd. 1989—), Nat. Speleolog. Soc., Internat. Soc. Biometerology, Real Sociedad Espanola de Historia Natural, Nat. Sci. Tchrs. Assn. Achievements include description of new subspecies of Pleistocene walrus; correlation of solar activity cycle MHz radiation to human behavior; first use of x-rays to study internal structure of stalactites; used polymer resin to make molds of underground nests of parasitic wasps; research in solar cycle-Earth relationships, especially climate and human behavior. Office: Laboratoriq de Investigaciones, Sobre Biorritmos Humanos Huerta Bicho 27, 38350 Tacoronte Spain

KARDES, FRANK ROBERT, marketing educator; b. Olean, N.Y., June 3, 1958; s. Frank Robert Kardes and Dorothy Marie (Chap) Rittberg; m. Perilou Goddard, May 26, 1985. MA in Psychology, U. Dayton, 1982; PhD in Psychology, Ind. U., 1986. Asst. prof. mgt. sci. MIT, Cambridge, 1986-89; assoc. prof. mktg. U. Cin., 1989-93, prof. mktg., 1993—; mem. edit. bd. Jour. Consumer Rsch., Champaign, Ill., 1989--, Jour. Consumer Psychology, Champaign, 1992--; mem. exec. com. Soc. for Consumer Psychology, Washington, 1989--. Contbr. articles to profl. jours. Named Reviewer of Yr. Jour. Consumer Rsch., 1990, 91, 92. Mem. AAAS, APA, The Inst. Mgmt. Scis., Soc. for Judgment and Decision Making, Soc. for Consumer Psychology. Office: U Cin Coll Bus Adminstn Cincinnati OH 45221-0145

KARDOMATEAS, GEORGE ALEXANDER, aerospace engineering educator; b. Athens, Greece, Aug. 5, 1958; came to U.S., 1981; s. Alexander and Katherine (Bechraki) K.; m. Maria Kalou-Papadatou, Aug. 31, 1982; 1 child, Alexander. MScME, MIT, 1982, PhD in Mech. Engring., 1985. Sr. rsch. engr. GM Rsch. Lab., Warren, Mich., 1985-89; asst. prof. aerospace engring. Ga. Inst. Tech., Atlanta, 1989-92, assoc. prof. aerospace engring., 1992—. Contbr. articles to profl. jours. Bodosakis Found. scholar, 1981. Mem. AIAA (sr., cert. of appreciation 1992), ASME. Office: Ga Inst Tech Dept Aerospace Engring Atlanta GA 30332-0150

KARDOUSLY, GEORGE J., chemical engineer; b. Jan. 2, 1964; s. John and Jacqueline (Arslan) K. BSChemE, Calif. State U., Long Beach, 1986. Cons. Factory Mut. Engring., Orange, Calif., 1986-88, AIG Consultants, L.A., 1988-89; acct. engr. AIG/Starr Tech. Risks, L.A., 1989—. Co-author: The Book of Apple Software Review, 1983. Vol. Union Rescue Mission, L.A., 1988—; asst. minister Ctrl. City Community Ch., L.A., 1989—; vol. L.A. Olympics Organizing Com., 1984. Mem. AICE, Soc. Fire Protection Engrs., Am. Internat. Cos. Achievements include involvement in student chem. engring. rsch. assisting profs. at Calif. State U. Home: 169 E 234 Pl Carson CA 90745 Office: Starr Tech Risks Agy 3699 Wilshire Blvd Los Angeles CA 90010

KARIGL, GÜNTHER, mathematician; b. Vienna, Austria, Sept. 5, 1953; s. Hermann and Elfriede (Grohmann) K.; m. Maria-Theresia Florian, May 10, 1980; children: Florian, Stephan, Benedikt, Dominik. PhD, U. Vienna, 1976; habilitation for biomath., Tech. U. Vienna, 1983. Tchr., high sch. Bundesgymnasium Wien 15, Vienna, 1974; asst. prof. Tech. U. Vienna, 1977-82, lectr., 1983—; reviewer in field. Author: Basic Course in Statistics, 1986, Mathematics in Computer Science and Economics, 1988; contbr. articles to profl. jours. Mem. Austrian Math. Soc., Austrian Stats. Soc. Roman Catholic. Avocations: family, house, gardening, skiing. Office: Tech U Vienna, Wiedner Hauptstr 8-10, A-1040 Vienna Austria

KARIM, AMIN H., cardiologist; b. Karachi, Pakistan, Oct. 10, 1951; came to U.S., 1981; MD, Dow Med. Coll., Karachi, Pakistan, 1977. Diplomate Am. Bd. Cardiology. Intern Sinai Hosp., Balt., 1981-84; resident Baylor Coll. Medicine, Houston, 1984-87, asst. prof., 1991—; pres. Angiocardiac Care of Tex., Houston, 1987—. Fellow Am. Coll. Physicians, Am. Coll. Cardiologists; mem. AMA. Office: Angiocardiac Care of Tex 6560 Fannin # 1532 Houston TX 77030

KARIN, SIDNEY, research and development executive; b. Balt., July 8, 1943. BSME, CCNY, 1966; MS in Nuclear Engring., U. Mich., 1967, PhD in Nuclear Engring., 1973. Registered profl. engr., Mich. Computer programmer, nuclear engr. ESZ Assocs., Inc., Ann Arbor, Mich., 1968-72; sr. engr., sect. leader Gen. Atomics (formerly GA Techs., Inc.), San Diego, 1973-75, mgr. fusion div. Computer Ctr., 1975-82; dir. info. systems div. GA Techs., Inc., 1982-85; dir. San Diego Supercomputer Ctr., 1985—; mem. computer sci. and telecomms. bd. NRC, 1988-93; adj. prof. computer sci. and engring. U. Calif., San Diego, 1985—; mem. adv. com. Fed. Networking Coun., 1992—. Contbr. articles to profl. jours. NDEA fellow, AEC fellow. Mem. AAAS, Assn. for Computing Machinery, IEEE Computer Soc. Avocations: flying, technical rock climbing, motorcycle riding, alpine skiing, reading. Home: 748 Avocado Ct Del Mar CA 92014 Office: San Diego Supercomputer Ctr PO Box 85608 San Diego CA 92186-5608

KARIS, THOMAS EDWARD, chemical engineer; b. Scotia, N.Y., Aug. 10, 1954; s. Raymond Thomas and Adrienne Viola (Aubry) K.; m. Elizabeth Myers, Dec. 31, 1986. BSEE, Union Coll., 1976; MSChE, Rensselaer Poly. Inst., 1979; PhDChE, Carnegie Mellon U., 1982. Rsch. staff mem. IBM Corp., San Jose, Calif., 1982—. Office: IBM ARC K93/801 650 Harry Rd San Jose CA 95120-6099

KARKHECK, JOHN PETER, physics educator, researcher; b. N.Y.C., Apr. 26, 1945; s. John Henry and Dorothy Cecilia (Riebling) K.; m. Kathleen Mary Shiels, Nov. 8, 1969; children: Lorraine, Michelle, Eric. BS, LeMoyne Coll., 1966; MA, SUNY, Buffalo, 1972; PhD, SUNY, Stony Brook, 1978. Various positions Grumman Corp., Bethpage, N.Y., 1964-68; grad. asst. SUNY, Buffalo, 1968-70; tchr. secondary schs. Mattituck (N.Y.) Sch. Dist., 1970-71, Shelter Island (N.Y.) Sch. Dist., 1971-73; grad. asst. SUNY, Stony Brook, 1973-78, postdoctoral fellow, 1978-79; rsch. assoc. SUNY, Stony Brook, N.Y., 1979-81; asst. prof. physics GMI Engring. and Mgmt. Inst., Flint, Mich., 1981-84, assoc. prof., 1984, prof., dir. physics, 1988—, head. dept. sci. and math., 1989-93; prof., chmn. dept. physics Marquette U., Milw., 1993—; physics assoc. Brookhaven Nat. Lab., Upton, N.Y., 1975-79, cons., 1979-85, STS, Hauppauge, N.Y., 1983, BID Ctr., Flint, 1985-90; acad. assoc. Mich. State U., 1988, 90, vis. scholar, 1989, vis. scientist, 1991; reviewer Addison-Wesley Pub., 1990, 93; regional dir. Mich. Sci. Olympiad, 1991-92, 92-93. Contbr. numerous articles to profl. jours. Den leader Cub Scouts Am., Flint, 1987-91; leader Boy Scouts Am., 1991-93; bd. dirs. Flint Area Sci. Fair, 1991-93; judge local sci. fairs. Dept. Energy rsch. grantee, 1977-79, NATO travel grantee, 1983-86, 89. Mem. Am. Phys. Soc., AAAS. Roman Catholic. Avocations: swimming, reading, travel, learning German. Home: 6592 N Bethmaur Ln Glendale WI 53233 Office: Marquette U Dept Physics 1700 W 3d Ave Milwaukee WI 53233

KARKIA, MOHAMMAD REZA, energy engineer, educator; b. Tabriz, Iran, Sept. 12, 1949; came to the U.S., 1979; s. Mohammad Taghi and Safieh M. Karkia; m. Mehran Karkia, June 20, 1981 (div. 1988); 1 child, Leila; m. Mojgan M. Karkia, Sept. 12, 1990; 1 child, Amir-Ali. BS in Engring., Sussex U., 1976; MS, U. Pitts., 1981; DBA, Newport U., 1990. V.p. W.F. Ryan & Assocs., Inc., Pitts., 1981-87; systemwide energy and utilities engr. Chancellor's Office Calif. State U., Los Alamitos, 1987—; mem. Calif. State Energy Policy Adv. Com., Sacramento, 1989—. Contbr. articles to Critical Issues in Facilities Mgmt. Mem. ASHRAE, Assn. Energy Engrs. (sr., Energy Engr. of Yr. 1985-86), Assn. Phys. Plant Adminstrs. Home: 12781 Chase St Garden Grove CA 92645 Office: Calif State U 4665 Lampson Ave Los Alamitos CA 90720

KARKUT, RICHARD THEODORE, clinical psychologist; b. Derby, Conn., Apr. 28, 1948; s. Harry Chester and Mary (Katz) K. AB, William Jewell Coll., 1971; MA, U. Mo., Kansas City, 1976; D Psychology, Forest Inst. Profl. Psychology, 1988. Lic. psychologist, Ohio, Ind.; cert. in biofeedback. Psychology intern Burrell Mental Health Ctr., Springfield, Mo., 1987-88; clin. psychologist Wabash Valley Hosp., Lafayette, Ind., 1989-91, Quinco Cons., North Vernon, Ind., 1991-93; CEO Adkar Assocs., Inc., North Vernon, Ind.; cons. Div. Family Svcs., Lafayette, 1989-90. Guest editor jour. Ind. Psychologist; contbr. articles to profl. jours. Mem. Am. Psychol. Assn., Soc. Behavioral Medicine, Assn. Applied Psychophysiology and Biofeedback, Am. Pain Soc., Am. Soc. Clin. Hypnosis, Am. Orthopsychiat. Assn., Am. Assn. Counseling and Devel., Am. Mental Health Counselor's Assn., Ind. Psychol. Assn., Ill. Psychol. Assn., Ind. Biofeedback Soc. Anglican. Home: 112 Hoosier St North Vernon IN 47265-1103

KARL, DANIEL WILLIAM, biochemist, researcher, consultant; b. Ill., Oct. 19, 1951; s. W.A. and Mildred A.; m. Jane A. Schmidt, 1981. AB cum laude, Monmouth Coll., 1973; PhD in Biochemistry, Iowa State U., 1981. Rsch. fellow Temple U., Phila., 1981-85, rsch. assoc., 1985-86; analysis lab. mgr., rsch. specialist U. Minn., Bioprocess Rsch. Inst., St. Paul, 1986-89, scientist, project mgr., 1989—. Contbr. rsch. publs. to profl. jours. Facilities chmn. St. Paul Open Sch., 1990-92. Recipient fellowship NIH, 1981-84, Am. Heart Assn., 1984-85. Mem. Am. Chem. Soc., Minn. Chromatography Forum. Achievements include rsch. in protein chemistry separation processes, dispersive processes in preparative chromatography, biopolymers and degradable plastics. Home: 430 Saratoga St S Saint Paul MN 55105 Office: U Minn/BPTI 1479 Gortner Ave Saint Paul MN 55108

KARL, GABRIEL, physics educator; b. Cluj, Romania, Apr. 30, 1937; came to Can., 1960; s. Alexander and Frida (Izsak) K.; m. Dorothy Rose Searle, Apr. 10, 1965; 1 child, Alexandra. Ph.D., U. Toronto, Ont., Can., 1964. Research assoc. Oxford U., Eng., 1966-69; prof. physics U. Guelph, Ont., Can., 1969—. Contbr. articles to profl. jours. German-Canadian Research Prize (Deutsch-Kanadischer Forschungspreis), 1992. Fellow Royal Soc. Can.; mem. Am. Phys. Soc., Can. Assn. Physicists (CAP medal 1991). Office: Univ Guelph, Dept Physics, Guelph, ON Canada N1G 2W1

KARLE, ISABELLA, chemist; b. Detroit, Dec. 2, 1921; d. Zygmunt Apolonaris and Elizabeth (Graczyk) Lugoski; m. Jerome Karle, June 4, 1942; children: Louise Hanson, Jean Marianne, Madeleine Tawney. BS in Chemistry, U. Mich., 1941, MS in Chemistry, 1942, PhD, 1944; DSc (hon.), U. Mich., 1976, Wayne State U., 1979, U. Md., 1986; LHD (hon.), Georgetown U., 1984. Assoc. chemist U. Chgo., 1944; instr. chemistry U. Mich., Ann Arbor, 1944-46; physicist Naval Rsch. Lab., Washington, 1946—; Paul Ehrlich lectr. NIH, 1991; mem. exec. com. Am. Peptide Symposium, 1975-81, adv. bd. Chem. and Engring. News, 1986-89. Mem. editorial bd. Biopolymers Jour., 1975—, Internat. Jour. Peptide Protein Rsch., 1981—; contbr. articles to profl. jours. Recipient Superior Civilian Service award USN, 1965, Fed. Women's award U.S. Govt., 1973, Annual Achievement award Soc. Women Engrs., 1968, Annual Achievement award U. Mich., 1987, Dexter Conrad award Office Naval Rsch., 1980, WISE Lifetime Achievement award Women in Sci. and Engring., 1986, award for disting. achievement in sci. Sec. of Navy, 1987, Gregori Aminoff prize Swedish Royal Acad. Scis., 1988, Adm. Parsons award Navy League U.S., 1988, Ann. Achievement award CCNY, 1989; Bijvoet medal U. Utrecht, The Netherlands, 1990, Vincent du Vigneaud award Gordon Conf. (Peptides), 1992; named to Michigan Women's Hall of Fame, 1989. Fellow Am. Acad. Arts Scis., Am. Inst. Chemists. (Chem. Pioneer award 1984); mem. NAS, Am. Crystallographic Assn. (pres. 1976), Am. Chem. Soc. (Garvan award 1976, Hillebrand award 1970), Am. Phys. Soc., Am. Philos. Soc., Biophys. Soc. Home: 6304 Lakeview Dr Falls Church VA 22041 Office: Naval Rsch Lab Code 6030 Washington DC 20375-5320

KARLE, JEAN MARIANNE, chemist; b. Washington, Nov. 14, 1950; d. Jerome and Isabella (Lugoski) K. BS in Chemistry, U. Mich., 1971; PhD, Duke U., 1976. Postdoctoral fellow Nat. Inst. Arthritis, Metabolic and Digestive Diseases NIH, Bethesda, Md., 1976-78, sr. staff fellow Nat. Cancer Inst., 1978-83; rsch. chemist Walter Reed Army Inst. Rsch., Washington, 1983—. Mem. Internat. Soc. for the Study of Xenobiotics, Am. Chem. Soc., Am. Crystallographic Assn., Am. Soc. for Tropical Medicine and Hygiene, Am. Assn. for Cancer Rsch. Office: Walter Reed Army Inst Rsch Dept Pharmacology Washington DC 20307-5100

KARLE, JEROME, research physicist; b. N.Y.C., June 18, 1918; married, 1942; 3 children. B.S., CCNY, 1937; A.M., Harvard U., 1938; M.S., U. Mich., 1942, Ph.D. in Phys. Chemistry, 1943. Rsch. assoc. Manhattan project, Chgo., 1943-44, U.S. Navy Project, Mich., 1944-46; head electron diffraction sect. Naval Rsch. Lab., Washington, 1946-58, head diffraction br., 1958-68, now head lab. for structure matter, 1968—; mem. NRC, 1954-56, 67-75, 78-87; chmn. U.S. Nat. Com. for Crystallography, 1973-75. Recipient Nobel prize in chemistry, 1985. Fellow Am. Phys. Soc.; mem. NAS (chairperson chemistry sect. 1988-91), AAAS, Am. Chem. Soc., Am. Math. Soc., Crystallograph Assn. (treas. 1950-52, pres. 1972), Internat. Union Crystallography (mem. exec. com. 1978-87, pres. 1981-84). Office: US Naval Rsch Lab Structure Matter Code 6030 Washington DC 20375

KARLIN, SAMUEL, mathematics educator, researcher; b. Yonova, Poland, June 8, 1924; s. Morris K.; m. Elsie (div.); children—Kenneth, Manuel, Anna. B.S. in Math., Ill. Inst. Tech., 1944; Ph.D. in Math., Princeton U., 1947; D.Sc. (hon.), Technion-Israel Inst. Tech., Haifa, 1985. Instr. math. Calif. Inst. Tech., Pasadena, 1948-49; asst. prof. Calif. Inst. Tech., 1949-52, assoc. prof., 1952-55, prof., 1955-56; vis. asst. prof. Princeton U., N.J., 1950-51; prof. Stanford U., Calif., 1956—; Andrew D. White prof.-at-large Cornell U., 1975-81; Wilks lectr. Princeton U.., 1977; pres. Inst. Math. Stats., 1978-79; Commonwealth lectr. U. Mass., 1980; lst Mahalanobis meml. lectr. Indian Statis. Inst., 1983, prin. invited speaker XII Internat. Biometics Meeting, Japan; prin. lectr. Que. Math. Soc., 1984; adv. dean math. dept. Weizmann Inst. Sci., Israel, 1970-77; Britton lectr. McMaster U., Hamilton, Ont., Can., 1990. Author: Mathematical Methods and Theory in Games, Programming, Economics, Vol. I: Matrix Games, Programming and Mathematical Economics, 1959, Mathematical Methods and Theory in Games, Programming, Economics, Vol. II: The Theory of Infinite Games, 1959, A First Course in Stochastic Processes, 1966, Total Positivity Vol. I, 1968; (with K. Arrow and H. Scarf) Studies in the Mathematical Theory of Inventory and Production, 1958; (with W.J. Sudden) Tchebycheff Systems: With Applications in Analysis and Statistics, 1966; (with H. Taylor) A First Course in Stochastic Processes, 2d edit., 1975, A Second Course in Stochastic Processes, 1980, An Introduction to Stochastic Modeling, 1984; (with C.A. Micchelli, A. Pinkus, I.I. Schoenberg) Studies in Spline Functions and Approximation Theory, 1976; editor: (with E. Nevo) Population Genetics and Ecology, 1976; (with T. Amemiya and L.A. Goodman) Studies in Econometric, Time Series, and Multivariate Statistics, 1983; (with K. Arrow and P. Suppes) Contributions to Mathematical Methods in the Social Sciences, 1960; (with K. Arrow and H. Scarf) Studies in Applied Probability and Management Sciences, 1962; (with S. Lessard) Theoretical Studies on Sex Ratio Evolution, 1986; editor: (with E. Nevo) Evolutionary Processes and Theory, 1986; sr. editor Theoretical Population Biology, Jour. D'Analyse; assoc. editor Jour. Math. Analysis, Lecture Notes in Biomath., Jour. Applied Probability, Jour. Multivariate Analysis, Jour. Approximation Theory, SIAM Jour. Math. Analysis, Jour. Linear Algebra, Computers and Math. with Applications, Ency. of Math. and Its Applications, Advanced in Applied Math.; contbr. articles to profl. jours. Recipient Lester R. Ford award Am. Math. Monthly, 1973, Robert Grimmett Chair Math., Stanford U., 1978, The John Von Neumann Theory prize, 1987, U.S. Nat. Medal Sci., 1989; Proctor fellow, 1945, Bateman Research fellow, 1947-48; fellow Guggenheim Found., 1959-60, NSF, 1960-61; Wald lectr., 1957. Fellow AAAS, Internat. Statis. Inst., Inst. Math. Stats.; mem. NAS (award in applied math. 1973), Am. Math. Soc., Am. Acad. Arts and Scis., Am. Soc. Human Genetics, Genetic Soc. Am., Am. Naturalist Soc., Human Genome Orgn., London Math. Soc. (elected hon. 1991). Office: Stanford U Bldg 380 Stanford CA 94305

KARLOW, EDWIN ANTHONY, physicist, educator; b. Glendale, Calif., May 13, 1942; s. Milton Anthony and Vera Marie (Cornwell) K.; m. Marilyn Edna Cross, Sept. 8, 1964; children: Marvin Anthony, Norman Edward. BS, Walla Walla Coll., 1966; MS, Wash. State U., 1968, PhD, 1971. Assoc. prof. math., physics Columbia Union Coll., Takoma Park, Md., 1974-78; prof. physics Loma Linda U., Riverside, Calif., 1978-90, La Sierra U., Riverside, 1990—; cons. Merlan Sci., Ltd., Georgetown, Ont., Can., 1986-90. Author: SIGPRO-Studies in Signal Processing, 1986. Chmn. bd. trustees La Sierra Acad. and Elem. Sch., Riverside, 1982-86. Mem. IEEE, Am. Assn. Physics Tchrs., Am. Phys. Soc., Audio Engring. Soc., Nat. Sci. Tchrs. Assn., Am. Sci. Affiliation. Seventh-day Adventist. Achievements include development of computer interfaces Champ, Nuclear Champ. Office: La Sierra Univ 4700 Pierce St Riverside CA 92515

KARLSON, KEVIN WADE, trial, forensic, and clinical psychologist, consultant; b. Madison, S.D., Sept. 23, 1952; s. Howard Earl and Merna Eunice (Pearson) K. BS in Psychology, S.D. State U., 1974; MA in Clin. Psychology, Tex. Christian U., 1976; PhD in Clin. Psychology, U. Tex. Southwestern Med. Ctr., Dallas, 1983; JD, So. Meth. U., 1984. Lic. psychologist; cert. health svc. provider, Tex.; mem. Nat. Register Health Svc. Providers in Psychology. Counseling psychologist Tex. Christian U., Ft. Worth, 1975-77; psychology intern U. Tex. Southwestern Med. Ctr., 1978-80, Fed. Correctional Instr., Ft. Worth, 1980-81; psychol. assoc. Charles A. Kluge P.C., Dallas, 1981-84; pvt. practice forensic & clin. psychology Dallas, 1984-91; prin., trial cons. Trial Psychology Inst., Dallas, 1991—; chair Tex. Psychology Polit. Action Com., 1989-91; adj. faculty U. Tex. Southwestern Med. Ctr. 1985—. Author: Loving Your Children Better, 991; co-author: Child Care Screening System, 1993, Uniform Child Custody Evaluation System, 1993; contbg. author: Psychiatric and Psychological Evidence, 1987; contbr. chpt. to book. Pres. Children's Arts and Ideas Found., Dallas, 1989-90; bd. dirs. Sammons Ctr. for the Arts, Dallas.; mem. Dallas County Mental Health Assn., 1984—. Mem. Am. Psychol. Assn., Tex. Psychol. Assn. (chair legis. com. 986-91), Dallas Psychol. Assn. (chair ethics com. 1990-92), Brookhaven Country Club, Phi Alpha Delta. Republican. Office: Trial Psychology Inst Ste 101 4000 Spring Valley Rd Dallas TX 75244

KARLSSON, INGEMAR HARRY, engineer; b. Stockholm, Oct. 7, 1944; s. Harry Edgar Gustav and Sanny Kristina (Lundgren) K.; m. Birgitta Britt Annie Gustavsson, Mar. 28, 1983; children: Krister, Karin, Robert, David. Degree in engring., Tekniskt Làroverk, Södertàlje, 1964; BBA, Norra Real, Stockholm, 1975. Engr. Stockholm Harbor, 1964-65, State Agriculture Bd., Solna, 1967-72; engr., owner Bröd. Karlsson Rörsvets, AB, Östhammar, Sweden, 1972—. Lutheran. Avocations: canoeing, skiing, long-distance skating. Home and Office: Bröd Karlsson Rorsvets, Sturegatan 3, 74231 Östhammar Sweden

KARNEY, JAMES LYNN, physicist, optical engineering consultant; b. San Antonio, July 3, 1941; s. Clyde Everett and Irene Jane (Rock) K.; m. Sarah Dana Abraham, Sept. 2, 1962 (div. 1975); children: Melissa Lynn, Michael James. BS in Physics, Tex. Tech U., 1963; MS, U. Okla., 1969. Physicist U.S. Dept. Def., Point Mugu, Calif., 1963-79; sr. physicist Xerox Electro-Optical Systems, Pasadena, Calif., 1979-85; v.p. engring. Copyguard, Inc., Sunland, Calif., 1988—; cons. in optical engring. Sunland, 1986—; cons., advisor Teltech, Inc., Mpls., 1989—; mem. adv. bd. Veritec, Inc., Chatsworth, Calif., 1988—; tchr., lectr. Calif. Jr. Coll. System, Pasadena, 1985—. Contbr. articles to profl jours. Mem. Soc. of Photo-optical Instrumentation Engrs., Nat. Mgmt. Assn., Enterprise Forum. Achievements include development of optical signature method for anticounterfeit deterrent; patents for absorption gas for laser countermeasure, optical photocopy prevention method, hand held optical symbology reader. Home: 11360 Alethea Dr Sunland CA 91040 Office: 11360 Alethea Dr Sunland CA 91040

KARNOVSKY, MORRIS JOHN, pathologist, biologist; b. Johannesburg, South Africa, June 28, 1926; came to U.S., 1955; s. Herman Louis and Florence (Rosenberg) K.; m. Shirley Esther Katz, Aug. 26, 1952; children; David Mark, Nina Jane. BS, U. Witwatersrand, Johannesburg, 1946, MB, BCh, 1950, DSc, 1984; diploma clin. pathology, U. London, 1954; M.A. (hon.), Harvard U., 1965. Prof. pathology Harvard U. Med. Sch., Boston, 1968-72, Shattuck prof., 1972—, chmn. program in cell and devel. biology, 1975-90, chmn. pathology dept., 1991-93. Recipient E.B. Wilson award The Am. Soc. for Cell Biology; hon. mem. German Soc. for Cell Biology. Fellow Royal Microscopic Soc.; mem. NAS Inst. of Medicine, Am. Soc. Cell Biology (pres. 1983-84), Am. Assn. Pathologists (co-pres. 1978-79 Rous-Whipple award). Office: Harvard Med Sch 25 Shattuck St Boston MA 02115-6092

KAROL, REUBEN HIRSH, civil engineer, sculptor; b. Toms River, N.J., Aug. 25, 1922; s. Joel Benjamin and Molly Karol; m. Sylvia Gross, Sept. 3, 1943 (dec. Oct. 1991); children: Diane, Leslee, Michael; m. Joan B. Baker, Feb. 6, 1993. B.S. in Civil Engring., Rutgers U., 1947, M.S., 1949. Lic. profl. engr., N.J. Asst. prof. civil engring. Rutgers U., New Brunswick, N.J., 1947-51, dir. Rutgers Ctr. Continuing Engring. Studies, 1967-85, prof. civil engring., 1980-85; prof. emeritus Rutgers U., 1985—; cons. engr. chem. grouting, design engr. Standard Oil Devel. Co., Linden, 1951-56; dir. Engring. Chem. Research Ctr. Am. Cyanamic Co., Princeton, 1956-67; pres. Karol-Warner, Inc., mfr. sci. instruments, 1952-85. Author four coll. texts including Chemical Grouting, 2d edit., 1990; contbr. numerous articles to profl. jours.; U.S. and fgn. patentee in field; exhibited wood sculpture in 9 one-man shows, 7 group shows; commd. wood sculpture, Busch Student Ctr., outdoor concrete sculpture Civil Engring. Lab., Rutgers U.; represented in permanent collections in galleries in N.J., Pa., Fla. Served to 1st lt. Signal Corps U.S. Army, 1943-46. Mem. ASCE (chmn. grouting com. 1976-82 Robert Ridgway award), ASTM (chmn. grouting com. 1979—, Outstanding Achievement award), Am. Soc. Engring. Edn., Nat. Soc. Profl. Engrs. Home and Office: 491 Sayre Dr Princeton NJ 08540-5851

KAROUNA, KIR GEORGE, chemist, consultant; b. N.Y.C., Feb. 2, 1929; s. George K. and Olga (Lebedinskaya) K.; m. Nina L. Paul, June 7, 1964; children: Maria, Natalie. AB, Hiram Coll., 1951. Chemist Ameco, N.Y., 1953-56; asst. chemist Allied Chem., Morristown, N.J., 1956-57; chemist Sapolin Paints, Bklyn., 1957-63; tech. dir. Hempels Marine Paints, Bklyn., 1963-73; sr. chemist Greenpoint Paint, Bklyn., 1973-76; tech. dir. Seagrave Corp., Bklyn., 1977-83; sr. chemist Fyn Paint and Lacquer, Bklyn., 1984-88; cons. E.A.O., N.Y.C., 1959—. Bd. dirs. Slavic Heritage Coun., N.Y.C., 1985—, Russian O. Theol. Fund, N.Y.C., 1982—, Russian Choral Soc., N.Y.C., 1984—. Sgt. U.S. Army, 1951-53. Mem. Soc. Paint Tech., Societé de Chimie Industrielle. Republican. Achievements include research in elemental fluorine, paints and coatings, corrosion control, pollution abatement and market research. Office: EAO New York Grand Central PO Box 1545 New York NY 10163-1545

KARP, WARREN BILL, medical and dental educator, researcher; b. Bklyn., Feb. 12, 1944; m. Nancy Virginia Blanchard, Jan. 4, 1976; children: Heather Anna, Michael Aaron. BS in Chemistry, Pace U., 1965; PhD, Ohio State U., 1970; DMD, Med. Coll. Ga., 1977. Accredited by Coll. of Am. Pathologists. Grad. teaching asst. physiol. chemistry, 1966-68; grad. rsch. assoc. physiol. chemistry Ohio State U., 1968-70, postdoctoral rsch. assoc. pediatrics, 1970-71; with Med. Coll. Ga., Augusta, 1971—; prof. Sch. Grad. Studies Med. Coll. Ga., 1988—, prof. oral diagnosis and patient care svcs., 1988—, prof. oral biology, 1988—, prof. pediatrics, 1988—, prof. biochemistry and molecular biology, 1991—; dir. clin. perinatal lab. Med. Coll. Ga. Hosps. and Clinics, 1977—; dir. nutritional cons. svcs. Med. Coll. Ga. Dental Sch., 1987—; rsch. chmn. Ga. Nutrition Coun., 1989-90; lectr. and speaker in field. Author: Cadmium in the Environment: Part 2, Health Effects, 1981, Vitamin E in Neonatology, 1986; peer reviewer Pediatrics, 1984—, Jour. Pediatrics, 1985—; contbr. articles to profl. jours. Grantee EPA, 1971-75, Biomed. Rsch. Support, 1973, 80-81, Roche Labs., 1985-86, Am. Diabetes Assn., 1986, Am. Cyanamid Co., 1989-90, Dept. Pediatrics Rsch. Support, 1992-93. Mem. AAAS, Am. Chem. Soc., Am. Assn. Dental Rsch., Am. Heart Assn. (grantee 1981-82, nutrition com. 1988-91, health site com. 1990-91), N.Y. Acad. Scis., Ga. Inst. Human Nutrition, Ga. Perinatal Assn., Ga. Nutrition Coun. (rsch. chmn. 1989-90, nominating com. 1990-91), Internat. Assn. Dental Rsch., So. Soc. Pediatric Rsch., Sigma Xi (membership chmn. 1971-75, 78-80, awards com. 1979-80, treas. 1980-84, audtior 1985-86, pres. 1987-88). Achievements include rsch. in placental enzyumatic activity profiles, cardiac congenital malformations induced by mercury and neonatal nutrition. Home: 402 Hastings Pl Augusta GA 30907 Office: Med Coll Ga Pediatrics Dept Bldg 114 Augusta GA 30912

KARPATI, GEORGE, neurologist; b. Hungary, May 17, 1934; m. Shira Tannor, July 31, 1966; children: Adam, Joshua. MD, Dalhousie U., Halifax, N.S., 1960; Doctorate (hon.), U. Marseille, 1993. Postdoctoral tng. Mont. Neurol. Inst., Can., 1960-64, neuroscientist, 1967—; rsch. tng. NIH, Bethesda, Md., 1965-67; staff neurologist Mont. Neurol. Hosp., 1967—; prof. neurology McGill U., Mont., 1978—, Isaac Walton Killam chmn. neurology, 1984. Co-author: Pathology of Skeletal Muscle, 1984; contbr. 160 articles to sci. jours. Recipient 125th Commemorative medal of Can., Gov. Gen. of Can., 1993. Fellow Royal Coll. Physicians and Surgeons Can., Am. Acad. Neurology; mem. Am. Neurol. Assn. Achievements include research in neuromuscular diseases and nerve-muscle biology. Office: Mont Neurol Inst, 3801 University St, Montreal, PQ Canada H3A 2B4

KARPINSKI, JACEK, computer company executive; b. Torino, Italy, Apr. 9, 1927; s. Adam and Wanda (Cumft) K.; m. Eulalia Gryniecka, Mar. 1, 1955 (div. 1975); children: Dorota, Ewa; m. Ewa Stepien, July 11, 1978; children: Adam, Daniel, Sylvan. Spl. student Harvard U., 1961-62; M.Sc.E.E., Politechnika Warsaw, 1951. System engr. Electronic Systems Mfg., Warsaw, 1951-54; adj. prof. Polish Acad. Scis., Warsaw, 1955-65; head computer lab. Warsaw U., 1965-70; mng. dir. Minicomputer R & D & Prodn., Warsaw, 1970-73; adj. prof. Warsaw Politechnic U., 1973-81; mng. dir. Karpinski Computer Systems, Le Mont, Switzerland, 1983-85; cons. in field of artificial intelligence and computer systems, 1985—; founder, pres. NEWAY Computer SA., Switzerland, JK Electronics Ltd., Warsaw, Poland. Author numerous computer systems. Served to lt. Polish Armed Forces, 1941-44. Decorated Cross of Valor (3), Polish Underground Army, 1944, AK Cross, 1944. Mem. IEEE, Polish Acad. of Engring. Roman Catholic. Home: Clos du Village, 1421 Grandevent Switzerland

KARPLUS, HENRY BERTHOLD, physicist, research engineer; b. Berlin, Feb. 9, 1926; came to U.S. 1949; s. Sigmar and Rose Erna (Anker) K.; m. Jean Avril Clarke, Sept. 6, 1951; children: Lester Clive, Kevin John, Wayne Eliot, Nadine Thea. BS, King's Coll. U. London, 1945; MS, IIT, 1953. Rsch. physicist AVIMO Ltd., Taunton, Eng., 1945-48, Telephone Mfg. Co., Sidcup Kent, Eng., 1948, IITRI Armour Rsch. Found., Chgo., 1949-71; rsch. engr. Argonne (Ill.) Nat. Lab., 1971-84, Karplus Consultants, Hindsdale, Ill., 1984—. Contbr. articles to profl. jours. Mem. Sigma Xi. Achievements include patents for Void/Particulate Detector, Doppler Flowmeter, Digital Pressure Transducer for use at High Temperature, Apparatus for Checking the Direction of Polarization of Shear Wave Transducers, Ultrasonic Transducer with Laminated Coupling Wedge, Transition Section for Acoustic Waveguide; discovery of coincidence of neuron delay with basilar membrane delay in the inner ear relating the neuron impulse sequence to the autocorrelation function of the input; development of ultrasonic flowmeters for liquid flow measurement without pipe penetration, time difference techniques for pure fluids and doppler methods for slurries; pulse-echo and through-transmission techniques for I-beam welds; ultrasonic imaging for inspection, noise flow diagrams for ranking noise paths for cost effective control strategies with application to design of jet-engine test-cells, quieting of appliances and for building components; measurement and prediction of sound propagation in various mixtures; environmental noise impact studies for projected Heliport facilities; measurements of aircraft noise propagation as a function of weather conditions. Home and Office: 5605 Monroe St Hinsdale IL 60521-5158

KARPLUS, MARTIN, chemistry educator; b. Vienna, Austria, Mar. 15, 1930; came to U.S., 1938; s. Hans and Isabella (Goldstern) K.; m. Marci Anne Hazard. BA, Harvard U., 1950; PhD, Calif. Inst. of Tech., 1953. NSF fellow Oxford (Eng.) U., 1953-55; asst. prof. chemistry U. Ill., 1957-60, assoc. prof., 1960; prof. Columbia U., N.Y.C., 1960-66; prof. Harvard U., Cambridge, Mass., 1966—, Theodore William Richard prof. chemistry, 1979—; prof. U. Paris VII, 1974-75, Coll. de France, Paris, 1980; prof. associé U. Paris-Sud, 1980-81, U. Louis Pasteur, Strasbourg, France, spring 1992. Author: (with R.N. Porter) Atoms and Molecules: An Introduction for Students of Physical Chemistry, 1970, (with C.L. Brooks III and B.M. Pettitt) Proteins: A Theoretical Perspective of Dynamics, Structure and Thermodynamics, 1988; also articles. Recipient Fresenius award Phi Lambda Epsilon, 1965, Harrison Howe award Am. Chem. Soc., 1967, Outstanding Contbn. award Internat. Soc. Quantum Biology, 1979, Disting. Alumni award Calif. Inst. Tech., 1986, Irving Langmuir award Am. Phys. Soc., 1987, Theoretical Chemistry award Am. Chem. Soc., 1993; Westinghouse scholar, 1947; nat. lectr. Biophys. Soc., 1991. Mem. NAS, Am. Acad. Arts and Scis., Internat. Acad. Quantum Molecular Sci., Netherlands Acad. Art and Scis. (fgn.). Office: Harvard U Dept Chemistry 12 Oxford St Cambridge MA 02138-2900

KARR, JAMES RICHARD, ecologist, researcher, educator; b. Shelby, Ohio, Dec. 26, 1943; s. Rodney Joll and Marjorie Ladonna (Copeland) K.; m. Kathleen Ann Reynolds, Mar. 23, 1963 (div. Nov. 1982); children: Elizabeth Ann, Eric Leigh; m. Helen Marie Herbst Serrano, Dec. 22, 1984. BS, Iowa State U., 1965; MS, U. Ill., 1967, PhD, 1970. Postdoctoral fellow in biology Princeton (N.J.) U., 1970-71; postdoctoral fellow in biology Smithsonian Tropical Rsch. Inst., Balboa, Panama, 1971-72, dep. dir., 1984-87; acting dir. Smithsonian Tropical Rsch. Inst., Balboa, 1987-88; asst. prof. biology Purdue U., Lafayette, Ind., 1972-75; assoc. prof. U. Ill., Urbana, 1975-80, prof., 1980-84; Harold H. Bailey prof. biology Va. Poly. Inst. and State U., Blacksburg, 1988-91; prof. zoology and fisheries, dir. Inst. Environ. Studies, U. Wash., Seattle, 1991—; cons. U.S. EPA on Water Resources, 1978—, Orgn. Am. States, Washington, 1980, South Fla. Water Mgmt. Dist., West Palm Beach, 1989-91. Grantee EPA, 1972-85, U.S. Forest Svc., 1980-81, 90-91, U.S. Fish and Wildlife Svc., 1979-82, NSF, 1982-84, TVA, 1990-93. Fellow AAAS, Am. Ornithologists Union; mem. Ecol. Soc. Am., Am. Soc. Naturalists. Achievements include development of Index of Biotic Integrity, now widely used in North America and Europe to assess directly the quality of water resources. Office: Inst for Environ Studies FM-12 U Wash Seattle WA 98195

KARR, JOSEPH PETER, podiatrist; b. Chgo., Sept. 7, 1925; s. Vendelin Stephan and Irene (Bielik) Karkoška; m. Marilyn Isabelle Calder, Sept. 1, 1951; children: Joseph Jr., Michael, Paul, Kenneth. D of Surgical Chiropody, Chgo. Coll. Chiropody and Pedic Surgery, 1951; D of Podiatric Medicine (hon.), Ill. Coll. Podiatric Med., 1973. Lic. podiatrist, Ill. Podiatrist Chgo., 1951-91; ret., ret., 1991; alumni advisor Dr. William M. Scholl Coll. of Podiatric Medicine, Chgo., 1986—. Author: (pamphlet) A Method To Alleviate and Cure the Painful Heel Syndrome, 1978. Leader, transp. chmn. Boy Scouts Am., 1968-72. With USCGR, 1943-46. Mem. Am. Podiatric Med. Assn., Ill. Podiatry Soc., Ill. Podiatry Edn. Group, Am. Legion. Roman Catholic. Avocations: philosophy, travel, home movies, family. Home: 10624 S Kildare Ave Oak Lawn IL 60453-5304

KARRI, SURYA B. REDDY, chemical engineer; b. Narsapur, India, Nov. 2, 1959; came to U.S. 1982.; s. Venkata R. and Saraswati (Satti) K.; m. Roopa Kala Neelapu, Feb. 7, 1992. B. Tech. in Chem. Engr., Andhra Univ., India, 1976; M.S. in Chem. Eng., U. N.H., 1984, PhD in Chem. Engring., 1988. Sr. rsch. engr. Particulate Solid Rsch., Inc/Inst. Gas Tech., Chgo., 1988-. Contbr. articles to profl. jours. Mem. Am. Inst. Chem. Engrs., Sigma Xi. Achievements included development of design procedures and models to design and operate indsl. fluid particle systems. Office: Particulate Solid Rsch Inc 3424 S State St Chicago IL 60616

KAR ROY, ARJUN, electrical engineer, educator; b. Ranchi, Bihar, India, Feb. 14, 1967; came to U.S., 1988; s. Satya Ranjan and Kalyani (Ghosh) Kar Roy. B in Tech with honors, Indian Inst. Tech., Kharagpur, India, 1988; MS, U. Calif., Irvine, 1990, PhD, 1993. Rsch. engr. Jenson and Nicholson Ltd., Ranchi, India, 1988; chancellors fellow U. Calif., Irvine, 1988-92, Regents fellow, 1993, teaching assoc., 1990—, rsch. scientist, 1989—. Contbr. articles to tech. jours. Recipient Nat. Talent Search scholarship Nat. Coun. Ednl. Rsch. and Tng., Indian, 1982-88. Mem. IEEE, Laser and Electro-Optic Soc., Optical Soc. Am., Internat. Soc. for Optical Engrs. Hindu. Achievements include research in integrated photonic circuits, guided-wave acousto-optics, acousto-optic signal processing, optical communications, and optical computing; made significant contributions in guided-wave acousto-optics. Home: 6294 Adobe Cir Rd S Irvine CA 92715 Office: U Calif Dept Elec Engring Irvine CA 92717

KARRS, STANLEY RICHARD, chemical engineer; b. New Kensington, Pa., Apr. 7, 1949; s. Stanley S. and Rose I. (Perriello) K.; m. Constance H. Mac, Nov. 24, 1979; children: Evan Jordan, Emily Jillian. BS in Chem. Engring., U. Pitts., 1971. Registered profl. engr., Pa., Mich., Ohio, Fla., S.C. Design engr. Lancy Labs., Zelienople, Pa., 1971-77; regional sales mgr. Oxy Metal Ind., Troy, Mich., 1977-79; project mgr. Lancy Internat., Zelienople, 1979-83; dir. ops. Alcoa Separations Tech., Warrendale, Pa., 1983-89; dir. process engring. U.S. Filter Inc., Warrendale, Pa., 1990—. Contbr. articles to profl. jours. Pack asst. Boy Scouts Am., Wildwood, Pa., 1992—; renew leader St. Catherines of Sweden, Hampton, Pa., 1990; vol. March of Dimes, 1989. Mem. Am. Electroplaters Soc. (program chmn. 1990-91, chmn. air environ. subcom., 1989-92), Hazardous Materials Control Resources Inst. Achievements include research in recycling air used in biological treatment processes to reduce VOC air emissions; patent pending. Home: 613 Westland Dr Gibsonia PA 15044 Office: US Filter Co 181 Thornhill Rd Warrendale PA 15044

KARSON, CATHERINE JUNE, computer programmer, consultant; b. Salt Lake City, Jan. 26, 1956; d. Gary George and Sylvia June (Naylor) Anderson; m. Mitchell Reed Karson, June 14, 1987. A in Gen. Studies, Pima Community Coll., Tucson, 1989, AAS in Computer Sci., 1990. Night supr. F.G. Ferre & Son, Inc., Salt Lake City, 1973-76, exec. sec., 1977-79; operating room technician Cottonwood Hosp., Salt Lake City, 1976-77; customer svc. rep., System One rep. Ea. Airlines, Inc., Salt Lake City and Tucson, 1979-88; edn. specialist Radio Shack Computer Ctr., Tucson, 1988-89; programmer/analyst Pinal County DPIS, Florence, Ariz., 1989-90; systems analyst Carondelet Health Svcs., Tucson, 1990; programmer/analyst Sunquest Info. Systems, Tucson, 1990—; cons. Pinal County Pub. Fiduciary, Florence, 1990, UBET, Barbados, W.I., 1990—, numerous clients, Tucson, 1990—. Mem. bus. adv. coun. Portable Practical Ednl. Preparation, Inc., Tucson, 1990-91. Mem. Nat. Systems Programmer Assn. Republican. Jewish. Avocations: reading, painting, music, light opera performance, dance classes. Home: 4287 N River Grove Cir # 104 Tucson AZ 85719

KARVELAS, DENNIS E., computers and information science educator; b. Athens, Attiki, Greece, Sept. 10, 1958; came to U.S., 1989; Diploma Elec. Engr., Nat. Tech. U. Athens, 1982; M Applied Sci., U. Toronto, 1984, PhD in Elec. Engring., 1990. Rsch. asst. dept. elec. engring. U. Toronto, Ont., Can., 1982-89, teaching asst. depts. elec. engring., computer sci., 1982-89; asst. prof. dept. computer and info. sci. N.J. Inst. Tech., Newark, 1989—. Contbr. articles to sci. jours. and major internat. confs. U. Toronto open doctoral fellow, 1986-87; AT&T grantee, 1990-92. Mem. IEEE, TEch. Chamber Greece, Profl. Elec. Engrs. Union (Greece). Achievements include research to show the relations between various service disciplines in token passing networks and apply them to derive exact or accurate approximate delay expressions for the ordinary, limited and Timed-Token Service (TTS) systems; investigate the voice/data performance of the Timed-Token Protocol (TTP) and Multiple-Priority Cycles (MPC) protocol under silence transmission and under silence suppression; investigate the performance of a hybrid-switching system in which the stations interconnected can communicate by means of both packet switching and circuit switching; introduce and analyze the performance of medium access control mechanisms which can provide fair bandwidth allocation in high speed Metropolitan Area Networks (MANs) without requiring bandwidth loss; propose and investigate effective priority mechanisms on large latency MANs, which can respond fast to traffic overloads and enable real time applications to meet their delay constraints; provide useful insight into the behavior of a large variety of network protocol. Home: 45 Beverly Rd Bloomfield NJ 07003 Office: NJ Inst Tech 323 Martin Luther King Blvd Newark NJ 07102

KASABACK, RONALD LAWRENCE, mechanical engineer; b. Sheffield, Pa., Dec. 3, 1935; s. Anthony John and Helen (Antal) K.; m. Carlotta Mary Wattsjer, Sept. 28, 1963; children: Ronald II, Carla Marie. BSME, U. Notre Dame, 1957; M Engring. Sci., Pa. State U., 1980. Registered profl. engr., Pa. Engr. Jack and Heintz, Inc., Cleve., 1957-59; product engr. Sylvania Co., Warren, Pa., 1959-63; design engr. Lord Mfg. Co., Erie, Pa., 1963-68; product engr. Erie Press Systems, 1968-78; design mgr. Weber Knapp, Jamestown, N.Y., 1978-79; chief engr. Mohawk Indsl.Design, Ridgway, Pa., 1979-80; equipemnt engr. Keystone Carbon Co., St. Mary's, Pa., 1980—. Chmn. Green Twp. Zoning Bd., Erie, 1975; mem., pres. Football Boosters Club, Ridgway, 1980-84; chmn. Mcpl. Authority, Ridgway, 1985—. Mem. Soc. Mfg. Engrs. Achievements include patent on torsionally resilient drive, engineering of largest mechanical forging press in world. Home: RD 1 Box 9A Ridgway PA 15853 Office: Keystone Carbon Co 1935 State St Saint Marys PA 15857

KASAI, PAUL HARUO, chemist; b. Osaka, Japan, Jan. 30, 1932; came to the U.S., 1950; s. Shunki and Chiyo (Kobayashi) K.; m. Toko Masako Hatori, Nov. 22, 1959; children: Yumi, Miki. BS in Chemistry, U. Denver, 1955; PhD, U. Calif., Berkeley, 1959. Mem. rsch. staff Hitachi Cen. Rsch. Lab., Tokyo, 1959-62, Union Carbide Rsch. Inst., Tarrytown, N.Y., 1962-66, 67-79; assoc. prof. U. Calif., Santa Cruz, 1966-67; tech. mgr. IBM Instruments, Danbury, Conn., 1979-86; mem. rsch. staff IBM Almaden Rsch. Ctr., San Jose, Calif., 1986—. Contbr. articles to Phys. Rev. Letter, Jour. Chem. Physics, Jour. Am. Chem. Soc., Jour. Phys. Chemistry, Macromolecules. Mem. Am. Chem. Soc. Democrat. Achievements include research in luminescent centers in phosphors by electron spin resonance, electrolytic property of zeolites, electron spin resonance study of isolated atoms, atom clusters, and atom-organic molecular complexes. Home: 18645 Castle Lake Dr Morgan Hill CA 95037 Office: IBM Almaden Rsch Ctr 650 Harry Rd San Jose CA 95120

KASH, JEFFREY ALAN, physicist, researcher; b. Whittier, Calif., Oct. 14, 1953; s. Sidney William and Rose (Eisenberg) K.; m. Judith Cooper. BA, U. Calif., Berkeley, 1975, PhD, 1981. Post-doctoral rschr. IBM Rsch. Div., Yorktown Heights, N.Y., 1981-83; rsch. staff member IBM Rsch. Div., Yorktown Heights, 1983—. Contbr. chpt. to Light Scattering and Other Secondary Emission Studies of Dynamic Processes in Semiconductors, 1991; contbr. articles to profl. jours. Sch. bd. mem. Pleasantville (N.Y.) Union Free Sch. Dist., 1992. Mem. Am. Phys. Soc. Achievements include 3 patents in field. Office: IBM Rsch PO Box 218 Yorktown Heights NY 10598

KASHAR, LAWRENCE JOSEPH, metallurgical engineer, consultant; b. Brooklyn, Calif., June 1, 1933; s. Samuel and Ethel (Dumanis) K.; m. Gail Ann Carleton Gray, May 1, 1965 (div. Sept. 1971); children: Evan Carleton, Summerlea Joan; m. Barbara Carleton Fasiska, Jan. 3, 1981; 1 stepchild, Desa. B in Metall. Engring., Rensselaer Poly. Inst., 1955; MSc, Stevens Inst. Tech., 1959, Carnegie-Mellon U., 1961; PhD, Carnegie-Mellon U., 1970. Registered profl. engr., Calif. Assoc. metallurgist Am. Metal Climax R & D Lab., Carteret, N.J., 1955-59; sr. rsch. metallurgist U.S. Steel Rsch. Lab., Monroeville, Pa., 1964-70; mem. tech. staff Rockwell Internat., L.A., 1971-73; staff engr. Martin Marietta, Orlando, Calif., 1973; v.p. Scanning Electron Analysis Labs., Inc., El Segundo, Calif., 1973-91; pres. Kashar Tech. Svcs. Inc., L.A., 1991—; lectr. grad. metallurgy course U. So. Calif., 1975-81; tchr. undergrad. metallurgy Calif. State U., Long Beach, 1973. Author conf. procs.; contbr. articles to profl. jours. Dow Chem. Co. fellow. Mem. ASTM, IEEE, Am. Soc. for Metals Internat. (tchr. short course 1983, chmn. symposium Westec conf. 1983, chmn. L.A. chpt. 1991), Electron Microscope Soc. Am., Internat. Soc. for Testing and Failure Analysis (vice chmn. 1979-87), Sigma Xi, Tau Beta Pi. Office: Kashar Tech Svcs Inc 250 N Nash St El Segundo CA 90245-4529

KASHDAN, DAVID STUART, chemist; b. N.Y.C., Oct. 21, 1950; s. Louis and Sylvia K.; m. Letitia Power Jones, May 28, 1983; children: Lee Harris, Benjamin Thomas. BS, Stevens Inst. Tech., 1972; PhD, U. Vt., 1977. From rsch. chemist to rsch. assoc. & lab. head Eastman Chem. Co., Kingsport, Tenn., 1979—. Post-doctoral Rsch. fellow U. Calif., Berkeley, 1977-79. Mem. Am. Chem. Soc. Achievements include several patents on chlorination reactions, polymer design for enteric coatings applications of cellulose esters for drug delivery and coating technologies. Home: 2064 Canterbury Rd Kingsport TN 37660 Office: Eastman Chem Co PO Box 1972 Bldg 150B Kingsport TN 37662

KASIMOS, JOHN NICHOLAS, pathologist; b. Chgo., Jan. 26, 1955; s. Nicholas John and Mia (Panos) K. BS in Biology, Loyola U., Chgo., 1978; MS in Biology, Ill. Inst. Tech., 1980; DO, Chgo. Coll. Osteopathic Med., 1984. Diplomate Nat. Bd. Examiners for Osteo. Physicians and Surgeons, Am. Osteo. Bd. Pathologists, Anatomic Pathology and Lab. Medicine. Intern Chgo. Osteo. Health Systems, 1984-85, resident pathology, 1985-89, pathologist, 1989—; asst. prof. pathology Chgo. Coll. Osteo. Medicine, 1989-93; assoc. prof. pathology Midwest U., 1993—; acad. mentor, advisor Chgo. Coll. Osteo. Medicine, 1989—, dir. residency tng., dept. pathology. Fellow Coll. Am. Pathologists, Am. Soc. Clin. Pathologists; mem. Am. Osteopathic Assn., Am. Osteopathic Coll. Pathologists, U.S. Acad. Pathologists, Can. Acad. Pathologists, Ill. Assn. Osteopathic Physicians and Surgeons, Ill. Pathology Soc., Chgo. Pathology Soc. Greek Orthodox. Achievements include research in nuclear magnetic resonance spectroscopy of tumors and pathophysiologic development of disease. Office: Olympia Fields Osteo Med Ct Med Ctr Dept Pathology 20201 Crawford Ave Olympia Fields IL 60461-1010

KASPER, HORST MANFRED, lawyer; b. Dusseldorf, Germany, June 3, 1939; s. Rudolf Ferdinand and Lilli Helene (Krieger) K.; 1 child, Olaf Jan. Diplom-Chemiker, U. Bonn., 1963, Dr. rer. nat., 1965; J.D., Seton Hall U., 1978. Bar admittee: N.J. 1978, U.S. Patent Office, 1977. Mem. staff Lincoln Lab., M.I.T., Lexington, 1967-69; mem. tech. staff Bell Telephone Labs., Murray Hill, N.J., 1970-76; assoc. Kirschstein, Kirschstein, Ottinger & Frank, N.Y.C., 1976-77; patent atty. Allied Chem. Corp., Morristown, N.J., 1977-79; sole practice, Warren, N.J., 1980-83; with Kasper and Weick, Warren, 1983-85, Kasper and Laughlin, 1985—. Mem. ABA, N.J. Bar Assn., Internat. Patent and Trademark Assn., Am. Patent Law Assn., N.J. Patent Law Assn., Am. Chem. Soc., Electrochem. Soc., Am. Phys. Soc., AAAS, N.Y. Acad. Scis. Contbr. numerous articles to profl. jours.; patentee semicondr. field.

KASS, LEON RICHARD, life sciences educator; b. Chgo., Feb. 12, 1939; s. Samuel and Anna (Shoichet) K.; m. Amy Judith Apfel, June 22, 1961; children—Sarah, Miriam. B.S., U. Chgo., 1958, M.D., 1962; Ph.D. in Biochemistry, Harvard U., 1967. Intern Beth Israel Hosp., Boston, 1962-63; staff assoc. Lab. Molecular Biology, Nat. Inst. Arthritis and Metabolic Diseases, NIH, Bethesda, Md., 1967-69; staff fellow Lab. Molecular Biology, Nat. Inst. Arthritis and Metabolic Diseases, NIH, 1969-70, sr. staff fellow, 1970; exec. sec. com. on life scis. and social policy NRC-NAS, Washington, 1970-72; tutor St. John's Coll., Annapolis, Md., 1972-76; Joseph P. Kennedy Sr. research prof. in bioethics Kennedy Inst., Georgetown U., 1974-76; Henry R. Luce prof. liberal arts of human biology in coll. U. Chgo., 1976-84, prof. com. on social thought, 1984-90, Addie Clark Harding prof. in coll. and com. on social thought, 1990—; founding fellow, bd. dirs. Hastings Ctr., 1969—; bd. govs. U.S.-Israel Binat. Sci. Found., 1982-88; mem. coun. Nat. Humanities Coun., 1984-91, vice chmn. 1987-89. Author: Toward a More Natural Science: Biology and Human Affairs, 1985; contbr. articles to profl. jours. Served with USPHS, 1967-69. NIH postdoctoral fellow, 1963-67, John Simon Guggenheim Meml. Found. fellow, 1972-73, Nat. Humanities Ctr. fellow, 1984-85, W.H. Brady, Jr. Disting. fellow Am. Enterprise Inst., 1991-92; NEH grantee, 1973-74. Mem. AAAS, Phi Beta Kappa, Alpha Omega Alpha. Jewish. Office: 1116 E 59th St Chicago IL 60637-1513

KASSAM, AMIRALI HASSANALI, agricultural scientist; b. Zanzibar, Tanzania, June 30, 1943; s. Hassanali Mohamed Saleh and Roshan Fazal (Bhanji) K.; m. Parin Suleman, May 3, 1968; children: Zahra, Shireen, Laila, Salman. BSc in Agri. Sci., Reading U., 1966, PhD in Agrl. Botany, 1971; MS in Irrigation Sci., U. Calif., Davis, 1967. Rsch. demonstrator Reading (Eng.) U., 1968-71; rsch. fellow Inst. for Agrl. Rsch., Ahmadu Bello U., Samaru, Nigeria, 1971-74; internat. scientist Internat. Crops Rsch. Inst. for Semi-Arid Tropics, Hyderabad, India, 1974-76; dir. Echemess Ltd., mgmt. and devel. cons., London, 1978-90; sr. cons. FAO, Rome, 1977-90, sr. agrl. rsch. officer, rsch. and tech. dept. div., 1990—; mem. WHO/FAO/UNEP panel of experts on Environ. Mgmt. of Vector Control, 1983-85; chmn. Aga Khan Found. (U.K.), 1985-89; mem. adv. com. Overseas Devel. Inst., London, 1988-91; referee Directorate Gen. for Sci., Rsch. and Tech., Commn. for European Community, Brussels, 1984—. Author: Agricultural Ecology of Savanna, 1978, Yield Response to Water, 1979, Agroecological Zones Project Reports, 1978-81, Potential Population Supporting Capacities of Lands in the Developing World, 1982, Assessment of Land Resources for Rainfed Crop Production in Mozambique, 1982, Land Resources Appraisal of Bangladesh for Developing Planning, 1988, Land Resource Inventory and Productivity Evaluation for National Development Planning, 1990, Agroecological Land Resources Appraisal for Development Planning in Kenya, 1991; co-editor Irrigation Sci. Jour., 1976—. Mem. Internat. Found. for Emergency Relief and Devel. New Zealand (life). English Speaking Union King George VI Meml. fellow, 1966. Fellow Inst. Biology (London). Ismaili Muslim. Home: 88 Gunnersbury Ave Ealing, London W5 4HA, England

KASSAN, STUART S., rheumatologist; b. White Plains, N.Y., Nov. 19, 1946; s. Robert Jacob and Rosalind (Suchin) K.; m. Gail Karesh, Apr. 4, 1971; children: Michael Andrew, Merrill Alissa. BA, Case Western Res., 1968; MD, George Washington U., 1972. Diplomate Am. Bd. Internal Medicine, Am. Bd. Rheumatology, Am. Bd. Geriatrics. Intern and resident Grady Meml. Hosp., Altanta, 1972-74; clin. fellow NIH, Bethesda, Md., 1974-76; fellow Hosp. for Spl. Surgery, Cornell Med. Ctr., N.Y.C., 1976-78; head rheumatology clinic VA Med. Ctr., Denver, 1978-80; asst. clin. prof. medicine U. Colo. Health Scis. Ctr., Denver, 1978-84, assoc. clin. prof. medicine, 1984—; med. dir. rehab unit. Luth. Med. Ctr., Wheatridge, Colo.,

1983-87; med. dir. rehab. unit St. Anthony Hosp., Denver, 1987—; cons. Annals Internal Medicine, Phila., 1986—; vis. alumni scholar George Washington U. Sch. Medicine, 1986; nat. med. adv. bd. Sjögren's Found., Port Washington, N.Y., 1987—. Co-editor: Sjögren's Syndrome, 1987; contbr. over 25 articles to profl. jours. Bd. dirs. Rocky Mountain chpt. Arthritis Found., Denver, 1978-90 (Polachek fellow, 1976-77). With USPHS, 1974-76. Fellow Am. Coll. Physicians, Am. Coll. Rheumatology (network physician 1989—); mem. Harvey Soc. Jewish. Office: Colo Arthritis Assoc 4200 W Conejos Pl Ste 314 Denver CO 80204-1311

KASSAS, MOHAMED, desert ecologist, environmental consultant; b. Egypt, July 6, 1921; s. Abdel-Fattah El Kassas and Rateeba Shabana; m. Freda Kamel Hosny, March, 21, 1957; children: Sherif, Aida. BSc, U. Cairo, Giza, Egypt, 1944, MSc, 1947; PhD, U. Cambridge, Eng., 1950; D in Forestry (hon.), Swedish Agrl. U., Uppsala, 1985; LLD (hon.), Am. U., Cairo, 1986. Prof. botany U. Cairo, 1965-81, prof. emeritus, 1981—; sr. advisor UN Environ. Program, 1978-92; mem. Shoura Coun., Cairo, 1980—. Pres. Internat. Union for Conservation of Nature, 1978-84; hon. v.p. World Wildlife Fund Internat., 1979-84. Recipient Environment prize UN, 1978, State Sci. prize, Egyptian govt., 1981. Fellow Egyptian Acad. Sci., Nat. Indian Acad. Sci., World Acad. Sci. and Arts; mem. Club of Rome. Achievements include development of the environmental movement in Egypt; research in plant ecology of Egyptian and Middle East deserts. Home: 41 Dokki St, Giza Arab Republic of Egypt Office: Cairo U, Faculty Sci, Giza 12613, Egypt

KASTEN, PAUL RUDOLPH, nuclear engineer, educator; b. Jackson, Mo., Dec. 10, 1923; s. Arthur John and Hattie L. (Krueger) K.; m. Eileen Alma Kiehne, Dec. 28, 1947; children: Susan (Mrs. Robert M. Goebbert), Kim Patrick, Jennifer. BSChemE, U. Mo., Rolla, 1944, M.S., 1947; Ph.D. in Chem. Engring., U. Minn., 1950. Registered profl. engr., Tenn. Staff mem. Oak Ridge Nat. Lab., 1950-88, dir. gas-cooled reactor and thorium utilization programs, 1970-78, dir. HTGR and GCFR programs, 1978-86, tech. dir. gas cooled reactor programs, 1986-88; cons., 1988—; guest dir. Inst. Reactor Devel., Nuclear Research Center, Jülich, Fed. Republic Germany, 1963-64; mem. faculty U. Tenn., Knoxville, 1953—, part-time prof. nuclear engring., 1965—. Fellow AAAS, Am. Nuclear Soc.; mem. Sigma Xi, Tau Beta Pi, Phi Lambda Upsilon. Lutheran. Rsch. and publ. in role of thorium in power reactor devel. high temperature gas-cooled reactors and modular gas reactors for gas-turbine and process-heat applications. Office: 341 Louisiana Ave Oak Ridge TN 37830-8514 also: U Tenn Dept Nuclear Engring Knoxville TN 37996-2300

KASTENBAUM, MARVIN AARON, statistician; b. N.Y.C., Jan. 16, 1926; s. Harry and Sarah (Strahl) K.; m. Helen Ganz, Dec. 22, 1955; children: Joan Kastenbaum Jackson, Robert H. BS, CCNY, 1948; MS, N.C. State U., 1950, PhD, 1956. Statistician Dun & Bradstreet, N.Y.C., 1952; biostatistician Atomic Bomb Casualty Com., Hiroshima, Japan, 1953-54; biometrician Oak Ridge (Tenn.) Nat. Lab., 1956-70; dir. stats. Tobacco Inst., Washington, 1970-87; vis. prof. U. Wis., Madison, 1965-66, Stanford (Calif.) U., 1969. Fellow AAAS, Am. Stat. Assn.; mem. N.Y. Acad. Scis. Jewish. Home: 16933 Timberlakes Dr Fort Myers FL 33908

KATAKKAR, SURESH BALAJI, hematologist, oncologist; b. Poona, India, Feb. 9, 1944; s. Balaji Vasudeo Katakkar and Padmavati (Gangadhar) Varavandkar; m. Sunila Moghe; children: Smita, Sucheta, Swati. MB, BS, Poona U., India, 1969. Diplomate Am. Bd. Internal Medicine, Am. Bd. Oncology, Am. Bd. Quality Assurance and Utilization Rev. Intern the resident St. Paul's Hosp., Saskatoon, 1969-71; resident U. Hosp., Saskatoon, 1971-72; resident clin. hematology Gen. Hosp., Ottawa, 1973-74; fellow in med. oncology W.W. Cross Cancer Inst., Edmonton, Can., 1974-75; sr. cancer clin. assoc. Sasketchewan Cancer Commn., 1975-78; clin. investigator NCI, USA, 1975—; med. oncologist Madigan Army Med. Ctr., 1978-80; pvt. practice Tucson, Ariz., 1980—; med. dir. N.W. Cancer Ctr., 1991—; chmn. tumor bd. St. Mary's Hosp., Tucson, 1981-83, chmn. transfusion com., 1982—; chmn. dept. med. Northwest Hosp., 1983-84, chief of staff, 1984-86, bd. trustees, 1984—. Contbr. articles to profl. jours. W.W. Cross Cancer Inst. fellow, 1974-75. Fellow ACP, Royal Coll. Physicians Can.; mem. AMA, Am. Soc. Clin. Oncology, Internat. Soc. Preventative Oncology, Am. Geriatrics Soc., Am. Hosp. Assn., Am. Assn. Blood Banks, Am. Bd. Med. Dirs., N.Y. Acad. Scis. Home: 1391 E Placita Mapache Tucson AZ 85718-3929 Office: NW Cancer Ctr 1845 W Orange Grove Rd Bldg 2 Tucson AZ 85704

KATAYAMA, TETSUYA, geologist, materials research petrographer; b. Odawara, Kanagawa, Japan, Jan. 20, 1953; s. Tsutomu and Noriko (Ishizaki) K.; m. Tomoko Ueda, Oct. 10, 1984; 1 child, Chiaki. BS in Geology, Chiba (Japan) U., 1977; MS in Geology, U. Tokyo, 1979. Registered profl. engr. in geology and civil engring., Japan. Researcher Cen. Rsch. Lab. Sumitomo Cement Co., Ltd., Funabashi, Japan, 1979-85, researcher Cement/Concrete Tech. Devel. Ctr., 1986-88, asst. sr. researcher Cement/Concrete Tech. Devel. Ctr., 1989-90, sr. researcher Cement/Concrete Rsch. Lab., 1991—. Recipient prize Cement Assn. Japan, 1982. Mem. Am. Concrete Inst., Am. Soc. Testing and Materials, Mineral. Soc. Am., Geol. Soc. Japan, Japan Soc. Civil Engrs. (tech. com. Tokyo chpt. 1989—), Limestone Assn. Japan (technical com. Tokyo chpt. 1988-89, 91—), Japan Concrete Inst. (pub. project Tokyo chpt. 1989—), Japan Mining Industry Assn. (tech. com. Tokyo chpt. 1990-91), Archtl. Inst. Japan (tech. com. Tokyo chpt. 1991—), Japan Cons. Engrs. Assn. (divisional com. Tokyo chpt. 1991-92), Found. Promotion Indsl. Sci. (tech. com. U. Tokyo 1993-). Achievements include research on alkali-aggregate reactions, concrete, durability of concrete, petrography of cement clinkers, concretes, metallurgical slags and other industrial by-products, geology of carbonate rocks, cement raw materials and concrete aggregates. Home: 2-13-16 Miyanodai, Sakura 285, Japan Office: Sumitomo Cement Co Ltd, 585 Toyotomi, Funabashi 274, Japan

KATCHUR, MARLENE MARTHA, nursing administrator; b. Belleville, Ill., Dec. 20, 1946; d. Elmer E. and Hilda B. (Gutherz) Wilde; m. Raymond J. Katchur, Feb. 22, 1969; 1 child, Nickolas Phillip. BSN, So. Ill. U., 1968; MS in Health Care Adminstrn., Calif. State U., L.A., 1982. RN; cert. critical care nurse. Staff nurse, head nurse, nursing supr. U. So. Calif Med. Ctr. LA County, 1968-81, assoc. dir. nursing, internal medicine nursing, 1981-83, mem. staff internal medicine nursing, info. systems dept., 1983-89, patient-centered info. systems cons., 1989-90, nursing info. systems cons. for pediatrics, psychiatry and ICU, 1990-92, psychiat. nursing svcs. human resources and info. systems, 1992—. Mem. Sheriff's Relief Assn. Mem. AACCN, NAFE, AAUW, Nat. Critical Care Inst. Edn., Am. Heart Assn., So. Ill. U. Alumni Assn. (life), Health Svcs. Mgmt. Forum, Orgn. Nurse Execs. Calif. (membership com.), Am. Soc. Profl. and Exec. Women, Soc. Clin. Data Mgmt. Systems (bd. dirs. 1990-91), Soc. Med. Computer Observers (charter), Am. Legion Aux., Nat. Hist. Soc., Job's Daus. (past honor queen). Avocations: reading, crocheting, embroidering, gardening, travel. Office: U So Calif Med Ctr LA County 1200 N State St Los Angeles CA 90033-4525

KATES, ROBERT WILLIAM, geographer, educator; b. Brooklyn, N.Y., Jan. 31, 1929; m. Eleanor Hackman, Feb. 9, 1948. Student, NYU, 1946-48, U. Ind., 1957; A.M., U. Chgo., 1960, Ph.D., 1962; Doctorate (hon.), Clark U. Mem. faculty grad. sch. geography Clark U., Worcester, Mass., 1962—, prof., 1968-92, univ. prof., 1974-88; univ. prof., dir. Alan Shawn Feinstein World Hunger Program, Brown U., Providence, 1986-92, univ. prof. emeritus, 1992—; dir. Bur. Resource Assessment and Land Use Planning, U. Coll., Dar es Salaam, Tanzania, 1967-69; hon. research prof. U. Dar es Salaam, 1970-71. Author: Risk Assessment of environmental Hazard, 1978, (with Ian Burton and Gilbert F. White) The Environment as Hazard, 1978; co-editor: Climact Impact Assessment, 1985, Hunger in History, 1990, The Earth as Transformed by Human Action, 1990; also author, editor or co-editor 15 other books, monographs; exec. editor: Environment; contbr. numerous articles to profl. jours. Prize fellow MacArthur Found; recipient Nat. Medal of Sci. President Bush NSF, 1991. Mem. NAS, AAAS, Assn. Am. Geographers (pres.), Academia Europaea, Tarzania Soc., Am. Acad. Arts and Scis. Home: PO Box 8075 Ellsworth ME 04605 Office: Brown U Alan Shawn Feinstein World Hunger Progra PO Box 1831 Providence RI 02912-0001

KATGERMAN, LAURENS, materials scientist; b. Deventer, The Netherlands, Jan. 29, 1945; came to U.K. 1984; s. Laurens and Jannetje Hendrika (Nap) K.; m. Evelien Zuiderhoek, Jan. 20, 1971; children: Nienke, Titia. Degree, Groningen U., 1967, doctoral degree, 1974. Lectr. Delft U., The Netherlands, 1974-79, sr. lectr., 1980-84; vis. scientist Aluminum Co. Am., Pitts., 1979-80; sr. scientist Alcan Internat., Banbury, Eng., 1984-86, prin. scientist, 1986-92; prof. metallurgy Delft U., The Netherlands, 1992—; spl. prof. Nottingham U., Eng., 1991—. Contbr. articles to profl. jours. Mem. Minerals, Metals and Materials Soc. Avocations: violin, classical music. Office: Delft U of Tech, Rotterdamsweg 137, Delft 2628AL, The Netherlands

KATHIRGAMANATHAN, POOPATHY, chemist; b. Inuvil, United Kingdom, Aug. 27, 1952; s. Muthuthamby and Sellappa (Luxumy) Poopathy; m. Jayanthy Kathirgamanathan, Oct. 20, 1980; children: Janany, Ganesh. BS with honors, U. Colombo, 1976; PhD, U. Exeter, 1980. Postdoctoral fellow U. Exeter, United Kingdom, 1980-82; sr. demonstrator U. Newcastle, United Kingdom, 1982-85; teaching fellow U. Southampton, United Kingdom, 1985-86; prin. scientist Cookson Group plc, Oxford, United Kingdom, 1986-92; mgr. USEP Univ. Coll., London, 1992—; sr. lectr. South Bank U., London, 1993—; com. mem. British Standards Inst., London, 1990—. Contbr. articles to profl. jours. Recipient Sir Monty Finniston award Eureka Engring. Materials and Designs, 1991. Mem. Royal Soc. Chemistry (exec.), Electrochem. Soc. Inc. Achievements include numerous patents in field; inventor microwave welding of plastics. Office: U Coll London, 20 Gordon St, London WC1H 0AJ, England

KATHMAN, R. DEEDEE, biologist, educator; b. Stamford, N.Y., Apr. 16, 1948; d. Frederick Henry and Elizabeth Kathman; m. Ralph O. Brinkhurst, Oct. 13, 1985. BS, SUNY, Oswego, 1970; MS, Tenn. Tech. U., 1981; PhD, U. Victoria, B.C., Can., 1989. Biol. technician LMS Engring., Oswego, 1973-76; biologist AWARE, Inc., Franklin, Tenn., 1978-81; project mgr. E.V.S. Cons., Sidney, B.C., Can., 1981-85; sole proprietor R. D. Kathman Cons., Sidney, 1987-90; dir. ecol. svcs. Woodward-Clyde Cons., Franklin, 1990-93; asst. dir. Aquatic Resources Ctr., Franklin, 1993—; adj. prof. Memphis State U., 1991—. Contbr. articles to profl. jours. Rsch. grantee Cities Svc. Corp., 1976-78; recipient Grad. Rsch. Engring. and Tech. award B.C. Rsch. Coun., 1985-87. Mem. AAAS, Internat. Assn. Meiobenthologists, N.Am. Benthological Soc., Ecol. Soc. Am., Sigma Xi.

KATO, DAISUKE, electronics researcher; b. Gifu, Japan, Aug. 7, 1942; s. Kichijiro and Chiya Kato; m. Michiko Tomita, Apr. 28, 1971; children: Atsuko, Yasuyo. BS, U. Tokyo, 1966, MS, 1968, PhD, 1971. Researcher Eletrotech. Lab., Tanashi, Japan, 1971—; sr. researcher Eletrotech. Lab., Tsukuba. Patentee optical fiber. Home: 3-11-10 Namiki, Ibaraki Tsukuba 305, Japan Office: Eletrotech Lab, 1-1-4 Umezono, Ibaraki Tsukuba 305, Japan

KATO, MICHINOBU, chemistry educator; b. Hiroshima, Japan, July 26, 1926; s. Sadao and Kosen (Yamada) K.; m. Michiko Itahashi, 1953. BS, Hiroshima U., 1950, DSc, 1960. Asst. Aichi (Japan) Gakugei U., Nagoya, 1950-60; lectr. Aichi Prefectural U., 1960-61, asst. prof., 1961-67, prof., 1967-92, emeritus prof., 1992—; chief kindergarten Aichi Kenritsu U., 1982-85. Contbg. author: Chelate Chemistry (Magnetism), 1976; contbr. articles to profl. jours. Avocations: writing poetry and Japanese verse (Tanka). Home: 6-5-603 Kamezaki 4 chome, Asakita-ku, Hiroshima 739-17, Japan Office: Aichi Prefectural U, Mizuho-ku, Nagoya Aichi 467, Japan

KATO, WALTER YONEO, physicist; b. Chgo., Aug. 19, 1924; s. Naotaro and Hideko (Kondo) K.; m. Anna Chieko Kurata, June 26, 1953; children—Norman, Cathryn, Barbara. B.S., Haverford (Pa.) Coll., 1946; M.S., U. Ill., 1949; Ph.D., Pa. State U., University Park, 1954. Research asso. Ordnance Research Lab., Pa. State U., 1949-52; research asso. Brookhaven Nat. Lab., Upton, N.Y., 1952-53; sr. nuclear engr., asso. chmn. dept. applied sci. Brookhaven Nat. Lab., 1975-77, asso. chmn. dept. nuclear energy, 1977-80, dep. chmn., 1980-88, chmn., 1988-91, sr. nuclear engr., 1991—; sr. physicist Argonne (Ill.) Nat. Lab., 1953-75; vis. prof. dept. nuclear engring. U. Mich., Ann Arbor, 1974-75; cons. Office Nuclear Regulatory Research, U.S. Nuclear Regulatory Commn., 1974—. Contbr. numerous articles to profl. jours. Bd. dirs. Naperville (Ill.) YMCA, 1966-74; mem. Order of Sacred Treasure Japanese Govt., 1992. Served with Ordnance Corps AUS, 1946-47. Fulbright Research fellow, 1958-59. Fellow Am. Nuclear Soc. (dir.), Argonne Univ. Assn. (Distinguished Appt. award 1974); mem. Am. Phys. Soc., A.A.A.S., Sigma Xi. Methodist. Home: 3 Chips Ct Port Jefferson NY 11777-1101 Office: Brookhaven Nat Lab Dept Nuclear Energy Upton NY 11973

KATRINAK, THOMAS PAUL, analytical chemist; b. Allentown, Pa., July 24, 1954; s. John P. and Fay C. (Moll.) K.; m. Judith A. Armstrong, Jan. 3, 1976; children: Christina S., Victoria F. BS in Chemistry, West Chester U., 1976. Lab. technician Air Products and Chems., Inc., Allentown, 1976-78; product devel. chemist Pennwalt Corp., King of Prussia, Pa., 1978-81; sr. analytical chemist Merck & Co., West Point, Pa., 1981-86, devel. chemist, 1986-89, supr., 1989-92, lead supr., 1992—; co-author, actor Chem. Magic Show for area high schs., 1973. Youth advisor Messiah Luth. Ch., Newtown Square, Pa., 1978-87, Christ Luth. Ch., Kulpsville, Pa., 1989-92; asst. softball coach Towamewcin Youth Assn., Kupsville, 1991, 92. Mem. Pa. Assn. Accredited Environ. Labs. Office: Merck and Co Sumneytown Pike West Point PA 19486

KATSIKADELIS, JOHN, civil engineering educator; b. Piraeus, Attica, Greece, Dec. 15, 1937; s. Theodore and Irene (Lautari) K.; m. Paraskevi-Eftichia Buyuka, Sept. 6, 1970; 1 child, Christina. Degree in civil engring., Nat. Tech. U., Athens, 1962, ED, 1973; MSc in Applied Mech., Poly. U. N.Y., 1975, PhD in Applied Mech., 1982. Lic. civil engr., Greece. Civil engr. J.T. Katsikadelis & Z. Bakalis Inc., Athens, 1962-74; lectr. engring. Nat. Tech. U., Athens, 1970-82, prof., 1982—, head dept., 1988-90, 93-94, dir. Inst. Structural Analysis and Aseismic Rsch., 1984—; prof. structural analysis Sch. Corp. Engrs., Hellenic Army, Athens, 1976—; dir. gen. Earthquake Planning and Protection Orgn., Greece, 1989-92, European Ctr. for Prevention and Forecasting of Earthquakes of Coun. Europe, Athens, 1989—; corr. Coun. Europe, 1989—, EEC, 1990—; sci. com. European Ctr. Nonlinear Dynamics and Aseismic Risks/Moscow, 1991—; expert of EEC in seismic hazard rsch., 1993. Co-author: The Boundary Element Method for Plates and Shells, 1991; contbr. articles on structural dynamics and analysis of plates by the boundary element method to scientific jours. Warrant officer Greek air force, 1962-65. Recipient honor Gen. Staff Greek Army, 1987, Assn. Alumni, Piraeus, Greece, 1990; grantee Ministry Labour, Greece, 1958-61, Poly. U. N.Y., 1974-75; Fulbright Rsch. scholar, 1974-75. Mem. Greek Soc. Civil Engrs., Hellenic Soc. Theoretical and Applied Mechanics (treas. 1986—), Greek Assn. Computational Mechanics (v.p.), Internat. Soc. Boundary Elements, Tech. Chamber Creece. Avocations: mountain climbing, skiing, cycling. Office: Nat Tech U, Zografou Campus, GR-157 73, Athens Greece

KATSIKAS, SOKRATIS KONSTANTINE, electrical engineer, educator; b. Athens, Attiki, Greece, Mar. 19, 1960; s. Konstantinos and Elisabeth Vasili (Vouzi) K.; m. Anna Nikolaos Papastamopoulou, July 21; 1 child, Konstantinos S. Diploma in Elec. Engring., U. Patras, Greece, 1982; PhD in Engring., U. Patras, 1987; MS in Elec. and Computer Engring., U. Mass., 1984. Registered electrical engr., Greece. Teaching assoc. Univ. Mass., Amherst, 1982-84; rsch. assoc. Univ. Patras, 1984-87; postdoctoral fellow Computer Tech. Inst., Patras, 1987-88; assoc. prof. Technol. Edn. Inst. of Athens, 1988—; asst. prof. Univ. Aegean, Karlovassi, Greece, 1990—; sr. cons. Indecon Advanced Technology, Athens, 1988—; assessor Ministry of Industry, Energy and Technology, Athens, 1989—; bd. dirs. Div. of Systems Analysis, Tech. Edn. Inst. of Athens. Author: (monograph) Data Security in the National Health System, 1990; contbr. articles to profl. jours. Rsch. grantee Commn. of European Communities, 1989—, Hellenic Ministry of Industry, 1989—, Hellenic Ministry Fgn. Affairs, 1991—. Mem. IEEE, Tech. Chamber of Greece, Greek Computer Soc., N.Y. Acad. Scis. Greek Orthodox. Avocations: cinema, theater, chess, swimming. Home: 1 Meg Alexandrou St, GR-15122 Marousi Greece Office: U Aegean, GR-83200 Karlovassi Greece

KATSUKI, HIROHIKO, biochemistry educator; b. Fukuoka, Japan, May 26, 1921; s. Kaisaku and Tamaki (Shigeto) K.; m. Fumiko K.; 1 child, Tsuneo. BS, Kyoto Imperial U., 1944; PhD, Kyoto U., 1959. From lectr. to prof. biochemistry Kyoto U., 1950-85, prof. emeritus, 1985—; prof. biochemistry Kinki U., Higashi-Osaka, 1985—. Author: Metabolic Regulation, 1975; contbr. articles to profl. jours. Home: Shikanodai Nishi 2-3-9, Ikoma 630-01, Japan Office: Kinki U, Kowakae 3-4-1, Higashi Osaka 577, Japan

KATZ, SIR BERNARD, physiologist; b. Leipzig, Germany, Mar. 26, 1911; s. Max and Eugenie (Rabinowitz) K.; m. Marguerite Penly, Oct. 27, 1945; children: David, Jonathan. MD, U. Leipzig, Germany, 1934; MD (hon.), U. Leipzig, German Dem. Republic, 1990; PhD, U. London, 1938, DSc, 1943; DSc (hon.), U. Southampton, 1971, U. Melbourne, 1971, Cambridge U., 1980; PhD (hon.), Weizmann Inst. Sci., 1979. Beit Meml. Research fellow, 1938-39; Carnegie Research fellow Sydney, Australia, 1939-42; asst. dir. biophys. research U. Coll., London, 1946-50, reader, 1950, prof., head biophysics dept., 1952-78; lectr. univs., socs. Author: Electric Excitation of Nerve, 1939; Nerve, Muscle and Synapse, 1966; The Release of Neural Transmitter Substances, 1969; also articles. Mem. Agrl. Research Council, 1967-77. Recipient Feldberg award, 1965, Copley medal Royal Soc., 1967, Nobel prize in medicine-physiology, 1970, Cothenius medal Deutsche Akademie der Naturforscher Leopoldina, 1989; created knight, 1969. Fellow Royal Soc. (council 1964-65, v.p. 1965, biol. sec. 1968-76), Royal Coll. Physicians (Baly medal 1967); fgn. mem. Royal Danish Acad. Scis. and Letters, Acad. Nat. Lincei, Am. Acad. Arts and Sci., Nat. Acad. Scis. U.S. (fgn. assoc.), Order Pour le Mérite für Wissenschaften und Künste (fgn.). Research on nerve and muscle function especially transmission of impulses from nerve to muscle fibers. Office: U Coll Dept Physiology, Gower St, London WC1E 6BT, England

KATZ, DAVID YALE, electronics engineering technology educator; b. Augusta, Ga., Mar. 3, 1950; s. Joseph and Miriam (Stadiem) K. BSEE, N.C. State U., 1974, MSEE, 1977. Instr. in electronics engring. tech. Augusta Tech. Inst., 1985—. Home: 2415 Bowdoin Dr Augusta GA 30909 Office: Augusta Tech Inst 3116 Deans Bridge Rd Augusta GA 30906

KATZ, IRVIN RONALD, research scientist; b. Bronx, N.Y., July 24, 1963; s. David H. and Felicia Hope (Weinberger) K. BS in Computer Sci. summa cum laude, Rensselaer Poly. Inst., 1984; PhD in Cognitive Psychology, Carnegie Mellon U., 1988. Rsch. asst. and programmer GE Corp. R&D, Schenectady, N.Y., 1984; vis. rsch. fellow Dept. Computer Sci., Keio U., Yokohama, Japan, 1988-90; rsch. scientist Ednl. Testing Svc., Princeton, N.J., 1990—; reviewer Human Computer Interaction, 1988, Jour. of Ednl. Measurement, 1993, Psychol. Sci. Author: (with others) Advanced Research on Computers and Education, 1990; contbr. articles to profl. jours. Grad. fellowship Carnegie Mellon U., 1984-85, 1985-88, postdoctoral fellowship Japan Soc. for the Promotion of Sci., 1988-90. Mem. Psychonomic Soc. (assoc.), Cognitive Sci. Soc., Am. Psychol. Soc. Jewish. Office: Ednl Testing Svc Rosedale Rd Princeton NJ 08541

KATZ, ISRAEL, engineering educator, retired; b. N.Y.C., Nov. 30, 1917; s. Morris and Sarah (Schwarts) K.; m. Betty Steigman, Mar. 29, 1942; children: Susan Raisner, Judith Kessar, Ruth Babai. BSME, Northeastern U., 1941; grad. in naval architecture, MIT, 1942; MME, Cornell U., 1944. Registered profl. engr., N.Y., Mass. Test engr. GE Co., Lynn, Mass., Schenectady, N.Y., 1938-42; sr. engr. USN diesel engring. lab. Cornell U., 1942-46, asst. prof. grad. sch. aero. engring., 1946-48, assoc. prof. Sibley Sch. Mech. Engring., 1948-56; mgr. cons. engring. GE Advanced Elec. Ctr., 1956-63; prof., dir. advanced engring. programs Northeastern U., Boston, 1963-88, dean continuing edn., 1967-74, ret., 1988; cons. engr. in pvt. practice, 1946-92; examiner Engrs. for Appointments and Promotions Com. Mass., 1964-88; engring. cons. NAS, Washington, 1971-88; chmn. seminars Nat. Engrs. Week, 1975-91, dir. New England observance, 1991—; ednl. advisor NSF, Washington, 1980-88; chmn. Mass. Engring. Coun., Boston, 1981-83. Author: (textbook) Principles of Aircraft Propulsion Machinery, 1949; editor various publs. Benwill Publ. Corp., Boston, 1950-88; contbr. articles to profl. jours. Recipient Outstanding Alumnus award Northeastern U., 1992; New Eng. award Engring. Socs. New Eng., 1993; various citations. Mem. Shriners, Bnai Brith. Democrat. Jewish. Achievements include patents for endodontic pressure syringe, and other classified projects. Home: 40 Auburn St Brookline MA 02146

KATZ, JOSE, cardiologist, theoretical physicist; b. Havana, Cuba, June 6, 1944; s. Lipa and Victoria (Masson) K.; m. Anke Ebsen; children: Susan, David, Rachel, Hannah. BS, U. Ill., 1963, MS, 1964, PhD, 1967; MD, F.U., Berlin, 1980. Rsch. assoc. physicist U. Hamburg, Fed. Republic Germany, 1967-69; instr. physics Purdue U., Lafayette, Ind., 1969-71; asst. prof. physics Free U. of Berlin, West Berlin, Fed. Republic Germany, 1971-74; prof. physics F.U., West Berlin, Fed. Republic Germany, 1974-82; resident in internal medicine Cleve. Met. Gen. Hosp., Mt. Sinai Med. Ctr., Cleve., 1982-85; cardiology fellow Southwestern Med. Sch., Dallas, 1985-88; asst. prof. medicine and radiology Columbia U., Coll. of Physicians and Surgeons, N.Y.C., 1988—; co-dir. cardiovascular magnetic resonance Columbia U. Coll. of Physicians and Surgeons, Presbyn. Hosp., N.Y.C., 1988—. Contbr. articles to profl. jours., chpts. to books. Fellow ACP, Am. Coll. Cardiology, Am. Coll. Chest Physicians, Am. Coll. Angiology, Am. Heart Assn. (coun. clin. cardiology, coun. on cardiovascular radiology, coun. on basic scis.), Soc. Magnetic Resonance Imaging, Soc. Magnetic Resonance in Medicine; mem. AMA, Radiol. Soc. N.Am., Soc. Nuclear Medicine, N.Am. Soc. Cardiac Imaging, Sigma Xi, Phi Kappa Phi, Sigma Tau, Pi Mu Epsilon, Tau Beta Pi. Office: Columbia U Div Cardiology 630 W 168th St New York NY 10032

KATZ, JOSEPH JACOB, chemist, educator; b. Apr. 19, 1912; s. Abraham and Stella (Asnin) K.; m. Celia S. Weiner, Oct. 1, 1944; children: Anna, Elizabeth, Mary, Abram. BSc, Wayne U., 1932; PhD, U. Chgo., 1942. Research asso. chemistry U. Chgo., 1942-43, asso. chemist metall. lab., 1943-45; sr. chemist Argonne Nat. Lab., Ill., 1945—; Tech. adviser U.S. delegation UN Conf. on Peaceful Uses Atomic Energy, Geneva, Switzerland, 1955; chmn. AAAS Gordon Research Conf. on Inorganic Chemistry, 1953-54. Am. editor Jour. Inorganic and Nuclear Chemistry, 1955-82. Recipient Distinguished Alumnus award Wayne U., 1955, Profl. Achievement award U. Chgo. Alumni Assn., 1983, Rumford Premium Am. Acad. Arts & Scis., 1992; Guggenheim fellow, 1956-57. Mem. Am. Chem. Soc. (award for nuclear applications in chemistry 1961, sec.-treas. div. phys. chemistry 1966-76), Nat. Acad. Scis., Phi Beta Kappa, Sigma Xi. Home: 1700 E 56th St Chicago IL 60637 Office: Argonne Nat Lab 9700 Cass Ave Lemont IL 60439-4801

KATZ, JULIAN, gastroenterologist, educator; b. N.Y.C., Apr. 3, 1937; s. Abraham M. and Fay (Sher) K.; m. Sheila Moriber, Aug. 18, 1963; children—Jonathan Peter, Sara Katherine. A.B., Columbia U., 1958; M.D., U. Chgo., 1962. Diplomate: Am. Bd. Internal Medicine. Intern U. Chgo. Hosps., 1962-63; resident in medicine Duke U., 1963-65; fellow in gastroenterology Yale U., 1965-67; practice medicine specializing in gastroenterology, internal medicine and geriatrics Phila., 1969—; prof. medicine, lectr. in physiology and biochemistry Med. Coll. Pa., 1970—; prof. medicine Jefferson Med. Coll., 1988—; also lectr. local and nat. groups; chief clin. gastroenterology Med. Coll. Pa. Editor profl. jours.; Contbr. articles to profl. jours. and books. Served with USN, 1967-69. Fellow ACP, Am. Coll. Gastroenterology; mem. Am. Soc. Gastrointestinal Endoscopy, Am. Soc. Study Liver Disease, Am. Gastroenterological Assn., others. Home: 701 Dodds Ln Gladwyne PA 19035-1516 Office: Gastrointestinal Specialists 2 Bala Pla Bala Cynwyd PA 19004

KATZ, LORI SUSAN, toxicologist; b. Framingham, Mass., Aug. 17, 1965; d. Philip Harvey Katz and Sandra Helene (Myers) Berardino. BA, Wheaton Coll., 1987; PhD, Boston U., 1991. Toxicologist Arthur D. Little, Cambridge, Mass., 1988-90; postdoctoral fellow Emory U., Atlanta, Ga., 1990-91; toxicologist Ctr. for Disease Control/Agy. for Toxic Substances and Disease Registry, Atlanta, 1991-92, Pfizer Ctrl. Rsch., Groton, Conn., 1992—. Author book chpt., 1990; contbr. articles to profl. jours. Recipient Merit Rsch. award Boston U. Med. Sch., 1990. Mem. AAAS, Soc. Toxicology, Am. Soc. Pharmacology and Exptl. Therapeutics. Home: 378 25-K Meridian Groton CT 06340 Office: Pfizer Ctrl Rsch Eastern Point Rd Groton CT 06340

KATZ, STEVEN EDWARD, psychiatrist, state health official; b. Phila., Aug. 10, 1937; s. Benjamin R. and Charlotte (Tomkins) K.; m. Marjorie A. Billstein, June 12, 1960; children: Barri L. Stryer, Stacey J. Herron. BA, Cornell U., 1959; MD, Hahnemann U., 1963. Cert. psychoanalyst, Columbia U., 1972. Diplomate Am. Bd. Psychiatry and Neurology. Intern Montefiore Hosp., Bronx, N.Y., 1963-64; resident Columbia U. N.Y. Psychiat. Inst., N.Y.C., 1966-69; dir. edn. dept. psychiatry Roosevelt Hosp., N.Y.C., 1971-74, assoc. dir., 1974-78; med. dir. dept. psychiatry Bellevue Hosp., N.Y.C., 1979-83; vice-chmn. dept. psychiatry NYU Med. Ctr., 1980-83, prof., exec. vice-chmn. dept. psychiatry, 1987—; commr. N.Y. State Office of Mental Health, Albany, 1983-87; dir. psychiatry Bellevue Hosp., 1987-91, dir. health policy NYU Med. Ctr., 1987-93, med. dir. dept. psychiatry, 1992—; clin. prof. psychiatry Albany Med. Coll., 1984-87. Contbr. articles to profl. jours., publs. and book chpts. Bd. dirs. Facilities Devel. Corp., Albany, 1983-87, Am. Mental Health Fund, 1983—, League Ctr., N.Y.C., 1988—, Vis. Nurse Svc., N.Y.C., 1992—. Capt. U.S. Army, 1964-66. Recipient Pub. Service award N.Y. Psychol. Soc., 1984, Pub. Service award Suffolk County Mental Health Assn., 1984, Exceptional Achievement award N.Y. State Office Mental Health, 1985, Governing Bd. award Crotona Park Community Mental Health Ctr., 1985, Pub. Svc. award N.Y. State Psychol. Assn., 1986, Pub. Svc. award for outstanding achievement Am. Assn. for Affirmative Action, 1986, Alexander P. Braile award, 1986, Horace M. Kallen Disting. Community Svc. award Am. Jewish Congress, 1987, William E. Byron award N.Y. State chpt. Assn. Mental Health Adminstrs., 1987; Cert. of Recognition Hosp. Assn. N.Y., 1991. Fellow Am. Psychiat. Assn. (commendation 1983), Am. Coll. Psychiatry; mem. AMA, Group for Advancement of Psychiatry, Am. Assn. Psychiat. Adminstrs. (Disting. Psychiat. Adminstr. award N.Y. regional chpt. 1990), Hosp. Assn. of N.Y. State (cert. of recognition 1991). Democrat. Office: NYU Med Ctr Dept Psychiatry 240 E 27th St New York NY 10016-9277

KATZENELLENBOGEN, JOHN ALBERT, chemistry educator; b. Poughkeepsie, N.Y., May 10, 1944; s. Adolph Edmund Max and Elisabeth (Holzheu) K.; m. Benita Schulman, June 7, 1967; children: Deborah Joyce, Rachel Adria. MA, Harvard U., 1967, PhD, 1969. Asst. prof. chemistry U. Ill., Urbana, 1969-75, assoc. prof. chemistry, 1975-79, prof. chemistry, 1979—, prof. Beckman Inst., 1988—, Roger Adams prof. chemistry, 1992—; chmn. BNP study section NIH, 1987-91; mem. adv. com. AUI Brookhaven, 1986-90. Mem. editorial Biochemistry, Jour. Med. Chem., Steroids; contbr. articles to profl. jours. Recipient Berson Yalow award Soc. Nuclear Medicine, 1988; Camille and Henry Dreyfus Tchr. scholar, 1974-79; fellow Alfred P. Sloan Found., 1974-76, John Simon Guggenheim Found., 1977-78; Univ. scholar, 1987-90. Fellow AAAS, Am. Acad. Arts and Scis.; mem. Am. Chem. Soc., Chem. Soc. (London). Office: U Ill 600 S Mathews Ave Box 37-5 Urbana IL 61801

KATZUNG, BERTRAM GEORGE, pharmacologist; b. Mineola, N.Y., June 11, 1932; m. Alice V. Camp; children: Katharine Blanche, Brian Lee. BA, Syracuse U., 1953; MD, SUNY, Syracuse, 1957; PhD, U. Calif., San Francisco, 1962. Prof. U. Calif., San Francisco, 1958—, v. chmn. Dept. Pharmacology, 1983—. Author: Drug Therapy, 1991, Basic and Clinical Pharmacology, 1992, Pharmacology, Examination and Board Review, 1993; contbr. articles to profl. jours. Markle scholar. Mem. AAAS, AAUP, Am. Soc. Pharmacology and Experimental Therapeutics, Biophysical Soc., Fed. Am. Scientists, Internat. Soc. Heart Rsch., Soc. Gen. Physiologists, Western Pharmacology Soc., Phi Beta Kappa, Alpha Omega Alpha, Golden Gate Computer Soc. Office: UCSF Dept Pharmacology PO Box 0450 San Francisco CA 94143-0450

KAUFFMAN, CHARLES WILLIAM, aerospace engineer; b. Waynesboro, Pa., Dec. 6, 1939; s. Charles Edgar and Florence Evelyn (Neibert) K.; m. Carol Ann Dussinger, Sept. 12, 1964. MS, Pa. State U., 1963; PhD, U. Mich., 1971. Engr. Martin Aircraft Co., Balt., 1961-62, HRB-Singer Inc., State College, Pa., 1963-65; asst. prof. U. Cin., 1971-75, assoc. prof., 1975-77; rsch. scientist U. Mich., Ann Arbor, 1977-85, assoc. prof., 1986—; pres. Explosion Rsch., Whitmore Lake, Mich., 1981—; cons. OSHA, Washington, 1979—. Contbr. articles to profl. jours. Recipient Smolenski medal Polish Acad. of Sci., 1988. Mem. ASME, AIAA, Combustion Inst. (pres. cen. state sect. 1991—), Soc. of Automotive Engrs. Democrat. Episcopalian. Home: 9669 Hermitage Way Whitmore Lake MI 48189 Office: U Mich Dept Aerospace Engring Ann Arbor MI 48109

KAUFFMAN, ERLE GALEN, geologist, paleontologist; b. Washington, Feb. 9, 1933; s. Erle Benton and Paula Virginia (Graff) K.; m. Claudia C. Johnson, Sept. 1989; children from previous marriage: Donald Erle, Robin Lyn, Erica Jean. BS, U. Mich., 1955, MS, 1956, PhD, 1961; MSc (hon.), Oxford (Eng.) U., 1970; DHC, U. Göttingen, Germany, 1987. Teaching fellow, instr. U. Mich., Ann Arbor, 1956-60; from asst. to full curator dept. paleobiology Nat. Mus. Natural History Smithsonian Instn., Washington, 1960-80; prof. geology U. Colo., Boulder, 1980—, chmn. dept. geol. scis., 1980-84, interim dir. Energy, Minerals Applied Rsch. Ctr., 1989-91; adj. prof. geology George Washington U., Washington, 1962-80; cons. geologist, Boulder, 1980—. Author, editor: Cretaceous Facies, Faunas and Paleoenvironments Across the Cretaceous Western Interior Basin, 1977; contbg. editor: Concepts and Methods of Biostratigraphy, 1977, Fine-grained Deposits and Biofacies of The Cretaceous Western Interior Seaway, 1985, High Resolution Event Stratigraphy, 1988, Paleontology and Evolution: Extinction Events, 1988, Extinction Events in Earth History, 1990, Evolution of the Western Interior Basin, 1993; also jour. articles. Recipient U.S. Govt. Spl. Svc. award, 1969, NSF Best Tchr. award U. Colo., 1985 R.C. Moore medal Soc. Sedimentary Geology, 1991; named Disting. Lectr. Am. Geol. Inst., 1963-64, Am. Assn. Petroleum Geologists, 1984, 85, 91, 92; Fulbright fellow, Australia, 1986. Fellow Geol. Soc. Am., AAAS; mem. Paleontol. Soc. (councilor under 40, pres. elect 1981, pres. 1982, past pres. 1983, chmn. 5 coms.); mem. NRC (rep.), Palaeontol. Assn., Internat. Paleontol. Assn. (v.p. 1982-88), Paleontol. Research Instn., Am. Assn. Petroleum Geol., Soc. Sedimentary Geology (com. mem., Spl. Svc. award 1985, Best Paper award 1985, Raymond C. Moore Paleontology medal 1991), Rocky Mountain Assn. Geologists (project chief) (Scientist of Yr. 1977), Paleontol. Soc. Wash. (pres., sec., treas.), Geol. Soc. Wash. (councilor), Md. Acad. Scis. (hon. Paleontology sect.), Sigma Xi, Phi Kappa Phi, Sigma Gamma Epsilon. Democrat. Avocations: music, fishing, climbing, photography. Home: Flagstaff Star Rte 3555 Bison Dr Boulder CO 80302 Office: U Colo Dept Geol Scis Campus Box 250 Boulder CO 80309

KAUFFMAN, JOEL MERVIN, chemistry educator, researcher, consultant; b. Phila., Jan. 3, 1937; s. David and Mathilde (Goldstein) K.; m. Thea Barbara Feldman, June 20, 1977 (div. Mar. 1980); m. Helen Ehrlich Plotkin, June 6, 1981; children: Michael, Alec. BS in Chemistry, Phila. Coll. Pharmacy and Sci., 1958; PhD in Organic Chemistry, MIT, 1963. Cert. profl. chemist, Am. Inst. Chemists. Sr. develop. chemist I.C.I. Organics Inc., Dighton, Mass., 1964-66; rsch. assoc. Mass. Coll. Pharmacy and Sci., Boston, 1966-67, 77-79, from asst. to assoc. prof. chemistry, 1979-92; dir. R & D div. pilot chems. New England Nuclear Corp., Watertown, Mass., 1969-76; prof. chemistry Phila. Coll. Pharmacy and Sci., 1992—; cons. Franklin Rsch. Ctr., Phila., 1982-90. Contbr. chpts. to books, articles to profl. jours. including Jour. Organic Chemistry, Jour. Pharm. Scis., Jour. Chem. Engring. Data, Optics Communus, Analyst, Jour. Chem. Edn., Apothecary, Jour. Computers in Math. and Sci. Teaching, Laser Chemistry, numerous others; pub. papers in field. Mem. Ams. for Legal Reform, Washington, 1985—; assoc. Consumers Union, Yonkers, N.Y., 1991—. Recipient Am. Inst. Chemists medal, 1958, Merck Chemistry award, Alumni medal Phila. Coll. Pharmacy and Sci.; grantee NSF. Mem. Am. Chem. Soc. (award), Nat. Motorists Assn., Pocono Environ. Ednl. Ctr., Phila. Organic Chemists Club. Achievements include patents in Process of Preparing Nitrosocarborane Monomers, Process for Preparing a Thiodiacyl Halide, Compositions and Process for Liquid Scintillation Counting, o,o-Bridged Oligophenylene Laser Dyes, and Dyestuff Lasers, and Methods of lasing Therewith, Radiation Hard Plastic Scintillator; research in antineoplastic drugs, direct synthesis of heterocyclic thiols, bridged quarterphenyls as flashlamp-pumpable laser dyes, new high efficiency fluors for liquid scintillation counting, glycosides and pseudoglycosides of 1,2,4-triazines as potential immunogenetic anticancer drugs, design of radiation-hard flours, development of oligophenylene laser dyes, photophysical properties of some new pyrazolines, others. Home: 65 Meadowbrook Rd Wayne PA 19087 Office: Phila Coll Pharmacy and Sci 600 S 43d St Philadelphia PA 19104-4495

KAUFFMAN, RAYMOND FRANCIS, biochemical pharmacologist; b. Dayton, Ohio, Aug. 20, 1952; s. Lloyd Benjamin and Margaret Elizabeth (Clark) K.; m. Jane Marie Clark, Apr. 20, 1974; children: Christopher Raymond, Carolyn Renee. BS in Chemistry, U. Dayton, 1973; PhD in Biochemistry, U. Wis., 1978. Postdoctoral fellow Enzyme Inst./U. Wis., Madison, 1978, Hormel Inst./U. Minn., Austin, 1979-81; sr. pharmacologist Cardiovascular Div./Eli Lilly & Co., Indpls., 1981-86, rsch. scientist, 1987-92, sr. rsch. scientist, 1993—; chmn. atherosclerosis com. Eli Lilly & Co., Indpls., 1988-91, vascular biology com., 1991—, myocardial contractility subcom., 1987-88. Patentee in field; contbr. articles to profl. jours. Dir. of mentor program Sycamore Sch. of Gifted Children, Indpls., 1986-88; den leader Cub Scouts, Crossroads of Am. Coun., Indpls., 1986-90. Grantee NIH, U. Wis., 1974-78; recipient Presdl. scholarship U. Dayton, 1970-74. Mem. Am. Soc. Pharmacology and Exptl. Therapeutics, Am. Chem. Soc., Am. Heart Assn., Sigma Xi (admissions chmn. 1988-90). Roman Catholic. Achievements include elucidation of biochem. mechanism of novel cardiac stimulants as isozyme - selective inhibitors of cyclic - AMP phosphodiesterase located in sarcoplasmic reticulum; rsch. interests include vascular biology and vascular occlusive disorders. Office: Eli Lilly and Co Lilly Corporate Ctr Indianapolis IN 46285

KAUFMAN, DENISE NORMA, psychologist, addictions counselor, educator; b. Trenton, N.J., Feb. 7, 1954; d. Charles Edwin and Luella (Barcroft) Farr; m. Peter Alan Kaufman, May 15, 1986 (div. Nov. 1989). BS, Trenton State Coll., 1976, MEd, 1977; EdD, Temple U., 1983. Cert. tchr. health, driver edn., spl. edn., N.J., cert. sch. psychologist, cert. addictions counselor, cert. in student pers. svcs., N.J. Health edn. tchr., dept. dir. Haddon Heights (N.J.) Pub. Schs., 1976-81; tchr. educationally handicapped adolescents Haddon Twp. (N.J.) High Sch., 1984-85; tchr. educationally handicapped adolescents, psychologist Archway Programs, Atco, N.J., 1984-90; tchr., psychologist Ferris Sch. for Boys, Dept. Children, Youth and Families, Wilmington, Del., 1991-92; tchr., cons. psychologist Willingboro (N.J.) Twp. Pub. Schs., 1991-92; pvt. practice psychology, addictions counselor Haddon Heights, 1979—; psychologist Atlantic County Spl. Svcs. Sch. Dist., Mays Landing, N.J., 1992-93; prof., supr. student interns Rowan Coll. of N.J., Glassboro, 1993—; mem. Gov. Brendan Byrne's Smoking and Health Com., 1978-80; assoc. prof. health edn. Mercer County Community Coll., Trenton, 1981; program dir. Phila. (Pa.) Health Mgmt. Corp., 1982; cons. Clearview Regional High Sch., Jr. High Sch. Pub. Sch. Dist., 1986—, Lower Camden County Regional Sch. Dist., 1986—; lectr. Assn. Schs. and Agys. for the Handicapped, 1986—; cons., lectr. Charter Fairmont Inst., Phila., 1991—. instr. Am. Red Cross Camden County Chpt.; instr., trainer Am. Red Cross Phila. Chpt., Am. Heart Assn. S.E. Pa. Chpt.; lectr., cons. Haddon Heights (N.J.) Rotary, 1988—; adj. prof. psychology Camden County Coll., Blackwood, N.J., 1993—. Mem. Eta Sigma Gamma, Kappa Delta Pi. Republican. Jewish. Avocations: reading, phys. fitness. Home: 1604 Chestnut Ave Haddon Heights NJ 08035 Office: Rowan Coll of NJ Dept Spl Edn Svcs and Instrn Robinson Hall Glassboro NJ 08028

KAUFMAN, DONALD WAYNE, research ecologist; b. Abilene, Tex., June 7, 1943; s. Leo Fred and Marcella Genevieve (Hobbie) K.; m. Glennis Ann Schroeder, Aug. 5, 1967; 1 child, Dawn. BS, Ft. Hays Kans. State Coll., 1965, MS, 1967; PhD, U. Ga., 1972. Postdoctoral fellow U. Tex., Austin, 1971-73; asst. prof. U. Ark., Fayetteville, 1974-75, SUNY, Binghamton, 1975-77; assoc. program dir. Population Biology, NSF, Washington, 1977-80; asst. prof. biology Kans. State U., Manhattan, 1980-84, assoc. prof. biology, 1984-91, prof. biology, 1991—; vis. scientist Savannah River Ecology Lab., Aiken, S.C., 1973-74; grant rev. panelist EPA, 1981-85; cons. NSF, 1984. Contbr. articles to profl. jours. NDEA fellow, 1967-69; NSF grantee, 1981—. Mem. AAAS, Am. Soc. Mammalogists (award 1972, bd. dirs. 1989-92), Ecol. Soc. Am., Am. Inst. Biol. Scis., Soc. for the Study Evolution. Office: Kans State U Div Biology Ackert Hall Manhattan KS 66506

KAUFMAN, SEYMOUR, biochemist; b. Bklyn., Mar. 13, 1924; s. Charles and Anna Kaufman; m. Elaine Elkins, Feb. 6, 1948; children: Allan, Emily, Leslie. BS, U. Ill., 1945, MS, 1946; PhD, Duke U., 1949. Fellow Dept. Pharmacology, NYU Med. Sch., 1949-50, instr., 1950-53, asst. prof., 1953-54; biochemist Lab. Cellular Pharmacology NIMH, Bethesda, Md., 1954-56, chief sect. on cellular regulatory mechanisms Lab. of Gen. and Comparative Biochemistry, 1956-68, acting chief Lab. Neurochemistry, 1968-71, chief Lab. Neurochemistry, 1971—. Contbr. articles to profl. jours. U.S. Pub. Health fellow Duke U., 1949. Mem. Am. Soc. Biol. Chemists, Am. Chem. Soc., Am. Acad. Arts and Scis., Internat. Soc. for Neurochemistry, Am. Soc. for Neurochemistry, Nat. Acad. Sci. Home: 10300 Rossmore Ct Bethesda MD 20814-2226 Office: Mental Health Intramural Res Div Lab Neurochem Rm 3D30 NIH Bldg 36 Bethesda MD 20892

KAUFMAN, TINA MARIE, physician recruiter, research consultant, presentation graphics consultant; b. Pasadena, Calif., Jan. 13, 1959; d. Thomas Edward and Joan Agnus (Trudeau) K. BA in Exercise Physiology, Humboldt State U., 1982; PhD in Physiology, U. Calif., San Francisco, 1987. Rsch. fellow dept. pediatric cardiology U. Tex. Southwestern Med. Ctr., Dallas, 1987-89, rsch. fellow dept. surgery, 1989-92; physician recruiter Roth Young Dallas, 1992—; graphics, rsch. cons., owner PHD Graphics, Dallas, 1990—. Author, presentor (poster, slide presentations) various sci. meetings and confs.; contbr. articles to profl. jours. Recipient Vera K. Woolford Premed. scholarship Humboldt State U., 1977-78, Regents scholarship, U. Calif. San Francisco, 1982-83. Mem. AAAS, Am. Heart Assn. (coun. on basic sci., coun. on cardiovascular disease in the young). Home: 2615 Bonnywood Ln Dallas TX 75233 Office: Roth Young 5344 Alpha Rd Dallas TX 75240

KAUFMAN, WILLIAM, internist; b. Dec. 31, 1910; s. Leo and Marie Kaufman; m. Charlotte R. Schnee, May 9, 1940. BA, U. Pa., 1931; MA in Chemistry, U. Mich., 1932, PhD in Physiology, 1937, MD cum laude, 1938. Diplomate Am. Bd. Internal Medicine. Intern Barnes Hosp., Washington U. Sch. Medicine, St. Louis, 1938-39; asst. resident, then resident Mt. Sinai Hosp., N.Y.C., 1939-40; Emanuel Libman fellow Yale U. Sch. Medicine, New Haven, 1940-41, Dazian Found. fellow, clin. asst., 1940-42; pvt. practice Bridgeport, Conn., 1940-65; courtesy staff mem. Bridgeport Hosp., 1941-65; courtesy staff St. Vincent's Hosp., Bridgeport, 1941-65; assoc. med. dir. L.W. Frohlich and Co./Intercon Internat. Inc., N.Y.C., 1964-65, med. dir., 1965-67, dir. med. affairs, 1967-68; assoc. med. dir. Klemtner Casey, Inc., N.Y.C., 1969-70, dir. med. affairs, 1970-71; v.p., dir. med. affairs Klemtner Advt., Inc., N.Y.C., 1971, sr. v.p., dir. sci. and med. affairs, 1971-81; pres., chmn. program and pub. edn. coms. Acad. Psychosomatic Medicine, 1953-55; founder fellow, mem. governing coun. Collegium Internationale Allergologicum, 1955-62; chmn. psychosomatic sect. 3d Internat. Congress Allergology, 1958; D. C. Y. Moore Meml. lectr. Manchester (Conn.) Med. Soc., 1967; cons. Family Life Film Ctr. Conn. Inc., 1967-74; mem. screening jury med. edn. sect. Am. Film Festival, 1975. Author: The Common Form of Niacinamide Deficiency Disease (Aniacinamidosis), 1943, The Common Form of Joint Dysfunction: Its Incidence and Treatment, 1949; (drawings) Kaufman's Kritters, 1990; (play) People Like Us; contbg. editor Internat. Archives Allergy and Applied Immunology, 1952-54, 67-69, Am. editor in chief, 1954-67; mem. bd. editorial collaborators Psychotherapeutica, Psychosomatica et Orthopaedagogica, 1955-62; contbg. editor Quar. Rev. Allergy and Applied Immunology, 1955; contbr. numerous articles to profl. and popular jours.; exhibited drawings and paintings at Housatonic Community Coll., New Canaan Art Show, others. Mem. adv. bd. Huxley Inst. So. Conn. for Biosocial Rsch., 1980-84. Recipient citation Internat. Assn. Gerontology, 1983, 1st Pl. award Faculty of Fine Arts, Housatonic Community Coll., 1983, Sci. and Math. medal Rensselaer Poly. Inst., 1928, Sternberg Meml. Gold medal, 1938. Hon. fellow Internat. Acad. Preventive Medicine (Tom Spies Meml. award and lectr. 1978); fellow AAAS, Am. Coll. Allergy and Immunology (emeritus, chmn. pub. edn. com. 1951-55, chmn. com. on allergy of nervous system 1962, Award of Merit 1981), Am. Coll. Nutrition, ACP (life), Gerontol. Soc. Am. (mem. various sects.), N.Y. Acad. Medicine, Royal Soc. Medicine (London); mem. AMA (life), Am Psychosomatic Soc. (emeritus), Conn. State Med. Soc. (life), Fairfield County Med. Assn. (life), Nat. Assn. Sci. Writers, N.Y. Acad. Scis., Dramatists Guild, Sigma Xi, Alpha Omega Alpha, Phi Sigma, Phi Lambda Upsilon, Phi Kappa Phi. Achievements include clin. rsch. using niacinamide (alone or with other vitamins) to reverse certain concomitants of aging, with resultant improvement in osteoarthritis, muscle strength, muscle working capacity and balance sense. Home: 3180 Grady St Winston Salem NC 27104-4008

KAUFMAN, WILLIAM CARL, biophysicist; b. Appleton, Minn., Jan. 21, 1923; s. William Carl and Octavia Marie (Weissbrodt) K.; m. Patricia Hurley, Nov. 23, 1946; children: Jane, William Carl III. BA, U. Minn., 1948; MS, U. Ill., 1952; PhD, U. Wash., 1961. Commd. 2d lt. USAF, 1944, advanced through grades to lt. col., ret., 1968, pilot, scientist, 1950-66; assoc. prof. Preventative Medicine Ohio State U., 1962-67; mem. nuclear effects rsch. com. Defence Atomic Support Agy., 1965-68; chief Biodynamics Aeromed. Rsch. Lab, Hollaman AFB, 1966-68; prof. human biology U. Wis., Green Bay, 1969-87, prof. emeritus, 1987—. Contbr. over 150 articles to profl. jours. NIH fellow, 1968-69. Mem. Am. Physiol. Soc., Aerospace Med. Soc., Sigma Xi. Achievements include research in human reponse to protection against environmental stress. Home: 19228 NE 202 St Woodinville WA 98072

KAUFMAN, WILLIAM MORRIS, research institute director, engineer; b. Pitts., Dec. 31, 1931; s. Nathan and Sarah M. (Paper) K.; m. Iris F. Picovsky, June 21, 1953; children: Nathan E., Marjorie L., D. Emily. BSEE, Carnegie Inst. Tech., 1953, MSEE, PhD in EE. Supr. Westinghouse Electric Corp., Pitts., 1955-62; dir. rsch. Gen. Instrument Corp., Newark, 1962-65; cons. engr. GE, Valley Forge, Pa., 1965-66; mgr. med. engr. dept. Hittman Assocs. Inc., Columbia, Md., 1966-71; v.p. engring. ENSCO, Springfield, Va., 1971-83; v.p. Ocean Data Systems Inc., Rockville, Md., 1984-85; v.p. applied rsch., dir. Carnegie Mellon Rsch. Inst. Carnegie Mellon U., Pitts., 1985—, mem. tech. transfer bd., 1989—, mem. employee retirement and welfare benefit plan com., 1988—; chmn. tech. adv. group Fostin Capital, Pitts., 1986—; mem. adv. bd. Pitts. Seed Fund, 1986—; bd. dirs. Mellon Pitt Carnegie Corp., Ben Franklin Tech. Ctr. of Western Pa., Maglev, Inc., Tech. Devel. and Edn. corp. Patentee in field. Mem. adv. coun. on regional development U. Pitts., 1986. Mem. IEEE (sr.), Sigma Xi, Tau Beta Pi, Eta Kappa Nu. Office: Carnegie Mellon Rsch Inst 4400 5th Ave Pittsburgh PA 15213-2683

KAUFMANN, GARY BRYAN, chemist; b. Akron, Ohio, Jan. 19, 1956; s. Marion Kenneth and Stella Miriam (Butcher) K.; m. Faith Margaret Wollenweber, Aug. 18, 1979; 1 child, Matthew Bryan Michael. AB in Polit. Sci., Greenville Coll., 1978; MS in Chemistry, U. Ky., 1987. Rsch. analyst U. Ky., Lexington, 1986-87; instr. Greenville (Ill.) Coll., 1988; chemist I Am. Bottoms Regional Waste Water Facility, Sauget, Ill., 1988-89, chemist II, 1989; rsch. technician PET, Inc., Greenville, 1989—, assoc. chemist, 1993—. Contbr. articles to profl. jours. Thomas B. Nantz scholar dept. chemistry U. Ky., 1984. Mem. Am. Chem. Soc. Methodist. Office: PET Inc 4649 Le Bourget Dr Saint Louis MO 63134-3120

KAUFMANN, PETER G., research biomedical scientist, psychologist, educator; b. Lauenburg, Fed. Republic of Germany, Feb. 5, 1942; came to U.S., 1952; m. Aukse J. Liulevicius, June 19, 1965; children: Viktoras Peter, Arius Vincentas, Vyga Genoveva. BS (hon.), Loyola U., 1964, MA, 1966; PhD, U. Chgo., 1970. Asst. prof. dept. psychology Emory Coll., Va., 1970-72, Henry Coll., Va., 1970-72; NIMH neurosciences rsch. fellow dept. physiology and pharmacology Duke U., Durham, N.C., 1972-75, asst. prof. dept. anesthesiology, 1975-83; health scientist adminstr. Behavior Medicine Br., Nat. Heart, Lung and Blood Inst., Bethesda, Md., 1983-91, acting chief, 1991-92, chief, 1992—; lectr. dept. psychology Loyola U., 1966-67; adj. assoc. prof. psychology George Mason U., 1992—. Editor: Research Methods Cardiovascular Behavior Medicine, 1989, Stress, Neuropeptides, Systemic Disease, 1991. Trustee Baltic Inst., Rockville, Md., 1991-92; pres. Assn. Lithuanian Cath. Grads., 1987-92. Fellow Soc. Behavioral Medicine; mem. APA, Internat. Soc. Behavioral Medicine, Am. Psychosomatic Soc., Am. Psychol. Soc., Acad. Behavioral Med. Rsch., Soc. Neuroscience, Undersea Med. Soc. Office: Nat Heart Lung and Blood Inst 7550 Wisconsin Ave Fed 216 Bethesda MD 20892

KAUFMANN, WALTER ERWIN, neurologist, neuropathologist, educator; b. Santiago, Chile, Feb. 19, 1959; came to U.S. 1986.; s. Walter Francisco and Maria Elisa (Rocco) K. AB, U. Chile, 1981, MD, 1982, PhD, 1986. Resident and fellow pediatrics/pediatric neurology U. Chile Med. Sch., Santiago, 1982-85, instr. in neurology and physiology, 1984-86; fellow in neurology Beth Israel Hosp. Harvard Med. Sch., Boston, 1986-88, rsch. assoc. neurology, 1988-89; resident and fellow Children's Hosp./Harvard Med. Sch., Boston, 1988-89; fellow in neuropathology Johns Hopkins U. Sch. Medicine, Balt., 1989-92, instr. neurology, 1992—; grant reviewer March of Dimes, Balt., 1990; rsch. cons. E.K. Shriver Mental Retardation Ctr., Boston, 1988-89. Contbr. articles to profl. jours. Recipient Clinician Scientist Merck award, 1993—, NICHD CIA award, 1993-98; Beth Israel Hosp. grantee, 1989; Internat. Brain Rsch. Orgn. fellow, 1986. Mem. Mem. Coll. Am. Pathologists, Internat. Brain Rsch. Orgn., N.Y. Acad. Scis., Soc. for Neurosci., Chilean Soc. of Child Neurology and Psychiatry (dir. 1985-86). Achievements include research in structural anomalies in dyslexic individuals and related animal models; cerebral anomalies in cocaine-exposed infants and experimental models of nutritional-environmental deprivation, and regulatory genes in cerebrocortical development.

KAUL, DAVID GLENN, civil engineer; b. Port Washington, Wis., Nov. 23, 1953; s. Robert William and Shirley Rae (Dorman) K.; m. Lois Ann Hanson, Sept. 30, 1978; children: Matthew, Steven, Eric. BSCE, U. Wis., 1978. Registered profl. engr., Wis. Project engr. Donohue & Assocs., Elkhorn, Wis., 1978-80; city engr. City of Stevens Point, Wis., 1980-85; dir. pub. works, city engr. City of Prairie, Wis., 1985-91; civil engr. State of Wis./ Dept. Adminstrn., Madison, 1991—. Mem. Am. Pub. Works Assn., Am. Soc. Civil Engrs., So. Wis. Assn. Pub. Works. Office: Dept Adminstrn State of Wis PO Box 7866 Madison WI 53707-7866

KAUNITZ, HANS, physician, pathologist; b. Vienna, Austria, Oct. 20, 1905; came to U.S., 1941; s. Arpad and Elsa (Hohenberg) K.; m. Esther Beckwith, Apr. 7, 1943. MD, U. Vienna, Austria, 1930. Lic. physician, N.Y. Supervising physician Vienna U. Hosp., 1932-38; assoc. prof. of medicine U. of Phillipines, Manila, 1938-40, from asst. prof. to clin. prof. Columbia U., N.Y.C., 1956-75. Contbr. numerous articles to profl. jours., sci. papers to meetings, confs. Recipient Presidential Merit medal Pres. of Phillipines, Presidential Hon. medal, Austria, 1961. Mem. Pirquet Soc. (Disting. Mem. award 1970), Am. Oil Chem. Soc. (Achievement award 1971, Alton Bailley medal 1981, Hans Kaunitz Student award 1988). Home: 152 E 94th St New York NY 10128-2510

KAUS, EDWARD GUY, occupational health consultant; b. Elizabeth, N.J., June 3, 1952; s. Joseph Andrew and Maria K. BA in Earth Sci., Kean Coll. N.J., 1980. Asst. environ. technician N.J. Dept. Transp., Trenton, 1983, environ. technician, 1984, environ. specialist, 1985-88, sr. environ. specialist, 1988-90, occupational health cons., 1990—; instr., searcher Med., Chem. and Toxicol. Database, N.J. Dept. Transp., Trenton, 1986—; cons. pvt. practice, Linden, N.J., 1987—; mem. Coun. Cert. Occupational Hearing Conservationist. Scoutmaster, unit commr., asst. dist. commr. Boy Scouts Am., Union County, N.J., 1978—, camping rep. Watchung coun., 1985-91; judge N.Y. City Sci. Fair, 1987-93; local rep. Am. Heart Assn., Linden, 1989;. Recipient Disting. Merit award Boy Scouts Am. Watchung Coun., 1981, Unit Scouters award Lenape Disting. Watchung Coun. Boy Scouts Am., 1990. Mem. N.Y. Acad. Scis., Am. Conf. of Govt. Indsl. Hygenists, Nu Delta Pi Alumni Assn. (Alumni of Yr. 1983), State Micro-Computer Users Group. Achievements include study of particulate concentrations in an urban-industrial area. Office: NJ Dept Trans 1035 Parkway Ave Trenton NJ 08618-2309

KAUSER, FAZAL BAKHSH, aerospace engineer, educator; b. Multan, Panjab, Pakistan, Nov. 15, 1943; came to U.S., 1980; s. Haji Khuda Bakhsh and Bagh Begum; m. Qamar, May 29, 1969; children: Hina Kauser, Shella Kauser. BSc in Physics and Math., Panjab U., 1961; BS with honors, D.I.S. in Aero. Engring., Loughborough U., Eng., 1966; MS in Aeronautical Engring., Air Force Inst. Tech., Wright Patterson AFB, Ohio, 1976. Registered profl. engr., aerospace engr., Fla., mech. engr., Calif. Dir. wind tunnels labs. Pakistan Air Force Engring. Acad., Karachi, 1966-76, head aerodynamic divs., 1977-80; rsch. asst. Pa. State U., College Park, Pa., 1980-82; asst. prof. Embry Riddle Aeronautical U., Daytona Beach, Fla., 1982-86; assoc. prof. Calif. State Polytechnic U., Pomona, 1986—; coord. aerospace Calif. Poly. Inst., Pomona, 1986-90; cons. Lockheed Aeronautical Systems Co., Burbank, Calif., 1988, NASA Jet Propulsion Lab, Pasadena; reviewer Delmar Pub. Co., N.Y., 1989-90; prin. investigator rsch. projects in field, 1978-90.

Advisor Pakistan Students Assn., Cal Poly State U., 1988-90. Squadron leader Pakistan AF, 1966-80. Mem. AIAA (assoc. fellow), ASME (mem. tech. com. propulsion 1992), Am. Soc. Engring. Edn., Tau Alpha Pi, Tau Beta Pi. Achievements include establishment of dept. aerospace engring. and setting up subsonic/supersonic wind tunnel test facilities at Engring. Acad., Pakistan Air Force, Karachi; designer and builder of low turbulence level annular wind tunnel for grad. research in turbulence at department of aerospace engineering of Pa. State University. Office: Calif State Poly Tech U 3801 W Temple Ave Pomona CA 91768

KAUTH, BENJAMIN, podiatric consultant; b. N.Y.C., Oct. 20, 1913; m. Bertha Locke. Student, CCNY, 1936-39; D in Podiatric Medicine, N.Y. Coll. Podiatric Medicine, 1939, postgrad., 1944-45, HHD (hon.), 1981. Pvt. practice N.Y.C., 1939-78; podiatric cons., 1960—; co-chief podiatry staff St. Clare's Hosp., N.Y.C.; chief of staff podiatry Jewish Home and Hosp. for Aged, Village Nursing Home of St. Vincents Hosp.; mem. staff French Polyclinic; chief podiatry panel 1199 Nat. Fund; coord. podiatry panel 32 B-J Health Ctr.; mem. med. panel Med. Malpractice Bronx County; trustee, mem. exec. coun. N.Y. Coll. Podiatric Medicine; cons. Podiatrist Local 1199 Health Fund, Equitable Life Assurance Co., various other third-party insurers, pub. rels. firms. Editorial asst. N.Y. Podiatrist Del. to Nat. Conv.; contbr. articles to profl. jours. Bd. dirs. Adams Sch. for Retarded Children, Am. Jewish Distbn. Com. Fellow Nat. Assn. of Professions; mem. Am. Coll. Foot Surgeons (assoc.), Am. Podiatric Med. Assn. (pub. affairs com., editorial asst.), Podiatry soc. of the State N.Y. (spl. asst. to pres., editorial asst. ann. meeting), N.Y. County Podiatry Soc. (sec., exec. bd.), Fair Harbor Yacht Club (sec.), Friars. Home and Office: 302 W 12th St New York NY 10014-1945

KAUTZ, FREDERICK ALTON, II, aerospace engineer; b. Knoxville, Aug. 27, 1950; s. Frederick Alton and Ora Irene (Wattenbarger) K.; m. Carol Ann Messere, Sept. 11, 1977; children: Catherine Anne, Elizabeth Jane. BSc, U. Tenn., 1972; SM, MIT, 1983, Nuclear Engr., 1983. Staff mem. Oak Ridge (Tenn.) Nat. Lab., 1972-73; nuclear engr. Combustion Engring., Windsor, Conn., 1973; rsch. asst. MIT, Cambridge, 1973-77; staff scientist CIA, Washington, 1977-86; staff mem. MIT Lincoln Lab., Lexington, 1986—. Contbr. articles to profl. jours. Mem. AIAA (sr. mem., sec. New Eng. sect. 1990-91, vice chair 1991-92, chmn. 1992-93, sec. thermophysics tech. com. 1987-90), IEEE, AAAS, Soc. Indsl. and Applied Math., Am. Phys. Soc., Am. Geophys. Union, N.Y. Acad. Sci., Sigma Xi, Tau Beta Pi, Phi Kappa Phi. Achievements include development of techniques for viscous, nonequilibrium supersonic/hypersonic flowfield solutions and Monte Carlo simulation of rarefied high speed external/internal flows and plume impingement. Office: MIT Lincoln Lab 244 Wood St Lexington MA 02173-6499

KAUZMANN, WALTER JOSEPH, chemistry educator; b. Mt. Vernon, N.Y., Aug. 18, 1916; s. Albert and Julia Maria (Kahle) K.; m. Elizabeth Alice Flagler, Apr. 1, 1951; children: Charles Peter, Eric Flagler, Katherine Elizabeth Julia Kauzmann Pacala. B.A., Cornell U., 1937; Ph.D., Princeton U., 1940; PhD (hon.), U. Stockholm, 1992. Westinghouse research fellow Westinghouse Mfg. Co., E. Pittsburgh, Pa., 1940-42; mem. staff Explosives Research Lab., Bruceton, Pa., 1942-44, Los Alamos Lab., 1944-46; asst. prof. Princeton U., 1946-51, asso. prof., 1951-60, prof. chemistry, 1960-82, chmn. dept., 1964-68, David B. Jones prof. chemistry, 1963-82, chmn. biochem. sci. dept., 1980-81; vis. scientist Atlantic Research Lab., NRC Can., 1983; vis. lectr. Kyoto U., 1974; vis. prof. U. Ibadan, 1975. Author: Quantum Chemistry, 1957, Kinetic Theory of Gases, 1966, Thermal Properties of Matter, 1967, (with D. Eisenberg) Structure and Properties of Water, 1969. Recipient Linderstrom-Lang medal, 1966, Stein and Moore award, 1993; Jr. fellow Soc. Fellows, Harvard U., 1942. Fellow AAAS, Am. Acad. Arts and Scis., Am. Phys. Soc.; mem. Nat. Acad. Scis., Am. Soc. Biochemistry and Molecular Biology, Am. Geophys. Union, Am. Chem. Soc.,, Fedn. Am. Scientists, Astron. Soc. Pacific, Royal Astron. Soc. Can., Math. Assn. Am., Protein Soc., Sigma Xi. Office: 301 N Harrison St Ste 152 Princeton NJ 08540-3512

KAVANAGH, YVONNE MARIE, physicist; b. Dublin, Ireland, Feb. 7, 1964; came to U.S., 1989; d. Sean Lydon and Maureen (Burke) Forde; m. William John Kavanagh, Aug. 8, 1986; children: Kevin Sean, Liam Adam. BSc, Univ. Coll. Galway, Ireland, 1984; M Applied Sci., Univ. Coll. Dublin, 1989. Tchr. physics County Dublin Vocat. Edn. Com., 1986-89; entr. Biorad, Boston, 1989-90; mem. tech. staff Hughes Aircraft/Light Valve Products, Inc., Carlsbad, Calif., 1990-93; lectr. in physics Regional Tech. Coll., Carlow, Ireland, 1993—. Mem. Internat. Soc. Optical Engring., Soc. for Info. Display, Cumann Ceimethe na Gallaimle. Roman Catholic. Office: Regional Tech Coll, Kilkenny Rd, Carlow Ireland

KAVANAUGH, HOWARD VAN ZANT, medical physicist; b. New Orleans, Sept. 20, 1931; s. Henry and Carmelita (Van zant) K.; m. Mary Wright, June 20, 1950; children: Terri, Howard, Robyn. BS, Southeastern U., 1960; MS, U. Fla., 1971. Engr. Westinghouse Elec. X-Ray, Baton Rouge, 1956-58; rsch. assoc./instr. La. State U., New Orleans, 1960-63, 65-68; engr. Chrysler Space Div., New Orleans, 1963-65; instr. U. Fla., Gainesville, 1970-72, 74-76; sr. physicist Ochsner Found. Hosp., Jefferson, La., 1972-74, Touro Infirmary, New Orleans, 1979-92, Romagosa Radiation Oncology, Lafayette, La., 1972—; clin. instr. Tulane U. Sch. Medicine, New Orleans, 1974—; rsch. assoc. Clark R&D, Folsom, La., 1992—; founder New Orleans Med. Physics, 1972—. Author: Dosimetry System for Radiation Therapy, 1971, Radiation Physics in Radiation Oncology, 1982, Theoretical Depth Dose Data, 1971, Health Physics in the Healing Arts, 1975. Mem. La. Air Mus., Patterson, 1987—. With U.S. Army, 1953-55. USPHS-Hew fellow, 1971; Radiol. Soc. Chgo. Sci. awardee, 1973. Mem. Am. Assn. Med. Physics, La. Assn. Radiation Oncology, New Orleans Radiation Oncology Assn., La. Assn. Med. Physics. Episcopalian. Home: 4908 Purdue Dr Metairie LA 70003 Office: Romagosa Radiation Ctr 917 Gen Mouton St Lafayette LA 70501-8511

KAVASOGLU, ABDULKADIR YEKTA, civil engineer; b. Urfa, Turkey, Jan. 1, 1952; came to U.S. 1983; s. Suleyman and Semiha (Yasar) K. BSCE, Istanbul Tech. U., 1975; MSCE, Fla. Internat. U., 1987. Registered profl. engr., Fla., Mass. Engr./project mgr. various firms, Turkey, 1975-83; engr. Tecton Engring., Boca Raton, Fla., 1984-85, THN Cons. Engrs., Miami, Fla., 1986; chief engr. D.E. Britt Assocs., Ft. Lauderdale, Fla., 1987-89; pres. GKA Cons. Engrs., Inc., Oakland Park, Fla., 1989-91, YKA Cons. Engrs., Inc., Pompano Beach, Fla., 1991—; adj. prof. Fla. Internat. U., Miami, 1989-91; cons. Fla. Quality Truss, 1990—, Horner Cons. Engrs., Ft. Lauderdale, 1989—. Treas. Turkish Am. Businessmen's Assn., Miami, 1991—. Mem. NSPE, ASCE, Assn. Turkish-Am. Scientists, Am. Concrete Inst., Fla. Turkish Am. Assn. Office: YKA Cons Engrs Inc 3120-B NW 16 Ter Pompano Beach FL 33064

KAVESH, SHELDON, chemical engineer; b. N.Y.C., Jan. 15, 1933; m. Shirley Kavesh; children: Deborah, Neal. BSChemE, MIT, 1957; MSChemE, Poly. Inst. Bklyn., 1960; PhD in Chem. Engring., U. Del., 1968. Rsch. engr. Celanese Corp., Summit, N.J., 1957-60, Foster Grant Co., Leominster, Mass., 1960-62, Avisun Corp., Marcus Hook, Pa., 1962-65; project leader Packaging Div. Union Carbide Corp., Chgo., 1968-70; sr. rsch. scientist Allied Signal Inc., Morristown, N.J., 1970—. Mem. AIChE, Am. Chem. Soc., Am. Phys. Soc. Achievements include 45 U.S. patents in High Strength Polyethylene Fibers and Amorphous Metals. Office: Allied Signal Inc 101 Columbia Rd Morristown NJ 07960

KAVIANY, MASSOUD, mechanical engineer educator; b. Tehran, Iran, July 1, 1948; came to U.S., 1968; s. Morad and Farideh (Etebaii) K.; m. MItra Kaviany, Jan., 1985; children: Saara, Parisa. BS in Mech. Engring., U. Ill., Chgo., 1973, MS in Mech. Engring., 1974; PhD in Mech. Engring., U. Calif., 1979. Staff scientist Lawrence Berkeley (Calif.) Lab., 1979-80; asst. prof. U. Wis., Milw., 1981-85, assoc. prof., 1985-86; assoc. prof. U. Mich., Ann Arbor, 1986-92, prof., 1992—; cons. Allen Bradley Co., Milw., 1985, Ford Motor Co., Dearborn, Mich., 1988—, Dow Chem. Co., Midland, Mich., 1992—, S.W. Rsch. Inst., San Antonio, 1992—. Author: Principles of Heat Transfer in Porous Media, 1991; contbr. articles to Internat. Jour. Heat & Mass Transfer, Jour. Heat Transfer, 1981—. Grantee NSF, 1982—, NASA, 1987—. Fellow ASME; mem. AAAS, Am. Assn. of Combustion Synthesis, Am. Phys. Soc., Am. Soc. Chem. Engrs. Home: 3048 Cedarbrook Ann Arbor MI 48105 Office: U Mich Ann Arbor 2350 Hayward Ann Arbor MI 48109-2125

KAWABATA, NARIYOSHI, chemistry educator; b. Yokohama, Kanagawa, Japan, July 13, 1935; s. Naotaro and Koyuki (Araki) K.; m. Akiko Usutani, Apr. 15, 1962; children: Etsuko, Haruko Ogawa, Hiromi. B in Engring., Kyoto (Japan) U., 1958, M in Engring., 1960, D in Engring., 1963. Staff asst. Kyoto U., 1963-69; assoc. prof. Kyoto Inst. Tech., 1969-76, prof., 1976—; dir. student affairs office Kyoto Inst. Tech., 1992-94. Active amateur symphony orch. Avocation: viola. Home: 1-2-24 Shioji, Nishinari-ku, Osaka 557, Japan Office: Kyoto Inst Tech, Matsugasaki, Sakyo-ku, Kyoto 606, Japan

KAWAHARA, FRED KATSUMI, research chemist; b. Penngrove, Calif., Feb. 26, 1921; s. Kentaro and Kiku (Seo) K.; m. Sumiko Hayami, May 5, 1952; children: Robert Katsumi, Kiku Seo, Richard Hojo; m. Andrea L. Eary, June 29, 1991. BS with honors, U. Tex., 1944; PhD, U. Wis., 1948. Assoc. chemist USDA, Peoria, Ill., 1948-51; postdoctoral fellow U. Chgo., 1951-53; sr. rsch. scientist Amoco Corp. (formerly Standard Oil of Ind.), Whiting, 1953-65; rsch. chemist EPA, Cin., 1965—; cons., expert witness U.S. Dept. Def., U.S. Dept. Air Force, U.S. Dept. Justice, State of Pa., State of N.J.; mentor EPA, others, 1965—. Author: Fossil Energy Extraction, 1983; contbr. chpts. to books, numerous articles to profl. jours. Fellow Am. Inst. Chemists. Achievements include 19 U.S. patents and 1 British patent; first to develop mechanism for soybean oil deterioration; first to develop additive for lubrication of ball bearings operating at 600 degrees Fahrenheit and at 10,000 R.P.M. with 50 pounds axial load for 180 hours to meet requirements for flight and space systems; characterized and identified oil spills with 98% certainty via linear discriminant function analyses and solved 48 oil spills throughout the U.S.; developed a suitable substitute for Freon 113 in order to help curtail ozone deterioration in the stratosphere; involved in hydrodechlorination of organic chlorine compounds which are toxic, carcinogenic to humans and deleterious to the atmosphere; involved in research to detoxify organic chemicals used in warfare. Home: 7751 E Bend Rd Burlington KY 41005-9631 Office: US EPA 26 W Martin Luther King Dr Cincinnati OH 45268-0001

KAWAHARA, MUTSUTO, civil engineer, educator; b. Tokyo, Aug. 31, 1942; s. Mutsuo and Ayako K.; m. Sumiko, Sept. 29, 1972; children: Ryuto, Kiyoto, Shigeto. BS, Waseda U., Tokyo, Japan, 1966, MS, 1968, PhD, 1973. Asst. prof. Chuo U., Tokyo, 1972-73, assoc. prof., 1974-82, prof., 1983—. Regional editor: Internat. Jour. Computational Fluid Dynamics, 1992—; contbr. articles to profl. jours. Recipient Incentive awards Japan Soc. Civil Engrs., 1978. Home: 8-3-22 Todoroki Setagaya-ku, Tokyo 158, Japan Office: 1-13-27 Kasuga, Bunkyo-ku Tokyo 112, Japan

KAWAKAMI, YUTAKA, biomedical researcher, hematologist; b. Osaka, Japan, Jan. 27, 1956; came to U.S., 1985; s. Kiichi and Etsuko Kawakami; m. Masako Abe, Apr. 29, 1982; children: Yuko, Tadashi, Kazuki. MD, Keio U, Tokyo, 1980. Resident Keio U. Hosp., 1980-82; mem. med. staff Nat. Okura Hosp., Tokyo, 1982-84; instr. med. div. hematology Keio U., 1984-85; rsch. assoc. dept. med. microbiology U. South Fla., Tampa, 1985-87; vis. fellow surgery br. Nat. Cancer Inst., NIH, Bethesda, Md., 1987-90, vis. assoc. surgery br., 1990-92, vis. scientist surgery br., 1992—. Contbr. articles to sci. jours. Keio U. Rsch. grantee, 1985; Fogarty Internat. fellow NIH, 1987-90. Mem. AAAS, N.Y. Acad. Sci., Am. Assn. Immunologists. Office: Nat Cancer Inst NIH Rm 4B50 9000 Rockville Pike Bldg 10 Bethesda MD 20892-0001

KAWAMATA, MOTOO, chemical company executive; b. Tokyo, Dec. 13, 1936; s. Takeo and Sanae (Takeda) K.; m. Michiko, Oct. 25, 1966; children: Nobuo, Toshio. BS, Tokyo Inst. Tech., 1960; MS, Baylor U., 1965. Registered chem. engr. Researcher Mitsui Toatsu Chems., Inc., Tokyo, 1960-68, rsch. specialist, 1968-72, rsch. assoc., 1972-77, sr. rsch. assoc., 1977-86, asst. dir., 1986—. Patentee in field. Avocation: tennis. Home: 2261-24 Kamigocho Sakae-ku, Yokohama 247, Japan Office: Mitsui Toatsu Chems Inc, 3-2-5 Kasumigaseki, Chiyoda-ku, Tokyo 100, Japan

KAWAMURA, MITSUNORI, material scientist, civil engineering educator; b. Shinkyo, China, Mar. 2, 1939; arrived in Japan, 1944; s. Tohgoro and Yoshino K.; m. Nobuko Yamaji; children: Kyoko, Yumi, Yoshiko. BS, Kyoto U., 1962, MS, 1964, DSc, 1971. Rsch. assoc. Dept. Civil Engring., Kyoto U., Kyoto, Japan, 1964-66; lectr. Dept. Civil Engring., Kanazawa U., Kanazawa, Japan, 1966-71; assoc. prof. Dept. Civil Engring., Kanazawa U., Kanazawa, 1972-79, prof. civil engring., 1979—; postdoctoral fellow Dept. Civil Engring., Purdue U., Lafayette, Ind., 1972-73; hon. vis. prof. Dept. Civil Engring. U. NSW, Canberra, Australia, 1986-87; vis. rsch. assoc. Sch. Civil Engring., Purdue U., 1991. Author: Construction Materials Reference Book, 1991; editor: Proceedings Alkali-Aggregate Reaction, 1989; author more than 100 papers concerning cement, concrete and other materials. Recipient Japan Concrete Inst. prize, 1984, Cement Assn. of Japan prize, 1987, The Sandberg prize, 1992. Home: Kodatsuno 1-36-18, Ishikawa, Kanazawa 920, Japan Office: Kanazawa U Faculty Engring, Kodatsuno 2-40-20, Kanazawa 920, Japan

KAWANO, HIROSHI, computer science and aesthetics educator; b. Fushun, Manchuria, Japan, Apr. 14, 1925; s. Seiji and Chiyo (Nakayabu) K.; m. Taeko Unoki, May 18, 1954. BA, U. Tokyo, 1951; PhD, Osaka (Japan) U., 1986. Asst. prof. U. Tokyo, 1955-61; assoc. prof. Tokyo Met. Coll. of Air Tech., 1961-72; prof. Met. Coll. of Tech., Tokyo, 1972-86, Tokyo Met. Inst. Tech., 1986-90, Nagano U., Ueda, Japan, 1990—; dir. Inst. of Computer Culture, Tokyo, 1990—; advisor Nippon Computer Graphics Assn., Tokyo, 1983—. Author: Information Science of Art, 1972, Art, Sign and Information, 1982, Logic of Art, 1983, Computer and Aesthetics, 1984. With Japanese mil., 1945. Mem. Assn. for Computing Machinery, Am. Soc. Aesthetics (emeritus), Soc. for Ethno-Art (bd. dirs. 1967—), Philosophy Sci. Soc. of Japan (bd. dirs. 1984—), Japanese Cognitive Sci. Soc., Info. Processing Soc. Japan, Japanese Soc. Artificial Intelligence. Avocations: travel, folk art. Office: 4-1-2-103 Toyogaoka, Tama-shi Tokyo 206, Japan

KAWAUCHI, HIROSHI, hormone science educator; b. Onomichi, Hiroshima, Japan, July 18, 1940; s. Shoichi and Kimiko (Kanenaga) K.; m. Sachiko Kitagawa, Mar. 8, 1971 (dec. May 1983); children: Shimako, Shigehiro, Naohiro; m. Yoriko Kitajima, Feb. 20, 1984. BSc, Tohoku U., Sendai, Japan, 1966, MSc, 1968, PhD, 1971. Postdoctoral fellow hormone rsch. lab. U. Calif., San Francisco, 1971-74; assoc. prof. Sch. Fisheries Scis. Kitasato U., Sanriku, Iwate, Japan, 1974-81, prof., 1981—, head libr., 1983-88, head of dept., 1992—; affiliate prof. Kitasato Inst., Tokyo, 1988—; mem. com. Ministry Internat. Trade and Industry, Tokyo, 1988—. Discovered hormones, 1983, 88, 90 (award 1988, 93). Mem. com. Ofunato City, Iwate, 1988. Grantee Mitubishi Found., 1990; recipient 1st Kitasato Sibasaburo Meml. award Kitasato Inst., 1988. Mem. AAAS, Japanese Agrl. Chem. Soc. (award 1979), Am. Chem. Soc., Japanese Soc. Comparative Endocrinology, Internat. Comparative Endocrinology Soc. (award 1993), Japanese Chem. Soc., N.Y. Acad. Scis., Nippon Suisan Gakkai (award 1987). Buddhist. Avocations: travel, fishing. Home: 5-51-608 Syowa-Machi, Sendai Miyagi 981, Japan Office: Kitasato U Sch Fishery Sci, 160-4 Okirai, Sanriku Iwate 022-01, Japan

KAY, ALAN, computer scientist; b. 1940. Formerly chief scientist Atari Inc.; Apple fellow Apple Computer Inc., Brentwood, Calif., 1984—. Office: Apple Computer Inc 131 S Barrington Pl # 200 Los Angeles CA 90049-3305

KAY, CYRIL MAX, biochemist; b. Calgary, Alta., Can., Oct. 3, 1931; s. Louis and Fanny (Pearlmutter) K.; m. Faye Bloomenthal, Dec. 30, 1953; children: Lewis Edward, Lisa Franci. B.Sc. in Biochemistry with honors (J.W. McConnell Meml. scholar), McGill U., 1952; Ph.D. in Biochemistry (Life Ins. Med. Research Fund fellow), Harvard U., 1956; postgrad., Cambridge (Eng.) U., 1956-57. Phys. biochemist Eli Lilly & Co., Indpls., 1957-58; asst. prof. biochemistry U. Alta., Edmonton, 1958-61; assoc. prof. U. Alta., 1961-67, prof., 1967—, co-dir. Med. Rsch. Coun. Group on Protein Structure and Function, 1974—, mem. protein engring. network Centre of Excellence, 1990—; Med. Rsch. Coun. vis. scientist in biophysics Weizmann Inst., Israel, 1969-70, summer vis. prof. biophysics, 1975, summer vis. prof. chem. physics, 1977, 80; mem. biochemistry grants com. Med. Research Council, 1970-73; mem. Med. Rsch. Coun. Can., 1982-88; Can. rep. Pan Am. Assn. Biochem. Socs., 1971-76; mem. exec. planning com. XI Internat. Congress Biochemistry, Toronto, Ont., Can., 1979; mem. med. adv. bd. Gairdner Found. for Internat. awards in Med. Sci., 1980-89. Contbr. numerous articles to profl. publs.; asso. editor Can. Jour. Biochemistry, 1968-82; editor-in-chief Pan Am. Assn. Biochem. Socs. Revista, 1971-76. Recipient Ayerst award in biochemistry Can. Biochem. Soc., 1970, Disting. Scientist award U. Alberta Med. Sch., 1988. Fellow N.Y. Acad. Scis., Royal Soc. Can.; mem. Can. Biochem. Soc. (council 1971—, v.p. 1976-77, pres. 1978-79). Home: 9408-143d St, Edmonton, AB Canada T5R 0P7 Office: U Alta Dept Biochemistry, Med Scis Bldg, Edmonton, AB Canada T6G 2H7

KAY, ELIZABETH ALISON, zoology educator; b. Kauai, Hawaii, Sept. 27, 1928; d. Robert Buttercase and Jessie Dowie (McConnachie) K. BA, Mills Coll., 1950, Cambridge U., Eng., 1952; MA, Cambridge U., Eng., 1956; PhD, U. Hawaii, 1957. From asst. prof. to prof. zoology U. Hawaii, Honolulu, 1957-62, assoc. prof., 1962-67, prof., 1967—; research assoc. Bishop Mus., Honolulu, 1968—. Author: Hawaiian Marine Mollusks, 1979, Shells of Hawaii, 1991; editor: A Natural History of The Hawaiian Islands, 1972. Chmn. Animal Species Adv. Commn., Honolulu, 1983-87; v.p. Save Diamond Head Assn., Honolulu, 1968-87, pres., 1987—; trustee B.P. Bishop Mus., Honolulu, 1983-88. Fellow Linnean Soc., AAAS; mem. Marine Biol. Assn. (Eng.), Australian Malacol. Soc. Episcopalian. Office: U Hawaii Manoa Dept Zoology 2538 The Mall Honolulu HI 96822-2233

KAYE, KENNETH MARC, physician, educator; b. N.Y.C., Feb. 5, 1960; s. Donald and Janet Kaye; m. Elaine Tracy, July 4, 1985; 1 child, Alexander James. AB summa cum laude, Harvard U., 1982, MD, 1986. Diplomate Am. Bd. Internal Medicine, also sub-bd. Infectious Disease. Resident in internal medicine Mass. Gen. Hosp., Boston, 1986-89; fellow in infectious disease Dana Farber Cancer Inst. Brigham & Women's Hosp., Beth Israel Hosp., Boston, 1989-91; instr. Med. Sch. Harvard U., Boston, 1991—; assoc. physician Brigham & Women's Hosp., Boston, 1991—. Contbr. articles to profl. jours. Recipient Edward H. Kass award for Clin. Excellence, Mass. Infectious Diseases Soc., 1991, Howard Hughes Med. Inst. Postdoctoral Fellowship for Physicians, 1991-92, Physician Scientist award NIH, 1992—. Fellow ACP; mem. AAAS, Phi Beta Kappa. Home: 33 Neillian Crescent Jamaica Plain MA 02130 Office: Brigham & Womens Hosp Divsn Infectious Diseases 75 Francis St Boston MA 02115

KAYE, NEIL S., psychiatry educator; s. Jesse J. Kaye and Shirley (Poskanzer) K.; m. Susan M. Donnelly. BA, Skidmore Coll., 1980; MD, Albany Med. Coll., 1984. Diplomate Nat. Bd. Med. Examiners, Am. Bd. Psychiatry and Neurology, Am. Acad. Pain Mgmt., Am. Bd. Geriatric Psychiatry. Resident Dept. Psychiatry, Albany Med. Ctr. Hosp., 1984-87; forensic fellowship Dept. Psychiatry, SUNY, 1987-88; asst. prof. psychiatry Sch. Medicine, U. Mass., 1988-90; spl. guest instr. Widener U. Sch. Law, 1991—; asst. prof. Dept. Psychiatry, Thomas Jefferson Sch. Medicine, 1991—; spl. investigation unit, hosp. expert reviewer U.S. Dept. of Justice, 1991—. Jour. reviewer Jour. of Forensic Scis., 1989—, Am. Jour. of Drug and Alcohol Abuse, 1991—, Hosp. and Community Psychiatry, 1991—, Bull. of the Am. Acad. of Psychiatry and the Law, 1991—; assoc. editor Am. Acad. of Psychiatry and the Law Newsletter, 1991—; contbr. numerous articles to profl. jours. including Psychiatry Digest, The Sciences, Am. Jour. Psychiatry, Am. Jour. on Addictions, Anglo-American Law Rev., Jour. of Forensic Scis., Behavioral Scis. and the Law. Mem. Riverfront Devel. Commn., Albany, N.Y., 1984-90, Downtown Redevel. Commn., Albany, 1984-90; pres. Empire State Regatta Fund, Inc., 1984-91, founder, co-chmn. 1984-87; ofcl. U.S. Rowing Assn., 1976—, rep. to sports medicine, judge referee com. 1984-88; instr. in Hebrew Saratoga Jewish Community Ctr., 1979; dir. Camp Shelley Day Camp, New Scotland, N.Y., 1977-82, many other activities. Mem. AMA, Am. Psychiat. Assn., Am. Acad. of Psychiatry and the Law (com. on ethics 1988—, com. on AAPL/APA rels. 1988-90), Nat. Assn. of Sports Ofcls., Am. Acad. of Forensic Scis., N.Y. Acad. Scis., Internat. Wine and Food Soc., Am. Acad. Psychiat. Adminstrs., Am. Acad. of Psychiatrists in Alcoholism and Addiction, Assn. for Convulsive Therapy, Am. Assn. for Geriatric Psychiatry, New Castle County Med. Soc., Med. Soc. of Del. (pub. laws com. 1990—), numerous others. Achievements include rsch. on paroxetine in obsessive compulsive disorder. Office: 1601 Concord Pike Ste 92-100 Wilmington DE 19803

KAYLOR, JEFFERSON DANIEL, JR., electronics executive; b. Birmingham, Ala., Dec. 10, 1947; s. Jefferson Daniel and Mary Charlye (Montague) K.; m. Terry Frances Hill, June 13, 1970; children: Christopher Robert, Laure Danielle. BS, USN Acad., 1970; MBA, Fla. Inst. Tech., 1981, MS, 1983. Sr. electronic systems engr. E-Systems, Garland, Tex., 1977-79; sr. staff engr. Sperry, Clearwater, Fla., 1979-83; program mgr. Amecom div. Litton Systems, College Park, Md., 1983-85, program dir., 1985-87, dir. program devel., 1989—; v.p. ops. Micro-Tel div. Adams-Russell, Hunt Valley, Md., 1987-88, div. mgr., 1988-89; cons. and lectr. in field. Lt. USN, 1970-77, capt. USNR. Mem. Armed Forces Comm. and Electronics Assn., Naval Res. Assn., Navy League, Assn. Old Crows (club pres. 1981-82). Republican. Methodist. Avocations: tennis, golf. Home: 4040 Firefly Way Ellicott City MD 21042 Office: Litton Systems Amecom Div 5115 Calvert Rd College Park MD 20740

KAYS, STANLEY J., horticulturist, educator; b. Stillwater, Okla., Feb. 3, 1945. BS, Okla. State U., 1968; MS, Mich. State U., 1971, PhD, 1971. Postdoctorate dept. biology Tex. A&M U., College Station, 1971; postdoctorate Sch. Plant Biology U.C. North Wales, Bangor, 1971-72; asst. prof. U. Ga., Tifton, 1973-76; assoc. prof. U. Ark., Fayetteville, 1976-77; assoc. prof. U. Ga., Athens, 1977-84, prof. horticulture, 1984—; Author: Postharvest Physiology of Perishable Plant Products, 1991, also over 115 rsch. papers in field. Coach Little League Baseball, Athens, 1986-89; officer high sch. athletics, Athens, 1991-92. NSF trainee, 1968; vis. scholar Wolfson Coll., 1985. Mem. AAAS, Am. Soc. Hort. Sci. (Gourley Rsch. award 1983), Econ. Botany, Internat. Soc. Tropical Root Crops. Achievements include identification of critical chemical components modulating the flavor and insect resistance of the sweet potato, contributions to understanding of use of ethylene in agriculture and mode of action of the hormone in plants.

KAZAN, BENJAMIN, research engineer; b. N.Y.C., May 8, 1917; s. Abraham Eli and Esther (Bookbinder) K.; m. Gerda B. Mosse, Nov. 4, 1988; 1 child from previous marriage, David Louis. BS in Physics, Calif. Inst. Tech., 1938; MA in Physics, Columbia U., 1940; PhD in Physics, Tech. U. Munich, 1961. Radio engr. Dept. Def., Ft. Monmouth, N.J., 1940-50; rsch. engr. RCA Labs., Princeton, N.J., 1950-58; head solid state display group Hughes Rsch. Lab., Malibu, Calif., 1958-61; head imaging sect. Electro-Optical Systems, Pasadena, Calif., 1961-68; head exploratory display group T.J. Watson Rsch. Ctr., Yorktown Heights, N.Y., 1968-74; prin. scientist Xerox Rsch. Ctr., Palo Alto, Calif., 1974-85; cons. display and imaging tech., 1985—; cons. Advisory Group Electron Devices, Dept. Def., 1973-82; adj. prof. U. R.I., Kingston, 1970-74. Author: (with others) Storage Tubes, 1952; Electronic Image Storage, 1968. Editor: Advances in Image Pickup and Display series, 1972-84; assoc. editor Advances in Electronics and Electron Physics series, 1984—; contbr. articles to profl. jours.; patentee in field. Recipient Silver medal Am. Roentgen Ray Soc., 1957. Fellow IEEE (assoc. editor Jour. Electron Devices 1979-83), Soc. Info. Display (editor jour. 1974-78); mem. Am. Phys. Soc., Sigma Xi, Tau Beta Pi. Home: 557 Tyndall St Los Altos CA 94022-3920 Office: Xerox Rsch Ctr 3333 Coyote Hill Rd Palo Alto CA 94304-1314

KAZAN, ROBERT PETER, neurosurgeon; b. Chgo., Mar. 29, 1947; s. Peter Joseph and Genevieve (Pauga) K.; m. Janet Rae Hoiland, June 21, 1975. BS, Loyola U., Chgo., 1969, MD, 1973. Diplomate Am. Bd. Neurol. Surgeons; lic. physician Ill., Minn. Intern in surgery Mayo Clinic, Rochester, Minn., 1973-74, resident in neurosurgery, 1974-78; neurosurg. cons. West Suburban Neurosurg. Assocs., Hinsdale, Ill., 1978-92; med. dir. neurosci. dept. Hinsdale Hosp., 1992; clin. asst. prof. neurosurgery U. Ill., Chgo., 1983—; various teaching appointments West Suburban Hosp. Dept. Surgery, Chgo. Med. Soc. Midwest Conf., Northwestern U.; staff neurosurgeon Hinsdale Hosp.; vice chmn. surgery Hinsdale Hosp., 1988-90, chmn. dept. surgery, 1990—. Contbr. articles to profl. jours. Fellow ACS; mem. AMA, DuPage County Med. Soc., Ill. Med.Soc., Mayo Clin. Neurosurg. Soc., Congress of Neurosurg. Surgeons, Am. Assn. Neurol.

Surgeons, Cen. Neurosurg. Soc., Soc. Med. Cons. Armed Forces of U.S., Am. Assn. Neurol. Surgeons (joint sec. trauma and disorders of spine and peripheal nerves), Congress Neurol. Surgeons (joint sect. trauma and disorders of spine and peripheral nerves), Internat. Skullbase Soc. Republican. Roman Catholic. Home: 120 Lakewood Cir Hinsdale IL 60521-6339 Office: West Suburban Neurosurg Assocs 20 E Ogden Ave Hinsdale IL 60521-3543

KAZEM, ISMAIL, radiation oncologist, educator, health science facility administrator; b. Cairo, Feb. 28, 1931; came to U.S., 1966; s. Mohamed and Khadiga A. (Abou-Hadid) K.; m. Barbara Jean Whitelock; children: Farid, Mohamed, Karen, Ramsey. MB, ChB, Ain Shams U., Cairo, 1955; diploma in radiotherapy, Royal Coll. Radiologists, London, 1960. Diplomate Am. Bd. Nuclear Medicine, Am. Bd. Radiology. Rotating intern Demerdach U. Hosp., Cairo, 1955-56; clin. demonstrator radiology dept. Ein Shams U. Faculty Medicine, 1956-59; trainee Meyerstein Inst. Radiotherapy, Middlesex Hosp., London, 1959, 60; IAEA fellow Strahlen Klinik, Czerny Krankenhaus, U. Heidelberg (Fed. Republic Germany), 1959; sr. registrar dept. radiotherapy St. Bartholomew's Hosp., London, 1960-61; lectr., then asst. prof. radiation therapy U. Alexandria (Egypt), 1962-65; sr. researcher Inst. Nuclear Medicine, German Cancer Rsch. Ctr., Heidelberg, 1965-66; instr., then asst. prof. radiology Hahnemann Med. Coll. and Hosp., Phila., 1966-70; prof., chmn. dept. radiation therapy and nuclear medicine Sint Radboud Acad. Hosp., Cath. U., Nijmegen, The Netherlands, 1970-83; dir. dept. radiation therapy and Regional Cancer Ctr. Mercer Med. Ctr., Trenton, N.J., 1983-92; dir. divsn. radiation oncology U. Medicine Dentistry-NJ Univ. Hosp., Newark, 1992—; clin. prof. radiation oncology Temple U., Phila., 1985-91; prof. clin. radiology U. Medicine and Dentistry N.J., Newark; mem. courtesy staff Helen Fuld Med. Ctr., Trenton; mem. cons. staff Freehold (N.J.) Area Hosp.; pres. Mercer County Med. Soc., 1993-94; presenter in field to sci. meetings. Author: (poetry) An Anthology of My Own Thing, 1975, Reflections and Definitions, 1978, Conversations with My Thoughts, 1992, Introduction to Oncology (in Dutch), 1983; mem. editorial bd. N.J. Medicine; editor Mercer County Medicine. Mem. exec. com. Mercer County unit Am. Cancer Soc., pres., 1992—; mem. pilot project task force for breast cancer screening in Mercer County, N.J. Dept. Health, Trenton, also mem. reaction group licensure reform project; mem. adv. coun. N.J. Office Pub. Guardian for Elderly. WHO fellow, 1963; Disting. fellow Am. Coll. Nuclear Medicine, 1993. Fellow Royal Soc. Medicine (London), Royal Coll. Radiologists (London), Acad. Medicine, N.J., Am. Coll. Nuclear Medicine (disting., charter); mem. AMA, Soc. Nuclear Medicine, Am. Coll. Radiology, Am. Soc. for Therapeutic Radiology and Oncology, Netherlands Soc. Radiotherapy, European Soc. Therapeutic Radiology and Oncology, Am. Assn. Cancer Edn., Am. Soc. Clin. Oncology, Pan Am. Med. Assn., World Med. Assn. (assoc.), Am. Endocurietherapy Soc., Pa. Med. Soc., N.J. Med. Soc., N.Y. Acad. Scis., Mercer County Med. Soc. (pres. 1993—). Office: U Medicine & Dentistry NJ Divsn Radiation Oncology 195 S Orange Ave U Heights Newark NJ 07103-2739

KAZEMI, HOMAYOUN, physician, medical educator; b. Teheran, Iran, Sept. 28, 1934; came to U.S., 1953, naturalized, 1970; s. Parviz and Irandokht K.; m. Katheryne McNulty, June 7, 1958; children: Paul, Laili. BA, Lafayette Coll., 1954; MD, Columbia U., 1958; MSc (hon.), Harvard U., 1990. Diplomate: Am. Bd. Internal Medicine. Intern M.I. Bassett Hosp., Cooperstown, N.Y., 1958-59; resident in medicine Mass. Gen. Hosp., Boston, 1963; chief pulmonary unit Mass. Gen. Hosp., 1967-89, chief pulmonary and critical care unit, 1989—; assoc. prof. medicine Harvard U., 1971-78, prof., 1979—, prof. medicine, Harvard/MIT Program in Health Sci. and Tech., 1980—; bd. dirs. Boston Tb Assn.; vis. prof. U. Ghent, 1975-76, Peking Union Med. Coll., China, 1992; dir. U.S. Beryllium Case Registry, 1968-78; vis. fellow Hammersmith Hosp., London, 1965; cons. Fed. Aviation Agy., 1987. Author: Disorders of the Respiratory System, 1976, (with L.G. Miller) Manual of Pulmonary Medicine, 1982—, Acute Lung Injury, 1986; mem. editorial bd. New Eng. JOur. Medicine, 1981-90, Respiratory Mgmt., 1989—. Bd. trustees Dublin (N.H.) Sch., 1987—. Am. Heart Assn. fellow, 1961-63; recipient Chadwick medal Mass. Thoracic Soc., 1988. Fellow ACP; mem. Am. Fedn. Clin. Rsch., Am. Thoracic Soc. (pres. Ea. sect. 1974-75), Am. Lung Assn. Boston (dir.), Mass. Med. Soc., Am. Physiol. Soc., Am. Heart Assn. Cardiopulmonary Coun. (exec. com. 1979—, v.p. 1985-87, pres. 1987-89), Am. Soc. Clin. Investigation, Soc. Occupational and Environ. Health, Am. Heart Assn. Rsch. review com.. Office: Mass Gen Hosp Boston MA 02114

KAZIMIR, DONALD JOSEPH, industrial engineer; b. Ossining, N.Y., July 8, 1934; s. Joseph Frances and Jean Rita (Sikorski) K.; m. Leila Elaine Caliendo, June 6, 1964; children: Donna Marie Kazimir King, Alicia Jean. BA, Columbia U., 1956, BS in Indsl. Engring., 1957. Commd. ensign USN, 1958, advanced through grades to lt., 1961, resigned, 1967; submarine capt. Ben Franklin, Bethpage, N.Y., 1967-71; mgr. ops. Access Ltd., Toronto, Ont., Can., 1973-74; founder, ptnr. Solar Devel., Inc., Riviera Beach, Fla., 1974—; mem. bd. Solar Rating and Cert. Corp., Washington, 1982-85; leader Gulfstream Drift Mission. Councilman Village of North Palm Beach, Fla., 1976-80; bd. dirs. Palm Beach County Right to Life, 1976—. Recipient Pub. Svc. Through Pub. Rels. award Advt. Club West Palm Beach, Fla., 1969, Guadelupe award Roman Cath. Diocese Miami, 1982, Solar Innovation award State of Fla., 1992. Mem. Fla. Solar Energy Industries Assn. (bd. dirs. 1986-91). Republican. Roman Catholic. Achievements include patent for solar water heater with bottom-return tank. Home: 106 Gulfstream Rd North Palm Beach FL 33408 Office: Solar Devel Inc 3607 A Prospect Ave Riviera Beach FL 33404

KAZMANN, RAPHAEL GABRIEL, civil and hydrologic engineer; b. Bklyn., Oct. 16, 1916; s. Boris and Elisabeth Anna (Maruchess) K.; m. Mary Caroline Beem, June 27, 1942; children: Elisabeth Paige, Hollis Beem, William McKee. BS in Civil Engring., Carnegie Inst. Tech., Pitts., 1939. Registered profl. engr., Ark., La., Ohio, Iowa, Ind., Tenn. Jr. hydrologic engr., assoc. hydrologic engr. U.S. Geol. Survey, Washington, Ohio, Tenn., 1940-45; chief hydrologic engr. Ranney Method Water Supplies, Columbus, Ohio, 1946-50; pvt. practice cons. engr. Stuttgart, Ark., 1951-63; assoc. prof., prof. civil engring. La. State U., Baton Rouge, 1963-82; cons. Baton Rouge, 1982—; assoc. dir. La. Water Resources Rsch. Inst., Baton Rouge, 1965-81. Author: Modern Hydrology, 1965, 72, 88; contbr. over 40 articles to profl. jours. Pres. Citizens Against Unnecessary Sewage Treatment, Baton Rouge, 1985—. Fellow ASCE; mem. AIME (life, chmn. com. 1968). Achievements include research on various aspects of surface and groundwater. Home: 231 Duplantier Blvd Baton Rouge LA 70808

KAZOR, WALTER ROBERT, statistical process control and quality assurance consultant; b. Avonmore, Pa., Apr. 16, 1922; s. Steven Stanley and Josephine (Lestic) K.; m. Gloria Rosalind Roma, Aug. 10, 1946; children: Steven Edward, Christopher Paul, Kathleen Mary Jo. BS in Mech. Engring., Pa. State U., 1943; MS, U. Pitts., 1953, M Letters in Econs. and Indsl. Mgmt., 1957. Research engr. Gulf Oil Corp., Pitts., 1946-57; with Westinghouse Electric Corp., 1957-84, quality assurance mgr. breeder reactor components project, Tampa, Fla., 1977-81, mgr. nuclear svc. ctr., Tampa, 1981-84; pres. Integrated Quality Systems Corp., Mgmt. Quality Assurance Cons., St. Petersburg, Fla., 1984-86; quality assurance specialist in nuclear waste mgmt. Sci. Applications Internat. Corp., Las Vegas, 1986-88; sr. cons. statis. process control and quality assurance Fischbach Tech. Svcs., Inc., Dallas, 1988-89; sr. cons. Gen. Physics Corp., Savannah River Site, Aiken, S.C., 1990-91; chmn., bd. dirs. ABB Fed. Tech. Svcs., 1990-91; prin. engr. Applied Statistics Reynolds Electrical Engring. Co., Las Vegas, 1992—; cons., guest lectr. in field. Bd. dirs. New Kensington (Pa.) council Boy Scouts Am., 1958-62. With USNR, 1944-46. Registered profl. engr., Pa. Mem. ASME, Am. Soc. Quality Control. Republican. Roman Catholic. Lodge: Lions (past pres. clubs). Author, patentee in field. Home: 1120 88th Ave N Saint Petersburg FL 33702-2966

KEAN, JAMES ALLEN, petroleum engineer; b. Oberlin, Ohio, Feb. 19, 1949; s. James Oscar and Mildred Lucille (Hess) K.; m. Constance Marie Chadwick, June 7, 1986; children; Laurence Chadwick, Victoria Marie. B-SChemE, Ohio State U., 1975. Registered profl. engr., Colo. Sr. staff process engr. Cities Svc. Co., Tulsa, 1975-82; supr. engring. Dome Petroleum Corp., Denver, 1982-84; staff process engr. Tex. Oil and Gas Co., Dallas, 1984-85, Atlantic Richfield Co., Dallas, 1985—. Contbr. articles to profl. jours. Mem. AICE. Achievements include research in structured packing in gas dehydration; in use of computational fluid dynamics in gas/liquid separation. Home: Simpruk Golf 13/#15, Simpruk Jakarta Selatan Indonesia Office: ARCO Indonesia PO Box 260888 Plano TX 75026

KEANE, WILLIAM FRANCIS, nephrology educator, research foundation executive; b. N.Y.C., Sept. 21, 1942; s. William F. and Theresa (Crotty) K.; m. Stephanie M. Gaherin, June 10, 1967; children: Alicia Anne, Elizabeth Gaherin. BS, Fordham U., 1964; MD, Yale U., 1968. Intern, resident, chief med. resident Cornell N.Y. Hosp. Med. Ctr., 1968-73; asst. prof. medicine U. Minn., Mpls., 1976-82, assoc. prof., 1982-87, prof., 1987—; pres. Minn. Med. Rsch. Found., Mpls., 1989—; chmn. dept. medicine Hennepin County Med. Ctr., 1992—. Office: Hennepin County Med Ctr Div Nephrology 701 Park Ave Minneapolis MN 55415*

KEAR, BERNARD HENRY, materials scientist; b. Port Talbot, South Wales, July 5, 1931; came to U.S., 1959, naturalized, 1965; s. Herbert and Catherine Ann (Rees) K.; m. Jacqueline Margaret Smith, Aug. 22, 1959; children: Andrew, Gareth, Edward, Gwyneth. B.Sc., U. Birmingham, 1954, Ph.D., 1957, D.Sc., 1970. With Tube Investments Ltd., Eng., 1957-59; staff scientist Franklin Inst., Phila., 1959-63; with United Technologies Corp., East Hartford, Conn., 1963-81; sr. cons. scientist United Technologies Corp., 1977-81; sci. adv. Exxon Research and Engring. Co., 1981-86; prof., chmn. dept mechs. and materials sci., dir. ctr. for materials synthesis Rutgers U., N.J., 1986—; John Dorn Meml. lectr., 1980, Henry Krumb lectr., 1983; mem. assessment panel Nat. Inst. Stds. & Tech.-Materials Sci. & Engring. Lab. Program, chmn., 1990; bd. dirs. Acta Metallurgica, Inc., chmn., 1989. Editor 8 books in field; contbr. 160 articles to profl. jours.; holder 25 patents. Bd. dirs., pres. Interfaith Housing for Elderly Project, Madison, Conn., 1974-79. Recipient Mathewson gold medal Am. Inst. Metall. Engrs., 1971. Fellow Am. Soc. Metals (Howe medal 1970); mem. Nat. Acad. Engring., Nat. Materials Adv. Bd. (chmn. 1986), Metall. Soc., Am. Soc. Metals. Office: Rutgers U Dept Mechs & Matl Sci PO Box 909 Piscataway NJ 08855-0909

KEARNEY, PHILIP CHARLES, biochemist; b. Balt., Dec. 31, 1932; s. Cyrus James and Nola Gertrude (Massengill) K.; m. Rita Anne Rogers, Sept. 4, 1955; children: James Douglas, Kathryn Ellen. BS, U. Md., 1955, MS, 1957; PhD, Cornell U., 1960. NSF postdoctoral fellow dept. biochemistry Mich. State U., East Lansing, 1960-62; rsch. chemist Agrl. Rsch. Svc., USDA, Beltsville, Md., 1962-65, rsch. leader, 1965-88, dep. area dir. Natural Resources Inst., 1988—; mem. 4 panels NAS; mem. White Ho. Com. on Agt. Orange. Contbr. over 250 articles to profl. jours.; editor 4 vols.: Chemistry of Herbicides, 1975-86. Named Disting. Scientist, Agrl. Rsch. Svc., USDA, 1986; recipient Rsch. award Weed Sci. Soc., 1974. Mem. Am. Chem. Soc. (chmn. divsn. pesticide chemistry 1972, Internat. Award for Rsch. in Pesticide Chemistry 1982), Internat. Union of Pure and Applied Chemistry (pres. applied chemistry divsn. 1985). Achievements include patent for pesticide wastewater destruction unit. Home: 8416 Shears Ct Laurel MD 20723-1016 Office: Natural Resources Inst Rm 208 Bldg 003 Beltsville MD 20705

KEARNS, DAVID RICHARD, chemistry educator; b. Urbana, Ill., Mar. 20, 1935; s. Clyde W. and Camille V. (French) K.; m. Alice Chen, July 5, 1958; children: Jennifer, Michael. BS in Chem. Engring., U. Ill., 1956; PhD., U. Calif., Berkeley, 1960. USAF doctoral fellow U. Chgo., 1960-61, MIT, Cambridge, 1961-62; asst. prof. chemistry U. Calif., Riverside, 1962-63, assoc. prof., 1964-67, prof., 1968-75; prof. U. Calif., San Diego, 1975—. Assoc. editor Molecular Photochemistry, 1969-75, Photochemistry and Photobiology, 1971-75, Chem. Revs., 1974; assoc. editor Biopolymers, 1975-78, editorial bd., 1978—. Sloan Found. fellow, 1965-67; Guggenheim fellow, 1969-70. Mem. Am. Chem. Soc. (Calif. sect. award 1973), Am. Phys. Soc., Am. Soc. Photobiology. Home: 8422 Sugarman Dr La Jolla CA 92037-2225 Office: U Calif San Diego Dept Chemistry La Jolla CA 92093

KEARNS, DAVID TODD, federal agency administrator; b. Rochester, N.Y., Aug. 11, 1930; s. Wilfrid M. and Margaret May (Todd) K.; m. Shirley Virginia Cox, June 1954; children—Katherine, Elizabeth, Anne, Susan, David Todd, Andrew. B.S., U. Rochester, 1952. With IBM, 1954-71, v.p. mktg. ops., data processing div., until 1971; with Xerox Corp., Stamford, Conn., 1971-91; group v.p. for info. systems Xerox Corp., 1972-75; group v.p. charge Rank Xerox and Fuji Xerox, 1975-77, exec. v.p. internat. ops., 1977; pres., chief exec. officer Xerox Corp., 1977-85, also dir., pres., chief operating officer, 1977-82, pres., chief exec. officer, 1982-85, chmn., chief exec. officer, 1985-90, chmn., 1990-91, also chmn. exec. com.; ret., 1991; dep. sec. edn. U.S. Dept. Edn., Washington, 1991—; bd. dirs. Rank Xerox Ltd., Time Warner, Inc., Fuji Xerox, Chase Manhattan Corp., Dayton Hudson Corp., Ryder Systems; chmn. Tri-State United Way; bd. trustees Ford Found. Bd. visitors Grad. Sch. Bus., Duke U.; bd. dirs. U. Rochester; trustee Nat. Urban League, Ford Found.; chmn. United Way Tri State; mem. Pres'. Edn. Policy Adv. Com. With USNR, 1952-54. Recipient Chairman's award Am. Assn. Engring. Socs., 1992. Mem. Am. Philos. Assn., Bus. Roundtable, Coun. on Fgn. Rels. Office: US Dept Edn 400 Maryland Ave SW Ste 4015 Washington DC 20202-0002

KEATING, CAROLE JOANNA, biotechnologist; b. Willington Quay, England, Dec. 26, 1958; came to the U.S., 1988; d. Fredrick and Joan Dorothy Mabel (Pardoe) Prain; m. Harold Joseph Keating III, June 1, 1990; 1 child Celina Louise. BS with honors, U. Aberdeen, 1981; PhD, U. Birmingham, 1985. Postdoctoral fellow Ctr. for Tropical Vet. Medicine, Edinburgh, U.K., 1985-88; postdoctoral fellow New Eng. Biolab., Beverly, Mass., 1988-92, product mgr., 1992—. Contbr. articles to Jour. Gen. Microbioogy, Jour. Bacteriology, Parasitology. Office: New Eng Biolabs 32 Tozer Rd Beverly MA 01915

KEATING, LARRY GRANT, electrical engineer, educator; b. Omaha, Jan. 15, 1944; s. Grant Morris and Dorothy Ann (Kauffold) K.; m. Barbara Jean Merley, Dec. 21, 1968. LLB, Blackstone Sch. Law, 1968, BS, U. Nebr., 1969; BS summa cum laude, Met. State Coll., 1971; MS, U. Colo., Denver, 1978. Chief engr. broadcoast electronics 3 radio stas., 1965-69; coord. engring. reliability Cobe Labs., Lakewood, Colo., 1972-74; quality engr. Statitrol Corp., Lakewood, Colo., 1974-76; instr. electrical engring. U. Colo., Denver, 1976-78; from asst. prof. to prof. Met. State Coll., Denver, 1978—, chmn. dept., 1984—; cons. Transplan Assocs., Boulder, Colo., 1983-84. 1st lt. U.S. Army, 1962-70. Recipient Outstanding Faculty award U. Colo., Denver, 1980, Outstanding Alumnus award Met. State Coll., 1985. Mem. IEEE (sr.), Instrument Soc. Am. (sr.), Robotics Internat. (sr.), Am. Soc. Engring. Edn., Nat. Assn. Radio and Telecommunications Engrs. (cert. engr.), Order of the Engr., Eta Kappa Nu, Tau Alpha Pi, Chi Epsilon. Avocations: skiing, astronomy. Home: 6455 E Bates Ave # 4108 Denver CO 80222-7135 Office: Met State Coll PO Box 173362 Campus Box 29 Denver CO 80217-3362

KEATING, TRISTAN JACK, retired aeronautical engineer; b. Fassett, Quebec, Can., Jan. 25, 1917; came to U.S., 1919; s. John Julian and Laure (Lalonde) K.; m. Helen Angela Condron, Dec. 27, 1947 (dec.); children: Timothy J., Jeffrey J., Jerome P., Daniel B., Sarah L.; m. Mary Ellen Hagan, Dec. 21, 1991. BS in Mech. Engring., U. Ark., 1939; MS, Ohio State U., 1948. Registered profl. engr., Ohio. Various tech. positions power plant and propulsion labs. Wright-Patterson AFB, Dayton, Ohio, 1939-53; chief nonrotating engines propulsion lab. Wright Patterson AFB, Dayton, 1953-56, chief nuclear propulsion div., 1956-57, chief rocket div., 1957-59, dep. dir. engring. x20 (Dynasoar) program office, 1959-63, dir. F-111 engring. system program office, 1963-73, dir. engring. F-15 system program office, 1973-76; cons. Dayton, 1976-87; sr. program mgr. Universal Tech. Corp., Dayton, 1987-93; cons. to aero. industry various orgns. including Battelle, Columbus Labs., U. Dayton Rsch. Inst., Pneumo Corp., USAF, 1976-87; briefer sec. def. navy, air force aero engring., 1960-76, speaker Tech. Socs. on hypersonic vehicle issues, 1950-63. Mem. Sch. Bd. St. Albert the Great Parish Sch., 1971-72; pres. St. Albert the Great Sch. PTA, 1972-73; mem. St. Albert the Great Parish Coun., Dayton, 1973-75; pres. Miami Valley Alzheimer's Assn., Dayton, 1983-86. Lt. U.S. Navy, 1943-46, PTO. Recipient Outstanding Fed. Profl. Employee award Dayton C. of C., 1965; named Outstanding Dayton Engr. Dayton profl. Socs., 1976, Air Force Decoration for Exceptional Civilian Svc., 1975, Air Force Systems Command Meritorious Civilian Svc. award, 1975. Assoc. fellow AIAA (pres. local chpt. 1950-51, mgmt. tech. com., 1977-78); mem. Am. Soc. Metals, Pi Mu Epsilon. Home: 7348 Hartcrest Ln Centerville OH 45459

KEATON, LAWRENCE CLUER, engineer, consultant; b. Gainesville, Tex., Nov. 24, 1924; s. William Lenard and Lettie (Phipps) K.; m. Emalee Prichard, Feb. 22, 1947; children: Lawrel Larsen, L.C. Jr., T.E. BSME, U. Okla., 1945; MS in Safety Mgmt. (hon.), Western States U., 1989, PhD in Bus. Adminstrn. (hon.), 1989. Registered profl. engr., Tex.; cert. lightning protection inspector; diplomate Coun. of Engring. Specialty Bds. In various engring. positions Phillips Petroleum Co., Borger, Tex., 1946-65; project devel. engr. Phillips Petroleum Co., N.Y.C., 1964-65; mng. dir. Nordisk Philback AB, Malmo, Sweden, 1965-73; dir. carbon black ops. Europe and Africa Phillips Petroleum Co., 1973-74, world-wide dir. carbon black ops., 1974-76; mng. dir. Sevalco Ltd., Bristol, Eng., 1976-81; ind. cons., 1981-85; mng. ptnr. System Engring. and Labs. Northwest Tex., Amarillo, 1985—. 5 patents in petrochem. processes. Lt. (j.g.) USN, 1943-45, PTO. Mem. ASME, Am. Soc. Safety Engrs., Lightning Protection Inst., Nat. Assn. Corrosion Engrs., Nat. Acad. Forensic Engrs., Nat. Assn. Fire Investigators, Nat. Assn. Profl. Accident Reconstruction Specialists, Nat. Soc. Profl. Engrs., Soc. Am. Mil. Engrs., Tex. Soc. Profl. Engrs., Amarillo Rotary, Shriners, Masons, Amarillo Club, Am. Legion, Tenn. Squires. Methodist. Avocations: gourmet cooking, gardening. Home: 1610 S Hughes St Amarillo TX 79102-2647 Office: System Engring and Labs NW Tex PO Box 1506 Amarillo TX 79105-1506

KEBBLISH, JOHN BASIL, retired coal company executive, consultant; b. Gray, Pa., Jan. 14, 1925; s. Joseph and Catherine (Benya) K.; m. Ruth L. Mueller, Oct. 14, 1955; children: John J., Heather R. BS in Mining Engring., Pa. State U., 1947, BS in Elec. Engring., 1948. With Consol. Coal Co. (and subs. cos.), various locations, 1948-71; pres. Consol. Coal Co. (Pocahontas Fuel Co. div.), Bluefield, W. Va., 1966-70; v.p. Consol. Coal Co., Pitts., 1970-71; exec. v.p. The Pittston Co., N.Y.C., 1971-73; pres., chief exec. officer Ashland Coal, Inc. (Ky.) (subs. Ashland Oil, Inc.), 1974-87; v.p., exec. officer Ashland Oil, Inc., 1976-87, also bd. dirs.; cons. in field. Served with AUS, 1944-46.

KECECIOGLU, DIMITRI BASIL, mechanical engineering educator; b. Istanbul, Turkey, Dec. 26, 1922; came to U.S., 1946, naturalized, 1956; s. Basil C. and Mary (Melayios) K.; m. Lorene June Legan, Dec. 22, 1951; children: Zoe Diana Kececioglu Draelos, John Dimitri. B.S., Robert Coll., Istanbul, 1942; M.S., Purdue U., 1948, Ph.D., 1953. Asst. instr. Purdue U., Lafayette, Ind., 1943-47; instr. Purdue U., 1947-52; engring. scientist in charge mech. research labs. Allis-Chalmers Mfg. Co., Milw., 1952-57; asst. to dir. mech. engring. industries group Allis-Chalmers Mfg. Co., 1957-60, cons. engr. industries group, 1960-63, dir. corp. reliability program, 1960-63; prof. aerospace and mech. engring. U. Ariz., Tucson, 1963—; reliability and maintainability engring. cons., Tucson, 1963—; dir. Reliability Engring. and Mgmt. Inst., 1963—, Reliability Testing Inst., 1975—; reliability cons. Northrop Space Labs., Gen. Elec. Co., Center for Mgmt. and Indsl. Devel., Rotterdam, Netherlands, Delco Radio div. Gen. Motors Corp., Aerojet-Gen. Corp., Westinghouse Elec. Co., U.S. Army Mgmt. Engring. Tng. Agy., Allied Signal, Data General, Polaroid, Storage Tek, Motorola, Digital Equipment, ITT, B.F. Goodrich, Gen. Dynamics, Xerox, Ford, JPL, Bendix, Cummins Engine, MOOG, Copeland, Eastman Kodak, Allied Chem.; Fulbright lectr. Nat. Tech. U., Athens, 1971-72; sr. extension tchr. UCLA, 1983; hon. prof. Shanghai U. Tech., 1984. Author: Bibliography on Plasticity, 1950, Introduction to Probabilistic Design for Reliability, 1975, Manual of Product Assurance Films and Videotapes, 1980, Reliability Engineering Handbook, Vols. 1-2, 1991, The 1992-94 Reliability Maintainability and Availability Software Handbook, 1992, Reliability and Life Testing Handbook, Vol. 1, 1993; contbr. over 122 sci. papers to profl. jours. Founder, fund raiser Dr. Dimitri Basil Kececioglu Reliability Engring. Rsch. Fellowships Endowment Fund, 1987. Recipient Presidency award Milw. Tech. Coun., 1962, Automotive Industries Author award, 1963, Ralph E. Teetor Outstanding Engring. Educator award Soc. Automotive Engrs., 1977, Anderson prize U. Ariz., 1983, U. Ariz. Scholarship Devel. Office award, 1991, Acad. of Achievement award in edn. Am. Hellenic Ednl. Progressive Assn., 1991-92. Mem. ASME (chmn. Milw. sect. 1960), IEEE, Soc. Exptl. Stress Analysis (chmn. Milw. sect. 1957), Am. Hellenic Ednl. Progressive Assn. (Acad. Achievement award in edn., 1992), Am. Soc. Engring. Edn., Am. Soc. Quality Control (Reliability Edn. Advancement award 1980, Allen Chop award for outstanding contbns. to reliability 1981), Soc. Reliability Engrs. (founder, pres. Tucson chpt. 1974-77), Hellenic Ops. Research Soc. Greece, Phi Beta Kappa, Sigma Xi, Tau Beta Pi, Phi Kappa Phi, Nat. Golden Key Soc. Patentee in field. Home: 7340 N La Oesta Ave Tucson AZ 85704-3119

KECK, DONALD BRUCE, physicist; b. Lansing, Mich., Jan. 2, 1941; s. William G. and Zelda D. Keck; m. Ruth A. Moilanen, July 10, 1965; children: Lynne Ann, Brian William. BS, Mich. State U., 1962, MS, 1964, PhD, 1967. With Corning (N.Y.) Glass Works, 1968-76, mgr. applied physics, 1976-86; dir. optoelectronics Corning, Inc., 1986—; bd. dirs. PCO, Inc., L.A.; lectr. in field. Co-author 4 books on optical fibers; editor Jour. Lightwave Tech., 1989; contbr. over 85 articles to profl. jours.; holder 25 patents. Chmn. troop com. Boy Scouts Am., Corning, 1968-71; pres. Civic Music Assn., Corning, 1971-75; moderator 1st Congl. Ch., Corning, 1986-87, 91-92; chmn. planning bd. Town of Corning, 1990—. Recipient Tech. Achievement award Internat. Soc. Optical Engring., 1981, IR-100 award Indsl. Rsch., 1981, Engring. Achievement award Am. Soc. Metals, 1983, John Tyndall award IEEE/Optical Soc. Am., 1992. Fellow IEEE, Optical Soc. Am., Nat. Inventors Hall of Fame, Nat. Acad. Engring. Avocations: water and snow skiing, music, woodworking. Home: 2877 Chequers Cir Big Flats NY 14814 Office: Corning Inc Sullivan Pk Corning NY 14831

KECK, JAMES COLLYER, physicist, educator; b. N.Y.C., June 11, 1924; s. Charles and Anne (Collyer) K.; m. Margaret Ramsey, Sept. 6, 1947; children: Robert Lyon, Patricia Anne. B.A., Cornell U., 1947, Ph.D., 1951. Research asst. Cornell U., 1951-52; sr. research fellow Calif. Inst. Tech., 1952-55; prin. scientist Avco-Everett Research Lab., Everett, Mass., 1955-65; dep. dir. Avco-Everett Research Lab., 1960-64; Ford prof. engring. MIT, Cambridge, 1965—. Served with AUS, 1944-46. Fellow Am. Acad. Sci.; mem. Am. Phys. Soc., Combustion Inst., Phi Beta Kappa, Sigma Xi, Phi Kappa Phi. Research high energy photonuclear reactions, theory of chem. reaction rates, high temperature gas dynamics, combustion, air pollution, thermionics. Office: MIT RM-3-342 Cambridge MA 02139

KEEFE, DEBORAH LYNN, cardiologist, educator; b. Oklahoma City, Nov. 23, 1950; d. Stanley William and Gloria Jean (Kelsoe) Denton; m. Richard Alan Keefe, June 14, 1971; children: Jennifer, Colin, Corwin. BA, Rice U., 1973; MD, N.Y. Med. Coll., 1976; MPH, Columbia U., 1990. Diplomate Am. Bd. Internal Medicine, Am. Bd. Cardiovascular Disease, Am. Bd. Critical Care, Am. Bd. Clin. Pharmacology. Intern and resident St. Vincent's Hosp., N.Y.C., 1976-79; fellow in cardiology Stanford (Calif.) Univ. Hosp., 1979-81; dir. CCU Bronx (N.Y.) Mcpl. Hosp., 1981-87; assoc. dir. Am. Cyanamid, Pearl River, N.Y., 1987-88; assoc. mem. Sloan-Kettering Meml. Hosp., N.Y.C., 1988—; asst. prof. medicine Albert Einstein Coll. Medicine, Bronx, 1981-87; assoc. prof. medicine Cornell U., N.Y.C., 1988—. Assoc. editor Jour. Clin. Pharmacology, 1985—; contbr. articles to Clin. Pharm. Therapeutics, Am. Heart Jour., Jour. Pharmacology and Exptl. Therapeutics, Jour. Cardiovascular Pharmacology, Am. Jour. Cardiology. Fellow Am. Coll. Cardiology, Am. Coll. Chest Physicians, Am. Coll. Angiology, Am. Coll. Clin. Pharmacology (regent 1985-89, 92—, treas. 1992-94); mem. Am. Bd. Clin. Pharmacology, Inc. Office: Sloan-Kettering Meml Hosp 1275 York Ave New York NY 10021-6094

KEEFE, FRANCIS JOSEPH, psychology educator; b. Framingham, Mass., May 7, 1949; s. Francis Joseph and Jeanne (Landry) K.; m. Delia Ware, Sept. 2, 1972; children: Daniel, Anne, John. BA, Bowdoin Coll., 1971; PhD, Ohio U., 1975. Psychologist Learning Therapies, Inc., Newton, Mass., 1975-76; postdoctoral fellow psychology lab. Med. Sch. Harvard U., Boston, 1976-78; asst. prof. Duke U. Med. Ctr., Durham, N.C., 1978-84, assoc. prof., 1984—; assoc. prof. social and health scis. Duke U., Durham, 1991—; cons. dept. labor Nat. Inst. Occupational Safety and Health, Washington, 1987-91; mem. behavioral medicine study sect. NIH, Washington, 1991—. Author: A Practical Guide to Behavioral Assessment, 1978, Behavioral Medicine in General Medical Practice, 1982, Assessment Strategies in Behavioral Medicine, 1982; editor Annals of Behavioral Medicine, 1990—. Fellow APA; mem. Internat. Assn. for Study of Pain, Assn. Advancement of Behavior Therapy, Soc. of Behavioral Medicine. Achievements include devel.

of reliable and valid method for recording pain-related behaviors in chronic pain patients; demonstration that cognitive-behavioral group treatment programs can reduce pain and improve functioning in osteopathic patients. Office: Duke Med Ctr Pain Mgmt Program Box 3159 Durham NC 27710

KEEFE, WILLIAM ROBERT, mechanical engineer; b. Austin, Minn., Feb. 10, 1965; s. William Robert Sr. and Dorothy Cecile (Speltz) F.; m. Sally Ann Sumner, June 15, 1991. BS in Aeroengring. & Mechanics, U. Minn., 1987; MSME, Washington U., St. Louis, 1990. Structural design engr. McDonnell Douglas Corp., St. Louis, 1987-91; product mgr. Command Corp. Internat., Mpls., 1991—. Mem. Soc. Mfg. Engrs., Soc. Automotive Engrs. Office: Command Corp Intenrat 11501 Eagle St NW Minneapolis MN 55448

KEELER, JILL ROLF, pharmacologist, army officer, consultant; b. Biloxi, Miss., Jan. 22, 1950; d. Curtis Holden and Caryl Jean (Anderson) K. BSN, U. Md., 1971; PhD, Uniformed Svcs. U. Health Scis., 1985. Cert. RN anesthetist. Enlisted U.S. Army, 1967, advanced through grades col., 1993; staff anesthetist 121st Evacuation Hosp., Seoul, 1976-77; staff anesthetist, instr. Madigan Army Med. Ctr., Tacoma, 1977-80; instr. pharmacology Acad. Health Scis., Ft. Sam Houston, Tex., 1985-87; prin. investigator U.S. Army Med. Rsch. Inst. Chem. Def., Aberdeen Proving Ground, Md., 1987-88, chief physiology br., 1989-92, chief chem. casualty care, 1992-93, cons. chem. casualty care issues, 1987-93, cons. respiratory threat agts., 1989-92; dir. U.S. Army nurse anesthesia program Army Med. Dept. Ctr. and Sch., Ft. Sam Houston, 1993—. Contbr. articles to profl. jours. Staff officer USCG Aux., Havre de Grace, Md., 1989-93. Decorated Bronze Star, Meritorious Svc. medal; grantee Uniformed Svcs. U. Health Scis., 1982, 83, 84. Mem. Am. Assn. Nurse Anesthetists (bd. dirs. Edn. and Rsch. Found. 1990—), Tex. Assn. Nurse Anesthetists, Sigma Xi. Achievements include establishment of inhalation pharmacology section as lead U.S. laboratory for medical research against chemical warfare agents. Office: AMEDD C&S AHS San Antonio TX 78234

KEELER, LYNNE LIVINGSTON MILLS, psychologist, educator, consultant; b. Detroit, Sept. 18, 1934; d. Robert Livingston Mills Staples and Lyda Charlotte (Diehr) Staples; m. Lee Edward Burmeister, July 16, 1955 (div. 1982); children: Benjamin Lee, Lynne Ann; m. Robert Gordon Keeler, Oct. 26, 1986. BS with honors, Cen. Mich. U., 1957; MA, U. Mich., 1965; student, Marygrove Coll., Cen. Mich. U., 1971-74. Ltd. lic. psychologist, sch. psychologist; cert. social worker, elem. permanent cons. and tchr. for mentally handicapped. First grade tchr. Shepherd (Mich) Schs., 1957-59; tchr. Kingston (Mich.) Schs., 1959-65; tchr. educationally handicapped Rialto (Calif.) Unified Sch. Dist., 1965-66; tchr., cons. Tuscola Int. Sch. Dist., Caro, Mich., 1966-71; sch. psychologist Huron Int. Sch. Dist., Bad Axe, Mich., 1971-74, Tuscola Int. Sch. Dist., Caro, 1974-89; instr. Delta Coll., University Center, Mich., 1976-88; tchr. spl. day classes Victorville (Calif.) High Sch., 1989; sch. psychologist Bedford (Ind.) Schs., 1990-91; clin. psychologist ACT team and outpatient therapy Sanilac County Mental Health Svcs., Sandusky, Mich., 1991—; cons. sch. psychologist Marlette (Mich.) Schs., 1982-86, Bartholomew Pub. Schs., Columbus, Ind., 1989, Johnson County Schs., Franklin, Ind., 1990; clin. psychologist Thumb Family Counseling, Caro, 1985-88. Conf. presenter in field. Del. NEA-Mich. Edn. Assn. Rep. Assemblies, 1970-89; pres., auction chmn. Altrusa Club, Marlett, 1982-88; style show chmn. Marlette Band Boosters, 1983; mem. exec. bd. Lawrence County Tchrs. Assn., Bedford, 1991; mem. Sanliac Symphonic Band, 1993. Fed. govt. grantee Wayne State U., 1968. Mem. Am. Federated State and Mcpl. Employees (chairperson #219 1993), Ind. State Tchrs. Assn. (rep. assembly del. 1991), Ind. Assn. Sch. Psychologists (pub. rels. bd. 1990-91). Democrat. Methodist. Avocations: antiques, swimming, gardening, pets, traveling. Home: 6726 Clothier Rd Clifford MI 48727-9501 Office: Sanilac County Community Mental Health 190 Delaware St Sandusky MI 48471

KEELER, ROGER NORRIS, physicist; b. Houston, Aug. 12, 1930; s. Roger Maurice and Alice Marie (Tangeman) K.; m. Ethel Miriam Hill, Dec. 6, 1987; children by previous marriage: Catherine Ann, John Allen, Roger David, Carolyn Elizabeth. BA, Rice Inst., 1951, BS, 1952; MS, U. Colo., 1958; PhD, U. Calif., Berkeley, 1962. Staff U. Calif., Lawrence Livermore Lab., 1963-68, dep. div. leader, 1968-69, div. leader, 1969-71; dep. head physics dept. Lawrence Livermore Nat. Lab., 1971-72, head physics dept., 1972-75, mem. staff of dir., 1978-80; dir. Navy Tech., Washington, 1975-78; pvt. cons. Washington, 1980-86; prin. sci. advisor Kaman Aerospace Corp., Washington, 1987-88; dir. tech. mktg. Kaman Diversified Techs. Corp., Washington, 1988—; lectr. dept. applied sci. U. Calif., Davis, 1967-75, Nat. Strategic Info. Ctr., Washington, 1987-88; adj. prof. physics and chemistry U.S. Naval Postgrad. Sch., Monterey, Calif., 1979-81; cons. Def. Advanced Projects Rsch. Agy., 1987-90, Energy Conversion Devices, Troy, Mich., 1987, The Valeron Corp., Detroit, 1978-83, ANSER, Washington, 1986-87, Los Alamos Nat. Lab., 1984—, Applied Physics Lab., Johns Hopkins U., 1986—, others; expert witness in fires, explosions and chem. effects, 1970—. Mem. editorial bd. Rev. of Sci. Instruments, 1967-71, High Pressure, 1987—; contbg. editor Am. Inst. Physics Handbook, 3rd edit., 1968; tech. advisor Signal Mag.; contbr. articles and book revs. to profl. jours. Capt. USNR, 1954-57; ret. Fellow Am. Phys. Soc., Am. Inst. Chemists, Washington Acad. Sci.; mem. Am. Soc. Naval Engrs. (life, Gold Medal award 1992), Internat. High Pressure Assn. (exec. com. 1975-79, v.p. 1979-85, pres. 1985-89), Capitol Hill Club, Armed Forces Comm. and Electronics Assn. (assoc. dir. 1992-93, dir. 1993—, Gold medal for engring. 1993). Achievements include over 20 patents in the area of laser ranging and underwater comm. Office: Kaman Diversified Tech Corp Ste 700 1111 Jefferson Davis Hwy Arlington VA 22202

KEEN, ALAN ROBERT, flavor scientist; b. Oamaru, New Zealand, Feb. 26, 1941; s. Alan Devine and Anzac Harriett Anne K.; m. Helen Elizabeth Nash, Feb. 12, 1982. BSc with honors, U. Otago, New Zealand, 1962, PhD, 1965. Cert. organic chemist. Head continuous culture sect. New Zealand Dairy Rsch. Inst., Palmerston North, 1965-70, rsch. officer flavor sect., 1970-76, sr. rsch. officer milkfat sect., 1976-80, head flavor sect., internat. arbitrator dairy flavor, 1980—; internat. arbitrator dairy product flavor New Zealand Dairy Rsch. Inst., 1980—. Contbr. numerous articles to profl. jours.; patentee in field. Mem. New Zealand Inst. Chemistry, New Zealand Soc. Dairy Sci. and Tech. Avocations: fishing, hunting, dogs, race horse betting systems, music. Office: New Zealand Dairy Rsch Inst, Private Bag, Palmerston North New Zealand

KEENE, CLIFFORD HENRY, medical administrator; b. Buffalo, Jan. 28, 1910; s. George Samuel and Henrietta Hedwig (Yeager) K.; m. Mildred Jean Kramer, Mar. 3, 1934; children: Patricia Ann (Mrs. William S. Kneedler), Martha Jane (Mrs. William R. Sproule), Diane Eve (Mrs. Gordon D. Simonds). AB, U. Mich., 1931, MD, 1934, MS in Surgery, 1938; DSc, Hahnemann Med. Coll., 1973; LLD, Golden Gate U., 1974. Diplomate Am. Bd. Surgery, Am. Bd. Preventive Medicine (occupational medicine). Resident surgeon, instr. surgery U. Mich., 1934-39; cons. surgery of cancer Mich. Med. Soc. and Mich. Dept. Health, 1939-40; pvt. practice surgery Wyandotte, Mich., 1940-41; med. dir. Kaiser-Frazer Corp., 1946-53; instr. surgery U. Mich., 1946-54; med. adminstrv. positions with Kaiser Industries and Kaiser Found., 1954-75, v.p., 1960-75; v.p., gen. mgr. Kaiser Found. Hosps. and Kaiser Found. Health Plan, 1960-67; med. dir. Kaiser Found. Sch. Nursing, 1954-67; dir. Kaiser Found. Research Inst., 1958-75; pres. Kaiser Found. Hosps. Health Plan, Sch. Nursing, 1968-75, dir., 1960-80; chmn. editorial bd. Kaiser Found. Med. Bull., 1954-65; lectr. med. econs. U. Calif.-Berkeley, 1956-75; mem. vis. com. Med. Sch., Stanford U., 1966-72, Harvard U., 1967-71, 79-85, U. Mich., 1973-78; Mem. Presdl. Panel Fgn. Med. Grads. (Nat. Manpower Commn.), 1966-69. Contbr. papers to profl. lit. Bd. visitors Harvard Bus. Adv. Council, 1972, Charles R. Drew Postgrad. Med. Sch., 1972-79; trustee Amman Civil Hosp., Jordan, 1973, Community Hosp. of Monterey Peninsula, 1983-92. Lt. col. M.C. AUS, 1942-46. Recipient Disting. Service award Group Health Assn. Am., 1974; Disting. Alumnus award U. Mich. Med. Center, 1976; Disting. Alumnus Service award U. Mich., 1985. Fellow ACS; mem. Am. Assn. Indsl. Physicians and Surgeons, Nat. Acad. Scis., Inst. Medicine, Calif. Acad. Medicine, Frederick A. Coller Surg. Soc., Calif., Am. med. assns., Alpha Omega Alpha (editorial bd., contbr. to Pharos mag. 1977—). Home: 3978 Ronda Rd PO Box 961 Pebble Beach CA 93953

KEENEY, DENNIS RAYMOND, soil science educator; b. Osceola, Iowa, July 2, 1937; s. Paul N. and Evelyn L. (Beck) K.; m. Betty Ann Goodhue, June 20, 1959; children: Marcia, Susan. BS, Iowa State U., 1959; MS, U. Wis., 1961; PhD, Iowa State U., 1965. Postdoctoral research assoc. Iowa State U., Ames, 1965-66; prof. U. Wis., 1966-88, Romnes research prof., 1975—; chmn. dept. soil sci. U. Wis., Madison, 1978-83; chmn. land resources program Inst. Environ. Studies, Madison, 1985-88; prof. dept. agronomy Iowa State U., Ames, 1988—, dir. Leopold Ctr. for Sustainable Agr., 1988—; dir. Iowa State Water Resources Inst., 1991—; sr. research scientist grasslands Dept. Sci. and Indsl. Research, Palmerston North, N.Z., 1975-76. Fellow Am. Soc Agronomy (rsch. grantee 1986, pres. 1992-93), Soil Sci. Soc. Am. (pres. 1987-88, rsch. grantee 1981). Office: Iowa State U 126 Soil Tilth Ames IA 50011

KEEPIN, GEORGE ROBERT, JR., physicist; b. Oak Park, Ill., Dec. 5, 1923; s. George Robert and Erlene Marie (Bennett) K.; m. Madge Mary Twomey, June 13, 1948; children: Robert, William, Ardis, Mavis, Denice. Ph.B, U. Chgo., 1943; B.S., MIT, 1946, M.S., 1947; Ph.D. in Physics, Northwestern U., 1949. Teaching fellow dept. physics MIT, Cambridge, Mass., 1947; postdoctoral fellow U. Calif.-Berkeley, 1950-52; research physicist Los Alamos Sci. Lab., 1952-63, group leader nuclear safeguards research, 1966-76, dir. nuclear safeguards program, 1976-80; head physics div. IAEA, Vienna, Austria, 1963-65, spl. adviser to dep. dir. gen. nuclear safeguards, 1982-85; fellow Los Alamos Nat. Lab., 1985—; mem. U.S. del. UN Atoms-for-Peace Conf., Geneva, 1955, 71, IAEA tech. adviser, 1964. Author: Progress in Nuclear Energy-Delayed Neutrons, 1956, Physics of Nuclear Kinetics, 1965; Arms Control Verification: The Technologies That Make It Possible, 1986; editor: Nuclear Analysis R and D; patentee in field. Fellow Los Alamos Nat. Lab, Am. Phys. Soc., Am. Nuclear Soc. (exec. com. 1967-69); mem. Inst. Nuclear Materials Mgmt. (nat. chmn. 1978-80, Disting. Service award 1984), N.Y. Acad. Scis., Sigma Xi. Office: Los Alamos Nat Lab MS-550 Los Alamos NM 87545

KEER, LEON MORRIS, engineering educator; b. Los Angeles, Sept. 13, 1934; s. William and Sophia (Bookman) K.; m. Barbara Sara Davis, Aug. 18, 1956; children: Patricia Renee, Jacqueline Saundra, Harold Neal, Michael Derek. B.S., Calif. Inst. Tech., 1956, M.S., 1958; Ph.D., U. Minn., 1962. Registered Profl. Engr., Calif. Mem. tech. staff Hughes Aircraft Co., Culver City, Calif., 1956-59; research fellow, instr. U. Minn., Mpls., 1959-62; asst. prof. Northwestern U., Evanston, Ill., 1964-66, assoc. prof., 1966-70, prof. engring., 1970—; assoc. dean research and grad. studies Northwestern U., 1985-92, chmn. dept. civil engring., 1992—; preceptor Columbia U., N.Y.C., 1963-64. Co-editor: monograph Solid Contact and Lubrication, 1980; contbr. articles to profl. jours. NATO fellow, 1962; Guggenheim Found. fellow, 1972; JSPS fellow, 1986. Fellow ASCE (chmn. engring. mech. divsn. 1992-93), Am. Acad. Mechanics (sec. 1981-88, pres.-elect 1987-88, pres. 1988-89), ASME (tech. editor Jour. Applied Mechanics 1988-92); mem. Acoustical Soc. Am., Sigma Xi, Tau Beta Pi. Home: 2601 Marian Ln Wilmette IL 60091-2207 Office: Northwestern U Dept Civil Engring 2145 Sheridan Rd Evanston IL 60208

KEEVIL, NORMAN BELL, mining executive; b. Cambridge, Mass., Feb. 28, 1938; s. Norman Bell and Verna Ruth (Bond) K.; m. Joan E. Macdonald, Dec. 1990; children: Scott, Laura, Jill, Norman Bell III. BA in Sci., U. Toronto, Ont., Can., 1959; PhD, U. Calif., Berkeley, 1964; LLD (hon.), U. B.C., 1993. Registered profl. engr., Ont. V.p. exploration Teck Corp., Vancouver, B.C., Can., 1962-68; exec. v.p. Teck Corp., Vancouver, 1968-81, pres., chief exec. officer, 1981-89, chmn., pres., chief exec. officer, 1989—; chmn. Cominco Ltd., Vancouver, 1986—. Named Mining Man of Yr. No. Miner, 1979. Mem. Can. Inst. Mining and Metallurgy (Selwyn G. Blaylock medal 1990), Prospectors and Developers Assn. (Disting. Svc. award 1990), Soc. Exploration Geophysicists, Vancouver Club, Shaughnessy Golf and Country Club (Vancouver). Office: Teck Corp, 200 Burrard St # 700, Vancouver, BC Canada V6C 3L9

KEFELI, VALENTIN ILICH, biologist; b. Moscow, Russia, July 12, 1937; s. Ilia Josef Kefeli and Alisa Michailovna Kefeli-Tongur; m. Galina Michailovna Mzen, Jan. 9, 1932; 1 child, Maria Valentinovna. Student, Agrl. Acad., Moscow, 1954-59; cand. of sci., Inst. Plant Physiology, Moscow, 1965, DSc, 1971. Asst. Inst. Phytopathology, Moscow region, 1959-61; sci. jr. Inst. Plant Physiology, Moscow, 1961-69, sci. sr., 1969-88, head lab., prof. biology, 1988—; dir. Inst. of Soil Sci. and Photosynthesis, Moscow region, 1988—. Author: Natural Growth Inhibitors, 1978; editor: Development of Acetabularia, 1979. V.p. Presidium of Pushchino Biol. Ctr., Moscow region, 1989. Recipient prize Russian Chek Acad., Moscow, 1979. Mem. Plant Physiol. Soc. (pres. 1993—), Soc. of Photobiologists (pres. 1992—). Home: Inst Soil Sci and Photosynthesis, 34B AB 74, 142292 Pushchino Russia 142292 Office: Pushchino Rsch Ctr, 142292 Moscow Oblast, Moscow Russia

KEGELES, S. STEPHEN, behavioral science educator; b. Manchester, N.H., June 2, 1925; s. Alex and Jennie (Wilder) K.; m. Jane Ainsworth, Jan. 3, 1948; children: Susan, Martha, Nancy, Robert, Dorothy. BA, Drake U., 1949; MA, Boston U., 1951, PhD, 1955. Rsch. psychologist Boston Psychopathic Hosp., 1950-52, USPHS, Washington, 1954-56; chief social psychol. studies sect. Div. Pub. Health and Resources, USPHS, 1957-62, chief social studies br., 1960-62; rsch. assoc., lectr. U. Mich., 1962-65, assoc. prof. pub. health, asst. dir. pub. health practice, 1965-66, co-dir., doctoral tng. program in pub. health adminstrn., 1966-69, prof. dept. behavioral scis. and community health, 1969—; sr. rsch. scientist Ctr. for the Environ. and Man, Inc., Hartford, Conn., 1970-71; assoc. dir. Conn. Cancer Control Rsch. Unit at Yale, 1986-89; head, behavioral scis., 1986-89; expert cons. WHO, 1969-79; tech. advisor surgeon gens. adv. com.; adv. com. Conn. State Dept. Health Svcs. Contbr. articles to profl. jours., chpts. to books. With USN, 1943-45. Recipient Disting. Sr. Sci. award Behavioral Scientists in Dental Rsch., 1988, numerous grants from various orgns. Fellow Acad. Behavioral Medicine Rsch.; mem. Internat. Assn. Dental Rsch., Am. Assn. Dental Rsch., Am. Pub. Health Assn., Am. Psychol. Assn. Home: 114 N Main St West Hartford CT 06107-1209 Office: U Conn Health Ctr Dept Behavioral Scis Farmington CT 06032

KEHR, AUGUST ERNEST, geneticist, researcher; b. Frankfort, Ky., Mar. 2, 1914; s. Carl Frederick August and Anna Esther (Heller) K.; m. Mary Louise Coon, Dec. 26, 1942; 1 child, Janet Marie Kehr Flick. BS, Cornell U., 1937, MS, 1947, PhD, 1950. Assoc. prof. La. State U., Baton Rouge, 1950-54; prof. Iowa State U., Ames, 1954-58; br. chief Agrl. Rsch. Svc. USDA, Beltsville, Md., 1958-72, mem. nat. program staff Agrl. Rsch. Svc., 1972-78. Author: (with others) Encyclopedia of Plant Physiology, 1965; author invitational papers, book chpts.; contbr. articles to profl. jours. Mem. team to study agr. in Egypt, 1975, 80; mem. team sponsored by Acad. Sci. to study vegetable farming in China, 1977. Recipient Gold medal Rhododendron Soc., 1977, Superior Svc. award USDA, 1975, B.Y. Morrison Lectureship award and medal, 1982. Fellow Am. Soc. for Hort. Sci.; mem. Am. Genetics Assn. (pres. 1963-65), Magnolia Soc. (nat. meetings planning com. 1989-91, D. Todd Gresham award 1992), Sigma Xi, Phi Kappa Phi, Gamma Sigma Delta. Achievements include several patents pending for plants; development of many varieties of azaleas, rhododendrons, and magnolias. Home: 240 Tranquility Pl Hendersonvlle NC 28739-9336

KEICHER, WILLIAM EUGENE, electrical engineer; b. Pitts., Dec. 28, 1947; s. William John and Gina Rina (Magrini) K.; m. Barbara Marie Gurgacz, Aug. 12, 1972; children: Lisa Anne, Kathy Marie, William Michael. BSEE, Carnegie-Mellon U., 1969, MSEE, 1970, PhD in Elec. Engring., 1974. Sr. elec. engr. CBS Labs., Stamford, Conn., 1974-75; mem. tech. staff Lincoln Lab., MIT, Lexington, Mass., 1975-83, asst. group leader, 1983-85, group leader, 1985—; cons. Sci. and Engring. Support Group for Strategic Def. Initiative, Arlington, Va., 1988; co-chair for numerous confs. in field. Editor: Millimeter Wave Technology, 1982; contbr. articles to profl. publs.; patentee spatial filter system. Capt. U.S. Army, 1974. Mem. IEEE (sr.), Optical Soc. Am., Assn. Old Crows. Roman Catholic. Avocations: history, snorkeling, travel, microcomputers. Home: 6 Winn Valley Dr Burlington MA 01803-4727 Office: MIT Lincoln Lab 244 Wood St Lexington MA 02173-6499

KEIFER, ORION PAUL, naval officer, mechanical engineer; b. Hiawatha, Kans., Oct. 21, 1951; s. Everett Wayne and Mary Keturah (Carper) K.; m. Patricia Ann Balaban, June 9, 1973; children: Rachel Keturah, Megan Denee, Dorion Amanda, Orion Paul Jr. BSME, U.S. Naval Acad., 1973; MSME, Naval Postgrad. Sch., 1984. Registered profl. engr., Calif., Fla. Enlisted USN, 1969, served in Vietnam, commd. ensign, 1973, advanced through grades to comdr., 1988; exec. officer USS Knox USN, Yokosuka, Japan, 1986-88; assoc. chmn. dept. mech. engring. U.S. Naval Acad., Annapolis, Md., 1988-92; exec. officer Naval ROTC unit Jacksonville (Fla.) U., 1992—; cons. Forensics Tech. Internat., Annapolis, 1989-92, Trident Engring., Annapolis, 1991-92; owner, cons. Orion Engring., Jacksonville Beach, Fla., 1992—. Mem. ASME, Soc. Automotive Engrs., Sigma Xi. Office: Jacksonville U Naval ROTC Bldg Jacksonville FL 32211

KEIGLER, JOHN E., aerospace engineer; b. Baltimore, Md., July 10, 1929; s. Arthur L. and Eliese E. (Doering) K.; m. Irene Tanis, 1955; children: Eliese A., Arthur L., John E. Jr., Elizabeth I., Janice M., James T. BE, Johns Hopkins U., 1950, MS in Elec. Engring., 1951; PhD, Stanford U., 1958. Aerospace rsch. scientist Ames Lab., NACA, Moffett Field, Calif., 1956; rsch. assoc. Stanford (Calif.) Electronics Labs., 1956-57; mgr. satellite system engring. RCA Astro Electronics, Princeton, N.J., 1958-71, mgr. comml. satellites, 1972-83; chief scientist GE Astro Space Div., Princeton, 1984-91; ret., 1991; mem. communications adv. com. NASA, 1980-83; cons. NRC, 1984-89; mem. Nat. Assn. for Search and Rescue, 1985-90. Contbr. articles to aerospace jours.; satellite design patentee. Trustee Dutch Neck (N.J.) Ch., 1965-68; councilman Boy Scouts Am., Princeton, 1968-70; vestry All Saints Ch., Princeton, 1982-85; Rep. committeeman, Princeton, 1984-86. Active duty USNR, 1951-55; comdr. USNR, ret. Recipient David Sarnoff medal RCA, 1976. Fellow IEEE (chmn. subcom. 1986-89), AIAA (tech. chmn. 1976-77, Aerospace Communications award 1990); mem. Internat. Acad. Astronautics, Electronic Industries Assn. (chmn. satellite telecom. com. 1975-79). Episcopalian. Avocation: light plane pilot.

KEIL, KLAUS, geology educator, consultant; b. Hamburg, Germany, Nov. 15, 1934; s. Walter and Elsbeth K.; m. Rosemarie, Mar. 30, 1961; children: Kathrin R., Mark K.; m. Linde, Jan. 28, 1984. M.S., Schiller U., Jena, Germany, 1958; Ph.D., Gutenberg U., Mainz, Fed. Republic Germany, 1961. Rsch. assoc. Mineral. Inst., Jena, 1958-60, Max Planck-Inst. Chemistry, Mainz, 1961, U. Calif., San Diego, 1961-63; rsch. scientist Ames Rsch. Ctr. NASA, Moffett Field, Calif., 1963-68; prof. geology, dir. Inst. Meteoritics, U. N.Mex., Albuquerque, 1968-90; pres., prof. U. N.Mex., 1985-90; chmn. dept. of geology U. N.Mex., Albuquerque, 1986-89; prof. geology U. Hawaii, Honolulu, 1990—, rsch. prof., head planetary geoscis. div., 1990—; cons. Sandia Labs., others. Contbr. over 450 articles to sci. jours. Recipient Apollo Achievement award NASA, 1970; recipient George P. Merrill medal Nat. Acad. Scis., 1970, Exceptional Sci. Achievement medal NASA, 1977, Regents Meritorious Service medal U. N.Mex., 1983, Leonard medal Meteoritical Soc., 1988, Zimmerman award U. N.Mex., 1988, numerous others. Fellow Meteoritical Soc., AAAS, Mineral. Soc. Am.; mem. Am. Geophys. Union, German Mineral. Soc., others. Office: U Hawaii at Manoa Planetary Geoscis Div Honolulu HI 96822

KEITH, DALE MARTIN, utilities consultant; b. Kansas City, Mo., Oct. 22, 1940; s. Floyd LeRoy and Pauline Constance (Brown) K.; m. Judith Ann Reynolds, May 8, 1964; children: Stephanie Deanna, Kirsten Michelle. BSBA in Indsl. Mgmt., U. Mo., 1965. Cert. mgmt. cons. Staff analyst Black & Veatch, Kansas City, Mo., 1965-68, asst. project mgr., 1968-75, adminstrv. coord., 1975-77, project mgr., 1977-88, mktg. dir., 1988-90, project dir., 1990-92; pres. Cert. Mgt. Cons., 1992—, Keith and Assocs., Ltd., Kansas City, 1993—; advisor Coun. of Econ. Regulation, Washington, 1988—. Mem. Eggs & Issues Forum, Kansas City, 1988—. Mem. Am. Mgmt. Assn., Assn. Energy Engrs., Inst. Mgmt. Cons. (mem. Coll. of Firm Prins., bd. dirs., v.p. Kansas City chpt., founding bd. mem. LAWSIG spl. interest group), Assn. Mgmt. Cons., Nat. Trust for Scotland (Edinburgh), St. Andrews Soc., Menninger, Kansas City C. of C. (quality com. 1988—), U.S. Energy Assn. (tech. collaboration com.), Inst. of the Ams., Internat. Platform Assn., Optimist Internat. (bd. dirs. Blue Valley chpt., Optimist Youth Homes). Republican. Presbyterian. Avocations: photography, sterophile, golf; computers, electric on-line world. Home: 17101 Canterbury Dr Stilwell KS 66085-9035 Office: 410 Archibald St Kansas City MO 64111

KEITH, FREDERICK W., JR., retired chemical engineer; b. Chgo., Jan. 20, 1921; s. Frederick W. and Elizabeth S. (Cummings) K.; m. Sidney P. Meeker, Oct. 23, 1943; 1 child, Katharine. BE, Yale U., 1942; PhD, U. Pa., 1951. Registered profl. engr., Pa. Rsch. engr. Du Pont, Wilmington, Del., 1942-44, Sharples Corp., Phila., 1944-50; teaching asst. U. Pa., Phila., 1948; dir. environ. tech. Pennwalt Corp., Phila., 1951-84. Author: Environmental Engineer's Handbook, 1974, Techniques of Chemistry: Separation, 1978, Kirk and Other Chemical Engineers. 2nd edit. Mem. Am. Chem. Soc., Am. Inst. Chem. Engrs. Achievements include 10 patents covering fundamental flow patterns in centrifuges. Home: The Quadrangle Apt 5105 3300 Darby Rd Haverford PA 19041

KEITH, JERRY M., molecular biologist; b. Salt Lake City, Oct. 22, 1940; s. Max Lewis Rose and Lois Lavon Watkins; m. Marla Sorensen, Mar. 24, 1971 (div. July 1991); children: Stephanie Daniels, Marlowe Jerry Dazley, Jonathan David Keith; m. Kim Yarborough Green, June 19, 1992. BA, U. Calif., Berkeley, 1973, PhD, 1976. Assoc. prof. NYU, N.Y.C., 1978-83; sect. head Rocky Mountain Labs./NIH, Hamilton, Mont., 1984-89; staff fellow NIH/NIAID, Bethesda, Md., 1976-78, sr. staff fellow, 1983, lab. chief, 1989—. Editor: Genetically Engineered Vaccines, 1992. Mem. Fedn. of Am. Socs. for Exptl. Biology, Am. Soc. for Virology, Am. Soc. for Microbiology. Achievements include patent for pertussis toxingene: cloning and expression of protective antigen. Office: NIH Bldg 30 Rm 316 Bethesda MD 20892

KEITH, ROGER HORN, chemical engineer; b. Phila., Apr. 30, 1933; s. Conrad G. and Cecilia A. (Horn) K.; m. Carol A. Burkhardt, Aug. 3, 1963; children: Paula, Thomas, Amy. BChE, U. Dayton, 1954; postgrad., U. Cin., 1954-56. Registered profl. engr., Ohio, Minn. Rsch. engr. Morris Bean & Co., Yellow Springs, Ohio, 1956-59, Dayton (Ohio) Rsch. Inst., 1960-70; engring. instr. dept. chemistry U. Dayton, 1961-67, asst. prof., 1967-70; sr. rsch. engr. indsl. tape/indsl. spltys. div. 3M, St. Paul, 1970-73, sr. rsch. engr. structural products dept., 1973-77, specialist, 1977-78, specialist indsl. spltys. div., 1978-80; specialist communications network and installation dept. 3M, St. Paul, Tex., 1980-82; sr. specialist TelComm Products Labs 3M, St. Paul, 1982-86; sr. specialist materials group TelComm Products Labs 3M, Austin, Tex., 1986-88, sr. specialist advanced systems, 1988-93, fiber optic terminating specialist, 1993—; cons. State of Minn. Workman's Compensation Bd., St. Paul, 1970-85; mem. aerospace free floaters Wright Patterson AFB Zero Gravity Expts. Editor: Part E Handbook of Instructions for Aerospace Vehicle Designators, 1962; contbr. articles to profl. jours. With U.S. Army, 1957-58. Recipient Gold medal Internat. Welding Expo, Brno, Czechoslovakia, 1978. Fellow Am. Inst. Chem. Engrs. (chmn. Balcones fault sect. 1989-90, oak leaf cluster); mem. Am. Welding Soc. (com. on safety and health/ventilation 1976-93), Instrument Soc. Am. (dir. telecom div. 1989-91), Sigma Xi (rsch. hon., chmn. 1969-70), Tau Beta Pi (Feldman award com. 1968). Achievements include 10 patents and 3 patents pending. Office: 3 M Bldg A147-2S-01 6810 River Pl Blvd Austin TX 78726-9000

KEITH, STEPHEN ERNEST, acoustical researcher; b. Toronto, Ont., Can., July 1, 1958; s. Ernest Everett and Joan Helen (Norton) K.; m. Renette Violette Sasouni, July 13, 1991. BSc in Aerospce Engring., U. Toronto, 1981; MSc in Aerospace Engring., U. Toronto Inst. Aerospace, 1983. Rschr. NRC/FERIC, Ottawa, Ont., 1983-86, NRC/Carleton U., Ottawa, 1986—. Mem. Acoustical Soc. Am., Can. Acoustical Assn. Achievements include rsch. on measurement methods and characterization of small engine noise and vibration. Home: Apt 1011, 2759 Carousel Crescent, Ottawa, ON Canada K1T 2N5 Office: Nat Rsch Coun, Bldg M-36, Montreal Rd, Ottawa, ON Canada K1A OR6

KEITH, THEO GORDON, JR., mechanical engineering educator; b. Cleve., July 2, 1939; s. Theo Gordon and Dorothy (Meech) K.; m. Sandra Jean Finzel, Aug. 20, 1960; children: Robin Lynne, Nicole Heather. BME, Fenn Coll., 1964; MS, U. Md., 1968, PhD, 1971. Mech. engr. Naval Ship R & D Ctr., Annapolis, Md., 1964-71; mem. faculty U. Toledo, 1971—, disting. univ. prof., 1990—; dir. resident programs Ohio Aerospace Inst.; prin. investigator, arc-jet microthruster modeling grantee NASA Lewis Rsch. Ctr., Cleve., 1986, prin. investigator, aircraft component de-icer modeling grantee,

1982—. Assoc. editor STLE. Recipient Outstanding Teaching award U. Toledo, 1977; Phi Kappa Phi scholar, 1968. Fellow ASME, AIAA (assoc., author Jour. of Aircraft); mem. Soc. Automotive Engrs. (Ralph Teetor award 1978), Am. Soc. Engring. Edn., Sigma Xi Rsch. award 1989), Pi Tau Sigma. Home: 3866 Laplante Rd Monclova OH 43542-9728 Office: Univ Toledo Dept Mech Engring 2801 W Bancroft St Toledo OH 43606

KEITH-LUCAS, DAVID, retired aeronautical engineer; b. Cambridge, England, Mar. 25, 1911; s. Keith and Alys (Hubbard) Lucas; m. Dorothy De Bauduy Robertson, Apr. 25, 1942 (dec. 1979); children; Mary, Michael, Christopher; m. Phyllis Marion Whurr, July 11, 1981. MA, Cambridge U., 1933; DSc (hon.), Queens U., 1968; DSc, Cranfield Inst. Tech., 1975. Chief designer Short Bros. & Harland Ltd., Belfast, Northern Ireland, 1949-58, tech. dir., 1958-65; prof. aircraft design Cranfield (England) Inst. Tech., 1965-72, prof. aeronautics, 1972-76, prof. emeritus, 1976—; bd. dirs. John Brown & Co., London, 1970-77, Civil Aviation Authority, London, 1972-81; chmn. Air Registration Bd., London, 1972-81. Author: The Shape of Wings to Come, 1952. Comdr. Order Brit. Empire. Fellow AIAA, Royal Aero. Soc. (pres. 1968-69, Gold medal 1975), Fellowship of Engring, Royal Acad. Engring. Achievements include design of Britain's first vertical take-off, jet lift aircraft. Home: Manor Close Emberton, Olney MK46 5BX, England

KEKATOS, DEPPIE-TINNY Z., microbiologist, researcher, lab technologist; b. Buffalo, Oct. 16, 1960; d. Soter Spyros and Mary Soter (Kassimis) Z. BS, CUNY, 1983; MS, St. John's U., Jamaica, N.Y., 1986. Lic. lab. technologist, N.Y. Clin. lab. technologist trainee Booth Meml. Hosp., Flushing, N.Y., 1986-87; clin. lab. technologist L.I. Jewish Hosp., New Hyde Park, N.Y., 1988-89, Elmhurst (N.Y.) Hosp., 1990—. Mem. Am. Pharm. Assn., St. John's U. Alumni Fedn. Home: 25-34 Crescent St Apt 5K Long Island City NY 11102 Office: Elmhurst Hosp 79-01 Broadway Elmhurst NY 11373

KELADA, NABIH PHILOBBOS, chemist, consultant; b. Cairo, Egypt, Nov. 12, 1930; came to U.S., 1969, naturalized, 1975; s. Philobbos Bey and Fahima (Takla) K.; B.Sc. in Chemistry and Biology, Cairo U., 1951; gen. dip. in sci. edn. Ain-Shams U., Cairo, 1952, spl. diploma in edn., 1957; M.Sc. in Chemistry, Am. U. in Cairo, 1968; grad. fellow Tech. U. Denmark, 1969; M.S. in Environ. Chemistry, U. Mich., 1971, Ph.D., 1972, postgrad., 1973; children—Reda, Samir. Tchr. sci. high schs., Egypt, 1952-64; sci. rep. and head public relations Wyeth Internat. Sci. Office, Cairo, 1964-68; research asso. Tech. U. of Denmark, Lyngby, 1968-69, Nat. Sanitation Found., Ann Arbor, Mich., 1969-71; prin. investigator environ. chemistry U. Mich., Ann Arbor, 1971-72, research asso. Sch. of Public Health, 1972-73; chief Environ. Labs., Ohio State Public Health Lab., Columbus, 1973-74; head methodology, instrumentation and toxic substances research and devel. Met. Water Reclamation Dist. Greater Chgo., 1974—; cons. to industry govt. agys. and USAID, 1973—. Mem. Am. Chem. Soc., ASTM, WPCF, Sigma Xi. Mem. Christian Coptic Orthodox Ch. Club: Glen Ellyn Tennis. Contbr. articles to profl. jours; patentee in cyanide systems. Address: 609 Midway Park Glen Ellyn IL 60137

KELDYSH, LEONID VENIAMINOVICH, physics educator; b. Moscow, 1931. Grad., Moscow U., 1954. With Moscow Physics Inst., 1954—, prof., 1969—. Mem. Acad. Scis. (corr.). Office: Lebedev Institute of Physics, Leninskiy Prospekt 53, 117924 Moscow Russia*

KELEMEN, DENIS GEORGE, physical chemist, consultant; b. Budapest, Hungary, June 18; came to U.S., 1947; s. Francis N. and Sarah (Koller) K.; m. Barbara Boushall, Feb. 3, 1951 (dec. Jan. 1991); 1 child, Peter B. Diploma in chem. engring., Tech. U. Budapest, 1947; PhD in Phys. Chemistry, Princeton U., 1951. Rsch. chemist E.I. Du Pont de Nemours & Co., Wilmington, Del., 1951-57, rsch. supr., 1957-66, rsch. mgr., 1967-70, devel. mgr., 1970-80, prin. cons., 1980-87; prin. cons. Tech. Appraisals, Lyme, N.H., 1988—. Contbr. articles to profl. jours. Past pres. Delawareans for Energy Conservation, Wilmington; sec. Lyme Conservation Commn., 1989-91, Lyme Hill and Valley Assn., 1991-92; recording sec. Lyme Zoning Bd. Adjustment, 1993—. Mem. IEEE, Am. Chem. Soc., Internat. Soc. Hybrid Microelectronics. Achievements include patents in silicon single crystal growing method, tape-automated semiconductor device packaging method. Home and Office: Franklin Hill Rd Lyme NH 03768

KELLAWAY, PETER, neurophysiologist, researcher; b. Johannesburg, Republic of South Africa, Oct. 20, 1920; s. Cecil John Rhodes and Doreen Elizabeth (Joubert) K.; m. Josephine Anne Barbieri, Apr. 1957; children: David, Judianne, Kevin, Christina, Jaime. BA, Occidental Coll., 1942, MA, 1943; PhD, McGill U., 1947; MD (hon.), U. Gothenberg, Sweden, 1977. Diplomate Am. Bd. Clin. Neurophysiology. Lectr. physiology McGill U., Montreal, Que., Can., 1946-47, asst. prof. physiology, 1947-48; assoc. prof. Baylor U. Coll. Medicine, Houston, 1948-61, prof., 1961-78, prof. neurology, 1978—, prof. div. neurosci., 1990—, dir. lab. clin. electrophysiology, 1948-65; dir. dept. clin. neurophysiology The Meth. Hosp., Houston, 1948-71, mem. attending staff, 1948—, chief, sr. attending physician Neurology Svc., 1971—; cons., neurophysiologist Hermann Hosp., Houston, 1949-73, dir. dept. electroencephalography, 1955-73; dir. electroencephalography lab. Ben Taub Gen. Hosp., Houston, 1965-79; mem. cons. staff, chief neurophysiology svc. Dept. Medicine Tex. Children's Hosp., Houston, 1972—; mem. cons. staff neurology St. Luke's Episc. Hosp., Houston, 1971-73, mem. cons. staff neurophysiology, chief neurophysiology svc., 1973—; dir. Blue Bird Circle Children's Clinic Neurol. Disorders The Meth. Hosp., 1949-60, dir. Blue Bird Circle Rsch. Labs., 1960-79, chmn. Instnl. Rev. Bd. Human Rsch., 1974-90, dir. Epilepsy Rsch. Ctr., 1975—; chmn. appointment and promotions com. Baylor U. Coll. Medicine, 1968-71, dir. Epilepsy Rsch. Ctr., 1975—, chief sect. neurophysiology Dept. Neurology, 1977—; cons., electrophysiologist VA Hosp., Houston, 1949—; cons. electroencephalography So. Pacific Hosp. Assn., Houston, 1949-57; cons., neurophysiologist M.D. Anderson Hosp. and Tumor Inst., Houston, 1953-62; mem. coun. adminstrs. Tex. Med. Ctr., Houston, 1954-60; cons. electroencephalography sect. NIH, 1961-62; hon. pres. Internat. Congress Clin. Neurophysiology, 1993. Author numerous books; editor Electroencephalography and Clin. Neurophysiology, an Internat. Jour., 1968-71, cons. editor, 1972-75, hon. cons. editor, 1989; contbr. more than 180 articles to profl. jours. Recipient Sir William Olsen medal Am. Assn. History of Medicine, 1946; grantee NIH, NASA; named Grass lectr. Am. Soc. EEG Technologists, 1989; Berger lectr., 1982, 92. Fellow Am. Acad. Pediatrics (hon.), Am. Electroencephalographic Soc. (hon., Jasper award 1991, coun. 1954, 64-66, treas. 1956-58, pres.-elect, 1962-63, pres. 1963-64); mem. Am. Epilepsy Soc. (sec.-treas. 1955-58, pres.-elect 1959, pres. 1960, Lennox lectr. 1981, Disting. Clin. Investigator award 1989), Am. Physiol. Soc., Am. Physiol. Assn., Am. Acad. Neurology, Am. Neurol. Assn., Can. Physiol. Soc., Internat. Fedn. Clin. Neurophysiology (hon. pres. internat. congress), Internat. League Against Epilepsy (Am. br.), So. Electroencephalographic Soc. (coun. 1953, v.p. 1954, pres. 1955), Ea. Assn. Electroencephalographic Soc., Cen. Encephalographic Soc., Houston Neurol. Soc. (v.p. 1957, pres. 1967, chmn. bd. trustees 1970-73), Soc. Neurosci., Child Neurology Soc., Epilepsy Assn. Houston/Gulf Coast (profl. adv. bd. 1985-92). Avocations: scuba diving, photography. Home: 627 E Friar Tuck Ln Houston TX 77024-5706 Office: Baylor Coll Medicine 1 Baylor Plz Houston TX 77030

KELLEHER, WILLIAM JOSEPH, pharmaceutical consultant; b. Hartford, Conn., July 18, 1929; s. Richard Francis and Julia Veronica (Bogash) K. BS in Pharmacy, U. Conn., 1951, MS in Pharmacy, 1953; PhD in Biochemistry, U. Wis., 1960. Registered pharmacist, Conn. Asst. prof. pharmacognosy U. Conn., Storrs, 1960-67, assoc. prof., 1967-70, prof., 1970-88, dept. chmn., 1971-76, asst. dean, 1976-81, interim dean, 1981; cons. Kelleher Cons. Svc., Storrs, 1988—. Contbr. chpts. to books; assoc. editor Jour. Natural Products, 1971-76. 1st lt. USMC, 1953-55, Korea. Grantee NIH, 1960-65, Nat. Cancer Inst., 1985-88. Mem. Am. Chem. Soc., Biochem. Soc., Am. Soc. Pharmacognosy (pres. 1973-74), Sigma Xi. Achievements include patents on pharmaceutical compositions and on sustained-release drug-delivery systems. Home and Office: PO Box 205 Storrs Mansfield CT 06268

KELLER, BEN ROBERT, JR., gynecologist; b. Big Spring, Tex., July 9, 1936; s. Ben Robert and Rowena Ward (Gibson) K.; children: Gwenyth Sue Keller Wood, Jennifer Lynn, Amy Jo Hightower Keller, Ben R. III, Destry S.L. BA, U. Tex., 1959; MD, U. Tex., Dallas, 1961. Diplomate Am. Bd. Obstetrics and Gynecology. Intern Hermann Hosp., Houston, 1961-62, ob-gyn resident, 1962-65; pvt. practice Arlington, Tex., 1967-79, 87—, Glenwood Springs, Colo., 1979-87, Arlington, 1987—; clin. instr. U. Tex., Dallas, 1975-79; assoc. clin. prof. U. Colo., Denver, 1983-86; mem. active staff Arlington Meml. Hosp., 1989—; courtesy staff South Arlington Med. Ctr., 1990—. Bd. dirs. Planned Parenthood North Tex., 1990—; chmn. speakers bur. Am. Cancer Soc., Arlington, 1968-73; mem. Arlington Drug Abuse Com., 1969-72, Glenwood Springs (Colo.) Coun. on Drug Abuse, 1984-876; chmn. bd. elders 1st Christian Ch., Arlington, 1975-76; chmn. Texpac com. Tarrant County, Ft. Worth, 1972-75. Capt. M.C., USAF, 1965-67. Fellow Am. Coll. Ob-Gyn.; mem. Tex. Med. Assn. (del. 1972-79, treas. 1974-79), Tarrant County Med. Soc., Rotary Internat., Sunlight Ski Club (chmn. bd. dirs. 1985-86). Republican. Mem. Christian Ch. (Disciples of Christ). Avocations: creative writing, music, golf, tennis, hunting. Office: 109 W Randol Mill Rd Ste 101 Arlington TX 76011-5855

KELLER, GEORGE HENRY, research administrator, consulting biochemist; b. Harrisburg, Pa., Oct. 21, 1950; s. George Henry and Agnes Marie (D'Antonio) K.; m. Nancy Jane Rossi, Aug. 17, 1974; children: Laura, Emily. BS in Biochemistry, U. Md., 1972; PhD in Biochemistry, Pa. State U., Hershey, 1978. Rsch. assoc. M. S. Hershey Med. Ctr., 1978-81; staff fellow NIH, Bethesda, Md., 1981-83; cons. biochemist Keller Rsch. Svcs., Palmyra, Pa., 1983-89; dir. comml. devel. Biotech Rsch. Labs. Inc., Rockville, Md., 1989-90; dir. project mgmt. Cambridge (Mass.) Biotech Corp., 1990—. Co-author: DNA Probes, 1989; contbr. articles to profl. jours. Am. Lung Assn. fellow, 1979. Mem. Am. Soc. for Microbiology, Assn. of Cons. Chemists and Chem. Engrs.

KELLER, JOSEPH BISHOP, mathematician, educator; b. Paterson, N.J., July 31, 1923; s. Isaac and Sally (Bishop) K.; m. Evelyn Fox, Aug. 29, 1963 (div. Nov. 17, 1976); children—Jeffrey M., Sarah N. B.A., N.Y.U., 1943, M.S., 1946, Ph.D., 1948. Prof. math. Courant Inst. Math. Scis., NYU, 1948-79; chmn. dept. math. Univ. Coll. Arts and Scis. and Grad. Sch. Engring. and Sci., 1967-73; prof. math. and mech. engring. Stanford U., 1979—; hon. prof. math. scis. Cambridge U., 1990—. Contbr. articles to profl. jours. Recipient U.S. Nat. Medal of Sci., 1988. Mem. Royal Soc. (fgn.), NAS, Am. Acad. Arts and Scis., Am. Math. Soc., Am. Phys. Soc., Soc. Indsl. and Applied Math. Home: 820 Sonoma Ter Stanford CA 94305-1024 Office: Stanford U Dept Math Stanford CA 94305-2060

KELLER, LAURENT, biologist, researcher; b. Lausanne, Switzerland, Feb. 28, 1961; s. Victor and Yvette (Desmeules) K. BA, U. Lausanne, 1983, MS, 1985, PhD, 1989. Postdoctoral rsch. fellow Mus. Zoology, Lausanne, 1989-90, Harvard U., Cambridge, Mass., 1990-92; rschr. dept. zoology Bern U., Switzerland, 1993—. Editor: Queen Number and Sociality in Insects; contbr. articles to profl. jours. Recipient Brunner prize U. Lausanne, 1990, Swiss Talent for Acad. Rsch. & Teaching award Swiss Nat. Sci. Found., 1992. Fellow French Sect. for Study of Social Insects; mem. Assn. for Study of Animal Behaviour. Achievements include contributions to understanding factors underlying partitioning of reproduction in animal societies; demonstration that social environment can affect gene expression; finding of single gene influencing the phenotype and reproductive success of an animal species; contribution to the comprehension of the social organization of insects. Home: rte du Mont 13, Prilly 1008, Switzerland Office: U Bern Dept Zool, Wohlenstrasse 50a, 3032 Hinterkappelen Switzerland

KELLER, MARK, medical educator; b. Austria, Feb. 21, 1907; came to U.S., 1913; s. Judah and Hannah (Fortgang) K.; m. Sarah Vivienne Hirsh, Dec. 30, 1930; 1 child, Ita Naomi. Med. editor med. sch. NYU, 1933-40; documentalist lab. applied physiology Yale U., New Haven, Conn., 1941-60; editor Internat. Bibliography of Studies on Alcohol, New Haven, Conn., 1966-82; lectr. applied physiology Yale U., New Haven, Conn., 1960-62; prof. documentation Rutgers U., New Brunswick, N.J., 1962-77; vis. prof. Ctr. Alcohol Studies Rutgers U., Piscataway, N.J., 1983—; assoc. prof. seminar on drugs and soc. Columbia U., 1971-77; adj. prof. Brandeis U., Waltham, Mass., 1980-82; adv. bd. grad. libr. sch. Rutgers U., 1970-74; chmn. adv. bd. Addiction Studies Found., Jerusalem, 1984—; dir. Alcohol Rsch. Documentation, New Brunswick, 1984—. Author: The Alcohol Language, 1958, CAAAL Manual, 1965, Dictionary of Words About Alcohol, 1968; 1982; editorial bd. Med. Communications, 1978-80; contbr. Ency. Americana, 1954, Ency. Judaica, 1971, Ency. Britannica, 1974-87, Ency. of Bioethics, 1978; editor Jour. of Studies on Alcohol, 1959-77; contbr. numerous articles to profl. jours. Recipient Jellinek Meml. award Jellinek Meml. Found., 1977. Mem. APHA, AAAS, Am. Med. Writers Assn. (Hammond award for disting. med. journalism 1976), Brit. Soc. for Study Addiction, Coun. Biology Editors. Jewish. Avocations: chess, archaeological digs, Bible translation, hiking. Home: 125 Stedman St Brookline MA 02146-3070 Office: Rutgers U Ctr Alcohol Studies Smithers Hall Piscataway NJ 08855

KELLER, STEPHEN, chemist educator; b. N.Y.C., July 21, 1932; s. Stefan and Mary (Mettendorfer) K.; m. Rosemarie Nancy Grumm, May 7, 1960. BS, CCNY, 1953; PhD, Rutgers U., 1959. Rsch. assoc. Columbia U., N.Y.C., 1960-79, asst. prof., 1980—; rsch. assoc. St. Luke's-Roosevelt Hosp. Ctr., N.Y.C., 1986—. Contbr. articles to profl. jours. Grantee John Policheck Found. 1972, NIH 1987, Stony Wold Herbert Fund 1990. Mem. AAAS, Sigma Xi. Republican. Presbyterian. Achievements include development of method to purify the enzyme collagenase which is now widely used in medical research, both clinical and basic; found relationship between lung elastin degradation and pulmonary emphysema; research includes developing diagnostic tests and treatments for pulmonary emphysema and asthma. Home: 7 Gifford Rd Somerset NJ 08873 Office: St Luke's-Roosevelt Hosp 428 W 59th St New York NY 10019

KELLERMANN, KENNETH IRWIN, astronomer; b. N.Y.C., July 1, 1937; s. Alexander Samuel and Rae (Goodstein) K.; m. Michele Kellermann; 1 child, Sarah. SB, MIT, 1959; PhD, Calif. Inst. Tech., 1963. Rsch. scientist CSIRO, Sydney, Australia, 1963-65; asst. scientist Nat. Radio Astronomy Obs., Green Bank, W.Va., 1965-67, assoc. scientist, 1967-69, scientist, 1978; asst. dir. Max Plank Inst. for Radio Astronomy, Bonn, Fed. Republic of Germany, 1978, dir., 1978-79, outside sci. mem., 1980—; sr. scientist Nat. Radio Astronomy Obs., Charlottesville, W.Va., 1980—; adj. prof. U. Ariz, Tucson, 1970-72; rsch. prof. U. Va., Charlottesville, 1985—. NSF fellow, Washington, 1965-66; recipient Rumford prize Am. Acad. Arts Scis., 1970, Warner prize Am. Astron. Soc., 1971, Gould prize NAS, 1973. Mem. NAS, Internat. Astron. Union (pres. com. 40 1982-85, pres. U.S. nat. com. 1990-920, Am. Astron. Soc., Internat. Radio Sci. Union, Am. Acad. Arts and Scis, Soviet Astron.Soc., Australian Astron. Soc. Avocation: amateur radio. Office: Nat Radio Astron Obs Edgemont Rd Charlottesville VA 22903

KELLETT, WILLIAM HIRAM, JR., retired architect, engineer, educator; b. Bryan, Tex., Oct. 15, 1930; s. William Hiram and Elizabeth (Minsky) K.; m. Christiane Maria Binsch, Feb. 2, 1962 (div.); children: Elizabeth Julia, Rene Janine, Kira Lorraine; m. Ann Robertson Wilkins, Dec. 11, 1971; children: Robert Lynn, Patricia Ann. AA, Victoria Coll., 1954; BArch, Tex. A&M U., 1960, MArch, 1967; postgrad., La. State U., 1986-87, Tex. Tech U., 1990-91. Registered architect, engr., Tex., La., Okla., N.Mex., Kans., Ala.; NCARB cert. 1990. Elec. technician W.E. Kutzschbach Co., Bryan, 1950-51; engring. technologist Johnston & Davis, Victoria, Tex., 1952-54; mech. elec. systems designer Hall Engring. Co., Bryan, 1955-62, Environments, Inc., Bryan, 1962-74; pres. Mech. & Elec Cons., Bryan, 1974-76; owner, operator William H. Kellett, Cons. Engrs. and Architects, Bryan, 1976—; prof. environ. design, architecture and bldg. constrn., constrn. sci. Tex. A&M U., College Station, 1962-88, ret., 1988; bd. dirs. Geranium Junction. Staff editor: Arch. Plus. Vice chmn. City Charter Com., Bryan, 1969; chmn. Bd. Equalization, 1969-70; vice chmn. City Floodwater Mgmt. Commn.; pres. Mayor's Com. on Spl. People. Named Outstanding State Handicapped Employee. Mem. AIA, Illuminating Engrs. Soc., Constrn. Specifications Inst., AAUP, ASHRAE, Refrigeration Engrs. and Tech. Assn., Nat. Soc. Profl. Engrs., Assn. Plumbing Engrs., Tex. Soc. Profl. Engrs., Phi Theta Kappa, Tau Beta Pi, Tau Sigma Delta. Home: 1000 Esther Blvd Bryan TX 77802-1821 Office: Cons Engrs and Architects 806 Oak St Bryan TX 77802-5321

KELLEY, ALBERT JOSEPH, management educator, executive consultant; b. Boston, July 27, 1924; s. Albert Joseph and Josephine Christine (Sullivan) K.; m. Virginia Marie Riley, June 7, 1945 (dec. Aug. 1988); children: Mark, Shaun, David; m. JoAnn Veronica Palmer, Dec. 14, 1991. BS, U.S. Naval Acad., 1945; BSEE, MIT, 1948, ScD, 1956; postgrad., U. Minn., 1954, Carnegie-Mellon U., 1974. Commd. ensign USN, 1945, advanced through grades to comdr., 1961; carrier pilot USN, Korea, 1950-51; exptl. test pilot Naval Air Test Ctr. USN, 1951-53, program engr. F-4 aircraft Bur. Aeros., 1956-58, mgr. Eagle missile program Bur. Weapons, 1958-60; mgr. Agena program NASA, 1960-61, dir. electronics and control, 1961-64, dep. dir. Electronics Research Ctr., 1964-67; dean sch. mgmt. Boston Coll., 1967-77; pres. Arthur D. Little program Systems Mgmt. Co., Cambridge, Mass., 1977-85, chmn., 1985-88; sr. group v.p. Arthur D. Little Inc., Cambridge, Mass., 1985-88; sr. v.p. strategic planning United Techs. Corp., Hartford, Conn., 1988-90; dep. undersec. of def. internat. programs Pentagon, Washington, 1990-93; fellow Kennedy Bus. Sch. Harvard U., Cambridge, Mass., 1993—; chmn. Bd. Econ. Advisors Commonwealth of Mass., 1970-74; chmn. bd. dirs. Arthur D. Little Valuation Inc., 1985-86; corp. mem. C.S. Draper Lab. Corp., Cambridge, 1975-90; cons. The White House; mem. NRC Space Applications Bd., 1976-82. Author: Venture Capital, 1977, New Dimensions of Project Management, 1982; contbr. articles to profl. jours. Trustee Milton (Mass.) Acad., 1975-83; bd. dirs. Mass. Bus. Devel. Corp., Boston, 1969-78, State Street Boston Co., State Street Bank and Trust, 1975-93, Mass. Tech. Devel. Corp., Boston, 1979-82, Am. Assembly Collegiate Schs. Bus., 1970-76. Recipient Exceptional Svc. medal NASA, 1967, Sec. Def. award U.S. Dept. Def., 1993. Fellow IEEE, AIAA (assoc.); mem. Internat. Acad. Astronautics, Armed Forces Communications and Electronics Assn. (v.p. 1962-65), Algonquin Club, Army Navy Country Club, Wollaston Golf Club, Milton-Hoosic Club, Sigma Xi, Tau Beta Pi. Avocations: golf, hiking, travel. Home: 4 Carberry Ln Milton MA 02186 Office: Harvard U Ctr for Bus and Govt Kennedy Sch 79 JFK St Cambridge MA 02138

KELLEY, CHARLES RAY, psychologist; b. Enid, Okla., Sept. 25, 1922; s. Charles G. and Estella F. (Hogan) K.; m. Barbara Crane, Sept. 1954 (div. 1966); children: Kathleen, John Tim; m. Erica Jean Smith, Mar. 1966; children: Eric, Kevin. BA, U. Hawaii, 1949; MA, Ohio State U., 1950; PhD, New Sch. for Social Research, 1958. Diplomate Am. Bd. Sexology. Asst. prof. N.C. State U., Raleigh, 1952-57; chief scientist Dunlap & Assocs., Santa Monica, Calif., 1957-70; dir., founder Radix Inst., Ojai, Calif., 1970-86; freelance tchr., writer Vancouver, Wash., 1987—; vis. scientist NATO div. Sci. Affairs, European Univs. and Rsch. Ctrs., 1962-70; vis. prof. U. Ill., Urbana, 1970. Author (books) Manual and Automatic Control, 1967, Education in Feeling and Purpose, 1970, The Science of Radix Processes, 1992, (course study) Science and the Life Force, 1990; contbr. articles to profl. jours. Sgt. USAF, 1942-46. Fellow APA, Am. Psychol. Soc. Unitarian. Home and Office: Steamboat Landing 13715 SE 36th St Vancouver WA 98684

KELLEY, FRANK NICHOLAS, university dean; b. Akron, Ohio, Jan. 19, 1935; s. John William Kelley and Rose (Hadinger) Bates; m. Judith Carol Lowe, Jan. 1, 1960; children: Katherine Rose Bruno, Frank Michael, Christopher Patrick. BS, U. Akron, 1958, MS, 1959, PhD, 1961. Br. chief propellant devel. Air Force Rocket Propulsion Lab., Edwards AFB, Calif., 1965-69, chief of plans, 1969-70, chief scientist, 1970-73; chief scientist Air Force Materials Lab., Wright-Patterson AFB, Ohio, 1973-77, dir., 1977-78; dir. Inst. Polymer Sci. U. Akron, 1978-88, dean Coll. Polymer Sci. and Engring., 1988—; bd. dirs. Premix, Inc., North Kingsville, Ohio, 1986—; cons. USAF, Thiokol Corp., others. Editor: Polymers in Space Research, 1965; contbr. articles to profl. jours. Capt. USAF, 1961-64. Named Outstanding Alumnus, Tau Kappa Epsilon, 1991. Mem. Am. Chem. Soc. Mem. Disciples of Christ Ch. Avocation: woodworking. Office: U Akron Coll Polymer Sci Engring Akron OH 44325-3909

KELLEY, GAYNOR NATHANIEL, instrumentation manufacturing company executive; b. New Canaan, Conn., May 12, 1931; s. James Thomas and Mabel Virginia (Seaf) K.; m. Diane Curio, Mar. 16, 1974; children: Gaynor Jr., Russell, Theodore, Ronald, Victoria. BSME, Delehanty Inst., 1951; grad. advanced mgmt. program, Northeastern U., Boston, 1965. Various mgmt. positions Perkin-Elmer Corp., Norwalk, Conn., 1951—, pres., chief oper. officer, 1985—; chmn. Concurrent Computer Corp., Tinton Falls, N.J., 1986-88; bd. dirs. Clark Equipment Corp., Gateway Bank, Hercules, Inc. Chmn. Waveny Health Care, New Canaan, 1983-86. Roman Catholic. Avocations: golf, tennis. Home: 1801 Ponus Rdg New Canaan CT 06840-2524 Office: Perkin Elmer Corp 761 Main Ave Norwalk CT 06859-0001*

KELLEY, MAURICE LESLIE, JR., gastroenterologist, educator; b. Indpls., June 29, 1924; s. Maurice Leslie and Martha (Daniel) K.; m. Carol J. Povec, Feb. 11, 1967; children: Elizabeth Ann, Mary Sarah. Student, U. Vt., Va. Ply. Inst., Princeton U., 1943-45; M.D., U. Rochester, 1949. Intern, resident Strong Meml. Hosp., Rochester, N.Y., 1949-51; Bixby fellow in medicine Strong Meml. Hosp., 1953-56; fellow in gastroenterology Mayo Clinic, Rochester, Minn., 1957-59; asst. prof. medicine U. Rochester, 1959-64, assoc. prof., 1964-67; practice medicine specializing in gastroenterology Rochester, N.Y., 1959-67; assoc. prof. clin. medicine Dartmouth Med. Sch., 1967-74, prof. clin. medicine, 1974-88; chmn. sect. internal medicine Hitchcock Clinic, 1972-74, chmn. sect. gastroenterology, 1974, 88; prof. medicine emeritus Dartmouth Med. Sch., 1988—; mem. staff Strong Meml. Hosp., Hitchcock Clinic, Mary Hitchcock Meml. Hosp.; cons. Canandaigua VA, Rochester Gen., Genesee hosps., VA. Med. Ctr., White River Junction. Contbr. articles to profl. jours., chpts. to books. Served with AUS, 1942-45; M.C. USAF, 1951-53. Fellow ACP (gov. for N.H. 1974-78, Laureate award 1993), Am. Gastroenterol. Assn.; mem. Am. Soc. Gastrointestinal Endoscopy, AMA (chmn. sect. gastroenterology 1970-71), Am. Physiol. Soc., Alpha Omega Alpha. Home: 15 Ledge Rd Hanover NH 03755-1612 Office: Dartmouth-Hitchcock Med Ctr 1 Medical Center Dr Lebanon NH 03756-0001

KELLEY, ROBERT FRANKLIN, systems analyst, consultant; b. Chgo., July 2, 1961; s. Jerry Dean and Jean (Laine) K. BA in Philosophy, Western Md. Coll., 1985; MBA in MIS, Ind. U., 1989. Human resource specialist Marriott Corp., Gaithersburg, Md., 1985-86; mgr. in tng. Courtyard by Marriott, Fairfax, Va., 1986-87; software applications specialist Hewlett-Packard, Palo Alto, Calif., 1989—. counselor Camp Allen for the Physically Handicapped, Manchester, N.H., 1977; track coach for disadvantaged youth Rockville (Md.) Recreation, 1980. Avocations: endurance horseback riding, bicycling, racquetball. Home: 408 Grant Ave Apt 309 Palo Alto CA 94306 Office: Hewlett-Packard 3000 Hanover St # 20bj Palo Alto CA 94304-1112

KELLEY, WILLIAM NIMMONS, physician, educator; b. Atlanta, June 23, 1939; s. Oscar Lee and Will Nimmons (Allen) K.; m. Lois Faville, Aug. 1, 1959; children: Margaret Paige, Virginia Lynn, Lori Ann, William Mark. MD, Emory U., 1963; MA (hon.), U. Pa., 1989. Diplomate Am. Bd. Internal Medicine (mem. bd. govs. 1978-86, chmn. 1985-86), Am. Bd. Med. Spltys. (mem. exec. com. 1980-82). Intern in medicine Parkland Meml. Hosp., Dallas, 1963-64, resident, 1964-65; sr. resident medicine Mass. Gen. Hosp., Boston, 1967-68; clin. asso., sect. on human biochem. genetics NIH, 1965-67; teaching fellow medicine Harvard U. Med. Sch., 1967-68; asst. prof. to prof. medicine, asst. prof. to asso. prof. biochemistry, chief div. rheumatic and genetic diseases Duke U. Sch. Medicine, 1968-75; Macy faculty scholar Oxford U., 1974-75; prof., chmn. dept. internal medicine, prof. dept. biol. chemistry U. Mich. Med. Sch., Ann Arbor, 1975-89; Robert G. Dunlop prof. medicine, biochemistry and biophysics U. Pa., Phila., 1989—, dean Sch. Medicine, 1989—; exec. v.p., chief exec. officer U. Pa. Med. Ctr., Phila., 1989—; mem. metabolism study sect. NIH, 1978-81; mem. adv. coun. Nat. Inst. Arthritis, Diabetes and Digestive and Kidney Diseases, 1984-86, Nat. Inst. Diabetes and Digestive and Kidney Diseases, 1986-87, mem. human gene therapy subcom., 1986-92, mem. recombinant DNA com., 1988-92, mem. dirs. adv. com., 1992—; bd. dirs. Merck & Co., Emory U., Woodruff Health Sci. Ctr. Author: (with J.B. Wyngaarden) Gout and Hyperuricemia, 1976, (with I.M. Weiner) Uric Acid, 1979, (with Harris, Ruddy and Sledge) Textbook of Rheumatology, 1981, 4th edit., 1993; editor-in-chief Textbook of Internal Medicine, 1989, 2d edit., 1992; also articles. Trustee Emory U., Woodruff Health Scis. Ctr., Wistar Inst., Leonard Davis Inst. Recipient C.V. Mosby award, 1963, John D. Lane award USPHS, 1969, Geigy Internat. prize rheumatology, 1969, Research Career Devel. award USPHS, 1972-75, Heinz Karger Meml. Found. prize, 1973, John Phillips Meml. award and medal Am. Coll. Phys., 1990; Disting. Med. Achievement award Emory U., 1985; Mead Johnson scholar, 1967; Clin. scholar Am. Rheumatism Assn., 1969-72; Josiah Macy Found. scholar, 1974-75. Mem. AAAS, ACP (master), Inst. Medicine of NAS (chmn. sect. 4, 1988-90, chmn. membership com. 1990—), Am. Soc. Clin. Investigation (editorial bd. 1974-79, pres. 1983-84), So. Soc. Clin. Investigation, Central Soc. for Clin. Research (pres. 1986-87), Am. Soc. Biochemistry and Molecular Biology (editorial bd. 1976-81), Am. Fedn. Clin. Rsch. (nat. coun. 1975-80, exec. com. 1975-80, nat. sec.-treas. 1976-78, pres. 1979-80), Assn. Am. Physicians, Assn. Profs. Medicine (nominating com. 1978-79, sec.-treas. 1987-89), Am. Coll. Rheumatology (chmn. membership com., program com., rsch. com., dir. 1975-76, exec. com. 1976-87, editorial bd. 1972-77, sec.-treas. 1982-85, pres. 1986-87, residency rev. com., 1986-93, com. chmn. 1990-93), Am. Soc. Human Genetics, Am. Soc. Nephrology, Am. Soc. Internal Medicine, Central Rheumatism Soc. (pres. 1978-79), Sigma Xi, Alpha Omega Alpha. Home: 768 Woodleave Rd Bryn Mawr PA 19010 Office: U Pa Sch Medicine 36th and Hamilton Walk Philadelphia PA 19104-6015

KELLNER, RICHARD GEORGE, mathematician, computer scientist; b. Cleve., July 10, 1943; s. George Ernest and Wanda Julia (Lapinski) K.; BS, Case Inst. Tech., 1965; MS, Stanford U., 1968, PhD, 1969; m. Charlene Ann Zajc, June 26, 1965; children: Michael Richard, David George. Staff mem. Los Alamos (N.M.) Scientific Lab., 1969-79, Los Alamos Nat. Lab., 1983-88; co-owner, dir. software devel. KMP Computer Systems, Inc., Los Alamos, 1979-84; mgr. spl. projects KMP Computer Systems div. 1st Data Resources Inc., Los Alamos, 1984-87; with microcomputer div., 1988; owner CompuSpeed, 1986—; co-owner Computer-Aided Communications, 1982-84; v.p. Applied Computing Systems Inc., 1988—; cons., 1979—. Recipient Commendation award for outstanding support of operation Desert Storm. Mem. IEEE, Assn. Computing Machinery, Math. Assn. Am., Soc. Indsl. and Applied Math., Am. Math. Soc. Home: 4496 Ridgeway Dr Los Alamos NM 87544-1960 Office: Applied Computing Systems Inc 120 Longview Dr Los Alamos NM 87544-3093

KELLOGG, CHARLES GARY, civil engineer; b. Des Moines, July 14, 1948; s. Charles Leonard and Patricia (Johnson) K.; m. Linda Louise Pries, Sept. 5, 1970; children: Karees, Tait. BS in Civil Engring. with distinction, Iowa State U., 1970, MS in Soil Engring., 1972, postgrad., U. Fla., 1986-87. Registered profl. engr., N.J., Fla., Md. Grad. asst. Iowa State U., Ames, 1970-72; structural designer George L. Levin, Mpls., 1973; engr. in tng. Iowa Dept. Transp., Ames, 1973-76; personal and mgmt. cons., Ames, Iowa and New Brunswick, N.J., 1976-80; civil engr. Epstein-Johnson, Plainfield, N.J., 1980-81; structural engr. Herbert A. Wiener, Newark, 1981; sr. engr. Jones, Edmunds & Assocs., Gainesville, Fla., 1981-83; regional engr. Universal Engring. and Testing, Gainesville, 1983-87; sr. structural engr., Spiegel and Zamecnik and Shah, Inc., Washington, 1987-93; v.p. structural engring. Delon Hampton and Assocs., 1987-93; pvt. engring. cons., 1987—; adj. lectr. civil engring., U. Fla., 1986. Recipient Donald T. Davidson award Iowa State U., 1972; Gibbs Cook scholar, 1968, 69. Mem. ASCE (dir. nat. capital sect., accreditation team, ASCE/Accreditation Bd. Engring. and Tech. visitor 1985—), Tau Beta Pi. Republican. Office: Delon Hampton and Assocs 720 N Lobby 800 K St NW Washington DC 20001

KELLOGG, KARL STUART, geologist; b. Grand Rapids, Mich., Apr. 3, 1943; s. William Welch and Elizabeth (Thorson) Kellogg; m. Nancy Jo Rader, Mar. 21, 1980; children: Kristoffer, Leah. BA, U. Calif., 1966; PhD, U. Colo., 1973. Geologist U.S. Geol. Survey, Denver, 1972-75, 77—; lectr. Sonoma State U., Rohnert Park, Calif., 1975-77. Contbr. articles to Geol. Soc. of Am. Bulletin, Jour. Geophys. Rsch., U.S. Geol. Survey Profl. Paper, African Jour. of Earth Sci., Geophys. Jour. of the Royal Astron. Soc. Recipient Svc. medal NSF, 1979; named Geographic Feature in Antarctica, 1984. Mem. Geol. Soc. of Am., Am. Geophys. Union, Colo. Sci. Soc., Tobacco Root Geol. Soc. Office: US Geol Survey Mail Stop 913 Denver Fed Ctr Denver CO 80225

KELLOGG, WILLIAM WELCH, meteorologist; b. New York Mills, N.Y., Feb. 14, 1917; s. Frederick S. and Elizabeth (Walcott) K.; m. Elizabeth Thorson, Feb. 14, 1942; children: Karl S., Judith K. Liebert, Joseph W., Jane K. Holien, Thomas W. BA, Yale U., 1939; MA, UCLA, 1942, PhD, 1949. With Inst. Geophysics UCLA, L.A., 1946-52, asst. prof., 1950-52; scientist Rand Corp., Santa Monica, Calif., 1947-59, head planetary scis. dept., 1959-64; assoc. dir. Nat. Ctr. Atmospheric Research, Boulder, Colo., also dir. lab. atmospheric scis., 1964-73, sr. scientist, 1973-87; mem. earth satellite panel IGY, 1956-59; mem. space sci. bd. Nat. Acad. Scis., 1959-68, mem. com. meteorol. aspects of effects of atomic radiation, 1956-58, mem. com. atmospheric scis., 1966-72, mem. polar research bd., 1972-77; mem. Rocket and Satellite Research Panel, 1957-62; mem. adv. group supporting tech. for operational meteorol. satellites NASA-NOAA, 1964-72; rapporteur meteorology of high atmosphere, commn. aerology World Meteorol. Orgn., 1965-71; chmn. internat. commn. meteorology upper atmosphere Internat. Union Geodesy and Geophysics, 1960-67, mem., 1967-75; mem. internat. com. climate Internat. Assn. Meteorology and Atmospheric Physics, 1978-87; mem. sci. adv. bd. USAF, 1956-65; chmn. meteorol. satellite com. Advanced Research Projects Agy., 1958-59; mem. panel on environment President's Sci. Adv. Com., 1968-72; mem. space program adv. council NASA, 1976-77; chmn. meteorol. adv. com. EPA, 1970-74, mem. nat. air quality criteria adv. com., 1975-76, air pollution transport and transformation adv. com., 1976-78; mem. council on carbon dioxide environ. assessment Dept. Energy, 1976-78; adv. to sec. gen. on World Climate Program, World Meteorol. Orgn., 1978-79; dir. research Naval Environ. Prediction Research Facility, Monterey, Calif., 1983-84; chmn. adv. com. Div. Polar Programs NSF, 1983-86; researcher on meteorology, dynamics and turbulence of upper atmosphere, prediction radioactive fallout and dispersal, applications of infrared techniques, atmospheres of Mars and Venus, theory of climate and causes of climate change. Served as pilot-weather officer USAAF, 1941-46. Co-recipient spl. award pioneering work in planning meteorol. satellite Am. Meteorol. Soc., 1961; recipient Risseca award contbn. human relations in scis. Jewish War Vets. U.S.A., 1962-63, Exceptional Civilian Service award Dept. Air Force, 1966, Spl. award for pioneering meteorol. satellites Dept. Commerce, 1985, Spl. Citation award for atmospheric conservation Garden Club of Am., 1988. Fellow Am. Geophys. Union (pres. meteorol. sect. 1972-74), Am. Meteorol. Soc. (council 1960-63, pres. 1973-74), AAAS (chmn. atmospheric and hydrospheric sect. 1984); mem. Sigma Xi. Home: 445 College Ave Boulder CO 80302-7131

KELLS, J. A., civil engineer. Recipient T. C. Keefer medal Can. Soc. Civil Engring., 1991. Home: 223 Christopher Cres, Saskatoon, SK Canada S7J 3R5*

KELLY, ALEXANDER JOSEPH, electrical engineer; b. Bklyn., Mar. 25, 1941; s. Patrick Joseph and Amy (Irvine) K.; m. Marion Siarkowski, June 16, 1962; children: Kathleen Righi, Deborah Kearns, Daniel Kelly. BSEE, Manhattan Coll., 1962; MSEE, Bklyn. Poly. Inst., 1964. Registered profl. engr., N.Y. Devel. engr. Wheeler Labs., Gt. Neck, N.Y., 1962-65; engr. AIL, Melville, N.Y., 1965-68; head microwave dept. LEL Div. Varian, Copiague, N.Y., 1968-71; prin. engr. Cardion Electronics, Woodbury, N.Y., 1971-73; leader rsch. and devel. group LNR Inc, Hauppauge, N.Y., 1973-77; dir. rsch. Hazeltine Corp., Greenlawn, N.Y., 1977-85; v.p. Satellite Transmission Systems, Hauppauge, 1985—. Contbr. articles to profl. jours. Mem. IEEE (chmn. L.I. sect. 1980-81, mem. exec. com. 1975-83, editorial rev. bd. microwave transactions 1977-85), Sigma Xi, Eta Kappa Nu. Achievements include establishment of basis of fundamental limits on conversion loss in microwave mixers and applied results to devel. of practical image-recovery mixer products; practical phased array antennas for airborne satellite communications. Office: Satellite Transmission Syst 125 Kennedy Dr Hauppauge NY 11788-4072

KELLY, A(LLAN) JAMES, environmental engineer; b. Malden, Mo., Aug. 12, 1951; s. Allan James and Janet Ruth (Girard) K.; m. Marlene Sue Rankinen, June 5, 1976; 1 child, Erin Lynn. BS in Geol. Engring., Mich. Tech. U., 1976, MS in Civil Engring., 1989. Registered profl. engr., Wis. Geol. engr. Tenneco Oil Co., Oklahoma City, 1976-78; staff devel. geologist Petro-Lewis Corp., Denver, 1978-81; sr. geologist Anadarko Prodn. Co., Englewood, Colo., 1981-83, Mesa Ltd. Partnership, Amarillo, Tex., 1983-88; project engr. Applied Techs. Inc., Brookfield, Wis., 1990-91; project environ. engr. Layne Geoscis. Inc., Pewaukee, Wis., 1991—. With USNR, 1970-72, Vietnam. Mem. ASCE, Assn. Groundwater Scientists and Engrs. Home: 251 Hillside Dr Oconomowoc WI 53066 Office: Layne Geoscis Inc N 4140 Duplainville Rd Pewaukee WI 53072

KELLY, CAROL JOHNSON, chemist; b. Detroit, Aug. 16, 1938; d. Ole J. and Grace (Kohlberg)Korsmo; m. Douglas E. Kelly, Mar. 23, 1961; 1 child, Elizabeth. BS Chemistry-Math., Wis. State U., River Falls, 1960; student, Northwestern U., 1961, Ill. Inst. Tech., Chgo., 1963; MS in Mgmt., U. Mich., Dearborn, 1981. Tchr. chemistry and phys. sci. Dearfield (Ill.) High Sch., 1960-61; rsch. chemist Universal Oil Products Co., Des Plaines, Ill., 1962-68; rsch. scientist sr., sci. rsch. staff Ford Co., Dearborn, 1968-75, mgr. metallurgy dept. ctrl. lab., 1975-88, mgr. ctrl. lab., 1988—; lectr. quality in the test lab. Maccomb Community Coll. Contbr. articles to profl. jours. Mem. Am. Chem. Soc., Am. Assn. Lab. Accreditation (bd. dirs. 1990-92), Soc. Automotive Engrs. (bright trim com. and reader com.), Motor Vehicle Mfrs. Assn. (accreditation com.). Achievements include research in the dark side of bright metal coatings, microscopy techniques for characterizing polymers, the use of measurement assurance checks in accrediting labs. and microscopy techniques in polymer rsch. Office: Ctrl Lab 15000 Century Dr Dearborn MI 48120

KELLY, DANIEL JOHN, physician; b. Binghamton, N.Y., June 23, 1940; s. William James and Mary Elizabeth (Schmitt) K.; m. Lois Ann Lanshe, Aug. 21, 1965; children: Britton James, Jeffrey Daniel, Reid William, Piper Ann. AB in History, Yale U., 1962; MD, Jefferson Med. Coll., 1966. Diplomate in Pathology, Nuclear Medicine, Dermatopathology. Intern Naval Hosp., Boston, 1966-67; resident Naval Hosp., Oakland, Calif., 1966-71; asst. chief lab. Naval Hosp., Great Lakes, Ill., 1971-73, chief lab. svcs., 1973-75; co-dir. lab. Highland Park (Ill.) Hosp., 1992—, dir. lab., 1980-89; co-dir. lab. Lake Forest (Ill.) Hosp., 1992—, dir. lab., 1989-91; with Dean, Hoffman & Clark Pathologists S.C., Lake Forest, 1975—; chief of staff elect Highland Park (Ill.) Hosp., 1992—, also bd. dirs.; mem. med. exec. com. Highland Park Hosp., 1992—, Lake Forest Hosp., 1989-91. Bd. dirs. Lake Forest Hist. Preservation Soc., 1979-89; mem. bldg. rev. bd. City Govt., Lake Forest, 1989-93; mem. clin. lab. and blood bank adv. bd. Ill. Dept. Pub. Health, 1990—; mem. Am. Pathology Found. Comdr. USNR, 1966-75. Fellow Coll. Am. Pathology, Am. Soc. Clin. Pathology, Internat. Acad. Pathologists, Am. Assn. Clin. Scientists; mem. AMA, Am. Soc. Nuclear Medicine, Ill. Soc. Pathologists, Am. Soc. Microbiology, Am. Soc. Dermatopathology, Internat. Soc. Dermatopathology, Am. Acad. Dermatology, Assn. Military Surgeons. Roman Catholic. Avocations: reading, art, music, fishing. Home: 499 E Illinois Rd Lake Forest IL 60045-2364 Office: Pathology and Nuclear Medicine Assocs 101 Waukegan Rd Ste 1250 Lake Bluff IL 60044

KELLY, DAVID REID, pathologist; b. Morganton, N.C., Nov. 21, 1947; s. Everett Oree and Paris Dreama (Keever) K.; m. Donna Marie Neale, May 30, 1970; children: Marie Keever, David Reid Jr. AB, U. N.C., 1970; MD, U. Tenn. Ctr. Health Scis., 1974. Diplomate Am. Bd. Pathology, Clin. and Anatomic Pathology, Am. Bd. Pediatric Pathology. Intern, then resident U. Ala., Birmingham, 1974-78; Am. Cancer Soc. fellow dept. pathology and lab. medicine U. Minn., Mpls., 1978-79; pathologist-in-chief, med. dir. labs. Children's Hosp. of Ala., Birmingham, 1979—; clin. prof. of pathology U. Ala.-Birmingham, 1985—; pathology cons. Pediatric Oncology Group, Chgo., 1979—. Author: Comprehensive Textbook of Oncology, 1986, Clinical Pediatric Oncology, 1991; contbr. articles to Cancer, Pediatric Pathology, Am. Jour. Clin. Pathology, Med. and Pediatric Oncology, Kidney Internat. Leader Cub Scouts and Boy Scouts Am., Birmingham, 1988—. Fellow Am. Soc. Clin. Pathologists, Coll. Am. Pathologists; mem. AMA (physician recognition award 1980, 91, 93), Soc. for Pediatric Pathology, Soc. Hematopathology, U.S. and Can. Acad. Pathology, Phi Alpha Theta, Phi Kappa Phi. Methodist. Office: Children's Hosp of Ala 1600 7th Ave S Birmingham AL 35233-1711

KELLY, DOUGLAS ELLIOTT, biomedical researcher, association administrator; b. Cheyenne, Wyo., Nov. 13, 1932; s. Raymond Douglas and Enid (McCaslin) K.; m. Louise Marie Webster, June 13, 1954 (div. 1984); children: Brian D., Alan D., Erin A., Megan L.; m. Joan Alyce Gudger, July 21, 1984; 1 child, Alison McGregor. BS in Zoology, Colo. State U., 1954; PhD in Biol. Sci., Stanford U., 1958. Asst. prof. biology U. Colo., Boulder, 1958-62; assoc. prof., vice chmn. biol. struc. U. Wash., Seattle, 1963-70; prof., chmn. biol. struc. U. Miami, Fla., 1970-74; prof., chmn. anatomy, cell biology U. So. Calif., L.A., 1974-89; assoc. v.p. biomed. rsch. Assn. Am. Med. Colls., Washington, 1989—; mem. human embryology devel. study sec. NIH, Bethesda, Md., 1978-85, chmn. 1983-85; adj. prof. anatomy George Washington U., Washington, 1989—. Author: (with others) Bailey's Textbook of Histology, 17th edit., 1978, Bailey's Textbook of Microscopic Anatomy, 18th edt., 1984; contbr. articles to profl. jours. Recipient Sigma Xi award Am. Scientist, 1962; Citation medal Japan Assn. Anatomists, Sendai, Japan, 1984; named Honor Alumnus, Colo. State U., 1978. Mem. Am. Assn. Anatomists (pres. 1986-87), Assn. Anatomy Chmn. (pres. 1977-78), Coun. Academic Socs. (chmn. 1987-88). Office: Assn Am Med Coll 2450 N St NW Washington DC 20037-1167

KELLY, EMERY LEONARD, bioenvironmental engineer; b. Endicott, N.Y., Sept. 13, 1960; s. Claud Harold and Jean (Hunsinger) K.; m. Nora Maureen Hackler, Nov. 14, 1981; children: Christopher Scott, Bridget Marie, Meaghan Cathleen. BSCE, Calif. State U., Sacramento, 1986, MSCE, 1991. Commd. 2nd lt. USAF, 1986—, advanced through grades to capt., 1990; design engr. 42nd Civil Engring. Squadron, Loring AFB, Maine, 1986-88, environ. engr., 1988-89; civil engr. 24 Civil Engring. Squadron, Howard AFB, Panama, 1989-90; environ. engr. HQ Air Force Space Command/Civil Engring., Peterson AFB, Colo., 1991-92; bioenviron. engr. 21st Med. Group, Peterson AFB, Colo., 1992—. Decorated Air Force Commendation medals (3); recipient Sr. Achievement award Alumni Assn. of Calif. State U., 1986. Mem. ASCE, Am. Soc. Mil. Engrs., Tau Beta Pi. Home: 8026 Scarborough Dr Colorado Springs CO 80920

KELLY, GERALD WAYNE, chemical coatings company executive; b. Charleston, W.Va., May 21, 1944; s. Wayne Woodside and Darrah (Myers) K.; m. Nancy Butenhoff, Sept. 15, 1965 (div. June 1983); children: Scott Wayne, Lauren Melissa (dec.); m. Elizabeth Long, Nov. 18, 1983. BS, W.Va. U., 1966. From sales corr. to regional mgr. duPont Corp., various locations, 1966-83; bus. mgr. Decatur (Ala.) div. Whittaker Corp., 1983-85; v.p. Decatur div. Morton Internat., 1985-86, pres. Decatur div., 1986-93; v.p. industrial coatings div. Morton Internat., Chgo., 1988-93; v.p. Morton Indsl. Coatings, Morton Internat., Chgo., 1993—. Bd. dirs. Ind. Cystic Fibrosis Found., Indpls., 1971-73. Mem. Nat. Coil Coaters Assn., Nat. Paint and Coatings Assn., Beta Theta Pi. Republican. Methodist. Avocation: automobiles. Home: 14 Tartan Ridge Hinsdale IL 60521 Office: Morton Internat 100 N Riverside Plaza Chicago IL 60606 also: IRP Inc PO Box 670 Falkville AL 35622

KELLY, JEFFREY JENNINGS, mechanical engineer; b. Columbia, S.C., July 28, 1947; s. Jesse Jennings and Mary Katherine (Hawkins) K.; m. Betty Jane Vickers, June 15, 1980 (div. May 1984). BS in Math., Va. Poly. Inst. and State U., 1970, PhD in Engring. Mechanics, 1981. Mathematician Naval Surface Weapons Ctr., Dahlgren, Va., 1970-76; rsch. assoc. Va. Poly. Inst. and State U., Blacksburg, 1979-81; mech. engr. David Taylor Rsch. Ctr., Bethesda, Md., 1981-82; asst. prof. Old Dominion U., Norfolk, Va., 1983-85; rsch. fellow U. Southampton, Eng., 1985-87; rsch. engr. Northrop Corp., L.A., 1987-90; staff engr. Lockheed Corp., Hampton, Va., 1990—; cons., Southampton, 1986, Hampton, 1991. Referee Jour. of Acoustical Soc. Am., 1981-82. NASA/Am. Soc. Engring. Edn. fellow NASA, Langley, Va., 1983-84. Mem. AIAA (sr.), Sigma Xi, Pi Mu Epsilon, Phi Kappa Phi. Achievements include development of signal processing schemes and codes for aircraft acoustic signals; performance of basic rsch. in aeroacoustics, duct acoustics, cavitation, structural-acoustic interaction and sound intensity. Home: 260 Marcella Rd # 1003 Hampton VA 23666 Office: Lockheed Engring & Scis Co 144 Research Dr Hampton VA 23666

KELLY, RICHARD LEE WOODS, civil engineer; b. Portsmouth, Va.; s. Charles W. and Mary H. Kelly. BS in Civil Engring., Old Dominion U., 1983, M Engring., 1988. Registered profl. engr., Va. Project engr. William C. Overman Assocs., Virginia Beach, Va., 1983-89; city engr. City of Suffolk, Va., 1989-90, acting dir. pub. utilities, 1989-90; v.p. constrn. and design Wake Med. Ctr., Raleigh, N.C., 1990—; cons. freelance structural inspections, Va., 1987-88, freelance structural design, 1989-90. Mem. property and grounds com. A.E. Finley YMCA, Raleigh, 1991—. Mem. NSPE, ASCE, N.C. Hosp. Engrs. Assn. Achievements include first city engineer for consolidated city of Suffolk, Va., first vice president in charge of construction and design at Wake Medical Center, Raleigh. Office: Wake Med Ctr 3000 New Bern Ave Raleigh NC 27610

KELLY, RICHARD MICHAEL, metallurgical engineer; b. Queens, N.Y., Oct. 29, 1955; s. Richard Lawrence and Patricia Mary (Walsh) K.; m. Jean Marie Barrett, June 6, 1986; children: Regina Michelle, Richard Mason. BS in Metallurgical Engring., Polytech. Inst. N.Y., 1977. Metallurgical tech. Lucius Pitkin Inc., N.Y.C., 1976-77; staff metallurgist Alumminum Assn. Inc., Washington, 1977-82; plant metallurgist R.D. Werner Co. Inc., Greenville, Pa., 1982-85; sr. metallurgist R.D. Werner Co. Inc., 1985-88, corp. engring. mgr. materials, tech. and quality assurance, 1988-92, dir. engring., 1992—. Recipient Best Paper award Internat. Extrusion Tech. Conf., 1984, 92. Mem. Am. Soc. Metals, Am. Soc. Testing Materials (Noah A. Kahn award 1977), Am. Soc. Assn. Execs., Am. Soc. Quality Control, SPI Composites Inst., Sigma Xi. Republican. Roman Catholic. Office: RD Werner Co Inc 93 Werner Rd Greenville PA 16125-9499

KELLY, SARAH ELIZABETH, chemist; b. Madison, Wis., May 12, 1959; d. Joseph R. and Joan (Wallenstein) K.; m. Eric R. Larson, Aug. 19, 1984; children: Erin I., Timothy C. BA, Carleton Coll., 1981; MS, PhD, Yale U., 1987. Rsch. scientist Pfizer Cen. Rsch., Groton, Conn., 1986-90, sr. rsch. scientist, 1990-93; sr. rsch. investigator, 1993—. Author: Comprehensive Organic Synthesis, Vol. 1, 1991; contbr. articles to profl. publs. NSF undergrad. rsch. fellow U. Minn., 1980, Am. Cancer Soc. Pre-doctoral fellow Yale U., 1982. Mem. Am. Chem. Soc. Achievements include patent for N-Trichloroacetyl-2-Oxindole-1-Carboxamides Intermediates for analgesic and antiinflammatory agents; development of a commercial process to Tenidap, of commercial process to ACAT inhibitor. Office: Pfizer Cen Rsch Eastern Point Rd Groton CT 06340

KELLY, TERRY LEE, computer systems analyst; b. Miami, Fla., Apr. 7, 1953; s. Stanley W. and Rose Kelly; m. Carol Monahan, Nov. 12, 1978. AS, Miami Dade Jr. Coll., 1973; BS, Fla. Internat. U., 1977. Rsch. scientist Dade div. AHSC, Miami, 1976-79; computer con. CSC, N.C., 1981-84; rsch. engr. Organon Teknika, 1986, 87; tech. svc. analyst Burroughs Wellcome, 1988--. Home: 2509 Whistling Quail Run Apex NC 27502 Office: Burroughs Wellcome 3030 Cornwallis Rd Biscoe NC 27209

KELLY, VINCENT MICHAEL, JR., orthodontist; b. Tulsa, Mar. 15, 1933 s. Vincent Michael and Ivy Maria (Phelps) K.; m. Donna Deane Amis, June 1955 (div. 1972); children: Kevin Marie, Leslie Rene, Karen Elizabeth, Carolyn Michelle, Kathleen Ann; m. Aleatha Wilkinson Kelly. BA in Biology, U. Mo., Kansas City, 1954, DDS, 1957, Orthodontist, 1962, MS in Oral Histology, 1963. Diplomate Am. Bd. Orthodontics. Pvt. practice Tulsa, 1963—; asst. prof. orthodontics St. Louis U., 1972-76; assoc. prof. U. Okla., Oklahoma City, 1980-90; bd. dirs. State Bank NA, Tulsa. Patentee in field. Capt. USAF, 1957-59. Mem. Tweed Found. Orthodontic Rsch. (instr. 1966-79), Am. Assn. Orthodontics, S.W. Soc. Orthodontics, Okla. Orthodontic Assn. (pres. 1976), Tulsa County Dental Soc., Coll. European Orthodontie (hon. life mem.), Soc. Panamena de Orthodoncia Panama (hon. life), Soc. Colombiana de Orthodoncia Colombia (hon. life), Assn. Mex. Orthodoncia Med. (hon.), European Soc. Lingual Orthodontists (hon.), So. Hills Country Club, Arrowhead Yacht Club. Republican. Roman Catholic. Avocations: flying, skiing, cooking, big-game hunting, gardening. Home: 2300 Riverside Dr Apt 15E Tulsa OK 74114-2404 Office: 1406 N Sioux Claremore OK 74017-3126

KELLY, WILLIAM CROWLEY, geological sciences educator; b. Phila., May 10, 1929; m. Anna Zauner; children—Geroge, Ted. A.B. in Geology, Columbia U., 1951, M.A. in Geology, 1953, Ph.D. in Geology, 1954. Map clk. Royal Liverpool Ins. Co., N.Y.C., 1946-47; geologist Naica mine Eagle Picher Co., Chihuahua, 1950; asst. in econ. geology Columbia U., N.Y.C., 1951-53; instr. geology Hunter Coll., N.Y.C., 1954; instr. Dept. Geol. Scis. U. Mich., Ann Arbor, 1956-58, asst. prof., 1958-62, assoc. prof., 1962-67, prof., 1967-83, C. Scott Turner prof., 1984—, chmn. dept., 1978-81; interim dir. Inst. Sci. and Tech., U. Mich., 1986-87, asst. v.p. for rsch., 1989—; acting dir. Mich. Sea Grant Coll. Program U. Mich.; vis. prof. geology U. Toronto, 1973, U. Tex., Arlington, 1978; Disting. vis. prof. U. Alta., 1982; mem. adv. panel in earth scis. NSF, 1977-80, mem. adv. panel on problem focused research, 1980, mem. oversight com. earth scis. div., 1983; mem. task force on submarine polymetallic sulfide deposits NOAA, 1982; dir. Econ. Geology Publishing Co.; lectr. and cons. in field. Mem. bd. assoc. editors: Jour. Econ. Geology, 1966-73, The Canadian Mineralogist, 76-80; contbr. chpts. to books and articles, and revs. to profl. jours. Trustee, Cranbrook Inst. Sci., 1964-71, chmn. personnel compensation, 1971. N.Mex. Bur. Mines fellow, 1953; grantee in field. Fellow Geol. Soc. Am., Mineralogical Soc. Am. (chmn. publications com. 1976); mem. Internat. Soc. Econ. Geologists (councilor 1973-76, v.p. 1978, pres. 1984), Geochem. Soc. Am. (co-editor jour. 1962, editor, 1963-64), Econ. Geology Found. (trustee 1983-85), Geol. Assn. Can., Mich. Acad. Sci., Phi Kappa Phi. Home: 17 Northwick Ct Ann Arbor MI 48105-1408 Office: U Mich Dept Geol Scis Ann Arbor MI 48109

KELLY, WILLIAM HAROLD, physicist, physics educator; b. Rich Hill, Mo., July 2, 1926; s. George Samuel and Ola Lorena (Ayers) K.; m. Altabelle Dougherty, Sept. 1, 1950; children: Douglas Scott, Linda Sue, Brian Patrick. A.A., Graceland Coll., 1948; B.S.E., U. Mich., 1950, M.S., 1951, Ph.D., 1955. Eastman Kodak predoctoral fellow in physics U. Mich., 1954-55; asst. prof. physics and astronomy Mich. State U., 1955-61; physicist U.S. Naval Research Lab., Washington, 1956, Lawrence Radiation Lab., Berkeley, Cal., 1961-62; asso. prof. physics Mich. State U., 1961-67, prof. physics, 1967-79, asso. chmn. undergrad. programs, 1968-76, chmn. dept. physics, 1976-79; dean Coll. Letters and Sci., prof. physics Mont. State U., Bozeman, 1979-83; dean Coll. Scis. and Humanities, dir. Scis. and Humanities Research Inst. Iowa State U., Ames, 1983-89, prof. physics, 1983—; summer research participant Oak Ridge (Tenn.) Nat. Lab., 1964; physicist Lawrence Radiation Lab., Berkeley, 1967-68; guest scientist Lawrence Berkeley Lab., 1989—; participating guest Lawrence Livermore (Calif.) Nat. Lab., summer, 1991; program dir. physics div. for undergraduate edn. Nat. Sci. Found., 1991-93. Trustee Graceland Coll., 1978-90, Univ.'s Research Assn., 1985—89, Fermilab Bd. of Overseers, 1987-89; elder Reorganized Ch. of Jesus Christ of Latter-day Saints, 1956—, high priest, 1971—. Served with USNR, 1944-46. Fellow Am. Phys. Soc.; mem. AAAS, Am. Assn. Physics Tchrs. (pres. 1981-82), Mich. Physics Tchrs. (sec.-treas. 1975-79), Am. Soc. Engring. Edn. (sec.-treas. physics div. 1978-80), Am. Inst. Physics (governing bd. 1980-83, exec. com. 1981-83), Tau Beta Pi, Sigma Xi, Phi Kappa Phi. Rsch., articles on nuclear structure physics, superdeformed nuclei, gamma ray spectroscopy, physics pedagogy, nuclear physics instrumentation. Home: 2021 Indian Grass Ames IA 50014 Office: Iowa State U Physics Dept Ames IA 50011

KELMAN, BRUCE JERRY, toxicologist, consultant; b. Chgo., July 1, 1947; s. LeRoy Rayfield and Louise (Rosen) K.; m. Jacqueline Anne Clark, Feb. 5, 1972; children: Aaron Wayne, Diantha Renee, Coreyanne Louise. BS, U. Ill., 1969, MS, 1971, PhD, 1975. Diplomate Am. Bd. Toxicology. Postdoctoral rsch. assoc. U. Tenn., Oak Ridge, 1974-76, asst. prof., leader prenatal toxicology group, 1976-79; mgr. devel. toxicology sect. Battelle NW, Richland, Wash., 1980-84, assoc. mgr. biology and chemistry dept., 1984-85, mgr., 1985-89, mgr. new products devel. Life Scis. Ctr., 1989-90; mgr. Internat. Toxicology Office, Battelle Meml. Inst., Richland, 1986-89; mng. scientist, mgr. toxicology dept. Failure Analysis Assocs., Inc., Menlo Park, Calif., 1990-93; mgr. toxicology and risk assessment Golder Assocs. Inc., Redmond, Wash., 1993—; mem. Nation Rsch. Coun. com. on possible effects of electromagnetic fields on biologic systems, 1993-94; adj. prof. N.Mex. State U., Las Cruces, 1983—. Co-editor: Interactions of Biological Systems with Static and ELF Electric and Magnetic Fields, 1987; mem. editorial bd. Trophoblast Rsch., 1983—, Biological Effects of Heavy Metals, 1990. Mem. adv. coun. Seattle Fire Dept., 1988-90; mem. Wash. Gov.'s Biotech. Targeted Sector Adv. Com., 1989-90. Fellow Am. Acad. Vet. and Comparative Toxicology; mem. Soc. Toxicology (founding pres. molecular biology splty. sect. 1988-89, pres. metals splty. sect. 1985-86, cert. of recognition 1989), Am. Soc. for Exptl. Pharmacology and Therapeutics,

Soc. for Exptl. Biology and Medicine (award of merit 1980), Teratology Soc., Wash. State Biotech. Assn. (bd. dirs. 1989-90). Office: Golder Assocs Inc 149 Commonwealth Dr 4104-148th Ave NE Redmond WA 98052

KELMAN, CHARLES D., ophthalmologist; b. Bklyn., May 23, 1930; s. David and Eva K.; m. Ann Gur-Arie; 1 child: Evan Ari Kelman; children from previous marriage: David, Lesley, Jennifer. B.S., Tufts U., 1950; B.M.S., U. Geneva, Switzerland, 1952, M.D., 1956. Diplomate Am. Bd. Ophthalmology. Intern, Kings County Hosp., N.Y.C., 1956-57; resident, Wills Eye Hosp., Pa., 1956-60; with Manhattan Eye, Ear, Nose and Throat Hosp., N.Y.C., 1967—, N.Y. Eye and Ear Infirmary, N.Y.C., 1983—; clin. prof. N.Y. Med. Coll., Valhalla, 1980—; Arthur J. Bedell Meml. lectr., 1991; hon. pres. elect 1994 World Congress on Lens Implant Surgery. Author: Cataracts-What You Must Know About Them, 1982; Atlas of Cryosurgical Techniques in Ophthalmology, 1966; Phacoemulsification Aspiration-The Kelman Technique of Cataract Extraction, 1975; Through My Eyes, 1985. Contbr. numerous articles to profl. jours. Recipient Gold Plate award Am. Acad. Achievement, 1969, 1st prize for sci. exhibit, Am. Acad. Ophthalmology, 1970, 1st Outstanding Achievement award Am. Soc. Contemporary Ophthalmology, 1981, Physicians Recognition award AMA, Can. Implant Assn. award, 1982, Congl. Salute, U.S. Senate 97th Congress, 1983, 1st Ann. Innovators award Am. Internat. Intraocular Lens Congress, 1985, Am. Acad. Ophthalmology Sr. Honor award, 1986, Binkhorst medal Am. Soc. Cataract and Refractive Surgery, 1989, Ridley medal Internat. Congress Ophthalmology, 1990, Special recognition award Am. Acad. Ophthalmology, Nat. Medal Tech., 1992, Pres. of U.S. Inventor of the Yr. award, 1992, Disting. Svc. award Tufts U., 1992. Fellow Am. Acad. Ophthalmology; mem. AMA, Internat. Assn. Ocular Surgeons, Am. Intraocular Implant Soc., Can. Implant Assn., N.Y. Implant Soc. (pres. elect), Am. Soc. for Contemporary Ophthalmology, Am. Soc. for Cataracts and Refractive Surgery (pres.-elect), N.Y. State Soc. Ophthalmology, N.Y. Acad. Medicine (sec. sect. on ophthalmology), European Phaco-Cataract Soc. (hon. life pres.), Soc. for Phacoemulsification and Related Techniques (hon. life pres.). Jewish. Avocations: golf, saxophone, composing, flying. Office: Empire State Bldg 350 5th Ave New York NY 10118-0110

KELMAN, DONALD BRIAN, neurosurgeon; b. Brandon, Manitoba, Can., Apr. 3, 1942; came to U.S., 1979; s. Alexander and May Marguerite (Ronayne) K.; m. Joan Ann Thompson, July 10, 1966 (div. Sept. 1985); children: Carl Michael, Melanie Catherine, Leslie Jane, Brian Andrew; m. Cynthia Marie Esser, Mar. 21, 1986; 1 child, Craig Richard. BA in Biology, U. Sask., 1964, MD, 1968. Diplomate Am. Bd. Neurol. Surgery. Jr. rotating intern St. Joseph's Hosp., Victoria, B.C., Can., 1968-69; pvt. practice Victoria, 1969-70; resident in neurosurgery Mayo Grad. Sch. of Medicine, Rochester, Minn., 1970-76; pvt. practice Prince George, B.C., Can., 1976-79; neurosurgeon Marshfield (Wis.) Clinic/St. Joseph's Hosp., 1979—; chmn. neurosurgery dept. Marshfield Clinic, 1984—. Wis. rep. Joint Coun. of State Neurol. Socs., 1990, 91, 92, 93. Mem. AMA, Am. Assn. Neurol. Surgeons, Wis. Med. Soc., Wis. Neurosurg. Soc. (sec., treas. 1987, pres. 1989). Avocations: rock and mineral collector, lapidary art, painting, computers, scuba diving. Home: 1403 N Broadway Ave Marshfield WI 54449-1321 Office: Marshfield Clinic 1000 N Oak Ave Marshfield WI 54449-5703

KELTON, KENNETH FRANKLIN, physicist, educator; b. Hot Springs, Ark., Aug. 3, 1954; s. John Franklin and Helen Marie (McClard) K.; m. Emily Brown, May 23, 1976; children: Franklin Immanuel, James Winfield. MS, Harvard U., 1980, PhD, 1983. Asst. prof. physics Wash. U., St. Louis, 1985—, assoc. prof. physics, 1990—, assoc. prof. metall. and materials sci., 1991—; co-organized 4th Internat. Conf. on Quasicrystals. Author: (with others) Solid State Physics; contbr. articles to profl. jours. Jour. of Chem. Physics and Phys. Rev. B. Mem. Am. Phys. Soc., Materials Rsch. Soc., Am. Ceramic Soc., Sigma Xi. Achievements include discovery of several titanium based quasicrystals, new complex crystalliche intermetallic phases. Home: 6800 Kingsbury Blvd Saint Louis MO 63130 Office: Wash U Dept Physics One Brookings Dr Saint Louis MO 63130

KEMNITZ, JOSEF BLAZEK, pathologist, scientist; b. Laun, Bohemia, Czechoslovakia, Nov. 14, 1943; s. Winnfried Heinrich and Anna Sarah (Klein) Blazek; m. Renée Edith Basche, Aug. 20, 1982; children: Martina, Kai-Joseph. MD, Charles U., Prague, 1967, PhD, 1975. Fellow Acad. Sci., Prague, 1967-70; postgrad. Weizman Inst., Rehovoth, Israel, 1969; resident med. faculty Charles U., Prague, 1970-72, asst. prof. med. faculty, 1972-75, assoc. prof. med. faculty, 1975-79; assoc. prof. Hannover (Germany) Med. Sch., 1982-89, full prof. pathology, 1989—; dir. Inst. of Pathology, Charles U., Prague, 1975-78; assoc. med. dir. Inst. Pathology, dir. Lab. of Immunopathology, Hannover Med. Sch., 1982; med. dir. Inst. Pathology, Bremen, 1991-92, Inst. Pathology, Essen, 1993; mem. Coun. Hannover Med. Sch., 1987-88; mem. Heart Rejection Study Group of Internat. Soc. for Heart Transplantation, Stanford, 1990. Editor, contbr. Diagnosis of Rejection in Biopsy Material of Caridac Allografts, 1992; contbr. articles to profl. jours. Mem. Internat. Soc. Heart Transplantation, Internat. Assn. Comparative Rsch. on Leukemia and Related Diseases, Internat. Acad. Pathology, Transplantation Soc., Immunocytochemistry Club (London), Deutsche Gesellschaft für Pathologie. Avocations: dogs, tennis.

KEMNITZ, JOSEPH WILLIAM, physiologist, researcher; b. Balt., Mar. 15, 1947; s. Harold Clarence Kemnitz and Alice Mae (Ziebarth) Delwiche; m. Amanda Marye Tuttle, Jan. 5, 1991; 1 child, Julia Ellen. BA, U. Wis., 1969, PhD, 1976. Rsch. assoc. Wis. Regional Primate Rsch. Ctr., Madison, 1976-79, asst. scientist, 1979-84, assoc. scientist, 1984—; assoc. scientist dept. medicine U. Wis., Madison, 1991—; cons. NIH, Bethesda, Md., 1981—, U. So. Calif. Med. Sch., L.A., 1988—; mem. Children's Diabetes Ctr., Madison, Wis., 1990—; steering com. Inst. on Aging, Madison, 1989—. Assoc. editor Hormones and Behavior, 1986—; contbr. articles to profl. jours. Grantee (various) NIH, 1977—. Mem. Am. Physiol. Soc., Am. Inst. Nutrition, Am. Diabetes Assn., Am. Soc. Primatologists, Gerontol. Soc. Am., Soc. Neurosci., N.Am. Assn. Study of Obesity. Office: Primate Rsch Ctr UW 1223 Capitol Ct Madison WI 53715

KEMP, BRUCE E., protein chemist; b. Sydney, Australia, Dec. 15, 1946; s. Norman B. and Mary F. Kemp; m. Alison Virginia Sanders, Jan. 23, 1970; children: Robert E.S., William E.B., Charles E.F. B Agrl. Sci., Adelaide U., Australia, 1970, B Agrl. Sci. with honors, 1971; PhD, Flinders U., Adelaide, Australia, 1975. Postdoctoral fellow U. Calif., Davis, 1974-76; heart found. fellow Flinders Med. Ctr., Adelaide, Australia, 1977-78; Queen Elizabeth II fellow Howard Florey Inst., Melbourne, Victoria, Australia, 1979-84; NH & MRC fellow U. Melbourne, 1985-88; deputy dir. St. Vincent's Inst. Med. Rsch., Melbourne, 1988—. Inventor synthetic substrates, 1976, autologous red cell agglutination test, 1989, synthetic peptide hormones, 1988. Recipient Selwyn Smith prize, 1988, Newman award, 1989, AIDS Trust award, 1989, Pharmacia-LKB medal, 1990, The Wellcome Australia medal, 1990, The Wellcome Rapid Diagnostics award, 1991. Mem. Australian Soc. Biochemistry and Molecular Biology, Australian Soc. Med. Rsch., Australian Soc. Microbiology. Avocations: tennis, aboriginal stone implements, art. Office: St Vincents Inst Med Rsch, 41 Victoria Parade Fitzroy, 3065 Melbourne Australia

KEMP, EJVIND, internist; b. Copenhagen, July 21, 1929; s. Gerhardt and Kirsten (Pedersen) K.; m. Grethe Serup Jorgensen, July 1, 1956; children: Michael, Helene, Kaare. MD, U. Copenhagen, 1956, D in Med. Sci., 1962. Sci. asst. Inst. Exptl. Medicine U. Copenhagen, 1957-61; registrar Bispebjerg Hosp., 1961-64; hon. clin. asst. dept. urology Leeds (Eng.) Gen. Infirmary, 1966-67; sr. registrar Copenhagen U. Hosp., 1964-68; cons. nephrology Odense U. Hosp., 1968, chmn. Lab. of Nephropatology, 1968, prof. internal medicine, 1972—; mem. group for rsch. in transplantation of organs across species EEC, 1991. Author: Hypertension in Poliomyelitis, 1957, To Be or Not To Be, 1975, Xenotransplantation, 1991. With Royal Danish Navy, 1956. Decorated Knight of Denmark; recipient U. Copenhagen Gold Medal, 1955, Odd Fellow prize, 1977, Spies grant of honor, 1990. Mem. Danish Soc. Nephrology, Transplant Soc., Internat. Soc. Nephrology. Home: 8 Hannerupvaenget, DK-5230 Odense M, Denmark Office: Odense U Hosp, Dept Nephrology, DK-5000 Odense C, Denmark

KEMP, EUGENE THOMAS, veterinarian; b. MacDonough, N.Y., Mar. 22, 1930; s. Oswald Milton and Almira Dorothy (Allen) K.; m. Ruth Emer Stoll, Sept. 29, 1951 (dec. Sept. 1977); 1 child, William Allen; m. Margaret Atenna Rowland, Dec. 27, 1980. BS, Cornell U., 1951, DVM, 1957. Sr. ptnr. Day Hollow Animal Clinic, Owego, N.Y., 1957—. Contbr. articles to profl. jours. Bd. dirs. 1st Ch. of Nazarene, Owego, 1991-93; v.p. Tioga County Bd. Health, 1988-93; mem. Owego-Apalachin Bd. Edn., 1961-71; mem. Broome-Tioga Bd. Coop. Edn. Svcs., Binghampton, N.Y., 1969-83, pres., 1971-76; founding pres. Broome Tioga Coun. Sch. Bd. Pres., 1973. Mem. So. Tier Vet. Med. Assn. (pres.-elect 1991-92, pres. 1992), Am. Vet. Med. Assn., N.Y. State Vet. Med. Assn., Kiwanis (pres. Owego chpt. 1968). Republican. Avocations: jazz piano, creative writing. Home and Office: 345 Day Hollow Rd Owego NY 13827-5307

KEMP, JOHN DANIEL, biochemist, educator; b. Mpls., Jan. 20, 1940; s. Dean Dudley and Catherine Georgie (Treleven) K.; children: Todd, Christine, Laura; m. Sharilyn May Buchanan, Feb. 14, 1985. B.A. in chemistry, UCLA, 1962, Ph.D, 1965. NIH postdoctoral fellow U. Wash., Seattle, 1965-68; prof. plant pathology U. Wis., Madison, 1968-81; assoc. dir. Agrigenetics Advanced Research Labs., Madison, 1981-85; prof., dir. plant genetic engrng. lab. N.Mex. State U., Las Cruces, 1985—. Author papers on plant molecular genetics. Grantee NSF; grantee Dept. Agr. Mem. Sigma Xi. Office: N Mex State U Genetic Engring Lab PO Box 3gl Las Cruces NM 88003

KEMP, SARAH (SALLY LEECH), neurodevelopment specialist; b. Bryn Mawr, Pa., Sept. 13, 1940; d. Thomas Bailey and Mary Elizabeth (Veasey) Leech; m. G. Philip Fritz, June 18, 1960 (dec. May 1968); 1 child, Mary Elizabeth Fritz Fitch; m. Garry Colquohoun Kemp, July 25, 1970; children: Sarah Katherine, Hannah Michelle. BA, Calif. State U., Sacramento, 1963; MA, U. Tulsa, 1970; EdM, Columbia U., 1989, PhD, 1990. English tchr. Holland Hall Sch., Tulsa, 1968-70; counselor, learning disabilities tchr. U.S. Dependents Sch., Greenham Common, England, 1971-75; coord. vols. Little Lighthouse Sch. for the Blind, Tulsa, 1974-75; tchr. Holland Hall Sch., Tulsa, 1976, middle sch. psychometrist, tchr., 1977-83, upper sch. resource rm. supr., psychometrist, tchr., 1983-84; program asst. for neuroscience and edn. program Tchrs. Coll., Columbia U., 1985-88; neurodevelopmental specialist Tulsa Devel. Pediatrics and Ctr. for Family Psychology, 1988—; instr. ednl. and devel. neuroscience Tchrs. Coll., Columbia U., fall 1987, summer 1988; adj. asst. prof. pediatrics U. Okla. Med. Sch., Tulsa, 1991—; presenter in field. Sec. Maple Ridge Assn., 1976-77, pres., 1977-83; lay reader Trinity Episc. Ch., 1990—; adv. bd. mem. Children With Attention Deficit Disorder of Green Country, 1991-92; bd. mem. Magic Empire Coun. Girl Scouts U.S.A., 1993—. Mem. Internat. Neuropsychological Soc., Assn. for Children and Adults With Learning Difficulties, Rodin Remediation Soc., Orton Soc. Episcopalian. Achievements include research on attention deficit disorder, autism, and school-related problems. Office: Tulsa Devel Pediatrics 4520 S Harvard Tulsa OK 74119

KEMPE, ROBERT ARON, venture management executive; b. Mpls., Mar. 6, 1922; s. Walter A. and Madge (Stoker) K.; m. Virginia Lou Wiseman, June 21, 1946; children: Mark A., Katherine A. BS in Chem. Engring., U. Minn., 1943; postgrad. metallurgy, bus. adminstrn., Case Western Res. U., 1946-49. Various positions TRW, Inc., Cleve., 1946-53, div. sales mgr., 1953; v.p. Metalphoto Corp., Cleve., 1954-63, pres., 1963-71, Allied Decals, Inc., affiliate, Cleve., 1963-68; v.p., treas. Horizons Rsch. Inc., 1970-71; pres. Reuter-Stokes, Inc. (now subs. of GE Co.), 1971-87; pres. Kempe Everest Co., Hudson, Ohio, 1987—; assoc. Paul Williams & Assocs., Medina, Ohio, 1987—; mem. adv. bd. Horizons Inc.; bd. dirs. Bicron Corp., 1987-90, TGM Detectors, Inc., 1988-92, Chagrin Valley Enterprises. Contbr. articles to profl. jours. Lt. (j.g.) USNR, 1944-46, PTO. Mem. Am. Nuclear Soc. (exec. officer, past chmn. No. Ohio sect.), Am. Soc. Metals, Ohio Citizens Adv. Coun. on Radiological Safety, Chemists Club (N.Y.C.), Country Club of Hudson, Sigma Chi. Achievements include patents in Method of and Apparatus for Making Poppet Valves, Method of Making Hollow Valves, Method of Making Hollow Castings, Method of Coating of Molybdenum Articles; vitreous coated refractory metals, method for producing the same and vitreous enamel composition, coated refractory body, aluminum plate with plural images and method of making same, process for developing photosensitized anodized aluminum plates. Home: 242 E Streetsboro St Hudson OH 44236-3474 Office: Kempe Everest Co 10 W Streetsboro St Hudson OH 44236-2850

KEMPER, KIRBY WAYNE, physics educator; b. N.Y.C., Apr. 13, 1940; s. Alfred Andrew and Anna (Bobetsky) K.; m. Margaret Ray Thurman, Aug. 24, 1964; children: Margaret, Andrew, Ann. BS, Va. Tech. U., 1962; PhD, Ind. U., 1968. Assoc. prof. Fla. State U., Tallahassee, 1975-79, prof., 1979--, dir. grad. physics program, 1982-88, assoc. chmn., 1985-88, dir. accelerator lab., 1990--; vis. fellow Australian Nat. U., Canberra, 1977, 81; bd. trustees SURA, Washington, 1984-88. Recipient Rsch. Support award Nat. Sci. Found., 1974--, Coll. Teaching award Fla. State U., 1990. Mem. Am. Phys. Soc. Democrat. Roman Catholic. Home: 550 Litchfield Rd Tallahassee FL 32312 Office: Fla State U Dept Physics Tallahassee FL 32306

KENAT, THOMAS ARTHUR, chemical engineer, consultant; b. Cleve., Aug. 6, 1942; s. Arthur Brian and Frances Lillian (Kuenzli) K.; m. Wynne Irene Kalvesmaki, June 13, 1964; children: Steven Thomas, Lisa Marie. BSChemE, Carnegie Inst. Tech., 1964, MSChemE, 1965; PhD in Chem. Engring., Carnegie-Mellon U., 1968. Registered profl. engr., Ohio. Rsch. engr. Chemstrand Rsch. Ctr., Durham, N.C., 1968-69; rsch. engr. B. F. Goodrich Co., Brecksville, Ohio, 1969-74, sr. rsch. engr., 1974-80, sr. engring. scientist, 1981-83, sr. R & D assoc., 1983-88; sr. R & D assoc. Camet Co., Hiram, Ohio, 1988-89; sr. project mgr. Quantum Techs., Inc., Twinsburg, Ohio, 1989-92, ind. cons. U.S. Cons., Inc., Cleve., 1992—, cons. Ameripol-Synpol Co., Port Neches, Tex., 1988, French Oil Mill Machinery Co., Piqua, Ohio. Contbr. articles to profl. jours. Elder Prince of Peace Luth. Ch., Medina, Ohio, 1975-90; mem. Medina Community Band, 1983—. Mem. AICE, Am. Chem. Soc., Am. Guild Organists, Nat. Assn. Corrosion Engrs. Republican. Lutheran. Avocations: music, sailing, pipe organ restoration. Home: 745 Falling Oaks Dr Medina OH 44256-2779 Office: US Cons Inc 4807 Rockside Rd #400 Cleveland OH 44131

KENDALL, HENRY WAY, physicist; b. Boston, Dec. 9, 1926; s. Henry P. and Evelyn Louise (Way) K. BA, Amherst Coll., 1950; PhD in Nuclear Physics, MIT, 1955; DSc (hon.), Amherst Coll., 1975. NSF fellow MIT, Cambridge, 1954-56, from asst. to assoc. prof., 1961-67, prof. physics, 1967—, J.A. Stratton prof., 1991—; rsch. assoc. High Energy Lab. Stanford U., 1956-57, lectr. physics, 1957-58, asst. prof., 1958-61. Recipient Nobel prize in physics, 1990. Fellow AAAS, NAS, Am. Acad. Arts and Scis., Am. Phys. Soc. (co-recipient Panofsky prize 1989). Office: MIT Dept of Physics 24-514 77 Massachusetts Ave Cambridge MA 02139

KENDALL, JILLIAN D., information systems specialist, program developer, educator; b. Catskill, N.Y., July 27, 1949; d. John S. and Patricia (Murphy) Rogers; m. Michael F. Kendall, Mar. 29, 1972 (div. June 1984). Graduated with honors, ITC Tech. Coll., 1972; BA in Bus. and Humanities, Golden Gate U., 1986, Cert. in Info. Systems, 1987; MA with honors in Humanities and Computers, U. San Francisco, 1989; PhD in Humanities, Computers, and Human Factors, U. Tex., 1991. Cert. instructor, Tex. Program developer ITC Tech. Coll., San Francisco, 1968-71; asst. programmer Crown Zellerbach, San Francisco, 1971-72; owner Alpine Sewing Emporium, Alburn, Calif., 1972-75, Crystal Milk Delivery, Sacramento, 1975-80; data processor State of Oreg., Covallis, 1980-82; asst. mgr. inventory control, computer operator Arneson Products, Corte Madera, Calif., 1982-85; with commodities and inventory control Hill Bros. Coffees, San Francisco, 1985-88; computer program developer IBM, Westlake, Tex., 1989-92; mgr., dir. rsch. Am. Church Lists, Arlington, Tex., 1992—; cons. pub. rels., instructor operating systems, tech. writing, human factors U. Tex., Arlington, 1989—; ednl. cons. IBM, 1989—; mentor, founder Student Found., 1990; prof. Dallas Community Coll., 1991—. Writer, producer, photographer (videos) The Industrial Revolution in China, 1988, Star Wars, 1989, Computers & Man, 1989; author The Computer is the Prime Symbol of a New Age: The Age of Synthesis; contbr. articles to profl. jours. Charter mem. U. Tex. Student Found., Arlington, Arlington, 1989-91; vol. ARC, Dallas, Ft. Worth, 1989-91; mem. Kimbel Art Mus., Hist. Soc. Scholar U. Calif., San Francisco, 1988, CIT Tech. Coll., 1967, Buck Found., 1988-89; grantee Golden Gate U., 1987. Mem. Assn. Interdisciplinary Studies, Western Social Sci. Assn., Golden Gate U. Alumni Assn., Toastmasters Internat. CTM (v.p. 1988). Avocations: computer graphics, travel, computers, videos. Home: 1514 Sherman St Arlington TX 76012

KENDALL, KAY LYNN, interior designer; b. Cadillac, Mich., Aug. 20, 1950; d. Robert Llewellyn and Betty Louise (Powers) K.; 1 child, Anna Renee Easter. BFA, U. Mich., 1973. Draftsman, interior designer store planning dept. Jacobson Stores, Inc., Jackson, Mich., 1974-79; sr. interior designer store planning dept. Jacobson Stores, Inc., Jackson, 1981—; prin. Kay Kendall Designs, Jackson, 1979—; cons. in field. Bd. dirs. Big Bros./Big Sisters of Jackson County. Mem. Am. Soc. Interior Designers (profl. mem., assoc. Cen. Mich. chpt.). Avocations: tennis, golf, sailing, skiing, martial arts. Home: 722 Beverly Park Pl Jackson MI 49203-3974 Office: Jacobson Stores Inc 3333 Sargent Rd Jackson MI 49201-8800

KENDIG, EDWIN LAWRENCE, JR., physician, educator; b. Victoria, Va., Nov. 12, 1911; s. Edwin Lawrence and Mary McGuire (Yates) K.; m. Emily Virginia Parker, Mar. 22, 1941; children: Anne Randolph (Mrs. R.F. Young), Mary Emily Corbin (Mrs. T.T. Rankin). B.A. magna cum laude, Hampden-Sydney Coll., 1932, B.S. magna cum laude, 1933, D.Sc. hon., 1971; M.D., U. Va., 1936. House officer Med. Coll. Va. Hosp., Richmond, Bellevue Hosp., N.Y.C., Babies Hosp., Wilmington, N.C., Johns Hopkins Hosp., Balt., 1936-40; instr. pediatrics Johns Hopkins U., 1944; practice medicine specializing in pediatrics Richmond, 1940—; dir. child chest clinic Med. Coll. Va., 1944—, prof. pediatrics, 1958—; chief of staff St. Mary's Hosp., Richmond, 1966-67; cons. diseases of chest in children; William P. Buffum orator Brown U., 1979; Abraham Finkelstein Meml. lectr. U. Md., 1983, Derwin Cooper lectr. Duke U. 1984, Renato Ma Guerrero lectr. Santo Tomas U., 1984, Bakwin Meml. lectr., NYU, Bellevue, 1986. Lectr. worldwide; contbr. numerous articles on diseases of chest in children to profl. publs.; editor: Disorders of Respiratory Tract in Children, 1967, 72, 77; co-editor: (with V. Chernick) Disorders of Respiratory Tract in children, 4th edit., 1983, (cons. editor to V. Chernick) 5th edit. pub. as Kendig's Disorders of the Respiratory Tract in Children, 1990, (with C.F. Ferguson) Pediatric Otolaryngology, 1972; contbg. editor: books Gellis and Kagan Current Pediatric Therapy, 12 edits., Antimicrobial Therapy, Kagan, 3 edits., Practice of Pediatrics, Kelley, Practice of Pediatrics, Maurer, Allergic Diseases of Infancy, Childhood and Adolescence, Bierman and Pearlman; mem. editorial bd. Pediatric Pulmonology; editorial adv. bd. Pediatric Annals; former mem. editorial bd. Pediatrics; mem. editorial bd. Alumnews U. Va, 1988. Chmn. Richmond Bd. Health, 1961-63; bd. visitors U. Va., 1961-72; former mem. bd. dirs. Va. Hosp. Svc. Assn.; former ofcl. examiner Am. Bd. Pediatrics; mem. White House Conf. on Children and Youth, 1960; dir. emeritus Dominion Nat. Bank; pres. alumni adv. com. U. Va. Sch. Medicine, Charlottesville, 1974-75; past bd. dirs. Maymont Found., Richmond.; bd. dirs. Children's Hosp., Sheltering Arms Hosp.; former mem. adv. bd. Ctr. for Study of Mind and Human Interaction, U. Va. Sch. Medicine, 1988; mem. steering com. One Hundred Twenty Fifth Anniversary, Med. Coll. of VA Hosps., 1986; bd. dirs. St. Mary's Health Care Found., 1990. Recipient resolution of recognition Va. Health Commr., 1978, Obici award Louise Obici Hosp., 1979, Bon Secours award St. Mary's Hosp., 1986, Keating award Hampden-Sydney Coll., 1989; named an Outstanding Alumnus Sch. Medicine U. Va., 1986; The Edwin Lawrence Kendig Jr. Disting Profesorship in Pediatric Pulmonary medicine named in honor Med. Coll. Va. Commonwealth U. Mem. Am. Acad. Pediatrics (past pres. Va. sect., chmn. sect. on diseases of chest, mem. exec. bd. 1971-78, nat. pres. 1978-79, Abraham Jacobi Meml. award with AMA, 1987, cons. com. on internat. child health), Am. Acad. Pediatrics for Latin Am. (ofcl. adv. to exec. bd. 1988), Va. Bd. Medicine (former pres.), Richmond Acad. Medicine (pres. 1962, chmn. bd. trustees 1963), Va. Pediatric Soc. (past pres.), Am. Pediatric Soc., AMA (pediatric residency rev. com.), So. Med. Assn., So. Soc. Pediatric Research, Internat. Pediatric Assn. (cons., standing com., medal 1986), Med. Soc. Va. (editor Va. Quarterly Jour. 1982, resolution of recognition), Soc. of Cincinnati, Raven, Phi Beta Kappa, Alpha Omega Alpha, Tau Kappa Alpha, Kappa Sigma, Omicron Delta Kappa. Episcopalian. Clubs: Commonwealth, Country of Va; Farmington (Charlottesville). Home: 5008 Cary Street Rd Richmond VA 23226-1643 Office: St Mary's Hosp 5801 Bremo Rd Richmond VA 23226-1900

KENDLER, TRACY SEEDMAN, psychology educator; b. N.Y.C., Aug. 4, 1918; d. Harry and Elizabeth (Goldfinger) Seedman; m. Howard Harvard Kendler, Sept. 20, 1941; children: Joel Harlan, Kenneth Seedman. BA, Bklyn. Coll., 1940; MA, U. Iowa, 1942, PhD, 1943. Statistician USAF, Washington, 1944-45; instr. U. Colo., Boulder, 1946-48; rsch. assoc. NYU, 1951-54; assoc. prof. Barnard Coll., Columbia U., N.Y.C., 1959-64; prof. U. Calif., Santa Barbara, 1965-89, prof. emeritus, 1990—; vis. prof. Hebrew U., Jerusalem, Israel, 1974-75, Tel Aviv (Israel) U., 1990. Co-author: Basic Psychology, 1971; contbr. chpts. to books and articles to profl. jours. Rsch. grantee NSF, 1953-76, Pub. Health Svc., 1965-69; Guggenheim fellow, Jerusalem, 1974; Fulbright scholar, Tel Aviv, 1990. Democrat. Jewish. Home: 4596 Camino Molinero Santa Barbara CA 93110

KENDREW, JOHN COWDERY, molecular biologist, former college president; b. Oxford, Eng., Mar. 24, 1917; s. Wilfrid George and Evelyn May Graham (Sandberg) K. B.A., Trinity Coll., Cambridge U., 1939, M.A., 1943, Ph.D., 1949; Sc.D., 1962, 1962. With Ministry Aircraft Prodn., 1940-45; sci. adv. allied air comdr. in chief S.E. Asia, 1944; dep. chmn. Med. Rsch. Coun. Lab. for Molecular Biology Cambridge (Eng.) U., 1947-75; fellow of Peterhouse, Cambridge U., 1947-75 (hon. fellow 1975); reader Davy-Faraday Lab., Royal Instn., London, 1954-68; dir.-gen. European Molecular Biology Lab., Heidelberg, Germany, 1975-82; pres. St. John's Coll., Oxford U., 1981-87 (hon. fellow 1987). Editor in chief Jour. Molecular Biology, 1959-87. Mem. council UN U., 1980-86, chmn., 1983-85. Decorated knight bachelor and comdr. Order Brit. Empire; recipient (with Max Perutz) Nobel prize in chemistry, 1962; Trinity Coll. Cambridge U. hon. fellow, 1972. Fellow Royal Soc., 1960; fgn. asso. Nat. Acad. Scis. (U.S.); fgn. hon. mem. Am. Acad. Arts and Scis.; hon. mem. Am. Soc. Biol. Chemists; mem. Brit., Am. biophys. socs., Internat. Orgn. Pure and Applied Biophysics (pres. 1969), Internat. Council Sci. Unions (sec. gen. 1974-80, pres. 1983-88). Achievements include determination, in work with myoglobin, of structure of a protein in general outline and atomic detail; observation of alpha-helix arrangement of the polypeptide chain, thereby confirming Pauling's earlier description. Home: Guildhall, 4 Church Ln, Linton Cambridge CB1 6JX, England

KENDRICK, PAMELA ANN, mathematics educator; b. Joplin, Mo., July 6, 1943; d. Laymon Harl and Margaret Alice (Stiers) Morrison; m. Anthony Eugene Kendrick , June 9, 1963. EdB, Pittsburg (Kans.) State U., 1965, MS, 1969. Cert. tchr., Mo., Kans. Computer programmer RCA Missile Test Project, Cape Canaveral Air Force Sta., Fla., 1969-72; statistician NASA Pub. Health Service, Cape Canaveral Air Force Sta., Fla., 1972-73; engr., computer analyst Jet Propulsion Lab., Cape Canaveral Air Force Sta., Fla., 1973-74; instr. math Fla. Inst. Tech., Melbourne, 1974-81; asst. prof. Brevard Community Coll., Cocoa, Fla., 1982—; cons. to various textbook pubs.; dir. computer calculus project in conjunction with Fla. Programs in Excellence, 1988; developer TV stats. course Brevard Community Coll. Recipient Outstanding Alumni award Pittsburg State U., 1975-76; named Fla. Outstanding Young Woman, 1976-77, One of Top 20 Outstanding Young Women Am., 1976-77. Mem. AAUW (sec. Brevard County 1983-85, v.p. programming Brevard County 1986-88), Math. Assn. Am. Democrat. Office: Brevard Community Coll 1519 Clearlake Rd Cocoa FL 32922-6597

KENIG, NOE, electronics company executive; b. Warsaw, Poland, June 5, 1923; came to U.S., 1974; naturalized, 1980; s. Lazaro Hersz and Felisa (Elenbogen) K.; m. Ida Melnik, Apr. 17, 1948; children: Jorge Alberto, Carlos Eduardo; Diploma mech. technologist, Nat. Indsl. Sch. Luis M. Huergo, Buenos Aires, 1941; diploma mech. and elec. engring., Nat. U. La Plata, Buenos Aires, Argentina, 1951. Licensee, Westinghouse Electric Corp., Argentina, 1941-49, Bendix Home Appliance Corp., Argentina, 1949-67; dir. Philco Argentina Corp., 1959-62; asst. pres., group gen. mgr. subs. Nat. Distillers and Chem. Corp., Argentina, 1968-72; with Motorola Inc., Schaumburg, Ill., 1972—, v.p., dir. corp. multinat. ops., 1983, dir. subs., pres. Mex. subs; pres. Motorola Spain, 1989; corp. v.p. Motorola, Inc., Dir. Latin Am. Ops., 1990; chmn. Motorola Internat. Inc., 1993—; mem. world bus. adv. coun. Am. Grad. Sch. of Internat. Mgmt., 1985; mem. internat. bus. devel. adv. bd. Northwestern U., Evanston, Ill., 1989 Office: 1303 E Algonquin Rd Schaumburg IL 60196

KENKEL, JAMES LAWRENCE, economics educator; b. Cin., Mar. 25, 1944; s. Lawrence J. and Mildred (Schmidt) K.; children: Julie, Tim. BA, Xavier U., Cin., 1966; MA, Purdue U., 1968, PhD, 1969. Prof. econs. U. Pitts., 1969—; cons. Fed. Home Loan Bank Bd, Washington, 1971-72, Jones & Laughlin Steel, Pitts., U.S. Steel, Pitts., Sony Corp., Nat. Steel, EPA, Mellon Bank, Westinghouse. Author: Risk in Mortgage Lending, 1973, Linear Dynamic Economic Models, 1974; Statistics for Management, 1989. Mem. Am. Econ. Assn., Am. Statis. Assn., Econometric Soc. Avocations: tennis, skiing, baseball. Home: 807 Academy Pl Pittsburgh PA 15243-2003 Office: U Pitts Forbes Quad Pittsburgh PA 15260

KENNEDY, D. J. LAURIE, civil engineering educator. Prof. civil engring. U. Alberta, Edmonton, Can. Recipient Sir Casimir Gzowski medal Can. Soc. Civil Engring., 1992, Le Prix P.L. Pratley award, 1992. Office: Univ of Alberta, Dept of Civil Engirning, Edmonton, AB Canada T6G 2G7*

KENNEDY, DOUGLAS WAYNE, physicist; b. La Mesa, Calif., Feb. 24, 1971; s. James W. and Geneva L. (Salsman) K. BA in Physics and Math., Southeastern State U., 1992. Co-owner Multimedia Prodns., Tishomingo, Okla., 1991—. Presbyterian. Home: 1500 E Main Tishomingo OK 73460 Office: Dept Physics Okla State U Stillwater OK 74075

KENNEDY, EUGENE PATRICK, biochemist, educator; b. Chgo., Sept. 4, 1919; s. Michael and Catherine (Frawley) K.; m. Adelaide Majewski, Oct. 27, 1943; children—Lisa Kennedy Helprin, Sheila Kennedy Violich, Katherine Kennedy Diller. BSc, DePaul U., 1941; PhD (Nutrition Found. fellow), U. Chgo., 1949, ScD (hon.), 1977; AM (hon.), Harvard U., 1960. Rsch. chemist chem. rsch. dept. Armour & Co., 1941-47; postdoctoral fellow Am. Cancer Soc., U. Calif., Berkeley, 1949-50; with Ben May Lab. Cancer Rsch., dept. biochemistry U. Chgo., 1950-56, prof. biochemistry, 1956-60; sr. postdoctoral fellow NSF, Oxford (Eng.) U., 1959-60; Hamilton Kuhn prof. biol. chemistry Harvard Med. Sch., 1960—, head dept., 1960-65; Macy scholar Cambridge U., 1976. Recipient Glycerine rsch. award, 1955; Am. Oil Chemist Soc. Lipid Rsch. award, 1970; Gairdner Found. award, 1976; Ledlie prize, 1976, Alexander von Humboldt prize, 1984; Passano Found. award, 1986, Heinrich Wieland Prize, 1986, William C. Rose Award in biochemistry. Am. Soc. Biochem. and Molecular Biology, 1992. Mem. NAS, Am. Chem. Soc. (Paul Lewis award 1958), Am. Soc. Biol. Chemists (pres. 1970-71), Am. Acad. Arts and Scis., Am. Philos. Soc. Home: 221 Mt Auburn St Cambridge MA 02138-4848 Office: Harvard Med Sch Dept Biol Chemistry Boston MA 02115

KENNEDY, JOANNE PATRICIA, technical service representative; b. Bronx, N.Y., Mar. 16, 1963. BS in Food Sci., Cornell U., 1984. Sales rep. TIC Gums, Belcamp, Md., 1984-85; mfg. specialist Estee Corp., Parsippany, N.J., 1985-86; food technologist Hercules, Inc., Middletown, N.Y., 1986-89; tech. svc. rep. Stepan Co., Maywood, N.J., 1989—. Contbr. articles to profl. jours. Mem. Inst. Food Technologists (newsletter editor 1986-89, sec. 1989-91, exec. mem-at-large 1992—), Am. Assn. Cereal Chemists, Am. Oil Chemists Soc., Am. Soc. Parenteral and Enteral Nutrition. Office: Stepan Co 100 W Hunter Ave Maywood NJ 07607

KENNEDY, JOHN WILLIAM, manufacturing company executive; b. Summit, N.J., May 20, 1956; s. William John and Jean Mary (Krutisia) K.; m. Cecelia Marie Hamrock, Dec. 26, 1981; 1 child, Sean Michael. BS with honors, North Adams State Coll., 1978; MBA with honors, Columbia Pacific U., 1987, BS in Indsl. Engring., 1988. Cert. tchr., N.J. Tchr. Mountainside (N.J.) Sch. Dist., 1979-82, Chatham (N.J.) Boro Sch. Dist., 1982-83; plant mgr. The Chatham Club Recreation Ctr., 1982-85; ops. mgr. Coleman Equipment, Inc., Irvington, N.J., 1985-91; acct. mgr., project mgr. automated sorting systems div. Sandvik Process Systems, Totowa, N.J., 1991—; plant mgr., ops. mgr., cons., Madison (N.J.) Community Pool., 1971-87. Co-patentee, vacuum lifter, air logic weightless circuit; contbr. tech. articles to trade publs. Active Denville (N.J.) area Boy Scouts Am., 1984—, chmn. dist. advancement com., 1990—; mem. area com. Spl. Olympics, Flanders, N.J., 1987—, event dir., Morris, Sussex and Warren counties, 1988—. Named Eagle Scout Boy Scouts Am., 1970. Mem. Am. Mgmt. Assn., Inst. Indsl. Engring., Am. Soc. for Quality Control. Republican. Roman Catholic. Avocations: camping, biking, racquetball, softball, coins. Home: 2 Burnet Rd Madison NJ 07940-1206 Office: Sandvik Seamco Systems 29 Commerce Way Totowa NJ 07512-1154

KENNEDY, KEN, computer science educator; b. Washington, Aug. 12, 1945; s. Kenneth Wade and Audrey Ruth K. BA in Math. summa cum laude, Rice U., 1967; MS in Math., NYU, 1969, PhD in Computer Sci., 1971. Asst. prof. dept. math. scis. Rice U., Houston, 1971-76, assoc. prof., 1976-80, prof., 1980-84, Noah Harding prof. dept. computer sci., 1985—, chmn. computer sci. program com., 1982-85, chmn. dept. computer sci., 1984-88, 90-92, dir. Computer and Info. Tech. Inst., 1986-92, dir. Ctr. for Rsch. on Parallel Computation, 1989—; vis. prof. computer sci. dept. Stanford U., 1985-86; v.p. R.M. Thrall and Assocs., Inc., 1974-81, pres., 1981-93; mem. programming langs. and implementation sub-area panel computer sci. and engring. rsch. Div. Computer Rsch. NSF, 1975-77, mem. adv. com. for computer rsch., 1984-88, chmn., 1985-87; vis. scientist Space Shuttle Program Lead Office NASA, 1975, Dept. Computer Sci. IBM Thomas J. Watson Rsch. Ctr., Yorktown Heights, N.Y., 1978-79, cons., 1979—, Lawrence Livermore Nat. Lab., 1985—; vis. staff mem. computer div. Los Alamos Sci. Lab., 1977—; mem. exec. com. CSNET, 1984-86, Computer Sci. and Telecom. Bd., NRC, 1992—, presdl. adv. com. on sci. and tech., High Performance and Comm. subpanel, Office Sci. and Tech. Policy, White House, 1992-93; presenter numerous profl. meetings; dir. numerous masters theses, PhD dissertations. Mem. editorial bd. Jour. Parallel and Distributed Computing, 1988—, Concurrency: Practice and Experience, ACM Transactions on Software Engring. and Methodology, 1989—; sect. editor langs. and programming Jour. Supercomputing, 1986—; contbr. numerous chpts. to books, articles to profl. jours. Bd. dirs. Houston Soc. Performing Arts, 1986—, v.p. artistic adv., 1987—. Grantee NSF, 1973—, IBM Corp., 1979—, DARPA, 1987—, W.M. Keck Found., 1990—, Office of Gov. State of Tex., 1990—, ONR, 1993—, NASA, 1993—; Woodrow Wilson Nat. fellow, 1967-68; NSF grad. fellow, 1968-71; recipient NYU Founders Day award for Acad. Achievement, 1972, Nat. Acad. Engring., 1990. Mem. AAAS, IEEE (sr.), Assn. Computing Machinery (program com. SIGPLAN nat. conf. 1982, 84, chmn. program com. principles of programming langs. conf. 1983, mem. software system award com. 1983-85, chmn. 1984, chmn. program com. Supercomputing 1991, chmn. Internat. Conf. Supercomputing 1992), Soc. Indsl. and Applied Math., Nat. Acad. Engring., Phi Beta Kappa, Sigma Xi. Office: Rice U Computer Info Tech Inst PO Box 1892 Houston TX 77251-1892

KENNEDY, PATRICK MICHAEL, fire analyst; b. Chgo., Jan. 13, 1947; s. John and Dorothy Jane (Petry) K.; m. Susan Ellen Baylis, Aug. 9, 1969; children: Christine, Heather, Kathryn. BS in Communications, St. Joseph's Coll., 1969; BS in Math. and Physics, SUNY, Albany, 1980; MS in Forensic Engring. (hon.), Pacific Western U., 1982, PhD in Fire Engring. (hon.), 1984; BS Fire Engring. Tech. summa cum laude, U. Cin., 1991. Cert. fire and explosion investigator; fire investigation instr. Police arson investigator Arlington Heights (Ill.) Police Dept., 1969-81; police commr. Algonquin (Ill.) Police and Fire Commn., 1981-83; sr. fire and explosion analyst John A. Kennedy & Assocs., Chgo., 1981—. Author: Fire and Arson Investigation, 1972, Fire-Arson-Explosion Investigation, 1978, Fires and Explosions - Determining Cause and Origin, 1985, Explosion Investigation and Analysis - Kennedy on Explosions, 1990; co-author: National Fire Code Guide on Fire & Explosion Investigations, 1991. With USN, 1969-71. Recipient Disting. Svc. award for Fire/Safety Engring. Tech. Coll. Law Enforcement, Ea. Ky. U., 1986. Mem. ASTM (Fire Standards com., 1988—, coord. com. on flash point, 1989—), Nat. Fire Protection Assn. (chmn. Fire Sci. and Tech. Educators Sect. 1991—, Fire Investigation tech. com. 1986—), Nat. Assn. Fire Investigators (bd. dirs. chmn.). Office: John A Kennedy Assocs 2155 Stonington Ave Ste 118 Schaumburg IL 60195-2057

KENNEDY, ROBERT SAMUEL, experimental psychologist, consultant; b. Bronxville, N.Y., Jan. 10, 1936; s. Robert and Helen (Marshall) K.; m. Margaret Draper, Aug. 23, 1964 (div. Aug. 1976); children: Kathryn Jeanette, Robert Carpenter, Richard Marshall, Kristyne Elizabeth. BA in English and Philosophy, Iona Coll., 1957; MA in Experimental Psychology, Fordham U., 1959; PhD in Sensation and Perception, U. Rochester, 1972. Commd. ensign USN, 1959, advanced through grades to commdr., ret., 1981; rsch. psychologist divsn. psychology Naval Sch. Aviation Medicine, Pensacola, Fla., 1959-65; head diver divsn. evaluations Naval Med. Rsch. Inst., Pensacola, Fla., 1968-70; head br. human factors engring. Naval Missile Ctr., Point Mugu, Calif., 1972-76; head human divsn. human factors Naval Air Devel. Ctr., Warminster, Pa., 1976; officer-in-charge dept. bioengring. scis. Naval Aerospace Med. Rsch. Lab. Detachment, New Orleans, 1976-79, head dept. human performance, 1977-79; head dept. human performance Naval Biomed. Lab., New Orleans, 1979-81; faculty dir. Essex Corp., Columbia, Md., 1981—; v.p. Essex Corp., Orlando, Fla., 1987—; prof. U. Ctrl. Fla., Orlando, 1987—; bd. dirs. Aviation, Space, and Environ. Medicine; lectr. grad. dept. Psychology Laverne Coll., Point Mugu, 1973-76, dept. Systems Mgmt. U. So. Calif., 1975-76; cons. NASA/Johnson Space Ctr., Houston, 1985—, Systems Tech., Inc., Univs. Space Rsch. Assn., Monterey Techs., Inc., Battelle, Am . Inst. Biol. Scis., Performance Metrics, Bolt, Beranek & Newman, NAS/Nat. Sci. Rsch. Coun., NASA/Ames Rsch. Ctr., U.S. Navy Med. Rsch. and Devel. Command; expert witness in human factors, Ala., Fla., Ga., Miss., N.Mex.; prin., assoc. investigator numerous projects; presenter numerous confs. Consulting editor, Aviation, Space, and Environmental Medicine, editorial adv. bd., 1991—, Behavior Research Methods, Instruments and Computers, Jour. Experimental Psychology: General, Perceptual and Motor Skills, Military Psychology, Ergonomics, Pediatrics, Perception and Psychophysics; co-author numerous tech. reports; contbr. chpts. to books, articles to profl. jours. Fellow APA (military divsn., pres. applied exptl. and engring. divsn. 1989-90), Am. Psychol. Soc., Aerospace Med. Assn. (Raymond F. Longacre awd., 1993); mem. AAAS, Am. Soc. Safety Engrs., Aerospace Human Factors Assn. (exec. com. 1991, individual differences tutorial group), Behavioral Toxicology Soc., Human Factors Soc. (forensics profl. group, visual performance tech. group, consulting editor), Soc. Neurosci., Behavioral Toxicology Soc., Illuminating Engring. Soc., Undersea Med. Soc. (individual differences tech. group), Barany Soc., N.Y. Acad. Scis., Psychonomic Soc., So. Soc. Philosophy and Psychology, SAFE Assn. Achievements include development of microcomputer based test battery, motion sickness data base and prediction tools. Office: Essex Corp Ste 227 1040 Woodcock Rd Orlando FL 32803

KENNEDY, STEPHEN DANDRIDGE, economist, researcher; b. N.Y.C., Feb. 25, 1942; s. Joseph Conrad and Frances (Midlam) K.; m. Joanna Court Bartlett, Nov. 27, 1965; children: Julia Paca, Benjamin Bartlett. AB, Harvard U., 1963; PhD, MIT, 1972. Mem. staff com. on banking and currency U.S. Ho. of Reps., Washington, 1964-66; adminstrv. asst. The Fed. Home Loan Bank Bd., Washington, 1966-67; analyst Abt Assocs., Inc., Cambridge, Mass., 1970, v.p., 1975, chief scientist, 1988—. Episcopalian. Avocations: gardening, sailing. Office: Abt Assocs Inc 55 Wheeler St Cambridge MA 02138-1125

KENNEDY, WILLIAM JAMES, pharmaceutical company executive; b. Troy, N.Y., Dec. 4, 1944; s. James Francis and Marjorie (Albrecht) K.; m. Mary Monika Silasz, July 22, 1967; children: Susan M., John R., Morgan E. BS, Siena Coll., Loudonville, N.Y., 1966; MA, Clark U., Worcester, Mass., 1969; PhD, SUNY, Buffalo, 1975. Assoc. dir. drug regulatory affairs Pfizer Pharms., N.Y.C., 1977-80; asst. dir. drug regulatory affairs Berlex Labs., Morristown, N.J., 1980-81; dir. drug regulatory affairs Kali Pharma, Elizabeth, N.J., 1981-82, GD Searle & Co., Skokie, Ill., 1982-86; v.p. drug regulatory affairs ICI Pharms. Group, Wilmington, Del., 1986-93, Zeneca Pharms. Group, Wilmington, 1993—; cons. various pharm. cos., 1981-86. Contbr. articles to profl. jours. and chpts. to books. NIH fellow, 1971-75. Mem. Pharm. Mfrs. Assn. (chmn. drug regulatory affairs com. 1991), Del. Valley Regulatory Affairs Forum (chmn. 1988), Nat. Acad. Sci. Home: 116 Marcella Rd Wilmington DE 19803-3411 Office: Zeneca Pharms Group Concord Pike & New Murphy Wilmington DE 19897

KENNEL, CHARLES FREDERICK, physicist, educator; b. Cambridge, Mass., Aug. 20, 1939; s. Archie Clarence and Elizabeth Ann (Fitzpatrick) K.; m. Ellen Lehman; children: Matthew Bochner, Sarah Alexandra. A.B. (Nat. scholar 1955-59), Harvard U., 1959; Ph.D. in Astrophys. Scis. (W.C. Peyton Advanced fellow 1962-63), Princeton U., 1964. Prin. research scientist Avco-Everett Research Lab., Mass., 1960-61, 64-67; vis. scientist Internat. Center Theoretical Physics, Trieste, Italy, 1965; mem. faculty U. Calif., Los Angeles, 1967—; prof. physics U. Calif., 1971—, chmn. dept., 1983-86; mem. Inst. Geophysics and Planetary Physics, 1972—, acting assoc. dir. inst., 1976-77; mem. space sci. bd. NRC, 1977-80, chmn. com. space physics, 1977-80; Fairchild prof. Calif. Inst. Tech., 1987; mem. space and earth scis. adv. com. NASA, 1986-89; mem. NRC Bd. Physics and Astronomy, 1987—, chmn., 1992—; chmn. plasma sci. NRC, mem. DOE fusion policy adv. com., 1990; Fulbright lectr. Brazil; visitor U.S.-USSR Acads. Exch., 1988-90; disting. vis. prof. U. Alaska, 1988-89, 90—; advisor U.S. Arctic Commn., 1993—; cons. in field. Co-author: Matter in Motion, The Spirit and Evolution of Physics, 1977; co-editor: Solar System Plasma Physics, 1978. Bd. dirs. Los Angeles Jr. Ballet Co., 1977-83, pres., 1979-80; bd. dirs. Inst. for Theoretical Physics, Santa Barbara, Calif., 1986-90. NSF postdoctoral fellow, 1965-66, Sloan fellow, 1968-70, Fulbright scholar, 1985, Guggenheim fellow, 1987. Fellow Am. Geophys. Union, Am. Phys. Soc. (pres. div. plasma physics 1989), AAAS; m. NAS, Am. Astron. Soc., Internat. Union Radio Sci., Internat. Acad. Astronautics. Office: U Calif Dept Physics Los Angeles CA 90024

KENNEL, ELLIOT BYRON, nuclear engineer; b. East Cleveland, June 15, 1957; s. Byron E. and Sook Cha (Lee) K.; m. Jonelle Blair, Nov. 17, 1984 (div. Feb. 1, 1992); 1 child, David Jeehyuk. BS, Miami U., 1979; MS, Ohio State U., 1982. Nuclear rsch. officer, capt. USAF, WPAFB, Ohio, 1980-84, nuclear engr., 1985-90; nuclear engr. Applied Scis., Inc., Cederville, Ohio, 1990—, Space Exploration Assocs., Cederville, Ohio, 1992—. Contbr. sci. articles on thermionic conversion, nuclear power sources for space exploration to profl. jours. Recipient Air Force Significant Achievement award, 1988, Air Force Commendation medal, 1984. Mem. ASME (v.p. direct conversion subcom.), AIAA, Am. Nuclear Soc., Planetary Soc. Achievements include co-founding of Space Exploration Assocs., joint U.S./Russian Co. for research on space exploration; research in thermionic conversion; patents for thermionic converters and related inventions. Home: PO Box 503 Yellow Springs OH 45387 Office: Space Exploration Assn 141 W Xenia Ave PO Box 579 Cederville OH 45314

KENNELLY, KEVIN JOSEPH, psychology educator; b. Wichita Falls, Tex., July 17, 1934; s. Edward Joseph and Lucille Marie (Hund) K.; m. Larua Leah Ballard, Aug. 26, 1961; children: Laura Kathryn, Kevin Garrett, Patrick Joseph, Daniel Thomas, Brendan Christopher. BA, Le Moyne Coll., 1956; MA, North Tex. State U., 1962; PhD, Syracuse U., 1967. Psychologist Whitfield State Hosp., Jackson, Miss., 1961-63; prof. psychology U. North Tex., Denton, 1967—; editorial cons. Holt, Rinehart & Winston, Inc., Harcourt, Brace, Jovanovich, Inc., Prentice-Hall, Inc., West Pub. Co.; referee grant proposal NSF, 1977; referee rsch. articles for many jours., 1977—. Contbr. articles to profl. publs. With U.S. Army, 1957-59. NDEA predoctoral fellow, 1963-66; NICHHD predoctoral trainee, 1966-67. Mem. Am. Psychol. Soc. Republican. Roman Catholic. Achievements include research findings in learned helplessness, test anxiety, hemispheric integration and lateralization of function. Home: 2105 Houston Pl Denton TX 76201 Office: U North Tex Dept Psychology Denton TX 76203

KENNELLY, WILLIAM JAMES, chemist; b. Cleve., Aug. 22, 1948; s. William James and Evelyn Ann (Hanson) K.; m. Maureen A. Sullivan, July 9, 1977; children: Margaret M., William R. BS, MIT, 1970; PhD, Northwestern U., 1975; MBA, Temple U., 1986. Postdoctoral fellow U. N.D., Grand Forks, 1975-76, MIT, Cambridge, 1976-77; sr. scientist Rohm and Haas Co., Spring House, Pa., 1977-84, Amax Inc., Ann Arbor, Mich., 1984-87; mgr. tech. svcs. Climax Molybdenum Co., Ypsilanti, Mich., 1987—. Contbr. articles to profl. jours. Mem. ASTM, Am. Chem. Soc., Soc. Plastics Engrs., Rotary. Home: 447 Marlpool Dr Saline MI 48176 Office: Climax Molybdenum Co P O Box 407 Ypsilanti MI 48197

KENNETT, JAMES PETER, geology and zoology educator; b. Wellington, New Zealand, Sept. 3, 1940; s. Stanley William and Muriel Jean K.; m. Diana Margaret Dawes, Dec. 12, 1964; children: Douglas, Mary. PhD, Wellington, 1965, Dsc, 1976. Sci. officer New Zealand Ocean Inst., 1965-66; rsch. assoc. Allan Hancock Fedn. U. So. Calif., L.A., 1966-68; assoc. prof. Fla. State U., Tallahassee, 1968-70; assoc. prof. U. R.I., Kingston, 1970-74, prof., 1974-87; prof. U. Calif., Santa Barbara, 1987—; mem. Antarctic drilling adv. com. JOIDES, 1970-75; mem. planning com. Internat. Program Ocean Drilling, 1975-79, 81-83. Author: Marine Geology, 1982, (with Srinivasan) Neogene Planktonic Foraminifera: A Phylogenic Atlas, 1983; editor: The Miocene Ocean: Paleoceanography & Biogeography, 1985, (with Warnke) The Antarctic Paleoenvironment, A Perspective on Global Change, 1992; contbr. articles to profl. jours. Recipient McKay Hammer award New Zealand Geol. Soc., 1968. Fellow Am. Geophys. Union (found. editor paleoceanography 1985-87), Geol. Soc. Am.; mem. Royal Soc. New Zealand (hon.). Episcopalian. Achievements include rsch. in paleoceanography, and in the importance of Antarctica and its surrounding ocean in earth system science. Office: U Calif Marine Sci Inst Santa Barbara CA 93106

KENNEY, TIMOTHY P., computer scientist; b. Burlington, Vt., Nov. 18, 1963; s. Thomas J. and Pauline L. (St. Hilaire) K.; m. Jennifer L. Moran, June 14, 1986; children: Maries C., Nathaniel J. BS in Computer Sci., St. Michael's Coll., 1986; MS in Computer Sci., U. Wis., 1987. Software engr., researcher IBM, Bethesda, Md., 1987; artificial intelligence engr. Computer Vision, Bedford, Mass., 1987-88; staff sr. engr. IDX, Burlington, 1989-. Mem., asst. treas. Dem. Com. Richmond, 1991-92; mem. Chittenden County DeM. Com., 1991-92. Recipient Bausch and Laumb Hon. Sci. award Rice Meml. Sci. Staff, 1986. Mem. IEEE, Am. Assn. for Artificial Intelligence. Democrat. Roman Catholic. Home: Box 275 Main St Richmond VT 05477 Office: IDX 1400 Shelburne Rd South Burlington VT 05403

KENNEY-WALLACE, GERALDINE, chemistry and physics educator; b. London, Mar. 29, 1943. Assoc., Royal Inst. Chemistry, 1965; MS, U. B.C., 1968, PhD in Chemistry, 1970. Research assoc. biophysics Oxford U., 1964-66; chemistry fellow U. B.C., 1970-71; assoc. Radiation Lab. U. Notre Dame, 1971-72; from instr. to asst. prof. Yale U., 1972-74, asst. prof. chemistry, 1974-78, assoc. prof., 1978-80; prof. chemistry, physics U. Toronto, from 1980; chmn. sci. Council of Can., Ottawa, Ont., 1987-90; pres., vice-chancellor McMaster U., Ont., 1990—; vis. scientist chemistry Argonne Nat. Lab., 1973—, Poly. Sch. of Paris, 1981; vis. prof. Stanford U., 1985-86. Recipient Corday-Morgan medal, 1979, Noranda award, 1984, Montreal medal, Chem. Inst. of Canada, 1991; Alfred P. Sloan fellow, 1977-79; Killam Research fellow, 1979-81; Guggenheim fellow, 1983; E.W.R. Stealil fellow, 1984. Mem. The Chem. Soc., Am. Chem. Soc., Am. Phys. Soc., Optical Soc. Am., InterAm Photochem Soc. Office: McMaster U, Office of Pres, Hamilton, ON Canada L8S 4L8

KENNY, DOUGLAS TIMOTHY, psychology educator, former university president; b. Victoria, B.C., Can., Oct. 20, 1923; s. John Ernest and Margaret Julia (Collins) K.; m. Lucille Rabowski, Apr. 18, 1950 (dec.); children—John Douglas, Kathleen Margaret; m. Margaret Lindsay Little, June 5, 1976. Student, Victoria Coll., 1941-43; BA, U. B.C., 1945, MA, 1947; PhD, U. Wash., 1950; LLD (hon.), U. B.C., 1983. Lectr. U. B.C., Vancouver, 1950-54, asst. prof. psychology, 1954-57, assoc. prof., 1957-64, prof., 1965-89, head dept. psychology, 1965-69, acting dean faculty of arts, 1969-70, dean faculty of arts, 1970-75, pres., vice chancellor, 1975-83, pres. emeritus, 1989—; vis. asso. prof. Harvard., 1963-65; trustee Can. Council, 1975-78, Social Scis. and Humanities Research Council, 1978-83, Monterey Inst. Internat. Studies, 1980-83, Discovery Found., B.C., 1979-83. Contbr. articles to profl. jours. Trustee Vancouver Gen. Hosp., 1976-78; founding mem. bd. govs. Arts, Scis. and Tech. Ctr., Vancouver, 1980; hon. patron Internat. Found. of Learning, 1983—. Recipient Queen's Silver Jubilee medal, 1977, Park O. Davidson Meml. award, 1984. Mem. Can. Psychol. Assn., B.C. Psychol. Assn. (pres. 1951-52), Am. Psychol. Assn., Am. Psychol. Soc., Vancouver Inst. (pres. 1973-74), U. B.C. Faculty Assn. (pres. 1961-62, hon. pres. 1975-83), B.C. Rsch. Coun. (trustee 1975-89), Vancouver Club, U. B.C. Faculty Club.

KENNY-WALLACE, G. A., chemical engineer. Pres. McMaster U., Hamilton, Ont., Can. Recipient Montreal medal Chem. Inst. Can., 1992. Office: McMaster U, GH-Rm 238 1280 Main St W, Hamilton, ON Canada L8S 4L8*

KENT, BARTIS MILTON, physician; b. Terrell, Tex., June 23, 1925; s. Bartis William and Annie (Smalley) K.; student So. Meth. U., 1942-44; M.D., Baylor U., 1948; m. Ann L. Kiel, July 6, 1954; children—Susan Ruth, Martha Lucille, Bartis Michael. Intern, Jefferson Davis Hosp., Houston, 1948-49; resident pathology Mass. Meml. Hosps., Boston, 1951; resident in internal medicine Baylor U., 1953-56; indsl. physician Humble Oil Co., Houston, 1949-51; instr. dept. medicine U. Iowa, 1956-58; staff physician Iowa City VA Hosp., 1956-58; practice medicine specializing in internal medicine, Muskogee, Okla., 1958—; cons. Muskogee VA Hosp.; clin. asst. prof. medicine U. Okla. Sch. Medicine, 1975—. Chmn., Muskogee County chpt. Am. Nat. Red Cross, 1963-65. Served with USAF, 1951-53. Decorated Air medal. Diplomate Am. Bd. Internal Medicine. Mem. A.C.P., Indsl. Med. Assn., Soc. Nuclear Medicine, Am. Fedn. Clin. Research, Am. Heart Assn., Aerospace Medicine Assn., Am., Okla. socs. internal medicine, Muskogee C. of C. Methodist. Mason (Shriner). Home: 800 N 45th St Muskogee OK 74401-1505 Office: 211 S 36th St Muskogee OK 74401

KENT, D. RANDALL, JR., engineerng company executive. V.p. Gen. Dynamics Corp., Fort Worth, Tex. Recipient Aircraft Design award Am. Inst. Aeronautics and Astronautics, 1992. Office: Gen Dynamics Corp Ft Worth Div PO Box 748 Fort Worth TX 76101-0748*

KENT, DENNIS VLADIMIR, geophysicist, researcher; b. Prague, Czechloslovakia, Nov. 4, 1946; s. Frank D. and Olga (Pospicil) K.; m. Carolyn Ann Cook, Dec. 18, 1971; 1 child, Amanda Grace. BS, CCNY, 1968; PhD, Columbia U., 1974. Rsch. assoc. Lamont-Doherty Earth Obs., Palisades, N.Y., 1974-79; sr. rsch. assoc. Lamont-Doherty Geol. Obs., Palisades, N.Y., 1979-84, Doherty sr. rsch. scientist, 1984—, assoc. dir., 1987-89, interim dir., 1989-90; dir. rsch. Lamont-Doherty Earth Obs., Palisades, N.Y., 1993—; adj. prof. dept. geol. scis. Columbia U., N.Y.C., 1987—; mem. ocean history panel JOIDES, 1987-90, mem. exec. com., 1989-90, 93—; mem. bd. govs. Joint Oceanographic Inst., Washington, 1989-90, 93—. Contbr. over 120 refereed articles to profl. jours. Grantee NSF, 1974—; named Conoco Disting. lectr. Woods Hole Oceanographic Inst., 1983, Turner/Conoco Disting. Lectr. U. Mich., Ann Arbor, 1985. Fellow Am. Geophys. Union (pres.-elect Geomagnetism/Paleomagnetism sect. 1992—), Geol. Soc. Am., AAAS. Office: Lamont-Doherty Earth Obs Rte 9W Palisades NY 10964

KENT, HOWARD LEES, obstetrician/gynecologist; b. Norristown, Pa., Nov. 27, 1930; s. Howard Linnaeus and Margaret (Cairns) K.; m. Margaret Louise Hermanutz, Oct. 17, 1959; children: Howard Lees Jr., Lisanne, Margaret, Kristyn. AB in Zoology, U. Pa., 1953; MD, Hahnemann U., 1958. Diplomate Am. Bd. Ob-Gyn. Intern Misericordia div. Mercy Cath. Med. Ctr., Phila., 1958-59; resident Hahnemann U., 1959-62; pvt. practice Hammonton, N.J., 1964-81; prof. Thomas Jefferson U., Phila., 1982—. Author: (with others) Vaginitis/Vaginosis, 1991; editor: Proceedings-Obstetrical Society of Philadelphia, 1980-82; contbr. articles to profl. jours. Fellow Internat. Soc. for Study of Vulvar Disease, Royal Soc. Medicine, Am. Coll. Ob-gyn. (key contact), Am. Coll. Surgeons, Coll. Physicians of Phila. (libr. com. 1980-89), Obstet. Soc. Phila. (asst. sec. 1978-81); mem. N.J. Ob-gyn. Soc., Vesper Club, U. Pa. Alumni. Republican. Avocations: travel, photography.

KENT, JAN GEORG, computer consultant; b. Oslo, Norway, Nov. 23, 1942; s. Rolf and Ragna Katarina (Kent) Nielssen; m. Elisabet Bigset, Mar. 20, 1973; 1 child, William. MS, U. Oslo, 1966, PhD, 1979. Lectr. U. Waterloo, Ont., Can., 1967; system programmer Stanford (Calif.) U., 1967; researcher Norwegian Def. Rsch. Est., Norway, 1968-69; stipendiat IBM Norway, Oslo, 1969-72, systems engr., 1977-79; researcher Norwegian Computing Ctr., Oslo, 1972-77; chief cons. Tandberg Data, Oslo, 1979-81; chief engr. Norwegian Def. Command, Norway, 1981-85; chief cons. Christiania Bank, Oslo, 1985-92; sr. quality assurance engr. Sci. Project Contractors, Oslo, 1992-93; cons. dept. EDP and Organl. Devel. Directorate of Customs and Excise, Oslo, 1993—. Contbr. articles to profl. publs. Mem. IEEE, ACM (Norwegian chpt. chair 1973—). Lutheran.

KENT, THEODORE CHARLES, psychologist; m. Shirley, June 7, 1948; children: Donald, Susan, Steven. BA, Yale U., 1935, MA, Columbia U., 1940, MA, Mills Coll., 1953, PhD, U. So. Calif., 1951; Dr. Rerum Naturalium, Johannes Gutenberg U., Mainz, Germany, 1960. Diplomate in clin. psychology. Clin. psychologist, behavioral scientist USAF, 1951-65, chief psychologist, Europe, 1956-60; head dept. behavioral sci. U. So. Colo., Pueblo, 1965-78, emeritus, 1978—; staff psychologist Yuma Behavioral Health, Ariz., 1978-82, chief profl. svcs., 1982-83; dir. psychol. svcs. Rio Colo. Health Systems, Yuma, 1983-85; clin psychologist, dir. mental health Ft. Yuma (Calif.) Indian Health Svc., USPHS, 1985-88; exec. dir. Human Sci. Ctr., San Diego, 1982—. Columnist Yuma Daily Sun, 1982-86. Author (tests) symbol arrangementtest, 1952, internat. culture free non-verbal intelligence, 1957, self-other location chart, 1970, test of suffering, 1982; (books) Skills in Living Together, 1983, Conflict Resolution, 1986, A Psychologist Answers Your Questions, 1987, Behind The Therapist's Notes, 1993; plays and video Three Warriors Against Substance Abuse. Named Outstanding prof. U. So. Colo., 1977. Fellow Am. Psychol. Assn. (disting. visitor undergrad. edn. program); mem. AAAS, Deutsche Gesellschaft fur Antropologie, Internat. Assn. Study of Symbols (founder, 1st pres. 1957-61), Japanese Soc. Study KTSA (hon. pres.), Home and Office: PO Box 270169 San Diego CA 92198-2169

KENTFIELD, JOHN ALAN, mechanical engineering educator; b. Hitchin, Eng., Mar. 4, 1930; s. William George and Cecile Lillian (Blackmore) K.; m. Amelia Elizabeth Emmerson, July 9, 1966. BS, U. Southhampton (Eng.), 1959; PhD, U. London, 1963. Registered prof. engr., Alta., Can. Trainee CVA-Kearney and Trecker, Ltd., Brighton, Eng., 1950-52; asst. tester Ricardo and Co., Shoreham, Eng., 1952-56; asst. lectr. Imperial Coll., U. London, 1962-63; project engr. Curtiss-Wright Corp., Woodridge, N.J., 1963-66; lectr. Imperial Coll., U. London, 1966-70; assoc. prof. U. Calgary (Alta., Can.), 1970-78, prof., 1978—; cons. several U.S. and Can. corps., 1976—; mem. assoc. com. on propulsion Nat. Rsch. Coun., Ottawa, Ont., Can., 1983-90; mem. wind-energy tech. adv. com. Energy Mines and Resources, Ottawa, 1985-87. Author: (reference/textbook) Nonsteady 1D, Internal Compressible Flows, 1992; author: (with others) Canadian Encyclopedia, 1985; contbr. 150 articles to profl. jours. including SAE Transactions Jour. of Engines, Transactions of ASME Jour. of Engring., AIAA Jour. of Aircraft. Killiam Resident fellow U. Calgary, 1980; rsch. operating grantee Nat. Scis. and Engring. Rsch. Coun., Ottawa, 1990; recipient Ordinary Nat. Cert. prize Brighton (Eng.) Tech. Coll., 1953, Higher Nat. Cert. prize Brighton Tech. Coll., 1955. Mem. ASME, AIAA, Am. Wind Energy Assn., Can. Wind Energy Assn. (R.J. Templin award 1992). Achievements include patents in field; research demonstrating what is believed to be the world's first gas-turbine equipped with a valveless, pulse, pressure-gain combustor; co-originator of novel form of water-pumping wind-turbine. Home: # 301 1222 Bowness Rd NW, Calgary, AB Canada T2N 3J7 Office: Dept Mech Engring, U Calgary Faculty Engring, Calgary, AB Canada T2N 1N4

KEON, WILBERT JOSEPH, cardiologist, surgeon, educator; b. Sheenboro, Que., Can., May 17, 1935; m. Anne Jennings, July 1, 1960; children—Claudia, Ryan, Neal. B.Sc., St. Patrick's Coll., Ottawa, Ont., Can., 1957; M.D., U. Ottawa, 1961; M.Sc. in Exptl. Surgery, McGill U., Montreal, Que., 1963. Jr. rotating intern Ottawa Civic Hosp., 1961-62, mem. staff, 1969—, chief div. cardiothoracic surgery, 1969—, surgeon-in-chief, 1977-84; gen. surg. resident Montreal Gen. Hosp., 1962-65, chief surg. resident, 1965-66; sr. cardiovascular surgery resident Toronto Gen. Hosp., Ont., 1966-67; sr. resident in cardiovascular surgery Toronto Hosp. for Sick Children, 1967-68; research and clin. assoc. Peter Bent Brigham Hosp. and Harvard Med. Ctr., Boston, 1968-69; practice medicine specializing in cardiac surgery, Ottawa, 1969—; assoc. prof. surgery U. Ottawa, 1969-76, prof., chmn. dept. surgery, 1976—, chmn. div. cardiovascular and thoracic surgery, 1969—, dir. Heart Inst., 1969—, bd. govs., 1979-82; mem. staff Ottawa Gen. Hosp., Children's Hosp. Eastern Ont.; cons., presenter in field; active Can. Heart Found., including vice chmn. med. adv. com. 1979-80, chmn. med. adv. com., 1981-83, bd. dirs., 1981—, med. v.p., 1983-85; mem. Cardiothoracic Manpower Study Group, 1972-83, subcom. chmn., 1979-83; mem. adv. med. bd. Ont. Cancer Treatment and Research Found., 1976—. Editorial adv. bd. Ont. Med. Jour., 1982—; mem. editorial bd. Can. Jour. Surgery, 1983-86. Contbr. chpts., numerous papers, abstracts to profl. pubs. Named Man of Yr., Ottawa Knockers Club, 1973; recipient Staff Research award U. Ottawa, 1975, Outstanding Alumnus award Carleton U., 1977; B'nai B'rith Man of Yr. award, 1985; officer Order of Can., 1985; McLaughlin fellow, 1968; Ont. Heart Found. sr. fellow, 1970-76; James IV Surg. Assn. travelling fellow, 1979. Fellow Royal Coll. Surgeons (Can.), Royal Coll. Physicians and Surgeons Can., ACS, Am. Coll. Cardiology, Council on Clin. Cardiology of Am. Heart Assn.; mem. Ont. Med. Assn., Can. Med. Assn., Acad. Medicine (Ottawa), Can. Cardiovascular Soc. (council 1977—), Am. Heart Assn., Royal Coll. Medicine (London) (affiliate), Internat. Cardiovascular Soc., Am. Assn. Thoracic Surgery, Soc. Thoracic Surgeons, Can. Assn. Clin. Surgeons (v.p. 1980-81, pres. 1981-82), Can. Soc. Clin. Investigation, Soc. Vascular Surgery, Can. Assn. Gen. Surgeons (founding), N. Am. Assn. Surg. Chairmen, Can. Assn. Vascular Surgeons, Assn. Surg. Edn., Internat. Soc. and Fedn. Cardiology, Interam. Soc. Cardiology, Can. Soc. Artificial Organs, Can. Assn. Univ. Tchrs., Am. Coll. Chest Physicians, Can. Assn. Cardiovascular and Thoracic Surgery, Pan Am. Med. Assn. (cardiovascular surgery council), Can. Atherosclerosis Soc. (chartered), Can. Assn. Trauma Surgery, Am. Surg. Soc., Alpha Omega Alpha (hon.). Home: 2298 Bowman Rd, Ottawa, ON Canada K0A 2T0 Office: U Ottawa Heart Inst Ottawa Civic Hosp, 1053 Carling Ave, Ottawa, ON Canada K1Y 4E9*

KEPNER, ROBERT ALLEN, agricultural engineering researcher, educator; b. Los Angeles, May 18, 1915; s. Louis Gilbert and Gertrude (Kennedy) K.; m. Denzil McBride, Sept. 27, 1941; children—Gilbert, Dorothy, Ronald, Harold. Student, Riverside Jr. Coll., 1932-34; B.S. in Agrl. Engring., U. Calif.-Davis, 1937. Registered profl. engr., Calif. Heater test engr. Stewart-Warner Corp., Chgo. and Indpls., 1942-47; assoc. agrl. engr. U. Calif.-Davis, 1937-42, prof., 1947-81, prof. emeritus, 1981—. Author: (with others) Principles of Farm Machinery, 1955, 72, 78. Contbr. articles to profl. jours. Fellow Am. Soc. Agrl. Engrs.; mem. Am. Soc. Engring. Edn., Phi Beta Kappa. Presbyterian. Home: 630 Miller Dr Davis CA 95616-3619

KERAMAS, JAMES GEORGE, engineering educator; b. Athens, Attica, Greece, Oct. 13, 1928; came to U.S., 1955; s. George Anthony and Irene (Poulios) K.; m. Virginia Krea, June 23, 1952; children: George, Renita Keramas Johnson. BSME, Athens Poly. Inst., 1950; MSME, Athens Poly. Inst., Greece, 1952; MEd in Occupational Edn., Fitchburg State Coll., 1978; EdD in Occupational Edn., U. Mass., 1990. Lic. engr. Mass.; cert. vocat. edn. instr., Mass. Design engr. Simplex Wire & Cable Co., Cambridge, Mass., 1955-60; sr. project engr. W.R. Grace Co., Woburn, Mass., 1960-62; dir. engring. Fibersearch Corp., Lawrence, Mass., 1962-63; pres. Alliance Engrs. & Rsch. Corp., Woburn, 1963-71; dir. rsch. Crompton and Knowles Corp., Agawan, Mass., 1971-73; sr. cons. engr. Foster-Miller Assocs., Waltham, Mass., 1973-76, Teledyne Corp., Woburn, 1976-77; assoc. prof. Daniel Webster Coll., Nashua, N.H., 1975-77, Middlesex Community Coll., Bedford, Mass., 1976-85; prof. U. Lowell, Mass., 1984-92, MIT, Cambridge, 1992—; cons. Concord Control, Inc., Boston, 1981-86, Abrasive Products Inc., Braintree, Mass., 1983—. Author: Curriculum Development for High Technology Programs, 1990; patentee High Pile Machine, Rechargeable Battery, Shrink Wrap Machine. Supt. Sun. Sch. Greek Orthodox Ch., Woburn, 1980-85. Lt. Greek Royal Navy, 1953-55. Fellow IEEE; mem. Masons (past master), Am. Hellenic Assn. (named Industrialist 1967). Avocations: writing, travel, swimming, boating, basketball. Office: MIT Dept Engring Rm E32-105 77 Massachusetts Ave Cambridge MA 02139-4307

KERBEL, ROBERT STEPHEN, cell biologist, cancer researcher; b. Toronto, Ont., Can., Apr. 5, 1945; s. Philip and Anne Gertrude (Feldman) K.; B.Sc., U. Toronto, 1968; m. Diane Barbara Smith, Nov. 15, 1970; 1 child, Alyssa Jane. Ph.D., Queen's U., 1972. Nat. Cancer Inst. Can. King George VI Silver Jubilee research fellow Inst. for Cancer Research, London, Eng., 1972-74, Sir Alexander Haddow vis. fellow, 1982; asst. prof. dept. pathology Queen's U., Kingston, Ont., 1975-80, assoc. prof., head cancer research div., 1981-85; head div. cancer research Mt. Sinai Hosp. Research Inst., Toronto, 1985—; head cancer rsch. div., Sunnybrook Health Sci. Ctr., Toronto, 1991—; prof. med. biophysics and med. genetics U. Toronto, 1985—, Tory Family prof. med. oncology, 1993—; research scholar Nat. Cancer Inst. Can., 1975-81, research assoc., 1981—, mem. grants panel B, 1978-81, Terry Fox career scientist, 1984—; mem. pathol. B. study sect. NIH, 1981-87, mem. nat. reviewers registery, 1990—. Chmn. Gordon Rsch. Conf. Cancer, 1991. Recipient Wildleitz award, 1981; Basmajian award Queen's U. Mem. Am. Assn. Immunologists, Can. Assn. Immunologists, Am. Assn. for Cancer Research. Contbr. articles on cancer biology, immunology to profl. jours; editor-in-chief Cancer Metastasis Revs., 1990—, Cancer Research, 1986—, Molecular and Cellular Biology, 1986-89, Invasion and Metastasis, 1981—, Clin. and Exptl. Metastasis, Am. Jour. Pathology, 1993—, Molecular Cell Differentiation, 1993—. Home: 48 Bennington Hts Dr, Toronto, ON Canada M4G 1A9 Office: Sunnybrook Health Sci Centre, Reichmann Rsch Bldg, 2075 Bayview Ave, Toronto, ON Canada M4N 3M5

KERBY, R. C., earth scientist. Recipient Sherritt Hydrometallurgy award Can. Inst. Mining and Metallurgy, 1990. Office: Can Inst Mining and Metallurgy, 3400 de Maisonneuve Blvd W, Montreal, PQ Canada H3Z 3B8*

KERCHNER, HAROLD RICHARD, physicist, researcher; b. Lewistown, Pa., Mar. 5, 1946; s. Harold Frey and Helen Grace (Burris) K.; m. Ruth Diane Fisher, June 16, 1974; children: Geoffrey Allen, Nichole Diane. AB, Harvard U., 1968; MS, U. Ill., 1972, PhD, 1974. Mem. rsch. staff Oak Ridge (Tenn.) Nat. Lab., 1974—; dir. Low Temperature Neutron Irradiation Facility, Oak Ridge, 1986-87. Contbr. papers on superconductivity and radiation effects in materials to sci. publs. With U.S. Army, 1969-71. Mem. Am. Phys. Soc., Materials Rsch. Soc. Lutheran. Office: Oak Ridge Nat Lab Solid State Div Mail Stop 6061 Oak Ridge TN 37831-6061

KERLEY, GERALD IRWIN, physicist; b. Houston, Mar. 23, 1941; s. James Gregory and Eulalia (McGee) K.; m. Donna Carrol Rice, Oct. 20, 1990. BS, Ohio U., 1963; PhD, U. Ill., 1966. Teaching asst. U. Ill., Urbana, 1963-67; physicist Los Alamos (N.Mex.) Nat. Lab., 1969-84, Sandia Nat. Labs., Albuquerque, 1984—. Contbr. over 40 articles to sci. publs. Capt. U.S. Army, 1967-69. Mem. Am. Phys. Soc., Am. Chem. Soc., Am. Trans. Assn. Achievements include developing equations of state for materials, especially at high pressures and temperatures, theories of liquids, chemical equilibrium, phase transitions, explosive and detonation phenomena; developed new techniques for use of material models in hydrodynamic codes and applications to shock wave phenomena in condensed matter. Home: PO Box 13835 Albuquerque NM 87192 Office: Sandia Nat Labs PO Box 5800 Albuquerque NM 87185

KERMAN, ARTHUR KENT, physicist, educator; b. Montreal, May 3, 1929; s. Samuel and Ida (Birn) K.; m. Enid Ehrlich, Dec. 21, 1952; children: Ben, Daniel, Elizabeth, Melissa, James. B.Sc., McGill U., 1950; Ph.D., MIT, 1953. Mem. faculty dept. physics MIT, Cambridge, 1956, prof., 1964—, dir. Ctr. Theoretical Physics, 1976-83, dir. lab. nuclear scis., 1983-92; vis. prof. SUNY-Stony Brook, 1970-71; adj. prof. Brklyn. Coll., 1971-75; cons. Argonne Nat. Lab., 1961-83, mem. sci. and tech. adv. com., 1984-90; cons. Brookhaven Nat. Lab., 1965-81, mem. relativistic heavy ion collider policy com., 1985—, vis. com. 1973-78, chmn. 1977; cons. Lawrence Berkeley Lab., 1975-80, mem. vis. com., 1980-83, chmn. 1981; cons. Lawrence Livermore Lab., 1964—, chmn. physical sci. advisory com. 1992—; cons., Los Almos Sci. Lab, 1961—, mem. physics div. adv. com., 1984—, mem. theo. div. adv. com. 1972—; cons. Nat. Bur. Standards, 1980-81, Oak Ridge Nat. Lab., 1979-85; mem. U. Calif. Pres.'s Sci. and Academic Advisory Com. 1981-92; mem. adv. com. to Office Sci. and Tech. White House Sci. Council, 1982-85, panel on sci. and tech. in govt., 1985, fed. lab. rev. panel, 1982-83; mem. adv. com. Woods Hole Sub-panel of U.S. Dept. Energy, 1982, com. on sci., engring. and pub. policy research briefing panel on sci. frontiers and superconducting super collider Nat. Research Council, 1985, nuclear sci. adv. com. Dept. Energy and NSF, 1982-85; mem. U.S. Dept. Energy Fusion Policy Advisory Com., 1990, mem. U.S. Dept. Energy Inertial Confinement Fusion Advisory Com. 1992—; mem. vis. com. Stanford U. Physics Dept., 1984, Yale U. Physics Dept., 1984, FONDS F.C.A.C. Comite des centres de Recherches pour le Laboratoire de Physique Nucleaire U. Montreal, 1982. Assoc. editor: Rev. Modern Physics, 1968-71. NRC fellow Calif. Inst. Tech., 1953-54, Niels Bohr Inst., Copenhagen, 1954-56; Guggenheim fellow U. Paris, 1961-62. Fellow Am. Phys. Soc. (program com. 1978-79, exec. com. div. nuclear physics 1970-72, pub. com. div. nuclear physics, Tom W. Bonner prize com. 1982-83), Am. Acad. Arts and Scis.; mem. N.Y. Acad. Scis. Office: MIT Dept Physics 6-305 77 Mass Ave Cambridge MA 02139

KERN, HARRY, developmental engineer; b. Velten, Berlin, Germany, Dec. 31, 1942; s. Walter and Gertrud (Stark) K.; m. Ching-Pu Tsai, Dec. 28, 1985; children: Sandra-Dee, Arthur-Tobias. Degree mech. engring., Gewerbeschule, Rheinfelden, Switzerland, 1962. Exec. v.p. Agietron Inc., Addison, Ill., 1973-79; pres. E.D.M. Tech. Inc., Central Islip, N.Y., 1980—. Contbr. articles to profl. jours. Recipient numerous award SME, 1975-79. Achievements include patents on multi wire electrical discharge machine technology, and others; developed world's first multiple fast hole electrical discharge machine technology, other special applications. Office: EDM Tech Inc 405-C Central Ave Bohemia NY 11716

KERNS, ALLEN DENNIS, energy and environmental manager; b. Asheboro, N.C., July 20, 1953. BS in Indsl. Arts. Edn., N.C. State U., Raleigh, 1975. Pre-vocat. tchr. Randolph County Schs., Asheboro, 1976-88, energy and environ. mgr., 1988—; mem. adv. com. for Solid Waste, Asheboro, 1990. Fin. officer 141st Composite Squadron CAP, Ramseur, N.C., 1989—. Mem. Nat. Asbestos Coun., Am. Water Works Assn., Assn. Energy Engrs. (assoc.), N.C. Pub. Sch. Maintenance Assn. (adv. coun. 1990—).

KERNS, JAMES ALBERT, structural engineer; b. Phila., Jan. 3, 1957. BS, Pa. State U., 1981; postgrad., U. Washington, 1982. Profl. engr. Pa., N.Y., Okla. Structural engr. United Engrs. and Constructors, Phila., 1981-84, Martin Marietta Corp., Orlando, Fla., 1984-86; assoc. Smith, Miller and Assocs., Inc., Kingston, Pa., 1986-91; prin. QPROQ Engring., Inc., Wilkes-Barre, Pa., 1991—; panel arbitrator Am. Arbitration Assn., Wilkes-Barre, 1992--. Bd. dirs. Pa. State Alumni Soc., Wilkes-Barre, 1986--; den leader Cub Scouts Am., Kingston, Pa., 1991--. With USN Res., 1975-77. Mem. Pa. Soc. Profl. Engrs. (Disting. Svc. award 1991-92), Am. Soc. Civil Engrs., Am. Legion. Office: QPROQ Engring Inc 33 Beekman St Wilkes Barre PA 18702

KERR, DONALD MACLEAN, JR., physicist; b. Phila., Apr. 8, 1939; s. Donald MacLean and Harriet (Fell) K.; m. Alison Richards Kyle, June 10, 1961; 1 dau., Margot Kyle. B.E.E. (Nat. Merit scholar), Cornell U., 1963, M.S., 1964, Ph.D. (Ford Found. fellow, 1964-65, James Clerk Maxwell fellow 1965-66), 1966. Staff Los Alamos Nat. Lab., 1966-76, group leader, 1971-72, asst. div. leader, 1972-73; exec. v.p., dir. Sci. Applications Internat. Corp., 1973-75; alt. div. leader Los Alamos Nat. Lab., 1975-76; dep. mgr. Nev. ops. office Dept. Energy, Las Vegas, 1976-77; acting asst. sec. def. programs Dept. Energy, Washington, 1978; dep. asst. sec. def. programs Dept. Energy, 1977-79, dep. asst. sec. energy tech., 1979; dir Los Alamos Nat. Lab., 1979-85; sr. v.p. EG&G, Inc., Wellesley, Mass., 1985-88, exec. v.p., 1988-89, pres., 1989-92; exec. v.p. Sci. Applications Internat. Corps., San Diego, 1993—; mem. Navajo Sci. Com., 1974-77; mem. sci. adv. panel U.S. Army, 1975-78; mem. engring. adv. bd. U. Nev., Las Vegas, 1976-78, Cornell U., 1985—; chmn. com. R&D Internat. Energy Agy., 1979-85; mem. nat. security adv. coun. SRI Internat., 1980-89; mem. adv. bd. U. Alaska Geophys. Inst., 1980-85; mem. sci. adv. group Joint Strategic Planning Staff, 1981-91; mem. adv. com. Naval Rsch., 1982-85; mem. corp. Draper Lab., 1982—; mem. adv. bd. Georgetown U. Ctr. Strategic Internat. Studies, 1981-87; bd. dirs. Mirage Systems, Sunnyvale, Calif., Resources for the Future, Washington. Published research on plasma physics, microwave electronics, ionospheric physics, energy and nat. security. Trustee New Eng. Aquarium, 1989—. Fellow AAAS; mem. Am. Phys. Soc., Am. Geophys. Union, Nat. Assn. Mfrs. (bd. dirs. 1986-92), Southwestern Assn. Indian Affairs, World Affairs Coun. Boston (bd. dirs. 1988—), Atlantic Coun. (bd. dirs. 1991—), Cosmos Club (Washington), Sigma Xi, Tau Beta Pi, Eta Kappa Nu. Office: Sci Applications Internat Corp 1241 Cave St La Jolla CA 92037

KERR, FRANK FLOYD, retired electrical engineer; b. Mayfield, Ky., June 28, 1924; s. William Floyd and Ruby Doyle (Trout) K.; m. Minnie Lee Crump, Nov. 28, 1949; 1 child, William Floyd. BSEE, U. Mo., Rolla, 1947. Registered profl. engr., Mo. Engr. elec. substation design Empire Dist. Electric Co., Joplin, Mo., 1947-58, system protection engr., 1958-89; Empire Dist. Electric Co. rep. South Ctrl. Electric Cos. Study Task Force, Mo.-Kans. Power Pool Study Task Force. S.W. Power Pool Transmission Planning Task Force, 1958-89. Mem. IEEE (life), NSPE (life), Mo. Soc. Profl. Engrs. (life). Republican. Methodist. Achievements include conducting of system planning and protection studies; designed and specified protection facilities for EDE Company and their interconnection with other electrical systems. Home: 109 N Connor Joplin MO 64801

KERR, FRANK JOHN, astronomer, educator; b. St. Albans, Eng., Jan. 8, 1918; s. Frank Robison and Myrtle Constance (McMeekin) K.; m. Maureen Parnell, Jan. 7, 1966; children: Gillian Wheeler (dec.), Ian Kerr, Robin Lowry. B.Sc., U. Melbourne, Australia, 1938, M.Sc., 1940, D.Sc., 1962; M.S., Harvard U., 1951. Rsch. scholar U. Melbourne, 1939-40; mem. staff radiophysics lab. Commonwealth Sci. and Indsl. Rsch. Orgn., Sydney, Australia, 1940-68; vis. prof. U. Md., 1966-68, prof., 1968-87, prof. emeritus, 1987—, dir. astronomy program, 1973-78, acting provost div. math. phys. scis. and engring., 1978-79, provost, 1979-85; vis. scientist Leiden U., 1957; vis. prof. U. Tex., 1964, U. Tokyo, 1967; Mem. NSF Adv. Panel Astronomy, 1969-72, chmn., 1971-72. Co-editor: Procs. Internat. Astron. Union Symposia, 1963, 73; Contbr. numerous articles to profl. jours. Trustee Assoc. Univs., Inc., 1981-84; dir. Univs. Space Rsch. Assn. Astronomy Program, 1984-. Fulbright travel grantee, 1950-51; Leverhulme fellow, 1967; NSF research grantee, 1967-83; Guggenheim fellow, 1974-75. Mem. Internat. Astron. Union (pres. commn. 33 1976-79), Am. Astron. Soc. (councillor 1972-75, v.p. 1980-82). Club: Cosmos (Washington). Home: 12601 Davan Dr Silver Spring MD 20904-3504 Office: U Md Astronomy Dept College Park MD 20742

KERR, JAMES WILSON, engineer; b. Balt., May 21, 1921; s. James W. and Laura Virginia (Wright) K.; m. Mary Thomas Montgomery, Feb. 25, 1945 (div., dec.); children: April Kerr Miller, Catherine Kerr Wood (dec.), Wilson, Andrew; m. June Walker, Dec. 27, 1977 (div.); m. Jance White Bain, Jan. 19, 1985. BS with honors, Davidson Coll., 1942; MS, NYU., 1948; postgrad. Freiburg U., 1957-60, Brookings Inst., 1970, 75, Fed. Exec. Inst., 1982; PhD, Kennedy Western U., 1989. Commd. 2d lt. U.S. Army, 1942, advanced through grades to lt. col., 1964; with inf., World War II, Korea; electronic staff, Ft. Bragg, N.C., 1948-51; weapons rsch., N.M., 1953-57; adviser French Army, 1957-60; staff electronics, Ft. Monroe, Va., 1960-62; rsch. mgr., div. dir. CD, Pentagon, 1962-64, as civilian, 1964-81, asst. assoc. dir. Fed. Energy Mgmt. Agy. for Rsch., 1981-85; sr. staff Michael Rogers, Inc., Winter Park, Fla., 1986—; dir. Mt. St. Helen's Tech. Office, 1980; v.p. Latherow & Co., Arlington, Va., 1965-86. Advanced English instr. French Army, 1957-60; cons. Am. Nat. Red Cross Mus., 1968-85, Smithsonian Instn. Dept. Postal History, 1966-85, NSF, 1976-85. Vol. fireman N.Y. State, 1946-48, Fairfax County, Va., 1969—; fire commr. Fairfax County, 1975-81, chmn., 1977-81, Orange County, Fla., 1986—, pres., 1987-90; active Boy Scouts Am., in U.S., Asia and Europe, 1933—; chmn. library bd., Orangeburg, N.Y., 1946-48. Decorated Bronze Star with three oak leaf clusters, Purple Heart; recipient Silver Beaver award Boy Scouts Am., 1956; Fulbright selectee, Japan, 1986; registered profl. engr., Calif. Fellow AAAS, Explorers Club; mem. Nat. Acad. Sci. (various coms. 1962-87), Internat. Assn. Fire Chiefs (chmn. rsch. com. 1969-88, chief sci. adviser 1982-86), Fed. Fire Council, Nat. Fire Protection Assn. (chmn. hosp. disaster com. 1973—), Presdl. Nat. Def. Exec., SAR, Black Forest Mardi Gras (Germany), Nat. Communications Club, Pentagon Officers Athletic Club, IEEE (sr.), Elks, Phi Beta Kappa, Gamma Sigma Epsilon, Delta Phi Alpha. Presbyn. (elder 1963—). Author: Korean-English Phrase Book, 1951; 19th Century Korea Postal Handbook, 1965, 2d edit., 1990. Editor Korean Philately mag., 1971-80, 85—. Contbr. articles to profl. jours. Club: University (Fla.). Home: PO Box 366 Winter Park FL 32790-0366 Office: MR Inc 199 E Welbourne Ave Winter Park FL 32789

KERR, NORBERT LEE, experimental social psychologist, educator; b. Lebanon, Mo., Dec. 10, 1948; s. Otis Leland and Martha Gertrude (Ellmer) K.; m. Jeanne Carol Wald, Aug. 21, 1971; children: Benjamin, Gabriel, Joshua. BA in Physics summa cum laude, Washington U., St. Louis, 1970; MA in Psychology, U. Ill., 1973, PhD, 1974. Asst. prof. psychology U. Calif. San Diego, La Jolla, 1974-79; asst. prof. psychology Mich. State U., East Lansing, 1979-81, assoc. prof., 1981-85, prof., 1985—. Co-author: Group Process, Group Decision, Group Action, 1992; co-editor: The Psychology of the Courtroom, 1982; contbr. articles to Jour. of Personality and Social Psychology, Orgnl. Behavior and Human Performance, Personality and Social Psychology Bull., Jour. Experimental Social Psychology, Law and and Human Behavior, others. Recipient grants from NIMH, NSF, others. Fellow APA, Am. Psychol. Soc., Soc. Personality and Social Psychology; mem Am. Psychology-Law Soc., European Assn. Exptl. Social Psychology (affiliate), Midwestern Psychol. Soc., Soc. Exptl. Social Psychology, Soc. for Psychol. Study of Social Issues, Phi Beta Kappa. Office: Mich Stae U Dept Psychology East Lansing MI 48824

KERR, PETER DONALD, geography educator emeritus; b. Toronto, Apr. 19, 1920. BA in Geography, U. British Columbia, 1941; MA in Geography, U. Toronto, 1943; postgrad., U. Calif., Berkeley, 1945-46; PhD in Geography, U. Toronto, 1950. Teaching asst. U. Calif., Berkeley, 1945-46; lectr. dept. geography U. Toronto, 1946-51, from asst. prof. to assoc. prof. dept. geography, 1951-62, prof. dept. geography, 1962-85, prof. emeritus dept. geography, 1985—; assoc. dean Sch. Grad. Studies U. Toronto, 1976-79, chmn. dept. geography, 1968-73; participant Pugwash Internat. Conf. chem. and biol. warfare, 1959; mem. Nat. Adv. Com. Geog. Rsch., 1067-70, publs. com. Social Sci. Rsch. Coun. Can., 1974-77; mem. exec. com. Hist. Atlas of Can., 1980—. Author: (with W.G. Kendrew) The Climate of British Columbia and the Yukon Territory, 1955, (with D.F. Putman) A Regional Geography of Canada, 1956, (with Jacob Spelt) The Changing Face of Toronto - A Study in Urban Geography, 1965, (with N.C. Field) Geographical Aspects of Industrial Growth in the Metropolitan Toronto Region, 1968; author: (with others) Canadian Regions, 1952, Toronto, 1973, Essays on World Urbanization, 1975, The Settlement of the West, 1977, Heartland and Hinterland: A Geography of Canada, 1987; sr. editor: Historical Atlas of Canada, 1990; contbr. articles to profl. jours. Meteorol. officer RCAF, 1943-45. Grantee Def. Rsch. Bd., 1946-49, Dept. Mines and Tech. Surveys, 1955-61, Emergency Measures Orgn., 1962-64, Dept. Treasury and Econs., 1966-68, Can. Coun., 1973-75, Hist. Atlas of Can. Project, 1980-89. Mem. Assn. Am. Geographers (chmn. local arrangements Toronto meetings 1966, honors com. 1976-77), Can. Assn. Geographers (councillor 1955-58, v.p. 1959, pres. 1960, CAG award for Service to the Profession of Geography, 1992), Internat. Geog. Union (Can. com. 1961-68, processes and patterns urbanization com. 1968-74), Sigma Xi. Achievements include research on climatology of British Columbia, industrial and urban geography of southern Ont., especially Metro Toronto, land use and industrial change in Metropolitan Toronto, changing wholesale trade in Winnipeg, long distance telephone communications, 1880-1930, changing status of Canadian ports, 1891-1961, various aspects of wholesale and retail trade in Canada, 1891-1961, resource development in Canada, 1946-61. Office: U of Toronto/Dept of Geography, 100 St George St, Toronto, ON Canada M5S 1A1

KERR, SANDRA LEE, psychology educator; b. Erie, Pa., Nov. 11, 1952; d. Elmer Norbert and Josephine Marie (Leone) Yacobozzi; m. Andre Stephen Kerr, Aug. 22, 1988. BA, Boston Coll., 1981; MA, SUNY, Stonybrook, 1989, PhD, 1991. Asst. prof. Alvernia Coll., Reading, Pa., 1991-92, No. Ill. U., DeKalb, 1992—. Contbr. articles to Jour. Abnormal Psychology, Psychotherapy Rsch. Jour. Mem. APA, Am. Psychol. Soc., Midwestern Psychol. Assn. Achievements include rsch. in emotion perception in schizophrenia, interpersonal and intrapersonal focus in cognitive therapy and psychodynamic interpersonal theories. Office: No Ill U De Kalb IL 60115

KERR, THOMAS ANDREW, senior program engineer; b. Seattle, Feb. 26, 1953; s. Gerald Dale and Maureen Eilish (Doherty) K.; m. Susan Myra Kellogg, May 18, 1980. Student, N.Mex. Tech., 1970-71, U. N.C., 1986-87. Tng. supr. Chem-Nuclear Systems, Inc., Barnwell, S.C., 1977-84; assoc. instr. Duke Power Co., Charlotte, N.C., 1984-87; chief low level radioactive waste mgmt. Ill. Dept. Nuclear Safety, Springfield, 1987-90; sr. program engr. nat. low level radioactive waste mgmt. EG&G Idaho, Inc., Idaho Falls, 1990—. Contbr. articles to N.Y. Acad. Medicine Bulletin, 1988, Waste Mgmt. '89, ASME Internat. Waste Mgmt. Conf., 1991. With USN, 1973-77. Mem. Am. Nuclear Soc., Nat. Environ. Tng. Assn. (cert. environ. trainer).

Achievements include design and implementation of program to site new low-level radioactive waste disposal facility in Illinois. Home: 9673 S Ammon Rd Idaho Falls ID 83406-8311 Office: EG&G Idaho Inc PO Box 1625 Idaho Falls ID 83415-2420

KERSEY, TERRY L(EE), astronautical engineer; b. San Francisco, June 9, 1947; s. Ida Helen (Schmeichel) K. Houseman, orderly Mills Meml. Hosp., San Mateo, Calif., 1965-68; security guard Lawrence Security, San Francisco, 1973-74; electronic engr. and technician engring. research and devel. dept. McCulloch Corp., Los Angeles, 1977; warehouseman C.C.H. Computax Co., Redondo Beach, Calif., 1977-78; with material ops. and planning customer support dept. Allied-Signal Aerospace Co., Torrance, Calif., 1978-91; security guard Guardsmark Inc., L.A., 1993; electronic technician J. W. Griffin, Venice, Calif., 1993—. Participant 9th Space Simulation conf., Los Angeles, 1977, 31st Internat. Astronautical Fedn. Congress., Tokyo, 1980, Unispace 1982 for the U.N., Vienna. Served to sgt. USAF, 1968-72, Vietnam. Decorated Vietnam Service medal with 2 bronze stars, Republic of Vietnam Campaign medal, Air Force commendation medal for Vietnam campaign Service. Mem. AAAS, AIAA (mem. space systems tech. com. 1981—, mem. aerodynamics com. 1980—, Wright Flyer Project Aerodynamics com. 1980—, pub. policy com. 1989—), Nat. Space Inst., Am. Astronautical Soc., The Planetary Soc., Internat. L5 Soc., Ind. Space Rsch. Group, IEEE Computer Soc., Space Studies Inst. (sr. assoc.). Zen Buddhist. Avocations: computers, sports, astronomy, science fiction literature.

KERSHAW, CAROL JEAN, psychologist; b. New Orleans, Apr. 11, 1947; d. Neal Howard and Gloria Jackson (Moss) Perkins; m. John William Wade, Aug. 20, 1983; stepchildren: Chris Wade, Stephen Wade, Tiffany Wade. BS in Secondary Edn., U. Tex., 1969; MS in Speech Communication, North Tex. State U., 1971, MEd in Counseling, 1976; EdD in Counseling, East Tex. State U., 1979. Lic. psychologist, Tex. Assoc. prof. DeVry Inst., Dallas, 1971-73; instr., counseling psychologist East Tex. State U., Commerce, 1976-78; counselor, instr. Tarrant County Jr. Coll., Hurst, Tex., 1971-74; dir. spl. svcs. Goodwill Industries, Dallas, 1974-76; marriage and family therapist, cons. mental health clinic Tex. Dept. Mental Health and Retardation, Greenville, 1977-79; asst. prof., dir. grad. program in marriage & family therapy Tex. Woman's U., Denton, 1980-83; coord. child devel. dept. Tex. Woman's U., Houston, 1983-88; pvt. practice Inst. for Family Psychology, Houston, 1986—; co-dir. Milton H. Erickson Inst. Houston, 1986—; bd. dirs. Milton H. Erickson Inst. Tex., Houston, 1986—; internat. presenter in field. Author: Therapeutic Metaphor in the Treatment of Childhood Asthma: A Systemic Approach, Ericksonian Monographs, Vol. 2, 1986, The Couple's Hypnotic Dance, 1991; co-author: Psychotherapeutic Techniques in School Psychology, 1984, Learning to Think for an Organ, Bridges of the Bodymind, 1980. Sec. Tex. Assn. for Marriage and Family Therapy, 1978-80. Recipient Visionary award, Meritorious Svc. award Tex. Assn. for Marriage & Family Therapy, 1980. Mem. Am. Psychol. Assn, Am. Assn. for Marriage and Family Therapy (clin., approved supr.), Soc. for Exptl. & Clin. Hypnosis, Am. Soc. for Clin. Hypnosis (assoc.), Internat. Soc. for Clin. & Exptl. Hypnosis, Psi Chi. Democrat. Methodist. Avocations: painting, reading, exercise, writing, singing. Office: Inst for Family Psychology 2012 Bissonnet St Houston TX 77005-1647

KERSTEN, ROBERT DONAVON, engineering educator, consultant; b. Carlinville, Ill., Jan. 30, 1927; s. Frederick Wilhelm and Buelah Louise (Surber) K.; m. Bonita Sue McCool, May 13, 1950; children: Susan, John. BS, MS, Okla. State U., 1956; PhD, Northwestern U., 1961. Profl. engr., Ariz., Fla., Okla. Asst. prof. engring. Okla. State U., Stillwater, 1953-56, 57-58; asst. prof. engring. Ariz. State U., Tempe, 1956-57, assoc. prof., then prof., chmn. civil engring., 1958-68; dir. univ. rsch. U. Cen. Fla., Orlando, 1968-69, dean engring., 1968-87, prof. engring., 1987—; mem., chmn. Fla. Bd. Profl. Engrs., Tallahassee, 1980-87, Engring. Accreditation Commn., N.Y.C., 1982-89; mem. Flood Control Adv. Bd., Phoenix, 1965-66. Author: Engineering Differential Systems, 1969; contbr. over 50 articles to profl. jours. Trustee Scottsdale (Ariz.) Bapt. Hosp., 1967-68, Polit. Action Com. NSPE, Washington, 1984-85; mem. Gov.'s Solar Energy Task Force, Tallahassee, 1975-77. With USNR, 1945-47. Faculty fellow Standard Oil Found., 1959-60; Royal E. Cabell fellow Northwestern U., 1960. Fellow ASCE, AAAS, Fla. Engring. Soc. (v.p.); mem. Pan Am. Union Engring. Socs. (engring. edn. com., v.p. 1991—), U.S. Coun. Internat. Engring. Practice (v.p. 1991—, bd. dirs.), Nat. Coun. Engring. Examiners, Rotary (pres. Orange County club 1975-76, Paul Harris fellow 1979), Sigma Xi, Tau Beta Pi, Phi Kappa Phi. Republican. Baptist. Home: 590 Dommerich Dr Maitland FL 32751 Office: U Cen Fla PO Box 25000 Orlando FL 32816-2450

KERWIN, COURTNEY MICHAEL, public health scientist, administrator; b. Kokomo, Ind., Dec. 12, 1944; s. Courtney M. and Helen Marie (Hartley) K.; divorced; children: Lisa Marie, Courtney Matthew. BS in Chemistry, Ill. Inst. Tech., 1967; PhD in Organic Chemistry, Mich. State U., 1972; MPH in Health Policy, Johns Hopkins U., 1993. Analytical chemist OSHA, Salt Lake City, 1973-74; med. quality assurance chemist Dept. Def., Richmond, Va., 1974-76; field investigator FDA, Balt., 1976-77; consumer safety officer FDA, Rockville, Md., 1977-84; health sci. adminstr., sci. rev. adminstr. Nat. Cancer Inst. NIH, Rockville, Md., 1984—. Editor Drug Shortage Newsletter. Pres. St. Paul's Luth. Ch. Coun., Walkersville, Md. chmn. evangel. com., del. to synod meetings, 1980-85; pres. Frederick (Md.) Area Computer Enthusiasts, 1987—, editor newsletter Perceptions of Frederick, 1983-86; chmn. archtl. control com. Discovery Devel., Walkersville, MD., 1980-82. Mem. Am. Chem. Soc., Internat. Assn. for Vitamin and Nutritional Oncology, Am. Volksports Assn. Mem. Evang. Luth. Ch. Am. Office: Nat Cancer Inst NIH Exec Pla N Rockville MD 20892

KERWIN, JOSEPH PETER, physician, former astronaut; b. Oak Park, Ill., Feb. 19, 1932; m. Shirley Ann Good; children: Sharon, Joanna, Kristina. B.A., Coll. Holy Cross, 1953; M.D., Northwestern U., 1957. Flight surgeon USN, 1959, aviator, 1962; astronaut, 1965; mem. Skylab 2 Crew, 1973; rep. Australia NASA, 1982-83; dir. Space-Life Sci., 1984-87; mgr. EVA systems Lockheed Missiles & Space Co., 1987—, mgr. Houston Manned Programs, 1990—. Address: Lockheed Missiles & Space Co 1150 Gemini A23 Houston TX 77058

KESKA, JERRY KAZIMIERZ, mechanical engineering educator; b. Wilkowisko, Poland, Feb. 14, 1945; came to U.S., 1984, naturalized, 1990; s. Jan and Stefania (Kawecka) K.; m. Jadwiga T. Sadowy, Sept. 11, 1966; children: Agnieszka, Marek. BSME, MS, State U. Krakow (Poland), 1970, PhD, 1974. Rsch. and teaching asst. State U. Krakow, 1965-70, lectr., 1970-74, asst. prof. Coll. Engring., 1974-75, assoc. prof. Lab. Multiphase Flow and Instrumentation, 1975-81; staff sr. scientist dept. mech. engring. State U. Karlsruhe (Germany), 1981-84; sr. project engr. rsch. dept. Copeland Corp., Sidney, Ohio, 1984-87; sr. systems engr. rsch. div. Technicon Instruments Corp., Tarrytown, N.Y., 1987-88; assoc. prof. mech. engring. dept. U. Nebr., Lincoln, 1988-92; sr. rsch. engr. Pacific N.W. Labs., Battelle, Richland, Wash., 1992—; vis. prof. dept. mech. engring. State U. Karlsruhe, 1975-76, Franzius Inst., State U. Hannover (Germany), 1980-81; lab. head Lab. Multiphase Flow and Instrumentation, Krakow, 1975-81; prin. investigator U. Nebr., Lincoln, 1988-92; chmn. internat. symposium Slurry Transport, Krakow, 1979. Author: (with others) Slurry Pipelines, 1976, Creation of Underground Cavern, 1978, Engineering Handbook, Sect. 8, Vol. 2, 1980; contbr. monographs and articles to profl. jours.; patentee in field. Deutsches Akademische Austauschdienst fellow, Bad Goesberg, Germany, 1975, Mina-James-Heinemann fellow, Hannover, Germany, 1980; recipient Gold medal Internat. Invention Exhbn., Brno, Czechoslovakia, 1980, Summer Faculty rsch. award Argonne Nat. Lab., 1990. Mem. ASME, Am. Soc. Engring. Edn., Am. Inst. Chem. Engrs., Am. Soc. Civil Engrs. Avocations: 5 languages, mycology, fly fishing, skiing, tennis. Home: PO Box 573 Richland WA 99352 also: 2348 Hood Ave # 3 Richland WA 99352 Office: Battelle PNL Battelle Blvd Richland WA 99352

KESKINER, ALI, psychiatrist; b. Kirsiher, Turkey, Mar. 10, 1929; came to the U.S., 1963; s. Mustafa and Ayse (Memis) K.; m. Lynne E. Hirz, Oct. 18, 1968 (div. 1982); children: Murad A., Aydin D. MD, Istanbul U., 1955; diploma in psychiatry, McGill U., 1962. Sr. intern psychiatry St. Anne's sect. Queen Mary Vets. Hosp., Montreal, Que., Can., 1958-59, resident psychiatry, 1959-60, sr. asst. resident psychiatry, 1960-61; sr. asst. resident in psychiatry Montreal Children's Hosp., 1961-62; asst. resident medicine-neurology Univ. Hosp., U. Sask., Saksatoon, Sask., Can., 1962-63; sr. rsch. scientist Mo. Inst. Psychiatry U. Mo., St. Louis, 1963-68, prof. psychiatry, 1964-76, chief behavioral rsch., 1968-75; staff psychiatrist, dir. rsch. Anclote Manor Hosp., Tarpon Springs, Fla., 1975-83; dir. rsch., staff psychiatrist VA Med. Ctr., Bay Pines, Fla., 1984-87, chief psychiatry svcs., 1988—; prof. psychiatry U. South Fla. Coll. Medicine, Tampa, 1984—; mem. rsch. com. U. South Fla. Dept. Psychiatry, Tampa, 1989—; mem. exec. com. VA Med. Ctr., 1992—; mem. sr. adv. com. Mo. Inst. Psychiatry, St. Louis, 1970-75. Contbr. chpt. to Therapeutic Studies in Therapy Resistant Schizophrenia, 1966-74; contbr. articles to profl. jours. Capt. Turkish Army, 1956-58. Grantee NIH, 1970-74; recipient Golden Eagle award Coun. on Internat. Non-Theatrical Events, 1975, 1st Place award in community mental health Nat. Mental Health Assn., 1977. Fellow Am. Psychiat. Assn.; mem. Fla. Psychiat. Soc. (chair various coms. 1984—), Turkish Am. Neuropsychiat. Assn. (pres. 1981-82). Office: VA Med Ctr Bay Pines FL 33504

KESSEL, BRINA, educator, ornithologist; b. Ithaca, N.Y., Nov. 20, 1925; d. Marcel and Quinta (Cattell) K.; m. Raymond B. Roof, June 19, 1957 (dec. 1968). B.S. (Albert R. Brand Bird Song Found. scholar), Cornell U., 1947, Ph.D., 1951; M.S. (Wis. Alumni Research Found. fellow), U. Wis.-Madison, 1949. Student asst. Patuxent Research Refuge, 1946; student teaching asst. Cornell U., 1945-47, grad. asst., 1947-48, 49-51; instr. biol. sci. U. Alaska, summer 1951, asst. prof. biol. sci., 1951-54, assoc. prof. zoology, 1954-59, prof. zoology, 1959—, head dept. biol. scis., 1957-66; dean U. Alaska (Coll. Biol. Scis. and Renewable Resources), 1961-72, curator terrestrial vertebrate mus. collections, 1972-90, curator ornithology collection, 1990—, adminstrv. asso. for acad. programs, grad. and undergrad., dir. acad. advising, office of chancellor, 1973-80; project dir. U. Alaska ecol. investigation for AEC Project Chariot, 1959-63; ornithol. investigations NW Alaska pipeline, 1976-81, Susitna Hydroelectric Project, 1980-83. Author book, monographs; contbr. articles to profl. jours. Fellow AAAS, Am. Ornithologists' Union (v.p. 1977, pres.-elect 1990-92, pres. 1992—), Arctic Inst. N.Am.; mem. Wilson, Cooper ornith. socs., Soc. for Northwestern Vertebrate Biology, Pacific Seabird Group, Assn. Field Ornithologists, Sigma Xi (pres. U. Alaska 1957), Phi Kappa Phi, Sigma Delta Epsilon. Office: U Alaska Mus PO Box 80211 Fairbanks AK 99708-0211

KESSENICH, KARL OTTO, aerospace engineer; b. Waco, Tex., Oct. 26, 1963; s. Jerome Otto and Geraldine Edna (Snydes) K.; m. Cynthia Elizabeth Strange, Oct. 15, 1988. BS in Aerospace Engring., U. Va., 1985. Sr. tech. staff Veda Inc., Arlington, Va., 1985—. Pres. Lafayette Forest Homeowners Assn., Annandale, Va., 1987-91. Mem. AIAA, Am. Def. Preparedness Assn., Tex. State Soc. Home: 306 N Nelson St Arlington VA 22201

KESSLER, DAVID A., health services commissioner; b. N.Y.C., May 31, 1951; married; 2 children. BA, Amherst Coll., 1973; JD, U. Chgo., 1978; MD, Harvard U., 1979; APC, NYU Sch. Bus. Food and drug law Columbia U. Sch. of Law; med. dir. Einstein-Montefiore Hosp., N.Y.C.; commr. FDA Dept. Health and Human Svcs., Rockville, Md., 1990—; mem. adv. commn. FDA; cons. U.S. Senate Labor and Human Resources com.; assoc. prof. pediatrics, Einstein Med. Sch., lectr. Office: Dept Health and Human Svcs FDA 5600 Fishers Ln Rockville MD 20857-0001

KESSLER, DIETRICH, biology educator; b. Hamilton, N.Y., May 28, 1936; s. William Conrad and Helga Martha Elizabeth (Wolfram) K.; m. Johanna Winterwerp Prins, Apr. 14, 1990; children from previous marriage: Jonathan Farley, Melissa Beth. BA with high honors, Swarthmore Coll., 1958; MS, U. Wis., 1960, PhD, 1964. Asst. prof. biology Haverford (Pa.) Coll., 1964-70, assoc. prof., 1970-77, prof. biology, 1977-84; prof., chmn. biology Colgate U., Hamilton, N.Y., 1984-90, prof., 1990—; vis. rsch. fellow dept. genetics U. Leicester, Eng., 1990-91; vis. prof. McArdle Lab. for Cancer Rsch. U. Wis., Madison, 1988; mem. rev. panel NSF Faculty Enhancement Program Proposals, 1990. Recipient Fulbright Research award U. Bonn, 1982-83; postdoctoral fellow Am. Cancer Soc. Brandeis U., 1966-67; NSF sci. faculty fellow Swiss Inst. for Cancer Research, Lausanne, 1971-72. Mem. Am. Soc. for Cell Biology, Mid. State Assn. Coll. and Secondary Schs. (evaluation teams for commn. on higher edn., various locations 1971—), Coun. for Internat. Exch. of Scholars (area adv. com. for We. Europe, subcom. for Austria/Federal Republic of Germany 1988-90), Phi Beta Kappa. Office: Colgate Univ Dept of Biology Hamilton NY 13346

KESSLER, DONNA KAY ENS, mathematics educator; b. St. Louis, Nov. 16, 1953; d. Milton A. and Anna Mae (Kerner) Ens; m. William D. Kessler, June 16, 1984. BA in Math., Cen. Meth. Coll., 1975; MEd, Webster U., 1979. Math. tchr. Ferguson-Florissant (Mo.) Sch. Dist., 1975-78, Francis Howell Sch. Dist., St. Charles, Mo., 1978-88, No. Callaway Sch. Dist., Kingdom City, Mo., 1988-92; instr. Moberly (Mo.) Area C.C., 1992—; instr. Webster U. Continuing Edn., St. Louis, 1988; cons. Parkway S. Jr. High Sch. Math Dept., Chesterfield, Mo., 1977. Bd. dirs. Ptnrs. in Edn. com., Kingdom City; sponsor Nat. Honor Soc., St. Charles, 1986-88, Kingdom City, 1988-92; mem. Presbyn. Women, Mexico, Mo., 1989—. Mem. NEA, Nat. Coun. Tchrs. Math., Francis Howell Community Tchrs. Assn. (chairperson profl. devel. com., 1983-84). Avocations: reading, crafts, Christian music. Home: RR 3 Box 293A Mexico MO 65265-9803

KESSLER, JOHN OTTO, physicist, educator; b. Vienna, Austria, Nov. 26, 1928; came to U.S., 1940, naturalized, 1946; s. Jacques and Alice Blanca (Neuhut) K.; m. Eva M. Bondy, Sept. 9, 1950; children: Helen J., Steven J. A.B., Columbia U., 1949, Ph.D., 1953. With RCA Corp., Princeton, N.J., 1952-66; sr. mem. tech. staff RCA Corp., 1960-66, mgr. grad. recruiting, 1964-66; prof. physics U. Ariz., Tucson, 1966—; Vis. research asso. Princeton, 1962-64; sr. vis. fellow, vis. prof. physics U. Leeds, Eng., 1972-73, sr. vis. fellow, 1990-91; vis. prof. Technische Hogeschool Delft, Netherlands, spring 1979; Fulbright fellow dept. applied math. and theoretical physics Cambridge U., Eng., 1983-84. Contbr. articles to tech. jours.; patentee bioconvection and consumption patterns of micro-organism population. Fellow AAAS; mem. Phycological Soc. Am., Am. Phys. Soc., Am. Soc. for Gravitational and Space Biology, European Low Gravity Rsch. Assn. Home: 2740 E Camino La Zorrela Tucson AZ 85718-3126 Office: U Ariz Physics Dept Bldg 81 Tucson AZ 85721

KESSLER, WAYNE VINCENT, health sciences educator, researcher, consultant; b. Milo, Iowa, Jan. 10, 1933; s. Joseph Edward and Genevieve (Frueh) K.; m. Olive Beatrice Buremaster, Sept., 10, 1953; children: Katherine Marie, Karl Matthew. B.S., N.D. State U., 1955, M.S., 1956; Ph.D., Purdue U., 1959. Asst. prof. pharm. chemistry N.D. State U., Fargo, 1959-60; asst. prof. health physics Purdue U., W. Lafayette, Ind., 1960-64, assoc. prof., 1964-68, prof. bionucleonics, 1968-79, prof. health scis., 1979—; cons. Mead Johnson Inc., Evansville, Ind., 1968-69, Miles Labs., Elkart, Ind., 1970. Author: Cadmium Toxicity, 1974. Purdue Research Found. fellow, 1957. Fellow AAAS, Acad. Pharm. Scis., Phi Kappa Phi; mem. Health Physics Soc. Presbyterian. Avocations: woodworking; traveling. Home: 2825 Forest Ln Lafayette IN 47904-2427 Office: Purdue U Sch Health Scis 1338 Civil Engring Bldg West Lafayette IN 47907-1338

KESTER, DALE EMMERT, pomologist, educator; b. Audubon, Iowa, July 28, 1922; s. Raymond and Fannie (Ditzenberger) K.; m. Daphne Dougherty; children: William Raymond, Nancy Inman. BS in Horticulture, Iowa State Coll., 1947; MS in Horticulture, U. Calif., Davis, 1949, PhD in Plant Physiology, 1951. Rsch. asst. dept pomology U. Calif., Davis, 1947-51, lectr., jr. pomologist, 1951-53, asst. prof., asst. pomologist, 1953-60, assoc. prof., associate pomologist, 1960-69, prof., pomologist, 1969-91, prof. emeritus, 1991—; vis. scholar dept. genetics U. Wis., Madison, 1962-63, Volcanic Rsch. Inst., Bet Dagan, Israel, 1975. Author: (with H. Hartmann) Plant Propagation: Principles and Practices, 1959, 5th revised edit., 1990; contbr. numerous articles to profl. and popular publs. 1st lt. USAF, 1943-45, ETO. Fellow Am. Soc. Hort. Sci. (Stark award 1980); mem. Internat. Plant Propagators Soc. (sec. 1961), Phytopath. Soc., Alpha Zeta, Gamma Sigma Delta, Phi Beta Kappa, Pi Alpha Xi. Republican. Presbyn. Achievements include introduction of 5 almond cultivars and 2 almond rootstocks. Home: 750 Anderson Rd Davis CA 95616-3511 Office: U Calif Dept Pomology Davis CA 95616

KESTIGIAN, MICHAEL, scientist; b. Charlton, Mass., Sept. 1, 1928; s. Vartan and Vartanoush Rose (Topelian) K.; m. Jean A. French, Aug. 21, 1949; children: Michael Brian, Mark Craig. BS with honors, U. Mass., 1952; MS, U. Conn., 1954, PhD, 1956. Teaching asst. U. Mass., Amherst, 1951-52; teaching asst. U. Conn., Storrs, 1952-53, rsch. asst., 1953-56; chemist E.I. du Pont de Nemours Inc., Wilmington, Del., 1956-58; tech. staff RCA Labs., Princeton, N.J., 1958-62; dept. mgr. Sperry Rsch. Ctr., Sudbury, Mass., 1962-83; sr. scientist Loral Infrared and Imaging Systems, Lexington, Mass., 1989—; pres. HYE Tech. Assn., Charlton, 1983—. Contbr. articles to Jour. Electrochem. Soc., Jour. Crystal Growth, numerous others. Lay min. Meth. Ch., Charlton, 1985—, lay leader, 1990-91; mem. Hopewell (N.J.) Twp. Bd. Edn., 1959-62; mem. sch. bd. Nashoba Regional High Sch., Bolton, Mass., 1969-72. Sgt. U.S. Army, 1946-48, Korea. Recipient David Sarnoff Lab. Achievement award, 1960, Engring. award Honeywell Corp., 1991; Lotta Crabtree scholar U. Mass., 1951. Fellow Am. Ceramic Soc. (exec. com. 1968-80), Am. Inst. Chemists; mem. Am. Crystal Growth Assn. (charter, exec. com. 1970-90), Electrochem. Soc. (treas. 1980s). Achievements include patents for single crystal growth of solid state lasers, magneto-optic, magnetic non-linear optic, Q switch and ferro electric materials; first to grow garnet oxide laser materials from direct melt by the Czochralski technique. Home: 28 Curtis Hill Rd Charlton Depot MA 01509 Office: Loral Infrared and Imaging 2 Forbes Rd MS 146 Lexington MA 02173

KESTLE, WENDELL RUSSELL, cost and economic analyst, consultant; b. Casper, Wyo., July 23, 1935; s. Philip Clayton and Ruby Maxine (Clifton) K.; m. Anne Marie Joujon-Roche, Nov. 18, 1961; children: Martha Anne, Joan Marie, Wendell Russell Jr. BA in Econs., Calif. State U., Northridge, 1961. Cost engr. Rand Corp., Santa Monica, Calif., 1961-63; mem. ops. rsch. staff Lockheed Aircraft Co., Burbank, Calif., 1963-65; mem. tech. staff Hughes Aircraft Co., El Segundo, Calif., 1965-66; sr. rsch. staff assoc. Northrop Corp., Hawthorne, Calif., 1966-69; sr. scientist Booz-Allen Applied Rsch., Inglewood, Calif., 1969-70; mgr. TRW, Inc., Redondo Beach, Calif., 1970-71; pvt. practice Carson, Calif., 1971-72; project mgr. PRC, Inc., Huntsville, Ala., 1972—; cons. in field, Washington. Author, editor sci. documents (4) 1987—. With USN, 1952-56, Korea. Recipient Letters of Commendation, NASA, 1976, 80, 92. Mem. AAAS, Am. Econ. Assn., Ops. Rsch. Soc. of Am., Nat. Space Soc., Planetary Soc., Bot. Garden Soc. of Huntsville, Historic Huntsville Soc., Friends of the Library, Am. Legion. Roman Catholic. Home and Office: 7904 Seville Dr Huntsville AL 35802

KETCHERSID, WAYNE LESTER, JR., medical technologist; b. Seattle, Oct. 16, 1946; s. Wayne Lester and Hazel May (Greene) K.; m. Wilette LaVerne Mautz, Oct. 6, 1972; 1 son, William Les. BS in Biology, Pacific Luth. U., 1976, BS in Med. Tech., 1978; MS in Adminstrn., Cen. Mich. U., 1990. Cert. med. technologist; cert. clin. lab. dir. Nat. Cert. Agy. for Med. Lab. Pers. Staff technologist Tacoma Gen. Hosp., 1978-79, chemistry supr., 1979-81, head chemistry, 1981-83; head chemistry Multicare Med. Ctr., 1984-86, mgr., 1986—. Mem. Nat. Rep. Com.; bd. trustees Polit. Action Com., 1991—. Served with U.S. Army, 1966-68. William E. Slaughter Found. scholar, 1975-76. Mem. Am. Assn. Clin. Chemistry, Am. Hosp. Assn., Am. Soc. Med. Tech. (cert., chmn. region IX adminstrn. 1984—, nat. del. 1984—, vice chmn. govt. affairs com. 1991-92, chmn. 1992—, nominee Mem. of Yr. 1992), Wash. State Soc. Med. Tech. (chmn. biochemistry sect. 1983-86, dist. pres. 1986—, cert. merit 1983, 84, 86, 88, pres. 1988-89, 89-90, mem. of the yr., 1990, chmn. govt. affairs com. 1991-92, chmn. 1992—), Am. Coll. Health Care Execs., Am. Soc. Clin. Pathologists (med. technolgist), N.W. Med. lab. Symposium (chmn. 1986-88, 90, 92). Lutheran. Contbr. articles to profl. jours. Office: Multicare Med Ctr 315 S K St Tacoma WA 98405-4234

KEVERN, NILES RUSSELL, aquatic ecologist, educator; b. Elizabeth, Ill., May 15, 1931; s. Russell William and Mary Alice (Cook) K.; m. Kathryn Kay Williams, Aug. 20, 1955; children: Michael, Tamara, JoAnn. BS, U. Mont., 1958; MS, Mich. State U., 1961, PhD, 1963. Radiation ecologist Oak Ridge (Tenn.) Nat. Labs., 1962-66; asst. prof., dept. fisheries and wildlife Mich. State U., East Lansing, 1966-67, assoc. prof., 1967-69, prof., 1969—, chmn. dept., 1969-92; assoc. dir. Inst. Water Research, Mich. State U., East Lansing, 1967-69, Mich. Sea Grant Coll. Program, 1977—. Contbr. numerous articles to profl. jours. Served to sgt. USAF, 1951-55. Mem. Internat. Assn. Great Lakes Research, Am. Fisheries Soc., Wildlife Soc., Mich. Acad. Sci., Sigma Xi, Phi Kappa Phi. Roman Catholic. Avocations: fishing, hunting, boating. Home: 1733 Ann St East Lansing MI 48823-3705 Office: Mich State U Dept Fisheries & Wildlife Nat Resources Bldg East Lansing MI 48824-1222

KEVREKIDIS, YANNIS GEORGE, chemical engineer; b. Athens, Greece, Mar. 26, 1959; came to U.S., 1981; s. George Yannis and Despoina (Bastea) K.; m. Fotini A. Zervaki, Mar. 1983. MA in Math., U. Minn., 1986, PhD in Chem. Engring., 1986. Registered profl. engr., Greece. Postdoctoral fellow Los Alamos (N.Mex.) Nat. Lab., 1985-86; asst. prof. Princeton (N.J.) Univ., 1986-91; assoc. prof. Princeton U., 1991—; sr. faculty applied and computational math. program Princeton U., 1993—; cons. Los Alamos Nat. Lab., 1988—. Assoc. editor (jours.) Nonlinear Sci., 1989—, Internat. Jour. Bifurcations & Chaos, 1990—. Named Packard Found. fellow David & Lucile Packard Foun., 1988; recipient NSF Presdl. Young Investigator award, 1989. Mem. Am. Inst. Chem. Engrs., Am. Phys. Soc. Orthodox. Achievements include research in reaction dynamics and bifurcation theory. Office: Chem Engring Princeton U Olden St Princeton NJ 08544

KEYES, MARION ALVAH, IV, manufacturing company executive; b. Bellingham, Wash., May 11, 1938; s. Marion Alvah and Winnefred Agnes (Nolte) K.; BS in Chem. Engring., Stanford U., 1960; MS in Elec. Engring., U. Ill., 1968; MBA, Baldwin Wallace Coll., 1981; m. Loretta Jean Mattson, Nov. 17, 1962; children—Marion A., Zachary Leigh, Richard. Registered profl. engr., Calif., Wis., N.Y., Ill., Ohio. Teaching asst. dept. math. Stanford U., 1958-59, technician Stanford Aerosol Labs., 1957-59; chem. engr. Ketchikan (Alaska) Pulp Co., 1960-63; dir. engring. Control Systems div. Beloit Corp. (Wis.), 1963-70; gen. mgr. digital systems div. Taylor Instrument Co., Rochester, N.Y., 1970-75; sr. v.p., group exec. Indsl. Products and Svcs. Group, mem. exec. operative bd. McDermott Internat. Inc., 1985-89; v.p. engring., pres. Bailey Controls Co., Wickliffe, Ohio, 1975-85, pres., chief exec. officer, 1989-90; chmn. Dcom Corp., Eastlake, Ohio, 1990-93; pres., chief exec. officer Trice Engrs., Chagrin Falls, Ohio, 1990—. Past bd. advisors Fenn Coll. Engring. Cleve. State U.; bd. dirs. Fact Inc., Baldwin Coll., United Cerebral Palsy, Cleve.; past pres., mem. exec. bd. N.E. Ohio coun. Boy Scouts Am.; past pres. Area 5 Boy Scouts Am. Holder 50 U.S. and over 50 fgn. patents; author over 100 tech. papers. Fellow ISA (hon. life mem.), Tech. Assn. Pulp & Paper Industry (pioneer award honoree), Am. Inst. Chemists, IEEE; mem. Ohio Acad. Scis. (bd. dirs., life, Named Centennial Honoree 1991), Cleve. Engring. Soc. (bd. dirs.), Am. Assn. Artificial Intelligence, Am. Mgmt. Assn., IEEE, U.S. Automation Rsch. Coun., Am. Automatic Control Coun. (past sec. and bd. dirs.) Instrument Soc. Am., Am. Inst. Chem. Engrs., Am. Chem. Soc., Wis. Acad. Arts, Scis. and Letters, Cleve. World Trade Assn. (Man of Yr. 1984). Republican. Lutheran. Clubs: Canterbury Golf. Author: Offshore Platform Automation, 1990; editor: A Glossary Of Automatic Control Terminology, 1970. Contbr. articles to profl. jours. Patentee in field. Home: 120 River Stone Dr Chagrin Falls OH 44022-1155 Office: Trice Inc Chagrin Falls OH 44022

KEYWORTH, DONALD ARTHUR, technical development executive; b. Flint, Mich., Apr. 21, 1930; s. Vern and Lillian May (Holcomb) K.; m. Dale Louise Bowlen, Oct. 12, 1979. MS, Mich. State U., 1954, PhD, 1958. Chief analytical control Lapaco, Lansing, Mich., 1951-52; analytical chemist Wyandotte (Mich.) Chem. Co., 1958-60; asst. rsch. dir. Universal Oil Products, Des Plaines, Ill., 1960-67; mgr. rsch. Tenneco Oil, Houston, 1968-85; tech. devel. mgr. Ako Chems., Bayport, Tex., 1985—; v.p., tech. dir. Sci. and Ednl. Svcs., Houston, 1967-68. Contbr. articles to profl. jours. Indsl. adv. bd. Tex. State Tech. Inst., Waco. With U.S. Army, 1954-56. Mem. Am. Chem. Soc. (CEDS steering com. 1954—), S.W. Catalysis, S.W. Sci. Forum, Detroit Chemist. Achievements include patents for Polygas, Isobutylene (high purity), ESEP (ethlene recovery), Cosorb (carbon monoxide recovery), HBB-2 (fire retardant), FAS (foulant control acetylene manufacture). Home: 5320 Dora Houston TX 77005 Office: Akzo Chems 13000 Bay Park Rd Pasadena TX 77507

KHACHATOURIANS, GEORGE GHARADAGHI, microbiology educator; b. Nov. 21, 1940; s. Sumbat and Mariam (Ghazarian) K.; m. Lorraine M. McGrath, Oct. 14, 1974; 1 child, Ariane K. BA, Calif. State U., San

Francisco, 1966, MA, 1969; PhD, U. B.C., Vancouver, 1971. Postdoctoral fellow Biol. Div. Oak Ridge (Tenn.) Nat. Lab., 1971-73; rsch. assoc. U. Mass. Med. Sch., Worcester, 1973-74; asst. prof. microbiology dept. U. Saskatchewan, Sask., Can., 1974-77, assoc. prof., 1977-80, prof., 1980-81, prof. applied microbiology and food sci., 1981—; resident prof. U. B.C., Vancouver, 1992; mem. Gov. of Can. Fed. Task Force on Biotech., Ottawa, Ont., Can., 1980-81, Operating Grants Panel, Can. Agr., 1981-84, Biomed. Grants Panel, Sask. Health Rsch., 1988—; bd. dirs. PhilomBios Inc.; pres. Khachatourians Enterprises Inc.; founding dir. BioInsecticide Rsch. Labs., U. Sask, 1982—; vis. prof. U. B.C., Vancouver, 1992. Contbg. author ency. chpts.; co-editor: (book series) Food Biotechnology-Microorganisms, 1993. Recipient grants Nat. Sci. Engrng. Coun., Ottawa, Can., 1974—, Sask. Agr. Rsch. Found., 1981-85, Nat. Rsch. Coun., 1977-88, Agrl. Devel. Found., Regina, Sask., 1985—. Mem. Am. Soc. Microbiology, Can. Soc. Microbiology, Soc. Indsl. Microbiol., Am. Entomol. Soc. Achievements include U.S. and Can. patents in anucleated live E. coli vaccines. Home: 1125 13th St East, Saskatoon, SK Canada S7H 0C1 Office: Applied Microb & Food Sci, U Saskatchewan, Saskatoon, SK Canada S7N 0W0

KHAIR, ABUL, chemistry educator; b. Noakhali, Bangladesh, Jan. 1, 1944; came to Omah, 1988; s. Mobarak Ullah and Ambia Khatoon; m. Montaz Khanom Lily, Dec. 16, 1963; children: Shaheen, Lina, Miti. BS, Dhaka U., Bangladesh, 1965, MS, 1966; PhD, Cambridge U., 1976. Lectr. chemistry Dhaka U., Bangladesh, 1967-76, asst. prof., 1976-78, assoc. prof., 1978-87, prof., 1987-88; faculty mem. Sultan Qaboos U., Oman, 1988—; mem. Dhaka U. Senate, Bangladesh, 1981-84. Compiler: Zinc in Nutrition, 1988; contbr. 18 articles to profl. jours. Named Best Chemist Pakistan Petroleum Ltd., 1968. Fellow Royal Soc. Chemistry (cert. chemist); mem. Am. Chem. Soc., Dhaka U. Tchrs. Assn. (sec.-gen. 1987). Home: vill Bhimpur PO Chatkhil, Dist Noakhali Bangladesh Office: Dept Chem Coll Sci Sultan Qaboos U, PO Box 36 Al Khod Postal Code 123, Muscat Oman

KHALATNIKOV, ISAAC MARKOVICH, theoretical physicist, educator; b. Dniepropetrovsk, USSR, Oct. 17, 1919; m. Valentina Nikolaevna Shchors; 2 children. Student, Dniepropetrovsk State U., USSR Acad. Scis. From jr. rschr. to sr. rschr. to head divsn., Inst. Physical Problems USSR Acad. Scis., 1945-65; dir. L. D. Landau Inst. Theoretical Physics USSR (now Russian) Acad. Scis., 1965-92, hon. dir. L. D. Landau Inst., adviser, 1993—; prof. Moscow Inst. Physics and Tech., 1954—. Recipient USSR State prize 1953, Landau prize in Physics, 1976, Alexander von Humboldt prize, 1992. Office: Landau Inst of Theor Physics, Ulitsa Kosygina 2, 117940 Moscow Russia*

KHALIMSKY, EFIM, mathematics and computer science educator; b. Odessa, USSR, June 23, 1938; came to U.S., 1978; s. David Khalimsky and Olga Weizman; m. Elena Merems, May 19, 1962; 1 child, Olga. MS in Math. with honors, Pedagogical Inst., Odessa, 1960; PhD in Math., Pedagogical Inst., Moscow, 1969. Tchr. high sch. Odessa, 1960-66; assoc. prof. Pedagogical Inst., Magnitogorsk, USSR, 1969-72; sr. research scientist Research and Prodn. Inst. for Food Industry, Odessa, 1972-73, Econs. Inst. Acad. Sciences, Odessa, 1973-77; asst. prof. Manhattan Coll., Riverdale, N.Y., 1980-85; assoc. prof. CUNY, 1979-80, Coll. of Staten Island (N.Y.), 1985-89; prof. Cen. State U., Wilberforce, Ohio, 1989—. Author: Ordered Topological Spaces, 1977, (with others) The Planning of Economic and Ecological Research at Sea Basins, 1976, (with others) Economical and Ecological Management of Water Resources, 1976, (with others) Methodological Foundations on Developing MIS System for Water Resources, 1976; area editor Jour. Applied Math. and Simulation, 1987—; contbr. numerous articles to profl. jours. Named Best Scientist USSR Acad. Sciences, 1986. Mem. IEEE, Am. Math. Soc., Assn. Computing Machinery, Soc. Indsl. and Applied Math., Ops. Research Soc. Am. Home: 1260 Brentwood Dr Dayton OH 45406-5713

KHALIQ, MUHAMMAD ABDUL, electrical engineer; b. Jhang, Punjab, Pakistan, May 4, 1948; came to U.S., 1984; s. Muhammad Abdul Ghani and Ayasha (Begum) Choudhry; m. Rasheeda B. Khaliq, Sept. 5, 1975; children: Mahmooda, Mansoora, Muhammad Abdul Ali. MSc in Electronics, U. Punjab, Pakistan, 1969; PhD in Elec. Engring., U. Ark., 1987. Lectr. Govt. Coll., Rabwah, Pakistan, 1970-83; assoc. prof. Elec. Engring. Mankato (Minn.) State U., 1988—. Patentee in field; contbr. articles to profl. jours. mem. IEEE (devices soc. 1986—, edn. soc. 1989—). Office: Mankato State U Dept Elec Engring Box 215 Mankato MN 56002

KHAN, ABDUL QUASIM, chemistry researcher; b. Karachi, Sindh, Pakistan, Dec. 29, 1959; s. Abdul Matin and Nadra Matin Khan. MSc, Karachi U.; PhD, HEJ Rsch. Inst. Chemistry, Krachi. Cert. organic chemistry rschr. Jr. rsch. fellow HEJ Rsch. Inst. Chemistry, 1985-87; rsch. fellow U. Karachi, 1987-91, postdoctoral rsch. fellow, 1991—; postdoctoral rsch. fellow dept. chemistry U. Alberta, Can. Contbr. articles to Jour. Natural Prodn., Phytochemistry, Heterocycles, Phytochemistry, Planta Medica, Tetrahedron, Jour. Syntheic Comm., Prakt. Chemie. Mem. Am. Chem. Soc. Achievements include research in isolation and structures elucidation of natural products like steroids, alkaloids from various indigenous medicinal plants; development of synthetic strategies for achieving novel syntheses of amino-sigras and alkaloidal glycoside. Home: B/2 Phase I UK Apts, 74800 Karachi Pakistan Office: U Alberta, Dept Chemistry, Edmonton, AB Canada T6G 2G2

KHAN, HAMID RAZA, physicist; b. Rampur, India, May 15, 1942; s. Hashmat Raza and Azghari (Begum) K. MS in Physics, Aligarh Muslim U., India, 1962; PhD in Physics, La. State U., 1970. Lectr. physics dept. Mahanand Mission Hindu Coll., Ghaziabad, India, 1962-63; researcher physics dept. Aligarh Muslim U., India, 1963-65; asst. instr. La. State U., Baton Rouge, 1965-70; rsch. scientist F.E.M., Schwaebish Gmuend, Germany, 1970-82, sr. scientist, 1982—, head solid state rsch. group; adj. prof. dept. physics and astronomy U. Tenn., Knoxville, 1984—. Co-author: Landau-Bernstein Superconductors, Vol. 21, 1990, Superconductivity of Noble Metal Alloys, 1985, Low Temperature Technology, 1981; contbr. over 100 sci. papers to profl. publs. Advisor Third World Acad. Internat. Ctr. for Theoretical Physics, Trieste, Italy, 1986—. Recipient Gold medal Am. Plating Surface & Finishing Soc., 1988. Mem. Am. Phys. Soc., German Phys. Soc., Am. Materials Rsch., European Material Rsch. Soc., European Phys. Soc., N.Y. Acad. Scis. Avocations: tennis, jogging. Office: F E M, Katharinenstrasse 17, 7070 Schwaebisch Germany

KHAN, LATIF AKBAR, mineral engineer; b. Ghazni Khel, Pakistan, Aug. 21, 1943; came to U.S., 1974; s. Ghulam Akbar and Habiba (Shahnawaz) K.; m. Riffat Begum, Dec. 25, 1973; children: Bilal, Muhammed, Habiba. Diplom ingenieur, Tech. U. Clausthal, Germany, 1968, Doktor Ingenieur, 1971. Sci. co-worker Tech. U. Clausthal, 1968; efficiency contr. Sudrohrbau, Ingoldstadt, Germany, 1971-72; gen. mgr., dir. Sarhad Devel. Authority, Peshawar, Pakistan, 1972-75; cons. Depaki Mineral Cons., Peshawar, 1975-76; rsch. engr. U. Calif., Berkeley, 1977-78; assoc. profl. scientist Ill. State Geol. Survey, Champaign, 1980—; organizing com. Nat. Symposium on Mining, Hydrology, Sedimentation and Reclamation; presenter short courses, symposia on mining; cons. in field. Vice-pres. Afghan Relief Fund, Champaign, 1988; pres. Bosnian Relief Com., Champaign, 1993. Mem. Soc. Mining, Metallurgy and Exploration, Am. Chem. Soc. (fellow div. colloid and surface chemistry). Islam. Achievements include invention of fine coal beneficiation method and advanced kinematically balanced and rocking akbar mill. Home: 1912 Bellamy Dr Champaign IL 61821 Office: Ill State Geol Survey 615 E Peabody Dr Champaign IL 61820

KHAN, MAHBUB R., physicist; b. Dhaka, Bangladesh, Sept. 11, 1949; came to U.S., 1974; s. Ebarat A. Khan and Zahura Khatun; m. Reena Yasmeen, Sept. 10, 1976; children: Madhury, Kamal, Jamal, Monika. MSc in Physics, Dhaka U., 1972; PhD in Physics, Boston Coll., 1979. Postdoctoral staff U. Nebr., Lincoln, 1979-81, Argonne (Ill.) Nat. Lab., 1981-83; sr. scientist MPI/CDC, Mpls., 1983-85, Stolle Corp./Alcoa, Sidney, Ohio, 1985-86; mgr. Seagate Magnetics, Fremont, Calif., 1986—. Contbr. articles to profl. jours. Recipient Gold medal Secondary Edn. Bd., Bangladesh, 1964. Mem. IEEE, Am. Phys. Soc. Islam. Home: 3463 Sagewood Ln San Jose CA 95132 Office: Seagate Magnetics 47001 Benicia St Fremont CA 94538

KHAN, MOHAMMAD ASAD, geophysicist, educator, former energy minister and senator of Pakistan; b. Aima, Lahore, Pakistan, Aug. 13, 1940; came to U.S., 1964; s. Ghulam Qadir and Hajira (Karim) K.; m. Tahera Pathan, Jan. 4, 1974; 1 dau., Shehzi Samira. B.S., U. Punjab, Lahore, Pakistan, 1957, M.S., 1963; postgrad., Harvard U., 1964-65; Ph.D. (East West Center scholar), U. Hawaii, 1967. Lectr. in geophysics U. Punjab, 1963-64; asst. prof. geophysics and geodesy U. Hawaii, 1967-71, assoc. prof., 1971-74, prof., 1974—; minister of petroleum and natural resources Govt. Pakistan, 1983-86, senator, 1984-86; chmn. internat. advisors, 1987—; chmn. Hydrocarbon Devel. Inst., Pakistan, 1984-86, Attock Oil Refinery, Pakistan, 1984-86; cabinet mem. Nat. Econ. Council, Govt. Pakistan, 1984-86; NSF and NASA fellow Summer Inst. Dynamical Astronomy at MIT, 1968-69; sr. vis. scientist geodynamics Goddard Space Flight Ctr., NASA, Greenbelt, Md., 1972-74; sr. scientist Computer Scis. Corp., Silver Spring, Md., 1974-76, sr. cons., 1976-77; diplomatic minister/adviser Resource Survey and Devel. Pakistan, 1974-76; sr. resident assoc. Nat. Acad. Scis., 1972-74; leader Am. Asian Studies and Contemporary Social Problems Seminar Series, Honolulu, 1968-69. Contbr. articles to profl. publs. Chmn. East and West: A Perspective for the 80's; mem. Hawaii Environ. Council, 1979-83, chmn. exec. com., 1979-83, vice chmn., 1981-83; chmn. Pakistan Relief Fund, Honolulu, 1971. Recipient Gold medal Rawalpindi Union of Journalists, 1985, Pakistan Engrng. Coun., 1985, Pakistan Assn. of Minorities, 1984, 85. Fellow Explorers Club; mem. Geol. Soc. U. Punjab (pres. 1962-63), Am. Geophys. Union, Pakistan Assn. Advancement Sci., Am. Geol. Inst., Am. Geophys. Union, East West Ctr. Alumni Assn. (dir. 1976-80, Disting. Alumnus award for career achievement and leadership 1984), Internat. Alumni of East West Ctr. (exec. com., chmn. 1977-80). Achievements include research in geophysics, geodetic and oceanographic applications of satellites, geodynamics, planetary interiors, global tectonics, global correlations, core-mantle boundary problems, equilibrium figures, gravity, isostasy, satellite altimetry, geodesy, earth models, geophysical exploration, ocean dynamics. Office: U Hawaii Hawaii Inst Geophysics 2525 Correa Rd Honolulu HI 96822-2285

KHAN, SAEED REHMAN, cell biologist, researcher; b. Rampur, India, Dec. 25, 1943; came to U.S., 1970; s. Habib R. and Gori Begum K.; m. Patricia Jo-Ann Clark; children: Omar S., Ameena Y. MS, Peshawar u., Pakistan, 1964; PhD, U. Fla., 1972. Asst. prof. King Abdulaziz U., Jeddah, Saudi Arabia, 1976-78; postdoctoral fellow U. Fla., Gainesville, 1978-79, asst. in surgery, 1979-82, rsch. scientist, 1982-85, asst. prof., 1987-91, assoc. prof., 1991—; dir. grad. program dept. pathology U. Fla., Gainesville, 1992—. Mem. editorial bd. Scanning Microscopy jour., 1985-92; contbr. over 123 articles to profl. jours. Recipient Rsch. award NIH, 1979, 90, 92; Fullbright-Hays scholar, 1970-73. Mem. N.Y. Acad. Scis., Fla. Soc. Electron Microscopy (pres. 1988-90), Microscopy Soc. Am., Am. Soc. Nephrology, Am. Urological Assn., Am. Soc. for Investigative Biology, Soc. for Biomaterials, Phi Kappa Phi. Muslim. Home: 3504 SW 1st Way Gainesville FL 32601 Office: U Fla Dept Pathology Box 100275 Gainesville FL 32610

KHAN, ZEYAUR RAHMAN, entomologist; b. Aurangabad, Bihar, India, Feb. 7, 1955; s. Tauquirur Rahman and Zinatun (Nisa) K.; m. Tamanna, Feb. 7, 1979; children: Zeba, Rizwan, Anam. MS in Entomology, Indian Agrl. Rsch. Inst., New Delhi, India, 1977, PhD in Insect Physiology, 1980. Asst. prof. Rajendra Agrl. U., Pusa, Bihar, India, 1980-83; postdoctoral scientist Internat. Rice Rsch. Inst., Los Banos, Philippines, 1983-85; rsch. assoc. U. Wis., Madison, 1985-86; assoc. entomologist Internat. Rice Rsch. Inst., Los Banos, 1986-91; sr. rsch. scientist Internat. Ctr. Insect Physiology and Ecology, Nairobi, Kenya, 1986—; vis. asst. prof. U. Philippines, Los Banos, 1987-91; coord. for deepwater rice for Asia Internat. Rice Rsch. Inst., 1988-91; referee Insect Sci. and Its Application, Kenya, Crop Protection, U.K., TAG, Germany; vis. sr. scientist Kans. State U., Manhattan, 1991-93. Author: World Bibliography of Rice Stem Borers, 1991, Insect Pests of Rice, 1993, Techniques for Evaluating Insect Resistance in Plants, 1993; contbr. over 50 articles to internat. jours. and chpts. to 9 books. Recipient Best Rsch. Paper award Pest Control Coun. of Philippines, 1984, 85, 88, 90; grantee Asian Devel. Bank, 1988. Mem. Entomol. Soc. of Am., Philippine Assn. Entomologists, Crop Sci. Soc. of Philippines, Kans. Entomol. Soc.Sigma Xi. Achievements include determination of allelochem. mechanisms of host plant resistance for major insect pests of rice, soybean and wheat; development of improved methods to evaluate crop plants for insect resistance; identification of rice insect pheromones. Office: Internat Ctr Insect Physiology & Ecology, PO Box 30, Mbita Nyanza, Kenya

KHANNA, FAQIR CHAND, physics educator; b. India, Jan. 23, 1935; came to U.S., 1958; s. Ram S. D. Khanna and Ram Ditti Malhotra; m. Swaraj Mukul, Jan. 16, 1966; children: Shrawan F., Varun F. BSc with honors, Panjab U., 1955, MSc with honors, 1956; PhD, Fla. State U., 1962. Sr. rsch. officer Chalk River Nuclear Labs., 1966-84; prof. physics U. Alta., Edmonton, Alta., Can., 1984—. Fellow Am. Phys. Soc. Achievements include research in subatomic physics; nuclear and particle physics; many-body physics. Office: U Alta, Theortical Physics Institute, Edmonton, AB Canada T6G 5G6

KHARDORI, NANCY, infectious disease specialist; b. Srinagar, Kashmir, India, Apr. 1, 1949; came to U.S., 1977; d. Moti Lal and Prabha (Burbuzoo) Misri; m. Romesh Khardori, Oct. 11, 1973; children: Amitabh, Ankush. Grad. premed., U. Jammu & Kashmir, India, 1966; MD, Govt. Med. Coll., Sringar, India, 1972; MD in Microbiology and Immunology, All India Inst. Med. Schs., New Delhi, 1977. Diplomate Am. Bd. Internal Medicine, Subbd. Infectious Disease. Instr. in microbiology, immunology All India Inst. Med. Scis., New Delhi, 1977-79; rsch. assoc. microbiology and immunology So. Ill. U. Sch. Medicine, Springfield, 1979-82; asst. internist, asst. prof. U. Tex. M.D. Anderson Cancer Ctr., Houston, 1986-89; asst. prof. microbiology, immunology Grad. Sch. Biomed. Scis., Houston, 1986-89; assoc. prof. medicine, microbiology So. Ill. U. Sch. of Medicine, Springfield, ., 1989—; cons. Inst. for Immunol. Disorders, Houston, 1986-88, Carlinville (Ill.) Area Hosp., 1991—; adj. faculty Grad. Sch. of Biomed Scis., Houston, 1989-91; adj. assoc. prof. U. Tex. M.D. Anderson Cancer Ctr., Houston, 1989—. Hindu. Avocation: cooking. Office: So Ill Sch Medicine PO Box 19230 Springfield IL 62794-9230

KHARE, MOHAN, chemist; b. Varanasi, India, May 15, 1942; came to U.S., 1967, naturalized, 1971; s. Dwarka Nath and Rampyari Devi Khare Srivastava; m. Meena K., Nov. 20, 1973; 1 child, Rohit. BSc, Banaras Hindu U., 1961, MSc, 1963, PhD, 1967. Rsch. assoc. U. Md., College Park, 1967-69, Oreg. State U., Corvallis, 1969-70; sr. rsch. assoc. Cornell U., Ithaca, N.Y., 1970-78; analytical specialist Hydroscience Inc., (subsidiary of Dow Chem. Co.), Knoxville, Tenn., 1978-80; tech. specialist IT Enviroscience subs. IT Corp., Knoxville, 1980-82; rsch. prof. chemistry U. Nev., Las Vegas, 1982-84, mgr. organic div. quality assurance lab. under coop. agreement with EPA, 1982-84; mgr. organic analysis lab. Environ. Monitoring Svcs. Rockwell Internat., Thousand Oaks, Calif., 1984-85; dir. environ. analytical lab. EA Engring, Sci., and Tech., Inc., Sparks, Md., 1985-87; sr. v.p. Recra Environ., Inc., Columbia, Md., 1987-88; pres., chief exec. officer Envirosystems, Inc., Columbia, 1989—. Contbr. articles to profl. jours. including protocols and standard oper. procedures for hazardous waste analytical program. Mem. AAAS, Am. Chem. Soc., Am. Soc. Mass Spectrometry, Am. Water Works Assn., Internat. Union Pure and Applied Chemistry. Home: 10189 Maxine St Ellicott City MD 21042-6316 Office: Envirosystems Inc 9200 Rumsey Rd Ste 102B Columbia MD 21045-1900

KHATIB, NAZIH MAHMOUD, mechanical engineer; b. Bireh, West Bank, Palestine, Dec. 22, 1956; came to U.S. 1975; s. Mahmoud Ahmad and Mariam Omar Khatib; m. Abeer Nazih, July 28, 1985; children: Hend, Leen. MSc, U. Ill., Chgo., 1980; PhDME, U. Mich., 1991. Instr. mech. engring. U. Jordan, Amman, 1980-83; lectr. U. Birzeit, West Bank, Palestine, 1983-86; sr. project engr. GM Co., Warren, Mich., 1990-93; tech. svcs. mgr. ICI Americas, New Castle, Del., 1993—. Mem. ASME, ASHRAE, SAE, Assn. Energy Engrs. Achievements include research in alternative/environmentally safe refrigeration and air conditioning methods, thermodynamics and energy analysis of refrigeration systems, global warming and ozone depletion issues. Home: 1667 McIntyre Dr 228 Sulky Way Chaddsford PA 19317 Office: ICI Americas Engring Bldg E2-HVAC 30200 Mound Rd 213 Cherrylane Bldg L-21 New Castle DE 19720

KHATTAK, CHANDRA PRAKASH, materials scientist; b. Rawalpindi, Pakistan, May 19, 1944; s. Diwan Chand and Krishnawanti (Vasdev) K.; m. Veena Khattak, July 11, 1970; children: Payal, Gautam. B. in Tech. with Hons., Indian Inst. Tech., Bombay, 1965; MS, SUNY, Stonybrook, 1971, PhD, 1973. Scientist D.M.R. Lab., Hyderabad, India, 1965-68; grad. asst. SUNY, Stonybrook, 1968-74; assoc. physicist Brookhaven (N.Y.) Nat. Lab., 1974-77; dir. rsch. and devel. div. Crystal Systems, Inc., Salem, Mass., 1977-80, sr. v.p., 1980-90, exec. v.p., 1990--. Contbr. articles to profl. jours. Recipient IR-100 award Indsl. R&D Mag. Mem. Am. Ceramic Soc., Inc., Am. Assn. Crystal Growth (sec. N.Eng. sect. 1979-80, vice chmn. N.Eng. sect. 1980-81, chmn. N.Eng. sect. 1981-82), The Electrochemical Soc., Inc., The Am. Phys. Soc., ASM Internat., Materials Rsch. Soc. Achievements include patents for castin an ingot and making a silica container, crystal growing, process of forming a plated wirepack with abrasive particles only in the cutting surface with a controlled kerf, multi-wafer slicing with a fixed abrasive, gallium arsenide crystal growth, method of growing silicon ingots using a rotating melt; development of a crucible to prevent cracking of silicon ingots solidified in silica crucibles; vacuum processing of silicon; silicon crystal growth by Heat Exchanger Method; growth of laser, scintillation, detector and compound semiconductor crystals. Home: 16 Delaware Ave Danvers MA 01923 Office: Crystal Systems Inc 27 Congress St Salem MA 01970

KHAZAN, NAIM, pharmacology educator; b. Baghdad, Iraq, Feb. 15, 1921; came to U.S. 1966, naturalized 1973; s. Rahamim adn Tova (Eliezer) K.; m. Evelyn Muallem, Nov. 12, 1952; children: Uri, Ron. PhC in Pharm. Chemistry, Sch. Pharmacy, Baghdad, 1943; PhD in Pharmacology (hon.), Hebrew U., Jerusalem, 1960. Asst. prof. pharmacology Coll. Pharm. Sci., Columbia U., N.Y.C., 1967-68, Mt. Sinai Sch. Med., N.Y.C., 1968-72; head dept. pharmocology Merrell Nat. Labs., Cin., 1972-73; prof., chmn. dept. pharmacology Sch. Pharmacy, U. Md., Balt., 1974-86, Emerson prof., 1976-86, prof. pharmacology, 1986-91, prof. emeritus, 1991—; sr. pharmacologist VA Med. Ctr., Washington, 1991—; clin. prof. affiliate psychiatry Georgetown U. Med. Ctr., Washington, 1992—. Mem. adv. editorial bd. Neuropharmacology Jour., 1977-88, Progress in Neuropsychopharmacology and Biol. Scis.; contbr. more than 150 articles on pharmacodynamics of addictive agts. to Jour. Pharmacology, Exptl. Therapeutics in Psychopharmacology, Neuropharmacology, Neuroscis., Life Sci., others. USPHS Internat. postdoctoral fellow, Washington, 1968; grantee NIMH, NIDA; apptd. eminent scholar Commn. on Higher Edn., State of Md., Annapolis, 1989. Mem. Am. Pharm. Assn., Am. Soc. Pharmacology and Exptl. Therapeutics (nominee Otto Krayer lectr. award 1990), Soc. Neurosci, N.Y. Acad. Sci. Achievements include research on the pharmacology of substance abuse and on EEG characteristics of opioids and other CNS active agents. Home: 2126 Caves Rd Owings Mills MD 21117-2326 Office: U Md Dept Pharmacology Sch Pharmacy 20 N Pine St Baltimore MD 21201-1180

KHAZEI, AMIR MOHSEN, surgeon, oncologist; b. Teheran, Iran, July 21, 1928; came to U.S., 1957; s. Abol Khasem and Esmat (Khaligh-Azam) K.; m. Carmeline Victoria Grace Picardi; children: Alan, Darla, Mia, Lance. BS, U. Lausanne, Switzerland, 1952, MD and Cert. d'Etudes Medicale, 1957. Diplomate Mass. Bd. Medicine, N.H. Bd. Medicine; qualified Am. Bd. Surgery. Intern Mercy Hosp., Pitts., 1957-58, resident in gen. surgery, 1957-62; fellow in surgery Lahey Clinic Found., Boston, 1962-63, assoc. staff mem. surgery and chemotherapy, 1963-67, assoc. dir. surg. rsch. lab., 1967-68; attending surgeon VA Hosp., Manchester, N.H., 1968-70; staff surgeon Cath. Med. Ctr./Elliot Hosp., Manchester, 1971—; pres., chmn. exec. com. of med. staff Cath. Med. Ctr., 1981-82. Mem. editorial bd. Living With Cancer; co-author book; contbr. articles to profl. jours. Bd. dirs. Incorporator Cath. Med. Ctr., 1981—, trustee, 1981-90; trustee Fidelity Health Alliance Bd., Manchester, 1991—, N.H. Nurses Found., 1991—; pres. N.H. div. Am. Cancer Soc., 1977-79, nat. del. dir., 1982-91. Recipient St. George medal Am. Cancer Soc., 1982, Golden Apple award Cath. Med. Ctr., 1990, Med. Staff award N.H. Hosp. Assn., 1990. Fellow Am. Coll. Angiology, Internat. Coll. Angiology, Inter-Am. Coll. Physicians and Surgeons; mem. AMA, Am. Fed. Clin. Rsch., N.H. Med. Soc. (pres., chmn. exec. com. 1986-87), N.Y. Acad. Scis., Orgn. State Med. Assn. Pres.'s (life), Transplantation Soc., Hillsborough County Med. Soc. (pres. 1981-82), Am. Assn. for Cancer Edn., Soc. Laparo-endoscopic Surgeons. Office: 88 McGregor St Ste # 304 Manchester NH 03102

KHO, EUSEBIO, surgeon; b. Philippines, Dec. 16, 1933; s. Joaquin and Francisca (Chua) K.: came to U.S., 1964; AA, Silliman U., Philippines, 1955; MD, State U. Philippines, 1960; fellow in surgery, Johns Hopkins, 1965-67; m. Grace Casas Lim, May 24, 1964: children: Michelle Mae, April Tiffany, Bradley Jude, Jaclyn Ashley, Matthew Ryan. Rotating intern Philippine Gen. Hosp., U. Philippines, 1959-60; resident gen. practice Silliman U. Med. Ctr., 1960-63; virology researcher Van Howelling Lab. Silliman U., 1963-64; intern in surgery Balt. City Hosp., 1964-65, resident in gen. surgery, 1965-67; rsch. assoc. pediatric surgery U. Chgo. Hosps., 1967-68; resident in gen. surgery, then chief resident U. Tex. Hosp., San Antonio, 1968-70; hosp. surgeon St. Anthony Hosp., Louisville, 1970-72; practice medicine specializing in surgery, Scottsburg, Ind., 1972—; chmn. dept. surgery Scott County Meml. Hosp., 1973—; cons. surgeon Washington County Meml. Hosp., Salem, Ind., also Clark County Meml. Hosp., Jeffersonville, Ind., 1973—; courtesy surgeon Suburban Hosp., Louisville, 1973—; gen. surgeon 5010 U.S. Army Hosp., Louisville, 1980—. Bd. dirs. Make-A-Wish Found. Ind., 1992—. Served to col. M.C., USAR, 1980—, Operation Desert Storm, 1990-91. Named to Hon. Order Ky. Cols., 1991. Diplomate Am. Bd. Surgery. Fellow A.C.S., Am. Soc. Abdominal Surgeons, Am. Coll. Emergency Physicians; mem. Am. Coll. Internat. Physicians (founding mem., trustee 1974—), AMA (Physician's Recognition award 1969, 72), Ind. State Med. Assn., Ky. Med. Assn., Philippines Med. Assn. of Ind. and Ky., Internat. Coll. Surgeons, Soc. Philippine Surgeons in Am. (life), Assn. Philippine Practicing Physicians in Am. (life), Assn. Mil. Surgeons of U.S. (life), Res. Officers Assn. of U.S. (life), Mark Ravitch Surg. Assn., Bradley Aust Surg. Soc., N.Y. Acad. Scis. Presbyterian. Clubs: Optimists, Masons. Home: 14 Carla Ln Scottsburg IN 47170-9707 Office: 137 E McClain Ave Scottsburg IN 47170

KHOJASTEH, ALI, medical oncologist, hematologist; b. Shiraz, Fars, Iran, Oct. 10, 1947; came to U.S., 1974; s. Mostafa and Pari Jan (Azimi) K.; children: Artemis, Amitis. Degree, Pahlavi U., Shiraz, 1968, MD, 1974. Vice dean Sch. Medicine Shiraz U., 1980-82, chmn. med. dept. Sch. Medicine, 1982-83; chief med. oncology Ellis Fischel Cancer Ctr., Columbia, Mo., 1983-87, chmn. med. dpet., 1987-90, med. dir. Meml. Community Cancer Ctr., Jefferson City, Mo., 1990—; pres. Columbia (Mo.) Comprehensive Cancer Care Clinic, 1990—; assoc. prof. U. Mo., Columbia, 1989—; prin. investigator Ellis Fischel CCOP, Columbia, 1988-90; chmn. Mo. Cancer Pain Initiative, 1991. Contbr. articles to New Eng. Jour. Medicine, Cancer, Am. Jour. Medicine; author: (with others) Immunoproliferative Small Intestinal Disease, 1988. Rsch. grantee Purdue Fredrick Co., Conn., 1984—, Adria Lab., Columbus, 1988—, Glaxo Rsch. Lab., Research Triangle Park, N.C., 1988-91, Ciba-Geigy Co., 1990—, Merril Dow Co., 1991. Fellow AAAS, Am. Coll. Physicians; mem. Am. Soc. Clin. Oncology, Am. Soc. Internat. Medicine, Smithsonian Soc., N.Y. Acad. Sci., Mo. Acad. Scis. (chmn. oncology sect. 1988-89), So. Med. Assn. Zoroastrian. Home: 2801 Greenbriar Dr Columbia MO 65203-3663 Office: Columbia Comprehensive Cancer Care Clinic 500 Keene St Ste 202 Columbia MO 65201-8159

KHOMAMI, BAMIN, chemical engineer, educator; b. Tehran, Iran, Nov. 3, 1962; came to U.S., 1978; s. Hedayat and Frough (Niknejad) K.; m. Faranak B. Jamshidi, July 1, 1989. BSChemE, Ohio State U., 1983; MS, U. Ill., 1985, PhD, 1987. Teaching and rsch. asst. U. Ill., Urbana, 1983-87; asst. prof. Chem. Engr. Washington U., St. Louis, 1987-92, assoc. prof. Chem. Engr., 1992—; cons. SACMI Corp., Imola, Italy, 1992—, LCI Corp., Rockhill, S.C., 1992, MEMC Electronic Materials, Inc., St. Louis, 1990-92. Author book chpt.; contbr. over 30 articles to profl. jours. Grantee NSF, 1988—, Naval Office Sci. Rsch., 1992—, Def. Advanced Rsch. Project Agy., 1993—, Nat. Supercomputer Ctr., 1990—. Mem. AIChE, Internat. Polymer Processing Soc., Soc. Rheology, Am. Phys. Soc., St. Louis Acad. Sci. (life), Tau Beta Pi, Phi Kappa Phi, Phi Lambda Upsilon. Achievements include research on applications of fundamental principles of rheology and fluid mechanics to polymeric and composite materials processing; translational flow stability of viscoelastic fluids, steady and transient viscoelastic flows in complex geometrics; mathematical modeling and simulation, numerical methods in heat and mass transfer. Office: Washington U Dept Chem Engr 1 Brookings Dr Box 1198 Saint Louis MO 63130

KHONSARI, MICHAEL M., engineering educator; b. Aug. 17, 1957; m. Karen Sue Troy, Sept. 1, 1990. BS in Mech. Engring. with honors, U. Tex., 1978, MS in Mech. Engring., 1979, PhD in Mech. Engring., 1983. Rsch. and teaching asst. U. Tex., Austin, 1978-83; asst. prof. Ohio State U., Columbus, 1984-87; asst. prof. U. Pitts., 1988-90, assoc. prof., 1990—; mem. mech. engring. grad. com. U. Pitts., 1988-90, design interest group, 1988—; mem. faculty ctr. motion control U. Pitts., 1989—; organizing tech. program dir., session chmn. Am. Soc. Engring. Edn. Mid-year Conf., 1985; paper solicitation vice chmn., session chmn. Am. Soc. Lubrication Engrs. Ann. Conf., 1986; paper solicitation chmn., session chmn. Am. Soc. Lubrication Engrs. Ann. Conf., 1987, Soc. Tribologists and Lubrication Engrs. Ann. Conf., 1988, vice chmn., 1989; organizing chmn. ASME/Am. Soc. Lubrication Engrs. Joint Conf. on Tribology, 1986, 1987, ASME/Soc. Tribologist and Lubrication Engrs. Joint Conf. on Tribology, 1988, 1989; session chmn. Soc. Tribologist and Lubrication Engrs. Ann. Conf. on Tribology, 1990, ASME/Soc. Tribologist and Lubrication Engrs. Ann. Conf. on Tribology, 1991, 92; reviewer NSF, NASA, Am. Chem. Soc. Books, McGraw Hill Books, Addison Wesley Books, Prentice-Hall Books, Holt Rinehart and Winston Books; cons. NASA-Lewis Rsch. Ctr., Gen. Motors Rsch. Labs.; lectr. in field. Mem. editorial bd., reviewer Jour. Engring. Design Graphics, 1987—; contbr., reviewer, mem. editorial bd. adv. com. CRC Handbook of Lubrication, vol. III, 1991-93; reviewer Lubrication Engring. Jour., Wear Jour.; pub. abstracts and reports; referee various jours.; contbr. articles to profl. jours. Recipient Found. award ALCOA, 1990, 91. Mem. ASME (conf. planning com. 1989—, reviewer Jour. Tribology and conf. papers, Burt L. Newkirk award 1990), Soc. Tribology and Lubrication Engrs. (bearings com. 1985—, chmn. 1988-91, assoc. editor, reviewer Tribology Transactions 1990—, Presdl. Rsch. Coun. award 1993). Achievements include research in thermal effects in hydrodome bearings, multi-phase flows in bearings, friction associated with instrument painting mechanisms operating under ultra low speeds. Home: 4305 Centre Ave Pittsburgh PA 15213 Office: Univ of Pittsburgh Dept of Mech Engring Pittsburgh PA 15261

KHOPKAR, SHRIPAD MORESHWER, chemistry educator; b. Nasik, Bombay, July 28, 1932; s. Manuorama and Moreshwav K.; m. D. Sucheta Suman, May 27, 1962; children: Samir, Supriya. BSc with honors, U. Bombay, 1954, MSc, 1956; PhD, Jadovpur U., 1960. Lectr. Dept. Chemistry, IIT Bombay, 1962-68, asst. prof., 1968-75, prof., 1975-92; emeritus scientist CSIR U. Bombay, 1993—; past mem. IUPAC commn. on microchemistry and trace analysis. Author: Solvent Extraction of Metals,1 970, Basic Concepts of Analytical Chemistry, 1985, Environmental Pollution Analysis, 1992, Crown Ethers and Cryptands in Solvent Extraction, 1993; editorial adv. com. Indian Jour. Chemistry, Jour. Indian Chem. Soc., Jour. Chem. Sci., Chem. and Environ. Rsch., Jour. Pollution Free Environ.; internat. adv. bd. Ency. Analytical Scis. Recipient Internat. award Can. Soc. Chem. Engrs. in Separation Scis., 1989. Mem. Am. Chem. Soc., Indian Chem. Soc. (pres. Bombay br. 1988-90), Indian Soc. for Analytical Sci. (v.p. 1984-90), Indian Coun. of Chemist (v.p. 1992—). Achievements include research in solvent extraction ion exchange, environmental monitoring. Home: Indian Inst Tech Powai, B 44 Lakeside Quarter, Bombay 400076, India Office: Dept Chemistry, Indian Inst of Tech, Bombay 400076, India

KHORANA, HAR GOBIND, chemist, educator; b. Raipur, India, Jan. 9, 1922; s. Shri Ganpat Rai and Shrimati Krishna (Devi) K.; m. Esther Elizabeth Sibler, 1952; children: Julia, Emilie, Dave Roy. BS, Punjab U., 1943, MS, 1945; PhD, Liverpool (Eng.) U., 1948; DSc, U. Chgo., 1967. Head organic chemistry group B.C. Rsch. Coun., 1952-60; vis. prof. Rockefeller Inst., N.Y.C., 1958—; prof. co-dir. Inst. Enzyme Rsch. U. Wis., Madison, 1960-70, prof. dept. biochemistry, 1962-70, Conrad A. Elvehjem prof. life scis., 1964-70; Alfred P. Sloan prof. biology and chemistry MIT, Cambridge, 1970—; vis. prof. Stanford U., 1964; mem. adv. bd. Biopolymers; researcher chem. methods for synthesis of nucleotides, coenzymes and nucleic acids, elucidation on the genetic code, lab. synthesis of genes, biol. membrane and light-transducing pigments. Author: Some Recent Developments in the Chemistry of Phosphate Esters of Biological Interests, 1961; mem. editorial bd.: Jour. Am. Chem. Soc, 1963—; contbr. numerous articles to profl. jours. Recipient Merck award Chem. Inst. Can., 1958, Gold medal Profl. Inst. Pub. Service Can., 1960, Dannie-Heinneman Preiz Göttingen, Germany, 1967, Remsen award Johns Hopkins U., 1968, Am. Chem. Soc. award for creative work in synthetic organic chemistry, 1968, Louisa Gross Horwitz prize, 1968, Lasker Found. award for basic med. research, 1968, Nobel prize in medicine, 1968; elected to Deutsche Akademie der Naturforscher Leopoldina HalleSaale, Germany, 1968; Overseas fellow Churchill Coll., Cambridge, Eng., 1967. Fellow Chem. Inst. Can., Am. Acad. Arts and Scis.; mem. NAS. Office: MIT Dept Biol/Chem Rm 18-511 77 Massachusetts Ave Cambridge MA 02139

KHOSLA, VED MITTER, oral and maxillofacial surgeon, educator; b. Nairobi, Kenya, Jan. 13, 1926; s. Jagdish Rai and Tara V. K.; m. Santosh Ved Chabra, Oct. 11, 1952; children: Ashok M., Siddarth M. Student, U. Cambridge, 1945; L.D.S., Edinburgh Dental Hosp. and Sch., 1950, Coll. Dental Surgeons, Sask., Can., 1962. Prof. oral surgery, dir. postdoctoral studies in oral surgery Sch. Dentistry U. Calif., San Francisco, 1968—; chief oral surgery San Francisco Gen. Hosp.; lectr. oral surgery U. of Pacific, VA Hosp.; vis. cons. Fresno County Hosp. Dental Clinic.; Mem. planning com., exec. med. com. San Francisco Gen. Hosp. Contbr. articles to profl. jours. Examiner in photography and gardening Boy Scouts Am., 1971-73, Guatemala Clinic, 1972. Granted personal coat of arms by H.M. Queen Elizabeth II, 1959. Fellow Royal Coll. Surgeons (Edinburgh), Internat. Assn. Oral Surgeons, Internat. Coll. Applied Nutrition, Internat. Coll. Dentists, Royal Soc. Health, AAAS, Am. Coll. Dentists; mem. Brit. Assn. Oral Surgeons, Am. Soc. Oral Surgeons, Am. Dental Soc. Anesthesiology, Am. Acad. Dental Radiology, Omicron Kappa Upsilon. Club: Masons. Home: 1525 Lakeview Dr Burlingame CA 94010-7330 Office: U Calif Sch Dentistry Oral Surgery Div 3D Parnassus Ave San Francisco CA 94117-4342

KHOUW, BOEN TIE, biochemist; b. Tegal, Java, Indonesia, Sept. 4, 1934; came to Can., 1957; s. Bian Hin and Swan Nio (Liem) K.; m. Eugenia Yuen-Chi Yu, Sept. 29, 1967; children: Charlotte, Vivian. BSc, Mt. Allison U., 1960; MSc, U. Windsor, 1965, PhD, 1968. Technician, Fisheries Rsch. Bd., Ellerslie, P.E.I., 1959-62; rsch. scientist Can. Packers, Inc., Toronto, Ont., 1967-73, sr. scientist, 1973-80, tech. group mgr. pharms., 1980-87 , sect. leader biochem. rsch., 1986-87 ; tech. mgr. Waitaki Internat. Biosciences, Toronto, 1987-92; tech. mgr. Intergen BioMfg. Corp., Toronto, 1992—; guest lectr. chem. engring. U. Toronto, 1980—. Contbr. articles to profl. jours.; patentee in field. Office: Intergen BioMfg Corp, 55 Glen Scarlett Rd, Toronto, ON Canada M6N 1P5

KHRISTOV, KHRISTO YANKOV, physicist; b. Varna, Bulgaria, June 12, 1915. Grad. physics with honors, grad. math., Sofia U., 1938; postgrad., U. Paris Sorbonne, Moscow U., Nuclear Rsch. Inst., Dubna, Soviet Union, 1966. Asst. chmn. nuclear physics and meteorology Sofia U., 1942, asst. prof. theoretical physics, 1946, prof. Physics and Math. Faculty, 1951, deputy head Physics and Math. Faculty, 1952, head Physics and Math. Faculty, 1956-58, chmn. nuclear physics, 1958-67, dep. rector Kl. Ohridski, 1958, rector Kl. Ohridski, 1972-73, vice chancellor Kl. Ohridski, 1974; corr. mem. Bulgarian Acad. Scis., 1951, full mem., 1961, dep. dir. Physics Inst., 1964-67, dir. Physics Inst., 1972; dep. dir. Nuclear Rsch. Inst., Dubna, 1967-69; head theory of elem. particles sect. Bulgarian Acad. Scis., 1968, head Cosmic Rays Lab., 1968, dep. chmn., 1973-77, founder, dir. Nuclear Rsch. and Nuclear Energy Inst., 1973, mem. Presidium, 1977-82; Bulgarian Govt. rep. Nuclear Rsch. Inst., 1970, mem. sci. coun.; chmn. Soc. Bulgarian Physicists, 1971. Recipient Dimitrov prize, 1952; named Honoured Sci. Worker, 1969, People's Sci. Worker, 1972; named to Order of People's Republic of Bulgaria 1st Cl., 1975, Order of Georgi Dimitrov, 1985. Office: Inst Nuclear Rsch & Energy, Bulgarian Acad Scis, Blvd Lenin 72, 1784 Sofia Bulgaria*

KHUDENKO, BORIS MIKHAIL, environmental engineer; b. Moscow, Dec. 23, 1939; s. Mikhail Alexei and Sophia Don (Charno) K. BS, MS, Belorussian Poly. Inst., Minsk, Belarus, 1961; PhD, Moscow Civil Engring. Inst., 1961. Registered profl. engr., Ga., Russia. Asst. prof. Belorussian Poly. Inst., Minsk, Belarus, 1961-63; jr. rsch. scientist Inst. Water Problems, Moscow, Russia, 1968-72, sr. rsch. scientist, 1972-75; cons. engr. Ralph Stone & Co., L.A., 1976-78; asst. prof. Wayne State U., Detroit, 1978-80; assoc. prof. Ga. Inst. Tech., Atlanta, 1980-87; v.p. Am. Combustion, Atlanta, 1987-88; pres. khudenko Engring., Atlanta, 1988; reviewer of papers for major profl. publs. Contbr. numerous articles and reports to profl. publs. Recipient Samuel Arnold Greely award, 1989, Best PhD Publ. award, 1967. Mem. ASCE, AICE, Internat. Assn. on Water Pollution Rsch., Sigma Xi. Achievements include one Russian patent and four U.S. patents in the field. Office: Khudenko Engring Inc 744 Moores Mill Rd Atlanta GA 30327

KHUHRO, SHAFIQ AHMED, biomedical engineer; b. Hyderabad, Sind, Pakistan, Dec. 5, 1943; came to U.S., 1967; s. Mohammed Murad and Zubeda (Panwhar) K.; m. Ramona S. Meaker, July 16, 1969; children: Shireen, Rafiq. BS, U. Sind, Hyderabad, 1964, MS, 1967; MS, U. N.Mex., 1969. Analytical chemist Rexall Drug Co., St. Louis, 1969-77; sr. material scientist Sherwood Med. Co., St. Louis, Deland, Fla., 1977-81; sr. scientist Critikon, Inc., Tampa, Fla., 1981-84; sr. biomed. engr. Storz Instrument Co., St. Louis, 1984—. Mem. Am. Inst. Chemists, Soc. Mfg. Engrs. Achievements include research and development of implantable hearing devices and intravenous catheters. Office: Storz Instrument Co 3365 Tree Court Industrial Blvd Saint Louis MO 63122

KHURI, NICOLA NAJIB, physicist, educator; b. Beirut, Lebanon, May 27, 1933; came to U.S., 1959, naturalized, 1970; s. Najib N. and Odette (Joujou) K.; m. Elizabeth Anne Tyson, Dec. 9, 1955; children: Suzanne Odette, Najib Nicholas. B.A with high distinction, Am. U. Beirut, 1952; Ph.D., Princeton U., 1957. Asst. prof. Am. U. Beirut, 1957-58, 60-61, assoc. prof., 1961-62; mem. Inst. Advanced Study, Princeton U., 1959-60, 62-63; vis. assoc. prof. Columbia, 1963-64; assoc. prof. Rockefeller U., 1964-68, prof., 1968—; cons. Brookhaven Nat. Lab., 1963-73; mem. Carnegie Panel on U.S. Security and Arms Control, 1981-83; vis. scientist European Ctr. for Nuclear Research, Geneva, Centre d'Etudes Nucléaires, Saclay, France, Max Planck Inst. für Physik, Munich, Fed. Republic Germany. Contbr. articles to profl. jours. Trustee Am. U. Beirut. Fellow Am. Phys. Soc.; mem. Council on Fgn. Relations. Club: Century (N.Y.C.). Home: 4715 Iselin Ave Bronx NY 10471-3323 Office: Rockefeller U New York NY 10021

KHUSH, GURDEV SINGH, geneticist; b. Rurkee, Punjab, India, Aug. 22, 1935; arrived in Philippines, 1967; s. Kartar Singh and Pritam Kaur (Dosanjh) Kooner; m. Harwant Kaur Grewal, Dec. 31, 1961; children: Ranjiv, Manjeev, Sonia, Kiran. BS in Agr., Punjab U., India, 1955; PhD, U. Calif., Davis, 1960; PhD (hon.), Punjab Agr. U., 1987. Rsch. asst. U. Calif., Davis, 1957-60, asst. geneticist, 1960-67; plant breeder Internat. Rice Rsch. Inst., Manila, 1967-72, plant breeder, head dept. plant breeding, 1972-85, prin. plant breeder, head dept. plant breeding, 1986—; cons. rice breeding programs Burma, Bangladesh, China, India, Indonesia, Iraq, Egypt, Sri Lanka, Bhutan, Kampuchea, Vietnam, Korea, Australia. Author: Cytogenetics of Aneuploids, 1973; editor: Rice Genetics Newsletter; contbr. articles to books and profl. jours. Recipient Borlaug award Coromandal Fertilizeers Ltd., Delhi, India, 1977, Japan prize Sci. and Tech. Found., Tokyo, 1987, Internat. Agronomy award Am. Soc. Agronomy, 1989. Fellow Rice Genetics Coop. (elected, sec. 1985—); mem. Genetic Soc. Am., Am. Soc. Agronomy (fellows award 1987), Indian Soc. Genetics and Plant Breeding (fellows award 1988), Crop Sci. Soc. Philippines (fellows award 1986), Indian Nat. Sci. Acad., U.S. NAS (fgn. assoc.), Third World Acad. Scis. Avocations: reading world history, jogging. Office: care Internat Rice Rsch Inst, PO Box 933, Manila The Philippines

KIA, SHEILA FARROKHALAEE, chemical engineer, researcher; b. Tehran, Iran, June 22, 1951; came to the U.S., 1981; d. Mohamadghasem and Touba (Naser) Farrokhalaee; m. Hamid G. Kia, Sept. 20, 1972; children: Michael, Kevin. MS, U. London, 1976; PhD, Cambridge U., 1980. Asst. prof. Arya-Mehr U., Iran, 1980-81; rsch. assoc. Case Western Res. U., Cleve., 1982, U. Mich., Ann Arbor, 1984-85; staff rsch. engr. GM Rsch. & Devel. Ctr., Warren, Mich., 1985—. Mem. editorial bd. Jour. Coating Tech., 1991—; contbr. articles to Water Rsch., Jour. Coating Tech., Colloid and Interface Jour. Mem. Am. Geophys. Union (com. mem. 1990—), Sigma Xi. Office: GM R&D Ctr 30500 Mound Rd Warren MI 48090-9055

KIBILOSKI, FLOYD TERRY, business and computer consultant, editor; b. Coldwater, Mich., Dec. 24, 1946; s. Floyd Benedict and Lucille Henrietta (Cholaj) K.; m. Peggy J. Foreman, Apr. 20, 1974; children: Sean, Angie. BBA in Acctg., Western Mich. U., 1978; MA in Computer Resource Mgmt., Webster U., 1989. CPA; cert. data processor. Enlisted USAF, 1966, advanced through grades to capt., ret., 1990, acct., 1966-70, computer technician, 1970-77, acct., computer cons. Bristol Leisenring and Co., 1977-80; computer cons. USAF, U.K., 1980-85, St. Louis, 1985-90; owner, editor Louisville (Ky.) Computer Times, 1990—; adj. prof. McKendree Coll., Lebanon, Ill., 1985-90, City Colls. of Chgo., 1982-84, Embry-Riddle Aero. U., 1983-84, U. Md., 1984; chair computer sci. dept. Sullivan Coll., Louisville, 1993—, adj. prof., 1990—. Author: Computer in the Audit, 1980, (human system seminar) Get the Most from Yourself, 1993. Christian youth leader Scott AFB, Belleville, Ill., 1985-90. Mem. Air Force Assn., Ret. Officers Assn. Republican. Roman Catholic. Avocations: traveling, photography. Home and Office: 3135 Sunfield Cir Louisville KY 40241-6527

KICE, JOHN EDWARD, engineering executive; b. Wichita, Kans., Sept. 11, 1949; s. Jack Wilbur and Anna Ruth (Jones) K.; m. Barbara Louise Svoboda, Oct. 27, 1973; children: Adam Wesley, Jason Mathew. BSBA and BS in Flour Milling Sci., Kans. State U., 1972; BS in Engring., Wichita State U., 1980. Registered profl. engr., Kans. Design engr. Kice Industries, Wichita, 1973-84, v.p. engring., 1984—; lectr. Wichita State U., 1980-86. Recipient Disting. Svc. award Assn. Operative Millers, 1988-90. Republican. Presbyterian. Achievements include patents for Positive Displacement Air Pump, Reciprocating Airlock Valve, Rotary Mixing Damper, Blade Type Mixing Damper, Conveying Air Velocity Control, Pneumatic Conveying Injector. Office: Kice Industries Inc 2040 S Mead St Wichita KS 67211-5017

KIDA, SIGEO, chemistry educator; b. Osaka, Japan, Aug. 28, 1927; s. Masaji and Sumie Kida; m. Yoshiko Wada, Mar. 8, 1953; children: Yoichi, Michiko Itasato. BSc, Osaka U., 1950, DSc, 1960. Rsch. assoc. Wakayama (Japan) U., 1951-56, lectr., 1956-61, assoc. prof., 1961-68, prof., 1968; prof. Kyushu U., Fukuoka, Japan, 1969-91; prof. chemistry Kumamoto (Japan) Inst. Tech., 1991—; mem. adv. bd. Inorganica Chimica Acta, Amsterdam. Mem. editorial bd. Coord. Chemistry Revs., Amsterdam, 1989—; author: Inorganic Chemistry (in Japanese), 1989. Mem. Chem. Soc. Japan (v.p. 1990-91, award 1986), Am. Chem. Soc. Achievements include studies on binuclear metal complexes. Home: 4-25 Kasumigaoka 3, Higashi-ku Fukuoka 813, Japan Office: Kumamoto Inst Tech, 22-1 Ikeda 4, Kumamoto 860, Japan

KIDD, JANICE LEE, nutritionist, consultant; b. Maryville, Tenn., Oct. 10, 1953; d. Lowery R. and Edna (Talbott) K.; m. Richard E. Woods, Jr., July 29, 1977; children: David, Jonathan. BS in Coordinated Program in Dietetics, U. Tenn., 1975, MBA in Mgmt. and Mktg., 1980, PhD in Nutrition, 1991. Registered dietitian; lic. dietitian-nutritionist, Tenn.; cert. home economist. Chpt. cons. Alpha Xi Delta Fraternity, Indpls., 1975-76; adminstrv. dietitian ARA Svcs. at Ft. Sanders Hosp., Knoxville, Tenn., 1976-79; nutrition cons. Y-12 Plant of Union Carbide Corp., Oak Ridge, Tenn., 1981-82, free standing dialysis ctrs., Tenn., 1979-88; owner Kidd and Co., Knoxville, 1981—; mem. adv. com. U. Tenn. Coordinated Program in Dietetics, 1984-86; mem. Nutrition and Health Adv. Com. to Food Policy Coun., Knoxville, 1983—; head Calorie Conscious Consumer Catering Project, 1987-88. Author: Tennessee Dietitian's Counseling Practicies Regarding the Role of Excercise in the Management of Non-Insulin Dependent Diabetics, 1991, Professional Guide to Stepping Out - A Guide for Exercise Educators Working with Persons with Diabetes, 1992. R.S. Tucker fellow U. Tenn. Coll. Bus., 1979; recipient Dorothy Nichols Sci. award Alpha Xi Delta, Indpls., 1988. Mem. Am. Dietetic Assn. (chmn., pub. rels. chmn. Cons. Nutritionists Practice Group 1993, Recognized Young Dietitian 1981), Am. Home Econs. Assn., Am. Diabetes Assn. (bd. dirs. Knoxville area chpt. 1983-87), Tenn. Dietetic Assn. (media rep. 1988-91), Knoxville Dist. Dietetic Assn. (Food Style TV host 1986—, Dial-A-Dietitian 1976—, Ask-A-Dietitian Column 1985—, Recognized Young Dietitian 1981), Soc. Nutrition Edn., Knoxville Home Econs. Assn. (pub. rels. com. chmn.), Golden Key, Kappa Omicron Nu. Office: Kidd and Co PO Box 11243 Knoxville TN 37939-1243

KIDD, JULIE JOHNSON, museum director. Chmn. Nat. Mus. Am. Indians, N.Y.C. Recipient Smithson medal Smithsonian Instn., 1991. Office: National Mus of Am Indians 3753 Broadway New York NY 10032*

KIDDER, RAY EDWARD, physicist, consultant; b. N.Y.C., Nov. 12, 1923; s. Harry Alvin and Laura Augusta (Wagner) K.; m. Marcia Loring Sprague, June 12, 1947 (div. Aug. 1975); children: Sandra Laura, David Ray, Matthew Sprague. BS, Ohio State U., 1947, MS, 1948, PhD, 1950. Physicist Calif. Rsch. Corp., La Habra, 1950-56, Lawrence Livermore Nat. Lab., Livermore, Calif., 1956—; mem. adv. bd. Inst. for Quantum Optics, Garching, Germany, 1976-90; bd. editors Nuclear Fusion IAEA, Vienna, 1979-84; cons. Sci. Applications Internat. Corp., San Diego, 1991—; mem. hon. adv. bd. Inst. for Advanced Physics Studies, La Jolla, Calif., 1991—. Contbr. chpts. to books. With USN, 1944-46. Recipient Humboldt award Alexander von Humboldt Found., 1988. Fellow Am. Phys. Soc. (Szilard award 1993); mem. AAAS, Sigma Xi. Achievements include research in physics of nuclear weapons, inertial confinement fusion, megagauss magnetic fields, laser isotope enrichment, containment of low-yield nuclear explosions. Home: 637 E Angela St Pleasanton CA 94566 Office: Lawrence Livermore Nat Lab PO Box 808 Livermore CA 94550

KIDDOO, RICHARD CLYDE, retired oil company executive; b. Wilmington, Del., Aug. 31, 1927; s. William Richard and Nellie Louise (Bounds) K.; m. Catherine Schumann, June 25, 1950; children: Jean L., William R., Scott F., David B. BSChemE, U. Del., 1948. With Esso Standard Oil and Esso Internat. Inc., Md., N.J., N.Y., 1948-66; internat. sales mgr. Esso Europe Inc., London, 1966-67; mng. dir., chief exec. officer Esso Pappas Indsl. Co., Athens, Greece, 1967-71; pres. Esso Africa Inc., London, 1971-72; v.p. Esso Europe Inc., London, 1973-81; v.p. mktg. Exxon Co., U.S.A., Houston, 1981-83; pres. Exxon Coal Internat., Coral Gables, Fla., 1983-86, ret.; vice chmn. mktg. com. Am. Petroleum, Washington, 1983. With USMC, 1945-46. Decorated Cross of King George I of Greece; recipient medal of distinction U. Del. Mem. AICE, Am. Petroleum Inst., Gibson Island Club, Circumnavigators Club. Home: Broadwater Way Gibson Island MD 21056-0224

KIDWELL, MICHAEL EADES, engineering executive; b. Nashville, Mar. 7, 1950; s. Leslie Elwin Kidwell and Juanita Phair (Reeves) Green; Lynn Warren Shaver, Dec. 27, 1977 (div. 1980); m. Marlene Celia Cohen, Aug. 2, 1986; 1 child, Alexander. BA, Emory U., 1972; MA, U. Tenn., 1976. Assoc. prof. Seton Hall U., South Orange, N.J., 1974-75; dir. forensics U. Ala., Tuscaloosa, 1975-80, Montgomery Bell Acad., Nashville, 1980-81; gen. mgr. Prince Analysis Inc., Washington, 1981-82; telecommunications mgr. Automated Scis. Group Inc., Silver Spring, Md., 1982-90; engring. cons. Perkom Sdn. Bhd., Kuala Lumpur, Malaysia, 1990-91; dir. systems integration Automated Scis. Group Inc., 1991-93; dir. bus. systems div. Synetics Corp., Columbia, Md., 1993—; cons. Ind. U. MUCIA Program, Shah Alam, Malaysia, 1990, Malaysian Stas. Inst., Shah Alam, 1991. Author/editor: Handbook of Issues on United States Education Policy, 1981, United States Foreign Military Sales, 1982, Uniform Criminal Justice Procedures, 1983, (with M. Cohen) Begining Debate, 1981. Treas. Potomac Valley Shetland Sheepdog Club, Washington, 1987. Mem. IEEE, Armed Forces Communications Assn., Eastern Communications Assn., Stonegate Citizens Assn., Emory U. Alumni Assn., Delta Sigma Rho, Tau Kappa Alpha (v.p. 1971-72). Democrat. Home: 508 Stone House Ln Silver Spring MD 20905

KIEFER, JOHN HAROLD, chemical engineering educator; b. New Ulm, Minn., Aug. 27, 1932; s. Harold Lyle and Margaret Olivia (Bentdahl) K.; m. Helen Murelle Chilton (div.); 1 child, Steven; m. Barbara June Berg, Dec. 30, 1971; children: Amy, Andrew. BS, U. Minn., 1954; PhD, Cornell U., 1961. Postdoctoral fellow Cornell U., Ithaca, N.Y., 1959-61; staff mem. Los Alamos (N.Mex.) Sci. Lab., 1961-66; assoc. prof. chem. engring. U. Ill., Chgo., 1966-72, prof. chem. engring., 1972—; acting head, 1989-91; joint appointee Argonne (Ill.) Nat. Lab., 1985-91. Grantee Dept. of Energy, 1978—, NSF, 1983, U.S. Israel Binat. Sci. Found., 1988. Office: U Ill 810 S Clinton Chicago IL 60607

KIEFFER, SUSAN WERNER, geology educator; b. Warren, Pa., Nov. 17, 1942. BS in Physics and Math., Allegheny Coll., 1964; MS in Geol. Scis., Calif. Inst. Tech., 1967, PhD in Planetary Scis., 1971; DSc (hon.), Allegheny Coll., 1987. Postdoctoral research geochemist UCLA, 1971-73, asst. prof. geology, 1973-79; geologist U.S. Geol. Survey, Flagstaff, Ariz., 1979-90; prof. geology Ariz. State U., Tempe, 1988—, Regents prof., 1991-93; prof., head dept. geol. sci. U. B.C., Vancouver, Can., 1993—. Co-editor: (with A. Navrotsky) Microscopic to Macroscopic: Atomic Environments to Mineral Thermodynamics, 1985. Alfred P. Sloan Found. fellow, 1977-79; W.H. Mendenhall lectr., U.S. Geol. Survey, 1980; recipient Disting. Alumnus award Calif. Inst. Tech., 1982, Meritorious Svc. award Dept. Interior, 1986, Spendiarov award Soviet Acad. of Scis., 1990. Fellow Am. Geophys. Union, Am. Acad. Arts and Scis., Mineral. Soc. Am. (award 1980), Geol. Soc. Am. (Arthur L. Day medal 1992), Meteoritical Soc.; mem. NAS. Avocations: athletics, music. Office: U BC, Dept Geol Sci, 6339 Storres Rd, Vancouver, BC Canada V6T 1Z4

KIEFL, ROBERT FRANCES, physicist, educator; b. Oct. 28, 1953. BASc with honors, Carleton U., 1976; MSc, U. B.C., 1978, PhD, 1982. Rsch. assoc. dept. physics U. B.C., 1982, NSERC Univ. rsch. fellow, 1987-90, asst. prof., 1990-92, assoc. prof., 1992—; NSERC postdoctoral fellow dept. physics U. Zurich, 1982-84; rsch. scientist II TRIUMF, 1984-87; mem. exptl. evaluation com. TRIUMF, 1984—; assoc. in superconductivity Can. Inst. Advanced Rsch., 1990; chair program com. 6th Internat. Conf. on Muon Spin Rotation, Maui, 1993; invited lectr. in field. Contbr. over 112 papers, notes, and comm. to refereed jours., 4 chpts. to books. Recipient Gerhard Herzberg medal Can. Assn. Physicists, 1992, Killiam Rsch. proze, 1992, McDowell medal, 1993; grantee NSERC, 1986 (two grants), 90-91, 92—, URF/NSERC, 1987-89, U. B.C., 1987. Office: Univ of BC, Physics Dept, Vancouver, BC Canada V6T 1Z1*

KIEFT, THOMAS LAMAR, biology educator; b. Feb. 6, 1951; married; 1 child. BA in Biology, Carleton Coll., 1973; MS in Biology, N.Mex. Highlands U., 1978; PhD in Biology, U. N.Mex., 1983. Lab. scientist serology dept. Scientific Lab. Div., Albuquerque, 1978-80; teaching asst. biology dept. U. N.Mex., Albuquerque, 1980-81; asst. curator microbiology collection Mus. Southwestern Biology, Dept. Biology, U. N.Mex., Albuquerque, 1981-82; asst. prof. Div. Sci. and Math., N.Mex. Highlands U., Las Vegas, 1982-83; vis. asst. rsch. microbiologist Dept. Plan and Soil Biology, U. Calif., Berkeley, 1983-85; asst. prof. dept. biology N.Mex. Inst. of Mining and Tech., Socorro, 1985-89, assoc. prof., 1989-93, prof., 1993—. Reviewer Applied and Environ. Microbiology, Biology and Fertility of Soils, Idaho DOE EPSCOR Program, Jour. Environ. Quality, NSF, N.Mex. Water Resources Rsch. Inst., Rsch. Found., U.S. EPA, U.S. Dept. Energy, Wyo. Abandoned Coal Mine Lands Rsch. Program; contbr. articles to profl. jours. including Cryobiology, Microbiology Ecology, Soil Biology and Biochemistry, Jour. Bacteriology, The Lichenologist, Applied and Environmental Microbiology, Current Microbiology, Geomicrobiology Jour., others. Numerous rsch. grants. Mem. AAAS, Am. Soc. Microbiology (N.Mex. br. sec.-treas. 1988—, br. rep. to the bd. edn. and tng. 1985-88, nominating com. div. N 1988, host state meeting at N.Mex. Tech, 1989), Sigma Xi (local membership com. 1985-88, treas. 1989, v.p. 1990-91). Achievements include rsch. on physiology and ecology of water-stressed microbes, soil microbiology, groundwater microbiology, chemolithotrophic microorganisms, biological ice nucleation, microbial ecology, environmental biology, biogeochemistry. Office: NMex Inst Mining/Tech Dept Biology Socorro NM 87801

KIEL, WILLIAM FREDERICK, architectural specifications consultant; b. Woodstock, Ill., Mar. 5, 1935; s. Hadwin Karl Martin and Laura Viola (Gile) K.; m. Greta Ann Stassen, Apr. 28, 1956 (div. 1969); children: Julie, Fred, Susan; me. Diane Helen Gumanowski, June 24, 1986; 1 child, William Colwell. BArch, Chgo. Tech. Coll., 1955. Archtl. draftsman Holabird & Root, Chgo., 1959-61; specificatin writer Schmidt, Garden & Erikson, Chgo., 1961-78; sr. specification writer Lester B. Knight Assocs., Chgo., 1978-81; chief of specifications Matthei & Colin, Chgo., 1981-82, Bertrand Goldberg

& Assocs., Chgo., 1982-86, Shayman & Salk, Northbrook, Ill., 1986-90; owner, pres. Con Spec Inc., Chgo., 1990—. Mem. Constrn. Specifications Inst. (cert., pres. 1972-73, 81-82, 82-83), Am. Arbitration Assn. Home: 4704 W Waveland Ave Chicago IL 60641 Office: Con Spec Inc 4704 W Waveland Ave Chicago IL 60641

KIELB, ROBERT EVANS, propulsion engineer; b. Youngstown, Ohio, May 11, 1949; s. John Joseph and Florence June (Evans) K.; m. Renee Marie Helle, Aug. 23, 1969; children: Eric Robert, Jason Joseph. BS in Aero. Engring., Purdue U., 1971; MSME, Ohio State U., 1975, PhD in Mech. Engring., 1981. Registered profl. engr., Ohio. Aerospace engr. USAF, Wright-Patterson AFB, Ohio, 1971-78; rsch. engr. NASA-Lewis Rsch. Ctr., Cleve., 1979-84, br. dep. mgr., 1984-87, br. mgr., 1987-89; sub-sect. mgr. GE Aircraft Engines, Cin., 1989—; advisor Coll. of Engring., Southern U., Baton Rouge, 1992—, Dept. Engring. Mechanics, Ohio State U., Columbus, 1990—. Editor: Bladed Disk Assemblies, 1985, 2d edit., 1987; contbr. over 25 articles ot profl. jours. Mem. ASME (com. chair 1988-92), AIAA. Methodist. Office: GE Aircraft Engines One Neumann Way MDA334 Cincinnati OH 45215

KIER, WILLIAM MCKEE, biologist, educator; b. Orleans, France, Apr. 25, 1956; s. Porter Martin and Mary (Lavely) K.; m. Kathleen K. Smith, June 9, 1984. BA, Colgate U., 1978; PhD, Duke U., 1983. Postdoctoral scholar Woods Hole (Mass.) Oceanographic Instn., 1983-84; postdoctoral fellow NATO, Marine Biol. Assn. U.K., Plymouth, Eng., 1984-85; vis. rsch. fellow U. Sheffield, Eng., 1985; assist. prof. U. N.C., Chapel Hill, 1985-91, assoc. prof. biology, 1991—; guest faculty Orgn. Tropical Studies, Monteverde, Costa Rica, 1986; panelist Sigma Xi Grants-in-Aid of Rsch., 1987—; course leader Program for Minority Advancement in Biomolecular Scis., Chapel Hill, 1991—; adv. panelist Life in a Phys. World exhibit N.C. Mus. Life and Sci., Durham, 1991. Contbr. sci. articles to profl. publs. Mentor N.C. Sch. Sci. and Math., Durham, 1990, 92. Recipient Presdl. Young Investigator Award NSF, 1987-93. Mem. AAAS, Am. Malacological Union, Am. Soc. Zoologists, Sigma Xi. Achievements include analysis of biomechanics of musculo-skeletal systems that rely on muscle for support and movement. Office: U NC Dept Biology CB 3280 Coker Hall Chapel Hill NC 27599-3280

KIERONSKA, DOROTA HELENA, computer science educator; b. Krakow, Poland, July 5, 1965; d. Jerzy Stefan and Anna Maria (Syrek) K. BSc with honors, U. Western Australia, 1986, PhD, 1991. Tutor U. Western Australia, Perth, 1988-90; lectr. Curtin U., Perth, 1990—; co-chair 3d Ann. Conf. on AI, Simulation and Planning, Perth, 1992. Contbr. articles to profl. jours. Vol. World Vision, Perth, 1991—. Mitsui Ednl. Found. studentship, 1986. Mem. IEEE (treas. Western Australia chpt. 1992—), Am. Assn. Artificial Intelligence, Indsl. Computing Soc., Planetary Soc. Avocations: reading, walking, badminton. Office: Curtin U Sch Computer Sci, PO Box U 1987, Perth 6001, Australia

KIERSCH, GEORGE ALFRED, geological consultant, educator emeritus; b. Lodi, Calif., Apr. 15, 1918; s. Adolph Theodore and Viola Elizabeth (Bahmeier) K.; m. Jane J. Keith, Nov. 29, 1942; children—Dana Elizabeth Kiersch Haycock, Mary Annan, George Keith, Nancy McCandless Kiersch Bohnett. Student, Modesto Jr. Coll., 1936-37; B.S. in Geol. Engring., Colo. Sch. Mines, 1942; Ph.D. in Geology, U. Ariz., 1947. Geologist 79 Mining Co., Ariz., 1946-47; geologist underground explosion tests and Folsom Dam-Reservoir Project U.S. C.E., Calif., 1948-50; supervising geologist Internat. Boundary and Water Commn., U.S.-Mex., 1950-51; asst. prof. geology, asst. prof., dir. mineral survey U. Ariz., Tucson, 1951-55, dir. Mineral Resources Survey Navajo-Hopi Indian Reservation, 1952-55; exploration mgr. resources survey So. Pacific Co., San Francisco, 1955-60; assoc. prof. geol. sci. Cornell U., Ithaca, N.Y., 1960-63, prof., 1963-78, prof. emeritus, 1978—, chmn. dept. geol. scis., 1965-71; geol. cons., Ithaca, 1960-78, Tucson, 1978—; chmn. coordinating com. on environment and natural hazards, Internat. Lithosphere Program, 1986-1991. Author: Engineering Geology, 1955, Mineral Resources of Navajo-Hopi Indian Reservations, 3 vols., 1955, Geothermal Steam-A World Wide Assessment, 1964; author: (with others) Advanced Dam Engineering, 1988; editor/author: Heritage of Engineering Geology--First Hundred Years 1888-1988 (vol. of Geol. Soc. Am.), 1991; editor: Case Histories in Engineering Geology, 4 vols., 1963-69; mem. editorial bd. Engring. Geology/Amsterdam. Mem. adv. council to bd. trustees Colo. Sch. Mines, 1962-71; mem. nine coms. NAE/NAS, 1966-88; reporter coordinating com. 1 CC1 Nat. Hazards U.S. GeoDynamics Com., 1985-90. Capt. C.E., U.S. Army, 1942-45. Recipient award for best articles Indsl. Mktg. Mag., 1964; NSF sr. postdoctoral fellow Tech. U. Vienna, 1963-64. Fellow ASCE, Geol. Soc. Am. (chmn. div. engring. geology 1960-61, mem. U.S. nat. com. on rock mechanics 1980-86, Disting. Practice award 1986, Burwell award 1992); mem. Soc. Econ. Geologists, U.S. Com. on Large Dams, Internat. Soc. Rock Mechanics, Internat. Assn. Engring. Geologists (U.S. com. 1980-86, chmn. com. 1983-87, v.p. N.Am. 1986-90), Assn. Engring. Geologists (1st recipient Claire P. Holdredge award 1965, hon. mem. 1985). Republican. Episcopalian. Clubs: Cornell (N.Y.C.); Statler, Tower (Ithaca); Mining of Southwest (Tucson). Home and Office: 4750 N Camino Luz Tucson AZ 85718-5819

KIERSTEAD, JAMES ALLAN, computer scientist; b. Orange, N.J., July 29, 1969; s. Allan Martin and Patricia Adele (Miller) K. BS in Computer Sci., Ramapo Coll. N.J.; postgrad., Poly. U., 1991—. Data processing mgr. Kanebridge Corp., Mahwah, N.J., 1988—; cons. J.K. Cons., Wayne, N.J., 1990—. Officer Bergen County SPCA Law Enforcement, South Hackensack, N.J., 1991—. Mem. Fraternal Order Police, ASPCA. Republican. Home: 9 Jefferson Pl Wayne NJ 07470 Office: Kanebridge Corp 360 Franklin Turnpike Mahwah NJ 07430

KIESELMANN, GERHARD MARIA, data processing executive; b. Saarbrücken, Germany, Dec. 8, 1956; s. Hermann and Gertrud (Weyer) K. Dipl., U. Karlsruhe, 1980; PhD in Physics, U. Bayreuth, 1985. Asst. scientist U. Bayreuth, 1982-86; computer systems developer Siemens AG, Munich, 1986-90; lab. head computer comm. Siemens Nixdorf Infos. Systems, Munich, 1990—; contbr. to X/Open Stds., Reading, U.K., 1988—; mem. IEEE/POSIX working group, 1990—. Contbr. articles to profl. jours.; co-author sci. tables: Landolt-Börnstein Vol. III/21a: Superconductors, 1990. Studienstiftung des deutschen Volkes scholar, 1975-80. Mem. IEEE Computer Soc., Am. Phys. Soc. Avocations: films, classical and jazz music, dancing.

KIESLING, ERNST WILLIE, civil engineering educator; b. Eola, Tex., Apr. 8, 1934; s. Alfred William and Louise (Kern) K.; m. Juanita Haseloff, Aug. 25, 1956; children: Carol, Chris, Max. B.S. in Mech. Engring, Tex. Tech. Coll., 1955; M.S. in Applied Mechanics, Mich. State U., 1959, Ph.D., 1966. Registered profl. engr. Asst. prof. Tex. Tech. Coll., 1959-63; sr. research engr. S.W. Research Inst., San Antonio, 1966-69; prof. civil engring. Tex. Tech U., Lubbock, 1969—, chmn. dept. civil engring., 1969-88, assoc. dean engring., 1988-93; prof. civil engring. Tex. Tech. U., Lubbock, 1993—. NSF faculty fellow, 1963-64. Mem. ASCE, Am. Soc. Engring. Edn., Tex. Soc. Profl. Engrs., Am. Underground Space Assn., Sigma Xi, Chi Epsilon, Tau Beta Pi. Home: 4912 94th St Lubbock TX 79424-4812 Office: Tex Tech U Dept Civil Engring Lubbock TX 79409

KIGOSHI, KUNIHIKO, geochemistry educator; b. Tokyo, July 7, 1919; s. Senpachi and Misao (Ito) K.; m. Noriko Hayashi, Oct. 14, 1944; children: Masako, Ikuko. MSc, U. Tokyo, 1942, DSc, 1954. Rsch. asst. Physics and Chemistry Rsch. Inst., Tokyo, 1942-46, Meteorol. Rsch. Inst., Tokyo, 1946-50; asst. prof. geochemistry Gakushuin U., Tokyo, 1950-54, prof., 1954-90, dir. Radiocarbon Lab., 1959—, dean Faculty Sci., 1969-7l, 82-84, prof. emeritus, 1990—. Author: Radiochemistry, 1956, Age Determination, 1965. Recipient award Nishina Meml. Found., Tokyo, 1970. Mem. Chem. Soc. Japan, Geochem. Soc. Japan, Am. Geophys. Union, Japanese Assn. for Quaternary Rsch. Home: Shibuya-ku Higashi 3-8-4, Tokyo 150, Japan Office: Gakushuin U, Toshima-ku Mejiro 1-5-l, Tokyo 171, Japan

KIHM, KYUNG D. (KEN KIHM), mechanical engineering educator; b. Seoul, Korea, Jan. 27, 1957; came to U.S., 1981; s. Hong-Chul and Yang-Ja (Park) K.; m. Hyeong-Ja Cha, Sept. 18, 1988; children: Grace, Christina. BS, Seoul Nat. U., 1979, MS, 1981; PhD, Stanford U., 1987. Registered profl. engr., Tex. Postdoctoral fellow mech. engring. Stanford U., Palo Alto, Calif., 1987; rsch. scientist mech. engring. Carnegie-Mellon U., Pitts., 1987-88; asst. prof. mech. engring. Tex. A&M U., College Station, 1989—. Contbr. more than 45 tech. articles to profl. jours. Recipient Select Young Faculty Fellow award Tex. Engring. Experiment Sta., 1990. Mem. AIAA, ASME, Inst. of Liquid Atomization and Spray System, Sigma Xi. Achievements include devel. of laser specklegram technique for thermal/fluid field measurements; synchronization of droplet sizing technique for intermittent fuel sprays; devel. dynamic property measurement technique for coal-water slurry mixtures. Home: 3013 Cortez St College Station TX 77845 Office: Tex A&M U Mechanical Engring Dept College Station TX 77843-3123

KIKUCHI, SHINYA, transportation engineer; b. Kobe, Japan, May 12, 1943; came to U.S., 1969; s. Saburo and Hideko (Tsukimoto) K.; m. Laura Velasco, Dec. 17, 1975. MS, Hokkaido U., Sapporo, Japan, 1969; PhD, U. Pa., 1974. Registered profl. engr., Mich. Assoc. Transp. Devel. Assn., Seattle, 1974-77; sr. project engr. GM Corp., Detroit, 1977-80, staff asst., 1980-82; asst. prof. civil engring. dept. U. Del., Newark, 1982-87, assoc. prof. civil engring. dept., 1987-93; prof., 1993—; dir. Del. Transp. Ctr., Newark, 1988—; cons. Chodai Co. Ltd., Tokyo, 1982—, Korea Transport Inst., Seoul, 1992. Contbr. articles to sci. jours. Active Ptnrs. of the Ams., Washington, 1985—, Del. Rail Passenger Assn., Wilmington, 1985—. Recipient Best Paper award Internat. Road Union, Lyon, France, 1992; rsch. grantee Fed. Transit Adminstrn., 1992. Mem. ASCE, Inst. Transp. Engrs., Soc. Logistics Engrs. Achievements include devel. of fuzzy set theory to transp. engring./planning process including schedule making, evaluation of transport svc. and modeling of transport investment decision process. Home: 5423 Crestline Rd Wilmington DE 19808 Office: U Del Civil Engring Dept Newark DE 19716

KIL, BONG-SEOP, biology educator; b. Muju, South Korea, Feb. 7, 1938; s. Sang-Man Kil and Soon-Im Song; m. Tae-Ok Lee; children: In-Sook, Ji-Hyun, Soo-Hyun, Joon-Il Kil. BS in Biology, Kongju Tchrs. Coll., Republic of Korea, 1960; MS in Plant Breeding, Wonkwang U., Iri, Republic of Korea, 1974; PhD in Plant Ecology, Chung-ang U., Iri, Republic of Korea, 1982. Tchr. Chonju Girl's High Sch., Chonju, Republic of Korea, 1960-76; prof. Biology Wonkwang U., Iri, Korea, 1976—; chair dept. Biology Wonkwang U., Iri, 1979-80, chief inst. natural sci., 1985-87, head dept. grad. sch. Biology, 1986-87; researcher Rural Dept. Agr., Iri, 1989-92; vice dean office Acad. Affairs Wonkwang U., Iri, 1989-91; vis. prof. Biology U. Mass., Amherst, 1980-81. Author ecology lab. manual, 1983; editor Korean Jour. Ecology and Botany, 1982—; contbr. articles to various handbooks and jours. With Korean Mil., 1960-61. Mem. Korean Soc. Ecology, Ecological Soc. Am., Botanical Soc. Am., Internat. Soc. Chem. Ecology, Internat. Assn. for Ecology, Korean Soc. Botany, Korean Soc. Plant Taxonomy. Methodist. Home: 102-1 Dongseohak-dong, Chonju Chonbuk, Republic of Korea 560-120 Office: Wonkwang U, 344-2 Shinyong-dong, Iri Chonbuk, Republic of Korea 570-749

KILBURN, PENELOPE WHITE, data processing executive; b. Freeport, N.Y., June 25, 1940; d. William Prescott and Marian (Churchill) White; m. Edwin Allen Kilburn, Feb. 7, 1964; children: Penelope Allen, Nancy Kitchen. BA, Barnard Coll., 1962. Elem. sch. tchr. Holmdel (N.J.) Bd. Edn., 1975-78; tech. writer Continental Data Ctr., Neptune, N.J., 1983-86; with Johnson & Higgins, N.Y.C., 1986-89; asst. v.p., 1989-91; v.p. Johnson & Higgins, N.Y.C., 1991—. Active mem. Jr. League, Monmouth County, 1973-80, sustaining mem., 1980—; chmn. St. Georges refugee com., Rumson, N.J., 1981-83; mem. St. Georges By the River Altar Guild, Rumson. Mem. Soc. for Tech. Communication. Episcopalian. Avocation: gardening. Office: Johnson & Higgins 125 Broad St New York NY 10004

KILBY, JACK ST. CLAIR, electrical engineer; b. Jefferson City, Mo., Nov. 8, 1923; s. Hubert St. Clair and Vina (Freitag) K.; m. Barbara Annegers, June 27, 1948; children: Ann, Janet Lee. BEE, U. Ill., 1947; MS, U. Wis., 1950; DEng (hon.), U. Miami, 1982; DSc (hon.), U. Wis., 1990; DEng (hon.), Rochester Inst. Tech., 1986; DSc (hon.), U. Ill., 1988; DSc, Rensselaer Poly. Inst., 1990. Program mgr. Globe-Union, Inc., Milw., 1948-58; asst. v.p. Tex. Instruments, Inc., Dallas, 1958-70; self-employed inventor Dallas, 1970—; disting. prof. elec. engring. Tex. A&M U., 1978-85; inventor monolithic integrated circuit, others; cons. to govt. and industry. Served with AUS, 1943-45. Recipient Nat. Medal of Sci., 1969, 90, Ballentine medal Franklin Inst., 1967, Alumni Achievement award U. Ill., 1974, Holley medal ASME, 1982, 89; inducted into Nat. Inventors Hall of Fame, U.S. Patent Office, 1981. Fellow IEEE (Sarnoff medal 1966, Brunetti award 1978, Medal of honor 1986); mem. NAE (Zworykin medal 1975, co-recipient Charles Stark Draper prize 1990). Home: 7723 Midbury Dr Dallas TX 75230-3211 Office: 6600 LBJ Fwy Ste 4155 Dallas TX 75240-6514

KILGORE, DONALD GIBSON, JR., pathologist; b. Dallas, Nov. 21, 1927; s. Donald Gibson and Gladys (Watson) K.; m. Jean Upchurch Augur, Aug. 23, 1952; children: Michael Augur, Stephen Bassett, Phillip Arthur, Geoffrey Scott, Sharon Louise. Student, So. Meth. U., 1943-45; MD, Southwestern Med. Coll., U. Tex., 1949. Diplomate Am. Bd. Pathology, Am. Bd. Dermatopathology, Am. Bd. Blood Banking. Intern Parkland Meml. Hosp., Dallas, 1949-50; resident in pathology Charity Hosp. La., New Orleans, 1950-54, asst. pathologist, 1952-54; pathologist Greenville (S.C.) Hosp. System, 1956—, dir. labs., 1985—; dir. labs. Greenville Meml. Hosp., 1972—; cons. pathologist St. Francis Hosp., Shriners Hosp., Greenville, Easley Baptist. Hosp.; vis. lectr. Clemson U., 1963—; asst. prof. pathology Med. U. S.C., 1968—; pres. Pathology Assocs. of Greenville, 1983—. Bd. dirs. Greenville County United Fund, 1966-74, 91—, Greenville Community Coun., 1968-71, Friends of Greenville County Libr., 1966-74; trustee Sch. Dist. Greenville County, 1970-90; bd. govs. S.C. Patient Compensation Fund, 1977—; patron Greenville Mus. Art, Greenville Little Theatre, 1956—; all state sch. bd. mem. Sc.C. Sch. Bd. Assn., 1990. Recipient Disting. Svc. award S.C. Hosp. Assn., 1976. Fellow Coll. Am. Pathologists (life, assemblyman S.C. 1968-71), Am. Soc. Clin. Pathologists (councilor S.C. 1959-62), Am. Soc. Dermatopathology; mem. Am. Assn. Blood Banks (life, adv. council 1962-67, insp. committeeman Southeast dist. 1965—), AMA (ho. of dels. 1978—), So. Med. Assn., S.C. Med. Assn., (exec. council 1969-76, 1978—, pres. 1974-75; A.H. Robins award for Outstanding Community Service 1985), Am. Soc. Cytology, Am. Coll. Nuclear Medicine, Nat. Assn. Med. Examiners, S.C. Inst. Med. Edn. and Rsch. (pres. 1974-80), S.C. Soc. Pathologists (pres. 1969-72), Richard III Soc. (co-chmn. Am. 1966-75), Am. Numis. Soc., Soc. Ancient Numismatics (life), Am. Numis. Assn. (life.), Blue Ridge Numis. Assn. (life), Royal Numis. Soc. (life), S.C. Numis. Assn. (life), Mensa (life), S.C. Congress Parents and Tchrs. (life), Greenville County Dental Soc. (hon. life), Greater Greenville C. of C., Greenville County Hist. Soc. (life), Preservation Soc. of Charleston (life), S.C. Hist. Soc. (life), Brookgreen Gardens Found. (life), Friends of Tewkesbury Abbey (life), Canterbury Cathedral Trust in Am. (life), Assn. Friends of Lincoln Cathedral (life), U.S. Power Squadron, Confrerie des Chevaliers du Tastevin (chevalier Atlanta chpt.), Soc. Med. Friends of Wine, Wine Acad. Am. (life), Soc. Wine Educators , Les Amis du Vin (life), Confrerie de la Chaine des Rotisseurs (bailli and conseiller Greenville chpt., L'Ordre de Mondial), Clan MacDuff Soc. Am. (exec. council 1980—), St. Andrews Soc. Upper S.C. (bd. govs. 1991-93), Phi Eta Sigma, Phi Chi. Democrat. Presbyterian (ruling elder 1969—). Clubs: Commerce (life), Poinsett (life), Torch (pres. 1964-65), Greenville Country (life), Thirty-Nine (pres. 1981-82), Chandon. Lodge: Rotary (Paul Harris fellow 1988). Home: 129 Rockingham Rd Greenville SC 29607-3620 Office: 8 Memorial Medical Ct Greenville SC 29605-4485

KILIAN, ROBERT JOSEPH, chemist; b. Chgo., Mar. 19, 1942; s. Edward Joseph and Irene Antoinette (Czemski) K.; m. Kathleen Mary Slayton, Apr. 15, 1989. BS in Chem. Engring., Northwestern U., 1964; PhD in Organic Chemistry, U. Wis., 1970. Sr. scientist Johnson & Johnson Products, North Brunswick, N.J., 1971-73; sr. scientist Johnson & Johnson Dental, East Windsor, N.J., 1973-78, mgr. devel., 1978-82; sr. group leader Johnson & Johnson Baby Products, Skillman, N.J., 1982-86; pres. Princeton (N.J.) ChemGroup, Inc., 1986—. Contbr. articles to profl. jours., chpts. to books. Recipient Harry McCormack award Am. Inst. Chem. Engrs., 1964. Mem. Am. Chem. Soc. (treas. 1985—), Assn. of Cons. Chemists (councilor 1992—), Sigma Xi, Pi Mu Epsilon, Tau Beta Pi. Achievements include 1 patent; developed and commercialized major new dental materials; developed new absorbent materials. Office: Princeton ChemGroup Inc 79 Marion Rd Princeton NJ 08540

KILKELLY, BRIAN HOLTEN, lighting company executive, real estate partner; b. East Orange, N.J., June 20, 1943; s. Daniel Joeseph and Mary Lorretta (Brown) K.; m. Judith Louise Kroger, May 21, 1966; children: Christopher, James. BS in Mktg., Fairleigh Dickinson U., 1968; MBA, Ga. State U., 1986. Sales rep. Thomas Lighting Div., Northern, N.J., 1965-68; mktg. svcs. Globe Inc., Hazelton, Pa., 1968-70; manpower devel./product mgr. Lithonia Lighting Div., Conyers, Ga., 1970-75; nat. market devel./southeastern mgr. Cooper Lighting Div., Atlanta, 1975-88; prin. Kilkelly Mgmt. Cons. Group, Conyers, 1988-89; partner Landmark Commercial & Investment Real Estate Inc., Conyers, 1988—; CEO Peachtree Lighting Inc., Covington, Ga., 1988—; Bd. dirs. Tech Able Handicapped Tech. Access. Contbr. articles to profl. jours. Active Kiwanis Internat. Conyers, 1988—. With USNR, 1961-67. Mem. Nat. Assn. Realtors (Ga. chpt., comml. coun., strategic planning com.), Nat. Fire Protection Assn. (joint 101/70 com.), Illuminating Engring. Soc. (chmn. tech. com. 1975—), Japan Am. Soc., Ga. Assn. Real Estate Exchangers, KC (grand knight, 1st degree team, chmn. com., Cert. of Merit 1990), EMBA Alumni Assn. (steering com., fund raising). Republican. Roman Catholic. Avocations: walking, U.G.A. football group, teaching, youth work, church work. Home: 2377 County Club Dr Conyers GA 30208

KILKSON, REIN, physics educator; b. Tartu, Estonia, Aug. 1, 1927; came to U.S., 1950; s. Ernst and Salme (Lehman) K. BS, Yale U., 1953, MS, 1954, PhD, 1956. Mem. tech. staff Bell Labs., Murray Hill, N.J., 1956-58; asst. prof. physics Wayne State U., Detroit, 1958-59; asst. prof. biophysics Yale U., New Haven, Conn., 1959-66; guest rschr. Karolinska Inst., Stockholm, Sweden, 1964-70; prof. physics and microbiology immunology U. Ariz., Tucson, 1970—. Contbr. 50 articles to Molecular Biophysics. Mem. AAAS, Am. Physical Soc., Am. Soc. Microbiology, Biophys. Soc., Sigma Xi. Achievements include research in the physical theory of the structure and evolution of the biological state of matter. Office: U Ariz Dept Physics Tucson AZ 85721

KILLDAY, K. BRIAN, organic chemist; b. Joliet, Ill., Jan. 24, 1961; s. Norbert Leo and Elsie Katherine (Prebe) K.; m. Shari Jo Keifer; June 8, 1991. BS in Chemistry, N. Mo. State U., 1983; MS in Chemistry, Mo. U., 1986. Tchr., researcher U. Mo., Columbia, 1983-88; chemist USDA, Gulfport, Miss., 1989, Harbor Br. Oceanographic, Ft. Pierce, Fla., 1990—; mem. safety com. Harbor Br. Oceanographic, Ft. Pierce, 1990-92, mem. recreation club 1989-92. Mem. city improvement com., White City, Fla., 1992. MFA scholar, 1979, Dugdale scholar, 1987. Mem. Am. Chem. Soc., Am. Soc. Pharmacognosy, Ins.t Food Technologists. Achievements include isolation of over sixteen new bioactive compunds from marine organisms. First to prove conclusively the structure of the cooked, cured meat pigment, Nitrosylhemochromogen. Office: Harbor Br Oceanographic 5600 Old Dixie Hwy Fort Pierce FL 34946

KILLEBREW, CHARLES JOSEPH, biologist; b. New Iberia, La., Dec. 15, 1941; s. Joseph and Rosamond (Lockett) K.; m. Winnie Hayman, May 25, 1974; children: Alyssa, Derek. BS in Zoology, Southeastern La. U., 1971, MS in Biology, 1973; postgrad., La. State U., 1987—. Rsch. biologist Gulf Coast Rsch. Lab., Ocean Springs, Miss., 1971-72; biologist La. Dept. Wildlife and Fisheries, Baton Rouge, 1975-89; tech. asst. Gov.'s Office Coastal Activities, Baton Rouge, 1989-90; exec. asst. to sec. La. Dept. Environ. Quality, Baton Rouge, 1992—; chmn. legis. task force on state natural and scenic rivers, Baton Rouge, 1987-88; exec. dir. Gov.'s Adv. Task Force on Environ. Quality, Baton Rouge, 1992—; curriculum adv. com. biol. scis. Southeastern La. U., Hammond, 1984-85. Contbr. to profl. publs. Mem. Am. Fisheries Soc. (chmn. tech. com. So. div. 1984-85), Sigma Xi, Phi Kappa Phi, Lambda Chi Alpha. Republican. Methodist. Achievements include contributions to design and devleopment of Louisiana's wetlands restoration and management program, research in structure and functioning of freshwater ecosystems. Office: La Dept Environ Quality PO Box 82263 Baton Rouge LA 70884-2263

KILLEBREW, ELLEN JANE (MRS. EDWARD S. GRAVES), cardiologist; b. Tiffin, Ohio, Oct. 8, 1937; d. Joseph Arthur and Stephanie (Beriont) K.; B.S. in Biology, Bucknell U., 1959; M.D., N.J. Coll. Medicine, 1965; m. Edward S. Graves, Sept. 12, 1970. Intern, U. Colo., 1965-66, resident 1966-68; cardiology fellow Pacific Med. Center, San Francisco, 1968-70; dir. coronary care, Permanent Med. Group, Richmond, Calif., 1970-83; asst. prof. U. Calif. Med. Center, San Francisco, 1970-83, assoc. prof., 1983-93, clin. prof. medicine, 1992—. Contbr. chpt. to book. Robert C. Kirkwood Meml. scholar in cardiology, 1970; recipient Physician's Recognition award continuing med. edn., Lowell Beal award excellence in teaching, Permanente Med. Group/House Staff Assn., 1992. Diplomate in cardiovascular disease Am. Bd. Internal Medicine. Fellow ACP, Am. Coll. Cardiology: mem. Fedn. Clin. Research, Am. Heart Assn. (research chmn. Contra Costa chpt. 1975—, v.p. 1980, pres. chpt. 1981-82, chm. CPR com. Alameda chpt. 1984). Home: 30 Redding Ct Belvedere Tiburon CA 94920-1318 Office: 280 W MacArthur Blvd Oakland CA 94611

KILLGORE, MARK WILLIAM, civil engineer; b. Vancouver, Wash., Jan. 1, 1956; s. Charles Roy and Barbara May (Coullahan) K.; m. Maria Teresa Preciadohopez, Aug. 13, 1993; children: Shannon Rose, Jason Philip. BCE, BA in Spanish, Seattle U., 1978; MS in Civil Engring., U. Wash., 1984. Registered profl. engr., Wash. Asst. engr. CRS Group, Seattle, 1978-81; assoc. engr. Ebasco Svcs. Inc., Bellevue, Wash., 1981-82, sr. assoc. engr., 1982-84, engr., 1985-88, sr. engr., 1988-91, prin. engr., 1991—. Author: Applying GIS to PMF Analysis in Microcomputer Environment, 1990. Cubmaster Boy Scouts Am., Bellevue, 1988-91. Mem. ASCE (pres. Seattle sect. 1992-93, Outstanding Young Mem. Zone IV), Am. Geophys. Union, Bellevue C. of C. (co. rep. 1985), Tau Beta Pi. Achievements include research on the hydraulic jump in a small rectangular channel. Home: 10824 158th Ct NE Redmond WA 98052 Office: Ebasco Svcs Inc 10900 NE 8th St Bellevue WA 98004-4405

KILLIAN, RUTH SELVEY, home economist; b. Rose Hill, Va., Sept. 8, 1921; d. James Robert and Mary Frances (Nolan) Selvey; m. Earl Willard Killian, Aug. 31, 1946. BS, Ea. Ky. State U., 1943; MA, Columbia U., 1952; student, Towson State U., 1974-75, Johns Hopkins U., 1957, 60. Tchr. Kings Mills (Ohio) High Sch., 1943-44, Oak Ridge (Tenn.) Sch. System, 1944-46; food cons. Girl's Clubs Am., Waterbury, Conn., 1946-47; food supr. Waterbury (Conn.) Hosp., 1947-49; tchr. home econs. Balt. County Schs., Towson, Md., 1950-62, county supr., 1962-68; coord. Cen. Balt. County Schs., Towson, Md., 1968-78; pvt. cons. Sebring, Fla., 1978—. Recipient Disting. Svc. award ARC, 1970, Cert. of Disting. Citizenship, State of Md., 1978. Mem. NEA, Am. Home Econs. Assn. (past mem. internat. scholarship com.), Md. Tchrs. Assn. (rep. 1950-92), Md. Home Econs. Assn. (past chmn. scholarship com.), Federated Woman's Club (chmn. publicity 1980-90, chmn. internat. affairs 1981-90), Federated Garden Club (chmn. publicity, cir. chmn. 1980-90), Golden Hills Garden Club (sec. 1992). Republican. Methodist. Avocations: reading, travel, golf, bridge, antiques. Home: 5145 NW 80th Ave Rd Ocala FL 34482

KILLORIN, EDWARD WYLLY, lawyer, tree farmer; b. Savannah, Ga., Oct. 16, 1928; s. Joseph Ignatius and Myrtle (Bell) K.; m. Virginia Melson Ware, June 15, 1957; children: Robert Ware, Edward Wylly, Joseph Rigdon. BS, Spring Hill Coll., Mobile, 1952; LLB magna cum laude, U. Ga., 1957. Bar: Ga. 1956. Pvt. practice in Atlanta, 1957—; ptnr. firm Gambrell, Russell, Killorin & Forbes, 1964-78; sr. ptnr. firm Killorin & Killorin, 1978—; lectr. Inst. Continuing Legal Edn. Ga., 1967—. Adj. prof. law Ga. State U., 1984-87. Chmn., Gov.'s Adv. Com. on Coordination State and Local Govt., 1973, Gov.'s Legal Adv. Council for Workmen's Compensation, 1974-76; bd. regents Spring Hill Coll., 1975-82, trustee, 1981-91. Served with AUS, 1946-47, 52-54. Recipient Disting. Alumnus award Spring Hill Coll., 1972. Mem. ABA, Internat., Ga. (chmn. jud. compensation com. 1976-77, chmn. legis. com. 1977-78), Atlanta (editor Atlanta Lawyer 1967-70, exec. com. 1971-74, chmn. legislation com. 1978-80) bar assns., Am. Judicature Soc., Lawyers Club Atlanta, Atlanta Legal Aid Soc. (adv. com. 1966-70, dir. 1971-74), Nat. Legal Aid and Defender Assn., Internat. Assn. Ins. Counsel (chmn. environ. law com. 1976-78), Atlanta Lawyers Found., Ga. Bar Found. (life), Ga. Def. Lawyers Assn. (dir. 1972-80), Ga. C. of C. (chmn. govtl. dept. 1970-75, chmn. workmen's compensation com. 1979—, Disting. Svc. award 1970-75), Def. Research Inst. (Ga. chmn. 1970-71), Spring Hill Coll. Alumni Assn. (nat. pres. 1972-74), U. Ga. Law Sch. Assn.

(nat. pres. 1986-87, Disting. Svc. Scroll 1989) Ga. Forestry Assn. (life, bd. dirs. 1969—, pres. 1977-79, chmn. bd. 1979-81), Am. Forestry Assn., Demosthenian Lit. Soc. (pres. 1957), Sphinx, Blue Key, Gridiron, Phi Beta Kappa, Phi Beta Kappa Assos., Phi Kappa Phi, Phi Delta Phi, Phi Omega. Clubs: Capital City, Peachtree Golf, Commerce, Buckhead (Atlanta); Oglethorpe (Savannah). Roman Catholic. Contbr. articles to legal jours. Home: 436 Blackland Rd NW Atlanta GA 30342-4005 Office: Killorin & Killorin 11 Piedmont Ctr NE Atlanta GA 30305-1733

KILMER, NEAL HAROLD, physical scientist; b. Orange, Tex., Apr. 24, 1943; s. Harold Norval and Luella Alice (Sharp) K. BS in Chemistry and Math., Northwestern Okla. State U., 1964; MS in Chemistry, Okla. State U., 1971; PhD in Chemistry, Mich. State U., 1979. Rsch. assoc. N.Mex. Petroleum Recovery Rsch. Ctr. N.Mex. Inst. Mining & Tech., Socorro, 1979-81, rsch. chemist, 1981-85, lectr. I geol. engring., 1984, asst. prof. mining engring., 1985-86; phys. scientist Phys. Sci. Lab. N.Mex. State U., Las Cruces, 1986—. Contbr. articles to profl. jours. Mem. Am. Chem. Soc., Am. Inst. Physics, Soc. Photo-Optical Instrumentation Engrs., Optical Soc. Am., Sigma Xi, Pi Mu Epsilon, Phi Lambda Upsilon. Presbyterian. Achievements include development of a preliminary screening procedure for testing polymers for suitability for use in enhanced oil recovery; major contributor in development of mathematical model of very low stratus clouds and sub-cloud regions. Home: 2200 Corley Dr Apt 14G Las Cruces NM 88001 Office: Phys Sci Lab PO Box 30002 Las Cruces NM 88003-0002

KILTIE, RICHARD ALAN, zoology educator; b. Camden, N.J., May 17, 1951; s. Thomas and Bertha Elizabeth (Bacon) K.; m. Grace Melvin Russell, May 8, 1981. AB, Princeton U., 1973, PhD, 1980; M Forestry Sci., Yale U., 1975. Lic. falconer. Postdoctoral fellow U. Fla., Gainesville, 1980-81, asst. prof. zoology, 1981-86, assoc. prof., 1986—; referee numerous biol. jours. and granting agys., 1981—; cons. Cross-Fla. Barge Canal Adv. Com., 1992. Contbg. author: Evoluation of Life Histories of Mammals, 1988, Great Cats, 1991; contbr. over 35 articles to Trends in Ecology and Evolution, Biol. Jour. Linnean Soc., Jour. Mammalogy. Rsch. grantee Whitehall Found., 1987-90. Mem. Am. Soc. Zoologists, Am. Soc. Naturalists, Am. Soc. Mammalogists, Soc. for Study Evolution. Achievements include research on ecology and morphology of mammals in remote neotropical areas, evolution of display organs and reproductive strategies, competition among carnivorans, psychophysics, computer vision and evolution of animal coloration. Office: U Fla Dept Zoology Gainesville FL 32611

KIM, BYUNG KYU, hematologist, consultant; b. Yang-San, Korea, Apr. 19, 1931; came to U.S., 1964; s. Yang Gil and Bong Sool (Cho) K.; m. Kildea Shim; children: Mihya, Norman, Noel. MD, Yonsei U., Seoul, Korea, 1959; MA, Brown U., 1983. Intern Yonsei U., Severance Hosp., Seoul, Korea, 1959-60, resident in clin. pathology, 1960-61, resident in internal medicine, 1961-64; rsch. assoc. in pharmacology and physiology SUNY, Upstate Med. Ctr., Syracuse, 1964-65; rsch. assoc. in biochem. pharmacology Brown U., Providence, R.I., 1965-67; assoc. staff, sr. investigator hematology Meml. Hosp., Pawtucket, R.I., 1967-82; asst. prof., then assoc. prof. Brown U. Med. Sch., Providence, 1975-83; sr. investigator sect. hematology New Eng. Deaconess Hosp., Boston, 1982-84; investigator Ctr. Blood Rsch., Harvard U., Boston, 1984-87; pre-clin. dir. Platelet Rsch. Product, Inc., Watertown, Mass., 1987-92; pre-clin. sr. dir., 1992—; cons. biol. rsch. Thomas J. Watson Rsch. Ctr., IBM, Yorktown Heights, N.Y., 1981-84, P-Z Lab., Worcester Found. Exptl. Biology, Shrewsbury, Mass., 1985—, Blood Bank Children's Hosp., Boston, 1985—; presenter at profl. confs. Contbr. articles to profl. jours., chpts. to books. Grantee NIH, 1971-76, U.S. AEC, 1971-80, R.I. Heart Assn., 1973-76, Meml. Hosp., Pawtucket, 1981, HHS, 1985-88. Mem. Am. Assn. Blood Bank, N.Y. Acad. Sci., Internat. Thrombosis and Hemostasis, Soc. Cryobiology, Boston Blood Club, New Eng. Korean Med. Assn. Achievements include development of platelet response to hypotonic stress test, method for preservation of blood platelets by freezing, research in therapeutic agent for thrombocytopenic patients from out-dated blood bank platelets. Home: 8 Morpheus Dr Cumberland RI 02864 Office: Platelet Rsch Products Inc 313 Pleasant St Watertown MA 02172

KIM, BYUNG-DONG, molecular biology educator; b. Chunan, Choongnam, Republic of Korea, Dec. 10, 1943; s. Boong-Han and Eul-Soon Kim; m. Il-Young Yoo, Apr. 1, 1972; children: Jihyun Jenifer, Soohyun Sarah. BS, Seoul (Rep. of Korea) Nat. U., 1966, MS, 1970; PhD, U. Fla., 1974. Postdoctoral fellow Sch. Medicine, U. Fla., Gainesville, 1975-76, asst. rsch. scientist, 1978-80; rsch. assoc. Sch. Medicine, W.Va. U., Morgantown, 1976-78; assoc. in rsch. Fla. State U., Tallahassee, 1980-83; asst. prof. U. R.I., Kingston, 1983-87; assoc. prof. molecular biology Seoul Nat. U., 1987—. Contbr. articles to profl. publs. Fellow AAAS, Biochem. Soc. Republic of Korea, Korean Soc. Horticultural Sci. (editor 1991—), Genetics Soc. of Korea, Korean Soc. Molecular Biology (assoc. editor 1990—). Baptist. Avocations: swimming, tennis, hiking, classical music. Office: Seoul Nat U Dept Horticultr, 103 Seodoon-dong, Suwon 441-744, Republic of Korea

KIM, HEESOOK PARK, chemist; b. Seoul, Dec. 13, 1958; came to U.S., 1981; d. Jaehoon and Youngsoon (Kim) Park; m. Sangsoo Kim, July 17, 1982. BS, Seoul Nat. U., 1981; PhD, Iowa State U., 1986. Sr. rschr. Agy. for Def. Devel., Daejeon, Korea, 1990—. Contbr. articles to Advances in Organometallics, Jour. Am. Chem. Soc., Inorganic Chemistry, Organometallics. Baptist. Office: Agy for Def Devel, PO Box 35 Yuseong, Daejon Republic of Korea 305-600

KIM, IH CHIN, pediatrician; b. Seoul, Korea, Aug. 6, 1925; s. Young Whan and Young Ho (Cho) K.; came to U.S., 1953, naturalized, 1965; MD, Seoul Nat. U., 1950; student Yon Sei U., 1944-46; postgrad. U. Pa., 1954-55; m. Helen Fern Wagner, Mar. 15, 1957; children: Catherine Joy Kim Smith, Stephen Thomas. Intern, Transp. Hosps., Seoul and Pusan, Korea, 1950-51; resident in pediatrics Pusan Children's Charity Hosp., 1951-53, Children's Hosp. Phila., 1953-55, fellow in pediatric gastroenterology, 1955-58, research assoc., 1958-67, med. staff, 1963-67; practice medicine, specializing in pediatrics, Easton, Pa., 1965—, Phillipsburg, N.J., 1971—; staff dept. pediatrics Hahnemann Med. Coll. and Hosp., Phila., 1967—, Easton Hosp., 1965—, Warren Hosp., Phillipsburg, N.J., 1966—, chief dept. pediatrics, 1978—; clin. asst. prof. pediatrics Hahnemann Med. Coll., Phila., 1971—. Diplomate Am. Bd. Pediatrics. Fellow Am. Acad. Pediatrics; mem. AMA. Presbyterian. Club: Country of Northampton County. Contbr. articles to med. jours. Address: 6 Ivy Court Easton PA 18042 Office: 545 Heckman St Phillipsburg NJ 08865

KIM, INN SEOCK, nuclear engineer; b. Busan, Korea, Jan. 2, 1954; s. Bongyee Kim and Soyeon Cha; m. Kyungmy Kim Moon, Dec. 18, 1983; children: Pio Moon, Jerome Moon. BS, Hanyang U., 1980; MS, Pa. State U., 1983; postgrad., U. Va., 1984-85; PhD, U. Md., 1988. Rsch. asst. Pa. State U., University Park, 1982-83, U. Va., Charlottesville, 1984-85; part-time cons. Rsch. Found. U. Md., College Park, 1985, rsch. fellow Systems Rsch. Ctr., 1985-88; mem. sci. staff Brookhaven Nat. Lab., Upton, N.Y., 1988—; cons. OECD Halden Reactor Project, 1993. Contbr. articles to profl. jours. With Korean Army, 1975-77. Mem. Am. Nuclear Soc., Assn. for Computing Machinery, Spl. Interest Group on Artificial Intelligence, Alpha Nu Sigma. Roman Catholic. Achievements include research in on-line process surveillance and diagnostics. Home: 5 Purdy Ave East Northport NY 11731-4501 Office: Brookhaven Nat Lab DNE Bldg # 130 Upton NY 11973

KIM, JAE NYOUNG, chemist; b. Seoul, Korea, Sept. 9, 1960; s. Mahn Hee and Sun Im (Park) K.; m. Mi Kyoung Choi, Apr. 13, 1986; children: Min Kyoung Kim, Min Seok Kim. BSc, Seoul Nat. U., 1984; PhD, Kaist, Taejeon, 1992. Rschr. Korea Rsch. Inst. of Chem. Tech., Taejon, 1984-91, sr. rschr., 1991—. Contbr. articles to profl. jours. including Jour. of Organic Chemistry, Bioorganic and Medicinal Chemistry Letters, Tetrhedron Letters, Chemistry Letters, Synthetic Comms. Mem. N.Y. Acad. Sci., Am. Chem. Soc., Korea Chem. Soc. Office: Korea Rsch Inst Chem Tech, Daedeong-Danji PO Box 9, Taejon 305-606, Republic of Korea

KIM, JAI SOO, physics educator; b. Taegu, Korea, Nov. 1, 1925; came to U.S., 1958, naturalized, 1963; s. Wan Sup and Chanam (Whang) K.; m. Hai Kyou Kim, Nov. 2, 1952; children: Kami, Tomi, Kihyun, Himi. B.Sc. in Physics, Seoul Nat. U., Korea, 1949; M.S. in Physics, U. Sask., Can., 1957, Ph.D., 1958. Asst. prof. physics Clarkson Coll. Tech., Potsdam, N.Y., 1958-59; asst. prof. physics U. Idaho, Moscow, 1959-62, assoc. prof., 1962-65, prof., 1965-67; prof. atmospheric sci. and physics SUNY, Albany, 1967—, chmn. dept. atmospheric sci., 1969-76, rep. Univ. Corp. for Atmospheric Research, 1970-76; cons. Korean Studies Program SUNY, Stony Brook, 1983-85; vis. prof. Advanced Inst. Sci. and Tech., Seoul, Korea, 1983; cons. U.S. Army Research Office, 1978-79, Battelle Meml. Inst., 1978-81, Environ. One Corp., 1978-84, N.Y. State Environ. Conservation Dept., 1976-82, Norlite Corp., 1982-84, Korean Antarctic Program, 1988—. Contbr. articles to profl. jours. Mem. Am. Inst. Physics, Am. Geophys. Union, Sigma Xi. Home: 33 Folmsbee Dr Albany NY 12204-1205 Office: 1400 Washington Ave Albany NY 12222-0001

KIM, JIN-KEUN, engineering educator; b. Milyang, Kyungnam, Republic of Korea, May 21, 1952; s. Yong-Sul Kim and Kye-Ah Ha; m. Min-Hee Choi, Dec. 26, 1976; children: Kil-Soo, Jung-Soo, Che-Young. BS, Seoul Nat. U., Seoul, Republic of Korea, 1975, MS, 1978; PhD, Northwestern U., 1985. Cert. profl. engr. Lectr. Ulsan U., Korea, 1979-81; asst. prof. Korea Advanced Inst. Sci. and Tech., Seoul, Republic of Korea, 1985-89; vis. scholar Tohoku U., Sendai, Japan, 1991; assoc. prof. Korea Advanced Inst. Sci. and Tech., Seoul, 1989—; cons. Dongyang Engring. Co., Seoul, 1989—, Daewoo Constrn. Co., Seoul, 1989-90, Rsch. Inst. of Korea Electric Power Co., Daejeon, Republic of Korea, 1990. Postdoctoral scholar Korea Sci. and Engring. Found., 1991. Mem. ASCE, ASTM, Soc. Exptl. Mechanics, Am. Concrete Inst. (bd. dirs., sec. Korea chpt. 1988-89), Prestressed Concrete Inst., Korea Concrete Inst. (sec. 1989-90, editor 1993—), Japan Concrete Inst., Réunion Internationale des Laboratoires d'Essais et de Recherches. Home: Hanbit Apt 132-1003 Eoun 99, Yusung Daejeon, Republic of Korea Office: Korea Advanced Inst Sci and Tech, Kusung 373-1, Yusung Daejeon, Republic of Korea

KIM, JONG SOO, polymer scientist; b. Korea, June 28, 1954. PhD, Poly. U., 1987. Dir. polymer rsch. lab. Oriental Chem. Industries, Inchon, Korea, 1990—. Achievements include development of polyvinyl alcohol and its derivatives. Office: Oriental Chem Industries, 587-102 Hak-Ik Dong, Inchon 402-040, Republic of Korea

KIM, KWANG HO, electrical engineer; b. Choonchun, Kangwon, Korea, July 25, 1946; came to U.S., 1971; s. Byung Hak and Sook Jung (Min) K.; m. Seyun Sohn, Sept. 23, 1972; children: Sarah, Rebecca. MSEE, Ohio State U.; PhDEE, Northeastern U., Boston. Project engr. Toledo Scale Co., Westerville, Ohio, 1974-81; engring. scientist RCA, Burlington, Mass., 1981-84; mem. tech. staff MITRE Co., Bedford, Mass., 1984—. Mem. IEEE (program mem., session chairs 1992, 93), AIAA, Internat. Neural Network Soc., Soc. Photo-Optical Instrument Engrs. Achievements include research in neural tracking, finite impulse response estimator, integrated kalman filter, proportional integral estimator, others. Home: 20 Lakin St Pepperell MA 01463 Office: The MITRE Corp 202 Burlington Rd Bedford MA 01730-1420

KIM, KYONG-MIN, semiconductor materials scientist; b. Seoul, Korea, Oct. 25, 1935; came to U.S. 1969; s. Yi-Oh and Yung-Shin K.; m. Heaja, Apr. 14, 1963; children: David H., Michelle. BSc, Seoul Nat. U., 1958; MSc, Tech. U. Braunschweig, Germany, 1962, PhD, 1965. Rsch. scientist Siemens Rsch. Ctr., Erlangen, Germany, 1965-67; vis. scientist N.S. Tech. Coll., Halifax, 1967-69; rsch. assoc. Ctr. for Material Sci., MIT, Boston, 1969-73; mem. tech. staff David Sarnoff Rsch. Ctr., RCA, Princeton, N.J., 1973-78; adv. scientist IBM, East Fishkill, N.Y., 1978-92; sr. mem. tech. staff MEMC Electronic Materials, Inc., St. Peters, Mo., 1992—; mem. rev. com. NASA Materials Processing in Space, Huntsville, Ala., 1970-72. Contbr. 35 articles to profl. jours. Recipient Skylab Achievement award NASA, 1974; Korean Atomic Energy Agy. scholar, 1959, West German Govt. scholar, 1963-65. Mem. Am. Phys. Soc., Electrochem. Soc., Am. Assn. Crystal growth. Achievements include 15 patents in area of semiconductor crystal growth, especially silicon crystal growth; convections in semiconductor crystal growth and study of silicon materials; others. Home: 1199 Whitmoore Dr Saint Charles MO 63304

KIM, MOON-IL, metallurgical engineering educator; b. Seoul, Republic of Korea, Sept. 11, 1929; s. Chang-Kyoo and Soon-Bok (Koh) K.; m. Yong Ok Hong, Oct. 24, 1968; children: Mee-Hye, Seong-Woong. B in Engring., Seoul Nat. U., 1956, DSc, 1970. Head researcher Sci. Rsch. Inst., Seoul, 1957-67; assoc. prof. Coll. Engring. Hanyang U., Seoul, 1966-70; prof. Coll. Engring. Yonsei U., Seoul, 1970—, chief Engring. Rsch. Inst., 1989-91; v.p. Korean Inst. Metals, Seoul, 1984-87; pres. Korean Soc. Heat Treatment, Seoul, 1988—. Author: Principles of Phase Diagram, 1976, Introduction to Mechanical Materials, 1982, Introduction to Physical Metallurgy, 1990, Engineering of Heat Treatment, 1990. Fellow Korean Soc. Heat Treatment; mem. Korean Inst. Metals, Korean Soc. Non-Destructive Testing, Korean Crystallographic Assn., Japan Inst. Metals, Japan Soc. Heat Treatment. Avocations: golf, baduk. Home: 408-6 Hongeun-Dong, Seodaemun-Ku Seoul 120-100, Republic of Korea Office: Yonsei U, 134 Shinchon-Dong, Seodaemun-Ku Seoul 120-749, Republic of Korea

KIM, MYUNG SOO, chemist, educator; b. Seoul, Korea, Oct. 28, 1948; s. Chonge Se and Young Soon (Lee) K.; m. Umi Kim, Oct. 17, 1979; children: Nuri, Ari, Sulki. BS, Seoul Nat. U., 1971; PhD, U. Chgo., 1976. Postdoctoral fellow Cornell U., Ithaca, N.Y., 1976-77; rsch. asst. Case Western Res. U., Cleve., 1977-78; rsch. assoc. prof. U. Utah, Salt Lake City, 1978-79; prof. dept. chemistry Seoul Nat. U., 1979—; vis. prof. U. Coll. Swansea, U.K., 1983; prin. researcher Korea Basic Sci. Rsch. Ctr., Seoul, 1989-91; coord. Korean Mass Sepctrometry Group, Seoul, 1990—. Author: Mass Spectrometry, 1987; editorial bd. Rapid Communication in Mass Spectrometry, 1990—; contbr. articles to sci. publs. and refereed jours. Grantee Ministry of Edn., Korea, 1980—, Korea Sci. and Engring. Found., 1980—. Mem. Am. Soc. Mass Spectrometry, Am. Chem. Soc., Korean Chem. Soc. (sec. phys. chem. div. 1988-89), Japanese Soc. Mass Spectrometry. Achievements include research in physical chemistry, spectroscopy and mass spectrometry. Office: Seoul Nat U Dept Chemistry, Kwanak Ku Shinlim Dong San 56-1, Seoul 151-742, Republic of Korea

KIM, MYUNGHEE, psychiatrist, child psychiatrist, psychoanalyst; b. Pusan, Korea, Nov. 8, 1932; came to U.S. 1959; d. Too Soo and Boo Sil (Kim) K.; m. Peter Reimann, June 28, 1962; children: Kim, Hannah. MD, Seoul Nat. U., Korea, 1959; Psychoanalyst, NYU, 1981. Intern Hackensack (N.J.) Hosp., 1959-60; resident in psychiatry Grassland Hosp., Valhalla, N.Y., 1960-62, Bronx Mcpl. Med. Ctr., 1962-63; staff psychiatrist Roosevelt Hosp., N.Y.C., 1964-65; fellow in child psychiatry Union County Psychiat. Clinic, Plainfield, N.J., 1968-70; child psychiatrist Child Guidance and Family Svc., Orange, N.J., 1970-72; pvt. practice child and adult psychiatry, psychoanalysis Springfield, N.J., 1972—; cons. Headstart Nursery Sch., Orange, 1971-72; faculty in psychiatry Bergen Pane Hosp., Paramus, N.J., 1977-83; instr. psychiatry N.J. Med. Sch., Newark, 1988—; clin. asst. prof. UMDNJ-Robert W. Johnson Med. Sch., 1991—. Editor N.J. Psychoanalytic Soc. Bull., 1989—; contbr. articles to profl. jours. Mem. Am. Psychoanalytic Assn., Internat. Psychoanalytic Assn., Pychoanalytic Assn. N.Y., Am. Psychiatric Assn., N.J. Psychoanalytic Soc., Seoul Psychoanalytic Study Group. Avocations: travel, gardening, art collecting, trekking. Home and Office: 272 Short Hills Ave Springfield NJ 07081-1029

KIM, PETER SUNG-BAI, biochemistry educator; b. Atlanta, Apr. 27, 1958; s. Mi Heh (Ryu) K.; m. Kathryn H. Spitzer; children: Michael, Jeremy. AB magna cum laude with distinction, Cornell U., 1979; PhD, Stanford U., 1985. Med. scientist tng. program fellow Stanford (Calif.) U., 1979-85; Whitehead fellow Whitehead Inst., Cambridge, 1985-88; asst. prof. biology MIT, Cambridge, 1988-92, assoc. prof., 1992—; asst. investigator Howard Hughes Med. Inst., Cambridge, 1990—. Recipient Stuart award for excellence in chemistry ICI Pharms. Group, 1989, Walter J. Johnson prize in molecular biology Jour. Molecular Biology, 1989, award in molecular biology NAS, 1993; scholar Rita Allen Found., 1990-92, Pew Charitable Trust, 1990-94. Mem. Whitehead Inst. Office: MIT/Whitehead Inst 9 Cambridge Ctr Cambridge MA 02142-1479

KIM, SAMUEL HOMER, pediatric surgeon; b. Boston, Sept. 11, 1936; s. Homer Tai-Sool and Ruth Aigyung (Choo) K.; m. Barbara Kim, Mar. 12, 1967 (dec. Nov. 1988); children: Susan, Stephen, Jeffrey, Jennifer. BA, Harvard Coll., 1958; MD, Harvard U., 1962. Diplomate Am. Bd. Surgery. Intern, 5th surg. svc. Boston City Hosp., 1962-63, resident in surgery, 1962-69; sr. registrar pediatric surgery Alday Children's Hosp., Liverpool, Eng., 1969-70; from asst. surgeon to assoc. surgeon, vis. surgeon Mass. Gen. Hosp., Boston, 1970—; pres. Mass. Gen. Physicians Corp., Boston, 1993—. Contbr. 8 chpts to books and 30 articles to profl. jours. Capt. USAF, 1964-66. Fellow ACS; mem. AAAS, AMA, Am. Acad. Pediatrics (surg. sect.), Am. Pediatric Surg. Assn., N.E. Med. Assn., New Eng. Surg. Soc., New Eng. Pediatric Soc., New Eng. Pediatric Surg. Soc., Mass. Med. Soc., Pan Am. Med. Assn., Boston Surg. Soc., Harvard Med. Alumni Orgn. (treas.). Achievements include research in pediatric surgery and pediatric urology. Home: 68 Whits End Rd Concord MA 01742 Office: Mass Gen Hosp Divsn Pediatric Surgery Boston MA 02114

KIM, SANGTAE, chemical engineering educator; b. Seoul, Korea, Aug. 2, 1958; naturlized, 1990; married; 2 children. BSChemE, Calif. Inst. Tech., 1979, MS, 1979; PhD, Princeton U., 1983. From asst. prof. to assoc. prof. chem. engring. U. Wis., Madison, 1983-90, prof., 1990—, prof. computer scis., 1990—; dist. prof. chem. engring. U. Wisconsin, Madison, Wisc., 1991—; cons. Amoco Oil Corp., Ill., 1983; process engr. Intel Corp., Santa Clara, Calif.; George A. Miller vis. scholar U. Ill., Urbana-Champaign, 1987; disting. vis. scholar U. Mass., Amherst, 1989; vis. prof. Pohang Inst. Sci. Tech., Korea, 1991. Author: (with S.J. Karrila) Microhydrodynamics: Principles and Select Applications, 1991; contbr. articles to profl. jours. Recipient Presdl. Young Investigator award NAS, 1985, NAS Award for Initiatives in Rsch., NAS, 1992, Allan P. Colburn award AICE, 1993; named Allan P. Colburn Meml. lectr. U. Del., 1989, Robert W. Vaughan lectr. Calif. Inst. Tech., 1991, Plenary lectr. Korean Inst. Chem. Engrs., 1991; Romnes Faculty fellow U. Wis., 1990; Petroleum Rsch. Fund grant Am. Chem. Soc., 1992—; grantee Office Naval Rsch., 1993—, NSF, 1993—. Achievements include research in dynamics of particulate suspensions; protein dynamics and simulations; computational methods on high-performance computers. Office: U of Wisc Dept Of Chem Engring Madison WI 53706*

KIM, SUNG CHUL, polymer engineering educator; b. Seoul, Korea, Jan. 1, 1945; s. Bok Sum and Kyung Ak (Park) K.; m. Myung Ja Rho, Sept. 29, 1968; children: Young Jin, Hyun Jin, Jae Jin. BS, Seoul Nat. U., 1967; D Engring., U. Detroit, 1975. Engr. Taekwang Industries, Ulsan, Korea, 1966-67, Korea Plastics Industries Co., Chin Hae, Korea, 1968; head lab. Korea Inst. Sci. & Tech., Seoul, 1975-78; prof. polymer engring. Korea Advanced Inst. Sci. Tech., Taejon, 1979—, dir., 1990—, dean, 1991—; sec. gen. Internat. Union Pure Applied Chem. Symposium, Seoul, 1988-89, Polymer Processing Soc. Symposium, 1989-90. Mem. Polymer Soc. Korea (dir. 1978—, acad. achievement award 1988), Am. Chem. Soc., Soc. Polymer Sci. Japan, Polymer Processing Soc., Korean Soc. Rheology (dir.1989—). Achievements include patents for IPN membrane materials for the separation of ethanol-water mixture; research in polymer alloys, reactive processing, polymers in membrane application. Office: Korea Advanced Inst Sci & Tech, 373-1 Kusongdong Yusongku, Taejon 305-701, Republic of Korea

KIM, SUNG WAN, pharmacology educator; b. Pusan, South Korea, Aug. 21, 1940; came to U.S., 1966; BS, Seoul U., MS; PhD, U. Utah. Asst. rsch. prof. U. Utah, Salt Lake City, 1971-73, asst. prof., 1974-76, assoc. prof., 1977-79, prof., 1980—, dir. Ctr. Controlled Chemical Delivery, 1986—; mem. study section SGY13 NIH, Bethesda, Md., 1985-89. Editor numerous books; contbr. articles to profl. jours. Recipient Clemson Basic Rsch. award Biomaterials Soc., 1987, Gov.'s medal for sci., State of Utah, 1989. Fellow Am. Assn. Pharm. Sci., Am. Inst. Med. Bioengring. Home: 4512 Juniper Dr Salt Lake City UT 84124 Office: U Utah Ctr Controlled Chem Delivery 421 Warkar Way Ste 318 Salt Lake City UT 84108

KIM, SUNG-HOU, chemistry educator, biophysical and biological chemist; b. Taegu, Korea, Dec. 12, 1937; s. Yong-Tai and Ok-Kum (Choi) K.; m. Rosalind Yuan, July 27, 1968; children: Christopher Sang Jai, Jonathan Sang-Joon. B.S., Seoul Nat. U., 1960, M.S., 1962; Ph.D., U. Pitts., 1966. Teaching asst. in chemistry Seoul Nat. U., 1960-62; lectr. chemistry Kun-Kook U., Seoul, 1960-62; research asst. dept. crystallography U. Pitts., 1963-66; research assoc. MIT, Cambridge, 1966-70, sr. research scientist, 1970-72; asst. prof. Duke U., Durham, N.C., 1972-73, assoc. prof., 1974-78; prof. chemistry U. Calif.-Berkeley, 1978—, Miller research prof., 1983-84; faculty sr. scientist Lawrence Berkeley Lab., 1979—, dir. div. structural biology, 1989—; exch. prof. Peking U., 1982; vis. prof. U. Paris, 1986; mem. adv. group biophysics and biophys. chemistry A Study sect. NIH, 1976-80; co-chmn. nucleic acids Gordon Research Conf., 1983; chmn. curriculum planning com. U.S. Nat. Com. for Crystallography, 1983-84. Contbr. numerous articles to sci. jours.; mem. editorial bd. Jour. Biol. Chemistry, 1979-83, Nucleic Acid Research, 1983-85. Awarded Presdl. Svc. Merit medal (Republic of Korea), 1985; recipient Sidhu award Pitts. Diffraction Conf., 1970, Rsch. Career Devel. award NIH, 1976-79, E.O. Lawrence award, 1987, Javits Neurosci. Investigator award HHS, 1988, Princess Takamatsu Cancer Found. award, 1989; Woo-Nam scholar Woo-Nam Found., Korea, 1959; Fulbright fellow, 1962; Guggenheim fellow, 1985-86; recipient Korean Overseas Compatriot's prize, 1993. Mem. Am. Soc. Biol. Chemists, Am. Chem. Soc., Am. Crystallographic Assn., AAAS, Korean Scientists and Engrs. in Am. Home: 1080 Country Club Dr Moraga CA 94556 Office: U Calif Dept Chemistry Berkeley CA 94720

KIM, TAE-CHUL, foundation engineer; b. Seoul, Korea, Mar. 13, 1943; arrived in Can., 1976; s. Woo-Jong and Chung-Sook (Lee) K.; m. Suzie Sook-Hee, May 13, 1972; children: Helen Sun-Young, Daniel Jun-Sik. BSc, Korea Mil. Acad., Seoul, 1966; MESc, U. Western Ont., London, Can., 1979. Commd. capt. Korean Army, 1966, advanced through grades to maj., 1975; liaison officer U.S. Forces Korean Army, Seoul, 1975-76; resigned Korean Army, 1976; rsch. assoc. U. Western Ont., London, Can., 1976-79; geotech. engr. Golder Assocs., Toronto, Ont., Can., 1979-82, sr. geotech. engr., 1982-87; found. design engr. Ont. Ministry Transp., Toronto, 1987-90, sr. found. engr., 1990—. Contbr. articles to profl. publs. Mem. ASTM, ASCE, EIC, ISSMFE, Can. Geotech. Soc. (exec. 1988-90), Korean-Can. Scientists and Engrs. Assn. (exec. 1990—), Assn. Profl. Engrs. Ont., CSCE, CGS. Mem. Liberal Party Can. Mem. United Ch. Can. Home: 4257 Camaro Ct, Mississauga, ON Canada L4W 3R1 Office: Ministry Transp Ont, CB Rm 315, 1201 Wilson Ave, Downsview, ON Canada M3M 1J8

KIM, WAN JOO, medicinal chemist; b. Kurye-Kun, ChonNam, Republic of Korea, Apr. 17, 1942; m. Bong Ae Chung, July 20, 1972; children: Mi-Jung, Jin, Mi-Kyoung. MS, U. Mainz, Fed. Republic Germany, 1972; PhD, U. Hamburg, Fed. Republic Germany, 1975. Sr. rschr. Schering Ag, Fed. Republic Germany, 1975-76; postdoctoral rschr. U. Cin., 1976-77; prin. rschr. Korea Inst. of Sci. and Tech., Seoul, 1977-85; prof. Songkyunkwan U., Seoul, 1985-86; head Dept. of Pharm. Chem., Korea Rsch. Inst. Chem. Tech., DaeJeon, Korea, 1986—; cons. DaeWoong Pharm. Ind., Seoul, 1988—; prof. Chungnam U., Daejeon, 1990—. Author: Steroid Chemistry, 1992; editor Korea Jour. of Medicinal Chemistry, 1991—. Recipient Presdl. award Korea Govt., 1983, Mog Ryun Chang award 1984, Ministry of Sci. and Tech. award, 1989, 1st Award of Comm. and Culture Chung Chin Ki Comm. and Culture Soc., 1986. Mem. Korea Chem. Soc., Korea Medicinal Soc., Am. Chem. Soc., West Germany Medicinal Soc. Roman Catholic. Achievements include 20 patents for cephalosporin antibiotics, 10 patents for quinolone antimicrobials. Home: Yuseong-ku, 38307 Doryong-Dong, Daejeon 305340, Republic of Korea Office: Korea Rsch Inst Chem Tech, 100 Jang-Dong, Yuseong-ku, Daejeon 305-606, Republic of Korea

KIM, YONG-DAL, physicist; b. Seoul, Korea, Dec. 23, 1957; came to U.S., 1986; m. Young-Eun Huh, June 25, 1988; 1 child, Jung-Ha. BS, Sung Kyun Kwan U., Seoul, 1980; MS, Tex. A&M U., 1990. Researcher Agy. Def. Devel., Seoul, 1979-85; rsch. asst. in physics Tex. A&M U., College Station, 1986-90, U. North Tex., Denton, 1990—. Co-author: Nuclear Instrumentation and Methods, 1992. Recipient fellowship Yuk Young Soo Found., 1978, Robert A. Welch Found., 1987. Mem. Am. Phys. Soc., Sigma Xi. Home: 502 Ave E Apt 208 Denton TX 76201 Office: Univ North Tex Dept Physics Denton TX 76203

KIM, YONGMIN, electrical engineering educator; b. Cheju, Korea, May 19, 1953, came to U.S., 1976; s. Ki-Whan and Yang-Whi (Kim) K.; m. Eunai Yoo, May 21, 1976; children: Janice, Christine, Daniel. BEE, Seoul Nat. U., Republic of Korea, 1975; MEE, U. Wis., Madison, 1979, PhD, 1982. Asst.

prof. U. Wash., Seattle, 1982-86, assoc. prof., 1986-90, prof., 1990—; bd. dirs. Optimedx; cons. MITRE Corp., McLean, Va., 1990, Lotte-Canon, Seoul, 1991, Seattle Silicon, Bellevue, Wash., 1990—, U.S. Army, 1989—, Neopath, Inc., Bellevue, Wash., 1989-90, Trinius Ptnrs., Seattle, 1989-91, Samsung Advanced Inst. Tech., Suwon, Republic of Korea, 1989-92, Daewoo Telecom Co., Seoul, 1989-91, Aptec Systems, Portland, Oreg., 1992—, Optimedx, Seattle, 1992—; bd. dirs. Image Computing Systems Lab., 1984—, Ctr. for Imaging Systems Optimization, 1991. Contbr. numerous articles to profl. jours., chpts. in books; editor Proceedings of the Annual International Conference of the IEEE EMBS, vol. 11, 1989, Proceedings of the SPIE Medical Imaging Conferences, vol. 1232, 1990, vol. 1444, 1991, vol. 1653, 1992, vol. 1897, 1993. Mem. various nat. coms.; chmn. numerous confs. Recipient Career Devel. award Physio Control Corp., 1982; grantee NIH, 1984—, NSF, 1984—, U.S. Army, 1986—, USN, 1986—; Whitaker Found. biomed. engring. grantee, 1986. Mem. IEEE (sr., Early Career Achievement award 1988, Disting. Speaker 1991), Assn. Computing Machinery, Soc Photo-Optical Instrumentation Engrs., Tau Beta Pi, Eta Kappa Nu. Presbyterian. Subspecialties: computer engring., high-performance image computing workstations, image processing, computer graphics, medical imaging, and multimedia workstations. Home: 4431 NE 189th Pl Seattle WA 98155

KIM, YOON BERM, immunologist, educator; b. Soon Chun, Korea, Apr. 25, 1929; s. Sang Sun and Yang Rang (Lee) K.; m. Soon Cha Kim, Feb. 23, 1959; children: John, Jean, Paul. M.D., Seoul Nat. U., 1958; Ph.D., U. Minn., 1965. Intern Univ. Hosp. Seoul Nat. U., 1958-59; mem. faculty U. Minn., Mpls., 1960-73; assoc. prof. microbiology U. Minn., 1970-73; mem., head lab. ontogeny of immune system Sloan Kettering Inst. Cancer Research, Rye, N.Y., 1973-83; prof. immunology Cornell U. Grad. Sch. Med. Scis., N.Y.C., 1973-83; chmn. immunology unit Cornell U. Grad. Sch. Med. Scis., 1980-82; prof. microbiology, immumology and medicine, chmn. dept. micorbiology and immunology U. Health Scis., Chgo. Med. Sch., 1983—; mem. Lobund adv. bd. U. Notre Dame, 1977-88. Contbr. numerous articles on immunology to profl. jours. Recipient research career devel. award USPHS, 1968-73. Mem. AAAS, Assn. Gnotobiotics (pres.), Internat. Assn. for Gnotobiology (founding), Am. Assn. Immunologists, Am. Soc. Microbiology, Am. Assn. Pathologists, Korean Med. Assn. Am., N.Y. Acad. Scis., Soc. for Leucocyte Biology, Internat. Soc. Devel. Comparative Immunology, Harvey Soc., Internat. Soc. Interferon and Cytokine Rsch., Chgo. Assn. Immunologists (pres.), Assn. Med. Soc. Microbiology Chmn., Internat. Endotoxin Soc. (charter), Soc. Natural Immunity (charter), Sigma Xi, Alpha Omega Alpha. Achievements include discovery of the unique germfree dolostrum-deprived immunologically "virgin" piglet model used to investigate ontogenic development and regulation of the immune system including T/B lymphocytes, natural killer/killer cells, and macrophages; research on ontogeny and regulation of immune system, immunochemistry and biology of bacterial toxins, host-parasite relationships and gnotobiology. Home: 313 Weatherford Ct Lake Bluff IL 60044-1905 Office: 3333 Green Bay Rd North Chicago IL 60064-3095

KIM, YOUDAN, aerospace engineer; b. Incheon, Korea, May 5, 1960; s. Yong-Duk and Choon Ja Kim; m. Hyekyung Park, June 15, 1986; children: Albert S., Christina S. MS, Seoul Nat. U., 1985; PhD, Tex. A&M U., 1990. Rsch. asst. Tex. A&M U., College Station, 1987-90, rsch. assoc., 1990-91; asst. prof. aerospace engirng. Seoul Nat. U., 1992—. Contbr. articles to Jour. of Guidance, Control and Dynamics, AIAA Progress in Astronautics and Aeronautics, Jour. Astronautical Scis. Mem. AIAA, Korean Soc. for Aero. and Space Scis. Roman Catholic. Office: Seoul Nat U, Dept Aerospace Engring, Seoul Republic of Korea 151-742

KIM, YOUNG KIL, aerospace engineer; b. Pusan, Korea, June 18, 1956; came to U.S., 1984; naturalized, 1988; s. Tae Hyun and Myong Ok (Shin) K.; m. Susan Katherine Hong, July 16, 1981; children: Steven Charles, Christina Kay. BS, Seoul Nat. U., Rep. of Korea, 1979; MS, Ga. Inst. Tech., 1985, PhD in Aerospace Engring., 1991. Rsch. engr. Korean Inst. Aero. Tech. Korean Air Lines, Seoul, 1978-84; vis. rsch. engr. Agy. for Def. Devel., Daedog, Republic of Korea, 1981-82; rsch. assoc. Univs. Space Rsch. Assn., Huntsville, Ala., 1991-93; rsch. engr. U. Ala. Rsch. Inst., Huntsville, 1993—. Mem. AIAA, Am. Helicopter Soc. Roman Catholic. Avocations: tennis, golf. Home: 1211 Willowbrook Dr SE Apt 2 Huntsville AL 35802-3827 Office: U Ala in Huntsville Rsch Inst Huntsville AL 35899

KIM, ZAEZEUNG, allergist, immunologist, educator; b. Hamhung, Korea, Feb. 21, 1929; came to U.S., 1967; s. Suh and Suyeo (Hahn) K.; m. Youngju Kim, June 2, 1961; children: Keungsuk, Maria. Student, Hamhung Med. Coll., Korea, 1946-50; MD, Seoul U., Korea, 1960; PhD in Immunology, U. Cologne, Fed. Republic of Germany, 1968. Diplomate Am. Bd. Allergy and Immunology. Intern Seoul Nat. U. Hosp., 1960-61, resident in medicine, 1961-63; resident in medicine Heidelberg U. Hosp., Fed. Republic of Germany, 1963-64; research fellow Max-Planck Inst., Cologne, 1965-67; fellow in hematology U. Tex., Houston, 1967-68; resident in allergy and immunology Temple U. Hosp., Phila., 1968-69; fellow in medicine Ohio State U., Columbus, 1969-71; instr. medicine Med. Coll. Wis., Milw., 1972-75, asst. prof., 1975-78, assoc. clin. prof., 1978—; practice medicine specializing in allergy and immunology Racine, Wis. Contbr. articles to profl. jours. Fellow Am. Acad. Allergy and Immunology, Am. Coll. Allergists; mem. AMA. Home: 4521 N Wildwood Ave Milwaukee WI 53211-1410 Office: 1300 S Green Bay Rd Racine WI 53406-4469

KIMBERLEY, BARRY PAULL, ear surgeon; b. Ont., Can.; s. Arthur and Maureen (Gibney) K.; m. Grace Khouri, May 13, 1989; 1 child, Caleigh. MD, Queen's U., Kingston, Ont., Can., 1983; PhD, U. Minn., 1990. Registered profl. engr., Can.; diplomate Am. Bd. Otolorynology. Scientific staff Bell No. Rsch., Ottawa, Ont., 1979-81; intern, resident U. Minn., 1983-88; asst. prof. dept. surgery U. Calgary, Alta., Can., 1990-92, Campbell McLaurin chair in hearing deficiencies, 1990—; Contbr. articles to profl. jours. including Jour. of Speech and Hearing Reseach, 1992, Acoustical Soc. Am., 1989, Laryngoscope, 1992. Recipient Clin. Investigator award Alta. Heritage Found. for Med. Rsch., 1991. Fellow Am. Acad. Otolaryngology, Royal Coll. of Physicians and Surgeons; mem. Acoustical Soc. Am., Assn. for Rsch. in Otolaryngology. Office: U Calgary, 3300 Hospital Dr NW, Calgary, AB Canada T2N 4N1

KIMBERLEY, JOHN A., mechanical engineer, consultant. BS in Mech. Engring., Lehigh U., 1942. Devel. engr. Curtis Wright; mgr. advanced products group Am. Bosch (name now Ambac); cons. Am. Bosch (name now Ambac), East Granby, Conn., 1986—; with United Techs. Mem. ASME (Internal Combustion engine award 1992), Soc. Automotive Engrs. Achievments include 15 patents in field and the development of a gas carburator, an electric governor for generator sets, a hydraulically-powered, electronically controlled unit injector research diesel fuel system, an optical start to injection timing, an automobile diesel injection pump with variable injection rate, and a system to control injection pressure in a pumpline-injection system independent of engine speed, a fuel system for injecting a slurry of coal and water, and a pilot diesel injector for gas/diesel engines. Home: 68 New Gate Rd East Granby CT 06026*

KIMBLER, DELBERT LEE, JR., industrial engineering educator; b. Whitman, W.Va., Sept. 8, 1945; s. Delbert and Jewell (Browning) K.; m. Elisabeth Moore Davidson, May 18, 1967. BS Engring. with distinction, U. South Fla., 1976; PhD in Indsl. Engring., Ops. Rsch., Va. Poly. Inst. and State U., 1980. Registered profl. engr., S.C., Fla. Asst. prof. dept. indsl. and mgmt. systems engring. U. South Fla., Tampa, 1980-84, assoc. prof. dept. IMSE, 1984-86; assoc. prof. dept. indsl. engring. Clemson (S.C.) U., 1986-90, head dept. indsl. engring., 1989-90, prof. dept. indsl. engring., 1990—; acad. adviser Systems Modeling Corp., State College, Pa., 1983-86; cons. engr. CIBA Vision Corp., Ga., 1992-93; coun. mem. Coll. Industry Coun. for Material Handling Edn., Charlotte, N.C., 1984-87. Editor: (proceedings) 19th Annual Simulation Symposium, 1986, (standard) ANSI Z94.17 in Industrial Engineering Terminology, 1990; editor (newsletter) Communications of SIM-IIE, 1989-90; sr. editor Jour. Mfg. Systems, 1991—. Mem., chmn. Zoning Bd. Adjustment, Clemson, 1989-92; unit commr. Boy Scouts Am., Clemson, 1990-92. With U.S. Army, 1966-70. Grantee 12 different sponsors, 1980-92; named Engring. Educator of Yr., S.C. Soc. Profl. Engrs., Piedmont chpt., 1992. Mem. Inst. Indsl. Engrs. (sr., pres. SIM 1988-90, Mfg. System award 1988), Am. Soc. for Quality Control (sr.), NSPE, Am. Soc. for Engring. Edn., Soc. Mfg. Engrs., Sigma Xi, Tau Beta Pi, Alpha Pi Mu. Democrat. Achievements include research in quality and the I.E. function in research and development. Office: Clemson U 104 Freeman Hall Clemson SC 29634-0920

KIMEL, WILLIAM ROBERT, engineering educator, university dean; b. Cunningham, Kans., May 2, 1922; s. Chester LeRoy and Klonda Florence (Hart) K.; m. Mila D. Brown, Aug. 14, 1952. B.S.M.E., Kans. State U., 1944, M.S.M.E., 1949; Ph.D. in Engring. Mechanics (WARF fellow), U. Wis., 1956. Registered profl. engr., Kans., Mo. Research asso. Argonne (Ill.) Nat. Lab., 1957-58; engr. Boeing Airplane Co., summer 1953, Westinghouse Electric Co., summer 1954, U.S. Forest Products Lab., 1955-56; from instr. to asso. prof. mech. engring. Kans. State U., Manhattan, 1946-58; prof., head nuclear engring. dept., 1958-68; dean, dir. engring. expt. sta., prof. nuclear engring. U. Mo., Columbia, 1968-86; Mem. Kans. Gov.'s Adv. Com. on Atomic Energy, 1961-68, chmn., 1966-68; mem. Mo. AEC, 1974-79; mem. Argonne Nuclear Engring. Edn. Com., 1959-84, chmn., 1966-67; AEC rep. UN Conf. on Peaceful Uses of Atomic Energy, 1964, Am. Nuclear Soc. rep., 1971; cons. N.Y. Regents External Degree Program for Nuclear Tech. Contbr. articles to profl. jours. Mem. Columbia Area Indsl. Devel. Commn., 1970—, chmn., 1973-76; mem. adv. com. to Office of Civil Def., 1964-70, chmn., 1969-70; bd. dirs. Engring. Colls. Consortium for Minorities, 1974-80, v.p., 1974-75; bd. dirs. Jr. Engring. Tech. Soc., 1976—, chpt. pub. relations com. chmn., 1976-78, pres., 1980-81; mem. Gen. Public Utilities Rev. Com. to evaluate operator-accelerated retraining program, TMI, 1979-80; mem. task force to establish training, edn. and accreditation requirements for nuclear plant operators Inst. Nuclear Power Ops., 1981-82, mem. accreditation bd., 1982—; mem. Mo. Gov.'s Task Force on Low-Level Radioactive Waste, 1981-85. Recipient Disting. Service award in engring. Kans. State U., 1972; Faculty Alumni award U. Mo., Columbia, 1979; Recipient Bliss award Am. Soc. Mil. Engrs., 1982; Disting. Service citation U. Wis. Coll. Engring., 1982; award of merit Engrs. Club St. Louis, 1983; Boss of Yr. award High Noon chpt. Am. Bus. Women's Assn., 1985. Fellow Am. Nuclear Soc. (chmn. edn. devel. com. 1963-66, 67-69, mem. exec. com. for tech. group edn. 1964-66, mem. planning com. 1966-75, chmn. 1970-74, mem. nominating com. 1969-70, 72-73, chmn. 79-80, exec. com. edn. div. 1968, chmn. edn. div. 1970-71, vice chmn. edn. div. 1969-70, dir. 1973-76, 77-81, exec. com. 1974-76, 77-79, mem. honors and awards com. 1975-78, engring. edn. and accreditation com. 1970-76, chmn. 1973, Governance award 1976, 77, 78, 79, 80, v.p. and pres.-elect 1977-78, pres. 1978-79, mem. public policy com. 1979-80, mem. blue ribbon coms. 1976-78, vice chmn. European Nuclear Conf. 1979), ASME; mem. Am. Soc. Engring. Edn. (chmn. nuclear engring. div. 1963-64, chmn. awards policy com. 1977-78, energy com. 1981-84, mem. steering com. council of tech. divs. 1964-65, mem. com. on relations with AEC 1964, 67—, chmn. nuclear engring. brochure com. 1966-68, engring. coll.'s council 1975-77), Engrs. Council for Profl. Devel. (dir. 1971-77, chmn. admissions com. 1973-76, mem. com. to evaluate advanced level accreditation 1975-77, exec. com. 1972, program chmn. 1972), NSPE (student profl. devel. com. 1973-75, edn. com. 1983-84, mem. legis. and govt. affairs com. 1984—, Ednl. Found. 1985—, task force on liability 1985—, NSPE award 1993), Profl. Engrs. in Edn. (chmn. N.C. region 1984-85, v.p.-elect 1985-86), Mo. Soc. Profl. Engrs. (mem. edn. adv. bd. 1968—, chmn. 1969-70, 73-74, 76-78, treas. 1978-79, sec. 1979-80, v.p. 1981-82, pres. 1983-84, profl. engrs. in edn. rep. to Nat. Soc. Profl. Engrs. polit. action com.), Columbia C. of C. (dir. 1968—, v.p. econ. devel. 1977—). Lodge: Rotary (v.p. pres.-elect Columbia 1977-79, pres. 1981-82). Home: 900 Yale Columbia MO 65203-1870

KIMERER, NEIL BANARD, SR., psychiatrist, educator; b. Wauseon, Ohio, Jan. 13, 1918; s. William and Ruby (Upp) K.; m. Ellen Jane Scott, May 22, 1943; children: Susan Leigh, Neil Banard, Brian Scott, Sandra Lynn. B.S., U. Toledo, 1941; M.D., U. Chgo., 1944; postgrad. (fellow), Menninger Sch., 1947-50. Diplomate Am. Bd. Psychiatry and Neurology. Intern Emanuel Hosp., Portland, Oreg., 1944; resident psychiatry Winter VA Hosp., Topeka, 1947-50; asst. physician Central State Hosp., Norman, Okla., 1950; cons. Central State Hosp., 1955—; chief out-patient psychiat. clinic U. Okla. Sch. Medicine, Oklahoma City, 1951-53; instr. dept. psychiatry, neurology and behavioral scis. U. Okla. Sch. Medicine, 1953-61, assoc. prof., 1961-69, clin. prof., 1969-85, clin. prof. emeritus, 1985—; practice medicine specializing in psychiatry Oklahoma City, 1953—; med. dir. Oklahoma City Mental Health Clinic, 1953-68; chmn. dept. psychiatry Bapt. Med. Ctr. Okla., 1979-83. Author: To Get and Beget, 1971; Contbr. articles in field to profl. jours. Mem. exec. com. Okla. Family Life Assn., 1958-60; bd. dirs. Oklahoma City Jr. Symphony Soc., 1959. Served as pfc ASTP, 1943-44; to capt. M.C. AUS, 1945-47. Fellow Am. Psychiat. Assn. (life); mem. AMA (life), Okla. Med. Assn., Oklahoma County Med. Soc., Oklahoma City Clin. Soc., AAAS, Alpha Kappa Kappa (pres. Nu chpt. 1943). Lodge: Rotary. Home: 2800 NW 25th St Oklahoma City OK 73107-2228 Office: 2600 NW Hwy Oklahoma City OK 73112

KIMES, MARK EDWARD, civil engineer; b. Youngstown, Ohio, May 16, 1965; s. Francis Dominic and Mary Rose (Ramsey) K.; m. Sharon K. Seman, Apr. 24, 1993. B Engring., Youngstown State U., 1991. Engr. in tng. State Registration Bd. for Profl. Engrs., Harrisburg, Pa., 1992—. Mem. NSPE (assoc.), ASPE, ASCE (assoc.). Republican. Roman Catholic. Home: 133 3rd St Pittsburgh PA 15225 Office: Baker Environ Inc 420 Rouser Rd AOP Bldg 3 Coraopolis PA 15108

KIMMEL, MAREK, biomathematician, educator; b. Gliwice, Poland, Sept. 17, 1953; came to U.S., 1982; s. Zbigniew and Janina (Rybicka) K.; m. Barbara Stankiewicz, June 27, 1981; children: Jan, Katarzyna. MS, Silesian Tech. U., Gliwice, 1977, PhD, 1980. Asst. prof. Silesian Tech. U., 1977-82, Sloan-Kettering Inst., N.Y.C., 1982-90; assoc. prof. dept. statistics Rice U., Houston, 1990—; cons. rsch. div. IBM, Yorktown Heights, N.Y., 1989—. Co-editor: Mathematical Population Dynamics, 1991; contbr. articles to profl. jours., including Jour. of Theoretical Biology, Biometrics, Genetics. Grantee Nat. Cancer Inst., 1985, NSF, 1989. Mem. Am. Statis. Assn., Inst. Math. Statistics, Am. Math. Soc., Cell Kinetics Soc. Roman Catholic. Achievements include research on the statistical model of natural history of lung cancer; mathematical model of unequal division of cells; methods of estimation of cell cycle kinetics; mathematical model of gene amplification. Office: Rice U Dept Stats PO Box 1892 Houston TX 77251-1892

KIMMEL, MELVIN JOEL, psychologist; b. Boston, July 9, 1944; s. Irving and Zelda (Gilinsky) K. MA, U. Mo., 1968; PhD, Wayne State U., 1974. Lic. psychologist, Md. NIH postdoctoral fellow SUNY, Buffalo, 1974-77; asst. prof. St. Ambrose Coll., Davenport, Iowa, 1977-80; rsch. psychologist U.S. Army Rsch. Inst., Alexandria, Va., 1980-89; sr. psychologist Balt. Gas & Electric Co., 1989—. Contbr. chpts. to books and articles to Jour. Personality and Social Psychology, Mil. Psychology. Named one of Outstanding Young Men Am., 1979. Mem. Am. Psychol. Assn., Am. Psychol. Soc., Soc. for Psychol. Study Social Issues, Soc. Indsl. and Organizational Psychology. Democrat. Home: 304 Kearney Dr Owings Mills MD 21117 Office: Balt Gas & Electric Po Box 1475 Baltimore MD 21203

KIMMEL, RICHARD JOHN, engineer; b. Brockport, N.Y., Feb. 21, 1942; s. John August and Verna Caroline (Richer) K.; m. Joy Darleen Isaac, May 23, 1964; children: Bret Richard, Amy Joy. BS in Engring., Ariz. State U., 1969. Registered profl. engr., Ariz. Engr. Ariz. Pub. Svc. Co., Phoenix, 1969-90, system engr., to 1990; sr. mech. engr. Volt Tech. Svcs., Garden Grove, N.Y., 1990-91; gen. engr. Dept. Energy, Idaho Falls, Idaho, 1991—. Hunter safety instr. Ariz. Dept. Game and Fish, 1970-90, Idaho Dept. Fish and Game, 1990—. With U.S. Army, 1964-66, Vietnam. Mem. ASME, ASCE, IEEE. Home: 867 N 800 E Shelley ID 83274

KIMMEL, ROBERT IRVING, communication systems design consultant, former state government official; b. Uniontown, Pa., Jan. 28, 1922; s. Andrew Filson and Dorothy Jean (Walker) K.; student Bucknell U., Lewisburg, Pa., 1940-41, 43-44, Washington U., St. Louis, 1942, Pa. State U., 1972; children—Donna Jean, Robert Filson, LuAnna Pat, Kevin Normaine, Gregory Paul. Self-employed entertainer, 1944; mgr. Cassiday Theaters, Midland, Mich., 1945-56; engring. illustrator Dow Chem. Co., Midland, 1956-59; engring. mgr. Radio Communications Co., Bloomsburg, Pa., 1959-64; chief electronics Pa. State Police, Harrisburg, 1964-74, dir. communications div., 1974-79; chmn. Pa. Law Enforcement Telecommunications Planning Com., 1976-79; design cons. Communications Systems Design Assocs., Harrisburg, 1979—; v.p. Partnership, Inc., 1980—; mgr. Paxton Herald and Paxton Herald West newspapers, 1981—; cons., lectr. in field. Mem. task force Cultural Center, Harrisburg, 1975-76; head coach Lakevue Midget Baseball Assn., 1976-78; pres. council St. Mark's Lutheran Ch., Harrisburg, 1975-79; bd. dirs. Harrisburg Performing Arts Co., 79, Emergency Health Services Fedn., 1984-86; v.p., bd. dirs. Am. Lung Assn. Cen. Pa., 1987—, treas., 1989—; bd. dirs. Am. Lung Assn. Pa., 1988-92; chmn. LP Customer Postal Coun., 1992—; bd. dirs. Salvation Army Rehab., 1992—; instr. Dancers Workshop, 1979—; sec.-treas. Susquehanna Valley Assn., 1984—; co-chmn. customer adv. coun. U.S. Postal Svc., 1991—. Served with USAAF, 1942-43. Recipient various pub. service awards, certs. of merit. Fellow Radio Club Am.; mem. Assn. Pub.-Safety Communications Officers (pres. 1978-79), Pa. Chiefs Police Assn. (life; chmn. frequency adv. com. 1967-79), Engrs. Soc. Pa. (pres. 1978-79), Nat. Assn. Dance and Affiliated Artists (past v.p.), Greater Harrisburg Arts Council (dir.), Internat. Platform Assn. Author papers in field. Developer vehicle location system, elec. security systems. Home: 880 Scenery Pl Harrisburg PA 17109-5323 Office: 101 Lincoln St Harrisburg PA 17112-2599

KIMURA, HARUO, aeronautical engineering educator; b. Kyuragi Town, Japan, Mar. 15, 1925; s. Seiichi and Asae (Hori) K.; m. Kazuko Kizuki, Dec. 20, 1956; 1 child, Yuko. M. Engring., Kyushu Imperial U., Fukuoka, Japan, 1947; D. Engring., Kyushu U., Fukuoka, 1967. Lectr. Kyushu U., Fukuoka, 1959-63, assoc. prof., 1964-70, prof., 1971-88, prof. emeritus, 1988—; prof. Nippon Bunri U., Oita, Japan, 1988—. Editor and author (with others): Handbook of Aerospace Engineering, enlarged edit. (Japanese), 1983; contbr. articles to Procs. of 7th World Congress IFAC, Memoirs of the Faculty of Engring. Kyushu U. Mem. Japan Soc. Mech. Engrs., Japan Soc. for Aeronautical & Space Scis., Japan Buoyant Flight Assn., AIAA. Home: 101 Green Hill Takajo, 29-09 Takajo Nishi Machi, Oita 870-01, Japan Office: Nippon Bunri U, 1727 Ichigi, Oita 870-03, Japan

KIMURA, ROBERT SHIGETSUGU, otologic researcher; b. Long Beach, Calif., June 5, 1920; s. Kumazo and Toku (Matsushita) K.; m. Ayako Iimura, Jan. 29, 1966; children: Tomomi Jane, Ibuki Anne. BA, Stanford U., 1948; PhD, Tokyo Med. & Dental U., 1976. Dir. histopathology lab. otolaryngology U. Chgo., 1950-61; rsch. assoc. Karolinska Inst., Stockholm, 1961-63; rsch. asst. Harvard Med. Sch., Boston, 1961-65; dir. electron microscope lab. Mass. Eye & Ear Infirmary, Boston, 1961—; rsch. assoc. Harvard Med. Sch., 1965-70, lectr., 1970-91, assoc. prof., 1991—; mem. subcom. human studies Mass. Eye & Ear Infirmary, 1967-85, mem. libr. com., 1970-80; mem. external adv. com. Baylor Coll. Medicine, Houston, 1981-84. Co-editor: Meniere's Disease, 1981, contbr.: Ultrastructural Atlas of the Inner Ear, 1984; contbr. articles to profl. jours. Named Javitz Neurosci. Investigator NIH, 1984. Mem. Assn. for Rsch. in Otolaryngology (Award of Merit 1990), Collegium Oto-Rhinolaryngology Amicitiae Sacrum (Shambaugh Prize in Otology 1988), Am. Otolog. Soc., Am. Assn. Anatomists. Achievements include development of animal model of endolymphatic hydrops; research in microcirculation of inner ear, pathology of inner ear due to vascular disorders, secretory cell distribution in vestibular labyrinth, electron microscopy of inner ear. Home: 21 Woodchester Dr Weston MA 02193 Office: Mass Eye & Ear Infirmary 243 Charles St Boston MA 02114

KIMZEY, JOHN HOWARD, chemical and aerospace engineer; b. New Orleans, Mar. 21, 1922; s. John Harrison and Mary Howard (Mallett) K.; m. Mary Eugenia Knight, Sept. 27, 1944; children: Carol Eugenia, Kaye Ellen, Mary Virginia, William Howard. AA, William and Mary Coll., Norfolk, Va., 1956; BS, Va. Poly., 1958. Registered profl. engr., Tex. Chem. engr. Nval Mine Engring. Facility, Yorktown, Va., 1958-62; aero. engr., mem. various publ. revs. Lyndon B. Johnson Space Ctr., Houston, 1962-86, subsystem mgr. spacecraft materials, 1985-86; engring. cons. Eagle Engring., Houston, 1986—. Author: Flammability as Related to Spacecraft Design and Operations, 1990, Material Selections for Spacecraft Applications, 1993; contbr. articles to profl. jours. Charter pres. Bay Area Lions Club, Houston, 1964. Lt. USN, 1943-53. Mem. NFPA (hyperbaric chambers com.). Baptist. Achievements include research in Skylab experiment M-479, Zero Gravity Flammability. Home: 1414 Basilan Ln Nassau Bay TX 77058

KINCHELOE, WILLIAM LADD, mechanical engineer; b. St. Louis, Oct. 18, 1925; s. Oscar Ladd and Merle (Jones) K.; m. Jean Margharita Sutherland, Apr. 1, 1951; 1 child, Kathryn. BSME, Washington U., St. Louis, 1949; MS, U. So. Calif., 1976; PhD, Calif. Coast U., 1993. Registered profl. engr. Mo, Calif., Nev. Jr. engr. Emerson Electric, St. Louis, 1949-51; project engr. Aerojet Gen., Azusa, Calif., 1955-65; program mgr. FMC Corp., San Jose, Calif., 1965-82; sr. assoc. D.R. Kennedy & Assocs., 1982—; program mgr. Physics Internat., San Leandro, Calif., 1983-87; proprietor Jayanbee Engring., Magalia, Calif., 1987—; program mgr. FMC Corp., San Jose, 1975-78, Aeroject Ordnance, Downey, Calif., 1982-83; system engr. Physics Internat., Lucerne, 1983-85; cons. Jayanbee Engring., 1987—. Author tech. reports, seminars in field. Asst. dir. Disaster Comms., Los Angeles County, 1960-65. Cpl. USAAF, 1944-46. Recipient award Ancient and Hon. Order of St. Barbara Artillery Bd., 1991. Mem. Am. Def. Preparedness Assn. (life), Acad. of Model Aeronautics, Pi Tau Sigma, Sigma Xi. Episcopalian. Achievements include patents for fuze delay timer, multipellet cartridge. Office: Jayanbee Engring Box 1444 5984 Pilgrim Ln Magalia CA 95954-1444

KINDT, THOMAS JAMES, chemist; b. Cin., May 18, 1939; s. James Michael and Barbara Katherine (Mayer) K.; m. Marie Louise Robinson, Sept. 5, 1964; children: Rachel Mary, James Thomas. BA cum laude, Thomas More Coll., Covington, Ky., 1963; PhD, U. Ill., 1967. Asst. rsch. scientist City of Hope Med. Ctr., Duarte, Calif., 1967-70; asst. prof. Rockefeller U., N.Y.C., 1970-73; assoc. prof. Rockefeller U., 1973-77, acting head lab. immunology/immunochemistry, 1975-78; vis. scientist Institut Pasteur, Paris, 1982-83; chief lab. immunogenetics NIH, Bethesda, Md., 1977—; assoc. dir. div. intramural rsch. Nat. Inst. Allergy and Infectious Diseases, Rockville, 1990—; adj. prof. Cornell U. Med. Coll., N.Y.C., 1973-78, Georgetown U., Washington, 1982—; sci. adv. Oncor Inc., Gaithersburg, Md., 1984—; mem. sci. adv. bd. Innovir Inc. N.Y.C. Dep. editor Jour. Immunology, 1987-92; assoc. editor FASEB Jour., 1992—; N.Am. editor Rsch. in Immunology, 1990—; co-author: Antibody Enigma, 1984; contbr. articles to profl. jours.; patentee in field. Adv. com. Multiple Sclerosis Found., N.Y.C., 1984—. With USN, 1957-59. Recipient Awd. for Exceptional Achievement, Asst. Sec. for Health, 1985. Mem. Am. Heart Assn., Am. Soc. Biol. Chemists, Am. Assn. Immunology, Am. Chem. Soc., Harvey Soc. Democrat. Roman Catholic. Home: 8313 Still Spring Ct Bethesda MD 20817-2727 Office: NIAID Twinbrook II Facility 12441 Parklawn Dr Rockville MD 20852

KING, ALBERT I., bioengineering educator; b. Tokyo, June 12, 1934; U.S. citizen; married; 2 children. BSc, U. Hong Kong, 1955; MS, Wayne State U., 1960; PhD in Engring. Mechanics, 1966. Demonstrator civil engring. Hong Kong, 1955-58; asst., instr. engring. mechanics Wayne State U., 1958-60, from instr. to assoc. prof., 1960-76, assoc. neurosurgery Sch. Medicine, 1971—, prof. bioengring., 1976—, Disting. Prof. mech. engring., 1990—. Recipient NIH Career Devel. award, Volvo award, 1984. Mem. Am. Soc. Engring. Edn., Am. Soc. Mech. Engrs. (Charles Russ Richards Meml. award 1980), Am. Acad. Orthopaedic Surgeons, Sigma Xi. Achievements include rsch. in human response to acceleration and vibration, automotive and aircraft safety, biomechanics of the spine, mathematical modelling of impact events, low back pain rsch. Office: Wayne State University Bioengineering Ctr 818 W Hancock St Detroit MI 48202*

KING, ALISON BETH, pharmaceutical company executive; b. Stamford, Conn., July 27, 1957; d. Stewart Alan and Marilyn Eugenia (Pearson) K.; m. Todd Stephan Miller, Sept. 23, 1989; 1 child, Travis King Miller. AB, Stanford U., 1979; PhD, Cornell U., 1987. Staff scientist Norwich (N.Y.) Eaton Pharms. div. of Procter & Gamble Co., 1987-91, group leader nutrition rsch., 1990-91, study dir. drug safety assessment, 1992; mgr. global health policy analysis Procter & Gamble Pharms., Norwich, 1992—; mem. tech./scis. com. Enteral Nutrition Coun., Atlanta, 1987-91; nutrition rev. bd. Physicians Assn. AIDS Care, Chgo., 1989. Contbr. to profl. publs. English tchr., Vols. in Asia, Taiwan, 1978; bd. dirs. Chenango County unit Am. Cancer Soc., Norwich, 1989-90; adv. com. The Children's Ctr., Norwich, 1992—. Sage grad. fellow Cornell U., 1981. Mem. Pharm. Mfrs. Assn. (pub. policy group 1992—), Nat. Pharm. Coun. (scientific affairs com.

1993—), Internat. AIDS Soc., Sigma Xi, Phi Kappa Phi. Office: Procter & Gamble Pharms 17 Eaton Ave Norwich NY 13815

KING, CHARLES HERBERT, JR., civil engineer; b. Hopkinsville, Ky., June 5, 1927; s. Charles Herbert and Emma Elizabeth (Pierce) K.; m. Audrey Ann Remmers, Dec. 31, 1955. BCE, U. Ky., 1957. Registered profl. engr., Ala., Ga., Fla., S.C., N.C., Va., W.Va., Md., Tenn., Ky., Ohio, Ill., Ariz., Colo., Nev., Miss. Constrn. engr. McDowell Co., Inc., Cleve., 1957-59; civil engr. USDA, Lexington, Ky., 1959-66; planning engr. U. Ky., Lexington, 1966-70; chief engr. GAC Properties, Inc., Miami, Fla., 1970-73; asst. v.p. Greiner Engring. Sci., Inc., Tampa, Fla., 1973-85; exec. v.p. W.K. Daugherty Consulting Engrs., Clearwater, Fla., 1985-91; ret., 1991. Editor: Proceedings of Urban and Regional Conflict in Water Related Issues Symposium, 1991. With U.S. Army, 1950-57, Korea. Recipient Cert. of Merit Gov. of Ky., 1966. Fellow ASCE (chmn. ops. design and maintenance com. 1989-91, pub. comm. irrigation and drainage jour. 1987-89); mem. Nat. Soc. Profl. Engrs. (sr.), Fla. Engring. Soc. (sr.), Tau Beta Pi. Democrat. Methodist. Achievements include patent for interlocking joint for precast concrete member. Home: 2249 Bascom Way Clearwater FL 34624

KING, DAVID STEVEN, quality control executive; b. Easton, Pa., May 16, 1960; s. Carl Stanley and Verna Marilyn (Frey) K. BS in Stats., Va. Poly. Inst. & State U., 1982. Cert. statistician. Quality and product design engr. SIECOR Corp., Hickory, N.C., 1982-86; quality supr. Alcatel Cable Systems, Fordyce, Ark., 1987-89; quality assurance mgr. Aeroquip Corp., Heber Springs, Ark., 1989-90; quality control mgr. Progress Lighting, Cowpens, S.C., 1990-92; quality mgr. Dana Corp., Greenville, S.C., 1993—. Mem. Am. Soc. for Quality Control, Am. Stats. Assn., ASTM. Home: 11 Vale St Spartanburg SC 29301-1224 Office: Dana Corp 500 Garlington Rd Greenville SC 29615

KING, DON E., air transportation executive; b. Indpls., May 11, 1939; 3 children. BA, Purdue U., 1966; postgrad., Highline C.C., 1990-92. Bus. (computer) systems analyst Boeing Co., Seattle, 1979—. Mem. Am. Rhododendron Soc. (sec., treas. Great Rivers chpt. 1977-79, v.p. 1979), Seattle Rhododendron Soc. (bd. dirs., pres. 1991-93), Rhododendron Species Found. (bd. dirs. 1984—, pres. 1991—), Magnolia Soc. Office: RSF PO Box 3798 Federal Way WA 98063

KING, FREDERICK ALEXANDER, neuroscientist, educator; b. Paterson, N.J., Oct. 3, 1925; s. James Aloysius and Louise Bisset (Gallant) K.; children: Alexander Karell, Elizabeth Gallant. A.B., Stanford, 1953; A.M. (John Carrol Fulton scholar 1953-55), Johns Hopkins, 1955, Ph.D., 1956. Instr. psychology Johns Hopkins U., 1954-56; asst. prof. psychiatry Ohio State U., 1957-59; mem. faculty Coll. Medicine, U. Fla., 1959-78, asst., then asso. prof., then prof. neurosurgery, 1965-69, prof., chmn. dept. neurosci., 1969-78; dir./co-dir. Center Neurobiol. Scis., 1964-78; dir. Yerkes Regional Primate Rsch. Ctr., rsch. prof. neurobiology, prof. anatomy and cell biology, psychology, assoc. dean Sch. Medicine Emory U., Atlanta, 1978—; adj. prof. psychology Emory and Ga. Inst. Tech.; mem. adv. com. Primate Research Centers, NIH, 1969-73; mem. psychobiology adv. panel, biol. and med. scis. div. NSF, 1963-67, cons. med. and biol. scis., 1967-70; chmn. research scientist devel. rev. com. NIMH, 1969-70, 75-78, chmn. com. for coordination and communication for dirs. in biol. research tng. programs, 1972—; sec.-treas. Fla. Anat. Bd., 1969-71; vice chmn. bd. sci. advisers Yerkes Regional Primate Research Center, 1974-78; mem. brain sci. com. NRC-Nat. Acad. Scis., 1974-78; mem. internat. sci. adv. bd. Nat. Mus. Kenya, 1983—; mem. Nat. Acad. Scis. com. for nat. survey of lab. animals, 1985-88, ILAR Com., 1986—. Gen. editor: Handbook of Behavioral Neurobiology, 7 vols, 1972-85; Contbr. articles to profl. jours. Served with USNR, 1943-46, 51. Research fellow NIH, 1955-56; spl. fellow NIMH, Inst. Physiology, U. Pisa (Italy) Faculty Medicine, 1961-62. Mem. Internat. Neuropsychology Soc. (sec.-treas.), Soc. Neurosci. (chmn. com. on edn., chair com. on animals in rsch. 1987, chair subcom. primates in rsch. 1987—), Am. Psychol. Assn. (chmn. membership com. div. physiol. and comparative psychology, chmn. com. for animal research and experimentation, bd. sci. affairs, spl. citation for leadership in sci. psychology 1984, pres. div. comparative and physiol. psychology 1989-90), Am. Assn. for Accrediation of Lab. Animal Care (bd. trustees 1987-90), Nat. Assn. Biomed. Rsch. (bd. dirs. 1987-89), Incurably Ill for Animal Rsch. (bd. advisors 1987—), Americans for Med. Progress (bd. dirs.), NIH (adv. com. to dir. 1989—). Home: 2681 Galahad Dr NE Atlanta GA 30345-3626 Office: Emory U Yerkes Regional Primate Rsch Ctr Atlanta GA 30322

KING, JERRY WAYNE, research chemist; b. Indpls., Feb. 19, 1942; s. Ernest E. and Miriam (Sanders) K.; m. Bettie Maria Dunbar, Aug. 8, 1965; children: Ronald Sean, Valerie Raquel, Diana Lynn. BS, Butler U., 1965; PhD, Northeastern U., 1973; fellow, Georgetown U., 1973-74. Research chemist Union Carbide Corp., Bound Brook, N.J., 1968-70; asst. prof. dept. chemistry Va. Commonwealth U., Richmond, 1974-76; research scientist Arthur D. Little, Inc., Cambridge, Mass., 1976-77; research assoc. Am. Can Co., Barrington, Ill., 1977-79; research scientist CPC Internat., Summit-Argo, Ill., 1979-86; lead scientist NCAUR-ARS div. USDA, Peoria, Ill., 1986—; guest lectr. various sci. groups, meetings, 1964—; organizer internat. symposia. Contbr. articles to profl. jours. Rsch. Corp. grantee, 1975-77; NSF fellow, 1973-74. Mem. Assn. Ofcl. Analytical Chemists, Inst. Food Technologists, Am. Oil Chemists Soc., Am. Chem. Soc. Home: 1820 W Sunnyview Dr Peoria IL 61614-4662 Office: Nat Ctr Agrl Utilization Rsch ARS/USDA 1815 N University St Peoria IL 61604-3999

KING, JOHN LAVERNE, III, facilities engineer, energy management specialist; b. Detroit, May 4, 1946; s. John LaVerne Jr. and Evelyn Ruth (Spires) K.; m. Sharon Lee Coryell, Oct. 3, 1975 (div. 1985); 1 child, John LaVerne IV; m. Christopher Ann Munford, July 3, 1988. BSME, U. Mich., 1968; BA in Math., Wayne State U., 1970. Registered profl. engr., Mich., Ariz. Engr., then project engr. Snell Environ. Group, Lansing, Mich., 1973-77; project. engr., then head mech. dept. Ayres, Lewis, Norris & May, Inc., Ann Arbor, Mich., 1977-81, head energy dept., 1981-85; facilities engr. World Bank, Washington, 1985-87, chief plant ops., 1987—. Author conf. procs. Recipient Honor award Am. Cons. Engrs. Coun., 1988. Mem. Assn. Profl. Energy Mgrs. (treas. 1990). Office: World Bank 1818 H St NW Washington DC 20433

KING, JONATHAN ALAN, molecular biology educator; b. Bklyn., Aug. 20, 1941; m. Jacqueline Dee. B.S. in Zoology, magna cum laude with high honors, Yale U., 1962; Ph.D., Calif. Inst. Tech., 1968. With NSF Antarctic Service, 1969; Brit. Med. Research Council postdoctoral fellow Cambridge (Eng.) U., 1970; asst. prof. MIT, Cambridge, 1971-73; assoc. prof. MIT, 1974-78, prof. molecular biology, 1979—, dir. biology electron microscope facility, 1971—; chmn. microbial physiology study sect. NIH, 1982-83. Contbr. numerous articles to sci. jours. Chmn. Nat. Jobs with Peace Campaign. Gen. Motors Nat. scholar, 1958-62; Jane Coffin Shields Fund fellow, 1968-70; recipient U.S. Antarctic Service medal, 1968; Woodrow Wilson fellow, 1962-63; NIH fellow, 1963-67. Fellow AAAS; mem. Genetics Soc. Am., Am. Soc. Microbiology, Biophysics Soc., Teratology Soc., Am. Soc. Biol. Chemists, Am. Pub. Health Assn. Home: 40 Essex St Cambridge MA 02139-2645 Office: MIT Dept Biology 77 Massachusetts Ave Cambridge MA 02139

KING, JOSEPH WILLET, child psychiatrist; b. Springfield, Mo., Aug. 26, 1934; m. Doris Ann Toby; children: Pamela Renee, Timothy Wells, Michael Brian, Bradley Christopher. BA, So. Meth. U., 1956; MD, U. Tex. Southwestern, 1962. Intern Baylor U. Med. Ctr., Dallas, 1962-63; resident in gen. psychiatry Timberlawn Psychiat. Hosp., 1963-64, Lisbon VA Hosp., 1965; fellow in child psychiatry U. Tex. Southwestern Med. Sch., 1965-67, Hillside Hosp., Glen Oaks, N.Y., 1967; staff child psychiatrist, dir. child and adolescent svcs. Timberlawn Psychiat. Ctr., Inc., Dallas, 1967-78; assoc. attending child psychiatrist dept. psychiatry Baylor U. Med. Ctr., Dallas, 1967-78; active attending child psychiatrist Children's Med. Ctr., Dallas, 1967-78; attending staff Dallas County Hosp. Dist./Parkland Meml. Hosp., 1967-78; cons. child psychiatry Girls Day Care Rehab. Ctr. Dallas County, Dallas, 1970-73; cons. child psychiatry and adminstrn. Meridell Achievement Ctr., Austin, Tex., 1971-73; dir. adolescent svcs. Portsmouth (Va.) Psychiat. Ctr., 1978-79; active attending child psychiatrist Maryview Hosp., Portsmouth, 1978-80; med. dir., chief exec. officer Psychiat. Inst. Richmond, Va., 1980-86; chief exec. officer, psychiatrist-in-chief Shadow Mountain Inst., Tulsa, 1987-90; v.p. Century Healthcare, Tulsa, 1987-90; assoc. clin. prof. Med. Coll. Va., Va. Commonwealth U., 1980—, Med. Sch. U. Okla., Tulsa, 1987—. Contbr. articles to profl. jours. Fellow Am. Psychiat. Assn. (Okla. dist. br.), Am. Soc. Adolescent Psychiatry (nat. pres. 1975-76), Am. Orthopsychiat. Assn., Am. Coll. Psychiatrists; mem. AMA, Tex. Med. Assn. (various coms.), Dallas County Med. Soc. (various coms.), Am. Acad. Child and Adolescent Psychiatry (ins. com. 1981-86, pres. Okla. coun. 1991-92, state del. to nat. coun.), Tex. Soc. Child Psychiatry (past officer), Nat. Assn. Pvt. Psychiat. Hosps. (chmn. adolescent care com. 1971-81, multiple com./task force functions, pres. ind. for profit sect. 1991-92, trustee 1992—), Okla. Psychiat. Assn., Okla. Med. Soc., Christian Med. Soc., Tulsa County Med. Soc., Tulsa Psychiat. Assn., Alumni Assn. U. Tex. Southwestern Med. Sch. (pres. 1991—). Office: Ste 9200 5555 E 71st St South Tulsa OK 74136-6557

KING, KENNETH VERNON, JR., pharmacist; b. Lexington, Miss., Dec. 17, 1950; s. Kenneth Vernon Sr. and Louise (Jordan) K.; m. Janis Marie Guynes, June 12, 1976; children: Kenneth V. King III, Nanette Marie King, Jason Guynes King. AA, Holmes Jr. Coll., 1971; BS in Pharmacy, U. Miss., 1973. Registered pharmacist Miss., Pa. Pharmacist Barretts Drug Store, Greenwood, Miss., 1973-74; registered pharmacist Eckerd Drugs, Greenwood, 1974-76, 77-88, Medi-Save Drugs Ellis Isle, Jackson, Miss., 1976; registered pharmacist Eckerd Drugs, Pearl, Miss., 1988-90, Jackson, 1990-92; compounding pharmacist Marty's Discount Drugs, Flowood, Miss., 1992—; cons. Whispering Pines Hospice, Hospice of Ctrl. Miss., 1992—, Sta-Home Hospice care of Miss., Grace House of Jackson, 1992—; contractor, researcher, cons. Profl. Pharm. Svcs, Inc. of Miss., Jackson; clin. pharmacy instr. U. Miss. Sch. Pharmacy, Oxford, 1985—; tchr. environ. illness VA Hosp.; Hospice pharmacist, cons., 1992—. 4-H advisor Leflore County 4-H, Greenwood, 1974-76; aux. patrolman Greenwood Police Dept., 1982-86; drug identification specialist Greenwood Aux. Police Dept., 1984-85; pres., founder Human Ecology Action League of Miss., Inc., 1988—; coord. Environ. Assocs. of Jack Eckerd, Inc., 1990—; mem. Rainbow Whole Food Co-op, 1989—; coord. Regional Support Svcs. for HEAL Inc., 1989—; bd. dirs. Nat. Human Ecology Action League, Inc., 1991—. Mem. Environ. Coalition of Miss. (co-founder), Environ. Assocs. of Jack Eckerd Inc. (coord.). Mem. Family Life Ch. Achievements include research in experimental, investigational dosage forms for hospice patients. Home: 97 Long Meadow Rd Brandon MS 39042-2182 Office: HEAL of Miss Inc 97 Long Meadow Rd Brandon MS 39042-2182

KING, K(IMBERLY) N(ELSON), computer science educator; b. Columbus, Ohio, Apr. 28, 1953; s. Paul Ellsworth and Marcelia Jeannette (Huston) K.; m. Cynthia Ann Stormes, Sept. 5, 1981 (div. Nov. 1991). BS with highest honors, Case Western Res. U., 1975; MS, Yale U., 1976; PhD, U. Calif., Berkeley, 1980. Asst. prof. info. and computer sci. Ga. Inst. Tech., Atlanta, 1980-86, rsch. scientist, 1986-87; assoc. prof. math. and computer sci. Ga. State U., Atlanta, 1987—; cons. Norfolk So. Rwy., 1991. Author: Modula-2: A Complete Guide, 1988; columnist Jour. Pascal, Ada, and Modula-2, 1989-90; contbr. articles to profl. jours. Vol. Ga. Radio Reading Svc., Atlanta, 1989—. Grad. fellow NSF, 1975-78; NSF grantee, 1981-84. Mem. AAUP, IEEE Computer Soc., Assn. for Computing Machinery, Tau Beta Pi. Home: 2661 Havermill Way NE Atlanta GA 30345-1427 Office: Ga State U Dept Math and Computer Sci Atlanta GA 30303

KING, LIONEL DETLEV PERCIVAL, retired nuclear physicist; b. Williamstown, Mass., Dec. 29, 1906; s. James Percival and Edith Marianne (Seyerlen) K.; m. Edith Marie Bork, Dec. 25, 1925 (dec. 1969); children: Nicholas S. P., Lidian King Watson; m. Jacqueline Cyzmoure, Jan. 1, 1970 (dec. 1973); m. Edna Johnson, Apr. 22, 1974 (dec. 1977); m. Elizabeth Lewis Sprang, Apr. 30, 1978. BSME, U. Rochester, 1930; PhD in Physics, U. Wis., 1937. Asst. in physics U. Rochester, N.Y., 1930-31, MIT, Cambridge, 1931-33, U. Wis., Madison, 1935-37; instr. physics Purdue U., Lafayette, Ind., 1937-42; fellow U.S. Corps of Engrs., 1942-43; group leader Los Alamos (N.Mex.) Sci. Lab., 1943-57, asst. div. leader, 1958-59, chmn. rover flight safety, 1960-69, rsch. advisor, 1969-73, retired, 1973; lectr. Atoms for Peace Conf., Geneva, 1955, U.S. tech. dir., 1958; advisor U.S. del. to UN, 1958; cons. Los Alamos Sci. Lab., 1973-78. Active Oppenheimer Meml. Com., 1969— (chmn. 1974-76); chmn., treas. and sec. El Cajon Grande Tesuque Ditch Assn., 1975-79, mayordomo, 1979—. Fellow AAAS, Am. Nuclear Soc. (chmn. prog. com. 1954-57, honors and awards com. 1954-57, bd. dirs. 1954-59), Am. Phys. Soc., Sigma Xi. Achievements include numerous patents, among them reactor design; member team achieving first critical assembly using enriched uranium; development of solutions to flight safety problems for nuclear rocket reactors. Home: RR 4 Box 16B Santa Fe NM 87501-9804

KING, MARY-CLAIRE, epidemiologist, educator, geneticist; b. Evanston, Ill., Feb. 27, 1946; m. 1973; 1 child, Emily King Colwell. BA in Math., Carleton Coll., 1966; PhD in Genetics, U. Calif., Berkeley, 1973. Asst. prof. U. Calif., Berkeley, 1976-80, assoc. prof., 1980-84, prof., 1984—; mem. bd. sci. counselors Nat. Cancer Inst.; cons. Com. for Investigation of Disappearance of Persons, Govt. Argentina, Buenos Aires, 1984—. Contbr. more than 80 articles to profl. jours. Recipient Alumni Achievement award Carleton Coll. Mem. AAAS, Am. Soc. Human Genetics, Soc. Epidemiologic Research, Phi Beta Kappa, Sigma Xi. Office: U Calif BEHS Sch Pub Health Berkeley CA 94720

KING, PATRICIA ANN, law educator; b. Norfolk, Va., June 12, 1942; d. Addison A. and Grayce (Wood) K.; m. Roger W. Wilkins, Feb. 21, 1981; 1 child, Elizabeth. BA, Wheaton Coll., 1963; JD, Harvard U., 1969. Bar: D.C. 1969, U.S. Supreme Ct. 1980. Spl. asst. to chair EEOC, Washington, 1969-71; dep. dir. civil rights office HEW, Washington, 1971-73; prof. law Georgetown Law Ctr., Washington, 1973—; cons. Women Judges' Fund for Justice, Washington, 1988—; mem. adv. com., health scis. policy bd. NAS, 1989—; mem. adv. com. NIH, Rockville, Md., 1990—; adj. prof. Sch. Hygiene and Pub. Health Johns Hopkins U., 1990—. Co-author: Law, Science and Medicine, 1984; contbr. articles to profl. jours. Chmn. Redevelopment Land Agy., Washington, 1976-80. Fellow Hastings Ctr.; mem. Am. Soc. Law and Medicine (bd. dirs. 1990—), Am. Law Inst.

KING, RANDALL KENT, mechanical engineer; b. Kansas City, Mo., Feb. 7, 1950; s. Joseph Edwin and Dorotha Velma (Johnson) K.; m. Deborah Diane Salmon, July 27, 1974; 1 child, Kristi M. BSME, Okla. State U., 1973, M. Engring. in Mech. Engring., 1974, PhDME, 1985. Registered profl. engr., Okla. Mfg. R&D engr. Boeing Co., Wichita, Kans., 1974-75; rsch. project leader Fluid Power Rsch. Ctr., Stillwater, Okla., 1975-79; instr. in materials Okla. State U., Stillwater, 1983-84; prin. engr. Halliburton Energy Svcs., Duncan, Okla., 1984—. Contbr. articles to ASME Jour. Mech. Design, SAE Transactions, Nat. Conf. on Fluid Power, ASM Internat. Conf. on Fatigue, Corrosion Cracking, Fracture Mechanics, and Failure Analysis. Bd. dirs. Duncan Community Residence, 1989—. 1st lt. Army N.G., 1973-79. Mem. ASME, NSPE. Achievements include patent for front discharge fluid end for reciprocating pump; research in embrittlement and fatigue of metals; design of high pressure equipment and high performance pumps. Home: 525 Allen Duncan OK 73533 Office: Halliburton Energy Svcs PO Drawer 1431-0448 Duncan OK 73536-0448

KING, ROGER LEE, electrical engineer, consultant; b. South Charleston, W.Va., Jan. 17, 1952; s. Christopher L. and Margaret M. (Barker) K.; m. Donna Marie Rogers, May 12, 1973; 1 child, Jason Scott. BSEE, W.Va. U., 1973; MSEE, U. Pitts., 1977; PhD, U. Wales, Cardiff, U.K., 1988. Elec. engr. Bettis Atomic Power Lab., Pitts., 1973-74; elec. engr. U.S. Bur. Mines, Pitts., 1974-80, supervisory elec. engr., 1980-81, rsch. supr., 1981-88; assoc. prof. elec. engring. Miss. State U., Mississippi State, 1988-93; prof. elec. engring., 1993—; cons. Southwire, Inc., Carrollton, Ga., 1990-92, St. Paul Fire and Marine Ins. Co., 1991; presenter workshops. Contbr. to profl. publs. Pres. Henderson Complex PTA, Starkville, Miss., 1991; trustee Capital Improvement Task Force, Starkville, 1991; mem. Supt.'s Fin. Com., Starkville, 1991. Mem. IEEE, Am. Assn. Artificial Intelligence, Internat. Neural Network Soc. Achievements include 3 patents, development of expert systems for mining industry, introduction of expert system technology to cable/wire manufacture, development of concept for intelligent drill for mapping roof of an underground coal mine. Home: 1109 Yorkshire Rd Starkville MS 39759 Office: Miss State U PO Drawer EE Mississippi State MS 39762

KING, RON GLEN, neuroscientist; b. Oakland, Calif., July 27, 1957; s. Eugene and Gladys Alberta K.; m. Camille Marion, Sept. 21, 1991. BA, Calif. State U., Hayward, 1979; PhD, U. Calif., Riverside, 1986. Rsch. fellow Merck Sharp & Dohme, Rahway, N.J., 1986-89; asst. prof. Howard U. Coll. of Medicine, Washington, 1992—; sr. staff fellow NIH, Bethesda, Md., 1989—; chmn. Equal Employment Opportunity, Bethesda, Md., 1990-91; cons. East Orange (N.J.) Sch. Dist., 1989-90. Mentor Montgomery County Sch. Dist., Md., 1992. Recipient Spl. Svc. award NIH, 1992, Intramural Rschs. Tng. award NIH, 1989, Merck Rsch. fellowship, 1986, Dissertation Rsch. award, 1985, Chancellor's Patent Fund award U. Calif. Riverside, 1983. Mem. Soc. for Neurosci. Achievements include cloning and expression of the Kainate binding protein; identification of a glutathione transferase response element. Home: 18212 Swiss Circle Germantown MD 20874 Office: NIH Bldg 36/3D02 9000 Rockville Pike Bethesda MD 20892

KING, RONOLD WYETH PERCIVAL, physics educator; b. Williamstown, Mass., Sept. 19, 1905; s. James Percival and Edith Marianne Beate (Seyerlen) K.; m. Justine Merrell, June 22, 1937 (dec. Aug. 1990); 1 son, Christopher Merrell; m. Mary M. Govoni, June 1, 1991. A.B., U. Rochester, 1927, S.M., 1929; Ph.D., U. Wis., 1932; student, U. Munich, Germany, 1928-29, Cornell U., 1929-30. Asst. in physics U. Rochester, 1927-28; Am.-German exchange student, 1929-30; White fellow in physics Cornell U., 1929-30; U. fellow in elec. engring. U. Wis., 1930-32, research asst., 1932-34; instr. physics Lafayette Coll., 1934-36, asst. prof., 1936-37; Guggenheim fellow Berlin, Germany, 1937-38; with Harvard U., 1938—, successively instr., asst. prof., assoc. prof., 1938-46, prof. applied physics, 1946-72, prof. emeritus, 1972—; cons. electromagnetics and antennas, 1972—. Author: Electromagnetic Engineering, Vol. 1, 1945, 2d edit, Fundamental Electromagnetic Theory, 1963, Transmission Lines, Antennas and Wave Guides, (with A.H. Wing and H.R. Mimmo), 1945, 2d edit., 1965, Transmission-Line Theory, 1955, 2d edit., 1965, Theory of Linear Antennas, 1956, (with T.T. Wu) Scattering and Diffraction of Waves, 1959, (with R.B. Mack and S.S. Sandler) Arrays of Cylindrical Dipoles, 1968, (with C.W. Harrison, Jr.) Antennas and Waves: A Modern Approach, 1969, Tables of Antenna Characteristics, 1971, (with G.S. Smith et al) Antennas in Matter, 1981 (with S. Prasad) Fundamental Electromagnetic Theory and Applications, 1986, (with M. Owens and T.T. Wu) Lateral Electromagnetic Waves Theory and Applications to Communications, Geophysical Exploration and Remote Sensing, 1992; also articles in field. Guggenheim fellow Europe, 1937, 58, IBM scholar Northeastern U., 1985; recipient Disting. Service citation U. Wis., 1973, Pender award U. Pa., 1986. Fellow IEEE (Centennial medal 1984), AAAS, Am. Acad. Arts and Scis., Am. Phys. Soc.; mem. IEEE Antennas and Propagation Soc. (Disting. Achievement award 1991), AAUP, Internat. Sci. Radio Union, Bavarian Acad. Sci. (contbg. mem.), Phi Beta Kappa, Sigma Xi. Home: 92 Hillcrest Pky Winchester MA 01890-1440 Office: Gordon McKay Lab 9 Oxford St Cambridge MA 02138

KING, STEVEN HAROLD, health physicist; b. Stamford, Conn., June 27, 1959; s. Richard H. and Joan W. (Weaver) K.; m. Kim Yvonne Bulmer, June 18, 1983; children: Christopher K., Brandon S., Arielle L. BA, SUNY, Buffalo, 1981, MA, 1983. Asst. health physicist Milton S. Hershey Med. Ctr. Pa. State U., Hershey, 1983-86, assoc. health physicist, 1986—, laser safety officer, 1989—; grad. program com. mem. Pa. State U., Harrisburg, Pa., 1989—. Author: (with others) Handbook of Management of Radiation Protection Programs, 1991; contbr. articles to profl. jours. Mem. Am. Assn. Physicists in Medicine, Am. Nuclear Soc., Health Physics Soc. (chair admissions com. 1989-91, pres. Susquehanna Valley chpt. 1989-90). Office: Milton S Hershey Med Ctr 500 University Dr Hershey PA 17033-2360

KING, THEODORE M., obstetrician, gynecologist, educator; b. Quincy, Ill., Feb. 13, 1931; married; 2 children. BS, Quincy Coll., 1950; MS, U. Ill., 1952, MD, 1959; PhD in Physiology, Mich. State U., 1959. Lab. asst. physiology U. Ill., Urbana, 1959; intern surgery Presbyn. Hosp., N.Y., 1959-60; from resident to chief resident ob-gyn Sloane Hosp. Women, 1960-65; asst. prof. physiology, ob-gyn Sch. Medicine U. Mo., 1965-68, assoc. prof., 1968; prof., chmn. dept. Albany Med. Coll., 1968-71; prof. ob-gyn, dir. dept. Sch. Medicine Johns Hopkins U., Balt., 1971. Macy fellow Sloane Hosp. Women, 1960-64, Macy faculty fellow obstetrics, 1966-67, Nat. Inst. Child Health & Human Devel. fellow, 1965-68. Achievements include rsch. in study of uterine contractile protein, influence of enzyme induction on animal reproduction. Office: Family Health Intern PO Box 13950 Research Triangle Park NC 27709*

KING, WALTER WING-KEUNG, surgeon, head and neck surgery consultant; b. Hong Kong, Jan. 27, 1950; s. Albert Cheng and Josephine Shou-Fan (Chao) K.; m. May Kam-Wei Poon, June 8, 1985; children: Kenneth S. F., Spencer S. W. BA with honors, U. Wis., 1971; MD, Vanderbilt U., 1975. Diplomate Am. Bd. Surgery. Intern Vanderbilt U. Hosp., Nashville, 1975-76; resident surgeon SUNY, Stony Brook, 1976-80; clin. asst. in surgery Mass. Gen. Hosp., Boston, 1980-82; clin. fellow in surgery Harvard U., Boston, 1980-82; fellow, head and neck svc. Meml. Sloan-Kettering Cancer Ctr., N.Y.C., 1982-83; asst. prof. surgery SUNY, Stony Brook, 1983-84; vis. fellow otolaryngology Stanford (Calif.) U. Med. Ctr., 1984; lectr. head and neck surgery Chinese U., Hong Kong, 1984-88, sr. lectr. surgery, 1988-93, reader surgery, 1993—; asst. dir. Nutritional Support Unit, Mass. Gen. Hosp., Boston, 1981-82; cons., chief head and neck unit, burns unit, Prince of Wales Hosp., Sha Tin, Hong Kong, 1988—. Contbr. articles to profl. jours. Shriners Burn Inst. rsch. fellow, 1980-82; Am. Soc. for Head and Neck Surgery fellow, 1984, Royal Coll. Surgeons Edinburgh fellow, 1991. Fellow Am. Coll. Surgeons, Royal Coll. Surgeons Can.; mem. Soc. Head and Neck Surgeons, Am. Burn Assn., Am. Soc. for Head and Neck Surgery, Internat. Assn. Endocrine Surgeons, Am. Acad. Facial Plastic and Reconstructive Surgery, Coll. Surgeons Hong Kong. Avocations: tennis, swimming, squash, computers. Office: Prince of Wales Hosp, Dept Surgery, Sha Tin Hong Kong

KINGERY, WILLIAM DAVID, ceramics and anthropology educator; b. N.Y.C., July 7, 1926; s. Lisle Byron and Margaret (Reynolds) K.; children: William, Rebekah, Andrew. SB, MIT, 1948, ScD, 1950; PhD (hon.), Tokyo Inst. Tech.; ScD (hon.), Ecole Poly. Federale de Lausanne. From instr. to assoc. prof. MIT, Cambridge, Mass., 1951-62, prof., 1962—, Kyocera prof. ceramics, 1984-88; prof. materials sci. and anthropology U. Ariz., Tucson, 1988—, Regents prof., 1992—. Author: (text) Introduction to Ceramics, 1960, 2d edit. 1976 (translated into 3 languages), Ceramic Masterpieces, 1986 (Hon. Mention, Pub. Inst.), others; editor: Ceramic Fabrication Processes, Property Measurements at High Temperatures, Kinetics of High Temperature Processes, Ceramics and Civilization I: Ancient Thechnology to Modern Science, 1985, Ceramics and Civilization II: Technology and Style, 1986, Ceramics and Civilization III: High Tech Ceramics-Past, Present and Future, 1987, Ceramics and Civilization, 1990, Technolo-gical Innovation, 1991; editor in chief Ceramics Internat. Chmn. bd. trustees Acad. Ceramics, 1989—. Named Wagener lectr. Tokyo Inst. Tech., 1976, Kurtz lectr. Technion, Haifa, Israel, 1978, Nelson W. Taylor lectr. Pa. State U., 1982; recipient Albert V. Bleininger award, 1977, F.H. Norton award, 1977; Regents fellow Smithsonian Instn., Washington, 1988. Fellow Am. Acad. Arts and Scis.; mem. Am. Ceramic Soc. (life, disting., Ross Coffin Purdy award, John Jeppson award 1958, Robert Sosman Meml. Lecture award 1973, Hobart M. Kraner award 1985), Nat. Acad. Engring. Clubs: Cosmos (Washington); Blue Water Sailing (Boston); Royal Hamilton Amateur Dinghy (Bermuda). Office: U Ariz 338 Mines Bldg Tucson AZ 85721

KINGREA, JAMES IRVIN, JR., electromechanical engineer; b. Phila., Mar. 10, 1928; s. James Irvin and Beatrice Mary (Sellers) K.; div.; children: James Irvin III, Kathleen Virginia. BSEE, Pa. Mil. Coll., 1950. Registered profl. engr., Pa. Design engr. Westinghouse Elec. Corp., Lester, Pa., 1952-56, mfg. engr., 1956-67, mgr. quality assurance, 1967-86; mgr. quality assurance Advanced Devel. and Engring. Ctr., Swarthmore, Pa., 1986-89; dir. quality Engineered Systems div. Datron, Aston, Pa., 1989-93; ind. cons., Springfield, Pa. With U.S. Army, 1945-47. Mem. ASME, Am. Soc. Quality Control. Achievements include development of method and instrumentation for high-accuracy steam turbine rotor bore size and contour measurement. Home: 65 Worrell Dr Springfield PA 19064 Office: Engineered Systems 2550 Market St Aston PA 19014

KINGSBURY, JOHN MERRIAM, botanist, educator; b. Boston, July 4, 1928; s. Willis Albert and Constance Elizabeth (Merriam) K.; m. Louise

Arnold Gerken, June 6, 1956; 1 dau., Joanna Merriam. B.S., U. Mass., 1950; A.M., Harvard U., 1952, Ph.D., 1954; Sc.D. (hon.), Dickinson Coll., 1985. Instr. Brandeis U., Waltham, Mass., 1953-54; mem. faculty N.Y. State Coll. Agr. and Life Scis., Cornell U., Ithaca, N.Y., 1954—, prof. botany, 1970-83; prof. botany emeritus N.Y. State Coll. Agr. and Life Scis., Cornell U., 1983—; prof. clin. scis. Coll. Vet. Medicine, Cornell U., 1978-83, dir. arboretum and bot. garden, 1982-83; instr. Marine Biol. Lab., Woods Hole, Mass., summers 1958-61; founding dir. Shoals Marine Lab., 1972-79; adj. prof. U. N.H., 1976-78; cons. Upstate Med. Ctr., Syracuse, N.Y., 1977-86; instr. Aquavet course Cornell U.-U. Pa., 1978—; lectr. Cornell U. Adult U., 1978—; proprietor Bullbrier Press, 1983—; lectr. Columbus project Sta. WGBH/Pub. Broadcasting Svc., Boston, 1990; chmn. Shoals Marine Lab. Endowment Com. Author: Poisonous Plants of the United States and Canada, 1964, Deadly Harvest—A Guide to Common Poisonous Plants, 1965, Seaweeds of Cape Cod and the Islands, 1969, The Rocky Shore, 1970, Oil and Water: The New Hampshire Story, 1975, 200 Conspicuous, Unusual, or Economically Important Tropical Plants of the Caribbean, 1988, Here's How We'll Do It—An Informal History of the Construction of the Shoals Marine Laboratory, 1991; mem. editorial bd. Cornell U. Press, 1985-86. Chmn. Shoals Marine Lab. endowment com., 1992—. NSF faculty fellow, 1958; Fulbright sr. scholar, 1980. Fellow Am. Acad. Vet. and Comparative Toxicology (hon.); mem. Bullard Meml. Farm Assn. (clk. 1978—, pres. 1990—), Sea Edn. Assn. (trustee 1977-92, pres. 1982-87), Marine Biol. Lab. (life), Nature Conservancy (trustee N.Y. state bd. 1983-90), Audubon Soc. (lectr. Mass. chpt. 1987-89), Sigma Xi, Phi Zeta. Office: Cornell U Plant Sci Bldg Ithaca NY 14853

KINGSLAKE, RUDOLF, retired optical designer; b. London, Aug. 28, 1903; came to U.S., 1929; s. Martin and Margaret (Higham) K.; m. Hilda G. Conrady, Sept. 14, 1929; children: David C., Alan H. (dec.). BSc, Imperial Coll., London, 1924, MSc, 1926, DSc, 1950; DSc (hon.), U. Rochester, 1986. Prof. U. Rochester, N.Y., 1929-37, 68-83; optical designer Eastman Kodak Co., Rochester, 1937-68. Author: Lenses in Photography, 1951, 2d edit., 1963, Lens Design Fundamentals, 1978, Optical Systems Design, 1983, A History of the Photographic Lens, 1989, Optics in Photography, 1992; also numerous articles. Named Engr. of Yr., Rochester Engring. Soc., 1978. Fellow Soc. Motion Picture and TV Engrs. (Progress medal 1964), Soc. Photographic Scientists and Engrs.; mem. Optical Soc. Am. (hon., pres. 1947-49, Ives medal 1973), Soc. Photog. Instrumentation Engrs. (life). Address: 56 Westland Ave Rochester NY 14618

KINGWOOD, ALFRED E., geologist, educator. Prof. geology McMaster U., Hamilton, Ont., Can. Recipient V. M. Goldschmidt award Geochemical Soc., 1991. Office: McMaster Univ-Dept of Geology, 1280 Main St W, Hamilton, ON Canada L8S 4M2*

KINI, ARAVINDA MATTAR, materials chemist; b. Karkala, Karnataka, India, Jan. 14, 1951; came to U.S., 1972; s. Rathnakar and Sushila (Pai) K.; m. Mridula Sharma, Feb. 28, 1979; children: Rohini, Nutan, Seema, Ashvin. MS, IIT Madras, India; PhD, U. Hawaii, 1979. Postdoctoral fellow The Johns Hopkins U., Balt., 1980-84; rsch. scientist LTV Aerospace and Def. Co., Dallas, 1984-86; asst. chemist Argonne (Ill.) Nat. Lab., 1986-91, chemist, 1991—. Pres. AMKA, Chgo., 1988. Named DOE Outstanding Accomplishment in Materials Chemistry, 1990. Fellow Acad. Gen. Edn.; mem. Am. Chem. Soc., Am. Phys. Soc. Achievements include discovery of ambient-pressure organic superconductor with the highest transition temperature known to date. Office: Argonne Nat Lab CHM Div 200/A125 9700 S Cass Ave Argonne IL 60439

KINIGAKIS, PANAGIOTIS, research scientist, engineer, author; b. Chanea, Greece, July 11, 1949; s. John and Evangelia (Vozinakis) K.; m. Kalliopi Paleologos, July 31, 1977; children: Evangelia, Maria Anna. BS, Superior Agrl. Sch., Athens, Greece, 1971, MS, 1973; MS in Food Sci., Rutgers U., 1979. Packaging devel. specialist Am. Cyanamid Co., Clifton, N.J., 1979-81; sr. packaging engr. Warner Lambert Co., Morris Plains, N.J., 1981-83; tech. services supr. M&M Mars Inc., Hackettstown, N.J., 1983-87; packaging engring. mgr. Gen. Foods Corp., White Plains, N.Y., 1987—; agrl. engr. Food Agrl. Orgn. div. of UN, Chanea, 1975-77. Patentee pkg. equipment and mfg. systems; contbr. articles to profl. jours. Advisor Greek Orthodox Youth Assn., Randolph, N.J., 1986, Hamilton, N.J., 1990. Mem. Inst. Food Tech., Tech. Assn. Pulp and Paper Industry, Inst. Packaging Profls., Soc. Plastics Engrs., N.Y. Acad. Sci. Greek Orthodox. Avocations: volleyball, soccer, tennis. Home: 65 Amherst Way Princeton Junction NJ 08550-1836 Office: Gen Foods USA 555 S Broadway Tarrytown NY 10591-6399

KINNEY, ANTHONY JOHN, biochemist, researcher; b. Ilkeston, Derbyshire, Eng., Sept. 9, 1958; came to U.S., 1983; s. John Frederick and Sheila (Baker) K.; m. Alison Jo Mack. BS, Sussex U., Falmer, U.K., 1980; PhD, Oxford (U.K.) U., 1985. Rsch. assoc. dept. botany La. State U., Baton Rouge, 1983-87; rsch. fellow Rutgers U., New Brunswick, N.J., 1987-89; prin. investigator Dupont Exptl. Sta., Wilmington, Del., 1989—. Mem. Am. Soc. Biochemistry and Molecular Biology, Am. Soc. Plant Physiologists, Sigma Xi. Achievements include research in pathways of phospholipid biosynthesis in higher plants; in molecular regulation of phospholipid metabolism in yeast; in modification of fatty-acid metabolism in oil seed plants by biotechnology. Office: Dupont Exptl Sta PO Box 80402 Wilmington DE 19880-0402

KINNEY, DONALD GREGORY, civil engineer; b. Anchorage, Mar. 22, 1957; s. James and Agatha Bernice (Caviezel) K.; m. Suzanne Lombardi, June 20, 1992. BS in Civil Engring., U. Alaska, Fairbanks, 1982; M in Civil Engring., U. Alaska, Anchorage, 1990. Registered profl. engr., Alaska. Warehouseman A&W Wholesale, Fairbanks, Alaska, 1975; oiler to grade checker H.C. Price, Fairbanks, 1975-78; asst. engr. City of Valdez (Alaska), 1980-81; project engr. H&H Contractors, Fairbanks, 1982; asst. city engr./project mgr. City of Homer (Alaska), 1983-85; civil engr. DOWL Engrs., Anchorage, 1985-87, Arctic Slope Consulting Engrs., Anchorage, 1987-89; sr. civil engr. Alyeska Pipeline Svc. Co., Anchorage, 1989—. Named Civil Engring. Student of the Year, U. Alaska, Fairbanks, 1982. Mem. Inst. Transp. Engrs. (chpt. sec. 1992-93), ASCE, ASME, Am. Soc. Profl. Engrs., Am. Water Works Assn. Republican. Roman Catholic. Avocations: white water kayaking, climbing, running, reading, skiing. Home: 2200 Sonstrom Dr Anchorage AK 99517-1018 Office: Alyeska Pipeline Svc Co 1835 S Bragaw St MS 556 Anchorage AK 99512

KINOE, YOSUKE, computer company researcher, engineer; b. Suginami-Ku, Tokyo, Japan, Feb. 2, 1961; s. Jiroh and Yoko Kinoe; m. Kazuko Kinoe. B. Engring., Keio U., Yokohama, Japan, 1984, M. Engring., 1986. Cert. engring. Researcher IBM Rsch., Tokyo Rsch. Lab., 1986-89; researcher product assurance lab. IBM Japan, Ltd., Kanagawa, Japan, 1989—. Contbr. articles to profl. jours. Mem. IEEE, Assn. for Computing Machinery, Japanese Soc. for Artificial Intelligence, N.Y. Acad. Scis. Office: IBM Japan Ltd, 1623-14 Shimotsuruma, Yamato-shi Kanagawa 242, Japan

KINOSHITA, SHIGERU, ophthalmologist; b. Osaka, Japan, Mar. 14, 1950; s. Kanezo and Masako K.; m. Junko Mikawa, Apr. 29, 1974; children: Manabu, Makoto. MD, Osaka U., 1974, DSc, 1983. Resident Osaka U. Hosp., 1974-79; rsch. fellow Harvard Med. Sch., Boston, 1979-82; instr. Med. Sch. Osaka U., 1982-84, asst. prof., 1984-92; dir. eye div. Osaka Rosai Hosp., 1984-88; dir. cornea svc. Osaka U. Hosp., 1988-92; assoc. chief ophthalmology Osaka U., 1991-92; prof., chmn. ophthalmology Kyoto Pref. Univ. Medicine, 1992—. Home: 1-3-11 Kitabatake Abenoku, Osaka 545, Japan

KINOSHITA, TOICHIRO, physicist; b. Tokyo, Japan, Jan. 23, 1925; came to U.S., 1952; s. Tsutomu and Fumi (Ueda) K.; m. Masako Matsuoka, Oct. 14, 1951; children: Kay, June, Ray. BS, Tokyo U., 1947, PhD, 1952. Mem. Inst. for Advanced Study, Princeton, N.J., 1952-54; postdoctoral fellow Columbia U., N.Y.C., 1954-55; rsch. assoc. Cornell U., Ithaca, N.Y., 1955-58, asst. prof., 1958-60, assoc. prof., 1960-63, prof., 1963-92, Goldwin Smith prof., 1992—; mem. tech. adv. panel U.S. Dept. Energy, Washington, 1982-83; com. fundamental constants Nat. Rsch. Coun., Washington, 1984-86. Author: Quantum Electrodynamics, 1990; contbr. over 100 articles to profl. jours. Guggenheim fellow, 1973-74. Fellow NAS, AAAS, Am. Physical Soc. (Recipient J.J. Sakurai prize 1990). Democrat. Home: 5 Winthrop Pl Ithaca NY 14850-1740 Office: Cornell U Newman Lab Ithaca NY 14853

KINOSHITA, TOMIO, economics educator; b. Shanghai, China, Dec. 29, 1944; s. Torashichi and Kazue (Nishio); m. Ikuko Ohkuma, July 9, 1982; 1 child, Taku. BA, U. Tokyo, 1967, MA, 1971, postgrad., 1972-74. Researcher Tokyo Ctr. for Econ. Rsch., 1974-76; asst. prof. econs. Musashi U., Tokyo, 1975-76, assoc. prof., 1977-90, prof., 1990—. Author: (in Japanese) Economics of Working Hours and Wages, 1990. Mem. Am. Econ. Assn., Japan Assn. Econs. and Econometrics. Avocation: igo. Home: Sekimae 2-chome 22-9, Musashino-shi, Tokyo 180, Japan Office: Musashi U Dept Econs, 26 Toyotama-kami 1-chome, Nerima-ku, Tokyo 176, Japan

KINOSZ, DONALD LEE, quality manager; b. Pitts., Dec. 7, 1940; s. Michael and Pearl (Buckner) K.; m. Deborah Michele Reed, June 2, 1978; children: Brigitte, Brenda, Wayne Casey. BS, U. Pitts., 1966. Process engr. Alcoa Tech. Ctr., Alcoa Center, Pa., 1966-76, Alcoa, Tenn., 1976-79; mgr. Alcoa Tech. Ctr., Alcoa Center, Pa., 1979-81, pers. mgr., 1981-83, chem. mgr., 1983-85, ceramics mgr., 1985-88, mgr. quality, 1988—; prodn. mgr. Anderson County Works, Palestine, Tex., 1978-79; adv. bd. U. Pitts., 1988—. City councilman City of Lower Burrell, Pa., 1980-88. Fellow Am. Inst. Chemists; mem. Am. Soc. Quality Control, Indsl. Rsch. Inst. (chmn. quality dirs. network), Strongland C. of C. (bd. dirs. 1985—), Sigma Xi. Achievements include patents in anti-pollution method, electrolytic production of magnesium, production of magnesium chloride, regeneration of activated carbon having materials adsorbed thereon, flow control baffles for molton salt electrolysis, metal production, disposal of waste gasses from production of aluminum chloride, electrolytic furnace lining, method of preparing an electrolytic cell for operation, situ cleaning of electrolytic cells, treatment of offgas from aluminum chloride production. Home: 491 Dakota Dr Lower Burrell PA 15068 Office: Alcoa Labs 100 Technical Dr Alcoa Center PA 15069

KINSBOURNE, MARCEL, neurologist, behavioral neuroscientist; b. Vienna, Austria, Nov. 3, 1931; came to U.S., 1967; s. David and Matilda (Gaster) Kinsbrunner; children: David, Daniel, Jeremy, Emily. BA, Oxford (Eng.) U., 1952, BM, BCh, 1955. Intern Lewisham & Brook Hosp., 1955-56; resident Hosp. for Sick Children, Great Ormond St., London and Bellevue Hosp., N.Y.C.; lectr. psychology Oxford U., 1964-67; assoc. prof. pediatrics Duke U., Durham, N.C., 1967-74; prof. pediatrics U. Toronto, 1974-80, prof. psychology, 1975-85; lectr. neurology Harvard Med. Sch., Boston, 1980-91; dir. behavioral neurology Shriver Ctr., Waltham, Mass., 1980-91; rsch. prof. cognitive sci. Tufts U., Medford, Mass., 1992—; policy advisor Nat. Inst. Neurol. Disease, Bethesda, Md., 1975-79; cons. neurologist VA Med. Ctr., Boston, 1989—; mem. neuropsychology com. Max Planck Gesellschaft, Germany, 1992—. Author: Children's Learning Attention Problems, 1979; editor: Hemispheric Disconnections and Cerebral Function, 1974, Asymmetrical Function of the Brain, 1978, Hemisphere Function in Depression, 1988, Assessment of Cognitive Function in Epilepsy, 1990. Fulbright scholar, 1958-59; recipient prize in neurology Nat. Hosp., London, 1961; Rockefeller Found. resident scholar, Bellagio, Italy, 1990; NIMH grantee, 1982—. Fellow APA, Gerontol. Soc.; mem. Internat. Neuropsychology Assn. (pres. 1977-78), Am. Neurol. Assn., Child Neurology Soc. Achievements include attentional model for human laterality; orientational model for unilateral neglect; right hemisphere compensation in aphasia; behavioral laboratory demonstrations of attention deficit disorder and stimulant effects. Office: Tufts U Medford MA 02155

KINSELLA, DANIEL JOHN, electrical engineer; b. Rochester, N.Y., Dec. 14, 1952; s. John Joseph and Lucille Kate (Taylor) K. BSEE, Rensselaer Poly. Inst., 1975, MSEE, 1976. Gen. radio telephone lic. FCC. Elec. design engr. Harris Corp., Rochester, N.Y., 1976-77; sr. elec. design engr. Raytheon Co., Sudbury, Mass., 1977-84; sr. field applications engr. Monolithic Memories, Inc., Framingham, Mass., 1984-86, dist. sales mgr., 1986-87; sr. field application engr. Advanced Micro Devices, Burlington, Mass., 1987-88; sr. field applications engr. WSI, Stow, Mass., 1988-92, Zilog, Nashua, N.H., 1992—. Mem. Boston Computer Soc., Mensa. Avocations: ham radio, computers,hockey, reading, study of Irish language. Home: 210 Barton Rd Stow MA 01775

KINSEY, JAMES LLOYD, chemist, educator; b. Paris, Tex., Oct. 15, 1934; s. Lloyd King and Elaine Mills K.; m. Berma McDowell, July 28, 1962; children: Victoria, Samuel, Adam. B.A., Rice U., 1956, Ph.D., 1959; NSF fellow, U. Uppsala, Sweden, 1959-60; postdoctoral fellow, U. Calif., Berkeley, 1960-62. Asst. prof. dept. chemistry M.I.T., 1962-67, asso. prof., 1967-74, prof., 1974-88, chmn. dept., 1977-82; dean natural scis., D.R. Bullard Welch Found. prof. sci. Rice U., Houston, 1988—, interim provost, 1993—; cons. Los Alamos Nat. Labs., external rev. com. chemistry and laser sci. div., 1983-89; Miller rsch. fellow, 1960-62; mem. NAS-NRC Bd. Chem. Scis., 1980-83, co-chmn., 1981-83; mem. steering com. U.S. Army Basic Sci. Rsch.-NRC, 1981-86; mem. oversight rev. com. chemistry div., NSF, 1989. Assoc. editor Jour. Chem. Physics, 1981-84; mem. editorial adv. bd. Jour. Phys. Chemistry, 1984-88, Ann. Rev. Phys. Chemistry, 1985-89; mem. adv. editorial bd. Chem. Physics Letters, 1992—; contbr. articles to profl. jours. Recipient E.O. Lawrence award U.S. Dept. Energy, 1987; Alfred P. Sloan fellow, 1964-68, Guggenheim fellow, 1969-70. Fellow AAAS, Am. Phys. Soc. (exec. com. div. chem. physics 1985-88), Am. Acad. Arts and Scis.; mem. NAS, Am. Chem. Soc. (chmn. div. phys. chemistry 1985, Nobel Laureate Signature award for grad. edn. 1990), Sigma Xi. Office: Rice U Office Dean Natural Scis PO Box 1892 Houston TX 77251-1892

KINSEY, JOHN SCOTT, environmental scientist; b. St. Louis, Jan. 8, 1949; s. Ralph Walter and Lorraine Lydia (Thomure) K.; m. Mary Sue Partney, June 13, 1969; children: Lynnette Marie, John Scott II. BS in Biology, Phys. Sci., Cen. Mo. State U., 1970; postgrad., U. Kans., 1973-74, U. Colo., 1975-78. Lab. technician Haver-Lockhart Labs., Shawnee Mission, Kans., 1970-71; sanitarian City of Kansas City, Mo., 1971-74; environ. technician R.W. Beck and Assocs., Denver, 1974; air pollution specialist Dept. Health State Colo., Denver, 1974-78; air quality engr. AeroVironment Inc., Pasadena, Calif., 1978-80; assoc. environ. scientist Midwest Rsch. Inst., Kansas City, Mo., 1980-85, sr. environ. scientist, 1985-86, sect. head air quality assessment sect., 1986-89, prin. environ. scientist, 1989—. Contbr. numerous articles to profl. jours. Unit commr. Boy Scouts Am., 1992—. Mem. Am. Assn. Aerosol Rsch., Fine Particle Soc. (exec. coun. 1986-89), Air and Waste Mgmt. Assn. (bd. dirs. 1990-93, numerous coms. including AE-1 Particulate Control Com., AS-5 Fugitive Emissions Com., Ann. meeting planning com., ann. meeting tech. com., etc.), Chi Epsilon, Beta Beta Beta. Achievements include patents for apparatus and method to produce charged fog, method and apparatus for smoke suppression. Office: Midwest Rsch Inst 425 Volker Blvd Kansas City MO 64110-2299

KINSLEY, CRAIG HOWARD, neuroscientist; b. Pico Rivera, Calif., Jan. 30, 1954; s. Howard Junior and Dorothy Geraldine (Sulenta) K.; m. Nancy Ellen Lustig, Nov. 8, 1987; 1 child, Devon Alyse. BA, Calif. State U., Sonoma, 1979; PhD, SUNY, Albany, 1985. Rsch. assoc. KOBA Assocs., Washington, 1980-81; lab. supr. SUNY, Albany, 1983-85; postdoctoral fellow Harvard Med. Sch., Boston, 1985-87, rsch. assoc., 1987-88, instr., 1988-89; asst. prof. biol. psychology U. Richmond, Va., 1989—; chmn. med. scis. sect. Va. Acad. Scis.; cons. Randolph-Macon Coll. IACUC, Ashland, Va., 1991—. Editorial adv. bd. Physiology and Behavior, San Antonio, 1989—; contbr. articles to profl. jours., chpts. to books. Scientist Va. Sci. Mus. "Sci. by Mail", Richmond, 1991—, Westinghouse-AAAS Middle Sch. Project, Richmond, 1992; judge Va. Jr. Acad. Sci., 1992. Grantee NSF, 1987, 88, 90, 92, Va. Found. for Ind. Colls., 1990, NIH, 1986, U. Richmond, 1989—, Keck Found., 1993, Va. Acad. Scis., 1993. Mem. AAAS, Soc. for Neurosci., Endocrine Soc., Am. Soc. Zoologists (animal care com.). Achievements include discovery of effects of reproductive experience on female physiology, prenatal stress effects on endogenous opioid, system prenatal stress effects on immune function and structure, opiate-induced modifications of olfaction in females. Home: 11221 Church Grove Ct Richmond VA 23233 Office: U Richmond Dept Psychology 116 Richmond Hall Richmond VA 23173

KINSLEY, HOMAN BENJAMIN, JR., chemist, chemical engineer; b. Balt., Dec. 31, 1940; s. Homan Benjamin and Mary Helen (DePriest) K.; m. Patricia Dorothy Harr, Dec. 20, 1962; children: Anne Elizabeth, Christine Dawn, Benjamin James. BA, Western Md. Coll., 1962; MS, Lawrence Univ., 1964, PhD, 1967. Rsch. assoc. Ethyl Corp., Richmond, Va., 1966-74; rsch. chemist DuPont, Richmond, Va., 1974-75; rsch. assoc. James River Corp., Richmond, Va., 1975-77, tech. dir., 1977-79, dir. tech., 1979-84, sr. rsch. fellow, 1984-90; sr. rsch. fellow Custom Papers Group, Inc., Richmond, Va., 1990—; project adv. com. The Inst. Paper Chemistry, Atlanta, 1983-86. Contbr. articles to profl. jours. Pres. Powhatan Village Pool, 1980; alumni councilman The Inst. Paper Chemistry, Appleton, Wis., 1981-89. Recipient Trustee Alumni award Western Md. Coll., 1991, Best Paper 1989 Conf. award Tech. Assn. Pulp & Paper Industry, 1989. Mem. Am. Chemical Soc., Am. Assn. Advancement Sci., Tech. Assn. Pulp & Paper Industry. Methodist. Achievements include numerous patents in field. Office: Custom Papers Group Inc 110 Tredegar St Richmond VA 23219

KINSMAN, FRANK ELLWOOD, engineering executive; b. Westfield, Pa., Oct. 2, 1932; s. Ellwood L. and Josephine I. (Champney) K. m. Ednamae J. Reuter, June 12, 1954; children: Patricia Scott, Beverly Armstrong, Cheryl Beezley, Lora Moriconi. BSEE, John Brown U., 1958. Cert. energy mgr. Tech. staff Cornell Aero. Lab., Buffalo, 1958-61; sr. engr. Tex. Instruments, Dallas, 1961-79; v.p. Bywaters & Assocs., Cons. Engrs., Dallas, 1980-86; pres. Kinsman & Assocs., Cons. Engrs., Dallas, 1986—; ops. rsch. Cornell Aero. Lab., Buffalo, 1958-61; energy resources mgr. Tex. Instruments, Dallas, 1974-79; energy system analysis and design Bywaters and Assoc. & Kinsman & Assoc., Dallas, 1980—; engr. adv. bd. John Brown U., Siloam Springs, Ark., 1968-71; vis. lectr. So. Meth. U., Dallas, 1984-85. Contbr. articles to profl. jours. Bd. chmn. Grace Bible Ch., Dallas, 1990. Mem. ASHRAE, NSPE, Assn. of Energy Engrs. (sr.), Tex. Soc. Profl. Engrs. Achievements include devel. of material signatures at long infrared wavelenghts; rsch. include airborne and satellite data interpretations. Office: Kinsman & Assocs Ste 600 1701 Greenville Ave Richardson TX 75081

KINSMAN, ROBERT PRESTON, biomedical plastics engineer; b. Cambridge, Mass., July 25, 1949; s. Fred Nelson and Myra Roxanne (Preston) K. BS in Plastics Engring., U. of Mass. at Lowell, 1971; MBA, Pepperdine U., 1982. Cert. biomed. engr., Calif.; lic. real estate sales person, Calif. Product devel. engr., plastics divsn. Gen. Tire Corp., Lawrence, Mass., 1976-77; mfg. engr. Am. Edwards Labs. divsn. Am. Hosp. Supply Corp., Irvine, Calif., 1978-80, sr. engr., 1981-82; mfg. engring. mgr. Edwards Labs., Inc. subs. Am. Hosp. Supply Corp., Añasco, P.R., 1983; project mgr. Baxter Edwards Critical Care divsn. Baxter Healthcare Corp., Irvine, 1984-87, engring. and prodn. mgr., 1987—; mem. mgmt. adv. panel Modern Plastics mag., N.Y.C., 1979-80. Instr. first aid ARC, N.D., Mass., Calif., 1971—; vol. worker VA, Bedford, Mass., 1967-71; pres., bd. dirs. Lakes Homeowners Assn., Irvine, 1985-91; bd. dirs., newsletter editor Paradise Park Owners Assn., Las Vegas, Nev., 1988—; bd. dirs. Orange County, Calif., divsn. Am. Heart Assn., 1991—, v.p. bd. dirs., 1993—, mem. steering com. Heart and Sole Classic fundraiser, 1988—, subcom. chmn., 1989, event chmn., 1991-92, mem. devel. com. Calif. affiliate. Capt. USAF, 1971-75. Recipient Cert. of Appreciation, VA, 1971, Am. Heart Assn., 1991, 92, 93. Mem. Soc. Plastics Engrs. (sr., Mem. of Month So. Calif. sect. 1989), Am. Mgmt. Assn., Arnold Air Soc. (comptr. 1969, pledge tng. officer 1970), Plastics Acad., Demolay, Profl. Ski Instrs. Am., Mensa, Am. Legion, Elks, Phi Gamma Psi. Avocations: skiing, scuba diving, marathon running, golfing, music. Office: Baxter Edwards Critical-Care 17221 Red Hill Ave Irvine CA 92714-5686

KINSTLE, JAMES FRANCIS, polymer scientist; b. Lima, Ohio, Nov. 23, 1938; s. Herbert W. and Marie (Naylor) K.; m. Alice Newman, Nov. 8, 1980. BA in Chemistry, Bowling Green (Ohio) State U., 1966, MA in Chemistry, 1967; PhD in Polymer Sci., U. Akron, 1970. Sr. scientist Ford Motor Co., Dearborn, Mich., 1969-72; prof. U. Tenn., Knoxville, 1972-83; corp. R & D mgr. Polaroid Corp., Cambridge, Mass., 1983-87; corp. R & D mgr. James River Corp., Neenah, Wis., 1987-92, Cin., 1992—; cons. Oak Ridge (Tenn.) Nat. Lab., also numerous cos., 1972-83; mem. adv. bd. polymer programs U. Conn., Storrs, 1989—. Co-editor: Chemical Reactions on Polymers, 1988, Radiation Curing of Polymers, 1989; editor-in-chief Jour. Radiation Curing, 1974-79; contbr. numerous articles on environ. impact of polymers, radiation curing, reactions of polymers, and polymer synthesis and characterization to profl. jours. Fellow NDEA and NSF, 1966-69. Mem. Am. Chem. Soc. (div. polymeric materials: sci. and engring., vice chmn. 1988, chmn.-elect 1989, chmn. 1990, A.K. Doolittle award 1977). Achievements include several patents and commercialized products. Office: James River Corp 1 Better Way Rd Milford OH 45150

KINTNER, ELISABETH TURNER, chemistry educator; b. Greensboro, N.C., Oct. 16, 1957; d. Holley Mack and Clara Murphy (Bond) Bell; 1 child, Robert Holley. BS in Chemistry, U. N.C., 1980; PhD, U. S.C., 1988. Chemist Trans World Chems., Washington, 1981; rsch. asst. U. S.C., Columbia, 1982-88, rsch. assoc., instr., 1988-90; lectr. U. S.C., Aiken, 1990; asst. prof. Winona (Minn.) State U., 1990-91, U. Pitts., Johnstown, 1991—. Author: Frontiers in Bioinorganic Chemistry, 1986; contbr. articles to profl. jours. Recipient Joseph W. BouKnight Teaching award U. S.C., 1983. Mem. Am. Chem. Soc. (bioinorganic subdiv., inorganic div., chem. edn. div.), Sigma Xi (Rsch. award 1984). Office: U Pitts Dept Chemistry Johnstown PA 15904

KINZER, ROBERT LEE, astrophysicist; b. Grandfield, Okla., June 23, 1941; s. Henry and Clara (Grant) K.; m. Lee Ann Johns, Feb. 4, 1967; children: Kirsten Lee, Robin Ann. PhD, Univ. Okla., 1967. Astrophysicist Naval Rsch. Lab., Washington, 1967—. Mem. Am. Physical Soc., Am. Astronomical Soc., Am. Assn. Advancement Scis. Office: Naval Rsch Lab Washington DC 20375

KINZIE, DANIEL JOSEPH, superconducting technology and mechanical engineer; b. Chapel Hill, N.C., June 28, 1966; s. Joseph Lee and Jeannie Lillian (Jones) K. BS in Mech. Engring. and Materials Sci., MIT, 1988; MS, Stanford U., 1989. Rsch. asst. MIT Ceramics Processing Lab., Cambridge, 1987-88, coop engr. Newport News (Va.) Shipbldg. and Drydock Co., 1988; sr. engr. space systems div. Gen. Dynamics, San Diego, 1989—. Contbr. articles to profl. jours. MacDonald fellow Stanford U., 1988, Guggenheim fellow Princeton U., 1988, Fgn. Lang. fellow NSF, 1989. Mem. AIAA, ASME, Pi Tau Sigma, Sigma Xi. Lutheran. Achievements include development of new sol-gel processing technique for Barium-Yttrium-Cuprate high temperature superconductor powder, quench heaters for SSC superconducting dipole magnets, superconducting x-ray lithography light source magnet.

KIPP, CARL ROBERT, mechanical engineer; b. Dayton, Ohio, June 15, 1959; s. Harvey Elwood and Janice Louise (Wright) K.; m. Paula Ann Lewis, Aug. 23, 1986; children: Stephanie Elizabeth, Valerie Catherine. BSME, GMI Engring./Mgmt. Inst., Flint; MSME, Purdue U., 1985. Jr. draftsman Systems Rsch. Labs., Inc., Dayton, Ohio, 1977-78; assoc. engr. Delco Products div. GM Corp., Dayton, 1983; teaching asst., rsch. asst. Purdue U., West Lafayette, Ind., 1983-85; mem. tech. staff AT&T Bell Labs., Whippany, N.J., 1985-93; mem. tech. staff, project mgr. AT&T Network Systems, Columbus, Ohio, 1993—. Contbr. articles to profl. jours. Mem. Adoptive Parents for Open Records, Morristown, N.J. Mem. ASME (chair subsect. 1988-91, reviewer tech. papers 1989, Membership Devel. award 1990), Soc. Automotive Engrs., Acoustical Soc. Am., Sigma Xi. Office: AT&T Network Systems 67 whippany Rd 6200 E Broad St Columbus OH 43213-1569

KIRA, GERALD GLENN, engineering executive; b. Pitts., Apr. 3, 1951; s. Michael and Dorothy (Kalzur) K.; m. Connie E. Sheets, June 23, 1973; children: Christopher G., Steven G. BSCE, U. Pitts., 1973. Registered profl. engr., lic. surveyor, Pa. Sr. estimator Mellon Stuart Co., Pitts., 1973-85, mgr. spl. projects, 1985-87; lead estimator Mellon Stuart Co., Orlando, Fla., 1987-88; project mgr. Boraj Craig Barber, Naples, Fla., 1988-90; v.p. Associated Cost Engrs., Orlando, 1990—. Com. mem. troop 749 Boy Scouts Am., Naples, 1989-92;mem. adv. bd. Collier 2000 (Community Study for Mcpl. Devel.), Naples, 1991. Mem. NSPE, Pa. Soc. Profl. Engrs. (scholarship com. Pitts. chpt. 1978-90), Chi Epsilon (hon.), Sigma Tau (hon.). Office: Associated Cost Engrs Ste I 12 4201 Vineland Rd Orlando FL 32811

KIRAN, ERDOGAN, chemical engineering educator; b. Malatya, Turkey, Nov. 10, 1946; s. Hasan and Fatma (Ozcicek) K.; m. Gunin Akkor; children: Levent, Dilara. BSc, MIT, 1969; MS, Cornell U., 1971; PhD, Princeton (N.J.) U., 1974. Rsch. leader Turkish Pulp and Paper Mfrs. Cen. Lab.,

Izmit, Turkey, 1974-76; dir. SEKA Cen. Lab., Izmit, Turkey, 1976-79; assoc. prof. U. Maine, Orono, 1981-86, Gottesman prof., 1986—; adj. prof. Bogazici U., Istanbul, Turkey, 1977-81, Istanbul Tech. U., 1979-81; cons. prof. Bilkent U., Ankara, 1991—; organizer, dir. NATO Advanced Study Inst. on Supercritical Fluids, Fundamentals for Application, 1993. Editor: Supercritical Fluid Engineering Science: Fundamentals and Applications, 1993; mem. editorial bd. Nature, Turkish Jour. Engring. Environ. Sci., 1989—; editor, founder Jour. Supercritical Fluids, 1987—; contbr. numerous articles to prof. jours. Recipient numerous rsch. grants. Mem. AICE (high pressure com. 1986—), Am. Chem. Soc., Tech. Assn. of the Pulp and Paper Industry, Soc. of Plastics Engrs. (sr.). Achievements include rsch. on the properties and utilization of supercritical fluids, applications in the polymers, pulp and paper and forest products industries. Office: U Maine 5737 Jenness Hall Orono ME 04469-5737

KIRBY, GARY NEIL, metallurgical and materials engineer, consultant; b. Neptune, N.J., Apr. 25, 1935; s. Edward Lewin and Margaret Lillian (Banks) K.; m. Carol Marie Chester, Apr. 29, 1967; children: John Edward, Kristin Marie. BAMetE, Cornell U., 1957; MAMetE, Lehigh U., 1961; PhD in Metall. Engring., U. Mich., 1972. Rsch. asst. metall. Union Carbide Corp., Niagara Falls, N.Y., 1957-59; rsch. asst. Rutgers U., New Brunswick, N.J., 1961-62; rsch. assoc. Internat. Nickel Co., Bayonne, N.J., Suffern, N.Y., 1962-66, Climax Molybdenum Co., Ann Arbor, Mich., 1969-71; dept. head Crawford & Russell E & C, Stamford, Conn., 1974-79; corp. materials engr. Ciba-Geigy Corp., Toms River, N.J., 1979-87; supr. EBASCO Svcs., Inc., N.Y.C., 1987-89; pres. Kirby Corrosion Control, Inc., Brielle, N.J., 1990—; cons. AlliedSignal Corp., Morristown, N.J., 1990-93. Editor: Dust Explosions, 1989; contbr. articles to profl. jours. Bd. dir. Garden State Philharm. Symphony Soc., Toms River, 1985-87, mem. Garden State Philharm. Chorus, 1985—. McMullen Regional scholar Cornell U., 1952, Grad. fellow Internat. Nickel Co., U. Mich., 1966. Mem. Nat. Assn. Corrosion Engrs. Achievements include co-inventing patents for copper-base alloy die casting and low-chromium age-hardenable cupronickels; lectured for industrial corrosion course for center for professional advancement for 12 years in U.S., Holland and Switzerland. Home and Office: 917 Teaberry Lane Brielle NJ 08730

KIRBY, JAMES THORNTON, JR., civil engineering educator; b. Easton, Md., Dec. 25, 1952; s. James Thornton and Pearl (Ross) K.; m. Barbara Ann Leventry, June 18, 1983; children: Joshua Thornton, Nicholas Ross, Benjamin Naylor. ScB, Brown U., 1975, ScM, 1976; PhD, U. Del., 1983. Asst. prof. Oceanography SUNY, Stony Brook, 1983-84; asst. prof. Ocean Engring. U. Fla., Gainesville, 1984-88; assoc. prof. Civil Engring. U. Del., Newark, 1989—; cons. Exxon Prodn. Rsch. Inc., Bechtel, Chevron Oil Field, U.S. Army Corp of Engrs., govt. labs. in U.S., Korea, Australia, The Netherlands, 1983—. Contbr. 37 articles to profl. jours. NSF Rsch. grantee, 1991—, ONR grantee, 1983—, Sea grant (NOAA), 1988—. Mem. ASCE (Walter L. Huber Civil Engring. Rsch. prize 1992, mem. publ. com. 1987—, mem. fluid mechanics com. 1991—), Am. Geophys. Union, Soc. for Indsl. and Applied Math., Sigma Xi. Democrat. Achievements include research in models for water wave propagation; developed numerical codes used in consulting industry and government laboratories worldwide. Home: 6 Pagoda Ln Newark DE 19711 Office: U Del Dept Civil Engring Newark DE 19716

KIRBY, KEVIN ANDREW, utilities company executive; b. Northampton, Mass., Aug. 25, 1950; s. Donald B. and Eileen B. (Boek) K.; m. Catherine Mary Faherty, Mar. 13, 1971; children: Theresa, Michael. BSEE, Northeastern U., Boston, 1973. Registered profl. engr., Mass., Calif., N.Y., R.I., N.H., N.J. Elec. engr. Stone & Webster Engring. Corp., Boston, 1973-79, United Engrs. and Constructors, Boston, 1979-80; mgr. elec. engring. Fern Engring., Inc., Bourne, Mass., 1980-84; cons. Kirby Engring. Assocs., Inc., Bourne, 1984-86; planning engr. EUA Svc. Corp., West Bridgewater, Mass., 1986-88, sr. project engr., 1988-89, dir. integrated resource mgmt., 1989—. V.p. Bourne Pigskin Club, 1990-92, pres. Swish Youth Basketball League, Bourne, 1987-88; sec. Bourne Babe Ruth, 1989-91. Mem. IEEE. Home: 61 Monument Neck Rd Bourne MA 02532 Office: EUA Svc Corp PO Box 543 750 W Center St West Bridgewater MA 02379

KIRBY, SHAUN KEVEN, physicist; b. Silver Spring, Md., Sept. 11, 1967; s. Marvin Devon and Doris Ann (Fink) K. BSEE, Princeton U., 1989; MS, Calif. Inst. Tech., 1991. Office of Naval Rsch. fellow Calif. Inst. Tech., Pasadena, 1989—; praktikant Asea Brown Boveri, Baden, Switzerland, summer 1988. Contbr. articles to profl. jours. Mem. Phi Beta Kappa, Sigma Xi, Tau Beta Pi. Methodist. Home: 446 S Catalina Ave Apt 203 Pasadena CA 91106 Office: Calif Inst Tech Mail Stop 128-95 Pasadena CA 91125

KIRCHHOFF, WILLIAM HAYES, chemical physicist; b. Chgo., Nov. 27, 1936; s. Paul August and Pauline (Swinehart) K.; m. Ann Elizabeth Rogers, Aug. 24, 1958; children Margaret Anne Kirchhoff Shearer, Daniel Rogers, David Paul, Jennifer Anne. BSc, U. Ill., 1958; MA, Harvard U., 1961, PhD, 1963. Physicist, chief office environ. measurements Nat. Bur. of Standards, Gaithersburg, Md., 1964-87; mgr. chem. physics program U.S. Dept. of Energy, Germantown, Md., 1987—. Contbr. articles to Jour. Molecular Spectroscopy. Recipient Silver medal U.S. Dept. of Commerce, 1977; NATO fellow Oxford (Eng.) U., 1963. Mem. ASTM (chair com. biol. effects and environ. fate 1980-85). Presbyterian. Achievements include research in molecular spectroscopy, thermodynamics and environmental policy. Office: US Dept of Energy Washington DC 20585

KIRCHMAYER-HILPRECHT, MARTIN, geologist; b. Bad Hall, Austria, Feb. 1, 1923; s. Peter and Katharina (Zäuninger) K.; 2 children. PhD, U. Vienna; diploma in edn., U. Heidelberg, PhD (hon.); DSc (hon.). Sci. asst. univ. and T.H., Vienna; cons. applied geology, geologist Bolidens Gruv AB, Sweden; ptnr. of rsch. Columbia U., N.Y.C.; lectr. U. Md.; collaborator Preussag, Hannover, Fed. Republic Germany; assoc. of rsch. Spl. Sch. of the Mines, Clausthal; geologist Prakla Gesellschaft für praktische Lagerstättenforschung; ptnr. of rsch., prof. edn. in mineralogy and petrography U. Heidelburg (Fed. Republic Germany). Contbr. articles to profl. jours. Recipient Prix M-Neumayr, U. Vienna, Cert. Appreciation U.S. Forces Support Dist. Baden-Wuerttemberg, 1974, Albert Einstein Bronze Medal for Peace, 1984, Cert. Disting. Svc. U. Md., 1986, Albert Einstein Internat. Acad. Cross of Merit 3d class with ribbon, 1988. Mem. Schlaraffia Heidelberg, Austrian Soc. Mineralogy, Group of Work of Mine Geology of the Coal-Mining Region of West Germany Bochum, Swedish Soc. Geology, Geol. Soc. Am., Am. Nat. Geog. Soc., U. Md. Alumni Assn. Internat., AAAS (life). Avocation: music. Home: Michael Gerber Strasse 22, D 69151 Neckargemuend Germany

KIRCHNER, JAMES WILLIAM, electrical engineer; b. Cleve., Oct. 17, 1920; s. William Sebastian and Marcella Louise (Stuart) K.; m. Eda Christene Landfear, June 11, 1950 (dec. May 1977); children—Kathleen Ann Kirchner Duda, Susan Lynn Kirchner Buonpane. B.S. in Elec. Engring., Ohio U., 1950, M.S., 1951. Registered profl. engr., Ohio. Instr. elec. engring. Ohio U., Athens, 1950-52; mgr. liaison engring. Lear Siegler Inc., Maple Heights, Ohio, 1952-64; coordinator engring. services Case Western Res. U., Cleve., 1964-72, gen. mgr. Med. Ctr. Co. (CWRU), 1972-91; ret., 1991—; sec. of corp. Thermagon, Inc., Cleve., 1992. Mem. Portage County Republican Exec. Com., 1961-62; treas. PTA, Aurora, Ohio, 1963-65, v.p., 1965-66; mem. The Ch. in Aurora, 1956—. Served with USAAF, 1942-45, PTO. Mem. Nat. Soc. Profl. Engrs., Ohio Soc. Profl. Engrs. (life mem.), Cleve. Engring. Soc. (chmn. environ. com. 1976), IEEE (life mem.), Am. Soc. Engring. Edn. (life mem.). Home: 200 Laurel Lake Dr W 306 Hudson OH 44236

KIRICK, DANIEL JOHN, agronomist; b. Port Jervis, N.Y., Nov. 8, 1953; s. Daniel and Mary Theresa K.; m. Jean Marie Guse, Sept. 27, 1986; 1 child, Nicholas. BA in Biology, History, U. Minn., Duluth, 1976; BS in Agronomy, U. Minn., St. Paul, 1977. Cert. profl. agronomist. Agronomist Delft (Minn.) Farm Chems., 1978, Skelly Fertilizer, Trimont, Minn., 1978-80, Mower County Svc. Co., Sargeant, Minn., 1980-86, Cenex Supply, Ellis, S.D., 1986-88, Rice (Minn.) Farm Supply, 1988-91, Kirick Agronomy Svcs., St. Cloud, Minn., 1992—. Mem. Community Edn. Devel. Adv. Coun., Sauk Rapids, Minn., 1990—, Youth Devel. Bd., Sauk Rapids, 1990, Benton County Ext. Com., 1993—. Mem. AAAS, Weed Sci. Soc. Am., Soil Sci. Soc. Am., Crop Sci. Soc. Am., Am. Soc. Agronomy. Roman Catholic. Home: PO Box 206 Rice MN 56367-0206 Office: Kirick Agronomy Svcs 3105 2d St SE Saint Cloud MN 56304

KIRK, THOMAS KENT, research scientist; b. Minden, La., Oct. 13, 1940; s. William Thomas and Wilda Inez (Gilstrap) K.; m. Celeste Hanson; children by previous marriage: Sharon Denise, Deborah Katherine, Sandra Kay. BS, La. Tech. U., 1962; MS, N.C. State U., 1964, PhD, 1968; postdoctoral, Chalmers Inst. Tech., Sweden, 1968-70. Rsch. microbiologist Forest Products Lab., USDA, Madison, Wis., 1970-80, project leader, 1980-85, dir. Inst. for Microbiol. and Biochem. Tech., 1985—; prof. dept. bacteriology U. Wis., 1982—; adj. prof. dept. wood and paper sci. N.C. State U., Raleigh, 1975—; vis. prof. Wood Rsch. Inst., Kyoto U., Uji, Japan, 1979-80. Contbr. more than 170 sci. papers and articles to profl. publs.; patentee in field. Recipient Marcus Wallenberg prize, Falun, Sweden, 1985. Fellow Internat. Acad. Wood Sci. (sec.-treas. 1984-89); mem. NAS, TAPPI (prize and medal R&D div. 1986), Am. Soc. Microbiology, Am. Chem. Soc. (Marvin J. Johnson Microbial and Biochemical Technology award 1993). Home: 3145 Timber Ln Verona WI 53593-9057 Office: USDA Forest Products Lab One Gifford Pinchot Dr Madison WI 53705-2398

KIRK, WILEY PRICE, JR., physics and electrical engineering educator; b. Joplin, Mo., July 24, 1942; s. Wiley Price Sr. and Inez Isabel (Watson) K.; m. Sally Ann Stoots, June 13, 1964; children: Camille Maura, Alexander Price. BS, Washington U., St. Louis, 1964; MS, SUNY, Stony Brook, 1967, PhD, 1970. Tech. collaborator Brookhaven Nat. Lab., Upton, N.Y., 1969-70; postdoctoral fellow U. Fla., Gainesville, 1970-72, asst. prof., 1972-75; asst. prof. Tex. A&M U., College Sta., 1975-77, assoc. prof., 1978-83, prof., 1984—, also bd. dirs.; dir. NanoFAB ctr. Tex. A&M U., 1990—. Editor: Nanostructure Physics and Fabrication, 1989, Nanostructures and Mesoscopic Systems, 1992; assoc. editor: Superlattices and Microstructures Acad. Press, Scottsdale, Ariz., 1991—; contbr. over 70 articles and papers to profl. jours.. Grantee NSF, 1973—, Rsch. Corp,. 1974; recipient Nat. Bur. Standards Precision Measurements award Nat. Inst. Standards and Tech., 1987, Tex. Engring. Experiment Sta. Rsch. Fellow award Tex. Engring. Exptl. Sta., 1992. Mem. AAAS, Am. Phys. Soc., Am. Vacuum Soc., Materials Rsch. Soc. Achievements include patent for gate adjusted resonant tunnel diode device and method of manufacture. Office: Tex A&M U Engring Physics Bldg # 423 College Station TX 77843-4242

KIRKBRIDE, CHALMER GATLIN, chemical engineer; b. nr. Tyrone, Okla., Dec. 27, 1906; s. Zachariah Martin and Georgia Anna (Gatlin) K.; m. Billie Lucille Skains, Apr. 13, 1939; 1 son, Chalmer Gatlin Jr. BSE and MSE, U. Mich., 1930; ScD (hon.), Beaver Coll., 1959; Eng. D. (hon.), Drexel U., 1960, Widener U., 1970. Chem. engr. rsch. dept. Standard Oil Co., Whiting, Ind., 1930-34; dir. tech. svc. Amoco, Texas City, Tex., 1934-41; chief chem. engring. devel. Mobil Oil Co., Dallas, 1942-44; Disting. prof. Tex. A&M U., 1944-47, cons. chem. engr., 1944-47; sci. cons. to sec. of War Bikini atomic bomb tests, 1946; v.p. charge R & D Houdry Process Corp., Marcus Hook, Pa., 1947-52; pres. and chmn. bd. Houdry Process Corp., Phila., 1952-56, dir., 1948-62; dir. Catalytic Constrn. Co., 1952-56; exec. dir. comml. devel., rsch., engring. and patent depts. Sun Oil Co., Phila., 1956-60, v.p. comml. devel., rsch., engring. and patents, 1960-70, corp. dir., 1963-70; pres. Avisun, Phila., 1959-60, dir., 1959-68; dir. Sunolin Chem. Co., 1957-68; exec. producer motion picture The Seeds of Evil, 1972; dir. Coordinating Rsch. Coun., 1958-70, pres., 1965-67; Mem. Pres. Nixon's Task Force on Oceanography, 1969; mem. adv. panel on sea grant programs NOAA Dept. Commerce, 1970-74; petroleum specialist Fed. Energy Agy., 1974-75; sci. adviser to adminstr. ERDA, 1975-77, cons. engr., 1978—; pres. Kirkbride Assocs. Inc., 1979—. Author: Chemical Engineering Fundamentals, 1947; contbr. articles to profl. jours.; patentee in field. Trustee Widener U., Chester, Pa., 1956-72, hon., 1972—, vice chmn. bd. trustees, 1959-71; chmn. bd. dirs. Riddle Meml. Hosp., 1965-67, dir., 1965-71. Served as 2d lt. Chem. Warfare Res., 1935-40. Recipient Disting. Pub. Service award U.S. Navy, 1968; Engring. Centennial medal Widener U., 1970; George Washington award Phila. Engring. Club, 1971; Dedicated Kirkbride Hall of Sci. and Engring. Widener U., 1965; elected to Nat. Acad. Engring., 1967. Fellow Am. Inst. Chem. Engrs. (pres. 1954, Profl. Progress award 1951, Founders award 1967, Fuels and Petrochem. award 1976, named Eminent Chem. Engr. 1983); mem. Am. Chem. Soc., Am. Petroleum Inst., Alpha Chi Sigma, Phi Lambda Upsilon, Tau Beta Pi. Clubs: Army-Navy (Washington). Home and Office: 4000 Massachussetts Ave NW Ste 1415 Washington DC 20016

KIRKLAND, MATTHEW CARL, nuclear engineer; b. Stamford, Conn., Apr. 28, 1959; s. Herbert John and Josephine (Schnellen) K.; m. Patricia Ann Walden, Jan. 26, 1985; children: Matthew Herbert, Ashley Rae. BS in Nuclear Engring., SUNY, Buffalo, 1981. Sr. reactor operator Farley Nuclear Plant-Ala. Power Co., Ashford, 1985-87, GE, 1987-89; nuclear engr. Detroit Edison Co.-Fermi Power Plant, Newport, Mich., 1989—. Achievements include coordination of complete control rod blade changeout for cobalt reduction, redesign of high density fuel storage rack surveillance, planning refueling outage offload and reload strategy. Home: 3829 Douglas Rd Ida MI 48140 Office: Detroit Edison Co Fermi 2 Power Plant 220AIB 6400 N Dixie Hwy Newport MI 48166

KIRKLAND, NED MATTHEWS, chemical engineering manager; b. Sheffield, Ala., Oct. 22, 1953; s. Ned Woodrow and Ona Mae (Matthews) K.; m. Vicky Jo Morrow, Aug. 12, 1978; children: William, Benjamin, Molly. BS in Chemistry, U. Ala., 1976; MSCE, U. Ky., 1979. Registered profl. engr., W.Va. Process engr. 3M Co., Guin, Ala., 1976-78; mass transfer specialist Union Carbide, South Charleston, W.Va., 1979-84, process engr., 1984-87; process/project engr. Amoco Performance Products, Greenville, S.C., 1987-90; engring. supr. Amoco Performance Products, Marietta, Ohio, 1990—. Mem. AICE. Home: 5110 Glenbrook Dr Vienna WV 26105 Office: Amoco Performance Products PO Box 446 Marietta OH 45750

KIRKLAND, SHARI LYNN, clinical psychologist; b. Sacramento, Calif., Sept. 15, 1961; d. Daniel Edward and Sumatra E. (Mulholland) K. BA, UCLA, 1983; MA, U. Ariz., 1985, PhD, 1988. Lic. clin. psychologist, Calif. Psychology extern Ariz. Dept. Corrections, Tucson, 1985-87; predoctoral intern Kaiser Permanente, L.A., 1987-88; postdoctoral intern Kaiser Permanente, Oakland, Calif., 1989-90; staff psychologist Kaiser Permanente, Hayward, Calif., 1990—; crisis therapist Kino County Hosp., Tucson, 1988-89; cons. U. Calif. San Francisco Dental Clinic, 1990; guest speaker ethnic-minority clin. issues and dangerous patient mgmt. Author: (with others) Discourse and Discrimination, 1988; contbr. articles to profl. publs., including Personality and Psychology Bull., Jour. Personality and Social Psychology. Vol. therapist Child Protective Svcs., Tucson, 1985-87. Recipient Grad. Tuition schoarship U. Ariz, Tucson, 1983-86, Minority Acad. scholarship, 1987. Mem. APA, Assn. Black Psychologists (charter, v.p. Tucson chpt. 1984-86). Office: Kaiser Permanente Dept Psychiatry 27400 Hesperian Blvd Hayward CA 94545

KIRKWOOD, ROBERT KEITH, applied physicist; b. Santa Monica, Calif., Mar. 10, 1961; s. Robert Lord and Patricia Cathrine (Keith) K.; m. Kimberly DeNeve Saunders, May 2, 1991. BS, UCLA, 1982, MS, 1984; PhD, MIT, 1989. Rsch. asst. dept. elec. engring. UCLA, 1982-84; mem. tech. staff TRW Space and Tech. Group, Redondo Beach, Calif., 1984-85; rsch. asst. MIT, Cambridge, 1985-89, vis. scientist Plasma Fusion Ctr., 1992—; postdoctoral fellow Calif. Inst. Tech., Pasadena, 1989-91; rsch. assoc. geophysics div. Air Force Phillips Lab., Hanscom AFB, Mass., 1991-92, physicist, 1992—. Contbr. articles to Nuclear Fusion, Physics of Fluids B, Rev. Sci. Instruments, Physics Letters A. Recipient Rsch. Associateship award NRC, 1991; postdoctoral fellow TRW Space and Tech. Group, 1985, Dept. Energy, 1989. Mem. Am. Phys. Soc. (Simon Ramo award in plasma physics 1991), Am. Nuclear Soc., Am. Geophys. Union. Achievements include development of cyclotron absorbtion diagnostics for plasmas, current drive in plasmas with low frequency waves. Office: Phillips Lab/PHP Hanscom AFB MA 01731

KIRMAN, LYLE EDWARD, chemist, engineer, consultant; b. N.Y.C., Sept. 12, 1946; s. Edward Irah and Arlene Margerite (Claybaker) K.; m. Cathleen Mary Dillon, July 18, 1966; children: Christopher, Matthew, Micah. BA in Chemistry, Case Western Res. U., 1972, MS in Chemistry, 1975. Lab. technician McGean Chem. Co., Cleve., 1966-68; process devel. engr., then sr. process devel. engr. Gould, Inc., Cleve., 1968-75, plant engr., 1975-83, environ. engr., 1983-86; mgr. systems engring. Kinetico Engineered Systems, Inc., Newbury, Ohio, 1987-92, v.p. tech., 1992—; presenter at profl. confs. Contbr. articles to profl. publs. Vice-pres. Coventry Neighbors, Inc., Cleveland Heights, Ohio, 1971-72. Mem. Am. Chem. Soc., Am. Electroplates Soc. Lutheran. Achievements include patents for process for tin and tin alloy plating; ion exchange process; deionization apparatus and control method. Office: Kinetico Engineered Systems 10845 Kinsman Rd Newbury OH 44065

KIRSCH, ROBERT, director analytical research and development; b. Bklyn., Feb. 27, 1952; s. Leonard and Judith (Spitz) K.; m. Elisa Zeitlin, Nov. 29, 1981; children: Rachel, Jessica. BS in chem., U. Md., 1974; MS, PhD, NYU, 1983. Sr. scientist Am. Cyanamid, Stamford, Conn., 1980-87; dir. analytical R&D Barr Labs., Pomona, N.Y., 1987-90; assoc. dir. analytical devel. DuPont Merck Pharm. Co., Garden City, N.Y., 1990—. Author: (chpt. in book) Laboratory Robotics, 1984, 85; contbr. articles to profl. jours. Campaign worker Dem. party Suffolk County, N.Y., 1990-93. Recipient Excellence award Am. Cyanamid, Stamford, Conn., 1984, Bus. Excellence award, DuPont Merck Pharm, Co., Garden City, N.Y., 1992. Mem. Am. Chem. Soc. (pres. 1975), Am. Assn. Pharm. Scientists (pres. 1989), Assn. Official Analytical Chemists (pres. 1992). Achievements include establishing new analytical, pioneer in laboratory Robotics, established first system for American Cyanamid and Barr Labs. Home: 224 Cedrus Ave East Northport NY 11731 Office: DuPont Merck Pharm Co 1000 Stewart Ave Garden City NY 11530

KIRSCH, TED MICHAEL, pathobiologist; b. Phila., Jan. 26, 1950; s. Joseph and Marjorie (Golden) K.; m. Leslie Ann Bryan, Feb. 5, 1970; children: Alys, Olivia. BA in Biology, Rider Coll., 1973; PhD, U. Pa., 1983. Rsch. fellowship Fels Rsch. Inst., Phila., 1984-86; clin. program mgr. Smith Kline Beckman Corp., Phila., 1986-88; sr. clin. scientist Wyeth-Ayerst Rsch., Phila., 1988-90, asst. dir., 1990—; AIDS researcher; cons. Franklin Inst., Phila., 1983. Contbr. articles to profl. jours. Vol. Recycling Com., Phila., 1984—, Spl. Olympics. Achievements include patent in Unique Column Configuration of Three Resins for Rapid Purification of Labile Proteins (potential use for mfg. recombinant proteins). Home: 427 Catharine St Philadelphia PA 19147-3105 Office: Wyeth-Ayerst Rsch PO Box 8299 Philadelphia PA 19101

KIRSCHNER, MARC WALLACE, biochemist, cell biologist; b. Chgo., Feb. 28, 1945. BA, Northwestern U., 1966; PhD in Biochemistry, U. Calif., Berkeley, 1971. Asst. prof. Princeton U., 1972-77, prof. biochemistry, 1977-78; prof. dept. biochemistry and biophysics U. Calif., San Francisco, 1978—. Recipient Rsch. Career Devel. award NIH, 1975-80; NSF fellow U. Calif., 1971-72. Mem. NAS (Richard Lounsberg award 1991), Am. Soc. Biol. Chemists, Am. Soc. Cell Biology, Am. Acad. Arts and Sci. Achievements include research in mechanism of microtubule assembly, regulation of mitosis and cell division in amphibian eggs, biophysical studies of macromolecules, embryonic induction. Office: U Calif Dept Biochemistry & Biophysics San Francisco CA 94143*

KIRSCHNER, RONALD ALLEN, osteopathic plastic surgeon, otolaryngologist, educator; b. N.Y.C., Jan. 18, 1942; s. Hyman C. and Eleanor (Pinkus) K.; m. Olivia Barbara Schlesinger, June 27, 1964; children: Andrew Scott, Julie Renee. AB, NYU, 1962; DO, Phila. Coll. Osteo. Medicine, 1966, MS in Otolaryngology, 1972. Diplomate Am. Osteo. Bd. Otolaryngology. Intern Le Roy Hosp., N.Y.C., 1966-67; resident Grandview Hosp., Dayton, Ohio, 1967-68; resident Phila. Coll. Osteo. Medicine, 1970-72, asst. prof., 1972-74, assoc. prof., 1974-76, clin. assoc. prof., 1976-85, clin. prof., 1985-90, prof., chmn. dept. otolaryngology, bronchoesophagology and facial plastic surgery, 1990-92, dir. emerging tech., 1992—; dir. neurosensory unit, 1973-76; NIH fellow Armed Forces Inst. Pathology, Washington, 1971; practice medicine specializing in plastic, otolaryngology and laser surgery, Bala Cynwyd, Pa., 1976—; attending physician Grad. Hosp., 1991—; attending physician, cons. Presbyn.-U. Pa. Med. Ctr., 1987—, Hosp. of Phila. Coll. Osteo. Medicine, chmn. laser and endoscopy com., 1987-89, 91—, mem. exec. com., 1990-92; attending physician Suburban Gen. Hosp., chief ear, nose and throat and plastic surgery, 1976—, chmn. div. surgery, 1983-89, exec. com., 1983-89; attending physician, cons. Del. Valley Med. Ctr., 1985—; v.p., chief med. adv. Courtlandt Group, 1979-85, exec. v.p., 1985-86, also dir. rsch. and edn., 1986; otolaryngologist Pa. Hearing Assn., 1986—; preceptor Xanar Laser Div., Johnson & Johnson, 1982; design cons. Pilling, Inc., 1982-87, Inframed Inc., 1985—, Sigma Dynamics Inc., Rhein Med., Inc., 1988—; otologic cons. Nat. Childrens' Hearing Aid Bank; pres. Kirschner Design Group, Inc., 1987—; bd. dirs. KDG-Rotem U.S.A., Pa. Acad. Cosmetic Surgery; dir. head and neck YAG laser protocol Cooper Lasersonics, 1983-88; chmn. med. symposium Internat. Conf. on Applied Laser Electro Optics, 1986, 87, 91; session chair Medtech '89, Freie Univ., Berlin, 1989; vis. prof. internat. sch. for quantam electronics Etore Majorana Nato, Erice, Sicily, 1990; cons. Bur. Vocat. Rehab., Imunodiagnostics Lab., Allergy Mgmt. Systems Inc.; dir. 1st World Congress on Cosmetic Laser Surgery, 1992. Served with M.C., USN, 1968-70; lt. comdr. Res. Recipient award for disting. teaching Lindbach Found., 1973, Legion of Honor, Chapel of Four Chaplains, 1982; Survivor of Yr. award, 1984; named Disting. Practitioner Am. Acads. of Practice. Fellow Pan Am. Allergy Assn., Phila. Acad. Facial Plastic Surgery, Phila. Laryngologic Soc., Phila. Coll. Physicians, Am. Soc. Lasers in Medicine and Surgery, Am. Auditory Soc., Am. Acad. Otolaryngology-Head and Neck Surgery, Soc. Ear, Nose, and Throat Advances in Children, Am. Acad. Facial Plastic Surgery (assoc.), Soc. Photo Optical Engrs., Osteo. Coll. Ophthalmology and Otorhinolaryngology, Am. Acad. Cosmetic Surgery; mem. Am. Osteo. Assn. (editorial cons. Jour. 1977—, editorial referee 1980—), Pa. Med. Soc., Pa. Acad. Otolaryngology, Pa. Acad. Cosmetic Surgery (bd. dirs. 1990—), Internat. Soc. Cosmetic Plastic Surgeons (bd. dirs.), Philadelphia County Osteo. Med. Assn. (chair laser com.), Centurian Club of Deafness Rsch. Found., Internat. Assn. Logopedics and Phoniatrics, Midwestern Biolaser Inst., Inst. for Applied Laser Surgery (pres.), Pa. Osteo. Med. Assn. (chmn. com. otolaryngology 1984-88, 90—, chmn. com. promotion of rsch. 1985-88), Am. Acad. Osteopathy, Survivors Club of Phila. Coll. Osteo. Medicine (pres. 1981-82), Internat. Soc. for Optical Engring., AAAS, AMA, Acad. Surgical Rsch., N.Y. Acad. Scis., Am. Soc. Liposuction Surgery, Laser Assn. Am. (sec. 1985-88), Laser and Electro Optics Mfrs. Assn., Am. Assn. Advancement Med. Instrumentation, Am. Soc. Cosmetic Surgeons, Pa. Hearing Aid Soc. (otologist), Pan Am. Assn. Otolaryngology and Bronchoesophagology, Pa. Acad. Opthalmology and Otolaryngology, Pa. Osteo. Med. Soc., Del Valley Tinnitus Assn. (chmn. com. otolarynology 1984-88, 90—, chmn. med. adv. bd.), Laser Inst. Am. (sr. Outstandi86, chmn. lasers 1987-89, bd. dirs. 1989—, dir., chmn. com. on biology and medicine 1989—), Pa. Acad. Cosmetic Surgery (bd. dir.), Am. Acad. Cosmetic Laser Surgery (bd. dirs. 1991—), Pa. Med. Soc., Montgomery County Med. Soc., Sigma Xi, Sigma Chi, Lambda Omicron Gamma (pres. 1981-82, Disting. Service award Caduceus chpt. 1982), Variety Club, NYU Club, Vesper Club, Pickwick Club of Phila., Masons, Shriners. Jewish. Med. editor Med. Portfolio, 1980-85; guest editor Surg. Clinics of N.Am., 1984; monthly columnist Photonics Spectra, 1987-91; contbg. editor Photonics Spectra, 1988—; mem. editorial bd. Pa. Osteo. Med. Jour., Laurin Publs., 1987—, Laser Applications; contbr. articles to med. jours., chpts. in med. texts; developer various med. instruments. Office: 2 Bala Cynwyd Plz Ste 17l Bala Cynwyd PA 19004

KIRSCHSTEIN, RUTH LILLIAN, physician; b. Bklyn., Oct. 12, 1926; d. Julius and Elizabeth (Berm) K.; m. Alan S. Rabson, June 11, 1950; 1 child, Arnold. B.A. magna cum laude, L.I. U., 1947; M.D., Tulane U., 1951; D.Sc. (hon.), Mt. Sinai Sch. Medicine, 1984; LL.D. (hon.), Atlanta U., 1985; DSc (hon.), Med. Coll. Ohio, 1986; LHD (hon.), L.I. Univ., 1991. Intern Kings County Hosp., Bklyn., 1951-52; resident pathology VA Hosp., Atlanta, Providence Hosp., Detroit, Clin. Ctr., NIH, Bethesda, Md., 1952-57; fellow Nat. Heart Inst. Tulane U., 1953-54; mem. staff NIH, Bethesda, 1957-72, 74—; asst. dir. div. biologics standards NIH, 1971-72; dep. dir. Bur. Biologics, FDA, 1972-73, dep. assoc. commr. sci., 1973-74; dir. Nat. Inst. Gen. Med. Scis., 1974—, acting assoc. dir. women's health, 1990-91; mem. Found. Advanced Edn. Scis.; chmn. grants peer rev. study team NIH; mem. Inst. Medicine, NAS, 1982—; co-chair, sec. Spl. Emphasis Oversight Com. on Sci. and Tech., 1989—; co-chair PHS Coordinating Com. on Women's Health Issues, 1990—; mem. Office of Tech. Assessment Adv. Com. on Basic Rsch., 1989—. Recipient Superior Service award, 1980, Presdl. Disting. Exec. Rank award, 1985,Pub. Svc. award Fedn. of Am. Socs. for Exptl. Biology, 1993. Mem. AMA (Dr. Nathan Davis award 1990), Am. Assn. Immunologists, Am. Assn. Pathologists, Am. Soc. Microbiology, Am. Acad.

Arts and Scis. Home: 6 West Dr Bethesda MD 20814-1510 Office: Nat Inst Gen Med Scis 5333 Westbard Ave Bethesda MD 20892-0001

KIRSHEN, PAUL HOWARD, water resources consultant; b. Summit, N.J., June 7, 1948; s. Howard Robert and Berys Margaret (Horrocks) K.; m. Bettina Burbank; 1 child, Andrew Taylor. SB, Brown U., 1970; MS, MIT, 1972, PhD, 1975. Rsch. assoc. MIT, Cambridge, 1975-76; sr. engr. Environ. Rsch. and Tech., Inc., Concord, Mass., 1976-78; rsch. prof. Dartmouth Coll., Hanover, N.H., 1978-80; asst. prof. Va. Tech., Blacksburg, 1980-81; pvt. cons. Mass., 1981—; reviewer Jour. Water Resources Planning and Mgmt. Div., N.Y.C., 1980-92. Contbr. articles to profl. jours. Mem. Solid Waste Com., Groton, Mass., 1987. Mem. Am. Soc. Civil Engrs. (water resources systems com. 1980-93, ops. mgmt. com. 1988-93). Democrat. Achievements include research in climate change water resources, hydroelectric scheduling systems for electric utilities, stream flow forecasting models for electric utilities, and planning and software tools for sewage mgmt. Home and Office: 90 Farmers Row Groton MA 01450

KIRSHENBAUM, RICHARD IRVING, public health physician; b. Bklyn., Aug. 19, 1933; s. Joseph and Anne (Hantman) K.; m. Jean Shicher, Aug. 17, 1957; children: Miriam, Susan, Rachel. AB, Temple U., 1955; DO, Phila. Coll. Osteo. Medicine, 1959; MPH, Columbia U., 1971. Diplomate Am. Bd. Preventive Medicine. Resident intern Met. Hosp., Phila., 1959-60; pvt. practice medicine Bklyn., 1960-70; resident in pub. health N.Y.C. Dept. Health, 1970-73, pub. health physician, 1973-81, regional health dir. for Queens County, 1977-80, chief epidemiologist for Manhattan Borough, 1980-81; pub. health physician N.Y. State Dept. Health, N.Y.C., 1981—. Contbr. articles to profl. jours. Lt. col. Med. Corps N.Y. Army NG, 1981-91, USAR, 1991—. Recipient Physician's Recognition award AMA, 1973, 76, 79, 82, 85, 88, 90. Fellow Am. Coll. Preventive Medicine; mem. APHA, N.Y. Acad. Scis. Home: 313 Whitman Dr Brooklyn NY 11234-6935 Office: NY State Dept Health 5 Penn Plz Fl 5 New York NY 10001-1810

KIRSHNER, ROBERT P., astrophysicist, educator; b. Long Branch, N.J., Aug. 15, 1949; s. D.R. and Virginia (Klarman) K.; m. Lucy Rand Herman, June 15, 1970; children: Rebecca Rand K., Matthew Klarman K. AB, Harvard Coll., 1970; PhD, Calif. Inst. Tech., 1975. Research assoc. Kitt Peak Nat. Observatory, Tucson, 1974-76; asst. prof. astronomy U. Mich., Ann Arbor, 1976-80, assoc. prof. astronomy, 1980-82, prof. astronomy, 1982-85; dir. McGraw-Hill Observatory, 1980-85; prof. astronomy Harvard U., Cambridge, Mass., 1985—, chmn. astronomy dept., 1990—; vis. com. Mt. Wilson and Las Campanas Observatory, Pasadena, 1986-89, Space Telescope Sci. Inst., 1988-89; chmn. observatory vis. com. Associated Univs. for Rsch. in Astronomy, Washington, 1983-86, bd. dirs., exec. com., 1989—; mem. sci. adv. com. Nat. New Tech. Telescope, Tucson, 1983-84, Com. on Space Astronomy and Astrophysics, 1982-85, Space Telescope Users Com., 1990—; Marc Aaronson Meml. lectr. U. Ariz., 1989; Grubb Parson lectr. U. of Durham, 1990; Delphasus lectr. U. Calif., Santa Cruz. Contbr. over 180 articles to scientific jours. and mags. including Nat. Geographic, Nat. History, Scientific Am. NSF fellow, 1970, Alfred P. Sloan Found. fellow, 1979. Fellow Am. Phys. Soc., Am. Acad. Arts and Scis.; mem. Am. Astron. Soc. (councilor 1986-88), Internat. Astron. Union, Astron. Soc. Pacific. Home: 174 Walden St Concord MA 01742-3623 Office: Ctr for Astrophysics MS-19 Harvard University 60 Garden St Cambridge MA 02138-1596

KIRSTEN, HENDRIK ALBERTUS, geotechnical engineer; b. Johannesburg, Republic of South Africa, Mar. 9, 1942; s. Louis Pieter and Laura (Odendaal) K.; m. Caroline Cecile Yvonne Lubbe, Dec. 24, 1966; children: Louis, Anton, Riette, Etienne. BS in Engring., Witwatersrand U., Johannesburg, Republic of South Africa, 1963, MSc in Engring., 1966, PhD in Engring., 1986. Registered profl. engr., Republic of South Africa. Jr. engr. Ove Arup & Ptnrs., Johannesburg, 1965; lectr. in structural engring. Witwatersrand U., 1966; lectr. in rock mechs. Ove Arup & Ptnrs., 1967-73; ptnr., founder Steffen Robertson & Kirsten, 1974—. Contbr. articles to profl. publs. Founding mem., div. chmn. Dem. Party, Randburg, Republic of South Africa, 1989-90. Fellow South African Inst. Mining and Metallurgy, South African Instn. Civil Engrs.; mem. South African Assn. Cons. Engrs. Mem. Dutch Reformed Ch. Home: 13 Yvette St Robin Hills, Randburg 2194, South Africa Office: Steffen Robertson & Kirsten, 265 Oxford Rd Illovo, Johannesburg 2196, South Africa

KIRWAN, GAYLE M., physics educator; b. Chapel Hill, N.C., Nov. 6, 1950; d. Caesar B. and Irene (Fitzer) Moody; m. Donald F. Kirwan, Aug. 31, 1991. BS, Northwestern La. U., 1972; MEd, La. State U., 1975. Tchr. East Baton Rouge Schs., 1972-77; instr. La. State U., Baton Rouge, 1977-89; spl. projects coord. Am. Inst. Physics, Washington, 1989—; cons. Argonne (Ill.) Nat. Lab., 1986—, Edn. & Tng. Assoc., Narragansett, R.I., 1985—; editorial adv. bd. Edn. Digest, 1983-86; instr. Smithsonian Instn., Washington, 1991—. Author: Explore the World for Elementary Teachers, 1991; assoc. editor: WonderScience, 1989—; column editor: The Physics Teacher, 1986-89. NSF grantee; recipient Presdl. award for excellence in sci. teaching Pres. U.S., 1983. Mem. AAAS (sect. Q rep. 1992), NSTA (Nat. Sci. Tchr. of Yr. 1983), Am. Assn. Physics Tchrs. (physics teaching resource agt.), Sigma Pi Sigma, Delta Kappa Gamma. Episcopalian. Office: American Inst Physics 1825 Connecticut Ave NW Ste 213 Washington DC 20009-5708

KISAK, PAUL FRANCIS, engineering company executive; b. Pitts., July 15, 1956; s. Paul F. and Catherine M. (Svaranowic) K. BSE in Nuclear Engring., Engring. Physics and Engring. Sci., U. Mich., 1982; MBA, Ea. Mich. U., 1984; postgrad., U. Va., 1986—. Intelligence officer, engr. CIA, Langley, Va., 1983-86; engr. U.S. Dept. of State, Washington, 1985-86; engr., mem. tech. staff Space Applications Corp., Vienna, Va., 1986-87; founder, pres. KKI, Inc., Fairfax, Va., 1986—; sr. scientist, program mgr. Info. Tech. & Application Corp., Reston, 1987-89; cons. devel. PFK Enterprises, Washington, 1986—; mem. working group Strategic Def. Initiative, 1986—. Holder software copyrights. Caseworker U.S. Sen. John Glenn, Columbus, Ohio, 1979; del. Loudoun County Rep. Nat. Party, 1988—. Mem. AIAA, ASME, Am. Phys. Soc., Am. Math. Soc., Am. Nuclear Soc., Am. Astronautical Soc., Am. Mgmt. Soc., Assn. MBA Execs., Bioengring. Soc., Mensa, Intertel, Texnikoi, Beta Gamma Sigma, Pi Mu Epsilon. Avocations: flying, reading, sports, movies, woodworking. Home: 1011 Chapel Rd Middletown VA 22645

KISCHER, CLAYTON WARD, embryologist, educator; b. Des Moines, Mar. 2, 1930; s. Frank August and Bessie Erma (Sawtell) K.; m.Linda Sese Espejo, Nov. 7. 1964; children: Eric Armine, Frank Henry. BS in Edn., U. Omaha, 1953; MS, Iowa State U., 1960, PhD, 1962. Asst. prof. biology Ill. State U., 1962-63; rsch. assoc. Argonne (Ill.) Nat. Lab., 1963; asst. prof. zoology Iowa State U., 1963-64; NIH postdoctoral fellow in biochemistry M.D. Anderson Hosp, Houston, 1964-66; chief sect. electron microscopy S.W. Found. Rsch. and Edn., San Antonio, 1966-67; assoc. prof. anatomy U. Tex. Med. Br., Galveston, 1967-77; assoc. prof. anatomy U. Ariz. Coll. Medicine, Tucson, 1977-92, prof. emeritus, 1993—; cons., dir. Scanning electron microscopy lab. Shrine Burns Inst., Galveston, 1969-73. Contbr. articles to profl. jours. Cubmaster pack 107 Island Dist., Galveston, 1974-76; bd. dirs. YMCA. With USN, 1947-49. NIH Rsch. grantee, 1968-89; Morrison Trust grantee, 1975-76. Mem. SAR, AAAS, Galveston Rsch. Soc. (pres. 1971-72), Am. Soc. Cell Biology, Electron Microscopy Soc. Am., Soc. Developmental Biology, Am. Assn. Anatomists, Tex. Soc. Electron Microscopy (hon.) (editor newsletter 1969-73, pres. 1975-76), Ariz. Soc. Electron Microscopy (pres. 1980-81), Gamma Pi Sigma. Home: 6249 N Camino Miraval Tucson AZ 85718-3024 Office: U Ariz Coll Medicine Dept Anatomy Tucson AZ 85724

KISER, CLYDE VERNON, retired demographer; b. Bessemer City, N.C., July 22, 1904; s. Augustus Burton and Minnie May (Carpenter) K.; m. Louise Venable Kennedy, Feb. 24, 1934 (dec. Mar. 1954). AB (Mangum medal 1925), U. N.C., 1925, AM, 1927; PhD (Richard W. Gilder fellow 1930-31), Columbia U., 1932. Rsch. fellow Milbank Meml. Fund, N.Y.C., 1931-33, rsch. assoc., mem. tech. staff, 1933-62, sr. mem. tech. staff, 1962-69, v.p. for tech. affairs, 1969-70; statis. cons. USPHS, 1936; vis. rsch. assoc., sr. rsch. demographer Office Population Rsch., Princeton, 1942-75; adj. prof. sociology N.Y. U., 1945-56; cons. Pan Am. Health, 1967; Mem. Census Adv. Com., 1965-71, Nat. Com. on Vital and Health Statistics, 1965-69; chmn. local arrangements com. N.Y.C., Internat. Population Conf., 1961; mem. standing com. Pub. Health Conf. Records and Statistics, Div. Vital Statistics, Dept. Health, Edn. and Welfare, 1958- 64, chmn. subcom. fertility measurement, 1963, chmn. subcom. on population dynamics, 1968-69. Author: Sea Island to City: A Study of St. Helena Islanders in Harlem and Other Urban Centers, 1932, Group Differences in Urban Fertility, 1942, (with Grabill and Whelpton) The Fertility of American Women, 1958, (with Grabill and Campbell) Trends and Variations in Fertility in the United States, 1968, The Milbank Memorial Fund: Its Leaders and Its Work, 1905-74, 1975; Editor: Research in Family Planning, 1962, (with Whelpton) Social and Psychological Factors Affecting Fertility, vols. I- V, 1946-58, Estudios de Demografia, 1967, Forty Years of Research in Human Fertility, 1971, (with A. L. Kiser) Kiser-Carpenter Chronicle, 1983. Bd. dirs. Gallaudet Coll., 1968-81. Recipient Grant Squires prize Columbia, 1940. Fellow Am. Statis. Assn.; mem. Am. Eugenics Soc. (pres. 1963-69), Population Assn. Am. (pres. 1952-53), Eastern Sociol. Soc. (v.p. 1959-60, chmn. com. social statistics 1960-68), Internat. Union Sci. Study Population (chmn. U.S. nat. com. 1958-61), Am. Pub. Health Assn. (com. on vital and health statistics monographs), Am. Sociol. Assn. Democrat. Lutheran. Home: Courtland Ter Apt 70 2300 Aberdeen Blvd Gastonia NC 28054

KISER, DON CURTIS, immunologist; b. Fitzpatrick, Ala., Oct. 2, 1953; s. Willie Lee and Lois (Elder) K. BS, Howard U., 1975; MD, Morehouse U., 1985. Intern, resident Grady Hosp., Atlanta; staff emergency rm. physician South Fulton Hosp., Atlanta, 1987-89; research scientist Ctrs. Disease Control, Atlanta, 1989-92, E-K-R, Inc., Atlanta, 1992—; med. cons. area health fairs, 1989-91. Vol. MLK Ctr. for Non-Violent Social Change, Atlanta, 1988; mem. Atlanta Philharmonic Chorale, 1992. Mem. NAACP, Howard U. Alumni Assn. Democrat. Office: EKR Inc 3114 Finance Sta Atlanta GA 30302

KISER, GLENN AUGUSTUS, pediatrician; b. Bessemer City, N.C., July 13, 1917; s. Augustus B. and May (Carpenter) K.; m. Katherine Parham, June 13, 1941 (dec. 1972); m. Muriel Coykendall, Feb. 4, 1973. BS, Duke U., 1941, MD, 1941. Diplomate Nat. Bd. Med. Examiners. Resident physician Duke Hosp., Durham, N.C., 1946-48; resident Johns Hopkins U., Balt., 1946; pvt. practice Salisbury, N.C., 1947-55; ret. Salisbury, 1955—; founder stockholder Food Lion, Inc.; med. cons. State of N.C., Raleigh, 1961-64, 75-76, New River Mental Health Ctr., Boone, N.C., 1976-77; chief pediatric dept. Rowan Meml. Hosp., Salisbury, 1947-55, chief of staff, 1951-52. Bd. advisors Chowan Coll., Murfreesboro, N.C., 1977-78. Surgeon USPHS, 1941-46. Mem. Pinnacle Club Duke Med. Ctr. (charter), Duke Med. Ctr. Alumni Assn. (coun. 1988), Lions (dep. dist. gov. N.C. chpt. 1959, pres. Milford Hills chpt. 1959, zone chmn. 1959, dep. dist. gov. 1960, internat. amb. 1961), Toastmasters (pres. Salisbury chpt. 1959), Salisbury Country Club. Republican. Presbyterian. Avocations: photography, music, motorcycling, alumni activities. Home: 728 Klumac Rd Ste 138C Salisbury NC 28144

KISER, KAREN MAUREEN, medical technologist, educator; b. St. Louis, Sept. 28, 1951; d. Arthur John and Elizabeth M. (Boyer) Meier; m. Winston Kiser, July 21, 1973; children: Cynthia Kay, Jessica Lea. BS in Med. Tech., S.E. Mo. State U., 1973; MA in Health Care Edn., Cen. Mich. U., 1984. Part-time lab. asst. Luth. Med. Ctr., St. Louis, 1970-71; part-time lab. technician Jewish Hosp., St. Louis, 1972-73, med. technologist, 1973-77; assoc. prof., edn. coord. St. Louis C.C. at Forest Park, 1977—; on-site supr. Nat. Accrediting Agy. for Clin. Lab. Sci., Chgo., 1986; speaker on continuing edn. St. Louis U., 1987; speaker Mo. Soc. for Med. Tech., St. Louis, 1987, 89; reviewer W.B. Saunders Co., Phila., 1986-91; capt. United Way, St. Louis, 1989, 90. Leader Girl Scouts U.S., 1986-92, co-leader, 1990-93; assoc. advisor Explorers Scouts, 1978-81. Mem. Am. Soc. for Med. Tech., Am. Soc. for Microbiology, NEA, Mo. Soc. for Med. Tech., Mo. Edn. Assn., Mo. Assn. Community and Jr. Colls., Am. Soc. Clin. Pathologists. Office: St Louis Community Coll 5600 Oakland Ave Saint Louis MO 63110-1393

KISER, KENNETH M(AYNARD), academic dean, chemical engineering educator; b. Detroit, Nov. 28, 1929; s. Kenneth Chapman and Emma (Kutkuhn) K.; m. Florence Mary Sclafani, June 26, 1954; children: David, Thomas, James, John, Melissa. BS in Chem. Engring., Lawrence Tech. U., 1951; MS in Chem. Engring., U. Cin., 1952; D.Engring., Johns Hopkins U., 1956. Registered profl. engr., N.Y. Chem. engr. Gen. Electric Co., Schenectady, N.Y., 1956-64; asst. prof. chem. engring. SUNY-Buffalo, 1964-65, assoc. prof., 1965-80, prof., 1980—, acting chair, 1977, 78, assoc. dean engring., 1978—; adj. prof. Rensselaer Poly. Inst., Troy, N.Y., 1962-64; chem. engring. cons., 1965—. Contbr. articles to profl. jours., 1957—. Patentee in field. Recipient Chancellor's award SUNY-Buffalo, 1974; grantee NSF, Heart Assn., others, 1965-80. Mem. Am. Inst. Chem. Engrs., Am. Soc. Engring. Edn., Alpha Chi Sigma, Sigma Xi, Tau Beta Pi (Tchr. award 1973). Office: SUNY Engring Dean's Office 412 Bonner Hall Buffalo NY 14260

KISER, THELMA KAY, analytical chemist; b. Oakridge, Tenn., Oct. 9, 1944; d. Lawrence T. and Sally Lura (Clay) K.; m. Robert Louis Klonis, Mar. 19, 1969 (div. Mar. 1978); 1 child, Melissa (dec.). BA in Biology, Emory & Henry Coll., 1968; BS in Chemistry, East Tenn. State U., 1979; postgrad., Ohio State U., 1981, 83, U. Cin., 1986, 87. Cert. collegeate profl. tchr. Tchr. Dickenson County Schs., Clintwood, Va., 1975-78; lab. mgr., rsch. scientist Mead Imaging div. Mead Corp., Dayton, Ohio, 1979-90; group leader, chem. control Marion Merrell Dow, Cin., 1990—. Contbr. chpt. to book on photopolymers; patentee in field. Vol. counselor Scioto-Paint Valley Crisis Ctr., Chillicothe, Ohio, 1980-84; founding mem. Mead Imaging Hiking Club, Chillicothe, Dayton, 1982-86. Mem. AAAS, Am. Chem. Soc., Ohio Acad. Sci., Soc. for Tech. Communications. Avocations: playing classical piano, hiking, writing non-fiction/poetry, natural history collecting. Office: Marion Merrell Dow 2110 E Galbraith Rd Cincinnati OH 45237-1625

KISH, MICHAEL STEPHEN, safety professional; b. Perth Amboy, N.J., Mar. 7, 1961; s. Stephen Joseph and Marie Josephine (Peleszak) K.; m. Mary Catherine Ruggeri, Apr. 25, 1964. BE in Chem. Engring., Stevens Inst. Tech., 1983, ME in Chem. Engring., 1989. Registered profl. engr., Mass.; cert. safety profl., cert. hazardous materials supt., cert. safety exec. Process engr. United Techs. div. Inmont Corp., Bound Brook, N.J., 1983-85; engring./maintenance supr. BASF div. Inmont Corp., Bound Brook, 1985-86; safety engr. Merck & Co., Inc., Rahway, N.J., 1986-90, asst. mgr. safety and indsl. hygiene, 1986-90, mgr. toll mfg., 1990-92; asst. mgr. safety and indsl. hygiene Merck & Co., Inc., West Point, Pa., 1992—. Mem. AICE, NSPE, Nat. Acad. Forensic Engrs., Am. Soc. Safety Engrs., Nat. Fire Protection Agy., World Safety Orgn., Nat. Forensic Soc. Avocations: reading, chess, computers and programming. Home: 1601 Bergey Rd Hatfield PA 19440-2875

KISHIMOTO, KAZUO, mathematical engineering educator; b. Yamada, Okayama, Japan, Sept. 9, 1952; s. Chikao and Keiko (Senō) K.; m. Mio Nakayama, Apr. 3, 1982; 1 child, Yuko. D. of Engring., U. Tokyo, 1980. Rsch. assoc. Hiroshima (Japan) U., 1980-86; asst. prof. U. Tsukuba, Japan, 1987-90, assoc. prof., 1990—. Mem. Japan Soc. Indsl. and Applied Math., Soc. Math. Biology, Ops. Rsch. Soc. Japan, Inst. Electronics, Info. and Communications Engrs. Japan. Home: 207-1-3-6 Asahi-cho, Funabashi Chiba 273, Japan Office: U Tsukuba Inst Socio-Econ, Tenno-dai, Tsukuba Ibaraki 305, Japan

KISHIMOTO, UICHIRO, biophysicist; b. Osaka, Japan, May 13, 1922; s. Usaburo and Kiku K.; m. Soyoko, May 3, 1946; children: Ryutaro, Kenjiro. BS, Osaka U., 1945, PhD, 1959. Rsch. assoc. Osaka U., 1953-55, lectr., 1955-62, assoc. prof., 1962-68, prof., 1968-86, prof., emeritus, 1986—; Rockefeller Fellow Darmouth Med. Sch./NIH, Hanover, N.H. and Bethesda, Md., 1959-60; vis. scientist NIH, Bethesda, 1960-61. Contbr. articles to profl. jours. Mem. Rotary. Democrat. Buddhist. Avocations: lawn tennis, music, microcomputer programming, travel. Home: Shinsenri Kitamachi 2-26-5, Osaka, 565 Toyonaka Japan

KISIELOWSKI, EUGENE, engineering executive; b. Leczowka, Tarnopol, Poland, May 30, 1932; arrived in U.S., 1959; s. Wladyslaw and Helena (Berezanko) K.; m. Rosemary Booth, July 31, 1957; children: Richard Julian, Helen Teresa, Eugene Stephen. BSME, Huddersfield (Eng.) Tech. Coll, 1955; MSc in Aero. Engring., Coll. Aeronautics, Cranfield, Eng., 1972. Aero. engr. Avro Aircraft Co., Malton, Ont., Can., 1957-59; asst. project engr. Kellett Aircraft Corp., Willow Grove, Pa., 1959-60; sr. analytical engr. Plasecki Aircraft Corp., Phila., 1960-61; supr. II Vertol div. Boeing Co., Ridley Park, Pa., 1961-65; dir. aero. engring. Dynasciences Corp., Blue Bell, Pa., 1965-71; pres., chief exec. officer United Terex, Inc., Fairview Village, Pa., 1971—. Contbr. articles to profl. publs. Vol. various local polit. campaigns. Mem. AIAA, Am. Helicopter Soc., Cranfield Students Soc. Republican. Avocations: swimming, tennis, chess. Home: 3020 Oak Dr Norristown PA 19401-1543 Office: United Terex Inc 2579 Industry Ln Fairviewill PA 19403

KISLOVSKI, ANDRE SERGE, electronics engineer; b. Dec. 6, 1933; s. Serge A. and Vera S. (Klimentov) K. Diploma in telecommunications engring., U. Belgrade, Yugoslavia, 1958. Asst., electrotech. U. Belgrade, 1959-61; devel. engr. Co. Electro-Mecanique, Paris, 1962-66, Soc. Electrotechnique, Paris, 1966-70; devel. engr., product engr. Ascom (Hasler) Energy Systems, Bern, Switzerland, 1970—. Author: Introduction to Dynamical Analysis of Switching DC-DC Converters, 1985, Dynamic Analysis of Switching DC-DC Converters (with others), 1991; contbr. papers to tech. publs. Mem. IEEE (sr.). Achievements include European and U.S. patents for switched-mode power electronics. Office: Ascom Energy Systems, Murtenstrasse 133, 3005 Bern Switzerland

KISSA, ERIK, chemist; b. Abja, Estonia, Apr. 7, 1923; came to U.S., 1951, naturalized, 1956; s. Mats and Selma (Wilson) K.; m. Selma Alide Tamm, Sept. 6, 1952; children: Erik Harold, Karl Martin. MS, Tech. U. Karlsruhe (Germany), 1951; PhD, U. Del., 1956. Rsch. chemist E. I. du Pont de Nemours & Co. Inc., Wilmington, Del., 1951-67, sr. rsch. chemist, 1967-74, rsch. assoc. Jackson Lab., 1974-86, sr. rsch. assoc., 1986-90, rsch. fellow, 1990—; UN tech. expert, India, 1978, 79, China, 1982, Korea, 1986, 87, 88. Recipient Soap and Detergent Assn. award, 1991. Fellow Am. Inst. Chemists; mem. AAAS, Am. Oil Chem. Soc., Am. Chem. Soc., Internat. Assn. Colloid and Interface Scientists, Fiber Soc., Du Pont Country Club, Del. Camera Club. Lutheran. Author: Fluorinated Surfactants, 1993; editor: (with W. G. Cutler) Detergency Theory and Technology, 1987; contbr. numerous articles, chpts. on surface chemistry of textiles, surfactants, dispersions, dyes, and analytical chemistry to profl. publs.; U.S., fgn. patentee in field. Home: 1436 Fresno Rd Wilmington DE 19803-5122 Office: EI DuPont de Nemours & Co Jackson Lab Wilmington DE 19898

KISSELL, KENNETH EUGENE, astronomer; b. Columbiana, Ohio, June 28, 1928; s. Thomas Franklin and Grace Jemima (Messersmith) K.; m. Theodora Julia, Apr. 3, 1953 (div. June 1975); children: Kevin Douglas, Bradley Thomas; m. Judith Lynn Lee, Aug. 25, 1988. BSc in Physics and Astronomy cum laude, Ohio State U., 1949, MSc in Physics, 1958, PhD in Astronomy, 1969. Rsch. assoc. Ohio State U., Columbus, 1948-51; rsch. physicist Aerospace Rsch. Labs., USAF, Wright-Patterson AFB, Ohio, 1951-59, lab. dir., 1959-72; sr. scientist AF Avionics Lab., Wright-Patterson AFB, 1972-80; sr. staff scientist Rocketdyne div. Rockwell Internat., Canoga Park, Calif., 1980-83; sr. scientist optics BDM Corp., McLean, Va., 1983-87; rsch. assoc. physics and astronomy U. Md., College Park, 1987-90, rsch. assoc. physics, 1992—; chief scientist AMOS/MOTIF Observatories, Rockwell Internat., Kihei, Hawaii, 1990-93; mem. Air Force studies bd. NAS, Washington, 1986-88. Editor ISA Transactions, 1961-66; contbr. articles to profl. jours. Team leader MOONWATCH Project IGY Vol. Support for Earth Satellites, Dayton, Ohio, 1957-60. Fellow Royal Astron. Soc.; mem. AIAA, Am. Astron. Soc., Am. Geophys. Union, Internat. Astron. Union, Optical Soc. Am., Soc. Photog. Instrumentation Engrs., Phi Beta Kappa, Sigma Xi, Sigma Pi Sigma. Achievements include 4 patents for Optical Devices/Space Sciences; research in Optical Space Object Identification; in 10 Solar Eclipse Expeditions. Office: U Md Dept Physics College Park MD 20742-0001

KISSICK, WILLIAM LEE, physician, educator; b. Detroit, July 29, 1932; s. William Leslie and Florence (Rock) K.; m. Priscilla Harriet Dillingham, June 16, 1956; children: William, Robert-John, Jonathan, Elizabeth. B.A., Yale U., 1953, M.D., 1957, M.P.H., 1959, Dr.P.H., 1961. Intern Yale-New Haven Med. Center, 1957-58; resident Montefiore Hosp. and Med. Center, N.Y.C., 1961-62; Div. Community Health Service, 1962-63; spl. asst. to asst. sec. for health U.S. Dept. HEW, 1964-65; dir. Office Program Planning Evaluation, Office of Surgeon Gen., USPHS, 1966-68; exec. dir., nat. adv. commn. health facilities The White House, Washington, 1968; prof., chmn. dept. community medicine Sch. Medicine U. Pa., 1968-71, George S. Pepper prof. public health and preventive medicine, 1969—, prof. research medicine, 1976—, prof. health care systems Wharton Sch., 1971—, prof. health policy and mgmt. Sch. Nursing, 1978—, dir. Center for Health Policy, 1981—; fellow, mem. exec. com. Nat. Ctr. for Health Care Mgmt.; dir. Health policy, chmn. bd. of govs. Leonard Davis Inst. Health Econs., 1989—; vis. prof. community medicine Guy's Hosp. Med. Sch.; vis. prof. dept. social sci. and adminstrn. London Sch. Econs. and Polit. Scis.; vis. prof. Inst. European Health Svcs. Rsch., Leuven U., 1974-75; cons. Nat. Ctr. Health Svcs. Rsch., Health Resources Adminstrn., Benedum Found., WHO, Appalachian Regional Commn., Smith Kline-Beckman, Pew Meml. Trust, Colonial Penn Group, Ctr. Disease Control; mem. Accrediting Commn. on Edn. for Health Svcs. Adminstrn., 1980-86; chmn. com. on med. affairs coun. Yale U., 1980-86, fellow Yale Corp., 1987—; mem. Mayor's Commn., 1981-83, coun. Coll. Physicians of Phila., 1983-88, coun. med. socs. Am. Coll. Physicians, 1983-88. Editor: Dimensions and Determinants of Health Policy, 1968. Contbr. articles to profl. jours. Bd. dirs. Met. Collegiate Ctr. Germantown; chmn. Yale U. Alumni Assn. Fund, 1988-90; trustee Appalachian Regional Hosps., 1969-76. With USPHS, 1962-68. Mem. AAAS, Am. Coll. Preventive Medicine, Am. Pub. Health Assn., Pa. Phila. Coll. Physicians, Assn. Health Svcs. Rsch. Assn. Tchrs. Preventive Medicine, Am. Coll. Physician Execs., Physicians for Social Responsibility, Nat. Assn. Pub. Health Policy. Home: Ellet Ln Philadelphia PA 19119

KISSILEFF, ALFRED, veterinarian, researcher; b. Camden, N.J., Aug. 2, 1908; s. Isaac and Dora (Shenitz) K.; m. Julia Reisman, Sept. 3, 1933; children: Harry Reisman, David Douglas. VMD, U. Pa., 1933. Pvt. practice vet. medicine Flourtown, Pa., 1933-42; owner, pres. Breeders Equipment Co., Flourtown, 1946—; prof. Delaware Valley Coll., Doylestown, Pa., 1950-52; hosp. dir. Ritter Animal Hosp., Phila., 1972-76; rsch. scientist William Cooper Nephews, Chgo., 1940-41. Contbr. articles to profl. jours. Lt. col. U.S. Army, 1942-46. Recipient commendation for design of instrnl. aids U.S. Army Q.M. Gen., 1944, citation U.S. Govt./U.S. Dept. Commerce, Washington, 1957, Alumni award of Merit U. Pa., 1993. Mem. AAAS, Am. Vet. Med. Assn. (practitioner rsch. award 1974), Entomol. Soc. Am., Am. Assn. Vet. Parasitologists. Achievements include discovery of cause of summer dermatitis in dogs; development of successful method of artificial insemination of dairy cattle; patent for Artificial Inseminator for Bovines. Office: Breeders Equipment Co 1232 Bethlehem Pike Flourtown PA 19031

KISSLINGER, CARL, geophysicist, educator; b. St. Louis, Aug. 30, 1926; s. Fred and Emma (Tobias) K.; m. Millicent Ann Thorson, Mar. 27, 1948; children: Susan, Karen, Ellen, Pamela, Jerome. B.S., St. Louis U., 1947, M.S., 1949, Ph.D., 1952. Faculty St. Louis U., 1949-72, prof. geophysics, geophys. engring., 1961-72, chmn. dept. earth and atmostpheric scis., 1963-72, prof. geophysics, 1972—; dir. Coop. Inst. Research in Environ. Scis., U. Colo., Boulder, 1972-79; UNESCO expert in seismology, chief tech. adviser Internat. Inst. Seismology and Earthquake Engring., Tokyo, 1966-67; chmn. com. seismology NRC-Nat. Acad. Scis., 1970-72; mem. U.S. Geodynamics Com., 1975-78; U.S. nat. corr. Internat. Assn. Seismology and Physics of Earth's Interior, 1970-72; mem. Internat. Union Geodesy and Geophysics, bur., 1975-83, v.p., 1983-91; mem. Gov.'s Sci. Adv. Council, State of Colo., 1973-77, com. on scholarly communication with People's Republic of China, Nat. Acad. Scis., 1977-81, NRC/Nat. Acad. Scis. adv. com. to U.S. Geol. Survey, 1983-88; governing bd. Am. Inst. Physics, 1989—. Recipient Alumni Merit award St. Louis U., 1976, Alexander von Humboldt Found. Sr. U.S. Scientist award, 1979, U.S. Geol. Survey's John Wesley Powell award, 1992, Disting. Svc. award U. Colo., 1993, Commemorative medal USSR Acad. Scis., 1985. Fellow Am. Geophys. Union (bd. dirs. sect. seismology 1970-72, fgn. sec. 1974-84), Geol. Soc. Am., Assn. Exploration Geophysicists (India), AAAS; mem. Soc. Exploration Geophysicists, Seismol. Soc. Am. (dir. 1968-74, pres. 1972-73), Austrian Acad. Sci. (corr.), Phi Beta Kappa, Sigma Xi. Club: Cosmos. Home: 4165 Caddo Pky Boulder CO 80303-3602

KIS-TAMÁS, ATTILA, chemist; b. Mór, Hungary, June 6, 1939; s. János and ERzsébet (Nándory) K.-T.; m. Éva Kovács (div. 1973); children: Attila, Gábor; m. Ágnes Kovács, Mar. 6, 1976; children: Barbara, Melinda. Degree in chem. engring., Tech. U., Budapest, Hungary, 1962, Pharma c.eng., 1970, PHD, 1970, postgrad., 1985. Chem. engr. Chinion Pharmaceuticals Ltd., Budapest, 1962-63, head rsch. dept., 1964-73; head rsch. dept. Egis Pharmaceuticals Ltd., Budapest, 1973-86; dir. Szeviki R & D Inst. J.V., Budapest, 1986-90, pres., dir., 1990—. Contbr. articles to profl. publs. Mem. Hungarian Chems. Prodn. Assn. (bd. dirs. 1991—), Hungarian Acad. Sci. Orgn. (chem. com., Acad. prize 1984), Hungarian Chamber of Innovation (bd. dirs. 1991—), Hungarian Assn. Plant Prot. Mfg. Roman Catholic. Achievements include over 50 patents in pharmaceuticals, plant protecting agents, fine chemicals, products hpetopargil, feuitropan. Home: 24 Szentendrei, H-1035 Budapest Hungary Office: Szeviki Organic Chem Rsch, 13 Stahly St, H-1085 Budapest Hungary

KISTER, HENRY Z., chemical engineer; s. J. M. and H. Kister. B in Engring., U. NSW, Kensington, Australia, 1973, M in Engring., 1977. Chartered engr., U.K., Australia, fuel technologist, U.K. From develop. engr. to sr. develop. engr. ICI Australia Ltd., Botany, NSW, Australia, 1974-77, startup supt., 1977-80; rsch. engr. Fractionation Rsch. Inc., South Pasadena, Calif., 1980-81; sr. process engr. C.F. Braun Inc., Alhambra, Calif., 1981-84; prin. process engr. Brown & Root Braun, Alhambra, 1984-90, engring. advisor, 1990—; mem. tech. adv. com. Fractionation Rsch. Inc. (FRI), Stillwater, Okla., 1982—, mem. design practices com., 1985—, mem. tech. com., 1987-93; leader seminar Practical Distillation Tech., 1983—. Author: Distillation Operation, 1990, Distillation Design, 1992; contbr. over 40 articles to profl. jours. Mem. Am. Inst. Chem. Engrs. (publ. com. 1991—, Outstanding Paper award 1992), Instn. Chem. Engrs. (U.K. and Australia, Humphry and Glasgow medal 1980), Inst. Energy (U.K.). Achievements include 2 U.S. patents in field; development of several published distillation/absorption design and troubleshooting methods used in the industry. Office: Brown & Root Braun 1000 S Fremont Alhambra CA 91802

KISTLER, ALAN LEWIS, engineering executive; b. Wilkesbarre, Pa., Dec. 9, 1943; s. Alan C. and Ruth (Lewis) K.; m. Jean Rowe (div. Nov. 1987); children: Alan, Michael; m. Joanne Gentry KEck, Dec. 26, 1987. BA, Tex. Christian U., 1965; MBA, Golden Gate U., 1972. Commd. 2d. lt. USAF, 1965, advanced through grades to capt., 1968, retired, 1975; engring. mgr. Aeronautic Ford, Palo Alto, Calif., 1975-80; mgr. exec. staff TRW, Sunnyvale, Calif., 1980-81; engring. mgr. Ford Aerospace Corp., Hanover, Md., 1981-89; ops. dir. Computer Scis. Corp., Falls Church, Va., 1990-92; program mgr. Martin Marietta, Reston, Va., 1992—. Mem. IEEE, Armed Forces Communications & Elec. Assn., Security Affairs Support Assn. Home: 1117 Kalmia Ct Crownsville MD 21032 Office: Martin Marietta 7115 Standard Dr Hanover MD 21076

KISTNER, DAVID HAROLD, biology educator; b. Cin., July 30, 1931; s. Harold Adolf and Hilda (Gick) K.; m. Alzada A. Carlisle, Aug. 8, 1957; children—Alzada H., Kymry Marie Carlisle. A.B., U. Chgo., 1952, B.S., 1956, Ph.D., 1957. Instr. U. Rochester, 1957-59; instr., asst. prof. biology Calif. State U., Chico, 1959-64, assoc. prof., 1964-67, prof., 1967-92, prof. emeritus, 1992—, sr. rschr. univ. found.; rsch. assoc. Calif. State U. Found., Chico, 1992—, Field Mus. Natural History, 1967—, Atlantica Ecol. Rsch. Sta., Salisbury, Zimbabwe, 1970—; CEO Kistner family Trust, 1982—; dir. Shinner Inst. Study Interrelated Insects, 1968-75. Author: (with others) Social Insects, Vols. 1-3; editor Sociobiology, 1975-82; contbr. articles to profl. jours. Patron Am. Mus. Natural History; life mem. Republican Nat. Com., 1980—. Recipient Outstanding Prof. award Calif. State Univs. and Colls., L.A., 1976; John Simon Guggenheim Meml. Found. fellow, 1965-66; grantee NSF, 1960—, Am. Philos. Soc., 1972, Nat. Geog. Soc., 1988. Fellow Explorers Club, Calif. Acad. Scis.; mem. AAUP, AAAS, Entomol. Soc. Am., Pacific Coast Entomol. Soc., Kans. Entomol. Soc., Am. Soc. Naturalists, Am. Soc. Zoologists, Soc. Study of Systematic Zoology, Internat. Soc. Study of Social Insects, Mus. Nat. Hist. (life), Chico State Coll. Assocs. (charter). Home: 3 Canterbury Cir Chico CA 95926-2411

KITADA, SHINICHI, biochemist; b. Osaka, Japan, Dec. 9, 1948; came to U.S., 1975; s. Koichi and Asako Kitada. MD, Kyoto U., 1973; MS in Biol. Chemistry, UCLA, 1973, PhD, 1979. Intern Kyoto U. Hosp., Japan, 1973-74; resident physician Chest Disease Research Inst., 1974-75; rsch. scholar lab. nuclear medicine and radiation biology UCLA, 1979-87, rsch. scholar Jules Stein Eye Inst., 1988-91; rsch. biochemist La Jolla (Calif.) Cancer Rsch. Found., 1992—. Author papers in field. Japan Soc. Promotion Sci. fellow 1975-76. Mem. Am. Oil Chemists Soc., N.Y. Acad. Scis., Sigma Xi. Home: 920 Kline St Apt 301 La Jolla CA 92037 Office: La Jolla Cancer Rsch Found 10901 N Torrey Pines Rd La Jolla CA 92037

KITAGAWA, TOSHIKAZU, chemistry educator; b. Hikone, Shiga, Japan, Mar. 9, 1958; s. Izo and Hatsue (Tanaka) K.; m. Kaori Matsushima, Jan. 14, 1990. B. in Engring., Kyoto (Japan) U., 1980, M. in Engring., 1982, D. in Engring., 1986. Japan Soc. for the Promotion of Sci. rsch. fellow dept. chemistry Kyoto U., 1985-86; postdoctoral fellow Harvard U., Cambridge, Mass., 1986-88; lectr. Inst. for Molecular Sci., Okazaki, Japan, 1989; instr. dept. chemistry Kyoto U., 1990—. Contbr. articles to profl. jours. Home: Shugakuin, Sakyo-ku, 1-3 Tsujinota-cho,, Kyoto 606, Japan Office: Kyoto U Dept Engring, Sakyo-ku, Kyoto 606, Japan

KITAHARA, SHIZUO, allergist; b. Tokyo, Nov. 20, 1922; s. Buntaro and Tamiko K.; m. Yoko Kitahara, April 4, 1959; 1 child, Taeko. MD, Tokyo U., 1945. Intern and resident Tokyo U. Hosp., 1946-56; chief allergist Doai Meml. Hosp., Tokyo, 1956-69; dir. Allergy Clinic, Tokyo, 1969—. Author: Anti-Allergics, 1970, Manual of Allergic Diseases, 1975, Treatment of Bronchial Asthma, 1985. Fellow Japan Allergy Soc.; mem. Am. Coll. Allergists, Am. Acad. Allergists, N.Y. Acad. Sci. Avocation: music. Home: 32-20 Minami-Ogikubo 4, Suginami-ku, Tokyo 167, Japan Office: Allergy Clinic, 11-18 Kita-Otsuka 1, Toshima-ku, Tokyo 170, Japan

KITAMURA, TOSHINORI, psychiatrist; b. Yokohama, Kanagawa, Japan, Oct. 16, 1947; s. Masanori and Kikuko (Matsumoto) K.; m. Fusako Oami, Mar. 17, 1973. M.D., Keio Gijuku U. Sch. Medicine, Tokyo, 1972. Psychiatrist, Inst. Psychiatry, Tokyo, 1973-76; hon. research fellow U. Birmingham (U.K.), 1976-80; clin. instr. Keio Gijuku U., Tokyo, 1980-83, lectr., 1983; chief sect. mental health for elderly NIMH, Ichikawa, Japan, 1983-91, dir. dept. sociocultural environ. rsch., 1991—; vis. lectr. Keio Gijuku U., 1986—; head Group for Research Assessment in Psychiatry, Tokyo, 1981—; mem. com. Med. Selection Japanese Astronauts, 1987-88, com. Psychiatric Diagnostic Criteria Japan, 1987-89, com. Guideline for Psychiatric Treatment, 1987-88. Editor-in-chief Archives of Psychiat. Diagnostics and Clin. Evaluation, Tokyo, 1989—. Contbr. articles to profl. jours. Fellow Royal Coll. Psychiatrists, British Council Japan Assn., Japanese Assn. Psychiatry and Neurology (coun. 1991—). Home: 8-12-4-305 Akasaka, Minato-ku, Tokyo 170, Japan Office: NIMH, 1-7-3 Konodai, Ichikawa, Chiba 272, Japan

KITANI, OSAMU, agriculture educator; b. Tokyo, Apr. 1, 1935; s. Tsuneyuki and Kyou (Hosotani) K.; m. Shigeko Tanaka, July 12, 1964; children: Yukiko, Mariko. BAgr, U. Tokyo, 1959, MAgr, 1961, DAgr, 1964; PhD, Mich. State U., 1966. Rsch. asst. Mich. State U., 1964-66; assoc. prof. Mie U., Tsu, Japan, 1966-78; prof. U. Tokyo, 1978—; guest prof. Tech. U. Munich, 1972-73; chmn. dept. agrl. engring. U. Tokyo, 1980, 88, chmn. grad. course agrl. engring., 1989, 90, dir. libr. of faculty, 1991-93. Author: Energy in Agriculture, 1983, Agricultural Machinery, 1984; editor: Biomass, 1981, Biomass Handbook, 1989, Bioproduction Machinery, 1993. Expert mem. Sci. and Tech. Agy., Tokyo, 1976-87, Sci. Coun., Ministry Edn., Tokyo, 1979-89, Coun. for Sci. and Tech., Prime Minister's Office, Tokyo, 1981—; rsch. mem. Sci. Coun. Japan, Rsch. Com., Tokyo, 1983—. Recipient Gov.'s award, Gov. Prefecture, Kagawa Prefecture, Japan, 1950, acad. award, Japanese Soc. Agrl. Machinery, 1976; fellowship Alexander von Humboldt Found., Bonn, Federal Republic Germany, 1972; grantee Fulbright Commn., Tokyo, 1964. Mem. Japanese Soc. Agrl. Machinery (dir. 1980-85, 89-91, pres. 1992—), Japanese Soc. Irrigation, Drainage and Reclamation (bd. dir. 1978-82), Soc. Agrl. Structures, Am. Soc. Agrl. Engrs., Am. Japan Soc., Japan Soc. Energy and Resources (dir. 1992—), Japan Agrl. Systems Soc. (pres. 1989-91, v.p. and dir. 1985-88), Japan Fedn. Agrl. Engring. (v.p. 1983-84), Japan Assn. Internat. Commn. of Agrl. Engring. (dir. and gen. sec. 1990—, rep. 1992—). Buddhist. Avocations: swimming, art, music. Home: Kataseyama 3-3-10, Fujisawa-shi Kanagaw 251, Japan Office: U Tokyo/Faculty Agr, Yayoi 1-1-1 Bunkyo-ku, Tokyo 113, Japan

KITANO, KAZUAKI, microbiologist, researcher; b. Kawachinagano, Osaka, Japan, May 5, 1939; s. Jiichi and Chiyo (Nishibata) K.; m. Naoko Miyoshi, Oct. 10, 1967; children: Yoko, Seiko, Yuko, Mutsuko. BA, Osaka U., 1962, PhD, 1977. Rsch. scientist Takeda Chem. Industries, Ltd., Takasago, Hyogo, Japan, 1962-71; rsch. assoc. Takeda Chem. Industries, Ltd., Osaka, 1972-77, rsch. head, 1980-86, sr. rsch. head, 1987-91, dir. discovery rsch. labs. II, 1992—; rsch. assoc. Rockefeller U., N.Y.C., 1978, asst. prof., 1979; chmn. biotech. com. Japan Pharm. Mfrs. Assn., Tokyo, 1988—. Author: (books) Practical Methods in Monoclonal Antibody, 1987, Animal Cell Bioreactors, 1991; contbr. articles to Jour. of Takeda Rsch. Labs., Progress in Indsl. Microbiology. Mem. Am. Soc. Microbiology, Japan Soc. Biosci., Biotech. and Agrochemistry, Soc. of Fermentation and Bioengring. (Saito prize 1975), Japanese Assn. Animal Cell Tech. (councillor). Achievements include discovery of single beta-lactam antibiotic of bacterial origin; establishment of a novel fermentation process for L-glutamate from acetate, novel screening system for beta-lactam antibiotics, effective production process for interferons by using recombinant E. coli, effective serum-free media for mammalian cells. Home: 4-15-4 Akasaka-dai, Sakai Osaka Japan 590-01 Office: Takeda Chem Industries Ltd, 17-85 Jusohonmachi 2-chome, Yodogawa-ku, Osaka 532, Japan

KITAZAWA, KOICHI, materials science educator; b. Nagano, Japan, Apr. 17, 1943; s. Yoshimi and Shigeko (Murakami) K.; m. Kuniko Tajima, Aug. 31, 1969; children: Tetsuya, Kay. BS in Chemistry, U. Tokyo, 1966; DSc in Materials Sci., MIT, 1972. Rsch. assoc. in chemistry U. Tokyo, 1973-79, lectr. in physics, 1979-81, assoc. prof. physics, 1981-87, prof. chemistry, 1987—. Recipient Grand Prize, Japan Ceramic Soc., 1987, Paper of Yr. award Applied Physics Soc. Japan, 1988, prize in physics IBM Japan, 1988. Mem. Am. Ceramic Soc. (Fulrath award 1988), Am. Phys. Soc., Chem. Soc. Japan, Materials Rsch. Soc. Achievements include world's first patent on high temperature superconductor. Office: U Tokyo Dept Ind Chemistry, 7-3-1 Hongo, Bunkyo-ku Tokyo 113, Japan

KITCHELL, SHAWN RAY, plant engineer, educator; b. Bonne Terre, Mo., Feb. 12, 1962; s. John Ray and Georgia Faye (Burch) K.; m. Jamie Elaine Ives, Dec. 8, 1984; children: Jonni Elizabeth, Ian Albert. AS, Mineral Area Coll., 1982; BSME, U. Mo., Rolla, 1984; MS in Mgmt., Cardinal Stritch Coll., 1991. Registered profl. engr. Wis. Mech. engr. 1st dist. U.S. Coast Guard, Boston, 1985-86; progect engr., mgr. Lee Mech. Contractors, Inc., Bismarck, Mo., 1986-87; project engr. Weyerhaeuser Co., Marshfield, Wis., 1987-88, chief engr., 1988-91, plant engr., 1991-92, process improvement engr., 1992—; adj. instr. Cardinal Stritch Coll., Milw., 1991—, Lakeland Coll., Sheboygan, Wis., 1992—; cons. Kitchell and Assocs., Marshfield, 1991—. Treas. Marshfield Bapt. Ch., 1990—, trustee, 1992. Mem. NSPE, ASME, ASTM, Am. Inst. Plant Engrs., Nat. Inst. Engring. Mgmt. and Systems. Democrat. Office: Weyerhaeuser Co 1401 E 4th St Marshfield WI 54449

KITCHING, PETER, physics educator; b. Leeds, Eng., Apr. 4, 1938; s. Jack Mayne and Bridget Catherine (Dolan) K.; m. Josephine Mary Rodgers, July 22, 1964; children: John Edward, Andrew James, Matthew Stephen. BA, Oxford (Eng.) U., 1960; MSc, Yale U., 1962, PhD, 1966. Rsch. scientist Nat. Inst. for Rsch. in Nuclear Sci., Daresbury, Eng., 1966-69; rsch. assoc. U. Alta., Edmonton, Can., 1969-71, asst. prof. physics, 1971-76, assoc. prof., 1976-82, prof., 1982—, dir. Ctr. for Subatomic Rsch., 1988—; assoc. dir. Triumf Lab., Vancouver, B.C., Can., 1983-88. Joint editor: Particle Physics—The Factory Era, 1991; contbr. over 60 articles to profl. jours. Mem. Can. Assn. Physicists. Home: 11135 83d Ave Apt 2301, Edmonton, AB Canada T6G 2C6 Office: U Alta, Ctr for Subatomic Rsch, Edmonton, AB Canada Z6G 2H5

KITSOPOULOS, SOTIRIOS C., electrical engineer, consultant; b. Athens, Greece, Feb. 12, 1930; came to U.S., 1958; s. Constantine S. and Maria (Lymberea) K.; m. Antonia Mitsakou, Dec. 7, 1957; children: Constantine A., Nicholas, Maria, Andrew T. Diploma, Swiss Fed. Inst. Tech., Zurich, 1952, PhD, 1955. Instr. U. Md., overseas, 1956-57; asst. prof. Rensselaer Poly. Inst., Troy, N.Y., 1958-59; supr. AT&T Bell Labs., Murray Hill, N.J., 1959-89; sr. ptnr. InterConsult, Internat. Mgmt. Consul, Summit, N.J., Athens, Greece and Zurich, Switzerland, 1989—. Author: Mesanastasis, 1986; editor (textbook) Einfuehrung in die Fernmeldetechnik, 1954; contbr. articles to profl. jours. Mem. IEEE (sr.), N.Y. Acad. Scis., Swiss Electrotech. Soc. (life), Inst. Mgmt. Consul. Achievements include 12 patents. Home and Office: 11 Robin Hood Rd Summit NJ 07901-3719

KITTAKA, ATSUSHI, chemist; b. Haibara, Shizuoka, Japan, Mar. 30, 1959; s. Shoji and Yoko (Seo) K. B, U. Tokyo, 1982, M, 1984, D of Pharmacology, 1987. Cert. organic chemist, pharmacist. Postdoctoral fellow Swiss Fed. Inst. Tech., Zurich, 1987-89; asst. Showa U., Tokyo, 1989—. Contbr. articles to profl. jours. Rsch. fellow Yamada Sci. Found. 1987, Takeda Sci. Found. 1991; grantee Ministry of Edn. 1990, 92. Mem. Pharm. Soc. Japan (editorial staff mem. 1990—), Chem. Soc. Japan, Soc. Synthetic Organic Chemistry Japan, Am. Chem. Soc., Internat. Soc. Antiviral Rsch. Avocations: skiing, tennis, driving. Home: Hongo 5-29-12-305, Tokyo Bunkyo-ku 113, Japan Office: Showa U, Hatanodai 1-5-8, Tokyo Shinagawa-ku 142, Japan

KITTO, JOHN BUCK, JR., mechanical engineer; b. Evanston, Ill., Dec. 22, 1952; s. John Buck and Marie (Comstock) K.; m. Cecilia Higgins, Aug. 17, 1974; children: Christopher Daniel, Andrew Comstock. BSME, Lehigh U., 1975; MBA, U. Akron, 1980. Reg. profl. engr., Ohio, Pa. Sr. engr. Babcock & Wilcox Co., Alliance, Ohio, 1975-80, research engr., 1980-81, program mgr., 1981—. Editor: Heat Exchangers for Two Phase Flow, 1983, Two-Phase Heat Exchanger, 1985, Maldistribution of Flow, 1987, Steam: Its Generation and Use, 1992; author and patentee in field. Active St. Marks Episc. Ch., North Canton, Ohio, 1987. Mem. ASME (chmn. local chpt. 1983-84, chmn. exec. com. of heat transfer divsn. 1992-93, v.p. region V 1992—, officer bd. comms. 1991—, Prime Movers award 1992, Dedicated Svc. award 1992, George Westinghouse Silver medal 1991), NSPE (Young Engr. of Yr. award 1986), Am. Inst. Chem. Engrs., Air Waste Mgmt. Assn., Tau Beta Pi, Pi Tau Sigma, Beta Gamma Sigma, Sigma Iota Epsilon. Republican. Avocations: reading, hiking, board games, coaching soccer. Home: 1150 7th St NE Canton OH 44720-2172 Office: Babcock & Wilcox Co Rsch and Devel Div 1562 Beeson St NE Alliance OH 44601-2165

KITTRELL, BENJAMIN UPCHURCH, agronomist; b. Kittrell, N.C., Oct. 25, 1937; s. Willie Arthur and Ethel Gladys (Hays) K.; m. Nancy Louise Lassiter, June 21, 1958; children: Benjamin Jr., Jan Elizabeth. BS, N.C. State U., 1960, PhD, 1975. Tchr. vocat. agr. Vance County Schs., Kittrell, 1960-63, Wake County Schs., Fuquay, N.C., 1963-65; supt. Lower Coastal Plain Rsch. Sta., Kinston, N.C., 1965-68; extension specialist N.C. State U., Raleigh, 1968-75, U. Ga., Athens, 1975-78; extension specialist Clemson U., Florence, S.C., 1978-87, dir. Pee Dee Rsch. and Edn. Ctr., 1987—. Contbr. articles to prof. jours. including Tobacco Sci., Jour. Agronomic Edn. Sgt. U.S. Army N.G., 1954-62. Recipient Disting. Svc. award Nat. Assn. County Agrl. Agts., 1988. Mem. Am. Agronomy Soc. (sect. chmn. 1991-92). Democrat. Methodist. Office: Clemson U Pee Dee Rsch and Edn Ctr RR 1 Box 531 Florence SC 29506-9801

KITZIS, GARY DAVID, periodontics educator, periodontist; b. Bklyn., Apr. 2, 1953. BA in Biology, Adelphi U., 1975; DMD, U. Fla., 1979; C.A.G.S. in Periodonics, Boston U., 1984, C.A.G.S. in Prosthodontics, 1986. Diplomate Am. Bd. Periodontology. Pvt. practice gen. dentistry Atlanta, 1980-82; pvt. practice periodontist, prosthodontist Woodbury, N.Y., 1986—; assoc. prof. periodontics SUNY Sch. Dental Medicine, Stony Brook, 1986—. Contbr. articles to profl. jours. Fellow Acad. Gen. Dentistry, Suffolk Acad. Medicine, Am. Coll. Prosthodontists (assoc.); mem. Am. Acad. Periodontology, Acad. Osseointegration, N.Y. Acad. Scis., ADA, Suffolk County Dental Soc. (bd. dels. 1988—), Long Island Soc. for Osseointegration (membership com. 1988-91, v.p. 1991-93), Am. Acad. Gold Foil Operators, Northeastern Soc. Periodontists. Achievements include determination of the accuracy of dental articulator interchange ability. Office: 156 Plainview Rd Woodbury NY 11797-2807

KITZMILLER, KARL WILLIAM, dermatologist; b. Cin., Sept. 23, 1931; m. Alice Ann Meehan, Jan. 29, 1955; children: Sue, John, Dan, Sarah, Brian. BS, U. Cin., 1953, MD, 1960. Diplomate, Am. Bd. Dermatology. Intern Cin. Gen. Hosp., 1960-61; fellow in dermatology Mayo Clinic, Rochester, Minn., 1961-64, asst. clin. prof. for family medicine, assoc. clin. prof. for dermatology; pvt. practice Cin., 1964—; asst. clin. prof. Dept. of Family Medicine U. of Cin. Med. Ctr., assoc. clin. prof. Dept. of Surgery Div. of Plastic Reconstructive and Hand Surgery, clin. prof. Dept. Dermatology; attending staff, chief dermatology Good Samaritan Hosp., Deaconess Hosp.; cons. Wright Patterson AFB, Dayton, Ohio; courtesy staff Our Lady of Mercy Hosp., Holmes Hosp., Margaret Mary Community Hosp., Batesville, Ind.; staff Children's Hosp., Bethesda Hosp.; attending physician Univ. Hosp., 1966—, Jewish Hosp., 1966—, dir. dept. dermatology, 1977—; attending staff Christ Hosp., 1976—, sec. dept. dermatology, 1977—; mem. Choice Care Physician Leadership and Mgmt. Edn. program Xaiver U., 1990-91. Contbr. articles to med. jours. Lt. USAF, 1954-56. Recipient Neil McElroy award, United Way, 1991. Fellow ACP; mem. AMA (alt. del., recognition award 1969-89, 90), Soc. Dermatol. Surgery, Am. Acad. Dermatology (CME award 1990), Ohio State Med. Assn. (sec. dermatology sect. 1974—, del., 1st dist. councilor 1991—, recognition award 1990), Acad. Medicine Cin. (pres. 1987-89), Chgo. Dermatol. Soc., Cin. Dermatol. Soc. (pres. 1973-74, 89-90), Noah Worcester Dermatol. Soc. (sec.-treas. 1985-90, pres. elect 1990-91, pres. 1992-93), The Cincinnatus Assn., Leadership Cin. (Class XII), Assn. Ohio Commodores, Kidney Found., Cin. C. of C., Cin. Tennis Club, Rotary, Cin. Country Club. Roman Catholic. Office: Towne Pl 9500 Kenwood Rd Cincinnati OH 45242-6174

KIVELSON, MARGARET GALLAND, physicist; b. N.Y.C., Oct. 21, 1928; d. Walter Isaac and Madeleine (Wiener) Galland; m. Daniel Kivelson, Aug. 15, 1949; children: Steven Allan, Valerie Ann. AB, Radcliffe Coll., 1950, AM, 1951, PhD, 1957. Cons. Rand Corp., Santa Monica, Calif., 1956-69; asst. to geophysicist UCLA, 1967-83, prof., 1983—, also chmn. dept. earth and space scis., 1984-87; prin. investigator of magnetometer, Galileo Mission, Jet Propulsion Lab., Pasadena, Calif., 1977—; overseer Harvard Coll., 1977-83; mem. adv. coun. NASA, 1987—; chair atmospheric adv. com. NSF, 1986-89, Com. Solar and Space Physics, 1977-86, com. planetary exploration, 1986-87, com. solar terrestial phys., 1989-92. Editor: The Solar System: Observations and Interpretations, 1986; contbr. articles to profl. jours. Named Woman of Yr., L.A. Mus. Sci. and Industry, 1979, Woman of Sci., UCLA, 1984; recipient Grad. Soc. medal Radcliffe Coll., 1983, 350th Anniversary Alumni medal Harvard U. Fellow AAAS, Am. Geophysics Union; mem. Am. Phys. Soc., Am. Astron. Soc. Office: UCLA Dept Earth & Space Scis 6843 Slichter Los Angeles CA 90024-1567

KIVIKAS, TÖIVELEMB, physicist, executive; b. Tallinn, Estonia, July 14, 1937; arrived in Sweden, 1944; s. Albert and Anna (Varik) K.; m. Tiiu, Dec. 31, 1965; children: Mart, Triinu, Malle. Ph lic, U. Lund (Sweden), 1964, PhD, 1971; postgrad., CERN, Geneva, 1964-67, INSEAD, Fontainebleau, France, 1980. Lectr. U. Lund, 1958-72, asst. prof., 1973-74; devel. mgr. Alfa-Laval AB, Lund, 1974-76; devel. mgr. AGA AB, Stockholm, 1976-78, bus. area mgr., 1978-81; v.p. Esab AB, Göteborg, Sweden, 1981-84; pres. Innocap AB, Stockholm, 1985-87; sr. mgmt. cons. PA Cons. Group, Stockholm and Cambridge, Eng., 1987-88; gen. mgr. nuclear div. Studsvik, Nyköping, Sweden, 1988-89; pres. Studsvik AB, Nyköping, 1990—; bd. dirs. Pacific Nuclear, Seattle. Author: (textbook) University Physics, 1971; patentee heat exchangers; contbr. articles to profl. jours. Mem. Swedish Phys. Soc. (bd. dirs. 1982), Royal Acad. Sci. (bd. dirs. swedish nat. com. physics sect. 1982), European Phys. Soc., Masons. Avocations: sailing, golf, tennis. Home: Wirsens Väg 10 B, 182 63 Djursholm Sweden Office: Studsvik AB, 611 82 Nykoping Sweden

KIVINEN, SEPPO TAPIO, obstetrician/gynecologist, hospital administrator; b. Saarijärvi, Finland, Dec. 30, 1946; s. Eino and Lempi Ihanelma (Nyman) K.; m. Ritva Leena Saari, June 13, 1974; children: Ilkka Tapio, Hanna-Leena, Jukka Tapio. MD, Oulu (Finland) U., 1972, spl. competence in obstetrics/gynecology, 1978, PhD, 1981. Gen. practitioner Pub. Health Svcs., Finland, 1972-75; resident dept. obstetrics and gynecology Oulu U. Cen. Hosp., 1975-80, asst. prof. dept. obstetrics and gynecology, 1981-83, dept. head div. gynecol. oncology, 1983-85; physician-in-chief dept. obstetrics and gynecology Kanta-Häme Cen. Hosp., Hämeenlinna, Finland, 1985—, med. dir.; med. dir. Kanta-Häme Cen. Hosp., 1991—; sec. Finnish Soc. Clin. Cytology, 1984-91; mgr. Sytogyn Lab., Finland, 1986—, Jaarli Pvt. Med. Ctr., Finland, 1987-90. Contbr. articles to profl. jours. Scholar Rotary Internat., 1981-82. Mem. European Assn. Gynecologists and Obstetricians, European Assn. Cancer Rsch., Internat. Club Physicians, Internat. Coll. Surgeons, Scandinavian Soc. Gynaecol. Oncology, N.Am. Menopause Soc., Internat. Editorial Bd. Menopause Digest, Finnish Med. Assn., Med. Assn. Duodecim, Finnish Assn. Gynaecologists and Obstetricians, Finnish Soc. Clin. Cytology (sec. 1984—), Soc. Clin. Endocrinology, Finnish Soc. Colposcopy, Soc. Social Medicine in Finland, Soc. Hosp. Physicians in Finland. Mem. Centre Party. Lutheran. Avocations: music, slalom, golf, politics, reading and writing. Home: Ansionmäentie 19, SF-13100 Hämeenlinna Finland Office: Kanta-Häme Cen Hosp, Ahvenisto, SF-13530 Hämeenlinna Finland

KIWAN, ABDUL MAGEED METWALLY, chemistry educator; b. Shibin El-Kom, Menofiya, Egypt, Apr. 19, 1934; s. Metwally and Fatima (Al Khateeb) K.; m. Angela Mina Griffiths, Sept. 11, 1965; children: Dina, Sarah, Nadia. BS, Alexandria U., Egypt, 1956; MS, Ain Shams U., Egypt, 1961; PhD, Leeds U., U.K., 1965. Researcher Nat. Rsch. Ctr., Cairo, Egypt, 1965-69; rsch. assoc. Leeds U., U.K., 1969-71; assoc. prof. Kuwait U., 1971-78, prof. chemistry, 1978-90; vis. scientist MIT, Cambridge, Mass., 1982-83; prof. chemistry U.A.E. Univ., Al-Ain, United Arab Emirate, 1991—; reviewer, translator Arabic Version Jour. of Scientific American, Kuwait, 1985—; referee Jour. Kuwait U. Faculty of Science, 1978-90, Gulf Sci. Jour., Riyadh, Saudi Arabia, 1980—. Contbr. articles to Jour. Chem. Soc., 1977, 81, Analytica Chimica Acta, 1982, Bulletin Chem. Soc. Japan, 1989, Jour. Coord. Chemistry, 1990. Grantee Kuwait U., 1979, 84, 88, grantee United Arab Emirate U., 1992. Fellow Royal Soc. Chemistry; mem. Am. Chem. Soc. Achievements include preparation and characterization ofmore than 25 dithizone analogues, investigation of their reactions with metals; studies on mesoionic compounds; elucidation of kinetics of cyclodehydrofluorination of 1, 5 Bis-(2-Flurorphenyl, 3-Mercaptoformazan). Home: Little Haven Golden Acre, Angmering On Sea BN16 1QP, United Kingdom Office: United Arab Emirate U, Al-Ain United Arab Emirates

KIXMILLER, RICHARD WOOD, chemist, corporate executive; b. Evanston, Ill., May 13, 1920; s. William and May (Wood) KixM.; m. Josephine Wallace, 1943 (div. 1984); children: Richard Farwell, John William, Daniel Walter; m. JoAnn Crise, 1985. AB, Princeton U., 1942. Rsch. chemist Standard Oil of Ind., 1942-43; asst. rsch. dir. with office of rubber dir. W.P.B. and Rubber Reserve Co., 1943-46; with Celanese Corp., 1946-82, v.p., gen. mgr. chem. div., 1955-59, also bd. dirs., exec. v.p., 1960-65, vice-chmn., 1965-66; pres. Celanese Chem. Co.-Celanese Plastics Co., 1959-60, in charge of internat. ops., 1960-64; chmn., dir. Wiltek, Inc., 1968-79; bd. dirs. Summit Bancorp., 1962-82, Keuffel & Esser Co., 1968-82, Seaboard Surety Co., 1967-63, Wallace Computer Svcs., Hanover Rsch. Corp. Trustee, sec. Drew U.; trustee, chmn. emeritus N.J. Symphony Orch.; pres., bd. dirs. Prospect Found.; bd. dirs. Hyde and Watson Found., Vt. Hist. Soc.; former mem. adv. coun. Princeton Sch. Engring., Princeton Alumni Coun.; former trustee Pingry Sch. Lt. (j.g.) USNR, 1944-46. Mem. AAAS, Am. Chem. Soc., Am. Inst. Chem. Engrs., Mfg. Chemists Assn. (dir. 1960-63), Comml. Chem. Devel. Assn., Chem. Market Rsch. Assn., Princeton Club, N.Y. Yacht Club, Beacon Hill Club (Summit, N.J. trustee 1962-65), Baltusrol Club (Springfield, N.J.), Phi Beta Kappa, Sigma Xi. Methodist. Home: 210 Fairmount Ave Chatham NJ 07928-1825 Office: The Hyde and Watson Found 437 Southern Blvd Chatham NJ 07928-1454

KLAPOETKE, THOMAS MATTHIAS, chemist; b. Gottingen, Germany, Feb. 24, 1961; s. Eberhard and Christa Brigitte F. (Pflug) K. BSC, Technische U. Berlin, 1982, MSc, 1984, PhD, 1986, Habil., 1990. Sci. asst. Technisch U. Berlin, 1984-86, instr., 1986-87; vis. scholar U. New Brunswick, Fredericton, 1987-88; rsch. group supr. and faculty mem. Technische U. Berlin, 1989—, asst. prof. chemistry (oxidizers, explosives), 1990—; cons. in chem. and pharm. fields; co-researcher, chemist Med. Dept., Berlin, 1988—; lectr. in field. Contbr. articles to profl. jours. textbooks; referee Elsevier Sci. Pub., Angew. Chem., 1989—. Recipient Schering prize, Schering AG, Berlin, 1987, Feodor-Lynen Stipend, Humboldt Found., Bonn,

1987, others. Mem. Royal Soc. Chemistry, German Chem. Soc., Fluorine div. of Am. Chem. Soc., Freunde der TU. Avocations: ballet, opera, modern art, music. Office: Tech U Berlin, Str d 17 Juni 135, D-10623 Berlin Germany

KLARMAN, KARL JOSEPH, electromechanical engineer; b. Scotia, N.Y., Mar. 18, 1922; m. Edith O. Wood, Sept. 27, 1952; children: Nancy Lee, James Douglas. BSEE, Union Coll., 1944; MSEE, Columbia U., 1947. Registered profl. engr., N.H. Lectr. Union Coll., Schenectady, N.Y., 1945-46; project engr. Eclipse-Pioneer divsn. Bendix, Teterboro, N.J., 1947-51; v.p. engrng. Electro Tec Corp., Hackensack, N.J., 1951-57; sect. head Sanders Assoc., Nashua, N.H., 1957-59; chief prodn. engr. Northrop, Norwood, Mass., 1959-61; engr., scientist RCA, Burlington, Mass., 1961-74; pvt. cons. Nashua, 1974—. With U.S. Army, 1944-45, ETO. Mem. IEEE (sr., gyro and accelerometer panel 1969—), Sigma Xi (life). Achievements include patents for Gyroscope Caging Device, Damping Mechanism. Home and Office: 20 Kipling St Nashua NH 03062

KLARREICH, SUSAN RAE, chemistry educator; b. N.Y.C., July 31, 1942; d. Benjamin and Evelyn (Bregstein) K.; m. David Nevin, June 20, 1963 (div. Mar. 1981); m. Brian Weiss, July 3, 1983. BS, CUNY, 1963, MA in Chemistry, 1967, PhD, 1974; MS in Computer Sci., Stevens Inst., 1986. Tchr. Bronx (N.Y.) High Sch. Sci., 1965, James Monroe High Sch., Bronx, 1965-66; from lectr. to adj. asst. prof. CUNY, N.Y.C., 1967-75; from instr. to asst. prof. Bergen Community Coll., Paramus, N.J., 1975-83, assoc. prof., 1983-93, prof., 1993—, coord. phys. scis., 1992—. Trustee Temple Israel Cliffside Park, N.J., 1988-89; pres. Sisterhood of Israel, Cliffside Park, 1989-91. Mem. Am. Chem. Soc., Internat. Union Pure and Applied Chemistry, N.Y. Acad. Scis., Amateur Astronomers Assn. N.Y. (bd. dirs. 1977-93). Democrat. Office: Bergen Community Coll 400 Paramus Rd Paramus NJ 07652-1508

KLASING, SUSAN ALLEN, environmental toxicologist, consultant; b. San Antonio, Sept. 10, 1957; d. Jesse Milton and Thelma Ida (Tucker) Allen; m. Kirk Charles Klasing, Mar. 3, 1984; children: Samantha Nicole, Jillian Paige. BS, U. Ill., 1979, MS, 1981, PhD, 1984. Staff scientist Life Scis. Rsch. Office, Fedn. Am. Socs. Exptl. Biology, Bethesda, Md., 1984-85; assoc. dir. Alliance for Food and Fiber, Sacramento, 1986; postgrad. researcher U. Calif., Davis, 1986-87; project dir. Health Officers Assn. Calif., Sacramento, 1987-89; cons. Klasing and Assocs., Davis, Calif., 1989—; mem. expert com. for substances-of-concern San Joaquin Valley Drainage Program, Sacramento, 1987, follow-up task force, 1990-91, drainage oversight com., 1992—. Author: (chpt.) Consideration of the Public Health Impacts of Agricultural Drainage Water Contamination, 1991. Mem. AAAS, Soc. Environ. Toxicology and Chemistry. Office: Klasing and Assocs 515 Flicker Ave Davis CA 95616

KLASSEN, JANE FRANCES, electrical engineer; b. St. Joseph, Mich., June 11, 1956; d. John Burkle and Carol Mae (Beebe) K.; m. Jay Russell Handy, June 12, 1976 (div. 1987); 1 child, Nicholas Anthony. AAS, Lake Mich. Coll., 1976; BSEE, Western Mich. U., 1990. Lab. rsch. engr. Argonne (Ill.) Nat. Lab., 1990-91; cons. engr. ABB Impell, Downers Grove, Ill., 1991—; Contbr. to profl. publs. Mem. IEEE, Am. Nuclear Soc., Soc. Women Engrs. Office: ABB Impell 1333 Butterfield Rd Downers Grove IL 60515

KLASSEN, LYNELL W., rheumatologist, transplant immunologist; b. Gossel, Kans., Jan. 24, 1947; married; 4 children. AB, Tabor Coll., 1969; MD, U. Kans., 1973. Resident in internal medicine U. Iowa Hosps. and Clins., 1973-75, chief resident internal medicine, 1977-78, asst. prof., 1978-82, assoc. prof. rheumatology & immunology, 1982-90; prof., vice chmn. internal medicine U. Nebr. Med. Ctr., 1990—; rsch. assoc. immunology Arthritis & Rheumatism Br., NIH, 1975-77; chief arthritis svc. rheumatology Omaha VA, 1982—; chmn. science rev. com. Nat. Inst. Alcohol Abuse, Alcoholism, 1989—. Mem. Edn. Coun., Am. Coll. Rheumatology, Am. Coll. Physicians, Am. Assn. Immunology. Achievements include rsch. in mechanisms of hematopoietic allograft rejection, pathophysiology of graft-versus-host disease, use of cytotoxic therapy in non-malignant diseases. Office: Omaha Sept of Vet Affair Med Ctr 4101 Woolworth Ave Omaha NE 68105*

KLATTE, DIETHARD W., mathematician; b. Berlin, Dec. 18, 1950; s. Gerhard and Herta (Matheja) K. Diploma in math., Humboldt U., Berlin, 1974, D Natural Sci., 1977, D habil. Natural Sci., 1984. Asst. Humboldt U., Berlin, 1976-82, asst. prof., 1982-85; assoc. prof. Pedagogical Inst., Halle, Germany, 1985-87, full prof. math., 1987-92; asst. prof. OR U. Zurich, Switzerland, 1992—; rsch. stay Leningrad State U., USSR, 1978-79, Internat. Inst. Applied Systems Analysis, Laxenburg, Austria, 1983. Co-author: Non-Linear Parametric Optimization, 1982; co-editor: Advances in Mathematics Optimization, 1988; contbr. articles to profl. jours. Mem. Math. Soc. Germany. Avocations: music, theatre, mountain-rambling, jogging, swimming. Home: Berliner Strasse 4, O-1100 Berlin Germany Office: U Zurich Inst OR, Moussonstr 15, CH-8044 Zurich Switzerland

KLAUBERT, EARL CHRISTIAN, chemical and general engineering administrator, consultant; b. Manchester, N.H., June 20, 1930; s. Charles Christian and Irene Adeline (Blanchette) k.; m. Charlotte Elizabeth Allen, July 30, 1955; children: Brian Douglas, Jeffrey Scott. BSChemE, Worcester Poly. Inst., 1952. Explosives engr. Hercules Powder Co., various locations, 1952-57; solid rocket propellant engr. Reaction Motors div. Thiokol Chem. Corp., Denville, N.J., 1957-65; radiometric instrumentation engr. Block Engrng., Inc., Cambridge, Mass., 1965-67; aerospace technologist NASA Elec. Rsch. Ctr., Cambridge, Mass., 1967-70; sr. staff engr. U.S. Dept. Transp., Cambridge, Mass., 1970—; automotive test engr. car and light truck fuel economy com. Soc. Automotive Engrs., Detroit, 1974-81. Webelos den leader Boy Scouts Am., Lexington, Mass., 1972-73, mem. troop com., 1972-76; Met. Coun. Ednl. Opportunity host parent Lexington Pub. Schs., 1974-78. With U.S. Army, 1953-55. Achievements include patents in solid propellant burning rate control; in flamethrower; in automotive exhaust system leak test; in time varying identification badge. Home: 12 Minute Man Ln Lexington MA 02173 Office: US Dept of Transp Transp Systems Ctr Kendall Square DTS-75 Cambridge MA 02142

KLAUSMEYER, DAVID MICHAEL, scientific instruments manufacturing company executive; b. Indpls., Aug. 29, 1934; s. David M. and V. Jane (Donellan) K.; m. Julie Ann Johnson, Oct. 29, 1955; children: Kathleen M., Kevin M., Gregory J. BSS, Georgetown U., 1955. Asst. to pres. White Cons. Ind., Cleve., 1957; auditor Ernst & Ernst, Cleve., 1957-59; pres. Photopipe, Inc., Cleve., 1960-63; v.p. McGregor & Werner Internat., Inc., Washington, 1964-70; internat. cons. Stratford of Tex., Houston, 1971-72; pres. FLR Corp., Houston, 1972-74, Southwest Cons., Houston, 1981-86, Imaging Products, Houston, 1987-90; pres. Nanodyanmics, Inc., Houston, 1988—, also bd. dirs.; pres. Corp. Devel., Houston, 1974-81; ptnr. Klausmeyer & Assoc., Houston; bd. dirs. S.W. Venture Reification, Houston, Imaging Products, Inc., Houston, Pharmaceutical Labs., Inc., Bedford, Tex., Chemblend Products, Inc., Houston, SI Diamond Tech., Inc., Houston, TWK Techs., Charlotte, N.C. With USCG, 1955-57. Republican. Roman Catholic. Home: 288 Litchfield Ln Houston TX 77024 Office: Nanodynamics Inc 10878 Westheimer Rd # 178 Houston TX 77042-3292

KLAVANO, PAUL ARTHUR, veterinary pharmacologist, educator, anesthesiologist; b. Valley, Wash., Nov. 30, 1919; s. Peter and Florence Caroline (Meyer) K.; m. Martha Emma Havighurst, June 2, 1945; children: Robert, Ruth, Beth, Ann. BS, Wash. State U., 1941, DVM, 1944; postgrad. in Philosophy, U. Minn., 1958. Lic. veterinarian, Wash. Chemist Wash. Horse Racing Commn., Pullman, 1942-45; instr. vet. anatomy Wash. State U., Pullman, 1944-45, instr. vet. physiology and pharmacology, 1945-48, asst. prof. physiology and pharmacology, 1948-52, assoc. prof., chmn., 1952-62, prof., chmn., 1962-72, prof. dept. vet. physiolgy and pharmacology, 1972-83, prof. emeritus, 1983—. Contbr. articles to profl. jours. With U.S. Army, 1943-44. Mem. AVMA, Wash. State Vet. Med. Assn., N.Y. Acad. Scis., Am. Acad. Vet. and Comparative Toxicology, Am. Soc. Vet. Physiology and Pharmacology. Democrat. Lutheran. Avocations: hunting, fishing, shop work. Home: SE 1125 Kamiaken Pullman WA 99163

KLAVER, MARTIN ARNOLD, JR., business administrator; b. Detroit, Mar. 9, 1932; s. Martin A. and Thelma (Wiegand) K.; m. Doris Ann Dukes, May 27, 1967; 1 child, Carol Ann. BA, Haverford (Pa.) Coll., 1954; MA, U. Wis., 1956. Indsl. engr. Joseph Bancroft & Sons Co., Wilmington, Del., 1961-68; project engr. Lenox China, 1968-89; mgr. ops. analysis Lenox China, Pomona, N.J., 1989-93; dir. MIS NET, Phila., 1993—. Editoral adv. bd. Speech Tech. Jour., N.Y.C., 1988—; contbr. articles to profl. jours. VonSchleinitz Found. fellowship U. Wis., 1955, Goethe Centennial fellowship, 1954. Democrat. Achievements include rsch. on combined use of industrial engineering and linguistics concepts for development of applications software for use of computer voice recognition for process and manufacturing control. Home: 108 Eastmont Ln Sickerville NJ 08081 Office: NET 2205 Bridge St Philadelphia PA 19137

KLEBAN, MORTON HAROLD, psychologist; b. Bklyn., Oct. 23, 1931; s. Samuel and Rose (Urbow) K.; m. Ferne Berman, Feb. 13, 1956; children: Karen, Joan, Ira. BBA, CCNY, 1953; MA, U. Iowa, 1955; PhD, U. N.D., 1960. Clin. psychology intern Nebr. Psychiat. Inst., Omaha, 1959-60; clin. psychologist Norristown (Pa.) State Hosp., 1960-64, med. rsch. scientis, 1966-90, sr. statistician, 1990—; rsch. cons. Office of Mental Health State of Pa., Harrisburg, 1964-66; sr. researcher, dir. psychometrics Phila. Geriatric Ctr., 1966—. Editorial staff: Pennsylvania Psychiatric Quarterly, 1965-69; cons. editor: Journal of Gerontology, 1985-88; editorial bd.: Experimental Professional Journal, 1960—; contbr. articles to profl. jours. Fellow APA (Pa. chpt. chmn. membership com. 1966-67, pub. affairs com. 1967-69), Am. Psychol. Soc., Gerontol. Soc. Am., Am. Statistical Assn., Pa. Psychol. Assn. Jewish. Home: 14 Colton Dr Norristown PA 19401 Office: Phila Geriatric Ctr 1501 Old York Rd Philadelphia PA 19141

KLEBANOFF, SEYMOUR JOSEPH, medical educator; b. Toronto, Feb. 3, 1927; s. Eli Samuel and Ann Klebanoff; m. Evelyn Norma Silver, June 3, 1951; children: Carolyn, Mark. MD, U. Toronto, 1951; PhD in Biochemistry, U. London, 1954. Intern Toronto Gen. Hosp., 1951-52; postdoctoral fellow Dept. Path. Chemistry, U. Toronto, 1954-57, Rockefeller U., N.Y.C., 1957-62; assoc. prof. medicine U. Washington, Seattle, 1962-68, prof., 1968—; mem. adv. coun. Nat. Inst. Allergy and Infectious Diseases, NIH, 1987-90. Author: The Neutrophil, 1978, also 201 jour. articles. Recipient Merit award NIH, 1988, Mayo Soley award Western Soc. for Clin. Investigation, 1991. Fellow AAAS; mem. NAS, Am. Soc. Clin. Investigation, Am. Soc. Biol. Chemists, Assn. Am. Physicians, Infectious Diseases Soc. Am. (Bristol award 1993), Endocrine Soc., Reticuloendothelial Soc. (Marie T. Bonazinga rsch. award 1985), Inst. of Medicine. Home: 509 Mcgilvra Blvd E Seattle WA 98112-5047 Office: U Wash Dept Medicine SJ-10 Div of Allergy & Infectious Dis Seattle WA 98195

KLECK, ROBERT ELDON, psychology educator; b. Archbold, Ohio, Aug. 3, 1937. AB in Philosophy, Denison U., 1959; PhD in Social Psychology, Stanford (Calif.) U., 1963. Postdoctoral fellow Stanford U., 1963-64; asst. prof. Williams Coll., Williamstown, Mass., 1964-66; asst. to assoc. prof. Dartmouth Coll., Hanover, N.H., 1966-75, prof. psychology, 1975—, John Sloan Dickey Third Century Prof. of Social Scis., 1985—; vis. research prof. Boy's Town Ctr. Study of Youth Devel., Stanford U., 1974-75; cons. VA Stroke Project, 1983—, Disadvantaged Children in N.H., 1974, Bur. Devel. Disabilities, Concord, N.H., 1975-80, Crotchet Mountain Rehab. Ctr., 1973, Abilities, Inc., Albertson, N.Y., 1979-81, Can. Research Council, NSF, USPHS; faculty sponsor USPHS Post-doctoral fellowship, 1977-78. Cons. editor Jour. Personality and Social Psychology, 1974-78, assoc. editor 1971-72; mem. editorial bd. Jour. Nonverbal Behavior, 1990-93; mem. editorial adv. bd. Action for Children's TV, 1975-79; editorial cons.various jours.; contbr. articles to profl. jours. Danforth fellow, 1959-63; Gen. Motors scholar, 1955-59. Mem. Am. Psychol. Assn., Am. Psychol. Soc., Internat. Soc. Research on Emotion, Soc. Experimental Social Psychology, New Eng. Psychol. Assn., New Eng. Soc. Psychol. Assn., Soc. Kent and Danforth Fellows, Sigma Xi, Phi Beta Kappa. Home: 28 Low Rd Hanover NH 03755-2207 Office: Dartmouth Coll Dept of Psychology Hanover NH 03755

KLECKNER, DEAN RALPH, trade association executive; b. Riceville, Iowa, Oct. 7, 1932; s. Ralph Burton and Grace Mary (Lenth) K.; m. Natalie Leone Kitzmann, June 7, 1953; children: Mark, Scott, Kirk, Rhonda, Lisa. LLD (hon.), Wartburg Coll., 1986. Sec. Floyd County (Iowa) Farm Bur., 1959, county orgn. dir., 1959-60, pres., 1960-62, voting del., 1962, dist. II dir., 1963-66; v.p. Iowa Farm Bur. Fedn., Des Moines, 1966-75, pres., 1975-86, also bd. dirs.; pres. Am. Farm Bur. Fedn., Park Ridge, Ill., 1986—; bd. dirs. U.S. Meat Export Fedn., Denver, 1980—, mem. food and agr. com. and adv. com. for trade negotiations, Washington, 1987—; mem. Nat. Econ. Commn., Washington, 1988-89; bd. dirs. First Interstate of Iowa. Mem. Iowans Right to Work, 1978—, Nat. Inst. for Rural Health, Des Moines, 1986—; mem. adv. com. dean's coun. Iowa State U., Ames, 1987—. Named Outstanding Young Farmer Iowa Jaycees, 1967. Office: Am Farm Bur Fedn 225 W Touhy Ave Park Ridge IL 60068-5874

KLECKNER, MARLIN DALLAS, veterinarian; b. Dubuque, Iowa, Oct. 21, 1932; s. Roland Reuben and Viola Marie (Kitchen) K.; m. Sheron Virginia Patterson, Jan. 28, 1956; children: Timothy Dallas, Daniel Aaron, Matthew David. BS, U. Ill., 1954, DVM, 1956. Epidemiologist USPHS, Communicable Disease Ctr., Epidemic Intelligence Svc., Atlanta, 1956-58; vet. practitioner Pearl City, Ill., 1958-63; researcher The Upjohn Co., Kalamazoo, Mich., 1963-90; cons. Plainwell, Mich., 1990—; bd. dirs. BallistiVet, Inc., Mpls., 1990—; com. mem. Animal Health Inst. Task Force, Alexandria, Va., 1970-90, Am. Feed Ingredients Assn. Animal Welfare, Roslyn, Va., 1985-89. Pres. Lake Doster Homeowners Assn., Plainwell, 1980's, Lake Doster Men's Golf Assn., Plainwell, 1985, 91; treas. Gun Plain-Plainwell Fire Bd., 1991-92. Capt. USPHS, 1956-58. Mem. Am. Vet. Med. Assn., Assn. Indsl. Veterinarians, Mich. Vet. Med. Assn. Republican. Achievements include management of animal drug registration of first prostaglandin for animals, lincomycin for multiples uses, amitraz for dogs and ceftiofur for cattle. Home and Office: 187 S Lake Doster Rd Plainwell MI 49080

KLEE, VICTOR LA RUE, mathematician, educator; b. San Francisco, Sept. 18, 1925; s. Victor La Rue and Mildred (Muller) K.; m. Elizabeth Bliss; children—Wendy Pamela, Barbara Christine, Susan Lisette, Heidi Elizabeth; m. Joann Polack, Mar. 17, 1985. B.A., Pomona Coll., 1945, D.Sc. (hon.), 1965; Ph.D., U. Va., 1949; Dr. honoris causa, U. Liège, Belgium, 1984. Asst. prof. U. Va., 1949-53; NRC fellow Inst. for Advanced Study, 1951-52; asst. prof. U. Wash., Seattle, 1953-54, asso. prof., 1954-57, prof. math., 1957—, adj. prof. computer sci., 1974—, prof. applied math., 1976-84; vis. asso. prof. UCLA, 1955-56; vis. prof. U. Colo., 1971, U. Victoria, 1975, U. Western Australia, 1979; cons. IBM Watson Research Center, 1972; cons. to industry; mem. Math. Scis. Research Inst., 1985-86; sr. fellow Inst. for Math. and its Applications, 1987. Author: (with H. Hadwiger and H. Debrunner) Combinatorial Geometry in the Plane, 1964, (with S. Wagon) Old and New Unsolved Problems in Plane Geometry and Number Theory, 1991; contbr. more than 200 articles to profl. jours. Recipient Research prize U. Va., 1952, Vollum award for disting. accomplishment in sci. and tech. Reed Coll., 1982, David Prescott Burrows Outstanding Disting. Achievment award Pomona Coll., 1988, Max Planck rsch. prize, 1992; NSF sr. postdoctoral fellow; Sloan Found. fellow U. Copenhagen, 1958-60; fellow Center Advanced Study in Behavioral Scis., 1975-76; Guggenheim fellow, Humboldt award U. Erlangen-Nürnberg, 1980-81, Fulbright award U. Trier, 1992; sr. fellow Inst. Mathematics and Its Applications, 1987. Fellow AAAS (chmn. sect. A 1975); mem. Am. Math. Soc. (asso. sec. 1955-58, mem. exec. com. 1969-70), Math. Assn. Am. (pres. 1971-73, L.R. Ford award 1972, Disting. Service award 1977, C. B. Allendoerfer award 1980), Soc. Indsl. and Applied Math. (mem. council 1966-68), Assn. Computing Machinery, Math. Programming Soc., Internat. Linear Algebra Soc., Phi Beta Kappa, Sigma Xi (nat. lectr. 1969). Home: 13706 39th Ave NE Seattle WA 98125-3810 Office: U Wash Dept Mathematics GN 50 Seattle WA 98195-0001

KLEENE, STEPHEN COLE, retired mathematician, educator; b. Hartford, Conn., Jan. 5, 1909; s. Gustav Adolph and Alice Lena (Cole) K.; m. Nancy Elliott, Sept. 2, 1937 (dec.); children: Paul Elliott and Kenneth Cole (twins), Bruce Metcalf, Pamela Lee; m. Jeanne M. Kleene, Mar. 17, 1988. AB, Amherst Coll., 1930; PhD, Princeton U., 1934. Instr. math., researcher Princeton U., 1930-35; from instr. math to assoc. prof. U. Wis., Madison, 1935-48, prof., 1948-64, Cyrus C. MacDuffee prof. math., then prof. emeritus, chmn. dept. math., 1957-58, 60-62; vis. prof. Princeton U., 1956-57; mem. Inst. Advanced Study, 1939-40,div. math. NRC, 1956-58; pres. Internat. Union History and Philosophy Sci., 1961. Author: Introduction to Metamathematics, 1952, (with Richard E. Vesley) The Foundations of Intuitionistic Mathematics, 1965; Logic, 1980; 2 other books; cons. editor Jour. Symbolic Logic, 1936-42, 46-49, editor, 1950-62; contbr. articles to profl. jours. Lt. comdr. USNR, 1942-46. Recipient Nat. medal Sci. NSF, 1990; Guggenheim fellow U. Amsterdam, 1950; NSF grantee U. Marburg, 1958-59. Fellow AAAS; mem. Phi Beta Kappa, Sigma Xi (pres. Wis. chpt. 1951-52). Achievements include climbing Mt. Everest. Home: 1514 Wood Ln Madison WI 53705-1457

KLEIMAN, DEVRA GAIL, zoologist, zoological park administrator; b. N.Y.C., Nov. 15, 1942. BS in Biopsychology, U. Chgo., 1964; PhD in Zoology, U. London, 1969. Rsch. asst. Wellcome Inst. Comparative Physiology, Zool. Soc. London, 1965-69; NIMH postdoctoral fellow Inst. Animal Behavior, Rutgers U., N.J., 1970-71; rsch. assoc. Smithsonian Instn., 1970-72; reproduction zoologist Nat. Zool. Pk., Smithsonian Instn., Washington, 1972-79, acting head dept. zool. rsch., 1979-81, head dept. zool. rsch., 1981—, acting asst. dir. animal programs, 1983-84, asst. dir. rsch. ednl. activities, 1984-85; adj. asst. prof. dept. Psychology George Washington U., 1974-77, dept. Zoology U. Md., 1979-83; studbook keeper International Studbook for Leontopithecus rosalia, 1974-84; grant reviewer NIMH, 1977, 81, NSF, 1978, 79; adj. prof. Biology George Mason U., Fairfax, Va., 1980-82; U.S. del. com. Internat. Ethnological Conf., 1980-86; mem. bd. fellowships and grants Smithsonian Instn., 1982-84, chair rsch. policy com., 1984-86; mem. species survival plan mgmt. com. L. r. chrysomelas, 1985—, L. r. chrysopygus, 1986—; scientific adv. com. Jersey Wildlife Preservation Trust, 1986—; co-studbook keeper Giant Panda Ailuropoda melanoleuca, 1988—; ad hoc reviewer behavioral and neuroscis. studies sect. NIH, 1988; adv. bd. program on zoos Sta. WQED, 1990—; presenter numerous confs. Mem. editorial bd. International Zoo Yearbook, 1977—, Carnivore, 1977-81, Zoo Biology, 1982—; consulting editor Am. Jour. Primatology, 1983-91; chief editorial advisor Wild Mammals in Captivity, 1983—; field editor Jour. Soc. conservation Biology, 1986—; contbr. articles to profl. jours. Bd. dirs. Scientists Ctr. Animal Welfare, 1984-86; trustee The Digit Fund, 1990—. Recipient Women in Sci. and Engring. award NSF, 1987, award for Disting. Achievement Soc. Conservation Biology, 1988. Fellow AAAS, Animal Behavior Soc. (sec. 1977-80, pres. 1983—); mem. Am. Assn. Zool. Pks. and Aquariums (species coord., internal mgmt. com. L. r. rosalia, species survival plan subcom. 1986-90, vice chair Giant Panda task force 1988—, chair 1992—, mem New World Primate TAG, Cheetah SSP, Brazil FIG, Reintro. adv. group, rsch. coord. group 1991—, chair behavior and husbandry adv. group 1992—), Internat. Union Conservation of Nature and Natural Resources (mem. SSC primate specialist group 1983—, SSC reintro. adv. group, rsch. coords. group 1991—), World Conservation Union (vice-chair primates 1989—), Consortium Aquariums, Univs. and Zoos (adv. com. 1986—), Internat. Soc. Endangered Cats (rsch. adv. bd. 1988-91), Sigma Xi. Office: National Zoological Park MS-551 3000 Connecticut Ave Washington DC 20008

KLEIMAN, NORMAN JAY, molecular biologist, biochemist; b. N.Y.C., Mar. 2, 1959; s. Irving Kleiman and Phyllis Lenore (Berger) Handler; m. Laurin Darcy Blumenthal, Aug. 21, 1983; children: Daryl, Charles. BS, Rensselaer Poly. Inst., 1979; PhD, Vanderbilt U., 1985. Post-doctoral fellow Columbia U., N.Y.C., 1985-89, assoc. rsch. scientist, 1989-92, asst. prof., 1992—; co-dir. basic sci. course ophthalmology Edward S. Harkness Eye Inst., Columbia U., N.Y.C., 1992—. Contbr. articles to profl. jours. Pres. Battery Park Synagogue, N.Y.C., 1990—. Recipient rsch. grant Nat. Eye Inst., 1991—, Individual Nat. Rsch. Svc. award, 1986-89, Rockefeller Clin. Rsch. award, 1989-90. Mem. Assn. for Rsch. in Vision and Ophthalmology, Internat. Soc. for Eye Rsch., Am. Soc. for Biochemistry and Molecular Biology, N.Y. Acad. Scis., Sigma Xi. Jewish. Office: Columbia U Dept Ophthalmology 630 W 168th St New York NY 10032

KLEIN, ARNOLD WILLIAM, dermatologist; b. Mt. Clemens, Mich., Feb. 27, 1945; m. Malvina Kraemer. BA, U. Pa., 1967, MD, 1971. Intern Cedars-Sinai Med. Ctr., Los Angeles, 1971-72; resident in dermatology Hosp. U. Pa., Phila., 1972-73, U. Calif., Los Angeles, 1973-75; pvt. practice dermatology Berverly Hills, Calif., 1975—; assoc. clin. prof. dermatology/medicine U. Calif. Ctr. for Health Scis; mem. med. staff Cedars-Sinai Med. Ctr.; asst. clin. prof. dermatology Stanford U., 1982-89; asst. clin. prof. to assoc. clin. prof. dermatology/medicine, UCLA; Calif. state commr., 1983-89; med. adv. bd. Skin Cancer Found., Lupus Found. Am., Collagen Corp.; presenter seminars in field. Reviewer Jour. Dermatologic Surgery and Oncology, Jour. Sexually Transmitted Diseases, Jour. Am. Acad. Dermatology; mem. editorial bd. Men's Fitness mag., Shape mag., Jour. Dermatologic Surgery and Oncology; contbr. numerous articles to med. jours. Mem. AMA, Calif. Med. Assn., Am. Soc. Dermatologic Surgery, Internat. Soc. Dermatologic Surgery, Calif. Soc. Specialty Plastic Surgery, Am. Assn. Cosmetic Surgeons, Assn. Sci. Advosors, Los Angeles Med. Assn., Am. Coll. Chemosurgery, Met. Dermatology Soc., Am. Acad. Dermatology, Dermatology Found., Scleroderma Found., Internat. Psoriasis Found., Lupus Found., Am. Venereal Disease Assn., Soc. Cosmetic Chemists, AFTRA, Los Angeles Mus. Contemporary Art (founder), Dance Gallery Los Angeles (founder), Am. Found. AIDS Research (founder, dir.), Friars Club, Phi Beta Kappa, Sigma Tau Sigma, Delphos. Office: 435 N Roxbury Dr Ste 204 Beverly Hills CA 90210-5087

KLEIN, CHRISTOPHER CARNAHAN, economist; b. Anniston, Ala., July 5, 1953; s. Wallace Carnahan and Frances Luvona (Meaders) K.; m. Vicki Lynn Brown, May 7, 1983; children: Hannah Marie Brown, Colin Christopher Brown. BA in Econs., U. Ala., 1976; PhD in Econs., U. N.C., 1980. Economist FTC, Washington, 1980-86; economist Tenn. Pub. Svc. Commn., Nashville, 1986—, rsch. dir., 1993—; adj. faculty Middle Tenn. State U. Murfreesboro, 1990—; mem. rsch. adv. com. Nat. Regulatory Rsch. Inst., Columbus, Ohio, 1990—, chmn., 1993-94; mem. staff subcom. on gas Nat. Assn. Regulatory Utility Commrs., 1990—. Contbr. articles to profl. jurs. Recipient Cert. of Commendation, FTC, 1985. Mem. Nat. Assn. Bus. Economists, Am. Econ. Assn., So. Econ. Assn., Indsl. Orgn. Soc., Mid-South Acad. Econs. and Fin., Alpha Pi Mu. Avocations: reading, writing poetry, photography. Office: Tenn Pub Svc Commn 460 James Robertson Pky Nashville TN 37243-0505

KLEIN, DALE EDWARD, nuclear engineering educator; b. Cooper County, Mo., July 6, 1947. BS, U. Mo., 1970, MS, 1971, PhD in Nuclear Engring., 1977. Design engr. Procter & Gamble Co., 1970-72; teaching and rsch. asst. nuclear engring. U. Mo., Columbia, 1973-77; asst. prof. U. Tex., Austin, 1977-82, assoc. program mech. engring. 82-G program, 1978—, dir. nuclear engring. teaching program, 1978—, assoc. dean rsch. coll. engring.; engr. Gen. Atomic Co., 1974; dep. dir. Ctr. Energy Studies, 1986—. Named Young Engr. of Yr., Travis chpt. Tex. Soc. Profl. Engring., 1982. Mem. ASME (Gustus L. Larson Meml. award 1990), NSPE, Am. Nuclear Soc. Achievements include research in thermal analysis of nuclear shipping containers, heat transfer augmentation for flow over rough surfaces, liquid metal flows through a packed bed under the influence of a transverse magnetic field. Office: U Tex Office Tech Devel/Transfer Coll Engring Cockrell Hall 10.340 Austin TX 78712-1080*

KLEIN, EDWARD ROBERT, pharmacist; b. New Orleans, Apr. 17, 1950; s. William Alfred Jr. and Evelyn Marie (Radabaugh) K.; m. Stephany Rhea Knower, July 21, 1973; 1 child, Kandace Rhea. BS, N.E. La. U., 1973. Pharmacy intern Osco Drugs, Springfield, Mo., 1972; pharmacist Eckerd Drugs, Lake Charles, La., 1973-74, Skillern Drugs, Dallas, 1974-80, Doctor's Hosp., Dallas, 1980; pres., pharmacist Klein's Pharmacy, Dallas, 1980—; cons. pharmacist Autumn Leaves Nursing Home, Dallas, 1986-93; mem. HIV medication program adv. com. Tex. Dept. Health, Austin, 1990-93; vis. profs. program Wyeth-Ayerst, 1991; community svcs. bd. v.p. Doctor's Hosp., 1990-91; adv. coun. squibb-Novo, 1991; co-chmn. Behren's Family Value Adv. Coun., 1989-93; mem. Novo Nordisk Speaker's Bur., 1992-93, U. Tex. Coll. Pharmacy Preceptor Program, 1992-93. Recipient Merck, Sharp & Dohme Leadership award, 1991. Mem. Nat. Assn. Retail Druggists, Am. Pharm. Assn., Dallas County Pharm. Soc. (editor directory 1988, 90, asst. editor newsletter 1989-91, Outstanding Pharmacist 1988-92, treas. 1981-91), N.E. La. U. Alumni Assn., U. Tex. Alumni Assn. of the Coll. of Pharmacy, Phi Lambda Sigma, Phi Delta Chi. Democrat. Roman Catholic. Avocations: golf, volleyball, travel, collecting pharmacy memorabilia. Home: 1207 E Berkeley Dr Richardson TX 75081-5808 Office: Kleins Pharmacy 9043 Garland Rd Dallas TX 75218-3920

KLEIN, GEORGE D., geologist, business executive; b. Den Haag, Netherlands, Jan. 21, 1933; came to U.S., 1947, naturalized, 1955; s. Alfred and Doris (deVries) K. BA, Wesleyan U., 1954; MA, U. Kans., 1957; Ph.D., Yale U., 1960. Research sedimentologist Sinclair Research Inc., 1960-61; asst. prof. geology U. Pitts., 1961-63; asst. prof. to assoc. prof. U. Pa., 1963-69; prof. U. Ill., Urbana, 1970-93; pres. N.J. Marine Scis. Consortium, Ft. Hancock, N.J., 1993—; dir. N.J. Sea Grant Coll.; vis. fellow Wolfson Coll. Oxford U., 1969; vis. prof. geology U. Calif., Berkeley, 1970; vis. prof. oceanography Oreg. State U., 1974, Seoul Nat. U., 1980, U. Tokyo, 1983; CIC vis. exchange prof. geophys. sci. U. Chgo., 1979-80; vis. prof. geophysics U. Utrecht, 1988; vis. prof. sedimentary geology Vrije U. of Amsterdam, The Netherlands, 1989; chief scientist Deep Sea Drilling Project Leg 58, 1977-78; continuing edn. lectr.; assoc. Ctr. Advanced Studies U. Ill., 1974, 83; co-dir. project Pangea of Global Sedimentary Geology program, 1991-93. Author: Sandstone Depositional Models for Exploration for Fossil Fuels, 3d edit, 1985, Clastic Tidal Facies, 1977, Holocene Tidal Sedimentation, 1976; assoc. editor Geol. Soc. Am. Bull., 1975-81, Jour. Geodynamics, 1992-93; cons. editor: McGraw-Hill Ency. of Sci. and Yearbook, 1977-89; chief cons. adv. editor: CEPCO div. Burgess Pub. Co, 1979-81; series editor: Geol. Sci. Monographs, Internat. Human Resources Devel. Corp. Press, Inc., 1981-87, Sedimentary Geology, Prentice-Hall Inc., 1988—; mem. editorial bd. Geology, 1973-74, 89-91, Sedimentary Geology, 1985-89; mng. editor Sedimentology, Earth Scis. Revs., 1987-92. Recipient Outstanding Paper award Jour. Sedimentary Petrology, 1970, Erasmus Haworth Disting. Alumnus award in geology U. Kans.,1980, Citation of Recognition, Ill. Ho. of Reps., 1980, Outstanding Geology Faculty Mem. award U. Ill. Geology Grad. Student Assn., 1983; sr. rsch. fellow Japan Soc. for Promotion of Sci., 1983, Fulbright Found., The Netherlands, 1989. Fellow Geol. Soc. Am. (chmn. div. sedimentary geology 1985-86); mem. Am. Geophys. Union, Am. Inst. Profl. Geologists, Soc. Exploration Geophysicists, Soc. for Sedimentary Geology, Internat. Assn. Sedimentologists, Am. Assn. Petroleum Geologists, Sigma Xi. Office: NJ Marine Sci Consortium Bldg 22 Fort Hancock NJ 07732

KLEIN, HAROLD PAUL, microbiologist; b. N.Y.C., Apr. 1, 1921; Alexander and Lillyan (Pal) K.; m. Gloria Nancy Dolgov, Nov. 14, 1942; children—Susan Ann, Judith Ellen. B.A., Bklyn. Coll., 1942; Ph.D., U. Calif., Berkeley, 1950. Am. Cancer Soc. fellow Mass. Gen. Hosp., Boston, 1950-51; instr. microbiology U. Wash., Seattle, 1951-54; asst. prof. U. Wash., 1954-55; asst. prof. biology Brandeis U., Waltham, Mass., 1955-56; assoc. prof. Brandeis U., 1956-60, prof., 1960-63, chmn. dept. biology, 1956-63; vis. prof. bacteriology U. Calif., Berkeley, 1960-61; div. chief exobiology, dir. life scis. Ames Research Center, NASA, Mountain View, Calif., 1963-84; scientist-in-residence Santa Clara U., Calif., 1984—; mem. U.S.-USSR Working Group in Space Biology and Medicine, 1971-84; leader biology team Viking Mars Mission, 1976; mem. space sci. bd. NAS, 1984-89; investigator US/USSR Cosmos 936 flight, 1975, Cosmos 1129 flight, 1975; participating scientist USSR Mars 1996 flight. Mem. editorial bd. Origins of Life, 1970-89. Served with U.S. Army, 1943-46. NSF Sr. Postdoctoral fellow, 1963; grantee NIH, 1955-63; NSF, 1957-63. Mem. Internat. Soc. Study Origin of Life, Am. Soc. Biol. Chemists, Internat. Astronautical Fedn., Phi Beta Kappa. Home: 1022 N California Ave Palo Alto CA 94303-3123 Office: Santa Clara U Dept Biology Santa Clara CA 95053

KLEIN, LEONARD, chemist. Sec. Am. Microchemical Soc., Princeton, N.J.; sr. rsch. chemist, analytical dept. Agrl. Chem. group FMC Corp. Office: Am Microchem Soc care FMC Corp PO Box 8 Princeton NJ 08543-0008

KLEIN, MARCIA SCHNEIDERMAN, analyst, programmer, consultant; b. N.Y.C., Dec. 16, 1940; d. Harry and Shirley (Bardack) Schneiderman; m. Robert M. Klein, June 8, 1963; children: Deborah, David. BS in Math., Syracuse U., 1962; MS in Computer Sci., NYU, 1984. Sci. programmer Rockefeller U., N.Y.C., 1984-87; sr. programmer Coopers & Lybrand, N.Y.C., 1987-88; computer application engr. AT&T, Valhalla, N.Y., 1988-90; analyst, programmer MONY Fin. Svcs., Teaneck, N.J., 1990-92; programmer, analyst Software Options, Englewood Cliffs, N.J., 1992—. Recipient award Solution Systems, 1985, Arity Corp., 1986; rsch. grantee Rockefeller U. 1977-87, 82—. Mem. IEEE Computer Soc., Assn. for Computing Machinery, Math. Assn. Am. (contest judge 1990). Achievements include research on multi-processor operating system AMPOS. Home: 15 Roundabend Rd Tarrytown NY 10591

KLEIN, MARSHALL S., health facility administrator; b. N.Y.C., Mar. 19, 1926; s. Charles Elias and Bertha Helen (Marshall) K.; m. Barbara Janet Cohen; children: Marcia Jill, Geoffrey Lee. BS, Fairleigh Dickinson U., 1966, MBA, 1980. Owner, dir. Styertowne Youth Ctr., Clifton, N.J., 1952-72; adminstr. Eye Inst. N.J., Newark, 1972—. Arbitrator Better Bus. Bur., Newark, 1983—. Served with USN, 1943-46. Mem. Contact Lens Mfg. Assn. (exec. dir. 1982-84), Eye Bank Found. N.J. (exec. dir. 1972—), N.J. Acad. Ophthalmology and Otolaryngology (exec. dirs. 1972—), Eye Bank Assn. Am. (chmn. pub. relations 1980-82), Soc. Fund Raising Execs. (bd. dirs. N.J. chpt. 1978—), Ophthalmology Polit. Action Council (treas. 1979—), Eye Screening Coordinating Council (treas. 1986—), Community Bus. Assn. (pres. 1980-81). Republican. Jewish. Club: Twin Brooks (Warren, N.J.) (pres. 1987-88). Lodge: Lions (pres. 1963-64, dist. gov. 1969-70, legis. agt. 1973—). Avocation: golf. Office: Eye Inst NJ 90 Bergen St Newark NJ 07103*

KLEIN, MARTIN, ocean engineering consultant; b. N.Y.C.; s. Allen and Muriel (Seidman) K.; children: Allen Jameson, Robyn Marie. SBEE, MIT, 1962. Program mgr. sonar systems EG&G Internat., Bedford, Mass., 1962-67; pres. Klein Assocs., Inc., Salem, N.H., 1968-89; cons. Andover, Mass., 1989—; mem. mgmt. coun. Project Urquhart (Loch Ness), London, 1992—; mem. bd. advisors B.Engring. Tech. program U. N.H., Durham, 1988—, Acad. Applied Sci., Concord, N.H., 1982—; mem. adv. bd. MIT Sea Grant, Cambridge, 1989—; bd. dirs. Marine Archaeol. and Hist. Rsch. Inst., Elliot, Maine, 1990—. Contbr. articles to mags. Minister search com., publicity dir. Unitarian Universalist Ch., Andover, 1990-91, chair publicity com., 1991—; trustee Andover Pub. Libr., 1992—. Recipient Small Bus. Person of Yr. award Small Bus. Adminstrn., 1983. Fellow Marine Tech. Soc. (dir budget and finance 1991—), Explorers Club; mem. IEEE. Achievements include patents in field; development of first commercially successful side scan sonar.

KLEIN, MARTIN JESSE, physicist, educator, historian of science; b. N.Y.C., June 25, 1924; s. Adolph and Mary (Neuman) K.; m. Miriam June Levin, Oct. 28, 1945 (div. 1973); children: Rona F., Sarah M. (Mrs. Joseph Zaino), Nancy R.; m. Linda L. Booz, Oct. 8, 1980; 1 child, Abigail M. A.B., Columbia U., 1942, M.A., 1944; Ph.D., Mass. Inst. Tech., 1948. With OSRD for USN, 1944-45; research asso. physics Mass. Inst. Tech., 1946-49; instr. physics Case Inst. Tech., 1949-51, asst. prof., 1951-55, asso. prof., 1955-60, prof., 1960-67, acting dept. head, 1966-67; prof. history physics Yale U., 1967-74, Eugene Higgins prof. history physics and prof. physics, 1974-91, Bass prof. history sci., prof. physics, 1991—, chmn. dept. history sci., 1971-74; William Clyde De Vane prof., 1978-81; Van der Waals guest prof. U. Amsterdam, 1974; vis. prof. Rockefeller U., 1975, adj. prof., 1976-79; vis. prof. Harvard U., 1989-90. Author: Paul Ehrenfest, Vol. I: The Making of a Theoretical Physicist, 1970; editor: Collected Scientific Papers of Paul Ehrenfest, 1959; sr. editor The Collected Papers of Albert Einstein, 1988—; editorial adviser Ency. Brit, 1956-76; translator: Letters on Wave Mechanics, 1967; contbr. articles to profl. jours. NRC fellow Dublin (Ireland) Inst. Advanced Studies, 1952-53; Guggenheim fellow Leyden, Netherlands, 1958-59; Guggenheim fellow Yale, 1967-68. Fellow Am. Acad. Arts and Scis., Am. Phys. Soc., AAAS; mem. NAS, History of Sci. Soc., Am. Assn. Physics Tchrs., Am. Hist. Assn., AAUP, Académie Internationale d'Histoire des Sciences, Phi Beta Kappa, Sigma Xi. Office: History of Science Box 2036 Yale U Station New Haven CT 06520

KLEIN, MICHAEL TULLY, chemical engineering educator, consultant; b. Wilmington, Del., Mar. 15, 1955; s. Donald Michael and Nancy (Tully) K.; m. Elizabeth Thompson, Aug. 7, 1976; children: Jennifer, Michael, Lisa. B-SChemE, U. Del., 1977; ScD, MIT, 1981. Asst. prof. chem. engring. U. Del., Newark, 1981-85, assoc. prof., 1985-89, prof., 1989—, dept. chmn., 1991—, assoc. dean Coll. Engring., 1987-88, dir. Catalysis Ctr., 1988-91. Contbr. over 90 articles to profl. jours. Named Presdl. Young Investigator NSF, 1985. Achievements include development of Monte Carlo reaction modeling software. Office: U Del Chem Engring Dept Newark DE 19716

KLEIN, MILES VINCENT, physics educator; b. Cleve., Mar. 9, 1933; s. Max Ralph and Isabelle (Benjamin) K.; m. Barbara Judith Pincus, Sept. 2, 1956; children—Cynthia Klein Banay, Gail. B.S., Northwestern U., 1954; Ph.D., Cornell U., 1961. NSF postdoctoral fellow Max Planck Inst., Stuttgart, Germany, 1961; prof. U. Ill. Urbana, 1962—. Co-author: Optics, 1986. Contbr. articles to profl. jours. A.P. Sloan Found. fellow, 1963. Fellow AAAS, Am. Phys. Soc.; mem. IEEE (sr.), Optical Soc. Am., Materials Rsch. Soc. Office: U Ill Sci & Tech 104 S Goodwin Ave Urbana IL 61801-2902

KLEIN, MORTON, industrial engineer, educator; b. N.Y.C., Aug. 9, 1925; s. Norbert and Lottie (Wigdor) K.; m. Gloria Ritterband, July 31, 1949; children: Lisa, Melanie. B.S.M.E., Duke U., 1946; M.S., Columbia U., 1952, D.Engring. Sci., 1957. Engr. Picatinny Arsenal, Dover, N.J., 1950-54; instr. Sch. Engring and Applied Sci., Columbia U., N.Y.C., 1956; asst. prof. Sch. Engring and Applied Sci., Columbia U., 1957-61, asso. prof., 1961-69, prof. ops. research, 1969—, chmn. dept. indsl. engring. and ops. research, 1982-85; cons. to industry, govt. Author: (with Cyrus Derman) Probability and Statistical Inference for Engineers, 1959; editor: Management Science, 1960-77; research and pubs. on prodn. planning, scheduling early cancer detection examinations, network flows, statistical quality control and ops. research. Served with USN, 1943-46. Mem. Ops. Research Soc. Am. Inst. Mgmt. Scis., Am. Inst. Indsl. Engrs., Pi Tau Sigma, Alpha Pi Mu, Omega Rho. Office: Columbia U 301 A SW Mudd New York NY 10027

KLEIN, PHILIPP HILLEL, physical chemistry consultant; b. N.Y.C., Sept. 14, 1926; s. Raphael and Lillian Rae (Wald) K.; m. Charlotte Feuerstein, June 21, 1953; children: Joshua David, Daniel William, Jonathan Henry. BS in Chemistry, Syracuse U., 1948, MS in Phys. Chemistry, 1951, PhD in Phys. Chemistry, 1953. Rsch. assoc. Knolls Atomic Power Lab., Schenectady, N.Y., 1952-56; phys. chemist GE Electronics Lab., Syracuse, N.Y., 1956-61; mem. rsch. staff Sperry Rand Rsch. Ctr., Sudbury, Mass., 1961-66; rsch. chemist NASA Electronics Rsch. Ctr., Cambridge, Mass., 1966-70; sect. head U.S. Naval Rsch. Lab., Washington, 1970-87, rsch. cons., 1987-90; prin. Philipp Klein Cons., Washington, 1990—. Assoc. editor Materials Letters, 1985-89; editor Advanced Energy Conversion, 1962; contbr. articles to profl. jours. With USNR 1945-46. Fellow Am. Inst. Chemists; mem. IEEE (chmn. com. on solid state devices 1962-63), Am. Ceramic Soc. (electronics com. 1968-70), Am. Assn. for Crystal Growth (program chmn. 1985-87), Am. Phys. Soc., Sigma Xi. Achievements include patents on the purification of fluorides, preparation of laser hosts, and deposition of silicon carbide shapes. Office: Philipp Klein Cons 2017 Hillyer Pl NW Washington DC 20009-1005

KLEIN, RONALD DON, molecular biologist; b. Mt. Clemens, Mich., July 30, 1948; s. Donald Howard and Virginia F. (Oberliesen) K.; m. Katherine May Keith, Apr. 26, 1968; children: Deanna Marie, Robert James. BA in Liberal Arts, Western Mich. U., 1970, MA in Social Sci., 1971; postgrad, Kalamazoo Coll., 1973-75; PhD in Molecular Biology, U. Wis., 1981. Grad. rsch. fellow dept. biochemistry Wayne State U., Detroit, 1975-76; rsch. assoc. U. Wis., Madison, 1976-81; postdoctoral fellow dept. chemistry MIT, Cambridge, 1981-82; rsch. scientist Phillips Petroleum Co., Bartlesville, Okla., 1982-84; sr. scientist Upjohn Co., Kalamazoo, Mich., 1984—; mem. peer rev. panel Food Prodn. Systems and Nutrition Ala. Rsch. Inst., 1985-87; co-chmn. Internat. Yeast Symposium, Kalamazoo, 1985. Editor: Biochemistry and Molecular Biology of Yeast, 3 vols., 1987; contbr. revs. and articles to profl. jours. Marvin scholar Kalamazoo Coll., 1974-75; recipient Nat. Svc. award Nat. Inst. Gen. Med. Sci., 1982. Mem. AAAS, Am. Soc. Microbiology, Genetics Soc. Am., Am. Soc. Parasitologists, Am. Assn. Vet. Parasitologists, Wis. Acad. Sci., Arts and Letters, Woodrow Wilson Ctr. Internat. Scholarship, Sigma Xi. Achievements include patents for methods of identifying antiparasitic drugs; devel. of first system indsl. yeast Schwanniomyces; rsch. in gene cloning/encoding drug targets from parasitic nematodes, cancer, cardiovascular; pharmacology of parasites. Home: 9721 S 6th St Schoolcraft MI 49087 Office: Upjohn Co Molecular Biology Rsch Kalamazoo MI 49001

KLEINPETER, JOSEPH ANDREW, computer operations specialist; b. Amite, La., Sept. 24, 1943; s. James Moore and Beulah (Gill) K.; m. Carolyn Furca, Oct. 16, 1965; children: Shawn Joseph, Shane Andrew, Shannon René. BSChemE, La. State U., 1965; PhD in Chem. Engring., Tulane U., 1969. Rsch. engr. Conoco, Inc., Ponca City, Okla., 1968-72, rsch. assoc., 1972-75; coal rsch. mgr. Conoco, Inc., Pitts., 1975-79; project mgr. Conoco, Inc., Stamford, Conn., 1979-82; rsch. mgr. DuPont Co., Wilmington, Del., 1982-85, telecomm. mgr., 1985-88, computer ops. mgr., 1988—. Co-author: Coal Liquefaction Products, 1983; contbr. articles to profl. jours. Named as Best Presentation at Nat. Meeting Soc. Mining Engrs., 1978. Mem. AIChE, Tulane Soc. Engrs., Sigma Xi, Tau Beta Pi (chpt. pres. 1965). Republican. Roman Catholic. Achievements include 6 patents in field. Home: 121 Shadow Ln Chadds Ford PA 19317 Office: DuPont Co Wilmington DE 19898

KLEINPOPPEN, HANS JOHANN WILLI, physics educator, researcher; b. Duisburg, Germany, Sept. 30, 1928; s. Gerhard and Emmi (Maas) K. Diploma in physics, U. Giessen, Germany, 1955; D in Physics, U. Tübingen, Germany, 1961, Habilitation, 1967. Pvt. dozent U. Tübingen, 1968—; prof. experimental physics U. Stirling, Scotland, 1968—; vis. assoc. prof. Columbia U., N.Y.C., 1968; vis. fellow U. Colo., Boulder, 1968; vis. prof. U. Bielefeld, Germany, 1978; vis. scientist Fritz-Haber Inst. of Max-Planck Gesellschaft, Berlin, 1991; chmn. several nat. and internat. conf. and summer schs. in atomic physics, Germany, Eng., Scotland, Italy, 1968, 74, 78, 80-82, 84, 87, 93. Co-editor (with P.G. Burke) of 11 books on atomic physics and a monograph series on physics of atoms and molecules. Fellow Am. Phys. Soc., Royal Astronomical Soc., Inst. Physics London, Royal Soc. of Edinburgh, Royal Soc. Arts. Home: 27 Kenningknowes, Stirling FK7 9JF, Scotland Office: U Stirling, Stirling Scotland

KLEMANN, LAWRENCE PAUL, chemical scientist; b. Cin., Aug. 13, 1943; s. Theodore A. and Frieda E. (Knoerzer) K.; m. Diane Wirsing, June 8, 1963; children: Lauren, Eric. BS, U. Mass., 1965, MS, 1968, PhD, 1969. Rsch. chemist Exxon Rsch. & Engring. Co., Clinton, N.J., 1969-86; sr. prin. scientist Nabisco Brands, Inc., East Hanover, N.J., 1986—. Contbr. over 20 articles to profl. jours. Mem. Am. Chem. Soc. (vice-chmn. div. food and nutritional biochemistry 1992), Am. Oil Chemists Soc. Achievements include patents for reducing calories from fat, new class of electrolytes for lithium batteries. Office: Nabisco Brands Inc 200 Deforest Ave East Hanover NJ 07936

KLEMENT, HAIM, electronics engineer; b. Trenčin, Czechoslovakia, Jan. 17, 1935; came to U.S., 1968; s. Joseph Knopflmacher and Elizabeth (Politzer) Klement; m. Shalva Karwasser, Apr. 3, 1963; children: Joseph, Orna. BSEE, Technion, Haifa, Israel, 1964, MSEE, 1967; MS in Mgmt., Poly. Inst. N.Y., 1986. Mgr. electronics lab. Aero. Labs. Technion, 1966-68; sr. prin. engr. EforM/Honeywell, Pleasantville, N.Y., 1968-83; mgr. advanced devel. Datascope Corp., Paramus, N.J., 1983-92; mgr. vital signs techs. Criticare Systems, Inc., Waukesha, Wis., 1993—. Mem. IEEE, Assn. for Advancement Med. Instrumentation, Internat. Soc. for Optical Engring. Achievements include patent for High Accuracy Delta Modulator, fiber-optic blood detection; developer numerous medical electronic instruments used worldwide in hospitals; manager research and development of Datascope pulse oximetry technology.

KLEMPERER, WALTER GEORGE, chemistry educator, researcher; b. Saranac Lake, N.Y., Apr. 2, 1947; s. Friedrich Wilhelm and Ingeborg Eveline (Klink) K.; m. Diane Risser, Aug. 20, 1977; children: Peter, Alexander. BA, Harvard U., 1968; PhD, MIT, 1973. Postdoctoral fellow NSF, 1968, 71-73; asst. prof. chemistry Columbia U., N.Y.C., 1973-78, assoc. prof., 1978-79, prof., 1979-81; prof. U. Ill., Urbana, 1981—, Beckman Inst. prof., 1989—. Contbr. articles to Jour. Am. Chem. Soc., Materials Rsch. Soc. Symposium, Science. Alfred P. Sloan Found. fellow, 1976-78; Guggenheim fellow, 1980; Camille and Henry Dreyfus tchr.-scholar, 1978-83. Fellow AAAS; mem. Royal Soc. Chemistry, Am. Chem. Soc. Office: U Ill 505 S Mathews Ave Urbana IL 61801-3664

KLEMPNER, LARRY BRIAN, network engineer, income tax preparer; b. Newark, Feb. 21, 1956; s. Elliot Klempner and Leila Irene (Sachs) Friedman; m. Ilene Nita Levine, Aug. 17, 1980; children: Jason Sean, Jeremy Lyle. BS, N.J. Inst. Tech., 1982. Draftsman Union Carbide Corp., Boundbrook, N.J., 1977-79, E-COM Corp., Stirling, N.J., 1979-80; network engineer AT&T Network Systems, Newark, 1980-85; bldg. studies engr. Northern Telecom, Inc., Tarrytown, N.Y., 1985-86; network engr. N.Y. Tel. Co., N.Y.C., 1986—; owner Klempner Income Tax, Edison, N.J., 1990—. Mgr. Reber Softball Little League, Union, N.J., 1976-88, Midtown & Edison (N.J.)Boys Little Leagues, 1988—; coach Edison Jets Pop Warner Football, 1990; v.p. and bd. dirs. Woodedge Tenants Corp., Edison, 1988—. Mem. IEEE. Office: NY Tel Co 140 West St Rm 1490 New York NY 10007

KLEPONIS, JEROME ALBERT, dentist; b. Ashland, Pa., July 26, 1955; s. Albert Francis and Anna Mae Catherine (Burns) K. BS in Biology summa cum laude, Allentown Coll. St. Francis de Sales, 1977; DMD, U. Pa., 1981. Resident in gen. dentistry Geisinger Med. Ctr., Danville, Pa., 1981-82; assoc. Office of Dr. Stephen D. Eingorn, Bethlehem, Pa., 1982-83; dir. dental svcs. Lock Haven (Pa.) Hosp., 1983-86; sr. staff dentist Tri-Town Med. Ctr., Williamstown, Pa., 1986-87; dir. dental svcs. Embreeville Ctr., Coatesville, Pa., 1987-92, Danville (Pa.) State Hosp., 1992—. Vol. Cheater County Buddies, West Chester, Pa., 1990-92; CPR instr. Am. Heart Assn., 1989—. Named to Outstanding Young Men Am., 1988. Fellow Am. Assn. Hosp. Dentists, Acad. Gen. Dentistry; mem. ADA, Acad. Dentistry for Handicapped, Am. Soc. Geriatric Dentistry, Pa. Dental Assn., Tri-County Dental Soc., Am. Soc. Dentistry for Children, Am. Soc. Forensic Odontology, Am. Assn. Mental Retardation, Elks, Am. Hose Co., Am. Legion Sons of Vets., Allentown Coll. Alumni Assn. (bd. dirs. 1983—, sec. bd. 1985-87, pres. bd. 1991—), Psi Omega (editor Zeta chpt. 1979-81). Roman Catholic. Avocations: traveling, photography, journalism, hiking. Home: 1201 Arch St Ashland PA 17921-1213 Office: Danville State Hosp PO Box 700 Danville PA 17821-0700

KLEPPER, ELIZABETH LEE, physiologist; b. Memphis, Mar. 8, 1936; d. George Madden and Margaret Elizabeth (Lee) K. BA, Vanderbilt U., 1958; MA, Duke U., 1963, PhD, 1966. Research scientist Commonwealth Sci. and Indsl. Research Orgn., Griffith, Australia, 1966-68, Battelle Northwest Lab., Richland, Wash., 1972-76; asst. prof. Auburn (Ala.) U., 1968-72; Plant physiologist USDA Agrl. Research Service, Pendleton, Oreg., 1976-85, research leader, 1985—. Assoc. editor Crop Sci., 1977-80, 88-90, tech. editor, 1990-92, editor, 1992—; mem. editorial bd. Plant Physiology, 1977-92; mem. editorial adv. bd. Field Crops Rsch., 1983-91; mem. editorial bd. Irrigation Sci., 1987-92; contbr. articles to profl. jours., chpts. to books. Marshall scholar British Govt., 1958-59; NSF fellow, 1964-66. Fellow AAAS, Crop Sci. Soc. Am. (fellows com. 1989-91), Soil Sci. Soc. Am. (fellows com. 1986-88), Am. Soc. Agronomy (monograph com. 1983-90); mem. Sigma Xi. Home: 1454 SW 45th Pendleton OR 98701 Office: USDA Argl Rsch Svc PO Box 370 Pendleton OR 98701

KLEPPER, JOHN RICHARD, physiologist, educator; b. Dayton, Ohio, Sept. 20, 1947. BS, Ohio State U., 1969; MA, Wash. U., 1975, PhD in Physics, 1980. Rsch. asst. Biomed. Computer Lab. Wash. U., 1977-80, rsch. bioengr., 1980-81; dir. dept. physical sciences, Inst. Applied Physiology & Medicine, Wash. U., 1981—, exec. dir., 1987—; affiliated asst. prof. elec. engring. U. Wash., 1982-87. Mem. IEEE, Engring. in Medicine & Biology Soc., Am. Inst. Ultrasound Medicine. Achievements include rsch. in devel. computer aided ultrasonic imaging systems for use in medical diagnosis, ultrasonic tissue characterization, through application of computer tomographic techniques, blood flow analysis, through Doppler shift measurements. Office: Inst Applied Physiology Medicine 701 16th Ave Seattle WA 98122-4599*

KLEPPNER, DANIEL, physicist, educator; b. N.Y.C., Dec. 16, 1932; s. Otto and Beatrice (Taub) K.; m. Beatrice Spencer; children: Paul, Sofie, Andrew. BS, Williams Coll., 1953; BA, Cambridge (Eng.) U., 1955; PhD, Harvard U., 1959. Asst. prof. physics Harvard U., Cambridge, Mass., 1962-66; assoc. prof. MIT, Cambridge, 1966-73, prof., 1974—, Lester Wolfe prof. physics, 1986—, assoc. dir. Rsch. Lab. of Electronics, 1987—. Author: Introduction to Mechanics, 1973, Quick Calculus, 1986. Fellow Am. Phys. Soc. (Davisson-Germer prize 1986, Julius Edgar Lilienfeld prize 1991), AAAS, Optical Soc. Am. (William F. Meggars award 1991), Am. Acad. Arts and Scis.; mem. NAS. Office: MIT Dept Physics 77 Massachusetts Ave Rm 26237 Cambridge MA 02139-4307

KLETZKINE, PHILIPPE, aerospace engineer; b. Paris, Dec. 30, 1957; came to U.S., 1981; s. Leon and Germaine (Moser) K.; m. Wilma C. Hindriks, June 20, 1991; 1 child, Jonathan. Degree in engring., ENS Des Mines, St. Etienne, France, 1981; BS in Econs., U. St. Etienne, 1980; MS, MIT, 1983. Space analyst French Air Force, Paris, 1983-84; launch facility engr. Ctr. Nat. D'Etudes Spatiales, Kourou, French Guiana, 1984-86; space propulsion engr. European Space Agy., Noordwijk, The Netherlands, 1986-93, mgr. XMM sci. spacecraft assembly integratio and launch, 1993—; mem. examination com. Tech. U. Delft, The Netherlands, 1990—; cons. Inmarsat, London, 1990; space launch commentator Arianespace, Kourou, 1985—. Contbr. articles to ESA Jour. With French Air Force, 1982-83. Scholar French Ministry Fgn. Affairs, 1981. Mem. AIAA, MIT Club Paris. Jewish. Achievements include development of European capability for production of high-purity spacecraft propellant and components for spacecraft propulsion. Office: European Space Agy, ESTEC-PXA, Keplerlaan 1 Postbus 299, 2200 AG Noordwijk The Netherlands

KLEVATT, STEVE, production software developer. MBA in Mktg., UCLA, 1990. Formerly with Robert Abel & Assocs., Wavefront Techs.; now in charge of prodn. software MetroLight Studios, L.A. Office: MetroLight Studios Ste 400 5724 W 3rd St Los Angeles CA 90036-3078

KLEYMAN, HENRY SEMYON, electrical engineer; b. Odessa, Ukrain, Russia, Dec. 2, 1926; s. Semyon G. and Hina M. (Schwartz) K.; m. Maya G. Haskelsky, Apr. 21, 1962 (dec. Sept. 1975); 1 child, Inga Hedjazi. MSEE, Poly. Inst., Kiev, Russia, 1949. Registered profl. engr., Calif. Engr. Movie Studio, Kiev, Russia, 1949-51; sr. engr. Engring. Inst., Kiev, Russia, 1951-60; group leader, chief engr. Engring. Indsl. Co., Kiev, Russia, 1960-78; engr. Bechtel Power Corp., Norwalk, Calif., 1980-87, Stone & Webster Engring. Corp., Cherry Hills, N.J., 1987-89; sr. engr. PTS Tech. Svcs., Hurst, Tex., 1989-91, The Atlantic Group, Norfolk, Va., 1991—. Mem. IEEE. Republican. Home: 28741 Vista Santiago Rd Trabuco Canyon CA 92679 Office: The Atlantic Group Mail Station 1596 411 N Central Ave Phoenix AZ 85004

KLIBANOV, ALEXANDER MAXIM, chemistry and biotechnology educator, researcher; b. Moscow, July 15, 1949; U.S., 1977, naturalized, 1983; s. Maxim and Eugenia (Tomas) K.; m. Margarita Romanycheva, Apr. 21, 1972; 1 child, Tanya. MS, Moscow U., 1971, PhD, 1974. Rsch. chemist Moscow U., 1974-77; postgrad. rsch. chemist U. Calif.-San Diego, 1978-79; asst. prof. applied biological sci. dept. MIT, Cambridge, 1979-83, H.L. Doherty prof., 1981-83, assoc. prof., 1983-87, prof., 1987-88, prof. chemistry dept., 1988—; cons. in field. Contbr. over 160 articles to profl. jours.; mem. editorial bd. Applied Biochemistry and Biotech, 1981—, Advances Biochem. Engring./Biotechnol., 1985—, Chimicaoggi, 1986—, Biocatalysis, 1988—, Applied Biocatalysis, 1989—, Biotechnology Progress, 1990—. Recipient Internat. Enzyme Engring. prize Engring. Found., 1993; grantee numerous orgns. Fellow Am. Inst. Med. and Biol. Engring. (founding); mem. NAE, Am. Chem. Soc. (Leo Friend award 1986, Ipatieff prize 1989, Marvin J. Johnson award 1991, Arthur C. Cope Scholar award, 1993), Am. Soc. Biochemistry and Molecular Biol. Jewish. Research on protein stability and stblzn., immobilized enzymes and cells, enzymes as catalysts in organic chemistry, nonaqueous biochemistry, enzymes in extreme environments. Home: 61 W Boulevard Rd Newton MA 02159-1218 Office: MIT Bldg 16-209 77 Massachusetts Ave Cambridge MA 02139

KLIEFOTH, A(RTHUR) BERNHARD, III, neurosurgeon; b. San Antonio, Nov. 26, 1942; s. Arthur Bernhard, Jr. and Pauline (Grey) K.; m. Ingrid R. Kunde, Apr. 22, 1968; children: Karena, Tanya. AB in Chemistry, Princeton U., 1965; M.D. U. Tex., 1970. Diplomate Am. Bd. Neurol. Surgery. Intern, Naval Hosp., Oakland, Calif., 1970-71; resident gen. surgery Naval Hosp., San Diego, 1972-73; neurosurg. tng. Washington U., St. Louis, 1973-78;

research fellow dept. radiation scis. Washington U., 1977-78; commd. ensign U.S. Navy, 1969, advanced through grades to comdr., 1977; staff neurosurgeon Naval Regional Med. Ctr., Oakland, 1978-81; resigned, 1981; capt. USNR, 1985; practice medicine specializing in neurosurgery, Knoxville, Tenn., 1981—; mem. staff U. Tenn. Hosp., St. Mary's Hosp., chmn. dept. surgery, 1989-90; clin. assoc. prof. surgery, U. Tenn.; bd. dirs. Tenn. Donor Svcs. Bd. dirs. Cole Neurosci. Found., Knoxville Donor Svcs.; pres. Princeton Alumni Assn. Knoxville and Eastern Tenn. Fellow ACS, Stroke Council Am. Heart Assn.; mem. AMA, Am. Assn. Neurol. Surgeons, Am. Soc. Stereotactic and Functional Neurosurgery, World Soc. Stereotactic, and Functional Neurosurgery, Congress Neurol. Surgeons, So. Neurosurgical Soc., So. Med. Assn., Tenn. Med. Assn., Knoxville Acad. Medicine, San Francisco Neurol. Soc., Soc. Med. Cons. to Armed Forces, Assn. Mil. Surgeons U.S., Soc. for Neurosci. Office: 1932 Alcoa Hwy Ste 550 Knoxville TN 37920-1573

KLIESCH, WILLIAM FRANK, physician; b. Franklinton, La., Nov. 4, 1928; s. Edward Granville and Elsie Jeni (Sylvest) K.; m. May Virginia Reid, Dec. 17, 1955; children: Thomas Karl, William August, John Francis. BS, La. State U., 1949, MD, 1953. Intern Valley Forge Hosp., Phoenixville, Pa., 1953-54; intern in med. rsch. Charity Hosp., New Orleans, 1956-57; resident, fellow in internal medicine Ochsner Found. Hosp., New Orleans, 1957-59; pvt. practice New Orleans, 1959-69, Jackson, Miss., 1969—; dir. spinal injury svc. Miss. Meth. Rehab. Ctr., Jackson, 1980—. Capt. U.S Air Force, 1953-56. Fellow Am. Coll. Emergency Physicians; mem. Am. Spinal Injury Assn., Internat. Paraplegia Soc. Episcopalian. Avocations: gardening, farming. Home: 8892 Gary Rd Jackson MS 39212-9732 Office: Miss Meth Rehab Vocat 1350 E Woodrow Wilson Ave Jackson MS 39216-5198

KLIMA, JON EDWARD, civil engineering educator, consultant; b. Cedar Rapids, Iowa, Nov. 10, 1939; s. William and Anna Marie Klima; m. Nancy Ann Neibergall, Nov. 27, 1962; children: Larry Jon, Tara Lynn. BSEE, Iowa State U., 1962; MSCE, U. Colo., 1987. Sr. staff engr. Martin Marietta Corp., Denver, 1962-80; prof. Red Rocks Community Coll., Lakewood, Colo., 1980—. Author: The Solar Controls Book, 1982; contbr. articles to profl. jours. Mem. Colo. Solar Energy Industries Assn. (sec. 1990—, Sunny award 1992). Office: Red Rocks Community Coll 13300 W 6th Ave Lakewood CO 80401

KLIMENT, ROBERT MICHAEL, architect; b. Prague, Czechoslovakia, June 9, 1933; came to U.S., 1950; s. Felix and Sophie (Baltinester) K.; m. Janet McClure, Sept. 12, 1959 (div. 1968); 1 child, Nicholas McClure; m. Frances Halsband, May 1, 1971; 1 child, Alexander Halsband. BA, Yale U., 1954, MArch, 1959. Registered architect Penn., N.Y., N.J., Mass., Conn., Ohio, Va., D.C., N.C., N.H., Md.; cert. Nat. Coun. Archtl. Registration Bds. Architect Mitchell/Giurgola Architects, Phila., 1961-66; architect, assoc. Mitchell/Giurgola Architects, N.Y.C., 1967-71; ptnr. R.M. Kliment Architect, N.Y.C., 1972-78, R.M. Kliment & Frances Halsband Architects, N.Y.C., 1978—; instr. U. Pa., Phila., 1963-66, vis. prof., 1972-73; asst. prof. Columbia U., N.Y.C., 1966-70, vis. prof., 1977, 84; vis. prof. MIT, Cambridge, Mass., 1970, Yale U., New Haven, 1972-74, N.C. State U., Raleigh, 1978, Rice U., Houston, 1979, U. Va., Charlottesville, 1979-80, Harvard U., Cambridge, 1980-81. Works include Computer Sci. Bldg. Princeton U., Salisbury (Conn.) Town Hall, U. Va. Life Scis. Bldg., Columbia U. Computer Scis. Bldg. (award NYSAA 1985, Tucker award Bldg. Stone Inst. 1985, other awards), Burke Chemistry Bldg., Dartmouth Coll., Adelbert Adminstrn. Bldg., Case Western Res. U, hdqrs. Marsh & McLennan Co., McCarter & English Law Office; works exhbtd. at Bklyn. Mus., 1977, The Drawing Ctr., 1977, Cooper Hewitt Mus., 1977, 78, Mus. Finnish Architecture, Helsinki, Finland, 1980, Harvard Grad. Sch. Design, 1981, NAD, 1981, 87, Smith Coll. Mus. Art, 1981, Rice U. Farrish Hall Gallery, 1983, Columbia U. Low Libr., 1986, Parrish Art Mus., 1987, German Architecture Mus., Frankfurt, 1989. With U.S. Army, 1955-57. Fulbright scholar, Italy, 1959-60; recipient Bard award for excellence in architecture City Club N.Y., 1989. Fellow AIA (Residential Design award N.Y.C. chpt. 1979, Disting. Architecture award 1985, Nat. Honor award 1987), Century Assn. Office: R M Kliment & Frances Halsband Architects 255 W 26th St New York NY 10001-6735

KLIMKOWSKI, ROBERT JOHN, photo reproduction process technical executive; b. Chicago, Mar. 11, 1930; s. Bruno Joseph and Harriett Marie (Lachowicz) K.; m. Dolores Loretta Ptasinski, Dec. 30, 1950; children: Mark, Doriane, Roberta, Paul, James. BS in Chemistry, De Paul Univ., 1951, MS in Chemistry, 1958. Pres. Canadian Thermal Images, Schiller Park, Ill., 1969-74, Reproduction Products, Schiller Park, Ill., 1969-74; dir. R&D Dietzgen Corp., Des Plaines, Ill., 1974-80, St. Regis Paper Co., Troy, Ohio, 1980-84; cons. Dietzgen Corp., Chgo., 1984-86, sr. chemist, 1986-89, dir. environ. affairs, 1989-92; cons., 1993—. Achievements include 20 patents in the fields of copying and duplicating. Home: 1285 Sequois Ct Apt A Tipp City OH 45371

KLIMOWICZ, THOMAS F., metallurgist; b. Bayonne, N.J., Sept. 7, 1955; s. Stanley Walter and Lorraine (Runkewicz) K.; m. Gabriele M. Gerbautz, June 9, 1986; 1 child, Jennifer. BS, MIT, 1977; MS, U. Calif., Berkeley, 1979, MBA, 1983. Mem. tech. staff Sandia Nat. Labs., Livermore, Calif., 1979-82; prodn. mgr. Raychem Corp., Livermore, 1982-87; mgr. quality control Duralcan U.S.A., San Diego, 1988—. Contbr. articles to profl. jours. Mem. Am. Soc. Metals, Minerals-Metals and Materials Soc., Tau Beta Pi. Office: Duralcan USA 10326 Roselle St San Diego CA 92121

KLINE, ARTHUR JONATHAN, electronics engineer; b. Prescott, Ariz., Feb. 29, 1928; s. Sarah Ann (Odell) Kline; m. Marilyn Sue Lane, Nov. 14, 1959; children: Lee Ann, Jonathan Lane. BS in Engring. Physics, U. Calif., Berkeley, 1953; MS in Applied Physics, UCLA, 1956. Electronics engr. Radio Corp. Am., Moorestown, N.J., 1953-55; electronics engr. Motorola Inc., Scottsdale, Ariz., 1956-93, retired, 1993; assoc. mem. Motorola Sci. Adv. Bd., 1972. With USN, 1946-52. Dan Noble fellow, 1975. Fellow AIAA (assoc., chmn. Phoenix chpt. 1968-69). Home: 6453 E Montcrosa St Scottsdale AZ 85251-3136

KLINE, DAVID GELLINGER, neurosurgery educator; b. Phila., Oct. 13, 1934; s. David Francis and Lois Ann (Gellinger) K.; m. Carol Anne Loewen, Mar. 1, 1958 (div.); children: Susan, Robert, Nancy. AB in Chemistry, U. Pa., 1956, MD, 1960. Diplomate Am. Bd. Neurol. Surgery (sec.-treas. 1978-83, chmn. 1983-84, adv. bd. 1984-90, chmn. 50th anniversary celebration 1990). Intern and resident in gen. surgery U. Mich., Ann Arbor, 1960-62; research investigator Walter Reed Army Inst. Research and Walter Reed Gen. Hosp., 1962-64; resident in neurosurgery and teaching instr. U. Mich., Ann Arbor, 1964-67; instr. neurosurgery and surgery Sch. Medicine La. State U., New Orleans, 1967-68, asst. prof., 1968-70, assoc. prof., 1970-75, prof., 1975—, head sect. of neurosurgery, 1971, chmn. dept. neurosurgery, 1976—; cons. USPHS Health Center Hosp., New Orleans VA Hosp., Kessler AFB Hosp.-Lederle Labs.; vis. investigator Delta Regional Primate Center, Covington; mem. Am. Bd. Med. Specialists, 1978-86, mem. residency rev. com., 1977-84; lectr. in field. Contbr. articles to sci. jours., also mem. numerous editorial bds. Capt. M.C. AUS, 1962-64. Recipient Frederick Coller Surg. prize, 1967; numerous grants. Mem. AMA, ACS, Orleans Parish Med. Society, La. State Med. Soc., New Orleans Neurol. Soc., Am. Acad. Neurol. Surgery, Soc. Neurol. Surgeons (treas. 1986-91), So. Neurol. Surgery Soc. (sec. 1976-79, pres. 1985-86), Am. Assn. Neurol. Surgeons (bd. dirs. 1985-89), Soc. Univ. Neurosurgeons, Congress Neurol. Surgeons, Assn. Acad. Surgery, Surg. Biol. Club II, Soc. Univ. Surgeons, Sunderland Club (pres. 1981), Phi Beta Kappa, Kappa Sigma, Phi Chi. Episcopalian (vestry and lay reader). Home: 307 Fairway Dr New Orleans LA 70124-1020 Office: La State U Med Ctr 1542 Tulane Ave New Orleans LA 70112-2865

KLINE, GORDON MABEY, chemist, editor; b. Trenton, N.J., Feb. 9, 1903; s. Manuel Kuhl and Florence (Campbell) K.; m. Dorothy Beard, Mar. 15, 1926; 1 child, Ann Linthicum (Mrs. Robert True Cook). A.B., Colgate U., 1925; M.S., George Washington U., 1926; Ph.D., U. Md., 1934. Research chemist N.Y. State Dept. Health, 1926-27; research chemist Picatinny Arsenal, 1928-29; chemist, phys. sci. adminstr. Nat. Bur. Standards, Washington, 1929-69; chief organic plastics sect. Nat. Bur. Standards, 1935-51, chief div. polymers, 1951-63, cons., 1964-69; tech. editor Modern Plastics Mag., 1936-90, tech. editor emeritus, 1991—; editorial dir., cons. Modern Plastics Ency., 1936-90; tech. investigator with U.S. Army, ETO, 1945; chmn. tech. com. on plastics Internat. Standardization Orgn., ann. meetings, N.Y.C., 1951, Turin, Italy, 1952, Stockholm, 1953, Brighton, Eng., 1954, Paris, 1955, The Hague, Netherlands, 1956, Burgenstock, Switzerland, 1957, Washington, 1958, U.S. del., Munich, 1959, Prague, 1960, 84, Turin, 1961, Warsaw, 1962, London, 1963, Budapest, 1964, Bucharest, 1965, Stockholm, 1966, 86, Phila., 1968, Prague, 1969, Paris, 1970, Moscow, 1971, Baden-Baden, 1972, Montreux, 1973, Tokyo, 1974, Pugnochiuso, Italy, 1975, Ottawa, 1976, London, 1977, Madrid, 1978, Budapest, 1980, Orlando, Fla., 1981, 90, The Hague, 1983, Warsaw, 1984; sec. div. plastics and polymers Internat. Union Pure and Applied Chemistry, 1951-59, vice chmn., 1959-63, chmn., 1963-67, mem. macro-molecules div., 1967-75; observer for ISO, 1976-86; plastics adv. council, Princeton, 1957-65. Editor: Analytical Chemistry of Polymers, Part I, 1959, Parts II and III, 1961; Contbr. articles profl. jours. Recipient Honor award Am. Inst. Chemists, Washington sect., 1952, Exceptional Service Gold medal Dept. Commerce, 1953, award Standards Engrs. Soc., 1964, Rosa award Nat. Bur. Standards, 1965, Meritorious Service award Am. Nat. Standards Inst., 1987; charter mem. Plastics Hall of Fame, 1973. Mem. ASTM (award of merit 1954; D-20 award of excellence 1986), Am. Chem. Soc., Am. Inst. Chemists, Soc. Plastics Engrs., Soc. Plastics Industry, Soc. Plastics Pioneers, Phi Beta Kappa, Sigma Xi. Clubs: Cosmos (Washington); Chemists (N.Y.C.). Home: 3063 Donnelly Dr Apt 318C Lake Worth FL 33462

KLINE, MILTON VANCE, psychologist, educator; b. Bklyn., Mar. 25, 1923; s. Joseph and Elizabeth (Zimmerman) K.; B.A., Pa. State U., 1944; M.A., Columbia U., 1945; Ph.D., Western U., 1952; m. Dorothy Weller, Feb. 25, 1952; 1 dau., Jill. Chief psychologist Westchester Health Dept., White Plains, N.Y., 1948; dir. research dept. psychology L.I.U., 1948, research project dir. grad. sch., 1953; research cons. VA, N.Y.C., 1950; pvt. practice hypnoanalysis N.Y.C., 1950—, professorial lectr. Seton Hall Med. Sch., 1960-62; pres. Inst. for Research in Hypnosis, N.Y.C., 1954—; lectr. Fairleigh Dickinson U., Rutherford, N.J., 1964—; dir. Morton Prince Clinic for Hypnotherapy; pres. Morton Prince Services, N.Y.C. and Balt., dir. substance abuse research lab., N.Y.C. and Mt. Kisco, 1987—, dir. employee asstance program consultation service, Mt. Kisco, 1987—; prof. med. psychology and hypnosis U. Milan New Sch. Medicine, Italy, 1988—; dir. Forensic Hypnosis Research and Consultation Center, Inst. for Research in Hypnosis, N.Y.C. and Mt. Kisco, N.Y.; co-dir. Internat. Soc. for Med. and Psychological Hypnosis; cons. NBC, Nat. Assn. Broadcasters; pres. Internat. Grad. U., Switzerland. Chmn. council sci. and profl. advs. Internat. Grad. Sch. Behavioral Scis., Switzerland; bd. advs. Am. Bd. Psychotherapy, 1982. Served with AUS, 1942-44. Recipient award for best book in hypnosis Soc. Clin. and Exptl. Hypnosis, 1958, 67, award for Best Paper on clin. hypnosis, 1973, Roy M. Dorcus award, 1971. Fellow Am. Med. Writers Assn., Acad. Psychosomatic Medicine; mem. Soc. Clin. and Exptl. Hypnosis (pres. 1961-63), Am. Psychol. Assn. (council reps.), Internat. Soc. Med. and Psychol. Hypnosis (bd. advisers), Authors Guild, Author: Hypnodynamic Psychology, 1955, A Scientific Report on The Search for Bridey Murphy, 1953; Freud and Hypnosis, 1958; The Nature of Hypnosis, 1962; Clinical Correlations of Experimental Hypnosis, 1965; Psychodynamics and Hypnosis, 1967; Forensic Hypnosis, 1983, Short Term Hypnotherapy and Hypnoanalysis, 1992; editor: Obesity: Etiology, Treatment and Research (C.C. Thomas), 1975. Editor emeritus Internat. Jour. Clin. and Exptl. Hypnosis. Home: 15 Kerry Ln Chappaqua NY 10514-1606 Office: 1991 Broadway New York NY 10023

KLINEBERG, JOHN MICHAEL, federal agency administrator, aerospace researcher; b. N.Y.C., Oct. 16, 1938; s. Otto and Selma Klineberg; m. Anne-Marie Michelle Mellet, Feb. 3, 1967; children: Eric, Arnaud, Logan. BS in Engring., Princeton U., 1960; MS, Calif. Inst. Tech., 1962, PhD, 1968. Engr. Douglas Aircraft Co., Santa Monica, Calif., 1960-62; grad. rsch. asst. Calif. Inst. Tech., Pasadena, 1962-68, rsch. engr., 1968-70; aerospace engr. NASA Ames Rsch Ctr., Moffett Field, Calif., 1970-74; aerospace engr. NASA Hdqrs., Washington, 1974-78, dep. assoc. adminstr. aeronautics and space technology, 1978-79; dep. dir. NASA Lewis Rsch. Ctr., Cleve., 1979-86, acting dir., 1986-87, dir., 1987—. Mem. corp. adv. coun. U. Cin., 1988—; bd. dirs. Greater Cleve. Growth Assn., 1987—, mem. air svc. devel. com., 1981—; mem. Coun. Great Lakes Govs.' Econ. Devel. Commn., 1988—, Ohio Gov.'s R&D Adv. Group, 1987—; mem. tech. leadership coun. Cleve. Tomorrow, 1988—; mem. Leadership Cleve., 1984; mem/ policy com. Cleve. Fed. Exec. Bd., 1986—. Recipient Outstanding Leadership medal NASA, 1982; named Meritorious Exec., U.S. Govt., 1986. Fellow AIAA; mem. Cleve. Engring. Soc., Nat. Space Club, (bd. govs. 1987—), Sigma Xi. Avocations: music, economic development. Home: 8900 Clifford Ave Bethesda MD 20815-4743 Office: NASA Lewis Research Ctr 21000 Brookpark Rd Cleveland OH 44135-3191

KLINGBIEL, PAUL HERMAN, information science consultant; b. Watertown, Wis., Nov. 3, 1919; s. Herman Carl and Elsa Helen (Zilisch) K.; Ph.B., U. Chgo., 1948, B.S., 1950; M.A., Am. U., 1966; m. Mildred Louise Wells, Nov. 30, 1968; stepchildren—Alice J. Blessley, Jo Ann Grayson. Abstractor, Armed Services Tech. Info. Agy., Dept. Def., Washington, 1953-58, editor Tech. Abstract Bull., 1958-60, dir. Office of Lexicography, 1960-66; phys. sci. adminstr., linguistics research Def. Documentation Center, 1966-79; sr. cons. Aspen Systems Corp., 1979-81; systems analyst PRC Data Services Co., Linthicum Heights, Md., 1981-82; lectr. Am. U., Washington, 1966-69; cons. div. med. scis. Nat. Acad. Scis., 1969-70. Served with AUS, 1943-46. Recipient Meritorious Civilian Service award, 1974, Disting. Career award, 1979. Fellow AAAS; mem. Assn. Computational Linguistics. Lutheran. Contbr. articles to profl. jours. Research in field of computational linguistics. Home: 2435 Sumatran Way Clearwater FL 34623-1824

KLINGEN, LEO H., computer science educator; b. Viersen, Germany, Mar. 30, 1926; s. Michael and Emma (Hertzer) K.; m. Ingrid F. Watzka, Oct. 29, 1960; children: Carmen C., Christa C., Claudia C., Karin M. State exam, U. Göttingen, Fed. Republic Germany, 1951, PhD, 1955. Tchr. grammer sch., Gummersbach, Fed. Republic Germany, 1953-56, Colegio Alemán, La Paz, Bolivia, 1957-61; catedrático titular U. San Andres, La Paz, 1958-61; dep. Helmholtz Gymnasium, Bonn, Fed. Republic Germany, 1962-76, prin., 1976-90; ret., 1990; prof. computer sci. in edn. U. Cologne, Fed. Republic Germany, 1972—; prof. computer sci. U. Bonn, 1990—; developer math. curriculum North Rhine Westphalian Ministry Edn., Düsseldorf, Fed. Republic Germany, 1972-80. Author, editor books on computer sci., 1980—. Home: Billrothstrasse 2, 5300 Bonn 1, Germany

KLINK, PAUL L., computer company executive; b. Auburn, N.Y., July 28, 1965; s. Charles Lawrence and Regina Joyce (Maniscalco) K. Student, SUNY, Cayuga, 1979-84. Pres., chief exec. officer Info. Tech., Honolulu, 1979—; v.p. direct mktg. mgrs. divsn. Milici Valenti Gabriel DDB Needham, Honolulu, 1979—; 2d v.p. external AD/steering com. Japanese C. of C., 1992-93. Contbr. and edited articles for profl. jours. Co-chmn. Aloha United Way, Honolulu, 1989—; active computer affairs Friends of Rep. Paul O'Shiro, Ewa Beach, Hawaii, 1988—, Friends of Gov. John Waihee, Honolulu, 1988—, Friends of Mayor Frank Fasi, Honolulu, 1988—; bd. dirs. Postal Customers Com., 1992-93. Mem. Info. Industry Assn. (dir. membership 1989—), Direct Mktg. and Advt. Assn. Hawaii (bd. dirs.), Database Mgmt. Assn. Hawaii (direct mktg. com.), Puualoa Rifle and Pistol Club, Mensa, Honolulu Club, La Marianas Sailing Club. Avocations: skiing, target shooting, reading, hiking, photography. Office: Info Tech USA Inc 330 Saratoga Rd # 88817 330 Saratoga Rd Box 88817 Honolulu HI 96815-9998

KLINKMULLER, ERICH, economist; b. Berlin, Oct. 21, 1928; m. Christa Edelbauer, June 11, 1964; children: Andreas, Veronika. Dr. rer. pol., Free U. Berlin, 1959. Rsch. fellow Harvard U., 1959-60; asst. prof. econs. Free U. Berlin, 1960-66; lectr. econs. U. Calif., Santa Barbara, 1966-67; prof. econs. U. Ariz., Tucson, 1967-68; assoc. prof. econs. St. Louis U., 1968-70; prof. econs. Free U. Berlin, 1970—. Author: Economic Cooperation Between CMEA Countries, 1960, Interdisciplinary Research Among the Social Sciences in the U.S. and Elsewhere, 1986, Soviet Russia Is No Super-power, 1988, Econ. History of Russia, 1845-88, 1992; contbr. articles to profl. jours. Mem. Am. Econ. Assn., Verein fur Socialpolitik, Berliner Wissensenaftliche Gesellschaft. Home: 34 A Bahnhofstrasse, H 12 207 Berlin Germany Office: Free U, 55 Garystrasse, H 14 195 Berlin Germany

KLIOUEV, VLADIMIR VLADIMIROVITCH, control systems scientist; b. Moscow, Russia, Jan. 2, 1937; s. Vladimir Matveevitch and Anna Danilovna K.; m. Larisa Mikhailovna Degtereva, July 23, 1960; children: Serguei, Zakhar. Degree in engring., Moscow State Tech. U., 1960, MSc in Engring., 1964, DSc, 1973. Engr., sci. worker Moscow State Tech. U., 1960-64, sr. sci. worker, head lab., head dept. Inst. Introscopy, 1964-70, dir. Inst. Introscopy, 1970—; gen. dir. Moscow Sci. Indsl. Assn. Spectrum, 1976—; chmn. ISO/TC 135 Non-Destructive Testing, Geneva, 1980-92; mem. presidium Znanie, 1978—, Highest Certifying Com. Russia, 1989—. Author: Test Equipment, 1982, Equipment for NDT of Materials, 1986, Technical Means for Diagnostics, 1989, X-Ray Engineering, 1992; mem. editorial bd. Defectoscopia, 1970—, European Jour. Non-Destructive Testing, 1990—. Mem. Russian Acad. Scis. (corr.), Academia Europena, Russian Soc. for Non-Destructive Testing and Tech. Diagnostics (pres. 1990—), Internat. Com. for Non-Destrictive Testing, European Com. for Non-Destructive Testing, Sci. Coun. on Automated Systems Diagnostics. Achievements include 75 Russian patents in field. Office: Moscow Sci Indls Assn Spectrum, 35 St Usacheva, 119048 Moscow Russia

KLITZING, KLAUS VON, institute administrator, physicist; b. Schroda, June 28, 1943; s. Bogislav and Anny (Ulbrich) von K.; m. Renate Falkenberg, May 27, 1971; children: Andreas, Christine, Thomas. Diploma, Tech. U. Braunschweig, 1969; Ph.D., U. Wuerzburg, 1972; Habilitation, 1978. Faculty mem. Tech. U., Munich, 1980-84; dir. Max Planck Inst. for Festkörperforschung, Stuttgart, Fed. Republic Germany, 1985—. Recipient Schottky prize Deutsche Phys. Gesellschaft, 1981, Hewlett Packard prize European Phys. Soc., 1982, Nobel prize in physics Royal Swedish Acad. Sci., 1985. Office: Max Planck Inst fur Festkörperforschung, Heisenbergstr 1, D-7000 D-70569 Germany

KLOEHN, RALPH ANTHONY, plastic surgeon; b. Milw., Dec. 18, 1932; s. Ralph Charles and Virginia Mary (Kosak) K.; m. Mary Theresa Landers, Nov. 4, 1961; Children: Colleen, Gregory, Kristine, Patricia, Timothy, Philip, Michelle. BS, Marquette U., 1954, MD, 1958. Diplomate Am. Bd. Plastic Surgery. Rotating intern Charity Hosp. La., New Orleans, 1958-59; gen. surgery resident Marquette U. Hosps., Milw., 1961-65; resident in plastic and maxillofacial surgery U. Tex. Med. Br., Galveston, 1965-68; fellowship in plastic and reconstructive surgery African Med. Rsch. Found., Nairobi, Kenya, 1968-69; pvt. practice medicine specializing in plastic surgery Milw., 1969—; med. cons. Surgitron Internat., Inc., Poway, Calif. Contbr. articles to profl. jours. Lt. USNR, 1959-61. Fellow ACS, Internat. Coll. Surgeons; mem. AMA, Am. Soc. Aesthetic Plastic Surgery, Am. Soc. Plastic and Reconstructive Surgery, Singleton Surgical Soc., Am. Soc. Maxillofacial Surgeons, Can. Soc. Aesthetic for (Cosmetic) Plastic Surgery. Republican. Roman Catholic. Avocations: photography, sports fishing. Home: N14 W 30082 High Ridge Rd # 5 Pewaukee WI 53072 Office: Affiliated Cosmetic and Plastic Surgeons 2323 N Mayfair Rd Ste 503 Milwaukee WI 53226-1507

KLOHS, MURLE WILLIAM, chemist, consultant; b. Aberdeen, S.D., Dec. 24, 1920; s. William Henry and Lowell (Lewis) K.; m. Dolores Catherine Borm, June 16, 1946; children: Wendy C., Linda L. Student Westmar Coll., 1938-40; BSc, U. Notre Dame, 1947. Jr. chemist Harrower Lab., Glendale, Calif., 1947, Rexall Drug Co., L.A., 1947-49; sr. chemist Riker Labs., Inc., L.A., 1949-57, dir. medicinal chemistry, Northridge, Calif., 1957-69, mgr. chem. rsch. dept., 1969-72, mgr. pharm. devel. dept., 1972-73, mgr. tech. liaison and comml. devel., 1973-82; cons. chemist, 1982—. Contbr. articles to profl. jours. Served to lt. USNR, 1943-46. Riker fellow Harvard U., 1950. Mem. Am. Chem. Soc., Am. Pharm. Assn., Adventures Club (L.A.). Home and Office: 19831 Echo Blue Dr Lake Wildwood Penn Valley CA 95946

KLÖPFFER, WALTER, chemist, educator; b. Graz, Austria, June 6, 1938; arrived in Germany, 1964; s. Otto and Josefine (Lammer) K.; m. Hedwig Gerlinde, Dec. 16, 1964; 1 child, Astrid Eva. PhD, U. Graz, 1964; prof., U. Mainz, 1975. Researcher Battelle Inst., Frankfurt, Germany, 1964-81, mgr., 1981-85, rsch. leader, 1985-91; rsch. leader Soc. for Environ. Cons. and Analysis Ltd., Frankfurt, 1992—; prof. U. Mainz, 1975—; expert OECD Chem. Group, Paris, 1978-82; steering com. on life cycle analysis Setac Europe, 1992-93. Author: Polymer Spectroscopy, 1984; contbr. articles to profl. jours. Mem. Ges. Oesterr. Chemiker, Ges. Deutsch. Chemiker, Am. Chem. Soc., Bunsenges. Phys. Chem., European Photochemistry Assn., Soc. Environ. Toxicology and Chemistry. Roman Catholic. Achievements include discovery of excitons in polymers; development of testing and evaluation methods for environmental chemicals, life cycle analysis. Home: Am Dachsberg 56E, D-60435 Frankfurt Germany Office: CAU GmbH, Am Roemerhof 35, D-60486 Frankfurt Germany

KLOSINSKI, DEANNA DUPREE, medical laboratory sciences educator; b. Goshen, Ind., Dec. 28, 1941; d. George C. and Gertrude (Todd) Dupree; m. Michael A. Klosinski, Jan. 30, 1965; children: Elizabeth, John, Robert, Lara. BS, Ind. State U., 1964; MS, Purdue U., 1972; PhD, Wayne State U., 1990. Diplomate Am. Soc. Clin. Pathologists; cert. med. technologist. Med. technologist South Bend (Ind.) Med. Found., 1959-68; lab. specialist Home Hosp., Lafayette, Ind., 1968-74; program dir. Ind. Vocat. Tech. Coll., Lafayette, 1968-75; clin. asst. prof. Oakland U., Rochester, Mich., 1985—; adj. asst. prof. Wayne State U., Detroit, 1991—; program dir. William Beaumont Hosp., Royal Oak, Mich., 1979—. Author: (videotape, monograph) Blood Collection: The Difficult Draw, 1992; co-author: (videotape, monograph) Blood Collection: The Routine Venipuncture, 1989 (chpt.) Molecular Biology and Pathology, 1993. Mem. pastoral coun. St. Hugo Cath. Ch., Bloomfield, Mich., 1991--. Named Outstanding Bus. Person Mich. Coun. on Vocat., 1992, Mich. Clinical Lab. Scientist, 1993; rsch. grantee William Beaumont Hosp. 1989-90. Mem. Am. Soc. Clin. Pathologists (chmn. tech. sample. 1986-93), Am. Soc. for Med. Technology (edn. sci. assembly), Mich. Soc. for Med. Tech. (treas. 1984-86, 88-92), Assn. Women in Sci., Sigma Xi, Alpha Mu Tau (scholarship award 1985, 87, 90), Delta Gamma Alumnae (v.p. 1991-93, pres. 1993—). Home: 90 Devon Rd Bloomfield Hills MI 48302 Office: William Beaumont Hosp 3601 W 13 Mile Rd Royal Oak MI 48073-6769

KLOSTERMAN, ALBERT LEONARD, technical development business executive, mechanical engineer; b. Cin., Oct. 22, 1942; s. Albert Clement and Mary J. Klosterman; m. Lynne Marie Gabelein, Jan. 4, 1964; children: Scott, Lance, Kimberly, Brad. BSMechE, U. Cin., 1965, MSMechE, 1968, PhD, 1971. Instr. U. Cin., 1966-70, adj. assoc. prof., 1974—; project mgr. Structural Dynamics Rsch. Corp., Milford, Ohio, 1970-72, mem. tech. staff, 1972-73, dir. tech. staff, 1973-78, v.p., gen. mgr., 1978-83, sr. v.p., chief tech. officer, gen. mgr., 1983—; mem. exec. steering com. Initial Graphics Exchange System/Product Data Exch. Speification of Nat. Standards Bd., Gaithersburg, Md., 1984—. Mem. editorial bd. Internat. Jour. Vehicle Design, 1979—. Recipient Disting. Alumnus award U. Cin., 1988. Mem. Assn. Computing Machinery, ASME (assoc.), Phi Kappa Theta. Republican. Roman Catholic. Home: 5444 Forest Ridge Cir Milford OH 45150-2821 Office: Structural Dynamics Rsch Corp 2000 Eastman Dr Milford OH 45150-2740

KLOTZ, IRVING MYRON, chemist, educator; b. Chgo., Jan. 22, 1916; s. Frank and Mollie (Nasatir) K.; m. Mary Sue Hanlon, Aug. 7, 1966; children: Edward, Audie Jeanne, David. B.S., U. Chgo., 1937, Ph.D., 1940. Rsch. assoc. in chemistry Northwestern U., 1940-42, instr., 1942- 46, asst. prof., 1946-47, assoc. prof., 1947-50, prof., 1950-63, Morrison prof. chemistry, 1963-86, prof. emeritus, 1986—; Lalor fellow Marine Biol. Lab., Woods Hole, Mass., 1947-48, corp. mem., 1947—, trustee, 1957-65. Author: Chemical Thermodynamics, 4th rev. edit., 1986, Energies in Biochemical Reactions, rev. edit., 1967, Introduction to Biomolecular Energetics, 1986; Diamond Dealers, Feather Merchants, 1986; articles sci. jours. Recipient Army-Navy cert. of appreciation for wartime research, 1948, William C. Rose award biochem. Am. Soc. Biochem. and Molecular Biology, 1993. Fellow Royal Soc. Medicine, Am. Acad. Arts and Scis., AAAS; mem. Nat. Acad. Scis., Am. Soc. Biol. Chemists, Am. Chem. Soc. (Eli Lilly award 1949, Midwest award 1970), Phi Beta Kappa, Sigma Xi, Phi Lambda Upsilon, Alpha Chi Sigma. Home: 2515 Pioneer Rd Evanston IL 60201-2203

KLOTZ, LOUIS HERMAN, structural engineer, educator, consultant; b. Elizabeth, N.J., May 21, 1928; s. Herman Martin and Edna Theresa (Kloepfer) K.; m. Virginia Helen Roll, Apr. 3, 1966; Emily Louise, Jennifer-

Claire Virginia. BSCE, Pa. State U., 1951; MCE, N.Y.U., 1956; PhD, Rutgers U., 1967. Registered profl. engr., N.J., N.H. Structural engr. various firms, N.Y., N.J. metro area, 1951-65; asst. prof. civil engring. dept. civil engring U. N.H., Durham, 1965-69, assoc. prof., 1969-86, chmn., 1971-74; spl. projects dir. ASCE, N.Y.C., 1986-87; cons. Klotz Assocs., Inc., New Castle, N.H., 1987-88; project mgr. Universal Engring. Corp., Boston, 1988-91; exec. dir. New Eng. States Earthquake Consurtium, 1991—; cons., evaluator Office of Energy Related Inventions, Gaithersburg, Md., 1978—; mem. energy policy adv. group N.H. Ho. of Reps., Concord, 1979-82; founding mem. N.H. Legis. Acad. Sci. & Tech., Concord, 1980-83. Editor: Energy Sources, The Promises and Problems, 1980; author: Users Manual Small Hydroelectric Financial/Economic Analysis, 1983; (monograph) Water Power, Its Promises and Problems; contbr. articles to Procs. of 1st Internat. Conf. on Computing in Civil Engring., Hydro Rev. Advisor Environ. Protection div. N.H. State Atty. Gen.'s Office, Concord, 1972-76; mem. New Castle (N.H.) Budget Com., 1977-79; tech. reviewer N.E. Appropriate Tech. Small Grants program Dept. Energy, Boston, 1979-80. Recipient Ford Found fellowship, 1962-65, Ford Found. grant, 1968, Systems Design fellowship, NASA, Assn. for Engring. Edn., Houston, 1975; named Gen. Acctg. Office Faculty Fellow, U.S. Gen. Acctg. Office, Washington, 1975-76. Mem. AAAS, ASCE (com. on coordination outside ASCE 1978-86), Am. Assn. Engring. Edn., N.Y. Acad. Scis. Republican. Episcopalian. Home: 90 Mainmast Cir New Castle NH 03854-9999 Office: New Eng States Earthquake Consortium 501 Islington St Portsmouth NH 03801

KLOTZ, RICHARD LAWRENCE, biology educator; b. Phila., Jan. 4, 1950; s. Richard Leidy and Eleanor (Evans) K.; m. Laurie Kraft, Aug. 23, 1975; children: Leidy Evans, Carrie Rose, Richard Lawrence Jr. BS, Denison U., 1972; MS, U. Conn., 1975, PhD, 1979. Asst. prof. SUNY, Cortland, 1979-82, assoc. prof., 1982-89, full prof., 1989—. Contbr. articles to Can. Jour. Fish Aquatic Sci., Limnology and Oceanography, Jour. Freshwater Ecology. League dir. Greater Homer (N.Y.) Youth Soccer Assn., 1990—. NSF grantee 1985, 92. Mem. AAAS, Am. Soc. Limnology and Oceanography, North Am. Benthological Soc., Sigma Xi. Achievements include rsch. on cycling of phosphorus in streams. Home: 2173 E Homer Rd Cortland NY 13045 Office: SUNY Dept Biology Box 2000 Cortland NY 13045

KLOTZ, WENDY LYNNETT, analytical chemist; b. Lebanon, Pa., June 15, 1966; d. William Lewis and Helen Irene (Schrader) Unger; m. Brian Lee Klotz, Sept. 29, 1990. BS, Delaware Valley Coll., Doylestown, Pa., 1988; postgrad., LaSalle U., Phila., 1989-92, Villanova U., 1992—. Cert. chemist. Chemist Rohm and Haas Co., Bristol, Pa., 1987-89; scientist Rohm and Haas Co., Spring House, Pa., 1989—. Active advertising The Hunger Project, Bucks County, Pa., 1986. Mem. AAAS, Am. Inst. Chemists (Outstanding Senior award 1988), Am. Chem. Soc. (Analytical Chemistry Undergraduate award 1987). Office: Rohm and Haas Co PO Box 904 727 Norristown Rd Spring House PA 19477-0904

KLØVE, TORLEIV, informatics educator; b. Bergen, Norway, Oct. 17, 1943; s. Anders and Borghild (Løken) K.; m. Marit Helene Hauge, Dec. 29, 1966; children: Unn, Tom, Adele. Cand Mag, U. Bergen, 1966, Cand Real, 1967, PhD, 1971. Rsch. fellow U. Bergen, 1968-70, sr. lectr. dept. math., 1971-82, prof. dept. informatics, 1982—, chmn. dept., 1986-87; lectr. Rogaland Distriktshøgskole, Stavanger, Norway, 1970-71; vis. colleague U. Hawaii at Manoa, Honolulu, 1975-76, 90-91; vis. prof. U. Hawaii at Manoa, Honolulu, 1981-82; mem. bd. info. tech. Norwegian Rsch. Coun. for Sci. and Tech., Oslo, 1989—. Contbr. articles on math. and computer sci. to profl. jours. Mem. bd. edn., Bergen, 1981-88. Mem. IEEE (sr.), Norwegian Math. Coun. (chmn. 1977-79), Norwegian Informatics Coun., Norwegian AI Soc. (mem. bd. 1985-88). Home: Djupedalen 23, N-5085 Morvik Norway Office: U Bergen Dept Informatics, Høgteknologisenteret, N-5020 Bergen Norway

KLUEPFEL, DIETER, microbiologist; b. Darmstadt, Fed. Republic Germany, Oct. 7, 1930; s. Max and Ute Maria (Paqué) K.; m. Jane Macmillan, Dec. 26, 1959 (div. 1975); children: Alexandra, Mark; m. Hélène Sasseville, June 25, 1983. Diploma sci. nat., ETH, Zurich, Switzerland, 1954, D. Natural Sci., 1957. Postdoctoral fellow Nat. Rsch. Coun. Can., Ottawa, 1957-59; rsch. scientist Lepetit Spa, Milan, Italy, 1959-61, head lab., 1961-65; sr. rsch. scientist Ayerst Labs., Montreal, Que., Can., 1965-70, rsch. assoc., 1970-75; rsch. prof. Inst. A. Frappier U. du Que., Laval, 1975—, head of dept., 1989-94. Assoc. editor Can. Jour. Microbiology, 1978-88; contbr. articles to profl. jours. Mem. Can. Soc. Microbiologist (pres. appt. sect. 1980-82), Am. Soc. Microbiology, Soc. Indsl. Microbiology. Achievements include patents in industrial microbiology and biotechnology; in process development of biobleaching of paper pulp. Home: 2314 Grand Blvd, Montreal, PQ Canada H4B 2W9 Office: Inst Armand Frappier, 531 Blvd des Prairies, Laval, PQ Canada 47N 4Z3

KLUG, AARON, molecular biologist; b. Aug. 11, 1926; s. Lazar and Bella (Silin) K.; m. Liebe Bobrow, 1948; 2 children. B.Sc., U. Witwatersrand; M.Sc., U. Cape Town; PhD, DSc, Cambridge U.; DSc (hon.), U. Chgo., 1978, Columbia U., 1978; D (hon.), U. Strasbourg, 1978; DSc (hon.), Stockholm U., 1980, U. Witwatersrand, 1984, Hebrew U., Jerusalem, 1984, Hull U., 1985, U. St. Andrews, 1987, U. Western Ont., 1991. Jr. lectr., 1947-48; rsch. student Cavendish Lab. Cambridge (Eng.) U., 1949-52; Rouse-Ball rsch. student Trinity Coll., 1949-52; Colloid Sci. dept., 1953; Nuffield rsch. fellow Birkbeck Coll., London, 1954-57, dir. virus structure rsch. group, 1958-61; mem. staff Med. Rsch. Coun. Lab. Molecular Biology, Cambridge U., 1962—, joint head div. structural studies, 1978-86, dir., 1986—; Leeuwenhoek lectr. Royal Soc., 1973; Dunham lectr. Harvard U. Med. Sch., 1975; Harvey lectr., N.Y.C., 1979, Lane lectr. Stanford U., 1983; Silliman lectr. Yale U., 1985; Cetus lectr. Berkeley U., 1986; Pauli lectr., Zürich, 1986; Nishina Meml. lectr., Tokyo, 1986; J. T. Baker lectr. Cornell U., 1987; Jean Weigle lectr., Geneva, 1989, Steenbock lectr. U. Wis., Madison, 1989; Innovators in Biochem. lectr. U. Va., Richmond, 1990; Calbiochem. lectr. U. Calif., San Diego, 1991. Contbr. articles to sci. jours. Recipient Heineken prize Royal Netherlands Acad. Sci., 1979, Louisa Gross Horwitz prize Columbia U., 1981, Nobel prize in chemistry, 1982; Gold medal of Merit, U. Cape Town, 1983; Copley medal Royal Soc., 1985; Knight, 1988. Fellow Royal Soc., Peterhouse (Coll., Cambridge), Royal Coll. Physicians (hon.), Royal Coll. Pathologists (hon.), Trinity Coll. (Cambridge, hon.); mem. Am. Acad. Arts and Scis. (fgn. hon.), French Acad. Scis. (fgn. assoc.), Max-Planck-Gesellschaft, Fed. Republic of Germany (fgn. assoc.), NAS (fgn. assoc.). Office: Med Rsch Coun, Lab Molecular Biology, Cambridge CB2 2QH, England

KLUGER, MATTHEW JAY, physiologist, educator; b. Bklyn., Dec. 14, 1946; s. Morris and Gladys (Feit) K.; m. Susan Lepold, Sept. 3, 1967; children: Sharon, Hilary. BS, Cornell U., 1967; MS, U. Ill., 1969, PhD, 1970. Postdoctoral fellow Yale U., New Haven, 1970-72; asst. prof. U. Mich. Med. Sch., Ann Arbor, 1972-76, assoc. prof., 1976-81, prof. physiology, 1981—; vis. prof. St. Thomas' Hosp., London, 1979, U. Witwatersrand, South Africa, 1992; vis. scientist Cetus Corp., Palo Alto, Calif., 1986-87. Author: Fever: Its Biology, Evolution and Function, 1979; editorial bd. Jour. Thermal Biology, 1991—, Cytokine, 1991—, Am. Jour. Physiology, 1992—; author workbooks; co-editor text books. Grantee NIH, other agys. Mem. Am. Physiol. Soc., Am. Assn. Immunologists, Am. Zool. Soc. Achievements include demonstration that fever has a long evolutionary history, studies of cytokines and other inflammatory mediators in fever, study of the role of cytokines in loss of food appetite during infection. Home: 2012 Vinewood Blvd Ann Arbor MI 48104 Office: U Mich Med Sch 7620 Medical Science II Ann Arbor MI 48109

KLUMP, WOLFGANG MANFRED, biochemist; b. Berlin, Germany, Dec. 28, 1954; came to U.S., 1987; s. Gerhard Wilhelm and Christa (Uhlig) K.; m. Gabriele Roland, Oct. 31, 1984 (div. 1989); 1 child, Susanne; m. Rosalind L. Marinou, May 2, 1993. Diploma in Chemistry, Ludwig-Maximilian Univ., Muenchen, Germany, 1981, PhD in Biochemistry, 1986. Postdoctoral scientist Max-Planck Soc., Muenchen, Germany, 1986-87, U. Calif. San Diego, La Jolla, Calif., 1987-92; rsch. scientist Viagene, Inc., San Diego, 1992—. Contbr. articles to profl. jours. Recipient Otto-Hahn medal Max Planck Soc., 1984, stipend Max Planck Soc., 1987, 88, 89, German Acad. Exch. Svc., 1989, 90. Mem. AAAS. Office: Viagene Inc 11075 Roselle St San Diego CA 92121

KLUN, JEROME ANTHONY, entomologist, researcher; b. Ely, Minn., May 4, 1939; s. Anton Dominic and Julia (Pishler) K.; m. Phyllis Ruth VanRiper, Mar. 18, 1989 (div.); children: Curt Anthony, Eric Leslie, Toinette Marie; m. Harriet Lee Rosenfeld, May 30, 1993. BA, U. Minn., 1961; PhD, Iowa State U., 1965. Rsch. assoc. entomology dept. Iowa State U., Ames, 1961-65, assoc. prof. entomology, 1968-77; rsch. entomologist agrl. rsch. USDA, Ankeny, Iowa, 1965-77, Beltsville, Md., 1977—; panel chmn. Agy. for Internat. Devel., Washington, 1990-92. Contbr. 92 articles to profl. jours. Mem. AAAS, Am. Chem. Soc., Entomol. Soc. Am. Achievements include patents for defined chemistry of many important insect, insect and plant, insect interaction. Home: 11621 Spring Ridge Rd Potomac MD 20854 Office: USDA Beltsville Agrl Rsch Ctr 10300 Baltimore Ave Beltsville MD 20705-2350

KLUTZ, ANTHONY ALOYSIUS, JR., safety and environmental manager; b. Wilkes-Barre, Pa., Dec. 2, 1954; s. Anthony A. Klutz and Matilda (Konopka) Weigand; m. LetaMarie A. Rydzewski, July 15, 1978; children: Athena Marie, Anthony A. III. BS, Kings Coll., Wilkes-Barre, 1976; MS, Rensselaer Poly. Inst., 1978; MBA, Clemson U., 1988. Material devel. engr. Sangamo Capacitor-Schlumberger, Pickens, S.C., 1978-87, product devel. engr., 1986-87; mgr. process engring. Sangamo Weston-Schlumberger, West Union, S.C., 1987-90; safety and environ. mgr. Schlumberger Industries, West Union, S.C., 1990—. Vice chmn. Oconee County Local Emergency Planning Com., Walhalla, S.C., 1988—; mem. coun. Holy Cross Parish, Pickens, 1986-87. Mem. Am. Vacuum Soc., Electro Chem. Soc., Am. Soc. Materials, S.C. C. of C. (tech. com.), Mgmt. Club (pres. 1985, 92, v.p. 1991), KC (Knight of Mo. award Pickens 1988). Avocations: reading, computing, travel. Home: 398 Chinquapin Rd Easley SC 29640-7053 Office: Sangamo Weston-Schlumberger Hwy 11 West Union SC 29696-9610

KLUTZOW, FRIEDRICH WILHELM, neuropathologist; b. Bandoeng, Preanger, Indonesia, Aug. 6, 1923; came to U.S., 1953; s. Rudolph F.W. and Pauline (Van Thiel) K.; m. Apr. 2, 1954; children: Judith A., Michael J. MD, U. Utrecht, Netherlands, 1951. Diplomate Am. Bd. Neuropathology and Anatomic Pathology. Chief staff Community Meml. Hosp., Oconto Falls, Wis., 1965-68; pathology resident U. Wis., Madison, 1968-72; neuropathologist VA Hosp., Mpls., 1972-75; dir. pathology dept. VA Hosp., Brockton, Mass., 1975-83, Wichita, Kans., 1983-87; chief staff VA Hosp., Bath, N.Y., 1987-90; neuropathologist VA Hosp., Bay Pines, Fla., 1991—; clin. assoc. prof. pathology U. Rochester (N.Y.) Sch. Medicine. Contbr. articles to profl. jours. Col. USAR, 1979-85. Paul Harris fellow Rotary Internat., Bath, 1990; recipient Outstanding Career award Dept. Vet. Affairs, Bath, 1990. Fellow Coll. Am. Pathologists; mem. Am. Assn. Neuropathologists. Republican. Methodist. Achievements include research in the practical approach to lesions and clin. diagnoses. Home: PO Box 7846 Sarasota FL 34278

KMETZ, CHRISTOPHER PAUL, lubrication systems development engineer; b. Bridgeport, Conn., Oct. 1, 1969; s. David John and Barbara (Fotta) K.; m. Jeanne Marie Leavey. BS in Aero. Engring., U. Notre Dame, 1991. Lubrication systems devel. engr. Textron Lycoming, Stratford, Conn., 1991—. Mem. AIAA. Home: 11 Sharon Rd Trumbull CT 06611 Office: Textron Lycoming Dept LSB 14 550 Main St Stratford CT 06497

KMIEĆ, BOGUMIL LEON, embryologist, educator, histologist; b. Radomsko, Poland, May 18, 1944; s. Leon Ignacy and Wladyslawa (Kaszuwara) K.; m. Krystyna Maria Hajduk, Jan. 6, 1979; two children. Grad., Med. Acad., Łódź, Poland, 1968, MD, 1977, Habilitation, 1989. Cert. assoc. prof. in medicine. Asst. dept. histology and embryology Inst. Biology and Morphology Med. Acad., Łódź, 1968-69, sr. asst., 1969-78, asst. prof., 1978-89; assoc. prof. dept. histology and embryology Med. U. of Łódź, Sch. of Medicine, Łódź, 1989—. Author: Morphdischemical Studies on Adrenalin Secretion by Adrenal Medulla of Guinea Pigs in Experimental Anaphylaxis, 1988 (Dept. Health and Welfare award 1989); contbr. numerous articles to profl. jours. Recipient 25th anniversary medal Med. Acad. Łódź, 1970, Ministry of Health and Welfare awards, 1976, '78. '84, '86, '88, '89, Hon. medal City of Łódź, 1989, Golden Cross of svc., Warsaw, 1989, Exemplary Work for Health medal, Warsaw, 1989. Mem. Iterasma, Polish Anatomical Soc., Polish Allergological Soc., Polish Biochem. Soc., Polish Anatomopathol. Soc. (electron microscopy section) Polish Histochem. and Cytochem. Soc. (pres. Łódź sect. 1968—). Roman Catholic. Avocations: bibliophile, travel, motoring, fishing. Home: Wierzbowa 42/13, 90-133 Łódź Poland Office: Dept Histology & Embryology, Narutowicza 60, Lodz Poland

KNAPOWSKI, JAN BOLESLAW, pathophysiology educator, physician, researcher; b. Poznań, Poland, July 18, 1933; s. Roch and Zofia (Krysiewicz) K.; m. Elzbieta M. Breborowicz, Aug. 8, 1962; children: Michal, Magdalena. Physician, U. Poznań, 1957, MD, 1962, PhD, 1967. Physician County Hosp., Zary, Poland, 1957-58; insp. Regional Sta. Epidemiology, Poland, 1959-61; asst. dept. pathophysiology U. Poznań Med. Sch., 1961-68, reader, 1968—, head dept., 1973—; trustee Biomed. Scis. Exch. Program between N.Am. and Europe, Hannover, Germany, 1991—. Editor, coauthor, co-editor handbooks and manuals for med. students, 1968-90; also over 200 articles. Capt. Polish Army, 1957-59. Grantee Govt. of Denmark, 1967-68, Max-Planck Soc., 1972, 74, 81. Mem. Internat. Nephrology Soc., Polish Nephrology Soc., European Dialysis and Transplant Assn., Polish Physiol. Soc. (chmn. audit commn. 1989), N.Y. Acad. Scis. Roman Catholic. Office: Med U Dept Pathophysiology, ul Swiecickiego 6, 60-781 Poznan Poland

KNAPP, EDWARD ALAN, scientist, government administrator; b. Salem, Oreg., Mar. 7, 1932; s. Gardner and Lucille (Moore) K.; m. Jean Elaine Hartwell, June 27, 1954; children: Sandra, David, Robert, Mary. A.B., Pomona Coll., 1954; Ph.D., U. Calif., Berkeley, 1958; D.Sc. (hon.), Pomona Coll., 1984, Bucknell U., 1984. With Los Alamos Sci. Lab., U. Calif., 1958-82, dir. accelerator tech. div., 1977-82; asst. dir., then dir. NSF, Washington, 1982-84; sr. fellow Los Alamos Nat. Lab., 1984; pres. Univs. Rsch. Assn., Washington, 1985-89; sr. fellow Los Alamos Nat. Lab., 1990, dir. Los Alamos meson physics facility, 1990-91; pres. Santa Fe Inst., 1991—; cons. in field. Contbr. articles to profl. jours. Fellow AAAS, Am. Phys. Soc.; mem. IEEE, Sigma Xi. Methodist. Office: Santa Fe Inst 1660 Old Pecos Trl # A Santa Fe NM 87501-4768

KNAPP, FREDERICK WHITON, entomologist, educator; b. Danbury, Conn., Mar. 19, 1915. BS, U. Calif., Davis, 1935, MS, 1956; PhD in Agrl. Chemistry, U. Calif., 1960. Asst. biochemist U. Fla., 1960-67, assoc. prof. food sci., assoc. biochemist inst. food & agrl. sci., 1967-77; prof. U. Ky., Lexington. Recipient CIBA-GEIGY/Entomol. Soc. Am. award CIBA-GEIGY Corp., 1992. Mem. Inst. Food Technologists. Achievements include research in use and control of enzymes in food processing, protein recovery from animal by-products. Office: U Kentucky Dept Entomology Lexington KY 40546*

KNAPP, MARK ISRAEL, industrial and management engineer; b. N.Y.C., Sept. 11, 1923; s. Dewey and Ialene (Rottner) K.; m. Dorothy Estelle Lipkin, May 26, 1944 (dec. 1963); children: A. Michael, Richard J., Carla E., Jeanine D.; m. Constance Ann Saunderson, Nov. 28, 1986. B Adminstrv. Engring., NYU, 1951; postgrad., Am. U., 1966. Sr. test engr. Wright Aero. Corp., Woodridge, N.J., 1945-49; project liaison engr. Fairchild Engine Div., Deer Park, N.Y., 1953-56; supr. planning and control Turbomotor div. Curtiss Wright Corp., Princeton, N.J., 1956-58; mgr. programming Flight Propulsion Lab. GE Co., Evandale, Ohio, 1958-59; div. chief fgn. tech. div. USAF, Dayton, Ohio, 1959-64; mem sr. rsch. staff Inst. Def. Analyses, Alexandria, Va., 1964-87, cons., 1987—; presenter at profl. confs. Author reports. Mem. Comprehensive Health Planning Adv. Coun., Prince Georges County, Md., 1974-76; chmn. governing body plan devel. com. So. Md. Health Systems Agy., Clinton, 1976-78. 1st lt. USAAC, 1943-45, ETO; capt. USAF, 1951-52. Decorated Air medal with 3 oak leaf clusters. Fellow AAAS; mem. Inst. Cost Analysis (cert.), Ops. Rsch. Soc. Am. Achievements include development of analytical method to decrease vulnerability of military aircraft to enemy ground fire, plan for balancing health care resources with health needs of residents of 4-county area of Maryland.

KNAPP, WILLIAM BERNARD, cardiologist; b. Paterson, N.J., Oct. 26, 1921; s. Joseph and Mary (Cannon) K.; m. Jeannette C. Zarnowiecki, Jan. 31, 1948; children: William, Thomas, Bernadette, Richard, Suzanne. Degree in Chemistry, Seton Hall U., 1942; MD, Loyola U., 1946. Diplomate Am. Bd. Internal Medicine. Intern St. Joseph Hosp., Paterson, N.J., 1946-47; resident in pathology, Goldwater Meml. Hosp., N.Y.C., 1947-48; resident in medicine Mercy Hosp., Chgo., 1948-49; sr. resident, Hines (Ill.) VA Hosp., 1949-53; attending physician Cook County Hosp., Chgo.; assoc. clin. prof. medicine Loyola U., Chgo.; chmn. medicine Little Co. of Mary Hosp., 1960-80; chmn. Holy Cross Hosp., Suburban Hosp., Hinsdale, Ill., 1976-81; practice medicine specializing in cardiology; chmn. S.W. Hosp. Planning, Chgo.; established 1st coronary care unit in Ill., 1965; lectr. Moscow Inst. Cardiology, 1969; mem. U.S. Bd. Examiners. Bd. dirs. Retirement Village, Civic Assn., Geneva Lake, Wis., 1970—; dir. water safety patrol, Geneva Lake, 1970—. 2d lt. U.S. Army, 1943-46. Recipient Rsch. award Ill. Inst. Medicine; 1st Professorial Chair Cardiology named in honor Loyola U., Chgo., 1985 . Fellow Am. Coll. Cardiology, Am. Coll. Chest Physicians, Am. Coll. Angiology, Chgo. Inst. Medicine; mem. N.Am. Soc. Pacing and Electrophysiology, AMA, ACP, Ill. Med. Soc., Chgo. Med. Soc., Inst. Medicine Chgo., Blue Key Honor Soc. Roman Catholic. Clubs: Butterfield Country (Oak Brook, Ill.); Big Foot Country; Northshore Country (Glenview, Ill.); Jupiter Hills (Fla.), Tracer.

KNASEL, THOMAS LOWELL, information systems consultant; b. Urbana, Ohio, Apr. 18, 1959; s. Richard Lowell and Edwina Francis (Williams) K.; m. Alisia Monyak, July 2, 1987. BS in Mfg. Engring., Milw. Sch. Engring., 1993. CNC programmer Grimes div. Midland-Ross Corp., Urbana, Ohio, 1977-80; NC systems analyst Cin. Milacron Inc., 1980-81; robotics systems analyst Caterpillar Tractor Co., Peoria, Ill., 1982-85; systems engr. GM, Detroit, 1985-87; CAM systems analyst ground systems div. FMC Corp., San Jose, Calif., 1987-89; sr. systems analyst A.O. Smith Corp. Automotive Products Co., Milw., 1989—; chmn. McAuto Robotics Users Group, St. Louis, 1983, 84. Mem. IEEE, Computer and Automated Systems Assn., Soc. Mfg. Engrs., Robotics Internat., Soc. Mfg. Engrs. Republican. Lutheran. Achievements include research in computer integrated manufacturing applications research and development, factory information systems technology development, development of robotic workcell simulation and offline programming technology. Office: A O Smith Automotive Products Co 3533 N 27th St Milwaukee WI 53216

KNAUSS, JOHN ATKINSON, federal agency administrator, oceanographer, educator, former university dean; b. Detroit, Sept. 1, 1925; s. Karl Ernst and Loise (Atkinson) K.; m. Marilyn Mattson, Sept. 6, 1954; children: Karl, William. BS, MIT, 1946; MS, U. Mich., 1949; PhD, U. Calif., 1959. Oceanographer Navy Electronics Lab, San Diego, 1947, Office Naval Rsch., 1949-51, Scripps Instn. Oceanography, 1951-52, 55-62; prof. Grad. Sch. Oceanography, U. R.I., Narragansett, 1962-90, dean, 1962-87, provost for marine affairs, 1969-82, v.p. marine programs, 1982-87; undersecretary for oceans and atmosphere Dept. Commerce, Washington, 1989-93; adminstr. Nat. Oceanic and Atmospheric Adminstrn., Washington, 1989-93; U.S. commr. Internat. Whaling Commn., 1991-93; leader 10 oceanographic expdns to study oceanic circulation, 1955-65; chair U.S. phys.-chem. panel Internat. Indian Ocean Expdn., 1959-62; mem. Pres's. Commn. on Marine Scis., Engring. and Resources, 1967-68; mem. State Dept. Pub. Adv. Com. on Law of Sea, 1970-82; chair sr. adv. com. on environ. scis. Ctr. for Energy and Environ. Rsch., U. P.R., 1977-80; mem. Nat. Adv. Com. on Oceans and Atmosphere, 1978-85, vice chair, 1979-81, chair 1981-85; chair bd. govs. Joint Oceanographic Instns., Inc., 1978-80; co-founder Law of Sea Inst., mem. exec. bd. 1965-76, 82-87; bd. dirs. Coun. for Ocean Law, 1983-89; chair Joint Oceanographic Instns. for Deep Earth Sampling, 1984-86; bd. dirs. Harbor Br. Oceanographic Instn., 1987-89; 1st vice chmn. Intergovernmental Oceanographic Commn., 1991-93. U.S. Congress renamed its Sea Grant fellowship the Dean John A. Knauss Fellowship program in 1987. With USNR, 1943-46. Named to R.I. Heritage Hall of Fame, 1983; recipient Albatross award Am. Miscellaneous Soc., 1959, Nat. Sea Grant award, 1974. Mem. AAAS (v.p. 1972-73), Am. Geophysics Assn. (pres. oceanography sect. 1965-67, Ocean Sci. award 1988), Nat. Assn. State U. and Land Grant Coll. (chair marine div. 1984-85), Am. Meteorol. Soc. (coun. 1980-82). Home: 126 Willett Rd Saunderstown RI 02874-3810

KNEEN, GEOFFREY, chemist; b. Ulverston, Cumbria, Eng., Aug. 7, 1949; s. Wilfred Cowper and Dorothy (Blackburn) K.; m. Clare Octavia Hampson, July 7, 1983; children: Paul Geoffrey, Sarah-Jane. BSc, U. Leeds, 1970, PhD, 1974. Postdoctoral rsch. fellow U. Nottingham, Eng., 1973-75; rsch. chemist Wellcome Rsch. Labs., Beckenham, Kent, 1975-82, sect. head, 1982-85; sect. head G.D. Searle & Co., Ltd., High Wycombe, Bucks., 1985-86; head synthetic chemistry Schering Agrochemicals, Ltd., Saffron Walden, Essex, 1986-89, head chem. rsch., 1989-90, dir. rsch., 1990—; also bd. dirs. Schering Agrochemicals, Ltd.; vis. prof. U. Reading, 1991—. Contbr. articles to profl. jours., chpts. to books; patentee in field of pharmacy. Fellow Royal Soc. Chemistry; mem. Soc. Chem. Industry. Avocations: stamp collecting, tennis, music. Office: Schering Agrochemicals Ltd, Chesterford Pk, Saffron Walden Essex CB10 1XL, England

KNELLER, ECKART FRIEDRICH, materials scientist, electrical engineer, educator, researcher; b. Magdeburg, Fed. Republic Germany, Jan. 1, 1928; s. Friedrich Karl Christian and Annamarie Erna (Mathiszig) K.; m. Brigitte Quass, 1968; children: Jo Hanna, Don Eckart. Diploma in Physics, Tech. U., Stuttgart, Fed. Republic Germany, 1951; D of Natural Scis., Tech. U., Stuttgart, Federal Republic of Germany, 1953, Habilitation, 1960. Staff mem. Max-Planck-Inst. Metallforschung, Stuttgart, 1953-59, 63-67; cons. IBM Rsch. Ctr., Yorktown Heights, N.Y., 1960-62; lectr., prof. Tech. U., Stuttgart, 1962-67; full prof. materials sci. Ruhr U., Bochum, Fed. Republic Germany, 1967-93, emeritus, 1993—, dean faculty elec. engring., 1968, 69, 87; guest lectr. T.H. Wien, Austria, U. Göttingen, Fed. Republic Germany, 1958, 59; dir. Inst. Elec. Engring. Materials, Bochum, 1967—. Author: Ferromagnetism, 1962; author (with others): Encyclopedia of Physics, Vol. XVIII, 1966; co-author, co-editor: Magnetism and Metallurgy, 1969; contbr. articles to profl. jours. Fellow Inst. Physics; mem. IEEE, Deutsche Gesellschaft für Metallkunde (Masing Preis 1962), Rheinisch-Westfälische Acad. Scis. (sec. 1988-91, v.p. 1990-91, sec.-gen. 1992-96). Avocations: oil and water color painting, graphic arts, history, mountaineering. Home: Vossegge 23, D-58456 Witten Germany Office: Ruhr U, Postfach 102148, D-4630 Bochum Germany

KNIBBS, DAVID RALPH, electron microscopist; b. Bristol, Conn., Feb. 4, 1953; s. John Milton and Dorothy (Brothwell) K.; children: Christopher John, Tabitha Jeanne. BS, U. Conn., 1977; MA, Cen. Conn. State U., 1986; postgrad., U. Conn., 1989—. Animal health technician Chippens Hill Vet. Hosp., Bristol, 1976-79; rsch. asst. U. Conn. Health Ctr., Farmington, 1979-82; electron microscopist Hartford (Conn.) Hosp., 1982—; adj. faculty U. Hartford, 1988—; presenter at profl. confs. Contbr. articles to Archives Otolaryngology, Jour. Histotechnology, Human Pathology, Ultrastructural Pathology, others. Adult leader Bristol area Boy Scouts Am., 1989—. Mem. Microscopy Soc. Am. (cert.), New Eng. Soc. Electron Microscopy, Conn. Microscopy Soc. (treas. 1985-87, v.p. 1987-88, pres. 1988-89), Sigma Xi. Office: Hartford Hosp EM Div 80 Seymour St Hartford CT 06115

KNIGHT, ALAN EDWARD WHITMARSH, physical chemistry educator; b. Kanpur, India, Nov. 17, 1946; arrived in Australia, 1956; s. Aubrey Edward and Ghislaine Gertrude (Simpson) Whitmarsh-Knight; m. Lea-Ann, Jan. 17, 1973; children: Iain Francis, Jane Elizabeth, Yvonne Michelle. BSc in Chemistry with honors, Australian Nat. U., Canberra, 1969, PhD in Phys. Chemistry, 1972. Rsch. assoc. Ind. U., Bloomington, Ind., 1973-75; rsch. fellow Melbourne (Australia) U., 1976; lectr. Griffith U., Brisbane, Australia, 1977-83, sr. lectr., 1983-87, assoc. prof., 1987-91, prof., 1991—; vis. prof. Ind. U., Bloomington, 1982-83; Alexander von Humboldt guest prof. Tech. U. Munich, 1987; Jila vis. fellow U. Colo., Boulder, 1991; Hill prof. U. Minn., Mpls., 1992. Contbr. articles to profl. jours. Fellow Am. Phys. Soc., Am. Chem. Soc., Royal Australian Chem. Inst. (Rennie medal for chem. rsch. 1979, H.G. Smith Meml. medal for chem. rsch. 1990). Office: Griffith U Div Sci & Tech, Nathan, Brisbane 4111, Australia

KNIGHT, GARY CHARLES, mechanical engineer; b. Bartlesville, Okla., Nov. 15, 1950; s. Charles Robert and Elizabeth India (Brown) K.; m. Lisa Jo Martin, Sept. 12, 1981; children: Amanda Joann, Gary Michael. BS in Mech. Engring., U. Tulsa, 1981. Registered profl. engr., Okla.; commd. inspector Nat. Bd. Boiler and Pressure Vessel Inspectors. Results engr. I Pub. Svc. Co. Okla., Tulsa, 1981-83, results engr. II, 1983-87, results engr.

III, 1987-89, maintenance supr., 1989-90, maintenance supt., 1990-92, asst. sta. mgr., 1992, sta. mgr., 1992—. With U.S. Army, 1970-73, Vietnam. Mem. ASME, NSPE. Republican. Home: 6848 E 59th St Tulsa OK 74145 Office: Pub Svc Co Okla PO Box 220 Oologah OK 74053

KNIGHT, HAROLD EDWIN HOLM, JR., utility company executive; b. Bklyn., Mar. 23, 1930; s. Harold Edwin Holm and Dorothy (Brown) K.; m. Janet Luft, Feb. 16, 1953. B.S., Yale U., 1952. Test engr. GE, Pittsfield, Mass., 1952; with Bklyn. Union Gas, 1954—, dir. ins., 1971-74, asst. sec., asst. treas., 1974-78, sec., 1978-84, v.p., sec., 1984—. Past pres. Bklyn. coun. Boy Scouts Am.; past pres. Estates Property Owners Assn., Garden City, N.Y., Garden City Community Fund. Served to capt. USAF, 1952-54. Mem. Bklyn. Club (bd. dirs., pres. 1989-91), North Fork Country Club. Home: Calves Neck Rd Southold NY 11971-0938

KNIGHT, KEVIN KYLE, research psychologist; b. Ridgecrest, Calif., Mar. 14, 1963; s. Perry John and Shirley Rae (Landa) K.; m. Danica Lee Kalling, May 23, 1987. BA, So. Meth. U., 1985, MA, 1988; PhD, Tex. Christian U., 1991. Assoc. rsch. scientist Inst. Behavioral Rsch. Tex. Christian U., Ft. Worth, 1991—. Contbr. articles to profl. jours. Ida M. Green fellow, 1988. Mem. APA, Am. Psychol. Soc., Soc. for Psychologists in Addictive Behaviors, Sigma Xi. Methodist. Home: 5908 Wimbleton Way Fort Worth TX 76133 Office: Tex Christian U Inst Behavioral Rsch Box 32880 Fort Worth TX 76129

KNIPLING, EDWARD FRED, retired research entomologist, agricultural administrator; b. Port Lavaca, Tex., Mar. 20, 1909; s. Henry John and Hulda Lena (Rasch) K.; m. Phoebe Rebecca Hall; children—Edwina, Anita, Edward B., Gary D., Ronald R. B.S., Tex. A&M U., 1930; M.S., Iowa State U., 1932, Ph.D., 1947; D.Sc. (hon.), Catawba Coll., 1962, N.D. State U., 1970, Clemson U., 1972. With USDA, various locations, 1931-73; dir. entomology research div. USDA, Beltsville, Md., 1953-73. Author: Principles of Insect Population Suppression, 1979; contbr. over 200 articles to profl. jours. Recipient Merit award Iowa State U., 1958; Disting. Service award USDA, 1960; Disting. Alumnus award Tex. A&M U., 1962; Nat. medal Sci., 1966; Rockefeller Pub. Service award, 1966; Pres.'s award for Disting. Fed. Civilian Service, 1971; World Food prize, World Food Prize Foundation, 1992. Fellow Entomol. Soc. Am. (pres. 1952); mem. Nat. Acad. Scis., Am. Acad. Arts and Scis. Club: Cosmos. Avocations: bow and arrow hunting; fishing; hiking. Home: 2623 Military Rd Arlington VA 22207-5117

KNIPPER, ANDREI LVOVICH, geologist, administrator, researcher; b. Moscow, Russia, Feb. 24, 1931; s. Lev Konstantinovich Knipper and Lubov Sergeevna Zalesskaya; m. Galina Fedorovna Karevna, Dec. 30, 1957 (dec. 1988); 1 child, Olga A.; m. Nataliya Ivanova Komkova, Aug. 18, 1991; 1 child, Kirill A. Grad., Moscow State U., 1954; MSc, USSR Acad. Sci., 1962, PhD, 1972. Jr. rschr. geol. inst. Russian Acad. Scis., Moscow, 1954-63, sr. rschr., 1963-85, head lab., chief. dept. tectonics, 1985-88, dep. dir., 1986-88, dir. geol. inst., 1988—; lectr. Moscow State U., 1989—; vis. lectr. Naccetepe U., Ankara, Turkey, 1992-93. Author: Oceanic Crust in the Structure of Alpine Fold Belt, 1975; co-author: Tectonics of Northern Eurasia, 1980; contbr. sci. papers. Recipient medal Bulgarian Acad. Scis., 1983, Labour Veteran medal Soviet Govt., 1984, Labour Distinction medal, 1986, medal Czechoslovakian Acad. Scis., 1985. Mem. USSR Acad. Sci. (corr.), Russian Acad. Scis. (academician), Nat. Com. Geologists, Moscow Scientists Club. Achievements include discovery of alumophosphates in Kazakhstan and blue asbestos. Office: Russian Acad Sci Geol Inst, Pyzhevskiy Pereulok 7, 109017 Moscow Russia

KNISBACHER, JEFFREY MARK, computer scientist; b. Balt., Oct. 6, 1941; s. Max and Rea (Lehman) K.; m. Anita Marshall, Aug. 29, 1965; children: Juliett A., Alden J. BA, Johns Hopkins U., 1961; MA, Brown U., 1963, PhD, 1971; BHL, Balt. Hebrew U., 1964. With rsch. dept. IBM T.J. Watson Rsch. Ctr., Yorktown Heights, N.Y., 1963, U. Tex. Linguistics Rsch. Ctr., Austin, 1965-66; instr. U. Pitts., 1967-70; asst. prof. U. Md., Balt., 1970-76; analyst DOD, Fort Meade, Md., 1976—; translator, computer reseller, integrator K&K Enterprises, Owings Mills, Md., 1980—. Author: M(A)T: Method from Madness, 1983, Some Theoretical Aspects of a Sentence Analysis Grammar, 1966; contbr. articles to profl. jours. Organizer Woods Homeowners Assn., Owings Mills, 1992; Sunday sch. tchr. Beth El Schs., Balt., 1982-87. Mem. Computer Assisted Lang. Instrn. Consortium. Democrat. Jewish. Home and Office: 2122 Harmony Woods Rd Owings Mills MD 21117-1642

KNISELY, RALPH FRANKLIN, retired microbiologist; b. Altoona, Pa., Mar. 30, 1927; s. Calvin Ross and Frieda Pauline (Neher) K.; m. Joan Marie Fitzgerald, Jan. 29, 1949 (div. 1955); 1 child, Patricia Ann; m. Ann Martin, May 21, 1960. BS, Pa. State U., 1953, postgrad., 1953. Bacteriologist Altoona Hosp., 1953-56, adminstrv. asst. to pathologist, 1957-59; microbiologist Chem. Corps Dept. Army, Ft. Detrick, Md., 1959-72; rsch. microbiologist Edgewood Arsenal, Aberdeen Proving Ground, Md., 1972-86. Contbg. author: Rapid Identification of Biological Agents, 1966; contbr. articles to Jour. Bacteriology, European Jour. Microbiology. Pres. Eastview Civic Assn., Frederick, Md., 1968-69. With USN, 1945-46, 50-51; capt. Res. ret., 1945-87. Mem. N.Y. Acad. Sci. (emeritus), Am. Soc. for Microbiology (emeritus), Rsch. Soc. Am., Assn. Mil. Surgeons U.S., Ret. Officer's Assn. (chpt. v.p. 1969-70), Am. Assn. Ret. Persons (bd. dirs. chpt. 636 1989-90, chpt. v.p. 1993—), Nat. Assn. Ret. Fed. Employees (v.p. chpt 409 1993—), Nat. Sojourners (pres. chpt. 354 1965, 81, sec. 1986—), George Washington Masonic Stamp Club (pres. 1978-80, sec. 1988—), Am. Legion, Elks, Masons, Scottish Rite, Jaffa Shrine Temple, Am. Philatelic Soc. (life), Masonic Rsch. Soc., Philalethes Soc., Sampson WWII Vets. (Md. state rep., exec. dir.). Republican. Lutheran. Avocations: family genealogist, historian. Home: 7400 Skyline Dr Frederick MD 21702-3652

KNIZE, RANDALL JAMES, physics educator; b. Tacoma, Feb. 4, 1953; s. Howard James and Nathalie (Gage) K. BA, MS, U. Chgo., 1975; MA, Harvard U., 1976, PhD, 1981. Staff physicist Princeton (N.J.) U., 1980-88; asst. prof. physics U. So. Calif., L.A., 1988—. Patentee hydrogen in metals; contbr. papers to sci. jours. NSF fellow, 1975-79. Mem. Am. Phys. Soc. Office: U So Calif Dept Physics MC 0484 Los Angeles CA 90089

KNOEDLER, ANDREW JAMES, aerospace engineer; b. Houston, Jan. 18, 1968; s. James Lawrence and Britta Eileen (Hurley) K.; m. Audra Ann Garner, June 20, 1992. SB in Aeros. and Astronautics, MIT, 1990, SM in Aeros. and Astronautics, 1991. Commd. 1st lt. USAF, 1991; engr. USAF Aero. Systems Ctr., Wright-Patterson AFB, Ohio, 1991—. Contbr. articles to profl. jours. Asst. scoutmaster troop 162 Boy Scouts Am., Wright-Patterson AFB, 1991—. Mem. AIAA, Air Force Assn., Sigma Xi. Office: Nat Aerospace Plane Joint Program Office Wright Patterson AFB OH 45433

KNOEPFLER, NESTOR BEYER, chemical engineer; b. New Orleans, Oct. 1, 1918; s. Adolphe Jean and Frieda June (Beyer) K.; m. Mary Janet Bierhorst, May 1, 1943; children: Mary Noel Knoepfler Banks, Nancy Ann Knoepfler Phillips, Jannes Jean Knoepfler Rost. B in Chem. Engring., Tulane U., 1940. Registered profl. engr., La. Lab. technician Sou Cotton Oil Co., Gretna, La., 1940-41; asst. mgr. Picture Box Mfg., New Orleans, 1945-50; rsch. fellow Nat. Cottonseed Products Assn., New Orleans, 1950-53; chem. engr., researcher USDA Sou Regional Rsch. Ctr., New Orleans, 1953-79; pvt. practice New Orleans, 1979—; guest lectr. Clemson U., Sam Houston State U. Contbr. articles to profl. jours. Lt. comdr. USNR, 1941-45. Recipient Cotton Batting award Nat. Cotton Batting Inst., Memphis, 1960, 78. Mem. Am. Oil Chemists Soc. (chmn. nat. conv. 1967, 69), Soc. Tulane Engrs. (pres. 1963), Sigma Xi (chpt. pres. 1970). Republican. Roman Catholic. Achievements include patents in field; research in cottonseed oil processing, naval stores processing, cotton batting, flame retardance and smolder resistance of cellulosics. Home and Office: 17 Jennifer Ct Mandeville LA 70448-6321

KNOLL, FLORENCE SCHUST, architect, designer; b. Saginaw, Mich., May 24, 1917; d. Frederick E. and M. Haisting Schust; m. Hans G. Knoll, July 1, 1946 (dec. 1955); m. Harry Hood Bassett, June 22, 1958 (dec. 1991). Student, Cranbrook Art Acad., Bloomfield Hills, Mich., 1935-37, Archtl. Assn., London, 1938-39; B.Arch., Ill. Inst. Tech., Chgo., 1941; D.F.A. (hon.), Parsons Sch. Design, 1979. Archtl. draftsman, designer Gropius & Breuer, Boston, 1941; design dir. Knoll Planning Unit, 1942-55; pres. Knoll Internat., N.Y.C., 1955-65; pvt. practice architecture and designer Coconut Grove, Fla., 1965—. Recipient Ill. Inst. Tech. Hall of Fame award, 1982; recipient Athena award R.I. Sch. Design, 1982, others. Mem. AIA (recipient Gold medal for indsl. arts 1961), Indsl. Designers Am. (hon.).

KNOOP, VERN THOMAS, civil engineer, consultant; b. Paola, Kans., Nov. 19, 1932; s. Vernon Thomas and Nancy Alice (Christian) K. Student, Kans. U., 1953-54; BSCE, Kans. State U., 1959. Registered profl. engr., Calif. Surveyor James L. Bell, Surveyors and Engrs., Overland Park, Kans., 1954; engr. asst. to county engr. Miami County Hwy. Dept., Paola, 1955; engr. State of Calif. Dept. Water Resources, L.A., 1959-85, sr. engr., 1986-88; chief, water supply evaluations sect. State of Calif. Dept. Water Resources, L.A., Glendale, 1989—; hydrology tchr. State of Calif. Dept. Water Resources, L.A., 1984; mem. Interagency Drought Task Force, Sacramento, 1988-91. Mem. Jefferson Ednl. Found., Washington, 1988-91, Heritage Found., Washington, 1988-91, Nat. Rep. Senatorial Com., Washington, 1990-91, Rep. Presdl. Task Force, Washington, 1990-91. With U.S. Army, 1956-57. Decorated Good Conduct medal U.S. Army, Germany, 1957. Mem. ASCE (dir. L.A. sect. hydraulics/water resources mgmt. tech. group 1985-86, chmn. 1984-85), Profl. Engrs. in Calif. Govt. (dist. suprs. rep. 1986—), Singles Internat. Baptist. Home: 116 N Berendo St Los Angeles CA 90004-4785 Office: State of Calif Dept Water Resources 770 Fairmont Ave Glendale CA 91203-1035

KNOPF, PAUL MARK, immunoparasitologist; b. Trenton, N.J., Apr. 4, 1936; s. Chiam David and Beatrice (Safir) K.; m. Carol Lois Harrison, June 29, 1958; children: Jeffrey William, Steven Harrison, Rachel Analiese. BSc, MIT, 1958, PhD, 1962. Postdoctoral fellow MRC Lab. Molecular Biology, Cambridge, Eng., 1962-64; spl. research assoc. Salk Inst., La Jolla, Calif., 1964-72; prof. med. sci. Brown U., Providence, 1972—, Charles A. and Helen B. Stuart prof. med. sci., 1992—, chmn. sect. molecular, cellular and devel. biology, 1990—; mem. study sect. on parasitic disease NIH, 1985-87. Recipient Career Devel. award NIH, 1966-72; grantee NIH, 1966-76, 84-88, 91—, Rockefeller Found., 1972-8o, Edna McConnell Clark Found., 1976-85, WHO, 1979—, MS Soc., 1989-90; Fulbright-Hays sr. fellow, 1978-79, Fogarty sr. internat. fellow, 1986-87. Mem. AAAS, Am. Assn. Immunologists, Am. Soc. Tropical Medicine and Hygiene, Soc. Neuroscience. Home: 2 Dana Rd Barrington RI 02806-4614 Office: Brown U Div Biology and Medicine Providence RI 02912

KNOSPE, WILLIAM HERBERT, medical educator; b. Oak Park, Ill., May 26, 1929; s. Herbert Henry and Dora Isabel (Spruce) K.; m. Adris M. Nelson, June 19, 1954. B.A., U. Ill., Chgo. and Urbana, 1951; B.S., U. Ill., 1952; M.D., U. Ill., Chgo., 1954; M.S. in Radiation Biology, U. Rochester, 1962. Diplomate Am. Bd. Internal Medicine and Subspecialty Bd. on Hematology. Rotating intern Upstate Med. Ctr. Hosps-SUNY-Syracuse, 1954-55; resident in medicine Ill. Central Hosp., Chgo., 1955-56, VA Research Hosp-Northwestern U. Med. Sch., Chgo., 1956-58; investigator radiation biology Walter Reed Army Inst. Research, Washington, 1962-64; investigator hematology, asst. chief dept. hematology Walter Reed Army Inst. Research, 1964-66; attending physician med. service Walter Reed Gen. Hosp., Washington, 1963-64, fellow in hematology, 1964-65, asst. chief hematology service, chief hematology clinic, 1964-66; asst. attending staff physician Presbyn. St. Luke's Hosp., Chgo., 1967-68, asst. dir. hematology radiohematology lab., 1967-74, assoc. attending staff physician, 1968-74, sr. attending staff physician, 1974—; asst. prof. medicine U. Ill.-Chgo., 1967-69, assoc. prof., 1969-72; assoc. prof. medicine Rush Med. Coll., Chgo., 1971-74, prof. medicine, 1974—; dir. sect. hematology Rush-Presbyn.-St. Luke's Med. Ctr., Chgo., 1974—; Elodia Kehm prof. hematology Rush-Med. Coll., Chgo., 1986—; speaker at profl. confs. U.S. and abroad; vis. prof. medicine dept. hematology U. Basel, Switzerland, 1980-81. Contbr. numerous articles to profl. publs. Trustee Ill. chpt. Leukemia Soc. Am., 1977—, v.p., 1979-80; trustee Bishop Anderson House (Rush-Presbyn.-St. Luke's Med. Ctr.), 1980—. Served to capt. M.C., USAR, 1958-61, to lt. col., U.S. Army, 1961-66. Fellow ACP; mem. Am. Fedn. Clin. Research, AMA, Am. Soc. Hematology, Am. Soc. Clin. Oncology, Central Soc. Clin. Research, Chgo. Med. Soc., Inst. Medicine Chgo., Internat. Soc. Exptl. Hematology, Radiation Research Soc., Southeastern Cancer Study Group, Polycythemia Vera Study Group, Eastern Coop. Oncology Group, Ill. State Med. Soc., Assn. Hematology-Oncology Program Dirs., Sigma Xi. Club: Chgo. Literary. Office: 1653 W Congress Pky Chicago IL 60612-3833

KNOTH, RUSSELL LAINE, psychologist, educator; b. San Jose, Calif., Oct. 14, 1957; s. Robert Joel and Dianne Louise (Turner) K. MA, U. N.H., 1986, PhD, 1988. Asst. prof. psychology Chapman U., Orange, Calif., 1988—; faculty advisor Psi Chi Honor Soc. Chapman U., 1989—, Lacrosse Club, 1989—. Contbr. articles to Behavioral Neurosci. Rsch. Methods, Cognitive Neurosci., Cognitive Psychology. Mem. Soc. for Neurosci., Scientific Rsch. Soc., Am. Psychol. Soc. Democrat. Office: Chapman U Dept Psychology Orange CA 92606

KNOTT, JOHN ROBERT, mathematics educator; b. Dale, Ind., June 24, 1937; s. Hilary Francis and Eunice Meriba (Heichelbech) K.; m. Mary Elizabeth Bockting, July 26, 1958; children: Susan, Lisa, Thomas. AB, Depauw U., 1959; MS, So. Ill. U., 1963; EdD, Ind. U., 1973. Tchr. math. North High Sch., Evansville, Ind., 1959-68; prof. math. U. Evansville, Evansville, 1968-70, Ind. State U., Evansville, 1966-68; assoc. instr. math. Ind. U., Bloomington, 1970-71; prof. math. U. Evansville, Evansville, 1971—; faculty athletics rep. NCAA, 1977—; math. dept. chmn. U. Evansville, 1983—. Mem. Nat. Coun. Tchrs. of Math., Math. Assn. Am., Phi Delta Kappa (chpt. pres. 1981-82), Phi Kappa Phi. Democrat. Roman Catholic. Office: U Evansville 1800 Lincoln Ave Evansville IN 47722-0002

KNOTTENBELT, HANS JORGEN, economist; b. Bandung, Netherlands, July 19, 1934; s. Anthony and Gertrud Annemarie (Rafflorze) K.; D.Econs., Erasmus U., Rotterdam, Netherlands, 1963; m. Marianne van Berkel, May 21, 1963; children—Karen A.E., Alexander. Researcher, Netherlands Econ. Inst., Rotterdam, 1960-61; sr. cons. Bakkenist, Spits & Co., Rotterdam, 1963-66; product mgr. J. van Nelle Co., Rotterdam, 1966-69; v.p. Netherlands Nat. Tourist Office, The Hague, 1969-74; ptnr. Custom Mgmt. B.V., Utrecht, Netherlands, 1974-85, 88—; pres. Koninklijke Vereenigde Tapijtfabrieken N.V., Moordrecht, Netherlands 1985-88; cons. and lectr. in field. With Royal Dutch Marines, 1955-57. Named knight Italian Order of Merit, 1972. Mem. Alumni Assn. Erasmus U. (pres. 1976-79), Netherlands Mgmt. Assn., Netherlands Inst. Mktg., Mars and Mercurius. Club: Rotary. Liberal.

KNOWLES, CHRISTOPHER ALLAN, healthcare executive; b. Washington, Oct. 24, 1949; s. Charles Edward and Eleanor Patricia (Murphy) K.; m. Mary Margaret O'Loughlin, Feb. 14, 1988; children: Sean Christopher, James Charles, Thomas Patrick. BA, U. Nebr., 1975; MPA, Drake U., Des Moines, 1982; postgrad., Fordham U., 1987—. Adminstrv. asst. to dir. Nebr. Dept. Water Resources, Lincoln, 1976-78; environ. planner Iowa Natural Resources Council, Des Moines, 1978-81; Md. Environ. Trust, Balt., 1982; fin. analyst Norwest Corp., Des Moines, 1982-83; asst. dir., dir. fin. Hospice of Cen. Iowa, Des Moines, 1983-85; assoc. dir. home health svcs. dept. Hackensack Med. Ctr., N.J., 1985-86; fiscal mgr. Family Health Ctr, Montefiore Med. Ctr., N.Y.C., 1986-87; assoc. dir. and adminstr. Comprehensive Family Care Ctr., Albert Einstein Coll., N.Y.C., 1987-88; exec. dir. Hospice Care of L.I., 1988; assoc. dir. Bronx-Lebanon Hosp., N.Y.C., 1989; chmn., chief exec. officer Knowles Econometrics, Inc., Pelham Manor, N.Y., 1990-91; dir. Vis. Nurse Svc., Martha's Vineyard Community Svcs., Oak Bluffs, Mass., 1991—; bd. dirs., treas. AIDS Alliance of Martha's Vineyard, Inc., Mass., 1992—. U.S. Dept. Edn. grantee, 1981-82. Mem. Pi Sigma Alpha, Pi Alpha Alpha. Democrat. Episcopalian. Avocation: sailing. Office: PO Box 369 Vineyard Haven MA 02568-0369

KNOWLES, DAVID EUGENE, chemist; b. Salt Lake City, May 30, 1959; s. Eugene Arthur and DeeAnn (Hancock) K.; m. Diana Lynn Blackham, Oct. 24, 1986; children: Karianne, Matthew David. BS, U. Utah, 1984. Chemist Dugway (Utah) Proving Grounds, 1985, Lee Sci., Salt Lake City, 1986-88; leader applications group Lee Sci. divsn. Dionex, Salt Lake City, 1988-90, tech. support mgr., 1990—. Co-contbr. chpt. to: Supercritical Fluid Extraction: Chromatography, 1988; contbr. chpt. to: Supercritical Fluids in Analytical Chemistry, 1993; author or co-author on eight other publs. Mem. ASTM, Assn. Ofcl. Analytical Chemists.

KNOWLES, JEREMY RANDALL, chemist, educator; b. Rugby, England, Apr. 28, 1935; came to U.S., 1974; s. Kenneth Guy Jack Charles and Dorothy Helen (Swingler) K.; m. Jane Sheldon Davis, July 30, 1960; children: Sebastian David Guy, Julius John Sheldon, Timothy Fenton Charles. B.A., Balliol Coll., Oxford (Eng.) U., 1958; M.A., D.Phil., Christ Ch., 1961; Doctor honoris causa, U. Edinburgh, 1992. Research fellow Calif. Inst. Tech., 1961-62; fellow Wadham Coll., Oxford U., 1962-74, univ. lectr., 1966-74; vis. prof. Yale U., 1969, 71; Sloan vis. prof. Harvard U., 1973; prof. chemistry Harvard U., 1974—, Amory Houghton prof. chemistry and biochemistry, 1979—, dean faculty of arts and scis., 1991—; Newton-Abraham vis. prof. Oxford U., 1983-84; hon. fellow Balliol Coll., Oxford U., Wadham Coll., Oxford U. Author papers, revs. bioorganic chemistry. Served as pilot officer RAF, 1953-55. Recipient Prelog medal ETH, Switzerland, CBE award (Queen's Birthday Honours), England, 1993. Fellow Am. Acad. Arts and Scis., Royal Soc. (Davy medal 1991), Royal Chem. Soc. London, (hon., Charmian medal); mem. NAS (fgn. assoc.), Biochem. Soc. London, Am. Chem. Soc. (Bader award, Cope Scholar award, Repligen Corp. award 1993), Am. Soc. Biol. Chemists, Am. Philos. Soc. Home: 7 Bryant St Cambridge MA 02138 Office: Harvard U Dean Faculty Arts & Scis University Hall 5 Cambridge MA 02138

KNOWLES, RICHARD JAMES ROBERT, medical physicist, educator, consultant; b. McPherson, Kans., Aug. 2, 1943; s. Richard E. and Pauline H. (Worland) K.; m. Stephanie R. Closter, May 14, 1970; 1 child, Guenevere Regina. BS, St. Louis U., 1965; MS, Cornell U., 1969; PhD, Poly. U., N.Y., 1979. Diplomate Am. Bd. Sci. in Nuclear Medicine, Am. Bd. Radiology. Chief med. physicist L.I. Coll. Hosp., Bklyn., 1977-81; dir. radiation physics lab. Downstate Med. Ctr., Bklyn., 1981-82; sr. med. physicist N.Y. Hosp. Cornell U. Med. Ctr., N.Y.C., 1982—; assoc. prof. physics in radiology Cornell U. Med. Coll., N.Y.C., 1989—. Author: Quality Assurance and Image Artifacts in Magnetic Resonance Imaging, 1988; contbr. articles to profl. jours. Mem. Am. Phys. Soc., Soc. Nuclear Medicine, Health Physics Soc., Am. Assn. Physicists in Medicine, N.Y. Acad. Scis., Soc. for Computer Applications in Radiology, Soc. Magnetic Resonance in Medicine, Sigma Xi. Office: NY Hosp-Cornell Med Ctr 525 E 68th St New York NY 10021-4873

KNOWLES, RICHARD NORRIS, chemist; b. Wilmington, Del., Aug. 8, 1935; s. Francis and Dorothy Edith K.; m. Alice Keith Pfohl, Aug. 30, 1957 (div. May 1987); children: Elizabeth Nelson, Dorothy Lawrence, Cynthia Norris; m. Claire Elaine Frerichs, Dec. 31, 1988. BS, Oberlin Coll., 1957; PhD, U. Rochester, 1961. With DuPont Co., Wilmington, 1960—, asst. works mgr. Chambers Works, N.J., 1980-83; mgr. Niagara Falls (N.Y.) plant, 1983, mgr. Belle (W.Va.) plant, 1987—; co. rep. to community awareness and emergency response com. Chem. Mfrs. Assn. Elder, Westminster Presbyn. Ch.; bd. dirs. Nat. Inst. of Chem. Studies; comsnr. W.V. State Emergency Comsn. Holder 40 patents in field. Kodak fellow, 1959-60. Mem. Am. Chem. Soc., W.Va. Mfrs. Assn. (bd. dirs.), Assn. for Quality and Participation, Audubon Soc., Rotary, Sierra Club, Nature Conservancy (DuPont Agrl. Products Crystal award 1991), Almost Heaven Hammered Dulcimer Soc.

KNOWLES, STEPHEN HOWARD, space scientist; b. N.Y.C., Feb. 28, 1940; s. Howard Nesmith and Emily Aurelia (Kent) K.; m. Joan Elizabeth Brame, June 5, 1965; children: Jennifer, Katherine. BA, Amherst Coll., 1961; PhD, Yale U., 1968. Space scientist Naval Rsch. Lab., Washington, 1961-86; tech. dir. Naval Space Surveillance Ctr., Dahlgren, Va., 1986—; vis. scientist Commonwealth Sci. and Indsl. Rsch. Orgn. Radiophysics Div., Sydney, Australia, 1974-76. Contbr. 75 articles to profl. jours. Mem. AIAA, Internat. Astron. Union, Am. Geophysical Union, Sigma Xi. Episcopalian. Achievements include rsch. in radar astronomy, spectral line radio astronomy, interferometry ionospheric physics, astrodynamics. Home: 12107 Harbor Dr Lakeridge VA 22192-2201 Office: Naval Space Surveillance Ctr Dahlgren VA 22448

KNOWLTON, NANCY, biologist; b. Evanston, Ill., May 30, 1949; d. Archa Osborn and Aline (Mahnken) K.; m. Jeremy Bradford Cook Jackson; 1 child, Rebecca Knowlton. AB, Harvard U., 1971; PhD, U. Calif., Berkeley, 1978. Asst. prof. biology Yale U., New Haven, 1979-84, assoc. prof., 1984; biologist Smithsonian Tropical Rsch. Inst., Panama, Republic of Panama, 1985—; panelist animal learning and behavior NSF, Washington, 1989-92; vis. scholar Wolfson Coll., Oxford (Eng.) U., 1990-91. Editor Am. Scientist, 1981-90. NATO postdoctoral fellow NSF, Liverpool, Cambridge, Eng., 1978-79. Mem. AAAS, Ecol. Soc. Am., Soc. Study Evolution. Office: Smithsonian Tropical Rsch Inst Unit 0948 APO AA 34002-0948

KNOX, DAVID LALONDE, ophthalmologist; b. Chgo., Sept. 3, 1930; s. Harry Jobes and Grace Elizabeth (LaLonde) K.; m. Linda Denny, 1958 (div. 1981); children: Benjamin, Mary Elspeth, Lucinda. MD, Baylor U., 1955. Diplomate Am. Bd. Ophthalmology. Assoc. prof. ophthalmology Johns Hopkins U. Sch. Medicine, Balt., 1962—; chmn. med. sch. admissions com. Johns Hopkins U. Sch. Medicine, 1976-79; cons. Montebello State Hosp., Balt., 1964—. Contbr. articles to profl. publs. Organizer Frank B. Walsh Neuroophthalmol. Soc., 1967. Capt. U.S. Army, 1956-58. Mem. Am. Ophthalmol. Soc. Office: Johns Hopkins U Sch Medicine 601 N Broadway Baltimore MD 21287-9013

KNOX, ERIC, botanist, educator. Prof. dept. botany U. Mich., Ann Arbor. Recipient George R. Cooley award Am. Soc. Plant Taxonomists, 1992. Office: Univ of Michigan Dept of Botany Ann Arbor MI 48108*

KNOX, RALPH DAVID, physicist; b. Wichita Falls, Tex., Nov. 16, 1961; s. Ralph Benner and Ruth Elaine (Aubol) K.; m. Anne Gray Chase, July 13, 1985. BS, U. Minn., 1983; PhD, Iowa State U., 1988. Scientist Iowa State U., Ames, 1988—. Contbr. articles to profl. jours. Mem. AAAS, IEEE, Am. Phys. Soc., Am. Vacuum Soc., Material Rsch. Soc., Sigma Pi Sigma. Home: 2101 Oakwood Rd # 327 Ames IA 50010 Office: Iowa State U 1925 Scholl Rd Ames IA 50011

KNOX, ROBERT ARTHUR, oceanographer, academic director; b. Washington, Jan. 15, 1943; s. James Milton and Virginia Matilda (Ernst) K.; m. Dorothy Chapin Hall, June 18, 1966; children: Leila Elizabeth, James Chapin. AB, Amherst Coll., 1964; PhD, MIT/Woods Hole Oceanographic Inst, 1971. Rsch. assoc. MIT, Cambridge, 1971-73; asst. rsch. oceanographer Scripps Inst. Oceanography/U. Calif. San Diego, La Jolla, 1973-81, acad. adminstr., 1980-90, assoc. rsch. oceanographer, 1981-86, rsch. oceanographer, 1986—, acting chair, chair oceanographic rsch. divsn., 1988-89, dir. physical oceanographic rsch. divsn., 1989-91, assoc. dir., 1992—; sr. vis. fellow Wolfson Coll., Oxford U., England, 1983. Contbr. articles to profl. jours. Coach, coord., v.p., pres. Del Mar (Calif.) Youth Soccer Club, 1981-90. Ford Found. fellow MIT, 1964-65. Mem. Am. Geophysical Union, Am. Meteorological Soc., Oceanography Soc., Phi Beta Kappa, Sigma Xi. Avocations: sailing, soccer. Home: 13019 Long Boat Way Del Mar CA 92014 Office: Scripps Inst Oceanography UCSD 0230 9500 Gilman Dr La Jolla CA 92093-0230

KNOX, WILLIAM JORDAN, physicist; b. Pomona, Calif., Mar. 21, 1921; s. Reginald Langenberger and Kate (Jordan) K.; m. Barbara Louise House, Sept. 27, 1948; children: William Jordan Jr., Margaret Louise, Sarah Ann, Reginald Vernon. BS in Chemistry, U. Calif., 1942, PhD in Physics, 1951. Rsch. asst. U. Chgo., 1942-43; jr. chemist Clinton Labs., Oak Ridge, Tenn., 1943-44; jr. technologist Hanford (Wash.) Engring. Works, 1944-45; chemist Clinton Labs., Oak Ridge, 1945-46; chemist, physicist Radiation Lab., U. Calif., Berkeley, 1946-51; rsch. physicist, asst. prof. Yale U., New Haven, 1951-60; assoc. prof., prof. physics U. Calif., Davis, 1960—; physicist U.S. Atomic Energy Commn., Div. of Rsch., Washington, 1953-55, cons., 1956; cons. Lawrence Berkeley Lab., 1980-85. Contbr. articles to profl. jours. Named Fulbright Rsch. fellow, 1973-74, Vis. Physicist, European Ctr. for Nuclear Rsch., 1973-74, Lawrence Berkeley Lab., 1981-82. Fellow Am. Phys. Soc. Achievements include patent in Method of Purifying Plutonium. Home: 731 Elmwood Dr Davis CA 95616 Office: U Calif Dept Physics Davis CA 95616

KNUDSEN, KNUD-ENDRE, civil engineer, consultant; b. Oslo, Norway, June 29, 1921; s. Edmund and Therese Marie (Ruud) K.; m. Kari Gulbrandsen, Aug. 14, 1951 (div. Nov. 1981); children: Trond, Lars, Per. MS in C.E., Norwegian Inst. Tech., Trondheim, 1946; PhD in Civl Engring., Lehigh U., 1949. Rsch. engr., Lehigh U., 1946-49, asst. prof., 1951-53; mng. dir. Norconsult Ethiopia, Addis Ababa, 1955-59, Norconsult A.S., Oslo, 1961-69, chmn., 1969-71; expert NATO Internat. Staff, Paris, 1959-61; cons. engr. K.E. Knudsen, Oslo, 1969—; pres. Saga Petroleum A.S., Oslo, 1973-79, mem., 1971-82, chmn., 1982-86; bd. dirs. Den norske Creditbank, Oslo; chmn. bd. reps. Factoring Finans A.S., Oslo, 1980-86, Norsk Skibs Hypthekbank, Oslo, 1980-87; mem. bd. reps. Norsk A/S Philips, 1980-90, Eksportfinans A.S., Oslo, 1978-85. Contbr. articles to profl. jours. Mem. Royal Norwegian Council for Sci. and Indsl. Resarch, 1976-79; bd. dirs. Fedn. Norwegian Industries, 1976-79; chmn. Commn on Orgn. of Norwegian State Petroleum Affairs, 1970-71; chmn. Norwegian Rsch. & Development Office for Continental Shelf, 1969-73. Hon. consul of Norway to Ethiopia, 1956-59; capt. Norwegian Army, 1944-45. Decorated knight 1st class Royal Norwegian Order St. Olav. Mem. Norwegian Acad. Tech. Scis. Conservative. Club: Norske Selskab (Oslo). Home: Stjerneveien 18, Oslo 3, Norway

KNUDSON, ALFRED GEORGE, JR., medical geneticist; b. Los Angeles, Aug. 9, 1922; s. Alfred George and Mary Gladys (Galvin) K.; m. Anna T. Meadows, June 20, 1977; children by previous marriage: Linda, Nancy, Dorene. B.S., Calif. Inst. Tech., 1944, Ph.D., 1956; M.D., Columbia U., 1947. Chmn. dept. pediatrics City of Hope Med. Center, Duarte, Calif., 1956-62; chmn. dept. biology City of Hope Med. Center, 1962-66; assoc. dean Health Sci. Center, SUNY, Stony Brook, 1966-69; dean Grad. Sch. Biomed. Scis., U. Tex. Health Sci. Center, Houston, 1970-76; dir. Inst. Cancer Research, Fox Chase Cancer Center, Phila., 1976-83, sr. mem., 1976—, disting. sci., 1992—, pres., 1980-82; mem. Assembly Life Scis., NRC, 1975-81. Author: Genetics and Disease, 1965; contbr. articles to profl. jours. Recipient Charles S. Mott prize Gen. Motors Cancer Research Found., 1988, medal of honor Am. Cancer Soc., 1989, Allan award Am. Soc. Human Genetics, 1991. Fellow AAAS; mem. NAS, Am. Philos. Soc., Am. Acad. Arts and Scis., Internat. Soc. Pediatric Oncology, Am. Soc. Human Genetics (pres. 1978), Assn. Am. Physicians, Am. Pediatrics Soc., Am. Assn. Cancer Rsch. Achievements include research in genetics of human cancer. Office: Inst Cancer Rsch 7701 Burholme Ave Philadelphia PA 19111-2497

KNUPFER, NANCY NELSON, computer science educator; b. Madison, Wis., Feb. 4, 1950; d. Alvie Charles and Cecelia Florence (Loisel) Nelson; m. Peter Bradley Knupfer, Oct. 27, 1984; children: Jason Scott Cartwright, Rebecca Lee, Kelly Marie, Sarah Jean. BS, U. Wis., LaCrosse, 1972; MA, U. Wis., 1984, PhD, 1987. Cert. tchr., Wis., Alaska. Instr. K-12, 1973-82; instr. sch. edn. U. Wis.-Madison, 1982-84; computer assisted instrl. devel. specialist Wis. Dept. of Health and Social Svcs., 1984-86; coord. tng., mgmt. informational specialist U. Wis., Madison, 1986-87; instrnl. designer CBT and IVD McDonnell Douglas Tng. Systems Inc., Litchfield Park, Ariz., 1989-90; asst. prof. Ariz. State U., Tempe, 1987-91; dir. multimedia prodn., asst. prof. Kans. State U., Manhattan, 1991—. Author: (with Robert Muffoletto) Educational Computing: Social Perspectives, 1993, (with others) My Computer and I: A Workbook for Secondary Students Using the IBM PC, 1984, (with others) Computer Related Activities Handbook: A Teacher Guide for Elementary and Intermediate Students, 1983; contbr. numerous articles to ednl. tech. jours. Meml. scholar Assn. Ednl. Communications and Tech., 1986; various grants; Conf. Internship award AECT, 1986. Mem. ASCD, ASTD, Internat. Visual Literacy Assn., Profs. Instrnl. Design and Tech., Assn. Ednl. Communications and Tech., Assn. Devel. Computer-Based Instrnl. Systems, Phi Kappa Phi. Home: 2804 Nevada St Manhattan KS 66502-2330 Office: Kans State U Coll of Edn 333 Bluemont Hall Manhattan KS 66506

KNUTH, DONALD ERVIN, computer sciences educator; b. Milw., Jan. 10, 1938; s. Ervin Henry and Louise Marie (Bohning) K.; m. Jill Carter, June 24, 1961; children: John Martin, Jennifer Sierra. BS, MS, Case Inst. Tech., 1960; PhD, Calif. Inst. Tech., 1963; DSc (hon.), Case Western Res. U., 1980, Luther Coll., Decorah, Iowa, 1985, Lawrence U., 1985, Muhlenberg Coll., 1986, U. Pa., 1986, U. Rochester, 1986, SUNY, Stony Brook, 1987, Valparaiso U., 1988, Oxford (Eng.) U., 1988, Brown U., 1988, Grinnell Coll., 1989, Dartmouth Coll., 1990, Concordia U., Montréal, 1991, Adelphi U., 1993; Docteur, U. Paris-Sud, Orsay, 1986, Marne-la-Valée, 1993; D Tech., Royal Inst. Tech., Stockholm, 1991; Pochetnogo Doktora, St. Petersburg U., Russia, 1992. Asst. prof. Calif. Inst. Tech., Pasadena, 1963-66, assoc. prof., 1966-68; prof. Stanford (Calif.) U., 1968-92, prof. emeritus, 1993—; cons. Burroughs Corp., Pasadena, 1960-68. Author: The Art of Computer Programming, 1968 (Steele prize 1987), Computers and Typesetting, 1986. Guggenheim Found. fellow, 1972-73; recipient Nat. Medal of Sci., Pres. James Carter, 1979, Disting. Alumni award, Calif. Inst. Tech., 1978, Priestly award, Dickinson Coll., 1981, Franklin medal, 1988, J.D. Warnier prize, 1989. Fellow Am. Acad. Arts and Scis.; mem. IEEE (hon., McDowell award 1980, Computer Pioneer award 1982), Nat. Acad. Scis., Nat. Acad. Engring., Assn. for Computing Machinery (Grace Murray Hopper award 1971, Alan M. Turing award 1974, Computer Sci. Edn. award 1986, Software Systems award 1986), Acad. Sci. (fgn. assoc. Paris and Oslo). Lutheran. Avocation: playing pipe organ. Office: Stanford Univ Computer Scis Dept Stanford CA 94305

KNUTSON, LYNN DOUGLAS, physics educator; b. Red Wing, Minn., Aug. 22, 1946; s. Olaf Leonard and Evelyn Julia (Frogum) K.; m. Joyce Clark, Aug. 24, 1968 (div. 1992). BA, St. Olaf Coll., 1968; MA, U. Wis., 1970, PhD, 1973. Rsch. assoc. U. Wis., Madison, 1973-74, asst. prof., 1977-80, assoc. prof., 1980-85, prof., 1985—; rsch. assoc. U. Wash., Seattle, 1974-76. Contbr. 44 articles to sci. jours. Fellow Am. Phys. Soc. Avocations: running, gardening, woodworking. Home: 3738 Ross St Madison WI 53705 Office: U Wis Dept Physics 1150 Univesity Ave Madison WI 53706

KO, MYOUNG-SAM, control engineering educator; b. Hamhung, Korea, Jan. 1, 1930; s. Chi-Young and Sun-hee (Kim) K.; m. Won-nam Lo, Apr. 13, 1962; 1 child, Woo-Sung. BS, Seoul (Korea) Nat. U., 1955, MS, 1960, PhD, 1972. Instr. to prof. control engring. Coll. Engring. Seoul Nat. U., 1962—, chmn. elec. engring. dept., 1976-78, chmn. control engring. dept., 1979-92; pres. Korean Inst. Elec. Engrs., Seoul, 1987-88, Korean Assn. Automatic Control, Seoul, 1989-90; dir. Automation & Sytems Rsch. Inst., Seoul, 1988—. Contbr. articles to profl. jours. Cpl. Republic of Korea Army, 1955-57, Korea. Recipient Order of Nat. Svc. Merit Pres. Republic of Korea, 1989. Fellow Japanese Soc. Instrumentation and Control Engrs. (internat. award 1991), Korean Inst. Elec. Engrs. (paper award 1973); sr. mem. IEEE (centenary award 1984). Achievements include development of industrial robot controller, automatic monitoring system for textile industry and FMS pilot plant at the inst. Home: Bangwi-dong Songpa-ku, Olympic Apt 310-503, Seoul 138-150, Korea Office: Seoul Nat U Coll Engring, 56-1 Shinlim-dong Kwanak-ku, Seoul 151-742, Republic of Korea

KOBAYASHI, MITSUE, chemistry educator; b. Tokyo, July 13, 1933; d. Tatsuo and Nobu (Haruyama) Nakazato; m. Masamichi Kobayashi, June 10, 1957. B in Engring., Waseda U., Tokyo, 1956; DSc, Osaka City U., 1962. Researcher Osaka City U., 1962-65, lectr., 1980—; rsch. assoc. Kyoto (Japan) U., Osaka, 1965-86, asst. prof. chemistry, 1986—. Author: Analytical Chemistry of Phosphorus Compounds, 1972. Mem. Chem. Soc. Japan, Atomic Energy Soc. Japan, Japanese Soc. Radiation Chemistry, Am. Chem. Soc., Internat. Soc. Neutron Capture Therapy. Home: Ohokubo 920-194 Kumatori, Sennan-gun, Osaka 590-04, Japan Office: Kyoto U Rsch Reactor Inst, Kumatori Sennan-gun, Osaka 590-04, Japan

KOBAYASHI, NAOMASA, biology educator; b. Tokyo, Mar. 4, 1929; s. Teiji and Aiko (Shimizu) K.; m. Eiko Kimura, Nov. 23, 1962; children: Norio, Sumiko. MSc, Kyoto (Japan) U., 1955, DSc, 1967. Lectr. Doshisha U., Kyoto, 1962-65, assoc. prof., 1965-71, prof., 1971—. Author: Water Pollution Bioassay, 1985. Mem. Am. Soc. Zoologists, N.Y. Acad. Scis., Zool. Soc. of Japan, Japanese Soc. Devel. Biologists, Japanese Soc. Sci. Fisheries. Avocations: philately, gardening, walking, touring. Home: 26, Kamitakano-morokicho,, Sakyo-ku, Kyoto 606, Japan Office: Doshisha U, Karasuma-Imadegawa,, Kamikyo-ku, Kyoto 602, Japan

KOBAYASHI, NOBUHISA, civil and coastal engineer, educator; b. Osaka, Japan, May 4, 1950; came to U.S., 1976; s. Kazunobu and Shigeko Kobayashi; m. Sharon G. Fisher, Aug. 13, 1983; children: Sachi C., Orion A. BCE, Kyoto (Japan) U., 1974, MCE, 1976; PhD, MIT, 1979. Sr. cons. engr. Brian Watt Assocs., Inc., Houston, 1979-81; asst. prof. U. Del., Newark, 1981-86, assoc. prof., 1986-91, prof. dept. civil engring., 1991—. Mem. editorial bd. Jour. Coastal Rsch., 1987—; assoc. editor Jour. Waterway, Port, Coast and Ocean Engring., 1990-92, editor, 1992—; editor Rational Design of Mound Structures, 1990; contbr. articles to profl. jours. Rsch. grantee NSF, 1984-92, NOAA, 1983—, U.S. Army Coastal Engring. Rsch. Ctr., 1988-92, U.S. Army Cold Regions Rsch. and Engring. Lab., 1986-87, Army Rsch. Office, 1992—. Mem. ASCE, Internat. Assn. Hydraulic Rsch., Am. Geophys. Union, Japan Soc. Civil Engrs. Achievements include development of new numerical methods for rational design of breakwaters and revetments used to protect harbors and beaches, arctic engring., coastal sediment transport, coastal hydrodynamics, marina design and oil spills. Office: U Del Dept Civil Engring Newark DE 19716

KOBAYASHI, SUSUMU, data processing executive, super computer consultant; b. Kumamoto, Japan, Apr. 3, 1939; s. Senkichiro and Michiko Kobayashi. BS, Tokyo Inst. Tech., 1963. Programmer Osaka (Japan) Gas Co., Ltd., 1963-65, C. Itoh Computing Services Co., Ltd., Tokyo, 1965-67; applications analyst, systems engr. Control Data Far East, Inc., Tokyo, 1967-75; asst. gen. mgr. systems dept. JMA Systems, Inc., Tokyo, 1975-79; dir. Nuclear Data Corp., Tokyo, 1979-89, Yokogawa Supertek Corp., Tokyo, 1989-90; tech. advisor sales div. Yokogawa Cray ELS Ltd., Tokyo, 1990-92; tech. advisor Cray Rsch. Japan Ltd., Tokyo, 1990—. Translator, editor: Fortran 4 (D.D. McCracken), 1968, Lisp 1.5 Primer (C. Weissman), 1970, A Few Good Men from Univac, (D.E. Lundstrom), 1992, The Official Computer Widow's (and Widower's) Handbook (by Experts on Computer Widow/Widowerhood), 1992, Future Computer Opportunities (Jack Dunning), 1993; contbr. articles to electronics mags. Mem. Assn. Computing Machinery, IEEE, Inc., Japan Math. Soc., Japan Info. Processing Soc., Am. Assn. for Artificial Intelligence. Avocations: motoring, audio/visual. Home: 85-2-206 Migawa 2-chome, Mito-shi, Ibaraki-ken 310, Japan Office: Cray Rsch Japan Ltd, 6-4 Ichiban-cho, Chiyoda-ku 7 Fl, Tokyo 102, Japan

KOBAYASHI, TOSHIRO, materials science educator; b. Sapporo, Hokkaido, Japan, May 20, 1939; s. Koichi and Tomoe (Yoshizumi) K.; m. Fumiko Fukuda, May 1, 1969; children: Chiharu, Toshiya. B in Engring., Hokkaido U., Sapporo, 1961, D in Engring., 1971. Registered profl. engr., Japan. Researcher Fuji Electric Corp. R&D Ltd., Yokosuka, Kanagawa, Japan, 1961-72; assoc. prof. Nagoya (Aichi, Japan) U., 1972-81; prof. Toyohashi (Aichi, Japan) U. Tech., 1982—, chmn. material system engr., 1990—. Inventor high resistivity al alloy, tekko-zairyo-Kogaku, CAI impact testing system; contbr. numerous articles to profl. jours. Recipient Japan Invention award Japan Invention Soc., 1981, Nishiyama Meml. prize Japan Inst. Iron & Steel, 1989, Iidaka prize Japan Foundrymen's Soc., 1993. Mem. ASTM, Minerals, Metals and Materials Soc., Inst. of Metals. Avocations: classical music, golf. Home: 21-16 Kodare, Ogasaki-cho, Toyohashi 441, Japan Office: Toyohashi U Tech, 1-1 Hibarigaoka, Tempakucho, Toyohashi 441, Japan

KOBAYASHI, YOSHINARI, polymer chemist; b. Tokyo, Aug. 19, 1934; s. Kinshiroh and Toku (Kashiwagura) K.; m. Kazue Okazaki, Nov. 15, 1969; children: Hiromi, Terue. Diploma in engring., Keio U., Tokyo, 1958; D. of Engring. (hon.), Keio U., 1974; D. of Agr. (hon.), Kyoto (Japan) U., 1990. Mem. staff Toray Co., Ltd., Tokyo, 1958; mem. tech. staff Toray Co., Ltd., Nagoya, Japan, 1958-63, researcher Nagoya Lab., 1963-65; dir. chem. fiber paper sect. Govt. Ind. Rsch. Inst., Shikoku, Takamatsu, Japan, 1975-81, dir. polymer resources sect., 1981-86, dir. planning sect., 1986-87, dir. system engring. dept., 1987-90, dir. material sci. dept., 1990-93; dir. tech. ctr. inform, exch. Govt. Ind. Rsch. Inst., Shikoku, Takamatsu, 1993—. Author: Round Trips for Washi, 1988, Papers From Seaweeds, 1990, Kenaf as an Earth-friendly Papermaking Resource, 1991, Papers from the Sea, 1993; editor: From Chemical Fiber Paper to High Performance Paper, 1988, High Performance Paper Reviews, 1993. Recipient prize Agy. of Sci. and Tech., Tokyo, 1990. Mem. High Performance Paper Soc. (exec. mng. dir. 1984—). Liberal Democrat. Buddist. Home: 2298-26 Yashimanish-machi, Takamatsu, Kagawa 761-01, Japan Office: Govt Indsl Rsch Inst, Shikoku #3 3-ban 2-chome Hananaomiya, Takamatsu Kagawa 761, Japan

KOBELSKI, ROBERT JOHN, analytical chemist; b. N.Y.C., Sept. 30, 1948; s. Frank William and Helen Clara (Himmel) K.; m. Pamela Gail Weed, June 15, 1974. MS, U. Vt., 1974; PhD, SUNY, Buffalo, 1986. Chemist E.I. du Pont Marshall Labs., Phila., 1973-75; sr. analytical chemist Buffalo Color Corp., 1980-85; prin. scientist Personal Products Co./Johnson & Johnson, Milltown, N.J., 1985-88; applications engr. Hewlett-Packard Analytical Edn., Atlanta, 1988—; governing bd. dirs. N.E. Regional Chromatography Discussion Group, Buffalo, 1983-85. Mem. Am. Chem. Soc., Am. Soc. for Mass Spectrometry. Home: 3349 Somerset Ter 3349 Somerset Tr Marietta GA 30067 Office: Hewlett-Packard C-03 2000 S Park Pl Atlanta GA 30339

KOBSA, HENRY, research physicist; b. Vienna, Austria, May 4, 1929; came to U.S., 1956; s. Rudolf and Johanna Maria (Urbanitzky) K.; m. Mary Jane Liston, Sept. 19, 1980. PhD in Chemistry, U. Vienna, 1955. Various positions DuPont Co., 1956-68; rsch. mgr. DuPont Co., Kinston, N.C., 1968-73, tech. supr., 1973-77; rsch. fellow DuPont Co., Wilmington, Del., 1977-82, sr. fellow, 1982-89, fellow, 1989—. Contbr. articles to profl. jours. Mem. AICE, SPIE, Am. Phys. Soc., Am. Optical Soc., Am. Chem. Soc., Laser Inst. Am. Libertarian. Achievements include patents for synthetic fiber products processes, laser treatment of polymers, micromachining with lasers. Office: DuPont PO Box 80715 Wilmington DE 19880-0715

KOBUS, DAVID ALLAN, research psychologist, consultant; b. Syracuse, N.Y., July 8, 1952; s. William Emil and Rita Evelyn (Brown) K.; m. Nancy Winifred Raner, Aug. 8, 1970; children: David, Amy, Jason. BA, San Diego State U., 1980; MS, Syracuse U., 1982, PhD, 1984. Enlisted USN, 1971, advanced through grades to lt. comdr.; psychologist Naval Submarine Med. Rsch., Groton, Conn., 1983-85; dept. head cognitive performance and psychophysiology dept. Naval Health Rsch. Ctr., San Diego, 1985—; asst. prof. psychology San Diego State U., 1986—. Contbr. articles to profl. jours. Mem. APS, Soc. Psychophysiol. Rsch., Western Psychol. Assn., Human Factors Soc. Achievements include developing filter for night-time ambient illumination used in Navy and aviation community. It allows enough ambient lighting to perform tasks at night while maintaining dark adaptation. Office: Naval Health Rsch Ctr PO Box 85122 San Diego CA 92186-5122

KOBUS, RICHARD LAWRENCE, architectural company executive; b. Chgo., Nov. 19, 1952. BS in Architecture, U. Ill., 1974; MArch, Harvard U., 1978. Registered architect, Mass., N.H., Maine, Ill., Pa. Designer Metz, Train, Olsen & Youngren, Chgo., 1974-75, Sheply, Bulfinch, Richardson & Abbott, Boston, 1978-79; assoc. Skidmore, Owings and Merrill, Boston, 1979-83; pres., prin., founder Tsoi/Kobus & Assocs., Inc., Cambridge, Mass., 1983—. Mem. AIA, U. L.I. NAIOP, Boston Soc. Architects, 1980, Soc. Brigham and Women's Hosp., Billings Soc., U. Chgo. Avocations: sailing, rowing, photography, auto racing. Office: Tsoi Kobus & Assocs Inc One Brattle Square Cambridge MA 02138-3726

KOCAN, KATHERINE MAUTZ, veterinary educator, researcher; b. Cleve., Mar. 27, 1946; d. Frederick Robert and Bertha (Specht) Mautz; m. Andrew Alan Kocan, June 15, 1968; children: Andrew James, Jonathan Michael. MSPH, U. N.C., 1971; PhD, Okla. State U., 1979. From rsch. assoc. to asst. prof. Coll. Vet. Medicine Okla. State U., Stillwater, 1974-1988, assoc. prof., 1988-93, prof., 1993—. Contbr. articles to profl. jours. Bd. dirs. Friends of Music, Stillwater, 1992—. Mem. Am. Soc. Tropical Vet. Medicine (pres.-elect 1991-93, pres. 1993-95, conf. chair 1991-93). Office: Okla State U Coll Vet Medicine Dept Vet Pathology Stillwater OK 74078

KOCAOGLU, DUNDAR F., engineering executive, industrial and civil engineering educator; b. Turkey, June 1, 1939; came to U.S., 1960; s. Irfan and Meliha (Uzay) K.; m. Alev Baysak, Oct. 17, 1968; 1 child, Timur. BSCE, Robert Coll., Istanbul, Turkey, 1960; MSCE, Lehigh U., 1962; MS in Indsl. Engring., U. Pitts., 1972, PhD in Ops. Rsch., 1976. Registered profl. engr., Pa., Oreg. Design engr. Modjeski & Masters, Harrisburg, Pa., 1962-64; ptnr. TEKSER Engring. Co., Istanbul, 1966-69; project engr. United Engrs., Phila., 1964-71; rsch. asst. U. Pitts., 1972-74, vis. asst. prof., 1974-76, assoc. prof. indsl. engring., dir. engring. mgmt., 1976-87; prof., dir. engring. mgmt. program, Portland State U., 1987—; pres. TMA-Tech. Mgmt. Assocs., Portland, Oreg., 1973—, pres. Portland Internat. Conf. Mgmt. Engring. and Tech., 1990—. Author: Engineering Management, 1981; editor: Management of R&D and Engineering, 1992; co-editor: Technology Management—The New International Language, 1991; series editor Wiley Series in Engring. and Tech. Mgmt.; contbr. articles on tech. mgmt. to profl. jours. Lt. C.E., Turkish Army, 1966-68. Fellow IEEE (Centennial medal 1984, editor-in-chief trans. on engring. mgmt. 1986—); mem. Inst. Mgmt. Scis. (chmn. Coll. Engring. Mgmt. 1979-81), Am. Soc. Engring. Edn. (chmn. engring. mgmt. div. 1982-83), IEEE Engring. Mgmt. Soc. (fellow, publs. dir. 1982-85), ASCE (mem. engring. mgmt. adminstrv. com. 1988—), Muhendis, Ilim Adamlari ve Mimarlar Dernegi Soc. Turkish Engrs. and Scientists (hon.), Am. Soc. Engring. Mgmt. (dir. 1981-86), Omega Rho (pres. 1984-86).

KOCH, ALISA ERIKA, biomedical researcher, rheumatologist; b. Jerusalem, Feb. 26, 1956; came to the U.S., 1958; d. Walter and Dora Deborah (Kaplan) K.; m. Howard Stein, June 19, 1988; 1 child, Joshua Stein. BS, Northwestern U., 1978, MD, 1980. Intern, then resident Loyola U. Med. Ctr., Maywood, Ill., 1980-83; instr. in medicine Northwestern U. Med. Sch., Chgo., 1983-86, asst. prof., 1986-92, assoc. prof., 1992—; staff physician Northwestern Meml. Hosp., Chgo.; assoc. investigator rsch. svcs. VA Lakeside Med. Ctr., Chgo., 1986-89, staff physician, 1989—, chief sect. rheumatology, 1991—; cons. physician Rehab. Inst. Chgo.; vis. scholar U. Mich., Ann Arbor, 1990-91; mem. spl. study sect. NIH, Bethesda, 1991, ad hoc mem. pathology study sect., 1992; mem. med. and sci. affairs com. Arthritis Found. Chgo., 1989—. Contbr. articles to Sci., Arthritis Rheumatism, Lab. Investigation, Am. Jour. Pathology, Jour. Immunology, Jour. Clin. Investigation. Arthritis Found. fellow, 1986-89; recipient Daniel and Ada Rice Found. Rsch. award Arthritis Found., 1989-90, Robert M. Kark prize for meritorious rsch. Chgo. Soc. Internal Medicine, 1991, Henry Christian award Am. Fedn. Clin. Rsch., 1992, Ralph and Marion Falk Challenge prize Arthritis Found. Fellow ACP; mem. Cen. Soc. for Clin. Rsch., Am. Coll. Rheumatology, Am. Fedn. Clin. Rsch. Achievements include patents for development of monoclonal antibodies detecting macrophage and endothelial antigens; research in immunopathogenesis of rheumatoid arthritis in regard to macrophage and endothelial cell surface antigens and function. Home: 1527 William St River Forest IL 60305 Office: Northwestern U Med Sch Ward 3-315 303 E Chicago Ave Chicago IL 60611

KOCH, DONALD LEROY, geologist, state agency administrator; b. Dubuque, Iowa, June 3, 1937; s. Gregory John and Josephine Elizabeth (Young) K.; m. Celia Jean Swede, July 5, 1962; children: Kyle Benjamin, Amy Suzanne, Nathan Gregory. BS, U. Iowa, 1959, MS in Geology, 1967, postgrad., 1971-73. Research geologist Iowa Geol. Survey, Iowa City, 1959-71, chief subsurface geology, 1971-75, asst. state geologist, 1975-80, state geologist and dir., 1980-86; state geologist and bur. chief Geol. Survey Bur., Iowa City, 1986. Contbr. articles to profl. jours. Fellow Iowa Acad. Sci. (bd. dirs. 1986-89); mem. Geol. Soc. Iowa (pres. 1969, 86), Iowa Groundwater Assn. (pres. 1986), Sigma Xi. Lodge: Rotary. Avocations: bicycling, camping, chess, numismatics. Home: 1431 Prairie Du Chien Rd Iowa City IA 52245-5615 Office: Geol Survey Bur 109 Trowbridge Hall Iowa City IA 52242-1319

KOCH, EVAMARIA WYSK, oceanographer, educator; b. Porto Alegre, Brazil, May 11, 1961; came to U.S., 1985; d. Walter and Eva Margarethe Elsa Anna (Wysk) K. BS in Oceanography, U. Rio Grande, Brazil, 1984; MS in Botany, U. South Fla., 1988, PhD in Marine Sci., 1993. Rsch. asst. U. Rio Grande, 1980-85, U. South Fla., Tampa, 1985-89; biol. scientist Fla. Marine Rsch. Inst., St. Petersburg, 1988-89; with NOAA, Milford, Conn. Contbr. to profl. publs. Mem. Brazilian Assn. Oceanography, Brazilian Phycological Soc., Estuarine Rsch. Fedn., Sigma Xi. Achievements include development of culture media for tissue culture of seagrasses. Office: NOAA 212 Rogers Ave Milford CT 06460

KOCH, KEVIN ROBERT, metallurgical engineer; b. Evansville, Ind., Oct. 22, 1967; s. Robert Louis and Cynthia (Ross) K. BS, U. Notre Dame, 1990. Assoc. engr. Westinghouse Electric Corp., Orlando, Fla., 1990—. Mem. ASTM. Office: Westinghouse Electric Corp 4400 N Alafaya Trail # 303 Orlando FL 32826-2399

KOCH, ROBERT MICHAEL, research scientist, consultant; b. Mineola, N.Y., Apr. 19, 1964; s. Roy Arthur and Ellen Anne (Trimble) K.; m. Laureen Theresa Chase, July 6, 1991. BSME, Poly. U., Bklyn., 1986, PhD in Applied Mechanics, 1991. Profl. engr., N.Y. Mech. engr. Vernitech Corp., Deer Park, N.Y., 1983-85; instr. Poly. U., Bklyn., 1986-91; rsch. scientist Naval Undersea Warfare Ctr., Newport, R.I., 1991—; cons. Beltran, Inc., Bklyn., 1988-91. Teaching fellow Poly. U., 1986-90, rsch. fellow, 1987, 90. Mem. AIAA, ASME, Sigma Xi. Republican. Roman Catholic. Achievements include research in underwater structural acoustics, adaptive procedures in h- and p- version finite element analysis, rapid prototyping with sterolithography, probabilistic structural mechanics, ultrasonic wave propagation in elastic solids. Home: 18 McIntosh Dr Portsmouth RI 02871 Office: Naval Undersea Warfare Ctr Code 8233 Bldg 108/2 Newport RI 02841-5047

KOCHAK, GREGORY MICHAEL, biophysical pharmacy/pharmacokinetics researcher; b. Warren, Ohio, May 8, 1953; s. Michael and Helen (Tymochko) K.; m. Jacqueline Lee White; children: Natalya, Thomas, Joseph, Emma Jane, Amanda. BS in Pharmacy, Ohio State U., 1976; PhD, U. Mo., 1981. Registered pharmacist, Ohio. Sr. scientist Ciba-Geigy Corp., Ardsley, N.Y., 1981-84, sr. rsch. scientist, 1985-86, mgr., 1986—; adj. prof. U. Mo., Kansas City, 1987—. Contbr. articles to profl. jours. Fellow N.Y. Acad. Sci.; mem. Am. Assn. Pharm. Scientists (chmn. ea. regional chpt. 1987, 88, pres. Hudson Valley pharmaceutics 1987). Achievements include pioneering of quantitative estimation of source components of variance in biological systems, in particular, pharmacokinetic systems. Home: 2743 Hyatt St Yorktown Heights NY 10598 Office: Ciba-Geigy Corp 444 Saw Mill River Rd Ardsley NY 10502

KOCHERLAKOTA, SREEDHAR, manufacturing engineer; b. Hyderabad, India, July 1, 1965; came to U.S., 1988; s. Ramana Murthy and Laxmi Kocherlakota. BE, Osmania U., Hyderabad, India, 1988; MS, N.J. Inst. Tech., 1991. Mfg. indsl. engr. Teledyne Adams, Union, N.J., 1990-91; mfg. quality engr. Mini-Circuit Labs., Bklyn., 1992—. Home: 515 W 59th St # 31E New York NY 10019

KOCHI, JAY KAZUO, chemist, educator; b. Los Angeles, May 17, 1927; s. Tsuruzo and Shizuko (Moriya) K.; m. Marion Kiyono, Mar. 1, 1959; children—Sims, Ariel, Julia. Student, Cornell U., 1945; B.S., UCLA, 1949; Ph.D., Iowa State U., 1952. Faculty Harvard U., 1952-55; NIH fellow Cambridge U., Eng., 1956; mem. faculty Iowa State U., 1956; with Shell Devel. Co., 1957-61; mem. faculty dept. chemistry Case Western Res. U., Cleve., 1962-69; prof. Case Western Res. U., 1966-69; prof. chemistry Ind. U., Bloomington, 1969-74; Earl Blough prof. chemistry Ind. U., 1974-84; Robert A. Welch Disting. prof. chemistry U. Houston, 1984—; cons. chemist, 1964—. Mem. Am. Chem. Soc., Chem. Soc. (London), Nat. Acad. Scis., Sigma Xi. Achievements include research on mechanism of catalysis of organic reactions, organometallics, electrochemistry and photochemistry, time-resolved spectroscopy of reactive intermediates. Home: 4372 Faculty Ln Houston TX 77004-6601

KOCI, HENRY JAMES, manufacturing company executive; b. Berwyn, Ill., Jan. 29, 1952; s. Henry Harold and Florence Vera (Svoboda) K.; m. Arlene Frances Banas, June 16, 1979; 1 child, Sarah. BS, Washington and Lee U., 1974; BS in Econ., DePaul U., Chgo., 1975, MBA, 1977, MS in Econ., 1977. Inventory analyst Ace Hardware, Oak Brook, Ill., 1978; dir. purchasing Medline Industries, Northbrook, Ill., 1979; commodity mgr. Navistar Internat., Schaumburg, Ill., 1979-89, Reliable Electric Div. Reliance Electric, Franklin Park, Ill., 1990-91, A.B. Dick Co., Niles, Ill., 1991—. Precinct capt. Lyons Twp. Reps., Western Springs, Ill., 1976-81; solicitor bldg. fund Christ Ch. of Oak Brook, Ill., 1986-88, mem. Breakfast Club. Mem. Chgo. Purchasing Assn. Avocations: Civil War historian, theater, golf, football, gardening. Home: 105 Heath Pl Westmont IL 60559-2645 Office: A B Dick Co 5700 Touhy Ave Niles IL 60714

KOCIVAR, BEN, aviation specialist, journalist; b. N.Y.C., Apr. 13, 1916; s. Izidor and Rebecca (Ladman) K.; m. Thelma Levine, Aug. 30, 1941; children: Karl, Carol, Jane. BS in Social Sci., CCNY, 1938; MA in Journalist, U. Mo., 1941. Info. specialist War Dept., Washington, 1941-45; sr. editor Look Mag., N.Y.C., 1945-69; pub. rels. mgr. N.Y. Airways, N.Y.C., 1971-79, Crossair Airlines, Zurich, 1985-88; cons. Pan Am. Airlines, N.Y.C., 1969-72, Agusta Helicopter USA, Houston, 1981-85, SAAB Aircraft, Washington, 1984-88; aviation editor Popular Sci., N.Y.C., 1971-88; aviation tech. specialist FAA, Washington, 1986—. Contbr. articles to profl. publs. Candidate North Castle (N.Y.) Town Bd., 1983. Recipient Aviation/Space Writers award All Am. Cities, 1975. Mem. AIAA, Nat. Sci. Writers Assn., N.Y. Airline Pub. Rels. Assn. (chmn.), Am. Soc. Journalists, Wings Club (N.Y.C.), Aero Club (Washington). Home: Rt 4 Box 330 Bedford NY 10506 Office: FAA JFK Airport Jamaica NY 11430

KOCK, LARS ANDERS WOLFRAM, physician, educator; b. Stockholm, Aug. 29, 1913; s. Gosta and Elsa (Wik) K.; m. Marianne Segerstéen, May 18, 1950; 1 child, Lars Kock. License medicine, Karolinska Institutet, 1941, MD, 1952. Intern and resident in Stockholm and Stocksund, 1940-54; pvt. practice medicine, specializing in gen. medicine and surgery, Stockholm; house surgeon several hosps., Stockholm, 1940-54; asst. prof. med. history rsch. Karolinska Institutet, Stockholm, 1953-79, prof., 1982, head physician Swedish Philips Co., Stockholm, 1954-78; dir. Mus. Med. History, Stockholm, 1955-88; sr. med. officer Defence Forces Med. Adminstr. Svc., Stockholm, 1955-76; fiduciary physician Social Ins. Office, 1976-82. Served with M.C., Swedish Navy, 1945-55. Decorated knight Vasa Order, knight Order North Pole Star, King's Golden Medal with Seraphimer ribbon; recipient Gold Medal for merit Mus. Med. History, Stockholm, 1977; scholarship of honor Stockholm County Coun., 1989. Mem./hon. mem. numerous socs. med. history; mem. Swedish Med. Soc. (150 Silver Jubilee medal 1958, Jubilee prize 1978, Carl Trafvenfelt Gold medal 1988, hon. sec. sect. med. history 1947-75, pres. 1975-89), Friends Mus. Med. History Stockholm (founder, hon. sec. 1952-75, pres. 1975-89), Société Internationale d'Histoire de la Médecine (asst. gen. sec. 1964-70), Internat. Acad. History Medicine, Scandinavian Soc. Med. History (pres. 1976-78), Finnish Soc. Med. Scis. Author: Resa till Rio, 1947; Resa till Kap, 1949; Kungl, Serafimerlasarettet 1752-1952, 1952; Medicinhistoriens Grunddrag, 1955; Svenska Konungars Sjukdomar, 1963; Kirurgminnen fran Karl Johanstiden, 1964; Olof af Acrel, 1967; Svenska Lakare som Vitterlekare, 1970; Svensk kirurgi-historisk rapsodi, 1978; (memoirs) Läkare och Lekman, 1982; Medicinhistorisk exposé, 1989; editor: Medicinalväsendeti Sverige 1813-1962, 1963; Dan Vincent Lundberg: Mina Minnen, 1983; editor-in-chief, founder Nordisk Medicinhistorisk Arsbok, 1953-90; contbr. articles to profl. publs. Home: Villa Walhall, Parkgatan 12, 15132 Södertälje Sweden

KOČKA, JAN VILÉM, physicist; b. Mladá Boleslav, Czechoslovakia, Jan. 19, 1946; s. Vilém V. and Jana A. (Vokálová) K.; m. Sylva Vodrážková, May 8, 1970; children: Viktor, Tomáš. D Natural Scis., Charles U., Prague, Czechoslovakia, 1969; PhD, Acad. Scis., Prague, 1975; DSc, Charles U., Prague, Czechoslovakia, 1990. Rsch. worker Inst. Solid State Physics, Acad. Scis., Prague, 1974-77; postdoctoral fellow Cavendish Lab. U. Cambridge, Eng., 1977-78; rsch. worker Inst. Physics, Acad. Scis., Prague, 1978—, head dept., 1990—; vis. prof. Institut für Physikalische Elektronik, U. Stuttgart, 1990, Tokyo Inst. Tech., 1993; mem. steering com. Internat. Conf. on Amorphous Semiconductors, 1991—; mem. adv. bd. Charles U., Prague, 1991—; mem. sci. bd. Inst. Physics, Prague, 1992—. With Czechoslovakia mil., 1969-70. Mem. Union of Czechoslovak Physicists and Mathematicians, Solidus Sporting Club. Office: Inst Physics, Cukrovarnická 10, 162 00 Prague 6, Czech Republic

KOCKA, THOMAS JOHN, mechanical engineer; b. Cleve., Sept. 10, 1957; s. Elmer Edward and Lillian Mary (Stedronsky) K. B in Mech. Engring., Cleve. State U., 1980. Registered profl. engr., Ohio. Mech. engr. Davy McKee, Independence, Ohio, 1980-83, Colpetzer Thomas, Willoughby, Ohio, 1983-87, East Ohio Gas, Cleve., 1987—. Author: Strategies for Reducing Natural Gas, Electric and Oil Costs, 1989. Mem. ASHRAE (energy com. 1989-), Assn. Energy Engrs., Cleve. Engring. Soc. Roman Catholic. Office: East Ohio Gas 1201 East 55th Cleveland OH 44101

KODALI, HARI PRASAD, electrical engineer; b. Guntur, India, July 14, 1949; came to U.S., 1982; s. Appiah Chowdary and Raja Ratnamma (Thottempudi) K.; m. Vijaya Lakshmi Tummala, Aug. 16, 1978; children: Sireesha, Deepa. BSEE, Sri Venkateswara U., Tirupathi, India, 1977. Jr. engr. Maharashtra State Electricity Bd., Nagpur, India, 1978-81; telecommunications engr. Electronic Data Systems, Southfield, Mich., 1985; engr. Bechtel Power Corp., Gaithersburg, Md., 1986-88; assoc. elec. engr. Niagara Mohawk Power Corp., Syracuse, N.Y., 1988-90, elec. engr., 1990—. Mem. IEEE. Hindu. Achievements include development of design basis to calculate thermal overload protection of safety equipment, development of fuse control program for nuclear power station. Office: Niagara Mohawk Power Corp 301 Plainfield Rd Syracuse NY 13212

KODYM, MILOSLAV, psychologist, researcher; b. Sobeslav, Czechoslovakia, Aug. 1, 1930; s. Ruzena Kodymova; m. Miroslava Lintnerova, Sept. 8, 1951; children: Miloslava, Roman. PhD, U. Prague, 1967, CSc, 1971. Tchr. Pedagogic Sch., Ceske Budejovice, Czechoslovakia, 1953-59; mem. faculty Faculty Pedagogy, Ceske Budejovice, 1959-79; sci. worker, dir. Inst. Psychology, Prague, Czechoslovakia, 1979-91; sci. worker, dep. dir. Inst. Edn. Fed. Ministry Interior, 1991—; chief editor Chechoslovak Psychology, Czechoslovak Acad. Scis., 1979-91; pres. sci. bd. pedagogy and psychology, 1982-91; chief psychology dept. Faculty Pedagogy, Prague, 1985-91. Author: Problem Solving and Performance, 1972, Selection of Talents, 1978, On the Theory of Abilities, 1987; author, editor: Psychological Aspects of Personality Development, 1987. Recipient silver medal Czechoslovk Acad. Scis., 1980, state honors for excellent work, 1985, laureat Acads. Scis. and Social Scis., Moscow-Prague, 1985. Mem. (corresponding) Extranjero de la Soc. Cubana de la Salud, Czechoslovak Psychol. Assn. (com. 1982—). Avocations: sports, music. Home: Machuldova 596/21, 142 00 Prague 4, Czech Republic Office: Ministry Interior, Inst of Edn, PO Box 7063, 170 21 Prague 7, Czech Republic

KOEKOEK, ROELOF, mathematics educator; b. Leiden, The Netherlands, Oct. 25, 1963; s. Albert and Jantje (Eilders) K.; m. Caroline Kroes, Jan. 22, 1988. D in Math., Groningen U., The Netherlands, 1986; D in Math., Delft (The Netherlands) U. of Tech., 1990. Asst. in opleiding Delft U. Tech., 1986-90, math. instr., 1990—; math. tchr. Hogesch. Rotterdam (The Netherlands) en Omstreken, 1987-90. Contbr. articles to profl. publs. Mem. Wiskundig Genootschap, Am. Math. Soc. Home: Bikolaan 154, 2622 EM Delft The Netherlands Office: Delft U Tech, Mekelweg 4, 2628 CD Delft The Netherlands

KOELMEL, LORNA LEE, data processing executive; b. Denver, May 15, 1936; d. George Bannister and Gladys Lee (Henshall) Steuart; m. Herbert Howard Nelson, Sept. 9, 1956 (div. Mar. 1967); children: Karen Dianne, Phillip Dean, Lois Lynn; m. Robert Darrel Koelmel, May 12, 1981; stepchildren: Kim, Cheryl, Dawn, Debbie. BA in English, U. Colo., 1967. Cert. secondary English tchr. Substitute English tchr. Jefferson County Schs., Lakewood, Colo., 1967-68; sec. specialist IBM Corp., Denver, 1968-75, pers. administr., 1975-82, asst. ctr. coord., 1982-85, office systems specialist, 1985-87, backup computer operator, 1987—; computer instr. Barnes Bus. Coll., Denver, 1987-92; owner, mgr. Lorna's Precision Word Processing and Desktop Pub., Denver, 1987-89; computer cons. Denver, 1990—. Editor newsletter Colo. Nat. Campers and Hikers Assn., 1992—. Organist Christian Sci. Soc., Buena Vista, Colo., 1963-66, chmn. bd. dirs., Thornton, Colo., 1979-80. Mem. NAFE, Nat. Secs. Assn. (retirement ctr. chair 1977-78, newsletter chair 1979-80, v.p. 1980-81), U. Colo. Alumni Assn., Alpha Chi Omega (publicity com. 1986-88). Republican. Club: Nat. Writers. Lodge: Job's Daus. (recorder 1953-54). Avocations: needlepoint, piano, bridge, reading, golf.

KOENIG, GOTTLIEB, mechanical engineering educator; b. Gottschee, Yugoslavia, Apr. 14, 1940; came to U.S., 1952; s. Ernst and Aloisia (Kump) K.; m. Berta Poje, June 25, 1966; children: Robert G., Elizabeth A. BSME, The Cooper Union, 1967; MSME, NYU, 1968, PhD, 1976. Lic. profl. engr., N.Y. Prof. aircraft design Acad. Aeronautics, Flushing, N.Y., 1960-67; chmn. aircraft design dept. Acad. Aeronautics, Flushing, 1970-73, chmn. techs. dept., 1973-76, assoc. dean acad. affairs, 1976-82; prof., chmn. dept. mech. and indsl. engring. N.Y. Inst. Tech., Old Westbury, N.Y., 1982—; adm. asst. prof. SUNY-Maritime Coll., Bronx, 1976-82; accreditation visitor Accrediting Bd. for Engring. and Tech. for AIAA Tech., 1989—. Contbr. articles to Profl. Jour. Proceedings. Recipient Cooper Union Alumni award The Cooper Union, 1967, Nat. Def. Edn. Act fellowship NYU, 1967-70. Mem. AIAA, ASME, Am. Soc. for Engring. Edn., Soc. Automotive Engrs. Roman Catholic. Office: N Y Inst Tech Northern Blvd & Wheatley Rd Old Westbury NY 11568

KOENIG, JACK L., chemist, educator; b. Cody, Nebr., Feb. 12, 1933; s. John and Lucille (Ewart) K.; m. Jeanus Brosz, July 5, 1953; children: John, Robert, Stan, Lori. BS, Yankton Coll., 1955; MS, U. Nebr., 1957, PhD, 1959. Chemist E. I. DuPont, Wilmington, Del., 1959-63; prof. Case Western Res. U., Cleve., 1963—; program officer NSF, Washington, 1972-74. Author: Chemical Microstructure of Polymer Chains, 1982, Spectroscopy of Ploymers, 1992; co-author: Physical Chemistry of Polymers, 1985, Theory of Vibrational Spectroscopy of Polymers, 1987. With U.S. Army, 1953-55. Recipient Disting. Lectr. award BASF, 1990, Internat. Rsch. award Soc. Plastics Engrs., 1991, Disting. Svc. award Cleve. Tech. Socs. Coun., 1991, Pioneer in Polymert Sci. award Polymer New Mag., 1991. Fellow Am. Physics Soc.; mem. Am. Chem. Soc., Soc. Applied Spectroscopy. Achievements include research in characterization of polymers by spectroscopic methods. Office: Case Western Res U 10900 Euclid Ave 7202 Cleveland OH 44106-7202

KOEPPE, EUGENE CHARLES, JR., electrical engineer; b. Chgo., Sept. 15, 1955; s. Eugene Charles and Lucille (Luczak) K. BSEE, Ill. Inst. Tech., 1977, MSEE, 1984. Registered profl. engr.-in-tng., Ill. R & D engr. Teletype Corp., Skokie, Ill., 1977-85; mem. tech. staff AT&T Bell Labs., Skokie, 1985-90, Naperville, Ill., 1990—; cert. TEMPEST engr., 1986-88. Mem. IEEE, NSPE, Am. Radio Relay League, Mensa, Tau Beta Pi. Office: AT&T Bell Labs 2000 N Naperville Rd PO Box 3033 Naperville IL 60566-7033

KOEPPE, PATSY PODUSKA, internist, educator; b. Memphis, Nov. 18, 1932; d. Ben F. and Lilly Mae (Reid) Poduska; m. Douglas F. Koeppe sr., Sept. 8, 1967; 1 child, Douglas F. Jr. BA, Tex. Women's U., 1954; MD, U. Tenn., 1957. Intern Roanoke (Va.) Meml. Hosp., 1960-61; resident in internal medicine VA Teaching Group Hosp., Memphis, 1961-62, Lahey Clinic, Boston, 1962-63; fellow in endocrinology and metabolism U. Tex. Med. Br., Galveston, 1963-65; pvt. practice Kingsville, Tex., 1972-73; dir. Women's Health Care Ctr., College Park, Md., 1974-77; instr. internal medicine and endocrinology Med. Br., U. Tex., Galveston, 1965-69, asst. prof. endocrinology, 1969-72, asst. prof. internal medicine, 1969-72, 78-87, assoc. prof., 1987—, mem. grad. faculty biomed. sci., 1983—, acting dir. div. geriatrics, 1991-92. Mem. Am. Geriatric Soc., Tex. Med. Assn., Tex. Med. Found., So. Assn. Geriatric Medicine, Galveston County Med. Soc. Presbyterian. Home: 323 Brookdale Dr League City TX 77573-1668 Office: Univ Tex Med Br 3.325 Jennie Sealy Hosp D60 Galveston TX 77550

KOEPSEL, WELLINGTON WESLEY, electrical engineering educator; b. McQueeney, Tex., Dec. 5, 1921; s. Wesley Wellington and Hulda (Nagel) K.; m. Dorothy Helen Adams, June 25, 1950; children: Kirsten Marta, Gretchen Lisa, Wellington Lief. BS in Elec. Engring., U. Tex., 1944, MS, 1951; PhD, Okla. State U., 1960. Engr. City Pub. Service Bd., San Antonio, 1946-47; research sci. Mil. Physics Research Lab., U. Tex., 1948-51; research engr. North Am. Aviation, Downey, Calif., 1951; asst. prof. So. Methodist U., 1951-59; assoc. prof. U. N.Mex., Albuquerque, 1960-63, Duke U., 1963-64; prof., head dept. elec. engring. Kans. State U., Manhattan, 1964-76; prof. elec. engring. Kans. State U., 1976-84, prof. emeritus, 1984—; pres., owner, chief engr. Mutronic Systems, Port Aransas, Tex.; vis. prof. Prairie View A&M U. Contbr. articles profl. jours. Served from ensign to lt. (j.g.) USNR, 1944-46. Mem. IEEE, Sigma Xi, Eta Kappa Nu. Achievements include research on microcomputer simulation and modeling of electromagnetic (microwave) sensor systems; digital signal processing; development of computer software for systems simulation. Address: Mutronic Systems PO Box 459 Port Aransas TX 78373

KOESTER, J. ANTHONY, science publication editor; b. Hampton, Iowa, Oct. 9, 1942; s. Raymond Harold and Margaret (Bell) K.; m. Judith Earle Ellaby, July 10, 1961; children: Deborah, David, Susan, John. Student, Purdue U., 1961-65. Office mgr. Vester and Assocs. Cons. Engrs., Lafayette, Ind., 1965-69; editor, mng. editor Railroad Model Craftsman mag., Newton, N.J., 1969-81; sr. tech. writer AT&T Bell Labs., Piscataway, N.J., 1981-84; mgr. tech. publs. Bellcore, Piscataway, 1984-85; mng. editor Bellcore Exch. sci. mag., Livingston, N.J., 1985—. Columnist Model Railroader mag., 1985—. Mem. Nickel Plate Rd. Hist. and Tech. Soc. (founder, nat. dir. 1966-68), Nat. Model RR Assn. (Disting. Svc. award 1980). Republican. Office: Bellcore 290 W Mt Pleasant Ave 1B-113 Livingston NJ 07039

KOESTLER, ROBERT JOHN, conservation research scientist, biologist; b. Hackensack, N.J., Nov. 23, 1950; s. James Alfred and Marie Antoinette (Stephenson) K.; m. Victoria Honoré-Maxine Riba, Aug. 13, 1972; children: Marina Riba, Daniel Lowell. BS, SUNY, Stony Brook, 1972; MA, Hunter Coll., CUNY, 1977; PhD in Biology, CUNY, 1985. Rsch. asst. Lamont-Doherty Geol. Obs. Columbia U., Palisades, N.Y., 1972-73; tech. specialist Am. Mus. Natural History, N.Y.C., 1973-81; rsch. scientist The Met. Mus. Art, N.Y.C., 1981—; chmn., session coord. internat. biodeterioration confs., 1989, 90, internat. electron microscopy confs., 1986, 87; asst. coord. ICOM Biology Working Group, 1992—; cons. to Am. Rsch. Ctr., Eqypt, NOVA, World Monument Fund, ICCROM, Rome; chmn. Am. Tech. Com. of the Corpus Vitrearum of Medeii, 1991—; adj. prof. Inst. Fine Arts NYU. Editor: Biodeterioration of Cultural Property, 1991; participating editor Jour. Internat. Biodeterioration and Biodegradation, 1987—, contbr. over 60 articles to profl. jours. Mem. Internat. Biodeterioration Soc., Electron Microscopy Soc. Am., Sigma Xi. Achievements include hundreds of scanning electron micrographs in a variety of publications. Office: Met Mus Art 1000 Fifth Ave New York NY 10028-0198

KOETSIER, JOHAN CAREL, clinical neurology educator; b. Amsterdam, the Netherlands, June 21, 1936; s. Hendrik and Johanna Carolina (Oosterhoff) K.; m. Nelly Adriana Bonne, July 31, 1938; children: Caroline, Annette. MD, Free U., Amsterdam, 1962; PhD, Free U., 1972. Cert. neurologist. Neurologist Valeriusclinic, Amsterdam, 1966; clin. neurophysiologist Valeriusclinic, 1967, staff mem. neurology svc., 1967-70, temp. chmn. neurology svc., 1970-75, chmn. neurology svc., 1975-85; chmn. neurology dept. Free U. Hosp., Amsterdam, 1985—; asst. lectr. Free U., Amsterdam, 1970-75; lectr. clin. neurology Free U., 1975-80, prof. clin. neurology, 1980—; chmn. neurology dept. Free U. Hosp., Amsterdam.; chmn. med. adv. bd. Dutch Multiple Sclerosis Soc., 1986. Editor: Handbook of Clinical Neurology, Vol. 47, 1985; adv. editor: Jour. Neurol. Sciences; contbr. articles in Neuroimmunology, Clin. Neurology, Jour. Neurology, and Jour. Neurol. Scis. Mem. N.Y. Acad. Scis., Soc. Neuroimmunology, Internat. Fedn. Multiple Sclerosis Soc. (internat. med. adv. bd.), Ned. Ver. Neurologie, Amsterdamsche Neurologen Vereeniging, Ned. Ver. Kinderneurologie, Studieclub voor Neurochirurgie. Christian Democrat. Mem. Dutch Reformed Ch. Avocations: piano, bird watching, reading. Home: Koningsvaren 50, 1391 NL Abcoude The Netherlands Office: Free U Hosp, De Boelelaan 1117, PO Box 7057, 1007 MB Amsterdam The Netherlands

KOFF, ANDREW, microbiologist, biomedical researcher; b. Jamaica, N.Y., Sept. 7, 1962; s. Morton Macy and Barbara Lee (Weiss) K.; m. Elizabeth Ann Moffitt, Sept. 16, 1990. BS, SUNY, Stonybrook, 1984, PhD, 1990. Rsch. fellow SUNY, Stonybrook, 1986-90, Fred Hutchinson Cancer Rsch. Ctr., 1991—; cons. Pharmingen, San Diego, 1992—. Contbr. articles to Cell, Science, Jour. Virology, Molecular Cell Biology. Recipient Rsch. fellowship Nat. Cancer Inst., 1987-90, '91-92; postdoctoral fellowship NIH, Washington, 1992-94. Achievements include patent on human cyclin E.; first to show an active cyclin-Kinase complex during G1 phase of cell cycle; first to show that initiation of herpes simplex virus replication requires assembly of multi-protein unwound DNA complex and viral initiation protein is not sufficient in itself to perform replication initiation. Office: Fred Hutchinson Cancer Rsch Ctr M 421 1124 Columbia St Seattle WA 98104

KOFF, BERNARD L., engineering executive; b. Huntington, N.Y., Mar. 24, 1927. BS in Mech. Engring., Clarkson U., 1951; MS in Mech. Engring., NYU, 1958; ScD honors cause, Clarkson U., 1993. Cert. profl. engr., Ohio, Fla. Test engr. Gen. Electric, 1951-52; design engr. Fairchild, Curtiss Wright, Gen. Electric, 1952-62; from sr. engr. to mgr. turbo machinery design & devel. Gen. Electric, 1962-65, mgr. compressor design & devel., 1965-68, mgr. preliminary design, 1968-89, mgr. advanced design engring., 1969-73, gen. mgr. devel., production engring., 1973-75, chief engr. aircraft engine engring. divsn., 1975-80; sr. v.p. engring. govt. products divsn. Pratt & Whitney, 1980-1983, sr. v.p. engring. divsn., 1983-87, sr. v.p. govt. engring. govt. engine bus., 1987-90; exec. v.p engring & tech. Pratt & Whitney, West Palm Beach, Fla., 1990—. Author numerous papers in field; symposia speaker; holder 13 patents. Mem. USAF Sci. Adv. Bd., 1986-90. Recipient Golden Knight award Clarkson U., 1982, Theodore von Karman award Air Force Assn., 1988, Daniel Guggenheim Medal award AIAA/ASME/SAE, 1992, Littlewood Meml. Lectr. award AIAA/SAE, 1991. Fellow AIAA (Disting. Lectr. program 1991-92, Elmer A. Sperry Bd. award 1991—, Reed Aeronautics award 1990, Speaker & Paper award 1989, Engr. of Yr. award 1989, Air Breathing Propulsion award), SAE (tech. bd., gen. fund devel. com., vice chmn. aerotech 1991, chmn. aerotech 1992); mem. ASME (hon., R. Tom Sawyer award 1988, Ind. Adv. Bd. 1987, Outstanding Paper award 29th internat. conf. 1985), AIA (past chmn. propulsion com.), Nat. Acad. Engring., Tau Beta Pi, Pi Tau Sigma. Office: Pratt & Whitney Mail Stop 711-25 PO Box 109600 West Palm Beach FL 33410-9600

KOGA, TATSUZO, aerospace engineer; b. Taipei, Taiwan, Apr. 14, 1935; s. Iwazo and Yoshi (Matsumura) Koga; m. Tamiko Hamano, May 21, 1977; children: Jun, Kei. B in Engring., U. Tokyo, 1961; PhD, Stanford U., 1968. Rsch. assoc. Stanford (Calif.) U., 1968; prin. scientist Nat. Aerospace Lab., Tokyo, 1974-78; head Thermo-Structural Lab. Nat. Aerospace Lab., Tokyo, 1978-80; prof. U. Tsukuba, Japan, 1980—, dean coll. engring., 1988-90, provost, 1993—; tech. advisor Nat. Personnel Authority, Tokyo, 1973-74; mem. Com. Indsl. Standards, 1975-80. Contbr. articles to profl. jours. and chpts. to books. Mem. AIAA, Japan Soc. Mech. Engrs. (award 1980), Japan Soc. Aero. Space Sci., Sigma Xi. Achievements include research in buckling criterion for pressurized spherical shells, effects of boundary conditions on vibrations and buckling of cyclindrical shells, bending and torsional rigidities of orthotropic laminates, isometric membrane valv. Home: 3-19-2 Yakushidai, Moriya 302-01, Japan Office: U Tsukuba, 1-1-1 Tennodai, Tsukuba 305, Japan

KOGELNIK, HERWIG WERNER, electronics company executive; b. Graz, Austria, June 2, 1932; came to U.S., 1960; naturalized, Jan. 1992; s. Sepp and Siglinde K.; m. Christa Muller, Mar. 7, 1964; children—Christoph N., Florian A., Andreas M. Dipl.-Ing., Tech. U. Vienna, 1955, Dr.techn., 1958; D.phil., Oxford U., 1960. Mem. research staff Bell Labs., Murray Hill, N.J., 1961-67; head coherent optics research dept. Bell Labs., Holmdel, N.J., 1967-76; dir. electronics research lab. Bell Labs., 1976-83, dir. photonics research lab., 1983—. Contbr. articles in field to profl. jours. Chmn. Monmouth (N.J.) Arts Found., 1973-76; past trustee N.Y. Mus. Holography. Recipient Johann Joseph Ritter von Prechtl medal Tech. U., Vienna, Austria, 1990; hon. fellow St. Peter's Coll., Oxford U., 1992. Fellow Optical Soc. Am. (pres. 1989, recipient Frederic IVES medal 1984), IEEE (David Sarnoff award 1989, Quantum Electronics award 1991), NAE, St. Peter's Coll., Oxford U. (hon.); mem. AAAS, Am. Phys. Soc., Am. Inst. Physics (gov.). Patentee in field of lasers, holography, electronics and optical comm. Home: 27 N Ward Ave Rumson NJ 07760-1913 Office: AT&T Bell Labs Holmdel NJ 07733

KOGER, MILDRED EMMELENE NICHOLS, educational psychologist; b. Jacksonville, Fla., Jan. 11, 1928; d. Hugh Huntley and Edna Wilhelmina (Snell) Nichols; m. Gerald Lee Gamache, Dec. 14, 1974. BA in Social Studies, Fla. State U., 1949; MusB, B Mus. Edn., Jacksonville U., 1955; MEd in Counseling, U. Fla., 1966, EdD in Counselor Edn., 1970. Lic. profl. counselor, Va.; nat. cert. counselor, Fla.; edn. cert. psychology, adminstr. supervision social sci., counseling, music; cert. Performance Systems Internat. Tchr. Jacksonville, Fla., 1949-65; counselor psychology dept. Fla. Jr. Coll., Jacksonville, 1967-68; staff psychologist, chief psychologist Duval County Sch. Bd., Jacksonville, 1968-70, sch. psychologist, counselor, 1972-81; resident coord. student teaching Fla. State U., Jacksonville, 1970-71; prof. ednl. psychology Edward Waters Coll., Jacksonville, 1971-72; pvt. practice profl. counseling Norfolk, Va., 1983-86; mgmt. and orgnl. analyst Orgn. Rsch. Group Tidewater, Inc., Norfolk, 1984-86; asst. prof. counseling and psychology Sch. Edn., Troy State U., Dothan, Ala., 1989-90; sch. psychology mental health dept. Fla. Sch. for Deaf and Blind, St. Augustine, 1990—; cons. Psychol. Rsch. Ctr., St. Augustine, 1991—; cons. Performance Systems Instrnl., Mpls., 1982—; staff developer Duval County Sch. Bd., Jacksonville, 1972-81; adj. tchr. U. Fla., 1966-70, Jacksonville U., 1970-72, Calhoun Ctr., 1970-72, New Drug Rehab. Ctr., 1972, U. North Fla., 1973-76, N.E. Fla. Counseling Svcs., 1975-76, Response and Assocs., 1975-76, Old Domionon U. Counseling Ctr., 1978-79, Golden Gate U., 1982-84, U. So. Calif., 1988. Recipient Outstanding Contbn. to Sch. award Rotary Club, Jacksonville, 1980, Leadership award Duval County Pers. and Guidance Assn., 1980. Mem. APA, ACA, Assn. Specialists in Group Work, Am. Mental Health Counselors Assn., Nat. Assn. Sch. Psychologists (expert witness Emotionally Handicapped, Symposium on Assessment 1993), Am. Choral Dirs. Assn., Ala. Assn. Counseling and Devel. (initiator, advisor 1989-90), Fla. Assn. Sch. Psychologists (stds. and practices com.), Pilot Club of St. Augustine (leadership chair, bd. dirs.), Sigma Alpha Iota (past dir. local groups), Phi Delta Kappa (bd. dirs Jacksonville chpt. 1993—), Phi Kappa Phi, Pi Lambda Theta. Achievements include research in the goals approach to performance objectives. Home: 8 Althea St Saint Augustine FL 32095 Office: Fla Sch Deaf and Blind 207 N San Marco Saint Augustine FL 32084

KOGUT, KENNETH JOSEPH, consulting engineer; b. Chgo., Dec. 3, 1947; s. Joseph Henry and Estelle Theresa (Swiercz) K.; student Lewis Coll., 1966-68; B.M.E., U. Detroit, 1971, M.E., 1972, postgrad, 1972—, m. Darlene Agnes Jedlicka, June 15, 1974. Mech. engr. Fluor Pioneer Inc., Chgo., 1972-73, cons. engr., 1973-75; project mgr. Engring. Corp. Am., Chgo., 1976-77; sr. cons. pub. utilities DeLoitte, Haskins & Sells, Chgo., 1977-79; individual practice as energy and mgmt. cons., 1979—. Registered profl. engr., Ill.; cert. energy mgr. Sloan fellow, 1971-73; recipient award Pres.'s Program for Energy Efficiency, Corporate Energy Mgmt. award, 1981, Regional Energy Profl. Devel. award, 1984, Regional Energy Engr. of Yr. award, 1987, Ill. Energy award, 1988, Illiana Energy Mgmt. Exec. of Yr. award 1992. Mem. Am. Nuclear Soc., Nat., Ill. socs. profl. engrs., Assn. Energy Engrs. (regional v.p. 1993), Environ. Engrs. and Mgrs. Inst., Demand-Side Mgmt. Soc., Exec. Hosp. Engrs. Soc. Ill., Blue Key, Tau Beta Pi, Pi Tau Sigma, Polish Nat. Alliance. Author: Energy Management for the Community Bank. Address: 5232 W 170th Pl Oak Forest IL 60452

KOH, CAROLYN ANN, chemical engineer; b. Hants, U.K., May 15, 1965; came to U.S., 1991; d. Eric P. and Ann C. (Tan) K. BSc, U. Brunel, Uxbridge, U.K., 1987; PhD, West London U., Uxbridge, U.K., 1990. Chartered chemist. Rsch. asst. Permutit-boby Ltd., Houslow, U.K., 1984, Soton Gen. Hosp., Princess Anne Hosp., Southampton, Hants, U.K., 1985, Smithkline-Beechams, Betchworth, Surrey, U.K., 1986, Brunel U., Uxbridge, U.K., 1983-87, Cornell U., Ithaca, N.Y., 1991—; mentor com. mem. Cornell Univ., Ithaca, N.Y., 1991-92; cons. Kyoto (Japan) Univ., 1991-92. Contbr. articles to profl. jours. Mem. N.Y. Acad. Scis., Assn. Chem. Engrs., Royal Soc. Chemistry. Roman Catholic. Achievements include developing novel catalyst for alkene selective hydrogenation, aluminum complexation work, perturbation equation of state and integration of this with molecular simulation and spectroscopy. Spectroscopic studies of gas hydrates. Office: Cornell U Olin Hall Ithaca NY 14853

KOH, PUN KIEN, retired educator, metallurgist, consultant; b. Shanghai, People's Republic China, Jan. 31, 1914; came to U.S., 1936, naturalized, 1949; s. Tse-Zan and Shun-Pao (Wang) K.; m. Jean Sie, Jan. 24, 1940; children: Robert, Jessica Koh Lewis. BSME, Nat. Chiao-Tung U., Shanghai, 1935; DSc in Phys. Metallurgy, MIT, 1939. Rsch. fellow Engring. Inst., Academia Sinica, Kumming, Yunan, People's Republic China, 1940-43; rsch. assoc. with rank asst. prof. dept. metallurgy MIT, Cambridge, 1943-45; head materials rsch. Engring. Rsch. Labs. Standard Oil Co. (Ind.), Whiting and Chgo., 1945-60; assoc. dir. rsch. Allegheny Ludlum Steel Corp., Brackenridge, Pa., 1960-66; rsch. engr. Homer Labs., Bethlehem (Pa.) Steel Corp.,

1966-81; prof. mech. engring. Tex. Tech U., Lubbock, 1966-81; tech. advisor Metal Industries Rsch. Inst. Ministry Econ. Affairs, Taiwan, Republic of China, 1981-89; cons. nuclear power dept. Taiwan Power Co., 1981-85; metall. cons. cen. elec. works Ministry Econ. Affairs, Kumming, 1971-73; adj. prof. mech. engring. Tsing-Hua U., Taiwan, 1941-43; cons. on noise pollution control China Petroleum Corp., Taiwan, 1978-85; cons. China Caprolactum (Nylon 6) Corp., Maoli, Rep. of China, 1973-78, China Copper and Gold Mining and Refining Corp., Taiwan, 1976-80. Contbr. articles to profl. jours. Fellow Am. Inst. Chemists, N.Y. Acad. Scis.; mem. AIME, ASTM, Am. Soc. Metals, Electron Microscopic Soc. Am. (sec. Phila. chpt. 1964), Sigma Xi. Republican. Achievements include successfully setting the pilot production of popular hydrogenation catalyst, namely, pyrophoric and non-pyrophoric grades of powdered Raney Nickel catalysts; started to recycle pig waste by using E-coli bacteria in the activated sludge arobic process and found that this processfor treating biomass can be further accomplished through future improvement in the better choice of bacteria for the mitosis and the additional use of proper enzymes and catalysts. Home: 3318 24th St Lubbock TX 79410-2131

KOHEL, RUSSELL JAMES, geneticist; b. Omaha, Nov. 30, 1934; married; 3 children. BS, Iowa State U., 1956; MS, Purdue U., 1958, PhD, 1959. Supervisory rsch. geneticist Argrl. Rsch. Svc. USDA, College Station, Tex., 1959—. Fellow Am. Soc. Agronomy; mem. Am. Soc. Plant Physiologists, Am. Genetic Assn., Genetics Soc. Am. Office: USDA So Crops Rsch Lab RR 5 Box 805 College Station TX 77845-9593

KOHLER, DYLAN WHITAKER, software engineer, animator; b. Provo, Utah, Oct. 10, 1966; s. Bryan Earl and Susan (Whitaker) K.; m. Lureline Weatherly, Dec. 22, 1990. BS in Computer Sci., Stanford U., 1987, BA in English, 1987. Computer cons. IRIS Stanford (Calif.) U., 1985-87; sr. software engr. Walt Disney Feature Animation, Glendale, Calif., 1986-90, screen cartoonist, 1990-91; owner Angst Animation Post Prodn., Glendale, 1991—. Recipient Acad. award for sci. and engring. Motion Picture Acad. Arts and Scis., 1992. Achievements includes U.S. patent for computer image production system, foreign patent for computer animation production system. Home: 1308 Truitt St Glendale CA 91201-2341 Office: Angst Animation Post Prodn 1308 Truitt St Glendale CA 91201-2341

KÖHLER, GEORGES J. F., scientist, immunologist; b. Apr. 17, 1946. Scientist, immunologist Max-Planck Institut fur Immunologie, Stubeweg, Fed. Republic Germany. Recipient Nobel prize in Medicine and Physiology, 1984, Albert Lasker Med. Research award, 1984.

KOHLI, TEJBANS SINGH, chemical engineer; b. Jullundur, Panjab, India, Jan. 19, 1958; came to U.S. 1980; s. Harbans Singh and Joginder Kaur (Anand) K.; m. Dolores Darlene Unrue, Jan. 24, 1983; 1 child, Tara. BSc in Chem. Engring., Panjab U., 1980; MS in Chem. Engring., U. Akron, 1983; MBA, Case Western Reserve U., 1993. Lab. mgr. Polymerics Inc., Akron, Ohio, 1983-84; materials engr. Clevite Industries, Milan, Ohio, 1984-88; materials engr. Lord Aerospace, Erie, Pa., 1988-90, mgr. materials engring., 1991-93, staff engr., 1993—. Contbr. articles to profl. jours. Mem. Am. Chem. Soc. (rubber div.), Am. Inst. Chem. Engrs. Achievements include co-invention of high temperature fluid mounts; development of high performance elastomers for aerospace products. Home: 3968 Stellar Dr Erie PA 16506 Office: Lord Aerospace 1635 W 12th St Erie PA 16506

KOHLMEIER, SHARON LOUISE, medical laboratory administrator, medical technologist; b. Brighton, Colo., Jan. 13, 1944; d. Lewis Herman and Emma Lou (Chambers) Hiller; m. Ronald Arthur Kohlmeier, July 30, 1966; children: Lisa Ann, Karin Sue. BA, U. No. Colo., 1966. Cert. medical tech. Med. technologist No. Colo. Med. Ctr., Greeley, Colo., 1966-67, Brighton (Colo.) Community Hosp., 1967-84; lab. dir. Platte Valley Med. Ctr., Brighton, Colo., 1984—; mem. quality assurance com., infection control com., safety com. Platte Valley Med. Ctr., 1984—. Analysis coord. health fair blood drive Platte Valley Med. Ctr., 1984-92. Mem. Am. Soc. Clin. Pathologists, Am. Hosp. Assn., Colo. Assn. for Continuing Med. Lab. Edn. Office: Platte Valley Med Ctr 1850 Egbert St Brighton CO 80601

KOHN, BARBARA ANN, veterinarian; b. Milw., Sept. 15, 1954; d. Willard Karl and Beth Elaine (Maule) K. BS, U. Wis., Green Bay, 1976; postgrad., MIT, 1976-79; BS, U. Minn., St. Paul, 1982, DVM, 1984. Water chemist City of Milw., 1980; sr. large animal health technician U. Minn. Coll. Vet. Medicine, St. Paul, 1981-83; rsch. veterinarian Wendt Labs., Belle Plaine, Minn., 1984-85; postdoctoral fellow Inst. Biomed. Aquatic Studies U. Fla., Gainesville, 1985-86; vet. clinician Norwell (Mass.) Vet. Hosp., 1986-92, Wildwood Animal Hosp., Marshfield, Wis., 1992; vet. med. officer U.S. Dept. Agr./Animal and Plant Health Inspection, Springfield, Ill., 1992-93, USDA/Animal and Plant Health Inspection, Hyattsville, Md., 1993—; mem. Institutional Animal Care and Use Com., U. Mass., Boston, 1988-92. Recipient Certs. of Appreciation ARC, Milw., 1972, Altrusa Internat., 1988. Mem. AAAS, Am. Vet. Med. Assn., Internat. Assn. Aquatic Animal Medicine, N.Y. Acad. Scis., Altrusa Internat., Sweet Adelines Inc. Roman Catholic. Office: USDA REAC Fed Bldg Federal Building 6505 Belcrest Rd Hyattsville MD 20782

KOHN, MICHAEL CHARLES, theoretical biochemistry professional; b. Bklyn., July 29, 1941; s. Mordecai and Rose (Teich) K.; m. Lynn Breeden Mitchell, Oct. 23, 1970. BS, MIT, 1964; PhD, U. S.C., 1970. NRC fellow Naval Underseas Ctr., Pasadena, Calif., 1971-73; rsch. prof. U. Pa., Phila., 1974-84, Duke U., Durham, N.C., 1984-91; expert Nat. Inst. Environ. Health Scis., NIH, Research Triangle Park, N.C., 1991—. Author: Practical Numerical Methods: Algorithms and Programs, 1987, 89; guest editor Bull. of Math. Biology, 1986, Jour. of Theoretical Biology, 1991; contbr. numerous articles to profl. jours. Grantee NIH, U. Pa., 1974-84, Duke U., 1984-91; fellowship Nat. Rsch. Coun., 1971-73. Mem. AAAS, Soc. for Math. Biology, Soc. for Computer Simulation, Sigma Xi. Achievements include invention of MetaNets, a graphical method for identification of regulatory properties of complex biochemical networks. Office: NIEHS PO Box 12233 Research Triangle Park NC 27709

KOHN, WALTER, educator, physicist; b. Vienna, Austria, Mar. 9, 1923; m. Mara Schiff; children: J. Marilyn , Ingrid E. Kohn Katz, E. Rosalind. BA, U. Toronto, Ont., Can., 1945, MA, 1946, LLD (hon.), 1967; PhD in Physics, Harvard U., 1948; Docteur es Sciences honoris causa, U. Paris, 1980; Sc (hon.), Brandeis U., 1981; PhD (hon.), Hebrew U. Jerusalem, 1981; DSc (hon.) honoris causa, Queens U., Kingston, Can., 1986. Indsl. physicist Sutton Horsley Co., Can., 1941-43; geophysicist Koulomzine, Que., 1944-46; instr. physics Harvard U., 1948-50; asst. prof. Carnegie Inst. Tech., 1950-53, assoc. prof., 1953-57, prof., 1957-60; prof. physics U. Calif., San Diego, 1960-79, chmn. dept., 1961-63; dir. Inst. for Theoretical Physics, U. Calif., Santa Barbara, 1979-84, prof. dept. physics, 1984-91, prof. emeritus, 1991—; rsch. physicist Ctr. for Quantized Electronic Structures, U. Calif., Santa Barbara, 1991—; vis. scholar U. Pa., U., Mich., U. Washington, Seattle, U. Paris., U., Copenhagen, U. Jersalem, Imperial Coll. London, ETH, Zurich, Switzerland; cons. Gen. Atomic, 1960-72, Westinghouse Rsch. Lab., 1953-57, Bell Telephone Labs., 1953-66, IBM, 1978; mem. or chmn. rev. coms. Brookhaven Nat. Labs., Argonne Nat. Labs., Oka Ridge Nat. Labs., Ames Nat. Labs.; mem. nat. adv. com. Michelson-Morley Centennial Celebration. Contbr. approximately 170 sci. articles and revs. to profl. jours. With inf., Can. Army, 1944-45. Recipient Oliver Buckley prize, 1960, Davisson-Germer prize, 1977, Nat. Medal of Sci., 1988; NRC fellow, 1951, NSF fellow, 1958, Guggenheim fellow, 1963, NSF sr. postdoctoral fellow, 1967. Fellow AAAS, Am. Phys. Soc. (counselor-at-large 1968-72), Am. Acad. Arts and Scis.; mem. NAS, Internat. Acad. Quantum Molecular Scis. (Feenberg medal 1991). Rsch. on electron theory of solids and solid surfaces. Office: U Calif Dept Physics Santa Barbara CA 93106

KOHNE, RICHARD EDWARD, retired engineering executive; b. Tientsin, China, May 16, 1924; s. Ernest E. and Elizabeth I. (Antonenko) K.; m. Gabrielle H. Vernaudon; children—Robert, Phillip, Daniel, Paul, Renee. B.S., U. Calif., Berkeley, 1948. Structural engr. hydro projects Pacific Gas & Electric Co., San Francisco, 1948-55; cons. engr. Morrison-Knudsen Engrs., Inc., San Francisco, 1955—, regional mgr. for Latin Am., then v.p., 1965-71, exec. v.p. world-wide ops. in engring. and project mgmt., 1971-79, pres., chmn., chief exec. officer, 1979-88; chmn., chief exec. officer Morrison-Knudsen Internat. Co., Inc., San Francisco, 1988-90, chmn. emeritus, 1990—. Decorated Chevalier Nat. Order of Leopold (Zaire). Mem. ASCE, U.S. Com. Large Dams, Cons. Engrs. Assn. Calif., World Trade Club (San Francisco). Democrat. Roman Catholic. Home and Office: 1827 Doris Dr Menlo Park CA 94025-6101

KOHNHORST, EARL EUGENE, tobacco company executive; b. Louisville, Apr. 15, 1947; s. Robert L. and Lali May (Bratton) K.; m. Mary Lou Pierce, Mar. 4, 1972; 1 child, Lauren Renea. BS, U. Louisville, 1970, MS, 1971. Chem. engr. Brown & Williamson Tobacco Corp., Louisville, 1971-74, supr. engring., 1974-77, mgr. devel. ctr., 1977-78, div. head, 1978-80, dir. mfg. planning and engring., 1980-83, v.p. R & D and engring., 1983-87; v.p. planning BATUS Inc, Louisville, 1987—. Patentee in field. Mem. U. Louisville Speed Sch. Adv. Com., 1988—. Mem. Am. Inst. Chem. Engrs., Sigma Pi Sigma. Republican. Methodist. Avocations: skiing, bicycle touring, running. Office: BATUS Inc 2000 Citizens Plz Louisville KY 40202-2875

KOHOUTEK, RICHARD, civil engineer; b. Prague, Czechoslovakia, May 14, 1943; s. Karel and Marie (Vencovsky) K.; m. Jaromira, Mar. 3, 1968; children: Claire Caroline, Lucie, Martina. Dipl. Ingenieur, Czech Tech. U., Prague, Czechoslovakia, 1966; PhD, Melbourne U., Australia, 1986. Registered engr., Prague. Asst. chief engr. Bur. Railways & Freeways, Prague, Czechoslovakia, 1967-68; supr. engr. Bur. of Roads, Prague, Czechoslovakia, 1968; asst. design engr. Hwys. Dept., Adelaide, South Australia, 1969-73, systems engr., 1973-79; lectr. U. Melbourne, Victoria, Australia, 1979-85; sr. lectr. U. Wollongong, NSW, Australia, 1985—; team leader NAASRA, Melbourne, 1976-77; corrd. group exptl. and analytical stress analysis. Author: Analysis and Design of Foundations for Vibrations, 1985; contbr. articles to profl. jours. Recipient Merit scholarship Czech Tech. U., 1960-66, Award for Excellence, Australian Concrete Inst., 1975, Writing Up award U. Melbourne, 1986. Mem. ASME, ASCE, ASTM, Instn. Engrs. (chmn. Australian chpt.), Internat. Assn. for Bridge and Structures, Soc. for Exptl. Mechanics, Assn. for Computing Machinery, Sigma Xi (sec. Australian chpt.). Avocations: sailing, skiing, photography. Office: U Wollongong, Dept Civil Enging, Wollongong 2500, Australia

KOHR, ROLAND ELLSWORTH, hospital administrator; b. Middletown, Ohio, Dec. 22, 1931; s. Roland Meredith and Mildred (Brandeberry) K.; m. Hilda Scherz, Sept. 6, 1952; children: Linda Kohr Harper, Roland Meredith, Jeffrey Stuart. BS, U. Cin., 1954; MS in Health Adminstrn., Northwestern U., 1959. Asst. adminstr. Bethesda Hosp., Cin., 1958-60; adminstr. William S. Major Hosp., Shelbyville, Ind., 1961-66; pres. Bloomington (Ind.) Hosp., 1966—; asst. prof. Sch. Pub. and Environ. Affairs, Ind. U.; bd. dirs. Precision Healthcare, Inc., Bank One, Bloomington, Vol. Hosps. Am.-Tri State Inc., Indpls.; chmn. bd. dirs. Bloomington Convalescent Ctr., Inc. Contbr. hosp. adminstrn. articles to profl. jours. Named for Disting. Svc., Shelbyville C. of C., 1966, Ind. Hosp. Assn., 1987; Paul Harris fellow Rotary Internat. Fellow Am. Colls. Healthcare Execs.; mem. Bloomington C. of C. (bd. dirs.), Rotary (Bloomington chpt., pres. 1987-88, bd. dirs. Bloomington Rotary Found. 1988—), Masons. Avocations: photography, scuba. Home: 2989 N Bankers Dr Bloomington IN 47408-1021 Office: Bloomington Hosp PO Box 1149 625 W 2d St Bloomington IN 47402

KOIZUMI, SHUNZO, surgeon; b. Kyoto, Japan, Mar. 14, 1946; s. Haruo and Chieko (Ushioda) K.; m. Yoko Yokoe, Aug. 19, 1971; children: Miyuu (dec.), Mitsuteru, Arei. MD, Kyoto U., 1971. Diplomate Am. Bd. Surgery. House staff in medicine Yamagami (Japan) Hosp., Yamato-Takada (Japan) City Hosp.; house staff in anesthesiology Osaka (Japan) Red Cross Hosp., 1972-74; resident Youngstown (Ohio) Hosp. Assn., 1975-76; gen. surg. resident St. Vincent's Med. Ctr., Bridgeport, Conn., 1976-80, chief surg. resident, 1979-80; staff first surg. dept. Kyoto U. Hosp., 1980; surg. cons. for residency dept. abdominal surgery Tenri (Japan) Hosp., 1980—; dir. Ichijoji Ctr. Meta-Med. Studies and Transnat. Lodge. Editor resident's manual, 1988. Mem. ACS, Japan Surg. Soc., Japan Soc. Med. Edn. Home: 16-1 Ichijoji-Iorino-cho, Sakyo-ku Kyoto 606, Japan Office: Tenri Hosp Dept Abdominal Surgery, 200 Mishima-cho, Tenri Nara 632, Japan

KOJIĆ-PRODIĆ, BISERKA, chemist; b. Aug. 29, 1938; married, Apr. 19, 1968; 1 child, Kristina. Degree in sci., U. Zagreb, Croatia, 1960. Technician dept. structural and inorganic chemistry U. Uppsala, Sweden, 1960-61, postdoctoral rsch. assoc. chemistry, 1972; rsch. asst. Ruder Bošović Inst., Zagreb, 1961-68, rsch. assoc., 1968-76, sr. rsch. assoc., 1976-81, sr. scientist dept. material scis., head of X-ray lab., 1981—, mem. faculty natural scis.; vis. scientist U. Utrecht, The Netherlands, 1983-84, Synchrotron Sta. DESY, Hamburg, 1988; vis. scientist Med. Found. Buffalo, 1976-77, Tex. Christian U., Ft. Worth, 1976-77. Contbr. sci. articles to internat. jours. Recipient Nat. Sci. award Nat. Sci. Adv. Bd., Zagreb, 1971. Mem. Internat. Union Crystallography Commn. on Small Molecules. Office: Rudjer Bošković Inst, Bijenićka 54 POB 1016, 41001 Zagreb Croatia

KOJIMA, RYUICHI O., computer educator; b. Gamagori, Aichi Pref, Japan, Apr. 19, 1949; s. Masao and Kimiyo (Ozawa) K.; m. Kumiko Adachi, Nov. 27, 1979; children: Makiko, Itsuko. B in Agrl., Kyoto U., 1974. Staff mem. Info. and Math. Rsch., Tokyo, 1975-80; exec. Ratoc System Engring., Tokyo, 1980-85; computer cons. pvt. practice, Gifu, Japan, 1985-88; ednl. staff mem. Gifu Tech. Coll., 1988-92; asst. prof. Aichi Coll. Tech., 1991—; researcher Computer Edn. Ctr., Tokyo, 1990; lectr., Gifu U., 1991. Author: C Language Training, 1986, Smalltalk Training, 1988, Smalltalk Programming, 1989, MS-DOS to UNIX, 1991, Interactive Multimedia, 1993. Mem. IEEE Computer Soc., Assn. Computing Machinery, Info. Processing Soc., Inst. of Elec. Info. and Communication Engrs. Avocations: camping, hiking, swimming, fishing. Home: 135-2 Saigo, Azumada-cho, Toyohashi-shi 440, Japan Office: Aichi Coll Tech, 50-2 Umanori, Hishihazama-cho, Gamagouri-shi 443, Japan

KOKINI, KLOD, mechanical engineer; b. Izmir, Turkey, Jan. 9, 1954; came to U.S., 1976; s. Moris and Maria Ana (Pavlovic) K.; m. Kathryn Essom, June 25, 1984. BS, Bogazici U., Istanbul, Turkey, 1976; PhD, Syracuse U., 1982. Asst. prof. U. Pitts., 1983-85; asst. prof. Purdue U., West Lafayette, Ind., 1985-90, assoc. prof. mech. engring., 1990—; internat. adv. bd. Functionally Gradient Material Forum, Sendai, Japan, 1990—; cons. Cummins Engine Co., Columbus, Ind. Contbr. articles to profl. publs. NSF grantee, 1987, 90, 92. Mem. ASME, Am. Ceramic Soc., Am. Acad. Mechanics, Soc. Automotive Engrs. Achievements include study of the effect of transient thermal loads on interface cracks in dissimilar materials; determination of thermal crack initiation in multilayer ceramic/metal systems. Office: Purdue U 1288 Mechanical Engr Bldg West Lafayette IN 47907-1288

KOKKALIS, ANASTASIOS, aeronautical engineer, consultant; b. Pireas, Greece, Nov. 28, 1957; s. John and Kiriaki (Gionis) K.; m. Renna Kliwis, Oct. 1, 1978; children: Joanna, Kiriakos. BSc, U. Salford, Manchester, Eng., 1980; PhD, U. Glasgow, Eng., 1988. Engr. Ruston Gas Turbines, Lincoln, 1980-81; rsch. asst. U. Glasgow, 1981-88; sr. engr. British Aerospace Dynamics Ltd., Stevenage, 1988-89; cons. engr. CASA, Madrid, Spain, 1989-90, Westland Helicopters Ltd., Yeovil, U.K., 1990—; mng. dir. Planability Ltd., Yeovil, 1990—. Contbr. articles to profl. jours. Gov. Priory Sch., St. Neots, 1988-90. Mem. AIAA, Royal Aero. Soc. (named Herbert Le Seur 1986), Engring. Coun. G.B. Achievements include rsch. in exptl. and theoretical helicopter rotor aerodynamics, stability and control of free and guided flying configurations, suppression of flow induced vibrations of bluff bodies. Office: Westland Helicopters Ltd, Lysander Rd, Yeovil England

KOKKINAKIS, DEMETRIUS MICHAEL, biochemist, researcher; b. Heraklion, Crete, Greece, Mar. 6, 1950; came to U.S., 1973; s. Michael-Byron and Kyriaki (Zeakis) K.; m. Jannie Tong, June 10, 1985; children: Michael Ross, Alexius Manuel. BS in Chemistry, Athens (Greece) Natl. U., 1973; MS, Pa. State U., 1975; PhD, W.Va. U., 1977. Postdoctoral fellow Tex. Tech U. Med. Sch., Lubbock, 1978-80; fellow Northwestern U. Med. Sch., Chgo., 1980-85, rsch. assoc., 1985-87, asst. prof., 1987-93; assoc. prof. U. Tex. Southwestern Med. Ctr., Dallas, 1993—; reviewer Merit Rev. Bd., Washington, 1988—, NIH, Bethesda, Md., 1990—. Author: (with others) Experimental Pancreatic Carcinogenesis, 1987; also articles. Active Animal Rights group, Evanston, Ill., 1990-93; chair Northwestern U. IACUC, 1989-93. Grantee Nat. Cancer Inst., 1987, Am. Cancer Soc., 1991. Mem. AAAS, Am. Assn. Cancer Rsch., N.Y. Acad. Sci. Republican. Orthodox. Achievements include determination of chemical structure and properties of nitrosamine pancreatic carcinogens, of nature of DNA damage induced by pancreatic carcinogens; introduction of evidence on promotion of pancreas, lung and kidney cancer by orotic acid; establishment of carcinogenic regimens to demonstrate the role of dietary protein on pancreatic carcinogenesis in experimental animal models. Home: 9677 Fallbrook Dr Dallas TX 75243 Office: Univ Tex Southwestern Med Ctr 5323 Harry Hines Blvd Dallas TX 75235

KOKOPELI, PETER HEINE, space designer; b. Chgo., Feb. 17, 1954; s. Ralph W. and Patricia Anne (Johns) H.; children: Eva M., Marc P. Student, North Ariz. U., 1991; BA cum laude, U. Wash., 1992. Ptnr. The Kokopeli Partnership, Seattle, 1980-86; owner, designer Settings Design, Seattle, 1986—; instr. Seattle C.C., 1989—; program asst. Inst. for Global Security Studies, Seattle, 1992. Herbert Sedville, Jr. Peace fellow Union Concerned Scientists, 1993. Mem. Phi Beta Kappa.

KOKOT, FRANCISZEK JÓZEF, physician; b. Olesno, Silesia, Poland, Nov. 24, 1929; s. Franciszek and Franciszka (Kostka) K.; m. Malgorza Skrzypczyk, Dec. 26, 1955; children: Stefan, Klaudiusz, Jan, Tomasz. Physician diploma summa cum laude, Silesian Sch. Medicine, Katowice, Poland, 1953, MD, 1957; Dr. h.c. (hon.), Med. Acad. Katowice, 1993. Technician dept. chemistry Silesian Sch. Medicine, Katowice, 1949-50, asst. and sr. asst. dept. pharmacology, 1950-57, asst. prof. dept. internal medicine, 1957-62, assoc. prof. dept. internal medicine, 1962-69, extraordinary prof. dept. internal medicine, 1969-74, extraordinary prof. dept. nephrology, 1974-82, ordinary prof. dept. nephrology, 1982—; rsch. fellow Clinique Therapeutique, Geneva, 1958-59; WHO fellow Middlesex Hosp., London, 1970. Mem. European Dialysis and Transplant Assn., Polish Acad. Arts and Scis.; hon. mem. Bulgarian Soc. Nephrology, German Soc. Nephrology, Yugoslavian Soc. Nephrology, Hungarian Soc. Nephrology, Macedonian Soc. Nephrology, Italian Soc. Nephrology. Roman Catholic. Home: Korfanty 8/162, 40-008 Katowice Poland Office: Dept Nephrology, Francŭska 20, 40-027 Katowice Poland

KOLATTUKUDY, PAPPACHAN ETTOOP, biochemist, educator; b. Cochin, Kerala, India, Aug. 27, 1937; came to the U.S., 1960; m. Marie M. Paul. BS, U. Madras, 1957; B in Edn., U. Kerala, 1959; PhD, Oreg. State U., 1964. Prin. jr. high sch. India, 1957-58, high sch. chemistry tchr., 1959-60; asst. biochemist Conn. Agrl. Experiment Sta., New Haven, 1964-69; assoc. prof. Wash. State U., Pullman, 1969-73, prof. biochemistry, 1973-80, dir. inst. biol. chemistry, 1980-86; dir. Ohio State Biotech. Ctr., Columbus, 1986—; cons. Analabs, New Haven, Allied Chem. Corp., Solvay, N.Y., Genencor Corp., South San Francisco, Calif., Monsanto Co., St. Louis; appointed to Overseas Adv. Com., India; mem. Edison BioTech. Ctr., Cleve., trustee; mem. adv. com. to MUCIA on Sci. and Tech, Nat. Agrl. Biotech. Consortium; Ohio state rep. to Midwest Plant Biotech. Consortium; external assessor faculty sci. environ. studies U. Agriculture Malaysia. Contbr. over 260 articles to profl. jours. Recipient Golden Apple award Wash. State Apple Commn., President's Faculty Excellence award Wash. State U.; grantee NIH, NSF, Am. Heart Assn., Am. Cancer Soc., DOE. Mem. Fedn. Am. Socs. for Exptl. Biology, Am. Soc. Plant Physiologists. Achievements include patent for cutinase for cleaning applications, two patents pending. Home: 2301 Hoxton Ct Columbus OH 43220 Office: Ohio State Biotech Ctr 1060 Carmack Rd Columbus OH 43210

KOLB, CHARLES CHESTER, humanities administrator; b. Erie, Pa., Sept. 4, 1940; s. John Christian and Edna Lucille (Church) K.; m. Joy Bilharz, June 3, 1972 (div. Mar. 1991); 1 child, Nancy Gwenyth; m. P. Jean Drew, July 20, 1991; 1 child, Catherine Claire Fraley. BA in History, Pa. State U., 1962, PhD in Archaeology and Anthropology, 1979. Instr. anthropology Pa. State U., University Park, 1966-69, Bryn Mawr (Pa.) Coll., 1969-73; instr. to asst. prof. anthropology Pa. State U., Erie, 1973-84; dir. rsch. and grants Mercyhurst Coll., Erie, 1984-89, asst. dir. Hammermill Libr., 1989; humanities adminstr., sci. officer div. state programs NEH, Washington, 1989-91, sci. officer div. preservation and access, 1991—; bd. mem., officer Pa. Humanities Coun., Phila., 1979-83, 86-89; grant proposal reviewer NSF, 1982-90, NEH, 1981-89, Wenner-Gren Found. for Anthropol. Rsch., 1987-89; manuscript reviewer Holt, Rinehart and Winston, Inc., 1977-89, Prentice-Hall Inc., 1979-85, William C. Brown, Pubs., 1982-85, U. Tex. Press, 1988—. Author: Marine Shell Trade and Classic Teotihuacan, 1987; editor: A Pot for All Reasons, 1988, Ceramic Ecology, 1988, 89, 93; contbr. articles to profl. jours., chpts. to books; film reviewer Sci. Books and Films, 1977—; manuscript reviewer Am. Antiquity, 1978—, Current Anthropology, 1979—, Ancient Mesoamerica, 1990—, Ceramic Abstracts, 1990—; regional editor La Tinaja: Newsletter of Archaeol. Ceramics, 1991—; anthropology reviewer Choice: Current Reviews for Acad. Librs., 1992—. Mem. Commonwealth Pa., Gov.'s Conf. on Librs. and Info. Systems, 1989. Mem. Am. Ceramic Soc., Am. Ethnological Soc., Am. Soc. Ethnohistory, Archaeol. Inst. Am., Assn. Field Archaeology, Coun. Mus. Anthropology, Materials Rsch. Soc.,Prehistoric Ceramic Rsch. Group, Soc. Am. Archaeology, Soc. Archaeol. Scis. (life), Soc. Historical Archaeology, Soc. Med. Anthropology, Soc. Am. Archivists, Soc. Profl. Archaeologists, U.S. Naval Inst., Nat. Rwy. Hist. Soc., N.Y. State Archaeol. Assn., Paleopathology Assn., Pearl Harbor History Assocs. (life). Achievements include tech. and cultural interpretations of archaeol. ceramics by using physiochem. analyses and petrographic microscopy; rsch. on ceramics from Afghanistan, Ctrl. Asia, Mexico, Guatemala, East Africa, Great Lakes Basin. Home: 1005 Pruitt Ct SW Vienna VA 22180-6429 Office: NEH Div Preservation & Access 1100 Pennsylvania Ave NW Washington DC 20506

KOLB, JAMES A., science association director, writer; b. Berkeley, Calif., May 31, 1947; s. James DeBruler and Evelyn (Thomas) K.; m. Mary Catherine Eames; children: Thomas, Catherine Mary. BA in Zoology, U. Calif., Berkeley, 1970, BA in Biol. Sci., Ecology, 1970, MS in Wildland Resource Sci., 1972. Rsch. asst. Sagehen Creek Rsch. Sta. U. Calif., Berkeley, 1970, teaching asst. dept. wildlife & fisheries, 1970-71; rsch. assoc. air pollution resource ctr. U. Calif., Berkeley, Riverside, 1971; tchr. secondary sci. Hayward (Calif.) Unified Sch. Dist., 1972-77; dir. Marine Sci. Ctr., Poulsbo, Wash., 1981-92; exec. dir. Marine Sci. Soc. Pacific Northwest, Poulsbo, Wash., 1992—; project dir. Marine Sci. Project FOR SEA, Poulsbo, 1978-81; mem. Wash. State Environ. Edn. Task Force, Olympia, 1986—, Puget Sound Water Quality Authority Edn. & Pub. Involvement, Olympia, 1987-91, Marine Plastics Debris Task Force, Olympia, 1987; cons./tchr., trainer Hood Canal Wetlands Project, Hoodsport, Wash., 1990. Author: Marine Science Activities, 1979 (NSTA award 1986), Marine Biology and Oceanography, 1979, 80, 81 (NSTA award 1985, 86), Marine Science Career Awareness, 1984 (NSTA award 1985), The Changing Sound, 1990, Puget Soundbook, 1991; co-author: A Salmon in the Sound, 1991, Discovering Puget Sound, 1991. Mem. NSTA, ASCD, Internat. Reading Assn., Nat. Marine Educators Assn. (bd. dirs.), Northwest Assn. Marine Educators (pres.), Wildlife Soc. Office: Marine Sci Soc Pacific NW 18743 Front St NE PO Box 2079 Poulsbo WA 98370

KOLB, MARK ANDREW, aerospace engineer; b. Hopewell, Va., Aug. 30, 1962; s. David Lee and Sanna Sue (Buckner) K.; m. Jean Marie Pettit, July 14, 1984; 1 child, Andrew Ryan. BS in Aeronautics and Astronautics, MIT, 1984, PhD in Aeronautics and Astronautics, 1990. Aerospace engr. Orbital Scis. Corp., Fairfax, Va., 1987; aero. engr. R & D Ctr. GE Corp., Schenectady, N.Y., 1990—; cons. Lockheed-Ga. Co., Marietta, 1985-86. Author Rubber Airplane computer program. Gardner fellow MIT, 1984. Mem. AIAA, Am. Assn. for Artificial Intelligence. Episcopalian. Achievements include rsch. in computer modeling, artificial intelligence for engring. conceptual and preliminary design. Office: GE Corp R & D Ctr PO Box 8 K1-ES 204 Schenectady NY 12301

KOLB, NOEL JOSEPH, research scientist; b. Bangalore, Mysore, India, Dec. 22, 1930; s. George Alfred and Blanche Cecilia (Temasfieldt) K. Cert. in radio isotopes, Chelsea Coll. Tech.; cert. in semiconductor tech., West Ham Coll. Cost acct. Binny's Mills, Bangalore, 1949-55; technologist ultrahigh vacuum, planning, materials Bharat Electronics Ltd., Bangalore, 1956-61; sr. technician, sr. rsch. technologist microwave, electronics, hydrogen fuel cell, chromatography Queen Mary and Westfield Coll., U. London, 1961—. Roman Catholic. Avocations: music, literature, poetry, electronics, computing. Home: Elms Acre Leaden Roding, Near Dunmow, Essex CM6 1QG, England Office: U London Queen Mary and Westfield Coll, Mile End Rd, London E1 4NS, England

KOLBAS, ROBERT MICHAEL, electrical engineering educator; b. Syracuse, N.Y., Nov. 13, 1953; s. John Michael and Frances C. (Woityra) K.; children: Michael Thomas, Daniel Robert, Sarah Anne. BS in Engring., Cornell U., 1975; MS in Physics, U. Ill., 1975, PhD, 1979. Rsch., teaching asst. U. Ill., Urbana, 1975-79; prin. rsch. scientist Honeywell, Inc., Bloomington, Minn., 1979-83, sr. prin. rsch. scientist, 1983-85; assoc. prof. N.C. State U., Raleigh, 1985-90, prof. elec. and computer engring., 1990—; chmn. edn. com. Lasers and Electro Optics Soc. IEEE, 1991—. Contbr. articles to profl. publs.; patentee in field. Mentor to high sch. students, N.C. Sch. Sci. and Math., Durham, 1988-91. Kodak Doctoral fellow, U. Ill./Kodak, 1978. Mem. Am. Phys. Soc., IEEE, Tau Beta Pia, Sigma Xi. Office: N C State U Box 7911 Raleigh NC 27695-7911

KOLBE, HELLMUTH WALTER, acoustical engineer, sound recording engineer; b. Wallisellen, Switzerland, Aug. 28, 1926; s. Walter Karl Kolbe and Editha Berta (Ehrbar) Kamm; m. Ursula Charlotte Delabro, Feb. 25, 1966; children: Martin, Christian, Daniel. Diploma, Conservatory Vienna (Austria), 1953. Editor Universal Edit., Vienna, Austria, 1947-50; producer RWR-Radio Sta. (U.S.), Vienna, Austria, 1948-59; recording engr., dir. Mastertone Rec., Vienna, Austria, 1951-63; chief rec. engr., producer CBS Masterworks Internat., N.Y./London, Winterthur, Switzerland, 1961-75; producer, owner Phonag Schallplatten AG & Rec. Studios, Winterthur, Switzerland, 1959-88; acoustic cons. Ing. Büro für Akustik, Wallisellen, Switzerland, 1970—; rschr. on acoustics, model-acoustics, auralization and "In the Ear" recording process. Grammy award nominee, 1967; recipient Diplome Grand Prix du Disque Acad. Charles Cross, Paris, 1967. Mem. Audio Engring. Soc. U.S., Acoustical Soc. Am. Avocations: model railroading, ski instructor. Office: Ing Buro fur Akustik, Schafligrabenstrasse 32, CH-8304 Wallisellen Switzerland

KOLDITZ, LOTHAR, chemistry educator; b. Albernau, Saxony, Germany, Sept. 30, 1929; s. Paul and Ella (Bauer) K.; m. Ruth Schramm, Aug. 25, 1952; children: Christiane, Martina. D. in Natural Scis., U. Berlin, 1954, D. in Natural Scis. Habilitation, 1957; D. (hon.), Bergakademie, Freiberg, Fed. Republic Germany, 1983. Asst. Humboldt U., Berlin, 1952-54, oberasst., 1954-57, lectr., 1957; prof. Tech. Hochschule, Leuna-Merseburg, Fed. Republic Germany, 1957-59, Friedrich-Schiller U. Jena (Fed. Republic Germany), 1959-62, Humboldt U., Berlin, 1956-80, Acad. of Scis., Berlin, 1980-91; vice dir. Tech. Hochschule, Leuna-Merseburg, 1957-59; bd. dirs. Friedrich-Schiller-U. Jena, 1959-62, Humboldt U., Berlin, 1962-80, Acad. of Scis., Berlin, 1980-90. Editor, author (with others): Anorganische Chemie, 1978; editor: Anorganikum, 1967; editor (jours.) Zeitschrift für Chemie, 1960-90, Zeitschrift für Anorg. Allgem. Chem., 1989-90. Hon. mem. States Coun. German Dem. Republic, 1982. Recipient Nationalpreis, Govt. German Dem. Republic, 1972, Clemens-Winkler-Medaille, Chem. Gesellschaft German Dem. Republic, 1976, Goldene Heyrowski-Medaille, Tshechosl. Akademie der Wissensch., 1989. Mem. Acad. of Scis. Berlin, Acad. of Scis. Soviet Union, Tschechoslovak. Chem. Gesellschaft (hon.). Avocation: hunting. Home: Dorfstrasse 16, D-16798 Steinfôrde Germany

KOLIATSOS, VASSILIS ELEFTHERIOS, neurobiologist; b. Athens, Mar. 31, 1957; came to U.S., 1985; s. Eleftherios Vassilios and Alexandra J. (Doukas) K.; m. Nefeli Efthymia Massia, July 5, 1987. MD, U. Athens, 1982. Intern Crete Naval Hosp., Chania, Crete, Greece, 1982-83; attending physician Greek Naval Acad., Piraeus, Greece, 1983-84; rsch. fellow U. Athens, 1984-85; rsch. fellow Johns Hopkins U. Sch. Medicine, Balt., 1985-87, rsch. assoc., 1987-89, instr., 1989-90, asst. prof. dept. pathology and neurology, 1990—, asst. prof. dept. neuroscience, 1991—. Contbg. author books in field; contbr. articles to profl. jours. Recipient scholarship Greek Govt., Athens, 1985, award for Leadership and Excellency in Alzheimers Disease, Nat. Inst. Aging, Bethesda, Md., 1991. Mem. AAAS, Soc. Neurosci., Internat. Brain Rsch. Orgn., Internat. Basal Ganglia Soc. Greek Orthodox. Achievements include demonstration of the effects of nerve growth factor on degenerating neurons in the inmate brain and the discovery of peptide growth factors that promote the survival of motor neurons in vivo. Home: 310 W Lanvale St Baltimore MD 21217-3608

KOLLROS, PETER RICHARD, child neurology educator, researcher; b. Iowa City, Dec. 30, 1953; s. Jerry J. and Catharine (Lutherman) K.; m. Barbara A. Konkle, Aug. 5, 1978; children: Daniel, Catharine. BA, Northwestern U., 1975; MD with honors, PhD, U. Chgo., 1982. Diplomate Am. Bd. Pediatrics, Am. Bd. Psychiatry and Neurology with spl. qualification in child neurology. Resident Children's Meml. Hosp., Chgo., 1982-84; resident, fellow U. Mich. Hosp., Ann Arbor, 1984-88; asst. prof. child neurology Thomas Jefferson U., Phila., 1988—. Mem. Soc. of Friends. Achievements include discovery that cyclic AMP inhibits collagen production in stretch smooth muscle cells, elevated glucose inhibits endothelial cell transport of myo-inositol; description of tic douloureux in a 13-month child, 3rd patient with glucose transporter dysfunction and 1st candidate mutation of this disease; correction of CSF copper in Menkes kinky hair disease. Office: Thomas Jefferson U Dept Pediatrics 1025 Walnut St Philadelphia PA 19107-5083

KOLOR, MICHAEL GARRETT, research chemist; b. Bklyn., May 1, 1934; s. Michael Austin and Frances (Nugent) K.; B.S. in Chemistry, Queens Coll. CUNY, 1956; postgrad. Adelphi U., 1958-60; m. Agnes Theresa Fitzpatrick, June 29, 1957; children: Mary Catherine, Michael Francis, Agnes Theresa, Johanna Margaret. Chemist, Nat. Dairy Corp., Oakdale, N.Y., 1956-59; assoc. chemist Gen. Foods Corp., Tarrytown, N.Y., 1959-62, rsch. chemist, 1962-65, sr. rsch. chemist, 1965-70, rsch. specialist, 1970-72, sr. rsch. specialist, 1972-79, rsch. scientist, head mass spectrometry lab., 1972-87; sr. chemist Champion Internat. Corp., West Nyack, N.Y., 1989—. Mem. Am. Soc. for Mass Spectrometry, N.J. Am. Chem. Soc. Mass Spectrometry Group (program chmn. 1968-69). Co-author: Biochemical Applications of Mass Spectrometry, 1972; Supplementary Volume of Biochemical Application of Mass Spectrometry, 1980; Mass. Spectrometry (practical spectroscopy/series, 1979), patentee in field, contbr. articles on chemistry to profl. jours. Roman Catholic. Home: 71 Margaret Keahon Dr Pearl River NY 10965-1040

KOLSTAD, GEORGE ANDREW, physicist, geoscientist; b. Elmira, N.Y., Dec. 10, 1919; s. Charles Andrew and Rose Catherine (Haesloop) K.; m. Christine Joyce Stillman, July 22, 1944; children: Charles Durgin, Martha Rae Kolstad Wilhelm, Peter Kenneth. BS magna cum laude, Bates Coll., 1943; PhD in Physics, Yale U., 1948. Asst. in rsch. Eastman Kodak Co., Rochester, N.Y., 1938-39; rsch. assoc. Harvard U., Cambridge, Mass., 1944-45; instr. physics Yale U., New Haven, 1948-50; physicist physics and math. rsch. AEC, Washington, 1950-52; head physics and math. rsch. Dept. Energy/AEC, Washington, 1952-73; head geosci. program Dept. Energy, Washington, 1973-90; ret.; vis. scientist Niels Bohr Inst., Copenhagen, 1956-57; chmn., organizer 3 internat. neutron data coms., Washington, 1956-75 and IAEA, Vienna, Austria, 1956-75; Dept. Energy, mem. Interagy. Coord. Group on Continental Sci. Drilling, Washington, 1982-90. Contbr. articles to Physics Today, others. Trustee Bates Coll., Lewiston, Maine, 1958-63, Sandy Spring (Md.) Friends Sch., 1965-68; bd. dirs. Friends Retirement Community, Sandy Springs, 1992—. Recipient award Drilling Observation and Sampling of Earth's Continental Crust Consortium, Houston, 1990. Fellow Am. Phys. Soc.; mem. Am. Geophys. Union (life), Cosmos Club, Lions (pres. Laytonsville club 1992-93), Sigma Xi (pres. 1965). Achievements include advanced devel. of gas centrifuge method of isotope separation. Home: 7920 Brink Rd Gaithersburg MD 20882-1618

KOLTAI, STEPHEN MIKLOS, mechanical engineer, consultant, economist; b. Ujpest, Hungary, Nov. 5, 1922; came to U.S., 1963; s. Maximilian and Elisabeth (Rado) K.; m. Franciska Gabor, Sept. 14, 1948; children: Eva, Susy. MS in Mech. Engring., U. Budapest, Hungary, 1948, MS in Econs., MS, BA, 1955. Engr. Hungarian Govt., 1943-49; cons. engr. and diplomatic service various European countries, 1950-62; cons. engr. Pan Bus. Cons. Corp., Switzerland and U.S., 1963-77, Palm Springs, Calif., 1977—. Patentee in field. Charter mem. Rep. Presdl. task force, Washington, 1984—. Avocations: tennis, golf.

KOMDAT, JOHN RAYMOND, data processing consultant; b. Brownsville, Tex., Apr. 29, 1943; s. John William and Sara Grace (Williams) K.; m. Linda Jean Garrette, Aug. 26, 1965 (div.); m. Barbara Milroy O'Cain, Sept. 27, 1986; children: Philip August, John William. Student U. Tex., 1961-65. Sr. systems analyst Mass. Blue Cross, Boston, 1970-74; pvt. practice data processing cons., San Francisco, 1974-80, Denver, 1981—; prin. systems analyst mgmt. info. svcs. div. Dept. of Revenue, State of Colo., 1986-89; prin. systems analyst Info. Mgmt. Commn. Staff Dept. Adminstrn. State Colo., 1989—; mem. Mus. Modern Art, CODASYL End User Facilities Com., 1974-76, allocation com. Mile High United Way. Served with U.S. Army, 1966-70. Mem. IEEE, AAAS, Assn. Computing Machinery, Denver Downtown Dem. Forum (mem. exec. com.), Denver Art Mus., Friend of Pub. Radio, Friend of Denver Pub. Libr., Colo. State Mgrs. Assn, Nature Conservancy. Democrat. Office: PO Box 9757 Denver CO 80209

KOMISAR, ARNOLD, otolaryngologist, educator; b. N.Y.C., Nov. 27, 1947; s. Samuel and Sonia (Schwartz) K.; m. Lenora I. Felderman, Dec. 23, 1984; children: Alexandra Danielle, Jonathan Reed. BS, Bradley U., 1968; DDS, NYU, 1972; MD, Hahnemann Med. Coll., 1975. Diplomate Am. Bd. Otolaryngology. Resident in surgery Beth Israel Med. Ctr., N.Y.C., 1975-76; resident in Otolaryngology Mt. Sinai Med. Sch., N.Y.C., 1976-79; asst. prof. otolaryngology Albert Einstein Coll. Medicine, N.Y.C., 1979-85, assoc. prof., 1985-86, assoc. clin. prof., 1986-90; assoc. dir. head and neck surgery Albert Einstein Affiliated Hosps., N.Y.C., 1982-86; attending otolaryngologist Montefiore Hosp. and Med. Ctr., N.Y.C., 1979-90, Bronx Mcpl. Hosp. Ctr., N.Y.C., 1979-90, N. Cen. Bronx Hosp., N.Y.C., 1979-90; clin. assoc. prof. otolaryngology Cornell U. Med. Coll., N.Y.C., 1990—; otolaryngologist Lenox Hill Hosp., N.Y.C., 1986—, asst. to dir. resident edn. dept. otolaryngology, 1986—, adj. otolaryngologist, 1987—, attending otolaryngologist, 1989—, assoc. dir. otolaryngologist, 1990—; cons. otolaryngology N.Y. Eye and Ear Infirmary, N.Y.C., 1986-89; couresty staff surgery-otolaryngology Doctors Hosp., N.Y.C., 1986-90; presenter in field. Contbr. articles to profl. jours. Fellow Am. Coll. Surgeons, Am. Soc. Head and Neck Surgery, Am. Acad. Facial Plastic and Reconstructive Surgery, Am. Acad. Otolaryngology/Head and Neck Surgery (Honor award), Triological Soc. (Mosher award), Am. Bronchoesophagological Soc., N.Y. Acad. Medicine, Am. Laryngol. Assn.; mem. AMA, Pan-Am. Soc. Brocho-esophagology, Soc. Univ. Otolaryngologists, N.Y. Head and Neck Soc., Med. Soc. N.Y., N.Y. County Med. Soc. Avocations: reading, travel. Office: 1317 3d Ave New York NY 10021-2995

KOMMEDAHL, THOR, plant pathology educator; b. Mpls., Apr. 1, 1920; s. Thorbjorn and Martha (Blegen) K.; m. Faye Lillian Jensen, June 2, 1924; children—Kris Alan, Siri Lynn, Lori Anne. B.S., U. Minn., 1945, M.S., 1947, Ph.D., 1951. Instr. U. Minn., St. Paul, 1946-51, asst. prof. plant pathology, 1953-57, assoc. prof., 1957-63, prof., 1963-90, prof. emeritus, 1990—; asst. prof. plant pathology Ohio Agrl. Research and Devel. Ctr., Wooster, 1951-53, Ohio State U., Columbus, 1951-53; prof. continuing edn. and extension U. Minn., St. Paul, 1990—; cons. botanist and taxonomist Minn. Dept. Agr., 1954-60, Sci. Mus. Minn., 1990—; 7th A.W. Dimock lectr. Cornell U., 1979. Author: Pesky Plants, 1989; cons. editor McGraw-Hill Ency. Sci. and Tech., 1972-78; editor-in-chief Phytopathology, 1964-67; editor: Procs. IX Internat. Congress Plant Protection, 2 vols., 1981, corn disease newsletter, 1970-76; sr. editor: Challenging Problems in Plant Health, 1982, Plant Disease Reporter, 1979; contbr. articles to profl. jours. Recipient Elvin Charles Stakman award, 1990; Guggenheim fellow, 1961, Fulbright scholar, 1968. Fellow AAAS, Am. Phytopathol. Soc. (councilor 1958-60, pres. 1971, publs. coord. 1978-84, Disting. Svc. award 1984, 93, sci. adv. 1984—, mem. adv. bd. office internat. programs 1987-93); mem. Am. Inst. Biol. Scis., Bot. Soc. Am., Coun. Biology Editors, Internat. Soc. Plant Pathology (councilor 1971-78, sec.-gen. and treas. 1983-88, treas. 1988-93, editor newsletter 1983-93), Mycol. Soc. Am., Minn. Acad. Sci., N.Y. Acad. Scis., Weed Sci. Soc. Am. (award of excellence 1968). Baptist. Home: 1666 Coffman St # 322 Saint Paul MN 55108-1326 Office: U Minn 496 Borlaug Hall 1991 Upper Buford Circle Saint Paul MN 55108-6030

KOMORITA, SAMUEL SHOZO, psychology educator; b. Seattle, Apr. 7, 1927; s. Ken and Kaga Komorita; m. Nori Ishimoto, Dec. 23, 1948; children: Paul, John, Lorene. BS, U. Wash., 1950, MS, 1952; PhD, U. Mich., 1957. Asst. prof. psychology Vanderbilt U., Nashville, 1957-61; prof. psychology Wayne State U., Detroit, 1961-69, Ind. U., Bloomington, 1969-74, U. Ill., Champaign, 1974—. Author: Social Dilemmas, 1993; contbr. articles to profl. jours. With U.S. Army, 1944-47. Fellow Am. Psychol. Soc. Home: 2112 Seaton Ct Champaign IL 61821 Office: U Ill Psychology Dept 603 E Daniel Champaign IL 61820

KOMVOPOULOS, KYRIAKOS, mechanical engineer, educator; b. Pireas, Greece, Nov. 3, 1955; came to the U.S., 1979; s. Konstantinos and Evagelia (Karoussos) K.; m. Eleni Dadi, July 19, 1986; children: Joanna, Marios. BS in Civil Engring., Nat. Tech. U., 1979; MS in Civil Engring., MIT, 1981, MS in Aeronautics and Astronautics, 1981, PhD in Mech. Engring., 1986. Rsch. assoc. Nat. Tech. U., Athens, 1978-79; teaching asst. MIT, Cambridge, 1980-81, rsch. asst., 1981-86; asst. prof. U. Ill., Urbana, 1986-89; asst. prof. U. Calif., Berkeley, 1989-91, assoc. prof., 1991—; instr. Boston Archtl. Ctr., 1982; cons. Kaiser Aluminum and Chem. Co., Pleasanton, Calif., 1991-92, IBM, San Jose, Calif., 1990, IBM, Rochester, Minn., 1993, Zimmer Inc., Warsaw, Inc., 1988. Contbr. articles to ASME Jour. Tribology, Jour. of Wear, ASME Jour. Engring. Materials and Tech., Tribology Transactions, ASME Jour. Engring. for Industry, Metallurgical Transactions, Jour. of Applied Physics, ASME Jour. Applied Mechanics. Named Presdl. Young Investigator NSF, 1989—, recipient Engring. Initiation award, 1987-89. Mem. ASME (assoc. editor Jour. Tribology, Burt L. Newkirk award 1988, rsch. com. on tribology), Soc. Tribologists and Lubrication Engrs. (chmn. ceramics and composites com. 1989—). Achievements include discovery of new theories of friction and wear phenomena; development of new materials for thin and thick surface films, experimental techniques to examine microstructural characteristics and physical properties of metals, ceramics and laser-processed materials; analysis of tribological interactions in magnetic recording devices and microelectromechanical systems. Office: U Calif Dept Mech Engring Berkeley CA 94720

KONCEL, JAMES E., electrical engineer; b. Chgo., Sept. 29, 1929; s. James and Frances (Chesney) K.; m. Shirley A. Cermak, Oct. 11, 1958; children: Scott J., Jill T. BSEE, Purdue U., 1951; MBA, Loyola U., 1972. Elec. engr. Sandia Corp., Albuquerque, 1953-54; v.p. engring. & devel. Bodine Electric, Chgo., 1956—. Contbr. articles to profl. jours. 1st lt. USAF, 1951-52. Eta Kappu Nu scholar, 1950. Mem. IEEE (sr.). Office: Bodine Electric 2500 W Bradley Pl Chicago IL 60618-4798

KONDO, JUN, physicist; b. Tokyo, Feb. 6, 1930; s. Yasuo and Nanae (Miyagawa) K.; m. Emiko Tashiro, May 24, 1959; children: Natsuko, Yuko. BS, U. Tokyo, 1954, PhD, 1959. Asst. Inst. Solid State Physics, Tokyo, 1960-63; rschr. Electrotech. Lab., Tokyo, 1963-80; rschr. Electrotech. Lab., Tsukuba, Japan, 1980-87, fellow, 1987-90; prof. Toho U., Chiba, Japan, 1990—. Recipient Emperor's prize Japan Acad. Sci., 1973, Fritz London Meml. award IUPAP, 1987, Acta Metallurgica Gold medal, Acta Metallurgica, 1991. Achievements include discovery of the Kondo effect. Home: Kamitakaido 2-2-29, Tokyo 168, Japan Office: Toho Univ Faculty of Science, Miyama 2-2-1, Chiba 274, Japan

KONECKY, MILTON STUART, chemist; b. Omaha, July 29, 1922; s. Eugene Max and Eve (Lipp) K.; m. Naomi Marie Schipporeit, Oct. 9, 1948; children: Mark, Chad. BS, Creighton U., 1944, MS, 1947; PhD, U. Ill., 1958. Chemist Entomology Rsch. div. USDA, Beltsville, Md., 1950-54; sr. chemist Exxon Rsch. & Engring. Co., Linden, N.J., 1957-61, rsch. assoc., 1961-63, sect. head, 1963-69, sr. staff advisor, 1969-85, sr. rsch. assoc., 1985-91; pres. Konecky Assocs Enterprises, Pottersville, N.J., 1991—. Contbr. dozens of reports and publs. to profl. jours. Advisor sci. curriculum Old Turnpike Sch., Tewksbury Twp., N.J., 1978-80; advisor mock trial team Voorhees High Sch., Glen Gardner, N.J., 1982-84. With USN, 1944-46, PTO. Texaco fellow, 1956-57. Mem. AAAS, Am. Chem. Soc., N.Y. Acad. Sci., Sigma Xi, Phi Lambda Upsilon. Achievements include patents for Synthesis of Decahydroacenaphthene as High Energy Fuel, Preparation of Succinic Acid from Acrolein, Preparation Biodegradable Detergetns, Polyphosphoric Acid as Cyclization Agent, Preparation Lubricating Compositions. Home: PO Box 307 Dryden Rd Pottersville NJ 07979 Office: Konecky Assocs Enterprises PO Box 307 Pottersville NJ 07979

KÖNIG, HEINZ JOHANNES ERDMANN, mathematics educator; b. Stettin, Germany, May 16, 1929; s. Josef and Meta (Bognitz) K.; m. Helga, Oct. 2, 1954, (wid. Feb. 1979); m. Karin Grewin, Nov. 21, 1980; 1 child, Daniel. D degree, U. Kiel, Fed. Republic Germany, 1952; Habilitation, U. Würzburg, Fed. Republic Germany, 1956; D degree (hon.), U. Karlsruhe, Fed. Republic Germany, 1979. Dozent Tech. U. Aachen, Fed. Republic Germany, 1957-60, assoc. prof., 1960-62; prof. U. Köln, Fed. Republic Germany, 1962-65, U. Saarbrücken, Fed. Republic Germany, 1965—; vis. prof. CALTECH, Pasadena, Calif., 1967-68, U. Washington, Seattle, 1970, U. Witwatersrand, Johannesburg, South Africa, 1992; vis. scholar, U. Tex., Austin, 1978; vis. fellow CALTECH, 1982-83, Australian nat. U., Canberra, 1988-89. Author: (with K. Barbey) Abstract Analytic Function Theory, 1977, Analysis I, 1984, (with M.M. Neumann) Mathematische Wirtschaftstheorie, 1986; contbr. numerous articles to profl. jours. Founding pres. Assn. Friends Hebrew U. in the Saar State, 1977—. Decorated officier Ordre Grand Ducal Couronne de Chêne (Luxembourg). Mem. Royal Soc. Sci. (Liège, Belgium) (corr.), German Math. Soc., Am. Math. Soc., French Math. Soc., Gesellschaft Angewandte Math. Mechanik, Soc. Math., Econs. and Ops. Rsch. (founding pres. 1977-81), Lions (pres. Saarbrücken-St. Johann 1977-78). Home: Auf Gierspel 36, D-66132 Saarbrücken Germany Office: U Saarlandes, D-66041 Saarbrücken Germany

KONIGSBERG, WILLIAM HENRY, molecular biophysics and biochemistry educator, administrator; b. N.Y.C., Apr. 5, 1930; s. Joseph and Jennie (Schneider) K.; m. Diane Danielson, Mar. 3, 1956 (div. 1975); 1 child, Jessica; m. Paulette Cohen, July 3, 1982; 1 child, Rachel. B.S. in Chemistry, Rensselaer Poly. Inst., 1952; Ph.D. in Organic Chemistry, Columbia U., 1956. NSF fellow Rockefeller Inst., 1956-57, research assoc., 1957-59, asst. prof., 1959-64; assoc. prof. biochemistry Yale U., New Haven, 1964-76, prof. molecular biophysics and biochemistry, 1976—, chmn. dept. molecular biophysics and biochemistry, 1984—; NSF panel mem.; mem. biochemistry study sect. NIH, 1970-74, physiol. chemistry study sect., 1970-74, adv. council minority career opportunity sect., 1976—; chmn. Gordon Conf. on Proteins, 1976-77; ad hoc cons. NSF, Am. Cancer Soc., Heart and Lung Inst. Mem. editorial bd. Archives of Biochemistry, 1968-72, Biochem. Biophys. Acta, 1969-73, Proteins: Structure, Function and Genetics, 1986; contbr. numerous articles to profl. jours. Mem. Am. Chem. Soc., Am. Soc. Biol. Chemists (membership com. 1969-70), Minority Biomed. Rev. Council, U.S.-Israel Binational Sci. Found. Office: Yale U Dept Molecular Biophysics and Biochemistry PO Box 3333 333 Cedar St Rm C-113 SHM New Haven CT 06510

KONISHI, KENJI, geology educator; b. Tokyo, Feb. 18, 1929; s. Kin'ichi and Hisa Konishi; m. Hiroko Hirakawa, Apr. 23, 1961; 1 child, Lisa Imamura. BS, U. Tokyo, 1951, DSc, 1961; MS, Colo. Sch. Mines, 1959. Fulbright rsch. fellow Colo. Sch. Mines, Golden, 1955-58; geologist Ohio Oil Co. Rsch. Ctr., Littleton, Colo., 1959; paleontologist U.S. Geol. Survey, Denver, 1959-60; asst. prof. U. Tokyo, 1960; lectr. Kanazawa (Japan) U., 1960-64, assoc. prof., 1964-68, prof., 1968—, dean, 1992—; vis. prof. U. Calif., Riverside, 1969-70. V.p. Hokuriku Japan-Am. Cultural Soc., Kanazawa, 1980—. Recipient Dist. rsch. awards Paleontol. Soc. Japan, 1974, Coral Reefs Rsch. awards Ishikawa TV Corps., 1990. Mem. Hokuriku Garioa-Fulbright Alumni Assn. (pres. 1992—). Home: 6-28-501 Tamagawa-cho, Ishikawa Kanazawa 920, Japan Office: Kanazawa U, Kakuma-machi, Kanazawa Ishikawa 920-11, Japan

KONKEL, R(ICHARD) STEVEN, environmental and social science consultant; b. Denver, June 27, 1950; s. E. Vernon and Rojean (Templeman) K.; m. Jane Frances Ohlert, July 14, 1984; children: Kaitlin Brooke and Britt Edward (twins). BS in Archtl. Engring., U. Colo., 1972; M in City Planning, Harvard U., 1975; PhD in Urban and Environ. Planning, MIT, 1991. Economist, planner Edward C. Jordan Co., Portland, Maine, 1975-77; cost-benefit analyst Oak Ridge (Tenn.) Nat. Lab., 1977-79; prin. economist Konkel Environ. Cons., San Francisco, 1980-82; policy analyst State of Alaska Office of Gov. and Dept. Commerce and Econ. Devel., Juneau, 1982-84; pres. Konkel & Co., Cambridge, Mass., 1984-91; sr. rsch. sci. energy environ. policy and dispute resolution Battelle Pacific Northwest Lab., 1992—; Coord. MIT faculty seminar on risk mgmt., 1988-89. Author: Environmental Impact Assessment Rev., 1987; co-editor: MIT Faculty Seminar on Risk Management, 1989. Co-chair Juneau Energy Adv. Com., 1984. Research grantee Nat. Inst. Dispute Resolution, Washington, 1986-87. Mem. Am. Inst. Cert. Planners (charter), Am. Planning Assn., Am. Econ. Assn., Internat. Assn. Energy Econs., Assn. Environ. and Resource Economists. Avocations: skiing, photography, backpacking, softball, fly fishing. Home: 8508 W Entiat Pl Kennewick WA 99336 Office: Battelle Pacific Northwest Lab Energy and Eviron Scis Bldgs PO Box 999 Richland WA 99352

KONRAD, WILLIAM LAWRENCE, electrical engineer; b. Pitts., Dec. 15, 1921; s. William Anton and Catherine Mary (Lawrence) K.; m. Dorothy Marie Weyman, Nov. 27, 1948; children: William W., Barbara J. BSEE, Carnegie Mellon U., 1948. Registered profl. engr., Conn. Engr. Union switch & Signal, Swissvale, Pa., 1948-59; sr. engr. Avco Corp., New London, Conn., 1959-67; supervisory engr. Raytheon Co., New London, Conn., 1967-70; engr. Naval Underwater Systems Ctr., New London, Conn., 1970-73, br. head, 1973-86, sr. engr., 1986—; head Navy Wide Nonlinear Acoustics Group, 1977-79. Contbr. numerous articles to profl. jours. With USN, 1944-46. Recipient Navy Engring. award Naval Underwater System Ctr., 1987, Decibel award, 1986. Fellow Acoustical Soc. Am.; mem. IEEE (sr.), Niantic Bay Yacht Club, Tri-City Amateur Radio Club. Achievements include 16 patents. Home: 54 Laurel Hill Dr Niantic CT 06357

KONSOWA, MOKHTAR HASSAN, mathematics educator; b. Tanta, Gharbia, Egypt, Aug. 25, 1953; s. Hassan Al Husieny and Zakia Muhammad (Khalil) K.; m. Amira Sayed Ahmed Al Hefny, May 5, 1985; children: Alaa, Nuha. BSc, Faculty Sci., Tanta, 1975, MSc, 1982; MSc, U. Cin., 1986, PhD, 1988. Lectr. Menouf Inst., 1976-82; teaching asst., lectr. U. Cin., 1982-88, lectr. Evening Coll., 1986-88; assoc. prof. math., mem. quantitative methods dept. King Saud U., Unizah, Saudi Arabia, 1988-93, assoc. prof., 1993—. Contbr. articles to math. jours. Univ. grad. scholar U. Cin., 1983-88, Univ. Rsch. Coun. fellow, 1987; grantee U. Wis., 1987. Mem. Egyptian Math. Soc., Am. Math. Soc. Avocations: soccer player, ping pong, swimming. Office: King Saud, Coll Bus and Econs, Unizah 505, Saudi Arabia

KONTNY, VINCENT, engineering and construction company executive; b. Chappell, Nebr., July 19, 1937; s. Edward James and Ruth Regina (Schumann) K.; m. Joan Dashwood FitzGibbon, Feb. 20, 1970; children: Natascha Marie, Michael Christian, Amber Brooke. BSCE, U. Colo., 1958, DSc honoris causa, 1991. Operator heavy equipment, grade foreman Peter Kiewit Son's Co., Denver, 1958-59; project mgr. Utah Constrn. and Mining Co., Western Australia, 1965-69, Fluor Australia, Queensland, Australia, 1969-72; sr. project mgr. Fluor Utah, San Mateo, Calif., 1972-73; sr. v.p. Holmes & Narver, Inc., Orange, Calif., 1973-79; mng. dir. Fluor Australia, Melbourne, 1979-82; group v.p. Fluor Engrs., Inc., Irvine, Calif., 1982-85, pres., chief exec. officer, 1985-87; group pres. Fluor Daniel, Irvine, Calif., 1987-88, pres., 1988—; pres. Fluor Corp., Irvine, 1990—. Contbr. articles to profl. jours. Mem. engring. devel. coun., U. Colo.; mem. engring. adv. coun., Stanford U. Lt. USN, 1959-65. Mem. Am. Assn. Cost Engrs., Australian Assn. Engrs., Am. Petroleum Inst. Republican. Roman Catholic. Club: Cet. (Costa Mesa, Calif.). Avocation: snow skiing. Office: Fluor Corp 3333 Michelson Dr Irvine CA 92730

KONTOS, EMMANUEL GEORGE, polymer chemist; b. Thessaloniki, Greece, Mar. 23, 1932; came to U.S., 1955; s. George and Aikaterini (Emmanuel) K.; m. Clare Wright, Feb. 22, 1958; children: Nina J., Leila M. MS in Chem. Engring., Nat. Tech. U., Greece, 1955; PhD in Chemistry, Columbia U., 1959. Rsch. chemist Uniroyal Chem. Div., Naugatuck, Conn., 1959-64; sr. group leader Uniroyal Chem. Div., Naugarung, Conn., 1964-67; mgr. polymer physics Uniroyal Corp. Rsch. Ctr., Wayne, N.J., 1967-73; mgr. materials and process rsch. Uniroyal Tire Co., Detroit, 1973-76; sr. rsch. asst. Uniroyal Chem. Div., 1976-79, mgr. tech., 1980—. Contbr. articles to profl. jours. Democrat. Unitarian. Home: 90 North St Trumbull CT 06611

KONUMA, MITSUHARU, materials scientist; b. Ashibetsu, Hokkaido, Japan, Jan. 19, 1950; s. Tatsuro and Michiko (Nasu) K.; m. Sumiko Takemasa, Nov. 4, 1976; 1 child, Akira. BS, Aoyama Gakuin U., Tokyo, 1975, MS, 1977, DSc, 1980. Supr. ANELVA Corp., Tokyo, 1980-83, 85-86; vis. scientist U. Zürich, Switzerland, 1983-85; materials scientist Max-Planck-

Inst. für Festkörperforschung, Stuttgart, Germany, 1986—. Author: Fundamentals of Plasma and Its Application to Film Deposition, 1986, Film Deposition by Plasma Techniques, 1992, (with G. Zhang) Fundamentals of Plasma Thin Film Technology, 1993; contbr. articles to profl. jours. Mem. Japanese Soc. Applied Physics; Deutsche Gesellschaft für Kristallwachstum und Kristallzüchtung. Home: Sommerhaldenstrasse 10A, 70195 Stuttgart Germany Office: Max-Planck Inst für FKF, Heisenbergstrasse 1, 70569 Stuttgart Germany

KOOMANOFF, FREDERICK ALAN, systems management engineer, researcher; b. N.Y.C., Sept. 2, 1926; s. Alexander Theodore and Margaret Theresa (McKendry) K.; m. Lora Gahimer, May 23, 1955; children: Vivre, Heather, Elena. BS in Indsl. Engring., NYU, 1952, MS in Indusl Engring., 1953. Dir. system scis. Battelle Meml. Inst., Columbus, Ohio, 1958-67; sr. assoc. Planning Rsch., Inc., Washington, 1967-70; dir. diversification Ea. Airlines, Washington, 1970-72; pres. Stentran Systems, Inc., Vienna, Va., 1972-76; dir. SPS program Dept. of Energy, Office Basic Energy Scis., Washington, 1978-81, dir. CO2 program, 1981-89, sci. facility mgr., 1989—; cons. Com. on Mgmt. Improvement in Govt., Washington, 1970-72; mem. at large R&D Aerospace Policy IEEE, Washington, 1989—. Author: Cybernetics on the Railways, 1962, Solar Power Via Satellite, 1993; also pubs. on environ. concerns, transp. and comm. systems. Cadet USAF, 1944-45. Recipient Calling of an Engr. award, Engring. Soc. Can., 1957. Mem. Sigma Xi. Home: 10700 Montrose Ave POB 221 Garrett Park MD 20896 Office: US Dept Energy Office of Basic Energy Scis Washington DC 20585

KOONS, LAWRENCE FRANKLIN, chemistry educator; b. Columbus, Ohio, Sept. 14, 1927; m. Benjamin Franklin and Ruth Elizabeth (Betsch) K.; m. Dorothy Helen Baesmann, Sept. 5, 1952 (dec. Dec. 1982); m. Helen Curtis Campbell, Mar. 30, 1991. BS in Chemistry, Ohio State U., 1949, PhD in Chemistry, 1956. Police chemist City of Columbus, 1952-56; instr. to asst. prof. DePaul U., Chgo., 1956-59; asst. prof. to prof. Tuskegee (Ala.) Inst., 1959—; Fulbright lectr. U. Dakar, Senegal, 1977-78; cons. Inst. at U. of Allahabad, Burdwan and Lucknow, 1965-67; abstractor of Russian articles Chem. Abstracts, 1960-87. Editor, translator: The Troubadours, 1965. Mem., sec. Airport Adv. Commn., Tuskegee, 1992—. Mem. AAAS, Am. Chem. Soc. (sect. chair 1972-73), Aircraft Owners and Pilots Assn., Sigma Xi. Home: PO Box 947 Tuskegee Institute AL 36087 Office: Tuskegee U Chemistry Dept Tuskegee AL 36088

KOONTZ, JAMES L., manufacturing executive; b. Dayton, Ohio, 1934. Various mgmt. positions in mfg. and engring. Gen. Motors, Chrysler, Bendix Corp.; with Micromatic, Inc., subs. Ex-Cell-O Corp., Detroit, Mich., 1966-68; gen. mgr. XLO Parker, subs. Ex-Cell-O corp., 1968; gen. mgr. divsn. machine tools Ex-Cell-O Corp., 1968, group v.p., 1969-78; exec. v.p. Kingsbury Corp., Keene, N.H., 1978-82, pres., CEO, 1982—; chmn. emeritus Nat. Ctr. Mfg. Scis.; mem. mfg. sci. bd. Dept. Defense. Recipient Gold medal Soc. Mfg. Engrs., 1990. Mem. Nat. Machine Tool Builders' Assn. (past chmn.). Office: Kingsbury Corp 80 Laurel St Keene NH 03431-4207*

KOOP, CHARLES EVERETT, surgeon, government official; b. Brooklyn, N.Y., Oct. 14, 1916; s. John Everett and Helen (Apel) K.; m. Elizabeth Flanagan, Sept. 19, 1938; children: Allen van Benschoten, Norman Apel, David Charles Everett, Elizabeth. AB, Dartmouth Coll., 1937; MD, Cornell U., 1941; DSc in Medicine, U. Pa., 1947; LLD (hon.), Ea. Bapt. Coll., 1960; MD (hon.), U. Liverpool, Eng., 1968; LHD (hon.), Wheaton Coll., 1973; DSc (hon.), Gwynedd Mercy Coll., 1978, Washington and Jefferson Coll., 1979; LLD (hon.), Phila. Coll. Osteo. Medicine, 1979; LHD (hon.), Phila. Theol. Sem., 1980; LLD (hon.), LaSalle Coll., 1983; DSc (hon.), Marquette U., 1983, Ea. Mich. U., 1985, N.Y. Med. Coll., 1985, Ball State U., 1987; LHD (hon.), Chgo. Med. Sch., 1988; DSc (hon.), Kirksville Coll. Osteo. Med., 1988; LLD (hon.), Colby-Sawyer Coll., 1988; DSc (hon.), Albany Med. Coll., 1988, Colby Coll., 1988, Yeshiva U., 1988, Phila. Coll. Pharmacy and Sci., 1988, Baylor Coll. Medicine, 1988, Dartmouth Coll., 1989, U. Mass., Boston, 1989, Brandeis U., 1990, Brown U., 1990, Northwestern U., 1990, U. pa., 1990; LLD (hon.), U. Miami, 1991, U. Cin., 1991; D. Pub. Svc. (hon.), George Washington U., 1991. Diplomate Am. Bd. Surgery, Nat. Bd. Med. Examiners. Intern Pa. Hosp., Phila., 1941-42; fellow in surgery Boston Children's Hosp., 1946; surgeon-in-chief Children's Hosp. of Phila., 1948-81; with U. Pa. Sch. Medicine, 1942-85, prof., 1959-85; former dep. asst. sec. for health HHS; surg. gen. of U.S., 1981-89; former dir. internat. health USPHS, from 1982; chair Safe Kids Nat. Campaign, Washington; dir. Elizabeth De Camp McInery prof. surgery C. Everett Koop Inst. Dartmouth-Hitchcock Med. Ctr., Hanover, N.H., 1993—; cons. USN, 1964-81; sr. scholar The C. Everett Koop Inst. at Dartmouth; dir. Room to Learn Program Carnegie Found., 1993—. Author: Visible and Palpable Lesions in Children, 1976, The Right to Live, The Right to Die, 1976, rev. edit., 1980, Smoking: The New Book of Knowledge, 1989; (with E. Koop) Sometimes Mountains Move, 1979; (with F. A. Schaeffer) Whatever Happened to the Human Race?, 1979, (with T. Johnson) Koop: The Memoirs of America's Family Docotr, 1991, Let's Talk, 1992; editor surgery sect. Jour. Clin. Pediatrics, 1961-64; mem. editorial bd. Zeitschrift fur Kinderchirurgie and Grenzqebiete, 1964-81; editor in chief: Jour. Pediatric Surgery, 1965-77; editorial cons. Japanese Jour. Pediatric Surgery and Medicine, 1970-81; chmn. editorial bd. PHS Reports, 1982-89; mem. editorial adv. bd. Tobacco Control: An Internat. Jour.; contbr. pubs. in surg. physiology, biomed. ethics, physiology of surg. neonate, tech. advances in pediatric surgery. Bd. dirs. Med. Assistance Programs, Inc., Brunswick, Ga., Friends Nat. Libr. of Medicine, Nat. Mus. of Health and Medicine Found. Inc. (pres.). Decorated chevalier Legion of Honor (France); Order Duarte, Sanchez and Mella (Dominican Republic); recipient medal City of Marseille, Presbyn. Man of Yr. award Presbyn. Social Union Phila., 1975, Super Achiever of Yr. award Phila. chpt. Juvenile Diabetes Found., 1975, Man of Yr. award Jewish Community Chaplaincy Svc. Phila., 1975, Kopernicus medal Polish Surg. Soc., 1977, Gold medal Children's hosp. Phila., 1981, Sec. of Health of Commonwealth of Pa. award, 1981, Thomas Linacre award Nat. Fedn. Cath. Physicians Guild, 1981, Award of Distinction Alumni Assn. Cornell U. Med. Coll., 1988, Humanitarian Svc. award City of Boston, 1989, Harry S. Truman award City of Independence, Mo., 1990, Daniel Webster award Dartmouth Coll., 1990, John Wiley Jones Disting. Lectr. award Rochester Inst. Tech., 1990, Tyler prize U. So. Calif., 1991, Albert Schweitzer prize Johns Hopkins U., 1991, Person of Yr. award Nat. Hosp. Orgn., 1991, others; recipient key to City of St. Louis, 1985; named an hon. citizen, City of Balt., 1985; C. Everett Koop Hon. Lectr. medal named in his honor Anchor & Caduceus Soc., 1991, C. Everett Koop Health Adv. award named in his honor Am. Soc. for Health Care Mktg. and P. Rels, Gustav O. Lienhard award Inst. Medicine, 1992; Disting. scholar to Carnegie Found. for advancement of teaching. Fellow ACS, Am. Acad. Pediatrics (William E. Ladd Gold medal), Royal Coll. Surgeons Eng. (hon.), Royal Coll. Physicians and Surgeons of Glasgow (hon.); mem. Am. Surg. Assn., Soc. U. Surgeons, Brit. Assn. Pediatric Surgeons (Dennis Browne Gold medal), Internat. Soc. Surgery, Assn. Mil. Surgeons U.S. (pres. 1982, 87, Founders medal), Societe Francaise de Chirugie Infantile, AMA, Deutschen Gesselschaft für Kinderchirugi, Societé Suisse De Chirurgie Infantile, Sigma Xi. Office: Dartmouth Coll Dartmouth Hitchkock Med Ctr C Everett Koop Inst Hanover NH 03755

KOOPMAN, WILLIAM JAMES, medical educator, internist, immunologist; b. Lafayette, Ind., Aug. 19, 1945; s. William James and Barbara Mary (Morehouse) K.; m. Lilliane Kathryn Desimone, June 15, 1968; children: Benjamin, Anna, Rebecca, Steven. BA, Washington and Jefferson U., 1967; MD, Harvard U., 1972. Diplomate Am. Bd. Internal Medicine. Intern/resident in medicine Mass. Gen. Hosp., Boston, 1972-74; rsch. fellow NIH, Bethesda, Md., 1974-77; from asst. prof., assoc. prof. to prof. medicine specializing in rheumatology and clin. immunology U. Ala., Birmingham, 1977—, Howard L. Holley prof. medicine, 1988—, dir. Multipurpose Arthritis Ctr., 1983—; mem. nat. adv. coun. Nat. Inst. Arthritis, 1987-90. Musculo-skeletal and Skin Diseases; chmn. bd. sci. counselors, NIH, NIAMS, 1991—. Editor Arthritis and Rheumatism jour., 1985-90; contbr. over 210 articles to profl jours. Recipient Carol Nachman Rsch. prize Fed. Republic Germany, 1982. Fellow ACP, Am. Coll. Rheumatology (pres. Southeastern region 1986-87, treas. 1992—); mem. Am. Soc. Clin. Investigation (pres. 1990-91), Assn. Am. Physicians, Am. Assn. Immunologists, Birmingham Area C. 0f C. Presbyterian. Avocations: fishing, gardening. Office: U Ala Tinsley Harrison Tower 429A Univ Station Birmingham AL 35294

KOOPMANN, GARY HUGO, educational center administrator, mechanical engineering educator; b. Howells, Nebr., May 8, 1939; s. Hugo Martin and Elsie (Hledik) K.; m. Barbara Bogue, May 26, 1972; children: Hannah, Eve. BS, U. Nebr., 1962; MS, Cath. U., 1966, PhD, 1969. Cert. engr., Tex. Rsch. scientist U.S. Naval Rsch. Lab., Washington, 1962-66; postdoctoral fellow Inst. Sound and Vibration, Southampton, Eng., 1969-70, univ. lectr., 1970-76; prof. U. Houston, 1976-87; dir. Ctr. Acoustics & Vibrations Pa. State U., State Coll., 1988—, prof. mechanical engring., 1988—; vis. prof. DFVLR, Berlin, 1982-83. Patentee noise reduction system, TRC suspension sim. Fellow Am. Soc. Mechanical Engrs. (editor jour.), Acoustical Soc. Am. Quaker. Avocation: music. Office: Pa State U Ctr Acoustics & Vibration 157 Hammond Bldg University Park PA 16802

KOOSER, ROBERT GALEN, chemical educator; b. Mankato, Minn., July 23, 1941; s. Galen Frances and Marion Standish (Drake) K.; m. Karen E. Bangs, June 5, 1966 (div. Nov. 1982); children: Ara S., Amanda C.; m. Patricia L. Carlson, June 8, 1985 (div. Nov. 1992). BA, St. Olaf Coll., 1963; PhD, Cornell U., 1968. Asst. prof. Knox Coll., Galesburg, Ill., 1968-76; assoc. prof. Knox Coll., Galesburg, 1976-84, prof., 1984—; postdoctoral assoc. Varian Assoc., Palo Alto, Calif., 1976-77; vis. assoc. prof. Dartmouth Coll., Hanover, N.H., 1983-84. Reviewer: Jour. Chem. Edn., 1968—, NSF, 1990—; contbr. chpts. to Chemically Modified Surfaces, vol. 3, 1990, ACS Symposium Series, 1992; contbr. articles to Jour. Phys. Chemistry, Analytical Chemistry, Concepts in Magnetic Resonance. Bd. mem. Civic Chorus, Galesburg, 1991-92. Mem. Am. Chem. Soc., Am. Phys. Soc., Sigma Xi, Phi Beta Kappa (local pres. 1973). Achievements include application of EPR to bonded chromatographic phases; research in role of value presuppositions in chemical education. Home: 1258 N Cherry St Galesburg IL 61401 Office: Knox College K46 2 E South St Galesburg IL 61401-4999

KOOYOOMJIAN, K. JACK, environmental engineer; b. Boston, Apr. 5, 1942; s. Haroutune and Mary (Sarafian) K.; m. Geraldine A. Smith, Nov. 3, 1973; children: Jennifer, Melissa, Jessica. BSME, U. Mass., 1965; MS in Mgmt. Sci., Rensselaer Poly. Inst., 1967, PhD in Environ. Engring., 1974. Mech. engr./aide Materials Rsch. Agy. and Ordnance Corps, U.S. Army, Watertown, Mass., 1963-65; grad. asst. to chairperson Rensselaer Poly. Inst., Troy, N.Y., 1966-68, grad. asst. Fresh Water Inst., 1967-74, instr. environ. engring., 1972; with U.S. EPA, Washington, 1974—, engr. Superfund program Emergency Response divsn., 1979-88, designated fed. ofcl. environ. engring. com. Sci. Adv. Bd., 1988—; sponsor, creator studies to promote better understanding of EPA and Dept. of Transp. hazardous materials/hazardous substance programs, 1983-86; mem. environ. and applied sci. prize panel Found. for Advancement of Sci., Kuwait, 1982. Contbr./co-contbr. articles to profl. publs. Chmn. PELT (planning, environ., land-use and transp.) com. Lake Ridge-Occoquan (Va.) Civic Assn., 1979—; mem. citizen's adv. com. to amend comprehensive plan Prince William County, 1986-90, chmn., 1986-89; mem. Environ. Clean-Up Task Force, Prince William County, 1986-92; bd. dirs. Tacketts Mill Found., Lake Ridge, Va., 1985—. Recipient Svc. awards Prince William County, 1988, 90, Prince William County Voluntary Action Ctr. cert. of appreciation, 1991, Gov.'s award for vol. excellence, State of Va., 1991. Mem. Water Pollution Control Fedn. (dir. bd. of control 1986-89, mem. policy adv. com. 1988-89, co-chmn. local arrangements com. for 1990 conf., Svc. awards 1982, 86, 89, 90, Arthur Sidney Biddell award 1988), Fed. Water Quality Assn. of Water Pollution Control Fedn. (pres. 1981-82, chmn. mgmt. com. 1983-89, Svc. awards), Sigma Xi, Omicron Delta Epsilon (charter, Rensselaer Poly. Inst. chpt.), Chi Epsilon, Adelphia (U. Mass. chpt.), Alpha Phi Omega (life, v.p. svc. U. Mass chpt. 1961-62). Avocations: civic activities, environ. protection activities, water pollution control through profl. assns., land-use planning. Home: 12453 Skipper Cir Woodbridge VA 22192 Office: US EPA 401 M St # 508 Washington DC 20460

KOP, TIM M., psychologist; b. Aug. 3, 1946; s. Michael and Antoinette Wanda (Stahurski) K.; m. Yoshino Fujita, Aug. 9, 1975; children: Maile K., Geoffrey M. BA in Psychology, U. Hawaii, 1972; MA in Edn., Mich. State U., 1976; MS in Psychology, Columbia Pacific U., 1989, PhD in Psychology, 1991. Air traffic controller FAA, Honolulu, 1968-74; with U.S. Dept. Def., 1974—; pres. PAOA, Inc., Honolulu, 1986—; cons. Tripler Army Med. Ctr., Honolulu, 1990-92, State of Hawaii, 1992—. Author: Neural Programming, 1991, Normal Language Learning, 1989, Normal Language Learning and Aphasia, 1988; editor: North Korean Military Forces, 1979; author manuscript: Counterinsurgency along the Thai-Malaysian Border, 1982. Vice pres. Waiau Gardens Community Assn., Pearl City, Hawaii, 1986-88. Capt. U.S. Army, 1965-68, 78. Recipient Sec. of the Navy Award for group achievement U.S. Sec. of Navy, 1982. Fellow Am. Orthopsychiat. Assn.; mem. APA, Am. Psychological Soc., Am. Assn. Artificial Intelligence, U. Hawaii Alumni Assn. (life), Assn. of Mil. Surgeons of U.S. (life). Democrat. Office: Century Ctr Ste 3-520 1750 Kalakaua Ave Honolulu HI 96826-3766

KOPALA, PETER STEVEN, mechanical engineer; b. Chgo., Oct. 28, 1946; s. Theodore Stanley and Eva Bernice (Kearney) K.; m. Ruth Ann Schacht, Dec. 19, 1971; children: Brian Steven, David Anthony, Jeffery Vernon. BS in Mech. Engring., U. Ill., 1969. Registered profl. engr., Ill., Wis., Minn., Mich., Ind. Mech. engr. Electro Motive Div. GM, LaGrange, Ill., 1969-74, sr. mech. engr., 1974-87; project mgr. BCM Engrs., Chgo., 1987-88, Amoco Corp., Naperville, Ill., 1988—; v.p. engring. Tri-Logic Inc., Oak Park, Ill., 1980-85; cons. in field. Scout leader Cub Scouts Am., Oak Park, 1981. Mem. NSPE, Ill. Soc. Profl. Engrs., Project Mgmt. Inst., ASME (alternate rep. B30 com. 1985-89). Achievements include patent for mass calculating and indicating means for weighing moving vehicles. Office: Amoco Rsch Ctr 150 W Warrenville Rd Naperville IL 60563-8460

KOPECKO, DENNIS JON, microbiologist, researcher; b. Ironwood, Mich., Jan. 14, 1947; s. Norbert Robert and Dorothy E. (La Chapelle) K.; m. Patricia Guerry, Dec. 10, 1977; 1 child, Jennifer Kristen. BS in Biology, Va. Mil. Inst., 1968; PhD in Microbiology, Va. Commonwealth U., 1973. Postdoctoral research fellow Stanford U. Med. Sch., Palo Alto, Calif., 1972-76; research scientist Walter Reed Army Inst. Research, Washington, 1976-80, sr. research scientist, 1980—, head molecular genetics and R-DNA unit, 1983-86, asst. chief dept. bacterial immunology, 1986—; rsch. adv. Nat. Rsch. Coun. Fellowships, Washington, 1980—; instr. NIH Grad. Sch., Bethesda, Md., 1980—; sci. rsch. cons. IGI Biotechnology Inc., Columbia, 1981-82; mem. panels sci. cons. at NIH, EPA, FDA, DOD, AID, 1979—; adj. prof. U. Md., College Park, 1984—. Author (chpt.) Burrows Textbook of Microbiology, 21st, 22nd edits., 1980, 86; contbr. articles to sci. jours and books.; editor, reviewer jours.; patentee in field. Vol. Heart Assn. Montgomery County, Rockville, Md., 1980—, Am. Cancer Soc., 1980—, Children's Hosp., Washington, 1980—; trustee Barrie Sch., Silver Spring, Md., 1992—. Served to capt. U.S. Army, 1976-79. Recipient Paul A. Siple award U.S. Army Sci. Conf., 1984. Mem. Am. Soc. Microbiology (chmn. microbial pathogenesis div 1990), Genetics Soc. Am., Fed. Exec. and Profl. Assn., Mid-Atlantic Regional Extrachromosomal Genetic Elements Group (co-founder), Sigma Xi. Democrat. Roman Catholic. Achievements include discovery of one of the first large transposable genetic elements in bacteria; patent for first oral live salmonella carrier vaccine, method for rapid detection of typhoid fever; patent pending for nucleic acid probes for detection bacterial dysentery; definition of genes involved in bacterial invasion of gut epithelial cells. Home: 4601 Flower Valley Dr Rockville MD 20853-1734 Office: Walter Reed Army Inst Rsch Dept Bacterial Immunology Washington DC 20307-5100

KOPELMAN, ARTHUR HAROLD, biology educator, population ecologist; b. N.Y.C., June 3, 1952; s. Joseph and Edith K. BA, CUNY (Queens Coll.), 1975; PhD, CUNY, 1982. adj. instr. dept. sci. and math. Fashion Inst. Tech., N.Y.C., 1981-84; adj. asst. prof. dept. biology, Queens Coll., 1982-86; vol. coord. OKeano Ocean Rsch. Found., Hampton Bays, N.Y., 1987—, naturalist, 1991—. Grad. fellow dept. biology Queens Coll., Flushing, N.Y., 1976-82; instr. dept. sci. and math. Fashion Inst. Tech., N.Y.C., 1984-86, asst. prof., 1986-92, assoc. prof., 1992—; adj. instr. dept. sci. and math. Fashion Inst. Tech., N.Y.C., 1981-84, adj. asst. prof. dept. biology Queens Coll., 1982-86; vol. coord. Okeanos Ocean Rsch. Found., Hempton Bays, N.Y., 1987—, naturalist, 1991—. Contbr. articles to profl. jours. Mem. AAAS, Ecol. Soc. Am., Entomol. Soc. Am., Am. Soc. Naturalists, N.Y. Acad. Scis., Soc. Conservation Biology, Sigma Xi. Office: Dept Sci and Math Fashion Inst Tech 227 W 27th St New York NY 10001

KOPKE, MONTE FORD, engineering executive; b. Great Bend, Kans., Feb. 22, 1949; s. Clifford Franklin ad Margaret Dorothy (Brack) K.; m. Sandra Jean Bachman, June 24, 1968 (dec. Sept. 1992); children: Melissa A., Chad F. BS, Fort Hays State U., 1972; M in Engring., Colo. U., 1990. Instrumentation engr. Dresser Industries, Houston and Great Bend, 1972-78; project engr. TRW, Denver, 1978-81; mgr. of engring. Martin Marietta, Denver, 1981—; instr. electronics Barton County Coll., Great Bend. Grantee NSF, 1973. Mem. Soc. of Profl. Well Log Analysts, Sigma Pi Sigma. Home: 6935 W Rowland Ave Littleton CO 80123 Office: Martin Marietta MS 0532 PO Box 179 Denver CO 80201

KOPLASKI, JOHN, structural engineer; b. Cheyenne, Wyo., Nov. 17, 1964; s. Walter Joseph and Haruko (Tsutsumi) K.; m. Lynn Margaret Haberlein, Apr. 20, 1991. BS in Civil Engring., U. Va., 1986. Registered profl. engr., Va. Structural engr. Bengtson, DeBell, Elkin & Titus, Ltd., Centreville, Va., 1986-89; project structural engr./asst. transp. dept. head Hayes, Seay, Mattern & Mattern, Inc., Rockville, Md., 1989—. Mem. ASCE (v.p. student chpt. 1985-86, Daniel W. Meade prize 1986), U. Va. Alumni Assn. (life). Democrat. Roman Catholic. Achievements include research on Gtstrudl computer software, an engring. case study U. Va. Replacement Hosp., 1986. Home: 10801 Hunt Club Rd Reston VA 22090 Office: Hayes Seay Mattern & Mattern Inc Ste 205 1801 Rockville Pike Rockville MD 20852

KOPP, DEBRA LYNN, manufacturing engineer, consultant; b. Bunker Hill AFB, Ind., Aug. 24, 1964; d. Dennis Frank and Elaine Mary (Mayer) Mathis; m. Bruce Alan Kopp, Sept. 28, 1964 (div. 1993). BS in Indsl. Engring., Ariz. State U., 1986; MS in Indsl. Engring., Stanford U., 1989. Cert. Am. Prodn. and Inventory Control Soc. Mfg. engr. Amdahl Corp., Sunnyvale, Calif., 1986-90; engr. Applied Physics Lab., Johns Hopkins U., Laurel, Md., 1991-93; prin. mem. technical staff Amecom dvsn. Litton Systems, Inc., College Park, Md., 1993—; cons., assoc. Synergistek Assocs., Round Lake Beach, Ill., 1992; cons., co-founder Integrated Techs. Group, Durham, N.H., 1993—. Contbr. articles to conf. procs. Mem. Inst. Indsl. Engrs. (sr.), Surface Mount Tech. Assn. (arrangements chmn. Capital chpt. 1992, v.p. 1993), Inst. for Interconnective and Packaging Electronic Cirs. (reliability coms. 1992—). Republican. Roman Catholic. Achievements include research on alternatives to tin-lead solder in electronic assemblies; surface mount technology design and processing with expertise in reflow technologies, especially vapor phase soldering, solder joint reliability studies to determine void and grain structure effects. Office: Litton Systems Inc Amecom Dvsn 5115 Calvert Rd College Park MD 20740

KOPP, MONICA, biologist, educator; b. East Lansing, Mich., Apr. 3, 1957; d. Francis A. and Diana (Reese) K.; m. Edward Cheever, Sept. 9, 1975 (div. 1986); children: Allan Ross Cheever, Cynthia Francine Cheever. BS in Biology, Mich. State U., 1979, MS in Biology, 1981. Postgrad. rsch. asst. Mich. State U., East Lansing, 1981-83, rsch. biologist, 1983-86, asst. prof. biol. scis., 1986-91, assoc. prof., 1991—; asst. scientist biol. rsch. divsn. Werik Ctr., Detroit, 1986-91, assoc. scientist, 1991—; mem. biol. rsch. fund com. U. Mich., 1988—, evaluation com. biol. sci. cirriculum, 1990—, biol. tech. forum, 1990—; mem. program. com. 3rd Annual Conf. Plant and Animal Life, Detroit, 1991, 92; head task force Mich. for Biol. Understanding, 1993—; cons. Ralston-Purina Co., 1989-90, Gruber Animal Techs., Inc., 1991—. Co-Author: Animal Biology and the Human Link, 1992; mem. editorial bd. Jour. Animal Sci., Vertebrates, Animal Sci. Letters; contbr. over 30 articles to sci. jours. Active PTA Dist. 43, Detroit, 1979-83; mem. pub. awareness com. SPCA, Detroit, 1980—, head., 1985-92. Recipient Humanitarian award SPCA, 1985. Mem. AAAS, AAUW (bd. dirs.), NOW, LWV, Am. Fedn. Biology (vice pres. 1990-91, Animal Biology award 1992), Am. Assn. Against Animal Testing (corr.), Am. Fedn. Humanity (bd. dirs.), Nat. Animal Rights League (sec. 1992-93), N.Y. Acad. Scis., Mich. Acad. Scis., Mich. Biology League (founder 1987—), Mensa, Sigma Xi. Office: Werik Ctr 645 Edward Ave Ste 972 Detroit MI 48226-4016

KOPPA, RODGER JOSEPH, industrial engineering educator; b. Oak Park, Ill., June 23, 1936; s. Thaddeus Marion and Edna (Gabrycowitz) K.; m. Patricia Dana Ford, June 28, 1957; children: Virginia K. Tipton, Cynthia A. BA, U. Tex., 1958, MA, 1960; PhD, Tex. A&M U., 1979. Registered profl. engr., Tex.; lic. psychologist, Tex. Human factors engr. LTV Aerospace Corp., Dallas, 1961-67; crew performance specialist GE Co., Houston, 1967-72; asst. rsch. psychologist Tex. Transp. Inst., Tex. A&M U., College Station, Tex., 1973-79, assoc. rsch. psychologist, 1979-82, assoc. rsch. engr., 1982—, head human factors div., 1977-91; assoc. prof. Indsl. Engring. Dept., Tex. A&M U., College Station, Tex., 1982—; cons. Rodger Koppa and Assoc., College Station, 1979—; chair evaluation com. Intelligent Vehicle and Hwy. Systems Am., Washington, 1991—. Author: (book chpt.) Human Factors in MIS, 1988, Automotive Engineering and Litigation, 1988; co-author: (manual) Heavy Truck & Bus Safety Inspection, 1982. Mem., co-chair Tex. Based Task Force on Older Drivers, Austin, 1991-92. Recipient Apolloneer award GE, 1969. Fellow Human Factors Soc.; mem. Inst. Indsl. Engrs., Soc. Automotive Engrs., Transp. Rsch. Bd., Lions. Episcopalian. Achievements include development of first comprehensive standard for automotive adaptive equipment for disabled drivers; design and evaluation of adaptive controls; design of documentation; trained Apollo astronauts for lunar surface operations. Home: 1214 N Ridgefield College Station TX 77840 Office: Tex A & M U Indsl Engring Dept College Station TX 77843

KOPPANY, CHARLES ROBERT, chemical engineer; b. Alhambra, Calif., Oct. 1, 1941; s. Charles Louis and Anne Marie (Donagrechia) K. MS, U. So. Calif., 1965, PhD, 1972. Registered profl. engr., Calif. Rsch. engr. C.F. Braun & Co., Alhambra, Calif., 1965-80; assoc. prof. Calif. Poly. U., Pomona, 1980-81; rsch. engr. Santa Fe Braun, Alhambra, 1981-86; process engr. Brown & Root Braun, Alhambra, 1987-91, R&E engr., 1991—. Mem. Am. Inst. Chem. Engrs. Democrat. Achievements include research in hydrocarbon processing and chem. engring.

KOPROWSKI, HILARY, microbiology educator, medical scientist; b. Warsaw, Poland; s. Pawel and Sarah (Berland) K.; m. Irena Grasberg; children: Claude Eugene, Christopher Dorian. BA, Nikolaj Rej Gymnasium of Luth. Congregation, Warsaw; MD, U. Warsaw; grad., Warsaw Conservatory Music and Santa Cecilia Acad., Rome; DSc (hon.), Ludwig-Maximilian U., Munich, Widener Coll.; D of Medicine & Surgery (hon.), U. Helsinki, Finland; D of Medicine (hon.), U. Uppsala, Sweden; LittD (hon.), Thomas Jefferson U.; D of Med. Sci. (hon.), U. Lublin, Poland. Rsch. asst. dept. exptl. and gen. pathology U. Warsaw, 1936-39; staff Yellow Fever Rsch. Svc., Rio de Janeiro, 1940-44; staff rsch. divsn. Am. Cyanamid Co., 1944-46; asst. dir. viral and rickettsial rsch. Lederle Lab., Pearl River, N.Y., 1946-57; dir. Wistar Inst., Phila., 1957-91, prof., 1957—; prof. microbiology Faculty Arts and Scis. U. Pa., Phila., 1957—, prof. microbiology and immunology, 1992—; prof. microbiology and immunology Thomas Jefferson U., Phila., 1992—; prof. Ctr. Neurovirology; cons. WHO, 1950—; mem. microbiology study sect. NIH, 1956-60; mem. PAHO; mem. adv. com. Nat. Multiple Sclerosis Soc., 1970-78; mem. immunobiology adv. com. NIH, USPHS, 1975-76; mem. bd. sci. counselors div. cancer etiology Nat. Cancer Inst., 1982-86, chmn., 1987-90; mem. biol. response modifiers program decision network com. NIH, 1985-87; mem. immunobiol. adv. com. NIH, USPHS, 1975-76. Co-editor: Methods in Virology, Viruses and Immunity, Current Topics in Microbiology and Immunology, 1965—, Cancer Research, Viral Immunology, Hybridoma. Decorated commandeur Ordre du Mérite pour la Recherche et l'Invention; chevalier Order Royal De Lion Belgium; recipient Alvarenga prize. Coll. Physicians Phila., 1959, Alfred Jurzykowski Found Polish Millenium prize, 1966, Felix Wankel Tierschutz prize, 1979, Alexander Von Humboldt Sr. U.S. Scientist award, Phila. Cancer Rsch. award Phila. Cancer Club, 1989, San Marino award, 1989, John Scott award, Nicolaus Copernicus medal Polish Acad. Scis., 1989, The Phila. award, 1990, John Scott award, 1990; Fulbright scholar Max Planck Inst. für Verhaltensphysiologie, Seewiesen, Fed. Republic Germany, 1971. Fellow AAAS, N.Y. Acad. Medicine, Phila. Coll. Physicians; mem. Nat. Acad. Scis., N.Y. Acad. Scis. (pres. 1959, trustee 1960-72), Yugoslavian Acad. Scis., Polish Acad. Scis., Russian Acad. Med. Scis. Achievements include co-development of genetically engineered oral rabies vaccine which is highly effective in preventing rabies in wild animals; development of monoclonal antibodies which were responsible for the first known cure of pancreatic cancer; discovery of possible causative link between multiple sclerosis and a specific retrovirus; development of first oral polio vaccine which prevented an outbreak of the disease in the Belgian Congo; research on AIDS. Office: The

Wistar Inst 3601 Spruce St Philadelphia PA 19104-4265 also: Thomas Jefferson U 462 Jefferson Alumni Hall 1020 Locust St Philadelphia PA 19107

KORAYEM, ESSAM ALI, computer company executive; b. Cairo, Egypt, Oct. 8, 1941; s. Ali Mohamed Korayem and Aida (Hidar) Shishini; m. Hanaa Mohamed, Nov. 23, 1967 (div. Nov. 1980); 1 child, Tamer Essam; m. Salwa Mokhtar, Mar. 17, 1981. BSc in Elec. Engring., Cairo U., 1963, postgrad. Diploma in Optemisation Tech., 1965. Rsch. engr. Missile Factory 333, Cairo, 1963-65; I/S mgr. The Arab Contractors, Cairo, 1965-69; head, mgmt. sci. NCR Mid. East Regional Support Co., Cairo, 1969-73; tech. mgr. Engring. Cons. Co., Cairo, 1969-78; from mng. dir. to vice-chmn. T.E.A. Computers, S.A., Cairo, 1974-79; ptnr. and dir. Electric and Electronic Works, Cairo, 1979-81; ptnr., mng. dir. Data Design Co., Cairo, 1982—; cons. Am. U. Cairo, 1971-73, Royal Sci. Soc., Amman, Jordan, 1972-73, ELDA-Pros. Assn., Geneva, 1970-72, Nat. Oil Distbn. Co., Umm Said, Qatar, 1987-91. Author: Pert & CPM, 1974, Pert/Time, 1975. Mem. Egyptian Engrs. Syndicate, Egyptian Computer Soc., Operation Rsch. Soc., Acad. Internat. Bus., Rotary (v.p. Giza 1986; sec. Maadi 1984-85). Islam. Avocations: reading, fishing, camping. Home: 85 Road No 9 PO Box 548 Maadi, Cairo Egypt Office: Data Design Co, 8 Mahmoud Sadek St Box 173, Heliopolis Cairo Egypt

KORBAN, SCHUYLER SAFI, plant geneticist; b. Beirut, Lebanon, July 9, 1954; came to U.S., 1976; s. Salim Tarraf and Lily (Fakhoury) K.; m. Tamra Cheryl Smith, Dec. 21, 1986; children: Christian Miles, Charles Martin. MS, Am. U., 1976; PhD, U. Nebr., 1980. Asst. prof. U. Ill., Urbana, 1982-88, assoc. prof., 1988—; mem. rsch. policy com. Coll. Agr., U. Ill., 1992—; chair Apple Crop Adv. Com., 1989-92; liaison officer U. Ill. Inst. Nat. de la Recherche Agronomique (France), Urbana, 1987-93. Contbr. articles to profl. jours., chpts. to books. Grantee Biology Rsch. and Devel. Corp., 1991, U. Ill. Biotech. Ctr., 1992, U. Ill. Campus Rsch. Bd., 1992. Mem. AAAS, Am. Soc. Horticultural Sci., Tissue Culture Assn., Sigma Xi, Gamma Sigma Delta. Achievements include patents for new apple cultivars released having disease resistance and high fruit quality, Goldrush, Enterprise, Dayton, Williams' Pride; isolation and cloning of coat protein gene of apple mosaic virus, gene from apple genomic libr., LCHP-II gene; gene from peach genomic Libr., LHCP-II; regeneration and gene transfer systems for various perennial plants. Office: U Ill 310 PABL 1201 W Gregory Urbana IL 61801

KORBITZ, BERNARD CARL, oncologist, hematologist, educator, consultant; b. Lewistown, Mont., Feb. 18, 1935; s. Fredrick William and Rose Eleanore (Ackmann) K.; m. Constance Kay Bolz, June 22, 1957; children: Paul Bernard, Guy Karl. B.S. in Med. Sci., U. Wis.-Madison, 1957, M.D., 1960, M.S. in Oncology, 1962; LL.B., LaSalle U., 1972. Asst. prof. medicine and clin. oncology, U. Wis. Med. Sch., Madison, 1967-71; dir. medicine Presbyn. Med. Ctr., Denver, 1971-73; practice medicine specializing in oncology, hematology, Madison, 1973-76; med. oncologist, hematologist Radiologic Ctr. Meth. Hosp., Omaha, 1976-82; practice medicine specializing in oncology, hematology, Omaha, 1982—; sci. advisor Citizen's Environ. Com., Denver, 1972-73; mem. Meth. Hosp., Omaha, 1977—; dir. Bernard C. Korbitz, P.C., Omaha, 1983—; bd. dirs., pres. B.C. Korbitz P.C. Contbr. articles to profl. jours. Webelos leader Denver area Council, Mid. Am. Council of Nebr. Boy Scouts Am.; bd. elders King of Kings Luth. Ch., Omaha, 1979-80; mem. People to People Del. Cancer Update to People's Republic China, 1986, Eastern Europe and USSR, 1987; mem. U.S. Senatorial Club, 1984, Republican Presdl. Task Force, 1984. Served to capt. USAF, 1962-64. Fellow ACP, Royal Soc. Health; mem. Am. Soc. Clin. Oncology, Am. Coll. Legal-Medicine, Am. Soc. Internal Medicine, AMA, Nebr. Med. Assn., Omaha Med. Society, Omaha Clin. Soc., Phi Eta Sigma, Phi Beta Kappa, Phi Kappa Phi, Alpha Omega Alpha. Avocations: photography, fishing, travel. Home: 9024 Leavenworth St Omaha NE 68114-5150 Office: 8300 Dodge St Ste 306 Omaha NE 68114-4145

KORCHMAROS, GABOR GABRIELE, mathematics educator; b. Mako, Csongrad, Hungary, Mar. 24, 1948; s. Pál and Jolán (Leskó) K.; m. Adriana Di Trana, May 28, 1986; 1 child, Annachiara. MSc in Math., Eötvös U., Budapest, Hungary, 1971; PhD in Math., EÖTVÖS U, Budapest, Hungary, 1972; Candidate in Math., Hungarian Acad. Scis., 1980. Teaching asst. Tech. U. Budapest, 1971-72, asst. prof., 1974-80; vis. prof. U. Bari, Italy, 1981; asst. prof. U. Calabria, Cosenza, Italy, 1982-83; asst. prof. math. U. Basilicata, Potenza, Italy, 1984-86, prof., 1987—, head dept., 1990—. Scholar Acad. Nat. Dei Lincei, Rome, 1973. Fellow Inst. Combinatorics and Applications (Can.); mem. Am. Math. Soc., Bolyai Janos Math. Soc. (Grünwaldy prize 1976), Unione Math. Italy, Australian Math. Soc., London Math. Soc. Roman Catholic. Office: U Basilicata, Dept Math, N Sauro 85, I-85100 Potenza Italy

KORCHYNSKY, MICHAEL, metallurgical engineer; b. Kiev, Ukraine, Apr. 11, 1918; came to U.S., 1950, naturalized, 1956; s. Michael and Jadwiga (Zdanowicz) K.; m. Taisija Lapin, Nov. 22, 1951; children—Michael, Marina, Roksana. Dipl. Ing. in Metals Tech., Tech. U. Lviv, 1942. Lectr. Tech. U. Lviv, 1942-44; chief engr. C.E., U.S. Army, Fed. Republic Germany, 1945-50; research metallurgist Union Carbide Co., Niagara Falls, N.Y., 1951-61; research supr. Jones & Laughlin Steel Corp., Pitts., 1962-68; dir. product research Jones & Laughlin Steel Corp., 1969-72; dir. alloy devel. metals div. Union Carbide Co., N.Y.C., 1973-77, Pitts., 1978-86; cons., prin. Korchynsky and Assocs., Pitts., 1986—; lectr. Niagara U., 1957-58. Author, patentee in field. Union Carbide sr. fellow, 1979. Fellow Am. Soc. Metals (Andrew Carnegie lectr. 1973, W.H. Eisenman medal 1984, F.C. Bain award 1986); mem. AIME (Howe Meml. lectr. 1983, Robert Earll McConnell Engring. Achievement award 1991), SAE Internat., Am. Iron and Steel Inst. (medalist), Inst. Metals, Am. Soc. Metals Internat., Acad. Engring. Scis. of Ukraine, Wire Assn. Internat. Home: 2770 Milford Dr Bethel Park PA 15102-1763

KORDENBROCK, DOUGLAS WILLIAM, biomedical electronics technician; b. Covington, Ky., June 8, 1964; s. Richard George and Mary Joyce K. A in Biomed. Electronics Tech. cum laude, Cin. Tech. Coll., 1983. Cert. biomed. electronics tech. Driver Esterkamps Auto Parts, Cin., 1980-84; biomed. electronics tech. II Mercy Hosp., Hamilton, Ohio, 1982-86; biomed. electronics tech. I Novare Biomed. Svcs., Cin., 1986-89, lead biomed. electronics tech., 1989-92, sr. biomed. electronics tech., account mgr., 1993—. Mem. Delta Mu Delta. Republican. Avocations: snow skiing, motorcycling, rock climbing. Office: Novare Biomed Svcs 4841 Business Center Way Cincinnati OH 45246-1319

KORDES, HAGEN, education researcher, author; b. Luebeck, Germany, Feb. 11, 1942; s. Hubert and Aenne (Schnittker) K.; m. Margret Guenther (div. 1978); m. Padmini Darmalengam, July 6, 1983; children: Khamini-Lise, Kamalla-Lilly, Kanita-Lilo. Diploma in philosophy, Philosophische Hochschule, Frankfurt, Fed. Republic Germany, 1963; diploma in social scis., U. Sorbonne, Paris, 1967; PhD in Philosophy, Westfälische U., Münster, Fed. Republic Germany, 1974. Prof. philosophy Lycée Condorcet, Paris, 1964-67; action researcher Community Devel., People's Republic of Benin, 1967-69; evaluator Ministry Edn., Düsseldorf, Fed. Republic Germany, 1971-78; prof. ednl. scis. Westfalische U., 1978—; animation evaluator German Svc. for Devel., People's Republic Benin, 1974-80, World Peace Svc., Ivory Coast, 1976—; edn. evaluator German Soc. for Tech. Coop., People's Republic Benin, 1980-82. Author, editor, and/or co-editor more than 25 books in German, French, English and Tamil, including: Curriculum Evaluation in abhängigen Gesellschaften, 1974, Animation politique et économique, 1976, Methoden der Erziehungs- und Bilungforschung, 1984, Aus Fehlern lernen, Von Fremden Lernen, 1986, Aufruhr unterm Kopftuch, 1988, Didaktik und Bildungsgang, 1989,The Go Between, 1990, Apprentissage Interculturel, 1991, Laberfach Erziehungswissenschaft?, 1991, Processing Experiences of Estrangaement, 1992, The Go Beyond, 1992, Der Gang der Bildung, 1993, Einubung in interkulturelles Lernen, 1993; contbr. more than 120 articles to profl. jours. Founder, bd. dirs. Citizen's Initiative for Refugee Asylum, Fed. Republic Germany; German counterpart Third World Network, Penang, Malaysia. Mem. German Soc. for Ednl. Scis. Home: Brochterbecker Strasse 7, D-4542 Tecklenburg Germany Office: Georgskommende 33, D-4400 Münster Germany

KORDESCH, MARTIN ERIC, physicist, educator; b. Lakewood, Ohio, July 22, 1956; m. Elizabeth Gierlowski, Dec. 30, 1978; 1 child, Alina. AB in Physics, U. Chgo., 1978; MS in Physics, Case We. Res. U., 1980, PhD in Physics, 1984. Scientist Fritz Haber Inst., Berlin, Germany, 1984-89; asst. prof. Ohio U., Athens, 1989-93, assoc. prof., 1993—. Mem. Am Vacuum Soc., Am. Phys. Soc., Microscope Soc. Am., German Phys. Soc. Office: Ohio U Physics Dept Athens OH 45701

KORENBERG, JACOB, mechanical engineer; b. Odessa, Ukraine, Sept. 23, 1930; came to U.S., 1978; s. Gregory and Fania (Feldman) K.; m. Rachel Vazeman, Oct. 16, 1954; 1 child, Marina Lasch. BS, Odessa Marine Inst., 1952; MS with honors, Moscow Energy U., 1961; PhD, Moscow Sci. Inst. Chemistry, 1965. Lab. mgr. Moscow Sci. Inst., 1954-79; v.p. Donlee Techs. (formerly York-Shipley), York, Pa., 1979—. Author: Hand-book of Sulphuric Acid Production, 1971, Fluidized Bed Pyrite Roasting, 1971; contbr. over 40 articles to profl. jours. Mem. ASME. Achievements include 37 patents. Office: Donlee Techs Inc 693 N Hills Rd York PA 17402

KORHONEN, ANTTI SAMULI, metallurgist, educator; b. Heinola, Finland, July 14, 1950; s. Vilho Veikko and Aura Inkeri (Kiesila) K.; m. Maire Annikki Niskanen, June 14, 1975; 1 child, Sanna Marita. MSc in Engring., Helsinki U. Tech., Espoo, Finland, 1974, lic. technology, 1977, D of Tech., 1981. Teaching asst. Helsinki U. Tech., Espoo, 1974-81; jr. fellow Acad. Finland, Espoo, 1981-86, sr. fellow, 1986-87; acting prof. Helsinki U. Tech., Espoo, 1987-88, prof., dir. lab., 1988—; teaching asst. Duke U., Durham, N.C., 1979-80; vicechmn. bd. Plasmatekniikka Oy, Inc., Helsinki, 1982-86. Contbr. articles to profl. jours. 2d lt. Finland Inf., 1976-77. Recipient R.F. Bunshah award Am. Vacuum Soc. 1987. Mem. ASME, ASM, Internat. Instn. for Prodn. Engring. Rsch., Materials Rsch. Soc., Surface and Coatings Tech. (mem. editorial bd.), Internat. Union Vacuum Sci., Techniques and Applications (vacuum metallurgy divsn. sect. 1992—). Office: Helsinki U Tech, Vuorimiehentie 2A, 02150 Espoo Finland

KORMILEV, NICHOLAS ALEXANDER, retired entomologist; b. Yalta, Crimea, Russia, Jan. 29, 1901; came to U.S., 1957; s. Alexander Nicholas and Catherine (Sakulin) K.; widower; 1 child, Alexander Nicholas. Diploma in engring. agronomy, State U., Zabreb, Yugoslavia, 1926. Prof. Agrl. Sch., Tetovo, Yugoslavia, 1931-32; adj. Directory of Agr., Skoplje, Yugoslavia, 1932-33, superior adj., 1933-40, councellor, 1940-41; sec. Ministry of Agr., Beograd, Yugoslavia, 1941-43; rsch. entomologist Nat. Mus. of Natural History, Buenos Aires, 1948-52, Inst. de Ciencias Naturales, San Miguel, Argentina, 1952-56; rsch. assoc. in entomology Bishop Mus., Honolulu, 1968-90. Author: Phymatidae Argentinas (Hemipteras), 1951, Revision of Phymatinae (Hemiptera), 1962, Mezirinae of Southeast Asia and Southern Pacific, 1971; (with Froeschner) Flat Bugs of the World, 1987, Phymatidae or Ambush Bugs of the World, 1989; contbr. 230 articles to profl. jours. With Yugoslavia Mil., 1939-40. Decorated Knight Cross of St. Sava, King of Yugoslavia, 1940. Mem. N.Y. Acad. Scis., Pacific Coast Entomol. Soc. Achievements include description of one new family, one new subfamily, scores of new genera, and more than a thousand new species. Home: 2930 54th St S Saint Petersburg FL 33707-5530

KORN, EDWARD DAVID, biochemist; b. Phila., Aug. 3, 1928; s. Joel and Carrie (Goldman) K.; m. Muriel Evelyn Fisher, June 23, 1950; children: Elizabeth Gail Korn Schoenherr, Sarah Harris Korn Gilchrist. BA, U. Pa., 1949, PhD, 1954. Scientist Nat. Heart Inst., Bethesda, Md., 1954-69; vis. scientist Cambridge (Eng.) U., 1958-59; prof. FAES Grad. Program, Bethesda, 1966-76; head sect. on cell biology Nat. Heart Lung and Blood Inst., Bethesda, 1969—, chief lab. of cell biology, 1974—, sci. dir., 1989—. Editor: (book series) Methods in Membrane Biology, 1974-79; assoc. editor Jour. Biol. Chemistry, 1977-93; contbr. over 260 sci. articles to jours. in field, 1953—. Recipient Superior Svc. award USPHS, 1980, Presdl. Meritorious Exec. Rank award, 1987; Mider lectr. NIH, 1985. Mem. NAS, Am. Soc. for Biochemistry and Molecular Biology, Biophys. Soc., Am. Soc. Cell Biology, Found. Advanced Edn. in Sics. (bd. dirs. 1977-92). Office: NIH Bldg 10 Rm 7N-214 Rm 7B-214 Bethesda MD 20892

KORNADT, HANS-JOACHIM KURT, psychologist, researcher; b. Stargard, Ger., June 16, 1927; s. Kurt Karl and Katharina (Bodenburg) K.; m. Gisela Trommsdorff; children from previous marriage: Claus-Ulrich, Tilmann, Nikola, Oliver. Diplom Psychol., U. Marburg, 1952, Ph.D., 1956. Research asst. U. Marburg, 1957; wissenschaftlicher asst. Wü rzburg, 1957-61; dozent Tchr. Tng. Coll., Saarbrü cken, 1961-64, prof., 1964-68; prof. ednl. psychology U. Saar, Saarbrucken, 1968—, dep. dir. Social-Psychol. Research Centre Devel. Planning, 1968—; research in E. Africa, 1965; lectr. Ruhr-U. Bochum, 1968. Mem. wissenschaflicher beirat Fed. Ministry Econ. Coop. and Devel., 1968—; exec. com. Wissenschaftsrat, 1975-81; chmn. Beirat Hochschulzugangstest der Kultus Minister Konferenz, 1976-86, mem. Kuratorium, 1976—, chmn., 1992—, Beirat Deutsches Inst. Japan-Studien, Tokyo, 1988—, chmn., 1991; vice chmn. Landes-Hochschulstruktur-Kommission Sachsen-Anhalt, 1991-92. Author: Thematische Apperzeptions Verfahren, 2d edit., 1979; Situation und Entwickungsprobleme des Schulsystems in Kenya, vol. 1, 1968, vol. 2, 1970; Toward a Motivation Theory of Aggression and Aggression Inhibition, 1974; Lehrziele, Schulleistung und Leistungs beurteilung, 1975; Cross-cultural Research on Motivation, 1980; Aggression and Frustration, Vol 1, 1981, Vol. 2, 1991; Aggressionsmotiv und Aggressions-Hemmung, 1982; Zur Lage Der Psychologie, 1985, Developmental Conditions of Aggression in Eastern and Western Cultures, 1991. Acad. stipende VW Found., 1977-78; research fellow Japan Soc. Promotion Sci., 1979, Japanese-German Research award, 1988. Mem. German Assn. Psychology (pres. 1982-84), Internat. Council Psychologists, Internat. Soc. Research on Aggression, Internat. Assn. Cross Cultural Psychology, Japanese-German Soc. for Social Scis. (pres. 1989—). Mem. Free Democratic Party. Home: PO Box 129, D-67142 Deidesheim Germany Office: U Saar, D-66041 Saarbrücken Germany

KORNBERG, ARTHUR, biochemist; b. N.Y.C., N.Y., Mar. 3, 1918; s. Joseph and Lena (Katz) K.; m. Sylvy R. Levy, Nov. 21, 1943 (dec. 1986); children: Roger, Thomas Bill, Kenneth Andrew; m. Charlene Walsh Levering, 1988. BS, CCNY, 1937, LLD (hon.), 1960; MD, U. Rochester, 1941, DSc (hon.), 1962; DSc (hon.), U. Pa., U. Notre Dame, 1965, Washington U., 1968, Princeton U., 1970, Colby Coll., 1970; LHD (hon.), Yeshiva U., 1963; MD honoris causa, U. Barcelona, Spain, 1970. Intern in medicine Strong Meml. Hosp., Rochester, N.Y., 1941-42; commd. officer USPHS, 1942, advanced through grades to med. dir., 1951; mem. staff NIH, Bethesda, Md., 1942-52, nutrition sect., div. physiology, 1942-45; chief sect. enzymes and metabolism Nat. Inst. Arthritis and Metabolic Diseases, 1947-52; guest research worker depts. chemistry and pharmacology coll. medicine NYU, 1946; dept. biol. chemistry med. sch. Washington U., 1947; dept. plant biochemistry U. Calif., 1951; prof., head dept. microbiology, med. sch. Washington U., St. Louis, 1953-59; prof. biochemistry Stanford U. Sch. Medicine, 1959—, chmn. dept., 1959-69; Mem. sci. adv. bd. Mass. Gen. Hosp., 1964-67; bd. govs. Weizmann Inst., Israel. Author: For the Love of Enzymes, 1989; contbr. sci. articles to profl. jours. Served lt. (j.g.), med. officer USCGR, 1942. Recipient Paul-Lewis award in enzyme chemistry, 1951; co-recipient of Nobel prize in medicine, 1959; recipient Max Berg award prolonging human life, 1968, Sci. Achievement award AMA, 1968, Lucy Wortham James award James Ewing Soc., 1968, Borden award Am. Assn. Med. Colls., 1968, Nat. medal of sci., 1979. Mem. Am. Soc. Biol. Chemists (pres. 1965), Am. Chem. Soc., Harvey Soc., Am. Acad. Arts and Scis., Royal Soc., Nat. Acad. Scis. (mem. council 1963-66), Am. Philos. Soc., Phi Beta Kappa, Sigma Xi, Alpha Omega Alpha. Office: Stanford U Med Ctr Dept Biochemistry Stanford CA 94305-5425

KORNBERG, SIR HANS LEO, biochemist; b. Herford, Germany, Jan. 14, 1928; s. Max and Margarete (Silberbach) K.; m. Monica Mary King, Oct. 6, 1956 (dec. June 1989); children: Julia Margaret, Rachel Elizabeth, Jonathan Paul, Simon Alexander; m. Donna Haber, July 28, 1991. B.Sc., U. Sheffield, 1949, Ph.D., 1953, D.Sc. (hon.), 1979; M.A., Oxford U., 1958, D.Sc., 1961; Sc.D. (hon.), U. Cin., 1974; Sc.D., Cambridge U., 1975; D.Sc. (hon.), Warwick U., 1975, Leicester U., 1979, Bath U., 1980, Strathclyde U., 1985; D.U. (hon.), Essex U., 1979; M.D. (hon.), Leipzig U., 1984. John Stokes research fellow U. Sheffield, 1951-53; Commonwealth Fund fellow Yale U., U. Calif., Berkeley, Pub. Health Research Inst., N.Y., 1953-55; mem. sci. staff M.R.C. cell metabolism rsch. unit, Oxford, 1955-60; prof. biochemistry U. Leicester, 1960-75; Sir William Dunn prof. biochemistry Cambridge (Eng.) U., 1975—, fellow Christ's Coll., 1975—, Master, 1982—; lectr. Worcester Coll., Oxford, 1958-60; Leeuwenhoek lectr. Royal Soc., 1972; Weizmann Meml. lectr., Rehovot, 1975; mem. Sci. Rsch. Coun., 1967-72, chmn. sci. bd., 1969-72; mem. U.G.C. Biol. Sci. Com., 1967-76; U.K. rep. NATO-ASI Panel, 1970-76, chmn., 1974-75; chmn. Royal Commn. on Environ. Pollution, 1976-81; mem. Agrl. Rsch. Coun., 1981-84; mem. Priorities Bd. for Rsch. and Devel. in Agr., 1984-90; chmn. adv. com. on Genetic Modification, 1986—. Mng. trustee Nuffield Found., 1972-93; gov. Hebrew U. Jerusalem, 1976—; sci. gov. Weizmann Inst. Sci., Rehovot, Israel, 1981-90, emeritus gov., 1990—; trustee Marine Biol. Lab., Woods Hole, Mass., 1982-87, 88-93, Wellcome Trust, 1990-92; gov. Wellcome Trust Ltd., 1992—; bd. dir. U.K. Nirex Ltd., 1986—; pres. elect Internatl Union of Biochemistry, 1988-91; pres. Biochem. Soc. U.K., 1990—, Assn. Sci. Edn., 1991-92; pres. Internat. Union of Biochemstry and Molecular Biology, 1991—. Recipient Colworth medal Biochem. Soc., 1963, Otto Warburg medal German Biochem. Soc., 1973; created knight bachelor, 1978; hon. fellow Worcester Coll., Oxford, 1981, Brasenose Coll., Oxford, 1982, Wolfson Coll., Cambridge, 1990. Fellow Royal Soc. (council 1975-77), Inst. Biology (v.p. 1970-72), Royal Soc. Arts, Royal Coll Physicians (London) (hon.), Am. Acad. Microbiology; hon. mem. Am. Soc. Biochemistry and Microbiology, Am. Acad. Arts & Scis., German Soc. Biol. Chemists, Japanese Biochem. Soc.; mem. NAS (fgn. assoc.), Am. Philos. Soc., German Acad. Scis. (Leopoldina). Author: (with Hans Krebs) Energy Transformations in Living Matter, 1957; contbr. articles to profl. jours. Office: U Cambridge, Dept Biochemistry, Cambridge CB2 1QW, England

KORNBREKKE, RALPH ERIK, colloid chemist; b. Bklyn., Nov. 22, 1951; s. Henning Norman and Esther (Pedersen) K.; m. Annette Elizabeth Kingman, Aug. 17, 1974. BS, Rensselaer Poly. Inst., 1974, PhD, 1981. Chemist Petroleum Action Inc., Rensselaer, N.Y., 1974-75, Rensselaer Rsch. Corp. Internat., Latham, N.Y., 1975-76; sr. rsch. chemist The 3M Corp., St. Paul, 1980-84; project leader Std. Oil of Ohio, Warrensville Hts., 1984-87; rsch. chemist IV The Lubrizol Corp., Wickliffe, Ohio, 1987-90, sr. rsch. chemist, 1990-91, rsch. scientist, 1991—; session chmn. Am. Chem. Soc. Nat. Meeting Colloid Div., N.Y.C., 1986; chmn. the Interface Sci. chpt. of 3M Tech. Forum, St. Paul, 1982-84; staff mem. NBS Molton Salts Data Ctr., Troy, 1975-76. Contbr. articles to profl. jours. Pres. Oakwood Lustre Townhome Assn., Oakdale, Minn., 1981-84; judge Reg. Sci. Fair, Mpls., Cleve., 1981—; team capt. Cleve. Orch. Campaign Fund Raising, 1988-90. N.Y. State Regents scholar 1970; named J. Willard Gibbs Rsch. fellow, 1979-80. Fellow Am. Inst. Chemists; mem. AAAS, Internat. Assn. Colloid and Interface Scientists, Am. Chem. Soc., Sigma Xi, Phi Lambda Epsilon. Achievements include discovery of stochastic nature of emulsion-type inversion process, complex nature of wetting near the critical point, special expertise surfactant interactions at solid-liquid interfaces, nonaqueous colloidal properties regarding dispersions and lubrication. Home: 8340 Tulip Ln Chagrin Falls OH 44023 Office: The Lubrizol Corp 29400 Lakeland Blvd Wickliffe OH 44092-2298

KORNEL, LUDWIG, medical educator, physician, scientist; b. Jaslo, Poland, Feb. 27, 1923; came to U.S., 1958, naturalized, 1970; s. Ezriel Edward and Ernestine (Karpf) K.; m. Esther Muller, May 27, 1952; children—Ezriel Edward, Amiel Mark. Student, U. Kazan Med. Inst., USSR, 1943-45; M.D., Wroclaw (Poland) Med. Acad., 1950; Ph.D., U. Birmingham, Eng., 1958. Intern Univ. Hosp., Wroclaw, 1949-50, Hadassah-Hebrew U. Hosp., Jerusalem, 1950-51; resident medicine Hadassah-Hebrew U. Hosp., 1952-55; Brit. Council scholar, Univ. research fellow endocrinology U. Birmingham, 1955-57, lectr. medicine, 1956-57; fellow endocrinology U. Ala. Med. Ctr., 1958-59, successively asst. prof., assoc. prof., prof. medicine, 1961-67; dir. steroid sect. U. Ala. Med. Center, 1962-67, assoc. prof. biochemistry, 1965-67; postdoctoral trainee in steroid biochemistry U. Utah, 1959-61; prof. medicine U. Ill. Coll. Medicine, Chgo., 1967-71; dir. steroid unit Presbyn.-St. Lukes Hosp., Chgo., 1967—; assoc. biochemist Presbyn.-St. Lukes Hosp., 1967-70, sr. biochemist on sci. staff, 1970-71, attending physician, 1967-71; prof. medicine and biochemistry Rush Med. Coll., 1970—; sr. attending physician, sr. scientist Rush-Presbyn.-St. Lukes Med. Cen., 1971—; hon. guest lectr. Polish Acad. Sci., Warsaw, 1965; vis. prof. Kanazawa (Japan) U., 1973, 82, 88, 93. Mem. editorial bd. Clin. Physiol. Biochemistry, 1984, Endocrinology, 1993—; co-editor: Yearbook of Endocrinology, 1986-90; coauthor: Encyclopedia of Human Biology, 1991; contbr. articles on endocrinology and steroid biochemistry to profl. jours.; contbr. chpts. to textbooks. Recipient Physicians Recognition award AMA, 1969, 73, 76, 81, Outstanding New Citizen award Citzenship Council Met. Chgo., 1970. Fellow Am. Coll. Clin. Pharmacology and Chemotherapy, Nat. Acad. Clin. Biochemistry (bd. dirs. 1982-86), Royal Soc. Health; mem. AMA, AAAS, AAUP, Endocrine Soc., Am. Fedn. Clin. Research, N.Y. Acad. Scis., Am. Physiol. Soc., Cen. Soc. Clin. Research, Am. Acad. Polit. and Social Scis., Fedn. Am. Socs. for Exptl. Biology (nat. corr. 1975—), Sigma Xi. Home: 6757 N Leroy Ave Lincolnwood IL 60646-3203 Office: Rush Presbyn St Lukes M C 1653 W Congress Pky Chicago IL 60612-3833

KORNER, ANNELIESE F., psychology research scientist; b. Munich, Germany; d. Leopold and Jenny (Deutsch) Friedsam; widowed; 1 child, Sue S. Kalman. MA, Columbis U., 1940, PhD, 1948. Cert. psychologist, Calif. Chief psychologist Mt. Zion Psychiat. Clinic, San Francisco, 1948-61; from rsch. assoc. to prof. dept. psychiatry Stanford (Calif.) U., 1964—; mem. Stanford Ctr. for Study of Families, Children & Youth, 1974—; bd. dirs. Zero to Three: Nat. Ctr. Clin. Infant Programs, Washington, 1984—; grant application reveiwer NIH-W.T. Grant Found., Nat. Found., March of Dimes, 1976—. Guest editor for jours. Child Devel., Devel. Psychology, Sleep, Pediatrics Sci., Infant Behavior and Devel., 1970—; author: Hostility in Young Children, 1949, Neurobehavioral Assessment of the Preterm Infant, 1990; contbr. articles to profl. publs., chpts. to sci. books. Rsch. grantee NICHD, Grant Found., Maternal and Child Health, HEW, NIMH, 1961-90. Fellow APA, Am. Psychol. Soc., Am. Orthopsychiat. Assn.; mem. Am. Men and Women of Sci., Soc. for Rsch. in Child Devel., Perinatal Rsch. Soc., San Francisco Psychoanalytic Inst., World Assn. for Infant Mental Health. Achievements include patents for oscillating incubator waterbed, method of treating perterm infants; findings that infants experienceing apnea of permaturity have significant reduction in apnea while placed on waterbed, preterm infant sleep is enhanced and irratibility is reduced while on the waterbed, neurobehavioral development of preterm infants was improved as a function of the waterbed treatment. Home: 2299 Tasso St Palo Alto CA 94301 Office: Stanford U Sch Medicine Psychiatry/Behavioral Scis 101 Quarry Rd Stanford CA 94305

KORNFELD, PETER, internist; b. Vienna, Austria, Mar. 16, 1925; came to U.S., 1939; s. Otto and Rosa (Weitzmann) K. BA summa cum laude, U. Buffalo, 1948; MD, Columbia U., 1952. Diplomate Am. Bd. Internal Medicine. Intern Mt. Sinai Hosp., N.Y.C., 1952-53; asst. resident, then chief resident in internal medicine Mt. Sinai Hosp., 1955-56; postdoctoral fellow cardiovascular physiology, physician Nat. Heart Inst. at Columbia U./Presbyn. Hosp., N.Y.C., 1953-54; pvt. practice, N.Y., N.J., 1956-88; clin. prof. medicine Stanford U. Sch. Medicine, Univ. Hosp., 1991—; cons. physician N.Y. State Bur. Disability Determination, 1960-87, Hackensack (N.J.) Hosp. Med. Ctr., 1988-91; dir. Myasthenia Gravis Clinic, Englewood (N.J.) Hosp., 1965-91; mem. nat. med. adv. bd. Myasthenia Gravis Found., 1970-91; attending physician Englewood Hosp., Mt. Sinai Hosp.; clin. prof. Mt. Sinai Sch. Medicine, CUNY, 1968-92. Contbr. numerous articles to med. jours. Grantee, NIH, 1966-70, Hoffman-LaRoche, Inc., 1966-73, Muscular Dystrophy Assn., 1978-81, 81-82, Rosenstiel Found., 1979-82; recipient Globus award, Mt. Sinai Jour. Medicine, 1976-77. Fellow ACP, Am. Coll. Cardiology (assoc.), N.Y. Acad. Sci., N.Y. Acad. Medicine; mem. AMA, Am. Fedn. Exptl. Biology, Am. Fedn. Clin. Rsch., Harvey Soc., Am. Diabetes Assn, Am. Heart Assn., Phi Beta Kappa, Alpha Omega Alpha, Sigma Xi. Avocations: numismatics, philately, photography, music, travel.

KORNFELD, STUART A., hematology educator; b. St. Louis, Mo., Oct. 4, 1936. AB, Dartmouth Coll., 1958; MD, Washington U., 1962. Rsch. asst. biochemistry dept. sch. medicine Washington U., St. Louis, 1958-62, from instr. to asst. prof. medicine, 1966-70, from asst. to assoc. prof. biochemistry, 1968-72, prof. medicine dept. internal medicine, 1972—, prof. biochemistry, co-dir. divsn. hematology and oncology, 1976—, dir. divsn. oncology, 1973-76; intern med. ward Barnes Hosp., 1962-63, asst. resident, 1965-66; rsch. assoc. nat. inst. arthritis and metabolic disease NIH, 1963-65; faculty rsch. assoc. Am. Cancer Soc., 1966-71; mem. cell biology study sect. NIH, 1974-77; mem. bd. sci. counselors Nat. Inst. Arthritis, Diabetes & Digestive & Kidney Disease, 1983-87; mem. sci. rev. bd. Howard Hughes Med. Inst., 1986—; mem. bd. sci. advisers Jane Coffin Childs Meml. Fund. Res., 1987—; Jubilee lectr. Biochemistry Soc., 1989. Assoc. editor Jour. Clin. Investigation, 1977-81, editor, 1981-82; assoc. editor Jour. Biol. Chemistry, 1982-87; author 145 publs. Recipient Borden award, 1962, Rsch. Career Devel. award NIH, 1971-76; named Harden Medallist, Biochemistry Soc., 1989, Passano Found. laureate, 1991. Mem. NAS (mem. inst. medicine), Am. Soc. Clin. Investigation (counselor 1972-75), Am. Soc. Hematology, Am. Soc. Biol. Chemists, Assn. Am. Physicians (sec. 1986—), Am. Acad. Arts and Sci., Am. Chem. Soc., Sigma Xi. Achievements include research in the structure, biosynthesis and function of glycoproteins, especially those which are found on the surface of normal and malignant cells, targeting of newly synthesized acid hydroloses to lysosomes. Office: Washington U Sch Med Dept Internal Medicine 660 S Euclid Ave Saint Louis MO 63110-1093*

KORNHAUSER, ALAN ABRAM, mechanical engineer; b. Washington, Dec. 1, 1950; s. Bernard and Sara Lea (Galtz) K.; m. Rhea Sue Epstein, Sept. 24, 1989; 1 child, Madelyn Lila. BS, Rensselaer Poly. Inst., 1973, ME, 1973; ScD, MIT, 1989. Registered profl. engr., N.Y. Engr. Gibbs & Cox, Inc., N.Y.C., 1973-74, Lockwood Greene Engrs., Inc., N.Y.C., 1974-79, Coca-Cola Co., Foods div., Leesburg, Fla., 1980-81; postdoctoral assoc. MIT, Cambridge, 1989; asst. prof. mech. engring. Va. Poly. Inst. and State U., Blacksburg, 1989—; cons. Saunders Coll. Pub., Phila., 1990-91, Steamsphere, Inc., Jerseyville, Ill., 1992. Contbr. articles to profl. jours. With U.S. Army, 1983-85. Mem. ASME (vice chair Stirling Engring. com. 1992-93), SAE (reviewer 1991-92), Sigma Xi. Democrat. Jewish. Achievements include research interests in transient heat transfer, two-phase flows, advanced refrigeration systems; advanced energy conversion systems. Home: 1430 Jefferson Forest Ln Blacksburg VA 24060 Office: Va Poly Inst and State U Dept Mech Engring Blacksburg VA 24061-0238

KORNSTEIN, MICHAEL JEFFREY, pathologist; b. Woonsocket, R.I., Mar. 14, 1955; s. Arnold Irving and Esta (Strong) K.; m. Ann Louise Kaplan, June 21, 1981; children: Sara Ellen, Joanna Leda. AB, Cornell U., 1977; MD, SUNY, 1980. Diplomate Am. Bd. Pathology; lic. physician, Va. Intern pediatrics Children's Hosp. Phila., 1980-81; resident pathology Hosp. of U. Pa., Phila., 1981-86, fellow surg. pathology, 1985-86; asst. prof. pathology Med. Coll. Va., Richmond, 1986-91, assoc. prof., 1991—. Contbr. over 30 articles to med. jours. Vol. Am. Cancer Soc., Richmond, 1989—. Fellow Coll. Am. Pathologists, Am. Soc. Clin. Pathologists; mem. U.S.-Can. Acad. Pathology, Phi Beta Kappa, Alpha Omega Alpha. Achievements include description of important prognostic factors in thymic tumors; description of immunopathology of the thymus in myasthenia gravis; promotion of fine needle aspiration of lymph nodes. Office: Med Coll Va Box 662 Richmond VA 23298

KORNYLAK, HAROLD JOHN, osteopathic physician; b. Jersey City, Feb. 16, 1950; s. Andrew Thomas and Lucille Bertha (Reilly) K.; children: Laura, Michael. BS in Physics with honors, Stevens Inst. Tech., 1971; MA, Maharishi Internat. U., 1977; MS, Maharishi European Rsch. U., 1977; DO, U. New Eng., 1983. Mem. indsl. R & D staff Kornylak Corp., Hamilton, Ohio, 1971-73, mgr. data processing, 1974-79; researcher Maharishi European Rsch. U., Weggis, Switzerland, 1973-74; intern Mich. Osteo. Med. Ctr., Detroit, 1983-84; staff physician Indian Health Svc., USPHS, San Carlos, Ariz., 1984-87, St. Louis Orthopedic Sports Medicine Clinic, 1987-88; pvt. practice Virginia Beach, Va., 1989—; cons. in systems analysis; instr. Atlantic U., Virginia Beach, 1989—, Harold J. Reilly Sch. Massotherapy, Virginia Beach, 1989—. Mem. Am. Osteo. Assn., Am. Acad. Osteopathy, Cranial Acad., Va. Osteo. Med. Assn. Avocations: sailboarding, backpacking, yoga, meditation. Home and Office: 1432 E Bay Shore Dr Virginia Beach VA 23451-3760

KOROLKOVAS, ANDREJUS, pharmaceutical chemistry educator; b. Siauliei, Lithuania, Aug. 27, 1923; arrived in Brazil, 1927; naturalized, 1961; s. Vasilius and Agafija (Semenova) K.; m. Ruzena Maglovsky, Aug. 17, 1944; children: Sonia (dec.), Miriam Mirna. B in Pharmacy-Biochemistry, U. São Paulo, 1961, PhD in Pharmacy-Biochemistry, 1966; postdoctoral, U. Mich., 1969-70. Asst. prof. U. São Paulo, 1962-66, asst. prof. doctor, 1966-70, free-dozent prof., 1970-73, assoc. prof., 1973-80, prof., 1980—; prof. grad. courses 4 Brazilian U., 1971-90, 2 Latin-Am. Univs., 1974-88; cons. FAPESP, CNPq, FINEP, CAPES, 1978-92; mem. Brazilian cons. Pharmacopeia Four Brazilian Ministries, 1978—. Author: Essentials of Molecular Pharmacology, 1970, (published Japanese and Portuguese) Grundlagen der Molekularen Pharmakologie, 1974, Esentials of Medicinal Chemistry, 1976 (published in Japanese, Spanish, Portuguese, Taiwan), Pharmaceutical Analysis, 1984, Essentials of Medicinal Chemistry, 2d edit., 1988; contbr. over 200 articles to profl. jours.; referee 4 Brazilian jours., 1972—. Vital Brazil medal Instituto Butantan, 1965, John R. Reitemeyer Prize Interam. Press Soc., 1967, ABIFARMA Prize, 1975. Mem. Am. Chem. Soc. Achievements include synthesis of potential drugs for schistosomiasis, malaria, and Chagas' disease, synthesis of prodrugs of schistosomicidal, antimalarial, and anti-Chagas' disease agents, molecular orbital calculations to elucidate the mechanism of action of some antiparasitic drugs, pharmaceutical analyses of new drugs. Home: Rua Edson 1272 Campo Belo, 04618-035 São Paulo Brazil Office: U São Paulo, Caixa Postal 66355, 05389 São Paulo Brazil

KOROS, WILLIAM JOHN, chemical engineering educator; b. Omaha, Aug. 31, 1947; s. William Alexander and Mary Ellen (Roth) K.; m. Ann Marie Teahan, Dec. 19, 1970. BSChemE, U. Tex., 1969, MSChemE, 1975, PhDChemE, 1977. Registered profl. engr., Tex. Chem. engr. E.I. DuPont, Wilmington, Del., 1969-71, cons., 1982—; engr. E.I. DuPont, Camden, S.C., 1971-73; research asst. U. Tex., Austin, 1973-77, prof., 1983—, B.F. Goodrich prof. in Materials Engring., 19910—; asst. prof. chem. engring. N.C. State U., Raleigh, 1977-80, assoc. prof. chem. engring., 1980-83. Editor in chief Jour. Membrane Sci.; mng. editor Membrane Quar. Recipient Sigma Xi Research award, 1980, Young Investigators award NSF, 1983, Alcoa Found Research award N.C. State U., 1983. Mem. Am. Chem. Soc., Am. Inst. Chem. Engrs. Office: U Tex Dept Chem Engring CPE Bldg Austin TX 78712-1104

KORSAH, KOFI, nuclear engineer; b. Accra, Ghana, Jan. 20, 1950; came to U.S., 1977; s. Isaac Francis and Ekua (Taakoa) K.; m. Hanna Howard-Turkson, Dec. 18, 1976; children: Ato, Kweku. BS, U. of Sci. and Tech., Ghana, 1973; M in Nuclear Engring., U. Mo., 1980, PhD, 1983. Asst. prof. elect. engring., nuclear physics U. Maine, Orono, 1982-84; asst. prof. U. Ghana, Accra, 1985-88, Tenn. Wesleyan Coll., Athens, 1988-90; rsch. staff mem. Oak Ridge (Tenn.) Nat. Lab., 1990—. Contbr. articles to IEEE Trans. in Nuclear Sci., Nuclear Instrn. and Methods, Am. Chem. Soc. jour., Trans. of Am. Nuclear Soc. Mem. IEEE, Am. Nuclear Soc. Baptist. Home: 6520 Ellesmere Dr Knoxville TN 37921 Office: Oak Ridge Nat Lab PO Box 2008 Oak Ridge TN 37831-2008

KORSMEYER, DAVID JEROME, aerospace engineer; b. Pitts., July 6, 1964; s. Jerome Daniel and Mary Abigail (Drake) K.; m. Katy Kuo, Nov. 5, 1988. BS in Aerospace Engring., Pa. State U., 1986; MS in Aerospace Engring., U. Tex., 1988, PhD in Aerospace Engring., 1991. Rsch. fellow Large Scale Programs Inst., Austin, 1987-91; staff engr. KDT Industries, Inc., Austin, 1991; staff scientist NASA Ames Rsch. Ctr., Moffett Field, Calif., 1991—; cons. in field, 1987-91. Author: (with others) Advances in the Astronautical Sciences, 1992, Modeling and Simulation of Advanced Space Programs, 1991; contbr. articles to profl. jours. Mem. AIAA (Best Student Speaker Mid-Atlantic region 1985). Republican. Presbyterian. Achievements include research on guidance and control of cislunar low-thrust trajectories, computational and information sciences, systems analysis and life cycle costing methodologies. Office: NASA Ames Rsch Ctr Mail Stop 269-1 Moffett Field CA 94035-1000

KORTE, BERNHARD HERMANN, mathematician, educator; b. Bottrop, Germany, Nov. 3, 1938; s. Bernhard F. and Agnes (Schmidt) K.; m. Sabeth Tensholter, Aug. 1, 1966; 1 child, Dagmar. PhD in Math., U. Bonn., 1968, Habilitation, 1970; PhD (hon.) U. Rome, 1987. Rsch. assoc. U. Bonn, Fed. Republic Germany, 1965-70, dir. Institut fur Gellschafts und Wirtschiftswissenschaften, 1972—; prof. U. Regensburg, Fed. Republic Germany, 1971, U. Bielefeld, Fed. Republic Germany, 1971; prof. Ops. Rsch. U Bonn., 1972—, dir. Inst. Ops. Rsch., 1972—, dep. univ. coun., vice rector, 1980-88, dean, 1984-87; disting. sr. fellow RUTCOR Rutgers U., New Brunswick, N.J., 1985—, dir. rsch. Inst. Discrete Math., 1987—; hon. prof. applied math. Acad. Sinica, Beijing, 1988—, U. Ponteficia Cath. Rio de Janeiro, 1988—. Recipient Grand Officier Cross of the Order of Merit of the Italian Republic, 1986; Prix Alexandre de Humboldt of the French Min. Rsch., 1990. Contbr. numerous articles to sci. jours. Fellow Inst. Combinatorics and Its Applications; mem. Rhenisch Westfalian Acad. Scis., Am. Math. Soc., Ops. Rsch. Soc. Am., Math. Programming Soc., Deutsche Mathematiker Vereinigung, N.Y. Acad. Scis. Home: Im Erlengrund 26, 53547 Impekoven Bonn Germany Office: Rsch Inst Discrete Math, Nassestrasse 2, 5300 Bonn 1, Germany

KOSÁRY, DOMOKOS, historian; b. Selmecbánya, Hungary, July 31, 1913; s. János and Lola (Réz) K.; m. Klára Huszti, Dec. 15, 1937 (widowed 1978). Degree, U. Budapest, Hungary, PhD, 1936; postgrad., Sorbonne U., Paris, 1936-37, Inst. Hist. Rsch., London, 1938-39. Prof. U. Budapest, 1937-50; dir. inst. history Teleki Inst., 1945-49; archivist; scientific researcher, scientific counsellor Inst. History Hungarian Acad. Scis., Budapest, corresponding mem., 1982, ordinary mem., 1985; pres. Nat. Com. Hungarian Historians, Budapest, 1985-90, Hungarian Acad. Scis., 1990—; founder, editor-in-chief Revue d'Histoire Comparée, Budapest, 1943-48; prof. U. Budapest, 1946-49; mem. European Acad., London, Paris, Brit. Acad., London. Author: Introduction to the Sources and Literature of Hungarian History, vols. 1-3, 1951-58, book on Artur Görgey, 1936, on Lajos Kossuth, 1946, studies on history of Hungary's international relations from the Middle Ages up to the 20th century; editor (with others) History of City of Budapest, vols. II-III, History of Hungarian Press, vols. I-II, 1979-85, Culture in 18th Century Hungary, 1980; contbr. articles to numerous profl. publs. Pres. Revolutionary Coun. Historians, 1956. Recipient Laureate of Hungarian State prize, 1988, Grand Cross award Hungarian Republic, 1993; named Officer of Ordre des Palmes Académiques de la République Française, 1988. Office: Magyar Tudomanyos Akademia, Roosevelt-tër 9, 1051 Budapest Hungary

KOSASKY, HAROLD JACK, gynecologist; b. Winnipeg, Man., Can., Oct. 19, 1927; s. Jack and Lillian (Resnick) K.; m. Shirley Anne Johnston, Sept. 3, 1955; children: Julia, Leah, Robert. BA, U. Manitoba, Can., 1948; MD, Licentiate, U. Manitoba, 1953. Diplomate Am. Bd. of Ob-gyn.; lic. Coll. of Physicians and Surgeons of Can. (8521), Ky. State Bd. of Health, Idaho State Bd. of Health (M2704), Mass. Bd. of Registration in Med. (29042), Med. Coun. of Can. Intern Deer Lodge VA and Grace Hosps., Winnipeg, Man., Can., 1952-53; resident in gen. surgery Col. Belcher Hosp., Calgary, Alta., Can., 1953-54; resident in psychiatry Warren (Pa.) State Hosp., 1955-56; jr. asst. resident, asst. resident, sr. resident in ob-gyn. Chgo. Lying-In Hosp., 1956-59; asst. and assoc. prof. U. Louisville Sch. Med., 1961-65; asst. and assoc. in Ob-gyn. various hosps., Boston, 1966-81; gynecologist and obstetrician Boston Hosp. for Women, 1965-81; gynecologist Brigham & Women's Hosp., Boston, 1981—; instr. ob-gyn. Harvard U., 1965—; cons. Ovutime, Inc., Boston, 1972—, Jordan Hosp., Plymouth, Mass., 1969—; pres. Saltime Co.; asst. visiting surgeon, Boston City Hosp., 1967-69; mem. Ky. Govs. Task Force on Mental Retardation, 1964-65, Com. on Malignancy (chmn.), 1963-65. Contbr. numerous articles to profl. jours.; co-inventor Ovutime Ovulation group of instruments. Fellow Royal Coll. of Surgeons of Can. (cert. FRCS), Royal Soc. of Health, Boston Obstet. Soc. (emeritus), Am. Coll. Obstetricians and Gynecologists (cert. FACOG); mem. W.Va. Obstet. and Gynecol. Soc. (hon.), Am. Fertility Soc., AAAS, FRCS (Can.), ACS (lic. FACS), Gen. Med. Council of Great Britain (lic. C5086), Royal Coll. of Obstetricians and Gynecologists of Eng. (cert. MRCOG), Assn. of Profs. of Ob-gyn, Louisville Obstet. and Gynecol. Soc. (sec., treas. 1962-65), Louisville Med. Forum (v.p.). Episcopalian. Club: Harvard. Office: 25 Boylston St Chestnut Hill MA 02167-1710

KOSCELNICK, JEANNE, computer scientist; b. Somerville, N.J., Nov. 11, 1960; d. Carl and Bernice (Stout) K. AAS in Computer Sci., Somerset County Coll., North Branch, N.J., 1980; BS in Computer Sci., Kean Coll. N.J., 1990. Programmer/analyst Western Electric, Piscataway, N.J., 1980-84; sr. programmer/analyst Bell Communications Rsch., Piscataway, N.J., 1984—. Mem. Lambda Alpha Sigma. Home: 227 N 8th Ave Manville NJ 08835-1232

KOSCHNY, THERESA MARY, environmental biologist; b. Washington, Feb. 28, 1954; d. William Simon and Bertha Margaret (Clarkin) K. BS, George Mason U., Fairfax, Va., 1977, MS, 1982. Sales clk. Montgomery Ward Co., Falls Church, Va., 1977; clk. of investigation FBI, Washington, 1977-83; environ. biologist I-95 Compost Facility, Lorton, Va., 1983-85, Lower Potomac Pollution Control Plant, Lorton, Va., 1985—; mem. stream survey Water Control Bd. of Va., 1983. Reviewer wastewater manual Water Pollution Control Fedn. Treatment Plant Manual, 1990, 91. Mem. Am. Inst. Biol. Sci., Ecol. Soc. Am., No. Va. Sci. Ctr., Washington Ind. Writers. Achievements include research on microbiological organisms in wastewater returned studies; determination of species diversity ratios as applicable to age of sludge. Home: 5704 Robinwood Ln Falls Church VA 22041-2606 Office: Lower Potomac Pollution Plt 9399 Richmond Hwy Lorton VA 22079-1825

KOSCIELAK, JERZY, scientist, science administrator; b. Lodz, Poland, Sept. 6, 1930; s. Jozef and Regina (Pokrzywa) K.; m. Anna Kitaszewska, 1969 (div. 1974); 1 child, Katarzyna. MB, Med. Acad., Warsaw, Poland, 1953, MD, 1960, DrSci, 1966. Asst. dept. physiol. chemistry Med. Acad., Warsaw, 1950-51, prof., 1973—; asst. and sr. asst. dept. biochemistry Inst. of Hematology, Warsaw, 1951-67; rsch. fellow Harvard U., Cambridge, 1964-65; head immunochem. lab. Inst. of Hematology, Warsaw, 1968-69, head dept. biochemistry, 1969—; sci. sec. Inst. of Hematology, Warsaw, 1969—. Editor-in-chief Acta Haematologica Polonica jour., 1976-85; contbr. articles to profl. jours. Mem. Polish Biochem. Soc. (chmn. Warsaw div. 1967-69), Polish Acad. Sci., Internat. Soc. Hematology, Internat. Glycoconjugate Orgn. (Polish rep. 1988—, pres. 1993—), N.Y. Acad. Scis. Avocation: history. Office: Inst of Hematology, Chocimska 5, 00957 Warsaw Poland

KOSHELEV, YURIY GRIGORYEVICH, mathematics educator; b. Novosibirsk, USSR, Oct. 24, 1947; s. Kapitolina Ivanovna Kosheleva; m. Zoya Evgenyevna, June 20, 1968 (div. 1975); 1 child, Irene; m. Olga Pavlovna Shevtsova, May 24, 1975; 1 child, Tatyana. Cand. of Math., Herzen Pedagogical Inst., Leningrad, 1976; Docent of Algebra Chair, Pedagogical Inst., Novosibirsk, 1985. Asst. Pedagogical Inst., Novosibirsk, 1970-72, 1975-76, sr. tchr., 1976-80, docent of the Algebra chair, 1980—. Contbr. articles to profl. jours. Mem. Am. Math. Soc., Siberian Math. Soc. Avocation: music. Home and Office: Pedagogical Inst, Vilyuiskaya 28, 630126 Novosibirsk Russia

KOSHI, MASAKI, engineering educator; b. Tokyo, Nov. 16, 1934; s. Taro and Fumiko (Komuro) K.; m. Yoko Sato, Oct. 10, 1959. BS in Civil Engring., U. Tokyo, 1957, MS in Civil Engring., 1959, D in Engring., 1969. Researcher Ministry of Construction, Tokyo, 1959-64; from lectr. to prof. in engring. U. Tokyo, 1964—. Author several books on road traffic engring.; contbr. articles to profl. jours. Chmn. road rsch. group Orgn. Econ. Cooperation and Devel.; advocator, leader R&D Automated Underground Freight Transport System. Recipient Article award, Internat. Assn. Traffic and Scis., Tokyo, 1984. Mem. Japan Soc. Civil Engrs., Japan Soc. Traffic Engrs. (v.p.), Internat. Assn. Traffic and Safety Scis. (v.p.). Avocations: skiing, sailing. Office: U Tokyo, 7-3-1 Hongo Bunkyoku, Tokyo 113, Japan

KOSHIYA, NAIHIRO, neuroscientist; b. Choshi, Chiba, Japan, Apr. 8, 1960; came to U.S., 1990; s. Yasuji and Kimi K.; m. Keiko Matsuda, Feb. 13, 1988; 1 child, Hitoshi Gene. BLA, Internat. Christian U., 1984; M of Med. Sci., U. Tsukuba, 1986, PhD, 1990. Lectr. Inst. Asian and African Langs., Mitaka, Tokyo, 1982-84, Matsudo City Hosp. Sch. Nursing, 1988-90; rsch. assoc. U. Va., Charlottesville, 1990—; vis. rsch. asst. U. Tokyo, 1989-90. Contbr. chpts. to books and articles to profl. jours. Spl. scholar Nihon Ikueikai, 1980-84, 84-90. Mem. Soc. Neurosci., Japanese Soc. Physiology, N.Y. Acad. Sci. Achievements include research on arterial chemoreceptor reflex, sympathetic nervous system, central adrenergic noradrenergic neurons, arterial baroreceptor reflex, central endothelin, respiratory neurons. Office: U Va Health Sci Ctr Box 448 Dept Pharmacology Charlottesville VA 22908

KOSHLAND, DANIEL EDWARD, JR., biochemist, educator; b. N.Y.C., Mar. 30, 1920; s. Daniel Edward and Eleanor (Haas) K.; m. Marian Elliott, May 25, 1945; children: Ellen, Phyllis, James, Gail, Douglas. BS, U. Calif., Berkeley, 1941; PhD, U. Chgo., 1949; PhD (hon.), Weizmann Inst. Sci., 1984; ScD (hon.), Carnegie Mellon U., 1985; LLD (hon.), Simon Fraser U., 1986; LHD (hon.), Mt. Sinai U.; LLD (hon.), U. Chgo., 1992, U. Mass., 1992. Chemist Shell Chem. Co., Martinez, 1941-42; research asso. Manhattan Dist. U. Chgo., 1942-44; group leader Oak Ridge Nat. Labs., 1944-46; postdoctoral fellow Harvard, 1949-51; staff Brookhaven Nat. Lab., Upton, N.Y., 1951-65; affiliate Rockefeller Inst., N.Y.C., 1958-65; prof. biochemistry U. Calif., Berkeley, 1965—, chmn. dept., 1973-78; fellow All Souls, Oxford U., 1972; Phi Beta Kappa lectr., 1976; John Edsall lectr. Harvard U., 1980; William H. Stein lectr. Rockefeller U., 1985; Robert Woodward vis. prof. Harvard U., 1986. Author: Bacterial Chemotaxis as A Model Behavioral System, 1980; mem. editorial bd. jours. Accounts Chem. Rsch., Jour. Biol. Chemistry, Jour. Biology, Biochemistry; editor jour. Procs. NAS, 1980-85; editor Sci. mag., 1985—. Recipient T. Duckett Jones award Helen Hay Whitney Found., 1977, Nat. Medal of Sci. NSF, 1990, Merck award Am. Soc. Biochemistry and Molecular Biology, 1991; Guggenheim fellow, 1972. Mem. NAS, Am. Chem. Soc. (Edgar Fahs Smith award 1979, Pauling award 1979, Rosentiel award 1984, Waterford prize 1984), Am. Philos. Soc., Am. Soc. Biol. Chemists (pres.), Am. Acad. Arts and Scis. (coun.), Acad. Forum (chmn.), Japanese Biochem. Soc. (hon.), Royal Swedish Acad. Scis. (hon.), Alpha Omega Alpha. Home: 3991 Happy Valley Rd Lafayette CA 94549-2423 Office: U Calif Dept Molecular & Cell Biology Berkeley CA 94720-0001

KOSHLAND, MARIAN ELLIOTT, immunologist, educator; b. New Haven, Oct. 25, 1921; d. Waller Watkins and Margaret Ann (Smith) Elliott; m. Daniel Edward Koshland, Jr., May 25, 1945; children—Ellen R., Phyllis A., James M., Gail F., Douglas E. B.A., Vassar Coll., 1942, M.S., 1943; Ph.D., U. Chgo., 1949. Research asst. Manhattan Dist. Atomic Bomb Project, 1945-46; fellow dept. bacteriology Harvard Med. Sch., 1949-51; asso. bacteriologist biology dept. Brookhaven Nat. Lab., 1952-62, bacteriologist, 1963-65; assoc. research immunologist virus lab. U. Calif., Berkeley, 1965-69, lectr. dept. molecular biology, 1966-70, prof. dept. microbiology and immunology, 1970-89, chmn. dept., 1982-89, prof. dept. molecular and cell biology, 1989—; mem. Nat. Sci. Bd., 1976-82; mem. adv. com. to dir. NIH, 1972-75; mem. coun. Nat. Inst. Allergy and Infectious Diseases NIH, 1991—. Contbr. articles to profl. jours. Mem. NAS, Nat. Acad. Arts and Scis., Am. Acad. Microbiology, Am. Assn. Immunologists (pres. 1982-1983), Am. Soc. Biol. Chemists. Home: 3991 Happy Valley Rd Lafayette CA 94549 Office: U Calif Dept Molecular/Cell Biology 439 LSA Berkeley CA 94720

KOSHY, VETTITHARA CHERIAN, chemistry educator, technical director/formulator; b. Kumbanad, Kerala, India, Jan. 5, 1952; came to U.S., 1984; s. Vettithara and Mariamma Cherian; m. Valsamma Koshy, Jan. 31, 1983; children: Rincy Mary, John Cherian. BSc in Chemistry, Kerala U., India, 1973; MSc in Chemistry, Ravishankar U., India, 1975, PhD in Chemistry, 1983; MS in Econ. Aspects of Chemistry, U. Detroit, 1992. Rsch. fellow chemistry Ravishankar U., Raipur, 1976-81; lectr., head dept. chemistry M.B. Patel Coll., 1981-83; lectr. dept. chemistry D.B. Sci. Coll., Gondia (India) Edn. Soc., India, 1983-84; group leader and evening supr. in R & D Widger Chem. Corp., Warren, Mich., 1984-87; mgr. automotive div., rsch. and devel. Croda Caourep Corp., Westland, Mich., 1987-89; dir. rsch. and devel., quality control and mfg. Autotek, inc., Farmington Hills, Mich., 1989—. Contbr. articles to Jour. Chem. Engring., Croatica Chemica Acta, Indian Acad. Scis., Nat. Acad. Scis. Sci. Letters, among others. Pres. sci. assn. J.M. Patel Coll., Bhandara, India, 1981-82; pres. chem. soc. Ravishankar U., Raipur, 1977-78, pres. rsch. scholars assn., 1979-81. Recipient numerous grants. Mem. Am. Chem. Soc., Am. Inst. Chemists, Soc. Automotive Engrs. (assoc.). Achievements include development of a formula for a universal sealer for automotive application. Home: 7030 White Pine Dr Bloomfield Hills MI 48301-3715 Office: Autotek Inc 23163 Commerce Dr Farmington MI 48335-2723

KOSKER, LEON KEVIN, electrical engineer; b. Denver, Oct. 23, 1954; s. Leon George Kosker and Barbara Jean (Schmidt) Charlton; m. Pamela Jean Fehr, June 13, 1981; children: Leah Jean, Leon Lucas. BEE, Carnegie Mellon U., 1979; MBA, Indiana U. of Pa., 1990. Assoc. engr. MSI Data Corp., Costa Mesa, Calif., 1976-79; mfg. engr. Fisher Sci. Co., Indiana, Pa., 1979-84, product design engr., 1984-87, quality assurance and test engr., 1987-89; gen. mgr. Quintech Electronics & Comm., Indiana, 1989; project engr. Concurrent Techs. Corp., Johnstown, Pa., 1990—. Home: 134 Euclid Ave Johnstown PA 15904 Office: Concurrent Techs Corp 1450 Scalp Ave Johnstown PA 15904

KOSKINEN, ARI MAURI PETRI, organic chemistry educator; b. Hyvinkää, Finland, Sept. 22, 1956; s. Veikko Aubust and Kirsti Kyllikki (Vannila) K.; m. Päivi Marketta Kataja, Dec. 10, 1988; children: Tiina Eerika, Joanna Päivikki, Heidi Petriina. MS, Helsinki U. Tech., Finland, 1979, lic. in Tech., 1982, D in Tech., 1983. Rsch. asst. Helsinki U. Tech., 1980-83; postdoctoral rsch. asst. U. Calif., Berkeley, 1983-87; group leader Orion Corp. Ltd.-Fermion, Finland, 1985-87; visiting scholar U. Calif., Berkeley, 1987-88; sr. scientist Kinnunen, Schroeder, Virtanen Lipids, Helsinki, 1989; lectr. U. Surrey, Eng., 1989-91; prof. U. Oulu, Finland, 1991—. Author 2 books; contbr. articles to profl. jours. Recipient grant Finnish Cultural Found., 1983, Finnish Acad. Sci., 1992, 93. Mem. AAAS, Finnish Chem. Soc., Royal Soc. Chemistry, Am. Chem. Soc., Ministry of Edn., Technology Devel. Ctr. Avocations: lit., tennis, motor sports. Home: Lepikkotie 2A1, Oulunsalo Finland Office: U Oulu, Linnanmaa, Oulu FIN 90570, Finland

KOSLOVER, ROBERT AVNER, physicist; b. Melrose, Mass., June 9, 1959; s. Monty and Lillian (Benjamin) K.; m. Deborah Anne Traynor, Aug. 16, 1981; children: Daniel, Rebecca. BS in Physics cum laude, U. Calif., Irvine, 1981, PhD in Physics, 1987. Scientist Sci. and Engring. Assocs., Albuquerque, 1987-88; sr. scientist Voss Sci., Albuquerque, 1988-92; rsch. physicist Phillips Lab./Weapons and Survivability, Kirtland AFB, N.Mex., 1993—. Mem. Am. Phys. Soc., Sigma Pi Sigma. Achievements include patent for Circular TM01 to TE11 Waveguide Mode Converter; patent pending for Compact, High-Gain, Ultra-Wideband Transverse Electromagnetic Planar Transmission-Line Array Horn Antenna; invention of optical tomography plasma diagnostic. Office: PL/WST Bldg 323 3550 Aberdeen Ave SE Kirtland AFB NM 87117-5776

KOSMATKA, JOHN BENEDICT, aerospace engineer, educator; b. Milw., Aug. 24, 1956; s. Benedict and Grace (Marciniak) K.; m. Ellen Pecchia, Aug. 13, 1988. BSME, U. Wis., 1978; MSME, U. Mich., 1980; PhD in Aerospace Engring., UCLA, 1986. Profl. engr., Wis. Machinist Svc. Tool & Die Co., Inc., West Hills, Wis., 1972-78; rsch. engr. Aerospace Corp., El Segundo, Calif., 1980-82, TRW Corp., Redondo Beach, Calif., 1982-86; asst. prof. mech. engring. Va. Poly. Inst. and State U., Blacksburg, 1986-89; asst. prof. dept. applied mechanics U. Calif., San Diego, 1989—. Contbr. over 50 articles to jours. and confs.; author over 15 govt. and indsl. reports; reviewer 12 tech. jours. Rsch. fellow NASA/Am. Soc. Engring. Edn., 1988-90; faculty fellow Newport News Shipbuilding, 1989-91; PhD fellow TRW Corp., 1984-85; recipient Outstanding Paper award ASME/AIAA, 1991. Mem. AIAA (faculty advisor 1991—), ASME, Am. Helicopter Soc. Office: U Calif Dept Applied Mechanics San Diego CA 92093-0411

KOSS, PETER, research administrator; b. Vienna, Austria, Mar. 21, 1932; s. Paul and Hilde (Fiedler) K.; m. Elsa Vedra, May 26, 1956; children: Michael, Christoph, Stephan. Ph.D. in Physics, U. Vienna, 1958; Univ. Dozent venia legendi, U. Vienna, 1967, univ. prof., 1981. Research asst. U. Vienna, 1955-58; research fellow MIT, Cambridge, 1959; with Atomics Internat., Canoga Park, Calif., 1959; head dept. metallurgy Austrian Research Ctr., Seibersdorf, Vienna, 1963-81, tech. sci. mng. dir., 1981—. Recipient Decoration of Honour for Merit in Silver (Austria), 1969, Cross of Honour for Sci. and Arts, 1975. Mem. Austrian Phys. Soc., Internat. Plansee Soc. Power Metallurgy, Chem. Phys. Soc. Office: Österreiches Forschungszentruti Seibersdorf GmbH, 2444 Seibersdorf, A-1010 Vienna Kramergasse 1, Austria*

KOSSMANN, CHARLES EDWARD, cardiologist; b. Brooklyn, N.Y., Apr. 20, 1909; s. Edward and Anna (Seidel) K.; m. Margaret Musgrave, Dec. 28, 1946; children: Michael Musgrave Kossmann, Margaret Olive Kossmann Dunklin. BS, NYU, 1928, MD, 1931, DMS, 1938. Asst. in medicine U. Mich. Sch. of Medicine, Ann Arbor, 1934; asst. in medicine to prof. medicine

NYU Sch. of Medicine, N.Y.C., 1934-67; prof. medicine U. Tenn. Coll. of Medicine, Memphis, 1967-76, prof. medicine emeritus, 1976—; hon. physician to vis. physician Bellevue Hosp. 3rd Med. Div., N.Y.C., 1931-67, cons. physician, 1968—, NY VA Hosp., 1964-67, Memphis VA Hosp., 1968-88; assoc. attending physician NYU Hosp., 1949-59, attending physician, 1955-67, staff physician, City of Memphis Hosp., 1968-88, cons. physician 1988—, staff physician U. Tenn. Hosp., Memphis, 1975-91, chief cons. in cardiology Cen. Office VA, Washington, 1951-55; mem. Sci. Adv. bd. Chief Staff USAF, 1952-56; cons. Nat. Heart and Lung Inst. Div. of Regional Med. Programs, Bethesda, 1966-73. Author: Flight Surgeon's Handbook, 1943, History of Electocardiographic Leads, 1988, Long Q-T Interval and Syndromes, 1987, Pericardiocentesis, 1980, Changing Views of Coronary Disease, 1976, Intraventricular Block, 1973, Electrocardiography Standards for Computers, 1970, Heart and Circulation, 1965, Vector Analysis in Acute Intarction, 1963, many others; editor: Flight Surgeon's Reference File, 1945, (with others) Advances in Electrocardiography, 1958; contbr. articles to profl. jours. Col. USAF, 1941-46. Named Disting. scholar St. Andrews Acad., 1980; numerous tng. grants Nat. Heart and Lung Inst., 1952-76. Fellow AMA, Am. Coll. Cardiology, N.Y. Acad. Scis., N.Y. Acad. Medicine; mem. ACP (master 1981), Assn. of Am. Physicians, Am. Heart Assn. (ctrl. com. 1958-60), Am. Soc. for Clin. Investigation, Soc. for Biology and Exptl. Medicine, Sociedad Mexicana de Cardiologica, Assn. U. Cardiologists, Sigma Xi. Office: U Tenn Coll Medicine 951 Court Ave Memphis TN 38163

KOSTERMAN, RICHARD JAY, political psychologist, political consultant; b. Boise, Idaho, Nov. 2, 1959; s. James Richard and Doris Jean (Soliday) K. BS, U. Wash., 1983; PhD, UCLA, 1991. Methodological cons. Fu-Gen, Inc. Rsch. and Investigation, Beverly Hills, Calif., 1988-89; teaching assoc. UCLA, 1984-89, rsch. assoc., 1983-92; adj. prof. U. Wash., Bothell, 1992—; dep. campaign mgr. Maria Cantwell for State Rep., Seattle, 1986; campaign mgr. Jim Kosterman for Port Commr., Vancouver, Wash., 1985, 91; issues dir. Maria Cantwell for Congress, Seattle, 1992. Contbr. chpts. to books, articles to profl. jours. Mem. APA, Am. Assn. for Pub. Opinion Rsch., Am. Polit. Sci. Assn., Internat. Soc. Polit. Psychology, Soc. for Psychol. Study of Social Issues. Democrat. Home: 8524 Phinney Ave N # 8 Seattle WA 98103 Office: U Wash 22011 26th Ave SE Bothell WA 98021

KOSTER VAN GROOS, AUGUST FERDINAND, geology educator; b. Leeuwarden, Friesland, The Netherlands, Jan. 9, 1938; came to U.S., 1962; s. Willem Hubertus and Gerardina Cornelia (Westphal) Koster van G.; m. Elisabeth Maria Theresa, Dec. 3, 1971; children: Sebastian Aernout, Paul Gijsbert. MS, Leiden U., 1962, PhD, 1966. Lectr. Utrecht U., The Netherlands, 1968-70; asst. prof. U. Ill., Chgo., 1970-77, assoc. prof., 1977-90, prof., 1990—, head prof. dept. geol. sci., 1992—; cons. in field. Contbr. articles to profl. jours. Mem. AAUP, AAAS, Am. Geophys. Union. Achievements include rsch. in liquid immiscibility between silicate and carbonate melts; high pressure differential thermal analysis of clay-water and salt-water systems. Office: U Ill Dept Geol Scis Chicago IL 60680

KOSTI, CARL MICHAEL, dentist, researcher; b. Wilmerding, Pa., Apr. 1, 1929; s. Michael John and Elizabeth Donna (Sotirovich) K.; m. Eileen Ann Peters, May 15, 1963; children: Stephen Michael, Peter Lawrence. BA in Chemistry, Wayne State U., 1957; DDS, U. Detroit, 1960. Pvt. practice Troy, Mich., 1960-86; owner, pres. Christephen Co., Troy, 1982-84, Am. Self Care Co., Inc., Fairfield, Ohio, 1986-88, Bio-Medica Concepts, West Chester, Ohio, 1988—. Cpl. U.S. Army, 1951-53. Achievements include 22 patents in medical, dental, home self-diagnosis products; research in relationship between use of Dilantin Sodium in treatment of adolescent seizures and resultant givgival hypertrophy, vasoconstrictor treatment of such tissue. Home: 4503 Williamsburg Rd NW Cincinnati OH 45215 Office: Bio Medica Concepts 8645 Cincinnati-Columbus Rd Westchester OH 48065

KOSTIN, VLADIMIR ALEXEEVICH, mathematics educator; b. Voronezh, USSR, Mar. 19, 1939; s. Alexey and Alexandra (Kortunova) K.; m. Alexandra Popova, Mar. 10, 1962 (div. 1975); children: Andrey, Veronika; m. Valentina Chernych, Jan. 9, 1976; children: Alexey, Dmitry. BSc, Voronezh State U., 1967, MSc, 1970. Sr. researcher Rsch. Inst. of Math., Voronezh, 1970-72; sr. lectr. math. Voronezh State U., 1972—, chief sci. Rsch. Centre Kentavr, 1991—. Contbr. articles to profl. jours. Grantee GSoros Fund, 1993. Avocations: chess, painting, football. Home: 11a/21 Festivalny Blvd, Voronezh 394062, Russia Office: Voronezh State U, 1 University Sq, Voronezh 394693, Russia

KOSTKA, MADONNA LOU (DONNA), naturalist, environmental scientist, ecologist; b. Hillsboro, Ill., July 15, 1934; d. John F. and Mabel L. (Hieronymus) Miller; m. Ronald W. Kostka, June 8, 1957 (div. 1980); children: Paul, Daniel, Jane; m. Floyd Reichman, Nov. 26, 1983. BS, U. Ill., 1956, MS, 1958; PhD, U. Minn., 1975. Camp dir. Citizens Club Settlement House, Mpls., 1957-59; ind. environ. cons. Minnetonka, Minn., 1975-76; asst. prof. U. Wis., River Falls, 1975-77; team leader environ. impact statement U.S. Fish and Wildlife Svc., Mpls., 1983-85; mgr. devel. program rotating positions Dept. Interior, Washington, 1985-87; environ. scientist U.S. EPA, Washington, 1987-88; project dir. Labat-Anderson, Inc., Arlington, Va., 1988-89; environ. scientist U.S. Dept. Energy, Washington, 1990—; chair Minnetonka Environ. Quality Commn., 1972-76; chair, bd. dirs. Minn. Assn. Environ. and Outdoor Edn., 1981-85; bd. dirs. Environ. Learning Ctr., 1983-85; del. Siberian People to People Exch. on Biodiversity, 1992. Editor: Great Lakes Fisherman, 1983; contbr. to profl. publs. Chair recreation adv. coms. City of Minnetonka, 1964-67, vice-chair Park Bd., 1968-70; chair coms. LWV, Minnetonka, 1967-69. Environ. Conservation fellow Nat. Wildlife Fedn., 1975. Mem. Met. Washington Environ. Profls. (bd. dirs. 1991—), Ecol. Soc. Am. (cert. ecologist), Nat. Assn. Environ. Profls. Home: 4510 48th St NW Washington DC 20016

KOSTYNIAK, PAUL JOHN, toxicology educator; b. Schenectady, Apr. 8, 1947; s. Theodore John and Veronica Anne (Wojnarowski) K.; m. Carol Ann Kusak, Aug. 22, 1970; children: Douglas, Gregory, Laura. BS in Biology, St. John Fisher Coll., 1970; PhD in Toxicology, U. Rochester, 1975. Diplomate Am. Bd. Toxicologists. Asst. prof. pharmacology SUNY, Buffalo, 1977-84, assoc. prof. pharmacology, 1984—; dir. Toxicology Rsch. Ctr., Buffalo, 1985—; toxicology cons. for exposure and risk assessment, Buffalo, 1981—; bd. advisors Union Occupational Health Ctr., Buffalo. Contbr. articles to profl. publs., chpt. to book. Fellow NIH, NIEHS; grantee NIEHS, Pharm. Mfrs. Assn. Mem. ASTM, APHA, AAAS, Am. Chem. Soc., N.Y. Acad. Scis., Soc. Toxicology (past pres. metals specialty sect.), Niagara Frontier Assn. for Rsch. Dirs., Tissue Culture Assn., Internat. Assn. Great Lakes Rsch., Internat. Zenobiotics, Western N.Y. Coun. on Occupational Safety and Health (bd. advisors). Office: SUNY 111 Farber Hall 3435 Main St Buffalo NY 14214-3000

KOSTYUK, PLATON GRIGOREVICH, physiologist; b. Kiev, USSR, Aug. 20, 1924. Grad. in biology, Kiev U., 1946, D Biol. Sci., 1956. Prof. biology Kiev U., 1960, head dept. physiology Inst. Animal Physiology, 1946-58; head lab. for gen. physiology Bogomoletz Inst. Physiology, Ukrainian Acad. Scis., 1958-66; dir. Bogomolets Inst. Physiology, Ukrainian Acad. Scis., 1966—. Contbr. articles to profl. jours. Mem. Ukrainian Acad. Scis., USSR Acad. Scis., Acad. Europea, European Fedn. Physiol. Soc. (v.p. 1991—). Office: Inst Physiology, Bogomoletz Str 4, Kiev 252024, Ukraine

KOSZEWSKI, BOHDAN JULIUS, internist, medical educator; b. Warsaw, Poland, Dec. 17, 1918; Came to U.S., 1952; s. Mikolaj and Helen (Lubienski) K.; children Mikolaj, Joseph, Wanda Marie, Andrzej Bohdan. MD, U. Zurich, Switzerland, 1946; MS, Creighton U., 1956. Resident in pathology U. Zurich, 1944-46, resident in internal medicine, 1946-50, assoc. in medicine, 1950-52; intern St. Mary's Hosp., Hoboken, N.J., 1953; practice medicine specializing in internal medicine Omaha, 1953—; mem. staff St. Joseph's Hosp., Mercy and Meth. Hosps.; instr. internal medicine Creighton U., 1956-57, asst. prof., 1957-65, assoc. prof. internal medicine, 1965-90; cons. hematology Omaha VA Hosp., 1957-90. Author: Prognosis in Diabetic Coma, 1952; contbr. numerous articles to profl. jours. Served with Polish Army, 1940-45. Fellow ACP, Am. Coll. Angiology; mem. AAAS, Am. Fedn. Clin. Research, Internat. Soc. Hematology, Polish-Am. Congress Nebr. (pres. 1960-68, 82-92). Home: 2901 Park Pl Dr Lincoln NE 68506 Office: Lincoln Ctr Bldg Lincoln NE 68542

KOSZI, LOUIS A., electronics engineer; b. Bethlehem, Pa., Oct. 7, 1944; s. Vilma E. Koszi; m. Lorraine F. Koszi, Oct. 17, 1970; children: Laura Lyn, Lacene Fay. BS in Physics, Moravian Coll., 1970. Sr. tech. aide AT&T Bell Labs., Murray Hill, N.J., 1970-73, assoc. mem. tech. staff, 1973-80, mem. tech. staff, 1980-88, Disting. mem. tech. staff, 1988—; indsl. advisor Union County C.C., Scotch Plains, N.J., 1983-92. Contbr. 35 papers to profl. jours. Youth soccer coach Scotch Plains-Fanwood Soccer Assn., 1986—. Recipient Engring. Excellence award Optical Soc. Am., 1992. Achievements include 17 U.S. and 5 European patents related to electro-optical components including fabrication processes, fundamental semiconductor designs, and applicable transmission systems. Home: 2097 Newark Ave Scotch Plains NJ 07076 Office: A T&T Bell Research Lab 600 Mountain Ave New Providence NJ 07974

KOTHA, SUBBARAMAIAH, biochemist; b. Kurnool, A.P., India, Aug. 14, 1960; came to U.S., 1989; s. Kotha and Kotha (Lalitha) Subramanyam; m. Kotha Sudharani, May 7, 1989. MS, Ctrl. U., Hyderabad, India, 1982, PhD, 1988. Postdoctoral fellow City Coll., N.Y.C., 1989—. Mem. AAAS, Am. Soc. Biochemistry and Molecular Biology, Sigma Xi. Achievements include research in biochemistry.

KOTHERA, LYNNE M., clinical neuropsychologist; b. Cleve., Dec. 18, 1938; d. Leonard Frank and Lillian (Shackleton) Kothera; m. Richard Litwin, Oct. 24, 1965. BA with hons., Denison U., Granville, Ohio, 1960; MA, NYU, 1983; PhD, L.I. U., Bklyn., 1989; postgrad. psychotherapy/psychoanalysis, NYU, 1992—. Dancer Martha Graham Dance Co., N.Y.C., 1961-62, Carmen DeLavallade Dance Co., N.Y.C., 1965-68, Glen Tetley Dance Co., N.Y.C., 1965-69; prin. dancer John Butler's, N.Y.C., 1971; artist-in-residence Boston High Schs. - Title III, 1969-71, Hobart-Smith Coll./Denison U., 1973; auditor N.Y. State Council of the Arts, N.Y.C., 1974-78; predoctoral fellow clin. psychology Yale-New Haven Hosp., 1987-88; postdoctoral fellow neuropsychology Inst. of Living, Hartford, Conn., 1989-91; with dept. rehab. medicine Mt. Sinai Med. Ctr., N.Y.C., 1991—. Mem. APA (divsn. 39, 40 and 42), Internat. Neuropsychol. Soc. Democrat. Avocation: the arts. Home: 23 E 11th St New York NY 10003 Office: Mt Sinai Med Ctr/Rehab Medicine 1 Gustave Levy Pl Box 1241 New Guggenheim Pavilion New York NY 10029

KOTKOV, BENJAMIN, clinical psychologist; b. Boston, Apr. 8, 1910; s. Moses and Annie (Hopner) K.; m. Sally B., Jan. 28, 1941; children: Ralph, Frank. AB, Cornell U., 1929; MA, Harvard U., 1934; PhD, Ottawa U., Ont., Can., 1954. Diplomate Am. Bd. Clin. Psychology, Am. Bd. Med. Psychotherapists, Am. Bd. Disability Cons. Staff to chief Nerve Clinic, New Eng. Med. Ctr., Boston, 1934-42, VA Mental Hygiene Unit, Boston, 1946-52; chief psychologist Mental Hygiene Clinics, State of Del., 1952-53; clin. exec. Child Guidance Ctr., Brattleboro, Vt., 1954-64; staff to prof. and faculty head Windham Coll., Putney, Vt., 1964-76; psychologist in pvt. practice Brattleboro, 1976—; internat. adv. bd. Acad. of Psychoanalysis, Germany, 1969—. Contbr. numerous articles to profl. jours. Lt. U.S. Army, 1942-46. NSF grantee, 1956, 57; recipient Editor's award Internat. Jour. Profl. Hypnosis, 1977, Medallion Acad. Psychosomatic Medicine, 1979. Fellow Am. Psychol. Soc., Internat. Soc. Profl. Hypnosis, Acad. Sci. Hypnotherapy; mem. AAUP (emeritus), APA (life), Soc. Personality Assessment (life), Vt. Psychol. Assn. (pres. 1968-69, chmn. certification bd. 1974-76), New England Soc. Clin. Hypnosis (pres. 1988), Diabled Am. Vets. (life), Lions (pres. 1973-74, 84-85). Home: RD 2 Box 95 Brattleboro VT 05301

KOTLYAKOV, VLADIMIR MICHAILOVICH, geographer, glaciologist researcher; b. Lobnya, USSR, Nov. 6, 1931; s. Michail Vasil'evich and Elena Alexandrovna (Abramova) K.; m. Kotlyakova Eleonora Maximovna Sheveleva; 1 child, Michail; m. Basanova Valentina Alexeevna; 1 child, Andrei. Diploma, Moscow State U., USSR, 1954; Kandidat of Sci., Inst. of Permafrost, USSR, 1961; PhD Sci., Inst. of Geography, USSR, 1967. Researcher Inst. Geography, USSR Acad. Scis., Moscow, USSR, 1954-68, head dept., 1968-86, prof., 1971, dir., 1986—; academician Russian Acad. Scis., Moscow, Russia, 1991; people's dep. USSR Supreme Coun., Moscow, 1989-91; pres. Internat. Commn. of Snow and Ice, 1987-91; sci. com. mem. Stockholm Internat. Geosphere Biosphere Program, 1987-93; steering com. Barselona Human Dimensions Global Environ. Change Program, 1990; mem. Earth Coun., San Jose, Costa Rica, 1993. Editor in chief: World Atlas of Snow and Ice Resources, 1976—, Data Of Glaciological Studies, 1961—, Izvestiya Academii Nauk, 1986—; Author: Antartic Snow Cover And Its Role In The Glaciation Of The Continent, 1961, Snow Cover Of The Earth And Glaciers, 1968, Mountains, Ice And Hypotethis, 1977, An Isotope and Geochemical Glaciology, 1982, Glaciological Glossary, 1984 (Litke Gold medal Russian Geographical Soc.), The Role Of Snow And Ice In The Earth Nature, 1986, Elsevier's Dictionary On Glaciology, 1991. Fellow Am. Geographical Soc.; mem. Mexican Geographical Soc., Academia Europeae, Acad. Scis. USSR, Internat. Geographical Union (v.p., 1988—), Internat. Assn. Hydrological Scis., (v.p. 1983-87), Russian Geographical Soc. St. Petersburg (v.p. 1980—). Achievements include contributions to the study of the earth's snow cover, and of the past and present regime of the Antarctic ice sheet, and the synthesis of socio-economoic and natural resource information on the Soviet Union, Russia and the World as a whole. Office: Inst Geography Russian Acad Sci, Staromonetny St 29, 109017 Moscow Russia

KOTOVSKY, KENNETH, psychology educator; b. Pitts., July 5, 1939; s. Jacob and Dorothy (Friedland) K.; m. Avis Brenda Lovit, June 10, 1962; children: Laura Lovit, Jack. BS in Econs./Polit. Sci., MIT, 1961; MS in Psychology, Carnegie Mellon U., 1970, PhD In Psychology, 1983. Systems analyst Stanford Rsch. Inst.-Control Systems Lab., Menlo Park, Calif., 1962-64 summers; USPHS trainee Biophys. Lab. Eye & Ear Hosp. of Pitts., Pitts., 1962-64; instr. biology Community Coll. of Allegheny County, Pitts., 1966-70, asst. prof. biology and psychology, 1970-75, assoc. prof., chmn. psychology, 1976-81, prof. psychology, 1981-89; assoc. prof., dir. undergrad. studies Carnegie Mellon U., Pitts., 1989—; rsch. assoc. psychology Carnegie Mellon U., Pitts., 1983-88, adj. prof. psychology, 1988-89. Editor: (with David Klahr) Complex Information Processing: The Impact of Herbert A. Simon, 1989, (with B. Fischhoft, H. Tuma and J. Bielak) A Two State Solution in the Middle East: Prospects and Possibilities, 1993; contbr. articles to profl. jours. Bd. dirs. Mid. East Forum, Pitts., 1987—; chmn. Troop com. Boy Scouts Am., Pitts., 1984-86; mem. Reizenstein Consortium Community Orgns., Pitts., 1979-82; com. on planning and allocation Pvt. Industry Coun. Allegheny County, 1988-91. Fellow NSF, 1964-66, grantee, 1981, 85-87. Mem. APA, Am. Psychol. Soc., Cognitive Sci. Soc. Democrat. Jewish. Avocations: skiing, bicycling, canoeing. Home: 1310 Murray Ave Pittsburgh PA 15217-1223 Office: Carnegie Mellon U Psychology Dept Pittsburgh PA 15213

KOTSUKI, HIYOSHIZO, chemist, educator; b. Kyoto, Japan, Jan. 2, 1951; s. Seigoro and Seiko (Takaya) K.; m. Noriko Hino; children: Chieko, Shunji, Kenji. BS, Kochi U., 1973; MS, Osaka City U., 1975, PhD, 1984. Rsch. asst. Kochi (Japan) U., 1975-85, asst. prof., 1985-87, assoc. prof., 1987—. Contbr. articles to Jour. Organic Chemistry, Tetrahedron Letters, Chemistry Letters, Bull. Chem. Soc. Japan. Office: Kochi U Dept Chemistry, Akebonocho, Kochi 780, Japan

KOTTAMASU, MOHAN RAO, physician; b. Gudivada, India, Jan. 13, 1947; came to U.S., 1973; s. Janardana Rao and Kantharatnamma (Maddi) K. MBBS, Gulbarga Med. Coll., 1972. House surgeon Govt. Gen. Hosp., Gulbarga, India, 1971-72; intern St. Vincent's Med. Ctr. of Richmond, S.I., N.Y., 1973-74, resident, 1974-76, chief resident, 1976-77; assoc. Valley Pulmonary and Med. Assocs., Springfield, Mass., 1979-81, ptnr., v.p., 1981—; adj. asst. prof. clin. pharmacy Mass. Coll. Pharmacy and Allied Health Scis., 1984—. Pres. house staff St. Vincent's Med. Ctr., 1976; founding pres. Indian Assn. Greater Springfield, 1985-86; pres. med. staff Mercy Hosp., Springfield, 1989-91. Pulmonary Diseases fellow Deaconess Hosp., Boston, 1977-79, Clin. fellow Harvard Med. Sch., Boston, 1978-79. Fellow Am. Coll. Physicians, Am. Coll. Chest Physicians; mem. Am. Thoracic Soc., Mass. Med. Soc. Hindu. Avocations: tennis, chess. Home: 112 Twin Hills Dr Longmeadow MA 01106 Office: Valley Pulmonary Med Assocs 222 Carew St Springfield MA 01104

KOTTLOWSKI, FRANK EDWARD, geologist; b. Indpls., Apr. 11, 1921; s. Frank Charles and Adella (Markworth) K.; m. Florence Jean Chriscoe, Sept. 15, 1945; children: Karen, Janet, Diane. Student, Butler U., 1939-42; A.B., Ind. U., 1947, M.A., 1949, Ph.D., 1951. Party chief Ind. Geology Survey, Bloomington, summers 1948-50; fellow Ind. U., 1947-51, instr. geology, 1950; adj. prof. N.Mex. Inst. Mining and Tech., Socorro, 1970—; econ. geologist N.Mex. Bur. Mines and Mineral Resources, 1951-66, asst. dir., 1966-68, 70-74, acting dir., 1968-70, dir., 1974-91, state geologist, 1989-91, emeritust dir., state geologist, 1991—; geologic cons. Sandia Corp., 1966-72. Contbr. articles on mineral resources, stratigraphy and areal geology to tech. jours. Mem. Planning Commn. Socorro, 1960-68, 71-78, chmn., 86-90; mem. N.Mex. Energy Resources Bd.; chmn. N.Mex. Coal Surface Mining Commn.; sec. Socorro County Democratic Party, 1964-68. Served to 1st lt. USAAF, 1942-45. Decorated D.F.C.; decorated Air medal; recipient Richard Owen Disting. Alumni award in Govt. and Industry U. Ind., 1987. Fellow AAAS, Geol. Soc. Am. (councilor 1979-82, exec. com. 1981-82); mem. AIME, Am. Assn. Petroleum Geologists (hon. mem., dist. rep. 1965-68, Disting. Svc. award, editor 1971-75, pres. energy minerals divsn. 1987-88), Assn. Am. State Geologists (pres. 1985-86), Soc. Econ. Geologists, Am. Inst. Profl. Geologists (Pub. Svc. award 1986), Am. Commn. Stratigraphic Nomenclature (past sec., chmn.), Cosmos Club, Sigma Xi. Home: 703 Sunset St Socorro NM 87801-4657 Office: NMex Bur Mines NMex Tech Campus Sta Socorro NM 87801

KOTWAL, GIRISH JAYANT, biochemist, educator; b. Bombay, India, Nov. 6, 1954; s. Jayant Vishwanath and Kumundini Jayant (Vaidya) K.; m. Archana Gupte, Feb. 13, 1986; 1 child, Mihir Girish. MSc, Bombay U., 1978; PhD, McMaster U., Hamilton, Ont., Can., 1985. Jr. rsch. fellow Hindustan Lever Rsch. Ctr., Bombay, 1978; sr. rsch. fellow Found. for Med. Rsch., Bombay, 1978-79; lectr. Sophia Coll., Bombay, 1979; teaching asst. McMaster U., 1979-84; vis. fellow NIH, Bethesda, Md., 1984-87, vis. assoc., 1987-90; asst. mem. staff James N. Gamble Inst. Med. Rsch., Cin., 1990-93; asst. prof. U. Louisville, 1993—; presenter in field; speaker Internat. Symposium on Viral Hepatitis, Madrid, 1992. Contbr. articles to profl. publs. Recipient 1st prize Indian Chem. Soc., 1977, Travel award Am. Soc. Virology, 1983, 90; J.N. Tata scholar, 1979; vis. fellow Fogarty Internat. Ctr., 1984-87. Mem. AAAS. Achievements include patents for synthetic anti-complement protein and gene encoding same, novel serine protease inhibitors and genes encoding same, HCV peptides, antibodies, olingonucleotides; rsch. in viral proteins mediating evasion of host defense. Home: 4664 Shenandoah Dr Louisville KY 40241 Office: U Louisville Dept Microbiology and Immunology Louisville KY 40292

KOTYNEK, GEORGE ROY, mechanical engineer, educator, marketing executive; b. Lake Forest, Ill., Apr. 18, 1938; s. Anton Joseph and Zdenka K.; m. Virginia Jean Hyde, Sept. 4, 1965 (div. 1973); children: John Anton, Joseph George. BSME, Ill. Inst. Tech., 1960. Registered profl. engr., Ill. Efficiency engr. Commonwealth Edison Co., Chgo., 1959-63; instr. physics Glenbard East High Sch., Lombard, Ill., 1963-67; systems engr. Sargent and Lundy, Chgo., 1967-77; prin. engr. Fluor Corp., Chgo., 1977-85; mgr. fossil tech. Stearns Catalytic World Corp., Oak Brook, Ill., 1985-86; mgr. mktg. Volund USA Ltd., Oak Brook, 1986—; mem. hazardous materials adv. com. Waubonsee C.C., Sugar Grove, Ill., 1992—. Contbr. articles to profl. publs. Mem. People to People Internat. Conventional and Nuclear Power Engring. Delegation to People's Republic of China, 1987. Mem. ASME (newsletter editor 1980-82, vice chmn. membership 1982-83, vice chmn. programs 1983-84). Achievements include design of electric generating station for cyclic service. Office: Volund USA Ltd Mountain Heights Ctr 430 Mountain Ave New Providence NJ 07974

KOTZABASSIS, CONSTANTINOS, agricultural engineering educator; b. Athens, Attiki, Greece, Oct. 20, 1958; came to U. S., 1986; s. George and Chrysanthi (Georganta) K.; m. Fabiana F. Bezerra May 19, 1990. BSc, Agrl. U. of Athens, Greece, 1982; MSC., U. Alberta, Edmonton, Can., 1986; PhD, Tex. A&M U., 1991. Crop damage inspector Hellenic Ministry Social Svcs., Athens, Greece, 1982-83; grad. rsch asst. U. Alberta, 1983-85; grad. rsch. and teaching asst. Tex. A&M U., College Station, Tex., 1986-91; instr. and extension agt. assoc. for Energy & Safety Tex. A&M U., College Station, 1991—; dir. internat. radio hour KAMV-FM, 1993—. Contbr. over 11 titles to Am. Soc. Agrl. Engrs. Pubd. Mem. Judicial Bd. Student Govt. Tex. A&M U., College Station, 1989-90, pub. rels. com. U. Apt. Coun. Tex. Aggies, Sigma Xi. Mem. Am. Soc Agrl. Engrs. (chmn. PM 46 1989-90, sec. PM 47 1992-93. assoc. editor newsletter, 1992—), Internat. Students Assn. Tex. A&M U. (mini olympics chmn. 1989-90), Nat. Inst. Farm Safety (assoc.), Tex. Safety Assn., Geotech. Chamber of Hellenic Republic, Agriculturists Assn. Of Attiki, Hellenic Student Assn. Tex. A&M U., Brazilian Student Assn., Tex. A&M U., Sigma Xi. Mem. Greek Orthodox Ch. Achievements include development of Farm Machinery Management Expert System (software). Office: Tex A&M Univ System Agrl Engring Dept College Station TX 77843

KOUPPARIS, MICHAEL ANDREAS, analytical chemistry educator; b. Polis Chr., Paphos, Cyprus, July 16, 1950; arrived in Greece, 1969; s. Andreas Michael Koupparis and Maria Christ. Nittis; m. Stella Spyros Kavvadia, Aug. 19, 1979; children: Andreas, Constantina. BS in Pharmacy, U. Athens (Greece), 1973, PhD in Chemistry, 1978. Cert. pharmacist. Prof. U. Athens, 1978—; postdoctoral rsch. assoc. U. Ill., Urbana, 1978-81. Coauthor: Instrumental Analysis, 1990, Quantitative Calculations in Pharmacy Practice, 1992; contbr. 70 articles to profl. jours.; author 50 conf. proceedings. 2d lt. Cypriot Army, 1968-69. Fellow Greek Found. Fellowships, Athens, 1970-73; rsch. grantee Gen. Secretariat for Rsch., Athens, 1988-92. Mem. Nat. Drug Orgn. (Athens, pharmacopeia com. 1987—). Achievements include research in new analytical methods for drug analysis, in new ion-selective electrodes, in construction of new dyalytical instruments. Home: Peloponisou 18, 15235 Vrilissia Attiki, Greece Office: U Athens Chemistry Dept, Panepistimiopolis, 15771 Athens Greece

KOURI, GUSTAVO PEDRO, virologist; b. Habana, Cuba, Jan. 11, 1936; s. Pedro and Mercedes (Flores) K.; m. Lidia Cardella (div. 1979); children: Lilliam, Vivian, Gustavo; m. Maria G. Guzman, Nov. 25, 1980; 1 child, Pedro. MD, Havana U., 1962; PhD, Nat. Ctr. Sci. Rsch., Havana, 1973; ScD, Charles U., Prague, Czechoslovakia, 1990. Chief virology dept. Nat. Ctr. for Sci. Rsch., Havana, 1965-70, dep. dir., 1965-70; vice-dean med. faculty Havana U., 1970-73, vice rector, 1973-76; nat. dir. for sci. Ministry of Higher Edn., Havana, 1976-78; dir. gen. Tropical Medicine Inst. "Pedro Kouri", Havana, 1979—; dir. WHO Collaborating Ctr. for Biol. Vector Control, Havana, 1990—; temp. advisor Pan Am. and WHO to present; lectr. in field; cons. in field. Contbr. numerous articles to profl. jours. Grantee TDR, 1979, 81, 82, 83, 84, 85, 86, IDRC, 1983, 86, French Govt., 1989; recipient Carlos Finlay Nat. Order and medal Cuban State Coun., 1990, Silver medal Charles U., 1988, Cesar Uribe medal NIH, Colombia, 1991. Mem. AAAS, N.Y. Acad. Sci., Cuban Acad. Sci. (presidium mem. 1978), Real Academia de Medicina y Cirujia de Galicia, Academia de Medicina y Cirijia de Galicia, Royal Soc. Tropical Medicine Hygiene, Latin Am. Soc. Tropical Medicine (pres.), Cuban Soc. Microbiology and Parasitology (pres. 1980), Latin Am. Soc. Parasitology (v.p. 1992), others. Achievements include research on Dengue Hemorrhagic Fever. Office: Inst Medicine Tropical, Pedro Kouri, Autopista Novia del Mediod, Havana Cuba

KOUTRAS, DEMETRIOS A., physician, endocrinology investigator, educator; b. Athens, Attica, Greece, Jan. 25, 1930; s. Anastasios and Phaidra Koutras; m. Maria Flouri, Dec. 1981. Med. diploma, Athens U., 1955. Trainee in medicine Alexandra Hosp., Athens, 1955-59; Brit. Coun. scholar Glasgow (Scotland) U., 1959-60, tem. lectr., 1961; assoc. physician Alexandra Hosp., Athens, 1962-63; vis. scientist NIH, Bethesda, Md., 1963-64; dir. thyroid unit Alexandra Hosp., Athens, 1964-76; assoc. prof. Athens U., 1976-90, prof. medicine, 1990—; dir. endocrinology unit Evgenidion Hosp., Athens, 1990—; researcher on iodine Greek Ministry of Health, Athens, 1959—. Author: Clinical Aspects of Iodine Metabolism, 1964, Introduction to Endocrinology, 1983; contbr. articles on medicine to profl. jours. Past treas. and v.p. European Thyroid Assn.; past pres. Hellenic Endocrine Soc.; past v.p. Soc. for Med. Studies. Lt. Med. Corps Greek mil. Metabolism and thyroid diseases grantee NIH, 1963-66; recipient Leo Oliner award, Washington, 1964, Empiricion prize, Athens, 1971, Choremion prize, Athens, 1973, Malamos prize, Athens, 1981. Fellow Royal Soc. Medicine; mem. Am.

Thyroid Assn. (assoc.), N.Y. Acad. Scis. Orthodox. Avocations: hunting, shooting, bridge. Home: 35 Vas Sofias Ave, GR 10675 Athens Greece

KOUTROUVELIS, PANOS GEORGE, radiologist; b. Feneos, Corinth, Greece, May 12, 1928; came to U.S., 1956; s. George Spiros and Dimitra (Revis) K.; m. Maria Chaconas, Sept. 25, 1960; children: Aristides, Alexander, Harris, Dimitri. MD, Med. Sch., Athens, 1956. Intern Trinity Luth. Hosp., Kansas City, 1956-57; resident St. Lukes Hosp., Kansas City, 1957-58; attending radiologist VA Hosp., Washington, 1961-63; chmn. dept. radiology Potomac Hosp., Woodbridge, Va., 1973-83; founder, pres. No. Va. Radiology and Nuclear Medicine, Falls Church, 1961—. Editor-in-chief, founder: (mag.) Imager Med. Imaging Technique and Achievements. Recipient Achievement in Medicine award Am. Hellenic Ednl. Progressive Assn. Fellow Am. Coll. Nuclear Medicine (Disting. Nuclear Medicine award 1991); mem. Am. Coll. Radiology, N.Y. Acad. Sci., Radiol. Soc. N.Am., Soc. Nuclear Medicine, Fairfax County Med. Soc., Hellenic Am. Soc. for the Health Scis. (pres. 1992-93). Achievements include patent for 3-D stereotactic device for needle guidance; development of stereotactic technique for percutaneous lumbar discectomy, technique for needle biopsies, cyst aspirations, and laser procedures. Office: 8320 Old Courthouse Rd 150 Vienna VA 22182

KOVAC, JEFFREY DEAN, chemistry educator; b. Cleve., May 29, 1948; s. Stanley Joseph and Lee (Chojnacki) K.; m. Susan Davis Kovac, June 9, 1973; children: Peter Jeffrey, Rachel Susan. BA, Reed Coll., 1970; MPhil, Yale U., 1972, PhD, 1974. Rsch. assoc. MIT/Dept. Chemistry, Cambridge, Mass., 1974-76; asst. prof. dept. chemistry U. Tenn., Knoxville, 1976-83, assoc. prof. dept. chemistry, 1983-91, prof. dept. chemistry, 1991—. Contbr. over 40 articles to profl. jours. on statis. mechanics of polymers and liquids on structure of coal. Co-chair Alliance for a Better Tomorrow, Knoxville, 1992-93; state dir. instrn. Tenn. State Soccer Assn., 1990—. Fellow Woodrow Wilson Found., 1966. Mem. AAAS, Am. Phys. Soc., Am. Chem. Soc. Unitarian Universalist. Office: U Tenn Dept Chemistry Knoxville TN 37996

KOVACH, JAMES MICHAEL, engineering executive; b. Phila., Nov. 26, 1955; s. James Joseph and Irene Teresa (kostovcik) K.; m. Margaret Marie Ashton, May 15, 1976 (div. July 1990); children: James, Thomas; m. Margaret Ann Stewart, May 19, 1991; children: Michael, Melissa, Melanie. A in Architecture, Am. Inst. Design, 1975; A in Mechanics, Scranton U., 1988. Draftsman Dynamic Svcs., Ivyland, Pa., 1973-75; plant layout designer Crown Cork and Seal Co., Inc., Phila., 1975-90; project mgr. Simplimatic Engring. Co., Lynchburg, Va., 1990—. Mem. ASTM, Guard Soc. Republican. Roman Catholic. Home: 313 Village Dr Lynchburg VA 24502 Office: Simplimatic Engring Co 1320 Wards Ferry Rd Lynchburg VA 24502

KOVACH, JOSEPH WILLIAM, management consultant, psychologist, educator; b. Hammond, Ind., Oct. 4, 1946; s. William Charles and Florence (Miotke) K. BA, St. Joseph Coll., Whiting, Ind., 1969; MA, Roosevelt U., 1974; PsyD, Chgo. Sch. Profl. Psychology, 1986. Lic. sch. psychologist, Ill., Ind., Mo. Chmn. depts. addictionology and psychology Calumet Coll. St. Joseph, Chgo.; exec. dir. Ednl. Rsch. Exch., Calumet City, Ill.; mgmt. cons. Joseph W. Kovach and Assocs., Ltd., Calumet City, 1990—; sr. ptnr. The Empowerment Group. Mem. APA, Midwest Psychol. Assn., Ill. Sch. Psychologists Assn., Nat. Guild Hypnosis. Office: PO Box 113 Calumet City IL 60409-0113

KOVACS, AUSTIN, research engineer; b. Woodside, N.Y., June 18, 1938; s. Nandor Semotan and Theresa (Werany) Kovacs; m. Betty Ann Roberts. BSCE, New Eng. Coll., 1960. Asst. constrn. supt. B. J. Lucarelli Co., 1960-61; field engr. IT&T Fed. Electric Corp., Paramus, N.J., 1961; rsch. engr. U.S.A. Cold Regions Rsch. and Engring. Lab., Hanover, N.H., 1964—; cons. in field; mem. internat. steering com. Port and Ocean Engring. Under Arctic Conditions, 1987—; pres. Kovacs Enterprise, 1992—. Contbr. articles to profl. publs. Chmn. Enfield (N.H.) Zoning Bd., 1973-75, Enfield Land Use Com., 1988; mem. Enfield Zoning Bd., 1989—. With U.S. Army, 1962-64. Recipient Antarctic Svc. medal NSF, 1979, 81, Spl. Rsch. and Devel. Achievement award U.S. Army Corps Engrs., 1990. Mem. ASCE, Internat. Glaciological Soc., Arctic Inst. N.Am.,. Achievements include patent and patent pending in field. Home: HC-63 Box 46 Lebanon NH 03766 Office: USA Cold Regions Rsch & Engring Lab 72 Lyme Rd Hanover NH 03755

KOVACS, ELIZABETH J., medical educator; b. N.Y.C., Apr. 5, 1957; d. George Joseph and Barbara (Toll) Kovacs; 1 child, Catherine Ann Kelley. BA, Reed Coll., 1978; PhD, U. Vt., 1984. Postdoctoral fellow Nat. Cancer Inst., Frederick, Md., 1984-87; asst. prof. dept. cell biology, neurobiology and anatomy Loyola U. Med. Ctr., Maywood, Ill., 1987-93; assoc. prof. dept. cell biology, neurobiology and anatomy Loyola U. Med. Ctr., Maywood, 1993—. Mem. Soc. for Leukocyte Biology (coun. mem. 1990—, presdl. adv. bd. 1991—). Office: Loyola U Med Sch Dept Cell Biology/Neurobiology/Anat 2160 S 1st Ave Maywood IL 60153

KOVAR, DAN RADA, pharmacist; b. Uniontown, Pa., Oct. 5, 1934; s. Daniel Rada and Annamabel (Craig) K.; m. Sandra Lynn Kregar, Aug. 10, 1963; children: Jennifer Lynn Kovar Blandford, Catherine Elizabeth Kovar. BS in Pharmacy, U. Pitts., 1958. Registered pharmacist, Pa., Ky. Pharmacist, mgr. Thrift Drug Co., Ashland, Ky., 1959-76, pharmacist, 1978-79; pharmacist, owner Laynes Pharmacy, Ashland, 1976—. Vol. faculty, bd. dirs. U. Ky. Coll. of Pharmacy Alumni Assn.; re-organizational pres. Greater Ashland Area Acad. of Pharmacy 1967; adminstrv. bd. Christ United Meth. Ch., Sunday sch. tchr.; advisor Ashland Area Hospice, Pathways Mental Health. Named Ky. Col. Mem. Nat. Assn. Retail Druggists, Nat. Assn. Bds. of Pharmacy, Am. Coll. Apothecaries, Christian Pharmacists Fellowship Internat. (founding pres. Ky. chpt. 1988-93), Ky. Pharmacists Assn. (Bowl of Hygeia award 1988, vice speaker of ho. of dels. 1992), speaker ho. of dels. 1993—), Ky. Pharmacists Svc. Assn. (div. operation group). Office: Laynes Pharmacy 2312 13th St Ashland KY 41101-3576

KOVENKLIOGLU, SUPHAN REMZI, chemical engineering educator; b. Istanbul, Turkey, Jan. 3, 1947; came to U.S., 1969; s. Mulhem and Melda (Sarpel) K.; m. Grace Ling-Yuh Lee, Oct. 20, 1972; children: Eileen, John. BS, Robert Coll., 1969; PhD, Stevens Inst. Tech., 1976. Asst. prof. Stevens Inst. Tech., Hoboken, N.J., 1977-82; assoc. prof. Stevens Inst. Tech., Hoboken, 1982-91, prof., 1991—. Contbr. articles to profl. jours. Bd. dirs., fin. chmn. Friends of Am. Schs. in Turkey, N.Y.C., 1991-92. Mem. Am. Inst. Chem. Engring. (N.J. sect. sec.-treas. 1984-86, vice chmn. 1986-87, chmn. 1987-88), Am. Chem. Soc., Sigma Xi (Stevens chpt. sec.-treas. 1989-92). Achievements include patent for Method of Hydrodehalogenating Halogenated ORganic Compounds in Aqueous Environmental Sources. Office: Stevens Inst Tech Chem Engring Dept Hoboken NJ 07030

KOVTYNOVICH, DAN, civil engineer; b. Eugene, Oreg., May 17, 1952; s. John and Elva Lano (Robie) K. BCE, Oreg. State U., 1975, BBA, 1976. Registered profl. engr., Calif., Oreg. V.p. Kovtynovich, Inc., Contractors and Engrs., Eugene, 1976-80, pres., chief exec. officer, 1980—. Fellow ASCE; mem. Am. Arbitration Assn. (arbitrator 1979—), N.W. China Coun., Navy League of U.S., Eugene Asian Coun. Republican. Avocations: flying, skiing, fishing, hunting. Office: Kovtynovich Inc 1595 Skyline Park Loop Eugene OR 97405-4466

KOWAL, CHARLES THOMAS, astronomer; b. Buffalo, Nov. 8, 1940; s. Charles Joseph and Rose (Myszkowiak) K.; m. Maria Antonietta Ruffino, Oct. 17, 1968; 1 dau., Loretta. B.A., U. So. Calif., 1963. Research asst. Mt. Wilson and Palomar obs., 1961-63, Calif. Inst. Tech., Pasadena, 1963-65, 66-75, U. Hawaii, 1965-66; asso. scientist Calif. Inst. Tech., 1976-78, scientist, 1978-81, mem. profl. staff, 1981-85; staff scientist Computer Sci Corp., 1986—; staff assoc. Hale Obs., 1979-80; lectr. in field. Author: Asteroids, Their Nature and Utilization, 1988. Recipient James Craig Watson award Nat. Acad. Scis., 1979. Mem. Am. Astron. Soc., Internat. Astron. Union. Discovered bright supernova, 1972, 13th satellite of Jupiter, 1974, large planetoid between orbits of Saturn and Uranus, 1977, also asteroids and comets; recovered lost comets and asteroids. Office: Space Telescope Sci Inst CSC Homewood Campus Baltimore MD 21218

KOWALCZYK, MACIEJ STANISLAW, obstetrician/gynecologist; b. Kraków, Poland, June 8, 1956; s. Bogumil Wieslaw and Teresa Maria (Matowska) K. MD, Med. Acad., Kraków, 1984; postgrad., Polish Acad. Sci., Kraków, 1984, Inst. Gyn.-Ob, 1988-91. Intern Narutowicz Hosp., Kraków, 1984-85; gen. practice medicine ambulatory Kraków-Srödmiescie, Kraków, 1984-85; gen. practice ambulatory medicine First Aid Svc., 1985-87; asst. in ob-gyn Szpital Polozniczy, Kraków, 1986—, Maternity Amb. for Sch. Tchrs., Kraków, 1987-92; tchr. Cathedral Normal Anatomy, Med. Acad., Kraków, 1984-86; prof. Med. Coll. for Midwives, 1991; mem. commn. in Soc. Ins. Inst., 1986-88. Contbr. articles to profl. jours. Mem. Polish Gynecol. Soc., vol. Life Saver's Assn., Polish Androl. Soc., Polish Sexology Soc.; gen. mem. Am. Inst. Ultrasound; mem. Polish Radiol. Soc. (ultrasound sect.), Am. Soc. for Colposcopy and Cervial Pathology, Polish Sonographic Soc., Internat. Soc. Ultrasound in Ob-Gyn, European Soc. Contraception, European Soc. Human Reproduction and Embriology, European Tourist Club. Roman Catholic. Avocation: numismatics. Home: Odroważa 22/7, Kraków 30009, Poland Office: Szpital Polozniczo-Ginekologiczny, ul Siemiradzkiego 1, Kraków 31 137, Poland

KOWALSKI, LYNN MARY, podiatrist; b. Passaic, N.J., Aug. 15, 1955; d. George J. and Gladys L. (Kucera) K.; m. Donald Storbeck, Feb. 9, 1975 (div. Mar. 1982); children: Jason, Jessica. BSN, William Paterson Coll., 1984; DPM, N.Y. Coll. Podiatric Medicine, 1988. RN, N.J.; diplomate Am. Bd. Podiatric Med. Examiners; lic. physician N.Y, N.J. Resident in podiatric surgery N.Y. Coll. Podiatric Medicine & Affiliated Hosps., 1988-89; pvt. duty nurse Bergen County, N.J., 1989; pvt. practice podiatrist Brick, N.J., 1990—; guest speaker Eldermed, Brick, 1991, Diabetes Support Group, Point Pleasant, N.J., 1991, 92, Arthritis Support Group, 1991, Garden State Rehab. Hosp., Toms River, N.J., Community Svcs., Toms River and Brick, 1992, Laurelton Village Community Edn., Brick, 1992, Lions Head North, 1992, Family Wellness Fair, Toms River, 1992, Green briar 2, 1992, Post Polio Support Group-Garden State Rehab. Hosp., Toms River, 1992, Treat Your Feet-Med. Ctr. Ocean County Health Edn. Network, 1992, Parkinsons Support Group-Med. Ctr. Ocean County, 1992. Contbr. articles to profl. jours. Mem. Toms River-Ocean (N.J.) County C. of C., 1990, Brick, 1991, Community Svcs., Toms River, 1992; vol. women's health day Med. Ctr. Ocean County, 1992, elder med. screening Sr. Citizen Villages, 1993. Mem. Am Diabetes Assn. (Tour de Cure 1992), Am. Coll. Foot Surgeons (assoc.), Am. Running and Fitness Assn., Am. Podiatric Med. Assn., N.J. Podiatric Med. Soc.,Kiwanis, Manasquan Elks (health fair 1992), Sigma Theta Tau (charter), Psi Chi. Office: 1608 Rte # 88 W Ste # 118 Brick NJ 08724

KOWLESSAR, MURIEL, retired pediatric educator; b. Bklyn., Jan. 2, 1926; d. John Henry and Arene (Driver) Chevious; m. O. Dhodanand Kowlessar, Dec. 27, 1952; 1 child, Indrani. AB, Barnard Coll., 1947; MD, Columbia U., 1951. Diplomate Am. Bd. Pediatrics. Instr. Downstate Med. Ctr., Bklyn., 1958-64, asst. prof., 1965-66; asst. prof. clin. pediatrics Temple U., Phila., 1967-70; assoc. prof. Med. Coll. Pa., Phila., 1971-83, dir. pediatric group svcs., 1975-90, acting chmn. pediatrics dept., 1981-83, vice chair pediatrics dept., 1982-91, prof., 1983-91, prof. emeritus, 1991—. Contbr. articles to med. jours. Vol. Phila. Com. for Homeless, 1991-92, Gateway Literacy Program, YMCA, Germantown Bridge, Pa., 1992—; mem. Pa. Gov.'s Task Force on Spl. Supplemental Food Program for Women, Infants and Children, Harrisburg, 1981-83, Phila. Bd. Health, 1982-86. Fellow Am. Acad. Pediatrics (emeritus); mem. Phila. Pediatric Soc., Phi Beta Kappa. Democrat. Avocations: opera, skiing, traveling.

KOYYALAMUDI, SUNDARRAO, chemistry educator; b. Kurukurn, India, June 5, 1956; s. Ammiraju and Nagaratnam Koyyalamudi; m. Manikyamma Rani, Dec. 8, 1984; children: Ratnatulas, Lakshmi. BS in Chemistry, Andhra U., 1978; MS in Chemistry, Roorkee U., 1980, PhD, 1985. Jr. rsch. fellow Regional Rsch. Lab., Hyderabad, India, 1981-83, sr. rsch. fellow, 1983-85; postdoctoral rsch. fellow Deakin U., Geelong, Australia, 1987-89; lectr. U. Papua New Guina, 1989–; vis. scientist Seoul (Korea) Nat. U., 1989-93, Tech. U. Nova Scotia, Halifax, Can., 1991-92. Mem. Am. Chem. Soc., Royal Soc. Chemistry, Am. Oil Chemists Soc., Oil Tech. Assn. India. Office: U Papua New Guinea, PO Box 320, Port Moresby Papua New Guinea

KOZAI, YOSHIHIDE, astronomer; b. Tokyo, Apr. 1, 1928; s. Yoshimasa and Sumie (Shidehara) K.; children: Mine Nagai, Aya K., Kay Okuyama. BS, U. Tokyo, 1951, DSc, 1958. Rsch. asst. Tokyo Astron. Observatory, U. Tokyo, 1952-58, assoc. prof., 1963-66, prof., 1966-81, dir., 1981-88; astronomer Smithsonian Astrophy. Observatory, Cambridge, Mass., 1958-62; dir. Nat. Astron. Observatory, Tokyo, 1988—; rsch. assoc. Coll. Observatory Harvard U., 1959-62. Recipient Asahi award Asahi Press, Tokyo, 1963, Dirk Brouwer award Am. Astron. Soc., 1988. Mem. Internat. Astron. Union (pres. 1988-91), Royal Astron. Soc. (assoc. 1969—), Astron. Soc. of Japan (pres. 1982-84), Japan Acad. (Imperial prize 1980). Home: 3-6-27 Takaido-Nishi, 168 Tokyo Japan Office: Nat Astron Obervatory, 2-21-1 Osawa, 181 Tokyo Japan

KOZAK, ROBERT WILLIAM, immunologist; b. N.Y.C., May 3, 1953; s. William and Leona (Donohue) K.; m. Elena Maria Grau; children: Sophia, Jessica, Robert. BS in Biology, Eckerd Coll., 1975; PhD in Immunology and Microbiology, Cornell U., 1982. Postdoctoral fellow in geriatrics and gerontology Cornell U. Med. Coll., N.Y.C., 1982-83; postdoctoral fellow Metabolism br., NIH, Bethesda, Md., 1983-87; sr. staff fellow in cytokine rsch. div. Cytokine Biology, FDA, Bethesda, 1987-90; sr. investigator Lab. Cellular Immunology, divsn. cytokine biology, FDA, Bethesda, 1990-92; acting chief Immunoconjugates Lab., div. Monoclonal Antibodies, FDA, Bethesda, 1992—. Contbr. articles to profl. jours. Recipient Nat. Rsch. Svc. award, 1986-87; fellow Juvenile Diabetes Found., 1986-87, ADA, 1985-86, Cancer Rsch. Inst., 1983-85. Mem. AAAS, Am. Assn. Immunologists, Am. Diabetes Assn. Office: FDA Bldg 29A 8800 Rockville Pike Rm 2A11 Bethesda MD 20892

KOZHEVNIKOVA, IRINA N., physicist; b. Moscow, Russia, Jan. 16, 1959; d. Nikolai N. and Luisa B. Kozhevnikova; 1 child, Elena V. MS in Physics, Moscow State U., 1984; PhD in Physics, Gen. Physics Inst., Moscow, 1987. Scientist Gen. Physics Inst. Acad. Sci. of USSR, Moscow, 1987-91, sr. scientist, 1991-92, mem. sci. com., 1989-92; group leader Tech. U. Denmark, 1993—; assoc. prof. Underwater Acoustics, 1992—. Author: (with other) Nonlinear Optics and Acoustics of Fluids, 1987; contbr. articles to profl. jours. Mem. sci. commn. Central Com. of Youth Communist Orgn., Moscow, 1988; mem. working group on interaction of atmosphere and ocean interface Govt. Com. on Sci. and Tech., 1991-92. Mem. Acoustical Soc. Am., Inst. of Acoustics U.K. Avocations: piano playing, badminton, books in history. Office: Tech U Denmark, Dept Indsl Acoustics Bldg 425, Lyngby 2800, Denmark

KOZHUHAROV, CHRISTOPHOR, physicist; b. Plovdiv, Bulgaria, Jan. 7, 1946; arrived in Fed. Republic Germany, 1970.; s. Vassil and Nadejda (Aceva) Kojouharov; m. Molly Sue Affleck, Aug. 5, 1983. Mgr inz., Politechnika Slaska, Gliwice, Poland, 1969; Dr. rer. nat., Technische U., Munich, 1974. With Technische U., Munich, 1974-78, GSI, Darmstadt, Germany, 1979—. Home: Dieburger Str 77, 64287 Darmstadt Germany Office: GSI, Planckstr, 64291 Darmstadt Germany

KOZICK, RICHARD JAMES, electrical engineering educator; b. Wilkes-Barre, Pa., June 1, 1964; s. Elmer George and Dolores Catherine (Simonovich) K.; m. Patricia Ann Atherton, June 14, 1986; 1 child, Justin Richard. BSEE, Bucknell u., 1986; MSEE, Stanford U., 1988; PhD in Elec. Engring., U. Pa., 1992. Cert. engr.-in-tng. Mem. tech. staff AT&T Bell Lab., Whippany, N.J., 1986-89, cons., 1989-90; rsch. fellow U. Pa., Phila., 1989-92; mem. tech. staff AT&T Bell Lab., Holmdel, N.J., 1992-93; asst. prof. elec. engring. Bucknell U., Lewisburg, Pa., 1993—; cons. AT&T Bell Lab., Whippany, 1989-90. Contbr. articles to profl. jours. Recipient Grad. fellowship USAF, 1989-92, Oliver J. Decker prize Bucknell U., 1986, George A. Irland prize Bucknell U., 1986, George Phillips award Bucknell U., 1983. Mem. IEEE, Sigma Xi, Tau Beta Pi. Achievements include development of new methods for imaging with sparse sensor arrays; unified framework for decorrelating coherent signals in adaptive beam forming. Home: 108 Miner St Wilkes-Barre PA 18702 Office: Bucknell U Elec Engring Dept Lewisburg PA 17837

KOZIOŁ, MICHAEL JOHN, biochemist, researcher; b. Southington, Conn., July 12, 1951; s. Stanley Michael and Janina Mary (Piela) K. BA in Chemistry Biology summa cum laude, St. Michael's Coll., Winooski, Vt., 1973; PhD, U. Oxford, Eng., 1980. Air pollution researcher Botany Sch. U. Oxford, Eng., 1980-85; head sci. support labs. Latinreco, S.A. (Nestle R&D Co.), Quito, Ecuador, 1985—; mem. botanical sci. delegation to China, People to People, 1988, to USSR, 1990, Spokane, Wash.; invited lectr. U. San Francisco de Quito, 1990—. Author: ABC de la Nutrición; editor: Gaseous Air Pollutants and Plant Metabolism. Recipient Rhodes scholarship Oxford, Eng., 1973, Conservation fellowship U.S. Nat. Wildlife Fedn., Am. Petroleum Inst., Oxford, 1976, EPA Cephalosporin Rsch. fellowship, Oxford U., 1983. Mem. N.Y. Acad. Scis., Am. Chem. Soc., Am. Soc. Plant Physiologists, Assn. Official Analytical Chemists, Inst. Biology. Achievements include rsch. in plant stress metabolism, especially in response to atmospheric pollutants; nutritional evaluation and reintroduction of native lost crops; ethnobotany and the identification of naturally ocurring antimicrobial, antitox. Office: Latinreco SA, Casilla 17 110 6053, Quito Ecuador

KOZLOWSKI, STEVE W.J., organizational psychologist; b. Providence, R.I., June 25, 1952; s. Stephen S. and Helena (Szczepan) K.; m. Georgia T. Chao, May 1, 1982. BA in Psychology, U. R.I., 1976; MS in Psychology, Pa. State U., 1979, PhD in Psychology, 1982. Asst. prof. Mich. State U., Lansing, Mich., 1982-87; assoc. prof. Mich. State U., Lansing, 1987—; cons. various corps., military rsch. labs., 1986—. Contbr. 22 articles and book chpts. Recipient fellowship Naval Personnel Rsch. & Devel. Ctr., San Diego, 1989, Naval Tng. System Ctr., Orlando, Fla., 1990. Mem. Acad. Mgmt., Soc. Indsl. and Orgnl. Psychology (editor TIP 1989-92), Mich. Assn. Indsl. and Orgnl. Psychology (mem. at large 1988-92). Office: Mich State U Dept Psychology Lansing MI 48824-1117

KOZMETSKY, GEORGE, computer science educator; b. Seattle, Oct. 5, 1917; s. George and Nadya (Omelan) K.; m. Ronya Keosiff, Nov. 5, 1943; children: Gregory Allen, Nadya Anne (Mrs. Michael Scott). B.A., U. Wash., 1938; M.B.A., Harvard U., 1947, D.C.S., 1957. Instr. Harvard U., 1947-50; asst. prof. Carnegie-Mellon U., Pitts., 1950-52; mem. tech. staff Hughes Aircraft Co., Los Angeles, 1952-54; dir. computer, controls lab. Litton Co., Los Angeles, 1954-59; v.p., asst. gen. mgr. electronic equipment div. Litton Co., 1959-60; exec. v.p. Teledyne Corp., Beverly Hills, Calif., 1960-66; prof. mgmt. and computer sci., dean Coll. Bus. Adminstrn. and Grad. Sch. Bus., U. Tex. at Austin, 1966-82, exec. assoc. for econ. affairs univ. system, 1966—; bd. dirs. LaQuinta, Hydril, Teledyne Corp., Dell Computer Corp.; Leatherbee lectr. Harvard U., 1967; vis. scholar U. Wash., 1968, Walker-Ames prof., 1970. Author: Financial Reports of Labor Unions, 1950, (with Simon and Guetzkow) Centralization Versus Decentralization in Organizing the Controller's Department, 1954, (with Paul Kircher) Electronic Computers and Management Control) 1956, (with Ronya Kozmetsky) Making It Together, 1981, Transformational Management, 1985; (with Gill and Smilor) Financing and Managing Fast-Growth Companies, 1985, Creating the Technopolis, 1988; (with Matsumoto and Smilor) Pacific Cooperation and Development, 1988, (with Peterson and Albaum) Modern American Capitalism, 1990. With AUS, 1942-45. Decorated Silver Star, Bronze Star with oak leaf cluster, Purple Heart; recipient of Nat. Medal of Tech., Nat. Sci. Found., 1993. Fellow AAAS; mem. AICPA, Inst. Mgmt. Sci. (chmn. bd., pres.), Assn. Advancement of Med. Instrumentation, Brit. Interplanetary Soc., Am. Soc. Oceanography. Home: PO Box 2253 Austin TX 78768-2253

KRAFT, ROBERT PAUL, astronomer, educator; b. Seattle, June 16, 1927; s. Victor Paul and Viola Eunice (Ellis) K.; m. Rosalie Ann Reichmuth, Aug. 28, 1949; children—Kenneth, Kevin. B.S., U. Wash., 1947, M.S., 1949; Ph.D., U. Calif.-Berkeley, 1955. Postdoctoral fellow Mt. Wilson Obs., Carnegie Inst., Pasadena, Calif., 1955-56; asst. prof. astronomy Ind. U., Bloomington, 1956-58, Yerkes Obs., U. Chgo., Williams Bay, Wis., 1958-59; staff Hale Obs., Pasadena, 1960-67; prof., astronomer Lick Obs., U. Calif., Santa Cruz, 1967-92; astromoner, prof. emeritus, 1992—; acting dir. Lick Obs., 1968-70, 71-73, dir., 1981-91; dir. U. Calif. Observatories, 1988-91; chmn. Fachbeirat, Max-Planck-Inst., Munich, Fed. Republic Germany, 1978-88; bd. dirs. Cara corp. (Keck Obs.), Pasadena, 1985-91; bd. dirs. AURA. Contbr. articles to profl. jours. Jila vis. fellow U. Colo., Nat. Bur. Standards, Boulder, 1970; Fairchild scholar Calif. Inst. Tech., Pasadena, 1980, Tinsley prof. U. Tex., 1991-92. Mem. Nat. Acad. Scis., Am. Acad. Arts and Scis., Am. Astron. Soc. (pres. 1974-76, Warner prize 1962), Internat. Astron. Union (v.p. 1982-88, pres.-elect 1994-97, pres. 1997-2000), Astron. Soc. Pacific (bd. dirs. 1981-87). Democrat. Unitarian. Avocations: contract bridge; art appreciation; classical music; opera; eonology. Office: U Calif Lick Observatory Santa Cruz CA 95064

KRAINER, EDWARD FRANK, engineering executive; b. Milw., Apr. 22, 1939; s. Edward Frank Sr. and Vera Ann (Leggett) K.; m. Louise Mary Protzmann, Aug. 21, 1965; children: Laura Ann, Brian Edward. BSChemE, U. Wis., 1962. Sales account rep. Union Carbide Corp., Cleve., 1965-75; supply mgr. hydrocarbons Union Carbide Corp., N.Y.C., 1976-79; with dept. chem. purchasing Union Carbide Corp., Danbury, Conn., 1979-84; dir. sales indsl. resins div. Borden Inc., Columbus, Ohio, 1984-86; gen. mgr. indsl. resins div. Borden Inc., Louisville, 1986—. Vestryman, warden Christ Ch., Redding, Conn., 1983; tchr. Christian edn. St. Mark's Ch., Columbus, 1985-86; vestryman, asst. treas. St. Luke's Ch., Louisville, 1990-92. 1st lt. U.S. Army, 1962-65. Mem. Jefferson Club, Valhalla Golf Club. Republican. Episcopalian. Home: 513 Woodlake Dr Louisville KY 40245-5120 Office: Borden Inc Indsl Resins 610 Meidinger Tower Louisville KY 40202-3443

KRAKAUER, HENRY, physics educator; b. Regensberg, Germany, Feb. 14, 1947; came to U.S., 1953; s. Mark and Sara K.; m. Sarah Yael, Oct. 16, 1971; children; Ilana, Mark, Benjamin. BA, Rutgers U., 1969; PhD, Brandeis U., 1975. Instr. W.Va. U., Morganstown, W.Va., 1975-77; postdoctoral fellow Northwestern U., Evanston, Ill., 1977-80; asst. prof. Coll. of William and Mary, Williamsburg, Va., 1980-84; assoc. prof. Coll. of William and Mary, Williamsburg, 1984-90, prof., 1990—. Contbr. numerous articles to profl. publs. Mem. Am. Phys. Soc., Materials Rsch. Soc. Office: Coll of William and Mary Dept Physics Williamsburg VA 23187

KRALEWSKI, JOHN EDWARD, health service administration educator; b. Durand, Wis., May 20, 1932; s. Joseph and Esther (Hetrick) K.; m. Marjorie L. Gustafson; Apr. 22, 1957; children: Judy, Ann, Sara. BS in Pharmacy, U. Minn., 1956, MHA, 1962, PhD, 1969. Asst. prof. U. Minn., Mpls., 1965-69, prof., 1978—; prof. U. Colo., Denver, 1969-78. Contbr. articles to profl. jours. 1st lt. USAF, 1957-60. Kellogg fellow Kellogg Found., 1962-65, Valencia (Spain) Acad. Medicine fellow, 1993. Mem. APHA, Assn. Health Svcs. Rsch. Avocation: oenology. Office: U Minn 420 Delaware St SE 15-205 Minneapolis MN 55455

KRAMAN, STEVE SETH, physician, educator; b. Chgo., Aug. 30, 1944; s. Julius and Ruth (Glassner) K.; m. Lillian Virginia Casanova, May 29, 1972 (div. Apr. 1991); children: Theresa, Pilar, Laura. BS, U. P.R., 1968, MD, 1973. Asst. prof. U. Ky., Lexington, 1978-84, assoc. prof., 1984-90, prof., 1990—; chief of staff VA Med. Ctr., Lexington, 1986—. Contbr. articles to profl. jours. Mem. Am. Coll. Chest Physicians. Achievements include patent for simple capsule pnenmograph. Office: VA Med Ctr 2250 Leestown Rd Lexington KY 40511

KRAMER, DAVID ALAN, biomathematician; b. Bklyn., Sept. 14, 1962; s. Murray and Zelda Reneé (Rabinowitz) K.; m. Janne Abullarade, Aug. 14, 1988. BS, Cornell U., 1984; MS, N.C. State U., 1987, PhD, 1993. Instr. Dept. Statis., N.C. State U., Raleigh, 1990-93; postdoctoral fellow Chem. Industry Inst. Toxicology, 1993—. Contbr. articles to profl. jours. Mem. Entomol. Soc. Am., Soc. for Risk Analysis, Biometric Soc., Sigma Xi, Phi Kappa Phi. Home: C-1 Fenway Ct Carrboro NC 27510 Office: CIIT PO Box 12137 6 Davis Dr Research Triangle Park NC 27709

KRAMER, GEOFFREY PHILIP, psychology educator; b. Bloomington, Ind., Oct. 23, 1953; s. Richard Niel and Josephine Angela (Defrank) K.; m. Michelle M. King, Apr. 27, 1985; children: Rachel Devin, Evan Robert. MA, Ctrl. Mich. U., 1977; PhD, Mich. State U., 1991. Staff psychologist State Prison of So. Mich., Jackson, 1978-84; ltd. lic. psychologist Ctr. for Health Psychology, Okemos, Mich., 1985-87; grad. rsch. asst.

Mich. State U., East Lansing, 1985-90; asst. prof. psychology Ind. U., Kokomo, 1990—. Contbr. articles to profl. jours. Rsch. grantee Mich. Criminal Jury Instrns. Com., Lansing, 1988. Mem. Am. Psychol. Soc., Am. Psychology-Law Soc., Midwest Psychol. Assn. Avocations: backpacking, guitar, tennis, home remodeling. Office: Ind U Kokomo 2300 S Washington Kokomo IN 46904

KRAMER, GORDON, mechanical engineer; b. Bklyn., Aug. 1937; s. Joseph and Etta (Grossberg) K.; m. Ruth Ellen Harter, Mar. 5, 1967 (div. June 1986); children: Samuel Maurice, Leah Marie; m. Eve Burstein, Dec. 17, 1988. BS Cooper Union, 1959; MS, Calif. Inst. Tech., 1960. With Hughes Aircraft Co., Malibu, Calif., 1959-63; sr. scientist Avco Corp., Norman, Okla., 1963-64; asst. div. head Batelle Meml. Inst., Columbus, Ohio, 1964-67; sr. scientist Aerojet Electrosystems, Azusa, Calif., 1967-75; chief engr. Beckman Instrument Co., Fullerton, Calif., 1975-82; prin. scientist McDonnell Douglas Microelectronics Co., 1982-83, Kramer and Assocs., 1983-85; program mgr. Hughes Aircraft Co., 1985—; cons. Korea Inst. Tech. NSF fellow, 1959-60. Mem. IEEE. Democrat. Jewish. Home: 153 Lakeshore Dr Rancho Mirage CA 92270 Office: 2000 E El Segundo Blvd El Segundo CA 90245-4599

KRAMER, HORST EMIL ADOLF, physical chemist; b. Friedrichshafen, Baden-Württemberg, Germany, Apr. 20, 1936; s. Max and Maria Magdalena (Kaufmann) K.; m. Ingeborg Maria Giesen, July 30, 1964; 1 child, Boris W.W. Diploma chemistry, U. Stuttgart, Baden-Wuerttemberg, 1961, D Natural Scis., 1964, Habilitation, 1970. Rsch. asst. Inst. Phys. Chemistry U. Stuttgart, 1966-70, lectr., 1970-74; prof. phys. chemistry Inst. Phys. Chemistry, Stuttgart, 1974—; dean faculty chemistry U. Stuttgart, 1988-90,. Assoc. editor Photochemistry and Photobiology, 1978-80; mem. editorial bd. Photobiochemistry and Photobiophysics, 1983-86; contbr. articles to profl. jours., chpts. to books. Mem. Gesellschaft Deutscher Chemiker, Fachgruppe Photochemie der Gesellschaft Deutscher Chemiker. Avocations: history, tennis, sailing, model railway. Home: Falkenweg 8, D-71126 Gaeufelden Germany Office: U Stuttgart Inst Phys Chemistry, Pfaffenwaldring 55, D-70569 Stuttgart Germany

KRAMER, JAY HARLAN, physiologist, biochemist, researcher, educator; b. Bklyn., Dec. 26, 1952; s. Albert and Blossom K.; m. Aisar Atrakchi, Apr. 18, 1993. BA with honors, Northeastern U., 1976; MS, Lehigh U., 1979, PhD, 1982. Clin. lab. technician Boston Med. Lab., Waltham, Mass., 1974-75; rsch. asst. Lehigh U., Bethlehem, Pa., 1979-81; rsch. assoc. Med. Coll. Va., Richmond, 1982-83; sr. rsch. assoc. Okla. Med. Rsch. Found., Oklahoma City, 1983-85; rsch. assoc. George Washington U., Washington, 1985-86, asst. rsch. prof. medicine, 1986-90, assoc. rsch. prof., 1990—, adj. assoc. prof. physiology, 1991—; lectr. physiology George Washington U., Washington, 1987-89; cons. Squibb & Sons, Princeton, N.J., 1989, mem. George Washington U. Instl. Animal Care and Use Com., 1988—. Contbr. more than 36 articles to profl. jours.; article referee profl. jours. Mem. basic sci. faculty assembly coun. George Washington U., 1992—. Grad. sch. scholar Lehigh U., 1980; named one of Outstanding Young Men of Am., Jaycees, 1981, 82. Mem. Am. Heart Assn., Am. Physiol. Soc., N.Y. Acad. Scis. (invited speaker 1993, presenter various nat. scientific meetings), Internat. Soc. for Heart Rsch., Internat. Soc. for Free Radical Rsch., Soc. for Exptl. Biology and Medicine, Acad. Honor Soc., Phi Sigma. Achievements include first to demonstrate relationship between toxic free radical prodn. and severity of ischemia in heart; first to demonstrate superoxide anion prodn. in postischemic heart using ESR spin trapping; first to demonstrate free radical prodn. in regionally ischemic canine and post-ischemic swine heart models; developed non-invasive ESR spin trapping technique for free radical detection; demonstrated occurrance of potentially toxic free radicals in human heart following open heart surgery. Office: George Washington U Dept Medicine 2300 I St NW Washington DC 20037

KRAMER, REX W., JR., former naval officer, business executive; b. L.A., June 22, 1934; s. Rex W. and Ruth (Roseberry) K.; m. Karen Blanchard, May 16, 1959; children: Timothy, E. Cecil. AB, Stanford U., 1956; MA, Cath. U. of Am., 1977. CPA. Div. officer, comdr. USN, 1956-60; student U.S. Navy Nuclear Power Sch., New London, Conn., 1960-61, instr., 1961-62; engring. adminstr. U.S. Navy Nuclear Submarine, 1962-71; commanding officer U.S. Navy Nuclear Submarine, Pearl Harbor, Hawaii, 1971-74; ops. analyst Staff Chief of Naval Ops., Washington, 1975-78; licensing supr. Palo Verde Nuclear Gen. Sta., Wintersburg, Ariz., 1978-84; pres. Southwestern Vactor Svc., Phoenix, 1986—; mem. Ariz. Hazwaste Soc., Phoenix, 1988—. Author: (textbook) Nuclear Reactor Kinetics, 1962. Pres. Kiwanis Club of Litchfield, Litchfield Park, Ariz., 1984-85; vestryman St. Peter's Episc. Ch., Litchfield Park, 1990-92. Achievements include development of teaching methodology and textbook used by USN to teach non-steady state light water moderated reactor behavior; teaching of Nuclear Reactor Theory at graduate level at U.S. Naval Nuclear Power Sch. Office: Southwestern Vactor Svc Inc PO Box 520 Litchfield Park AZ 85340

KRAMER, RICHARD HARRY, JR., biomedical engineer; b. Reading, Pa., Jan. 28, 1953; s. Richard H. Sr. and Arlene M. (Loy) K.; m. P. Lynn Kreidler, Oct. 28, 1978; children: Thane M., Garrett E. BSEE, Pa. State U., 1974, M in Engring. Sci., 1979. Engr. GE Space Ctr., Valley Forge, Pa., 1974-77; sr. elec. engr. Chemcut Corp., State College, Pa., 1978-84; sr. and chief elec. engr. E.R. Squibb and Sons Inc., Princeton, N.J., 1984-89; prin. devel. engr. Bristol-Myers Squibb Co., Princeton, 1989-90, assoc. dir. instrument devel., 1990—. Mem. IEEE, Assn. for Advancement of Medical Instrumentation, Sigma Nu. Achievements include developing Cardiogen radiodiagnostic infusion system design; researched biomedical electronics, psysiological sensing, application of smart materials, and lab robotics. Home: 158 Crest Rd Wrightstown PA 18940 Office: Bristol-Myers Squibb Co Pharma Rsch Inst PO Box 4000 MS-K 1704 PO Box 4000 MS-K 1703A Princeton NJ 08943-4000

KRAMER, THOMAS ROLLIN, automation researcher; b. N.Y.C., Sept. 3, 1943; s. Edward David Kramer and Esther Gertrude (Kraatz) Reddrop; m. Anne K. Welsh, Sept. 12, 1970 (div. 1982); 1 child, Jeanne Elizabeth; m. Teresa Rae Sessions, Oct. 27, 1983; children: Jennifer Katherine, Kimberly Jane. BA, Swarthmore Coll., 1965; MA, Duke U., 1968, PhD, 1971. Tchr. U.S. Peace Corps., Ghana, 1965-67; exec. sec. math. div. NRC, Washington, 1971-75; staff dir., subcom. Com. on Sci. & Tech. U.S. Ho. of Reps., Washington, 1975-84; guest researcher Nat. Inst. Stds. & Tech., Gaithersburg, Md., 1984—. Contbr. articles to profl. jours. Mem. AAAS, Am. Math. Soc. Achievements include devel. of integrated software system for computer-aided design and mfg. using feature-based design; rsch. in math., artificial intelligence, automated mfg. Office: Nat Inst Stds & Tech Bldg 220 Rm B-124 Gaithersburg MD 20899

KRAMPITS, MARK WILLIAM, power management software specialist; b. Waltham, Mass., Aug. 27, 1955; s. William George and Eleanore Julia (Foley) K.; m. Myong S. Hong, June 7, 1986; children: Jessica Ann, Jennifer Myong. BSEE, Portland State U., 1990. Engr. Siemens/Nixdorf Computer Corp., Denver, 1977-80, Seismograph Svc. Corp., Denver, 1980-83, Northrop Corp., Hawthorne, Calif., 1983-86, Intel Corp., Hillsboro, Oreg., 1990—. Mem. IEEE (vice-chmn. Portland State U. chpt. 1989-90), Eta Kappa Nu (chpt. sec. 1989-90). Democrat. Roman Catholic. Home: 6916 SE 68th Ct Hillsboro OR 97123-6132 Office: Intel Corp 5200 NE Elam Young Pky Hillsboro OR 97124

KRANTZ, GERALD WILLIAM, entomology educator; b. Pitts., Mar. 12, 1928; s. Harry Clifford and Anna (Schrager) K.; m. Vida June Kersch, Mar. 1, 1955; children: Wayne M., Georgia L., Valerie J. BS, U. Pitts., 1951; PhD, Cornell U., 1955. From asst. prof. to prof. Oreg. State U., Corvallis, 1955—, chmn., 1991—; exec. com. Internat. Congress Acarology, 1982-90; program officer NSF, Washington, 1984-85, Associé du Muséum Nat. d'Hist., Paris, 1984. Author: Manual of Acarology, 1970, 86; contbr. articles to profl. jours.; assoc. editor Internat. Jour. Acarology, 1992—; mem. bd. editors Exptl. & Appl. Acarology, 1984—, Acarologia, 1980—, Rev. Zool. Africa, 1984—. With USMC, 1946-48. Recipient Berlese award, 1979, Gloeckner Found. Rsch. award, 1979-82, Gilfillan award, 1988; NSF, NIH, USDA grantee, 1961-89. Mem. Entomological Soc. Am., Acarological Soc. Am. (bd. govs., chmn. 1971-73), Sigma Xi. Office: Oreg State U Corvallis OR 97331-2907

KRANTZ, JOHN HOWELL, psychology educator; b. Columbus, Ohio, Feb. 18, 1960; s. Albert Russell and Bargara Sewell (Hoyt) K.; m. Margaret Katharine Edsell, Nov. 16, 1959; children: Michael, Jennifer, Daniel. BA, St. Andrews Coll., 1982; MS, U. Fla., 1985, PhD, 1988. Rsch. assoc. Honeywell, Inc., Mpls., 1987-88; sr. rsch. scientist Honeywell, Inc., Phoenix, 1988-90; asst. prof. Hanover (Ind.) Coll., 1990—. Contbr. articles to Human Factors, Jour. of Optical Soc., Investigative Opthalmology and Visual Sci., Jour. Social Psychology. Fellow U. Fla., 1982, NSF, 1983. Mem. Am. Psychol. Soc., Assn. for Rsch. in Vision and Opthalmology, Soc. for Info. Display, Ind. Acad. Sci. (vice-chmn. psychology sect. 1992, chmn. 1993), Phi Beta Kappa. Achievements include development of model of display visibility for displays under dynamic viewing conditions; rsch. in optimum selection of number of gray sides vs. pixel density for color matrix display, rsch. demonstrating readjustments of local sign during saccadic eye movements. Office: Hanover Coll PO Box 108 Hanover IN 47243

KRAPE, PHILIP JOSEPH, chemist; b. Buffalo, Aug. 13, 1948; s. Philip R. and Doris H. (Moore) K.; m. Deborah Washburn, Mar. 18, 1978; children: Chase, Darren, Evan, Whitney. BA in Chemistry, U. Pa., 1970. Sr. bus. analyst fin. DuPont Merck Pharm. Co., Wilmington, Del., 1984-89, sr. mgr. new product devel., 1989-91, assoc. dir. bus. devel., 1991—. Office: DuPont Merck Pharm Co Barley Mill Plz Wilmington DE 19880-0025

KRAPPINGER, HERBERT ERNST, economist; b. St. Veit, Kärnten, Austria, Jan. 27, 1950; s. Herbert and Gerharda (Tomaschitz) K.; m. Doris Evelyn Meller, May 16, 1987; children: Julia, Marion. M in Econs., U. Vienna, Austria, 1972; D in Econs., Sch. Econs., Vienna, Austria, 1975; Fulbright scholar postgrad., U. Oregon and U. Chgo., 1976. Rsch assoc. U. Chgo., 1976-77; sec. Minister of State for Econ. Affairs, Vienna, 1978; dep. dir. for foreign econ. policy coord. Fed. Chancellery Dept. for Econ., Vienna, 1987-91; dir. for multilateral econ. affairs Fed. Chancellery Dept. for Econs., Vienna, 1991—; lectr. econs. Vienna Sch. Econs. and Bus. Adminstrn., 1987—; sr. econ. advisor UNO-Econ. Commn. for Europe, Geneva, 1989—. Recipient various scholarships and grants, 1969-75, Fulbright Scholarship, 1975. Mem. Austrian Econ. Assn., Austrian Assn. for Foreign Policy and Internat. Relations., Austrian Assn. for Pub. Fin. #D. Avocations: piano, literature, sports, philosophy, foreign cultures, painting.

KRASNER, PAUL R., psychologist; b. N.Y.C., May 10, 1951; s. Ernest and Elisa A. (Abrams) K.; m. Trudi Robin Klarman, June 17, 1973; children: Lori Alison, Ami Aileen, Michelle Janie. MA, Adelphi U., 1976, PhD, 1979. Staff psychologist N.C. Dept. Human Resources, Butner, 1978-81; dir. psychology Thoms Hosp., Ashville, N.C., 1981-85; clin. dir. Cary (N.C.) Psychology, 1985—. Contbr. articles to profl. jours. Pres. am. Diabetes Assn., N.C., Raleigh, 1985; bd. dirs. Beth Shalom, 1987—. Office: Cary Psychol Inc 1145A Executive Circle Cary NC 27511

KRASNEY, ETHEL LEVIN, research chemist; b. Norfolk, Va., June 30, 1961; m. Philip Albert Krasney, Aug. 7, 1983. BA in Chemistry, U. Va., 1982; MS in Chemistry, Vanderbilt U., 1985. Rsch. intern U. Va., Charlottesville, 1980-82; chemistry tchr. Harpeth Hall Sch., Nashville, 1985-88; assoc. chemist SmithKline Beecham, King of Prussia, Pa., 1988-91; sr. rsch. chemist Sterling Winthrop, Rensselaer, N.Y., 1991—; vis. scienitst educator vis. scientist teaching program SmithKline Beecham, Phila., 1988-91; presenter in field. Contbr. articles to profl. jours. Inorganica Chimica Acta, Jour. Cell Biology. B'nai Brith Undergrad. scholar, 1978, Bayley Mus. Undergrad. scholar U. Va., 1978-82; grad. fellow Vanderbilt U., 1983-85. Mem. Am. Assn. Pharm. Scientists, Am. Chem. Soc. (continuing edn. com. Phila. chpt. 1990-91), Nat. Sci. Tchrs. Assn., Sigma Delta Tau, Alpha Chi Sigma. Democrat. Jewish. Achievements include patent for four melphalan/cis-platinum drug complexes; research in the development of pharmaceutical procedures by HPLC, CZE and similar chromatographic techniques, synthesis of various cis-platinum and melphalan anti-cancer drugs; patentee for four melphalan/CIS platinum drug complexes. Home: 113 Woodin Rd # B Clifton Park NY 12065-6104 Office: Sterling Winthrop 81 Columbia Turnpike Rensselaer NY 12144

KRASNOW, MAURICE, psychoanalyst, educator; b. N.Y.C., Jan. 3, 1944; s. Hershel and Sylvia (Lerner) K.; m. Susan D. Godden, Feb. 28, 1969 (div. 1989); 1 child, Joshua Samuel. BA, U. Ariz., 1968; MA, New Sch. for Social Rsch., 1971; PhD, Columbia Pacific U., 1983. Cert. psychoanalyst. Psychotherapist Payne Whitney Psychiat. Clinic, N.Y.C., 1964-67; child psychotherapist So. Ariz. Mental Health Clinic, Tucson, 1967-68; psychotherapist, group therapist Mt. Sinai Hosp., N.Y.C., 1969-72, Fifth Ave. Ctr. for Psychotherapy, N.Y.C., 1971-74; psychotherapist C.G. Jung Inst. Clinic, N.Y.C., 1976-79; pvt. practice Jungian analyst N.Y.C. and Wilton, Conn., 1972—; mem. exec. com. C.G. Jung Found. N.Y., N.Y.C., 1984-88, chmn. publs. com. 1984-88, trustee, 1984-89. Chmn. editorial bd. Quadrant jour. 1983-89. 2d lt. U.S. Army, 1968-71. Mem. APA (founder psychologist-psychoanalysts' forum sect. 5, div. psychoanalysis 1986, chmn. bylaws com. psychologist-psychoanalysts' forum 1986-88), APGA, C.G. Jung Inst. N.Y. (instr., curriculum and faculty 1984-91, supervising analyst 1982—, faculty 1983—), N.Y. Assn. for Analytical Psychology (chmn. standing com. ethics and profl. practices 1981-86), Internat. Assn. Analytical Psychology, Nat. Assn. for Advancement of Psychoanalysis (v.p. 1983-87, pres. 1987-88, legis. com. 1984-87, profl. practices com. 1985-87, Profl. Achievement award 1990), N.Y. Acad. Scis., Joint Coun. for Mental Health Svcs. Office: 334 W 86th St Ste 1A New York NY 10024

KRASNY, HARVEY CHARLES, research biochemist, medical research scientist; b. High Point, N.C., July 27, 1945; s. Morris Theodore and Elizabeth (Nurkin) K. BS, Lynchburg (Va.) Coll., 1967; MS, U. N.C., 1969, PhD, 1976. Rsch. biochemist Burroughs Wellcome Co., Research Triangle Park, N.C., 1969-83, sr. rsch. biochemist, 1983—. Contbr. articles to profl. jours. Mem. Am. Soc. for Clin. Pharmacology and Therapeutics, Am. Soc. for Pharmacology and Exptl. Therapeutics, Soc. Toxicology, Sigma Xi (pres. chpt. 1993—). Achievements include study of pharmacokinetics, bioavailability and drug metabolism of the antiviral agent acyclovir (Zovirax) in humans; pre-clinical development of AZT (Retrovir). Home: 120 Woodbridge Ln Chapel Hill NC 27514

KRASSER, HANS WOLFGANG, physicist; b. Coburg, Bavaria, Germany, May 13, 1937; s. Johann Georg and Irmgard Klara (v. Schultes) K.; m. Ingeborg Helga Döller; children: Heike, Jürgen, Inge. Grad., U. Würzburg, 1964; D, U. Bonn., 1969. Group leader Chem. Inst. KFA, Jülich, Germany, 1970-74, Inst. Neutron Scattering KFA, Jülich, Germany, 1974-86, Inst. Electronic Properties KFA, Jülich, Germany, 1986—. Inventor/patentee in field. Avocations: tennis, badminton, classical and jazz music. Home: An der Lünette 4c, Jülich Germany Office: Rsch Ctr Jülich, Leo Brandt Strasse, 517 Jülich Germany

KRASZEWSKI, ANDRZEJ WOJCIECH, electrical engineer, researcher; b. Poznan, Poland, Apr. 22, 1933; arrived in Can., 1980.; s. Tadeusz Jozef and Waleria Barbara (Pietrzyk) K.; m. Janina Wiktoria Okula, June 24, 1956; children: Marcin Jan, Andrzej Maria (dec.). BSEE, Tech. U., Warsaw, Poland, 1954, MSEE, 1958; DSc in Tech. Sci., Polish Acad. Scis., Warsaw, 1973. Rsch. engr. Indsl. Inst. for Telecoms., Warsaw, 1953-58; sr. rsch. engr. Lamina Works, Iwiczna, Poland, 1958-60; asst. prof., dept. electronics Tech. U., 1960-62; microwave labs. mgr. UNIPAN Scientific Instruments, Warsaw, 1963-72; co-founder, mgr. microwaves WILMER Instruments, Warsaw, 1972-80; assoc. prof., dept. electrical engring. U. Manitoba, Winnipeg, Manitoba, Can., 1976-77; vis. prof., dept. electrical engring. U. Ottawa, Ontario, Can., 1980-86; rsch. electronics engr., Agrl. Rsch. Svc. U.S. Dept. Agriculture, Athens, Ga., 1987—. Author: Microwave Switching Circuits, 1966, Microwave Gas Discharge Devices, 1967; contbr. chpts. to books, articles to profl. jours.; pub. numerous papers. Recipient 3 Gold medals Internat. Fair, Leipzig, Germany, 1964, 65, 77, Indsl. Rsch. award Indsl. Rsch. Inst., Chgo., 1976, Scientific award Ministry Sci. and Tech., Warsaw, 1977, Sci. State Prize State Coun., Warsaw, 1980; Rsch. in Engring. grantee Natural Sci. and Engring. Rsch. Coun., Ottawa, 1985, Rsch. grantee Binational Agrl. Rsch. and Devel., Washington, 1989. Mem. IEEE (sr. 1989—, editorial bd. 1992—), Internat. Microwave Power Inst. (editorial bd. 1975-84), Material Rsch. Soc., N.Y. Acad. Scis., Sigma Xi. Roman Catholic. Achievements include 19 patents on moisture content determination using microwave parameter measurements; pioneer works in microwave aquametry, experimental RF&MW dosimetry, permittivity of tissues in-vivo, DNA solutions, applications of microwave techniques for measurement of nonelectrical quantities; research in plant structure and composition using electromagnetic waves, multi-parameter microwave measurements for density-independent moisture content determination, microwave switching circuits, microwave gas discharge devices. Office: US Dept Agriculture Russell Rsch Ctr 950 Coll Station Rd Athens GA 30613

KRATZ, RUEDIGER, neurologist, researcher; b. Rudolstadt, Germany, Jan. 12, 1943; came to U.S., 1970; s. Robert Hermann Wolfgang and Hildegard Marie (Kruger) K.; m. Ai-Wen Josephine Wu; children: Stefan Theodor Sklode, Johannes Ruediger Ulrich. MD, Univ. Chgo., 1973; MDiv, Friedrich Wilhelm Universitat, Bonn, Germany, 1967. Diplomate Am. Bd. Psychiatry, Am. Bd. Neurology. Internship in medicine Univ. Chgo., 1973-74; residency in neurology Washington Univ., St. Louis, 1974-77; fellow neuromuscular diseases Washington Univ, Nat Inst. Health, St. Louis, Bethesda, 1977-80; clinical asst. prof. medicine Georgetown Univ., Washington, 1980-89, clinical assoc. prof. neurology, 1989—. Contbr. articles to profl. jours.; author chapter in book. Dir. Muscular Dystrophy Clinic, Georgetown, 1988—; medical adv. Muscular Dystrophy Assn. Recipient clinic grant Muscular Dystrohpy Assn., 1992. Fellow Am. Acad. Neurology; mem. Medical Soc. Lutheran. Office: Neuroscience Assn PO Box 924 Mc Lean VA 22101-0924

KRÄTZEL, EKKEHARD, mathematics educator; b. Stassfurt-Leopoldshall, Germany, June 27, 1935; s. Karl and Herta (Wilke) K.; m. Ilse Wille, Dec. 13, 1958; children: Carsten, Katrin, Andreas. D, Friedrich-Schiller U., Jena, Germany, 1963, Qualification Univ. Lectr., 1965. Sci. asst. Friedrich-Schiller U., Jena, 1959-66, lectr., 1966-69, prof. number theory, 1969-92; vis. prof. U. Freiburg and Vienna, 1993. Author: Zahlentheorie, 1981, Lattice Points, 1988, (with Bernulf Weissbach) H. Minkowski: Ausgewählte Arbeiten zur Zahlentheorie und zur Goemetrie, 1989. Avocations: philately, table tennis. Home: Am Friedensberg 4, 07745 Jena Germany Office: Friedrich Schiller U Faculty Math, Universitäts hochhaus, 17 OG 07245 Jena Germany

KRATZER, GUY LIVINGSTON, surgeon; b. Gratz, Pa., Apr. 24, 1911; s. Clarence U. and Carrie E. (Schwalm) K.; m. Kathryn H. Miller, Jan. 27, 1940; 1 son, Guy Miller. Student, Muhlenberg Coll., 1928-31; M.D., Temple U., 1935; M.S., U. Minn., 1945. Diplomate: Am. Bd. Proctology. Intern Harrisburg Hosp., 1935-36; fellow proctology, surgery Mayo Clinic, 1942-46, fellow surgery, 1949-50; asso. surgeon Pottsville Hosp., 1936-41; asso. proctologist Allentown (Pa.) Hosp., 1946—, mem. tumor clinic, 1955—, chief, dept. proctology, 1958—; mem. cons. staff Sacred Heart Hosp., 1946—, chief dept. colon and rectal surgery, 1974—; clin. asso. prof. surgery Milton S. Hershey Med. Center, Pa. State U., 1972-75, clin. prof., 1975—, cons., 1975—; mem. Pa. Bd. Med. Edn. and Licensure, 1984—. Author: Disease of the Colon and Rectum, 1985; contbr. numerous articles to med. jours. Pres. Lehigh Valley chpt., bd. dirs. Am. Cancer Soc. Recipient Award for Exceptional Svc. and Significant Contbns. Am. Soc. of Colon and Rectal Surgeons, 1982. Mem. Fellow ACS (pres. S.E. Pa. 1965-66), Am. Proctologic Soc., Internat. Coll. Surgeons; mem. Shelter House Soc., Am. Med. Writers Assn., Pa. Proctologic Soc. (past pres.), Pa. Med. Soc., Am. Med. Authors, Lehigh Valley Med. Soc. (past pres.), Allentown C. of C. (gov.), Lions, Union League. Republican. Evangelical. Achievements include rsch. in local anesthesia in anorectal surgery and study with collaboration of Johns Hopkins U. on 4 generations of a family with congenital polyps of colon and rectum. Address: 1447 Hamilton St Allentown PA 18102

KRAUSE, JOSEPH LEE, JR., electrical engineer; b. Williams Bay, Wis., Feb. 14, 1958; s. Joseph L. Sr. and Janet Ruth (Boettcher) K.; m. Anne E. Carey; children: Courtney Carey, Rachel Anne. A.S.E.T. with distinction, Southeast C.C., Milford, Nebr., 1978; BSEE with distinction, U. Nebr., 1985, MS in Mfg. Systems Engring., 1992. Registered profl. engr., Nebr.; lic. comml. pilot. Field svc. technician Westinghouse Electric, Kansas City, Kans., 1978-80, inside salesman, 1980-82; devel. engr. Goodyear Tire and Rubber, Lincoln, Nebr., 1986-89; staff engr. Goodyear Tire and Rubber, Lincoln, 1989—. Regents scholar U. Nebr., 1982. Mem. IEEE, NSPE. Republican. Achievements include patent pending in field. Home: PO Box 80322 Lincoln NE 68501 Office: Goodyear Tire and Rubber 4021 N 56th St Lincoln NE 68501

KRAUSE, KURT LAMONT, biochemistry educator; b El Paso, Tex., Jan. 19, 1956; s. Elmer L. and Caroline L. (Garrison) K. BA in Chemistry summa cum laude, Trinity U., San Antonio, 1977; MD cum laude, Baylor U., 1980; MA in Chemistry, Harvard U., 1983, PhD in Chemistry, 1986. Diplomate Am. Bd. Internal Medicine. Staff physician student health ctr. U. Houston, 1979-80; staff physician geriatrics program Met. State Hosp., 1981-83; asst. physician dept. internal medicine McLean Hosp., 1981-86; affiliated physician Cambridge Psychiat. Assocs., 1984-86; teaching asst., rsch. asst. Harvard U., 1980-85, postdoctoral fellow in chemistry, 1985-86; resident in internal medicine Baylor Coll. of Medicine, Houston, 1986-89, asst. prof. dept. medicine and cell biology, 1989—; asst. prof. biochemistry and chemistry U. Houston, 1989—; attending physician dept. medicine Ben Taub Gen. Hosp., 1989—. Exxon fellow, 1981-82, Kleberg Found. fellow, 1976; Pres.' scholar, 1974-76, Nat. Merit scholar, 1974. Mem. AAAS, Am. Chem. Soc., Am. Crystallographic Assn., Phi Beta Kappa. Office: U Houston Dept Biochem & Biophys Scis 3201 Cullen BCHS Rm 402 Houston TX 77004-5934 also: Baylor Coll Medicine Div Atherosclerosis & Liproprot 6565 Fannin Houston TX 77030

KRAUSE, LOIS RUTH BREUR, chemistry educator, engineer; b. Paterson, N.J., Mar. 26, 1946; d. George L. and Ruth Margaret (Farquhar) Breur; m. Bruce N. Pritchard, Aug. 10, 1968 (div. May 1982); children: John Douglas, Tiffany Anne; m. Robert H. Krause, June 16, 1990. Student, Keuka Coll., 1964-65; BS in Chemistry cum laude, Fairleigh Dickinson U., 1980; post-grad., Stevens Inst. Tech. With dept. R & D UniRoyal, Wayne, N.J., 1966-68, Jersey State Chem. Co., North Haledon, 1968-69, Inmont, Clifton, N.J., 1969; from chemist to sr. analyst Lever Bros., Edgewater, N.J., 1976-80; process engr. Bell Telephone Labs., Murray Hill, N.J., 1980-84, RCA, Somerville, N.J., 1984-86; sr. engr. electron beam lithography ops. Gain Electronics Corp., Somerville, 1986-88; ind. tech. cons. Pritchard Assocs., Budd Lake, N.J., 1988-92; tchr. of math. and scis. Mt. Olive Bd. Edn. (temporary assignments), 1990-92; tchr. chemistry Morris Hills Regional Dist., 1992—; presenter profl. papers for profl. confs. Patentee package design. Troop leader, trainer, cons. Bergen County council Girl Scouts U.S., 1969-80, troop leader Morris Area council, 1980-83, head com. Mt. Olive twp., 1980-81; den leader, den leader coach, trainer Boy Scouts Am., 1973-76. Fellow AICE; mem. IEEE (sr.), NEA, NRA (life), Components, Hybrids and Mfg. Tech. Soc. (semicondr. tech. subcom. Electronic Components Conf. program com. 1981-86), Am. Soc. Quality Control, Soc. Women Engrs., Am. Chem. Soc., Assn. Women in Sci., Nat. Woodlot Owners Assn., N.J. Edn. Assn., Arbor Day Found., N.J. Forestry Assn., Mensa, Marine Corps League Aux., Phi Omega Epsilon. Republican. Episcopalian. Achievements include work in electron beam lithography, statistical process control. Avocations: photography, writing, teaching, forestry and wildlife management. Home and Office: 80 Lozier Rd Budd Lake NJ 07828-1210

KRAUSE, RICHARD MICHAEL, medical scientist, government official, educator; b. Marietta, Ohio, Jan. 4, 1925; s. Ellis L. and Jennie Mae (Waterman) K. B.A., Marietta Coll., 1947, D.Sc. (hon.), 1978; M.D., Case Western Res. U., 1952, D.Sc. (hon.), U. Rochester, 1979, Med. Coll. Ohio, Toledo, 1981, Hahnemann Med. Coll. and Hosp., 1982; LLD (hon.), Thomas Jefferson U., 1982. Rsch. fellow dept. preventive medicine Case Western Res. U., 1950-51; intern Ward Med. Service, Barnes Hosp., St. Louis, 1952-53; asst. resident Ward Med. Service, Barnes Hosp., 1953-54; asst. physician to hosp. Rockefeller Inst., 1954-57, asst. prof., assoc. physician to hosp., 1957-61; prof. epidemiology Sch. Medicine, Washington U., St. Louis, 1962-66; assoc. prof. medicine Sch. Medicine, Washington U., 1962-65, prof. medicine, 1965-66; assoc. prof., physician to hosp. Rockefeller U., 1966-68, prof., sr. physician, 1968-75; dir. Rockefeller U. (Animal Rsch. Ctr.), 1974-75, Nat. Inst. Allergy and Infectious Diseases, NIH, HEW, Bethesda, Md., 1975-84; USPHS surgeon, 1975-77, asst. surgeon gen., 1977-84; dean Emory U. Sch. Medicine, Atlanta, 1984-89, Robert W. Woodruff prof. medicine, 1984-89; mem. program com. Inst. Medicine, 1986-87; sr. sci. adv. Fogerty Internat. Ctr. NIH, Bethesda, 1989—; Bd. dirs. Mo.-St. Louis

Heart Assn., 1962-66, mem. research com., 1963-66; mem. exec. com. council on rheumatic fever and congenital heart disease Am. Heart Assn., 1963-66, chmn. council research study com., 1963-66, mem. assn. research com., 1963-66, mem. policy com., 1966-70; mem. commn. streptococcal and staphylococcal diseases U.S. Armed Forces Epidemiol. Bd., 1963-72, dep. dir., 1968-72; bd. dirs. N.Y. Heart Assn., 1967-73, chmn. adv. council on research, 1969-71, mem. dirs. council, 1973-75; cons. mem. coccal expert com. WHO, 1967—; mem. steering com. Biomed. Sci. Scientific Working Group, WHO, 1978; mem. infectious disease adv. com. Nat. Inst. Allergy and Infectious Disease, NIH, 1970-74; bd. dirs. Royal Soc. Medicine Found., Inc., 1971-77, treas., 1973-75; bd. dirs. Allergy and Asthma Found. Am., 1976-77, Lupus Found. Am., 1977-79. Assoc. editor: Jour. Immunology, 1963-71; sect. editor: Viral and Microbial Immunology, 1974-75; editor: Jour. Exptl. Medicine, 1973-75; adv. editor, 1976-84; mem. editorial bd. Bacteriological Revs, 1969-73, Infection and Immunity, 1970-78, Immunochemistry, 1973-80, Clin. Immunology and Immunopathology, 1976-78; contbr. numerous articles to profl. jours. Served with U.S. Army, 1944-46. Decorated Gumhuria medal Egypt; recipient Disting. Service medal HEW, 1979; C. William O'Neal Disting. Am. Service award; Robert Koch Medal in Gold, Berlin, 1985; Sr. U.S. Scientist award Alexander Von Humboldt Found., Fed. Republic Germany, 1986. Mem. U.S. Nat. Acad. Scis., Inst. Medicine, Assn. Am. Physicians, Am. Acad. Allergy, Am. Soc. Biol. Chemists, Am. Soc. Clin. Investigation, Am. Assn. Immunologists, Am. Soc. Microbiology, Harvey Soc., Am. Coll. Allergists, AAAS, Infectious Diseases Soc. Am., Royal Soc. Medicine, Practitioner's Soc. N.Y., Am. Epidemiol. Soc. Clubs: Century Assn. (N.Y.C.); Cosmos (Washington). Rsch. on pathogenesis and epidemiology of streptococcal diseases; immunochem. studies on streptococcal antigens; immunogenetics; recognition of rabbit antibodies with molecular uniformity, genetics of immune response. Home: 4000 Cathedral Ave NW Apt 413B Washington DC 20016-5249 Office: Fogerty Internat Ctr NIH Rm B2C02 Bldg 31 Bethesda MD 20892

KRAUSE, SONJA, chemistry educator, researcher; b. St. Gall, Switzerland, Aug. 10, 1933; came to U.S., 1939; d. Friedrich and Rita (Maas) K.; m. Walter Walls Goodwin, Nov. 27, 1970. B.S., Rensselaer Poly. Inst., 1954; Ph.D., U. Calif.-Berkeley, 1957. Sr. phys. chemist Rohm & Haas Co., Phila., 1957-64; asst. lectr. Lagos U., Nigeria, 1964-65; asst. prof. Gondar Health Coll., Ethiopia, 1965-66; vis. asst. prof. U. So. Calif., Los Angeles, 1966-67; chemistry faculty Rensselaer Poly. Inst., Troy, N.Y., 1967—; prof. Rensselaer Poly. Inst., 1978—; mem. coun. Gordon Rsch. Conf., 1981-83; sabbatical Inst. Charles Sadron, Ctr. de Recherches sur les Macromolécules, Strasbourg, France, 1987. Author: (with others) Chemistry of Environment, 1978; editor: Molecular Electro-Optics, 1981; mem. editorial adv. bd. Macromolecules, 1982-84. Fellow Am. Phys. Soc. (coun. div. biol. physics 1980-83); mem. NRC (com. on polymer sci. and engring. 1992—), IUPAC (assoc. mem. 1991, Am. Chem. Soc. (councillor ea. N.Y. sect. 1991—, adv. bd. petroleum rsch. fund 1979-81, chmn. ea. N.Y. sect. 1981-82), Biophys. Soc. (coun. 1977), N.Y. Acad. Scis., Sigma Xi (pres. Rensselaer chpt. 1984-85). Office: Rensselaer Poly Inst Dept Chemistry Troy NY 12180

KRAUSMAN, JOHN ANTHONY, mechanical engineer; b. Balt., Oct. 11, 1948; s. Joseph A. and Sonya E. (Mellon) K.; m. Diana M. Rizzo, Jan. 20, 1979; children: Mark E., Russell A., Paul A. BSME, Johns Hopkins U., 1971; MS, Clarkson U., 1976. Engring. aide Westinghouse Electric Corp., Hunt Valley, Md., 1967-68, sr. engr., 1976-89; liaison engr. C.M. Kemp Mfg. Co., Glen Burnie, Md., 1968-69; product engr. Western Electric Co., Balt., 1969-76; fellow engr. TCOM, L.P., Columbia, Md., 1976—. Contbr. articles to profl. jours. Fellow AIAA (assoc., chmn. lighter than air tech. com.). Republican. Home: 10202 Castle Hill Ct Ellicott City MD 21042-5858 Office: TCOM LP 7115 Thomas Edison Dr Columbia MD 21046-2113

KRAUSS, HANS LUDWIG, emeritus educator; b. Halle, Fed. Republic Germany, June 4, 1927; s. Ludwig and Lore (Schletterer) K. Dipl. chem., Tech. Hochschule, München, 1951, Dr.rer.nat., 1955; habilitation, T.U., München, 1959. Lectr., curator T.U., 1959—, prof., 1964—; vis. prof. George Washington U., 1968—; prof. F.U., Berlin, 1968-93, U. Bayreuth, 1976-93. Contbr. articles to profl. jours. Fellow Rotary Internat.; mem. Ges. deutscher Chemiker Bunsengesellschaft Phys. Chem. Achievements include patent for surface chemistry. Home: 5b Heunischstr, 96049 Bamberg Bavaria, Germany

KRAUSS, JONATHAN SETH, pathologist; b. Bklyn., May 25, 1945; s. Maurice Daniel and Rose (Halpern) K.; m. Robin Livingston, June 18, 1972; children: Timothy, Rachel. AB, Cornell U., 1966; MD, U. Fla., 1970. Intern Med. Coll. Va., Richmond, 1970-71; resident U. N.C., Chapel Hill, 1973-78; asst. prof. Med. Coll. Ga., Augusta, 1978-83, assoc. prof., 1983-93; prof. Med. Coll. Ga., 1993—; cons. VA, Augusta, 1978—, Ga. Dept. Corrections, Augusta, 1983—. Contbr. articles to Am. Jour. Clin. Pathology, Thrombosis Rsch., Clin. Chemistry, Ann. Club. Lab. Sci., Modern Pathology, ObGyn. Pres. Augusta Authors Club, 1989-90; treas. Firends of Libr., Augusta, 1992-93. Lt. USNR, 1971-73. Grantee NIH, 1991, So. Med. Assn., 1980. Fellow Coll. Am. Pathologists, Coll. Am. Physicians; mem. Am. Soc. Hematology, Torch Club Augusta. Democrat. Hebrew. Achievements include research in crossed affinity electrophoresis of von Wilebrand factor antigen, granulocytic fragments in sepsis suggestive of irreversible shock, Factor VII deficient individuals in Georgia, secondary neoplasia in ovarian cancer. Home: 2407 McDowell St Augusta GA 30904-4635 Office: Med Coll Ga BIH 222B 1120 15th St Augusta GA 30912-3620

KRAUSS, ROBERT WALLFAR, botanist, university dean; b. Cleve., Dec. 27, 1921; s. Wallfar Gradifer and Emma Eleanor (Mueller) K.; m. Wilberta Tucker Bunker, Aug. 29, 1947 (div. 1969); children: Robert Geoffrey, Douglas Andrew; m. Marilyn J. Marsh, Dec. 19, 1986. BA, Oberlin Coll., 1947; MS, U. Hawaii, 1949; PhD, U. Md., 1951. Research fellow Carnegie Instn., 1951-54; research assoc. U. Md., College Park, 1951-54, mem. faculty, 1955-72, prof. plant physiology, 1959-72, head dept., 1964-73; dean Coll. Sci. Oreg. State U., Corvallis, 1973-79; exec. dir. Fedn. Am. Socs. Exptl. Biology, Bethesda, Md., 1979-90; sr. vis. sci. NASA, Washington, 1990—; staff mem. Marine Biol. Lab., Woods Hole, Mass., 1955, 56, 57; cons. USAF Sch. Aviation Medicine, 1961—; spl. advisor on U.S./Soviet relations to administr. NASA, 1964—; sr. research affiliate Chesapeake Biol. Lab., 1968—. Contbr. numerous articles to tech. jours., chpts. to books. Served to 2d lt. AUS, 1943-46. Recipient Achievement award in biology Washington Acad. Scis., 1961, Controlled Ecol. Life Support System Achievement award NASA, 1989. Fellow AAAS (council mem.); mem. Am. Soc. Plant Physiologists (trustee 1964-70), Am. Inst. Biol. Scis. (spl. award 1974, sec.-treas. 1963-68, pres. 1973), Bot. Soc. Am. (Darbaker award 1956), Bot. Soc. Washington (pres. 1964), Phycological Soc. Am. (pres. 1963—), Phi Beta Kappa, Sigma Xi. Club: Cosmos (Washington). Office: NASA Code SBM Capital Gallery Bldg Ste 510 600 Maryland Ave SW Washington DC 20024-2520

KRAUSS, SUE ELIZABETH, radiological medical management technologist; b. Poplar Bluff, Mo., Oct. 29, 1951; d. Raymond Harry and Wanda Elizabeth (Randol) Gibson; 1 child, Emily Sue. AS in Radiol. Tech., Santa Fe Jr. Coll., 1971. Radiol. technologist U. Fla., Gainesville, 1971-73; sect. chief Mt. Sinai Hosp., Miami Beach, Fla., 1973-76; asst. chief Miami Heart Inst., 1976-80; radiol. technologist Heights Hosp., Houston, 1981-82, Casa Grande (Ariz.) Regional Med. Ctr., 1983—; adminstrv. mgr. AMMAN, Inc., Casa Grande, 1986—. Mem. Am. Soc. Radiol. Technologists, Am. Registry Radiol. Technologists, Radiology Bus. Mgrs. Assn., Casa Grande Regional Med. Aux., Am. Cancer Soc., Am. Mgmt. Assn., Nat. Parks and Conservation Assn., Audubon Soc., World Wildlife Fedn., World Wildlife Fund, Sierra Club. Republican. Baptist. Avocations: reading, education. Office: AMMAN Inc 900 E Florence Blvd Ste D Casa Grande AZ 85222-4673

KRAUT, JOANNE LENORA, computer programmer, analyst; b. Watertown, Wis., Oct. 29, 1949; d. Gilbert Arthur and Dorothy Ann (Gebel) K.; BA in Russian, U. Wis., Madison, 1971, MS in Computer Sci., 1973. Computer programmer U. Wis. Sch. Bus., Madison, 1969-72, Milw. Ins. Co., 1973-74; tech. coord. Wis. Dept. Justice, Madison, 1974-83; tech. svcs. supr. CRC Telecommunications (formerly Benchmark Criminal Justice Systems), New Berlin, Wis., 1983-89; sr. programmer/analyst Info. Communications Corp., Pub. Safety Software, Inc., 1989-91; advanced systems engr. EDS, 1991-93; sr. programmer/analyst Time Ins., Milw., 1993—. Mem. Lakewood Gardens Assn. (dir. 1981-83), Dundee Terrs. Condominium Assn. (officer 1983—). Mem. Phi Beta Kappa. Home: 609 Dundee Ln Hartland WI 53029-2722 Office: Time Ins 501 W Michigan Milwaukee WI 53201

KRAUT, JOEL ARTHUR, ophthalmologist; b. Jersey City, July 21, 1937; s. Alan and Lillian Betty (Kravitz) K.; m. Cathy Jane Kleven, June 30, 1963; children: David Terence, Amy Melissa. AB cum laude, Princeton U., 1958; MD, Columbia U., 1962. Diplomate Am. Bd. Ophthalmology. Intern Boston U. Med. Ctr., 1962-63; resident in ophthalmology NYU-Bellevue Med. Ctr., N.Y.C., 1963-66; chief ophthalmology USAF Hosp., Tachikawa, Japan, 1966-68; pvt. practice specializing in ophthalmology Brookline, Mass., 1968—; clin. assoc., clin. instr. ophthalmology Harvard U. Med. Sch.; clin. instr. ophthalmology Tufts U. Sch. Medicine, 1968-91, clin. assoc. prof. ophthalmology, 1991—, assoc. surgeon ophthalmology, 1981-91, surgeon in ophthalmology, 1991—; dir. Low Vision Ctr., Mass. Eye & Ear Infirmary, 1968—, med. dir. Rehab. Ctr.; bd. dirs. physiol. optics dept. ophthalmology Tufts-New Eng. Med. Ctr., 1968-73; bd. surgeons Mass. Eye and Ear Infirmary, 1993—; cons. U.S. 5th Air Force, Japan, 1966-68. Contbr. articles to med. and profl. jours. Chmn. United Way campaign, 1973; bd. dirs. Boston Aid to Blind, 1987—; mem. adv. bd. Mass. Commn. for Blind, 1988—; mem. adv. bd. Nat. Assn. of Visually Handicapped, 1991—. Cane scholar, 1954-59, St. John-Princeton scholar, 1958-62; U. Calif. rsch. fellow, 1960. Fellow ACS; mem. Royal Soc. Medicine, Am. Acad. Ophthalmology (honor award 1991), New Eng. Ophthal. Soc., Mass. Ophthal. Soc., Soc. Geriatric Ophthalmology, Intraocular Lens Soc., New Eng. Implant Soc. (sec. 1979-81, pres. 1981-83), Mass. Med. Soc. Greater Boston Med. Soc., Mass. Soc. Eye Physicians and Surgeons (exec. bd., 1988—, recorder 1991—), Hazel Hotchkiss Wightman Tennis Club, du Bailliage de la Chaine des Rotissurs, Princeton U. Club (spl. gifts com. 1992-93), Phi Beta Kappa, Sigma Xi. Office: 16 Webster St Brookline MA 02146-4938

KRAUTER, STEFAN CHRISTOF WERNER, electrical engineer; b. Goeppingen, Fed. Republic Germany, Apr. 5, 1963; s. Werner and Charlotte (Rau) K. Student, Max Planck Soc. Plasma Physics, 1984; Dipl.-Ing., Tech. U. Munich, Fed. Republic Germany, 1988; PhD, Tech. U. Berlin, 1993; postgrad., Prof. M. Green U., NSW, Australia, 1994. Postgrad. tchr., rsch. Inst. Elec. Machines, Tech. U. Berlin, 1989—; fellow, advisor, tchr. project-orientated edn. bd. Tech. U. Berlin, 1989-91; mem. adv. bd. Internat. Solar Ctr. Berlin, 1992—; trustee symposia in field, 1991-92. Author procs. in field, 1990—. Mem. Internat. Solar Energy Soc., German Soc. for Solar Energy, Eurosolar. Achievements include research on optical and thermal optimization of photovoltaic modules, influence of skylight polarization on reflection, incidence angles of diffuse light, optical interaction of module encapsulation layers, partly structured surfaces, optical dispersion, spectral mismatch, measurement of free convective heat transfer coefficient, thermal simulation, cooling strategies, systems optimization. Office: Tech U Berlin Inst Electrial Maschines, Sec EM4 Einsteinufer 11, D-10587 Berlin Germany

KRÄUTLER, BERNHARD, chemistry educator; b. Dornbirn, Austria, Nov. 2, 1946; s. Alfons H. and Margarete (Ganser) K.; m. Rita Doll, Oct. 26, 1973; children: Raphael, Vincent, Nike. Diploma in chemistry, Eidgenoessische Tech. Sch., Zurich, 1970; Dr.sc.nat., Eidgenoessische Tech. Sch., 1977. Guest prof. U. Ill., 1985; dozent Eidgenoessische Technische Hochschule, 1986-91; prof. U. Innsbruck, Austria, 1991—. Author/editor: Vitamine II, 1989. Mem. Am. Chem. Soc., Swiss Chem. Soc. (Werner prize 1987), Gesellschaft Deutscher Chemiker, Gesellschaft Österr. Chemiker. Office: U Innsbruck, Innrain 52a, Innsbruck A-6020, Austria

KRAUZE, TADEUSZ KAROL, sociologist, educator; b. Warsaw, Poland, Feb. 27, 1934; came to U.S., 1958; s. Alfons W. and Kazimiera (Plazewska) K.; m. Sharon Robin Reland, Mar. 17, 1967; 1 child, Andrzej. MA, U. Lodz, 1955; PhD, NYU, 1974. Lectr. sociology Rutgers U., Newark, 1966-69; lectr. polit. sci. Haverford (Pa.) Coll., 1970-72; rsch. assoc. sociology Cornell U., Ithaca, N.Y., 1975-78; asst. prof. sociology Hofstra U., Hempstead, N.Y., 1971-75, assoc. prof., 1978-85, prof., 1985—; cons. Com. for Sci. Rsch., Warsaw, 1992. Editor: (with K. Slomczynski) Class Structure and Social Mobility in Poland, 1978, Social Stratification in Poland: Eight Empirical Studies, 1986; contbr. articles to Am. Sociol. Rev., European Sociol. Rev., Social Forces. Grantee Internat. Rsch. and Exch. Bd., 1976, 83, 84; Fulbright fellow U.S. Info. Agy., 1981-82. Achievements include rsch. in social stratification, sociology of sci. and quantitative methodology. Home: 4 Stuyvesant Oval 9A New York NY 10009 Office: Hofstra U Dept Sociology Hempstead NY 11550

KRAVATH, RICHARD ELLIOT, pediatrician, educator; b. N.Y.C., May 25, 1935; s. Reuben and Fannie Kravath; m. Pauline Sara Hauser, Aug. 27, 1960; children: Robert, Peter, Caroline. AB, Columbia U., 1956; MD, SUNY, Bklyn., 1960. Diplomate Am. Bd. Pediatrics. Intern Montefiore Med. Ctr., 1960-61, pediatric resident, chief resident, 1964-66, pediatric pulmonary fellowship, 1966-68; dir. div. intensive care pediatrics Albert Einstein Coll. Medicine, Bronx, 1981-82, prof. pediatrics, 1982; dir. in-patient pediatrics King's County Hosp. Ctr., Bklyn., 1982—; prof. clin. pediatrics SUNY, Bklyn., 1982—. Author, co-author: Pediatrics: Pretest, 1987, 89, 91; co-author: Water and Electrolytes in Pediatrics, 1982, 2d edit., 1993; contbr. articles to profl. jours. Capt. USAF, 1961-64. Mem. Alumni Assn. SUNY-Bklyn. (pres. 1992). Achievements include patents for monitoring of stress and a partitioning device for pools. Home: 6 Scott St Dobbs Ferry NY 10522-2614 Office: SUNY Bklyn Dept Pediatrics 450 Clarkson Ave Brooklyn NY 11203-2098

KRAVEC, CYNTHIA VALLEN, microbiologist; b. Newark, Sept. 8, 1951; d. William George and Elizabeth Irene (VanAllen) K. BS, Syracuse (N.Y.) U., 1974; MS, Seton Hall U., S. Orange, N.J., 1980; MBA, Monmouth Coll., W. Long Branch, N.J., 1986. Registered microbiologist. Sr. technician GIBCO/Invenex, Millburn, N.J., 1974-79; rsch. scientist Wampole Labs. div. Carter-Wallace Inc., East Windsor, N.J., 1979-90; scientist Roche Diagnostic Systems subsidiary Hoffmann-LaRoche, Inc., Nutley, N.J., 1990—. Contbr. articles to profl. jours. Mem. Am. Soc. Microbiology, Tissue Culture Assn., Soc. of Indsl. Microbiology. Home: 1006 Coolidge St Westfield NJ 07090-1215 Office: Roche Diagnostic Systems 1080 US Hwy 202 North Branch NJ 08876-1760

KRAVITZ, RUBIN, chemist; b. Framingham, Mass., Mar. 22, 1928; s. Abe and Lillian (Cohen) K. m. Geraldine Pudaim, Aug. 20, 1950 (dec.); children: Richard Alan, Steven Jay, Stuart Paul; m. Annabelle S. Durieux, July 16, 1978; 1 child, Michelle Pearl. BS, Northeastern U., 1952, D in Pharm, 1982. Analytical chemist FDA, HEW, Boston, 1956-61; analytical chemist Alcohol and Tobacco div. U.S. Treasury Dept., Boston, 1961-65; supr. phys. testing lab. plastic div. Am. Hoechst Corp., Leominster, Mass., 1967-78; rsch. chemist plastic div. Am. Hoechst Corp., Leominster, 1978-83; sr. devel. engr. EPS, 1983-85; pres. Nat. Plastics Mus. Inc., 1981-85; dir., pres. T.H.E. Hypnosis Ctr., Virginia Beach, Va., 1986-89; staff pharmacist MacDonald Army Hosp., Ft. Eustis, Va., 1987-89; chief pharmacist U.S. Army Health Clin., Fort Monroe, Va.; pres., chief exec. officer Cadet Labs., Virginia Beach, 1984—; chief pharmacist U.S. Army Health Clinic, Ft. Monroe, 1989—; del. Va. Pharm. Assn., 1988; mem. Mid-Atlantic Cholesterol Coun. Cubmaster Boy Scouts Am., Worcester, Mass., 1967-68; trustee, founding pres. Nat. Plastics Ctr. and Mus., 1985—. With USAAF, 1946-48. Mem. Assn. Mil. Surgeons U.S., Soc. Plastic Engrs. (newsletter editor 1969-71, treas. Pioneer Valley sect. 1972-73, v.p. 1973-74, chmn. tech. com. 1973, pres. Pioneer Valley sect. 1975-76, chmn. sect. museum 1979-85, achievement award 1981), ASTM (chmn. compression molding 1969-70, vice chmn. publicity and papers com. D-20 on plastics 1972-76, chmn. subcom. specimen preparation, chmn. sect. plastic furniture, chmn. specimen preparation 1976, chmn. task group Kravitz impact test method 1976, chmn. D 20.12 Olefin Plastics com., mem. exec. com. 1982-85), Assn. Analytical Chemists, Assn. to Advance Ethical Hypnosis, Am. Soc. Rsch. and Clin. Hypnosis, K.P. (chancellor comdr. 1963-64).

KRAWETZ, ARTHUR ALTSHULER, chemist, science administrator; b. Chgo., Oct. 30, 1932; s. John and Grace (Altshuler) K. BS in Chemistry, Northwestern U., 1952; MS in Phys. Chemistry, U. Chgo., 1953, PhD in Phys. Chemistry, 1955. V.p. Phoenix Chem. Lab., Inc., Chgo., 1950-73, tech. dir., 1958—, pres., 1974—. Contbr. articles to profl. jours. Mem. Nat. Safety Coun. 1st Lt. USAF, 1956-58, capt. Res. Fellow Am. Inst. Chemists (life); mem. ASTM (chmn. sub-com. XI engring. scis., sub-com. N-VI Fire Resistance 1974-84, sub-com IX-D oxidation 1974-81, task force on Precautionary Statements for Hazardous Material and Lab. Ops. 1976-84, mem. various coms.), Am. Indsl. Hygiene Assn. (Chgo. sect.), Am. Chem. Soc. (div. phys. chemistry, div. analytical chemistry, div. petroleum chemistry, div. indsl. chemistry), Instrument Soc. Am., Air Pollution Control Assn., Soc. for Applied Spectroscopy, Soc. Automotive Engrs., Am. Soc. Lubrication Engrs., Nat. Lubricating Grease Inst., The Coblentz Soc., Nat. Fire Protection Assn. (com. on classification and properties of hazardous chemical data), Assn. Official Analytical Chemists, Chgo. Gas Chromatograph Discussion Group, Internat. Assn. Stability and Handling Liquid Fuels (hon.), Royal Soc. Chemistry (chartered chemist), Phi Beta Kappa, Sigma Xi, Phi Lambda Upsilon, Pi Mu Epsilon. Achievements include patents for temperature control apparatus and method, for method of determining acid content of oil sample, for automatic oxygen measuring system, and for viscometers. Office: Phoenix Chem Lab Inc 3953 W Shakespeare Ave Chicago IL 60647-3497

KRAWETZ, STEPHEN ANDREW, molecular biology and genetics educator; b. Fort Frances, Ont., Can., Sept. 17, 1955; s. Stephen and Michaelene (Medynski) K.; m. Lorraine Ruth St. John, Aug. 19, 1977; 1 child, Rhochelle Tairaesa. BS, U. Toronto, Ont., 1977, PhD, 1983. Tchr. Scarborough Bd. Edn., Ont., 1976-77; AHFMR postdoctoral fellow in med. biochemistry U. Calgary, Alta., Can., 1983-89; asst. prof. rsch. Ctr. for Molecular Biology Wayne State U., Detroit, 1989, asst. prof. molecular biology and genetics, 1989-92, asst. prof. obstetrics and gynecology and molecular biology and genetics, 1992—; biotechnology cons. Calgary, 1985-89, Grosse Pointe Woods, Mich., 1989—; co-founder Genetic Imaging, Inc., 1988. Contbr. numerous articles to scholarly jours.; mem. editorial bd. BioTechniques, Ag Biotech News and Info. Recipient B.C. Childrens Hosp. Rsch. award, Vancouver, 1984, Computer Applications in Molecular Biology award, IntelliGenetics Inc., Mountain View, Calif., 1988; Alta. Heritage Found. Med. Rsch. fellow, 1985-88. Mem. Can. Biochem. Soc., Am. Soc. Human Genetics, N.Y. Acad. Scis., AAAS. Achievements include development of splinkers for sequencing DNA, of a computer-based imaging system for biological data, of VPCS cloning vectors, of the basis of biological sequence alignment algorithm; one of the first to describe overlapping reading frames in eucaryotes; first detailed analysis of a mammalian protamine gene; first definition of sequence interpretation errors in the GenBank database. Home: 805 Canterbury Rd Grosse Pointe Woods MI 48236-1417 Office: Wayne State U CS Mott Ctr Human Growth/Devel Depts Ob-Gyn/Mol Biol/Genet Detroit MI 48201

KREBS, EDWIN GERHARD, biochemistry educator; b. Lansing, Iowa, June 6, 1918; s. William Carl and Louise Helena (Stegeman) K.; m. Virginia Frech, Mar. 10, 1945; children: Sally, Robert, Martha. AB in Chemistry, U. Ill., 1940; MD, Wash. U., St. Louis, 1943; hon. degree, U. Geneva, 1979, Med. Coll. Ohio, 1993. Intern, asst. resident Barnes Hosp., St. Louis, 1944-45; rsch. fellow biol. chemistry Wash. U., St. Louis, 1946-48; asst. prof. biochemistry U. Wash., Seattle, 1948-52, assoc. prof. biochemistry, 1952-57, prof. biochemistry, 1957-66, prof. biochemistry, dean for planning Sch. Medicine, 1966-68; prof., chmn. dept. biol. chemistry, Sch. Medicine U. Calif., Davis, 1968-76; prof., chmn. dept. pharmacology U. Wash., Seattle, 1977-83; investigator, sr. investigator Howard Hughes Med. Inst., Seattle, 1983-90, sr. investigator emeritus, 1991—; prof. biochemistry and pharmacology U. Wash., Seattle, 1984—; mem. Phys. Chemistry Study Sect. NIH, 1963-68, Biochemistry Test Com. Nat. Bd. Med. Examiners, 1968-71, rsch. com. Am. Heart Assn., 1970-74, bd. sci. counselors Nat. Inst. Arthritis, Metabolism and Digestive Diseases, NIH, 1979-84, Internat. Bd. Rev., Alberta Heritage Found. for Med. Rsch., 1986, External Adv. Com. Weis Ctr. for Rsch., 1987-91. Mem. editorial bd. Jour. Biol. Chemistry, 1965-70; mem. editorial adv. bd. Biochemistry, 1971-76; mem. editorial and adv. bd. Molecular Pharmacology, 1972-77; assoc. editor Jour. Biol. Chemistry, 1971—; mem. internat. adv. bd. Advances in Cyclic Nucleotide Rsch., 1972—; editorial advisor Molecular and Cellular Biochemistry, 1987—. Recipient Nobel Prize in Medicine or Physiology, 1992, Disting. lectureship award Internat. Soc. Endocrinology, 1972, Gairdner Found. award, Toronto, Ont., Can., 1978, J.J. Berzelius lectureship, Karolinska Institutet, 1982, George W. Thorn award for sci. excellence, 1983, Sir Frederick Hopkins Meml. lectureship, London, 1984, Rsch. Achievement award Am. Heart Assn., Anaheim, Calif., 1987, 3M Life Scis. award FASEB, New Orleans, 1989, Albert Lasker Basic Med. Rsch. award, 1989, CIBA-GEIGY-Drew award Drew U., 1991, Steven C. Beering award, Ind. U., 1991, Welch award in chemistry Welch Found., 1991, Louisa Gross Horwitz award Columbia U., 1989; John Simon Guggenheim fellow, 1959, 66. Mem. NAS, Am. Soc. Biol. Chemists (pres. 1986, ednl. affairs com. 1965-68, councillor 1975-78), Am. Acad. Arts and Scis., Am. Soc. Pharmacology and Exptl. Therapeutics. Achievements include life-long study of the metabolism of glycogen and its ATP/ADP cycle. Office: U Wash Dept Pharmacology Howard Hughes Med Inst SL-15 Seattle WA 98195

KREBS, JAMES NORTON, retired electric power industry executive; b. Sauk Center, Minn., Apr. 20, 1924. BS, Northwestern U., 1945. Mgr. design devel., markets Gen. Elec. Co., Fairfield, Conn., 1946-78, v.p. mil. engring. programs, 1978-82, v.p. LYNN engring. ops., 1982-84, v.p. tech., mgmt. assessment, aircraft engring. group, 1984-85. Mem. Nat. Acad. Engring., Am. Inst. Aeronautics and Astronautics (Sylvanus A. Reed Aeronautics award 1992). Address: 84 Harbor Ave Marblehead MA 01945*

KREEK, MARY JEANNE, physician; b. Washington; d. Louis Francis and Esperance (Agee) Kreek; BA, Wellesley Coll., 1958; MD, Columbia, 1962; m. Robert A. Schaefer, Jan. 24, 1970; children: Robert A., Esperance Anne. Med. researcher NIH, Bethesda, Md., 1957-62; intern, resident Cornell N.Y. Hosp. Med. Ctr., N.Y.C., 1962-65, fellow, 1965-67; instr. medicine Cornell Med. Coll., 1966-67; acad. medicine specializing in internal medicine, endocrinology, gastroenterology, clin. pharmacology, N.Y.C., 1966—; mem. staff N.Y. Hosp.-Cornell U., 1968-77, clin. asst. prof., asst. attending physician, now assoc. attending physician, adj. assoc. prof.; asst. prof. Rockefeller U., 1967-72, sr. rsch. assoc., physician, 1972-83, assoc. prof., physician, 1983—; head Ind. Lab. on Biology of Addictive Diseases, 1975—; mem. gen. medicine study sect. NIH, 1973-77; co-chmn. John E. Fogarty (NIH) Internat. Conf. Hepatotoxicity Due to Drugs and Chems., 1977; vis. prof. Pahlavi U., Shiraz, Iran, summer 1977; spl. adv. Nat. Inst. Drug Abuse, 1976-86, mem. Nat. Adv. Coun., 1991—, prin. investigator Rsch. Ctr. Biol. Bass Addictive Diseases, 1987—; mem. gastroenterology adv. com. FDA, 1975-79, 92-96, NIH Gen. Clin. Recipient Borden Rsch. award, 1962; Career Scientist award Health Rsch. Council City N.Y., 1974—; Dole/Nyswander award; Outstanding Rsch. Svc. in Addictive Diseases award,NIH-NIDA, 1984; Rsch. Scientist award, NIH, 1978—. Rsch. Ctr. Study Sect., 1979-83, chmn., 1982-83; mem. exec. com. Problems Drug Dependence, 1982-87, 91, chmn. exec. com., 1985-87, chair sci.program com., 1991—; dir. NIH-NIDA Rsch. Ctr., 1987—. Mem. Am. Fedn. for Clin. Rsch., Shakespeare Soc. of Wellesley, Am. Gastroent. Assn., N.Y. Gastroent. Assn. (pres. 1987), Endocrine Soc., Am. Assn. Study Liver Diseases, Internat. Assn. Study Liver, Internat. Narcotic Research Conf. Group, Research Soc. on Alcoholism, Am. Coll. Neuropsychopharmacology, Soc. on Neurosciences, Phi Beta Kappa, Sigma Xi. Home: 1175 York Ave New York NY 10021-7169 Office: Rockefeller U New York NY 10021

KREHBIEL, DARREN DAVID, civil engineer; b. Columbia, Mo., Oct. 7, 1963; s. David Gordon and Dana Kay (Hatcher) K.; m. Christine Marie McKinney, May 25, 1988. BS in Civil Engring., U. Mo., 1987. Registered profl. engr., Mo., Ark., Kans., Okla. Assoc. transp. engr. Caltrans, San Francisco, 1987-89; pres. Krehbiel Engring., Inc., Camdenton, Mo., 1989—. Mem. NSPE, Mo. Soc. Profl. Engrs., Mo. Assn. Registered Land Surveyors, Rotary Internat. (Paul Harris fellow 1989, Youth Exch. chair 1990—). Christian. Home: PO Box 562 Camdenton MO 65020 Office: Krehbiel Engring Inc 109 Blair Ave Camdenton MO 65020

KREHBIEL, DAVID KENT, chemical engineer; b. McPherson, Kans., May 4, 1966; s. Darrell Kent and Maria Louisa (Gutierrez) K.; m. Leesa Lamia Lobban, June 7, 1986; 1 child, Andrew David. BS in Chemistry, McPherson (Kans.) Coll., 1988; MS in Chem. Engring., U. Okla., 1991. Engr., cons. Univ. Technologists, Norman, Okla., 1991-92; engr. E.I. duPont de Nemours & Co., Beaumont, Tex., 1992—. Ford Found. fellow NRC, 1988, NSF grad. fellow, 1988-91; recipient State of Excellence award Gov. of Okla., 1990. Mem. Am. Chem. Soc., Am. Inst. Chem. Engrs., Phi Kappa Phi. Achievements include development of technique for chromatographic separation of enkephalins extracted from rat brain tissue, development of an ultrafiltration technique to remove chromates from waste water using polyelectrolytes. Office: E I duPont de Nemours & Co PO Box 3269 Beaumont TX 77704

KREISMAN, NORMAN RICHARD, physiologist; b. Chgo., June 26, 1943; s. Albert and Florence (Tweer) K.; m. Jane Ellen Schexnayder, May 18, 1975; 1 child, Anne E. MS, U. Mich., 1968; PhD, Med. Coll. Pa., 1971. From instr. tp asst. prof. dept. physiology Med. Sch. Tulane U., New Orleans, 1971-73, assoc. prof., 1979—; mem. neurology B study sect. NIH, Bethesda, Md., 1992; co-founder, dir. interdisciplinary program in neurosci., Tulane U. Contbr. articles to Jour. Neurophysiology, Jour. Cerebral Blood Flow and Metabolism, Brain Rsch., Am. Jour. Physiology, Jour. Applied Physiology. Resource vol. New Orleans Pub. Schs., 1973—. Grantee Schleider Found., 1973-75, NIH, 1975-78, 83-86, Epilepsy Found. Am., 1980. Mem. Am. Heart Assn. (peer rev. com. La. affiliate 1982-92, grantee 1989-92), Am. Physiol. Soc., Soc. for Neurosci. (chpt. pres. 1977), Internat. Soc. Cerebral Blood Flow and Metabolism. Office: Tulane Med Sch Dept Physiology 1430 Tulane Ave New Orleans LA 70112-2699

KREITHEN, MELVIN LOUIS, biologist, educator; b. Phila., July 18, 1941; s. Alexander and Rose (Greenberg) K.; m. Marian Evans Miller, Aug. 14, 1963; children: David, Amy. BA, U. Md., 1968; PhD, Cornell U., 1974. Rsch. assoc. Cornell U., Ithaca, N.Y., 1974-80; assoc. prof. biol. scis. U. Pitts., 1980—; mem. avian adv. panel U.S. Windpower, Oakland, Calif., 1992—; mem. adv. bd. Am. Racing Pigeon Union, Oklahoma City, 1992—. Achievements include research in avian special senses: infrasounds, ultraviolet light, polarized light, barometric pressure detection. Office: U Pitts Dept Biol Scis Pittsburgh PA 15260

KREITZBERG, FRED CHARLES, construction management company executive; b. Paterson, N.J., June 1, 1934; s. William and Ella (Bohen) K.; m. Barbara Braun, June 9, 1957; children: Kim, Caroline, Allison, Bruce, Catherine. BSCE, Norwich U., 1957. Registered profl. engr., Ala., Alaska, Ariz., Ark., Calif., Colo., Conn., Del., D.C., Fla., Ga., Idaho, Ill., Ind., Kans., Ky., Md., Mass., Minn., Miss., Mo., Nebr., Nev., N.C., N.H., N.J., N.Mex., N.Y., Ohio, Okla., Oreg., Pa., S.D., Tenn., Va., Vt., Wash., W.Va., Wis., Wyo. Asst. supt. Turner Constrn. Co., N.Y.C., 1957; project mgr. Project Mercury RCA, N.J., 1958-63; schedule cost mgr. Catalytic Constrn. Co., Pa., 1963-65, 65—; cons. Meridien Engring., 1965-68; prin. MDC Systems Corp., 1968-72; owner, pres., chief exec. officer, bd. dirs. O'Brien-Kreitzberg and Assocs. Inc., San Francisco, 1972—; lectr. Stanford (Calif.) U., U. Calif., Berkeley. Author: Crit. Path Method Scheduling for Contractor's Mgmt. Handbook, 1971; tech. editor Constrn. Inspection Handbook, 1972; contbr. articles to profl. jours. bd. dirs. Partridge Soc.; trustee Norwich U. 2d lt. C.E., U.S. Army, 1957-58. Recipient Disting. Alumnus award Norwich U., 1987; named Boss of Yr., Nat. Assn. Women in Constrn., 1987; named in his honor Kreitzberg Amphitheatre, 1987, Kreitzberg Libr. at Norwich U., 1992; Bay Area Discovery Mus.-Birthday rm. and snack bar named in honor of Kreitzberg family, 1989. Mem. ASCE (Constrn. Mgr. of Yr. 1982); mem. Am. Arbitration Assn., Constrn. Mgmt. Assn. Am. (founding, bd. dirs.), Soc. Am. Value Engrs., Community Field Assn., Ross Hist. Soc., N.J. Soc. Civil Engrs., N.J. Soc. Profl. Planners, Project Mgmt. Inst. Constrn. Industry Pres. Forum. Avocations: running, bicycling, tropical fish. Home: 19 Spring Rd Box 1200 Ross CA 94957 Office: O'Brien-Kreitzberg & Assocs Inc 188 The Embarcadero San Francisco CA 94105-1231

KRENKE, FREDERICK WILLIAM, electrical engineer; b. Port Huron, Mich., Aug. 27, 1946; s. William Frederick and Marian Mable (Schaffer) K.; m. Constance Caye Aylesworth, Nov. 3, 1967 (div. 1979); children: Timothy, Kimberlee; m. Bonnie Jean Feakes, Sept. 7, 1979; children: Kalla, Ashley, Nicole. BSEE, U. Mich., 1973. Elec. constrn. engr. Bechtel Power Corp., various locations, 1973-84; elec. engr. Farley Nuclear Generating Sta. Delcon, Dothan, Ala., 1975-76, Jelco, Inc., Castledale, Utah, 1976-77, Ebasco Svcs. Inc., Cohasset, Minn., 1978-79, Matsco, Reform, Mo., 1982; start-up engr. Bechtel Power Corp.-Ariz. Pub. Svcs., Palo Verde, 1982-83; elec. engr. nuclear generating sta. Ill. Power Co., Clinton, 1984-85; lead elec. and instrument and control engr. Fla. Power and Light Co., Miami, 1985-87; start-up engr. Limerick Generating Sta. Gen. Physics Corp., Pottstown, Pa., 1987-89; elec. design engr. nuclear unit Brand Utility Svcs., Russellville, Ark., 1989—. With USN, 1968-69. Home: 3114 Alaskan Tr Russellville AR 72801

KRESH, J. YASHA, cardiovascular researcher, educator; b. Russia, July 13, 1948; came to U.S., 1967; m. Myrna Blickman Masucci. BSEE, N.J. Inst. Tech., 1971; MSBME, Rutgers U., 1973, PhD, 1976. Rsch. assoc. Beth Israel Med. Ctr., Newark, 1976-79; dir. rsch. Jefferson Med. Coll., Phila., 1979-86; prof. medicine, dir. cardiovascular biophysics and computing Likoff Cardiovascular Inst., Phila., 1986—, prof., dir. rsch. cardiothoracic surgery, 1986—; lectr. in field. Author over 100 publs. in physiol. cardiology and bioengring. jours.; patentee in field. Fellow Am. Coll. Cardiology; sr. mem. IEEE, Biomed. Engring. Soc.; mem. Am. Heart Assn., Am. Soc. Artificial Internal Organs, Sigma Xi, Tau Beta Pi, Eta Kappa Nu. Avocations: fractal art, swimming, sailing, computing. Office: Likoff Cardiovascular Inst Mail Stop 110 Broad & Vine Sts Philadelphia PA 19102

KRESINA, THOMAS FRANCIS, immunologist, educator; b. Balt., June 18, 1954; s. Thomas Francis and Bertha (Miller) K.; m. Marilee Keim, June 10, 1979 (div. 1991); children: Rachel Ann, Jennifer Lynn, Rebecca Marie; m. Laura Cheever June 12, 1993. BS, Cath. U. Am., 1975; PhD, U. Ala., Birmingham, 1979; MA, Brown U., 1987. Postdoctoral fellow Brandeis U., Waltham, Mass., 1980-82; asst. prof. Case Western Res. U., Cleve., 1982-87; assoc. prof. dept. medicine Brown U., Providence, R.I., 1987-93; dir. Biliary and Pancreas Disease Programs Nat. Inst. of Diabetes and Digestive and Kidney Diseases, NIH, 1993—; mem. spl. study sect. NIH, Washington, 1989, 91, 92. Editor: Monoclonal Antibodies, Cytokine and Arthritis, 1991; contbr. chpts. to books, articles to jours. Asst. troop leader R.I. unit Girl Scouts U.S., 1989-91; sci. adviser George Peters Elem. Sch., Cranston, R.I., 1990-91; mem. bd. dirs. PTO, Cranston, 1989-93. Grantee NIH, 1983-93, various pvt. founds. Mem. Am. Assn. Pathologists, Am. Assn. Immunologists, N.Y. Acad. Sci., Sigma Xi. Achievements include discovery of idiotypie immune network in schistosomiasis, cytokine antagonists in experimental arthritis, anti-idiotypie antibody vaccine in schistosomiasis, human monoclonal antibodies in schistosomiasis. Office: NIDDK NIH Rm 3A17 Westwood Bethesda MD 20892

KRESS, ALBERT OTTO, JR., polymer chemist; b. Cullman, Ala., June 15, 1950; s. Albert Otto and Odell Pearl (Norris) K.; m. Ruth Jeanette Beach, Dec. 30, 1972 (div. Aug. 1978); children: Adrian Konrad, Katyna Ileana; m. Roby Lynn Rice, Apr. 14, 1984; 1 child, Ashley Alan Rice Kress. BS, U. Montevallo, 1972; PhD, U. Ala., 1979. Rsch. scientist Hercules Chem. Corp., Wilmington, N.C., 1979-83; rsch. assoc. Clemson (S.C.) U., 1983-84; assoc. prof. U. Montevallo, Ala., 1984-86; rsch. assoc. U. So. Miss., Hattiesburg, 1986-88; sr. scientist Schering-Plough HealthCare Products Corp., Memphis, 1988—. Contbr. articles to Jour. Organic Chemistry, Dissertation Abstracts Internat. B., Jour. Chem. Soc., Jour. Applied Polymer Sci. Recipient Dean's scholarship U. Ala., 1975. Mem. AAAS, SPI, SPE, Am. Chem. Soc., Moose. Republican. Lutheran. Home: 9596 Pigeon Roost Rd Olive Branch MS 38654-2611 Office: Schering-Plough HealthCare Products Corp 3030 Jackson Ave Memphis TN 38151-0001

KRESS, GERARD CLAYTON, JR., psychologist, educator; b. Buffalo, N.Y., July 10, 1934; s. Gerard Clayton and Eleanor Amelia (Rupp) K.; m. Suzanne Ardys Raloff, May 4, 1957 (div. 1980); children: Timothy, Peter, Jennifer. AB, U. Rochester, 1956; PhD in Psychology, SUNY, Buffalo, 1962. Assoc. rsch. scientist Am. Insts. for Rsch., Pitts., 1961-64, rsch. scientist, 1964-68; asst. prof. U. Pitts., 1968-71; asst. dean Harvard Sch. Dental Medicine, Boston, 1971-80, assoc. dean, 1980-88; prof., dir. behavioral scis. Baylor Coll. Dentistry, Dallas, 1988—; cons. U.S. VA, Washington, 1982—, Boston U., 1980-82, U. Minn., Mpls., 1973, 88-90, 91. Author: Behavior Management in the Classroom, 1969; contbr. chpt.: Dental Clinics of North America, 1988; contbr. articles to profl. jours. With USAR, 1957-60. Mem. Am. Psychol. Assn., Am. Assn. Dental Schs. (chair behavioral sci. sect. 1992), Internat. Assn. Dental Rsch. (pres. behavioral scis. group 1974), Am. Ednl. Rsch. Assn. Democrat. Home: 7714 Arborside Dr Dallas TX 75231 Office: Baylor Coll Dentistry 3302 Gaston Ave Dallas TX 75246

KRESSEL, NEIL JEFFREY, psychologist; b. Newark, N.J., Aug. 28, 1957; s. Morris Israel and Betty (Weiss) K.; m. Dorit Fuchs, Aug. 11, 1991. BA, MA, Brandeis U., 1978; MA, Harvard U., 1981, PhD, 1983. Sophomore tutor Harvard U., Cambridge, 1979-83; asst. prof. William Paterson Coll., Wayne, N.J., 1984-93, chmn. dept. psychology, 1992—, assoc. prof., 1993—; adj. assoc. prof. mgmt. Stevens Inst. Tech., Hoboken, N.J., 1989--; adj. asst. prof. NYU, 1990-91; manuscript cons. Political Psychology, N.Y.C., 1990-91, social, clin. and indsl. psychology, Leonia, N.Y., 1983—. Editor: Political Psychology, 1993; contbr. articles to profl. jours. Mem. Internat. Soc. Political Psychology, Am. Assn. for Pub. Opinion Rsch., Zionist Orgn. Am. Jewish. Office: William Paterson Coll Dept Psychology Wayne NJ 07470

KRETSCH, MICHAEL GENE, civil engineer; b. Newark, Dec. 4, 1959; s. Albert and Veronica Ann (Gruenberg) K.; m. Mary Noreen Cheney, Feb. 14, 1981; children: Allen Glen, Emma Michelle. BCE, Ga. Tech., 1983. Registered profl. engr., Ga; land surveyor-in-tng. Instrument man Gordon Story & Assoc., Inc., Snellville, Ga., 1984-85; crew chief Cordes Quintana & Assoc., Inc., Cumming, Ga., 1985-86; project engr. Precision Planning, Inc., lawrenceville, Ga., 1986-91; prin. engr. Dept. Pub. Utilities Gwinnett County, Ga., 1991—. Mem. Ga. Soc. Profl. Engrs. (treas. 1990-91, sec. 1991-92), Survey & Mapping Soc. Ga., Nat. Rifle Assn. Libertarian. Presbyterian. Office: Dept Pub Utilities 75 Langley Dr Lawrenceville GA 30245

KREUTER, KONRAD FRANZ, software engineer; b. Nuremberg, Germany, Mar. 20, 1939; s. Kurt E. and Eva Friedel Kreuter; m. Gerit D. Mayer, 1966; children: Fei Silke, May Frauke. Diploma, U. Mainz, Fed. Republic Germany, 1965, Dr rer. nat., 1968. With Siemens AG, Karlsruhe, Fed. Republic Germany, 1969-72; software engr. ESG, Munich, 1972-76; sr. cons. Sesa Deutschland GmbH, Frankfurt, Fed. Republic Germany, 1977-81; software engr. BBC Brown Boveri & Cie AG, Mannheim, Germany, 1981—, ABB Asea Brown Boveri SAN/RDL, Mannheim, 1988—. Co-designer (programming language) PEARL. Mem. Landeselternbeirat Baden Württemberg, 1990-93. Recipient Ehrennadel honor Land Baden-Württemberg, 1990. Mem. GI Gesellschaft für Informatik, Assn. Computer Machinery. Achievements include patents on electrical switchyard interlocking. Home: Leutershausen Schlehdornweg 2, D-69493 Hirschberg Germany

KREUZ, ROGER JAMES, psychology educator; b. Toledo, Aug. 18, 1961; s. Paul Lawrence and Ila Mae (Noe) K. BA, U. Toledo, 1983; MA, Princeton U., 1985, PhD, 1987. Rsch. assoc. Duke U., Durham, N.C., 1987-88; asst. prof. Memphis State U., 1988-93, assoc. prof., 1993—. Editor: (with M.S. MacNealy) Empirical Approaches to Literature and Aesthetics, 1994; contbr. articles to profl. jours. Mem. Am. Psychol. Soc., Psychonomic Soc., Sigma Xi (assoc.). Democrat. Home: 1351 Linden Ave Apt 4 Memphis TN 38104-3648 Office: Memphis State U Dept Psychology Memphis TN 38152

KRICHEVER, MARK, optical engineer; b. Penza, Russia, Nov. 16, 1941; came to U.S., 1980; s. Jacob and Zoia (Rubashkina) K.; m. Inessa Trilio, Nov. 10, 1962 (div. Feb. 1989); 1 child, Maria. MS in Optical Engring., LITMO, St. Petersburg, Russia, 1964; A. in Elec. Engring., Popov Inst., St. Petersburg, 1967. Shop engr. Lomo, St. Petersburg, 1964-67; group leader Inst. for Precision Instrumentation, St. Petersburg, 1967-74; sr. design engr. Tadiran Electronics Ltd., Holon, Israel, 1976-80; sr. engring. mgr. SymBol Technologies Inc., Bohemia, N.Y., 1980—; cons. Inst. of Geology, St. Petersburg, 1971-72. Author: (publs.) Spectrophotometer with Automatic Processing of Measuring Data, 1974, Double Channel Atomic-Absorption Spectrophotometer, 1973. Recipient Best Sci. Instrument Design award USSR Com. on Exhbn. of Indsl. Achievements, 1972. Mem. IEEE, Am. Mgmt. Assn., Mgmt. Roundtable. Jewish. Achievements include inventions of conceptually new types of optical laser scanners; more than 20 patents in field. Office: Symbol Technologies 110 Wilbur Pl Bohemia NY 11716

KRICK, IRVING PARKHURST, meteorologist; b. San Francisco, Dec. 20, 1906; s. H. I. and Mabel (Royal) K.; m. Jane Clark, May 23, 1930; 1 dau., Marllynn; m. Marie Spiro, Nov. 18, 1946; 1 son, Irving Parkhurst II. B.A., U. Calif., 1928; M.S., Cal. Inst. Tech., 1933, Ph.D., 1934. Asst. mgr. radio sta. KTAB, 1928-29; meteorologist, 1930—; became mem. staff Calif. Inst. Tech., 1933, asst. prof. meteorology, 1935-38, assoc. prof., prof. and head dept., 1938-48; organizer, pres. Am. Inst. Aerological Rsch. and Water Resources Devel. Corp., 1950; pres. Irving P. Krick Assocs., Inc., Irving P. Krick, Inc., Tex., Irving P. Krick Assocs. Can. Ltd.; now, chmn. emeritus, sr. cons. strategic Weather Svc. Krick Ctr. Weather R & D, Palm Springs, Calif.; established meteorology dept. Am. Air Lines, Inc., 1935, established Internat. Meteorol. Cons. Svcs., 1946, mng. dir.; cons. in field, 1935-36; mem. sci. adv. group Von Kármán Army Air Force, 1945-46. Pianist in concert and radio work, 1929-30; Co-author: Sun, Sea and Sky, 1954; Writer numerous articles on weather analysis, weather modification and forecasting and its application to agrl. and bus. industries. Served as lt. Coast Arty. Corps U.S. Army, 1928-36; commd. ensign USNR, 1938; maj., then lt. col. USAAF, 1943; Weather Directorate, Weather Central Div. unit comdr. of Long Range Forecast Unit A 1942-43; dep. dir. weather sect. U.S. Strategic Air Forces Europe, 1944; chief weather information sect. SHAEF 1945. Decorated Legion of Merit, Bronze Star with Oak leaf cluster U.S.; Croix de Guerre France; recipient Distinguished Service award Jr. C. of C.; chosen one of 10 outstanding men under age 35 by U.S. Jr. C. of C. Fellow AIAA (assoc.), Royal Soc. Arts; mem. AAAS, Royal Meteorol. Soc., Am. Meteorol. Soc., Am. Geophys. Union, Sigma Xi. Republican. Achievements include creation of 1st modern airline weather forecasting service for Western Air Express, 1932; development of private weather forecasting service, supplying information to various cos.; development of long-range weather forecasting method covering periods up to 25 years; creation of applications rain increase and hail control; perfected comml. five year weather forecasts. Home: Apt 13 1200 S Orange Grove Blvd Pasadena CA 91105-3353 Office: Krick Ctr Weather R & D 610 S Belardo Rd Palm Springs CA 92264

KRIDER, E. PHILIP, atmospheric scientist, educator; b. Chgo., Mar. 22, 1940; s. Edmund Arthur and Ruth (Abbott) K.; m. Barbara A. Reed, June 13, 1964 (div. Mar. 1983); children: Ruth Ellen, Philip Reed. BA in Physics, Carleton Coll., 1962; MS in Physics, U. Ariz., 1964, PhD in Physics, 1969. Resident rsch. assoc. NASA Manned Spacecraft Ctr. NAS, Houston, 1969-71; asst. rsch. prof. Inst. Atmospheric Physics U. Ariz., Tucson, 1971-75, asst. prof. dept. atmospheric scis., 1973-75, assoc. prof. dept. atmospheric Scis. and Inst. Atmospheric Scis., 1975-80; exec. v.p., part-time chmn. Lightning Location and Protection, Inc., Tucson, 1976-83; adj. prof. dept. elec. engring. U. Fla., Gainesville, 1988—; prof. dept. atmospheric scis. Inst. Atmospheric Physics U. Ariz., 1980—, dir. Inst. Atmospheric Physics, head dept. atmospheric scis., 1986—; pres. Internat. Commn. Atmospheric Electricity, 1992—; co-chmn. panel Earth's electrical environment geophysics study com. NAS, 1982-86; mem. panel weather support for space ops. NAS, 1987-88, geostationary platform sci. steering com. NASA, 1987—; chmn. mems. nominating com. Univ. Corp. for Atmospheric Rsch., 1986—; mems. rep. Univ. Corp. for Atmospheric Rsch., 1986—, mem. sci. program evaluation com., 1988-89; mem. U.S. nat. com. Internat. Sci. Radio Union; mem. lightning and sferics subcom. Internat. Commn. on Atmospheric Electricity, 1976—; mem. aerospace corp. adv. team USAF Launch Vehicle Lightning/Atmospheric Electrical Constraints, Post Atlas/Centaur 67 Incident, 1987-89; sci. advisor Air Force Geophys. Lab., 1988; mem. lightning adv. com. U.S. Army Missile Command, 1986-87; co-chief editor Jour. of Atmospheric Scis., 1980-91, editor, 1991-92; assoc. editor Jour. Geophys. Rsch., 1977-79; referee Jour. Geophys. Rsch., Geophys. Rsch. Letters, Jour. of Atmospheric Scis., Planetary and Space Sci.; lectr. in field. Author: (with others) Thunderstorms, 1983, Lightning Electromagnetics, 1990, Benjamin Franklin des Lumieres á nos Jours, 1991; contbr. numerous articles to profl. jours. Fellow Am. Meteorol. Soc. (Outstanding Contbn. to Advance Applied Meteorology award 1985, chmn. com. on atmospheric electricity 1987-89); mem. IEEE (Transactions Prize Paper award EMC Soc. 1982), Am. Assn. Physics Tchrs., Am. Geophys. Union (Smith medal selection com. 1992—, com. on atmospheric and space electricity 1990—), Sigma Xi, Sigma Pi Sigma. Achievements include patents for All-Sky camera apparatus for time-resolved lightning photography, photoelectric lightning detector apparatus, transient event data acquisition apparatus for use with radar systems and the like, lightning detection system utilizing triangulation and field amplitude comparison techniques, thunderstorm sensor and method of identifying and locating thunderstorms. Office: U of Arizona Inst Atmospheric Physics PAS Bldg Tucson AZ 85721

KRIEGER, IRVIN MITCHELL, chemistry educator, consultant; b. Cleve., May 14, 1923; s. William I. and Rose (Brodsky) K.; m. Theresa Melamed, June 9, 1965; 1 dau., Laura. B.S., Case Inst. Tech., 1944, M.S., 1948; Ph.D., Cornell, 1951. Rsch. asst. Case Inst. Tech., Cleve., 1946-47; teaching fellow Cornell U., Ithaca, N.Y., 1947-49; instr. Case Western Res. U., 1949-51, asst. prof., 1951-55, assoc. prof., 1955-68, prof., 1968-88, prof. emeritus, 1988—; dir. Center for Adhesives, Sealants and Coatings, 1983-88; vis. prof. U. Bristol, 1977-78; cons. for chem. firms; prof. invité Ecole Nat. Supérieure de Chimie de Mulhouse, 1987, Louis Pasteur U., Strasbourg, France, 1989. Contbr. articles to profl. jours. Served as ensign USNR, 1943-46. NSF fellow Université Libre De Bruxelles, 1959-60; sr. fellow Weizmann Inst., 1970. Mem. Am. Chem. Soc., Am. Inst. Chem. Engrs., AAUP, Soc. Rheology (pres. 1977-79, Bingham medalist 1989). Home: 15691 Fenemore Rd Cleveland OH 44112-4010 Office: Case Western Res U Cleveland OH 44106

KRIKOS, GEORGE ALEXANDER, pathologist, educator; b. Old Phaleron, Greece, Sept. 17, 1922; came to U.S., 1946; s. Alexios and Helen (Spyropoulou) K.; m. Aspasia Manoni, June 22, 1949; children: Helen, Alexandra, Alexios. D.D.S., U. Pa., 1949; Ph.D., U. Rochester, 1959; Ph.D. hon. doctorate, U. Athens, Greece, 1981. Asst. prof. pathology U. Pa. Sch. Dentistry, 1958-61, assoc. prof., 1961-67, prof., 1967-68, chmn. dept., 1964-68; assoc. prof. oral pathology U. Pa. Grad. Sch., 1962-68, prof. oral pathology, 1968; prof. pathobiology Sch. Dentistry, U. Colo., Denver, 1968-75, chmn. dept. pathobiology, 1968-73, prof. oral biology, 1975-86, clin. prof. oral biology, 1986-91, prof. oral biology emeritus, 1991—; asst. dean basic sci. affairs Sch. Dentistry, U. Colo., 1973-75, asso. dean oral biology affairs, 1975-76; vis. prof. Sch. Dentistry, U. Athens, 1980-81; mem. dental study sect. NIH, 1966-70; mem. cancer com. Colo.-Wyo. Regional Med. Program, 1970-72; cons. oral pathology Denver VA Hosp., 1970-72. Served with AUS, 1949-54. Mem. Am. Soc. Investigative Pathology, Internat. Assn. Dental Rsch., Sigma Xi. Home: 350 Ivy St Denver CO 80220-5855 Office: U Colo Sch Dentistry 4200 E 9th Ave Denver CO 80262-0001

KRIMIGIS, STAMATIOS MIKE, physicist, researcher, space science/engineering manager, consultant; b. Chios, Greece, Sept. 10, 1938; s. Michael and Angeliki (Tsetseris) K.; children: Michael, John. B.S., U. Minn., 1961; M.S., U. Iowa, 1963, Ph.D., 1965. Research assoc. and asst. prof. physics U. Iowa, Iowa City, 1965-68; supr. space physics sect. Applied Physics Lab., Johns Hopkins U., Laurel, 1968-74; supr. space physics and instrumentation group Applied Physics Lab., Johns Hopkins U., 1974-81, chief scientist space dept., 1980-90, head space dept., 1991—; mem. Space Sci. Bd., Nat. Acad. Scis. NRC, 1983-86; chmn. com. on solar and space physics, 1983-86, cons.; mem. steering com. space sci. working group Assn. Am. Univs., 1983-85; mem. NASA space sci. and applications adv. com., 1987-91. Contbr. over 265 articles to sci. jours.; author book chpts. on solar, interplanetary and magnetospheric plasma physics, cosmic rays, magnetospheres of Earth, Jupiter, Saturn, Uranus and Neptune. Recipient Exceptional Sci. Achievement medal NASA, 1981, 86. Fellow Am. Geophys. Union, Am. Phys. Soc.; mem. AAAS. Greek Orthodox. Home: 613 Cobblestone Ct Silver Spring MD 20905-5806 Office: Johns Hopkins U Applied Physics Lab Laurel MD 20723

KRIMM, MARTIN CHRISTIAN, electrical engineer, educator; b. Shively, Ky., Dec. 15, 1921; s. Martin C. Sr. and Effie Verlee (Boling) K.; m. Helenora Magdelyn Schenk, Feb. 11, 1946; children: Marsha Lee, Sharon Cecilia,David Leslie, Timothy Wilson. MSEE, U. Ky., 1962. Contractor Louisville, 1945-50, 56; design engr., cons. Am. Standard Rsch., Louisville, 1954-56; asst. prof. elec. engring. U. Ky., Lexington, 1957-90, VPA prof. emeritus, 1991—; areas of interest include electrobiology and psychobioybernetics. Contbr. numerous broad-based articles for engring. edn. to profl. jours. With USN, 1942-45. MEm. AAUP, Tau Beta Pi, Eta Kappa Nu.

KRISHNAMACHARI, SADAGOPA IYENGAR, mechanical engineer, consultant; b. Chidambaram, Tamil Nadu, India, Sept. 14, 1944; came to U.S., 1982; s. Renga Iyengar and Alamelu Sadagopan; m. Lalitha Ramanujam, June 2, 1969; children: Sriram, Parashar. BS in Math., U. Madras, India, 1963, BSME, 1966; MS in Mechanics, Ill. Inst. Tech., 1984. Engr. Bharat Heavy Elecs. Ltd., Tiruchi, Tamil Nadu, 1967-73, sr. engr., 1973-77, mgr. nuclear engring., 1977-82; sr. stress analyst L.J. Broutman & Assocs., Chgo., 1984-91, mgr. computational svcs., 1991—; industry rep. Indian Boiler Regulatory Bd., 1976-77; mem. vis. faculty dept. mech. engring. Ill. Inst. Tech., Chgo., 1985—. Author: Applied Stress Analysis of Plastics--A Mechanical Engineering Approach, 1992. Founder classical music sch. for children, Tiruchi, 1978. Mem. ASME, Soc. Plastics Engrs., Soc. Exptl. Mechanics. Hindu. Achievements include pioneered capabilities for stress analysis of pressure vessels, piping, thermal and mechanical design of nuclear heat exchangers, components made of non-metallic materials; development of stress analysis of non-metallic materials. Office: LJ Broutman & Assocs 3424 S State St Chicago IL 60616-3896

KRISHNAMOORTHI, RAMASWAMY, biochemistry educator; b. Ooty, Tamilnadu, India, June 1, 1952; came to U.S., 1978; s. Ramaswamy V. and Rajammal Kaliyur; m. Shyamala Rajagopalan, July 4, 1983; children: Sindhuja, Bharath. BSc, Madras U., 1973, MSc, 1975; PhD, U. Calif., Davis, 1983. Postdoctoral fellow Purdue U., West Lafayette, Ind., 1983-85, U. Wis., Madison, 1985-86; asst. prof. dept. biochemistry Kans. State U., Manhattan, 1986-92, assoc. prof., 1992—. Contbr. articles to profl. jours. Recipient Grant-in-Aid Am. Heart Assn., 1987—; rsch. grantee Wesley Found., 1990—, NIH, 1993—. Mem. AAAS, Am. Assn. Molecular Biologists (assoc.), Sigma Psi. Home: 3317 Newbury Manhattan KS 66502 Office: Kansas State U Dept Biochemistry Manhattan KS 66506

KRISHNAMURTHY, RAMACHANDRAN (KRISH), chemical engineer, researcher; b. Madras, India, Feb. 4, 1957; came to U.S., 1980; s. Marathukudi Gopalasastry and Yogambal Ramachandran; m. Girija Ganapathisubramanian, Oct. 11, 1981; children: Vivek, Vyas, Verna. B of Tech., U. Madras, 1979; MS, Clarkson U., 1983, PhD, 1984. Registered profl. engr., N.J. Process engr. mgmt. trainee Engrs. India Ltd., New Delhi, 1979-80; sr. engr. Boc Group Inc., Murray Hill, N.J., 1985-88, lead engr., 1988-90, sect. mgr., 1990—. Contbr. articles to profl. jours. Recipient Ted Peterson Student Paper award AICE, 1990, Tech. Innovation award Airco Indsl. Gases, 1991. Mem. AICE (chmn. N.J. local sect. 1990), AAAS, Phi Kappa Phi. Democrat. Hindu. Achievements include 19 patents in the field of separation processes for purification of hydrogen, carbon monoxide, carbon dioxide and argon. Home: 13 Ross Ave Chestnut Ridge NY 10977-6909 Office: BOC Group Inc Tech Ctr 100 Mountain Ave Murray Hill NJ 07974

KRISHNAMURTHY, SURESH KUMAR, chemist, researcher; b. Palghat, Kerala, India, Sept. 24, 1966; s. Krishnamurthy and Rajeswari (Krishnamurthy) Subramania. BS, Bombay U., 1987; MS, Indian Inst. Tech., 1989. Jr. rsch. fellow Indian Inst. Tech., Bombay, 1989-91, sr. rsch. fellow, 1991—. Contbr. articles to profl. jours. including Jour. Applied Polymer Sci. Home: B-8 Sarvodaya Bldgs RC Marg, Chembur, 400-071 Bombay India Office: Indian Inst Tech Powai, Dept Chemistry, 400-076 Bombay India

KRISHNAN, ANANTHA, mechanical engineer; b. Ranchi, Bihar, India, Feb. 3, 1962; came to U.S., 1984; s. Venkatraman and Seetha (Lakshmi) Krishnaswamy. MS, Marquette U., 1986; DSc, MIT, 1989. Teaching asst. Marquette U., Milw., 1984-86; rsch. asst. MIT, Cambridge, Mass., 1986-89; project engr. CFD Rsch. Corp., Huntsville, Ala., 1989-91, sr. project engr., 1991-93, group leader, 1993—; cons. NASA, Huntsville, 1989-91, Sematech, Austin, Tex., 1992. Contbr. over 30 publs. to internat. jours. Mem. AIAA. Home: 12001 Chimney Hollow Trail Huntsville AL 35803 Office: CFD Rsch Corp 3325-D Triana Blvd Huntsville AL 35805

KRISHNAPPAN, BOMMANNA GOUNDER, fluid mechanics engineer; b. Madras, India, Jan. 15, 1943. BE, Madras U., 1966; MSc, U. Calgary, 1968;

PhD in Civil Engring., Queen's U., Ont., 1972. Rsch. scientist Can. Ctr. Inland Waters, Ont., 1972-77, rsch. scientist hydraulics, nat. water rsch. inst., 1978—; flow system engr. Ont. Hydro, 1977-78; asst. Nat. Rsch. Coun. Can., 1966-72. Recipient Les Prix Camille A. Dagenais award Can. Soc. Civil Engring., 1991. Mem. Internat. Assn. Hydraulic Rsch. Achievements include research in sediment transport in open channel flows, dispersion of mass in open channels, mathematical models for river morphology, thermal models. Office: Hydraulics Div Ntl Water Res, 867 Lakeshore Rd, Burlington, ON Canada L7R 4A6*

KRISS, JOSEPH JAMES, data processing executive; b. Pitts., Nov. 29, 1942; s. Joseph Frank and Dorothy (Crowley) K.; m. Lilyan Marie Hart, Mar. 25, 1972; 1 stepchild, James W. Wile II. AS in Bus. Mgmt., Westmoreland Community Coll., Youngwood, Pa., 1975; BSBA, Kennedy-Western U., 1990, MBA, 1993. Data processing mgr. Housewares, Inc., Pitts., 1964-67; data processing mgr. Stuart's Drug & Surg. Supply, Inc., Greensburg, Pa., 1967-72, dir. mgmt. info. systems, 1972-86, v.p. mgmt. info. systems, 1986-89, chief info. officer, 1989—. Democrat. Avocations: golf, walking, racquet ball, tennis. Office: Stuart Med Inc Supply Inc Donahue and Luxor Rds Greensburg PA 15601

KRITCHEVSKY, DAVID, biochemist, educator; b. Kharkov, Russia, Jan. 25, 1920; came to U.S., 1923, naturalized, 1929; s. Jacob and Leah (Kritchevsky) K.; m. Evelyn Sholtes, Dec. 21, 1947; children—Barbara Ann, Janice Eileen, Stephen Bennett. B.S., U. Chgo., 1939, M.S., 1942; Ph.D., Northwestern U., 1948. Chemist Ninol Labs., Chgo., 1939-46; postdoctoral fellow Fed. Inst. Tech., Zurich, Switzerland, 1948-49; biochemist Radiation Lab., U. Calif. at Berkeley, 1950-52, Lederle Lab., Pearl River, N.Y., 1952-57, Wistar Inst., Phila., 1957—; prof. biochemistry Sch. Vet. Medicine U. Pa., Phila., 1965—; prof. biochemistry Sch. Medicine U. Pa., 1970—, chmn. grad. group molecular biology, 1972-84; Mem. USPHS study sect. Nat. Heart Inst., 1964-68, 72-76; chmn. research com. Spl. Dairy Industry Bd., 1963-70; mem. food and nutrition bd. Nat. Acad. Sci., 1976-82. Author: Cholesterol, 1958, also numerous articles.; editor: (with G. Litwack) Actions of Hormones on Molecular Processes, 1964; co-editor: (with R. Paoletti) Advances in Lipid Research, 1963-89, (with P. Nair) 1973, Bile Acids, 1971; Western Hemisphere editor Atherosclerosis, 1978-90, cons. editor, 1990—. Recipient Research Career award Nat. Heart Inst., 1962, award Am. Coll. Nutrition, 1978, Herman award Am. Soc. Clin. Nutrition, 1992; research on role vehicle when cholesterol and fat produces atherosclerosis in rabbits, effects of saturated and unsaturated fat, deposition of orally administered cholesterol in aorta of man and rabbit, caloric restriction and cancer. Fellow Am. Inst. Nutrition (Borden award 1974, pres. 1979); mem. AAAS, Am. Soc. Biol. Chemists, Am. Chem. Soc. (award Phila. sect. 1977), Soc. Exptl. Biology and Medicine (pres. 1985-87), Arteriosclerosis Coun., Am. Heart Assn., Am. Soc. Oil Chemists (chmn. methods com. 1963-64), Internat. Soc. Fat Rsch. Home: 136 Lee Cir Bryn Mawr PA 19010-3724 Office: Wistar Inst 36th and Spruce Sts Philadelphia PA 19104

KRIZ, GEORGE JAMES, agricultural research administrator, educator; b. Brainard, Nebr., Sept. 20, 1936; s. George Jacob and Rose Agnes Kriz; m. Patricia Elizabeth Kelly (div. Feb. 1989); children: Rosalie Sue, Richard Patrick, Thomas George; m. Rhoda Mae Whitacre, June 23, 1989. BS in Agrl. Engring., Iowa State U., 1960, MS in Agrl. Engring., 1962; PhD, U. Calif., Davis, 1965. Lectr. U. Calif., Davis, 1965; asst. prof. agrl. engring. N.C. State U., Raleigh, 1965-69, assoc. prof., 1968-69, prof., 1972—, assoc. dept. head, 1969-73, asst. rsch. dir., 1973-81, assoc. rsch. dir., 1981—. Fellow Am. Soc. Agrl. Engring. (bd. dirs. 1983-85, found. trustee 1986—, Presdl. citation 1988, 91); mem. Coun. Agrl. Scis. and Tech. Avocation: swimming. Office: NC State U Box 7643 100 Patterson Hall Raleigh NC 27695-7643

KROCK, HANS-JURGEN, civil engineer; b. Krakow, Poland, Aug. 30, 1942; came to U.S. 1952; s. Gunther and Lucie Natalie (Prietz) K.; m. Lynda Gibson, Sept. 1966 (div. 1972). BS in Civil Engring. magna cum laude, Ariz. State U., 1965; MS in Sanitary Engring., U. Calif., Berkeley, 1967, PhD in Environ. Engring., 1972. Registered profl. engr. Hawaii. Engring. trainee BAgdad (Ariz.) Copper Corp., 1958-62; researcher Salt River Project, Tempe, Ariz., 1963-64; pub. health engr. Ariz. State Health Dept., Phoenix, 1965-66; rsch. engr. L.A. County Sanitation Dist., Pomona, Calif., 1967-68; researcher SERL/U. Calif., Berkeley, 1970-72; sr. engr., lab. dir., chief diver M&E Pacific Inc., Honolulu, a1972-80; pres. OCEES, Internat., Inc., Honolulu, 1988—; dir., assoc. prof. J.K.K. Look Lab., U. Hawaii, Honolulu, 1980—; chmn. dept. Ocean Engring. U. Hawaii, Honolulu, 1992—; mem. Water Quality Stds. Bd., Dept. Health, Hawaii, 1978—; chmn. Water Quality Bd., EPA, Pago Pago, Am. Samoa, 1986—. Editor: ICOER '89, 1990; patentee in field; contbr. articles to profl. jours. Recipient Commendation, Gov. Hawaii, 1979. Fellow Am. Inst. Chemists; mem. NSPE, ASCE (control mem. nat. com. ocean energy), Engring. Assn. Hawaii (pres. 1982-83), Am. Chem. Soc., Marine Tech. Soc., Water Pollution Control Fedn. Avocations: diving, kayaking. Office: U Hawaii JKK Look Lab 811 Olomehani St Honolulu HI 96813-5513

KROCKOVER, GERALD HOWARD, science educator; b. Sioux City, Iowa, Nov. 12, 1942; s. Marvin H. and Rose (Holdowsky) K.; m. Sharon D. Shulkin, Jan. 30, 1965; children: Mark A., Chad B. BA, U. Iowa, 1964, MA, 1966, PhD, 1970. Cert. sci. tchr., Iowa, 1968. Mid. sch. sci. tchr. Bettendorf (Iowa) Community Sch. Dist., 1964-67; chemistry tchr. Univ. High Sch., Iowa City, 1967-70; asst. prof., assoc. prof. Purdue U., West Lafayette, Ind., 1970-80; prof. Purdue U., West Lafayette, 1980—; bd. dirs. Coun. for Elem. Sci. Internat., Washington, 1990-93; vis. scholar U. Tex., Austin, 1976. Author: Creative Sciencing: A Practical Approach, 1980, Activities Handbook for Energy Education, 1981, Creative Museum Methods and Educational Techniques, 1982, Creative Sciencing: Ideas and Activities for Teachers and Children, 1991, Creative Teaching: A Practical Approach, 1993. Grants com. mem. Pub. Sch. Found., Lafayette, Ind., 1986—; mem. Golden Apple awards com., Lafayette, 1988—, Tchr. Tng. and Licensing Adv. com., Indpls., 1991-92. Grants to improve insvc. sci. tchr. edn. 1971—. Fellow AAAS; mem. NSTA (chair internat. com. 1990-93), Sch. Sci. and Math. (jour. reviewer), Nat. Assn. for Rsch. in Sci. Teaching (jour. reviewer), Kappa Delta Pi, Phi Delta Kappa. Office: Purdue U 1440 Liberal Arts/Edn Bldg West Lafayette IN 47907-1440

KROESEN, GERRIT MARIA WILHELMUS, physicist, educator; b. Heerlen, Limburg, Netherlands, May 10, 1958; s. Albertus F. and Wilhelmina M. (Hegeman) K. Ingenieur, Eindhoven U., 1983, PhD, 1988. Rsch. staff U. Tech., Eindhoven, 1986-88, asst. prof. physics, 1988—; vis. scientist IBM Rsch., Yorktown Heights, N.Y., 1990, Ecole Polytechnique, Paris, 1984; mem. internat. sci. com. European Sectional Conf. on Atomic and Molecular Physics of Ionized Gases, 1990—. Contbr. articles to profl. jours. Organist Cath. Ch., Eindhoven, 1990—, Eindhoven U., 1990—. Mem. Dutch Phys. Soc., Am. Vacuum Soc., Electrochem. Soc. Roman Catholic. Achievements include 2 patents. Home: Boschdijk 123C, Eindhoven The Netherlands 5612 HB Office: Univ Technology, Dept Physics PO Box 513, 5600 MB Eindhoven The Netherlands

KROFTA, MILOS, engineer; b. Ljubljana, Slovenia, July 23, 1912; came to U.S., 1951; s. Hanus and Minka (Jelačin) K. m. Maria Hybler, Aug. 3, 1937; children: Tjasa, Hanka. BS in Engring., U. Ljubljana, 1932; MSME, U. Prague, 1934; PhD in Papermaking, U. Darmstadt, Germany, 1937. Profl. engr., N.Y., Mass. Chemist Paper Mills V.G.M., Ljubljana, Yugoslavia, 1935-37, tech. mgr., 1937-45; prin. Krofta Engring. Co., Milan, Italy, 1945-51, Lenox, Mass., 1951—; owner, mgr. Krofta Apparatebau, Karlsruhe, Germany, 1960—, Manchester, U.K., Krofta U.K., Manchester, U.K., 1960—, Krofta Swizterland, France, Japan, Taiwan, Korea, Mexico, Brazil, Argentina, 1970—. Achievements include 30 patents on flotation technology for water-waste water clarification. Office: Lenox Inst of Water Tech Inc 101 Yokun Ave PO Box 1639 Lenox MA 01240

KROGER, MANFRED, food science educator; b. Bad Oeynhausen, Germany, May 19, 1933; arrived in Can., 1955; came to U.S., 1961; s. Alfred and Adele (Wolf) K.; m. Goldie Laris, Dec. 27, 1962; children: Hans, Erika, Steven. BS, U. Manitoba, 1961; MS, Pa. State U., 1963, PhD, 1966. Dairy quality control officer Office of Provincial Govt., Winnipeg, Manitoba, Can., 1959-60; prof. food sci. Pa. State U., University Park, 1963—; bd. trustees Drinking Water Rsch. Found., Alexandria, Va., 1989—; edit. cons. Rodale Press-Prevention Mag., Emmaus, Pa., 1988—; cons. in field. Co-editor: Nutrition Forum; tech. editor: Prometheus Books; assoc. editor: Jour. Food Sci.; author: Ency. Fermented Fresh Milk Products, 1992; co-author 4 books, several chpts.; contbr. articles to profl. jours. Mem. Am. Cultured Dairy Products Inst., Am. Coun. Sci. and Health, Inst. Food Techs. (food sci./food issues communicator). Home: 711 McKee St State College PA 16803 Office: Pa State U 104 Burland Lab University Park PA 16802

KROGH-JESPERSEN, MARY-BETH, chemist, educator; b. Schenectady, N.Y., Aug. 10, 1949; parents: George Henry and Barbara Veronica (Norton) Baillie; m. Karsten Krogh-Jespersen, Dec. 20, 1975; children: Erik, Sheila Ann, Michelle Grace. BA in Chemistry, Northeastern U., 1972; PhD in Chemistry, NYU, 1976; MBA in Mgmt., Pace U., 1990. Postdoctoral fellow U. Maine, Orono, 1978-79; lectr. Rutgers U., New Brunswick, N.J., 1979-81; prof. chemistry Pace U., N.Y.C., 1981-92, chair chemistry dept., 1990-92; dean coll. of sci., prof. chemistry Rochester (N.Y.) Inst. Tech., 1992—; speaker on women in sci. Contbr. over 20 articles to profl. publs. Vis. sci. tchr., St. Matthew's Sch., Edison, N.J., 1991-92. NSF grantee, 1988; Dyson Coll., Pace U., fellow, 1984—. Mem. AAUW, Am. Chem. Soc., Delta Mu Delta. Achievements include studies of spectroscopy of platinum anti-cancer agents. Office: Rochester Inst Tech PO Box 9887 Rochester NY 14623-0887

KROHA, JOHANN, physicist; b. Weilheim, Bavaria, Federal Republic of Germany, Mar. 14, 1961; s. Wilhelm and Hedwig (Hell) K. BA in Physics, Tech. U. Munich, 1983; M Physics, 1988; PhD in Physics, U. Karlsruhe (Germany), 1993. Teaching asst. Tech. U. Munich, 1984-85; rsch. asst. U. Fla., Gainesville, 1986-88; rsch. assoc. U. Karlsruhe, 1988, theoretical physicist, 1989—; cons. U. Karlsruhe, 1990-93; rsch. assoc. Cornell U., 1993—. Contbr. articles to profl. jours. Govt. scholar, Bavaria, Germany, 1981-86, Studienstiftung des Deutschen Volkes scholar German NSF, 1984-87, Alexander von Humboldt-Stiftung scholar, 1993—. Mem. Am. Phys. Soc., Deutsche Physikalische Gesellschaft. Office: Cornell U Lab of Atomic and Solid State Physics Ithaca NY 14853

KROHN, KARSTEN, chemistry educator; b. Hademarschen, Germany, Apr. 20, 1944; s. Hans-Peter and Ingeborg (Weber) K.; m. Odile Cornic, Mar. 31, 1972; children: Nicolas, Caroline. PhD in Chemistry, U. Kiel, 1971. Asst. U. Kiel (Fed. Republic Germany), 1969-71; postdoctoral fellow U. Kiel, Fed. Republic Germany, 1971-73, U. Hannover, Fed. Republic Germany, 1973-74; asst. prof. U. Hamburg, Fed. Republic Germany, 1975-81; assoc. prof. U. Braunschweig, Fed. Republic Germany, 1981-91; prof. U. Paderborn, Fed. Republic Germany, 1991—; vis. prof. U. Wis., Madison, 1984. Contbr. articles to profl. jours. Home: Brauerskamp 34, D 3300 Braunschweig Germany Office: Fachbereich Chemie, Warburger Str 100, D 33095 Paderborn Germany

KROHN, KENNETH ALBERT, radiology educator; b. Stevens Point, Wis., June 19, 1945; s. Albert William and Erma Belle (Cornwell) K.; m. Marijane Alberta Wideman, July 14, 1968; 1 child, Galen. BA in Chemistry, Andrews U., 1966; PhD in Chemistry, U. Calif., 1971. Acting assoc. prof. U. Wash., Seattle, 1981-84, assoc. prof. radiology, 1984-86, prof. radiology and radiation oncology, 1986—, adj. prof. chemistry, 1986—; guest scientist Donner Lab. Lawrence Berkeley (Calif.) Lab., 1980-81; radiochemist, VA Med. Ctr., Seattle, 1982—. Contbr. numerous articles to profl. jours.; patentee in field. NDEA fellow. Fellow AAAS; mem. Am. Chem. Soc., Radiation Rsch. Soc., Soc. Nuclear Medicine, Acad. Coun., Sigma Xi. Home: 11322 23d Ave NE Seattle WA 98125 Office: Imaging Rsch Lab RC-05 U Washington Seattle WA 98195

KROLL, CASIMER V., engineering executive; b. Buffalo, May 29, 1947; s. Casimer Walter and Marie (Matarese) K.; m. Rebecca Little, Sept. 6, 1969; children: Casimer J., Kimberly. BS in Aerospace Engring., SUNY, Buffalo, 1968; MBA, SUNY, 1982; M in Applied Sci., U. Toronto, 1970. Project engr. Cornell Aero. Lab., Cheektowaga, N.Y., 1969-70; dept. head Buffalo Bd. Edn., 1971-75; product mgr., dir. tech. svcs. Sentrol Systems Ltd./Inc., Toronto, Atlanta, Can., Ga., 1975-78; gen. mgr. AIRCO, Inc., Montvale, N.J., 1978-84; v.p. ops. DeVry Div. Bell & Howell, Inc., Evanston, Ill., 1984-87; sr. v.p. API Systems, Culver City, Calif., 1987-90; pres., CEO Thorn Automated Systems, Inc., Westlake, Ohio, 1990—. Mem. ASIS (fellow). Home: 408 Windward Way Avon Lake OH 44012 Office: Thorn Automated Systems Inc 835 Sharon Dr Westlake OH 44145

KROLL, MARK WILLIAM, electrical engineer; b. Mpls., July 11, 1952; s. William H.O. and Irene Claudia Kroll; m. Lori Carolyn Palm, Sept. 6, 1975; children: Braden, Mollie, Ryan, Chase. BS in Math., U. Minn., 1975, MSEE, 1983, PhDEE, 1987; MBA, U. St. Thomas, St. Paul, 1990. Circuit designer Medtronic, Fridley, Minn., 1970-72; teaching asst. U. Minn., Mpls., 1973-78; v.p. R & D Intercomp, Plymouth, Minn., 1978-84; v.p. of rsch. and devel. Cherne Med. Co., Edina, Minn., 1985-90; v.p. rsch. Angemed, Plymouth, 1991—; bd. dirs. BioDimensions, Inc., Houston, Surviva-Link, Plymouth, Minn.-Heart-Tech, St. Louis Park, Minn. Co-editor Implantable Cardioverter Defibrilator Therapy, 1993; contbr. over 34 papers to profl. publs. Bd. dirs. St. Peter's Luth. Sch., Edina, 1990—. Alfred P. Sloan fellow, 1971. Mem. IEEE, Am. Heart Assn. (coun. on clin. cardiology). Achievements include 57 patents in field which include patents for method for defibrillation, heart sound sensor, lung sound cancellation method and apparatus, wheel scale assembly, medical current limiting circuit, flexible and disposable electrode belt device (U.S., U.K.), terrain biased dynamic multiple threshold synchronization method and apparatus, load cell assembly, medical current limiter, electrode belt adaptor, optical fiber transmissive signal modulation system, bioelectric noise cancellation system, optical fiber reflective signal modulation system, flexible wegwerfbare elektrodenbandvorrichtung (Germany). Office: AngeMed 3650 Annapolis Ln N Minneapolis MN 55447-5434

KROMIDAS, LAMBROS, cell biologist, physical scientist; b. Chios, Greece, Nov. 25, 1956; came to U.S., 1969.; s. George and Maria (Moutafi) K.; m. Marie-Louise Caloustian, June 28, 1987. BA, NYU, 1980; MS, St. John's U., 1984, PhD, 1990. Teaching asst. St. John's U., Jamaica, N.Y., 1981-87; rsch. fellow St. John's U., Jamaica, 1988-90; postdoctoral fellow dept. physiology Med. Coll. Cornell U., N.Y.C., 1990-92; physical scientist U.S. Dept. Energy, N.Y.C., 1992—; environ. field worker U.S. Dept. Energy, N.Y.C., 1991. Author: Jour. Toxicology Letters, Vol. 51, 1990, Vol. 60, 1992. Rsch. vol. Mt. Sinai Sch. of Medicine, N.Y.C., 1983-84. Valergakis grantee, 1986. Mem. Internat. Soc. for Study of Xenobiotics, Soc. Environ. Toxicology and Chemistry, N.Y. Acad. Scis., N.Y. Soc. Electron Microscopists, Soc. Toxicology (Neurotoxicology Splty. award 1990), Tissue Culture Assn., Hellenic U. Club of N.Y. Achievements include demonstration of methyl mercury disruption of the cytoskeleton and protection of glutathione against toxicity of methyl mercury. Office: US Dept Energy Environ Measurements Lab 376 Hudson St New York NY 10014

KRON, RICHARD G., astrophysicist, educator. Prof. dept. astronmy and astrophysics U. Chgo.; dir. Yerkes Obs., Williams Bay, Wis. Office: Yerkes Obs U Chgo PO Box 258 Williams Bay WI 53191*

KRONENBERG, STANLEY, research physicist; b. Krosno, Poland, May 3, 1927; came to U.S. 1953; s. Ferdinand and Eugenie (Hutny) K.; m. Eva Maria Kroupa, Apr. 1, 1953; children: Eric, Olga. PhD, U. Vienna, 1952. Rsch. physicist U.S. Army, Ft. Monmouth, N.J., 1953—; cons. Fed. Emergency Mgmt. Agy., Washington, 1960—. Author: High Intensity Radiation Dosimetry with Semirad, 1966; contbr. articles to profl jours. Recipient Outstanding Pub. Svc. award Fed. Emergency Mgmt. Agy., 1986. Mem. Am. Phys Soc., Am. Philatelic Soc., Polonus Philatelc and Numismatic Soc. (hon.). Achievements include 30 patents in the area of radiology. Office: US Army CECOM NVESD Bldg 9401 (Evans) Fort Monmouth NJ 07703

KROPOTOFF, GEORGE ALEX, civil engineer; b. Sofia, Bulgaria, Dec. 6, 1921; s. Alex S. and Anna A. (Kurat) K.; came to Brazil, 1948, to U.S., 1952, naturalized, 1958; BS in Engring., Inst. Tech., Sofia, 1941; postgrad. in computer sci. U. Calif., 1968; Registered profl. engr., Calif.; m. Helen P., July 23, 1972. With Standard Eletrica S.A., Rio de Janeiro, 1948-52, Pacific Car & Foundry Co., Seattle, 1952-64, T.G. Atkinson Assocs., Structural Engrs., San Diego, 1960-62, Tucker, Sadler & Bennett A-E, San Diego, 1964-74, Gen. Dynamics-Astronautics, San Diego, 1967-68, Engring. Sci., Inc., Arcadia, Calif., 1975-76, Incomtel, Rio de Janeiro, Brazil, 1976, Bennett Engrs., structural cons., San Diego, 1976-82; project structural engr. Hope Cons. Group, San Diego and Saudi Arabia, 1982-84; cons. structural engr. Pioneered engring. computer software. With U.S. Army, 1941-45. Fellow ASCE; mem. Structural Engrs. Assn. San Diego (assoc.), Soc. Am. Mil. Engrs., Soc. Profl. Engrs. Brazil. Republican. Russian Orthodox. Home: 9285 Edgewood Dr La Mesa CA 91941-5612

KROTKI, KAROL JOZEF, sociology educator, demographer; b. Cieszyn, Poland, May 15, 1922; emigrated to Can., 1964; s. Karol Stanislaw and Anna Elzbieta (Skrzywanek) K.; m. Joanna Patkowski, July 12, 1947; children—Karol Peter, Jan Jozef, Filip Karol. B.A. (hons.), Cambridge (Eng.) U., 1948, M.A., 1952; M.A., Princeton U., 1959, Ph.D., 1960. Civil ser. Eng., 1948-49; dep. dir. stats. Sudan, 1949-58; vis. fellow Princeton U., 1958-60; research adviser Pakistan Inst. Devel. Econs., 1960-64; asst. dir. census research Dominion Bur. Stats., Can., 1964-68; prof. sociology U. Alta., 1968-83, univ. prof., 1983-91, univ. prof. emeritus, 1991—; vis. prof. U. Calif., Berkeley, 1967, U. N.C., 1970-73, U. Mich., 1975; coord. program socio-econ. rsch. Province Alta., 1969-71; dir. Can. Futures Rsch. Inst., Edmonton, 1970—; cons. in field. Author 11 books and monographs; contbr. numerous articles to profl. jours. Served with Polish, French and Brit. Armed Forces, 1939-46. Recipient Achievement award Province of Alta., 1970, Commemorative medal for 125th Anniversary of Can., 1992; grantee in field. Fellow Am. Statis. Assn., Royal Soc. Can. (v.p. 1986-88), Acad. Humanities and Social Scis. (v.p. 1984-86, pres. 1986-88); mem. Fedn. Can. Demographers (v.p. 1977-82, pres. 1982-84), Can. Population Soc., Association des Demographes du Quebec, Soc. Edmonton Demographers (founder, pres. 1990—), Cen. and E. European Studies Soc. (pres. 1986-88), Population Assn. Am., Internat. Union Sci. Study Population, Internat. Statis. Inst., Royal Statis. Soc. Roman Catholic. Home: 10137 Clifton Pl, Edmonton, AB Canada T5N 3H9 Office: U Alta, Dept Sociology, Edmonton, AB Canada T6G 2H4

KROTO, HAROLD WALTER, science educator; b. Oct. 7, 1939; s. Heinz and Edith K.; m. Margaret Henrietta Hunter, 1963; 2 children. Student, U. Sheffield, 1961-64. Rsch. scientist Bell Tel. Labs., N.J., 1966-67; lectr. U. Sussex, Brighton Sussex, Eng., 1968-77, reader, 1977-85, prof. chemistry, 1985-91, Royal Soc. Rsch. prof., 1991—; vis. prof. U. B.C., 1973, U. So. Calif., 1981, UCLA, 1988-90; mem. phys. chemistry com. Sussex European Rsch. Ctr., 1987—, synchrotron com., 1987—, chemistry com., 1988—. Contbr. articles to profl. jours. Postdoctoral fellow NRCC, 1964-66; recipient Internat. New Materials prize Am. Phys. Soc., 1992. Office: Univ Sussex, Dept Molecular Sciences Falmer, Brighton Sussex BN1 9RH, England*

KRUEGER, DARRELL GEORGE, nuclear power industry consultant; b. Brainerd, Minn., Oct. 26, 1948; s. George William and Dorthea Viola (Kjera) K.; m. Deborah Ann Blake, May 31, 1986; 1 child, Sarah Kirsten. BS in Edn., St. Cloud State Coll., 1972. Classroom instr. Westinghouse Nuclear Tng. Ctr., Zion, Ill., 1981-82, simulator instr., 1982-83, program devel. specialist, 1983-84; program devel. specialist Richardson-Estes, Ltd., Gurnee, Ill., 1984-88, project mgr., 1988-90, mgr. Zion sta. projects, 1990-92, dir. quality assurance, 1992—. Co-founder Living With Infertility and Experimentation Support Group, Evanston, Ill., 1991. With USN, 1972-80. Office: Richardson-Estes Ltd 4215 Grove Ave Gurnee IL 60031

KRUEGER, KURT EDWARD, environmental management company official; b. Santa Monica, Calif., June 24, 1952; s. Richard L. and Peggy J. (Cisler) K.; m. Maureen S. Catland, Aug. 4, 1973; children: Corey Edward, Brendan Kurt, Alyssa Marie. BA in Biology, Calif. State U., Northridge, 1978; MS in Environ. and Occupational Health & Safety, Calif. State U., 1980; MBA, Pepperdine U., 1988. Registered environ. health specialist. Regional health and safety coord. Internat. Tech. Corp., Wilmington, Calif., 1979-82, mgr. emergency response program, 1982-85, ops. mgr., 1985-88, gen. mgr., 1988-89; dir. health and safety Internat. Tech. Corp., Torrance, Calif., 1989—. Mem. Am. Indsl. Hygiene Assn. (cert.), Masons. Office: Internat Tech Corp 23456 Hawthorne Blvd Ste 300 Torrance CA 90505-4738

KRUEGER, RONALD, aerospace engineer; b. Calw, Ger., Nov. 28, 1958; s. Margit Evelin Nassi Krueger. Abitur, Kepler Gymnasium, Pforzheim, 1978; Diplom Ingenieur, U. Stuttgart, 1989. Asst. U. Stuttgart, 1989—. With German Army, 1978-79. Mem. AIAA, Deutsche Gesellschaft für Luft- und Raumfahrt Lilienthal-Oberth e.V. DGLR. Home: Glasbronnenstr 10, 75449 Wurmberg 70550, Federal Republic of Germany Office: Inst for Statics & Dyn, Pfaffenwaldring 27, Stuttgart 70569, Germany

KRUG, EDWARD CHARLES, environmental scientist; b. New Brunswick, N.J., Aug. 24, 1947; s. Edward and Regina (Bartkoviak) K.: m. Nancy Wegner, July 19, 1988. BS in Environ. Sci with highest honors, Rutgers U., 1975, PhD in Soil Sci., 1981. Asst. scientist Conn. Agrl. Expt. Sta., New Haven, 1980-85; assoc. scientist Ill. State Water Survey U. Ill., Champaign, 1985-90; dir. environ. projects CFACT, Washington, 1991—; advisor CFACT, 1989-90, Conn. Tech. Lakes Com., Hartford, 1980-84. Author: (chpt.) Encyclopedia for Earth System Science, 1992; contbr. articles to Jour. Pollution Control Assn., Jour. Air and Waste Mgmt. Assn., Jour. Hydrology. Mem. N.J. Ad Hoc Water Quality Control Com., New Brunswick, 1972-73; reviewer, tech. advisor N.J. PIRG, New Brunswick, 1972-75; chmn. ch. and soc. coms. United Meth. Ch., Winona, Minn., 1990-91. With USN, 1967-69. Recipient Frank G. Helyar award Rutgers U., 1973, Excellence in Rev. award Jour. Environ. Quality, 1991. Mem. AAAS, Am. Geophys. Union, Internat. Soc. Soil Sci., Soil Sci. Soc. Am. Achievements include development of organic acid buffering theory; generalization of Rosenquist land-use theory to include naturally increased acidity of watershed from accelerated loss of bases; unified theory of acid/base biogeochemistry. Office: CFACT PO Box 65722 Washington DC 20035-5722

KRUG, MAURICE F., engineering company executive; b. 1929. BS in Mech. Engring., U. Dayton, 1955. Rschr. U. Dayton, 1955-60; founder, chmn., CEO Krug Internat. Corp., 1959—. With U.S. Army, 1950-52. Office: Krug Internat Corp 6 Gem Plaza Ste 500 Dayton OH 45402*

KRUGER, JEROME, materials science educator, consultant; b. Atlanta, Feb. 7, 1927; s. Isaac and Sarah (Stein) K.; m. Mollee Coppel, Feb. 20, 1955; children: Lennard, Joseph. B.S., Ga. Inst. Tech., 1948, M.S., 1949; Ph.D., U. Va., 1952. With Naval Research Lab., Washington, 1952-55; with Nat. Bur. Standards, Commerce Dept., Washington, 1955-83; group leader Corrosion and Electrodeposition Nat. Bur. Standards, Commerce Dept., 1966-83; prof. Johns Hopkins U., 1984—, chmn. materials sci. and engring., 1986-88; cons. Argonne Nat. Lab., Lockheed, Balt. Gas & Electric, Teletech Thompson, Dalton & DeRose, Mueller Brass. Divisional editor Jour. Electrochem. Soc., 1966-83; subject area editor: Ency. of Materials Sci. and Engring.; also editor books; contbr. articles to tech. jours., chpts. to book. DuPont fellow U. Va., 1951-52; recipient Silver medal Commerce Dept., 1962, Gold medal, 1972; Blum award Nat. Capitol sect. Electrochem. Soc., 1966; Samuel Wesley Stratton award Nat. Bur. Standards, 1982; Presdl. rank of Meritorious Exec. of Sr. Exec. Svc., 1982; U.R. Evans award Brit. Inst. Corrosion, 1991. Fellow Electrochem. Soc. (treas. 1982-86, hon. mem. 1987, Outstanding Achievement award 1977), Nat. Assn. Corrosion Engrs. (bd. dirs. 1983-86, W.R. Whitney award 1976); mem. AIME, Am. Soc. Materials, Am. Inst. Conservation, Internat. Corrosion Coun. (1st v.p. 1984-87, pres. 1987-90), Metall. Soc., Fedn. Materials Socs. (pres. 1977), Sigma Xi, Tau Beta Pi. Jewish (bd. dirs. 1966-69). Home: 619 Warfield Dr Rockville MD 20850-1921 Office: Johns Hopkins U Dept Materials Sci and Engring Baltimore MD 21218

KRUGER, PAUL, nuclear civil engineering educator; b. Jersey City, June 7, 1925; s. Louis and Sarah (Jacobs) K.; m. Claudia Mathis, May 19, 1972; children: Sharon, Kenneth, Louis. BS, MIT, 1950; PhD, U. Chgo., 1954. Registered profl. engr., Pa. Rsch. physicist GM, Detroit, 1954-55; mgr. dept. chemistry Nuclear Sci. and Engring. Corp., Pitts., 1955-60; v.p. Hazleton Nuclear Sci. Corp., Palo Alto, Calif., 1960-62; prof. civil engring. Stanford (Calif.) U., 1962-87, prof. emeritus, 1987—; cons. Elec. Power Rsch. Inst., Palo Alto, 1975—, Los Alamos (N.Mex.) Nat. Lab., 1985—. Author: Principles of Activation Analysis, 1973, Geothermal Energy, 1972. 1st lt. USAF, 1943-46, PTO. Recipient achievement cert. U.S. Energy R &

D Adminstrn., 1975. Fellow Am. Nuclear Soc.; mem. ASCE (divsn. chmn. 1978-79). Home: 819 Allardice Way Stanford CA 94305-1050 Office: Stanford U Civil Engring Dept Stanford CA 94305

KRUGER, WILLIAM ARNOLD, consulting civil engineer; b. St. Louis, June 13, 1937; s. Reynold and Olinda (Siefker) K.; m. Carole Ann Hofer, Oct. 17, 1959. BCE, U. Mo.-Rolla, 1959; MS, U. Ill., 1968. Registered profl. engr., Ill., Mo., Fla., Miss., N.Y., Iowa, Del., Ohio, Ind. Civil engr. City of St. Louis, 1959; with Clark, Dietz & Assocs., and predecessors, Urbana, Ill., 1961-79, sr. design engr., 1963-67, dir. transp. div., 1968-79; civil engr. div. hwys. Ill. Dept. Transp., Paris, 1979-83; part-owner ESCA Cons., Inc., Urbana, 1983-88; civil engr. Zurheide-Herrmann, Inc., Champaign, 1988—; instr. Parkland Coll., Champaign, 1972; mem. Ill. Profl. Engrs. Examining Com., 1982-89, Ill. State Bd. of Profl. Engrs., 1990—. With C.E., AUS, 1959-61. Mem. ASTM, ASCE (br. pres. 1982-83, sect. pres. 1988-89, chmn. dist. coun. 1989), NSPE, Ill. Soc. Profl. Engrs. (chpt. pres. 1974, state chmn. registration laws com. 1973, 78), Ill. Assn. Hwy. Engrs., Am. Pub. Works Assn. (sect. dir. 1974-77, 80), Inst. Transp. Engrs., Ill. Profl. Land Surveyors Assn., Soc. Am. Mil. Engrs., Nat. Coun. Examiners for Engring. and Surveying (Disting. Svc. award 1992), U. Mo.-Rolla Acad. Civil Engrs., U. Mo.-Rolla Alumni Assn., Champaign Ski Club, Theta Tau, Tau Beta Pi, Chi Epsilon, Pi Kappa Alpha. Home: 1811 Coventry Dr Champaign IL 61821-5239 Office: 2 Henson Pl Ste 1 Champaign IL 61820-7805

KRUGMAN, STANLEY LIEBERT, science administrator, geneticist; b. St. Louis, June 8, 1932; s. Bernard and Della (Goldberg) K.; m. Judith Rachel Alfend, June 28, 1958; children: Mark Bernard, Jeffrey Jon. BS in Forestry, U. Mo., 1955; MF, U. Calif., Berkeley, 1956, PhD in Plant Physiology, 1961. Rsch. aide U. Calif., 1956-61, rsch. assoc., 1961-62; rsch. physiologist U.S. Forest Svc., 1962-64, project leader, 1964-71; staff geneticist U.S. Forest Svc., Washington, 1971-80, staff dir., 1980—; cons. in field. Editor: Seeds of Woody Plants, 1974, Advances in Reproductive Biology, 1974, Management Biosphere Reserves, 1979, Advances in Forest Physiology, 1980. Fellow AAAS, Soc. Am. Foresters (William Schlich medal 1990); mem. Internat. Union Forestry Orgn. Jewish. Home: 6515 Dryden Dr Mc Lean VA 22101-4627 Office: USDA Forest Svc 201 14th St SW Washington DC 20090-6090

KRULL, EDWARD ALEXANDER, dermatologist; b. Oakville, Conn., Oct. 25, 1929; s. Alexander and Marian (Ruppert) K.; m. Joan Marie Adams, Sept. 7, 1955; children: Alisa M., Lael Adams, Edward A., Jr. Student Yale U., 1948-51, MD, 1955. Intern San Francisco City-County Hosp., 1955-56; with Madigan Gen. Hosp., 1959-60; resident Henry Ford Hosp., Detroit, 1960-63, staff physician dept. dermatology, 1965-76, chmn. 1976—; practice medicine specializing in dermatology, Grand Rapids, Mich., 1963-65; bd. dirs. Skin Cancer Found., 1977-80, Found. Internat. Dermatologic Edn., 1980-82; mem. residency rev. com. in dermatology, 1984—, chmn. residency rev. com., 1987—. Bd. govs. Henry Ford Hosp., trustee 1986—, exec. com. bd. trustees, 1986—. Editorial bd. Jour. Dermatol. Surg. and Oncology, 1976-79. Capt. M.C., U.S. Army, 1957-59, Iran. Fellow Am. Dermatol Assn., Am. Coll. Chemosurgery, Am. Acad. Dermatology (editorial bd. jour. 1979-84, chmn. task force on surgery 1978-84, presdl. blue ribbon com. 1984, exec. com. adv. bd. 1974-76, chmn. task force surgery Coun. and Lab. Svcs., 1978-80, bd. dirs. 1982-86, exec. com. bd. dirs. 1984-86, v.p. 1986-87, Bronze award exhibit, 1969), Am. Bd. Dermotology (diplomate, bd. dirs. 1984—, v.p. 1992-93), Am. Bd. Med. Splts. (diplomate, chmn. dermotology sect. 1992—), Mich. Dermatol. Soc. (chmn. liaison Blue Cross Blue Shield 1971-72, sec. treas. 1973-75, pres. 1976-77), AMA, Mich. State Med. Soc. (sec. dermatology sect. 1972-73, pres. 1973-74), Wayne County Med. Soc., Am. Soc. Dermatologic Surg. (pres. elect. 1980-81, pres. 1982, bd. dirs., 1973-76, 79-82, chmn. resident com. 1979-84, scholarship com., 1976-81, cons. bd. dirs. 1976-79, chmn. educ. coordinating com., 1978—, Leon Goldman Achievement award 1988), Assn. Professors of Dermatology (bd. dirs. 1988-89), Assn. Acad. Dermatologic Surgeons (pres. 1988-89). Episcopalian. Avocations: tennis, trout fishing, golf. Home: 422 University St Grosse Pointe MI 48230 Office: Henry Ford Hosp Dept Dermatology 2799 W Grand Blvd Detroit MI 48202-2689

KRUMHANSL, CAROL LYNNE, psychologist, educator; b. Providence, Sept. 17, 1947; d. James A. Krumhansl and Barbara Bayer. BA in Math., Wellesley Coll., 1969; AM in Math., Brown U., 1973; PhD in Psychology, Stanford U., 1978. Asst. prof. Rockefeller U., 1978-79, Harvard U., Cambridge, Mass., 1979-80; asst. prof. Cornell U., Ithaca, N.Y., 1980-84, assoc. prof., 1984-90, prof., 1990—. Mem. editorial bd. Jour. of Exptl. Psychology: Human Learning and Memory, 1980-81, Jour. of Exptl. Psychology: Human Perception and Performance, 1981-88, Music Perception, 1981—; contbr. numerous articles to profl. jours. Recipient NIMH traineeship, 1975-78, Milton Fund award Harvard U., 1979-80, NSF grant, 1981-84, APA Disting. Sci. award, 1983, NIMH grant, 1984-87, NSF grant, 1987-88; fellow Ctr. for Advanced Study in Behavioral Scis., 1983-84, Inst. for Advanced Studies, Ind. U., 1986; named Vis. scientist I.R.C.A.M., 1987-88. Mem. APS, Psychonomic Soc., Soc. for Math. Psychology, Soc. for Rsch. in Psychology of Music and Music Edn., European Soc. for Cognition of Music. Office: Cornell U Dept Psychology Uris Hall Ithaca NY 14853-7601

KRUPOWICZ, JOHN JOSEPH, metallurgical engineer; b. New Brighton, Pa., July 7, 1946. BS, U. Cin., 1969, MS, 1971, PhD, 1974. Metall. trainee Crucible Steel Co., Midland, Pa., 1965-68; prin. engr. Westinghouse-Bettis Atomic Power Lab., West Mifflin, Pa., 1975-79; cons. engr. Combustion Engring. Inc., Windsor, Conn., 1979-85; engring. assoc. Mobil R&D Corp., Pennington, N.J., 1985—. Contbr. articles to numerous profl. jours. Mem. ASM, Nat. Assn. Corrosion Engrs., Alpha Sigma Mu, Sigma Xi, Phi Lambda Upsilon. Home: 2058 Hawthorne Pl Paoli PA 19301 Office: Mobil R & D Corp Box 1026 Pennington NJ 08543-1026

KRUPP, EDWIN CHARLES, astronomer; b. Chgo., Nov. 18, 1944; s. Edwin Frederick and Florence Ann (Olander) K.; m. Robin Suzanne Rector, Dec. 31, 1968; 1 son, Ethan Hembree. B.A., Pomona Coll., 1966; M.A., UCLA, 1968, Ph.D. (NDEA fellow, 1970-71), 1972. Astronomer Griffith Obs., Los Angeles Dept. Recreation and Parks, 1972—, dir., 1976—; mem. faculty El Camino Coll., U. So. Calif., extension divs. U. Calif.; cons. in edni. TV Community Colls. Consortium; host teleseries Project: Universe. Author: Echoes of the Ancient Skies, 1983, The Comet and You, 1986 (Best Sci. Writing award Am. Inst. Physics 1986), The Big Dipper and You, 1989, Beyond the Blue Horizon, 1991, The Moon and You, 1993; editor, co-author: In Search of Ancient Astronomies, 1978 (Am. Inst. Physics-U.S. Steel Found. award for Best Sci. Writing 1978), Archaeoastronomy and the Roots of Science; editor-in-chief Griffith Obs., 1984—. Mem. Am. Astron. Soc. (past chmn. hist. astronomy div.), Astron. Soc. Pacific (past dir., recipient Klumpke-Roberts outstanding contbns. to the public understanding and appreciation of astronomy award 1989), Internat. Astron. Union, Explorers Club, Sigma Xi. Office: Griffith Observatory 2800 E Observatory Rd Los Angeles CA 90027-1299

KRUSE, OLAN E., physics educator; b. Coupland, Tex., Sept. 6, 1921; s. Max Edward and Irma Pauline (Miller) K.; m. Lucille Thomas, Sept. 4, 1942; children: John E., James L. BS, Tex. A&I, 1942; MA, Univ. Tex., 1949, PhD, 1951. Asst. prof., assoc. prof., prof. and chmn. physics Stephen F. Austin State, Nacogdoches, Tex., 1951-56; prof., chmn. physics Tex. A&M Univ., Kingsville, Tex., 1956-87, prof. physics, 1987—; instr. pre-radar sch., Harvard, 1943-44, Bowdoin, Brunswick, Me., 1944-45. Contbr. articles to profl. jours. Lt. USNR, 1942-45. Recipient Disting. Teaching award, Tex. A&I, 1980. Fellow Tex. Acad. of Sci. (v.p. 1989-90); mem. Am. Assn. Physics Teachers, Tex. Assn. Coll. Teachers., AAPT (pres. 1986-87). Achievements include patent for an educational apparatus. Home: 325 Seale Kingsville TX 78363 Office: Tex A&M U West Santa Gertrudes Kingsville TX 78363

KRUSHAT, WILLIAM MARK, mathematical statistician; b. Kenfield, Calif., Sept. 5, 1949; s. Lester W. and Jane (White) K.; m. Cheryl Barbara Sweet, Aug. 30, 1980; children: John Arthur, Joshua Parker, Daniel Marin. BA, U. Calif., Riverside, 1972; MPH, U. Calif., Berkeley, 1975; DSc, Johns Hopkins Univ., 1992. Commd. lt. (jg) USPHS, 1975, advanced through grades to comdr., 1987; math.-statistican Dept. Health and Human Svcs. Divsn. of Hosp. and Clinics USPHS, San Francisco, 1975-80; math.-statistican Dept. Health and Human Svcs. FDA, USPHS, Rockville, Md., 1981-83; math.-statistican Office of Inspector Gen. Dept. Health and Human Svcs., Balt., 1983—. Author Dept. Health and Human Svcs. report leading to Fed. law banning physician ownership of clin. labs; contbr. articles to profl. jours. Mem. Am. Pub. Health Assn., Am. Statis. Assn., Soc. for Epedimecol. Rsch. Achievements include presentation of FDA position on association between Reyes syndrome and aspirin, eventually led to labeling of aspirin. Office: Dept Health and Human Svcs Office Inspect Gen RM 1-D-16 OM 6325 Security Blvd Rm 116D Baltimore MD 21207-5187

KRUSKAL, MARTIN DAVID, mathematical physicist, educator; b. N.Y.C., Sept. 28, 1925; m. 1950; 3 children. BS, U. Chgo., 1945; MS, NYU, 1948, PhD in Math., 1952. Rsch. scientist Plasma Phys. Lab., Princeton U., 1951-61, prof. astrophys. sci., 1961—, prof. math., 1981—, emeritus, 1989—; David Hilbert prof. math. Rutgers U., New Brunswick, N.J., 1989—; trustee Soc. for Indsl. & Applied Math., 1985-91, Math. Scis. Edn. Bd. of NRC, 1986-89, Ext. Adv. Com. (curriculum chmn.), Ctr. for Nonlinear Studies, Los Alamos Nat. Lab., 1980—. Sr. fellow NSF, 1959-60, sr. fellow Weizmann Inst. of Sci., 1973-74, Gibbs Lecturer Am. Math. Soc., 1979, fellow Jap. Soc. for Promotion of Sci., 1979, recipient Dannie Heineman prize in math. phys., 1983, Potts gold medal of Franklin Inst., 1986, award in appl. math. and num. anal., Nat. Acad. Scis., 1989, Nat. Medal of Sci., Nat. Sci. Found., 1993. Mem. AAS, Am. Math. Soc., Math. Assn. Am., Am. Phys. Soc. (fellow), Nat. Acad. Scis. (chmn. sect. of applied math. scis. 1990—). Home: 60 Littlebrook Rd N Princeton NJ 08540-4062 Office: Rutgers U Dept Math Hill Ctr Busch Campus New Brunswick NJ 08903

KRUTZ, RONALD L., computer engineer; b. McKeesport, Pa., Aug. 27, 1938; s. Louis Joseph and Rose Marie Krutz; m. Hilda Mae Napolitano, Apr. 29, 1961; children: Sheri Rose, Lisa Maria. BSEE, U. Pitts., 1960, MSEE, 1967, PhD in Elec. Engring., 1972. Registered profl. engr., Pa. Sr. project engr. Gulf Rsch. and Devel. Co., 1964-74; mgr. computer scis. dept., Rsch. and Devel. Ctr. Singer Corp., 1974-75; faculty mem. dept. elec. engring. Carnegie Mellon U., Pitts., 1975-78, assoc. dir. Carnegie Mellon Rsch. Inst., 1978—, dir., founder computer engring. ctr., 1978—; rsch. coun. Carnegie Mellon U., COEST coun., Greater Pitts. High Tech. Coun.; served as technical reference N.Y. Times. Author: (text book) Microprocessors and Logic Design, 1980, Microprocessors for Managers, 1983, Interfacing Techniques, 1988; devel. video tape course on microprocessoes and software; contbr. articles to profl. jours. 1st. lt. U.S. Army, 1961-64. Mem. IEEE (sr. mem.), Am. Assn. for Artificial Intelligence, ADPA, Univ. Club (bd. dirs.), Eta Kappa Nu, Sigma Tau. Achievements include seven patents for computer and digital systems. Office: Carnegie Mellon Rsch Inst 4616 Henry St Pittsburgh PA 15213

KRUUS, HARRI KULLERVO, physician; b. Kuusankoski, Finland, Nov. 13, 1950; s. Ilmari and Vuokko Maire (Poutiainen) K.; m. Raija Marjatta Majlander, Dec. 31, 1971; children: Joanna Marietta, Jesse Christian. MD, Univ. Helsinki, 1977. Gen. practice Cen. Hosp., Vasteras, Sweden, 1978-79; occupational physician ASEA AB, Vasteras, Sweden, 1979-80; gen. practice Tawam Hosp., United Arab Emirates, 1981-84; occupational physician Personnel Adminstrn., Helsinki, 1984-85; resident Tammiharju Hosp., Tammisaari, Finland, 1986-87; gen. practice Vantaa Health Ctr., Vantaa, Finland, 1988-90; resident Rheumatism Found. Hosp., Heinola, Finland, 1990-92; gen. practice Vantaa (Finland) Health Ctr., 1992—. Chmn. Finnish-English Bible Soc., Helsinki 1972-76. Mem. Finnish Med. Balint Soc. Lutheran. Avocations: literature, swimming, tennis, skiing.

KRYSTAL, ANDREW DARRELL, psychiatrist, biomedical engineer; b. Detroit, June 17, 1960; s. Henry and Esther Rose (Reichstein) K. BS and MS, MIT, 1983; MD, Duke U., 1987. Bd. cert. in medicine. Rsch. assoc. Rsch. Lab. of Electronics, MIT, Cambridge, 1981-83, Duke U., Durham, N.C., 1986-91; resident in psychiatry Duke U., Durham, 1987-91, chief resident in psychiatry, 1990-91, fellow in rsch. methodology, 1991—, clin. assoc. psychiatrist, 1991—. Contbr. articles to profl. jours. Bd. mem. Durham (N.C.) Companions Big Bro./Sister, 1991—. Recipient Outstanding Young Investigator award UpJohn Corp., 1990; named Laughlin fellow Am. Coll. Psychiatrists, 1990. Mem. AMA, AAAS, Am. Psychiat. Assn., Sigma Xi. Achievements include development of EEG markers of electroconvulsive therapy treatment efficiency and side-effects. Office: Duke U Med Ctr Box 3309 Durham NC 27710

KU, JENTUNG, mechanical and aerospace engineer; b. Hsinchu, Republic of China; came to U.S., 1974; BS, Tsing Hua U., Hsinchu, 1972; MS, Purdue U., 1976, PhD, 1980. Rsch. asst. Purdue U., West Lafayette, Ind., 1976-80; mem. tech. staff Advanced Tech. Ctr., Bendix Corp., Columbia, Md., 1980-83; section head/program mgr. OAO Corp., Greenbelt, Md., 1983-91; sr. engr. NASA Goddard Space Flight Ctr., Greenbelt, Md., 1991—. Contbr. articles on heat transfer and thermal control systems to profl. jours.; pioneer in developing capillary pumped loop heat transport systems (2 Tech. Innovation awards NASA). Bd. dirs. Columbia Chinese Bapt. Ch., 1991—. Mem. ASME, AIAA, Am. Nuclear Soc., Tau Beta Pi, Phi Tau Phi. Baptist. Avocation: tennis. Home: 14208 Bradshaw Dr Silver Spring MD 20905 Office: NASA Goddard Space Flight Ctr Greenbelt MD 20771

KU, THOMAS HSIU-HENG, biochemical and specialty chemical company executive; b. Nanking, China, Sept. 1, 1948; s. Cheng-Kang and May (Wang) K.; m. Elsa Yeou-Ying Lin, Jan. 7, 1979; children—Catherine Yeou-Jung, James Yeou-Liang. B.S. in Chemistry, Nat. Taiwan U., 1970; M.S. in Chemistry, Carnegie-Mellon U., 1973, Ph.D. in Chemistry, 1975. Research scientist Brookhaven Nat. Lab., Upton, N.Y., 1975-79; assoc. prof. Nat. Tsing-Hwa U., Hsinchu, Taiwan, 1981-82; tech. dir., dir. mfg., dir. bus. planning and devel., pharm. div. mgr., from dep. mng. dir. to mng. dir., gen. mgr. Cyanamid Taiwan Corp., Taipei, 1980—; mem. chem. sub-com. ind. devel. adv. com. Ministry Econ. Affairs, 1989-91. Contbr. 18 articles to profl. jours. Dir. Chinese-Am. Acad. Assn., N.Y., 1977-78; com. mem. Taipei City Basketball Com., 1983-85; dir. 4-H Assn., 1989—. Mem. Am. Chem. Soc., Chinese Chem. Soc., Chinese Mgmt. Assn., Chinese Assn. Advancement Mgmt. (chmn. mfg. com., prodn. com. 1989-90, outstanding mgr. assn. 1991, dep. sec. gen. 1991, bd. dirs. 1992—, Outstanding Profl. Mgr. award 1984), ROC Bioindustry Devel. Assn. (bd. dirs. 1989—), Chinese Multinational Industry Rsch. Assn. (supr. 1992—), Am. Club in China.

KU, Y. H., engineering educator; b. Wusih, Kiangsu, China, Dec. 24, 1902; came to U.S., 1950; s. Ken Ming Ku and Ching-Su Wang; m. Wei-zing Wang, Apr. 1, 1929; children: Wei-Lien, Wei-Ching, Wei-Wen (Mrs. Chi-Liang Hsieh), Walter, John, Victor, Anna (Mrs. Yuk-Kai Lau). S.B., MIT, 1925, S.M., 1926, Sc.D., 1928; M.A., LL.D., U. Pa., 1972. Prof. elec. engring., head dept. Chekiang U., China, 1929-30; dean engring. Cen. U., China, 1931-32; pres. Central U., 1944-45; dean engring. Tsing Hua U., China, 1932-37; vice minister Ministry Edn., Republic of China, 1938-44; edn. commr. Shanghai, 1945-47; pres. Nat. Chengchi U., Nanking, 1947-49; vis. prof. MIT, 1950-52; prof. U. Pa., 1952-71, prof. emeritus, 1972—; hon. prof. Jiao-Tong U., Shanghai, 1979—, Xi'an, Southwestern and Northern, 1985—, Northeastern U. Tech. and NW Inst. Telecommunications, 1986—, S.E. U. Nanjing, 1988—; cons. Gen. Electric Co., Univac, RCA. Author: Analysis and Control of Nonlinear Systems, 1958, Electric Energy Conversion, 1959, Transient Circuit Analysis, 1961, Analysis and Control of Linear Systems, 1962, Collected Scientific Papers, 1971; poems, plays, novels, essays in Chinese Collected Works, 1961; Woodcutter's Song, 1963, Pine Wind, 1964, Lotus Song, 1966, Lofty Mountains, 1968, The Liang River, 1970, The Hui Spring, 1971, The Si Mountain, 1972, 500 Irregular Poems, 1972, The Great Lake, 1973, 1000 Regular Poems, 1973, 360 Recent Poems, 1976, The Tide Sound, 1980, History of Chan (Zen) Masters, 1976, History of Japanese Zen Masters, 1977, History of Zen (in English), 1979, The Long Life, 1981, One Family-Two Worlds (in English), 1982, Poems after Chin Kuan, 1983, Poems after Tao Chien, 1984, 303 Poems after Tang Poets, 1986, Flying Clouds and Flowing Water, 1987, Poems After Wu Wen-Ying, 1988, Selected Plays, 1990, Eyebrows, 1991, Scientific Papers, 1992, Old Age, 1993, Clearn Water and Beautiful Flowers, 1994. Recipient Gold medal Ministry Edn., Republic of China; Pro Mundi Beneficio Gold medal Brazilian Acad. Humanities, 1975; Gold medal Chinese Inst. Elec. Engrs., 1972. Fellow Academia Sinica, IEEE (Lamme medal 1972), Instn. Elec. Engrs. (London); mem. Am. Soc. Engring. Edn., Internat. Union Theoretical and Applied Mechanics (mem. gen. assembly), U.S. Nat. Com. on Theoretical and Applied Mechanics, Sigma Xi, Eta Kappa Nu, Phi Tau Phi. Home: 1420 Locust St 22G Philadelphia PA 19102-4223 Office: 200 S 33d St Philadelphia PA 19104

KUAN, PUI, nuclear engineer; b. Szechuan, People's Republic of China, Nov. 4, 1945; came to U.S., 1966; s. Chao-Hsiang and Ya-Chih (Wei) K.; m. Habibah Hassan, Apr. 17, 1986. BS in Astronomy, Calif. Inst. Tech., 1970; MA in Astronomy, U. Calif., Berkeley, 1972, PhD in Astronomy, 1973; MBA, Columbia U., 1991. Postdoctoral fellow Kitt Peak Nat. Obs., Tucson, 1974-76; scientist Energy Inc., Idaho Falls, Idaho, 1977-80; sr. engring. specialist Idaho Nat. Engring. Lab., Idaho Falls, 1980—. Mem. Am. Nuclear Soc., Am. Astron. Soc. Achievements include research in stellar atmospheres, safety of nuclear reactors, nuclear fuel cycle; development of accident scenario for the Three Mile Island Unit 2 reactor incident. Home: 1699 Laguna Dr Idaho Falls ID 83404-7416 Office: Idaho Nat Engring Lab PO Box 1625 Idaho Falls ID 83415-3890

KUBACKI, KRZYSZTOF STEFAN, mathematics educator; b. Lublin, Poland, Mar. 23, 1953; s. Stefan and Adela (Wieczorek) K.; m. Urszula Dorota Surma, Feb. 7, 1981; children: Kamila, Mateusz. MS in Math., Maria Curie-Sklodowska U., Lublin, Poland, 1977, MS in Math. Edn., 1979; PhD in Math. Scis., Math. Inst. Polish Acad. Scis., Wroclaw, 1986. Lectr. Agrl. Acad., Lublin, 1977-81, sr. lectr., 1981-86, adj., 1986-90, asst. prof., 1990—; asst. Math. Inst. Polish Acad. Scis., Wroclaw, 1979-81; sr. asst. Maria Curie Sklodowska U., Lublin, 1984-85; vis. asst. prof. Johns Hopkins U., Balt., 1990. Contbr. articles to profl. jours. Recipient prize Math. Inst. Polish Acad. Scis., 1986, prize Min. Nat. Edn., 1989; Kosciuszko Found. scholar, 1989-90; grantee Computer Sci. Inst. of the Polish Acad. Scis., 1979, Math. Inst. of the Polish Acad. Scis., 1981-82, 83-85, 86-89, 86-88, Ministry of the Nat. Edn., 1990, 91. Mem. Polish Biometrical Soc., Bernoulli Soc. for Math. Stats. and Probability, Am. Math. Soc., Inst. Math. Stats., Polish Math. Soc. Home: Szwajcarska 7 M 31, 20-861 Lublin Poland Office: Agrl Acad Inst Applied Math, Akademicka 13, PO Box 158, 20-950 Lublin Poland

KUBALE, MAREK EDWARD, computer scientist, educator; b. Gdynia, Poland, July 2, 1946; s. Witold Ludwik and Henryka Zofia (Lewicka) K.; m. Anna Eleonora Techman, Dec. 26, 1970; 1 child, Marcin. MSEE, Tech. U., Gdańsk, 1969, PhD, 1975, DS, 1989. Asst. prof. Tech. U., 1969-89, assoc. prof., 1989-91; assoc. prof. U. Gdansk, 1989-91, prof., 1991—; prof. Tech. U., 1991—; rector's proxy Tech. U., 1984-87, head dept., 1990—, chair rector bd., 1991—; reviewer Math. Reviews, 1989—, Zentralblatt für Math., 1990—. Assoc. editor NETWORKS, 1993—; contbr. articles to profl. jours. Erskine fellow, 1993; recipient Min. Nat. Edn. award, 1990. Mem. Polish Cybermetical Soc. (chair auditing bd. 1984-88), Com. Sci. Rschrs. (sec. Informatics 1992-93, grantee 1991, 92), Polish Acad. Scis. (com. computer sci. 1993—), Am. Math. Soc., Inst. of Combinatorics and its Applications. Avocations: tennis, birdge. Home: Pilotów 8B, 80-270 Gdansk Poland Office: Tech U, Narutowicza 11/12, 80-952 Gdańsk Poland

KUBANEK, GEORGE R., chemical engineer; b. Opeln, Germany, Oct. 26, 1937; arrived in Canada, 1956; s. Joseph and Julie (Vecerova) K.; m. Anne-Marie Weidler, Sept. 2, 1967; children: Julia, Ian. B in Engring., McGill U., 1961, PhD, 1966; MBA, Rutgers U., 1976. Sr. rsch. engr. Allied Chem. Corp., Morristown, N.J., 1966-69; asst. mgr. Johnson & Johnson Rsch., New Brunswick, N.J., 1969-73; head dept. chem. engring. Noranda Rsch. Ctr., Pointe-Claire, Quebec, Canada, 1973-79, mgr. process engring. div., 1979-88; mgr. forest tech. lab. Noranda Tech. Ctr., Pointe-Claire, Quebec, Canada, 1988—; mem. chem. engring. adv. com. McGill U., 1986-93; mem. rsch. program com. Pulp and Paper Rsch. Inst. Canada, 1989-92. Contbr. articles to profl. jours. Fellow Chem. Inst. Canada and Canadian Soc. Chem. Engring. (treas. 1982-85, jour. editorial bd. 1979-85), World Congress Chem. Engrs. (v.p. 1979-81); mem. AIChE, Canadian Pulp and Paper Assn., Tech. Assn. Pulp and Paper Industires, Order of Engrs. Quebec, Indsl. Rsch. Inst. Achievements include patents for arc plasma technology. Office: Noranda Tech Ctr, 240 Blvd Hymus, Pointe Claire, PQ Canada H9R 1G5

KUBAS, GREGORY JOSEPH, research chemist; b. Cleve., Mar. 12, 1945; s. Joseph Arthur and Esther (Polcyn) K.; m. Jeanne Henry, Dec. 22, 1973; children: Kelly Richmond, Sherry Richmond. BS, Case Inst. Tech., 1966; PhD, Northwestern U., 1970. Postdoctoral fellow Princeton (N.J.) U., 1971-72; postdoctoral fellow Los Alamos (N.Mex.) Nat. Lab., 1972-74, mem. staff, 1974—. Contbr. articles to profl. jours. Mem. Am. Chem. Soc. (Inorganic Chemistry award 1993). Clubs: Los Alamos Tennis (pres. 1983-84), Los Alamos Duplicate Bridge. Home: 345 Valle Del Sol Rd Los Alamos NM 87544-3563 Office: Los Alamos Nat Lab MS-C346 Los Alamos NM 87545

KUBIDA, WILLIAM JOSEPH, patent lawyer; b. Newark, Apr. 3, 1949; s. William and Catherine (Gilchrist) K.; m. Mary Jane Hamilton, Feb. 4, 1984; children: Sara Gilchrist, Kathleen Hamilton. B.S.E.E., U.S. Air Force Acad., 1971; JD, Wake Forest U., 1979. Bar: N.C. 1979, U.S. Patent Office 1979, Ind. 1980, U.S. Dist. Ct. (no. dist.) Ind. 1980, U.S. Dist. Ct. (so. dist.) Ind. 1980, U.S. Ct. Appeals (7th cir.) 1981, U.S. Dist. Ct. (Ariz.) 1982, U.S. Ct. Appeals (9th cirs. and fed.) 1982, Ariz. 1982, Colo. 1990, U.S. Dist. Ct. Co., 1990, U.S. Ct. Appeals (10th cir.) 1990. Patent and trademark lawyer Lundy and Assocs., Ft. Wayne, Ind., 1979-81; patent atty. Motorola, Inc., Phoenix, 1981-85; Intellectual Property Counsel Nippon Motorola, Ltd., Tokyo, 1985-87; ptnr. Lisa & Kubida, P.C., Phoenix, 1987-89; engring. law counsel Digital Equipment Corp., Colorado Springs, Colo., 1989-92; of counsel Holland & Hart, Denver and Colorado Springs, 1992—. 1st lt. USAF, 1971-76. Mem. Am. Intellectual Property Law Assn. (computer software sect.), Am. C. of C. (patents, trademarks and lic. sect., Japan), Licensing Exec. Soc. (Pacific Rim subcom.), Country Club Colo., Mensa, Phi Delta Phi. Republican. Presbyterian. Home: 4165 Regency Dr Colorado Springs CO 80906-4368

KUBISTA, THEODORE PAUL, surgeon; b. N.Y.C., July 20, 1937; s. Theodore Anton and Antonette (Balasch) K.; m. Alice Elizabeth Maris, Dec. 26, 1963; children: Theodore Stephen, Christian Gregory. BS, Pa. State U., 1959; MD, U. Pa., Phila., 1963. Diplomate Am. Bd. Surgery. Intern U. Pa. Hosp., 1963-64; resident in surgery Mayo Grad. Sch. Medicine, Rochester, Minn., 1964-69; gen. surgeon Duluth (Minn.) Clinic, 1971—. Contbr. articles to profl. jours. Lt. comdr. USN, 1969-71. Decorated Bronze star with combat V. Fellow ACS (Minn. sect. com. on trauma 1971—); mem. AMA, Minn. Surg. Soc. (pres. 1986-87), Duluth Surg. Soc. (pres. 1976-77), Soc. Clin. Vascular Surgery. Presbyterian. Achievements include improvement of survival in cancer of rectum with combined surgery, chemotherapy and radiation therapy. Office: Duluth Clinic 400 E 3d St Duluth MN 55805

KUBLER-ROSS, ELISABETH, physician; b. Zurich, Switzerland, July 8, 1926; came to U.S., 1958, naturalized, 1961; d. Ernst and Emma (Villiger) K.; children: Kenneth Lawrence, Barbara Lee. M.D., U. Zurich, 1957; D.Sc. (hon.), Albany (N.Y.) Med. Coll., 1974, Smith Coll., 1975, Molloy Coll., Rockville Centre, N.Y., 1976, Regis Coll., Weston, Mass., 1977, Fairleigh Dickinson U., 1979; LL.D., U. Notre Dame, 1974, Hamline U., 1975; hon. degree, Med. Coll. Pa., 1975, Anna Maria Coll., Paxton, Mass., 1978; Litt.D. (hon.), St. Mary's Coll., Notre Dame, Ind., 1975, Hood Coll., 1976, Rosary Coll., River Forest, Ill., 1976; L.H.D. (hon.), Amherst Coll., 1975, Loyola U., Chgo., 1975, Bard Coll., Annandale-on-Hudson, N.Y., 1977, Union Coll., Schenectady, 1978, D'Youville Coll., Buffalo, 1979, U. Miami, Fla., 1976; D.Pedagogy, Keuka Coll., Keuka Park, N.Y., 1976. Rotating intern Community Hosp., Glen Cove, N.Y., 1958-59; rsch. fellow Manhattan State Hosp., 1959-62; resident Montefiore Hosp., N.Y.C., 1961-62; fellow psychiatry Psychopathic Hosp., U. Colo. Med. Sch., 1962-63; instr. psychiatry Colo. Gen. Hosp., U. Colo. Med. Sch., 1962-65; mem. staff LaRabida Children's Hosp. and Rsch. Ctr., Chgo., 1965-70; chief cons. and rsch. liaison sect. LaRabida Children's Hosp. and Rsch. Ctr., 1969-70; asst. prof. psychiatry Billings Hosp., U. Chgo., 1965-70; med. dir. Family Service and Mental Health Ctr. S. Cook County, Chicago Heights, Ill., 1970-73; pres. Ross Med. Assos. (S.C.), Flossmoor, Ill., 1973-77; pres., chmn. bd. Shanti Nilaya Growth and Health Ctr., Escondido, Calif., 1977—; mem. numerous adv., cons. bds. in field. Author: On Death and Dying, 1969, Questions and Answers on Death and Dying, 1974, Death-The Final Stages of Growth, 1975, To Live Until We Say Goodbye, 1978, Working It Through, 1981, Living With Death and Dying, 1981, Remember The Secret, 1981, On Children and Death, 1985, AIDS: The Ultimate Challenge, 1988, On Life after Death, 1991; contbr. chpts. to books, articles to profl. jours. Recipient

Teilhard prize Teilhard Found., 1981; Golden Plate award Am. Acad. Achievement, 1980; Modern Samaritan award Elk Grove Village, Ill., 1976; named Woman of the Decade Ladies Home Jour., 1979; numerous others. Mem. AAAS, Am. Holistic Med. Assn. (a founder), Am. Med. Women's Assn., Am. Psychiat. Assn., Am. Psychosomatic Soc., Assn. Cancer Victims and Friends, Ill. Psychiat. Soc., Soc. Swiss Physicians, Soc. Psychophysiol. Research, Second Attempt at Living. Address: care Celestial Arts Pub PO Box 7327 Berkeley CA 94707

KUBO, ISOROKU, mechanical engineer; b. Tokyo, May 16, 1942; came to U.S., 1977; s. Shogo and Sono (Ito) K.; m. Mary Ann Stone, Mar. 17, 1974; children: Tomiko J., Yukari J., Kiyokaz J., Yuri J. PhD in Aero./Mech., Cornell U., 1974; MBA, Ind. U., 1987. Engr. Komatsu Ltd., Tokyo, 1965-73, asst. chief engr., 1973-77; rsch. assoc. Applied Rsch. Lab., Pa. State U., State College, 1977-79; group leader Cummins Engine Co., Columbus, Ind., 1979-82, tech. advisor, 1982-88, dir., 1988—. Recipient Spl. Scholarship Japan Scholarship Assn., 1961-65; Cornell U. fellow, 1969. Mem. ASME (assoc.), Soc. Automotive Engrs. (assoc.), Sigma Xi, Beta Gamma Sigma. Achievements include patents for gas lubricated piston ring assembly, a Nox reduction method and integrated diesel-rankine system; patent pending for solar energy system. Home: 3405 Putter Pl Columbus IN 47203 Office: Cummins Engine Co PO Box 3005 Columbus IN 47202-3005

KUĆAN, ŽELJKO, biochemistry educator; b. Zagreb, Croatia, May 24, 1934; s. Vilim and Aleksandra (Fischer) K.; children: Iva, Maja, Tanja. BSc, U. Zagreb, 1958, PhD, 1964. Grad. student Rudjer Bošković Inst., Zagreb, 1958-64, chmn. dept. organic chemistry and biochemistry, 1972-77; rsch. assoc. Rockefeller U., N.Y.C., 1961-63; vis. scientist NYU Sch. Medicine, N.Y.C., 1969-72, vis. prof., 1977-79; prof. biochemistry U. Zagreb, 1982—, dean, faculty sci., 1990—. Contbr. over 50 articles to major internat. profl. jours., 1961—. Recipient fellowship Internat. Atomic Energy Agy, Vienna, 1961, Rudjer Bošković award Republic of Croatia, 1991. Mem. Croatian Acad. Scis. and Arts, Croatian Biochem. Soc. (pres. 1991—). Achievements include contributions to radiobiology; and to nucleic acids biochemistry. Home: A Augustinćića 11, Zagreb 41000, Croatia Office: Faculty of Sci, Strossmayerov trg 14, Zagreb 41000, Croatia

KUCHIBHOTLA, SUDHAKAR, environmental engineer, consultant; b. Machilipatnam, India, May 7, 1964; came to U.S., 1989; s. Sarma Sivarama Krishna and Satyabhama (Peddi) K.; m. Madhuri Tamirisa, Aug. 12, 1990. BE, Andhra U., 1986; MS, Miami U., Oxford, Ohio, 1991, M.Environ. Engring., 1993. Jr. engr. 3S Fabricators, Visakhapatnam, India, 1986-88; project engr. Instrumentation Ltd., Visakhapatnam, 1988-89; rsch. asst. Miami U., 1989-92; san. engr. Ind. Dept. Environ. Mgmt., Indpls., 1992; environ. engr. Environ. Quality Mgmt., Inc., Cin., 1992—. Contbr. articles to profl. jours. Recipient scholarships and assistantship. Mem. TAPPI, Instrument Soc. Am., Sigma Xi. Home: 11755 Norbourne Dr Apt 1302 Cincinnati OH 45240-4430 Office: Environ Quality Mgmt Inc 1310 Kemper Meadow Dr # 100 Cincinnati OH 45240

KUCHNER, EUGENE FREDERICK, neurosurgeon, educator; b. N.Y.C., Nov. 19, 1945; s. Morton H. and Edna Estelle (Marks) K. m. Joan Ruth Freedman, Sept. 2, 1968; children: Marc Jason, Eric Benjamin. AB, Johns Hopkins U., 1967; MD, U. Chgo., 1971. Resident in surgery Yale U. Sch. Medicine, New Haven, 1971-72; resident in neurosurgery Montreal (Que., Can.) Neurol. Inst., McGill U., 1972-76, spine fellow, 1976; neurosurgeon Sch. Medicine, SUNY, Downstate, 1976-79, Stony Brook, 1979—; mem. staff North Shore U. Hosp.-Cornell U. Med. Ctr., Univ. Hosp., Stony Brook, Nassau County Med. Ctr., St. John's Hosp.; cons. in field. Contbr. articles to profl. publs.; specialist in microsurgery, magnetic resonance imaging, spinal trauma, pituitary surgery. Recipient K.G. McKenzie Meml. award Royal Coll. Physicians and Surgeons Can., 1976, Open Scholarship award Johns Hopkins U., yearly, 1963-66, Scholarship award U. Chgo., yearly, 1967-70; NSF fellow, 1968, Blackman-Hoffman Found. fellow, 1969-70, USPHS fellow, 1969. Mem. ACS, AMA, Am. Assn. Neurol. Surgeons, Congress Neurol. Surgeons, N.Y. Acad. Scis., L.I. Neurosci. Acad., Suffolk Acad. Medicine, Montreal Neurol. Inst. Fellows Soc., N.Y. State Neurosurg. Soc., N.Y. State Med. Soc., N.Y. State Soc. Surgeons, Am. Epilepsy Soc., Am. Soc. Neuroimaging, Internat. Platform Assn., Nat. Alumni Schs. (chmn. com. Johns Hopkins U.), Assn. Yale Alumni in Medicine, Princeton Club N.Y., Johns Hopkins Club, Sigma Xi. Office: Stony Brook Med Ctr PO Box 721 Stony Brook NY 11790-0721

KUCHTA, STEVEN JERRY, psychologist; b. Chgo., Mar. 29, 1961; s. Ronald James and Joan Rita (Farnum) K.; m. Susan Kay Bartow, Oct. 17, 1987. BA, Blackburn Coll., 1983; MA, Ea. Ill. U., 1984; PhD, Ind. State U., Terre Haute, 1990. Instr. Ea. Ill. U., Charleston, 1987-88; psychologist Heartland Hosp., Nevada, Mo., 1988—; therapist Family Counseling Ctr., Nevada, 1990-92; psychologist Family Counseling Ctr., Springfield, 1993—; student Forest Inst. Profl. Psychology, Springfield, Mo., 1992. Bd. dirs. Am. Cancer Soc., Vernon County, Mo., 1990-91. Mem. Am. Psychol. Assn. Democrat. Roman Catholic. Office: Heartland Hosp 1500 W Ashland Nevada MO 64772

KUCIC, JOSEPH, management consultant, industrial engineer; b. Mali Losing, Croatia, Yugoslavia, Dec. 21, 1964; came to U.S., 1967, naturalized, 1974; s. Roman Kucic and Esterina (Karcic) Milevoj; m. Gia Michelle Bonavisa, Sept. 11, 1992. AAS, Coll. of Aeronautics, 1984; BS, Thomas A. Edison State Coll., 1986; B in Tech., N.Y. Inst. Tech., 1986; MBA, St. John's U., Jamaica, N.Y., 1989. Workload planner Butler Aviation-Newark, Inc., 1984-85; tech. planner N.Y. Airlines, Flushing, N.Y., 1985-86; product support engr. United Techs.-Pratt & Whitney, East Hartford, Conn., 1986; indsl. engr. Montefiore Med. Ctr., Bronx, 1986-88; sr. work mgmt. analyst Bank Leumi Trust Co., N.Y.C., 1988-89; sr. methods analyst Salomon Bros., Inc., N.Y.C., 1989-92; mgmt. cons. United Mgmt. Techs., N.Y.C., 1992-93; sr. sys. analyst N.Y.C. Health & Hosp. Corp. Metro. Hosp. Ctr., 1993—. Contbr. articles to profl. jours. Mem. AIAA, IEEE (assoc.), SAE (affiliate), Inst. Indsl. Engrs. (chpt. pres. 1988-89, chmn. bd. N.Y.C. chpt. 1989-90, Cert. of Recognition 1988, bd. govs. 1988—), MBA Execs., Coll. of Aeronautics Alumni Assn. (pres. 1990-92), St. John's U. Coll. of Bus. Adminstrn. Alumni Assn. (bd. dirs. 1991—), The Wings Club (N.Y.C.), Tau Alpha Pi. Democrat. Roman Catholic. Avocation: tennis. Home: 1542 Silver St Bronx NY 10461 Office: Met Hosp Ctr 1901 1st Ave and 97th St New York NY 10029

KUCK, DAVID JEROME, computer system researcher, administrator; b. Muskegon, Mich., Oct. 3, 1937; s. Oscar Ferdinand and Alyce (Brems) K.; m. Sharon McCure, July 16, 1977; children: Julianne, Jonathan. BSEE, U. Mich., 1959; MS, Northwestern U., 1960, PhD, 1963. Asst. prof. elec. engring. MIT, Cambridge, 1963-65; asst. prof. computer sci. U. Ill., Urbana, 1965-68, assoc. prof., 1968-72, prof. computer science, elec. and computer engring., 1972-93, dir. Ctr. for Supercomputing R & D, 1984-92; chmn. Kuck and Assocs., Inc., Champaign, Ill., 1979—; cons. Alliant Computer Systems, Littleton, Mass., 1982-87, Burroughs Corp., Paoli, Pa., 1972-80; mem. adv. bd. Sequent Computer Systems, Portland, Oreg., 1985-93; mem. computer sci. and tech. bd. NRC, Washington, 1986-90. Author: Structure of Computers and Computations, vol. 1, 1978; also numerous papers; editor 12 jours.; patentee supercomputer, 1979. Recipient Alumni Merit award Northwestern U., 1989, Eckert-Mauchly award ACM-IEEE, 1993. Fellow AAAS, IEEE (editor Transactions on Computers 1973-75, Emanuel R. Piore award 1987, Outstanding Paper award Internat. Conf. Parallel Processing 1986, ACM); mem. NAE, Assn. for Computing Machinery (editor jour. 1980-83). Home: 405 Yankee Ridge Ln Urbana IL 61801-7113 Office: kuck and Assocs 1906 Fox Dr Champaign IL 61820

KUCK, MARIE ELIZABETH BUKOVSKY, retired pharmacist; b. Milw., Aug. 3, 1910; d. Frank Joseph and Marie (Nozina) Bukovsky; Ph.C., U. Ill., 1933; m. John A. Kuck, Sept. 20, 1945 (div. Nov. 1954). Pharmacist, tchr. Am. Hosp., Chgo., 1936-38, St. Joseph Hosp., Chgo., 1938-40, Ill. Masonic Hosp., Chgo., 1940-45; chief pharmacist St. Vincent Hosp., Los Angeles, 1946-48, St. Joseph Hosp., Santa Fe, 1949-51; dir. pharm. services St. Luke's Hosp., San Francisco, 1951-76; pharmacist Mission Neighborhood Health Center, San Francisco, 1968-72; docent Calif. Acad. Sci., 1977—, DeYoung Mus., 1989—; mem. peer rev. com. Drug Utilization Com., Blue Shield Calif. and Pharm. Soc. San Francisco. Recipient Bowl of Hygeia award Calif. Pharm. Assn., 1966. Mem. No., Calif. (legis. chmn. aux. 1967-69, chmn. fund raising luncheon 1953-71, pres. San Francisco aux. 1974), Nat., Am., No. Calif. (pres. 1955-56, pres. San Francisco aux. 1965-66, editor ofcl. publ. 1967-70), San Francisco (sec. 1977-79, treas. 1979-80, pres. 1982-83; Pharmacist of Yr. award 1978) pharm. socs., Am. Pharm. Assn. (pres. No. Calif. br. 1956-57, nat. sec. women's aux. 1970-72, hon. pres. aux. 1975—), Calif. Council Hosp. Pharmacists (organizer 1962, sec.-treas. 1962-66), Am. Soc. Hosp. Pharmacists, Assn. Western Hosps. (gen. chmn. hosp. pharmacy sect. conv. San Francisco 1958), Internat. Pharmacy Congress (U.S. del. Brussels 1958, Copenhagen 1960), Fedn. Internationale Pharmaceutique, Lambda Kappa Sigma. Home: 2261 33d Ave San Francisco CA 94116

KUDRNA, FRANK LOUIS, JR., civil engineer, consultant; b. Chgo., Sept. 11, 1943; s. Frank Louis Sr. and Helen Georgiana (Malcik) K.; m. Joann Helen Danca, May 3, 1964; children: Karen, Matthew, David. BS in Archtl. Engring., Chgo. Tech. Inst., 1964; MS, Ill. Inst. Tech., 1973, PhD, 1975; MBA, U. Chgo., 1985. Engr. State Dept. Transp., Chgo., 1966-68; supervising engr. Metro. Water Reclamation Dist., Chgo., 1968-76; dir. water resources State of Ill., Chgo., 1976-82; pres. Epstein Civil Engring., Chgo., 1982-86; pres., CEO Kudrna and Assoc., Chgo., 1986—; commr. Great Lakes Commn., Ann Arbor, Mich., 1976—; panel mem. Sea Grant Panel NOAA, Washington, 1992—. Contbr. articles to profl. jours. Mem. Union League Club, Chgo., 1986. Office: Kudrna and Assoc Ltd 203 N Cass Westmont IL 60559

KUEBELER, GLENN CHARLES, engineering executive; b. Sandusky, Ohio, Aug. 1, 1935; s. William Louis and Helen Amelia (Johnson) K.; m. Iris Elaine Parker, Sept. 7, 1957; children: Mark Kenton, Gregory Glen. BSChemE, Case Inst. Tech., 1957. Project engr. Hercules Inc., Cumberland, Md., 1957-61, program mgr., 1962-69; field resident rep. Hercules Inc., Sunnyvale, Calif., 1961-62; sr. tech. specialist Hercules Inc., Wilmington, Del., 1969-77, aero mktg. mgr., 1977-80, sr. sales rep., 1980-87, quality mgr. purchasing, 1987-88, mgr. purchasing divsn., 1988-91, dir. safety, 1991—. Co-author: Composite Materials: Testing and Design, 1974, Beer Advertising Openers, 1978, Commercial Opportunities for Advanced Composites, 1980. Bd. dirs. Suppliers Advanced Composite Materials Assn., Arlington, Va., 1985-87, Del. Safety Coun., Sci. Alliance Del., Wilmington, 1991—, vice chairperson, 1991—; adv. bd. dirs. ctr. for Composite Materials, Newark, Del., 1982-87, del. Aerospace Edn. Found., Wilmington, 1991—; chair Sci. Alliance Bd., 1993—; active Firelands Hist. Soc., Hist. Lyme Village, Heymann Hist. Soc. Mem. ASTM, NRA, Soc. Advancement Material and Process Engring. (chpt. chmn. 1983-84), Delaware County Field and Stream Assn., Palatines to Am., Nat. Assn. Breweriana Collecting, East Coast Breweriana Assn., Hercules Country Club, Johnny Appleseed Postcard Club, Ohio Bottle Club, Case Alumni Assn., Zeta Psi, Alpha Chi Sigma. Republican. Avocations: genealogy, breweriana collecting, tennis, shooting, history. Home: 2410 Dacia Dr Wilmington DE 19810 Office: Hercules Inc 1313 N Market St Wilmington DE 19894

KUEHLER, JACK DWYER, computer company executive; b. Grand Island, Nebr., Aug. 29, 1932; s. August C. and Theresa (Dwyer) K.; m. Carmen Ann Kubas, July 16, 1955; children—Cynthia Marie, Daniel Scott, Christina L., David D., Michael P. BSME, U. Santa Clara. Design engr. jet engines dept. Gen. Electric Co., Evandale, Ohio, 1954-55; with IBM, 1958—; dir. IBM Raleigh Comunications Lab., 1967-70, IBM San Jose and Menlo Park Labs., 1970-72; v.p. devel. gen. products div. IBM, 1972-77, asst. group exec. data processing product group, 1977-78, pres. system products div., 1978-80, corp. v.p., from 1980; pres. gen. tech. div. IBM, White Plains, N.Y., 1980-81, info. systems and tech. group exec., 1981-82, sr. v.p., 1982-88, vice chmn., 1982-88, pres., 1988—; bd. dirs. Olin Corp. Patentee in field. Trustee U. Santa Clara (Calif.). Served as 1st lt. U.S. Army, 1955-57. Mem. IEEE (sr.), Nat. Acad. Engring., Am. Electronics Assn. Office: IBM Old Orchard Rd Armonk NY 10504-1709 also: IBM Info Systems & Tech Group 1000 Westchester Ave White Plains NY 10604

KUEHNERT, DEBORAH ANNE, medical center administrator; b. Raleigh, N.C., Nov. 21, 1949; d. Eldor Paul and Lila Catherine (Gilbert) K. Student, Valparaiso (Ind.) U., 1967-69; BS in Biology, Lenior Rhyne Coll., Hickory, N.C., 1977. Cert. med. technologist. Rsch. asst. Strong Meml. Hosp., Rochester, N.Y., 1967-68; lab. technician Richard Baker Hosp., Hickory, N.C., 1969-76; med. technician, shift supr. Glenn R. Frye Hosp., Hickory, 1977-83; lab. tech. dir. Frye Regional Med. Ctr., Hickory, 1983-85, adminstrv. dir. lab. svcs., 1986-92; sr. tech. dir. lab. svcs. Al-Fanateer Hosp., Jubail, Saudia Arabia, 1993—; instr. microbiology Catawba Valley Tech. Coll., Hickory, 1977—, Lenior Rhyne Coll., Hickory, 1978—; cons. Frye Physicians, Hickory, 1985—; lab. cons. Am. Med. Internat., New Orleans, 1986, Lake City, Fla., 1984-85; cons. Med. Lab. Observer, Chgo., 1989. Mem. Am. Soc. Clin. Pathologists, N.C. Soc. Blood Bankers. Lutheran. Avocations: guitar, piano, hiking, traveling, reading. Home: 34 Penny Ln Hickory NC 28601-9341

KUENZEL, WAYNE JOHN, avian physiologist, neuroscientist; b. Phila., Jan. 22, 1942; s. E. Gustav and Minerva (Beckmann) K.; m. Nancy T. Kuenzel, June 8, 1969 (div. 1978); m. Kimberley Ann Kuenzel, Nov. 10, 1985; children: Lauren Margaret, Carolyn April, Jonathan Patrick. BS in Biology, Bucknell U., 1964, MS, 1966; PhD in Zoology, U. Ga., 1969. NIH postdoctoral fellow Cornell U., Ithaca, N.Y., 1971-73, rsch. assoc., 1973-74; asst. prof. U. Md., College Park, 1974-78, assoc. prof., 1978-84, prof. physiology, 1984—. Author: A Stereotaxic Atlas of the Brain of the Chick, Gallus domesticus, 1988; contbr. over 60 articles to profl. jours. Capt. U.S. Army, 1969-71. Fulbright-Hays sr. rsch. fellow Coun. Internat. Exch. of Scholars, Scotland, 1980-81, Germany, 1988-89. Achievements include research in cell and tissue, comparative neurology, developmental brain rsch., physiology and behavior and poultry sci. Home: 6829 Pineway University Park MD 20782 Office: U Md Poultry Sci Dept College Park MD 20742

KUENZLER, EDWARD JULIAN, ecologist and environmental biologist; b. West Palm Beach, Fla., Nov. 11, 1929; s. Edward and Flora Caroline (Jeske) K.; m. Jutta Gertraud Koslowski, Sept. 4, 1965; children—Doreen Friederika, Dirk Edward. B.S., U. Fla., 1951; M.S., U. Ga., 1954, Ph.D., 1959. Assoc. scientist Woods Hole (Mass.) Oceanographic Inst., 1959-65; assoc. prof. environ. scis. and engring. U. N.C., Chapel Hill, 1965-71, prof. environ. biology, 1972-92; prof. emeritus U. N.C., 1992—; program dir. environ. chemistry and biology U. N.C., Chapel Hill, 1980-83, program dir. aquatic and atmospheric scis., 1990-92, dep. chmn. environ. scis. and engring., 1984-87, chmn. curriculum in marine scis., 1968-71, 72-73; program dir. for biol. oceanography NSF, 1971-72; mem. panel Nat. Acad. Scis., 1974-75; mem. N.C. Gov.'s Tech. Coordinating Com., 1968-70, N.C. Comml. and Sports Fisheries Adv. Com., 1975-77; cons. and mem. adv. panels in field; tchr. grad. courses phytoplankton ecol., wetland ecol., 1967-92; dir. grad. rsch., 1965-92. Contbr. articles on phytoplankton ecology, water quality, elemental cycling in estuarine, freshwater and wetland ecosystems to profl. jours. Served to capt. USAF, 1954-57. Named to U. Fla. Hall of Fame, 1951; grantee AEC, 1962-70, NOAA Office Sea Grants, 1971-76, Office Water Rsch. and Tech., U.S. Geol. Survey-N.C. Water Resources Rsch. Inst., 1970-89, EPA, 1987-90. Mem. Ecol. Soc. Am., Am. Soc. Limnology and Oceanography, Estuarine Rsch. Fedn., N.C. Acad. Sci. (treas. 1982-85), Elisha Mitchell Sci. Soc. (pres. 1979-81), Soc. Wetland Scientists, Nat. Assn. Scholars, Phi Beta Kappa, Sigma Xi. Republican. Methodist. Home: 6015 Old Greensboro Rd Chapel Hill NC 27516-8516

KUESEL, THOMAS ROBERT, civil engineer; b. Richmond Hill, N.Y., July 30, 1926; s. Henry N. and Marie D. (Butt) K.; m. Lucia Elodia Fisher, Jan. 31, 1959; children—Robert Livingston, William Baldwin. B. Engring. with highest honors, Yale U., 1946, M. Engring., 1947. With Parsons, Brinckerhoff, Quade & Douglas, 1947-90; project mgr. Parsons, Brinckerhoff, Quade & Douglas, San Francisco, 1967-68; ptnr., sr. v.p. Parsons, Brinckerhoff, Quade & Douglas, N.Y.C., 1968-83, chmn. bd., dir., 1983-90; cons. engr., 1990—; vice chmn. OECD Tunneling Conf., Washington, 1970; mem U.S. Nat. Com. on Tunneling Tech., 1972-74; chmn. Geotech. bd. Nat. Rsch. Coun., 1988-89. Contbr. 60 articles to profl. jours.; designer more than 120 bridges, 135 tunnels and numerous other structures in 36 states and 20 fgn. countries, most recent SSC physics rsch. tunnel Tex., 1990—, Geo Coleman Bridge Replacement, Yorktown, Va., 1991-93, Boston Ocean Outfall Tunnel, 1988-90, Cumberland Gap Tunnel, Ky. and Tenn., 1986-90, Jamuna River Bridge, Bangladesh, 1985—, Trans Koolau Tunnel, Hawaii, 1985-90, Ft. McHenry Tunnel, Balt., 1978-85, Rogers Pass Rwy. Tunnel, B.C., 1981-85, Glenwood Canyon Tunnel, Colo., 1981-88, subways Boston, N.Y., Balt., Wash., Atlanta, Pitts., Seattle, L.A., Singapore and Taipei. Fellow ASCE; mem. Nat. Acad. Engring., Internat. Assn. for Bridge and Structural Engring. Brit. Tunnelling Soc., Yale Sci. and Engring. Assn., Am. Underground Space Assn. (hon. mem.), Yale Club N.Y.C., Wee Burn Club, The Moles, Farmington Club, Tau Beta Pi.

KUGELMAN, IRWIN JAY, civil engineering educator; b. Bklyn., Feb. 15, 1937; s. Samuel Solomon and Sylvia (Habas) K.; m. Ruth Lillian Cariski, Aug. 28, 1958; children: Sylvia E., Harold M., Elizabeth A., Maura J. BSCE, Cooper Union U., 1958; SM in Sanitary Engring., MIT, 1960, ScDCE, 1963. Cert. engr.-in-tng., N.Y. Jr. and asst. civil engr. City of N.Y., 1958-59; rsch. asst. MIT, Cambridge, Mass., 1960-61; asst. prof. civil engring. NYU, N.Y.C., 1962-65; rsch. scientist Am. Standard Corp., Piscataway, N.J., 1965-70; rsch. sanitary engr. U.S. EPA, Cin., 1970-82; dir. environ. studies, prof. civil engring. Lehigh U., Bethlehem, Pa., 1982-88, prof., chair dept. civil engring., 1985—; cons. Alexander Potter Assoc., N.Y.C., 1963-64, Hydrotechnic Corp., N.Y.C., 1964-65; bd. dirs. ACT Corp., Allentown, Pa.; mem. environ. engring. peer panel U.S. EPA, 1979—. Author numerous engring. reports, design manuals; contbr. over 100 articles to tech. jours. Fin. sec., v.p. No. Hills Synagogue, Cin., 1971-81; v.p. Bur. Jewish Edn., Cin., 1975-81. Mem. ASCE (dept. chair coun. 1989-91), Water Environ. Fedn. (asst. chair rsch. com. 1986-91), Am. Water Works Assn., Am. Chem. Soc. Achievements include patents in field. Home: 524 Kevin Dr Bethlehem PA 18017 Office: Lehigh U Civil Engring Dept 13 E Packer Ave Bethlehem PA 18015

KUHI, LEONARD VELLO, astronomer, university administrator; b. Hamilton, Ont., Can., Oct. 22, 1936; came to U.S., 1958; s. John and Sinaida (Rose) K.; m. Patricia Suzanne Brown, Sept. 3, 1958 (div.); children: Alison Diane, Christopher Paul; m. Mary Ellen Murphy, July 15, 1989. B.S., U. Toronto, Can., 1958; Ph.D., U. Calif., Berkeley, 1964. Carnegie postdoctoral fellow Hale Obs., Pasadena, Calif., 1963-65; asst. prof. U. Calif., Berkeley, 1965-69, assoc. prof., 1969-74, prof., 1974-89, chmn. dept. astronomy, 1975-76, dean phys. scis. Coll. Letters and Sci., 1976-81, provost, 1983-89; prof. astronomy, sr. v.p. for acad. affairs, provost U. Minn., Mpls., 1989-91; vis. prof. U. Colo., 1969, Coll. de France, Paris, 1972-73, U. Heidelberg, 1978, 80-81. Contbr. articles to profl. jours. Recipient Alexander von Humboldt Sr. Scientist award, 1980-81; NSF research grantee, 1966—. Fellow AAAS; mem. Am. Astron. Soc. (treas. 1987), Astron. Soc. Pacific (pres. 1978-80), Royal Astron. Soc. (Can.), Internat. Astron. Union, Sigma Xi. Office: U Minn Dept Astronomy 116 Church St SE Minneapolis MN 55455-0149

KUHL, DAVID EDMUND, physician, radiology educator; b. St. Louis, Oct. 27, 1929; s. Robert Joseph and Caroline Bertha (Waldemar) K.; m. Eleanor Dell Kasales, Aug. 7, 1954; 1 son, David Stephen. AB, Temple U., Phila., 1951; MD, U. Pa., 1955; LHD (hon.), Loyola U. Chgo., 1992. Diplomate: Am. Bd. Radiology, Am. Bd. Nuclear Medicine (a founder; life trustee 1977—). Intern, then resident in radiology Sch. Medicine and Hosp. U. Pa., 1955-56, 58-63, mem. faculty, 1963-76, prof. radiology, 1970-76, vice chmn. dept., 1975-76, chief div. nuclear medicine, 1963-76; prof. radiol. scis. UCLA Sch. Medicine and Hosp., 1976-86, chief div. nuclear medicine, 1976-84, vice-chmn. dept., 1977-86; prof. internal medicine and radiology, chief div. nuclear medicine U. Mich. Sch. Medicine, Ann Arbor, 1986—; Disting. Faculty lectr. in biomed. rsch. U. Mich. Med. Sch., 1992; mem. adv. com. Dept. Energy, NIH, Internat. Commn. on Radiation Units and Measures; mem. sci. adv. bd. Max Planck Inst., Cologne, Fed. Republic of Germany, John Douglas French Found. for Alzheimer's Disease. Mem. editorial bd. various jours.; contbr. articles to med. jours. Served as officer M.C. USNR, 1956-58. Recipient Research Career Devel. award USPHS, 1961-71, Ernst Jung prize for medicine Jung Found., Hamburg, 1981, Emil H. Grubbe gold medal Chgo. Med. Soc., 1983, Berman Found. award peaceful uses atomic energy, 1985, Steven C. Beering award for advancement med. sci. Ind. U., 1987, Disting. Grad. award U. Pa. Sch. Medicine, 1988, William C. Menninger Meml. award ACP, 1989, Javits Neuroscience Investigator award NIH, 1989. Fellow Am. Coll. Radiology, Am. Coll. Nuclear Physicians; mem. Assn. Am. Physicians, Am. Epilepsy Soc., Assn. Univ. Radiologists, Radiol. Soc. N.Am., Soc. Nuclear Medicine (Nuclear Pioneer citation 1976, Herman L. Blumgart, M.D., Pioneer award 1979, Disting. Scientist award 1981, ann. lectr. 1991), Am. Heart Assn. (fellow council circulation), Am. Neurol. Assn., Rocky Mountain Radiol. Soc., Soc. Neurosci., Inst. Medicine of Nat. Acad. Scis., Sigma Xi, Alpha Omega Alpha. Office: U Mich Hosp Div Nuclear Medicine 1500 E Medical Ctr Dr Ann Arbor MI 48109-0028

KUHLKE, WILLIAM CHARLES, plastics engineer; b. Plainfield, N.J., Oct. 13, 1930; s. William H. and Bertha V. (Link) K.; m. Carol E. Kuhlke, May 10, 1956; children: William C. Jr., Stephen M., Susan M., Patricia A. BSChemE, Lehigh U., 1952. With Shell Chem., 1952-86; v.p. Plasyer Fun House, 1984—, DeWitt & Co., 1986-91; pres. Kuhlke & Assocs., Houston, 1991—. Author: DeWitt Polymer Books, 1986-91; co-author: Concise Encyclopedia of Polymer Science and Engineering, 1990. Mem. Soc. Plastic Engrs. (pres. 1984-85), Chem. Mgmt. Resource Assn., European Assn. for Bus. Res. Home: 14519 Cindywood Houston TX 77079 Office: Kuhlke & Assocs 14519 Cindywood Houston TX 77079

KUHLMAN, ROBERT E., orthopedic surgeon; b. St. Louis, Nov. 7, 1932; s. Carl W. and Irma V. (Huetteman) K.; m. Ana Marie, Aug. 25, 1958; children: Lisa Anne, Marcella, Robert E. Jr., Caroline, Richard. AB, Washington U., St. Louis, 1953; MS cum laude, Washington U. Sch. Medicine, St. Louis, 1956. Diplomate Am. Bd. Orthopedic Surgeons. Intern Barnes Hosp., St. Louis, 1956-57; sr. asst. surgeon USPHS, NIH, 1957-59; with various hosps., 1959-63; asst. in clin. orthopedic surgery Washington U. Sch. Medicine, St. Louis, 1963-64, instr. clin. orthopedic surgery, 1964-68, asst. prof., 1968—; speaker and presenter in field. Contbr. articles to profl. jours. Fellow Am. Coll. Surgeons; mem. AMA, AAAS, Am. Acad. Orthopedic Surgeons, Am. Bd. Orthopedic Surgery, Mid-Ctrl. State Orthopedic Assn., Mo. Chpt. Am. Coll. Surgeons, Mo. State Med. Soc., St. Louis Med. Soc. St. Louis Orthopedic Soc., St. Louis Surg. Soc., Southern Med. Assn., Orthopedic Rsch. Soc., Assn. Bone and Joint Surgeons, Clin. Orthopedic Soc., Soc. Internat. De Chirurgie ORthopedic Et De Traumatology, Sigma Xi,. Achievements include application of Quantitatice Microchemistry to epiphyseal bone formation molecular biology of bone formation. Office: Orthopedic Cons Inc Ste 16301 W Pavilion Barnes Hosp Pl Saint Louis MO 63110

KUHN, LEIGH ANN, air pollution control engineer; b. Wichita, Kans., May 14, 1967; d. Judy Stites Kuhn; m. Jeff A. Carroll, Feb. 20, 1993. BS in Chem. Engring., Kans. State U., 1990. Registered engr.-in-tng. Kans. Air pollution control engr. Burns & McDonnell, Kansas City, Mo., 1990—. Contbr. articles to profl. jours. Coord. Burns & McDonnell - S.E. Middle Magnet Sch. Community Partnership, Kansas City, 1991-92; judge Greater Kansas City Sci. Fair, 1991-92. Mem. NSPE, Mo. Soc. Profl. Engrs., Am. Inst. Chem. Engrs., Air and Waste Mgmt. Assn. (tech. tours com. 85th Ann. Exhbn. 1991-92, tech. rev. com., 1991-92). Office: Burns & McDonnell Engring 4800 E 63d St Kansas City MO 64130

KUHN, MATTHEW, engineering company executive; b. Sacalaz, Banat, Romania, Mar. 19, 1936; came to U.S., 1967; s. Peter and Katherine (Gerres) K.; m. Betty Jane Ritchie, Aug. 20, 1966; children: Andrew Jason, Andrea Suzanne. BASc in Engring. Physics, Queen's U., Kingston, Ont., Can., 1962; MASc, U. Waterloo, Ont., 1963, PhD in Elect. Engring., 1967, D of Engring. (hon.), 1985. Fellow Brown U., Providence, 1967-68; supr. MTS Bell Telephone Labs., Murray Hill, N.J., 1968-73; mgr. adv. tech. BNR Ltd., Ottawa, Ont., 1973-80, dir. corp. devel., 1980-85; asst. v.p. BNR Inc., Research Triangle Park, N.C., 1985-89; pres. Microelectronics Ctr. of N.C., Research Triangle Park, 1989—; presenter numerous profl. meetings. Contbr. articles to profl. jours. Mem. N.C. Bd. Sci. & Tech., 1991—; chmn. adv. coun. Queen's U., 1983-84; chmn. engring. adv. coun. Duke U., Durham, N.C., 1989—. Fellow IEEE (editor spl. issue Electron Devices Jour. Optoelectronics 1975). Roman Catholic. Achievements include discovery of quasi-static method measurement technique for integrated circuit development; co-development first generation fiber optics technology. Home: 2 Whisper Ln Chapel Hill NC 27514-1635 Office: MCNC Inc 3021 W Cornwallis Rd Research Triangle Park NC 27709

KUHN, ROBERT HERMAN, public works and utilities executive, engineer; b. Canton, Ill., Apr. 10, 1946; s. Orval Jesse Sr. and J. Nellie (Gallien) K.; m. Marlene Elizabeth Shuffer, May 29, 1971; children: Jesse Lee, Regina Marie. BS, U. Ill., 1969, MS, 1975. Registered profl. engr., Mich., Ill. Engr. Crawford, Murphy & Tilley, Springfield, Ill., 1965-68; cons. Grt. Basin Engring., Ogden, Utah, 1973; pub. works engr. City of Bloomington, Ill., 1974; asst. city engr. City of Muskegon, Mich., 1975-78; asst. dir. pub. works County of Muskegon, 1978-85, dir. pub. works and utilities, 1985—; pres. K&S Component Cars, Muskegon, 1978-82; owner BoMar Commodities, Muskegon, 1982—; mem. Solid Waste Planning Bd., 1988-89, Muskegon County Dept. Pub. Works Bd., 1989—. Speaker, facilitator PreCana Marriage Preparation Seminars, 1972—; mem. exec. bd. St. Francis deSales Parish Coun., 1985-88, Muskegon County Cooperating Chs., 1989-92; mem. Charter Study Co., 1979; guitarist ecumenical religious retreats Muskegon Correctional Facilities. With USAF, 1969-73. Decorated Meritorious Svc. medal; recipient Govt. award Am. City and County Mag., 1987; named One of Outstanding Young Men Am., 1971, Recycler of Yr. Mich. Recycling Coalition, 1989; Harry H. Gunther scholar U. Ill., 1965-69. Mem. ACSE, Am. Pub. Works Assn., Am. Water Works Assn., Am. Soc. Pub. Adminstrs., Nat. Assn. Fleet Mgrs., Govtl. Refuse Collection and Disposal Assn., Naturist Soc. Roman Catholic. Avocations: contemporary gospel guitar, basketball, coaching, motivational speaker. Home: 3080 W Sherman Blvd Muskegon MI 49441-1154 Office: City of Muskegon Pub Svc 1350 E Keating Ave Muskegon MI 49442

KUHNS, JAMES HOWARD, communications engineer; b. Uniontown, Pa., Dec. 10, 1953; s. Howard Hill and Barbara Jenn (Beard) K.; m. Ernestine Yvonne Wiseman, Nov. 15, 1978; children: James Bryan, Sherry Katherine, Jonathan Michael, Rebecca Marie. Cert. in electronic tech., Cleve. Inst. Electronics, 1985, Nat. Cable TV Inst., Denver, 1988. Line technician Storer Cable Communications, North Charleston, S.C., 1982-84; chief technician Mills Communications, Kiawah Island, S.C., 1984-85; sr. technician Summit Communications, Woodstock, Ga., 1985-87; engring. mgr. Summit Communications, Statesville, N.C., 1987-88; dist. tech. tng. mgr. Continental Cablevision, Oak Park, Mich., 1989-92; regional staff engr. Comcast, Warren, Mich., 1992—. Contbr. articles to tech. publs. Mem. Soc. Broadcast Engrs. (sr.), Soc. Cable TV Engrs. (sr. chpt. officer 1986—). Republican. Presbyterian. Achievements include copyrights for development of technical software for broadcast, cable TV and satellite industries. Office: Comcast 5700 Enterprise Ct Warren MI 48092

KUHR, RONALD JOHN, entomology and toxicology educator; b. Appleton, Wis., Dec. 29, 1939; m. Mary Margaret Zeh, Sept. 9, 1961; children: Christopher, Katharine, Matthew. BS, U. Wis., 1963; PhD, U. Calif., Berkely, 1966. Asst. assoc. prof. Cornell U., Geneva, N.Y., 1968-77; prof. Cornell U., Ithaca, N.Y., 1978-80, assoc. dir. agrl. expt. sta., 1977-80; dept. head N.C. State U., Raleigh, 1980-86, dir. agrl. expt. sta., 1987-91, prof., 1992--. Author: Carbamate Insecticides, 1976; co-editor: Linking Research to Crop Production, 1980, Minimizing the Risk Associated with Pesticide Use, 1987, Safer Inseticides: Development and Use, 1990. Office: NC State U Entomology Box 7613 Raleigh NC 27695-7613

KUIPERI, HANS CORNELIS, chemical trading company executive; b. Hague, Netherlands, May 15, 1939; arrived in Belgium, 1970; s. Henri C.C.H. and Wilhelmina (Limburg) K.; m. Martine Lebeau, Sept. 19, 1980 (div. Apr. 1991; children: Marcel, Cuno. Degree in chemistry, U. Leiden, Netherlands, 1970. Mgr. V.G.L., Utrecht, Netherlands, 1964-67; sr. rep. Conoco Chem. Europe, Brussels, 1967-73; exec. Fallek SA, Brussels, 1973-78; mng. dir. Gill & Duffus, Brussels, 1978-82; mgr. Kaiser Internat., Dusseldorf, 1982-85; asst. gen. mgr. Mitsui Benelux, Brussels, 1985-88; mng. dir. Joss Belgium, Brussels, 1988—. 1st lt. Air Force, 1962-64. Avocation: golf.

KUKULL, WALTER ANTHONY, epidemiologist, educator; b. Everett, Wash., June 30, 1945; s. Anthony and Winnifred (Rayner) K.; m. Diane Neal, June 20, 1975; children: Megan J., Anthony T., Benjamin J. BA, U. Wash., 1971, PhD in Epidemiology, 1983; MS, We. Wash. State Coll., 1974. Project dir. VA Merit Review project VA Med. Ctr., Seattle, 1983-86; rsch. assoc. dept. community health care systems U. Wash., Seattle, 1983-85, epidemiologist alzheimer's disease clin rsch. ctr., NIMH, dept. psychiatry and behavioral scis., 1985-86, rsch. assoc. dept. epidemiology, 1985-88; co-investigator alzheimer's disease rsch. ctr. depts. pathology/medicine, co-prin. investigator patient registry, Seattle, 1986—; prin. investigator genetic differences in alzheimer's cases and controls dept. epidemiology U. Wash., Seattle, 1988—, sr. rsch. assoc. dept. epidemiology, 1989-91, assoc. prof. dept. epidemiology, 1991—, adj. assoc. prof. dept. health scis., 1992—; faculty senate and group VII rep. for faculty senate exec. com. U. Wash., 1991-92, 92-93, mem. faculty coun. sch. pub. health and community medicine, 1993—; mem. northwest health svcs. rsch. and devel. steering com. VA Med. Ctr., Seattle, 1983-89; mem. ad hoc study sect. Nat. Inst. on Aging, 1992, NIH. Contbr. articles and abstracts to sci. jours., chpts. to books; invited presenter in field. With USN, 1966-70. VA Health Svcs. Rsch. and Devel. predoctoral fellow, 1980-82, Nat. Inst. on Aging grantee, 1985-90, 86-89, 88-91, 89-91, 89—, 90—, 93—. Fellow Am. Coll. Angiology (sci. coun.); mem. APHA, Soc. for Epidemiologic Rsch., Gerontol. Soc. Am. Achievements include research in alzheimer's disease/neuroepidemiology, epidemiologic methods, health services research. Office: U Wash Dept Epidemiology SC-36 Seattle WA 98195

KUKUSHKIN, VLADIMIR IVANOVICH, aviation engineer, educator; b. Yarzoslavl, Russia, July 23, 1931; arrived in The Ukraine, 1955.; s. Ivan Nikolayevich and Alexandra Alexandrovna (Tushina) K.; m. Valentina Mikhailovna Egorova, Feb. 26, 1955; 1 child, Sergei. Diploma in engring., Moscow Aviation Inst., 1955; D in Engring. Scis., Dniepropetrovsk, 1984; academician, Acad. Engring. Scis., Kiev, 1991. Engr. Yuzhnoye Design Office, Dniepropetrovsk, 1955-57, from chief of group to chief of dept., 1957-67, chief of design office, chief designer, 1967—; prof. Dniepropetrovsk State U., 1986—. Author: Design of Rocket Engines, 1969; Thrust Control Systems, 1989. Recipient Order of the Red Banner of Labour Presidium of the USSR, 1959, Order of Lenin, Supreme Soviet, 1961, Order of the Red Banner of Labour, Supreme Soviet, 1971, Lenin prize winner, USSR Com., 1976. Mem. Trade Union of Defense Industry Profl. Workers. Home: Karl Marks Ave 20 flat 89, 320027 Dniepropetrovsk Ukraine Office: Yuzhnoye Design Office, Kzivorozhskaya Str 3, 320008 Dniepzopetrovsk Ukraine

KULANDER, KENNETH CHARLES, physicist; b. St. Paul, Nov. 26, 1943; s. Merrill Alton and Nellie Ava (Brownjohn) K.; m. Monica Ann Romanovski, Dec. 27, 1969. BS, Cornell Coll., 1965; PhD, U. Minn., 1972. Postdoctoral fellow U. Minn., Mpls., 1972-75; sr. rsch. assoc. Daresbury Lab., Warrington, U.K., 1975-78; chemist Lawrence Livermore (Calif.) Nat. Lab., 1978-82, physicist, 1983-86, group leader, 1986—; vis. scientist Max Planck Inst. Quantum Optics, Garching, Germany, 1982-83; mem. rev. panels Dept. Energy, Washington, 1990—. Editor: Time Dependent Methods for Quantum Dynamics, 1991; editorial Bd. Internat. Jour. Nonlinear Optical Physics, Singapore, 1991—; contbr. to profl. publs. Fellow Am. Phys. Soc. (fellowship com. 1990-91); mem. Inst. Physics, Optical Soc. Am. Achievements include theoretical studies of dynamics of atomic and molecular systems and their interactions with laser fields. Office: Lawrence Livermore Nat Lab PO Box 808 Livermore CA 94550

KULASIRI, GAMALATHGE DON, computer educator, mechanical and agricultural engineer; b. Horana, Sri Lanka, Apr. 7, 1957; arrived in New Zealand, 1990; s. David Gamalath and Jayawathie Sinhabahu; m. Sandhya Samarasinghe, May 16, 1985. BSc in Engring., U. Peradeniya, Sri Lanka, 1980; MS, Va. Poly Inst. & State U., 1988, PhD, 1990. From instr. dept. mech. engring. to rsch. engr. U. Peradeniya, 1981-84; project engr. Ceylon Tobacco Co. Ltd., Colombo, Sri Lanka, 1984-85; grad. rschr. dept. agrl. engring. Va. Poly Inst. & State U., Blacksburg, 1985-90; lectr. Lincoln U., Canterbury, New Zealand, 1990—. Mem. Am. Soc. Agrl. Engrs., Soc. for Computer Simulation, Sigma Xi, Gamma Sigma Delta. Avocations: cricket, bowling. Office: Lincoln U Ctr Computing Biometrics, PO Box 84, Canterbury New Zealand

KULESZA, FRANK WILLIAM, chemical engineer; b. Cambridge, Mass., Apr. 6, 1920. BSChemE, Northeastern U., 1950. Prodn. chemist Synthon Inc., Cambridge, Mass., 1950-51; R & D chemist Borden Chem., Bainbridge, N.Y., 1951-53; assoc. engr. IBM, Poughkeepsie, N.Y., 1953-66; pres. Epoxy Tech. Inc., Billerica, Mass., 1966—; chmn. Poly-Organics Inc., Newburyport, Mass., 1981—. Sgt. USAF, 1941-45, CBI. Mem. Am. Chem. Soc., Internat. Soc. Hybrid Mfrs., Semi-Conductor Equipment and Materials Inst. Avocations: fishing, boating, swimming, music, reading. Home: 3 Grant Rd Winchester MA 01890-1016 Office: Epoxy Tech Inc 14 Fortune Dr Billerica MA 01821-3922

KULICK, RICHARD JOHN, computer scientist, researcher; b. New Kensington, Pa., Mar. 27, 1949; s. John Anthony and Anna Teresa (Tuzik) K. BS, Pa. State U., 1971; MBA, U. Md., 1973. Project acct. PPG Industries, Inc., Ford City, Pa., 1973-75; programmer analyst Allegheny Ludlum Steel Corp., Brackenridge, Pa., 1975-77, systems analyst, 1977-82, sr. systems analyst, 1982, sr. MIS planner, 1982-86; system design specialist Allegheny Ludlum Corp., Brackenridge, 1986-91; mgmt. info. systems assoc. Allegheny Ludlum Corp., Vandergrift, Pa., 1991—. Author: Heuristic Coil Slitting Optimization, 1986, (manual) Data Modeling Standards, 1988, Information Systems Integration Strategy, 1989. Mem. Computer Soc. of IEEE, Nat. Systems Programmers Assn., Assn. for Computing Machinery (voting), Datamation High Tech. Panel, Smithsonian Assocs., U.S. Tennis Assn., The Racquet Club Pitts., Pa. State U. Club Alle-Kiski Valley. Avocations: music, reading, stamp collecting, running, fine art. Home: 483 Lillian Rd Leechburg PA 15656 Office: Allegheny Ludlum Corp 132 Lincoln Ave Vandergrift PA 15690-1232

KULIKOWSKI, CASIMIR A., computer science educator, research program director; b. Hertford, Herts, Eng., May 4, 1944; s. Victor A. and Isabel S. (Tuckett) K.; m. Christine A. Wilk, May 31, 1969; children: Michael Edward, Victoria Anne. BE with honors, Yale U., 1965, MS, 1966; PhD, U. Hawaii, 1970. From asst. prof. to assoc. prof. Rutgers U., New Brunswick, N.J., 1970-77, prof., 1977—, chmn. dept. computer sci., 1984-90, dir. Lab. Computer Sci. Rsch., 1985—; mem. editorial bd. Jour. Am. Med. Informatics Assn., Bethesda, Md., 1993—, Computers in Biology and Medicine, Washington, 1980—, Experts Systems: Rsch. and Applications, Washington, 1985—; mem. bd. sci. counselors Nat. Libr. Medicine, Bethesda, 1984-87. Co-author: A Practical Guide to Designing Expert Systems, 1984, Computer Systems That Learn, 1992; editor: Artificial Intelligence Expert Systems and Languages in Modeling & Sinolation, 1988. Pres. Highland Park (N.J.) Residents Assn., 1983-88. Mem. AAAS, Am. Assn. Artificial Intelligence, Am. Coll. Med. Informatics, Nat. Acad. Scis. Inst. Medicine. Office: Rutgers U Detp Computer Sci Hill Ctr Busch Campus New Brunswick NJ 08903

KULKARNI, RAVI SHRIPAD, mathematics educator, researcher; b. Solapur, India, May 22, 1942; came to U.S., 1962; s. Shripad D. and Neela S. (Page) K. BA in Math., Poona U., 1962; PhD in Math., Harvard U., 1967. Mem. Inst. Henri Poincaré, Paris, 1967; rsch. assoc. Harvard U., Cambridge, Mass., 1967-68; asst. prof. Johns Hopkins U., Balt., 1968-70; rsch. fellow U. Bonn, Germany, 1970-71; mem. Inst. for Advanced Study, Princeton, N.J., 1971-73; Ritt asst. prof. Columbia U., N.Y.C., 1973-76; vis. assoc. prof. Rutgers U., New Brunswick, N.J., 1976-77; assoc. prof. dept. math. Ind. U., Bloomington, 1977-79, prof., 1979-87; prof. Queens Coll. and Grad. Ctr. CUNY, N.Y.C., 1986—; vis. prof. Princeton U., 1980, U. Colo., Boulder, 1981, Columbia U., N.Y.C., 1981, U. Mich., Ann Arbor, 1982, Max-Planck-Inst., Bonn, 1984, Harvard U., 1985, Mittag-Leffler Inst., Stockholm, 1989-90; prin. speaker Internat. Mtg. on Differential Geometry and Math. Physics, Brno, Czechoslovakia, 1986; lectr. in field. Author: Index theorems and curvature invariants, 1972; (with U. Pinkall) Conformal Geometry, Aspects of Mathematics: E, Vol. 12, 1988; (manuscript) (with M. Wang) Geometries for 3-and 4-Dimensional Manifolds; contbr. articles to profl. jours. Vice pres. Maharashtra Found., N.J., 1988. Grantee NSF, 1968—, recipient Interntat. Program award, 1985, 89; Guggenheim fellow, 1981-82. Mem. Am. Math. Soc. (invited speaker 1986), Math. Assn. Am. Achievements include research in differential geometry, Riemann surfaces, and discontinuous groups. Office: CUNY Grad Ctr 33 W 42d St New York NY 10036

KULKARNI, SANJEEV RAMESH, electrical engineering educator; b. Bombay, India, Sept. 21, 1963; came to U.S., 1966; s. Ramesh Madhav and Pushpa Ramesh (Nene) K.; m. Marian Jane Fisher, June 22, 1985; children: Mykel Jay, Kristina Marie, Lauryn Nicole. BS in Math., Clarkson U., Potsdam, N.Y., 1983, BS in Elec. Engring., 1984, MS in Math., 1985; MS in Elec. Engring., Stanford U., 1985; PhD, MIT, 1991. Mem. tech. staff MIT Lincoln Lab., Lexington, Mass., 1985-91; asst. prof. dept. elec. engring. Princeton (N.J.) U., 1991—; part-time faculty U. Mass., Boston, 1986. Contbr. articles to profl. jours. Mem. IEEE. Office: Princeton U Dept Elec Engring Princeton NJ 08544

KULKARNI, SHRINIVAS R., astronomy educator. Prof. dept. radio astronomy Calif. Inst. Tech., Pasadena. Recipient Alan T. Waterman award NSF, 1992. Office: Calif Inst of Technology Radio Astronomy Pasadena CA 91125*

KUMAR, ANIL, nuclear engineer; b. Agra, India, Aug. 3, 1952; came to U.S., 1988; s. Vedprakash and Satyawati (Sudhir) Parashar; m. Geeta Sharma, Nov. 29, 1979; 1 child, Amitabh. MSc in Physics, Agra U., 1973; PhD in Nuclear Engring., U. Bombay, India, 1981. Sci. officer Bhabha Atomic Rsch. Ctr., Bombay, 1974-81; sr. researcher Ecole Poly. Fed. Lausanne, Switzerland, 1982-88; devel. engr. UCLA, 1988-90, sr. devel. engr., 1990—. Contbr. articles to Jour. Fusion Energy, Nuclear Sci. and Engring., Fusion Tech., Fusion Engring. and Design, Atom Kern Energie, proc. internat. confs. and symposia. Mem. Am. Nuclear Soc., Am. Phys. Soc., Soc. Indsl. and Applied Math. Achievements include research in modified wigner rational approximation in neutronics, Boltzmann Fokker Planck transport equation, measurements of induced radioactivity and nuclear heating in fusion neutron environment, muon catalyzed fusion. Office: UCLA 43-133 Eng IV 405 Hilgard Ave Los Angeles CA 90024

KUMAR, BHAGAVATULA VIJAYA, electrical and computer engineering educator; b. Porumamilla, India, Aug. 15, 1953; came to U.S., 1977; s. Ramamurthy and Saradamba (Kanchinadam) Bhagavatula; m. Latha Bayya, July 1, 1982; children: Ramamurthy, Madhusudan, Chandrasekhar. BSEE, Indian Inst. Tech., Kanpur, India, 1975; MEE, Indian Inst. Tech., Kanpur, 1977; PhD in Elec. Engring., Carnegie Mellon U., 1980. Asst. prof. elec. and computer engring. Carnegie Mellon U., Pitts., 1982-87, assoc. prof., 1987-91, prof., 1991—; cons. Unicorn Systems, Inc., Pitts., 1984-91, Two-Six, Inc., 1989-90, U.S. Army, Huntsville, 1990-92. contbr. articles to profl. jours. Pres. S.V. Temple, Pitts., 1991-92. Grantee in field. Fellow SPIE, Optical Soc. Am.; mem. IEEE, Sigma Xi. Achievements include development of synthetic discriminent functions, optical processor for pattern recognition, optical moment calculators, advances in optical computing. Home: 2668 Rossmoor Dr Upper Saint Clair PA 15241 Office: Carnegie Mellon U ECE CMU Pittsburgh PA 15213

KUMAR, BINOD, materials engineer, educator; b. Jamalpur, Bihar, India, Jan. 13, 1946; came to U.S., 1971; s. Rambaran and Ramdulari (Rai) Singh; m. Shyama Thakur, May 23, 1969; children: Vineet, Sunita. MS, Pa. State U., 1973, PhD, 1976. Glass technologist Seraikella Glass Works, Konnagar, India, 1968-71; rsch. engr. Anchor Hocking Corp., Lancaster, Ohio, 1976-79; sr. rsch. engr. U. Dayton, 1980—, prof., 1992; cons. Zimmer, Inc., Warsaw, Ind., 1987—, Mead, Inc., Dayton, 1988, JAFE, Inc., Greenville, Ohio, 1987-90. Editor Newsletter of Ideas, Inc., Cleve., 1990—; contbr. articles to profl. jours. Trustee Centerville (Ohio) Soccer League, 1988-91, India Found., Dayton, 1991—. Mem. Am. Ceramic Soc., Electrochem. Soc., Indian Ceramic Soc. (life). Achievements include patents for segmented YAG laser rods and method, bioabsorbable glass fiber, fast ionic conductors. Office: U Dayton 300 College Park Ave Dayton OH 45469-0170

KUMAR, KRISHNA, physics educator; b. Meerut, India, July 14, 1936; came to U.S., 1956, naturalized, 1966; s. Rangi and Susheila (Devi) Lal; m. Katharine Johnson, May 1, 1960; children—Jai Robert, Raj David. B.Sc. in Physics, Chemistry and Math., Agra U., 1953, M.Sc. in Physics, 1955; M.S. in Physics, Carnegie Mellon U., 1959, Ph.D. in Physics, Carnegie Mellon U., 1964. Research assoc. Mich. State U., 1963-66, MIT, 1966-67, rsch. fellow Niels Bohr Inst., Copenhagen, 1967-69; physicist Oak Ridge Nat. Lab., 1969-71; assoc. prof. Vanderbilt U., Nashville, 1971-77; fgn. collaborator AEC of France, Paris, 1977-79; Nordita prof. U. Bergen, Norway, 1979-80; prof. physics Tenn. Tech. U., Cookeville, 1980-83, univ. prof. physics, 1983—; lectr. in field; cons. various rsch. labs. Author: Nuclear Models and the Search for Unity in Nuclear Physics, 1984, Superheavey Elements, 1989; contbr. articles to profl. jours, books. Sec. India Assn., Pitts., 1958-59. Faculty advisor, assoc. mem. Triangle Fraternity, 1990—; deacon Presbyn. Ch., 1991—. Recipient gold medal Agra U., 1955; NSF rsch. grantee, 1972-75. Mem. Indian Phys. Soc., Am. Phys. Soc., Tenn. Acad. Scis., Internat. Community Hospitality Assn. (pres. 1992—), Planetary Soc., Rotary (Paul Harris fellow, bd. dirs. internat. coms. 1991-92), Sigma Pi Sigma, Sigma Xi (bd. dirs., 1992-93). Democrat. Home: 1248 N Franklin Ave Cookeville TN 38501-1677 Office: Tenn Tech U Cookeville TN 38505-5051

KUMAR, MANISH, engineer; b. Bhopal, Madhya, India, May 21, 1963; came to U.S., 1987; s. Murari Lai and Maya Saxena; m. Charu Srivastava, Apr. 17, 1992. B of Tech., Indian Inst. Tech., Bombay, 1985; MS, Auburn U., 1990. Registered geotech. engr. in-tng., Ala. Trainee engr. Tata Electric Cos., Bombay, 1985-86; grad. researcher, teaching asst. Auburn (Ala.) U., 1987-88; project engr. Law Engring. Inc., Birmingham, Ala., 1988—; cons. NASA advanced solid rocket motor facility, Iuka, Miss., 1990-92, Auburn U. slope stability studies, Auburn, 1992, Inland Dam Renovation, Birmingham, Ala., 1992. Donor United Way, Birmingham, 1991-92; adopt-a-sch. participant Birmingham Pub. Schs., 1992. Nat. Merit cum Means scholar Govt. of India, Bombay, 1983-85. Mem. ASCE, Indian Cultural Assn. Birmingham. Achievements include pioneering research work to develop a three-dimensional, finite element based computer model to simulate deep foundations subjected to lateral loads which could eventually be used to generate P-Y curves. Home: 4010 Heatherbrooke Rd Birmingham AL 35242 Office: Law Engring Inc 800 Concourse Pkwy Birmingham AL 35244

KUMAR, RAJENDRA, electrical engineering educator; b. Amroha, India, Aug. 22, 1948; came to U.S., 1980; s. Satya Pal Agarwal and Kailash Vati Agarwal; m. Pushpa Agarwal, Feb. 16, 1971; children: Anshu, Shipra. BS in Math. and Sci., Meerut Coll., 1964; BEE, Indian Inst. Tech., Kanpur, 1969, MEE, 1977; PhD in Electrical Engring., U. New Castle, NSW, Australia, 1981. Mem. tech. staff Electronis and Radar Devel., Bangalore, India, 1969-72; rsch. engr. Indian Inst. Tech., Kanpur, 1972-77; asst. prof. Calif. State U., Fullerton, 1981-83, Brown U., Providence, 1980-81; prof. Calif. State U., Long Beach, 1983—; cons. Jet Propulsion Lab., Pasadena, Calif., 1984-91. Contbr. numerous articles to profl. jours.; patentee; efficient detection and signal parameter estimation with applications to high dynamic GPS receivers; multistage estimation of received carrier signal parameters under very high dynamic conditions of the receiver; fast frequency acquisition via adaptive least squares algorithms. Recipient Best Paper award Internat. Telemetering Conf., Las Vegas, 1986, 10 New Technology awards NASA, Washington, 1987-91. Mem. IEEE (sr.), NEA, AAUP, Calif. Faculty Assn., Auto Club So. Calif. (Cerritos), Sigma Xi, Eta Kappa Nu, Tau Beta Pi (eminent mem.). Avocations: gardening, walking, hiking, reading. Home: 13910 Rose St Cerritas CA 90701-5044 Office: Calif State U 1250 N Bellflower Blvd Long Beach CA 90840-0001

KUMAR, SANJAY, systems engineer; b. New Delhi, Feb. 18, 1961; came to U.S., 1985; s. Krishan and Jagdish (Sachdev) K.; m. Devyani Khanna, Mar. 25, 1986; children: Swati, Nikhil Shiv. BTech in Chem. Engring., Indian Inst. Tech., New Delhi, 1983. Process engr. Oil & Natural Gas Commn., Mehsana, India, 1983-85; engring. systems engr. EDS Corp., Dayton, Ohio, 1985-86; engring. systems engr. EDS Corp., Lansing, Mich., 1985-89, advanced engring. systems engr., 1989—. Mem. DEC User Group (treas. Lansing chpt. 1990—, bd. dirs. 1990—). Achievements include design of tool change system for all power train plants in General Motors, integrated CIM facility for Ngee Ann Polytechnic, Singapore, internal pull system for GM North America, PIC and non-PIC based; design and implementation of real time cellular phone billing system for National Car Rental; development of technical infrastructure for Regional Support Center. Office: EDS Plant Automation Div 905 Southland MS 1027 Lansing MI 48910

KUMAR, SURINDER, food company executive; b. Lahore, Panjab, India, Apr. 2, 1944; came to U.S., 1966; s. Kanshi Ram and Kailash (Wanti) Arora; m. Janet Lauer, July 21, 1973; 1 child, Daven Arora. BS, Nat. Dairy Research Inst., Karnal, India, 1965; MS, Ohio State U., 1968, PhD, 1971; MBA, U. Chgo., 1979. Group leader Quaker Oats Co., Barrington, Ill., 1972-75, mgr., 1975-81, assoc. dir., 1981-82; dir. Frito Lay, Dallas, 1982-84; v.p. Pepsi Cola Co., Valhalla, N.Y., 1984—; mem. nutritional adv. bd. Monell Chem. Senses Ctr.; mem. mgmt. bd. Food Update. Patentee in field; contbr. articles to profl. jours. Pres. India Assn. Ohio State U., 1969; mem. tech. adv. bd. Rutgers U. Named Outstanding Grad., Indian Agriculture Research Inst., 1966; recipient Disting. Alumni award Ohio State U., 1987. Mem. Indsl. Research Inst., Inst. Food Technologists, Sigma Xi, Gamma Tau Delta. Home: PO Box 771 Goldens Bridge NY 10526-0771 Office: Warner-Lambert Co 201 Tabor Rd Morris Plains NJ 07950

KUNC, JOSEPH ANTHONY, physics and engineering educator, consultant; b. Baranowicze, Poland, Nov. 1, 1943; came to U.S., 1978; s. Stefan and Helena (Kozakiewicz) K.; m. Mary Eva Smolska, May 24, 1979; 1 child, Robert. PhD, Warsaw Tech. U., 1974. Assoc. prof. Warsaw Tech. U., 1974-79; rsch. assoc. prof. U. So. Calif., L.A., 1980-84, assoc. prof., 1985-89, prof. dept. aerospace engring., dept. physics, 1990—; rsch. affiliate Jet Propulsion Lab., Calif. Inst. Tech., 1982-83; vis. scholar Inst. Theoretical Atomic and Molecular Physics, Harvard U., Cambridge, Mass., 1991; vis. scholar dept. high-temperature plasma Nat. Inst. for Nuclear Studies, Warsaw, 1991; vis. scholar atomic and plasma radiation div. Nat. Bur. Standards, Washington, 1979; cons. Nat. Tech. Systems, L.A., 1984-86, Phys. Optics Corp., Torrance, Calif., 1988—, Wolfsdorf and Assocs., L.A., 1991—; mem. com. on arcs and flames Nat. Rsch. Coun., 1985-86; chmn. numerous sci. workshops and symposia. Author: (with others) Advances in Pulsed Power Technology, 1991, Progress in Astronautics and Aeronautics, vol. 116, 1989; contbr. over 150 articles to profl. jours., confs., symposia. Recipient Award of Merit, U.S. Congl. Adv. Bd., 1986, Nat. Bur. Standards, 1979; fellow Nat. Bur. Standards, 1978. Fellow AIAA (assoc.), Am. Phys. Soc.; mem. IEEE (sr.), Phi Beta Delta (co-founder Beta Kappa chpt.). Achievements include patent for heat release in micromechanical actuators and engines; principal investigator numerous government-sponsored research programs. Office: U So Calif University Park MC-1191 Los Angeles CA 90089-1191

KUNDEL, HAROLD LOUIS, radiologist, educator; b. N.Y.C., Aug. 15, 1933; s. John A. and Emma E. (Tolle) K.; m. Alice Marie Pape, Mar. 28, 1958; children—Jean, Catherine, Peter. A.B., Columbia U., 1955, M.D., 1959; M.S., Temple U., 1963; M.A. (hon.), U. Pa., 1980. Diplomate Am. Bd. Radiology. Asst. to assoc. prof. Temple U., Phila., 1967-73, prof. radiology, 1973-80; Matthew J. Wilson prof. research radiology U. Pa., Phila., 1980—; dir. Pendergrass Diagnostic Imaging Labs., U. Pa., Phila., 1980—. Contbr. articles to profl. jours. Com. mem. Nat. Council Radiation Protection and Measurements. Served to capt. USAF, 1963-65. Fellow Am. Coll. Radiology; mem. Assn. Univ. Radiologists (Meml. award 1963, Stauffer award 1982), Radiol. Soc. N.Am. Honor award 1978), Am. Roentgen Ray Soc., Soc. Med. Decision Making, Soc. Thoracic Radiology, Alpha Omega Alpha. Lutheran. Office: U Pa Hosp 3400 Spruce St Philadelphia PA 19104-4220

KÜNDIG, ERNST PETER, chemist, researcher, educator; b. Weinfelden, Thurgau, Switzerland, Sept. 2, 1946. Dipl.chem.ETH, ETH Zurich, 1971; PhD in Chemistry, U. Toronto, Can., 1975. Rsch. assoc. U. Bristol, U.K., 1975-77; chargé de Recherche U. Geneva, Switzerland, 1977-83, maitre de'Enseignement et de Recherche, 1983-86, prof. adjoint, 1986-90, prof., 1990—, head of organic chemistry dept., 1989-90, 92—. Contbr. articles to profl. jours. including Jour. Am. Chem. Soc., Organometallics, Jour. Chem. Soc., Synlett, Tetrahedron, Helv. Chim Acta. Recipient Werner Prize Swiss Chem. Soc., 1986. Mem. New Swiss Chem. Soc., Royal Soc. of Chemistry (U.K.), Am. Chem. Soc. Achievements include research on asymmetric transformation of areues into alicyclic compounds, organometallic reaction mechanisms. Office: U Geneva, Dept Organic Chemistry, 30 quai Ernest Ansermet, 1204 Geneva Switzerland

KUNDU, DEBABRATA, mechanical engineer; b. Calcutta, W. Bengal, India, Sept. 11, 1957; came to U.S., 1979; s. Charu Chandra and Chapala

Bala (Mullick) K.; m. Madhuchhanda Ray, Feb.. 21, 1990. MME, U. Houston, 1984; PhD, U. Tex., Arlington, 1989. Registered profl. engr., Tex. Adj. instr. Fla. Inst. Tech., Melbourne, 1981; teaching fellow U. Houston, 1982-84; teaching asst. mech. engring. U. Tex., Arlington, 1985-89, teaching assoc. mech. engring., 1989; plant engr. IMCOA, Hatltom City, 1989-91, dir. rsch. and devel., 1991—; cons. IMCOA, Halton City, 1986-89. Contbr. articles to profl. jours. including Numerical Heat Transfer, Jour. Hear Transfer, proceedings annual meeting ASME, Soc. Plastic Engrs. Mem. Nat. Wildlife Found., 1991—. Recipient Nat. scholarship Bd. of Secondary Edn., Calcutta, India, 1974, Competitive scholarship U. Tex. at Arlington, 1985, Watamull scholarship, 1987. Mem. ASME, ASTM, Soc. Plastic Engrs., Sigma Xi. Hinduism. Achievements include establishment of correlations for heat transfer over a row of in-line cylinders placed between the parallel plates for different aspect ratios and Reynolds number; establishment of a correlation showing effects of shear strain, temperature of the polymer and the concentration of blowing agent on the viscosity of polythylene; determination of the diffusion rate of isobutane from polythylene foam and the effect of the tempurature on the diffusion rate. Home: 6468 G Industrial Pk Blvd North Richland Hills TX 76180-6033 Office: IMCOA 4325 Murray Ave Haltom City TX 76117

KUNG, H. T., computer science and engineering educator, consultant; b. Shanghai, China, Nov. 9, 1945; came to U.S., 1969; s. F.K. Kung and D.Z. Hsu; m. Ling-Ling Chang, Nov. 6, 1970; children—Allen, Angela. B.S., Nat. Tsing Hua U., 1968; M.A., U. N.Mex., 1970; Ph.D., Carnegie-Mellon U., 1974. Research assoc. Carnegie Mellon U., Pitts., 1973-74, asst. prof. computer sci., 1974-78, assoc. prof. computer sci., 1978-82, prof. computer sci., 1982—; Shell Disting. chair Carnegie Mellon U., 1985-92; Gordon McKay prof. elec. engring. and computer sci. Harvard U., Cambridge, 1992—; cons. GE Co., Schenectady, N.Y., ESL, Inc., Sunnyvale, Calif.; staff Bell-No. Rsch., 1991—. Mem. Nat. Acad. Engring., Academia Sinica (Taiwan). Home: One Joseph Comee Rd Lexington MA 02173 Office: Harvard U Div of Applied Scis Cambridge MA 02138

KUNG, LING-YANG, electronics engineer, educator; b. Kaoyaw, Canton, Peoples Republic of China, Oct. 31, 1944; s. Cheung-Kote and May-Siu (Lee) K. PhD/Electronic Engring., summa cum laude, Poly. di Torino, Italy, 1971. Researcher Istituto Nazionale Eletrotecnico Galileo Ferraris, Turin, Italy, 1970-71; assoc. prof. Nat. Cheng Kung U., Tainan, Taiwan, 1972-79, prof., 1981-90, prof. elect. engring. dept., 1990—; computer system analyst DB Data Processing, New Rochelle, N.Y., 1979-80; vis. expert Nat. Sci. Coun., Taipei, 1981—; founder dir. Inst. Info. Engring., Tainan, 1987—; cons. Teco Electric, Taipei, 1977-78, Digital Equipment Corp., Taoyuan, Taiwan, 1988-92, China Steel Cooperation, Kaoshiung, Taiwan, 1989-90. Patentee in field; contbr. articles to profl. jours. Grantee Nat. Sci. Coun., 1981—; named one of top ten talent in info. technology of Taiwan, 1990. Mem. IEEE, Assn. for Computing Machinery, Digital Equipment Cooperation User Group, Don Bosco Alumni Club (advisor 1985—). Avocations: hiking, biking. Office: Nat Cheng Kung U, University Rd 1, Tainan 700, Taiwan

KUNG, PANG-JEN, materials scientist, electrical engineer; b. I-Lan, Taiwan, May 13, 1959; s. Ching-Yu and A-Se (Yu) K.; m. Tzyy-Yun Tzeng, May 18, 1986. MS in Chem. Engring., Nat. Tsing Hua U., 1983; MS in Elec. Engring., Auburn U., 1988; ME in Metall. Engring., Carnegie Mellon U., 1991, PhD in Materials Sci, 1993. Jr. engr. Tatung Co., Taipei, Taiwan, 1979-80; teaching asst. Nat. Tsing Hua U., Hsin-Chu, Taiwan, 1981-82, rsch. asst., 1982-83; assoc. scientist Indsl. Tech. Res. Inst., Hsin-Chu, 1985-86; teaching and rsch. asst. Auburn (Ala.) U., 1986-89; rsch. asst. Carnegie Mellon U., Pitts., 1989-91; staff rsch. asst. Los Alamos (N.Mex.) Nat. Lab., 1991-92, postdoctoral rsch. fellow, 1993—; chmn. acad. affairs Tatung Inst. Tech., Taipei, 1979-80; tech. info. editor Indsl. Tech. Rsch. Inst., Hsin-Chu, 1985-86; translator tech. articles Super Tech. Books Co., Taipei, 1986. Author, editor: Unit Operation in Chemical Engineering, 1986; contbr. articles to Jour. Applied Physics, Applied Physics Letter, Physical Review B, Jour. Materials Rsch., Jour. Vacuum Sci. & Tech. 2d lt. Chinese Air Force, 1983-85. Recipient Editor's Choice award Nat. Poetry Assn., 1989, 90; Am.-Chinese Engr. scholar Am.-Chinese Assn. Engrs., 1980; Liang Ji-Duan fellow Carnegie Mellon U., 1991. Mem. AAAS, IEEE, SPIE, ASM, Materials Rsch. Soc., Am. Vacuum Soc., Am. Ceramic Soc., Soc. for Applied Spectroscopy,. Internat. Soc. for Hybrid Microelectronics, N.Y. Acad. Scis. Achievements include research on diamond thin film deposition and high Tc superconductors; superconducting quantum interference devices and biomagnetism; surface characterization and microstructural analysis; flux pinning and critical current density; biomagnetic systems; Ag-sheathed superconducting tapes; thin-film devices. Home: 3974B Alabama Ave Los Alamos NM 87544-1639 Office: Los Alamos Nat Lab PO Box 1663 Los Alamos NM 87545-0001

KUNG, SHAIN-DOW, molecular biologist, educator; b. China, Mar. 14, 1935, came to U.S., 1971, naturalized, 1977; s. Chao-tzen and Chih (Zhu) K.; grad. Chung-Hsing U., Taiwan, China, 1958; Ph.D., U. Toronto, Can., 1968; m. Helen C.C. Kung, Sept. 5, 1964; children: Grace, David, Andrew. Research fellow Hosp. for Sick Children, Toronto, 1968-70; biologist UCLA, 1971-74; asst. prof. biology U. Md., Baltimore County, 1974-77, assoc. prof., 1977-82, prof., 1982-86, acting chmn. dept., 1982-84, assoc. dean arts and sci., 1985-86, prof. botany U. Md., College Park, 1986-93; hon. prof. Fudan U., 1986, Beijing Agrl. U., 1987; acting dir. Ctr. for Agri. Biotech, 1986-88, dir. 1988-93; acting provost Md. Biotech. Inst., 1989-91, dean Sch. Sci. Hong Kong U. Sci. and Tech., 1991-93; pro vice chancellor for Academic Affairs, Hong Kong U. Sci. and Tech., 1992—. Author 1 book; editor 8 books; contbr. chpts. to books, articles to profl. jours. Recipient Philip Morris award for disting. achievement in tobacco sci., 1979, Outstanding Alumni award, 1990, Outstanding Svc. award, 1990; named Disting. Scholar, Nat. Acad. Sci., 1981; Fulbright grantee, 1982-83, grantee NSF, NIH. Mem. Am. Soc. Plant Physiologists, AAAS. Office: U Md Ctr for Agricultural Biotech 2111 Agricultural Life Sciences College Park MD 20742

KUNHARDT, ERICH ENRIQUE, physicist, educator; b. Montecristy, Dominican Republic, May 31, 1949; came to U.S., 1961; s. Juan Enrique and Irma Mercedes (Grullon) K.; m. Christine Ann Koza, Oct. 23, 1976. BS, NYU, 1969; PhD, Poly. U., Bklyn., 1976. Asst. prof. Tex. Tech U., Lubbock, 1976-80, assoc. prof., 1980-83, prof., 1983-85; prof. Poly. U., 1985-91; George Mead Bond prof. physics Stevens Inst. Tech., Hoboken, N.J., 1991—; dir. Weber Rsch. Inst., Poly, U., 1986-91. Editor: Breakdown and Discharges in Gases, 1983, The Liquid State and its Electrical Properties, 1985; mem. adv. bd. Jour. Transport Theory and Statis. Physics, 1984—; contbr. articles to profl. jours. Recipient Citation for Excellence in Rsch. Nassau County, 1988. Mem. IEEE, Am. Phys. Soc. Achievements include observation of plasma wavepacket bifurcation; research on kinetic behavior of a streamer, and on method for closure of fluid equations. Office: Stevens Inst Tech Castle Point Hoboken NJ 07030

KUNICKI, JAN IRENEUSZ, physicist; b. Wroclaw, Poland, Mar. 25, 1950; came to U.S., 1983; s. Miroslaw and Janina (Staszkiewicz) K.; m. Jolanta Elzbieta Drobik, June 28, 1975; 1 child, Marcin Szymon. MS,BS in Physics, U. Wroclaw, 1972. Lectr., lab. instr. dept. physics U. Wroclaw, 1972-80; instr., cons. Inst. Low Temperatures/Structure Rsch., Polish Acad. Scis., Wroclaw, 1980-83; cons. Meml. Sloan-Kettering Cancer Ctr., N.Y.C., 1983-84, clin. lab. technologist, 1984-87; researcher, teaching specialist U. Medicine-Dentistry N.J., Piscataway, 1987-88; rsch. assoc., lab. mgr. Blood Rsch. Inst., Newark, 1988—. Contbr. articles to profl. pubs. Active Solidarity Workers Union, Wroclaw, 1980-83, Polish Boy Scout Orgn., Perth Amboy, N.J., 1989—. Mem. Internat. Soc. Analytical Cytology. Roman Catholic. Home: 234 Burnside Pl Ridgewood NJ 07450 Office: Blood Rsch Inst 268 Martin Luther King Blvd Newark NJ 07102

KUNIEDA, HIRONOBU, physical chemistry educator; b. Shinagawa, Tokyo, Nov. 17, 1948; s. Toshihiro and Nobuko (Mikami) K.; m. Akemi Wakabayashi, Jan. 7, 1973; children: Yukiko, Naoko. BS, Yokohama Nat. U., 1971, MS, 1973; D Engring., Tokyo U., 1980. Rsch. assoc. Yokohama (Japan) Nat. U., 1973-80, postdoctoral assoc., 1980-88, assoc. prof. phys. chemistry, 1988—; postdoctoral fellow U. Mo., Rolla, 1984-85. Recipient Progressing award Japan Oil Chemists' Soc., 1984. Buddist. Avocations: music, travel, badmington. Office: Yokohama Nat U, Hodogaya-ku, Tokiwadai 156, Kanagawa, Yokohama 240, Japan

KUNKEL, LOUIS MARTENS, research scientist, educator; b. N.Y.C., Oct. 13, 1949; s. Henry George and Betty (Martens) K.; m. Susan Manter, Aug. 24, 1985; children: Sarah, Johanna, Ellen. BA, Gettysburg Coll., 1971; PhD, Johns Hopkins U., 1978. Rsch. fellow U. Calif., San Francisco, 1978-80; rsch. fellow Children's Hosp., Boston, 1980-82, chief div. genetics, 1989—; instr. in pediatrics Harvard Med. Sch., Boston, 1982-84, asst. prof., 1984-86, assoc. prof., 1987-90, prof., 1990—; assoc. investigator Howard Hughes Med. Inst., 1987-91, investigator, 1991—; George Cotzias Meml. lectr. Am. Acad. Neurology, Cin., 1988; Pruzansky lectr. March of Dimes Birth Defects Found., 1989; lectr. Harvard Med. Sch., 1983-89, Harvard U., 1984; tutor Harvard Med. Sch., 1987-89. Mem. editorial bds. Cytogenetics and Cell Genetics, 1986—, Genomics, 1987—, Muscle & Nerve, 1989—, Jour. Molecular Neurosci., 1989—; contbr. articles to profl. jours., chpts. to textbooks. Recipient Wellcome Found. prize Royal Soc. London, 1988, Young Scientist award Passano Found., Balt., 1989, Nat. Med. Rsch. award Nat. Health Coun., Washington, 1989, Internat. award Gairdner Found., Toronto, 1989, E. Mead Johnson award, 1991, Silvio O. Conte Decade of Brain award, 1991, Sanremo Internat. award for genetic rsch., Italy, 1991. Mem. NAS, Am. Soc. Human Genetics. Republican. Lutheran. Avocations: gardening, fishing, sailing, tennis, running. Office: Children's Hosp 300 Longwood Ave Boston MA 02115-5737

KUNKLE, DONALD EDWARD, physicist; b. New Kensington, Pa., Mar. 9, 1928; s. James Laine and Orpha May (Black) K.; m. Joan Stewart Hoagland, June 10, 1950 (div. 1985); children: John, Merry, Shelley; m. Claire Regina Maskel, July 20, 1990. BS in Physics, Lafayette Coll., Easton, Pa., 1950. Rsch. physicist Alcoa Rsch. Labs., New Kensington, 1950-55, group leader non-destructive testing, 1955-62, group leader process control, 1962-69; mgr. product analysis/control Kaiser Aluminum & Chem., Pleasanton, Calif., 1969-75, rsch. physicist instrumentation, 1975-80, staff engr. process tech., 1980-85; nuclear engr. U.S. Navy, Mare Island, Calif., 1986-90; physicist/cons. Analyte Corp., Medford, Oreg., 1991—; Contbr. articles to profl. jours. Mem. Soc. for Nondestructive Testing (chmn. 1959-60), Kiwanis (pres. 1993-94), Knife and Fork Club (bd. dirs. 1992—). Republican. Achievements include design of rolling mill gages for accurately controlling the thickness of aluminum can stock sheet; develped first zone purification technique; developed first eddy current techniques for quality control of aluminum, others. Home: 1419 Village Center Dr Medford OR 97504

KUNOV, HANS, biomedical engineering educator, electrical engineering educator; b. Copenhagen, Mar. 14, 1938; arrived in Can., 1967; s. Jens Christian and Ruth (Valeur) K.; m. Helle H.D. Jorgensen, Sept. 12, 1964 (div. 1972); children Mark Jacob, Niels Peter; m. D. Clare Lamb, Aug. 1, 1977. MASc, Tech. U. Denmark, Copenhagen, 1963, PhD, 1966. Registered profl. engr., Ont. Postdoctoral fellow Tech. U. Denmark, 1966-67; asst. prof. U. Toronto, Ont., Can., 1967-73, assoc. prof., 1973-82, prof., 1982—, dir. Inst. Biomed. Engring., 1989—; dir. Elec. Engring. Consociates, Toronto, 1972—; pres. Artel Engring., 1975—; dir. rsch., co-founder Paul Madsen Med. Svcs. Ltd., Toronto, 1992—; mem. grant selection com. Natural Scis. and Engring. Rsch. Coun., Ottawa, Ont., 1990-93. Contbr. numerous sci. papers and publs. Chmn. United Way, U. Toronto, 1991-92; mem. Big Bros. Met. Toronto, 1980—, dir., 1988-92. Recipient Big Brother of Yr. award Big Bros. Met. Toronto, 1985, 86, Irving Pomerantz award Big Bros. Met. Toronto, 1989. Mem. IEEE (assoc. editor BME Trans. 1991-93), Acoustical Soc. Am., Can. Med. Biol. Engring. Soc., Danish Engring. Soc. Achievements include development of novel audiometric techniques, of accurate mechano-acoustic models of human hearing and speech apparatus. Home: 4 Princeton Rd, Etobicoke, ON Canada M8X 2E2 Office: U Toronto, 4 Taddle Creek Rd, Toronto, ON Canada M5S 1A4

KUNTZ, HAL GOGGAN, petroleum exploration company executive; b. San Antonio, Dec. 29, 1937; s. Peter A. and Jean M. (Goggan) K.; m. Vesta McClain, Oct. 7, 1983; children: Hal Goggan, Peter, Michael B., Vesta. BS in Engring., Princeton U., 1960; MBA, Oklahoma City U., 1972. Line, staff positions Mobil Oil Corp., Dallas, Oklahoma City, New Orleans, 1963-74; co-founder, pres. CLK Corp., New Orleans, Houston, 1974—, IPEX Co., New Orleans, 1975—; pres. Gulf Coast Exploration Co., New Orleans, 1979—, pres. CLK Investments I, II, III and IV, 1979—, CLK Producing, CLK Oil and Gas Co., CLK Exploration Co., 1980—. Mem. Mus. Fine Arts, Houston, 1978—; mem. condrs. circle Houston Symphony, 1980; governing bd. Houston Opera. Served with AUS, 1960-63. Mem. Am. Mgmt. Assn., Nat. Small Bus. Assn., Inter-Am. Soc., Soc. Exploration Geophysics, Am. Assn. Petroleum Geologists, Aircraft Owners and Pilots Assn., Houston C. of C., River Oaks C. of C. Republican. Roman Catholic. Clubs: Petroleum of Houston, University of Houston; Argyle, Order of Alamo (San Antonio), Brae-Burn Country, The Coronado, Princeton of N.Y., River Oaks Country. Avocations: golf, skiing, birdshooting, flying. Office: CLK Co 1001 Fannin St Ste 1400 Houston TX 77002-6708

KUNTZE, HERBERT KURT ERWIN, aeronautical engineer; b. Neidenburg, Germany, Dec. 22, 1920; s. Erwin and Doris (Gast) K.; m. Christa Tourneau, Feb. 2, 1957; children: Juergen, Doris. Diplom Ingenieur, Technische U. Hannover, 1952. With Continental Rubber, Hannover, 1952-56; sales staff Bosch, Stuttgart, 1956-57; with Westinghouse, Hannover, 1957-58; with German Air Force, 1959-78, advanced through ranks to lt. col., 1970; project mgr. Fed. Office for Mil. Technics and Procurements, various locations, 1970-78; ret., 1978. Mem. AIAA. Achievements include patents on controlled air-suspension for automobiles. Home: Weimarer Str 30, D56075 Koblenz Germany

KUNUGI, SHIGERU, scientist, researcher, educator; b. Kyoto, Japan, May 25, 1949; s. Masanaga and Yuriko (Nanasawa) K.; m. Yukiko Uehara, May 18, 1979; 1 child, Motoshi. BS, Kyoto U., 1972, M in Engring., 1974, D in Engring., 1978. Rsch. assoc. Kyoto U., 1977-83; assoc. prof. Fukui (Japan) U., 1983-90; prof. Kyoto Inst. of Tech., 1990—. Buddist. Office: Kyoto Inst of Tech, Matsugasaki, Kyoto 606, Japan

KUNZ, DONALD LEE, aeronautical engineer; b. Geneva, N.Y., Oct. 19, 1949; s. Clarence E. and Mildred M. (Kerr) K.; m. Christine W. Bickel, Aug. 30, 1986; 1 child, Shannon. BS, Syracuse (N.Y.) U., 1971; MS, Ga. Inst. Tech., 1972, PhD, 1976. Rsch. scientist, aeroflightdynamics dir. U.S. Army, Moffett Field, Calif., 1976-89; group leader U.S. Army Aeroflightdynamics dir., Moffett Field, Calif., 1989; mem. tech. staff V McDonnell Douglas Helicopter Co., Mesa, Ariz., 1989-91, rsch. and engr. specialist, 1991—. Contbr. articles to profl. jours. Fellow AIAA (assoc., tech. program chmn. dynamics specialists conf. 1992, mem. structural dynamics tech. com. 1989—); mem. Am. Helicopter Soc. (dynamics com. 1990—). Home: 183 W Stacey Ln Tempe AZ 85284 Office: McDonnell Douglas Helicopter Co 5000 E McDowell Rd Mesa AZ 85205

KUNZ, SIDNEY, entomologist; b. Fredericksburg, Tex., Dec. 24, 1935. BS, Tex. A&M U., 1958, MS, 1962; PhD in Entomology, Okla. State U., 1967. Survey entomologist Okla. State U., 1961-64, ext. entomologist, 1964-67; rsch. entomologist agr. rsch. svc. USDA, Kerrville, Tex., 1967-69; rsch. entomologist USDA, College Station, 1969-77; rsch. leader, rsch. entomologist sci. and edn. USDA, 1977-86; lab. dir. U.S. Livestock Insects Lab., Kerrville, 1986—; entomology cons. food & agr. orgn. UN Devel. Prog., Mauritius, 1973-74, USAID, Tanzania, IAEA, Somalia, 1982. Recipient CIBA-GEIGY/Entomol. Soc. Am. award CIBA-GEIGY Corp., 1991. Mem. Entomol. Soc. Am., Am. Registry Profl. Entomologists, Sigma Xi. Achievements include research in biology, ecology and area integrated pest management control of biting flies of cattle, horn flies and stable flies. Office: US Livestock Insects Lab PO Box 232 Kerrville TX 78028*

KUNZE, ERIC, physical oceanographer, educator; b. Nelson, B.C., Can., June 13, 1956; came to U.S., 1979; s. Junior Otto and Ann (Orth) K. Bsc in Math. and Physics, U. B.C., Vancouver, 1979; MS in Oceanography, U. Wash., 1982, PhD in Oceanography, 1985. Rsch. asst. dept. physics U. B.C.; rsch. asst. applied physics lab. U. Wash., Seattle, 1979-85, rsch. asst. prof. sch. oceanography, 1987-93, assoc. prof. sch. oceanography, 1993—, teaching asst. sch. oceanography, 1981; summer fellow Woods Hole (Mass.) Oceanographic Inst., 1983, postdoctoral fellow, 1985-86, postdoctoral rsch. scientist, 1986-87; seminar organizer Phys. Oceanography Lunch Seminar, 1987-89. Asst. editor Jour. Marine Rsch., 1987—; contbr. articles to profl. jours. Vernon C. of C. scholar Okanagan Coll., 1975; Grad. Student Tuition Waiver scholar U. Wash., 1984. Fellow Am. Geophysical Union (Father James B. Macelwane Young Investigators medal 1993, Sverdrup lectr. 1992); mem. ACLU, Am. Meteorological Soc., Oceanography Soc., Can. Meteorological and Oceanographic Soc., Nature Conservancy, Greenpeace. Achievements include research in interaction of meso- to microscale oceanic phenomena including fronts, eddies, internat waves, turbulence, double diffusion, bottom topography and surface forcing and their effects, through mixing and water-mass modification, on larger scales; demonstration that internal waves are sensitive to rotation in the ocean as well as that of earth, can be trapped by negative veloicty, dominate finescale vertical wavelengths of about 10m, and are the principal agents for turbulence production in the pycnocline, vorticies co-exist on internal-wave scales. Office: Univ of Washington Dept Oceanography WB-10 Seattle WA 98195

KUNZE, HANS-JOACHIM DIETER, physics educator; b. Kauffung, Fed. Republic Germany, Mar. 19, 1935; s. Wilhelm and Elsa (Hoffmann) K.; m. Regina Erna Zellerer, May 5, 1937; children: Stefanie, Martina. Diploma in physics, T.H., Munich, 1961, D. in Natural Scis., 1964. Scientist Inst. for Plasmaphysics, Garching, Fed. Republic Germany, 1961-65; rsch. assoc. U. Md., College Park, 1965-67; asst. prof. U. Md., 1967-70, assoc. prof., 1970-72; prof. physics Ruhr U., Bochum, 1972—. Author: Physikalische Messmethoden, 1986; contbr. articles on plasma spectroscopy, laser diagnostics, plasma physics to profl. jours. Mem. German Physical Soc. (chmn. div. plasma physics 1986-88), European Physical Soc., Am. Physical Soc. Lutheran. Home: Wagenfeldstr 12, 58456 Witten Germany Office: Ruhr Univ, Universitätsstr 150, 44780 Bochum Germany

KUNZE, OTTO ROBERT, retired agricultural engineering educator; b. Warda, Tex., May 27, 1925; s. John Paul and Hermine Amanda (Moerbe) K.; m. Alice Ruth Eifert, Aug. 5, 1951; children: Glenn, Allen, Charles, Karen. BS, Tex. A&M U., 1950; MS, Iowa State U., 1951; PhD, Mich. State U., 1964. Registered profl. engr., Tex. Agrl. and indsl. engr. Ctrl. Power and Light Co., San Benito, Tex., 1951-56; rsch. asst. agrl. engring. dept. Mich. State U., East Lansing, 1961-64; assoc. prof. agrl. engring. dept. Tex. A&M U., College Station, 1956-61, 64-69, prof. agrl. engring. dept., 1969-90, prof. emeritus agrl. engring. dept., 1990—; engring. cons. Farmers Rice Coop., Sacramento, 1992, Post Harvest Process and Food Engring. Ctr., G.B. Pant U., Pantnagar, India, Rice Process Engring. Ctr., Indian Inst. Tech., Khangput, India; lectr. on harvesting, Taichung, Taiwan, 1985, 87; publ. coord. Rice Tech. Working Group, 1976-90. Contbr. chpts. to 3 books, more than 70 publs. in field of post harvest rice technology. Mem. A&M Consol. Bd. Equalization, College Station, 1969-71, Tex. A&M Control Bd., Austin, 1979-90, Pediatric Scholarship Com., M.D. Anderson Cancer Ctr., Houston, 1990—. With U.S. Army, 1944-46, ETO. Recipient Outstanding Svc. award Rice Tech. Working Group, 1990; NSF faculty fellow, 1961-62. Fellow Am. Soc. Agrl. Engrs. (tech. dir., numerous coms.), Am. Assn. Cereal Chemists (assoc. editor), Sigma Xi (sec. 1969-70, chmn. 1970-71), Phi Kappa Phi (pub. rels. officer 1984-85). Lutheran. Home: 1002 Milner College Station TX 77840 Office: Tex A&M U Agrl Engring Dept College Station TX 77843

KUO, CHANG KENG, electrical engineer; b. Tachia, Taiwan, Republic of China, Mar. 14, 1935; s. Chin Quen and Chin Chiaw (Lee) K.; m. Chung Yu, Jan. 22, 1962; children: Jane May, John Hawell, Alfred Chung. BSEE, Taiwan U., Taipei, 1957; MSEE, U. Tenn., 1963; postgrad., U. Calif., Berkeley, 1968-73. Lic. profl. elec. engr., Calif., Oreg. Jr. elec. engr. Taiwan Sugar Co., Taichung, 1959-62; elec. designer Kocher Cons. Engrs., L.A., 1964-66; asst. elec. engr. State Dept. of Water Resources, Sacramento, Calif., 1966-68; engr. designer Pacific Gas & Elec. Co., San Francisco, 1972-74; elec. engr. Bechtel Power Corp., San Francisco, 1974-83, engr. specialist, 1983-91, project engr., 1991—. 2d Lt. Taiwan Air Force, 1957-59. Democrat. Buddhist. Achievements include design of Pebble Spring Nuclear Power Plant in Oregon, and participation in the design and modificiations of other nuclear power plants. Home: 2836 Buckskin Rd Pinole CA 94564 Office: Bechtel Power Corp 50 Beale St San Francisco CA 94105

KUO, PETER TE, cardiologist; b. Fukien, Peoples Republic of China, Mar. 21, 1916; came to the U.S., 1946; s. Lang Shan and Su Dzen (Liu) K.; m. Nancy N. Huang, Dec. 25, 1949; children: Larry, Kathy. BS, St. John's U., 1936, MS in Medicine, 1939; DSc in Medicine, U. Pa., 1949. Intern, resident St. Luke Hosp., Shanghai, 1939-43; resident, rsch. fellow in medicine and cardiology Pa. Hosp. and Hosp. U. of Pa., Phila., 1947-50; rsch. fellow Am. Heart Assn., 1955-56, established investigator, 1956-61; from instr. to asst. prof. of medicine St. John's U., Shanghai, 1939-46; from instr. to prof. of medicine U. Pa., Phila., 1952-72; prof. medicine, chief cardiovascular diseases Robert Wood Johnson Med. Sch., New Brunswick, N.J., 1973-82; cons. medicine and cardiovascular diseases Princeton U., New Brunswick, N.J., 1973—; dir. hyperlipidemia VA Med. Ctr., Houston, 1987—; prof. medicine Baylor Coll. Medicine, 1973—; cons. NHLBI Health Svcs. and Mental Health Adminstrn., Bethesda, Md., 1982-87, dir. atherosclerosis; John G. Detwiler prof. cardiology Robert Wood Johnson Med. Sch., New Brunswick, N.J., 1982-87. Mem. editorial bd. Chest, Cardiopulmonary Jour., Angiology, Am. Jour. Clin. Nutrition; contbr. 112 articles and 20 abstracts to profl. jours. and 14 chpts. to books. Recipient U.S. Pub. Health Rsch. Career Devel. award, 1961-66. Fellow ACP, AAAS; mem. AMA, Am. Heart Assn., Am. Coll. Cardiologists, Am. Coll. Angiology, Am. Assn. Chest Physicians, Am. Inst. for Nutrition, Am. Fedn. for Clin. Rsch., Gerontol. Soc., Assn. Univ. Cardiologists, Alpha Omega Alpha. Achievements include research in dyslipidemias and hypercholesterolemia to prevent and control coronary artery disease. Home: 4215 Milton Houston TX 77005 Office: VA Med Ctr 2002 Holcombe Blvd Houston TX 77030

KUO, TUNG-YAO, industrial engineering educator; b. Tou Liou, Taiwan, Feb. 25, 1926; s. Chau-Cheng and Yen (Hsiao) K.; m. Hsueh-Yun Chi, Apr. 7, 1953; children: Howard, Joanne, Joseph. BS, Nat. Taiwan U., Taipei, 1951; MS, Stanford U., 1973. Chmn. indsl. engring. dept. Tunghai U., Taichung, Taiwan, 1968-86, dean Engring. Coll., 1974-75, dean of students, 1975-76, prof. indsl. engring., 1978—, dir. ctr. for extensional edn., 1979—; part-time faculty mem. Old Dominion U., Norfolk, Va., 1984-85, Okinawa (Japan) Internat. U., 1990; cons. Taiwan Power Corp., Taipei, 1986, Funai Elec. Corp., Taichung, Sanfu Automobile Coirp., Taichung; acting pres. Tainan Sem., 1992. Chmn. bd. dirs. YMCA, Taichung, 1987—. Recipient Excellent Tchr. award Ministry of Edn., 1993. Mem. Chinese Inst. Engrs. (Outstanding Engring. Prof. award 1993), China Inst. Indsl. Engring. (past bd. dirs.), China Inst. Mgmt. Sci., Chinese Assn. Standardization (past bd. dirs.), Indsl. Safety and Health Assn. of Republic of China, Internat. Cultural Interchange Assn. of Republic of China (bd. dirs.), Family Wellness Assn. of Republic of China (bd. dirs.), Assn. Mgmt. Cons. Japan (pres. 1989-91), Far East Econ. Assn. (pres. 1989-91), Rotary (pres. Tatong club 1990-91). Presbyterian. Avocations: music, travel, tennis, golf. Office: Tunghai U, Taichung Taiwan

KUPER, GEORGE HENRY, research and development institute executive; b. Washington, Oct. 16, 1940; s. James B. Horner and Mariette (Kovesi) K.; m. Danielle E. Pienaar, Feb. 1982; children: James A.H., E. Andrew. BA, Johns Hopkins U., 1963; MS in Econs., London Sch. Econ., 1964; MBA, Harvard U., 1970. Mktg. analyst Morgan Guaranty Trust Co., N.Y.C. and Paris, 1968-70; exec. v.p. Boston Venture Mgmt. Co., 1970-71; dep. dir. Mayor's Office Justice Adminstrn., Boston, 1971-72; exec. dir. Nat. Ctr. Productivity, Washington, 1972-78; staff assoc. Gen. Electric Co., Fairfield, Conn., 1978-83; exec. dir. Mfg. Studies Bd. Nat. Rsch. Coun., Washington, 1983-88; pres. Indsl. Tech. Inst., Ann Arbor, Mich., 1988—; mem. council on trends and perspective U.S. C. of C., 1979-91; founder Cambridge Parallel Processors, Inc., 1986—, Nat. Ctr. for Mfg. Scis., 1986—; bd. dirs. Tech. Mgmt. Ctr., Phila., Access Computer Products, 1991—; cons., Boston, 1970-72, Washington, 1978, 83. Mem. editorial adv. bd. John Wiley & Sons, N.Y.C., 1984-86; mem. editorial bd. Internat. Jour. Tech. Mgmt., 1987—, Mfg. Rev., 1987—; contbr. articles to profl. jours. Served to lt. USNR, 1964-68. Mem. Am. Productivity Mgmt. Assn. (founder, v.p. 1979—), Delta Phi (pres. 1960-61, 62-63, scholarship award 1961-62), N.Y. Yacht Club, Pequot Yacht Club (Conn.), Waterloo Hunt Club (Mich.), Fairfax Hunt Club (Va.), Pi Sigma Alpha.

KUPFER, CARL, ophthalmologist, science administrator; b. N.Y.C., Feb. 9, 1928; s. James and Hannah Kupfer; m. Muriel I. Kaiser, Dec. 9, 1969; children: Charles, Sarah. AB, Yale U., 1948; MD, Johns Hopkins U., 1952;

DSc (hon.), U. Pa., 1982, SUNY, 1992. Diplomate Am. Bd. Ophthalmology. Intern, resident Johns Hopkins U., 1952-55, 57-58; asst. prof. Harvard U. Med. Sch., Boston, 1960-66; prof., chmn. dept. ophthalmology U. Wash. Sch. Med., Seattle, 1966-69; dir. Nat. Eye Inst. NIH, Bethesda, Md, 1970—. Recipient Migel award Am. Found. for the Blind, 1976, Pisart award Lighthouse for the Blind, N.Y.C., 1984, Presdl. Rank award, 1991. Recipient Humanitarian award Lions Club Internat., 1992. Mem. Johns Hopkins Soc. Scholars, Insts. Medicine, NAS. Office: HHS 9000 Rockville Pike Bethesda MD 20892-0001*

KUPPER, PHILIP LLOYD, chemist; b. Louisville, Ky., Sept. 7, 1940; s. Louis James and Mary Sylvia (Noonan) K.; m. Lynn Anne Abbinanti, Dec. 28, 1968; children: Nicole Marie, Rachel Ann Mary. BS in Chemistry, U. Md., 1963; MBA, Ga. State U., 1970. Tech. counselor Nat. Soft Drink Assn., Washington, 1964-68; dir., rsch. & devel. Moxie Monarch NuGrape, Atlanta, 1968-69; chief chemist GRAF/S Beverages, Milw., 1970-75; rsch. mgr. Crush Internat., Evanston, Ill., 1975-81; flavor chemist Procter & Gamble, Cin., 1981—. Patentee in field. With USAR, 1963-69. Mem. Soc. Flavor Chemists (cert.), Soc. Soft Drink Techs., Inst. Food Tech. Roman Catholic. Avocations: golf, sports, reading. Home: 1332 Cryer Ave Cincinnati OH 45208-2807 Office: Procter & Gamble 11450 Grooms Rd Cincinnati OH 45242

KUREK, DOLORES BODNAR, physical science and mathematics educator; b. Toledo, Dec. 14, 1935; d. James J. and Veronica Clara (Gorajewski) Bodnar; m. Arnold John Kurek, Aug. 30, 1958; children: Kerry, Darrah, Michele, James, Ursula. BS, Mary Manse Coll., 1958; MEd, U. Toledo, 1968, doctoral candidate. Chemistry, physics and math. tchr. St. Ursula Acad., Toledo, 1961-75; sci. tchr. McAuley High Sch., Toledo, 1975-78; chemistry tchr. St. Francis de Sales High Sch., Toledo, 1978-83; instr. math. and chemistry Owens Tech. Coll., Toledo, 1980-86; instr. chemistry, physics and astronomy Lourdes Coll., Sylvania, Ohio, 1983-86, assoc. prof. phys. sci., 1986—; instr. math. U. Toledo, 1986—; pres., chmn. judging, co-dir. N.W. Dist. Sci. Day, Toledo, 1975—; regional dir. Women of Sci., Toledo, 1987—; bd. dirs. Toledo Jr. Sci. Humanities Symposium, Toledo; dir. Copernicus Planetarium, Lourdes Coll., 1990—; v.p. edn. Tech. Soc. Toledo, 1989—; pres. Tech. Found. Toledo, 1991—; dir. field-based earth sci. program for mid. sch. tchrs. grant Ohio Bd. Regents, 1992. Inventee in field; contbr. articles to profl. jours. Mem. Toledo Mus. of Art, 1987—, Toledo Zool. Soc., 1988—. Named One of 100 Women Sci. Exemplars in Ohio, Women in Sci., Engring. and Math. Consortium Ohio, 1988, Woman of Toledo, St. Vincent Med. Ctr., 1988; recipient award for teaching excellence and campus leadership Sears Roebuck Found., 1991; Mary Manse scholar, 1954-58. Mem. Am. Chem. Soc. (James Conant Bryant award 1980, 81), Ohio Acad. Sci. (Acker award 1980), Nat. Sci. Tchrs. Assn., Soc. for Coll. Sci. Tchrs., Am. Assn. Physics Tchrs., Mensa, Astron. Soc. Pacific, Gt. Lakes Planetarium Assn., Phi Delta Kappa (newsletter editor 1984-86). Roman Catholic. Avocations: aerobics, running, crossword puzzles, swimming. Home: 624 Arcadia Ave Toledo OH 43610-1108 Office: Lourdes Coll 6832 Convent Blvd Sylvania OH 43560-2891

KUREPA, ALEXANDRA, mathematician; b. Zagreb, Yugoslavia, Dec. 31, 1956; came to U.S., 1985; d. Svetozar and Zora (Lopac) K.; m. Rodney Anthony Waschka II, June 24, 1988; 1 child, Andre Kurepa Waschka. BS, U. Zagreb, 1978, MS, 1982; PhD, U. North Tex., 1987. Asst. prof. math. U. Zagreb, 1987-88, Tex. Christian U., Ft. Worth, 1988-93, N.C. A&T State U., Greensboro, 1993—. Author: Matematica 2, 1989; contbr. articles to profl. jours. Rsch. grantee UNESCO, 1988, 89. Mem. Am. Math. Soc., Math. Assn. Am., Assn. for Women in Math. Office: NC A&T State U Dept Math Greensboro NC 27411

KURIAN, PIUS, physician; b. Arpookara, Kerala, India, May 9, 1959; came to U.S., 1986; s. Pylo and Mariamma Kurian; m. Sally Kurian, May 11, 1986; children: Michelle Maria, Matthew Paul. BSci, Kuriakose (India) Elias Coll., 1979; MB, BS, Kottayam (India) Med. Coll., India, 1986. Diplomate Am. Bd. Internal Medicine, 1991. Sr. staff physician Sacred Heart Med. Ctr., Kottayam, 1986; resident physician Nassau County Med. Ctr., East Meadow, N.Y., 1988—; fellow in nephrology Nassau County Med. Ctr., East Meadow, 1991—. Mem. ACP, AAAS, AMA. Roman Catholic. Office: 200 Carman Ave Apt 8A East Meadow NY 11554-1148

KURITA, CHUSHIRO, writer, engineering educator, researcher; b. Nakanegishi machi, Tokyo, Japan, Apr. 16, 1910; s. Kamejiro and Shige Kurita; m. Chieko Hagiwara, Dec. 19, 1946; 1 child, Hidemi. B in Engring., Waseda U., Tokyo, 1935. Researcher lab. sch. sci. Fusidenkiseizo Co., Waseda U., Tokyo, 1945-81; asst. prof. elec. dept. 2d Sch. Sci. and Engring. Waseda U., Tokyo, 1959; ret. Waseda U., 1981; lectr. elect. dept. Kokushikan U. Sch. Engring., 1977-89; pres. Gakujutsu Bunken Shuppankai, Tokyo, 1977—. Author: Thermo-Electrical Engineering, 1973, Principle of Contact Relating to Solids, 1979, Electronic Contact Properties Relating to Solid States, 1979, The Principles of Electron Theory (The True Values of Elementary Charge e, IC and IA and the others), 1987; patentee thermoelectric refrigerator, 1960. Avocation: Noh Song. Home: Kohinata 2-Chome, Bunkyo-Ku, Tokyo 112, Japan Office: Gakujutsu Bunken Shuppankai, Tokyo 112 Bunkyo-ku, Japan

KURLAN, MARVIN ZEFT, surgeon; b. Wilkes-Barre, Pa., Feb. 20, 1934; s. Ephraigm Joseph and Fannye Lillian (Rosenbluth) Kurlancheek; m. Eleanor Frank, June 21, 1964; 1 child, Todd. BA, Wilkes Coll., 1957; MS, U. Ill., 1958; MD, SUNY, Buffalo, 1964. Diplomate Nat. Bd. Med. Examiners, Am. Bd. Surgery. Intern then resident in surgery Millard Fillmore Hosp., Buffalo, 1964-69, dir. trauma svcs., 1974-82, sr. attending surgeon, 1984—, plant surgeon Bethlehem (Pa.) Steel Corp., 1969-74; med. dir. Bros. of Mercy Health Facilities, Clarence, N.Y., 1976-80; assoc. examiner Am. Bd. Surgery, Phila., 1987—; chmn. James Platt White Soc., Sch. Medicine and Biomed. Scis., SUNY, Buffalo. Contbr. articles to profl. jours. Vol. Empire State Games, Buffalo, 1986; mem. Jack Kemp Forum, Buffalo, 1985-91; bd. dirs. Jewish Fedn. Allentown, Pa., 1972-74. Fellow Am. Coll. Gastroenterology, Am. Trauma Soc. (founder); mem. Assn. Mil. Surgeons, Buffalo Surg. Soc. (v.p. 1988-89, pres. 1989-90, sec. 1986-88), ACS (life fellow leadership soc.), Am. Biog. Inst. (dep. gov.), Internat. Biog. Assn. (dep. dir. for the Americas), Grand Coun. World Parliament, Confederation of Chivalry, Knight of Humanity, Order White Cross Internat. (dist. comdr. N.Y., U.S.A.), Chevalier Grand Cross, Ordre Souverain et Militaire de la Milice du Saint Sepulcre. Republican. Club: Sci. Progress Research (Buffalo) (v.p. 1983-84). Lodges: Masons, Shriners. Avocation: world travel. Home and Office: 413 Dan Troy Dr Buffalo NY 14221-3558

KURLANDER, ROGER JAY, medical educator, researcher; b. Bklyn., Feb. 21, 1947; m. Marion Danis; m. Cara Beth, Jacob Eli, Rachel. BA, Bklyn. Coll., 1967; MD, U. Chgo., 1971. Med. intern U. Chgo. Hosp., 1971-72, med. resident, 1972-74; fellow in hematology Duke U. Med. Ctr., Durham, N.C., 1974-76, assoc. in medicine, 1976-80, asst. prof. medicine, 1980-86, assoc. prof. medicine, 1986—. Author: Methods in Hematology: Immune Hemolytic Anemias, 1985; contbr. to med. book. Mem. Am. Soc. Hematology, Am. Fedn. Clin. Rsch., Am. Soc. Clin. Investigation, Am. Assn. Immunologists. Office: Duke U Med Ctr 330 Sands Blvd Box 3486 Durham NC 27710

KURLINSKI, JOHN PARKER, physician; b. Buchanon, W.Va., Jan. 17, 1948; s. John Peter and Jean (Holloway) K.; m. Claire Sawyer, June 12, 1971; children: Joshua John, Ryan Edward, Seth Parker. AB cum laude, Williams Coll., 1970; MD, Johns Hopkins Sch. Medicine, 1974. Intern, then resident Johns Hopkins Hosp., Baltimore, 1974-77; fellowship neonatal/perinatal medicine U. Calif., San Diego, 1977-79; chief resident pediatrician Johns Hopkins Hosp., 1979-80; asst. prof. of pediatrics-obstetrics U. Nev. Sch. Medicine, Reno, 1980–; vice chief of staff Sunrise Children's Hosp., Las Vegas, 1989-90, chief of staff, 1990—; pediatrician, co-dir. neonatology S.W. Regional Neonatal Ctr. at Sunrise Hosp. and Med. Ctr., Las Vegas, 1980—; vice chief pediatrics Humana Hosp. Sunrise, Las Vegas, 1983-90; bd. dirs. S.W. Regional Neonatal Ctr. Edn. Found.; mem. Med.-Legal Screening Panel, Nev., 1986—; many hosp. coms., 1980—. Bd. dirs. So Nev. chpt. March of Dimes, Las. Vegas, 1984—. Mem. AMA, Am. Acad. Pediatrics (v.p. Nev. chpt. 1987-90, pres. 1990—, coun. mem. dist. VIII sect. on perinatal pediatrics), Clark County Med. Soc., Las Vegas Pediatric Soc. (founding), Phi Beta Kappa. Avocations: rugby, skiing, hiking, camping. Home: 3322 Beam Dr Las Vegas NV 89118-5902 Office: Sunrise Childrens Hosp 3186 S Maryland Pky Las Vegas NV 89109-2306

KURNIT, DAVID MARTIN, pediatrician, educator; b. Bklyn., Dec. 24, 1947; s. Victor and Helen (Oxhandler) K.; m. Kristine Kurnit, May 1, 1993; children: Katherine, Jennifer. BA, CUNY, 1968; PhD, Albert Einstein Coll. Medicine, 1974, MD, 1975. Diplomate Am. Bd. Pediatrics, Am. Bd. Medical Genetics. Intern Childrens Hosp., Pitts., Pa., 1975-76, resident, 1976-77; fellow med. genetics U. Washington, Seattle, 1977-79; asst. prof. pediatrics Harvard Med. Sch., Boston, 1979-84, assoc. prof., 1985-86; investigator Howard Hughes Med. Inst., Ann Arbor, 1986—; prof. pediatrics and human genetics U. Mich. Med. Ctr., Ann Arbor, 1986—. Contbr. articles to Procs. NAS, Am. Jour. Human Genetics, Nature Genetics. Bd. dirs. Ann Arbor Ctr. for Ind. Living, 1992; chmn. adv. com. Ann Arbor Transp. Authority, 1992. Grantee NIH-SUPHS, 1968-75, NIH, 1977-79, 1982—. Mem. Am. Soc. Human Genetics, Phi Beta Kappa. Achievements include development of recombination-based assay to isolate genes; elaboration of stochastic mechanism that underlies phenotypic defects in a plurality of subjects. Office: Howard Hughes Med Inst 3520 MSRB I Box 0650 1150 W Medical Center Dr Ann Arbor MI 48109-0650

KUROBANE, ITSUO, chemical company executive; b. Karasuyama, Tochigi, Japan, Dec. 23, 1944; s. Takeo and Kiyono K.; m. Sachiko Ishimori, Sept. 23, 1973; 1 child, Emie Laura. BSc, U. Tokyo, 1970, MSc, 1972; postgrad., Wash. State U., 1973-74; D of Agrl. Sci., U. Tokyo, 1975. Rsch. assoc. Dalhousie U., Halifax, Can., 1975-79; guest scientist NRC; sr. rsch. assoc. Wis. U., Madison, 1979-80; assoc. faculty Northwestern U., Chgo., 1980-82; dep. mgr. Hoechst Japan Rsch. Lab., Kawagoe, 1982-88; project mgr. Rhone-Poulenc Agrochimie, Lyon, France, 1988-90; rsch. dir. Rhone-Poulenc Agro, Tsukuba, Japan, 1990—. Grantee Chgo. Heart Assn. 1981-82. Mem. Am. Chem. Soc., Am. Soc. Microbiology, Chem. Inst. Can., Japan Soc. Agrl. Chemistry, Japan Biochem. Soc., Japan Bioindustry Assn., Japan Antibiotics Rsch. Assn. Avocations: gardening, tennis, golf. Home: 109-5 Furushiro, Makabe Ibaraki 300-44, Japan Office: Rhone-Poulenc Agr Rsch Ctr, 1500-3 Mukoueno, Akeno Ibaraki 300-45, Japan

KURODA, ROKURO, chemist, educator; b. Tokyo, Oct. 7, 1926; s. Washizo and Sato (Yamada) K.; m. Sachiko Furubayashi, Nov. 30, 1958; children: Yuzo, Keiko. BS, Tokyo U., 1950, PhD, 1957. Asst. prof. Tokyo Kyoiku U., 1961-65; prof. U. Chiba, Japan, 1965-93; prof. emeritus U. Chiba, 1993—; prof. Univ. Entrance Exam. Ctr., Tokyo, 1984-85; editor Bunseki Kagaku, Tokyo, 1981-83, Analytical Scis., Tokyo, 1988-90, IUPAC-ICAS, Chiba, 1991. Contbr. articles to profl. jours. Mem. Chem. Soc. Japan, Am. Chem. Soc., Japan Soc. Analytical Chemistry (v.p. 1984-85, award 1976). Home: 9-13 Miyamae 4-Chome, Tokyo Suginami-ku 168, Japan Office: U Chiba Faculty of Engring, Yayoi-cho, Inage-ku Chiba 263, Japan

KUROKAWA, KISHO, architect; b. Aichi Prefecture, Japan, Apr. 8, 1934; s. Miki and Ineko K.; m. Ayako Wakao; 2 children. B.Arch., Kyoto U., 1957; M.Arch., Tokyo U., 1964. Pres. Kisho Kurokawa Architect & Assocs.; chmn. Urban Design Cons., Inc.; prin. Inst. Social Engring., Inc.; pres. Kurokawa Internat., Inc.; analyst Japan Broadcasting Corp., 1974—; hon. prof. U. Buenos Aires, Argentina, 1985—; vis. prof. Tsinghua U., Beijing, People's Rep. of China, 1986—; advisor Internat. Design Conf. Aspen, U.S.A., 1974—; chmn. Japan Com. on Bicentennial of French Revolution, 1988—; commr. World Architecture Triennale, Nara, 1992; gen. producer World Architecture Exposition 1998, Nara, 1991—. Author: Prefabricated House, Meatbolism, 1960, Urban Design, 1965, Action Architecture, 1967, Homo-Movens, 1969, Architectural Creation, 1969, The Work of Kisho Kurokawa, 1970, Creating Contemporary Architecture, 1971, Conception of Metabolism, In the Realm of the Future, 1972, The Archipelago of Information: The Future Japan, 1972, Introduction to Urbanism, 1973, Metabolism in Architecture, 1977, A Culture of Grays, 1977, Concept of Space, 1977, Concept of Cities, 1977, Architecture et Design, 1982, Towards Japanese Space, 1982, Thesis on Architecture, 1982, A Cross Section of Japan, 1983, Architecture of the Street, 1983, Under the Road: Landscape under Roads, 1984, Drawing Collection of World Architecture, 1984, Kisho Kurokawa: Il Futuro Nella Tradizione, 1984, Prospective Dialogues for the 21st Century, 1985, Philosophy of Symbiosis, 1987, New Tokyo Plan, 2025, 1987, Kisho KUROKAWA Architecture of Symbiosis, 1988, Rediscovering Japanese Space, 1989, Era of Nomad, 1989, Thesis on Architecture II, 1990, Hanasuki, 1991, Intercultural Architecture—The Philosophy of Symbiosis, 1991, Kisho Kurosawa-From Metabolism to Symbiosis, 1992, Poem of Architecture, 1992, New Wave Japanese Architecture, 1993; exhbns. include: Heinz Gallery, Royal Inst. Brit. Architects, 1981, Institut Francais d'Architecture, 1982, Construma by Ministry of Bldg. and Urban Devel., Budapest, Hungary, 1984, Central House for Architects, Moscow, 1984, Mus. Finnish Architecture, Helsinki, 1985, Buenos Aires Biennale of Architecture, Argentina, 1985, Mus. Architecture in Wroclaw and Warsaw, 1986, Calif. Mus. Sci. and Industry, 1987, U. Calif., 1987, Columbia U., 1992, Sackler Galleries Royal Acad. Arts, 1993. Active Urban Culture Coun., 1981—, Coun. Urban Landscaping Nagoya City, 1981—, Rd. and Environ. Coun., Japan Hwy. Corp., 1973—, Sci. and Tech. Agy., 1980—, Rsch. Soc. Creation Cultural Environ. in Schs., Ministry of Edn., 1981—, Ministry Internat. Trade and Industry, 1981—, Ministry of Constrn., 1983—, Japanese-Chinese Friendship for 21st Century Com., 1985—, Sympo. of the French Revolution, 1989; Recipient Takamura Kotaro Design award, 1965; Hiroba prize, 1977, Japan S.D.A. award silver prize Sony Tower, 1977; Chubu Archtl. Award Ishikawa Cultural Ctr., 1978, Store Front Competition Silver prize for head Offices Chubu Gas Group, 1978; Hon. Citizenship, Sofia, Bulgaria, 1979; B.C.S. award Nat. Ethnological Mus., 1979; decorated comdr. Order of Lion (Finland); Recipient Gold medal Acad. Architecture, France, Richard Neutra award State Poly. U., Calif., 1988, Grand Prix with Gold medal Hiroshima Mus., Sofia Biennale, 1989; Chevalier de l'Ordre des Arts et des Letteres Ministry of Culture, France, 1989, prize Japan Art Acad., 1992, others. Fellow AIA (hon., award of excellence for book Intercultural Architecture), Royal Inst. Brit. Architects (hon.), Union Architects Bulgaria (hon.), Royal Soc. Arts U.K. (life); mem. Japan Inst. Architects, Archtl. Inst. Japan, City Planning Inst. Japan, Japan Soc. Futurology, Japan Soc. Ethnology. Avocation: photography. Address: 11F Aoyama Bldg, 1-2-3 Kita Aoyama, Minato-ku, Tokyo Japan

KURT, THOMAS LEE, medical toxicologist; b. Wichita, Kans., Sept. 7, 1938; s. Lester John and Dorothea Rosemary (Gufler) K.; m. Carol Sue Fox, Sept. 4, 1971; children: Johanna, Gretchen, Erika. BS, U. Notre Dame, 1960; MD, U. Kans., 1964; MPH, Harvard U., 1974. Diplomate Am. Bd. Preventive Medicine, Am. Bd. Med. Toxicology. Intern Georgetown Med. div./D.C. Gen. Hosp., 1965; resident in internal medicine U. Colo. Med. Ctr., Denver, 1965-67, Nat. Heart and Lung Inst. fellow in cardiology, 1967-69, from instr. medicine to assoc. prof., 1969-78, dir. health svc., 1969-72, dep. dir. emergency rm., 1976-78; assoc. prof. U. Tex./S.W. Med. Ctr., Dallas, 1979-89, prof., 1989—; regional med. officer FDA, Dallas, 1989-91; acting dir., prof. North Tex. Poison Ctr./U. Tex. S.W. Med. Ctr., Dallas, 1991—; cons. Airline Pilots Assn., Denver, 1975-78, Internat. Assn. Fire Fighters, 1974—, Forensic Inst., Dallas, 1979—, many others; staff Parkland Meml. Hosp., Dallas, 1978—, St. Paul Med. Ctr., Dallas, 1978—, others in past. Asst. editor PoisindexR; editorial bd. Occupational Health and Safety, Jour. of Toxicology/Clinical Toxicology; contbr. articles to profl. jours. Maj. USAFR, 1965-69, Wyo. Air N.G., 1969-72. Recipient numerous grants. Fellow Am. Acad. Clin. Toxicology, Am. Coll. Clin. Pharmacology, Am. Coll. Preventive Medicine, Am. Coll. Occupational and Environ. Medicine, Coun. on Epidemiology/Am. Heart Assn.; mem. AAAS, Am. Assn. Pub. Health Physicians, Am. Bd. Med. Toxicology, Am. Coll. Epidemiology, Am. Coll. Cin. Pharmacology, Am. Indsl. Hygiene Assn., Tex. Med. Assn., Soc. for Epidemiologic Rsch., Dallas County Med. Soc., many others. Achievements include discovery that firefighter exposure to ambient carbon monoxide caused EKG changes and health effects; odor-triggered panic attacks respond to tricyclics; neuropathy associated with new plastic foaming agent, others. Home: 3645 Stratford Ave Dallas TX 75205-2810 Office: North Tex Poison Ctr PO Box 36032 Dallas TX 75235-1032

KURTH, PAUL DUWAYNE, biotechnical services executive; b. Milw., Nov. 12, 1946; s. Paul John and Estelle Kurth; m. Joan Elizabeth Dunn, June 6, 1970. BA, North Cen. Coll., 1969; MS, U. Wis., 1972; PhD, Johns Hopkin's U., 1979. Supr. molecular biology Ortho Diagnostic Systems, Carpinteria, Calif., 1984-85; corp. sci. advisor Am. Hosp. Supply Corp., Valencia, Calif., 1985-86; dir. devel., quality assurance and regulatory affairs Nunc, Inc., Naperville, Ill., 1986-90; pres. Internat. Biotech. Svcs., Naperville, 1990—. Contbr. articles to Jour. Cell Biology. Mem. AAAS, N.Y. Acad. Scis., Broadcast Music Inc. Achievements include development of 1st FDA approved DNA probe diagnostic kits for viral Herpes virus infection; design and development of flask for producing human skin grafts; design of device for removal of tissue culture cells from tissue culture flasks. Office: Internat Biotech Svcs PO Box 566 Naperville IL 60566-7604

KURTZ, ALFRED BERNARD, radiologist; b. Albany, N.Y., May 1, 1944; s. Leonard David and Esther (Lederman) K.; m. Barbara Ellen, July 3, 1973; children: Dana, Liza, Amy. BA, NYU, 1966; MD, Stanford U., 1972. Diplomate Am. Bd. Radiology. Internal medicine intern Montefiore Hosp. and Med. Ctr., Bronx, N.Y., 1972-73, resident in internal medicine, 1973-74, resident in diagnostic radiology, 1974-77; assoc. prof. ob/gyn Jefferson Med. Coll. Thomas Jefferson Univ. Hosp., Phila., 1982-85; assoc. dir. Div. of U.S. and Radiol. Imaging, Phila., 1982-86, Body Computed Tomography, Thomas Jefferson U. Hosp., Phila., 1986-89, Div. Diagnostic U.S., Thomas Jefferson U. Hosp., Phila., 1986-89; prof. radiology Jefferson Med. Coll. Thomas Jefferson U. Hosp., Phila., 1983—, prof. ob/gyn., 1985—, vice chmn. dept. radiology, 1989—; med. advisor Blue Shield of Pa., Phila., 1983—; mem. adv. com. Ctr. of Excellence in Biomed. Imaging, Phila., 1987—. Author: Obstetrical Measurements in Ultrasound: A Reference Manual, 1988; editor: Atlas of Ultrasound Measurements, 1990; contbr. articles to profl. jours. Fellow Coll. Physicians of Phila., Am. Inst. of Ultrasound in Medicine (bd. govs. 1990—), Am. Coll. Radiology (chmn. com. on edn. and tng. of commn. 1987—, commn. on ultrasound 1987—); mem. Soc. Radiologists in Ultrasound (pres.-elect). Achievements include advancement of the ability of ultrasound to establish an accurate fetal age; establishment of ultrasound patterns for analysis of diffuse liver disease; advancement of ultrasound in further evaluation of obstetrical and gynecologic problems and analysis of the prostate by intravaginal scanning transrectal approach. Home: 1050 Indian Creek Rd Wynnewood PA 19096-3407 Office: Thomas Jefferson U Hosp 111 S 11th St Philadelphia PA 19107-5084

KURTZ, ANDREW DALLAS, chemical engineer; b. Phila., Apr. 28, 1954; s. William L. K. and Constance H. (Dallas) Millet; m. Melissa J. Wilkins, June 16, 1979; children: Amanda L., Jennifer W. BSChE, Princeton U., 1976; M in Chem. Engring., Cornell U., 1977. Registered profl. engr., N.J. Rsch. engr. FMC Corp., Princeton, N.J., 1977-86; sr. engr. Church & Dwight Co., Princeton, 1986-89, mgr., 1989-92, dir. process, 1992—. Contbr. articles to Jour. Fire and Flammability. Deacon Princeton Fellowship Ch., 1980—. Achievements include 4 patents. Office: Church & Dwight Co 469 N Harrison St Princeton NJ 08543

KURTZ, EDWIN BERNARD, biology educator, researcher; b. Wichita, Kans., Aug. 11, 1926; s. Edwin B. and Florence (Warner) K.; m. Lois L. Leecing, June 12, 1952 (dec. July 1984); children: Kathryn, Jane. B.S., U. Ariz., 1948, M.S., 1949; Ph.D., Calif. Inst. Tech., 1952. From asst. prof. to prof. U. Ariz., Tucson, 1951-68; prof., chmn. dept. Kans. State Tchrs. Coll., Emporia, 1968-72; prof. biology, chmn. dept. biology U. Tex.-Permian Basin, Odessa, 1972-89, prof. emeritus, 1989—. Author books including: Adventures in Living Plants, 1965; Modules for Contemporary Natural Science I and II, 1978; also articles. Pres. Planned Parenthood of Permian Basin, Odessa, 1982-83, bd. dirs., Phoenix, 1991—. Recipient Amoco Found. Outstanding Tchr. award, 1982, 1st Pres.'s award for excellence in teaching U. Tex., 1986; NRC-AEC fellow Calif. Inst. Tech., Pasadena, 1949-51. Fellow AAAS (asst. dir. edn. 1964-67), Sigma Xi; mem. Ariz. Acad. Sci. (pres. 1962-63), Phi Beta Kappa, Phi Kappa Phi. Unitarian. Avocation: gardening. Home: 1620 N Kutch Dr Flagstaff AZ 86001-1229 Office: U Tex Permian Basin Odessa TX 79762

KURTZ, MAX, civil engineer, consultant; b. Bklyn., Mar. 25, 1920; s. Samuel and Ida (Malkin) K.; B.B.A., CCNY, 1940; postgrad. Rutgers U., 1943-44; m. Ruth Ingraham, Sept. 9, 1967. Structural engr. Kurtz Steel Constrn. Corp., Mineola, N.Y., 1946-56; pvt. practice cons. engring., Flushing, N.Y., 1956—; condr. seminars on ops. research; instr. review courses for profl. engrs. licensing examinations. Served with U.S. Army, 1943-45. Registered profl. engr., N.Y. Mem. N.Y. State Soc. Profl. Engrs. (Honor award Kings County chpt. 1970), Nat. Soc. Profl. Engrs. Author: Structural Engineering for P.E. Examinations, 3d edit., 1978; Engineering Economics for P.E. Examinations, 3d edit., 1985; Comprehensive Structural Design Guide, 1968; Handbook of Engineering Economics, 1984; Handbook of Applied Mathematics for Engineers and Scientists, 1991; editor Kings County Profl. Engr., 1967-71; project editor Civil Engineering Reference Guide, 1986. Home and Office: 33-47 91st St Flushing NY 11372

KURTZ, MICHAEL E., medical educator; b. St. Louis, Mar. 30, 1952; m. Karen Rodefeld, Apr. 16, 1983; children: Brian, Alison. BA in Biology, U. Mo., St. Louis, 1976, MS in Biology, 1980. Cert. histocompatibility technologist. Sr. rsch. technician cancer biology sect. Washington U., St. Louis, 1976-81; rsch. biologist John Cochran VA Hosp., St. Louis, 1981-85, mgr. histocompatibility lab., 1985-89; rsch. assoc. surgery and transplant svcs. St. Louis U., 1989-90, mgr., assoc. dir. Surg. Rsch. Inst., 1990—, cons. HLA svcs., 1986—, instr. rsch. methodology, 1990—; med. computing cons. 9th Wave Computing Inc., Chapel Hill, N.C., 1991—; mem. lab. adv. com. HLA, 1992, 93. Bd. dirs. Washington Luth. Sch., St. Louis, 1992—. Mem. Soc. Rsch. Adminstrs., Assn. Acad. Surg. Adminstrs., Am. Soc. Histocompatibility and Immunogenetics (mem. regional edn. com. 1986-88), Am. Assn. Lab. Animal Sci., Nat. Coun. U. Rsch. Adminstrs. Republican. Lutheran. Home: 5833 Morning Field Pl Saint Louis MO 63128 Office: St Louis U Surg Rsch Inst PO Box 15250 3635 Vista Ave at Grand Blvd Saint Louis MO 63110-0250

KURTZ, RUSSELL MARC, laser/optics engineer; b. Lincoln, Nebr., Aug. 11, 1959; s. Richard Allen and Patricia Jane (Hazen) K.; m. Susan Irene Miller, Dec. 30, 1983 (div. June 1991). BS, MIT, 1981; PhD, U. Southern Calif., 1991. Cert. elec. engr. Calif. Engr. Hughes Aircraft, El Segundo, Calif., 1981-85; pres. TMS, L.A., 1985—; cons. Allied-Signal, Westlake Village, Calif., 1985-87, LIWA and Co., Monterey Park, Calif., 1990—. Author: (with others) Solid State Lasers, 1987, Advanced Solid State Lssers, 1989; contbr. articles to profl. jours. Speaker Youth Motivation Task Force, L.A., 1989, Voter Registration, L.A., 1992. Mem. IEEE, Toastmasters Internat., Sigma Xi. Achievements include early work in multiple-wavelength solid-state lasers; pioneering studies of degenerate phase conjugation in active media; study of laser system improvement through preventive maintenance. Home: PO Box 2049 Artesia CA 90702

KURTZ, STEWART KENDALL, physics educator, researcher; b. Bryn Mawr, Pa., June 9, 1931; s. Stewart S. Jr. and Ellen (Chase) K.; m. Dora Grandinetti, July 1, 1951; children: Philip, David, Timothy, John. BSc in Physics, Ohio State U., 1955, MS in Physics, 1956, PhD in Physics, 1960. Mem. tech. staff Bell Telephone Labs., Murray Hill, N.J., 1960-69; dir. exptl. rsch. Philips Labs., Briarcliff Manor, N.Y., 1969-78; v.p. engring. Clairol Bristol Meyers Co., Stamford, Conn., 1978-84, v.p. tech. Clairol, 1984-85; sr. scientist Clairol Rsch. Labs., Stamford, 1985-87; prof. elec. engring. Pa. State U., University Park, 1987—, assoc. dir. Materials Rsch. Lab., 1988-89, dir. Materials Rsch. Lab., Murata prof. materials rsch., 1989-91, vice chair, exec. adminstr.Materials Rsch. Inst., 1992—; cons. Clairol Rsch., Stamford, 1987, Alcoa Packaging, Pitts., San Diego, 1989-90. Co-author: Landolt-Bornstein, Vol. 2, 1969, Landolt-Bornstein, Vol. 18, 1984; author: (with others) Quantum Electronics, A Treatise, Vol. 1, 1975, Systematic Materials Analysis, Vol. IV, 1978. ITT fellow, 1957, NSF teaching fellow, 1958. Mem. IEEE (sr.), Am. Phys. Soc., Am. Ceramic Soc., European Pigment Cell Soc., Materials Rsch. Soc., Phi Beta Kappa. Republican. Roman Catholic. Achievements include patents for electro-optic memory plane, for backward wave oscillator, for nonlinear optical iodates, for photoelectric switching device, for second harmonic analyzer, for roller, for high efficiency frequency doubler. Home: 1614 Woodledge Cir State College PA 16803-1874 Office: Pa State U Materials Rsch Inst 117 Barbara Bldg II University Park PA 16802-1013

KURTZMAN, CLETUS PAUL, microbiologist; b. Mansfield, Ohio, July 19, 1938; s. Paul A. and Marjorie M. (Gartner) K.; m. Mary Ann Dombrink,

Aug. 4, 1962; children: Mary, Mark, Michael. BS, Ohio U., 1960; MS, Purdue U., 1962; PhD, W.Va. U., 1967. Microbiologist Nat. Ctr. Agrl. Utilization Rsch./USDA, Peoria, Ill., 1967-85, rsch. leader, 1985—; U.S. rep. Internat. Commn. on Yeasts, 1980—, World Fedn. Culture Collections, 1988—. Editor: Yeasts in Biotechnology, 1988; contbr. papers to sci. jours. 1st lt. U.S. Army, 1962-64. Named Midwest Area Outstanding Scientist USDA, 1986; recipient Medal of Merit award Ohio U., 1992. Fellow Am. Acad. Microbiology; mem. AAAS, Internat. Mycol. Assn. (sec.-gen. 1990—), Mycol. Soc. Am., Am. Soc. Microbiology (div. chair 1991—, J. Roger Porter award 1990), U.S. Fedn. Culture Collections (pres. 1976-78), Soc. Gen. Microbiology. Achievements include patent for Xylose Fermentation in Yeasts; research in the correlation of DNA relatedness and fertility in yeasts, correlation of ribosomal RNA divergence. Office: Nat Ctr Agrl Utilization Rsch 1815 N University St Peoria IL 61604-3999

KURUSU, YASUHIKO, chemistry educator; b. Wakayama, Japan, Dec. 27, 1938; s. Yasuichi and Kimiko Kurusu; married, Apr. 29, 1969; children: Tamami, Kazuhiko. M of Engring., Osaka Pref. U., Osaka-Sakai, 1963; D of Engring., Tokyo Inst. of Tech., 1970. Assoc. rschr. Tokyo Inst. of Tech., 1965-69; assoc. prof. Sophia U., Tokyo, 1969-79, prof., 1979—. Author: Hydrocarbon Chemistry, 1991; editorial staff mem. The Soc. of Synthetic Organic Chemistry Japan, 1970-71, Chem. and Edn., 1989-90; steering com. mem. Macromolecular Metal Complex Rsch. Group, Tokyo, 1986-92. Recipient Encouragement award Asahi Glass Industry, 1990. Home: Hiyoshi hon cho 3-24-14, Koho ku-ku 223, Japan Office: Sophia U, Kioicho 7-1, Chiyoda-ku Tokyo 102, Japan

KURYLA, WILLIAM COLLIER, chemist, consultant; b. Akron, Ohio, Sept. 3, 1934; s. Vladimir Alexander and Helen (Lynn) K.; m. Arlene Batten, June 22, 1957; children: Paul T., Matthew L. BSc, Kent State U., 1956; MSc, U. Minn., 1958, PhD, 1960. Rsch. chemist Union Carbide Corp., Charleston, W.Va., 1960-67, rsch. scientist, 1967-79, tech. mgr., 1974-76, toxicology adminstrv. mgr., 1979-84; sr. mgr. Union Carbide Corp., Danbury, Conn., 1984-86, dir. prodn. safety, 1986-87, assoc. dir. prodn. safety, 1987—; v.p. Sci. Horizons, Danbury, 1985-92. Sr. editor book series Flame Retardancy...Polymers, 1973-79; contbr. articles to sci. publs. Adult vol. Boy Scouts Am., Charleston and Danbury, 1960—. NIH fellow, 1959-60; recipient Silver Beaver award, Silver Antelope award, Disting. Eagle award Boy Scouts Am., 1968-92. Fellow AAAS; mem. Am. Chem. Soc. (sect. chmn. 1958-92), Am. Coll. Toxicology, W.Va. Acad. Sci. (life, pres. 1985-86), N.Y. Acad. Sci., Sigma Xi. Achievements include patents on polymer polyols. Home: 4 Peaceful Dr New Fairfield CT 06812 Office: Union Carbide Corp 39 Old Ridgebury Rd Danbury CT 06817

KURYS, JURIJ-GEORGIUS, environmental engineer, scientist, consultant; b. Cerkovna-Dolyna, Ukraine, USSR, May 3, 1926; arrived in Can., 1952; s. Nicolas and Maria-Sofia (Uzela) K.; m. Irene-Alexandra Martyniuk, May 11, 1962; children: Oksana, Natalie, Christine. Diploma in engring., Tech. U. Regensburg, Fed. Republic of Germany, 1951; DSc, Tech. U. Munich, 1972. Registered profl. engr., Ont., Can. Rsch. scientist Ministry of Health, Toronto, Ont., 1963-67; sr. rsch. scientist Ministry of Environment, Toronto, 1967-70, tech. advisor, 1970-76; sci. advisor Ministry of Industry and Trade, Toronto, 1976-80; sr. policy advisor Ministry of Industry, Trade and Tech., Toronto, 1980-90; pres. J.G. Kurys & Assocs. Inc., Environ. Mgmt., Toronto, 1990—; mem. adv. bd. St. Clair Coll. Tech., Windsor, Ont., Can., 1978—; rektor Ukrainisches Technisch-Wirtschaftliches Institut, Muenchen, 1989—; pro-rektor Poly. U. Lviv, Ukraine, 1991—. Author: Methyl Bromide Fumigation, Procedure, 1966, Industrial Toxicology, 1977, (book) Health Aspect of Pesticides, 1977, (corr. course) How to Use Pesticides, 1968, (govt. study) Petrochemical Strategy in Ontario, 1980. Fellow Chem. Inst. Can., Schevchenko Sci. Soc.; mem. Assn. Profl. Engrs. Ont., Assn. Chem. Profls. Ont., Can. Pub. Health Assn., World Fedn. Ukrainian Engring. Socs. (pres. 1988—). Home and Office: 27 Newel Ct, Toronto-Etobicoke, ON Canada M9A 4T9

KURYU, MASAO, economics educator; b. Oita, Japan, Feb. 18, 1920; d. Sugitaro and Chiyo (Shigaki) Fujino; m. Yoshiko Kuryu, Mar. 14, 1947; 1 child, Norio. B. of Commerce, Toyko Comml. Coll., 1943. Lectr. Oita U., 1952-54, asst. prof., 1955-62, prof., 1963-83, dean faculty of econs., 1971, prof. emeritus, 1983—; prof. Kyushu Kyoritsu U., Kitakyushu, Japan, 1983-92, dean faculty of econs., 1983-88; prof. Miyazaki Sangiokeiei U., Miyakonojo, Japan, 1992—. Author: The Analysis of Econimic Models, 1978. Chmn. Oita Minimum Wage Coun., 1981—. Fellow Oita Regional Econ. Rsch. Inst.; mem. Japan Assn. Econs. and Econometrics. Avocations: music, travel. Home: 2-12 Minami Kasugamachi, Oita 870, Japan Office: Kyushu Kyoritsu U, Faculty Econs, Jiyugaoka Yahatanishi-ku, Kitakyushu 807, Japan

KURZWEG, ULRICH HERMANN, engineering science educator; b. Jena, Germany, Sept. 16, 1936; came to U.S., 1947, naturalized, 1952; s. Hermann Herbert and Erna Herta (Michaelis) K.; m. Sophia Speth, Dec. 21, 1963; 1 dau., Tina. B.S., U. Md., 1958; M.A. (Woodrow Wilson fellow 1958-59), Princeton U., 1959, Ph.D. in Physics, 1961. Sr. theoretical physicist United Tech. Research Labs., East Hartford, Conn., 1962-68; adj. assoc. prof. math. Rensselaer Poly. Inst., Hartford (Conn.) Grad. Center, 1964-68; mem. faculty U. Fla., Gainesville, 1968; prof. engring. scis. U. Fla., 1968—. Contbr. numerous articles to sci. and tech. publs. Fulbright grantee, 1961-62; recipient Cert. of Recognition, NASA, 1984, award for excellence in undergrad. teaching U. Fla., 1991. Mem. AAAS, Am. Phys. Soc., N.Y. Acad. Scis., Sigma Xi. Home: 8407 NW 4th Pl Gainesville FL 32607-1414 Office: U Fla Dept Aerospace Engring Mechanics and Engring Sci Gainesville FL 32607

KUSHLAN, JAMES A., biology educator; b. Cleve., Oct. 11, 1947; m. Paula Frohring; children: Kristin, Philip. BS in Biology and Chemistry cum laude, U. Miami, 1969, MS in Biology, 1972, PhD in Biology, 1974. Rsch. biologist U.S. Dept. of Interior, 1975-84; assoc. prof. biology East Tex. State U., Commerce, 1984-87, prof. biology, 1987-88; prof. biology, chair dept. biology U. Miss., 1988—; adj. assoc. prof. biology U. Miami, 1980-86; dir. ctr. water resources studies East Tex. State U., 1986-88; hydrology technician US Geol. Survey, 1972-73; biol. technician Nat. Park Svc., 1973-75; mem. Fed. Fla. Panther Recovery Team, 1975-82, Fed. Am. Crocodile Recovery Team, 1975-83, Crocodile Specialist Group, 1981—; vis. scientist Darwin Rsch. Sta., Galapagos, 1978; leader Fed. Cape Sable Sparrow Recovery Team, 1979-83; rsch. assoc. Ctrl. U. Venezuela, 1979-83; mem. tech. adv. com. EPA 208 Water Quality Program, 1976-78, tech. adv. bd. Miss. Nature Conservancy, 1991—; mem. Miss. Environ. Edn. Consortium, 1992—, Miss. Rsch. Consortium, 1992—, Miss. Pub. Edn. Forum, 1993—, Internat. Waterfowl and Wetlands Rsch. Bur., 1986—, exec. bd., steering com., 1992—; mem. various coms. East Tex. State U., 1986-88, U. Miss., 1989—; tchr. workshops; lectr. in field. Author: (with J. Hancock) The Herons Handbook, 1984, (with W.F. Loftus) Freshwater Fishes of Southern Florida, 1987, (with J. Hancock and M.P. Kahl) Storks, Ibises, and Spoonbills of the World, 1992, (with others) Environments of South Florida, Present and Past, 1974, Carrying Capacity for Man Nature in South Florida, 1976, Wading Birds, 1978, Rare and Endangered Biota of Florida, 1979, Crocodiles, 1982, Status and Management of Osprey and Eagles, 1983, Dictionary of Birds, 1985, Encyclopedia of Birds, 1985, Managing Cumulative Effects in Florida Wetlands, 1986, Ecosystems of Florida, 1990, The Rivers of Florida, 1991; editor Fla. Field Naturalist, 1981-86; mem. editorial bd. Colonial Waterbirds, 1981-84, editor, 1985-88; mem. editorial bd. Wetlands, 1982, assoc. editor, 1993—; pub. papers, revs., commentaries; contbr. articles to profl. jours. Com. chair Maloy Community Improvement Assn., 1987; active United Way Planning Coun., University, Miss., 1991-92; trustee John Cabot U., 1992—. Grantee Nat. Park Svc., 1975-83, U.S. Fish & Wildlife Svc., 1975-76, EPA, 1978-80, 1991-92, Am. Mus. Natural History, 1978, Ctrl. U. Venezuela, 1979, Tex. Edn. Agy., 1986-87, Brehm Fond, 1987-89, U.S. Geol. Survey and Tex. Water Resource Inst., 1988-89, Insts. Higher Learning and U.S. Dept. Edn., 1990, Soil Conservation Svc., 1990-91, Internat. Coun. Bird Protection, 1993; Paul Harris fellow Rotary Internat., 1989; recipient Citizen award WIOD Radio, Miami, 1980. Fellow Am. Ornithologists' Union (chair membership com., asst. to sec., coord. sci. program, biography and membership coms.); mem. AAAS, Am. Soc. Naturalists, Ecol. Soc. Am., Miss. Acad. Scis., Soc. Rsch. Adminstrs., Soc. Wetland Scientists (life, assoc. editor, editorial bd.), Colonial Waterbird Soc. (editor jour., elec.-counselor, editorial bd., local com. chair, mem. various coms.), Rotary Club (pres.-elect, bd. dirs., dist. environ. chair), Sigma Xi (nominations com., chpt. pres. 1987—). Achievements include research in adaptation and accommodation of animal populations to fluctuating water conditions, the influence of environmental fluctuationson population and community dynamics, international conservation and management of biodiversity, nature reserves, wetland ecosystems and their typical organisms. Office: University of Mississippi Dept of Biology University MS 38677

KUSHNER, BRIAN HARRIS, pediatric oncologist; b. N.Y.C., July 8, 1951; s. William Isidore and Sheila Elaine (Kasselbranar) K.; m. Phyllis Debra Levinberg, Feb. 22, 1986; children: Sarah Lynn, Carolyn Joy. AB, Harvard U., 1972; MD, Johns Hopkins U., 1976. Diplomate Am. Bd. Pediatrics, Am. Bd. Pediatric Hematology-Oncology. Pediatric intern and resident Babies Hosp. of Columbia-Presbyn. Med. Ctr., N.Y.C., 1976-78; pediatric sr. resident N.Y. Hosp., N.Y.C., 1978-79; clin. fellow in pediatric hematology-oncology Children's Hosp., Boston, 1979-80; staff pediatrician Boston City Med. Clinics, North End Community Health Ctr., 1980-81; coord. in-patient svc., dir. ICU dept. pediatrics Lincoln Hosp., N.Y. Med. Coll., Bronx, 1982-83; clin. rsch. fellow in pediatric hematology-oncology Meml. Sloan-Kettering Cancer Ctr., N.Y.C., 1983-86, chief fellow dept. pediatrics, 1985-86, spl. fellow dept. pediatrics, 1986-87, clin. asst. pediatrician, 1987-92; asst. attending pediatrician dept. pediatrics N.Y. Hosp., N.Y.C., 1988—; asst. attending pediatrician Meml. Sloan-Kettering Cancer Ctr., N.Y.C., 1992—; staff physician Internat. Rescue Com., Khao-I-Dang Refugee Camp, Thailand, 1981; staff physician Oxfam Relief, Khiam, Lebanon, 1983; instr. in pediatrics Cornell U. Med. Coll., N.Y.C., 1987—. Contbr. articles to Jour. Clin. Oncology, Jour. Pediatrics, Cancer, Blood, Leukemia, Med. Pediatric Oncology, Exptl. Cell Biology, others. Recipient Clin. Scholars Nat. Rsch. Svc. award, 1988-90, Career Devel. award Am. Cancer Soc., 1990—. Mem. AMA, AAAS, Am. Acad. Pediatrics, Am. Assn. Cancer Rsch., Am. Soc. Clin. Oncology, Am. Soc. Hematology, Am. Soc. Pediatric Hematology-Oncology, N.Y. Acad. Scis. Office: Meml Sloan Kettering Cancer Ctr 1275 York Ave # 299 New York NY 10021-6094

KUSHNER, MICHAEL JAMES, neurologist, consultant; b. Hackensack, N.J., July 18, 1951; s. Samuel and Ruth Ellen (Paul) K.; m. Sarah Joan Warden, Aug. 14, 1976; 1 child, Hunter Paul. BA in Physics, Yale U., 1973; MD, NYU, 1977. Diplomate Am. Bd. Psychiatry, Am. Bd. Med. Examiners. Intern Parkland Meml. Hosp., U. Tex., Dallas, 1977-78; resident in neurology Neurol. Inst., Columbia-Presbyn. Med. Ctr., N.Y.C., 1978-81; rsch. assoc. U. Pa., Phila., 1981-83, asst. prof. neurology, 1983-90; attending physician Hosp. of U. Pa., Phila., 1983-90; with Wilson (N.C.) Neurology Ctr., 1992—; dir. SPECT facility Hosp. of U. Pa., 1986-90, asst. dir. neurovascular lab., 1987-90; mem. sensory disorders and lang. study sect. NIH, Bethesda, Md., 1988-90; cons. Dupont Med. Products Div., Billerica, Mass., 1987—; staff neurologist Wilson (N.C.) Neurology Ctr. Contbr. numerous articles to profl. jours. Interviewer alumni schs. com. Yale U., Phila., 1984—. Fellow Am. Acad. Neurology, Am. Heart Assn. (stroke coun.); mem. AMA, Internat. Soc. for Blood Flow and Metabolism, Yale of N.Y.C., Yale of Cen. N.C., Yale of Phila. Republican. Episcopalian. Avocations: oenology, travel, swimming, golf. Home: 1110 Salem St NW Wilson NC 27893-2125 Office: Wilson Neurology Ctr PO Box 3148 Wilson NC 27895-3148

KUST, ROGER NAYLAND, chemist; b. Berwyn, Ill., Apr. 20, 1935; s. Victor Nathan and Millicent Rose (Crabtree) K.; m. Mary Angela Black, Oct. 1, 1957 (dec. Oct. 15, 1976); children: Paul, Peter; m. Claudette Ann Bouten, June 15, 1979. BS, Purdue U., 1957; PhD, Iowa State U., 1963. Asst. prof. chemistry U. Utah, Salt Lake City, 1965-71; sect. head chemistry Kennecott Copper Corp., Lexington, Mass., 1971-79; div. mgr. Exxon Minerals Co., Houston, 1979-87; v.p. R&D Tetra Techs. Inc., Houston, 1987—. Contbr. articles to profl. jours. Lt. (j.g.) USN, 1957-60. Grad. fellow NSF, 1957; NROTC scholar USN, 1953. Fellow Am. Inst. Chemists (life); mem. Am. Chem. Soc. (student award Ind. sect. 1957), AAAS, Am. Acad. Arts & Scis., N.Y. Acad. Scis. Achievements include 6 patents in inorganic process chemistry; development process to recover metal values from manganese nodules, and process to reclaim copper and zinc from brass dust. Home: 8408 Crescent Wood Ln Spring TX 77379-8711 Office: Tetra Techs 9391 Grogans Mill Rd Spring TX 77380-3627

KUSTIN, KENNETH, chemist; b. Bronx, N.Y., Jan. 6, 1934; s. Alex and Mae (Marvisch) K.; m. Myrna May Jacobson, June 24, 1956; children—Brenda Jayne, Franklin Daniel, Michael Thorpe. B.Sc., Queens Coll., 1955; Ph.D., U. Minn., 1959. Postdoctoral fellow Max Planck Inst. for Phys. Chemistry, Gottingen, W. Ger., 1959-61; asst. prof. chemistry Brandeis U., 1961-66, asso. prof., 1966-72, prof., 1972—, chmn. dept. chemistry, 1974-77; vis. prof. pharmacology Harvard U. Med. Sch., 1977-78; Fulbright-Hays lectr., 1978; program dir. NSF, 1985-86. Editor: Fast Reactions, vol. 16 of Methods in Enzymology, 1969; bd. editors Internat. Jour. Chem. Kinetics, 1983-90, Inorganic Chemistry, 1993—; rsch. and publs. in field. Mem. Am. Chem. Soc. (councilor 1983-85), Phi Beta Kappa. Office: Brandeis U Dept Chemistry PO Box 9110 Waltham MA 02254

KUSUMI, AKIHIRO, scientist, educator; b. Kyoto, Japan, Oct. 17, 1952; came to U.S., 1979; s. Shigeru and Fumi (Kato) K.; m. Taeko Tanaka, Nov. 6, 1986; children: Natsuko, Masahiro. BS, Kyoto U., 1975, PhD, 1980. Rsch. assoc. Med. Coll. Wis., Milw., 1979-82; rsch. fellow Princeton (N.J.) U., 1982-84; asst. prof. Kyoto U., 1984-88; prof. U. Tokyo, 1988—; vis. prof. Med. Coll. Wis., Milw., 1984—; exec. com. mem. Sci. and Tech. Agy. of Japanese Govt., Tokyo, 1989-91; prin., co-prin. investigator U.S. NIH, Japanese govt., various internat. agys. Contbr. articles to profl. jours. Grantee NIH, 1984—, Japanese Ministry of Edn., Sci. and Culture, Tokyo, 1984—, Human Frontier Sci. Program, Grenoble, 1991—; recipient various awards pvt. founds. in field, Japan, 1988—. Achievements include discovery of domain structures of the cell membrane; invention of time-resolved fluorescence microscopy; first measurement of oxygen permeability across the cell membrane; research on fluorescence lifetime imaging microscopy. Office: Med Coll Wis Nat Biomed ESR Ctr 8701 W Watertown Plank Rd Milwaukee WI 53226-4801

KUWATA, KAZUHIRO, chemist; b. Antone-yu, Antone, Korea, Mar. 2, 1942; arrived in Japan, 1945; s. Kazumi and Hisayo (Matsumoto) K. B in Engring., Hiroshima (Japan) U., 1965, M in Engring., 1967, D of Engring., 1983. Chemist Osaka (Japan) Prefectural Govt., 1967-68; chemist Environ. Pollution Control Ctr., Osaka, 1968-86, chemist exec. staff, 1986-93; exec. staff air pollution ctrl. divsn. Environ. Bur. Osaka Prefectural Govt., Oska, 1993—. Contbr. articles to profl. jours. Mem. Am. Chem. Soc., Air & Waste Mgmt. Assn., Air Pollution Control Assn. Japan, Analytical Chem. Soc. Japan, Chem. Soc. Japan. Avocations: Japanese art swords and fine arts, go game, music, sports. Home: 316 2-11-53 Tamagushimotomachi, Higashi Osaka 578, Japan Office: Environ Bur Osaka Perfectural Govt, 2 Ohtemae Chuo-Ku, Osaka 540, Japan

KUYATT, BRIAN LEE, toxicologist, scientific systems analyst; b. Lincoln, Mar. 8, 1952; s. Chris Ernie Earl and Patricia Lou (Peirce) K.; m. Susan Lynn Farber, June 5, 1976; children: Kristen M., Jill E., Laura E. and Leslie E. (twins). BA in Biology, Cath. U. of Am., 1974; MS in Human Anatomy, U. Md., 1976, postgrad., 1976-79; postgrad., U. Balt., 1986-89. Instr. anatomy and physiology Catonsville (Md.) C.C., 1977-90; rsch. biologist NIH, Nat. Inst. Aging, Balt., 1978-82, NIH, Nat. Inst. on Dental Rsch., Bethesda, 1982-85, NIH, Addiction Rsch. Ctr., Balt., 1985-87; dir. customer svcs. Loats Assocs., Inc., Westminster, Md., 1987-90; toxicologist, image analyst morphometrics resource ctr. Eli Lilly & Co., Greenfield, Ind., 1990—; rsch. collaboration Dr. Marian DeMyer, IUPUI, Indpls., 1990-91; cons. Dr. Keith March-Krannert Heart Inst., Indpls., 1991-93, Dr. John Duguid, VA Hosp., Indpls., 1991-92, Dr. John Russ, N.C. State U., 1992-93. Contbr. articles to profl. jours. Bd. dirs. Good News Mission, Indpls., 1992; Sunday sch. supt. Halethorpe Community Ch., Balt., 1985-90, clk., 1982-85. Mem. AAAS, N.Y. Acad. Scis., Sigma Xi, Phi Kappa Phi. Achievements include patent for semi-automated point counting macro for the MacIntosh computer, semi-automated image analysis method for evaluating centrilobular hepatocellular hypetrophy, artificial intelligence method for evaluating color images of vaginal smears; development of fully automated method for counting micronuclei in RBC's, fully automated counting of BrdU cells as a marker of cell proliferation. Home: 3594 S Eland Dr New Palestine IN 46163 Office: Eli Lilly & Co PO Box 708 Greenfield IN 46140

KUYK, WILLEM, mathematics educator; b. Amsterdam, Nov. 19, 1934; arrived in Belgium, 1969; m. Minke Zuidema; children—Egbert, Sijtze, Maarten. DSc in Math., Free U. Amsterdam, 1960. Asst. prof., fellow math. ctr. U. Amsterdam, 1961-63; postdoctoral research fellow Nat. Research Council, Ottawa, Ont., Can., 1963-64; assoc. prof. McGill U., Montreal, Que., Can., 1964-68; prof. math. U. Antwerp, Belgium, 1968—, dean sci. 1980-88. Author: Complementarity in Mathematics, 1977, Cardiovascular System Simulation, 1993; co-editor math. books. Mem. Dutch Math. Soc. (pres. 1963), N.Y. Acad. Sci., Belgian Math. Soc., Am. Math. Soc., Can. Math Soc. Avocation: sculpture. Home: Fruithoflaan 116-13, 2600 Berchem, Antwerp Belgium Office: R U C A, Groenenborgerlaan 171, 2020 Antwerp Belgium

KUYKENDALL, TERRY ALLEN, environmental and chemical engineer; b. Atlanta, Sept. 11, 1953; s. Gen Lee and Elsie Lee (Morris) K. BS, Ga. State U., 1977; MS, Columbia Pacific U., 1991. Cert. environ. profl.; registered environ. assessor. Plant chemist Ga. Power Co., Taylorsville, 1977-79; radiochemist Environ. Sci. and Engring. Inc., Gainesville, Fla., 1979-80; dir. radiochem. scis. PBS&J Environ. Lab., Orlando, Fla., 1980-81; nuclear chemist Fla. Power & Light Corp., Miami, 1981-84; exec. cons. engr. NUS Corp., Inc., Aiken, S.C., 1984-88; mgr. S.C. office Peer Cons., P.C., Aiken, 1988-90; mgr. systems engring. Parsons Environ. Svcs., Inc., Charlotte, N.C., 1990—; mem. ind. rev. and adv. panel Battelle-Pacific N.W. Lab. Analytical Support Svcs., Richland, Wash., 1992—; mem. ind. rev. panel U.S. Dept. Energy-Savannah River, Aiken, 1986. Mem. N.C. Assn. Environ. Profls. (treas. 1991-92, chmn. waste com.), Nat. Assn. Environ. Profls., Am. Chem. Soc., Health Physics Soc., ASTM. Achievements include devel. of programmatic environ. appraisal program and safety analysis report preparation guide and devel. of DDE ES&H engring. programs. Home: 8759 W Cornell Ave #22-5 Lakewood CO 80227 Office: Parsons Environ Svcs Inc 4701 Hedgemore Dr Charlotte NC 28209

KUYPER, PAUL, physicist; b. Lochem, Gelderland, The Netherlands, May 11, 1932; s. Abraham Kuyper and Rachel de Wilde; m. Johanna M. Loen, Dec. 13, 1962 (div. Mar. 1973); children Saskia, Laura N., Jessica S. M. Doctorandus, State Oldest U., Leiden, Holland, The Netherlands, 1963; Doctor, U. Amsterdam, Holland, 1969. Cert. clin. psychologist, audiology scientist. Docent of physics City of Leiden, 1959-62, City of Haarlem, Holland, 1963-64; asst. prof. U. Amsterdam, 1964—; docent of physics Sch. for Logo Paedicts, Amsterdam, 1975-87; sr. audiologist Communical Audiol Ctr., Amsterdam, 1975—; scientific advisor Sch. for Hard Of Hearing and Specially Impared Children, Amsterdam, 1977-85. Author: Hearing With Two Ears, 1969, Possibilities Modern Hearing Aids, 1975. Mem. Internat. Soc. Audiology, Acoustic Soc. Am., Netherlands Soc. Audiology, Netherlands Soc. Ear Nose Throat Doctors, Dutch Acoustical Soc. Home: 29 Dotter, Abcoude 1391 SK, The Netherlands Office: U Amsterdam Audiol C, AMC-D2 Meibeydreef 9, Amsterdam 1105 AZ, The Netherlands

KUZMANOVIĆ, BOGDAN OGNJAN, structural engineer; b. Beograd, Serbia, Yugoslavia, July 16, 1914; came to U.S., 1965; s. Ognjan and Savka (Praporčetovic) K.; m. Galina V. Movčanjuk, Dec. 19, 1952; 1 child, Natalija Natasha. BS and MS in Civil Engring., U. Beograd, Yugoslavia, 1937; DSc in Tech., Serbian Sci. Acad., Beograd, Yugoslavia, 1956. Registered profl. engr., Kans., Fla. Design engr. Yugoslav Railways, Beograd, 1938-41, sr. design engr., 1945-52; prof. U. Sarajevo, Yugoslavia, 1952-58; dean, head dep.t U. Khartoum, Sudan, 1959-65; prof. U. Kans., Lawrence, 1965-81; v.p., dir. Beiswenger, Hoch & Assocs., Inc., North Miami Beach, Fla., 1982—; cons. Howard, Needles, Tammen & Bergendoff, Kansas City, Mo., 1970-74. Author: Steel Design for Structural Engineers, 1977, 2d edit., 1983; contbr. 39 articles to ASCE Structural Jour., Periodica, Internat. Assn. for Bridge and Structural Engring., Jour. Computers and Structures, Engring. Fracture Mechanics. 2d lt. Yugoslav Army C.E., 1941-45. Fellow ASCE (pres. Kans. chpt. 1978); mem. Am. Assn. Civil Engrs. (life), Internat. Assn. Bridge and Structural Engring. (sr.). Orthodox. Home: 1528 Wiley St Hollywood FL 33020

KUZMIC, PETR, chemist; b. Ostrava, Czechoslovakia, Mar. 10, 1955; came to U.S. 1987; s. Petr and Anna (Javurkova) K.; m. Eva Porjes, June 30, 1979 (div. 1991); children: Anezka, Michal. MS, Tech. Inst. Prague, 1979; PhD, Acad. of Scis., Prague, 1984. Rsch. scientist Sch. Pharmacy U. Wis., Madison, Wis., 1987—; staff scientist Czechoslovakia Acad. Scis., Praque, 1984-87; pres. Biokin Ltd., Madison, 1991—; vis. scientist Nat. Acad. Scis., Washington, 1987. Contbr. articles to profl. jours. Am. Cancer Soc. instl. rsch. grantee, 1991; named Disting. Young Scientist Czechoslovakian Acad. Scis., 1985. Mem. AAAS, Am. Chem. Soc. Unitarian Universalist. Office: Univ of Wis 425 N Charter St Madison WI 53706

KVALSETH, TARALD ODDVAR, mechanical engineer, educator; b. Brunkeberg, Telemark, Norway, Nov. 7, 1938; married; 3 children. B.S., U. Durham, King's Coll., Eng., 1963; M.S., U. Calif.-Berkeley, 1966, Ph.D., 1971. Research asst. engring. expt. sta. U. Colo., Boulder, 1963-64, teaching asst. dept. mech. engring., %; mech. engr. Williams & Lane Inc., Berkeley, 1964-65; research asst. dept. indsl. engring. and ops. research U. Calif.-Berkeley, 1965-71, research fellow, 1973; asst. prof. Sch. Indsl. and Systems Engring. Ga. Inst. Tech., Atlanta, 1971-74; sr. lectr. indsl. mgmt. div. Norwegian Inst. Tech. U. Trondheim, 1974-79, head indsl. mgmt. div., 1975-79; assoc. prof. dept. mech. engring. U. Minn., Mpls., 1979-82, prof., 1982—; guest worker NASA Ames Research Ctr., Calif., 1973; mem. organizing com. 1st Berkeley-Monterey Conf. Timespan, Pay and Discretionary Capacity, 1973; mem. steering com. Internat. Conf. Human Factors in Design and Op. Ships, Gothenburg, Sweden, 1977; mem. bd. Norwegian Ergonomics Com., 1977-80; gen. session chmn. Conf. Work Place Design and Work Environ. Problems, Trondheim, 1978. Author book chpts., articles, presentations, reports in field; editor text books; mem. editorial bd., reviewer for numerous profl. jours. Mem. Am. Inst. Indsl. Engrs. (sr.), IEEE, Human Factors Soc. (pres. upper Midwest chpt.), Nordic Ergonomics Soc. (council 1977-80), Internat. Ergonomics Assn. (gen. council 1977-80, v.p. 1982-85), Ergonomics Soc., Psychonomic Soc., Sigma Xi. Lutheran. Club: Campus (U. Minn.). Home: 108 Turnpike Rd Minneapolis MN 55416-1149 Office: U Minn Dept Mech Engring Minneapolis MN 55455

KVAMME, JOHN PEDER, electric technology company executive; b. Scanlon, Minn., July 1, 1918; s. John Conrad and Nellie (Larson) K.; m. Vera Joyce Fox, July 28, 1943; children: John, David, Karene, Jonell. BBA, U. Minn., 1943. Sr. acct. U. Minn., Mpls., 1946-49; pres. Glabman and Assocs., Muscada, Wis., 1949-52; comptroller and treas. Kato Engring. Co., Mankato, Minn., 1953-80, pres., 1974-80; pres. Minn. Elec. Tech., Mankato, 1980—. Mem. Sch. Dist. Ind. Dist. #77, Mankato. Lt. USNR, 1943-46, PTO. Office: Minn Elec Tech 1507 1st Ave Mankato MN 56001

KVARAN, AGUST, chemistry educator, research scientist; b. Husavik, Thingey, Iceland, Aug. 19, 1952; s. Axel K. and Hanna Hofsdal; m. Edda Sigfrid Jonasdottir, Aug. 19, 1975; children: Melkorka Arny, Bergthora. BS in Chemistry, U. Iceland, Reykjavik, 1975; PhD in Phys. Chemistry, U. Edinburgh, Scotland, 1979. Postdoctoral U. Pitts., 1978-79; rsch. scientist Sci. Inst., U. Iceland, Reykjavik, 1979-86; sr. rsch. scientist U. Iceland, Reykjavik, 1986-91, prof. phys. chemistry, 1991—; part-time lectr. chemistry dept. U. Iceland, 1979-91, adj. prof., 1982-91, vice-chmn., 1983-91, chmn., 1991—; vis. assoc. prof. chemistry dept. U. Utah, Salt Lake City, 1988; adj. rep. sic. faculty U. Iceland, 1984-86, union vice rep. univ. coun., 1988—. Contbr. articles to profl. jours. Mem. bd. Physicists Against Nuclear Armament, Reykjavik, 1985-87. SERC fellow, 1982, Fulbright fellow, 1988; Icelandic NSF grantee, 1980-91, NATO grantee, 1983-86; Brit. Coun. scholar, 1977-78. Fellow Acad. Sci. Iceland; mem. Am. Photochem. Soc., Am. Chem. Soc., Am. Physics Soc., European Photochem. Assn., Icelandic Physics Soc. (mem. bd. 1985-87), Royal Chem. Soc. Britain, N.Y. Acad. Scis. Lutheran. Avocations: photography, jogging, badminton, reading. Home: Hadarstigur 2, 101 Reykjavik Iceland Office: U Iceland Sci Inst, Dunhaga 3, 107 Reykjavik Iceland

KVINT, VLADIMIR LEV, mining engineer, economist, educator; b. Krasnoyarsk, Siberia, Russia, Feb. 21, 1949; s. Lev V. Kvint and Lidia E. Fridland Adamskia; m. Natalia Darialova, Apr. 30, 1981; children: Liza, Valeria. MS in Mining Engring., Inst. Non-Ferrous Metals, Krasnoyarsk, 1972; PhD in Econs., Inst. Nat. Economy, Moscow, 1975; D of Econs., Inst. Economy, Acad. of Scis., Moscow, 1988, life-title: Prof. Pol. Economy, 1989.

Diplomate in mining engring. Asst. prof. Inst. of Non-Ferrous Metals, 1972; chief of dept. non-ferrous metals co., Norilsk, Russia, 1975-76; dep. chair Automation of non-ferrous metals com., 1976-78; chief dept. sci.-tech. progress Siberian br. Acad. of Scis., Novosibirsk, 1978-82; leading rschr. Inst. of Economy, Acad. of Scis., Moscow, 1982-89; prof. Grad. Sch. Fordham U., N.Y.C., 1990—; cons. Gen. Electric, N.Y.C., 1989—; vis. prof. Vienna (Austria) Econ. U., 1989-90; Disting. lectr. Babson Coll., Babson Park, Mass., 1991; sr. cons. Arthur Andersen & Co., N.Y.C., 1992—. Author: The Introduction and Use of Automation Systems, 1981, The Krasnoyarsk Experiment, 1982, Management of Scientific-Technical Progress, 1986, The Scientific-Technical Information, 1987, The Barefoot Shoemaker: Capitalizing on the New Russia, 1993; contbr. articles to Forbes, Instl. Investor, others. Bd. dirs. USSR Exporters Assn., Moscow, 1988-90; mem. internat. com. Muhlenberg Coll., Allentown, Pa. Recipient Silver medal for achievements in nat. economy USSR Main Nat. Com., Moscow, 1986. Mem. N.Y. Acad Scis., Philos. Soc., Am. Econ. Assn., AAUP, Russian Acad. Natural Scis. (life). Jewish. Achievements include development of theory of regionalization of scientific technical progress; evaluation of role of scientific-technical policy in development of regional economy; development of regional programs in science and technology. Home: 404 E 66 St Apt 4E New York NY 10021 Office: Fordham U 113 W 60 St 6th Fl New York NY 10023

KWAK, NOWHAN, physics educator; b. Seoul, Korea, Sept. 16, 1928; came to U.S., 1954; m. Changsook Oh, May 14, 1958; children: Larry, Sarah. BS, Seoul Nat. U., 1952; PhD, Tufts U., 1962. Asst. prof. physics U. Kans., Lawrence, 1965-68, assoc. prof. physics, 1969-77, prof. physics, 1978—; sr. physicist Austrian Acad. Scis., Vienna, 1973-74; vis. physicist CERN, Geneva, 1974-76; vis. fellow Deutsches Electonen Synchrotron, Hamburg, Fed. Republic Germany, 1980-81. Mem. Am. Phys. Soc., Assn. Korean Physicists in Am. Home: 2105 Inverness Dr Lawrence KS 66047 Office: U Kans Dept Physics Lawrence KS 66045

KWAN, HENRY KING-HONG, pharmaceutical chemist; b. Hong Kong, Feb. 19, 1950; came to U.S., 1967; s. Cho-Yiu and Wai-Fun (Chow) K.; m. Katherine Kit-Ling Ip, Jan. 18, 1973; children: Herbert, Raymond. BSc. in Chemistry, Marquette U., 1971; PhD in Pharm. Chemistry, U. Mich., 1977. Sr. scientist Schering-Plough Rsch., Kenilworth, N.J., 1978-80, prin. scientist, 1980-82, sr. prin. scientist, 1982-87, mgr., 1987-90, assoc. dir. pharm. product devel., 1991-92; sr. assoc. dir. Schering-Plough Rsch. Inst., 1992—; mem. pharm. subcom. "Bilateral Forum on Biotech.-Derived Products," U.S.-Japan Bus. Coun., Washington, 1988-90. Contbr. articles to profl. jours. State league rep. and registrar Summit (N.J.) Soccer Club, 1984-88. Mem. AAAS, N.Y. Acad. Scis., Am. Assn. Pharm. Scientists, Parenteral Drug Assn. (chmn. awards com. 1989—), Phi Lambda Upsilon, Rho Chi. Achievements include patents on stabilization of interferon compositions for injectable and nasal administration (U.S., European).

KWITEROVICH, PETER OSCAR, JR., medical science educator, researcher, physician; b. Danville, Pa., June 24, 1940; s. Peter O. Sr. and Mary E. (Marks) K.; m. Kathleen Ann Justin, Aug. 14, 1965; children: Kris Ann, Peter III, Karen Ann. AB, Holy Cross Coll., Worcester, Mass., 1962; B in Med. Sci., Dartmouth Coll., 1964; MD, Johns Hopkins U., 1966. Intern Boston Children's Meml. Hosp., 1966-67; staff assoc. NIH, Bethesda, Md., 1967-70; resident Johns Hopkins Hosp., Balt., 1970-72; from asst. prof. to assoc. prof. in med. sci. Johns Hopkins U., Balt., 1972-84, prof. in med. sci., 1984—; bd. dirs. various clinics Johns Hopkins Hosp., 1971—. Author: Beyond Cholesterol, 1989 (Blakeslee award 1991); contbr. articles to profl. jours. Platt rep., Roland Park Civic League, Balt., 1986-88, v.p., 1988-89, pres., 1989-91. Surgeon USPHS, 1967-70. Fellow Coun. Arteriosclerosis; mem. Soc. Pediatrics Rsch., Am. Soc. Clin. Investigation. Republican. Roman Catholic. Avocations: boating, running, fishing, wildlife, reading. Office: Johns Hopkins Hosp 600 N Wolfe St Baltimore MD 21205

KWOK, LANCE STEPHEN, optometry educator, vision researcher; b. Moree, Australia, Jan. 22, 1954; came to U.S., 1987; s. Layton and Sunny (Ng) K. BEE, U. New South Wales, Sydney, Australia, 1976, B in Optometry, 1981, PhD, 1986. Scientist Eye Rsch. Inst., Sydney, Australia, 1986-87; postdoctoral rsch. assoc. La. State U. Eye Ctr., New Orleans, 1987-90; asst. prof. U. Houston, Coll. of Optometry, 1990-; reveiwer Optometry & Vision Sci., Washington, 1987—, Investigative Ophthalmology and Visual Sci., Balt., 1992—, Exptl. Eye Rsch., N.Y., 1992—; mem. tech. program com. Ophthalmic and Visual Optics meeting Optical Soc. Am., 1993; mem. tech. program subcom. Vision Sci. and Its Applications meeting Ophthalmic and Visual Optics, 1994. Author: (book chpt.) A Textbook of Contact Lens Practice, 1992; contbr. articles to Jour. of Theoretical Biology, Proceedings of the Royal Society London, Biophys. Jour. Recipient Internat. Soc. for Eye Rsch. travel award, San Francisco, 1988, Internat. Union for Pure Appl. Biophysics travel fellowship, Vancouver, Canada, 1990, Biophysical Soc. award, Bethesda, Md., 1990, Contact Lens Assn. of Ophthalmologists Fison travel award fellowship, New Orleans, 1991, Gordon Rsch. Conf. travel award, 1990. Fellow Am. Acad. Optometry; mem. IEEE, Biophys. Soc., Assn. for Rsch. in Vision and Ophthalmology, Internat. Soc. for Eye Rsch. Roman Catholic. Office: Univ of Houston Coll of Optometry Houston TX 77204-6052

KWOK, RAYMOND HUNG FAI, economics educator; b. Hong Kong, Dec. 23, 1963; s. Yuk Cheung and Wai Mui (Lok) K.; m. Siu Ha Anna Chung, Nov. 26, 1989; 1 child, Deborah Shun Ling. BA in Econs., York U., Can., 1985, BA in Econs. and Bus. with honors, 1986, MA in Econs., 1987; postgrad., City U. London. Grad. asst., teaching asst. York U., 1986-87; mktg. officer Sun Hung Kai Commodities Ltd., Hong Kong, 1987-88, lectr. City Poly. of Hong Kong, 1988—; lectr. U. East Asia, Macao, 1990-91, Henley Distance Learning, Eng., 1991; rschr. econ. social commn. Asia and Pacific, Bangkok, 1991. Mem. Am. Econ. Assn., Ctr. for Pacific Basin Monetary and Econ. Studies, Fed. Reserve Bank of San Francisco, Hong Kong Inst. Econ. Sci., Acad. Internat. Bus. Avocations: squash, tennis, basketball, bowling, table tennis. Home: Block 36 20/F Unit G, Laguna City, Kwan Tong, Hong Kong Hong Kong Office: City Polytechnic Hong Kong, 83 Tat Chee Ave, Hong Kong Hong Kong

KWOK, RUSSELL CHI-YAN, retail company executive; b. Hong Kong, Mar. 9, 1935; s. Lansing and Corinne (Jue) K.; m. Linda Ling-Hui Cheng, Dec. 18, 1965; children: Linnet, Kurt. BS in Pharmacy, U. San Francisco, 1959, PhD, 1963. Sr. organic chemist Eli Lilly & Co., Indpls., 1964-69; rsch. chemist Cutter Labs, Berkeley, Calif., 1969-72; asst. mgr. The Wing on Co. Ltd., Hong Kong, 1972-73, mgr., 1973-82, exec. dir., 1982-84, gen. mgr., 1984-89; gen. mgr. The Wing on Dept. Stores (HK) Ltd., Hong Kong, 1989-92, mng. dir., 1993—. Contbr. articles to profl. jours. Mem. Pharmacy and Poison Appeal Tribunal, Hong Kong, 1984-92; supr. Lingnan Dr CWK Meml. Secondary Sch., Hong Kong, 1984-88, mgr., 1985—; acting vice chmn. coll. coun. Lingnan Coll., Hong Kong, 1991-92. Mem. AAAS, Am. Chem. Soc., N.Y. Acad. of Sci., Pharm. Soc. of Hong Kong. Achievements include patents in field. Home: 15B Jardine's Lookout, 7 Boyce Rd, Hong Kong Hong Kong Office: Wing on Dept Stores HK, 7/F Wing on Centre, 211 Des Voeux Rd C, Hong Kong Hong Kong

KWOK, SUN, astronomer; b. Hong Kong, Sept. 15, 1949; arrived in Canada, 1967.; s. Chuen-Poon and Pui-Ling (Chan) K.; m. Shiu-Tseng Emily Yu, June 16, 1973; children: Roberta Wing-Yue, Kelly Wing-Hang. BSc, McMaster U., Hamilton, Ont., Can., 1970; MSc, U. Minn., Mpls., 1972, PhD, 1974. Postdoctoral fellow U. British Columbia, Vancouver, Can., 1974-76; asst. prof. U. Minn., Duluth, 1976-77; rsch. assoc. Centre for Rsch. in Exptl. Space Sci., Toronto, Ont., Can., 1977-78, Herzberg Inst. of Astrophysics, Ottawa, Ont., Can., 1978-83; asst. prof. U. Calgary, Calgary, Alta., Can., 1983-85, assoc. prof., 1985-88, prof., 1988--; vis. fellow Joint Inst. Lab. Astrophysics, Boulder, Colo., 1989-90; project specialist Internat. Adv. Panel, World Bank, 1984; mem. grant selection com. Natural Sci. and Engring. Rsch. Coun., Ottawa, Can., 1985-88; mem. Nat. Facilities Bd. Nat. Rsch. Coun., Ottawa, Can., 1986-89. Editor (books) Late Stages of Stellar Evolution, 1987, Astronomical Infrared Spectroscopy, 1993, Future Observational Directions, 1993; contbr. over 130 articles to scholarly and profl. jours. Natural Sci. and Engring. Rsch. Coun. grantee, 1984--, NASA grantee, 1990; Nat. Inst. for Standards and Tech. vis. fellowship, 1989-90. Mem. Internat. Astron. Union, Can. Astron. Soc., Am. Astron. Soc. Achievements include formulation of interacting winds theory of planetary nebulae formation. Home: 139 Edgeland Rd NW, Calgary, AB Canada T3A 2Y3 Office: Univ of Calgary, Dept of Physics & Astronomy, Calgary, AB Canada T2N 1N4

KWON, BYOUNG SE, geneticist, educator; b. Songak, Dangjin, Korea, Dec. 17, 1947; came to U.S., 1978; s. Won S. and Hyung Gap (Lee) K.; m. Myung Hee Han, Feb. 24, 1973; children: David Hyungjoong, Edwin Eujoong, Patrick Myungjoong. PhD in Cell and Molecular Biology, Med. Coll. Ga., Augusta, 1981; postgrad., Yale U., 1981-83. Assoc. scientist dept. human genetics Yale U., New Haven, 1983-84; scientist Guthrie Rsch. Inst., Sayre, Pa., 1984-88, Walther Oncology Ctr., Indpls., 1988—; assoc. prof. dept. microbiology and immunology Sch. Medicine Ind. U., Indpls., 1988-93, prof., 1993—. Contbr. articles to Jour. Exptl. Medicine, Jour. Immunology, Procs. NAS. Capt. Korean Army, 1974-77. Grantee NIH, March of Dimes Birth Defects Found., Am. Heart Assn. Mem. AAAS, AAI, Nat. Orgn. Albinism (bd. dirs.), Gene Cloning and Expression (bd. dirs.), Am. Soc. Microbiology, Sigma Xi. Achievements include invention of DNA clone for human tyrosinase and silvergene, genetic basis of human albinism, new treatment modalities for cancer and autoimmune diseases, DNA clones for new lymphokines and receptors, and isolation of immune-enhancing tumor suppressing agents from blood cells. Office: Ind U Sch of Medicine 635 Barnhill Dr # 255 Indianapolis IN 46202-5126

KWON, GLENN S., pharmaceutical scientist; b. Republic of Korea, July 8, 1963; came to U.S., 1967; s. Young Jin and Susanne (Kim) K.; m. Grace H. Oh, June 3, 1989; 1 child, Aaron. BS in Chemistry cum laude, U. Utah, 1985, PhD in Pharms., 1991. Rsch. assoc. Ctr. for Controlled Chem. Delivery, Salt Lake City, 1986—; postdoctoral fellow Tokyo Women's Med. Coll., 1991—; presenter in field. Contbr. articles to Jour. Controlled Release, Jour. Colloid Interface Sci., Internat. Jour. Pharmaceutics, others. AMAX scholar U. Utah, 1983; Osco/Skaggs Pharms. fellow U. Utah, 1987, NSF/JSPS Postdoctoral fellow Japan Soc. for the Promotion of Sci., 1991-93. Mem. A.A.P.S., Am. Chem. Soc., Japan Polymer Soc. Achievements include research on polymeric and colloidal drug delivery systems, on surface and interfacial phenomena with emphasis on biomedical applications, on protein and polypeptide delivery. Office: Sci U Tokyo ICBS Inst Biosc, Yamazaki 2669, Noda 278, Japan

KWON, JOON TAEK, chemistry researcher; b. Kimpo, Kyunggi Do, Republic of Korea, Mar. 10, 1935; came to U.S., 1955; s. Young Tae and Byoung Soon (Kim) K.; m. Moon Ja You, Aug. 15, 1964; children: Howard Albert, Daphne Elsa. BS in Chemistry, U. Ill., 1957; MS in Chemistry, Cornell U., 1959, PhD in Chemistry, 1962; postdoctoral fellow, U. B.C., Vancouver, Can., 1962-64. Instr. II dept. chemistry U. B.C., 1964-65; assoc. rsch. chemist Chemcell Ltd., Edmonton, Alta., Can., 1965-66; rsch. chemist Celanese Corp., Summit, N.J., 1967-70; sr. rsch. chemist Lummus Co., Bloomfield, N.J., 1970-78; prin. rsch. chemist Lummus Crest Inc., Bloomfield, N.J., 1978—; lead tech. catalysis specialist Combustion Engring., Stamford, Conn., 1985—. Co-author: Handbook of Chemical Production Process, 1986; mem. mgmnt. advs. panel Chemical Week, 1980—; contbr. over 13 articles to profl. jours. Mem. Silver Beaver lodge Monmouth council 347 Boy Scouts Am., vigil mem. Order of the Arrow. Indsl. matching grantee Nat. Rsch. Coun., Ottawa, Can., 1966-67. Fellow Am. Inst. Chemists; mem. Royal Soc. Chemistry, Soc. Chem. Industry (N.Am. sect.), N.Am. Thermal Analysis Soc., Am. Chem. Soc., Korean Chem. Soc. (life, rec. sec. N.Am. 1975-91), Korean Scientists and Engrs. in Am. (pres. N.J. chpt. 1976-77), Catalysis Soc. Met. N.Y.C., U. Ill. Alumni Assn. (life), Cornell U. Alumni Assn. Methodist. Achievements include patent for prodn. process for propylene oxide and 12 other patents in field of organometallic chemistry and process rsch. Home: 142 Derby Dr Freehold NJ 07728-2767 Office: ABB Lummus Crest Inc Dept 379 1515 Broad St Bloomfield NJ 07003-3099

KWONG, JAMES KIN-PING, geological engineer; b. Kowloon, Hong Kong, Sept. 12, 1954; came to U.S., 1985; s. Joseph and Mary (Sung) K.; m. Annie May-Ching Loh, June 7, 1980; 1 child, John Richard. BSc with honors, U. London, Eng., 1977; MSc, U. Leeds, Eng., 1978, PhD, 1985. Registered profl. civil engr., Hawaii; chartered engr., Eng. Engring. geologist Maunsell Cons., Hong Kong, 1978-80, Palmer & Turner Geotechnics, Hong Kong, 1980-82; cons., rsch. assoc. U. Guelph, Ont., Can., 1982-85; project mgr. Geolabs Hawaii, Honolulu, 1985-87, Dames & Moore, Honolulu, 1987-91; v.p. Pacific Geotech. Engrs., Inc., Honolulu, 1991—; guest lectr. dept. civil engring. U. Hawaii, Honolulu, 1990—. Contbr. articles to profl. jours. Recipient J.F. Kirkaldy's prize U. London, 1977. Mem. ASCE (exec. bd., pres.-elect Hawaii sect. 1989—), NSPE, Cons. Engrs. Coun. Hawaii (bd. dirs.), Can. Geotech. Soc., Instn. Mining and Metallurgy. Avocations: fencing, hiking, swimming, golfing. Office: Pacific Geotech Engrs Inc # 101 1030 Kohou St Honolulu HI 96817-4434

KYBETT, BRIAN DAVID, chemist; b. Oxford, U.K., May 10, 1938; arrived in Can., 1965; s. Henry and Gwenllian (Williams) K.; m. Gaynor Margaret Davies, Aug. 31, 1963; children: Gareth, Spencer. BSc, UCW, 1960; PhD, U. Wales, 1963. PDF Rice U., Houston, 1963-65; prof. U. Regina, Sask., Can., 1965—; dir. Energy Rsch. Unit, Regina, 1987—. Author over 200 papers and reports on chemistry, fossil fuel and renewable energy sources. Office: Univ of Regina, Energy Research Unit, Regina, SK Canada S4S 0A2

KYIN, SAW WILLIAM, chemist, consultant; b. Rangoon, Burma, Aug. 6, 1954; came to U.S., 1981; s. U. Shin Nga and Daw (Swa) Khin; m. Myint Oo, Jan. 30, 1975; children: Tim, Moe. BS, Rangoon Arts and Sci. U., 1977; MS, Western Ill. U., 1984. Dir. biotech. ctr. Genetic Engring. Facility, U. Ill., Urbana Champaign, 1981 92, dir. molecular biology dept. Synthesis/Sequencing Facility, Princeton, N.J., 1992—. Mem. AAAS, Am. Chem. Soc., Am. Peptide Soc., Assn. Biomolecular Resource Facilities, Protein Soc. Office: Princeton Univ Dept Molecular Biology Washington Rd Princeton NJ 08544

KYLE, BENJAMIN GAYLE, chemical engineer, educator; b. Atlanta, Dec. 4, 1927; s. John Curtis and Kathryn (Greer) K.; m. Patricia Lee Brugler, Dec. 6, 1952; children: Mary Kathryn, Shirley Jean, John Hobart, Benjamin Gayle Jr. BS in Chem. Engring., Ga. Inst. Tech., 1950; MS, U. Fla., 1955, PhD, 1958. Engr. Monsanto Chem. Co., Dayton, Ohio, 1950-52; from asst. to assoc. prof. chem. engring. Kansas State U., Manhattan, 1958-64; prof. chem. engring. Kans. State U., Manhattan, 1964—. Author: Chemical and Process Thermodynamics, 1984, 2d edit., 1992. With U.S. Army, 1946-47. Mem. AICE, AAAS, Am. Chem. Soc., Am. Soc. Engring. Edn. Democrat. Presbyterian. Achievements include patent on low-energy process of producing gasoline-ethanol mixtures. Office: Kans State U Durland Hall Dept Chem Engring Manhattan KS 66506

KYLE, CHESTER RICHARD, mechanical engineer; b. L.A., Nov. 18, 1927; s. Chester Raymond and Lorena Dale (Olson) K.; m. Joyce Sylvia; children: Scott D., Kelley L., Cova-Lee, Chester W. BSME, U. Ariz., 1951; MS in Engring., UCLA, 1964, PhD, 1969. Registered profl. engr., Calif. Prodn. engr. Shell Oil Co., Long Beach, Calif., 1951-57; prodn. supt. Internat. Petroleum Co., Talara, Peru, 1957-59; prof., mech. engr. Calif. State U., Long Beach, 1959-84; dir. Sports Equipment Rsch. Assocs., Weed, Calif., 1989—; pub., editor Cycling Sci., Mt. Shasta, Calif., 1989—; mem. sports equipment and tech. com. U.S. Olympic Com., Colorado Springs, Colo., 1984-88; cons. on solar cars U.S. Dept. Energy, 1993. Sci. editor Bicycling, 1987-91; mem. editorial bd. Internat. Jour. Sports Biomechanics, 1988-92; contbr. articles to Sci. Am., Smithsonian, other profl. jours.; contbg. author/author books and publs. in field. Named Faculty fellow in sci. NSF, UCLA, 1967, Rsch. fellow U.S. Olympic Com., Long Beach, 1982-88, Fulbright prof., Lima, Peru, 1964-65; recipient Paul Dudley White award League Am. Wheelman, 1993. Fellow Explorers Club (N.Y.); mem. Internat. Human Powered Vehicle Assn. (founder, bd. dirs. 1975—), L.A. Adventurer's Club (pres. 1986). Achievements include design of bicycles for medal winning U.S. Olympic team, clothing for U.S. track team, parts for World Solar Challenge winner Sunrayer, of world record setting U.S. streamlined bicycle, of Paul MacCready's Kramer Prize winning Gossamer Condor; research in human powered vehicles. Home and Office: 9539 N Old Stage Rd Weed CA 96094-9714

KYLE, ROBERT ARTHUR, medical educator, oncologist; b. Bottineau, N.D., Mar. 17, 1928; s. Arthur Nichol and Mabel Caroline (Crandall) K.; m. Charlene Mae Showalter, Sept. 11, 1954; children: John, Mary, Barbara, Jean. AA, N.D Sch. Forestry, 1946; BS, U. N.D., 1948; MD, Northwestern U., 1952; MS, U. Minn., 1958. Diplomate Am. Bd. Internal Medicine; subsplty. Hematology. Fellow Mayo Grad. Sch., Rochester, Minn., 1953-59; clin. asst. Tufts U. Sch. Medicine, Boston, 1960-61; cons. internal medicine Mayo Clinic, Rochester, 1961—; prof. medicine and lab. medicine Mayo Med. Sch., Rochester, 1975—; pres. med. subjects unit Am. Topical Assn., Johnstown, Pa., 1976-81; chmn. standards, ethics and peer rev. orgn. Cancer & Acute Leukemia Group B, Scarsdale, N.Y., 1978-82; Robert A. Hettig lectr. in hematology Baylor U. Coll. of Medicine, Houston, 1984; Waldenström lectr., Stockholm, 1988; Redlich Meml. lectr Cedars-Sinai Med. Ctr., U. Calif., L.A. Author: The Monoclonal Gammopathies, 1976, Medicine and Stamps, Vols. 1 and 2, 1980; author/editor: Neoplastic Diseases of the Blood, vols. 1 and 2, 1985, 2d edit., 1990. Chmn. bd. trustees First Presbyn. Ch., Rochester, Minn., 1967; chmn. Rochester Med. Ctr. Ministry, 1979-86. Capt. USAF, 1955-57. Named Disting. Topical Philatelist, Am. Topical Soc., 1982; recipient Waldenström award Internat. Workshop for Myeloma, Italy, 1991. Fellow ACP; mem. N.Y. Acad. Scis., Am. Soc. Hematology, Internat. Soc. Hematology (sec. gen. Inter-Am. div. 1990), Am. Assn. Cancer Rsch., Phi Beta Kappa. Republican. Avocation: philately. Home: 1207 6th St SW Rochester MN 55902-1918 Office: Mayo Clinic & Hosps 200 1st St SW Rochester MN 55905-0001

KYRIACOU, DEMETRIOS, electro-organic chemist, researcher, consultant; b. Spilaion, Greece, Oct. 16, 1922; came to U.S., 1946; s. Kyriacos and Eleni (Manos) K.; m. Eleni Sarantitis, July 19, 1953; children: Damon, Tula. BS in Chemistry, Northeastern U., 1953. Chemist Abbott Labs., North Chicago, Ill., 1953-56; rsch. chemist Pacific Vegetable Oil Corp., Richmond, Calif., 1956-58, Calif. Rsch. Corp., Richmond, 1961-64; chemist A Aerojet-Gen. Corp., Sacramento, 1958-61; rsch. specialist Dow Chem. Co., Pittsburg, Calif., 1964-81; vis. specialist scientist U. Thessaloniki, Greece, 1981-84; adj. prof. of chemistry U. Lowell, Mass., 1985—. Author: Basics of Electroorganic Synthesis, 1981, Electrocatalysis for Organic Synthesis, 1986; contbr. articles to Am. Chem. Soc. and internat. jours. Cpl. U.S. Army, 1946-48. Mem. Am. Chem. Soc., N.Y. Acad. Scis. Achievements include 13 patents (U.S. and foreign); invention of chemical and electroorganic processes including one for manufacture of Lontrel herbicide; research in electrosynthesis, biomass electrochemistry, bioelectrochemistry. Home: 81 4th Ave Lowell MA 01854-2729

LAALY, HESHMAT OLLAH, research chemist, roofing consultant, author; b. Kermanshah, Iran, June 23, 1927; came to Germany, 1951, Can., 1967, U.S., 1984; s. Jacob and Saltanat (Afshani) L.; m. Parvaneh Modarai, Oct. 7, 1963; (div. 1971); children: Ramesh, Edmond S.; m. Parivash M. Farahmand, Feb. 7, 1982. BS in Chemistry, U. Stuttgart, Germany, 1955; MS in Chemistry, U. Stuttgart, Republic of Germany, 1958, PhD in Chemistry, 1962. Chief chemist Kress Sohne, Krefeld, Germany, 1963-67; analytical chemist Gulf Oil Research Ctr., Montreal, Que., Can., 1967-70; material scientist Bell-Northern Research, Ottawa, Ont., Can., 1970-71; research officer NRC of Can., Ottawa, 1972-84; pres. Roofing Materials Sci. and Tech., L.A., 1984—; Patentee in field. Author: The Science and Technology of Traditional and Modern Roofing Systems, 1992; patentee bifunctional photovoltaic single-ply roofing membrane. Mem. AAAS (Can. chpt.), ASTM, Inst. Roofing and Waterproofing Cons., Single-Ply Roofing Inst., Assn. Profl. Engrs. Ontario, Am. Chem. Soc., Internat. Union of Testing and Rsch. Labs. for Material and Structures (tech. com. 75), Constrn. Specifications Inst., Nat. Roofing Contractors Assn., UN Indsl. Devels. Orgn., Internat. Conf. Bldg. Ofcls., Roofing Cons. Inst., Can. Standard Assn., Can. Gen. Standards Bd. Office: Roofing Materials Sci & Tech 9037 Monte Mar Dr Los Angeles CA 90035-4235

LAANANEN, DAVID HORTON, mechanical engineer, educator; b. Winchester, Mass., Nov. 11, 1942; s. Joseph and Helen Katherine (Horton) L.; m. Mary Ellen Storck, Sept. 9, 1967 (div. 1981); children: Gregg David, Robin Kaye; m. Delores Ann Talbert, May 21, 1988. BS in Mech. Engring., Worcester Poly. Inst., 1964; MS, Northeastern U., 1965, PhD, 1968. Project engr. Dynamic Sci., Phoenix, 1972-74; asst. prof. Pa. State U., State College, 1974-78; mgr. R&D Simula Inc., Phoenix, 1978-83; assoc. prof. Ariz. State U., Tempe, 1983—, dir. aerospace rsch. ctr., 1992—. Referee: Jour. Aircraft, Jour. Mech. Design; contbr. articles to Jour. Aircraft, Jour. Am. Helicopter Soc., Jour. Safety Rsch., Jour. Thermoplastic Composite Materials, Composites Sci. and Tech. Fellow AIAA (assoc., mem. structures tech. com., design engring. tech. com.); mem. ASME, Am. Helicopter Soc., Sigma Gamma Tau, Sigma Xi, Pi Tau Sigma. Democrat. Achievements include research in aircraft crash survivability, composite structures. Office: Ariz State U Dept Mech Aerospace Engring Tempe AZ 85287-6106

LAANO, ARCHIE BIENVENIDO MAAÑO, cardiologist; b. Tayabas, Quezon, Philippines, Aug. 10, 1939; naturalized U.S. citizen; s. Francisco M. and Iluminada (Maaño) L.; m. Maria Esmeralda Eleazar, May 2, 1964; 1 child, Sylvia Marie. A.A., U. Philippines, 1958, B.S., 1959, M.D., 1963. Diplomate Am. Bd. Internal Medicine. Rotating intern Hosp. St. Raphael, New Haven, 1963-64; resident internal medicine, 1964-65; rotating resident pulmonary diseases Laurel Heights Hosp., Shelton, Conn., 1965; affiliated rotating resident Yale-New Haven Med. Ctr., 1965; resident internal medicine Westchester County Med. Ctr., Valhalla, N.Y., 1965-66, resident cardiology, 1966-67; resident fellow cardiology Maimonides Med. Ctr., Bklyn., 1967-68; rotating sr. resident cardiology Coney Island Hosp., Bklyn., 1967-68; fellow internal medicine Mercy Hosp., Rockville Centre, N.Y., 1968-70; med. dir. 54 Main St. Med. Ctr., Hempstead, N.Y., 1971-76, Bloomingdale's, Garden City, N.Y., 1972—, Esselte Pendaflex Corp., Garden City, 1976—; attending staff Nassau County (N.Y.) Med. Ctr., Hempstead Gen. Hosp.; practice medicine specializing in cardiology, internal medicine, Nassau County, 1971—; chief med. svcs., chief profl. svcs. U.S. Army 808th Sta. Hosp., Hempstead, N.Y., 1979—; brig. gen. 1st U.S. Army AMEDD Augmentation Detachment, Ft. Meade, Md., 1989—, M.C., chief of staff, chief profl. svcs. U.S. Army Meddac Hosp., Ft. Dix, N.J., 1990—; med. dir. Cities Svc. Oil Co. (CITGO), L.I. div., 1972— ; mem. adv. bd. Guardian Bank, Hempstead, chmn. adv. coun., 1973-89; clin. prof. medicine SUNY at Stony Brook, 1979—; professorial lectr. medicine (cardiology) U.S. Mil. Acad.-Keller Army Med. Ctr., West Point, N.Y., 1979; affiliated teaching hosp. Harvard Med. Sch, 1979; vis. prof. Harvard U., 1979—; cons. physician ICC, Citgo, Liberty Mut. Ins. Co. Boston, 1972—, U.S. Dept. Transp.; counsel White House Commn. on Mil. Medicine, 1988—. Decorated Silver Star, Bronze Star, Legion of Merit, Soldiers medal, Joint Svc. Command medal, Army Meritorious Svc. medal, Dept. Def. Joint Svc. Achievement award, Southwest Asia Svc. award-Desert Storm, others. Fellow Internat. Coll. Angiology, Am. Coll. Angiology, Am. Coll. Internat. Physicians, Internat. Coll. Applied Nutrition, Am. Soc. Contemporary Medicine and Surgery, Acad. Preventive Medicine, Internat. Acad. Med. Preventives, Philippine Coll. Physicians, Am. Coll. Acupuncture, N.Y. Acad. of Sci.; mem. AMA, Am. Coll. Cardiology, N.Y. Med. Soc., Nassau County Med. Soc., Am. Heart Assn., N.Y. Cardiol. Soc., World Med. Assn., Royal Soc. Medicine (overseas, London), Nassau Acad. Medicine, Am., N.Y. State, Nassau Soc. Internal Medicine, N.Y. Soc. Acupuncture for Physicians, Am. Geriatrics Soc., Nassau Physicians Guild, Res. Officers Assn. U.S., Assn. Mil. Surgeons, Assn. Philippine Physicians Am. (bd. govs. rep. N.Y. State 1984-86, v.p. 1988-89, chmn. com. nominations and election 1987-88. spl. counsel to pres. 1986-87), Philippine Med. Assn. Am. (spl. counsel 1988-89, bd. dirs, 1989-90, spl. counsel to pres, 1989-90, dir. continuing med. edn. 1990—, chmn. scholarship com. 1989—), Assn. Philippine Physicians of N.Y. (founding v.p., pres. 1985-87, pres. emeritus 1988—, chmn. com. on constitution and by-laws, nominations and election, med. coord. Internat. Games for Disabled Olympics 1984), Soc. Philippine Surgeons Am. (Medallion of Honor 1991), U.S. Knights of Rizal, U. Philippines Med. Alumni Soc. (pres. class of 1963, 1981—), U. Philippines Med. Alumni Soc. Am. (chmn. bd. 1985—), Royal Soc. Medicine Club, The Oxford Club, Rolls Royce Club L.I., N.Y. Club (chmn. 1987—), West Point OfficersClub, Garden City Country Club, Phi Kappa Mu (overseas coord. U. Philippines 1985—), Beta Sigma (coun. adivsers Ea. U.S. 1990—)program chmn. 1975—, chmn. bd. 1978—, pres. 1978-79), Lions (Garden City program chmn. 1975—, chmn. bd. 1978—, pres. 1978-79). Home: 80 Stratford Ave Garden City NY 11530-2531 Office: 200 Hilton Ave Ste 106 Plaza 230 Profl Condo Hemps

LABAN, SAAD LOTFY, medicinal chemist; b. Cairo, Egypt, Jan. 13, 1960; came to U.S., 1983; s. Lofty and LouLou (Wanis) L.; m. Mona Nosshy Beshay Saleib, Aug. 22, 1987; 1 child, MacCary. BS in Pharmacy, Cairo U.,

1981; cert. radiopharmacist, U. N.Mex., 1984; MS in Medicianl Chemistry, U. Iowa, 1985, PhD in Medicinal Chemistry, 1988. Lic. pharmacist, Egypt, Ill. Pharmacist Lofty Pharmacy, Cairo, 1981-83, cons., 1983—; resident in radiopharmacy Coll. Pharmacy U. N.Mex., Albuquerque, 1983-84; rsch. asst. Coll. Pharmacy U. Iowa, Iowa City, 1984-88, postdoctoral asst., cons. Coll. Medicine, 1988-90; scientist III Sandoz Crop Protection, Des Plaines, Ill., 1990; sr. scientist Sandoz Agro, Inc., Des Plaines, Ill., 1991—. Mem. AAAS, Am. Chem. Soc., Am. Soc. Pharmacognosy, Quantum Chemistry Program Exch. Achievements include research in spin labeling of spiperone for MRI potential use, synthesis of topical carbonic anhydrase inhibitors, synthesis of tropical ocular inflammatories, conformational analysis of ganglioside oligosaccharides; invention of matrix derivatization of soil and corn for residue analysis. Office: Sandoz Agro Inc 1300 E Touhy Ave Des Plaines IL 60018

LABATH, OCTAVE AARON, mechanical engineer; b. Milw., Sept. 22, 1941; s. Octave Adrain and Bertha Jane (Johnson) LaB.; m. Carole Marion Clay, Jan. 23, 1965; children—Melissa, Michelle, Mark. B.S. in Mech. Engring., U. Cin., 1964, M.S., 1969. Registered profl. engr., Ohio. v.p. engring. Cin. Gear Co., 1972—. Contbr. articles to profl. jours. Mem. Am. Gear Mfg. Assn., ASME. Methodist. Men's Club. Home: 5105 Kenridge Dr Cincinnati OH 45242-4831 Office: Cin Gear 5657 Wooster Pike Cincinnati OH 45227-4199

LABBE, ARMAND JOSEPH, museum curator, anthropologist; b. Lawrence, Mass., June 13, 1944; s. Armand Henri and Gertrude Marie (Martineau) L.; m. Denise Marie Scott, Jan. 17, 1969 (div. 1972). BA in Anthropology, Univ. Mass., 1969; MA in Anthropology, Calif. State U., 1986; lifetime instr. credential in anthropology, State Calif. Curator collections Bowers Mus., Santa Ana, Calif., 1978-79, curator anthropology, 1979-86, chief curator, 1986-91, dir. rsch. and collections, 1991—; tchr. Santa Ana Coll., 1981-86, Calif. State U., Fullerton, 1982, 83, 88, U. Calif., Irvine, 1983, 87, 91, 93; trustee Balboa Arts Conservation Ctr., San Diego, 1989—, Ams. Found., Greenfield, Mass., 1985—, Quintcentenary Festival Discovery, Orange County, Calif., 1990-91, Mingei Internat. Mus., La Jolla, Calif., 1993—; mem. adv. bd. Élan Internat., Newport Beach, Calif., 1992—; inaugural guest lectr. Friends of Ethnic Art, San Francisco, 1988. Author: Man and Cosmos, 1982, Ban Chiang, 1985, Colombia Before Columbus, 1986 (1st prize 1987), Leigh Wiener: Portraits, 1987, Colombia Antes de Colón, 1988 (honored at Gold Mus. Bogotá, Colombia, 1988), Images of Power: Master Works of the Bowers Museum of Cultural Art, 1992; co-author Tribute to The Gods: Treasures of the Museo del Oro, Bogotá, 1992. Cons. Orange County Coun. on History and Art, Santa Ana, 1981-85; mem. Task Force on County Cultural Resources, Orange County, 1979; cons., interviewer TV prodn. The Human Journey, Fullerton, 1986-89. With USAF, 1963-67. Recipient cert. of Recognition Orange County Bd. Suprs., 1982, award for outstanding scholarship Colombian Community, 1987; honored for authorship Friends of Libr., 1987, 88. Fellow Am. Anthrop. Assn.; mem. AAAS, Am. Assn. Mus., N.Y. Acad. Scis., S.W. Anthrop. Assn. Avocations: photography, travel. Office: Bowers Mus 2002 N Main St Santa Ana CA 92706-2731

L'ABBE, GERRIT KAREL, chemist; b. Oostende, Belgium, Dec. 13, 1940; s. Bertram L'abbé and Marie-Rose Coenegrachts; m. Christianne Platteau, Aug. 17, 1967; children: Annick, Caroline. Lic. in Chemistry, U. Leuven, Belgium, 1963; Dr. in Sci., U. Leuven, 1966. Postdoctoral fellow A. von Humboldt, Erlangen, Fed. Republic Germany, 1967-68; Fulbright fellow Boulder, Colo., 1969-70; docent U. Leuven, 1972-75, prof. chemistry, 1975-77, prof. ordinarius, 1977—; rsch. dir. U. Leuven, 1972—. Contbr. 200 articles to profl. jours. Alexander von Humboldt fellow, 1967-68, Fulbright Rsch. fellow, 1969-70, Nat. Fund for Scientific Rsch. fellow, 1963-72; recipient Belgian BP-prize 1959, P. Bruylants prize Chemici Lovanienses, 1966, J.S. Stas medal Belgian Royal Acad., Brussels, 1966, Biennial prize of the Jour. Industrie Chimique Belge, Brussels, 1967-68, Laureate Belgian Royal Acad. Scis., 1972, Breckpot prize Flemish Chem. Soc., 1983. Fellow The British Royal Soc. Chemistry (chartered chemist); mem. AAAS, Am. Chem. Soc., Internat. Soc. Heterocyclic Chem., Belgian Royal Acad. Sci., Royal Flemish Chem. Soc. Home: Merellaan 33, 3210 Linden Belgium Office: U Leuven, Dept Chemistry, Celestijnenlaan 200F, 3001 Heverlee Belgium

LABEN, DOROTHY LOBB, volunteer nutrition educator, consultant; b. Yonkers, N.Y., Mar. 7, 1914; d. John David and Edna Lyall (Klein) Lobb; m. Robert C. Laben, Nov. 29, 1946; children: John Victor, Robert James, Elizaeth Jean Cunningham, Catherine Lyda. AB, Wellesley Coll., 1935; MS, Conn. Coll. for Women, 1937. Cert. med. technologist Am. Soc. Clin. Pathologists. Med. technologist Carl H. Wies M.D., New London, Conn., 1937-40; rsch. chemist Hixon lab. U. Kans. Med. Sch., Kans. City, 1940-41; med. technologist chemistry Blodgett Meml. Hosp., Grand Rapids, Mich., 1941-43; grad. researcher chemistry Cornell U. (Nutrition), Ithaca, N.Y., 1943-45; researcher U.S. Plant, Soil and Nutrition Lab. Cornell U., Ithaca, 1944-45; researcher food composition Bur. Human Nutrition and Home Econs., USDA, Washington; lab. asst. and instr. Okla. A&M Chemistry Dept., Stillwater, 1947; cons. Yolo County Coalition Against Hunger, Davis, Calif., 1970—. Bd. dirs. Econ. Opportunity Commn. of Yolo County Calif., 1970-75. Recipient Brinley award City of Davis, 1985; named Woman of Yr. Calif. State Senate, Sen. Pat Johnston, Sacramento, 1991. Mem. LWV, UN Assn. (bd. mem. 1986-88). Presbyterian. Home: 502 Oak Ave Davis CA 95616

LABLANCHE, JEAN-MARC ANDRE, cardiologist, educator; b. Langres, Haute-Marne, France, June 4, 1946; s. Marcel Georges and Paulette (Renard) L.; m. Brigitte Laurence Lespagnol, Nov. 7, 1969; children: Christelle, Catherine. MD, U. Medicine, Lille, France, 1975. Intern, resident U. Hosp. U. Medicine, 1973-77, from asst. prof. cardiology to assoc. prof., 1977—; lectr. in field. Author books; contbr. articles to med. jours. Served with French Health Corps, 1972-73. Fellow Am. Coll. Cardiology, European Soc.; mem. French Soc. Cardiology. Roman Catholic. Office: Hopital Cardiologique, Lille, 59037 Nord France

LABRIE, TERESA KATHLEEN, research scientist; b. Albany, N.Y., Aug. 8, 1960; d. Edwin F. and Bridget (Griffin) LaB. BA, Coll. St. Rose, 1982; MS, Union Coll., 1993. Cert. lab. animal technologist. Technician Albany (N.Y.) Med. Ctr. Lab., 1982-85, 85—; biol. lab. technician VA Med. Ctr. Hosp., Albany, 1985-86; rsch. biol. Sterling Winthrop Rsch. Inst., Rensselaer, N.Y., 1986-88; rsch. scientist Sterling Winthrop Pharm. Rsch. Div., Rensselaer, 1989—. Contbr. articles to profl. jours. Owner Wysiwyg Great! Danes, East Nassau, N.Y., 1983—; mem. Sterling Winthrop Rescue Squad, Rensselaer, 1986—. Mem. AAAS, Soc. Toxicology, Internat. Soc. Analytical Cytology, Drug Info. Assn., Soc. Leukocyte Biology. Roman Catholic. Office: Sterling Winthrop Pharm Rsh 81 Columbia Turnpike Rensselaer NY 12144

LABRIOLA, JOSEPH ARTHUR, environmental biologist; b. Paterson, N.J., Jan. 25, 1949; s. Joseph and Sophie (Andrew) L. BS in Environ. Sci., Rutgers U., 1970; postgrad. in Environ. Edn., Glassboro State Coll., 1971-72; MS in Biology, W.Va. U., 1974. Cert. ecologist Ecol. Soc. Am., wetlands profl. N.J. Assn. Environ. Profls. Prin. biologist Edwards and Kelcey, Inc., Livingston, N.J., 1976-89; prin. environ. scientist Hayden/Wegman-McCumsey, Parsippany, N.J., 1990—; adj. faculty biology Bergen C.C., Paramus, N.J., 1975-79, 85-86, 90—, Fairleigh Dickinson U., Teaneck-Hackensack, N.J., 1993—; mem. adv. com. grad. program in environ. systems sci. Fairleigh Dickinson U., Teaneck, 1991—. Co-author (primary) Proceedings of W.Va. Acad. of Sci., 1974, Proceedings of Annual Hackensack River Symposium of Fairleigh Dickinson Univ. (1991), 1992; author other abstracts. Vol. Wildlife Conservation Corps of N.J. Divsn. Fish, Game and Wildlife, Pequest Natural Resource Edn. Ctr., Oxford, N.J., 1990—. Grad. Student Rsch. grantee Sigma Xi, 1974; recipient Cert. of Appreciation, N.J. Divsn. Fish, Game and Wildlife, 1991. Mem. N.J. Assn. Environ. Profls. (founding mem., bd. trustees), Soc. Wetland Scientists (edn. com. North Atlantic chpt. 1990—), Nature Conservancy, N.J. Acad. Sci., Ecol. Soc. Am., Am. Inst. Biol. Scis., Soil and Water Conservation Soc. Am., Torrey Bot. Club. Democrat. Roman Catholic. Home: 863 Allwood Rd Apt D-1 Clifton NJ 07012 Office: Hayden Wegman McCumsey 1055 Parsippany Blvd Ste 110 Parsippany NJ 07054

LABUZ, RONALD MATTHEW, design educator; b. Utica, N.Y., Nov. 17, 1953; s. Emil John and Elsie (Pritchard) L.; m. Carol Ann Altimonte, Sept. 5, 1975. BA, SUNY, Oswego, 1975; MA, Ohio State U., 1977; MPhil, Syracuse U., 1993. Acquisition dir. Collegiate Pub., Columbus, Ohio, 1977-78; pres. Advt., Pub. and Avatar Media Advt. Agy., Columbus, 1978-80; prodn. specialist Am. Ceramic Soc., Columbus, 1980-81; assoc. prof. advt. Mohawk Valley Community Coll., Utica, 1981-85, prof., dept. head advt. design, 1985—. Author: Contemporary Graphic Design, Typography and Typesetting, The Computer in Graphic Design, 5 other books; contbr. articles and revs. to profl. jours. Recipient Chancellor's award for excellence, SUNY, 1989, faculty exch. scholar, 1990. Mem. Printing History Soc., Am. Printing History Assn., Graphic Design Educators Assn. (bd. dirs. 1989—, treas. exec. bd. 1992—), Am. Ctr. for Design, Internat. Graphic Arts Educators Assn., Am. Inst. Graphic Arts, N.Y. State Assn. Two-Yr. Colls. (bd. dirs. 1990-91). Office: Mohawk Valley CC 1101 Sherman Dr Utica NY 13501-5308

LA CELLE, PAUL LOUIS, biophysics educator; b. Syracuse, N.Y., July 4, 1929; s. George Clarke and Marguerite Ellen (Waggoner) La C. A.B., Houghton Coll., 1951; M.D., U. Rochester, 1959. Resident U. Rochester Med. Center-Strong Meml. Hosp., 1960-62; asst. prof. medicine U. Rochester, 1967-70, asso. prof., 1970-74, prof., 1974—, chmn. dept. biophysics, 1977—; cons. to govt. Mem. Gates-Chili Sch. Bd., Rochester, 1964-72; trustee Houghton Coll., 1976—. Served to lt. USNR, 1952-55. NIH spl. fellow, 1965-66; recipient von Humboldt Sr. Scientist award, 1982-83. Mem. Biophys. Soc., Microcirculation Soc., European Microcirculation Soc., Alpha Omega Alpha. Achievements include research in biophysics of blood cells, physiology of microcirculation. Office: U Rochester Dept Biophysics 601 Elmwood Ave Rochester NY 14642-8408

LACERNA, LEOCADIO VALDERRAMA, research physician; b. Lucena, Quezon, The Philippines, Nov. 16, 1958; came to U.S., 1984; s. Leocadio Saragoza and Margarita (Valderrama) L.; m. Wendy Hernandez, June 28, 1984; 1 child, Jason Anthony. BS, U. Santo Thomas, Manila, 1978, MD, 1982. Intern Jose Reyes Meml. Hosp. and Med. Ctr., Manila, 1982-83; resident Peter Paul Hosp., Manila, 1983-85; fellow clin. rsch. br. div. cancer treatment Nat. Cancer Inst., NIH, Bethesda, Md., 1986-88, clin. trials monitor cancer treatment evaluation program, 1987-89; sr. scientist sci. info. svcs. dept. Hoffmann-La Roche, Inc., Nutley, N.J., 1989-92; assoc. mgr. profl. svcs. Amgen, Inc., Thousand Oaks, Calif., 1992—. Contbr. articles to profl. publs. Mem. Am. Assn. Cancer Rsch., Am. Fedn. Clin. Rsch., N.Y. Acad. Scis., Drug Info. Assn., NIH Alumni Assn.

LACHEL, DENNIS JOHN, geologist, engineer, consultant; b. Chgo., Sept. 28, 1939; s. Raymond J. and Evelyn (Gore) L. BA, Monmouth Coll., 1961; MS, U. Calif., 1974. Cert. engring. geologist. Mgmt. trainee Humble Oil & Refinery Co., Pelham, N.Y., 1961-64; geologist engr. Alaska Dist. U.S. Army Corps Engrs., Anchorage, 1964-70; geologist engr. Mo. River Divsn. U.S. Army Corps Engrs., Omaha, 1970-75; mgr. Underground Divsn. Woodward Clyde Cons., Clifton, N.Y., 1975-76; exec. v.p. LACHEL HANSEN & Assocs., Inc., South Orange, N.J., 1976-80, Golden, Colo., 1980-85; pres. LACHEL & Assocs., Inc., Golden, 1985—; Nuclear Waste Rev. Bds., U.S. Dept. Energy, Washington, 1980-84; industry rep. U.S. Nat. Commn. on Tunneling Tech., 1978-82, vice chmn. 1982-83, chmn. 1983-84; U.S. rep. Internat. Tunneling Soc., 1982-86. Editor: Alaska Mining Journal, 1969; contbr. articles to profl. jours.; lectr. on geotechnics and underground structures at over 20 universities and colleges. Explorer advisor Boy Scouts Am.. Tenafly, N.J., 1961-64; coun. advisor Boy Scouts Am., Anchorage. Recipient James O'Daniel award Monmouth Coll., 1989. Mem. ASCE. Achievements include direct involvement in more than 200 tunneling projects worldwide. Office: LACHEL & Assocs Inc PO Box 5266 Golden CO 80401

LACKEY, JAMES FRANKLIN, JR., civil engineer; b. Tatum, N.Mex., Jan. 3, 1932; s. James Franklin and Florence Estelle (Morris) L.; m. Sue Linda Shockley Lackey, Sept. 22, 1953; children: James Franklin III, Dale Robert, John Patrick, Patricia Marie. BS in Civil Engring., Okla. State U., 1962. Registered Profl. Engr. Calif., Okla., Tex., N.Mex., Ariz., Nebr., Wash., Utah, Idaho, Colo. Engr. Morrison-Knudsen Sedalia, Mo., 1962-63; asst. civil engr. Internat. Engring. Co., Inc., San Francisco, 1963-64, assoc. civil engr., 1964-67, sr. civil engr., 1969-73, prin. engr., 1973-74, chief civil engr., 1974-77, chief engr., 1978-82, design mgr., 1983-86; mgr. spl. projects Morrison-Knudson Engrs., San Francisco, 1986-90; mgr. projects, infrastructure Morrison-Knudson Engrs., Dallas, 1991—. Capt. U.S. Army, 1952-59. Recipient Commendation medal with oak leaf, U.S. Army; named Kavanaugh Community Bldg. award scholar, 1961. Mem. ASCE, Nat. Soc. Profl. Engrs., Tex. Soc. Profl. Engrs. Home: 1316 Mosslake Dr DeSoto TX 75115 Office: PB/MK Team 5510 S Westmoreland Rd Dallas TX 75237

LACKEY, MARY MICHELE, physician assistant; b. Johnson City, N.Y., Dec. 22, 1955; d. Joseph Charles and Jane Ann (Weston) Reardon; m. Donald V. Lackey Jr., Oct. 27, 1979. AAS in Nursing, Broome Community Coll., Binghamton, N.Y., 1978; cert. family nurse practitioner, Albany Med. Coll., 1982; BS in Psychology and Sociology, U. State of N.Y., Albany, 1989. Cert. physician asst., family nurse practitioner, nurse midwife; RN, N.Y., Conn. Physician asst. Streit, Hickey & Lasky M.C., P.C., Saratoga Springs, N.Y., 1982-85, Litchfield Hills Ob/Gyn., Sharon, Conn., 1986-89, Foothills Family Health Ctr., Amenia, N.Y., 1991—; physican asst. Vassar Coll. Health Svcs., Poughkeepise, N.Y., 1990—. Leader, instr. Girl Scouts U.S.A., Dutchess County, N.Y., 1990—. Lt. col. U.S. Army, 1975—. Fellow Am. Acad. Physician Assts., Am. Coll. Nurse Midwives; mem. Nat. Guard Assn. U.S., Malitia Assn. N.Y., Phi Theta Kappa. Roman Catholic. Avocations: breeding exhbn. poultry and geese, gourmet cooking, collect early med. books and equipment. Home: RR 1 Box 222 Salt Point NY 12578-9733

LACKNER, JAMES ROBERT, aerospace medicine educator; b. Virginia, Minn., Nov. 11, 1940; s. William and Lillian Mae (Galbraith) L.; m. Ann Martin Graybiel, Aug. 26, 1970. BSc, MIT, 1966, PhD, 1970. Asst. prof. psychology Brandeis U., Waltham, Mass., 1970-74, assoc. prof. psychology, 1974-79, Riklis prof. physiology dept. psychology, 1977—, chmn. dept. psychology, 1975-83, provost, dean faculty, 1986-89, dir. Ashton Graybiel Spatial Orientation Lab., 1982—; research assoc. dept. psychology and clin. research ctr. MIT, Cambridge, 1970-80; sci. adv. bd. Space Biomed. Research Inst., Houston, 1982—, Aphasia Research Ctr. Boston U. Sch. Med., 1977-82, Eunice Kennedy Shriver Ctr. Harvard U. Med. Sch., Cambridge, 1980—; sci. adv. panel astronaut longitudinal health program Johnson Space Ctr., NASA, 1983, exec. sec. space adaptation syndrome steering com., 1982-84, pre-adaption trainer working group, 1986—, artificial gravity working group, 1987—; fabricant com. life scis. experiments for a space sta., 1982; space scis. bd. sensory motor panel NAS, 1984-86; com. on hearing, bioacoustics and biomechanics NRC, 1985-89, com. on vision, 1987-92, com. on space, biology and medicine, 1991—. Contbr. over 150 articles to sci. jours. Mem. Am. Soc. for Gravitational and Space Biology, Aerospace Med. Assn. (Arnold B. Tuttle award), Soc. for Neurosci., Psychonomics Soc., Internat. Brain Research Orgn., Barany Soc. (hon.), Internat. Acad. Astronautics (hon.). Rsch. in human sensory-motor coordination and spatial orientation. Home: Boyce Farm Rd Lincoln MA 01773-4813 Office: Brandeis Univ Dept of Psychology POB 9110 Waltham MA 02154-2700

LACKOFF, MARTIN ROBERT, engineer, physical scientist, researcher; b. Queens, N.Y., Oct. 6, 1946; s. Samuel K. and Esther (Soifer) L.; m. Mara Lee Feinstein, Oct. 29, 1989; 1 child, Blythe Ann. BSEE, U. R.I., 1969, MS in Ocean Engring., 1971, PhD in Ocean Engring., 1974. With tech. staff Stein Assocs., Inc., Waltham, Mass., 1974; tech. staff to dir. of tech. Naval Underwater Systems Ctr., New London, Conn., 1975-77, tech. staff advanced systems tech., 1977-80; staff to mgr. Computing Systems div. Analogic Corp., Wakefield, Mass., 1980-83; staff to mgr. signal processing dept. Equipment div. Raytheon Co., Wayland, Mass., 1983-85; lead engr. The Mitre Corp., Bedford, Mass., 1985—. Mem. IEEE, Am. Geophys. Union, Am. Meterol. Soc., Tau Beta Pi, Phi Kappa Phi. Office: The Mitre Corp Burlington Rd Bedford MA 01730-1306

LACOMBE, RONALD DEAN, mechanical engineer; b. Kansas City, Mo., July 28, 1962; s. Marvin L.D. and Twila Kay (Allen) LaC.; m. Joan Marie Sweaney, Sept. 20, 1986; children: Jacob Dean, Veronica Marie, Scott James. BSME, U. Mo., 1985. Registered profl. engr., Mo. Mich., Pa., Calif., Ohio. Engr. AT&T Tech. Systems, Lee's Summit, Mo., 1985-86; survey engr. Viron Corp., North Kansas City, Mo., 1986-87; project engr. Viron Corp., North Kansas City, 1987-91; project mgr. Viron Corp., Riverside, Mo., 1991—. Mem. bldgs. and ground com. St. Regis Cath. Ch., Kansas City, 1992—. Mem. Assn. of Energy Engrs. Home: 622 Village Dr Lee's Summit MO 64063

LACY, CLAUD H. SANDBERG, astronomer; b. Shawnee, Okla., June 5, 1947; s. Lester Claud and Leola Chrstine (Hinton) L.; m. Patricia Kathryn McCoy, Apr. 1, 1971 (div. 1984); m. Patricia Alison Sandberg, Dec. 19, 1988; children: Adrian R., Kathryn Mia Rose. MS in Physics, U. Okla., Norman, 1971; PhD in Astronomy, U. Tex., 1978. Vis. asst. prof. Tex. A&M U., College Station, Tex., 1978-80; asst. prof. astronomy U. Ark., Fayetteville, 1980-86, assoc. prof. astronomy, 1986—. Author: Astronomy Laboratory Exercises, 1981; contbr. articles to Astron. Jour. With U.S. Army, 1971-73. NSF grantee, 1981-84. Mem. Am. Astron. Soc., Internat. Astron. Union. Achievements include determination of accurate absolute properties for stars in 11 eclipsing binary star systems; discovery of 40 new double-lined eclipsing binaries. Office: U Ark Dept Physics Fayetteville AR 72701

LACY, MARK EDWARD, computational biologist, systems scientist; b. Cookeville, Tenn., Oct. 2, 1955; s. Edward Arnold and Rita Sharon (Webb) L.; m. Valerie Ann Sedor, June 21, 1975 (div. Dec. 1990); 1 child, Kendra Lynn; m. Barbara May Hickling, June 27, 1992. BS in Chemistry, Fla. Atlantic U., 1976; MS in Biochemistry, Med. U. S.C., 1979, PhD in Biometry, 1982. Biometrician Procter and Gamble Pharms., Norwich, N.Y., 1982-85, group leader, 1985-86, sect. head, 1986-93, prin. scientist, 1993—. Guest editor Tetrahedron Computer Methodology, 1990. Deacon First Bapt. Ch., Norwich, 1989-92. Fla. Atlantic U. faculty scholar, 1973, State of S.C. grad. trainee, 1976. Mem. AAAS, Am. Chem. Soc. (symposium chmn. 1989), Mu Sigma Rho. Democrat. Office: Procter & Gamble Pharms Miami Valley Labs PO Box 398707 Cincinnati OH 45239-8707

LADA, ELIZABETH A., astronomer. With Harvard Smithsonian Ctr., Cambridge, Mass. Recipient Annie Jump Cannon award Am. Astonomical Soc., 1992. Office: Harvard Smithsonian Ct 60 Garden St MS 42 Cambridge MA 02138*

LADD, CHARLES CUSHING, III, civil engineering educator; b. Bklyn., Nov. 23, 1932; s. Charles Cushing and Elizabeth (Swan) L.; m. Carol Lee Ballou, June 11, 1954; children: Melissa, Charles IV, Ruth, Matthew. A.B., Bowdoin Coll., 1955; S.B., MIT, 1955, S.M., 1957, Sc.D., 1961. Asst. prof. civil engring. MIT, 1961-64, assoc. prof., 1964-70, prof., 1970—, dir. Ctr. Sci. Excellence in Offshore Engring., 1983—; gen. reporter 9th Internat. Conf. Soil Mechanics and Found. Engring., Tokyo, 1977; co-gen. reporter 11th Internat. Conf. Soil Mechanics and Found. Engring., San Francisco, 1985; mem. geotech. bd. NRC, 1992—. Contbr. articles to profl. jours. Mem. Concord (Mass.) Republican Town Com., 1968-82; commr. Concord Dept. Pub. Works, 1965-78, chmn., 1972-74. Recipient Civil Engring. Effective Teaching award MIT, 1980. Fellow ASCE (rsch. prize 1969, Croes medal 1973, Norman medal 1976, Terzaghi lectr. 1986); mem. NAE, ASTM (Hogentogler award 1990), NSPE, Boston Soc. Civil Engrs. (bd. govt. 1972-81, pres. 1977-78), Transp. Rsch. Bd., Internat. Soc. Soil Mechanics and Found. Engring., Am. Soc. Engring. Edn., Assn. of Engring. Firms Practicing in the Geosciences. Home: 7 Thornton Ln Concord MA 01742-4107 Office: MIT Dept Civil & Environ Engrng Cambridge MA 02139

LADEROUTE, CHARLES DAVID, engineer, economist, consultant; b. Helena, Mo., Aug. 2, 1948; s. Estel and Anna Maude (Stuart) L.; m. Linda Dodd, June 8, 1985; 1 child, Lindsay; 1 stepchild, Erik. BS in Engring. Mgmt., U. Mo., Rolla, 1971, BS in Econs., 1972; MA in Econs., Ea. Mich. U., 1980; postgrad., Harvard U., 1979-81. Sr. rate analyst Consumers Power Co., Jackson, Mich., 1972-79; prin. cons. Chas. T. Main, Inc., Boston, 1979-81; pres., chief exec. officer Charles D. Laderoute, Ltd., Topsfield, Mass., 1981—, also chmn. bd. dirs.; mem. supplemental faculty Jackson Community Coll., 1974-78; course dir. Ctr. for Profl. Advancement, East Brunswick, N.J., 1981—; ptnr. Knowledge Applications Software, LP, Acton, Mass., 1986—; lectr. in field. Contbr. articles to profl. jours. Mem. Nat. Rep. Senatorial Com., 1990-91, Rep. Nat. Com., 1990—; charter mem. Rep. Campaign Coun., 1991. Mem. Am. Econs. Assn., Am. Meteorol. Soc., Nat. Assn. Bus. Economists, Nat. Soc. Rate of Return Analysts, Internat. Assn. Energy Economists (charter, pres. N.E. chpt. 1984-86), Am. Soc. Engring. Mgmt. (charter, life), Planning Engrs. Desktop Computer Users Group (charter, pres. 1987-88), ABA (assoc.), Am. Gas. Assn. (assoc.), Can. Gas. Assn. (assoc.), Assn. Energy Engrs., Demand-Side Mgmt. Soc. (charter), Assn. Demand-Site Mgmt. Profls. (charter), Omicron Delta Epsilon. Republican. Achievements include creation of Relative System Utilization method used in utility industry for allocation of demand related costs; research in effect of weather and possible global warming on energy sales and peak demands. Office: Charles D Laderoute Ltd PO Box 328 Boxford MA 01921

LADINO, CYNTHIA ANNE, cell biologist; b. New Bedford, Mass., Nov. 8, 1962; m. Mark Stephen Saunders, June 13, 1992. BA, Wheaton Coll., Norton, Mass., 1984; PhD, U. Mass., Lowell, 1988. Postdoctoral fellow Worcester Found. for Exptl. Biology, Shrewsbury, Mass., 1989-92; rsch. scientist Immuno Gen, Inc., Cambridge, Mass., 1992—. Contbr. articles to sci. jours. Vol. pediatric unit U. Mass. Med. Ctr., Worcester, 1989-92. Grantee NIH, 1989-92. Mem. AAAS, Am. Soc. for Cell Biology, Am. Chem. Soc., N.Y. Acad. of Scis., Sigma Xi, Sigma Delta Epsilon. Achievements include research on reduction of adenine nucleotide content of clone 4 molck cells, protein carboxyl methylation reactions in density-fractionated human erythrocytes, regulation of protein carboxyl methylation in Hela cells during stress response. Office: ImmunoGen Inc 148 Sidney St Cambridge MA 02139

LADISCH, THOMAS PETER, chemical engineer; b. Yeadon, Pa., Dec. 31, 1953; s. Rolf Karl and Brigitte Maria (Gareis) L.; m. Sonia Roxanne Painter, May 31, 1980; children: Jennifer L., Nikolaus P. BS in Chem. Engring., U. Pa., 1976. Engr. ABEC, Bath, Pa., 1977-79, Valley Instrument Co., Exton, Pa., 1979-81; pres. Penna Polymeric, Inc., Gilbertsville, Pa., 1981-83; v.p. Associated Bioengineers, Bethlehem, Pa., 1983-85; pres. Thomas Ladisch Assoc., Alburtis, Pa., 1985--. Active United We Stand, Reading, Pa., 1992. Mem. Am. Inst. Chem. Engrs. Achievements include patent for sanitary valve design. Home: Box 558 RR 1 Alburtis PA 18011 Office: 500 County Line Rd Gilbertsville PA 19525

LADOUCEUR, HAROLD ABEL, mechanical engineer, consultant; b. Center Line, Mich., Aug. 2, 1927; s. Abel and Mary Elisabeth (Macklem) L.; m. Lucille Jugon, Aug. 24, 1950; 1 child, Marcia Ann Kolle. BSME, Lawrence Technol. U., 1963. Engr., sect. chief Vickers Inc., Detroit, 1950-66; engring. mgr. Multifastener Corp., Detroit, 1966-79, 82-92; dir. engring. Alpha Industries, Novi, Mich., 1979-82; v.p., cons. Halco Inc., Livonia, Mich., 1992—. Campaign worker Livonia mayoral candidate, 1987, 91. With U.S. Army, 1945-47, ETO. Named Eminent Engr., Tau Beta Pi, 1993. Mem. ASME, Engring. Soc. Detroit, Soc. Automotive Engrs., Am. Soc. Metals. Republican. Achievements include over 35 U.S. patents, additional foreign patents. Home: 30123 Bentley St Livonia MI 48154 Office: Halco Inc 30123 Bentley St Livonia MI 48154

LA DU, BERT NICHOLS, JR., physician, educator; b. Lansing, Mich., Nov. 13, 1920; s. Bert Nichols and Natalie (Kerr) La D.; m. Catherine Shilson, June 14, 1947; children: Elizabeth, Mary, Anne, Jane. B.S., Mich. State Coll., 1943; M.D., U. Mich., 1945; Ph.D. in Biochemistry, U. Calif., Berkeley, 1952. Intern Rochester (N.Y.) Gen. Hosp., 1945-46; research asso. N.Y.U. Research Service, Goldwater Meml. Hosp., N.Y.C., 1950-53; sr. asst. surgeon USPHS, Nat. Heart Inst., 1954-57; surgeon, later sr. surgeon, med. dir. Nat. Inst. Arthritis and Metabolic Disease, 1957-63; prof., chmn. dept. pharmacology N.Y.U. Med. Sch., 1963-74; prof. pharmacology U. Mich. Med. Sch., Ann Arbor, 1974-89, prof. emeritus, 1989—, chmn. dept., 1974-81. Contbr. articles to profl. jours. Served with AUS, 1943-45. Mem. AAAS, Am. Chem. Soc., N.Y. Acad. Sci. (pres.), Am. Soc. Biol. Chemistry, Am. Soc. Pharmacol. Therapeutics (pres.), Am. Soc. Human Genetics, Biochem. Soc. (Gt. Britain). Home: 817 Berkshire Rd Ann Arbor MI 48104-2630 Office: U Mich Med Sch 6322 Med Sci I Ann Arbor MI 48109-0626

LADUE, ROBIN ANNETTE, psychologist, educator; b. Seattle, Sept. 8, 1954; d. Charles Vernon and Thalia Kathleen (Collias) La.; m. Marshall Bruce Harmon, Nov. 24, 1984. MS, Wash. State U., 1980, PhD, 1982. Licensed clin. psychologist, Wash. Asst. prof. U. Mo., Kansas City, 1982-83; clin. psychologist VA, Honolulu, 1983-84, Muckleshoot Indian Tribe, Auburn, Wash., 1984-86; pvt. practice Renton, Wash., 1986—; postdoctoral fellow U. Wash., Seattle, 1986-89, clin. asst. prof., 1989—; cons. VA Adv. Com. on Am. Indian Vets., Washingotn, 1986-88, VA Working Group on Am. Indian Vietnam Vets., Washington, 1983-87. Author: Adolescents and Adults with Fetal Alcohol Syndrome, 1986, (with others) Perinatal Substance Abuse, 1992; contbr. chpt.: Coyote Returns: Survival in the 1990s, 1991; contbr. articles to Jour. AMA; assoc. editor: FOCUS Newsletter, 1990-92. Mem. Rsch. Soc. on Alcohol, Am. Psychol. Assn. (mem. adv. com., com. on ethnic minority affairs, minority fellowship adv. com.), Soc. Indian Psychologists. Achievements include research in fetal alcohol syndrome, American Indian mental health. Office: 1500 Benson Rd S # 201 Renton WA 98055

LAESSIG, RONALD HAROLD, pathology educator, state official; b. Marshfield, Wis., Apr. 4, 1940; s. Harold John and Ella Louise (Gumz) L.; m. Joan Margaret Spreda, Jan. 29, 1966; 1 child, Elizabeth Susan. B.S., U. Wis.-Stevens Point, 1962; Ph.D., U. Wis.-Madison, 1965. Jr. faculty Princeton (N.J.) U., 1966; chief clin. chemistry Wis. State Lab. Hygiene, Madison, 1966-80, dir., 1980—; asst. prof. preventive medicine U. Wis.-Madison, 1966-72, assoc. prof., 1972-76, prof., 1976—, prof. pathology, 1980—; cons. Ctr. Disease Control, Atlanta; dir. Nat. Com. for Clin. Lab. Standards, Villanova, Pa., 1977-80; chmn. invitro diagnostic products adv. com. FDA, 1974-75; mem. rev. com. Nat. Bur. Standards, 1983-86. Mem. editorial bd. Med. Electronics, 1970—, Analytical Chemistry, 1970-76, Health Lab. Sci., 1970—; contbr. articles to profl. jours. Mem. State of Wis. Tech. Com. Alcohol and Traffic Safety, 1970-88. Sloan Found. grantee, 1966; recipient numerous grants. Mem. Am. Assn. Clin. Chemistry (chmn. safety com. 1984-86, bd. dirs. 1986-89, Natelson award 1989, Contbns. Svc. to Profession award 1990), Am. Pub. Health Assn. (Difco award 1974), Am. Soc. for Med. Tech., Nat. Com. Clin. Lab. Standards (pres. 1980-82, bd. dirs. 1984-87), Sigma Xi. Avocation: Woodworking. Office: State Lab Hygiene 465 Henry Mall Madison WI 53706-1578

LA FARGE, TIMOTHY, plant geneticist; b. N.Y.C., Mar. 14, 1930; s. Louis Bancel and Hester Alida (Emmet) La F.; m. Anne Blackstone, Oct. 16, 1960 (div. Mar. 1964); m. Frances Madelyne Holst, Aug. 6, 1966 (dec. 1992); 1 child, Jason Emmet. BSc in Forestry, U. Maine, 1964; M Forestry, Yale U., 1965; PhD, Mich. State U., 1971. Forestry aid Forest Svc., Orono, Maine, 1962-64; lab. technician geology dept. Yale U., New Haven, 1965; rsch. forester USDA Forest Svc., Macon, Ga., 1965-69, plant geneticist Southeastern Sta., 1970-82; plant geneticist Nat. Forest System USDA Forest Svc., Atlanta, 1982—. Contbr. articles to profl. jours. Recipient Certs. of Merit, USDA Forest Svc., Atlanta, 1986, 88. Mem. AAAS, Soc. Am. Foresters. Democrat. Achievements include demonstration that backcrossing and hybridization between shortleaf pine and loblolly pine can effectively produce fast-growing back-cross hybrids that are resistant to fusiform rust; application of Best Linear Prediction to analysis of unbalanced or messy progeny test data. Home: 2679 Tritt Springs Terr Marietta GA 30062-5266 Office: USDA Forest Svc 1720 Peachtree Rd NW Atlanta GA 30367

LAFONTAINE, THOMAS E., chemical engineer; b. North Adams, Mass., Jan. 2, 1952; s. Omer Charles and Isabella Frances (Luczinski) LaF.; m. Catherine Hackett, July 9, 1981; children: Christina, Colima. Dir. tech. internat. Cameron-Yakima (Wash.), Inc., 1981-87; dir. engring. Agro Industries de Tecuman S.A., Tecuman, Mexico, 1987-88; pres., owner Teltech Engring., Yakima, 1988-91; tech. dir., v.p. Intercon Pacific, Inc., Portland, 1989—; cons. with Phillippine govt. to amend regulatory laws for pollution control. Contbr. articles to profl. jours. Mem. Am. Water Works Assn., Air Pollution Control Assn., Internat. Carbon Soc. Achievements include development of several specially treated activated carbons for specific use; designed installation and startup of activates carbon manufacturing plants in Philippines, China, Mexico and U.S. Office: Intercon Pacific Inc PO Box 2784 Yakima WA 98907

LAGALLY, MAX GUNTER, physics educator; b. Darmstadt, Germany, May 23, 1942; came to U.S., 1953, naturalized, 1960; s. Paul and Herta (Rudow) L.; m. Shelley Meserow, Feb. 15, 1969; children—Eric, Douglas, Karsten. B.S. in Physics, Pa. State U., 1963; M.S. in Physics, U. Wis.-Madison, 1965, Ph.D. in Physics, 1968. Registered profl. engr., Wis. Instr. physics U. Wis., Madison, 1970-71, asst. prof. materials sci., 1971-74, assoc. prof., 1974-77, prof. materials sci. and physics, 1977—, dir. thin-film deposition and applications ctr., 1982—, John Bascom Prof. materials sci., 1986—, E.W. Mueller Prof. materials sci. and physics, 1993—; Gordon Godfrey vis. prof. physics, U. New South Wales, Sydney, Australia, 1987; cons. in thin films, 1977—; vis. scientist Sandia Nat. Lab., Albuquerque, 1975. Editor: (with others) Methods of Experimental Physics, 1985; editor: Kinetics of Ordering and Growth at Surfaces, 1990; mem. editorial bd., also editor spl. issue Jour. Vacuum Sci. and Tech., 1978-81; prin. editor Jour. Materials Rsch., 1990—; contbr. articles to profl. jours.; patentee in field. Max Planck Gesellschaft fellow, 1968, Alfred P. Sloan Found. fellow, 1972, H.I. Romnes fellow, 1976; Humboldt Sr. Rsch. fellow, 1992; grantee fed. agys. and industry. Fellow Am. Phys. Soc., Australian Inst. Physics; mem. AAAS, Am. Vacuum Soc. (program, exec. coms. 1974—, M. W. Welch prize 1991), Am. Soc. Metals Internat., Am. Chem. Soc. (colloid and surface chem. div.), Materials Research Soc. Home: 5110 Juneau Rd Madison WI 53705-4744 Office: U Wis Materials Sci & Engring 1509 University Ave Madison WI 53706-1595

LAGANÁ, ANTONIO, chemical kinetics educator; b. Como, Italy, Nov. 4, 1944; s. Vincenzo and Giuseppa (Vinci) L.; m. Giovanna Lepri, June 23, 1945; 1 child, Leonardo. Laurea in Chemistry, U. Perugia (Italy), 1969. Tchr. physics and math. Istituto Magistrale A. Pieralli, Gubbio, Italy, 1971-75; lab. demonstrator U. Perugia (Italy), 1972-75; rsch. contractor U. Perugia, 1975-77; rsch. asst. U. Manchester (Eng.), 1977-78; lectr. chem. kinetics U. Perugia, 1979-85, prof., 1985—; collaborator Los Alamos (N.Mex.) Nat. Lab., 1987—, Centro Nazionale Universitario Calcolo Elettronico, Pisa, Italy, 1988—; NATO fellow Calif. Tech., Pasadena, 1982; vis. prof. U. Cambridge, 1985. Contbr. more than 120 articles to profl. jours. Cpl. Italian mil., 1970-71. Fellow Royal Soc. Chemistry; mem. Soc. Chimica Italiana, Assn. Italy Chimica Fisica, Assn. Italy Calcolo Automatico (rep. at Co-Operation in Sci. and Tech. Rsch. field (EC)). Avocation: economics. Office: Dept Chemistry U Perugia, Via Elce di Sotto 8, I-06123 Perugia Italy

LAGERLOF, RONALD STEPHEN, sound recording engineer; b. Long Beach, Calif., Mar. 24, 1956; s. Paul Richard and Harriett Jane Lagerlof; m. Janice Martin, June 7, 1989; 1 child, Jackson Martin. Chief rec. engr. Wishbone Rec., Muscle Shoals, Ala., 1978-79; tech. engr. TM Communications, Dallas, 1980-81; gen. mgr. Studio Svcs., Inc., Dallas, 1984-85; head tech. engr. Dallas Sound Lab., 1985-87; studio mgr. Motown/Hitsville Studios, West Hollywood, Calif., 1987-89; dir. tech. ops. Soundworks West, West Hollywood, 1989-90; ops. mgr. Skywalker Sound divsn. Lucasfilm, Marin County, Calif., 1990-92; owner Visioneering Design Co., Woodland Hills, Calif., 1992—; cons. digital audio-for-video installations Pacific Ocen Post and Record Plant studios. Engr. LP rec. Family Tradition, 1979 (gold record 1980); re-rec. engr. (TV spl.) Motown Merry Christmas, 1987 (Emmy nomination 1988); songwriter copyrighted and pub. works. Mem. Audio Engring. Soc., Soc. Motion Picture and TV Engrs. Avocations: sailing, skiing, scuba diving.

LAHIRI, DEBOMOY KUMAR, molecular neurobiologist, educator; b. Varanasi, Utta, Pradesh, Sept. 9, 1955; came to U.S. 1983.; s. Benoy Kumar and Nilima Rani (Moitra) L. MS, Benaras Hindu U., India, 1975, PhD, 1980. Rsch. fellow Benaras Hindu U., Varanasi, 1975-79; jr. scientist Indian Coun. of Agrl. Rsch., New Delhi, India, 1979-81; postdoctoral fellow McMaster U. Sch. Medicine, Hamilton, Ont., Can., 1982; asst. rsch. scientist NYU, N.Y.C., 1983-86; rsch. assoc. N.Y. State Inst. for Basic Rsch., Staten Island, N.Y., 1987; asst. prof. Mt. Sinai Sch. Medicine, N.Y.C., 1988-90; asst. prof., chief molecular neurogenetics lab. Inst. Psychiat. Rsch. Ind. U. Sch. Medicine, Indpls., 1990—. Contbr. articles to profl. jours, made presentations to confs. and sci. meetings. U.P. Govt. Merit scholar, 1970-75; Univ. Grants Commn. New Delhi jr. rsch. fellow, 1975-79; grantee NIH, 1991—. Mem. AAAS, Soc. for Neurosci., N.Y. Acad. Scis. Democrat. Hindu. Achievements include the molecular cloning and sequencing a cDNA for a major hnRNP (heterogenous nuclear ribonucleoprotein particle) core protein determination of the presence of beta amyloid precursor protein (APP) in different regions of human brain, and alternatively spliced APP transcripts in different tissues and various cell types; demonstration of a relationship between cholinergic agonists and the processing of APP; first characterization of the beta amyloid gene promoter; first to show the presence of enhancer like element in the beta amyloid gene promoter; research related to the origin and biogenesis of Alzheimer amyloid plaque and the general areas of gene regulation and genetics of Alzheimer's Disease, development of rapid, economical, non-enzymatic and non-organic method for DNA extraction, elucidation of the genetic basis of neuropsychiatric disorders by the linkage studies using molecular genetic methods and PCR (polymerase chain reaction) based genotyping, RFLP (restriction fragment length ploymorphism) and candidate gene studies in families ascertained through the NIMH Genetics Initiative in order to confirm association between the inheritance of an RFLP with the member of the family sharing the illness, development of a sensitive radioimmunoassay to measure melatonin in human plasma samples, bipolar patients have an increased sensitivity to the effects of light on the circadian rhythm of melatonin secretion, and the risk of mood disorder seems to be related to this hypersensitivity to light, the suppression of melatonin by light may be a trait marker for bipolar affective disorder. Home: 5731 Arabian Run Indianapolis IN 46208 Office: Inst Psychiat Rsch Ind Univ 791 Union Dr Indianapolis IN 46202-4887

LAHR, BRIAN SCOTT, pilot; b. Atlanta, Feb. 22, 1961; s. Robert Scott and Jane Anne (Towns) L.; m. Karen Ann Cook, Mar. 25, 1989; 1 child, Robert Scott. BS in Physics, North Ga. Coll., 1982; MS in Aerospace Engring., Auburn (Ala.) U., 1986. Cert. airline transport pilot-learjet type rating. Staff flight instr. Auburn U., 1984-86; flight instr. Epps Air Svc., Chamblee, Ga., 1986-87; line pilot Flight Internat. Inc., Newport News, Va., 1987-88; line pilot Kalitta Flying Svc., Morristown, Tenn., 1988-90, asst. chief pilot, 1990-92; flight mgr. Flight Internat. Inc., Naples, ITaly, 1992—. Mem. AIAA, Sigma Gamma Tau. Republican. Baptist. Office: PSC 817 Box 33 FPO AE 09622-0033

LAI, GEORGE YING-DEAN, metallurgist; b. Chutung, Taiwan, Republic of China, July 18, 1940; came to U.S., 1964; s. Jung-Lai and Chin-Mei (Chien) L.; m. Mei-Huei Tsai, Dec. 28, 1969; 1 child, Christine Wenly. MS in Metall. Engring., Va. Poly. Inst., 1968; PhD in Materials Engring., N.C. State U., 1972. Metallurgist Techalloy Co., Inc., Rahns, Pa., 1968-70; postdoctoral rsch. assoc. U. Calif., Berkeley, 1972-74; sr. engr. Gen. Atomic Co., San Diego, 1974-77, staff engr., 1977-80; sr. engring. assoc. Haynes Internat., Inc., Kokomo, Ind., 1980-83; mem. tech. staff haynes Internat., Inc., Kokomo, Ind., 1983-85, group leader, 1985—. Author: High Temperature Corrosion of Engineering Alloys, 1990; co-editor: Materials Performance in Waste Incineration Systems, 1992; contbr. numerous articles to profl. jours. Mem. Nat. Assn. Corrosion Engrs. (co-chmn. task group on materials problems in waste incinerator fireside and air pollution equipment 1985—, chmn. unit com. on high temperature mateirals performance 1992-94, symposium chmn. annual meetings 1987, 91). Achievements include inventions of two commercial superalloys. Office: Haynes Internat Inc 1020 W Park Ave Kokomo IN 46902

LAI, MING-CHIA, mechanical engineering educator, researcher; b. Taipei, Taiwan, Republic of China, July 28, 1957; came to U.S., 1982; s. Wu-Sing and Pet-Chin (Ho) L.; m. Wen-Haw Jeng, Jan. 1, 1985; children: Deborah J., Elizabeth K., Ian B. BSME, Nat. Taiwan U., 1979; MS, PhD, Pa. State U., 1985. Instr. Nat. Taiwan U., Taipei, 1981-82; grad. rsch. asst. Pa. State U., University Park, 1982-85; postdoctoral fellow U. Mich., Ann Arbor, 1985-86; rsch. assoc. MIT, Cambridge, 1986-87; asst. prof. Wayne State U., Detroit, 1987-92, assoc. prof., 1992—; cons. Ford Motor Co., Dearborn, Mich., 1990-92, Inst. Gas Tech., Chgo., 1992—, CFD Rsch. Corp., Huntsville, Ala., 1990; summer faculty fellow NASA Lewis Rsch. Ctr., Cleve., 1988-89. Contbr. articles to Jour. Fluid and Structure, Jour. Materials and Mfg., Jour. Engring. in Gas Turbine and Power, Phoenics Jour. of CFD App., Jour. Inst. of Energy, AIAA Jour., Jour. Heat Transfer, Combustion Sci. and Tech., Jour. Fluid Engring., Jour. Engines, among others. Recipient Best Paper award 23rd Nat. Heat Transfer Conf., ASME and AIChE, Pitts., 1987, Young Engr. of Yr. award ASME, 1989, Facility Rsch. award Wayne State U., 1988, Am. Natural Resources Rsch. award, 1989, NASA/Case Summer Faculty fellow, 1988, 89; grantee NASA Lewis Rsch. Ctr., 1988-93, NSF, 1988-93, Dept. Def., 1988-93, Ford Motor Co., 1989-93, Chrysler Corp., 1992-93, Army Rsch. Office, 1992-95. Mem. AIAA, ASME, Soc. Automotive Engrs., The Combustion Inst. Achievements include research in thermal/fluid sciences, laser-based optical diagnostics and multidimensional numerical models as applied to energy systems involving turbulent mixing, combustion, heat transfer, and multiphase flow processes. Office: Wayne State U Dept Mech Engring Detroit MI 48202

LAI, RICHARD JEAN, research administrator; b. Poet-Celard, Drome, France, May 18, 1940; s. Efisio Armand and Jeanne Marguerite (Peysson) L.; m. Marie Odile Trensz, July 10, 1965 (div. 1980); 1 child, Emmanuelle; m. Lucette Scamaroni, Apr. 23, 1983. Maitrise chimie, Fac. Sciences St. Charles, Marseille, France, 1967, doctorat ès sciences, 1974. Rschr. CNRS, Marseille, 1969—. Contbr. articles to profl. jours. Recipient award ITERG, 1970, PHIRAMA, 1973. Fellow Royal Soc. Chemistry; mem. Am. Chem. Soc., Soc. Francaise de Chimie. Home: La Caroline Bd Delestrade, 13190 Allauch B du R, France

LAI, SHU TIM, physicist; b. Hong Kong, May 23, 1938; came to U.S., 1963; s. Shing Y. Lai and Choi K. Chan; m. Dorcas W. Y. Yee, Jan. 9, 1972. MA, Brandeis U., 1967, PhD, 1971. Mem. tech. staff Logicon, Lexington, Mass., 1973-78; mem. tech. staff Lincoln Lab. MIT, Lexington, 1978-79; sr. physicist Boston Coll., Chestnut Hill, Mass., 1980-81; rsch. physicist Air Force Geophysics Lab. (name now Phillips Lab.), Hanscom AFB, Mass., 1981—. Contbr. numerous articles to profl. jours., including Jour. Geophys. Rsch., Annals of Physics, Geophys. Rsch. Letters, others. Fellow Inst. Physics; mem. AIAA (sr., mem. tech. com. 1989-92), IEEE (chair Boston sect. nuclear and plasma sci. chpt. 1993—, sr.), Am. Geophys. Union (life), Am. Phys. Soc. Office: Phillips Lab Mail Stop WSSI Hanscom AFB MA 01731

LAIDLAW, HARRY HYDE, JR., entomology educator; b. Houston, Apr. 12, 1907; s. Harry Hyde and Elizabeth Louisa (Quinn) L.; BS, La. State U., 1933, MS, 1934; PhD (Univ. fellow, Genetics fellow, Wis. Dormitory fellow, Wis. Alumni Rsch. Found. fellow), U. Wis., 1939; m. Ruth Grant Collins, Oct. 26, 1946; 1 child, Barbara Scott Laidlaw Murphy. Teaching asst. La. State U., 1933-34, rsch. asst., 1934-35; prof. biol. sci. Oakland City (Ind.) Coll., 1939-41; state apiarist Ala. Dept. Agr. and Industries, Montgomery, 1941-42; entomologist First Army, N.Y.C., 1946-47; asst. prof. entomology, asst. apiculturist U. Calif.-Davis, 1947-53, assoc. prof. entomology, assoc. apiculturist, 1953-59, prof. entomology, apiculturist, 1959-74, assoc. dean Coll. Agr., 1960-64, chair agr. faculty, staff, 1965-66, prof. entomology emeritus, apiculturist emeritus, 1974—; coord. U. Calif.-Egypt Agrl. Devel. Program, AID, 1979-83. Rockefeller Found. grantee, Brazil, 1954-55, Sudan, 1967; honored guest Tamagawa U., Tokyo, 1980. Trustee, Yolo County (Calif.) Med. Soc. Scholarship Com., 1965-83. Served to capt. AUS, 1942-46. Recipient Cert. of Merit Am. Bee Jour., 1957, Spl. Merit award U. Calif.-Davis, 1959, Merit award Calif. Central Valley Bee Club, 1974, Merit award Western Apicultural Soc., 1980, Gold Merit award Internat. Fedn. Beekeepers' Assns., 1986; recipient Disting. Svc. award Ariz. Beekeepers Assn., 1988. Cert. of Appreciation Calif. State Beekeepers' Assn., 1987, award Alan Clemson Meml. Found., 1989: NIH grantee, 1963-66; NSF grantee, 1966-74. Fellow AAAS, Entomol. Soc. Am. (C.W. Woodworth award Pacific br. 1981, honoree spl. symposium 1990); mem. Am. Inst. Biol. Scis., Am. Soc. Naturalists, Am. Soc. Zoologists, Internat. Bee Rsch. Assn., Nat. Assn. Uniformed Svcs., Ret. Officers Assn. (2d v.p. Sacramento chpt. 1984-86), Scabbard and Blade, Sigma Xi (treas. Davis chpt. 1959-60, v.p. chpt. 1966-67), Alpha Gamma Rho (pres. La. chpt. 1933-34, counsellor Western Province 1960-66). Democrat. Presbyterian. Author books, the most recent being: Instrumental Insemination of Honey Bee Queens, 1977; Contemporary Queen Rearing, 1979; author slide set: Instrumental Insemination of Queen Honey Bees, 1976. Achievements include determination of cause of failure of attempts to artificially inseminate queen honey bees; invention of instruments and procedures to consistently accomplish same; elucidation of genetic relationships of individuals of polyandrous honey bee colonies; design of genetic procedures for behavioral study and breeding of honey bees for general and specific uses. Home: 761 Sycamore Ln Davis CA 95616-3432 Office: U Calif Dept Entomology Davis CA 95616

LAILA, ABDUHAMEED ABDELRAHMAN, chemistry educator; b. Kafermaitk, West Bank, Oct. 11, 1943; s. Abdelrahman Abdulhameed and Zaniah (Eid) L.; m. Maysoon Muhamad, Sep. 14, 1978; children: Aia, Majd, Saher, Shahd. BS, Damascus U., 1966; PhD in Chemistry, Reading U., 1976. Tchr. High Schs., Abha, Saudi Arabia, 1966-73; teaching asst. Reading (Eng.) U., 1974-75; asst. prof. Birzeit (West Bank) U., 1972-80, chmn., 1980-84, assoc. prof., 1984--. Contbr. articles to profl. jours. Recipient Abdes Sallam award. Home: Kafrmalik, Ramallah Israel Office: Birzett U, Chemistry Dept, Ramallar Israel

LAIN, ALLEN WARREN, reliability and maintainability engineer; b. Kansas City, Mo., July 15, 1953; s. Franklin Kenneth and Virginia Ervine (Aurig) L.; m. Karen Lee Burke, Sept. 8, 1973. Student, U. Bridgeport (Conn.), 1984-88, Bridgeport Engring. Inst., Fairfield, Conn., 1990-93. Engrs. aide Sikorsky Aircraft, Stratford, Conn., 1980-88, reliability and maintainability engr., 1988—; mem. Aviation Week Rsch. Adv. Panel, N.Y.C., 1990-92. Staff sgt. USMC, 1973-80. Mem. Am. Helicopter Soc. Achievements include development of methodology used to determine structural reliability of partially failed aircraft structure, of battle damage repair methodology for helicopter systems, reliability analysis of composite structure, component reliability analysis of fluid systems. Office: Sikorsky Aircraft Z109A 6900 Main St Stratford CT 06601-1381

LAIN, DAVID CORNELIUS, health scientist, researcher; b. Savannah, Ga., May 17, 1955; s. Marion Cornelius and Sandra (Weatherly) L.; m. Brenda Kay Gastin, May 24, 1980; children: Candace, Heather. BS, Columbia Pacific U., 1985, MS, 1985, PhD, 1987. Spl. chemistry lic., Ga.; lic. respiratory care practitioner. Instr. dept. continuing edn. Ga. So. U., Statesboro, 1983; rsch. devel. coord. Meml. Med. Ctr. Inc., Savannah, Ga., 1983-87; rsch. coord., clin. instr. dept. allied health sci. Med. Coll. Ga., Augusta, 1987-90; clin. mgr. Ohmeda Respiratory Care, Columba, Md., 1990-91, Ohmeda Critical Care, Columbia, 1991—; bd. dirs. Ga. Soc. Cardiopulmonary Tech., Atlanta, 1987; adv. bd. Respiratory Therapy Adv. Com., Augusta, 1987-90; cons. Aero-Med. Internat., 1987; rsch. affiliate Siemen. Elem., Schaumburg, Ill., 1986; manuscript reviewer Am. Assn. Respiration Therapy, Dallas, 1988. Contbr. articles to profl. jours. Recipient Appreciation award Am. Heart Assn., 1985, Outstanding Achievement award Calif. Coll. Health Sci., 1986. Mem. AAAS, So. Med. Assn., N.Y. Acad. Sci., Am. Assn. Respiratory Care, Nat. Bd. Respiratory Care (registered respiratory therapist). Democrat. Achievements include 9 patents pending; research on reduction of peak inspiratory pressure during acute lung injury to reduce iatrogenic progression of lung pathology. Home: 1861 Norhurst Way N Baltimore MD 21228-4123 Office: Ohmeda 9065 Guilford Rd Columbia MD 21046-1836

LAING, MALCOLM BRIAN, geologist, consultant; b. Toronto, Ont., Can., Apr. 4, 1955; s. Alexander Duncan and Joan (Dawson) L.; m. Vicki Lynne; 1 child, Megan Jené. BS in Geology, Tex. Christian U., 1978. Geologist Electro-Seise, Inc., Ft. Worth, 1978-79, Exploration Logging Co., Houston, 1979-80, Thomas-Powell Royalty Co., Ft. Worth, 1980-82, Lentex Petroleum Inc., Abilene, Tex., 1982-84; cons., 1984-90, Tex. Dept. Health, 1990-92, Tex. Water Commn., 1992—; bd. dirs. Tailwheel First, Inc. Mem. Am. Assn. Petroleum Geologists, Soc. Petroleum Engrs., Tex. Air Mus., Panhandle Squadron CAF (past leader), Phantom Squadron, West Tex. Wing CAF (past fin. officer, CAF check pilot). Republican. Baptist. Office: 4630 50th St Ste 603 Lubbock TX 79424

LAIRE, HOWARD GEORGE, chemist; b. Waterbury, Conn., Nov. 14, 1938; s. Howard George and Adele Melvina (St. Marie) L.; m. Wanda Carolyn Curtis, Aug. 13, 1966; children: Danny Wayne, Nicole Michele, David Tatanka, Carolyn Lorraine. Grad., U. Miami, 1970; MS, Nova U., 1977. Tchr. Dade County Schs., Miami, Fla., 1970-80; environ. specialist, chemist Colo. UTE Electric Co., Montrose, 1981-90; sr. chemist A/C Power, Trona, Calif, 1990--; participant Colo. Wildlife Consortium, Boulder, Colo., 1984. Chmn. Am. Indian Urban Ctr., Miami, Fla., 1977-80; bd. dirs. Region 10 Econ. Devel., Montrose, 1983-86, Chpt. Two Ednl. Grants, Denver, 1987-90. Mem. Nat. Assn. Corrosion Engrs. Home: 713 N Inyo Ridgecrest CA 93555 Office: A/C Power 12801 Makiposa St Trona CA 93592

LAITONE, EDMUND VICTOR, mechanical engineer; b. San Francisco, Sept. 6, 1915; s. Victor S. L.; m. Dorothy Bishop, Sept. 1, 1951; children: Victoria, Jonathan A. BSME, U. Calif., Berkeley, 1938; PhD in Applied Mechanics, Stanford U., 1960. Aero. engr. Nat. Adv. Com. for Aeros., Langley Field, Va., 1938-45; sect. head, flight engr. Cornell Aero. Lab., Buffalo, 1945-47; prof. U. Calif., Berkeley, 1947—; cons. aero. engr. Hughes Aircraft & Douglas Aircraft, 1948-78; U.S. acad. rep. to flight mechanics AGARD/NATO, 1984-88; chmn. engring. dept. U. Calif. Extension, Berkeley, 1979—. Author: Surface Waves, 1960; author, editor: Integrated Design of Advanced Fighter Aircraft, 1987; contbr. articles to Jour. Aero. Scis., Aircraft and Math. Jour. Named Miller Rsch. prof., 1960, U.S. Exch. prof., Moscow, 1964; vis. fellow Balliol Coll., 1968; vis. prof. Northwestern Poly. Inst., Xian, China, 1980. Fellow AIAA (San Francisco region chmn. 1960-61, assoc. fellow 1964-88); mem. Am. Math Soc., Am. Soc. for Engring. Edn. Achievements include discovery of effect of acceleration on longitudinal dynamic stability of a missile, nonlinear dynamic stability of space vehicles entering or leaving atmosphere; higher approximations to nonlinear water waves. Home: 6915 Wilson Way El Cerrito CA 94530-1853 Office: U of Calif Dept Mech Engring Berkeley CA 94720

LAJEUNESSE, ROBERT PAUL, design engineer; b. Toledo, Ohio, Oct. 31, 1950; s. Wilfrid Joseph and Ruth Agatha (Seger) LaJ. BEE, U. Detroit, 1974; MS in Engring., U. Mich., 1976. Project engr. Chrysler Corp., Detroit, 1974-80, TRW Transp. Elec., Farmington Hills, Mich., 1980-82; design engr. Com-2 Inc., Saline, Mich., 1982-88, Nematron Corp., Ann Arbor, Mich., 1988—; cons. Programmable Designs, Dexter, Mich., 1990—. V.p. Traver Lakes Community Maintenence Assn., Ann Arbor, 1991-92. Home: 2553 Meade Ct Ann Arbor MI 48105 Office: Nematron Corp 5840 Interface Dr Ann Arbor MI 48103

LAJTHA, ABEL, biochemist; b. Budapest, Hungary, Sept. 22, 1922; naturalized; married; 2 children. PhD in Chemistry, Eotvos Lorand U., Budapest, 1945; D (hon.), U. Padua. Asst. prof. biochemistry Eotvos Lorand U., 1945-47; asst. prof. Inst. Muscle Rsch., Mass., 1949-50; sr. rsch. scientist N.Y. State Psychiat. Inst., 1950-57, assoc. rsch. scientist, 1957-62, prin. rsch. scientist, 1962-66; dir. N.Y. State Rsch. Inst. Neurochemistry, 1966—; prof. exptl. psychiatry Sch. Medicine NYU, 1971—; asst. prof. Coll. Physicians & Surgeons, Columbia U., 1956-69. Zoology Station fellow Italy, 1947-48, Rsch. fellow Royal Inst. Great Britain, 1948-49. Mem. Internat. Brain Rsch. Orgn., Am. Soc. Biol. Chemists, Am. Acad. Neurol., Am. Coll. Neuropsychopharmacol., Internat. Soc. Neurochem. (pres.), Am. Chem. Soc., Am. Soc. Neurochem. (pres.). Achievements include rsch. in neurochemistry, amino acid and protein metabolism of the brain and the brain barrier system. Office: Dept Neurochem Nathan S. Kline Inst for Psy Rsc Orangeburg NY 10962*

LAKDAWALA, VISHNU KESHAVLAL, electrical and computer engineering educator; b. Bangalore, India, Sept. 11, 1951; came to U.S., 1982; s. Keshavlal M. and Shantaben (Patel) L.; m. Rita V. Chhatiawala, Nov. 30, 1976; children: Mayuri, Anushka. MSEE, Indian Inst. Sci., 1974; PhD in Elec. Engring., U. Liverpool, 1980. Sr. rsch. assoc. U. Liverpool (England) Dept. Elec. Engring., 1980-82, U. Tenn./Oak Ridge Nat. Lab., 1982-83; asst. prof. Old Dominion U. Dept. Elec. Engirng., Norfolk, Va., 1983-89, assoc. prof., 1989—, grad. program dir., 1990—. Contbr. articles to profl. jours. Chmn. Asian Indians of Tidewater, Norfolk, 1985-88; sec. Hindu Temple of Hampton Rds., Norfolk, 1989-92, treas., 1993—. Recipient Indian Inst. Sci. NR Khambati Meml. award, 1974. Mem. IEEE (sr. mem., chmn. Va. coun. region 3 1990-92), Eta Kappa Nu (faculty advisor 1987-90, Leadership award 1986). Achievements include 3 patents on optically controlled and electron beam controlled semiconductor; developed a novel switch which can be turned on and off optically without needing external source to maintain the photoconductivity. Home: 4112 Cheswick Ln Virginia Beach VA 23455 Office: Old Dominion U Dept Elec Engring Norfolk VA 23529

LAKE, LARRY WAYNE, petroleum engineer; b. Del Norte, Colo., Jan. 31, 1946; s. Ralph Wayne and Ina Belle (Card) L.; m. Carole Sue Holmes, Mar. 22, 1975; children: Leslie Sue, Jeffrey Wayne. B.S., Ariz. State U., 1967; Ph.D., Rice U., 1973. Registered profl. engr., Tex. Prodn. engr. Motorola Co., Phoenix, 1968-70; sr. research engr. Shell Devel. Co., Houston, 1973-78; W.A. "Tex" Moncrief Centennial chair prof. U. Tex., Austin, 1978—; chmn. dept. petroleum engring.; cons. enhanced oil recovery. Recipient U. Tex. Engring. Found. award, 1979, Disting. Faculty Adv. award U. Tex., 1981, Grad. Engring. Council award, U. Tex., 1987. Mem. AICE, Soc. Mining Engrs., Soc. Petroleum Engrs. (mem. editorial rev. com., author video, Disting. Achievement award 1981), Soc. of Petroleum Engrs. (disting. lectr., bd. dirs., Reservoir Engr. award 1991). Baptist. Home: 4003 Edgefield Ct Austin TX 78731-2902 Office: U Tex Dept Petroleum Engring Austin TX 78712

LAKES, STEPHEN CHARLES, research chemist, educator; b. Hamilton, Ohio, Jan. 1, 1951; s. Carl and Evelyn Lakes; m. Gayle Ann Bruce; 1 child, Elizabeth Ann. BS, Bowling Green State U., 1973; MS, U. Cin., 1977. Process chemist Emery Industries, Cin., 1973-87; sr. rsch. chemist Henkel Corp./Emery Group, Cin., 1987—; instr. No. Ky. U., Highland Heights, 1980-81, OCAS/U. Cin., 1978-80; instr. synthetic lubricants Soc. Tribologists and Lubricating Engrs., Park Ridge, Ill., 1990—. Mem. Soc. Tribologist and Lubricating Engrs., Nat. Lubricating Grease Inst., Soc. Auto. Engrs., Am. Chem. Soc. Office: Henkel Corp Emery Group 4900 Este Ave Cincinnati OH 45232

LAKKARAJU, HARINARAYANA SARMA, physics educator, consultant; b. Bapatla, India, Sept. 20, 1946; came to U.S., 1971; PhD, SUNY, Amherst, 1979. Vis. asst. prof. dept. physics Tex. A&M U., College Station, 1979-81; asst. prof. dept. physics San Jose (Calif.) State U., 1981-83, assoc. prof., 1984-89, prof., 1990—; cons. laser and electro-optics. Contbr. articles to profl. jours. Mem. Am. Phys. Soc., Optical Soc. Am., Soc. Photo-Optical Instrumentation Engrs. Office: San Jose State Univ Dept Physics One Washington Square San Jose CA 95192-0106

LAKSHMANAN, MARK CHANDRAKANT, physiologist, physician; b. Cheverly, Md., Oct. 14, 1953; s. Sitarama and Florence (Lazicki) L.; m. Shelley Anne Boyer, Nov. 7, 1973; children: Damaris Victoria, Anastasia Sitarama. BS, Univ. Md., 1977, MD, 1981. Intern, resident, chief resident Cleve. Met. Gen. Hosp., 1981-85; medical staff fellow NIDDK, NIH, Bethesda, Md., 1985-88; sr. instr. dept. medicine, dept. physiology and biophysics CWRU, Cleve., 1988-90; physician MetroHealth Medical Ctr., Cleve., 1988—; asst. prof. dept. of medicine CWRU, Cleve., 1990—; cons. Brookhaven Nat. Labs., Long Island, NY., 1986—. Author: (chpt.) Thyroid Hormone Transport from Blood Into Brain, 1989, Thyroid Pharmacology, 1993. With U.S. Army, 1973-75. Recipient Clinical Investigator award NIH, 1989, ATA Travel award, Americal Throid Assn., 1988, Peter J. Adams award Cleve. Met. General Hosp., 1985. Mem. ACP, Am. Fedn. Clinical Rsch., Am. Physiological Soc. Democrat. Office: MetroHealth Med Ctr 2500 MetroHealth Dr Cleveland OH 44109-1998

LAKSHMINARAYANA, BUDUGUR, aerospace engineering educator; b. Shimoga, India, Feb. 15, 1935; came to U.S., 1963, naturalized, 1971; m. Saroja L.; children—Anita, Arvind. B.E. in Mech. Engring., Mysore U., India, 1958; Ph.D., U. Liverpool, Eng., 1963, D.Eng., 1981. Grad. trainee Steel Constrn. Co., Bangalore, India, 1957; asst. mech. engr. Kolar Gold Mining Undertakings, Kolar, India, 1958-60; researcher in mech. engring. Liverpool U., Eng., 1960-63; vis. asst. prof. aerospace engring. Pa. State U., University Park, 1963-65, asst. prof. aerospace engring., 1965-69, assoc. prof. aerospace engring., 1969-74, prof. aerospace engring., 1974-85, dir. computational fluid dynamics studies, 1980-87, Disting. Alumni Prof. aerospace engring., 1985-86, Evan Pugh prof. Aerospace Engring., 1986—; vis. fellow scientist Cambridge U., St. John's Coll., Eng., 1971-72; vis. assoc. prof. aeros. and astronautics MIT, 1972; vis. prof. dept. mech. engring. Indian Inst. Sci., 1979; aerospace engr. computational fluid mechanics group NASA Ames Research Ctr., Moffett Field, Calif., 1979; CNRS vis. prof. Laboratoire de Mecanique des Fluides at d'Acoustique, Ecole Centrale de Lyon, France, 1987-88; vis. prof. Tech. U. of Aachen, Fed. Republic of Germany, 1988; cons. Pratt & Whitney Aircraft, GE Aircraft Engine Div., Garrett Turbines, Teledyne CAE, Inc.; UN, NATO/AGARD lectr.; Gen. Motors, Rolls Royce (England), European Sponice Agy.; mem. NASA adv. group on computational fluid dynamics, 1980; lectr. in field. Editor 2 books; contbr. numerous papers on fluid dynamics, turbomachinery, computational fluid dynamics, turbulence modelling and acoustics to profl. publs. Recipient award for outstanding achievement in rsch. Pa. State U., 1977, Henry R. Worthington N.Am. tech. award, 1977, outstanding achievement in rsch. award Pa. State U. Coll. Engring., 1983, sr. prof. Fulbright award, 1988, Arch T. Colwell merit award SAE, 1992; Mysore U. merit scholar, 1953-57; Leverhulme fellow Liverpool U., 1962-63; grantee NSF, numerous others. Fellow AIAA (chmn. Central Pa. chpt. 1970, Pendrey lit. award 1989), ASME (Freeman Scholar award 1990). Office: Pa State U Coll Engring 153 Hammond Bldg University Park PA 16802

LALL, B. KENT, civil engineering educator; b. Feb. 4, 1939; m. Margaret Vivienne Boult, Nov. 30, 1970; 1 child, Niren Nicolaus. BS in Civil Engring., Panjab Engring. Coll., Chandigarh, India, 1961; ME in Hwy. Engring., U. Roorkee, India, 1964; PhD in Transp., U. Birmingham, Eng., 1969. Registered profl. engr. Commonwealth scholar U. Birmingham, 1966-69; lectr. Indian Inst. Tech., New Delhi, 1964-72, asst. prof., 1972-75; assoc. prof. U. Man., Winnipeg, Can., 1975-77; assoc. prof. civil engring. Portland (Oreg.) State U., 1977-84, prof., 1984—; vis. prof. U. Adelaide, South Australia, 1985; cons. Nat. Rds. Bd., Ministry of Works, Wellington, New Zealand, 1986. Contbr. articles to profl. jours. Vol. Meals on Wheels, Portland, 1991-93. Fellow ASCE (exec. com. urban transp. divsn., chair pub. transp. com. 1988-91, mem. high speed ground transport com.), S.W. Portland Rotary (bd. dirs. 1990-91). Office: Portland State U Dept Civil Engring PO Box 751 Portland OR 97207-0751

LALONDE, ROBERT THOMAS, chemistry educator; b. Bemidji, Minn., May 7, 1931; s. Clarence A. and Barbara B. (Rafferty) L.; m. Suzanne Denniston, Dec. 28, 1957; children: Robert J., Judith M., Mary C., Jane F., Suzanne, Jerome V., Thomas A. BA, St. John's U., Collegeville, Minn., 1953; PhD, U. Colo., 1957. Sr. rsch. engr. Jet Propulsion Lab., Pasadena, Calif., 1957-58; postdoctoral rsch. assoc. Univ. Ill., Urbana, 1958-59, Coll. Environ. Sci. and Forestry, SUNY, Syracuse, 1959-60; asst. prof. chemistry Coll. Environ. Sci. and Forestry, SUNY, 1960-64, assoc. prof. chemistry, 1964-69, prof. chemistry, 1969—; vis. scholar Stanford Univ., Palo Alto, Calif., 1965-66; vis. scientist NRC, Ottawa, Can., 1973; exch. prof. Hanover (Germany) Univ., 1980. Contbr. articles to profl. jours. Rsch. grantee Rsch. Corp., NSF, Nat. Inst. for Allergy and Infectious Diseases, Nat. Inst. of Environ. Health Scis., N.Y. State Sea Grant, Petroleum Rsch. Fund, U.S. Geol. Survey. Mem. Am. Chem. Soc. (grantee, Syracuse sect. chmn. 1978-79), Am. Soc. Pharmacognosy, Environ. and Molecular Mutagenesis Soc., Sigma Xi. Democrat. Roman Catholic. Achievements include patents for nuphar syntheses and antifungal products; discovery of structure and syntheses of antifungal nuphar alkaloids, elucidation of the molecular structure features responsible for the activiy of chlorinated drinking water genotoxins; recognized role of alkali polysulfides in the formation of organosulfur compounds in the geosphere. Office: Coll Environ Sci and Forestry 1 Forestry Dr Syracuse NY 13210-2786

LAM, BING KIT, pharmacologist; b. Fukien, People's Republic of China, Mar. 18, 1957; s. Man Yuk and Yuk Ying (Cheng) L.; m. Siu Ming Tan, Sept. 18, 1986; children: Jonathan Chun-Ming, Stephanie Mei-Fung. PhD, N.Y. Med. Coll., 1987. Rsch. fellow Harvard Med. Sch., Boston, 1987-89, instr., 1989-91, asst. prof., 1991—. Mem. Soc. Chinese Bioscientists Am. Achievements include research in mechanism of release of leukotrienes from human leukocytes. Office: Harvard Med Sch 250 Longwood Ave Boston MA 02115-5719

LAM, DANIEL HAW, chemical engineer; b. Manila, Philippines, Dec. 27, 1961; s. Ah Sian Haw. MChemE, U. Tulsa, 1989, PhDChemE, 1990. Registered profl. engr., Tex. Dept. supr. Nation Alcohol Inc., Philippines, 1985-87; process engr. ABB Lummus Crest Inc., Houston, 1990—. Contbr. articles to profl. jours. Mem. AICE, Sigma Xi, Tau Beta Pi. Achievements include research on multiphase equilibrium behavior of the mixture ethane methanol n-decanol liquid-liquid-vapor phase equilibrium behavior of certain binary ethane n-alkanol mixtures. Home: Apt #1509 11900 Wickchester Ln Houston TX 77043

LAM, SIMON SHIN-SING, computer science educator; b. Macao, July 31, 1947; came to U.S., 1966; s. Chak Han and Kit Ying (Tang) L.; m. Amy Leung, Mar. 29, 1971; 1 child, Eric. B.S.E.E. with distinction, Wash. State U., 1969; M.S. in Engring., UCLA, 1970, Ph.D., 1974. Research engr. ARPA Network Measurement Ctr., UCLA, Los Angeles, 1971-74; research staff mem. IBM Watson Research Ctr, Yorktown Heights, N.Y., 1974-77; asst. prof. U. Tex.-Austin, Austin, 1977-79, assoc. prof., 1979-83, prof. computer sci., 1983—; David S. Bruton Centennial prof. U. Tex., Austin, 1985-88, anonymous prof., 1988—; chmn. dept. computer sci. U. Tex.-Austin, Austin, 1992—. Contbr. articles to profl. jours., conf. procs.; editor: Principles of Communicaton and Networking Protocols. NSF grantee, 1978—; Chancellor's Teaching fellow UCLA, 1969-73. Fellow IEEE (Leonard G. Abraham prize 1975); mem. Assn. for Computing Machinery (program chmn. symposium 1983). Avocations: tennis, swimming, skiing, travel. Office: Univ Tex Dept of Computer Sci Austin TX 78712

LAMAGNA, JOHN THOMAS, chemical engineer; b. N.Y.C., Dec. 21, 1961; s. Liborio Joseph and Anne M. (Calzi) LaM.; m. Tammy L. Koster, Oct. 24, 1992. BChEng, Manhattan Coll., 1983, MChEng, 1984; MBA, Iona Coll., 1988. Plant mgr. Superior Wall Products, N.Y.C., 1984-86; safety engr. Pfizer Inc., N.Y.C., 1986-88, prodn. supr., 1988-89, aseptic projects supr., 1989-91; mfg. mgr. Pfizer Inc., Terre Haute, Ind., 1991-93, mgr. projects and maintenance, 1993—. Bus. mgr. Jr. Achievement, Terre Haute, 1992, 93. Mem. Am. Inst. Chem. Engrs. (treas.), Parenteral Drug Assn. (mem. cleaning validation task force, Twenty-First Century Securities Investment Club (pres.). Home: PO Box 6133 Terre Haute IN 47802

LAMAN, JERRY THOMAS, mining company executive; b. Muskogee, Okla., Mar. 1, 1947; s. Thomas J. and Juanita J. (Pittman) L.; m. Lenora J. Laman, July 1, 1972; children: Troy T., Brian D. Silver Diploma, Colo. Sch. Mines, 1969. Refinery engr. ARCO, Torrance, Calif., 1969-71; chem. engr. Cleveland-Cliffs Iron Co., Mountain City, Nev., 1971-73, asst. mine supt., 1973-77; chief uranium metallurgist Cleveland-Cliffs Iron Co., Casper, Wyo., 1977-83; project mgr. In-Situ, Inc., Laramie, Wyo., 1983-85, v.p., 1985—; also bd. dirs. In-Situ, Inc.; pres. Solution Mining Corp., Laramie, Wyo., 1990—, also bd. dirs., 1990—. Mem. Soc. for Mining, Metallurgy and Exploration, Optimist (pres. Laramie club 1989). Avocations: golf, fishing. Home: 1085 Colina Laramie WY 82070

LA MANTIA, CHARLES ROBERT, management consulting company executive; b. N.Y.C., June 12, 1939; s. Joseph Ferdinand and Catherine (Perniciaro) LaM.; m. Ann Christine Carmody, Sept. 16, 1961; children: Elise, Matthew. BA, Columbia U., 1960, BS, 1961, MS, 1962, ScD, 1965; grad. advanced mgmt. program, Harvard Bus. Sch., 1979. Cons. staff Arthur D. Little, Inc., Cambridge, Mass., 1967-77, v.p., 1977-81, pres., chief oper. officer, 1987-88, pres., chief exec. officer, 1988—, also bd. dirs.; pres., chief exec. officer Koch Process Systems, Westboro, Mass., 1981-86; mem. engr. coun. Columbia U.; mem. bd. govs. New England Med. Ctr., 1989—; bd. dirs. State Street Bank, State Street Boston Corp.; trustee Meml. Dr. Trust, 1988—. Mem. bd. overseers Mus. Sci., Boston, 1986—; overseer, Sta. WGBH pub. broadcasting, 1991—, mem. Conf. Bd., 1989—; mem. Mass. Bus. Roundtable, 1992. Served to lt. USN, 1965-67. NSF fellow, 1965; Sloan Found. fellow, 1962. Mem. Am. Inst. Chem. Engrs., Soc. Chem. Industries. Office: Arthur D Little Inc 25 Acorn Park Cambridge MA 02140-2301

LAMAR, JAMES LEWIS, JR., chemical engineer; b. Antlers, Okla., June 13, 1959; s. James Lewis and Priscilla (Henderson) L.; m. Carol Horton, May 16, 1982; children: Joy Loree, Amanda Beth. AS, Bee County Coll., 1979; BSChemE, Tex. A&I U., 1982. Cert. quality engr., 1991, cert. quality auditor, 1992. Prodn. engr. Union Carbide Corp., Soutn Charleston, W.Va., 1982-86; diagnostic engr. Union Carbide Chems. & Plastics, South Charleston, W.Va., 1986-90; sr. quality engr. Union Carbide Chems. & Plastics, League City, Tex., 1990—; instr. Tex. Dept. Commerce, Houston, 1992. Author/programmer software, 1989, 91, 92. Mem. Pine Dr. Babt. Sch. Bd., Dickinson, Tex., 1992. Named Jaycees Outstanding Young Men Am., 1983. Mem. Am. Soc. Quality Control, Cowboys for Christ. Republican. Office: Union Carbide 2525 S Shore Blvd Ste 300 League City TX 77539

LAMARQUE, MAURICE PATRICK JEAN, health products executive; b. Boston, Feb. 6, 1948; s. Jean Maurice and Mary Frances (Mahon) L.; m. Michelle Marie Tocci, July 22, 1979; children: Renee, Jennifer, John. BS in Chemistry, Boston State Coll., 1969. With tech. svc. dept. New Eng. Nuclear, Boston, 1969-71, tech. rep., 1972-75, sr. tech. rep., 1976-79; sales mgr. Collaborative Rsch., Inc., Waltham, Mass., 1979-80; dir., co-founder KOR Biochems. Inc., Cambridge, Mass., 1980-81; founder, chief exec. officer Biomed. Techs. Inc. Co., Stoughton, Mass., 1981—; mktg. cons. KOR Isotypes, Cambridge, 1981. Coach little league baseball, Walpole, Mass., 1989; pres. bd. trustees Deerfield Corp. Ctr., Stoughton, 1988—. Roman Catholic. Office: Biomed Techs Inc 378 Page St Stoughton MA 02072-1141

LAMARRE, BERNARD, engineering, contracting and manufacturing advisor; b. Chicoutimi, Que., Can., Aug. 6, 1931; s. Emile J. and Blanche M. (Gagnon) L.; m. Louise Lalonde, Aug. 30, 1952; children: Jean, Christine, Lucie, Monique, Michèle, Philippe, Mireille. BSc, Ecole Poly., Montreal, Que., Can., 1952; MSc, Imperial Coll., U. London, 1955; LLD, St. Francis Xavier U., N.S., Can., 1980; D in Engring. (hon.), U. Waterloo, Ont., 1984; LLD (hon.), U. Concordia, Montreal, 1985; D in Engring. (hon.), U. Montreal, 1985; D in Applied Sci. (hon.), U. Sherbrooke, Que., 1986; D in Bus. Adminstrn. (hon.), U. Chicoutimi, Que., 1987; D in Sci. (hon.), Queen's U., Kingston, Ont., 1987; D in Engring. (hon.), U. Ottawa, Ont., 1988, Tech. U. N.S., 1989, Royal Mil. Coll., Kingston, 1990. Structural and founds. engr. Lalonde-Valois, Montreal, 1955-60, chief engr., 1960-62; ptnr., gen. mgr., pres. Valois, Lalonde, Valois, Lamarre, Montreal, 1962-72; chmn., chief exec. officer Lavalin Group, 1972-91; sr. advisor SNC-Lavalin Inc., 1991—; adv. Publicité Martin Inc. Contbr. articles to profl. jours. Bd. dirs. Télésystéme Inc.; chmn. Bellechasse Santé Inc., Santra Inc.; hon. chmn. Montreal Mus. Fine Arts, Coll. Stanislas; chmn. Can. cultural property export rev. bd. Decorated officer Ordre nat. du Québec, Order of Can.; Athlone fellow, 1952. Fellow Engring. Inst. Can., Can. Soc. Civil Engring.; mem. ASCE, Order Engrs. of Que. (chmn.), Inst. Design Montreal (chmn.), Mont-Royal Club, Laval-sur-le Lac Club. Roman Catholic. Home: 4850 Cedar Crescent, Montreal, PQ Canada H3W 2H9

LAMB, EDWARD ALLEN, JR., business owner; b. Elgin, Ill., Oct. 5, 1957; s. Edward Allen and Sara Dina (Mirs) L. Sr.; m. Alice Craig, Sept. 19, 1977 (div. Sept. 1980); 1 child, Jeremy Edward; m. Cynthia Lee Gray, Mar. 24, 1990; children: Brittany Mary Sarah, Benjamin Allen Douglas. Student, Devry Inst., Chgo., 1975-77. Svc. mgr. Stereo Studio, Arlington Heights, Ill., 1977-79; co-founder, svc. mgr. Ariz. Instrument Inc., Tempe, 1979-90; prin. owner Lamb Tech. Svcs., Gilbert, Ariz., 1991—; bd. dirs. Odor Control Com./Water Environment Fedn., Alexandria, Va., 1987—. Contbr. articles to profl. jours. Mem. Water Environment Fedn. Achievements include use of monitoring instrumentation to reduce public nuisance/How to Complaints. Office: Lamb Tech Svcs 133 S Pueblo St Gilbert AZ 85234

LAMB, HENRY GRODON, safety engineer; b. Saco, Maine, Feb. 9, 1906; s. Charles Barnard and Fannie Mabel (Prentiss) L.; 1 child, Patricia Miller Lamb Steen. BS, Dartmouth Coll., 1926; BS in CE, MIT, 1928. Cert. profl. safety engr.; registered profl. engr., Calif. Civil engr. Stone & Webster Engring., Boston, 1928-31; safety engr. Liberty Mutual Ins., Boston, 1931-38, Scott Paper Co., Chester, Pa., 1938-42, Am. Nat. Stds. Inst., N.Y.C., 1942-71; ret. Mem. Sch. Bd., Freedom, N.H., 1983-89. Mem. Am. Soc. Safety Engrs. (pres. 1968-69), Am. Indsl. Hygiene Assn., ASME, Masons (worshipful master 1986-88). Republican. Congregationalist. Avocations: hiking, snowshoeing, chopping wood. Address: Scarboro Rd PO Box 211 Freedom NH 03836

LAMB, LOWELL DAVID, physicist; b. Abilene, Tex., Nov. 8, 1955; s. Donald Wayne Lamb and Gaylon (Jordan) Monteverde; m. Victoria Irene Russo, May 8, 1985. BS in Physics, BS in Math., Abilene Christian U., 1977; MA in Physics, Johns Hopkins U., 1983; PhD in Physics, U. Ariz., 1991. Systems programmer Fed. Res. Bank of N.Y., N.Y.C., 1981-85; asst. dir. Ariz. Fullerene Consortium, U. Ariz., Tucson, 1991—. Contbr. articles to profl. jours. Fellow ARCS Found., 1988, 89; faculty scholar U. Ariz., 1985. Mem. Am. Phys. Soc. Achievements include development of method for large-scale production of the fullerenes including C60 and buckminsterfullerene or buckyball; discovery of third crystalline form of pure carbon, fullerite. Home: 905B N Venice Ave Tucson AZ 85711 Office: Dept Physics U Ariz Tucson AZ 85721

LAMB, MARGARET WELDON, lawyer; b. Arlington, Mass., June 26, 1935; d. Hubert Weldon and Lydia Cazneau (Baker) L. BA, U. Denver, 1959; JD, Boston Coll., 1964. Bar: Mass. 1964, N.M. 1969. Pvt. practice Taos County, N.Mex., 1971-76; dist. atty. N.Mex. 8th Jud. Dist., Taos, 1978-80; pvt. practice specializing in aviation adminstrv. law, 1981—; air safety investigator, writer, flight instr., 1981—. Contbr. articles on aviation safety to profl. publs. Mem. AIAA, Am. Meteorol. Soc., Aerospace Med. Assn., Lawyer-Pilots Bar Assn., NTSB Bar Assn. Achievements include research into microscale mountain weather and aircraft crashes. Home and Office: Box 650 Questa NM 87556

LAMB, PETER JAMES, meteorology educator, researcher, consultant; b. Nelson, New Zealand, June 21, 1947; came to U.S., 1971; s. George Swan and Dorothy Elizabeth (Smith) L.; m. Barbara Helen Harrison, Aug. 29, 1970; children: Karen Deborah, Brett Timothy. BA, U. Canterbury, Christ Ch., New Zealand, 1969, MA with honours, 1971; PhD, U. Wis., 1976. Asst. lectr. U. Canterbury, 1971; rsch. asst. U. Wis., Madison, 1971-76, rsch. assoc., 1976; lectr. U. Adelaide, Australia, 1976-79; sr. sci. Ill. State Water Survey, Champaign, 1979-91, section head, 1984-90; prof. U. Okla., Norman, 1991—; vis. rsch. assoc. U. Miami, Fla., 1978-79; adj. prof. U. Ill., Urbana, 1983—; dir. Coop. Inst. Mesoscale Meteorological Studies, Norman, 1991—; cons. Dept. State, Dept. Energy, Agy. Internat. Devel., Nat. Oceanic & Atmospheric Adminstrn., NSF, World Meteorological Orgn., Kingdom of Morocco, U. Wis., U. Adelaide, Univs. Space Rsch. Assn., EPA, 1983—; site sci. atmospheric radiation measurement program, Dept. Energy, 1992—. Co-author rsch. monographs, book chpt., numerous sci. papers. Coach Champaign Youth Soccer Orgn., 1983-91. Recipient 29 rsch. grants U.S. Fed. Agencies including NSF, EPA, Dept. Energy, Nat. Atmospheric & Oceanic Administrn.; MacArthur Found. grantee. Fellow Am. Meteorological Soc. (chief editor Jour. Climate 1989—); mem. Royal Meteorological Soc. (Margary lectr. 1991), Am. Assn. State Climatologists, Sigma Xi. Achievements include research on heat transport by the Atlantic Ocean; investigations into the role of the ocean in causing droughts in Sahelian Africa. Home: 3616 Burlington Dr Norman OK 73072 Office: Univ of Oklahoma CIMMS-Sarkeys Energy Ctr 100 E Boyd Rm 1110 Norman OK 73019

LAMB, WILLIS EUGENE, JR., physicist, educator; b. Los Angeles, Calif., July 12, 1913; s. Willis Eugene and Marie Helen (Metcalf) L.; m. Ursula Schaefer, June 5, 1939. BS, U. Calif., 1934, PhD, 1938; DSc (hon.), U. Pa., 1953, Gustavus Adolphus Coll., 1975, Columbia U., 1990; MA (hon.), Oxford (Eng.) U., 1956, Yale, 1961; LHD (hon.), Yeshiva U., 1965. Mem. faculty Columbia U., 1938-52, prof. physics, 1948-52; prof. physics Stanford U., 1951-56; Wykeham prof. physics and fellow New Coll., Oxford U., 1956-62; Henry Ford 2d prof. physics Yale U., 1962-72, J. Willard Gibbs prof. physics, 1972-74; prof. physics and optical scis. U. Ariz., Tucson, 1974—, Regents prof., 1990—; Morris Loeb lectr. Harvard U., 1953-54; Gordon Shrum lectr. Simon Fraser U., 1972; cons. Philips Labs., Bell Telephone Labs., Perkin-Elmer, NASA; vis. com. Brookhaven Nat. Lab. Recipient (with Dr. P. Kusch) Nobel prize in physics, 1955, Rumford premium Am. Acad. Arts and Scis., 1953; Rsch. Corp. award, 1954, Yeshiva award, 1962; Guggenheim fellow, 1960-61, Sr. Alexander von Humboldt fellow, 1992-93. Fellow Am. Phys. Soc., N.Y. Acad. Scis.; hon. fellow Inst. Physics and Phys. Soc. (Guthrie lectr. 1958), Royal Soc. Edinburgh (fgn. mem.); mem. Nat. Acad. Scis., Phi Beta Kappa, Sigma Xi. Office: U Ariz Optical Scis Ctr Tucson AZ 85721

LAMBERG, LYNNE FRIEDMAN, medical journalist; b. St. Louis, Jan. 13, 1942; d. Ralph Maurice and Fay Geraldine (Bialick) Friedman; m. Stanford Irwin Lamberg, Aug. 19, 1962; children: Nicole, Ryan. BA, Washington U., St. Louis, 1963; MA, U. Pa., 1967. Mem. pub. rels. staff Jewish Hosp., St. Louis, 1962-64; dir. pub. rels. Hosp. U. Pa., Phila., 1964-67; freelance mental health writer, 1968—. Author: The American Medical Guide to Better Sleep, 1984, Drugs and Sleep, 1988, Skin Disorders, 1989; co-author: Crisis Dreaming, 1992; contbr. articles to profl. jours. Editor newsletter Mt. Washington Improvement Assn., Balt., 1976-88, various coms. Balt. City Commn. for Women, 1980; writer Md. chpt. Nat. Multiple Sclerosis Assn., Balt., 1980. Recipient numerous journalism awards. Mem. Nat. Assn. Sci. Writers. Home and Office: 3704 Gardenview Rd Baltimore MD 21208

LAMBERG-KARLOVSKY, CLIFFORD CHARLES, anthropologist, archaeologist; b. Prague, Czechoslovakia, Oct. 2, 1937; came to U.S., 1939; s. Carl Othmar von Lamberg and Bettina Karlovsky; m. Martha Louise Veale, Sept. 12, 1959; children—Karl Emil Othmar, Christopher William. A.B., Dartmouth Coll., 1959; M.A. (Wenner-Gren fellow), U. Pa., 1964, Ph.D., 1965; M.A. (hon.), Harvard U., 1970. Asst. prof. sociology and anthropology Franklin and Marshall Coll., 1964-65; asst. prof. anthropology Harvard U., 1965-69, prof., 1969-90, Stephen Phillips prof. archaeology, 1991—; curator Near Eastern archaeology Peabody Museum Archaeology and Ethnology, 1969—, mus. dir., 1977-90; assoc. Columbia U., 1969—; trustee Am. Inst. Iranian Studies, 1968—, Am. Inst. Yemeni Studies, 1976-77; dir. rsch. Am. Sch. Prehist. Rsch., 1974-79, 86—; trustee Am. Sch. Oriental Rsch., 1969-71, 86—, Centro di Richerche Ligabue, 1984; Reckitt archaeol. lectr. Brit. Acad., 1973; dir. archaeol. surveys in Syria, 1965, excavation projects at Tepe Yahya, Iran, 1967-75, Sarazm, Tadjikistan, USSR, 1985, archaeol. surveys in Saudia Arabia, 1977-80, USSR, 1990-91; corresponding fellow Inst. Medio and Extremo Orient, Italy; mem. UNESCO com. for sci. study of mankind, 1989—. Author: (with J. Sabloff) Ancient Civilizations: The Near East and Mesoamerica, 1979; editor: (with J. Sabloff) The Rise and Fall of Civilizations, 1973, Ancient Civilizations and Trade, 1975, Hunters, Farmers and Civilization, 1979, Archaeological Thought In America, 1988; author, gen. editor Tepe Yahya: The Early Periods, 1986. Recipient medal Iran-Am. Soc., 1972; NSF grantee, 1966-75, 78-80; Nat. Endowment for Arts grantee, 1977—; Nat. Endowment for Humanities grantee, 1977—. Fellow Soc. Antiquaries Gt. Brit. and Ireland (sec. N.Am. chpt.), Am. Anthrop. Assn., AAAS (chmn. USA/USSR archaeol. exchange program), N.Y. Acad. Sci., USSR Acad. Sci., Soc. Am. Archaeology, Archeol. Inst. Am., Instituto para Medio et Extrema Oriente; mem. German Archaeol. Inst., Danish Archaeol. Inst., Brit. Archaeol. Inst., Tavern Club (Boston). Club: Tavern (Boston). Office: Peabody Mus Archaeology & Ethnology 11 Divinity Ave Cambridge MA 02138-2096

LAMBERT, DAVID L., astronomy educator. BS, Univ. Coll., Oxford, Eng.; PhD, Balliol Coll., Oxford. Research fellow Calif. Inst. Tech., Pasadena, Mount Wilson Palomar Observatories; with dept. astronomy U. Tex., Austin, 1969, prof., 1974—, now Isabel McCutcheon Harte Centennial prof. astronomy. Recipient Dannie Heineman Prize for Astrophysics, Am. Astron. Soc., 1987; Guggenheim fellow, vis. Erskine fellow Univ. Canterbury, New Zealand, 1985. Fellow Royal Astron. Soc.; mem. Am. Astron. Soc., Internat. Astron. Union. Office: U Tex Austin Dept Astronomy Austin TX 78712

LAMBERT, DENNIS MICHAEL, virologist, researcher; b. Chichamauga, Ga., May 12, 1947; s. Marshall Tarver and Mildred Louise (Dennis) L.; m. Andrea Lee Spencer, Aug. 24, 1968; children: Angela Lynn, Debra Megan. BA, Ind. Cen. Coll., 1969; MA, Ind. State U., 1972, PhD, 1977. Postdoctoral fellow Pub. Health Rsch. Inst., N.Y.C., 1976-77, Christ Hosp. Inst. Med. Rsch., Cin., 1977-79; asst. mem. James N. Gamble Inst. Med. Rsch., Cin., 1979-88; sr. investigator Smith Kline Beckman Pharm. Rsch. and Devel., King of Prussia, Pa., 1988-90; asst. dir. dept. virology and host def. SmithKline Beechman Pharm. Rsch. and Devel., King of Prussia, Pa., 1990-93; dir. virology Trimeris, Inc., Research Triangle Park, N.C., 1993—;

mem. antiviral rev. bd. U.K. SmithKline Beechman Pharm. Rsch. and Devel., Great Burgh, Eng., 1990-92. Author: Design of Anti-AIDS Drugs, 1990. Rsch. grantee NIAID/NIH, 1983, Thrasher Rsch. Fund, 1983, WHO, 1988. Mem. AAAS, Am. Soc. for Virology, Internat. Soc. for Antiviral Rsch., Am. Soc. for Microbiology. Achievements include first to demonstrate antiviral activity of HIV-1 protease inhibitors Nature 1990. Home: 101 Centerville Ct Cary NC 27513 Office: Trimeris Inc 1000 Park Forty Plz Ste 300 Durham NC 27713

LAMBERT, JAMES ALLEN, industrial electrician; b. Tanapeg, Saipan, Northern Mariana Islands, Aug. 30, 1956; came to U.S. 1958; s. Romulus Dewey and Frances Anajean (Moore) L.; m. Nancy Jean Lubniewski, Sept. 10, 1977; children: Kristin Noelle, James Allen Jr. Student, Gateway Tech. Inst., Kenosha, Wis., 1985-86, Coll. Lake County, Grayslake, Ill., 1987-91. Maint. group leader Anchor Hocking Corp., Gurnee, Ill., 1974-78; maint. tech. Cherry Elec. Products, Waukegan, Ill., 1978—. Mem. Phi Theta Kappa. Achievements include development of adjustable frequency drives and piece part inspection systems for high speed metal stamping presses. Home: 23615 82nd Pl Salem WI 53168

LAMBERT, JAMES LEBEAU, chemistry educator; b. Sanford, Fla., Feb. 11, 1934; s. Edward Arthur and Marcella (Labadie) L. BS, Spring Hill Coll., 1959; PhD, Johns Hopkins U., 1963. Chemistry educator Springs Hill Coll., Mobile, Ala., 1968—, acad. dean, 1976-78, prof. chmn. chemistry, 1982-93. Home: 4000 Dauphin St Mobile AL 36608 Office: Springs Hill Coll Dept Chemistry Mobile AL 36608

LAMBERT, JAMES MICHAEL, chemist, researcher; b. Charleston, W.Va., Oct. 20, 1959; s. Damon Keith and Bettie Lou (Mays) L.; m. Jeanne Elizabeth Barrett, Apr. 21, 1990. BS summa cum laude, W.Va. State Coll., Institute, 1981; PhD, Va. Poly. Inst., 1986. Scientist Rsch. Ctr. Becton Dickinson, Research Triangle Park, N.C., 1985-88; sr. scientist polymer rsch. Becton Dickinson, Dayton, Ohio, 1988-92; sr. scientist polymer surfaces Becton Dickinson Vascular Access, Sandy, Utah, 1993—. Author: (with others) Synthesis of Segmented Poly (Arylene Ether Sulfone)-Poly Arylene Terephalate) Compolymers, 1985; contbr. articles to Jour. Vynyl Tech., Polymer Preprints. Trustee Meml. Bapt. Ch., Dayton, 1991-92. Faculty scholar W.Va. State Coll., 1977-78, 79-80. 80-81. Mem. Am. Chem. Soc., Soc. Plastics Engrs. (med. plastics div., bd. dirs. 1991-92, chmn. edn. com. 1991-93). Baptist. Achievements include patents for liquid crystalline catheter, expandable catheter having hydrophobic surfaces. Office: Becton Dickinson 9450 S State St PO Box 1285 Sandy UT 84070

LAMBERT, JAMES MORRISON, physics educator; b. Chgo., Feb. 18, 1928; s. Robert D. and Elizabeth M. (Morrison) L.; m. Margaret E. Arnold, June 13, 1953; children: Karen E. Hartman, Diane M. Thomas, Byron J. AB, Johns Hopkins U., 1955, PhD, 1961. Instr. U. Mich., Ann Arbor, 1961-63, asst. prof., 1963-64; asst. prof. Georgetown U., Washington, 1964-68, assoc. prof., 1968-74, prof., 1974—, chmn. physics dept., 1984-90; cons. Navy Dept., Washington, 1966—; vis. scientist SLAC Stanford U., Palo Alto, Calif., 1985-90, CEBAF, Newport News, Va., 1991; vis. prof. Duke U., Durham, 1990-91. Mem. Southeastern U. Rsch. Assn. (trustee), Phi Beta Kappa (pres. 1960), Sigma Xi (sec. 1960). Home: 5312 Augusta St Bethesda MD 20816 Office: Georgetown U 37th O St NW Washington DC 20057

LAMBERT, JOHN BOYD, chemical engineer, consultant; b. Billings, Mont., July 5, 1929; s. Jean Arthur and Gail (Boyd) L.; m. Jean Wilson Bullard, June 20, 1953 (dec. 1958); children: William, Thomas, Patricia, Cathy, Karen; m. Ilse Crager, Sept. 20, 1980. BS in Engring., Princeton U., 1951; PhD, U. Wis., 1956. Rsch. engr. E.I. DuPont de Nemours Co., Wilmington, Del., 1956-69; sr. rsch. engr. Fansteel, Inc., Balt., 1969, mktg. mgr./plant mgr., North Chicago, Ill., 1970-73, mgr. mfg. engring., Waukegan, Ill., 1974-80, corp. tech. dir., North Chicago, 1980-86, gen. mgr. metals, 1987-91, v.p., corp. tech. dir., 1990-91; ind. cons., Lake Forest, Ill., 1991—; bd. dirs. Lake Forest Grad. Sch. Mgmt., 1984-91. Cont. to profl. publs. Sec. Del. Jr. C. of C., Wilmington, 1972-74. Recipient Charles Hatchett medal Inst. Metals, London, 1986. Mem. AICE, Am. Chem. Soc., Am. Soc. Metals, Sigma Xi. Episcopalian. Achievements include patents in field of dispersion-strengthened metals, refractory metals, chemical vapor deposition, both products and processes. Home and Office: 617 E Greenbriar Ln Lake Forest IL 60045-3214

LAMBERT, JON KELLY, engineer; b. Seattle, Nov. 4, 1954; s. William Edward and Irene Myrtle (Paulson) L.; m. Linda Lenore LeMere, July 18, 1980; 1 child, Kelly Renee. Cert. nuclear lead auditor. Nuclear quality control inspector various orgns., 1981-87; dir. quality Tanco, Inc., Houston, 1987; welding engr. Joy Technologies, Inc., Thompson, Tex., 1987; quality inspector Townsend & Bottum Svcs. Group, 1987-88; welding engr. M.K. Ferguson Co., Bridgman, Mich., 1988; quality engr. M.K. Ferguson Co., Aiken, S.C., 1988-90; welding engr. Westinghouse Savannah River Co., Aiken, 1990—. Mem. Am. Welding Soc. (cert. welding insp., mem. sub-com. structural welding code-steel, charter com. stainless steel welding code 1990—). Republican. Achievements include work on safety of nuclear power plants and defense nuclear facilities.

LAMBERT, JOSEPH BUCKLEY, chemistry educator; b. Ft. Sheridan, Ill., July 4, 1940; s. Joseph Idus and Elizabeth Dorothy (Kirwan) L.; m. Mary Wakefield Pulliam, June 27, 1967; children: Laura Kirwan, Alice Pulliam, Joseph Cannon. B.S., Yale U., 1962; Ph.D. (Woodrow Wilson fellow 1962-63, NSF fellow 1962-65), Calif. Inst. Tech., 1965. Asst. prof. chemistry Northwestern U., Evanston, Ill., 1965-69, assoc. prof., 1969-74, prof. chemistry, 1974-91, Clare Hamilton Hall prof. chemistry, 1991—; chmn. dept. Northwestern U., 1986-89, dir. integrated sci. program, 1982-85; vis. assoc. Brit. Mus., 1973, Polish Acad. Scis., 1981, Chinese Acad. Scis., 1988. Author: Organic Structural Analysis, 1976, Physical Organic Chemistry through Solved Problems, 1978, The Multinuclear Approach to NMR Spectroscopy, 1983, Archaeological Chemistry III, 1984, Introduction to Organic Spectroscopy, 1987, Recent Advances in Organic NMR Spectroscopy, 1987, Acyclic Organonitrogen Stereodynamics, 1992, Cyclic Organonitrogen Stereodynamics, 1992; audio course Intermediate NMR Spectroscopy, 1973; editor in chief Journal of Physical Organic Chemistry; editorial bd. Reson. Chem. mag.; contbr. articles to sci. jours. Recipient Nat. Fresenius award, 1976, James Flack Norris award, 1987, Fryxell award, 1989, Nat. Catalyst award, 1993; Alfred P. Sloan fellow, 1968-70, Guggenheim fellow, 1973, Interacad. exch. fellow (U.S.-Poland), 1985, Air Force Office sci. rsch. fellow, 1990. Fellow AAAS, Japan Soc. for Promotion of Sci., Brit. Interplanetary Soc., Ill. Acad. Sci. (life); mem. Am. Chem. Soc., Royal Soc. Chemistry, Soc. Archaeol. Scis. (pres. 1986-87, assoc. editor Bull.), Phi Beta Kappa, Sigma Xi. Home: 1956 Linneman St Glenview IL 60025-4264 Office: Northwestern U Dept Chemistry 2145 Sheridan Rd Evanston IL 60208-3113

LAMBERT, MICHAEL IRVING, systems engineer; b. Montpelier, Vt., Aug. 13, 1958; s. Gordon Rice and Barbara Ann (Lamson) L.; m. Judith Kietzmann, June 28, 1980; children: Christopher David, Anna Michelle. BA, U. Vt., 1980. Pres., sr. systems engr. Creative Mgmt. Cons., Williston, Vt., 1982—; sr. systems engr. McAuliffe Office Products, Burlington, Vt., 1986-92. Fellow IEEE; mem. Waterbury Area Lions Club (pres. 1987-89), Ctrl. Vt. PC Users Group (pres. 1984-86). Home: PO Box 477 Waterbury VT 05676-0477 Office: Creative Mgmt Cons Inc 2 Winter Sport Ln Williston VT 05495

LAMBERT, RALPH WILLIAM, civil engineer; b. Salem, Oreg., July 12, 1946; s. Chester William and Ethyl May (Butterfield) L.; m. Linda Lee Neuschwander, July 14, 1967. BSCE, Oreg. State U., 1969. Profl. Engr. Civil and Traffic, Oreg. Hwy. engr. I Washington Dept Transp., Bellevue, 1969-70, hwy. engr. II, 1970-72; civil engr. II City of Salem (Oreg.) Pub. Works, 1972-80, civil engr. III, 1980—. Pres. Mid. Williamette Coun. Square Dance, Salem, Oreg., 1990, Oreg. Fed. Square Dance Clubs, pres. 1991. Recipient Disting. Svc. award Mid Williamette Valley Coun. Govts., Salem. Oreg., 1985. Mem. Am. Sc. Civil Engrs., Inst. Transp. Engrs., Am. Pub. Works Assn. Home: 419 Idylwood Dr SE Salem OR 97302 Office: City of Salem Pub Works 555 Liberty St SE Salem OR 97301

LAMBERT, RICHARD BOWLES, JR, science foundation program director, oceanographer; b. Clinton, Mass., Apr. 20, 1939; s. Richard Bowles and Dorothy Elisabeth (Peck) L.; m. Sherrill Faye Smith, July 4, 1964; 1 child, Lisa Beth Lauren. AB in Physics, Lehigh U., 1961; ScM in Physics, Brown U., 1964, PhD in Physics, 1966; postgrad., Goethe Institut, Germany, 1966, NATO Internat. Sch., Germany, 1966, Max Planck Inst. for Physics & Astrophysics, Germany, 1966. Fulbright Postdoctoral fellow Institut fur Stromungsmechanik Technische Hochschule, Munich, Germany, 1966-67; asst. prof. Grad. Sch. Oceanography U. R.I., 1968-74, assoc. prof. Grad. Sch. Oceanography, 1974; program dir. physical oceanography program NSF, Washington, 1975-77; rsch. oceanography Sci. Applications Internat. Corp., 1977-79, mgr. ocean physics divsn., 1979-83, asst. v.p., 1980-83, sr. rsch. oceanographer, 1983-84; assoc. program dir. physical oceanography program NSF, Washington, 1984-91, program dir. physical oceanography program, 1991—; adv. com. NOAA; assoc. dir. U.S. TOGA Project Office 1985-91; delegate Intergovernmental TOGA Bd.; delegation head Intergovernmental WOCE Panel; co-investigator R/V Trident Oceanographic Cruises, Feb. 1971 (21 days), Dec. 1971 (6 days), April 1974 (14 days), Nov. 1974 (19 days); chief scientist R/V Trident Oceangraphic Cruises, Oct. 1971 (17 days), Nov. 1971 (8 days), April 1973 (15 days); co-investigator R/V Atlantis II Oceanographic Cruises, July 1973 (18 days). Contbr. articles to Jour. Fluid Mech. and other profl. jours. Bd. advisors Christian Performing Artist's Fellowship, Fairfax, Va., 1988—. Mem. Am. Assn. Advancement Sci., Am. Geophysical Union, Am. Physical Soc., The Oceanography Soc. (life mem.), Phi Beta Kappa, Sigma Xi. Office: NSF Phys Oceanography Program 4201 Wilson Blvd Arlington VA 22203

LAMBERT, RICHARD DALE, JR., structural engineer, consultant; b. Lansing, Mich., Dec. 7, 1952; s. Richard Dale Sr. and Ruth (Mittelstadter) L.; m. Cynthia Sue Johnston, June 2, 1973; children: Thomas, Rebecca, Russell. BSCE, N.C. State U., Raleigh, 1975. Cert. structural engr.; cert. land surveyor. Structural engr. U.S. Army C.E., Charleston, S.C., 1979—; structural engr., prin. Lambert Engring. Co., Charleston, S.C., 1983—. Elder Grace Alliance Ch., North Charleston, S.C., 1990—. Mem. ASCE, Soc. Am. Mil. Engrs., Charleston Civil Engrs. Club. Home and Office: 820 Creekside Dr Mount Pleasant SC 29464

LAMBERT, RICHARD WILLIAM, mathematics educator; b. Gettysburg, Pa., May 1, 1928; s. Allen Clay and Orpha Rose (Hoppert) L.; m. Phyllis Jean Bain, Sept. 2, 1949 (div. May 1982); children: James Harold, Dean Richard; m. Kathleen Ann Waring, Aug. 30, 1982; stepchildren: Gregory Scott Gibbs, LeAnn Marie Gibbs. BS, Oreg. State U., 1952; MA in Teaching Math., Reed Coll., 1962. Instr. Siuslaw High Sch., Florence, Oreg., 1954-55, David Douglas High Sch., Portland, Oreg., 1955-67; instr. Mt. Hood Community Coll., Gresham, Oreg., 1967-87, ret., 1987. NSF grantee, 1959, 60, 62. Mem. Nat. Coun. Tchrs. of Maths., Am. Math. Assn., Am. Math. Assn. of Two Yr. Colls., Oreg. Coun. Tchrs. of Maths. Democrat. Methodist. Avocations: travel, camping, home improvements, reading. Home: 11621 SE Lexington St Portland OR 97266-5933

LAMBIRTH, THOMAS THURMAN, clinical psychologist; b. Decatur, Ill., Dec. 24, 1949; s. Thurman Albert and Frances Hester (Ritchey) L. BA in Psychology, Millikin U., 1974; MA in Clin. Psychology, Ea. Ill. U., 1982; PhD in Counseling Psychology, U. So. Miss., 1988. Lic. clin. psychologist, Ill. Psychology intern VA Med. Ctr., Ann Arbor, Mich., 1987-88; postdoctoral fellow Office of Naval Tech. Naval Aerospace Med. Rsch. Lab., Pensacola, Fla., 1988-89; staff psychologist Ill. Dept. of Mental Health, Decatur, 1990—. Contbr. articles to profl. jours. Mem. APA (div. 36), Human Factors Soc. (chair, program co-chair tech. group on personality and individual differences in human performance 1988-89). Achievements include rsch. on perceptual and cognitive factors in obsessive-compulsive behavior and the devel. of a computer-mounted personality instrument. Home: 2510 Twin Oaks Ct Apt 17 Decatur IL 62526 Office: Adolf Meyer Ctr 2310 E Mound Rd Decatur IL 62526

LAMBOS, WILLIAM ANDREW, computer consultant, social science educator; b. N.Y.C., Sept. 7, 1956; s. Constantine Peter and Theodora (Kaganis) L.; m. Barbara Ann O'Brien, Feb. 25, 1989. BA in Psychology, Vassar Coll., 1979; postgrad., McMaster U., Hamilton, Ont., Can., 1980-85, PhD in Psychology, 1986; postgrad., Kennedy Western U., 1991—. Mgr. info. svcs. The Eagles, Ltd., Odessa, Fla., 1986; ind. computer cons. Tampa, Fla., 1986—; adj. prof. social sci. St. Petersburg Jr. Coll., Tarpon Springs, Fla., 1986—; Campus Lands, Inc., Gainesville, Fla., 1980—. Contbr. articles to profl. jours. Mem. APA, Pinellas IBM Users Group. Democrat. Greek Orthodox. Avocations: racquetball, classical guitar, photography, pro-football. Office: 16407 Birkdale Dr Odessa FL 33556-2806

LAMBREMONT, EDWARD NELSON, JR., nuclear science educator; b. New Orleans, July 29, 1928; s. Edward Nelson and Caroline Josephine (Joachim) L.; m. Mary Chris Bittle, May 30, 1981 (dec. Jan. 1987); m. Carol Jane Annis, June 16, 1951; children: Carol, Suzanne, John, Barbara; m. Janice P. Savoy, Apr. 6, 1990. B.S., Tulane U., 1949, M.S., 1951; Ph.D., Ohio State U., 1958. Research entomologist U.S. Dept. Agr., Baton Rouge, 1958-66; assoc. prof. nuclear sci. La. State U., Baton Rouge, 1966-73, prof., 1973—, dir. Nuclear Sci. Ctr., 1974—; councilor Oak Ridge Assoc. Univs., 1971-79, bd. dirs., 1979-84, vis. scientist med. div., 1967—; vis. scientist Internat. Atomic Energy Agy. Lab., Seibersdorf, Austria, 1988-90; cons. nuclear-related corps., 1978—. Contbr. articles to sci. jours. Served with U.S. Army, 1951-54, col. Res. (ret.). Grantee NIH, 1964-71; grantee NSF, 1971-77, U.S. Dept. Agr., 1979-83. Mem. AAAS, Entomol. Soc. Am., Health Physics Soc., Sigma Xi (S.E. regional dir. 1983-90, bd. dirs. 1983-90). Home: 2913 Calanne Ave Baton Rouge LA 70820-5408 Office: La State U Nuclear Sci Ctr Baton Rouge LA 70803

LAMEIRO, GERARD FRANCIS, research institute director; b. Paterson, N.J., Oct. 3, 1949; s. Frank Raymond and Beatrice Cecilia (Donley) L.; BS, Colo. State U., 1971, MS, 1973, PhD, 1977. Sr. scientist Solar Energy Rsch. Inst., Golden, Colo., 1977-78; asst. prof. mgmt. sci. and info. systems Colo. State U., Fort Collins, 1978-83, mem. editorial bd. energy engring., 1978-82, editorial bd. energy econs. policy and mgmt., 1981-82, lectr. dept. computer sci., 1983, lectr. dept. mgmt., 1983; pres. Successful Automated Office Systems, Inc., Fort Collins, 1982-84; product mgr. Hewlett Packard, 1984-88; computer networking cons., 1988-89, Ft. Collins; mem. editorial bd. The HP Chronicle, 1986-88, columnist, 1988, mgmt. strategist, 1988-91; dir. Lameiro Rsch. Inst., 1991—. Mem. editorial bd. Hp Chronicle, 1986-88, Energy Engring., Policy and Mgmt., 1981-82, Energy Engring., 1978-82. Mem. P-resdl. Electoral Coll., 1980. Recipient nat. Disting. Svc. award Assn. Energy Engrs., 1981, Honors Prof. award Colo. State U., 1982; Colo. Energy Rsch. Inst. fellow 1976; NSF fellow 1977. Mem. Assn. for Computing Machinery, Assn. Energy Engrs. (pres. 1980, Nat. Distinguish Service award 1981, internat. bd. dirs. 1980-81), Am. Mgmt. Assn., Am. Soc. for Tng. and Devel., Am. Mktg. Assn. (exec.), Am. Soc. For Quality Control, IEEE Computer Soc., Inst. Indsl. Engrs., U.S. C. of C., Crystal Cathedral Golden Eagles Club, The Heritage Found., Sigma Xi, Phi Kappa Phi, Beta Gamma Sigma, Kappa Mu Epsilon. Roman Catholic. Contbr. articles in mgmt. and tech. areas to profl. jours. Home: PO Box 9580 Fort Collins CO 80525-0500 Office: 3313 Downing Ct Fort Collins CO 80526-2315

LAMIRANDE, ARTHUR GORDON, editor; b. Holyoke, Mass., July 19, 1936; s. Joseph Armand Arthur and Marion Gordon (Beaton) L. AA, Holyoke Community Coll., 1956; BA, Am. Internat. Coll., 1959. Editorial asst. Merriam-Webster Dictionary, Springfield, Mass., 1959-61; assoc. editor N.Y. Acad. Scis., N.Y.C., 1966-67; mng. editor Grune & Stratton Pub. Co., N.Y.C., 1967-68; assoc. editor H.S. Stuttman Co., N.Y.C., 1968-70, Sci. and Medicine Pub. Co., N.Y.C., 1971-73; editorial cons. N.Y.C., 1973-81; dir. editorial dept. Profl. Exam. Svc., N.Y.C., 1981-92, exec. editor, 1992—. Contbr. articles to The Diapason, PES News. Organist Ch. of Holy Name of Jesus, N.Y.C., 1973-82. Home: 461 Fort Washington Ave New York NY 10033 Office: Profl Exam Svc 475 Riverside Dr New York NY 10115 also: 461 Fort Washington Ave New York NY 10033

LAMM, DONALD LEE, urologist; b. L.A., Oct. 24, 1943; m. Wanda Kay Hartgrove; children: Donald A., Loreen M., Theresa D., Kelly Jo Anne. BA in Zoology, UCLA, 1965, MD, 1968. Diplomate Am. Bd. Urology. Intern U. Oreg. Med. Sch., Portland, 1968-69; resident in gen. surgery UCLA Med. Sch., 1971-72; resident in urology U. Calif., San Diego, 1972-76; active staff Bexar County Hosp., San Antonio, 1976-85; chief urology sect. surg. svc. Audie L. Murphy Meml., VA Hosp., San Antonio, 1976-85; asst. prof. div. urology, dept. surgery to assoc. prof. U. Tex. Health Sci. Ctr., San Antonio, 1976-85, acting head div. urology dept. surgery, 1984-85, prof. div. urology, 1983-85; prof., chmn. Dept. Urology, W.Va. U. Med. Ctr., Morgantown, 1985—. Contbr. numerous articles to profl. jours. With USPHS, 1969-71. Recipient numerous grants, 1976—. Mem. AAAS, ACS, AMA, Am. Soc. Clin. Oncology, Am. Urol. Assn. (program com. 1988—, chmn. ad hoc com. for FDA approval of BCG, mid-Atlantic sect. editorial com. 1988—), Assn. for Acad. Surgery, Ea. Coop. Oncology Group, Monongalia County Med. Soc., San Antonio Urol. Soc. (sec., treas. 1983-84, pres. 1984-85), Soc. of U. Urology Residents (chmn. 1975-76), Soc. Urologic Oncology (chmn. bylaws com.), So. Med. Assn. (assoc. councilor for W.Va. 1988, pres. Urology sect., 1992), others. Achievements include initial controlled trial demonstration of superiority of Bacillus Calmette-Guerin (BCG) immunotherapy over surgery alone for superficial transitional cell carcinoma of the bladder; subsequential demonstration of superiority of BCG over doxorubicin and mitomycin C chemotherapy and the superiority of maintenance BCG immunotherapy. Home: 312 Jackson Ave Morgantown WV 26505 Office: Dept Urology WVa U Morgantown WV 26506

LAMMERS, THOMAS GERARD, museum curator; b. Burlington, Iowa, Sept. 20, 1955; s. Robert William and Shirley Anne (Lohman) L.; m. Diane Lynn Seebach, June 19, 1976; children: Valerie Andrea, Michael Thomas. BS in Botany, Iowa State U., 1977; MA in Biology, U. No. Iowa, 1981; PhD in Botany, Ohio State U., 1988. Asst. curator dept. botany Field Mus. Natural History, Chgo., 1990—; vis. asst. prof. dept. botany Miami U., Oxford, Ohio, 1988-90. NSF Rsch. grantee, 1984; recipient Alumni Rsch. award Ohio State U., 1987. Mem. Am. Soc. Plant Taxonomists (Herbarium Travel award 1987), Botanical Soc. Am., Internat. Assn. Plant Taxonomy, Sigma Xi (Rsch. grantee 1983). Office: Field Mus Natural History Roosevelt Rd/ Lake Shore Dr Chicago IL 60605-2496

LAMMERTSMA, KOOP, chemist, consultant; b. Makkum, Frysland, The Netherlands, Aug. 29, 1949. MS, U. Groningen, The Netherlands, 1975; PhD, U. Amsterdam, 1979. Postdoctoral fellow U. Coll. London, Eng., 1980, U. Erlangen, Nürnberg, Germany, 1980-81; rsch. fellow U. So. Calif., L.A., 1981-83; asst. prof. U. Ala., Birmingham, 1983-87, assoc. prof., 1987-92, prof., 1992—. Contbr. numerous articles to profl. jours. Mem. Am. Chem. Soc. Office: U Ala Uab # 221 Birmingham AL 35294-0001

LAMMING, ROBERT LOVE, retired surgeon; b. Hull, Yorkshire, Eng., Dec. 6, 1910; s. Robert Sydney and Clara Gertrude (Redfearn) L.; m. Olive Mary Callow, Sept. 23, 1939 (dec. May 1970); children: Alison Joan, Roberta Mary. BSc in Physiology, U. Leeds, Eng., 1932, MSc in Physiology, 1933, MS ChB, 1936; FRCS, Royal Coll. Surgeons, Eng., 1939. Intern, then resident Gen. Infirmary at Leeds, 1939-45; cons. Isle of Man Health Svc., Douglas, 1946-76; ret., 1976; hon. surgeon to gov.'s household Isle of Man, 1975-85. Maj. Royal Army M.C., 1939-46. Decorated Territorial Decoration, Order of Brit. Empire, knight Order St. Larazus of Jerusalem. Mem. Masons. Avocations: shooting, fishing. Home and Office: Kantara Little Switzerland, Douglas Isle of Man

LAMOREAUX, PHILIP ELMER, geologist, hydrogeologist, consultant; b. Chardon, Ohio, May 12, 1920; s. Elmer I. and Gladys (Rhodes) L.; m. Ura Mae Munro, Nov. 11, 1943; children: Philip E Jr., James W., Karen L. BA, Denison U., 1943, PhD (hon.), 1972; MS, U. Ala., 1949. Registered profl. geologist, Ga., N.C., S.C., Tenn., Ind., Ariz., Ark., Fla., Wyo. Geologist U.S. Geol. Survey, Tuscaloosa, Ala., 1943-45, dist. geologist Groundwater Office, 1945-57, div. hydrologist Water Resources Programs, 1957-59; chief Ground Water Br. U.S. Geol. Survey, Washington, 1959-61; state geologist, oil and gas supr. Ala. Geol. Survey, Tuscaloosa, 1961-76; pres. P.E. LaMoreaux & Assocs. Inc., Tuscaloosa, 1976-87, chmn. bd., 1987-90, sr. hydrologist, 1990—; mem. Karst Commn., 1961—; chmn. Hydrology Hazardous Waste Commn., 1983-91; lectr. Am. Geol. Inst. College Program, 1961-71, Am. Geophys. Union Coll. Program, 1961—, NSF, Ala. Acad. Sci. High Sch. Program, 1961—, No. Engring. and Testing , Salt Lake City, 1985, Ga. State U., Fla. State U., Vanderbilt U., Denison U., Auburn U., U. of Montpellier, France, U. Christ Church, New Zealand, University of Praetoria, Republic of South Africa; hydrogeology cons. to 30 fgn. countries. Editor in chief Jour. Environ. Geology and Water Scis., 1984-92; editor in chief: Annotated Bibliography Carbonate Rocks, vols. 1-5; author numerous published articles in field. Mem. Nat. Drinking Water Adv. Coun. EPA, 1984-88, mem. Tech. Rev. Group Oak Ridge Nat. Lab., 1984-88; trustee Denison U. Recipient Comdrs. medal C.E., 1990. Mem. NAS (nat. rsch. coun. geotech. bd. 1990-92, water sci. and tech. bd. 1990-92, bd. on earth scis. and resources 1992—), AAAS, Ala. Acad. Sci., Ala. Geol. Soc., Am. Assn. Petroleum Geologists (acad. liaison com., del. to Ho. of Dels. 1970-72, com. for preservation samples and cores, chmn. divsn. geosci. hydrogeology com.), Am. Geol. Inst. (chmn. com. on publs. 1968-70, pres. 1971-72, Ian Campbell award 1990), Am. Geophys. Union, Am. Inst. Hydrology, AIME, Am. Inst. Profl. Geologists (chmn. com. on rels. with govtl. agencies 1967-70, bd. dirs. 1969-70), Am. Soc. Testing and Materials, Assn. Am. State Geologists (statistician 1966-69, chmn. liaison com. with fed. agencies 1968-70, pres.), Geol. Soc. Am. (1st chmn. hydrogeology group 1963, chmn. O.E. Meinzer award com. 1965, cons. on membership S.E. Sect. 1967-68, chmn. nominating com., bd. dirs., bd. trustees, publs. com.), Geol. Soc. London, Internat. Assn. Hydrogeologists (pres. 1977-80, v.p. 1973-77, mem. com. on water rsch., 1978-80), Internat. Water Resources Assn., Interstate Oil Compact Commn. (vice chmn. 1963, chmn. rsch. com.), Miss. Geol. Soc., NAE, Nat. Assn. Geology Tchrs., Nat. Rivers and Harbors Congress, Nat. Speleological Soc., Nat. Water Resources Assn., Nat. Water Well Assn., Soc. Econ. Geologists, Soc. Econ. Paleontologists and Mineralogists, Soil Conservation Soc. Am., Southeastern Geol. Soc., Ala. C. of C. (Pres.'s adv. com., Rep. of Energy, 1980), Geol. Soc. Am. (chmn.), numerous other orgns. and offices. Republican. Presbyterian. Avocations: golfing, photography, stamp collecting, coin collecting, gardening. Office: P E LaMoreaux & Assocs Inc 2610 University Blvd Tuscaloosa AL 35401-1566

LAMP, BENSON J., tractor company executive; b. Cardington, Ohio, Oct. 7, 1925; m. Martha Jane Motz, Aug. 21, 1958; children: Elaine, Marlene, Linda, David. BS in Agrl. and B in Agrl. Engring., Ohio State U., 1949, MS in Agrl. Engring., 1952; PhD in Agrl. Engring., Mich. State U., 1960. Registered profl. engr., Ohio. Prof. agrl. engring. Ohio State U., Columbus, 1949-61, 87-91, prof. emeritus, 1991—; product mgr. Massey Ferguson Ltd., Toronto, Can., 1961-66; product planning mgr. Ford Tractor Ops. div. Ford Motor Co., Troy, Mich., 1966-71, mktg. mgr., 1971-76, bus. planning mgr., 1978-87; v.p. mktg. and devel. Ford Aerospace div. Ford Motor Co., Dearborn, Mich., 1976-78. Author: Corn Harvesting, 1962. Served to 2d lt. USAF, 1943-45. Fellow Am. Soc. Agrl. Engrs. (pres. 1985-86, Gold medal 1993), Country Club at Muirfield Village (Dublin, Ohio). Avocations: golf, tennis, bridge. Office: BJM Company Inc 6128 Inverurie Dr E Dublin OH 43017

LAMPERT, LEONARD FRANKLIN, mechanical engineer; b. Mpls., Nov. 13, 1919; s. Arthur John Lampert and Irma (Potter) Smith. BME, U. Minn., 1943, B in Chem. Engring., 1959, MS in Biochemistry, 1964, PhD in Biochemistry, 1969. Registered profl. engr., Minn. With flight measurement rsch. dept. Douglas Aircraft Corp., El Segundo, Calif., 1943-47; researcher, tchr. U. Minn., Mpls., 1947-83; with rsch. engring. dept. Mpls. Honeywell Corp., 1950-55; info. scientist Control Data Corp., Mpls., 1982-88; mech. engr. Leonard Lampert Co., White Bear Lake, Minn., 1988—; scientist Eurasion Watermilfoil Control, White Bear Lake, 1989—; stockholder rep. Lampert Lumber Co., St. Paul, 1988—. Contbr. articles to profl. jours. Mem. Am. Inst. Chem. Engrs. (award 1959), Am. Chem. Soc., U. Minn. Alumni Assn. (advisor) MIT Alumni Assn. (advisor), Phi Gamma Delta (advisor), Gamma Alpha, Phi Lambda Upsilon. Republican. Avocations: Ballroom dancing competitions, snow and water skiing, biking, geography, travel. Home and Office: 2467 S Shore Blvd Saint Paul MN 55110-3820

LAMPSON, BUTLER WRIGHT, computer scientist; b. Washington, Dec. 23, 1943; s. Edward Tudor and Mary Caroline (Wright) L.; m. Lois Helen Alterman, Sept. 23, 1967; children—Michael Alterman, David Wright. A.B., Harvard U., 1964; Ph.D., U. Calif.-Berkeley, 1967; D.Sc. (hon.), Eidgenossiche Technische Hochschule, Zurich, 1986. Asst. prof. U. Calif.-Berkeley, 1967-70, assoc. prof., 1970-71; dir. system devel. Berkeley Computer Corp., 1969-71; prin. scientist Xerox Research Ctr., Palo Alto, Calif., 1971-75, sr. research fellow, 1975-84; sr. cons. engr. Digital Equipment Corp., Palo Alto, 1984-86, corp. cons. engr., 1986—. Contbr. articles to profl. jours. Patentee in field. Mem. NAE, Assn. Computing Machinery (Software System award 1984, A.M. Turnig award 1992). Office: Digital Equipment Corp 1 Kendall Sq Cambridge MA 02139-1562

LAMSON, EVONNE VIOLA, computer software company executive, computer consultant, pastor, Christian education administrator; b. Ithaca, Mich., July 8, 1946; d. Donald and Mildred (Perdew) Guild; m. James E. Lamson, Nov. 2, 1968; 1 child, Lillie D. Assoc. in Math., Washtenaw C.C., Ypsilanti, Mich., 1977; BS, Ea. Mich. U., 1989; MA in Pastoral Counseling Ashland (Ohio) Theol. Sem., 1993. Data base mgr. ERIM, Ann Arbor, Mich., 1978-91; mgr. product svcs. Comshare, Ann Arbor, 1981-90, project leader, tng. course designer info. techs., 1991-93; founder, pres. G & L Consultants, Brighton, Mich., 1982—; tng. specialist ComShare, Ann Arbor, 1990-93, tng. course designer, 1991-93; Assoc. Pastor, dir. Christian edn. Keystone Community Ch., Saline, Mich., 1993—; founder Living Water Ministries, 1993—. Study leader Brighton Wesleyan Ch., 1981-93; lic. minister Weseleyan Ch. Am., 1993—; program dir. Wesleyan Womens Assn. of Brighton, 1983-91; clin. staff counselor Women's Resource Ctr., Howell, Mich., 1991—. Mem. AACD, NAFE, Am. Mgmt. Assn., Fairbanks Family of Am., Internat. Platform Assn. Avocations: skiing, motivational speaking, reading. Home: 6708 Calfhill Ct Brighton MI 48116 Office: Keystone Community Ch 567 Eastlook Saline MI 48176

LAMUNYON, CRAIG WILLIS, biology researcher; b. Fullerton, Calif., Feb. 2, 1960; s. Harold Lee and Bernardine Marie (Schutten) LaM.; m. Cynthia Ann Munson, Aug. 20, 1983; children: Diana Nicole, Kyle Clark. BA in Biology, Calif. State U., Fullerton, 1984; PhD in Chem. Ecology, Cornell U., 1992. Rsch. assoc. Cornell U., Ithaca, N.Y., 1992-93; rsch. assoc. dept. molecular and cellular biology U. Ariz., Tucson, 1993—; predoctoral trainee NIMH, Cornell U., 1990-92. Mem. AAAS, Entomol. Soc. Am., Animal Behavior Soc., Sigma Xi. Achievements include research in multiple mating and its implications in an arctiid moth, use and effect of an anal defensive substance in larval chrysopidae.

LANA, PHILIP K., emergency response educator, consultant. BS in Biology, Abilene Christian Univ., 1973, MS in Microbiology, 1976. Cert. hazards mgr., hazardous materials supr., registered environ. profl. Grad. teaching asst. Biology dept. Abilene (Tex.) Christian Univ., 1973-76, biology tchr., 1976-81; training specialist Tex. A&M Univ., College Station, 1981-85; hazardous materials instr. Assn. Am. Railroads, Pueblo, Colo., 1985-86, mgr., 1986-90; pres. Philip Lana Assoc., Inc., Denton, Tex., 1990—; adv. Pueblo Colo. Hazardous Materials Response Authority, 1986-90; exec. subcom. Am. Soc. Testing and Materials, Phila., 1990—. Editorial adv. bd. Environmental Waste Management, 1991—; contbr. articles to profl. jours. Reserve deputy Brazos County, Tex. Sheriff's Office, Bryan, Tex., 1984-85; deputy Kiowa County, Colo. Sheriff's Office, Eads, Colo., 1987—; mem. Denton County Extension Environ. Com., 1992—. Mem. Spill Control Assn. Am., Nat. Environ. Training Assn., Am. Soc. Testing and Materials. Office: Philip Lana Assocs Inc 1807 N Elm Ste 122 Denton TX 76201-3023

LANCASTER, FRANCINE ELAINE, neurobiologist; b. Belton, Tex., Feb. 25, 1947; d. James Henry and Doris (Foster) L.; children: Jennifer, Stacy, Noelle. BA in Psychology, Tex. Woman's U., 1973, MS in Biology, 1975, PhD in Molecular Biology, 1977. Lectr. biology Tex. Woman's U., Denton, 1976-77; asst. prof. biology Tex. Woman's U., Houston, 1977, coord. biol. dept., 1978-91, assoc. prof. biology, 1983-89, prof. biology, 1989—; health scientist adminstr. Nat. Inst. Alcohol Abuse and Alcoholism, Rockville, Md., 1991-93. Contbr. chpts. to books in field; contbr. articles to profl. jours. Clayton Found. researcher; rsch. grantee NIAAA. Mem. Soc. Neurosci., Am. Soc. Neurochemistry, Am. Women in Sci., Rsch. Soc. Alcoholism, Sigma Xi. Office: Tex Woman's U Dept Biology PO Box 23971 Denton TX 76204

LANCASTER, JOHN HOWARD, civil engineer; b. Bklyn., July 3, 1917; s. George York and Alice Eliot (Littlejohn) L.; m. Phyllis Elaine Metcalf, June 1, 1938; children: Judith Ann, Barbara Jean, Marylin Sharon, Kathryn Joy, Debra Elizabeth. B.S., Worcester (Mass.) Poly. Inst., 1939. Lic. master mariner USCG. Engr. Austin Co., N.Y.C., 1939-40; engr. C.E., 1940-42; asst. to div. engr. C.E., N.Y.C., 1942-43; chief engring. and constrn. AEC, Upton, N.Y., 1946-54; chief project engr. Brookhaven Nat. Lab., Upton, 1954-72; asst. dir. Nat. Radio Astronomy Obs. and program mgr. very large array radiotelescope program, Socorro, N.Mex., 1972-81; propr. John H. Lancaster & Assos. (cons. engrs.), 1950-72; cons. NRAO/Associated Univs. Inc., 1981—; cons. in field, 1970—; mem. comml. panel Am. Arbitration Assn. Served with USNR, 1943-46. Recipient Meritorious Service award NSF, 1976. Mem. NSPE, N.Y. Soc. Profl. Engrs., N.Mex. Soc. Profl. Engrs., Rotary, Masons, Sigma Xi, Alpha Tau Omega. Home: PO Box 981 1400 Hilton Pl Socorro NM 87801

LANCE, JAMES WALDO, neurologist; b. Wollongong, Australia, Oct. 29, 1926; s. Waldo Garland and Jessie Forsyth (Stewart) L.; m. Judith Lilian Logan, July 6, 1957; children: Fiona, Sarah, Jennifer, Robert, Sophie. MB, BS, U. of Sydney, Australia, 1950, MD, 1955; DSc (hon.), U. NSW, 1992. Resident med. officer Royal Prince Alfred Hosp., Sydney, Australia, 1950-51; rsch. fellow U. Sydney, 1952-53; house physician Hammersmith Hosp., London, 1954; asst. house physician Nat. Hosp. for Nervous Diseases, London, 1955; supt. Northcott Neurological Ctr., Sydney, Australia, 1956-60; rsch. fellow Mass. Gen. Hosp., Boston, Mass., 1960-61; neurologist The Prince Henry Hosp., Sydney, 1961-91; prof. of neurology U. of New South Wales, Sydney, 1975-91; cons. neurologist Inst. Neurol. Scis., Randwick, Australia, 1991—. Author: Mechanism and Management of Headaches, 1969, 5th edit. 1993, Migraine and Other Headaches, 1986; co-author: A Physiological Approach to Neurology, 3rd edit. 1981, Introductory Neurology, 1983, 2nd edit. 1989. Decorated with comdr. Order of the Brit. Empire, 1978, officer Order of Australia, 1991. Fellow Royal Coll. Physicians (London), Royal Australasian Coll. Physicians, Australian Acad. of Sci. (v.p. 1984-85); mem. Australian Assn. of Neurologists (pres. 1978-81), Internat. Headache Soc. (pres. 1987-89), World Fed. of Neurology (v.p. 1989—). Avocations: skiing, trout fishing. Home: 15 Coolong Rd, Vaucluse 2030, Australia Office: Inst Neurol Scis Prince of Wales Hosp, High St, Randwick 2031, Australia

LAND, GEOFFREY ALLISON, science administrator; b. Jeannette, Pa., July 9, 1942; s. Albert E. Jr. and Helene (Matthews) L.; m. Maxine McCluskey, Jan. 22, 1966; children: Kevin Jeffrey, Melissa Allison, Kyle Robert. MS in Biology (Biochemistry), Tex. Christian U., 1970; PhD in Microbiology/Immunology, Tulane U., 1973. Cert. clin. lab. dir. Am. Bd. Bioanalysis. Dir. mycology Wadley Institutes Molecular Medicine, Dallas, 1974-78; dir. mycology, assc. dir. microbiology U. Cin. Med. Ctr., 1978-81; dir. microbiology/immunology Meth. Med. Ctr., Dallas, 1981—, assoc. adminstrv. dir. pathology, 1990—; dir. histocompatibility Stewart Blood Ctr., Tyler, Tex., 1987—; adj. full prof. biology Tex. Christian U., Ft. Worth, 1982—. Mem. rev. bd. Jour. Clin. Microbiology, 1980—; author: Pictorial Handbook of Medically Important Fungi, 1982, (with others) The Dermatophytes, 1992, Handbook of Applied Mycoses, 1991, Manual of Clinical Microbiology, 1992. Coach Denton (Tex.) Soccer Assn., 1981—; minister Tioga (Tex.) Ch. of Christ, 1985—. Recipient Svc. Above Self award Rotary Club. Mem. Mycol. Soc. Am. (pres. 1990-93), Billy H. Cooper-Maridian award 1992), Tex. Soc. Clin. Microbiology (pres. 1977-79, 81-83), N.Y. Acad. Sci., Am. Soc. Histocompatibility and Immunogenetics, Am. Soc. Microbiology. Mem. Ch. of Christ. Office: Meth Med Ctr 1441 N Beckley Ave Dallas TX 75203

LANDAU, WILLIAM MILTON, neurologist; b. St. Louis, Oct. 10, 1924; s. Milton S. and Amelia (Rich) L.; m. Roberta Anne Hornbein, Apr. 3, 1947; children: David, John, Julia, George. Student, U. Chgo., 1941-43; M.D. cum laude, Washington U., St. Louis, 1947. Diplomate: Am. Bd. Psychiatry and Neurology (dir. 1967, pres. 1975). Intern U. Chgo. Clinics, 1947; resident St. Louis City Hosp., 1948; fellow Washington U., St. Louis, 1949-52, NIH, Bethesda, Md., 1952-54; instr. neurology Washington U., 1952-54, asst. prof., 1954-58, assoc. prof., 1958-63, prof., 1963—, dept. head, 1970-91, co-head dept. neurology and neur. surgery, 1975; chmn. Nat. Com. for Research in Neurol. and Communicative Disorders, 1980. Editorial bd.: Neurology, 1963, A.M.A. Archives Neurology, 1965, Annals Neurology, 1977. Mem. ACLU (trustee East Mo. 1956—), Am. Neurol. Assn. (pres. 1977), Am. Acad. Neurology, Assn. U. Profs. Neurology (pres. 1978), Soc. Neurosci., Am. Physiol. Soc., Am. Electro-encephalography Soc. Rsch. in neurophysiology. Office: 660 S Euclid Ave Saint Louis MO 63110-1093

LANDAUER, ROLF WILLIAM, physicist; b. Stuttgart, Germany, Feb. 4, 1927; came to U.S., 1938, naturalized, 1944; s. Karl and Anna (Dannhauser) L.; m. Muriel Jussim, Feb. 26, 1950; children—Karen, Carl, Thomas. SB, Harvard U., 1945, AM, 1947, PhD, 1950; DSc (hon.), Technion, 1991. Solid state physicist Lewis Lab., NACA (now NASA), Cleve., 1950-52; with IBM Research (and antecedent groups), 1952—; asst. dir. research IBM Research (T. J. Watson Research Center), Yorktown Heights, N.Y., 1966-69; IBM fellow IBM Research (T. J. Watson Research Center), 1969—; Scott lectr. Cavendish Lab., 1991. Contbr. articles on solid state theory, computing devices, statis. mechanics of computational process to profl. jours. Served with USNR, 1945-46. Recipient Stewart Ballantine medal Franklin Inst., 1992, Centennial medal Harvard U., 1993. Fellow IEEE, AAAS, Am. Phys. Soc.; mem. NAE, NAS, European Acad. Scis. and Arts, Am. Acad. Arts and Scis. Achievements include initiating IBM programs leading to injection laser, large scale integration. Office: IBM Rsch Ct PO Box 218 Yorktown Heights NY 10598-0218

LANDE, ALEXANDER, physicist, educator; b. Hilversum, The Netherlands, Jan. 5, 1936; s. Leo and Bella (Berlin) L. BA, Cornell U., 1957; PhD, MIT, 1963. Instr. Princeton (N.J.) U., 1963-66; asst. prof. physics Niels Bohr Inst., Copenhagen, 1968-70; lectr. in physics Groningen (The Netherlands) U., 1972-79, prof. physics, 1979—, chmn. Inst. Theoretical Physics, 1976-83. Author and co-author over 40 papers in field. NSF fellow, 1966-67. Mem. Am. Phys. Soc., European Phys. Soc., Netherlands Phys. Soc., AAAS, Sigma Xi, Phi Beta Kappa.

LANDECK, ROBIN ELAINE, environmental engineer; b. Teaneck, N.J., Sept. 20, 1966; d. George Arthur and Jacqueline (Favara) L.; m. Robert P. Miller, June 12, 1993. BS, Manhattan Coll., 1988, MS, 1990. Rsch. engr. Manhattan Coll., Bronx, 1988-90; cons. engr. Hydro Qual, Inc., Mahwah, N.J., 1990—; Contbr. chpt. to book: Modeling Carbon Utilization by Bacteria in Natural Water Systems, 1992. Mem. Water Environ. Fedn., Sigma Xi. Roman Catholic. Home: 4 Chelsea Ct Ramsey NJ 07446 Office: Hydro Qual Inc 1 Lethbridge Plaza Mahwah NJ 07430

LANDER, DEBORAH ROSEMARY, chemist; b. Capetown, South Africa, Aug. 22, 1962; came to U.S., 1967; d. Gerald Heath Lander and Jean Christine (Oswald) Pleshar; m. Christopher John Kendig, Oct. 15, 1988. BS in Chemistry, McGill U., 1984; MA and PhD in Chemistry, Rice U., 1990. Lab. technician U. Rochester (N.Y.) Med. Ctr., 1984-85; plant chemist Exxon Chemical Co., Baton Rouge, 1989-91; sr. lab chemist Exxon Chemical Co., Baytown, Tex., 1991—. Contbr. articles to Jour. Chem. Physics, Jour. Phys. Chemistry, article to book Laser Spectroscopy, vol. III. Robert A. Welch predoctoral fellow Rice U., 1985-89; Harry B. Weiser scholar Rice U., 1987. Mem. Am. Chem. Soc., Am. Soc. Testing and Materials, Baytown Profl. Forum (co-chair events 1992-93). Democrat. Office: Exxon Chemical Co 3525 Decker Dr Baytown TX 77520

LANDERSMAN, STUART DAVID, engineer; b. Bklyn., May 26, 1930; s. Joseph David and Thelma (Domes) L.; m. Martha Britt Morehead, Sept. 2, 1955; children: David Wesley, Mark Stuart. BA, Dakota Wesleyan U., Mitchell, S.D., 1953; MS, George Washington U., 1967. Commd. ens. USN, 1953, advanced through grades to capt., 1974, retired, 1982; engr. Applied Physics Lab., Johns Hopkins U., Laurel, Md., 1982—; convoy commodore USN, Royal Navy, Can. Armed Forces, 1984-92. Author books on shiphandling, naval tactics, principles of naval warfare; contbr. articles to mags. Decorated Bronze Star, (3) Legion of Merit. Home: 13220 Cooperage Ct Poway CA 92064-1213 Office: JHU/APL Rep COMNAVSURFPAC NAB Coronado San Diego CA 92155-5035

LANDIS, FRED, mechanical engineering educator; b. Munich, Germany, Mar. 21, 1923; came to U.S., 1947, naturalized, 1954; s. Julius and Elsie (Schulhoff) L.; m. Billie H. Schiff, Aug. 26, 1951 (dec. Jan. 10, 1985); children—John David, Deborah Ellen, Mark Edward. B.Eng., McGill U., 1945; S.M., MIT, 1949, Sc.D., 1950. Design engr. Canadian Vickers, Ltd., Montreal, Can., 1945-47; asst. prof. mech. engring. Stanford U., 1950-52; research engr. Northrop Aircraft, Inc., Hawthorne, Calif., 1952-53; asst. prof. NYU, 1953-56, assoc. prof., 1956-61, prof., 1961-73, chmn. dept. mech. engring., 1963-73; dean, prof. mech. engring. Poly. U., Bklyn., 1973-74; dean Coll. Engring. and Applied Sci., U. Wis., Milw., 1974-83, prof. mech. engring., 1984—; staff cons. Pratt & Whitney Aircraft Co., 1957-88. Cons. editor, Macmillan Co., 1960-68; cons. editorial bd.: Funk & Wagnalls Ency., 1969—, Compton's Ency., 1984—; contbr. numerous rsch. articles to profl. jours. and encys., including Ency. Britannica. Mem. Bd. Edn., Dobbs Ferry, N.Y., 1965-71, v.p., 1966-67, 70-71, pres., 1967-68; bd. dirs. Westchester County, Sch. Bds. Assn., 1969-70, v.p., 1970, pres., 1970-71; bd. dirs. Engring. Found., 1986—. Fellow AIAA (assoc.), Am. Soc. Engring. Edn., ASME (hon. mem., div. exec. com. 1965-73, policy bd. 1973-89, v.p. 1985-89, 92—, bd. govs. 1989-91); mem. Sigma Tau, Tau Beta Pi, Pi Tau Sigma. Home: 2420 W Acacia Rd Milwaukee WI 53209-3306

LANDIS, PAMELA ANN YOUNGMAN, tribologist; b. Lancaster, Pa., Nov. 14, 1941; d. George Baldwin and Hazel Elizabeth (Ray) Youngman; m. Carrol Paul Landis, Aug. 21, 1971. BA in Chemistry, Millersville (Pa.) State U., 1973. Chief chemist Howmet Aluminum Corp., Lancaster, 1974-87; corp. inter plant tribologist, supr. chem. lab. Alumax Mill Products, Inc., Lancaster, 1987-89; sr. lubrication engr. Alumax, Inc., Hinsdale, Ill., 1989—; presenter papers in field. Patentee in field. Advisor Jr. Achievement of Lancaster County, 1979, exec. advisor, 1980-83; judge Lancaster Sci. and Engring. Fair, Quarryville, Pa., 1983; vol. Easter Seal Soc., Lancaster County, 1983-85. Recipient Company of Yr. award Jr. Achievement of Lancaster County, 1983. Mem. Soc. Tribologists and Lubrication Engrs. (program com. 1985-91, sec. 1988, vice chmn. 1989, chmn. program com. 1990, ann. meeting program com.), Non Ferrous Metals Coun. (session chmn. 1980—, sec.-treas. exec. com. 1983-84, vice chmn. 1984-85, chmn. 1985-86, asst. indsl. coord. 1991-93, dir. 1993—). Republican. Presbyterian. Avocations: reading, cooking. Home and Office: 240 Salem Rd Pound Ridge NY 10576

LANDIS, WAYNE G., environmental toxicologist; b. Washington, Jan. 20, 1952; s. James G. and Harriet E. L.; m. Linda S.; 1 child, Margaret Evelyn. BA in Biology, Wake Forest U., 1974; MA in Biology, Ind. U., 1978, PhD in Zoology, 1979. Document mgr. Franklin Rsch. Ctr., Silver Spring, Md., 1979-81; rsch. biologist U.S. Army Chem. Rsch., Devel. & Engring. Ctr., Aberdeen Proving Ground, Md., 1982-89; dir., prof. Inst. Environ. Toxicology & Chemistry, Western Wash. U., Bellingham, 1989—. Contbr. numerous articles to profl. jours. Recipient Hankins scholarship Wake Forest U., 1972-74, U.S. Army Rsch. and Devel. Achievement award, 1984, Spl. Act award, 1985, Exceptional Performance award, 1986; named to Outstanding ILIR Rsch. Program, 1983, 84. Mem. AAAS, ASTM, SETAL, PNWSETAL, Genetics Soc. Am., Sigma Xi. Achievements include evaluation of aquatic toxicology of smoke, riot control materials and binary system compounds; devel. theory to predict response of biol. communities to chem. insults using resource competition models, of new methods for ecosystem anal. and applications of chaos theory to chem. impacts. Office: Western Wash U Huxley Coll Inst Environ Toxicology 516 High St Bellingham WA 98225-5946

LANDON, SUSAN MELINDA, petroleum geologist; b. Mattoon, Ill., July 2, 1950; d. Albert Leroy and Nancy (Wallace) L.; m. Richard D. Dietz, Jan. 24, 1993. BA, Knox Coll., 1972; MA, SUNY, Binghamton, 1975. Cert. profl. geologist; cert. petroleum geologist. Petroleum geologist Amoco Prodn. Co., Denver, 1974-87; mgr. exploration tng. Amoco, Houston, 1987-89; ind. petroleum geologist Denver, 1990—; instr. petroleum geology & exploration Bur. of Land Mgmt., U.S. Forest Svc., Nat. Park Svc., 1978-86. Editor: Interior Rift Basins, 1993. Mem., chmn. Colo. Geol. Survey Adv. Com., Denver, 1991-93; mem. Bd. on Earth Sci. and Resources-NRC, 1992—. Recipient Disting. Alumni award Knox Coll., 1986, Disting. Svc. award Rocky Mountain Assn. Geologists, 1986. Mem. Am. Assn. Petroleum Geologists (treas.), Am. Inst. Profl. Geologists (pres., Martin Van Couvering award 1991, Ben H. Parker Meml. medal 1991). Achievements include frontier exploration for hydrocarbons in U.S., especially in Midcontinent Rift System. Home: 790 Ballantine Rd Golden CO 80401 Office: Thomasson Ptnr Assocs 718 17th St Ste 2300 Denver CO 80202

LANDRENEAU, RODNEY EDMUND, JR., physician; b. Mamou, La., Jan. 17, 1929; s. Rodney Edmund and Blanche (Savoy) L.; M.D., La. State U., 1951; m. Colleen Fraser, June 4, 1952; children—Rodney Jerome, Michael Douglas, Denise Margaret, Melany Patricia, Fraser Edmund, Edythe Blanche. Intern, Charity Hosp., New Orleans, 1951-52, resident, 1952-54, 56-58; practice medicine specializing in surgery, Eunice, La., 1958—; pres., dir. Eunice Med. Center, Inc., 1960—; mem. staff Moosa Meml. Hosp., Eunice, 1958—, chief med. staff; vis. staff Opelousas Gen. Hosp., 1958—; assoc. faculty La. State U.-Eunice, chmn. community adv. com., v.p.; cons., staff Lafayette (La.) Charity Hosp.; cons. staff surgery Savoy Meml. Hosp., Mamou; pres. Eunice Med. Center, Inc.; mem. La. State Hosp. Bd., 1972—. Mem. Evangeline council Boy Scouts Am.; bd. dirs. Moosa Meml. Hosp., Eunice, 1986—, bd. govs., 1985-91, vice chmn. bd. dirs. Served with M.C., AUS, 1954-56. Recipient Physician's Recognition award AMA, 1978-85. Diplomate Am. Bd. Surgery. Fellow Internat. Coll. Surgeons (regional dir.), ACS (local chmn. com. trauma, instr.), Southeastern Surg. Congress, Pan Pacific Surg. Congress; mem. Am. Bd. Abdominal Surgeons, Am. Geriatrics Soc., St. Edmunds Athletic Assn., St. Landry Hist. Soc. (v.p. chpt.), St. Landry Parish Med. Soc. (pres. 1969-71, 85-87), Am. Legion, SCV, SAR, Alpha Omega Alpha. Democrat. Roman Catholic. Club: St. Edmund's High Sch. Scholastic Booster (pres. 1986—). Home: 1113 Williams Ave Eunice LA 70535-4939

LANDRUM, HUGH LINSON, JR., engineering executive; b. Texas City, Tex., Oct. 11, 1963; s. Hugh Linson Sr. and Bootsey (Taylor) L. Mech. Engr., Tex. A&M U., 1988; MBA, U. St. Thomas, Houston, 1992. Registered profl. engr., Tex.; registered profl. appraiser. Plant engr. Igloo Products Corp., Houston, 1988-90; valuation engr. Hugh L. Landrum & Assocs., Houston, 1990-92; v.p. Hugh L. Landrum & Assocs., Inc., Houston, 1992—. Mem. Pol. Action Com. for Engrs. Mem. NSPE, Internat. Assn. Assessing Officers, Tex. Soc. Profl. Engrs., Tex. Assn. Assessing Officers. Achievements include research in fabrication, evaluation and application of evacuated panel insultation. Office: Hugh L Landrum & Assocs Inc 1320 S Loop W Houston TX 77054

LANDRUM, LARRY JAMES, computer engineer; b. Santa Rita, N.Mex., May 29, 1943; s. Floyd Joseph and Jewel Helen (Andreska) L.; m. Ann Marie Hartman, Aug. 25, 1963 (div.); children: Larry James, David Wayne, Andrei Mikhail, Donal Wymore; m. 2d, Mary Kathleen Turner, July 27, 1980. Student N.Mex. Inst. Mining and Tech., 1961-62, N. Mex. State U., 1963-65; AA in Data Processing, Eastern Ariz. Coll., 1971; BA in Computer Sci., U. Tex., 1978. Tech. svc. rep. Nat. Cash Register, 1966-73; with ASC super-computer project Tex. Instruments, Austin, 1973-80, computer technician, 1973-75, tech. instr., 1975-76, product engr., 1976-78, operating system programmer, 1978-80; computer engr. Ariz. Pub. Svc., Phoenix, 1980-84, sr. computer engr., 1984-87, lead computer engr., 1987-88, sr. computer engr., 1988-90, sr. control systems engr., 1990—; pres., chmn. bd. dirs. Glendale Community Housing Devel. Orgn., 1993—; instr. computer fundamentals Eastern Ariz. Coll., 1972-73, Rio Salado C.C., Phoenix, 1985-86; mem. bd. trustees Epworth United Meth. Ch., 1987-89, chmn. 1988; mem. community devel. adv. com. City of Glendale (Ariz.), 1988-90, chmn., 1991-92; local arrangements chmn. Conf. on Software Maintenance, 1988. Mem. IEEE Computer Soc., Assn. Computing Machinery, Mensa, Phi Kappa Phi. Methodist. Home: 6025 W Medlock Dr Glendale AZ 85301-7321 Office: Ariz Nuclear Power Project PO Box 52034 Phoenix AZ 85072-2034

LANDSBERG, DENNIS ROBERT, engineering executive, consultant; b. Bklyn., Dec. 27, 1948; s. Nathan and Marilyn (Scherling) L.; m. Marjorie Ina Sahr, Aug. 18, 1974; children: Rebecca, Nancy, Jeniffer. BS in Aero. Engring., Poly. U., N.Y., 1969, MS, 1971; PhD, SUNY, Albany, 1975. Rsch. asst. Poly. Inst. N.Y., Bklyn., 1969-71; rsch. assoc. SUNY, Albany, 1971-75, Atmospheric Sci. Rsch. Ctr., Albany, 1976-79; pres. Assoc. Weather Svcs., Albany, 1979-85; v.p. W.S. Fleming & Assocs., Albany, 1977-85; pres. The Fleming Group, Syracuse, N.Y., 1986—. Author: (with R. Stewart) Improving Energy Efficiency in Buildings, 1980. Recipient Energy Conservation Design award Owens Corning Fiberglas, 1982; Soc. for Tech. Communications Mohawk Regional Mgmt. award, 1984. Mem. ASHRAE, AIAA, Am. Meteorol. Soc., Sigma Xi. Democrat. Jewish. Avocations: swimming, fishing, bowling, golf. Home: 6733 Harmony Dr Fayetteville NY 13066-1782 Office: The Fleming Group 6310 Fly Rd East Syracuse NY 13057-9370

LANDSBERG, LEWIS, endocrinologist, medical researcher; b. N.Y.C., Nov. 23, 1938. AB, Williams Coll., 1960; MD, Yale U., 1964. From instr. to asst. prof. medicine Sch. Medicine Yale U., 1969-72; from asst. prof. to prof. medicine Harvard Med. Sch., 1972-90; Irving S. Cutter prof., chmn. dept. medicine Med. Sch. Northwestern U., 1990—, dir. Ctr. Endocrinology, Metabolism & Nutrition, 1990—; assoc. physician Yale-New Haven Hosp., 1969-71, attending physician, 1971-72, Beth Israel Hosp., 1974-79, physician, 1979-88, sr. physician, 1988-90; attending physician West Haven VA Hosp., 1970-72; assisting physician Boston City Hosp., 1972-73, assoc. vis. physician, 1973-74; physician-in-chief dept. medicine Northwestern Meml. Hosp., 1990—. Fellow Am. Coll. Physicians; mem. AAAS, Am. Fedn. Clin. Rsch., Endocrine Soc., N.Y. Acad. Scis., AHA, Am. Soc. Pharmacology & Exptl. Therapeutics, Am. Physiology Soc., Am. Soc. Clin. Investigators, Assn. Am. Physicians. Achievements include rsch. in catecholamines and the sympathoadrendal system, nutrition and the sympathetic nervous system, obesity and hypertension. Office: Ctr Endocrinology Metabolism & Nutrition 10-555 Searle 303 Chicago Ave Chicago IL 60611*

LANE, ADELAIDE IRENE, computer systems specialist, researcher; b. Bronx, N.Y., Sept. 27, 1939; d. Anton John and Constance Mary (Fogle) Pospisil; m. Robert Walton Lane, Sept. 26, 1964; children: Frank Anton, Miriam Helen, Robin Ann. BS in Edn. cum laude, SUNY, Oneonta, 1961; MS in Edn., Hofstra U., 1963; MS in Computer Sci., Rensselaer Poly. Inst., 1983. Cert. tchr., N.Y., Vt. Tchr. Island Trees Jr. High Sch., Levittown, N.Y., 1961-64; copy editor, typesetter Pennysaver & Press, Bennington, Vt., 1976-77; tchr. Mt. Anthony Jr./Sr. High Sch., Bennington, 1977-80; computer operator Rensselaer Poly. Inst., Troy, N.Y., 1981-82, graphics application programmer, 1982-87, sr. graphics application programmer, 1987-92, mgr. instrnl. computing, 1992—; tchr. computer sci. Russell Sage Coll., Troy, 1984-85; cons. Union Coll., Schenectady, 1990. Editor: The Rock Ribs of Bennington Town, 1977; contbr. articles to profl. jours. Asst. coach Bennington Swim Team, 1972-75; troop leader Girl Scouts U.S., Hoosick Falls, N.Y., 1973-79; tchr. Hoosick Falls Ice Skating Club, 1975-80. Interuniv. Consortium for Ednl. Computing fellow, 1984. Mem. IEEE (affiliate), Nat. Computer Graphics Assn., Ednl. Uses of Info. Tech./EDUCOM (mem. Joe Wyatt Challenge selection com. 1990-91). Republican. Roman Catholic. Avocations: multimedia computing, weaving, watercolor painting. Office: Rensselaer Poly Inst ECS-CII 3111 Troy NY 12180

LANE, JOSEPH M., orthopaedic surgeon, oncologist; b. N.Y.C., Oct. 27, 1939; s. Frederick and Madelaine Lane; m. Barbara Greenhouse, June 23, 1963; children: Debra, Jennifer. AB in Chemistry, Columbia U., 1957; MD, Harvard U., 1965. Surg. intern Hosp. U. Pa., Phila., 1965-66, resident in gen. surgery, 1966-67, resident, 1969-72, chief resident, 1972-73, chief MBD sect., 1973-76; rsch. assoc. NIH, NIDR, Bethesda, Md., 1967-69; rsch. fellow Phila. Gen. Hosp., 1969-70; chief MBD unit Hosp. Spl. Surgery, N.Y.C., 1976—, dir. rsch. div., 1990—; assoc. dir. MultiPurpose Arthritis Ctr., N.Y.C., 1988—; cons., collaborator Collagen Corp., Warsaw, Ind., Genetics Inst., Andover, Mass., EBI, Fairfield, N.J. Recipient N.Y. Mayoral Proclamation, 1988. Fellow Am. Acad. Orthopaedic Surgeons; mem. AMA, Am. Soc. Bone and Mineral Rsch., Med. Soc. State N.Y., Internat. Soc. Frac. Rep. (bd. dirs.), Musculoskeletal Tumor Soc. (pres. 1982-83), Orthopaedic Rsch. Soc. (pres. 1984-85). Office: Hosp Spl Surgery 535 E 70th St New York NY 10021-4898

LANE, LEONARD J., hydrologist; b. Tucson, Apr. 25, 1945; s. Edward M. and Audrey M. (Cutshaw) L.; m. Rose Mary Toon; children: Brian James, Leonard Brett. BS with distinction, U. Ariz., 1970, MS in Systems & Indsl. Engring., 1972; PhD in Hydrology & Water Resources, Colo. State U., 1975. Cert. erosion control specialist. Staff mem. Los Alamos Nat. Lab., 1981-82,

83-84; systems engr., coord. natural resource modeling activities ARS, 1989-91; hydrologist USDA-ARS, Tucson, 1970-72, 75-89, rsch. leader S.W. Watershed Rsch. Ctr., 1991—; cons. City of Tucson, Ariz. State U., San Diego State U., Los Alamos Nat. Lab., Washington State Dept. Ecology, Sabol Engring.; adj. assoc. prof. U. Ariz. Author or co-author over 140 publs. Recipient Outstanding Performance award, 1970, 77, 87, 88, 89, 92, Arthur S. Flemming award, 1982, Outstanding Performance award WEPP, 1989; named Mountain States Area Scientist of Yr., 1986. Mem. ASCE, AGU, ASAE (co-author outstanding paper award 1981, 90), AWRA, BGRG, SRM. Office: S W Watershed Rsch Ctr 2000 E Allen Rd Tucson AZ 85719

LANE, MEREDITH ANNE, botany educator, museum director; b. Mesa, Ariz., Aug. 4, 1951; d. Robert Ernest and Elva Jewell (Shilling) L.; m. Donald W. Longstreth, Apr. 6, 1974 (div. Feb. 1985). BS, Ariz. State U., 1974, MS, 1976; PhD, U. Tex., 1980. Asst. prof. U. Colo., Boulder, 1980-88, assoc. prof., 1988-89; assoc. prof., dir. McGregor Herbarium U. Kans., Lawrence, 1989—; vis. asst. prof. U. Wyo., Laramie, 1985-86; vis. scholar U. Conn., Storrs, 1989; cons. editor McGraw-Hill Ency. of Sci. and Tech., N.Y.C., 1985-92. Contbr. 30 articles to profl. jours. Mem. Am. Soc. Plant Taxonomists (sec. 1986-88, program dir. 1986-90, councillor 1993-96, Cooley award 1982), Bot. Soc. Am. (editor 1990—, sect. chair 1984-86, sect. sec. 1986-90), Internat. Orgn. for Plant Biosystematics (councillor 1989-92), Internat. Assn. Plant Taxonomists, Calif. Bot. Soc. Democrat. Lutheran. Avocations: reading, conversation, country dance, hiking, furniture refinishing. Office: R L McGregor Herbarium 2045 Constant Ave Lawrence KS 66047-3279

LANE, MICHAEL JOSEPH, microbiology educator; b. St. Paul, Oct. 10, 1956; s. J. Edgar and M. Cecile (Moore) L.; m. Leslie E. Mills, Nov. 14, 1986; 1 child, Christopher. BS, Southeastern Mass. U., 1979; PhD, Syracuse U., 1984. Postdoctoral fellow genetics and devel. Columbia U., N.Y.C., 1984-86; asst. prof. dept. microbiology SUNY Health Sci. Ctr., Syracuse, 1987-92, asst. prof. dept. medicine, 1986-92, assoc. prof. dept. microbiology, 1992—, assoc. prof. dept. medicine, 1992—; scientific founder Genmap, Inc., New Haven, Conn., 1989-91; mem. grant review panel U.S. Dept. Energy, 1991—. Contbr. articles to profl. jours. Recipient Am. Chem. Soc. undergraduate rsch. award for Raman spectral studies of mononucleotides, 1979, NIH postdoctoral cancer tng. grant, 1985-86. Achievements include patents pending for a method for physically mapping genomic material, a multimegabase cloning vehicle, a strategy for rapidly mapping chromosomes, a process by which the sequence dependence of DNA reactivity can be determined. Office: SUNY Health Sci Ctr 750 E Adams St Syracuse NY 13210

LANE, NEAL FRANCIS, university provost, physics researcher; b. Oklahoma City, Aug. 22, 1938; s. Walter Patrick and Harietta (Hattie) Charlotta (Hollander) L.; m. Joni Sue Williams, June 11, 1960; children: Christy Lynn Lane Saydjari, John Patrick. B.S., U. Okla., 1960, MS, 1962, Ph.D., 1964. NSF postdoctoral fellow, 1964-65; asst. prof. physics Rice U., Houston, 1966-69, assoc. prof., 1969-72, prof., 1972—; dir. div. physics NSF, Washington, 1979-80 (on leave from Rice U.); chancellor U. Colo., Colorado Springs, 1984-86; provost Rice U., 1986—; dir. Nat. Science Foundation, Washington, D.C., 1993—; non-resident fellow Joint Inst. for Lab. Astrophysics U. Colo., Boulder, 1984—, vis. fellow, 1965-66, 75-76; mem. commn. on phys. sci., math. and applications NRC, 1989—; bd. overseers Superconducting Super Collider (SSC) Univs. Rsch. Assn., 1985-93; disting. Karcher lectr. U. Okla., Norman, 1983; disting. vis. scientist U. Ky., Lexington, 1980; mem. adv. com. math. and phys. sci. NSF, 1992—. Co-author: Quantum States of Atoms, Molecules and Solids, Understanding more Quantum Physics; contbr. articles to profl. jours. Active Cath. Commn. Intellectual and Cultural Affairs, 1991. Recipient George Brown prize for superior teaching Rice U., 1973-74, 76-77; Sloan Found. fellow, 1967-71. Fellow Am. Phys. Soc. (councilor-at-large 1981-84, chmn. panel on pub. affairs 1983, exec. com. 1981-83, chmn. div. of electron and atomic physics 1977-78), AAAS; mem. Am. Inst. Physics (gov. bd. 1984-87), Am. Assn. Physics Tchrs., Phi Beta Kappa, Sigma Xi (pres. elect 1992, pres. 1993). Roman Catholic. Avocations: tennis; squash. Office: Rice U Office of the Provost PO Box 1892 Houston TX 77251-1892

LANE, ORRIS JOHN, JR., engineer; b. Sigourney, Iowa, Apr. 21, 1932; s. Orris John and Hester Hanna (Hazen) L.; m. Joan Joyce Nelson, June 19, 1954; children: Jerry, Beth, Dona, Seth. BS in Engring., Iowa State U., 1957. Registered profl. engr., Iowa. Spl. investigations engr. Iowa Dept. Transp., Ames, 1962-64, Portland cement concrete engr., 1964-72, 83-87; dist. materials engr. Iowa Dept. Transp., Atlantic, 1972-83; testing engr. Iowa Dept. Transp., Ames, 1987—. Achievements include development of bridge floor repair procedure using bonded concrete, and development of fast-track procedure for concrete pavement opened to early traffic. Home: 1111 Garfield St Ames IA 50010 Office: Iowa Dept Transp 800 Lincoln Way Ames IA 50010

LANE, PETER, ornithologist; m. Jacqueline LaBrecque. Author: Les Oiseaux du Québec, 1980, L'Alimentation des Oiseaux, 1985. Recipient Snowy Owl Conservation award Québec Zoological Gardens, 1992. Mem. Ornithologist's Club Québec, Cercle des Mycologues Québec. Office: 210 Chemin de l'Eperon, Lac Beauport, PQ Canada G0A 2C0*

LANE, SUSAN NANCY, structural engineer; b. Dayton, Ohio, Aug. 20, 1960; d. Richard Evenson and Nancy (Schmidt) L. BSCE, Pa. State U., 1982, MSCE, 1984. Registered profl. engr., Va. Asst. engr. Parsons Brinckerhoff FG, Inc., Trenton, N.J., 1981-85; structural engr. Doyle Engring. Corp., Landover, Md., 1985; engr. I Parsons Brinckerhoff Quade & Douglas, Inc., Herndon, Va., 1985-88; rsch. structural engr. U.S. Dept. Transp., Fed. Hwy. Adminstrn., McLean, Va., 1988—; technician Ocean County (N.J.) Engring. Dept., Toms River, summers 1978-81; mem. prestressing steel com. Precast/Prestressed Concrete Inst., 1989—, R&D com., 1990—, bridge com., 1992—. Contbr. articles to Pub. Rds., PCI (Precast/Prestressed Concrete Inst.) Jour. Mem. McLean Bible Ch., 1986—. Mem. ASCE (mem. reinforced concrete rsch. coun.), Am. Concrete Inst. (mem. prestressed concrete com. 1992—). Office: Fed Hwy Adminstrn 6300 Georgetown Pike Mc Lean VA 22101-2200

LANE, WILLIAM KENNETH, physician; b. Butte, Mont., Nov. 5, 1922; s. John Patrick and Elizabeth Marie (Murphy) L.; m. Gilda Antoinette Parision, Aug. 21, 1954; children: William S., Francine Deirdre. Student, U. Mont., 1940-41, Mt. St. Charles Coll., 1941-43; MD, Marquette U., 1946. Intern Queen of Angels Hosp., L.A., 1946-47, resident physician, 1954-56; pvt. practice internal medicine San Francisco, 1947-51; resident in urology VA Hosp., Long Beach, Calif., 1956-58; physician VA Hosp., Long Beach, Oakland and Palo Alto, Calif., 1958—; lectr. on psychology of the elderly Foothill Coll., Los Altos, 1972-74; rschr. in field. Bd. dirs., mem. No. Cheyenne Indian Sch.; mem. Josef Meier's Black Hills Theatrical Group, S.D., 1940. With U.S. Army, 1943-46, ETO, lt. USN, 1951-54, Korea. Mem. AMA, Am. Geriatrics Soc., Nat. Assn. VA Physicians, San Francisco County Med. Soc., Woodrow Wilson Ctr. (assoc.), St. Vincent de Paul Soc., Cupertino Landscape Artists (past pres.), Audubon Soc., Stanford Hist. Soc., San Jose Camera Club. Roman Catholic. Avocations: oil and watercolor painting, hiking, mountain climbing, outdoor video camcorder photography. Home: 18926 Sara Park Cir Saratoga CA 95070-4164 Office: Stanford VA Med Ctr 3801 Miranda Ave # 171 Palo Alto CA 94304-1207

LANEY, LEROY OLAN, economist, banker; b. Atlanta, Mar. 20, 1943; s. Lee Edwin and Paula Izlar (Bishop) L.; m. Sandra Elaine Prescott, Sept. 3, 1966; children: Prescott Edwin, Lee Olan III. B Indsl. Engring., Ga. Inst. Tech., 1965; MBA in Fin., Emory U., 1967; MA in Econs., U. Colo., 1974, PhD in Econs., 1976. Budget analyst Martin-Marietta Corp., Denver, 1971-72; economist Coun. Econ. Advisers, Washington, 1974-75; internat. economist U.S. Treasury Dept., Washington, 1975-78; sr. economist Fed. Res. Bank Dallas, 1978-88; prof. econs., chmn. dept. Butler U., Indpls., 1989-90; v.p., chief economist 1st Hawaiian Bank, Honolulu, 1990—; chmn. Fed. Res. Com. on Internat. Rsch., Washington, 1981-83; vis. prof. U. Tex., Arlington and Dallas, 1978-85; adj. prof. So. Meth. U., Dallas, 1982-85. Editor bank periodicals, 1975-88; contbr. articles to profl. jours. Mem. Internat. Fin. Symposium, Dallas, 1982-85, Overland Stage Neighborhood Assn., Arlington, 1983-88, Hawaii Coun. on Revenues. Lt. USN, 1967-71. Scholar Ga. Inst. Tech., 1961; rsch. fellow Emory U., 1965-67, teaching fellow U. Colo., 1972-73; rsch. grantee Butler U., 1989-90. Mem. Am. Econ. Assn., Western Econ. Assn., Indpls. Econ. Forum, Plaza Club, Omicron Delta Epsilon, Kappa Sigma. Avocations: sailing, skiing, reading, tennis, jogging. Office: 1st Hawaiian Bank PO Box 3200 Honolulu HI 96847-0001

LANFLISI, RAYMOND ROBERT, aeronautical engineer; b. N.Y.C., Dec. 24, 1929; s. Joseph and Louisa (Pentecoste) L.; m. Gloria Faith Cerrato, June 21, 1952; children: Raymond Robert II, Robert Alan, James Dwight. BS in Aero. Engring., Princeton U., 1952; MS in Aero. Engring., Cornell U., 1953. Aero. engr. Consol. Vultee Aircraft Co., San Diego, 1953-54; design specialist Conair div. Gen. Dynamics, San Diego, 1954-61, chief aeroballistics astronautics div., 1961-65, chief aero mechanics aerospace div., 1965-79, sr. engring. specialist Convair div., 1979—; cons. contbr. space shuttle tech. steering group NASA, 1974-80. Author: Flow Separation From Delta Wings at Yaw, 1954. Curtis-Wright fellow, 1952-53; recipient Pub. Svc. Achievement award NASA, 1981. Fellow AIAA (assoc.); mem. Nat. Mgmt. Assn., Sigma Xi. Roman Catholic. Achievements include development of aerospace vehicle synthesis design program, wave drag and flight path optimization techniques, flight software to prevent missile engine surge, performance analyses leading to upgrade of UC2 Tomahawk missile. Home: 11408 Rolling Hills Dr El Cajon CA 92020

LANG, DEREK EDWARD, aerospace engineer; b. Oakland, Calif., Apr. 12, 1966; s. Eddie Yee and Arlene (Lum) L. BS in Aero. Engring., Calif. Poly. State U., 1988; cert., Internat. Space U., Cambridge, Mass., 1988; MS in Aero.-Astro. Engring., Stanford U., 1989. Engring. aide USN/Navy Facilities Engring. Command, San Bruno, Calif., 1985; engring. clk., cons. United Technologies, San Jose, 1986-87; project adminstr. Calpoly Space Systems, San Luis Obispo, Calif., 1985-88; rsch. asst. Stanford (Calif.) U., 1989; aero. engr. Office Comml. Space Transp., U.S. Dept. Transp., Washington, 1989—. Bd. dirs. Conf. Asian Pacific Am. Leadership, Washington, 1991—. Mem. AIAA (mem. space launch systems com. on standards 1991—, mem. orbital debris com. on standards 1992—), Women in Aero., Orgn. Chinese-Ams., Tau Beta Pi, Sigma Gamma Tau.

LANG, JAMES DOUGLAS, aerospace engineer, educator; b. Chgo., Mar. 22, 1942; s. John and Marjorie Virginia (Marti) L.; m. Elaine Marie Carter, June 18, 1966; children: Nina Marie, Andrew Douglas. BS in Engring., U.S. Mil. Acad., 1963; MS in Aeronautics and Astronautics, Stanford U., 1969; PhD in Aerodynamics, Cranfield (Eng.) Inst. Tech., 1975. Registered profl. engr., Colo. Commd. 2d lt. USAF, 1963, advanced through grades to col., 1983; ret., 1987; forward air contr. USAF, Republic of Vietnam, 1966-67; assoc. prof. USAF Acad., Colorado Springs, 1969-78; dep. dir. flight test engring., engring. test pilot 4950 Test Wing USAF, Wright-Patterson, Ohio, 1978-82; chief flight control divsn. Wright Labs., Wright-Patterson, 1983-84; dir. avionics lab. Wright-Patterson, 1984-85; dep. commdr. for engring. Aeronautical Systems Divsn., Wright-Patterson, 1985-87; dir. flight scis. McDonnell Douglas Corp., St. Louis, 1988-90, dir. engring. nat. aerospace plane, 1990-92, dir. engring.F-15 and F-4, 1992—; mem. adv. bd. U. Mo., Columbia, 1989—, Aerospace Ill., Urbana, 1989—; affiliate prof. Washington U., St. Louis, 1990—. Author, editor: Aircraft Performance, Stability and Control, 1974; author 24 tech. papers on engring. and mgmt. Asst. scoutmaster Boy Scouts Am., Chesterfield, Mo., 1988-90. Decorated DFC (2), 1967, Purple Heart, 1967, Legion of Merit, 1987. Fellow AIAA (bd. dirs. 1988-91, 93—). Achievements include rsch. in control of separating flows being applied to fighter aircraft to improve maneuverability. Home: 1745 Horseshoe Ridge Rd Chesterfield MO 63005

LANG, OTTO E., business executive, former Canadian cabinet minister; b. Handel, Sask., Can., May 14, 1932; s. Otto T. and Maria (Wurm) L.; m. Adrian Ann Merchant, 1963-88; children: Maria (dec.), Timothy, Gregory, Andrew, Elisabeth, Amanda, Adrian; m. Deborah McCawley, 1989; stepchildren: Andrew, Rebecca. BA, U. Sask., 1951, LLB, 1953; BCL (Rhodes scholar), Oxford (Eng.) U., 1955; LLD (hon.), U. Man., 1987. Bar: Sask. 1956, Ont., Yukon and N.W.T 1972, Man. 1988; created Queen's counsel 1972. Mem. faculty Law Sch., U. Sask., 1956-68, asso. prof. law, 1958-61, prof., dean law, 1961-68; M.P. for Sask.-Humboldt, 1968-79; Canadian minister without portfolio, 1968-69, minister for energy and water, 1969, minister of manpower and immigration, 1970-72, minister of justice, 1972-75, 78-79, minister transport, 1975-79; minister-in-charge Canadian Wheat Bd., 1969-79; exec. v.p. Pioneer Grain Co. Ltd., James Richardson & Sons Ltd., Winnipeg, Man., Can., 1979-88; chmn. Transp. Inst., U. Man., Winnipeg, 1983-93; mng. dir. Winnipeg Airpots Authority, Inc., 1992-93; pres., CEO Ctrl. Gas Manitoba Inc., 1993—; mem. Queen's Privy Council for Can. Editor: Contemporary Problems in Public Law, 1967. Vice pres. Sask. Liberal Assn., 1956-62, fed. campaign chmn., Sask., 1963-64; campaign chmn. Winnipeg United Way, 1983. Roman Catholic. Clubs: Knights of Sovreign Order Malta; Bridge; Golf (St. Charles). Office: U Manitoba, 444 St Mary Ave, Winnipeg, MB Canada R3T 2N2

LANG, PATRICIA LOUISE, chemistry educator, vibrational spectroscopist; b. Kingston, R.I., Oct. 21, 1954; d. Ralph Edgington and Margaret Evelyn (Hartman) Murphy; m. Terry Howard Lang, Nov. 22, 1972; 1 child, Jill Rae. BS in Chemistry, Ball State U., 1983; PhD in Phys. Chemistry, Miami U., 1987. Asst. prof. dept. chemistry Ball State U., Muncie, Ind., 1987-91, assoc. prof. dept. chemistry, 1991—. Contbr. chpts. in books on infrared and raman microspectroscopy; contbr. articles to profl. jours. Recipient Student award Fedn. Analytical Chemistry and Spectroscopy Socs., 1986. Mem. Am. Chem. Soc., Soc. Applied Spectroscopy, Coblentz Soc., Sigma Xi. Office: Ball State Univ Dept Chemistry Muncie IN 47306

LANG, WILLIAM EDWARD, mathematics educator; b. Salisbury, Md., Oct. 22, 1952; s. Woodrow Wilson and Clara T. L. BA, Carleton Coll., 1974; MS, Yale U., 1975; PhD, Harvard U., 1978. Vis. mem. Inst. for Advanced Study, Princeton, N.J., 1978-79; exch. prof. Universite de Paris, Orsay, 1980; C.L.E. Moore instr. MIT, Cambridge, 1980-82; asst. prof. U. Minn., Mpls., 1982-83, assoc. prof., 1983-89; vis. assoc. prof. Brigham Young U., Provo, Utah, 1988-89, prof., 1989—. Contbr. articles to profl. jours. Fellow NSF 1974-77, 79-80. Mem. Am. Math. Soc., Math. Assn. Am., Math. Scis. Rsch. Inst., Sigma Xi. Republican. Office: Brigham Young Univ Dept Math Provo UT 84602

LANGDON, GEOFFREY MOORE, architect, educator; b. Englewood, N.J., Dec. 23, 1955; s. Palmer Hull and Anne (Moore) L.; m. Cindy Muller, Aug. 22, 1992. BS in Bldg. Sci., Rensselaer Poly. Inst., 1977, BArch, 1987, MS, 1986. Archtl. designer Cybul and Cybul Architects, Edgewater, N.J., 1978-80, Donal Simpson Assocs., Newport, R.I., 1980-81; project architect Garofolo Assocs., Providence, 1981; solar energy cons. Solar Design Specialists, Troy, N.Y., 1982-87; microcomputer cons. Info. Tech. Svcs., Rensselaer Poly. Inst., Troy, 1984-87; CADD cons. Archtl. CADD Cons., Beverly, Mass., 1987—; prof. architecture Wentworth Inst. Tech., Boston, 1989—; system mgr. Architects Online Electronic Network, Boston Soc. Architects, 1990—; vis. lectr. Cons. for Architects, 1990-92. Author: CADD and the Small Firm (versions 1 through 4.1), 1988-92, CADD and the Small Firm '93 (version 5.2), 1993; contbr. to profl. publs. Fellow Am. Coll. Schs. Architecture, 1990;p grantee CADKEY, Inc., 1990, 91, Graphisoft, Inc., 1991, Abvent, 1992. Mem. AIA, Am. Solar Energy Soc., Boston Soc. Architects, Storm Trysail Club. Achievements include software development, anthropological fieldwork in Mahina district of Tahiti documenting that Tahitian women speak French while the men speak Tahitian. Home: 6 Devon Ave Beverly MA 01915 Office: Wentworth Inst Tech Dept Architecture 550 Huntington Ave Boston MA 02115

LANGDON, GEORGE DORLAND, JR., museum administrator; b. Putnam, Conn., May 20, 1933; s. George Dorland and Anne Claggett (Zell) L.; m. Agnes Domandi; children—George Dorland, III, Campbell Brewster. A.B. cum laude (Coolidge scholar), Harvard Coll., 1954; M.A., Amherst Coll., 1957; Ph.D. (Coe fellow), Yale U., 1961; LHD (hon.), Colgate U. Instr. in history and Am. studies Yale U., 1959-62, asst. provost, lectr. in history, 1969-70, asso. provost, lectr. in history, 1970-72, dep. provost, lectr. in history, 1972-78; asst. prof. history Calif. Inst. Tech., 1962-64; asst. prof. Vassar Coll., 1964-67, asso. prof., 1967-68, spl. asst. to pres., 1968; pres., trustee Colgate U., 1978-88, pres. emeritus, 1988—; pres. Am. Mus. Natural History, N.Y.C., 1988-93; bd. dirs. Quest for Value Dual Purpose Fund, Inc., Custodial Trust Co. Author: A History of New Plymouth 1620-1691, 1966 (State and Local Assn. of History award of Merit). Trustee The Kresge Found., St. Lukes/Roosevelt Hosp. Ctr., Wenner-Gron Found. for Anthrop. Rsch. Lt. U.S. Army, 1954-56. Fellow Pilgrim Soc. Office: Am Mus Natural History Cen Pk W 79th St New York NY 10024-5192

LANGDON, PAUL EUGENE, JR., consulting civil engineer; b. Evanston, Ill., Nov. 9, 1931; s. Paul Eugene and Phyllis (Kehling) L.; m. Carol Anne Porter, Oct. 16, 1955; 1 child, Theresa Anne. BS in Engring., Calif. Inst. Tech., 1953, MSCE, 1954. Registered profl. engr., Ill., Mich., S.D., Fla., Ga., Tex., Okla., Ariz., Oreg., Wash., D.C.; diplomate Am. Acad. Environ. Engrs. San. engr. USPHS, Cin., 1954-57; engr. Greeley and Hansen, Chgo., 1957-68, assoc., 1968-71, ptnr., 1971—. Mem. ASCE, NSPE, Am. Cons. Engrs. Coun., Water Environ. Fedn., Am. Water Works Assn., Univ. Club Chgo., Tower Club Chgo. Office: Greeley and Hansen 100 S Wacker Dr Chicago IL 60606

LANGE, GARRETT WARREN, psychology educator; b. Mineola, N.Y., Mar. 26, 1942; s. Warren Frederick Lange and Marjorie (MacLeod) Tudor; m. Carol Marie Relyea, Dec. 27, 1964 (div. 1987); children: Michelle, Scott; m. Marilyn Margo Melton, Oct. 3, 1992; children: Corey, Garrett. BA, W.Va. Wesleyan Coll., 1965; MA, U. N.H., 1967; PhD, Pa. State U., 1970. Asst. prof. Vassar Coll., Poughkeepsie, N.Y., 1970-73; assoc. prof. U. N.C., Greensboro, 1973-76, prof. human devel., 1981—; prof. Purdue U., West Lafayette, Ind., 1976-81; cons., reviewer book manuscripts Allyn & Bacon, Prentice-Hall, Wadsworth, W.C. Brown, Addison-Wesley, South-Western Pubs., 1977—. Reviewer Child Devel., Devel. Psychology, Jour. Exptl. Child Psychology, Merrill-Palmer Quar.; contbr. numerous articles to profl. jours. With USMC, 1960-68. Mem. Am. Ednl. Rsch. Assn., Am. Psychol. Soc., Soc. for Rsch. in Child Devel., Psi Chi, Omicron Delta Kappa, Alpha Kappa Delta, Sigma Xi. Home: 5400 Chatfield Sq Greensboro NC 27410 Office: U NC Stone Bldg Greensboro NC 27412

LANGE, NICHOLAS THEODORE, statistician; b. Valparaiso, Ind., Mar. 18, 1952; s. Lester Henry and Beverley Jane (Brown) L.; m. Dorothy Cresswell, Sept. 6, 1976 (div. 1982); children: Sarah Elisabeth, Nicholas Cresswell; m. Louise Marie Ryan, Dec. 15, 1984. ScB, Northeastern U., 1976; ScM, U. Mass., 1981; ScD, Harvard U., 1986. Instr. applied math. MIT, Cambridge, 1986-87; asst. prof. med. statistics Brown U., Providence, 1987-93; specialist in neuroimaging NIH, Bethesda, Md., 1993—; expert witness McGovern, Noel & Benik Providence, 1991-93, M. Scherzer, N.Y.C., 1993; cons. in field. Editor: Case Studies in Biometry, 1991—; assoc. editor: Jour. Am. Statis. Assn. Revs., 1993-95; guest editor: Statistics in Medicine, 1992. Am. Cancer Soc. grantee, 1991-93; recipient Robert Reed prize Harvard U., 1986. Mem. Am. Statis. Assn., Biometric Soc., Bernoulli Soc., Soc. Optical Enging. (hon.). Home: 243 Concord Ave #11 Cambridge MA 02138 Office: NIH Fed Bldg 7550 Wisconsin Ave Bethesda MD 20892

LANGE, ROBERT WILLIAM, immunotoxicologist; b. Chgo., Nov. 28, 1960; s. Robert Frederick and Joan Jeanette (Schmidt) L.; m. Marcia Greer, Oct. 1, 1988. BS, Western Ill. U., 1982; MS, Ill. Inst. Tech., 1991; postgrad., N.C. State U., 1991. Assoc. biologist Rsch. Inst. Ill. Inst. Tech., Chgo., 1984-91; assoc. scientist Mantech Environ. Techs., Inc., Research Triangle Park, N.C., 1991—; vol. scientist Mus. of Sci.: Sci. By Mail, Boston, 1992—. Contbr. articles to profl. publs. Mem. Internat. Soc. Chronobiology, Immunotoxicology Discussion Group, Sigma Xi. Lutheran. Home: 3112 Julian Dr Raleigh NC 27604 Office: Mantech Environ Tech Inc 2 Triangle Dr Research Triangle Park NC 27709

LANGENEGGER, OTTO, hydrogeologist; b. Arbon, Switzerland, Apr. 22, 1938; s. Robert and Anna (Suhner) L.; m. Dorothea Kloss, June 24, 1967; children: Urs, Thomas. Diploma, Engring. Sch., Winterthur, Switzerland, 1964; PhD, U. Bern, Switzerland, 1973. Hydrogeologist Christoffel Mission, Addis Ababa, Ethiopia, 1974-76; high sch. tchr. Gymnasium Untere Waid, Mörschwil, Switzerland, 1976-79; cons. Electrowatt Engring. Ltd., Accra, Ghana, 1979-81; project officer World Bank (UNDP), Washington, 1981-82; regional project officer World Bank UN Devel. Program, Abidjan, Ivory Coast, 1983-88; lectr. Engring. Sch., St. Gallen, Switzerland, 1989—; pvt. practice cons. Gais, Switzerland, 1989—; asst. U. Bern, 1971-72; rsch. asst. Axel Heiberg expedition, Can. Arctic, McGill U., Montreal, Can., 1968. Author: World Bank: GW Quality and Handpump Corrosion in West Africa, 1993; co-author: World Bank: Community Water Supply-The Handpump Option, 1987; contbr. articles to profl. publs. Mem. Am. Chem. Soc., Am. Water Works Assn., Internat. Assn. Hydrogeologists, Internat. Water Resources Assn., Nat. Groundwater Assn., Swiss Soc. Engrs. and Architects, Assn. Groundwater Scientists and Engrs., Swiss Gas and Water Industry Assn., Swiss Acad. Natural Scis. Achievements include study of impact of handpump corrosion on water quality, guidelines for application of galvanized equipment for handpumps, mapping of groundwater quality, mapping air pollution by means of snow samples. Home and Office: Riesern 1561, CH-9056 Gais Switzerland

LANGER, ARTHUR MARK, mineralogist; b. N.Y.C., Feb. 18, 1936; s. Morton Livingston and Ruth Regina (Lewitz) L.; m. Catherine Chilcott Josi, apr. 11, 1977; children: Erica Margaret, Andrew Michael, Elliott Mark, Christopher Morton. BA, Hunter Coll., 1956; MA, Columbia U., 1962, PhD, 1965. Exploration geologist Rosario Exploration Chibougamau Mining and Smelting, 1956; field asst. in geology Beartooth Mountains, Mont. Columbia U., N.Y.C., 1957-58, teaching asst. dept. geology, 1958-59, cons. mineralogist, 1960-65, rsch. asst. dept. geology, 1961-64; lectr. CUNY, 1964-65, mem. grad. faculty, 1982—; rsch. assoc. environ. medicine, dept. medicine Mt. Sinai Hosp., N.Y.C., 1965-67, asst. prof. dept. community medicine, 1967-68; assoc. prof. mineralogy, dept. community medicine Mt. Sinai Sch. Medicine, N.Y.C., 1968-86, 87-88, head phys. scis. sect., assoc. dir. Environ. Scis. Lab., 1969-86, sci. adminstr. Environ. Scis. Lab., 1983-84, assoc. prof. Ctr. for Polypeptide and Membrane Rsch., 1986-88; rsch. assoc. dept. mineral scis. Am. Mus. Natural History, N.Y.C., 1979—; dir. Environ. Scis. Lab. Inst. of Applied Scis. CUNY, 1988—; adj. assoc. prof. mineralogy grad. div. CUNY, 1968-69; expert cons. NIH, 1974—, WHO, 1975—, Nat. Heart, Lung and Blood Inst., 1975—, EPA, 1975—, Nat. Inst. for Environ. Health Scis., 1975—, Nat. Inst. for Occupational Safety and Health, 1975—, EPA Superfund cases, 1985—, other regional consultations; cons. Inst. Pub. Health, Norway, 1977, Ministry of Mines, South Africa, 1977, Internat. Agy. Rsch. Cancer, 1976, 86, Internat. Program Chem. Safety (WHO), 1985, Internat. Fedn. Bldg. Wood Workers, 1989; mem. internat. coms. on pollution and health. Assoc. editor Environ. Rsch., 1978-85, adv. editor, 1985-87; asst. editor Am. Jour. Indsl. Medicine, 1980-85, assoc. editor, 1985-86; mem. editorial rev. bd. Journ. Environ. Pathology and Toxicology, 1978-82, Jour. Environmental Pathology, Toxicology and Oncology, 1983—; mem. editorial adv. bd. Advances in Modern Eenviron. Toxicology, 1981-82; manuscript reviewer many jours.; reviewer, author fed. and industry documents, 1978—; contbr. chpts. to books, articles and abstracts to profl. jours. and symposia proc. Recipient award Dept. Geology, Hunter Coll., 1956, Dust Rsch. award Polachek Found., 1965-67, Career Scientist award Nat. Inst. Environ. Health Scis., 1969-74; named to Hall of Fame, Hunter College-CUNY, 1978; grantee Health Rsch. Coun., 1966-67, 75-77, Polacheck Found., 1966-68, NIH, 1967-78, Johns-Manville Corp., 1968-73, Am. Cancer Soc., 1971-74, 80-81, EPA, 1973, Ford Motor Co., 1973-75, Nat. Inst. Occupational Safety and Health, 1976-79, 82-84, Nat. Inst. Environ. Health Scis, 1978-86, Nat. Cancer Inst., 1979-81, Mobil Found., 1979—, Vanderbilt Talc Co., 1988, Battelle, Columbus, 1988—, Consumer Products Safety Commn., 1987-90, Ga. Pacific, 1989-91, others. Fellow Collegium Ramazzini, Geol. Soc. Am., Mineral. Soc. Am., N.Y. Acad. Scis.; mem. Phi Beta Kappa, Sigma Xi. Home: 6 Rochambeau Dr Hartsdale NY 10530 Office: Bklyn Coll of CUNY Brooklyn NY 11210

LANGER, GLENN ARTHUR, cellular physiologist, educator; b. Nyack, N.Y., May 5, 1928; s. Adolph Arthur and Marie Catherine (Doscher) L.; m. Beverly Joyce Brawley, June 5, 1954 (dec. Nov. 1976); 1 child, Andrea; m. Marianne Phister, Oct. 12, 1977. BA, Colgate U., 1950; MD, Columbia U., N.Y.C., 1954. Diplomate Am. Bd. Internal Medicine. Asst. prof. medicine Columbia U. Coll. Physicians and Surgeons, N.Y.C., 1963-66; assoc. prof. medicine and physiology UCLA Sch. Medicine, 1966-69, prof., 1969—, Castera prof. of cardiology, 1978—, assoc. dean rsch., 1986-91, dir. cardiovascular rsch. lab., 1987—; Griffith vis. prof. Am. Heart Assn., L.A., 1979; cons. Acad. Press, N.Y.C., 1989—. Editor: The Mammalian Myocardium,

1974, Calcium and the Heart, 1990; mem. editorial bd. Circulation Rsch., 1971-76, Am. Jour. Physiology, 1971-76, Jour. Molecular Cell Cardiology, 1974—; contbr. over 170 articles to profl. jours. Capt. U.S. Army, 1955-57. Recipient Disting. Achievement award Am. Heart Assn. Sci. Coun., 1982, Heart of Gold award, 1984, Cybulski medal Polish Physiol. Soc., Krakow, 1990, Pasarow Found. award for Cardiovascular Sci., 1993; Macy scholar Josiah Macy Found., 1979-80. Fellow AAAS, Am. Coll. Cardiology; mem. Am. Soc. Clin. Investigation, Am. Assn. Physicians. Achievements include research on control of cardiac contraction. Office: UCLA Sch Medicine Los Angeles CA 90024

LANGER, JERK WANG, physician, medical journalist; b. Copenhagen, Denmark, Aug. 26, 1960; s. Ulf W. and Elsebeth (Mollerup) L.; m. Karen Lyager Horve. MD, U. Copenhagen, 1987. Med. cons. several cos. Author: Nutrition, Health and Disease, 1990, Vitamins and Minerals, 1992, Hair, Skin and Nails, 1992; editor: Complete Family Health Encyclopedia, 1990; translator several popular med. books; contbr. articles to profl. jours. Mem. Assn. Danish Sci. Journalists (mem. com.), N.Y. Acad. Scis. Home and Office: Mediconsult, Sommervej 13, DK-2920 Charlottenlund Denmark

LANGER, ROBERT SAMUEL, chemical, biochemical engineering educator; b. Albany, N.Y., Aug. 29, 1948; s. Robert Samuel Sr. and Mary (Swartz) L.; m. Laura Feigenbaum, July 31, 1988; children: Michael David, Susan Katherine. BS, Cornell U., 1970; ScD, MIT, 1974. Rsch. assoc. Children's Hosp. Med. Ctr., Boston, 1974—; asst. prof. MIT, Cambridge, Mass., 1978-81, assoc. prof., 1981-85, prof., 1985-89, Germeshausen prof., 1989—; bd. dirs. Alkermes, Cambridge, Omniquest, N.H.; tchr. Group Sch., Cambridge, 1971-73; endowed lectr. U. P.R., 1983, Case Western Res. U., 1986, U. Mich., 1987, U. Wash., 1988, U. Kans., 1989, U. Calif., San Francisco, 1991, U. Wis., 1991, Ga. Inst. Tech., 1991, Ohio State U., 1991, U.. Pitts., 1992, Purdue U., 1992, U. Del., 1993, Pa. State U., 1993; cons. to numerous cos. including Genentech, San Francisco, 1981—, Merck Sharpe and Dohme, 1981-85; sci. advisor Cygnus, Redwood City, Calif., 1987—, Perseptive Biosystems, Cambridge, 1991—. Author: (with D. Cincotta and K. Cole) Group School Chemistry Curriculum, 1972, (with W. Thilly) Laboratory in Applied Biology, 1978, Analytical Practices in Biochemistry, 1979, (with W. Hrushewsky and F. Theeuwes) Temporal Control of Drug Delivery, 1991; editor: (with M. Chasin) Biodegradable Polymers in Drug Delivery, 1990, (with D. Wise), Medical Applications on Controlled Release, Vols. I and II, 1984; editor (with R. Steiner, P. Weisz) Angiogenesis, 1992; contbr. more than 500 articles to profl. publs.; patentee in field. Union Oil fellow 1970-71, Chevron fellow 1971-72; cited for Outstanding Patent in Mass., Intellectual Property Owners Inc., 1989. Mem. NAS, Nat. Acad. Engring., Inst. Medicine of NAS, Am. Inst. of Med. and Biol. Engineers (founding fellow), AICE (Food, Pharm. & Bioengring. award, 1986, Profl. Progress award 1990, Charles M. Shine Materials Sci. and Engring. award 1991), Soc. for Biomaterials (Clemson award 1990), Am. Chem. Soc. (recipient Creative Polymer award, 1989, Phillips Applied Polymer Sci. award 1992, Pearlman Meml. lectr. award 1992), Internat. Soc. Artificial Internal Organs (Organon-Teknika award 1991), Biomed. Engring. Soc. (bd. dirs. 1991—), Controlled Release Soc. (bd. govs. 1981-85, chmn. regulatory affairs com. 1985-89, pres. 1991-92, recipient Founders award 1989, Outstanding Pharm. Paper award 1992), Am. Soc. Artificial Internal Organs (mem. program com. 1984-87), Internat. Soc. Artificial Internal Organs. Avocations: magic, jogging. Office: MIT Dept Chem Engring 77 Massachusetts Ave Cambridge MA 02139

LANGER, SEPPO WANG, physician, registrar, medical journalist; b. Esbönderup, Denmark, Sept. 6, 1963; s. Ulf Wang and Elsebeth (Mollerup) L.; m. Marianne Hemicke Hansen, June 2, 1990; 1 child, Natasha Hemicke. MD, U. Copenhagen, 1989. Sr. house officer dept. oncology Herlev U. Hosp., Copenhagen, Denmark, 1989-90; sr. house officer dept. medicine Bispebjerg U. Hosp., Copenhagen, 1990-91, 91—; sr. house officer dept. surgery Nat. U. Hosp., Copenhagen, 1991; med. cons. several publ. cos., 1986—. Editor: Home Medical Library, 16 vols., 1990—; co-editor: Complete Family Health Ency., 1990, World of Science Ency., 1990; contbr. articles to profl. jours. Mem. Danish Med. Assn., Danish Soc. Young Oncologists, Danish Soc. Med. Oncology (mem. edn. com.). Home: Strandvejen 110 C, DK-2900 Hellerup Copenhagen Denmark

LANGER, STEVE GERHARDT, biomedical physicist, consultant; b. Mauston, Wis., Apr. 10, 1963; s. Calvin Lloyd and Betty Marie (Neustader) L. BS, U. Wis., 1986; MS, Mich. State U., 1988; postgrad., Oakland U., 1989—. Rsch. asst. Nat. Superconducting Cyclotron, East Lansing, Mich., 1986-88, William Beaumont Hosp., Royal Oak, Mich., 1988—; v.p. physics adv. com. Mich. State U., East Lansing, 1987-88; pres. Prof. Physicists and Students Assn. Oakland U., Rochester Hills, Mich., 1989-92. Editor News from Detroit jour., 1991—. Mem. Am. Phys. Soc., N.Am. Hyperthermia Group, Radiation Rsch. Soc., U. Wis. Physics Soc. (pres. Madison chpt. 1985-86). Achievements include development of pulse-shape method of neutron detection; of precessing magnetic induction method for non-invasive cancer treatments; of device independant software for medical device control and communication. Office: William Beaumont Hosp 3601 W 13 Mile Rd Royal Oak MI 48073

LANGERMAN, SCOTT MILES, geotechnical engineer; b. Oakland, Calif., Oct. 1, 1968; s. Duane Lee and Linda Ruth (Wilson) L. BSCE, U. Tex., 1992, postgrad., 1992—. Geotechnical engr. MLA Labs., Austin, Tex., 1992—. Mem. ASCE. Democrat. Presbyterian. Home: 506 Bellevue Pl Austin TX 78705

LANGLAIS, CATHERINE RENEE, science administrator; b. Paris, Apr. 9, 1955; d. Jean-Jacques and Anne-Marie (Pavie) Brissot; m. Gerard Michel Langlais, Nov. 12, 1977; children: Nathalie, Gilles. Engr. degree, Mines, Nancy, France, 1976; MSME, Stanford U., 1977. Rsch. engr. CRIR Isover St. Gobain, Rantigny, France, 1978-83, head thermal sci. lab., 1984-87, head physics and new products rsch. dept., 1988—. Mem. editorial bd. Jour. Thermal Insulation; contbr. articles to profl. jours. Recipient Aniuta Winter Klein prize Acad. des Scis., Paris, 1990. Office: CRIR Isover St Gobain, BP19, 60290 Rantigny France

LANGLAIS, PHILIP JOSEPH, psychologist, educator; b. Salem, Mass., Apr. 20, 1946; s. Robert Joseph and Denise (Le Brun) L.; m. Lynne Sanborn, May 24, 1969; children: Sean, Scott, Sharna, Eric, Stephanie, Lisa. MA, U. Tex., 1974; PhD, Northeastern U., Boston, 1986. Health care scientist VA Med. Ctr., Brockton, Mass., 1985-87, La Jolla, Calif., 1988—; instr. psychiatry, Harvard U., Boston, 1985-87; assoc. prof. neuroscis., U. Calif., San Diego, 1993—; assoc. prof. psychology San Diego State U., 1988-91, prof. 1991—; reviewer NIH Study Sect., 1992. Contbr. articles to profl. jours. Presenter, coord. World Wide Marriage Encounter, San Diego, 1989—. Grantee NIH, NIA 1987-95, VA Merit Rev. 1987-96; named Outstanding Program Vol. Am. Heart Assn. 1982. Mem. APS, Soc. Neurosci., Brit. Brain Rsch. Assn., Rsch. Soc. Alcoholism. Roman Catholic. Achievements include findings of identification of glutamate excitotoxicity as a mechanism(s) involved with thiamine deficiency induced Wernicke's encephalopathy, importance of damage to specific thalamic structures in the learning and memory impairments of diencephalic amnesia, development and implementation of series electrochemical detectors in high performance liquid chromatography for the analyses of biologically active substances in mammalian nervous system. Office: Dept Psychology San Diego State Univ San Diego CA 92182-0350

LANGLANDS, ROBERT PHELAN, mathematician; b. New Westminster, Can., Oct. 6, 1936; came to U.S., 1960; s. Robert and Kathleen (Phelan) L.; m. Charlotte Lorraine Cheverie, Aug. 13, 1956; children: William, Sarah, Robert, Thomasin. BA, U. B.C., 1957, MA, 1958, DS honoris causa, 1985; PhD, Yale U., 1960; DSc (hon.), McMaster U., 1985, CUNY, 1985; D in Math. (hon.), U. Waterloo, 1988; DSc (hon.), U. Paris, 1989, McGill U., 1991, Toronto U., 1993. From instr. to asso. prof. Princeton (N.J.) U., 1960-67; prof. math. Yale U., New Haven, 1968-72, Inst. Advanced Study, Princeton, 1972—. Author: Euler Products, 1971, (with H. Jacquet) Automorphic Forms on GL (2), 1970, On the Functional Equations Satisfied by Eisenstein Series, 1976, Base Change for GL (2), 1980, Les Débuts d'une Formule des Traces Stable, 1983. Recipient Wilbur Lucius Cross medal Yale U., 1975, Common Wealth award Sigma Xi, 1984, Mathematics award Nat. Acad. Sci., 1988. Fellow Royal Soc. London, Royal Soc. Can.; mem. NAS, Am. Math. Soc. (Cole prize 1982), Nat. Acad. Sci., Can. Math. Soc. Office: Inst Advanced Study Sch Mathematics Olden Ln Princeton NJ 08540

LANGLEY, PATRICIA COFFROTH, psychiatric social worker; b. Pitts., Mar. 1, 1924; d. John Kimmel and Anna (McDonald) Coffroth; m. George J. Langley, May 1, 1946; children: George Julius III, Mary Patricia, Kelly Joan; stepchildren: Robin Spencer, Veronica Bell. BA, Empire State Coll., 1976; MSW, Hunter Coll., 1980. Diplomate Clin. Social Worker; cert. ind. social worker; cert. Conn. Psychiat. rehab. worker. Credentialed alcoholism treatment counselor, supervisor, Bronx Mcpl. Hosp. Center, Albert Einstein Med. Coll., 1970-74, case worker, comprehensive alcoholism treatment center, dept. psychiatry, 1974-80; asst. coordinator outpatient psychiat. alcoholism Meridian Ctr., Stamford, Conn., 1980-83; dir. family treatment Meridian Ctr.; pvt. practice and consultation. Vol., DuBois Day Clinic, Stamford, 1966-67, Greenwich Hosp., 1966-67. Mem. NASW, Conn. Soc. for Clin. Social Workers. Home and Office: 25 W Elm St Greenwich CT 06830-6465

LANGLEY, ROLLAND AMENT, JR., construction and engineering company executive; b. San Francisco, Aug. 22, 1931; s. Rolland Ament and Kathryn Lee (Beals) L.; m. Pamela Winston, May, 15, 1954 (div. 1978); children: Owen C., Cynthia, James R.; m. Chiara Bini-Sexton, Apr. 12, 1978. BS in Engring., Physics, U. Calif., Berkeley, 1953; MME, U. Pitts., 1961; MBA, Golden Gate U., 1973. Engr. Bettis Atomic Power Lab. of Westinghouse Electric Corp., Pitts., 1957-62; with Bechtel Corp., San Francisco, 1962-71; mgr. refinery and chem. nuclear fuel ops. Bechtel Inc., San Francisco, 1977-78; mgr. projects nuclear fuel ops. Bechtel Nat. Inc., San Francisco, 1979-80, mgr. decontamination and restoration nuclear fuel ops., 1980-81; v.p., mgr. nuclear fuels ops. Bechtel Nat. Inc., Oak Ridge, Tenn., 1981-84; sr. v.p., mgr. div. ops., research and devel. ops. Bechtel Nat. Inc., San Francisco, 1985-89; dep. mgr. Uranium Enrichment Assocs., San Francisco, 1972-76; v.p. Uranium Enrichment Tech. Inc., San Francisco, 1976-77; pres., dir. Bechtel Systems Mgmt. Inc., 1988-90; pres., chief exec. officer BNFL Inc., 1990—; Mem. Nat. Acad. Panel on Separations Tech., 1991—; bd. dirs. No. Ireland Partnership-U.S.A. Contbr. numerous articles to profl. jours. Capt. USNR. Recipient Bausch and Lomb Sci. award, 1948. Mem. Naval Res Assn. (past pres. Golden Gate chpt.). Achievements include patents in nuclear fuel and reactor systems design; research on uranium enrichment, nuclear waste disposal, fast breeder reactors, and engineering management. Home: PO Box 208 Middleburg VA 22117-0208 Office: BNFL Inc 1776 Eye St NW Ste 750 Washington DC 20006-0037

LANGLEY, TEDDY LEE, engineering executive; b. Ligon, Ky., Oct. 14, 1943; s. Raymond Matthew and Hazel Marie (Henson) L.; m. Rhoda Catherine McCloud, Jan. 26, 1963; children: Cynthia, Matthew, Scott. BSEE, U. Ky., 1968; MBA, Ohio State U., 1982. Sr. design engr. Radiation, Inc., Melbourne, Fla., 1968-71; devel. programs mgr. AccuRay Corp., Columbus, Ohio, 1971-83; program mgr. Picker Internat., Highland Heights, Ohio, 1983-84; mgr. product engring. Anaconda Advanced Tech., Dublin, Ohio, 1984-85; v.p. engring. Gould Electronics, Inc., Valley View, Ohio, 1986-90; pres. Profl. Success Group, Columbus, Ohio, 1991—. Head usher Cornerstone Chapel, Foursquare Gospel Ch., Medina, Ohio, 1990. With U.S. Army, 1963-65. Mem. IEEE, Found. Christian Living, Full Gospel Businessmen's Fellowship Internat., Tau Beta Phi, Eta Kappa Nu. Democrat. Avocations: hunting, racquetball. Home: 7750 Chickory Hollow Ct Worthington OH 43085-5801 Office: Profl Success Group 30 Dilmont Dr Ste 277 Columbus OH 43235

LANGMORE, JOHN PRESTON, biophysicist, educator; b. Santa Monica, Calif., Jan. 2, 1947; s. Herbert C. and Elizabeth (Harmon) L.; children: Ian, Katherine. BS in Physics, Stanford U., 1968; PhD in Biophysics, U. Chgo., 1975. Asst. prof. biology U. Mich., Ann Arbor, 1979-85, assoc. prof. biology, 1985-91, chmn. dept. biophysics, 1989—, prof. biology, 1991—.

LANGRALL, HARRISON MORTON, JR., internist; b. Balt., Mar. 24, 1922; s. Harrison Morton and Hazel Lucille (Clarke) L.; m. Mary Ann Saviano; children: Lucille Clarke, Kate Kidwell, Hazel Marie Langrall. BA, Johns Hopkins U., 1949; MD, U. Md., 1953; MS in Medicine, U. Minn. Mayo Found., 1958. Lic. MD, Md., Minn., Ind., N.J., Pa.; diplomate Am. Bd. Internat Medicine. Rotating internship Winchester (Va.) Meml. Hosp., 1953-54; fellow internal medicine Mayo Found., Mayo Clinic, Rochester, Minn., 1954-66; pvt. practice in internal medicine, endocrinology The Davis Clinic, Marion, Ind., 1957-66; attending staff Marion (Ind.) Gen. Hosp., 1964-66, chief of medicine, 1966-77; assoc. med. dir. Hoffmann LaRoche, Inc., Nutley, N.J., 1966-77; v.p., med. rsch. Sandoz, Inc., East Hanover, N.J., 1977-82; pres. Langrall Assocs., Inc., Paoli, Pa., 1982-92; v.p. med. rsch. Photofrin Med., Inc., Buffalo, 1984-85, Great Valley Pharms., Inc., Malvern, Pa., 1992—; clin. instr. in medicine Ind. U. Sch. of Medicine, Marion County Gen. Hosp., Indpls., 1957-66. Contbr. articles to profl. jours. 1st lt. Med. Adm. Corps., U.S. Army, 1942-46. Recipient Physician's Recognition award, 1982, 85, 88, 91. Fellow Am. Coll. Clin. Pharmacology; mem. AMA, N.J. Acad. Medicine, Chester County Med. Soc., Pa. State Med. Soc. Office: Great Valley Pharms 10 Great Valley Pkwy Malvern PA 19355

LANGSTON, WANN, JR., paleontologist, educator, researcher emeritus; b. Oklahoma City, July 10, 1921; married; 2 children. B.S., U. Okla., 1943, M.S., 1947; Ph.D. in Paleontology, U. Calif., 1951. Instr. geology Tex. Tech. Coll., 1946-48; preparator Mus. Paleontology-U. Calif., 1949-54; lectr. Mus. Paleontology, U. Calif., 1951-52; paleontologist Nat. Mus. Can., 1954-62; research scientist Tex. Meml. Mus., Austin, 1962-86; dir. vertebrate paleontology lab. U. Tex., Austin, 1969-86, prof. dept. geol. sci., 1975-85; ret., 1986; 1st Mr. and Mrs. Charles E. Yager prof. geol. scis. U. Tex., Austin, 1986-90, 1st Mr. and Mrs. Charles E. Yager prof. emeritus, 1990—; research assoc. Cleve. Mus. Natural History, 1974-79. Mem. Geol. Soc. Am., Soc. Vertebrate Paleontology (v.p. 1973-74, pres. 1974-75), Am. Soc. Ichtvol. and Herpetology, Am. Assn. Petroleum Geologists. Office: U Tex Dept Geol Sci UT Sta Austin TX 78712

LANGTON, MAURICE C., marketing and business services entrepreneur, consultant; b. Manchester, Eng., Feb. 5, 1933; s. Joseph L. and Elizabeth (Carr) L. BSc Tech. with honors, Manchester U., 1955, PhD, 1958. Head microbiology dept. Simon Carves Ltd., Manchester, 1958-64; cons. hydraulics dept. H.A. Simons Internat. Ltd., Vancouver, B.C., Can., 1967-69; chief process engr. Air Products Ltd., North Wales, 1969-70; prin. environ. engr. F.M.C. Corp., N.Y.C., 1970-74; cons., indsl. div. Acres Cons. Svcs. Ltd., Toronto, 1974-77; sr. ptnr. Langton Cons. Svcs., Vancouver, 1978—; gen. mgr. Laser In-Vitro Testing Ltd., Waterloo, Ont., 1987-90; v.p. Altech Tech. Systems Inc., Toronto, 1990-92; pres. Langton Svcs., Toronto, 1992—; liason coord. Toronto New Bus. Devel. Ctr., 1993—. Recipient Manchester (U.K.) City scholarship, 1952-55, Manchester (U.K.) City Rsch. scholarship, 1955-56, Univ. Manchester (U.K.) Teaching scholarship, 1956-58. Mem. Toastmasters Internat. (sec. 1979-82). Achievements include patent on Flour Milling Alternative. Office: Langton Svcs, 1460 Bayview Ave Ste 404, Toronto, ON Canada M4G 3B3

LANGTON, MICHAEL JOHN, mechanical engineer, consultant; b. Kingston, N.Y., July 29, 1957; s. Richard John and Marilyn (Thaler) L.; m. Jill Ann Salque, Aug. 8, 1981 (div. Dec. 1985); m. Deanna Joyce Dial, July 22, 1989 (div. Jan. 1993). BSME, Calif. Polytech. Inst., 1983; MBA, So. Meth. U., 1987. Registered profl. engr., cert. energy mgr., cert. lighting efficiency profl. Sr. assoc. IBM Corp., Armonk, N.Y., 1983-88; assoc. Ccrd Partners, Dallas, 1989-90; pres. Langton Enterprises, Inc., Dallas, 1990—. Recipient Cert. of Appreciation Nat. Multiple Sclerosis, Dallas, 1988, Energy Profl. of Yr. award Energy Odyssey, 1993. Mem. ASHRAE, Assn. Energy Engrs. (treas. 1988, v.p. 1989, pres. 1990). Home: 3229 Sugarbush Carrollton TX 75007 Office: Langton Enterprises Inc 3710 Rawlins Ste 1100 Dallas TX 75219

LANGWORTHY, HAROLD FREDERICK, manufacturing company executive; b. White Plains, N.Y., Aug. 1, 1940; s. B. Fred and Helen W. (Studwell) L.; m. Nikki N. Smith, Sept. 4, 1965; children: Katherine, Kristen, Thomas. BS in Math., Rensselaer Poly. Inst.; PhD in Math., U. Minn. Lab. head Kodak Rsch. Labs., Rochester, N.Y., 1979-81, div. dir., 1981-85; group lab. dir. Eastman Kodak Co., Rochester, 1985-90, asst. gen. mgr. printer products div., 1990-92, assoc. dir. mfg. rsch. and engring., 1992-93, dir. mfg. rsch. and engring., 1993—; mem. adv. bd. Ctr. for Indsl. Innovation, Troy, N.Y., 1985-87, Microelectronics and Computer Tech. Corp., Austin, Tex., 1985-88. Mem. AAAS, Optical Soc. Am. Achievements include patents, research and publications on imaging in the 21st century. Home: 732 Hightower Way Webster NY 14580 Office: Eastman Kodak Co Rochester NY 14650

LANHER, BERTRAND SIMON, biological spectroscopist; b. Charleville, France, May 27, 1965; came to U.S., 1990; s. Gilbert and Micheline Juliette (Constant) L.; m. Debra Mary Ubbelohde, July 28, 1991. MSc, U. des Scis. et Tech. de Lille, France, 1987; PHD, U. Bourgogne, Dijon, France, 1991. Chemist Helio Jean Didier, Hellemnes, France, 1985; quality assurance/quality control mgr. SRBG-Coca Cola, Fâches Thumesnil, France, 1986-87; rsch. engr. INRA, Poligny, France, 1988-91; product mgr. Nicolet Instrument Corp., Madison, Wis., 1991-92; gen. mgr. Chemometrics Cons., Cottage Grove, Wis., 1992—; pres., CEO Anadis Instruments USA, Inc., Madison, Wis., 1993—; speaker at confs. in field. Contbr. articles to profl. publs. Musician Dijon Jazz Band, 1987-90, Am. Jazz Express, 1992—. French Govt. grantee, 1990. Mem. AAAS, Assn. Ofcl. Analytical Chemists, Am. Musicians Union, N.Y. Acad. Scis. Roman Catholic. Achievements include patent for monitoring of the kinetics of milk enzymic coagulation usint FT-IR spectroscopy; finding of implementation of algorithms for the automated calibration of FT-IR spectrometers. Office: Anadis Instruments USA Inc 2009 S Stoughton Rd Madison WI 53716

LANKFORD, MARY ANGELINE GRUVER, pharmacist; b. Homerville, Ga., Jan. 17, 1964; d. Charles Laney Jr. and Elizabeth Pamela (Waters) Gruver; m. Leo LaFel Lankford, July 27, 1985; 1 child, Charles Jacob (Jake). PharmD, Mercer U., 1989. Pharmacist South Fulton Drug Shoppe, East Point, Ga., 1989; pharmacist in charge Given's Pharmacy, East Point, 1989; co-dir. of pharmacy 5 stores Barnes Drugs, Valdosta, Ga., 1989-91; nursing home cons. Health Care Resources, Valdosta, 1990—; pharmacist in charge Barnes Drugs Downtown, Valdosta, 1990—. Mem. adminstrv. com., women's com. Homerville United Meth. Ch., 1990—; young adult mem. Homerville United Meth. Ch. Adminstrv. Com., 1992—. Mem. Am. Pharm. Assn., Ga. Pharm. Assn., South Ga. Pharm. Assn., Delta Zeta (v.p. 1984-85, scholarship chair 1984-85, Best Sister award 1985). Avocations: wildlife, geriatric interests, law enforcement. Home: Rte 1 Box 83-C Homerville GA 31634 Office: Barnes Drugs 200 S Patterson St Valdosta GA 31601-5621

LANNER, MICHAEL, research administrator, consultant; b. Montreal, Que., Can., Sept. 14, 1943; came to U.S., 1969; s. Hyman Alter and Anne P. (Rasnikopf) L.; m. Bluma Pauline Weiskopf, Dec. 27, 1946; children: Brian, Jennifer, Lisa. BS, Loyola U., Montreal, Que., Can., 1968. Sr. rsch. technician McGill U., Montreal, 1964-69; sr. rsch. assoc. Beth Israel Hosp., Boston, 1969-80, rsch. mgr., 1980-84, deputy dir. rsch., 1984-88, dir. rsch., 1988—; pvt. practice Boston, 1987—; instr. in medicine Harvard Medical Sch., 1992—. Contbr. articles to Biochemistry, Diabetes, Nature. Active Stoughton (Mass.) Town Meeting, 1985—, Stoughton Youth Commn., 1988—; chmn. Stoughton Sch. Com., 1989—. Grantee NIH, 1980—, 1990. mem. Soc. Rsch. Adminstrs. (bd. dirs., sec. 1990-91), Mass. Soc. Med. Rsch. (mem. policy bd. 1989—). Achievements include patents for Convertible Animal Cage, Insert for Animal Cage. Office: Beth Israel Hosp 330 Brookline Ave Boston MA 02215-5491

LANNI, JOSEPH ANTHONY, military officer; b. Niskayuna, N.Y., Oct. 9, 1958; s. Ralph and Gertrude Olivia (Seiffert) L.; m. Pamela Amy Morgan, July 16, 1983; children: Nicole, Kyle, Caitlin. BS in Engring., Mech. and Math., USAF Acad., 1980; MS in Aerospace Engring., U. Fla., 1991. Commd. 2d lt. USAF, 1980, advanced through grades to major, 1992. Contbr. articles to profl. jours. Decorated Air medal, 1990; recipient JABARA Airmanship award USAF, 1991. Mem. AIAA, Soc. Exptl. Test Pilots. Achievements include research in flight testing on F-16 aircraft which defined structural aircraft limits. Home: 4504 Whelk Pl Las Vegas NV 89031

LANNIN, JEFFREY S., physicist, educator; b. N.Y.C., Aug. 21, 1940; s. Sidney and Ann Lannin; m. Delores M. Slawinski, Apr. 7, 1971; 1 child, Joshua. BS, Purdue U., 1962; MS, U. Ill., 1964; PhD, Stanford U., 1971. Physicist Lockheed Rsch. Lab., Palo Alto, Calif., 1967-68, Max Planck Inst., Stuttgart, Germany, 1971-74, Argonne (Ill.) Nat. Lab., 1974-75; vis. prof. U. Del., Newark, 1975-76; asst. prof. Pa. State U., State College, 1976-81, assoc. prof., 1981-86, prof., 1986—. Contbr. articles to profl. jours. and chpts. to books. Mem. Am. Phys. Soc., Am. Vacuum Soc., Materials Rsch. Soc. Achievements include research in basic structure and dynamics of amorphous solids and fullerene materials. Office: Pa State U Dept Physics 104 Davey Lab State College PA 16802

LANOUE, ALCIDE MOODIE, medical corps officer, health care administrator; b. Tonawanda, N.Y., Nov. 2, 1934; s. Alcide and Isabelle LaNoue; children: Claire L., Alcide J., George E., Michelle; m. Beth Gortner, Dec. 20, 1986. B.A., Harvard U., 1956; M.D., Yale U., 1960. Diplomate Am. Bd. Orthopedic Surgery. Commd. lt. U.S. Army Med. Corps, 1966, advanced through grades to maj. gen., 1988; intern Brooke Army Med. Ctr., Ft. Sam Houston, Tex., 1960-61, resident in orthopedic surgery, 1963-66; mem. staff 2d Surg. Hosp., Republic of Vietnam, 1966-67, 24th Evacuation Hosp., Republic of Vietnam, 1967, Valley Forge Gen. Hosp., Phoenixville, Pa., 1967-70, 71-73, Command & Gen. Staff Coll., Ft. Leavenworth, Kans., 1970-71, Walter Reed Army Med. Ctr., Washington, 1973-77, Hanau Clinic, Fed. Republic Germany, 1977-80; comdr. U.S. Army MEDDAC, Ft. Stewart, Ga., 1980-82; mem. staff U.S. Army MEDDAC, Ft. Benning, Ga., 1982-84; comdg. gen. Dwight David Eisenhower Army Med. Ctr., Ft. Gordon, Ga., 1984-86; comdt. Acad. Health Scis., U.S. Army, Ft. Sam Houston, Tex., 1986—; now dep. surgeon gen. Surgeon Gen.'s Office, Falls Church, Va. Contbr. chpts. to books, articles to profl. jours. Decorated Bronze Star, Legion of Merit, Meritorious Service medal with 3 oak leaf clusters. Fellow ACS; mem. Am. Acad. Orthopedic Surgery. Office: Surgeon Gen's Office 5109 Leesburg Pike # 6 Falls Church VA 22041-3208*

LANTERMAN, WILLIAM STANLEY, III, plant pathologist, researcher, administrator; b. Portsmouth, Va., Sept. 8, 1947; s. William Stanley Lanterman Jr. and Elisabeth Chesley Rasch; m. Denise Marie MacDonald, Aug. 16, 1975; children: Samuel MacDonald, Ian Kennedy. Diploma in plant sci., Nova Scotia Agrl. Coll., Truro, 1978; BS in Plant Sci., U. Maine, Orono, 1980; M Plant Sci. in Agriculture, Cornell U., 1982, PhD, 1985. Head plant quarantine sect. Agriculture Can. Rsch. and Quarantine Sta., Sidney, B.C., Can., 1985-87; dir. Ctr. for Plant Health (formerly Saanichton Plant Quarantine Sta.), Sidney, 1987—; chmn. com. on cert. standards North Am. Plant Protection Orgn., Ottawa, Ont. Can., 1989—. Dir., v.p. Sidney and North Saanich Community Hall Assn., Sidney, 1986-92; dir. Meml. Park Soc., Sidney, 1990-92; dir., exec. mem. coop. adv. bd. U. Victoria, B.C., 1988—. Office: Agriculture Can Ctr Plant Health, 8801 E Saanich Rd, Sidney, BC Canada V8L 1H3

LANTOS, PETER R(ICHARD), industrial consultant, chemical engineer; b. Budapest, Hungary, July 18, 1924; came to U.S. 1939; naturalized citizen; s. Ernest and Bertha (Wigner) L.; m. Janice Kirchner, Dec. 20, 1947 (div. 1982); children: Geoffrey P., Greggory P., Gabrielle, Giselle. BChemE, Cornell U., 1945, PhD in Chem. Engring., 1950. Devel. chemist GE, Pittsfield, Mass., 1946-47; rsch. engr. E. I. Du Pont Nemours & Co., Wilmington, Del., 1950-55, supr. rsch., 1955-60; mgr. application and product devel. Celanese Plastics Co., Clark, N.J., 1961-63, mgr. R & D, 1964-69; dir. devel. Sun Chem. Corp., Carlstadt, N.J., 1969-70, v.p. R & D, 1970-75; gen. mgr. div. plastics Rhodia Inc., N.Y.C., 1975-76; dir. R & D Arco Polymers, Inc., Phila., 1976-77, v.p. R & D, 1978-79; pres. Target Group, Inc., Phila., 1980—. Contbr. over 30 articles to profl. jours. Pres. Bd. Health, Kennett Square, Pa., 1955-59. Mem. Am. Chem. Soc. (chmn. div. chem. mktg. and econs. 1988-89), Am. Inst. Chem. Engrs., Assn. Cons. Chemists and Chem. Engrs., Soc. Plastics Engrs., Soc. Plastics Industry. Home: PO Box 27247 Philadelphia PA 19118-0247 Office: Target Group Inc 1000 Harston Ln Philadelphia PA 19118-1037

LANTZ, CHARLES ALAN, chiropractor, researcher; b. Haywood County, N.C., Dec. 15, 1947; s. Harry Adrian and Betty Elizabeth (Stewart) L.; m. Susan Ruppalt (div. June 1971); children: Heather Leigh; m. Judith Daye Wagner (div. May 1990); children: Sky Micah, Spring Laurel, Miriah

Lee. Cert. in French, U. Lyon, France, 1969; BS in Zoology, U. N.C., 1971, PhD in Pharmacology, 1978; D Chiropractic magna cum laude, Life Chropractic Coll., Marietta, Ga., 1987. Postdoctoral fellow Johns Hopkins U., Balt., 1977-81; assoc. prof. biochemistry Life Chiropractic Coll., 1982-87, assoc. rsch. prof., 1987-90; dir. rsch. Life Chiropractic Coll.-West, San Lorenzo, Calif., 1990—; mem. adv. bd. Chiropractic Rsch. Jour., 1988—; mem. rsch. adv. com. Found. for Chiropractic Edn. and Rsch., Arlington, Va., 1990—; mem. exec. com. Consortium for Chiropractic Rsch., Sunnyvale, Calif., 1990—; mem. commn. Consensus Conf. To Establish Guidelines for Chiropractic Quality Assurance and Practice Parameters. Mem. Am. Chiropractic Assn. (cons. on technique 1992—), Calif. Chiropractic Assn., Mich. Chiropractic Coun. (hon.), Internat. Chiropractors Assn. Achievements include development of mathematical theory for thin-film dialysis, developing a theoretical basis for chiropractice—the vertebral subluxation complex. Home: 492 Scenic Rd Fairfax CA 94930 Office: Life Chiropractic Coll West 2005 Via Barrett San Lorenzo CA 94580

LANTZ, NORMAN FOSTER, electrical engineer; b. Pekin, Ill., June 8, 1937; s. Norman Gough and Lenore (Elsbury) L.; m. Donnis Maureen Ballinger, Sept. 7, 1958 (div. Aug. 1991); children: Katherine, Deborah, Norman Daniel; m. Judith Eliane Peach, Dec. 7, 1991. BSEE, Purdue U., 1959, MSEE, 1961. System engr. GE Co., Phila., 1961-72; mem. tech. staff The Aerospace Corp., El Segundo, Calif., 1972-75, mgr., 1975-79, dir., 1979-83, prin. dir., 1990, sr. project engr., 1991—; dir. Internat. Found. for Telemetering, Woodland Hills, Calif., 1985—. 2d lt. U.S. Army, 1960-61. Mem. AIAA (sr.), IEEE, Internat. Test and Evaluation Assn., Am. Mgmt. Assn. Office: The Aerospace Corp El Segundo CA 90245-4691

LANZAFAME, RAYMOND JOSEPH, surgeon, researcher; b. Rochester, N.Y., Sept. 30, 1952; s. Ray J. and Mary Vera (DeMeis) L.; m. Patricia Marie Volkmar, Apr. 26, 1980; children: Mark Raymond, Karen Elizabeth. BS with honors and distinction, Cornell U., 1974; MD, George Washington U., 1978. Diplomate Nat. Bd. Med. Examiners, Am. Bd. Surgery. Clin. asst. prof. U. Rochester, N.Y., 1983-87, asst. prof., 1987-92, assoc. prof., 1992—; mem. laser task force N.Y. State Dept. of Health, 1990; dir. laser tng., chmn. laser usage com., Rochester Gen. Hosp. 1983—, dir. surg. laser rsch. lab., 1988—, dir. Laser Ctr., 1984—; bd. dirs. Rochester Gen. Hosp. Found., 1990—. Sr. editor Jour. Clin. Laser Surgery and Medicine, co-editor-in-chief; mem. editorial bd. Laser Surgery and Medicine, 1987-93, referee, 1987—; mem. editorial bd. Laser Medicine and Surgery News and Advances, 1988-90, Jour. of Laparoendoscopic Surgery, 1991—, Surgery Alert, 1991—; consulting editor Biomedical Optics, 1992—. Grantee Am. Cancer Soc. Fellow ACS (young surgeon rep. 1992—, councillor upstate N.Y. chpt.); mem. AMA, Southern Med. Assn., Med. Soc. State of N.Y., Monroe County Med. Soc., Rochester Surg. Soc., Rochester Acad. Medicine, Am. Soc. for Laser Medicine and Surgery, Assn. for Acad. Surgery, Collegium Internationale Chirugie Digestive, N.Y. Acad. Sci., Acad. Surg. Rsch., Internat. Soc. for Lasers in Surgery and Medicine, Internat. Soc. Surgery, Soc. for Surgery of Alimentary Tract, Cen. Surg. Assn., Cen. N.Y. Surg. Soc., Soc. Photo-Optical Instrumentation Engrs., Biomedical Optics Soc. Achievements include research on comparison of the effects of cholecystectomy incisions made with the CO2 laser on pulmonary functions and atelectasis vs. scalpel incisions, on comparison of modified radical mastectomies done with CO2 laser vs. scapel as regards the amount of hematocrit of drainage, length of hospital stay, frequency of seroma formation, on hematoporphyrin derivative as primary cancer therapy, on hematoporphyrin derivative fluorescence imaging systems, on experimental surgical techniques with various lasers, on testing of prototype lasers and laser accessories, on development of photosensitizing agents as cancer therapies, on mechanisms of reduction of tumor recurrence by laser surgery, on laser tissue interaction, on the host-tumor response and its effect on local tumor recurrence, on the influence of hyperthermia (laser) on local recurrence, on wound factors/cytokines on host-tumor response, on development of laparoscopic surgical techniques. Office: 1445 Portland Ave # 202 Rochester NY 14621-3008

LANZANO, RALPH EUGENE, civil engineer; b. N.Y.C., Dec. 26, 1926; s. Ralph and Frances (Giuliano) L.; B.C.E., NYU, 1959. Engring. aide Seelye, Stevenson, Value, Knecht, N.Y.C., 1957-58; jr. civil engr. N.Y.C. Dept. Public Works (name changed to N.Y.C. Dept. Water Resources), 1960-63, asst. civil engr., 1963-68; civil engr., 1968-71; sr. san. engr. Parsons, Brinckerhoff, Quade & Douglas, N.Y.C., 1971-72; civil engr. N.Y.C. Dept. Water Resources, 1972-77, N.Y.C. Dept. Environ. Protection, 1978-90; ret., 1990. Registered profl. engr., N.Y. Mem. NRA (life), ASCE (life), ASTM, Nat., N.Y. socs. profl. engrs., Water Environ. Fedn., Am. Water Works Assn., Am. Public Health Assn., Am. Fedn. Arts (contbg.), U.S. Inst. Theatre Tech., NYU Alumni Assn., Am. Nat. Theatre and Acad., Lincoln Center Performing Arts, Film Soc. Lincoln Ctr., N.Y.C. Ballet Guild, Asia Soc., Nat. Fire Protection Assn., N.Y. Pub. Libr. (friend), Sta. WNET-TV (Thirteen), U.S. Lawn Tennis Assn. (life), Nat., Internat. wildlife fedns., Bible-A-Month-Club, Nat. Parks and Conservation Assn., Nat. Geog. Soc., Nat. Audubon Soc., Am. Automobile Assn., Bklyn. Bot. Garden, Am. Mus. Natural History, Chi Epsilon. Home: 17 Cottage Ct Huntington Station NY 11746-1104

LAOR, HERZEL, physicist; b. Tel-Aviv, Israel, Aug. 19, 1947; came to U.S., 1988; s. Haim and Hana (Tahoreq) Lieberman; m. Ita Laudon, Sept. 9, 1968; children: Sharon, Yuval, Elad. BS in Physics, Tel-Aviv U., 1974. Physicist Rehovot Instruments, Rehovot, Israel, 1973-76; capt. Israel Defence Force, 1976-78; cons. Rehovot, 1980-86; mng. dir. Shoval Optronics, Rehovot, 1980-86, Shoval Internat., Rotterdam, The Netherlands, 1982-88; cons. Time Warner Cable, Denver, 1988-89; engr. Pragmatronics, Inc., Boulder, Colo., 1990-91; founder, chief tech. officer Astarte Fiber Networks, Inc., Boulder, Colo., 1992—. Contbr. articles to profl. jours. Mem. IEEE, Optical Soc. of Am. Achievements include 6 U.S. patents in field and 17 world-wide patents in field.

LAPERRIERE, FRANCIS WILLIAM, electrical engineer, audio recording engineer; b. Grosse Pointe Farms, Mich., Apr. 6, 1943; s. Edgar Daniel and Louise Georgiana (Merz) L.; m. Sharon Judene Kowalewski, Dec. 20, 1975; children: Anne Ruth, Neil Ross. BS in Physics, Wayne State U., 1976, MS in Physics, 1979. Electronics instr. Radio Electronic-TV Schs., Detroit, 1965-69; studio engr. Sta. WWJ Radio, Detroit, 1969-73; rsch. asst. physics Wayne State U ., Detroit, 1976-79; mem. tech. staff Energy Conversion Devices, Troy, Mich., 1980-82; elec. engr. Gen. Dynamics Land Systems, Center Line, Mich., 1982-85, U.S. Army TACOM Rsch. and Devel. Ctr., Warren, Mich., 1985-86, AMC & Chrysler Corp., Detroit, 1986—; mem. SAE Electromagnetic Radiation Tech. com., Detroit, 1986-92, Std. and Test Methods Tech. Com., 1986-90. Contbr. articles to profl. jours. Mem. Am. Phys. Soc., Detroit Theater Organ Soc. (bd. dirs. 1991—). Home: 1360 Buckingham Ave Grosse Pointe Park MI 48230 Office: Chrysler Corp 800 Chrysler Dr E Auburn Hills MI 48326

LAPIDUS, MORRIS, retired architect, interior designer; b. Odessa, Russia, Nov. 25, 1902; came to U.S., 1903, naturalized, 1914; s. Leon and Eva (Sherman) L.; m. Beatrice Perlman, Feb. 22, 1929 (dec. 1992); children: Richard L., Alan H. B. Arch., Columbia, 1927. With Warren & Wetmore, N.Y.C., 1926-28, Arthur Weisner, N.Y.C., 1928-30; asso. architect Ross-Frankel, Inc., N.Y.C., 1930-42; prin. Morris Lapidus Assos., 1942-86. Author: Architecture-A Profession and a Business, 1967, Architecture of Joy, 1979, Man's Three Million Odyssey, 1988, A Pyramid in Brooklyn, 1989, Morris Lapidus: The Architect of the American Dream, 1992 (English and German edits. by Martina Duttmann); architect-designer: Fontainebleau Hotel, 1954, Eden Roc Hotel, 1955, Americana Hotel, N.Y.C., 1962, Sheraton Motor Inn, N.Y.C., 1962, Internat. Inn, Washington, 1964, Fairfield Towers, Bklyn., 1966, Summit Hotel, N.Y.C., 1966, Paradise Island Hotel, Nassau, 1967, Paradise Island Casino, Nassau, 1968, Out-Patient and Rehab. Center, continuing care wing Mt. Sinai Hosp, Miami Beach, Fla., 1967, Research Bldg, 1981, congregation Beth Tfiloh, Pikesville, Md., 1967, Internat. Hdqrs. of Jr. Chamber Internat., Coral Gables, Fla., 1968, Americana Hotel, N.Y.C., 1968, Cadman Plaza Urban Redevel., Bklyn., 1969, Miami Internat. Airport, 1969-74, Penn-Wortman Housing Project, Bklyn., 1971, Bedford-Stuyvesant Swimming Pool and Park, Bklyn., 1970, El Conquistador Hotel, P.R., 1969, Trelawney Beach Hotel, Jamaica, W.I., 1973, Greater Miami Jewish Fedn. Hdgrs, 1970, Hertz Skycenter Hotel, Jacksonville, Fla., 1971, Aventura, Miami, 1971, U. Miami (Fla.) Concert Hall, 1972, Griffin Sq. Office Bldg, Dallas, 1972, U. Miami Law Library, 1975, Citizens Fed. Bank Bldg, Miami, 1975, Miami Beach Theater of Performing Arts, 1976, Ogun State Hotel, Nigeria, 1977, Exhbn. Designers, Forum Design, Linz, Austria, 1980; others; assoc. architect: Keys Community Coll, Key West, Fla., 1977, Churchill Hotel London, Grandview Apt. Complex, 1980, La Union Ins. Bldg., Guayaquil, Ecuador, 1983, Daniel Tower Hotel, Herzlea, Israel, 1983, Colony Performing Arts Ctr., 1983, Jabita Hotel, Nigeria, 1984; lectr. store, hotel design; one-man exhibit 40 Yrs. Art and Architecture, Lowe Gallery, Miami U., 1967, Fedn. Arts and Archtl. League N.Y., 1970, Weiner Galleries, 1972, Exhibit 55 Yrs. Architecture, Rotterdam, The Netherlands, 1992. Mem. Miami Beach Devel. Commn., 1966-67. Winner nat. competition S.W. Urban Renewal Program in Wash., internat. competition for trade center on The Portal Site in Washington; recipient Justin P. Allman award Wallcovering Wholesaler's Assn., 1963; Outstanding Specifications award Gypsum Drywall Contractors Internat., 1968; certificate merit N.Y. Soc. Architects, 1971; others. Mem. Miami Beach C. of C. (gov.), Kiwanis. Achievements include initial use of modern in merchandising field; areas of work include housing, hosps., hotels, shopping ctrs., office bldgs., religious instns. Home: 3 Island Ave Miami FL 33139-1363

LAPINSKI, JOHN RALPH, JR., aerospace engineer; b. Detroit, Sept. 4, 1952; s. John Ralph and Maxine La Verne (Friend) L.; m. Ann Maureen Buettner, Aug. 1, 1986; 1 child, Jonathan Ralph. BSME, U. Mich., 1980, MSME, 1981. Part time instr. U. Mich., Dearborn, 1980-81, St. Louis C.C., Florissant, Mo., 1982-86, U. Mo., St. Louis, 1984-85; engr. DSP Laser Systems Thermal Design, McDonnell Douglas, St. Charles, Mo.; sr. engr. Pilot Laser System Thermal Design, McDonnell Douglas, St. Charles, Mo., 1982-87; lead engr. Modular Power Subsystem Thermal Design, McDonnell Douglas, St. Charles, Mo., 1987-89; tech. specialist Space Sta. Freedom, McDonnell Douglas, St. Charles, Mo., 1989—. Contbr. articles to profl. jours. With U.S. Army, 1976-79. Recipient Cert. of Achievement, McDonnell Douglas, 1991, Army Commendation medal, 1979, Good Conduct medal, 1979; named Teammate of Distinction, McDonnell Douglas, 1988, Battalion Soldier of the Yr., U.S. Army, 1978. Mem. AIAA, Tau Beta Pi Hon. Soc., Whitmoor Country Club. Achievements include patent for Isothermal Multi-passage Cooler. Home: 3844 Chablis Saint Charles MO 63304 Office: McDonnell Douglas Systems PO Box 426 Saint Charles MO 63302

LAPLANTE, MARK JOSEPH, laser and electro-optics engineer; b. Troy, N.Y., Mar. 10, 1957; s. William L. and Janet A. (Gagnon) LaP.; m. Diana D. Duclos, May 21, 1977; children: Erica, Keith. AAS in Elec. Engring., Hudson Valley Community Coll., 1984. Laborer Del. and Hudson Railway, Colonie, N.Y., 1976; electrician Del. and Hudson Railway, Mechanicsville, N.Y., 1980-81; sr. engring. specialist IBM, Hopewell Junction, N.Y., 1984—. Sgt. USAF, 1976-80. Mem. Tau Alpha Pi. Achievements include patents for formation of high quality patterns for substrates and apparatus therefor; for method and apparatus for an increased pulse repetition rate for a CW pumped laser; patent pending for method and apparatus for an increased output for a pumped laser using a moving aperture. Office: IBM East Fishkill Rt 52 M/S 81K Hopewell Junction NY 12533

LAPP, ROGER JAMES, consulting pharmacist; b. Buffalo, Jan. 29, 1933; s. Roger Vincent and Georgia James (Saemenes) L.; student Mich. State U., 1952-53; BS in Pharmacy, U. Buffalo, 1957; MA, Trinity Theol. Sem., 1993; m. Judith Bure, Mar. 30, 1956; children: Eric Roger, Mark Frederick. Pharmacist intern Nobb Hill Pharmacy, Buffalo, 1956-57; pharm. intern Buffalo Gen. Hosp., 1957, pharm. resident, 1958; pharmacy mgr. Morton Plant Hosp., Clearwater, Fla., 1960-84, dir. profl. services, 1984-86; cons. pharmacist Basic Am. Med. Co., 1986-88; pharmacist, Healthcare Prescription Services, Gainesville, Fla., 1986-89, mgr., cons., 1986-91; clin. instr. Sch. of Pharmacy U. Fla., 1990—; tchr. profl. seminars.; cons. pharmacist several nursing homes, Ocala., Fla. Mem. Human Rights Advocacy Com. for Pinellas and Pasco Counties (Fla.), 1973-82, chmn., 1973-81; pres. Upper Pinellas Assn. Mental Retardation Assns., 1970-72, bd. dirs., 1969-78; pres. Am. Cancer Soc., Pinellas County, 1979-82, life bd. dirs.; pres. Pinellas Epilepsy Found., 1978-79; v.p. Fla. Assn. Retarded, 1971-78; exec. v.p., sr. v.p.; bd. dirs. Christian Corp Found. for Mentally Disabled, 1983-86, pres., 1985-86; bd. dirs. Bethel Bethany Homes for the Mentally Disabled, 1986-89, Isaiah Found. for the Disabled, 1990—, vice chmn., 1992—. With U.S. Army, 1958-60. Named Man of Yr., Upper Pinellas Assn. Retarded, 1970, Fla. Cons. Pharmacist of Yr., 1987; recipient Nat. Bowl of Hygeia, Fla. Pharm. Assn. and A.H. Robins Co., 1975, Smith award for helping retarded Kiwanis Club, Clearwater Beach, 1978, cert. of merit for public edn. Am. Cancer Soc., 1978, Citizen Health award Clearwater Sun, 1981. Fellow Am. Soc. Cons. Pharmacists, Fla. Soc. Hosp. Pharmacists (pres. 1972-73, chmn. bd. 1972-74, dir. 1970-78, 79-81; Fla. Hosp. Pharmacist of Yr. 1975, Fla. Cons. Pharmacist of Yr. 1987), Fla. Pham. Assn. (award for public rels. 1981, futuristic com. 1989—, chmn. 1992-93), Pinellas Soc. Pharmacists (pres. 1982, exec. sec. 1983-86, dir. 1979-86), S.W. Fla. Soc. Hosp. Pharmacists (pres., sec., dir., President's award 1982), Christian Pharmacists Fellowship Internat. (dir. 1984—, pres. 1989-92), Fla. Assn. Retarded Citizens (v.p. 1971-79, Brotherhood award 1975, Pres.'s award 1978, sr. v.p. 1979-81, exec. v.p. 1982), Upper Pinellas Assn. Retarded Citizens, Gideons Internat. Republican. Baptist. Author: Antibiotics, 1974, 5th rev. edit., 1984; contbr. articles to profl. jours. Home: 3527 E Lazy River Ct Dunnellon FL 34434 Office: RJL Consulting 3527 E Lazy River Dr Dunnellon FL 34434

LAQUAGLIA, MICHAEL PATRICK, pediatric surgeon, neuroblastoma researcher; b. Newark, Aug. 6, 1950; s. Michael and Dorothy Theresa (Livsey) LaQ.; m. Joanne Drako, June 26, 1982; children: Michael Joseph, Catherine Elizabeth. BS, N.J. Inst., 1972, MD, 1976. Diplomate Am. Bd. Surgery; Cert. Spl. Competence Pediatric Surgery. From intern to chief resident in gen. surgery Mass. Gen. Hosp., Boston, 1976-83, clin. fellow in transplantation, 1980-81, clin. fellow in vascular surgery, 1984; hon. sr. registrar in surgery Broadgreen Regional Chest Ctr., Liverpool, Eng., U.K., 1982; assoc. chief resident in pediatric surgery Children's Hosp. Med. Ctr., Boston, 1983-86, chief resident in pediatric surgery, 1986-87, asst. attending surgeon and pediatrician Meml. Sloan-Kettering Cancer Ctr., N.Y.C., 1987—; asst. attending surgeon N.Y. Cornell U. Med. Ctr., N.Y.C., 1989—; asst. prof. surgery Med. Sch. Cornell U., N.Y.C., 1989—; asst. attending surgeon, pediatrician Meml. Sloan-Kettering Cancer Ctr., N.Y.C.; asst. attending surgeon N.Y. Cornell U. Med. Ctr., N.Y.C.; asst. prof. surgery Cornell U., N.Y.C. Mem. AAAS, Am. Assn. Cancer Rsch., Am. Pediatric Surg. Assn., Soc. Surg. Oncology. Office: Meml Sloan Kettering Cancer Ctr 1275 York Ave New York NY 10021

LARAGH, JOHN HENRY, physician, scientist, educator; b. Yonkers, N.Y., Nov. 18, 1924; s. Harry Joseph and Grace Catherine (Coyne) L.; m. Adonia Kennedy, Apr. 28, 1949; children: John Henry, Peter Christian, Robert Sealey; m. Jean E. Sealey, Sept. 22, 1974. MD, Cornell U., 1948. Diplomate Am. Bd. Internal Medicine. Intern medicine Presbyn. Hosp., N.Y.C., 1948-49; asst. resident Presbyn. Hosp., 1949-50; fellow cardiology, trainee Nat. Heart Inst., 1950-51; rsch. fellow N.Y. Heart Assn., 1951-52; asst. physician Presbyn. Hosp., 1950-55, asst. attending, 1954-61, assoc. attending, 1961-69, attending physician, 1969-75, pres. elect med. bd., 1972-74; dir. cardiology Delafield Hosp., N.Y.C., 1954-55; mem. faculty Coll. Physicians and Surgeons Columbia U., 1950-75, prof. clin. medicine Physician and Suregons, 1967-75, spokesman exec. com. faculty coun. Physicians and Surgeons, 1971-73; vice chmn. bd. trustees for profl. and sci. affairs Presbyn. Hosp., 1974-75; dir. Hypertension Ctr. Columbia-Presbyn. Med. Ctr., 1971-75, chief Nephrology div., 1968-75; Hilda Altschul Master prof. medicine, dir. Hypertension and Cardiovascular Ctr., N.Y. Hosp.-Cornell Med. Ctr., 1975—, chief cardiology div., 1976—, dir.; cons. USPHS, 1964—. Editor-in-chief Am. Jour. Hypertension, Cardiovascular Reviews and Reports; Editor: Hypertension Manual, 1974, Topics in Hypertension, 1980, Frontiers in Hypertension Rsch., 1981; editor Hypertension: Pathophysiology, Diagnosis, and Management, 1990; editorial bd.: Am. Jour. Medicine, Am. Jour. Cardiology, Kidney Internat., Jour. Clin. Endocrinology and Metabolism, Hypertension, Jour. Hypertension, Circulation, Am. Heart Jour., Procs. of Soc. Exptl. Biology and Medicine, Heart and Vessels. Mem. policy adv. bd. detection and follow-up program Nat. Heart and Lung Inst., 1971, bd. sci. counselors, 1974-79; chmn. U.S.A.-USSR Joint Program in Hypertension, 1977—. With AUS, 1943-46. Recipient Stouffer prize med. rsch., 1969. Fellow ACP, Am. Coll. Cardiology; mem. Am. Heart Assn. (chmn. med. adv. bd. council high blood pressure rsch. 1968-72), Am. Soc. Clin. Investigation, Assn. Am. Physicians., Am. Soc. Contemporary Medicine and Surgery (adv. bd.), Am. Soc. Hypertension (founder, 1st pres. 1986-88), Internat. Soc. Hypertension (pres. 1986-88, v.p. sci. coun.), Harvey Soc., Kappa Sigma, Nu Sigma Nu, Alpha Omega Alpha. Clubs: Winged Foot (Mamaroneck, N.Y.); Shinnecock Hills Golf (Southampton). Research on hormones and electrolyte metabolism and renal physiology, on mechanisms of edema formation and on causes and treatment of high blood pressure. Home: 435 E 70th St New York NY 10021-5342 Office: NY Hosp-Cornell Med Ctr 525 E 68th St Starr 4 New York NY 10021

LAREW, HIRAM GORDON, III, research coordinator; b. Lafayette, Ind., Jan. 30, 1953; s. H. Gordon and Mary Jo (Thompson) L. BS summa cum laude, U. Ga., 1975; MS in Botany, Oreg. State U., 1977, PhD, 1982. Rsch. entomologist Agr. Rsch. Svc. USDA, Belthsville, Md., 1982-89; biotechnology specialist Rsch. Office USAID, Washington, 1989-92; rsch. coord. Office Strategic Planning U.S. Agy. for Internat. Devel., Washington, 1992—. Editor: Insect Caused Galls, 1986; contbr. articles to profl. jours. Pre-Doctoral fellow nat. Sci. Found., 1976; Travel grantee Nat. Geographic Soc., 1987, Rsch. grantee USDA, 1985. mem. Entomological Soc. Am. (internat. affairs com. 1990--), Am. Assn. for the Advancement Sci. (Diplomacy fellow 1987), Soc. for Internat. Devel., Entomological Soc. Washington. Achievements include discovery of sex reversal in Atlantic Sea Bass, staining reaction in cancer cells, cytochemical demonstration of viruses. Home: 3312 Gumwood Dr Hyattsville MD 20783 Office: USAID Rm 3889 NS Washington DC 20523

LARGMAN, KENNETH, strategic analyst, strategic defense analysis company executive; b. Phila., Apr. 7, 1949; s. Franklin Spencer and Roselynd Marjorie (Golden) L.; m. Suzanna Forest, Nov. 7, 1970 (div. Nov. 1978); 1 child, Jezra. Student, SUNY-Old Westbury, 1969-70. Ind. strategic analyst, 1970-80; chmn., chief exec. officer World Security Council, San Francisco, 1980—, dir. joint project with Apple Computer to develop improved techniques for decision analysis; dir. US/Soviet Nuclear Weapons and Strategic Def. Experiment: Discovery of Unanticipated Dangers and Possible Solutions. Author: (research documents) Space Peacekeeping, 1978, Preventing Nuclear Conflict: An International Beam Weaponry Agreement, 1979, Space Weaponry: Effects on the International Balance of Power and the Prevention of Nuclear War, 1981, Defense Against Nuclear Attack: U.S./Soviet Interactions, Moves, and Countermoves, 1985, 2 vols. on U.S./Soviet options in strategic def. race, 1986. Mem. World Affairs Council. Achievements include research in experimental decision analysis, in development of new methodologies of policy analysis and decision making for use in summit negotiations, in exposure of decision making to extremely rigorous and methodical scientific examination.

LA RIVIÈRE, JAN WILLEM MAURITS, environmental biology educator; b. Rotterdam, The Netherlands, Dec. 24, 1923; m. Louise A. Kleijn, 1958; 3 children. Student, Rotterdam and Delft U. Tech. Postdoctoral Rockefeller fellow Stanford U.; mem. scientific staff microbiology dept. Delft U. Tech., 1953-63; prof. environ. biology, dep. dir. Internat. Inst. Infrastructural, Hydraulic and Environ. Engring., Delft, 1963-88, now prof. emeritus; vis. prof. Harvard U., 1967-68; sec.-gen. ICSU, 1988—; mem. biol. coun. Royal Netherlands Acad. Arts and Scis., numerous internat. and nat. coms.; del. U.N. confs., adviser, lectr.; hon. mem. coun. Internat. Cell Rsch. Orgn. Author: Microbiology of Liquid Waste Treatment, 1977, Biotechnology in Development Cooperation, 1983, Water Quality: Present Status, Future Trends, 1987, Threats to the World's Water, 1989, Co-operation between Natural and Social Scientists in Global Change Research: Imperatives, Realities, Opportunities, 1991. Fellow World Acad. Art and Sci., Ky. Order Lion of Netherlands. •

LARK, RONALD EDWIN, logistics support engineer; b. Rochester, Minn., Oct. 14, 1934; s. Harley Edward and Greta Margret (Pearson) L.; m. Janean Ilene Musolf, May 26, 1978; children: Robert Allen, David Wayne, Dana Lynn. Diploma, Capitol Radio Engr. Inst., 1972. Sr. tech. writer McDonnell Aircraft Co., St. Louis, 1963-67, 70-77; work dir. tech. writing Honeywell, St. Louis Park, Minn., 1978-79; mgr. product support Control Data, Bloomington, Minn., 1979-91; program mgr., tech. writing Weiser Scott & Assocs., Shoreview, Minn., 1991-93; pres. Lark Data, Jordan, Minn., 1991—; mem. edit. rev. bd. Mag. of Svc. Mgmt., 1992—, Logistics Spectrum, 1991-92, reader opinion panel NASA Tech. Briefs, 1992-93. Mem. Jordan City Coun., Minn., 1992; coach Little League, Jordan, 1979-83. With USAF, 1952-62. Mem. Soc. Logistics Engrs. (v.p. adminstrn. 1992-93). Home and Office: Lark Data 266 Valley Green Park Jordan MN 55352

LARKIN, PETER ANTHONY, zoology educator, university dean and official; b. Auckland, New Zealand, Dec. 11, 1924; s. Frank Wilfrid and Caroline Jane (Knapp) L.; m. Lois Boughton Rayner, Aug. 21, 1948; children: Barbara, Kathleen, Patricia, Margaret, Gillian. BA, MA, U. Sask., 1946; DPhil (Rhodes scholar), Oxford U., 1948; LLD (hon.), U. Sask., 1989; DSc (hon.), U. B.C., 1992. Bubonic plague survey Govt. of Sask., 1942-43; fisheries investigator Fisheries Research Bd. of Can., 1944-46; chief fisheries biologist B.C. Game Commn., 1948-55; asst. prof. U. B.C., 1948-55, prof. dept. zoology, 1959-63, 66-69; prof. Inst. Animal Resource Ecology, 1969—, also dir. fisheries, 1955-63, 66-69, head dept. zoology, 1972-75, dean grad. studies, 1975-84, assoc. v.p. research, 1980-86, v.p. research, 1986-88, univ. prof., 1988—; hon. life gov. Vancouver Pub. Aquarium; mem. Canadian nat. com. Spl. Com. on Problems Environment, 1971-72; mem. Killam selection com. Can. Council, 1974-77; mem. Sci. Council Can., 1971-76; govtl. research coms.; mem. Nat. Research Council Can., 1981-84; mem. Can. Com. on Seals and Sealing, 1981-86, Can. Inst. Advanced Research, 1982-85; mem. Internat. Ctr. for Living Aquatic Resources Mgmt., 1977, chmn. bd., 1991; dir. B.C. Packers, 1980—; bd. govs. Internat. Devel. Research Ctr., 1986-92; mem. Nat. Sci. Eng. Research Council Can., 1987—; pres. Rawson Acad., 1988-91; mem. interim gov. coun. U. No. B.C., 1991—. Contbr. articles to profl. jours. Pres. B.C. Conservation Found., 1987-90. Recipient Centennial medal Govt. of Can., 1967, Master Tchr. award U. B.C., 1970, Silver Jubilee medal, 1977, award Can. Sport Fishing Inst., 1979; Nuffield Found. fellow, 1961-62. Fellow Royal Soc. Can.; mem. Internat. Limnological Assn., Am. Fisheries Soc. (award of excellence 1983), B.C. Natural Resources Conf. (pres. 1954), B.C. Wildlife Fedn., Canadian Soc. Zoologists (pres. 1972, Fry medal 1978), Canadian Assn. Univ. Research Adminstrn. (pres. 1979), Am. Inst. Fisheries Biologists (Outstanding Achievement award 1986). Home: 4166 Crown Crescent, Vancouver, BC Canada V6R 2A9 Office: U BC Fisheries Ctr, 2204 Main Mall, Vancouver, BC Canada V6T 1Z4

LARKINS, BRIAN ALLEN, botany educator; b. Bellville, Kans., Aug. 14, 1946. BSEd, U. Nebr., 1969, PhD in botany, 1974. Rsch. assoc. biochem. genetics U. Nebr., Lincoln, 1975-76, from asst. prof. to assoc. prof., 1976-83; Houde prof. genetics Purdue U., W. Lafayette, Ind., 1984—. Mem. AAAS, Am. Soc. Plant Physiologists (Charles Albert Schull award 1983), Sigma Xi. Research in protein and nucleic acid biosynthesis, seed storage protein, metabolism, regulation of gene Activity during seed formation. Office: Purdue U Dept of Botany West Lafayette IN 47907

LARKY, STEVEN PHILIP, electrical engineer; b. Allentown, Pa., June 18, 1962; s. Arthur Irving and E. Georgia (Stonehill) L.; m. Alexis Eve Solomon, Sept. 8, 1985. BSEE, MIT, 1984. Staff engr. rsch. div. IBM, Yorktown Heights, N.Y., 1984-88; adv. engr. work sta. div. IBM, Austin, Tex., 1988-90; OEM mgr. Infotronic Am., Inc., Austin, 1990—. Mem. Assn. for Computing Machinery, Sigma Xi (assoc.), Tau Beta Pi, Eta Kappa Nu. Achievements include patents for virtual display adapter, display using ordered dither, bit gating for efficient use of RAMS in variable plane displays. Home: 8911 Scotsman Dr Austin TX 78750 Office: Infotronic Am Inc 8834 N Capital of Texas Hwy Austin TX 78759

LA ROCCA, ALDO VITTORIO, mechanical engineer; b. Caserta, Italy, Apr. 22, 1926; came to U.S., 1951, naturalized, 1959; s. Vincenzo and Anna (Casagrande) La R.; m. Elizabeth Müller, Aug. 31, 1955; children: Renato, Dario, Marcello. Dottore Ingegneria Meccanica cum laude, Univ. and Poly. Naples, 1950; PhD in Applied Mechanics, Poly. Inst. Bklyn., 1955. Rsch. assoc. Poly. Inst. Bklyn., 1951-55; with Gen. Electric Co., 1955-72, advanced propulsion specialist Flight Propulsion Lab., Evendale, Ohio, 1955-60, mgr., cons. engring. advanced propulsion, missile and space div., Phila., also Valley Forge (Pa.) Space Tech. Ctr., 1960-72; advanced tech. mgr. Fiat S.p.A. Rsch.

Ctr., Turin, Italy, 1972-77, mem. central staff, advanced tech. planner, 1977-86, head auto high power laser program, 1979-86; founder, pres. Lara Cons. S.r.l., 1986—; co-founder, pres. ALTEC S.r.l., 1990-92; mem. space and energy com. Internat. Acad. Astronautics, 1978-86; mem. sci. com. Italian Nat. Research Council, 1976-84; sci. cons. high power laser program lectr. seminars Royal Swedish Acad. Scis., Soc. German Engrs., Philips Rsch. Ctr., Eindhoven, Netherlands; invited speaker numerous internat. sci. confs. Fulbright fellow, 1951-53; recipient Wallenberg Found. award Royal Swedish Acad. Scis., 1984. Assoc. fellow AIAA; mem. Italian Industry Mgrs. Assn., Sigma Xi. Unitarian. Author, patentee in field; contbr. articles to profl. jours. including V.D.I., Laser Focus, Sci. Am., Expansion. Home: Viale dei Castagni 4bis, 10020 Moncalieri Italy

LAROCCA, PATRICIA DARLENE MCALEER, mathematics educator; b. Aurora, Ill., July 12, 1951; d. Theodore Austin and Lorraine Mae (Robbins) McAleer; m. Edward Daniel LaRocca, June 28, 1975; children: Elizabeth S., Mark E. BS in Edn./Math., No. Ill. U., 1973, postgrad., 1975. Tchr. elem. sch. Roselle (Ill) Sch. Dist., 1973-80; instr. math. Coll. DuPage, Glen Ellyn, Ill., 1980—; pvt. cons., Downers Grove, Ill. Bd. dirs. PTA, Hillcrest Elem. Sch., Downers Grove; active Boy Scouts Am.; mem. 1st United Meth. Ch. Ill. teaching scholar, 1969. Methodist. Avocations: antiques, softball, organ, dance. Home and Office: 5648 Dunham Rd Downers Grove IL 60516-1246

LA ROCCA, RENATO V., oncologist, researcher; b. Cin., June 16, 1957; m. Margaret Carolyn Cauthron, Sept. 5, 1987; children: Alessandra, Marcello, Victoria, Chac. MS, Liceo Sci. Statale, Turin, Italy, 1976; postgrad., U. Padua, Italy, 1976-80; MD, Cornell U., 1982. Diplomate Nat. Bd. Med. Examiners, Am. Bd. Internal Medicine, Am. Bd. Oncology. Resident in internal medicine N.Y. Hosp.-Cornell Med. Ctr., N.Y.C., 1982-85; med. oncology fellow medicine br. Nat. Cancer Inst., Bethesda, Md., 1985-88, sr. investigator medicine br., 1988-90; pvt. practice Kentuckiana Med. Oncology Assocs., PSC, Louisville, 1990—; cons. UpJohn Pharms.; researcher med. br. Nat. Cancer Inst., NIH, Bethesda; mem. steering com. Ky. Cancer Pain Initiative; mem. cancer coms. Jewish Hosp., St. Anthony Med. Ctr., Louisville. Author: (chpts. in books) Molecular and Cellular Biology of Prostate Cancer, Molecular Foundations Oncology; contbr. articles to profl. jours.; patentee in field. Recipient USPHS Commendation medal, 1990. Mem. Am. Soc. Clin. Oncology, Am. Assn. Cancer Rsch., Am. Cancer Soc. (Ky. divsn.), Am. Coll. Physicians and Inventors, Am. Pain Soc., Jefferson County Med. Soc., Ky. Oncology Soc. (bd. dirs.), Ky. Med. Assn., Ind. Med. Assn., Alpha Omega Alpha. Avocations: sailing, computers, astronomy, skiing, political science. Office: Kentuckiana Med Oncology Assn 250 E Liberty St Ste 802 Louisville KY 40202-1507

LAROSA, JOHN CHARLES, internist, educator, researcher; b. Pitts., Feb. 17, 1941; s. Henry Gaetano and Sue Mary (Davis) LaR.; m. Judith Frances Hoag, Mar. 28, 1964; children: Christopher Henry, Jennifer Ann. BS, U. Pitts., 1961, MD, 1963. Resident internal medicine Peter Bent Brigham Hosp., Boston, 1963-67; clin. resident NIH, Bethesda, Md., 1967-70; mem. faculty George Washington Univ. Med. Coll., Washington, 1970-85, dean clin. affairs, 1985-90, dean rsch., 1990—; chmn. nutrition com. Am. Heart Assn., Dallas, 1986-89, task force on cholesterol issue Am. Heart Assn., Dallas, 1989—. Contbr. numerous sci. papers, some book chpts. to field. Sr. asst. surgeon USPHS, 1967-70. Mem. Am. Heart Assn., Am. Coll. Physicians. Democrat. Presbyterian. Home: 1958 Dundee Rd Rockville MD 20850 Office: George Washington U Med Coll 2300 Eye St NW Ste 713 Washington DC 20037

LAROUSSI, MOUNIR, electrical engineer; b. Sfax, Tunisia, Aug. 9, 1955; came to U.S., 1981; s. Habib and Manana (Jeloul) L.; m. Nicole Christine Mache, Dec. 30, 1966; children: Alexander Habib, Alyssa Jehan. BS in Elec. Engring., Tech. Faculty Sfax, 1979; MS in Elec. Engring., Nat. Sch. Radio and Elec., Bordeaux, France, 1981; PhD in Elec. and Computer Engring., U. Tenn., 1988. Grad. teaching asst. dept. elec. and computer engring. U. Tenn., Knoxville, 1983-85, rsch. asst. plasma sci. lab., 1984-88; asst. prof. Nat. Sch. Engring., Sfax, 1988-89; assoc. prof. Faculty Scis., Sfax, 1989-90; rsch. assoc. Plasma Sci. Lab. U. Tenn., Knoxville, 1990—. Contbr. articles to profl. jours. Recipient award Air Force Office Sci. Rsch., Washington, 1991. Mem. IEEE (Cir. and Systems Soc., Nuclear and Plasma Scis. Soc., Antennas and Propagation Soc.), Sigma Xi (rsch. award 1987). Achievements include new method to heat plasmas with magnetic pumping; invention of new type of microwave tunable filters; contbn. to new technique to hide aerospace objects from radar detection using plasma cloaking. Office: U Tenn Knoxville TN 37996

LARRAZÁBAL ANTEZANA, ERIK, economics educator, organization executive; b. Oruro, Bolivia, Dec. 23, 1958; s. Arturo and Ruth (Antezana) Larrazabal; m. Carolina Melgar, Apr. 26, 1985; children: Camila, Diego. BA, Bolivian Cath. U., La Paz, 1984; MA in Econs., U. Colo., 1989. Jr. economist Nat. Coun. of Salary, La Paz, 1983, Cen. Bank Bolivia, La Paz, 1983-85; jr. economist Unity of Analysis of Econ. Policy, La Paz, 1985-89, sr. economist, 1989-90; prof. econs. U. La Paz, 1984—; treas. Fin. Fund for Plata Basin Devel. (FONPLATA), Sucre, Bolivia, 1990—. Contbr. articles to profl. jours. Scholar AID, 1987. Office: FONPLATA, España # 74, PO Box 47, Sucre Bolivia

LARSEN, ALVIN HENRY, chemical engineer; b. Salt Lake City, June 22, 1939; s. Henry Victor and Marie (Bartholomew) L.; m. Sharon Leila Branch, Aug. 20, 1966; children: Reed Henry, Adrian Platt, Cherry Ann, Elliott Kimball, Cameron McKay, Katrina Joy. BSChE with honors, BA, U. Utah, 1965; PhD, Calif. Inst. Tech., 1969. Registered profl. engr., Mo. Sr. mathematician Monsanto Co., St. Louis, 1968-70, sr. engr., 1970-72, engring. specialist, 1972-77, engring. supt., 1979-85, prin. engring. specialist, 1977-79, 86—; chmn. Tech. Community of Monsanto, St. Louis, 1982; v.p. Fluid Properties Rsch., Inc., Stillwater, Okla., 1977-78. Author: Physical Property Data Book, 1975, 77; contbr. articles to profl. jours. Fellow AIChE (chmn. data compilation 1979-85, Design Inst. Phys. Property Data Mgmt. award 1986); mem. Am. Chem. Soc., Sigma Xi. Mem. LDS Ch. Achievements include development of correlation library of physical property programs in Monsanto; of new liquid enthalpy model for process simulation. Office: Monsanto Co 800 N Lindbergh Blvd Saint Louis MO 63167

LARSEN, IB HYLDSTRUP, engineer; b. Horsholm, Denmark, Mar. 25, 1919. BS, Tech. U. Denmark, 1935, MSc in Elec. Engring., 1942. Engr. Danish Posts & Telegraphs, Denmark, 1943, staff engr., 1949; asst. chief engr. Telegraphs & Telephones, Denmark, 1960; chief engr. and dep. dir. telecom. Danish Posts & Telegraphs, Denmark, 1977—; dir. The Danish Energy Agency, Denmark. Decorated knight Order of Dannebrog, 1964, knight 1st class Order of Dannebrog. Mem. Danish Soc. Chem., Elec. and Mech. Engrs. Office: Danish Energy Agy, Landemaerket 11, 1119 Copenhagen Denmark*

LARSEN, JANINE LOUISE, electrical engineering educator; b. Long Beach, Calif., Apr. 23, 1959; d. Neil and Jannet May (Rawlings) L. BSEE, Calif. State U., Long Beach, 1981; MS in Bioengring., U. Mich., 1985, MSEE, 1987, PhD in Bioengring., 1988. Rsch. engr. Nimbus Inc., Rancho Cordova, Calif., 1981-83; rsch. asst. U. Mich., Ann Arbor, 1983-89; rsch. assoc. Pritzker Inst. Med. Engring., Chgo., 1987-88, asst. prof., 1988—; asst. prof. elec. engring. Ill. Inst. Tech., Chgo., 1988—. Active ACLU, Amnesty Internat., Art Inst. Chgo. Grantee Whitaker Found., 1989-92. Mem. AAUP, IEEE (sr. mem.), IEEE Engring. in Medicine and Biology Soc., IEEE Controls Soc., Internat. Fedn. Automatic Control (affiliate). Avocations: sailing, hiking, art collecting, antiques, sky diving. Office: Ill Inst Tech Dept Elec Engring Chicago IL 60616

LARSEN, JESPER KAMPMANN, applied mathematician, educator; b. Bogense, Denmark, Apr. 7, 1950; s. Knud and Hanne (Kampmann) L.; m. Julie Anna Maria H. Szabad, Oct. 21, 1971 (dec. July 1990); 1 child, Cecile Julie Kampmann Szabad. MSc in Civil Engring., Tech. U. Denmark, 1974, PhD in Applied Math., 1977. Founding owner Math-Tech, Denmark, 1977—; civil engr. DSB, Denmark, 1978-80; ext. instr. Roskilde U., Denmark, 1980-88; sr. hydraulic engr. Danish Hydraulic Inst., Denmark, 1982-86; dir. Math-Tech, 1986—; ext. prof. applied math. Rosklide U., 1989—; bd. dirs. Danish PARSIM Consortium, Denmark, 1988-93. Contbr. articles to profl. jours. Mem. Danish Soc. Applied Math. (founding mem., chmn. 1988—). Office: Math Tech, Kildeskovsvej 67, DK 2820 Gentofte Denmark

LARSEN, RALPH IRVING, environmental engineer; b. Corvallis, Oreg., Nov. 26, 1928; s. Walter Winfred and Nellie Lyle (Gellatly) L.; B.S. in Civil Engring., Oreg. State U., 1950; M.S., Harvard U., 1955, Ph.D. in Air Pollution and Indsl. Hygiene, 1957; m. Betty Lois Garner, Oct. 14, 1950 (dec. Feb. 1989); children: Karen Larsen Cleeton, Eric, Kristine Larsen Burns, Jan Alan; m. Anne Harmon King, Aug. 3, 1991; children: Vikki King Ball, Terri King Redding, Cindi King King. San. engr. div. water pollution control USPHS, Washington, 1950-54; chief tech. service state and community service sect. Nat. Air Pollution Control Adminstrn., Cin., 1957-61; with EPA and Nat. Air Pollution Control Adminstrn., 1961—, environ. engr. Atmospheric Rsch. and Exposure Assessment Lab., Rsch. Triangle Park, N.C., 1971—; air pollution cons. to Poland, 1973, 75, Brazil, 1978; condr. seminars for air pollution researchers, Paris, Vienna and Milan, 1975; adj. lectr. Inst. Air Pollution Tng., 1969—; Falls of Neuse community rep. City of Raleigh (N.C.), 1974—. Recipient Commendation medal USPHS, 1979. Mem. Air and Waste Mgmt. Assn. (mem. editorial bd. jour. 1971-88), Rsch. Soc. Am., Conf. Fed. Environ. Engrs., USPHS Commd. Officers Assn. (past br. pres.). Republican. Mem. Christian and Missionary Alliance Ch. (elder). Contbr. numerous articles to profl. jours. Home: 4012 Colby Dr Raleigh NC 27609-6045 Office: MD-76 EPA Research Triangle Park NC 27711

LARSON, BENNETT CHARLES, solid state physicist, researcher; b. Buffalo, N.D., Oct. 9, 1941; s. Floyd Everet and Gladys May (Hogen) L.; m. Piola Anne Taliaferro, June 6, 1969; children—Christopher Charles, Andrea Kay. B.A. in Physics, Concordia Coll., Moorhead, Minn., 1963; M.S. in Physics, U. N.D., 1965; Ph.D. in Physics, U. Mo., 1970. Rsch. physicist, group leader x-ray diffraction, sect. head thin films and microstructures solid state div. Oak Ridge Nat. Lab., Tenn., 1969—. Contbr. numerous articles to profl. jours. Recipient Sidhu award Pitts. Diffraction Soc., 1974. Fellow Am. Phys. Soc.; mem. Am. Crystallographic Assn. (Bertram E. Warren Diffraction Physics award 1985), Materials Research Soc. Office: Oak Ridge Nat Lab Solid State Div Oak Ridge TN 37831

LARSON, BETTY JEAN, dietitian, educator; b. Grand Forks, N.D., May 4, 1949; d. Harold and Eileen W. (Kelly) Hanson; m. Wayne E. Larson, Dec. 13, 1975; children: Brian, Robin. BS in Nutrition, U. N.D., 1971; MS in Nutrition, N.D. State U., 1973; EdD in Edn., U. N.D., 1989. Registered dietitian, Minn. Intern in dietitics St. Mary's Hosp., Rochester, Minn., 1971-72; instr. Concordia Coll., Moorhead, Minn., 1974-78; nutrition clinic dietitian VA Hosp., Fargo, N.D., 1978-80; assoc. prof. Concordia Coll., Moorhead, 1983—. Contbr. articles to profl. jours. Pres. Hope Luth. Ch. Woman, Fargo, 1989, Fargo/Moorhead Dietitic Assn., 1979. Recipient Collaborative Rsch. award Ohio State U., 1989; named Young Dietitian of Yr., N.D., 1979. Mem. Am. Dietetic Assn. (review panel Coun. of Edn. 1990-93), Soc. Nutrition Edn., Inst. Food Technologists, Sigma Xi. Home: 3017 Edgewood Dr Fargo ND 58102 Office: Concordia Coll 901 S 8th St Moorhead MN 56562

LARSON, CARL SHIPLEY, engineering educator, consultant; b. Chgo., Sept. 23, 1934; s. Carl Uno and Marion Jean (Woufel) L.; m. Vivian Phylis Peuckert, Dec. 28, 1957; children: Carl, Michael, Daniel. BSME, U. Ill., 1956, MSME, 1958, PhD, 1965. Registered profl. engr., Ill. Engr. Western Electric, Chgo., 1955-56; from instr. to asst. prof. U. Ill., Urbana, 1965-72, assoc. prof., 1972-91, asst. dean, 1974—, prof., 1991—; cons. Larson & Assocs., Urbana, 1975—; bd. dirs. Capsonic Corp., Elgin, Ill., 1989—. Contbr. articles to profl. jours. Bd. dirs. United Way, Urbana, 1987. Teaching Fellowship Nat. Sci. Found, Urbana, 1960-64. Mem. Am. Soc. Engring. Edn. (sect. chmn. 1990-91), Nat. Coun. Examiners Engring. (cons., vice-chmn. 1989-92). Office: 411 E Mumford Dr Urbana IL 61801-6230 also: U of Ill 207 Engring Hall 1308 W Green St Urbana IL 61801-2936

LARSON, ERIC HEATH, chemical engineer; b. Medford, Mass., Apr. 8, 1950; s. Frank Robert and Ann Louise (Huntington) L.; m. Maria Veronica Contreras, June 1, 1973; children: Richard, Peter. BS magna cum laude, Tufts U., 1973; MS, Yale U., 1974; PhD, Syracuse U., 1978. Rsch. engr. Allied Signal, Syracuse, 1978-81, sr. rsch. engr., 1981-84, rsch. assoc., 1984-91; group leader Rhone-Poulenc, Cranbury, N.J., 1991—. Contbr. articles to profl. jours. Mem. Tau Beta Pi. Achievements include 7 patents in electrochemistry and water treatment chemicals. Office: Rhone Poulenc Prospect Plains Rd Cranbury NJ 08512

LARSON, NANCY CELESTE, computer systems analyst, music educator; b. Chgo., July 17, 1951; d. Melvin Ellsworth and Ruth Margaret (Carlson) L. BS in Music Ed., U. Ill., 1973, MS in Music Edn., 1976; postgrad., Purdue Univ., 1982-86. Vocal music educator Consol. Sch. Dist., Gilman, Ill., 1975-77; elem. vocal music tchr. Sch. Dist. 161, Flossmoor, Ill., 1977-87; instr. Vander Cook Coll., Chgo., 1980-88; systems programmer analyst Sears, Roebuck & Co., Chgo., 1987-92, tech. instr., 1989-90, project leader, 1990-91, sr. systems analyst, 1991-92; sr. systems analyst Trans Union Corp., Chgo., 1992—; tchr. adult computer edn. Homewood-Flossmoor High Sch., 1986-90. Chmn. Faith Luth. Ch., 1982-87, pres. bd., 1988-91, vocal soloist and voice-over performer. Mem. Ill. Music Educators Assn., Music Educators Nat. Conf., Ill. Educators Assn., Nat. Educators Assn., Am. ORFF Schulwerk Assn., Flossmoor Edn. Assn. (negotiator 1983-86). Republican. Lutheran. Avocations: swimming, skiing, reading, antique hunting. Office: Trans Union Corp 555 W Adams Chicago IL 60606

LARSON, REED WILLIAM, psychologist, educator; b. Mpls., Nov. 17, 1950; s. Curtis Luverne and Miriam Rue (Johnson) L.; m. Sharon Lee Irish, Dec. 29, 1982; children: Miriam, Renner. BA, U. Minn., 1973; PhD, U. Chgo., 1979. Postdoctoral fellow Michael Reese Hosp., Chgo., 1979-80, coord. clin. rsch. program in adolescence, 1980-85, dir. lab. for study of adolescence, 1982-87; rsch. assoc. U. Chgo., 1980-85; mem. faculty human ecol. U. Ill., Urbana, 1985—. Co-author: Being Adolescent, 1984. NIMH grantee, 1985—; Nat. Inst. on Child Health and Devel. fellow, 1991-92. Home: 608 W Iowa Urbana IL 61801 Office: U Ill 1105 W Nevada Urbana IL 61801

LARSON, RONALD GARY, chemical engineer; b. Litchfield, Minn., Mar. 30, 1953; s. Roy Joseph and Louise Rose (Nistler) L.; m. Beatrice Lathrop, Apr. 11, 1987; children: Rachel, Emily, Andrew. BSChemE, U. Minn., 1975, MSChemE, 1977, PhDChemE, 1980. Cons. Gary Operating Co., Denver, 1976-77; mem. tech. staff AT&T Bell Labs., Murray Hill, N.J., 1980-87, disting. mem. tech. staff, 1987—; editor-in-chief book series on polymers and complex materials Am. Inst. Physics, 1992—. Author: Constitutive Equation for Polymer Melts and Solutions, 1988; contbr. articles to Jour. Fluid Mechanics, Nature, Macromolecules. Fellow NSF, 1975, German Acad. Exch. Svc., 1980. Mem. AIChE, AAAS, Am. Phys. Soc., Soc. Rheology (steering com. 1991—). Republican. Mem. Christian and Missionary Alliance. Achievements include co-discovery of a class of polymer elastic-flow instabilities, of universal viscoelastic behavior of layered liquids, discovery of instabilities in flows of liquid crystalline polymers; computer simulation of self-assembly of ordered surfactant phases. Office: AT&T Bell Labs 600 Mountain Ave Rm 6E-320 Murray Hill NJ 07974

LARSSON, ANDERS LARS, anesthesiologist; b. Motala, Sweden, Dec. 3, 1952; s. Arne Teofil and Ewa Sara (Edsner) L.; m. Elna-Marie Barup, Aug. 16, 1975; children: Mans, Elisabet, Hanna. MD, U. Lund, Sweden, 1977; PhD, U. Lund, 1988; DEAA, European Acad. Anaesthesiology, Stockholm, 1990. Intern Kalmar (Sweden) Hosp., 1976-78, resident in anesthesia, 1978-81; resident Univ. Hosp. U. Lund, 1981-83, asst. prof., 1983-89, assoc. prof., 1989—, co-dir. Intensive Care Unit, 1990-92, dir. Intensive Care Unit, 1992—; vis. assoc. prof. Dept. Anesthesiology Health Sci. Ctr. U. Tex., San Antonio, 1989-90. Grantee Swedish Nat. Bd. Tech. Devel., AGA Co. Mem. Swedish Med. Assn., Swedish Soc. Med. Sci. (grantee), Swedish Soc. Anesthesiology and Intensive Care, Scandinavian Soc. Anesthesiologists, Soc. Critical Care Medicine, European Soc. Intensive Care Medicine. Avocations: sailing, jogging. Home: Gilleskroken 3, Lund S-22647, Sweden Office: Univ Hosp Univ Lund, Dept Anesthesia & ICU, Lund S-22185, Sweden

LARUBIO, DANIEL PAUL, JR., civil engineer; b. Bklyn., Nov. 9, 1959; s. Daniel and Connie (Gentile) LaR. BSCE, U. Nev. Las Vegas, 1983. Registered profl. engr., Nev. Staff engr. M.E.A. Engrs., Las Vegas, 1984-86, V.T.N. Nev. Engrs., Las Vegas, 1986-87; project engr. Wojcik Engring., Las Vegas, 1988—. Mem. NSPE, ASCE. Democrat. Roman Catholic. Home: 4801 Spencer St #163 Las Vegas NV 89119

LARUSSO, NICHOLAS F., gastroenterologist, educator. Prof., chmn. divsn. gastroenterology Mayo Med. Sch. Clin. & Found., Rochester, Minn., 1977—, dir. Ctr. Basic Rsch. Digestive Disorders, 1977—. Office: Ctr for Basic Rsch in Diges Dis Guggenheim 17 Mayo Clinic Rochester MN 55905*

LASHLEE, JOLYNNE VAN MARSDON, army officer, nursing administrator; b. Asheville, N.C., May 22, 1948; d. William Reid and Frances (Furey) Van Marsdon. BS in Nursing, U. Fla., 1971; M Health Care Adminstrn., Baylor U., 1982. Team leader surg. specialties Shand Teaching Hosp., Gainesville, Fla., 1971; commd. lt. U.S. Army Nurses Corps, 1971, advanced through grades to lt. col., 1986; asst. head nurse organ transplant service unit Walter Reed Hosp., Washington, 1972; staff nurse surg. ICU, head nurse multi-service nursing unit Nurnberg, W. Ger., 1972-75; head nurse recovery room William Beaumont Army Med. Center, Ft. Bliss, El Paso, Tex., 1975-76, dep. dir. patient care specialist course, 1976-78; ednl. coordinator, project officer U.S. Lyster Hosp., Ft. Rucker, Ala., 1978; adminstrv. resident Madigan Army Med. Center, Tacoma, 1981-82; chief nurse methods div. Walter Reed Hosp., 1982-85; mem. Army Surgeon Gen.'s Task Force on Health Care, 1985-86; insp. gen. U.S. Army, 1986-89; asst. chief nurse Ireland Army Hosp., Ft. Knox, Ky., 1989-92; coord. composite health care system Tripler Army Med. Ctr., Hawaii, 1993—; regional coord. AMEDD, 1993—; pvt. image marketer Color 1 Assocs. Mem. Am. Hosp. Assn., Am. Coll. Healthcare Execs., Assn. Health Care Adminstrs. Nat. Capital Area, Am. Assn. Critical Care Nurses, Baylor U. Healthcare Adminstrs. Alumni, Honolulu Hawaii Healthcare Found. (elua ali'i). Author: Ireland Army Community Hospital Access Study. Home: 1221 Victoria # 2401 Honolulu HI 96814

LASHLEE, JON DAVID, physical scientist; b. Beacon, N.Y., Jan. 4, 1962; s. John Stephen Lashlee and Juanita Shirley (Fanning) Raines; m. Amy Elizabeth Bradley, Mar. 27, 1983; children: Megan Elizabeth, Emily Katherine. BA in Geology, Berea (Ky.) Coll., 1984; MS in Geography, Murray (Ky.) State U., 1987. Registered profl. geologist, Tenn. Program mgr. Hilton Systems, Inc., Vicksburg, Miss., 1987-88; phys. scientist U.S. Army Corps of Engrs./Waterways Experiment Sta., Vicksburg, 1988—; user's group mem. U.S. Army Engr. Mission to Planet Earth, 1993—; mem. Waterways Experiment Sta. GIS Users Group, 1992-93. Contbr. articles to profl. jours. Recipient 1993-94 U.S. Army Engr. Advance Studies Program. Mem. Am. Soc. for Photogrammetry and Remote Sensing (2d v.p. mid-south region 1993, cert. mapping scientist 1993, Ford Bartlett award 1992), Assn. of Engring. Geologists (vice-chmn. remote sensing com. 1991-92), Nat. Geographic Soc., Sigma Xi. Achievements include research on digital image processing, unsupervised cluster analysis of multivariate data, terrain analysis, land cover classification accuracy, and thematic mapping, the integration of remotely sensed data and geographic information systems. Office: US Army Engr Waterways Experiment Sta 3909 Halls Ferry Rd Vicksburg MS 39180-6199

LASHLEY, VIRGINIA STEPHENSON HUGHES, retired computer science educator; b. Wichita, Kans., Nov. 12, 1924; d. Herman H. and Edith M. (Wayland) Stephenson; m. Kenneth W. Hughes, June 4, 1946 (dec.); children: Kenneth W. Jr., Linda Hughes Tindall; m. Richard H. Lashley, Aug. 19, 1954; children: Robert H., Lisa Lashley Van Amberg, Diane Lashley Tan. BA, U. Kans., 1945; MA, Occidental Coll., 1966; PhD, U. So. Calif., 1983. Cert. info. processor, tchr. secondary and community coll., Calif. Tchr. math. La Canada (Calif.) High Sch., 1966-69; from instr. to prof. Glendale (Calif.) Coll., 1970—, chmn. bus. div., 1977-81, coord. instructional computing, 1974-84; 88—; sec., treas., dir. Victory Montessori Schs., Inc., Pasadena, Calif., 1980—; pres. The Computer Sch., Pasadena, 1983-92; ret., 1992; pres. San Gabriel Valley Data Processing Mgmt. Assn., 1977-79, San Gabriel Valley Assn. for Systems Mgmt., 1979-80; chmn. Western Ednl. Computing Conf., 1980, 84. Editor Jour. Calif. Ednl. Computing, 1980. Mem. DAR. NSF grantee, 1967-69, EDUCARE scholar U. So. Calif., 1980-82; John Randolph and Dora Haynes fellow, Occidental Coll., 1964-66; student computer ctr. renamed Dr. Virginia S. Lashley computer ctr., 1992. Mem. AAUP, AAUW, Data Processing Mgmt. Assn., Calif. Ednl. Computing Consortium (bd. dir. 1979—, v.p. 1983—, pres. 1985-87, ret. 1992), Orgn. Am. Historians, San Marino Women's Club, Phi Beta Kappa, Pi Mu Epsilon, Phi Alpha Theta, Phi Delta Kappa, Delta Phi Upsilon, Gamma Phi Beta. Republican. Congregationalist. Home: 1240 S San Marino Ave San Marino CA 91108-1227

LASRY, JEAN-MICHEL, mathematics educator; b. Paris, Oct. 29, 1947; m. Elisabeth du Boucher; children: Laura, Romain, Julien. M. in Econs., U. Paris-Assas, 1970; these d'etat in math., U. Paris IX, 1975. Rsch. fellow Ctr. Nat. Recherche Scientifique, Paris, 1971-78; prof. Paris-Dauphine U., 1978—, chmn. math. dept., 1980-83; cons. Compagnie Bancaire, Paris, 1988-91; mem. bd. Caisse des Depots DABF, 1991—. Contbr. articles in math., econ. and fin. to profl. jours. Mem. Soc. Mathematiques apliquies et industrielles (bd. dirs.), Soc. Mathematique de France, Assn. Francaise de Finance, Ecole de la Cause Freudienne.

LASSER, HOWARD GILBERT, chemical engineer, consultant; b. N.Y.C., Nov. 24, 1926; s. Milton and Tessie (Rosenthal) L.; m. Barbara Ann Katz, Aug. 24, 1950; children: Cathy, Ellen, Alan. BSChemE, Lehigh U., 1950; postgrad., Columbia U., 1951; Dr.Ing., Darmstadt Tech. Inst., Germany, 1956. Registered profl. engr., D.C., Va., Calif. Chem. engr. Belvoir Rsch. Engring. & Devel. Ctr., Ft. Belvoir, Va., 1951-55, 58-72; materials engr. Naval Sea Systems Command, Washington, 1956-57, chem. engr., 1955-56; materials engr. Naval Facilities Engring. Command, Alexandria, Va., 1972-82; chem. engr. Materials Rsch. Cons., Alexandria & Springfield, Va., 1982—. Author: Design of Electroplating Facilities, 1990; contbr. articles to profl. jours. Fellow AAAS, Oil and Colour Chemists Assn.; mem. Am. Electroplaters Soc., Am. Inst. Chemists, Am. Inst. Chem. Engrs., Nat. Assn. Corrosion Engrs. (cert.), Tau Beta Pi, Sigma Xi, Alpha Chi Sigma, Pi Delta Epsilon. Achievements include 6 patents in electroplating and metal finishing; description of thermodynamic properties of carbon dioxide; development of thermotropic dyes for aluminum oxides; over 500 publs. in materials and chemical engineering. Home: 5912 Camberly Ave Springfield VA 22150-2438 Office: Materials Rsch Cons 1121 King St Alexandria VA 22314-2924

LASSLO, ANDREW, medicinal chemist, educator; b. Mukacevo, Czechoslovakia, Aug. 24, 1922; came to U.S., 1946, naturalized, 1951; s. Vojtech Laszlo and Terezie (Herskovicova) L.; m. Wilma Ellen Reynolds, July 9, 1955; 1 child, Millicent Andrea. MS, U. Ill., 1948, PhD, 1952, MLS, 1961. Rsch. chemist organic chems. div. Monsanto Chem. Co., St. Louis, 1952-54; asst. prof. pharmacology, div. basic health scis. Emory U., 1954-60; prof. and chmn. dept. med. chemistry Coll. Pharmacy, U. Tenn. Health Sci. Ctr., 1960-90, Alumni Disting. Svc. prof. and chmn., dept. medicinal chemistry, 1989-90, professor emeritus, 1990—; cons. Geschickter Fund for Med. Research Inc., 1961-62; rsch. contractor U.S. Army Med. R & D Command, 1964-67; dir. postgrad. tng. program sci. librarians USPHS, 1966-72; chmn. edn. com. Drug Info. Assn., 1966-68, bd. dirs., 1968-69; dir. postgrad. tng. program organic medicinal chemistry for chemists FDA, 1971; exec. com. adv. council S.E. Regional Med. Library Program, Nat. Library of Medicine, 1969-71; chmn. regional med. library programs com. Med. Library Assn., 1971-72; mem. pres.'s faculty adv. council U. Tenn. System, 1970-72; chmn. energy authority U. Tenn. Center for Health Scis., 1975-77, chmn. council departmental chmn., 1977, 81; chmn. Internat. Symposium on Contemporary Trends in Tng. Pharmacologists, Helsinki, 1975. Producer, moderator (TV and radio series) Health Care Perspective, 1976-78; editor: Surface Chemistry and Dental Integuments, 1973, Blood Platelet Function and Medicinal Chemistry, 1984; contbr. numerous articles in sci. and profl. jours.; mem. editorial bd. Jour. Medicinal and Pharm. Chemistry, 1961, U. Tenn. Press, 1974-77; composer (work for piano) Synthesis in C Minor, 1968; patentee in field. Trustee First Bohemian Meth. Ch., Chgo., 1951-52, mem. bd. stewards, 1950-52; mem. ofcl. bd. Grace Meth. Ch., Atlanta, 1955-60; mem. adminstrv. bd. Christ United Meth. Ch., Memphis, 1973-75, 77-79, 81-83, 88-90, chmn. commn. on edn., 1965-67, chmn. bd. Day Sch., 1967-68. Served to capt. M.S.C., USAR, 1953-62. Recipient Research prize U. Ill. Med. Ctr. chpt. Sigma Xi, 1949, Honor Scroll Tenn. Inst. Chemists, 1976,

Americanism medal DAR, 1976; U. Ill. fellow, 1950-51; Geschickter Fund Med. Research grantee, 1959-65, USPHS Research and Tng. grantee, 1958-64, 66-72, 82-89, NSF research grantee, 1964-66, Pfeiffer Research Found. grantee, 1981-87. Fellow AAAS, Am. Assn. Pharm. Scientists, Am. Inst. Chemists (nat. councilor for Tenn. 1969-70), Acad. Pharm. Rsch. and Sci.; mem. ALA (life), Am. Chem. Soc. (sr.), Am. Pharm. Assn., Am. Soc. Pharmacology and Exptl. Therapeutics (chmn. subcom. pre and postdoctoral tng. 1974-78, exec. com. ednl. and profl. affairs 1974-78), Sigma Xi (pres. elect U. Tenn. Ctr. for Health Sci. chpt. 1975-76, pres. 1976-77, Excellence in Rsch. award 1989), Beta Phi Mu, Phi Lambda Sigma, Rho Chi. Methodist. Achievements include 7 U.S. and 10 fgn. patents in field; identification of platelet aggregation-inhibitory specific functions in synthetic organic molecules; design and synthesis of novel human blood platelet aggregation inhibitors, novel compound for mild stimulation of central nervous system activity; research on relationships between structural features of synthetic organic entities, their physicochemical properties and their effects on biologic activity. Home and Office: 5479 Timmons Ave Memphis TN 38119-6932

LAST, ROBERT LOUIS, plant geneticist; b. N.Y.C., Nov. 10, 1958; s. Jerry G. and Terry (Axelrod) L.; m. Jill M. Canny. BA, Ohio Wesleyan U., 1980; PhD, Carnegie Mellon U., 1986. Predoctoral fellow Carnegie Mellon U., Pitts., 1980-86; post doctoral Whitehead Inst. for Biomed. Rsch., Cambridge, Mass., 1986-89; assoc. plant molecular biologist Boyce Thompson Inst., Ithaca, N.Y., 1989—; adj. assoc. prof. Cornell U., Ithaca, 1990—. Co-author: Internat. Reviews of Cytology, 1993; contbr. articles to profl. jours. Mem. Am. Assn. for the Advancement Sci., Internat. Soc. Plant Molecular Biology, Am. Soc. Plant Physiologists. Achievements include the isolation of amino acid requiring mutants of plants. Office: Boyce Thompson Inst Tower Rd Ithaca NY 14853-1801

LASTER, BRENDA HOPE, radiobiologist; b. Pittston, Pa., June 24, 1941; d. Harold Russell and Sylvia Rita (Lustig) Wruble; m. Marvin Joseph Laster; children: Michelle, Jan, Karen, Jonathan. BSc, Stern Coll. for Women, N.Y.C., 1962; MS, Meml. Sloan-Kettering Inst., 1963; PhD, Union Inst., Cin., 1991. Diagnostic cytologist Meml. Sloan Kettering, N.Y.C., 1963-64, Plainview, N.Y., 1964-82; instr. SUNY Sch. Medicine, Stony Brook, 1984-91, asst. prof., investigator, 1991—; researcher Brookhaven Nat. Lab., Upton, N.Y., 1983-91, sci. mentor, 1989—; cons. Ben Gurion U., Beer Sheva, Israel, 1992-93. Contbr. 40 articles to profl. jours., chpt. to ency. Bd. dirs. Mid-Island Y, Plainview, N.Y., 1981-92; pres. Washington Ave. Civic Assn., Plainview, 1983—. Recipient Svc. awards Mid-Island Y., Plainview, 1985, Inst. for Gifted and Talented Youth, Suffolk County, N.Y., 1988; named to Hall of Fame Gallaudet U., Washington, 1991. Mem. AAAS, Internat. Soc. for Neutron Capture Therapy, Radiation Rsch. Soc., Brookhaven Women in Sci., Scientist's Ctr. for Animal Welfare. Achievements include research in SIRH669-Samarium-145 and its use as a radiation source. Home: 662 Washington Ave Plainview NY 11803 Office: Brookhaven Nat Lab Med Dept 30 Bell Ave Upton NY 11973

LASTER, RICHARD, biotechnology executive; b. Vienna, Austria, Nov. 10, 1923; came to U.S., 1940; naturalized, 1944; s. Alan and Caroline (Harband) L.; student U. Wash., 1941-42; BChE cum laude, Poly. Inst. Bklyn., 1943; postgrad. Stevens Inst. Tech., 1945-47; m. Liselotte Schneider, Oct. 17, 1948; children: Susan Laster Rubenstein, Thomas. With Gen. Foods Corp., 1944-82, corp. rsch. and devel., Hoboken, N.J., 1944-58, ops. mgr. Franklin Baker div., Hoboken, N.J., Atlantic Gelatin div., Woburn, Mass., 1958-64, mgr. rsch. devel. Jell-O div., White Plains, N.Y., 1958-64, corp. mgr. quality assurance, White Plains, 1964-67, ops. mgr. Maxwell House div., White Plains, 1967-68, exec. v.p. Maxwell House div., 1968-69, pres. Maxwell House div., 1969-71, corp. group v.p., White Plains, 1971-73, exec. v.p. Gen. Foods Corp., 1974-82, also dir., rsch., devel. and food-away-from-home, 1975-82; bd. dirs. DNA Plant Tech. Corp., 1982—, chmn., 1988—, CEO, 1982-92, pres. 1982-91; bd. dirs. RiceTec, Bowater Inc., Indsl. Biotechnology Assn., mem. sch. bd., Chappaqua, N.Y., 1971-74, pres., 1973-74; chmn., mem. bd., 1st v.p. United Way of Westchester, 1978; chmn. adv. com. Poly. Inst. Westchester, 1977; trustee Poly. Inst. N.Y.,1978—; mem. coll. coun. SUNY, Purchase; chmn. Purchase Coll. Found, 1986—; mem. corp. N.Y. Botanical Garden; mem. adv. panel to chancellor SUNY, Albany; v. chmn. Westchester Edn. Coalition, 1992—. Recipient Disting. Alumnus award. Fellow Poly. Inst. N.Y. Mem. AAAS, N.Y. Acad. Scis., Am. Inst. Chem. Engrs. (Food and Bioengring. award 1972), Am. Chem. Soc., Am. Inst. Chemists, Tau Beta Pi, Phi Lambda Upsilon. Contbr. articles on food sci. to profl. publs. Patentee in field. Home: 23 Round Hill Rd Chappaqua NY 10514-1622 Office: DNA Plant Tech Corp 103 S Bedford Rd Mount Kisco NY 10549-3440

LASTRA, JOSE RAMON, plant virologist; b. Orense, Spain, May 22, 1939; arrived in Venezuela, 1954, naturalized; s. Ramon and Maria (Rodriguez) L.; m. Ana Maria Mumm, Mar. 3, 1953; children, Ricardo, Daniel, Eduardo, Andres. Degree in Biology, U. Cent. Venezuela, 1966; MSc in Plant Pathology, U. Calif., Berkeley, 1970, PhD in Plant Pathology, 1974. Head plant virus lab IVIC, Caracas, 1974-85; prof. U. Cen. Venezuela, Caracas, 1976-85; head molecular biology lab. CATIE, 1980-87; coord. IPM project CATIE, Turrialba, Costa Rica, 1985-86; dir. ednl. programs CATIE, 1986—; cons. in field. Mem. Am. Phytopathological Soc. (pres. Caribbean div. 1978), Soc. Latinoamericana de Fitopatologos, Soc. Venezolana de Fitopatologia, Phi Beta Kappa. Roman Catholic. Avocations: tennis, philately. Office: CATIE, Turrialba 7170, Costa Rica

LATANISION, RONALD MICHAEL, materials science and engineering educator, consultant; b. Richmondale, Pa., July 2, 1942; s. Stephen and Mary (Kopach) L.; m. Carolyn Marie Domenig, June 27, 1964; children—Ivan, Sara. BS, Pa. State U., 1964; PhD in Metall. Engring., Ohio State U., 1968. Postdoctoral fellow Nat. Bur. Standards, Washington, 1968-69; research scientist Martin Marietta, Balt., 1969-73, acting head materials sci., 1973-74; dir. H.H. Uhlig Corrosion Lab. MIT, Cambridge, 1975—, Shell Disting. prof. materials sci. and engring., 1983-88, dir. Materials Processing Ctr., 1984-91; co-founder Altman Materials Engring. Corp., 1992—; mem. tech. adv. bd. Modell Devel. Corp., Framingham, Mass., 1987—; sci. advisor com. on sci. and tech. U.S. Ho. of Reps., 1982-83; chmn. ad hoc com. Mass. Advanced Materials Ctr., Boston, 1985—; mem. adv. bd. Mass. Office Sci. and Tech.; Co-PI, NSF/SSI project PALMS; co-founder ALTRAN Materials Engring. Corp., 1992. Editor: Surface Effects in Crystal Plasticity, 1977, Atomistics of Fracture, 1983, Chemistry and Physics of Fracture, 1987, Advances in Mechanics and Physics of Fracture, 1981, 83, 86; contbr. articles to profl. jours. Coach Winchester Soccer Club, Mass. Recipient sr. scientist award Humboldt Found., 1974-75, David Ford McFarland award Pa. State U., 1986; named Henry Krumb lectr. AIME, 1984, Disting. Alumnus, Ohio State U. Coll. Engring., 1991; hon. alumnus MIT, 1992. Fellow Am. Soc. Metals Internat.; mem. NAE, Am. Soc. Metals (govt. and pub. affairs com. 1984), Nat. Assn. Corrosion Engrs. (A.B. Campbell award 1971), New Eng. Sci. Tchrs. (founder, co-chmn.), Nat. Materials Adv. Bd. Roman Catholic. Office: MIT Materials Sci & Engring 77 Massachusetts Ave Rm 8202 Cambridge MA 02139-4307

LATHAM, ELEANOR RUTH EARTHROWL, neuropsychology therapist; b. Enfield, Conn., Jan. 12, 1924; d. Francis Henry and Ruth Mary (Harris) Earthrowl; m.Vaughan Milton Latham, July 20, 1946; children: Rebecca Ann, Carol Joan, Jennifer Howe, Vaughan Milton Jr. BA, Vassar Coll., 1945; MA, Smith Coll., 1947, Clark U., Worcester, Mass., 1974; EdD, Clark U., Worcester, Mass., 1979. Lic. psychologist, Mass. Guidance counselor Worcester Pub. Schs., 1967-74, sch. psychologist, 1975-80; pvt. practice neuropsychology Worcester, 1981—; postdoctoral trainee Children's Hosp.-Harvard Med. Sch., Boston, 1980-81; med. staff Hahnemann Hosp., Worcester, St. Vincent Hosp., Worcester, The Med. Ctr. Cen. Mass., Worcester; assoc. in pediatrics U. Mass. Med. Ctr. and Med. Sch., Worcester, 1982—. Author: Neuropsychological Impairment in Duchene Muscular Dystrophy, 1985, Motor Coordination and Visual-Motor Development in Duchenne Muscular Dystrophy; contbr. chpts. to books. Mem. Internat. Neuropsychology Soc., Am. Psychol. Assn. Republican. Unitarian. Avocations: chamber music, piano, travel, swimming. Home: 59 Berwick St Worcester MA 01602-1442 Office: Vernon Med Ctr 10 Winthrop St Worcester MA 01604-4435

LATHAM, PAUL WALKER, II, electrical engineer; b. Birmingham, Ala., July 18, 1957; s. Allen and Martha (Scharmann) L.; m. Colleen Mary O'Meara, July 11, 1981; children: Allen, Emily Agnes. BSEE, U. Minn., 1980; PhD in Elec. Engring., U. N.H., 1992. Design engr. Microcomputer Systems Corp., 1980-82, MPI div. Control Data, Mpls., 1982-84; group mgr. Data Gen., Durham, N.H., 1984-89; sect. mgr. Allegro Microsystems, Concord, N.H., 1989—. Mem. Sigma Xi. Achievements include several patents in field. Home: 30 Wheelwright Dr Lee NH 03824

LATHI, BHAGAWANDAS PANNALAL, electrical engineering educator; b. Bhokar, Maharashtr, India, Dec. 3, 1933; came to U.S., 1956; s. Pannalal Rupchand and Tapi Pannalal (Indani) L.; m. Rajani Damodardas Mundada, July 27, 1962; children: Anjali, Shishir. BEEE, Poona U., 1955; MSEE, U. Ill., 1957; PhD in Elec. Engring., Stanford U., 1961. Research asst. U. Ill., Urbana, 1956-57, Stanford (Calif.) U., 1957-60; research engr. Gen. Electric Co., Syracuse, N.Y., 1960-61; cons. to semicondr. industry India, 1961-62; assoc. prof. elec. engring. Bradley U., Peoria, Ill., 1962-69, U.S. Naval Acad., Annapolis, Md., 1969-72; prof. elec. engring. Campinas (Brazil) State U., 1972-78, Calif. State U., Sacramento, 1979—; vis. prof. U. Iowa, Iowa City, 1979. Author: Signals, Systems and Communication, 1965, Communication Systems, 1968 (transl. into Japanese 1977), Random Signals and Communication Theory, 1968, Teoria Signalow I Ukladow Telekomunikacyjnych, 1970, Sistemy Telekomunikacyjne, 1972, Signals, Systems and Controls, 1974, Sistemas de Comunicacion, 1974, 86, Sistemas de Comunicacao, 1978, Modern Digital and Analog Communication Systems, 1983, 89 (transl. into Japanese 1986, 90), Signals and Systems, 1987, Linear Systems and Signals, 1992; contbr. articles to profl. jours. Mem. IEEE (sr.). Avocations: swimming, poetry. Office: Calif State U 6000 J St Sacramento CA 95819-2605

LATHROP, KAYE DON, nuclear scientist, educator; b. Bryan, Ohio, Oct. 8, 1932; s. Arthur Quay and Helen Venita (Hoos) L.; m. Judith Marie Green, June 11, 1957; children: Braxton Landess, Scottfield Michael. BS, U.S. Mil. Acad., 1955; MS, Calif. Inst. Tech., 1959, PhD, 1962. Staff mem. Los Alamos Sci. Lab., 1962-67; group leader methods devel. Gen. Atomic Co., San Diego, 1967-68; with Los Alamos Sci. Lab., 1968-84, assoc. div. leader reactor safeguards and reactor safety and tech. div., 1975-77, alt. div. leader energy div., 1977-78, div. leader computer sci. and svcs. div., 1978-79, assoc. dir. for engring. scis., 1979-84; assoc. lab dir., prof. applied rsch. Stanford Linear Accelerator Ctr. Stanford U., 1984—; vis. prof. U. N.Mex., 1964-65, adj. prof., 1966-67; guest lectr. IAEA, 1969; mem. adv. com. reactor physics ERDA, 1973-77; mem. reactor physics vis. com. Argonne Nat. Lab., 1978-83; mem. mgmt. adv. com. y-12 div. Union Carbide Corp., 1979-82; mem. engring. nat. adv. com. U. Mich., 1983-92; mem. steering com. Joint MIT-Idaho Nat. Engring. Lab. Rsch. Program, 1985-89; mem. external adv. com. Nuclear Tech. and Engring. div. Los Alamos Sci. Lab., 1988-91, 92—; mem. com. on material control and acctg. for spl. nuclear materials NRC, 1988-89; mem. energy rsch. adv. bd. panel on new prodn. reactor tech. assessmemnt Dept. of Energy, 1988; cons. in field. Author reports, papers, chpts. to books; mem. editorial adv. bd. Progress in Nuclear Energy, 1983-85. Served to 1st lt. C.E. U.S. Army, 1955-58. Spl. fellow AEC, 1958-61; R.C. Baker Found. fellow, 1961-62; recipient E.O. Lawrence Meml. award ERDA, 1976; Disting. Svc. award Los Alamos Nat. Lab., 1984. Fellow Am. Nuclear Soc. (chmn. math. and computation div. 1970-71, nat. dir. 1973-76, 79-82, treas. 1977-79, Outstanding Performance award 1980); mem. Am. Phys. Soc., Nat. Acad. Engring. Republican. Episcopalian. Home: 672 Junipero Serra Blvd Stanford CA 94305-8444

LATHROP, LESTER WAYNE, mechanical engineer; b. Shawnee, Okla., Feb. 26, 1935; s. Wilmer Claude and Nellie Alma (Gray) L.; m. Virginia Lathrop, Aug. 10, 1957; children: Kenneth, Karen, Kay. BSME, U. Okla., 1962; MSME, U. N.Mex., 1964. Mem. tech. staff Sandia Labs., Albuquerque, 1962-76; test dir. Sandia Labs., Tonopah, Nev., 1976-80, Albuquerque and Kauai, 1980-81; range ops. supr. Sandia Labs., Tonopah, 1981-86; rocket systems supr. Sandia Labs., Albuquerque and Kauai, 1986-91; range ops. supr. Sandia Labs., Tonopah, 1991—; range safety officer Sandia Labs. Tonopah Test Range, 1981-86, 91-92. Staff sgt. USAF, 1955-58. Assoc. fellow AIAA. Home: 9633 Buckhorn Dr Las Vegas NV 89134 Office: Tonopah Test Range PO Box 871 Tonopah NV 89049

LATIFF-BOLET, LIGIA, psychologist; b. Colon, Panama, Sept. 10, 1953; came to the U.S., 1971; d. Antonio and Graciela (Whittaker) L.; m. Celso G. Bolet, Aug. 1991; 1 child, Ligia Alanna. BA, Loyola U., New Orleans, 1975; MA, Caribbean Ctr. Advanced Study, Santurce, Puerto Rico, 1979, PhD, 1982. Lic. clin. psychologist. Staff psychologist Puerto Rico State Psychiat. Hosp., Rio Piedras, 1980-82; project dir. Secretariat Aux. Mental Health, Rio Piedras, 1982-83; chief psychologist U.S. Army MEDDAC Panama, Panama City, 1983-88; dir. dept. psychology 1st Hosp. Panamericano, Cidra, Puerto Rico, 1988-91; coord. family preservation program Aiken-Barnwell Mental Health Ctr., 1993—; lectr. various agys., 1982—. Trainer Crisis Line Cath. Ch. Santuario Nacional, Panama City, 1987-88; coord. workshops program Un Mensaje al Corazon, Panama City, 1987-88. NIMH scholar, 1978-79. Mem. Am. Psychol. Assn., World Fedn. for Mental Health. Roman Catholic. Avocations: jogging, aerobic exercises, swimming, reading. Home: 9 Myer Dr Fort Gordon GA 30905

LATIMER, GEORGE WEBSTER, JR., chemist. BS in Chemistry, George Washington U., 1955; PhD in Phys./Analytical Chemistry, Princeton U., 1961; diploma in corr. with honors, Air War Coll., 1982. Group leader analytical svcs., sr. rsch. chemist PPG Industries, Corpus Christi, Tex., 1962-64, 67-73; chief nutritional quality control, sr. group leader pharm. and nutritional quality control Mead Johnson, Evansville, Ill., 1973-78; asst. mgr. aulity assurance Whitehall Labs., Elkhart, Ind., 1978-81; mgr. quality control bulk pharm. chems. SmithKline Chems., Conshohocken, Pa., 1981-84; mgr. corp. quality control Norwood Industries, Malvern, Pa., 1984-86; chief agrl. analytical systems Office Tex. State Chemist, College Station, Tex., 1986-92, acting state chemist, 1988-90, state chemist, 1990—; asst. prof. analytical chemistry., dir. analytical lab. U. Utah., Salt Lake City, 1965-67; asst. prof. sci. U. Evansville, 1974-76; admissions liason counselor Hdqs. Air Force Acad., 1978-86; lectr. Air U., Maxwell AFB, Ala., 1981-86; asst. prof. chemistry Bucks County C.C., Pa., 1985-86; mem. shrimp mariculture task force Tex. Agrl. Expt. Sta. safety com., 1987, 88; mem. Tex. State Legis. Com. on Naturally Occurring Radioactive Materials, San Antonio, 1989, Houston, 1990; rep. Infant Formoula Coun., 1976-78; mem. peer review com. Wright Labs., U.S. Dept. Def., 1992; cons. Kennecott Copper, 1966-67, Deseret Industries, 1966-67, FGIS/USDA CONASUPO Negotiations, 1990, USAF, 1992. Co-founder, editor Separation Sci., 1966-70; contbr. articles to peer reviewed sci. jours.; invited presenter in field. Col. USAFR, 1981-86. Grantee NSF. Mem. Am. Chem. Soc. (sec./treas. loacl sta. Corpus Christi 1968-72, divsn. profl. rels. 1986—), Assn. Am. Plant Food Control Ofcls. (Magruder check sample com. 1988—, good mfg. practices com. 1991—, bd. dirs. 1992—, investigator investigational allowances 1992—), Am. Assn. Feed Control. Ofcls. (parliamentarian, collaborative check sample com. 1988—, states-industry rels. com. 1989—), Am. Lab. Mgrs. Assn., Assn. Official Analytical Chemists assoc. referee iodine-containing compounds in feeds 1987—, com. on feeds., fertilizers and related topics 1989—, pres. Southwest regional sect., 1989-90, treas. Southwest regional sect. 1992—, regional selection com. 1990, ad hoc com. on analytical tng. 1990—, statis. methods sect. editor jour. 1990—, gen. referee analytical methids for feed and fertilizer 1991—, com. long range planning), Cound. for Sci. and Handicapped, Res. Officers Assn. (life), Air Force Assn. (life), Alpha Chi Sigma, Pi Beta Kappa, Sigma Xi. Achievements include research in method for determining dimethylsulfoxide in triampterene. Office: Tex Agrl Expt Sta PO Drawer 3160 321A Reed McDonald Bldg College Station TX 77841

LATIMER, JAMES HEARN, systems engineer; b. Texarkana, Ark., Oct. 27, 1941; s. James Hearn and Helen (Harris) L.; m. Michele Renee Halbeisen, July 5, 1969; children: Eric James, Veronique Claire. BS, MIT, 1963. Engr. Smithsonian Astrophys. Obs., Cambridge, Mass., 1963-71; applied mathematician Smithsonian Astrophys. Obs., 1971-77, mgr. satellite data, 1977-83; tech. staff The MITRE Corp., Bedford, Mass., 1983-90, lead engr., 1990—. Contbg. author: 1973 Smithsonian Standard Earth (III), 1973. Mem. ACM, AIAA (tech. com. on astrodynamics 1988-91). Republican. Methodist. Avocations: mountain climbing. Office: The Mitre Corp Burlington Rd Bedford MA 01730-1306

LATIMER, PAUL JERRY, non-destructive testing engineer; b. Springfield, Tenn., July 21, 1943; s. Paul Daniel and Juanita Inez (Richey) L.; m. Sylvia Susan Cole, June 6, 1966; children: Zachary Nathaniel, Matthew Jason. BS in Physics with honors, U. Tenn., 1966, MS in Physics, 1979, PhD in Physics, 1983. Devel. engr. Oak Ridge (Tenn.) Nat. Lab., 1980-81; faculty rsch. assoc. Ohio State U., Columbus, 1981; rsch. asst. U. Tenn., Knoxville, 1981-83; sr. rsch. engr. Babcock and Wilcox, Lynchburg, Va., 1983-91. Patentee in field. Co-leader cub pack Lynchburg area Boy Scouts Am., 1983-84. Mem. Sigma Pi Sigma. Avocations: martial arts, hiking, lapidary, mineral collecting. Home: 303 Juniper Dr Lynchburg VA 24502-5661 Office: Babcock and Wilcox Lynchburg Rsch Ctr Lynchburg VA 24506

LATORRE, ROBERT GEORGE, naval architecture and engineering educator; b. Toledo, Jan. 9, 1949; s. Robert James and Madge Violette (Roy) L.; m. Yuko Yoshino, Sept. 5, 1980. BS in Naval Architecture and Marine Engring. with honors, U. Mich., 1971, MS in Engring., 1972; MSE in Naval Architecture, U. Tokyo, 1975, PhD. in Naval Architecture, 1978. Asst. prof. U. Mich., Ann Arbor, 1979-83; assoc. prof. U. New Orleans, 1984-87, prof. naval architecture, marine engring., 1987, prof., chair naval architecture, marine engring., 1989—; assoc. prof. mech. engring., U. Tokyo, 1986-87; rsch. scientist, David Taylor Naval R & D Lab., Bethesda, Md., 1980, 81, Bassin d'Essais des Carenes, Paris, 1983; cons. in field. Contbr. to profl. publs. Mem. Soc. Naval Architects, Royal Instn. Naval Architects Gt. Britain, ASME, Soc. Naval Architects Japan, Am. Soc. Engring. Edn. (program chmn. ocean engring. div. 1989—), Japan Club New Orleans. Roman Catholic. Home: 300 Lake Marina Dr New Orleans LA 70124-1676 Office: U New Orleans 911 Engineering Bldg New Orleans LA 70148

LATTAL, KENNON ANDY, psychology educator; b. Selma, Ala., Jan. 7, 1943; s. Anton and Ella Gene (Agee) L.; m. Alice Darnell Hammer, Apr. 16, 1965; children: Kennon Matthew, Anna Rachel, Laura Ashley. BS, U. Ala., Tuscaloosa, 1964, PhD, 1969. Postdoctoral fellow U. Calif.-San Diego, LaJolla, 1971-72; asst. prof. W.Va. U., Morgantown, 1972-75, assoc. prof., 1975-79, prof. psychology, 1979—. Co-editor: Experimental Analysis of Behavior, 1991; contbr. over 60 articles to profl. jours.; assoc. editor Jour. Exptl. Analysis of Behavior, 1982-86. Capt. U.S. Army, 1969-71. Benedum Disting. scholar W.Va. U., 1989. Fellow APA (pres. elect div. 25, 1993, Outstanding Tchr. award 1987), Am. Psychol. Soc.; mem. Assn. Behavior Analysis (pres.-elect 1993), Animal Behavior Soc. Achievements include research in understanding of animal learning and behavior. Home: 848 Vandalia Rd Morgantown WV 26505 Office: W Va U Dept Psychology Morgantown WV 26506-6040

LATTES, ARMAND, educator, researcher; b. Toulouse, France, Mar. 31, 1934; s. André and Elise (Dirles) L.; m. Françoise Rouzet, Mar. 25, 1958 (div. June 1991); children: Jean-Michel, Bernard; m. Isabelle Rico, Dec. 18, 1991; 1 child, Julien. Ingenieur, E.N.S.C.T., Toulouse, 1955; Dr ès Sciences, Paris U., 1960; Dr en Pharmacie, Toulouse U., 1982. Asst. prof. Faculté Des Sciences, Toulouse, 1955-64, assoc. prof., 1964-67; assoc. prof Université Paul Sabatier, Toulouse, 1967-74, full prof., 1974—; pres. CNRS, Paris, 1987-91. Co-author: Introduction á la Chimie Structurale, 1967; editor, co-author: Du Cours aux Applications, 1967. Conseiller municipal Mairie de Ramonville, France, 1989. Recipient Médaille Berthelot, Academie des Sciences, Paris, 1989, Grammaticakis-Neumann award Academie des Sciences, Paris, 1989, Van T'Hoff award Ste Chimique des Paysbas, Holland, 1969. Achievements include discovery of olefines amination by aminomercuration, prototropy of allylic amines, photochemical rearrangement of spiro oxaziridines, nitrogen inversion of unsubstituted aziridines, reactivity in organized media, molecularaggregation in polar non-aqueous solvents. Office: Univ Paul Sabatier, 118 Route de Narbonne, 31062 Toulouse France

LATTIS, RICHARD LYNN, zoo director; b. Louisville, May 31, 1945; s. Albert Francis and Jean Elizabeth (Baker) L.; m. Sharon Louise Elkins, June 22, 1968; children Michael David, Robert Brian, Theodore James. BS in Biol., U. Louisville, 1967, MS in Ecol., 1970. Asst. curator edn. Bronx (N.Y.) Zoo, 1974-75, curator edn., 1975-78; chmn. edn. N.Y. Zoological Soc., 1978-80, dir. city zoos, 1980-93, v.p. conservation parks and aquariums, 1993—; lectr., cons. zoos, aquariums, nature ctrs.; past cons. Time-Life Wild Wild World Animals film series; appeared Who's Who in the Zoo WNBC-TV, N.Y.C. Sgt. USAR, 1970-76. Mem. Am. Assn. Advancement Sci., Am. Assn. Zoological Parks and Aquariums (vice chmn. honors and awards com., chmn. ISIS task force, vice chmn. regional task force, nom. com., govt. affairs com., accreditation review program), Nat. Hist. Soc., Soc. Conservation Biol., Zoo Biol, Sigma Xi. Avocations: fishing, photography, golf, gardening, bird watching. Home: 1650 Maxwell Dr Yorktown Heights NY 10598 Office: NY Zoological Soc 830 5th Ave New York NY 10021-7095

LATZ, JOHN PAUL, aerospace engineer; b. Indpls., Jan. 31, 1969; s. John Patrick and Pamela Sue (Ridgeway) L. BSE in Aero. Engring., Princeton U., 1991; MS in Aeronautics, Calif. Inst. Tech., 1993. Fellow Northrop Corp., Pico Rivera, Calif., 1991—. Mem. AIAA (student chair 1990), Sigma Xi, Tau Beta Pi.

LAU, HARRY HUNG-KWAN, acoustical and interior designer, consultant; b. Hong Kong, May 8, 1939; s. Kang Hoi and Yuk Jing (Chan) L. BArch, Ohio State U., 1965, M.Arch., 1966; postgrad. in archl. acoustics, MIT, 1967. Acoustical designer Bolt, Beranek & Newman, N.Y.C., 1967-69; archl. designer Marcel Breuer & Asscos., N.Y.C., 1969-70; archl. designer Edward L. Barnes & Assocs., N.Y.C., 1970-74; pres. MKC Design, N.Y.C., 1975-76; pres. Lau & Asscos., N.Y.C., 1977—; instr. of design N.Y. Inst. Tech., 1974. Summer grantee Harvard-Cornell Sardis Expdn., 1966. Mem. Acoustical Soc. Am., Nat. Council Interior Design, Am. Soc. Interior Design. Address: 30 E 95th St New York NY 10028

LAU, K(WAN) P(ANG), independent power developer; b. China, July 20, 1944; m. Myra Clontz, Aug. 19, 1972; children: Karen, Kimberly, Marc. BSEE, U. N.C., 1970. Registered profl. engr., N.C., S.C. Group head power plant design Duke Power Co., Charlotte, N.C., 1970-84, legis. liaison, 1987-88, asst. to v.p. design engring. dept., 1988-89; legis. asst. Senator Pete Domenici, Washington, 1984-85; profl. staff Senate Energy & Natural Resources Com., Washington, 1985-86; v.p. Am. Nuclear Energy Coun., Washington, 1989-92; CEO Internat. Project Mgmt., Inc., Charlotte, 1993—; cons. Energy Info. Adminstrn., Washington, 1987. Author conf. proceedings. Vol. tutor Mecklenbury Sch. System, Charlotte, 1980's, Toastmasters, Charlotte, 1984, Civitan Club, Charlotte, 1984. Mem. IEEE (chmn. engring. adv. com. 1987, congsl. fellow selection com. chmn. 1988-90, N.C. Coun. Svc. award 1983, Outstanding Svc. award 1985, U.S. Activity Bd. Pub. Svc. citation 1986, Prize Paper award 1987). Office: PO Box 7346 Alexandria VA 22307

LAU, YUN-TUNG, physicist, consultant; b. Macao, Jan. 10, 1964; m. Monica Yu, Aug. 15, 1990. BSc, Zhangshan U., Canton, China, 1984; PhD, MIT, 1988. Rsch. asst. Plasma Fusion Ctr. MIT, Cambridge, 1984-88; rsch. assoc. Lab. Plasma Rsch. U. Md., College Park, 1988-90; rsch. fellow NASA/Goddard Space Flight, Greenbelt, Md., 1991-93; rsch. assoc. Lab. Plasma Rsch. U. Md., College Park, 1993—; cons. Mulimax, Inc., Greenbelt, 1989-90. Contbr. articles to Astrophys. Jour. Nat. Rsch. Coun. fellow, 1991-93. Mem. Am. Phys. Soc., Am. Geophys. Union. Office: Univ MD Lab Plasma Rsch College Park MD 20742

LAUBE, THOMAS, chemist, researcher; b. Berlin, Fed. Republic of Germany, Sept. 20, 1952; arrived in Switzerland, 1981; s. Herbert Urbschat and Ingeborg Laube. Diploma in Chemistry, Free U., Berlin, 1980; dissertation, Eidgenössische Technische Hochschule, Zürich, Switzerland, 1984; habilitation, ETH, Zürich, Switzerland, 1988. Asst. ETH, Zürich, Switzerland, 1983-89, privatdozent, 1989—. Contbr. articles to profl. jours. Recipient Silver medal ETH, Zürich, Switzerland, 1984, Ruzicka prize Swiss Univ. Coun., 1988, dozentenstipendium Fund of the Chem. Industry, Frankfurt, Fed. Republic of Germany, 1988. Mem. Am. Crystallographic Assn. Avocation: mathematics. Office: Lab für Org Chemie ETH, Universitätstrasse 16, CH-8092 Zurich Switzerland

LAUBENHEIMER, JEFFREY JOHN, civil engineer; b. Westbend, Wis., Aug. 3, 1963; s. Henry Wilbur and Ann Elizabeth (Fassbender) L.; m. Mary Ellen Gebhardt, July 7, 1990; 1 child, Michael John. BS in Civil Engring., Marquette U., 1989. Civil technician J.C. Zimmerman Engring. Corp., Greenfield, Wis., 1987-89; project engr. Larsen Engrs., Milw., 1989—. Sec. Richfield Sportsman Club, 1992. Mem. Am. Soc. Civil Engrs., Nat. Soc.

Profl. Engrs., Wis. Soc. Profl. Engrs. Republican. Methodist. Office: Larsen Engrs 735 W Wisconsin Ave Ste 800 Milwaukee WI 53233

LAUBER, JOHN K., research psychologist; b. Archbold, Ohio, Dec. 13, 1942; s. Kenneth Floyd and Fern Elizabeth (Rupp) L.; m. Susan Elizabeth Myers, Sept. 16, 1967; 1 stepchild, Sarah H. BS, Ohio State U., 1965, MS, 1967, PhD, 1969. Rsch. psychologist U.S. Naval Tng. Equipment Ctr., Orlando, Fla., 1969-73; chief aero. human factors office NASA Ames Rsch. Ctr., Moffett Field, Calif., 1973-85; mem. Nat. Transp. Safety Bd., Washington, 1985—; mem. aeros. adv. com. NASA, 1987—, U.S. Air Force Studies Bd., NAS, Washington, 1987-89. Contbr. articles to profl. jours. Recipient Industry Svc. award Air Transport World, N.Y.C., Disting. Svc. award Flight Safety Found., Tokyo, 1987, Joseph T. Nall Meml. award Nat. Air Traffic Contrs. Assn., 1992, Paul T. Hansen Lectureship award, 1993. Fellow Aerospace Med. Assn. (chmn. aviation safety com. 1978-82, R. F. Longacre award 1990).; mem. Human Factors Soc. Democrat. Avocations: sailing, flying, amateur radio, cooking. Office: Nat Transp Safety Bd 490 L'Enfant Pla E SW Washington DC 20594

LAUCHLE, GERALD CLYDE, acoustics educator; b. Williamsport, Pa., Sept. 20, 1945; s. Clarence Walter and Helen (Borowski) L.; children: Keith, Paul. BS, Pa. State U., 1968, PhD, 1974. Rsch. asst. Pa. State U., State College, 1968-74, rsch. assoc. Applied Rsch. Lab, 1975-79, sr. rsch. assoc., 1979-85, sr. scientist, 1985-90, prof. Coll. Engring., 1990—; cons. Ford Motor Co., Dearborn, Mich., 1987, AT&T Bell Labs, Murray Hills, N.J., 1991-92, GM, Milford, Mich., 1993, Eastman Kodak, Rochester, N.Y., 1992, AMETEK Corp., Kent, Ohio, 1993. Contbg. author: Lecture Notes in Engineering, 1989; contbr. articles to profl. jours. Fellow Acoustical Soc. Am., Inst. Noise Control Engring. Republican. Roman Catholic. Achievements include model of high-frequency scattering by spheroids, extended laminar flow control on underwater bodies, model of sound radiation from boundary-layer transition, noise control for fans, blowers, benchmark data on wall pressure fluctuations under turbulent flows. Office: Pa State Univ Applied Rsch Lab PO Box 30 State College PA 16804

LAUGHLIN, DAVID EUGENE, materials science educator, metallurgical consultant; b. Phila., July 15, 1947; s. Eugene L. and Myrtle M. (Kramer) L.; m. Diane Rae Seamans, June 13, 1970; children—Jonathan, Elizabeth, Andrew, Daniel. B.Sc., Drexel U., 1969; Ph.D., MIT, 1973. Asst. prof. materials sci. Carnegie-Mellon U., Pitts., 1974-78, assoc. prof., 1978-82, prof., 1982—; dir. Magnetic Materials Rsch. Group; research scientist Oxford (Eng.) U., 1985. Editor: Solid-Solid Phase Transformations, 1982; catagory editor of copper: Am. Soc. Metals/Nat. Bur. Standards Phase Diagram, 1981—; assoc. editor Metallurgical Transactions, 1982-87, editor, 1987—; contbr. over 140 articles to profl. jours. Mem. sch. bd. Trinity Christian Sch., Pitts., 1976-85, 87—, pres., 1978-83, sec. 1988-91, pres., 1991—; ruling elder Covenant Presbyn. Ch., Pitts., 1982—; foster parent Children's Home of Pitts., 1984-90; bd. dirs. Christian Schs. Internat., 1991—. Recipient Ladd Teaching award Carnegie-Mellon U., 1975; postdoctoral fellow Nat. Acad. Scis., 1974. Fellow Am. Soc. Metals; mem. Metallurgical Soc. of AIME, Am. Sci. Affiliation, Materials Rsch. Soc. Orthodox Presbyterian. Avocations: sports; books. Home: 2357 Mcnary Blvd Pittsburgh PA 15235-2779 Office: Carnegie-Mellon U Dept Materials Sci and Engring Pittsburgh PA 15213

LAUGHLIN, ETHELREDA ROSS, chemistry educator; b. Cleve., Nov. 13, 1922; d. Edward Walter and Marie Cecilia (Solinski) Ross; m. James J. Laughlin III, June 14, 1951 (div. Jan. 1956); 1 child, J. Guy. AB, Flora Stone Mather U., 1942; MS, Western Res. U., 1944, PhD, 1962. Tchr., researcher Industry Med. Schs., 1942-49; instr. St. John's Coll., Cleve., 1949-51; chemistry tchr. Cleve. Heights High Sch., Ohio, 1954-62; assoc. prof. Ferris State Coll., Big Rapids, Mich., 1962-63; div. chmn. Cuyahoga Community Coll., Cleve. and Parma, Ohio, 1963-76; prof. chemistry Cuyahoga Community Coll., Parma, 1976-85, chemistry prof. eneritus, 1985--; exec. com. Two Yr. Coll. Chemistry, 1968-86; pres. Western Cuyahoga Audubon Soc., Cleve., 1977. Contbr. articles to profl. jours. Mem. Parma Dem. Orgn., 1987--, Cleve. Mus. Natural History, 1966--. Recipient Excellence in Chemistry Teaching award Manufacturing Chem. Assn., 1973, Nat. Catalyst award, 1979, Disting. Grad. Alumni Citation Case Western Res. U., 1968, Postdoctoral Vis. Prof. award Case Western Res. U., 1970-71; LOCI grantee Nat. Sci. Found., 1977-78. Mem. Two Yr. Coll. Chemistry Assn., Am. Chem. Soc. (chmn. 1974), Nat. Edn. Assn., Nat. Sci. Tchrs. Assn., Sigma Xi. Democrat. Home: 6486 State Rd #12 Parma OH 44134

LAUGHLIN, PATRICK RAY, psychologist; b. Mpls., May 25, 1934; s. William P. and Ada I. (Martin) L.; m. Rosemary Munch, Sept. 3, 1966; children: Gregory, Mark, Richard. BA, St. Louis U., 1959; PhD, Northwestern U., Evanston, Ill., 1964. Asst. to assoc. prof. Loyola Univ., Chgo., 1965-70; assoc. to full prof. U. Ill., Urbana-Champaign, 1970—. Contbr. articles to profl. jours. Recipient Rsch. grant NSF, 1992. Office: Univ Ill 603 E Daniel Champaign IL 61820

LAUKARAN, VIRGINIA HIGHT, epidemiologist; b. N.Y.C., Dec. 14, 1948; d. Jack R. and Charlotte (Painton) Hight; children: Anson A., Maya Jasodra. BS, Tufts U., 1970; MPH, UCLA, 1975, DrPH, 1978. Rsch. assoc. U. Calif., Berkeley, 1978-79; staff assoc. Population Coun., N.Y.C., 1979-84; sr. program officer nat. Acad. Scis., Washington, 1984-88; assoc. prof. pub. health St. George's U., Grenada, W.I., 1988-90; cons. Population Coun., Grenada, W.I., 1990; sr. staff fellow in clin. nutrition FDA, Washington, 1991; rsch. asst. prof. Sch. Medicine Georgetown U., Washington, 1991—, sr. assoc. for rsch. Inst. Reproductive Health, 1991—; mem. sterilization and hysterectomy adv. com. N.Y. HHC, 1982-84; mem. editorial adv. com. Studies in Family Planning, N.Y., 1982-84. Editor: Infant Feeding in Four Societies, 1988; contbr. numerous articles on breastfeeding, contraception and related policy issues to various publs. Vice-chair bd. dirs. Internat. Sch. Grenada, 1988-90. Mem. APHA (program com. 1987), Internat. Epidemiology Assn., Am. Coll. Epidemiology. Achievements include research on breast feeding, maternal and child health, and maternity care delivery in developing countries and Washington, and epidemiologic studies of women's health. Home: 6817 Old Chesterbrook Rd McLean VA 22101 Office: Georgetown Univ Inst Reproductive Health 3800 Reservoir Rd Washington DC 20007

LAUN, HANS MARTIN, rheology researcher; b. Heilbronn, Germany, Oct. 1, 1944; s. Kurt Karl and Hedwig (Ludwig) L.; m. Ursula Eichholz, Aug. 17, 1973; children: Thomas, Eva. Pre diploma Mathematik, U. Stuttgart, Germany, 1967, diploma Physik, 1971; Dr.rer.nat., U. Ulm, Germany, 1974. Researcher BASF AG, Ludwigshafen, Fed. Republic Germany, 1974-87, group leader, 1988—. Co-author: Die Kunststoffe, vol. 1, 1990, Praktische Rheologie der Kunststoffe und Elastomere, 1991, Kunststoff-Physik, 1991; mem. editorial bds. several jours. in field. Mem. IUPAC (sec. working party IV.2.1 1989—), Deutsche Rheologische Gesellschaft, Brit. Soc. Rheology (Ann. award 1981), Soc. Rheology, Polymer Processing Soc., GVC Fachausschuss Rheologie (vice chair 1987—). Avocations: music, opera. Office: BASF, Aktiengesellschaft ZKM-G 201, D-67056 Ludwigshafen am Rhein Germany

LAUNEY, GEORGE VOLNEY, III, economics educator; b. Ft. Worth, Feb. 8, 1942; s. George Volney and Harriet Louise (Pitts) L.; m. Sondra Ann Schwarz, May 29, 1965; children: George Volney IV, David Vincent. BBA, U. N. Tex., Denton, 1965, MBA, 1966; PhD, U. Ark., 1970. Asst. prof. econs. N.E. La. U., Monroe, 1968-70; asst. prof., assoc. prof. econs. Franklin (Ind.) Coll., 1970-83, chmn. econs. and bus. dept., 1971-81, prof. econs., Joyce and E. Don Tull prof. bus. and econs., 1983—, chmn. social sci. div., 1983—; pres. Econ. Evaluation, Inc., Franklin, 1985—; cons. Von Durpin, Div. Ingersol Rand, Bargersville (Ind.) State Bank, Ind. Dept. Ins., Med. Malpractice Bd., Indpls. Contbr. articles to profl. jours. Recipient Branigin award for teaching excellence Franklin Coll. Bd. Trustees, 1979. Mem. Am. Econ. Assn., Am. Assn. Forensic Economists, Am. Acad. Fin. and Econ. Experts (bd. editors 1988—). Avocation: coin collecting. Home: 1875 Hillside Dr Franklin IN 46131-9559 Office: Franklin Coll Dept Econs Franklin IN 46131

LAURA, ROBERT ANTHONY, coastal engineer, consultant; b. Syracuse, NY, June 4, 1955; s. John Emil and Rosemary (Ross) L.; m. Susan Ann Sieve, Dec. 30, 1978; children: Carolyn Ruth, Alyson Anne, Katy Marie. BS in Engring. Sci., SUNY, Buffalo, 1977; MS in Ocean Engring., U. Miami, Fla., 1980. Registered profl. engr., Fla. Sr. assoc. Post, Buckley, Schuh & Jernigan, Inc., Miami, Fla., 1979-88; sr. civil engr. Law Environ. Inc., Ft. Lauderdale, Fla., 1988-92; sr. civil engr. South Fla. Water Mgmt. Dist., West Palm Beach, Fla., 1992—. Contbr. articles to Jour. of Hydraulic Engring. and proceedings of sci. meetings. Recipient fellowship from Conoco to U. Miami, 1979. Mem. ASCE (control mem. coastal engring. tech. com. 1988—), Am. Water Resources Assn., Nat. Soc. Profl. Engrs., Fla. Engring. Soc., Am. Shore and Beach Preservation Assn. Democrat. Roman Catholic. Achievements include development of many special computer programs for hydrodynamics and water resources. Home: 1081 Northumberland Ct Wellington FL 33414 Office: South Fla Water Mgmt Dist PO Box 24680 West Palm Beach FL 33416

LAURENO, ROBERT, neurologist; b. Cleve., Mar. 2, 1945; s. Raymond Rudolph and Reva (Gelb) L.; m. Karen Jayne Knoller, July 13, 1969; children: Caroline, Rachel, Meredith. AB, Cornell U., 1967, MD, 1971. From asst. prof. to prof. George Washington U., Washington, 1977—; chmn. dept. neurology Washington Hosp. Ctr., 1977—. Contbr. articles to profl. jours. With USPHS, 1972-74. Grantee Medlantic Rsch. Found. Fellow Am. Assn. for Electrodiagnosis, Am. Acad. Neurology; mem. Am. Neurol. Assn. Achievements include experimental demonstration of cause of central pontine myelinolysis, a human brain disease due to rapid correction of hyponatremia. Office: Washington Hosp Ctr 110 Irving St NW Washington DC 20010

LAURENT, DUANE GILES, software engineer; b. New Orleans, Dec. 17, 1952; s. Ewell Joseph and Lorrie Marie (Montz) L.; m. Susan Marie Waguespack, Aug. 2, 1974; 1 child, Sarah Elizabeth. BS, Southeastern La. U., Hammond, 1974; PhD in Physics, La. State U., 1981. Sr. engr. UTC-Mostek Corp., Carrollton, Tex., 1981-85; staff engr. SGS-Thomson Microelectronics, Carrollton, 1985—. Contbr. articles to profl. jours. Mem. IEEE, Am. Phys. Soc., Am. Radio Relay League. Home: 726 Holly Oak Dr Lewisville TX 75067 Office: SGS-Thomson Microelectronic MS600 1310 Electronics Dr Carrollton TX 75006

LAURENT, TORVARD CLAUDE, biochemist, educator; b. Stockholm, Dec. 5, 1930; s. Torbern and Bertha E. (Svensson) L.; m. Ulla B. G. Hellsing., Oct. 9, 1953; children: Birgitta, Claes, Agneta. B Medicine, Karolinska Inst., Stockholm, 1950, MD, D Med. Scis., 1958; MD (hon.), Turku (Finland) U., 1993. Instr. histology and chemistry Karolinska Inst., 1949-52, 55-58; rsch. fellow., assoc. Retina Found., Boston, 1953-54, 59-61; assoc. prof. U. Uppsala, Sweden, 1961-66, prof. med. and physiol. chemistry, 1966—, chair dept. med. and physiol. chemistry, 1973-77, 87-91, dep. dean faculty medicine, 1969-72; dep. chmn. Biomedical Ctr., Uppsala, 1973-77, 87-91; vis. prof. biochemistry Monash U, Melbourne, Australia, 1979-80; mem. Swedish Natural Sci. Rsch. Coun., 1968-70, Swedish Med. Rsch. Coun., 1970-77. Contbr. 200 papers. Recipient Anders Jahre Med. prize U. Oslo, 1968, Pharmacia award Pharmacia, Inc., 1986, Eric K. Fernström Med. prize U. Lund, 1989, Björkén prize U. Uppsala, 1990. Mem. Royal Swedish Acad. Scis. (pres. 1991—), Swedish Biochemical Soc. (sec. 1967-70, chmn. 1972-76), Wenner-Gren Found. (scientific sec. 1993—). Achievements include research in chemistry of connective tissue, physical properties, physiological functions, turnover and medical applications of the polysaccharide hyaluronan (hyaluronic acid), ophthalmic biochemistry, physical chemistry of polysaccharide networks, transport processes in polysaccharide solutions, biochemical separation techniques (e.g. a theory of gel filtration) and methods for cell separation. Office: U Uppsala Inst Med/Physiol Chemistry, BMC Box 575, S-751 23 Uppsala Sweden

LAURER, TIMOTHY JAMES, electrical engineer, consultant; b. Fremont, Ohio, July 5, 1962; s. George William and Leona Mary (Bertsch) L.; m. Teresa Gay Poggemeyer, May 9, 1992. BS in elec. engr., Univ. Toledo, 1985. Electronic system designer Helm Instrument, Maumee, Ohio, 1985-88; devel. engr. Bay Controls, Toledo, 1988-89; systems analyst PSI, Fremont, 1989—. Mem. Moose. Republican. Roman Catholic. Office: Profl Supply Inc 504 Liberty St Fremont OH 43420

LAURSEN, BRETT PAUL, psychology educator; b. Lincoln, Nebr., Jan. 10, 1962; s. Paul H. and M. Gail (Thompson) L. BA, Nebr. Wesleyan U., 1984; MA, U. Minn., 1987, PhD, 1989. Asst. prof. psychology U. Maine, Orono, 1989-91, Fla. Atlantic U., Ft. Lauderdale, 1991—; cons. Brooks/Cole Textbooks, 1991—. Editorial bd. Child Devel., 1991—; contbr. articles to profl. jours. North Am. Young scholar Johann Jacobs Found., 1991, Nat. Merit scholar, 1984; NIMH traineeship, 1987-89. Mem. APA, Am. Psychol. Soc., Soc. for Rsch. on Adolescence, Soc. for Rsch. in Child Devel. Office: Fla Atlantic Univ Dept Psychology 2912 College Ave Fort Lauderdale FL 33314

LAUTERBACH, HANS, pharmaceutical company executive; b. 1934; married. With Bayer AG, Leverkusen, Germany, 1951-91; pres. diagnostic divsn. Miles Inc., Tarrytown, N.Y., 1992—. Office: Miles Inc 511 Benedict Ave Tarrytown NY 10591*

LAUTERBUR, PAUL C(HRISTIAN), chemistry educator; b. Sidney, Ohio, May 6, 1929. BS, Case Inst. Tech., 1951; PhD, U. Pitts., 1962; PhD (hon.), U. Liege, Belgium, 1984; DSc (hon.), Carnegie Mellon U., 1987; DSc (med. hon.), Corpernicus Med. Acad., Cracow, Poland, 1988; DSc (hon.), Wesleyan U., 1989, SUNY, Stony Brook, 1990; DEng (hon.), Rennselaer Poly. Inst., 1991. Rsch. asst. and assoc. Mellon Inst., Pitts., 1951-53, fellow, 1955-63; assoc. prof. chemistry SUNY, Stony Brook, 1963-69, prof. chemistry, 1969-84, with, 1963-85, rsch. prof. radiology, 1978-85, univ. prof., 1984-85; prof. (4) depts. U. Ill., Urbana, 1985—; Disting. Univ. prof. Coll. Medicine U. Ill., Chgo., 1990—. Contbr. articles to profl. jours.; mem. editorial bds.; mem. sci. couns. Cpl. U.S. Army, 1953-55. Recipient Clin. Rsch. award Lasker Found., 1984, Nat. Medal of Sci., U.S.A., 1987, Fiuggi Internat. prize Fondazione Fiuggi, 1987, Roentgen medal, 1987, Gold medal Radiol. Soc. N.Am., 1987, Nat. Medal of Tech., 1988, Gold medal Soc. Computed Body Tomography, 1989, The Amsterdam (Alfred Heineken) prize in medicine, 1989, Laufman-Greatbatch award Assn. for Advancement Med. Intrumentation, 1989, Leadership Tech. award Nat. Elec. Mfr. Assn., 1990, Bower award and prize for achievement in sci. Benjamin Franklin Nat. Meml. Commn. of the Franklin Inst., 1990. Fellow AAAS, Am. Phys. Soc.; mem. NAS, Am. Chem. Soc., Soc. Magnetic Resonance Medicine. Office: U Ill-Urbana-Champaign 1307 W Park St Urbana IL 61801-2332

LAUTERSTEIN, JOSEPH, cardiologist; b. Vienna, Austria, Dec. 1, 1934; came to U.S., 1940; s. Bernard and Hajnalka (Stern) L.; m. Erika Stein, Jan. 24, 1964 (dec. Aug. 1990); children: Deborah Ann, Brenda Rose. BA, Syracuse U., 1955; MD, U. Vienna, 1964. Lic. physician, N.Y. Intern, then resident in internal medicine The Bklyn. Cumberland Med. Ctr., 1964-66, 68-69, fellow in cardiology, 1969-70; attending physician, cons. internal medicine and cardiology Hamilton Ave. Hosp., Monticello, N.Y., 1970-78; attending physician, cons. internal medicine and cardiology Community Gen. Hosp. Sullivan County, Harris, N.Y., 1970—, chief cardiology, 1971—, chief of staff, 1981-82; mem. courtesy staff dept. internal medicine and cardiology The Bklyn. Hosp. Ctr., 1971—; clin. asst. dept. internal medicine and cardiology St. Vincent's Hosp. and Med. Ctr. N.Y., 1974-80, asst. attending physician, 1981-86, assoc. attending physician, 1987—; with Sullivan Internal Medicine Group, P.C., Monticello, 1970—; dir. ICU Community Gen. Hosp. Sullivan County, 1971-79, dir. CCU, 1978—, dir. spl. diagnostics, 1984—, pres. med. bd., 1981-82; mem. pacemaker task force Empire State Med. Sci. and Ednl. Found., 1985-89; med. dir. Sullivan County EMT-D Program, 1989—; police surgeon Village of Monticello, 1974—, Sullivan County, 1972—; med. advisor Monticello Vol. Ambulance Corps, 1970-80, 89—; mem. Sullivan County Emergency Svcs. Coun., 1990, 91. Co-contbr. articles to Jour. Cardiovascular Surgery, Annals of Thoracic Surgery, Angiology, Chest. Mem. bd. trustees Community Gen. Hosp. Sullivan County, 1981-82, 92—. Capt. USAF, 1966-68. Fellow Am. Coll. Cardiology (N.Y. State chpt., del. to N.Y. Med. Soc. Ho. Dels. 1991—, councilor 1991—, com. mem. 1990—), Am. Coll. Chest Physicians (assoc.), Am. Coll. Angiology, Internat. Coll. Angiology, N.Y. Cardiological Soc. (exec. bd. dirs. 1982—, mem. various coms.), N.Y. Acad. Medicine; mem. AMA, Am. Geriatrics Soc., Am. Soc. Internal Medicine, Soc. for Critical Care Medicine, N.Y. Acad. Scis., N.Am. Soc. for Pacing and Electrophysiology, Med. Soc. State of N.Y., others. Office: Sullivan Internal Medicine Group PC 370 Broadway Monticello NY 12701-1155

LAUTH, ROBERT EDWARD, geologist; b. St. Paul, Feb. 6, 1927; s. Joseph Louis and Gertrude (Stapleton) L.; student St. Thomas Coll., 1944; BA in Geology, U. Minn., 1952; m. Suzanne Janice Holmes, Apr. 21, 1947; children—Barbara Jo, Robert Edward II, Elizabeth Suzanne, Leslie Marie. Wellsite geologist Columbia Carbon Co., Houston, 1951-52; dist. geologist Witco Oil & Gas Corp., Amarillo, Tex., 1952-55; field geologist Reynolds Mining Co., Houston, 1955; cons. geologist, Durango, Colo., 1955—. Appraiser helium res. Lindley area Orange Free State, Republic of South Africa, 1988, remaining helium res. Odolanow Plant area Polish Lowlands, Poland, 1988. With USNR, 1944-45. Mem. N.Mex., Four Corners (treas., v.p., pres., symposium com.) geol. socs., Rocky Mountain Assn. Geologists, Am. Inst. Profl. Geologists, Am. Inst. Mining, Metall. and Petroleum Engrs., Am. Assn. Petroleum Geologists, Helium Soc., N.Y. Acad. Sci. Am. Assn. Petroleum Landman, Soc. Econ. Paleontologists and Mineralogists, The Explorers Club. Republican. Roman Catholic. K.C. Clubs: Durango Petroleum (dir.), Denver Petroleum, Elks. Author: Desert Creek Field, 1958; (with Silas C. Brown) Oil and Gas Potentialities of Northern Arizona, 1958, Northern Arizona Has Good Oil, Gas Prospects, 1960, Northeastern Arizona; Its Oil, Gas and Helium Prospects, 1961; contbr. papers on oil and gas fields to profl. symposia. Home: 2020 Crestview Dr PO Box 776 Durango CO 81302 Office: 555 S Camino Del Rio Durango CO 81301-6826

LAUVEN, PETER MICHAEL, anesthesiologist; b. Leverkusen, Fed. Republic of Germany, May 13, 1948; s. Peter Aloysius and Katharina (Oedekoven) L.; m. Anne-Kareen Wetje, Nov. 7, 1970; children: Anne-Laureen, Lars-Peter. Diploma in Chem., U. Bonn, Fed. Republic of Germany, 1970, Dr. rer. nat., 1974, Dr. med., 1979, priv.-dozent, 1985. Teaching asst. Inst. Organic Chem. U. Bonn, Fed. Republic of Germany, 1970-76, scientist Inst. Anaesthesiology, 1976-79, physician, 1979—, anaesthesiologist, 1983—, asst. dir., 1983-85, vice-chmn., 1985-92, prof. of anaesthesia, 1986—; mem. German Fed. Drug Admission Com., 1987—. Author, co-editor: Das Zentralanticholinergische Syndrom, 1985, Klinische Pharmakologie und rationale Arzneimitteltherapie, 1992; author, editor: Anasthesie und der Geriatrische Patient, 1989, Postoperative Schmetztherapie, 1991. Recipient scholarship Stipendien Fonds der Chemischen Inst., Frankfurt, 1970, Paul Martini award, Paul Martini Found., Bonn, 1988. Mem. Gesellschaft Deutscher Chemiker, Deutsche Gesellschaft für Anasthesiologie und Intensiv Medizin, Am. Soc. of Anaesthesiology (affiliate), Am. Soc. Regional Anaesthesia, European Acad. of Anaesthesiology, European Soc. of Regional Anaesthesia, European Soc. Intensive Care Medicine, European Soc. Anaesthesiology. Home: Niehausweg 22, D-33739 Bielefeld 3, Germany Office: Anaesthesia Klinik Staedt Krankenanstalten Mitte, Teutoburger Str 50, D-33604 Bielefeld Germany

LAUX, DAVID CHARLES, microbiologist, educator; b. Freeport, Pa., Jan. 1, 1945; s. Charles L. and Margaret K. Laux; m. Sara Ellen Pollen; 1 child, Benjamin. BA, Washington and Jefferson Coll., 1966; MS, Miami U., Oxford, Ohio, 1968; PhD, U. Ariz., 1971. Postdoctoral fellow Pa. State U. Sch. Medicine, Hershey, 1971-73; prof. dept. microbiology U. R.I., Kingston, 1973—, chmn. dept., 1988—. Contbr. articles to profl. jours. NIH rsch. grantee. Mem. AAAS, Am. Soc. Microbiology, Am. Assn. Immunology. Office: U RI Dept Microbiology Kingston RI 02881

LAVALLEE, H.-CLAUDE, chemical engineer, researcher; b. Cap-Santé, Que., Can., July 28, 1938; s. Henri Lavallée and Yvonne Lavallée-Légaré; m. Ginette Morissette, June 25, 1966. BScA, Univ. Laval, Que., 1964, MScA, 1965, DSc, 1970. Rschr. Def. Rsch. Establishment of Valcartier Govt. of Can., Que., 1965-67; prof. chem. engring. U. Que. at Trois-Rivières, 1970-74; sr. engr. pulp & paper industry Ministry of Environment-Govt. of Que., Quebec City, 1974-87, head pulp & paper industry, 1986-87; dir. Pulp & Paper Rsch. Ctr. U. Que. at Trois-Rivières, 1987—; pres. H.C. Lavallée Inc., Donnacona, Que., 1989—; cons. Roche Ltée, Québec City, 1988—; adminstr. John Meunier Inc., Montréal, 1991—, Centre des technologies du gaz naturel, Montréal, 1992—. Contbr. articles to profl. jours., chpts. to books. Recipient Prize Raimbeault de Montigny Conf. Technologique, Point-au-Pic, 1990, Prize of excellence SNC-Lavalin Assn. Qué. Technique de l'eau, 1993. Mem. TAPPI, Can. Pulp and Paper Assn., Ordre des Ingénieurs du Qué. Roman Catholic. Office: Ctr Rsch Pulp & Paper, 3351 Blvd des FOrges, Trois Rivières, PQ Canada G9A 5H7

LAVALLI, KARI LEE, marine biologist; b. Detroit, Feb. 23, 1960; d. Walter Ray and Lee (Kefauver) L. BA, Wells Coll., 1982; PhD, Boston U., 1992. Rsch. asst. U. Pitts., 1981, Wells Coll., Aurora, N.Y., 1982; field asst. Rocky Mountain Biol. Lab., Crested Butte, Colo., 1982; teaching fellow Boston U., 1983-89; lectr., staff MIT, Cambridge, 1988-91, 93; postdoctoral fellow Marine Sci. Ctr. Northeastern U., East Point, Mass., 1992—; vis. lectr. marine program Boston U., 1991, Northeastern U., 1993; freelance writer, 1991; grant reviewer Sigma Delta Epsilon, 1989-90, 92; vis. scientist Sea Edn. Assn., Woods Hole, Mass., 1990. Article reviewer: Jour. Exptl. Marine Biology and Ecology, 1991—, Marine Behavior and Physiology, 1992—; contbr. chpts. to books and articles to Marine Ecology Progress Series, Fishery Bull., Lobster Newsletter, Jour. Crustacean Biology, AWIS Mag. Del. Nat. Women's Conf. in Observance of Internat. Woman's Yr., Houston, 1977. Recipient Koch prize Wells Coll., 1982, Alumni award Boston U., 1987, Grad. Student Travel award, 1986, 89. Mem. AAUW, Am. Soc. Zoologists, Animal Behavior Soc., Assn. for Women in Sci., Crustacean Soc., Sigma Xi, Sigma Delta Epsilon. Democrat. Achievements include discovery that lobsters are capable of suspension feeding in their juvenile stages when they are highly susceptible to predation. Home: 32 Summer St Nahant MA 01908 Office: Northeastern U Marine Sci East Point Nahant MA 01908

LAVATELLI, LEO SILVIO, retired physicist, educator; b. Mackinac Island, Mich., Aug. 15, 1917; s. Silvio E. and Zella (Cunningham) L.; m. Anna Craig Henderson, June 14, 1941 (dec. Sept. 1966); children: Nancy Jack, Mark Leo; m. Celia Burns, Jan. 23, 1967 (dec. May 1976); 1 stepchild, Faith Stendler (dec.); m. Barbara Gow, Nov. 22, 1976 (div. Jan. 1979); stepchildren: Ann Deemer, Lindsay Deemer; m. Olwen Thomas, Mar. 4, 1982; stepchildren: Alice Ann Williamson (Mrs. Michael W. Cone), Caroline Hill Williamson (Mrs. Patrick Blessing), Thomas Holman Williamson, Hugh Stuart Williamson. BS, Calif. Inst. Tech., 1939; MA, Princeton U., 1943, Harvard U., 1949; PhD, Harvard U., 1951. Instr. physics, chemistry, algebra, calculus, symbolic logic Deep Springs (Calif.) Jr. Coll., 1939-41; instr. Princeton (N.J.) U., 1941; rsch. asst. Manhattan Dist. Office Sci. R&D, Nat. Def. Rsch. Coun., Princeton, 1942-43; jr. staff mem. Los Alamos (N.Mex.) Nat. Lab. (formerly Manhattan Dist. Site Y), 1943-46; rsch. asst. Harvard U., Cambridge, Mass., 1946-50; asst. prof. physics, staff mem. Control Systems Lab. U. Ill., Urbana, 1950-55, assoc. prof., 1955-58, prof., 1958-79, prof. emeritus, 1979—; witness of "Trinity," the 1st atomic bomb test; mem. design team orbit plotting/control circuit logic for new cyclotron project Harvard U., 1946; observer air/ground exercises U.S. Dept. Def., Waco, Tex., 1952; observer joint air exercises NATO, Fed. Republic Germany, 1955; mem. project quick-fix Control Systems Lab., 1953; cons. Ill. group Phys. Sci. Study Com., 1956-57, Sci. Teaching Ctr., MIT, 1966, Teheran (Iran) Rsch. Unit, U. Ill., 1970; participant info. theory in biology conf. U. Ill., 1952. Producer silent film cassettes on orbit graphing U. Ill., 1964; co-interviewee video tape Logical Thinking in Children and Science Education, Nat. Japanese TV, Tokyo, 1970; phys. sci. cons. The Macmillan Science Series, 1970 edit., The Macmillan Co., N.Y.C., 1967-70; contbr. articles and revs. to profl. publs. Co-moderator discussion Fedn. Atomic Scientists, 1945, 1947. John Simon Guggenheim Meml. fellow, U. Bologna, Italy, 1957. Fellow Am. Phys. Soc.; mem. Harvard Faculty Club (nonresident). Avocations: music, painting, art history, books, movies. Home: RFD 2 Box 510 Spring Hope NC 27882-9543

LAVELLE, SEÁN MARIUS, clinical informatics educator; b. County Mayo, Ireland, May 21, 1928; s. John and Eileen Veronica (Dempsey) L.; m. Frances Angela Roche-Kelly, Apr. 6, 1956; children: Sean, Connla, Diarmuid, Eimear, Eilin, Nuala, Malachy, Frances, Maria. MB, Nat. U. Ireland, Galway, 1952, MD, 1960. Diplomate Am. Bd. Pathology. Resident Cen. Hosp., Galway, 1952-56, Boston City Hosp., 1956-58; fellow U. Pitts., 1958-60; prof. exptl. medicine Univ. Coll., Galway, 1960—, bd. govs., 1980-86; cons. Regional Hosp., Galway, 1960-78; mem. Med. Rsch. Coun. Ireland, Dublin, 1968-71, 81-85, rsch. fellow, 1953-55; vice chmn. rsch. com. on bioengring. EEC, Brussels, 1984-91; vis. rsch. prof. SUNY, Bklyn., 1966; rsch. fellow Tufts U., Boston, 1957-58. Editor: Objective MEdical Decisions, 1983; also articles. Councillor Galway Hospice Movement, 1987—. Travel fellow Med. Rsch. Coun., 1956-57. Fellow Royal Coll. Physicians (Dublin); mem. Irish Med. Orgn., Assn. for Study Med. Edn. (councillor), Irish Assn. for Cancer Rsch. (chmn. 1979-81), European Assn. for Cancer Rsch. (councillor 1987—), Irish Fedn. Univ. Tchrs. (chmn. 1970-72), Galway Golf Club, County Club. Avocations: Irish lang. books, hill walking, golf, Beethoven, elements of thinking. Office: Univ Coll Galway, Dept Exptl Medicine, Galway Ireland

LAVEN, DAVID LAWRENCE, nuclear and radiologic pharmacist, consultant; b. Detroit, Jan. 31, 1953; s. Harold Sanford and Ada Rae (Blumenthal) L.; m. Maxine Frances Miller, May 14, 1977; 1 child, Ryan Stuart. BA in History, Biology, Albion Coll., 1975; BS in Pharmacy, U. N.Mex., 1981. Rsch. technologist, biodistbn. specialist U. N.Mex. Coll. Pharmacy, Albuquerque, 1978-81; asst. mgr. Syncor, Inc. (formerly Pharmatopes), Miami, Fla., 1981-84; instr. nuclear pharmacy U. Miami, 1982-85; pres., owner Ganmascan Cons., Bay Pines, Fla., 1982—; staff pharmacist Hollywood (Fla.) Med. Ctr., 1983-84; asst. mgr. Nuclear Pharmacy, Inc., Sunrise, Fla., 1984-85; mem. adv. panel on radiopharms. U.S. Pharmacopeial Conv. Inc., Rockville, Md., 1985-95; dir. nuclear pharmacy program VA Med. Ctr., Bay Pines, 1985—; cons. nuclear pharmacy Nat. Assn. Bds. Pharmacy, Chgo., 1987-92; adj. assoc. clin. prof. U. Fla. Coll. Pharmacy, Gainesville, 1986—, U. the Health Scis. Coll. Pharmacy Southeastern, North Miami Beach, Fla., 1990—; edn. cons. Nuclear Tech. Rev. Series Edn. Rev., Inc., 1988—; mem. specialty coun. on nuclear pharmacy Bd. Pharm. Specialities, 1988—. Editor, co-pub. Clini-Scan Monthly, 1982-84; co-guest editor Jour. Pharmacy Practice, 1989, mem. editorial bd., 1991—; guest editor Fla. Jour. Hosp. Pharmacy, 1990, cons. editor, 1986—; guest author In-Svc. Rev. in Nuclear Medicine, 1990-92; mem. editorial bd. New Perspectives in Cancer Diagnosis and Management, 1992—; nat. field editor ASHP Signal Newsletter, 1985-87. Mem. Henry Morgan chpt. B'nai B'rith, Southfield, Mich., 1975-77. Fellow Am. Soc. Hosp. Pharmacists (chmn.-designate specialized practice group on radiologic pharmacy 1993-94, edn. program assoc. 1988—), Am. Soc. Cons. Pharmacists; mem. Am. Pharm. Assn. (chmn. sect. on specialized pharm. svcs. 1989-90, chmn.-elect 1992-93, chmn. section on nuclear pharmacy 1993-94, Acad. Pharmacy Practice and Mgmt. Practitioner Merit award 1990, Acad. Pharmacy Practice and Mgmt. Presentation award 1990, 91, fellow 1992, del. 1986—, edn. cons. 1987—), Am. Assn. Colls. Pharmacy, Am. Soc. Pharmacy Law, Fla. Pharmacy Assn. (chmn. ednl. affairs coun. 1989-90, chmn. nuclear pharmacy section 1987-89, 91-93, chmn. acad. pharmacy practice 1988-90, chmn. orgnl. affairs coun. 1992-93, del. 1988—, edn. cons. 1987—, Number 1 Club 1990, Disting. Young Pharmacist award 1990, Acad. Pharmacy Practice Practitioner Merit award 1992, Sidney Simkowitz Pharmacy Involvement award 1992), Acad. Pharmacy Practice (chmn. 1988-90, chmn. nuclear pharmacy sect. 1987-89, 91—), Fla. Soc. Hosp. Pharmacists, Soc. Nuclear Medicine (edn. cons. 1989—), Pinellas Pharmacist Soc. (pres. elect 1991—, pres. 1992-93, Pharmacist of Yr. award 1992), Pasco-Hernando Pharmacy Assn. (treas. 1991—), Hillsborough County Pharmacy Assn. (sec. 1991-92, pres.-elect 1993-94), Polk County Pharmacy Assn., Ctrl. Fla. Pharmacy Assn., Kappa Psi,Psi Chi, Phi Alpha Theta, Beta Beta Beta. Avocations: art collecting, intramural sports, camping, traveling, writing. Home: 3955 Orchard Hill Cir Palm Harbor FL 34684-4141 Office: VA Med Ctr PO Box 636 Bay Pines FL 33504-0636

LAVENDA, BERNARD HOWARD, chemical physics educator, scientist; b. N.Y.C., Sept. 18, 1945; s. Nathan and Selma (Dubnow) L.; m. Fanny Malka, Mar. 6, 1973; children: Marlene Allyn, Jason Isaac. BSc, Clark U., 1966; MSc, Weizmann Inst., 1967; PhD, Free U. Brussels, 1970. Prof., U. Pisa, Italy, 1972-73, U. Naples, Italy, 1975-80, U. Camerino, Italy, 1980—; cons. TEMA, Bologna, Italy, 1978-84, Nuovo Pignone, Firenze, Italy, 1972-73; vis. prof. Inst. de Fisica U. Fed. de Rio Grande do Sul, Porto Alegre Brasil, 1986; assoc. scientist Internat. Ctr. Theoretical Biology, Venice, 1987-92; founding mem. Internat. Ctr. Thermodynamics, 1992—, pres., 1993—. Author: Thermodynamics of Irreversible Processes, 1978; Nonequilibrium Statistical Thermodynamics, 1985; Statistical Physics: A Probabilistic Approach, 1991; co-author (with E. Santamato) Introduzione Alla Fisica Atomica E Statistica, 1989. Contbr. articles to profl. jours. Fellow Royal Soc. Chemistry; mem. Am. Math. Soc. Avocation: chess. Home: Frazione Sentino 30/A, 62030 Camerino Italy Office: Internat Ctr Thermodynamics c/o Enea, CRE Casaccia via SP Anguillarse 301, 00060 South Maria Di Galeria Rome Italy

LAVENDA, NATHAN, physiology educator; b. N.Y.C., Dec. 10, 1918; s. Zukin and Etie Feige (Weinstein) L.; m. Selma Lavenda, Aug. 24, 1943 (div. June 1961); children: Bernard, Ronald; m. Harriet Rebecca Lavenda, June 24, 1961; children: Elaine, David. MS, NYU, 1947, PhD, 1952. Aquatic biologist U.S. Fish and Wildlife Svc., N.Y.C., 1944-47; biologist USPHS, S.I., N.Y., 1947-50; instr. physiology Howard U. Med. Sch., Washington, 1952-56; chmn. dept. biology North Adams (Mass.) State Coll., 1961-67; chmn. dept. physiology Ill. Coll. Podiatric Medicine, Chgo., 1967-69; rsch. assoc. Michael Reese Hosp., Chgo., 1970-82, U. Ill., Chgo., 1983—; chmn. physiology Nat. Bd. Podiatric Examiners, Chgo., 1967-69. Contbr. articles to profl. jours. With USAF, 1941-44. Recipient grant U.S. Atomic Energy Commn., 1953. Mem. Am. Soc. Zoologists, Soc. for Study Reproduction. Achievements include discovery of sex reversal in Atlantic Sea Bass, staining reaction in cancer cells, cytochemical demonstration of viruses. Home: 8338 Karlov Ave Skokie IL 60076

LAVER, MURRAY LANE, chemist, educator; b. Warkworth, Ont., Can., Mar. 7, 1932; s. Oscar Frederick and Clara Gertrude (Lane) L.; m. Mary Margaret Smolska, Feb. 9, 1963; children: Ann Margaret, Elizabeth Clara. BSA, Ont. Agrl. Coll., Guelph, Ont., Can., 1955; PhD, Ohio State U., 1959. Rsch. chemist Westreco Inc. (Nestlé Co.), Marysville, Ohio, 1959-63; rsch. scientist Rayonier Can., Vancouver, B.C., 1963; profl. specialist Weyerhaeuser Co., Seattle, 1964-67; rsch. fellow U. Wash., Seattle, 1968-69; assoc. prof. forest products dept. Oreg. State U., Corvallis, 1969—; rsch. fellow Harvard U., Cambridge, Mass., 1977-78; cons. Tooze Marshall Holloway & Duden, Portland, Oreg., 1988-90, Am. Cemwood, Albany, Oreg., 1991—, Black-Helterline, Portland, 1992-93, Pozzi Wilson Atchison O'Leary & Conboy, Portland, 1993—. Contbr. chpt. to Wood Structure and Composition, 1991; contbr. articles to profl. jours. Speaker local high schs., Alsea and Philomath, Oreg., 1988, 90. Grantee USDA, 1985—, 86-88, 92-93, Hill Family Found., 1990. Mem. AAAS, Am. Chem. Soc. (chair Oreg. sect. 1985), N.Y. Acad. Sci., Oreg. Acad. Sci. Achievements include consultancy on new resin system for particle board manufacture, research on Douglas Fir bark wax resulting in commercialization. Home: 1950 SW Whiteside Dr Corvallis OR 97333-1409 Office: Dept Forest Products Oreg State Univ Corvallis OR 97331-5703

LAVERNIA, ENRIQUE JOSE, materials science and engineering educator; b. Havana, Cuba, July 30, 1960; came to U.S., 1965; s. Carlos Manuel and Ana Margot (Borrego) L.; m. Julie M. Schoenung, Oct. 4, 1986. BS in Solid Mechanics, Brown U., 1982; MS in Metallurgy, MIT, 1984, PhD in Materials Engring., 1986. Rsch. asst. MIT, Cambridge, 1982-86, postdoctoral assoc., 1986, rsch. assoc., 1986-87; assoc. prof. materials sect. dept. mech. & aerospace engring. U. Calif., Irvine, 1987-91; assoc. prof. U. Calif., 1991—; mem. sci. adv. com. Ceracon Inc., Sacramento; cons. orgns. including Swedish Inst. Metals Rsch., Stockholm, 1987, Martin Marietta, Balt., 1988, Oak Ridge (Tenn.) Nat. Labs., 1988, Assn. Sci. Adivsors, L.A., 1990—, Flow Internat., Seattle, 1990—; book proposal reviewer CRC Press., Fla., Allyn & Bacon, Mass., Advanced Materials & Mfg. Processes; proposal reviewer U.S.-Israel binat. fund. NSF, Idaho State Bd. Edn., Univ. Tech. Transfer Inc., Army Rsch. Office; invited lectr. in field. Reviewer Internat. Jour. Powder Metallurgy, Jour. Advanced Materials and Mgd. Processes, Metallurgical Transactions: Atomization and Sprays. Rockwell Internat. fellow, 1982-84, ALCOA fellow, 1990-92; Rsch. grantee NSF, 1988—, Co-rsch. grantee; recipient Young Investigator award Office Naval Rsch., 1990-93; named Presdl. Young Investigator, NSF, 1989-94. Mem. Am. Soc. Metals (Internat. Bradley Stoughton award for Young Tchrs. 1993), Materials Rsch. Soc., The Metallurgy Soc. (sec. com. on synthesis and analysis in materials processing, organizer/chmn. various meeting sessions), Am. Powder Industries Fedn., Phi Delta Beta (nat. hon. soc. Alpha Kappa chpt.), Sigma Xi. Avocations: tennis, jogging, handball, scuba diving. Home: 13701 Wheeler Pl Tustin CA 92680-1931 Office: U Calif Dept Mech and Aerospace Engring 616 Engineering Office 516 F Irvine CA 92717

LAVEROV, NIKOLAI PAVLOVITCH, science foundation executive; b. Pozarische, Arkchangelsk, USSR, Jan. 12, 1930; s. Pavel Nicolaevitch and Klaudia (Savvatievna) L.; m. Valentina Leonidovna, Jan. 1, 1952; 2 children. D of Geology and Mineral Scis., Inst. Gold and Nonferrous Metals, Moscow, 1958, PhD in Geology and Mineral Scis., 1974. Exec. sec. Inst. Geology of Minerals, Petrology, Minerology and Geochemistry, Moscow, 1958-66; head dept. USSR Ministry Geology, Moscow, 1966-83; dep. rector USSR Acad. Nat. Econ., Moscow, 1983-87; pres. Acad. Sci. of Kirgiz Republic, Frunze, 1987-88; dep. prime minister, chmn. USSR State Com. for Sci. and Tech., Moscow, 1989—; dir. Inst. of Geology for Minerals, Petrology, Minerology and Geochemistry, 1988—. Author: USSR Ore Deposits, 1974, Geology of Hydrothermal Uranium Deposits, 1966, Basic Principles of Forecast for Uranium, 1976, others; contbr. over 190 articles to sci. jours. People's dep. USSR Supreme Soviet, Moscow, 1989. Mem. USSR Acad. Sci. (corr. mem. 1979, academician 1987, v.p. 1988—). Mem. Communist Party. Avocations: fishing, mountain climbing. Home: 11 Tverskajastr, 103905 Moscow USSR Office: Acad Scis USSR, Leninsky Pr 14, Moscow V-71, Russia*

LAVINE, LEROY STANLEY, orthopedist, surgeon, consultant; b. Jersey City, Oct. 28, 1918; s. Max and Katherine (Miner) L.; m. Dorothy Kopp, Feb. 14, 1946; children: Michael, Nancy. AB, NYU, 1940, MD, 1943. Diplomate Am. Bd. Orthopedic Surgery. Intern Morrisania City Hosp., Bronx, N.Y., 1944; resident Jewish Hosp., Bklyn., 1949-51, Ind. U. Med. Ctr., Indpls., 1951-52; dir., prof. orthopedic surgery Health Scis. Ctr., SUNY, Bklyn., 1970-80; vis. orthopedic surgeon Mass. Gen. Hosp., Boston, 1983—; dir. rehab. svcs. Spaulding Rehab. Hosp., Boston, 1983—; prof. emeritus Health Scis. Ctr., SUNY, Bklyn., 1980—; lectr. Harvard Med. Sch., Boston, 1983—; mem. adv. panel FDA, Rochville, MD, 1988-91, cons. 1992—; cons. Dept. Health and Human Svcs., NIH, Bethesda, Md., 1992—. Contbr. more than 100 articles to profl. jours. Mem. coun. MIT, Cambridge, 1991—. Capt. U.S. Army, 1946-48. Grantee NIH, 1968-80, AEC, 1968-80. Fellow AAAS, ACS, N.Y. Acad. Scis., Am. Acad. Orthopedic Surgeons; mem. Cosmos Club. Achievements include initial performance clinical case of elec. bone stimulation. Office: Mass Gen Hosp Dept Orthopedic Surgery Boston MA 02114

LAVKULICH, LESLIE MICHAEL, soil science educator; b. Coaldale, Alta., Can., Apr. 28, 1939; s. Michael Lavkulich and Mary Alexa; m. Mary Ann Elizabeth Olah, Sept. 15, 1962; children: Gregory Michael, Miles Andrew. BS with distinction, U. Alta., Edmonton, 1961, MS, 1963; PhD, Cornell U., 1967. From asst. prof. to assoc. prof. U. B.C., Vancouver, 1967-75, prof., 1975—, dept. head, 1980-90, chair rsch. mgmt., 1979—. Contbr. 2 chpts. to books, more than 40 tech. reports, more than 80 refereed sci. publs. Mem. Internat. Soc. Soil Sci. (sec. 1976-78), Am. Soc. Soil, Can. Soc. Soil Sci. (pres. 1980-81), Pacific Regional Soc. Soil Sci. (founder 1979). Office: U BC, 2206 East Mall, Vancouver, BC Canada V6T 1Z4

LAW, CHUNG KING, aerospace engineering educator, researcher; b. Shanghai, People's Republic of China, Sept. 21, 1947; came to U.S., 1969; s. Chun and Chin (To) Lo; m. Helen Kwan-Mei Chen, Aug. 11, 1973; children: Jonathan, Jennifer, Jeffrey. BS, U. Alberta, Edmonton, Can., 1968; MS, U. Toronto, Can., 1970; PhD, U. Calif., San Diego, 1973. Assoc. sr. rsch. engr. GM Rsch. Lab., Warren, Mich., 1973-75; from assoc. prof. to prof. Northwestern U., Evanston, Ill., 1976-84; prof. U. Calif., Davis, 1984-88; rsch. staff Princeton (N.J.) U., 1975-76, prof., 1988—. Recipient Curtis W. McGraw award Am. Soc. Engring. Edn., 1984. Fellow AIAA, ASME; mem. Combustion Inst. (Silver medal 1990). Office: Princeton U Dept Mech/Aerospace Engring Princeton NJ 08544

LAWFORD, G. ROSS, research and development company executive; b. Toronto, Ont., Can., Feb. 27, 1941; s. Frederick Hugh and Edith Mildred Lawford; m. Beatrice Lawford, Sept. 24, 1966; children: Grant, Janine. BSc, U. Toronto, 1963, PhD, 1966. Gen. mgr., tech. dir. George Weston Ltd., Toronto, 1977-89; pres. Ortech Internat., Mississauga, Ont., 1990—; mem. adv. bd. Inst. for Biol. Scis., NRC, Ottawa, Ont., 1991—. Mem. Can. Rsch. Mgmt. Assn. Mem. United Ch. Canada. Office: Ortech Internat, 2395 Speakman Dr, Mississauga, ON Canada L5K 1B3

LAWLER, JAMES EDWARD, physics educator; b. St. Louis, June 29, 1951; s. James Austin and Delores Catherine L.; m. Katherine Ann Moffatt, July 21, 1973; children: Emily Christine, Katie Marie. BS in Physics summa cum laude, U. Mo., Rolla, 1973; MS in Physics, U. Wis., 1974, PhD in Physics, 1978. Rsch. assoc. Stanford (Calif.) U., 1978-80; asst. prof. U. Wis., Madison, 1980-85, assoc. prof., 1985-89, prof., 1989—; product devel. cons. Nat. Rsch. Group, Inc., Madison, 1977-78; cons. GE, Schnectady, N.Y., 1985—, Teltech, Inc., 1990—; mem. exec. com. Gaseous Electronics Conf., 1987-89, treas., 1992-94. Editor: (with R.S. Stewart) Optogalvanic Spectroscopy, 1991; contbr. articles to profl. jours. Schumberger scholar U. Mo., 1971-72; Grad. fellow Wis. Alumni Rsch. Found., 1973-74, NSF, 1974-76, H.I. Romnes Faculty fellow U. Wis., 1987. Fellow Am. Physical Soc. (DAMOP program com. 1993—, Will Allis prize 1992); mem. Optical Soc. Am., Sigma Xi. Achievements include patent for Echelle Sine Bar for dye laser cavity; development of laser diagnostics for glow discharge plasmas, of methods for measuring accurate atomic transition probabilities and radioactive lifetimes. Office: U Wis Dept of Physics 1150 University Ave Madison WI 53706

LAWLESS, JOHN JOSEPH, anatomy educator; b. St. Paul, Nov. 21, 1908; s. Patrick Joseph and Christina Marie (Grantz) L.; m. Louise Rebecca Gatewood, June 3, 1937; children: Carolyn Christina, Laura Louise, Joan Elizabeth. BSc, U. Minn., 1930, PhD, 1933; MD, U. Chgo., 1939. Fellow anatomy U. Minn. Med. Sch., Mpls., 1930-33; instr. anatomy Georgetown Med., Washington, 1933-34; asst. prof. anatomy W.Va. Med. Sch., Morgantown, 1934-40, U. Tex. Med. Branch, Galveston, 1940-42; intern Greenbrier Hosp., Ronceverte, W.Va., 1942-43; physician White Sulphur Springs, W.Va., 1943-44; dir. student health, asst. prof. medicine W.Va. U., 1944-71; adj. prof. anatomy U. Fla. Med. Sch., 1979—. Home: 1421 NW 30th St Gainesville FL 32605

LAWN, IAN DAVID, marine biologist; b. Great Yarmouth, United Kingdom, Oct. 20, 1947; arrived in Australia, 1981; s. Leslie George and Catherine Maude (Sadler) L.; m. Myriam Preker; children: Pippi, Torkel. BSc with honors, U. Nottingham, Eng., 1969; PhD, U. St. Andrews, Scotland, 1973. Rsch. fellow U. Alberta, Edmonton, Canada, 1973-75; rsch. fellow Bamfield Marine Station, B.C., Canada, 1976-79, asst. dir., 1980-81; dir. and reader Heron Island Rsch. Station, Queensland, Australia, 1981—. Contbr. articles to profl. jours. Chmn. Heron Island Mgmt. Com., 1991, 88, 85. Mem. Australian Coral Reef Soc. Avocations: photography, music, travel, natural history, bush walking. Home and Office: U Queensland Heron Island Rsch Sta, Gt Barrier Reef via Gladstone, Queensland 4680, Australia

LAWRANCE, CHARLES H., civil and sanitary engineer; b. Augusta, Maine, Dec. 25, 1920; s. Charles William and Lois Lyford (Holway) L.; m. Mary Jane Hungerford, Nov. 22, 1947; children: Kenneth A., Lois R., Robert J. BS in Pub. Health Engring., MIT, 1942; MPH, Yale U., 1952. Registered profl. engr., Calif. Sr. san. engr. Conn. State Dept. Health, Hartford, 1946-53; assoc. san. engr. Calif. Dept. Pub. Health, L.A., 1953-55; chief san. engr. Koebig & Koebig, Inc., Cons. Engrs., L.A., 1955-75; engr., mgr. Santa Barbara County Water Agy., Santa Barbara, Calif., 1975-79; prin. engr. James M. Montgomery Cons. Engrs., Pasadena, Calif., 1979-83; v.p. Lawrance, Fisk & McFarland, Inc., Santa Barbara, 1983—. Author: The Death of the Dam, 1972; co-author: Ocean Outfall Design, 1958; contbr. articles to profl. jours. Bd. dirs. Pacific Unitarian Ch., Palos Verdes Peninsula, Calif., 1956-60, chmn. bd. 1st lt. USMCR, 1942-46, PTO. Fellow ASCE (life; Norman medal 1966); mem. Am. Water Works Assn. (life), Am. Acad. Environ. Engrs. (life diplomate), Water Environment Fedn. Republican. Unitarian. Home: 1340 Kenwood Rd Santa Barbara CA 93109

LAWRENCE, CAROLYN MARIE, engineer; b. Waterbury, Conn., Jan. 27, 1963; d. Donald Pasquale and Jacqueline Mary (Asselin) Salcito; m. Victor Braswell Lawrence, Sept. 4, 1987; 1 child, Brittany Elizabeth. BS in Bioengring., Tex. A&M U., 1987. Simulation Engr. Lockheed Engring. & Scis. Co., Houston, 1988-90; biomed. engr. KRUG Life Scis., Houston, 1990-91; crew procedures engr. Martin Marietta Svcs. Co., Houston, 1991—. Mem. Palmer Meml. Ch. Choir, 1992—. Acad. scholar Houston Northwest Med. Ctr., 1981. Mem. AIAA, Aerospace Med. Assn., Assn. Profl. Engrs. Republican. Epsicopalian. Achievements include development of automated graphics package for generating analytical plots using batch file systems; research, procurement and manfest of a quick donning emergency breathing system for use aboard NASA's space shuttle. Office: Martin Marietta Svcs Co 1050 Bay Area Blvd Houston TX 77058

LAWRENCE, JORDAN, psychologist; b. N.Y.C., May 10, 1929; s. Benjamin and Frances (Peck) L.; m. Betty Esther, July 13, 1956; 1 child, Richard. PhD, Cath. U., Washington, 1965; MLA, Johns Hopkins U., 1975. Lic. psychologist, Md. Intern psychologist State Mental Health System, N.Y.C., 1954-55; rsch. psychologist NIMH Grant, Md., 1958-60; sch. psychologist Dept. Edn., Balt. County, 1960—; tchr. Towson State U., Balt., Cath. U., Washington, 1964-65, Notre Dame Coll., Balt., 1965-78; psychologist Children's Guild, Balt., 1966-76, Xcell Drug Rehab., Balt., 1986-91; cons. psychologist, Balt., Washington, 1960—; lectr. assorted groups, Md. Designer numerous instrnl. formats for psychologists involving emotional and behavioral disorders in children. Mem. APA, N.Y. Acad. Scis. Achievements include research in developmental, emotional disorders, learning problems in children and adolescents, drug use in young adults.

LAWRENCE, MARGERY H(ULINGS), utilities executive; b. Harmarville, Pa., June 17, 1934; d. Richard Nuttall and Alva (Burns) Hulings; student Bethany Coll., 1951-52; B.S. in Mktg., Carnegie-Mellon U., 1955. Asst. mdse. buyer Joseph Horne Co., Pitts., 1955-57; home econs. editor Pitts. Group Cos. Columbia Gas System, Pitts., 1957-64, dir. home econs., 1968-72; home economist Columbia Gas Pa., Jeannette, 1964-68, dist. marketing mgr., 1972-87, div. mgr., 1987-91; dir. mktg. Columbia Gas Pa. & Columbia Gas Md., 1991—. Bd. dirs., treas. Ohio Valley Gen. Hosp. Mem. DAR, AmGas Assn. (Home Svc. Achievement award 1964), Mfrs. Assn. of Tri-Counties (pres., bd. dirs.), Pa. Economy League (bd. dirs. Beaver County chpt.), Beaver County C. of C. (bd. dirs.). Office: Columbia Gas Pa Inc 650 Washington Rd Pittsburgh PA 15228

LAWRENCE, MARTIN WILLIAM, physicist; b. Sydney, N.S.W., Australia, Nov. 3, 1943; s. James Joscelyn and Iris (Carder) L.; m. Claire Jeannette O'Loughlin, July 5, 1966; children: Jessica Jane, Bronwyn Claire. BSc in Physics, U. Sydney, 1965; MSc, U. N.S.W., 1970, PhD, 1974. Rsch. physicist AWA Rsch. Lab., Sydney, 1965-74, sr. rsch. physicist, 1975-78; rsch. scientist Def. Sci. and Tech. Orgn., Sydney, 1978-79, sr. rsch. scientist, 1979-89, prin. rsch. scientist, 1989—; mem. staff Woods Hole (Mass.) Oceanographic Instn., 1983-84. Contbr. articles to profl. publs., chpts. to books. Recipient medal Marconi Found., 1976. Fellow Australian Inst. Physics; mem. Acoustical Soc. Am., Australian Acoustical Soc., Am. Geophys. Union. Home: 29 Innes Rd, Greenwich NSW 2065, Australia Office: Def Sci and Tech Orgn, Wharf 17 Jones Bay Rd, Pyrmont NSW 2009, Australia

LAWRENCE, RAYMOND JEFFERY, physicist; b. Cornwall, N.Y., Feb. 25, 1939; s. Raymond J. and Hazel Lillian (Hamler) L.; m. Jane MacLean Whipple, June 29, 1963; children; Janet Beth Lawrence Sanchez, David Jeffery. BA, Lawrence Coll., 1961; MS, U. N.Mex., 1970. Rsch. physicist Air Force Weapons Lab., Albuquerque, 1963-67; staff mem. Sandia Nat. Labs., Albuquerque, 1967—. Contbr. articles and reports to profl. publs. Capt. USAF, 1963-67. Mem. Am. Phys. Soc. Republican. Presbyterian. Home: 1308 Kirby St NE Albuquerque NM 87112 Office: Sandia Nat Labs PO Box 5800 Albuquerque NM 87185

LAWRENCE, ROBERT EDWARD, electrical engineer; b. Boston, May 29, 1946; s. Jules P. and Gertrude (Lander) L.; m. Marjorie Alberta Holman; 1 child, Andrew Jon. BS, Rensselaer Poly. Inst., 1968, MS, 1969, PhD, 1972. Mem. tech. staff Bell Tel. Labs., Whippany, N.J., 1972-74; engr. Vitro Labs., Silver Spring, Md., 1974-77, sr. staff engr., 1978-80; asso. Booz Allen & Hamilton, Bethesda, Md., 1977-78; dir. Litton Amecom, College Park, Md., 1980-86; consulting engr. The MITRE Corp., McLean, Va., 1986—. Founder, bd. dirs. Vitro Fed. Credit Union, 1970; mem. Prince George's County Econ. Devel. Corp., 1985-89; cons. U.S. Army Sci. Bd., 1991; tech. chmn. Milcom 1991, bd. dirs., 1993. NSF fellow, 1970. Mem. IEEE (sr.), Nat. Security Indsl. Assn., Armed Forces Communications Electronics Assn. (past chpt. v.p., past chpt. treas., bd. dirs. 1993), Assn. Old Crows, Sigma Xi, Tau Beta Pi, Eta Kappa Nu. Contbr. articles to profl. jours. Home: 9011 Copenhaver Dr Potomac MD 20854-3012

LAWRENCE, ROBERT MICHAEL, research chemist; b. Windsor, N.C., July 6, 1961; s. George Robert and Gerti Gizelle Lawrence; m. Laura Jane Kirby, Aug. 20, 1988. BS in Chemistry, U. N.C., 1983; PhD in Organic Chemistry, U. Tex., 1988. Postdoctoral rsch. assoc. U. Utah, Salt Lake City, 1988-90; rsch. chemist Bristol-Myers Squibb, Princeton, N.J., 1990—. Contbr. articles to profl. jours. Recipient Grad. fellowship U. Tex., 1987. Mem. AAAS, Am. Chem. Soc. Home: 48 W Crown Ter Yardley PA 19067-7346 Office: Bristol Myers Squibb H2806A PO Box 4000 Princeton NJ 08543-4000

LAWRENCE, RODERICK JOHN, architect, social science educator, researcher, consultant; b. Adelaide, Australia, Aug. 30, 1949; s. Keith and Babette Naomi (Radford) L.; m. Clarisse Christine Gonet, Sept. 30, 1977; children: Xavier Gerard, Adrien Keith, Kevin John. BS with first class hons., Adelaide U., Australia, 1972; MS, Cambridge U., Eng., 1977; PhD, Ecole Poly., Lausanne, Switzerland, 1983. Architect Edwards, Madigan and Torzillo, Sydney, Australia, 1972-74, S. Australian Housing Trust, Adelaide, 1973-76; rsch. fellow St. John's Coll., Cambridge U., Eng., 1975-77; asst. prof. Ecole Poly. Fed., Lausanne, Switzerland, 1978-84; cons. Econ. Commn. Europe, Geneva, 1984—; master tchr. and rsch. U. Geneva, 1984—; vis. prof. U. Que., Montreal, Can., 1987; vis. fellow Flinders U., Adelaide, 1985; mem. editorial bd. Open House Internat., 1986; speaker, guest lectr. various European and Australian univs. Author: Le Seuil Franchi..., 1986, Housing, Dwellings, and Homes, 1987; contbr. articles to profl. jours. and chpts. to books; mem. editorial bd. Architecture and Behavior, 1980, Open House Internat., 1986; guest editor to jours. Nat. Sci Found. Switzerland fellow, 1984. Mem. Internat. Assn. Study of People and their Phys. Surroundings (bd. dirs. 1986—), Environ. Design Rsch. Assn., People and Phys. Environ. Rsch. Soc., Open House Internat. Assn., Internat. Sociological Assn. (regional editor for newsletter 1988—). Avocations: photography, bushwalking, skiing. Office: Univ of Geneva, 102 Blvd Carl Vogt, 1211 Geneva 4, Switzerland

LAWSER, JOHN JUTTEN, electrical engineer; b. Bryn Mawr, Pa., July 4, 1941; s. John Jacob and Audrey (Jutten) L.; m. Sarah Anne Hardin, June 12, 1965; children: Susan, Amy. MSE, U. Mich., 1964, PhD in Elec. Engring., 1970. Mem. staff AT&T Bell Labs, Holmdel, N.J., 1970-78; supr. switch planning AT&T Bell Labs, Holmdel, 1979-83, supr. signaling dept., 1983-91; svc. devel. mgr. AT&T Network Svcs. Div., Holmdel, 1991—; mgr. capital budgeting AT&T Gen. Depts., Basking Ridge, N.J., 1978-79. Author 20 papers on common channel signaling, intelligent network architecture, new telecommunications svcs. and fault tolerant networks. Commodore Monmouth Boat Club, Red Bank, N.J., 1990-92; pres. U. S. Albacore Assn., 1982-84; elder Lincroft (N.J.) Presbyn. Ch., 1974-77, 89-92. Bell Labs. fellow, 1991. Mem. IEEE (sr. mem.), Sigma Xi, Tau Beta Pi, Eta Kappa Nu. Office: AT&T Bell Labs Crawford Corners Rd Holmdel NJ 07733

LAWSON, LARRY DALE, environmental consultant; b. Galesburg, Ill., Jan. 10, 1954; s. Kenneth Dale and Darlene June (Peterson) L. BA in Math., BA in Chemistry, Ark. Coll., 1975; MS in Inorganic Chemistry, Iowa State U., 1978. Grad. teaching asst. Iowa State U., Ames, 1975-76, grad. rsch. asst., 1976-78; mem. staff Lawson Electric Co., Oneida, Ill., 1978-80; pvt. practice environ. cons. to mcpl. utilities Oneida, 1980—. Elder Oneida Presbyn. Ch.; active Wastewater Bd. Certification, 1993—. Mem. Heart Ill. Water Pollution Control Operators (pres. 1991), Ill. Water Environ. Fedn. (Ops. award 1985-86), Am. Chem. Soc., Am. Water Works Assn., Water Environ. Fedn., Ill. Assn. Water Pollution Control Operators. Home and Office: 208 S Sage PO Box 7 Oneida IL 61467

LAWSON, MERLIN PAUL, dean, climatologist; b. Jamestown, N.Y., Jan. 12, 1941; s. Merle Nelson and Cecile May (Post) L.; m. Nina Louise Rising, Jan. 25, 1964; children: Keith, Kenneth, Kristin. BA, SUNY, Buffalo, 1963; MA, Clark U., 1966, PhD, 1973. Prof., chair dept. geography U. Nebr., Lincoln, 1980-87; asst. dean grad. studies U. Nebr., 1986, assoc. dean, asst. vice chancellor, 1986-92, dean grad. studies, 1992—. Author: Reconstruction of Climate of the Western Interior United States, 1974; editor: Images of the Plains, 1975; contbr. articles on climatology to profl. jours. Mem. AAAS, Assn. Am. Geographers (nat. program chair 1983, nat. exec. coun. 1984-87), Am. Meteorol. Soc., Sigma Xi. Achievements include development of seasonal temperature forecasts as products of antecedent temperature arrays; of contingency analysis and seasonal forecast skill. Home: Rt 9 Lincoln NE 68506 Office: U Nebr Lincoln 301 Administration Lincoln NE 68588-0434

LAWSON, ROBERT DAVIS, theoretical nuclear physicist; b. Sydney, Australia, July 14, 1926; came to U.S., 1949; s. Carl Herman and Angeline Elizabeth (Davis) L.; m. Mary Grace Lunn, Dec. 16, 1950 (div. 1976); children—Dorothy, Katherine, Victoria; m. Sarah Virginia Roney, Mar. 13, 1976. B.S., U. B.C., Can., 1948; M.S., U. B.C., 1949; Ph.D., Stanford U., 1953. Research assoc. U. Calif., Berkeley, 1953-57; research assoc. Fermi Inst. U. Chgo., 1957-59; assoc. physicist Argonne Nat. Lab., Ill., 1959-65; sr. physicist Argonne Nat. Lab., 1965—; vis. scientist U.K. Atomic Energy Authority, Harwell, Eng., 1962-63, Oxford U., Eng., 1970, 85; vis. prof. SUNY, Stony Brook, 1972-73; vis. fellow Australian Nat. U., Canberra, 1982; vis. prof. U. Groningen, 1973, U. Utrecht, 1974, Technische Hochschule, Darmstadt, 1975, 78, Free U., Amsterdam, 1976, 81, others; TRIUMF, U. B.C., Vancouver, Can., 1984. Author: Theory of the Nuclear Shell Model, 1980. Contbr. articles to sci. jours. Fellow Weizmann Inst. Sci., 1967-68, Niels Bohr Inst., 1976-77; Sir Thomas Lyle fellow U. Melbourne, Australia, 1987. Fellow Am. Phys. Soc. Home: 1590 Raven Hill Dr Wheaton IL 60187 Office: Argonne Nat Lab Bldg 314 Argonne IL 60439

LAWSON, WILLIAM BRADFORD, psychiatrist; b. Richmond, Va., Nov. 27, 1945; s. Thomas Henry and Violet Serena (Roane) L.; m. Rosemary Jackson, Aug. 6, 1983; children: Robert, Anthony. BS, Howard U., 1966; MA, U. Va., 1968; PhD, U. N.H., 1971; MD, U. Chgo., 1978. Diplomate Am. Bd. Psychiatry and Neurology. Asst. prof. dept. psychology U. Ill., Urbana, 1971-74; intern, resident Stanford (Calif.) U. Med. ctr., 1979-82; clin. rsch. fellow NIMH, Washington, 1981-84; asst. prof. U. Calif., Irvine, 1984-86, Vanderbilt U. Med. Ctr., Nashville, 1986-91; assoc. prof. U. Ark. Med. Sch., Little Rock, 1991—; chief chronic mentally ill, psychiat. svc. McClellan VA Med. Ctr., Little Rock; dir. rsch. Met. State Hosp., Norwalk, Calif., 1984-86; cons. Meharry Med. Coll., Nashville, 1988-91; chief med. officer Tenn. Mental Health and Mental Retardation, 1988-91. Contbr. chpts. to books, articles to profl. publs. Bd. dirs. Ctr. Living and Learning, Nashville, 1991. Mem. ACLU (bd. dirs. 1990), NAACP (bd. dirs. 1991), Am. Psychiat. Assn. (rsch. coun., com. on under-represented minorities), Collegium Internat. Neuropsychopharmacologium (pres.-elect), Black Psychiatrists Am. (editor quar.), Omega Psi Phi (Gold medal 1992, 93). Democrat. Baptist. Home: 7 No 6 Hill Cove Little Rock AR 72205 Office: VA Med Ctr 116A NLR 2200 Fort Roots Dr North Little Rock AR 72114-1706

LAX, PETER DAVID, mathematician; b. Budapest, Hungary, May 1, 1926; came to U.S., 1941, naturalized, 1944; s. Henry and Klara (Kornfeld) L.; m. Anneli Cahn, 1948; children: John, James D. BA, N.Y. U., 1947, PhD, 1949; DSc (hon.), Kent State U., 1976; DSC (hon.), Brown U., 1993; Dr. honoris causa, U. Paris, 1979; D. Natural Scis. (hon.), Technische Hochschule Aachen, Germany, 1988; DSc (hon.), Herriot Walt U., 1990; D. (hon.), Leningrad State U., 1991, U. Md. Baltimore County, 1993; PhD (hon.), Tel Aviv U., 1992, Beijing U. Asst. prof. NYU, 1949-57, prof., 1957—; dir. Courant Inst. Math. Scis., 1972-80. Author: (with Ralph Phillips) Scattering Theory, 1967, Scattering Theory for Automorphic Functions, 1976, (with A. Lax and S.Z. Burstein) Calculus with applications and computing, 1976, Hyperbolic Systems of Conservation Laws and the Mathematical Theory of Shock Waves, 1973. Mem. Pres.'s Com. on Nat. Medal of Sci., 1976, Nat. Sci. Bd., 1980-86. Served with AUS, 1944-46. Recipient Semmelweis medal Semmelweis Med. Soc., 1975, Nat. Medal Sci., 1986, Wolf Prize, 1987. Mem. AAAS, Am. Math. Soc. (pres. 1979-80, Norbert Wiener prize, Leroy P. Steele prize 1993), Nat. Acad. Scis. (applied math. and numerical analysis award 1983), Math. Assn. Am. (bd. govs., Chauvenet prize), Soc. Indsl. and Applied Math., Académie des Scis. (fgn. assoc.), Russian Acad. Sci. (fgn. assoc.), Academia Sinica (hon.), Hungarian Acad. Sci. (hon.). Office: Courant Inst Math Scis Rm 912 251 Mercer St New York NY 10012-1185

LAY, PETER ANDREW, chemistry educator; b. Korumburra, Victoria, Australia, Aug. 27, 1955; s. Esbert William and Dulcie Joan (Hopwood) L.; m. Jill Lesley Gordon, Apr. 24, 1983; children: Carly Marie, Robert Michael, Timothy Dylan. BSc, U. Melbourne, 1977; PhD, Australian Nat. U., 1981. CSIRO postdoctoral fellow Stanford (Calif.) U., 1981-83, Div. of Applied Organic Chemistry, CSIRO, Melbourne, 1983-84; Queen Elizabeth II fellow Deakin U., Geelong, Australia, 1984-85; vis. fellow Australian Nat. U., Canberra, Australia, 1982-85; lectr. U. Sydney, 1985-88, sr. lectr., 1989—; guest prof. U. Berne, Switzerland, 1991; chair electochem. subdiv. plan com. Internat. Coordination Chem. Conf., Gold Coast, Australia, 1986-89; com. mem. Standards Australia, Sydney, 1987—. Author: (with others) Advances in Inorganic Chemistry, 1991; contbr. articles to profl. jours. Recipient Edgeworth David medal Royal Soc. of NSW, 1988; grantee Australian Rsch. Coun., 1986—, Cancer Rsch. fund, 1990—, Australian Inst. of Nuclear Sci. and Engring., 1988, 91—. Fellow Royal Australian Chem. Inst.; mem. Am. Chem. Soc., Soc. of Electroanalytical Chemistry, Royal Australian Chem. Inst. (Bloom-Gutmann Prize 1984, Rennie medal 1987), Soc. of Electroanalytical Chemistry. Achievements include patents for new materials and catalytic destruction of intractable organochlorine wastes; rsch. in conceptual advances in inorganic reaction mechanisms, devel. of reagents that prevent mutations induced by industrially important chromium chemicals; synthesis and characterization of structural models of vanadium proteins and nitrogenase; developments in microelectrode electrochemistry technology. Office: U Sydney, Dept Inorganic Chemistry, Sydney 2006, Australia

LAY, THORNE, geosciences educator; b. Casper, Wyo., Apr. 20, 1956; s. Johnny Gordon and Virginia Florence (Lee) L. BS, U. Rochester, 1978; MS, Calif. Inst. Tech., 1980, PhD, 1983. Rsch. assoc. Calif. Inst. Tech., Pasadena, 1983; asst. prof. geosciences U. Mich., Ann Arbor, 1984-88, assoc. prof., 1988-89; prof. U. Calif., Santa Cruz, 1989—; cons. Woodward Clyde Cons., Pasadena, 1982-84; dir. Inst. Tectonics, 1990—. Contbr. articles to scientific jours. NSF fellow, 1978-81, Guttenberg fellow Calif. Inst. Tech., 1978, Lilly fellow Eli Lilly Found., 1984, Sloan fellow, 1985-87, Presidential Young Investigator, 1985-90. Fellow Royal Astron. Soc., Am. Geophys. Union (Macelwane medal 1991), Soc. Exploration Geophysicist, Seismol. Soc. Am., AAAS. Home: 248 Meadow Rd Santa Cruz CA 95060-2040 Office: U Calif Santa Cruz Earth Sci Bd Santa Cruz CA 95064

LAYCOCK, ANITA SIMON, psychotherapist; b. Cheyenne, Wyo., Dec. 17, 1940; d. James Robert and Dorothy (Dearmin) Simon; m. Maurice Percy Laycock, June 18, 1965(dec. 1976); 1 child, (dec.). BA, U. Wyo., 1962, MA, 1971. Lic. counselor, Wyo., nationally cert. addiction specialist. Grad. student counselor, psychometrist Wyo. State Prison, Rawlins, 1971-73; counselor, trainer Dept. of Insts. State of Colo., Denver, 1973-75; counselor, tchr. supr. Jefferson County Evaluation-Diagnostic Ctr., Rawlins, 1975-78; psychometrist Wyo. State Penitentiary, Rawlins, 1978-79; counselor, therapist Rocky Mountain Arts and Scis., Cheyenne, 1979-81; counselor, therapist supr., dir. SWARA, Rock Springs, Wyo., 1981-85; therapy dir. St. Joseph Residential Treatment, Torrington, Wyo., 1985-88; dir. psychiatric unit Nat. Med. Enterprises Hill-Haven-Pk. Manor, Rawlins, 1988-89; chief exec. officer Simon-Laycock & Assocs., Rawlins, 1989—; cons. Kids in Distressed Situations, Rawlins, 1990-91, Child Devel. Ctr., Rawlins, 1991—; dir. Pub. Offender and Forensic Mental Health Program, Rawlins, 1988-91. Author: (programs) related to sex offenders. Pres. Cheyenne City Panhellenic, 1965-68. Named Miss Wyo.-Miss Universe, 1960; named Miss Wool of Wyo., 1965. Mem. ACA, Nat. Sex Offenders Counselors, Nat. Assn. Drug and Alcohol Counselors, Pub. Offenders Counselors Assn., Western Corrections Assn., Wyo. Assn. Addiction Specialists (pres. 1988—). Avocations: profl. animal trainer, artist. Home: PO Box 3027 Cheyenne WY 82009 Office: Simon Laycock & Assocs 1716 Yellowstone hwy Cheyenne WY 82009

LAYZER, DAVID, astrophysicist, educator; b. Ohio, Dec. 31, 1925; married; 6 children. A.B., Harvard U., 1947, Ph.D. in Theoretical Astrophysics, 1950. NRC fellow, 1950-51; lectr. astronomy U. Calif., Berkeley, 1951-52; research assoc. physics Princeton (N.J.) U., 1952-53; research assoc. Harvard U., Cambridge, Mass., 1953-55, research fellow, lectr., 1955-60, prof. astronomy, 1960-80, Donald H. Menzel prof. astrophysics, 1980—. Office: Harvard U Ctr for Astrophysics 60 Garden St Stop 31 Cambridge MA 02138-1596

LAZAR, ANNA, chemist; b. Budapest, Hungary, Jan. 10, 1931; came to U.S. 1956; d. Lajos and Maria (Grits) Varga; m. Jospeh Lazar, Apr. 11, 1955 (div. 1969); 1 child, Julie Anna. Diploma, Eötvös Lorand Sci. U., Budapest, 1955. Chemist Harvard U., Boston, 1957-59, Arthur D. Little, CAmbridge, Mass., 1959-61, Wilkens Instrument and Rsch., Lafayette, Calif., 1961-62, Stanford (Calif.) U., 1962-63, Hercules Inc., Wilmington, Del., 1964-69, Cancer Rsch. Inst., Fox Chase Phila., Pa., 1969-70, Food and Drug Administrn., Phila., 1970--. Roman Catholic. Home: 428 W Montgomery Ave Haverford PA 19041

LAZAR, JUDITH TOCKMAN, pharmaceutical company researcher; b. Denver, Sept. 10, 1947; d. Abe and Terry (Grass) Tockman; m. Robert J. Lazar, Dec. 30, 1979 (div. Oct., 1986). BA, U. Colo., 1969, MA, 1971, PhD, 1975. Postdoctoral rsch. fellow Scripps Clinic and Rsch. Found., La Jolla, Calif., 1975-76, Salk Inst., La Jolla, 1977; assoc. dir. rsch. Nutrition 21, La Jolla, 1978; info. rsch. scientist Abbott Labs., North Chgo., Ill., 1978-83; sr. clin. rsch. assoc. Abbott Internat. Ltd., North Chgo., 1983-84, clin. project mgr., 1985-86; asst. dir. sci. affairs and safety assurance Pfizer Internat. Inc., N.Y.C., 1986-89; assoc. dir. clin. rsch. Vestar, Inc., San Dimas, Calif., 1989-91, Liposome Tech., Inc., Menlo Park, Calif., 1991-93; dir. internat. drug devel. Liposome Co., Inc., Princeton, N.J., 1993—. Contbr. articles to sci. jours. Mem. Am. Soc. Microbiology, So. Calif. Skeptics, L.A. Sci. Fantasy Soc., Gustav Mahler Soc., Mensa, Phi Beta Kappa, Sigma Xi. Democrat. Jewish. Avocations: singing, music, playing percussion, sci. fiction, walking. Home: 615 11th St Brooklyn NY 11215-5202 Office: Pfizer Internat Inc 235 E 42nd St New York NY 10017-5703

LAZAR, MACIEJ ALAN, electronics executive, engineer; b. Cieszyn, Poland, Aug. 14, 1957; s. Gustaw and Janina (Michalak) L.; m. Magdalena Helena Newald, Apr. 24, 1982; children: Katarzyna, Witold. MSEE, Acad. Mining, Kracow, Poland, 1980. Design engr. Apena, Bielsko-Biala, Poland, 1981-87; dir. engring. Elpa, Rybarzowice, Poland, 1987-89; pres. Analog, Bielsko-Biala, 1989—. Contbr. articles to profl. jours.; patentee motor control system. Mem. Assn. Polish ELectricians. Avocations: skiing, tennis. Home: Miarki 15/7, 43-300 Bielsko-Biala Poland Office: Analog, Legionow 81, 43-300 Bielsko-Biala Poland

LAZAR, RANDE HARRIS, otolaryngologist; b. N.Y.C., Feb. 27, 1951; s. Irving and Dorothy (Tartasky) L.; m. Linda Zishuk, Aug. 11, 1974; 1 child, Lauren K. BA, Bklyn. Coll., 1973; MD, U. Autonoma de Guadalajara, Mexico, 1978; postgrad., N.Y. Med. Coll., 1978-79. Diplomate Am. Bd. Otolaryngology-Head and Neck Surgery; lic. physician, N.Y., Ohio, Tenn. Gen. surgery resident Cornell-North Shore Community Hosp., Manhasset, N.Y., 1979-80; gen. surgery resident Cleve. Clinic Found., 1980-81, otolaryngology-head and neck surgery resident, 1980-84, chief resident dept. otolaryngology & communicative disorder, 1984-84; physician Otolaryngology Cons. Memphis, 1984—; fellow pathology head and neck dept. otolaryngologic pathology Armed Forces Inst. Pathology, Washington, 1983, pediatric otolaryngology fellow LeBonheur Children's Med. Ctr., Memphis, 1984-85, 89—, chief surgery, 1989, chief staff East Surgery Ctr.; chmn. dept. otolaryngology head and neck surgery Meth. Health Systems, 1990-91; courtesy staff Bapt. Meml. Hosp., Bapt. Meml. Hosp.-East, Eastwood Med. Ctr., Meth. Hosp., Germantown, Tenn.; chief dept. otolaryngology Les Passees Rehab. Ctr., 1988—. Contbr. articles to profl. jours. Bd. dirs. Bklyn. Tech. Found. Recipient award of honor Am. Acad. Otolaryngology-Head and Neck Surgery, 1991. Fellow Internat. Coll. Surgeons; mem. AMA, Am. Acad. Otolaryngology-Head and Neck Surgery, Am. Acad. Facial Plastic and Reconstructive Surgery, Am. Acad. Otolaryngic Allergy, Centurions Deafness Rsch. Found., Am. Auditory Soc., Nat. Hearing Assn., Soc. Ear, Nose Throat Advances in Children, Am. Soc. Laser Medicine and Surgery, So. Med. Assn., N.Y. Acad. Scis., Tenn. Med. Soc., Tenn. Acad. Otolaryngology-Head and Neck Surgery, Memphis and Shelby County Med. Soc., Memphis/Mid South Soc. Pediatrics. Office: Otolaryngology Cons Memphis 777 Washington Ave Ste 240P Memphis TN 38105-4526

LAZAROFF, NORMAN, microbiologist, researcher; b. N.Y.C., Nov. 24, 1927; s. Samuel and Mollie (Cohen) L.; m. Sandra Mirel Nord, Aug. 24, 1958; children: Alan, Deborah. AB in Chemistry, Syracuse U., 1950, MS in Microbiology, 1952; PhD in Microbiology, Yale U., 1961. Biochemist Rsch. Found. SUNY, Bklyn., 1955; bacteriologist Schwarz Labs., Mt. Vernon, N.Y., 1955-56; project leader, cons. Evans R&D Corp., N.Y.C., 1956-59; microbiologist B.C. Rsch. Coun., Vancouver, Can., 1961-62; asst. prof. biology U. So. Calif., L.A., 1962-64; sr. rsch. scientist Syracuse (N.Y.) U. Rsch. Corp., 1964-66; assoc. prof. biol. scis. SUNY, Binghamton, 1966-90, assoc. prof. emeritus, 1990—; rsch. prof., founder pvt. lab. Micronostix, Vestal, N.Y., 1992—; cons IBM Corp., Endicott, N.Y., 1972-74, EG&G, Idaho Falls, Idaho, 1989—; mem. coun. Onondaga County Sci. Coun., Syracuse, 1965-66; initiator, coord. high sch. program Microbe Hunters, SUNY-Binghamton, 1988-89. Contbr. articles to profl. publs. including Jour. Gen. Microbiology, Jour. Bacteriology. With U.S. Army, 1946-47. NSF fellow, 1952-54; grantee NIH, 1963-66, 68-71, 63-65, Dept. of Energy, 1977-78. Mem. Am. Soc. Microbiology (sec.-treas. cen. N.Y. br. 1971-73); Am. Chem. Soc., Phycol. Soc. Am. Jewish. Achievements include co-invention and patent for photoinduced cyanobacterial screening for anti-adhesin antibiotics, discovery of photo-induction, photo-reversal of development in filamentous cyanobacteria, sulfate requirement for chemolithotrophic iron oxidation. Office: Micronostix 312 Front St Vestal NY 13850

LAZARUS, STEVEN S., management consultant, marketing consultant; b. Rochester, N.Y., June 16, 1943; s. Alfred and Ceal H. Lazarus; m. Elissa C. Lazarus, June 19, 1966; children: Michael, Stuart, Jean. BS, Cornell U., 1966; MS, Poly. U. N.Y., 1967; PhD, U. Rochester, 1974. Pres. Mgmt. Systems Analysis Corp., Denver, 1977—; dir. Sci. Application Intern Corp., Englewood, Colo., 1979-84; assoc. prof. Metro State Coll., Denver, 1983-84; sr. v.p. Pal Assocs. Inc., Denver, 1984-85; with strategic planning and mktg. McDonnell Douglas, Denver, 1985-86; mktg. cons. Clin. Reference Systems, Denver, 1986—; assoc. exec. dir. Ctr. Rsch. Ambulatory Health Care Adminstrn., Englewood; spl. cons. State of Colo., Denver, 1976-81; mktg. cons. IMX, Louisville, Ky., 1986-87; speaker Am. Hosp. Assn., Chgo., 1983—; speaker Med. Group Mgmt. Assn., 1975—. Author chpts. to books; patentee med. quality assurance. NDEA fellow U. Rochester, 1968-71. Mem. Inst. Indsl. Engring. (sr.), Med. Group Mgmt. Assn., Operations Research Soc. Am. Lodge: Optimists (program chmn. Denver club 1976-78). Home: 7023 E Eastman Ave Denver CO 80224-2845 Office: Ctr for Rsch in Ambulatory Hlth 104 Inverness Terr E Englewood CO 80112*

LAZICH, DANIEL, aerospace engineer; b. Galjipovci, Yugoslavia, Jan. 1, 1941; came to U.S., 1963; s. Stojan and Ljubica Lazic; m. Spomenka Krkljus, Aug. 11, 1968. BS in Engring., U. Ill., 1974; postgrad., U. Tex., Arlington, 1976-78. Analytical engr. Pratt & Whitney Aircraft, West Hartford, Conn., 1974-75; aircraft structures engr. Gen. Dynamics Corp., Ft. Worth, 1975-78; aerospace engr. Shrike missile Air Systems Commn. USN, Arlington, Va., 1978-81; sr. propulsion engr. Joint Cruise Missiles Project, Dept. Def., Arlington, 1981-85; prin. staff engr., tech. advisor kinetic energy weapons Strategic Def. Commn., Arlington, 1985—. Contbr. articles to profl. jours. Sgt. U.S. Army, 1966-68. Mem. AIAA, Aircraft Owners and Pilots Assn., No. Va. Astronomy Club. Avocations: flying, photography. Home: 739 19th St S Arlington VA 22202-2704

LE, KHANH TUONG, utility executive; b. Saigon, Vietnam, Feb. 25, 1936; parents Huy Bich and Thi Hop; m. Thi Thi Nguyen, Apr. 22, 1961; children: Tuong-Khanh, Tuong-Vi, Khang, Tuong-Van. BS in Mech. Engring., U. Montreal, 1960, MS in Mech. Engring., 1961. Cert. profl. engr. Project mgr. Saigon Met. Water Project Ministry Pub. Works, Saigon, 1961-66; dep. dir. gen. Cen. Logistics Agy. Prime Min. Office, Saigon, 1966-70; asst. dir., chief auditor Nat. Water Supply Agy. Min. Pub. Works, Saigon, 1970-75; mgr. Willows Water Dist., Englewood, Colo., 1975—; dean sch. mgmt. scis., asst. chancellor acad. affairs Hoa-Hao U., Long-Xuyen, Vietnam, 1973-75; bd. dirs. Asian Pacific Devel. Ctr.; adv. bd. Arapahoe County Utility Douglas County Water Resources Authority. Treas. Met. Denver Water Authority, 1989-92; mem. Arapahoe County Adv. Bd., Douglas County Water Resources Authority, 1993—. Recipient Merit medal Pres. Republic Vietnam, 1966, Pub. Health Svc. medal, 1970, Svc. award Asian Edn. Adv. Coun., 1989; named to Top Ten Pub. Works Leaders in Colo., Am. Pub. Works Assn., 1990. Mem. Am. Water Works Assn., Vietnamese Profl. Engrs. Soc. (founder), Amnesty Internat. Buddhist. Avocations: reading, swimming, tennis, hiking. Office: Willows Water Dist 6970 S Holly Cir Ste 200 Englewood CO 80112-1066

LE, QUANG NAM, engineering researcher; b. Quangngai, Vietnam, Apr. 11, 1953; s. Le Quang Mai and Nguyen Thi Tien; m. Le Nguyen Thi Ngoc Quynh, Nov. 22, 1978; children: Le Quynh Dung, Le Quynh Anh. MSc, Leningrad (Russia) U., 1977; PhD, U. Paris 7, 1992. Rschr. Centre of Solar Energy, Hochiminh City, Vietnam, 1977-83, Inst. Physics, Hochiminh City, Vietnam, 1984-88, CNRS-LPSB, Meudon, France, 1989-92; engring. rschr. Photowatt Internat. S.A., Bourgoin-Jallieu, France, 1992—. Contbr. articles to Internat. Jour. Solar Energy, Jour. Phys. III France; contbr. articles to confs. Office: Photowatt Internat S AZA Champfleuri, 33 Rue St-Honoré, 38300 Bourgoin Jallieu, France

LEACH, ROBERT ELLIS, physician, educator; b. Sanford, Maine, Nov. 25, 1931; s. Ellis and Estella (Tucker) L.; m. Laurine Seber, Aug. 20, 1955; children: Cathy, Brian, Michael, Craig, Karen, Diane. A.B., Princeton U., 1953; M.D., Columbia U., 1957. Resident orthopedic surgery U. Minn., 1957-62; orthopedic surgeon Lahey Clinic, Boston, 1964-68; chmn. dept. Lahey Clinic, 1968-70; prof., chmn. dept. Boston U. Med. Sch., 1970—; lectr. Tufts U. Med. Sch., Medford, Mass., 1971—; head physician U.S. Olympic Team, 1984; chmn. sports medicine council U.S. Olympic Com., 1984—. Editor-in-chief Am. J. Sports Med.; contbr. articles to profl. jours. Served to lt. comdr. USNR, 1962-64. Am., Brit., Canadian Orthopedic Travelling fellow, 1971; Sports Medicine Man of the Yr., 1988. Mem. Am. Acad. Orthopedic Surgeons, Continental Orthopedic Soc. (sec. 1966—), Am. Orthopedic Assn. (pres. elect 1993), Am. Orthopedic Soc. Sports Medicine (pres. 1983). Club: Longwood Cricket (Brookline). Home: 40 Rockport Rd Weston MA 02193-1428 Office: 75 East Newton St Boston MA 02118

LEADBETTER, MARK RENTON, JR., orthopaedic surgeon; b. Phila., Nov. 7, 1944; s. Mark Renton and Ruth (Protzeller) L.; m. Letitia Ashby, July 28, 1973 (div. June 1990); m. Jan Saker, 1991. BA, Gettysburg Coll., 1967; MSc in Hygiene, U. Pitts., 1970; MD, Temple U., 1974. Surg. intern Univ. Hosps., Boston, 1974-75, resident in surgery, 1975-76; emergency room physician Sturdy Meml. Hosp., Attleboro, Mass., 1976-78; resident in orthopaedics U. Pitts., 1978-81; orthopaedic physician Rockingham Meml. Hosp., Harrisonburg, Va., 1981-82, courtesy staff, 1982—; pvt. practice, Staunton, Va., 1982—; mem. active staff King's Daus. Hosp., Staunton, 1982—; active staff Samaritan Hosp., Moses Lake, Wash.; courtesy staff Columbia Basin Hosp., Ephrata, Wash. Contbr. articles to med. jours.; patentee safety syringes, safety cannulas. Mem. Am. Coll. Sports Medicine, So. Med. Assn., So. Orthopaedic Assn., County Med. Soc., Nat. Futures Assn. (assoc.). Republican. Avocations: flying, skiing, raising bird dogs. Home: 660 Coolidge St Moses Lake WA 98837-1877

LEAF, ALEXANDER, physician, educator; b. Yokohama, Japan, Apr. 10, 1920; came to U.S., 1922, naturalized, 1936; s. Aaron L. and Dora (Hural) L.; m. Barbara Louise Kincaid, Oct. 1943; children—Caroline Joan, Rebecca Louise, Tamara Jean. B.S., U. Wash., 1940; M.D., U. Mich., 1943; M.A., Harvard, 1961. Intern Mass. Gen. Hosp., Boston, 1943-44; mem. staff Mass. Gen. Hosp., 1949—, physician-in-chief, 1966-81; resident Mayo Found., Rochester, Minn., 1944-45; research fellow U. Mich., 1947-49; practice internal medicine Boston, 1949-90; faculty Med. Sch., Harvard, 1949—, Jackson prof. clin. medicine, 1966-81, Ridley Watts prof. preventive medicine, 1980-90, chmn. dept. preventive medicine and clin. epidemiology, 1980-90, Jackson prof. clin. medicine emeritus, 1990—; Disting. physician VA Medical Ctr. Brockton/W. Roxbury Hosps., Boston, 1992—. Served to capt. M.C. AUS, 1945-46. Recipient Outstanding Achievement award U. Minn., 1964; vis. fellow Balliol Coll., Oxford, 1971-72; Guggenheim fellow, 1971-72; named Disting. Physician, VA, 1991—. Fellow Am. Acad. Arts and Scis.; mem. NAS, Inst. Medicine, ACP (master), Am. Soc. Clin. Investigation (past pres.), Am. Physiol. Soc., Biophys. Soc., Assn. Am. Physicians. Home: 1 Curtis Cir Winchester MA 01890-1703 Office: Mass Gen Hosp Boston MA 02114

LEAKE, DONALD LEWIS, oral and maxillofacial surgeon, oboist; b. Cleveland, Okla., Nov. 6, 1931; s. Walter Wilson and Martha Lee (Crowe) L.; m. Rosemary Dobson, Aug. 20, 1964; children: John Andrew Dobson, Elizabeth, Catherine. AB, U. So. Calif., 1953, MA, 1957; DMD, Harvard U., 1962, MD, Stanford U., 1969. Diplomate Am. Bd. Oral and Maxillofacial Surgery. Intern Mass. Gen. Hosp., Boston, 1962-63; resident Mass. Gen. Hosp., 1963-64; postdoctoral fellow Harvard U., 1964-66; practice medicine specializing in oral and maxillofacial surgery; asso. prof. oral and maxillofacial surgery Harbor-UCLA Med. Ctr., Torrance, 1970-74, dental dir., chief oral and maxillofacial surgery, 1970—, prof., 1974—; assoc. dir. UCLA Dental Rsch. Inst., 1979-82, dir., 1982-86; prof. extranjero Escuela de Graduados, Asociacion Medica Argentina, 1990—; cons. to hosps.; dental dir. coastal health services region, Los Angeles County, 1974-81; oboist Robert Shaw Chorale, 1954-55; solo oboist San Diego Symphony, 1954-59. Contbr. articles to med. jours.; rec. artist: (albums on Columbia label) The Music of Heinrich Schütz, Stockhausen, Zeitmasse for 5 Winds, Schönberg, Orchestra Variations-Opus 31; freelance musician various film studio orchs., Carmel Bach Festival, 1949, 52-53, 67-81, numerous concerts with Coleman Chamber Music, The Cantata Singers, Boston, Garden St. Chamber Players, Cambridge, Baroque Consortium, L.A., Corona Del Mar Baroque Festival, others; world premieres (oboe works) by Darius Milhaud, William Kraft, Alice Parker, Mark Volkert, Eugene Zádor, Robert Linn. Mem. Commn. on the Future of Rose-Hulman Inst. Tech., Terre Haute, Ind., 1992-93. Recipient 1st prize with greatest distinction for oboe and chamber music Brussels Royal Conservatory Music Belgium, 1956. Fellow ACS; mem. AAAS, ASTM, Internat. Assn. Dental Rsch., Internat. Assn. Oral Surgeons, Soc. Biomaterials, Biomed. Engring. Soc. (sr. mem.), L.A. County Med. Assn., N.Y. Acad. Sci., L.A. Acad. Medicine, European Assn. Maxillofacial Surgeons, Brit. Assn. Oral and Maxillofacial Surgeons, Internat. Gesellschaft fur Kiefer-Gesichts-Chirurgie, Internat. Soc. Plastic, Aesthetic and Reconstructive Surgery, Harvard Club (Boston and N.Y.C.), Phi Beta Kappa, Phi Kappa Phi. Clubs: Harvard (Boston and N.Y.C.). Achievements include patents for work related to bone reconstruction, 1974, 82, 90. Home: 2 Crest Rd W Rolling Hills CA 90274 Office: Harbor-UCLA Med Ctr 1000 W Carson St Torrance CA 90502-2004 also: Harbor UCLA Profl Bldg 21840 S Normandie Ave Ste 700 Torrance CA 90502

LEAKEY, MARY DOUGLAS, archaeologist, anthropologist; b. Feb. 6, 1913; d. Erskine Edward and Cecilia Marion (Frere) Nicol; m. Louis Seymour Bazett Leakey, 1936 (dec. 1972); 3 sons. Ed. pvt. schs.; D.Sc. (hon.), U. Witwatersrand, 1968, Western Mich. U., 1980, U. Chgo., 1981, Yale U., 1976, U. Cambridge, 1987; D.Litt. (hon.), U. Oxford, 1981, Emory U., 1988, Mass. U., 1988, Brown U., 1990, Columbia U., 1991. Former dir. Olduvai Gorge Excavations. Author: Excavation in Beds I and II, 1971, Africa's Vanishing Art, Disclosing the Past; also articles; editor: (with J.M. Harris) Laetoli, A Pliocene Site in North Tanzania. Joint recipient (with L.S.B. Leakey) Prestwich medal of Geol. Soc. London; recipient Nat. Geog. Soc. Hubbard medal, Centennial award, 1988; Gold medal Soc. Women Geographers; Lineus medal Stockholm, 1978; Stopes medal Geol. Assn. London, 1980; Bradford Washburn prize Boston Mus. of Sci., 1980; Elizabeth Blackwell award Hobart and Smith Coll., 1980. Mem. Royal Swedish Acad. Scis. (hon.), Am. Acad. Arts and Scis. (assoc.), Nat. Acad. Sci. (fgn.). Office: National Museum of Kenya, PO Box 40658, Nairobi Kenya

LEAKEY, RICHARD ERSKINE, paleoanthropologist, museum director; b. Nairobi, Kenya, Dec. 19, 1944; s. Louis Seymour and Mary Douglas (Nicol)

L.; m. Meave Epps, 1970; children: Anna, Louise, Samira. Student, Lenana Sch., Nairobi; DSc (hon.), Wooster Coll., 1978, Rockford Coll., 1983; LittD, U. Kent, 1987. Dir. tour co. Kenya, 1965-66; asst. dir. Center for Prehistory and Palaeontology, 1966-67; adminstrv. dir. Nat. Museums of Kenya, Nairobi, 1968-74, dir., chief exec., from 1974; chmn. East African Wildlife Soc., 1985—; head Wildlife Conservation Dept., 1989—; leader expdn. to West Lake Baringo, Kenya, 1966, Internat. Omo River Expdn. in So. Ethiopia, 1967, East Rudolf Expdn., 1968; leader, coordinator Koobi Fora Research Project, Lake Turkana, 1969—; mem. Kenya del. UNESCO, 1972, 76; chmn. Found. Research into Origins of Man; trustee, vice chmn. Found. Social Habilitation; vice-chmn. Environ. Prep. Group, Kenya, 1972-74; mem. Nakali/Suguta Valley Expdn., 1978, West Turkana Research Project, 1982, 84, 85, 86, Buluk-Early Miocene Project, 1982, advisor TV series The Making of Mankind; numerous pub. and scholarly lectures, U.S., Can., U.K., Scandinavia, New Zealand, Kenya, China, 1968—. Author: (with R. Lewin) Origins, 1977, Koobi Fora Research Project: a Catalogue of Hominid Fossils, Vol. I, 1978, People of the Lake, 1978, Making of Mankind, 1981, One Life An Autobiography, 1984; films include The Ape that Stood Up, 1977; lecture films on prehistory; various sci. programs, talk shows and news interviews since 1968; contbr. (with R. Lewin) chpts. to books, articles to profl. jours. Trustee Nat. Fund for the Disabled in Kenya, trustee Rockford Coll., Nat. Kidney Found. Kenya, Agrl. Research Found., Kenya, Gallmann Meml. Found., Kenya. Recipient Franklin Burr prize, 1965, 73, Centennial award Nat. Geographic Soc., 1988. Fellow Royal Anthrop. Inst., AAAS, Kenya Acad. Scis. (founding fellow), Inst. Cultural Research U.K.; mem. Explorers Club, Wildlife Clubs of Kenya (trustee, hon. chmn. 1969-85), Kenya Exploration Soc. (chmn. 1969-72), East African Wildlife Soc. (hon. chmn. 1984—), Pan African Assn. Prehistoric Studies (sec.), Sigma Xi.

LEAL, GEORGE D., engineering company executive; b. 1934. B in Civil Engring., MA, Santa Clara U., 1959. With Dames & Moore, Inc., L.A., 1959—, CEO, 1981—. Office: Dames & Moore Inc 911 Wilshire Blvd Ste 700 Los Angeles CA 90017-3499*

LEAL, JOSEPH ROGERS, chemist; b. New Bedford, Mass., Sept. 14, 1918; s. Joaquim S. and Mary C. (Rogers) L.; m. Mary Desmond, Apr. 25, 1944; children: Joseph E., Michael J., Patricia M., Victoria A. Diploma, U. Mass., Dartmouth, 1940; BS summa cum laude, U. Mass., 1949; PhD, Ind. U., 1953. Asst. chemist CPC Internat., Edgewater, N.J., 1940-42, Revere Copper & Brass Co., New Bedford, 1942-43, 45-46; rsch. chemist Am. Cyanamid Co., Bound Brook, N.J., 1952-57; tech. rep. Am. Cyanamid Co., Washington, 1957-63; mgr. contract rels. Am. Cyanamid Co., Stamford, Conn., 1963-67; sr. staff assoc. Celanese Rsch. Co., Summit, N.J., 1967-83; pres. Crescent Cons., Maplewood, N.J., 1983—. Frederick Gardiner Cottrell fellow, 1950, Corn Industries Rsch. fellow, 1951-52. Mem. AAAS, Am. Chem. Soc., N.Y. Acad. Scis., SAMPE. Achievements include research in high temperature resistant polymers, nonflammable fibers, high strength high modulus reinforcement materials, fiber reinforced organic, ceramic and metal composites. Office: Crescent Consultants 10 S Crescent Maplewood NJ 07040

LEAMAN, GORDON JAMES, JR., chemical engineer; b. Tulsa, Mar. 7, 1951; s. Gordon James and Alberta M. (Ryder) L.; m. Marilyn Remmers, May 18, 1973; 1 child, Rebecca Jennifer. BSChemE, U. Okla., 1973. Registered profl. engr., Okla., Ariz. Chem. engr. Continental Oil Co., Ponca City, Okla., 1973-75, process engr., 1975-77; project coord. Conoco Coal Devel. Co., Stamford, Conn., 1977-81; sr. process engr. Conoco Inc., Ponca City, 1981-87, Billings, Mont., 1987-90; quality leader Conoco Inc., Ponca City, 1990-92; mgr. petrochem. M3 Engring. & Tech., Tucson, 1992—; presenter in field. Contbr. articles, reports to profl. publs. Officer USCG Aux., 1977-81. Mem. Am. Inst. Chem. Engrs. (sect. chair 1986, mem. nat. admissions com. 1978-81), Nat. Soc. Profl. Engrs., Okla. Soc. Profl. Engrs. (chair chpt. 1987, Okla. Young Engr. of Yr. 1987). Office: M3 Engring & Tech Corp 2440 W Ruthrauff Ste 170 Tucson AZ 85705

LEAMON, TOM B., industrial engineer, educator; b. Ossett, Yorkshire, U.K., Apr. 16, 1940; came to U.S., 1980; s. Harold and Dorothy (Wilde) L.; m. Geraldine Bland, July 1, 1967; children: Amanda Charlotte, Jonathan Barton, Genevieve Emma. BS in Chemistry, U. Manchester, U.K., 1961; MS in Indsl. Engring., Inst. Tech., Cranfield, U.K., 1964; MS in Applied Psychology, U. Aston, Birmingham, U.K., 1970; PhD in Indsl. Engring. Production Mgmt., Inst. Tech., Cranfield, 1982. Chartered engr.; chartered profl. ergonomist. Indsl. engr. Ilford (Eng.) Ltd., 1961-62; mgr. ergonomics Pilkington Ltd., St. Helens, Eng., 1964-71; dir. masters course Inst. Tech., Cranfield, 1970-75; head ergonomics br. Nat. Coal Bd., Burton-upon-Trent, U.K., 1975-80; dir. grad. program in occupational safety U. Ill., Chgo., 1981-82; prof., chmn. No. Ill. U., DeKalb, 1982-87, acting dean, 1985-86; prof., chmn. Tex. Tech. U., Lubbock, 1987-91; v.p., dir. rsch. ctr. Liberty Mut. Ins. Group, Hopkinton, Mass., 1991—. Trustee Am. Soc. Safety Engrs. Found. Named 1985 Disting. Lectr., Coll. Lake County. Fellow Inst. Prodn. Engrs. (Sir Ben Williams Silver medal 1969), Human Factors Soc. (chmn. indsl. ergonomics 1983-85), Ergonomics Soc.; mem. Inst. Indsl. Engrs. (chmn. ergonomics div. 1990-91, Spl. citation for outstanding contbns. to the enhancement of ergonomics), Am. Soc. Safety Engrs., Sigma Xi. Office: Liberty Mutual Rsch Ctr 71 Frankland Rd Hopkinton MA 01748

LEAP, DARRELL IVAN, hydrogeologist; b. Huntington, W.Va., Oct. 19, 1937; s. Willis Ivan and Bertie Estal (Messinger) L.; m. Wilma Ann Elkins, Feb. 24, 1962 (div. 1977); m. Myra Kay McCammon, May 14, 1983. BS, Marshall U., 1960; MA, Ind. U., 1966; PhD, Pa. State U., 1974. Registered profl. hydrogeologist; cert. profl. geologist. Geologist, hydrologist S.D. Geol. Survey, Vermillion, S.D., 1966-71; hydrologist U.S. Geol. Survey, Denver, 1974-80; assoc. prof. Purdue U., West Lafayette, Ind., 1980-93, prof., 1993—; pv. cons., West Lafayette, 1985—; cons. U.S. EPA, Washington, 1990—, U.S. Nuclear Regulatory Com., Washington, 1992—. Contbr. articles to profl. jours. Singer Bach Chorale Singers, Lafayette, 1981-92, bd. dirs., 1992, pres., 1993—; bd. dirs. Water Conservancy Dist., Battleground, Ind., 1992. Lt. USNR, 1960-65. Mem. Am. Geophys. Union, Geol. Soc. Am. (com. chair 1991-92), Nat. Assn. Ground Water Scientists and Engrs., Ind. Water Resources Assn. (pres. 1982-83). Republican. Presbyterian. Achievements include research development of two new ground water tracing methods, three new ground water modeling methods, method to test relative retardation of substances in the ground; first to report influence of pore pressure on dispersivity. Office: Purdue U Dept Earth and Atmospheric Scis West Lafayette IN 47907-1397

LEAR, ERWIN, anesthesiologist, educator; b. Bridgeport, Conn., Jan. 1, 1924; s. Samuel Joseph and Ida (Ruth) L.; m. Arlene Joyce Alexander, Feb. 15, 1953; children—Stephanie, Samuel. MD, SUNY, 1952. Diplomate Am. Bd. Anesthesiology, Nat. Bd. Med. Examiners. Intern L.I. Coll. Hosp., Bklyn., 1952-53; asst. resident anesthesiology Jewish Hosp., Bklyn., 1953-54; sr. resident Jewish Hosp., 1955, asst., 1955-56, adj., 1956-58, assoc. anesthesiologist, 1958-64; attending anesthesiologist Bklyn. VA Hosp., 1958-64, cons., 1977—; assoc. vis. anesthesiologist Kings County Hosp. Ctr., Bklyn., 1957-80; staff anesthesiologist Kings County Hosp. Ctr., 1980-81; vis. anesthesiologist Queens Gen. Hosp. Ctr., 1955-67; dir. anesthesiology Queens Hosp. Ctr. Jamaica, 1964-67, cons., 1968—; chmn. dept. anesthesiology Catholic Med. Ctr., Queens and Bklyn., 1968-80; dir. anesthesiology Beth Israel Med. Ctr., N.Y.C., 1981—; clin. instr. SUNY Coll. Medicine, Bklyn., 1955-58; clin. asst. prof. SUNY Coll. Medicine, 1958-64, clin. assoc. prof., 1964-71, clin. prof., 1971-80, prof., vice-chmn. clin. anesthesiology, 1980-81; prof. anesthesiology Mt. Sinai Sch. Medicine, 1981—. Author: Chemistry Applied Pharmacology of Tranquilizers; contbr. articles to profl. jours. Served with USNR, 1942-45. Fellow Am. Coll. Anesthesiologists, N.Y. Acad. Medicine (sec. sect. anesthesiology 1985-86, chmn. sect. anesthesiology 1986-87); mem. AMA, Am. Soc. Anesthesiologists (chmn. com. on by-laws 1982-83, dir. 1981—, ho. of dels. 1973—, editor newsletter 1984—, chmn. adminstrv. affairs com., 1987—), N.Y. State Bd. Profl. Med. Conduct, N.Y. State Soc. Anesthesiologists (chmn. pub. relations 1963-73, chmn. com. local arrangements 1968-73, dist. dir. 1972-73, v.p. 1974-75, pres. 1976, bd. dirs. 1972—, chmn. jud. com. 1977-81, assoc. editor Bulletin 1963-77, editor Sphere 1978-84), N.Y. State Med. Soc. (chmn. sect. anesthesiology 1966-67, sec. sect. 1977-81), N.Y. County Med. Soc., SUNY Coll. Medicine Alumni Assn. (pres. 1983, trustee alumni fund 1980), Alpha Omega Alpha. Address: Harriman Dr Sands Point NY 11050

LEATH, KENNETH THOMAS, research plant pathologist, educator; b. Providence, Apr. 29, 1931; s. Thomas and Elizabeth (Wootten) L.; m. Marie Andreozzi, Aug. 1955; children: Kenneth, Steven, Kevin, Maria Beth. BS, U. R.I., 1959; MS, PhD, U. Minn., 1966. Rsch. plant pathologist U.S. Regional Pasture Rsch. Lab. USDA-ARS, University Park, Pa., 1966—; prof. Pa. State U., University Park, 1966—; advisor numerous state and nat. orgns. Contbr. numerous articles to profl. jours. With USN, 1951-55. Mem. Elks. Achievements include research on root diseases and systemic wilts of forage species. Office: Pa State U US Regional Pasture Rsch La Curtin Rd University Park PA 16802

LEAVELL, MICHAEL RAY, computer programmer and analyst; b. Port St. Joe, Fla., Sept. 28, 1955; s. Ray Carl and Willodean (Griggs) L. AS in Electronics Tech., Gulf Coast Jr. Coll., Panama City, Fla., 1975; BS in Systems Sci., U. West Fla., 1979. Engr. Sta. WDTB-TV (now WMBB-TV), Panama City, 1976; radio announcer Sta. WJOE, Port St. Joe, 1979; computer programmer III, Fla. Dept. Labor, Tallahassee, 1979-80, computer programmer, analyst II, 1980—. Democrat. Office: Fla Dept Labor 246 Howard Bldg Tallahassee FL 32399-0692

LEBL, MICHAL, peptide chemist; b. Prague, Czechoslovakia, Aug. 21, 1951; came to U.S., 1991; s. Bedrich and Olga (Krystofova) Leblova; m. Zuzana Bucinova, June 9, 1973; children: Martin, George. MS, Inst. Chem. Tech., Prague, 1974; PhD, Inst. Organic Chemistry & Biochemistry, Prague, 1978; DSc in History and Biochemistry, Inst. Organic Chemistry, 1992. Scientist Inst. Organic Chemistry and Biochemistry, Prague, 1978-87, group leader, 1987-89, dept. head, 1989-91; dir., owner CSPS, Prague, Tucson, 1990—; dir. chemistry Selectide Corp., Tucson, 1991—; pres. Spyder Instruments Inc., 1993—; vis. prof. McMaster U., Hamilton, Can., 1982, U. Ariz., Tucson, 1983, 86, cons., 1989; mem. bd. chemistry Czechoslovak Acad. Sci., Prague, 1988-91; mem. program com. European Peptide Symposium, Interlaken, Switzerland, 1992; mem. Debiopharm Award Com., Bonmot, Switzerland, 1992. Author, editor: Handbook of Neurohypophysial Hormone Analogs, 1987, numerous sci. publs.; editor-in-chief jour. Collection of Czechoslovak Chem. Comm., 1987-91; mem. editorial bd. Internat. Jour. Peptide and Protein Rsch., 1989—, Peptide Rsch. jour., 1988—; contbr. 130 articles to profl. jours. Sci. grantee NIH, 1992-95, Czechoslovak Acad. Scis., 1991-93; recipient prize Czech Republic. Govt., 1990, Czechoslovak Acad. Scis., 1980, 86, 89. Mem. Am. Chem. Soc., Am. Peptide Soc., European Peptide Soc. (Zervas prize 1990), Czechoslovak Chem. Soc. Achievements include 25 patents in field; design of automatic peptide synthesizer utilizing cotton as a solid carrier, of several selectively acting analogs of neurohypophysial hormones, of peptide and non-peptide libraries. Office: Selectide Corp 1580 E Hanley Blvd Tucson AZ 85737

LEBLANC, ROGER MAURICE, chemistry educator; b. Trois Rivières, Que., Can., Jan. 5, 1942; s. Henri and Rita (Moreau) L.; m. Micheline D. Veillette, June 26, 1965; children: Daniel, Hughes, Marie-Jose, Nancy. B.Sc., U. Laval, 1964, Ph.D., 1968. NRC postdoctoral fellow Davy Faraday Rsch. Lab. Royal Inst. Great Britain, London, 1968-70; prof. phys. chemistry U. Que., Trois-Rivières, 1970—, chmn. dept., 1971-75, dir. Biophysics Rsch. Group, 1978-81, chmn. Photobiophysics Rsch. Ctr., 1981-91; hon. prof. Jilin U., Chang Chun, People's Republic of China, 1992. Recipient Barringer award Spectroscopy Soc. Can., 1983, Medaille du Merite Universitaire du Que. a Trois-Rivieres, 1987, John Labatt Ltd. award Can. Soc. Chemistry, 1992, Commemorative medal for 125th Anniversary of Confedn. Can., 1993. Fellow Chem. Inst. Can. (Noranda award 1982, John Labatt Ltd. award 1992); mem. Am. Chem. Soc., Assn. Canadienne Francaise pour l'Avancement des Sciences (Prix Vincent 1978), Am. Soc. Photobiology, Biophys. Soc., European Photochem. Assn. Roman Catholic. Home: 5539 Marseilles, Trois Rivières Ouest, PQ Canada G8Y 3Z6 Office: U Que, 3351 blvd des Forges CP 500, Trois Rivières, PQ Canada G9A 5H7

LEBOWITZ, JOEL LOUIS, physicist, educator; b. May 10, 1930; came to U.S., 1946, naturalized, 1951; m. Estelle Mandelbaum, June 21, 1953. BS, Bklyn. Coll., 1952; MS, Syracuse U., 1955, PhD, 1956; hon. doctorate, Ecole Poly. Federale, Lausanne, Switzerland, 1977. NSF postdoctoral fellow Yale U., New Haven, Conn., 1956-57; mem. faculty Stevens Inst. Tech., Hoboken, N.J., 1957-59; mem. faculty Yeshiva U., N.Y.C., 1959-77, prof. physics, 1965-77, acting chmn. Belfer Grad. Sch. Sci., 1964-67, chmn. dept., 1967-76; prof. math. and physics, dir. Ctr. for Math. Scis. Rutgers U., New Brunswick, N.J., 1977—. Co-editor: Phase Transitions and Critical Phenomena, 1980, editor Jour. Statis. Physics, 1975—, Studies in Statis. Mechanics, 1973—, Com. Math. Physics, 1973—; contbr. articles to profl. jours. Recipient Boltzmann medal Internat. Union of Pure and Applied Physics, 1992; Guggenheim fellow, 1976-77. Fellow Am. Phys. Soc.; mem. NAS, AAAS, AAUP, N.Y. Acad. Scis. (pres. 1979, A. Cressy Morrison award in natural scis. 1986), Am. Math. Soc., Phi Beta Kappa, Sigma Xi. Office: Rutgers U Ctr Math Sci Rsch Busch Campus-Hill Ctr New Brunswick NJ 08903

LECHEVALIER, MARY PFEIL, retired microbiologist, educator; b. Cleve., Jan. 27, 1928; d. Alfred Leslie Pfeil and Mary Edith Martin; m. Hubert Arthur Lechevalier, Apr. 7, 1950; children: Marc E.M., Paul R. BA in Physiology-Biochemistry, Mt. Holyoke Coll., 1949; MS in Microbiology, Rutgers U., 1951. Rsch. fellow Rutgers U., New Brunswick, N.J., 1949-51, rsch. assoc. inst. microbiology, 1962-74, from asst. to assoc. rsch. prof., 1974-85, rsch. prof. Waksman inst. microbiology, 1985-91, prof. emerita, 1991—; ind. rschr., 1955-59; microbiologist steroid preparative lab. E.R. Squibb and Sons, New Brunswick, 1960-61; vis. investigator Inst. Biology Czechoslovak Acad. Scis., Svc. de Mycologie Pasteur Inst., Prague, Paris, 1961-62, cons. in field. Contbr. over 100 chpts. to books and articles to rsch. jours.; mem. adv. com. actinomycetes Bergey's Manual of Determinative Bacteriology, 8th edit.; chair adv. com. muriform actinomycetes Bergey's Manual, 9th edit. Assoc. mem. Bergey's Trust, 1989-92. Recipient Charles Thom award Soc. Indsl. Microbiology, 1982, Waksman award Theobald Smith Soc., 1991. Mem. AAAS, Am. Soc. Microbiology (former mem. com. actinomycetales), U.S. Fedn. Culture Collections (exec. com. 1982-85, J. Roger Porter award nominating com. 1983-84, 87-88, chair 1989-90, J. Roger Porter award 1992), N.Am. Mycol. Assn., Soc. Gen. Microbiology, Sigma Xi (pres. Rutgers U. chpt. 1977-78). Achievements include patents for immunological adjuvant and process for preparing same, pharmaceutical composition and process, restriction endonuclease Fse I. Home: RR 2 Box 2235 Morrisville VT 05661

LECKMAN, JAMES FREDERICK, psychiatry and pediatrics educator; b. Albuquerque, Dec. 3, 1947; s. Frederick Arnold and Alberta Beatrice (Lane) L.; m. Hannah Jean Hone, Dec. 27, 1971; children: Emily Beth, Peter Edwin. BA, Coll. Wooster, 1969; MD, U. N.Mex., 1973; MA (hon.), Yale U., 1990. Diplomate Am. Bd. Psychiatry, Am. Bd. Child Psychiatry. Intern USPHS Marine Hosp., San Francisco, 1973-74; clin. assoc. NIMH, Bethesda, Md., 1974-76; adult and child psychiatric resident Yale U., New Haven, 1976-80, from asst. prof. to assoc. prof., 1980-90, Nelson Harris prof. child psychiatry and pediatrics, 1990—; mem. psychopathology and clin. biology initial rsch. rev. com. NIMH, 1985-90; Milton M. & Harriet H. Parker lectr. psychiatry and human genetics Ohio State U., 1985; cons. U.S. Army, Heidelberg, 1986, Nat. Adv. Mental Health Coun., 1989-90; chmn. steering com. study of rsch. on child and adolescent mental disorders Inst. Medicine, 1988-89, child psychology and treatment initial rsch. rev. com. NIMH, 1992—; sci. adv. bd. Sophia Found. Med. Rsch., Rotterdam, Netherlands, 1989—. Co-author: Tourette's Syndrome and Tic Disorders, 1988, Fragile X Syndrome, 1993; mem. editorial bd. Devel. and Psychopathology, 1988—, Acta Paedopsychiatrica, 1992—; contbr. over 200 articles to sci. and jours. Recipient Seymour L. Lustman Rsch. award, 1978, 79; fellow USPHS-AAMC, 1972; William T. Grant Found. Rsch. scholar, 1980-83, Merck Faculty scholar, 1982-91; grantee NIH, 1972-93, 92-95, Nat. Inst. Child Health and Human Development, 1970-95, Nat. Inst. Neurological Disease and Stroke, 1980-96, NIMH, 1980-83, 89-93, 92-95. Fellow APA, Am. Coll. Neuropsychology, Am. Acad. Child and Adolescent Psychiatry (editorial bd. jour. 1982-88, guest co-editor spl. sect. Tourette's syndrome 1984, sci. program com. 1983-87, Outstanding Mentor 1990, 92); mem. ACP (H.P. Laughlin fellow 1981), Tourette Syndrome Assn. (sci. adv. bd. 1991—), Conn. Coun. Child and Adolescent Psychiatrists (pres. 1991—), Phi Beta Kappa, Alpha Omega Alpha, Sigma Xi. Home and Office: 125 Spring Glen Ter Hamden CT 06517

LE COMTE, CORSTIAAN, radar system engineer; b. Sommelsdijk, The Netherlands, Sept. 4, 1911; s. Karel Thomas and Digna (Meijer) le C.; m. Eelktje Elisabeth Makkes; children: Charles Thomas, Eelktje Elisabeth, Adriaan. Lab. asst. Philips, Eindhoven, The Netherlands, 1929-32; chief engr. Philips Telecommunication, other cos., The Netherlands, 1932-72; cons. engr. Hollandse Signaal Apparaten, Hengelo, 1970-73; designer radar systems for airports, waterways, navy in Europe, The Netherlands. Mem. Koninklijk Inst van Ingenieurs (Speurwerkprijs van het 1970). Home: Willem Dreeslaan 6, Huizen 1272 DC, The Netherlands

LECOMTE, ROGER, physicist; b. St. Sebastien, Que., Can., July 11, 1951; s. Richard and Lucille (Lareau) L.; m. Danielle Potvin, May 26, 1976; 1 child, Noemie. MSc in Applied Physics, U. Montreal, 1977, PhD in Nuclear Physics, 1981. Postdoctoral fellow U. Sherbrooke, Que., 1981-83, rsch. assoc., 1983-84, asst. prof., 1984-88, assoc. prof., 1988—; peer rev. com. Med. Rsch. Coun. Can., Ottawa, Int., 1985-89; cons. Biocapital, Montreal, 1991; peer rev. panel U.S. Dept. Energy, L.A., 1992. Contbr. articles to profl. jours. Mem. Fondation quebecoise en environment, Montreal, 1990—. Med. Rsch. Coun. Can. fellow, 1980-83; Fonds de la recherche en sante du Que. scholar, 1984—; Med. Rsch. Coun. Can. grantee, 1984—; NSERC/RCA grantee, 1987-90. Mem. IEEE, Am. Phys. Soc., Can. Assn. Physicists, Assn. canadienne française pour l'avancement des sciences. Achievements include development of proton-induced x-ray emission (PIXE) for trace element analysis; identification and explanation of nuclear shape transition in doubly even germanium isotopes; development of avalanche photodiode detectors and high resolution camera in positron emission tomography, multispectral imaging in positron emission tomography; patent for Scintillation Detector for Tomographs. Office: U Sherbrooke, Dept Nuclear Medicine, Sherbrooke, PQ Canada J1H 5N4

LEDEEN, ROBERT WAGNER, neurochemist, educator; b. Denver, Aug. 19, 1928; s. Hyman and Olga (Wagner) L.; m. Lydia Rosen Hailparn, July 2, 1982. B.S., U. Calif., Berkeley, 1949; Ph.D., Oreg. State U., 1953. Postdoctoral fellow in chemistry U. Chgo., 1953-54; rsch. assoc. in chemistry Mt. Sinai Hosp., N.Y.C., 1956-59; rsch. fellow Albert Einstein Coll. Medicine, Bronx, N.Y., 1959; asst. prof. Albert Einstein Coll. Medicine, 1963-69, assoc. prof., 1969-75, prof., 1975-91; prof., dir. div. neurochemistry U. Medicine and Dentistry N.J., Newark, 1991—. Contbr. articles to profl. jours.; dep. chief editor Jour. Neurochemistry. Mem. neurol. scis. study sect. NIH; mem. study sect. Nat. Multiple Sclerosis Soc. NIH grantee, 1963—; Nat. Multiple Sclerosis Soc. grantee, 1967-74; recipient Humboldt prize, Javits Neurosci. Investigator award. Mem. Internat. Soc. Neurochemistry, Am. Soc. Neurochemistry, Am. Chem. Soc., Am. Soc. Biol. Chemists, N.Y. Acad. Sci. Jewish. Achievements include discoveries in the biochemistry of brain glycolipids and myelin. Home: 8 Donald Ct Wayne NJ 07470-4608 Office: U Medicine and Dentistry NJ Dept Neurosci 185 S Orange Ave Newark NJ 07103-2714

LEDER, FREDERIC, chemical engineer; b. N.Y.C., Nov. 1, 1939; s. Julian Mitchell and Dorothy (Turkel) L.; m. Barbara Zayne Alexander, Dec. 11, 1971; children: Helene Jill, Henry Alexander. BS in Chemistry, Queens Coll., 1961; BS, Columbia U., 1961; MS, Yale U., 1963, PhD, 1965. Planning assoc., group leader Exxon Rsch. and Engring. Co., Linden, N.J., 1965-76; dir. new ventures, exploratory/minerals rsch., energy rsch. Occidental Rsch. Corp., Irvine, Calif., 1976-83; rsch. dir., v.p. Cities Svc. R&D Corp., Tulsa, 1983-85; mgr. fracture R&D Dowell Schlumberger, Tulsa, 1985-88; mng. ptnr. Splty. Capital Group, Tulsa, 1988—; mem. task force on R&D needs in fossil fuels U.S. Dept. Energy, 1979. Contbr. numerous articles to profl. jours. Mem. AICE (founder rsch. com. on energy 1976, Tech. Achievement award 1982), Soc. Petroleum Engrs., Sigma Xi. Achievements include U.S. patents, invention of acid gas separations technology. Home: 4219 E 87 St Tulsa OK 74137 Office: Splty Capital Group 4219 E 87th St Tulsa OK 74137

LEDER, PHILIP, geneticist, educator; b. Washington, Nov. 19, 1934; married; 3 children. A.B., Harvard U., 1956, M.D., 1960. Research assoc. Nat. Heart Inst., Nat. Cancer Inst.; lab chief molecular genetics Nat. Inst. Child Health and Human Devel., NIH, 1972-80; prof. genetics Harvard U. Med. Sch., Boston, Mass., 1980—, now John Emory Andrus prof. genetics; sr. investigator Howard Hughes Med. Inst. Recipient Albert Lasker Med. Rsch. award, 1987, Nat. Medal of Sci., 1989. Mem. NAS, Inst. Medicine.

LEDERBERG, JOSHUA, geneticist, educator; b. Montclair, N.J., May 23, 1925; s. Zwi Hirsch and Esther (Goldenbaum) L.; m. Marguerite S. Kirsch, Apr. 5, 1968; children: David Kirsch, Anne. BA, Columbia U., 1944; PhD, Yale U., 1947. With U. Wis., 1947-58; prof. genetics Sch. Medicine, Stanford (Calif.) U., 1959-78; pres. Rockefeller U., N.Y.C., 1978-90, Univ. prof., 1990—; bd. dirs. Procter & Gamble Co., Cin., Ann. Revs., Inc., Palo Alto; adj. prof. Columbia U., 1990—; mem. adv. com. med. rsch. WHO, 1971-76; mem. bd. sci. advisors Affymax N.V., Palo Alto, Calif., Bellcore, Livingston, N.J., Aviron, Belmont, Calif.; cons. U.S. Def. Sci. Bd., U.S. Dept. Energy, NSF, NIH, NASA, ACDA. Trustee Camille and Henry Dreyfus Found.; bd. dirs. Chem. Industry Inst. Toxicology, N.C., Am. Type Culture Coll., Wash., OTA Assessment Adv. Coun. With USN, 1943-45. Recipient Nobel prize in physiology and medicine for rsch. in genetics of bacteria, 1958, U.S. Nat. Medal of Sci., 1989; Sackler Found. scholar. Fellow AAAS, Am. Philos. Soc., Am. Acad. Arts and Scis., N.Y. Acad. Medicine (hon.) Academie Universelle des Cultures (Paris); mem. NAS (Inst. of Medicine), Coun. Fgn. Rels., Royal Soc. London (fgn.), N.Y. Acad. Scis. (pres.), Ordre des Lettres et des Arts (fgn.). Office: Rockefeller U 1230 York Ave Ste 400 New York NY 10021-6399

LEDERER, JOHN MARTIN, aeronautical engineer; b. Solomon, Kans., May 12, 1930; s. George Martin and Angie Belle (Faubion) L.; m. Joan Elizabeth Patrick, June 15, 1963; children: Jeffrey Mark, Carol Elizabeth. BS in Aero. Engring., Kans. State U., 1953; MSEE, Air Force Inst. Tech., 1955; postgrad., U. N.Mex., 1962-65. Registered profl. aero. engr., Ohio. Project engr. Air Force Spl. Weapons Ctr., Albuquerque, 1955-63, chief project engring. div., 1963-67, chief electromagnetics div., 1967-70; tech. adviser Air Force Weapons Lab., Albuquerque, 1970-73, 76-87, chief nuclear systems surety div., 1988-91; dir. nuclear systems engring. aero. systems Nuclear Systems Engring. Directorate/USAF Materiel Command, Albuquerque, 1991-92; dir. nuclear systems engring. aero. systems ctr. Air Force Materiel Command, Albuquerque, 1992—; tech. dir. 4900th test group, Albuquerque, 1973-76; chmn. Dept. of Def. Design Rev. and Acceptance Group, Albuquerque, 1979-91; flying instr. airplanes, instruments. Co-inventor digital distance measuring instrument. Founder One of Ten Young Am. Football League, Albuquerque, 1964. Served to 1st lt. USAF, 1953-58. Recipient Outstanding Performance award Dept. Air Force, Albuquerque, 1965, 66, 68, 73, 74, 79, Sustained Superior Performance award Dept. Air Force, Albuquerque, 1961, 81, 83-86, 88-92. Mem. NSPE, FAA (cert. flight instr.), Inst. Aerospace Scis. Republican. Episcopalian. Avocations: archery, flying. Home: 3012 El Marta Ct NE Albuquerque NM 87111-5618 Office: Nuclear Systems Engring Directorate OL-NS/EN 1651 1st St SE Kirtland AFB NM 87117-5617

LEDERER, KLAUS, macromolecular chemistry educator; b. Kalwang, Austria, May 10, 1942; s. Helmuth and Elisabeth (Liehm) L.; m. Beatrix Perrelli, Apr. 2, 1976; children: Martin, Nina. PhD, U. Graz, Austria, 1968; habil., Tech. Hochschule, Darmstadt, Germany, 1974. Postdoctoral position SUNY, Albany, 1968-70; asst. Tech. Hochschule, Darmstadt, 1970-71, lectr., 1971-75; lectr. Montanuniversitaet Leoben, Austria, 1975-77; assoc. prof. Montanuniversitat Leoben, Austria, 1977-89, head inst., 1989--. Author: Kunststoffe, 1987; editor (jour.) Thrombosis Rsch., 1981-85; contbr. articles to profl. jours. Recipient Goldene Vok Ehrennadel award Vereinigg Osterr Kunstoff Verarbeiter, 1991. Mem. Verbolnd Leobener Kunstoff Techniker (pres. 1991—). Roman Catholic. Achievements include development of rheological instrumentation, elucidation of size and shape of fibrinogen; development of method for calibration of separation and peak broadening in size exclusion chromatography, elucidation of thermal degradation of polyesters and polyimides. Office: Inst Chemi der Kunstoffe, Franz Josef Str 18, A-8700 Leoben Austria

LEDERMAN, LEON MAX, physicist, educator; b. N.Y.C., July 15, 1922; s. Morris and Minna (Rosenberg) L.; m. Florence Gordon, Sept. 19, 1945; children: Rena S., Jesse A., Heidi R.; m. Ellen Carr, Sept. 17, 1981. B.S.,

CCNY, 1943, DSc (hon.), 1980; A.M., Columbia U., 1948, Ph.D., 1951; DSc (hon.), No. Ill. U., 1984, U. Chgo., 1985, Ill. Inst. Tech., 1987. Assoc. in physics Columbia, N.Y.C., 1951; asst. prof. Columbia, 1952-54, assoc. prof., 1954-58; prof. Columbia U., 1958-89, Eugene Higgins prof. physics, 1972-79; Frank L. Sulzberger prof. physics U. Chgo., 1989-92; dir. Fermi Nat. Accelerator Lab., Batavia, Ill., 1979-89, dir. emeritus, 1989—; Pritzker prof. physics Ill. Inst. Tech., Chgo., 1992—; dir. Nevis Labs., Irvington, N.Y., 1962-79; guest scientist Brookhaven Nat. Labs., 1955; cons. Nat. Accelerator Lab., European Orgn. for Nuclear Rsch. (CERN), 1970—; mem. high energy physics adv. panel AEC, 1966-70; mem. adv. com. to div. math. and phys. scis. NSF, 1970-72; sci. advisor to gov. State of Ill., 1989-92. Author: Quarks to the Cosmos, 1989, The God Particle, 1993; also articles. 1st lt. Signal Corps, AUS, 1943-46. Recipient Nat. Medal of Sci., 1965, Townsend Harris medal CUNY, 1973, Elliot Cresson medal Franklin Inst., 1976, Wolf prize, 1982, Nobel prize in physics, 1988, Enrico Fermi prize, 1992; Guggenheim fellow, 1958-59, Ford Found. fellow European Ctr. for Nuclear Rsch., Geneva, 1958-59, fellow NSF, 1967. Fellow AAAS (pres. 1990-91, chmn. 1992-93), Am. Phys. Soc.; mem. NAS, Italian Phys. Soc., Aspen Inst. Physics (pres. 1990-92), Ill. Math. Sci. Acad. (vice chmn. 1985—), Tchrs. Acad. for Math. and Sci. in Chgo. (co-chmn. 1990—). also: Ill Inst Tech Dept Physics 3300 S Federal St Chicago IL 60616

LEDFORD, RICHARD ALLISON, food science educator, food microbiologist; b. Charlotte, N.C., June 30, 1931; s. Travis Allison and Sarah (Moon) L.; m. Martha Ann Worley, Jan. 26, 1957; children: Richard Jr., Roeby, Ann, Jeanne, Robert. BS, N.C. State U., 1954, MS, 1958; PhD, Cornell U., 1961. Dir. food lab. N.Y. State Agr. & Markets, Albany, 1961-64; asst. prof. food sci. Cornell U., Ithaca, N.Y., 1964-70, assoc. prof., 1970-80, prof., 1980—, chmn. dept., 1972-77, 85—; dir. Inst. Food Sci., 1988—. Served to 1st lt. U.S. Army, 1954-56. Mem. Am. Soc. Microbiology, Inst. Food Technologists, Am. Dairy Sci. Assn. Lodge: Rotary. Office: Cornell U Dept Food Sci Ithaca NY 14853

LEDIC, MICHÈLE, economist; b. Zagreb, Croatia, 1951; d. Vitomir and Nelly (Juricev) L.; m. Zangwill Aubrey Silberston. BA in Econs., U. Zagreb, 1974, MS in Econs., 1981. Lectr., sr. lectr. econs. U. Zagreb, 1974-85; economist The World Bank, Washington, 1983, Shell Internat. Petroleum Co., London, 1985-86; rschr., cons. Birkbeck Coll., U. London, 1987-91; dir.-gen. L'Observatoire Européen du Textile et de L'Habillment (OETH), Brussels, 1991—; econ. cons. U.K. Dept. Trade and Industry, BP Solar Internat., The Common Law Inst. Intellectual Property, Oxford Analytica, 1987—; invitee The World Econ. Forum, Davos, Switzerland, 1991-93. Author: The Impact of the Net Flow of Foreign Capital on the Dynamics of Growth of the Yugoslav Economy until the Year 1990, 1984, (with Z.A. Silberston) The Future of the Multi-fibre Arrangement, Implications for the UK Economy, 1989; contbr. numerous articles to profl. jours. Mem. Royal Inst. Internat. Affairs-Chatham House (econ. cons.), Great Britain-China Centre. Office: Dir Gen OETH, Rue Belliard 197 Bte 9, 1040 Brussels Belgium

LEDLEY, ROBERT STEVEN, biophysicist; b. N.Y.C., June 28, 1928. DDS, NYU, 1948; MA, Columbia U., 1949. Rsch. physicist Columbia U. Radiation Labs., Columbia, 1948-50; instr. physics Columbia U., 1949-50; vis. scientist Nat. Bur. Standards, 1951-52; physicist, 1953-54; ops. rsch. analyst Johns Hopkins U., 1954-56; assoc. prof. elec. engring George Washington U., 1957-60; instr. pediatric. Johns Hopkins U., Sch. Medicine, 1960-63; prof. elec. engring. George Washington U., 1968-70; prof. physiology, biophysics & radiology Georgetown U., from 1970; pres., rsch. dir. Nat. Biomed. Rsch. Found., from 1960; pres. Digital Info. Sci. Corp., 1970-75. named to Nat. Inventor Hall of Fame, 1990. Office: Georgetown U Nat Biomed Rsch Found 3900 Reservoir Rd Washington DC 20057

LEDLEY, TAMARA SHAPIRO, earth system scientist, climatologist; b. Washington, May 18, 1954; d. Murray Daniel and Ina Harriet (Gordon) Shapiro; m. Fred David Ledley, June 6, 1976; children: Miriam Esther, Johanna Sharon. BS, U. Md., 1976; PhD, MIT, 1983. Rsch. assoc. Rice U., Houston, 1983-85, asst. rsch. scientist, 1985-90, sr. faculty fellow, 1990—; mem. Alaska SAR facility archive working team NASA, Pasadena, Calif., 1988; McMurdo SAR facility sci. working team, 1990; participant workshop of Arctic leads initiative Office Naval Rsch., Seattle, 1988, lst DeLange Conf. on Human Impact on Environ., Houston, 1991; cons. Houston Mus. Natural Sci., 1989-90, Ea. Rsch. Group Inc., Arlington, Mass., 1989—, Broader Perspectives, Houston, 1989; dir. weather project for tchr. tng. program George Obs., Rice U., 1990-92; co-dir. Rice Houston Mus. Natural Sci. Summer Solar Inst., 1993. Contbr. articles to profl. pubs. Spl. judge Houston Area Sci. and Engring. Fair, 1985; judge S.W. Tex. Region High Sch. Debates, 1986, Houston Area Sci. and Engring. Fair, 1990, 91; guest expert Great Decisions '88 Polit. Discussion Group, 1988. Fellow sci. computing Nat. Ctr. for Atmospheric Rsch., Boulder, Colo., 1978, Fed. Jr. fellow, 1972-74; senatorial scholar State of Mo., 1972-76; grantee NSF, 1985-87, 87-88, 89-92, 90-93, 92-94, Tex. Higher Edn. Coordinating Bd., 1988-90, 90-92, Univ. Space Rsch. Assn., NASA, 1990-93, 1991-92. Mem. Am. Geophys. Union, Am. Meteorol. Soc., AAAS, Oceanography Soc., Sigma Xi, Phi Beta Kappa, Phi Kappa Phi, Alpha Lambda Delta. Avocations: reading, tennis, aerobics. Office: Rice U Dept Space Physics & Astronomy 6100 Main St Houston TX 77005-1892

LEDUY, ANH, engineering educator; b. Vietnam, Feb. 6, 1946; came to Can., 1965; s. Thach and Tam (BuiThi) LeD.; m. Suzanne Roger, Sept. 24, 1977; children: Isabelle, Dominic. B.S. in Mech. Engring., U. Sherbrooke, Que., Can., 1969, M.S. in Chem. Engring., 1972; Ph.D. in Biochem. Engring., U. Western Ont., Can., 1975. Registered profl. engr., Que. Research asst. CNRC, Univ. Sherbrooke, Que., Can., 1975-77; asst. prof. chem. engring. Universite Laval, Sainte-Foy, Que., 1977-81, assoc. prof., 1981-85, prof., 1985—; mem. grant selection coms.; cons. in field. Presenter symposiums, confs. Contbr. numerous articles to profl. jours. Mem. Order of Engrs. of Que., Am. Soc. Microbiology, Chem. Inst. Can., Can. Aeronautics and Space Inst., Can. Soc. Chem. Engring., Am. Soc. Engring. Edn., Am. Inst. Chem. Engrs., Can. Soc. Microbiologists, N.Y. Acad. Scis., Genetics Soc. Can. Office: Universite Laval, Dept Chem Engring, Sainte-Foy, PQ Canada G1K 7P4

LEE, ALEXANDRA SAIMOVICI, civil engineer; b. Negrest, Vaslui, Romania, Nov. 6, 1932; came to U.S., 1969; d. Leonidas and Etlea (Schreibman) Saimovici; m. Jack Lee, July 14, 1972. Grad. in constrn. engring., Constrn. Inst., Bucharest, Romania, 1956. Registered profl. engr., S.C. Structural engr. Energo Constructia, Bucharest, 1956-61, Elcora Constrn. Metalicas, Buenos Aires, 1961-69, Walter Kidde, N.Y.C., 1969-70, John Kassner, N.Y.C., 1970-72; civil engr. I, City of Columbia, S.C., 1972-77, design engr., 1977-82, civil engr. II, 1982—. Mem. NSPE, Am. Pub. Works Assn. Home: 3733 Greenbriar Rd Columbia SC 29206 Office: City of Columbia PO Box 147 Columbia SC 29217

LEE, BERNARD SHING-SHU, research company executive; b. Nanking, People's Republic of China, Dec. 14, 1934; came to U.S., 1949; s. Wei-Kuo and Pei-fen (Tang) L.; m. Pauline Pan; children: Karen, Lesley, Tania. BSc, Poly. Inst. Bklyn., 1956, DSc in Chem. Engring., 1960. Registered profl. engr., N.Y., Ill. With Arthur D. Little, Inc., Cambridge, Mass., 1960-65; with Inst. Gas Tech., Chgo., 1965-78, pres., 1978—; mem. adv. bd. Ctr. Applied Energy Rsch. U. Ky.; chmn. M-C Power Corp., Burr Ridge, Ill.; bd. dirs. NUI Corp., Bedminster, N.J., Peerless Mfg. Co., Dallas, Energy BioSystems Corp., The Woodlands, Tex. Contbr. more than 60 articles to profl. jours. Recipient Outstanding Personal Achievement in Chem. Engring. award Chem. Engring. mag., 1978. Fellow AAAS, Am. Inst. Chem. Engrs. (33d inst. lectr. 1981); mem. AIME, Am. Chem. Soc., Am. Gas Assn. (Gas Industry Rsch. award 1984), Econ. Club Chgo. Office: Inst Gas Tech 3424 S State St Chicago IL 60616-3896

LEE, BRENDAN, geneticist; b. Hong Kong, Aug. 5, 1966. BS, CUNY, 1986; PhD, SUNY, Bklyn., 1991, postgrad. Rsch. assoc. Mt. Sinai, Med. Ctr., 1990-91; rsch. asst. prof. SUNY Health Sci. Ctr., Bklyn., 1991—. Contbr. articles to Sci., Nature, NEJM. Achievements include initial demonstration of gene defect (in Type II collagen) in chrodysplasia, first to clone novel class of matrix protein genes-fibrillin; cloning of gene encoding for protein defect in Marfan Syndrome. Office: SUNY Health Sci Ctr 450 Clarkson Ave Box 44 Brooklyn NY 11203

LEE, CALVIN K., aerospace engineer; b. Hong Kong, Oct. 18, 1943; came to U.S., 1961; s. Ning K. and King Y. (Yee) L.; m. Janyce D. Tow, Aug. 10, 1969; children: Jonathan, Alexandria. BS, Mich. U., 1968; MS, Brown U., 1970, PhD, 1973. Mech. engr. Pitts. Rsch. Ctr., U.S. Bur. Mines, 1975-81; rsch. aerospace engr. U.S. Army Natick (Mass.) RD&E Ctr., 1981—; cons. on aerodynamic decellerator systems. Contbr. over 50 rsch. articles to profl. publs. Recipient Army Materiel Command Fed. Engr. of Yr. award Nat. Soc. Profl. Engrs., 1985. Mem. AIAA (tech. mem. aerodynamic systems tech. com. 1992—). Achievements include patents on fire research and aerodynamic decellerators technology, low-altitude clustered parachutes; improved mine fire safety and advanced airdrop technology. Office: US Army Natick RD&E Ctr Kansas St Natick MA 02760

LEE, CARLTON K. K., clinical pharmacist, consultant, educator; b. Honolulu, June 17, 1962; s. Hsiang Tsing and Ngan Kar (Ching) Lee. PharmD, U. of the Pacific, 1985; postgrad., Johns Hopkins U., 1990—. Hosp. pharmacy resident Johns Hopkins Hosp., Balt., 1985-86, clin. staff pharmacist pediatrics dept. pharmacy, 1986-88, sr. clin. pharmacist pediatrics dept. pharmacy, 1988-90, clin. coord. pediatrics dept. pharmacy, 1990—; asst. prof. Sch. Pharmacy Howard U., Washington, 1987-88; clin. asst. prof. Sch. Pharmacy U. Md., Balt., 1989—; instr. pediatrics Sch. Medicine Johns Hopkins U., Balt., 1992—; cons. Home Intensive Care Inc., Hunt Valley, Md., 1992—. Contbg. author: Harriet Lane Handbook, 1990, 93, Newborn Nursery Handbook, 1992; investigational drug advisor Med. Sci. Bull., 1992—; contbr. articles to profl. jours.; author, co-author conf. papers. Mem. Am. Soc. Hosp. Pharmacists , Am. Coll. Clin. Pharmacy, Internat. Assn. Therapeutic Drug Monitoring and Clin. Toxicology. Office: Johns Hopkins Hosp Dept Pharmacy Svcs 600 N Wolfe St Baltimore MD 21287-6180

LEE, CHARLES C., physicist; b. Szechwan, China, Oct. 8, 1940; came to U.S., 1965; s. James K.C. and Floral (Han) L.; m. Dora Lee, Sept. 5, 1970; children: Andrew, Wendell. BS in Physics, Peking U., Beijing, 1964; MS in Solid State Physics, Purdue U., 1968, PhD in Solid State Physics, 1974. Sr. scientist cen. rsch. lab. 3M Co., St. Paul, 1974-78, rsch. specialist physics and material lab., 1978-82, sr. rsch. specialist engring. systems divsn., 1982-88, scientist engring. document systems divsn., 1988-93, scientist comml. graphics divsn., 1993—; tech. auditor 3M Co. Freelance writer Chinese newspaper, N.Y.C.; author: Multi-function CAD Printer, 1984; translator: Optical Electronics, 1980; contbr. articles on laser scanning tech. to profl. jours. Pres. Chinese Am. Assn. Minn., Mpls., 1982; treas. Citizen for S.B. Woo in Minn., Mpls., 1988; organizer Music Festival, St. Paul, Mpls., 1990; chair Friends of Esther Lee Yao in Minn., 1991. Mem. Soc. Photo-Optical Instrumentation Engrs., Soc. Info. Display, Soc. for Imaging Sci. and Tech., Assn. Info. and Image Mgmt. Achievements include patents for Laser Diode Printer; for Laser Scanning Apparatus using a Fan Style Grating Plate; patent pending for Infrared Sensitization of Photoconductor. Office: Comml Graphics Divsn 207 BW 09 3M Co Saint Paul MN 55144

LEE, CHI-WOO, chemistry educator; b. Pyungtaek, Kyungido, Republic of Korea, May 27, 1954; s. Jong-Hwan and Sang-Un Lee; m. Mi-Kyung Kim, June 10, 1980; children: Sang-Ho, Sang-Hwa. BS, Korea U., Seoul, 1976, MS, 1978; PhD, Calif. Inst. Tech., 1984. Lectr. Korea U., Seoul, 1978-80; asst. prof. Korea U., Jochiwon, 1988-90, assoc. prof. chemistry, 1990—, chmn. chemistry dept., 1989-91; postdoctoral assoc. U. Calif., Berkeley, 1983-85; R. A. Welch postdoctoral fellow U. Tex., Austin, 1985-88. Contbr. articles to profl. jours. Korean Govt. fellow, 1980-83. Mem. Am. Chem. Soc., Electrochem. Soc., Soc. for Electroanalytical Chemistry, Internat. Soc. Electrochemistry. Avocation: martial arts. Home: Samik Apt # 5-102, Seocho 4-Dong, Seoul 137, Republic of Korea Office: Korea U, Chemistry Dept, Jochiwon Choongnam 339, Republic of Korea

LEE, CHOOCHON, physics educator, researcher; b. Seoul, Korea, June 8, 1930; came to U.S., 1962.; s. Yoon Young and Soon Ye (Rhee) L.; m. Chung Sun Yun, Apr. 9, 1960; children: John Taihee, Jane Eun Kyoung, Carol Eunmee. BS in Physics, Seoul Nat. U., 1953, MS in Physics, 1957; PhD in Physics, U. Ill., 1968. From instr. to asst. prof. Seoul Nat. U., 1957-62; rsch. asoc. U. Ill., Urbana, 1968; rsch. physicist U. Montreal, Que., Can., 1968-75; assoc. prof. Korea Advanced Inst. Sci., Seoul, 1975-78; prof. Korea Advanced Inst. Sci. and Tech., Seoul, Taejon, Korea, 1978—; pres. Korea Advanced Inst. Sci. and Tech., Seoul, 1980-82; vis. scholar Harvard U., Cambridge, Mass., 1982-83, 87-88. Author: Physics of Semiconductor Materials and Applications, 1986; contbr. over 100 articles to profl. jours. Recipient Presdl. Sci. prize Pres. Korean Govt., 1985, Recipient Incheon prize in Academic Achievement, 1993. Mem. IEEE, Am. Phys. Soc., Korean Phys. Soc. (pres. 1991-93), Soc. Info. Display (bd. dirs. 1991-94). Methodist. Achievements include discovery of negative staerble-wronski experiment; first to confirm mechanism of Controversial State III annealing in gold, to explain persistent photocontercivity in a-Si:H and a-Si:H/a-SiN multilayers. Home: KAIST Apt 3-302, 383-2 Toryong-dong, Yusong-ku, Taejon Korea 305-340 Office: Korea Advanced Inst Sci and Tech, 373-1 Kusong-dong Yusong-ku, Taejon Republic of Korea 305-701

LEE, CHUNG KEEL, biologist; b. Seoul, Dec. 20, 1940; came to U.S., 1969; s. Kyung Sok and Shin Hee (Kyun) L.; m. Chung Hee Ryu, Sept. 5, 1970; children: Grace, Angela, Jennifer. BS, Seoul Nat. U., 1963, MS, 1967; PhD, U. Ill., Urbana, 1974. English instr. Adjutant Gen. Sch., Korean Army, Young-chon, 1963-64; teaching, rsch. asst. Seoul Nat. U., 1965-69, U. Ill., Champaign-Urbana, 1969-74; postdoctoral appointee Argonne (Ill.) Nat. Lab., 1974-76, asst. scientist, 1976-79; scientist The Salk Inst., Swiftwater, Pa., 1979-86; sr. scientist The Salk Inst., 1986—; cons. Schs. of Basic Med. Scis. and Vet. Medicine, Univ. Ill., Urbana, 1973, Rockefeller Found., N.Y., 1987, 92, 93, WHO, Switzerland, 1992-93, Pan Am. Health Orgn., Washington, 1991-93; judge Pa. Jr. Acad. Sci.Region II, Pa., 1980—. Contbr. articles to profl. jours. Deacon, ruling elder Korean Ch. of Lehigh Valley, Whitehall, Pa., 1980-88; ruling elder Korean Presbyn. Ch. of Pocono, Swiftwter, 1988—; treas. Presbytery of Phila., Korean-Am. Presbyn. Ch., 1991-93. Grantee Nat. Cancer Inst., NIH, 1976-79. Mem. Tissue Culture Assn., Am. Soc. Microbiology, Korean Scientists and Engrs. Assn. in Am. (pres. Lehigh chpt. 1990-91), Sigma Xi. Achievements include research on radiosensitivity of the choromsomes in cultured human cells, chromosome analysis of cultured uterine carlinoma, in vitro properties of FBR murine osteosarcoma virus, antiviral properties of polyinosinic acids containing thio and methyl substitutions, human vaccine development and approval process in the U.S.A. Home: HCR 1 Box 70 Swiftwater PA 18370-9711 Office: The Salk Inst PO Box 250 Swiftwater PA 18370-0250

LEE, DANIEL DIXON, JR., nutritionist, educator; b. Dillon, S.C., Sept. 27, 1935; s. Daniel Dixon and Mattie (McLemore) L.; m. Anne Moore, June 28, 1958; children: Rebecca, Sarah, Dixon III, John. BS, Clemson U., 1957, MS, 1964; PhD, N.C. State U., 1970. Owner, operator dairy farm Dillon, S.C., 1959-62; rsch. supr. dept. biochemistry N.C. State U., Raleigh, 1967-70; asst. prof. dept. animal industries So. Ill. U., Carbondale, 1970-73, assoc. prof., 1973-76, asst. dean rsch. Coll. Agr., 1976-82; head dept. dairy sci. Clemson (S.C.) U., 1986-90. Contbr. articles to Jour. Animal Sci., Jour. Dairy Sci., Jour. Nutrition, Procs. Soc. Exptl. Medicine and Biology. Capt. U.S. Army, 1958-62. Mem. Am. Registry Profl. Animal Scientists, Rotary, Masons, Shriners, Sigma Xi, Phi Kappa Phi, Gamma Sigma Delta, Alpha Zeta. Home: 208 University Dr Seneca SC 29678-9299 Office: Dept Animal Dairy Vet Scis Clemson Univ 119 P&A Bldg Clemson SC 29634-0363

LEE, DANIEL KUHN, economist; b. Kyoto, Japan, Dec. 18, 1946; came to U.S., 1977; s. Chu G. and Myung N. (Lee) L.; m. Kaye K.S. Kwon, Apr. 10, 1976; children: David, Alexander. BS, Kyoto U., Japan, 1970; MA, Seoul Nat. U., Seoul, Republic of Korea, 1973, SUNY, Stony Brook, 1979; PhD, Iowa State U., 1981. Postdoctoral rsch. assoc. Iowa State U., Ames, 1981-82, instr., 1982; sr. economist Miss. Rsch. and Devel. Ctr., Jackson, 1982-88; dir. of econs. Miss. Insts. of Higher Learning, Jackson, 1988—; adj. prof. Jackson State U., 1986-88; advisor Gov.'s Econ. Task Force, Jackson, 1982-84. Author: A Study of Mississippi Input-Output Model, 1986; contbr. articles to profl. jours. Exec. dir. So. Regional Assn., Washington, 1992—; elder Presbyn. Ch. USA, 1991—. Travel grantee UN Indsl. Devel. Orgn., Vienna, Austria, 1986. Mem. So. Regional Sci. Assn., North Am. Regional Sci. Assn., Am. Econ. Assn., So. Econ. Assn., Gamma Sigma Delta. Avocations: jogging, swimming. Home: 656 Old Agency Rd Jackson MS 39213 Office: Miss Insts Higher Learning 3825 Ridgewood Rd Jackson MS 39211-6463

LEE, DAVID WOON, chemist, lawyer; b. Hong Kong, July 14, 1949; came to U.S., 1967; s. Kwoon and Sau Yuen Lee; m. Helen Lee, May 23, 1970; children: Victor, Malinda. BS, U. Winnipeg, Can., 1970; BS with honors, U. Waterloo, Can., 1979; JD, Glendale (Calif.) U., 1991. Bar: Calif. 1992, U.S. Dist. Ct. (cen. dist.) Calif. 1992, U.S. Ct. Appeals (9th cir.) 1993, U.S. Patent & Trademark Office 1993. Rsch. chemist Atomic Energy of Can., Pinawa, 1970-80; supr. maj. facilities Atomic Energy of Can., Chalk River, 1980-87; indsl. waste specialist County of L.A., 1987-88; chemist City of L.A., 1988—; del. citizen amb. program nuclear waste mgmt. U.S.S.R., 1989. Contbr. articles to Can. Jour. of Chemistry, Jour. of Colloid and Interface Sci., Jour. Electrochem. Soc., Electrochimica Acta 22; contbr. over 20 articles to profl. jours. Br. rep. Atomic Energy of Can. Profl. Employees Assn., Chalk River, 1984-87; judge Pembroke (Can.) Regional Sci. Fair, 1985-86. Mem. Am. Nuclear Soc., People-to-People Internat. Home: 8636 Zerelda St Rosemead CA 91770-1249 Office: City of LA 2002 W Slauson Ave Los Angeles CA 90047-1019

LEE, EDWARD KING PANG, dermatologist, public health service officer; b. Canton, Kwangtung, China, Nov. 11, 1929; arrived in Hong Kong, 1948; s. Huan Hsin and Yau Lin (Chan) L.; m. Kwok Yung Liao, July 18, 1953; children: Jean, Stephen Siu Wing, Ruby Siu Yin. MD, Sun Yat Sen U. Med. Scis., Canton, China, 1951. Resident in medicine Kwong Hua Med. Coll. Hosps., Canton, 1951-53; lectr. in pharmacology South China Med. Coll., Canton, 1953-57; med. officer Med. and Health Svcs., Hong Kong, 1959-63; med. officer, social hygiene, dermatology, venereology Queen Elizabeth Hosp., Sai Ying Pun VD Clinic, among others, Hong Kong, 1963-89; ret., 1989; now studying wildlife and working with refugees on island Hei Ling Chau.; clin. demonstrator for med. student venerology Med. Sch. U. Hong Kong, 1963-85, lectr. for nurses, health nurses and midwives Sch. Nursing, 1966-85. Contbr. articles to Canton Med. Jour., WHO Adv. Studies, Brit. Jour. Dermatology. Group supr. Aux. Med. Svc., Hong Kong, 1965—. Recipient Meritorious Svc. certs. and Gold medal Govt. of Hong Kong, 1989, 90, 1st Clasp Def. medal Aux. Med. Svcs., 1993. Mem. Chinese Acad. Physiology, Hong Kong Soc. Dermatology (founding mem. 1974—), Hong Kong Soc. Dermatology and Venereology. Avocations: painting, poetry, elephant chess, classical music, photography. Home: 4-C Shiu Fai Terr 7/F, Stubbs Rd, Hong Kong Hong Kong

LEE, ELDON CHEN-HSIUNG, chemist; b. Kaohsiung, Taiwan, Mar. 1, 1944; came to U.S., 1971; s. Cheng-En and Yuen (Chen) L.; m. Lisa Tsai-Feng Chen, Nov. 25, 1971; 1 child, David Chen. MS, Kans. State U., 1973, PhD, 1976. Tech. mgr. He Sung Beverage Co., Ltd., Taiwan, 1968-71; chief chemist water dept. City of Kansas City, Mo., 1975-77; sr. rsch. scientist Westreco, Inc./Nestle S.A., New Milford, Conn., 1977—; mem. PhD com. Rutgers U., U. N.J., 1989-92. Contbr. articles to profl. pubs. 1st lt. Taiwan Navy, 1967-68. Mem. Am. Chem. Soc., Am. Assn. Cereal Chemists, Inst. Food Technologists. Achievements include 6 patents in field. Home: 10 High Trail New Milford CT 06776 Office: Westreco Inc/Nestle SA 201 Housatonic Ave New Milford CT 06776

LEE, ELHANG HOWARD, physicist, researcher, educator; b. Seoul, Republic of Korea, Dec. 19, 1947; came to U.S., 1972; s. Pyung Sup and Aeyoung Lee; m. Namsoo Chang, Oct. 19, 1974; children: David Hanseul, Jennifer Hanbyul. BSEE summa cum laude, Seoul Nat. U., 1970; MS, Yale U., 1973, MPhil, 1975, PhD, 1977. Applied scientist, rsch. staff Yale U., New Haven, 1977-78; rsch. scientist Princeton (N.J.) U., 1979-80, Monsanto Co., St. Louis, 1980-84; sr. rsch. scientist AT&T Bell Labs., Princeton and Murray Hill, N.J., 1984-90; exec. dir. rsch., fellow scientist Electronics and Telecom Rsch. Inst., Daeduk Science Town, Republic of Korea, 1990—; adj. prof. Chung-Nam Nat. U., 1992—; lectr. Seoul Nat. U., 1990, Korea Advanced Inst. Sci. and Tech., 1992. Contbr. more than 60 articles to profl. jours.; author, translator sci., philosophy, sci. history books, editorials. Choir condr. Protestant Christian Ch., 1974—. Scholar Yale U., 1972-77. Fellow Korean Inst. Telematics and Electronics (chmn. rsch. com. 1991-92), Korean Phys. Soc.; mem. AAAS, IEEE (sr.), Am. Phys. Soc., Optical Soc. Am., Optical Soc. Korea (bd. dirs.), Soc. Future Studies, SPIE of Korea (exec. dir.), Materials Rsch. Soc. (session chmn. symposium and conf. activities), N.Y. Acad. Scis., Sigma Xi. Achievements include research in solid-state/semiconductor physics and laser optical/photonic science/technology, quantum physics, crystal growth physics, epitaxial and bulk, artificial superlattices, optoelectronics, photonics, MBE. MOCVD, laser/atom/molecule solid interactions, light scattering, lightwave optical communication technology. Office: Electronics and Telecom Rsch Inst, PO Box 8 Daeduk Science Town, Daejon 305 606, Republic of Korea

LEE, E(UGENE) STANLEY, industrial engineer, mathematician, educator; b. Hopeh, China, Sept. 7, 1930; came to U.S., 1955, naturalized, 1961; s. Ing Yah and Lindy (Hsieng) L.; m. Mayanne Lee, Dec. 21, 1957 (dec. June 1980); children: Linda J., Margaret H.; m. Yuan Lee, Mar. 8, 1983; children—Lynn Hua Lee, Jin Hua Lee, Ming Hua Lee. BS, Chung Cheng Inst. Tech., Republic of China, 1953; MS, N.C. State U., 1957; PhD, Princeton U., 1962. Research engr. Phillips Petroleum Co., Bartlesville, Okla., 1960-66; asst. prof. Kans. State U., Manhattan, 1966-67; asso. prof. Kans. State U., 1967-69, prof. indsl. engring., 1969—; prof. U. So. Calif., 1972-76; cons. govt. and industry. Author: Quasilinearization and Invariant Imbedding, 1968, Coal Conversion Technology, 1979, Operations Research, 1981; editor: Energy Sci. and Tech., 1975—; assoc. editor Jour. Math. Analysis and Applications, 1974—, Computers and Mathematics with Applications, 1974—; editorial bd. Jour. Engring. Chemistry and Metallurgy, 1989—, Jour. of Nonlinear Differential Equations, 1992—. Grantee Dept. Def., 1967-72, Office Water Resources, 1968-75, EPA, 1969-71, NSF, 1971—, USDA, 1978-90, Dept. Energy, 1979-84, USAF, 1984-88. Mem. Soc. Indsl. and Applied Math., Ops. Rsch. Soc. Am., N. Am. Fuzzy Info. Processing Soc., Internat. Neural Network Soc., Sigma Xi, Tau Beta Pi, Phi Kappa Phi. Office: Kans State U Dept Indsl Engring Manhattan KS 66506

LEE, FANG-JEN SCOTT, biochemist, molecular biologist; b. Taipei, Taiwan, Republic of China, Apr. 20, 1957; came to U.S., 1982; s. Tien-Te Lee and Bao-Cheng Lin; m. Leewen Lin, Jan. 21, 1984; children: Alice Christina, Albert Alexander. BS, Nat. Taiwan U., 1980; PhD, N.C. State U., 1986. Rsch. asst. N.C. State U., Raleigh, 1982-86; rsch. fellow Harvard Med. Sch./Mass. Gen. Hosp., Boston, 1987-90; sr. staff mem. Nat. Heart, Lung and Blood Inst., NIH, Bethesda, Md., 1990—; cons. Yung Shin Pharm. Indsl. Co. Ltd., Taiwan, 1986—. Author: Advances in Gene Technology: Protein Engineering and Production, 1988, Methods in Protein Sequence Analysis, 1989; contbr. articles to Jour. Biol. Chemnistry, Jour. Bacteriology, others. Mem. AAAS, Am. Soc. Biochemistry and Molecular Biology (Keyston Symposium grantee 1992), Am. Soc. Microbiology, Protein Soc. (grantee 1988), Sigma Xi, Gamma Sigma Delta. Achievements include 3 patents for genes and enzymes involved in amino-terminal processing. Home: 6320 Windermere Cir North Bethesda MD 20852 Office: Nat Heart Lung Blood Inst NIH 9000 Rockville Pike Bethesda MD 20892

LEE, FU-MING, chemical engineer; b. Kwei-Lin, Peoples Republic of China, Dec. 27, 1943; came to the U.S., 1969; s. Soong-Lin and Kwei-Yin (Moon) L.; m. Mina Shang, Apr. 17, 1971; children: Timothy, Jeffrey. BSChE, Tunghai U., 1967; MS, U. Toledo, 1971, PhD, 1974. Devel. group leader Pan Am. Chem. Corp., Toledo, 1974-76; sr. process engr. Sherwin & Williams Co., Coffeyville, Kans., 1976-77; rsch. engr. Phillips Petroleum Co., Bartlesville, Okla., 1977-80, sr. rsch. engr., 1980-86, engring. assoc., 1986—; tech. rep. Fractionation Rsch. Inst., Stillwater, Okla., 1992—, Ctr. for Indsl. Rsch., Oslo, 1988—. Contbr. articles to 14 jours. and 5 tech. conf. publs. Mem. Am. Inst. Chem. Engrs., Sigma Xi. Mem. Wesleyan Ch. Achievements include 32 U.S. patents for inventions related to refining and petrochemical processes; invention of technology to recover high purity chemicals from natural gas liquid/petroleum streams; research in extractive distillation and heavy oil processing technology. Home: 645 SE Castle Rd Bartlesville OK 74006 Office: Phillips Petroleum Co 182 PDC Phillips Rsch Ctr Bartlesville OK 74006

LEE, GARY L., engineering executive; b. Canton, Ohio, Sept. 21, 1947; s. Clifford M. and Delores A. (Lones) L.; m. Rena N. Hudgens, Aug. 1, 1983; children: Jeffrey, Michael, Mathew. B in Mech. Engring. Tech., Akron U.,

1979. Asst. plant engr. Timken Co., Canton, 1970-79; grinding engr. SKF Industries, Glasgow, Ky., 1979-80; prodn., mech. engring. mgr. Brenco, Inc., Petersburg, Va., 1980-85; engr. mgr. J.P. Industries, Caldwell, Ohio, 1985-89; plant mgr. INA Bearing, Cheran, S.C., 1989-91; engr. mgr. Prym Dritz, Spartanburg, S.C., 1991—. Author: Grinding Study, 1979. Sgt. USAF, 1967-71. Mem. Internat. Orgn. Packaging Profls., Soc. Mech. Engrs., Plating Orgn., Jaycees (treas. 1977). Achievements include invention of O2 probe in carburizing laser welding of bimetal bushings; hi-speed packing machine; cellular manufacturing in a traditional environment. Home: 306 Shady Dr Inman SC 27344 Office: Prym Dritz 950 Bruisack Rd Spartanburg SC 29303

LEE, GEORGE C., civil engineer, university administrator; b. Peking, China, July 17, 1933; s. Shun C. and J. T. (Chang) L.; m. Grace S. Su, July 29, 1961; children—David S., Kelvin H. B.S., Taiwan U., 1955; M.S. in Civil Engring., Lehigh U., 1958, Ph.D., 1960. Research assoc. Lehigh U., 1960-61; mem. faculty dept. civil engring. SUNY, Buffalo, 1961—; prof. SUNY, 1967—, chmn. dept., 1974-77, dean sch. of engring. and applied scis., 1978—; head engring. mechanics sect. NSF, Washington, 1977-78; assoc. dir. Calspan-U. Buffalo Rsch. Ctr., 1985-89; acting dir. Nat. Ctr. for Earthquake Engring. Rsch., 1989-90, dir., 1992—; sci. cons. Nat. Heart Lung and Blood Inst., NSF. Author: Structural Analysis and Design, 1979, Design of Single Story Rigid Frames, 1981, Cold Region Structural Engineering, 1986, Stability and Ductility of Steel Structures Under Cyclic Loading, 1991; contbr. articles to profl. jours. in areas of structural design, nonlinear structural mechanics, biomed. engring. and cold region structural engring. Recipient Adams Meml. award Am. Welding Soc., 1974; Superior Accomplishment award NSF, 1977. Mem. ASCE, Am. Welding Soc., Welding Research Council, Structural Stability Research, Council, Am. Soc. Engring. Edn., AAAS, Sigma Xi, Chi Epsilon, Tau Beta Pi. Office: SUNY Buffalo 412 Bonner Hall Amherst NY 14260

LEE, GILBERT BROOKS, retired ophthalmology engineer; b. Cohasset, Mass., Sept. 10, 1913; s. John Alden and Charlotte Louise (Brooks) L.; m. Marion Corrine Rapp, Mar. 7, 1943 (div. Jan. 1969); children: Thomas Stearns, Jane Stanton, Frederick Cabot, Eliot Frazar. BA, Reed Coll., 1937; MA, New Sch. for Social Rsch., 1949. Asst. psychologist U.S. Naval Submarine Base Civil Svc., Psychophysics of Vision, New London, Conn., 1950-53; rsch. assoc. Project Mich., Vision Rsch. Labs., Willow Run, 1954-57; rsch. assoc. dept. ophthalmology U. Mich., Ann Arbor, 1958-72, sr. rsch. assoc., 1972-75, sr. engring. rsch. assoc. ophthalmology, 1975-82, part-time sr. engr. ophthalmology, 1982—; sec. internat. dept., 23d St. YMCA, N.Y.C.; cons. W.K. Kellogg Eye Ctr., Ann Arbor, 1968—. Local organizer, moderator (TV program) Union of Concerned Scientists' Internat. Satellite Symposium on Nuclear Arms Issues, 1986; producer (TV show) Steps for Peace, 1987; designer, builder portable tristimulus Colorimeter; (videotape) Pomerance Awards, UN.; broken lake ice rescue procedure rsch., by one person in a dry suit, all weather conditions, 1966, 89-93 (videotape). Precinct del. Dem. County Conv., Washtenaw County, 1970, 74; treas. Dem. Club, Ann Arbor, Mich., 1971-72, 74-79; vice chmn. nuclear arms control com., 1979; chmn. Precinct Election Inspectors, 1968-75; scoutmaster Portland (Oreg.) area coun. Boy Scouts Am., 1932-39. Capt. AUS, 1942-46, 61-62. Mem. AAAS, Optical Soc. Am., Fedn. Am. Scientists, N.Y. Acad. Sci., Nation Assocs., ACLU, Sierra Club, Amnesty Internat. Home: 23080 Guidotti Pl Salinas CA 93908-1022

LEE, GORDON MELVIN, electrical engineering consultant; b. Mpls., Jan. 3, 1917; s. Melvin and Alma Matilda (Lindahl) L.; m. Harriet Lily Malkerson; children: Theodore, James, David, Mary. BEE, U. Minn., 1938; MS, U. Mo., 1939; DSc, MIT, 1944. Rsch. and teaching asst. U. Mo., Columbia, 1938-39; rsch. asst. MIT, Cambridge, 1939-44, mem. staff, 1944-45; dir. elec. engring., sec.-treas. Ctrl. Rsch. Labs., Inc., Red Wing, Minn., 1945-73, pres., 1973-81, cons., 1981—; lectr. U. Minn., Mpls., 1946. Contbr. articles to Phys. Rev., Proceedings of Inst. of Radio Engring. Mem. Red Wing Bd. Edn., 1946-64, pres., 1951-65; bd. dirs., chmn. Interstate Rehab. Ctr., 1984-91. Recipient Browder J. Thompson prize Inst. Radio Engrs., 1946, Outstanding Sch. Bd. Mem of Yr. award Minn. Sch. Bds. Assn., 1962, B'nai B'rith award Anti Defamation League, 1964, Lay Ministry award Episcopalian Diocese Minn., 1989. Mem. AAAS, IEEE, Am. Nuclear Soc., Kiwanis Internat. Democrat. Episcopalian. Achievements include development of high-speed microoscillograph; patents for remote handling. Home and Office: Wacouta Beach Red Wing MN 55066

LEE, HUA, electrical engineering educator; b. Taipei, Taiwan, Sept. 30, 1952; came to U.S., 1976; s. Chi-Sun and Min-Eeh (Poon) L.; m. Rayshin Wang, June 5, 1976; children: Michelle, Michael. BS, Nat. Taiwan U., Taipei, 1974; MS, U. Calif., Santa Barbara, 1978, PhD, 1980. Asst. prof. U. Calif., Santa Barbara, 1980-83, prof., 1990—; asst. prof. U. Ill., Urbana, 1983-87, assoc. prof., 1987-90; adv. bd. mem. Acoustical Imaging Conf., 1988—; rev. panel mem. NSF, 1991—, NRC, 1991—. Author: Engineering Analysis, 1988; editor: Imaging Technology, 1986, Modern Acoustical Imaging, 1986, Acoustical Imaging, vol. 18, 1990; editor various jours. Recipient Presdl. Young Investigator award NSF, 1985; named Prof. of Yr. Mortar Bd. Honor Soc., U. Calif.-Santa Barbara, 1992. Fellow IEEE, Acoustical Soc. Am.; mem. Am. Soc. Engring. Edn., Eta Kappa Nu. Achievements include development of scanning laser tomographic microscope, of microwave subsurface imaging system for NDE of civil structures, of synthetic-aperture sonar imaging system, and of high-performance imaging techniques. Office: U Calif Santa Barbara Dept Elec Engring Santa Barbara CA 93106

LEE, JAMES KING, technology corporation executive; b. Nashville, July 31, 1940; s. James Fitzhugh Lee and Lucille (Charlton) McGivney; m. Victoria Marie Marani, Sept. 4, 1971; children: Gina Victoria, Patrick Fitzhugh. BS, Calif. State U., Pomona, 1964; MBA, U. So. Calif., 1966. Prodn. and methods engring. foreman Gen. Motors Corp., 1963-65; engring. adminstr. Douglas MSSD, Santa Monica, Calif., 1965-67; mgr. mgmt. systems TRW Systems, Redondo Beach, Calif., 1967-68; v.p. corp. devel. DataStation Corp., L.A., 1968-69; v.p., gen. mgr. Aved Systems Group, L.A., 1969-70; mng. ptnr. Corp. Growth Cons., L.A., 1970-81; chmn., pres. chief exec. officer Fail-Safe Tech. Corp., L.A., 1981-93. Author industry studies, 1973-79. Mem. L.A. Mayor's Community Adv. Com., 1962-72, aerospace task force L.A. County Econ. Devel. Commn., 1990-92; bd. dirs. USO of Greater L.A., 1990-92, v.p. personnel 1990-92, exec. v.p., 1992-93, pres. 1993—; asst. adminstr. SBA, Washington, 1974; vice chmn. Traffic Commn., Rancho Palos Verdes, Calif., 1975-78; chmn. Citizens for Property Tax Relief, Palos Verdes, 1976-80; mem. Town Hall Calif. Recipient Golden Scissors award Calif. Taxpayers' Congress, 1978. Mem. So. Calif. Tech. Execs. Network, Am. Electronics Assn. (chmn. L.A. coun. 1987-88, vice chmn. 1986-87, nat. bd. dirs. 1986-89), Nat. Security Industries Assn. Republican. Baptist. Home: 28874 Crestridge Rd Palos Verdes Peninsula CA 90274-5063 Office: Fail-Safe Tech Corp Ste 318 710 Silver Spur Rd Rolling Hills Estates CA 90274-3695

LEE, KENNETH, physicist; b. San Francisco, July 3, 1937; s. Kai Ming and Ah See Lee; A.B. with honors in Physics, U. Calif., Berkeley, 1959. Ph.D., 1963; m. Cynthia Ann Chu, June 28, 1959; children—Marcus Scott, Stephanie Denise. Research physicist Varian Assocs., Palo Alto, Calif., 1963-68; mem. research staff, mgr. IBM, San Jose, Calif., 1968-83; dir. memory techs. Southwall Techs., Palo Alto, Calif., 1983-84; sr. v.p. product devel. Domain Tech., Milpitas, Calif., 1984-89; chief tech. officer, exec. v.p. engring. Quantum Corp., Milpitas, Calif., 1989—. Fellow Am. Phys. Soc.; sr. mem. IEEE; mem. Phi Beta Kappa, Sigma Xi. Contbr. articles to profl. jours.; patentee in field. Home: 20587 Debbie Ln Saratoga CA 95070-4827 Office: Quantum Corp 500 McCarthy Blvd Milpitas CA 95035

LEE, KENNETH STUART, neurosurgeon; b. Raleigh, N.C., July 23, 1955; s. Kenneth Lloyd and Myrtie Lee (Turner) L.; m. Cynthia Jane Anderson, May 23, 1981; children: Robert Alexander, Evan Anderson. BA, Wake Forest U., 1977; MD, East Carolina U., 1981. Diplomate Nat. Bd. Med. Examiners, Am. Bd. Neurol. Surgeons; med. lic. N.C., Ariz. Intern then resident in neurosurgery Wake Forest U. Med. Ctr., Winston-Salem, N.C., 1981-88; fellow Barrow Neurol. Inst., Phoenix, 1988-89; clin. asst. prof. neurosurgery East Carolina U., Greenville, N.C., 1989—. Assoc. editor Current Surgery, 1990—; contbr. 30 articles to profl. jours. and 5 chpts. to books. Mem. Ethicon Neurosurgical Adv. Panel, 1989—. Bucy fellow, 1988. Fellow Am. Heart Assn. (stroke coun.); mem. AMA, N.C. Med. Soc., Am. Assn. Neurol. Surgeons (assoc.), Am. Acad. Neurology (assoc.), Am. Soc. Stereotactic and Functional Neurosurgery, So. Med. Assn., Congress Neurol. Surgeons, N.C. Neurosurg. Soc. (sec.-treas. 1991—). Democrat. Baptist. Achievements include research on the efficacy of certain surgical procedures, particularly carotid endarterectomy, in the prevention of strokes. Home: 3600 Baywood Ln Greenville NC 27834-7630 Office: Ea Carolina Neurosurgical 2325 Stantonsburg Rd Greenville NC 27834-7546

LEE, KI DONG, aeronautical engineer, educator; b. Seoul, Korea, June 5, 1944; came to U.S., 1971; s. Kun Sung and Boo Oak (Suh) L.;m. Jounghyoun Kim, July 31, 1971; children: Angie, Joyce. BS, Seoul Nat. U., 1967; PhD, U. Ill., 1976. Instr. Korea Air Force Acad., Seoul, 1967-71; prin. engr. The Boeing Co., Seattle, 1977-85; rsch. asst. U. Ill., Urbana, 1973-76, rsch. assoc., 1976-77, assoc. prof., 1985—; dir. Computational Fluid Dynamics Lab., Urbana, 1986—. Contbr. numerous tech. papers to profl. jours. Univ. fellow U. Ill., 1971-73. Fellow AIAA (assoc., chmn. Ill. sect. 1989-93), ASME (assoc.); mem. Soc. Automotive Engrs. Home: 403 Holmes St Urbana IL 61801 Office: U Ill 104 S Wright St Urbana IL 61801

LEE, KOTIK KAI, physicist; b. Chungking, Peoples Republic of China, May 30, 1941; came to the U.S., 1967; s. Shi-Shan and Wa-J (Hsia) L.; m. Lydia S.M. Rue, Sept. 8, 1967 (div. 1991); children: Jennifer M., Peter H. MS, U. Ottawa, 1967; PhD, Syracuse U., 1972. Asst. prof. Rio Grande (Ohio) Coll., 1973-74; vis. prof. U. Ottawa, Ontario, Canada, 1974-76; scientist U. Rochester, N.Y., 1977-82, TRW, Redondo Beach, Calif., 1982-83; sr. staff scientist Gen. Electric Co., Binghamton, N.Y., 1983-86, Perkin-Elmer Corp., Danbury, Conn., 1986-89; assoc. prof. U. Colo., Colorado Springs, 1989—. Author: Lectures on Dynamical Systems, Structural Stability and Their Applications, 1992; editor: Optical Bistability, Instability and Optical Computers, 1988; contbr. articles to profl. jours. Grad. fellow Nat. Rsch. Coun. Canada, 1966-67. Mem. Am. Phys. Soc. (edn. com. 1992—), Optical Soc. Am. (edn. coun. 1989-91), Am. Math. Soc. (reviewer), N.Y. Acad. Scis. Democrat. Roman Catholic. Achievements include patents and patents pending; research in semiconductor lasers, solid-state lasers, laser phase-locked coupling, non-linear optics. Office: U Colo Dept Elec and Computer Engr Colorado Springs CO 80933-7150

LEE, KWAN RIM, statistician; b. Seoul, Korea, Oct. 24, 1948; s. Seung Chang and Donghwa (Cho) L.; m. Young S. Rho, Nov. 9, 1974; children: James J., Elliot E. BS, Seoul Nat. U., 1971, MS, 1973; MS, So. Meth. U., 1980, PhD, 1981. Cert. quality engr., reliability engr. Asst. to assoc. prof. Lamar U., Beaumont, Tex., 1981-87; sr. tech. staff BMDP Stat Software Inc., L.A., 1984-85; statistical cons. DuPont de Nemours and Co., Beaumont, 1985-87; assoc. prof. E. Tenn. State U., Johnson City, 1987-89; sr. statistician Eastman Chem. Co., Kingsport, Tenn., 1988-91, prin. statistician, 1991--. Contbr. articles to profl. jours. Mem. Am. Statistical Assn., Am. Soc. for Quality Control, Biometric Soc. Home: 2213 Kipping St Johnson City TN 37601 Office: Eastman Chem Co Kingsport TN 37662

LEE, KYO RAK, radiologist; b. Seoul, Korea, Aug. 3, 1933; s. Ke Chang and Ok Hi (Um) L.; came to U.S., 1964, naturalized, 1976; M.D., Seoul Nat. U., 1959; m. Ke Sook Oh, July 22, 1964; children—Andrew, John. Intern, Franklin Sq. Hosp., Balt., 1964-65; resident U. Mo. Med. Center, Columbia, Mo., 1965-68; instr. dept. radiology U. Mo., Columbia, 1968-69, asst. prof., 1969-71; asst. prof. dept. radiology U. Kans., Kansas City, 1971-76, assoc. prof., 1976-81, prof., 1981—. Served with Republic of Korea Army, 1950-52. Diplomate Am. Bd. Radiology. Recipient Richard H. Marshak award Am. Coll. Gastroenterology, 1975. Fellow Am. Coll. Radiology; mem. Radiol. Soc. N.Am., Am. Roentgen Ray Soc., Assn. Univ. Radiologists, Kans. Radiol. Soc., Greater Kansas City Radiol. Soc., Wyandotte County Med. Soc. Presbyterian, Korean Radiol. Soc. N.Am., Soc. Cardiovascular & Internat. Radiology. Contbr. articles to med. jours. Home: 9800 Glenwood St Shawnee Mission KS 66212-1536 Office: U Kans 39th St and Rainbow Blvd Kansas City KS 66103

LEE, LAWRENCE CHO, commodities advisor; b. Ithaca, N.Y., Aug. 30, 1953; s. Tak Yan and Mary (Foo) L. BA, Cornell U., 1976; diploma, Cambridge U., 1976; MA, U. Mich., 1980. Chartered comodity analyst. Instr. world history Northfield (Mass.) Mount Hermon Sch., 1976-77; data counselor Lockheed Space Missiles, Sunnyvale, Calif., 1982-83; commodity analyst Hy & Ht Lee Bros. Ltd., Hong Kong, 1984—; cons. Wibaux, Mont., 1987—; cons. securities Hong Kong and Shanghai Bank, 1984—; advisor commodities First Nat. Bank & Trust, Wibaux, 1989—. Author: (poem) Evening Star, 1970; inventor in commodity field, 1990, 91. Fundraiser UNICEF, Ithaca, 1975; liaison officer Dems., Hartford, Conn., 1972; mem. Rep. Nat. Com. Fellow Neural Network Inst.; mem. Royal Econometric Soc. (HK pres. 1986-91), Artificial Intelligence Assn. Democrat. Avocations: opera singing, martial arts, poetry writing, astronomy, foreign languages.

LEE, LIHSYNG STANFORD, medical researcher, biotechnical consultant; b. Linsen, China, Oct. 28, 1945; came to U.S., 1969; s. Honping and Kuorung (Shea) L.; m. Alice S.F. Chang, Sept. 8, 1974; children: Jenny, Oriana. MS, Yale U., 1972, PhD, 1974. Postdoctoral fellow Roswell Park Meml. Cancer Inst., Buffalo, N.Y., 1974-76; staff assoc. Columbia U. Coll. Physicians and Surgeons, N.Y.C., 1976-79; staff mem. GE Rsch. Ctr., N.Y., 1979-84; prin. scientist Cytogen Corp., Princeton, N.J., 1984-88; prin. investigator Enzo Biochem., N.Y.C., 1988-90; sci. supr. Enzon, Inc., South Plainfield, N.J., 1990—; sci. adv. bd. Leels Biotech., Princeton, 1988-92. Contbr. papers on sci. discoveries to sci. jours., conf. procs., books. Pres. Growth Bus. Investment Club, Princeton, 1988—. Yale U. fellow, 1973. Mem. AAAS, Am. Assn. Cancer Rsch., Am. Soc. Pharmacology and Exptl. Therapeutics. Achievements include patents for protein drug to treat Gaucher's disease, protein drug to treat Fabry's disease, method of non-radioactive detection of HIV genome in AIDS patients. Home: 22 Van Wyck Dr Princeton Junction NJ 08550-1640

LEE, LILLIAN VANESSA, microbiologist; b. N.Y.C., June 1, 1951; d. Wenceslao and Ada (Otero) Cancel; B.S. in Biology, St. Johns U., 1972; M.S. in Microbiology, Wagner Coll., 1974; m. Thomas Christopher Lee, June 11, 1972; children—Tovan, John-Peter, Phillip-Michael. Grad. lab. asst. in microbiology Wagner Coll., S.I., N.Y., 1972-74; clin. microbiology technologist Queens Hosp. Center, Jamaica, N.Y., 1974-81, clin. microbiology supr., 1981-84; sect. head microbiology Nyack Hosp. (N.Y.), 1984-93, acting lab. mgr., 1992-93; tech. dir. Tb. lab. N.Y.C. Dept. Health, 1993—. Cert. registered microbiologist and specialist in microbiology, clin. lab. specialist. Mem. Am. Soc. Clin. Pathologists, Am. Soc. Microbiology, Am. Acad. Microbiology, Med. Mycology Soc., N.Y., N.Y. Acad. Scis., Nat. Cert. Agy. Med. Lab. Personnel, Synergists Soc. Home: 14 Continental Dr West Nyack NY 10994-2803 Office: Nyack Hosp N Midland Ave Nyack NY 10960

LEE, MARTIN YONGHO, mechanical engineer; b. Apr. 13, 1937; s. Yee Whan and Myo Ryun (Choi) L.; m. Su Ja Bang, Nov. 29, 1969; children: Mu Young, Tae Young. BSME, Han Yang U., Seoul, Republic of Korea, 1964. Lic. stationary engr., N.Y. Startup engr. power plant Korea Electric Power Co., Seoul, 1969-75; stationary engr. CUNY, N.Y.C., 1981-92, Queens, 1981—. Mem. Assn. Energy Engr., Co-Generation Assn. Home: 146-28 34th Flushing NY 11354 Office: Police Hdqs NYC 1 Police Plz New York NY 10038

LEE, MATHEW HUNG MUN, physiatrist; b. Hawaii, July 28, 1931; married; 3 children. AB, Johns Hopkins U, 1953; MD, U. Md., 1956; MPH, U. Calif., 1962. Diplomate Am. Bd. Physical Medicine & Rehab. Resident Inst. Physical Medicine & Rehab., NYU, 1962-64, assignee rehab. svc.N.Y. State Health Dept., 1964-65, from asst. prof. to assoc. prof. rehab. medicine, 1965-73, dir. edn. & training dept. rehab. medicine, 1966-68, assoc. dir., 1968, prof. rehab medicine, 1973—; dir. dept. rehab. medicine Goldwater Meml. Hosp., 1968—; assoc. vis. physician Goldwater Meml. Hosp., 1965-68, vis. physician, 1968—, chief electrodiagnosis unit, 1966—, v.p. med. bd., 1969-70, pres., 1971' asst. clin. prof. Coll. Dentistry NYU, 1966-69, clin. asst. prof., 1969-70, clin. assoc. prof., 1970—; cons. Daughters of Israel Hosp., N.Y., 1965-72, Bur. Adult Hygiene, 1965—, Human Resources Ctr., 1966—; asst. attending physician Hosp. NYU, 1968—; attending physician Bellevue Hosp. Ctr., 1971—; cons. World Rehab. Fund, Gordon Seagrave & Maryknoll Hosps., Korea, 1969, U.S. Dept. Interior. Fellow Am. Acad. Physical Medicine & Rehab., Am. Coll. Physicians, Am. Pub. Health Assn.; mem. AAAS, Pan-Am. Med. Assn. Office: Jerry Lewis Neuromuscular Dis Ctr Dept Rehab Medicine 400 E 34th St New York NY 10016*

LEE, NANCY FRANCINE, psychologist; b. L.A., Aug. 20, 1956; d. Aaron and Renee Anne (Ball) Kumetz; m. John Stanley Lee, July 8, 1984; children: David Alan, Michael Ethan. BA, UCLA, 1978; PhD, U. Tex. Southwestern Med. Ctr., Dallas, 1983. Lic. clin. psychologist. Postdoctoral fellow in behavioral medicine Harbor-UCLA Med. Ctr., 1984-85; pvt. practice Beverly Hills, Calif., 1986—; clin. psychologist St. John's Hosp. and Health Ctr., Santa Monica, Calif., 1988—; instr. The Maple Ctr., Beverly Hills, 1989—. Contbr. articles to profl. jours. Mem. APA, L.A. County Psycholog. Assn., Phi Beta Kappa. Avocations: swimming, tennis, reading. Office: 8500 Wilshire Blvd Beverly Hills CA 90211

LEE, PATRICK A., physics educator; b. Hong Kong, Sept. 8, 1946; m. Jeanne M. Tran, June 7, 1970; children: Eric, Brian. BS, MIT, 1966, PhD, 1970. Gibbs instr. Yale U., New Haven, 1970-72; asst. prof. U. Wash., Seattle, 1973-74; mem. tech. staff Bell Labs., Murray Hill, N.J., 1974-82; prof. physics MIT, Cambridge, 1982—. Fellow Am. Phys. Soc. (Oliver Buckley prize 1991); mem. NAS, Am. Acad. Arts and Scis. Office: MIT Dept of Physics Rm 12-117 Cambridge MA 02139

LEE, PAUL HUK-KAI, biomedical research scientist; b. Hong Kong, July 11, 1956; came to U.S., 1985; s. Fei Yin and Sau Wah (Chan) L.; m. Rachel Hau-Yin Chan, May 22, 1983; children: Carolyn, Stephanie. BSc, U. Liverpool, England, 1980; PhD, U. Hong Kong, 1985. Rsch. assoc. Nat. Inst. Environ. Health Sci., Research Triangle Park, N.C., 1985-89; rsch. scientist Burroughs Wellcome Co., Research Triangle Park, 1989—. Contbr. articles to profl. jours. Mem. AAAS, Soc. Neurosci., Internat. Brain Rsch. Orgn. Achievements include initial successful culture of Dynorphin containing hippocampal dentate granule cells, and first to show that injection of mu opioid agonists into the hippocampus induced convulsions in rats. Office: Burroughs Wellcome Co 3030 Cornwallis Rd Research Triangle Park NC 27709

LEE, PETER Y., electrical engineer, consultant; b. Taipei, Taiwan, May 20, 1959; s. Jack T. and Joanna C. (Chen) L. BSEE, U. So. Calif., MSEE. MTS TTI/Citicorp, Santa Monica, Calif., 1981-82; project mgr. Tomy Corp., Carson, Calif., 1982-83; program devel. officer City Net Rsch. and Devel., L.A., 1983-86; system mgr. Hughes Aircraft Co./EDSG, El Segundo, Calif., 1986-87; founder, pres. Hyper Systems, Walnut, Calif., 1985—. Mem. IEEE. Republican. Roman Catholic. Avocations: basketball, ham radio, flying, fine art. Home: 254 Viewpointe Ln Walnut CA 91789-2078 Office: Hyper Systems 1313 N Grand Ave # 453 Walnut CA 91789-1317

LEE, PHILIP RANDOLPH, medical educator; b. San Francisco, Apr. 17, 1924; married, 1953; 4 children. AB, Stanford U., 1945, MD, 1948; MS, U. Minn., 1956; DSc (hon.), MacMurray Coll., 1967. Diplomate Am. Bd. Internal Medicine. Asst. prof. clin. phys. medicine & rehab. NYU, 1955-56; clin. instr. medicine Stanford (Calif.) U., 1956-59, asst. clin. prof., 1959-67; asst. sec. health & sci. affairs U. Calif., San Francisco, 1967-69, chancellor, 1969-72, prof. social medicine, 1969—, dir. inst. health policy studies, 1972—; mem. dept. internal medicine Palo Alto Med. Clinic, Calif., 1956-65; cons. bur. pub. health svc. USPHS, 1958-63, adv. com., 1978, nat. commn. smoking & pub. policy, 1977-78; dir. health svc. office tech. cooperation & rsch. AID, 1963-65; dep. asst. sec. health & sci. affairs HEW, 1965, asst. sec., 65-69, mem. nat. coun. health planning & devel., 1978-80; co-dir. inst. health & aging, sch. nursing U. Calif., San Francisco, 1980—; pres. bd. dirs. World Inst. Disability, 1984—; mem. population com. Nat. Rsch. Coun.- Nat. Acad. Sci., 1983-86; mem. adv. bd. Scripps Clinic & Rsch. Found., 1980—. Author over 10 books; contbr. articles to profl. jours. Recipient Hugo Schaefer medal Am. Pharm. Assn., 1976. Mem. AAAS, AMA, ACP, Am. Pub. Health Assn., Am. Fedn. Clin. Rsch., Am. Geriatric Soc., Assn. Am. Med. Colls., Inst. Medicine-Nat. Acad. Sci. Achievements include research in arthritis and rheumatism, especially Rubella arthritis, cardiovascular rehabilitation, academic medical administration, health policy. Office: University of Cal Institute for Hlth Policy 1388 Sutter St 11th Fl San Francisco CA 94109*

LEE, RICHARD A., mechanical engineer, consultant; b. Spokane, Wash., Jan. 27, 1952; s. Parley Lewis and Shirley Laureen (Hastings) L.; m. Aura Esperanza Salguero, Oct. 18, 1974; children: Leslie, Consuelo Marie, Dolores Rose, Laura Ann, Richard Lisandro. Vocat.-tech. cer., Idaho State U., 1975, BS, 1986. Fitter-welder Bucyrus-Erie, Pocatello, Idaho, 1975-83; nuclear technician Westinghouse Electric Corp., Scoville, 1984-86, opers. engr., 1986-90; mech. engr. Alpha Engrs., Inc., Pocatello, 1990—; Registered profl. engr., Idaho. Mem. NSPE, Am. Welding Soc., Instrument Soc. Am., Idaho Soc. Profl. Engrs. Mem. Ch. Jesus Christ Latter-day Saints. Avocations: cabinet making, rabbit raising, gardening, hunting, fishing. Office: Alpha Engrs Inc 850 S Main Pocatello ID 83205-4849

LEE, ROBERT GUM HONG, chemical company executive; b. Montreal, Que., Can., May 22, 1924; s. Hai Chong Lee and Toy Kay Yip; m. Maude Toye; children: Peter, Patricia, Cathrine. BS in Engring., McGill U., Montreal, 1947. Research engr. Can. Liquid Air, Ltd., Montreal, 1947—. Co-inventor OBM/Q-BOP Oxygen Seel Refining Process, 1967. Bd. dirs. Montreal Chinese Hosp., 1975-82. Mem. Soc. for Crybiology, Can. Inst. of Mining and Metallurgy (Airey award 1974, Falconbridge Innovation awd., 1992), Am. Inst. of Mining and Metallurgy, Am. Chem. Soc. Office: Can Liquid Air Ltd, 1155 Sherbrooke St W, Montreal, PQ Canada H3A 1H8

LEE, ROBERT JEFFREY, municipal utility professional; b. Gary, Ind., Apr. 3, 1955; s. Charles Austin and Joan Jeanette (Julian) L.; m. Donna Kathleen Webb, Oct. 12, 1979. Cert., Ind. Vocat. Tech. Coll., 1985. Foreman City of Cown Point, Ind., 1975—; mem. Mayoral Ad Hoc Environ. Com., Crown Point, 1992--. Mem. March Dimes. Mem. Am. Water Works Assn. (cert.), Ind. Rural Water Assn., Am. Backflown Prevention Assn. Office: Crown Point Water Dept 1313 E North St Crown Point IN 46307

LEE, SANBOH, materials scientist; b. Chiayi, Taiwan, China, July 2, 1948; s. Ching-Shiang and Shioh-Yeh (Chang) L.; m. Hsiao-Fan Wang, June 21, 1986; children: I-Fan, Tau-Fan. BS, Fu Jen U., Taipei, Taiwan, 1970; MS, Tsing Hua U., Hsinchu, Taiwan, 1972; PhD, U. Rochester, 1980. Rsch. engr. Taiwan Power Co., Taipei, 1972-75; rsch. assoc. U. Rochester (N.Y.), 1982-83; assoc. prof. Nat. Tsing Hua U., Hsinchu, 1983-85; prof. Nat. Tsing Hua U., 1985—; vis. scholar Lehigh U., Bethlehem, Pa., 1987-88; head gen. affairs Materials Sci. Ctr., Hsinchu, 1985-87. Contbr. articles to profl. jours. Recipient Outstanding Rsch. award Nat. Sci. Coun., 1989, 91. Mem. AIAA, ASM Internat., Minerals, Metals and Materials Soc., Am. Phys. Soc. Avocations: bridge, go, Chinese chess, table tennis, tennis. Home: Nat Tsing Hua U, W Ct 60, Hsinchu 30043, Taiwan Office: Nat Tsing Hua U, Dept Materials Sci, Hsinchu 30043, Taiwan

LEE, SANG-GAK, astronomy educator; b. Seoul, Korea, Feb. 6, 1948; d. Chae-Ku and Auh-keum (Chun) L.; m. Young-June Lee, Sept. 27, 1983; 1 child, June-Hae. BS in Astronomy, Seoul Nat. U., 1971; MS in Astronomy, Case Western Res. U., 1974, PhD in Astronomy, 1978. Lectr. Seoul Nat. U./Yonsei U., Seoul, 1979-80; asst. prof. Seoul Nat. U., 1980-85, assoc. prof., 1985-91, prof., 1991—. Translator: (book) Journey to the Stars, 1990; editor: Korean Astron. Jour., 1988-90, Jour. of Korean Astron. Soc., 1988-90. Mem. Am. Astron. Soc., Astron. Soc. of the Pacific, Internat. Astron. Union, Korean Astron. Soc. Office: Seoul Nat U, san 56-1 ShinRim-Dong Kwanak-ku, 151-742 Seoul Republic of Korea

LEE, SHYAN JER, physical chemist; b. Kaohsiung, Taiwan, Oct. 21, 1961; came to U.S., 1987; s. Hao-Jye and Roung (Yang) L.; m. Lynn Farh, Dec. 14, 1989. BS, Chung Yang U., Taiwan, 1984; PhD, U. Iowa, 1992. Mgr. Isosceles Inc., Kaohsiung, Taiwan, 1986-87; teaching and rsch. asst. U. Iowa, Iowa City, 1988-92, rschr., 1992—. Contbr. articles to profl. jours. Mem. Am. Phys. Soc., Am. Vacuum Soc., Optical Soc. Am. Achievements include research in laser assisted partical removal - this novel technique is capable of removing micron or submicron particulate contamination from critical surfaces. Home: 265 Hawkeye Ct Iowa City IA 52246 Office: Univ Iowa 265 Hawkeye Ct Iowa City IA 52246

LEE, STEPHEN, chemist, educator. Prof. dept. chemistry U. Mich., Ann Arbor. MacArthur fellow John D. and Katherine T. MacArthur Found., 1993. Office: Univ of Michigan Dept Chemistry 1543 Chemistry Bldg Ann Arbor MI 48109*

LEE, SUN BOK, biochemical engineering educator; b. Cheong Ju, Chung Buk, Korea, Dec. 23, 1953; s. Won Ku and Moon G. (Yeon) L.; m. Young Ran Cho, Oct. 6, 1981; children: Jae Won, Ji Won. BS, Seoul (Korea) Nat. U., 1976; PhD, Korea Adv. Inst. Sci. Tech., Seoul, 1981. Cert. biochem. engring. Rsch. fellow Calif. Inst. Tech., Pasadena, 1981-83; asst. prof. Korea Adv. Inst. Sci. Tech., Seoul, 1983-88; head enzyme technol. lab. Genetic Engring. Rsch. Inst., Seoul, 1988-89; assoc. prof. biochem. engring. Pohang (Korea) Inst. Sci. Tech., 1989—; vis. prof. U. Calif., Davis, 1985. Author: Biocatalysis in Organic Solvent, 1991, Bioprocess Kinetics of Recombinant Culture, 1991. Mem. adv. bd. Pohang City Coun., 1991—. Recipient govt. fellowship Korea Adv. Inst. Sci. Tech., Seoul, 1976, postdoctoral fellowship Korea Sci. and Engring. Found., Seoul, 1981, Govt. award Ministry Sci. and Tech., Seoul, 1988. Mem. Am. Inst. Chem. Engrs., Am. Soc. Microbiolgoy, Korean Inst. Chem. Engrs., N.Y. Acad. Scis. Achievements include patents for separation of aminoglycoside antibiotics, enzymatic synthesis of cephalexin, L-trytophan fermentation, enzymatic synthesis beta-lactam antibiotics; findings on quantitative description of gene expression kinetics, enzyme kinetics in anhydrous media. Office: Pohang Inst Sci Tech, PO Box 125, Pohang 790 600, Republic of Korea

LEE, SUNG JAI, medicinal chemist; b. Seoul, Korea, Sept. 20, 1955; came to U.S., 1964; s. Hwan W. and Myung S. (Park) L.; m. Jane Bi Yi, Nov. 7, 1981; children: Samuel, Peter. BS, UCLA, 1977; PhD, U. Minn., 1983. Sr. rsch. scientist Bristol-Myers Pharm., Evansville, Ind., 1983-84, Ayerst Labs., Princeton, N.J., 1984-86; v.p. chemistry Biofor, Inc., Waverly, Pa., 1986—. Achievements include patents on novel oxazinone anti-inflammatory agents and pyrolidinone anti-inflammatory agents.

LEE, SUNG TAICK, aerospace engineer; b. Seoul, South Korea, Jan. 22, 1957; s. Yong Ho and Hee Kyung Lee; m. Sun Hwa Cho, Sept. 8, 1986. BSChemE, Hanyang U., 1979; MSChemE, Seol Nat. U., 1981; MSChem and Aerospace Engring., Ga. Inst. Tech., 1988, PhD in Aerospace Engring., 1991. Engr. Daewoo Engring. Co., Seoul, 1979-80; researcher Korea Explosives Co., Inchon, South Korea, 1980-86; sr. rschr. gen. mgr. Aerospace and Satellite Comm. Divsns. HAN WHA Co., Daejeon, South Korea, 1991—. Contbr. articles to Combustion and Propulsion. Mem. AIAA, Korean Inst. Aeronautics and Astronautics. Home: 6 Dong 1002 Ho, Hanyang Apt 6-1002 Wolpi-Dong447, Ansan Kyungki-Do 425-070, Republic of Korea Office: HAN WHA Co, Woisam Dong Yousung Gu, Daejeon 305-156, Republic of Korea

LEE, TERRY JAMES, environmental engineer, consultant; b. Easton, Pa., May 12, 1947; s. Frank Herman and Marguerite Helen (Weirbach) L.; m. Carolyn Ann Ranktis, Aug. 19, 1947; 1 child, Carrie Jean. BSChemE, Lafayette Coll., 1969; MSChemE, U. Mass., 1971. Profl. engr., Pa. Process engr. Pfizer Inc., Easton, 1971-78, mgr. environ. engring., 1979-90, sr. environ. engr., 1991-92; mgr. environ. engring. Harcros Pigments Inc., Easton, 1990; sr. environ. engr. Minerals Techs., Inc., Easton, 1992—; ind. cons., Easton, 1991—. Contbr. articles to profl. jours. Trustee Hugh Moore Park & Canal Mus., Easton, 1991—; commr. Joint Planning Commn. of Lehigh & Northampton Counties, Allentown, Pa., 1988—; bd. dirs. Bushkill Stream Conservancy, Easton, 1991—. Fellow Am. Prodn. & Inventory Control Soc.; mem. Tau Beta Pi. Republican. Mem. United Ch. of Christ. Achievements include U.S. and French patents for Sulfur Oxides Reduction. Home: 89 Old Well Rd Easton PA 18042-7077 Office: Minerals Techs Inc 640 N 13th St Easton PA 18042

LEE, THOMAS HENRY, electrical engineer, educator; b. Shanghai, China, May 11, 1923; came to U.S., 1948, naturalized, 1953; s. Y. C. and Nan Tien (Ho) L.; m. Kin Ping, June 12, 1948; children—William F., Thomas H. Jr., Richard T. B.S.M.E., Nat. Chiao Tung U., Shanghai, 1946; M.S.E.E., Union Coll., Schenectady, 1950; Ph.D., Rensselaer Poly. Inst., 1954. Registered profl. engr., Pa. Mgr. research and devel. Gen. Electric Co., Phila., 1959-74; mgr. strategic planning Fairfield, Conn., 1974-78, staff exec., 1978-80; prof. elec. engring. MIT, Cambridge, 1980-84, 87—; dir. Internat. Inst. for Applied Systems Analysis, Laxenburg, Austria, 1984-87; pres. Tech. Assessment Group, Schenectady, 1980-84, Ctr. for Quality Mgmt., Cambridge, Mass., 1990—. Author: Physics and Engineering of High Power Switching Devices, 1973, Energy Aftermath, 1989; patentee in field. Recipient Davis medal for outstanding engring. accomplishment Rensselaer Poly. Inst., 1987. Fellow IEEE (Power Life award 1980, Haraden Pratt award 1983), AAAS; mem. NAE, Swiss Acad. Engring. Sci., Power Engring. Soc. (pres. 1974-76). Office: MIT 77 Massacusetts Ave Cambridge MA 02139-4307

LEE, THOMAS J., aerospace scientist; b. Wedowee, Ala., 1935; m. Jean Gullatt; children: Kevin, Patrick. BS in Aero. Engring., U. Ala., 1958; grad. advanced mgmt. program, Harvard U., 1985; PhD (hon.), U. Ala., Huntsville, 1993. Registered profl. engr., Ala. Aero. rsch. engr. U.S. Army Ballistic Missle Agy., Redstone Arsenal, Ala., 1958-60; systems engr. Marshall Space Flight Ctr., Huntsville, Ala., 1960-69, tech. asst. to. tech. dep. dir., 1969-73, mgr. Sortie Lab task team, 1973-74, mgr., 1974-80, dep. dir., 1980-89, dir., 1989—. Recipient Medal for Exceptional Svc, Outstanding Leadership medal, Equal Opportunity award NASA, Meritorious Exec. award, 1988, Disting. Exec. award, 1991, Executive Excellence Disting. Svc. award Sr. Execs. Assn. Profl. Devel. League, 1992, Werner Von Braun Space Flight trophy Hunstville chpt. Nat. Space Club, 1993; named to Ala. Engring. Hall of Fame, 1993; Disting. Engring. fellow U. Ala. Fellow AIAA (Von Braun award for Excellence in Space Program Mgmt.). Office: NASA Marshall Space Flight Center Huntsville AL 35812*

LEE, TSU TIAN, electrical engineering educator; b. Taipei, Taiwan, Republic of China, Feb. 1, 1949; s. Chan sen and Mei (Pai) L.; m. Fay Suzanne Lue, Feb. 26, 1976; children: Stephen Yin-chen, George Mingchiao. BS, Nat. Chiao-Tung U., Hsinchu, Republic of China, 1970; MS, U. Okla., 1973, PhD, 1975. Assoc. prof. Nat. Chiao-Tung U., Hsinchu, 1975-78, prof., chmn., 1978-84; vis. prof. U. Ky., Lexington, 1986-87, prof. elec. engring., 1987-90; prof., chmn. elec. engring. Nat. Taiwan Inst. Tech., Taipei, 1991—; cons. Mech. Industry Rsch. Lab., Hsinchu, 1984-86; rsch. fellow Academia Sinica, Taipei, 1985-86. Author: Control Systems I &II, 1982; contbr. articles to profl. jours. Mem. IEEE (sr.). Achievements include invention of quadruped walking robot. Office: Nat Taiwan Inst Tech, 43 Keelung Rd Sec 4, Taipei 10772, Taiwan

LEE, TSUNG-DAO, physicist, educator; b. Shanghai, China, Nov. 25, 1926; s. Tsing-Kong L. and Ming-Chang (Chang); m. Jeannette Chin, June 3, 1950; children: James, Stephen. Student, Nat. Chekiang U., Kweichow, China, 1943-44, Nat. S.W. Assoc. U., Kunming, China, 1945-46; PhD, U. Chgo., 1950; DSc (hon.), Princeton U., 1958; LLD (hon.), Chinese U., Hong Kong, 1969; DSc (hon.), CCNY, 1978. Research assoc. in astronomy U. Chgo., 1950; research assoc., lectr. physics U. Calif., Berkeley, 1950-51; mem. Inst. for Advanced Study, Princeton (N.J.) U., 1951-53, prof. physics, 1960-63; asst. prof. Columbia U., N.Y.C., 1953-55, assoc. prof., 1955-56, prof., 1956-60, 63—, adj. prof., 1960-62, Enrico Fermi prof. physics, 1964—, also Univ. prof.; Loeb lectr. Harvard U., Cambridge, Mass., 1957, 64. Editor: Weak Interactions and High Energy Nutrino Physics, 1966, Particle Physics and Introduction to Field Theory, 1981. Recipient Albert Einstein Sci. award Yeshiva U., 1957, (with Chen Ning Yang) Nobel prize in physics, 1957. Mem. NAS, Acad. Sinica, Am. Acad. Arts and Scis., Am. Philos. Soc., Acad. Nazionale dei Lincei. Office: Columbia U Dept Physics Morningside Heights New York NY 10027

LEE, USIK, mechanical engineer, educator; b. Namwon, Korea, July 6, 1956; came to U.S., 1979; s. Suk-Bong and Jeong-Kyung (Koh) L. MS, Stanford U., 1982, PhD, 1985. Rsch. asst. Stanford (Calif.) U., 1981-85; rsch. specialsit Korea Aero. Inst. Tech., Seoul, 1985-89; prof. Inha U., Inchon, Korea, 1989—. Contbr. articles to profl. jours. Mem. AIAA, ASME. Office: Inha U, 253 Yong Hyun-Dong Nam-Ku, Inchon 402-751, Republic of Korea

LEE, WILLIAM JOHN, petroleum engineering educator, consultant; b. Lubbock, Tex., Jan. 16, 1936; s. William Preston and Bonnie Lee (Cook) L.; m. Phyllis Ann Bass, June 10, 1962; children—Anne Preston, Mary Denise. B.Chem. Engring., Ga. Inst. Tech., 1959, M.S. in Chem. Engring., 1961, Ph.D. in Chem. Engring., 1963, Nat. Acad. Engring., 1993. Registered profl. engr., Tex., Miss. Sr. research specialist Exxon Prodn. Research Co., Houston, 1962-68; assoc. prof. petroleum engring. Miss. State U., Starkville, 1968-71; tech. advisor Exxon Co., Houston, 1971-77; prof. petroleum engring. Tex. A&M U., College Station, 1977—, holder Noble chair in petroleum engring., 1985—; dir. Crisman Inst. for Petroleum Reservoir Mgmt. at Tex. A&M U., 1987—; exec. v.p. S.A. Holditch & Assocs., Inc., College Station, 1979—. Author: Well Testing, 1982. Recipient award of excellence Halliburton Edn. Found., 1982, Meritorious Engring. Teaching award Tenneco, Inc., 1982, Disting. Teaching award Assn. Former Students, Tex. A&M U., College Station, 1983; Tex. Engring. Experiment Sta. fellow, 1987-88, sr. fellow 1990. Mem. Soc. Petroleum Engrs. (disting., chmn. edn. and accreditation com. 1985-86, disting. lectr. 1980, disting. faculty achievement award 1982, Reservoir Engring. award 1986, Regional service award 1987, disting. svc. award 1992). Presbyterian. Avocation: investments. Home: 3100 Rolling Gln Bryan TX 77801-3209 Office: Tex A&M U Petroleum Engring Dept College Station TX 77843-3116

LEE, WILLIAM STATES, utility executive; b. Charlotte, N.C., June 23, 1929; s. William States and Sarah (Everett) L.; m. Janet Fleming Rumberger, Nov. 24, 1951; children—Lisa, States, Helen. B.S. in Engring. magna cum laude, Princeton U., 1951. Registered profl. engr., N.C., S.C. With Duke Power Co., Charlotte, 1955—, engring. mgr., 1962-65; v.p. engring. Duke Power Co., 1965-71, sr. v.p., 1971-75, exec. v.p., 1976-77, pres., chief operating officer, 1978-82, chmn., chief executive officer, 1982-89, chmn., chief exec. officer, 1989—, also dir., mem. mgmt. and fin. coms.; mem. U.S. Com. on Large Dams, 1963—; bd. dirs. Liberty Corp., J.P. Morgan Co., Morgan Guaranty Trust Co., Knight-Ridder, Tex. Instruments. Bd. dirs. United Community Svcs., Am. Nuclear Energy Coun., Edison Electric Inst., Found. of the Carolinas; mem. chmn. N.C. Gov.'s Bus. Coun.; trustee Queens Coll., U. N.C. Charlotte Found., Presbyn. Hosp. Found. With C.E. USNR, 1951-54. Named Outstanding Engr. N.C. Soc. Engrs., 1969. Fellow ASME (George Westinghouse gold medal 1972, James N. Landis medal 1991), ASCE; mem. Nat. Acad. Engring., Nat. Soc. Profl. Engrs. (Outstanding Engr. award 1980), Edison Electric Inst. (dir. econs. and fin. policy com., dir.), Charlotte C. of C. (chmn. 1979), Am. Nuclear Soc., Phi Beta Kappa, Tau Beta Pi. Presbyn. (ruling elder). Office: Duke Power Co 422 S Church St Charlotte NC 28202

LEE, WON JAY, radiologist; b. Seoul, Korea, Feb. 2, 1938; came to U.S., 1965; s. Kang Sei and Choon Ja (Park) L.; m. Moon Jung, Feb. 24, 1968; children: Julie, Lisa, Jennifer. Dr.med., Yonsei U., Seoul, 1962. Diplomate Am. Bd. Radiology, Am. Bd. Nuclear Medicine. Intern Wyckoff Heights Hosp., Bklyn., 1965-66; resident NYU Med. Ctr., N.Y.C., 1966-69; fellow, asst. radiologist L.I. Jewish Med. Ctr., New Hyde Park, N.Y., 1969-71, staff radiologist, 1975-82, chief uroradiology, 1983—; assoc. radiologist Binghamton (N.Y.) Gen. Hosp., 1971-75; asst. prof. SUNY, Stony Brook, 1975-86, assoc. prof. radiology, 1987-89; prof. radiology Albert Einstein Coll. Medicine, 1989—; cons. in field. Contbr. chpts. to books and articles to profl. jours. First lt. USMC, 1962-65, Korea. Recipient Clin. award Can. Assoc. Radiologists, 1979. Fellow Am. Coll. Radiology, Cardiovascular and Interventional Radiology; mem. Assn. Univ. Radiologists, Am. Roentgen Ray Soc., Radiol. Soc. N.Am., Soc. Uroradiology. Democratic. Methodist. Avocations: gardening, gofl, traveling. Home: 15 Lucille Ln Huntington Station NY 11746-5848 Office: LI Jewish Med Ctr 27005 76th Ave New Hyde Park NY 11040-1433

LEE, YOUNG KI, economist, researcher; b. Dae Gu, Republic of Korea, July 25, 1946; s. Chong Ho and Duck Hae (Chung) L.; m. Soo Jung Kang, Mar. 31, 1973; children: Eugene S., Sung Hwan. B in Engring., Seoul Nat. U., 1969, MBA summa cum laude, 1972; DBA, Boston U., 1983. Rsch. assoc. Korea Devel. Inst., Seoul, 1972-83, sr. rsch. fellow, 1983—, dir. rsch. planning and coord., 1992—; advisor Korea Ins. Commn., 1986; cons. Ministry of Fin., 1987-89; bd. dirs. Korea Stock Exchange, 1990—; mem. fin. devel. com. Ministry of Fin., 1988—. Author: Securities Market and Industries in Korea, 1990; contbr. articles to prof. jours. Fellow Korea Mgmt. Devel. Inst.; mem. Korea Fin. Assn. (bd. dirs. 1988-92), Korea Corp. Fin. Assn. (bd. dirs. 1988-91). Avocations: photography, tennis. Home: Ban Po Apt 69-105, Seocho Ku, Seoul Republic of Korea Office: Korea Devel Inst Cheong Ryang PO Box 113, Seoul Republic of Korea

LEE, YOUNG MOO, polymer chemist; b. Seoul, Republic of Korea, Oct. 23, 1954; s. Seung Nyong and Hong Wol (Kim) L.; m. Haewon Wang, Mar. 13, 1981; children: Bumkil, Juliet. BS, Hanyang U., Seoul, 1977, MS, 1979; PhD, N.C. State U., 1986. Rsch. Rensselaer Poly. Inst., Troy, N.Y., 1986-87; sr. polymer chemist 3M Corp., St. Paul, 1987-88; asst. prof. polymer sci. Hanyang U., Seoul, 1988-92, assoc. prof. polymer sci., indsl. chem., 1992—. Contbr. articles to profl. jours. Sgt. Korean Army, 1979-81. Mem. Am. Chem. Soc., N.Am. Membrane Soc., Korean Indsl. Chem. Soc. (bd. dirs., organizing com.). Avocation: tennis. Home: Samik Apt 201-1105, Song padong, Song pa-ku, Seoul 138-171, Republic of Korea Office: Hanyang U Dept Indst Chem, 17 Haengdang-dong, Seoul 133-791, Republic of Korea

LEE, YUAN T(SEH), chemistry educator; b. Hsinchu, Taiwan, China, Nov. 29, 1936; came to U.S., 1962, naturalized, 1974; s. Tsefan and Pei (Tasi) L.; m. Bernice Wu, June 28, 1963; children: Ted, Sidney, Charlotte. BS, Nat. Taiwan U., 1959; MS, Nat. Tsinghua U., Taiwan, 1961; PhD, U. Calif., Berkeley, 1965. From asst. prof. to prof. chemistry U. Chgo., 1968-74; prof. U. Calif., Berkeley, 1974—, also prin. investigator Lawrence Berkeley Lab. Contbr. numerous articles on chem. physics to profl. jours. Recipient Nobel Prize in Chemistry, 1986, Ernest O. Lawrence award Dept. Energy, 1981, Nat. Medal of Sci., 1986, Peter Debye award for Phys. Chemistry, 1986; fellow Alfred P. Sloan, 1969-71, John Simon Guggenheim, 1976-77; Camille and Henry Dreyfus Found. Tchr. scholar, 1971-74. Fellow Am. Phys. Soc.; mem. NAS, AAAS, Am. Acad. Arts and Scis., Am. Chem. Soc. Office: U Calif Dept Chemistry Berkeley CA 94720

LEE, YUEN FUNG, mechanical engineer; b. Kwangsi, China, Feb. 28, 1940; came to U.S., 1964; s. Ping S. L.; m. Rita H. Shih, Aug. 14, 1971; children: Kay, Jerry. BS in Structure Engring., Chung Yuan Coll., 1963; MS, U. Notre Dame, 1966; PhD, Cornell U., 1971. Sr. engr. Corning (N.Y.) Glass Works, 1972-75, Acurex Corp., Mt. View, Calif., 1975-77; lead engr. Westinghouse, Sunnyvale, Calif., 1977-81; adv. engr. Tandem Computers, Cupertino, Calif., 1981—. Contbr. articles to profl. jours. Recipient Invention award Westinghouse Electric, 1985. Mem. Inst. Packaging Profls. Achievements include patent in shock absorbent gas seal; establishment of dynamics lab and test specification for testing electronic equipment. Home: 10446 Avenida Ln Cupertino CA 95014-3947

LEE, YUNG-KEUN, physicist, educator; b. Seoul, Korea, Sept. 26, 1929; came to U.S., 1953, naturalized, 1968; s. Kwang-Soo and Young-Sook (Hur) L.; m. Ock-Kyung Pai, Oct. 25, 1958; children—Ann, Arnold, Sara, Sylvia, Clara. B.A., Johns Hopkins, 1956; M.S., U. Chgo., 1957; Ph.D., Columbia, 1961. Research scientist Columbia U., N.Y.C., 1961-64; prof. physics Johns Hopkins U., Balt., 1964—; vis. mem. staff Los Alamos Sci. Lab., 1971; vis. researcher Institut Scis. Nucléaires, Grenoble, France, 1975; cons. Idaho Nat. Engring. Lab., 1988—. Contbr. articles to profl. jours. Mem. Am. Phys. Soc. Democrat. Methodist. Club: Johns Hopkins. Home: 1318 Denby Rd Baltimore MD 21286 Office: Johns Hopkins U 34th and Charles Sts Baltimore MD 21218

LEE, YUNG-MING, educator; b. Taichung, Taiwan, Republic of China, June 1, 1957; s. Chung-Lin and Shih (Chen) L.; m. Mei-Yane Chung, Aug. 12, 1985; 1 child, Ho Lee. MS, U. Tex., Arlington, 1987, PhD, 1991. Assoc. prof. Feng Chia U., Taichung, 1991—. Contbr. articles to profl. jours. Mem. AIAA, ASME (assoc.). Office: Dept Aero Engring, Feng Chia U 100 Wen-Hwa Rd, Taichung 40724, Taiwan

LEE, ZUK-NAE, psychiatry educator, psychotherapist; b. Hab-Chun, Kyungnam, Republic of Korea, Feb. 5, 1940; s. Sang-Yong and Yeum-Chun Song-Lee; m. Young-Hee Kwon-Lee, May 7, 1968; children: Kyung-Im, Sung-Lim. MD, Kyungpook U., Taegu, Republic of Korea, 1965; Lic. in Philosophy, Zurich U., Switzerland, 1986, PhD, 1990. Lic. psychiatrist. Intern Korean First Army Hosp., Taegu, 1965-66, resident, 1966-70; dir. Non-San Army Hosp., 1970-71, 102 Korean Army Hosp., Natrang, Vietnam, 1971-72; psychiat. researcher HQ for Research, Seoul, Republic of Korea, 1972-73; teaching staff Med. Coll. Chungnam U., Taechun, Republic of Korea, 1973-74; tng. candidate Jung Inst., Zurich, 1974-78; asst. prof. Med. Coll. Kyungpook U., Taegu, 1978-82, assoc. prof. psychiatry, 1982-90, prof. psychiatry, 1990—. Chief editor: Shim-Song Yon-Gu, 1986—. Served to maj. Korean mil., 1982-83. Mem. Korean Acad. Psychotherapists (pres. Taegu br. 1988-89), Korean Assn. Psychotherapists (exec. com. 1984—), Korean Soc. Analytical Psychology (tng. analyst 1978—, v.p. 1986-88, pres. 1988-91), Inst. for Human Sci. (exec. com. 1987-89), Korean Neuropsychiatric Assn. (pres. Taegu br. 1988-89). Avocation: mountain climbing. Home: 50 Samduk-Dong 1-Ka Chung-Ku, Taegu 700-411, Republic of Korea Office: Kyungpook U Hosp, 52 Samduk-Dong Chung-Ku, Taegu 700-412, Republic of Korea

LEEB, CHARLES SAMUEL, clinical psychologist; b. San Francisco, July 18, 1945; s. Sidney Herbert and Dorothy Barbara (Fishstrom) L.; m. Storme Lynn Gilkey, Apr. 28, 1984; children: Morgan Evan, Spencer Douglas. BA in Psychology, U. Calif.-Davis, 1967; MS in Counseling and Guidance, San Diego State U., 1970; PhD in Edn. and Psychology, Claremont Grad. Sch., 1973. Assoc. So. Regional Dir. Mental Retardation Ctr., Las Vegas, Nev., 1976-79; pvt. practice, Las Vegas, 1978-79; dir. biofeedback and athletics Menninger Found., Topeka, 1979-82, dir. children's div. biofeedback and psychophysiology ctr. The Menninger Found., 1979-82; pvt. practice, Claremont, Calif., 1982—; dir. of psychol. svcs. Horizon Hosp., 1986-88; dir. adolescent chem. dependency and children's program Charter Oak Hosp., Covina, Calif., 1989-91; founder, chief exec. officer Rsch. and Treatment Inst., Claremont, 1991; lectr. in field. Contbr. articles to profl. jours. Mem. Am. Psychol. Assn., Calif. State Psychol. Assn. Office: 937 W Foothill Blvd Ste D Claremont CA 91711-3358

LEECH, SALLY See KEMP, SARAH

LEELAND, STEVEN BRIAN, electronics engineer; b. Tampa, Fla., Dec. 27, 1951; s. N. Stanford and Shirley Mae (Bahner) L.; m. Karen Frances Hayes, Dec. 20, 1980; children: Crystal Mary, April Marie. BSEE, MSEE magna cum laude, U. South Fla., 1976. Registered profl. engr., Ariz. Engr. Bendix Avionics, Ft. Lauderdale, Fla., 1976-77; prin. engr., instr. Sperry Avionics, Phoenix, 1977-84; prin. staff engr. Motorola Govt. Electronics Group, Scottsdale, Ariz., 1984-88; sr. staff engr. Fairchild Data Corp., Scottsdale, 1988—; cons. Motorola Govt. Electronics Group, 1991. Patentee systolic array, 1990; contbr. articles to profl. jours. Mem. IEEE (Phoenix chpt. Computer Soc. treas. 1978-79, sec. 1979-80, chmn. 1980-81, 81-82), Tau Beta Pi, Pi Mu Epsilon, Phi Kappa Phi, Omicron Delta Kappa, Themis. Republican. Adventist. Avocations: chess, computers, biking, exercise, health. Home: 10351 E Sharon Dr Scottsdale AZ 85260-9000 Office: Fairchild Data Corp 350 N Hayden Rd Scottsdale AZ 85257-4692

LEEMAN, SUSAN EPSTEIN, neuroscientist; b. Chgo., May 9, 1930; d. Samuel and Dora (Gubernikoff) Epstein; m. Cavin Leeman (div.); children: Eve, Raphael, Jennifer. BA, Goucher Coll., 1951; MA, Radcliffe Coll., 1954, PhD, 1958; DS (hon.), SUNY, Utica, 1992; hon. degree, Goucher Coll., 1993. Instr. Harvard Med. Sch., Boston, 1958-59; rsch. assoc., adj. asst. prof., asst. rsch. prof. Brandeis U., Waltham, Mass., 1966-68, 68-71; asst. prof. Harvard Med. Sch., 1972-73, assoc. prof., 1973-80; prof. U. Mass. Med. Ctr., Worcester, 1980-92, dir. interdept. neurosci. program, 1984-92; prof. Boston U. Sch. Medicine, 1992—. Postdoctoral fellow Brandies U., 1959-62, 62-66; recipient Burroughs Wellcome Vis. Professorship award U. Ky., 1992, Women's Excellence in Scis. award Fedn. Am. Socs. for Exptl. Biology, 1993. Mem. Nat. Acad. Scis. Office: Boston U Sch Medicine Dept Physiology 80 E Concord St Boston MA 02118

LEEMANS, WIM PIETER, physicist; b. Gent, Belgium, June 7, 1963. BS in Elec. Engring., Free U. Brussels, 1985; MS in Elec. Engring., UCLA, 1987, Ph.D. in Elec. Engring., 1991. Teaching asst. UCLA, 1986-87, rsch. asst., 1987-91; staff scientist Lawrence Berkeley Lab., Berkeley, Calif., 1991—; presenter numerous seminars. Contbr. articles to profl. jours. Recipient Simon Ramo awd., Am. Physical Soc., 1992; grad. scholar IEEE Nuclear and Plasma Soc., 1987. Fellow Belgian Am. Ednl. Found., Francqui Found.; mem. IEEE (Nuclear and Plasma scis. soc. grad scholar 1987), SPIE, Am. Phys. Soc. (Simon Ramo award 1992), Royal Flemish Engrs. Soc. Achievements include research in high-intensity laser-plasma interaction in pre-formed and laser-produced plasmas; plasma beat-wave acceleration of electrons, driven density fluctuations using collective Thomson scattering of visible laser probe beam, and analysis of forward and backward scattered spectra pf the pump beam and hole-coupled resonator mode analysis in support of LBL's infared FEL program. Office: Lawrence Berkeley Lab Divsn Accelerator Fusion Rsch 1 Cyclotron Road MS 71-259 Berkeley CA 94720*

LEES, RON MILNE, physicist, educator; b. Sutton, Eng., Oct. 28, 1939; married, 1962; 2 children. BS, U. B.C., 1961, MS, 1965; Phd in Physics, Bristol U., 1967. Assoc. prof. U. N.B., Fredericton, 1968-77, prof. physics, 1977—, dept. chmn., 1981-88; vis. assoc. prof. physics dept. U. B.C., Vancouver, 1974-75; prin. investigator Ctrs. of Excellence in Molecular & Interfacial Dynamics, Fed. Networks of Ctrs. Excellence Program. Nat. Rsch. Coun. Can. fellow Nat. Rsch. Coun., Ottawa, 1966-68. Mem. Can. Am. Physicists, Am. Assn. Physics Tchrs., Optical Soc. Am., Soc. Photo-Optical Instrumentation Engrs. Achievements include research in atomic and molecular physics, molecular spectroscopy. Office: Can Assn Physicists, 151 Slater St # 903, Ottawa, ON Canada K1P 5H3*

LEESER, DAVID O., materials engineer, metallurgist; b. El Paso, Tex., Aug. 3, 1917; s. Oscar D. and Rose R. (Goodman) L.; m. Marilyn Bachman Kalina, Mar. 18, 1945; children: Barbara H., Joyce N. BSc in Mining, U. Tex., El Paso, 1943; MSc in Materials Sci., Ohio State U., 1950; postgrad., U. Fla., 1962, U. Mich., 1985. Registered profl. engr., Ohio, Calif., Ariz. Mining engr. Bradley Mining Co., Stibnite, Idaho, 1943-44; rsch. engr. Battelle Meml. Inst., Columbus, Ohio, 1944-50; assoc. metallurgist Argonne Nat. Lab., Lemont, Ill., 1950-54; mgr., staff metallurgist Atomic Power Devel. Assn., Detroit, 1954-61; chief scientist, materials Chrysler Corp./Missiles, Cape Canaveral, Fla., 1961-68; chief metallurgist Chrysler Corp., Amplex div., Detroit, 1968-75; sr. staff engr. Burroughs Corp. (Unisys), Plymouth, Mich., 1975-86; prin., materials cons., forensic engr. D.O. Leeser, Profl. Egnr., Scottsdale, Ariz., 1987—; materials cons. fgn. tech. div. Wright-Patterson AFB, Dayton, Ohio, 1958-80; charter mem. Missile, Space and Range Pioneers, Cape Canaveral, 1967—; organizer Fla. Indsl. Exhbn., Orlando, 1968. Contbr. articles on nuclear radiation effects on structural materials, flame deflector materials for rockets, forensic investigations to profl. publs. Vice-pres. Manzanita Villas Home Owners Assn., Scottsdale, 1990—. Recipient citation War Manpower Commn., 1945, U.S. Sci. R&D Office, 1946, Appreciation award State of Fla., 1968; Engr. of Yr. Fla. Engring. Soc.-Canaveral Coun. Tech. Socs., 1967, Outstanding Alumnus of Yr., U. Tex. at El Paso, 1969; recipient Apollo Achievement award NASA, 1969, Exemplary Action award Burroughs Corp., 1986. Mem. Am. Soc. materials Internat., Sigma Xi. Achievements include collaboration on development of atomic bomb, Apollo space program, first nuclear-powered naval vessel; pioneering studies on effects of nuclear radiation on materials. Home: 11515 N 91st St #151 Scottsdale AZ 85260 Office: DO Leeser Profl Engr 11515 91st St Ste 151 Scottsdale AZ 85260

LE FAVE, GENE MARION, polymer amd chemical company executive; b. Green Bay, Wis., May 18, 1924; s. Thomas Paul and Marie Agnes (Young) Le F.; m. Rosemary Beatrice Sackinger, Aug. 28, 1948; children: Laura, Deborah, Michele, Mark, Camille, Jacques, Louis. BS, U. Notre Dame, 1948; MS, Butler U., 1950. Staff engr. P.R. Mallory & Co., Indpls., 1953-54; sr. staff engr. Lear, Inc., Santa Monica, Calif., 1954-56; chief engr. G.M. Giannini & Co., Pasadena, Calif., 1956; v.p., dir. Coast Pro Seal & Mfg. Co., Compton, Calif., 1956-64; cons. Input/Output, Whittier, Calif., 1964-67, Diamond Shamrock, Painesville, Ohio, 1967-71; mem. bd. cons. U.S. Army

Corps of Engrs., Mariemont, Ohio, 1964-71; cons. Joslyn Mfg. & Supply Co., Chgo., 1964-72, Arco Chem. Co., Phila., 1969-75; pres. Fluid Polymers, Inc., Las Vegas, Nev., 1970—; bd. dirs. Polimeros Flexibles de Monterrey (Mexico), SA, Desert Industries, Inc., Las Vegas. Contbr. articles to profl. jours. Bd. dirs. adv. com. Nat. Bus., Las Vegas, 1990—; mem. regents com. on sci. and tech. U. Nev., Las Vegas, 1990—; mem., chmn. Rep. Party, Whittier, 1965-67. Mem. Am. Inst. Chem. Engrs., Am. Inst. Chemists, Am. Concrete Inst. Byzantine Catholic. Avocations: golf, bridge. Home: 1568 Leatherleaf Dr Las Vegas NV 89123-1942

LEFCORT, HUGH GEORGE, zoologist; b. Boston, Sept. 7, 1962; s. Dorothy (Sedlezky) L. BA in Anthropology, U. Wash., 1985; PhD, Oreg. State U., 1993. Researcher U. Wash., Seattle, 1987-88; lab. coord. Oreg. State U., Corvallis, 1988-92, researcher, 1988-92; researcher U.S. Forest Svc., Corvallis, 1992; instr. Ctrl. Washington U., Corvallis, 1993; asst. prof. Ga. So. U., Stateboro, 1993—; cons. Wm. C. Brown Pub., Dubuque, Iowa, 1991. Contbr. articles to profl. jours. Recipient Bailey fellowship Oreg. State U., 1991, Sigma Xi Grad. award, 1991, Rsch. grant N.Y. Mus. Natural History, 1991, Grant-in-Aid of Rsch., Oreg. State U., 1991, 90. Mem. Animal Behavior Soc., Soc. for Study of Amphibians and Reptiles, Am. Soc. Naturalists, Am. Soc. Zoologists. Democrat. Achievements include discovery that part of the immune response can alter the behavior of parasitized animals. Office: Ga So U Landrum Box 8042 Statesboro GA 30460-8042

LEFEBURE, ALAIN PAUL, family physician; b. Paris, Nov. 17, 1946; s. Rene Julien and Ginette (Peradon) L.; m. Lucia Lefebure; children: Vincent, Benjamin. RN, Sch. Nursing, Paris, 1970; Dr. Medicine, U. Paris 7, 1978. Male nurse Tenon Hosp., Paris, 1970-71; intern Bichat Hosp., Paris, 1974-76; resident Bretonneau Hosp., Paris, 1976-78; lectr. in pediatrics Claude Bernard Hosp., Paris, 1976-81; practice medicine specializing in family medicine Paris, 1981—. Author six langs. med. dictionary Lexica Medica Polyglotta. Dep. Internat. Parliament for Safety and Peace; knight Templars of Jerusalem. Recipient Clarinetist Soloist prize Congratulations from the Jury, 1965. Mem. Comdr. des Lofsensischen Ursiniusordens, Internat. Order of Merit. Avocations: music, computing, langs. Office: 2 rue Pierre Mouillard, 75020 Paris France

LEFER, ALLAN MARK, physiologist; b. N.Y.C., Feb. 1, 1936; s. I. Judah and Lillian G. (Gastwirth) L.; m. Mary E. Indoe, Aug. 23, 1959; children—Debra Lynn, David Joseph, Barry Lee and Leslie Ann (twins). B.A., Adelphi Coll., 1957, Western Res. U., 1959; Ph.D. (NSF fellow), U. Ill., Urbana, 1962. Instr. physiology, USPHS-NIH fellow Western Res. U., 1962-64; asst. prof. physiology U. Va., 1964-69, asso. prof., 1969-71, prof., 1972-74; vis. prof. Hadassah Med. Sch., Jerusalem, 1971-72; prof., chmn. dept. physiology Jefferson Med. Coll., Thomas Jefferson U., Phila., 1974—; dir. Ischemia-Shock Research Inst., 1980—; cons. Merck & Co., Upjohn Co., Genentech Inc., Syntex, Inc., Ciba-Geigy, NIH, Smith, Kline and Beecham Labs., Bristol-Myers Squibb, Cytel Corp.; Wellcome Found. vis. prof., 1985-86; Nat. Bd. of Med. Examiners, Step 1, 1993—. Author: Pathophysiology and Therapeutics of Myocardial Ischemia, 1977, Prostaglandins in Cardiovascular and Renal Function, 1979, Cellular and Molecular Aspects of Shock and Trauma, 1983; Leukotrienes in Cardiovascular and Pulmonary Function, 1985; mng. editor: Eicosanoids, 1988-93; cons. editor Circulatory Shock, 1973-80; mem. editorial bd. Critical Care Medicine, Am. Jour. Physiology, Drug News and Perspectives Endothelium; contbr. to World Book Ency. Sci. Yearbook, 1979; contbr. over 500 sci. articles to profl. jours. Active Acad. Com. on Soviet Jewry, 1970—; chmn. United Jewish Appeal, 1973-74; coach basketball and baseball Huntingdon Valley Athletic Assn., 1975-78. Recipient Pres. and Visitor's prize in research U. Va., 1970. Fellow Am. Coll. Cardiology; mem. AAAS, Am. Physiol. Soc., Am. Soc. Pharmacology and Exptl. Therapeutics, Internat. Heart Research Soc., Am. Heart Assn. (established investigator 1968-73, fellow circulation coun., Nat. Grant Review Com., 1993—), Pa. Heart Assn. (research com.), Shock Soc. (chmn. membership com., pres. 1983-84, chmn. devel. com. 1985-89, chmn. internat. rels. com. 1993), Soc. Exptl. Biology and Medicine, Israel Soc. Physiology and Pharmacology, Phila. Physiol. Soc. (pres. 1978-79), Sierra Club, B'nai B'rith (Charlottesville chpt., v.p. 1967-68, chmn. Va. Hillel 1970-71), Sigma Xi. Democrat. Home: 3590 Walsh Ln Huntingdon Valley PA 19006-3226 Office: Thomas Jefferson Univ 1020 Locust St Philadelphia PA 19107-6799

LEFEVRE, ELBERT WALTER, JR., civil engineering educator; b. Eden, Tex., July 29, 1932; s. Elbert Walter Sr. and Hazie (Davis) LeF.; m. Joyce Ann Terry, Nov. 28, 1957; children: Terry Ann, Charmaine Rene, George Walter, John Philip. BS in Civil Engring, Tex. A&M U., 1957, MS in Civil Engring, 1961; PhD, Okla. State U., 1966. Registered profl. engr., Ark., Tex. Faculty Tex. A&M U., Bryan, 1958, Tex. Technol. Coll., Lubbock, 1959-63, Okla. State U., Stillwater, 1963-66, U. Ark., Fayetteville, 1966—; head dept. civil engring. U. Ark., 1971-82, dean engring., 1982-83; sr. v.p. Engring. Svcs., Inc., Springdale, Ark., 1973—; dir. Nat. Rural Transp. Study Ctr., 1992—; mem. Ark. State Bd. Registration for Profl. Engrs. and Land Surveyors, 1984—,Nat. Coun. Examiners for Engring. & Surveying, 1984—, v.p. So. Zone, 1991—, mem. accreditation bd. engring. and tech., 1985-91. Served to 1st lt. AUS, 1953-56. Fellow ASCE (pres. Mid-South sect. 1972, chmn. dist. 14 1977-80, dir. dist. 14 1983-86, nat. dir. 1981-83); mem. NSPE (v.p. 1984-85, 86-87, pres. 1988-89), Transp. Rsch. Bd., Am. Soc. Engring. Edn. (pres. Mid-West sect. 1976-77), Nat. Soc. Profl. Engrs. (v.p. profl. engrs. in edn. 1983, v.p. SW region 1984-86, pres. 1989-90), Ark. Soc. Profl. Engrs. (pres. 1979-80, Outstanding Ark. Engr. 1980), Sigma Xi, Chi Epsilon, Tau Beta Pi, Phi Beta Delta. Lodges: Masons, Rotary (pres. 1973). Home: 300 Paradise Ln Springdale AR 72762-3832 Office: Univ Ark Dept Civil Engring Fayetteville AR 72701

LEGATES, JOHN CREWS BOULTON, information scientist; b. Boston, Nov. 19, 1940; s. Eber Thomson and Sybil Rowe (Crews) LeG.; m. Nancy Elizabeth Boulton, Apr. 28, 1993. BA in Math., Harvard U., 1962. Edn. svcs. mgr. Telcomp Dept. Bolt Beranek & Newman, Cambridge, Mass., 1966-67; v.p. Washington Engring. Svcs., Cambridge, 1967-69; v.p., co-founder Cambridge Info. Systems, 1968-69; v.p., founder Computer Adv. Svc. to Edn., Wayland, Mass., 1966-72; exec. dir. Educom Interuniversity Communications Coun., Boston, 1969-72; founder, mng. dir. Program on Info. Resources Policy Harvard U., 1973—, founder, pres. Ctr. Info. Policy Rsch., 1978—; cons. in field. Contbr. articles to profl. jours. Bd. dirs. Nat. Telecommunications Conf., Washington, 1979. Kent fellow, 1964. Mem. NAS/NRC (telecommunications privacy, reliability and integrity panel), IEEE, Nat. Sci. Found., Soc. for Values in Higher Edn. Episcopalian. Club: Nashuba Valley Hunt (Pepperell, Mass.) (pres. 1974-80). Avocations: sailing, fox-hunting, mountaineering, classical music. Home: PO Box 331 Lincoln MA 01773-0012

LEGGETT, WILLIAM C., biology educator, educational administrator; b. Orangeville, Ont., Can., June 25, 1939; s. Frank William and Edna Irene (Wheeler) L.; m. Claire Holman, May 9, 1964; children: David, John. AB, Waterloo U. Coll., 1962; MSc, U. Waterloo, 1965; PhD, McGill U., 1969; Dsc, U. Waterloo, 1992. Don of men St. Pauls Coll. U. Waterloo, 1963-65; research scientist Essex (Conn.) Marine Lab., 1965-70, rsch. assoc., 1970-73; asst. prof. McGill U., Montreal, Que., Can., 1970-72, asso. prof., 1972-79, prof., 1979—, chmn. dept. biology, 1981-85, dean of sci., 1986-91, acad. v.p., 1991—; pres., chmn. bd. Huntsman Marine Lab., 1980-89; pres. Groupe Interuniversitaire de Recherche Oceanographique du Que., 1986-91; chmn. grant selection com. for population biology Natural Scis. and Engring. Research Council Can., 1980-81, chmn. grant selection com. for oceans, 1986-87. Mem. editorial bd.: Can. Jour. Fisheries and Aquatic Sciences, 1980-85, Le Naturaliste Canadien, 1980-91, Can. Jour. Zoology, 1982-86; contbr. in field. Recipient Dwight D. Webster award Am. Fisheries Soc., 1989, Award for Excellence for Fisheries Edn., 1990, Fry medal Can. Soc. Zoologists, 1990; grantee in field. Fellow Rawson Acad., Royal Soc. Can.; mem. Am. Fisheries Soc. (pres. North-East div. 1977-78), Can. Com. for Fishery Research, Can. Soc. Zoologists, Am. Soc. Limnology and Oceanography, Am. Soc. Naturalists. Office: 1205 Ave Docteur Penfield, Montreal, PQ Canada H3A 1B1

LEGGON, CHERYL BERNADETTE, sociologist, staff officer; b. Cleve., Aug. 19, 1948; d. Robert Winston and Bernice (Metcalfe) L.; m. Edward W. Gray Jr., July 18, 1970; 1 child, Robert. BA, Barnard Coll., 1970; MA, U. Chgo., 1973, PhD, 1975. Asst. prof. sociology Mount Holyoke Coll., South Hadley, Mass., 1975-77, U. Ill., Chgo., 1977-80; postdoctoral rsch. fellow U. Chgo., 1980-82, rsch. fellow ctr. for the study of indsl. socs., 1982-83; assoc. prof. sociology Chgo. State U., 1983-84; staff officer NRC, Washington, 1985—; adj. assoc. prof. sociology Georgetown U., 1991—; cons. James H. Lowry and Assocs., Chgo., 1982-83; adj. assoc. prof. sociology Georgetown U., 1991—. Contbr. chpts. to books. Recipient George F. Baker Fund scholarship, 1966-70, Ford Found. fellow, 1970-75. Mem. AAAS, Am. Sociol. Asns., N.Y. Acad. Scis., Soc. for Social Study of Sci., So. Sociol. Soc. (chair com. on sociol. practice). Baptist. Avocation: martial arts. Office: NRC 2101 Constitution Ave NW Washington DC 20418-0001

LEGOFFIC, FRANCOIS, biotechnology educator; b. Pluzunet, France, Nov. 10, 1936; s. Jacques and Marie LeGoffic; m. Marie-Thérèse Castel, Nov. 28, 1957; 1 child, Marc. Dr 3e Cycle, U. Paris, 1962, Dr Sci., 1963. Various positions Nat. Ctr. Sci. Rsch., France, 1962-74; prof. Ecole Nationale Supérieure de Chimie de Paris, 1975—; bd. dirs. UA 1389 du Centre Nat. de la Recherche Scientifique. Contbr. over 250 articles to Organic Chemistry, Biochemistry, Biotechnology and Microbiology; holder 20 patents. Achievements include research on total synthesis of natural compounds of therapeutic value, mechanism of resistance of bacteria to antibiotics, mechanism of action of antibiotics, enzyme inhibitors as drugs in antibacterial, antifungal and anticancer areas, and biotechnology, such as new membranes, immobilized enzymes and cells, enzymes in organic synthesis, valorization of molecules from marine origin. Home: 42 rue Jean Georget, Clamart France 92140 Office: ENSCP, 11 rue Pierre Marie Curie, Paris France 75231

LEGRAND, RONALD LYN, nuclear facility executive; b. Fort Rucker, Ala., Nov. 26, 1952; s. Marion Lynwood and Nettie Jean (Gardner) LeG.; m. Tammy Annette Cook, Mar. 24, 1979; children: Veronica Lyn, Vanessa Annette. BS, U. S.C., 1981. Chem-rad technician Plant Hatch, 1981-83, lab. foreman, 1983; shift foreman Plant Vogtle, Burke County, Ga., 1983-84, shift supr., 1985-87, ops. supr., 1987-89, chemistry mgr., 1989-91, mgr. ops., 1991—; shift supr. Plant Farley, 1984-85. Methodist. Office: Ga Power Co PO Box 1600 Waynesboro GA 30830

LEHBERGER, CHARLES WAYNE, mechanical engineer, engineering executive; b. Newark, Dec. 3, 1944; s. Arthur Norman and Lillian Beatrice (Grimshaw) L.; m. Roxann C. Roulin, Oct. 3, 1970; children: Jason C., Benjamin J., Andrea J. BSME, N.J. Inst. Tech., 1967; MBA, U. New Haven, 1977. Registered profl. engr., N.J., Conn. Contract adminstr. Combustion Engring., Inc., Windsor, Conn., 1967-73; dir. sales, mktg. The Lee Co., Westbrook, Conn., 1973-90; pres. Farmington Engring., Inc., Madison, Conn., 1990—. Mem. ASME, Soc. Automotive Engrs. Home: 37 Kimberly Ln Madison CT 06443-2080 Office: Farmington Engring Inc 7 Orchard Park Rd Madison CT 06443-3403

LEHÉ, JEAN ROBERT, mechanical engineer; b. Paris, Jan. 10, 1946; s. Jacqueline Lehé; m. Chantal O. Fayard, Jan. 20, 1973; children: Olivier J., Sandra A. Baccalauréat in Math., Lycée E. Pailleron, Paris, 1965; M in Mech. Engring., U. Nancy, France, 1970. Rsch. engr. European Soc. of Propulsion, Vernon, France, 1971-75; engines specialist European Soc. of Propulsion, Vernon, 1975-80, turbopump group mgr., 1980-91, mgr. quality suppliers and lab., 1991—. Recipient Bronze medal Centre Nat. d'Études Spatiales, 1981. Achievements include project engineering 3d stage HM7 cryogenic TP Ariane IV; conceptual design of VULCAIN turbopumps for Ariane V Launcher. Office: Soc Européenne de Propulsion, BP802 Forêt de Vernon, 27207 Vernon France

LEHMA, ALFRED BAKER, mathematician, educator; b. Cleve., Mar. 21, 1931. BS, Ohio U., 1950; PhD in Math., U. Fla., 1954. Instr. math. Tulane U., 1954; mem. staff acoustics lab and rsch. lab. electronics MIT, 1955-57; asst. prof. math. Case Inst. Tech., 1957-61; rsch. assoc. Rensselaer Polytech. Inst., 1963; rsch. mathematician Walter Reed Army Inst. Rsch., 1964-67; prof. math. and computer sci. U. Toronto, Can., 1967—; vis. mem. math. rsch. ctr. U. Wis., 1961-63; vis. prof. U. Toronto, 1965-67. Recipient award Delbert Ray Fulkerson Fund, Am. Math. Soc., 1992. Mem. Math. Assn. Am., Soc. Industrial and Applied Math. Achievements include research in combinatorics. Office: Univ of Toronto, 215 Huron St, Toronto, ON Canada M5S 1A1*

LEHMAN, THOMAS ALAN, chemistry educator; b. Berne, Ind., Jan. 12, 1939; s. Leslie Burkhart and Naomi Edith (Neuenschwander) L.; m. Mary Jo Diller, June 18, 1961; children: Alan, Kathryn. BS, Bluffton (Ohio) Coll., 1961; PhD, Purdue U., 1967. Asst. prof. Bluffton Coll., 1966-69; rsch. assoc. U. N.C., Chapel Hill, 1969-70; assoc. prof. Nat. U. Zaire, Kisangani, 1971-73; assoc. prof. Bethel Coll., North Newton, Kans., 1973-81, prof. chemistry, 1981—; chercheur libre U. Libre Brussels, 1971; vis. scientist Nat. Inst. Environ. Health Scis., Research Triangle Park, N.C., 1979-80; guest prof. Inst. Exptl. Physics, U. Innsbruck, 1985; vis. prof. Univs. Utah, Del., Kans. Author: Ion Cyclotron Resonance Spectrometry, 1976; contbr. articles to profl. jours. Mem. Am. Chem. Soc. (chair Wichita sect. 1988), Am. Soc. Mass Spectrometry. Mennonite. Office: Bethel Coll North Newton KS 67117

LEHMANN, JOHN CHARLES, neuropharmacologist; b. Kitchener, Ont., Can., Feb. 5, 1952. BS, Calif. Tech., Pasadena, 1974; PhD, U. B.C., Vancouver, 1980. Postdoctoral fellow Synthelabo, Paris, 1980-82, Johns Hopkins U., Balt., 1982-84; project leader Ciba Geigy, Summit, N.J., 1984-88, Servier Pharm., Paris, 1988-90; assoc. prof. Hahnemann U. Sch. Medicine, Phila., 1990—; scientific bd. mem. Drug News and Perspectives, Barcelona, Spain, 1990—. Achievements include discovery of first competitive NMDA antagonist to be used clinically to protect against stroke and traumatic brain injury. Home: 486 Church Rd Devon PA 19333 Office: Hahnemann Univ Mail Stop 409 Philadelphia PA 19102-1192

LEHN, JEAN-MARIE PIERRE, chemistry educator; b. Rosheim, Bas-Rhin, France, Sept. 30, 1939; s. Pierre and Marie (Salomon) L.; m. Sylvie Lederer, 1965; 2 children. Grad., U. Strasbourg, France, 1960, PhD, 1963; PhD (hon.), U. Jerusalem, 1984, U. Autonoma, Madrid, 1985, U. Göttingen, 1987, U. Bruxelles, 1987, U. Heraklion, Greece, 1989, U. Bologna, 1989, Charles U., Prague, 1990, U. Twente, 1991; postgrad., U. Sheffield, 1991. Various positions Nat. Ctr. Sci. Rsch., France, 1960-66; postdoctoral rsch. assoc. Harvard U., Cambridge, Mass., 1963-64; asst. prof. U. Strasbourg, France, 1966-69; assoc. prof. U. Louis Pasteur of Strasbourg, 1970, prof. of chemistry, 1970-79; prof. Coll. France, Paris, 1979—; vis. prof. chemistry Harvard U., 1972, 74, E.T.H., Zurich, Switzerland, 1977, Cambridge (Eng.) U., 1984, Barcelona (Spain) U., 1985, Frankfurt U. (Fed. Republic Germany), 1985-86. Contbr. about 400 articles to sci. publs. Recipient Bronze, Silver and Gold medals Ctr. Nat. Sci. Rsch., Gold medal Pontifical Acad. Sci., 1981, Paracelsus prize Swiss Chem. Soc., 1982, von Humboldt prize, 1983, Nobel Prize for Chemistry, 1987, Karl-Ziegler prize, 1989; decorated officier Légion d'Honneur; officier Ordre Nat. du Mérite, Ordre Pour le Mérite for Scis. and Arts, Fed. Republic Germany, 1990. Mem. AAAS (fgn. hon.), Inst. de France, Deutsche Acad. der Naturforscher Leopoldina, Acad. Nazionale dei Lincei, NAS (fgn. assoc.), Royal Netherlands Acad., Am. Philos. Soc. (Phila., fgn. mem.), Acad. Europaea, Acad. Wissenschaften Literalur-Mainz, Acad. Wissenschaften, Göttingen, Yougoslav Acad. Arts and Scis. Zagreb., Indian Acad. Scis., Polish Acad. Scis., Royal Acad. Scis. Letters & Fine Arts (Belgium), Acad. Arts & Scis. P.R., Acad. Scis. Ukraine, Inst. Grand Ducal (Luxembourg). Home: 21 Rue d'Oslo, 67000 Strasbourg France Office: Coll France, 11 Pl Marcelin Berthelot, 75005 Paris France also: U Louis Pasteur, 4 Rue Blaise Pascal, 67000 Strasbourg France

LEHNER, GERHARD HANS, engineer, consultant; b. Gera, Germany, Feb. 6, 1924; s. Hans Walther and Margarete (Leibinger) L.; divorced; m. Anne Freby. Degree in engring. telecommunications, Mil. Engring. Sch., Wetzlar, Fed. Republic Germany, 1942. Engr. Armed Forces Radio Service, Munich, 1945-48, Bavarian Broadcasting Network, Munich, 1948-52, Radio Free Europe, Munich, 1952-55; chief engr., tech. dir. Barclay Records, Paris, 1956-81; sound system installer, cons. Lido, Paris, Moulin Rouge, Paris, Crazy Horse Saloon, Paris. Served with German Army, 1943-45, Russia. Mem. Audio Engring. Soc. (v.p. 1979-80). Home: 63 Rue de Courcelles, 75008 Paris France

LEHRER, WILLIAM PETER, JR., animal scientist; b. Bklyn., Feb. 6, 1916; s. William Peter and Frances Reif (Muser) L.; m. Lois Lee Meister, Sept. 13, 1945; 1 child, Sharon Elizabeth. BS, Pa. State U., 1941; MS in Agr., MS in Range Mgmt., U. Idaho, 1946, 55; PhD in Animal Nutrition, Wash. State U., 1951; LLB, Blackstone Sch. Law, 1972; JD, U. Chgo., 1974; MBA, Pepperdine U., 1975. Mgmt. trainee Swift & Co., Charleston, W.Va., 1941-42; farm mgr. Maple Springs Farm, Middletown, N.Y., 1944-45; rsch. fellow U. Idaho, Moscow, 1945; asst. prof. to prof. U. Idaho, 1945-60; dir. nutrition Albers Milling Co., L.A., 1960-62; dir. nutrition and rsch. Albers Milling Co., 1962-74, Albers Milling Co. & John W. Eshelman & Sons, L.A., 1974-76, Carnation Co., L.A., 1976-81; ret.; cons. in field; speaker, lectr. more than 40 univs. in U.S. and abroad. Contbr. articles to profl. jours.; co-author: The Livestock Industry, 1950, Dog Nutrition, 1972; author weekly col., Dessert News, Salt Lake City. Mem. rsch. adv. co. U.S. Brewers Assn., 1969-81; mem. com. on dog nutrition, com. animal nutrition Nat. Rsch. Coun. NAS, 1970-76. With U.S. Army Air Corps, 1942-43. Named Disting. Alumnus, Pa. State U., 1963, 1983, Key Alumnus, 1985; named to U. Idaho Alumni Hall of Fame, 1985, others. Fellow AAAS, Am. Soc. Animal Sci.; mem. Am. Inst. Nutrition, Coun. for Agrl. Sci. & Tech., Am. Registry of Profl. Animal Scientists, Am. Inst. Food Technologists, Animal Nutrition Rsch. Coun., Am. Dairy Sci. Assn., Am. Soc. Agrl. Engrs., Am. Feed Mfrs. Assn. (life, nutrition coun. 1962-81, chmn. 1969-70), Calif. State Poly. U. (adv. coun. 1965-81, Meritorious Svc. award), The Nutrition Today Soc., Am. Soc. Animal Sci., Poultry Sci. Assn., Nat. Block & Bridle Club, Hayden Lake Country Club, Alpha Zeta, Sigma Xi, Gamma Sigma Delta (Alumni Award of Merit), Xi Sigma Pi. Republican. Avocations: river running, hunting, fishing, gardening, restoring furniture. Home: Rocking L Ranch 12180 Rimrock Rd Hayden Lake ID 83835

LEHRMAN, DAVID, orthopedic surgeon; b. N.Y.C., June 12, 1936; s. Irving and Bella (Goldfarb) L.; m. Sandra Rich, Dec. 18, 1958 (div. 1969); m. Linda Schwartz, Nov. 27, 1970; children: Richard, Michael, Steven, Robert. BA, Brandeis U., 1958; MD, U. Miami, 1962. Diplomate Am. Bd. Orthopaedic Surgery. Intern Lenox Hill Hosp., 1962-63; resident in gen. surgery Manhattan VA Hosp., N.Y.C., 1963-64; resident in orthopaedics Bklyn. Jewish Brookdale Med. Ctr., N.Y.C., 1964-67; mem. assisting attending med. staff Dept. Orthopaedics, Tulane U. Med. Sch., New Orleans, 1967-68; dep. chief orthopaedic surgery U.S. Pub. Health Hosp., New Orleans, 1968-69; chief orthopaedic surgery U.S. Pub. Health Hosp., Boston, 1968-69; orthopaedic surgeon Miami Beach, Fla., 1969—; chmn. dept. orthopedics South Shore Hosp., Miami Beach, 1975-78, pres. med. staff, 1978-81; chmn. dept. orthopedics St. Francis Hosp., Miami Beach, 1983—; chmn. Human Performance Technology, Miami Beach, 1988—, pres., med. dir. Lehrman Back Systems, Miami Beach, 1988—, Baxter Physiotherapy Ctr., Miami Beach, 1992—; pres. Backaid Systems, Miami Beach, 1984—; cons. Waterbed Health Orgn., L.A., 1987—, contbr. spine cons. Krames Communication, San Francisco, 1989—. Columnist Better Health and Living mag., 1987, creator videotape on back exercises, 1986, software Back Typing, 1989; inventor back support pillows Backaid. Mem. Mayor's Safety com. City of Miami Beach, 1975-78, Mayor's Health adv. bd., 1991—. With USPHS, 1967-69. Fellow ACS, Am. Acad. Orthopaedic Surgeons, Internat. Coll. Surgeons; mem. Am. Arthroscopy Assn., Internat. Arthroscopy Assn., Am. Pain Soc., Nat. Back Found., Am. Back Soc. (bd. dirs., treas.), N.Am. Spine Soc., Am. Occupational Med. Assn., Am. Coll. Sports Medicine. Republican. Jewish. Avocations: tennis, writing poetry. Office: 1680 Michigan Ave Miami Beach FL 33139-2519

LEIBACHER, JOHN WILLIAM, astronomer; b. Chgo., May 28, 1941; s. George W. and Irene (Novotney) L.; m. Lise H. Ouvarard, Dec. 21, 1976. AB, Harvard U., 1963, PhD, 1971. Postdoctoral fellow U. Colo., Boulder, 1970-71; scientist Laboratoire de Physique Stellaire et Planetaire, Paris, 1972-74, Lockheed Rsch. Lab., Palo Alto, Calif., 1975-81; astronomer Nat. Solar Obs., Tucson, 1982—, dir., 1988-93; dir. Gong Project, 1984—. Office: Nat Solar Observatory 950 N Cherry Ave Tucson AZ 85719-4933

LEIBETSEDER, JOSEF LEOPOLD, nutritionist, educator; b. Linz, Austria, Mar. 7, 1934; s. Josef and Pauline (Brandner) L.; m. Hedwig Vana, May 23, 1959; children: Florian, Veronika, Valentin, Sebastian. DVM, U. Vet. Medicine Vienna, 1958; PhD, U. Vet. Medicine, 1963. Asst. prof. U. Vet. Medicine, Vienna, Austria, 1958-72; assoc. prof. U. Calif., Davis, 1972-73; assoc. prof. U. Vet. Medicine, Vienna, 1974, full prof., head, 1975—. Co-author: Supplements to Lectures and Practice in Animal Nutrition, 1989, Textbook of Dog and Cat Nutrition, 1993; editor, co-author: Nutrition of Monogastric Animals, 1993; editor: (jours.) Ernährung Nutrition, 1992, Wiener Tierärztliche Monatsschrift, 1989—. Feedstuff com. Ministry of Agriculture, 1975—; chem. com. Ministry of Ecology Youth and Family, 1988—. Recipient Sandoz award, 1972, Gustav Fingerling award Deutsche Landwirtschafts-Gesellschaft, 1987. Mem. Austrian Nutrition Soc. (pres. 1991—), European Soc. Vet. and Comp. Nutrition (pres. 1992—), Vet. Assn. (v.p. 1986—). Roman Catholic. Office: Vet Med Univ, Inst Nutrition, Linke Bahngasse 11, A-1030 Vienna Austria

LEIBHARDT, EDWARD, optics industry professional; b. New Rome, Wis., Oct. 13, 1919; s. Stephan and Roza (Jilling) L.; m. Maidi Wiebe, June 3, 1961; children: Barbara, Leslie. BA, Northwestern U., 1954, PhD, 1959. Engraver R.R. Donnelley and Socs Co., Chgo., 1937-43; ptnr. Leibhardt Bros., Maywood, Ill., 1943-46, Leibhardt Engring., Maywood, 1946-51; pres. Diffraction Products Inc., Woodstock, Ill., 1951—; cons. in optics rsch., devel., 1951—. Mem. Optical Soc. Am., Optical Soc. Chgo., Sigma Xi. Achievements include development of diffraction grating ruling engines. Home: 9416 W Bull Valley Rd Woodstock IL 60098

LEIBOVICH, SIDNEY, engineering educator; b. Memphis, Apr. 2, 1939; s. Harry and Rebecca (Palant) L.; m. Gail Barbara Colin, Nov. 24, 1962; children: Bradley Colin, Adam Keith. BS, Calif. Inst. Tech., Pasadena, 1961; PhD, Cornell U., 1965. NATO postdoctoral fellow U. Coll., London, 1965-66; asst. prof. thermal engring. Cornell U., Ithaca, N.Y., 1966-70, assoc. prof. thermal engring., 1970-78, prof. mech. and aerospace engring., 1978-89, Samuel B. Eckert prof. mech. and aerospace engring., 1989—. Editor: Nonlinear Waves, 1974; assoc. editor: Jour. Fluid Mechanics, 1982—; co-editor: Acta Mechanica, 1986-92; mem. editorial bd. Ann. Revs. of Fluid Mechanics, 1989—. Disting. lectr. Naval Ocean Rsch. Devel. Activity, 1983. Recipient MacPherson prize Calif. Inst. Tech., 1961. Fellow ASME (chmn. applied mechanics div. 1987-88), Am. Phys. Soc. (chmn. div. fluid dynamics 1987-88), Am. Acad. Arts and Scis., U.S. Nat. Com. for Theoretical and Applied Mechanics (chair 1990-92.), Nat. Acad. Engring. Office: Cornell U Upson Hall Ithaca NY 14853

LEIBOWITZ, ALAN JAY, environmental scientist; b. Phila., Sept. 18, 1956; s. Arthur and Marlene (Bezar) L.; m. Amy Carol Polin, Aug. 23, 1981; children: Joshua Abel, Adam Samuel. BS, Drexel U., 1979, MS with honors, 1980. Toxicology researcher Drexel U., Phila., 1979-80; environ. coord. ITT Def. Communications Div., Nutley, N.J., 1980-83, Environ. Health and Safety specialist, 1983-85; mgr. Environ. Health and Safety ITT Defense Tech. Corp., Nutley, N.J., 1985-89; dir. Environ. Health and Safety ITT Def. & Electronics, Arlington, Va., 1989—; vice chmn. EHS coun., Aerospace Industries Assn., Washington, 1991-92, chmn., 1992-93. Pres. Hillel, Drexel U., 1978-79. Recipient Environ. Health grant, HEW, 1979-80. Mem. Am. Indsl. Hygiene Assn. (mgmt. com. 1989—), Am. Soc. Safety Engrs. Office: ITT Def and Electronics Inc 1000 Wilson Blvd Arlington VA 22209-3901

LEIBY, CRAIG DUANE, control systems engineer, electrical design engineer; b. Fairless Hills, Pa., Mar. 14, 1958; s. Dorothea Doris (Beidler) Yurcho; divorced; 1 child, Craig Duane Jr.; m. Lori Renee Kobrinski, Aug. 11; children: Christopher Edward, Clinton Daniel. Elec. cert., Burlington Co. Vo-Tech, Mt. Holly, N.J., 1976. Electro-mech. technician Nassau Chem., Trenton, N.J., 1984-87; controls mgr. Finmac Inc., Warminster, Pa., 1987—. Office: Finmac Inc 21 Bonair Dr Warminster PA 18974

LEIGH, GERALD GARRETT, research engineer; b. Burley, Idaho, Sept. 10, 1931; s. Wilbur Garrett and Iva Grace (Morehead) L.; m. Ruth Ann Kurtz, Dec. 28, 1957 (div. Feb. 1983); m. Ann Christen Galloway, March 5, 1983; children: Christi Leigh, Lauren Schieffer, Thomas Crawley (stepson). BS in Chem. Engr., U. Idaho, 1955; MS in Mechanical Engring., Ariz. State U., 1964, PhD in Engring. Mechanics, 1971. Commd. USAF, 1955, retired, 1976; sr. rsch. engr. N.Mex. Engring. Rsch. Inst., Albuquerque, 1976-93. Contbr. articles to profl. jours. Home: 528 Live Oak Pl Albuquerque NM 87122 Office: U N Mex M Mex Engring Rsch Inst Albuquerque NM 87131

LEIGH, HOYLE, psychiatrist, educator; b. Seoul, Korea, Mar. 25, 1942; came to U.S., 1965; m. Vincenta Masciandaro, Sept. 16, 1967; 1 child, Alexander Hoyle. MA, Yale U., 1982; MD, Yonsei U., Seoul, 1965. Diplomate Am. Bd. Psychiatry and Neurology. Asst. prof. Yale U., New Haven, 1971-75, assoc. prof., 1975-80, prof., 1980-89, lectr. in psychiatry, 1989—; dir. Behavioral Medicine Clinic, Yale U., 1980-89; dir. psychiat. cons. svc. Yale-New Haven Hosp., 1971-89; chief psychiatry VA Med. Ctr., Fresno, Calif., 1989—; prof., vice chmn. dept. psychiatry U. Calif., San Francisco, 1989—; head dept. psychiatry U. Calif., San Francisco, Fresno, 1989—; cons. Am. Jour. Psychiatry, Archives Internal Medicine. Author: The Patient, 1980, 2d edit., 1985, 3d edit., 1992; editor: Psychiatry in the Practice of Medicine, 1983. Fellow Am. Psychiat. Assn., ACP, Internat. Coll. Psychosomatic Medicine; mem. AMA, AAUP, World Psychiat. Assn. Avocations: reading, music, skiing. Office: U Calif 12535 Moffatt Ln Fresno CA 93703-2286

LEIGH, LINDA DIANE, psychologist, clinical neuropsychologist; b. Miami, Fla., June 12, 1946; d. Ernest Violan and Gladys May (Crisman) L.; children: Marcus, Noa, Maryse. BA, Brown U., 1977; MA, Mich. State U., 1984, PhD, 1987. Lic. psychologist, Conn., Fla., Mich. Postdoctoral fellow Fairfield Hills Hosp., Newtown, Conn., 1988-89; clin. neuropsychologist Datahr Rehab. Inst., Brookfield, Conn., 1989-90; pvt. practice psychologist East Lansing, Mich., 1990—; cons. neuropsychologist Tamarack, Inc., Sparrow Hosp., East Lansing, 1992. Mem. APA (div. clin. neuropsychology, div. ind. practice), Nat. Acad. Neuropsychologists, Coun. for Nat. Register Health Svc. Providers in Psychology. Democrat. Avocations: swimming, hiking, dancing, reading. Home: 1426 Meadow Rue East Lansing MI 48823

LEIGH, MICHAEL CHARLES, electrical engineer; b. Corry, Pa., Dec. 31, 1948; s. Michael Leech and Ann (Zawoiski) Parks; m. Emily Rose Drolz, Feb. 15, 1986; children: Jeremy Spencer Morony, Hannah Millicent, Piper Gillian. BSEE, Pa. State U., 1970. Elec. designer Jacob Engring., Inc., Pitts., 1974-76; product systems engr. Ingersoll-Rand Mining Div., Belle Vernon, Pa., 1976-82; instrumentation engr. Boeing Svcs. Internat., Pitts., 1982-89; lead instrument engr. Boeing Helicopters, Phila., 1989-90; elec. engr. Heyl & Patterson, Inc., Pitts., 1990—. Mem. IEEE, AIAA (assoc.), Soc. Automotive Engrs. Republican. Achievements include patent for cutting sound enhancement systems for mining machines.

LEIGHTON, CHARLES RAYMOND, construction inspector; b. Limestone, Maine, Mar. 22, 1921; s. Harold Raymond and Sarah Lydia (Phair) L.; m. Edith Jesse Smith, May 28, 1947 (div. 1975); children: Jean Elizabeth, Robert Scott; m. Sandra Barrett Smith, June 25, 1976. BS, U. Maine, 1975. Owner Bldg. Supply and Wood Mill, Limestone, Maine, 1946-51; mentor, designer Limestone Machine Co., 1951-53; maintenance supt. Loring AFB, Loring AFB, Limestone, 1953-72; constrn. inspector various archtl. firms, Maine, 1975-91. Town councilor Town of Limestone, Maine. 1966-67, planning bd., 1970-80. Lt. USN, 1943-46. Named for Outstanding Performance, USAF, 1971. Mem. NSPE, Am. Phys. Soc. Republican. Achievements include patents for Potato Harvester, Improved Harvester, Bag Tying Machine.

LEIKIN, SERGEY L., biophysicist, researcher; b. Moscow, July 13, 1961; came to U.S., 1989; m. Eugenia A. Kreer, June 29, 1983; 1 child, Alexander S. MS, Moscow Steel and Alloys Inst., 1984; PhD in Biophysics, Moscow State U., 1987. Rsch. engr. Frumkin Inst. Electrochemistry, Acad. Sci. USSR, Moscow, 1984-86, rsch. fellow, 1986-89, rsch. scientist, 1989; vis. fellow NIH, Bethesda, Md., 1989-92, vis. assoc., 1992—. Contbr. articles to Phys. Rev. A, Jour. Chem. Physics, Jour. Theoretical Biology. Named Hon. Citizen, State of Tex., 1989. Mem. Biophys. Soc., Am. Phys. Soc. Achievements include development of theory of structure factors for hydration forces in biomolecular systems; measured intermolecular forces between collagen triple helices. Office: NIH Bldg 12A Rm 2007 Bethesda MD 20892

LEINEN, MARGARET SANDRA, oceanographic researcher; b. Chgo., Sept. 20, 1946; d. Earl John and Ester (Louis) Leinen; m. Denzel Earl Gleason, 1984; 1 child, Daniel Glenn Whaley. BS, U. Ill., 1969; MS, Oreg. State U., 1975; PhD, U. R.I., Kingston, 1980. Marine scientist U. R.I., Kingston, 1980-82, asst. rsch. prof., 1982-86, assoc. prof., 1986-88, prof., 1988—, assoc. dean, 1988-92, dean and vice provost, 1992—. Office: U RI Grad Sch Oceanography Narragansett RI 02882-1197

LEISEY, APRIL LOUISE SNYDER, chemist; b. Williamsport, Pa., Sept. 14, 1955; d. Donald Clifford and Eva Juanita (Smith) S.; 1 child, Janel Jennifer Snyder. BS in Biology/Chemistry, Lock Haven U., 1986. Lab. technician Croda Mfg., Mill Hall, Pa., 1986-88; chemist technician Textron Lycoming, Williamsport, 1988-89, chemist, 1989—. Mem. Am. Chem. Soc., Am. Electroplaters and Surface Finishers. Office: Textron Lycoming 652 Oliver St Williamsport PA 17701

LEISSA, ARTHUR WILLIAM, mechanical engineering educator; b. Wilmington, Del., Nov. 16, 1931; s. Arthur Max and Marcella E. (Smith) L.; m. Gertrud E. Achenbach, Apr. 11, 1974; children: Celia Lynn, Bradley Glenn. BME, Ohio State U., 1954, MS, 1954, PhD, 1958. Engr., Sperry Gyroscope Co., Great Neck, N.Y., 1954-55; rsch. assoc. Ohio State U., 1955-56, instr. engring. mechanics, 1956-58, asst. prof., 1958-61, assoc. prof., 1961-64, prof., 1964—; vis. prof. Eidgenossische Technische Hochschule, Zurich, Switzerland, 1972-73, USAF Acad., Colorado Springs, Colo., 1985-86; Plenary lectr. 2nd Internat. Conf. on Recent Advances in Structural Dynamics, Southampton, Eng., 1984, 4th Internat. Conf. on Composite Structures, Paisley, Scotland, 1987, Dynamics and Design Conf., Japan Soc. Mech. Engrs., Kawasaki, 1990, Energy Sources and Tech. Conf., ASME, Houston, 1992; cons. in field. Performer Columbus Symphony Orch. Operas, 1971-79; author: Vibration of Plates, 1969, Vibration of Shells, 1973, Buckling of Laminated Composite Plates and Shell Panels, 1985; assoc. editor Applied Mechanics Revs., 1985-93, editor-in-chief, 1993—; assoc. editor Jour. Vibration and Acoustics, 1990—; mem. editorial bd. Jour. Sound and Vibration, 1971—, Internat. Jour. Mech. Sci., 1972—, Composite Structures, 1982—, Applied Mechanics Revs., 1988-93; contbr. over 150 articles to profl. jours. Gen. chmn. Pan Am. Congress Applied Mechanics, Rio de Janeiro, 1989; leader Ohio State U. Mt. McKinley Expdn., 1978. Recipient Recognition plaque Inst. de Mecanica Applicada, Argentina, 1977. Fellow ASME, Am. Acad. Mechanics (pres. 1987-88), Japan Soc. for Promotion of Sci.; mem. Am. Soc. for Engring. Edn., Am. Alpine Club. Home: 1294 Fountaine Dr Columbus OH 43221-1520 Office: 155 W Woodruff Ave Columbus OH 43210-1181

LEITCH, CRAIG H. B., earth scientist. Recipient Barlow medal Can. Inst. Mining and Metallurgy, 1991. Office: care Xerox Tower Ste 1210, 3400 de Maisonneuve Blvd W, Montreal, PQ Canada H3Z 3B8*

LEITE, CARLOS ALBERTO, physician; b. Rio de Janeiro, Feb. 2, 1939; s. Indayassu and Munira (Raed) L. BSc, Coleg. Ext. Sao Jose, Rio de Janeiro, 1956; MD, U. Brazil, Rio de Janeiro, 1962, PhD, 1972. Intern Rochester (N.Y.) Gen. Hosp., 1963-64; resident Henry Ford Hosp., Detroit, 1964-65; resident, fellow, researcher Jackson Meml. Hosp. and U. Miami, Fla., 1965-68; ltd. practice Nanticoke Meml. Hosp., Seaford, Del., 1968; prof. medicine U. Fed. Rio de Janeiro, 1972—; emeritus prof. medicine Faculty Medicine Soc. Ens. Sup. Nova Iguacu, Rio de Janeiro, 1986—; dir. Hosp. de Nova Iguacu-Posse, Rio de Janeiro, 1991; instr. medicine Fac. Nac. Med., U. Fed. Reio de Janeiro, 1963-72; chief in-patient ward Santa Casa da Misericordia Hosp., Rio de Janeiro, 1968-72, chief out-patient dept., 1968-76, cons. physician surg. unit, 1969—; chercheur visitant temporaire Inst. Pasteur, Paris, 1988. Med. writer Today's Medicine/Jour. Commerce, 1975—; editor: Metabolic Aspects of 95% Pancreatic Resection, 1971, Medicine, Logique and Reasoning, 1992, Limited Abduction of the Thumb-A New Physical Sign, 1992, Signs and Manoeuvers in Physical Diagnosis, 1992; contbr. articles to profl. jours. 2d lt. Brazilian Army, 1961-62. Recipient Carlos Chagas medal State of Guanabara, 1972, prize Argentine Meeting of Gastroenterology, 1971. Fellow ACP, Colegio Interamericano de Medicos y Cirurjanos; mem. N.Y. Acad. Scis., Am. Venereal Disease. Brazilian Coll. Surgeons, Clube Monte Libano (counsel mem. 1972—). Home: 70 Rua Redentor Apt 101, Ipanema, 22421 Rio de Janeiro Brazil Office: Ste 302, 595 Rua Visconde de Piraja, 22420 Rio de Janeiro Brazil

LEITHEISER, JAMES VICTOR, environmental specialist; b. Great Lakes NAS, Ill., July 31, 1962; s. William John and Hortense (Torres) L. BS in Chemistry, U. Ala., 1986. Environ. coord. SPPI, Columbus, Miss., 1987-89, mgr. environ. affairs, 1989-92; environ. supr. Ariz. Chem., Panama City, Fla., 1992—; mem. Local Emergency Planning Com., Columbus, 1988—; instr. seminars Miss. State U., U. Ala. Mem. Miss. Mfrs. Assn. (mem. environ. com. 1988—), Miss. Econ. Coun. (advisor com. on environment 1989—), Am. Chem. Soc., Am. Soc. Quality Control, Acad. Hazardous Materials Mgrs., Nat. Assn. Environ. Mgmt. Home: 7601 Yellowbluff Rd Panama City FL 32404 Office: Ariz Chem 2 Everitt Ave Panama City FL 32401

LEKBERG, ROBERT DAVID, chemist; b. Chgo., Feb. 2, 1920; s. Carl H. and Esther (Forsberg) L.; m. Sandra Sakal, Oct. 19, 1970; children by previous marriage: Terry Lee, Jerrald Dean, Roger Daryl, Kathleen Sue, Keith Robin. AA, North Park Coll., 1940; BS, Lewis Inst., 1943. Chemist Glidden Co., Hammond, Ind., 1940-43, Wilson Packing Co., Calumet City, Ill., 1943-45, Libby, McNeil & Libby, Blue Island, Ill., 1945-47; chief chemist Dawes Labs., Chgo., 1947-50; owner, pres. Chemlek Labs. Inc., Oak Forest, Ill., 1950—, Chemlek Labs. Can. Ltd., Windsor, Ont., 1960-66; cons. AL Labs. Internat. Editor: Indsl. Dist. Assn. Newsletter, 1975—; patentee chem. processes, pollution control used in Mex., Italy and Brazil. Violinist Chgo. Civic Symphony, 1940-42, N.W. Ind. Symphony, 1976—, Chicago Heights Symphony, 1959—, Southwest Symphony, 1970—; officer with U.S. Power Squadron. Mem. Ill. Mfrs. Assn., Ill. Indsl. Council, Chgo. Feed Club. Republican. Club: Dolton Yacht (past commodore). Home and Office: 6624 Linden Dr Oak Forest IL 60452-1556

LEKIM, DAC, chemist; b. Hanoi, Vietnam, Aug. 4, 1936; s. Kien and Duong Thi (An) L.; m. Ingrid Joisten, 1964; 1 child, Tanja. Dr. rer. nat., U. Cologne, 1969; Dr. med. habil., Med. Acad., Posen, Germany, 1989. Rsch. dept. head Nattermann & CIE, Cologne, Germany, 1970-77; rsch. dir. Fresenius AG, Bad Homburg, Germany, 1977-80, Pearson & Co., Kruft, Germany, 1980—; assoc. prof. Med. Acad., 1989—. Author: Germanium in Medicine, 1986, Phospholipid Advances, 1989; editor: Lecithin, 1983, 85, 87. Chmn. Neumann Found., Cologne, 1988-90, German/VN Assn., Bonn, 1969-79. Mem. Am. Chem. Soc., Am. Assn. Pharm. Scientists, N.Y. Acad. Scis. Roman Catholic. Achievements include more than 20 patents. Home: Girlitzweg 22, 50829 Cologne Germany Office: Pearson & Co, Bahnhofstrasse 20, 56642 Kruft Germany

LELAND, HAROLD ROBERT, research and development corporation executive, electronics engineer; b. Eau Claire, Wis., Apr. 18, 1931; s. Harold F. and Julia E. (Porter) L.; m. Loraine M. Weiss, Feb. 5, 1958; children: Jane Valorie, Robert William. BEE, U. Wis., 1954, MS, PhD, 1958. Registered profl. engr., N.Y. With Calspan Corp., Buffalo, 1958—, pres., 1983—; v.p. Arvin Industries, Inc., Columbus, Ind., 1984—; Vice chmn., dir. Calspan SUNY Buffalo Research Ctr., 1983—, The U. Tenn. Calspan Ctr. for Aerospace Research, 1985—. Served to capt. USAF, 1954-56. Recipient Alumni award U. Wis., 1981, Dean's award SUNY at Buffalo, 1987. Mem. AIAA, IEEE, Assn. Old Crows (Washington, Buffalo, N.Y.), Air Force Assn. (pres, sec. 1978—), Assn. U.S. Army, Am. Defense Preparedness Assn., Niagara Frontier Assn. (sec. 1974—), Sigma Xi, Tau Beta Pi, Sigma Phi (sec. 1952—). Unitarian. Office: Arvin Industries Inc PO Box 400 Buffalo NY 14225*

LELL, EBERHARD, retired inorganic chemist; b. Linz, Austria, Oct. 3, 1927; came to U.S., 1957; s. Wilhelm and Margarete (Sigel) L.; m. Edith Döller, May 26, 1962; children: Christoph, Bertrand, Claudia. Dipl.Ing., T.U., Graz, Austria, 1952; PhD, T.U., Stuttgart, Germany, 1956. Asst. prof. T.U., Stuttgart, 1956-57; sr. scientist Bausch & Lomb Inc., Rochester, N.Y., 1957-69, Itek Corp., Lexington, Mass., 1969-74; project mgr. Voest-Alpine, Linz, 1975-91, ret., 1991. Contbr. articles to Progress in Ceramic Sci., Can. Jour. Physics, Physics and Chemistry of Glasses, Jour. Am. Ceramic Soc., Applied Physics, Am. Ceramic Soc. Bull. Mem. Gesellschaft Österreichischer Chemiker, Export Club Linz, Austrian Sr. Expert Pool. Home: Knabenseminarstrasse 47, Linz A-4040, Austria

LEM, KWOK WAI, research scientist; b. Kaiping, Republic of China, July 14, 1952; came to Can., 1968; s. Man Oi and Lau Yuet (Wong) L.; m. Margaret Yun-Min Hsieh, Mar. 22, 1986; children: Paul C., Richard C. BS in Engring. Sci., U. Toronto, 1976; PhD, Poly. Inst. of N.Y., 1983. Registered profl. engr., Ont., Can. Chem. specialist Can. Hanson Ltd, Toronto, 1976-77; polymer chemist Schenectady Chems. Can., Toronto, 1977-78; rsch. engr. Allied Corp., Morristown, N.J., 1983-85; rsch. engr. Allied-Signal, Inc., Morristown, 1985-87, sr. rsch. engr., 1987—; adj. prof. dept. chem. engring. Poly. U., Bklyn., 1991—; mem. safety com. Allied-Signal, Inc., 1990—, ASPAC com., 1988-89. Contbr. numerous articles to profl. jours. Recipient Melvin M. Gerson award Soc. of Plastic Engrs., 1980, 81. Mem. Am. Men and Women of Sci., Assn. of Profl. Engrs. (Ont.), Soc. of Rheology, Sigma Xi, Tau Beta Pi. Achievements include 6 patents and 20 patents pending. Home: 11 Old Coach Rd Randolph NJ 07869

LEMAIRE, PAUL JOSEPH, scientist; b. Colchester, Vt., Aug. 11, 1953; s. Henry Paul and Theresa Mary (Leary) L.; m. Laura Catherine Pribuss, July 19, 1986; children: Caitlin, Paul; stepchildren: Sarah Lester, Danielle Lester. SB, MIT, 1975, PhD, 1980. Mem. tech. staff AT&T Bell Labs., Murray Hill, N.J., 1980—; speaker at confs. in field. Contbr. articles to profl. publs. Recipient R&D 100 award R&D Mag., 1988. Mem. Am. Ceramic Soc. (Ross Coffin Purdy award 1984), Materials Rsch. Soc., Soc. Photo-optical Instrumentation Engrs. Roman Catholic. Achievements include 5 patents for hermetically coated optical fibers, optical fiber reliability predictions, UV induced fiber phase gratings, and optical fiber design. Office: AT&T Bell Labs 600 Mountain Ave PO Box 636 Murray Hill NJ 07974-0636

LEMARBE, EDWARD STANLEY, engineering manager, engineer; b. Chicago Heights, Ill., June 30, 1952; s. Gerald Joseph and Irene Helen (Jelen) LeM.; m. Patricia Ann Czyz, May 28, 1977; children: Kyle Bradford, Randall Jered. BS in Mech. Tech., Purdue U., 1976; MBA, Lewis U., 1984. Field engr. Morrison Constrn. & Engring., Hammond, Ind., 1976-78; sr. engr. Miner Enterprises, Inc., Geneva, Ill., 1978-85; mgr. product devel. Alco Dispensing Systems div. Alco Standard, Torrington, Conn., 1985-88; v.p. engring. Jet Spray Corp., Norwood, Mass., 1988-92; sr. dir. Engring. Multiplex Co., 1992—; mem. pres.' staff Alco Dispensing/Selmix-Alco, Torrington, 1986-88; mem. exec. com. Jet Spray Corp., Norwood, 1988-92; mem. resource allocation com. Multiplex Co. Inc., 1992—. Mem. ASTM (subcom. 1988—), Am. Mgmt. Assn. (assoc.), Internat. Food Svc. Mfrs. Assn. (corp. mem.), Hickory Bend Condo Assn. (bd. dirs. 1984-85). Republican. Roman Catholic. Avocations: golf, tennis, scuba diving.

LEMASTER, ROBERT ALLEN, mechanical engineer; b. Troy, Ohio, Sept. 18, 1953; s. Robert Milton and Betty Jean (Reynolds) LeM.; m. Debbie Lou Hinton, Apr. 6, 1974; 1 child, Michael Allen. BSME, Akron U., 1976; MS in Engring. Mechanics, Ohio State U., 1978; PhD in Engring. Sci., U. Tenn., 1983. Engr. Hissong Cons., Mt. Vernon, Ohio, 1976-77; program engr. W.J. Schafer & Assocs., Dayton, Ohio, 1988-89; engr. AEDC Group Arnold Engring. Devel. Ctr. Group Sverdrup Tech. Inc., Tullahoma, Tenn., 1978-83; project engr. Tech. Group Sverdrup Tech. Inc., Tullahoma, Tenn., 1983-88; dept. mgr. MSFC Group Marshal Space Flight Ctr. Group Sverdrup Tech. Inc., Huntsville, Ala., 1989—. Contbr. tech. papers to profl. publs. Mem. budget com., fund raising com. Bob Jones High Sch. Athletic Booster Club, Madison, Ala., 1989—. Mem. ASME, AIAA (structures tech. com.), ASCE (advisor, electric transmission structures com.). Republican. Baptist. Avocations: jogging, scuba diving. Home: 116 Silver Creek Cir Madison AL 35758-7637 Office: Sverdrup Tech Inc MSFC Group 620 Discovery Dr NW Huntsville AL 35806-2802

LEMAY, GERALD J., electrical engineer, educator; b. Fall River, Mass., Mar. 11, 1949; s. Julien Alphonse and Lorraine Marie Louise (Remy) L.; m. Vilimaina Ranadi, June 30, 1974; children: Sam, Melanie. BS in Physics, U. Mass., 1971; MS in Physics, U. Mass., Dartmouth, 1978, BSEE, 1980; PhD in Electrical Engring., U. Rhode Island, 1988. Vol. tchr. U.S. Peace Corps, Labasa, Fiji Islands, 1971-74; rsch. assoc. Robotics Rsch. Ctr. U. R.I., Kingston, 1982-84; from vis. lectr. to asst. prof. U. Mass., Dartmouth, 1978-89, assoc. prof., 1989—; cons. Interval Corp., Framingham, Mass., 1985-86, Bose Speaker Corp., Framingham, 1985-86, Ariel Corp., N.J., 1990-91; prin. instr. Computer Aided Engring. of Analog Electronic Cirs., 1988, Digital Filters Workshop, 1990, 91. Author in field. Asst. scoutmaster Boy Scouts Am., Kingston, 1988-91, com. mem., 1991—. Recipient Small Coll. Opportunity award NSF, 1985, 86, 87. Mem. Internat. Electrical Engrs., Acoustic Soc. Am., Audio Engring. Soc., Boston Computer Soc. Home: 5 Westwind Rd Wakefield RI 02879-1502 Office: U Mass Elec and Computer Engring Dept North Dartmouth MA 02747

LEMAY, MARJORIE JEANNETTE, neuroradiologist; b. Medical Lake, Wash., May 6, 1917; d. Samuel and Grace (Lobingier) LeM.; m. Walter Eugene Knox (dec.); children: Tamsen Ann, Walter Eugene IV. AB, U. Kans., 1939, MD, 1942. Diplomate Am. Bd. Radiology. Instr. in radiology Presbyn. Hosp., N.Y.C., 1946-48, Columbia U., N.Y.C., 1951-52; asst. clin. prof. radiology Boston U. Sch. of Medicine, 1954-68; clin. instr. in radiology Tufts U., Boston, 1959-63; asst. clin. prof. radiology Harvard Med. Sch., Boston, 1964-72; asst. clin. prof. neurology Boston U. Sch. of Medicine, 1969-74; assoc. prof. radiology Harvard Med. Sch., 1972—; vis. scientist dept. radiotherapeutics Cambridge U., Eng., 1949-51; vis. prof., acting chair dept. radiology Am. U., Beirut, Lebanon, 1961-62. Contbr. articles to profl. jours. Home: 10 Garden Terr Cambridge MA 02138-1407 Office: Brigham & Women's Hosp Dept Radiology 75 Francis St Boston MA 02115

LEMBERG, HOWARD LEE, telecommunications network research executive; b. N.Y.C., July 29, 1949; s. Seymour Lemberg and Joan Frances (Schusterman) Hartstein; m. Christine Ann Van Ullen, Dec. 26, 1970; children: Kathryn M., Diana L. BA, Columbia U., 1969; PhD, U. Chgo., 1973. Mem. tech. staff Bell Labs., Murray Hill, N.J., 1973-75; asst. prof. chemistry U. N.C., Chapel Hill, 1975-78; mem. tech. staff Bell Labs., Naperville, Ill., 1978-81; supr. Bell Labs., Holmdel, N.J., 1981-83; dist. mgr. Bellcore, Morristown, N.J., 1984-91, dir., 1991—. Trustee Bernardsville Pub. Libr., 1989—. Mem. IEEE (sr., vice chmn. optical com. 1991—, tech. editor jour. 1990—). Office: Bellcore 445 South St Morristown NJ 07960-6438

LEMBONG, JOHANNES TARCICIUS, cardiologist; b. Tomini, Indonesia, June 10, 1930; s. Joseph and Maria (Yo) L.; m. Yetty Maria-Paula Tabanan, Sept. 7, 1958; children: Augustinus Budirahmat, Thomas Trikasih. MD, U. Indonesia Med. Sch., Jakarta, 1960; PhD, U. Dusseldorf Med. Sch., Fed. Republic of Germany, 1980. Cert. internist, cardiologist, otorhinolaryngologist, acupuncturist. Lectr. U. Indonesia Med. Sch., Jakarta, 1958-73; guest physician U. Heidelberg Med. Sch., Fed. Republic of Germany, 1973; resident physician Rhön Clin., Bad Kissingen, Fed. Republic of Germany, 1974-77, St. Elisabeth Hosp., Recklinghausen, Fed. Republic of Germany, 1977-79, Roderbirken Clin., Leichlingen, Fed. Republic of Germany, 1979-81; practice medicine specializing in cardiology, Otorhinolarngology Jakarta, 1981—; pres. commissary P.T. Pharos Indonesia Ltd. Pharm. Mfr., Jakarta, 1981—. V.p. Cath. Students Union, Jakarta, 1953; chmn. Students Cons. Council, Jakarta, 1954; dir. Jr. Econ. High Sch., Jakarta, 1953. Mem. Assn. Internists, Assn. Cardiologists, Assn. Otorhinolaryngologists. Roman Catholic. Avocations: reading, swimming. Home: Jalan Limo 39, 12210 Jakarta Indonesia Office: Jalan Tanah Abang Dua 67, 10160 Jakarta Indonesia

LEMELSON, JEROME H., inventor; b. 1923. MSc degrees in aeronautical, industrial engineering, New York U., N.Y.C. Holder more than 450 patents. Achievements include nearly 500 patents. Address: 48 Parkside Dr Princeton NJ 08540

LEMENTOWSKI, MICHAL, surgeon; b. Warsaw, Poland, Sept. 20, 1943; came to U.S., 1972; s. Wlodzimierz and Tamara (Bakurewicz) L.; m. Celeste Gallo, Mar. 29, 1983; children: Maria, Michal Jr., Jennifer, Lisa, Jason, Sean. MD, Acad. Medicine, Warsaw, 1968. Diplomate Am. Bd. Surgery. Surg. intern Kings County Hosp., Bklyn., 1974-75; surg. resident St. Francis Hosp. Med. Ctr., U. Conn., Hartford, 1975-79; asst. dept. urology Grochowski Hosp., Warsaw, 1970-72; clin. instr. surgery SUNY Downstate Med. Ctr., Bklyn., 1974-75, U. Conn., Hartford, 1978-79; pvt. practice Monessen, Pa.; chmn. dept. surgery Brownville (Pa.) Gen. Hosp., 1988, Monongahela (Pa.) Valley Hosp., 1989—. Fellow ACS, Internat. Coll. Surgeons; mem. AMA (physician recognition award 1989), Pa. Med. Soc., Fayette County Med. Soc., Monessen C. of C. Republican. Roman Catholic. Avocations: travel, chess. Home: 30 Heritage Hills Rd Uniontown PA 15401-5620 Office: 331 Schoonmaker Ave Monessen PA 15062-1210

LEMESSURIER, WILLIAM JAMES, structural engineer; b. Pontiac, Mich., June 12, 1926; s. William James and Bertha Emma (Sherman) LeM.; m. Dorothy Wright Judd, June 20, 1953; children: Claire Elizabeth, Irene Louise, Peter Wright. AB cum laude, Harvard U., 1947, postgrad., 1948; SM, MIT, 1953. Registered profl. engr., Mass., D.C., N.Y., Tenn., Colo. Ptnr. Goldberg, LeMessurier Assoc., Boston, 1952-61; pres. LeMessurier Assocs. Inc., Boston and Cambridge, 1961-73; chmn. Sippican Cons. Internat. Inc., Cambridge, 1973-85, LeMessurier Cons. Inc., Cambridge, 1985—; instr. dept. bldg. constrn. and engring. MIT, 1951-52, asst. prof., 1952-56, assoc. prof. dept. architecture, 1964-67, sr. lectr. dept. civil engring., 1976-77, lectr., 1983-86; assoc. prof. grad. sch. design Harvard U., 1956-61, adj. prof., 1973—; vis. lectr. Yale U., U. Mich., U. Ill., Chgo., Rice U., Washington U., St. Louis, Northeastern U., R.I. Sch. Design, Cornell U., U. Pa., Roger Williams Coll., U. Calif., Berkeley, U. Tex., Austin; speaker Assn. Collegiate Schs. of Architecture Constrn. Materials and Tech. Inst., Harvard U.; mem. sci. adv. com. Nat. Ctr. Earthquake Engring. Research. Coauthor: Structural Engineering Handbook, 1968, 2d edit., 1979; prin. works include New Boston City Hall, Shawmut Bank Boston, 1st Nat. Bank Boston, Fed. Res. Bank Boston, Citicorp Ctr., N.Y.C., Dallas-Ft. Worth Airport Terminal Bldgs., Ralston Purina Hdqrs., St. Louis, Nat. Air and Space Mus., Washington, Am. Inst. Architects Hdqrs., Washington, One Post Office Sq., Boston, Bank Southwest Tower, Houston, InterFirst Plaza Dallas Main Ctr., Treasury Bldg., Singapore, King Khalid Mil. City, Al Batin, Saudi Arabia, TVA Hdqrs., Chattanooga, Lafayette Place Hotel, Boston, academic bldgs. at Harvard U., Princeton U., Amherst Coll., Bowdoin Coll., Williams Coll., U. Mass., U. Wis., U. Ill., Colby Coll., U. N.H., Northeastern U., Kirkland Coll., Cornell U. Mem. Cambridge Experimentation Rev. Bd., 1977; juror Capitol area archtl. and planning bd. State Capitol Bldg. Extension, St. Paul, 1977, Am. Inst. Architects Regional Awards Program, No. Vt., 1980, Progressive Architecutre mag., Portland Cement Assn. Awards, 1986, Fazlur Rahman Kahn Internat. Fellowship, N.Y.C., 1987; tech. advisor to jury Boston Archtl. Ctr. Competition. Recipient Am. Inst. Architects' Allied Professions medal, 1968, Profl. Service award Engring. News-Record, 1978, Prestressed Concrete Inst. award, 1984. Fellow ASCE (hon.), Am. Concrete Inst., Am. Inst. Architects, Nat. Inst. Architects (hon.); mem. NAE, Boston Soc. Architects (hon.), Boston Soc. Civil Engrs. (hon.), Boston Assn. Structural Engrs. (past pres.), Am. Inst. Steel Constrn. (specifications adv. com. 1961—, Award of Excellence 1962, 66, 70, 77, 79, Spl. award 1972), VA (adv. bd. div. constrn.), Nat. Com. on Housing Tech., Structural Clay Products Inst. (bldg. code com.), Seismic Safety Council (task com.), Bldg. Code Com., Internat. Masonry Inst. (research council), Tau Beta Pi. Episcopalian. Club: Met. (N.Y.C.). Developed Mah-LeMessurier high-rise housing system; conceived and developed Staggered Truss System for use in high-rise steel structures; conceived, developed and applied Tuned Mass Damper System to reduce tall bldg. motion. Avocation: music. Office: LeMessurier Cons 1033 Massachusetts Ave Cambridge MA 02138-5319

LEMIEUX, RAYMOND URGEL, chemistry educator; b. Lac La Biche, Alta., Can., June 16, 1920; s. Octave L.; m. Virginia Marie McConaghie, 1948; children: 1 son, 5 daus. BS with honors, U. Alta., 1943; PhD in Chemistry, McGill U., 1946; DSc (hon.), U. N.B., 1967, Laval U., Quebec, 1970, U. Ottawa, 1975, U. Waterloo, 1980, Meml. U., Nfld., 1981, Université du Quebec, 1982, Queen's U., Kingston, 1983, McGill U., Montreal, 1984; Dr. honoris causa, Université de Provence, Marseille, France, 1972; LLD (hon.), U. Calgary, 1979; DSc (hon.), Université de Sherbrooke, 1986, McMaster U., 1986, U. Alta., 1991; PhD (hon.), U. Stockholm, 1988. Postdoctoral fellow Ohio State U., Columbus, 1946-47; asst. prof. U. Sask.,

Saskatoon, Can., 1947-49; sr. rsch. officer NRC of Can., Saskatoon, 1949-54; prof., chmn. chemistry dept. U. Ottawa, Can., 1954-61; vice dean, faculty of pure and applied sci. U. Ottawa, 1954-61; prof. organic chemistry U. Alta., Edmonton, Can., 1961-81, Univ. prof., 1981-85, prof. emeritus, 1985—. Author 243 published articles in sci. field. Decorated officer Order of Can.; recipient Louis Pariseau medal, Association Canadienne Francaise pour l'advancement des Sciences, 1961, Centennial of Can. medal, 1968, award of achievement Province of Alta., 1980, diplome d'Honneur Le Groupe Francais des Glucides, Lyon, France, 1981, Izaak Walton Killam award The Can. Coun., 1981, rsch. prize U. Alta., 1982, Sir Frederick Haultain prize, Govt. Alta., 1982, Tishler award lectr. Harvard U., 1983, Gairdner Foun. Internat. award, 1985, Rhone-Poulenc award Royal Soc. Chemistry, Eng., 1989, King Faisal Internat. prize in sci., 1990, Gold medal Nat. Scis. and Engring. Rsch. Coun. Can., 1991, Manning award of distinction, 1992, PMAC Health Rsch. Found. Medal of Honor, 1992, Albert Einstein award World Cultural Coun., 1992; inauguration of The Lemieux Lectures, U. Ottawa, 1972; inauguration of The Raymond U. Lemieux Lectures on Biotechnology, U. Alta., 1987; inauguration of The R.U. Lemieux Award for Organic Chemistry, Can. Soc. for Chemistry, 1992; inducted into Alta. Order of Excellence, 1990. Fellow Chem. Inst. Can. (1st div. award div. organic chemistry 1954, Palladium medal, 1964) Royal Soc. Can., Royal Soc. London, Am. Chem. Soc. (C.S. Hudson award 1966), The Chem. Soc. (Haworth medal 1978), medal of hon. Can. Med. Assn., 1985). Home: 7602 119th St, Edmonton, AB Canada T6G 1W3 Office: U Alta, Dept Chemistry, Edmonton, AB Canada T6G 2G2

LEMING, W(ILLIAM) VAUGHN, electronics engineer; b. Pawhuska, Okla., Dec. 11, 1945; s. William Dalton Leming and Mattie Cornelia (Hatfield) Kafer; m. Janis Diana Lee (div.); children: Heather Lynne, Hilary Ann; m. Donna Faye Sartor, May 18, 1975; 1 child, Chandra Paige. Student, U. Okla., 1964-67. 68-70, U. Tulsa, 1967-68; cert., diploma, DeVry Inst. Tech., Chgo, 1977; cert., diploma with highest honors, Nat. Radio Inst., Washington, 1981. FCC Gen. Radio telephone, Naber Cert., IHF Consumer Audio Assoc. Spl. instr. Tri County Vocat.-Tech. Sch., Bartlesville, Okla., 1975-76; pres., chief exec. officer Fantasia Sound Systems, Inc., Jenks, Okla., 1976-87; announcer KWON Radio, Bartlesville, 1980-81; chief electronics technician The Sound Centre, Bartlesville, 1978-81; electronic technician 1A Burlington No. R.R. Co., Tulsa, 1981-92; founder, CEO FeS2 Pictures, 1991—; founder FeS2 Pictures. Musician, actor, entertainer, 1966—; bassist The New Orleans Jazz Band, Jenks, Okla., 1979—; actor: (films), The Outsiders, 1982, Rumble Fish, 1983, Fandango, 1984, Schizophrenia, 1989; dir., editor (documentary) Adjuvant Nutrition in Cancer Treatment Symposium. Recipient Gold Medal in Electronics, U.S. Skill Olympics, Lawton, Okla., 1976. Mem. Internat. Brotherhood Elec. Workers, Assn. Comm. Technicians (sr.), Nat. Assn. Bus. and Ednl. Radio (charter), Internat. Platform Assn. Avocations: antique Brit. sports cars, motorcycle riding, collecting Disney artifacts. Home: 523 East E St Jenks OK 74037-3326 Office: FeS2 Pictures Box 772 Jenks OK 74037-0772

LEMKE, CINDY ANN, support center founder and administrator; b. Ft. Eustice, Va., Nov. 10, 1963; d. Richard Wilburt and Darla Ann (Smith) L. BS with honors, U. Iowa, 1985; MS, Nova U., 1992. Behavior specialist Assn. Retarded Citizens of Rock Island County, Ill., 1988-89, birth-to-three coord., 1989-90; founder, exec. dir. Quad Cities Child and Family Support Ctr. Inc., Davenport, Iowa, 1988—; speaker's bur. Coun. on Children and Risk, Moline, Ill., 1988-90. Bd. dirs. Dirs. of Vols. in Agys., Davenport, 1989-90. Avocations: weight lifting, ceramics, animals, aerobics, distance running. Home: 609 NE 10th Ave Fort Lauderdale FL 33304-4671 Office: Quad Cities Child & Family Support Ctr 718 Bridge Ave Davenport IA 52803-5620

LEMKE, HEINZ ULRICH, computer science educator; b. Stettin, Germany, Mar. 25, 1941; m. Jean Gladys Chilvers, Sept. 21, 1970; 1 child, Kono Heinz. PhD, Cambridge (Eng.) U., 1970. Systems analyst Plessey Telecommunications, Taplow, Eng., 1971-72; software team leader, then project mgr. Graphical Software Ltd., London, 1972-74; chmn., speaker Faculty Informatics/Tech. U. Berlin, 1979-81, prof., 1975—; vis. prof., U. Hangzhou, China, 1985, U. Cairo, 1986, U. Calif., Irvine, 1987, U. Osaka, Japan, 1992; steering com. European Soc. Picture Archiving and Communication System, 1987—. Editor: Computer Assisted Radiology, 1985, 87, 89, 91, 93. Mem. Assn. Computing Machinery, Brit. Inst. Radiology, German Roentgen Soc., Soc. Computer Applications in Radiology (dir.-at-large 1988—). Avocations: sports, music. Office: Tech U Berlin, Franklin Strasse 28/29, 10587 Berlin Germany

LEMKE, PAUL ARENZ, botany educator; b. New Orleans, July 14, 1937; s. Paul A. and Glory Ann (Schellinger) L.; m. Joy Faye Owens, 1963 (div. 1982); children: Paul Arenz, Anne Wellesley. B.S., Tulane U., 1960; M.A., U. Toronto, 1962; Ph.D., Harvard U., 1966. Instr. Tulane U., New Orleans, 1962-63; sr. scientist Eli Lilly & Co., Indpls., 1966-72; assoc. prof. Carnegie-Mellon U., Pitts., 1972-79, sr. fellow, 1972-79; prof. Auburn (Ala.) U., 1979—; head dept. botany, plant pathology and microbiology Auburn U., 1979-85; cons. Marcel Dekker Pub., 1974—, E.R. Squibb & Sons, 1975-76, Schering Corp., 1987-90; dir. Am. Genetics, Inc., Denver, 1982-86. Editor: Viruses and Plasmids in Fungi, 1979, Applied and Environmental Microbiology, 1982, Applied Microbiology and Biotechnology, 1985; author: (with others) Plasmids of Eukaryotes, 1986; patentee in field. Recipient Humboldt Found. award, 1978; recipient Charles Porter award Soc. Indsl. Microbiology, 1982; Woodrow Wilson fellow, 1960. Fellow Am. Acad. Microbiology; mem. Soc. for Indsl. Microbiology (pres. 1979-80), Gordon Research Conf. (council mem. 1980), Am. Inst. Biol. Sci. (governing bd.), Am. Soc. Microbiology (chmn. fermentation and biotech. div. 1982), Phi Beta Kappa. Republican. Roman Catholic. Club. University. Lodge. Lions. Home: 1309 Gatewood Dr Apt 1109 Auburn AL 36830-2836 Office: Auburn Univ Dept of Botany & Microbiology 129 Funchess Hall Auburn AL 36849-3501

LEMMON, HAL ELMONT, computer scientist; b. Salt Lake City, July 24, 1932; s. Claude C. and Virginia (Olson) L.; m. Mary June Faerber, Aug. 9, 1950; children: Paula, Cindy, Howard, Craig, Teresa, Julie. BS in Math., U. Utah, 1956, MS in Math., 1959, PhD in Chem. Engring., 1963. Computer scientist USDA, Salt Lake City, 1975—. Mem. AAAS, Sigma Xi. Mem. LDS Ch. Achievements include development of original expert system for cotton crop management. Office: USDA 800 Buchanan St Berkeley CA 94710

LEMON, DOUGLAS KARL, physicist; b. Logan, Utah, Aug. 16, 1950; s. Karl Alvin and Bessie (Kirkham) L.; m. Alice Mae Baugh, June 27, 1973; children: Brett, Derek, Karl, Thad, Nathan. BS, Utah State U., 1974, PhD, 1978. Rsch. scientist Pacific N.W. Lab. Battelle, Richland, Wash., 1978-85; sect. mgr. Battelle, Richland, 1985-90, tech. mgr., 1990—; nat. program mgr. The Amtex Partnership for Am. Textile Industry and Dept. of Energy, Washington, 1992—; mgr./leader PNL's advanced mfg. initiative. Contbr. articles to profl. jours. Councilman City of West Richland, Wash., 1980-84; scoutmaster Boy Scouts Am., Richland, 1982-85 (dist. award of merit 1988). Named Scholar of Yr. Utah State U., 1974; recipient Excellence in Tech. Transfer award Fed. Lab. Consortium, 1989. Sr. mem. Soc. Mfg. Engrs. Mormon. Achievements include patent in acoustic emission linear pulse holography. Office: Pacific N W Lab PO Box 999 K7-02 Richland WA 99352

LEMONICK, AARON, physicist, educator; b. Phila., Feb. 2, 1923; s. Samuel and Mary (Ferman) L.; m. Eleanor Leah Drutt, Feb. 12, 1950; children—Michael Drutt, David Morris. B.A., U. Pa., 1950; M.A., Princeton U., 1952, Ph.D., 1954. Asst. prof. physics Haverford Coll., Pa., 1954-57, assoc. prof., 1957-61; assoc. prof. Princeton U., N.J., 1961-64, prof., 1964—, assoc. chmn. dept. physics, 1967-69, dean grad. sch., 1969-73, dean faculty, 1973-89, dean of faculty emeritus, 1989—; assoc. dir. Princeton-Pa. Accelerator, 1961-67; v.p. Princeton U. Press, 1973—; dep. dir. Princeton Plasma Physics Lab., 1989-90. Trustee Bryn Mawr Coll., 1988—. Fellow AAAS, Am. Phys. Soc.; mem. AAUP, Am. Assn. Physics Tchrs., Phi Beta Kappa, Sigma Xi. Office: Princeton U Dept Physics Princeton NJ 08544

LENEY, GEORGE WILLARD, environmental administrator, consulting geologist; b. Wausau, Wis., Nov. 13, 1927; s. Bert and Iva Irene (Skoog) L.; m. Arax G. Tefankjian, June 25, 1955 (dec. Aug. 1983); children: Sara Ann, Janet Ellen, John Alan, Ruth Alison. BS, U. Mich., 1950, MS, 1952, MA, 1955. Teaching fellow U. Mich., 1951-53, 53-55; geophysicist Gulf Oil Co., Harmarville, Pa., 1955-56; chief geophysicist Hanna Mining Co., Cleve., 1956-64; staff geophysicist Shell Oil Co., Houston, 1964-66; chief geologist H.K. Porter Co., Inc., Pitts., 1966-76, cons., 1976-77, 81-86, regional geologist U.S. Dept. Energy, 1977-81; air pollution adminstr. Allegheny County Health Dept., Pa., 1986—; v.p., bd. dirs. Pacific Asbestos Corp., 1970-75. With USN, 1946-48. Mem. Soc. Econ. Geologists, Am. Inst. Mining Engrs., Soc. Exploration Geophysicists, Geologic Soc. Am., Pitts. Geol. Soc., Pa. Acad. Sci., Air and Waste Mgmt. Assn. Address: 5335 Tomfran Dr Pittsburgh PA 15236

LENFANT, CLAUDE JEAN-MARIE, physician; b. Paris, Oct. 12, 1928; came to U.S., 1960, naturalized, 1965; s. Robert and Jeanine (Leclerc) L.; children: Philipe, Bernard, Martine Lenfant Wayman, Brigitte Lenfant Martin, Christine. B.S., U. Rennes, France, 1948; M.D., U. Paris, 1956; D.Sc. (hon.), SUNY, 1988. Asst. prof. physiology U. Lille, France, 1959-60; from clin. instr. to prof. medicine physiology and biophysics U. Wash. Med. Sch., 1961-72; asso. dir. lung programs Nat. Heart, Lung and Blood Inst. NIH, Bethesda, Md., 1970-72; dir. div. lung diseases Nat. Heart, Lung and Blood Inst. NIH, 1972-80; dir. Fogarty Internat. Center NIH, 1980-82, assoc. dir. internat. research, 1980-82; dir. Nat. Heart, Lung and Blood Inst., 1982—. Assoc. editor: Jour. Applied Physiology, 1976-82, Am. Jour. Medicine, 1979-91 ; mem. editorial bd.: Undersea Biomed. Research, 1973-75, Respiration Physiology, 1971-78, Am. Jour. Physiology and Jour. Applied Physiology, 1970-76, Am. Rev. Respiratory Disease, 1973-79; editor-in-chief: Lung Biology in Health and Disease. Fellow Royal Coll. Physicians; mem. Assn. Am. Physicians, Am. Soc. Clin. Investigation, French Physiol. Soc., Am. Physiol. Soc., N.Y. Acad. Scis., Undersea Med. Soc., Inst. of Medicine of Nat. Acad. Sci., USSR Acad. Med. Scis., French Nat. Acad. Medicine. Home: 13201 Glen Rd Gaithersburg MD 20878-8855 Office: Nat Heart Lung & Blood Inst Bldg 31A Rm 5A52 Bethesda MD 20892

LENG, GERARD SIEW-BING, mechanical and production engineering educator; b. Singapore, Mar. 5, 1963; s. Bernard Kwok-Leong and Juliana (Lee) L. PhD, U. Ill., 1990. Rsch. engr. Def. Sci. Orgn., Ministry of Def., Singapore, 1990-92; lectr. mech. and prodn. engring. Nat. U. Singapore, 1992—; cons. Singapore Air Force, 1992—. Recipient Overseas Engring. Scholarship Econ. Devel. Bd. Singapore, 1984-88. Mem. AIAA. Office: Nat U Singapore, Mech & Prodn Engring, 10 Kent Ridge Crescent, Singapore 0511, Singapore

LENG, MARGUERITE LAMBERT, regulatory consultant, biochemist; b. Edmonton, Alberta, Canada, Sept. 25, 1926; came to the U.S., 1950; d. Joseph Edouard and Marie (Kiwit) Lambert; m. Douglas Ellis Leng, June 18, 1955; children: Ronald Bruce, Janet Elaine, Douglas Lambert. BSc in Chemistry with honors, U. Alberta, 1947; MSc, U. Saskatchewan, 1950; PhD, Purdue U., 1956. Rsch. asst. U. Mich. Med. Rsch. Inst., Ann Arbor, 1950-53; sr. rsch. chemist bioproducts Dow Chem. Co., Midland, Mich., 1956-59, sr. registration specialist, product registration mgr., 1966-73, rsch. assoc. for internat. registration agrochems., 1973-86, mgr. internat. regulatory affairs, 1986-90; pres., cons. Leng Assocs., Midland, 1991—. Editor: Pesticide Chemist and Modern Toxicology, 1981, Regulation of Agrochemicals Environmental Fate in the Nineties, 1994; contbr. articles to profl. jours. and chpts. to books. Life ins. med. rsch. fellow Equitable Life Assurance Co., 1949-50. Fellow Am. Inst. Chemists (bd. dirs. 1992—, vice chmn. bd. dirs., exec. com. 1993—), Am. Chem. Soc. (agrochems. div. fellow 1976, chmn. 1981, program chmn. 1980, alt. councilor 1984-91, councilor 1992—), N.Y. Acad. Scis., mem. Internat. Soc. for Study Xenobiotics, Sigma Xi. Avocations: sailing, curling, international travel, family activities, foreign languages. Home and Office: 1714 Sylvan Ln Midland MI 48640-2538

LENG, XIN-FU, chemist, educator; b. Hai Yang, Shandong Province, Peoples Republic of China, Sept. 18, 1927; s. Huan-Yu and Xian-Ying Leng; m. Cha-Yun Sha, Feb. 25, 1955; 1 child, Li-Xing. BS, Peking Agrl. U., 1953. Rsch. asst. Inst. Entomol. Acad. Sinica, Beijing, 1953-62; rsch. assoc. Inst. Zoology Acad. Sinica, Beijing, 1963-77, rsch. assoc. prof., 1978-80, prof., 1986—; vis. scholar SUNY, Syracuse, 1980-83. Editor: Principle and Application of Insecticide, 1993; contbr. articles to Bull. Environ. Contamination Toxicology, Pesticide Biochemistry and Physiology, Sci. Bull., Chinese Biochem. Jour. Recipient Prize for Sci. Chinese Acad. Sci., 1978, 80, Prize for Malaria Control Nat. Com. Sci. and Tech., 1981. Mem. Chinese Entomol. Soc., Am. Chem. Soc. (div. agrochemistry 1985). Achievements include Chinese patent for method of preparation for methoxyphenyl acetate. Office: Inst Zool Acad Sinica, 19 Zhongguancun Lu Haidien, Beijing 100080, China

LENGYEL, JOSEPH WILLIAM, engineering manager; b. Homestead, Pa., Feb. 6, 1941; s. Joseph William and Sophia (Nachylowski) L.; children from previous marriage: Carolyn Ann, Lori Kay, Maura Lynn, Matthew Joseph; m. Audrey Jean Smith, July 11, 1992. BS, U.S. Mil. Acad., 1963; MSCE, U. Pitts., 1972. Registered profl. engr., Pa., Ky., W.Va. Mgr. transp. projects Pullman-Swindall divsn. Pullman Inc., Pitts., 1969-73; group v.p. Green Internat., Inc., Sewickley, Pa., 1973-76, sr. v.p., 1981-84; dir. design engring. engring. constrn. div. Pa. Engring. Corp., Pitts., 1976-81; mgr. design engring. Trimark Engring. and Constrn., Inc., Pitts., 1984-85; dir. engring. and design SE Techs., Inc., Bridgeville, Pa., 1985-93; dir. engring. and constrn. Allegheny County Sanitary Authority, 1993—; mem. site evaluation com. Louisville (Ky.) and Jefferson County Air Bd., 1972-73. Author: (corp. manual) A Systems Approach for the Analysis and Design of Urban Traffic Networks; author corp. tech. manuals on engring. mgmt.; mem. adv. bd. New Concepts in Trans. Trans. and environ. advisor various local mcpl. govts., Pitts., 1970—. Capt. U.S. Army, 1963-69. Named Ky. Col., 1973. Mem. ASCE, Am. Inst. Iron and Steel Engrs., Am. Railway Engring. Assn., Engrs. Soc. Western Pa. Achievements include devel. of major relational database applications for use in mng., engring., devel. of significant software programs for use in engring. estimating and resource scheduling. Home: 1017 Valley Dr Pittsburgh PA 15237 Office: Allegheny County Sanitary 3300 Preble Ave Pittsburgh PA 15233-1092

LENNON, GERARD PATRICK, civil engineering educator, researcher; b. N.Y.C., Nov. 15, 1951; s. Eugene Francis and Monica (Burghardt) L.; m. Linda More, June 5, 1976; children: Elizabeth, Brian, Marianne. BS, Drexel U., 1975; MS, Cornell U., 1977, PhD, 1980. Rsch. assoc. Cornell U., Ithaca, N.Y., 1978-80; asst. prof. Lehigh U., Bethlehem, Pa., 1980-86, assoc. prof., 1986—, acting dir. Environ. Studies Ctr., 1989-91; cons. Woodward-Clyde Cons., Plymouth Meeting, Pa., 1985—. Editor Symposium on Groundwater, 1991; contbr. articles to profl. jours. including Jour. Hydraulic Engring., Marine Geology. Consistory 1st United Ch. of Christ, Hellertown, Pa., 1988—, budget chmn., 1990—. Mem. ASCE (tech. com. 1983-91). Republican. Achievements include design of (with Richard Weisman) fluidization systems for coastal applications; research in boundary element method for solving groundwater flow problems. Office: Lehigh U 13 E Packer Ave Bethlehem PA 18015

LENNOX BUCHTHAL, MARGARET AGNES, neurophysiologist; b. Denver, Dec. 28, 1913; d. William Gordon and Emma Stevenson (Buchtel) L.; m. Gerald Klastskin, 1941 (div. 1947); 1 child, Jane Herner; m. Fritz Buchtal, Aug. 19, 1957. BA, Vassar Coll., 1934; MD, Yale Sch. Medicine, 1939; D of Medicine, Copenhagen U., 1972. Intern pediatrics Strong Meml. Hosp., Rochester, N.Y., 1939-40; asst. resident pediatrics N.Y. Hosp., N.Y.C., 1941-42; instr. Yale Sch. Medicine, New Haven, Conn., 1942-44, asst. prof. dept. psychiatry, 1945-51; asst. prof. U. Copenhagen, Inst. Neurophysiology, Denmark, 1957-72; assoc. prof. U. Copenhagen, Inst. Neurophysiology, 1972-81; ret., 1981; head clin. electroencephalography Yale U. Sch. of Medicine, 1942-51, head clinic epileptology, 1942-51; chief editor Epilesia Pub. by Elsevire, 1967-73. Contbr. articles to profl. jours. Republican. Methodist. Home: 289 El Cielito Rd Santa Barbara CA 93105-2306

LENOIR, WILLIAM BENJAMIN, aeronautical scientist-astronaut; b. Miami, Fla., Mar. 14, 1939; s. Samuel S. Lenoir; m. Elizabeth Lenoir, 1964; children: William Benjamin, Samantha. SB, MIT, 1961, SM, 1962, PhD, 1965. Asst. elec. engr. MIT, Cambridge, 1962-64, instr., 1964-65, asst. prof., 1965-67, Ford fellow engring., 1965-66; scientist-astronaut NASA Johnson Space Ctr., Houston, 1967-84; mission specialist 5th Mission of Columbia, NASA, 1982; with Booz Allen & Hamilton, Inc., Arlington, Va., 1984—. Mem. AAAS, Am. Geophys. Union. also: NASA Space Flight 600 Independence Ave SW Washington DC 20546

LENTS, THOMAS ALAN, waste water treatment company executive; b. Jasper, Ind., July 3, 1946; s. John Lee and Frances Elizabeth (Schmitt) L.; m. Margaret Louise Haake, Aug. 24, 1968; children: Nick Christopher, Caley Thomas. BS, Purdue U., 1969. Plant operator Jasper Wastewater Facilities, 1973-74, lab technician, 1974-79, chief operator, 1979-87, plant mgr., 1987—. Pres. Little League Baseball, Dubois, Ind., 1986-87. With U.S. Army 1969-70. Decorated Bronze Star. Mem. S.W. Ind. Operators' Assn. (pres. 1983-86, sec., treas 1986--), Ind. Water Pollution Control Assn. (Resolution of Appreciation 1984-85). Home: 10910 E 475 N Dubois IN 47527 Office: Jasper Wastewater 110 US Highway 231 Jasper IN 47546

LENTZ, RICHARD DAVID, psychiatrist; b. Passaic, N.J., Jan. 27, 1942; s. Harold Arthur and Ruth (Bitterman) L.; m. Joan Ellen Sacks, June 25, 1983; children: Daniel Keith, Andrew Simon. Student, John Hopkins U., 1959-61; AB cum laude, NYU, 1964; MS in Pathology, U. Rochester, 1969, MD with distinction, 1969. Diplomate Am. Bd. Psychiatry and Neurology, Am. Bd. of Pediatrics, Am. Bd. of Pediatric Nephrology. Intern U. Minn. Hosps., Mpls., 1969-70, resident in pediatrics, 1970-71, fellow in pediatric nephrology, 1972-74; resident in neurology and pediatrics Washington U., St. Louis, 1971-72; resident in psychiatry, fellow consultation-liaison U. Minn. Hosps., Mpls., 1979-81; chief pediatric nephrology Walter Reed Army Med. Ctr., Washington, 1974-76; instr. dept. of pediatrics Georgetown Med. Ctr., Washington, 1975; asst. prof. U. Md., Balt., 1978; psychiatrist Park Nicollet Med. Ctr., St. Louis Park, Minn., 1981—, vice chmn. dept. psychiatry, 1981-85, chmn. patient rels., 1983—, mem. risk mgmt. com., ops. com., dir. Medctr. Health Plan, 1985-90; clin. asst. prof. U. Minn., Mpls., 1981-85, clin. assoc. prof., 1985-90; clin. prof. U. Minn., 1990—; chmn. psychiatry Abbott-Northwestern Hosp., Mpls., 1991-92; cons. Courage Ctr., Mpls., Comprehensive Epilepsy Ctr., Bill Kelly House, numerous others. Contbr. articles to profl. jours. Maj. U.S. Army, 1974-76. Mem. Am. Psychiat. Assn., Minn. Med. Assn., Hennepin County Med. Soc., Mpls. Acad. Medicine. Office: Park Nicollet Med Ctr 2001 Blaisdell Ave Minneapolis MN 55404-2414

LENTZ, THOMAS LAWRENCE, biomedical educator, dean; b. Toledo, Mar. 25, 1939; s. Lawrence Raymond and Kathryn (Heath) L.; m. Judith Ellen Pernaa, June 15, 1961; children: Stephen, Christopher, Sarah. Student, Cornell U., 1957-60; MD, Yale U., 1964. Instr. in anatomy Yale U. Sch. Medicine, New Haven, 1964-66, asst. prof. of anatomy, 1966-69, assoc. prof. of cytology, 1969-74, assoc. prof. of cell biology, 1974-85, prof. of cell biology, 1985—, asst. dean for admissions, 1976—, vice chmn. cell biology, 1992—; mem. cellular and molecular neurobiology panel NSF, 1987-88, mem. cellular neurosci. panel, 1988-90. Author: The Cell Biology of Hydra, 1966, Primitive Nervous Systems, 1968, Cell Fine Structure, 1971; contbr. over 80 articles to sci. publs. Chmn. Planning and Zoning Commn., Killingworth, Conn., 1979—, Killingworth Hist. Soc. Grantee NSF, 1968-92, Dept. Army, 1986, NIH, 1987—; fellow Trumbull Coll., Yale U. Mem. AAAS, Am. Soc. Cell Biology, Soc. for Neurosci., N.Y. Acad. Scis., Appalachian Mountain Club, Alpha Omega Alpha. Republican. Mem. United Ch. of Christ. Achievements include study of primitive nervous systems, identification of neurotoxin binding site on the acetylcholine receptor, identification of cellular receptor for rabies virus. Office: Yale U Sch Medicine Dept Cell Biol 333 Cedar St New Haven CT 06520-8002

LENZ, CHARLES ELDON, electrical engineering consultant; b. Omaha, Apr. 13, 1926; s. Charles Julius and Hattie Susan (Wageck) L. SB, MIT, 1951; MS, U. Calif., Irvine, 1971; SM, MIT, 1953; PhD, Cornell U., 1957. Registered profl. engr., Mass. Elec. engr. GE, Pittsfield, West Lynn, Mass., Syracuse, Ithaca, N.Y., 1949-56; sr. staff engr. Avco Corp., Wilmington, Mass., 1958-60; mem. tech. staff Armour Rsch. Found., Chgo., 1960-62; sr. staff engr. North Am. Aviation, Inc., Anaheim, Calif., 1962-69; prof. U. Hawaii, Honolulu, 1966-68, U. Nebr., Omaha, 1973-78; sr. engr. Control Data Corp., Omaha, 1978-80; elec. engr. USAF, Offutt AFB, Nebr., 1980-84; sr. rsch. engr. Union Pacific Railroad, Omaha, 1984-87; engring. cons., 1987—; guest lectr. Coll. of Aeronautics, Cranfield, Eng., 1969, Cornell U., Ithaca, N.Y., UCLA, U. Minn., Mpls. Contbr. articles to profl. jours. including IEEE Internat. Conv. Record, ISA Conf. Procs., ISA Transactions, Procs. of the IEEE Electronics Conf., Procs. of Internat. Aerospace Instrumentation Symposium, Procs. of Modeling and Simulation Conf., Procs. of IEEE Electronics Conf., Procs. of Nat. Aerospace Electronics Conf., Procs. of Nat. Congress of Applied Mechanics. Program com. mem. N.E. Rsch. and Engring. Meeting, Boston, 1959. With USNR, 1944-46. Recipient 1st prize for tech. papers Am. Inst. Elec. Engrs., 1949; tuition scholar MIT, 1948-51, John McMullen Grad. scholar Cornell U., 1955-56; Charles Bull Earl Meml. fellow Cornell U., 1956-57, postgrad. fellow, U. Calif., Irvine, 1971. Mem. ASCE, IEEE (life, sr., nat. feedback control com., nat. control subcom. on computers), Sci. Rsch. Soc. Am. (life), N.Y. Acad. Scis. (life), Inst. Soc. Am. (sr.), Tau Beta Pi (life), Eta Kappa Nu (life), Sigma Xi (life). Achievements include patents in the fields of computer safety, storage and testing, phase-angle measurement, radar-signal processing, and ultraprecise control. Home: 5016 Western Ave Omaha NE 68132

LEODORE, RICHARD ANTHONY, electronic manufacturing company executive; b. Phila., Dec. 10, 1954; s. Pasquale and Norma (DiMartino) L.; m. Mariaelena D'Antonio, Mar. 3, 1989; children: Lauren, Richard Jr., Jennifer. BS in Acctg., Econs., Rutgers U., 1980; MBA in Fin., Drexel U., 1982. Yield analyst Atlantic Richfield Co., Phila., 1978-82; cost analyst ChemLink, Inc., Newtown Square, Pa., 1982-83, mgr. bus. analysis, 1983-86, cost acctg. mgr., 1986-87; contr. Amplifonix, Inc., Phila., 1987-88, v.p. ops., 1988-92, v.p., gen. mgr., 1993—. Home: 209 10th Ave Haddon Heights NJ 08035 Office: Amplifonix Inc 2707 Black Lake Pl Philadelphia PA 19154

LEONARD, CONSTANCE JOANNE, civil, environmental engineer; b. Bethlehem, Pa., Sept. 6, 1963; d. Roy Junior and Charlotte Jean (Biela) L. BSCE, U. Calif., Berkeley, 1985; postgrad., Loyola Marymount U., 1992—. Civil, environ. engr. Black & Veatch, L.A., 1988—. Vol. Habitat for Humanity, L.A., 1992. Mem. ASCE, Calif. Water Pollution Control Assn., Mensa, Intertel. Home: 3526 1/2 Market St Glendale CA 91208

LEONARD, ELIZABETH ANN, veterinarian; b. Anderson, Ind., Dec. 15, 1950; d. Doanld W. and Mary Jane (Elsbury) L. Student, Ind. U., Purdue U., 1969-71; DVM, Purdue U., 1975. Assoc. veterinarian Shirley Animal Hosp., Pendleton, Ind., 1970—. Avocations: needlecrafts, dog obedience tng. Home: 137 John St Pendleton IN 46064-1009 Office: Shirley Animal Hosp RR 2 Box 364A Pendleton IN 46064-9802

LEONARD, GILBERT STANLEY, oil company executive; b. Kingsport, Tenn., Sept. 3, 1941; s. Robert Spencer and Hope (Palmer) L.; m. Barbara Ann Bell, June 12, 1965 (div. 1982); m. Linda Marie Gremillion, Oct. 27, 1984. BS in Indsl. Mgmt., Purdue U., 1964; MS in Bus., U. Kans., 1970. Summer trainee Tenn. Eastman Co., Kingsport, 1963, prodn. planner, 1966-68; mktg. analyst Exxon Co., USA, Houston, 1970-74, distbn. specialist, 1975-78, staff systems analyst, 1979-81, group supr. applications support, 1982-84, strategic systems planner, 1984-85, supr. applications devel., 1986-89, system supr. Exxon Card Ctr., 1990—; instr., facilitator team leadership forum Exxon Co., USA, Houston, 1988, instr., facilitator quality forum, 1990; advisor, cons. Nat. Jr. Achievement, Houston, 1972-74. Lay reader Episc. Ch. Good Shepherd, Kingwood, Tex., 1975-82; treas. Forrest Lake Townhome Assn., Houston, 1987-88, pres., 1988-89. Lt. USNR, 1964-66. Recognized for excellence in Naval ROTC Chgo. Tribune, 1963. Mem. Quarterdeck Soc. (pres. 1962-63), Beta Gamma Sigma. Avocations: private pilot, golf, running. Home: 15515 Township Glen Ln Cypress TX 77429-5505 Office: Exxon Co USA PO Box 2180 Houston TX 77252-2180

LEONARD, JOEL I., biomedical engineer; b. N.Y.C., May 10, 1939; s. Julius and Anne (Schenkman) L. BS, Worcester Poly. Inst., 1960; PhD, U. Mich., 1971. Rsch. scientist NASA-Ames Rsch. Ctr., Moffett Field, Calif., 1962-66; sr. engr. NASA-Johnson Space Ctr., Houston, 1973-85; sr. scientist Lockheed Engring. and Sci. Co., Washington, 1985—. Contbr. over 90 articles to profl. jours. Pres. Washington Balalaika Soc.; bd. dirs. No. Va. Folk Festival, 1992-93, Balalaika & Domra Assn. of Am., 1992-93. Grantee NASA, NIH, NSF. Mem. Aerospace Med. Assn., Biomed. Engring. Soc., Am. Soc. of Gravitational and Space Biology. Home: 400 Madison St #2103 Alexandria VA 22314 Office: Lockheed & Sci Co 500 E St SW Washington DC 20024

LEONARD, KATHLEEN MARY, environmental engineering educator; b. Grand Rapids, Mich., Aug. 14, 1954; d. Melvin Frank and Mary Ann (Adkins) L.; m. John Andrew Gilbert, Dec. 31, 1985; children: Brian J., Benjamin M., Rebecca M., Allison M. BSCE, U. Wis., Milw., 1983, MSCE, 1985; PhD in Engring., U. Huntsville, 1990. Registered engr.-in-tng., Ala. Asst. prof. U. Ala., Huntsville, 1991—; speaker and presenter workshops. Contbr. articles to profl. jours. NSF grantee, 1992-94. Mem. ASCE (treas. 1992-93), Soc. Women Engrs. (v.p. 1991-92, Engr. of Yr. award 1990), Water Environment Fedn., Assn. Environ. Engring. Profs. Achievements include research in utilizing optical fibers for contaminant monitoring in life support systems, remote chemical sensing using fiber optics, fiber optics and the environment. Office: U Ala Civil & Environ Engring Huntsville AL 35899

LEONARD, KURT JOHN, plant pathologist, university program director; b. Holstein, Iowa, Dec. 6, 1939; s. Elvin Elsworth and Irene Marie (Helkenn) L.; m. Maren Jane Simonsen, May 28, 1961; children: Maria Catherine, Mary Alice, Benjamin Andrew. BS, Iowa State U., 1962; PhD, Cornell U., 1968. Plant pathologist Agrl. Rsch. Svc. USDA, Raleigh, N.C., 1968-88; dir. Cereal Rust Lab. U. Minn. USDA, St. Paul, 1988—. Author: (with others) Annual Review of Phytopathology, 1980; co-editor: Plant Disease Epidemiology, vol. 1, 1986, vol. 2, 1989; editor-in-chief Phytopathology, 1981-84; contbr. 70 articles to profl. jours. Fellow Am. Phytopathol. Soc. (coun. 1981-84); mem. Am. Mycological Soc., Internat. Soc. Plant Pathology (councilor 1982-93), British Soc. Plant Pathology, Phi Kappa Phi, Sigma Xi, Gamma Sigma Delta. Achievements include description of new species and genera of plant pathogenic fungi; research on spread of disease through crop mixtures, on relationships between virulence and fitness in plant pathogenic fungi. Office: U Minn USDA ARS Cereal Rust Lab Saint Paul MN 55108

LEONARD, NELSON JORDAN, chemistry educator; b. Newark, Sept. 1, 1916; s. Harvey Nelson and Olga Pauline (Jordan) L.; m. Louise Cornelie Vermey, May 10, 1947 (dec. 1987); children: Kenneth Jan, Marcia Louise, James Nelson, David Anthony; m. Margaret Taylor Phelps, Nov. 14, 1992. B.S. in Chemistry, Lehigh U., 1937, Sc.D., 1963; B.Sc., Oxford (Eng.) U., 1940, D.Sc., 1983; Ph.D., Columbia U., 1942; D.h.c., Adam Mickiewicz U., Poland, 1980; D.Sc. (hon.), U. Ill., 1988. Fellow and rsch. asst. chemistry U. Ill., Urbana, 1942-43, instr., 1943-44, assoc., 1944-45, 46-47, asst. prof., 1947-49, assoc. prof., 1949-52, prof. organic chemistry, 1952-68, head div. organic chemistry, 1954-63, prof. biochemistry, 1973-86, R.C. Fuson prof. chemistry, mem. Ctr. for Advanced Study, 1981-86, R.C. Fuson prof. emeritus, 1986—; investigator antimalarial program Com. Med. Research, OSRD, 1944-46; sci. cons. and spl. investigator Field Intelligence Agy. Tech., U.S. Army and Dept. Commerce, 1945-46; mem. Can. NRC, summer 1950; Swiss-Am. Found. lectr., 1953, 70; vis. lectr. UCLA, summer 1953; Reilly lectr. U. Notre Dame, 1962; Stieglitz lectr. Chgo. sect. Am. Chem. Soc., 1962; Robert A. Welch Found. lectr., 1964; Disting. vis. lectr. U. Calif.-Davis, 1975; vis. lectr. Polish Acad. Scis., 1976; B.R. Baker Meml. lectr. U. Calif., Santa Barbara, 1976; Ritter Meml. lectr. Miami U., Oxford, Ohio; Werner E. Bachman Meml. lectr. U. Mich., Ann Arbor, 1977; vis. prof. Japan Soc. Promotion of Sci., 1978; Arapahoe lectr. U. Colo., 1979; mem. program com. in basic scis. Arthur P. Sloan, Jr. Found., 1961-66; Philips lectr. Haverford Coll., 1971; Backer lectr., Groningen, Netherlands, 1972; FMC lectr. Princeton U., 1973; plenary lectr. Laaxer Chemistry Conf., Laax, Switzerland, 1980, 82, 84, 88, 90, 92; Calbiochem-Behring Corp. U. Calif.-San Diego Found. lectr., 1981; Watkins vis. prof. Wichita State U. (Kans.), 1982; Ida Beam Disting. vis. prof. U. Iowa, 1983; Fogarty scholar-in-residence NIH, Bethesda, Md., 1989-90; Sherman Fairchild Disting. scholar Calif. Inst. Tech., 1991; Syntex. disting. lectr. U. Colo., 1992; faculty assoc. Calif. Inst. Tech., 1992—; mem. adv. com. Searle Scholars program Chgo. Community Trust, 1982-85; ednl. adv. bd. Guggenheim Found., 1969-88, mem. com. of selection, 1977-88. Editor: Organic Syntheses, 1951-58, mem. adv. bd., 1958—, bd. dirs., 1969—, v.p., 1976-80, pres., 1980-88; editorial bd. Jour. Organic Chemistry, 1957-61, Jour. Am. Chem. Soc., 1960-72; adv. bd. Biochemistry, 1973-78, Chemistry International, 1984-91, Pure and Applied Chemistry, 1984-91; contbr. articles to profl. jours. Recipient Am. Chem. Soc. award, 1963, medal Synthetic Organic Chem. Mfrs., 1970, Wheland award U. Chgo., 1991; named to Mt. Vernon High Sch. Hall of Fame, N.Y., 1985; Rockefeller Found. fellow, 1950, Guggenheim Meml. fellow, 1959, 67. Fellow Am. Acad. Arts and Scis. (v.p. 1991—); mem. NAS, AAAS, Polish Acad. Scis. (fgn.), Ill. Acad. Sci. (hon.), Am. Chem. Soc. (Edgar Fahs Smith award and lectureship Phila. sect. 1975, Centennial lectr. 1976, Roger Adams award 1981), Am. Soc. Biol. Chemists, Chem. Soc. London, Swiss Chem. Soc., Am.-Can. Soc. Plant Physiologists, Internat. Union Pure and Applied Chemistry (sec. organic chemistry div. 1989, v.p. 1989-91, pres. 1991-93), Pharm. Soc. Japan, (hon.), Phi Beta Kappa, Phi Lambda Upsilon (hon.), Tau Beta Pi, Alpha Chi Sigma. Achievements include patents on synthesis of sparteine; esters of pyridine dicarboxylic acid as insect repellents; fluorescent derivatives of adenine- and cytosine-containing compounds. Home: 389 California Ter Pasadena CA 91105

LEONARD, THOMAS ALLEN, physicist; b. Muskegon, Mich., July 23, 1941; s. Percy James and Verona Mae (Lukonic) L.; m. Ellen Marie Dullberg, Nov. 24, 1965 (div. 1974); m. Bonita C. Bentz, Apr. 11, 1976 (div. 1984); children: Christine Anne, Michael James; m. Marsha Kay Bishop, May 19, 1990. MS in Nuclear Engring., U. Mich., 1966, PhD in Plasma Physics, 1972. Physicist KMS Fusion Inc., Ann Arbor, Mich., 1972-78; program mgr. U. Dayton (Ohio) Rsch. Inst., 1978-85; sr. program mgr. Ball Corp., Dayton, 1985-89; sr. scientist, pres. Science Services, Inc., Dayton, 1989—; cons. USAF, Strategic Defense Initiative, 1986-92. Contbr. articles to profl. jours. AEC fellow, 1970. Mem. Am. Phys. Soc., Optical Soc. Am., Soc. Photographic Instrumentation Engrs., U.S. Space Found., Fusion Power Assoc., Mensa. Achievements include research on infrared ellipsometry, optical scatter measurement, plasma physics and laser fusion. Home: 93 Shelford Way Beavercreek OH 45440 Office: Sci Svcs Inc PO Box 340607 Beavercreek OH 45434-0607

LEONARDOS, GREGORY, chemist, odor consultant; b. Cambridge, Mass., Dec. 30, 1935; s. Nicholas C. and Evangeline (Niarchos) L.; m. Virginia Shinopoulos, May 23, 1965; children: Nicholas, Charles. AB in Biochem. Sci., Harvard U., 1957; MS in Chemistry, Northeastern U., 1964, MBA, 1969. Rsch. assoc. Protein Found., Boston, 1960-63; sr. project leader Arthur D. Little Inc., Cambridge, Mass., 1963-80; prin. G. Leonardos Cons., Arlington, Mass., 1980—; vis. lectr. Univ. Mass., Dartmouth, 1992—. Contbr. articles to profl. jours. Mem. Am. Chem. Soc., Air and Waste Mgmt. Assn. (chmn. TT-4 com. 1975-78). Greek Orthodox. Avocations: coaching youth soccer, golf. Home and Office: 43 Ronald Rd Arlington MA 02174-1421

LEON DUB, MARCELO, chemicals executive, entrepreneur; b. Quito, Ecuador, Mar. 30, 1946; arrived in Colombia, 1946; s. Jorge E. and Helen (Dub) Leon Fernandez-Salvador; m. Patricia Gomez, Dec. 19, 1974; children: Mateo, Nicolas, Manuela. Diploma in chem. engring., U. Nacional, Bogota, Colombia, 1968; MBA, Harvard U., Boston, 1972. Tecnico Dept. Nacional De Planeacion, Bogota, 1968-70; gen. mgr. Aknaz Colombiana, S.A., Bogota, 1972-74; co. gen. mgr. Quimica Comml. Andina, S.A., Bogota, 1974—, Distribuidora Andina S.A., Bogota, 1976—; bd. dirs. Quala, S.A., Bogota, Oxigenados Y Derivados Ltd., Medellin, Sociedad Agricola De Dibulla, Ltd., Bogota, Seguros Aurora, S.A., Bogota. Bd. dirs. Corporacion Sindrome de Down, Bogota, 1988-92, Fundacion Recreacion Y Cultura, Bogota, 1990—. Mem. Young Pres. Orgn. (membership chmn. 1988-89, chmn. 1993—), MIT and Harvard Club, Bogota Country Club (bd. dirs. 1990-91), Chicala. Office: Distribuidora Andina SA, Calle 12 A No. 68B-81, Bogota Colombia

LEONG, MANG SU, economist; b. Zhongshan, China, Jan. 17, 1960; s. Ka Kei Leong and Wan Sam Lei. BSSc in Econs., U. East Asia, Macau, 1989. Apprentice Seng Kong Antigos Electricos, Macau, Macao, 1979-80; roomboy Hotel Estoril, Macau, Macao, 1980-84; rsch. asst. China Econ. Rsch. Ctr. U. East Asia, 1990—. Mem. Macau Soc. Social Scis. Avocations: reading, swimming. Office: U Macau China Econ Rsch Ctr, PO Box 3001, Macau Macau

LEON-SANZ, MIGUEL, physician, endocrinologist, nutritionist; b. Valencia, Spain, June 29, 1957; s. Francisco Jose Leon-Tello and Virginia Sanz-Sanz. MD, Complutense U., Madrid, 1980; endocrinology and nutrition specialist, Insalud, Madrid, 1985; PhD, Complutense U., 1988. Resident tng. program in endocrinology and nutrition Hosp. 12 de Octubre, Madrid, 1982-85, endocrinologist, cons. in endocrinology and nutrition, 1986-88; endocrinologist Hosp. Virgen de Alarcos, Ciudad Real, Spain, 1989; dir. nutrition dept. Hosp. 12 de Octubre, Madrid, 1990—; assoc. prof. medicine Complutense U. Contbr. articles to profl. jours. Fellow N.Y. Acad. Scis.; mem. Soc. Española Endocrinologia, Assn. Nutrición C. y Dietetica (treas. 1988-91), Assn. Endocrinologia CAM (sec. 1989—), European Soc. Parenteral and Enteral Nutrition, Am. Soc. Parenteral and Enteral Nutrition. Avocations: piano, literature, history, tennis. Home: Fernando El Catolico 77, 28015 Madrid Spain Office: Hosp 12 de Octubre, Cra Andalucia KM 5.4, 28041 Madrid Spain

LEOPOLD, LUNA BERGERE, geology educator; b. Albuquerque, Oct. 8, 1915; s. Aldo and Estella (Bergere) L.; m. Barbara Beck Nelson, 1973; children: Bruce Carl, Madelyn Dennette. BS, U. Wis., 1936, DSc (hon.), 1980; M.S., UCLA, 1944; Ph.D., Harvard, 1950; D Geography (hon.), U. Ottawa, 1969; DSc (hon.), Iowa Wesleyan Coll., 1971, St. Andrews U., 1981, U. Murcia, Spain. With Soil Conservation Service, 1938-41, U.S. Engrs. Office, 1941-42, U.S. Bur. Reclamation, 1946; head meteorologist Pineapple Research Inst. of Hawaii, 1946-49; hydraulic engr. U.S. Geol. Survey, 1950-71, chief hydrologist, 1957-66, sr. research hydrologist, 1966-71; prof. geology U. Calif. at Berkeley, 1973—. Author: (with Thomas Maddock, Jr.) The Flood Control Controversy, 1954, Fluvial Processes in Geomorphology, 1964, Water, 1974, (with Thomas Dunne) Water in Environmental Planning, 1978; also tech. papers. Served as capt. air weather service USAAF, 1942-46. Recipient Disting. Svc. award Dept. of Interior, 1958; Veth medal Royal Netherlands Geog. Soc., 1963; Cullum Geog. medal Am. Geog. Soc., 1968; Rockefeller Pub. Service award, 1971; Busk medal Royal Geog. Soc., 1983, Berkeley citation U. Calif., David Linton award British Geomorphol. Research Group, 1986, Linsley Award Am. Inst. Hydrology, 1989, Caulfield medal Am. Water Resources Assn., 1991, Nat. Medal Sci. NSF, 1991. Mem. NAS (Warren prize), ASCE (Julian Hinds award), Geol. Soc. Am. (Kirk Bryan award 1958, pres. 1972, Disting. Career award geomorphological group 1991), Am. Geophys. Union (Robert E. Horton medal 1993), Am. Acad. Arts and Scis., Am. Philos. Soc., Sigma Xi, Tau Beta Pi, Phi Kappa Phi, Chi Epsilon. Club: Cosmos (Washington). Home: 400 Vermont Ave Berkeley CA 94707-1722

LEPAGE, WILBUR REED, electrical engineering educator; b. Kearny, N.J., Nov. 16, 1911; s. Wilbur Nicholas and Gertrude Elizabeth (Reedt) LeP.; m. Eveline Marie Jacobsen, June 9, 1936; 1 dau., Margaret Ann. E.E., Cornell U., 1933, Ph.D., 1941; M.S. in Physics, U. Rochester, 1939. Instr. engring. U. Rochester, 1933-38; grad. student, teaching asst. Cornell U., 1939-41; mem. staff advanced devel. lab. RCA, 1941; physicist radiation lab. Johns Hopkins, 1942-45; sr. research engr. Stromberg Carlson Co., 1946; prof. elec. engring. Syracuse U., 1947—, chmn. dept., 1956-74; initiated ann. Sagamore Conf. Elec. Engring. Edn., 1952; cons. to Signal Corps. U.S. Army, 1953; ednl. cons. UNESCO, 1967; guest lectr. Rumanian Assn. Scientists, 1977; mem. Nat. Com. on Elec. Engring. Films, 1964-71. Author: Analysis Alternating Current Circuits, 1952, (with S. Seely) General Network Analysis, 1952, Complex Variables and The Laplace Transform for Engineers, 1961, (with N. Balabanian) Electrical Science, Book I, 1970, Book II, 1972, Applied APL Programming, 1978. Recipient Gen. Electric award for teaching excellence, 1985. Fellow IEEE (chmn. com. basic scis. 1955-56), AAUP, N.Y. Acad. Scis., Sigma Xi. Home: 217 Dewitt Rd Syracuse NY 13214-2006 Office: Syracuse U Link Hall Syracuse NY 13244

LEPPLA, DAVID CHARLES, pathology educator; b. Denver, July 22, 1953; s. Charles Frederick and Lucille Josephine (Schneider) L. BS, Seattle U., 1975; MD, Colo. U., 1979. Diplomate Am. Bd. Pathology. Intern in internal medicine U. Tex. Health Sci. Ctr., Dallas, 1979-80, fellow in mineral metabolism and endocrinology, 1980-82, rsch. assoc., 1982-83; resident in pathology Marshall U. Sch. Medicine, Huntington, W.Va., 1984-87, chief resident in pathology, 1987-88, asst. prof., 1988—. Fellow Am. Soc. Clin. Pathology (alt. to adv. com. 1990); mem. AAAS, Alpha Omega Alpha. Office: Marshall U Sch Medicine 1542 Spring Valley Dr Huntington WV 25704-9388

LERMAN, GERALD STEVEN, software company executive; b. Bklyn., Nov. 13, 1951; s. Joseph and Selma Lillian Lerman; m. Kathleen A. Kania; children, Mar. 24, 1977; children: Candice, Alexander. BBA, Pace U., 1973. Sr. analyst Honeywell Info. Systems, Billerica, Mass., 1978-81; devel. mgr. Atex Systems, Bedford, Mass., 1981-83, Lotus Devel. Corp., Cambridge, Mass., 1984—; pres. Lerman Assocs., Westford, Mass., 1986—. Author software SeeMore (Best of '87 and '89 award PC mag.), Extra K (Star of 1989 award Reseller News). Mem. Software Pubs. Assn., Boston Computer Soc.

LERMAN, ISRAËL CÉSAR, data classification and processing researcher; b. Beirut, Jan. 2, 1940; arrived in France, 1963; s. Jacob and Sarina (Tawil) L.; m. Rollande Genet, Oct. 14, 1966; children: Sabine, Alix Léa, Judith. Lic. Math., U. Lyon, Beirut, 1963; diploma in Stats., U. Paris, 1966, D, 1966, DSc, 1971. Rsch. engr. Maison des Scis. de l'Homme, Paris, 1966, 73; prof. U. Rennes (France) 1, 1973—; project mgr. Rsch. Inst. Informatics and Random Systems, Rennes, 1981. Author: Les Bases de la Classification Automatique, 1970, Classification et Analyse Ordinale des Données, 1981; assoc. editor Rairo-Operations Research; mem. editorial bd. Applied Stochastic Models and Data Analysis, Math. Informatique and Scis. Humaines, Rairo: Automatic Control Production Systems. Avocations: painting, yoga. Home: Ave Sergent Maginot, 35000 Rennes Bretagne, France Office: Rsch Inst Informatics and, Random Systems, Ave du Gen Leclerc, 35042 Rennes Bretagne, France

LERNER, ARMAND, acoustical consultant; b. Antwerp, Belgium, Oct. 3, 1932; s. Jacques and Mia (Engel) L. BBA, Baruch Coll., 1954; postgrad., Columbia U., 1955. V.p. Jacques Lerner & Son, Inc., N.Y.C., 1951-60; mgr. Indsl. Acoustics Co. Inc., N.Y.C., 1956-61; pres. Lerner Equipment Co., Inc., 1961—, Controlled Acoustics Corp., New Rochelle. Contbr. articles to profl. jours. Bd. dirs. S.D.I.C., New Rochelle, 1986-92. Mem. Construction Specifications Inst., Inst. Bus. Designers. Achievements include patent in Specialty Noise Control Product and Systems.

LERNER, HENRY HYAM, chemist; b. Bklyn., Jan. 28, 1940; s. Samuel Ber and Lillian (Cohen) L.; m. Bella Kurtz, June 29, 1961; children; Solomon, Deborah, Emanuel, Riuka, Dena. BS, Bklyn. Coll., 1961; MS in Chemistry, Rutgers U., 1970. Dir. quality control Bristol-Myers Squibb Co., New Brunswick, N.J., 1961—. Author: Analytical Profile of Drug Substances, 1971, Vol. 2&3, 1973. Sec. Jewish Fedn. Greater Middlesex County, Edison, N.J., 1983—. Mem. Am. Chem. Soc., Am. Assn. Pharm. Scientists, Asn. Official Analyt. Chemists. Jewish. Home: 1 Celler Rd Edison NJ 08817 Office: Bristol-Myers Squibb Co PO Box 191 New Brunswick NJ 08903

LERNER, JAMES PETER, software engineer; b. N.Y.C., Aug. 7, 1956; s. Arnold Aaron and Rita (Guggenheim) L.; m. Anita Elisabeth Springer, Sept. 12, 1987. BS in Biology/Computer Sci. with honors, Union Coll., Schenectady, N.Y., 1978. Engr. Raytheon Co., Sudbury, Mass., 1978-81; cons. Paramin, Inc., Wellesley, Mass., 1981-83; engr. Kurzweil Applied Intelligence, Waltham, Mass., 1983-87; mgr. Sun Microsystems, Chelmsford, Mass., 1987—. Author software Portable Mail, 1989. Mem. IEEE, Sigma Xi. Achievements include software patents for linguistic expert system, mouse emulation, fast I/O. Home: 17 Tamarac Rd Newton MA 02164 Office: Sun Microsystems Inc 2 Elizabeth Dr Chelmsford MA 01824

LERNER, JOSEPH, dean; b. Wilkes-Barre, Pa., Jan. 16, 1942; s. Solomon Samuel and Dorothy Rose (Gromer) L.; m. Linda Dell, Aug. 28, 1963; children: Michael Jeffrey, Michele Ann. BS, Rutgers U., 1963, PhD, 1967. Rsch. biochemist U.S. Dept. Agrl., Phila., 1967-68; asst. prof. biochemistry U. Maine, Orono, 1968-71, assoc. prof., 1971-77; rsch. assoc. Avian scis U. Calif., Davis, 1974; dir. PhD program nutrition U. Maine, Orono, 1975-77, prof. biochemistry, 1977-83; chmn. biochemistry, 1978-83; fellow acad. adminstrn. U. N.H., Durham, 1982-83; dean coll. arts & scis. Tenn. Tech. U., Cookeville, 1984—; mem. adv. coun. Tenn. State Bd. Edn., Nashville, 1987-88; pres. Tenn. Coun. Arts& Scis. Deans, 1987. Author: A Review of Amino Acid Transport Processes in Animal Cells and Tissues, 1978; contbr. over 51 articles to profl. jours. Rsch. grantee NIH, 1974-83. Mem. AAAS, Am. Chem. Soc., Am. Inst. Nutrition, Tenn. Coun. Arts Scis. Deans, N.Y. Acad. Scis., Coun. Coll. Arts and Scis., Sigma Xi, Rotary. Home: 709 Sutton Pl Cookeville TN 38501 Office: Tenn Tech U Box 5065 Cookeville TN 38505

LERNER, RICHARD ALAN, chemistry educator, scientist; b. Chgo., Aug. 26, 1938; s. Peter Alex and Lily (Orlinsky) L.; m. Diana Lynn Pritchett, June 1966 (div. 1977); children: Danica, Arik, Edward; m. Nicola Green, Sept. 1, 1979. Student, Northwestern U., 1956-59; BS, MD, Stanford U., 1964; MD (hon.), Karolinska Inst., 1990. Intern Palo Alto (Calif.) Stanford Hosp., 1964-65, rsch. fellow, 1965-68; assoc. mem. Wistar Inst., Phila., 1968-70; assoc. mem. dept. exptl. pathology Scripps Clinic and Rsch. Found., La Jolla, Calif., 1970-72, mem., 1972-74, mem. dept. immunopathology, 1974-82; chmn. and mem. dept. molecular biology Rsch. Inst. Scripps Clinic, La Jolla, 1982-87, prof. dept. chemistry, 1988—, dir., 1987—; pres. The Scripps Rsch. Inst., La Jolla; cons. Johnson & Johnson, 1983—, PPG Industries, Inc., Pitts., 1987—; sci. advisor Igen Inc., Rockville, Md., 1986—; spl. advisor Genex Corp., Gaithersburg, Md., 1988—; bd. dirs. Cytel Corp.; chmn. Internat. Symposium on Molecular Basis Cell-Cell Interaction, 1977, 78, 79, 80; mem. organizing com. for Modern Approaches to Vaccines, Cold Spring Harbor, 1983-89. Contbr. over 250 sci. papers; mem. editorial bd. Jour. Virology, Molecular Biology and Medicine, Protein Engring., Vaccine, In Vivo, Peptide Rsch. Mem. sci. policy adv. com. Uppsala U. (Sweden), sci. adv. bd. Econ. Devel. Bd., Singapore. Decorated Oficial de La Orden de San Carlos (Colombia); recipient NIH AID Career Devel. award, 1970, Parke Davis award, 1978, John A. Muntz Meml. award, 1990, San Marino prize, 1990, Burroughs Wellcome Fund and FASEB Wellcome Vis. Prof. award, 1990-91, College de France award, 1991, 1oth Ann. Jeanette Piperno Meml. award, 1991, Arthur C. Cope scholar award in chemistry, 1991. Fellow ACS (screening com. Calif. div.); mem. NAS Inst. Med. (ad hoc com. new rsch. opportunities in immunology), Am. Soc. Virology (charter), Am. Soc. Nephrology, Am. Assn. Immunologists, Am. Soc. Exptl. Pathology, Am. Soc. Microbiology, N.Y. Acad. Scis., Biophys. Soc., Royal Swedish Acad. Sci, Nat. Cancer Inst. (cancer preclin. program project rev. com. 1985-88), Royal Swedish Acad. Scis. (fgn., Lita Annenberg Hazen prof. immunochemistry 1986), 1st Thursday Club, Phi Eta Sigma, Alpha Omega Alpha. Avocations: tennis, walking, skiing, polo. Home: 2630 Torrey Pines Rd Apt 21 La Jolla CA 92037-3447 Office: Scripps Rsch Inst 10666 N Torrey Pines Rd La Jolla CA 92037-1027*

LERNER, SHELDON, plastic surgeon; b. N.Y.C., Mar. 3, 1939; s. Louis and Lillian L.; AB with honors, Drew U., Madison, N.J., 1961; MD, U. Louisville, 1965. Intern, resident Albert Einstein Coll. Medicine, Bronx-Mcpl. Hosp. Center, 1965-73; practice medicine, specializing in plastic surgery Plastic Cosmetic and Reconstructive Surgery Center, San Diego, 1973—. Served with USPHS, 1968-70. Mem. AMA, Am. Soc. Plastic and Reconstructive Surgeons, Calif. Med. Soc., San Diego County Med. Soc., San Diego Internat. Plastic Surgery Assn. Clubs: Masons, Shriners. Office: 3399 First Ave San Diego CA 92103

LEROY, CLAUDE, physics educator, researcher; b. Charleroi, Hainaut, Belgium, Sept. 30, 1947; s. Bernard and Rénee (Jacobeus) L. Mathématique Spéciale, Faculté St. Louis, Brussels, 1967; Lic. en Sci., U. Louvain, Belgium, 1971, PhD, 1976. Rsch. assoc. McGill U., Montréal, 1977-80; attaché de rsch U. Montréal, 1978-80; rsch. assoc. Northwestern U., Evanston, Ill., 1980-81; chercheur du fonds du devel. scis. U. Louvain, 1981-83; rsch. scientist Inst. Particle Physics, Montréal, 1983-90; assoc. prof. physics McGill U., 1983-90; titular prof. physics U. Montréal, 1990—, dir. nuclear physics lab., 1991—; vis. rsch. fellow U. Southampton, Eng., 1976-77; sci. assoc.Ctr. European Rsch. Nuclear Physics, Geneva, Switzerland, 1980—. Contbr. over 200 sci. papers to profl. jours. Killam Rsch. fellow The Can. Coun., 1993. Fellow Royal Soc. Can. (Rutherford Prize for Physics, 1988); mem. Inst. Particle Physics Can., Can. Assn. Physicists. Roman Catholic. Avocations: Egyptian Hieroglyphics, history, fishing. Home: 5155 Boulevard Lasalle, Verdun, PQ Canada H4G 2C1 Office: U Montréal Nuclear Physics Lab, CP 6128, succursale A, Montreal, PQ Canada H3C 3J7

LEROY, EDWARD CARWILE, rheumatologist; b. Elizabeth City, N.C., Jan. 19, 1933; s. J. Henry and Grace Brown (Carwile) LeR.; m. Garnette DeFord Hughes, June 11, 1960; children: Garnette DeFord, Edward Carwile. B.S. in Sci. summa cum laude, Wake Forest Coll., 1955; M.S. in Pathology, U. N.C., 1958, M.D. with honors, 1960. Med. intern Presbyn. Hosp., N.Y.C., 1960-61; resident Presbyn. Hosp., 1961-62; clin. asso. Nat. Heart Inst., Bethesda, Md., 1962-65; fellow in rheumatology Columbia U., 1965-67; dir. Edward Daniels Faulkner Arthritis Clinic; asso. attending physician Presbyn. Hosp., N.Y.C., 1970-75; asso. prof. Columbia U. Coll. Phys. and Surg., 1970-75; prof. medicine, dir. div. rheumatology and immunology Med. U. S.C., Charleston, 1975—; bd. dirs. Arthritis Found. Contbr. articles med. jours. Mem. Am. Coll. Rheumatology (bd. dirs.), AAAS, Am. Fedn. Clin. Research, Harvey Soc., A.C.P., Am. Assn. Immunologists, Soc. Exptl. Biology and Medicine, Am. Soc. Clin. Investigation, Microvascular Soc., N.Y. Acad. Scis., So. Soc. Clin. Investigation, Assn. Am. Physicians, Orthopedic Research Soc. First Scots Presbyterian. Clubs: Yeamans Hall, Carolina Yacht. Office: Med Univ SC Arthritis Clin & Rsch Ctr 171 Ashley Ave Charleston SC 29425-2229

LESH, JAMES RICHARD, engineering manager; b. Greeley, Colo., May 31, 1944; s. Arthur James and Clare Marjorie (Parker) L.; m. Aja Maria Tulleners (div.); children: Richard, Jeffrey, Timothy. BS, UCLA, 1969, MS, 1971, PhD, 1976. Devel. engr. Bell & Howell Corp., Pasadena, Calif., 1966-69; systems engr. Leach Corp., Azusa, Calif., 1969-71; rsch. engr. Jet Propulsion Lab., Pasadena, Calif., 1971-77, rsch. supr., 1977—; lectr. Calif. Inst. Tech., Pasadena, 1978-79. Editor-in-chief IEEE Transactions on Comm., 1985-88, editor, 1975-84; contbr. over 70 articles to profl. jours. Dir., choir local ch., Pasadena, 1979—; cubmaster Cub Scouts, Pasadena, 1978-88. With USN, 1964-65. Fellow IEEE; mem. Soc. of Photo Instrument Engrs. Achievements include patents for multiple rate digital command detection system, means for phase-locking the outputs of a surface emitting laser diode array; prin. investigator for NASA GOPEX optical comms. demonstration with Galileo spacecraft. Office: Jet Propulsion Labs 4800 Oak Grove Dr Pasadena CA 91109

LESHKO, BRIAN JOSEPH, civil engineer; b. Charleston, S.C., Aug. 13, 1962; s. Francis Theodore and Carol Anne (Gravatt) L.; m. Debra Louise Bogen, Mar. 4, 1988. BSCE, USAF Acad., 1985; MS in Structural Engring., U. Conn., 1990. Registered profl. engr., Colo., Md. Design engr., project mgr. 820th Civil Engring. Squadron, Nellis AFB, Nev., 1985-88; quality assurance evaluator 4700th Ops. Support Squadron, Sondrestrom AB, Greenland, 1988-89; instr. civil engring. USAF Acad., Colo., 1990-92; grad. student, teaching and rsch. asst. The Johns Hopkins U., Balt., 1992-93, Eisenhower Transp. fellow, 1993—; soaring flight instr. USAF Acad. Soaring, 1991-92, Bay Soaring, Woodbine, Md., 1992—. Named one of Outstanding Young Men in Am., 1988, Outstanding Grad. in Civil Engring., 1985. Mem. ASCE, Soc. Am. Mil. Engrs. (sec. Nellis post 1987-88), Air Force Assn., Soaring Soc. of Am. Achievements include research to quantify the rate of permeability of air through a sample of simulated lunar concrete fabricated using a dry mold steam injection system. Office: The Johns Hopkins U Dept Civil Engring 3400 N Charles St Baltimore MD 21218

LESHNOWER, ALAN LEE, podiatrist; b. Bklyn., Apr. 9, 1938; s. William Julius Max and Florence (Liebowitz) L.; m. Tobi Thea Delofsky, Oct. 30, 1976. BS in Biology-Chemistry, Queens Coll., Flushing, N.Y., 1959; MA in Biology, Hofstra Coll., Hempstead, N.Y., 1961; D Podiatric Medicine, N.Y. Coll. Podiatric Medicine, N.Y.C., 1985. Bd. cert. in Podiatric Medicine and Surgery by Am. Inst. Foot Medicine. Tchr. sci. N.Y.C. Pub. Schs., 1959-69; fin. cons., 1969-82, pvt. practice, 1985—. With U.S. Army, 1961-62. Mem. Acad. Ambulatory Foot Surgery, Am. Inst. Foot Medicine. Republican. Jewish. Avocations: skiing, swimming. Home: 27127G Grand Central Pky Floral Park NY 11005-1200 Office: Park 89 Podiatry 17 E 89th St New York NY 10128

LESIKAR, JAMES DANIEL, II, physicist; b. Houston, Feb. 24, 1954; s. James Daniel and Ludine Luella (Kosel) L. BSME cum laude, Rice U., 1976, MME, 1978, MA, 1981, PhD in Physics, 1982. Registered profl. engr., Va., Md., Tex., Washington. Rsch. asst. T.W. Bonner Nuclear Labs. Rice U., Houston, 1976-81; asst. prof. physics U.S. Naval Acad., Annapolis, Md.,

1984-85; sr. analyst system scis. div. Computer Scis. Corp., Lanham, Md., 1986—. Author (with others) NASA reports, 1990; contbr. articles to Physics Letters, Phys. Rev., Phys. Rev. Letters. Co-moderator Math Counts Program, Annapolis, Md., 1991. Capt. U.S. Army, 1981-85; maj. USAR, 1988—. Recipient Bronze Cross for Achievement Legion of Valor, Inc., Army Parachutist badge; grad. fellow Rice U., Houston, 1976-77, Nettie S. Autrey Meml. fellow in sci., 1978-79. Mem. AIAA (sr. mem.), ASME, AAAS, IEEE, NSPE, Am. Phys. Soc. (life), Am. Def. Preparedness Assn. (life), Optimist Club (bd. dirs. 1983-84), Sigma Xi, Sigma Pi Sigma. Achievements include supervision and participation with flight dynamics support team for the Cosmic Background Explorer satellite; research in high energy spin-dependence of hadron interactions. Home: 1037 Oak Tree Ln Annapolis MD 21401-5011 Office: Computer Scis Corp 10110 Aerospace Rd Lanham Seabrook MD 20706-2262

LESKO, JOHN NICHOLAS, JR., research scientist; b. Mt. Lebanon, Pa., May 1, 1957; s. John Nicholas and Regina Mae (Emme) L.; m. Debra Ann Defreitas, Nov. 7, 1986; children: Jonathan, Kelsey. BS, U.S. Military Acad., 1979; MS, Boston Univ., 1989. Various command and staff positions U.S. Army, 1979-86; rsch.and devel. staff officer U.S. Army Lab. Command Material Tech. Lab., Watertown, Mass., 1986-90; rsch. scientist Battelle Meml. Inst., Arlington, Va., 1990—; pres., founder Lesko's Info Enterprises, Lake Ridge, Va., 1991—; lectr. Boston U., Air Force Inst. Tech. Contbr. articles to profl. jours. Vol. Springwood Sch. Adv. Bd., Lake Ridge Parks and Recreation Assn., 1992—. Capt. U.S. Army, 1979-90, maj. USAR, 1990—. Decorated Army Achievement medal with oak leaf cluster, U.S. Army, 1982, Army Commendation medal with oak leaf cluster, U.S. Army, 1985, Meritorious Svc. medal with oak leaf cluster, U.S. Army, 1990. Mem. Am. Def. Preparedness Assn., Soc. Logistics Engrs., World Future Soc. Office: Battelle 4001 N Fairfax Dr Ste 600 Arlington VA 22203

LESLIE, ROBERT FREMONT, mobile testing executive, non-destructive testing inspector; b. St. Louis, July 8, 1952; s. Robert Day and Dena (Lange) L. BA in Bus., Psychology, North Cen. Coll., 1975. Salesman Fairmount Hydraulics, Chgo., 1975-77; salesman, technician Torco Equip. Co., Louisville, 1978-81; pres., insp. Delta Mobile Testing, Inc., LaGrange, Ky., 1981—. Mem. ASTM (F18 com. 1988—), Nat. Fire Protection Assn. (stds. writing com. 1988—), Am. Welding Soc. Achievements include development of intensified fluoroscopic X-ray. Home and Office: Delta Mobile Testing Inc 2306 Running Brook Rd LaGrange KY 40031

LESSARD, ROGER A., physicist, educator; b. East Broughton, Que., Can., Sept. 11, 1944; s. J. Emilien and Gabrielle (Robert) L.; m. Nicole Doyon, July 15, 1967; children: Pascal, Guillaume. BA, Moncton U., Edmundston, N.B., 1965; BSc in Physics, Laval U., Que., 1969, DSc in Optics, 1973. Rschr. Gentec Inc., Sainte-Foy, Que., 1971-72; asst. prof. physics Laval U., University City, Que., 1972-73, adjoint prof., 1973-77, assoc. prof., 1977-82, full prof., 1982—; vice-dir. physics Laval U., 1985-88, dir. LROL, 1989, dir. Ctr. for Optics Photonics & Lasers, 1989—. Author over 80 publs.; contbr. chpt. to book. Fellow Internat. Soc. for Optical Engring., Optical Soc. Am.; sr. mem. IEEE; mem. Can. Assn. Physics (v.p. 1993—), Material Rsch. Soc. Roman Catholic. Office: Laval U-Ctr for Optics & Lasers, Cite Universitaire, Quebec, PQ Canada G1K 7P4

LESSER, RONALD PETER, neurologist; b. L.A., Jan. 17, 1946; m. Sara Elizabeth Roesler; 2 children. BA, Pomona Coll., 1966; MD, U. So. Calif., L.A., 1970. Diplomate Nat. Bd. Med. Examiners, Am. Bd. Psychiatry and Neurology, Am. Bd. of Qualification in Electroencephalography, Assn. Sleep Disorder Ctrs.; lic. physician N.Y., Calif., Ohio, Md. Intern in pediatrics Mayo Grad. Sch. Medicine, Rochester, Minn., 1970-71; resident in psychiatry N.Y. Psychiat. Inst./Columbia-Presbyn. Med. Ctr., N.Y.C., 1973-76; rsch. fellow neurophysiolgoy and electroencephalograpy Neurol. Inst., Columbia-Presbyn. Med. Ctr., N.Y.C., 1975-76, resident neurology, 1976-79; spl. fellow in electroencephalography, clin. assoc. Cleve. Clinic Found., 1979, staff neurologist, 1980-86; mem. Mind/Brain Inst. Mind/Brain Inst., Johns Hopkins U., Balt., 1987—; asst. clin. prof. neurology Case Western Res. U., Cleve., 1980-86; assoc. prof. neurosurgery Johns Hopkins U., Balt., 1987—, assoc. prof. neurology, 1986—; lectr. and cons. in field. Assoc. editor Epilepsy Advances, 1988—; cons. editor Electroencephalography and Clin. Neurophysiology, 1992—; editorial bd. Epilepsia, 1992—, Epilepsy Advances, 1985-88, Epilepsy Rsch., 1987—, Thermology, 1990—, Jour. neurosurg. Anesthesiology, 1990—, Therapeutic Drug Monitoring, 1987—, others; ad hoc reviewer various jours.; co-editor: Epilepsy: Electroclinical Syndromes, 1987; editor: The Diagnosis and Management of Seizure Disorders, 1991; contbr. numerous articles. With USPHS, 1971-73. Grantee NIH and numerous others. Fellow Am. Acad. Neurology, Am. Sleep Disorders Assn.; mem. AAAS, Am. Electroencephalographic Soc. (chmn. com. on psychophysiol. rsch. 1991-93, electrode position nomenclature com. 1988-90, treas. 1988-91, mem. coun. 1987-91), Ea. Assn. Electroencephalographers, Am. Epilepsy Soc. (rules com. 1984—, Lennox Trust com. 1991—, liaison to nat. assn. 1992—), Am. Neurol. Assn., N.Y. Acad. Scis., Soc. for Neurosci., Acad. of Aphasia. Office: Johns Hopkins Hosp 2-147 Meyer 600 N Wolfe St Baltimore MD 21287-7247

LETOKHOV, VLADILEN STEPANOVICH, physicist, educator; b. Irkutsk, USSR, Nov. 10, 1939; s. Stepan Grigorievich and Anna Vasilievna (Sevastianova) L; m. Tiina Karu, Nov. 30, 1979. Engr. degree, Moscow Phys. Tech. Inst., 1963; PhD, Lebedev Phys. Inst., Moscow, 1969, sci. degree, 1970. Researcher P.N. Lebedev Phys. Inst., Moscow, 1966-70; head laser spectroscopy dept. Inst. Spectroscopy, USSR Acad. Scis., Moscow, 1970—, vice-dir. for rsch., 1970-89; prof. Moscow Phys.-Tech. Inst., 1973—; dir. lab. Soviet Br. of World Lab., USSR, 1990—. Author: Nonlinear Laser Chemistry, 1983, Laser Photoionization Spectroscopy, 1982, more than 600 sci. papers; co-author: Nonlinear Laser Spectroscopy, 1977. Fellow Am. Optical Soc.; mem. N.Y. Acad. Scis., M. Planck Soc. Home: Puchkovo 66, Troitzk 142092, Russia Office: Inst Spectroscopy, Troitzk 142092, Russia

LETOWSKI, TOMASZ RAJMUND, acoustical engineer, educator; b. Warsaw, Oct. 12, 1942; came to the U.S., 1981; s. Henryk Konrad and Prakseda Emma (Swietochowska) L.; m. Anna Zofia Stolarek, Jan. 12, 1974; children: Szymon, Jan. MS, Warsaw Tech. U., 1965, DSc, 1986; PhD, Wroclaw Tech. U., 1973. Asst. prof. Chopin Acad. Music, Warsaw, 1965-81; rsch. assoc. prof. U. Tenn., Knoxville, 1981-89; assoc. prof. Pa. State U., University Park, 1989—. Contbr. more than 85 articles to profl. jours. Fulbright scholar, 1976-77. Mem. Acoustical Soc. Am., Audio Engring. Soc., Am. Speech, Lang. and Hearing Assn. Achievements include 4 patents.

LETSINGER, ROBERT LEWIS, chemistry educator; b. Bloomfield, Ind., July 31, 1921; s. Reed A. and Etna (Phillips) L.; m. Dorothy C. Thompson, Feb. 6, 1943; children: Louise, Reed, Sue. Student, Ind. U., 1939-41; B.S., Mass. Inst. Tech., 1943, Ph.D., 1945. Research assoc. MIT, 1945-46; research chemist Tenn. Eastman Corp., 1946; faculty Northwestern U., 1946—, prof. chemistry, 1959—, chmn. dept., 1972-75, joint prof. biochemistry and molecular biology, 1974—, Clare Hamilton Hall prof. chemistry, 1986-92, Clare Hamilton Hall prof. emeritus chemistry, 1992—; Mem. med. and organic chemistry fellowship panel NIH, 1966-69, medicinal chem. A study sect., 1971-75; bd. on chem. scis. and tech. Nat. Research Council, 1987-90. Mem. bd. editors Nucleic Acids Research, 1974-80; contbr. articles to profl. jours. Guggenheim fellow, 1956; JSPS fellow Japan, 1978; recipient Rosenstiel Medallion, 1985, Humboldt Sr. US Scientist award, 1988, NIH merit award, 1988, Arthur C. Cope scholar award, 1993. Fellow Am. Acad. Arts and Scis., Nat. Acad. Scis., Am. Assn. Arts and Scis.; mem. Am. Chem. Soc. (bd. editors 1969-72, bioconjugate chemistry 1992—, Arthur C. Cope scholar award 1993), Internat. Union Pure and Applied Chemistry, Sigma Xi, Phi Lambda Upsilon (hon. mem.). Home: 316 3d St Wilmette IL 60091 Office: Northwestern U Chemistry Dept 2145 Sheridan Rd Evanston IL 60208

LETSOU, GEORGE VASILIOS, cardiothoracic surgeon; b. Boston, 1958; s. Vasilios George and Helen (Valacellis) L.; m. Jane Elizabeth Carter, June 1, 1985; children: Christopher George, Philip Taylor. AB magna cum laude, Harvard U., 1979; MD, Columbia U., 1983. Diplomate Am. Bd. Surgery, Am. Bd. Thoracic Surgery. Resident in gen. surgery Yale-New Haven Hosp., 1983-88, chief resident, 1987-88, clin. fellow in cardiothoracic surgery, 1988-89, resident in cardiothoracic surgery, 1990-91, Cystic Fibrosis Found. fellow in cardiopulmonary transplantation, 1989-90, Winchester scholar in cardiothoracic surg. rsch., 1990-91, chief resident in cardiothoracic surgery, 1991-92, attending surgeon, 1992—; instr. surgery Yale U., New Haven, 1987-88, 91-92, asst. prof., 1992—. Mem. AMA, ACS, Am. Coll. Cardiology, Am. Coll. Chest Physicians, Soc. Thoracic Surgeons. Office: Yale U Sch Medicine Divsn Cardiothoracic Surgery 333 Cedar St New Haven CT 06510

LETTS, LINDSAY GORDON, pharmacologist, educator; b. Warragul, Victoria, Australia, Jan. 9, 1948; came to U.S., 1987; m. Barbara Dawn Hawkey, Sept. 13, 1969; children: Michelle Maree, Kathryn Jane, David Gordon. BS, Monash U., Australia, 1971; PhD, Sydney U., 1980. Tutor Sydney (Australia) U., 1976-80; rsch. scientist Royal Coll. Surgeons Eng., London, 1980-82; sr. rsch. fellow Merck Frosst Can. Inc., 1982-87; dir. pharmacology Boehringer Ingelheim Pharms., Inc., Ridgefield, Conn., 1987—; adj. assoc. prof. Yale U. Sch. Medicine, New Haven, 1991—. Editor Mediators of Inflammation, 1992—, Pulmonary Pharmacology, 1992—; sect. editor Prostaglandins, 1986—. Bd. dirs. Nat. Inst. for Community Health Edn., Quinnipiac Coll., Hamden, Conn., 1990—, Conn. United Rsch. Excellence, Wallingford, 1991—; dir. Inflammation Rsch. Assn., 1992—, Conn. Biomed. Rsch. Assn., 1992—. Office: Boehringer Ingelheim Pharms 90 E Ridge Rd PO Box 368 Ridgefield CT 06877

LETZRING, TRACY JOHN, civil engineer; b. Cavalier, N.D., July 2, 1962; s. Richard J. and Phoebe A. L.; m. Karen E. Vatnsdal, Dec. 24, 1987; 1 child, Tylor J. AS in Civil Engring. Tech., N.D. State Sch. Sci., Wahpeton, 1983; BSCE, U. N.D., 1987. Profl. engr., Calif. Field engr. Swingen Constrn., Grand Forks, N.D., 1983; field technician N.D. Hwy. Dept., Grand Forks, 1983-85; engr. 0-1 USPHS, Sioux City, Iowa, 1986; assoc. civil engr. L.A. County Dept. Pub. Works, Alhambra, Calif., 1987-88; dir. engring. Tait & Assocs., Inc., Orange, Calif., 1988—. Mem. ASCE, NSPE. Roman Catholic. Office: Tait & Assocs Inc 800 N Eckhoff St Orange CA 92668

LEUNG, CHARLES CHEUNG-WAN, technological company executive; b. Hong Kong, June 27, 1946; came to U.S., 1969; s. Mo-Fan and Lai-Ping (Tam) L.; m. Jessica Lan Lee, Sept. 1, 1972; children: Jennifer W., Cheryl E., Albert H. BS with spl. honors, U. Hong Kong, 1969; PhD, U. Chgo., 1976. Sr. staff scientist Corning (N.Y.) Glass Ctrl. Lab., 1975-79; sr. mem. tech. staff Motorola, Mesa, Ariz., 1979-81; engring. mgr. Avantek, Santa Clara, Calif., 1981-88; chmn., pres. Bipolarics Inc., Los Gatos, Calif., 1981—. Mem. IEEE, Am. Phys. Soc., Am. Vacuum Soc., Asian Am. Mfrs. Assn. Achievements include patent on wafer planarization technology; developed fastest silicon transistor, widest bandwidth microwave oscillator 2-20 GHz; inventor of silicon microwave monolithic integrated circuit technology for communication. Office: Bipolarics Inc 108 Albright Way Los Gatos CA 95030

LEUNG, KAM H., medical company executive; b. Hong Kong, Nov. 6, 1951; m. Katrina; 1 child, Anson. BS summa cum laude, Sam Houston State U., 1972; PhD, Cornell U., 1977. Rsch. assoc. U. Chgo. Sch. Medicine, 1976-78; scientist Procter and Gamble, Cin., 1978-80; sr. rsch. assoc. U. So. Calif./L.A. Med. Ctr., 1980-81; toxicology mgr. Pharmaseal Labs., Irwindale, Calif., 1981-83; project mgr. Am. Pharmaseal, Glendale, Calif., 1983-85; tech. advisor Baxter Pharmaseal, Valencia, Calif., 1985—. NSF fellow, 1971, NIH fellow, 1972-76. Mem. Chinese Am. Chemists Assn. (bd. dirs. 1987-89, chmn. So. Calif. chpt. 1985-88). Achievements include invention of wound bandages, hydrophillic dressings, sprays and washes; decubitus ulcer preventive wheelchair cushions and bed mattresses; pulmonary embolism preventive stockings, devices and air pumps. Home: 2260 Robles Ave San Marino CA 91108 Office: Baxter Healthcare Corp 27200 N Tourney Rd Valencia CA 91355

LEUNG, WOON-FONG, mechanical engineer, research scientist; b. Hong Kong, Jan. 25, 1954; s. Shing-Lam and So-Wan (Cheung) L.; m. Stella Po-Chun Cheng, June 25, 1978; children: Jessica, Jeffrey. SB, Cornell U., 1977; MSME, MIT, 1978, ScD, 1981. Cons. Water Purification Assoc., Cambridge, Mass., 1979-80; rsch. engr. Gulf Rsch. & Devel. Co., Harmarville, Pa., 1981-83; sr. engr. Gulf Rsch. & Devel. Co., Houston, 1983-84; project leader Schlumberger Tech., Sugarland, Tex., 1984-86; sr. rsch. scientist Bird Machine Co., South Walpole, Mass., 1986—. Editor: System Approach to Separation and Filtration Process Equipment, 1993; assoc. editor: Fluid-Particle Separation Jour., 1990-92; contbr. articles to profl. jours. Organizer MIT Chin. Students Club Concert, 1991. Recipient Cedric Ferguson medal AIME, Soc. Petroleum Engrs., Dallas, 1987, Baker Hughes Tech. Achievement award, Houston, 1992. Mem. Am. Men and Women of Sci., Am. Soc. Mech. Engrs., Am. Filtration Soc. (bd. dirs. 1992-93, chmn. centrifuge network 1990—, ann. meeting chair and organizer 1993, Engring. Merit award 1992), Soc. Petroleum Engrs., Soc. Rheology. Achievements include patents for centrifugal separation; notable findings on petroleum production and well "skin" testing, aquifer water flow, centrifuge (industrial scale), rotating flow, rheological behavior of non-Newtonian fluid, ultrafiltration and lamella sedimentation. Home: 11 Ames Dr Sherborn MA 01770 Office: Bird Machine Co 100 Neponset St South Walpole MA 02071

LEUSCH, MARK STEVEN, microbiologist; b. Cleve., June 8, 1961; s. Thomas Arthur and Elaine Margaret (Torma) L.; m. Cindy June Campos, Feb. 4, 1989; children: Steven Alexander, Kristen Denise. BA, U. Ariz., 1984, PhD, 1990. Postdoctoral assoc. Monsanto Co., St. Louis, 1990-92; scientist healthcare div. Procter and Gamble Co., Cin., 1992—. Contbr. articles to Infection and Immunity, Procs. Nat. Acad. Sci., Biochem. Biophys. Rsch. Communications. Mem. AAAS, Internat. Assn. Dental Rsch., Am. Soc. Microbiology. Roman Catholic. Achievments include patent pending on generation of infectious recombinant baculoviruses in E. coli. Office: Sharon Woods Tech Ctr 11450 Grooms Rd Cincinnati OH 45242-1434

LEUS MCFARLEN, PATRICIA CHERYL, water chemist; b. San Antonio, Mar. 12, 1954; d. Norman W. and Jacqueline S. (Deason) Leus; m. Randy N. McFarlen, June 28, 1986; 1 child, Kevin Bryant. AA, Highline Community Coll., 1974; BS in Chemistry, Eastern Wash. U., 1980. Cert. operator grade II water treatment & distbn., grade I wastewater & collection operator Ariz. Dept. Environ. Quality. Lab. technician, oil analyst D.A. Lubricant, Vancouver, Wash., 1982-83; plant chemist Navajo Generating Sta., Page, Ariz., 1983-92, chemist, 1992—. Sci. judge Page Schs. Sci. Project Fair, 1985, 91; chemist Navajo Generating Sta./Page Sch. Career Day, 1986, 89, 90; life mem. Girl Scouts Am. Mem. Sigma Kappa (life mem., treas. 1976-78). Methodist. Avocations: sewing, hiking, snow skiing, archery, flying. Office: Navajo Generating Sta Environ & Lab Svcs Dept PO Box W Page AZ 86040-1949

LEV, ALEXANDER SHULIM, mechanical engineer; b. Tselinograd, USSR, May 4, 1945; came to U.S., 1979; s. Borukh and Golda (Kopitman) L.; m. Polina Zhdanovskaja, Aug. 31, 1968; 1 child, Victoria. MSME, Lvov Polytech Inst., 1968. Project mgr. Glavspetsavtotrans, Lvov, USSR, 1968-78; sr. engr. Machine Plant, Lvov, USSR, 1978-79; metalurgist Ronson Metals Corp., Newark, N.J., 1980-82, foundry quality control mgr., 1982-83, reduction dept. mgr., 1986-86; project mgr. FMB Systems Inc., Harrison, N.J., 1986—. Patentee in field. Mem. Am. Metals Soc., Metallurgical Soc. Avocations: reading, art, metal work, tennis. Home: 16 Princeton St Maplewood NJ 07040

LE-VAN, NGO, research chemist; b. Hué, Vietnam, Jan. 1, 1949; came to U.S., 1978; s. Khoi and Thi Nhon (Ngo) L.; m. Cuong T.V. Pham, Feb. 6, 1976. MS, T.U., Berlin, 1975, PhD, 1977. Postdoctoral rsch. fellow La. State U., Baton Rouge, 1978, Tex. A&M U., College Station, 1979-80; sr. rsch. chemist Monsanto Agrl., St. Louis, 1980-84, rsch. specialist, 1984-87; rsch. specialist Sandoz Agrl. Ltd., Basle, Switzerland, 1987—. Contbr. articles to profl. jours. Doctoral fellowship City of West Berlin, 1975-77; postdoctoral fellowship Dentsch. Forschungs-gemeinschaft, 1977-78, Prof. Wigner of West Berlin, 1978-79, NIH, 1979-80. Mem. Am. Chem. Soc., Am. Soc. Pharmacology, Phytochem. Soc. N.Am. Achievements include patent in production of dehydrosinefungin by microorganisms and agricultural use. Home: 10 Rue Du Kai, F68220 Neuwiller France Office: Sandoz Agro Ltd, Lichtstrasse 35, CH 4002 Basel Switzerland

LEVEL, ALLISON VICKERS, science librarian; b. Cookeville, Ill.. BSPA, U. Ark., 1981; MEd, Kent State U., 1985; MLS, Emporia State U., 1990. Student svcs. Kent (Ohio) State U., 1983-85, Northern Ariz. U., Flagstaff, 1985-87; scholar project coord. U. Ark., Fayetteville, 1988; asst. dean, distance edn. Emporia (Kans.) State U., 1989-90; sci. & tech. libr. Libr. of Congress, Washington, 1990—; co-chmn. Libr. Congress Reference Forum. Vol. Smithsonian Inst., Washignton, 1992. Mem. ALA, D.C. Libr. Assn. Office: Sci & Tech Div Libr of Congress Washington DC 20540

LEVENSON, MARIA NIJOLE, medical technologist; b. Kaunas, Lithuania, Mar. 24, 1940; came to U.S., 1948; d. Zigmas and Monika (Galbuogis) Sabataitis; m. Coleman Levenson, Nov. 21, 1975. BA, Annhurst Coll., 1962. Sr. rsch. technician Case Western Res. U., Cleve., 1962-69; phys. sci. technician Nat. Oceanographic Data Ctr., Washington, 1969-70; biologist NIH, Bethesda, Md., 1970-76; nuclear medicine technologist VA Med. Ctr., New Orleans, 1977-79; paramed. examiner Hooper Industries, New Orleans, 1980-82; assoc. chemist Computer Scis. Corp., Stennis Space Ctr., La., 1982-83; med. technologist VA Med. Ctr., New Orleans, 1984—; sec. Lithuanian Cath. Youth Assn., Putnam, Conn., 1960-62, Lithuanian Club, Annhurst Coll., South Woodstock, Conn., 1960-62. Participant Freedom Movement for Baltic Independence, Slidell, La., 1990-91; counselor Life with Cancer, Slidell, 1989—. La. State Nursing Sch. scholar, 1989. Mem. Daus. of Lithuania. Avocations: reading, traveling, cooking, flying, family. Home: 126 Woodcrest Dr Slidell LA 70458-5130

LEVENTHAL, CARL M., neurologist; b. N.Y.C., July 28, 1933; s. Isidor and Anna (Semmel) L.; m. Brigid Penelope Nesburn Gray, Feb. 4, 1962; children: George Leon, Sarah Elizabeth Roark, Dinah Susan Lacefield, James Gray. A.B. cum laude, Harvard U., 1954; M.D., U. Rochester (N.Y.), 1959. Diplomate: Am. Bd. Psychiatry and Neurology. Fellow in anatomy U. Rochester, 1956-57; intern, then asst. resident in medicine Johns Hopkins Hosp., 1959-61; asst. resident, then resident in neurology Mass. Gen. Hosp., Boston, 1961-64; commd. officer USPHS, 1963—, asst. surgeon gen., 1979-83; asso. neuropathologist Nat. Inst. Neurol. Diseases and Blindness, 1964-66; neurologist Nat. Cancer Inst., 1966-68; asst. to dep. dir. sci., 1968-73; acting dep. dir. sci. NIH, 1973-74; dep. dir. bur. drugs FDA, Rockville, Md., 1974-77; dep. dir. Nat. Inst. Arthritis, Diabetes and Digestive and Kidney Diseases, 1977-81; div. dir. Nat. Inst. Neurol. Disorders and Stroke, 1981—; sr. policy analyst for life scis. Office of Sci. and Tech. Policy, Exec. Office of Pres., 1983; asst. clin. prof. neurology Georgetown U. Med. Sch., 1966-76. Recipient Commendation medal USPHS, 1970, Meritorious Svc. medal, 1974, 77, 91, Outstanding Svc. medal, 1988, dir's. award NIH, 1992. Fellow Am. Acad. Neurology; mem. Am. Assn. Neuropathologists, Am. Neurol. Assn., Alpha Omega Alpha. Home: 9254 Old Annapolis Rd Columbia MD 21045-1832 Office: Nat Inst Neur Disorders & Stroke Fed Bldg Rm 810 Bethesda MD 20892

LEVENTIS, NICHOLAS, chemist, consultant; b. Athens, Greece, Nov. 12, 1957; came to U.S., 1980; s. Spyro and Efrosine (Nenou) L.; m. Chariklia Sotiriou, Nov. 12, 1988. BS in Chemistry, U. Athens, Greece, 1980; PhD in Chemistry, Mich. State U., 1985; grad. cert. in adminstrn. and mgmt., Harvard U., 1992. Grad. asst. Mich. State Univ., East Lansing, 1980-85; rsch. assoc. MIT, Cambridge, Mass., 1985-88; project dir. Molecular Displays, Inc., Cambridge, 1988-90, v.p. of R&D, 1990—; cons. Igen, Inc., Rockville, Md., 1987—, Hyperion Catalysis Internat., Cambridge, 1988—, Delta F Corp., Woburn, Mass., 1992—. Contbr. articles on electrochromic phenomena and devices to Yearbook of Ency. of Sci. & Tech., Jour. Mat. Chem., Chem. of Materials, Jour. Electrochem. Soc. Recipient Greek Inst. State Scholarships awards Greek Govt. Dept. Edn., 1976-79, Katie Y. F. Yang prize Harvard U., Cambridge, 1992; named Ethyl Corp. fellow Mich. State U., East Lansing, 1983, Yates Meml. fellow Mich. State U., East Lansing, 1984. Mem. Am. Chem. Soc. (Arthur K. Doolittle award 1993), Am. Soc. for Photobiology, Internat. Union Pure & Applied Chemistry (affiliate mem.), Soc. for Info. Display. Greek Orthodox. Achievements include patents for Electrochromic, Electroluminescent and Electrochemiluminescent Displays; Electrically Conductive Polymer Composition, Method of Making Same and Device Incorporating Same; others. Home: 32 Skehan St Somerville MA 02143

LEVERE, JAMES GORDON, electrical engineer; b. Mineola, N.Y., July 23, 1957; s. Richard Craig and Marilyn Florence (Harms) L.; m. Susan Denise Hunter, Aug. 20, 1983; children: Matthew James, Gregory Hunter. BSEE, Met. State Coll., 1983. Sr. planning engr. Burlington No. RR, Overland Park, Kans., 1984-87; engr. methods and stds. Burlington No. RR, Overland Park, 1988—; supr. control systems Burlington No. RR, Wenatchee, Wash., 1987-88. Contbr. articles to profl. jours. Elder Grace Covenant Presbyn. Ch., Overland Park, 1985-87. Mem. Assn. of Am. RR (hwy. grade warning systems 1989—, signal systems, 1993—). Office: Burlington No RR 9401 Indian Creek Pkwy Overland Park KS 66201-9136

LEVERE, RICHARD DAVID, physician, academic administrator, educator; b. Bklyn., Dec. 13, 1931; s. Samuel and Mae (Fain) L.; m. Diane L. Gonchar, Jan. 15, 1978; children—Elyssa C., Corinne G., Scott M. Student, N.Y. U., 1949-52; M.D., SUNY, N.Y.C., 1956. Intern Bellevue Hosp., N.Y.C., 1956-57, resident, 1957-58; resident Kings County Hosp., 1960-61; asst. prof. medicine SUNY Downstate Med. Center, 1965-69, assoc. prof., 1969-73, prof., 1973-77, vice-chmn. dept. medicine, 1975-77, chief hematology/oncology div., 1970-77; asst. prof. Rockefeller U., 1964-65; prof., chmn. dept. medicine N.Y. Med. Coll., 1977-93, vice dean, 1991-93; med. dir. Westchester County Med. Ctr., 1991-92; v.p. med. affairs St. Agnes Hosp., 1991-93; sr. v.p. The Bklyn. Hosp. Ctr., 1994—; assoc. dean, prof. medicine NYU Sch. Medicine, 1994—, adj. prof. Rockefeller U., 1973—. Contbr. in field. Bd. dirs. Leukemia Soc. Am., 1970-85, Am. Heart Assn., 1978—; trustee Our Lady of Mercy Med. Ctr., 1993. Capt. USAR, 1958-60. NIH grantee, 1971-76, 65-86. Fellow ACP (Physician Recognition award 1986, gov. elect N.Y. State, Downstate I 1989, gov. 1990—, pres. N.Y. state chpt. 1992-93); mem. Harvey Soc., Am. Soc. Clin. Investigation, Soc. Study of Blood (pres. 1973-74), Soc. Devel. Biology, Am. Soc. Pharm. Exptl. Therapeutics, Den Tiroler Adler-Ordern of Austria, Alpha Omega Alpha. Home: 5 Seymour Pl W Armonk NY 10504-2516 Office: NY Med Coll Dept Medicine Munger Pavilion Valhalla NY 10595

LEVERENZ, HUMBOLDT WALTER, retired chemical research engineer; b. Chgo., July 11, 1909; s. Paul Frederick and Lydia (Humboldt) L.; m. Edith Ruggles Langmuir, Nov. 30, 1940; children: David, Edith, Julia, Ellen. BA in Chemistry, Stanford U., 1930; postgrad., U. Muenster, 1930-31. Rsch. engr. RCA Mfg. Co., Camden, Harrison, N.J., 1931-42; rsch. engr. RCA Labs., Princeton, N.J., 1942-54, dir. physics and chem. rsch. lab., 1954-57, asst. dir. rsch., 1957-59, dir. rsch., 1959-61, assoc. dir., 1961-68; staff v.p. RCA Corp., Princeton, 1968-74; mem. Materials Adv. Bd., Washington, 1964-68. Author: Luminescence of Solids, 1950, 70; contbr. articles to profl. publs. Named Modern Pioneer Nat. Assn. Manufacturers, 1940; recipient Frank P. Brown medal Franklin Inst., 1954. Fellow Am. Phys. Soc., Optical Soc. Am., IEEE; mem. Nat. Acad. Engring., Am. Chem. Soc., Sigma Xi. Achievements include 67 patents; devel. of phosphors and luminescent screens used in fluorescent lamps and picture tubes, ferrites for TV receivers. Home: 2240 Gulf Shore Blvd N Naples FL 33940

LEVESQUE, RENE JULES ALBERT, former physicist; b. St. Alexis, Que., Can., Oct. 30, 1926; s. Albert and Elmina Louisa (Veuilleux) L.; m. Alice Farnsworth, Apr. 6, 1956 (div.); children: Marc, Michel, Andre; m. Michile Robert, Feb., 1992. B.Sc., Sir George Williams U., 1952; Ph.D., Northwestern U., 1957. Research assoc. U. Md., 1957-59; asst. prof. U. Montreal, 1959-64, assoc. prof., 1964-67, prof., 1967-87, dir. nuclear physics lab., 1965-69, chmn. dept. physics, 1968-73, vice dean arts and scis., 1973-75, dean, 1975-78, v.p. research, 1978-85, v.p. research and planning, 1985-87, prof. emeritus, 1987; mem. Atomic Energy Control Bd., Ottawa, Can., 1985-87, pres., 1987-93; ret., 1993; mem. adv. com. ING project Atomic Energy of Can. Ltd., 1966-69; mem. adv. bd. physics NRC Can., 1972-74, pres. nuclear physics grant selection, 1973; mem. adv. bd. on TRIUMF, 1979-87; v.p. Commn. Higher Studies Que. Ministry Edn., 1976-77, Natural Scis. and Engring. Research Council Can., 1981-87; v.p. bd. dirs. Can.-France-Hawaii Telescope Corp., 1979-80, pres., 1980-81; pres. permanent research com. Conf. Rectors and Prins. Que. Univs., 1979-80; pres. Mouvement Laïc de Langue française, 1961. Mem. Can. Assn. Physicists (pres. 1976-77), U. Montreal Faculty Assn. (pres. 1971), Fedn. Que. Faculty Assn. (pres. 1971-

72), Interciencia Assn. (v.p. bd. dirs. 1979-80), Assn. Sci., Engring. and Technol. Community Can. (v.p. 1979-80, pres. 1980-81). Home: 190 Willowdale PH 1, Outremont, PQ Canada H3T 1G2

LEVEY, SANDRA COLLINS, civil engineer; b. Richmond Heights, Mo., July 3, 1944; d. Edward Milton Collins and Florence May (Godat) Berta; m. James Raymond Levey, Sept. 12, 1964 (div. Oct. 1986); children: Melinda Elisabeth, Cynthia Eileen. BS in Civil Engring., U. Ill., 1967; BS in Indsl. Engring., U. Wis., Platteville, 1979; MBA, U. Puget Sound, 1984. Registered profl. engr., Ill., Wash. Jr. sanitary engr. Jenkins, Merchant & Nankivil, Springfield, Ill., 1969-71; civil engr. I dist. 6 Ill. Dept. Transp., Springfield, 1971-74; gen. engr. Pan Am. World Svcs., Bermerton, Wash., 1979-80, facility engr., 1980-84; budget officer Pub. Utility Dist. 2, Grant County, Ephrata, Wash., 1985-87, civil engr., hydro engring., 1987—; mem. solid waste adv. com. Grant County, Ephrata, 1990—. Mem. Planning and Zoning Commn., Platteville, Wis., 1976-78; council person City of Soap Lake (Wash.), 1990—. Mem. NSPE (Bremerton sect., treas. 1984-85), ASCE, Soc. Women Engrs. (sr. mem., editor newsletter 1989-91, chair Ea. Wash. sect. 1989-91, pres. Ea. Wash. sect. 1992-93, Nat. Continuing Devel. award 1990, Nat. Career Guidance award 1991. Avocation: reading. Home: PO Box 487 Ephrata WA 98823-0487 Office: Pub Utility Dist 2 Grant County PO Box 878 Ephrata WA 98823-0878

LEVICH, ROBERT ALAN, geologist; b. Bklyn., Apr. 16, 1941; s. Leonard Walter and Dinah (Cohen) L.; m. Stella Araba Nkrumah, June 10, 1964; children: Alexander Kwamina, Walter Abraham, Leo Augustine. BS in Geology, CUNY, Bklyn., 1963; MA in Geol. Scis., U. Tex., 1973. Cert. profl. geologist. Geologist Ghana Geol. Survey, Sunyani, 1969-72, U.S. Atomic Energy Commn., Austin, Tex., Spokane, Wash., 1973-81; regional mgr. Apache Energy & Minerals Co., Spokane, 1981-82; cons. expert East Africa Internat. Atomic Energy Agy., Vienna, Austria, 1982-83; geologist U.S. Dept. Energy, Argonne, Ill., Las Vegas, Nev., 1984-88; chief tech. analysis br. U.S. Dept. Energy/YMP, Las Vegas, 1988-89, internat. programs mgr., 1989—; U.S. rep. Internat. Stripa Project, Stockholm, 1989-92; lead, coord. U.S. del. OECD/Nuclear Energy Agy. Site Evaluation and Design Experiments, Paris, 1990—; U.S. Dept. Energy rep. OECD/NEA Alligator Rivers Analogue Project, Sydney, NSW, Australia, 1990-92; project dir. USDOE/Atomic Energy, Can. Ltd. Subsidiary Agreement, 1991—; tech. coord. USDOE/Switzerland Nat. Co-op. for Disposal Radioactive Waste Project Agreement, 1991—; project dir. USDOE/Swedish Nuclear Fuel & Waste Mgmt. Co-Project Agreement, 1993—. Author, coauthor, editor reports. Fellow Geol. Soc. Am., Soc. Econ. Geologists; mem. Am. Nuclear Soc. (Internat. High Level Radioactive Waste Mgmt. Conf. program com., steering com.), Assn. Geoscientists for Internat. Devel., Am. Inst. Profl. Geologists (v.p. Nev. sect. 1993). Achievements include development of international radioactive waste natural analogue study. Office: US Dept Energy YMP 101 Convention Center Dr Las Vegas NV 89109-2003

LEVI-MONTALCINI, RITA, neurobiologist, researcher; b. Turin, Italy, Apr. 22, 1909; came to U.S., 1947; naturalized, 1956; d. Adamo Levi and Adele Montalcini. MD, U. Turin, 1936. Asst. in neurology Inst. Anatomy, Neurology Clinic, Turin Sch. Medicine, 1936-37; researcher Neurol. Inst. Brussels, 1939; with Allied health svc., Italy, 1944-45; resident, assoc. zoologist Washington U., 1947-51, assoc. prof., 1951-58, prof., 1958-81; prof. emeritus Washington U., St. Louis, 1977; dir. neurobiology rsch. program CNR (Nat. Rsch. Coun.), Rome, 1961-69, dir. cellular biology lab., 1969-79, guest prof. cellular biology lab., 1979-89, guest prof. inst. neurobiology, 1989—; pres. Ency. Italiana, 1993, Italian Nat. Commn. of United World Colls., 1993. Author: In Praise of Imperfection: My Life and Work, 1988. Recipient Albert Lasker Med. Rsch. award, 1986, Nobel prize Physiology-Medicine, (with Stanley Cohen) for work on chem. growth factors which control growth and devel. in humans and animals, 1986, Lewis S. Rosenstiel award, U.S. Nat. Medal of Sci. Mem. AAAS, Soc. Devel. Biology, Am. Assn. Anatomists, Tissue Culture Assn., NAS, Pontifical Acad., Nat. Acad. dei Lincei, Harvey Soc., Am. NAS, Belgian Royal Acad. Medicine, NAS of Italy, European Acad. Scis., Arts and Letters, Acad. Arts and Scis. of Florence. Office: Inst Neurobiol, CNR Viale Marx 15, 00156 Rome Italy

LEVIN, BERNARD H., psychologist educator; b. Phila., Dec. 19, 1942; s. Morris L. and Freda (Weiner) L.; children: Judith, Matthew; m. Karen T. Levin, Mar. 4, 1992. AB, Temple U., 1968; MS, N.C. State U., 1970; EdD, Va. Tech., 1985. Asst. prof. psychology Pembroke (N.C.) State U., 1972-73; prof. psychology Blue Ridge C.C., Weyers Cave, Va., 1973—. Contbr. articles to profl. jours. Bd. dirs. ACLU Va., Richmond. Mem. APA, AAAS, Va. Social Sci. Assn. (pres. 1992—), Waynesboro Police Res. (adv. bd. 1983—). Democrat. Unitarian. Achievements include designing and implementing first intentional culture in a maximum security prison. Office: Blue Ridge C C Box 80 Weyers Cave VA 24486

LEVIN, KEN, radiologist; b. Wilkes-Barre, Pa., June 21, 1951. MD, Jefferson Med. Coll., 1977; M of Computer Sci., U. Ill., 1981. Diplomate Am. Bd. Radiology, Am. Bd. Nuclear Medicine. Intern United Health & Hosp. Svcs., Wilkes-Barre, 1977-78; fellow in nuclear medicine George Washington U., Washington, 1982-84; resident in radiology Temple U. Hosp., Phila., 1984-87; asst. prof. nuclear medicine Hahnemann U. Hosp., Phila., 1987-88; asst. prof. radiology Georgetown Univ. Hosp., Washington, 1988-90; sect. head div. nuclear medicine dept. radiology St. Lukes Hosp., Bethlehem, Pa., 1990—. Home: 1834 W Hamilton St Allentown PA 18104-5658

LEVIN, PETER LAWRENCE, electrical engineering educator; b. N.Y.C., July 3, 1961; s. Myles Richard and Evelyn Ann (Meyerfield) L.; m. Klaudia Martha Lechner, June 1, 1991; 1 child, Jakob Mark. BS, Carnegie Mellon U., 1983, MS in Elec. and Computer Engring., 1984, PhD, 1987. Asst. prof. elec. engring. Worcester (Mass.) Poly. Inst., 1988-92, assoc. prof. elec. and computer engring., 1992, inst. assoc. prof. elec. engring., 1992—; reviewer IEEE, 1989—, NSF, 1991—; profl. translator Asea Brown Boveri, Mannheim Riley Stoker, Worchester, 1992; vis. scientist Deutscher kademischer Austauschdienst, Munich, 1988. Contbr. 25 articles to profl. jours. Presdl. Young Investigator, NSF, 1991. Satin disting. fellow Worcester Poly. Inst., 1990; grad. fellow Am. Electronics Assn., Pitts., 1983-87. Mem. IEEE, Electrostatics Soc. Am., Sigma Xi, Eta Kappa Nu. Democrat. Jewish. Achievements include foundation of Computational Field Lab.; creation of charge simulation program; research on high voltage engring., computer simulation, math. and sci. edn. Home: 8 Cascade Rd Worcester MA 01602 Office: Worcester Poly Inst 100 Institute Rd Worcester MA 01609

LEVIN, ROBERT BRUCE, industrial engineer, executive, consultant; b. Roanoke, Va., Sept. 24, 1918; s. Isaac and Belle (Weinkrantz) L.; m. Juanita Kahn, July 26, 1952; children: Susan Lee, Carol Lynn. BS with honors, Ga. Tech. U., 1940; postgrad. in Chemistry, U. Miami, 1946-52. Registered profl. engr., Fla., Calif. Exec. trainee RCA Mfg. Co., Camden, N.J., 1940-41; pres. Tripure Water Co., Miami, Fla., 1946-76, Mountain Valley Water, Miami, 1977—; bd. dirs. Mountain Valley Spring Co., Hot Springs, Ark., 1984-86. Editor: Bomb Loading Inspection Manual. Pres. Rotary Club of Miami, 1984, Disting. Rotarian, 1988. Col. U.S. Army, 1941-46. Mem. Inst. Indsl. Engrs. (sr. mem., pres. Miami chpt. 1966), Am. Chem. Soc., Am. Water Works Assn. (life), Greater Miami C. of C. (trustee 1986—), Internat. Bottled Water Assn. (founder, past pres., Hall of Fame), Miami Club. Office: Mountain Valley Water Co 3455 NW 73rd St Miami FL 33147-5830

LEVINE, ARNOLD JAY, molecular biology educator, researcher; b. Bklyn., July 30, 1939; s. Samuel and Marion (Wisot) L.; m. Linda Hirst, June 5, 1962; children—Samantha, Alison. B.A., Harper Coll. SUNY, 1961; Ph.D., U. Pa., 1966. Postdoctoral fellow Calif. Inst. Tech., Pasadena, 1968; asst. prof. Princeton U., N.J., 1968-73, assoc. prof., 1973-76, prof. biochemistry, 1976-79; prof. dept. molecular biology, chmn. dept. SUNY, Stony Brook, 1979-83; Harry C. Wiess prof. molecular biology, chmn. dept. Princeton U., N.J., 1984—; mem. human cell biology panel NSF, 1971-72, mem. genetics and biology panel, 1972-73. Assoc. editor Jour. Cellular and Molecular Biology, 1984—; editor-in-chief Jour. Virology, 1984—; editor Virology Jour., 1974-84; contbr. articles to profl. jours. Mem. cell biology panel Am. Heart Assn., 1978-80; mem. adv. com. virology and cell biology Am. Cancer Soc., 1974-78, vice chmn., 1976, chmn. 1977; mem. Nat. Bd. Med. Examiners. Dryfus fellow, recipient Lila Gruber Cancer Rsch award, Am. Acad. of Dermatology, 1992, Charles Rodolphe Brupbacher award, Switzerland, 1993, Katherine Berkan Judd award, Memorial Sloan-Kettering Cancer Ctr, 1993, Josef Steiner Prize, Switzerland, 1993. Mem. Am. Soc. Microbiology, AAAS, Nat. Acad. of Scis., 1991, Assn. Med. Sch. Microbiology (chmn. adv. and ednl. coms.), Sigma Xi. Office: Princeton U Dept Molecular Biology Princeton NJ 08544*

LEVINE, ARNOLD MILTON, retired electrical engineer, documentary filmmaker; b. Preston, Conn., Aug. 15, 1916; s. Samuel and Florence May (Clark) L.; m. Bernice Eleanor Levich, Aug. 31, 1941; children: Mark Jeffrey, Michael Norman, Kevin Lawrence. BS in Radio Engring., Tri-State U., Angola, Ind., 1939, DSc (hon.), 1960; MS, U. Iowa, 1940. Head sound lab. CBS, N.Y.C., 1940-42; asst. engr., div. head ITT, N.Y.C. and Nutley, N.J., 1942-65; lab. head, lab. dir. ITT, San Fernando, Calif., 1965-71; v.p. aerospace, gen. mgr., sr. scientist ITT, Van Nuys, Calif., 1971-86; ret., 1986. Patentee fiber optics, radar, communications and TV fields. Past mem. bd. dirs., v.p., pres. Am. Jewish Congress, L.A. Recipient San Fernando Valley Engr. of Yr. award, 1968; Profl. designation Motion Picture Art & Scis., UCLA, 1983. Fellow IEEE (life), Soc. Motion Picture and TV Engrs., USCG Aux. (vice comdr. 1990-91, flotilla cmdr. 1992-93). Avocations: sailing, amateur radio, filmmaking, swimming. Home: 10828 Fullbright Ave Chatsworth CA 91311-1737

LEVINE, ARTHUR SAMUEL, physician, scientist; b. Cleve., Nov. 1, 1936; s. David Alvin and Sarah Ethel (Rubinstein) L.; m. Ruth Eleanor Rubin, Oct. 14, 1959; children: Amy Elizabeth, Raleigh Hannah, Jennifer Leah. AB, Columbia U., 1958; MD, Chgo. Med. Sch., 1964. Diplomate Am. Bd. Pediatrics, Am. Bd. Pediatric Hematology-Oncology. Intern in pediatrics U. Minn., Mpls., 1964-65, resident in pediatrics, 1965-66, USPHS fellow in hematology and genetics, 1966-67; capt. USPHS, 1967-92, rear adm., asst. surgeon gen., 1992—; clin. assoc. div. cancer treatment Nat. Cancer Inst., Bethesda, Md., 1967-69, sr. staff fellow, 1969-70, sr. investigator, 1970-73, head sect. infectious disease, pediatric oncology br., 1973-75, chief pediatric oncology br., 1975-82; sci. dir. Nat. Inst. Child Health and Human Devel., Bethesda, 1982—; clin. prof. medicine and pediatrics Georgetown U., Washington, 1975—; clin. prof. pediatrics Uniformed Svcs. U. Health Scis., Bethesda, 1983—; vis. prof. Cold Spring Harbor Lab., N.Y., 1973, Benares Hindu U., India, 1975, U. Minn., 1974, Hebrew U., Israel, 1981, U. Bologna, 1989, Northwestern U., 1992; Karon meml. lectr. U. So. Calif., 1983; Scham lectr. U. Minn., 1983. Author: Cancer in the Young, 1982; editor-in-chief The New Biologist, 1989—; contbr. articles to profl. jours. Recipient Disting. Alumnus award Chgo. Med. Sch., 1972; NIH Dir.'s award, 1984, Meritorious Service award Pub. Health Service, 1987, Disting Svc. award, 1991. Mem. Am. Soc. Clin. Investigation, Soc. Pediatric Research, Am. Assn. Cancer Research, Am. Soc. Hematology, Am. Soc. Clin. Oncology, Am. Fedn. Clin. Research, AAAS, Am. Soc. Microbiology, Am. Soc. Pediatric Hematology/Oncology. Office: NIH Bldg 31 Room 2A-50 Nat Inst Child Health and Human Devel 9000 Rockville Pike Bethesda MD 20892-0001

LEVINE, DONALD JAY, electrical engineer; b. N.Y.C., Oct. 10, 1921; s. Samuel O. and Rose (Weiss) L.; m. Gloria Lerner, May 18, 1946; children: Judith Ellen LeVine Duvall, Nancy Beth LeVine Intrator. BEE, CCNY, 1943; MEE, Poly. Inst. Bklyn., 1952. Registered profl. engr., N.Y., N.J., Mass., Calif. Dir. communication systems The Aerospace Corp., Washington, 1975; v.p. engring. Kings Electronics, Tuckahoe, N.Y., 1976-83, Am. Nucleonics Corp., Westlake Village, Calif., 1983-84; chief broadcast systems div. Voice of Am., USIA, Washington, 1984-86; cons. Donald J. LeVine, P.E., Bethesda, Md., 1986-88, 88—; pres. Internat. Broadcast Systems, Inc., Arlington, Va., 1988; cons. Varian Corp., Radio Free Europe, Radio Liberty, Voice of Am.; vol. rsch. Nat. Air and Space Mus., Washington, 1990—. Editor IEEE Transactions, 1988, 89; contbr. articles to profl. jours. With Signal Corps U.S. Army, 1942-46; ETO. Recipient Scott Helt Meml. award IEEE Broadcast Tech. Soc., 1989. Mem. IEEE (sr. life), NSPE, Armed Forces Communications and Electronics Assn., Soc. Sr. Aerospace Execs. Achievements include 7 patents in field of microwave technology; profl. activity spanned mfg., systems engring., design, rsch. and devel. for microwave components, antennas and systems; rsch. work with short wave broadcast systems/station design. Home and Office: Apt 609 7420 Westlake Ter Bethesda MD 20817

LEVINE, LEON, chemical engineer; b. Bklyn., Aug. 24, 1946; s. David and Lee (Schulman) L.; m. Marlene Rosenblatt, June 8, 1968; children: Faye, Seth. BChemE, CCNY, 1967. Group leader Procter & Gamble, Cin., 1968-75; sr. rsch. scientist Pillsbury Co., Mpls., 1976-86; pres. Leon Levine & Assoc., Plymouth, Minn., 1986—. Author: Food Process Operations and Scale-Up, 1990, (with others) Extrusion Cooking, 1989, Dough Rheology and Baked Product Texture, 1990, Food Engineering Handbook, 1991, Cookie and Cracker Technology, 1993. Mem. AICE (Young Chem. Engr. of Yr. Minn. sect. 1978), Am. Assn. of Cereal Chemists, Inst. of Food Technologists. Achievements include patents for process for bleaching edible oils, method of forming rippled snack products, method for dewaxing vegetable oils, pea separating apparatus and methods of use, microwave expandable product and process for manufacture. Office: Leon Levine & Assocs 2665 Jewel Ln Plymouth MN 55447

LEVINE, LOUIS DAVID, museum director, archaeologist; b. N.Y.C., June 4, 1940; s. Moe Wolf and Jeanne (Greenwald) L.; m. Dorothy Abrams, Dec. 30, 1962 (div. 1991); children: Sarra L., Samuel E. Student, Brandeis U., 1960; BA with honours, U. Pa., 1962, PhD with distinction, 1969. Instr. of Hebrew U. Pa., Phila., 1966-69; asst. curator Royal Ont. Mus., Toronto, Can., 1969-75, assoc. curator, 1975-80, curator, 1981, assoc. dir., 1987-90; asst. commr., dir. N.Y. State Mus., Albany, 1990—; vis. sr. lectr. Hebrew U., Jerusalem, 1975-76; vis. prof. U. Copenhagen, 1985; asst. prof. U. Toronto, 1969-74, assoc. prof. U. Toronto, 1974-81, prof., 1981-90; dir. Seh Gabi Expdn., western Iran, 1971-73, dir. Mahidasht Project, western Iran, 1975-79. Author: Two Stelae from Iran, 1972, The Neo-Assyrian Zagros, 1974; editor: Mountains and Lowlands, 1977; contbr. articles to profl. jours. NDEA fellow U. Pa., 1962-65, Fulbright fellow, 1965, W.F. Albright fellow, Am. Schs. of Oriental Rsch., 1966, fellow Inst. for Advanced Studies, Hebrew U. Mem. Brit. Inst. of Persian Studies, Brit. Sch. of Archaeology in Iraq, Am. Assn. Mus., Am. Oriental Soc. Jewish. Avocations: woodworking, music. Office: NY State Mus Cultural Edn Ctr Rm 3099 Albany NY 12230

LEVINE, MARTIN DAVID, computer science and electrical engineering educator; b. Montreal, Que., Can., Mar. 30, 1938; s. Max and Ethel (Tauber) L.; m. Deborah Tiger, June 6, 1961; children: Jonathan, Barbara. B.Eng., McGill U., Montreal, 1960, M.Eng., 1963; Ph.D., Imperial Coll., U. London, 1965, also diploma. Part-time lectr. Imperial Coll., 1963-64; mem. faculty McGill U., 1965—, prof., 1977—; mem. tech. staff Jet Propulsion Lab., Pasadena, Calif., 1972-73; vis. prof. Hebrew U., Jerusalem, 1979-80; PRECARN/Can. Inst. Advanced Rsch. assoc., 1990; pres. Internat. Assn. Pattern Recognition, 1988-90; cons. computer vision, biomed. image processing, robotics; dir. McGill Rsch. Ctr. for Intelligent Machines, 1986—. Author: Vision in Man and Machine, 1985; co-author: (P. Noble) Computer-Assisted Analyses of Cell Locomotion and Chemotoxis; contbr. articles to profl. jours.; Assoc. editor Computer Vision, Graphics and Image Processing, Image Understanding; gen. editor Advances in Computer Vision and Machine Intelligence. Fellow Am. Soc. Engring. Edn. Ford Found., 1972-73. Fellow IEEE; mem. Computer Soc. of IEEE (assoc. editor Trans. on Pattern Analysis and Machine Intelligence), Pattern Recognition Soc. (assoc. editor Pattern Recognition), Order Engrs. Que., Can. Image Processing and Pattern Recognition Soc. (past founding pres.), Internat. Assn. Pattern Recognition (past pres.). Jewish. Office: McGill U Rsch Ctr for Intelligent, 3480 University St, Montreal, PQ Canada H3A 2A7

LEVINE, MARVIN, psychologist, author; b. N.Y.C., Mar. 16, 1928; s. Louis and Dora (Chaifetz) L.; m. Tillie Cascio, July 14, 1951 (dec. Mar., 1987); children: Laurie, Todd; m. Mara Sandler, Sept. 2, 1991. MA, Harvard U., 1951; PhD, U. Wis., 1959. Asst. prof. Ind. U., Bloomington, 1959-65; from asst. prof. to prof, to prof. emeritus SUNY, Stony Brook, 1965—. Author: A Cognitive Theory of Learning, 1979, Effective Problem Solving, 1987. Mediator Community Mediation Ctr., Suffolk County, N.Y., 1980-85; Sgt. U.S. Army, 1954-56. Grantee: NIMH, 1964-76, NSF, 1979-83; recipient NSF Rsch. fellowship, Brussels, 1961-62. Fellow Am. Psychol. Assn., Am. Psychol. Soc.; mem. Psychonomic Soc. Achievements include discovery of the blank-trial law, the map contra-alignment effect; development of hypothesis probes. Office: State Univ NY Dept of Psychol Stony Brook NY 11794

LEVINE, MICHAEL WILLIAM, vision researcher, educator; b. N.Y.C., Mar. 10, 1943; s. Lester and Beulah (Cirker) L.; m. Jane Freeman, May 25, 1969; children: Matthew S., Andrea E. BS, MIT, 1965, MS, 1967; PhD, Rockefeller U., 1972. Asst. prof. U. Ill. at Chgo., 1972-77, assoc. prof., 1977-85, prof., 1985—; project engr. Lion Rsch. Corp., Newton, Mass., 1967. Author: (with others) Fundamentals of Sensation and Perception, 1980, 91; contbr. articles to profl. jours. including Vision Rsch., Sci., Visual Neurosci., Biol. Cybernetics, Jour. of Physiology and others. Sr. Internat. fellow Fogarty Ctr., NIH, 1987; rsch. grant Eye Inst. of NIH, 1977-80. Mem. AAAS, Assn. for Rsch. in Vision and Ophthalmology, Soc. for Neuroscience, Sigma Xi (local treas.). Office: U Ill m/c 285 1007 W Harrison St Chicago IL 60607

LEVINE, PAUL ALLAN, cardiologist; b. Bklyn., Dec. 22, 1944; s. Lawrence Jerome and Shirely (Silverstein) L.; m. Lucille Doris Pohley, Aug. 6, 1971. BS, MD, Boston U., 1968. Intern Bronx (N.Y.) Mcpl. Hosp. Ctr., 1968-69, resident, 1969-70; resident Georgetown U. Hosp., Washington, 1972-73; cardiology fellow VA Med. Ctr., Washington, 1973-74; cardiology fellow Univ. Hosp. Boston U. Med. Ctr., 1974-76; assoc. prof. medicine Boston U., 1976-89; clin. assoc. prof. medicine UCLA, 1990—; clin. prof. medicine Loma Linda (Calif.) U., 1989—; v.p., med. dir. Siemens Pacesetter, Sylmar, Calif., 1989—; med. advisor Siemens Pacesetter, Sylmar, 1977-89. Author: Pacing Therapy: A Guide to Cardiac Pacing, 1983. Lt. comdr. USN, 1970-72. Decorated Bronze Star; teaching scholar Am. Heart Assn., 1980-83; hon. fellow Argentine Sco. Electrophysiology, 1989. Fellow ACP, Am. Coll. Cardiology (pacemaker com. 1987-89), Coun. Clin. Cardiology Am. Heart Assn.; mem. N.Am. Soc. Pacing and Electrophysiology (bd. trustees 1985-89). Jewish. Achievements include development of interaction ednl. programs in cardiac pacing; advancement in study of the art of pacing therapy. Office: Siemens Pacesetter Inc 15900 Valley View Ct Sylmar CA 91392

LEVINE, RAPHAEL DAVID, chemistry educator; b. Alexandria, Egypt, Mar. 29, 1938; brought to U.S., 1939; s. Chaim S. and Sofia (Greenberg) L.; m. Gillah T. Ephraty, June 13, 1962; 1 child, Ornah T. MSc, Hebrew U., Jerusalem, 1959; PhD, Nottingham (Eng.) U., 1964; DPhil, Oxford (Eng.) U., 1966; PhD honoris causa, U. Liege, Belgium, 1991. Vis. asst. prof. U. Wis., 1966-68; prof. theoretical chemistry Hebrew U., Jerusalem, 1969—, chmn. research ctr. molecular dynamics, 1981—, Max Born prof. natural philosophy, 1985—; Battelle prof. chemistry and math. Ohio State U., Columbus, 1970-74; Brittingham vis. prof. U. Wis., 1973; adj. prof. U. Tex., Austin, 1974-80, MIT, 1980-88, UCLA, 1989—; Arthur D. Little lectr. MIT, 1978; Miller rsch. prof. U. Calif., Berkeley, 1989, A.D. White prof. at large Cornell U., 1989—. Author: Quantum Mechanics of Molecular Rate Processes, 1969, Molecular Reaction Dynamics, 1974, Lasers and Chemical Change, 1981, Molecular Reaction Dynamics and Chemical Reactivity, 1986; mem. editorial bds. several well known scientific jours.; contbr. articles to profl. jours. Served with AUS, 1960-62. Recipient Ann. award Internat. Acad. Quantum Molecular Sci., 1968, Landau prize, 1972, Israel prize in Exact Scis., 1974, Weizman prize, 1979, Rothschild prize, 1992; co-recipient Chemistry prize Wolf Found., 1988; Ramsay Meml. fellow, 1964-66, Alfred P. Sloan fellow, 1970-72. Fellow Am. Phys. Soc.; mem. Israel Chem. Soc., Israel Acad. Scis., Max. Planck SOc. (fgn. mem.). Office: UCLA Dept of Chemistry Los Angeles CA 90024-1569 also: Hebrew U Jerusalem, 91904 Jerusalem Israel

LEVINE, SEYMOUR, psychology educator, researcher; b. Bklyn., Jan. 23, 1925; s. Joseph and Rose (Reines) L.; m. Barbara Lou McWilliams, Feb. 19, 1949; children: Robert Thomson, Leslie Ingrid, Alicia Margaret. B.A., U. Denver, 1948; M.A., NYU, 1950, Ph.D., 1952. Asst. prof. Boston U., 1952-53, Ohio State U., Columbus, 1956-60; assoc. prof. Stanford U., 1962-69, prof., 1969—, dir. Stanford Primate Facility, 1976—, dir. biol. scis. research tng. program, 1971—; cons. Found. Human Devel., Dublin, Ireland, 1973—. Editor: Hormones and Behavior, 1972, Psychobiology of Stress, 1978, Coping and Health, 1980. Served with U.S. Army, 1942-45. Recipient Hoffheimer Research award, 1961; recipient NIMH Career Devel. award, 1962, Research Scientist award NIMH, 1967—. Fellow Am. Psychol Assn., AAAS; mem. Internat. Soc. Devel. Psychobiology (pres. 1975-76), Internat. Soc. Psychoneuroendocrinology (pres. 1990-93, Am. Soc. Primatologists, Internat. Primatol. Soc. Home: 927 Valdez Pl Palo Alto CA 94305-1008 Office: Stanford U Sch Medicine Dept Psychiatry and Behavioral Scis Stanford CA 94305

LEVINE, ZACHARY HOWARD, physicist; b. Phila., Apr. 6, 1955; s. Murray and Adeline Francis (Gordon) L. BS, MIT, 1976; PhD, U. Pa., 1983. Mem. tech. staff AT&T Engring. Rsch. Ctr., Princeton, N.J., 1983-86, sr. mem. tech. staff, 1986-87; vis. scientist Cornell U., Ithaca, N.Y., 1987-89; sr. rsch. assoc. Ohio State U., Columbus, 1989-92, rsch. specialist, 1992--. Mem. Am. Phys. Soc. Achievements include patent for routing circuit boards. Office: Ohio State U 174 W 18th Ave Columbus OH 43210

LEVINGSTON, ERNEST LEE, engineering company executive; b. Pineville, La., Nov. 7, 1921; s. Vernon Lee and Adele (Miller) L.; m. Kathleen Bernice Bordelon, June 23, 1944; children: David Lewis, Jeanne Evelyn, James Lee. BME, La. State U., 1960. Gen. foreman T. Miller & Sons, Lake Charles, La., 1939-42; sr. engr., sect. head Cities Service Refining Corp., Lake Charles, 1946-57; group leader Bovay Engrs., Baton Rouge, 1957-59; chief engr. Augenstein Constrn. Co., Lake Charles, 1959-60; pres. Levingston Engrs., Inc., Lake Charles, 1961-85, gen. mgr. SW La., Austin Indsl., 1985-88; pres. Levingston Engrs., 1989—. Mem. Lake Charles Planning and Zoning Commn., 1965-70; mem. adv. bd. Sowela Tech. Inst., 1969—; mem. Regional Export Expansion Council, 1969-70, chmn. code com., 1966—; mem. La. Bd. Commerce and Industry, 1978—; bd. dirs. Lake Charles Meml. Hosp.; bd. dirs., regional chmn. La. Chem. Industry Alliance, 1990—. With USNR, 1942-46. Named Jaycee Boss of Year, 1972. Registered profl. engr., La., Tex., Miss., Ark., Tenn., Pa., Md., Del., N.J., D.C., Okla., Colo. Mem. La. Engring. Soc. (pres. 1967-68, state bd. dirs. 1967-68, 90-91), Nat. Inst. Cert. Engring. Technologists (past trustee, mem. exam. com.), La. Assn. Bus. and Industry, Lake Area Industries/McNeese Engring., Lake Charles C. of C. (dir. 1969-73). Baptist (deacon 1955—). Office: PO Box 1865 Lake Charles LA 70602-1865

LEVINSKY, NORMAN GEORGE, physician, educator; b. Boston, Apr. 27, 1929; s. Harry and Gertrude (Kipperman) L.; m. Elena Sartori, June 17, 1956; children—Harold, Andrew, Nancy. A.B. summa cum laude, Harvard U., 1950, M.D. cum laude, 1954. Diplomate Am. Bd. Internal Medicine. Intern Beth Israel Hosp., Boston, 1954-55; resident Beth Israel Hosp., 1955-56; commd. med. officer USPHS., 1956; clin. assoc. Nat. Heart Inst., Bethesda, Md., 1956-58; NIH fellow Boston U. Med. Center, 1958-60; practice medicine, specializing in internal medicine and nephrology Boston, 1960—; chief of medicine Boston City Hosp., 1968-72, 93—; physician-in-chief, dir. Evans dept. clin. research Univ. Hosp., Boston, 1972—; asst. prof., then assoc. prof. medicine Boston U., 1960-68, Wesselhoeft prof., 1968-72, Wade prof. medicine, 1972—, chmn. dept. medicine, 1972—; mem. drug efficacy panel NRC.; mem. nephrology test com.-Am. Bd. Internal Medicine, 1971-76; mem. gen. medicine B rev. group NIH; mem. comprehensive test com. Nat. Bd. Med. Examiners, 1986-89; chmn. com. to study end-stage renal disease program Nat. Acad. Scis./Inst. Medicine, 1988-90. Editor: (with R.W. Wilkins) Medicine: Essentials of Clinical Practice, 3d edit., 1983; contbr. chpts. to books, sci. articles to med. jours. Recipient Distinguished Teacher awd., Am. Coll. of Physicians, 1992. Master ACP (Disting. Tchr. award 1992); mem. AAAS, Am. Fedn. Clin. Rsch., Am. Soc. Clin. Investigation, Am. Heart Assn., Assn. Am. Physicians, Am. Physiol. Soc., Assn. Profs. Medicine (sec., treas. 1984-87, pres.-elect 1987-88, pres. 1988-89), Am. Soc. Nephrology, Inst. Medicine NAS, Interurban Clin. Club (pres. 1985-86), Phi Beta Kappa, Alpha Omega Alpha. Home: 20 Kenwood Ave Newton MA 02159-1439 Office: Boston U Med Ctr 75 E Newton St Boston MA 02118-2347

LEVINSON, ARTHUR DAVID, molecular biologist; b. Seattle, Mar. 31, 1950; s. Sol and Malvina (Lindsay) L.; m. Rita May Liff, Dec. 17, 1978; children: Jesse, Anya. BS, U. Wash., 1972; PhD, Princeton U., 1977.

Postdoctoral fellow U. Calif., San Francisco, 1977-80; sr. scientist Genentech, South San Francisco, 1980-84, staff scientist, 1984—, dir. cell genetics dept., 1988-89, v.p. rsch., 1990-93, sr. v.p. rcsh. and devel., 1993—. Mem. editorial bd. Virology, 1984-87, Molecular Biology and Medicine, 1986-90, Molecular and Cellular Biology, 1987—, Jour. of Virology, 1988-91. Mem. Am. Soc. Microbiology, Am. Soc. Biochemistry and Molecular Biology. Office: Genentech Inc 460 Point San Bruno Blvd South San Francisco CA 94080

LEVINTON, MICHAEL JAY, electrical engineer, energy consultant; b. Bklyn., Apr. 6, 1957; s. Morris Aaron and Marian (Bordman) L.; m. Patricia S. Herschtein, Aug. 25, 1988; children: Marlon A., Reuben U. BSEE magna cum laude, Poly. U., Bklyn., 1979, MSEE, 1981. Assoc. engr. Am. Electric Power Svc. Co., N.Y.C., 1980-84; dep. dir. design and constrn. SUNY Health Sci. Ctr., Bklyn., 1984—; cons. MJ Cons., Bklyn., 1987—. Contbr. articles to profl. publs. Pres. bd. dirs. Westport Real Estate Corp., Bklyn., 1987-91; del. N.Y.C. Bd. Elections, Bklyn., 1982-86. Mem. IEEE (sr.), Assn. Energy Engrs. (sr.), Hosp. Engring. Soc. Greater N.Y. Achievements include implementation of energy conservation measures for university campus. Office: SUNY Health Sci Ctr Bklyn Box 13 450 Clarkson Ave Brooklyn NY 11203

LEVIS, DONALD JAMES, psychologist, educator; b. Cleve., Sept. 19, 1936; s. William and Antoinette (Stejskal) L.; children: Brian, Katie. Ph.D., Emory U., 1964. Postdoctoral fellow clin. psychology Lafayette Clinic, Detroit, 1964-65; asst. prof. psychology U. Iowa, Iowa City, 1966-70, assoc. prof., dir. research and tng. clinic, 1970-72; prof. SUNY-Binghamton, 1972—. Author: Learning Approaches to Therapeutic Behavior Modification, 1970, Implosive Therapy, 1973; cons. editor: Jour. Abnormal Psychology, 1974-80, Jour. Exptl. Psychology, 1976-77, Behavior Moedification, 1977-81, Behavior Therapy, 1974-76, Clin. Behavior Therapy Rev., 1978—; contbr. articles to profl. jours. Served to capt. AUSR, 1958-66. Fellow Behavior and Therapy Research Soc. (charter, clin.), Am. Psychol. Assn.; mem. Assn. Advancement Behavior Therapy (publ. bd. 1979-82), AAAS, Psychonomic Soc., N.Y. State Psychol. Assn., Sigma Xi. Home: 48 Riverside Dr Binghamton NY 13905-4326 Office: SUNY Dept Psychology Binghamton NY 13901

LEVITT, GEORGE, retired chemist; b. Newburg, N.Y., Feb. 19, 1925; married; 4 children. BS, Duquesne U., 1950, MS, 1952; PhD, Mich. State U., 1957. Rsch. chemist Exptl. Sta. E.I. du Pont de Nemours & Co., Inc., 1956-63, rsch. chemist Stine Lab., 1963-66, rsch. chemist Exptl. Sta., 1966-68, sr. rsch. chemist, 1968-80, rsch. assoc., 1981-86; instr. Del. Tech. and C.C., 1975-80. Recipient Quadrennial award Swiss Soc. Chem. Industries, 1982, Nat. Agrl. award Nat. Agrl. Mktg. Assn., 1987, 88, Am. Chem. Soc. Creative Invention award Corp. Assocs. Am. Chem. Soc., 1989, Nat. Medal of Tech., NSF, 1993. Mem. AAAS, Am. Chem. Soc., Internat. Union Pure & Applied Chemistry, Sigma Xi. Achievements include research in organic syntheses, herbicides, fungicides, medicinals, pesticides, heterocyclic compounds, synthesis, characterization and identification of novel organic compounds for biological evaluation, developed programs to define and optimeze chemical structure- biological activity relationships, sulfonylurea herbicides, heteroyclics, exploratory process research. Office: E I du Pont de Nemours and Co Inc Agrl Products Wilmington DE 19898*

LEVITT, ISRAEL MONROE, astronomer; b. Phila., Dec. 19, 1908; s. Joseph and Jennie (Marriner) L.; m. Alice Gross, July 3, 1937; children: Peter Leighton, Nancy Bambino. B.S. in Mech. Engring., Drexel U., 1932, D.Sc., 1958; M.A., U. Pa., 1937, Ph.D., 1948; D.Sc., Temple U., 1958, Phila. Coll. Pharmacy and Sci., 1963. Astronomer, Fels Planetarium of The Franklin Inst., Phila., 1934-39; asst. asso. dir. Fels Planetarium of The Franklin Inst., 1939-49, dir., 1949-72, v.p. inst., 1970-72; exec. dir. Phila. Mayor's Sci. and Tech. Adv. Council, 1972-93; sr. lectr. astronomy U. Pa., 1977; astronomer The Flower Obs., 1946-48; dir. (Sci. Council), 1953—; sci. cons. to City of Phila., 1956-63; chmn. Air Pollution Control Bd. Phila., 1965—. Author: Precision Laboratory Manual, 1932, (with Roy K. Marshall) Star Maps for Beginners, 1942, Space Traveler's Guide to Mars, 1956, Target for Tomorrow, 1959, Exploring The Secrets of Space, 1963, (with Dandridge M. Cole) Beyond the Known Universe, 1974; developer NASA Spacemobile; inventor oxygen mask, pulse counting photoelectric photometer (with William Blitzstein); contbr. articles in jours., mags. on sci. subjects; author: internationally syndicated Space column for Gen. Features. Recipient USN Ordnance Devel. award, 1945; Henry Grier Bryant gold medal Geog. Soc. Phila., 1962; Joseph Priestley award Spring Garden Inst., Phila., 1963; Writing award Aviation/Space Writers Assn., 1965; Samuel S. Fels Medal award, 1970; cert. of recognition NASA, 1977. Fellow AAAS, Brit. Interplanetary Soc., Am. Astron. Soc.; mem. AIAA, Rittenhouse Astron. Soc. (past pres.), Acad. Scis. Phila. (v.p.), Aviation Writers Assn., Nat. Assn. Sci. WRiters, Acad. Scis. Phila. (v.p. 1993), Explorers Club, Pi Tau Sigma. Home: 3900 Ford Rd Apt 19D Philadelphia PA 19131-2039 Office: 1515 Market St Fl 17 Philadelphia PA 19102-1901

LEVY, BORIS, chemistry educator; b. N.Y.C., Nov. 24, 1927; s. Samuel and Frieda (Meisel) L.; m. Ann Dolores Reiter, July 15, 1956; children: Michael, Daniel, Jennifer. PhD, NYU, 1955. Sr. scientist Westinghouse Rsch. Lab., Pitts., 1956-60, Mobil Oil Co., Pennington, N.J., 1960-65; mgr. Polaroid Corp., Cambridge, Mass., 1965-89; rsch. prof. chemistry Boston U., 1989—; cons. in field. Editor/author: Symposium on Electronic and Ionic Properties of Silver Halides, 1991; contbr. articles to profl. jours.; assoc. editor Photographic Sci. Engring. Fellow Soc. for Imaging Sci., Sigma Xi. Achievements include 14 patents; development of time resolved technique for measurement of charge carrier dynamics in particulate photoconducting materials; discovered photographic diodes and transisters. Office: Boston U Dept Chemistry 590 Commonwealth Ave Boston MA 02215

LEVY, DANIEL, economics educator; b. Tschakaia, Georgian Republic, Georgia, Nov. 13, 1957; came to U.S., 1982; s. Shabtai and Simha (Levi ashvili) Leviashvili; m. Sarit Adler, Sept. 10, 1981; 1 child, Avihai. BA, Ben-Gurion U., Beer-Sheva, Israel, 1982; MA, U. Calif., Irvine, 1989, PhD, 1990. Lectr. U. Minn., Mpls., 1983-88, St. Olaf Coll., Northfield, Minn., 1986-88, The Coll. St. Catherine, St. Paul, 1987-88; prof. Pepperdine U., Irvine, 1989-90, U. Calif., Irvine, 1990-91, Union Coll., Schenectady, N.Y., 1991-92, Emory U., Atlanta, 1992—; computer software programmer Mac Cartuli, 1989. Contbr. articles to profl. jours. Treas. Minn. Student Orgn., 1984-85. Mem. Internat. Inst. Forecasters, Am. Econ. Assn., Soc. Econ. Dynamics and Control, Econometric Soc., Western Econ. Assn. Avocations: basketball, tennis, chess, computers. Office: Emory U Dept Economics Atlanta GA 30322

LEVY, DONALD HARRIS, chemistry educator; b. Youngstown, Ohio, June 30, 1939; s. Gabriel and Minnie (Lerner) L.; m. Susan Louise Miller, June 14, 1964; children—Jonathan G., Michael A., Alexander B. B.A., Harvard U., 1961; Ph.D., U. Calif.-Berkeley, 1965. Asst. prof. chemistry U. Chgo., 1967-74, assoc. prof., 1974-78, prof., 1978—, chmn. dept. chemistry, 1983-85; mem. chemistry adv. com. NSF. Assoc. editor Jour. Chem. Physics, 1983—. Fellow Am. Phys. Soc., AAAS; mem. Am. Chem. Soc., Optical Soc. Am., Am. Acad. Arts and Scis., Nat. Acad. Scis. Office: U Chgo Dept Chemistry 5640 S Ellis Ave Chicago IL 60637-1467

LEVY, EDWARD K., mechanical engineering educator. BS, U. Md., 1963; MS, Mass. Inst. Tech., 1964, ScD, 1967. Prof. mech. engring. Lehigh U., 1967—; assoc. prof. Nat. Acad. Engring. Mem. AICE, Am. Soc. Mech. Engrs., Am. Nuclear Soc. Acheivements include research in fluid mechanics, heat transfer and applied thermodynamic aspects of energy with emphasis on power generation systems. Office: Lehigh U Energy Rsch Ctr 117 Atlss Dr Bethlehem PA 18015*

LEVY, ETIENNE PAUL LOUIS, surgical department administrator; b. Paris, Feb. 17, 1922; s. Pierre-Paul and Jeanne (Dreyfus) L.; m. Suzanne Binvignat, Nov. 20, 1965; 1 child, Anne Cécile. Baccalaureat, Janson Sailly, Paris, 1935; engr., French Nat. Agronomic Inst., Paris, 1947; PhD in Engring., U., Paris, 1949, MD, 1956. Chief lab., dept. surgery Salpetriere Hosp., Paris, 1956-61; attaché rsch. INSERM, Paris, 1956-61; chief rsch. Insern, Paris, 1961-67, master rsch., 1967-87, dir. rsch., 1981-91; head dept. surg. gastrointestinal ICU Hosp. St. Antoine, Paris, 1960—; assoc. prof. Faculty Medicine, Paris, 1967—; cons. in field. Contbr. articles to profl. jours. Maj. French Mil., 1943-45, 56-57. Decorated officer Legion of Honour. Mem. N.Y. Acad. Scis., French Alpine Club. Home: 106 Ave Villiers, 75017 Paris France Office: Hosp St Antoine, 184 rue fb St Antoine, 75012 Paris France

LEVY, EUGENE HOWARD, planetary sciences educator, researcher; b. N.Y.C., May 6, 1944; s. Isaac Philip and Anita Harriet (Guttman) L.; m. Margaret Lyle Rader, Oct. 13, 1967; children: Roger P., Jonathan S., Benjamin H. AB in Physics with high honors, Rutgers U., 1966; PhD in Physics, U. Chgo., 1971. Teaching asst. dept. physics U. Chgo., 1966-69, rsch. asst. Enrico Fermi Inst., 1969-71; postdoctoral fellow dept. physics and astronomy U. Md., 1971-73; asst. prof. physics and astrophysics Bartol Rsch. Found., Franklin Inst., Swarthmore, Pa., 1973-75; asst., then assoc. prof. U. Ariz., Tucson, 1975-83, prof. planetary scis., 1983—, mem. faculty applied math. program, 1981—, head dept. planetary scis., dir. lunar and planetary lab., 1983—, mem. theoretical astrophysics program, 1985—, dir. NASA-Ariz. Spacegrant Coll. Consortium, 1989—; mem. com. on planetary and lunar exploration of space sci. bd., Nat. Acad. Scis., 1976-79, chmn., 1979-82, co-chair Space Sci. Bd. Study on Exploration Primitive Solar-System Bodies, 1978, mem. Space Sci. Bd., 1979-82, head U.S. del., co-chair Nat. Acad. Scis.-European Sci. Found. Joint Working Group on Cooperation in Planetary Exploration, 1982-84, mem. steering group com. on major directions for space sci. 1995-2015, 1984-86, chair adv. com. on internat. cooperation for Mars sample return, 1986-88; mem. Comet Halley Sci. Working Group, NASA, 1977, mem. spacelab phys. sci. rev. panel space sci. steering com., 1979, mem. rev. panel on origin plasmas in Earth's neighborhood, 1980, mem. solar system exploration com. of Adv. Coun., 1980-83, mem. Ames Rsch. Ctr. Planetary Detection Study, 1983, Solar System Exploration Mgmt. Coun., 1983-87, mem. com. on future space-sta. sci. projects, 1985, mem. Space Sta. Sci. Users' Working Group, 1985-86, Space and Earth Sci. Adv. Com., 1985-88, chair Comet Rendevous and Asteroid Flyby Rev. Panel, 1986, mem. Mars Exploration Strategy Adv. Group, 1986, Mars Rover Sample Return Sci. Working Group, 1987—; sci. cons. Rockwell Internat. Corp., 1980; mem. COSPAR Internat. Tech. Panel on Comets, 1980-82; U.S.-NASA del. to discussions on internat. cooperation investigations of Comet Halley, Padua, Italy, 1981, to U.S.-USSR Joint Working Group on Near-Earth Space, the Moon and Planets, 1981; mem. program adv. bd. Internat. Conf. on Cometary Exploration, Budapest, Hungary, 1982; mem. exec. com. univs.' space sci. working group Assn. Am. Univs., 1982-86; study panel U.S.-Soviet cooperation in space sci. U.S. Cong. Office of Tech. Assessment, 1984; chair planetary exploration panel Pacific Rim Nations Internat. Space Yr. Conf., Kona, Hawaii, 1987; mem. working group planetary systems sci. NASA, 1988—, rev. panel lunar and planetary, 1988-90, rev. panel origins solar systems programs, 1990-91; mem. astronomy and astrophysics survey com., sci. opportunities panel NAS, 1989-90; mem. study panel on robotic exploration of Moon and Mars, U.S. Cong. Office Tech. Assessment, 1991; chmn. coun. of instns., bd. dirs. U.S. Space Rsch. Assn., 1991—, vice-chmn. bd. dirs. 1993—; cons. and lectr. in field. Editor: Protostars and Planets III, 1993; contbr., author articles for gen. pub., adv. reports for Congl. Record, abstracts, book revs., others. Recipient Disting. Pub. Svc. medal NASA, 1983, Alexander von Humboldt-Stiftung Sr. Scientist award Fed. Republic Germany, 1989; Disting. vis. scientist Jet Propulsion Lab., Calif. Inst. Tech., 1985-91; NASA predoctoral fellow U. Chgo., 1966-69, fellow Ctr. for Theoretical Physics, U. Md., 1971-73; rsch. grantee NASA, NSF. Mem. AAAS, Am. Astron. Soc., Am. Geophys. Union, Am. Phys. Soc., Internat. Astron. Union, Univs. Space Rsch. Assn. (bd. dirs. 1991—), Coun. Instns. (pres. 1991-92, vice chmn. bd. dirs. 1993-94), Phi Beta Kappa, Sigma Xi. Achievements include research in theoretical cosmical physics, planetary geophysics, magnetohydrodynamics, space and solar physics, magnetic field generation, physical processes associated with the formation of stars any systems. Home: 5442 E Burns St Tucson AZ 85711-3126 Office: U Ariz Lunar and Planetary Lab Tucson AZ 85721

LEVY, HAROLD DAVID, psycholinguist; b. Rochester, N.Y., Aug. 25, 1938; s. Barnet Lewis and Ada Sylvia (Zimmerman) L.; m. Jan Patricia Schwartz, Mar. 3, 1959 (div. 1961); children: Marvin Lee; m. Natalie Miller, Nov. 27, 1969 (div. 1982); children: Benjamin Eli; m. Judy Weiner, Sept. 9, 1987. BS in Psychology, U. Rochester, 1969, MA in Edn., 1971. Permanent cert. to teach langs. (7-12). Sociotherapist Convalescent Hosp. for Children, Rochester, 1971-76; tutor City Sch. Dist., Rochester, 1973-83; editor, ednl. dir. Operaton Friendship, Rochester, 1983-88; pvt. tutor home and social agencies Rochester, 1982—; activity therapist Genesee Hosp., Rochester, 1983—. Author: (textbooks) Forced Categories: A Taxonomy for Languages, 1971, Languages: Their Common Elements, 1990, Linguistics: Theory of Names, 1990, Language Learning by Slices, 1990, Linguistics: The Binary System, 1990. Avocations: jazz piano, mental health education, nutrition. Home: 111 East Ave Apt 719 Rochester NY 14604-2542

LEVY, HAROLD JAMES, physician, psychiatrist; b. Buffalo, Feb. 15, 1925; s. Sidney Harold and Evelyn (Sperling) L.; m. Arlyne Adelstein, July 3, 1958; children: Sanford Harvey, Richard Alan, Kenneth Lee. MD, U. Buffalo, 1946. Diplomate in psychiatry Am. Bd. Neurology and Psychiatry. Intern Meyer Meml. Hosp., Buffalo, 1946-47, asst. resident in psychiatry, 1947-48; fellow in psychosomatic medicine Med. Sch. U. Buffalo, Meyer Meml. Hosp., 1950-53; psychiatrist Buffalo, 1950—; mem. courtesy staff Millard Fillmore Hosp., 1957, clin. asst., 1958, asst. attending physician, 1959-63, assoc. attending physician, 1963-64, attending physician, 1964-90, chmn. dept. psychiatry, 1968-90, cons., 1990—; attending psychiatrist BryLin Psychiat. Hosp. (formerly Linwood Bryant Hosp.), Buffalo, 1955—, clin. dir. psychiatry, 1966—; attending psychiatrist Meyer Meml. Hosp., 1953—, asst. chief psychiatry, 1953-59; staff psychiatrist Psychiat. Clinic, Family Ct. Erie County, N.Y., 1959-63, psychiat. dir. clinic, 1963-80; mem. courtesy staff in psychiatry St. Joseph's Intercommunity Hosp., Buffalo, 1969-71, cons. in psychiatry, 1971—; cons. in psychiatry St. Francis Hosp., 1972—, Sisters of Charity Hosp.; asst. in psychiatry Med. Sch. SUNY, Buffalo, 1950-52, asst. 1952-55, instr. 1955-61, assoc. 1955-70, clin. asst. prof. 1970-86, clin. assoc. prof., 1986—; mem. psychiat. staff Rosa Coplon Jewish Home and Infirmary, 1957-72, chmn. dept. psychiatry, 1969-72; staff psychiatrist Chronic Disease Rsch. Inst., sect. on alcoholism Med. Sch. SUNY, Buffalo, 1950-53; psychiat. cons. Dent Clinic Found. Millard Fillmore Hosp., 1967—, Sisters of Charity Hosp., Buffalo, 1987—, Lafayette Gen. Hosp., Buffalo, 1973-88. Pres. Lemezo Enterprises Inc., Buffalo, 1970, Sanricken Enterprises Inc., Buffalo, 1970—; mem. exec. com. Blue Shield Western N.Y. Served to capt. M.C., AUS, 1948-50. Fellow Am. Psychiat. Assn. (life, pres. Western N.Y. dist. br. 1969-70), Am. Soc. Psychoanalytic Physicians, Am. Soc. Advancement Electrotherapy; mem. AMA, Israel Med. Assn., N.Y. State Med. Soc., Erie County Med. Soc. (chmn. com. on mental health, econs. com., publ. com. for bull. 1969-70), Buffalo Acad. Medicine, Jerusalem Acad. Medicine, Maimonides Med. Soc. (pres. 1968-69), N.Y. State Soc. Med. Rsch., Western N.Y. Neuropsychiat. Soc. (pres.-elect 1965-66), Western N.Y. Psychiat. Assn. (past pres. 1974-75), Gen. Alumni Assn. SUNY, Buffalo (treas. exec. bd. 1967-69, numerous offices), SUNY-Buffalo Sch. Medicine Alumni Assn. (past pres., numerous offices), Med. Students' Aid Soc. (past. nat. pres., now -chmn. bd. dirs.), B'nai B'rith (exec. com. Anti Defamation League), Cherry Hill Golf and Country Culb, Alpha Omega Alpha, Phi Lambda Kappa (nat. dir., past nat. v.p., past nat. pres., now chmn. bd. dirs.), Beta Sigma Rho. Home: 47 Longleat Dr Buffalo NY 14226-4199 Office: Psychiat Assocs of Western NY 2740 Main St Buffalo NY 14214-1702

LEVY, MELVIN, tunnel engineer, union executive; b. N.Y.C., Jan. 27, 1932; s. Irving and Rose (Lichtenstein) L.; m. Sylvia Levine, May 31, 1953; children: Jeffrey David, Fran Ellen. B in Engring., CCNY, 1952; MSCE, Poly. Inst. Bklyn., 1964. Registered profl. engr., N.Y. Engr. Lockheed Air Craft, Burbank, Calif., 1952, Edwards Kelcey & Beck, Newark, 1952-53, Bklyn. Navy Yard, 1953-54, M.H. Treadwell, N.Y.C., 1956-57; sr. engr. Smillie & Griffin, N.Y.C., 1957-60, Sverdrup & Parcel, N.Y.C., 1960-64; sr. civil engr. N.Y.C. Transit Authority, Bklyn., 1964—; del. Dist. Coun. 37 AFSCME, N.Y.C., 1989—. Bd. dirs. Mill Island Civic Assn., Bklyn, 1962-63. With U.S. Army, 1954-56. Mem. ASCE, NSPE. Achievements include development of torsion box for support of monorail at New York's 1964 World's Fair; correction of tunnel roadway design formula for American Association of State Highway Officials. Office: NYC Transit Authority 370 Jay St Brooklyn NY 11201-3814

LEVY, MIGUEL, physicist; b. Lima, Peru, Sept. 24, 1951; s. Jose and Ethel (Goldfarb) L.; m. Anita Ackerman, Jan. 15, 1972; children: Diego Jose, Tania. MS in Physics, Cornell U., 1975; PhD in Physics, CUNY, 1988. Postdoctoral rsch. scientist City Coll. N.Y., N.Y.C., 1988-91; postdoctoral rsch. scientist Columbia U., N.Y.C., 1991-92, rsch. scientist, 1993—. Cpmtbr. articles to profl. jours. Office: Columbia U 500 W 120 St Rm 1312 New York NY 10027

LEVY, MOISES, physics educator; b. Concepcion, Panama, Apr. 8, 1930; came to U.S., 1948; s. Abood and Adela Levy; m. Patricia Dolores Hollingsworth, June 1959 (div. 1983); m. Janina Elizabeth Rozanska, Nov. 14, 1983; 1 child, Aleksandra Braginski Diebald. BS, Calif. Inst. Tech., 1952, MS, 1955; PhD, UCLA, 1963. Mem. staff Specialty Resins, Lynwood, Calif., 1952-53; fellow Hughes Aircraft, Inglewood, Calif., 1955-57; postdoctoral fellow ETH, Zurich, Switzerland, 1962-63; asst. prof. physics U. Pa., Phila., 1964-65, UCLA, 1965-70; asst. prof. physics U. Wis., Milw., 1971-73, assoc. prof. physics, 1971-73, prof. physics, 1973—, chmn. physics dept., 1975-78; cons. Northrup Corp., El Segundo, Calif., 1967-69, Energy Conversion Devices, Troy, Mich., 1981-82, Astronautics Corp., Milw., 1988—; speaker in field. Guest editor: Ultrasonics of High Tc and Other Unconventional Superconductors, 1992; contbr. over 170 articles to sci. jours. With U.S. Army, 1953-55. Fellow Hughes Aircraft, 1955-57, NATO, 1962-63, Lady Davis Found., 1983-84. Fellow Am. Phys. Soc.; mem. IEEE-Ultrasonics Ferroelectrics and Frequency Control Soc. (sr., Disting. Lectr. 1991-92, chmn. symposium 1974, 83, 90), Acoustical Soc. Am. Jewish. Office: U Wis Milw Physics Dept Milwaukee WI 53201

LEVY, NORMAN B., psychiatrist, educator; b. N.Y.C., 1931; s. Barnett Theodore and Lena (Gulnick) L.; m. Lya Weiss (dec.); children: Karen, Susan, Joanne; m. Carol Lois Spiegel, 1 son, Robert Barnett. B.A. cum laude, NYU, 1952; M.D., SUNY. Diplomate: Am. Bd. Psychiatry and Neurology (examiner). Intern Maimonides Med. Center, Bklyn.; resident physician in medicine U. Pitts.-Presbyn. Hosp.; resident in psychiatry Kings County Hosp. Center, Bklyn.; instr. psychiatry State U. N.Y. Downstate Med. Center Coll. Medicine, Bklyn.; asst. prof. State U. N.Y. Downstate Med. Center Coll. Medicine, asso. prof., prof.; presiding officer faculty State U. N.Y. Downstate Med. Center Coll. Medicine (Coll. Medicine), assoc. dir. med-psychiat. liaison service; prof. psychiatry, medicine, surgery and coordinator psychiat. liaison services N.Y. Med. Coll.; dir. liaison psychiatry div. Westchester County Med. Center, mem. exec. com. of med. staff, 1981-85, 89—; vis. prof. psychiatry and medicine So. Ill. U. Sch. Medicine; vis. prof. psychiatry John A. Burns Sch. Medicine, U. Hawaii, 1981; coordinator 1st Internat. Conf. Psychol. Factors in Hemodialysis and Transplantation, 1978, 2d-8th Internat. Confs. on Psychonephrology; cons. NIMH.; chief med. services USAF Hosp., Ashiya, Japan. Author: (with others), editor: Living or Dying: Adaptation to Hemodialysis, 1974, Psychonephrology I: Psychological Factors in Hemodialysis and Transplantation, 1981, Men in Transition: Theory and Therapy, 1982, Psychonephrology II: Psychological Problems in Kidney Failure and their Treatment, 1983; contbr. articles to jours., chpts. to textbooks in field.; assoc. editor: Gen. Hosp. Psychiatry; also sect. editor, 1978—; sect. editor: Internat. Jour. Psychiatry in Medicine, 1977-78; mem. editorial bd., book rev. editor Jour. Dialysis and Transplantation, 1979—; mem. editorial bd. Resident and Staff Physician, 1981-91, Internat. Jour. Artificial Internal. Organs, 1983—, Geriatric Nephrology and Urology, 1990—. Served to capt. M.C., USAF. Recipient Willaim A Consol Master Tchr. award, SUNY, Brooklyn, 1991; Thomas P. Hackett award Acad. Psychosomatic med., 1993. Fellow ACP, Am. Coll. Psychiatrists, Am. Psychiat. Assn. (pres. Kings County dist. br. 1981-82), Internat. Coll. Psychosomatic Medicine, Acad. Psychosomatic Medicine; mem. AAAS, Am. Psychosomatic Soc., N.Y. Acad. Scis., Psychonephrology Found. (pres. 1978—), Assn. Acad. Psychiatry, Internat. Soc. Nephrology, Am. Soc. Nephrology, Am. Assn. Artificial Internal Organs, Soc. Liaison Psychiatry (bd. dirs. 1979-80, sec. 1980-81, pres. elect 1991-92, pres. 1992—), Phi Beta Kappa, Sigma Xi. Home: 169 Westminster Rd Brooklyn NY 11218-3445 Office: NY Med Coll Westchester County Med Ctr Valhalla NY 10595

LEVY, ROBERT ISAAC, physician, educator, research director; b. Bronx, N.Y., May 3, 1937; s. George Gerson and Sarah (Levinson) L.; m. Ellen Marie Feis, 1958; children: Andrew, Joanne, Karen, Patricia. B.A. with high honors and distinction, Cornell U., 1957; M.D. cum laude, Yale U., 1961. Intern, then asst. resident in medicine Yale-New Haven Med. Ctr., 1961-63; clin. assoc. molecular diseases Nat. Heart, Lung and Blood Inst., Bethesda, Md., 1963-66, chief resident, 1965-66, attending physician molecular disease br., 1965-80, head sect. lipoproteins, 1966-80, dep. clin. dir. inst., 1968-69, chief clin. services molecular diseases br., 1969-73, chief lipid metabolism br., 1970-74, dir. div. heart and vascular diseases, 1973-75, dir. inst., 1975-81; v.p. health scis., dean Sch. Medicine Tufts U., Boston, 1981-83, prof. medicine, 1981-83; v.p. health scis. Columbia U., N.Y.C., 1983-84, prof., 1983—, sr. asst. v.p. health scis., 1985-87; pres. Sandoz Research Inst., East Hanover, N.J., 1988-92; pres. Wyeth-Ayerst Rsch. Wyeth-Ayerst Labs div. Am. Home Products, Phila., 1992—; attending physician Georgetown U. med. div. D.C. Gen. Hosp., 1966-68; spl. cons. anti-lipid drugs FDA. Editor: Jour. Lipid Rsch., 1972-80, Circulation, 1974-76, Am. Heart Jour., 1980-90; contbr. articles to profl. jours. Served as surgeon USPHS, 1963-66. Recipient Kees Thesis prize Yale U., 1961; Arthur S. Flemming award, 1975; Superior Service award HEW, 1975; Rsch. award and Van Slyke award Am. Soc. Clin. Chemists, 1980; Roger J. Williams award, 1985; award Humana Heart Found., 1988. Mem. Am. Heart Assn. (mem.-at-large exec. com. council basic sci., mem. exec. council on atherosclerosis), Am. Inst. Nutrition, Am. Fedn. Clin. Research, N.Y. Acad. Scis., Am. Soc. Clin. Nutrition, Am. Soc. Clin. Investigation, Am. Coll. Cardiology, Inst. Medicine of Nat. Acad. Scis., Am. Soc. Clin. Pharmacology and Therapeutics, Assn. Am. Physicians, Phi Beta Kappa, Sigma Xi, Alpha Omega Alpha, Alpha Epsilon Delta, Phi Kappa Phi. Office: Wyeth-Ayerst Labs PO Box 8299 Philadelphia PA 19101

LEVY, RONALD, medical educator, researcher; b. Carmel, Calif.. B.S., Harvard U.; M.D., Stanford U., 1968. Diplomate Am. Bd. Internal Medicine. Intern Mass. Gen. Hosp., Boston, 1968-69; researcher Mass. Gen. Hosp., 1969-70; Helen Hay Whitney Found. fellow in dept. chem. immunology Weizmann Inst. Sci., Rehovot, Israel, 1973-75; mem. faculty Stanford U., Calif., 1975—, now assoc. prof. dept. medicine-oncology. Co-recipient (with G. Telford) 1st award for cancer research Armand Hammer Found. Mem. ACP, Am. Soc. Clin. Oncology. Office: Stanford U Dept Medicine-Oncology Stanford CA 94305

LEVY, RONALD BARNETT, aeronautical engineer; b. Oceanside, N.Y., Apr. 10, 1951; s. Walter and Beatrice (Moskowitz) L.; m. Mary Frances Severn, Aug. 27, 1978; 1 child, Benjamin Julius. BS in Engring., U. Mich., 1972; M Aviation Mgmt., Embry-Riddle Aero. U., 1985. Pilot Ky. Flying Svc., Louisville, 1977-80; weapons systems officer USAF, Cannon AFB, N.Mex., 1980-82, RAF Upper Heyford, U.K., 1982-88; sr. analyst Survice Engring Co., Aberdeen, Md., 1988—. Contbr. articles to profl. publs. Cub scout leader Boy Scouts Am., Bel Air, Md., 1988-91; v.p. Temple Adas Shalon, Havre de Grace, Md., 1992—. With USN, 1973-77; capt. USAFR, 1978-88. Mem. AIAA (sr.), Am. Def. Preparedness Assn., Air Force Assn., B'nai B'rith (pres. local lodge 1992—). Jewish. Achievements include development of methodology for detailed vulnerability analysis of turbine aircraft engines, establishment of Air Force's first integrated command, control, intelligencs and mission planning network system. Office: Survice Engring Co 1003 Old Philadelphia Rd Aberdeen MD 21001

LEVY, SIDNEY, psychologist; b. Bklyn., June 5, 1909; s. Benjamin and Minnie (Zwickel) L.; BS, CCNY, 1932, MS, 1936; PhD, NYU, 1948; m. Estelle Turteltaub, Jan. 22, 1931; 1 son, Richard. Supr. social work N.Y.C. Dept. Welfare, 1941-45; chief psychologist Westover Air Force Hosp., 1944-45; rsch. assoc. Bur. Naval Rsch. , NYU, 1947; pvt. practice psychology, N.Y.C., 1947—; faculty NYU, 1947-67; staff psychologist Northport (N.Y.) Psychiatric Hosp., 1947-52; dir. profl. tng. VA Mental Hygiene Svc., 1950-52; exec. dir. Rsch. Inst. Personality, Psychotherapy and Edn., N.Y.C., 1956-75, Atlantic Inst. Internat. Conflict and Behavioral Analysis, N.Y.C., 1969-75; dir. Profl. Seminars in Diagnosis and Therapy, 1962—. Served in USAAF, 1944-45. Mem. Am. Psychol. Assn., N.Y. State Psychol. Assn., AAAS, AAUP, N.Y. Acad. Sci., Phi Delta Kappa. Originator: Levy Animal Symbol Test, 1944; Little Momsa Technique in Psychotherapy, 1952. Office: 9511 Shore Rd Brooklyn NY 11209-7506

LEVY, STANLEY BURTON, mechanical engineer, marketing executive; b. Providence, Mar. 18, 1939; s. Louis Abraham and Ida Sarah (Lipski) L.; m.

Rhea Lobel, Nov. 19, 1966; children: Faith Michelle, William Todd, Ellen Beth. BSME, U. R.I., 1960; MSME, U. Conn., 1963, PhD, 1966. Registered profl. engr., Del. Engr. GE Gas Turbine, West Lynn, Mass., summer 1963; auto. engr. GM Corp., Detroit, summers 1964-65; from rsch. engr. to sr. rsch. assoc. DuPont Devel. Dept., Wilmington, Del., 1966-90; mktg. mgr. DuPont High Performance Films, Wilmington, 1990-93; owner Clear Solutions, Wilmington, 1993—; bd. dirs. Burton Energy and Solar Tech., North Babylon, N.Y., 1985-86. Contbr. articles to profl. jours. Named Outstanding Am. Inventor Intellectual Property Owners Found., 1989. Mem. ASME, SAE, Soc. Plastics Engrs., Del. Assn. Profl. Engrs. Achievements include 10 patents. Home and Office: Clear Solutions 632 Kilburn Rd Wilmington DE 19803

LEVY, WILLIAM JOEL, endocrinologist; b. Pitts., Mar. 7, 1949; s. Millard Lester and Shirley (Lubovsky) L.; m. Tammey J. Naab; children: Nicole, Natalie, Adam, Alaina. BA, Case Western Res. U., 1971; MD, U. Pitts., 1975. Intern Cleve. Clinic, 1975-76, resident, 1976-78, fellow in endocrinology, 1978-90; chief endocrinology St. Thomas Hosp., Akron, Ohio, 1980-84; endocrinologist in pvt. practice Akron, 1980-84, No. Va., 1985, Suitland, Md., 1986-93; endocrinologist Silver Spring, Md., 1993—. Contbr. articles to profl. jours. Fellow ACP; mem. Am. Diabetes Assn., Am. Thyroid Assn. Office: Ste 328 344 University Blvd W Silver Spring MD 20901

LEW, LAWRENCE EDWARD, chemical engineer; b. Santa Monica, Calif., June 11, 1952; s. Robert Joseph and Evelyn (Joy) L.; m. Doris Schaefer, May 28, 1976; children: Katherine, Henry. BSChE and Petroleum Refining Engring., Colo. Sch. Mines, 1974. Registered profl. engr., Okla. Process/constrn. engr. Natural Resources Group Phillips, Teesside, Eng./Emden, Germany, 1975-78; process engr. corp. engring. Phillips, Bartlesville, Okla., 1978-80, process engr. refining div., 1980-84, prin. engr. catalytic processes, 1984-89, supr. residual oil rsch., 1989-90, tech. planning dir., 1990-91, mgr. process modeling and catalytic cracking hydrotreating br., 1991—. Mem. AICE, Nat. Petroleum Refiners Assn. (screening com. 1988-93). Democrat. Roman Catholic. Achievements include 2 patents. Office: Phillips Petroleum Co 333 PL Bartlesville OK 74004

LEWALD, PETER ANDREW, process engineer; b. Bklyn., Dec. 25, 1968; s. Gunnar and Elke (Lorenzen) L. BS in Chem. Engring., U. Tulsa, 1990. Chem. engr. Conoco, Inc., Ponca City, Okla., 1990-91, process engr., 1991—. Mem. Soc. Petroleum Engrs. (assoc.). Republican. Home: 3605 Bellflower Ave Ponca City OK 74604 Office: Conoco Inc 1000 S Pine Ponca City OK 74601

LEWALSKI, ELZBIETA EWA, mechanical engineer, consultant; b. Warsaw, Poland, Oct. 22, 1933; arrived in Can., 1975; came to U.S., 1983; d. Stefan and Klementyna (Dziewierska) Banaszczyk; m. Marian Zdzislaw, Apr. 6, 1957; 1 child, Wojciech (Bart). BSME, Tech. U. Warsaw, 1955, MS in Aero. Engring., 1957. Registered mech. engr. Mgr. testing lab. Fed. Inst. Aero. Tech., Poland, 1957-65; sr. project engr. Design Ctr. for Steel Constrn., Poland, 1965-69, Fed. Rsch. and Devel. Inst., Poland, 1969-75; designer Hawker Siddley Can. Ltd., 1976-79, Vacudyne Altair and M.H. Detrick Co., U.S., 1979-80; stds. engr. UTDC, Can., 1980-83; approvals engr. Bently Nev. Corp., U.S., 1984-89; pres. Transit Performance Engring., U.S., 1986—; exec. officer Polish Soc. of Mech. and Civil Engrs., 1965-75. Contbr. articles to profl. jours. Mem. Assn. of Profl. Engrs. of the Province of Ont. (exec. bd. 1980-83), State of Nev. Profl. Engrs. Roman Catholic.

LEWANDO, ALFRED GERARD, JR., oceanographer; b. Boston, Apr. 17, 1945; s. Alfred Gerard and Marie Helen (Coughlin) L.; m. Carol Ann Kologe, Nov. 8, 1969; children: Jennifer Ann, Christina Marie. BS in Earth Sci., State Coll. Boston, 1967; MBA, U. So. Miss., 1986, MS in Polit. Sci., 1989, MS in Pub. Rels., 1990, MEd in Adult Edn., 1991. Lic. real estate broker and notary pub., Miss. Staff oceanographer Naval Oceanographic Office, Washington, 1967-76, head fleet support br., 1976-80; dir. tactical analysis div. Naval Oceanographic Office, Bay St. Louis, Miss., 1980-86, dir. oceanographic programs div., 1986-88; dep. asst. chief of staff for ops. Naval Oceanography Command, Stennis Space Ctr., Miss., 1988—; mem. policy bd. Ctr. of Higher Learning, Stennis Space Ctr., Miss., 1990—; mem. adv. com. Cape Fear Jr. Coll., Wilmington, N.C., 1974—, Miss. State U. Rsch. Ctr., 1988—; mem. steering com. Summer Indsl. fellowships for Gulf Coast Tchrs., 1990—; mem. organizing com. 44th Internat. Sci. and Engring. Fair, 1993. Contbr. articles to profl. publs. Commr., City of Long Beach (Miss.) Port Authority, 1986-88. Mem. Miss. Acad. Scis., Gamma Theta Upsilon. Home: 553 Mockingbird Dr Long Beach MS 39560-3105 Office: Naval Oceanography Command Stennis Space Center MS 39529-5000

LEWANDOWSKI, ANDREW ANTHONY, utilities executive, consultant; b. Kiel, Fed. Republic of Germany, Nov. 29, 1946; came to U.S., 1949; s. Kazimierz and Emily (Lewandowski) L.; m. Mary Ann Zuza; 1 child, Adam Christopher. Student, Rutgers U., 1964-66; BS in Mech. Engring., N.J. Inst. Tech., 1969; postgrad., Pa. State U., 1969-70; MS in Mech. Engring., N.J. Inst. Tech., 1973. Registered profl. engr., N.J.; cert. profl. planner, N.J. NSF trainee N.J. Inst. Tech., 1970-72; Engr. I DeLeuw, Cather & Co., Newark, 1970; gas utilities engr. DeLeuw, Cather & Co. of N.Y., Inc., N.Y.C., 1972, specifications writer, 1972-74, chief specifications, 1974-75; supv. engr. Elizabethtown Gas Co., Iselin, N.J., 1976-79; mgr. planning, system improvement Elizabethtown Gas Co., Iselin, 1979-81, mgr. planning, budgets, 1981-86; internal cons., computer mgmt. Elizabethtown Gas Co., Elizabeth, N.J., 1986-87; internal cons. ops., engring. Elizabethtown Gas Co., Iselin, N.J., 1987-89; internal cons. engring., budgets Elizabethtown Gas Co., Union, N.J., 1989—. Editor Jaycee newsletter, 1979-80, local Rep. newsletter, 1986. Den leader, asst. cubmaster Cub Scouts Boy Scouts Am.; active various local govt., religious, polit. and charitable orgns. Recipient Dir. of Yr. award South Plainfield Jaycees, 1972, Disting. Svc. award, 1975, Outstanding Young Man of Yr. award N.J. Jaycees. 1975, South Plainfield Jaycees, 1976; named one of Outstanding Young Men of Am., 1977. Mem. NSPE, ASME, KC, Internat. Platform Assn., South Plainfield Polish Nat. Home. Republican. Roman Catholic. Home: 1910 Murray Ave South Plainfield NJ 07080-4713 Office: Elizabethtown Gas Co 1 Elizabethtown Plz Union NJ 07083-7138

LEWES, KENNETH ALLEN, clinical psychologist; b. Charleston, W.Va., June 8, 1943; s. Joseph and Anne L. BA, Cornell U., 1964; PhD in English, Harvard U., 1972; PhD in Psychology, U. Mich., 1983. Asst. prof. English Rutgers U., New Brunswick, N.J., 1967-75; psychotherapist Orchard Hills Psychiatric Ctr., Novi, Mich., 1983—. Author: The Psychoanalytic Theory of Male Homosexuality, 1988. Woodrow Wilson fellow, 1964-66. Home: 1913 Geddes Ave Ann Arbor MI 48104 Office: Orchard Hills Psychiat Ctr 42450 W Twelve Mile Rd Novi MI 48377

LEWEY, SCOT MICHAEL, gastroenterologist; b. Kansas City, Mo., Sept. 10, 1958; s. Hugh Gene and Janice Vivian (Arnold) L.; m. Julie Ann Williams, July 17, 1982; children: Joshua Michael, Aaron Scot, Rachel Anne. BA in Chemistry, William Jewell Coll., 1980; DO, U. Health Scis., 1984. Resident internal medicine and pediatrics William Beaumont Army Med. Ctr., El Paso, Tex., 1985-89; asst. chief pediatric svc. Irwin Army Hosp., Ft. Riley, Kans., 1989-90; asst. chief dept. medicine Irwin Army Hosp., Ft. Riley, 1990, chief emergency med. svcs., 1990; comdr. F co. 701st support bn. 1st inf. Operation Desert Shield Operation Desert Storm U.S. Army, Saudi Arabia, 1990-91; chief dept. pediatrics Munson Army Hosp., Ft. Leavenworth, Kans., 1991-92, chief dept. medicine, 1992-93; fellow gastroenterology svc. Fitzsimons Army Med. Ctr., Aurora, Colo., 1993—. Named Outstanding Young Man Am., 1982; decorated Bronze Star. Fellow ACP, Am. Acad. Pediatrics; mem. AMA (physician recognition award), Am. Osteo. Assn., Am. Soc. Internal Medicine, Assn. Mil. Osteo. Physicians and Surgeons. Republican. Mem. Christian Ch. Avocations: racquetball, golf, running, genealogy, reading. Office: Ftizsimons Army Med Ctr Gastroenterology Dept Svc 403 W Aurora CO 80045-5001

LEWIN, ANITA HANA, research chemist; b. Bucarest, Romania, Oct. 27, 1935; came to U.S., 1956; d. Efim Chaim and Fima (Schweitzer) Shapira; m. Arie Yehuda Lewin, Sept. 2, 1956; children: Tal Mia, Oren Chaim. BS, UCLA, 1959, PhD, 1963. Postdoctoral assoc. U. Pitts., 1963-64, rsch. asst. prof., 1964-66; from asst. prof. to assoc. prof. Poly. Inst. Bklyn., 1966-74; postdoctoral assoc. Rsch. Triangle Inst., Research Triangle Park, N.C., 1974-75, sr. rsch. chemist, 1975—. Contbr. articles to profl. jours. Democrat. Jewish. Achievements include patents for cocaine receptor binding ligands; isoprenoid phospholipose A2 inhibitors. Office: Rsch Triangle Inst PO Box 12194 Research Triangle Park NC 27709

LEWIN, KEITH FREDERIC, environmental science research associate; b. Riverhead, N.Y., July 20, 1952; s. Fred A. Lewin and Margaret (Kart) Boergesson; m. Barbara Gundersen, Feb. 15, 1975; children: Kimberly, Anita, Robert. BS, Cornell U., 1974. Rsch. assoc. Brookhaven Nat. Lab., Upton, N.Y., 1975—. Contbr. articles to Transactions ASAE, Jour. of the Air and Waste Mgmt. Assn., Environ. Pollution, Vegetatio; co-author: FACE: Free Air CO2 Enrichment for Plant Research, 1993. Charter mem. Riverhead Town Vol. Ambulance, 1978—, chief, 1988-89. Mem. Sigma Xi, Chi Phi. Congregationalist. Achievements include patent pending for Valve Design related to Free-Air Facility Design; chief designer for moveable rainfall exclusion shelters constructed to study acid rain effects on crop plants, free-air CO2 enrichment facilities developed for Dept. Energy funded research in U.S. and also for programs being developed in New Zealand and Switzerland. Office: Brookhaven Nat Lab Bldg 318 Upton NY 11973

LEWIN, KLAUS J., pathologist, educator; b. Jerusalem, Israel, Aug. 10, 1936; came to U.S., 1968; s. Bruno and Charlotte (Nawratzki) L.; m. Patricia Coutts Milne, Sept. 25, 1964; children: David, Nicola, Brumo. Attended, King's Coll. U. London, 1954-55; MB, BS, Westminster Med. Sch. London, Eng., 1959; MD, U. London, 1966. Diplomate Am. Bd. Pathology, Royal Coll. Pathologists (London), lic. Calif. Casualty officer Westminster Med. Hosp., 1960; resident Westminster Hosp. Med. Sch., London, 1960-68; pediatric house physician Westminster Hosp. Med. Sch., Westminster Children's Hosp., 1961; house physician St. James Hosp., Balham, London, 1961; asst. prof. pathology Stanford (Calif.) U., 1970-76; assoc. prof. pathology UCLA, L.A., 1977-80; attending physician Dept. Medicine Gastroenterology divsn. UCLA, Wadsworth, Va., 1978—; prof. pathology UCLA Med. Sch., L.A., 1980—, prof. dept. medicine divsn. gastroenterology, 1986—; resident pathologist clinical chemistry, bacteriology, hematology, blood transfusion, serology, Westminster Hosp. Med. Sch., 1961-62, registrar dept. morbid anatomy, 1962-64, rotating sr. registrar morbid anatomy, Royal Devon, Exeter Hosp., 1964-68; vis. asst. prof. pathology, Stanford U. Med. Sch., 1968-70; vice chmn. pathology UCLA, L.A., 1979-86; pres. L.A. Soc. Pathologists Inc., 1985-86; mem. curriculum com. U. Calif. Riverside, 1977-84; cons. Wadsworth VA Hosp., L.A., carcinoma of esophagus intervention study, Polyp Prevention study, Nat. Cancer Inst., Cancer Preservation Studies br., Bethesda, Md., Sepulveda VA Hosp.; mem. various coms. UCLA in field; rschr. structure, function, pathologic disorders of gastrointestinal tract and liver. Author: (with Riddel R., Weinstein W.) Gastrointestinal Pathology and Its Clinical Implications, 1992; edit. bd. Human Pathology, 1986—, Am. Jour. Surg. Pathology, 1990—; reviewer Gastroenterology and Archives of Pathology; contbr. papers, abstracts, review articles to profl. jours., chpts. in books; lectr., presenter in field. Dir. diagnostic Immonuhistochenistry Lab.; mem. diagnostic surg. Pathology svc. Recipient Chesterfield medal Inst. Dermatology, London, 1966; named Arris and Gale lectr. Royal Coll. Surgeons, London, 1968; Welcome Trust Rsch. grantee, 1968; fellow Found. Promotion Cancer Rsch., Tokyo, 1992. Fellow Royal Coll. Pathologists (Eng.); mem. Pathological Soc. Great Britain, Am. Gastroenterology Soc., Gastrointestinal Pathology Soc. (founder, pres. 1985-86, exec. com., edn. com. 1990—), U.S. Acad. Pathology, Can. Acad. Pathology, Assn. Clin. Pathologists, Pathological and Bacteriological Soc. Great Britain, Internat. Acad. Pathology, L.A. Pathology Soc. (bd. dirs.), Calif. Soc. Pathology (edn. com. 1983—), So. Calif. Soc. Gastrointestinal Endoscopy, Arthur Purdy Stout Soc., Gastrointestinal Pathology Soc. (pres., by-laws com., chmn. edn. com., exec. com.). Avocations: internat. travel, geographic pathology, hiking, swimming. Home: 333 Las Casas Ave Pacific Palisades CA 90272 Office: UCLA Sch Medicine Dept Pathology 10833 Le Conte Ave Los Angeles CA 90024

LEWIS, ALAN JAMES, pharmaceutical executive, pharmacologist; b. Newport Gwent, UK. BSc, Southampton U., Hampshire, 1967; PhD in Pharmacology, U. Wales, Cardiff, 1970. Postdoctoral fellow biomedical sci. U. Guelph, Ont., Can., 1970-72; rsch. assoc. lung rsch. ctr. Yale U., 1972-73; sr. pharmacologist Organon Labs., Ltd., Lanarkshire, Scotland, 1973-79; rsch. mgr. immunoinflammation Am. home products Wyeth-Ayerst Rsch., Princeton, N.J., 1979-82, assoc. dir. exptl. therapeutics, 1982-85, dir., 1985-87, asst. v.p., 1987-89, v.p. rsch., 1989—. Editor allergy sect. Agents & Actions & Internat. Archives Pharmacodynamics Therapy; reviewer Jour. Petroleum Tech., Biochemical Pharmacology, Can. Jour. Physiol. Pharmacology, European Jour. Pharmacology, Jour. Pharm. Sci. Mem. Am. Soc. Pharmacological and Exptl. Therapeutics, Am. Rheumatism Assn., Mid-Atlantic Pharmacology Soc. (v.p. 1991—), Pulmonary Rsch. Assn., Inflammation Rsch. Assn. (pres. 1986-88), Pharm. Mfrs. Assn. Achievements include research in mechanisms and treatment of inflammatory diseases including arthritis and asthma, cardiovascular pharmacology, metabolic disorders, central nervous system pharmacology, osteoporosis. Office: Wyeth-Ayerst Research PO Box CN-8000 Princeton NJ 08543*

LEWIS, BENJAMIN PERSHING, JR., pharmacist, public health service officer; b. Danville, Ky., June 2, 1942; s. Benjamin Pershing Lewis and Juanita Elizabeth (Garner) Applewhite; m. Patricia Marlene Glover, Aug. 7, 1968; children: Laura Denise, Jason Matthew. BS in Pharmacy, Auburn U., 1966, MS in Pharmacy, 1972; PhD in Health Svcs. Mgmt., Century U., L.A., 1989. Registered pharmacist, Ky., Ala. Instr. Auburn (Ala.) U. Sch. Pharmacy, 1972-73, now affiliate asst. prof.; commd. lt. comdr. USPHS, 1976, advanced through grades to capt., 1985; pharmacy officer Bur. Drugs FDA, Rockville, Md., 1976-82, health scientist adminstr. orphan products devel., 1982-87, AIDS coord., 1987-89; spl. asst. to assoc. dir. Ctr. Biologics Evaluation-Rsch. FDA, Bethesda, Md., 1989-92; pharmacist/dir. divsn. Congressional and pub. affairs FDA Ctr. Biologics Evaluation and Rsch., Rockville, Md., 1993—. Co-author: Veterinary Drug Index, 1982; editor: FDA Role in AIDS, 1988, The International Ramifications of Drug Development, 1988, Report of the Criticism Task Force on Career Development, 1989; co-editor: Poliovirus Attenuation: Molecular Mechanisms and Practical Aspects, 1993; contbr. articles to profl. jours. Officer U.S. Army, 1972-76. Recipient letter of commendation FDA, 1984. Mem. Am. Pharm. Assn., Am. Acad. Pharm. Rsch. and Sci., Am. Soc. for Pharmacy Law, Commd. Officers Assn. USPHS, Sigma Xi. Methodist. Home: 24137 Newbury Rd Gaithersburg MD 20882-4009 Office: FDA Ctr Biologics Eval and HFM 12 1401 Rockville Pike Rockville MD 20892-1448

LEWIS, BRIAN KREGLOW, computer consultant; b. Durban, Republic of South Africa, Sept. 2, 1932; s. Arthur Armington and Isabel (Kreglow) L.; m. Mary Helen Kidwell, July 14, 1953; children: Brian E., James A., Charles A., Carol J., Robert E., Sharon H. BS, Ohio State U., 1954; PhD, Tufts U., 1971. Biology tchr. Lincoln-Sudbury (Mass.) Regional High Sch., 1965-66; rsch. assoc. May Inst. for Med. Rsch., Cin., 1971-75; from asst. to assoc. prof. health sci. Grand Valley State U., Allendale, Mich., 1975-81; prin. Lewis Assocs., Sarasota, Fla., 1984—; adj. asst. prof. physiology Cin. Coll. Medicine, 1972-75; assoc. prof. Ponce (Puerto Rico) Sch. Medicine, 1981-84, prof., chmn. physiology, 1987-91. Contbr. revs. and articles to Computer Shopper, Proceedings Soc. Exptl. Biology Medicine, Am. Heart Jour., Atherosclerosis; developer business and ednl. software. Cubmaster, scoutmaster Boy Scouts Am., 1963-78; mem. fin. com., ch. choir St. Andrew Ch., Sarasota, 1984—; active Village Voices, Greenhills, Ohio, 1972-75. Lt. USN, 1954-62. NIH fellow, 1965-71. Mem. Endocrine Soc., Soc. for Study Reproduction, Soc. for Study Fertility, Assn. Ind. Computer Profls. (sec. 1992). Office: 6423 Caracara St Sarasota FL 34241

LEWIS, CEYLON SMITH, JR., physician; b. Muskogee, Okla., July 19, 1920; s. Ceylon Smith and Glenn (Ellis) L.; m. Marguerite Dearmont, Dec. 20, 1943; children: Sarah Lee Lewis Lorenz, Ceylon Smith III, Carol D. Lewis Kast. B.A., Washington U., 1942, M.D., 1945. Diplomate: Am. Bd. Internal Medicine, bd. govs., 1976-82. Intern Salt Lake Gen. Hosp., Salt Lake City, 1945-46; fellow in cardiovascular disease Salt Lake Gen. Hosp., 1950-51; resident in internal medicine VA Hosp., Salt Lake City, 1948-50; practice medicine specializing in internal medicine and cardiology Tulsa, 1951—; mem. staff St. John's Hosp., Tulsa, 1952—; chief of staff St. John's Hosp., 1963, chmn. dept. medicine, 1970-71, mem. teaching staff, 1952—; cons. in internal medicine and cardiology Indian Health Svc., Claremore (Okla.) Indian Hosp., 1956—; vis. staff Hillcrest Med. Ctr., Tulsa, 1953—, St. Francis Hosp., Tulsa, 1963—, Univ. Hosp., Oklahoma City, 1973—; med. dir. St. John's Med. Ctr., Tulsa, 1976-78; asst. clin. prof. medicine Okla. U., 1970-74, assoc. clin. prof., 1975-76; clin. prof. U. Okla. Coll. of Medicine, Tulsa, 1976—, dir. Internat. Studies in Medicine, 1990—; adj. clin. prof. med. scis. U. Tulsa Coll. Nursing, 1978—; mem. Okla. Phsician Manpower Tng. Commn. Contbr. articles to profl. jours. Trustee Coll. Ozarks, 1962-72, chmn., 1964-66; bd. dirs. United Way, Tulsa, 1973-81, St. John Med. Ctr., 1974-84, Okla. Found. Peer Rev., 1973-79, treas., 1973—; bd. dirs. Med. Benevolence Found., 1993—; pres. Tulsa Med. Edn. Found., Inc., 1973-77, 82-84, Presbyn. Med. Mission Fund, Inc., Woodville, Tex., 1973-77, 82-84, bd. dirs. 1972-89; elder 1st Presbyn. Ch., Tulsa, 1959—; chmn. joint com. on health affairs United Presbyn. Ch. in U.S.A., 1969-72. With M.C. AUS, 1946-48. Fellow Royal Australian Coll. of Physicians, 1986; recipient Heart of Yr. award Okla. Heart Assn., 1974, Disting. Svc. award Am. Heart Assn., 1976. Fellow ACP (chmn. bd. govs. 1978-79, bd. regents 1980—, treas. 1983-85, pres.-elect 1985-86, pres. 1986-87, immediate past pres 1987-88, pres. emeritus 1988—), Am. Coll. Cardiology, Am. Coll. Chest Physicians, Royal Australasian Coll. Physicians (hon.); mem. Tulsa County Med. Soc. (pres. 1971), Am. Clin. and Climatological Assn., Okla. Med. Assn. (pres. 1977-78), Inst. Medicine Nat. Acad. Scis. (1984—), AMA, Am. Fedn. Clin. Research, Okla. Soc. Internal Medicine (pres. 1971-72), Am. Soc. Internal Medicine, Tulsa County Heart Assn. (dir. 1952-74, pres. 1956-57), Okla. Heart Assn. (dir. 1952—, pres. 1959-60), Am. Heart Assn. (fellow council clin. cardiology, v.p. 1974-75), Royal Soc. Medicine (London), Action in Internat. Medicine (v.p. 1991—). Clubs: So. Hills Country, Tulsa. Office: 1725 E 19th St Tulsa OK 74104-5424

LEWIS, DANIEL LEE, mechanical engineer; b. Clarion, Pa., July 26, 1964; s. Robert Jon Nathaniel and Lydia Emogene (Brocious) L. BS in Mechanical Engring., Carnegie Mellon U., 1986; MS in Mechanical Engring., Stanford U., 1990. Profl. engr., Calif. Mechanical engr. Lawrence Livermore (Calif.) Lab., 1986—. Contbr. articles to profl. jours. Mem. Am. Soc. Mechanical Engrs. (assoc.). Achievements include development of an automated non-destructive x-ray gauging system for characterizing multiple-element materials using x-ray transmission. Office: Lawrence Livermore Lab 7000 East Ave L-125 Livermore CA 94550

LEWIS, DAVID LAMAR, research microbiologist; b. Tampa, Fla., May 28, 1948; s. Homer Willard and Thelma (Marion) L.; m. Kathy Ann Langford, Sept. 15, 1984; children: Joshua David, Jedediah Keith. BS, U. Ga., 1971, PhD, 1986. Analytical chemist Ciba-Geigy Corp., McIntosh, Ala., 1975-76; microbiologist U.S. EPA, Athens, Ga., 1971-74, rsch. microbiologist, 1977—; mem. grad. faculty and dept. ecology U. Ga., Athens, 1989—. Mem. editorial bd. Soc. Environ. Toxicology and Chemistry, 1991-93; contbr. articles to profl. jours. Mem. Sigma Xi (sec. 1991-93, pres. chpt. 1994-95), Phi Kappa Phi. Mannabaptist. Achievements include a patent for a dolomitic activated carbon filter that uses microbial competition to prevent colonization of the water filter by harmful microbes; the discovery that dental handpieces and their attachments could still transmit hepatitis and AIDS viruses after treatment with germicides , prompting upgrading of sterilization guidelines; development of theory that natural microbial processes, such as nutrient cycling, carbon fixation, and greenhouse gas production depend on interactions between quiescent and active microbial populations to maintain system-level stability. Office: US EPA College Station Rd Athens GA 30613

LEWIS, DON ALAN, civil, sanitary engineer, consultant; b. Louisville, Ky., July 11, 1952; s. Robert Henry and Gennetta (Smith) L.; m. Lidia Maria Scheffs, Apr. 27, 1991. BSCE, U. Ky., 1974. Registered profl. engr., Ky. Project engr. GRW Engrs., Inc., Lexington, Ky., 1974-78; assoc. PDR Engrs., Inc., Lexington, 1978-85; chief engring. svcs. ITT Fed. Svcs., Inc., Kaiserslautern, Germany, 1985-91; supervising engr. Internat. Airports Projects Ministry Def. and Aviation, Riyadh, Saudi Arabia, 1991—. Contbr. to profl. publs. Mem. ASCE, Water and Environ. Fedn. Achievements include design of innovative and alternative technology wastewater and sludge treatment systems throughout Kentucky. Home: 4910 Southern Pky Louisville KY 40214 Office: Internat Airports Projects, PO Box 12531, Riyadh 11483, Saudi Arabia

LEWIS, EDWARD B., biology educator; b. Wilkes-Barre, Pa., May 20, 1918; s. Edward B. and Laura (Histed) L.; m. Pamela Harrah, Sept. 26, 1946; children: Hugh, Glenn (dec.), Keith. B.A., U. Minn., 1939; Ph.D., Calif. Inst. Tech., 1942; Phil.D., U. Umea, Sweden, 1982. Instr. biology Calif. Inst. Tech., Pasadena, 1946-48, asst. prof., 1949-56, prof., 1956-66, Thomas Hunt Morgan prof., 1966-88, prof. emeritus, 1988—; Rockefeller Found. fellow Sch. Botany, Cambridge U., Eng., 1948-49; mem. Nat. Adv. Com. Radiation, 1958-61; vis. prof. U. Copenhagen, 1975-76, 82; researcher in developmental genetics, somatic effects of radiation. Editor: Genetics and Evolution, 1961. Served to capt. USAAF, 1942-46. Recipient Gairdner Found. Internat. award, 1987, Wolf Found. prize in medicine, 1989, Rosenstiel award, 1990, Nat. medal of sci. NSF, 1990, Albert Lasker Basic Med. Rsch. award, 1991, Louisa Gross Horowitz prize Columbia U., 1992. Fellow AAAS; mem. NAS, Genetics Soc. Am. (sec. 1962-64, pres. 1967-69, Thomas Hunt Morgan medal), Am. Acad. Arts and Scis., Royal Soc. (London) (fgn. mem.), Am. Philos. Soc., Genetical Soc. Great Britian (hon.). Home: 805 Winthrop Rd San Marino CA 91108-1709 Office: Calif Inst Tech Div Biology 1201 E California Blvd Pasadena CA 91125

LEWIS, FORBES DOWNER, computer science educator, researcher; b. New Haven, Apr. 15, 1942; s. Taylor Downer Lewis and Clara (Bartholow) Hall. B.S., Cornell U., 1967, M.S., 1969, Ph.D., 1970. Asst. prof. Harvard U., Cambridge, Mass., 1970-75; assoc. prof. SUNY-Albany, 1975-78; assoc. prof. U. Ky., Lexington, 1978-83, prof. computer sci., 1983—. Contbr. articles to profl. jours. Served with U.S. Army, 1960-63. Mem. Assn. for Computing Machinery, Assn. for Symbolic Logis, Soc. for Indsl. Applied Math., IEEE. Office: National Sci Found CISE/CDA Rm 436 1800 G st NW Washington DC 20550

LEWIS, GRAHAM THOMAS, analytical inorganic chemist; b. Broken Hill, NSW, Australia, May 28, 1944; s. Lawrence Michael and Jessica (Ripper) L.; m. Susanne Brook, May 10, 1969; children: Melissa Jane, Simon Matthew, Katrina Jane. BS, U. NSW. Cadet metallurgist North Broken Hill Ltd., 1962-69, asst. chief chemist, 1970, chief assayer, 1970, chief analyst, 1971-76, chief chemist, 1976-85, supt. metall. lab., 1985-89; supt. analytical svcs. Pasminco Mining, Broken Hill, 1989-93; mem. Broken Hill Lead Dust Monitoring Com., 1986-91. Fellow Australasian Inst. Mining and Metallurgy (past br. chmn., local com.); mem. Royal Australian Chem. Inst. (local committeeman 1978-93, br. chmn. 1991-93). Anglican.

LEWIS, GREGORY LEE, gastroenterologist; b. Moberly, Mo., Sept. 21, 1949; s. Claude and Aletha Avonell (Harrison) L.; m. Joan Mary Montello, Dec. 23, 1973; 1 child, Alexander Peter. AB in Chemistry, U. Pa., 1971; MD, Jefferson Med. Coll., 1975. Diplomate Am. Bd. Internal Medicine, Am. Bd. Gastroenterology. Intern in internal medicine Thomas Jefferson U. Hosp., Phila., 1975-76, resident in internal medicine, 1976-78; fellowship in gastroenterology Grad. Hosp., Phila., 1978-80; internist, gastroenterologist Masland Assocs., Carlisle, Pa., 1980-90; gastroenterologist Carlisle Digestive Disease Assocs., 1990—; head div. of gastroenterology Carlisle Hosp., 1981-87, chmn. dept. of medicine, 1986-88, chmn. biomed. ethics com., 1987—, chmn. quality assurance com., 1988-90. Co-chmn. AIDS Task Force, Carlisle, 1989-90; bd. dirs. Cumberland County Coalition for Shelter, Carlisle, 1988-90, Carlisle Hosp. Med. Care Found., 1992-93. Fellow ACP (presenter 1986), Am. Gastroenterology Assn., Phila. Coll. Physicians and Surgeons.; mem. AMA (Physician Recognition award 1986, 89, 92), N.Y. Acad. Scis., Cumberland County Med. Soc. (pres. 1986-88). Democrat. Roman Catholic. Office: Carlisle Digestive Disease 40 Brookwood Ave Carlisle PA 17013-9173

LEWIS, GREGORY WILLIAMS, neuroscientist; b. Seattle, Mar. 3, 1940; s. Delbert Srofe and Eileen Juliann (Williams) L.; m. Stephanie Marie Schwab, Sept. 18, 1966; children: Jeffrey Williams, Garrick Peterson. BS, Wash. State U., 1962, MA, 1965, PhD, 1970. Tchr., rsch. asst. Wash. State U., Pullman, 1965-69; prin. investigator U.S. Army Med. Rsch. Lab., Ft. Knox, Ky., 1970-74; prin. investigator USN Pers. R & D Ctr., San Diego, 1974—, head neurosci. lab., 1980—, leader security systems, 1981-83, head neurosci. projects office, 1987-89, div. head neuroscis., 1989—; cons. in field. Contbr. articles to profl. jours. Bd. dirs., pres. Mesa View Homeowners

Assn., Calif., 1980-82. Capt. U.S. Army, 1967-74. Fellow Internat. Orgn. Psychophysiology; mem. AAAS, Soc. Neurosci., Internat. Brain Rsch. Orgn., N.Y. Acad. Scis., Soc. Psychophysiol. Rsch., Sigma Xi, Alpha Kappa Delta, Delta Chi, Psi Chi. Achievements include neuroelectric research and development for improving prediction of job performance; neuromagnetic research directed toward individual differences and personnel performance; neuroelectric and neuromagnetic data acquisition and analysis; development of neuroelectric-neuromagnetic system for individual identification and impairment of function using artificial neural network analyses. Home: 410 Santa Cecelia Solana Beach CA 92075-1505 Office: US Navy Pers R&D Ctr 53335 Ryne Rd San Diego CA 92152

LEWIS, GWENDOLYN L., sociologist, policy analyst; b. Sweetwater, Tenn., July 26, 1943; d. Robert Martin and Glenna Louise (Parker) L.; m. David Carey Montgomery, July 18, 1987. BA in Math., Reed Coll., 1965; MS in Sociology, San Jose State Coll., 1968; PhD in Sociology, Princeton U., 1975. Asst. prof. sociology U. Pitts., 1973-80; sr. rsch. assoc. Cornell U., Ithaca, N.Y., 1980; dir. premed. edn. project Associated Colls. of Midwest, Chgo., 1981-84; data svcs. officer Nat. Rsch. Coun., Washington, 1984-86; sr. policy analyst Coll. Bd., Washington, 1986-89; assoc. dir. sociology program NSF, Washington, 1989-91; sr. edn. specialist coop. state rsch. svc. USDA, Washington, 1991—; cons. Am. Assn. State Colls. and Univs., Washington, 1991-92; mem. selection com. RJR Nabisco scholars Nat. Assn. State Univs. and Land-Grant Colls., Washington, 1987-89. Co-author: (chpts.) The Federal Government and Higher Education: Traditions, Trends, Stakes and Issues, 1987, Administrative Staff: Salaries and Issues, Higher Education in American Society, 1993; co-author (publ.) Trends in Student Aid: 1980-89, 1989; contbr. articles to profl. jours. Pres. Reed Coll. Alumni Bd. Mgmt., Portland, Oreg., 1992-94; mem. nat. adv. coun. Race and Ethnic Studies Inst., Tex. A&M U., College Station, 1991-92; mem. everyday sci. exhibit com. Mus. of Sci. and Industry, Chgo., 1982-84. Recipient Fulbright-Hayes Faculty Rsch. fellowship, 1976-77, grant NSF, 1986, Josiah Macy Jr. Found. Grant Renewal, 1982, N.Y. State Vocat. Edn. grant, 1980, Dissertation Rsch. grant NSF, 1973. Mem. AAAS, Am. Sociol. Assn. (chmn. status on women com. 1987-89, com. on coms. 1991-93), Assoc. Instl. Rsch., Am. Ednl. Rsch. Assn., Soc. Social Studies Sci. Office: USDA CSRS HEP Rm 350A 14th and Independence Aves Washington DC 20250-2250

LEWIS, JONATHAN JOSEPH, surgeon, molecular biologist; b. Johannesburg, South Africa, May 23, 1958; s. Myer Philip and Maisie (Bagg) L.; m. Nanci Lynn Vicedomini, May 20, 1990. MB BCH, Witwatersrand U., Johannesburg, 1982; PhD, Med. Sch., South Africa, 1990. Registrar in surgery Witwatersrand U. Sch. Medicine, 1982-87; postdoctoral assoc. Yale U. Sch. Medicine, New Haven, Conn., 1987-90; chief resident, surgery Yale U. Sch. Medicine, New Haven, 1990-92; fellow dept. surgery Meml. Sloan-Kettering Cancer Ctr., N.Y.C., 1992—. Contbr. articles to profl. jours. Recipient Sulliman medal Witwatersrand U., 1979, Abelheim medal, 1982, Trubshaw medal Coll. of Surgeons, Johannesburg, 1984, Ohse award Yale U., 1989, 90. Fellow Royal Coll. Surgeons; mem. Am. Soc. Cell Biology, Am. Assn. Cancer Rsch., N.Y. Acad. Scis. Jewish. Achievements include research in oncogenes, growth factors, signal transduction gene therapy. Home: 265 E 66th St Apt 28F New York NY 10021 Office: Meml Sloan-Kettering Cancer Ctr Dept Surgery 1275 York Ave New York NY 10021

LEWIS, LARRY, mathematics educator; b. Blakely, Ga., May 3, 1944; s. Julius and Edith (Brown) L.; m. Cynthia Troupe, July 5, 1986 (div. May 1988). BS, Ala. State U., 1965; MAT, Purdue U., 1969; EdD, Utah State U., 1975. Tchr. Excelsior High Sch., Rochelle, Ga., 1965-67, Sch. City of Gary, Ind., 1967-70, Shaw U., Raleigh, N.C., 1970-71, Ala. State U., Montgomery, 1975-83, Ft. Valley (Ga.) State Coll., 1985—. Martin Luther King fellow Utah State U., 1973-75; NSF grantee Purdue U., 1967-69. Mem. Nat. Coun. Tchrs. of Maths., Ga. Coun. Tchrs. of Maths., Phi Delta Kappa, Omega Psi Phi, Pi Mu Epsilon. Democrat. Mem. Ch. of Christ. Avocations: reading, bowling, pool, music, photography. Home: 701 Orange St Apt 8A Fort Valley GA 31030-3517 Office: Ft Valley State Coll State College Dr Fort Valley GA 31030-3302

LEWIS, NORMAN G., academic administrator, researcher, consultant; b. Irvine, Ayrshire, Scotland, Sept. 16, 1949; Came to U.S., 1985; s. William F. and Agnes H. O. L.; m. Christine I. (div. Oct. 1993); children: Fiona, Kathryn. BSc in Chemistry with honors, U. Strathclyde, Scotland, 1973; PhD in Chemistry 1st class, U. B.C., 1977. NRC postdoctoral fellow U. Cambridge, Can., 1978-80; rsch. assoc. chemistry dept. Nat. Rsch. Coun., Can., 1980; asst. scientist fundamental rsch. divsn. Pulp and Paper Rsch. Inst. Can., Montreal, 1980-82, group leader chemistry and biochemistry of woody plants, grad. rsch. chemistry divsn., 1982-85; assoc. prof. wood sci. and biochemistry Va. Poly. Inst. and State U., Blacksburg, 1985-90; dir. Inst. Biol. Chemistry, Wash. State U., Pullman, 1990—; Cons. NASA, DOE, USDA, NIH, NSF, other industries, 1985—. Mem. editorial bd. Holzforschung, 1986, TAPPI, 1986, 89, Jour. Wood Chemistry and Tech., 1987, Cellulose Chemistry and Tech., 1987—, Phytochemistry, 1992—, Polyphenols Actualities, 1992—; author or co-author over 80 publications: books, articles to profl. jours. Hon. mem. Russian Assn. Space and Mankind. Recipient ICI Merit awards Imperial Chem. Industries, 1968-69, 69-70, 70-71, 71-72, ICI scholar, 1971-73; recipient Chemistry awards Kilmarnock Coll., 1969-70, 70-71; NATO/SRC scholar U. B.C., 1974-77. Mem. Am. Chem. Soc. (at-large cellulose divsn., organizer symposia, programme sub-com. cellulose, paper and textile divsn. 1987-90, editorial bd.), Am. Soc. Plant Physiologists, Am. Soc. Gravitational and Space Biology, Phytochemical Soc. N.A. (phytochemical bank com. 1989—), Chem. Inst. Can. (treas. Montreal divsn. 1982-84, Am. Inst. Chemists and Chem. Inst. Can. Montreal conf. 1982-84), Can. Pulp and Paper Assn., Tech. Assn. of Pulp and Paper Industry, Societe de Groupe Polyphenole, Gordon Rsch. Conf. (vice-chmn. raenewable resources com. 1993—). Presbyterian. Achievements include 2 patents in field. Home: 1710 Upper Dr Pullman WA 99163 Office: Washington State U Inst Biol Chemistry Clark Hall Pullman WA 99164

LEWIS, PAUL LE ROY, pathology educator; b. Tamaqua, Pa., Aug. 30, 1925; s. Harry Earl and Rose Estella (Brobst) L.; m. Betty Jane Bixby, June 2, 1953; 1 child, Robert Harry. AB magna cum laude, Syracuse U., 1950; MD, SUNY, Syracuse, 1953. Diplomate Am. Bd. Pathology. Intern Temple U. Hosp., Phila., 1953-54; resident in pathology Hosp. of U. Pa., Phila., 1954-58, asst. instr., 1957-58; instr. pathology Thomas Jefferson U. Coll. Medicine, Phila., 1958-62, asst. prof., 1962-65, assoc. prof., 1965-75, prof., 1975—; pathologist Thomas Jefferson U. Hosp., 1958—; attending pathologist Meth. Hosp., Phila., 1975—, dir. clin. labs., chmn. dept. pathology, 1975—; pres. Penndel Labs. Inc., Ardmore, Pa., 1974-85; cons. VA Hosp., Coatesvillle, Pa., 1976-85; mem. med. adv. com. ARC Blood Bank, Phila., 1978—. Contbg. author: Atlas of Gastrointestinal Cytology, 1983; contbr. articles to med. jours. 2d lt. USAAF, 1943-46. Fellow Am. Soc. Clin. Pathologists, Coll. Am. Pathologists; mem. AMA, Pa. Med. Soc., Philadelphia County Med. Soc., Internat. Acad. Pathology, Am. Soc. Cytology, Masons, Phi Beta Kappa, Alpha Omega Alpha, Nu Sigma Nu. Republican. Methodist. Avocations: photography, hiking. Home: 521 Baird Rd Merion Station PA 19066 Office: Methodist Hosp Dept Path 2301 S Broad St Philadelphia PA 19148-3594

LEWIS, PAUL MARTIN, engineering psychologist; b. Washington, Apr. 7, 1943; s. Verne Bruce and Margaret (Fuglie) L. BA, Oberlin (Ohio) Coll., 1965; PhD, U. Calif., Berkeley, 1981. Instr. U. Calif., Berkeley, 1981; sr. rsch. scientist Battelle, Richland, Wash., 1981-89; engring. psychologist U.S. Nuclear Regulatory Commn., Rockville, Md., 1989—. Contbr. articles to profl. jours. Mem. ACM, Am. Nuclear Soc. Office: US Nuclear Regulatory Commn 5650 Nicholson Ln Rockville MD 20852

LEWIS, RICHARD VAN, chemist; b. Dallas, June 14, 1951; s. William Elton and Betty Jo (Lay) L.; m. Cathy Elaine Baker, Jan. 21, 1972 (div. Janu 1977); m. Karen Moorman, May 2, 1992; children: Eric Ryan Pampe, Paige Elizabeth Pampe. BS, U. Tex., 1979; MS, E. Tex. State U., 1986. Sr. engring. tech. Mostek Corp., Carrollton, Tex., 1979; assoc. rsch. chemist Arco Oil and Gas Co., Plano, Tex., 1979-83; assoc. rsearcher I U. N. Tex., Denton, 1986-89; staff scientist Radian Corp., Austin, Tex., 1989--. Contbr. articles to profl. jours. Mem. Am. Chem. Soc. Methodist. Achievements include development of methods for determining compunds in process stemming in site. Home: 8402 Candelaria Austin TX 78737 Office: Radian Corp 8501 MoPac Blvd Austin TX 78720

LEWIS, ROY ROOSEVELT, space physicist; b. Richmond, Va., Mar. 4, 1935; s. Jesse NMN and Elizabeth (Lewis) L.; m. Debra Blondell, Sept. 21, 1968 (div. Aug. 1974); 1 child, Roy Jr.; m. Linda Eleanor, Dec. 19, 1985. BS, Va. Union U., Richmond, 1958; MS, Howard U., 1962, UCLA, 1969; PhD, UCLA, 1972. Mem. tech. staff Hughes Rsch. Lab., Malibu, Calif., 1972-75, Aerospace Corp., El Segundo, Calif., 1977-81, TRW, Redondo Beach, Calif., 1981-82; dir. minority engring. Calif. State U., Long Beach, 1982-83; assoc. prof. Calif. State U., 1982-86, Calif. State Polytech. U., Pomona, Calif., 1986-89; faculty fellow Jet Propulsion Lab. Cal. Inst. Tech., Pasadena, Calif., 1987-89; mem. tech. staff Jet Propulsion Lab. Cal. Inst. Tech., 1989-93; pres. Roy Lewis & Assocs., Inglewood, Calif., 1993—. Author: LewLearns, Science Lessons For Children, 1977; contbr. articles to profl. jours. Mem. Am. Soc. Engring. Edn., Nat. Soc. Black Physicists, L.A. Coun. Black Engrs., IEEE, Inglewood Dem. Club, Sigma Xi, Alpha Phi Alpha, Sigma Phi Sigma. Episcopalian. Home: 1401 Overhill Dr Inglewood CA 90302-1346

LEWIS, STUART WESLIE, surgeon; b. Bellefield, Mandeville, Jamaica, Oct. 23, 1938; came to U.S., 1959; s. Phillip Augustus and Ivy Hyacinth (Glegg) L.; m. Cordia L. Beverley; children: Camille, Hope Louise, Denise, Hara. BA, NYU, 1965; MD, Harvard U., 1974. Dir. Youth in Action, Neighborhood Youth Corps, Bklyn., 1966-69; resident in surgery NYU Med. Ctr.-Bellevue Hosp., N.Y.C., 1974-76; resident in surgery SUNY Downstate Med. Sch.-Kings County Hosp., Bklyn., 1976-79, chief surg. resident, 1979-80; pres. Monad Med. Svcs., Bklyn., 1981—; med. dir. N.Y.C. Transit Authority, Bklyn., 1989—. Chair David Dodge Bedford Stuyvesant Neighborhood House, Bklyn., 1990—. Recipient Marcus Garvey award Com. to Honor Marcus Garvey, Bklyn., 1991, Recognition award Cen. Bklyn. Coord. Coun., 1992. Mem. Med. Execs., 100 Black Men. Avocations: reading, computer hacking. Home: 45 E 89th St New York NY 10128

LEWIS, WILLIAM SCHEER, electrical engineer; b. Mt. Vernon, N.Y., Feb. 7, 1927; s. Perley Linwood and Nellie Cora (Scheer) L.; m. Jane Alexander, Feb. 4, 1950 (div. 1972); children: Christopher A., Pamela Scheer Shaw, David Robert; m. Barbara Johnson, June 24, 1972. SB, MIT, 1950, SM, 1950. Registered profl. engr. Mass, N.Y. Sales engr. Gen. Electric Co., Erie, Pa., 1950-53, Morrissey Tractor Co., Burlington, Mass., 1953-56, Hubbs Engine Co., Cambridge, Mass., 1956-57; mgr. contract div. Payne Elevator Co., Cambridge, 1957-69; sales mgr. diversified systems Otis Elevator Co., N.Y.C., 1969-72, mktg. analyst/gen. sales, 1972-73; mgr. vertical transp. Jaros, Baum & Bolles, N.Y.C., 1973-91, ptnr., 1978-91, ret., 1991—. Author: (handbooks) Materials Handling, Freight Elevators, 1985, Building Structural Design-Vertical Transportation, 1987; editor: (monograph) Tall Buildings—Vertical and Horizontal Transportation, 1978. Mem. Wayland (Mass.) Bd. Assessors, 1954-69, chmn., 1963-69. Recipient 1st prize award N.Y. Assn. Consulting Engrs., 1987, Honor award Am. Consulting Engrs. Council, 1987. Mem. ASME. Republican. Unitarian. Avocations: white water canoeing, hunting, fishing, photography. Home: The Chase House RR2 Box 909 Cornish NH 03745-9743 Office: Jaros Baum & Bolles Cons Engrs 345 Park Ave New York NY 10154-0004

LEWITT, MILES MARTIN, computer engineering company executive; b. N.Y.C., July 14, 1952; s. George Herman and Barbara (Lin) L.; m. Susan Beth Orenstein, June 24, 1973; children: Melissa, Hannah. BS summa cum laude, CCNY Engring., 1973; MS, Ariz. State U., 1976. Software engr. Honeywell, Phoenix, 1973-78; architect iRMX line ops. systems, x86 line microprocessors Intel Corp., Santa Clara, Calif., 1978; engring. mgr. Intel, Hillsboro, Oreg., 1978-80, 1981-89, corp. strategic staff, 1981-82; engring. mgr. Intel, Israel, 1980-81; v.p. engring. Cadre Techs., Inc., Beaverton, Oreg., 1989-91; v.p. rsch. and devel. ADP, Portland, Oreg., 1991—; instr. Maricopa Tech. Coll., Phoenix, 1974-75. Contbr. articles to profl. jours. Recipient Engring. Alumni award CCNY, 1973, Eliza Ford Prize CCNY, 1973, Advanced Engring. Program award, Honeywell, 1976, Product of Yr. award Electronic Products Mag., 1980. Mem. IEEE (sr.), IEEE Computer Soc. (voting mem.), Assn. Computing Machinery (voting mem.), Am. Electronics Assn. (exec. com. Oreg. Coun.). Democrat. Avocations: photography, internat. travel, beach walking. Office: Automatic Data Processing 2525 SW 1st Ave Portland OR 97201-4753

LEX, BARBARA WENDY, medical anthropologist; b. Buffalo, N.Y., Mar. 12, 1941; d. Walter Albert and Ruth (Allgrim) L. PhD in Anthropology, Syracuse U., 1969; MPH in Pub. Health, Harvard U., 1982. Asst. prof. Lehigh U., Bethlehem, Pa., 1969-70, W. Mich. U., Kalamazoo, 1970-75; assoc. in psychiatry Harvard Med. Sch., Cambridge, Mass., 1975-87, asst. prof., 1987-90, assoc. prof., 1990-; assoc. in anthropology McLean Hosp. ADARC, Belmont, Mass., 1987-88; assoc. anthropologist McLean Hosp. ADARC, 5, Mass., 1989--; mem. edit. bd. Alcohol and Drug Rsch. Newsletter, 1990, Sci. Matters, 1990; mem. DAAR-1, IRG, Rockville, Md., 1991--. Co-author: The Neurobiology of Ritual Trance, 1979, Alcohol Problems in Special Population, 1992; contbr. articles to profl. jours. Expert witness U.S. Congress Senate Com., Washington, 1988, Ho. of Reps. Com. on Children, Washington, 1990. Recipient Dissertation Completion award Syracuse U., 1969, Faculty Rsch. award W. Mich. U., 1972; N.Y. State Regents scholar, 1957, Syracuse U. Trustee scholar, 1957. Fellow Soc. for Applied Anthropology, Am. Anthrop. Assn.; mem. Am. Pub. Health Assn., Soc. for Study Psychiatry and Culture. Achievements include development of the concept of ritual trance, autonomic t. Office: McLean Hosp ADARC 115 Mill St Belmont MA 02178

LEYBOVICH, ALEXANDER YEVGENY, electrophysicist, engineer; b. Leningrad, Russia, Oct. 31, 1948; came to U.S., 1989; s. Yevgeny Alexander and Maria (Pruslin) L.; m. Maina M. Paykin, Apr. 18, 1975; children: Benjamin A., Arnold A. MD in Electrophysics with honors, Poly. Inst., Leningrad, 1972. Sr. process engr. Granite Prodn. Corp., Leningrad, USSR, 1972-80; lead rsch. engr. Granite Rsch. Inst., Leningrad, USSR, 1980-84; lead process engr. Gosmetr Co., Leningrad, USSR, 1985-89; lab. engr. Tosoh Splty. Metals Divsn., Inc., Grove City, Ohio, 1990—; cons. Tech. Rsch. Corp., McLean, Va.,1 990-91. Editor, reviewer Granite Rsch. Inst., 1980-84; contbr. articles to profl. jours. Recipient Bronze medal Nat. Economy Achievements Exhbn., Moscow, 1976, Achievements cert., 1977. Achievements include 9 patents in microelectronics, electrochemistry, physics, over 50 technical innovations; research in thin-film technology, sputtering, electrical discharges. Office: Tosoh SMD Inc 3600 Gantz Rd Grove City OH 43123

LEYH, GEORGE FRANCIS, association executive; b. Utica, N.Y., Oct. 1, 1931; s. George Robert and Mary Kathleen (Haley) L.; m. Mary Alice Mosher, Sept. 17, 1955; children—Timothy George, Kristin Ann. B.C.E., Cornell U., 1954; M.S. (Univ. fellow), 1956. Structural engr. Eckerlin and Klepper, Syracuse, N.Y., 1956-59; asso. dir. engring. Martin Marietta Corp., Chgo., 1959-63; structural engr. Portland Cement Assn., Chgo., 1963-67; dir. mktg. Concrete Reinforcing Steel Inst., Chgo., 1967-75; exec. v.p. Am. Concrete Inst., Detroit, 1975—; editor jour. Am. Concrete Inst., 1975—. Mem. Planning Commn., Streamwood, Ill., 1960-68; chmn. Lake Bluff (Ill.) Citizens Com. for Conservation, 1972. Recipient Bloem Disting. Service award Am. Concrete Inst., 1972. Mem. ASCE, Am. Soc. Assn. Execs. (chmn. key profl. assns. com. 1989-90), Nat. Inst. Bldg. Scis., Am. Ry. Engring. Assn., Am. Nat. Standards Inst. (bd. dirs. 1986—), Am. Soc. for Concrete Constrn. (bd. dirs. 1984-86), Phi Kappa Phi. Clubs: North Cape Yacht, Lake Bluff Yacht (dir. 1969-74, commodore 1973). Home: 1327 Lone Pine Rd Bloomfield Hills MI 48302-2756 Office: Am Concrete Inst PO Box 19150 22400 W 7 Mile Rd Detroit MI 48219-1885*

LEYLEK, JAMES H., aerospace engineer. With Gen. Electric Aircraft Engrs., Cin. Recipient Gas Turbine award ASME, 1992. Office: Gen Electric Aircraft Engines Mail Drop A-322 1 Newmann Way Cincinnati OH 45215*

LEZAK, MURIEL DEUTSCH, psychology, neurology and psychiatry educator; b. Chgo., Aug. 26, 1927. PhB, U. Chgo., 1947, AM in Human Devel., 1949; PhD in Clin. Psychology, U. Portland. Diplomate Am. Bd. Profl. Psychology, Am. Bd. Clin. Neuropsychology; lic. psychologist, Oreg. Staff psychologist Community Child Guidance Clinic, Portland, Oreg., 1949-53; chief psychologist Clackamas County Child Guidance Clinic, Oregon City, Oreg., 1959-61; counselor, asst. prof. ednl. psychology Portland State Coll., 1961-63; lectr. psychology U. Portland, 1963-66; pvt. practice Portland, 1966-85; prof. psychology U. Oreg., Eugene, 1980-82; prof. neurology, psychiatry and neurosurgery Oreg. Health Scis. U., Portland, 1986—; clin. neuropsychologist VA Hosp., Portland, 1966-85; mem. faculty Am. Acad. Judicial Edn., Washington, 1974-78; hon. vis. prof. West China U. Med. Schs., Chengdu, 1987; lectr. in field. Author: Neuropsychological Assessment, 1976, rev. 2d edit., 1983; editor: Assessment of the Behavioral Consequences of Head Injury, 1989; adv. editor jour. Contemporary Psychology, 1987-93; mem. editorial bd. Arch. Clin. Neuropsychol. 1993—, Rehab. Psychology, 1981-84, Devel. Neuropsychology, 1984—, Neuropsychol. Rehab., 1990—, Neuropsychology, 1992—, Perceptual and Motor Skills, 1991—; contbr. numerous articles to profl. jours., chpts. to books. Mem. adv. coun. Family Survival Project for Brain Damaged Adults, San Francisco, 1982-88; mem. nat. adv. bd. Nat. Head Injury Found., 1982— (Clin. Svc. award 1989); mem. NIH coma data bank project, 1987—; mem. examination com. Am. Assn. State Psychology Bds., 1982-86; cons. Calif. State Athletic Commn., 1991-93; mem. neurosci. behavior and sociology of aging com. Nat. Inst. Aging, 1991—. Pfizer fellow, 1981; grantee VA Med. Ctr., 1972-81, Oreg. Health Scis. U., 1982-87, Nat. MS Soc. and NIH, 1990—. Fellow APA (div. 40, program com. 1982, 83); mem. Internat. Neuropsychol. Soc. (bd. dirs. 1978-81, 83-86, pres. 1987-88), Oreg. Psychol. Assn. (chmn. com. on incorp. 1961-62, chmn. ethics com. 1968-69), Portland Psychol. Assn. (sec.-treas. 1960-61, pres. 1972-73). Home: 1811 SW Boundary St Portland OR 97201-2172 Office: Oreg Health Scis U Dept Neurology L 226 3181 SW Sam Jackson Park Rd 3181 SW Sam Jackson Park Rd Portland OR 97201-3098

LI, HONG, physicist; b. Xiamen, Fujian, China, Feb. 25, 1962; s. Li and Lianbi (Wu) L.; m. Yun-chou Tan, Oct. 17, 1988; 1 child, Amy Yi. MS, U. Wis., Milw., 1984, PhD, 1988. Rsch. asst. Lab. for Surface Studies, U. Wis., Milw., 1984-88; postdoctoral researcher SUNY, Stony Brook, 1988-91; sr. scientist BOC Group Inc., Group Tech. Ctr., Murray Hill, N.J., 1991—. Contbr. over 40 articles to profl. jours. Mem. Am. Phys. Soc., Am. Vacuum Soc., Materials Rsch. Soc. Achievements include demonstration of the photoelectron and auger electron forward scattering effect experimentally; first use of the photoelectron and auger electron to identify metastable fcc and bcc cobalt films; research in surface science, materials physics, chemical and structural analysis, electron spectroscopy and scanning probe microscopy, and thin film technology. Office: Group Tech Ctr/BOC Group In 100 Mountain Ave New Providence NJ 07974-2005

LI, JAMES CHEN MIN, materials science educator; b. Nanking, China, Apr. 12, 1925; came to U.S., 1949; s. Vei Shao and In Shey (Mai) Li; m. Lily Y.C. Wang, Aug. 5, 1950; children—Conan, May, Edward. B.S., Nat. Central U., China, 1947; M.S., U. Wash., 1951, Ph.D., 1953. Research assoc. U. Calif.-Berkeley, 1953-55; supr. Mfg. Chemists Assn. project Carnegie Inst. Tech., Pitts., 1955-56; phys. chemist Westinghouse Electric Co., Pitts., 1956-57; sr. scientist U.S. Steel Corp., Monroeville, Pa., 1957-69; mgr. strength physics Allied Chem. Co., Morristown, N.J., 1969-71; A.A. Hopeman prof. engring. U. Rochester, N.Y., 1971—; vis. prof. Columbia U., N.Y.C., 1964-65, adj. prof., 1965-71; adj. prof. Stevens Inst. Tech., Newark, 1971-72; vis. prof. Ruhr U., Bochum, Fed. Republic Germany, 1978-79. Holder 5 patents:hemisphere Laue camera and pulsed annealing of amorphous metals with applied magnetic fields; author 2 books; contbr. 250 articles to profl. jours. Recipient Alexander von Humboldt award, 1978, Acta Metallurgica Gold medal, 1990. Fellow TMS/AIME (Robert F. Mehl medal and lectr. 1978, Champion H. Mathewson Gold medal 1972, Structural Materials Divsn. luncheon speaker, 1993), ASM Internat. (chmn. materials sci. div. 1982-84), Am. Phys. Soc.; mem. Chinese Soc. for Materials Sci. (Lu Tse-Hon medal, 1988). Office: U Rochester Dept Mech Engring Rochester NY 14627-0133

LI, JIANMING, molecular and cellular biologist; b. Xinxiang, Henan, China, Oct. 23, 1956; came to U.S., 1983; s. Zhonghe and Zhimei (Liang) L.; m. Wei Xiao, Sept. 27, 1982; children: Christina Bo, Tracy. BS, Wuhan U., 1982; PhD, CUNY, 1989. Assoc. rsch. scientist Yale U., New Haven, 1989—. Contbr. articles to Jour. Cell Biology, Gene, Jour. Bacteriology, others. Recipient Swebilius cancer rsch. award Yale U., 1991. Mem. AAAS, Am. Assn. Cancer Rsch., Am. Soc. Microbiology, N.Y. Acad. Scis. Achievements include demonstration of role of granulocyte colony-stimulating factor receptor in leukemia cell differentiation; characterization of induction effect of LiCl on leukemia differentiation, synergistic effects of G-CSF and retinoic acid. Office: Yale U Sch Medicine 333 Cedar St New Haven CT 06510

LI, KAM WU, mechanical engineer, educator; b. China, Feb. 16, 1934; came to U.S., 1959; s. Yang Chung and Oy Lan Li; M.S., Colo. State U., 1961; Ph.D., Okla. State U., 1965; m. Shui Mui Chan, Aug. 30, 1956; children—Christopher, Charles. Asst. prof. mech. engring. Tex. A&I U., 1965-67; assoc. prof. N.D. State U., Fargo, 1967-73, prof., 1973—, assoc. dean engring. and arch., 1989-91; cons. Charles T. Main Inc., Boston, 1973-80, Center for Profl. Advancement, East Brunswick, N.J., 1982—. Recipient cert. appreciation U.S. Navy, 1974; NSF fellow, 1966; Ford Found. fellow, 1972. Mem. ASME, N.Y. Acad Scis., Sigma Xi, Tau Beta Pi, Pi Tau Sigma, Kappa Mu Epsilon. Author: Power Plant System Design. Contbr. numerous articles to profl. jours.; govt. engring. research, 1965—. Home: 2516 18th St S Moorhead MN 56560-4811 Office: ND State U University Ave Fargo ND 58105

LI, LINXI, biomedical scientist; b. Renshou, Sichuan, People's Republic of China, Aug. 14, 1956; came to U.S., 1986; s. Maorong Li and Qingyan Zhu; m. Ming Lei, Dec. 18, 1982; children: Lily Yu, Kevin Guohong, Jonathan Guozhi. MD, WEst China U. Med. Scis., Chengdu, Sichuan, 1982; PhD, Bowman Gray Sch. Medicine, 1990. Instr. West China U. med. Scis., 1982-86; postdoctoral fellow Bowman Gray Sch. Medicine, Winston-Salem, N.C., 1986-87, rsch. asst. prof., 1991—. Travel fellow 3d Internat. Meeting Neurotransplantation, 1989. Mem. Assn. for Rsch. in Vision and Ophthalmology, Internat. Congress Eye Rsch., Soc. for Neurosci., Sigma Xi. Achievements include development surgical technique for transplantation of retinal pigment epithelial cells in an animal model which successfully rescue photoreceptor cells fro degeneration in an inherited retinal disease. Office: Bowmen Gray Sch Medicine Medical Ctr Rd Winston Salem NC 27157

LI, LIQIANG, physicist; b. Lai Yuan, HeBei, China, Jan. 17, 1963; s. Xi Ruo Li and Feng Mei (Xiaoshi) Zhang. BS, HeBei Engring. Inst., TianJin, China, 1982; MS, Va. Tech, 1990. Grad. diploma in electro-optics. Grad. student Electro. Tech. U. of China, Chengdu, China, 1982-85; rsch. engr. Guilin Inst. of Optical Communication, Guilin, China, 1985-88; PhD candidate U. Tenn. Space Inst., Tullahoma, 1990—. Home: 406 Brandywine Apts Tullahoma TN 37388 Office: Ctr for Laser Applications MS-35 U Tenn Space Inst Tullahoma TN 37388

LI, YAO, science educator; b. Beijing, China, July 19, 1958; came to U.S., 1982; s. Dao-Zeng Li and Qin (Shi) Shi; m. Hong-Bing Chen, Sept. 20, 1984; children: Calvin C., Monica C. ME, CCNY, 1984, PhD, 1987. Asst. prof. CCNY, 1987-90, assoc. prof., 1991—; cons. SCS Telecom Inc., Port Washington, N.Y., 1990—, Zybron, Inc., Dayton, Ohio, 1989—, NEC Rsch. Inst., 1991. Contbr. over 60 articles to profl. jours. Grantee NSF, Washington, 1990, AFOSR/USAF, Washington, 1987, 88, 90, 91, DARPA, 1990, 91. Mem. IEEE, Optical Soc. Am., Soc. for Photo-Optical Instrumentation Engrs. Achievements include 4 patents in field and 3 patents pending. Home: 32 Essex Dr South Brunswick NJ 08852 Office: CCNY Convent Ave & 138th St New York NY 10031-9198 also: NEC Rsch Inst 4 Independence Way Princeton NJ 08540

LI, YIPING (Y.P.), applied mathematics educator; b. Taishan, Guangdong, China, Sept. 30, 1943; s. Keung and Qiongzhen (Yu) L.; m. Shouzhuang Liang, Jan. 15, 1973; 1 child, Linning. Grad., Xian Jiaotong U., People's Republic China, 1966; MS, U. Washington, Seattle, 1983, PhD, 1987. Technician, engr. Chongqing Br. Sci. Rsch. Inst. Communications, Peoples Republic China, 1967-80; teaching asst., rsch. asst. U. Washington, Seattle, 1982-87; postdoctoral rsch. assoc. U. Wash., Seattle, 1987; lectr. Zhongshan U., Peoples Republic China, 1987-88; assoc. prof. Zhongshan U., 1988—, U. Macau, 1990—. Author: Applications of Computers on Structure Analysis,

1980; contbr. articles to profl. jours. Mem. Soc. for Indsl. and Applied Math., Am. Math. Soc., Guangzhou Soc. for Indsl. and Applied Math. (coun. 1990—). Home: 622-202 Puyuan Dist, Guangzhou 510275, China Office: U Macau Faculty Sci-Tech, University Dr, PO Box 3001, Taipa Macau

LI, YONGHONG, physics researcher; b. Hubel, China, Mar. 15, 1966; came to U.S., 1988; s. Chuanshen and Ben Wei (Liu) L.; m. Rui Zhang, Sept. 13, 1990; 1 child, Ruth. BS, Fudan (China) U., 1988; MS, Columbia U., 1990. Rsch. asst. Harvard U., Cambridge, Mass., 1991—. Mem. Am. Phys. Soc. Achievements include research on mechanism of high temperature superconductors. Office: Harvard U Chemistry Dept Box 182 Cambridge MA 02138

LI, ZILI, research physicist; b. Jinan, Shandong, China, Feb. 8, 1956; came to U.S., 1984; s. Luping and Yi (Liu) L.; m. Min Jiang, May 24, 1984; children: Jeanne, Ray. BS in Physics, Shandong U., Jinan, 1982; PhD in Physics, Case Western Res. U., 1991. Teaching asst. dept. phsycis Shandong U., 1982-84; rsch. asst. Case Western Res. U., Cleve., 1984-90; mem. rsch. staff Liquid Crystal Inst., Kent (Ohio) State U., 1991—. Contbr. articles to Applied Physics Letters, Molecular Crystal and Liquid Crystal, Phys. Rev., Jour. Chem. Physics, Phys. Rev. Letter. Recipient rsch. award Am. Lung Assn., 1985. Mem. Am. Phys. Soc., Internat. Liquid Crystal Soc. Achievements include discovery of novel nematic electroclinic effect and the first direct measurement of anchoring strength in smectic A phrase; research on magneto and electroptic studies of liquid crystals, liquid crystal-polymer dispersions, structural transition in liquid crystals, interfacial phenomena between liquid crystals and substrates; and ferroelectric liquid crystals. Office: Liquid Crystal Inst Kent State U Kent OH 44242

LIAKISHEV, NIKOLAI PAVLOVICH, metallurgist, materials scientist; b. Orel, Russia, Oct. 5, 1929; s. Pavel Vasilievich Liakishev and Dari Jakovlevna Liakisheva; m. Maija Leonidovna Diomkina, June 18, 1960; children: Natalia Nikolaevna, Tatiana Nikolaevna. Grad., Moscow Inst. Steel and Alloys, 1954, DSc, 1974; Candidate of Sci., Bardin Inst., Moscow, 1965. Cert. metallurgy engr. Sci. worker I.P. Bardin Ctrl. Res., Moscow, 1954-67; lab. head Inst. Ferrous Metallurgy, 1967-74, vice-dir., 1974-75; dir. A.A. Baikov Inst. Metallurgy, Moscow, 1975-87, Russian Acad. Scis., Moscow, 1987—; chmn. qualification coun. Russian Govt. Commn., Moscow. Author: Niobium in Steels and Alloys, Theory and Technology of Ferroalloys Production; editor: Metallurgy: Problems and Decisions, New Metallurgical Processes and Materials; editor-in-chief (jours.) Zavodskaia Laboratiria, Fizika i Khimia, 1979—, Obrabotki Mater., 1992—. Pres. USSR-Luxemburg Friendship Soc., Moscow, 1987-92. Recipient State Premium, USSR, 1970, Lenin Premium, 1976. Mem. Russian Acad. Scis. (chmn. coun. metallurgy and materials 1975—, mem. presidium 1993—), Internat. Union Metallurgists (mem. presidium 1992—). Achievements include patent for Process for Combined Production of Ferrosilicozirconium and Zirconium Corundum; invention of various methods of fabrication of steels, alloying with niobium, vanadium, chromium, zirconium and other elements. Office: A A Baikov Inst Metallurgy, Leninsky Prospect # 49, 117334 Moscow Russia

LIANG, JASON CHIA, research chemist; b. Beijing, Peoples Republic China, Feb. 24, 1935; came to U.S., 1978, naturalized 1984; s. Tsang Truan and Shulin (Tang) L.; m. Joan Chorng Chen, June 11, 1960; children: Cheryl, Chuck. BS in Pharm. Chemistry, U. Beijing, 1957; postgrad., Pharm. Research Instn., Beijing, 1961; MS in Organic Chemistry, U. Oreg., 1980. Chemist Beijing Chem. Factory, 1961-71; rsch. chemist Beijing Pharm. Factory, 1971-78; rsch. chemist Tektronix Inc., Beaverton, Oreg., 1980-85, sr. rsch. chemist, 1985-88; sr. rsch. chemist Kalama (Wash.) Chem. Inc., 1988—; presenter Internat. Pitts. Conf. on Analytical Chemistry and Applied Spectroscopy, 1988. Contbr. articles to profl. jours.; patentee in field. Fellow Am. Inst. Chemists; mem. Am. Chem. Soc. (organic chemistry divsn., paper presenter 1984-93), Internat. Union Pure and Applied Chemistry (affiliate). Office: Kalama Chem Inc 1296 NW 3D St Kalama WA 98625

LIANG, JEFFREY DER-SHING, retired electrical engineer, civil worker; b. Chungking, China, Oct. 25, 1915; came to U.S., 1944, naturalized, 1971; s. Tze-hsiang and Sou-yi (Wang) L.; m. Eva Yin Hwa Tang, Jan. 2, 1940; 1 child, Shouyu. BA, Nat. Chengchih U., Chungking, 1940; BAS, U. B.C., Vancouver, 1960. Office asst. Ministry of Fgn. Affairs, Chungking, 1940-43; vice consul Chinese consulate Ministry of Fgn. Affairs, Seattle, 1944-50; consulate-gen. Ministry of Fgn. Affairs, San Francisco, 1950-53; consul Chinese consulate-gen. Ministry of Fgn. Affairs, Vancouver, 1953-56; engr.-in-tng. Can. Broadcasting Corp., Vancouver, 1960-65; assoc. engr. Boeing Co., Seattle, 1965-67, rsch. engr., 1967-70, engr., 1970-73, sr. engr., 1973-75, specialist engr., 1975-78; cons. Seattle, 1979-81. Mem. chancelor's cir. Wesbrook Soc. U. B.C., Vancouver, 1986—, Seattle-King County Adv. Coun. on Aging, 1984-88, Gov.'s State Coun. on Aging, Olympia, 1986-88, Pres. Coun., Rep. Nat. Com.; permanent mem. Rep. Nat. Senatorial Com., Washington State Rep. Party, Seattle Art Mus.; life mem. Am. Assn. Individual Investors, Mutual Fund Investors Assn., Rep. Presdl. Task Force; sustaining mem. Rep. Nat. Congl. Com., Rep. Presdl. Adv. Com.,. Mem. IEEE (life), Heritage Found., Hwa Sheng Chinese Music Club (v.p. 1978-79, chmn. nomination com. 1981-88), Pacific West Clubs, Health and Tennis Corp. Am. Republican. Mem. Christian Ch. Avocations: Chinese calligraphy, opera, poetry, physical fitness. Home: 2428 158th Ave SE Bellevue WA 98008-5416

LIANG, JUNXIANG, aeronautics and astronautics engineer, educator; b. Hangzhou, Zhejiang, China, Aug. 17, 1932; s. Yigao and Yunruo (Yu) L.; m. Junxian Sun, Jan. 27, 1960; 1 child, Song Liang. Grad., Harbin Inst. Tech., 1960. Head control dept. Shenyang (Liaoning, China) Jet Engine R&D Inst., 1960-70, China Gas Turbine Establishment, Jiangyou, Sichuan, China, 1970-78; assoc. chief engr. China Gas Turbine Establishment, Jiangyou, 1978-83; vis. scientist MIT, Cambridge, Mass., 1984-86; prof. China Aerospace Inst. Systems Engring., Beijing, China, 1986—; grads. supr. Beijing U. Aero-Astronautics, Beijing, 1986—; chief engr. Full Authority Digital Elec. Engine Control China Aerospace Industry Ministry, Beijing, 1986—; mem. China Aerospace Sci. and Tech.Com., Beijing, 1983—, Aero-engine R&D Adv. Bd., Beijing, 1991—; bd. dirs. China Aviation Ency. Editorial Bd., Beijing, 1991—. Author: Nonlinear Control System Oscillation, 1964; contbr. articles to Jour. Aeronautics and Astronautics, Jour. Propulsion Tech., Internat. Aviation, Acta Aeronautica et Astronautica Sinica. Recipient Nat. Sci. and Tech. 2d award, China Nat. Sci. and Tech. Com., Beijing, 1965, Sic. and Tech. Progress award, China Aerospace Industry Ministry, 1991, Nat. Outstanding Sci. and Tech. Contbn. award, 1992. Mem. AIAA, Chinese Soc. of Aeronautical, Astronautical Engine Control (mem. commn. 1987—). Achievements include solution of oscillation problem on nonlinear control system; formulation of overall strategy, study and control of High Thrust/Weight Engine Rsch. Program. Office: China Aerospace Inst Systems Engring, Beiyuan-Dayuan #2 An-Wai, Beijing 100012, China

LIANG, NONG, chemist, researcher; b. Tianjin, Peoples Republic of China, Mar. 20, 1958; came to U.S., 1982; parents Shaoli Liang and Huiwen Wang; m. Cheng Liu, Nov. 8, 1985. BS, Nankai U., Tianjin, 1982; PhD, Northwestern U., 1988. Postdoctoral fellow Argonne (Ill.) Nat. Lab., 1988-91; rsch. staff scientist Procter & Gamble Co., Cin., 1991—. Contbr. articles to profl. jours. Mem. AAAS, Am. Chem. Soc. Achievements include first experimental demonstration of the quantum mechanical nature of Marcus' inverted effect; development of noval methodologies; pioneering research in understanding of long-range biological electron transfer reactions; research into effect of phylogenetically conserved residue-82 of Cc in long-range electron tunneling within protein complexes. Office: Procter & Gamble Co HC Box B38 SWTC Cincinnati OH 45241-9974

LIANG, VERA BEH-YUIN TSAI, psychiatrist, educator; b. Shanghai, China, July 29, 1946; came to U.S., 1970, naturalized, 1978; d. Ming Sang and Mea Ling Chu Tsai; m. Hanson Liang, Nov. 6, 1971; children: Eric G., Jason G. MBBS, U. Hong Kong, 1969. Diplomate Am. Bd. Psychiatry and Neurology. Intern Cambridge Hosp. (Mass.), 1970-71; resident Hillside div. L.I. Jewish Med. Ctr., New Hyde Park, N.Y., 1971-73; fellow Albert Einstein Coll. Medicine, Bronx, N.Y., 1973-75, asst. clin. prof., 1989—; instr. SUNY, Bklyn., 1975-79; asst. prof. SUNY, Stony Brook, 1979-89; med. dir. Hillside Ea. Queens Ctr., Queens Village, N.Y., 1987-90, 91-92; staff child psychiatrist Schneider Children's Hosp., New Hyde Park, N.Y., 1990-92; sr. psychiatrist South Oaks Hosp., Amityville, N.Y., 1992—; cons. in field. Contbr. articles to profl. jours. Mem. Am. Psychiat. Assn., Am. Acad. Child Psychiatry. Office: South Oaks Hosp 400 Sunrise Hwy Amityville NY 11701

LIAO, CHUNG MIN, agricultural engineer, educator; b. Taipei, Taiwan, China, Apr. 10, 1957; s. Kuo Jung and Soong May (Cheng) L.; m. Tehsin Kuan, Jan. 5, 1991. BS, Chung Yuan U., Chungli, 1980; MS, Nat. Taiwan U., Taipei, 1984; PhD, Iowa State U., 1988. Asst. scientist Agrl. Engring. Rsch. Ctr., Chungli, Taiwan, 1984-85; postdoctoral fellow U. Alberta, Edmonton, Can., 1988-89; assoc. fellow Nat. Taiwan U., Taipei, 1989-90; assoc. prof. Nat. I-Lan (Taiwan) Inst. Agr. and Tech., 1990-92, Nat. Taiwan U., 1992—. Contbr. articles to profl. jours. Recipient Academic Rsch. award Nat. Sci. Coun. 1990. Mem. ASHRAE, Am. Soc. Agrl. Engrs., Chinese Soc. Agrl. Engrs. Avocations: philosophy, science, mathematics, music, movies. Home: 6 Lane 20 Kuang Chow St, Hsinchu 30026, Taiwan Office: Nat Taiwan U, Dept Agrl Engring, Taipei 10264, Taiwan

LIAO, HSIANG PENG, chemist; b. Nanping, China, May 4, 1924; came to U.S., 1948, naturalized, 1953; s. Samuel and Chung (Chang) L.; B.S., Fukien Christian U., 1945; Ph.D., Northwestern U., 1952; m. Chen Hansing, Jan. 6, 1950; children—Jacob, Wesla Mildred, Michael Lawrence. Chemist, Standard Oil Co., Ind., 1952-60; research chemist FMC Corp., Balt., 1960—, research asso., Princeton, N.J., 1980—; lectr. Fukien Christian U., 1945-47; grad. seminar lectr. Johns Hopkins, W.Va. U., 1961. Mem. AAAS, Am. Chem. Soc. Sigma Xi, Alpha Chi Sigma, Phi Lamda Upsilon. Methodist. Patentee in field. Home: 260 Fisher Pl Princeton NJ 08540-6444 Office: FMC Corp PO Box 8 Princeton NJ 08540

LIAO, KEVIN CHII WEN, data processing executive; b. Taiwan, Republic of China, Feb. 2, 1937; s. Zuway-Ching and Ah-Chin Liao; m. Shirley Huay-Ling Ni, Jan. 1, 1966; children: Eliza Yuh-Yun, Sabrina Yuh-Chi. BA, Nat. Taiwan U., 1959. Acct. Taiwan Sugar Corp., Taipei, 1961-68; mgr. programming Taxation Reform Com. div Exec. Yuan Republic of China, Taipei, 1968-74; dir. info. systems dept. China Steel Corp., Taipei, 1974-81, Ministry of Econ. Affairs, Taipei, 1981-90; pres. China Data Processing Ctr., Taipei, 1990—. Named Outstanding Person in Info. Industry, Republic of China, 1985. Mem. Computer Soc. (bd. dirs.), Info. Mgrs. Soc. (bd. dirs.), Value Added Network Assn. (bd. dirs.), Chinese Open System Assn., Mgr. Sci. Assn. Home: 41 Ln 22 Kuo-Sing St, Shih-Chih, Taipei Taiwan Office: China Data Processing Ctr, 1 Roosevelt Rd Sec 1, Taipei 10757, Taiwan

LIAW, HAW-MING (CHARLES), physicist; b. Hua-lien, Taiwan, China, Jan. 20, 1942; came to the U.S., 1967; s. Tzu-Wei and Chu-Mei (Wang) L.; m. Susan Karwoski (div. 1972); m. Sandra J. Gabler, 1978. BS in Physics, Taiwan Nat. Normal U., 1965; MS, U. Lowell, 1971; PhD in Biophysics, Boston U., 1979. Researcher Boston U., 1975-80; postdoctoral scholar U. Calif., San Francisco, 1980-81; sr. scientist Lockheed Calif. Co., Burbank, 1981-82; systems engr. Hughes Aircraft Co., El Segundo, Calif., 1982-84; prin. engr., project mgr. Northrop Corp., Anaheim, Calif., 1984-86; project mgr. Jet Propulsion Lab., Pasadena, Calif., 1986-89; prin. scientist Rockwell Internat. Corp., El Segundo, 1989—; chair SPIE Symposia, L.A., 1987-89; conf. chair IRIS Confs., Washington, 1989-92. Contbr. articles to profl. jours. Achievements include development of several new sensor systems for military and civilian applications. Home: 2925 Poplar Blvd Alhambra CA 91803 Office: Rockwell Internat Corp 201 N Douglas El Segundo CA 90245

LIBBRECHT, KENNETH, astronomy educator. Prof. Calif. Inst. Tech., Pasadena. Recipient Newton Lacy Pierce prize Am. Astron. Soc., 1991. Office: California Inst Technology Pasadena CA 91125*

LIBCHABER, ALBERT JOSEPH, physics educator; b. Paris, Oct. 23, 1934; came to U.S., 1983; s. Charles and Cyrla (Markowska) L.; m. Irene Gelman, Sept. 11, 1955; children—Jacques, Remy, David. B.S., U. Paris, 1956; M.S., U. Ill., 1959; Ph.D., Ecole Normale Superieure, Paris, 1965. Matre de Recherche CNRS, Ecole Normale, Paris, 1967-74, dir. research, 1974-83; prof. physics U. Chgo., 1983-91, Paul Snowden Disting. Svc. prof., 1987—; prof Dept. Physics Princeton (N.J.) U., 1991—. Served as officer French Army, 1959-61. Recipient Wolf prize Wolf Found., Herzlia, Israel, 1986, MacArthur Found. Fellow, 1986-91. Fellow NEC Rsch. Inst. Princeton; mem. Am. Phys. Soc., French Phys. Soc. (Silver medal 1971, prix Ricard 1979), Am. Acad. Arts and Scis. Jewish. Home: 1640 E 50th St Chicago IL 60615-3161 Office: Princeton U Dept Physics Jadwin Hall PO Box 708 Princeton NJ 08544 also: NEC Rsch Inst Inc 4 Independence Way Cranbury NJ 08512

LIBERMAN, IRVING, optical engineer; b. N.Y.C., June 24, 1937; m. Lois Hrabar, May 9, 1970. BEE, CCNY, 1958; PhD, Northwestern U., 1965. Sr. engr. through program mgr. Westinghouse STC, Pitts., 1963—; indsl. staff mem. Los Alamos (N.Mex.) Nat. Lab., 1974-80. Contbr. articles to profl. jours., chpts. to books. Recipient IR100 award R&D Mag., 1969. Mem. Am. Inst. Physics, Optical Soc. Am., Sigma Xi, Tau Beta Pi, Eta Kappa Nu. Achievements include 5 patents on lasers and atomic clocks. Home: 125 Westland Dr Pittsburgh PA 15217 Office: Westinghouse STC 1310 Beulah Rd Pittsburgh PA 15235

LIBERMAN, MICHAEL, metallurgist, researcher; b. Gomel, USSR, June 24, 1938; came to U.S., 1980; s. David Liberman and Tsivya Goldberg; m. Alla Itkin, Feb. 23, 1963; 1 child, Vladimir. MS, Poly. Inst., Minsk, USSR, 1961; PhD, Poly. Inst., Leningrad, USSR, 1974. Process engr. Machinery Plant, Gomel, 1961-63; sr. scientist Inst. of Alumnium, Leningrad, 1963-79; sr. rsch. assoc. N.J. Inst. Tech., Newark, 1980-81; metall. engr. Exxon Enterprises, Greer, S.C., 1981; metallurgist Steel Heddle Mfg. Co., Greenville, S.C., 1981-82, tech. svcs. mgr., 1982-87, chief metallurgist, 1987-89; R&D mgr. Memtec Am. Corp., Deland, Fla., 1989—; cons. Siege Corp., Greenville, 1981-83. Contbr. 30 articles to profl. jours.; patentee in field. Mem. ASTM, Am. Soc. Metals (abstractor), The Metall. Soc. Avocations: chess, physical fitness, gardening. Home: 1505 Covered Bridge Dr De Land FL 32724-7931

LIBERMAN, ROBERT PAUL, psychiatry educator, researcher, writer; b. Newark, Aug. 16, 1937; s. Harry and Gertrude (Galowitz) L.; m. Janet Marilyn Brown, Feb. 16, 1973; children: Peter, Sarah, Danica, Nathaniel, Annalisa. AB summa cum laude, Dartmouth Coll., 1959, diploma in medicine with honors, 1960; MS in Pharmacology, U. Calif.-San Francisco, 1961; MD, Johns Hopkins U., 1963. Diplomate Nat. Bd. Med. Examiners, Am. Bd. Psychiatry and Neurology; cert. community coll. instr., Calif. Intern Bronx (N.Y.) Mcpl. Hosp.-Einstein Coll. Medicine, 1963-64; resident in psychiatry Mass. Mental Health Ctr., Boston, 1964-68; postdoctoral fellow in social psychiatry Harvard U., 1966-68, teaching fellow in psychiatry, 1964-68; mem. faculty group psychotherapy tng. program Washington Sch. Psychiatry, 1968-70; with Nat. Ctr. Mental Health Svc., Tng. and Rsch., St. Elizabeths Hosp., also mem. NIMH Clin. and Rsch. Assocs. Tng. Program, Washington, 1968-70; asst. clin. prof. psychiatry UCLA, 1970-72, assoc. clin. prof., 1972-73, assoc. rsch. psychiatrist, 1973-76, rsch. psychiatrist, 1976-77, prof. psychiatry, 1977—; dir. Camarillo-UCLA Clin. Rsch. Unit, 1970—, UCLA Clin. Rsch. Ctr. Schizophrenia and Psychiat. Rehab., 1977—; chief Rehab. Medicine Svc., West L.A. VA Med. Ctr., Brentwood divsn., 1980-92; cons. div. mental health and behavioral scis. edn. Sepulveda (Calif.) VA Hosp., 1975-80; practice medicine specializing in psychiatry, Reston, Va., 1968-70, Thousand Oaks, Calif., 1977—; staff psychiatrist Fairfax Hosp., Falls Church, Va., 1968-70, Ventura County Mental Health Dept., 1970-75; staff psychiatrist Ventura County Gen. Hosp.; mem. med. staff UCLA Neuropsychiatric Inst. and Hosp., Ventura Gen. Hosp., Camarillo State Hosp., West Los Angeles VA Med. Ctr.; dir. Rehab. Rsch. and Tng. Ctr. Mental Illness, 1980-85. Bd. dirs. Lake Sherwood Community Assn., 1978—, pres., 1979-81, 89-91; mem. Conejo Valley Citizens Adv. Bd., 1979-81. Served as surgeon USPHS, 1964-68. Research grantee. Mem. Assn. Advancement Behavior Therapy (exec. com. 1970-72, dir. 1972-79), Am. Psychiat. Assn., Assn. Clin. Psychosocial Research (exec. com. 1985—), Phi Beta Kappa. Author: (with King, DeRisi and McCann) Personal Effectiveness: Guiding People to Assert Their Feelings and Improve Their Social Skills, 1975; A Guide to Behavioral Analysis and Therapy, 1972; (with Wheeler, DeVisser, Kuehnel and Kuehnel) Handbook of Marital Therapy: An Educational Approach to Treating Troubled Relationships, 1980, Psychiatric Rehabilitation of Chronic Mental Patients, 1987, Social Skills Training for Psychiatric Patients, 1989, (with Kuehnel, Rose and Storzbach) Resource Book for Psychiatric Rehabilitation, 1990, Handbook of Psychiatric Rehabilitation, 1992, Stress in Psychiatric Disorders, 1993, Behovior Therapy in Psychiatric Hospitals, 1993; mem. editorial bd. Jour. Applied Behavior Analysis, 1972-78, Jour. Marriage and Family Counseling, 1974-78, Jour. Behavior Therapy and Exptl. Psychiatry, 1975—, Behavior Therapy, 1979-84, Assessment and Invervention in Devel. Disabilities, 1980-85; assoc. editor Jour. Applied Behavior Analysis, 1976-78, Schizophrenia Bull., 1981-87; internat. Rev. Jour. Psychiatry, 1988—; contbr. over 250 articles to profl. jours., chpts. to books. Home: 528 Lake Sherwood Dr Lake Sherwood CA 91361-5120 Office: 11301 Wilshire Blvd 116 AR Los Angeles CA 90073

LIBNOCH, JOSEPH ANTHONY, physician, educator; b. South Bend, Ind., Feb. 1, 1934; s. Casimir Louis and Regina (Kaczorowski) L.; m. Irma Lee Lewis, Jan. 14, 1967 (div. 1984); children: Mark Alan Conine, Michael A., Robert Keith, Sharon M., Andrea E. AB in Chemistry, U. Ill., 1955, MD, 1958. Diplomate Am. Bd. Internal Medicine. Rotating intern Cook County Hosp., Chgo., 1958-59; USPHS fellow in rheumatic disorders U. Ill. Rsch. & Ednl. Hosps., Chgo., 1959-60; resident in internal medicine VA West Side Hosp., Chgo., 1960-62, fellow in hematology, 1962-65; asst. chief hematology Clement J. Zablocki VA Med. Ctr., Milw., 1965-66, 68-75; chief hematology oncology sect. Clement D. Zablocki VA Med. Ctr., Milw., 1976—; acting dir. hematology, oncology sects. Med. Coll. Wis., Milw., 1977-78; instr. medicine Med. Coll. Wis., 1965-66, 68-71, asst. prof. medicine, 1972-88, assoc. prof. clin. medicine, 1988—; sr. attending staff Milw. County Med. Complex, 1968—. Contbr. over 43 articles to profl. jours. Lt. comdr. USNR, 1966-68. Mem. AAAS, Am. Soc. Hematology, Internat. Soc. for Exptl. Hematology, Milw. Hematology Oncology Club, Am. Coll. Rheumatology, Phi Beta Kappa, Phi Beta Phi, Omega Beta Pi, Phi Eta Sigma. Achievements include discovery of the high oxygen affinity hemoglobin variant (human) HB Wood Beta 97 (F64) His- Leu which causes familial erythrocytosis. Home and Office: Clement Zablocki VA Med Ctr 5000 W National Ave Milwaukee WI 53295-1000

LIBSHITZ, HERMAN I., radiologist, educator; b. Phila., May 12, 1939. AB, U. Pa., 1959; MD, Hahnemann Med. Coll., 1963. Intern Mt. Zion Hosp., San Francisco, 1963-64; resident Jefferson Hosp., Phila., 1966-70; assoc. prof. Thomas Jefferson U. Hosp., Phila., 1970-74, Duke U. Med. Ctr., Durham, N.C., 1974-76; prof. radiology U. Tex. M.D. Anderson Cancer Ctr., Houston, 1979—, sect. head thoracic, 1989—; cons. VA Hosp., Houston, 1977—, Brooke Army Med. Ctr., San Antonio, 1977—; adj. prof. Baylor U. Coll. Medicine, Houston, 1983—. Capt. USAF, 1964-66. Fellow Am. Coll. Radiology; mem. AMA, Am. Roentgen Ray Soc., Houston Radiol. Soc. (pres. 1985-86, v.p. 1983-85). Home: 2637 University Blvd Houston TX 77005 Office: U Tex MD Anderson Cancer Ct 1515 Holcombe Blvd Houston TX 77030

LICEAGA, CARLOS ARTURO, research computer engineer; b. San Juan, P.R., Nov. 20, 1958; s. Carlos and Eneida (López) L.; m. Marisol Piña, Sept. 27, 1980; children: Juan Carlos, Camil Enid, Mariel Deliz. BSEE, U. P.R., 1981; MS in Computer Sci., Coll. William and Mary, 1984; PhD in Elec. and Computer Engring., Carnegie Mellon U., 1992. Registered profl. engr., Va. Instr. U. P.R., Mayaquez, 1980, Electronics Coll. Computer Programming, Mayaquez, 1981; asst. instr. Carnegie Mellon U., Pitts., 1985; instr. Thomas Nelson C.C., Hampton, Va., 1987-88; computer engr. Compass Cons. Corp., Yorktown, Va., 1990; rsch. computer engr. Langley Rsch. Ctr. NASA, Hampton, 1981—; asst. prof. Coll. William and Mary, Williamsburg, Va., 1993—, Old Dominion U., Norfolk, 1993—. Contbr. tech. reports to sci. jours. Mem. IEEE (computer and reliability socs., tech. com. on fault-tolerant computing and software engring., subcom. on software reliability engring.). Methodist. Achievements include development of ARM - first computer program to specify a Markov reliability model from a graphical description of a fault-tolerant computer; rsch. in automatic specification of reliability models for life-critical processor-memory-switch structures. Office: NASA Langley Rsch Ctr Mail Stop 130 Hampton VA 23681-0001

LICHSTEIN, EDGAR, cardiologist; b. N.Y.C., Nov. 27, 1936; s. Joseph and Ruth (Weisner) L.; m. Marilyn Dorf, June 19, 1966; children: Adam Robert, Amy Ruth. AB, Columbia Coll., 1957; MD, SUNY, Bklyn., 1961. Diplomate Am. Bd. Internal Medicine, Am. Bd. Cardiovascular Disease. Intern Lenox Hill Hosp., N.Y.C., 1961-62, resident in medicine, 1962-63; resident in medicine NYU, N.Y.C., 1963-64; fellow in cardiology NYU-Nat. Heart Inst., 1964-66; chief cardiology Mt. Sinai Med. Services Elmhurst, N.Y.C., 1971-77; dir. cardiology Maimonides Med. Ctr., Bklyn., 1977-89, chmn. dept. medicine, 1989—; bd. dirs. Maimonides Rsch. and Devel. Found., Bklyn., N.Y. Heart Assn. Author: Hemodynamict's Reference File, 1971; contbr. articles to profl. jours. Mem. New Rochelle (N.Y.) Sch. Bd., 1977-81; bd. dirs. New Rochelle Youth Soccer LEague, 1976. Served to capt. USAF, 1966-68. Fellow Am. Coll. Cardiology, Am. Coll. Physicians, Am. Coll. Chest Physicians, Council Clin. Cardiology; mem. N.Y. Heart Assn. (chmn. council community programs, bd. dirs. 1983—). Jewish. Avocation: swimming. Office: Maimonides Med Ctr 4802 10th Ave Brooklyn NY 11219-2999

LICHSTEIN, HERMAN CARLTON, microbiology educator emeritus; b. N.Y.C., Jan. 14, 1918; s. Siegfried and Luba (Berson) L.; m. Shirley Riback, Jan. 24, 1942 (dec.); children: Michael, Peter. A.B., NYU, 1939; M.S. in Pub. Health, U. Mich., 1940, Sc.D. in Bacteriology, 1943. Diplomate: Am. Bd. Microbiology. Instr. U. Wis., 1943-46; assoc. prof., then prof. U. Tenn., 1947-50, U. Minn., 1950-61; prof. microbiology, dir. dept. U. Cin. Coll. Medicine, 1961-78, dir. grad. studies microbiology, 1962-78, 81-83, prof. microbiology, 1961-84, prof. emeritus, 1984—; mem. Linacre Coll., Oxford U., 1976; mem. microbiology tng. com. NIH, 1963-66, microbiology fellowships com., 1966-70; mem. sci. faculty fellowship panel NSF, 1960-63; Herman C. Lichstein disting. lectureship microbial physiology and genetics endowed, 1987. Author: (with Evelyn Oginsky) Experimental Microbial Physiology, 1965; editor: Bacterial Nutrition, 1983; also articles. Mem. Gov. Minn. Com. Edn. Exceptional Child, 1956-57; Bd. dirs. Walnut Hills (Ohio) High Sch. Assn., 1964-66. NRC fellow Cornell U., 1946-47; fellow U. Cin. Grad. Sch., 1965. Fellow AAAS, Am. Acad. Microbiology (gov. 1972-75, chmn. 1973); mem. Am. Soc. Microbiology (hon. 1986), Am. Soc. Biol. Chemistry, Soc. Gen. Microbiology, Assn. Med. Sch. Microbiology Chairmen (council 1970-72, trustee 1974-77), Soc. Exptl. Biology and Medicine, Am. Inst. Biol. Scis., Sigma Xi (pres. Cin. chpt. 1971-73), Phi Kappa Phi, Pi Kappa Epsilon.

LICHTENBERG, BYRON K., futurist, manufacturing executive, space flight consultant; b. Stroudsburg, Pa., Feb. 19, 1948; s. Glenn John and Georgianna (Bierei) L.; m. Lee Lombard, July 25, 1970 (divorced); children—Kristin, Kimberly. Sc.B., Brown U., 1969; M.S., MIT, 1975, Sc.D., 1979. Rsch. scientist MIT, Cambridge, 1978-84; pres. Payload Systems, Inc., Cambridge, 1984-89, chief scientist, 1989-91; pres., chief exec. officer Omega Aerospace Inc., Virginia Beach, Va., 1991—. Contbg. author NASA Payload Specialist, 1979—, Flew on Space Shuttle Mission #9, #45; contbr. articles to profl. jours. Served to lt. col. USAF, Mass. Air N.G., 1969-93. Recipient NASA Space Flight award, 1983, 92, Spaceflight award VFW, 1983, Haley Spaceflight award AIAA, 1983. Mem. Tau Beta Pi, Sigma Xi. Avocations: golf; racquetball; windsurfing; skiing. Office: Omega Aerospace Inc 728 Wolfsnare Cres Virginia Beach VA 23454

LICHTENHELD, FRANK ROBERT, physician, plastic and reconstructive surgeon, urologist; b. Weimar, Thuringia, Fed. Republic of Germany, Apr. 11, 1923; s. Georg Kaspar and Carolina Henriette (Schmidt) L.; m. Ingrid Lichtenheld (div. 1982); children: Mathias, Claudia; children: (adopted) Ana-Carolina, Fernando F., Michele C., Filipe T.; m. Eva-Maria Matiasek; 1 child, Maximilian Emmanuel. MD, U. Frankfurt, Fed. Republic Germany, 1951. Cert. surgeon, plastic surgeon, urologist. Intern Presbyterian Hosp., Newark, N.J., U.S.A., 1951-52; resident in surgery Presbyterian Hosp., Newark, 1952; fellow in surgery Mayo Clinic, Rochester, Minn., U.S.A., 1953-56; asst. chief urology USAF Hosp., Wiesbaden, Fed. Republic of Germany, 1956-64; med. researcher Winthrop Pharm. GmbH, Frankfurt, Fed. Republic of Germany, 1964-65, Hoechst AF, Pharma Forschung Ausland, Frankfurt, 1965-66; chief of clinic for plastic and reconstructive surgery Wiesbaden, 1966-88. Contbr. 45 articles to profl. jours. Lt. German Navy,

1941-45. Mem. Ventnor Alumni Assn. (pres. 20th anniversary Ventnor Found. 1971), Alumni Assn. of Mayo Found., Assn. of Mil. Surgeons of U.S., Deutsche Gesselschaft für Chirurgie, Deutsche Gesellschaft für Äsethetisch-Plastiche Chirurgie (pres. 1976, 82), Deutsche Gesellschaft für Plastische-und Wiederherstellungschururgie. Avocations: European and world politics, econs. Home: Vorder Brennberg 6, 8966 Altusried Frauenzell, Allgau Germany

LICHTENTHALER, FRIEDER WILHELM, chemist, educator; b. Heidelberg, Germany, Jan. 19, 1932; s. Wilhelm and Emma (Hick) L.; m. Evemaria von Infeld, Apr. 15, 1966; children: Matthias, Johannes, Kathrin. Diploma Chemistry, U. Heidelberg, 1956, D in Natural Scis., 1959; DSc (hon.), Kossuth Univ., Debrecen, Hungary, 1993. Rsch. fellow dept. biochemistry U. Calif., Berkeley, 1959-62; habilitation Technische Hochschule Darmstadt, Germany, 1963, assoc. prof., 1969-72, prof., 1972—; vis. prof. Keio U., Tokyo, 1973, Kyoto U., 1981, Tongji U., Shanghai, 1981, U. Calif., Berkeley, 1985, Yokohama U., 1992. Editor: (monograph) Carbohydrates as Organic Raw Materials, 1991; contbr. more than 200 articles to sci. jours. Fellow Royal Soc. Chem.; mem. Gesellschaft Deutscher Chemiker, Am. Chem. Soc., Chem. Soc. Japan, European Carbohydrate Orgn. (pres. 1985-87). Home: Am Willgraben 5, D-64367 Mühltal 4, Germany Office: Inst Organische Chemie, Petersenstrasse 22, D-64287 Darmstadt Germany

LICHTENWALNER, OWEN C., engineer, telecommunications executive; b. Seattle, Oct. 30, 1937; s. John J. and Irene (Haskell) L.; m. Susan Jenner, Nov. 22, 1958; children: David, Peter, Dan, Joseph, Sally, Matthew, Andrew. BSME, Seattle U., 1960; postgrad., U. Colo., 1965. Registered profl. engr., Ga., Mo., Ill. Sr. engr. Pacific NW Bell, Seattle, 1960-70; v.p. Contel-Cen. Region, St. Louis, 1970-83; gen. program mgr. Contel Bus. Products Inc., Atlanta, 1983-85; asst. v.p. engring. Contel, Atlanta, 1985-91; dir. Pactel Cellular, Atlanta, 1991—; mem. nat. svcs. adv. com. USTA, Washington, 1988-90. 1st lt. U.S. Army, 1960-62. Mem. NSPE, IEEE-Comm. Soc. (chpt. pres. 1983-85), Ga. Soc. Profl. Engrs. Achievements include implementation of new communication technologies digital switchingand transmission, fiber and wireless; research on integration of computing and communications. Office: Pactel Cellular Ste 300 4151 Ashford Dunwoody Rd Atlanta GA 30319

LICHTER, PAUL RICHARD, ophthalmology educator; b. Detroit, Mar. 7, 1939; s. Max L. and Buena (Epstein) L.; m. Carolyn Goode, 1960; children: Laurie, Susan. BA, U. Mich., 1960, MD, 1964, MS, 1968. Diplomate Am. Bd. Ophthalmology. Asst. to assoc. prof. ophthalmology U. Mich., Ann Arbor, 1971-78, prof., chmn. dept. ophthalmology, 1978—; chmn. Am. Bd. Ophthalmology, 1987. Editor in chief Ophthalmology jour., 1986—. Served to lt. comdr. USN, 1969-71. Fellow Am. Acad. Ophthalmology (bd. dirs. 1981—, Sr. Honor award 1986); mem. AMA, Pan Am. Assn. Ophthalmology (bd. dirs. 1988—, sec.-treas. English speaking countries 1991—), Mich. State Med. Soc., Washtenaw County Med. Soc., Mich. Ophthalmol. Soc. (pres. 1993—), Assn. Univ. Profs. Ophthalmology (trustee 1986-93, pres. 1991-92), Alpha Omega Alpha. Office: U Mich Med Sch Kellogg Eye Ctr 1000 Wall St Ann Arbor MI 48105-1912

LICHTER, ROBERT LOUIS, science foundation administrator, chemist; b. Cambridge, Mass., Oct. 26, 1941. AB, Harvard U., 1962; PhD, U. Wis., 1967. NIH postdoctoral fellow Technische U. Braunschweig, Fed. Republic of Germany, 1967-68; rsch. fellow Calif. Inst. Tech., Pasadena, 1968-70; from asst. prof. to prof. Hunter Coll. CUNY, N.Y.C., 1970-83; chair dept. chemistry Hunter Coll. CUNY, N.Y.C., 1977-81; regional dir. grants Rsch. Corp., Port Washington, N.Y., 1983-86; v.p. for rsch. and grad. studies SUNY, Stony Brook, 1986-89; exec. dir. Camille & Henry Dreyfus Found., N.Y.C., 1989—; vis. scientist Sandoz Rsch. Labs., 1981-82, Exxon Rsch. and Engring. Co., 1981-82; treas. Exptl. NMR Confs., Inc., 1981-82; cons. Rsch. Corp., 1990—; bd. govs. Nat. Conf. on Undergrad. Rsch., 1992—; councillor Coun. on Undergrad. Rsch., 1993—. Co-author: N-15 Nuclear Magnetic Resonance Spectroscopy, 1979, C-13 Nuclear Magnetic Resonance Spectroscopy, 1980, NMR Spectroscopy Techniques, 1987; editor Concepts in Magnetic Resonance Jour., 1989—; mem. editorial bd. Magnetic Resonance in Chemistry Jour., 1983-87; mem. edn. adv. bd. N.Y. Acad. Scis., 1991—; contbr. over 36 articles to profl. jours. Mem. Am. Assn. Advancement Sci., Am. Chem. Soc. (chair NMR topical group Northern Jersey chpt. 1982-83), N.Y. Acad. Scis. (chair sect. on chem. scis. 1987). Office: Camille & Henry Dreyfus Found 555 Madison Ave Ste 1305 New York NY 10022-3301

LICHTIG, LEO KENNETH, health economist; b. Bklyn., Oct. 20, 1953; s. Samuel and Alyne Norma (Strauss) L.; m. Susan Mary Walsh, May 15, 1977; children: Brielle Joy, Danica Jill. BS, MS, Rennselaer Poly. Inst., 1974, PhD, 1976. Asst. prof. SUNY, Albany, 1976-77; project specialist, econometrician N.J. State Dept. Health, Trenton, 1977-82; dir. utilization econs. and rsch. Empire Blue Cross/Blue Shield, Albany, 1982-90; v.p. rsch. and demonstration Health Care Rsch. Found., Albany, 1982-90; v.p. Network, Inc., Latham, N.Y., Randolph, N.J., 1990—; pvt. practice cons., Latham, 1982-90; mem. nat. diagnosis related group, steering com. health care fin. adminstrn. Yale U., Washington, 1979-81; mem. adj. faculty Russell Sage Grad. Sch. Health Adminstrn., Albany, 1986—, Union Coll. Grad. Mgmt. Inst., Schenectady, N.Y., 1991—; expert reviewer Health Care Financing Adminstrn., Washington, 1987, 89. Author: Hospital Information Systems for Case Mix Management, 1986; contbg. editor (newsletter) Nat. Report on Computers & Health, 1982-85; contbr. articles to profl. jours. Mem. tech. adv. com. Statewide Planning and Rsch. Coop. System, N.Y. State Dept. Health. Mem. Am. Mgmt. Assn., Assn. For Helath Svcs. Rsch., Am. Statis Assn (com. on privacy and confidentiality 1981-84, subcom. on quality and productivity measures 1988-90). Avocation: Arthurian legends. Office: Network Inc 1572 Sussex Turnpike Randolph NJ 07869

LICINI, JEROME CARL, physicist, educator. BA in Physics, Princeton U., 1980; PhD, MIT, 1987. Assoc. prof. Lehigh U., Bethlehem, Pa., 1987—. Office: Lehigh U Dept Physics 16 Memorial Dr E Bethlehem PA 18015

LICK, WILBERT JAMES, mechanical engineering educator; b. Cleve., June 12, 1933; s. Fred and Hulda (Sunntag) L.; children—James, Sarah. B.A.E., Rensselaer Poly. Inst., 1955, M.A.E., 1957, Ph.D., 1958. Asst. prof. Harvard, 1959-66; sr. research fellow Calif. Inst. Tech., 1966-67; mem. faculty Case Western Res. U., 1967-79, prof. earth scis., 1970-79, chmn. dept., 1973-76; prof. mech. engring. U. Calif.-Santa Barbara, 1979—, chmn. dept., 1982-84. Home: 1236 Camino Meleno Santa Barbara CA 93111-1007 Office: U Calif Dept Mech and Environ Engring Santa Barbara CA 93106

LIDDIARD, GLEN EDWIN, physicist; b. San Diego, Nov. 4, 1919; s. Thomas Phillip and Ruth Elizabeth (Durfee) L.; m. Roberta Gordon, June 21, 1942; children: Donald Eugene, Carol Anne. BA in Physics, San Diego State U., 1942. Physicist U.S. Naval Ordnance Lab., Washington, 1942-46, U.S. Navy Electronics Lab., San Diego, 1946-61; asst. chief engr. Bendix Corp., North Hollywood, Calif., 1961-63; supervisory physicist U.S. Naval Electronics Lab., San Diego, 1963-64; v.p. Ametek Corp., El Cajon, Calif., 1964-78; pres. Internat. Transduser Corp., Santa Barbara, Calif., 1978-86; cons. Liddiard Consultingq, Pine Valley, Calif., 1986—. Author: U.S. Underwater Acoustics Journal, 1959-60. Mem. IEEE, Acoustical Soc. of Am., Sigma Pi Sigma. Republican. Presbyterian. Achievements include devel. of unqiue underwater transducer designs for deep submergence. Home: PO Box 309 8761 Pine Creek Rd Pine Valley CA 91962 Office: Liddiard Consulting PO Box 309 8761 Pine Creek Rd Pine Valley CA 91962

LIDE, DAVID REYNOLDS, science editor; b. Gainesville, Ga., May 25, 1928; s. David Reynolds and Laura Kate (Simmons) L.; m. Mary Ruth Lomer, Nov. 5, 1955 (div. Dec. 1988); children: David Alston, Vanessa Grace, James Hugh, Quentin Robert; m. Bettijoyce Breen, 1988. BS, Carnegie Inst. Tech., 1949; PhD, Harvard U., 1952, AM, 1951. Physicist Nat. Bur. Standards, Washington, 1954-63, chief molecular spectroscopy sect., 1963-69; dir. standard reference data Nat. Bur. Standards, Gaithersburg, Md., 1969-88; editor-in-chief Handbook of Chemistry and Physics, CRC Press, 1988—; pres. Com. on Data for Sci. and Tech., Paris, 1986-90. Editor Jour. Phys. and Chem. Reference Data, 1972-92; contbr. more than 100 articles to profl. publs. Recipient Skolnik award for Chem. Info., Am. Chem. Soc., 1988, Patterson-Crane award, 1991, Presdl. Rank award in sr. exec. svc., 1986. Mem. Internat. Union Pure and Applied Chemistry (pres. phys. chemistry div. 1983-87). Achievements include use of microwave spectroscopy for studying hindered internal rotation, explanation of HCN laser. Home and Office: 13901 Riding Loop Dr Gaithersburg MD 20878-3879

LIDICKER, WILLIAM ZANDER, JR., zoologist, educator; b. Evanston, Ill., Aug. 19, 1932; s. William Zander and Frida (Schroeter) L.; m. Naomi Ishino, Aug. 18, 1956 (div. Oct., 1982); children: Jeffrey Roger, Kenneth Paul; m. Louise N. DeLonzor, June 5, 1989. B.S., Cornell U., 1953; M.S., U. Ill., 1954, Ph.D., 1957. Instr. zoology, asst. curator mammals U. Calif., Berkeley, 1957-59; asst. prof., asst. curator U. Calif., 1959-65, assoc. prof., assoc. curator, 1965-69; assoc. dir. Mus. Vertebrate Zoology, 1968-81, acting dir., 1974-75, prof. zoology, curator mammals, 1969-89, prof. integrative biology, curator of mammals, 1989—. Contbr. articles to profl. jours. Bd. dirs. No. Calif. Com. for Environ. Info., 1971-77; bd. trustees BIOSIS, 1987-92, chmn., 1992; N.Am. rep. steering com., sect. Mammalogy IUBS, UNESCO, 1978-89; chmn. rodent specialist group Species Survival Commn., IUCN, 1980-89; mem. sci. adv. bd. Marine World Found. at Marine World Africa USA, 1987—; pres. Dehnel-Petrusewicz Meml. Fund, 1985—. Fellow AAAS, Calif. Acad. Scis.; mem. Am. Soc. Mammalogists (dir., 2d v.p. 1974-76, pres. 1976-78, C.H. Merriam award 1986), Am. Soc. Naturalists, Berkeley Folk Dancers Club (pres. 1969, tchr. 1984—), others. Office: U Calif Mus Vertebrate Zoology Berkeley CA 94720

LIEB, ELLIOTT HERSHEL, physicist, mathematician, educator; b. Boston, July 31, 1932; s. Sinclair M. and Clara (Rosenstein) L.; m. Christiane Fellbaum; children—Alexander, Gregory. B.Sc., M.I.T., 1953; Ph.D., U. Birmingham, Eng., 1956; D.Sc.h.c., U. Copenhagen, 1979. With IBM Corp., 1960-63; sr. lectr. Fourah Bay Coll., Sierra Leone, 1961; mem. faculty Yeshiva U., 1963-66, Northeastern U., 1966-68, M.I.T., 1968-75; prof. physics, 1963-68, prof. math., 1968-73, prof. math. and physics, 1973—; prof. math. and physics Princeton U., 1975—. Author: (with D.C. Mattis) Mathematical Physics in One Dimension, 1966, (with B. Simon and A. Wightman) Studies in Mathematical Physics; also articles. Recipient Boris Pregel award chem. physics N.Y. Acad. Scis., 1970; Dannie Heineman prize for mathematical physics Am. Inst. Physics and Am. Phys. Soc., 1978; Prix Scientifique, Union des Assurances de Paris, 1985; Birkhoff prize Am. Math. Soc. and Soc. Indsl. Applied Math., 1988; Max-Planck medal German Phys. Soc., 1992; Guggenheim Found. fellow, 1972, 78. Fellow Am. Phys. Soc.; mem. NAS, Austrian Acad. Scis., Danish Royal Acad., Internat. Assn. Math. Physics (pres. 1982-84). Office: Princeton U Jadwin Hall-Physics Dept PO Box 708 Princeton NJ 08544-0708

LIEBENSON, GLORIA KRASNOW, interior design executive; b. Chgo., Apr. 6, 1922; d. Henry Randolph and margaret (Rivkin) Krasnow; m. Herbert Liebenson, Mar. 11, 1944; children: Lauren Ward, Lynn Liebenson Green. Student, Int. Inst. Interior Design, Washington, 1961; B Am. Studies, Dunbarton Coll., Washington, 1974. Lic. Interior Designer, D.C. Numerous positions Journalism, Advt., editing, 1942-62; interior design exec. Creative Interiors, 1962—; tchr. interior design YMCA, Washington, 1980-82. Mem. editorial staff Champlin Encyclopedia, 1945-47; journalist Shreveport Jour., 1944. Bd. Dirs. Jewish Social Svc. Agy., Washington, 1983-85, Nat. Coun. Jewish Women. 1982-84; pres. Friends Nat. Museum African Art, 1983-85, D.C. Mental Health Assn., 1986-88. Mem. Womens Nat. Dem. Club. Democrat. Jewish. Avocations: theater, concerts, scrabble, reading, travel. Home and office: 2703 Unicorn Ln NW Washington DC 20015-2233

LIEBERMAN, EDWARD MARVIN, biomedical scientist, educator; b. Lowell, Mass., Feb. 10, 1938; s. Irving and Lillian (Silverman) L.; m. Harriet Handman, June 12, 1960; children: Lynn Rebecca, Dana Beth, Kurt Michael. BS, Tufts U., 1959; MS, U. Mass., 1961; PhD, U. Fla., 1965. Rsch. fellow Swedish Med. Rsch. Coun. U. Uppsala, Sweden, 1966-68; asst., assoc. prof. Bowman Gray Sch. Medicine Wake Forest U., Winston-Salem, N.C., 1968-74; assoc. prof., prof. physiology sch. medicine East Carolina U., Greenville, N.C., 1974—; vis. prof. Venezuelan Inst. Sci. Rsch., Caracas, 1978-84, Marine Biol. Assn. of U.K., Plymouth, Eng., 1984-90; vis. rschr. marine lab. Duke U., Beaufort, N.C., 1990—; program officer NSF, Washington, 1991-92. Contbr. articles to profl. jours. Bd. dirs., mem. rsch. com. N.C. affiliate Am. Heart Assn., Chapel Hill, 1975-85. Rsch. grantee NSF, 1980-86, Army Rsch. Office, 1986—. Mem. Am. Physiol. Soc., Am. Soc. Molecular Biology and Biochemistry, Biophysical Soc., Soc. Neuroscience. Jewish. Achievements include research on neuron-glia interaction, nerve metabolism, neurochemistry, ion transport, blood brain barrier, comparative physiology, neurobiology. Office: East Carolina U Sch Medicine Dept Physiology Greenville NC 27858

LIEBERMAN, GERALD J., statistics educator; b. N.Y.C., Dec. 31, 1925; s. Joseph and Ida (Margolis) L.; m. Helen Herbert, Oct. 27, 1950; children—Janet, Joanne, Michael, Diana. B.S. in Mech. Engring., Cooper Union, 1948; A.M. in Math. Stats., Columbia U., 1949; Ph.D., Stanford U., 1953. Math. statistician Nat. Bur. Standards, 1949-50; mem. faculty Stanford U., 1953—, prof. statistics and indsl. engring., 1959-67, prof. statistics and operations research, 1967—, chmn. dept. operations research, 1967-75, assoc. dean Sch. Humanities and Scis., 1975-77, acting v.p. and provost, 1979, vice provost, 1977-85, dean research, 1977-80, dean grad. studies and research, 1980-85, provost, 1992-93; cons. to govt. and industry, 1953—. Author: (with A. H. Bowker) Engineering Statistics, 1959, 2d edit., 1972, (with F.S. Hillier) Introduction to Operations Research, 1967, 5th edit., 1990. Ctr. Advanced Studies in Behavioral Scis. fellow, 1985-86. Fellow Am. Statis. Assn., Inst. Math. Statistics, Am. Soc. Quality Control (Shewhart medal 1972), AAAS; mem. Nat. Acad. Engring., Inst. Mgmt. Sci. (pres. 1980-81), Ops. Research Soc. Am., Sigma Xi, Pi Tau Sigma. Home: 811 San Francisco Ter Stanford CA 94305-1021

LIEBERMAN, HARVEY MICHAEL, hepatologist, gastroenterologist, educator; b. N.Y.C., Feb. 24, 1949; s. Louis and Ellie (Miller) L.; m. Lewette Alexandra Fielding, Nov. 24, 1985. BA magna cum laude, NYU, 1972, MD, 1976. Intern Bronx (N.Y.) Mcpl. Hosp./Albert Einstein Coll. Medicine, N.Y.C., 1976-77, jr. and sr. resident, 1977-79; fellow in gastroenterology and liver disease Albert Einstein Coll. Medicine, 1979-81, rsch. assoc. Liver Rsch. Ctr., 1983; asst. prof. Albert Einstein Coll. Medicine, Bronx, N.Y., 1984-86; dir. gastroenterology Gouverneur Hosp., N.Y.C., 1986-90; asst. chief gastroenterology Lenox Hill Hosp., N.Y.C., 1992—, founding dir. liver clinic, 1992—; clin. asst. prof. Sch. Medicine NYU, 1986—; prin. investigator Liver Rsch. Ctr. Albert Einstein Coll. Medicine, 1984-87, vis. scientist, 1992—; med. adv. bd. Crohn's and Colitis Found. of Am., Am. Liver Found., N.Y. chpts., 1987—; researcher in molecular biology of hepatitis B virus and relationship to viral infection and liver cancer. Author: Relationship of Hepatitis B Viral Infection in Serum to Viral Replication, 1983. Recipient Clin. Investigator award NIH, 1984-87. Fellow ACP, Am. Coll. Gastroenterology, N.Y. Acad. Gastroenterology (pres. 1990-91). Achievements include development of assay to measure DNA of hepatitis B virus directly in serum; first to note its greater sensitivity in measuring active viral replication liver disease compared to conventional serological tests. Office: 345 E 37th St New York NY 10016-3217

LIEBERMAN, JAMES LANCE, chemistry educator; b. Newport News, Va., Nov. 26, 1952; s. Lawrence Lipman and Evelyn (Wilks) L.; married, 1992. BS, U. Richmond, 1975; MBA, U. Colo., 1988. Cert. indsl. hygienist. Lab. specialist U. Wis., Madison, 1976-78; chemist Vail (Colo.) Assocs., Inc., 1981-84, supr. sanitation ops., 1984-85; instr. celestial navigation Colo. Mountain Coll., Vail, 1981—; pres. Advanced Purification, Inc., Boulder, Colo., 1988-91, Environ. Info. Svcs., Inc., Boulder, 1992—; instr. skiing Vail/Beaver Creek Ski Sch., 1979-80; cons. staff scientist NFT Inc., Golden, Colo., 1989-92. Patentee in field. Mem. Vail Mountain Rescue Group, 1982—, local lic. bd., Vail, 1982-85. Named Instr. of Yr. Colo. Mountain Coll., 1984. Mem. Am. Chem. Soc., Internat. Ozone Assn., Am. Indsl. Hygiene Assn. Jewish. Avocations: skiing, sailing, mountaineering. Home: 4790 Shawnee Pl Boulder CO 80303-3818 Office: Environ Info Svcs Inc 4790 Shawnee Pl Ste 102 Boulder CO 80303

LIEBERMANN, HOWARD HORST, metallurgical engineering executive; b. Neustadt, Germany, Nov. 27, 1949; came to U.S., 1956; s. Hermann Arno and Gerda (Klamt) L.; m. Lynda Jeanne Kobuskie, Apr. 7, 1979; children: Daniel Howard, Amanda Jeanne. BS in Metall. Engring., Poly. Inst. Bklyn., 1972; MS in Material Sci. and Metallurgy, U. Pa., 1975, PhD in Material Sci. and Metallurgy, 1977. Staff metallurgist GE Corp R&D, Schenectady, N.Y., 1977-81; sr. metallurgist AlliedSignal Advanced Materials, Amorphous Alloys, Parsippany, N.J., 1982-85, rsch. assoc., 1985, mgr. R&D, 1985—. Editor: Rapidly Solified Alloys, 1993; contbr. 81 articles to profl. jours. Mem. IEEE (sr.), Metals Soc. of AIME, Am. Soc. Metals, Iron and Steel Soc. Achievements include 25 patents in field. Home: 11 Cynthia Dr Succasunna NJ 07876 Office: AlliedSignal Advanced Materials Amorphous Metals 6 Eastmans Rd Parsippany NJ 07054

LIEBERSTEIN, MELVIN, administrative engineer; b. N.Y.C., May 28, 1934; s. Morris Jacob and Gertrude (Udovin) L.; m. Gloria Schrier, July 29, 1961; children: Rochelle Beth, Barbra Lynn. BChE, CCNY, 1955; ME, The Cooper Union, 1978. Lic. profl. engr. N.Y., N.J. Rsch. scientist NACA, Cleve., 1955-57; analytical engr. Fairchild Engine Div., Deer Park, N.Y., 1957-59; project engr. Curtiss Wright Corp., Wood Ridge, N.J., 1959-66; design engr. Walter Kidde Co., Inc., Belleville, N.J., 1966-70; chief thermal engr. Mesco Tectonics Inc., Clifton, N.J., 1970-71; free lance engring. cons. Paramus, N.J., 1971-72; air pollution control engr. N.Y.C. Dept. Air Resources, 1972-81; admistrv. engr. N.Y.C. Dept. Sanitation, 1981--. Book reviewer: Water Air Soil Pollution Jour., 1975, 77. Mem. N.J. State Dem. Party, Paramus, 1992. Recipient Lion of Judah award State of Israel Bonds, 1991. Mem. Am. Inst. Chem. Engrs., Knights Pythias (sec. 1981--), Am. Geographic Soc. Jewish. Office: NYC Dept Sanitation 44 Beaver St New York NY 10004

LIEBLER, ELISABETH M., veterinary pathologist; b. Nuernberg, Germany, June 20, 1960; d. H. and M. (Andres) L. Fachtierarzt, Vet. Pathology, 1991; DVM, Vet. Sch. Hannover, Germany, 1985; MS in Vet. Pathology, Iowa State U., Ames, 1986. Rsch. asst. Iowa State U., Ames, 1985-86; rsch. asst. Vet. Sch. Hannover, 1986-89, akad. rat., 1990—. Author chpt. in book: Diseases of Swine, 7th edit. 1991-91. Office: TIHO Inst of Pathology, Buenteweg 17, Hannover Germany 3000

LIEBMAN, PAUL ARNO, biophysicist, educator; b. Pitts., Aug. 1, 1933; s. Arno Jack and M. Josephine (Schurr) L.; m. Elizabeth Loeffler, Nov. 6, 1982; 1 child, Erica. BS in Chemistry, U. Pitts., 1954; MD, Johns Hopkins U., 1958. Intern in medicine Barnes Hosp., St. Louis, 1958-59; postdoctoral fellow Johnson Rsch. Found., Phila., 1959-63; asst. prof. physiology U. Pa. Sch. Medicine, Phila., 1963-67, assoc. prof. anatomy, 1967-76, prof. anatomy, 1976-92, prof. ophthalmology, 1982—, prof. neurosci., prof. biophysics/biochemistry, 1992—; dir. U. Pa. Vision Rsch. Ctr., Phila., 1976—. Contbr. numerous articles to Jour. Biol. Chemistry, Science, Nature, others. Grantee NIH, 1962—. Mem. Biophys. Soc. for Neurosci., Assn. for Rsch. in Vision and Ophthalmology, Phila. Canoe Club (hon. life mem.). Achievements include invention of UV-VIS microspectrophotometers, ultrasensitive pH recording instruments; discovered color vision pigments, biochemical mechanism of visual transduction, membrane protein lateral diffusion. Office: U Pa Vision Rsch Ctr 36th & Hamilton Walk Philadelphia PA 19104

LIEBMAN, SHIRLEY ANNE, analytical research scientist; b. Boston, Sept. 4, 1934; d. John A. and Fay Glazier; m. Harmon L. Liebman, June 23, 1956; children: Robert C., David J. BS, Northeastern U., Boston, 1956; PhD, Temple U., Phila., 1969. Lab. technician MIT, 1953-55; jr. engr. Boeing Co., Seattle, 1956-58; rsch. chemist Monsanto Rsch. Corp., Everett, Mass., 1958-61; sr. rsch. scientist Armstrong World Ind., Lancaster, Pa., 1969-80; mgr. application and contract rsch. Chem. Data Systems, Oxford, Pa., 1980-83; sr. assoc. NRC-Ballistic Rsch. Lab., Aberdeen, Md., 1984-86; sr. scientist Geo-Ctrs., Inc., Aberdeen, 1986-91; v.p., dir. contract rsch. and applications CCS Instrument Systems, West Grove, Pa., 1991—; sr. adv. bd. Environ. Rsch. Ctr., U. Nev., Las Vegas, 1990—; cons. Computer Chem. Systems, Inc., Avondale, Pa., 1986-91, The CECON Group, Inc., Wilmington, Del., 1991—. Co-editor: Pyrolysis and GC in Polymer Analysis, 1985; contbr. over 170 articles to profl. jours., chpts. to books. Citizens adv. coun. Lancaster Housing and Redevel. Authority, 1986—. Mem. AAAS, Am. Chem. Soc., Soc. for Applied Spectroscopy, Royal Soc. Chemistry (London). Republican. Unitarian Universalist. Achievements include patents in field; research in major disciplines of physical/analytical organic chemistry, chemical engineering and computers, polymer characterization, materials and environmental sciences, and the integrated intelligent instrument approach to analytical sciences. Home: 91 Pinnacle Rd W Holtwood PA 17532-9641

LIEBOWITZ, HAROLD, aeronautical engineering educator, university dean; b. Bklyn., June 25, 1924; s. Samuel and Sarah (Kaplan) L.; m. Marilyn Iris Lampert, June 24, 1951; children: Alisa Lynn, Jay, Jill Denice. B. in Aero. Engring., Poly. Inst. Bklyn., 1944, M. in Aero. Engring., 1946, D. in Aero. Engring., 1948. With Office Naval Research, 1948-60; asst. dean Grad. Sch., exec. dir. engring. expt. sta. U. Colo., Boulder, 1960-61; also vis. prof. aero. engring. U. Colo.; head structural mechs. br. Office Naval Research, 1961-68, dir. Navy programs in solid propellant mechanics, 1962-68; dean Sch. Engring. and Applied Sci., George Washington U., Washington, 1968-90; L. Stanley Crane endowed prof., 1990—; rsch. prof., endowed prof. Cath. U., Washington, 1962-68; cons. AERDCO, 1960—, Office Naval Rsch., 1970—, Pratt & Whitney Aircraft Co., 1981-82, Pergamon Press, 1968—, Acad. Press, 1968—; mem. Israeli-Am. Materials Adv. Group, 1970—; sci. adviser Congl. Ad Hoc Com. on Environ. Quality, 1969—; co-dir. Joint Inst. for Advancement Flight Scis. NASA-Langley Rsch. Ctr., Hampton, Va., 1971—. Editor: Advanced Treatise on Fracture, 7 vols., 1969-72; founder, editor-in-chief Internat. Jour. Computers and Structures, 1971—; founder, editor Internat. Jour. Engring. Fracture Mechanics, 1968—; founder computer series jours., 1968—; contbr. articles to profl. jours. Bd. govs. U. Denver, 1987—. Recipient Outstanding award Office Naval Research, 1961, Research Accomplishment Superior award, 1961; Superior Civilian Service award USN, 1965, 67; Commendation Outstanding Contbns. sec. navy, 1966; Wash. Soc. Engrs. award, Ednl. Service award, Wash. Soc. Engrs., Fundacion Gran Mariscal de Avacucho, Service cert. Tech. Contbns. to Structures and Materials Panel, NATO, Disting. Alumnus award Polytechnic Inst. Bklyn., Tech. Achievement cert. Washington sect. ASME, Civilian Service award USN, Albert Einstein prize, 1991, Washington Acad. of Scis. award for Disting. career in Sci., 1990. Fellow AAAS, AIAA, Am. Soc. Metals; mem. Soc. Engring. Scis. (past pres.), Sci. Rsch. Soc. Am., Am. Technion Soc. (dir.), Am. Acad. Mechanics (founder, pres.), Internat. Coop. Fracture Inst. (founder, v.p.), Engrs. Coun. for Profl. Devel., nat. Acad. Engring. (home sec.), Engring. Acad. Japan, Argentina NAS, Hungarian Acad. Scis., Sigma Xi, Tau Beta Pi, Sigma Gamma Tau, Omega Rho, Pi Tau Sigma, Sigma Tau. Office: George Washington U Sch Engring & Applied Sci Dept Civil/Mech/Environ Eng Washington DC 20052

LIEF, HAROLD ISAIAH, psychiatrist; b. N.Y.C., Dec. 29, 1917; s. Jacob F. and Mollie (Filler) L.; m. Myrtis A. Brumfield, Mar. 3, 1961; Caleb B., Frederick V., Oliver F.; children from previous marriage: Polly Lief Goldberg, Jonathan F. BA, U. Mich., 1938; MD, N.Y. U., 1942; cert. in psychoanalysis, Columbia Coll. Physicians and Surgeons, 1950; MA (hon.), U. Pa., 1971. Intern Queens Gen. Hosp., Jamaica, N.Y., 1942-43; resident psychiatry L.I. Coll. Medicine, 1946-48; pvt. practice psychiatry N.Y.C., 1948-51; asst. physician Presbyn. Hosp., N.Y.C., 1949-51; asst. prof. Tulane U., New Orleans, 1951-54, asso. prof., 1954-60, prof. psychiatry, 1960-67; prof. psychiatry U. Pa., Phila., 1967-82, prof. emeritus, 1982—; dir. div. family study U. Pa., 1967-81; dir. Marriage Council of Phila., 1969-81, Ctr. for Study of Sex. Edn. in Medicine, 1968-82; mem. staff U. Pa. Hosp., 1967-81, Pa. Hosp., 1981—. Author: (with Daniel and William Thompson) The Eighth Generation, 1960; Editor: (with Victor and Nina Lief) Psychological Basis of Medical Practice, 1963, Medical Aspects of Human Sexuality, 1976, (with Arno Karlen) Sex Education in Medicine, 1976, Sexual Problems in Medical Practice, 1981, (with Zwi Hoch) Sexology: Sexual Biology, Behavior and Therapy, 1982, (with Zwi Hoch) International Research in Sexology, 1983, Human Sexuality With Respect to AIDS and HIV Infection, 1989; contbr. numerous articles to publs. Bd. dirs. Ctr. for Sexuality and Religion; mem. La. State Commn. Civil Rights, 1958-67. Maj. M.C. U.S. Army, 1943-46. Commonwealth Fund fellow, 1963-64; recipient Gold Medal award Mt. Airy Hosp., 1977, Lifetime Achievement award Phila. Psychiat. Soc., 1992. Fellow Phila. Coll. Physicians, Am. Psychiat. Assn. (life), N.Y. Acad. Scis., AAAS, Am. Acad. Psychoanalysis (charter, past pres.), Am. Coll. Psychiatrists (founding), Am. Coll. Psychoanalysts (charter); mem. Am. Assn. Marriage and Family Therapists, Sex Info. and Edn. Council U.S. (past pres.),

Group Advancement Psychiatry (life), Am. Soc. Adolescent Psychiatry, Am. Psychosomatic Soc., Assn. Psychoanalytic Medicine (life), Internat. Acad. Sex Rsch., Soc. Sci. Study of Sex, Am. Soc. Sex Educators, Counselors and Therapists, Soc. Sex Therapists and Researchers, World Assn. Sexology (past v.p.), Soc. Exploration of Psychotherapy Integration (adv. bd.), Univ. of World (health com.), Columbia Club, Mich. Club of Greater Phila., Sigma Xi, Alpha Omega Alpha, Phi Eta Sigma, Phi Kappa Phi. Home: 101 S Buck Ln Haverford PA 19041-1104 Office: 700 Spruce St Fl 503 Philadelphia PA 19106-4027

LIEGEL, LEON HERMAN, soil scientist, forester; b. Richland Center, Wis., Sept. 30, 1947; s. Luke Alois and Elizabeth Theresa (Dischler) L.; m. Beth Appleton, Mar. 26, 1978; children: Lea Noel, Lora Hope. BS, U. Wis., 1970; MS, SUNY, Syracuse, 1973; D, N.C. State U., 1981. Forester Soil Conservation Svc. USDA, Bayamon, P.R., 1970; rsch. forester Forest Svc. USDA, Rio Piedras, P.R., 1974-79, soil scientist, 1979-85; soil scientist Forest Svc. USDA, Corvallis, Oreg., 1985-91; rsch. forester Forest Svc. USDA, Corvallis, 1991—; natural resources specialist P.R. Dept. Natural Resources, San Juan, 1971-73; cons. U.S. Agy. for Internat. Devel., Haiti, 1982; forest soils scientist EPA, Corvallis, 1985-91. Author: (with others) Forest Nursery Management in the Caribbean, 1987. Grantee Forest Svc. USDA, Raleigh, N.C., 1976-78, Man and Biosphere, 1980, 93, U.S. Agy. for Internat. Devel., 1983-87. Mem. Soil Sci. Soc. Am., Soc. Am. Foresters (Mary's Peak chpt.), Am. Soc. Photogrammetry and Remote Sensing, Internat. Soc. Tropical Foresters, Sigma Xi. Office: Pacific NW Rsch Sta 3200 SW Jefferson Way Corvallis OR 97331-4401

LIEHR, JOACHIM GEORG, pharmacology educator, cancer researcher; b. Namslau, Silesia, Germany, June 20, 1942; came to U.S., 1965; s. Georg and Anna Maria (Skupin) L.; m. Katherine M. Liehr, Dec. 17, 1988; 1 child, Christopher Joachim. Diploma, U. Münster, Germany, 1965; PhD, U. Del., 1968. With space sci. ctr. U. Calif., Berkeley, 1968-69; with dept. chemistry Tech. U. Munich, Germany, 1970-72; with lipid rsch. inst. coll. medicine Baylor U., Houston, 1972-74; with dept. spectroscopy Ciba-Geigy Ltd., Basle, Switzerland, 1974-76; with dept. biochemistry med. sch. U. Tex., Houston, 1976-83, with dept. pharmacology med. sch., 1983-85; with dept. pharmacology med. sch. U. Tex., Galveston, 1985—; mem. chem. pathology study sect. NIH, NCI, Washington, 1986-90. Contbr. articles to sci. jours. Achievements include research in synthesis of estrogens with decreased carcinogenic activity, mechanism and prevention of estrogen-induced carcinogenesis. Home: 1712 Winnie Galveston TX 77550 Office: U Tex Dept Pharmacology Galveston TX 77550

LIEKHUS, KEVIN JAMES, chemical engineer; b. Omaha, June 10, 1960; s. Alvin Jacobs and Joan Frances (Knust) L. BS in Chem. Engring., Rose Hulman Inst. Tech., 1982, MS in Chem. Engring., 1984; PhD in Chem. Engring., Fla. State U., 1990. Registered profl. engr., Idaho. Adj. instr. Fla. State U., Tallahassee, 1985-89; engring. specialist EG&G Idaho, Inc., Idaho Falls, 1990—. Mem. Am. Inst. Chem. Engrs., Sigma Xi, Tau Beta Pi, Omega Chi Epsilon. Achievements include research in chem. engring.

LIEM, ANNIE, pediatrician; b. Kluang, Johore, Malaysia, May 26, 1941; d. Daniel and Ellen (Phuah) L. BA, Union Coll., 1966; MD, Loma Linda U., 1970. Diplomate Am. Bd. Pediatrics. Intern Glendale (Calif.) Adventist Hosp., 1970-71; resident in pediatrics Children's Hosp. of Los Angeles, 1971-73; pediatrician Children's Med. Group, Anaheim, Calif., 1973-75, Anaheim Pediatric Med. Group, 1975-79; practice medicine specializing in pediatrics Anaheim, 1979—. Fellow Am. Acad. Pediatrics; mem. Los Angeles Pediatric Soc., Orange County Pediatric Soc., Adventist Internat. Med. Soc., Chinese Adventist Physicians' Assn. Avocations: reading, gardening. Office: 1741 W Romneya Dr # D Anaheim CA 92801-1805

LIEM, KHIAN KIOE, medical entomologist; b. Semarang, Java, Indonesia, Jan. 11, 1942; came to U.S., 1969; s. Coen Ing T and Marie Soei-Nio (Goei) L.; m. Anita Tumewu, Apr. 3, 1980; children: Brian Dexter, Tiffany Marie, Jennifer Amanda, Ashley Elizabeth. BS, Bandung Inst. Tech., Bandung, Indonesia, 1964; MS, Bandung Inst. Tech., 1966, Eastern Ill. U., 1970; PhD, U. Ill., 1975. Registered profl. entomologist, vector ecologist. Grad. teaching asst. Bandung Inst. Tech., 1964-66, grad. instr., 1966-68; grad. rsch./teaching asst. Eastern Ill. U., 1969-70; grad. teaching asst. U. Ill., 1970-74; med. entomologist South Cook County Mosquito Abatement Dist., Harvey, Ill., 1974-76; mgr./dir.-med. entomologist South Cook County Mosquito Abatement Dist., Harvey, 1977—; cons. U.S. AID, Washington, 1979—. Recipient Community Svc. award Asian Am. Coalition, 1993. Mem. Am. Mosquito Control Assn. (chmn. resolution com. 1977-78, mem. editorial bd. 1980-83, mem. worldwide com. 1987—), Ill. Mosquito Control Assn. (pres. 1979-81), Entomol. Soc. Am. (com. on book revs.), Am. Tropical Medicine and Hygiene Assn., Am. Registry of Profl. Entomologists, Scientists Inst. Pub. Info., Soc. Vector Ecology, Sigma Xi, Phi Sigma. Roman Catholic. Avocations: soccer, tennis, martial arts, camping, classical music. Home: 7824 E Sequoia Ct Orland Park IL 60462-4241 Office: Mosquito Abatement Dist 15440 Dixie Hwy Harvey IL 60426-3402

LIEPKALNS, VIS ARGOTS, biochemist; b. Riga, Latvia, July 3, 1945; s. George and Alma (Pukulis) Cernaks; m. Christine Icard, Dec. 27, 1980; children: Justine, Jimmy. BA magna cum laude, Hiram Coll., 1966; MS, Johns Hopkins U., 1969; PhD, Loyola U., 1972. Postdoctoral fellow U. Iowa, Iowa City, 1974-75, rsch. scientist dept. medicine, 1975-76; rsch. assoc. dept. medicine Boston U., 1977-79; instr. dept. pathology Ohio State U., Columbus, 1979-80, asst. prof., 1980-82; guest scientist Assn. for Cancer Rsch., Paris, 1982-84; charge de recherche CNRS, Paris, 1984-92; vis. scientist McLean Hosp., Harvard U., Belmont, Mass., 1990-91; investigator, NATO, 1990-91; reviewer Internat. Jour. Cancer, Internat. Neurochemistry, Jour. Lipid Rsch., Molecular Immunology. Contbr. articles to profl. publs. Grantee Am. Heart Assn., 1974, Am. Cancer Soc., 1980, Assn. Cancer Rsch., 1982-89, NSF, 1966-69. Mem. AAAS, Am. Chem. Soc., Am. Soc. Cell Biology, N.Y. Acad. Scis., Soc. Complex Carbohydrates. Avocations: tennis, piano, photography. Office: U Paris Sud, Biochimie Bldg 432, 91405 Orsay France

LIEPMANN, HANS WOLFGANG, physicist, educator; b. Berlin, Germany, July 3, 1914; came to U.S. 1939, naturalized, 1945; s. Wilhelm and Emma (Leser) L.; m. Kate Kaschinsky, June 19, 1939 (div.); m. Dietlind Wegener Goldschmidt, 1954; 2 children. Student, U. Istanbul, 1933-35, U. Prague, 1935; Ph.D., U. Zurich, 1938; Dr. Engring. (hon.), Tech. U. Aachen, 1985. Research fellow U. Zurich, 1938-39; mem. faculty Calif. Inst. Tech., Pasadena, 1939, prof. aeronautics, 1949—, dir. Grad. Aeronautical Labs., 1972-85, Charles Lee Powell prof. fluid mechanics and thermodynamics, 1976-83, Theodore von Kármán prof. aeronautics, 1983-85, Theodore von Kármán prof. aeronautics emeritus, 1985—; mem. research and tech. adv. com. on basic research NASA. Co-author: (with A.E. Puckett) Aerodynamics of a Compressible Fluid, 1947; (with A. Roshko) Elements of Gas-dynamics, 1957. Contbr. articles to profl. jours. Recipient Physics prize U. Zurich, 1939, Prandtl Ring, German Soc. Aeros. and Astronautics, 1968, Worcester Reed Warner medal ASME, 1969, Nat. Medal of Sci., 1986, Guggenheim medal, 1986, Nat. Medal of Tech., Nat. Sci. Found., 1993. Fellow AIAA (Fluid and Plasmadynamics award 1990), Am. Acad. Arts and Scis., Am. Phys. Soc. (Fluid Dynamics prize 1980, Otto Laporte award 1985); hon. fellow Indian Acad. Scis., 1985; fgn. fellow Max-Planck Institut, 1988; mem. NAS (award 1971), NAE (award 1965), AAAS, ASME (hon.), Internat. Acad. Astronautics, Sigma Xi (Monie A. Ferst award 1978). Address: Calif Inst Tech Dept Aeronautics Pasadena CA 91125

LIEPSCH, DIETER WALTER, engineering educator; b. Weissenfels, Fed. Republic Germany, Aug. 30, 1940; s. Walter Paul and Johanna (Herold) L.; m. Joyce McLean; children: Eva-Beate, Stephan. Diploma in engring., Tech. U., Munich, 1966, Dr. in Engring., 1974, Habilitation, 1986. R&D engr. Siemens AG, Munich, 1967-69; researcher Gesellschaft für Strahlen- u. Umweltf., Neuherberg, Fed. Republic Germany, 1969-72; prof. Fachhochschule Munich, 1972—; rsch. scientist/dir. Eisenhower Med. Ctr., U.S., 1986-89; prof. Fachhochschule Munich, 1989—; guest prof. U. Houston, 1983-84; founder/pres. Inst. für Biotechnik, Munich, 1985; v.p. Ingenieure der Versorgungstechnick, Munich, 1975-86; cons. in field. Editor: Blood Flow in Large Arteries, 1989, Biofluid Mechanics, 1990, 1st Con. Versorgungs Ing., 1977. Mem. ASME, VDI, SPIE, Am. Phys. Soc., GAMM, European Soc. Biomechanics, European Soc. on Engring. and Medicine, European Soc. for Microcirculation, Internat. Soc. for Biorheology, Biomed. Engring. Soc. Home: Am Buchenwald 29, 82340 Feldafing Germany Office: Fachhochschule Munich, Lothstr 34, 8000 Munich Germany

LIEW, CHONG SHING, structural engineer; b. Malacca, Malaysia, Oct. 13, 1964; came to U.S., 1984; s. Ching H. Liew and Beng H. Tan. BS in Civil Engring., U. Southwestern La., 1987; MS in Civil Engring., Ga. Tech., 1991. Engr. in tng. Engring. technician The Burke Co., Atlanta, 1988-90; structural engr. Brockway & Assocs., Atlanta, 1989-90; chief engr. BPS Equipment, Atlanta, 1990—. Mem. ASCE, NSPE, Am. Concrete Inst. Home: 6402 Park Ave Atlanta GA 30342

LIFSCHITZ, KARL, sales executive; b. N.Y.C., Jan. 26, 1949. BA, Yeshiva U., 1970; MA, Columbia U., 1971. Founder, pres. GFI Advanced Techs., Inc., Forest Hills, N.Y., 1987—. Office: GFI Advanced Techs Inc 112-41 69th Ave Flushing NY 11375-3917

LIGGETT, MARK WILLIAM, mechanical engineer; b. Syracuse, N.Y., Apr. 28, 1957; s. William Edward and Joan Petrican (Hiser) L.; m. Elisabeth Rita Hergenhan, Apr. 4, 1992; 1 child, Christina Lynn. BSME, U. Minn., 1980. Registered profl. engr., Calif. Engr. U.S. Bur. Mines & Metalurgy, St. Paul, 1978-80; engring. specialist Gen. Dynamics Space Systems, San Diego, 1980—; energy specialist, ptnr. Energy Conservation Engring., San Diego, 1987-88. Contbr. ARticles to Cryogenics Jour.; author tech. articles. Missionary Christian Outreach Internat., Salzburg, Austria, 1990-91; hockey player North American Crusaders, East and West Europe, 1989-91; head coach Huntsville (Ala.) Amateur Hockey Assn., 1992. Recipient Cert. of Recognition award NASA, 1993. Mem. AIAA, ASME (assoc., Outstanding Svc. 1980). Achievements include development of Cryogenic On Orbit Liquid Analytical Tool (COOLANT) computer code for cryogenic storage behavior for NASA; research in heat transfer, thermodynamics analysis, of Atlas/Centaur space vehicle, aircraft, superconducting magnets; and prototype system testing. Office: Gen Dynamics Space Systems 620 Discovery Dr Huntsville AL 35806

LIGHT, WILLIAM ALLAN, mathematics educator; b. Chester, Cheshire, Eng., Apr. 19, 1950; s. Louis John and Mary Goodbrand (Findlay) L.; m. Anita Mary Edwards, July 24, 1971. BSc, Sussex U., Eng., 1971; MA with distinction, Lancaster U., Eng., 1973, PhD, 1976. Cert. in postgrad. edn., Wales. Tutorial fellow U. Lancaster, 1974-76, lectr. math., 1976-89, sr. lectr., 1989-91; prof., head dept. math. and computer sci. U. Leicester, Eng., 1991—; vis. prof. Tex. A&M U., College Station, 1982-83, U. Tex., Austin, 1987-88; guest prof. U. Kuwait; bd. dirs. NATO Rsch. Team, 1985-88, 92—; dir. summer sch. in numerical analysis Sci. Rsch. Coun., Lancaster, 1990; speaker in field. Author: Approximation in Tensor Product Spaces, 1986, Abstract Analysis; editor Advances in Numerical Analysis, vols. I and II, Jour. Approximation Theory; contbr. over 60 papers to profl. confs. and jours. Chmn. Lancaster U. Grad. Assn., 1972-73. Grantee Sci. Rsch. Coun.; Erskine fellow U. Canterbury, New Zealand, 1993. Mem. Am. Math. Soc., London Math. Soc. Avocations: skiing, private pilot. Office: U Leicester, Dept Math and Computer Sci, Leicester LE1 7RH, England

LIGHTFOOTE, MARILYN MADRY, molecular immunologist; b. Jacksonville, Fla.; d. Arthur Chester and Janie (Cowart) Madry; m. William Edward Lightfoote II, Oct. 23, 1971; 1 child, Lynne Jan-Maria. BA in Chemistry magna cum laude, Fisk U.; MS in Biochemistry, Georgetown U.; PhD in Microbiology and Immunology, U. Va., 1983. Staff fellow Lab. Immunogenetics Nat, Inst. Allergy Infectious Diseases, NIH, Bethesda, Md., 1983-85, staff fellow Lab. Immunoregulation, 1985-87; rsch. faculty dept. biochemistry George Washington U., 1987-90; molecular immunologist FDA, Rockville, Md., 1990—; Graves meml. lectr. biology dept. N.C. A&T Coll., Greensboro; NSF tng. fellow U. Va., 1979-83; presenter various orgns., workshops and symposiums. Author: Biology of Light, 1992; issue editor Immunomethods, 1992-93. Dir. fund raising Jack and Jill of Am., Reston, Va., 1986-90; dir. Project Lead, Links Inc., Reston, 1990; mem. vestry St. Paul's Episcopal Ch., Alexandria, Va., 1987-90. Mem. AAAS, Am. Assn. Immunology, Mortarboard, Sigma Xi. Achievements include development and characterization of cell lines for study of HIV virus; first isolation and publication of amino acid analysis of HIV Reverse Transcriptase. Home: 827 Swinks Mill Rd McLean VA 22102 Office: FDA 12709 Twinbrook Pwy Rockville MD 20857

LIGI, BARBARA JEAN, architectural and interior designer; b. Binghamton, N.Y., June 13, 1959; d. Robert Richard and Helen Margaret (Wagner) Taylor; m. Alan Joseph Ligi, July 24, 1982; children: Curtis John, Ryan Robert. AA, Mt. Ida Coll., 1979; BFA, Syracuse U., 1982. Cert. N.C.D.I.Q. Designer Norman Davies, Architect, Binghamton, 1982—; mem. adj. faculty design Broome C.C., Binghamton, 1987—. Active Nat. Trust for Hist. Preservation. Mem. Am. Soc. Interior Designers (profl.), Gold Key. Democrat. Roman Catholic. Avocation: travel photography. Home: 2803 Robins St Endwell NY 13760 Office: Norman J Davies Architect 783 Chenango St Binghamton NY 13901

LIGON, DAISY MATUTINA, systems engineer; b. Manila, Philippines, Jan. 18, 1947; came to U.S., 1971; d. Dominador P. and Antipolo (Villaruel) N.; m. Augusto B. Ligon, Sept. 4, 1976; 1 child, Kevin. BS in Chemistry, U. Philippines, Qezon City, 1968; MBA with distinction, Nat. U., San Diego, Calif., 1982; MS in Systems Engring., George Mason U., 1992. Instr. dept. chemistry U. Philippines, Quezen City, 1968-71; engr. G.A. Techs., San Diego, 1971-87; lead engr. Mitre Corp., McLean, Va., 1987—. Mem. membership com. Troop 1347 Boy Scouts Am., Burke, Va., 1991—. Mem. Am. Nuclear Soc. Achievements include rsch. on prodn. of citric acid from Philippine molasses by fermentation with different strains of aspergillus niger. Office: Mitre Corp 7525 Colshire Dr Mc Lean VA 22102

LIGOTTI, EUGENE FERDINAND, retired dentist; b. N.Y.C., June 10, 1936; s. Eugene A. and Lee (D'Agata) L.; m. Corbina Theresa Loscalzo, Nov. 21, 1959; children: Gina Maria Ligotti Aliperti, Lisa Anne. BA, Adelphi U., 1958; DDS, NYU, 1962. Pvt. practice Huntington, N.Y., 1962-90; instr. operative dentistry NYU, N.Y.C., 1962-65. Author historic fiction; freelance contbr. articles to profl. jours. and mags. Founder, pres. Upper Bay Civic Assn., Inc., Huntington, 1979—. Mem. ADA, N.Y. State Dental Soc., Suffolk County Dental Soc., German Shepherd Dog Club (pres. 1971-75), Xi Psi Phi (founder, pres. alumni chpt. 1981-82), Chi Sigma. Republican. Roman Catholic. Avocations: travel, writing, raising show dogs.

LI-KAM-WA, PATRICK, research optics scientist, consultant; b. Mauritius, Oct. 22, 1958; came to U.S., 1989; s. Joseph and Moy Mou (Wong Yan Man) Li Kam Wa; m. Micheline Chin-Chew, Sept. 8, 1982; children: Wendy Karen, Robert Patrick. B Engring., U. Sheffield, Eng., 1982, PhD, 1987. Tchr. Eden Coll., Mauritius, 1978-79; rsch. asst. U. Sheffield, 1985-87, rsch. assoc., 1987-89; rsch. scientist ctr rsch. in electro-optics and lasers U. Cen. Fla., Orlando, 1989—; cons. Novatec Laser Systems, San Diego, 1991—. Contbr. articles to Electronics Letters, IEEE Jour. Quantum Electronics, Optical and Quantum Electronics, Optics Letters. Recipient award Instn. Elec. Engrs., London, 1986. Mem. IEEE, Optical Soc. Am. Roman Catholic. Achievements include research on all-optical switching in waveguide, ultrashort pulse generation from chromium doped laser crystals. Office: U Cen Fla CREOL 12424 Research Pky Orlando FL 32826

LILJESTRAND, HOWARD MICHAEL, environmental engineering educator; b. Houston, July 29, 1953; s. Walter Emmanuel and Frances Newland (Lane) L.; m. Blinda Eve McClelland, Aug. 19, 1986; children: Emily Morgan, Frasier Lane. BA, Rice U., 1974; PhD, Calif. Inst. Tech., 1980. Registered profl. engr., Tex. Asst. prof. civil engring. Calif. State U., L.A., 1979-80; asst. prof. civil engring. U. Tex., Austin, 1980-85, assoc. prof., 1985-92, prof., 1992—; reviewer U.S. Nat. Acid Precepitation Assessment Program, 1983-90; mem. adv. bd. Alta. (Can.) Acid Deposition Rsch. Program, Calgary, 1987-88. Contbr. articles to Jour. Environ. Sci. and Tech., Atmospheric Environ., Water Sci. and Tech. Mem. ASCE, Am. Chem. Soc., Sigma Xi, Tau Beta Pi. Achievements include initial documented existence of acid rain in the western U.S., importance of nitric acid in acid rain in the west, and importance of dry deposition of acids in the west. Office: Civil Engring 8.6 ECJ U Tex Austin TX 78712

LILLER, KAREN DESAFEY, health education educator; b. Pitts., Nov. 18, 1956; d. Thomas and Irene (Cenderelli) DeSafey; m. David Allen Liller, Aug. 30, 1980; 1 child, Matthew Thomas Allen Liller. BS, W.Va. U., 1978; MA, U. South Fla., 1982, EdS, 1986, PhD, 1988. Med. technologist Fla. Hosp., Altamonte Springs, 1978-81; lab. instr. Tampa (Fla.) Med. Coll., 1982-83; edn. dir. Sch. of Med. Tech. Tampa Gen. Hosp., 1983-85; sci. advisor Mylan Pharms., Inc., Tampa, 1986-87; postdoctoral fellow Coll. of Pub. Health U. South Fla., Tampa, 1988-90, asst. prof. Coll. of Pub. Health, 1990—. Contbr. articles to profl. jours. Mem. Am. Soc. Clin. Pathologists, Assn. for the Advancement of Health Edn., APHA. Home: 16509 Cayman Dr Tampa FL 33624-1065 Office: U South Fla Coll Pub Health 16509 Cayman Dr Tampa FL 33624

LILLIEHÖÖK, J(OHAN) BJÖRN O(LOF), health science association administrator; b. Stockholm, Feb. 12, 1945; s. Nils-Olof and Barbro (Brandel) L.; m. E. Margareta Setterberg, Sept. 7, 1968; children: Veronica, Lovisa, Alexandra. LLB, U. Stockholm, 1969. Dist. judge's assessor Dist. Ct. Sollentuna and Farentuna, Solna, Stockholm, Sweden, 1969-72; prin. clk. Swedish Nat. Bd. Immigration, Stockholm, 1972-73; dir., v.p., head regional office Swedish Employers' Confederation, Stockholm, 1973-90; sec.-gen., chief exec. officer Swedish Heart Lung Found., Stockholm, 1990—; chmn. Stockholm New Enterprise Ctr., 1987-90; bd. dirs. F.O.F. Rsch. and Progress Found., Stockholm, Nordic Assn., Stockholm. Mem. Internat. Soc. for Labour Law and Social Security, Sallskapet, SvD Exec. Club, Rotary (bd. dirs. Stockholm club, pres. 1989-90). Lutheran. Avocations: badminton, tennis, skiing, photography. Office: Swedish Heart Lung Found, Kungsgatan 54, S11135 Stockholm Sweden

LILLY, LES J., micronautics engineer; b. Lewiston, Idaho, July 14, 1950; s. Walter Franklin and Mary Lee (Roberg) L.; m. Judith C. Aites, Mar. 1970 (div.Nov. 1977); children: Christopher Brian, Deserea Dawn; m. Kimberly Lynn Boniventura, Nov. 17, 1984; 1 child: Natasha Crystal; 1 child with Jennie Marie Warwick: Jerimiah Michael Warwick-Lilly; 1 child with Victoria Smith: Vanesa Ashley Smith-Lilly. Diploma EMT/paramedic, U.S. Army Sch., Denver, 1974; AS in Math./Sci., Pikes Peak Community Coll., Colorado Springs, Colo., 1979. Cert. level I res. police officer, Calif. Rsch. technician NCR-Plasma/LCD Display Rsch. and Devel. Ctr., Colorado Springs, 1979-81; sr. engring. technician Mostek/United Techs., Colorado Springs, 1981-83, Intel-Fab 7, Rio Rancho, Colo., 1983-84; asst. engr. Fairchild Advanced Rsch. Devel. Lab., Palo Alto, Calif., 1985-87, Fairchild/Nat. Semiconductor, Puyallup, Wash., 1987-89; sr. process engr. Silicon Systems, Santa Cruz, Calif., 1989—; cons. Semitool, Inc., San Jose, Calif., 1985-86, G&K Svcs., Santa Clara, Calif., 1989—, Cafe Bene, Santa Cruz, 1989—. Editor: NCR monthly newsletter, 1980; contbr. articles to publs. in field. Vol. Am. Wildlife Rescue Ctr., Scotts Valley, Calif., 1989-91; co-founder Coalition for the Homeless, Santa Cruz, 1990—. With U.S. Army, 1973-75, Korea, USANG, 1987-90, USAR, 1990—. Recipient Author of Merit award Nat. Semiconductor, Puyallup, 1989, Silicon Systems Pen & Quill award. Mem. Inst. Environ. Scis., Am. Chemists Soc., Electrochem. Soc., Fine Particle Soc. Republican. Buddhist. Achievements include patent pending for sealing process for LCD's displays; co-development team member of 1 micron Bicmos 256K SRAM memory chip which established world record for speed; development team member of world's first 6 inch wafer FAB-semiconductors (Intel 7AB7) and 5 inch wafer (FAB-mostek/UTI). Office: Silicon Systems 2300 Delaware Ave Santa Cruz CA 95060-5728

LIM, ALEXANDER RUFASTA, neurologist; b. Manila, Philippines, Feb. 20, 1942; s. Benito P. and Maria Lourdes (Cuyegkeng) L.; m. Norma Sue Hanks, June 1, 1968; children: Jeffrey Allen, Gregory Brian, Kevin Alexander, Melissa Gail. AA, U. Santo Tomas, Manila, Philippines, 1959, MD, 1964. Intern Bon Secours Hosp., Balt., 1964-65; resident internal medicine Scott and White Clinic, Temple, Tex., 1965-67; resident in neurology Cleve. Clinic, 1967-69, chief resident in neurology, 1969-70, fellow clin. neurophysiology, 1970-71, clin. assoc. neurologist, 1971-72; neurologist Neurol. Clinic, Corpus Christi, Tex., 1972—; co-founder, co-mng. ptnr., pres. Neurology, P.A., Corpus Christi, 1972-92; chief neurology Meml. Med. Ctr., Corpus Christi, 1975-76, 78, 85, 90, Spohn Hosp., Corpus Christi, 1974, 76, 86, 90, Reynolds Army Hosp., Ft. Sill, Okla., 1990-91; clin. assoc. prof. Sch. Medicine U. Tex. Health Sci. Ctr., San Antonio. Mem. editorial bd. Coastal Bend Medicine, 1988—. Lt. col. Med. Corps, 1990-91. Recipient Army Commendation medal, 1991, Nat. Def. medal U.S. Army, 1991. Mem. AMA, Tex. Med. Assn. (chmn. neurology 1985-86), Tex. Neurol. Soc. (sec. 1986-88, pres. 1989-90), Am. Acad. Neurology, Am. Epilepsy Soc., Am. Acad. Clin. Neurophysiology, Am. Electroencephalographic Soc., So. Electroencephalographic Soc., Am. Acad. Pain Mgmt., K.C. Republican. Roman Catholic. Avocations: tennis, philately, travel, snow skiing, bonsai. Home: 4821 Augusta Cir Corpus Christi TX 78413-2711 Office: The Neurological Clinic 3006 S Alameda St Corpus Christi TX 78404-2698

LIM, JOSEPH DY, oral surgeon; b. Manila, Nov. 2, 1948; s. Celestino Yu and Soledad (Dy) L.; m. Giok Leng Cua, Nov. 10, 1974; children: Joseph Oliver, Alistair Bryan, Kenneth Lester, Mark Andrew. DMD cum laude, U. of the East, Philippines, 1974. Pres. Filipino-Chinese Dental Found., Inc., 1982-83; mng. dir. Internat. Dental Supply, Inc., Manila, 1982—; pvt. practice Manila, 1975—; pres. Oral Implantology Ctr. of The Philippines, 1992—; chmn. trade exhibits Philippine Dental Assn., 1982-89; co-chmn. trade exhibits Fedn. Dentaire Internat., Philippines, 1986, Asian-Pacific Dental Congress, 1994; chief del. Philippine delegation, 14th Asian-Pacific Dental Congress, Seoul, South Korea, 1989; organizer CIVAC Team, Manila Dental Soc., 1982—; com. mem. Asian Oral Implant Acad.; Philippine del. 16th Asian Pacific Dental Congress, Kuala Lumpur, Malaysia, 1993. Editorial cons. MDS Digest, 1989; contbr. sci. articles to profl. jours. Chmn., team leader Civic Action, Filipino-Chinese Dental Found., Inc., 1981—. Recipient Disting. Svc. award Filipino-Chinese Dental Found., Inc., 1982. Fellow Philipine Coll. Oral Maxillo-Facial Surgeons, Internat. Coll. of Dentists, Internat. Assn. Oral & Maxillofacial Surgeons, Asian Oral Implant Acad. (hon.); mem. ADA (assoc.), Philippine Dental Assn. (auditor 1982-83, 92—, trustee 1988-89, Presdl. award of merit 1983), Manila Dental Soc. (pres. 1989-90), U. East Dental Alumni Inc. (dir. 1983—), Fedn. Dentaire Internat., Asian-Pacific Dental Fedn., Assn. Asian Oral Maxillo-Facial Surgeons, Asian Acad. Craniomandibular Disorders, Assn. for the Dental Practitioners of the Philippines (pres. 1993—), Internat. Congress of Oral Implantologists, Oral Implantology Cetr. Study Club (founding mem.). Avocations: reading, swimming, bowling, horseback riding, tennis. Office: Tytana Pla Ste 821, Pla Lorenzo Ruiz Binondo, Manila The Philippines also: Ortigas Ctr Complex, Jollibee Ctr Bldg Ste 202, Pasig, Metro Manila Philippines

LIM, KIERAN FERGUS, chemistry educator; b. Kuala Lumpur, Selangor, Malaya, June 15, 1962; arrived in Australia, 1970; s. Patrick and Elizabeth (Tang) L.; m. Jeanne Lee, July 11, 1992. BSc with honors, U. Sydney, 1985, PhD, 1989. Postdoctoral rsch. Stanford (Calif.) U., 1988-90; vis. scientist U. Göteborg, Sweden, 1990; lectr. in chemistry U. New England, Armidale, Australia, 1991-93, U. Melbourne, Australia, 1993—. Mem. Am. Chem. Soc., Royal Australian Chem. Inst. (sec. phys. chemistry div. 1991—). Achievements include rsch. on the theory of collisional energy transfer from highly vibrationally excited molecules, theory of chem. reaction dynamics. Office: U Melbourne, Sch Chemistry, Parkville Victoria 3052, Australia

LIM, RAMON (KHE-SIONG), neuroscience educator; b. Cebu City, Philippines, Feb. 5, 1933; came to U.S., 1959; s. Eng-Lian and Su (Yu) L.; m. Victoria K. Sy, June 21, 1961; children—Jennifer, Wendell, Caroline. A.B., U. Santo Tomas, Manila, 1953, M.D. cum laude, 1958; Ph.D. in Biochemistry, U. Pa., 1966. Research neurochemist U. Mich., Ann Arbor, 1966-69; asst. prof. biochemistry U. Chgo., 1969-76, assoc. prof. Brain Research Inst., 1976-81; prof. dept. neurology U. Iowa, Iowa City, 1981—, dir. div. neurochemistry and neurobiology, 1981—. Mem. editorial bd. Internat. Jour. Devel. Neurosci., 1984—. Contbr. articles to sci. jours. Discovered Glia Maturation Factor, 1972, sequencing and cloning of the same, 1990. Grantee NIH, 1971—, NSF, 1979—, VA, 1981—. Mem. Am. Soc. Biochem. Molecular Biology, Internat. Soc. Neurochemistry, Am. Soc. Neurochemistry, Soc. Neurosci., Am. Soc. Cell Biology. Avocations: calligraphy, painting, writing, music. Home: 118 Richards St Iowa City IA 52246-3516 Office: U Iowa Iowa City IA 52242

LIM, SHUN PING, cardiologist; b. Singapore, Dec. 1, 1947; came to U.S., 1980; s. Tay Boh and Si Moi (Foo) L.; m. Christine Sock Kian Ng; children: Corinne Xian-li, Damiel John Xian-ming, Justin David Xian-an. MBBS with honors, Monash U., Clayton, Australia, 1970, PhD, 1981; M in Medicine, Nat. U. Singapore, 1975; M, Royal Australasian Coll. Physicians, 1975. Rsch. scholar Australian Nat. Health and Med. Rsch. Coun., Canberra, 1978-79; fellow in cardiology Michael Reese Hosp., Chgo., 1980-82; chief noninvasive cardiovascular imaging Cin. V.A.M.C., 1982-86; asst. prof. U. Cin., 1982-86; cardiologist Quain and Ramstad Clinic, Bismarck, N.D., 1986-88; clin. asst. prof. U. N.D., Bismarck, 1986-90; pvt. practice cardiovascular diseases, 1988-91; assoc. prof. medicine U. N.D., 1991—; dir. catheterization lab. Marion (Ohio) Gen. Hosp., 1992—; pres. Inst. for Advanced Med. Tech., 1990-93; chmn. ICU com. VA Med. and Regional Office Ctr., Fargo, N.D., 1991-93, chief cardiology sect., 1992—. Contbr. articles to profl. jours.; catheter tip polarographic lactic acid and lactate sensor. Fellow ACP, Am. Coll. Cardiology, Internat. Coll. Angiology, Am. Coll. Angiology, Royal Australian Coll. Physicians, Am. Coll. Chest Physicians, Coun. on Clin. Cardiology of Am. Heart Assn.; mem. Am. Fedn. Clin. Rsch., Am. Soc. Echocardiography, Am. Heart Assn. (grantee 1984-85), Ohio Med. Assn., N.Y. Acad. Scis. (life). Methodist. Office: Cardiology Ste 4th Fl High McKinley Pk Dr Marion OH 43302-2498

LIM, YOUNGIL, economist; b. (Bakchon) Seoul, Korea, June 12, 1932; arrived in Austria, 1981; s. Suk-Moo and Anna (Lee) L.; m. Helen Kum-Hyun Shin, June 14, 1959; children: Mihae, Sunghae Anna, Michael. BA, Harvard U., 1958; PhD in Econs., UCLA, 1965. From asst. prof. to prof. U. Hawaii, Honolulu, 1964-81; rsch. economist UN Indsl. Devel. Orgn., Vienna, 1981—; cons. UN Economic and Social Commn. for Asia and Pacific, Bangkok, 1971-75; com. mem. Social Sci. Rsch. Coun.-Am. Coun. Learned Socs. on Korea, N.Y., 1973-79; cons. Korea Internat. Economic Inst., Seoul, 1977-81; sr. economist Global Issues and Policy Analysis Br. of UNIDO, 1981—, officer ann. series pub. Industry and Develop. Global Report. Author: Government Policy and Private Enterprise: Korean Experience in Industrialization, 1981; contbr. articles in profl. jours. Recipient Fulbright-Hays fellowship, U.S. Dept. Edn., Washington, 1968, 77, East-West Ctr. fellow, East-West Ctr., Honolulu, 1973; grantee Asia Found., San Francisco, 1970-71, Korea Traders Assn., Seoul, 1978-79. Avocations: music, tennis, reading. Home: 47 Haizingergasse, A-1180 Vienna Austria Office: UNIDO, Vienna Internat Ctr, A-1400 Vienna Austria

LIN, ALICE LEE LAN, physicist, researcher, educator; b. Shanghai, China, Oct. 28, 1937; came to U.S., 1960, naturalized, 1974; m. A. Marcus, Dec. 19, 1962 (div. Feb. 1972); 1 child, Peter A. AB in Physics, U. Calif., Berkeley, 1963; MA in Physics, George Washington U., 1974. Statis. asst. dept. math. U. Calif., Berkeley, 1962-63; rsch. asst. in radiation damage Cavendish Lab. Cambridge (Eng.) U., 1965-66; info. analysis specialist Nat. Acad. Scis., Washington, 1970-71; teaching fellow, rsch. asst. George Washington U., Catholic U. Am., Washington, 1971-75; physicist NASA/Goddard Space Flight Ctr., Greenbelt, Md., 1975-80, Army Materials Tech. Lab., Watertown, Mass., 1980—. Contbr. articles to profl. jours. Mencius Ednl. Found. grantee, 1959-60. Mem. AAAS, N.Y. Acad. Scis., Am. Phys. Soc., Am. Ceramics Soc., Am. Acoustical Soc., Am. Men and Women of Sci., Optical Soc. Am. Democrat. Avocations: rare stamp and coin collecting, art collectibles, home computers, opera, ballet. Home: 28 Hallett Hill Rd Weston MA 02193-1753 Office: Army Materials Tech Lab Mail Stop MRS Bldg 39 Watertown MA 02172

LIN, CHI-HUNG, chemical engineer; b. Taipei, Apr. 1, 1943; came to U.S., 1966; s. Chuan and Kin-Yu Lin; m. Betty Chu, Aug. 13, 1967; children: Sandra, Sherry, Sara. BS, Nat. Taiwan U., 1965; PhD, U. Ill., 1970. Rsch. engr. R&D, Amoco Oil, Naperville, Ill., 1970-75, various positions, 1975-90, rsch. assoc., 1990—. Contbr. articles to profl. jours. Mem. AICE (sect. chmn. Anaheim nat. meeting 1984). Achievements include 6 patents on polypropylene manufacturing processes; pioneer of gas-phase polypropylene process development which has led to one of the most preferred modern polypropylene manufacturing processes; rsch. on polypropylene copolymer. Office: Amoco Chem R&D PO box 3011 Naperville IL 60566

LIN, CHING-FANG, engineering executive; b. Taiwan, Mar. 23, 1954. BEE, Nat. Chiao-Tung U.; MS, PhD in Computer, Info. Control Engr., U. Mich. Rsch. engr. U. Va., Charlottesville, 1977-78; lectr. U. Mich., Ann Arbor, 1978-80, adj. prof. elec. and computer engring., 1980-82; sr. rsch. scientist Applied Dynamics Internat., 1980-82; mem. fauclty elec. and computer engring., dir. control engr. U. Wis., Madison, 1982-83; lead engr., project supr. flight controls tech. Boeing Mil. Aircraft Co., Seattle, 1983-86; pres. Am. GNC Corp., Chatsworth, Calif., 1986—. Contbr. 150 articles to profl. publications; author: Modern Navigation, Guidance and Control Processing, 1991, Advanced Control Systems Design, 1993. Recipient numerous grants. Fellow (assoc.) AIAA; mem. IEEE (sr.), SCS, IMACS, IFIP, Sigma Xi, Tau Beta Pi. Office: Am GNC Corp 9131 Mason Ave Chatsworth CA 91311

LIN, FEI-JANN, zoology educator, researcher; b. Keelung, Taiwan, May 23, 1934; s. A-Long and Haw (Li) L.; m. Mei Show Kao, Mar. 17, 1963; children: Chien-Zu, Peih-Rur, Chia-Zu. BS, Nat. Taiwan Normal U., 1957; postgrad., U. Tex., Austin, 1968-71. Rsch. sci. engr. U. Tex., 1968-71; prof. Nat. Chung-Hsing U., Tai-chung, Taiwan, 1974-85, Nat. Taiwan Ocean U., Keelung, 1975-76, Tung-Hai U., Tai-chung, 1982-83; prof. zoology Nat. Taiwan U., Taipei, 1986—; rsch. fellow Inst. Zoology Academia Sinica, Nankang, Taipei, Taiwan, 1974—; editor Bull. Inst. Zoology Academia Sinica, Nankang, Taipei, 1989—; guest prof. Tokyo Met. U., 1978-79; program dir. Nat. Sci. Coun., Taipei, 1986-88; cons. China Times, 1975 . Editor Sci. Monthly, 1974-84, Jour. Chinese Entomology, 1986-88; contbr. articles to profl. jours. Sci. attaché Assn. East Asian Rels., Tokyo, 1986-88. 2d lt. Chinese Army, 1958-60. Fellow Royal Entomol. Soc. London; mem. AAAS, Am. Naturalists, Chinese Entomol. Soc., Chinese Biol. Soc., Soc. Chinese Environ. Protection, Willi Hennig Soc., Soc. Systematic Zoology, Soc. for Study of Evolution, N.Y. Acad. Sci. Avocation: bridge. Home: 21 1 Alley 4 Ln 61, Yien-Chiu-Yuan Rd Sect 2, Taipei 11521, Taiwan Office: Academia Sinica, Inst Zoology, Nankang 11529, Taiwan

LIN, GUANG HAI, research scientist, consultant; b. Shanghai, China, Oct. 30, 1942; came to U.S., 1981; s. Zhong D. Lin and Xiang Y. Shi; m. Mu Zhi He, Jan. 25, 1968; children: Hong Lin, Jun Lin. BS in Chem. Physics, Chinese Sci. and Tech. U., Beijing, 1964; MS, U. Colo., 1983, PhD in Physics, 1985. Rsch. assoc. Inst. Mechanics, Academia Sinica, Beijing, 1964-77; asst. rsch. prof. Inst. Mechanics, Beijing, 1978-81; rsch. asst. dept. physics U. Colo., Boulder, 1982-85; rsch. assoc. JILA, U. Colo., Boulder, 1985-87; rsch. assoc. dept. chemistry Tex. A&M U., College Station, 1988-89; asst. rsch. scientist Ctr. Electrochem. Systems and Hydrogen Rsch., College Station, 1989-91; rsch. scientist dept. chemistry Tex. A&M U., College Station, 1991—; adv. Burstein Assocs., Boston, 1992—; cons. Phila. Project, Zurich, 1992—. Author: Introduction to Plasma Physics,1 980; contbr. articles to profl. jours. Mem. Material Rsch. Soc. Achievements include patent for novel hydrogenated amorphous silicon alloys, patent on amorphous silicon-based photovoltaic semiconductor materials free from Staebler-Wronski effects; finding low-energy nuclear reactions. Home: 2907 Indiana Ave Bryan TX 77803 Office: Tex A&M U Dept Chemistry College Station TX 77843

LIN, JUNG-CHUNG, microbiologist, researcher; b. Ping Tung, Taiwan, Republic of China, Nov. 15, 1939; came to U.S., 1970; s. Wan-Ho and Kwei-Tzu (Chen) L.; m. Shou-Huei, July 19, 1967; children: Melissa, Richard. PhD, Temple U., 1974. Rsch. asst. unit 2 U.S. Naval Med. Rsch., Taipei, Taiwan, 1965-66; asst. prof. Nat. Def. Med. Coll., Taipei, 1967-70; assoc. mem. Inst. of Zoology Academia Sinica, Taipei, 1967-74; rsch. assoc. U. N.C., Chapel Hill, 1977-80, rsch. asst. prof., 1980-84, rsch. assoc. prof., 1984-90; chief molecular biology sect. Ctrs. for Disease Control, Atlanta, 1990—; adj. prof. Emory U., Atlanta, 1992—. Author chpts. in books; contbr. over 74 articles to profl. jours. Grantee NCI, NIH, NIAID. Mem. AAAS, Am. Soc. for Microbiology, Am. Soc. for Virology, Am. Assn. for Cancer Rsch., Internat. Soc. for Antiviral Rsch., Internat. Assn. for Rsch. on Epstein-Barr Virus and Associated Diseases. Achievements include patent for design of antisense oligodeoxynucleotide to cure Epstein-Bar virus latent infection. Home: 3723 Toxaway Ct Atlanta GA 30341-4622 Office: Ctrs for Disease Control 1600 Clifton Rd NE Atlanta GA 30333-4046

LIN, JUSTIN YIFU, economist, educator; b. Yilan, Taiwan, China, Oct. 15, 1952; s. Huoshu and Jinhua (Li) L.; m. Yunying Chen, Jan. 21, 1976; children: Leon Xuchu, Lindsay Xi. MA, Zhengzhi U., Taibei, Taiwan, 1978, Peking U., Beijing, 1982; PhD, U. Chgo., 1986. Postdoctoral fellow Econ. Growth Ctr., Yale U., New Haven, Conn., 1986-87; head div. econ. growth studies, dep. dir. Devel. Inst., Rsch. Ctr. for Rural Devel. of the State Coun., Beijing, 1987-89; dep. dir. Dept. Rural Economy, Devel. Rsch. Ctr. of State Coun., Beijing, 1990—; assoc. prof. Peking U., Beijing, 1987—; cons. World Bank, Washington, 1987-90, 93—, Internat. Rice Rsch. Inst., 1992-93; vis. assoc. prof. UCLA, 1990—; adj. prof. Australian Nat. U., Canberra, 1991—. Author: Institutions, Technology and China's Rural Development, 1992 (Sun Yefang prize 1993); mem. bd. editors Agrl. Econs., China Social Scis., Jour. Comparative Econ. and Social Systems; contbr. articles to profl. jours. Mem. nat. com. Chinese People's Polit. Consultative Conf., Beijing, 1988—; dir. Nat. Assn. for Promoting Peaceful Unification of China, Beijing, 1989—. Named David Lam Economist, New Asia Coll., Chinese U. of Hong Kong, 1993; rsch. grantee Rockefeller Found., 1988-89, 91—, travel grantee Ford Found., 1990, Prince fellow Prince Found., 1982-86, postdoctoral fellow Rockefeller Found., 1986-87. Fellow Chinese Economists Soc.; mem. Am. Econ. Assn., Royal Econ. Soc., Econometric Soc., Econ. History Assn., Internat. Assn. Agr. Economists, Assn. Comparative Econ. Studies, Chinese Assn. Agrl. Economists (bd. dirs. 1990—, dep. sec.-gen. 1992—, mem. standing com. 1992—), Am. Agrl. Economist Assn. Office: Dept Rural Economy Devel Rsch Ctr, Xihuangchenggen Nanjie No 9, Beijing 100032, China

LIN, LI-MIN, dentistry educator; b. Sie-Chuang, Republic of China, Aug. 2, 1944; s. Shun-Tao and Cheng-Sing Lin; m. Mei-Jane Chen, Feb. 5, 1967; children: David, Sarah. DDS, Kaohsiung Med. Coll., Taiwan, Republic of China, 1969; MS, U. Chgo., 1975; PhD, U. Oriental Studies, 1985. Diplomate Acad. Oral Pathology, Republic of China. Instr. dental surgery U. Chgo., 1975-76; asst. prof. Loyola U., Chgo., 1976-79; assoc. prof. Kaohsiung Med. Coll., 1979-85, prof., 1985—; dean Sch. of Dentistry, Kaohsiung Med. Coll., 1979-82, dean of student affairs, 1992—, chmn. oral pathology dept., 1985—. Fellow Internat. Coll. Dentistry, Am. Acad. Oral Pathology, Pierre Fauchard Acad. Home: 202 Sun-Tong St, Kaohsiung Taiwan Office: Kaohsiung Med Coll Sch Dentistry, 100 Sie-Chung 1st Rd, Kaohsiung Taiwan

LIN, MING SHEK, allergist, immunologist; b. Taipei, Taiwan, Republic of China, Oct. 11, 1937; came to U.S., 1965; s. Joseph and Tong-Kai (Chan) Lynn; m. Mary Liao, Nov. 22, 1969; children: Jerry, Michael. MD, Nat. Taiwan U., 1964; PhD, U. Pitts., 1974. Diplomate Am. Acad. Allergy and Immunology, Am. Bd. Pediatrics. Asst. prof. U. Pitts. Grad. Sch. Pub. Health, 1976-80; asst. and assoc. prof. dept. pediatrics U. Pitts. Sch. Medicine, 1981—; chief sect. of allergy and immunology Forbes Health System, Pitts., 1987—. Contbr. articles to Jour. Allery and Immunology, Internat. Congress of Immunology, Jour. Allergy, Jour. Pediatrics, Jour. Cellular Immunology. Named Winklestan lectr. Hirsch Med. Sch., 1979. Fellow Am. Soc. for Microbiology; mem. AMA. Home: 81 Locksley Dr Pittsburgh PA 15235-5117 Office: 4099 William Penn Hwy Ste 805 Monroeville PA 15146-2518

LIN, OTTO CHUI CHAU, materials scientist, educator; b. Swatow, China, Aug. 8, 1938; s. Wei-min and Yen-Ching (Chang) L.; m. Ada Ma, Sept. 7, 1963; children: Ann, Gene, Dean. BS, Nat. Taiwan U., 1960; MA, Columbia U., 1963, Ph.D., 1967. Rsch. chemist E.I. duPont, Wilmington, 1967-69, sr. rsch. chemist, 1969-71, rsch. assoc., supr., 1971-78, 79-83; dean engring., prof. Nat. Tsing-Hua U., Hsinchu, Taiwan, 1978-79; dir. materials rsch. lab. Indsl. Tech. Rsch. Inst., Hsinchu, Taiwan, 1983-88, v.p., 1985-88, pres., 1988— . Patentee in polymers; contbr. numerous articles to profl. jours. Recipient Disting. Leadership award Drexel Univ., 1992—. Mem. Am. Chem. Soc., Royal Acad. Engring. Sci. (fgn.), Am. Inst. Physics, Chinese Soc. Materials Sci (bd. dirs. 1984—, pres. 1986—). Roman Catholic.

LIN, PING-WHA, educator; b. Canton, China, July 11, 1925; m. Sylvia Lin; children: Karl, Karen. BS, Chiao-Tung U., Shanghai, China, 1947; PhD, Purdue U., 1951. Engr. various, 1951-61; cons., engr. WHO, Geneva, 1962-66, 84, project mgr., 1980-82; Laurence L. Dresser chair, prof. Tri-State U., Angola, Ind., 1966—; pres. Lin Techs., Inc., Angola, 1989—. Contbr. articles to profl. jours. Grantee Dept. of Energy, 1983-84. Fellow ASCE (past pres. Ind. chpt.); mem. NSPE, Am. Chem. Soc., Am. Water Works Assn. (life), Sigma Xi. Achievements include patents in the fields. Home: 506 S Darling St Angola IN 46703 Office: Tri-State U Angola IN 46703

LIN, SHUNDAR, sanitary engineer; b. Tainan, Taiwan, July 24; came to U.S., 1963; s. Wen-Lee ind King-Yu (Yong) L.; m. Meiling Chang, Mar. 6, 1965; children: Luke, Lucy. BS, Nat. Taiwan U., Taipei, 1958; MS, U. Cin., 1964; PhD, Syracuse U., 1967. Teaching asst. Sch. Pub. Health Nat. Taiwan U., 1961-63; engr. Ill. State Water Survey, Peoria, 1967—. Mem. ASCE (com. mem.), Water Environ. Fedn., Am. Water Works Assn. (com. mem.), Rotary Club. Achievements include development of improved methods for total coliform, fecal coliform and fecal stryptococcus recoveries. Home: 3012 W Alan Ct Peoria IL 61615 Office: Ill State Water Survey PO Box 697 Peoria IL 61652

LIN, TAO, electronics engineering manager; b. Shanghai, People's Republic of China, Aug. 6, 1958; came to U.S., 1986; s. Zheng-hui Lin and Wei-jing Wu; m. Ping Kuo, Aug. 18, 1989; children: Jason, Jessie. BS, East China Normal U., Shanghai, 1982; MS, Tohoku U., Sendai, Japan, 1985; PhD, Tohoku U., 1990. Technician Dongtong Electronics Inc., Shanghai, 1977-78; rsch. asst. Electronics Rsch. Lab U. Calif., Berkeley, 1986-87, postgrad. researcher, 1987-88; applications engr. Integrated Device Technology Inc., Santa Clara, Calif., 1988-90; sr. applications engr. Sierra Semiconductor Corp., San Jose, Calif., 1990-91, applications mgr., 1991—. Contbr. articles to profl. jours. Mem. IEEE. Home: 3552 Rockett Dr Fremont CA 94538-3423 Office: Sierra Semiconductor Corp 2073 N Capitol Ave San Jose CA 95132-1000

LIN, TU, endocrinologist, educator, researcher, academic administrator; b. Fukien, China, Jan. 18, 1941; came to U.S., 1967; s. Tao Shing and Jan En (Chang) L.; m. Pai-Li, July 1, 1967; children: Vivian H., Alexander T., Margaret C. MD, Nat. Taiwan U., Taipei, 1966. Diplomate Am. Bd. Internal Medicine, Am. Bd. Endocrinology and Metabolism. Intern Episcopal Hosp.-Temple U., Phila., 1967-68; resident in medicine Berkshire Med. Ctr., Pittsfield, Mass., 1968-70; fellow in endocrinology Lahey Clinic, Boston, 1970-71, Roger Williams Gen. Hosp.-Brown U., Providence, 1971-73; rsch. fellow in med. sci. Brown U., 1971-73; chief, endocrine sect. WJB Dorn Vet. Hosp., Columbia, S.C., 1975—; asst. prof. U. S.C. Sch. Medicine, Columbia, 1976-80, assoc. prof., 1980-84, prof. medicine, 1984—, prof., dir. divsn. endocrinology, diabetes and metabolism, 1992—; mem. Merit Review Bd. of Endocrinology, Dept. Vet. Affairs, 1990-94. Author: (book chpt.) Disorders of Male Reproductive Function, 1990; mem. editorial bd. Biology of Reproduction, 1990, Jour. of Andrology, 1993—; contbr. articles to med. and profl. jours. Recipient Disting. Investigator award U. S.C. Sch. Medicine, 1981, 88. Fellow ACP; mem. Endocrine Soc., Am. Soc. Andrology (chmn. ann. meeting, coun. 1993-96), Soc. for the Study of Reproduction. Office: U S C Sch Medicine Med Library Bldg 3rd Fl Columbia SC 29208

LIN, TUNG YEN, civil engineer, educator; b. Foochow, China, Nov. 14, 1911; came to U.S., 1946, naturalized, 1951; s. Ting Chang and Feng Yi (Kuo) L.; m. Margaret Kao, July 20, 1941; children: Paul, Verna. BS in Civil Engring., Chiaotung U., Tangshan, Republic of China, 1931; MS, U. Calif., Berkeley, 1933; LLD, Chinese U. Hong Kong, 1972, Golden Gate U., San Francisco, 1982, Tongji U., Shanghai, 1987, Chiaotung U., Taiwan, 1987. Chief bridge engr., chief design engr. Chinese Govt. Rys., 1933-46; asst., then assoc. prof. U. Calif., 1946-55, prof., 1955-76, chmn. div. structural engring., 1960-63, dir. structural lab., 1960-63; chmn. bd. T.Y. Lin Internat., 1953-87, hon. chmn. bd., 1987-92; pres. Inter-Continental Peace Bridge, Inc., 1968—; cons. to State of Calif., Def. Dept., also to industry; chmn. World Conf. Prestressed Concrete, 1957, Western Conf. Prestressed Concrete Bldgs., 1960. Author: Design of Prestressed Concrete Structures, 1955, rev. edit., 1963, 3d edit. (with N.H. Burns), 1981, (with B. Bresler, Jack Scalzi) Design of Steel Structures, rev. edit., 1968, (with S.D. Statesbury) Structural Concepts and Systems, 1981, 2d edit., 1988; contbr. articles to profl. jours. Recipient Berkeley citation award, 1976, NRC Quarter Century award, 1977, AIA Honor award, 1984, Pres.'s Nat. Medal of Sci., 1986, Merit award Am. Cons. Engrs., Coun., 1987, John A. Roebling medal Bridge Engring., 1990, Am. Segmental Bridge Inst. Leadership award, 1992; named Outstanding Alumni of Yr., U. Calif. Engring. Alumni Assn., 1984, Hon. Prof., Chiaotung U., 1982, Hon. Prof., Tongji U., 1984, Hon. Prof., Shanghai Chiaotung U., 1985; fellow U. Calif. at Berkeley. Mem. ASCE (hon., life, Wellington award, Howard medal), Nat. Acad. Engring., Academia Sinica, Internat. Fedn. Prestressing (Freyssinet medal), Am. Concrete Inst. (hon.), Prestressed Concrete Inst. (medal of honor). Home: 8701 Don Carol Dr El Cerrito CA 94530-2734 Office: 315 Bay St San Francisco CA 94133

LIN, Y. K., engineer, educator; b. Foochow, Fukien, China, Oct. 30, 1923; came to U.S., 1954, naturalized, 1964; s. Fa Been and Chi Ying (Cheng) L.; m. Ying-yuh June Wang, Mar. 29, 1952; children: Jane, Della, Lucia, Winifred. B.S., Amoy U., 1946; M.S., Stanford U., 1955, Ph.D., 1957. Tchr. Amoy U., China, 1946-48, Imperial Coll. Engring., Ethiopia, 1957-58; engr. Vertol Aircraft Corp., Morton, Pa., 1956-57; research engr. Boeing Co., Renton, Wash., 1958-60; asst. prof. U. Ill., Urbana, 1960-62, assoc. prof., 1962-65, prof. aero. and astron. engring., 1965-83; Charles E. Schmidt Eminent scholar chair Coll. Engring., dir. Ctr. for Applied Stochastics Rsch. Fla. Atlantic U., Boca Raton, 1984—; vis. prof. mech. engring. M.I.T., 1967-68; sr. vis. fellow Inst. Sound and Vibration Research, U. Southampton, Eng., 1976; cons. Gen. Motors Corp., Boeing Co., Gen. Dynamics Corp., TRW Corp., Brookhaven Nat. Lab. Author: Probabilistic Theory of Structural Dynamics, 1967, Stochastic Structural Mechanics, 1987, Stochastic Approaches in Earthquake Engineering, 1987, Stochastic Structural Dynamics, 1990; contbr. articles to profl. jours. Sr. postdoctoral fellow NSF, 1967-68. Fellow ASCE (Alfred M. Freudenthal medal 1984), Am. Acad. Mechs., Acoustical Soc. Am., AIAA (assoc.); mem. Sigma Xi. Home: 2684 NW 27th Ter Boca Raton FL 33434-6001 Office: Fla Atlantic U Coll Engring Boca Raton FL 33431

LIN, YEOU-LIN, systems engineer, consultant; b. Taipei, Taiwan, July 2, 1957; came to U.S., 1981; s. Chuan and Chiao-Chen (Chang) L.; m. Ting-Ting Yao, June 25, 1983; children: Cheryl Chang, Calvin Yao. BSEE, Nat. Chiao-Tung U., Hsinchu, Taiwan, 1979; MSEE, U. Pitts., 1982; PhD in Elec. and Computer Engring., Carnegie Mellon U., 1987. Image processing analyst Internat. Robomation Intelligence, Carlsbad, Calif., 1987; computer scientist Four PI Systems Corp., San Diego, 1987-89, sr. vision scientist, 1991-92; sr. systems engr., program mgr. Taiwan Aerospace Corp., Taipei, Taiwan, 1992—; owner Global Linking Tech. Svcs., San Diego, 1991—. Home: 13175 Janetta Pl San Diego CA 92130 Office: Taiwan Aerospace Corp Ste 2901, 333 Keelung Rd Sec 1, Taipei Taiwan

LIN, YI-HUA, environmental chemist; b. Taipei, Taiwan, Oct. 8, 1953; s. T.S. and Y. (Tsai) L.; m. Jessica C. Yen, Dec. 30, 1985; 1 child, Calvin Y. MS, U. R.I., 1982, PhD, 1986. Sr. chemist ERCO/ENSECO, Cambridge, Mass., 1986-87, pesticide lab. mgr., 1987-88; sr. chemist Roy F. Weston, Inc., Edison, N.J., 1988-89, quality control coord., 1989-91, orgn. lab. mgr., 1991--. Home: 20 Magellan Way Franklin Park NJ 08823 Office: Roy F Weston Inc 2890 Woodbridge Ave Bldg 209 Edison NJ 08837

LIN, YING-CHIH, chemistry educator; b. Tainan, Republic of China, Dec. 21, 1949; s. Kun-Yu and Chen-Chih (Chen) L.; m. Shuang-Chyuan Ho, June 24, 1973. MS, Nat. Tsing Hua U., Hsin-Chu, Republic of China, 1973; PhD, UCLA, 1981. Vis. rsch. scientist CR&DD duPont de Nemours, Wilmington, Del., 1981-83; assoc. prof. chemistry Nat. Taiwan U., Taipei, 1983-86, prof. chemistry, 1986—. Contbr. articles to profl. publs. Mem. Am. Chem. Soc. Office: Nat Taiwan U, Chemistry Dept, Taipei 10764, Taiwan

LIN, YOU JU, physicist; b. Nanjing, Peoples Republic of China, Dec. 15, 1942; s. Chang Xing and Xue Zhen (Cheng) L.; m. Bao-Zhu Wang, Apr. 26, 1970; 1 child, Lin. BS in Nuclear Physics, Nanjing U., MS in Radio Physics. Radio engr. Nanjing Electron Tube Factory, 1968-83; radiotherapy physicist Nanjing 81st Hosp., 1983—; cons. bd. Chinese Assn. Jiang Su Radiation Measurement, 1987—. Contbr. articles to Jour. Radiation Physics, Chinese Jour. Radiation Oncology, Med. and Pharm Jour., Jiang Su Jour. Measurement. Mem. Chinese Nuclear Soc., Chinese Assn. Radiation Measurement, Chinese Assn. Measurement, Chinese Assn. Radiation Physics, Chinese Soc. Med. Physics, Am. Assn. Physicists in Medicine (corr.), Am. Inst. Ultrasound in Medicine, Stamp Collector Soc. China. Home: No 35 34 Biao Tai Ping S Rd, Nanjing 210002, Peoples Republic of China Office: 81st Hosp Dept Rad Oncology, No 34 34 Biao Tai Ping S Rd, Nanjing 210002, China

LIN, ZHEN-BIAO, acoustical and electrical engineer; b. Minhau, Fu-jian, China, Oct. 24, 1938; came to U.S., 1989; s. Shi-Xun and Sai-Ying (Chen) L.; m. Zhi-Rong Ma, Feb. 24, 1968; children: Jian-Quan, Jian-Jin. BSEE, Nanjing U. Tech., 1962. Lectr. South-China U. of Tech., Guangzhou, 1963-80, assoc. prof., 1985-87, prof., 1987-89; vis. scientist Superior Normal Sch., Lyon, France, 1981-83; vis. prof. U. Mo., Columbia, 1989; rsch. assoc. Purdue U., Indpls., 1990—. Contbr. articles to profl. jours. Recipient Nat. Sci. Congress prize China Nat. Sci. Congress, Beijing, 1978, 2nd prize Nat. Sci. and Tech. Progress, China Nat. Commn. Sci. and Tech., Beijing, 1987. Mem. IEEE, Electric and Electronic Soc. France, Acoustical Soc. Am., Acoustical Soc. China. Achievements include use of artificial neural networks and time frequency (scale) representation in modeling of bat sonar; development of side scan sonar systems. Office: Purdue Univ Dept Elec Engring 723 W Michigan St SL2167 Indianapolis IN 46202

LINCICOME, DAVID RICHARD, biomedical and animal scientist; b. Champaign, Ill., Jan. 17, 1914; s. David Rosebery and Olive Iola (Casper) L.; m. Dorothy Lucile Van Cleave, Sept. 1, 1941 (dec. Nov. 1952); children: David Van Cleave, Judith Ann; m. Margaret Stirewalt, Dec. 29, 1953. BS, MS, U. Ill., 1937; PhD, Tulane U., 1941. Diplomate (emeritus) Am. Bd. Microbiology; cert. animal scientist Am. Registry Profl. Animal Scientists. Asst. instr. U. Ill., 1937; asst. instr. tropical medicine Tulane U. Med. Sch., 1937-41; asst. prof. parasitology U. Ky., 1941-47; sr. rsch. parasitologist Du Pont Co., 1949-53; from asst. prof. to full prof. biol. Scis. Howard U., 1953-70; guest scientist USDA, 1978—; vis. scholar Nat. Agrl. Libr., USDA, 1990-92; registrar Jacob Sheep Conservancy, 1989—, bd. dirs., 1990—. Founder, editor Exptl. Parasitology, 1949-76; editor Transactions of the Ky. Acad. Sci., 1946-49, Transactions of the Am. Microscopical Soc., 1970-71, Internat. Rev. Tropical Medicine, 1953-63; founder Virology, 1950, Advances in Vet. Sci., 1952. Lt. col. Med. Svc. Corps, U.S. Army, World War II, PTO. Recipient Anniversary award Helminthological Soc., 1975; rsch. grantee NIH, 1958-68. Mem. Helminthological Soc. (pres. 1958, emeritus), Am. Physiol. Soc. (emeritus), Soc. Invertebrate Zoology (emeritus), Am. Soc. Zoologists (emeritus), Am. Soc. Parasitologists, Am. Microscopical Soc. (emeritus), Royal Soc. Tropical Medicine (emeritus), Am. Soc. Tropical Medicine (emeritus), Am. Goat Soc. (bd. dirs. 1992—), Am. Dairy Goat Assn. (bd. dirs. 1972-87), Nat. Pygmy Goat Assn. (bd. dirs. 1976-92, pres. 1979), Natural Colored Wool Growers Assn. (bd. dirs. 1990—), Jacob Sheep Breeders Assn., Jacob Sheep Soc. (Eng.), Nat. Tunis Sheep Registry (bd. dirs. 1991—, sec. 1991-92), Greater Washington D.C. Area Soft-Coated Wheaten Terrier Club (founder, pres. 1991-92). Achievements include breeding of two rare and endangered breeds of sheep, Jacob and Tunis. Home: Frogmoor Farm RR 1 Box 352 Midland VA 22728-9748 also: 4419 Cambria Ave Box 13 Garrett Park MD 20896 Office: USDA BARC East Bldg 467 Rm 207 Beltsville MD 20705

LINCOLN, JANET ELIZABETH, experimental psychologist, consultant; b. Dallas, Mar. 17, 1945; d. Frank Jackson and Ruth Louise (Boedeker) L.; m. Robert Jack Kessler, Oct. 15, 1977. BA, U. Ill., 1968; MA, NYU, 1980, PhD, 1986. Ind. cons., 1980—; cons. U. Dayton (Ohio) Rsch. Inst., 1980-86, 89—, MacAulay-Brown, Inc., Dayton, 1986-88, Search Tech., Inc., Atlanta, 1987-90. Co-editor: Engineering Data Compendium: Human Perception and Performance (4 vols.), 1988; contbg. author: Handbook of Perception and Human Performance, 1986; co-author conf. papers. Recipient James McKeen Cattell award N.Y. Acad. Scis., 1987. Mem. Human Factors Soc., Optical Soc. Am. Achievements include an award-winning study of human depth perception. Home and Office: PO Box 159 Stuyvesant NY 12173

LINCOLN, WALTER BUTLER, JR., marine engineer, educator; b. Phila., July 15, 1941; s. Walter Butler and Virginia Ruth (Callahan) L.; m. Sharon Platner, Oct. 13, 1979; children: Amelia Adams, Caleb Platner. BS in Math., U. N.C., 1963; Ocean Engr., MIT, 1975; MBA, Rensselaer Poly. Inst., 1982; postgrad., Naval War Coll. Registered profl. engr., N.H., Conn.; chartered engr., U.K. Ops. rsch. analyst Applied Physics Lab. Johns Hopkins U., Silver Spring, Md., 1968-70; grad. asst. MIT, Cambridge, 1971-75; ocean engr. USCG R&D Ctr., Groton, Conn., 1976-78, chief marine engring. br., 1983—; prin. engr. Sanders Assocs., Nashua, N.H., 1978-83; lectr. U. Conn., Avery Point, 1986—; master, U.S. Mcht. Marine. Contbr. articles to profl. jours. Mem. planning bd. Town of Brookline, N.H., 1982. Comdr. USNR, 1963—. Fellow MIT, 1971. Mem. SAR, Am. Soc. Naval Engrs., Am. Geophys. Union, Nat. Assn. Underwater Instrs. (instr. 1971—), Royal Inst. Naval Architects, Soc. Naval Architects & Marine Engrs., Marine Tech. Soc. (exec. bd. New Eng. sect. 1980—), Navy Sailing Assn. (ocean master), Pi Mu Epsilon. Achievements include rsch. in integrated systems modeling and engring. of deep ocean systesm; devel. of algorithms for simulation of hydromechs. of ocean systems and ships; engring. mgmt. of ship and marine system rsch., devel., test and evaluation. Office: USCG R&D Ctr Avery Point Groton CT 06340

LINDBERG, DONALD ALLAN BROR, library administrator, pathologist, educator; b. N.Y.C., Sept. 21, 1933; s. Harry B. and Frances Seeley (Little) L.; m. Mary Musick, June 8, 1957; children: Donald Allan Bror, Christopher Charles Seeley, Jonathan Edward Moyer. AB, Amherst Coll., 1954, ScD (hon.), 1979; MD, Columbia U., 1958; ScD (hon.), SUNY, 1987; LLD (hon.), U. Mo., Columbia, 1990. Diplomate Am. Bd. Pathology, Am. Bd. Med. Examiners (exec. bd. 1987-91). Rsch. asst. Amherst Coll., 1954-55; intern in pathology Columbia-Presbyn. Med. Ctr., 1958-59, asst. resident in pathology, 1959-60; asst. in pathology Coll. Physician and Surgeons Columbia U., N.Y.C., 1958-60; instr. pathology Sch. of Medicine U. Mo., 1962-63, asst. prof. Sch. of Medicine, 1963-66, assoc. prof. Sch. of Medicine, 1966-69, prof. Sch. of Medicine, 1969-84, dir. Diagnostic Microbiology Lab. Sch. of Medicine, 1960-63, dir. Med. Ctr. Computer Program Sch. of Medicine, 1962-70, staff, exec. dir. for health affairs Sch. of Medicine, 1968-70, prof., chmn. dept. info. sci. Sch. of Medicine, 1969-71; dir. Nat. Libr. of Medicine, Bethesda, Md., 1984—; dir. Nat. Coord. Office for High Performance Computing and Comms., exec. office of Pres., Office Sci. & Tech. Policy, 1992—; mem. computer sci. and engring. bd. Nat. Acad. Sci., 1971-74 chmn. Nat. Adv. Com. Artificial Intelligence in Medicine, Stanford U., 1975-84; U.S. rep. to Internat. Med. Informatics Assn. of Internat. Fedn. for Info. Processing, 1975-84, also trustee; mem. peer rev. group Dept. Def., 1979-84; bd. dir. Symposium on Computer Applications in Med. Care. Author: The Computer and Medical Care, 1968; The Growth of Medical Information Systems in the United States, 1979; editor: (with W. Siler) Computers in Life Science Research, 1975; (with others) Computer Applications in Medical Care, 1982; editor Methods of Info. in Medicine, 1970-83, assoc. editor, 1983—; editor Jour. Med. Systems, 1976—, Med. Informatics Jour., 1976—; chief editor procs. 3d World Conf. on Med. Informatics, 1980; contbr. articles to sci. jours., chpts. to books. Recipient Silver Core award Internat. Fedn. for Info. Processing, 1980, Walter C. Alvarez award Am. Med. Writers Assn., 1989, PHS Surgeon Gen.'s medallion, 1989, Nathan Davis award AMA, 1989, Presdl. Disting. Exec. Rank award Sr. Exec. Svc., Outstanding Svc. medal Uniformed Svcs. U. of the Health Scis., 1992; Simpson fellow Amherst Coll., 1954-55; Markle scholar in acad. medicine, 1964-69. Mem. Inst. Medicine of NAS, Coll. Am. Pathologists (commn. on computer policy and coordination 1981-84), Mo. State Med. Assn., Assn. for Computing Machines, Salutis Unitas (Am. v.p. 1981-91), Am. Assn. for Med. Systems and Informatics (internat. com. 1982-89, bd. dirs. 1982, editor conf. procs. 1983, 84), Gorgas Meml. Inst. Tropical and Preventive Medicine (bd. dirs. 1987—), Am. Med. Informatics Assn. (pres. 1988-91), Sigma Xi. Democrat. Club: Cosmos (Washington). Avocations: photography; riding. Home: 13601 Esworthy Rd Germantown MD 20874-3319 Office: Nat Libr of Medicine 8600 Rockville Pike Bethesda MD 20894-0001

LINDBERG, ROBERT E., JR., space systems engineer, controls researcher; b. Bklyn., May 31, 1953; s. Robert E. and Rose Winifred (Woodhead) L.; m. Nancy Kathleen Montalbine, June 12, 1976; children: Bethany Christine, Christian John, Sarah Elizabeth. BS in Physics with distinction, Worcester Poly. Inst., 1974; MS in Engring. Physics, U. Va., 1976; Engring. ScD in Mech. Engring., Columbia U., 1982. Rsch. physicist Naval Rsch. Lab., Washington, 1976-83, sect. head, 1983-86, br. head, 1986-87; dir. advanced projects Orbital Scis. Corp., Fairfax, Va., 1987-90, program mgr., 1990—; mem. adj. faculty George Washington U., Washington, 1983-87; mem. engring. adv. com. Worcester (Mass.) Poly. Inst., 1990—. Editor: Astrodynamics 1987, 1987; assoc. editor Jour. Astronautical Scis., 1984-87, guest editor spl. issue, 1990; also over 25 articles. Bd. dirs. Lake Ridge (Va.) Parks and Recreation Assn., 1991—, High Frequency Wavelengths, N.Y.C. and Virginia Beach, Va., 1992—. Recipient rsch. publ. award Naval Rsch. Lab., 1979, Alan Berman rsch. publ. award, 1983, 87, 91; Engr. of Yr. award Washington Coun. Engring. and Archtl. Socs., 1988, Ichabod Washburn award Worcester Poly. Inst., 1989. Mem. AIAA (assoc. fellow, founding chmn. No. Va. chpt. 1988-89, mem. astrodyn. tech. com. 1984-87, 88-90, coun. Nat. Capital sect. 1986-90, Outstanding Young Engr.-Scientist award Nat. Capital sect. 1985), Am. Astronautical Soc. (sr., v.p. membership 1992-93), Am. Geophys. Union, Sigma Xi (sec. Naval Rsch. Lab chpt. 1986-87), Sigma Pi Sigma. Methodist. Achievements include research in field small spacecraft and launch systems, developed PegaStar integrated spacecraft design, now used for Air Force APEX satellite and NASA SeaStar satellite; member original Pegasus launch vehicle design team. Office: Orbital Scis Corp 21700 Atlantic Blvd Dulles VA 20166

LINDE, HAROLD GEORGE, chemist; b. N.Y.C., Mar. 18, 1945; s. Harold George and Mildred Amanda (White) L.; m. Elizabeth Theresa Murphy, July 4, 1988; children: Kathleen, Michael, Daniel, Julia. BS, SUNY, Stony Brook, 1966; PhD, U. Vt., 1972. Chemist, criminalist Vt. State Police, Montpelier, 1971-80; chemist IBM Corp., Essex Junction, Vt., 1980—. Contbr. articles to profl. jours. Mem. Am. Chem. Soc., Am. Acad. Forensic Sci., Northeastern Assn. Forensic Scientists. Achievements include patents for forming reactive ion etch barriers, adhesion promoters for cured polyimide resins, polymers for enhanced polyimide adhesion, novel solvent for the removal of hardened photoresist, chemical stabilization of polymer formulations, use of oxidative catalysts in etching, planarizing silsesquioxane copolymers, improved anisotrophic silicon etchants, numerous others. Home: HCR Box 188 Richmond VT 05477 Office: IBM Corp River Rd Essex Junction VT 05452

LINDE, ROBERT HERMANN, economics educator; b. Schlewecke, Fed. Republic Germany, July 22, 1944; s. Robert and Emma (Lohmann) L.; m. Sabine Rinck, Mar. 12, 1976 (div. 1985); children: Niels Christian, Johanne Cornelia; m. Ingrid Windus, June 30, 1987. Diploma in Econs., U. Gottingen, Fed. Republic Germany, 1969, D in Polit. Sci., 1977, D in Habilitation, 1981. Asst. U. Gottingen, 1976-81, lectr., 1981-86, prof. econs., 1986-87; prof. econs. U. Luneburg, Fed. Republic Germany, 1987—. Author: Theory of Product Quality, 1977, Pay and Performance, 1984, Introduction to Microeconomics, 2d edit., 1992; co-author: Production Theory, 1976. Mem. Am. Econ. Assn., Verein für Socialpolitik. Office: U Lüneburg, Barckhausen Strasse 35, 21332 Lüneburg Germany

LINDELL, MICHAEL KEITH, psychology educator; b. West Point, N.Y., Oct. 14, 1946; s. Keith Gordon Lindell and Mary Alicia Alrich; m. Kerry Lee Bair, Aug. 1, 1971; children: Ashley Caye, Dennis Michael. BA, U. Colo., 1969, PhD, 1975. Rsch. scientist Battelle Meml. Inst., Seattle, 1974-89; assoc. prof. psychology Mich. State U., East Lansing, 1987—; adj. asst. prof. U. Wash., Seattle, 1981-87; vis. assoc. prof. Ga. Inst. Tech., Atlanta, 1986-87. Author: Evacuation Planning in Emergency Management, 1981, Living With Mt. St. Helens, 1990, Behavioral Foundations of Community Emergency Planning, 1992, (computer assisted instruction courseware) Hyperstat: Statistical Principles in Personnel Research, 1992. Office: Mich State U 129 Psychology Rsch Bldg East Lansing MI 48824-1117

LINDEN, BARNARD JAY, electrical engineer; b. Bklyn., May 18, 1943; s. Abraham and Beatrice (Westler) L.; m. Janet Sigrid Jaffe, Sept. 12, 1965; 1 child, Michelle. BS, Monmouth Coll., 1965; postgrad., Adelphi U., 1966, N.Y. Inst. Tech., 1970-71. Registered profl. engr., N.Y., N.J. Electronics engr. Grumman Aerospace, Bethpage, N.Y., 1965-72; sr. electronics engr. Litton Industries, Melville, N.Y., 1972-73; elec. engr. N.Y.C. Bd. Edn., 1973-75; chief elec. engr. N.Y. State Div. Housing, N.Y.C., 1975—; chmn. Profl. Engrs. in Govt., 1992-95; cons. in field, N.Y. Vol. Spl. Olympics, Patchogue, N.Y., 1991, Hauppage, Stony Brook, 1992. Mem. N.Y. State Soc. Profl. Engrs. (legis. chmn. 1986-88, chair membership 1988-90, treas. 1990-91), Nat. Soc. Profl. Engrs., Sigma Pi Sigma. Jewish. Avocations: skiing, amateur radio, golf, classical music. Office: NY State Div Housing One Fordham Plz Bronx NY 10458

LINDEN, HENRY ROBERT, chemical engineering research executive; b. Vienna, Austria, Feb. 21, 1922; came to U.S., 1939, naturalized, 1945; s. Fred and Edith (Lermer) L.; m. Natalie Govedarica, 1967; children by previous marriage: Robert, Debra. B.S., Ga. Inst. Tech., 1944; M.Chem. Engring., Poly. U., 1947; Ph.D., Ill. Inst. Tech., 1952. Chem. engr. Socony Vacuum Labs., 1944-47; with Inst. of Gas Tech., 1947-78, various rsch. mgmt. positions, 1947-61, dir., 1961-69, exec. v.p., dir., 1969-74, pres., trustee, 1974-78; various acad. appointments Ill. Inst. Tech., 1954-86, Frank W. Gunsaulus Disting. Prof. of Chem. Engring., 1987-90, McGraw prof. energy and power engring. and mgmt., 1990—, interim pres., chief exec. officer, 1989-90, interim chmn., chief exec. officer IIT Rsch. Inst., 1989-90; chief oper. officer Gas Devel. Corp. (name now GDC, Inc.) subs. Inst. of Gas Tech., Chgo., 1965-73, chief exec. officer, 1973-78, also bd. dirs.; pres., dir. Gas Rsch. Inst., Chgo., 1976-87, exec. advisor, 1987—; mem. energy engring. bd. NRC; dir. Sonat Inc., So. Natural Gas Co., Larimer & Co., AES Corp., Resources for the Future Inc. Author tech. articles; holder U.S. and fgn. patents in fuel tech. Recipient award of merit, oper. sect. Am. Gas Assn., Disting. Svc. award Am. Gas Assn., Gas Industry Rsch. award Am. Ga Assn., 1982, Nat. Energy Resources Orgn. R & D award, 1986, Homer H. Lowry award for excellence in Fossil Energy Rsch., U.S. Dept. of Energy, 1991, award U.S. Energy Assn., 1992, Walton Clark medal Franklin Inst., Bunsen-Pettenkofer-Ehrentafel medal Deutscher Verein des Gas-und Wasserfaches; named to IIT Hall of Fame, 1982. Fellow AICE, Inst. of Fuel; mem. NAE, Am. Chem. Soc. (recipient H.H. Storch award, chmn. div. fuel chemistry 1967, councilor 1969-77), So. Gas Assn. (hon. life). Office: Ill Inst Tech 10 W 33d St PH 135 Chicago IL 60616 also: Gas Rsch Inst 8600 W Bryn Mawr Ave Chicago IL 60631

LINDEN, LYNETTE LOIS, bioelectrical engineer; b. Cheyenne, Wyo., Feb. 5, 1951; d. Byron Nels and Mary Ann (Savage) L. BA with honors, U. Calif., Santa Cruz, 1972; MS, MIT, 1974, PhD, 1988. Asst. engr. Burroughs Corp., Pasadena, Calif., 1969-70; engr., cons. Burroughs Corp., La Jolla, Calif., 1971-73; teaching asst. U. Calif., Santa Cruz, 1974-76; teaching asst. MIT, Cambridge, Mass., 1973-75, tutor, 1976-79; engr. Lincoln Labs., Lexington, Mass., 1979-80; asst. prof. engring. Boston U., 1980-90; ind. rsch. scientist Somerville, Mass., 1990—. Contbr. articles to profl. jours. Mem. AAAS, Soc. Women Engrs., Sigma Xi. Achievements include research in dimensionality constraints on color perception, application of group theory to computational models of neurons, visual perception, sensory systems, living systems, and biophysics of sensory systems. Office: 1A Banks St Boston MA 02144-3104

LINDENBAUM, S(EYMOUR) J(OSEPH), physicist; b. N.Y.C., Feb. 3, 1925; s. Morris and Anne Lindenbaum; m. Leda Isaacs, June 29, 1958. A.B., Princeton U., 1945; M.A., Columbia U., 1949, Ph.D., 1951. With Brookhaven Nat. Lab., Upton, N.Y., 1951—; sr. physicist Brookhaven Nat. Lab., 1963—, group leader high energy physics research group, 1954—; vis. prof. U. Rochester, 1958-59; Mark W. Zemansky chair in physics CCNY, 1970—; cons. Centre de Etudes Nucleaire de Saclay, France, 1957, CERN, Geneva, 1962; dep. for sci. affairs ERDA, 1976-77. Author: Particle Interaction Physics at High Energies, 1973. Contbr. articles to profl. jours. Fellow Am. Phys. Soc.; mem. N.Y. Acad. Scis., AAAS. Achievements include discovering nucleon isobars dominated high energy particles interactions, isobar model; inventor on line computer technique in scientific experiments; proved experimentally that Einstein's special theory of relativity was correct down to subnuclear distances one hundredth the radius of a proton; discovered the glueball states predicted by quantum chromodynamics. Office: Brookhaven Nat Lab Dept Physics Upton NY 11973

LINDESMITH, LARRY ALAN, physician, administrator; b. Amarillo, Tex., July 27, 1938; s. Lyle J. and Imogene Agnes (Young) L.; m. Patricia Ann Brady, June 6, 1959 (div. Mar. 1973); children: Robert James, Lisa Ann; m. Diane Joyce Bakken, Nov. 22, 1973; children: Abigail Arleen, Nathan Lyle, David Alan. BA, U. Colo., 1959; MD, Bowman-Gray Sch. Medicine, Winston-Salem, N.C., 1963. Diplomate Am. Bd. Internal Medicine, Am. Bd. I.M.-Pulmonary Disease; Nat. Inst. Occupational Safety and Health B Reader; provider ACLS, advanced trauma life support. Medical intern U. Chgo. Hosps., Clinics, 1963-64; I.M. resident U. Colo. Med. Ctr., Denver, 1964-66; pulmonary disease fellowship U. Colo. Med. Ctr., Webb-Waring Lung Inst., Denver, 1966-67; asst. dir. infectious and pulmonary disease svc. Madigan Gen. Hosp., Tacoma, Wash., 1967-69; chief pulmonary disease Gundersen Clinic, Ltd., La Crosse, Wis., 1969-87, chief pulmonary and occupational medicine, 1979-89, chmn. dept. medicine, 1987-93; chief occupational health, preventive medicine, 1988—; bd. govs. Gundersen Clinic, Ltd., 1987-93; adj. prof. phys. therapy U. Wis., La Crosse, 1977-92; cons. VA Hosp., Tomah, Wis., 1977—; clin. asst. prof. internal medicine U. Wis., Madison, 1982-92, clin. assoc. prof., 1992—; med. dir. RESTOR U. Wis., La Crosse, 1986—, Svcs. to Bus. and Industry Gundersen/Luth. Med. Ctr., La Crosse, 1987—. Contbr. book chpts. and articles to profl. publs. Mem. Air Pollution Control Coun. State of Wis. Dept. Natural Resources, 1978-81; vice-chmn. Bd. Control Luther High Sch., Onalaska, Wis., 1990-93. Maj. USAR, 1968-69; chmn. bd. dirs. Greater La Crosse Area C. of C., 1991. Boettcher Found. scholar, 1955-59; named Pagliara Tchr. of Yr. Gundersen Med. Found., 1984; recipient Dist. Svc. award Am. Lung Assn. Wis., 1988. Fellow Am. Coll. Chest Physicians, Am. Coll. Occupational and Environ. Medicine (assoc., chmn. pvt. practice coun., chmn. occupational lung disorders com.); mem. AMA, Am. Assn. Respiratory Therapy, Clin. Sleep Soc., Am. Thoracic Soc. (Wis. counselor 1978-81), Cen. States Occupational Med. Assn. (bd. govs. 1984—, pres. 1991), Am. Lung Assn. Wis. (pres. 1975-77), Wis. Thoracic Soc. (gen. conf. chmn. 1987), State Med. Soc. Wis. (chmn. environ. and occupational health com. 1989-91). Republican. Lutheran. Avocation: photography. Home: 4965W Woodhaven Dr La Crosse WI 54601-2435 Office: Gundersen Clinic Ltd 1836 South Ave La Crosse WI 54601-5494

LINDH, ALLAN GODDARD, seismologist; b. Mason City, Wash., Mar. 18, 1943; s. Quentin Willis and Helma (Beagle) L.; m. Julie Gunda Pulver, Mar. 21, 1971; children: Briana Christine, Quentin William. BA in Geology and Physics, U. Calif., Santa Cruz, 1972; MS, Stanford U., 1974, PhD, 1980. Geophysicist U.S. Geol. Survey, Menlo Park, Calif., 1972—, chief scientist Parkfield prediction experiment, 1986—. Avocation: prevention of nuclear war. Office: US Geol Survey 345 Middlefield Rd MS 977 Menlo Park CA 94025

LINDHOLM, CLIFFORD FALSTROM, II, engineering executive, mayor; b. Passaic, N.J., Dec. 8, 1930; s. Albert William and Edith (Neandross) L.; m. Margery Nye (div.); children: Clifford, Elizabeth, John; m. Karen Cooper, Oct. 7, 1989. BS in Engring., Princeton U., 1953; M in Engring., Stevens Inst. Tech., 1957. Supr. prodn. GM, Linden, N.J., 1953-56; pres. Falstrom Co., Passaic, N.J., 1956—; Bd. dirs. N.J. Mfg. Ins. Co., Trenton. Mayor Twp. Montclair, N.J., 1988—; pres. Montclair Bd. Edn., 1968-72; bd. dirs. Albert Payson Terhune Found., N.J., 1976—. Mem. N.J. Bus. and Industry Assn. (bd. dirs. 1977—), Princeton Club N.Y., Upper Montclair Golf Club, Mantoloking Yacht Club. Republican. Mem. Ch. of Christ. Home: 10 Mountainside Park Ter Montclair NJ 07043-1209 Office: Falstrom Co 3 Falstrom Ct Passaic NJ 07055-4443

LINDLER, KEITH WILLIAM, marine engineering educator, consultant; b. Balt., Oct. 4, 1954; s. Jack Griffith and Mary Ellen (Boone) L.; m. Jean Ann Schmalenberger, May 21, 1977; children: Jason Edward, Jennifer Ann, Kathryn Michelle. BSME summa cum laude, U. Md., 1975, MSME, 1978, PhD in Mech. Engring., 1984. Assoc. design engr. Radiation Systems, Inc., McLean, Va., 1975; instr. mech. engring. U. Md., College Park, 1976-84; assoc. prof. marine engring. U.S. Naval Acad., Annapolis, Md., 1984—; cons. TPI, Inc., Bethesda, 1981-89; forensic engr. Severna Park, Md., 1989—; conf. presenter. Author: (handbook) NBS Passive Solar Test Building Handbook, 1982. Mem. ASME, Am. Soc. Naval Engrs., Am. Solar Energy Soc., Internat. Solar Energy Soc. Achievements include development of computer programs RANKINE and BRAYTON which are used for analysis of steam plants and gas turbines at universities throughout the U.S. Home: 11 St Andrews Crossover Severna Park MD 21146 Office: US Naval Acad Stop 11D 590 Holloway Rd Annapolis MD 21402-5042

LINDNER, ALBERT MICHAEL, structural engineer; b. Milw., Jan. 12, 1952; s. Albert A. and Sylvia V. (Budzinski) L.; m. Beth A. Blaske, Nov. 29, 1980; 1 child, Sarah F. BS, U. Wis., Milw., 1974, ME, 1977. Registered profl. engr., Wis., Ill. Design engr. P&H Corp., Milw., 1976-78, Inryco Corp., Milw., 1978-80; design engr. Graef Anhalt Schloemer, Milw., 1980-83, project engr., 1983-87, project mgr., 1987-91, ptnr., 1991—. Mem. Brookfield (Wis.) Bd. Appeals, 1991. Recipient Santa Monica Bridge award Wis. Ready Mix Concrete Assn., 1990. Mem. ASCE, Wis. Soc. Profl. Engrs. (chpt. pres. 1988-89), Concrete Reinforcing Steel Inst., Am. Inst. Timber Constrn. Office: Graef Anhalt Schloemer 345 N 95th St Milwaukee WI 53226

LINDNER, DUANE LEE, materials science management professional; b. Ft. Dodge, Iowa, May 7, 1950; s. Orville Richard and Dorothy Ann (Licht) L.; m. Deborah Georgine Ford, Sept. 10, 1977; children: John R., Nathaniel R., Elizabeth K. SB, MIT, 1972; PhD, U. Calif., Berkeley, 1977. Mem. tech. staff Sandia Nat. Labs., Livermore, Calif., 1977-86, supr. chemistry div., 1986-90; mgr. materials dept. Sandia Nat. Labs., Livermore, 1990-92, mgr. materials programs, 1992—. With U.S. Army, 1972-78. Mem. Am. Chem. Soc., Materials Rsch. Soc., Sigma Chi. Office: Sandia Nat Lab Dept 8701 Livermore CA 94551

LINDQUIST, ANDERS GUNNAR, applied mathematician, educator; b. Lund, Sweden, Nov. 21, 1942; s. Gunnar David and Gudrun Katarina (Dahl) L.; m. Kerstin Birgitta Rickander, Jan. 7, 1966; children: Johan, Martin; m. Galina Jurievna Degtyareva, Jan. 8, 1986; 1 child, Max Josef. MS, Royal Inst. Tech., 1967, PhD, 1972. Researcher Swedish Inst. Nat. Def., Stockholm, 1967-68; docent Royal Inst. Tech., Stockholm, 1972, prof., chmn., 1982—; asst. prof. U. Fla., Gainesville, 1972-73; assoc. prof. Brown U., Providence, 1973; assoc. prof., then prof. U. Ky., Lexington, 1974-84; affiliate prof. Washington U., St. Louis, 1989—. Communicating editor Jour. Math. Systems, Estimation and Control, 1991—; assoc. editor Systems and Control Letters, 1989-93; mem. editorial bd. Internat. Jour. Adaptive Control and Signal Processing, 1991—, (books) Progress in Systems and Control, 1991—, Systems and Control: Foundations and Applications, 1991. Fellow IEEE; mem. Soc. Indsl. and Applied Math.

LINDQUIST, DANA RAE, mechanical engineer; b. Corvallis, Oreg., Oct. 21, 1963; d. Norman Fred Lindquist and Lois (Allen) Lindquist Woll. BS, Cornell U., 1986; SM, MIT, 1988, PhD, 1992. Tech. support engr. Flomerics, Inc., Westborough, Mass., 1992—. Contbr. articles to profl. jours. AFRAPT fellow, 1988-91, Amelia Earhart/Zonta fellow, 1989-90. Mem. AIAA. Office: Flomerics Inc 33 Lyman St Westborough MA 01581

LINDQUIST, EDWARD LEE, biological scientist, ecologist; b. Hamilton, Mont., Aug. 2, 1942; s. Clarence Algut and Elizabeth May (Pigg) L.; m. Shirley Annette Johnson, Nov. 1, 1967; children: Lance Edward, Jason James. BS, U. Mont., 1966. Insect physiologist Agr. Rsch. Svc., Beltsville, Md., 1969-74; wildlife biologist USDA Forest Svc., Washington, 1974; wildlife ecologist Sawtooth Nat. Recreation Area, Katchum, Idaho, 1974-77; wildlife biologist Ashley Nat. Forest-USDA Forest Svc., Vernal, Utah, 1978-81, Superior Nat. Forest-USDA Forest Svc., Duluth, Minn., 1981-90; coord. Geog. Info. System, Duluth, Minn., 1990—; exec. bd. mem. N. Cen. Caribou Corp., Duluth, 1988—; cons. ecologist U.S. Fish and Wildlife Svc., Hawaii Nat. Park, 1991; chair Uinta Basin Biol. Conf., Vernal, Utah, 1979, N.Am. Moose Conf., Duluth, 1987. Editor: Information Needs Assessment-Superior National Forest, 1992; contbr. articles to profl. and popular publs. Pres. Walther League, Hamilton, 1960; v.p. Greater Laurel (Md.) Jaycees, 1972-73. With USAF, 1967-69. Named Jaycee of Yr. Md. Jaycees, 1973; recipient Conservation award Field and Stream mag., 1972, Order of Alces award N.Am. Moose Congress, 1987. Mem. Wildlife Soc. (pres. Minn. chpt. 1984, profl. devel. cert.), Nat. Wildlife Fedn., Ecol. Soc. Am. Achievements include reintroduction of Rocky Mountain Bighorn sheep in Utah, reintroduction of American Peregrine Falcoln to Minnesota, 1984-88. Office: Superior Nat Forest PO Box 338 515 W 1st St Duluth MN 55801

LINDQUIST, LOUIS WILLIAM, artist, writer; b. Boise, Idaho, June 26, 1944; s. Louis William and Bessie (Newman) L.; divorced; children: Jessica Ann Alexandra, Jason Ryan Louis. BS in Anthropology, U. Oreg., 1968; postgrad., Portland State U., 1974-78. Researcher, co-writer with Asher Lee, Portland, Oreg., 1977-80; freelance artist, painter, sculptor Oreg., 1980-91. Sgt. U.S. Army, 1968-71, Vietnam. Mem. AAAS, Internat. Platform Assn., N.Y. Acad. Scis. Democrat. Avocations: reading, beachcombing. Home and Office: PO Box 991 Bandon OR 97411

LINDQVIST, JENS HARRY, physician; b. Saltsjoebaden, Sweden, Feb. 3, 1967; s. Bert Alfred and Gun Gertrud Ingegerd Lindqvist. MD, Karolinska Inst., Stockholm, 1991. Clin. researcher dept. endocrinology Karolinska Inst. and Hosp., Stockholm, 1989-91; clin. researcher Kabipharmacia AB, Stockholm, 1989—; physician Huddinge U. Hosp., Stockholm, 1991-92; intern Oskarshamn (Sweden) Hosp., 1992—; speaker annual Swedish Med. Congress, Stockholm, 1989. Contbr. articles to Acta Endocrinologica. Moderata Samlingspartiet. Avocations: sailing, golf, business administration. Home: John Bergs Plan 2, 11250 Stockholm Sweden

LINDSAY, ROBERT, physicist, educator; b. New Haven, Mar. 3, 1924; s. Robert Bruce and Rachel (Easterbrooks) L.; m. Charlotte Rose Melton, May 3, 1952; children—Nancy Carolyn, William Melton, Ann Elizabeth. B.S., Brown U., 1947; M.A., Rice U., 1949, Ph.D., 1951. Physicist Nat. Bur. Standards, Washington, 1951-53, summer 1955; asst. prof. physics So. Meth. U., Dallas, 1953-56, Trinity Coll., Hartford, Conn., 1956-59; assoc. prof. Trinity Coll., 1959-65, prof., 1965-90, Brownell-Jarvis prof. natural philosophy and physics, 1978-90, prof. emeritus, 1990—, chmn. dept., 1985-87. Editor: Early Concepts of Energy in Atomic Physics, 1979; Contbr. to Acad. Am. Ency., 1987; Contbr. research articles on magnetism to profl. jours. Served to 1st. lt. USAAF, 1943-46. Mem. Am. Phys. Soc., Sigma Xi. Home: 62 Midwell Rd Wethersfield CT 06109-2839 Office: Trinity Coll Physics Dept Hartford CT 06106

LINDSETH, RICHARD EMIL, orthopaedic surgeon; b. Denver, Apr. 3, 1935; s. Emil Victor and Audrey Madera (Yeo) L.; m. Marilyn Martha Miller, July 7, 1959; children: Erik Lars, Ellen Sue. BA, Dartmouth Coll., 1957; MD, Harvard U., 1960. Diplomate Am. Bd. Orthopaedic Surgery. Intern, then resident in surgery SUNY, Syracuse, 1960-62, resident in orthopaedic surgery, 1964-67; prof. orthopaedic surgery Ind. U., Indpls., 1975—, acting chmn. dept. orthopaedic surgery, 1984-86; pres. Assoc. Orthopaedic Surgeons, Inc., Indpls., 1983-86. Mem. editorial bd. Jour. Pediatric Orthopaedics; contbr. articles to profl. jours. Fellow ACS, Am. Acad. Orthopaedic Surgeons, Am. Acad. Pediatrics, Scoliosis Rsch. Soc., Am. Orthopaedic Assn.; mem. Pediatric Orthopaedic Soc. N.Am. (pres. 1986-87), European Pediatric Orthopaedic Soc., Internat. Soc. for Rsch. into Hydrocephalus and Spina Bifida, Spina Bifida Assn. Am. (profl. adv. coun. 1978-88, chmn. 1987-88). Methodist. Office: Ind U Med Ctr Riley Hosp 702 Barnhill Dr Indianapolis IN 46202-5128

LINDSEY, ALFRED WALTER, federal agency official, environmental engineer; b. Camden, N.J., Jan. 10, 1942; s. Alfred Hazel Lindsey and May Marquerite (Ergood) Warrington; m. Kathleen Francis Leighton, Aug. 15, 1964; 1 child, Amy Elizabeth. BS in Pulp and Paper Tech., N.C. State U., 1964. Dep. dir. tech. div. Office of Solid Waste EPA, Washington, 1979-85, dep. dir. office environ. engring., Office R&D, 1985-88, dir. office environ. engring., 1988—. Contbr. articles to profl. jours. Bagpiper Prince Georges Police Pipes and Drums. Mem. Sr. Execs. Assn., Nat. Sanitation Found. (coun. pub. health cons. 1985-91), Nature Conservancy (Va. chpt.). Republican. Methodist. Office: EPA RD 681 401 M St SW Washington DC 20460-0002

LINDSEY, DOUGLAS, trauma surgeon; b. Oakdale, La., Oct. 26, 1919. MD, Yale U., 1943, D in Pub. Health, 1950; MSIA, George Wash.

U., 1965. Diplomate Am. Bd. Surgery, Am. Bd. Preventive Medicine, Am. Bd. Emergency Medicine. Col., med. officer U.S. Army, 1944-73; mem. surgery faculty U. Ariz., Tucson, 1973—. Author books, chpts., and articles on trauma and soft tissue infections. Fellow Am. Assn. for the Surgery of Trauma; mem. Surg. Infection Soc. Office: Half-S Enterprises Box 531 Cortaro AZ 85652

LINDSTROM, ERIC EVERETT, ophthalmologist; b. Helena, Mont., Nov. 28, 1936; s. Everett Harry and Nan Augusta (Johnson) L.; BS, Wheaton Coll., 1958; MD, U. Md., 1963; MPH, Harvard U., 1966; m. Nancy Jo Alexander, July 24, 1960; children: Laura Ann, Eric Everett. Intern, Madigan Army Med. Center, Tacoma, Wash., 1963-64; resident in aerospace medicine Sch. Aerospace Medicine, Brooks AFB, Tex., 1966-68, resident in ophthalmology Brooke Army Med. Ctr., Ft. Sam Houston, Tex., 1972-75; surgeon 12th combat aviation group U.S. Army, Vietnam, 1968-69, chief profl. svcs. and aviation medicine Beach Army Hosp., Ft. Wolters, Tex., 1969-72; asst. chief ophthalmology clinic Madigan Army Med. Center, Tacoma, 1975-76; now with Lindstrom Eye Clinic; med. dir. Palo Pinto County (Tex.) Mental Health Clinic., 1970-72; cons. Tex. State Rehab. Com., 1971-72; chmn. bd. trustees South Cen. Regional Med. Ctr.; sr. aviation med. examiner, FAA; flight surgeon Miss. ANG, ret. Deacon First Bapt. Ch., Laurel, 1978—; bd. dirs. Laurel Salvation Army. Decorated Bronze Star, Air medal with 2 oak leaf clusters, Meritorious Svc. medal. Diplomate Am. Bd. Preventive Medicine, Am. Bd. Ophthalmology. Fellow ACS, Am. Coll. Preventive Medicine, Aerospace Med. Assn. (assoc.), Am. Acad. Ophthalmology; mem. AMA, FAA (sr. aviation med. examiner), Miss. Med. Assn. (trustee), Miss. EENT Assn., South Miss. Med. Soc., Southern Med. Assn., La.-Miss. EENT Assn., Flying Physicians Assn., Soc. Mil. Ophthalmologists, Soc. of USAF Flight Surgeons, Alliance of Air N.G. Flight Surgeons, Aircraft Owners and Pilots Assn., Nu Sigma Nu. Club: Kiwanis. Home: 809 Cherry Ln Laurel MS 39440-1651 Office: Lindstrom Eye Clinic PO Box 407 Laurel MS 39441-0407

LINDSTROM, KRIS PETER, environmental consultant; b. Dumont, N.J., Oct. 18, 1948; s. Sven Rune and Moyra Hilda (Coughlan) L.; m. Annette Gail Chaplin, June 25, 1978; 1 child, Karl Pierce. MPH, U. Calif., Berkeley, 1973; MS in Ecology, U. Calif., Davis, 1983. Registered environ. health specialist, Calif. Sr. lab. analyst County Sanitation Dists. Orange County, Fountain Valley, Calif., 1970-72, environ. specialist, 1973-74; environ. specialist J.B. Gilbert and Assocs., Sacramento, 1974-78; prin. K.P. Lindstrom, Inc., Sacramento, 1978-84; pres. K.P. Lindstrom, Inc., Pacific Grove, Calif., 1985—; mem. rsch. adv. bd. Nat. Water Rsch. Inst., Fountain Valley, 1991—. Author: Design of Municipal Wastewater Treatment Plants, 1992; editor publs., 1989, 90. Chmn. City of Pacific Grove (Calif.) Mus. Bd., 1992, City of Seal Beach (Calif.) Environ. Bd., 1970. Mem. Water Environ. Fedn. (chmn. marine water quality com. 1987-90), Calif. Water Pollution Control Assn., Friends of Monarchs (bd. dirs.). Office: KP Lindstrom Inc PO Box 51008 Pacific Grove CA 93950-6008

LINDSTROM, TIMOTHY RHEA, chemist; b. Jonesville, Va., Nov. 8, 1952; m. Candice K. Brown, Apr. 26, 1975; 1 child, Jonathan P. BA, LeTourneau Coll., 1975; PhD, Iowa State U., 1980. Area chemist E.I. duPont, Parkersburg, W.Va., 1980-83, div. chemist, 1983-87, sr. chemist, 1987, 90-92, area supt. tech., 1987-90, area supt. lab., 1992—; owner, oper. Lindstrom Home Video Prodns., 1987—. Contbr. articles to profl. jours. Republican. Am. Baptist. Home: 27 Hazel St Washington WV 26181 Office: EI duPont de Nemours & Co PO Box 1217 Parkersburg WV 26102

LINDZEN, RICHARD SIEGMUND, meteorologist, educator; b. Webster, Mass., Feb. 8, 1940; s. Abe and Sara (Blachman) L.; m. Nadine Lucie Kalougine, Apr. 7, 1965; children: Eric, Nathaniel. A.B., Harvard U., 1960, S.M., 1961, Ph.D., 1964. Research assoc. U. Wash., Seattle, 1964-65; Research asso. U. Oslo, 1965-66; with Nat. Center Atmospheric Research, Boulder, Colo., 1966-68; mem. faculty U. Chgo., 1968-72; prof. meteorology Harvard U., 1972-83, dir. Center for Earth and Planetary Physics, 1980-83; Alfred P. Sloan prof. meteorology MIT, 1983—; Lady Davis vis. prof. Hebrew U., 1979; Sackler prof. Tel Aviv U., 1992; Vikram Sarabhai prof. Phys. Research Lab., Ahmedabad, India, 1985; Lansdowne lectr. U. Victoria, 1993; cons. NASA, Jet Propulsion Lab., others; mem. space studies bd. and bd. on atmospheric scis. and climate NRC. Author: Dynamics in Atmospheric Physics; co-author: Atmospheric Tides; contbr. to profl. jours. Recipient Macelwane award Am. Geophys. Union, 1968. Fellow NAS, AAAS, Am. Geophys. Union, Am. Meteorol. Soc. (Meisinger award 1969, councillor 1972-75, Charney award 1985), Am. Acad. Arts and Scis.; mem. Internat. Commn. Dynamic Meteorology, Woods Hole Oceanographic Instn. (corp.), Institut Mondial des Scis. (founding mem.). Jewish. Office: MIT 54-1720 Cambridge MA 02139

LINEBERGER, WILLIAM CARL, chemistry educator; b. Hamlet, N.C., Dec. 5, 1939; s. Caleb Henry and Evelyn (Cooper) L.; m. Katharine Wyman Edwards, July 31, 1979. BS, Ga. Inst. Tech., 1961, MSEE, 1963, PhD, 1965. Research physicist U.S. Army Ballistic Research Labs., Aberdeen, Md., 1967-68; postdoctoral assoc. Joint Inst. for Lab. Astrophysics U. Colo., Boulder, 1968-70, from asst. prof. to prof. chemistry, 1970-83, E.U. Condon prof. chemistry, 1983—; Phi Beta Kappa nat. lectr., 1989. Served to capt. U.S. Army, 1965-67. Fellow AAAS, Joint Inst. for Lab. Physics, Am. Phys. Soc. (H.P. Broida prize 1981, Bomem Michelson prize 1987, Optical Sci. Am. Meggers prize 1988, Plyler prize 1992); mem. NAS, Am. Chem. Soc., Sigma Xi. Office: U Colo Joint Inst Lab Astrophysics CB 440 Boulder CO 80309-0440

LINEBERRY, GENE THOMAS, mining engineering educator; b. Bluefield, W.Va., July 29, 1955; m. Diana C. Cooper, Dec. 23, 1988. AS, Bluefield Coll., 1975; BS, Va. Poly. Inst. and State U., 1977, MS, 1979; PhD, W.Va. U., 1982. Instr. dept. mining and minerals engring. Va. Poly. Inst. and State U., Blacksburg, 1979; lectr. dept. mining engring. W.Va. U., Morgantown, 1980-81, asst. prof. dept. mining engring., 1982-87; assoc. prof. dept. mining engring. U. Ky., Lexington, 1987—; cons. office of energy-related inventions U.S. Dept. Energy, Nat. Bur. Stds., Gaithersburg, Md., 1990—; rsch. assoc. U.S. Army Rsch. Inst., Ft. Rucker, Ala., summer 1990; rsch. scientist, mining engr. Pitts. Rsch. Ctr. U.S. Bur. Mines, 1992-93. Co-editor proceedings of multi-nat. conf. on mine planning and design, 1989; contbr. 50 articles, papers, conf. proceedings to profl. jours. Recipient Stephen McCann award for Excellence in Edn., Pitts. Coal Mining Inst. Am., 1991. Mem. Soc. for Mining, Metallurgy and Exploration (sect. coord. and author handbook, Stefanko award 1989), Tau Beta Pi, Phi Theta Kappa, Sigma Xi. Achievements include co-development of computer model for objective evaluation of secondary mining potential of abandoned and inactive highwalls in Appalachia; rsch. in prototype expert system for underground coal mine design. Office: U Ky 230 MMRB Lexington KY 40506-0107

LINEVSKY, MILTON JOSHUA, physical chemist; b. Glen Cove, N.Y., Apr. 20, 1928; s. David I. and Tillie (Ain) L.; m. Barbara Jody Rutenberg, June 29, 1958; children: Joanne, Richard. BS in Chemistry, Rensselaer Poly. Inst., 1949; MS, Pa. State U., 1950, PhD, 1953. Phys. chemist GE, King of Prussia, Pa., 1957-79, Johns Hopkins Applied Physics Lab., Laurel, Md., 1979—; with NSF, Washington, 1993—. With U.S. Army, 1955-57. Gotshall Powel scholar, 1946. Mem. Am. Chem. Soc. Achievements include invention and development of technique for matrix isolation applied to high-temperature materials, development of first flame laser using chemical pumping reactions. Home: 700 Hermleigh Rd Silver Spring MD 20902 Office: Johns Hopkins Applied Physics Lab Johns Hopkins Rd Laurel MD 20723 also: NSF 1800 G St NW Washington DC 20550

LINFORD, RULON KESLER, physicist, program director; b. Cambridge, Mass., Jan. 31, 1943; s. Leon Blood and Imogene (Kesler) L.; m. Cecile Tadje, Apr. 2, 1965; children: Rulon Scott, Laura, Hilary, Philip Leon. BSEE, U. Utah, 1966; MS in ElecE, Mass. Inst. Tech., 1969, PhD in ElecE, 1973. Staff CTR-7 Los Alamos (N.Mex) Nat. Lab., 1973-75, asst. group leader CTR-7, 1975-77, group leader CTR-11, 1977-79, program mgr., group leader compact toroid CTR-11, 1979-80, program mgr., asst. div. leader compact toroid CTR div., 1980-81, assoc. CTR div. leader, 1981-86, program dir. magnetic fusion energy, 1986-89, program dir., div. leader CTR div. office, 1989-91, program dir. nuclear systems, 1991—. Contbr. articles to profl. jours. Recipient E. O. Lawrence award Dept. of Energy, Washington, 1991. Fellow Am. Physical Soc. (exec. com. 1982, 90-91, program com. 1982, 85, award selection com. 1983, 84, fellowship com. 1986); mem. Sigma Xi. Office: Los Alamos Nat Lab P O Box 1663 MS H854 Los Alamos NM 87545

LING, EDWARD HUGO, mechanical engineer; b. Beeville, Tex., Nov. 7, 1933; s. Will and Tiny Blue (Hausenfluck) L.; m. Linda Nell McClure, Apr. 30, 1960; children: Kathy Suzanne, Lori Sharon, Edward Jesse. BSME, U. Tex., 1956. Registered profl. engr. Jr. field engr. Schlumberger Well Surveying Corp., Tex., N.Me, 1956-60; mech. engr. Dyess AFB, Abilene, Tex., 1960-66, Hdqs. Air Tng. Command, Randolph AFB, Tex., 1967—. Contbr. articles to profl. jours. Councilman City of Cibolo, Tex., 1970-78, mayor, 1981-83; deacon 1st Bapt. Ch., Schertz, Tex., 1968—. Staff sgt. USAFR, 1958-65. Mem. Assn. Energy Engrs. (EMCS coun. 1986—), Soc. Am. Mil. Engrs. Achievements include development of reorganization plan for Air Force civil engineering to implement the new technology of energy management and control systems; pioneered full application of energy management and control systems technology to Air Force facilities. Home: PO Box 246 Cibolo TX 78108-0246 Office: Hdqs ATC/DEME Randolph AFB TX 78150-5001

LING, JOSEPH TSO-TI, manufacturing company executive, environmental engineer; b. Peking, China, June 10, 1919; came to U.S., 1948, naturalized, 1963; s. Ping Sun and Chong Hung (Lee) L.; m. Rose Hsu, Feb. 1, 1944; children: Lois Ling Olson, Rosa-Mai Ling Ahlgren, Louis, Lorraine. B.C.E., Hangchow Christian Coll., Shanghai, China, 1944; M.S. in Civil Engring, U. Minn., 1950, Ph.D. in San. Engring, 1952. Registered profl. engr., Minn., Ala., N.J., Okla., W.Va., N.Y., Ill., Ind., Pa., Mich. Civil engr. Nanking-Shanghai R.R. System, 1944-47; research asst. san. engring. U. Minn., 1948-52; sr. staff san. engr. Gen. Mills, Inc., Mpls., 1953-55; dir. dept. san. engring. research Ministry Municipal Constrn., Peking, 1956-57; prof. civil engring. Bapt. U., Hong Kong, 1958-59; head dept. water and san. engring. Minn. Mining & Mfg. Co., St. Paul, 1960-66, mgr. environ. and civil engring., 1967-70, dir. environ. engring. and pollution control, 1970-74, v.p. environ. engring. and pollution control, 1975-84, community service exec., 1985—; adv. mem. on air pollution Minn. Bd. Health, 1964-66; mem. Minn. Gov.'s Adv. Com. on Air Resources, 1966-67; mem. adv. panel on environ. pollution U.S. C. of C., 1966-71; mem. chem. indsl. com., adv. to Ohio River Valley Water Sanitation Commn., 1962-76; mem. environ. quality panel Electronic Industries Assn., 1971-80; mem. environ. quality com. NAM, 1965-84; mem. Pres.'s Adv. Bd. on Air Quality, 1974-75; vice chmn. environ. com. U.S. Bus. and Industry Adv. Com. to OECD, 1975-84; mem. adv. subcom. on environ., health and safety regulations Pres.'s Domestic Policy Rev. of Indsl. Innovation, 1978-79; adv. panel indsl. innovation and health, safety and environ. regulation Office Tech. Assessment of U.S. Congress, 1978-80; exec. com. engring. assembly NRC, 1977-80; also environ. studies bd. Commn. Natural Resources, 1977-82; mem. staff svcs. subcom. of environ. com. Bus. Roundtable, 1975-84; vice chmn. environ. com. U.S. Coun., Internat. C. of C., 1978-89; adv. com. on rsch. applications policy NSF, 1976-80; mem. Sci. Adv. Bd. EPA, 1984-88, selection com. Pres.' Environment and Conservation Challenge Award, 1991-93. Contbr. articles to profl. jours. Trustee Belwin Outdoor Lab., St. Paul, 1970—; bd. dirs. Fresh Water Found., 1974—, Northwest Area Found., 1970-87, St. Paul Area YMCA, 1974-80, Midwest China Study Center, Minn. Environ. Sci. Found., 1970-78, Nat. Water Alliance, 1983-88, World Environ. Ctr., 1984—. Woodrow Wilson Sr. fellow, 1975—; recipient numerous awards, including Joan Hodges Queneau award Am. Assn. Engring. Socs., 1990; named Laureate, Global 500 Honor Role UN Environ. Program. Fellow ASCE; mem. NAE, Am. Acad. Environ. Engrs. (diplomate, chmn. examination update com. 1981-83), Minn. Assn. Commerce and Industry, Am. Water Works Assn. (life), Air and Waste Mgmt. Assn. (dir.), Chem. Mfg. Assn. Club: Rainbow (Mpls.). Home: 2090 Arcade St Saint Paul MN 55109-2564 *

LING, RUNG TAI, physicist; b. Taipei, Taiwan, July 28, 1943; came to U.S., 1966; s. Suei Yuen and Lan Shian (Su) L.; m. Carol Tih Wang, Dec. 16, 1972; children: Bianca, Fricka. BS, Nat. Taiwan U., 1965; PhD, U. Calif., San Diego, 1972. Lectr. San Diego State U., 1972-73; rsch. fellow Calif. Inst. Tech., Pasadena, Calif., 1973-76; rsch. scientist STD Rsch. Corp., Arcadia, Calif., 1976-78, R & D Assocs., Marina del Rey, Calif., 1978-81; rsch. specialist Lockheed Corp., Burbank, Calif., 1981-84; sr. tech. specialist Northrop Corp., Pico Riveria, Calif., 1984—. Contbr. article to Phys. Revs., AIAA Jour., Jour. Applied Physics, Jour. of E&M Waves and Application, Computer Physics Communications. Mem. AIAA, Am. Phys. Soc., Electromagnetics Acad. Achievements include work in the application of differential equation approach to electromagnetic scattering. Office: Northrop Dept T234/GK B-2 Div 8900 E Washington Blvd Pico Rivera CA 90660-3783

LING, VICTOR, oncologist, educator; b. Mar. 16, 1943. BS in Biochemistry, U. Toronto, 1966; PhD in Biochemistry, U. B.C., 1969. Staff scientist Ont. Cancer Inst., Toronto, 1971—, head divsn. molecular and structural biology, 1989—; prof. med. biophysics U. Toronto, 1983—, mem. coun. sch. grad. studies, 1984—, mem. faculty of medicine rsch. com., 1985—, vice-chmn., 1988—; mem. study sect. of experimental therapeutics Nat. Insts. of Health, USA, 1986—; bd. govs. Wellesly Hosp. Rsch. Inst., 1988-90, mem. MRC scholarship com. Med. Rsch. Coun. of Can., 1988—, bd. sci. advisors Hong Kong Inst. Biotech., 1989—, adv. bd. Internat. Jour. Anti-cancer Drugs, 1990—, external adv. com. U. Wis. Clin. Cancer Ctr., 1990—, bd. sci. counselors divsn. cancer treatment Nat. Insts. of Health, 1990—; bd. dirs. Hosp. for Sick Children Found., 1992—. Assoc. editor Cancer Rsch., 1986—, Jour. Cellular Physiology, 1989—, Jour. Cellular Pharmacology, 1989—, Jour. Molecular Pharmacology, 1992—, Jour. Biomed. Sci., 1992—. Victoria U. Alumni scholar in Life Scis., 1965, Centennial fellow MRC of Can., 1969-71; recipient C. Chester Stock award Meml. Sloan-Kettering Cancer Ctr., 1988, Cancer Rsch. award The Milken Family Med. Found., 1988, Merit award The FCCP (Ont.) Edn. Found., 1989, Internat. award Gairdner Found., 1990, Kettering prize Gen. Motors Cancer Rsch. Found., 1991. Fellow Royal Soc. Can.; mem. Am. Assn. Cancer Rsch. (bd. dirs. 1992—, Bruce F. Cain Meml. award 1993), Am. Soc. Cell Biology, Can. Cancer Soc. (bd. dirs. 1992—), Can. Soc. Cell Biology, Can. Biochem. Soc., Genetics Soc. Can. Office: Ontario Cancer Inst, 500 Sherbourne St, Toronto, ON Canada M4X 1K9*

LINK, FREDERICK CHARLES, systems engineer; b. Phila., Sept. 2, 1948; s. Frederick Adolph and Elizabeth Mary (Gabriel) L. BSEE, Drexel U., 1971; MSEE, Bucknell U., 1973. Mem. tech. staff AT&T Bell Labs., Holmdel, N.J., 1973-84; mem. tech. staff Bellcore, Red Bank, N.J., 1984-86, disting. mem. tech. staff, 1986—. Mem. IEEE, AIAA, Nat. Space Soc., The Planetary Soc., Space Studies Inst. Roman Catholic. Achievements include extensive internal Bell technical memoranda involving contributions to various telecommunications projects. Office: Bellcore NVC 2Z301 331 Newman Springs Rd Red Bank NJ 07701-5699

LINK, STEVEN OTTO, environmental scientist, statistician; b. St. Paul, Oct. 23, 1953; s. Carl August and Wanda Jean (Baldwin) L. BS in Biology, BM in Math., U. Minn., 1977; PhD in Botany, Ariz. State U., 1983. Postdoctoral fellow Utah State U., Logan, 1983-85; rsch. scientist Battelle Meml. Inst., Richland, Wash., 1985—; courtesy asst. prof. Wash. State U., Pullman, 1986—. Contbr. articles to profl. jours. Office: Pacific Northwest Lab Environ Sci P7-54 Richland WA 99352

LINKER, KERRIE LYNN, systems engineer; b. Poughkeepsie, N.Y., Dec. 12, 1966; d. William Landes and Charlotte Louise (Scofield) Linker. BS in Physics and Math., Susquehanna U., Selinsgrove, Pa., 1989; MEng in Ops. Rsch. and Indsl. Engring., Cornell U., 1992. Mem. tech. staff AT&T Bell Labs., Holmdel, NJ., 1989—. Mem. Inst. Indsl. Engrs., Ops. Rsch. Soc. Am., AT&T Bell Labs. Future Pioneers. Office: AT&T Bell Labs Rm 2F-512 101 Crawfords Corner Rd Holmdel NJ 07733-3030

LINKER, LEWIS CRAIG, environmental engineer; b. Balt., Mar. 6, 1956; s. Fredrick Lewis and Miriam Evelyn (Fleagle) L.; m. Julia Augusta White, May 28, 1983; children: Jason James, Christopher Lewis. BS cum laude, Towson U., 1980; MA, Johns Hopkins U., 1987. Rsch. asst. Chesapeake Bay Inst., Balt., 1977-81; dir. biochem. lab. Johns Hopkins U., Balt., 1981-86; environ. engr. Chesapeake Bay program U.S. EPA, Annapolis, Md., 1987—. Contbr. to profl. jours. Mem. Am. Chem. Soc., Water Pollution Control Fedn., Estuarine Rsch. Fedn. Achievements include research in the creation of wetlands for the improvement of water quality. Office: US EPA Chesapeake Bay Program 410 Severn Ave Annapolis MD 21403

LINN, CAROLE ANNE, dietitian; b. Portland, Oreg., Mar. 3, 1945; d. James Leslie and Alice Mae (Thorburn) L. Intern, U. Minn., 1967-68; BS, Oreg. State U., 1963-67. Nutrition cons. licensing and cert. sect. Oreg. State Bd. Health, Portland, 1968-70; chief clin. dietitian Rogue Valley Med. Ctr., Medford, Oreg., 1970—; cons. Hillhaven Health Care Ctr., Medford, 1971-83; lectr. Local Speakers Bur., Medford. Mem. ASPEN, Am. Dietetic Assn., Am. Diabetic Assn., Oreg. Dietetic Assn. (sec. 1973-75, nominating com. 1974-75, Young Dietitian of Yr. 1976), So. Oreg. Dietetic Assn., Alpha Lambda Delta, Omicron Nu. Democrat. Mem. Christ Unity Ch. Avocations: sewing, needlecrafts, cooking, swimming, skiing. Office: Rogue Valley Med Ctr 2825 E Barnett Rd Medford OR 97504-8332

LINN, PAUL ANTHONY, nuclear engineer; b. Pitts., July 17, 1958; s. James E. and Lois M. (Howe) L.; m. Marianne T. McGartland, May 11, 1985; children: Laura, Paul. BS in Chem. Engring., Carnegie Mellon U., 1980, MS Chem Engring., 1985. Sr. engr. Westinghouse Electric, Monroeville, Pa., 1980-85; sr. engr. Volian Enterprises, Inc., Murrysville, Pa., 1985-91, mgr. engring., 1991--. Contbr. articles to profl. jours. Pres. Alle-West Colt Assn., Pircairn, Pa., 1983-86. Recipient Renselear award Renselear U., 1975, 76. Mem. ANS, Am. Inst. Chem. Engrs. Home: 1135 Preston Dr North Versailles PA 15137 Office: Volian Enterprises Inc PO Box 410 Murrysville PA 15668-0410

LINNVILLE, STEVEN EMORY, psychologist, navy officer; b. Pensacola, Fla., Apr. 2, 1956; s. Emory Maxwell and Eve Helen (Szmulewicz) L. BS, U. Fla., 1978; MA, U. West Fla., 1983; PhD, So. Ill. U., 1988. Commd. lt. USNR, 1989; psychologist USN, San Diego, 1989—. Contbr. articles to profl. jours., chpts. to books. Vol. usher San Diego Symphony, 1989-92; vol. AIDS task force 1st Unitarian Ch., San Diego, 1990-91. NIMH postdoctoral fellow, 1988; recipient postdoctoral rsch. award U. Colo. Health Scis. Ctr., 1988. Mem. APA, AAAS, Sigma Xi. Democrat. Episcopalian. Achievements include identification of attentional processing differences in HIV patients using electrophysiology, sex-related differences in brain function using magnetoencephalography, brain function differences in schizophrenics using magnetoencephalography, effects of lead to brain development using electrophysiology in animal model. Office: Naval Health Rsch Ctr 271 Catalina Blvd San Diego CA 92186-5122

LINOWES, DAVID FRANCIS, political economist, educator; b. N.J., Mar. 16, 1917; m. Dorothy Lee Wolf, Mar. 25, 1946; children: Joanne Linowes Alinsky, Richard Gary, Susan Linowes Allen (dec.), Jonathan Scott. BS with honors, U. Ill., 1941. Founder, ptnr. Leopold & Linowes (name now BDO Siedman), Washington, 1946-62; cons. sr. ptnr. Leopold & Linowes (name now BDO Siedman), 1962-82; nat. founding ptnr. Laventhol & Horwath, 1965-76; chmn. bd, chief exec. officer Mickleberry Corp., 1970-73; dir. Horn & Hardart Co., 1971-77, Piper Aircraft, 1972-77, Saturday Rev./World Mag., Inc., 1972-77, Chris Craft Industries, Inc., 1958—, Work in Am. Inst., Inc.; prof. polit. economy, pub. policy, bus. adminstrn. U. Ill., Urbana, 1976—, Boeschenstein prof. emeritus, 1987—; cons. DATA Internat. Assistance Corps., 1962-68, U.S. Dept. State, UN, Sec. HEW, Dept. Interior; chmn. Fed. Privacy Protection Commn., Washington, 1975-77, U.S. Commn. Fair Market Value Policy for Fed. Coal Leasing, 1983-84, Pres.'s Commn. on Fiscal Accountability of Nation's Energy Resources, 1981-82; chmn. Pres.' Commn. on Privatization, 1987-88; mem. Council on Fgn. Relations; cons. panel GAO; adj. prof. mgmt. NYU, 1965-73; Disting. Arthur Young Prof. U. Ill., 1973-74; emeritus chmn. internat. adv. com. Tel Aviv U.; headed U.S. State Dept. Mission to Turkey, 1967, to India, 1970, to Pakistan, 1968, to Greece, 1971, to Yugoslavia, 1991; U.S. rep. on privacy to Orgn. Econ. Devel. Intergovtl. Bur. for Informatics, 1977-81, cons., N.Y.C., 1977-81; U.S. State Dept. mission to Chile, Argentina and Uruguay, July, 1988, Yugoslavia, May, 1991. Author: Managing Growth Through Acquistion, Strategies for Survival, Corporate Conscience; commn. report Personal Privacy in Information Society, Fiscal Accountablility of Nation's Energy Resources, The Privacy Crisis In Our Time; editor: The Impact of the Communcation and Computer Revolution on Society, Privacy in America, 1989; contbr. articles to profl. jours. Trustee Boy's Club Greater Washington, 1955-62, Am. Inst. Found., 1962-68; assoc. YM-YWHA's Greater N.Y., 1970-76; chmn. Charities Adv. Com. of D.C., 1958-62; emeritus bd. dirs. Religion in Am. Life, Inc.; former chmn. U.S. People for UN; chmn. citizens com. Combat Charity Rackets, 1953-58. Served to 1st lt. Signal Corps, AUS, 1942-46. Recipient 1970 Human Relations award Am. Jewish Com., U.S. Pub. Service award, 1982, Alumni Achievement award U. Ill., 1989, CPA Distinguished Pub. Svc. award, Washington, 1989. Mem. AICPA (v.p. 1962-63), U. Ill. Found. (emeritus bd. dirs. 1), Coun. Fgn. Rels., Cosmos Club (Washington), Phi Kappa Phi (nat. bd. dirs.), Beta Gamma Sigma. Home: 803 Fairway Dr Champaign IL 61820-6325 Office: U Ill 308 Lincoln Hall Urbana IL 61801 also: 9 Wayside Ln Scarsdale NY 10583

LINTON, WILLIAM SIDNEY, marketing research professional; b. Cullman, Ala., Sept. 18, 1950; s. James Hubert and Elaine (Ryan) L.; m. Phyllis Leigh Holland, July 13, 1972; children: Jodi K., Philip W. BA, Ambassador Coll., 1972; MBA, Furman/Clemson, 1983. Purchasing agt. Jamestown Corp., Birmingham, Ala., 1972-73; mktg. adminstr. J.M. Tull Metals, Greenville, S.C., 1974; mgr. mktg. rsch. Fluor Daniel, Greenville, 1975-83; v.p. Tecton Group, Greenville, 1983-87; pres. Linton & Assocs., Greenville, 1987—; cons. in field. Contbr. articles to profl. jours.

LINTZ, PAUL RODGERS, physicist, patent examiner; b. Dallas, Feb. 8, 1941; s. Norman Edmund and Sarah Kathleen (Powers) L.; m. Mary Grace Caggiano, Nov. 27, 1965; children: Matthew Thomas, Eileen Sarita, Jerome Peter, Elizabeth Irene. BA cum laude, U. Dallas, 1963; MS, Cath. U. Am., 1965, PhD, 1977. Mem. tech. staff Tex. Instruments, Dallas, 1965-67; rsch. physicist Teledyne Geotech. Co., Alexandria, Va., 1967-74; scientist Planning Systems Inc., McLean, Va., 1974-76; prin. investigator Sci. Applications Internat. Corp., McLean, 1976-84; systems engr. Mitre Corp., McLean, 1984-87; ind. cons. Vienna, Va., 1987-92; patent examiner U.S. Patent and Trademark Office, Washington, 1992—. Mem. Providence Dist. Dem. Com., Vienna, 1985; com. mem. Vienna area Boy Scouts Am., 1986-90; pres. Tysons Woods Civic Assn., Vienna, 1991. Mem. IEEE, Patent and Trademark Office Soc., Sigma Xi. Roman Catholic. Achievements include development of digital signal processing techniques for defense purposes in seismic detection of underground nuclear blasts, submarine sonar signal processing, passive bistatic radar signal processing, detection, tracking. Home: 2222 Craigo Ct Vienna VA 22182 Office: Patent and Trademark Office Crystal Park 2 Crystal City VA 22202

LINYARD, SAMUEL EDWARD GOLDSMITH, civil engineer; b. North Augusta, S.C., Jan. 7, 1937; s. David P. and Frieda (Goldsmith) L.; m. Margaret JoAnn Bell, Sept. 2, 1956 (dec. 1987); 1 child, Beth Louise; m. Sue Pardue, July 18, 1991; children: Susan Langley, David Rhett. Student, Augusta (Ga.) Coll., 1954-55, Clemson U., 1955-57. Registered profl. engr., S.C., Fla., N.C., N.H., Ky., W.Va. Engring. technician Patchen and Zimmerman, Augusta, 1957-58; engring. designer Michael Baker Jr., Columbus, Ohio, 1958-60, Rackoff Assocs., Columbus, 1960-61, Wilbur Smith Assocs., Inc., Columbia, S.C., 1961-69; assoc.-in-charge Wilbur Smith Assocs., Inc., Orlando, Fla., 1969-72, Raleigh, N.C., 1972-85; sr. v.p. Wilbur Smith Assocs., Inc., Columbia, 1985—. Mem. A. Pub. Works Assn., Cons. Engrs. S.C. (pres. 1987-89), S.C. Coun. Engring. Socs. (pres. 1989-91), S.C. Soc. Profl. Engrs. Episcopalian. Home: 230 King Charles Rd Columbia SC 29209 Office: Wilbur Smith Assocs Inc 4500 Jackson Blvd Columbia SC 29209

LINZ, ANTHONY JAMES, osteopathic physician, consultant, educator; b. Sandusky, Ohio, June 16, 1948; s. Anthony Joseph and Margaret Jane (Ballah) Linz; m. Kathleen Ann Kovach, Aug. 18, 1973; children: Anthony Scott, Sara Elizabeth. BS, Bowling Green State U., 1971; D.O., U. Osteo. Med. and Health Scis., 1974. Diplomate Nat. Bd. Osteo. Examiners; bd. cert., diplomate Am. Osteo. Bd. Internal Medicine, Internal Medicine, Med. Diseases of Chest and Critical Care Medicine. Intern Brentwood Hosp., Cleve., 1974-75, resident in internal medicine, 1975-78; subsplty. fellow in pulmonary diseases Riverside Meth. Hosp., Columbus, Ohio, 1978-80; med.

dir. pulmonary svcs. Sandusky (Ohio) Meml. Hosp., 1980-85; med. dir. cardio-pulmonary svcs. Firelands Community Hosp., Sandusky, 1985—; cons. pulmonary, critical care and internal medicine, active staff sect. internal medicine, chmn. dept. medicine, head div. pulmonary medicine Firelands Community Hosp., 1985—; cons. staff dept. medicine Good Samaritan Hosp., 1982-85, sect. internal medicine specializing pulmonary diseases; cons. pulmonary and internal medicine Providence Hosp., Sandusky, Mercy Hosp., Willard, Ohio; clin. prof. internal medicine Ohio U. Coll. Osteo. Medicine; clin. prof. medicine Univ. Health Scis. Coll. Osteo. Medicine, Kansas City, Mo.; clin. asst. prof. med. Med. Coll. of Ohio at Toledo; adj. prof. applied scis. Bowling Green State U.; mem. respiratory tech. adv. bd. Firelands Campus, Bowling Green State U., 1983—; med. dir. Respiratory Therapy program., Bowling Green State U., 1984—. Author, contbr. articles and abstracts to profl. jours. Water safety instr. ARC, 1965—; med. dir. Camp Superkid Asthma Camp, 1984—; bd. trustees Stein Hospice. Recipient Sr. award Excellence in Pharmacology UOMHS-COMS, 1974, Edward Ruff Community Svc. award Am. Lung Assn., 1985, Master Clinician award Ohio U. Coll. Osteopathic Medicine, 1987, Golden Rule award J.C. Penney, 1990. Fellow Am. Coll. Chest Physicians, Am. Coll. Critical Care Medicine, Am. Coll. Osteo. Internists; mem. AAAS, Am. Osteo. Assn., Ohio Osteo. Assn. (past. pres., past v.p., past. sec.-treas., acad. trustee 5th dist. acad.), Am. Heart Assn., Am. Thoracic Soc., Ohio Thoracic Soc., Am. Lung Assn. (pres., 1st v.p., med. adv. bd. chmn., exec. bd. dirs., bd. dirs. Ohio's So. Shore sect. 1984—), Nat. Assn. Med. Dirs. Respiratory Care, Ohio Soc. Respiratory Care (med. adviser/dir. 1982—), Soc. Critical Care Medicine, Found. Crit. Care (mem. Founder's Circle), Sandusky Yacht Club, Sandusky High Sch. Alumni Assn. (life), U. Osteo. Medicine and Health Scis.-Coll. Osteo. Medicine and Surgery Alumni Assn. (life), U.S. Power Squadron (Sandusky chpt.), Alpha Epsilon Delta (premed hon. soc.), Beta Beta Beta (biology hon. soc.), Pi Kappa Alpha, Atlas Med. Fraternity. Roman Catholic.

LIONS, JACQUES LOUIS, science educator, space studies center administrator; b. Grasse, France, May 2, 1928; s. Honore Antoine and Anne (Muller) L.; m. Andree Olivier, Aug. 21, 1950; 1 child, Pierre Louis. Agregation Math., Ecole Normale Superieure, Paris, 1950; Dr. es Sci., U. Paris, 1954; hon. dr., U. Liege, 1973, U. Madrid, 1976, U. Fudan, 1981, Goteborg U., 1984, Heriot Watt U., Edinburgh, 1982, Poly. U. Madrid, 1988, St. Jacques de Compostella, 1993, U. Malaga, U. Santiago-Chile. Faculty mem. U. Nancy, France, 1954-62, U. Paris, 1962-73; prof. Coll. de France, Paris, 1973—; prof. Ecole Polytechnique, Paris, 1967-86; pres. Institut National de Recherche en Informatique et en Automatique, 1980-84, Centre Nat. Etudes Spatiales, 1984-92; high scientific advisor DASSAULT Industry 1993—; scientific advisor E.D.F., E.L.F.; chmn. Scientific Advisory Com. Pechiney and GAZ of France. Author: Les Inequations en Mecanique et en Physique, 1969, Some Methods in the Mathematical Analysis of Systems and of their Control, 1981, Controle des Systemes distribues Singuliers, 1983, others. Decorated commandeur Ordre Nat. du Merite, 1988, Ordre Nat. de la Legion d'Honneur, 1993; recipient Japan prize, 1991, Harvey prize, 1991. Mem. Acad. des Scis., Pontifical Acad. Scis.; fgn. mem. Acad. Royale de Liege, Acad. Scis. Lombardie, Acad. Brasileira de Ciencias, Russian Acad. Scis., Ukraine Acad. Scis., Acad. Royale Belgium, Am. Acad. Arts and Scis., Internat. Acad. Astronautics, Acad. Sci. of Chile, Internat. Mathematical Union (pres. 1990—), Academia Europae. Office: College de France, 3 Rue d'Ulm, 75005 Paris France

LIOTTA, LANCE A., pathologist; b. Cleve., July 12, 1947; married; 2 children. BA in Gen. Sci. and Biology, Hiram Coll., 1969; PhD in Biomed. Engring. and Biomath., Case Western U., 1974, MD, 1976. Cert. basic life support Am. Heart Assn., advanced life support Am. Heart Assn. Instr. pathology for inhalation therapists dept. pathology St. Luke's Hosp., Cleve., 1972-74; sr. instr. pulmonary pathology Phase I and Phase II, Sch. Medicine Case Western Reserve U., 1973-74; USPHS resident physician Lab. Pathology, Nat. Cancer Inst. NIH, Bethesda, Md., 1976-78, pathologist, expert/cons. Lab. Pathophysiology, Nat. Cancer Inst., 1978-80, sr. investigator, pub. health svc. officer Lab. Pathophysiology and Pathology, Nat. Cancer Inst., 1980-82, chief tumor invasion and metastases sect. Lab. Pathology and Lab. Pathology, Nat. Cancer Inst., 1982—, dir. anatomic pathology residency program Lab. Pathology, Nat. Cancer Inst., 1982—, dep. dir. intramural rsch., 1992—; clin. prof. pathology Sch. Medicine George Wash. U.; mem. adj. faculty Sch. Medicine Georgetown U.; invited faculty mem. Rockefeller U., 1979; speaker in field. Author: (with others) Cancer Invasion and Metastasis, 1977, Pulmonary Metastasis, 1978, Metastatic Tumor Growth, 1980, Bone Metastasis, 1981, Cell Biology of Breast Cancer, 1980, New Trends in Basement Membrane Research, 1982, Tumor Invasion and Metastasis, 1982, Progress in Clinical and Biological Research, 1982, Growth of Cells in Hormonally Defined Media, 1982, Understanding Breast Cancer: Clinical and Laboratory Concepts, 1983, The Role of Extracellular Matrix in Development, 1984, Basic Mechanisms and Clinical Treatment of Tumor Metastasis, 1985, Hemostatic Mechanisms and Metastasis, 1984, Biological Responses in Cancer, vol. 4, 1985, The Cell in Contact: Adhesions and Junctions as Morphogenetic Determinants, 1985, Rheumatology, vol. 10, 1986, Progress in Neuropathology, vol. 6, 1986, Cancer Metastasis: Experimental and Clinical Strategies, 1986, Biochemistry and Molecular Genetics of Cancer Metastasis, 1986, Basement Membranes, 1985, 1986 Year Book of Cancer, New Concepts in Neoplasia as Applied to Diagnostic Pathology, 1986, Head and Neck Management of the Cancer Patient, 1986, Cancer Metastasis: Biological and Biochemical Mechanisms and Clinical Aspects, 1988, Important Advances in Oncology, 1988, Breast Cancer: Cellular and Molecular Biology, 1988, Cancer: Principles and Practice of Oncology, vol. 1, 3d edit., 1989, Molecular Mechanisms in Cellular Growth and Differentiation, 1991, Peptide Growth Factors and Their Receptors, 1990, Molecuar Genetics in Cancer Diagnosis, 1990, Cancer Surveys-Advances & Prospects in Clinical, Epidemiological and Laboratory Oncology, vol. 7, no. 4, 1988, Genetic Mechanisms in Carcinogenesis and Tumor Progression, 1990, Molecular and Cellular Biology, Host Immune Responses and Perspectives for Treatment, 1989, Origins of Human Cancer: A Comprehensive Review, 1991, Cancer and Metastasis Reviews, vol. 9, 1990, Comprehensive Textbook of Oncology, 1991, Textbook of Internal Medicine, 2d edit., vol. 2, 1992, Molecular Foundations in Oncology, 1991, Genes, Oncogenes, and Hormones: Advances in Cellular and Molecular Biology of Breast Cancer, 1991, Cell Motility Factors, 1991, Oncogenes and Tumor Suppressor Genes in Human Malignancies, 1993, Principles and Practice of Gnecologic Oncology, 1992, Cancer Medicine, 3d edit., 1993; contbr. articles to profl. jours. NIH Pre-doctoral fellow; recipient Arthur S. Flemming award, 1983, Flow award lectureship Soc. Cell Biology, 1983, Nat. award and lectureship Am. Assn. Clin. Chemistry, 1987, Rsch. award Susan G. Komen Found., 1987, Disting. Lectr. award Rush Cancer Ctr., 1987, George Hoyt Whipple award and lectureship Sch. Medicine U. Rochester, 1988, Karen Grunebaum Symposium award lectureship Hubert H. Humphrey Cancer Rsch. Ctr., 1988, Cancer Rsch. award Milken Family Med. Found., 1988, William M. Shelly Meml. award and lectureship Centennial Johns Hopkins Med.st., 1989, Josef Steiner Cancer Found. prize, 1989, Basic Rsch. award Am. Soc. Cytology, 1989, Officer's Recognition award Equal Employment Opportunity, 1990, John W. Cline Cancer Rsch. award and lectureship U. Calif., 1990, Herman Pinkus award lectureship Am. Soc. Dermatology, 1990, Simon M. Shubitz award U. Chgo. Cancer Ctr., 1991, Stanley Gore Rsch. award, 1991, Lila Gruber Cancer Rsch. award Am. Acad. Dermatology, 1991, Am.-Italian Found. Cancer Rsch. award, 1992. Mem. Am. Assn. Cancer Rsch. (bd. dirs., 6th Ann. Rhoads Meml. award 1985), Am. Assn. Pathologists (Warner-Lambert/Parke-Davis award 1984), Am. Soc. Cell Biology, Am. Soc. Clin. Investigation, Internat. Acad. Pathology, Internat. Assn. Metastasis Rsch. (pres.-elect 1990), Sigma Xi, Phi Beta Kappa. Achievements include patents for method and device for determining the concentration of a material in a liquid, method for isolating bacterial colonies, test method for separating and/or isolating bacteria and tissue cells., divice and method for detecting phenothiazine-type drugs in urine, in vitro assay for cell invasiveness, enzyme immunoassay with two-zoned device having bound antigens, metalloproteinase peptides, matrix receptors role in diagnosis and therapy of cancer, genetic method for predicting tumor aggressiveness, therapeutic application of an anti-invasive compound; patents pending for role of tumor motility factors in cancer diagnosis, role of tumor metalloproteinases in cancer diagnosis, peptide inhibitor of metalloproteinases, protein inhibitors of metalloproteinases, autotaxin motility stimulating proteins diagnosis and therapy, motility receptor protein and gene diagnosis and therapy. Office: Lab of Pathology Nat Cancer Inst 9000 Rockville Pike Bethesda MD 20892-0001

LIOU, JENN-CHORNG, electronics engineer; b. Hsinchu, Taiwan, May 1956; came to U.S., 1985; s. Hwan-Sern and Chun-Shieh Liou; m. Pi-Chi Huang, July 1, 1989. BSEE, Nat. Chiao Tung U., Hsinchu, Taiwan, 1978, MSEE, 1980; PhD, U. Mass., 1991. Instr. Nat. Chiao Tung U., Hsinchu, 1982-83; sr. engr. Microelectronics Tech., Inc., Hsinchu, 1983-85; sr. microwave engr. Lucas Epsco, Inc., Hopkinton, Mass., 1990—. Contbr. articles to profl. jours. Mem. IEEE. Achievements include research in tuner design with lossy filters, characteristics of transmission lines on semiconductors, microwave and RF, microwave theory technology and electron device. Office: Lucas Epsco Inc 99 South St Hopkinton MA 01748

LIPKIN, BERNICE SACKS, computer science educator; b. Boston, Dec. 21, 1927; d. Milton and Esther Miriam (Berchuck) Sacks; m. Lewis Edward Lipkin; children: Joel Arthur, Libbe Lipkin Englander. BS in Biology, Chemistry, Northeastern U., 1949; MA in Psychology, Boston U., 1950; PhD in Experimental Psychology, Columbia U., 1961. Rsch. and devel. scientist Directorate Sci. and Tech., CIA, Washington, 1964-70; scientist dept. computer sci. U. Md., Greenbelt, 1971-72; health sci. adminstr. NIH, Bethesda, Md., 1972-88; cons. computerized text analysis, data exploration L B and Co., Bethesda, 1989—. Editor: Picture Processing and Psychopictorics, 1970; contbr. articles on computer-based text searches and data analysis to profl. publs. Cerebral Palsy Soc. fellow in neurophysiology, 1961-62; NIH trainee, 1955-58. Mem. AAAS, IEEE, APA, Soc. Neurosci., Optical Soc. Am., Assn. Computing Machinery, Sigma Xi. Jewish. Achievements include design of system for manipulation and analysis of text data files, documentation and instruction manuals; teaching children computer concepts and programming. Office: 9913 Belhaven Rd Bethesda MD 20817

LIPKIN, GEORGE, dermatologist, researcher; b. N.Y.C., Dec. 31, 1930; s. Samuel and Celia (Greenfield) L.; m. Sari Berger, June 16, 1957; children: Michael, Lisa. AB, Columbia Coll., 1952; MD, SUNY, Bklyn., 1955. Diplomate Am. Bd. Med. Examiners, Am. Bd. Dermatology. From instr. to assoc. prof. dermatology NYU Sch. Med., N.Y.C., 1961-73, prof. dermatology, 1973—; bd. dirs. Berger Found. for Cancer Rsch., Clifton, N.J.; reviewer numerous sci. jours. and publ., 1961—. Author: (with others) Methods in Cancer Research, vol. 8, 1973, Pharmacological Effects of Lipids III-Lipids in Cancer Research, 1989, Proceedings of National Academy of Science, 1974, Experimental Cell Research, 1976, Cancer Research, 1978; contbr. articles to profl. jours. Capt. U.S. Army, 1957-59. Rsch. grantee Nat. Cancer Inst., 1967-78, Dermatology Found., 1970, Skin Cancer Found., 1980, Cancer Rsch. Corp. N.J., 1981-84, Bion Rsch. Corp., 1984-88. Mem. AAAS, Am. Acad. Dermatology, Soc. Investigative Dermatology, N.Y. Dermatolog. Soc., Harvey Soc. Achievements include patent (with other) for the improved method for purification of contact inhibitory factor; development of first cell-derived factor which reverses the malignant melanoma phenotype. Office: NYU Medical Ctr 530 First Ave New York NY 10016

LIPKIN, MARTIN, physician, scientist; b. N.Y.C., Apr. 30, 1926; s. Samuel S. and Celia (Greenfield) L.; m. Joan Schulein, Feb. 16, 1958; children—Richard Martin, Steven Monroe. A.B., NYU, 1946, M.D., 1950. Diplomate: Nat. Bd. Med. Examiners. Fellow Cornell U. Med. Coll., 1952; practice medicine specializing in internal medicine, gastroenterology and neoplastic diseases N.Y.C.; mem. staff N.Y. Hosp., Meml. Hosp. for Cancer and Allied Diseases; assoc. prof. medicine Cornell U. Med. Coll., N.Y.C., 1963-78; prof. medicine Cornell U. Med. Coll., 1978—, prof. Grad. Sch. Med. Scis., 1978—; attending physician, dir. Irving Weinstein Lab. for Gastrointestinal Cancer Prevention Meml. Sloan Kettering Cancer Ctr.; mem. Meml. Sloan-Kettering Cancer Ctr., 1985—; vis. physician Rockefeller U. Hosp., 1981—; hon. lectr. Israel Med. Assn. and Gastroenterology Soc., 1982; nominator Nobel Prize for Physiology and Medicine, 1982. Mem. editorial bd. Cancer Epidemiology, Biomarkers and Prevention, Cancer Research, Cancer Letters; editor: Gastrointestinal Tract Cancer, 1978, Inhibition of Tumor Induction and Development, 1981, Gastrointestinal Cancer: Endogenous Factors, 1981, Calcium, Vitamin D and Prevention of Colon Cancer, 1991, Cancer Chemoprevention, 1992; contbr. articles to Cancer Rsch.; contbr. articles to profl. jours. Served as officer USN, 1953-55. Recipient NIH career devel. award, 1962-71; Albert F.R. Andresen ann. award and lectureship N.Y. State Med. Soc., 1971. Fellow ACP, Am. Coll. Gastroenterology; mem. Med. Soc. State of N.Y. (assoc. chmn. sci. program com. 1977-90, chmn. sci. program com. 1990-91, chmn. edn. com. 1991—), Digestive Diseases Soc. (founder), Internat. Soc. Investigative Gastroenterology (founder), Am. Soc. Clin. Investigation, Am. Physiol. Soc., Am. Assn. Cancer Rsch., Am. Gastroenterol. Assn., Soc. for Exptl. Biology and Medicine, Harvey Soc. Office: 1275 York Ave New York NY 10021-6094

LIPKIN, MARY CASTLEMAN DAVIS (MRS. ARTHUR BENNETT LIPKIN), retired psychiatric social worker; b. Germantown, Pa., Mar. 4, 1907; d. Henry L. and Willie (Webb) Davis; student Acad. Fine Arts, Pa., 1924-28, grad. sch. social work U. Wash., 1946-48; m. William F. Cavenaugh, Nov. 8, 1930 (div.); children: Molly C. (Mrs. Gary Oberbillig), William A.; m. 2d, Arthur Bennett Lipkin, Sept. 15, 1961 (dec. June 1974). Nursery sch. tchr. Miquon (Pa.) Sch., 1940-45; caseworker Family Soc. Seattle, 1948-49, Jewish Family and Child Service, Seattle, 1951-56; psychiat. social worker Stockton (Calif.) State Hosp., 1957-58; supr. social service Mental Health Research Inst., Fort Steilacoom, Wash., 1958-59; engaged in pvt. practice, Bellevue, Wash., 1959-61. Former mem. Phila. Com. on City Policy. Former diplomate and bd. mem. Conf. Advancement of Pvt. Practice in Social Work; former mem. Chestnut Hill women's com. Phila. Orch; mem. Bellevue Art Mus., Wine Luke Mus. Mem. ACLU, Linus Paul Inst. Sci. and Medicine, Inst. Noetic Scis., Menninger Found., Union Concerned Scientists, Physicians for Social Responsibility, Center for Sci. in Pub. Interest, Asian Art Council, Seattle Art Mus., Nature Conservancy, Wilderness Soc., Sierra Club.Women's Univ. Club Seattle, Friday Harbor Yacht Club Washington). Home: 10022 Meydenbauer Way SE Bellevue WA 98004-6041

LIPKIN, RICHARD MARTIN, journalist, science writer; b. N.Y.C., June 3, 1961; s. Martin and Joan (Schulein) L. AB, Princeton U., 1983. Rschr. MacNeil/Lehrer Newshour, N.Y.C., 1983-84; reporter UPI, Washington, 1984-85; assoc. editor The Wilson Quar., Washington, 1985-87; staff writer Insight Mag., Washington, 1987-91; freelance sci. writer Washington, 1991-93; editor Sci. News, Washington, 1993—. Contbg. author: Mathematical Impressions, 1991. Mem. AAAS, Nat. Assn. Sci. Writers, D.C. Assn. Sci. Writers, Am. Math. Soc., N.Y. Acad. Scis., Philos. Soc. Washington. Office: Sci News 1719 N St NW Washington DC 20036

LIPMAN, DANIEL GORDON, neuropsychiatrist; b. London, July 22, 1912; s. Jacob Meyer and Chana Gitel (Halbmillion) L.; m. Ida Wolfson; children: Ralph I., Frances J. Lipman Yellin, Philip W., Leta P. Lipman Cotton, Gail R. Lipman Mandel. MD, Middlesex Med. Coll., Waltham, Mass., 1940; PhD, Columbia Pacific U., San Rafael, Calif., 1983. Intern Harbor Hosp., Bklyn., 1940-41; resident Wadsworth Hosp., N.Y.C., 1941-42; gen. practice medicine Lynn, Mass., 1942-43, 46-49; internal medicine practice/research in adaptation disease Lynn, 1949-55; dir Stress & Hypertension Clinic Dispensary, U.S. Naval Weapons Plant, Washington, 1955-59; psychiatry resident VA Hosp., Perry Point, Md., 1959-62; tng. officer NIMH, St. Elizabeth Hosp., Washington, 1966-70; dir. research and edn. Creative Research Inst., Gaithersburg, Jerusalem, 1975—; asst. prof. clin. psychiatry George Washington U., Washington, 1967-83, Howard U. Med. Sch., Washington, 1968-83; med. officer neuropsychiatry Bur. of Hearings and Appeals, med. adv. staff. Social Security Adminstrn., HEW, Washington, 1971-73; cons. in field. Originator Lipman Personality Image Projection Test indicating brain harmony, disharmony and hemispheric dominance; inventor med. 3 dimensional stethoscope; contbr. articles to profl. jours.; inventor in field. 1st lt. U.S. Army, 1943-46.c Music Assn., Lynn, 1952-54. Recipient Physician Recognition award, AMA, APA, 1969—. Fellow Am. Geriatric Soc.; mem. AMA, Am. Psychaitric Assn., Mass. Med. Soc., N.Y. Acad. Sci. Office: Creative Rsch Inst, PO Box 16236, Jerusalem Israel 91163

LIPMAN, RICHARD PAUL, pediatrician; b. Cambridge, Mass., Aug. 1, 1935; s. Hyman Zelig and Betty (Likovsky) L.; m. Mary Alice Wilcox, Aug. 25, 1963; children: Gregory, Susan. AB magna cum laude, Harvard U., 1957; MD cum laude, Tufts U., 1961. Diplomate Am. Bd. Pediatrics. Intern Boston Floating Hosp., 1961-62, jr. resident, 1962-63, sr. resident, 1963-64, chief resident, 1964; rsch. fellow infectious disease Med. Sch. U. N.C., Chapel Hill, 1967-69; practice pediatrics Peabody and Lynn, Mass., 1969—; mem. staff North Shore Children's Hosp., Salem, Mass., assoc. chief of staff, 1974-76, pres., chief of staff, 1976-79, chief of medicine, 1979-83, trustee, 1980-84, corporator, 1985-86, mem. staff Tufts-New Eng. Med. Ctr., Boston, Boston Children's Hosp. Med. Ctr., AtlantiCare Med. Ctr., Salem Hosp.; clin. instr. pediatrics Tufts U. Sch. Medicine, Boston, 1969-74, asst. clin. prof., 1974-78, assoc. clin. prof., 1978—; bd. dirs. Tufts Assoc. Health Maintanance Orgn., 1988—. Contbr. articles to profl. jours. Capt. M.C., AUS, 1964-66. Fellow Am. Acad. Pediatrics; mem. AMA, Am. Soc. Microbiology, New Eng. Pediatric Soc., Mass. Med. Soc., Tufts Alumni Assn., Nat. Assn. Watch and Clock Collectors. Office: 1 Roosevelt Ave Peabody MA 01960-2227 also: 225 Boston St Salem MA 01970

LIPOMI, MICHAEL JOSEPH, health facility administrator; b. Buffalo, Mar. 9, 1953; s. Dominic Joseph and Betty (Angelo) L.; m. Brenda H. Lipomi, Dec. 23, 1977; children: Jennifer, Barrett. BA, U. Ottawa, 1976. Mktg. dir. Am. Med. Internat. El Cajon Valley Hosp., Calif., 1980-83; dir. corp. devel. Med. Surg. Ctrs. Am., Calif., 1983-85; exec. dir. Stanislaus Surgery Ctr., Modesto, Calif., 1985—. Author: Complete Anatomy of Health Care Marketing, 1988; co-host med. TV talk show Health Talk Modesto. Bd. dirs. Am. Heart Assn., Modesto, 1988-89; pres. Modesto Community Hospice, 1987-88; active local govt.; sec.-treas. Modesto Industry and Edn. Council, 1989. Mem. Calif. Ambulatory Surgery Assn. (pres. 1988-89), No. Calif. Assn. Surgery Ctrs. (pres. 1986-88), Federated Ambulatory Surgery Assn. (govt. rels. com. 1988, bd. dirs. 1989—, chmn. govt. rels. com. 1990), C. of C. (bd. dirs. 1989-92), Rotary. Lodge: Rotary. Avocations: golf, tennis. Office: Stanislaus Surgery Ctr 1421 Oakdale Rd Modesto CA 95355-3359

LIPOWITZ, JONATHAN, materials scientist; b. Paterson, N.J., Apr. 25, 1937; s. Alex and Esther (Knoble) L.; m. Evelyn R. Jacobs, Dec. 1960; children: Robert A., Suzanne J. BS, Rutgers U., 1958; PhD, U. Pitts., 1964. Postdoctoral fellow Pa. State U., State College, 1964-65; mem. rsch. staff Dow Corning Corp., Midland, Mich., 1965—, rsch. scientist, 1987—. Contbr. articles to Jour. Fire and Flammability, other profl. jours. Predoctoral fellow NSF, 1962-64. Mem. AAAS, Am. Chem. Soc., Am. Ceramic Soc., N.Y. Acad. Scis. Achievements include initial study of flame chemistry/fire properties of silicones, comprehensive studies of structure of ceramic fibers derived from organosilicon polymers, preparation of stoichiometric, fine diameter continuous silicon carbide fiber. Office: Dow Corning Corp 3901 S Saginaw Rd Midland MI 48686-0995

LIPPARD, STEPHEN JAMES, chemist, educator; b. Pittsburgh, Pa., Oct. 12, 1940; s. Alvin I. and Ruth (Green) L.; m. Judith Ann Drezner, Aug. 16, 1964; children: Andrew (dec.), Joshua, Alexander. B.A., Haverford Coll., 1962; Ph.D., MIT, 1965. Postdoctoral research fellow chemistry MIT, Cambridge, 1965-66, prof. chemistry, 1983-89, Arthur Amos Noyes prof. chemistry, 1989—; asst. prof. chemistry Columbia U., N.Y.C., 1966-69; asso. prof. Columbia U., 1969-72, prof., 1972-82; mem. study sect. medicinal chemistry NIH, 1973-77. Editor: Progress in Inorganic Chemistry, 1967-92; mem. editorial bd. Inorganic Chemistry, 1981-83, 89-91, assoc. editor, 1983-88; mem. editorial bd. Account Chem. Res., 1986-88; contbr. articles to profl. jours. Coach Demarest Borough Soccer Team, 1975-82, league adminstr., 1979-82. NSF fellow, 1962-66; Alfred P. Sloan fellow, 1968-70; Guggenheim fellow, 1972; recipient Tchr.-Scholar award Camille and Henry Dreyfus Found., 1971-76, Henry J. Albert award Internat. Precious Metals Inst., 1985, Alexander von Humboldt U.S. Sr. Scientist award, 1988; sr. internat. fellow John E. Fogarty Internat. Center, 1979. Fellow AAAS; mem. NAS, Am. Acad. Arts and Sci., Am. Chem. Soc. (chmn. bioinorganic subdiv. 1987-88, Inorganic Chemistry award 1987, Remson award 1987, assoc. editor jour. 1989—, chmn. elect inorganic div. 1991, chmn. 1992), Am. Crystallographic Assn., Am. Soc. Biol. Chemists, Nat. Inst. Medicine, Chem. Soc. (London), Biophys. Soc., Phi Beta Kappa. Home: 450 Memorial Dr Cambridge MA 02139-4306 Office: MIT Dept Chemistry Rm 18-290 77 Massachusetts Ave Cambridge MA 02139

LIPPE, PHILIPP MARIA, neurosurgeon, educator; b. Vienna, Austria, May 17, 1929; s. Philipp and Maria (Goth) L.; came to U.S., 1938, naturalized, 1945; m. Virginia M. Wiltgen, 1953 (div. 1977); children: Patricia Ann Marie, Philip Eric Andrew, Laura Lynne Elizabeth, Kenneth Anthony Ernst; m. Gail B. Busch, Nov. 26, 1977. Student Loyola U., Chgo., 1947-50; BS in Medicine, U. Ill. Coll. Medicine, 1952, MD with high honors, 1954. Rotating intern St. Francis Hosp., Evanston, Ill., 1954-55; asst. resident gen. surgery VA Hosp., Hines, Ill., 1955, 58-59; asst. resident neurology and neurol. surgery Neuropsychiat. Inst., U. Ill. Rsch. and Ednl. Hosps., Chgo., 1959-60, chief resident, 1962-63, resident neuropathology, 1962, postgrad. trainee in electroencephalography, 1963; resident neurology and neurol. surgery Presbyn.-St. Luke's Hosp., Chgo., 1960-61; practice medicine, specializing in neurol. surgery, San Jose, Calif., 1963—; instr. neurology and neurol. surgery U. Ill., 1962-63; clin. instr. surgery and neurosurgery Stanford U., 1965-69, clin. asst. prof., 1969-74, clin. assoc. prof., 1974—; staff cons. in neurosurgery O'Connor Hosp., Santa Clara Valley Med. Ctr., San Jose Hosp., Los Gatos Community Hosp., El Camino Hosp. (all San Jose area); chmn. div. neurosugery Good Samaritan Hosp, 1989—; founder, exec. dir. Bay Area Pain Rehab. Center, San Jose, 1979—; clin. adviser to Joint Commn. on Accreditation of Hosps.; mem. dist. med. quality rev. com. Calif. Bd. Med. Quality Assurance, 1976-87, chmn., 1976-77. Served to capt. USAF, 1956-58. Diplomate Am. Bd. Neurol. Surgery, Nat. Bd. Med. Examiners. Fellow ACS, Am. Coll. Pain Medicine (pres. 1992, bd. dirs., v.p. 1991—); mem. AMA (Ho. of Dels. 1981—), Calif. Med. Assn. (Ho. of Dels. 1976-80, sci. bd., council 1979-87, sec. 1981-87, Outstanding Svc. award 1987), Santa Clara County Med. Soc. (coun. 1974-81, pres. 1978-79, Outstanding Contbn. award 1984, Benjamin J. Cory award 1987), Chgo. Med. Soc., Congress Neurol. Surgeons, Calif. Assn. Neurol. Surgeons (dir. 1974-82, v.p. 1975-76, pres. 1977-79), San Jose Surg. Soc., Am. Assn. Neurol. Surgeons (chm. sect. on pain 1987-90, dir. 1983-86, 87-90, Disting. Svc. award 1986, 90), Western Neurol. Soc., San Francisco Neurol. Soc., Santa Clara Valley Profl. Standards Rev. Orgn. (dir., v.p., dir. quality assurance 1975-83), Fedn. Western Socs. Neurol. Sci., Internat. Assn. for Study Pain, Am. Pain Soc. (founding mem.), Am. Acad. Pain Medicine (sec. 1983-86, pres. 1987-88), Alpha Omega Alpha, Phi Kappa Phi. Assoc. editor Clin. Jour. of Pain; contbr. articles to profl. jours. Pioneered med. application centrifugal force using flight simulator. Office: 2100 Forest Ave Ste 106 San Jose CA 95128-1496

LIPPMAN, LOUIS GROMBACHER, psychology educator; b. Whittier, Calif., Jan. 10, 1941; s. Robert Weiler and Ruth Major (Grombacher) L.; m. Marcia Zoe Luehrs, Dec. 21, 1965; children: Leah N., David R. BA, Stanford U., 1962; MA, Mich. State U., 1963, PhD, 1966. Teaching asst. Mich. State U., East Lansing, 1962-66; asst. prof. psychology Western Wash. U., Bellingham, 1966-69, assoc. prof., 1969-74, prof., 1974—; vis. prof. San Diego State U., 1978. Author children's piano accompaniment book; editorial bd. mem. Jour. Irreproducible Results, 1984—; contbr. sci. and sci. humor articles to profl. jours. Mem. Am. Psychol. Soc., Psychonomic Soc., Rocky Mountain Psychol. Assn., Midwestern Psychol. Assn., Behavioral and Brain Scis. (assoc.), N.Am. Soc. for Psychology of Sport and Physical Activity, Sigma Xi, Psi Chi. Republican. Jewish. Avocations: walking, photography, piano and organ music. Office: Western Wash U Psychology Dept Bellingham WA 98225-9089

LIPPMAN, MARC ESTES, pharmacology educator; b. Bklyn., Jan. 15, 1945. BA, Cornell U., 1964; MD, Yale U., 1968. Intern Osler med. svc. Johns Hopkins Hosp., Balt., 1968-69, asst. resident, 1969-70; clin. assoc. leukemia svc. Nat. Cancer Inst., NIH, Washington, 1970-71, clin. assoc. lab. biochemistry, 1971-73, sr. investigator med. br., 1974-88, head med. breast cancer sect., 1976-88; clin. prof. medicine & pharmacology, uniformed svc. U. Health Sci., 1978—; dir. Vincent T. Lombardi Cancer Ctr. Georgetown U., Washington, 1988—, prof. medicine & pharmacology, 1988—; mem. merit rev. bd. oncology Vet. Adminstrn. Med. Rsch. Svc., 1977-81, endocrine treatment com. Nat. Surg. Adjuvant Breast Project, 1977-86; cons. dept. pharmacology George Washington Sch. Medicine, 1978-89; co-chmn. Gordon Rsch. Conf. on Hormone Action, 1984, chmn., 1985; treas. Internat. Congress Hormones & Cancer, 1984—; mem. med. adv. bd. Nat. Alliance Breast Cancer Orgn., 1986—; mem. stage III monitoring com. Nat. Surg. Adjuvant Project Breast & Bowel Cancers, 1987-89; bd. trustees Am. Cancer Soc., Washington, 1989-92; mem. sci. adv. bd. Coordinated Coun. Cancer

Rsch., 1989—; hon. dir. Y-ME, Nat. Orgn. Breast Cancer Info. & Support, 1990—; Woodward vis. prof., mem. Sloan-Kettering, 1990; Sidney Sachs Meml. lectr. Case Western Reserve, 1985, D.R. Edwards lectr. Tenovus Inst., Wales, 1985, Gosse lectr. Dalhousie U., Halifax, N.S., 1987, Transatlantic lectr. Brit. Endocrine Socs., 1989, Barofsky lectr. Howard U., 1990, Rose Kushner Meml. lectr. Long Beach Meml. Med. Ctr., 1990, Constance Wood Meml. lectr. Hammersmith Hosp., Eng., 1991. Endocrinology fellow Yale Med. Sch., 1973-74; recipient Mallinckrodt award Clin. Radioassay Soc., 1978, D.R. Edwards medal Tenovus Inst., 1985, Transatlantic medal Brit. Endocrine Socs., 1989, Tiffany award of Distinction, Komen Found., 1989. Fellow ACP, Am. Fedn. Clin. Rsch., Am. Soc. Cell Biology, Am. Assn. Cancer Rsch. (program com. 1986), Am. Soc. Clin. Oncology (program com. 1987-89, chmn. local organizing com. 1989-90), Endocrine Soc. (pub. affairs com. 1980-81, Edward B. Astwood Lecture award 1991), Metastasis Rsch. Soc.; mem. Assn. Am. Physicians, Am. Soc. Clin. Investigators (program com. 1988), Am. Soc. Biol. Chemists. Achievements include research in growth regulation of cancer, breast cancer, cancer endocrinology, growth factor receptors. Office: Georgetown U Vincent T Lombardi Cancer 3800 Reservoir Rd NW Washington DC 20007-2196*

LIPPMAN, MURIEL MARIANNE, biomedical scientist; b. N.Y.C., Oct. 16, 1930; d. Louis George and Erna (Hirsch) L. BA, Syracuse U., 1951; MS, U. Pa., 1955; postgrad., Tufts U., 1965-66, Yale U., 1966-67; PhD, U. Chgo., 1970. Chmn. sci. dept. St. Agnes High Sch., Roshester, N.Y., 1957-59, Nazareth Acad., Rochester, 1959-63; asst. prof. biology, research dir. Nazareth Coll., Rochester, 1963-65; scientist Retina Found., Boston, 1965-66; vis. scientist Karolinska Inst., Stockholm, 1967; assoc. prof. biology Seton Hall U., South Orange, N.J., 1970-71; sr. staff fellow Nat. Cancer Inst., Bethesda, Md., 1971-76; sr. scientist Food and Drug Adminstrn. Bur. Med. Devices, Silver Spring, Md., 1976-77; sr. staff scientist Nat. Acad. Scis., Washington, 1977-78; dir. scientific planning and review Clement Assocs., Washington, 1978-79; pres. ERNACO, Inc., Silver Spring, 1979—; cons. EPA, Washington, 1979-84; adj. prof. biology Am. U., Washington, 1981-83; vis. prof. Cook Coll., Rutgers State U., N.J., 1985-86. Contbr. articles to profl. jours. Commr. Human Relations Commn. Montgomery County, Md., 1982-83. Recipient numerous grants and fellowships including Cancer Rsch. grantee Damon Runyon Found., 1964, Am. Cancer Soc. grantee, 1969-70, Biomedical rsch. grantee Evans Found., 1984-91, Nat. Heart, Lung and Blood Inst. NIH, 1986-87; U.S. Pub. Health fellow, 1965-66, KC Rsch. fellow, 1967, Danforth Teaching fellow U. Chgo., 1970; Teaching Excellence award Rochester Acad. Scis., 1963. Mem. N.Y. Acad. Scis., Soc. for Complex Carbohydrates, Soc. Toxicology Nat. Capital Br. Culture Assn., Sigma Xi. Home: 3740 Capulet Ter Silver Spring MD 20906-2644 Office: ERNACO Inc PO Box 6522 Silver Spring MD 20916-6522

LIPSCOMB, DAVID MILTON, audiologist, consultant; b. Morrill, Nebr., Aug. 4, 1935; s. Roy M. and Elsie M. (Schmidt) L.; m. Gail Frances Beck, Dec. 13, 1978; children: Clinton, Julia. BA, Redlands U., 1957, MA, 1959; PhD, U. Wash., 1966. Audiologist Amarillo (Tex.) Regional Hearing and Speech Found., 1961-63, Ea. Tenn. Regional Speech and Hearing Clinic, Knoxville, 1963-64; from asst. prof. to prof. audiology U. Tenn., Knoxville, 1966-87; pres. Correct Svc., Inc. Stanwood, Wash., 1987—; cons. to industry, govt. entities, 1966—. Author: Noise, The Unwanted Sounds, 1974, Introduction to Lab Methods in Study of the Ear, 1974; co-editor Noise Abatement and Control, 1978; editor Noise and Audiology, 1978; author, editor Hearing Conservation in Industry, 1988; sect. editor Audiology Today, 1990—. Recipient Disting. Tchr. award Beltone Corp., 1985. Fellow Am. Speech-Lang.-Hearing Assn.; mem. Am. Auditory Soc. (bd. dirs. 1984-86). Achievements include patent on rebreathing device. Office: Correct Svc Inc PO Box 1680 Stanwood WA 98292

LIPSCOMB, THOMAS HEBER, JR., civil engineer, consultant; b. Lexington, Miss., Dec. 11, 1912; s. Thomas Heber and Lutie (Scott) L.; m. Louise Buchanan Heiss, June 6, 1935; children: Thomas III, Peg, Jane. BS, U.S. Mil. Acad., 1934; MS in Engring., Cornell U., 1938. Commd. 2d lt. U.S. Army, 1934, advanced through grades to maj. gen., 1968, ret., 1968; engr. U.S. Army Corps Engrs., various, 1938-62; exec. dir. Del. River Port Authority, Camden, N.J., 1968-70; gen. mgr. Southeast Mich. Transp. Authority, Detroit, 1970-74; pvt. practice Moorestown, N.J., 1974—; cons. Met. Transp. Devel. Authority, San Diego, 1976-78. Fellow ASCE (S. Jersey br., N.J. sect. bd. dirs. 1984-86), Union League, Chi Epsilon. Achievements include constrn. of 3 dams on Columbian River and tributaries, 1951-54, serving as divsn. engr. N. Atlantic Divsn. U.S. Army Corps of Engrs., 1959-62, engr. U.S. Forces Far East, 1958-59, commdr. Ft. Leonard Wood, Mo., 1965-67. Home: 549 Chews Landing Rd Haddonfield NJ 08033 Office: 214 W Main St Ste 204-C Moorestown NJ 08057

LIPSCOMB, WILLIAM NUNN, JR., retired physical chemistry educator; b. Cleveland, Ohio, Dec. 9, 1919; s. William Nunn and Edna Patterson (Porter) L.; m. Mary Adele Sargent, May 20, 1944; children: Dorothy Jean, James Sargent; m. Jean Craig Evans, 1983. BS, U. Ky., 1941, DSc (hon.), 1963; PhD, Calif. Inst. Tech., 1946; DSc (hon.), U. Munich, 1976, L.I. U., 1977, Rutgers U., 1979, Gustavus Adolphus Coll., 1980, Marietta Coll., 1981, Miami U., 1983, U. Denver, 1985, Ohio State U., 1991; Transylvania U., 1992, Transylvania U., 1992. Phys. chemist Office of Sci. R&D, 1942-46; faculty U. Minn., Mpls., 1946-59, asst. prof., 1946-50, assoc. prof., 1950-54, acting chief phys. chemistry div., 1952-54, prof. and chief phys. chemistry div., 1954-59; prof. chemistry Harvard U., Cambridge, Mass., 1959-71, Abbott and James Lawrence prof., 1971-90, prof. emeritus, 1990—; mem. U.S. Nat. Commn. for Crystallography, 1954-59, 60-63, 65-67; chmn. program com. 4th Internat. Congress of crystallography, Montreal, 1957; mem. sci. adv. bd. Robert A. Welch Found.; mem. rsch. adv. bd. Mich. Molecular Biology Inst.; mem. adv. com. Inst. Amorphus Studies; mem. sci. adv. com. Nova Pharms., Daltex Med. Svc., Gensia Pharms., Binary Therapeutics. Author: The Boron Hydrides, 1963, (with G.R. Eaton) NMR Studies of Boron Hydrides and Related Compounds, 1969; assoc. editor: (with G.R. Eaton) Jour. Chemical Physics, 1955-57; contbr. articles to profl. jours.; clarinetist, mem.: Amateur Chamber Music Players. Guggenheim fellow Oxford U., Eng., 1954-55; Guggenheim fellow Cambridge U., Eng., 1972-73; NSF sr. postdoctoral fellow, 1965-66; Overseas fellow Churchill Coll., Cambridge, Eng., 1966, 73; Robert Welch Found. lectr., 1966, 71; Howard U. distinguished lecture series, 1966; George Fisher Baker lectr. Cornell U., 1969; centenary lectr. Chem. Soc., London, 1972; lectr. Weizmann Inst., Rehovoth, Israel, 1974; Evans award lectr. Ohio State U., 1974; Gilbert Newton Lewis Meml. lectr. U. Calif., Berkeley, 1974; also lectureships Mich. State U., 1975, U. Iowa, 1975, Ill. Inst. Tech., 1976, numerous others; also speaker confs.; Recipient Harrison Howe award in Chemistry, 1958; Distinguished Alumni Centennial award U. Ky., 1965; Distinguished Service in advancement inorganic chemistry Am. Chem. Soc., 1968; George Ledlie prize Harvard, 1971; Nobel prize in chemistry, 1976; Disting. Alumni award Calif. Inst. Tech., 1977; sr. U.S. scientist award Alexander von Humboldt-Stiftung, 1979; award lecture Internat. Acad. Quantum Molecular Sci., 1980. Fellow Am. Acad. Arts and Scis., Am. Phys. Soc.; mem. NAS, Am. Chem. Soc. (Peter Debye award phys. chemistry 1973, chmn. Minn. sect. 1949-50), Am. Crystallographic Assn. (pres. 1955), The Netherlands Acad. Arts and Scis. (fgn.), Math. Assn. Bioinorganic Scientists (hon.), Academie Europeenne des Sciences, des Arts et des Lettres, Royal Soc. Chemistry (hon.), Phi Beta Kappa, Sigma Xi, Alpha Chi Sigma, Phi Lambda Upsilon, Sigma Pi Sigma, Phi Mu Epsilon. Office: Harvard U Dept Chemistry 12 Oxford St Cambridge MA 02138

LIPSCOMBE, TREVOR CHARLES EDMUND, physical science editor, researcher; b. London, Feb. 14, 1962; s. Edmond George and Amy Joan (Jackson) L.; m. Kathleen Mary Zanella, May 26, 1990; 1 child, Mary Elizabeth Lipscombe. BS, U. London, 1983; PhD, Oxford U., 1986. Rsch. ast. CUNY, N.Y.C., 1986-88, Covenant House, N.Y.C., 1988-90; manuscript editor Physical Review, Ridge, N.Y., 1990-92; physical sci. editor Princeton (N.J.) U. Press, 1992—. Contbr. articles to profl. jours. Mem. Am. Math. Soc., Am. Phys. Soc., N.Y. Acad. Scis. Office: Princeton Univ Press 41 William St Princeton NJ 08540

LIPTON, JAMES ABBOTT, epidemiologist, researcher; b. N.Y.C., July 24, 1946; s. Benjamin M. and Ann (Rappaport) L.; m. Jill Friedman, Oct. 8, 1978; 1 child, Gordon. DDS, Columbia U., 1971, PhD, 1980. Region II cons. USPHS, N.Y.C., 1978-84; chief dental officer Bur. Health Care Delivery and Assistance, USPHS, Rockville, Md., 1984-85; chief office of planning, evaluation and data systems Nat. Inst. of Dental Rsch., NIH, Bethesda, Md., 1985-91; spl. asst. to dir. for sci. devel. Nat. Inst. for Dental Rsch., NIH, Bethesda, Md., 1991—. Author 10 book chpts.; contbr. articles to profl. jours. Capt. USPHS, 1976-92. Mem. over 20 profl. and sci. orgns. Achievements include creation of articles on the social and cultural aspects of orofacial pain; development of first Branch on Molecular Epidemiology at Nat. Inst. Dental Rsch., NIH; first articles on prevalence and sociodemographic variation of particular types of oral and craniofacial pain in U.S. Home: 4500 Saul Rd Kensington MD 20895 Office: Nat Inst Dental Rsch Westwood Bldg Bethesda MD 20892

LIPTON, LESTER, ophthalmologist, entrepreneur; b. N.Y.C., Mar. 14, 1936; s. George and Rita (Steinbaum) L.; m. Harriet Arfa, June 25, 1960; children: Sherri, Brandi, Shawn. BA, NYU, 1959; MD, Chgo. Med. Sch, 1964. Rsch. fellow Chgo. Med. Sch., 1959-60; intern Brookdale Hosp. Ctr., Bklyn., 1964-65; resident Harlem Eye and Ear Hosp., N.Y.C., 1965-68; assoc. attending Polyclinic French hosps., N.Y.C., 1968-75; asst. attending physician, ophthalmologist, surg. instr. St. Clare's Hosp., N.Y.C., 1975—; attending ophthalmologist Cabrini Med. Ctr., N.Y.C., 1982—; founder Lipton Eye Clinic, N.Y.C., 1981—; v.p. Van Arfa Realty, N.Y.C., 1984-88; pres. H&L Realty, Suffern, N.Y., 1981—. Mem. U.S. Congl. Adv. Bd.; mem. bd. deacons Congregationalist Ch. With AUS, 1956-58. Named Internat. Amigo, OAS; recipient Presdl. Citation for outstanding community svc., 1991. Mem. N.Y. Med. Soc., Am. Assn. Individual Investors, Bronx High Sch. Sci. Alumni Assn., United Shareholders Assn., Internat. Platform Assn., Wider Quaker Fellowship, Vanderbilt U. Cabinet Club. Republican. Home: Interlaken Estates Lakeville CT 06039 also: 1199 Park Ave New York NY 10128 Office: Lipton Eye Clinic 51 E 90th St New York NY 10128-1205

LISACK, JOHN, JR., cartographer, executive; b. May 22, 1945; married, 1972; 3 children. BSCE in Civil Engring., U. Mass., 1968, MBA in Mgmt. and Fin., 1970. Asst. adminstr. Prince William County, Va., 1971-72; mgr. land devel. and planning Foxvale Construction Co., Inc., Reston, Va., 1972-73; exec. dir., property adminstrn. Real Equity Investments, Inc., Washington, 1973-74; fin. analyst County of Fairfax, Va., 1974-75; dep. dir., membership, projects and fed. relations Am. Soc. Engring. Edn., Washington, 1975-86; exec. v.p. Nat. Assn. Personnel Cons., Alexandria, Va., 1986-90; exec. dir., Am. Congress Surveying and Mapping, Bethesda, Md., 1990—; bd. dirs., Greater Washington Bd. of Trade, Small Bus. Legis. Coun.; chmn. bd. Greater Washington Soc. Assn. Execs.; com. chmn., Coun. Engring and Sci. Soc. Execs.; fellowship cons. Fed. Hwy. Adminstr., Lincoln Arc Welding Found. Capt. USAR. Mem. Am. Soc. Assn. Execs. (coun. mem), Scabbard & Blade (disting. grad. 1968). Office: Am Congress On Surveying 5410 Grosvenor Ln Bethesda MD 20814

LISKOV, BARBARA HUBERMAN, software engineering educator; b. Los Angeles, Nov. 7, 1939. BA in Math., U. Calif., Berkeley, 1961; MS in Computer Sci., Stanford U., 1965, PhD, 1968. With applications programming sect. Mitre Corp., Bedford, Mass., 1961-62, mem. tech. staff, 1968-72; with Harvard U., Cambridge, Mass., 1962-63; grad. research asst. dept. computer sci. Stanford U., Palo Alto, Calif., 1963-68; prof. computer sci. and engring. MIT, Cambridge, 1972—, NEC prof. software sci. and engring., 1984—. Author: (with others) CLU Reference Manual, Lecture Notes in Computer Science 114, 1981; (with J. Guttag) Abstraction and Specification in Program Development, 1986; assoc. editor Transactions on Programming Langs. and Systems; contbr. articles to profl. jours. Mem. IEEE, Am. Acad. Arts and Scis., Assn. Computing Machinery (spl. interest groups on databases, oper. systems and programming langs.), Nat. Acad. Engring.

LISLE, MARTHA OGLESBY, mathematics educator; b. Charlottesville, Va., June 29, 1934; d. Earnest Jackson and Lucy Elizabeth (Berger) Oglesby; m. Leslie M. Lisle, June 18, 1955; children: Lucie A., Karen B., John D. BA, Randolph-Macon Woman's Coll., 1955; MA, Fla. State U., 1957. Instr. various univ., 1957-69; tchr. Am. Sch., Khartoum, Sudan, 1971-72, Holton-Arms, Bethesda, Md., 1974-78, Rabat Am. Sch., Morocco, 1978-81, Stone Ridge Sch., Bethesda, 1981-82; instr. part-time Montgomery Coll., Takoma Pk., Md., 1982-83; assoc. prof. Prince George's Community Coll., Largo, Md., 1983—. Mem. DAR, Nat. Coun. Tchrs. Math., Am. Math. Assn. Two Yr. Coll., Math. Assn. Am., Md. Math Assn. Two Yr. Coll., Assn. Women in Math., Md. Coun. Tchrs. Math., Pi Mu Epsilon. Democrat. Mem. Unitarian Ch. Avocations: sewing, working crafts, playing flute, singing in choir. Home: 11108 Woodson Ave Kensington MD 20895-1607 Office: Prince George's Community Coll 301 Largo Rd Upper Marlboro MD 20772-2199

LISSAUER, JACK JONATHAN, astronomy educator; b. San Francisco, Mar. 25, 1957; s. Alexander Lissauer and Ruth Spector. SB in Math., MIT, 1978; Phd in Applied Math., U. Calif., Berkeley, 1982. NAS-NRC resident rsch. assoc. NASA-Ames Rsch. Ctr., Moffett Field, Calif., 1983-85; asst. rsch. astronomer U. Calif., Berkeley, 1985; vis. postdoctoral researcher dept. physics Inst. for Theoretical Physics U. Calif., Santa Barbara, 1985-87; asst. prof. astronomy program, dept. earth and space scis. SUNY, Stony Brook, 1987—; rep. Univs. Space Rsch. Assn., SUNY, Stony Brook, 1987—; mem. organizing/edit. com. Protostars and Planets III conf. and book, 1990; vis. scholar dept. planetary scis. and lunar and planetary lab. U. Ariz., Tucson, 1990; professeur invité dept. de physique Université de Paris VII et Observatoire de Paris, Meudon, France, 1990; mem. Lunar and Planetary Geoscis. Rev. Panel, 1989, 91; vis. asst. rsch. physicist Inst. for Theoretical Physics, U. Calif., Santa Barbara, 1992, organizerProgram on Planet Formation, 1992. Contbr. numerous articles on planet and star formation, spiral density wave theory, rotation of planets and comets to jours. including Nature, Astron. Jour., Icarus, Science, Astrophys. Jour. Letters, Astrophys. Jour., Jour. Geophys. Rsch., others. NASA Grad. student fellow, 1981-82, Alfred P. Sloan Found. fellow, 1987-91. Mem. Am. Astronomical Assn. (divsn. planetary scis., divsn. dynamical astronomy, Harold C. Urey prize divsn. planetary scis. 1993), Internat. Astronomical Union, Am. Geophys. Union. Achievements include research in cratering, binary and multiple star systems, circumstellar disks, resonances and chaos, formation and rotational properties of giant molecular clouds, spacecraft imaging processing. Office: SUNY Dept Astonomy Stony Brook NY 11794 Office: SUNY Astronomy Program Dept Earth and Space Scis Stony Brook NY 11794

LIST, HANS C., manufacturing engineer; b. Graz, Austria, Apr. 30, 1896; s. Hugo and Anna (Raab von Rabenau) L.; m. Elfriede L. Wachter, 1937. Student, Tech. U., Graz, D in Tech. Scis. (hon.), 1963. Designer Grazer Wagon und Maschinenfabrik; prof. Tungchi U., Woosung, China, 1926-32, Tech. U., Graz, 1932-41, Tech U., Dresden, Austria, 1941-45; owner, pres. AVL (inst.) Prof. List Ges. mbH, Graz, 1945—. Co-author, editor: Die Verbrennungskraftmaschine. Recipient Gold medal Assn. Austrian Engrs. & Architects, 1954, Grand medal for Svc., Austrian Republic, 1958, Grand Silver medal for Svcs., 1967, Grand Gold medal for Svcs., 1976, ring of Honor, City of Graz, 1959, title of Citizen of Honor, 1976, Grand Gold medal with Star, Province of Styria, 1976, ring of Honor, 1981, Internat. Tech. Promotion prize Inst. Internat. de Promotion et de Prestige, 1971, Wilhelm Exner medal Austrian Trade Assn., 1971, Hon. Baurat h.c. title Pres. Aus. Republic, 1972. Mem. ASME (Soichiro Honda medal 1991), SAE, ÖAIV, Austrian Acad. of Sci. (Schrödinger prize 1980), Assn. German Engrs., Conseil Internat. des Machines a Combustion, Rotary Club, Styrian Golf Club. Home: Heinrichstrasse 112, Graz A-8010, Austria Office: AVL, Kleiststrasse 48, Graz A-8020, Austria*

LISTER, BRUCE ALCOTT, food scientist, consultant; b. Bklyn., Dec. 23, 1922; s. James Alan and Georgana Martha (Hunt) L.; m. Doris Ann Jonassen, May 20, 1990. BSChemE, Columbia U., 1943, MSChemE, 1947; cert. in Food Tech., MIT, 1952. From lab. asst. to div. rsch. dir. General Foods Corp., Hoboken, N.J. and White Plains, N.Y., 1943-62; from mgr. tech. svcs. to v.p. corp. affairs Nestle Foods Corp., White Plains and Purchase, N.Y., 1962-88; exec. dir. Tea Assn. U.S.A., N.Y.C., 1990-91; cons. food industry Baldwin, 1989—; pres. Internat. Hydrolyzed Protein Coun., Washington, 1979-90; trustee cacao biogenetic rsch. project Pa. State U., State College, 1988-89; chmn. stds. regulations com. Nat. Coffee Assn., N.Y.C., 1968-88, tech. com. Chocolate Mfrs. Assn., McLean, Va., 1972-88; U.S. del., chmn. tea com., mem. coffee com. Internat. Stds. Orgn.; U.S. del., full commn., chocolate, soup, dietary foods, labeling coms. Codex Alimentarius; U.S. del., tea com. UN Commn. on Trade and Devel. Bd. dirs., fin. chmn., v.p. Bethany Meth. Home, Bklyn., 1969—; treas., dir. Coun. Against Drug Abuse, Baldwin, 1969—; trustee First Ch. Baldwin, United Meth., 1972—; chmn. Nassau County Bd. Health, Mineola, N.Y., 1975—; bd. dirs. Nassau County Dept. Social Svcs., 1972-74; mem. adv. bd. Hofstra U., Hempstead, N.Y., 1982—; pres. Rep. Club, Baldwin, 1967-68. Lt. (j.g.) USNR, 1944-46. Named Man of Yr., Baldwin Rep. Club, 1968; recipient Hon. doctorate Nat. Assn. Food Equip. Mfrs., 1973. Fellow Am. Inst. Chemists; mem. AIChemE (profl.), Am. Chem. Soc. (emeritus), Inst. Food Technologists (emeritus, chmn. internat. div. 1985-88, liaison to Codex Alimentarius 1990—), Phi Tau Sigma (nat. pres. 1989-90), Tau Beta Pi, Sigma Xi, Phi Lambda Upsilon. Achievements include patents in food formulations; working with NASA to supply menu items for all space missions from Mercury thru Shuttle. Home: 1976 Oakmere Dr Baldwin NY 11510-2739

LISTER, CHARLES ALLAN, electrical engineer, consulting engineer; b. Trenton, N.J., Nov. 15, 1918; s. Lyman Llewellyn and Helen Cullens (Jones) L.; m. Janet Albin Dressler, Feb. 9, 1946; children: Joan Elizabeth, Judith Gail, Robert Charles. BS in EE summa cum laude, Tufts U., 1940; MS in EE, Case We. Res. U., 1951. Registered profl. engr., N.Y. Engr.-in-tng. Gen. Electric, Schenectady, 1940-43, application engr., 1946-47; asst. prof. Swarthmore (Pa.) Coll., 1947-49; engring. supr. Square D Co., Cleve., 1949-62; engring. mgr. Lockheed Aircraft Co., Sunnyvale and Ontario, Calif., 1962-65, Otis Elevator Co., N.Y.C. and Mahwah, N.Y., 1965-75, Square D Co., Columbia, S.C., 1975-84; cons. engr. Naples, Fla., 1984—. Contbr. articles to profl. jours. Lt. USNR, 1943-46, PTO. Mem. IEEE (life sr.), Sigma Xi, Tau Beta Pi. Achievements include patents on medium voltage air-break contactors with high interrupting ratings, adjustable torque electric brakes. Home and Office: # 511 3215 Gulf Shore Blvd N Naples FL 33940-3915

LISTER, E. EDWARD, animal science consultant; b. Harvey, N.B., Can., Apr. 14, 1934; s. Earle Edward and Elizabeth Hazel (Coburn) L.; m. Teresa Ann Moore, June 4, 1983. BSc in Agriculture, McGill U., Montreal, Can., 1955, MSc in Animal Nutrition, 1957; PhD in Animal Nutrition, Cornell U., 1960. Feed nutritionist Ogilvie Flour Mills, Montreal, 1960-65; rsch. scientist rsch. br. Animal Rsch. Ctr. Agriculture Can., Ottawa, 1965-74, dep. dir. rsch. br. Animal Rsch. Ctr., 1974-78, program specialist ctrl. region rsch. br., 1978-80; dir. gen. Atlantic region rsch. br. Agriculture Can., Halifax, N.S., 1980-85; dir. gen. plant health and plant products and pesticides, food prodn. and inspection br. Agriculture Can., Ottawa, 1985-87, dir. rsch. br. Animal Rsch. Ctr., 1987-91; dir. Ctr. Food and Animal Rsch., 1991-92; cons., 1992—; lectr. Ont. Ministry of Agriculture and Food.; mem. Ont. Agrl. Rsch. and Svc. Coms.; former chmn. Beef Rsch. Com.; invited presenter Atlantic Livestock Conf., 1968, Ea. CSAS meetin, 1969, Can. Com. Animal Nutrition, 1976, CSAS Symposium Laval U., 1974, Guelph Nutrition Conf., 1973, 74, Asst. Dep. Minister Dairy Review meeting, Toronto, 1985, U. Guelph ethics conf., 1991, Can. Consumers Assn., Saskatchewan, 1991. Contbr. 54 articles to profl. jours. co-chmn. United Way/Health Ptnrs. for Agriculture Can., Ottawa, 1991; former dir. N.S. Inst. Agrologists. McGill U. scholar, 1953-55; recipient Nat. Rsch. Coun. Post Grad. Spl. scholarship Cornell U., 1957-59. Mem. Am. Soc. Animal Sci., Am. Dairy Sci. Assn., Assn. Advancement Sci. in Can., Agrl. Inst. Can., Can. Soc. Animal Sci. (former dir.), Ont. Inst. Agrologists. Achievements include research in the determination of nutrient requirements of beef cows during winter pregnancy, determination of protein and energy levels and appropriate sources of nutrients fot young dairy calves for optimal growth; development of intensive feeding system for raising high quality beef from Holstein male calves. Home: 390 Hinton Ave, Ottawa, ON Canada K1Y 1B1

LISTERUD, MARK BOYD, surgeon; b. Wolf Point, Mont., Nov. 19, 1924; s. Morris B. and Grace (Montgomery) L.; m. Sarah C. Mooney, May 26, 1954; children: John, Mathew, Ann, Mark, Sarah, Richard. BA magna cum laude, U. Minn., 1949, BS, 1950, MB, 1952, MD, 1953. Diplomate Am. Bd. Surgery. Intern King County Hosp., Seattle, 1952-53; resident in surgery U. Wash., Seattle, 1953-57; practice medicine specializing in surgery Wolf Point, 1958-93; mem. admission com. U. Wash. Med. Sch., Seattle, 1983-88; instr. Dept. Rural and Community Health, U. N.D. Med. Sch., 1991. Contbr. articles to med. jours. Mem. Mont. State Health Coordinating Council, 1983, chmn. 1986—; bd. dirs. Blue Shield, Mont., 1985-87. Served with USN, 1943-46. Fellow Am. Coll. Surgeons, Royal Soc. Medicine; mem. N.E. Mont. Med. Soc. (pres.), Mont. Med. Assn. (pres. 1968-69), AMA (alt. del., del. 1970-84). Club: Montana. Lodge: Elks. Avocations: fishing, hunting. Home: Rodeo Rd Wolf Point MT 59201 Office: 100 Main St Wolf Point MT 59201-1530

LISTON, AARON IRVING, botanist; b. Cleve., Dec. 2, 1959; s. Herbert Liston and Seema Beatrice Feinstein; m. Sara Noelle Meury, Nov. 11, 1990. BSc in Biology, The Hebrew Univ., Jerusalem, 1982, MSc cum laude in Botony, 1984; PhD in Botany, Claremont Grad. Sch., 1990. Herbarium asst. The Hebrew Univ., Jerusalem, 1982-86; rsch. asst. Rancho Santa Ana Botanic Garden, Claremont, Calif., 1987-89; postdoctoral rschr. dept. Genetics U. Calif., Davis, 1990; asst. prof., dir. herbarium, dept. Botonay and Plant Pathology Oreg. State U., Corvallis, 1991—. Contbr articles to profl. jours. Grantee Claremont Grad. Sch., 1986, 87, 89, Sigma Xi, 1988, NSF, 1988, 92, Oreg. Dept. Agr., 1991, M.J. Murdock Charitable Trust, 1991-92, Hardman Found., 1993, Hardman Found. and Hoover Trust, 1993, NAS/Nat. Rsch. Coun., 1993; recipient G. Ledyard Stebbins award, Calif. Native Plants Soc., 1987. Mem. AAAS, Am. Soc. Plant Taxonomists, Assn. Systematics Collections, Botanical Soc. Am., Soc. Study Evolution, Soc. Systematic Biologists, Calif. Botanical Soc. Office: Oregon State University Cordley Hall 2082 Corvallis OR 97331

LISZCZAK, THEODORE MICHAEL, university administrator; b. Meriden, Conn., May 29, 1942; s. Michael Ambrose and Sophie (Laskowski) L.; m. Elizabeth Hazeltine Crocker, Sept. 12, 1964 (dec. Aug. 1978); m. Elizabeth Ann Young, Feb. 24, 1979; children: Kristin, Caitlin. BA, U. Conn., 1965; MA, Montclair State U., 1973; PhD, Tufts U., 1981; MBA, Suffolk U., 1986. Cert. tchr., N.J., Mass. Rsch. assoc. Yale U., New Haven, 1965-68; electronmicroscopist Pfizer Inc., Maywood, N.J., 1968-73; rsch. assoc. Mass. Gen. Hosp., Boston, 1973-81, dir. ultrastructural rsch., 1973-81, asst. neuroanatomist, 1981-86; rsch. analyst Cowen & Co., Boston, 1986-88; assoc. dir. sponsored program Tufts U., Medford, Mass., 1988—. Contbr. 22 publs. to profl. jours. and chpt. to book. Recording sec. Barnstable (Mass.) Village Civic Assn., 1982-86, pres., 1986-88, exec. bd., 1980-88. Mem. Electron Microscope Soc. Am., N.Y. Soc. Electron Microscopists, N.Y. Acad. Sci., Soc. for Neurosci., Am. Assn. Anatomists, Nat. Coun. Univ. Rsch. Adminstrs., Cruising Club Cape Cod, Barnstable Yacht Club. Republican. Episcopalian. Achievements include rsch. in cerebral arteries devoid of arterial supply. Home: 107 Mill Way Barnstable MA 02630 Office: Tufts U Packard Hall Medford MA 02155

LITKE, JOHN DAVID, computer scientist; b. Winchester, Mass., May 30, 1944; s. E. David and Clara Edna (Killenberg) L.; m. Louise May Lockier, June 11, 1966. BS in Physics, MIT, 1965; PhD in Physics, Johns Hopkins U., 1976. Instr. Johns Hopkins U., Balt., 1966-76; mem. tech. staff Bell Labs., Holmdel, N.J., 1976-80; prin. scientist Photocircuits, Glen Cove, N.Y., 1980-84; dep. dir. software tech. Grumman Data Systems, Woodbury, N.Y., 1984-88; dir. computing and controls rsch. Grumman Corp. Rsch. Ctr., Bethpage, 1989—; mfg. tech. cons., Huntington, N.Y., 1984—. Contbr. articles to computer sci. jours. Mem. AIAA, IEEE, Assn. for Computing Machinery, Am. Phys. Soc., N.Y. Acad. Scis., Inst. for Interconnecting and Packaging Electronic Circuits (computer standards com. 1982-84). Achievements include inventions in field of circuit technology. Office: Grumman Corp Rsch Ctr Mailstop A08-35 Bethpage NY 11714-3580

LITMAN, DIANE JUDITH, computer scientist; b. N.Y.C., Mar. 5, 1958; d. Philip and Freda Rae (Grumet) L.; m. Mark William Kahrs, May 17, 1987. BA, Coll. William and Mary, 1980; MS, U. Rochester, 1982, PhD, 1986. Mem. tech. staff AT&T Bell Labs., Murray Hill, N.J., 1985-90, 92—; asst. prof. Columbia U., N.Y.C., 1990-92. Editorial bd. Computational Linguistics, 1991—; contbr. articles to profl. jours. Mem. Assn. for Computational Linguistics, Am. Assn. of Artificial Intelligence, Phi Beta Kappa. Achievements include rsch. in artificial intelligence (natural language, plan recognition, knowledge representation). Office: AT&T Bell Labs 600 Mountain Ave Murray Hill NJ 07974

LITSTER, JAMES DAVID, physics educator, dean; b. Toronto, Ont., Can., June 19, 1938; s. James Creighton and Gladys May (Byers) L.; m. Cheryl Ella Schmidt, June 26, 1965; children: Robin Joyce, Heather Claire. B Engring., McMaster U., Hamilton, Ont., 1961; PhD, MIT, 1965. Instr. physics MIT, Cambridge, Mass., 1965-66, asst. prof. physics, 1966-71, assoc. prof. physics, 1971-75, prof. physics, 1975—, head div. atomic, condensed matter and plasma physics, dept. physics, 1979-83, dir. Ctr. for Materials Sci. and Engring., 1983-88, dir. Francis Bitter Nat. Magnet Lab., 1988-92, interim assoc. provost, v.p. for rsch., 1991, v.p., dean for rsch., 1991—; mem. rsch. staff Thomas J. Watson Rsch. Ctr. IBM, 1969, cons. to liquid crystal group Thomas J. Watson Rsch. Ctr., 1969-70; vis. prof. U. Paris (Orsay), 1971-72; lectr. in physics Harvard Med. Sch., 1974-75; mem. ad hoc oversight com., Solid State Chemistry, NSF, 1977-78; cons. N.Y. State Edn. Dept., 1978, vis. scientist Risø Nat. Lab., Denmark, 1978; mem. condensed matter scis. subcom. (chmn. 1980-81), materials rsch. adv. com. NSF, 1978-81; mem. solid state scis. panel, NRC (chmn. 1991-92), 1986—. Regional editor Molecular Crystals and Liquid Crystals, 1986—. Recipient Gold medal Assn. Profl. Engrs. Ont., 1961, Chancellor's Gold medal McMaster U., 1961, Irving Langmuir Chem. Physics prize Am. Chemical Soc., 1993; Kennecott Copper Co. fellow, 1964-65, John Simon Guggenheim Meml. fellow, 1971-72. Fellow Am. Phys. Soc., Am. Acad. Arts and Scis. Office: MIT Rm 3-240 77 Massachusetts Ave Cambridge MA 02139-4307

LITTLE, CHARLES GORDON, geophysicist; b. Liuyang, Hunan, China, Nov. 4, 1924; s. Charles Deane and Caroline Joan (Crawford) L.; m. Mary Zughaib, Aug. 21, 1954; children: Deane, Joan, Katherine, Margaret, Patricia. B.Sc. with honors in Physics, U. Manchester, Eng., 1948; Ph.D. in Radio Astronomy, U. Manchester, 1952. Jr. engr. Cosmos Mfg. Co. Ltd., Enfield, Middlesex, Eng., 1944-46; jr. physicist Ferranti Ltd., Manchester, Lancashire, Eng., 1946-47; asst. lectr. U. Manchester, 1952-53; prof. dept. geophysics U. Alaska, 1954-58, dep. dir. Geophys. Inst., 1954-58; cons. Ionosphere Radio Propagation Lab. U.S. Dept. Commerce Nat. Bur. Standards, Boulder, Colo., 1958-60, chief Upper Atmosphere and Space Physics div., 1960-62, dir. Central Radio Propagation Lab., 1962-65; dir. Inst. Telecommunication Sci. and Aeronomy, Environ. Sci. Services Adminstrn., Boulder, Colo., 1965-67; dir. Wave Propagation Lab. NOAA (formerly Environ. Sci. Services Adminstr.), Boulder, Colo., 1967-86; sr. UCAR fellow Naval Environ. Prediction Research Facility, Monterey, Calif., 1987-89; George J. Haltiner rsch. prof. Naval Postgrad. Sch., Monterey, 1989-90. Author numerous sci. articles. Recipient U.S. Dept. Commerce Gold medal, 1964, mgmt. and sci. research awards NOAA, 1969, 77, Presdl. Meritorious Exec. award, 1980. Fellow IEEE, Am. Meteorol. Soc. (Cleveland Abbe award 1984); mem. NAE, AIAA (R.M. Losey Atmos. Sci. award 1992). Address: 6949 Roaring Fork Trail Boulder CO 80301

LITTLE, DOROTHY MARION SHEILA, chemist; b. Glasgow, Scotland, Aug. 25, 1939; came to U.S. 1964; d. Theodore Christiaan and Janet Baird (Craig) Parsons.; m. Robert Little, Aug. 15, 1962; children: Alan, Colin, Fiona. BSc, U. Glasgow, 1961, PhD, 1964. Sr. prof. tech. staff Frick Chem. Lab. Princeton (N.J.) U., 1979--. Office: Frick Chem Lab Princeton U Washington Rd Princeton NJ 08544

LITTLE, JACK EDWARD, oil company executive; b. Dallas, Sept. 9, 1938. BS, Tex. A&M U., 1960, MS, 1961, PhD, 1966. Dir. prodn. research and exploration research Bellaire Research Ctr., Shell Devel. Co., Houston, 1977; div. prodn. mgr. onshore div. So. region Shell Oil Co., New Orleans, 1978-79; gen. mgr. prodn. western exploration and prodn. region Shell Oil Co., Houston, 1979-80, gen. mgr. Pacific div. western exploration and prodn. ops., 1980-81, v.p. corp. planning, 1981-82, sr. v.p. adminstrn., 1985-86, exec. v.p. exploration and prodn., 1986—, also bd. dirs.; head Southeast Asia div. Shell Internat. Petroleum Co., London, 1982-85; mem. equity adv. com. Gen. Electric Investment Corp., Stamford, Conn., 1987—; bd. dirs. Am. Petroleum Inst., Washington, 1986—. Trustee, mem. exec. com. United Way Tex. Gulf Coast, Houston, 1987—; mem. external adv. com. Tex. A&M Coll. Engring., 1988—. Mem. Soc. Petroleum Engrs. (chmn. nat. coms. on career guidance, investments and mgmt. and gen. interest 1967-81, nat. bd. dirs. 1977-78, bd. dirs. Bakersfield sect. 1970, bd. dirs. Houston sect. 1969), Nat. Ocean Industries Assn. (bd. dirs., mem. exec. com. 1986—), Mid-Continent Oil and Gas Assn. (exec. com. 1986—). Baptist. Clubs: Petroleum (bd. dirs. 1987-90), Forum of Houston (bd. govs. 1985—), Lakeside Country (Houston, bd. dirs. 1988-91). Avocation: golf. Office: Shell Oil Co 900 Louisiana St Houston TX 77002

LITTLE, JOHN BERTRAM, physician, radiobiology educator, researcher; b. Boston, Oct. 5, 1929; s. Bertram Kimball and Nina (Fletcher) L.; m. Francoise Cottereau, Aug. 4, 1960; children—John Bertram, Frederic Fletcher. A.B. in Physics, Harvard U., 1951; M.D., Boston U., 1955. Diplomate Am. Bd. Radiology. Intern in medicine Johns Hopkins Hosp., Balt., 1955-56; resident in radiology Mass. Gen. Hosp., Boston, 1958-61; fellow Harvard U., Cambridge, Mass., 1961-63; from instr. to assoc. prof. radiobiology Harvard Sch. Pub. Health, Boston, 1963-75; prof. Harvard Sch. Pub. Health, 1975—, chmn. dept. physiology, 1980-83, James Stevens Simmons prof. radiobiology, 1987—; dir. Kresge Ctr. Environ. Health, Boston, 1982—; cons. radiology Mass. Gen. Hosp., Boston, 1965—, Brigham and Women's Hosp., Boston, 1968—; chmn. bd. sci. counsellors, Nat. Inst. Environ. Health Sci., 1982-84; bd. sci. counsellors Nat. Toxicology Program, 1988-92. Mem. editorial bd. numerous nat. and internat. jours.; contbr. chpts. to books and articles to profl. jours. Served to capt. U.S. Army, 1956-58. Am. Cancer Soc. grantee, 1965-68; recipient numerous rsch. and grants NIH, 1968—; named one of Outstanding Investigator grantee Nat. Cancer Inst., 1988—. Mem. Radiation Rsch. Soc. N.Am. (pres.-elect 1985, pres. 1986-87), Am. Assn. Cancer Rsch., Am. Physiol. Soc., Health Physics Soc., Am. Soc. Photobiology, AAAS (coun. in med. scis. 1988-91). Avocations: music, architectural history. Office: Harvard U Dept Radiobiology 665 Huntington Ave Boston MA 02115-6021

LITTLE, RICHARD ALLEN, mathematics and computer science educator; b. Coshocton, Ohio, Jan. 12, 1939; s. Charles M. and Elsie Leanna (Smith) L.; m. Gail Louann Koons, June 12, 1960 (div. May 1989); children: Eric, J. Alice, Stephanie; m. Laura Ann Novosel, June 15, 1991. BS in Math. cum laude, Wittenberg U., 1960; MA in Edn., Johns Hopkins U., 1961; EdM in Math., Harvard U., 1965; PhD in Math. Edn., Kent State U., 1971. Tchr. Culver Acad., Ind., 1961-65; instr., curriculum cons. Harvard U., Cambridge, Mass. and Aiyetoro, Nigeria, 1965-67; from instr. to assoc. prof. Kent State U., Canton, Ohio, 1967-75; from assoc. prof. to prof. Baldwin-Wallace Coll., Berea, Ohio, 1975—, dept. chair, 1978-83; mathematician/educator Project Discovery Ohio Bd. Regents, 1992—, Ohio State U., Columbus, 1992—, vis. prof., math. , 1992—; cons. in field; vis. prof., math. Ohio State U., Columbus, 1987-88, 92—; lectr. various colls. and univs.; pres. Cleve. Collaborative on Math. Edn., 1986-87, bd. dirs. 1985—. Contbr. articles to profl. jours. Bd. dirs. Canton Symphony Orch., 1973-75; Catechism tchr. St. Paul Luth. Ch., Berea, 1976-90, lector, 1980-90; bd. deacons Holy Cross Luth. Ch., Canton, 1968-74, chmn., 1971-74. Mem. Nat. Coun. Tchrs. Math. (profl. devel. and status adv. com. 1987-90), Ohio Coun. Tchrs. Math. (pres. 1974-76, v.p. 1970-73, sec. 1982-84, dir. state math. contest, 1983-92, Christofferson-Fawcett award 1990), Ohio Math. Educators Leadership Coun. (pres. 1990-91, bd. dirs. 1988-92), Greater Canton Coun. Tchrs. Math. (pres. 1969-70), Math. Assn. Am. (pres. Ohio sect. 1983-84, editor 1978-83). Avocations: hiking, tennis, handball. Office: Baldwin-Wallace Coll Dept Math and Computer Sci Berea OH 44017-2088

LITTLE, ROBERT JOHN, JR., botanist, environmental consultant; b. Oceanside, Calif., July 8, 1946; s. Robert John and E. Ruth (Price) L.; m. Cynthia L. Haugen, June 11, 1968; children: Jeffery, Branden. BS in Botany, U. Utah, 1968; MA in Biology, Calif. State U., Fullerton, 1977; PhD in Botany, Claremont Grad. Sch., 1980. Cert. lifetime community coll. instr.: biology, botany and ornamental horticulture, Calif. Instr. biology Saddleback Coll., Mission Viejo, Calif., 1977-80; lectr. botany Calif. State U., Fullerton, 1977-80; dir. environ. svcs. Envirosphere Co., Sacramento, 1980-91, Sycamore Environ. Cons., Sacramento, 1991—; instr. biology Sierra Coll., Auburn, Calif., 1992—; dir.'s environ. com. Calif. Dept. Transp., 1992—; rsch. assoc. Herbarium, U. Calif., Berkeley, 1988—. Co-author: A Dictionary of Botany, 1980; co-editor: Handbook of Experimental Pollination Biology, 1983; author: (publs.) Violaceae of Calif., 1992, Violaceae of Ariz., 1992; contbr. articles and book revs. to profl. jours. Bd. dirs. So. Calif. Botanists, 1979-85; pres. Orange County chpt., Calif. Native Plant Soc., 1983-84. Lt. USN, 1968-72, Vietnam. Audubon Soc. rsch. grantee, 1975-77; recipient Naval ROTC scholarship, 1964. Mem. Bot. Soc. Am., Am. Soc. Plant Taxonomists, Ecol. Soc. Am., Am. Inst. Biol. Scis., Sigma Xi. Mem. Mennonite Ch. Home and Office: Sycamore Environ Cons 16 Pebble River Cir Sacramento CA 95831

LITTLE, W. KEN, JR., structural engineer; b. Charleston, S.C., Jan. 29, 1959; s. W. K. and Elizabeth Pearl (Brooks) L. Sr.; m. Kim B. Little; children: Jana, Aubree. BSCE, Clemson U., 1982. Registered profl. engr., S.C. Coop. student U.S. Naval Shipyard, Charleston, S.C., 1981-82, civil engr., 1983-86; civil engr. USMC Air Sta., Iwakuni, Japan, 1986-89, specifications chief, 1989-91; civil engr. chief USMC Base Camp Butler, Okinawa, Japan, 1991, engring. dir., 1991-92; structural engr. USCG Facilities Design, Norfolk, Va., 1992—. Author: Foreign Object Debris Prevention Manual for Airfield Pavement, 1988. Mem. ASCE, Clemson Club. Home: 525 Yorkshire Dr Chesapeake VA 23320

LITTLEFIELD, JOHN WALLEY, physiology educator, geneticist, cell biologist, pediatrician; b. Providence, Dec. 3, 1925; s. Ivory and Mary Russell (Walley) L.; m. Elizabeth Legge, Nov. 11, 1950; children: Peter P., John W., Elizabeth L. M.D., Harvard U., 1947; MHS, Johns Hopkins U, 1992. Diplomate: Am. Bd. Internal Medicine. Intern Mass. Gen. Hosp., Boston, 1947-48; resident in medicine Mass. Gen. Hosp., 1948-50, staff, 1956-74, chief genetics unit children's service, 1966-73; asso. in medicine Harvard U. Med. Sch., 1956-62, asst. prof. medicine, 1962-66, asst. prof. pediatrics, 1966-69, prof. pediatrics, 1970-73; prof., chmn. dept. pediatrics Johns Hopkins U. Sch. Medicine, Balt., 1974-85, prof., chmn. dept. physiology, 1985-92; pediatrician-in-chief Johns Hopkins U. Hosp., 1974-85. Author: Variation, Senescence and Neoplasia in Cultured Somatic Cells, 1976. Served with USNR, 1952-54. Guggenheim fellow, 1965-66; Josiah Macy Jr. Found. fellow Oxford U. Mem. NAS, Am. Acad. Arts and Scis., Am. Soc. Biol. Chemists, Am. Soc. Clin. Investigation, Tissue Culture Assn., Soc. Pediatric Rsch., Am. Soc. Human Genetics, Am. Pediatric Soc., Assn. Am. Physicians, Phi Beta Kappa, Alpha Omega Alpha, Delta Omega. Home: 304 Golf Course Rd Owings Mills MD 21117-4114 Office: Johns Hopkins U Sch Medicine Dept Physiology Baltimore MD 21205

LITTMAN, SUSAN JOY, physician; b. Oak Park, Ill., Dec. 13, 1957; d. Howard and Arline (Caruso) L.; m. Karsten Peppel, May 27, 1990; 1 child, Andreas Peppel. BS, Cornell U., 1980; MS, SUNY, Albany, 1985; MD, Albany Med. Coll., 1989. Lab. asst. Albany Med. Coll, 1981-82, fellow, 1989-90; teaching and rsch. asst. biology dept. SUNY, Albany, 1983-85; intern, resident Parkland Meml. Hosp., Dallas, 1990-93; fellow Duke U. Med. Ctr., Durham, N.C., 1993—. Contbr. articles to Biosci., Jour. Biol. Chemistry, Jour. Interferon Rsch. Pub. Health Svc. Oncology fellow Albany Med. Coll., 1986, Horoschak scholar, 1986-88; Irene M. Dreny scholar N.Y. State Realtors Assn., 1976. Mem. AAAS, Am. Med. Soc., N.Y. Acad. Scis., Tex. Med. Assn., Dallas County Med. Soc. Office: Duke Univ Med Ctr Dept Internal Medicine Divsn Hematology-Oncology Durham NC 27710

LITVINCHEV, IGOR SEMIONOVICH, mathematician, educator; b. Malahovka, Moscow, USSR, Dec. 9, 1956; s. Semion Danilovich and Larisa Nikolayevna L. MS in Automatic Control, Moscow Phys. Tech. Inst., 1979; DSc in Ops. Rsch./Computing Ctr., USSR Acad. Scis., 1984. Researcher Computing Ctr./USSR Acad. Sci., Moscow, 1979-83, sr. researcher, 1984-88, reading researcher, 1989—; asst. prof. Inst. Civil Aviation Engring., Moscow, 1979-83, Phys. Tech. Inst., Moscow, 1984-88; assoc. prof. Inst. Radioelectronics and Automatics, Moscow, 1989—; sr. cons. All-Union Inst. Mgmt. in Non-Indsl. Area, 1986—. Author: Decomposition in Interconnected Large-Scale Control Problems, 1991; contbr. numerous articles to profl. jours. Mem. USSR Operational Rsch. Soc., Script (bd. dirs. 1990-92), RADOS (gen. dir. 1992—). Avocations: painting, playing music, walking. Office: Computing Ctr/Russian Acad Sci, Vavilov Str 40, 117967 Moscow Russia

LITZSINGER, ORVILLE JACK, aerospace executive; b. Richmond Heights, Mo., Sept. 11, 1936; s. Orville Frank and Irma Helen (Krisay) L.; m. Frances Elaine Shepard, June 14, 1958; children: Karen Elaine, Cheryl Denise. BS, U. Mo., 1958; MBA, Auburn U., 1972. Commd. USAF, 1958-82, advanced through grades to col., ret. 1982; ICBM program dir. USAF, Ogden, Utah, 1979-82; pres., cons. OJL Inc., Alexandria, Va., 1982-85, 92—; dir. Aerojet, Washington, 192-85, 92-93; cons. Textron, Titan Systems, Aerojet, Washington, 1982-85, NASA HQ, 1992, Dept. Energy, USAF, 1993. Pres. Clermont Woods Community Assn., Fairfax County, Va., 1980-81. Recipient Key to City of Omaha, 1967, City of San Bernardino, Calif., 1970. Mem. AIAA, Am. Space Transp. Assn. (bd. dirs. 1991—, vice chmn. 1992-93), Washington Space Bus. Roundtable, Air Force Assn. Achievements include cooperation with inventors in the commercialization of energy saving ideas. Avocation: surf fishing. Home and Office: 5824 Wessex Ln Alexandria VA 22310-1437

LIU, ALAN FONG-CHING, mechanical engineer; b. Canton, China, Mar. 25, 1933; came to U.S., 1958; s. Gee Call and Shuk Hing (Chen) L.; m. Iris P. Chan, Sept. 2, 1962; children: Kent, Willy, Henry. BSME, U. Chiba, Japan, 1958; MSME, U. Bridgeport, 1965. Sr. structures engr. Lockheed Calif. Co., Burbank, Calif., 1968-73; sr. tech. specialist/project mgr. Rockwell Internat. Space div., Downey, Calif., 1973-76; sr. tech. specialist Northrop Corp. Aircraft div., Hawthorne, Calif., 1976-88, Rockwell Internat./N.Am. Aircraft, El Segundo, Calif., 1988—. Contbr. articles to Jour. of Aircraft, AIAA Jour., Res Mechanica, Jour. Engring. Materials and Tech., Engring. Fracture Mechanics, procs. nat. and internat. confs. and symposia. Fellow AIAA (assoc.); mem. ASTM, Am. Soc. Metals Internat. Achievements include research on fatigue and fracture of metallics and composites, on durability and damage tolerance of airframe structures.

LIU, BAI-XIN, materials scientist, educator; b. Shanghai, Peoples Republic of China, June 10, 1935; s. Yin and Yue-Li (Rong) L.; m. Hui-Ling Ni, Jan. 10, 1963; 1 child, Xiao-Yun. BS, Tsinghau U., Beijing, 1961. Asst. Tsinghau U., Beijing, 1961-77, lectr., 1978-80, 1983-84, assoc. prof., 1985-86, prof., 1987—; prof. Sichuan U., Chengdu, China, 1990—; vis. assoc. Calif. Inst. Tech., Pasadena, 1981-82; spl. mem. Ctr. Condensed Matter and Radiation Physics, China Ctr. of Advanced Sci. and Tech. (World Lab.), Beijing, 1988-89; prof. dep. dir. dept. Materials Sci. and Engring., Tsinghau U., Beijing, 1988-91; mem. adv. editorial bd. Internat. Jour. of Nuclear Materials, 1993—. Contbr. over 150 articles to internat. profl. jours. and conf. proceedings. Mem. Am. Phys. Soc. Chinese Phys. Soc., Am. Materials Rsch. Soc., Böhmische Phys. Soc. (scientific), N.Y. Acad. Sci. Avocations: music, sports, movies, reading, bridge. Office: Tsinghau U, Dept Materials Sci/Engring, 100084 Beijing China

LIU, CHANG YU, engineering educator; b. Potin, Hopei, People's Republic of China, June 21, 1935; s. Tien-fu and Cherish Liu; m. Peng Yun Wu, Aug. 13, 1967; children: Zeh Chen, Zeh Wen. BSc, Chinese Naval Coll. Tech., Tsoying, Republic of China, 1958; MSc, Colo. State U., 1965, PhD, 1967. Rsch. scientist Chung-shan Inst. Tech., Lungtan, Republic of China, 1967-73; prof. Nat. Taiwan U., Taipei, Republic of China, 1973-75; assoc. prof. Singapore U., 1975-78, State U. of Campinas (Brazil), Sao Paulo, 1978-84; assoc. prof. Nanyang Tech. Inst., Singapore, 1984-88, prof., 1988—; cons. Hydraulic Lab., Taipei, 1969-70, Indsl. Tech. Rsch. Inst., Hsinchu, Republic of China, 1989-90. Author: Principle of Naval Architecture, 1974, (publ.) Twin Wire Resistance Probe Rotameter, 1990. Recipient Merit of Publ., Ministry of Edn., Republic of China, 1976. Mem. AIAA, Aeronautical and Astronautical of the Republic of China. Achievements include patents for Improved Semi-Balance Rudder; for Twin Wire Resistance Proble Rotameter; for Manometer for Very Low Differential Pressure Measurement. Home: Head Counsellor Residence, Nanyang Crescent Hall 4, Singapore 2263, Singapore Office: Nanyang Tech U Sch MPE, Nanyang Ave, Singapore 2263, Singapore

LIU, CHAOQUN, staff scientist; b. Yizheng, Jiangsu, China, Apr. 8, 1945; came to U.S., 1986; s. Jixiang and Guiying (Han) L.; m. Weilan Jin, Dec. 25, 1972; children; Haiyan, Haifeng. MS, Tsinghua U., Beijing, China, 1981; PhD, U. Colo., 1989. Asst. prof. Nanjing Aero. Inst., Nanjing, China, 1981-86; vis. scholar Miss. State U., 1986; rsch. asst. U. Colo., Denver, 1986-89; staff scientist Ecodynamics Rsch. Assocs., Inc., Denver, 1990—; asst. prof. adj. U. Colo., Denver, 1990-93, assoc. prof. adj., 1993—. Contbr. articles to profl. jours. Grantee USN, 1992-93, NASA, 90-93, 92, 93—, USAF, 1991. Mem. AIAA, Am. Phys. Soc. Office: Univ of Colorado Denver Computational Math Group Campus Box 170 Denver CO 80217

LIU, CHARLES CHUNG-CHA, transportation engineer, consultant; b. Ping-Tung, Taiwan, Oct. 6, 1953; came to U.S., 1977; s. Lian-Chyan and Sheue-Er (Chien) L.; m. Ing-Ing Tsai, Aug. 19, 1979; children: Alexander Charles, Andrew Huey. BSCE, Nat. Taiwan U., 1975; MSCE, Purdue U., 1979, PhD, 1982. Registered profl. engr., Tenn. Instr. Chinese Army Engr. Sch., Taipei, Taiwan, 1975-77; grad. instr. Sch. Civil Engring. Purdue U., West Lafayette, Ind., 1977-82; asst. prof. civil engring. Tenn. State U., Nashville, 1982-84; prin. engr. SRA Techs., Inc., Alexandria, Va., 1984-89; on-site automatic data processing contract mgr. Turner-Fairbank Hwy. Research Ctr., Fed. Hwy. Adminstrn., McLean, Va., 1988—; dir. transp. engring. AEPCO, Inc., Rockville, Md., 1989—; pres., CEO LENDIS Corp., McLean, Va., 1992—; cons. Cumberland Tectonics, Inc., Nashville, 1984; Engring. Directions Internat., Leesburg, Va., 1986-92, UN Devel. Program People's Republic of China, 1991, Vicor Assoc., Manassas, Va., 1988— . Contbr. articles to profl. jours. Mem. ASCE (referee Jour. of Transp. Engring. 1989—), Inst. Transp. Engrs. (Past Pres.'s award 1988), Ops. Soc. Am., Sigma Xi, Omega Rho. Office: AEPCO Inc 15800 Crabbs Branch Way Ste 300 Rockville MD 20855-2604

LIU, EDWARD CHANG-KAI, biochemist; b. An-Hwei, China, Sept. 17, 1946; came to U.S. 1973; s. I-Chi and Ro-Lan (Chen) L.; m. Yuen-Hwa, Nov. 25, 1975; children: Jennifer, Carol, David. BS, Nat. Taiwan U., 1968; MS, U. Dayton, 1975. Sr. med. technologist St. Peter's Med. Ctr., New Brunswick, N.J., 1975-77; sr. rsch. scientist Bristol-Meyers-Squibb Inst. for Med. Rsch., Princeton, N.J., 1977--. Contbr. articles to profl. jours. Mem. Am. Chem. Soc., Am. Assn. for Advancement Sci. Achievements include first to find losartan an antihypertensive interacts with thromboxan receptors. Home: 2 Deer Run Ct East Brunswick NJ 08816 Office: Bristol-Myers-Squibb PO Box 4000 Princeton NJ 08543-4000

LIU, GUOSONG, neurobiologist; b. Chengdu, Peoples' Republic China, Jan. 18, 1956; came to U.S., 1985; s. Wanfu Liu and Guizhen Wang; m. Xiaoyun Sun, Aug. 13, 1981; 1 child, Robert S. MD, Chuanbei Med. Sch., Sichuan, China, 1980; PhD, UCLA, 1990. Intern Chuanbei Med. Sch., 1979-80; rsch. assoc. Acad. Traditional Chinese Medicine, Beijing, 1982-85; rsch. fellow U. Ill., Chgo., 1985-86; rsch. fellow UCLA, 1986-90, postgrad. researcher, 1990-93; rsch. fellow Stanford U., 1993—. Contbr. to profl. publs. Mem. Soc. Neurosci., Sigma Xi. Achievements include research on the quanta nature of excitatory synaptic transmission in central nervous system, presynaptic modulation of synaptic transmission in mammanian neuron, dendritic properties of neuron. Office: Stanford U Dept Mol & Cell Physiol Beckman 103 Stanford CA 94305

LIU, HAN-SHOU, space scientist, researcher; b. Hunan, China, Mar. 9, 1930; came to U.S., 1960, naturalized, 1972; s. Yu-Tin and Chun-Chen (Yeng) L.; m. Sun-Ling Yang Liu, May 2, 1957; children—Michael Fu-Yen, Peter Fu-Tze. Ph.D., Cornell U., 1963. Research asst. Cornell U., 1962-63; research assoc. Nat. Acad. Sci., Washington, 1963-65; scientist NASA Goddard Space Flight Center, Greenbelt, Md., 1965—; Pres. Mei-Hwa Chinese Sch., 1980-81. Contbr. articles to profl. jours. Fellow AAAS; mem. Am. Astron. Soc., Am. Geophys. Union, Planetary Soc., AIAA. Office: NASA Goddard Space Flight Ctr Code 921 Greenbelt MD 20771

LIU, JIANGUO, ecologist; b. Hengnan, Hunan, People's Republic China, Feb. 17, 1963; came to U.S., 1988; s. Xiaoyu and Changlian (Yi) L.; m. Qiuyun Wang, Nov. 28, 1987. BS in Agr., Hunan Agrl. Coll., 1983; MS in Ecology, Chinese Acad. Sci., Beijing, 1986; PhD in Ecology, U. Ga., 1992. Rsch. asst. Chinese Acad. Scis., Beijing, 1986-88; rsch. assoc. Harvard U., 1992—. Editor-in-chief: Advances in Modern Ecology; co-editor: Ecology and Social-Economic Development, 1989; contbr. chpts. to books, articles to profl. publs. Mem. AAAS, Internat. Assn. Ecology, Internat. Soc. Ecol. Modeling, Sigma Xi. Office: Harvard U HIID Cambridge MA 02138

LIU, PINGYU, physicist, educator; b. Shanghai, China, May 27, 1941; came to U.S., 1980; s. Shih-Tsan and Wen-Chang (Chen) L.; m. Jiagi Wang, Dec. 22, 1972 (dec. 1983); 1 child, Mei; m. Carolyn Hsu, Oct. 14, 1983; 1 child, Helen Wen. Diploma, Dalian (China) Poly. Inst., 1963; MS, U. Utah, 1982, PhD, 1984. Rsch. asst., postdoctoral rsch. assoc. U. Utah, Salt Lake City, 1981-85; devel. engr. Garrett AiResearch, L.A., 1985-86; sr. engr., engring. specialist, engr. III Allied-Signal Aero. Co., Tucson, Ariz., 1986-92; assoc. prof. imaging sci., dept. radiology Ind. U., Indpls., 1992—. Mem. IEEE, Am. Assn. Physicists in Medicine, Biomed. Optics Soc., SPIE. Achievements include patent on enhancement of fluoroscopically generated images. Office: Ind U Sch Medicine 541 Clinic Dr Indianapolis IN 46202

LIU, SHI JESSE, physiologist, researcher; b. Kaohsiung, Taiwan, July 17, 1953; came to U.S., 1981; s. Cherng-Chyuan and Chyo-Lian (Lin) L; m. Mei-Yueh Guo, Oct. 23, 1978; children: Daniel, Eden, Peter. BS, Nat. Taiwan Normal U., Taipei, 1976; MS, Nat. Taiwan U., Taipei, 1980; PhD, Duke U., 1986. Asst prof. dept. medicine cardiology divsn. U. Ark. for Med. Scis., Little Rock, 1990—. Home: 10607 Brazos Valley Dr Little Rock AR 72212 Office: U Ark for Med Scis 4301 W Markham St MS 532 Little Rock AR 72205

LIU, SI-KWANG, veterinary pathologist; b. Kwangsi, China; came to U.S., 1959; s. Yeeshao and Shinmei (Yeh) L.; m. Sing-ping Chueh, Dec. 20, 1961; children: Davis, Ernest, Diana, Phillip. DVM, Chinese Vet. Coll., Anshun Kweichow, 1950; PhD, U. Calif., Davis, 1964. Chief veterinarian Taitung Agrl. Rsch. Sta., Taiwan, 1951-56; instr., chief Nat. Taiwan U. Vet. Hosp., Taipei, 1956-59; rsch. asst. U. Calif. Sch. Vet. Med., Davis, 1959-64; pathologist, sci. fellow N.Y. Zool. Soc., Bronx, 1964-88, 88—; pathologist, chief, sr. staff mem. Animal Med. Ctr., N.Y.C., 1964—; from asst. assoc. prof. to prof. N.Y. Med. Coll., N.Y.C., 1966-90; cons. Pig Rsch. Inst., Taiwan, 1984—; vis. expert Nat. Sci. Coun., Taipei, 1976, 83, 88, 91; vis. prof. Nat. Taiwan U., Taipei, 1976, 88, 91, Nat. Chung Hsing U., Taichung, Taiwan, 1983; lectr., speaker in field. Author: An Atlas of Cardiovascular Pathology, 1989; contbr. over 145 articles to Jour. Vet. Med. Assn., Am. Jour. Pathology, others. Elder Presbyn. Ch. of Newtown, Elmhurst, N.Y., 1970-80. Recipient rsch. award Ralston Purina Co., 1982, Feline Disease award Cornation, 1984, Rsch. Excellence award Beecham, 1986, comparative pathology award Chinese Pathology Soc., 1989, Outstanding Svc. award N.Y.C. Vet. Assn., 1991, Outstanding Svc. award N.Y. State Vet. Medicine Soc., 1991, award Chinese Vet. Med. Assn., 1993. Mem. Internat. Acad. Pathology, Internat. Skeletal Soc., Internat. Cardiovascular Pathology Soc., N.Y. Acad. Scis., Am. Vet. Med. Assn., Vet. Med. Assn. N.Y.C. (hon.), N.Y. State Vet. Medicine Soc. Office: Animal Med Ctr 510 E 62d St New York NY 10021

LIU, SUYI, biophysicist; b. Dalian, Liaoning, People's Republic of China, May 23, 1955; came to U.S., 1983; s. Hanzhou and Yiying Liu; m. Wei Sun, July 27, 1985; 1 child, Andrew. BS, Tsinghua U., Beijing, 1982; PhD, U. Ill., 1990. Postdoctoral rsch. assoc. U. Ill., Urbana, 1990; dir. R&D World Precision Instruments, Sarasota, Fla., 1990—. Contbr. articles to Biophys. Jour., Photochem. Photobiol. Mem. Biophys. Soc. Achievements include research in photoelectric response of biological membranes. Office: World Precison Instruments 175 Sarasota Center Blvd Sarasota FL 34240-9258

LIU, TI LANG, physics educator; b. Harbin, China, Mar. 16, 1932; parents Bingshia and Minghsin (Fu) Lang; m. Jong-Diing Liu, Feb. 24, 1957. BS, Nat. Taiwan U., Taipei, Republic of China, 1954; MS, Nat. Tsing Hua U., Hsinchu, Republic of China, 1959; PhD, Temple U., 1968. Asst. prof. Ohio No. U., Ada, 1968-69; assoc. prof. Nat. Tsing Hua U., Hsinchu, 1969-76, prof., 1976—; head physics dept. Nat. Tsing Hua U., Hsinchu, 1970-72, 1981-83, dir. Inst. of Physics, 1981-83; dir. phys. rsch. ctr. Nat. Sci. Coun. Taiwan, 1981-83. Contbr. articles to profl. jours. Mem. Am. Phys. Soc., Chinese Phys. Soc., Chinese Physics Edn. Soc. Office: Nat Tsing Hua U Dept Physic, 101 Sect II Kung Fu Rd, Hsinchu 30043, Taiwan

LIU, YINSHI, metallurgical engineer; b. Shanghai, China, Dec. 6, 1960; came to U.S., 1986; s. Shikai Liu and Huixin Yang. M in Engring., Carnegie Mellon U., 1989, PhD, 1992. Lab. instr. Southwestern Jiaotong U., Emei, China, 1982-86; rsch. asst. U. Ky., Lexington, 1986-87; sr. rsch. engr. U.S.

Steel, Pitts., 1992—. Contbr. articles to profl. jours. Mem. Am. Soc. Metals, Minerals Metals and Materials Soc. Office: USS Tech Ctr Div 5 4000 Tech Center Dr Monroeville PA 15146

LIU, YONG-BIAO, entomologist; b. Ningxia, China, May 15, 1960. B Agr., Beijing Forestry U., 1982; MS, U. B.C., Vancouver, Can., 1987; PhD, U. Maine, 1990. Rsch. asst. U. Maine, Orono, 1987-90; postdoctoral scholar U. Ky., Lexington, 1990—. Recipient George F. Dow award, Fred Griffee Meml. award Maine Agrl. Experiment Sta., Orono, 1990. Mem. Entomol. Soc. Am., Sigma Xi. Office: Univ Ky Dept Entomology S 225 Agrl Scis N Lexington KY 40546

LIU, YOUNG KING, biomedical engineering educator; b. Nanjing, China, May 3, 1934; came to U.S., 1952; s. Yih Ling and Man Fun (Teng) L.; m. Nina Pauline Liu, Sept. 4, 1964 (July, 1986); children—Erik, Tania. B.S.M.E., Bradley U., 1955; M.S.M.E., U. Wis.-Madison, 1959; Ph.D., Wayne State U., 1963. Cert. acupuncturist, Calif. Asst. prof. Milw. Sch. of Engring., 1956-59; instr. Wayne State U., Detroit, 1960-63; lectr. then asst. prof. U. Mich., Ann Arbor, 1963-69; assoc. prof. then prof. Tulane U., New Orleans, 1969-78; prof. biomed. engring., dir. dept. U. Iowa, Iowa City, 1978—. Contbr. articles to profl. jours., chpts. to books. NIH spl. research fellow, 1968-69; recipient Research Career Devel. award NIH, 1971-76. Mem. Internat. Soc. Lumbar Spine (exec. com., central U.S. rep.), Orthopedic Research Soc., Am. Coll. Sports Medicine, Am. Soc. Engring. Edn., Sigma Xi. Democrat.

LIU, YUAN HSIUNG, drafting and design educator; b. Tainan, Taiwan, Feb. 24, 1938; came to U.S., 1970; s. Chun Chang and Kong (Wong) L.; m. Ho Pe Tung, July 27, 1973; children: Joan Anshen, Joseph Pinyang. BEd, Nat. Taiwan Normal U., Taipei, 1961; MEd, Nat. Chengchi U., Taipei, 1967, U. Alta., Edmonton, 1970; PhD, Iowa State U., 1975. Cert. tchr. Tchr. indsl. arts and math. Nan Ning Jr. High Sch., Tainan, Taiwan, 1961-62, 63-64; tech. math. instr. Chung-Cheng Inst. Tech., Taipei, 1967-68; drafter Sundstrand Hydro-Transmission Corp., Ames, Iowa, 1973-75; assoc. prof. Fairmont (W.Va.) State Coll., 1975-80; per course instr. Sinclair Community Coll., Dayton, Ohio, 1985; assoc. prof. Miami U., Hamilton, Ohio, 1980-85, Southwest Mo. State U., Springfield, 1985—; cons. Monarch Ind. Prevision Co., Springfield, 1986, Gen. Electric Co., Springfield, 1988, Fasco Industries, Inc., Ozark, Mo., 1989, Springfield Remfg. Corp., 1990, 92. 2d lt. R.O.C. Army, 1962-63. Recipient Excellent Teaching in Drafting award Charvoz-Carsen Corp., Fairfield, N.J., 1978. Mem. Am. Soc. Engring. Edn., Am. Design Drafting Assn., Days Inn Credible Card Club. Avocations: walking, watching TV. Office: Southwest Mo State U Tech Dept 901 S National Ave Springfield MO 65804-0094

LIU, YUANFANG, chemistry educator; b. Zhijiang, People's Republic of China, Feb. 1, 1931; s. Xiaokong Liu and Jingjuan Zhu; m. Xiaoyan Tang, Mar. 1, 1955; children: Zuoyi, Zuojian. Grad., Yinjing U., Peking, 1952. Asst. chem. dept. Peking U., Beijing, 1952-55, lectr., 1955-62, assoc. prof., 1962-83, prof., 1983—; titular mem. Commn. on Radiochemistry and Nuclear Techniques, Internat. Union Pure and Applied Chemistry, 1987—, vice-chmn., 1991—. Adv. bd. mem. jour. Raciochimica Acta, Strasburg, France, 1989—; author: Radiochemistry, 1988, Actinide Chemistry, 1990; contbr. articles to profl. jours. Recipient First Rank award for Progress of Sci. and Tech., State Commn. of Edn., Bejing, 1987. Mem. Chinese Acad. Scis., Chinese Nuclear and Radiochemistry Soc. (chmn. 1990—), Chinese Nuclear Soc. (standing dir. bd. 1990—). Achievements include first prototype ultracentrifuge for 235U enrichment in China, decay property of nuclide 251Bk, advanced separation procedure for Rd, Pd, Tc from high level radwaste, rare earth binding protein in vivo, hypothesis on non-essential but nutritious biological trace elements, 111 In-labeled monoclonal antibodies with very high specific activity of 30 mCi per g. protein. Address: Zhong Guanyuan, Bldg 47 Rm 105, Peking U, Bejing 100871, People's Republic of China Office: Dept Technical Physics, Peking U, Beijing 100871, China

LIU, YU-JIH, engineer; b. China, June 2, 1948; came to U.S., 1971; s. Tung-Shen and Cha-Ching (Hu) L.; m. Eau-Yin Wang, Nov. 20, 1976; 1 child, Eva. BS, Nat. Taiwan U., 1970; MS, U. Rochester, 1972; PhD, Ohio State U., 1983. Engr. I Lear Siegler Instrument Div., Grand Rapids, Mich., 1978-83; sr. staff engr. Nartron Corp., Reed City, Mich., 1983-85; sr. mem. tech. staff ITT Aerospace/Communications Div., Nutley, N.J., 1985—. Author: On Creating Averaging Templates, 1984, A High Quality Speech Coder at 400 bps, 1989, A High Quality Speech Coder at 600 bps, 1990, Advances in Speech Coding, 1990. Mem. IEEE. Achievements include development of isolated word recognition system in the environment of aircraft noise; development of speech coder at 400 bps and 600 bps with high intelligibility using the techniques of vector quantization, trellis coding. Home: 16 Ridge Rd Wharton NJ 07885-2828 Office: ITT Aerospace/Communication Div 492 River Rd Nutley NJ 07110-3696

LIU, ZI-CHAO, aerospace engineering educator; b. Canton, Republic of China, Jan. 12, 1933; came to U.S., 1988; s. Luk-Fung and Lai-Jing (Wun) L.; m. Li-Sheng Li, May 20, 1958; children: Yuan, Alice Yu. BS in Aeronautics, Inst. of Aeronautics, Beijing, 1953, PhD in Aeronautics, 1956. Asst. prof., head div. elec. propulsion Beijing Inst. Aeronautics/Astronautics, 1960-71, head of div. physics, 1977-79, assoc. prof. dept. Jet Propulsion, 1979-80; vis. fellow MAE dept. Princeton (N.Y.) U., 1979-80; vis. scientist A Dept., Stanford (Calif.) U., 1980-91; vis. prof. U. Essen, Germany, 1987-88; vis. prof. dept. theoretical and applied mechanics, dept. chem. engring. U. Ill., Urbana, 1988—; referee Nuclear Sci. and Engring., 1990, Experiments in Fluids, 1990. Author: Liquid Droplets Diagnostics Using Laser Techniques, 1988; contbr. articles to profl. jours. Achievements include rsch. on wall turbulence structure using particle image velocimetry, particle and droplet size and velocity distribution in two-phase flow, holography and holographic interferometry in fluid flow, electric propulsion. Office: U Ill 211 RAL Box C-3 1209 W California St Urbana IL 61801

LIVENGOOD, TIMOTHY AUSTIN, astronomer; b. Indpls., June 18, 1962; s. David Robert and Sue Elise (Vanderbeck) L.; m. Gwyn Francis Fireman, Oct. 8, 1989; 1 child, Haggis Thorndyke. BA, Washington U., St. Louis, 1984; PhD, Johns Hopkins U., 1991. Rsch. assoc. Nat. Rsch. Coun., Greenbelt, Md., 1991—; organizing & program coms. Internat. Workshop on Variable Phenomen in Jovian Planetary Systems, Annapolis, Md., 1992. Contbr. articles to profl. jours. Grad. Tuition fellow Johns Hopkins U., 1984-91. Mem. Am. Astron. Soc., Am. Geophys. Union, Internat. Jupiter Watch.

LIVIGNI, RUSSELL A., polymer chemist; b. Akron, Ohio, July 20, 1934. BS, U. Akron, 1956, PhD in Polymer Chemistry, 1960. Sr. rsch. chemist Gen. Tire & Rubber Co., 1961-62, group leader, 1962-63, sect. head, 1963-75, mgr., 1975-80, assoc. dir., 1980-87; v.p., dir. rsch. Gencorp, Akron, 1988—; chmn. Gordon Rsch. Conf. Elastomers, 1978; adv. mem. Akron Coun. Engring. and Sci. Soc., 1988—; mem. indsl. panel sci. and tech. NSF, 1990—; trustee Edison Polymer Innovation Corp. Mem. AAAS, Am. Chem. Soc. Achievements include 30 patents; research in polymer based technology. Office: Gencorp Rsch Div 2990 Gilchrist Rd Akron OH 44305*

LIVINGOOD, MARVIN DUANE, chemical engineer; b. Corning, Kans., Aug. 15, 1918; s. Aldo Morton and Gladys Orda (Peck) L.; m. Agnes Dyer, Apr. 17, 1947; children: Christopher, Winifred, Matthew, Abigail. BS in Chem. Engring., Okla. State U., 1938, MS in Chem. Engring., 1940; PhD in Chem. Engring., Mich. State U., 1952; postgrad., U. Louisville, 1980-81. Registered profl. engr., Ky., Ohio, Ind. Grad. asst. Okla. State U., Stillwater, 1939-40, Armour Inst., Chgo., 1940-41; instr. Mo. U., Rolla, 1941-45; asst. prof. Mich. State U., East Lansing, 1945-50; rsch. engr. Mich. Engring. Expt. Sta., East Lansing, 1950-52, duPont, Deerwater, N.J., 1950-58; sr. engr. cons. duPont, various locations, 1958-83; cons. profl. engr. RCI Ltd. Systems Cons., Louisville, 1983—. Contbr. articles to profl. jours. Fellow Am. Inst. Chem. Engrs. (chmn. profl. devel. com. 1979-82, sec. mgmt. div. 1979-84). Republican. Episcopalian. Achievements include patents in polyurethanes; correlation of experimental data to codify design of safe equipment for hazardous chemical manufacturing. Home and Office: 2603 Landor Ave Louisville KY 40205-2333

LIVINGSTON, MARY M., psychology educator; b. Birmingham, Ala., June 15, 1948; d. James Archebald Jr. and Margaret Morrow (Gresham) Livingston; m. Brian Ward Hindman, Nov. 28, 1987; children: James William Hindman, Margaret Livingston Hindman. BA, U. Mich., 1970; MA, U. Ala., Birmingham, 1972; PhD, U. Ala., Tuscaloosa, 1982. Rsch. asst. community dentistry-psychology U. Ala., Birmingham, 1970-72; asst. psychologist to project Children's Hosp. U. Ala. Med. Ctr., Birmingham, 1972-73; psychol. trainee McGuire VA Hosp., Richmond, Va., 1974-76; intern in psychology U. Ark. for Med. Scis., Little Rock, 1976-77; asst. prof. psychology behavioral scis. dept. La. Tech. U., Ruston, 1977-84, assoc. prof. psychology, 1984-92, prof., 1992—; mem. Regional Mental Health Bd., Monroe, La., 1985-91, gov.'s task force for Mental Health, La., 1986-87; chmn., mem. Ruston Mental Health Bd., 1985-92. Contbr. articles to sci. jours. Active sustainer Jr. League, Birmingham., 1970—; vol. instr., rschr. hunter safety La. Dept. Wildlife and Fisheries, Baton Rouge, 1986—. Rsch. grantee State La. Dept. Wildlife and Fisheries, 1990-91; Univ. Enhancement grantee LEQSF La. Bd. Regents, 1990-91. Mem. AAUP (v.p. 1986-87), Am. Psychol. Soc., Am. Fedn. Univ. Profs., Phi Delta Kappa, Sigma Xi. Episcopalian. Achievements include work on subclinical cytomagalovirus' behavioral effects; research on sex roles and stereotypic perceptions of infants and psychiatric patients, firearm ownership patterns. Home: 500 Hundred Oaks Ruston LA 71270 Office: La Tech U Tech Sta Dept Bahavioral Scis Ruston LA 71272

LIVINGSTON, ROBERT BURR, neuroscientist, educator; b. Boston, Oct. 9, 1918; s. William Kenneth and Ruth Forbes (Brown) L.; m. Mandana Beckner, Dec. 21, 1954 (div. 1977); children: Louise, Dana, Justyn. AB, Stanford U., 1940, MD, 1944. Intern, asst. resident internal medicine Stanford Hosp., San Francisco, 1943-44; instr. physiology Yale U., New Haven, Conn., 1946-48; asst. prof. Yale U., New Haven, 1950-52; rsch. asst. psychiatry Harvard U., Cambridge, Mass., 1947-48; NRC sr. fellow neurology Inst. Physiology, Geneva, Switzerland, 1948-49; Wilhelm Gruber fellow neurophysiology Switzerland, France, Eng., 1949-50; exec. asst., pres., chmn. Nat. Acad. Sci. and Nat. Rsch. Coun., 1950-52; assoc. prof. physiology and anatomy UCLA, 1952-56, prof., 1956-57; dir. basic rsch. NIMH and Nat. Inst. Neurol. Diseases and Blindness, 1956-61; chief neurobiology lab. NIMH, 1960-65; prof. dept. neurosci. U. Calif., San Diego, 1965-89, chmn. dept. neurosci., 1965-70, prof. dept. neurosci. emeritus, 1989—; Gäst prof. U. Zürich, Switzerland, 1971-72; Ernest Sachs lectr. Dartmouth Med. Sch., 1981; cons. NRC, VA, NASA, HEW, NSF, Dept. Def.; assoc. neurosci. rsch. program MIT, 1963-76, hon. assoc., 1976—; emissary to 6 Arab nations for Internat. Physicians for the Prevention of Nuclear War, Sept., 1986; del. Internat. Conflict Resolution in Cen. Am., Rust, Austria, Internat. Conf. The Cen. Am. Challenge, Costa Rica, 1988; neurosci. tutor to Dalai Lama, Dharamsala, India, 1987, 90, Newport Beach, Calif., 1989; sci. adv. Dalai Lama, 1991—. Mem. adv. editorial bd. Jour. Neurophysiology, 1959-65; mem. editorial bd. Internat. Jour. Psychobiology, 1970-80, Neurol. Rsch., 1979—; cons. editor Jour. Neurosci. Rsch., 1975-85. Bd. dirs. Foundations' Fund for Rsch. in Psychiatry, 1954-57; bd. incorporators Jour. History of Medicine and Allied Scis.; incorporator Inst. Policy Studies, 1963, Elmwood Inst., 1984. Lt. (j.g.) M.C. USNR, 1944-46. Decorated Bronze Star; recipient Award for Excellence Matrix: Midland Festival, 1981. Fellow AAAS (chmn. commn. sci. edn. 1968-71), Am. Acad. Arts and Scis.; mem. Am. Physiol. Soc., Am. Assn. Anatomists, Am. Neurol. Assn., Am. Acad. Neurology, Assn. for Rsch. in Nervous and Mental Diseases, Am. Assn. Neurol. Surgeons, Soc. for Neurosci.

LIVINGSTONE, DANIEL ARCHIBALD, zoology educator; b. Detroit, Aug. 3, 1927; s. Harrison Lincoln and Elizabeth Agnes (Matheson) L.; m. Bertha Griffin Ross, June 17, 1952 (div.); children: Laura Ross, Mary Lisa, John Malcolm, Christina Ann, Elizabeth; m. Patricia Greene Palmer, June 3, 1989. BS, Dalhousie U., Halifax, N.S., Can., 1948, MSc, 1950; PhD, Yale U., 1953. Postdoctoral fellow Cambridge U., Eng., 1953-54, Dalhousie U., Halifax, 1954-55; asst. prof. zoology U. Md., College Park, 1955-56; asst. prof. zoology Duke U., Durham, N.C., 1956-59, assoc. prof., 1959-66, prof., 1966—, James B. Duke prof. zoology, 1980—; prof. geology Duke U., Durham, 1990—; limnogist U.S. Geol. Survey, Washington, summers 1956-58; mem. adv. panel Environ. Biology br. NSF, 1964-67, Tundra Biome Project, 1974-76, mem. human origins panel, 1978, mem. river-ocean interaction panel, 1978; mem. U.S. Nat. Com. for Internat. Hydrological Decade, 1964; mem. adv. council for systematic and environ. biology Fgn. Currency program, Smithsonian Instn., 1977-80; mem. external rev. com. U. Minn. Dept. Ecology and Behavioral Biology, 1979; collaborator Nat. Park Service U.S. Dept. Interior, 1974; cons. on geochemistry to A.D. Little Inc., 1972-73, on study on mineral cycling in Volta Lake, Ghana, Smithsonian Instn., 1973, to South-East Consortium for Internat. Devel., Kakamega, Kenya, 1985; convenor various workshop sessions. Mem. editorial bd. Limnology and Oceanography, 1969-72, Ecology, 1970-72, Ann. Rev. Ecology and Systematics, 1975-80, African Archaeol. Rev., 1985—; assoc. editor Paleobiology, 1974-79; mem. editorial adv. bd. Tropical Freshwater Biology, 1987—; contbr. articles to profl. jours. NRC fellow, 1953, 54-55; John Simon Guggenheim Meml. Found. fellow, 1960-61. Fellow AAAS; mem. Nova Scotian Inst. Sci., Internat. Union for Quaternary Research (U.S. nat. com. 1970-78, sub.-commn. for African stratigraphy 1973-82), Am. Quaternary Assn. (council 1982, counselor 1971-73, 81—), Am. Soc. Ichthyologists and Herpetologists, Ecol. Soc. Am. (council 1961-66, ecology study com. 1964, chmn. weather working group 1965-66), Am. Soc. for Limnology and Oceanography (G.E. Hutchinson medal 1989), N.C. Acad. Scis., Freshwater Biol. Assn. U.K., Southeast Electron Microscopy Soc., Hydrobiol. Soc. East Africa, Can. Quaternary Assn., Am. Assn. Stratigraphic Palynologists, Internat. Assn. for Fundamental and Applied Limnology, Limnological Soc. So. Africa, Assn. Pour l'Etude Taxonomique de la Flore d'Afrique Tropicale, Assn. Sénégalaise Pour l'Étude du Quaternaire de l'Ouest Africain, Assn. des Palynologues de la Langue Française, Sigma Xi. Avocation: woodworking. Home: 3431 Cromwell Rd Durham NC 27705-5408 Office: Duke U Dept Zoology Durham NC 27706-2584

LIZAK, MARTIN JAMES, physicist; b. Chgo., June 5, 1964; s. Eugene Francis and Genevieve (Galkowski) L. BA, Washington U., St. Louis, 1986, PhD, 1991. Harvard postdoctoral fellow Mass. Gen. Hosp., Charlestown, 1991—. Mem. Am. Phys. Soc. Home: 188 Sherman St Cambridge MA 02140 Office: Mass General Hospital NMR Ctr Bldg 149 13th St Charlestown MA 02139

LIZOTTE, MICHAEL PETER, aquatic ecologist, researcher; b. Central Falls, R.I., May 18, 1960; s. Roland Peter and Estelle Aurora (Demeule) L.; m. Suzette Marie Quenneville, Aug. 5, 1989; children: Jolie Estelle, Sophie Michelle. BS, U. Mass., 1982; MS, Va. Polytech. Inst., 1984; PhD, U. So. Calif., L.A., 1989. Post-doct. fellow Mont. State U., Bozeman, 1989-92; sr. staff earth observing sci. BDM Internat., Inc., Washington, 1992—. Contbr. articles to profl. jours. Mem. AAAS, Am. Soc. Limnology and Oceanography, Sigma Xi. Office: BDM Internat Inc Ste 340 409 Third St NW Washington DC 20024

LJUBICIC, BLAZO, mechanical engineering researcher; b. Ivangrad, MonteNegro, Yugoslavia, Feb. 1, 1948; came to U.S., 1991; s. Raoloslav and Vera (Loncarevic) L.; m. Jasmina Jezdimirovic, Dec. 24, 1981; children: Nadja, Filip. MSc, U. Novi Sad, Yugoslavia, 1988, PhD, 1992. Rsch. assoc. Faculty Tech. Scis., Novi Sad, 1975-82, asst. prof., 1982-91; rsch. engr. Energy and Environ. Rsch. Ctr., Grand Folks, N.D., 1991—. Mem. ASME, Yugoslav Soc. Mech. Engrs. Achievements include research in hydraulic transport of coal, preparation and transport of coal, lugid lignite, energy alternative, power generation from Alaskan coal, pipeline transport of upgraded coal. Office: Energy Rsch Ctr PO Box 8213 Grand Forks ND 58202

LLENADO, RAMON, chemist, chemicals executive. BS in Chemistry, U. Santo Tomas, Philippines; PhD in Chemistry, SUNY. With Procter & Gamble Co., 1972-83; divisional v.p. household products rsch., dir. product devel., v.p. R & D L & F Products, Inc. (formerly Lehn & Fink Products Group a subsidiary of Eastman Kodak Co.); group v.p. Clorox Co., 1991—; bd. dirs. Soap & Detergent Assn., Chem. Specialties Mfg. Assn. Mem. adv. bd. U.C. Berkeley Lawrence Hall of Sci., Bay Area Sci. Fair, Inc. Office: The Clorox Co Tech Ctr 7200 Johnson Dr PO Box 493 Pleasanton CA 94566

LLOYD, DONALD GREY, JR., civil engineer; b. Raleigh, N.C., Mar. 8, 1961; s. Donald Grey Sr. and Shelby Jean (Adams) L.;. A in Marine Tech., Cape Fear C.C., Wilmington, N.C., 1984; B in Civil Engring., N.C. State U., 1990. Engr. in tng. Systems operator Geophysical Svcs., Inc., Dallas, 1984-86; project cons. R.V. Buric Construction Cons., Wilmington, 1990-. Home: 2020 White Rd Wilmington NC 28405 Office: RV Buric Construction 4009 Oleander Dr Wilmington NC 28403

LLOYD, JOSEPH WESLEY, physicist, researcher; b. N.Mex., Jan. 31, 1914; s. William Washington and Mattie May (Barber) L.; m. Lenora Lucille Hopkins, Jan. 24, 1944 (dec. June 1969); 3 children (dec.); m. Ruth Kathryn Newberry, Nov. 19, 1988; children: Kathryn Ruth Jordan, Mary Evelyn Jordan. Student, Pan Am. Coll., 1942. Plumber Pomona, Calif., 1951-57; plumber, pipefitter Marysville, Calif., 1957-79; ret., 1979; ind. researcher in physics and magnetism, Calif., 1944—. With CAP, 1944-45. Mem. AAAS, N.Y. Acad. Scis. Mem. Ch. of Christ.

LO, CHUN-LAU JOHN, aerospace engineer, consultant; b. Hong Kong, Hong Kong, June 22, 1954; came to U.S., 1973; s. Woodman Kang-Chi and Jenny Lin-Ying (Lai) L.; m. Po-See Betsy Yim, Aug. 8, 1981; children: Derek Kay, Darren Lun. BS in Aerospace Engring., Tex. A&M U., 1977; MS in Systems Mgmt., U. Denver, 1991. Analyst OAO Corp., Houston, 1979-80; facility supr. Northrop Svcs. Inc., Houston, 1980-86; sr. engr. Space Test Inc., Houston, 1986-87, Lockheed Engring. & Scis. Co., Houston, 1987-89; sr. systems safety engr. Hernandez Engring., Greenbelt, Md., 1989-90; project engr. systems safety NSI-Mantech, Greenbelt, 1990—; pres. The Shuttle Safety Co. Contbr. tech. paper to AIAA conf.; reviewer NASA document. Recipient 1st Shuttle Flight Achievement Team award NASA, 1981. Mem. AIAA (sr.). Achievements include research on space shuttle payload safety requirements. Office: The Shuttle Safety Co 14516 Dowling Dr Burtonsville MD 20866-1754

LO, KWOK-YUNG, astronomer; b. Nanking, Jiangsu, China, Oct. 19, 1947; came to U.S., 1965; s. Pao-Chi and Ju-Hwa (Hsu) Lu; m. Helen Bo Kwan Chen Lo, Jan. 1, 1973; children: Jan Hsin, Derek. BS in Physics, MIT, 1969, PhD in Physics, 1974. Rsch. fellow Calif. Inst. Tech., Pasadena, 1974-76, sr. rsch. fellow, 1978-80, asst. prof., 1980-86; prof. U. Ill., Urbana, 1986—, assoc. Ctr. for Advanced Study, 1991-92; chmn. Vis. Com. to Haystack Obs., Westford, Mass., 1991-92; assoc. Ctr. Advanced Study, U. Ill., 1991-92. Contbr. articles to Astrophysical Jour., Science, Nature. Grantee NSF, 1977-93; Miller fellow U. Calif., Berkeley, 1976-78, James Clerk Maxwell Telescope fellow U. Hawaii, 1991. Mem. Am. Astron. Soc., Internat. Astron. Union. Achievements include studies of center of the galaxy and of star formation in galaxies. Office: U Ill Astronomy Dept 1002 W Green St Urbana IL 61801

LO, SHUI-YIN, physicist; b. Canton, Peoples Republic of China, Oct. 20, 1941; came to the U.S., 1959; s. Long tin and Ty-Fong (Chow) L.; m. Angela Kwok-Kie Lau, Dec. 18, 1969; children: Alpha Wei-min, Fiona Ai-ming, Hao-min. BS, U. Ill., 1962; PhD, U. Chgo., 1966. Rsch. assoc. Rutherford High Energy Lab., Chilton, United Kingdom, 1966-69, Glasgow (United Kingdom) U., 1969-72; sr. lectr. U. Melbourne, Australia, 1972-89; pres. Inst. for Boson Studies, Pasadena, Calif., 1986—; dir. Sinotronic Co., Hong Kong, 1980; dir. rsch. Am. Environ. Tech. Group, Monrovia, Calif., 1993—. Author: Scientific Studies of Chinese Character, 1986; author, editor: Geometrical Picture of Hadron Scattering, 1986; contbr. over 100 articles to profl. jours. Prin. Chinese Sch. of Chinese Fellowship Victoria, Australia, 1977-84. Fellow Australian Inst. Physics; mem. Am. Phys. Soc. Achievements include patents for Chinese computer and BASER. Office: Am Environ Tech Group 425 E Huntington Dr Pasadena CA 91019-3762

LOBASHEV, VLADIMIR MIKHAILOVICH, physicist; b. Leningrad, USSR, July 29, 1934; s. Mikhail Yephimovich and Nina Vladimirovna (Y-evropeitseva) L. BS in Physics, State U., Leningrad, 1957; MS, Radium Inst. of the Acad. Scis., Leningrad, 1963; DSc, A.F. Ioffe Physico-Tech. Inst., Leningrad, 1968. Engr., researcher, head of lab. A.F. Ioffe Physico-Tech. Inst., Leningrad, 1957-70; head of lab. Leningrad Inst. Nuclear Physics, 1971—; head div. Inst. Nuclear Research USSR Acad. Scis., Moscow, 1972—. Contbr. numerous articles to profl. jours. Recipient Lenin Prize, 1974. Mem. USSR Acad. Sci. (corr.). Mem. Communist Party. Office: USSR Acad of Scis, Inst of Nuclear Rsch, 72 117312 Moscow Russia*

LOBAY, IVAN, mechanical engineering educator; b. Koltuny, Ukraine, Oct. 4, 1911; came to U.S., 1961, naturalized, 1968; s. Stephan and Clementina (Maret) Lobay; m. Halyna Makarenko, Apr. 25, 1943; children: Maria Ivanna, Halyna Blahoslava. Mech. Engr., Inst. Tech., Brno, Czechoslovakia, 1940, Cen. U. Venezuela, Caracas, 1956. Registered profl. engr., Conn. Engr., designer Erste Bruenner Maschinenfabriksgesellschaft, Brno, 1940-41; asst. prof. dept. mech. engring. Inst. Tech., Lviv, Ukraine, 1942-43; sci. asst. dept. mech. engring. Inst. Tech., Brno, 1943-45; engr. san. and civil engring. Ministry San. Affairs, Caracas, Venezuela, 1948-59; prof. dept. civil engring. U. Santa Maria, Caracas, 1957-60; prof. Mech. Engring. Sch., chmn. div. tech. machines & prodn Cen. U. Venezuela, Caracas, 1956-62; prof. dept. mech. engring. U. New Haven, West Haven, 1963-77, 83-84, prof. emeritus, 1984—; prof. gas sect. Inst. Algerien du Petrole, Boumerdes, Algeria, 1977-82. Author: Lecciones de Elementos de Maquinas, No. 3, 1960, No. 2, 1961, Estudio Sobre Descarga de Aguas de Lluvia, 1962. With U.S. Army, 1945-47. Mem. AAUP, AAAS, ASME, NSPE, Conn. Soc. Profl. Engrs., N.Y. Acad. Scis., Ukrainian Am. Assn. Univ. Profs., Ukrainian Engrs. Soc. Am., Colegio de Ingenieros de Venezuela, Asociacion de Profesores de la Universidad Central de Venezuela. Home: 873 Orange Center Rd Orange CT 06477-1712

LOBB, MICHAEL LOUIS, psychologist; b. Pine Bluff, Ark., Feb. 16, 1942; s. Grady Boyd and Mary Elizabeth (Baldus) L.; m. Mary J. Ferebee, Oct. 14, 1972 (div. 1990); children: Susan Kathryn, George Chapman; m. Nancy Gale Love (div. 1971); 1 child, William Boyd; m. Helen Jeanette Greenwood, Nov. 23, 1990. BA, U. Tex., Arlington, 1968, MA, 1974, PhD, 1978. Licensed psychologist, Tex.; cert. profl. sch. psychologist (life), Tex. Ednl. Agy. Rsch. assoc., asst. prof. U. Tex., Arlington, 1977-87; dir. dept. psychol. svcs. Ind. Sch. Dist., Fort Worth, 1985-87; adj. prof. Tex. Wesleyan U., Fort Worth, 1986-90; dir. rsch. Applied Behavioral Scis., Austin, Tex., 1986—; clin. trainee Mental Health Clinic USAF Regional Hosp., Fort Worth, 1980-82; summer faculty rsch. fellow USAF Sch. Aerospace Medicine, San Antonio, 1982, rsch. assoc., 1983-84. Author: (test) Psycho Orthopedic Pain Screen, 1988; contbr. articles to Internat. Jour. Neuro Sci., Animal Learning and Behavior, Zool. Record; author: Native American Youth and Alcohol, 1987. Lt. col. USAFR, 1967-90. Decorated Bronze star; grantee U.S. DOD, 1982., Air Force Office Scientific Rsch., 1983. Mem. Am. Psychol. Assn., Am. Coll. Forensic Psychology, Vietnam Vets. Am., Beta Phi. Achievements include demonstration of oculomotor indices of altered states of consciousness, eyelid velocities predictive of decision errors, stomach ulcer prevention through sectioning adrenal splanchnic, behavioral sink effect in population density control mechanisms, evaluation of performance effects of atropine sulfate; Psycho Orthopedic Pain Screen and Patient Health History Test. Office: Applied Behavioral Scis PO Box 27467 Austin TX 78755-2467

LOBEL, STEVEN ALAN, immunologist; b. N.Y.C., Feb. 8, 1952; s. Norman and Frances (Mandel) L.; m. Patricia Sebag, Dec. 27, 1978; children: Gilana, Liora, Nataniel. BA, U. Tex., 1973; PhD, SUNY, Buffalo, 1977. Diplomate Am. Bd. Med. Lab. Immunology. Lady Davis postdoctoral Hebrew U., Jerusalem, 1977-78; postdoctoral fellow U. Pitts., 1978-80; asst. prof. U. Ill., Chgo., 1980-84, Med. Coll. Ga., Augusta, 1984-88; lab. dir. Am. Med. Labs., Chantilly, Va., 1988--; gov. Am. Bd. Med. Lab. Immunology, Washington, 1992--. Contbr. articles to profl. jours. Sgt. U.S. Army, 1970-71. Vietnam. Mem. Am. Assn. Immunologists, Am. Assn. Pathologists, Am. Soc. Microbiology, Am. Soc. Clin. Pathology. Achievements include one patent for enhanced quick method for bacterial isolation. Office: Am Med Labs 14225 Newbrook Dr Chantilly VA 22021

LOBO, ROY FRANCIS, structural engineer; b. Calcutta, India, Nov. 1, 1963; came to U.S., 1987; s. Francis Leonard and Nora Clotilda (D'Souza) L. B of Tech. Civil Engring., Indian Inst. Tech., Madras, 1987; MS in Civil Engring., SUNY, Buffalo, 1989. Teaching asst. SUNY, Buffalo, N.Y., 1987-90, rsch. asst., 1990—. Contbr. articles to sci. jours. Roman Catholic.

Achievements include program for three dimensional analysis of reinforcement concrete structures subjected to static/dynamic (earthquake) loads for elastic as well as inelastic deformation of the structure. Home: 20 Englewood Ave Buffalo NY 14214

LOCHOVSKY, FREDERICK HORST, computer science educator; b. Neukirchen, Germany, Mar. 20, 1949; arrived in Hong Kong, 1991; s. Frank and Rosalia (Mesh) L.; m. Amelia Fong, May 16, 1983; children: Lucas, Conrad. BASc, U. Toronto, Ont., Can., 1972, MSc, 1973, PhD, 1978. Asst. prof. computer sci. U. Toronto, 1978-83, assoc. prof., 1983-88, Prof., 1988-90; prof. Hong Kong U. Sci. and Tech., 1991—; vis. scientist IBM Rsch. Lab., San Jose, Calif., 1983. Author: Data Base Management Systems, 1977, Data Models, 1982; editor: Object-Oriented Concepts, Databases and Applications, 1988. Mem. IEEE Computer Soc. (editor-in-chief Office Knowledge Engring. 1986—), Assn. for Computing Machinery (assoc. editor Trans. Info. Systems 1985-89), Internat. Fedn. for Info. Processing (vice chmn. WG 8.4, 1986-89). Avocations: reading, swimming, cycling, skiing. Home: Flat 16B, 8A Old Peak Rd, Hong Kong Hong Kong

LOCKARD, WALTER JUNIOR, petroleum company executive; b. El Dorado, Kans., Dec. 6, 1926; s. Walter Allen and Ida May (Akright) L. BS in Bus., Emporia (Kans.) State U., 1952. Pres. Lockard Petroleum Inc., Hamilton, Kans., 1958—. With USN, 1945-47. Mem. Soc. Vertebrate Paleontology, Paleontol. Soc., Kans. Acad. Sci. Achievements include discovering and collecting Hamilton Quarry fossils for Emporia State U. and other universities. Home and Office: HC-1 Box 68 Hamilton KS 66853-9746

LOCKETT, STEPHEN JOHN, medical biophysicist; b. Amersham, Bucks, Eng., Dec. 9, 1961; came to U.S., 1987; s. Alfred John and Margaret (Winterburn) L.; m. Karen Audrey MacMillan, Oct. 18, 1992. BS, Manchester U., 1983; MS, Birmingham U., 1984, PhD, 1987. Rsch. asst. prof. U. N.C., Chapel Hill, 1991—; cons. Am. Innovision Inc., San Diego, 1991—. Contbr. articles to profl. jours. U. N.C. Rsch. fellow, 1987-91; Rsch. grantee Lineberger Cancer Rsch. Ctr., 1990-91, Whitaker Found., 1991-94. Mem. Internat. Soc. Analytical Cytology, Internat. Soc. Optical Engring., European Soc. Analytical Cellular Pathology. Home: 903B Dawes St Chapel Hill NC 27516 Office: U NC 230B Taylor Hall CB # 7090 Chapel Hill NC 27599

LOCKHART, FRANK DAVID, healthcare company executive; b. Columbus, Miss., Aug. 17, 1944; s. Arvile Gray and Dorothy (Pennington) L.; m. Anita Gilmer, June 9, 1968; 1 child, David Franklin. BS, Miss. State U., 1968; MEd, Southeastern La. U., 1972. Cert. radiation safety officer. Tchr. Jefferson Parish Sch. System, Gretna, La., 1968-74; v.p. Boundary Healthcare, Columbus, Miss., 1974—. Mem. ASTM, Internat. Nonwovens Assn., Am. Soc. Quality Control. Democrat. Baptist. Achievements include penile implant drape. Home: 1696 Gatlin Rd Columbus MS 39701 Office: Boundary Healthcare Corp PO Box 2425 Columbus MS 39704-2425

LOCKHART, JOHN CAMPBELL, bioengineer, physiologist; b. Norfolk, Va., July 23, 1963; s. Norman Deroy Gale and Elizabeth Clark (Davidson) Lockhart. BSc, U. Glasgow, Scotland, 1985; PhD, U. Strathclyde, Scotland, 1988. Rsch. fellow U. So. Calif., L.A., 1989-90; head of lab. U. Heidelberg, Germany, 1990; rsch. fellow Mayo Clinic and Found., Rochester, Minn., 1991—. Contbr. chpt. to book, articles to profl. jours. Mem. Microcirculatory Soc., Sigma Xi. Office: Mayo Clinic Dept Physiology-Biophysics Rochester MN 55905-0001

LOCKHEAD, GREGORY ROGER, psychology educator; b. Boston, Aug. 8, 1931; s. John Roger and Ester Mae (Bixby) L.; m. Jeanne Marie Hutchinson, June 9, 1957; children: Diane, Elaine, John. B.S., Tufts U., 1958; Ph.D., Johns Hopkins, 1965. Psychologist research staff IBM Research, Yorktown Heights, N.Y., 1958-61; research assoc., instr. Johns Hopkins, 1961-65; asst. prof. psychology Duke, 1965-68, assoc. prof., 1968-71, prof., 1971-91; chmn. dept. exptl. psychology Duke U., Durham, N.C., 1991—; scholar Stanford U.; research asso. U. Calif., Berkeley, 1971-72; fellow Wolfson Coll., Oxford (Eng.) U., 1980-81; scholar Fla. Atlantic U., 1981; Cons. in human engring. Cons. editor: Perception and Psychophysics; contbr. articles to profl. jours., co-author, editor chpts. in book. Served with USN, 1951-55. NSF grantee, 1966-69, 79-84, USPHS grantee, 1963-69, 70-79, Air Force Office Sci. Rsch., 1983-91. Fellow Am. Psychol. Assn., Am. Psychol. Soc.; mem. Psychonomic Soc., Internat. Soc. Psychophysics, Sigma Xi, Phi Beta Kappa (hon.). Home: 2900 Montgomery St Durham NC 27705-5638 Office: Duke U Dept Exptl Psychology Durham NC 27706

LOCKRIDGE, ALICE ANN, secondary education educator; b. Gread Bend, Kans., Mar. 27, 1951; d. Richard Lee and Madeleine Francis McMillan; m. Patrick Henry Lockridge, Jan. 1, 1988. AS, Pratt (Kans.) Community Coll., 1971; BS, U. Kans., 1973; MS in Phys. Edn., U. Wash., 1977. Cert. fitness instr., phys. fitness specialist/trainer. Tchr. Kansas City (Kans.) Pub. Schs., 1973-74, Highline Pub. Sch. Dist., Seattle, 1974-76; fitness instr. Seattle Fire Dept., 1977-79; insvc. trainer various sch. dists., 1984—; prog. instr./health fitness technologist Renton (Wash.) Vocat. Tech. Inst., 1985-87; fitness instr. Apprenticeship and Non-Traditional Employment for Women, Renton, 1981-87; exercise physiologist Seattle City Light Apprenticeship, 1988—; owner PRO-FIT, Renton, 1983—; fitness cons. police dept., Seattle, 1991; testing cons. police, fire, electric and water depts., various cities, 1991; tchr. tng. lectr., various sch. dists., 1984—. Author: (book/study cards) PRO-FACTS, 1986, (edn. chart) Training Heart Rate Chart, 1983, (slide show series) Do It Right...Teach It Safe, 1985, (consumer edn. series) Never Exercise with a Jerk, 1990. Recipient Presdl. Sports awards, Presdl. Coun. on Phys. Fitness, 1978-86, Outstanding Support award Apprenticeship and Non-Traditional Employement for Women, 1988. Mem. IDEA, AAHPERD, Assn. for Fitness Profls. (com. mem.), Am. Coun. on Exercise (cert. com. 1986, cert. trainer of fitness instrs.), Wash. Alliance Health, Phys. Edn., Recreation and Dance, Nat. Speakers Assn. (Pacific N.W. chpt. bd. dirs. 1992-93), Nat. Dance Assn. (advocacy com.). Avocations: coaching rugby, weight lifting, walking, gardening. Office: PRO-FIT 12012 156th Ave SE Renton WA 98059-6317

LOCKSHIN, MICHAEL DAN, rheumatologist; b. Columbus, Ohio, Dec. 9, 1937; s. Samuel Dan and Florence (Levin) L.; m. Jane Toby Roberts, Sept. 2, 1965; 1 child, Amanda. AB, Harvard U., 1959, MD, 1963. Diplomate Am. Bd. Internal Medicine. From asst. prof. to prof. Cornell U. Med. Coll., N.Y.C., 1970-89; attending physician Hosp. for Spl. Surgery and N.Y. Hosp., N.Y.C., 1970-89; dir. extramural program Nat. Inst. Arthritis & Musculoskeletal Skin Diseases/NIH, Bethesda, Md., 1989—. Contbr. over 100 articles to jours., chpts. to books. Mem. Am. Rheumatism Assn. (2d v.p. 1984-85), La Sociedad Chilena de Reumatologica (hon.), Alpha Omega Alpha. Office: NIAMS/NIH Bldg 31 Rm 4 C32 Bethesda MD 20892

LOCKWOOD, DAVID JOHN, physicist; b. Christchurch, N.Z., Jan. 7, 1942; arrived Can., 1970, naturalized, 1981; s. Robert Keith and Nola Joyce (Radford) L.; B.Sc., U. Canterbury (N.Z.), 1964, M.Sc., 1966, Ph.D., 1969; D.Sc., U. Edinburgh (U.K.), 1978; m. Eugenia Dubovitskaya, Aug. 24, 1979; children: Alisa Nola, Ilana Emilia. Teaching fellow in physics U. Canterbury, 1965-69; postdoctoral fellow in chemistry U. Waterloo (Ont., Can.), 1970-71; research fellow in physics U. Edinburgh, 1972-78; prin. research officer in physics NRC of Can., Ottawa, Ont., 1978—, sect. head surface and interface physics, 1987-90, head phys. characterization group, 1990-92, head thin films group, 1991-92; cons. visitor Battelle Centre de Recherche, Switzerland, 1972-76; tutor Open U., Eng., 1977-78; cons. visitor U. Paul Sabatier, Toulouse, France, 1977-92; cons. visitor U. Essex, (Eng.), 1981-83; program cons., reviewer Natural Scis. and Engring. Research Council of Can., 1986—; disting. vis. prof. Centre Nat. de la Recherche Sci., France, 1987-88; mem. NATO Sci Panel, 1987-90, chmn. 1989-90 ; mem. Can. Adv. Group on NATO Sci. Com., 1990—; disting. visitor Chinese Acad. Scis., Beijing, 1992. Co-author: Light Scattering in Magnetic Solids, English edit. 1986, Russian edit., 1991; editor: P.L. Kapitsa-Letters to Mother, 1989; co-editor: Condensed Systems of Low Dimensionality, 1991, Light Scattering in Semiconductor Structures and Superlattices, 1991, Optical Phenomena in Semiconductor Structures of Reduced Demensions, 1993; mem. editorial bd. Oenophorum, 1989—, Solid State Comm., 1992—; co-dir. two advanced rsch. workshops on semicondrs. NATO, 1990; dir. advanced rsch. workshops on semiconductor optical properties NATO, 1992; contbr. over 200 articles to profl. jours., chpts. to books. NATO research grantee, 1980-82. Mem. Profl. Inst. Pub. Service of Can., Royal Commonwealth Soc., Am. Physical Soc., World Fedn. Scientists. Office: Microstructural Scis Inst, Nat Research Council, Ottawa, ON Canada K1A 0R6

LOCKWOOD, JOHN PATTERSON, electrical engineer; b. Bridgeton, N.J., Dec. 19, 1924; s. Lee John and Frances Mary (Patterson) L.; m. Evelyn May Anderson, June 9, 1951 (dec. 1980); m. Mary Denise Belcastro, Aug. 16, 1990; children: Donald, David, Darin. BS in Mech. Engring., Northwestern U., 1950. Project engr. Micro Switch div. Honeywell, Freeport, Ill., 1950-56, lab. supr., 1956-60, sr. engring. cons., 1960-72, ops. mgr. for D.C. control motors, 1972-74, dir. quality assurance, 1975-86; cons. engr. Freeport, 1986—; tchr., lectr. Bibl. archaeology, 1962—. Author: Applying Precision Switches, 1972; contbr. articles to profl. jurs. Mem. ASTM (subcom. on elec. contact materials 1965-72), Rockford Soc. Archaeol. Inst. Am. (pres. 1970-72). Republican. Presbyterian. Achievements include development of first precision switch for use at 1000 degrees F. Home: 528 Timber Hills Dr Freeport IL 61032

LOCKWOOD HOURANI, LAUREL LEE, epidemiologist; b. Carmel, Calif., Sept. 10, 1950; d. Eugene Franklin and Katherine Ruth (Miller) Betz; m. Ghazi Fayez Hourani, Feb. 28, 1984; children: Nathan, Danna, Lisa. BA, Chico State U., 1977; MPH, Am. Univ. Beirut, 1983; PhD, U. Pitts., 1990. Prog. evaluator Community Hosp. Monterey Peninsula, Carmel, Calif., 1978-81; instr./researcher Am. Univ. Beirut, 1981-85; predoctoral fellow U. Pitts., 1985-89; researcher, cons. V.A. Med. Ctr., Pitts., 1988-90; dir., tumor registry U. Calif. Irvine Med. Ctr., Orange, 1990-92; researcher Naval Health Rsch. Ctr., San Diego, Calif., 1993—; cons. Nat. Devel. Commn. South Lebanon, 1981-83. Author: No Water, No Peace, 1985; contbr. articles to profl. jours. Bd. dirs. Am. for Justice in Middle East, Beirut, 1982-85, Nat. Devel. Com., South Lebanon, 1983-85. Recipient grant V.A., Pitts., 1989, rsch. grant U. Rsch. Bd., Beirut, 1985. Mem. Am. Psychol. Assn., Am. Pub. Health Assn., Soc. for Epidemiologic Rsch. Office: Naval Health Rsch Ctr Divsn Epidemiology PO Box 85122 San Diego CA 92186-5122

LOCONTO, PAUL RALPH, chemist, consultant; b. Worcester, Mass., June 21, 1947; s. Peter Raphael and Rose Rita (Chivallatti) L.; m. Priscilla Ann Hamel, May 20, 1972; children: Jennifer, Michelle, Allison, Julia, Elizabeth. BS, Lowell Tech. Inst., 1970; MS, Ind. U., 1973; PhD, U. Lowell, 1986. Analytical chemist Am. Cynamid Co., Stamford, Conn., 1970, Dow Chem. Co., Midland, Mich., 1973-74; asst prof. Dutchess C.C., Poughkeepsie, N.Y., 1974-86; mgr. rsch. and devel. NANCO Labs., Wappingers Falls, N.Y., 1986-90; dir. analytical svcs. Mich. Biotech. Inst., Lansing, Mich., 1990-92; analytical lab dir. hazardous substance rsch. ctr. Mich. State U., East Lansing, 1992—; cons. NANCO Labs., 1981-86, Perkin Elmer, Norwalk, Conn., 1987-90, Toxi-Lab, Irvine, Calif., 1991-92, 3M Corp., 1993, Mich. Biotech. Inst., Lansing, Mich., 1993. Contbr. articles to Jour. Liquid Chromatography, Jour. Chromatographic Science, LC-GC, Am. Environ. Lab. Small Bus. Innovative Rsch. grant EPA, 1987, N.Y. State Sci. and Tech. Found., 1988. Mem. Am. Chem. Soc., K of C, Kappa Sigma. Achievements include development of extensive analytical method studies in trace environmental analysis. Home: 4300 Manitou Dr Okemos MI 48864

LODDER, ADRIANUS, physics educator; b. Oud Beijerland, The Netherlands, May 22, 1939; s. Adrianus and Lidewij (Visser) L.; m. Adriana Korteweg, July 9, 1963; children: Adriana Liduina, Adrianus Roelant, Lidewij Nathanja. MSc, Free U. Amsterdam, The Netherlands, 1963, PhD, 1966. Rsch. asst. Found. for Fundamental Rsch. Matter, Utrecht, The Netherlands, 1963-65; asst. prof. physics Free U. Amsterdam, 1966-73, assoc. prof., 1973-79, prof., 1980—; rsch. fellow Purdue U., West Lafayette, Ind., 1967; postdoctoral fellow SUNY, Buffalo, 1968; mem. Netherlands Solid State Physics Rsch. Com., 1969—; chmn. edn. com. for physics Netherlands Acad. Coun., The Hague, 1981-85. Contbr. numerous articles on nuclear and solid state physics and statis. mechanics to profl. jours. Rep. music students parents Netherlands Parents Coun., Amersfoort, 1979-81. Mem. Am. Phys. Soc., Netherlands Phys. Soc., Soc. for Advancement Natural and Health Scis., European Phys. Soc., T.H.O.R. Reunists (chmn. 1981—). Christian Democratic. Avocations: sailing, skiing, horseback riding, swimming, cycling. Home: Stromboli 13, 1186 CH Amstelveen The Netherlands Office: Free U Physics Lab, Faculty Physics & Astronomy, De Boelelaan 1081, 1081 HV Amsterdam The Netherlands

LODWICK, GWILYM SAVAGE, radiologist, educator; b. Mystic, Iowa, Aug. 30, 1917; s. Gwylim S. and Lucy A. (Fuller) L.; m. Maria Antonia De Brito Barata; children by previous marriage: Gwilym Savage III, Philip Galligan, Malcolm Kerr, Terry Ann. Student, Drake U., 1934-35; B.S., State U. Iowa, 1942, M.D., 1943. Resident pathology State U. Iowa, 1947-48, resident radiology, 1948-50; fellow, sr. fellow radiologic and orthopedic pathology Armed Forces Inst. Pathology, 1951; asst., then asso. prof. State U. Iowa Med. Sch., 1951-56; prof. radiology, chmn. dept. U. Mo. at Columbia Med. Sch., 1956-78, research prof. radiology, 1978-83, interim chmn. dept. radiology, 1980-81, chmn. dept. radiology, 1981-83, prof. bioengring., 1969-83, acting dean, 1959, assoc. dean, 1959-64; assoc. radiologist Mass. Gen. Hosp., 1983-88, radiologist, 1988-91; hon. radiologist Mass. Gen. Hosp., Boston, 1991—; vis. prof. dept. radiology Harvard Med. Sch., 1983-93; cons. in field; vis. prof. Keio U. Sch. Medicine, Tokyo, 1974; chmn. sci. program com. Internat. Conf. on Med. Info., Amsterdam, 1983; trustee Am. Registry Radiologic Technologists, 1961-69, pres., 1964-65, 68-69; mem. radiology tng. com. Nat. Inst. Gen. Med. Scis., NIH, 1966-70; com. radiology Nat. Acad. Scis.-NRC, 1970-75; chmn. com. computers Am. Coll. Radiology, 1965, Internat. Commn. Radiol. Edn. and Info., 1969-73; cons. to health care tech. div. Nat. Ctr. for Health Services, Research and Devel., 1971-76; dir. Mid-Am. Bone Tumor Diagnostic Ctr. and Registry, 1971-83; adv. com. mem. NIH Biomed. Image Processing Grant Jet Propulsion Lab., 1969-73; nat. chmn. MUMPS Users Group, 1973-75; mem. radiation study sect. div. research grants NIH, 1976-79, mem. study sect. on diagnostic radiology and nuclear medicine div. research grants, 1979-82, chmn. study sect. on diagnostic radiology div. research grants, 1980-82; mem. bd. sci. counselors Nat. Library of Medicine, 1985, chmn. 1987-89; dir. radiology Spaulding Rehab. Hosp., 1986-92. Adv. editorial bd. Radiology, 1965-86, cons. to editor, 1986-91; adv. editorial bd. Current/Clin. Practice, 1972-88; mem. editorial bd. Jour. Med. Systems, 1976—, Radiol. Sci. Update div. Biomedia, Inc., 1975-83, Critical Revs. in Linguistic Imaging, 1990; mem. cons. editorial bd. Skeletal Radiology, 1977-92, Contemporary Diagnostic Radiology, 1978-80; assoc. editor Jour. Med. Imaging, 1988—. Served to maj. AUS, 1943-46. Decorated Sakari Mustakallio medal Finland; named Most Disting. Alumnus in Radiology, State U. Ia. Centennial, 1970; recipient Sigma Xi Research award U. Mo., Columbia, 1972, Gold medal XIII Internat. Conf. Radiology, Madrid, 1973, Founder's Gold medal Internat. Skeletal Soc., 1990. Fellow AMA (radiology rev. bd. council med. edn., council rep. on residency rev. com. for radiology 1969-74), Am. Coll. Radiology (co-chmn. ACR-NEMA standardization com. 1983-90); mem. NAS Inst. Medicine, Am. Coll. Med. Informatics (founding), Nat. Acad. Practice in Medicine, Radiol. Soc. N.Am. (3d v.p. 1974-75, chmn. ad hoc com. representing assoc. scis. 1979-87, chmn. assoc. scis. com. 1981-87), Assn. Univ. Radiologists, Mo. Radiol. Soc. (1st pres. 1961-62), Salutis Unitas; hon. mem. Portuguese Soc. Radiology and Nuclear Medicine, Tex. Radiol. Soc., Ind. Roentgen Soc., Phila. Roentgen Ray Soc., Finnish Radiol. Soc. (hon.), Rotary, Alpha Omega Alpha. Clubs: Harvard of Boston, Cosmos. Home: 307 Playa del Mar 3900 Galt Ocean Dr Fort Lauderdale FL 33308-6631

LÖE, HARALD, dentist, educator, researcher; b. Steinkjer, Norway, July 19, 1926; s. Haakon and Anna (Bruem) L.; m. Inga Johansen, July 3, 1948; children: Haakon, Marianne. D.D.S., U. Oslo, 1952; Dr.Odont., 1961; hon. degrees, U. Gothenburg, 1973, Royal Dental Coll., 1980, U. Athens, 1980, Catholic U., Leuven, 1980, U. Lund, 1983, Georgetown U., 1983, U. Bergen, 1985, U. Md., 1986, U. N.J., 1987, Royal Dental Coll., Copenhagen, 1988, U. Toronto, 1989, U. Detroit, 1990, U. S.C., 1990, U. Helsinki, Finland, 1992, Pacific U., 1993. Instr. Sch. Dentistry, Oslo U., 1952-55; research assoc. Norwegian Inst. Dental Research, 1956-62; Fulbright research fellow, research asso. dept. oral pathology U. Ill., Chgo., 1957-58; Univ. research fellow Oslo U., 1959-62, asso. prof. dept. periodontology, 1960-61; prof. dentistry, chmn. dept. periodontology Royal Dental Coll., Aarhus, Denmark, 1962-72; asso. dean, dean-elect Royal Dental Coll., 1971-72; prof., dir. Dental Research Inst. U. Mich., Ann Arbor, 1972-74; dean, prof. periodontology Sch. Dental Medicine U. Conn., Farmington, 1974-82; dir. Nat. Inst. Dental Research, Bethesda, Md., 1983—; vis. prof. periodontics Hebrew U., Jerusalem, 1966-67; hon. prof. Med. Scis. U. Beijing, 1987; cons. WHO, NIH. Served with Norwegian Army, 1944-48. Recipient 75th Anniversary award Norwegian Dental Assn., 1958, Aalborg Dental Soc. prize Denmark, 1965, William J. Gies Periodontology award, 1978, Internat. prize Swedish Dental Soc., 1988, U.S. Surgeon Gen.'s Exemplary award, 1988; decorated Royal Norwegian Order of Merit. 1989. Mem. AAAS, ADA, Danish Dental Assn., Am. Acad. Periodontology, Am. Assn. Dental Research, Am. Soc. Preventive Dentistry (internat. award), Mass. Dental Soc. (internat. award), Internat. Assn. Dental Research (award for basic research in periodontology 1969, pres. 1980), Internat. Coll. Dentists, Scandinavian Assn. Dental Research, Scandinavian Soc. Periodontology. Office: Nat Inst Dental Rsch Bldg 31 9000 Rockville Pike Bethesda MD 20892-0001*

LOEB, BARBARA L., chemistry educator; b. Santiago, Chile, Dec. 23, 1950; d. Federico and Ursula L.; m. Eduardo Boys, Mar. 30, 1974; children: Michael, Maureen, Henry. Lic. in Chemistry, P. Cath. U. Chile, Santiago, 1975, MSc in Chemistry, 1976; DSc in Chemistry, U. Chile, Santiago, 1985. Asst. prof. chemistry DP. Cath. U. Chile, Santiago, 1976-91; adj. prof. Cath. U. Chile, Santiago, 1991—. Translator: Atkins Physical Chemistry, 1991; contbr. articles to profl. jours. including Jour. Phys. Chemistry. Mem. Chilean Chem. Soc., Am. Chem. Soc., Third World Orgn. of Women in Sci. Roman Catholic. Office: P Cath U Chile, Vicuna Mackenna 4860 PO Box 306, Santiago Met, Chile

LOEBLICH, HELEN NINA TAPPAN, paleontologist, educator; b. Norman, Okla., Oct. 12, 1917; d. Frank Girard and Mary (Jenks) Tappan; m. Alfred Richard Loeblich, Jr., June 18, 1939; children: Alfred Richard III, Karen Elizabeth Loeblich, Judith Anne Loeblich Covey, Daryl Louise Loeblich Valenzuela. BS, U. Okla., 1937, MS, 1939; PhD, U. Chgo., 1942. Instr. geology Tulane U., New Orleans, 1942-43; geologist U.S. Geol. Survey, Washington, 1943-45, 47-59; mem. faculty UCLA, 1958—, prof. geology, 1966-84, prof. emeritus, 1985—, vice chmn. dept. geology, 1973-75; research assoc. Smithsonian Instn., 1954-57; assoc. editor Cushman Found. Foraminiferal Research, 1950-51, incorporator, hon. dir., 1950—. Author: (with A.R. Loeblich, Jr.) Treatise on Invertebrate Paleontology, part C, Protista 2, Foraminiferida, 2 vols., 1964, Foraminiferal Genera and Their Classification, 2 vols., 1987; author: The Paleobiology of Plant Protists, 1980, also articles profl. jours., govt. publs., encys.; editorial bd.: Palaeoecology, 1972-82, Paleobiology, 1975-81. Recipient Joseph A. Cushman award Cushman Found., 1982; named Woman of Yr. in Sci. Palm Springs Desert Mus., 1987; Guggenheim fellow, 1953-54. Fellow Geol. Soc. Am. (sr., councilor 1979-81); mem. Paleontol. Soc. (pres. 1984-85, patron 1987, medal 1982), Soc. Sedimentary Geology (councilor 1975-77, hon. mem. 1978, Raymond C. Moore medal 1984), UCLA Med. Ctr. Aux. (Woman of Yr. medal), AAUP, Internat. Paleontological Assn., Paleontol. Rsch. Inst., Am. Microscopical Soc., Am. Inst. Biol. Scis., Phi Beta Kappa, Sigma Xi. Home: 11427 Albata St Los Angeles CA 90049-3403 Office: UCLA Dept Earth and Space Scis Los Angeles CA 90024

LOEHR, JOHANNES-MATTHIAS, physician; b. Frankfurt/Main, June 3, 1959; s. Eberhard R. and Magdalena (Wohldorf) L.; m. Isabel Andree, July 21, 1988. Colloquium theologicae, U. Kiel, Fed. Republic Germany, 1980; MD, U. Hamburg, Fed. Republic Germany, 1986, Edn. Com. for Fgn. Grad., Phila., 1986. Resident pathology U. Hamburg, 1986-87; postdoctoral fellow dept. immunology Scripps Clinic, La Jolla, Calif., 1987-88; resident dept. medicine U. Erlangen, Fed. Republic Germany, 1989-93; lab. dir. div. med. rsch. dept. medicine U. Erlangen,, Fed. Republic Germany, 1989-93; chief resident, lab dir. divsn. med. rsch. dept. medicine U. Rostock, Germany, 1993—. Mem. Am. Assn. Cancer Rsch. Mem. AAAS, European Assn. for Study Diabetes, Internat. Assn. Pancreatology, German Assn. Scientists and Physicians, Am. Fedn. Clin. Rsch., N.Y. Acad. Scis., European Pancreas Club, Royal Soc. Medicine (London). Office: U Rostock Dept Medicine, Ernst-Heydemannstr 6, D-18057 Rostock Germany

LOEHR, THOMAS MICHAEL, chemist, educator; b. Munchen, Germany, Oct. 2, 1939; came to the U.S., 1951; s. Max and Irmgard (Kistenfeger) L.; m. Joann Sanders, June 20, 1965. BS in Chemistry, U. Mich., 1963; PhD, Cornell U., 1967. Asst. prof. Cornell U., Ithaca, N.Y., 1967-68, Oreg. Grad. Ctr., Portland, 1968-74; assoc. prof. Oreg. Grad. Ctr., Beaverton, 1974-78, prof., 1978—, acting head dept. chem. and bio. scis., 1981-82; acting head dept. chem. and bio. scis. Oreg. Grad. Inst. Sci. and Tech., Beaverton, 1992-93; vis. prof. Portland State U., 1974-75, adj. prof., 1979—; vis. assoc. Calif. Inst. Tech., Pasadena, 1978-79; chmn. Metals in Biology/Gordon Rsch. Conf., 1987; mem. Metallobiochemistry study sect. NIH, 1978-82. Editor: Iron Carriers and Iron Proteins, 1989; contbr. more than 100 articles to profl. jours. Mem. Am. Chem. Soc. Office: Oreg Grad Inst Sci Tech PO Box 91000 20000 NW Walker Rd Portland OR 97291-1000

LOEW, FRANKLIN MARTIN, veterinary medical and biological scientist, university dean; b. Syracuse, N.Y., Sept. 8, 1939; s. David Franklin and Sarah (Adelaide) L.; m. Mary Moffatt, Sept. 9, 1964; children—Timothy, Andrew. B.S., Cornell U., 1961, D.V.M., 1965; Ph.D., U. Sask., 1971. Lic. veterinarian; diplomate Am. Coll. Lab. Animal Medicine. Research asst. R.J. Reynolds Co., Winston-Salem, N.C., 1965-66; research asst. Tulane U., New Orleans, 1966-67; prof. U. Sask., Saskatoon, Can., 1967-77; dir. comparative medicine Johns Hopkins U., Balt., 1977-82; dean Sch. Vet. Medicine, Tufts U., Boston, 1982—, Henry and Lois Foster prof. comparative medicine, 1985—; v.p. Tufts U. Devel. Corp. Inc., Boston, 1991—; pres. Tufts Biotech. Corp., Boston, 1993—; cons. Can. Council Animal Care, Ottawa, Ont., 1969-84; mem. life scis. com. Nat. Acad. Sci., Washington, 1981-88, chmn. inst. lab. animal resources, 1981-87; N.B. lectr. Am. Soc. Microbiology; mem. nat. adv. bd. Ctr. on Bioethics Lit., Kennedy Inst. Georgetown U., 1986—; Schofield lectr. U. Guelph, Can.; Smith lectr. U. Sask.; Schalm lectr. U. Calif.; dir. Mass. Biotech. Rsch. Inst., 1985—, Commonwealth BioVentures, Inc., TSI Corp., 1988—; mem. sci. adv. com. Harvard Primate Rsch. Ctr., 1988-92, Mass. Health Resources Inst.; sci. and tech. adv. com. State of Mass., 1988-92; mem. USDA Sec.'s Adv. Com. Nat. Rsch. Initiative, 1992—; pres. Tufts Biotech Corp., 1993—. Author: Vet in the Saddle, 1978; editor: Laboratory Animal Medicine, 1984; contbr. numerous articles to profl. jours. Chmn. bd. trustees Boston Zool. Soc., 1984-88; trustee Worcester Acad., 1984-90; mem. Nat. Ctr. Rsch. Resources adv. coun., NIH, 1988-92, Blue Ribbon adv. coun. USDA, 1987-91; bd. dirs. Ea. States Exhbn., 1988—. Decorated Queen Elizabeth II Jubilee medal Gov.-Gen. Can., 1977; Med. Rsch. Coun. Can. fellow, 1969-71; recipient Charles River prize Am. Vet. Med. Assn., 1988, named Vet. of Yr., 1989; recipient Disting. Svc. award Mass. Vet. Med. Assn., 1992. Mem. NAS/Inst. Medicine, AAAS, Am. Inst. Nutrition, Soc. Toxicology, Assn. Am. Vet. Med. Colls. (pres. 1985-86), Am. Coll. Lab. Medicine (bd. dirs. 1979-82), Nat. Acads. Practice, Fedn. Am. Socs. for Exptl. Biology, Am. Antiquarian Soc., Mass. Agrl. Club. Office: Tufts U Sch Vet Medicine 200 Westboro Rd North Grafton MA 01536-1895

LOEW, LESLIE MAX, biophysicist; b. N.Y.C., Sept. 2, 1947; s. Ernest and Selma (Sonneberg) L.; m. Helen Karen Jeremias, Jan. 20, 1970; children: Daniel, Rena, Aviva. BS, CCNY, N.Y.C., 1969; MS, Cornell U., 1972, PhD, 1974. Rsch. assoc. Harvard U., Cambridge, Mass., 1973-74; asst. prof. SUNY, Binghamton, 1974-79, assoc. prof., 1979-84; assoc. prof. U. Conn. Health Ctr., Farmington, 1984-86, prof., 1986—; vis. scientist Weizmann Inst., Rehovot, Israel, 1981, 83; vis. assoc. prof. Cornell U., 1984; adj. prof. med. sch U. Mass., 1991—. Editor: Spectroscopic Membrane Probes, 1988. Recipient Rsch. Career Devel. award NIH, 1979-84; Rsch. grantee NIH, 1976—, ONR, 1990-92. Mem. AAAS, Biophys. Soc., Am. Chem. Soc., Phi Beta Kappa. Achievements include development of reagents and methods for mapping membrane potential distbns. in living cells. Office: U Conn Health Ctr Dept Physiology Farmington CT 06030-3505

LOEWENSTEIN, GEORGE WOLFGANG, retired physician, UN consultant; b. Germany, Apr. 18, 1890; m. Johanna Sabath, Nov. 27, 1923; children: Peter F. Lansing (dec.) and Ruth Edith Gallagher (twins). Student, Royal William Coll., Germany, 1909, Friedrich William U., Germany, 1919, London Sch. Tropical Hygiene and Medicine, 1939. Dir. pub. health Neubabelsberg, 1920-24, Berlin, 1924-34; dir. pub. health and welfare City of Berlin, 1923-33; pvt. practice medicine, Chgo., 1940-46,

Chebeague and Dark Harbor, Maine, 1947-58; instr. Berlin Acad. Prevention of Infant Mortality, Postgrad. Acad. Physicians; permanent cons., v.p., rep. Internat. Abolitionists Fedn. at ECOSOC, UN, 1947-90; med. cons. German Gen. Consulate, Atlanta, Miami, Fla., 1963; lectr. Morton Plant Hosp., Clearwater, Fla., also Clearwater campus St. Petersburg Jr. Coll.; guest prof. U. Bremen, Berlin, 1981-82. Author: Public Health Between the Time of Imperium and National Socialism, The Destruction of Public Health Reforms of the First German Republic, 1985, others; transl. from the Japanese Origin of Syphilis in the Far East, Static Atony, Sexual Pedagogic; contbr. 300 articles to med. jours. and revs. to books. Served with German Army, 1914-18. Decorated Cross Merit I Class (Germany), 1965; recipient Commendation awards Pres. of U.S., 1945, 70, 65 Year Gold Service Pin, AMA and ARC, 1985, Service to Mankind award Sertoma, 1972-73, Sport award Pres. Carter, 1977, Musicologist award Richey Symphony, 1979, Reconciliation award Germany-U.S.A., 1983, Friendship award Fed. Republic of Germany, USA, 1985, Teaching award Morton Plant Hosp., 1988, 70 Yrs. Svc. Red Cross, others. Fellow Am. Acad. Family Physicians (charter, life, 40-Yr. Svc. award 1986), AAAS, Am. Coll. Sport Medicine (emeritus, charter, life), Am. Pub. Health Assn. (life, 40-Yr. Svc. award 1984), Brit. Soc. (emeritus); mem. World Med. Assn. (life), German Assn. History of Medicine (life), Acad. Mental Retardation (charter, life), Am. Pub. Health Assn. (life, 40 Yr. Svc. award), Fla. Health Assn. (life), Brit. Pub. Health Assn. (life), AMA (hon.), Am. Assn. Mil. Surgeons (life), Acad. Preventive Medicine (life), Steuben Soc., Richey Symphony Soc. (charter, Musicologist 1979), World Peace Through World Law Ctr. Clubs: City (Chgo. chmn. hygiene sect. 1944-46). Lodges: Rotary (life, Harris fellow 1980), Masons (32 deg.), Shriners (comdr., life v.p.). Home: 2470 Rhodesian Dr Apt 34 Clearwater FL 34623-1948

LOEWY, ROBERT GUSTAV, engineering educator, aeronautical engineering executive; b. Phila., Feb. 12, 1926; s. Samuel N. and Esther (Silverstein) L.; m. Lila Myrna Spinner, Jan. 16, 1955; children—David G., Esther Elizabeth, Joanne Victoria, Raymond M. B in Aero. Engring., Rensselaer Poly. Inst., 1947; MS, MIT, 1948; PhD, U. Pa., 1962. Sr. vibrations engr. Martin Co., Balt., 1948-49; assoc. research engr. Cornell Aero. Lab., Buffalo, 1949-52, prin. engr., 1953-55; staff stress engr. Piasecki Helicopter Co., Morton, Pa., 1952-53; chief dynamics engr., also chief tech. engr. Vertol div. Boeing Co., Essington, Pa., 1955-62; from assoc. prof. to prof. mech. and aerospace scis. U. Rochester, 1962-73, dean Coll. Engring. and Applied Sci., 1967-74; dir. Space Sci. Center, 1966-71; v.p., provost Rensselaer Poly. Inst., Troy, N.Y., 1974-78, Inst. prof., 1978-93; dir. Rotorcraft Tech. Ctr., 1982-93; dir. sch. aerospace engring. Ga. Inst. Tech., 1993—; chief scientist USAF, 1965-66; cons. govt. and industry, 1959—; mem. aircraft panel Pres.'s Sci. Adv. Coun., 1968-72; mem. Air Force Sci. Adv. Bd., 1966-75, 78-85, vice chmn., 1971, chmn., 1972-75, chmn. aero. systems div. adv. group, 1978-84; mem. Post Office Rsch. and Engring. Adv. Coun., 1966-68; mem. rsch. and tech. adv. com. on aeros. NASA, 1970-71, mem. rsch. and tech. adv. coun., 1976-77, chmn. aero. adv. com., 1978-83; mem. aerospace engring. bd. NRC, 1972-78, 1988-93, mem. bd. on army sci. and tech., 1986-90; mem. naval studies bd. NAS, 1979-82; chmn. tech. adv. com. FAA, 1976-77; bd. dirs. Vertical Flight Found. Contbr. articles to profl. jours. Served with USNR, 1944-46. Recipient NASA disting. pub. service award, 1983; Gotshall-Powell scholar Rensselaer Poly. Inst.; USAF Exceptional Civilian Service awards, 1966, 75, 85. Hon. fellow AAAS, AIAA (Lawrence Sperry award 1958), Am. Helicopter Soc. (tech. dir. 1963-64); mem. Am. Soc. Engring. Edn., Nat. Acad. Engring., Sigma Xi, Sigma Gamma Tau, Tau Beta Pi. Home: 9 Loudon Hts N Albany NY 12211-2011 Office: Rensselaer Poly Inst 110 8th St Troy NY 12180-3522

LOFASO, ANTHONY JULIUS, mechanical engineer; b. N.Y.C., May 1, 1923; s. Antonio and Angelina (Cirrincione) L.; m. Angelina Barbera, Dec. 6, 1944; children: Angela, Christine, Anthony, James, Richard. BS in Aeronautics, N.Y. U., 1946, MME, 1948. Dir. programs Sperry, Great Neck, N.Y., 1949-86; dir. engring. ops. Norden Systems Inc., Melville, N.Y., 1986—. Co-editor: (jour.) Organization for Productivity, 1985. Mem. Environ. Control Commn., Oysterbay, N.Y., 1979-84; lt. col. N.Y. Guard, N.Y.C., 1981-86; trustee, pres. Bd. Edn., Bethpage, N.Y., 1956-85. Sgt. U.S. Army Air Corp, 1942-46. Mem. AIAA. Roman Catholic. Home: 16 Dorothea St Plainview NY 11803 Office: Norden Systems Inc 75 Maxess Rd Melville NY 11747

LOFREDO, ANTONY, chemical engineer; b. Bellville, N.J., Sept. 28, 1926; s. Joseph Anthony and Emilia (Pomponio) L.; m. JoAnn Cangene; children: Carol Ann, Joseph Michael, James Anthony, Robert Francis, Mary Ellen, Richard Charles, Thomas Patrick. BChemE, NYU, 1952. Jr. engr. to sr. processing engr. Airco Indsl. Gases, Murray Hill, N.J., 1952-58, supr., mgr. dir. process engring., 1959-64, v.p. engring., 1965-68, gen. mgr. process systems, mktg. process plants, 1969-73, prodn. engring. energy control mgr., 1974-92, effiency plant improvements, ops. up-grading mgr., 1974-92; engring. assoc., 1992—. With U.S. Army, 1944-46. Mem. Am. Chem. Soc., Am. Inst. Chem. Engrs., Tau Beta Pi. Roman Catholic. Achievements include 10 process patents on crogenic refrigerators, liquefiers and gas separations for process and nuclear industries. Home: 38 Skylark Rd Springfield NJ 07081

LOFTIN, KARIN CHRISTIANE, biomedical specialist, researcher; b. Hamburg, Germany, Dec. 3, 1947; came to U.S., 1959; d. Joseph Jr. and Regine (Juhn) Cibula; m. Richard Bowen Loftin, Nov. 23, 1972; children: Elisabeth C., Benjamin B. BA in Biology, Oakland U., 1970; cert. med. tech., Baylor U., 1971; MS, Grad Sch. Biol Scis., U. Tex., Houston, 1973, PhD, 1979. Instr. biology U. Houston, Univ. Park, 1979-81; rsch. assoc. microbiology U. Tex., Galveston, 1982, rsch. asst. dept. pediatrics, 1982; Henry Holcomb postdoctoral fellow in clin. immunology U. Tex. M.D. Anderson Hosp. and Tumor Inst., Houston, 1983-84; instr. infectious diseases and clin. microbiology U. Tex. Med. Sch., Houston, 1984-86, sr. rsch. assoc. dept. ob/gyn. and reproductive scis., 1986-89; sr. rsch. scientist Krug Life Scis., Inc., Houston, 1989—; senator-at-large to faculty senate U. Tex. Med., 1985-86, course dir. for microbiology labs., 1984-86. Contbr. articles to Am. Jour. Reproductive Immunology, Leukemia Rsch., Blood, Jour. Dental Rsch.; contbr. abstracts to numerous confs., symposia. Judge dist. sci. fair Clear Creek Ind. Sch. Dist., judge sr. div., 1981-83, jr. div., 1977-78, judges coord. Regional Houston Sci. and Engring. Fair, 1984-92; chair blood drive Gulf Coast Blood Ctr.; coach basketball YMCA, 1991, 92, coach baseball, 1987; team mgr. Bay Area Youth Sports Baseball, 1988, Nasa Area Little League Baseball, 1989, Nasa Area Soccer Club, 1988, bd. dirs., 1989-92; judge stroke and turn Nassau Bay Swim Team, 1987-91, bd. dirs., 1989-90; pres. Women in Svc. Orgn., Lord of Life Luth. Ch., 1991-92, v.p. Gulf Coast chpt. Luth. Women Missionary Soc., 1992-93. Mem. Am. Soc. for Microbiology, Nat. Mgmt. Assn. (v.p. local chpt. 1993-94), AIAA, Am. Assn. for Artificial Intelligence, Sigma Xi Sci. Rsch. Soc.

LOGAN, BRUCE DAVID, physician; b. Salem, Mass., Nov. 17, 1945; s. Jack Merill and Miriam Jane (Buckley) L.; m. Ann Marie Viola, Mar. 24, 1983; children: Anna Leah, Jennifer Mary. AB, Colby Coll., 1967; MD, Columbia U., 1972. Dir. ambulatory care Beekman Hosp., N.Y.C., 1981-87; pvt. practice N.Y.C., 1987—; chief of medicine N.Y. Downtown Hosp., N.Y.C., 1991—. Mem. AAAS, AMA, Assn. Program Dirs. in Internal Medicine, Phi Beta Kappa. Avocation: skiing.

LOGAN, NORMAN, chemistry researcher, educator; b. Shipley, Yorkshire, Eng., Jan. 25, 1935; s. Samuel and Mary Emma (Paley) L.; m. Maureen McEvan, Aug. 3, 1957; children: Karen Louise, Timothy James, Alexandra Jane. BSc in Chemistry, U. Nottingham, U.K., 1956, PhD in Inorganic Chemistry, 1959, DSc in Inorganic Chemistry, 1989. Chartered chemist. Lectr. in inorganic chemistry U. Nottingham, 1961-70, sr. lectr., 1970-76, reader, 1976-92, rsch. fellow, 1991—; Miller Rsch. fellow chemistry dept. U. Calif., Berkeley, 1964-65; vis. prof. dept. chemistry Panjab U., Chandigarh, India, 1975-76, 79; editorial rep. for U.K. Am. Inst. Physics, 1968-70; cons. INMARSAT, Brit. Aerospace, Royal Ordnance Rocket Motors Div., European Space Agy. Co-author: Preparative Inorganic Reactions, 1964, Developments in Inorganic Nitrogen Chemistry, Nitrogen NMR, 1973; editorial adv. bd. mem. Inst. Sci. Info., Phila, 1967-70; contbr. articles to profl. jours. Chmn. Liberal Party, Keyworth, Nottinghamshire, Eng., 1975-86; councillor Parish Coun., Keyworth, 1979-83. Rsch. grantee USAF, 1970-80, Royal Ordnance Rocket Motors Div., 1981-88, Royal Aerospace Establishment, 1983—, U.S. Army, 1986-90, Ford Aerospace/INTELSAT, 1988-90. Fellow Royal Soc. Chemistry (sec./treas. East Midland sect. 1967-69); mem. AIAA, Internat. Bank Note Soc. Methodist. Achievements include research on corrosion chemistry of metals in nitrogen tetroxide and nitric acid liquid rocket propellant oxidisers including electrochemistry and surface analysis, on chemistry of anhydrous metal nitrates, nitrogen oxides and nitric acid, including discovery of the iron nitrate responsible for flow decay of propellant nitrogen tetroxide, on 14N and 15N NMR spectroscopy of inorganic nitrogen compounds. Office: Dept Chemistry U Nottingham, University Park, Nottingham NG7 2RD, England

LOGANI, KULBHUSHAN LAL, civil and structural engineer; b. Mardan, Panjab, India, Oct. 20, 1943; came to U.S., 1969; s. Sulakhan Mal and Shankri Devi L.; m. Suresh Logani, Jan. 24, 1965; children: Sanjay, Monica, Ronica. BSCE, Panjab U., 1961; ME in Structural Engring., Iowa State U., 1970, PhD, 1973. Registered profl. engr., lic. structural engr. Design engr. Bhakra & Beas Design Orgn., New Delhi, 1961-65; engring. cons. Ministry of Agr., Ghana, West Africa, 1965-69; rsch. asst. Iowa State U., Ames, 1969-73; consulting engr. Harza Engring. Co., Chgo., 1973-86; v.p., dir. Facilities Cons. Ltd., Chgo., 1986—; pres. KL Cons. Ltd., Glenview, Ill., 1991—; cons. dam design and constrn. Dept. Hydraulic Resources, San Juan, Argentina, 1979-83; cons. devel. of instrumentation under artesian conditions Reza Shah Kabir Dam, Iran, 1978-79. Author publs. in field including Proceedings of VII Pan Am. Conf. on Soil Mechanics and Found. Engr.-Can., 1983, Transaction of the 14th Internat. Congress on Large Dams - Brazil, Proceedings of the Internat. Conf. on Recent Advances in Geotechnical Earthquake Engring., U. Mo., Rolla, 1981, various others confs. in field; contbg. author ency. article, 1988; contbr. articles to profl. jours. Founding mem. Assn. of Indian in Am., Chgo. Rsch. grantee Def. Nuclear Agy., Washington, 1970-73. Fellow ASCE; mem. ASTM (com. on soil and rock 1984—), Internat. Soc. Soil Mechanics and Found. Engring., Instn. of Engrs. India. Achievements include math. model for rock creep and progressive failure, math. formulation to model rock failure in plane-strain, devel. of equipment and technique to install instruments under high pore water pressure condition and deep water. Home: 1144 Bette Ln Glenview IL 60025-2429

LOGOMARSINO, JACK, nutrition educator; b. Staten Island, N.Y., May 6, 1945; s.m. Luise, Nov. 28, 1970; 1 child, Lisa. BS, Cornell U., 1967; MS, Purdue U., 1969; PhD, Cornell U., 1973. Registered dietitian. Asst. prof. SUNY, Geneseo, 1973-77; assoc. prof. U. S.C., Columbia, 1977-89, U. Ill., Urbana, 1989-91; prof. nutrition Ctrl. Mich. U., Mt. Pleasant, 1991—. Mem. edit. bd. Jour. Coll. and Univ. Foodsci., 1991—; contbr. articles to profl. jours. Recipient USDA Rsch. award, 1989, Instructional award NSF, 1982. Mem. Am. Dietetic Assn. (state exec. bd. 1987), Sigma Xi (chpt. sec. 1983). Office: Ctrl Mich U Mount Pleasant MI 48859

LOGUE, JOHN JOSEPH, psychologist; b. Phila., Nov. 16, 1929; s. Edwin J. and Ellen V. (Mallon) L.; m. Evelyn Bortnick, Apr. 24, 1954; 1 child, Eileen Logue Handel. BS, Temple U., 1954, MEd, 1958, EdD, 1966. Lic. psychologist Pa., Md., N.J., Del. Ptnr., sr. cons. RHR Internat., Phila., 1966-88; mgmt. psychologist pvt. practice Phila, 1988—. With U.S. Army, 1954-56. Fellow Royal Soc. Health; mem. APA (indsl., orgn., cons., counseling, edn. divsns.), Vesper Club, Quaker City Yacht Club. Home: 710 Kenilworth Ave Philadelphia PA 19126-3715 Office: 205 Keith Valley Rd Horsham PA 19044-1499

LOHMANN, GEORGE YOUNG, JR., neurosurgeon, hospital executive; b. Scranton, Pa., Aug. 9, 1947; s. George Young Lohmann and Elizabeth (Nichols) Frantzen; m. Joette Calabrese, May 15, 1973 (div. 1981); m. Rosemary Ei-Ling Ma, Sept. 24, 1988; 1 child, Norelle Christa Victoria. AB in Chemistry with honors, Hobart Coll., 1968; MD, SUNY, Buffalo, 1972. Diplomate Am. Bd. Neurol. Surgeons, Am. Acad. Pain Specialists. Resident gen. surgery Wesley Meml. Hosp., Chgo., 1972-73; from jr. resident to chief resident Georgetown U. Hosp., Washington, 1975-79; asst. med. dir. West Side Orgn., Chgo., 1973-74; emergency physician St. James Hosp., Chicago Heights, Ill., 1973-74; pvt. practice Baton Rouge, 1979-81, 81-84; dir. dept. neurosurgery Brookdale Hosp. Med. Ctr., Bklyn., 1984—; pres. Bklyn. Neurosurg. Svcs., Inc., 1985—; mem. Med. Dir. Com., Risk Mgmt. Com., Exec. Quality Assurance Com., 1987—; mem. Med. Bd. Com., 1985—, Exec. Bd. Com., 1984—, Pain Mgmt. Com., 1988-91. Patentee in field; contbr. articles to profl. jours, poetry to lit. mags. Mem. adv. bd. Ctr. Latin Affairs, Baton Rouge, 1982-84; mem. Senatorial Inner Cir., 1988; bd. trustees Christian Victory Ctr., Hempstead, N.Y., 1986-88; fellow Am. Coll. Pain Mgmt. Fellow ACS; mem. AMA, Am. Assn. Neurologic Surgeons (spine sect.), N.Y. State Neurosurg. Soc., N.Y. Soc. Neurosurgery, Congress Neurologic Surgeons (spine sect.), So. Med. Soc., Presdl. Roundtable. Avocations: skiing, painting, poetry, music, cooking. Office: Brookdale Hosp Med Ctr 1 Brookdale Plz Brooklyn NY 11212

LOHMANN, JOHN J., quality assurance engineer; b. Detroit, Aug. 18, 1933; s. Peter and Elisabeth (Nielsen) L.; div.; children: Virginia, Judith, John. BA cum laude, U. of South, 1959, MDiv, 1962; MBA, Ohio U., 1981. Curate Episc. Ch., Lexington, Ky., 1962-64; vicar Episc. Ch., Hendersonville, Tenn., 1964-69; rector Episc. Ch., Detroit, 1969-76; pers. specialist Owens-Corning, Newark, Ohio, 1976-79, lab. supr., 1979-86, div. engr. quality assurance, 1986-89; div. engr. quality assurance Owens-Corning, Toledo, 1989-90; adv. div. engr. quality assurance Owens-Corning World Hdqrs., Toledo, 1990—, lead assessor qualified; bd. dirs. Owens Corning Federal Credit Union, Toledo. Sgt. USAF, 1952-56. Mem. Am. Soc. Quality Control (sr., cert. quality engr., reliability engr., quality auditor, quality technician; bd. dirs. Columbus and Toledo sects.), Mensa, Phi Beta Kappa. Republican. Home: 5311 Brookfield Ln Sylvania OH 43560 Office: Owens Corning World Hdqrs Fiberglas Tower Toledo OH 43659

LOIGNON, GERALD ARTHUR, JR., nuclear engineer; b. N.Y.C., June 25, 1950; s. Gerald Arthur Loignon Sr. and Nancy MacLean (Walker) Bucknell; m. Margaret Mary Hamburger, Aug. 7, 1971; children: Brian MacLean, Matthew Thomas, Teresa Marie. BS in Nuclear Engring., N.C. State U., 1976. Registerd profl. engr., S.C.; lic. sr. reactor operator Nuclear Regulatory Commn. Health physics technician nuclear fuel div. Westinghouse, Columbia, S.C., 1970-72; office equipment technician Cavin's Inc., Raleigh, N.C., 1973-76; quality assurance engr. Met. Edison Co., Reading, Pa., 1976-79; shift tech. advisor Met. Edison Co., Gen. Pub. Utility, Harrisburg, Pa., 1979-81; shift tech. advisor S.C. Electric and Gas Co., Jenkinsville, 1981-83, assoc. mgr. performance and results, 1983-88, shift engr., 1988-91, test unit supr., 1992—. Mem. choir, dir. religious edn. St. Marks Parish, Newberry, S.C., 1991—. Mem. Am. Nuclear Soc. (chair S.C. chpt. 1985-86, profl. engring. exam. com. 1987—, cert. appreciation 1985, cert. governence 1986), Profl. Reactor Operator Soc. Roman Catholic. Home: Rte 2 Box 183 Kinards SC 29355 Office: SC Electric and Gas Co VC Summer Nuclear Sta PO Box 88 Jenkinsville SC 29065-0088

LOLIS, ELIAS, biomedical researcher; b. N.Y.C., July 10, 1962. BA in Chemistry, Columbia Coll., 1984; PhD in Chemistry, MIT, 1989. Post doctoral assoc. Rockefeller U., N.Y.C., 1989-91; asst. prof. pharmacology dept. Yale U., New Haven, Conn., 1991--.

LOLLAR, JOHN SHERMAN, III (PETE LOLLAR), hematologist; b. St. Louis, Apr. 17, 1951; s. John Sherman and Constance Lilyan (Maggard); m. Carolyn Ann Strawn: children: John Robin, Stefan Scott, Ryan Douglas. BS, La. State U., 1973; MD, St. Louis U., 1977. Diplomate Am. Bd. Internal Medicine, Am. Bd. Hematology. Resident internal medicine U. Iowa Med. Ctr., Iowa City, 1977-78, 80-81, fellow hematology, 1978-80, chief resident internal medicine, 1981-82; fellow Mayo Clinic, Rochester, Minn., 1982-84; asst. prof. medicine U. Vt., Burlington, 1984-89, assoc. prof. medicine, 1989-90; assoc. prof. medicine Emory U., Atlanta, 1990—. Contbr. articles to Jour. Clin. Investigation, Blood, Biochemistry, Jour. Biol. Chemistry. Recipient Clin. Investigator award, NIH, 1982, Established Investigator award, Am. Heart Assn., 1987, Am. Soc. Clin. Investigation award, 1991. Mem. Am. Chem. Soc., Am. Soc. Hematology, Am. Soc. Biochemistry and Molecular Biology. Achievements include isolation and characterization of activated blood coagulation factor VIII; determination of mechanisms governing the stability of activated factor VIII. Office: Emory U Drawer AJ Atlanta GA 30322

LOMAS, LYLE WAYNE, agricultural research administrator, educator; b. Monett, Mo., June 8, 1953; s. John Junior and Helen Irene Lomas; m. Connie Gail Frey, Sept. 4, 1976; children: Amy Lynn, Eric Wayne. BS, U. Mo., 1975, MS, 1976; PhD, Mich. State U., 1979. Asst. prof., animal scientist Kans. State U. S.E. Expt. Sta., Parsons, 1979-85, assoc. prof., head, 1985-92, prof., head, 1992—. Contbr. articles to refereed sci. jours. Mem. Am. Soc. Animal Sci., Am. Registry Profl. Animal Scientists, Am. Forage and Grassland Coun., Rsch. Ctr. Adminstrs. Soc. (bd. dirs. 1993—), Parsons Rotary Club (bd. dirs. 1992—), Phi Kappa Phi, Gamma Sigma Delta. Presbyterian. Achievements include research in ruminant nutrition, forage utilization by grazing stocker cattle. Home: Rt 1 Box 31 Dennis KS 67341 Office: Kansas State Univ Southeast Kans Expt Sta PO Box 316 Parsons KS 67357

LOMBARDO, JANICE ELLEN, microbiologist; b. Chgo., Sept. 22, 1951; d. John Robert and Betty Jane (Westfall) Richardson; m. Peter Anthony Lombardo, Aug. 17, 1979; 1 child, Gina Ellen. BA in Biology, Northeastern Ill. U., 1972, MS in Biology, 1984. Tech. supr. St. Joseph's Hosp., Chgo., 1973-78; lab. leader, microbiologist Cabrini Hosp., Chgo., 1979-80, lab. mgr., 1980-84; med. technologist Stroink Pathology Lab., Bloomington, Ill., 1984-85; microbiology supr. Damon Clin. Labs., Berwyn, Ill., 1985-87; microbiology technologist Damon Clin. Labs., Berwyn, 1987-90; microbiology supr. Columbus-Cabrini Med. Ctr., Chgo., 1990-92, lab. adminstrv. dir., 1992—. Mem. social ministry com., Lutheran Meml. Ch., Chgo. Grantee Campaign for Human Devel., Des Moines, 1978. Mem. Am. Soc. Microbiology, Am. Soc. Clin. Pathologists, South Ctrl. Assn. Clin. Microbiology, N.Y. Acad. Sci., Clin. Lab. Mgmt. Assn., Nat. Cert. Agy. Achievements include environmental study on Chicago River, botanical study on bacterial pathogens, and microbiological analysis in botulism litigation, 1985. Home: 5605 N Nagle Chicago IL 60646

LOMET, DAVID BRUCE, computer scientist; b. Neptune, N.J., Aug. 2, 1939; s. Pierre and Helen (Foster) L.; m. Charlotte Jean Vandermark, Aug. 15, 1964; children: Bruce, Kevin. BS in Physics, Lafayette Coll., Easton, Pa., 1961; MS in Math., George Washington U., Washington, 1966; PhD in Computer Sci., U. Pa., Phila., 1969. Vis. researcher U. Newcastle (U.K.)-upon-Tyne, 1975-76; mem. rsch. staff IBM Corp., Yorktown Heights, N.Y., 1969-85; prof. computer sci. Wang Inst. Grad. Studies, Tyngsboro, Mass., 1985-87; sr. info. cons. Digital Equipment Corp., Nashua, N.H., 1987-89; sr. cons. engr. and mem. rsch. staff Digital Equipment Corp., Cambridge, Mass., 1989—; grant reviewer NSF, NASA, NRC (Can.); participant program coms. Editor Data Engring. Bulletin; contbr. articles to profl. jours. Mem., v.p. Bd. Edn., Yorktown Heights, N.Y., 1980-85. IBM resident grad. fellow, 1966. Mem. IEEE, AAAS, Assn. Computer Machinery, Phi Beta Kappa. Democrat. Achievements include twelve patent applications; research in database systems, programming languages, computer architecture, and distributed systems. Office: Digital Equipment Corp Cambridge Rsch Lab 1 Kendall Sq Bldg 700 Cambridge MA 02139

LOMHEIM, TERRENCE SCOTT, physicist; b. Mexico City, Apr. 29, 1953; came to U.S., 1953; s. James Henry and Darlene DeLoris (Hofer) L.; m. Katherine Bernice Crawford, Nov. 3, 1977; children: Jill Lynn, Justin Michael, Jason Ryan, Jordan Neal. BA in Physics, Calif. State U., Fullerton, 1973; MA in Physics, U. So. Calif., 1976, PhD in Physics, 1978. Mem. tech. staff The Aerospace Corp., El Segundo, Calif., 1978-82, staff engr., 1982-84, mgr. electro-optics, 1984—; lectr. dept. physics Calif. State U., Dominguez Hills, 1981—; instr. Internat. Optical Engring. symposium, San Diego, 1989, 92, 93; speaker workshops IEEE, Waterloo, Can., 1991, 93, So. Calif. Modern Physics Inst. for Tchrs., Fullerton, 1990. Co-author: (book chpt.) Electro-Optical Displays, 1992; contbr. 16 articles to profl. jours. Mem. Am. Phys. Soc., Optical Soc. Am., Soc. Photo-Optical Instrumentation Engrs. Republican. Home: 1724 N Peacock Ln Fullerton CA 92633 Office: The Aerospace Corp 2350 E El Segundo Blvd El Segundo CA 92633

LONDON, J. PHILLIP, professional services company executive; b. Oklahoma City, Apr. 30, 1937; s. Harry Riles and Laura Evalyn (Phillips) L.; separated; J. Phillip Jr., Laura McLain. BSc, U.S. Naval Acad., 1959; MSc, U.S. Naval Postgrad. Sch., 1967; D in Bus. Adminstrn., George Washington U., 1971. Commd. ensign USN, 1959, advanced through grades to capt., resigned, 1971; program mgr. Challenger Research Inc., 1971-72; mgr. CACI Internat. Inc., Arlington, Va., 1972-76, v.p., 1976-77, sr. v.p., 1977-79, exec. v.p., 1979-82, pres. operating div., 1982-84, pres., chief exec. officer, 1984-90, chmn. bd., 1990—. Recipient Alumni of Yr. award George Washington U. Sch. Govt. & Bus. Adminstrn., Washington, 1987. Mem. Profl. Services Council. Episcopalian. Club: George Town. Office: CACI Internat Inc 1100 N Glebe Rd Arlington VA 22201*

LONDON, RAY WILLIAM, clinical and forensic psychologist, consultant, researcher; b. Burley, Idaho, May 29, 1943; s. Loo Richard and Maycelle Jerry (Moore) L. AS, Weber State Coll., 1965, BS, 1967; MSW, U. So. Calif., 1973, PhD, 1976, Exec. MBA, 1989, postgrad, Pepperdine U. Law Sch., 1992—. Diplomate: Am. Bd. Psychol. Hypnosis (dir. 1984—, pres. 1989—), Am. Acad. Behavioral Medicine, Am. Bd. Psychotherapy, Am. Bd. Med. Psychotherapy, Internat. Acad. Medicine and Psychology., Am. Bd. Profl. Neuropsychology, Am. Bd. Adminstrv. Psychology, Am. Bd. Examiners Clin. Soc. Work, Am. Bd. Clin. Hypnosis in Social Work (pres. 1989-91), Am. Bd. Profl. Psychology, Am. Bd. Family Psychology (dir. 1993—), Am. Bd. Child and Adolescent Psychology (dir. 1992—), NASW Clin. Soc. Work; cert. Am. Assn. Sex Therapists, Soc. Med. Analysts; registered internat. cons., cert. mgmt. cons., congl. asst. U.S. Ho. of Reps., 1964-65; rsch. assoc. Bus. Advs., Inc., Ogden, Utah, 1965-67; dir. counseling and consultation svcs. Meaning Found., Riverside, Calif., 1966-69; mental health and mental retardation liaison San Bernardino County (Calif.) Social Svcs., 1968-72; clin. trainee VA Outpatient Clinic, L.A., 1971-72, Children's Hosp., 1972-73, clin. fellow, 1973-74; clin. trainee Reiss Davis Child Study Ctr., L.A., 1973-74, L.A. County-U. So. Calif. Med. Ctr., 1973; psychotherapist Benjamin Rush Neuropsychiat. Ctr., Orange, Calif., 1973-75; clin. psychology postdoctoral intern Orange County (Calif.) Mental Health, 1976-77; postdoctoral fellow U. Caif.-Irvine-Calif. Coll. Medicine, 1978; clin. psychologist Orange Police Dept., 1974-80; pvt. practice consultation and assessment, Santa Ana, Calif., 1974—; chief oper. officer London Assocs. Internat., 1974-80; cons. to public schs., agys., hosps., bus., nationally and internationally, 1973—; presenter nat. and internat. lectures, seminars and workshops; pres. bd. govs. Human Factor Programs, Ltd., 1976—; pres. Internat. Bd. Medicine and Psychology, 1980-84; chief exec. officer Human Studies Ctr., 1987—; pres., chief exec. officer London Assocs. Internat.; Organizational Behavior-Crisis-Devel. Cons., 1980—; research affil. Ctr. for Crisis Mgmt. U. So. Calif. Grad. Sch. Bus. Adminstrn., 1988-90; pres., chief exec. officer Am. Bd. Clin. Hypnosis, Inc.; mem. faculty UCLA, U. So. Calif., Calif. State U., U. Calif., Irvine, Calif. Coll. Medicine, Internat. Cong. of Psychosomatic Medicine, Internat. Coll.; research assoc. Nat. Commn. for Protection of Human Subjects of Biomed. and Behavioral Research, 1976; fellow Inst. for Social Scientists on Neurobiology and Mental Illness, 1978. Editor: Internat. Bull. Medicine and Psychology, 1980—, A.B.C.D. Report, 1988— behavioral medicine Australian Jour., 1980, adv. editor Internat. Jour. Clin. and Exptl. Hypnosis, 1981-92, mng. editor, 1991—, assoc. editor, 1992—; cons. editor Internat. Jour. Psychosomatics, 1984—; Experimentelle und Klinische Hypnose, 1987—, cons. Am. Jour. Forensic Psychology, 1986, Jour. Mgmt. Consulting, 1992—; pub.: London Behavioral Medicine Assessment, 1982, A Behavior-Cris-Development newsletter, ABCD Newsnote; producer: TV series Being Human, 1980; contbg. author World Book Ency. and books; contbr. articles to profl. jours. Recipient Congl. recognition U.S. Ho. of Reps., 1976; named scholar laureate Erickson Advanced Inst., 1980. Fellow Internat. Acad. Medicine and Psychology (dir. 1981—), Soc. Clin. Social Work (dir. 1979-80), Royal Soc. Health, Am. Coll. Forensic Psychology, Soc. Clin and Experimental Hypnosis (bd. dirs. 1985—, treas. 1987-89); mem. Acad. Psychosomatic Medicine, Am. Psychol. Assn., Am. Group Psychotherapy Assn., Am. Orthopsychiat. Assn., N.Y. Acad. Sci., Soc. Behavioral Medicine, Internat. Psychosomatic Inst., Australian Coll. Pvt. Consulting Psychologists, Australian Psychol. Soc., Internat. Coun. Psychologists, Acad. Mgmt., Assn. Profl. Cons., Inst. Mgmt. Cons., Internat. Forum Corp. Dirs., Nat. Assn. Corp. Dirs., Profl. and Tech.Indsl. and Orgnl. Psychology, Assn. Profl. Cons., So. Calif. Mediation Assn. (mem. law soc.), Am. Soc. Trial Cons., Am. Psychology Law Soc., Toastmasters, Phi Delta Kappa, Delta Sigma Rho, Tau Kappa Alpha, Pi RhoPhi, Lambda Iota Tau. Office: London Assocs Internat 1125 E 17th St Ste 209E Santa Ana CA 92701-2201

LONDON, WILLIAM THOMAS, internist; b. N.Y.C., Mar. 11, 1932; s. William Wolf and Lillian (Mann) L.; m. Linda Greenman, June 23, 1957; children: Barbara, Katharine, Emily, Nancy. BA, Oberlin Coll., 1953; M, Cornell U., 1957. Intern Bellevue Hosp., N.Y.C., 1957-58; resident in medicine Bellevue and Meml. Hosps., N.Y.C., 1960; sr. rsch. physician Fox Chase Cancer Ctr., Phila., 1978-90, sr. mem., 1990—; assoc. physician Jeanes Hosp., Phila., 1966—; cons. Am. Oncologic Hosp., Phila., 1985—; adj. prof. U. Pa. Sch. Medicine, Phila., 1978—; sr. mem. Fox Chase Cancer Ctr., 1990—; chmn. AIDS and Related Rsch. Study Sect., Bethesda, Md., 1990-92; chmn. med. adv. com. Am. Liver Found., Delaware Valley chpt., 1991—; mem. Bd. Sci. Counselors, Bethesda, 1985-89. Assoc. editor Jour. Acquired Immune Deficiency Syndrome, 1989-92, Cancer Epidemiology, Biomarkers and Prevention, 1991—; contbr. articles to profl. jours. Bd. dirs. Cheltenham Twp. Adult Sch., Wyncote, Pa., 1975—. With USPHS, 1964-66. Recipient Med. Excellence award Am. Liver Found., 1991, Award for Svc. Rendered to Chinese communities of Phila. Am. Assn. Ethnic Chinese, 1991. Mem. AMA, Am. Assn. Cancer Rsch., Am. Soc. Preventive Oncology (pres. 1989-91). Office: Fox Chase Cancer Ctr 7701 Burholme Ave Philadelphia PA 19111

LONERGAN, BRIAN JOSEPH, economist, planner; b. New London, Conn., May 1, 1951; s. James and Olga Lonergan; m. Melissa Ann Weaver, Sept. 28, 1985; children: Kate, Keah. BA, U. Conn., 1973; MPA, Columbia U., 1981. Adminstr. Town of Waterford, Conn., 1977-79; intern Brookings Instn., Washington, 1979-80; energy policy analyst Port Authority N.Y. and N.J., N.Y.C., 1981-88; lead planning analyst United Illuminating Co., New Haven, 1988—; mem. adv. bd. environ. policy adv. com. Electric Power Rsch. Inst., Palo Alto, Calif., 1991-92. Home: 45 Sunset Beach Rd Branford CT 06405 Office: United Illuminating Co 157 Church St New Haven CT 06506-0901

LONERGAN, THOMAS FRANCIS, III, criminal justice consultant; b. Bklyn., July 28, 1941; s. Thomas Francis and Katherine Josephine (Roth) L.; m. Irene L. Kaucher, Dec. 14, 1963; 1 son, Thomas F. BA, Calif. State U., Long Beach, 1966, MA, 1973; MPA, Pepperdine U., L.A., 1976; postgrad., U. So. Calif., L.A., 1976. Dep. sheriff Los Angeles County Sheriff's Dept., 1963-70; U.S. Govt. program analyst, 1968—; fgn. service officer USIA, Lima, Peru, 1970-71; dep. sheriff to lt. Los Angeles Sheriff's Office, 1971-76, aide lt. to div. chief, 1976-79; dir. Criminal Justice Cons., Downey, Calif., 1977—; cons. Public Adminstrv. Service, Chgo., 1972-75, Nat. Sheriff's Assn., 1978, 79; cons. Nat. Inst. Corrections, Washington, 1977—, coordinator jail ctr., 1981-82 ; tchr. N. Calif. Regional Criminal Justice Acad., 1977-79; lectr. Nat. Corrections Acad., 1980-83; spl. master Chancery Ct. Davidson County, Tenn., 1980-82, U.S. Dist. Ct. (no. dist.) Ohio, 1984-85, Santa Clara Superior Ct. (Calif.), 1983-89, U.S. Dist. Ct. Ga., Atlanta, 1986-87, U. S. Dist. Ct. (no. dist.) Calif., 1982—, U.S. Dist. Ct. (no. dist.) Idaho, 1986, U.S. Dist. Ct. Oreg. 1986, U.S. Dist. Ct. Portland 1987, U.S. Dist. (no. dist.) Calif. 1984-89; also ct. expert. Author: California-Past, Present & Future, 1968; Training-A Corrections Perspective, 1979; AIMS-Correctional Officer; Liability-A Correctional Perspective; Liability Law for Probation Administrators; Liability Reporter; Probation Liability Reporter; Study Guides by Aims Media. Mem. Am. Correctional Assn., Nat. Sheriff's Assn. Roman Catholic.

LONG, ALFRED B., former oil company executive, consultant; b. Galveston, Tex., Aug. 4, 1909; s. Jessie A. and Ada (Beckwith) L.; student S. Park Jr. Coll., 1928-29, Lamar State Coll. Tech., 1947-56, U. Tex., 1941; m. Sylvia V. Thomas, Oct. 29, 1932; 1 dau., Kathleen Sylvia (Mrs. E.A. Pearson, II). With Sun Oil Co., Beaumont, Tex., 1931-69, driller geophys. dept., surveyor engring. dept., engr. operating dept., engr. prodn. lab., 1931-59, regional supr., 1960-69, cons., 1969—. Mem. sr.'s bd. dirs. Bapt. Hosp., Beaumont, Tex.; mem. sr.'s vols. bd. dirs. S.E. Tex. Rehab. Hosp., Beaumont; chaplain Seniors-Lawmen Coun.; mem. Jefferson County Program Planning Com., 1964; mem. tech. adv. group Oil Well Drilling Inst., Lamar U., Beaumont. Mem. Soc. Petroleum Engrs., Am. Petroleum Inst., Am. Assn. Petroleum Geologists, IEEE, Houston Geol. Soc., Gulf Coast Engring. and Sci. Soc. (treas. 1962-65), U.S. Power Squadron, Soc. Wireless Pioneers. Recipient Nat. Jefferson award for Outstanding Pub. Svc. Am. Inst. for Pub. Svc., 1992. Inventor various oil well devices. Office: PO Box 7266 Beaumont TX 77726-7266

LONG, AUSTIN, geosciences educator; b. Olney, Tex., Dec. 21, 1936; s. Jesse Lee and Sara Louise (Taylor) L.; m. Virginia Lee Haldeman, 1962 (div. 1976), children: Kirsten, Tonya, Lara; m. Karen Anne Long, 1976; children: Kathy, Stephanie. BA, Midwestern State U., 1957; MA, Columbia U., 1959; PhD, U. Ariz., 1966. Sr. sci. Smithsonian Inst., Washington, 1963-68; prof. U. Ariz., Tucson, 1968—. Editor Radiocarbon, 1987—. Rsch. grantee NSF, Dept. Energy, Am. Chem. Soc. Mem. Am. Geophysical Soc., Geochemical Soc. Home: 2715 E Helen St Tucson AZ 85716 Office: Univ of Arizona Geosciences Dept Tucson AZ 85721

LONG, CARL FERDINAND, engineering educator; b. N.Y.C., Aug. 6, 1928; s. Carl and Marie Victoria (Wellnitz) L.; m. Joanna Margarida Tavares, July 23, 1955; children: Carl Ferdinand, Barbara Anne. S.B., MIT, 1950, S.M., 1952; D.Eng., Yale U., 1964; A.M. (hon.), Dartmouth Coll., 1971. Registered profl. engr., N.H. Instr. Thayer Sch. Engring., Dartmouth Coll., Hanover, N.H., 1954-57; asst. prof. Thayer Sch. Engring., Dartmouth Coll., 1957-64, assoc. prof., 1964-70, prof., 1970—, assoc. dean, 1972, dean, 1972-84, dean emeritus, 1984—, dir. Cook Design Ctr., 1984—; engr. Western Electric Co., Alaska, 1956-57; v.p. ops., dir. Controlled Environment, 1975-81; pres., dir. Q-S Oxygen Processes, Inc., 1979-87; N.H. Water Supply and Pollution Control Com., U.S. Army Small Arms Systems Agy.; mem. New Eng. Constrn. Edn. Adv. Council, 1971-74; mem. adv. com. U.S. Patent and Trademark Office, 1975-79; mem. ad hoc vis. com. Engrs. Council for Profl. Devel., 1973-81; pres., dir. Roan of Thayer, Inc., 1986-93; bd. dirs. Micro Tool Co., Inc., Micro Weighing Systems, Inc., 1986-91, Roan Ventures, Inc., 1987-91, Hanover Water Works Co., 1989—, pres., 1990—. Mem. Hanover Town Planning Bd., 1963-75, chmn., 1966-74; trustee Mt. Washington Obs., 1975-92; bd. dirs. Eastman Community Assn., 1977-80; mem. corp. Mary Hitchcock Meml. Hosp., 1974—. NSF Sci. Faculty fellow, 1961-62; recipient Robert Fletcher award Thayer Sch. of Engring., 1985, Fellow Members awd., Am. Soc. for Engineering Education, 1992. Fellow AAAS, ASCE, Am. Soc. Engring. Edn. (chmn. New Eng. sect. 1977-78, chmn. council of sects. Zone 1, dir. 1981-83); mem. Sigma Xi, Chi Epsilon, Tau Beta Pi. Republican. Baptist. Home: 25 Reservoir Rd Hanover NH 03755-1327

LONG, CEDRIC WILLIAM, health research facility executive; b. Mpls., Mar. 4, 1937; s. Tracy Steven and Clarice Cecilia (Robertson) L. BA, UCLA, 1960, MA, 1962; PhD, Princeton U., 1966. Postdoctoral fellow U. Calif., Berkeley, 1966-68; instr. NYU Med. Sch., N.Y.C., 1969-70; lab. chief Flow Labs., Rockville, Md., 1968-70, Litton Industries, Frederick, Md., 1976-80; preclin. chief NIH, Nat. Career Inst., DCT, Bethesda, Md., 1980-86; gen. mgr. Nat. Cancer Inst.- Frederick Cancer R&D Ctr., 1986—. Home: 2 Basildon Cir Rockville MD 20850-2724*

LONG, CHRISTOPHER, toxicologist; b. Woodside, N.Y., June 13, 1949; s. Walter Anthony and Jeanne (Bishop) L.; m. Maureen Ann Otremba, Jan. 2, 1978; children: Elizabeth Marie, Matthew Christopher. BA in Chemistry, Marist, 1971; MS in Biomedicine, L.I. U., 1974; MS in Pharm./Toxicology, St. John's U., 1979, PhD in Toxicology, 1981. Diplomate Am. Bd. Forensic Toxicology. Toxicologist, supr. Nat. Health Labs., East Meadow, N.Y., 1972-73, N.Y. Med. Labs., Great Neck, 1973-74; adj. faculty St. John's U., Jamaica, 1983-85; toxicologist Nassau County Med. Examiner, East Meadow, 1974-85; adj. faculty So. Ill. U. Med. Sch., Springfield, 1986-88; chief toxicologist Ill. State Police, Springfield, 1985-88; dir. toxicology, asst. prof. St. Louis U. Med. Sch., 1988—; expert toxicologist, cons. FDA, 1990—; chief toxicologist St. Louis County, 1991—. Contbr. articles to profl. jours. Leader Cub Scouts, 1991. Recipient Ednl. Rsch. award Soc. for Toxicology, 1981. Fellow Am. Acad. Forensic Scis.; mem. Soc. Forensic Toxicology, Internat. Soc. Forensic Toxicology, N.Y. Acad. Scis. Roman Catholic. Office: Saint Louis U Med Sch 1402 S Grand Blvd Saint Louis MO 63104-1004

LONG, ERIC CHARLES, biochemist; b. Reading, Pa., Nov. 20, 1962; s. Ronald Barry and Carole Kay (Mauger) L.; m. Maureen Theresa Walsh, May 25, 1985. BS in Chemistry, Albright Coll., 1984; PhD, U. Va., 1988. Rsch. fellow Columbia U., N.Y., 1988-89; fellow Jane Coffin Childs Meml. Fund for Med. Rsch. Calif. Inst. Tech., Pasadena, Calif., 1989-91; mem. rsch. faculty Calif. Inst. Tech., Pasadena, 1989-91; asst. prof. chemistry Ind. U., Purdue U., Indpls., 1991—. Contbr. articles to profl. jours. Recipient MDS Labs. award Albright Coll., 1984, Trustee Grant award, 1980-84. Mem. Am. Chem. Soc., AAAS. Office: Ind U Purdue U Indpls 402 N Blackford St Indianapolis IN 46202

LONG, FRANKLIN ASBURY, chemistry educator; b. Great Falls, Mont., July 27, 1910; s. F.A. and Ethel (Beck) L.; m. Marion Thomas, 1937; children: Franklin, Elizabeth. A.B., U. Mont., 1931, M.A., 1932; Ph.D., U. Calif., 1935. Instr. chemistry U. Calif., 1935-36, U. Chgo., 1936-37; instr. chemistry Cornell U., 1937-38, prof., 1939-79, prof. emeritus, 1979—, chmn. dept., 1950-60, v.p. research and advanced studies, 1963-69, Henry Luce prof. sci. and society, 1969-79, dir. program on sci., tech. and society, 1969-73, dir. peace studies program, 1976-79; adj. prof. U. Calif., Irvine, 1988—; dir. United Tech. Corp., Exxon Corp., 1969-81, cons., 1970-82; Mem. President's Sci. Adv. Com., 1961-66; asst. dir. U.S. Arms Control and Disarmament Agy., 1962-63, cons., 1963-73, 77-79 dir. Arms Control Assn., 1971-76; mem. adv. com. for planning and instnl. affairs NSF, chmn., 1973-74, mem. adv. panel for policy research analysis, 1976-80; co-chmn. Am. Pugwash Steering Com., 1974-79; mem. Indo-U.S. subcom. for ednl. and cultural affairs, 1974-82, co-chmn., 1977-82; bd. dirs. Assoc. Univs., Inc., 1947-74, hon. bd. dirs., 1975—; bd. dirs. Albert Einstein Peace Prize Found., 1979—, Fund for Peace, 1981—. Mem. editorial bd.: Am. Scientist, 1974-81, Bulletin of the Atomic Scientists, 1986—; Contbr. articles on chemistry, sci. policy and pub. affairs and arms control and disarmament to books, jours., encys. and reference works. Faculty Trustee Cornell U., 1956-57, Alfred P. Sloan Found., 1970-83, Fund for Peace, 1981—. Recipient U.S. Medal of Merit, 1948; (Korea) Dongbaeg medal, 1975; Guggenheim fellow, 1970. Fellow AAAS (v.p. 1976-80, Abelson prize 1989); mem. NAS, Coun. on Fgn. Rels., Am. Chem. Soc. (Charles Lathrop Parsons award 1985). Home: 446 Cambridge St Claremont CA 91711

LONG, GILBERT MORRIS, chemical engineer; b. Bellefonte, Pa., Dec. 15, 1947; s. Robert B. and Marie (Parker) L.; children: Erik, David, Robin; m. Tonimarie McGlynn, Sept. 3, 1993. BSChE, Lafayette Coll., 1970; MSChE, U. Mass., 1972. Cert. hazardous materials mgr. Unit engr. Exxon Chem. Co. USA, Bayway, N.J., 1974-77; asst. to pres. Publicker Chem. Corp., Phila., 1977-78; project engr. Handy & Harman, Fairfield, Conn., 1978-84; plant engr. Bel Ray Co., Farmingdale, N.J., 1985-87; project mgr. IT Corp., Edison, N.J., 1987-89; program mgr. ENSR Cons. Engring., Somerset, N.J., 1989-92, gen. mgr., 1992—. Co-author: Practical Environmental Bioremediation, 1992; contbr. articles to AWMA Jour., Chem. Engring. Progress. Active Henry Hudson Regional Sch. Bd. Edn., Highlands, N.J., 1986-89. Mem. AIChE, Air and Waste Mgmt. Assn. (steering com. satellite teleconferences 1991-92), Hazardous Materials Control Rsch. Inst., Inst. Hazardous Materials Mgmt. Office: ENSR Cons and Engring 1 Executive Dr Somerset NJ 08873

LONG, HARRY (ON-YUEN ENG), chemist, rubber science and technology consultant; b. Passaic, N.J., June 22, 1932; s. Eng Yick and Yue York (Ng) L.; m. Linda Lai-King Yu, Sept. 18, 1960; 1 child, Steven Eng Park-Ning. BS, N.J. Inst. Tech., Newark, 1959. Asst. devel. engr., belts and splty. products Uniroyal, Inc., Passaic, 1959-62, devel. engr. hose and expansion joints, 1962-67, sr. process engr., 1967-71; chief devel. engr. Raybestos-Manhattan, Inc., Passaic, 1971-72; chief chemist Goodall Rubber Co., Trenton, N.J., 1972-76, tech. mgr., 1976-90; v.p. tech. Pelmor Labs., Inc., Newtown, Pa., 1990—. Editor, author: Basic Compounding and Processing of Rubber, 1985. Mem. AAAS, ASTM, Am. Chem. Soc. (area dir. Rubber div. 1990-92, Spl. Svc. award Rubber div. 1985), Phila. Rubber Group (chmn. 1980). Achievements include development of the rubber technology course used by the subdivisions of the rubber division of American Chemical Society throughout the U.S., Canada, Mexico and Colombia; organization of national symposium on rubber compounding. Office: Pelmor Labs Inc 401 Lafayette St Newtown PA 18940-0309

LONG, JAMES ALVIN, exploration geophysicist; b. Porto Alegre, Brazil, July 13, 1917; s. Frank Millard and Eula (Kennedy) L.; m. Vivienne V. Peratt, Apr. 13, 1940 (dec. 1979); children: Frank, David, Susan, Kathryn. AB, U. Okla., 1937. With Stanolind Oil & Gas Co., 1937-46; mgr. United Geophys. Co., Venezuela and Brazil, 1946-61; tech. svcs. and divsn. mgr. United Geophys. Co., Pasadena, Calif., 1961-72; sr. geophysicist Tetratech, Houston, also Peru, 1973-74; geophys. adviser Yacimientos Petroliferos Fiscales Bolivianos, Santa Cruz, Bolivia, 1974-77; internat. geophys. cons., Australia, S.Am., U.S., China, 1978-86. Author, editor manuals for United Geophys. Co.; contbr. articles to profl. jours. Fellow The Explorers Club; mem. Geophys. Soc. Houston (emeritus), Soc. Exploration Geophysicists (emeritus), Naples North Rotary Club (dir. 1987-88), Country Club of Naples, Earthwatch (5 expeditions). Avocations: running, golf, languages, travel, exploration. Home: 3951 Gulf Shore Blvd N Apt 504 Naples FL 33940-3644

LONG, JOHN KELLEY, nuclear reactor physicist, consultant; b. New Rochelle, N.Y., Dec. 12, 1921; s. John K.H. and Alva Rae (Taylor) L.; m. Maye Louise Hampton, Dec. 30, 1948 (dec. June 1990); children: Iona, John, Brady. BSChemE, Columbia U., 1942; PhD in Physics, Ohio State U., 1953. Chemist Hercules Powder Co., Wilmington, Del., 1942-45; materials engr. Wright-Patterson AFB, Dayton, Ohio, 1946-49; heat transfer technician Battelle Meml. Inst., Columbus, Ohio, 1952-55; reactor physicist Argonne Nat. Lab., Idaho Falls, Idaho, 1955-73, Nuclear Regulatory Commn., Bethesda, Md., 1973-83; sr. cons. Halliburton NUS, Gaithersburg, Md., 1990—. With U.S. Army, 1945-46. Home: 227 S 35th W Idaho Falls ID 83402

LONG, KEVIN JAY, medicolegal consultant; b. Chgo., May 19, 1961. Student, Chgo. Med. Sch., 1983-86; BS in Math./Stats., Loyola U., Chgo., 1985; postgrad., John Marshall Law Sch., 1988-90. Researcher Cons. in Neurology, Ltd., Skokie, Ill., 1981-84, Assn. for Women's Health Care, Ltd., Chgo., 1982-83; researcher dept. neurology U. Ill., Chgo., 1985-86; law clk. Steven K. Jambois, Chgo., 1989; med. paralegal Hilfman & Fogel, P.C., Chgo., 1989-92; internal medicolegal cons. Robert A. Clifford & Assocs., Chgo., 1992; medicolegal cons. Chgo., 1992—. Contbr. articles to Current Problems in Obstetrics and Gynecology, Archives of Neurology, Archives of Internal Medicine, Pediatrics, Clin. Electroencephalography, Am. Jour. Medicine, Houston Medicine, Hosp. Pharmacy, Pediatric Emergency Care, Quality Management In Health Care. Mem. Nat. Hon. Soc. Secondary Schs., Assn. Trial Lawyers Am., Am. Med. Student Assn., N.Y. Acad. Scis., Blue Key Nat. Hon. Frat., Beta Beta Beta Biol. Honor Soc., Alpha Epsilon Delta Premed. Honor Soc. Jewish. Home: Ste 3-South 1325 W North Shore Ave Chicago IL 60626-4763

LONG, MICHAEL WILLIAM, cell/molecular biologist, educator; b. Detroit, Sept. 4, 1946; s. William Henry and Louise (Boren) L.; m. Sarah Ellen Clune, Jan. 19, 1983; children: Timothy, Christopher. BSc, Wayne State U., 1968, MSc, 1975, PhD, 1979. Instr. dept. anesthesia Wayne State U., Detroit, 1974-79; rsch. fellow Sloan Kettering Inst., N.Y.C., 1979-82; fellow Leukemia Soc. Am., N.Y.C., 1979-81; spl. fellow Leukemia Soc. Am., N.Y.C., Ann Arbor, Mich., 1981-83; asst. prof. dept. pediatrics U. Mich., Ann Arbor, 1983-88, assoc. prof. dept. pediatrics, 1988—; referee over 10 sci. jours.; ad hoc reviewer NIH, Bethesda, Md., 1983—; guest speaker various sci. socs., 1979—. Editor: (textbook) The Hematopoietic Microenvironment, 1992; assoc. editor Exptl. Hematology, 1991—; contbr. over 50 rsch. articles to various sci. jours. Scholar Leukemia Soc. Am., 1984-89, Stohlman Meml. scholar, 1987-88; recipient Established Investigator award Am. Heart Assn., 1991-96; grantee in field. Mem. AAAS, Internat. Soc. for Exptl. Hematology, Am. Soc. Hematology, Am. Fedn. for Clin. Rsch. Achievements include research in control of transplantable blood precursor cells; the development of bone-forming cells. Office: Univ Mich Rm 7510 MSRB-1 Box 0684 1150 W Medical Center Dr Ann Arbor MI 48109

LONG, SHARON RUGEL, molecular biologist, plant biology educator; b. San Marcos, Tex., Mar. 2, 1951; d. Harold Eugene and Florence Jean (Rugel) Long; m. Harold James McGee, July 7, 1979; 2 children. BS, Calif. Inst. Tech., 1973; PhD, Yale U., 1979. Rsch. fellow Harvard U., Cambridge, Mass., 1979-81; asst. prof. molecular biology Stanford U., Palo Alto, Calif., 1982-87, assoc. prof., 1987-92, prof., 1992—. Assoc. editor (jour.) Plant Physiology, 1992—; mem. editorial bd. Jour. Bacteriology, Molecular Plant-Microbe Interactions . Recipient postdoctoral award NSF, 1979, NIH, 1980, Shell Rsch. Found. award 1985, Presdl. Young Investigator award NSF, 1984-89, Faculty awards for women, 1991; rsch. grantee NIH, Dept. Energy, NSF; MacArthur fellow, 1992-97. Fellow AAAS; mem. NAS, Genetics Soc. Am., Am. Soc. Plant Physiology (Charles Albert Shull award 1989), Am. Soc. Microbiology, Soc. Devel. Biology, Am. Soc. for Microbiologists. Office: Stanford U Dept Biol Scis Stanford CA 94305-5020

LONG, TIMOTHY SCOTT, chemist, consultant; b. Racine, Wis., Dec. 20, 1937; s. Leslie Alexander and Esther (Sand) L.; m. Karen M. Koniarski, July 13, 1985; children by previous marriage: Corinne, Christine. BS in Chemistry, Winona State U., 1975. Staff chemist IBM, Rochester, Minn., 1962-77; adv. chemist IBM, Harrison, N.Y., 1977-80, IBM Instruments, Inc., Danbury, Conn., 1980-81; mgr. Midwest Instrument Ctr. IBM Instruments, Inc., Chgo., 1981-85; mgr. corp. environ. engring. IBM, Stamford, Conn., 1985-89; industry cons. IBM, White Plains, N.Y., 1989-92; environ. cons. Geraghty & Miller, Inc., Rochelle Park, N.J., 1992—; mem. World Environ. Ctr., N.Y.C., 1985-89; adv. bd. Coop. Ctr. Rsch. in Hazardous and Toxic Substances, Newark, 1985-89. Autohr: Testing for Prediction of Material Performance, 1972, Methods for Emissions Spectrochemical Analysis, 1977, 2d edit., 1982; contbr. articles to Applied Spectroscopy, Plating, Polymer Engring. and Sci. Mem. ASTM (com. emission spectroscopy), Soc. Applied Spectroscopy (chmn. Minn. sect. 1976-77), Assn. Am. Indian Affairs, Soc. Plastics Engrs. (bd. reviewers 1975-76). Achievements include demonstration of world's first application using ion chromatography in the analysis of indsl. waste water. Home: 362F Heritage Hills Dr Somers NY 10589-1781 Office: Geraghty & Miller Inc 201 W Passaic St Rochelle Park NJ 07662

LONG, WILLIAM MCMURRAY, physiology educator; b. Greenville, S.C., Nov. 9, 1948; s. William McMurray and Cecile Mae (Ariail) L.; m. Kathleen Webb, Mar. 18, 1971 (dec. Oct. 1990); m. Marianne Castrén, July 22, 1992. BA, Tulane U., 1970, BS, 1974; PhD, La. State U., 1980. Rsch. assoc. Med. Ctr. La. State U., New Orleans, 1974-75; pathology extern Charity Hosp. of La., New Orleans, 1975-80; Nat. Rsch. Svc. Award fellow Pa. State Med. Ctr., Hershey, 1980-82; rsch. assoc. Mt. Sinai Med. Ctr., Miami Beach, Fla., 1983-89; rsch. physiologist VA Med. Ctr., Miami, Fla., 1982-89; asst. prof. medicine U. Miami, 1982-89; asst. prof. physiology U. N.D., Grand Forks, 1989—; cons. VA Med. Ctr., Miami, 1991; ad hoc reviewer Am. Jour. Physiology, Bethesda, Md., 1990-91, VA Ctrl. Office, 1987-90; dir. Minority Access to Rsch. Careers, U. N.D., 1992—. Author: Non-Steriodal Agents in Sepsis Syndrom, 1989, (with others) Airways: Asthma, Bronchietasis and Emphysema, 1992; contbr. articles to profl. jours. Chmn. Nat. Letter-In Com., New Orleans, 1968, Cliff Solar Fund, New Orleans, 1973; cood. Spring Jazz Festival, New Orleans, 1970. Recipient Rsch. award Bush Found., 1990, Nat. Rsch. Svc. award NIH, 1980-82; grantee NIH, 1986-89, Fla. Lung Assn., 1984-85, VA, 1986-90, Am. Heart Assn. Dakota affiliate, 1991-93, Nat. Inst. Gen. Med. Scis., 1992—. Mem. Nat. Inst. Gen. Med. Scis., Am. Physiol. Soc., Am. Thoracic Soc., N.Y. Acad. Sci. Da Vinci Soc. (sec. 1987-88). Achievements include research in modification of cardiac proteolysis with amino acid methyl esters, in inefficacy of steroids in treatment of septic shock syndrome, in differentiation of histamine effects on bronchial flow and bronchomotor tone, on protein profiles in differentiating mechanisms of pulmonary edema, in role of bronchial blood flow in allergic airway disease and pharmacologic modification of that response; establishment of research program for minorities. Home: 1023 Reeves Dr Grand Forks ND 58201-5646 Office: U ND Sch Medicine Grand Forks ND 58202

LONGFELLOW, LAYNE, psychologist. D in Exptl. Psychology magna cum laude, U. Mich., 1967. Acad. v.p. Prescott (Ariz.) Coll., 1972-74; dir. exec. seminars Menninger Found., 1975-78; sr. assoc. Health Edn. Inst. Phoenix (Ariz.) Bapt. Hosp., 1980-81; dir. wilderness seminars Banff Ctr., Alberta, Canada, 1978-85; founder Lecture Theatre, Inc., 1978, Earthtrust, Inc., 1989. Author: (videos, audios) Beyond Success, Healthy, Wealthy, and Wise, Ethics and the Work Ethic, Earth is Alive. Fellow NSF, Woodrow Wilson Found., NIH. Mem. Phi Beta Kappa. Office: Lecture Theatre Inc 1134 Haining St Prescott AZ 86301

LONGLEY, GLENN, biology educator, research director; b. Del Rio, Tex., June 2, 1942; s. Glenn L. and Cleo M. (Tipton) L.; m. Francis Van Winkle, Aug. 5, 1961; children: Kelly Francis, Kristy Lee, Katherine Camille, Glenn C. BS in Biology, S.W. Tex. State U., 1964; MS in Zoology and Entomology, U. Utah, 1966, PhD in Environ. Biology, 1969. Prof. aquatic biology, dir. Edwards Aquifer Rsch. Ctr. S.W. Tex. State U., San Marcos, 1979—. Contbr. 138 papers to profl. jours. or meetings. Recipient more than 50 grants or contracts for water related studies; named Eminent Hydrologist, Am. Inst. Hydrology and Am. Water Resource Assn., 1993. Fellow Tex. Acad. Sci. (past pres.); mem. Am. Water Resource Assn., Tex. Orgn. for Endangered Species (past pres.), Tex. Water Conservation Assn., Tex. Water Pollution Control Assn. (past pres.), Assn. Groundwater Scientists and Engrs., Water Environment Fedn., Sigma Xi (life), Phi Sigma (former chpt. pres.). Achievements include research on redescription and assignment to the new Lircelous of the Texas troglobitic water slater, Asellus smithii, the larva of a new subterranean water beetle, Haideoporus texanus, watchlist of endangered, threatened and peripheral vertebrates of Texas, the generic status and distribution of Monodella texana Maguire, the only known North American Thermosbaeneacean, the Edwards Aquifer, Hadocerus taylori, a new genus and species of phreatic Hydrobiidae from South-Central Texas, Phreatoceras, a new name for Hadoceras Hershler and Longley, Phreatodrobia coronae, a new species of cavesnail from southwestern Texas, reproductive patterns of the subterranean shrimp Palaemonetes antrorum Benedict from Cental Texa, population size, distribution, and life history of Eurycea nana in the San Marcos River. Home: 814 Palomino Ln San Marcos TX 78666 Office: SW Tex State U Edwards Aquifer Rsch Ctr San Marcos TX 78666

LONGMUIR, ALAN GORDON, manufacturing executive; b. Vancouver, B.C., Can., Mar. 1, 1941; came to U.S., 1968; s. Gordon and Marjorie (Hobbs) L.; m. Barbara Ann Brown, Feb. 3, 1962 (div. Mar. 1987); children: Paul James, Kyle Michael; m. Linda Rae Peterson, Feb. 22, 1989. BS, U. B.C., 1964, PhD, 1968. Profl. elec. engr., Calif. From staff engr. to sr. staff engr. Kaiser Aluminum, Oakland, Calif., 1968-77; mgr. automated systems Kaiser Aluminum, Oakland, 1977-82, mgr. automated systems, elec. engr., 1982-85; dir. mfg. systems Kaiser Aluminum, Pleasanton, Calif., 1985-88; v.p. R&D Kaiser Aluminum, Pleasanton, 1988—; assoc. editor Automatica, Internat. Fedn. Automatic Control, 1975-82; rep. Indsl. Rsch. Inst., Aluminum Assn. tech. com. Achievements include development of indsl. computer control systems. Home: 5747 San Carlos Way Pleasanton CA 94566 Office: Kaiser Aluminum 6177 Sunol Blvd Pleasanton CA 94566

LONGO, JOSEPH THOMAS, electronics executive; b. Ferndale, Mich., Jan. 13, 1942; s. Thomas Alfred and Marie Ann (Wirmel) L.; m. Marjorie Louise Bower; children: Marjorie Louise, Joseph Thomas. BS in Physics, U. Detroit, 1964; MS in Physics, Mich. State U., 1966, PhD in Physics, 1968; exec. program, Stanford U., Palo Alto, 1975. Engr. U.S. Army Tank Automotive Ctr., Warren, Mich., summers 1963, 64; mem. tech. staff Rockwell Internat. Sci. Ctr., Thousand Oaks, Calif., 1968-72, mgr. narrow gap semiconductor devices, 1972-77, asst. dir. solid state electronics, 1977-78, dir. electro-optics, 1978-82, assoc. ctr. dir., 1982-85, v.p., gen. mgr., 1985—, v.p. rsch., 1992—. Contbr. over 30 articles to publs. to profl. jours.; patentee in field. Bd. dirs. Conejo Symphony Orch., Thousand Oaks, Calif., 1987—. Named Engr. of Yr., Rockwell Internat., Los Angeles, 1977; recipient Silver Knight award Nat. Mgmt. Assn., 1988, Gold Knight of Mgmt. award, 1990. Mem. IEEE, So. Calif. Tech. Execs., Infrared Info. Symposium (exec. com. 1986—, Levinstein award 1987), Am. Phys. Soc. Office: Rockwell Internat Sci Ctr 1049 Camino Dos Rios Thousand Oaks CA 91360-2398

LONGO, LAWRENCE DANIEL, physiologist, gynecologist; b. Los Angeles, Oct. 11, 1926; s. Frank Albert and Florine Azelia (Hall) L.; m. Betty Jeanne Mundall, Sept. 9, 1948; children: April Celeste, Lawrence Anthony, Elizabeth Lynn, Camilla Giselle. B.A., Pacific Union Coll., 1949; M.D., Coll. Med. Evangelists, Loma Linda, Calif., 1954. Diplomate: Am.

Bd. Ob-Gyn. Intern Los Angeles County Gen. Hosp., 1954-55, resident, 1955-58; asst. prof. ob-gyn UCLA, 1962-64; asst. prof. physiology and ob-gyn U. Pa., 1964-68; prof. physiology and ob-gyn Loma Linda U., 1968—; head div. perinatal biology Centre Loma Linda U. Sch. Medicine, 1974—; mem. perinatal biology com. Nat. Inst. Child Health, NIH, 1973-77; chmn. reprodn. scientist devel. program NIH; NATO prof. Consiglio Nat. delle Richerche, Italian Govt. Editor: Respiratory Gas Exchange and Blood Flow in the Placenta, 1972, Fetal and Newborn Cardiovascular Physiology, 1978, Charles White and A Treatise on the Management of Pregnant and Lying-in Women, 1987; co-editor: Landmarks in Perinatology, 1976-75; editor classic pages in ob-gyn. Am. Jour. Ob-Gyn., 1970-80; contbr. articles to profl. jours. Served with AUS, 1945-47. Recipient Research Career Devel. award NIH, 1967; NIH grantee, 1966—. Mem. Am. Assn. History Medicine (coun.), Am. Coll. Obstetricians and Gynecologists, Am. Osler Soc. (bd. govs., sec.-treas., Am. Physiol. Soc., Assn. Profs. Ob-Gyn., Perinatal Rsch. Soc., Soc. Gynecologic Investigation (past pres.), Neurosci. Soc. Adventist. Office: Loma Linda U Sch Medicine Divsn Perinatal Biology Loma Linda CA 92350

LONGUET, GREGORY ARTHUR, automation engineer, consultant; b. Pensacola, Fla., Nov. 1, 1945; s. Harry Charles and Gretchen (Gregory) L.; m. Elaine Gail Shuler, July 11, 1970; children: Ondreja N., Courtney E. BS, Ga. State U., 1974. Cert. mfg. specialist. Toolmaker GM, Doraville, Ga., 1970-74; mfg. engr. Gen. Dynamics Corp., Ft. Worth, 1974-79; automation engr., cons. IBM, Lexington, Ky., 1979-91, design team cons., rep., 1985-91; mfg. consulting engr. Ala., Miss. trading area IBM, Montgomery, Ala., 1991—. Elder Presbyn. Ch., Lexington, 1980—; mem. Habitat for Humanity, Girl Scouts U.S., South Ctrl. Ala. Capt. U.S. Army, 1966-70, Vietnam. Mem. Am. Prodn. and Inventory Control Soc., Soc. Mfg. Engrs., Nat. Mgrs. Assn., Masons, Beta Phi Gamma. Avocations: sports cars, aircraft, astronomy. Office: IBM 4525 Executive Park Dr Montgomery AL 36116-1600

LONKS, JOHN RICHARD, physician; b. Teaneck, N.J., Nov. 23, 1960; s. Harold and Sylvia Marie (Weisberg) L.; m. Maryann Katherine Vinton, Nov. 19, 1988; 1 child, Sara Anne. BS magna cum laude, U. Lowell, 1982; MD, U. Med. and Dentistry N.J., 1986. Diplomate Nat. Bd. Med. Examiners, Am. Bd. Internal Medicine; lic. physician, R.I. Intern Miriam Hosp./Brown U., Providence, 1986-87, med. resident, 1987-89; chief med. resident Providence VA Med. Ctr./Brown U., 1989-90; infectious diseases fellow Brown U./R.W.M.C., 1990-92; asst. instr. medicine Brown U., 1989-92, clin. instr. medicine, infectious diseases rsch. fellow, 1992—. Mem. AMA, ACP, Infectious Diseases Soc. Am., Am. Soc. for Microbiology, N.Y. Acad. Sci., Sigma Xi, Alpha Omega Alpha. Achievements include research in prevalence of erthromycin resistance among s. pneumoniae: design/development of absorbable internal fracture fixation devices; research in role of lipid unsaturation and peroxidation in cell aging; effect of malarial serum on graft versus host reaction. Office: Miriam Hosp 164 Summit Ave Providence RI 02906

LOOK, DWIGHT CHESTER, JR., mechanical engineering educator, researcher; b. Smith Center, Kans., Aug. 25, 1938; s. Dwight Chester and Margery Rae (Bash) L.; m. Patricia Ann Wellbaum, June 4, 1960; children: Dwight C. III, Douglas C. AB, Cntrl. Coll., Fayette, Mo., 1960; MS, U. Nebr., 1962; PhD, U. Okla., 1969. Teaching asst. U. Nebr., Lincoln, 1960-63; aerosystems engr. Gen. Dynamics, Ft. Worth, 1963-67; asst. prof. U. Mo., Rolla, 1969-73, assoc. prof., 1973-78, prof., 1978—. Co-author: Thermodynamics, 1982, Engineering Thermodynamics, 1986; contbr. over 75 papers to profl. jours. Recipient Ralph Teeter award Soc. Automotive Engring., 1978; rsch. grantee NSF, 1970—. Fellow AIAA (assoc.); mem. ASME, Am. Soc. Engring. Edn., Internat. Soc. Optical Engring. Office: U Mo 203 Mechanical Engring Bldg Rolla MO 65401

LOOMIS, EARL ALFRED, JR., psychiatrist; b. Mpls., May 21, 1921; s. Earl Alfred and Amy Louise (Shore) L.; m. Victoria Malkerson, June 2, 1944 (div.); children: Rebecca Marie Keith, Kathleen Victoria Rioja, Jennifer Lee; m. Lucile Meyer, July 1, 1962 (dec. 1967); 1 child, Amy Windeler; m. Anita Muriel Peabody, Mar. 22, 1969. MD, U. Minn., 1945. Diplomate Am. Bd. Psychiatry and Neurology. Intern in internal medicine, pediatrics Univ. Hosp., Boston, 1945-46; resident Western Psychiat. Hosp., Pitts., 1946-48, Hosp. U. Pa., Phila., 1948-50; assoc. prof. child psychiatry U. Pitts. Sch. Medicine, 1952-56; prof. psychiatry and religion Union Theol. Sem., N.Y.C., 1956-63; chief div. child psychiatry St. Luke's Hosp., N.Y.C., 1956-62; rsch. fellow U. Geneva (Switzerland) Inst. Jean-jacques Rousseau, 1962-63; med. dir. Blueberry Treatment Ctr./Severly Emotionally Ill Children, Bklyn., 1963-81; prof. psychiatry Med. Coll. Ga., Augusta, 1980-90; pvt. practice, cons. Vet. Hosp., Charter Hosp. of Augusta, 1985—; cons. Gracewood Sch. and Hosp., Augusta, 1983-89, Eisenhower Army Med. Ctr., Augusta, 1983—. Author: The Self in Pilgrimage, 1960; contbr. articles to profl. jours. Lt. (j.g.) USNR, 1950-52. Rsch. grant NIMH, 1956-63, travel grant, 1962-63, U.S. Info. Svcs., 1963. Fellow Am. Psychiat. Assn. (chair psychiatry and religion 1955-60); Group for the Advancement of Psychiatry (chair psychiatry and religion 1959-62), Am. Psychoanalytic Assn., Psychoanalytic Study Group of S.C. (founder, pres. 1981-88). Achievements include development of techniques for studying ego functions in psychotic, retarded and normal children via play pattern observations. Home and Office: 1002 Katherine St Ste 6 Augusta GA 30904-6106

LOOMIS, RONALD EARL, biophysicist; b. Elmira, N.Y., June 22, 1954; s. Donald Earl and Liselotta Hedwig (Hammer) L.; m. Pamela Marie Wingert, Dec. 28, 1974; children: Donald Arthur, Arthur Scott, Theodore Thom. BA in Chemistry and Biology, SUNY, Brockport, 1976; PhD in Biophysics, Roswell Park Cancer Inst., 1983. Postdoctoral rsch. fellow U. Oreg., Eugene, 1983-84; rsch. asst. prof. U. Buffalo, 1984-85, asst. prof. health scis., 1985-92, clin. assoc. prof. health scis., 1992—. Contbr. articles to Biopolymers, Biophys. Jour., Jour. Adhesive Sci. Tech., others. Grantee NIH, 1986—, Health Inst. Devices Inst., 1986-90. Mem. Am. Chem. Soc., Am. Phys. Soc., Soc. for Biomaterials, N.Y. Acad. Sci., Am. Assn. Dental Rsch., Internat. Soc. Magnetic Resonance, Biophys. Soc., Internat. Union Pure and Applied Chemistry (afffiliate), Trout Unltd., Fedn. of Fly Fishers, Sigma Xi. Presbyterian. Home: 53 Puritan Pl Orchard Park NY 14127-4615 Office: SUNY Buffalo 321-322 Foster Hall Buffalo NY 14214

LOONEY, NORMAN EARL, pomologist, plant physiologist; b. Adrian, Oreg., May 31, 1938; came to Can., 1966; s. Gaynor Parks and Lois Delilah (Francis) L.; m. Arlene Mae Willis, Oct. 4, 1957 (div. 1982); children: Pamela June, Patricia Lorene, Steven Paul; m. Norah Christine Keating, July 16, 1983. BSc in Agr. Edn., Washington State U., 1960, PhD in Horticulture, 1966. Rsch. scientist Agr. Can., Summerland, B.C., 1966-71, scientist, sect. head, 1972-81, sr. scientist, 1982-87, sr. scientist, sect. head, 1987-90, prin. scientist, sect. head, 1991—; vis. scientist food rsch. divsn. CSIRO, Sydney, Australia, 1971-72, East Malling Rsch. Sta., Maidstone, Kent, 1981-82, Dept. Horticulture U. Lincoln, Christchurch, New Zealand, 1990-91; mem. B.C. Plant Sci. Lead Com. and Can. Agr. Sci. Coordinating Com., Vancouver, 1968-90; sec. Expert Com. on Horticulture, Ottawa, Ont., Can., 1986-90; chmn. working group on growth regulators in fruit prodn. Internat. Soc. for Hort. Sci., Wageningen, The Netherlands, 1987—. Contbr. numerous articles and chpts. to sci. publs. Fellow Am. Soc. Hort. Sci. Achievements include patent for Promotion of Flowering in Fruit Trees (U.S. and foreign). Office: Agr Can Rsch Sta, Summerland, BC Canada V0H 1Z0

LOONEY, THOMAS ALBERT, psychologist, educator; b. Dallas, Feb. 8, 1947; s. Billy Albert and Helen Dorothy (Holland) L.; m. Margaret Ann Thomas, Aug., 1968 (div. 1982). BA, Tex. Tech. Coll., 1969; MS, Fla. State U., 1971, PhD, 1973. Instr. Northeastern U., Boston, 1973-75; asst. prof. Lynchburg (Va.) Coll., 1975-79, assoc. prof., 1979-86, prof., 1986—, chair dept. psychology, 1987—; referee NSF. Contbr. articles to jour. Comparative and Physiol. Psychology, Exptl. Analysis of Behavior, Bull. Psychonomic Soc., Worm Runner's Digest, Physiology and Behavior, Animal Learning and Behavior, Neurosci. and Biobehavioral Revs.; contbr. chpt. to book. Grantee HEW, HIMH, Gwathmey Trust; recipient Disting. Scholar award, 1986, Excellence in Teaching award, 1992. Mem. Am. Psychol. Soc., Am. Psychol. Assn., Psychonomic Soc., Eastern Psychol. Assn., Southwestern Psychol. Assn., AAAS, Gold Key, Phi Kappa Phi, Psi Chi, Sigma Xi. Home: 1205 Westridge Circle Lynchburg VA 24502 Office: Lynchburg Coll Dept Psychol 1501 Lakeside Dr Lynchburg VA 24501

LOPATIN, GEORGE, research chemist; b. N.Y.C., Feb. 13, 1929; s. Charles and Sylvia (Drell) L.; m. Carolyn Shaw, Aug. 17, 1952 (dec. Sept. 1990); m. Helene Berkowitz, May 3, 1992; children: Alan, Craig, Laurel, Andrew. BS, Brooklyn Coll., 1950; PhD, Bklyn. Poly. U., 1958; postgrad., U. Liverpool, Eng., 1960. Control chemist Nepera Chem. Co., Harriman, N.Y., 1950-54; rsch. chemist Shell Devel. Co., Emeryville, Calif., 1961-72; head dept. Israel Fiber Inst., Jerusalem, 1972-76; asst. dir. Albany (N.Y.) Internat., 1976-82; rsch. fellow Millipore Corp., Bedford, Mass., 1982—. Mem. ACS, N.Y. Acad. Scis. Jewish. Achievements include patents in fibers, membranes and plastic materials. Home: 494 Ward St Newton Centre MA 02159 Office: 80 Ashby Rd Bedford MA 01730

LOPER, CARL RICHARD, JR., metallurgical engineer, educator; b. Wauwatosa, Wis., July 3, 1932; s. Carl Richard Sr. and Valberg (Sundby) L.; m. Jane Louise Loehning, June 30, 1956; children: Cynthia Louise, Anne Elizabeth. BS in Metall. Engring., U. Wis., 1955, MS in Metall. Engring., 1958, PhD in Metall. Engring., 1961; postgrad., U. Mich., 1960. Metall. engr. Pelton Steel Casting Co., Milw., 1955-56; instr., rsch. assoc. U. Wis., Madison, 1956-61, asst. prof., 1961-64, assoc. prof., 1964-68, prof. metall. engring., 1968-88, prof. materials sci. and engring., 1988—; asso. chmn. dept. metall. and mineral engring. U. Wis., 1979—; rsch. metallurgist Allis Chalmers, Milw., 1961; cons., lectr. in field. Author: Principles of Metal Casting, 1965; contbr. over 250 articles to profl. jours. Chmn. 25th Anniversary of Ductile Iron Symposium, Montreal, 1973; sec. Yodrasil Literary Soc., 1986-87. Foundry Ednl. Found. fellow, 1953-55; Wheelabrator Corp. fellow, 1960; Ford Found. fellow, 1960; recipient Adams Meml. award Am. Welding Soc., 1963, Howard F. Taylor award, 1967, Service citation, 1969, 72, others; recipeint Silver medal award of Sci. Merit Portuguese Foundry Assn., 1978, medal Chinese Foundrymen's Assn., 1989. Fellow Am. Soc. Metals (chmn. 1969-70), AIM; mem. Am. Foundrymen's Soc. (dir. 1967-70, 76-79, best paper award 1966, 67, 85, John A. Penton gold medal 1972, Hoyt Meml. lectr. 1992), Am. Welding Soc., Foundry Ednl. Found., Torske Klubben (bd. dirs., co-founder 1978—), Blackhawk Country Club, Sigma Xi, Gamma Alpha, Alpha Sigma Mu, Tau Beta Pi. Lutheran. Achievements include significant contributions to understanding the solidification and metallurgy of ferrous and non-ferrous alloys; recognized authority on solidificaton and cast iron metallurgy, and on education in metallurgy and materials science. Office: U Wis Cast Metals Lab 1509 University Ave Madison WI 53706-1595

LOPER, WARREN EDWARD, computer scientist; b. Dallas, Aug. 2, 1929; s. Leon Edward and Belva (Fannin) L.; BS in Physics, U. Tex. at Austin, 1953, BA in Math. with honors, 1953; m. Ruth M. Wetzler, June 17, 1967; 1 child, Mary Katherine. Commd. ensign U.S. Navy, 1953, advanced through grades to lt., 1957; physicist U.S. Naval Ordnance Test Sta., China Lake, Calif., 1956-61; operational programmer U.S. Navy Electronics Lab., San Diego, 1962-64; project leader, systems programming br., digital computer staff U.S. Fleet Missile Systems Analysis and Evaluation Group, Corona, 1964-65, sr. systems analyst digital computer staff U.S. Naval Ordnance Lab., Corona, 1965-69; head systems programming br. Naval Weapons Center, Corona Labs, 1969; computer specialist compiler and operating systems devel., Naval Electronics Lab. Ctr., San Diego, 1969-76; project leader langs., operating systems and graphics Naval Ocean Systems Ctr., San Diego, 1977-90, employee emeritus, 1990—. Navy rep. on tech. subgroup Dept. Def. High Order Lang. Working Group, 1975-80. Recipient Disting. Svc. award Dept. Def., 1983. Democrat. Roman Catholic. Home: 6542 Alcala Knolls Dr San Diego CA 92111-6947

LOPEZ, CAROLYN CATHERINE, physician; b. Chgo., Oct. 13, 1951; d. Joseph Compean and Angela (Silva) L. BS, Loyola U., Chgo., 1973; MD, U. Ill., 1978. Diplomate Am. Bd. Family Practice. Intern, resident Rush/Christ Hosp., Chgo., 1978-81; med. dir. Wholistic Health Ctr., Oak Lawn, Ill., 1981-82; clin. dir. Anchor HMO, Oak Brook, Ill., 1982-84, assoc. med. dir., 1984-87; med. dir. Chgo. Pk. Dist., 1987-91; med. dir. Rush Access HMO, Chgo., 1991—, v.p., 1992—; asst. dean Rush Med. Coll., 1990-93; med. dir. Rush Access HMO, Chgo., 1991-93; v.p. Rush Anchor HMO, 1992-93, v.p. for profl. affairs, 1993; sr. v.p. and chief med. officer Rush-Prudential Health Plans, 1993—. Primary Care Policy fellow USPHS, 1993. Mem. AMA, APHA, Am. Acad. Family Physicians, Chgo. Med. Soc., Ill. State Med. Soc., Am. Coll. Physicians Execs., Ill. Acad. Family Physicians (bd. dirs. 1987-89, speaker 1990-91, bd. chair 1990-91, pres.-elect 1991-92, pres. 1992-93, alt. del. to Am. Acad. Family Physicians 1992—), Am. Med. Women's Assn. Roman Catholic. Avocations: swimming, cooking. Office: Rush Prudential Health Plan 33 E Congress Pkwy Chicago IL 60605-1288

LOPEZ, JORGE WASHINGTON, veterinary virologist; b. Valdivia, Chile, Jan. 19, 1948; came to U.S., 1978; s. Washington Eduardo Lopez and Anisia Lorca; m. Cecile Therese Angulo, Aug. 29, 1992. DVM, U. Austral, Valdivia, 1976; MSc, U. Ill., 1983, PhD, 1985. Prodn. mgr. Meat Packing Plant, Concepcion, Chile, 1977-78; rsch. assoc. vet. pathobiology dept. U. Ill., Urbana, 1985-86; postdoctoral assoc. Utah State U., Logan, 1986-88; asst. dir. virology, asst. prof. Diagnostic Lab., Vet. Coll., Cornell U., Ithaca, N.Y., 1988—. Mem. AAAS, Am. Vet. Med. Assn., Soc. Tropical Vet. Medicine, N.Y. Acad. Sci., N.Y. Vet. Med. Assn., Phi Zeta. Achievements include development of several diagnostic tests and techniques for detection of viruses and bacterias in livestock. Office: Diagnostic Lab NY Coll Vet Medicine Ithaca NY 14851

LOPEZ, JOSEPH JACK, oil company executive, consultant; b. N.Y.C., July 26, 1932; s. Florentino Estrada and Leah (Bodner) L.; m. June Elliott, June 20, 1953; children: Karen Marie Lopez Lynch, Debra Jo Lopez Newton, Laura Jean Lopez Berrell. Student, CCNY, 1955-59. Project estimator Chem. Constrn.-Engrs., N.Y.C., 1960-64, Dorr Oliver-Engrs., Stamford, Conn., 1964-66; chief estimator R.M. Parsons-Engrs., Frankfurt, Germany, 1966-74; mgr. project svcs. A.G. McKee-Engrs., Berkley Heights, N.J., 1974-76; mgr. tech. svcs. Rsch. Cottrell Corp., Sommerville, N.J., 1976-78; cons. Booz Allen & Hamilton, Abu Dhabi, United Arab Emirates, 1978-84; v.p. XL Tech. Corp., N.Y.C., 1984-87; cons. Qatar Gen. Pete Corp., Doha, 1987-90; pres. J. Lopez Cons., Babylon, N.Y., 1990—; estimator Combustion Engring. Co., N.Y.C., 1955-60. With USAF, 1950-54. Mem. Am. Assn. Cost Engrs., Project Mgmt. Inst. Republican. Roman Catholic. Home and Office: 15 Hinton Ave. Babylon NY 11702

LOPEZ, RAFAEL, pediatrician; b. Dominican Republic, Dec. 15, 1929; came to U.S., 1948; s. Ygnacio and Idalia (Santelices) L.; m. Noris Dijkhoff, June 7, 1956; children: Rafael Jr., Idalia Lopez de Diaz. MD, U. P.R., San Juan, 1956; BS, Seton Hall U., 1952. Diplomate Am. Bd. Pediatrics. Gen. rotating internship, 1956-57; residency pediatrics St. Joseph Mercy, Pontiac, Mich., 1959; assoc. rsch. U. Mich., 1959; pvt. practice, 1960-65, N.Y.C., 1960-65; from asst. to assoc. prof. N.Y. Med. Coll., N.Y.C., 1965—; assoc. dir. Our Lady of Mercy Med. Ctr., N.Y.C., 1980-85, dir., 1985—; mem. pediatric hematology-oncology bd. Am. Bd. Pediatrics, 1974—. Author: Handbook of Vitamin, 1991. Fellow AMA, Am. Acad. Pediatrics, N.Y. Acad. Sci., Am. Soc. Hematology; mem. Am. Coll. Med. Quality Assurance. Home: 140 Cabrini Blvd New York NY 10033 Office: Our Lady of Mercy Med Ctr 600 E 233d St Bronx NY 10466

LÓPEZ-CANDALES, ANGEL, cardiologist, researcher; b. Habana, Cuba, Sept. 3, 1960; came to U.S., 1963; s. Angel Lopez and Maria Dolores Candales; m. Hilda L. Lissette Lopez, May 10, 1986; children: Ashley Marie, Gabriel Alexander. BS magna cum laude, U. P.R., 1982, MD, 1986. Diplomate Am. Bd. Internal Medicine. Intern Vets. Hosp., San Juan, P.R., 1986-87, resident, 1987-89; rsch. fellow Jewish Hosp., St. Louis, 1989-91, clin. fellow, 1991-92; instr. medicine, rschr. Jewish Hosp. at Wash. U., St. Louis, 1992—. Cardiology fellowship award Hewlett-Packard, 1993-94. Fellow Am. Coll. Cardiology; mem. AAAS, Am. Fedn. Clin. Rsch. (Henry Christian award 1992), N.Y. Acad. Scis., Alpha Omega Alpha. Roman Catholic. Achievements include development of a drug compound to inhibit cholesterol absorption; first to demonstrate a dietary induction of pancreatic enzyme to stimulate cholesterol absorption; research to modify the intestinal absorption of cholesterol in the animal model; presently working on the molecular and acoustic characterization of vascular restenosis after angioplasty. Home: 578 Oaktree Crossing Ballwin MO 63021

LOPEZ DE MANTARAS, RAMON, computer scientist, researcher; b. Sant Vicenc de Castellet, Barcelona, Spain, May 8, 1952. M in Elec. Engring., U. Toulouse, France, 1974, PhD in Applied Physics, 1977; M in Computer Sci., U. Calif., Berkeley, 1979; PhD in Computer Sci., U. Polytech., Barcelona, 1981. Rsch. asst. Electronics Rsch. Lab. U. Calif., Berkeley, 1978-79; computer scientist Ikerlan, Mondragon, Spain, 1979-80; asst. prof. Polytech. U., Barcelona, 1980-85; rsch. scientist Spanish Nat. Rsch. Coun., Blanes, 1986-89, rsch. prof., 1989—, head, artificial intelligence, 1986—; vis. prof. U. Paris VI, 1986. Author: Approximate Reasoning, 1990; contbr. articles to profl. jours. Recipient award European Artificial Intelligence Rsch. Paper, 1987, City of Barcelona Rsch. prize, 1982, Bages Culture prize Omnium Cultural of Bages, 1987. Mem. Am. Assn. for Artificial Intelligence, Spanish Assn. for Artificial Intelligence (v.p. 1987-92), Computer Profls. for Social Responsibility. Avocation: jazz piano. Home: Auladell 14, 08190 Sant Cugat Spain Office: Centre of Advanced Studies, Cami de Santa Barbara, 17300 Blanes Spain

LÓPEZ DE VIÑASPRE URQUIOLA, TEODORO, engineering company executive; b. Bilbao, Vizcaya, Spain, Sept. 20, 1939; s. López de Viñaspre Ozaeta, Teodoro and Urquiola, Soledad; m. Maria Begona Maurolagoitia, May 9, 1975; 1 child, Alberto. BS, St. Apostolic Coll., Bilbao, 1956; MS in Indsl. Engring., Tech. Sch. Indsl. Engring., Bilbao, 1967; MBA, City of London Poly., 1979; PhD in Econs., Pacific Western U., 1987. Registered profl. engr., U.S. Project engr. Heinrich Koppers GmbH, Bilbao and Madrid, Spain, 1965-68; mng. dir. Hydraulic Krane, S.A., Alava, Spain, 1969-71; mgr. of bldgs. Butler Mfg. Co., Madrid, 1971-73; dir. of area Maquiobras-Maquipo, S.A., Bilbao, 1973-77; cons. auditor Caja Lab. Pop.-Mec. Alt. Precis., Bilbao and fgn., 1977-81; dir. of studies Ingenieros Consultores, S.A., Bilbao, 1982-90; gen. dir. CEI-Bus./Innovation Ctr., Vitoria, Spain, 1990—. Author: Electrical Oil Transformers in Industry, 1979, Diagnostic of Television Chair, 1984, Prospects to Electrical Firms, 1986; contbr. papers in field. Lt. Spanish Air Force, 1965. Mem. Am. Inst. Indsl. Engrs. (sr.), Am. Acctg. Assn., Nat. Assn. Bus. Econs., Brit. Bus. Grads. Soc., Sociedad Bilbaina Club, Fireproof Indsl. Assn. Avocations: walking, reading, swimming, traveling, music. Home: Avda de las Universidades 4, 48007 Bilbao Spain Office: CEI-Bus/Innovation Ctr, Vitoria Alava, Spain

LOPEZ-NAKAZONO, BENITO, chemical engineer; b. Nuevo Laredo, Tam., Mex., Oct. 26, 1946; came to U.S., 1973; s. Benito and Ayko (Nakazono) Lopez-Ramos; m. Anastacia Espinoza, June 22, 1981; children: Benito Keizo, Tanzy Keiko, Mayeli, Aiko Michelle. BSc in Chem. Engring., ITESM, Monterrey, Mexico, 1968; MS in Chem. Engring., U. Houston, 1991. Prof. chem. engring. ITESM, Monterrey, 1971-72; vessel analytical design engr./process engr. M.W. Kellogg, Houston, 1973-79; product mgr. Ind. Del Alcali, Monterrey, 1980-81; sr. process engr. Haldor Topsoe, Inc., Houston, 1982—. Pres. adminstrn. coun. United Meth. Ch., Houston, 1990-92. ITESM fellow, 1963, U. Houston fellow, 1968. Mem. Tex. Soc. Profl. Engrs., Am. Inst. Chem. Engrs., Sigma Xi. Achievements include development and design of hydrogen, ammonia, methanol, formalhyde and SNOx/WSA plants, supervision of ammonia plants in Mexico, Russia, Somalia and India. Home: 1805 Lanier Dr League City TX 77573 Office: Haldor Topsoe Inc 17629 El Camino Real #302 Houston TX 77058

LOPEZ-PORTILLO, JOSÉ RAMON, economist, international government representative; b. Mexico City, Feb. 2, 1954; s. Jose Lopez-Portillo Pacheco and Maria Del Carmen Romano Nolck; m. Maria Antonieta Garcia-Lopez Loreza, Mar. 21, 1980; three children. Degree in econs., Univ. Anahuac, 1976; postgrad., Oxford (Eng.) U., 1989—. Dir. gen. documentation and analysis Secretaria de Programmaciòn, Mexico City, 1976-80, vice minister of evaluation, 1980-81; gen. dir. info. and evaluation Revolutionary Instnl. Party, Mexico City, 1981-82; permanent rep. FAO, WFP, WFC, IFAD, Rome, 1983-88; rep. of Mexico program com. FAO, Rome, 1989—. Contbr. articles to profl. jours. Avocations: astronomy, painting, sports. Home: 61 Charlbury Rd, Oxford OX26UJ, England Office: FAO, Via Delle Termedi Caracalla, Rome Italy

LOPINTO, CHARLES ADAM, chemical engineer; b. N.Y.C., June 21, 1952; s. Anthony and Anna (Schiessmann) LoP.; m. Lidia Llamas, June 14, 1976; 1 child, Carla. BChE, Cooper Union, 1974; ME, Manhattan Coll., 1975. Process engr. Exxon Co., Florham Park, N.J., 1976-86; project mgr. Lotepro, Valhalla, N.Y., 1987-90, mktg. engr., 1990—. Editor Engring. Software Exchange Newsletter of CAE Cons., 1982-88. Mem. Am. Inst. Chem. Engrs., Tau beta Phi. Home: 41 Travers Ave Yonkers NY 10705

LOPINTO, ROBERT ANTHONY, environmental engineer; b. N.Y.C., Nov. 27, 1946; s. Anthony and Anna E. (Schliessman) LoPinto; m. Rosanne Vigna, May 9, 1970; children: Lisa-Ann, Diane, Anthony F. BSChemE, Poly. Inst. Bklyn., 1967; ME in Environ. Engring., Manhattan Coll., 1973. Registered profl. engr., N.Y. Commd. 2d lt. U.S. Army, 1967, advance through grades to lt. col., 1988; served in Vietnam, 1968-69; san. engr. North Atlantic Div., C.E., N.Y.C., 1973-76; co. comdr. Germany, 1976-78; insp. gen. U.S. Am. Armament Munitions and Chem. Command, 1983-86; prof. mil. sci. Hofstra Univ., N.Y., 1986-88; chief engr. Shapiro Engring., P.C., Bklyn., 1988—; chair Queens County Solid Waste Adv. Bd., N.Y.C., 1990—; mem. steering com. Citywide Recycling Adv. Bd., N.Y.C., 1990—. Chmn. environ. com. Community Bd., Queens, N.Y., 1991—. Decorated Bronze Stars. Mem. N.Y. State Soc. Profl. Engrs. (Queens County chpt.), Soc. Am. Mil. Engrs., Assn. U.S. Army, VFW. Roman Catholic. Home: 6-29 161 St Beechhurst NY 11357 Office: Shapiro Engring P C 6315 Mill Ln Brooklyn NY 11234

LOPKER, ANITA MAE, psychiatrist; b. San Diego, May 25, 1955; d. Louis Donald and Betty Jean (Sayman-Campbell) L. BA magna cum laude, U. Calif., San Diego, 1978; MD, U. Rochester, 1982. Diplomate Nat. Bd. Med. Examiners. Intern in internal medicine Yale U. Sch. Medicine-Greenwich Hosp., 1982-83; resident in psychiatry Yale U. Sch. of Medicine, 1983-86; postdoctoral fellow Yale U. Sch. Medicine, New Haven, Conn., 1982-86; clin. instr. Yale U. Sch. Medicine, New Haven, 1986-88; pvt. practice specializing in eating disorders Westport, Conn., 1987—; cons. psychiatrist Yale-New Haven Hosp Lyme Disease Study Clinic, 1987—, Yale U. Lyme Disease Rsch. Project, 1986—, Alcoholism and Drug Dependency Coun., Inc., 1989-90; internat. lectr. on Lyme psychiat. syndrome; nat. lectr. on eating disorders, substance abuse. Contbr. articles to profl. jours. Founding mem. Nat. Mus. for Women in the Arts, Washington, 1987; patron Menninger Found., Met. Opera. Mem. AAAS, N.Y. Acad. Scis., Am. Psychiat. Assn., Conn. Psychiat. Soc., World Fedn. Mental Health (life), Menninger Found., Alpha Omega Alpha, Phi Beta Kappa. Achievements include discovery of preventable neuropsychiatic disorders associated with Lyme disease. Home: 27 Strathmore Ln Westport CT 06880-4700 Office: 7 Whitney St Ext Westport CT 06880-3761

LOPPNOW, HARALD, biologist; b. Itzehoe, Fed. Republic Germany, Feb. 14, 1954; s. Bernhard and Helga (Fetscher) L. Diploma, Kiel U., Fed. Republic Germany, 1983; PhD, Kiel U., 1986. Lab. instr. Kiel U., 1981-83, rsch. assoc., 1983-88, 90—; rsch. assoc. Sclavo Rsch. Ctr., Siena, Mai, Italy, 1987, Tufts U., Boston, 1988-90. Reviewer circulation rsch. Am. Soc. Microbiology, Birmingham, Ala., 1990, Am. Soc. for Biochemistry and Molecular Biology; contbr. articles to profl. jours. and books. Cpl. German Army, 1974-76. Recipient travel award IV. Internat. grant 385/4-1 of Deutsche Forschungsgemeinschaft Conf. on Immunopharmacology, 1988. Mem. German Soc. Immunology, am. Assn. Immunologists, Internat. Endotoxin Soc., German Soc. for Cell Biology. Home: Rendsburger Landstr 87, Schl-Holstein, D-24113 Kiel Germany Office: Forschungsinstitut Borstel, Parkallee 22, D-23845 Borstel Germany

LOPREATO, JOSEPH, sociology educator, author; b. Stefanaconi, Italy, July 13, 1928; s. Frank and Marianna (Pavone) L.; m. Carolyn H. Prestopino, July 18, 1954; (div. 1971); children: Gregory F., Marisa S. Schmidt; m. Sally A. Cook, Aug. 24, 1972 (div. 1978). B.A. in Sociology, U. Conn., 1956; Ph.D. in Sociology, Yale U., 1960. Asst. prof. sociology U. Mass., Amherst, 1960-62; vis. lectr. U. Rome, 1962-64; assoc. prof. U. Conn., Storrs, 1964-66; prof. sociology U. Tex., Austin, 1968—; chmn. dept. sociology U. Tex., 1969-72; vis. prof. U. Catania, Italy, 1974, U. Calabria, Italy, 1980; mem. steering com. Council European Studies, Columbia U., 1977-80; chmn. sociology com. Council for Internat. Exchange of Scholars, 1977-79; mem. Internat. Com. Mezzogiorno, 1986—; Calabria Internat. Com., 1988—. Author: Italian Made Simple, 1959, Vilfredo Pareto, 1965, Peasants No More, 1967, Italian Americans, 1970, Class, Conflict and

Mobility, 1972, Social Stratification, 1974, The Sociology of Vilfredo Pareto, 1975, La Stratificazione Sociale negli Stati Uniti, 1945-1975, 1977, Human Nature and Biocultural Evolution, 1984, Evoluzione e Natura Umana, 1990, Mai Più Contadini, 1990; contbr. articles to profl. jours. Mem. Nat. Italian-Am. Com. for U.S.A. Bicentennial; mem. exec. com. Congress Italian Politics, 1977-80. Served to cpl. U.S. Army, 1952-54. Fulbright faculty research fellow, 1962-64, 73-74; Social Sci. Research Council faculty research fellow, 1963-64; NSF faculty research fellow, 1965-68; U. Tex. Austin research fellow, spring 1985, spring 1993; Guido Dorso award for U.S.A., Italy, 1992. Mem. AAAS (behavioral sci. rsch. prize com. 1992—), Internat. Sociol. Assn., Am. Sociol. Assn., European Sociobiolog. Soc., Evolution and Behavior Soc., So. Sociol. Soc. (assoc. editor Am. Sociol. Rev. 1970-72, Social Forces 1987-90, Jour. Polit. and Mil. Sociology 1980—), Internat. Soc. Human Ethology. Catholic-Episcopalian. Office: Univ of Tex Dept Sociology Austin TX 78712

LORANCE, ELMER DONALD, organic chemistry educator; b. Tupelo, Okla., Jan. 18, 1940; s. Elmer Dewey and Imogene (Triplett) L.; m. Phyllis Ilene Miller, Aug. 31, 1969; children: Edward Donald, Jonathan Andrew. BA, Okla. State U., 1962; MS, Kansas State U., 1967; PhD, U. Okla., 1977. NIH research trainee Okla. U., Norman, 1966-70; asst. prof. organic chemistry So. Calif. Coll., Costa Mesa, 1970-73, assoc. prof., 1973-80, prof., 1980—, chmn. div. natural scis. and math., 1985-89, chmn. chemistry dept., 1990-93, chmn. divsn. natural scis. and math., 1993—. Contbr. articles to profl. jours. Mem. AAAS, Am. Chem. Soc., Internat. Union Pure and Applied Chemistry (assoc.), Am. Inst. Chemists, Am. Sci. Affiliation, Phi Lambda Upsilon. Republican. Mem. Ch. Assembly of God. Avocations: reading, gardening, music. Office: So Calif Coll 55 Fair Dr Costa Mesa CA 92626-6597

LORBER, DANIEL LOUIS, endocrinologist, educator; b. N.Y.C., Sept. 21, 1946; s. Jerome Zachary Lorber and Ruth (Frank) Cook. AB, Columbia U., 1968; MD, Albert Einstein, 1972. Diplomate Am. Bd. Internal Medicine, Endocrinology and Metabolism. Intern medicine Bronx (N.Y.) Municipal Ctr., 1972-73; resident medicine Albert Enstein, Bronx, 1973-75; fellow in endocrinology Vanderbilt U., Nashville, 1975-77; from asst. prof. to asst. dean NYU Sch. Medicine, N.Y.C., 1977-84; from asst. clin. prof. to assoc. clin. prof. Albert Einstein Coll. Medicine, Bronx, 1984—; med. dir. Diabetes Control Found., Flushing, N.Y., 19865—, Diabetes Treatment Ctr., The N.Y. Hosp. Med. Ctr. of Queens. Editor in chief Practical Diabetology Magazine, 1987—. Fellow ACP; mem. Am. Diabetes Assn. (bd. dirs. N.Y. Downstate Affiliate 1982—), Endocrine Soc. Democrat. Jewish. Avocations: Skiing, tennis, sailing. Office: Diabetes Control Found 138-26 58th Ave Flushing NY 11355-5232

LORBER, MORTIMER, physiology educator, researcher; b. N.Y.C., Aug. 30, 1926; s. Albert and Frieda (Levin) L.; m. Eileen Segal, May 20, 1956; children: Kenneth, Stephanie. BS, NYU, 1945; DMD cum laude, Harvard U., 1950, MD cum laude, 1952. Diplomate Nat. Bd. Med. Examiners. Rotating intern A.M. Billings Hosp., 1952-53; resident in hematology Mt. Sinai Hosp., N.Y.C., 1953-54, asst. resident, 1957; asst. resident medicine Georgetown U. Hosp., Washington, 1958; instr., asst. prof. dept. physiology and biophysics Georgetown U., Washington, 1959-68, assoc. prof., 1968; lectr. physiology U.S. Naval Dental Sch., Bethesda, Md., 1962-70, Walter Reed Army Inst. Dental Rsch., Washington, 1963-70; guest scientist Naval Med. Rsch. Inst., Bethesda, 1978-83. Contbr.: The Merck Manual, 14th, 15th and 16th edit., 1982, 87, 92; contbr. articles to profl. jours. Lt. USNR, 1954-56. Recipient Lederle Med. Faculty award Lederle Co., Pearl River, N.Y., 1960-63, USPHS Rsch. Career Devel. award Nat. Inst. Dental Rsch., Bethesda, 1963-70; grantee Am. Cancer Soc., USPHS. Mem. Am. Physiol. Soc., Am. Soc. Hematology, Assn. Rsch. in Vision and Ophthalmology, Internat. Assn. Dental Rsch. Jewish. Achievements include discovery that the ground substance is masked but not lost in calcification, removal of spleen is followed by a reticulocytosis that is permanent in dogs, dogs have many more young reticulocytes in their blood than man, stretching of skin increases mitoses in the rat showing physical factors can modulate DNA and cell division, adult Gaucher cells contain iron secondary to erythrophagocytosis, the spleen protects against insecticide-induced hematoxicity, biological armature provides internal stability to exocrine glands. Home: 5823 Osceola Rd Bethesda MD 20816-2032 Office: Georgetown U Sch Medicine 3900 Reservoir Rd NW Washington DC 20007-2187

LORD, GUY RUSSELL, JR., psychiatry educator; b. Panama City, Panama, July 6, 1943; s. Guy Russell Sr. and Marie Loretta (Madden) L.; m. Patricia Lawlor, 1968 (div. 1974); m. Mary Margaret Steiner, May 21, 1977; 1 child, Katherine. BA, Reed Coll., 1966; MD, Duke U., 1970. Intern Baylor Affiliated Hosps., Houston, 1970-71; resident in psychiatry San Diego County Mental Health Svcs., 1971-74, staff psychiatrist, 1974-77; fellow in child psychiatry U. Wis., Madison, 1977-79; staff child psychiatrist Mendota Mental Health Inst., Madison, 1979-81; asst. prof. psychiatry U. Utah, Salt Lake City, 1981-83, Med. Coll. Wis., Milw., 1983-90; chief child and adolescent psychiatry St. Mary's Hill Hosp., Milw., 1990—; pres. Wis. Coun. Child and Adolescent Psychiatry, 1990-92. Mem. Am. Psychiatric. Assn., Am. Acad. Child Psychiatry, State Med. Soc. Wis. Democrat. Mem. Soc. of Friends. Office: St Mary's Hill Hosp 2350 N Lake Dr Milwaukee WI 53211

LORD, NORMAN WILLIAM, physicist, consultant; b. Bklyn., Feb. 28, 1925; s. Louis and Rebecca (Cohen) L.; m. Maxine Levin, Dec. 29, 1949 (div. Oct. 1961); children: Nancy Theresa, Susan Amy, Robert James; m. Estella Cahan, Nov. 15, 1961 (dec. Dec. 1989). BEE, Poly. Inst., Bklyn., 1944; PhD in Physics, Columbia U., 1953. Rsch. scientist Raytheon Rsch. Div., Waltham, Mass., 1952-54; sr. scientist Applied Physics Lab. of Johns Hopkins, Silver Spring, Md., 1954-60; sr. rsch. assoc. Hudson Labs. of Columbia U., Dobbs Ferry, N.Y., 1960-68; sr. rsch. scientist Travelers Rsch. Ctr., Hartford, Conn., 1968-72; sr. cons. Auerbach Assocs., Rosslyn, Va., 1972-76; lead engr. Mitre Corp., McLean, Va., 1976-89; pvt. cons. Washington, 1989—; chmn. underwater acoustic communication Nat. Security Indsl. Assn., Washington, 1968-70. Author: Heat Pump Technology, 1980, Heat Treatment of Metals, 1981, Coal-Oil Mixture Technology, 1982, Advanced Computers, 1983; contbr. articles to profl. jours. Chmn. study of high sch. student drug use Hartford Conn. Citizens Study Groups on Secondary Edn., Newington, 1969-70. Cpl. U.S. Army, 1944-46, ETO. Mem. IEEE (sr. mem.), Am. Phys. Soc. Achievements include exact analysis of molten-zone purification, patent for generating and analyzing high information signals, algorithmic construction of optimized waveforms, deep sound channel acoustic communication feasibility, alkali halide charged vacancy wave function determination. Home: 4501 Nebraska Ave NW Washington DC 20016-1849

LORD, THOMAS REEVES, biology educator; b. Alexandria, Va., Mar. 14, 1943; s. Arthur Roberts and Dorothy (Reeves) L.; m. Jane Tompkins, June 17, 1967; children: Erik Thomas, Andrea Margaret, Elizabeth Jane. BS in Biology, Rutgers U., 1965, D Sci. Edn., 1983; MS in Biology, Trenton State Coll., 1969. Tchr. biology Rancocas Valley Regional High Sch., Mt. Holly, N.J., 1965-69; instr. biology Frankfurt (Germany) Internat. Sch., 1969-73; prof. Burlington County Coll., Pemberton, N.J., 1973-89; prof. biology, sci. edn. Indiana U. Pa., 1989—; cons. Innovative Community Colls., Kansas City, Kans., 1989-91, Zero Population Growth, Inc., Washington, 1989—; fellow, cons. Dept. Higher Edn., Trenton,1985-89; fellow Birmingham (Eng.) U., 1980-81. Author: Stories of Lake George - Fact and Fancy; contbr. articles to profl. jours. Instr. swimming, life saving, canoeing, sailing, first aid, CPR ARC, Burlington County, 1975—. Grantee Biol. Sciences Curriculum Studies, 1990-92, Pa. Acad. for Profession of Teaching, 1991-92, NSF-Biology Sci. Curriculum Studies, 1991-92). Mem. Nat. Sci. Tchrs. Assn., Nat. Assn. for Rsch. in Sci. Teaching (jour. reviewer 1988-93), Nat. Assn. Biology Tchrs. (coll. com. 1989-92), Sigma Xi. Mem. Soc. of Friends. Avocations: jogging, swimming, canoeing. Home: 92 Valley Rd Indiana PA 15701 Office: Indiana U Pa Weyandt Hall Indiana PA 15705

LORENZ, BETH JUNE, mechanical engineer; b. Camden, N.J., June 2, 1961; d. Manuel Jack and Carolyn Ann (Miller) Lambersky; m. Michael Eric Lorenz, Feb. 14, 1992. BSME, Drexel U., 1984, postgrad. Registered profl. engr., Pa. Project engr. E.I. DuPont de Nemours & Co., Inc., Towanda, Pa., 1984-87; energy engr. RCA, Moorestown, N.J., 1987-88; HVAC staff engr. CUH2A, Princeton, N.J., 1988-90; project HVAC engr. The Kling Lindquist Partnership, Phila., 1990-91, 92—; constrn. coord. Lott Constructors, Harleysville, Pa., 1991-92. Mem. ASHRAE, Inst. Energy Engrs., Pi Lambda Phi. Office: Kling Lindquist Partnership 2301 Chestnut St Philadelphia PA 19103

LORENZ, EDWARD NORTON, meteorologist, educator; b. West Hartford, Conn., May 23, 1917; 3 children. AB, Dartmouth Coll., 1938; AM, Harvard U., 1940; SM, MIT, 1943, DSc, 1948. Asst. meteorologist MIT, Cambridge, 1946-48, mem. staff, 1948-54, asst. prof., then assoc. prof., 1955-62, prof. meteorology, 1962—; head dept. MIT, 1977-81; vis. assoc. prof. UCLA, 1954-55. Recipient Crafoord prize Swedish Acad. Scis., 1983, Roger Revelle medal Am. Geog. Union, 1993. Mem. Nat. Acad. Scis. Office: MIT Dept Earth Atmospheric & Planetary Sci Cambridge MA 02139

LORENZ, RUEDIGER, neurosurgeon; b. Niederfischbach, Fed. Republic Germany, Sept. 9, 1932; s. Johannes and Ena (Mueller) L. m. Gunde Hussmann, Feb. 13, 1959; children: Matthias, Mechthild. Study of medicine U. Bonn and Goettingen, 1951-1956; Dr. Med. U. Goettingen, 1956; Dr.h.c. U. Zaragoza, 1991. Med. tng. in internal medicine, gen. pathology, gen. surgery, ob-gyn, 1957-1959. Spl. tng. in gen. surgery 1959-1962, neurophysiology 1963, neurosurgery, 1962. Specialist in neurosurgery 1967. Habilitation for Neurosurgery U. Giessen 1971; prof. neurosurgery U. Giessen Faculty Medicine, 1973-80; prof. and chmn. neurosurgery, head dept. Univ. Hosp., Frankfurt, 1980—. Mem. Am. Soc. Neurological Surgeons (hon.), Med. Acad. Zaragoza (Spain), Med. Acad. Burma, Med. Acad. Argentina. Contbr. articles in field to profl. jours., chpts. in books, monographs. Home: Lerchesbergring 86a, D-60598 Frankfurt am Main Germany Office: Leiter der Klinik fuer, Neurochirurgie 2-16 Schleusenweg, D-60528 Frankfurt am Main Germany

LORENZEN, COBY, emeritus engineering educator; b. Oakland, Calif., Nov. 30, 1905; s. Coby and Catherine (O'Keefe) L.; m. Ina Voss, Aug. 7, 1937; children: Robert, Jacklyn, Donald, Kenneth. BSME, U. Calif., Berkeley, 1929, MSME, 1934. Profl. engr., Calif. Recipient John Scott medal City of Phila., 1976, Cyrus McCormick Gold medal Am. Soc. Agrl. Engrs., Chgo., 1981. Achievements include design of first commercial mechanical harvester for canning tomatoes. Home: Country Club Dr Carmel Valley CA 93924

LORENZEN, ROBERT FREDERICK, ophthalmologist; b. Toledo, Ohio, Mar. 20, 1924; s. Martin Robert and Pearl Adeline (Bush) L.; m. Lucy Logsdon, Feb. 14, 1970; children: Roberta Jo, Richard Martin, Elizabeth Anne. BS, Duke, 1948, MD, 1948; MS, Tulane U., 1953. Intern, Presbyn. Hosp., Chgo., 1948-49; resident Duke Med. Center, 1949-51, Tulane U. Grad. Sch., 1951-53; practice medicine specializing in ophthalmology, Phoenix, 1953—; mem. staff St. Joseph's Hosp., St. Luke's Hosp., Good Samaritan Hosp., Surg. Eye Ctr. of Ariz. Pres. Ophthalmic Scis. Found., 1970-73; chmn. bd. trustees Rockefeller and Abbe Prentice Eye Inst. of St. Luke's Hosp., 1975—. Recipient Gold Headed Cane award, 1974; named to Honorable Order of Ky. Cols. Fellow ACS, Internat. Coll. Surgeons, Am. Acad. Ophthalmology and Otolaryngology, Soc. Eye Surgeons; mem. Am. Assn. Ophthalmology (sec. of ho. of dels. 1972-73, trustee 1973-76), Ariz. Ophthal. Soc. (pres. 1966-67), Ariz. Med. Assn. (bd. dirs. 1963-66, 69-70), Royal Soc. Medicine, Rotary (pres. Phoenix 1984-85). Republican. Editor in chief Ariz. Medicine, 1963-66, 69-70. Office: 367 E Virginia Ave Phoenix AZ 85004-1275

LORENZINO, GERARDO AUGUSTO, linguist; b. Buenos Aires, Sept. 24, 1959; came to U.S., 1985; s. Gerardo Augusto and Maria Teresa (Julia) L.; m. Laura Sepp, Aug. 17, 1985. MS, NYU, 1987; MA, CUNY, 1992. Adj. lectr. Hostos C.C., N.Y.C., 1990-92; adj. lectr. Hunter Coll., N.Y.C., 1992, NYU, 1992. Contbr. articles to profl. jours. Recipient Am. Inst. Chemists' award, 1982, Social Sci. Rsch. Coun. fellowship, Africa, 1991. Mem. Am. Anthrop. Assn., Linguistic Soc. Am.

LORENZ-MEYER, WOLFGANG, aeronautical engineer; b. Wohltorf, Lauenburg, Germany, Mar. 6, 1935; s. Paul Lorenz and Martha Maria Bothilde (Russ) L-M.; m. Erika Baeumler, Oct. 12, 1962; children: Ulrike, Stefan Lorenz, Klaus. Diploma in Mech. Engring., Tech. High, Hannover, Germany, 1962; D in Engring., Tech. U., Braunschweig, Germany, 1968. Scientist Aerodynamische Versuchsanstalt, Göttingen, Germany, 1963-69; dep. head of inst. Deutsche Forchungs-und Vesuchsanstalt für Luft-und Raumfahrt, Göttingen, 1969-80, div. chief, 1973-86; cons. engr. pvt. practice, Göttingen, 1986—; rep. Deutsch Forschungs-und Vesuchsanrstalt für Luft-und Raumfahrt to Supersonic Tunnel Assn., U.S., 1971-86; German mem. NATO DRG AC 243 PG7/WG1, Brussels, 1973-76; rep Deutsche Forchungsaustalt für Luft-und Raumfahrt to European Transonic Windtunnel related working group, Amsterdam, 1976-86. Contbr. articles to Jahrbuch der Wissenschaftliche Gesellschaft für Luft-und Raumfahrt, Jahrbuch der Deutsch Gesellschaft für Luft-und Raumfahrt, Ing. Arch and others, 1966-79. Recipient Ernst-Mach Preis, Deutsche Forschungs-und Vesuchsanstalt für Luft-und Raumfahrt, Bonn, Germany, 1968. Mem. AIAA, Deutsche Gesellschaft f. Luft-und Raumfahrt, Verein Deutscher Ingenieure. Lutheran. Achievements include U.S., European patents for active suppression of flow-induced vibration on sting supported wind-tunnel models; contbrs. to strain-gauge balance devel. under cryogenic conditions; wind tunnel corrections and measuring accuracy in transonic flow. Home: Sudetenlandstrasse 4, 37085 Göttingen L Saxony, Germany

LORIJN, JOHANNES ALBERTUS, economist; b. Deventer, The Netherlands, May 17, 1925; s. Albertus and Antoinette (Dorgelo) L.; m. Margaretha Elizabeth Tennekes van Ouwerkerk, Jan. 18, 1950; children: Ronald, Harold, Yolande, Niels. BBA, Algemene Handelschool, Assen, The Netherlands, 1949; diploma in fiscal law, European Fiscal Coll., Brussels, 1964; LLD (hon.), Coll. Rome, 1964; D in Econs. (hon.), U. Coventry (Eng.), 1978. Acct. Cen. Acctg. Office, The Hague, The Netherlands, 1947-52, dir., 1952-57; sr. exec. mgmt. cons. Bus. Counselors GMBH, Frankfurt, Fed. Republic Germany, 1957-58; mng. dir. Bus. Counselors GMBH, Athens, Greece, 1958-59; pres., chief exec. officer I.E.E. S.A., Barcelona, Spain, 1959-68; pres. Internat. Devel. Cons., Lagos, Nigeria, 1969-76, Cam Cons. Corp., Wilmington, 1978—, Carelo S.A., Barcelona, 1980-84; pres., mng. dir. P.D. & M. Corp., Brussels, 1985—; adviser Dutch Govt. del. to Nigeria, 1969, World Bank Assn., N.Y., 1981; v.p. Tilapia Internat. Found., Brussels, 1961-89, Robert Sherower Group, N.Y., 1977-84. Decorated Knight Sovereign Order Constantine Magnificent, 1959; recipient Diploma de Gratitud Red Cross, Spain, 1962. Mem. ABA, Dutch Assn. Acct., Dutch Assn. Mgmt. Cons., Dutch Coll. Tax Cons., German Assn. Econ. Experts, Order des Juristes Fiscaux D'Europe, Am. Entrepreneurs Assn. Avocations: piano, model building. Home: Postbus 38097, 6503 AB Nijmegen The Netherlands Office: PD&M Corp, Terlindenvijverweg nr 5, 1700 Asse Belgium

LORING, DAVID WILLIAM, neuropsychologist, researcher; b. Richmond, Ind., July 13, 1956; s. Richard William and Janet (Teetor) L.; m. Debra Rogers, Jan. 5, 1980 (div. 1983); m. Sherrill Rabon, July 30, 1988; children: Jason Michael, Sarah Elizabeth, Rachel Erin. BA, Wittenberg U., 1978; PhD, U. Houston, 1982. Diplomate Am. Bd. Profl. Psychology. Postdoctoral fellow dept. neurology Coll. of Medicine Baylor U., Houston, 1982-83; instr. div. neurosurgery U. Tex. Med. Br., Galveston, 1983-85; asst. prof. dept. neurology Med. Coll. Ga., Augusta, 1985-89, assoc. prof. dept. neurology, 1989—; lectr. Parke-Davis Nat. Speakers Bur., 1991—. Author: Amobarbital Effects and Lateralized Brain Function, 1992; cons. editor Clin. Neuropsychologist Jour., 1991—, Jour. Epilepsy, 1991—, Jour. Clin. and Exptl. Neuropsychology, 1993—, Aging and Cognition, 1993—; contbr. articles to Jour. Clin. and Exptl. Neuropsychology, Neuropsychologia, Archives Neurology, Neurology. Mem. Internat. Neuropsychol. Soc. (bd. govs. 1991—), Am. Epilepsy Soc., Am. Psychol. Assn., Am. Acad. Neurology (assoc.). Achievements include research in neuropsychology. Home: 147 Savannah Pt North Augusta SC 29841-3568 Office: Med Coll Ga Dept Neurology Augusta GA 30912

LORING, THOMAS JOSEPH, forest ecologist; b. Haileybury, Ont., Can., May 27, 1921; s. Ernest Moore and Margaret Evangeline (Bacheller) L.; m. Beth Rogers McLaughlin, Oct. 29, 1966; children: John Francis, Christopher Thomas. BSc in Forestry, Mich. Tech. U., 1946; M Forestry, N.Y. State Coll. Forestry, 1951. Forester McCormick Estates, Champion, Mich., 1947; cons. Porteous and Co., Seattle, 1948-49; forester Penokee Veneer Co., Mellon, Wis., 1951-53; cons. E.M. Loring Consulting, Noranda, Que., Can., 1954-55; forester USDA Forest Svc., Albuquerque, 1956-81; cons. Tom Loring, Cons., Victoria, B.C., Can., 1986—; mem. Parks and Recreation Commn., Victoria, 1988-92, mem. environment adv. com., 1993—. Editor: Directory of the Timber Industry in Arizona and New Mexico,1 972; co-editor: Ecology, Uses and Management of Pinyon-Juniper Woodlands, 1977. Pres. Shawnigan Lake Residents and Rate Payers Assn., B.C., 1985-86. Mem. Soc. Am. Foresters (sect. chair 1960-62), Ecol. Soc. Am., Forest Products Rsch. Soc. (regional rep. 1980-81), Can. Inst. Forestry, Soc. Ecol. Restoration. Home: 59 Moss St, Victoria, BC Canada V8V 4M1

LOS, MARINUS, agrochemical researcher; b. Ridderkerk, The Netherlands, Sept. 18, 1933; came to U.S., 1960; s. Cornelis and Neeltje (Zoutewelle) L.; m. Lorraine Betty Lowe, May 11, 1957; children: Simon, Sija, Michael, Martin (dec.). BS, Edinburgh U., Scotland, 1955, PhD, 1957. Sr. rsch. chemist Am. Cyanamid Co., Princeton, N.J., 1960-71, group leader, 1971-84, sr. group leader, 1984-86, mgr. crop protection chems., 1986-88, assoc. dir. crop scis., 1988-92, rsch. dir. crop scis., 1992—. Recipient Disting. Inventor of 1990 award Intellectual Property Owners, Inc., Washington, 1990, Thomas Alva Edison Patent award, R&D Coun. of N.J., 1991, Nat. Medal of Tech., Nat. Sci. Found., 1993. Mem. AAAS, Am. Chem. Soc., Plant Growth Regulator Soc. Achievements include 53 patents. Office: Am Cyanamid Co PO Box 400 Princeton NJ 08543-0400

LOSADA, MARCIAL FRANCISCO, research scientist, psychologist; b. Rancagua, Chile, July 3, 1939; came to U.S., 1977; s. Marcial and Hortensia (Menendez) L.; m. Giovanna Morchio, Jan. 5, 1964; children: Ximena, Andrea. MA, U. Mich., 1975, PhD, 1977. Prof. Univ. Catolica, Chile, 1969-73; sr. study dir. U. Mich., Ann Arbor, 1977-80; dir. Coherent Forecasting Systems, Ann Arbor, 1981-84; sr. rsch. scientist Electronic Data Systems, Warren, Mich., 1985-86; sr. rsch. scientist EDS Ctr. for Machine Intelligence, Ann Arbor, 1986-88, program dir., 1989-90; dir. EDS Ctr. for Advanced Rsch., Ann Arbor, 1991—. Mem. Am. Psychol. Soc. Achievements include development of group analyzer, a system for dynamic analysis of group interaction. Office: EDS Ctr for Advanced Rsch 2001 Commonwealth Blvd Ann Arbor MI 48105

LOSS, JOHN C., architect, educator; b. Muskegon, Mich., Mar. 6, 1931; s. Alton A. and Dorothy Ann (DeMars) Forward; m. LaMyrna Lois Draggoo, June 7, 1958. B.Arch., U. Mich., 1954, M.Arch., 1960. Lic. architect Md., Va., Mich. Architect Eero Saarinen & Assocs., Bloomfield Hills, Mich., 1956-57; owner John Loss & Assocs, Detroit, 1960-75; prof., acting dean Sch. Architecture, U. Detroit, 1960-75; prof., head dept. architecture N.C. State U., Raleigh, 1975-79; assoc. dean. Sch. Architecture U. Md., College Park, 1981-83; prof. architecture U. Md., 1979—; dir. Architecture and Engring. Performance Info. Ctr., 1982—; pvt. practice architecture Annapolis and College Park, 1979—; mem. com. NRC-NAS, 1982—; mem. bldg. diagnostics com. Adv. Bd. on Build Environment, 1983—; mem. com. on earthquake engring. NRC, 1983—; survey team leader tornado damage in Pa. and Ohio, 1985. Author: Building Design for Natural Hazards in Eastern United States, 1981, Identification of Performance Failures in Large Structures and Buildings, 1987, Analysis of Performance Failures in Civil Structures and Large Buildings, 1990, Performance Failures in Buildings and Civil Works, 1991; works include med. clinic, N.C.; Aldersgate Multi Family Housing, Oscoda, Mich. Advisor Interfaith Housing Inc., Detroit, 1966-74; advisor Detroit Mayor's Office, 1967-69, Interim Housing Com. Mich. State Housing Devel. Authority, Lansing, 1969-71, Takoma Park Citizens for Schs. (Md.), 1981-82; advisor, cons. Hist. Preservation Commn., Prince George's County, Md. With U.S. Army, 1954-56. NSF grantee, 1978-81, 1982-84, 86-87, 88-90; named one of Men of Yr., Engring. News Record, 1984. Fellow AIA; mem. Earthquake Engring. Research Inst., Assn. Collegiate Sch. Arch., Md. Soc. Architects, Archtl. Research Ctrs. Consortium (bd. dirs. 1981-89). Democrat. Roman Catholic. Office: U Md Sch Architecture College Park MD 20742

LOTRICK, JOSEPH, aeronautical engineer; b. Plymouth, Pa.; s. Stephen and Catherine (Turpak) L.; m. Barbara Sue Vining; 1 child, Pegge Jo. Student, U. Pa., 1943; BS in Aero. Engring., Northrop U., 1962. Sr. engr. flight test N.Am. Aviation, L.A., 1952-86; project engr. Rockwell Internat., L.A., 1986—. With USN, 1943-46. Mem. AAAS, AIAA (mem. nat. tech. com. flight test 1984-86), Assn. Naval Aviation, Aircraft Owners and Pilots Assn., Elks. Republican. Achievements include flight testing of exotic airborne R&D avionics systems and sensors. Home: 2531 Highcliff Dr Torrance CA 90505 Office: Rockwell Internat 201 N Douglas St MC:GB21 El Segundo CA 90245

LOTT, IRA TOTZ, pediatric neurologist; b. Cin., Apr. 15, 1941; s. Maxwell and Jeneda (Totz) L.; m. Ruth J. Weiss, June 21, 1964; children: Lisa, David I. BA cum laude, Brandeis U., 1963; MD cum laude, Ohio State U., 1967. Intern Mass. Gen. Hosp., Boston, 1967, resident in pediatrics, 1967-69, resident in child neurology, 1971-74; clin. assoc. NIH, Bethesda, Md., 1969-71; from clin. rsch. fellow to asst. prof. Harvard Med. Sch., Boston, 1971-82; clin. dir. Eunice Kennedy Shriver Ctr. for Mental Retardation, Waltham, Mass., 1974-82; assoc. prof. U. Calif., Irvine, 1983-91, prof., 1992—; chmn. dept. pediatrics U. Calif., Irvine, 1992—, dir. pediatric neurology, 1983—; pres. Prof. Child Neurology, Mpls., 1992—. Editor: Down Syndrome-Medical Advances, 1991; contbr. articles to profl. jours. Sec., treas. Child Neurology Soc., Mpls., 1987-90. Lt. comdr. USPHS, 1969-71. NIH grantee, 1974—; recipient Career Devel. award Kennedy Found., 1976. Fellow Am. Acad. Neurology; mem. Am. Pediatric Soc., Am. Neurol. Assn., Nat. Down Syndrome Soc. (sci. acad. bd. 1985—), Western Soc. for Pediatric Rsch. (councillor 1989-91). Achievements include research in relationship of Down Syndrome to Alzheimer's disease, neurometabolic disease, extracorporeal membrane oxygenation in infants. Office: U Calif Irvine Med Ctr Dept Pediatrics Bldg 27 Rt 81 101 City Dr S Orange CA 92668

LOTTER, DONALD WILLARD, environmental educator; b. Davis, Calif., Sept. 17, 1952; s. Willard S. and Jane (Baker) L. BS in Agronomy, U. Calif., Davis, 1977; M in Profl. Studies, Cornell U., 1981; postgrad., U. Calif., Davis. Instr. Davis Experimental Coll. Experimental Coll. U. Calif., Davis, 1973—; bd. dirs. Yolo Environ. Resource Ctr. Author software program. Achievements include first to develop personal environmental accounting software (Enviro Account). Home and Office: 605 Sunset Ct Davis CA 95616

LOTZ, WILLIAM ALLEN, consulting engineer; b. Hartford, Conn., May 26, 1932; s. John and Ruby (Allen) L.; m. Barbara Hockert, Sept. 16, 1953 (div. 1970); children: Esther, Nelson, Seth, Valerie, Heather, Pat; m. Jacquelyne Ann Lotz, Dec. 24, 1974. BSME, U. Miami, Fla., 1956; postgrad., Ohio State U., 1956-60. Registered profl. engr., Ohio, Vt., Maine, Mass., R.I., Conn., N.H. Testing engr. Owens Corning Fiberglas, Granville, Ohio, 1956-64; area engr. Kaiser Aluminum, Newark, Ohio, 1964-67; facility engr. IBM Corp., Burlington, Vt., 1967-70; v.p. Brand Insulation, Chgo., 1971-72; pres. Darbron, Houston, 1973-74; prin. William A. Lotz, Engr., Acton, Maine, 1975—; vice chair Maine state Bd. Registration for Profl. Engrs. Contbr. articles to profl. publs.; syndicated columnist Lotz on Energy, 1978-81. Chmn. Acton Town Dem. Com., 1980—; chmn. Dem. State Platform Com., Maine, 1984, 86, 88; chmn. ministerial search com. 1st Parish Ch., 1976, chmn. worship and spl. events com., 1975-83, chmn. budget com. 1983; exec. com. So. Maine Regional Planning Commn., 1980—; pres. Unitarian Universalist Psi Symposium, 1977-87; sec. Main Dem. parthy, 1985-89, chmn. Dem. State Conv. Com., 1989; vice-chair York County Dem. Com., 1986. Fellow ASHRAE (chpt. pres., tech. com. on insulation and vapor barriers, mem. indoor air pollution task force, large bldg. air condition com.); mem. ASME, Am. Nuclear Soc, Air Pollution Control Assn. Democrat. Unitarian. Home and Office: Acton Ridge Rd Acton ME 04001

LOTZE, EVIE DANIEL, psychodramatist; b. Roswell, N.Mex., Mar. 6, 1943; d. Wadsworth Richard and Lee Ora (Norrell) Daniel; m. Christian Dieter Lotze, June 9, 1963; children: Conrad, Monica. BA cum laude, La. State U., 1964; MA, Goddard Coll., 1975; PhD, Union Inst., Cin., 1990. Dir. Casa Alegre, Hogares, Albuquerque, 1979-80; pvt. practice Riyadh, Saudi Arabia, 1980-83, Silver Spring, Md., 1983-85; dir. Gulf States Psychodrama Tng., Houston, 1986-88; founder, dir. Innerstages Psychodrama Tng., Houston, 1988—; supr. Houston area psychodramatists, 1988—; tng. cons. Assn. Applied Psychologists, Moscow, 1992; guest lectr.

New Univ. Bulgaria, Sofia, 1992—; cons. in field. Author: (tng. manual) Clinical Psychodrama Training Manual, 3 vols., 1990. Bd. dirs. Interact Theater, Houston, 1992. Fellow Am. Soc. for Group Psychotherapy and Psychodrama; mem. Am. Group Psychotherapy Assn., Internat. Coun. Psychologists. Democrat. Lutheran. Avocations: cross-country skiing, biking, hiking, camping, reading. Home: 11926 Riverview Way Houston TX 77077 Office: Innerstages Psychodrama Tng 11926 Riverview Way Houston TX 77077

LOTZOVÁ, EVA, immunologist, researcher, educator; b. Prague, Czech Republic; came to U.S., 1968; married. MS, Charles U., Prague, Czech Republic, 1965, PhD, 1968. Asst. prof. SUNY, Buffalo, 1972-73, Baylor Coll. Medicine, Houston, 1973-76; asst. prof./asst. immunologist M.D. Anderson Cancer Ctr. U. Tex., Houston, 1976-78, assoc. prof./assoc. immunologist, 1978-84, prof., immunologist, 1984—, chief sect. natural immunity dept. surg. oncology, 1989—; endowed prof. Florence Maude Thomas Cancer Rsch., Florence, Italy, 1988—; prof. affiliate Pitts. Cancer Inst., 1986—; mem. grant com. Dutch Cancer Soc., 1989—, Leukemia Rsch. Fund, 1992, Experimental Therapeutics NIH, 1991, U.S. Dept. Agrl. Office Grants, 1987—, Med. Rsch. Coun. Can., 1987—, U.S.-Israeli Binational Sci. Found., 1986—, Leukemia Found., 1986—; cons. chmn. organizing com. Soc. Natural Immunity, 1989; ad hoc mem. grant site visit teams NIH, 1984—, study sections, 1984—. Editor-in-chief Natural Immunity Mag., 1984—; editor: (book) NK Cell Mediated Cytotoxicity: Receptors, Signalling and Mechanisms, 1992, Interleukin-2 and Killer Cells in Cancer, 1990, Experimental Hematology Today, 1989, Immunobiology of Natural Killer Cells, vol. i, 1986, vol. ii, 1986, Natural Immunity, Cancer and Biological Response Modification, 1986; (jours.) Symposiua Publs., 1982-90, Symposia Abstracts, 1985-92, Folia Biologica, 1992-97; mem. editorial bd. Jour. Internat. Soc. Experimental Hemotology, 1977-81, Survey and Synthesis Path. Rsch., 1982-85; section editor: 1986, 87, 88 Yearbooks of Cancer; contbr. articles to profl. jours. Grantee NIH, 1976-79, 79-82, 85-91, 89-94, 92-95; recipient Ernesto Nuti Internat. prize Jour. Experimental Clin. Cancer Rsch., Rome, 1992. Mem. Internat. Soc. Experimental Hemotology, Internat. Soc. Natural Immunity (v.p.), Am. Assn. Immunologists, Am. Assn. Cancer Rsch., Transplantation Soc., Am. Soc. Hematology, N.Y. Acad. Scis., Clin. Immunology Soc. Achievements include leading scientific contributions to field of immunity against cancer, regulating natural killer (NK) cells by cytokines, immunotherapy and experimental bone marrow transplantation; bridged the scientific and clinical disciplines by translating the scientific data to the clinic by developing new approaches to cancer treatment. Office: U Tex MD Anderson Cancer Ctr 1515 Holcomb Blvd Box 18 Houston TX 77030

LOU, ZHENG (DAVID), mechanical engineer, biomedical engineer; b. Changshu, Jiangsu, Peoples Republic China, Apr. 25, 1959; came to U.S., 1982; s. Gui-Xin and Pei-Ling (Wang) L.; m. Min Yu, 1984; 1 child, Katherine Hua. BE, Zhejiang U., Hangzhou, China, 1982; PhD, U. Mich., 1990. Asst. rsch. scientist Transp. Rsch. Inst. U. Mich., Ann Arbor, 1990-93; mfg. tech. engr. Ford Motor Co., Ypsilanti, Mich., 1993—. Contbr. articles to Jour. Rheology, Jour. Biomechanics, others. Grantee NASA, 1992—, U.S. Army, 1992—. Mem. ASME, Soc. Rheology, Soc. Automotive Engring., Tau Beta Pi, Sigma Xi. Achievements include research in nonlinear dynamic interaction between an electrorheological fluid and a viscometer, dynamics of electrorheological valves and dampers, heat transfer model in hyperthermia as a tumor therapy. Home: 1613 Old Salem St Plymouth MI 48170 Office: Ford Motor Co EFHD PO Box 922 McKean and Textile Rds Ypsilanti MI 48197

LOUCKS, ERIC DAVID, water resources engineer, consultant; b. Madison, Wis., June 7, 1958; s. Orie Lipton and Elinor Jane (Bernstein) L. PhD, U. Wis., 1989. Instr. U. Wis., Madison, 1980-88; engr. RUST Environ. and Infrastructure, Schaumburg, Ill., 1988—. Co-author: Great Lake Diversion Impacts (2 vols.), 1985; designed Flood Control Facility, Elmhurst Quarry Flood Control Diversion Works, 1992; developer Great Lakes Hydrologic Response Model, 1983-85. Trustee Chi Epsilon, Madison, 1987-93. Mem. ASCE (vice chmn., sec., chmn. water resource div., Ill. sect. 1990-93), Internat. Assn. Great Lakes Rsch., Sigma Xi. Home: 425 South St # 1203 Honolulu HI 96813 Office: Rust E & I 1501 Woodfield Rd Ste 200E Schaumburg IL 60173

LOUCKS, VERNON R., JR., health care products and services company executive; b. Evanston, Ill., Oct. 24, 1934; s. Vernon Reece and Sue (Burton) L.; m. Linda Kay Olson, May 12, 1972; 6 children. B.A. in History, Yale U., 1957; M.B.A., Harvard U., 1963. Sr. mgmt. cons. George Fry & Assos., Chgo., 1963-65; with Baxter Travenol Labs., Inc. (now Baxter Internat. Inc.), Deerfield, Ill., 1966—, exec. v.p., 1976, also bd. dirs., pres., chief oper. officer, 1980, chief exec. officer, chmn., 1987—; bd. dirs. Dun & Bradstreet Corp., Emerson Electric Co., Quaker Oats Co., Anheuser-Busch Cos.; bd. advisors Nestlē U.S.A. Trustee Rush-Presbyn.-St. Lukes Med. Ctr.; assoc. Northwestern U.; sr. fellow Yale Corp.; chmn. Yale Devel. Bd. 1st lt. USMC, 1957-60. Recipient Citizen Fellowship award Chgo. Inst. Medicine, 1982, Nat. Health Care award B'nai B'rith Youth Svcs., 1986, William McCormick Blair award Yale U., 1989, Semper Fidelis award USMC, 1989, Disting. Humanitarian award St. Barnabas Found., 1992, Alexis de Tocqueville award for community svc. United Way Lake County, 1993; named 1983's Outstanding Exec. Officer in the healthcare industry Fin. World; elected to Chgo.'s Bus. Hall of Fame, Jr. Achievement, 1987. Mem. Health Industry Mfrs. Assn. (chmn. 1983), Bus. Roundtable (conf. bd., mem. policy com.), Bus. Coun. Clubs: Chgo. Commonwealth, Commercial, Mid-America; Links (N.Y.C.). Office: Baxter Internat Inc 1 Baxter Pky Deerfield IL 60015-4625

LOUDON, RODNEY, optics scientist, educator; b. Manchester, Eng., July 25, 1934; s. Albert and Doris Helen (Blane) L.; m. Mary A. Philips, 1960; 2 children. Student, Oxford U., U. Calif., Berkeley. Sci. civil servant RRE, Malvern, 1960-65; mem. tech. staff Bell Labs., Murray Hill, N.J., 1965-66, 70, RCA, Zürich, 1975, British Telecom Rsch. Labs., 1984, 89; prof. physics Essex U., Colchester, Eng., 1967—; vis. prof. Yale U., 1975, U. Calif., Irvine, 1980, Ecole Polytechnique, Lausanne, 1985, U. Rome, 1987. Author: The Quantum Theory of Light, 1973, 83, (with W. Hayes) Scattering of Light by Crystals, 1978 (with D. Barber) An Introduction to the Properties of Condensed Matter, 1989; editor: (with Agranovich) Surface Excitations, 1984. Recipient Thomas Young medal and prize Inst. of Physics, 1987, Max Born award Optical Soc. of Am., 1992. Avocations: classical music, choral singing, musical instrument making. Office: Univ of Essex Physics Dept, Wivenhoe Pk, Colchester C04 3SQ, England*

LOUGHEED, EVERETT CHARLES, retired horticulture educator, researcher; b. Thornbury, Ont., Can., July 16, 1927; s. John Henry and Charlotte Druscilla (Dobson) L.; m. Leslie Ross Burness, Aug. 15, 1959; children: Stephen, Katherine, Robert. BSc. in Agr., Ont. Agrl. Coll., 1958; MSc. in Agr., U. Toronto, 1960; PhD, Mich. State U., 1964. Cert. profl. agrologist. Orchard labourer L.A. Myles, Thornbury, 1945-50, 52-54; lectr. Ont. Agrl. Coll., Guelph, 1960-62; asst. prof. U. Guelph, Ont., 1964-67, assoc. prof., 1967-71, prof., 1971-92, retired and associated grad. faculty, 1992—; cons. FAO, 1985; mgr. aid project Can. Internat. Devel. Agy., Argentina, 1989-92. Mem. editorial bd. Sci. Host, 1982—, Postharvest Biology and Tech., 1990—; assoc. editor, editor Can. Jour. Plan Tsoc., 1982—. Fellow Am. Soc. Hort. Sci. (chair postharvest working group 1978-79, assoc. editor 1985-87), Agrl. Inst. Can.(chair editorial bd., Fellowship award 1992); Can. Soc. Hort. Sci. (hon. life, pres. 1973-74), Ont. Inst. Agrologists, Internat. Soc. Hort. Sci., Internat. Standards Orgn. (Can. subcom. horticultural crops 1985—), Sigma Xi. Home: 156 Palmer St, Guelph, ON Canada N1E 2R6 Office: Univ of Guelph, OAX Dept of Horticultural, Guelph, ON Canada N1G 2W1

LOUGHLIN, KEVIN RAYMOND, urologic surgeon; b. N.Y.C., Aug. 10, 1949; s. Raymond Gerard and Josephine (McGrath) L. AB, Princeton U., 1971; MD, N.Y. Med. Coll., 1975. Diplomate Nat. Bd. Med. Examiners, Am. Bd. Urology. Surgery instr. Harvard Med. Sch. Brigham & Women's Hosp., Boston, 1983-86, asst. prof. surgery Harvard Med. Sch., 1986-90, assoc. prof. surgery Harvard Med. Sch., 1991—; staff urologist Dana Farber Cancer Inst., Boston, 1991—. Contbr. over 60 articles to profl. jours. Fellow Am. Cancer Soc., 1982-83, Nat. Kidney Found., 1980-81. Fellow ACS; mem. AAAS, Am. Soc. Andrology, Am. Urologic Assn., Boston Surg. Soc. Achievements include patent pending in laparoscopic surgical instruments. Home: 30 Lime St Boston MA 02108 Office: Brigham & Womens Hosp 75 Francis St Boston MA 02115

LOUGHMILLER, GROVER CAMPBELL, psychologist, consultant; b. Dallas, July 5, 1937; s. George Campbell and Opal Lynn (Nicolaides) L.; m. Carol Kay Beckstead, June 3, 1963; children: Trelesa, Lark, Kishl, Marlette, Velora, Logan. PhD, U. Utah, 1970. Lic. psychologist, sch. psychologist, Tex. Postdoctoral fellow George Peabody Coll., Nashville, 1971, asst. prof., 1969-72; pvt. practice psychology Loughmiller Inst., Tyler, Tex., 1972—; applicant evaluator Tex. State Bd. Psychology, Austin. Contbr. articles to profl. jours.; author psychol. tests: Media Personnel Survey, 1987, Loughmiller Diagnostic Interview Schedule for Children and Adolescents, 1988, Loughmiller Quick Test, 1988, Law Enforcement Candidate Questionnaire, 1989, Self-Choosing Guide, 1989; co-author book: Measurement and Predictors of Physician Performance, 1971, Study Guide and Topical Index to Major Source Materials in Transactional Analysis, 1974. Pres. Camp Fire Bd., 1979-80; adv. bd. Tyler Big Bros./Big Sisters Program, 1991—. With USMC, 1958-59. Mem. Am. Assn. Marriage and Family Therapy (supr.), Rotary (bd. dirs. 1991—). Republican. LDS. Office: Loughmiller Inst 422 S Spring Tyler TX 75702

LOUGHRAN, GERARD ANDREW, chemistry consultant, polymer scientist; b. Mt. Vernon, N.Y., Sept. 10, 1918; s. George Andrew and Harriet Willhelmenia (Reiss) L.; m. Kathleen Pearse O'Connor, Aug. 11, 1945; children: Maura, Kathleen, Gerard Jr., Judith Ann. BS in Chemistry, Fordham U., 1941; MS in Chemistry, N.Y.U., 1948. Analytical chemist N.Y. Quinine & Chem. Works, Bklyn., 1941-43; grad. asst. chemistry Fordham U., Bronx, N.Y., 1943-44; rsch. chemist Am. Cyanamid Co., Stamford, Conn., 1946-56, R.T. Vanderbilt Co., East Norwalk, Conn., 1956-59; rsch. scientist USAF Materials Lab., Wright-Patterson AFB, Ohio, 1960-86; cons. Kettering, Ohio, 1986—; cons. Universal Tech. Corp., Dayton, Ohio, 1989; presenter in field. Contbr. 15 articles to profl. jours. With U.S. Navy, 1944-46. Fellow Am. Inst. Chemists (profl. chemist accredited); mem. AAAS, Am. Chem. Soc. (Dayton sect., polymeric materials div., sci. and engring. div., polymer chemistry div., rubber div.), Nat. Assn. Retired Fed. Employees, N.Y. Acad. Scis, Am. Legion. Achievements include 12 patents for Acetelyne Disubstituted Phenoxy-Sulfone Compositions, Reactive Plasticizers for Thermoplastice Polysulfone Resins, Thermally Stable Dioxobenzisoquinoline Compositions and Method Synthesizing Same, Ladder Polydisalicylides, Antiozonants for Rubber, Alkoxyphenyldithiophosphonamidates as Vulcanization Accelerators, Ester and Ester Salts of Olefin-Phosphorous Trisulfide Oxygen Condensation Products, Lubricating Oil Additives, O,O Diakyl Alkylthiophosphonates, Copolymers N-(Hydroxymethyl)acrylamide, and Corrosion Inhibitors for Lubricating Oil. Home: 4575 Irelan St Dayton OH 45440-1548

LOUISOT, PIERRE AUGUSTE ALPHONSE, biochemist, medical educator; b. Marnay, France, Sept. 9, 1933; s. Marcel and Paulette (Coupat) L.; MD, U. Lyon, 1958, PhD, 1969; D honoris causa U. Montevideo, Uruguay, 1982, U. Shanghai, People's Republic of China, 1986; m. Marie-Colette Raymond, May 5, 1959; children: Christine, Philippe, Alain, Jean. Prof. gen. biochemistry and medicine U. Lyon, 1966—; dir. biochem. service Cardiol. Hosp. Lyon, 1978—; bd. dirs. Internat. Life Scis. Inst. Europe; v.p. Nat. Agy. for AIDS Rsch.; mem. sci. coun. Nat. Inst. for Sci. Rsch.; dir. unity Institut Nat. de la Santé et de la Recherche Médicale; dir. lab. Centre National de la Recherche Scientifique; pres. de l'Inst. Francais pour la Nutrition; pres. Ctr. of Nat. Studies of the Recommandations of Nutrition and Nourishment, pres. nourishment and nutrition Coun. For The Good Public Health of France; Author book; contbr. articles to profl. jours. Decorated officer Legion of Honor, Chevalier Nat. Order of Merit; recipient 4 sci. and med. prizes, 1960, 66, 71, 79; named Laureate Nat. Acad. Medicine, 1982. Mem. Soc. Biochemistry and Molecular Biology (bd. dirs.), French Soc. Clin. Biology, European Cell Biology Orgn., French Soc. Microbiology, N.Y. Acad. Scis. Home: 152 Cours Gambetta, 69007 Lyons France Office: Lyon-Sud Med Sch, BP 12, 69921 Oullins France

LOVE, DANIEL MICHAEL, manufacturing engineer; b. Gary, Ind., July 4, 1954; s. John Paulus Love and Carol B. (Hattenbach) Dallman; m. Jawanda K. Fairchild, July 14, 1984; children: Jessica K., Jacob E. AS, Ind. State U., 1983, BS, 1985. Biomed. engr. St. Margaret Hosp., Hammond, Ind., 1984; sr. systems engr. Tridan Tool and Machines, Danville, Ill., 1986-87; process engr. Heil Quaker Inc., Lewisburg, Tenn., 1987-89; sr. advanced mfg. engr. Lennox Industries Inc., Columbus, Ohio, 1989—. Achievements include the design of computer controls for hairpin benders, expanders, 100 ton fin presses for coil manufacturing equipment. Home: 1387 Brentfield Pl Columbus OH 43228 Office: Lennox Industries Inc 1711 Olentangy River Rd Columbus OH 43212

LOVE, JAMES BREWSTER, pharmaceutical engineer; b. Evanston, Ill., Oct. 14, 1948; s. James S. and Elise (Longino) L.; divorced; children: James Randal, Michael Christopher; m. Denise Hansen, Jan. 3, 1993. BSME, Bradley U., 1971. Engr. Commonwealth Edison, Chgo., 1971-76; project mgr. Travenol Labs., Deerfield, Ill., 1976-81; dir. engring. G D Searle, Skokie, Ill., 1981-91, Synetx, Humacao, P.R., 1991—; cons. PharmaCon, Chgo., 1991. Mem. Internat. Soc. Pharm. Engring. Home: PO Box 879 Ste 192 Humacao PR 00792 Office: Syntex NC01 Box 16625 Humacao PR 00791-9731

LOVEJOY, THOMAS EUGENE, tropical and conservation biologist, association executive; b. N.Y.C., Aug. 22, 1941; s. Thomas Eugene and Audrey Helen (Paige) L.; m. Charlotte Seymour, 1966 (div. 1978); children: Elizabeth Paige and Katherine Seymour (twins), Anne Williams. BS, Yale U., 1964, PhD in Biology, 1971; DSc (hon.), Colo. State U., 1989, Williams Coll., 1990; LHD (hon.), Coll. Boca Raton, 1991. Rsch. assoc. in biology U. Pa., 1971-74; exec. asst. to sci. dir. Acad. Natural Scis., Phila., 1972-73, asst. to v.p. for resources and planning, 1972-73; program dir. World Wildlife Fund-U.S., Washington, 1973-78, v.p. for sci., 1978-85, exec. v.p., 1985-87; asst. sec. external affairs Smithsonian Instn., Washington, 1987—; sci. advisor to Sec. Interior, 1993; bd. dirs. Manhattan Life Ins. Co., N.Y.C., chmn. exec. com., 1982-86, chmn., 1986—; rsch. assoc. in ornithology Acad. Natural Scis., 1971—; chmn., bd. dirs. Wildlife Preservation Trust Internat., 1974—; treas. Pan Am. sect. Internat. Coun. for Bird Preservation, 1973-84; sci. fellow N.Y. Zool. Soc., 1978—; mem. adv. bd. Environ. Assessment Coun., 1980—; founder, advisor Nature series Sta. WNET, 1980—; vis. lectr. on tropical ecology Yale U. Sch. Forestry and Environ. Studies, 1982. Coauthor: Nearctic Avian Migrants in the Neotropics, 1983; co-editor: Key Environments: Amazonia, 1985, Conservation of Tropical Forest Birds, 1985, Global Warming and Biological Diversity, 1992; contbr. articles, chpts. to profl. publs. Mem. Smithsonian Coun., 1982-87; trustee Millbrook Sch., N.Y., 1971—, Rocky Mountain Biol. Lab., 1983—, Acad. Natural Scis. Phila., 1987—, The Ozone Soc., 1990—; mem. U.S.-Brazil panel White House Office of Sci. and Tech., Washington, 1986-87; past chmn. U.S. Man and Biosphere Com., 1987—; treas., mem. exec. com. Sci. Com. on Problems of the Environ., 1988-91; sec. J. Paul Getty Wildlife Conservation prize, Washington, 1974-87, jury mem. 1988—; mem. adv. & tech. bd. Fundacion Neotropica and de Parques Nacional, Costa Rica, 1987—; mem. White House Sci. Coun., exec. office Pres., 1988-90; mem. Pres.'s Coun. Advisors in Sci. & Tech., 1990-92; dir. Rainforest Alliance, 1988—; mem. sci. coun. FPCN (Conservation Found., Peru), 1988—; mem. adv. bd. Am. Soc. Protection Nature Israel, 1988—; co-prin. investigator World Wildlife Fund/INPA, North Manaus, Brazil, 1979—; bd. govs. N.Y. Botanical Garden, N.Y.C., 1986—; dir. Ctr. for Plant Conservation, 1987—; bd. dirs. Fundacion Maquipucuna, Ecuador, 1988—, Resources for the Future, 1989—, World Resources Inst., 1989—, Peruvian Cultural Ctr., 1989—. Grantee Nat. Geog. Soc., NIH, NSF, Mellon Found., Rockefeller Found.; recipient Ibero-Am. award II Ibero Am. Ornithological Congress, 1983, Cert. of Merit, Goeldi Mus., 1985, 50th Anniversary medal Brazilian Nat. Parks, 1987, Carr medal Fla. Mus. Natural History, 1990; named comdr. Order of Merit of Mato Grosso, 1987, comdr. Order of Rio Branco, Brazil, 1988, UN Environment Programme Global 500 Roll of Honour, 1992. Fellow AAAS (wildlife panel 1981), N.Y. Zool. Soc., Linnean Soc. London, Am. Ornithologists Union; mem. Am. Inst. Biol. Scis. (bd. dirs. 1989—, pres. 1991-92), Ecol. Soc. Am., Brit. Ecol. Soc., Brit. Ornithologists Union, Cooper Ornithol. Soc., Soc. Study of Evolution, Internat. Union for Conservation of Nature (species survival commn.), Soc. for Conservation Biology (gov. 1986-89, pres. 1989-91), Century Club, Cosmos Club, Knickerbocker Club, New York Lawn Club. Home: 8526 Georgetown Pike Mc Lean VA 22102-1206 Office: SI 317 Smithsonian Instn Washington DC 20560

LOVELACE, ALAN MATHIESON, aerospace company executive; b. St. Petersburg, Fla., Sept. 4, 1929; married; 2 children. BA, U. Fla., 1951, MA, 1952, PhD in Chemistry, 1954. Staff mem. Air Force Materials Lab., Wright-Paterson AFB, 1954-72; dir. sci. and tech. Andrews AFB, Washington, 1972-73; prin. dep. asst. sec., dept. R & D USAF, 1973-74; assoc. adminstr. Aerospace Tech. Office, 1974-76; dep. NASA Aerospace Tech. Office, 1976-81; v.p. sci. and engring., space systems div. Gen. Dynamics Corp., San Diego, 1981-82, corp. v.p. quality assurance and productivity, 1982-85, corp. v.p., gen. mgr., 1985—. Fellow AIAA (Goddard Astronautics award 1989, George M. Low Space Transp. award 1992); mem. NAE, Air Force Assn., Sigma Xi. Office: Gen Dynamics Corp Space Systems Div PO Box 85990 San Diego CA 92186-5990

LOVELACE, EUGENE ARTHUR, psychology educator; b. Montour Falls, N.Y., July 27, 1939; s. Ralph Andrew and Doris C. (Wellman) L.; m. Mary Jo Rittinger; children: Kristin Lee, Shelley Anne. BA, Harpur Coll., 1960; MA, U. Iowa, 1963, PhD, 1964. From asst. to assoc. prof. U. Va., Charlottesville, 1964-85; sr. researcher Duke U., Durham, N.C., 1981-82; prof. Alfred (N.Y.) U., 1985—; vis. prof. U. Colo., Boulder, 1974. Editor, co-author: Aging and Cognition, 1990; contbr. 40 articles to profl. jours. Recipient Nat. Rsch. Svc. award Nat. Inst. Aging, 1981-82; fellow NASA, 1962-64. Mem. Am. Psychol. Soc., Gerontol. Soc. Am., Psychonomic Soc. (cons. editor 1977—). Office: Alfred U Sci Ctr Psychology Dept Alfred NY 14802

LOVELL, CHARLES RICKEY, biologist, educator; b. Winter Haven, Fla., July 27, 1957; s. E. Hershell and Noma D. (Ridgway) L. BS, Fla. State U., 1979; PhD, Purdue U., 1984. Postdoctoral rsch. assoc. U. Ga., Athens, 1984-87; asst. prof. U. S.C., Columbia, 1987-93, assoc. prof., 1993—. Mem. editorial bd. Jour. Microbiol. Methods, 1991—; contbr. articles to Applied Environ. Microbiology, Microbial Ecology, Biochemistry, Jour. Biol. Chemistry, Marine Ecology Progress Series, Jour. of Molecular Biology. David Ross fellow Purdue U., 1981-83; grantee NSF, 1989-91, 92—, S.C. Rsch. & Edn. Found./Westinghouse Savannah River Co./Dept. Energy, 1990-91, 92—. Mem. AAAS, Am. Soc. for Microbiology, Estuarine Rsch. Fedn., Sigma Xi. Achievements include co-discovery of first know flavin-containing haloperoxidase (purified from a marine invertebrate); development of functional group specific DNA probe for acetogenic bacteria. Office: U SC Dept Biol Scis Columbia SC 29208

LOVELL, JAMES A., JR., business executive, former astronaut; b. Cleve., Mar. 25, 1928; s. James A. and Blanch L.; m. Marilyn Gerlach; children: Barbara Lynn, James Arthur, Susan Kay, Jeffrey C. Student, U. Wis., 1946-48; B.S., U.S. Naval Acad., 1952; grad., Aviation Safety Sch., U. So. Calif., 1961. Commd. in USN, advanced through grades to capt., 1965; test pilot Navy Air Test Center Patuxent River, Md., 1958-61; flight instr., safety officer Fighter Squadron 101, Naval Air Sta. Oceana, Va; became astronaut with Manned Spacecraft Center, NASA, 1962, dep. dir. sci. and applications directorate Manned Spacecraft Center, NASA, 1971-73, ret., 1973; pres. Fisk Telephone Systems, Inc., 1977-81; sr. v.p. adminstrn. Centel Corp., Chgo., 1980—. Recipient Disting. Service award NASA, 1965, Medal of Honor, 1970; recipient Robert J. Collier trophy, 1969, Grand Medallion award Aero Club France, 1972. Fellow Am. Astronautical Soc. Club: Toastmasters. Made 14 day orbital Gemini 7 flight, Dec. 1965, including rendezvous with Gemini 6, Gemini 12 (last of Gemini series) Nov. 1966, Apollo 8 1st journey to moon, Dec. 21-27, 1968, Apollo 13, aborted and returned to earth, Apr. 11-17, 1970. Office: Centel Corp 8725 W Higgins Rd Chicago IL 60631-2702

LOVELL, ROBERT R(OLAND), engineering executive; b. Gladwin, Mich., Feb. 22, 1937. BS, U. Mich., MS. Various tech. and tech. mgmt. positions Lewis Rsch. Ctr. NASA, Cleve., 1962-80; dir. satellite comm., advanced R & D programs NASA, 1980-87; corp. v.p., pres. divsn. space systems, mgr. Pegasus Program Orbital Sci. Corp., Fairfax, Va., 1987—. Author over 50 tech. publs. Recipient Nat. Medal Tech. U.S. Dept. Commerce Tech. Administrn., 1991, Yuri Gagarin medal USSR Acad. Cosmonautics. mem. Am. Inst. Aeronautics and Astronautics. Achievements include development of unmanned spacecraft technology, communications systems and research in rocket propulsion. Office: Orbital Science Corp 12500 Fair Lakes Circle Fairfax VA 22033*

LOVELL, THEODORE, electrical engineer, consultant; b. Paterson, N.J., May 10, 1928; s. George Whiting and Ethel Carol (Berner) L.; m. Wilma Syperda, May 8, 1948 (div. Oct. 1961); m. Joyce Smelik, July 15, 1962; children: Laurie, Dorothy Jane, Valerie, Cynthia, Karen, Barbara. BEE, Newark Coll. Engring., 1948; postgrad., Canadian Inst. Tech., 1950. Exec. dir. Lovell Electric Co., Franklin Lakes, N.J., 1955-82; ptnr., exec. dir. Lovell Design Services, Swedesboro, N.J., 1982—. Author engring. computer software, 1982. Bd. dirs., treas. Contact "Help" of Salem County, 1991—; pres. Bloomingdale Bd. Edn., N.J., 1970-82; mem. Mcpl. Planning Bd., Bloomingdale, 1980-82, Swedesboro/Woolwich Bd. Edn., 1987—, v.p., 1990-92, pres. 1993—; mayoral candidate Borough of Bloomingdale, 1982. Recipient Outstanding Service award Lake Iosco Co., Bloomingdale, 1985. Mem. Am. Soc. Engring. Technicians, Radio Club Am., Dickinson Theater Organ Soc. Republican. Presbyterian. Avocations: sailing, theater organ music. Home: 502 Liberty Ct Swedesboro NJ 08085-9416 Office: Lovell Design Svcs 530 Commerce St Franklin Lakes NJ 07417-1310

LOVELL, WALTER CARL, engineer, inventor; b. Springfield, Vt., May 7, 1934; s. John Vincent and Sophia Victoria (Klementowicz) L.; m. Patricia Ann Lawrence, May 6, 1951; children: Donna, Linda, Carol, Patricia, Diane, Walter Jr. B of Engring., Hillyer Coll., Hartford, Conn., 1959. Project engr. Hartford Machine Screw Co., Windsor, Conn., 1954-59, design engr. DeBell and Richardson Labs., Enfield, Conn., 1960-62; cons. engr. Longmeadow, Mass., 1962—; freelance inventor Wilbraham, Mass., 1965—. Numerous patents include Egg-Stir mixer, crown closure sealing gasket, circular unleakable bottle cap, sonic wave ram jet engine, solid state heating tapes, card key lock; composer over 50 country-and-Western songs. Achievements include patents for sonic wave ram jet engine, solis state heat & resistor tape, card key lock, security lock system. Office: 348 Mountain Rd Wilbraham MA 01095-1724

LOVRO, ISTVAŃ, aircraft engineer; b. Kisszékely, Tolna M, Hungary, Nov. 15, 1944; s. György and Julianna (Kiss) L.; m. Anna Garas, June 1, 1968; 1 child, Adrienn. Engr. degree, Polytech., Budapest, 1967, econ. degree, 1977. Cert. aircraft engr., economical engr., agrl. aviation specialist. Engr. Air Svc., Budapest, Hungary, 1967-68, 73-92; lectr. Agrl. High Sch., Nyiregyhaza, Hungary, 1968-73; engr., mgr. Aero Kft. Ltd., Budapest, 1992—. Author: Aerodynamics, Aircraft Structure, 1969, Agricultural Aviation, 1970; contbr. articles to profl. jours. Mem. AIAA, Royal Aeronaut. Soc. (U.K.), Sci. Soc. Mech. Engrs. (Hungary), Chamber of Engrs. (Hungary). Achievements include research in increased efficiency of Agricultural Aviation; improved agricultural equipment of aircraft; overhaul of helicopter. Home: Halmi ut 53, H-1115 Budapest Hungary Office: Aero Kft, Koerberki ut 36, H-1112 Budapest Hungary

LOW, BOON CHYE, physicist; b. Singapore, Feb. 13, 1946; came to U.S., 1968; s. Kuei Huat and Ah Tow (Tee) Lau; m. Daphne Nai-Ling Yip, Mar. 31, 1971; 1 child, Yi-Kai. BSc, U. London, Eng., 1968; PhD, U. Chgo., 1972. Scientist High Altitude Observatory Nat. Ctr. for Atmospheric Rsch., Boulder, Colo., 1981-87, head coronal interplanetary physics sect., 1987-90, acting dir., 1989-90, sr. scientist, 1987—; mem. mission operation working group for solar physics NASA, 1992-94. Mem. editorial bd. Solar Physics, 1991-94. Named Fellow Japan Soc. for Promotion of Sci., U. Tokyo, 1978, Sr. Rsch. Assoc., NASA Marshall Space Flight Ctr., 1980. Mem. Am. Physical Soc., Am. Astron. Soc., Am. Geophysical Union. Office: Nat Ctr for Atmosph Rsch PO Box 3000 Boulder CO 80307-3000

LOW, CHOW-ENG, chemistry educator; b. Selama, Perak, Malaysia, May 31, 1938; s. Ah-Cheow and Tiew-Tiah (Ng) L.; m. Teresa Lingchun Kao, June 26, 1966; children: Albert H. C., Cecilia C. C., Jasmine I. M . BS, Chinese U. Hong Kong, 1962; MS, Tex. So. U., 1966; PhD, U. Tex., 1970.

Vis. asst. prof. La. State U., Baton Rouge, 1970-71; postdoctoral rsch. scientist Ind. U., Bloomington, 1972-75; sr. rsch. scientist dept. human biol. chemistry/genetics U. Tex. Med. Br., Galveston, 1976-78; asst. prof. dept. biochemistry George Washington U Med. Ctr., Washington, 1978-83; prof. chemistry Nat. Cheng Kung U., Tainan, Taiwan, Republic of China, 1984—; chmn. dept. Nat. Cheng Kung U., Tainan, Republic of China, 1986-89, dir. Inst. Chem. Rsch., 1986-89; dir. mass spectrometry lab. So. Instrument Ctr., Tainan, 1985-89. Contbg. author: Mass Spectrometry, 1991; editor Jour. Chemistry, 1986—; contbr. articles to profl. jours. Examiner Govt. Employee's Highest Cert. Exam., Ministry of Exam., Taipei, 1989—. Nat. Sci. Coun. grantee, 1985. Mem. AAAS, Am. Chem. Soc., Chinese Chem. Soc., Phi Tau Phi, Sigma Xi. Avocations: badminton, bridge, swimming, hiking. Home: 68-4 Lane 93, Tung-Ning Rd, Tainan 70104, Taiwan Office: Nat Cheng Kung U Dept of Chemistry, 1 Ta-Hsieh Road, Tainan 70101, Taiwan

LOW, EMMET FRANCIS, JR., mathematics educator; b. Peoria, Ill., June 10, 1922; s. Charles Walter and Nettie Alys (Baker) Davis; m. Lana Carmen Wiles, Nov. 23, 1974. B.S. cum laude, Stetson U., 1948; M.S., U. Fla., 1950, Ph.D., 1953. Instr. physics U. Fla., 1950-54; aero. research scientist NACA, Langley Field, Va., 1954-55; asst. prof. math. U. Miami, Coral Gables, Fla., 1955-60; assoc. prof. U. Miami, 1960-67, prof., 1967-72, chmn. dept. math., 1961-66; acting dean U. Miami (Coll. Arts and Scis.), 1966-67, assoc. dean, 1967-68, assoc. dean faculties, 1968-72; prof. math. Clinch Valley Coll., U. Va., 1972-89, dean, 1972-86, chmn. dept. math. scis., 1986-89; emeritus prof. math., 1989—; vis. research scientist Courant Inst. Math. Scis., NYU, 1959-60. Contbr. articles to profl. jours. Served with USAAF, 1942-46. Recipient award for excellence in teaching Clinch Valley Coll., 1988. Mem. Am. Math. Soc., Math. Assn. Am., Soc. Indsl. and Applied Math., Nat. Council Tchrs. of Math., Southwest Va. Council Tchrs. of Math., AAUP, AAAS, Sigma Xi, Delta Theta Mu, Phi Delta Kappa, Phi Kappa Phi. Clubs: Univ. Yacht (Miami, Fla.); Kiwanis. Office: PO Box 3417 Wise VA 24293-3417

LOW, FRANK NORMAN, anatomist, educator; b. Bklyn., Feb. 9, 1911; s. William Wans and Hilda (Nelson) L. BA, Cornell U., 1932, PhD, 1936; DSc (hon.), U. N.D., 1983. Postdoctoral Charlton fellow Sch. Medicine Tufts Coll., Boston, 1936-37; instr. to asst. prof. U. N.C., Chapel Hill, 1937-45; assoc. Sch. Medicine U. Md., Balt., 1945-46; assoc. prof. U. W.Va., Morgantown, 1946; asst. prof. Johns Hopkins Med. Sch., Balt., 1946-49; assoc. prof. to prof. anatomy Sch. Medicine La. State U., New Orleans, 1949-64, vis. prof., 1981—; rsch. prof. anatomy U. N.D., Grand Forks, 1964-81, emeritus, 1981—, Chester Fritz Disting. prof., 1975-77; mem. regional rev. bd. Am. Heart Assn., Grand Forks, 1971-74. Author: (with J.A. Freeman) Electron Microscopic Atlas of Normal and Leukemic Human Blood, 1958; assoc. editor Am. Jour. Anatomy, 1971-91; contbr. over 100 rsch. articles to profl. jours. Participant People's Republic China-U.S. exchange program, People to People; citizen amb. Soviet Union, 1991; del. Anniversary Caravan '91, People to People Internat., Russia, Uzbekistan. Mem. Am. Assn. Anatomists (exec. com. 1976-80, Henry Gray award 1989), Am. Soc. Cell Biology, La. Soc. Electron Microscopy (chmn. 1962), Am. Assn. History of Medicine, World Trade Ctr. (New Orleans), Sigma Xi (pres. U. N.D. chpt. 1977). Avocation: travel, history of medicine. Office: La State Med Ctr 1901 Perdido St New Orleans LA 70112-1328

LOW, ROBERT B., physiology educator; b. Greenfield, Mass., Sept. 19, 1940; s. Merritt B. and Marian J. Low; m. Elizabeth S. Low, Sept. 2, 1967; children—Jonathan, Bronwyn. A.B., Princeton U., 1963; Ph.D., U. Chgo., 1968; postgrad., MIT, 1968-70. Asst. prof. physiology U. Vt., Burlington, 1970-74, assoc. prof., 1974-79, prof., 1979—, assoc. dean for research, 1984—; research fellow U. Geneva, 1979-80; dir. Vt. Pulmonary Specialized Ctr. Research, Burlington, 1976—. Contbr. articles to profl. jours. Recipient numerous grants. Mem. Am. Soc. Cell Biology, Am. Thoracic Soc., Reticuloendothelial Soc., Sierra Club (steering com. No. Vt. 1972-75). Avocations: gardening; hiking; camping; cross-country skiing. Home: RD 1 PO Box 203 Richmond VT 05477 Office: U Vt Dept Physiology E-211 Given Burlington VT 05405*

LOWE, ANGELA MARIA, civil engineer; b. Newark, Nov. 15, 1963; d. Eleanor Gugliocciello; m. Thomas Edward Lowe, Nov. 1, 1986; 1 child, Matthew Richard. BSCE, Pa. State U., 1985. Registered profl. engr., N.J. Engr. Greenhorne & O'Mara, Inc., Greenbelt, Md., 1985-86; structural engr. Goodkind & O'Dea, Inc., Rutherford, N.J., 1986-88; civil engr. Charles Mackie Assocs., Inc., Barnegat, N.J., 1988-90; planning engr. Naval Weapons Sta. Earle, Colts Neck, N.J., 1992-93; v.p., CEO Compro Techs., Inc., Barnegat, 1993—. Admissions vol. Pa. State Alumni Admissions Vol. Program, 1991. Lt. (j.g.) Civil Engring. Corps, USNR, 1990—. Mem. NSPE, Pa. State Alumni Assn., Naval Res. Assn. Home: 28 Deer Run Dr S Barnegat NJ 08005-2216

LOWE, CAMERON ANDERSON, dentist, endodontist, educator; b. Alcester, S.D., Dec. 19, 1932; s. Richard Barrett and Emma Louise (Anderson) L.; m. Doris Teresita Franquez, Dec. 23, 1957; children: Barrett, Steven, Leslie. Student, George Washington U., 1951-53, U. Va., 1955-56; DDS, Georgetown U., 1956-60; cert. residency in endodontics, U.S. Naval Dental Sch., 1969. Commd. lt. (j.g.) U.S. Navy Dental Corps, 1960, advanced through grades to capt., 1976, ret., 1978; pvt. practice endodontist Newport News, Va., 1978-81; assoc. prof. dentistry emeritus Old Dominion U., Norfolk, Va., 1991, asst. chair Sch. Dental Hygiene, 1985-89; adj. asst. prof. Med. Coll. Va.-Va. Commonwealth U. Sch. Dentistry, Richmond, 1979-81. Contbr. articles to profl. jours. Pack and troop chmn. Boy Scouts Am., Guam, 1969-72, Virginia Beach, Va., 1972-78. With USN, 1953-55. Mem. Assn. Mil. Surgeons of U.S., Am. Assn. Endodontists, Am. Acad. Oral Medicine, Am. Dental Assn., Va. Acad. Endodontics, Peninsula Dental Soc., Sigma Alpha Epsilon, Delta Sigma Delta, Sigma Phi Alpha Dental Hygiene Honor Soc. Republican. Methodist. Avocations: tennis, drawing, carving, reading, sculpting. Home: 1497 Wakefield Dr Virginia Beach VA 23455-4541

LOWE, DOUGLAS GEORGE, analytical chemist; b. Jacksonville, Fla., Nov. 18, 1940; s. George Henry and Ophelia Leola (Childres) L.; m. Mildred Arline Province, June 23, 1967; 1 child, Daniel Douglas. BS in Chemistry, Calif. State U., 1968. Analytical chemist Occidental Rsch., Irvine, Calif., 1969-80; chemist ICC Co., Claypool, Ariz., 1980-81; chief chemist Occidental Oil Shale, Grand Junction, Colo., 1981-82; chemist Water Food Rsch. Lab., Portland, Oreg., 1983-85, Cyprus Mineralss Miami, Claypool, 1985--. Republican. Baptist. Office: Cyprus Minerals Miami PO Box 4444 Claypool AZ 85532

LOWE, JOHN BURTON, molecular biology educator, pathologist; b. Sheridan, Wyo., June 13, 1953; s. Burton G. and Eunice D. Lowe. BA, U. Wyo., 1976; MD, U. Utah, 1980. Diplomate Am. Bd. Pathology. Asst. med. dir. Barnes Hosp. Blood Bank, St. Louis, 1985-86; instr. Sch. of Medicine Washington U., St. Louis, 1985, asst. prof. Sch. of Medicine, 1985-86; asst. investigator Howard Hughes Med. Inst., Ann Arbor, Mich., 1986-92; asst. prof. Med. Sch. U. Mich., Ann Arbor, 1986-91, assoc. prof. Med. Sch., 1991—. Contbr. articles to Jour. Biol. Chemistry, Genes and Devel., Nature, Cell. Office: U Mich Howard Hughes Med 1150 W Medical Center Dr Ann Arbor MI 48109-0650

LOWE, JOHN RAYMOND, JR., mechanical engineer; b. Cleve., May 4, 1922; s. John Raymond Lowe and Mildred Esther (Potter) Grover; m. Doris Jean Woolmington, Mar. 27, 1943; children: David B., Cynthia Ann. BSME, 1949. Registered profl. mech. engr., Ohio; cert. energy mgr. Assn. Energy Engrs. Machinist Warner & Swasey, Cleve., 1940-42; indsl. engr. GM, Elyria, Ohio, 1949-51; mgr. motor design Reliance Electric, Cleve., 1952-70; product mgr. Colt Indsl. Elec. Motors, Beloit, Wis., 1971-73; dir. Lau Industries, Dayton, Ohio, 1974-75; designer Andritz Sprout-Bauer, Muncy, Pa., 1976—. Contbr. articles to profl. jours. Pres. Chagrin Falls (Ohio) PTO, 1964-65, Cleve. St. Alumni Assn., 1965-66, Chagrin Falls Bd. of Edn., 1968-70; mem. Jr. Achievement, Cleve., 1966. With U.S. Army, 1944-45. Recipient Purple Heart, Bronze Star, Two Battle Stars, Cert. of Appreciation, 1961. Achievements include design improvements and increased performance that has reduced the cost and broadened the market for double disc pulp refiners that produce the stock for newsprint paper machines. Home: 107 James Rd Lewisburg PA 17837-8851 Office: Andritz Sprout-Bauer Sherman St Muncy PA 17756-1203

LOWE, RONALD DEAN, aerospace engineer, consultant; b. Poteau, Okla., Aug. 19, 1961; s. Dempsey Eugene and Helen Nadine (Lankford) L. BS in Aerospace Engring., U. Okla., 1983; MS in Aerospace Engring., U. Tex., Arlington, 1992. Engr. Gen. Dynamics, Ft. Worth, 1983-87; sr. engr. LTV Aircraft Products Group, Dallas, 1987-89; lead engr. LTV Aerospace and Def., Dallas, 1989-91, engring. specialist, 1991-92; owner Daytowa Enterprises, Arlington, 1985—, The Daytona Group, Arlington, 1992—. Contbr. articles to profl. jours. Recipient Outstanding Engr. Achievement award Soc. Profl. Engrs., 1983. Mem. AIAA, Nat. Mgmt. Assn., Sigma Gamma Tau. Achievements include development of Lowe's upwind scheme and Lowe's high accuracy scheme for the Euler MR Navie-Stokes Equations; co-developer of the Rankine-Hugonzot Flux Difference Splitting Scheme. Home and office: PO Box 501 Tecumseh OK 74873-0501

LOWENKAMP, WILLIAM CHARLES, JR., medical device engineer, researcher, consultant; b. N.Y.C., Oct. 9, 1941; s. William Charles Sr. and Margaret (Poll) L.; m. Martha Marie Engle, Aug. 1964 (div. 1968); 1 child, Kristina Monica; m. Deborah Diane Grimm, July 13, 1991; 1 child, Eric Thomas; 1 stepchild, Cory York. BA, Sussex Coll., Eng., 1977; MSc, City U. L.A., 1987, PhD, 1990. Sr. ops. supr. Johnson & Johnson, Inc., Southington, Conn., 1969-74; sr. mfg. supr. TSI, Inc., New Milford, Conn., 1974-76; sr. mfg.-ops. supr. U.S. Surg. Corp., Stamford, Conn., 1976-78; v.p. Lowenkamp Mfg. Co., Jackson, New Milford, Miss., 1980-87; pres. Lowenkamp Internat. Inc., Hazelhurst, Miss., 1987—; cons. WAH-SHEN, China, 1993—; cons. SAF-T-MED, Dallas, 1993—. Author: On Life and Love, 1982, The Country Poet on the Road Again, 1983, (manual) QA-GMP for Small Medical Device Manufacturers, 1988; contbr. articles to profl. jours. With U.S. Army, 1959-64. Mem. ASTM (voting), AAAS, Am. Soc. for Quality Control (sr.), Am. Chem. Soc., Inst. for Environ. Scis. (sr.), N.Y. Acad. Sci., Internat. Union Pure Analytical Chemists (Europe), Inst. Indsl. Engrs. Achievements include patents for cathether batch processing procedure, for blood collection device; patents pending for retractable needle-syringe, for bio-hazard culture dish, for cord clamp. Home: 13022 Monticello Rd Hazlehurst MS 39083 Office: Lowenkamp Internat Inc 13022 Monticello Rd Hazlehurst MS 39083

LOWENSTEIN, ALFRED SAMUEL, cardiologist; b. Frankfurt, Germany, Oct. 19, 1931; came to U.S., 1938; s. Ernst and Babette (Stern) L.; m. Mirjam Stern, June, 1957 (div. Feb. 1981); children: Esther, David, Eve; m. Lucy Zilbersweig, Nov. 1, 1981; children: Elie, Daniel, Ariel. BA cum laude with honors in German, NYU, 1953; MD, SUNY, Bklyn., 1957. Diplomate Am. Bd. Internal Medicine, Am. Bd. Cardiology. Intern Montefiore Hosp., Bronx, N.Y., 1957-58, resident, 1959-60, fellow in cardiology, 1960-61; resident Bklyn. VA Hosp., 1958-59; fellow in cardiology St. Vincent's Hosp., N.Y.C., 1970-72; internist Far Rockaway, N.Y., 1963-70; attending cardiologist Beilenson Hosp., Petah Tikva, Israel, 1972-75; pvt. practice Cedarhurst, N.Y., 1975—. Capt. U.S. Army, 1961-63. Fellow Am. Coll. Cardiology, N.Y. Cardiol. Soc.; mem. AMA, N.Y. State Nassau County Med. Soc., Phi Beta Kappa. Jewish. Office: 1490 Broadway Hewlett NY 11557

LOWENSTEIN, DEREK IRVING, physicist; b. Hampton Court, Eng., Apr. 26, 1943; came to U.S., 1946; s. Siegfried and Ilse (Mildenberg) L.; m. Elaine Hartmann, July 6, 1968; children: Jessica R., Peter D. BS, CCNY, 1964; MS, U. Pa., 1965, PhD, 1969. Postdoctoral fellow U. Pa., Phila., 1969-70; research assoc. U. Pitts., 1970-73; asst. physicist Brookhaven Nat. Lab., Upton, N.Y., 1973-75; assoc. physicist Brookhaven Nat. Lab., 1975-77, physicist, 1977-83, sr. physicist, 1983—, head Exptl. Planning and Support div., 1977-84, dep. chmn. accelerator dept., 1981-84, chmn. Alternating Gradient Synchrotron dept., 1984—; assoc. mem. U.S.-Russia Joint Coordinating Commn. on Fundamental Properties of Matter, 1983—, U.S.-Japan Commn. on High Energy Physics, 1984—; mem. DOE High Energy Physics Adv. Panel, 1993—. Contbr. articles on particle and accelerator physics to profl. jours. Fellow Am. Phys. Soc.; mem. AAAS, N.Y. Acad. Scis., Sigma Xi. Office: Brookhaven Nat Lab AGS Dept 35 Lawrence Dr Bldg 911 Upton NY 11973-9999

LOWERY, LEE LEON, JR., civil engineer; b. Corpus Christi, Tex., Dec. 26, 1938; s. Lee Leon and Blanche (Dietrich) L.; children: Kelli Lane, Christianne Lindsey. B.S. in Civil Engring, Tex. A&M U., 1960, M.E., 1961, Ph.D., 1965. Prof. dept. civil engring. Tex. A&M U., 1960; rsch. engr. Tex. A&M Rsch. Found., 1962—; pres. Pile Dynamics Found. Engring., Inc., Bryan, Tex., 1962—; pres. Tex. Measurements, Inc., College Station, 1965—; pres. Interface Engring. Assos., Inc., College Station, 1969—; dir. Braver Corp. Bd. dirs. Deep Found. Inst. Recipient Faculty Disting. Achievement award Tex. A&M U., 1979, Zachry award, 1989, 91, award of merit Tex. A&M Hon. Soc., 1991; NDEA fellow, 1960-63. Mem. ASCE, NSPE, Tex. Soc. Profl. Engrs., Sigma Xi, Phi Kappa Phi, Tau Beta Pi. Baptist. Achievements include patents in field. Home: 2905 S College Ave Bryan TX 77801-2510 Office: Tex A&M U Dept Civil Engring College Station TX 77843

LOWI, ALVIN, JR., mechanical engineer, consultant; b. Gadsden, Ala., July 21, 1929; s. Alvin R. and Janice (Haas) L.; m. Guillermina Gerardo Alverez, May 9, 1953; children: David Arthur, Rosamina, Edna Vivian, Alvin III. BME, Ga. Inst. Tech., 1951, MSME, 1955; PhD in Engring., UCLA, 1956-61. Registered prof. engr., Calif. Design engr. Garrett Corp., Los Angeles, 1956-58; mem. tech. staff TRW, El Segundo, Calif., 1958-60, Aerospace Corp., El Segundo, 1960-66; prin. Alvin Lowi and Assocs., San Pedro, 1966—; pres. Terraqua Inc., San Pedro, Calif., 1968-76; v.p. Daeco Fuels and Engring. Co., Wilmington, Calif., 1978—; also bd. dirs. Daeco Fuels and Engring. Co.; vis. research prof. U. Pa., Phila., 1972-74; sr. lectr. Free Enterprise Inst., Monterey Park, Calif., 1961-71; bd. dirs. So. Calif. Tissue Bank; research fellow Heather Found., San Pedro, 1966--. Contbr. articles to profl. jours.; patentee in field. Served to lt. USN, 1951-54, Korea. Fellow Inst. Humane Studies; mem. ASME, NSPE, Soc. Automotive Engrs., Soc. Am. Inventors, So. Bay Chamber Music Soc., Scabbard and Blade, Pi Tau Sigma. Jewish. Avocations: chamber music, jazz, photography, classic automobiles, motor sports, philosophy of science. Home and Office: 2146 W Toscanini Dr San Pedro CA 90732-1420

LOWKE, GEORGE E., biotechnology executive; b. Vernon, Tex., 1939. BS, Tex. A&M U., 1962; MS, North Tex. U., 1965; PhD, U. Ariz., 1969; postgrad., U. Tex., 1970. With Pfizer, 1970-79, Warner-Lambert, 1979-85, Johnson & Johnson, 1983-87; dir. R & D for diagnostic products IGEN, Inc., Rockville, Md., 1987-89; v.p. R & D Life Techs., Inc., Gaithersburg, Md., 1989—. Mem. AAAS, Am. Assn. Clin. Chemists, N.Y. Acad. Scis. Office: Life Techs Inc 8400 Helgerman Ct PO Box 6009 Gaithersburg MD 20884-9980*

LOWN, BERNARD, cardiologist, educator; b. Utena, Lithuania, June 7, 1921; came to U.S., 1935; s. Nisson and Bella (Grossband) L.; m. Louise Charlotte Lown, Dec. 29, 1946; children—Anne Lown Green, Frederick, Naomi Lown Lewiton. BS summa cum laude, U. Maine, 1942, DS (hon.), 1982; MD, Johns Hopkins U., 1945; DSc (hon.), Worcester State Coll., 1983, Charles U., Prague, 1987, Bowdoin Coll., 1988, SUNY, Syracuse, 1988, Columbia Coll., Chgo., 1989; LLD (hon.), Bates Coll., Lewiston, Maine, 1983, Queen's U., Kingston, Ont., Can., 1985; LHD (hon.), Colby Coll., 1986, Thomas Jefferson U., 1988; PhD (hon.), U. Buenos Aires, 1986; D honoris causa, Autonomous U. Barcelona, Spain, 1989; D Univ. (hon.), Hiroshima (Japan) Shudo U., 1989. Asst. in pathology Yale U.-New Haven Hosp., 1945-46; intern in medicine Jewish Hosp., N.Y.C., 1947-48; asst. resident in medicine Montefiore Hosp., N.Y.C., 1948-50; research fellow in cardiology Peter Bent Brigham Hosp., Boston, 1950-53, asst. in medicine, 1955-56, dir. Samuel A. Levine Cardiovascular Research Lab., 1956-58, jr. assoc. in medicine, 1956-62, research assoc. in medicine, 1958-59, assoc. in medicine, 1962-63, sr. assoc. in medicine, 1963-70, dir. Samuel A. Levine Coronary Care Unit, 1965-74, physician, 1973-81, sr. physician, 1982—; asst. in medicine Harvard U., Boston, 1955-58, asst. prof. medicine dept. nutrition Sch. Pub. Health, 1961-67, assoc. prof. cardiology, 1967-73, prof. cardiology, 1974—, dir. cardiovascular research lab. Sch. Pub. Health, 1961—; cons. in cardiology Newton-Wellesley Hosp., Mass., 1963-77, Beth Israel Hosp., Boston, 1963—, Children's Hosp. Med. Ctr., Boston, 1964—; spl. cons. WHO, Copenhagen, 1971; coordinator U.S.-USSR Coop. Study, 1973-81; mem. lipid metabolism adv. com. NIH, Bethesda, Md., 1975-79; vis. prof., lectr., guest speaker numerous univs., hosps., orgns. Author: (with Samuel A. Levine) Current Advances in Digitalis Therapy, 1954; (with Harold D. Levine) Atrial Arrhythmias, Digitalis and Potassium, 1958, (with A. Malliani) Neural Mechanisms and Cardiovascular Disease, 1986; mem. editorial bd. Circulation, Coeur et Medecine Interne, Jour. Electrocardiology; mem. editorial adv. bd. Jour. Soviet Research in Cardiovascular Diseases; contbr. numerous articles to profl. jours.; mem. internat. adv. bd. Internat. Med. Tribune, 1987—; inventor cardioverter; introduced Lidocaine as antiarrythmic drug. Recipient Modern Medicine award, 1972, Ray C. Fish award and Silver medal Tex. Heart Inst., Houston, 1978, A. Ross McIntyre award and Gold medal U. Nebr. Med. Ctr., Omaha, 1979, Richard and Hinda Rosenthal award Am. Heart Assn., 1980, George W. Thorn award Brigham and Women's Hosp., 1982, 1st Cardinal Medeiros Peace medallion, 1982, Nikolay Burdenko medal Acad. Med. Scis. USSR, 1983; co-recipient Peace Edn. award UN Edn., Sci. and Cultural Orgn., 1984, Beyond War award, 1984, Nobel Peace prize, 1985, Ghandi Peace award, 1985, New Priorities award, 1986, Andres Bello medal 1st class Ministry Edn. and Ministry Sci., Venezuela, 1986, Gold Shield, U. Havana, Cuba, 1986, Dr. Tomas Romay y Cahcon Medallion Acad. Sci., Havana, 1986, George F. Kennan award, 1986, Fritz Gietzelt Medaille Council of Medico-Sci. Socs. of German Democratic Republic, 1987; named hon. citizen City of New Orleans, 1978, Pasteur award Pasteur Inst., Leningrad, USSR, 1987, Alumni Humanitarian award U. Maine, Orono, 1988, Internat. Peace and Culture award Soka Gokkai, Tokyo, 1989, Golden Door award Internat. Inst. Boston, 1989; named Disting. Citizen and recipient Key to City Buenos Aires, 1986. Fellow Am. Coll. Cardiology; mem. Am. Soc. for Clin. Investigation, Am. Heart Assn., Assn. Am. Physicians, AAAS, Physicians for Social Responsibility (founder, 1st pres. 1960-70), U.S.-China Physicians Friendship Assn. (pres. 1974-78), Internat. Physicians for Prevention Nuclear War (pres. 1980—); corr. mem. Brit. Cardiac Soc., Cardiac Soc. Australia and New Zealand, Swiss Soc. Cardiology, Belgian Royal Acad. Medicine, Acad. Medicine of Columbia (hon.), Nat. Acad. Scis. (sr. mem. inst. medicine), Phi Beta Kappa, Alpha Omega Alpha. Club: Harvard (Boston). Avocations: photography; music; philosophy; bicycling. Office: Lown Cardiovascular Group PC PO Box 3800 Boston MA 02241-0001

LOWREY, D('ORVEY) PRESTON, III, mechanical engineer; b. Middletown, Conn., July 5, 1952; s. O. Preston Jr. and Silvia G. (Roberts) L.; m. Ying Liu, Apr. 19, 1987; 1 child, Kendall. BS in Geology, Zoology magna cum laude, Duke U., 1974; MSME, N.C. State U., 1981, PhD in Mech. Engring., 1985. Instr. N.C. State U., Raleigh, 1985; vis. prof. Ohio U., Athens, 1986; asst. prof. mech. engring. San Diego State U., 1986-90, assoc. prof., 1990—; summer faculty rsch. fellow Jet Propulsion Lab., Naval Ocean Systems Ctr. Contbr. articles to Solar Energy, other sci. publs. Grantee NSF, 1986, Pacific Gas and Electric Co., 1987-88, San Diego Gas and Electric Co., 1989, Calif. Inst. Energy Efficiency, 1990-91, Calif. Sea Grant, 1990-93. Mem. ASME, Soc. Mech. Engring. Achievements include patents on aquaculture in nonconvective solar ponds, new method of distilling fresh water from seawater without a conventional heat source. Office: San Diego State U Dept Mech Engring San Diego CA 92182-0191

LOWRIE, ALLEN, geologist, oceanographer; b. Washington, Dec. 30, 1937; s. Allen and Mary (Green) L.; m. Mildred C. McDaniel, Feb. 2, 1985; 1 child from previous marriage, Tanya Anne. B.A., Columbia U., 1962. Cert. Assn. Profl. Geol. Scientists, Ark. Bd. Registration Profl. Geologists. Geologist, Lamont Doherty Geol. Obs., Palisades, N.Y., 1963-68; oceanographer U.S. Naval Oceanographic Office, SSC/NASA, Miss., 1968-81, 83—; geologist Mobil Oil Corp., New Orleans, 1981-83; cons. Mobil Research & Devel. Corp., Dallas, Sci. Applications Inc., McLean, Va., Geo-Cons Internat., Inc., Kenner, La., Seagull Internat. Exploration Inc., Houston, Planning Systems, Inc., Slidell, La., Corporacion Miners de Cerro Colorado, Republic of Panama, Hotel Drotama, Santa Marta, Colombia, others. Contbr. articles to profl. jours. Mem. Am. Assn. Petroleum Geologists (cert.), Soc. Econ. Paleontologists and Mineralogists, N.Y. Acad. Scis., Am. Geophys. Union, Sierra Club, Sigma Xi. Episcopalian. Avocations: reading; hiking; ranching. Home: 230 FZ Goss Rd Picayune MS 39466-9707

LOWRY, ROBIN PEARCE, nephrologist, educator; b. Edinburgh, Scotland, Apr. 23, 1947; came to U.S., 1977; s. Edward Loxley-Wood and Catherine Eileen (McCall) L.; m. Barbara M. Glijer, Dec. 12, 1981; children: Philip Martin, Christine Isabelle. BSc, McGill U., Montreal, Can., 1967, MD, 1971. Diplomate Am. Bd. Internat. Medicine, Am. Bd. Nephrology. Intern, resident Royal Victoria Hosp., 1971-74; jr. lectr. faculty of medicine U. Nairobi, Kenya, 1974-75; renal fellow Royal Victoria Hosp. & McGill U., 1975-77; fellow in nephrology Peter Bent Brigham Hosp., Boston, 1977-80; rsch. fellow in medicine Harvard Med. Sch., Boston, 1977-80; asst. prof. McGill U., 1980-86, assoc. prof., 1986-89; assoc. physician Royal Victoria Hosp., Montral, Que., Can., 1983-89; assoc. prof. Sch. Medicine Emory U., Altanta, 1989—; vis. assoc. Calif. Inst. Tech., Pasadena, 1987-89. Mem. editorial bd. Transplantation, 1993—; contbr. chpt. to book. Med. Rsch. Coun. Can. scholar, 1981; Rsch. grantee NIH, 1991. Mem. Am. Soc. Nephrology, Am. Soc. Transplant Physicians (program and publ. com.), Internat Transplant Soc. Democrat. Roman Catholic. Achievements include research in cellular and molecular basis of allograft rejection and transplantation tolerance, identification of the lymphocyte subpopulations which mediate tolerance and rejection, and the soluble factors released by each subset. Office: Emory U Sch Medicine 1364 Clifton Rd NE Atlanta GA 30322

LOWY, ISRAEL, internist, educator; b. Bklyn., Jan. 5, 1957; s. Dow and Fela (Wagshal) L.;m. Julie Martha Ostwald, Aug. 14, 1988; 1 child, Jacob. AB, Princeton U., 1977; MD, Columbia U., 1984, PhD, 1991. Diplomate Am. Bd. Internal Medicine, Am. Bd. Infectious Disease. Med. resident Columbia Presbyn. Med. Ctr., N.Y.C., 1984-87, fellow infectious diseases, 1987-88; asst. prof. medicine Columbia U., N.Y.C., 1988—; mem. sci. bd. Progenics Pharms., Tarrytown, N.Y., 1989—, Nat. Eye Transplant Rsch. Found., N.Y.C., 1990—. Recipient Clinician-Scientist award Am. Heart Assn., Dallas, 1988-93. Mem. ACP, AAAS, Am. Assn. Clin. Rsch., N.Y. Acad. Scis. Achievements include early demonstration of method for gene isolation using DNA mediated gene transfer; evidence that mechanism of inhibition of HIV reverse transcriptuse by dideoxynucleosides is not by previous accepted modes.

LOY, RICHARD FRANKLIN, civil engineer; b. Dubuque, Iowa, July 6, 1950; s. Wayne Richard and Evelyn Mae (Dikeman) L.; m. Monica Lou Roberts, Sept. 2, 1972; children: Taneha Eve, Spencer Charles. BSCE, U. Wis., Platteville, 1973. Registered profl. engr., Wis., Ohio. Engr. aid Wis. Dept. of Transp., Superior, 1969; asst. assayer Am. Lead & Zinc Co., Shullsburg, Wis., 1970; asst. grade foreman Radandt Construction Co., Eau Claire, Wis., 1970; air quality technician U. Wis., Platteville, 1972-73; asst. city engr. City of Kaukauna, Wis., 1973-77; asst. city engr. City of Fairborn, Ohio, 1977-89, city engr., 1989-93; pub. works dir. City of Fairborn, 1993—. Bd. dirs. YMCA Fairborn, 1990—; mem. coun. Trinity United Ch. of Christ, Fairborn, 1989-92; chmn. Chillicothe dist. Tecumseh coun. Boy Scouts Am., 1991-93. Recipient Blue Coat award, 1983; named to Exec. Hall of Fame, N.Y., 1990. Mem. ASCE, NSPE, Am. Pub. Works Assn., Inst. Transp. Engrs.

LOYONNET, GEORGES-CLAUDE, engineer; b. Paris, July 24, 1924; came to U.S., 1962; s. Paul Louis and Edith (Tourn) L.; m. Julie Fara, 1948 (div. 1991); stepchildren: Julie Sullivan, William Sullivan; m. Jung-Ja Moon, July 24, 1991; 1 stepchild, Gina An. Diploma in Engring., Ecole Arts & Metiers, Lyon, France, 1947. Project engr. Air Liquide Engring., Montreal, Can., 1948-58; prodn. mgr. ea. Can. Canadian Liquid Air, Montreal, 1958-60, mgr. of projects, 1960-62; project mgr. Union Carbide - Linde Div., Tonawanda, N.Y., 1962-69; prodn. mgr. Chemetron Corp./Indsl. Gases Div., Chgo., 1969-79, Liquid Air Corp., San Francisco, 1980-86; sr. project mgr. Liquid Air Engring. Corp., Walnut Creek, Calif., 1986—. Recipient 5-yr. Svc. citation Chgo. Pub. Schs. Sci. Fair, 1978. Mem. Compressed Gas Assn. (com. chmn. 1986-88, Svc. Recognition award 1989), ASTM. Achievements include active participation in the large growth of the industrial gas industry and specialty supplies, 1948—; helped start the first large tonnage oxygen plant in the world, 1949-50; responsible for design and construction of first cryogenic methane separation plant and largest oxygen plant in the world. Home: 295 Pimlico Dr Walnut Creek CA 94596 Office: Liquid Air Engring Corp 2121 N California Blvd Walnut Creek CA 94596

LOZANSKY, EDWARD DMITRY, physicist, consultant; b. Kiev, Ukraine, Feb. 10, 1941; s. Dmitry R. and Dina M. (Chizhik) L.; m. Tatiana I. Yershov, Feb. 27, 1971; 1 child, Tania. MS, Moscow Phys. Engring. Inst., 1966; PhD, Inst. Atomic Energy, Moscow, 1969. Asst. prof. Moscow State U., 1969-71; assoc. prof. Mil. Tank Acad., Moscow, 1971-75; prof. U. Rochester, N.Y., 1977-80, Am. Univ., Washington, 1981-83, L.I. U., Bklyn., 1983-87; pres. Independent U., Washington, 1987—; exec. dir. Andrei Sakharov Inst., Washington, 1981-86; pres. Russia House, Inc., 1991—, Am. U. in Moscow, 1992—. Author: Theory of the Spark, 1976, For Tatiana, 1984, Andrei Sakharov, 1986, Mathematical Competitions, 1988. Republican. Jewish. Avocations: skiing, chess, lecturing on Russia. Office: Russia House 1800 Connecticut Ave NW Washington DC 20009

LOZOVATSKY, MIKHAIL, civil engineer, consultant; b. Kiev, Ukraine, USSR, Aug. 3, 1949; came to U.S., 1979; s. Leonid and Rebecca (Tolkatch) L.; children: Helen, Stan. MSCE, Inst. Civil Engring., Kiev, 1973; BSCE, John Hopkins U., 1985. Registered profl. engr., Md. Draftsman Communications Design Inst., Kiev, 1966-69, engr., 1969-72, group leader, 1972-75, project mgr., 1975-79; designer Johnson, Mirmiran & Thompson, Balt., 1980-82; group leader Md. Transp. Authority, Balt., 1982-89, chief design sect., 1989--. Mem. ASCE. Home: 609 Old Crossing Dr Baltimore MD 21208-3332 Office: Md Transp Authority PO Box 9088 Baltimore MD 21222-0788

LU, CATHERINE MEAN HOA, electrical engineer; b. Taipei, Taiwan, China, Oct. 27, 1963; d. Chin Yao and Mei Shan (May) L. BSChemE, Ill. Inst. Tech., 1984, MSEE, 1987. Student rsch. engr. Rockwell Internat., Downers Grove, Ill., 1985, summer intern, 1986; summer intern GM Corp., Rochester Hill, Mich., 1987; rsch. asst. Ill. Inst. Tech. (Advanced Telecomm. Rsch. Lab.), Chgo., 1987-89, Ill. Inst. Tech. (Advanced Digital Signal Processing Lab.), Chgo., 1989—; instr. Ill. Inst. Tech. (Elec. and Computer Engring. Dept.), Chgo., 1992—. Contbr. articles to profl. jours. Recipient Rsch. assistantship IEEE, 1987—; named to Dean's list. Mem. IEEE, Acoustical Soc. Am.. Home: 3140 S Michigan Ave Apt 510 Chicago IL 60616 Office: Dept Elec/Computer Engring Ill Inst Tech Chicago IL 60616

LU, CHENG-YI, mechanical engineer; b. Tainan, Taiwan, Republic of China, Sept. 15, 1954; came to U.S., 1978; s. Nai-Cheng and Yu-Min (Cheng) L.; m. Mei-June Lu; children: Peter, Angela. BSChemE, Tunghai (Taiwan) U., 1976; D of Engring., Cleve. State U., 1983. Sr. rsch. assoc. Cleve. State U., 1983-88; team mgr. Rochetdyne, Rockwell Internat., Canoga Park, Calif., 1988—; Contbr. articles to profl. jours. Fellow Inst. of Advancement of Engring.; mem. AIAA. Office: Rocketdyne 6633 Canoga Ave Canoga Park CA 91303

LU, HSIAO-MING, physicist; b. Ling Xian, Shan Dong, China, Dec. 30, 1956; s. Zhen Bang and Zhong (Chen) L.; m. Rui Qi, Jan. 17, 1982; children: Michael, Diana. BS, Nanjing U., 1982; PhD, Ariz. State U., 1988. Faculty rsch. assoc. U. Nebr. Physics Dept., Lincoln, 1988-91, asst. rsch. prof., 1992-93; instr. Harvard Med. Sch., Boston, 1993—. Mem. Am. Phys. Soc. Achievements include exact analytical theory of resonance Raman scattering including mode mixing and non-condon coupling; first principles simulation of ionic molecular solids. Office: Harvard Med Sch 50 Binney St Boston MA 02115

LU, KWANG-TZU, physicist; b. Chung-King, China, June 28, 1942; came to U.S., 1964; s. Fu-Ting and Sheng-Yung (Wang) L.; divorced. BS, Nat. Taiwan U., 1963; PhD, U. Chgo., 1971. Rsch. assoc. U. Ariz., Tucson, 1970-72; cons. U. Chgo., 1972-73; rsch. assoc. Imperial Coll., London, 1974-76; physicist Argonne (Ill.) Nat. Lab., 1976-84; vis. scientist NBS, Gaithersburg, Md., 1985-87; pres. Atomic Engring. Co., Gaithersburg, Md., 1987—; mem. Com. Line Spectra Element, Washington, 1979-85; vis. prof. Jilin U., China, 1981, Chinses Acad. Scis., Bejing, 1984; mem. faculty Watson Rsch. Ctr., IBM, Yorktown Heights, N.Y., summer, 1982; cons. devel. project World Bank, China, 1987. Author: Progress in Atomic Spectroscopy, Part C, 1984, Photophysics and Photochemistry in VUV, 1985; contbr. over 70 articles to profl. jours. Mem. Am. Optical Soc., Am. Phys. Soc. Achievements include research in Lu-Fane theory of atomic collison and spectroscopy, Rydberg atom in external field, gas lasers; copy rights on scientific data management software systems. Office: Atomic Engring Co PO Box 3342 Gaithersburg MD 20885-3342

LU, PAUL HAIHSING, mining engineer, geotechnical consultant; b. Hsinchu, Taiwan, Apr. 6, 1921; came to U.S., 1962; m. Sylvia Chin-Pi, May 5, 1951; children: Emily, Flora. BS in Mining Engring., Hokkaido U., Sapporo, Japan, 1945; PhD in Mining Engring., U. Ill., 1967. Sr. mining engr., br. chief Mining Dept. Taiwan Provincial Govt., Taipei, 1946-56; sr. indsl. specialist mining and geology U.S. State Dept./Agy. for Internat. Devel., Taipei, 1956-62; rsch. mining engr. Denver Rsch. Ctr. Bur. of Mines, U.S. Dept. Interior, 1967-90; geotech. cons. Lakewood, Colo., 1991—. Contbr. over 60 articles to profl. jours. Rsch. fellow Hokkaido U., 1945-46, Ill. Mining Inst., 1966-67. Mem. Internat. Soc. for Rock Mechanics, Soc. for Mining, Metallurgy, and Exploration (AIME), Mining and Materials Processing Inst. Japan, Chinese Inst. of Mining and Metall. Engrs. (dir., mining com. chair 1960-62, Tech. Achievement award 1962). Achievements include development of prestressed concrete mine supports; invention of new technologies of rock stress measurement with hydraulic borehole pressure cells and measurement of geomechanical properties of rock masses with borehole pressure cells; invention of integrity factor approach to mine structure design. Home and Office: 1001 S Foothill Dr Lakewood CO 80228-3404

LU, RUTH, structural engineer; b. Hong Kong, Mar. 5, 1961; d. Louis Y. and Linda (Lee) L.; m. Joseph Anthony Petrino, Oct. 4, 1986; 1 child, Benjamin V. BSCE, Union Coll., 1983; MSCE, Manhattan Coll., 1989. Registered profl. engr., N.Y. Structural engr. Goldrich, Page and Thropp, N.Y.C., 1983-85, Burns and Roe, Oradell, N.J., 1985-86; project mgr. Envirodyne Engrs., N.Y.C., 1986-89; sr. engr. Metro-North Commuter R.R., N.Y.C., 1990—. Mem. ASCE, Am. Railway Engring. Assn. Home: 517 E 87th St New York NY 10128 Office: Metro-North Commuter R R 11 W 42d St New York NY 10036

LU, TIAN-HUEY, physics educator; b. Kaohsiung, Taiwan, Mar. 5, 1939; s. Ding-Chuen and Shuang-Fang (Chern) Lu; m. Fang-Lan Liu, Apr. 18, 1968; 1 child, Biing-Jyh. BS, Taiwan Normal U., Taipei, 1963; MS, Nat. Tsing Hua U., Hsinchu, Taiwan, 1965; DEng, Nagoya (Japan) U., 1989. Asst. physics lab Nat. Tsing Hua U., 1965-66, instr. theoretical mechanics, physics, electromagnetism, 1966-73, assoc. prof. electromagnetism, physics, biophysics, 1973-79, prof. biophysics, physics, x-ray diffraction, applications, 1979—; vis. scholar dept. chemistry U. Calif., Berkeley, 1982-83; co-rsch. fellow dept. applied chemistry Nagoya U., 1988-89; vis. rsch. fellow Inst. Protein Rsch., Osaka (Japan) U., 1989; referee patent office Cen. Bur. Standards, Taipei, 1974-91; referee Acta Crystallographica, Nagoya, 1989—, Jour. Chinese Chem. Soc., Taipei, 1983—; vis. scholar dept. crystallography U. London, 1993—. Contbr. articles to profl. publs. 2nd lt. Chinese armed forces, 1965-66. Grantee Nat. Sci. Coun., Taipei, 1968—, Vet.'s Gen. Hosp., 1990-91. Fellow Chung-Shan Acad. Found. Achievements include solution of over 50 crystal structures; study of structures of powder, structure determination of biological molecule and material science; deduction of common features of metal compounds with quadridentate ligands. Office: Nat Tsing Hua U Dept Physics, No 101 Sec 2 Kuang Fu Rd, Hsinchu 300, Taiwan

LU, WUAN-TSUN, microbiologist, immunologist; b. Taichung, Taiwan, July 8, 1939; came to U.S., 1964; s. Yueh and Jinmien Lu; m. Rita Man Rom, July 25, 1970; children: Dorcia, Loretta. BS in Agrl. Econs., Nat. Taiwan U., 1960; MS in Microbiology, Brigham Young U., 1968; PhD, U. Okla., 1978. Microbiologist, chemist Murray Biol. Co., L.A., 1969-71; microbiologist Reference Lab., North Hollywood, Calif., 1971-73; rsch. assoc. U. Okla., Okla. City, 1973-78; lab. supr. Reference Med. Lab., San Jose, Calif., 1980; mng. dir. Anakem Labs., Los Gatos, Calif., 1981-85; toxicologist SmithKline Labs., San Jose, 1981; founder, pres. United Biotech, Inc., Mountain View, Calif., 1983—, dir., chmn., mng. dir., 1987—; bd. dirs. Sino-U.S. Hunan Bioengring. Co. Ltd., Internat. Biopharm. Inc., Sacramento. Mem. N.Y. Acad. Sci., Am. Soc. Clin. Pathologists, Am. Soc. Clin. Chemists, Delta Group. Office: United Biotech Inc 110 Pioneer Way # C Mountain View CA 94041-1517

LU, WUDU, mathematics educator; b. Guangzhou, Guangdong, China, Dec. 20, 1947; s. Shu Du and Meng Dan (Liang) L.; m. Yean Wang, Sept. 10, 1984. MS, Zhongshan U., Guangzhou, 1983; postgrad., Chinese U. of Hong Kong, 1992—. Lectr. South China Normal U., Guangzhou, 1983—. Contbr. articles to profl. jours. Avocations: TV, reading, walking, swimming. Office: Chinese U of Hong Kong, Dept Math, Sha Tin New Territories, Hong Kong

LU, YINGZHONG, nuclear engineer, educator, researcher; b. Jiangsu, People's Republic China, June 24, 1926; came to U.S., 1988; s. Zheng and Yi Lu; m. Huaiquing Chen; 1 child, Dan. BSc, Tsinghua U., Beijing, China, 1950. Teaching asst. Tsinghua U., Beijing, 1950-52, lectr., 1952-60, assoc. prof., 1960-79, prof., 1979—, dir. inst. nuclear energy tech., 1960-85, dir. inst. for techno-econs. and energy system analysis, 1985—; tech. advisor Profl. Analysis Inc., Oak Ridge, Tenn., 1990—; advisor Sages' group of energy and environment World Bank, Washington, 1990-91; advisor internat. panel energy program Office of Tech. Assessment, Washington, 1990-92; advisor sci. and tech. panel Global Environment Facility, World Bank, UN Devel. Program and UN Environ. Program, Washington and Nairobi, Kenya, 1991-93. Author: Fueling One Billion--Chinese Energy Policy, 1993, 14 other books and monographs; contbr. more than 60 articles to profl. jours. Recipient 1st Grade Sci. & Tech. Advance award Chinese Govt., 1988, 3d Grade Sci. & Tech. Advance award, 1987. Mem. Am. Nuclear Soc., China Nuclear Soc. Achievements include 2 foreign patents for Deep Pool Type Heating Reactor and Hydraulic Drive for Reactor; development of Chinese nuclear engineering technology and education system, of Chinese modern energy policy, of nuclear heat program in People's Republic China, of radical theory on Chinese economic reform.

LUBAN, MARSHALL, physicist; b. Seattle, May 29, 1936; s. Joseph D. and Sara R. (Gann) L.; m. Pnina Harpak, Oct. 4, 1982; children: Dekel, Aviv, Sarit. BA, Yeshiva U., 1957; MS, U. Chgo., 1958, PhD, 1962; postgrad., Inst. Advanced Study, 1962-63. Asst. prof. U. Pa., Phila., 1963-66; assoc. prof. Bar-Ilan U., Ramat Gan, Israel, 1967-74, prof., 1974-82, chmn. dept. physics, 1967-70, dean faculty natural scis., 1969-71; prof. Iowa State U., Ames, 1982—, chmn. dept. physics and astronomy, 1990—; trustee Jerusalem (Israel) Inst. Tech., 1979-81; vis. prof. Washington U., St. Louis, 1981-82; sr. physicist Ames Lab., 1982—. Contbr. 80 articles to profl. jours. Trustee Bar-Ilan U., 1975-81. Fellow NSF, 1959-62, J.S. Guggenheim, 1966-67. Mem. AAAS, Am. Phys. Soc., Israel Phys. Soc. (pres. 1980-82), Rotary Internat., Sigma Xi. Achievements include research in quantum many body systems; statistical mechanics and phase transitions; localization phenomena; quantum nanostructures. Home: 2801 Torrey Pines Rd Ames IA 50014 Office: Iowa State U Dept Physics and Astronomy Ames IA 50011

LUBCHENCO, JANE, marine biologist, educator; b. Denver, Dec. 4, 1947; married; 2 children. BA, Colo. Coll., 1969; MS, U. Wash., 1971; PhD in Ecology, Harvard U., 1975. Asst. prof. ecology Harvard U., Cambridge, Mass., 1975-77; from asst. prof. to assoc. prof. Oreg. State U., Corvallis, 1978-88, prof. zoology, 1988—; rsch. assoc. Smithsonian Inst., 1978—; prin. investigator NSF, 1976—; adv. panel long term ecol. rsch. programs, 1977; vis. asst. prof. Discovery Bay Marine Lab., 1976; sci. adv. Ocean Trust Found., 1978-84, West Quoddy Marine Sta., 1981-88; vis. assoc. prof. U. Antofagasta, Chile, 1985, Inst. Oceanography, Qingdao, China, 1987. Fellow John D. and Katherine T. MacArthur Found., 1993. Mem. Ecol. Soc. Am. (George Mercer award 1979, mem. coun. 1982-84, chair awards com. 1983-86, nominating com. 1986), Phycological Soc. Am. (nat. lectr. 1987-89), Am. Soc. Naturalists, Am. Soc. Zoologists, Am. Inst. Biol. Sci. Achievements include research in population and community ecology, plant-herbivore and predator-prey interactions, competition, marine ecology, algal ecology, agal life histories, biogeography and chemical ecology. Office: OR State U Dept of Zoology Corvallis OR 97331*

LUBECK, MARVIN JAY, ophthalmologist; b. Cleve., Mar. 20, 1929; s. Charles D. and Lillian (Jay) L. A.B., U. Mich., 1951, M.D., 1955, M.S., 1959. Diplomate Am. Bd. Opthamology; m. Arlene Sue Bitman, Dec. 28, 1955; children: David Mark, Daniel Jay, Robert Charles. Intern, U. Mich. Med. Ctr., 1955-56, resident ophthalmology, 1956-58, jr. clin. instr. ophthalmology, 1958-59; pvt. practice medicine, specializing in ophthalmology, Denver, 1961—; mem. staff Rose Hosp., Porter Hosp., Presbyn. Hosp., St. Luke's Hosp.; assoc. clin. prof. U. Colo. Med. Ctr.; cons. ophthalmologist State of Colo. With U.S. Army, 1959-61. Fellow ACS; mem. Am. Acad. Ophthalmology, Denver Med. Soc., Colo. Ophthalmol. Soc., Am. Soc. Cataract & Refractive Surgery. Home: 590 S Harrison Ln Denver CO 80209-3517 Office: 3865 Cherry Crk North Dr Denver CO 80209-3803

LUBINSKY, ANTHONY RICHARD, physicist; b. Cleve., Sept. 26, 1946; s. Anthony and Anne (Beskid) L.; children: Steven, Michael. BS, Case Western Res. U., 1968; MS, Northwestern U., 1969, PhD, 1975. Project mgr. Xerox Corp., Rochester, N.Y., 1975-83; rsch. scientist Eastman Kodak Co., Rochester, 1983—. Contbr. to profl. publs. Mem. Am. Phys. Soc., Soc. for Imaging Sci. and Tech., Am. Assn. Physicists in Medicine. Achievements include invention of systems for electrophotographic development used in various copiers and duplicators, liquid electrophotographic method for color proofing, method for computed radiography, other inventions in hard copy output. Home: 128 Country Manor Webster NY 14580 Office: Eastman Kodak Co Research Labs 182 Rochester NY 14650-2123

LUBKIN, VIRGINIA LEILA, ophthalmologist; b. N.Y.C., Oct. 26, 1914; d. Joseph and Anna Fredericka (Stern) L.; m. Arnold Malkan, June 6, 1944 (div. 1949); m. Martin Bernstein, Aug. 28, 1949; children: James Ernst, Ellen Henrietta, Roger Joel, John Conrad. BS, NYU, 1933; MD, Columbia U., 1937. Diplomate Am. Bd. Ophthalmology. Intern Harlem Hosp., N.Y.C., 1938-40; asst. resident neurology Montefiore Hosp., N.Y.C., 1940, asst. resident pathology, 1940-41, fellow in ophthalmology, 1941-42; resident ophthalmology Kings County Hosp., Bklyn., 1942-43, Mt. Sinai Hosp., N.Y.C., 1943-44; attending opthalmologist, assoc. clin. prof. emeritus Mt. Sinai Sch. Medicine, 1944—; pvt. practice N.Y.C., 1945-90; rsch. prof. N.Y. Med. Coll., 1986—; creator, chief of rsch. bioengineering lab. N.Y. Eye and Ear Infirmary (name now The Aborn), N.Y.C., 1978—; creator first grad. course in oculoplastics and bi-yearly symposia in devel. dyslexia Mt. Sinai Sch. Medicine; educator courses in psychosomatic ophthalmology Am. Acad. Ophthalmology, 1950-60, educator course in complications of blepharoplasty, 1980-90; bd. dirs. Jewish Guild for the Blind; tchr. surg. ophthalmology in French Cameroon, Presbyn. Mission, 1951; lectr. in numerous countries including India, 1976, 92, Pakistan, 1976, 84, China, 1978, Sri Lanka, 1979, South Africa, 1982, Singapore, 1984, Thailand, 1984, Argentina, 1986, Peru, 1987. Author: (with others) Ophthalmic Plastic and Reconstructive Surgery, 1989; contbr. articles to profl. jours. Grantee Intraocular Lens Implant Mfrs., 1989. Fellow AMA, AAAS, Am. Soc. Ophthalmic Plastic and Reconstructive Surgery (founding), Am. Coll. Surgeons, N.Y. Acad. Medicine, N.Y. Acad. Scis., Am. Acad. Ophthalmology, Am. Soc. Cataract and Refractive Surgery, PanAm. Soc. Ophthalmology, Soc. Light Treatment and Biol. Rhythms. Democrat. Home and Office: One Blackstone Pl Bronx NY 10471

LUBOWSKY, JACK, academic director; b. Bklyn., July 11, 1940; m. Marcelle Kaplan, Jan. 1, 1986. BEE, CCNY, 1962; MSEE, Polytech. Inst. Bklyn., 1966, PhD in Elec. Engring., 1973. Registered profl. engr., N.Y. Project engr. Airborne Instruments Lab., Melville, N.Y., 1962-66; dir. sci./acad. computing ctr. SUNY-Health Sci. Ctr., Bklyn., 1966—; chmn. SUNY-FACT com. Contbr. 40 articles to profl. jours.; patent on imaging in a random medium. Recipient Congl. Sci. fellowship AAAS, Washington, 1983. Mem. IEEE (sr. mem., chmn. U.S. govt. activities coun., chmn. fed. legis. agenda task force for 100th and 103d Congress, chmn. Washington internships for students of engring. program, mem. supercomputer com., congl. fellows com, computer soc., instrumentation soc., engring in biology and medicine soc., vice chmn. U.S. activities bd.), Eta Kappa Nu, Sigma Xi. Home: 2064 Beverly Way Merrick NY 11566-5416 Office: SUNY Health Sci Ctr Bklyn Sci-Acad Computing 450 Clarkson Ave # 7 Brooklyn NY 11203-2098

LUCA-MORETTI, MAURIZIO, nutrition scientist, researcher; b. Rome, June 2, 1945; s. Giuseppe and Elena (Moretti) L.; m. Anna Grandi, Jan. 2, 1974; 1 child, Elena. BS, Ministry of Edn., Caracas, Venezuela, 1969; MS in Allied Health Scis., Pacific Western U., 1990, PhD, 1990, DSc in Human Nutrition, 1990. Dir. rsch. Inst. Italiano di Terapia Fisica e Medicina Interna, Rome, 1974-80, Caracas, 1973-88; dir. rsch. human nutrition rsch. program, AIDS rsch. program InterAm. Med. and Health Assn., Boca Raton, Fla., 1989—; cons. clin. nutrition Civic Hosp. Pescara, Italy, 1990, invited scientist, 1991—; invited prof. sch. medicine U. Chieti, Italy, 1991—. Contbr. articles to Madrid Medico, Federazione Medica, JIMHA, Advances in Therapy; dep. editor: Internat. Jour. Immunopathology and Pharmacology, 1990—; editor in chief Jour. InterAm. Med. and Health Assn., 1991—. Mem. Royal Nat. Acad. Medicine, Nat. Acad. Medicine, N.Y. Acad. Scis., AAAS, Am. Pub. Health Assn., Internat. AIDS Soc., Am. Soc. for Parenteral and Enteral Nutrition, Am. Med. Writers Assn., Inventors Club Am. Achievements include discovery of synthetic chemical foods. Office: InterAm Med and Health Assn 3025 Saint James Dr Boca Raton FL 33434

LUCAS, BILLY JOE, philosophy educator; b. Houston, Jan. 7, 1942; s. Joseph Cuthel and Billie Louise (Smith) L.; m. Diana Stephens, July 13, 1965 (div. 1976); children: Lisa Ann, Deborah Lynn; m. Shelby Hearon, Apr. 19, 1981. BA, U. Houston, 1970; MA, McMaster U., 1972; PhD, U. Tex., 1981. Instr. Houston Community Coll., 1973-77; asst. instr. U. Tex., Austin, 1978-81; asst. prof. Manhattanville Coll., Purchase, N.Y., 1981-86, assoc. prof., 1986-89, prof., 1989—; edn. policy cons. to the pres. Manhattanville Coll., 1990-91, 92-93; invited del. Citizem Amb. Program Philosophy Del., People's Rep. China, 1993. Assoc. editor Internat. Jour. for Philosophy of Religion, 1990—; contbr. articles to profl. jours. Mem. AAAS, Soc. for Philosophy of Religion (exec. coun. 1986-93, v.p. 1989-90, pres. 1990-91), Assn. for Symbolic Logic, Computer Soc. of IEEE (tech. com. on multiple valued logic 1987—), Am. Acad. Religion, Am. Philos. Assn. (logic sect. adv. com. 1990-93), N.Am. Soc. for Social Philosophy (exec. coun. 1986-93, co-chair Eastern div. 1986-90), Am. Math. Soc., N.Y. Acad. Scis., Phi Kappa Phi. Achievements include complete axiomatization of the minimal modal logic in which each formula is provably equivalent to a formula of finite degree, related work on reduction laws in deontic and other modal logics, applications of formal logic to philosophy, and logic curriculum development. Office: Manhattanville Coll 2900 Purchase St Purchase NY 10577-2400

LUCAS, DONALD, physical chemist; b. Chgo., Sept. 29, 1951; s. Albert and Angelina (Seleskis) L.; m. Vicki Lynne Britt, Sept. 26, 1981; children: Scott, Kelsey. BS, IIT, 1972; PhD, U. Calif., Berkeley, 1977. Staff scientist Lawrence Berkeley (Calif.) Lab., 1980--; vis. rsch. engr. U. Calif., Berkeley, 1989--, lectr., 1986-91; vis. prof., rsch. assoc. Ind. U., Bloomington; exec. com. Western States sect. Combustion Inst., 1992--; tech. advisor CAlif. Inst. for Energy Efficiency, Berkeley, 1991--. Contbr. articles to profl. jours. Mem. Am. Chem. Soc., Combustion Inst., Sigma Xi. Achievements include development of new laser and non-laser detection methods for hazardous wastes. Office: Lawrence Berkely Lab B29C Berkeley CA 94720

LUCAS, JOHN STEWART, marine biologist; b. Perth, Australia, Dec. 25, 1940; s. John Stewart and Bertha Violet (Foote) L.; m. Helen Rhys Davies, Sept. 2, 1966; children: Jenny, Robyn, Wendy. BSc in Zoology with 1st class honours, U. Western Australia, 1964, PhD, 1969. Lectr. zoology Univ. Coll., Townsville, Australia, 1968-76; sr. lectr. James Cook U., 1977-85; joint project coord. Australian Ctr. for Internat. Agrl. Rsch., 1984-89, 89-92; assoc. prof. zoology James Cook U., 1986—; study leaves at Duke U. Marine Lab., N.C., 1974-75, CSIRO Western Marine Labs., Australia, U. Ryukyus, Okinawa, Japan, and Scripps Inst. Oceanography, Calif., 1981-82, Tonga Fisheries and Shellfish Culture Ltd., Tasmania, 1992. Author: (with C. Birkeland) Acanthaster planci: Major Management Problem of Coral Reefs, 1990; contbr. chpts. to books. Recipient Australian Marine Scis. Assn. prize, 1967; grantee Adv. Com. on Rsch. into Crown-of-thorns Starfish, 1973-75, Australian Rsch. Grants Com., 1977-79, Australian Marine Scis. & Tech. Adv. Com., 1981-83, Marine Scis. & Tech. Grants Com./Crown-of-thorns Starfish Adv. Com./Crown-of-thorns Starfish Rsch. Com., 1985-92, Fishing Industry Rsch. Com. of Australia, 1987-89, Australian Rsch. Coun., 1990-92. Fellow ISCast; mem. Australian Inst. Biology, Australian Marine Scis. Assn., Internat. Soc. for Reef Studies, Australian Coral Reef Soc., Australian Mariculture Assn., World Aquaculture Soc., Asian Fisheries Soc. Achievements include description of life-cycle of the Crown-of-thorns starfish and demonstration of potential importance of environmental influences on early developmental stages in triggering its population outbreaks.Research on the breeding of the giant clam Tridacna gigas. Office: J Cood Univ of N Queensland, Dept Zoology, Townsville 4811, Australia

LUCAS, MELINDA ANN, health facility director; b. Maryville, Tenn., June 27, 1953; d. Arthur Baldwin and Dorthy (Shields) L. BA, Maryville Coll., 1975; MS, U. Tenn., 1976, MD, 1981. Diplomate Am. Bd. Pediatrics; lic. dr. N.Y., Tenn. Intern in pediatrics U. Rochester, N.Y., 1981-82, resident in pediatrics, 1982-84; pvt. practice Maryville, Tenn., 1984-85; emergency room pediatrician U. Tenn. Med. Ctr., Knoxville, 1985-90, dir. child abuse clinic, 1987-90, pediatric intensivist, 1987—, acting dir. pediatric ICU, 1990—; mem. Pediatric Cons., Inc., Knoxville; physician rep. Project Search Working Symposium, 1990. Contbr. articles to profl. jours. Mem. Blount County Foster Care Rev. Bd., Maryville, Tenn., 1985—, Blount County Exec. Bd., Maryville Coll. Alumni Assn. United Presbyn. Ch. scholar, 1971, Mary Lou Braly scholar, 1971, 72, 73; U. Tenn. Med. Ctr. faculty, 1988-89, 91—. Fellow Am. Acad. Pediatrics; mem. AMA (Physician Recognition award 1984-87, 88-91, 91—), Am. Profl. Soc. on Abuse of Children, Knoxville Area Pediatric Soc., Soc. Critical Care Medicine (abstract reviewer 1991, 92, 93). Methodist. Avocations: basketball, tennis, piano. Home: 1608 Mcilvaine Dr Maryville TN 37801-6230

LUCCA, DON ANTHONY, mechanical engineer; b. Hammonton, N.J., Feb. 9, 1954; s. Domenic and Marie (Luca) L.; m. Joyce Ann Miller, Sept. 18, 1982. BS, Cornell U., 1976; MSE, Princeton U., 1979; PhD, Rensselaer Polytechnic Inst., 1982. Cert. mfg. engr. Mgr. advanced process design and devel. Luke Tool & Engring., Hammonton, 1982-88; asst. prof. indsl. and systems engring. Ohio State U., Columbus, 1988-89; assoc. prof. Sch. Mech. and Aerospace Engring. Okla. State U., Stillwater, 1990—; guest scientist Los Alamos (N.Mex.) Nat. Lab., 1989—. John McMullen scholar Cornell U., Ithaca, N.Y., 1972-76; rsch. grantee NSF, Washington, 1989—; recipient Wire Awareness award Wire Assn. Internat., Guilford, Conn., 1979. Mem. Soc. Mfg. Engrs. (sr. mem., Outstanding Young Mfg. Engr. award 1990, periodical publs. com. 1990—), Am. Soc. Precision Engring. (bd. dirs., chmn. edn. com), ASME, Soc. Tribologists and Lubrication Engrs., Am. Phys. Soc., Sigma Xi, Pi Tau Sigma. Home: 1417 S Fairfield Dr Stillwater OK 74074 Office: Okla State U Sch Mech and Aero Engring 218 Engineering N Stillwater OK 74078

LUCCHESI, ARSETE JOSEPH, mathematician, university dean; b. Bagni di Lucca, Italy, Aug. 1, 1933; came to U.S. 1935; s. Frank and Ormeda (Cinquini) L.; m. Agnes Ann Caminiti, Aug. 19, 1962; children: Arthur John, Lisa Claire. BS in Math., Queens Coll., 1956; MS in Math., NYU, 1959. Project engr. Bendix Aviation, Teterboro, N.J., 1956-59; instr. Sch. Engring., Cooper Union, N.Y.C., 1959-62, asst. prof. math., 1962-70, dir. computer ctr., 1969-78, assoc. prof., 1970-81, prof. math., 1981—, assoc. dean engring., 1988—; assoc. dir. Comprehensive Math & Sci. Program, N.Y.C., 1976-86; statis. cons. N.Y.C. Dept. Air Resources, 1981-82. Bd. trustees The Chad Sch. Founc., Newark, 1991—, Sch. for Physical City, N.Y.C.; adv. coun. N.Y.C. Transit Museum; mem. coun. City of North Plainfield (N.J.), 1977-84; mem. Somerset County Econ. Devel., 1980-82, North Plainfield Planning Bd., 1977-79. NSF fellow, 1965; recipient Bronze award Coun. for Advancement and Support of Edn., 1989. Mem. Am. Soc. Engring. Edn., Math. Assn. Am. Roman Catholic. Home: One Mali Dr North Plainfield NJ 07062 Office: Cooper Union 51 Astor Pl New York NY 10003-7139

LUCCHESI, JOHN C., biology educator. BA, La Grange Coll., 1955; MS, U. Ga., 1958; PhD, U. Calif., 1963. Cary C. Boshamer prof. biology U. N.C., Chapel Hill, 1965-90; Asa G. Candler prof., chmn. biology Emory U., Atlanta, 1990—. Office: Emory U Dept of Biology Atlanta GA 30322

LUCCHESI, LIONEL LOUIS, lawyer; b. St. Louis, Sept. 17, 1939; s. Lionel Louis and Theresa L.; m. Mary Ann Wheeler, July 30, 1966; children: Lionel Louis III, Marisa Pilar. BSEE, Ill. Inst. Tech., 1961; JD, St. Louis U., 1969. Bar: Mo. 1969. With Emerson Electric Co., 1965-69; assoc. Polster, Polster & Lucchesi, St. Louis, 1969-74, ptnr., 1974—; city atty. City of Ballwin (Mo.), 1979-85, 92—. Alderman City of Ballwin, 1977-79, mem. Zoning Commn., 1971-77. Served to lt. USNR, 1961-65. NROTC scholar, 1957-61; recipient Am. Jurisprudence award St. Louis U., 1968-69. Mem. ABA, Am. Patent Law Assn., Assn. Trial Lawyers Am., St. Louis Met. Bar Assn. (exec. com., pres.-elect 1984, pres. 1985-86), Newcomen Soc. N.Am. Republican. Roman Catholic. Clubs: Forest Hills, Rotary (St. Louis) (pres. elect 1991-92, pres. 1992-93). Office: 763 S New Ballas Rd Saint Louis MO 63141

LUCE, TIMOTHY CHARLES, physicist; b. Greenville, Ill., Sept. 28, 1960; s. Charles Howard and Jane Clair (Boehmer)L.; m. Sonja Lee Fischer, May 23, 1981; 1 child, Rachel Marie. BS, U. Ark., 1982; MA, Princeton U., 1984, PhD, 1987. Sr. scientist Gen. Atomics, San Diego, 1987-90, assoc. staff scientist, 1990—. Mem. Am. Phys. Soc. (divsn. plasma physics). Achievements include research in electron cyclotron heating and current drive, perturbative and steady-state transport studies in magnetic confinement controlled nuclear fusion plasmas, inward energy transport in tokamak plasmas, recent ECH results from DIII-D tokamak. Office: Gen Atomics PO Box 85608 San Diego CA 92186

LUCID, SHANNON W., biochemist, astronaut; b. Shanghai, China, Jan. 14, 1943; d. Joseph O. Wells; m. Michael F. Lucid; children: Kawai Dawn, Shandara Michelle, Michael Kermit. BS in Chemistry, U. Okla., 1963, MS in Biochemistry, 1970, PhD in Biochemistry, 1973. Sr. lab. technician Okla. Med. Rsch. Found., 1964-66, rsch. assoc., from 1974; chemist Kerr-McGee, Oklahoma City, 1966-68; astronaut NASA Lyndon B. Johnson Space Ctr., Houston, 1979—; mission specialist flights STS-51G and STS-34 NASA Lyndon B. Johnson Space Ctr., mission specialist on Shuttle Atlantis Flight, 1991. First woman to fly on the shuttle three times. Address: NASA Johnson Space Ctr Astronaut Office Houston TX 77058

LUCKE, JOHN EDWARD, mechanical engineer; b. Mpls., July 10, 1963; s. William John and Patricia Ann (Moran) L.; m. Joyce Joanna Freeman, Nov. 9, 1985. BSMET, Milw. Sch. Engring., 1985. Registered profl. engr., Wis. Mech. engr. Kohler (Wis.) Co., Engine Div., 1985-90, mech. project engr., 1990; sr. engr. Cummins Engine Co., Columbus, Ind., 1990—. Vol. Housing Partnership Inc., Columbus, 1991—. Mem. NSPE, Soc. Automotive Engrs. Office: Cummins Engine Co PO Box 3005 Columbus IN 47202-3005

LUCKENBACH, ALEXANDER HEINRICH, dentist; b. Heidenheim, Fed. Republic Germany, Aug. 1, 1956; s. Siegfried and Gertraud (Hirsch) L.; m. Christine O. Heckmann, June 13, 1987. DDS, U. Tuebingen, Fed. Republic Germany, 1981; DMD, U. Tuebingen, 1983. Clin. dentist U. Tuebingen, 1983-85, asst. prof. dept. prosthodontics, 1986—, v.p. R&D, cons., 1988—, staff ZMK-Klinik, 1989%; tech. design cons. Girrbach Dental Co., Pforzheim, Fed. Republic Germany, 1985—; pres. Advanced Dental Computer Techs. Co., Tuebingen, 1991—. Author software; patentee techniqes for dental appliances, computer aided 3D joint movement recording system. Fellow German Gesellschaft F. Zahn, Mund and Kieferheilkunde, German Rsch. Soc. (spl. rsch. unit implantology), Apple Programmers and Developers Assn. Avocations: travelling, hiking, music, photography. Office: Bus Park Praxis, Zettachring 4, D-70567 Stuttgart Germany

LUCKETT, JOHN PAUL, JR., medical radiation physicist; b. Baton Rouge, La., Dec. 28, 1962; s. Jack and Joan Leona (Herman) L.; m. Suzanne Shaye Smith, July 29, 1990; 1 child, John Paul III. BS in Zool., La. State U., 1985, MS in Nuclear Sci., 1990. Dosimetrist Mary Bird Perkins Cancer Ctr., Baton Rouge, 1989-90, med. radiation physicist, 1990–; assoc. mem. Med./Physics/Dental Staff, Baton Rouge, 1990–; mem. specified profl. staff Our Lady of the Lake Regional Med. Ctr., Baton Rouge, 1990–. Mem. Am. Assn. Med. Physicist. Republican. Roman Catholic. Office: Mary Bird Perkins Cancer Ctr 4950 Essen Ln Baton Rouge LA 70809-3432

LUCKEY, GEORGE WILLIAM, research chemist; b. Dayton, Apr. 17, 1925; s. George Paul and Olive (Lehmer) L.; m. Doris Waring, Mar. 29, 1958; children: Robert, Jana, John. BA in Chemistry, Oberlin Coll., 1947; PhD in Chemistry, U. Rochester, 1950. Rsch. and staff asst. Eastman Kodak Co., Rochester, N.Y., 1950-59, rsch. assoc., 1959-69, lab. mgr., rsch. fellow, 1969-86. Contbr. articles to profl. jours. Mem. Am. Chem. Soc., Am. Phys. Soc., The Electrochem. Soc., Royal Soc. Chemistry, Sigma Xi, Phi Beta Kappa. Achievements include U.S. and fgn. patents; participation in improvements in diagnostic imaging with x-rays by improvements in intensifying screens, films and processing systems; improvement of performance of systems for mammography, other diagnostic uses. Home: 240 Weymouth Dr Rochester NY 14625-1917

LUCKING, PETER STEPHEN, industrial engineering consultant; b. Kalamazoo, Oct. 11, 1945; s. Henry William, Sr., and Mary (Lynn) L.; m. Marilyn Barbara Jensen, Dec. 18, 1971. BA, Western Mich. U., 1968; BS in Indsl. Engring., 1973. Indsl. engr. Motorola, Phoenix, 1974, Revlon, Inc., Phoenix, 1974-75; indsl. engr. Hooker Chem. and Plastics Co., Niagara Falls, N.Y., 1975-76, sr. corp. indsl. engr., 1976-77; indsl. engr. Carborundum Co., Niagara Falls, 1977-78; cons. H.B. Maynard and Co., Pitts., 1978-85; mgr. indsl. engring. Carrier, Tyler, Tex., 1985-88; cons. H.B. Maynard and Co., Pitts., 1988-92; pres. MARPET Systems, Inc., 1992—; lectr. in field, 1989. Advisor, Jr. Achievement, Niagara Falls, 1977. Author: (chtp. in book) Handbook of Industrial Engineering. Served with U.S. Army, 1969-70, Vietnam. Mem. Inst. Indsl. Engrs. (sr. mem., region v.p. 1983-85), Inst. Indsl. Engrs. (pres. Niagara Frontier chpt. 1977-78, paper presented fall conf.), Soc. Mfg. Engrs. (sr.). Democrat. Roman Catholic. Home: 12826 Weatherstone Dr Florissant MO 63033-4045 Office: MARPET Systems Inc Ste 141 11200 W Flossant Saint Louis MO 63033

LUCKNER, HERMAN RICHARD, III, interior designer; b. Newark, Ohio, Mar. 14, 1933; s. Herman Richard and Helen (Friednour) L. BS, U. Cin., 1957. Cert. interior designer and appraiser. Interior designer Greiwe Inc., Cin., 1957-64; owner, internat. designer Designers Loft Interiors, Cin., 1964—; owner Designer Accents, Cin., 1991—. Mem. bd. adv. Ohio Valley Organ Procurement Ctr., Cin., 1987—, U. Cin. Fine Arts Collection and Hist. Southwest Ohio. Mem. Am. Soc. Interior Designers, Appraisers Assn. Am., Metropolitan Club. Republican. Avocations: needlepoint, collecting 18th century Chinese porcelain. Home and Office: 555 Compton Rd Cincinnati OH 45231-5005

LUDDEN, JOHN FRANKLIN, financial economist; b. Michigan City, Ind., May 6, 1930. BS in Econs., U. Wis., 1952, MS in Econs., 1955; postgrad., U. Mich., 1955-59. Wage and hour investigator U.S. Dept. Labor, 1960, mgmt. intern, 1960-61, labor economist, 1963; economist, instr. U.S. Bur. of Labor Statis., 1961-63; economist Office of Internat. Ops. IRS, 1963-68, fin. economist Audit div., 1968-86, fin. economist Office of the Asst. Commr. Internat., 1986—. With U.S. Army, 1952-54. Recipient Spl. Svc. award U.S. Dept. of Treasury, 1967, 68, Spl. Achievement award U.S., 1984, Spl. Svc. award, 1987, Spl. Act award, 1990. Mem. Am. Econ. Assn.

LUDORF, MARK ROBERT, cognitive psychologist, educator; b. Milw., Apr. 15, 1963; s. Wayne Francis and Charlotte Floris (Baltes) L.; m. Lynn Marie Frasheski, Dec. 22, 1989; 1 child, Katherine. BS, U. Wis., Stevens Point, 1986; MA, U. Kans., 1989, PhD, 1991. Asst. prof. dept. psychology Stephen F. Austin State U., Nacogdoches, Tex., 1990—. Mem. Am. Psychol. Soc., Midwestern Psychol. Assn., Southwestern Psychol. Assn. Roman Catholic. Office: Stephen F Austin State U PO Box 13046 SFA Sta Nacogdoches TX 75962

LUDWIG, GEORGE HARRY, physicist; b. Johnson County, Iowa, Nov. 13, 1927; s. George McKinley and Alice (Heim) L.; m. Rosalie F. Vickers, July 21, 1950; children: Barbara Rose, Sharon Lee, George Vickers, Kathy Ann Ramsay. BA in Physics cum laude, U. Iowa, 1956, MS, 1959, PhD in Elec. Engring., 1960. Head fields and particles instrumentation sect. Goddard Space Flight Center, NASA, 1960-65, chief info. processing div., 1965-71, assoc. dir. for data ops., 1971-72; dir. systems integration Nat. Environ. Satellite Service, NOAA, 1972-75, dir. ops., 1975-80, tech. dir., 1980; sr. scientist Environ. Rsch. Labs. NOAA, Boulder, Colo., 1980-81; dir. Environ. Rsch. Labs. NOAA, 1981-83; asst. to chief scientist NASA, 1983-84 ret., ind. cons. data mgmt. and space sta. design, 1983—; sr. rsch. assoc. Lab. for Atmospheric and Space Physics U. Colo., 1985-91; vis. sr. scientist NASA hdqrs., Calif. Inst. Tech., 1989-91; prin. designer radiation detection instrumentation for numerous sci. spacecraft including Explorer I, 1956-65; co-discoverer Van Allen radiation belts; expert on NASA sci. and applications research data processing; oversaw devel. and operation U.S. Nat. Environ. Satellite System, 1972-80; oversaw environ. research program Nat. Oceanic and Atmospheric Adminstrn., 1981-83. Served from pvt. to capt. USAF, 1946-52. Van Allen scholar, 1958; research fellow U.S. Steel Found., 1958-60; recipient Exceptional Service medal NASA, 1969, Program Adminstrn. and Mgmt. award NOAA, 1977, Exceptional Sci. Achievement medal NASA, 1984. Mem. IEEE (sr.), Am. Geophys. Union, Am. Meteorol. Soc., Phi Beta Kappa, Sigma Xi, Phi Eta Sigma, Eta Kappa Nu. Home: 215 Aspen Trail Winchester VA 22602

LUECKE, CONRAD JOHN, aerospace educator; b. Escanaba, Mich., Dec. 30, 1932; s. John Frederick and Rose Margaret (Jaeger) L.; m. Shirley Kerfoot, Dec. 30, 1969 (dec. 1978); m. Freida Belle Shell, Dec. 15, 1979; 1 child, B. Michael. BA, Albion Coll., 1954. Commd. 2d lt. USAF, 1954, advanced through grades to command pilot, 1969, ret., 1977; instr. orbiter systems Lockheed Space Ops. Co., Titusville, Fla., 1984—; instr. celestial navigation U.S. Power Squadron, Cocoa Beach, Fla., 1984—; trainer emergency egress shuttle astronauts NASA, Kennedy Space Ctr., Fla., 1984—. Decorated Silver star, Air medal with 5 oak leaves. Democrat. Baptist. Avocations: golf, sailing, flying, travel. Home: 655 Doral Ln Melbourne FL 32940-7601 Office: Lockheed Space Ops Co 1100 Lockheed Way Titusville FL 32780-7910

LUECKE, KENN ROBERT, software engineer; b. St. Charles, Mo., May 12, 1966; s. Robert Clarence and Louise Nora (Meers) L. BS in Computer Sci., U. Mo., St. Louis, 1989. Programmer-coop McDonnell Douglas, St. Louis, 1988, software engr., 1989—. Trustee, sec., chief usher Holy Cross Luth. Ch., St. Charles, 1992, high rep., 1991-92. Mem. AIAA, Am. Soc. for Quality Control, Am. Assn. Artificial Intelligence, Ops.Rsch. Soc. Am. Home: 1005 Olde Coventry Saint Charles MO 63301 Office: McDonnell Douglas PO Box 512 Saint Louis MO 63101

LUEDKE, PATRICIA GEORGIANNE, microbiologist; b. Milw., May 4, 1956; m. Michael Andrew Luedke, July 15, 1978; children: Christopher M., Sean P. BS, Marquette U., 1978, postgrad., 1981-82. Registered med. technologist; registered microbiologist. Med. technologist Med.-Surgical Clinic, Milw., 1978-79, Milw. County Hosp., 1979, Fort Atkinson (Wis.) Meml. Hosp., 1979-88, Franciscan Shared Lab., Wauwatosa, Wis., 1988—; cons. Forensic Rsch. Assocs., Oconomowoc, Wis., 1978—. Mem. Am. Soc. Microbiology, Am. Soc. Clin. Pathologists, N.Y. Acad. Scis. Home: 739 Elizabeth St Oconomowoc WI 53066-3703

LUEPKER, RUSSELL VINCENT, epidemiology educator; b. Chgo., Oct. 1, 1942; s. Fred Joeseph and Anita Louise (Thornton) L.; m. Ellen Louise Thompson, Dec. 22, 1966; children: Ian, Carl. BA, Grinnell Coll., 1964; MD with distinction, U. Rochester, 1969; MS, Harvard U., 1976. Intern U. Hosp. San Diego County; Intern U. Calif., San Diego, 1969-70; resident Peter Bent Brigham Hosp., Boston, 1973-74; from rsch. asst. to asst. resident Peter Bent Brigham Hosp./Med., Boston, 1971-74; dir. Lipid Clinic Peter Bent Brigham Hosp., Boston, 1975-76; asst. prof. div. epidemiology, med. lab. physiol. hygiene U. Minn., Mpls., 1976-80, assoc. prof., 1980-87, assoc. dir. div. epidemiology, 1986-91, prof. div. epidemiology and medicine, 1987—, dir. div. epidemiology, 1991—; cons. NIH, Bethesda, Md., 1980—, U. So. Calif., L.A., 1985—; vis. prof. U. Goteborg, Sweden, 1986. Lt. comdr. USPHS, 1970-73. Harvard U. fellow, 1974-76, Bush Leadership fellow, 1990; recipient Prize for Med. Rsch. Am. Coll. Chest Physicians, 1970, Nat. Rsch. Svc. award Nat. Heart, Lung and Blood Inst., Bethesda, 1975-77, Disting. Alumni award Grinnell Coll., 1989. Fellow ACP, Am. Coll. Cardiology, Am. Heart Assn. Coun. om Episemiology (chmn. 1992—), Am. Coll. Epidemiology; mem. Delta Omega Soc. (Nat. Merit award 1988). Office: Univ Minn Sch Pub Health Div Epidemiology 1300 S Second St Ste 300 Minneapolis MN 55454-1015

LUEPTOW, RICHARD MICHAEL, mechanical engineer, educator; b. Port Washington, Wis., July 8, 1956; s. Wayne Richard and Marguerite Anne (Veh) L.; m. Maiya Lee, Mar. 18, 1989; 1 child, Hannah. MS, MIT, 1980, ScD, 1986. Registered profl. engr., Mass. Rsch. engr. Haemonetics Corp., Braintree, Mass., 1980-83, sr. rsch. engr., 1986-87; asst. prof. mech. engring. Northwestern U., Evanston, Ill., 1988—. Methodist. Office: Northwestern Univ 2145 Sheridan Rd Evanston IL 60208

LUFFY, RONALD JON, mechanical engineer; b. Pitts., May 26, 1964; s. Thomas Craig and Marlene Mary (Pasquinelli) L.; m. Christine Maria Fleck, July 19, 1986. BSME summa cum laude, Ohio State U., 1986; MS in Engring., U. Cin., 1991; postgrad., Xavier U., 1991—. Registered profl. engr., Ohio. Co-op IBM, 1983-85; engr. GE Aircraft Engines, Cin., 1986—. Mem. AIAA, NSPE, ASME. Achievements include devel. of multi-hole cooling concept; rsch. on boosted cooling system, impact of shroud geometry on ejector pumping performance, exhaust system cooling methodology, transient 3D compressor clearance prediction methodology. Office: GE Aircraft Engines 1 Neumann Way MD A326 Cincinnati OH 45215

LUGG, MARLENE MARTHA, health information systems specialist; b. Wauwatosa, Wis., Mar. 6, 1938; d. Armand Werner and Elise (Kuehni) Heinrich; m. Richard S.W. Lugg, June 11, 1966 (div. Dec. 1976); children: Jennifer Elsie, William Thomas Armand. BS, U. Wis., 1960; MPH, U. Pitts., 1966, DrPH, 1981. Dep. chairperson Nat. Com. on Health and Vital Stats., Canberra, Australia, 1973-83; dir. State Ctr. for Health Stats. and Planning Health Dept. Western Australia, Perth, 1966-83; dir. health info. systems program UCLA, 1983-88; vis. prof. pub. health Calif. State U., Northridge, 1987—; health info. systems specialist Kaiser-Permanente-So. Calif., Pasadena, 1988—; cons. software applications, L.A., 1987—; civil svc. examiner L.A. Civil Svc. Commn., 1986-88; vis. prof. Pasadena City Coll., 1992—; mem. Calif. State Health Info. Policy Interagency Com., 1992—; bd. dirs. Pub. Health Found., L.A., 1991—. Author: Medical Manpower in Western Australia, 1978; contbr. articles on injury, health data systems, planning, injury control and Pub. Health Conf. stats. and records to profl. jours. Leader, trainer Girl Scouts U.S.A., Milw., Pitts., L.A., 1956—, Australian Girl Guides, Perth, Australia, 1966-82; explorer leader, trainer Boy Scouts Am., Western L.A. and Verdugo Hills, 1983—. Recipient Broughton award Izaak Walton League Am., Wis., 1966, Fisher award Am. Med. Technologists, 1971, Outstanding Young Person award Jaycees, Perth, Australia, 1977, Take Pride in Am. award U.S. Govt., Washington, 1990, Wm. T. Hornaday Gold medal Boy Scouts Am., 1991; named Career Woman of Yr., Daily News, 1983; Nat. Health and Med. Rsch. Coun. pub. health fellow, Australia, 1978. Fellow Am. Pub. Health Assn., Australian Coll. Health Execs. (state bd. dirs. 1977-82), Royal Soc. Health, London; mem. Internat. Epidmiological Assn., So. Calif. Pub. Health Assn. (bd. dirs. 1987—). Lutheran. Achievements include research in development of serial section microcinematography. Office: Kaiser-Permanente So Calif 393 E Walnut St Pasadena CA 91188-0001

LUGIATO, LUIGI ALBERTO, physics educator; b. Limbiate, Milano, Italy, Dec. 17, 1944; s. Pietro and Maria (Morteo) L.; m. Vilma Tagliabue, Apr. 7, 1969; 1 child, Paolo. Deg. in Physics, U. Milan, 1968, Postgrad. Dipl. in Atomic/Nuclear Physic, 1975. Supply asst. U. Milan, 1971-72; researcher Nat. Inst. Nuclear Physics, Milan, 1972-74; asst. in theoretical physics U. Milan, 1974-83, lectr. physics, 1977-83, assoc. prof. physics, 1983-86, prof. quantum electronics, 1990—; prof. atomic physics Turin Poly., 1987-89; mem. physics group Human Capital and Mobility programme Commn. of European Communities, Brussels, 1992—; hon. adj. prof. Drexel U., Phila., 1980-91. Mem. editorial bd. Phys. Rev. A of Am. Inst. Physics, 1991—, Optics Communications, 1988—, Quantum Optics, 1989—; co-editor: 4 vols. of conf. proceedings, 1987, 88, 89, 90; contbr. over 250 articles to profl. jours. Recipient Albert A. Michelson medal Franklin Inst., Phila., 1987; CNR and Ministry of Pub. Edn. scholar, 1969-71. Fellow Optical Soc. Am., Franklin Inst. (life), Am. Phys. Soc.; mem. Italian Phys. Soc. (Prize SIF-Laser Optronics 1991), European Phys. Soc. (bd. quantum electronics and optics div.). Avocation: classical music. Home: Via Cremona 17, Legnano Milano, Italy 20025 Office: Dept Physics Univ Milan, Via Celoria 16, Milan Italy 20133

LUGO, ARIEL E., botanist, federal agency administrator; b. Mayagüez, P.R., Apr. 28, 1943; m. Helen Nunci; 2 children. BS in Biology, U. P.R., 1963, MS in Biology, 1965; PhD in Ecology, U. N.C., 1969. Asst. prof. dept. Botany U. Fla., Gainesville, 1969-73, 75-76, assoc. prof., 1976-79, acting dir. ctr. for wetlands, 1977-78; asst. sec. planning and resource analysis P.R. Dept. Natural Resources, Puerta de Tierra, 1973-74, asst. sec. sci. and tech., 1974-75; staff mem. Coun. Environ. Quality Exec. Office Pres., Washington, 1978-79; head divsn. ctr. energy and environ. rsch. U. P.R., 1980-88; project. leader Internat. Inst. Tropical Forestry, USDA Forest Svc., Rio Piedras, P.R., 1980-92, dir., 1986-92, acting dir., supervisory rsch. ecologist, 1992—; cons. Save Our Bays Assn., 1970, Am. Oil Co., 1972, H.W. Lochner, Inc., 1972, U.S. Postal Svc., 1972, U.S. Forest Svc., 1972, U.S. Dept. Interior, 1972, 73, 76, 77, U.S. EPA, 1974, U.S. Justice Dept., 1974, 78, UNESCO, 1975-76, 78, 83, 85-86, P.R. Dept. Natural Resources, 1975, 76, S.W. Fla. Regional Planning Coun., 1976-77, Environ. Quality Bd., 1976, 77, Fla. Dept. Natural Resources, 1976, Rockefeller Found., 1976, Nat. Audubon Soc., 1977, County of Lee, Fla., 1977, Nat. Wildlife Fedn., 1978, World Bank, 1978, Orgn. Am. States, 1979, 80, Collier County Nature Conservancy, 1980; hon. prof., lectr. U. P.R., 1974-76, 80—, hon. assoc. prof., 1985-86; with dept. Environ. Engring. Scis., Ctr. Latin Am. Studies U. Fla., Gainesville, 1977, dept. Botany, 1980-92; co-chmn. Fed. Com. Ecol. Reserves, 1977-79; mem. Man and Biosphere directorate 7-B Caribbean Islands, 1978-79, chmn. directorate I Tropical Forest Ecosystems, 1980-89, chmn. directorate Tropical Ecosystems, 1990; mem. Endangered Species Scientific Authority, Interagy. Arctic Rsch. Coordinating Com., Interagy. Tropical Deforestation Task Force, 1978-79, Tropical Diversity Interagy. Com., 1985; chmn. rsch. com. Latin Am. Forestry Commn. UNESCO, 1980—; mem. commn. on ecology Internat. Union Conservation Nature and Natural Resources, 1980-86; nat. rsch. coun. com. ecol. problems associated with devel. humid tropics, 1980-82; exec. com. U.S. Man and Biosphere program, 1981-84, nat. com. 1990; sr. adv. com. to Pres. U. P.r., 1982-88, chmn., 1985-88; scientific advisor Inst. Energy Analysis, Oak Ridge, Tenn., 1982-84; active Decade of Tropics Internat. Union Biol. Scis., 1985-88; adv. com. Yale Tropical Resources Inst., 1985—; advisor coastal zone program tropical countries, R.I. U., 1985-88; review team dept. forestry, Sch. Forest Resources and Conservation U. Fla., 1989; nat. rsch. coun. com. sustainable agriculture and the environment in the humid tropics, 1990-93; mem. grad. faculty dept. Environ., Population and Organismic Biology, U. Colo., Boulder, 1990-91; apptd. Consejo Consultivo del Programa de Patrimonio Nacional, 1990-92; expert witness in field. Mem. editorial bd. Vegetatio, 1981-82, Jour. Litoral, 1982—, Acta Cientifica, 1987—, Jour. Sustainable Forestry, 1991—, Restoration Ecology, 1992—, assoc. editor; contbr. articles to profl. jours. Trustee Fla. Defenders of Environment, 1980-81; rsch. assoc. Islands Resources Found., U.S. Virgin Islands, 1980—; bd. overseers Harvard Coll., com. to visit dept. Organismic and Evolutionary Biology, 1989—; bd. govs. Soc. Conservation Biology, 1989-91; bd. dirs. P.R. Conservation Found., 1990—,. Recipient Disting. Scientist award Interam. U., San Juan, P.R., 1992; grantee U. Fla., 1969, 70, 71-72, U.S. Forest Svc., 1970, U.S. Dept. Interior, 1971, 72-75, Br. Sport Fisheries and Wildlife, 1971-73, 73-76, , Am. Oil Co., 1972, Fla. Dept. Natural Resources, 1972-75, H.W. Lochner, Inc., 1972, State of Fla., 1972-74, Inst. Food and Agrl. Scis., 1973, NSF, 1974, 81-83, 88—, U.S. EPA, 1977-79, Conservation Found., 1977, U.S. Dept. Energy, 1978-82, 83-87, 88-91, U.S. Man and the Biosphere Consortium, 1980-87, 90-91, U.S. AID, 1982-83, FAO, 1983, U.S. Oceanographic Adminstrn., 1984-85, Tech. Wetlan 1986, World Wildlife Fund, 1989; Fullbright-Hayes fellow U. La Plata, Argentina, 1978. Mem. AAAS (coun. Caribbean divsn. 1991-92), Ecology Soc. Am., Internat. Soc. Tropical Foresters (exec. com. 1990-93), Soc. Ecol. Restoration and Mgmt. (founding 1988, bd. dirs. 1992), Internat. Soc. Tropical Ecology (bd. bearers 1990-93), Internat. Assn. Ecology, Fla. Acad. Scis, Soc. Caribbean Ornithology, P.R. Acad. Arts and Scis., Sigma Xi, Beta Beta Beta (hon. Zeta Zeta chpt.). Achievements include research in assessment of role of tropical forests in the carbon cycle of the world; studies of tropical tree plantations in Puerto Rico; comparisons of plantations and natural forests; relations between forest management and soil and water quality in Caribbean forests; studies of tropical wetlands; study on rate of decomposition in the tropics. Home: 1528 Tamesis El Paraiso San Juan PR 00926 Office: USDA Forest Svc Internat Inst Tropical Forestry Call Box 25000 San Juan PR 00928-2500

LUGOWSKI, ANDRZEJ MIECZYSLAW, electrical engineer; b. Kepno, Poland, Dec. 28, 1952; s. Eugeniusz and Wanda (Bednarek) L. MASc, Tech. U., Wroclaw, Poland, 1977. Rsch. worker R&D div. DIORA, Wroclaw, 1981-84; sr. engr. DIORA Receiver Mfg., Wroclaw, 1984-87; product mgr. DIORA S.A. Mfg. Co., Wroclaw, 1987-91; R&D adviser Polish Telecom, Wroclaw, 1992—. Contbr. articles to profl. jours.; patentee broadcasting systems. Mem. IEEE, Comm. Soc., Aerospace Soc., Soc. Broadcast Engrs. Roman Catholic. Avocations: music, swimming, book collecting. Home: Mielecka 19-1, 53401 Wroclaw Poland

LUGTENBURG, JOHAN, chemistry educator; b. Nieuwenhoorn, The Netherlands, Apr. 20, 1942; s. J. and I. (Arkenbout) L.; m. J.E. van Dam, Oct. 21, 1988. PhD in Chemistry, Leiden (The Netherlands) U., 1970. Student asst. Leiden U., 1962-68, rsch. asst., 1968-72, sci. co-worker, 1972-76, sci. head co-worker, 1976-84, prof. orgahic chemistry, 1984—, head cen. NMR facility, 1980—; mem. com. sect. bio-organic chemistry Netherlands Found. Chem. Rsch., 1989-91; trustee IUPAC Congress on Photochemistry, 1978—. Mem. Royal Dutch Chem. Soc. (mem. com. sect. organic chemistry 1989-92). Office: Leiden U Gorlaeus Labs, PO Box 9502 Einsteinweg 55, 2300 RA Leiden The Netherlands

LUI, ERIC MUN, civil engineering educator, practitioner; b. Hong Kong, Feb. 2, 1958; came to U.S., 1977; s. Kui Leung and Yin Fong (Leung) L. BS in Civil and Environ. Engring., U. Wis., 1980; MS in Civil Engring., Purdue U., 1982, PhD, 1985. Teaching asst. Purdue U., W. Lafayette, Ind., 1981-82; rsch. asst. Purdue U., W. Lafayette, 1983-85, post-doctoral rsch. asst., 1985-86, lectr., 1985-86; asst. prof. Syracuse (N.Y.) U., 1986-91, assoc. prof., 1992; engring. cons. in field; advisor Grad. Student Orgn., Syracuse, N.Y., 1988—, ASCE Student Chpt., 1992—; coun. Tall Bldg. and Urban Habitat. Co-author: Structural Stability-Theory and Implementation, 1987, Stability Design of Steel Frames, 1991; mem. editorial bd. Jour. Singapore Structural Steel Soc.; contbr. over 50 articles to profl. jours., papers to sci. proceedings and chpts. to books and monographs. Recipient Bleyer scholarship U. Wis., 1979, Bates & Rogers Found. scholarship, 1980, David Ross fellowship Purdue U., 1982. Mem. ASCE (student chpt. faculty advisor Syracuse sect. 1992—), AAUP, Am. Concrete Inst., Am. Acad. Mechanics, Am. Inst. Steel Constrn., Am. Soc. Engring. Edn., Coun. Tall Bldgs. and Urban Habitat, Structural Stability Rsch. Coun., Tau Beta Pi, Phi Kappa Phi, Sigma Xi. Avocations: painting, classical music, piano playing. Office: Syracuse U 220 Hinds Hall Syracuse NY 13244-1190

LUI, MING WAH, electronics executive; b. China, Apr. 4, 1938; arrived in Can., 1968; m. Adeline Pli-Fong Cheng, 1970; children: Nana, Justin. MS, U. N.S.W., 1968; PhD, U. Saskatchewan, 1973. Chief metallurgist Chiap Hua Industry Co., Hong Kong, 1963-65; project metallurgist Comalco Aluminium Co., Ltd., Sydney, Australia, 1966-68; rsch. fellow U. Saskatchewan, Can., 1973-76; dep. gen. mgr. Meyer Aluminium, Ltd., 1976-81; mng. dir. Keystone Electronics Co., Ltd., 1981—, Nantin Enterprise, Ltd., 1985—; gen. mgr. Keystone Trading, Ltd., 1988—; bd. dirs. Keystone Instruments Ltd., Hong Kong., Autotech Ltd., China; Justice of the Peace, 1992. Mem. Labour Adv. Bd., Hong Kong, 1986—; lay assessor Magistrates Cts. of Hong Kong, 1983—; mem. coun. Sir Edward Youde Meml. Fund, Hong Kong, 1987—; chmn. Electronics Industry Tng. Bd., Hong Kong, 1988—; justice of peace, Hong Kong, 1992—. Mem. Inst. of Metallurgists, Profl. Engrs. of Camada, Australia Inst. Mining and Metallurgy (exec. com. mem.), Chines Mfrs. Assn. of Hong Kong (chmn. 1993—), Hong Kong Electronics Assn. (v.p. 1987—), Hong Kong Assn. for Advancement of Sci. and Tech. (v.p. 1987—), Royal Gulf Club, Royal Jockey Club, Pacific Club, Tower Club. Office: Keystone Electronics Co Ltd, 11/F 23 Hing Yip St, Hong Kong Hong Kong

LUI, NG YING BIK, engineering educator, consultant; b. Amoy, Fujian, China, Dec. 2, 1936; arrived in Hong Kong, 1975; d. Yu-Zhou and Kao (Li-

Tsing) Wu; m. Robert Sai-Boom Lui, Dec. 31, 1961; children: Rosan Zi-Lan, Simon Zi-Yin, Hong-Wu. BSc, Chejiang Univ., 1960; M.Phil., Univ. Hong Kong, 1983. Asst. lectr., lectr. Chejiang Univ., Hang Zhou, China, 1960-75; mech. engr. ELectronic Devices, Ltd., Hong Kong, 1975-76, Hong Kong Industrial Co., Ltd., Hong Kong, 1976-77; equipment design engr. Precision Mould Ltd., Hong Kong, 1977; lectr. Hong Kong Polytech., Hong Kong, 1977—; cons. Industrial Acoustic Co. Ltd., Hong Kong, 1991-92, Rising Advances Co., Ltd., Hong Kong, 1990-91, Tacko Devel. Ltd., Hong Kong, 1986-87. Contbr. articles to profl.jours. Recipient Honorary Cons. award Xiamen Scientific Assn., 1988. Mem. Am. Soc. Mech. Engrs., Am. Solar Energy Soc., Institution of Mech. Engrs., Institution of Engrs. Australia, The Engineering Coun. (chartered engr.). Achievements include development of model 165 F air-cooled small power diesel engine. Office: Hong Kong Polytechnic, Hung Hom, Kowloon Hong Kong

LUISO, ANTHONY, international food company executive; b. Bari, Italy, Jan. 6, 1944; s. John and Antonia (Giustino) L.; m. Nancy Louise Bassett, June 26, 1976. B.B.A., Iona Coll., 1967; M.B.A., U. Chgo., 1982. Audit sr. Arthur Andersen & Co., Chgo., 1966-71; supr. auditing Beatrice Foods Co., Chgo., 1971-74, adminstr. asst. to exec. v.p., 1974-75, v.p. ops. internat. div., dairy div., 1975-77, exec. v.p. internat. div., 1977-82, prof. internat. div., 1982-83, chief operating officer internat. food group, 1984-86, pres. U.S. Food segment, from 1986; group v.p. Internat. Multifoods Corp., Mpls., until 1988, pres., 1988—, chief oper. officer, 1988-89, chmn., chief exec. officer, 1989—; bd. dirs. Black & Decker Co. Mem. adv. council U. Chgo. Grad. Sch. of Bus. Served with USAR, 1968-74. Mem. AICPA. Republican. Roman Catholic. Clubs: Univ. (Chgo.), Internat. (Chgo.). Office: Internat Multifoods Corp Multifood Tower PO Box 2942 Minneapolis MN 55402-0942

LUK, KING SING, engineering company executive, educator; b. Canton, China; came to U.S., 1954, naturalized, 1964; s. Yau Kong and Liang Yu L.; m. Kit Ming Wong; children—Doris, Stephen, Eric, Marcus. B.S., Calif. State U., Los Angeles, 1957; M.S. in Civil Engring., U. So. Calif., 1960; Ph.D. with distinction, UCLA, 1971. Chief engr. R.E. Rule Inc., 1958-60; pres. King S. Luk & Assoc., Inc. Cons. Engrs., Los Angeles, 1960—; prof. civil engring. Calif. State U., Los Angeles, 1960-82, prof. emeritus, 1982—, chmn. civil engring. dept., 1968-74; commr. Calif. Seismic Safety Commn., 1979-83; dir., corp. dir., exec. officer Mechanics Nat. Bank, Cathay Pacific Inc., Luk & Luk, Inc. Author: Civil Engineering Reviews, 1964; contbr. articles in field to profl. jours.; research in reinforced concrete and earthquake engring. NSF fellow, 1966, 70. Fellow ASCE. Office: Luk & Luk Inc 55 S Raymond Ave Ste 302 Alhambra CA 91801-7106

LUKAS, ELSA VICTORIA, radiobiologist, radiobiochemist; b. Baden nr. Vienna, Austria, Feb. 28, 1927; d. Johann and Victoria (Hauer) L.; Degree for High Sch. Tchrs., U. Vienna, 1952, Ph.D., 1955; DSc in Physics, Biology and Physiology (hon.) Marquis Giuseppe Scicluna Internat. U., 1987; PhD in Physics (hon.), Albert Einstein Internat. Acad. Found., 1990. Researcher, Max Planck Inst. Biophysics, Frankfurt/Main, Federal Republic Germany, 1959-64, Path. Inst. Justus Liebig U., Giessen, Fed. Republic Germany, 1961-64, Oak Ridge Nat. Lab., U. Radiation Biology, U. Tenn., Knoxville, 1964-67; high sch. tchr., country insp. schs., Vienna, 1967—; research. Author numerous publs. on biochem. effects of ionizing radiation in living cells, especially in their nucleic acids. Recipient Dr. J. Kowarschick award, 1957, Dr. Karl Luick award 1957, Theodor Kö rner prize 1960, Alexander von Humboldt award, 1961, Vibert Douglas award Internat. Fedn. Univ. Women, 1962, medal of honor Am. Biog. Inst., 1987, Golden Acad. award, 1991, Profl. of Yr. nomination, 1991, Cert. of Merit as Foremost Woman of the 20th Century Internat. Biog. Centre, Cambridge, Eng., 1987; named hon. citizen State of Tenn., 1965, Woman of Yr. Am. Biog. Inst., 1990; Fulbright Hays scholar, 1964; inducted to Am. Biog. Inst's. 5,000 Personalities of the World Hall of Fame, N.C., 1989. Mem. Biophys. Soc., Radiation Research Soc., Soc. German Scientists and Physicians, Austrian Biochem. Soc., Am. Inst. Biol. Scis., Soc. Parapsychology, German Bot. Soc., Soc. German Biologists, Gregor Mendel Soc., Soc. Austrian Chemists, Univ. Assn. Alma Mater Rudolphina. Roman Catholic. Home: 60 Elisabethstrasse, Baden 2500, Austria

LUKE, SUNNY, medical geneticist, researcher; b. Neericad, Kerala, India, Nov. 8, 1951; came to U.S., 1985; d. Chacko and Sosamma Mattathilparambil; m. Philo Luke, Aug. 15, 1976; children: Sandeep, Sanoop. BSc in Biology, U. Kerala, 1972, MSc in Genetics, 1975; MS in Biology, Adelphi U., 1987; DSc (hon.), Somerset (Eng.) U., 1993. Lic. clin. cytogenetics supr., N.Y.C.; chartered biology Inst. Biology, London; cert. clin. lab. specialist in cytogenetics Nat. Cert. Agy. Secondary tchr. Nigerian Ministry of Edn., Plateau State, 1977-84, insp. for biology and agrl. scis., 1981-84; teaching asst. Adelphi U., Garden City, N.Y., 1985-87, U. Toledo, 1987; clin. cytogeneticist, rsch. scientist in med. genetics L.I. Coll. Hosp., Bklyn., 1988—; seminar presenter on clin. genetics and primate evolution to various univs. and clin. genetic labs., 1991—; rsch. mentor Westinghouse and N.Y. State talent search, 1992-93. Contbr. articles and abstracts to Am. Jour. Med. Genetics, Genomics, Nature Genetics, Jour. Cell Sci. Mem. Assn. Clin. Scientists, Sigma Xi. Roman Catholic. Achievements include display of research discoveries in Hall of Human Biology and Evolution in the American Museum of Natural History. Home: 3152 Kennedy Blvd Jersey City NJ 07306 Office: Div Genetics LI Coll Hosp Brooklyn NY 11201

LUKERIS, SPIRO, engineer, consultant; b. Bridgeport, Conn., Jan. 23, 1948; s. Gabriel Peter and Katherine (Contarini) L.; m. Sally Eva Campbell, June 26, 1976; children: Katherine Sophia, Cassandra Campbell, Rebecca Sarah. BSc, McGill U., Montreal, Que., Can., 1965; MS, U. Conn., 1973. Cons. Katie Engring., Sandy Hook, Conn., 1991—. Tutor, adviser computer coop., mem. enrichment acad. Newtown (Conn.) Sch. System, 1992—. Mem. IEEE, AAAS, Math. Assn. Am., Assn. Computing Machinery, Engrs. for Excellence. Home: PO Box 14 217 Berkshire Rd Sandy Hook CT 06482

LULL, WILLIAM PAUL, engineering consultant; b. Indpls., Nov. 5, 1954; s. William Roger and Florence Elizabeth (Morris) L.; m. Mary Ann Garrison, Dec. 22, 1989. Student, Ind. State U., 1973-75; BS in Arts & Design, MIT, 1978. Systems designer James Assocs., Architects, Engrs., Indpls., 1978-79; architect TVA, Knoxville, Tenn., 1980; mgr. energy mgmt. div. Dubin-Bloome, Engrs., N.Y.C., 1981; asst. chief of design Syska & Hennessy, Engrs., N.Y.C., 1982-83; prin. Garrison/Lull Inc., Princeton Junction, N.J., 1984—; adj. assoc. prof. NYU, 1983—; lectr., presenter cons. environ. field. Author: Conservation Environment Guidelines for Libraries and Archives, 1990; co-author: Criteria for Storage of Paper-Based Archival Records, 1984; contbr. articles to profl. publs. Mem. ASHRAE (affiliate, conf. presenter 1983), Am. Assn. Mus. (registrars' com., presenter 1988), Am. Inst. Conservation of Historic and Artistic Works (assoc.), Sigma Pi Sigma. Achievements include initial discipline of consulting on conservation environments for preservation of museum and archival collections. Home: 7 High St Allentown NJ 08501 Office: Garrison/Lull Inc PO Box 337 Princeton Junction NJ 08550-0337

LUM, PAUL, writer; b. Canton, Republic of China, Jan. 10, 1923; came to U.S., 1927; s. Hing and Lae Jun (Ng) L. BA in Sociology, U. Oreg., 1945. Sales rep., ptnr. Worchung, Portland, 1946-67; advt. mgr. Walnut Restaurant, 1972-73. Vol. reception Ambassador from Taiwan, Portland, 1973. Mem. AAAS, Am. Statis. Assn., Sci. and Engring. Group. Republican. Home: 16485 SW Pacific Hwy Portland OR 97224-3446

LUMBIGANON, PISAKE, obstetrician/gynecologist; b. Khon Kaen, Thailand, Aug. 28, 1953; s. Narong and Tasanee (Thirasas) L.; m. Pagakrong Teptanavatana, Jan. 4, 1979; children: Disajee, Supanat. MD, Ramathibodi, Bangkok, Thailand, 1976; MS, U. Pa., Phila., 1991. Diplomate Bd. Ob-Gyn. Thailand. Intern Ramathibodi Hosp., Bangkok, 1976-77, resident, 1977-80; lectr. Khon Kaen U., 1980-82, asst. prof., 1982-87, assoc. prof., 1987—, dept. chmn., 1987-91, asst. dean rsch., 1989-92; cons. WHO, Rangoon, 1984, Geneva, 1987, 91, 92, 93, Johns Hopkins Programme of Internat. Edn. for Gynecology and Obstetrics, Cairo, 1991, Jarkatar, 1992. Postdoctoral fellowship Rockefeller Found., 1982-83, rsch. grant, 1984, WHO, 1984, 90, IDRC, 1986. Mem. Internat. Epidemiol. Assn., Thai Med. Assn., Thai Coll. Ob-Gyn. Buddhist. Home: 6/3 Klangmuang Rd, Khon Kaen 40000, Thailand Office: Khon Kaen U, 123 Mitraparb Rd, Khon Kaen 40002, Thailand

LUMLEY, JOHN LEASK, physicist, educator; b. Detroit, Nov. 4, 1930; s. Charles S. and Jane Anderson Campbell (Leask) L.; m. Jane French, June 20, 1953; children: Katherine Leask, Jennifer French, John Christopher. B.A., Harvard, 1952; M.S. in Engring, Johns Hopkins, 1954, Ph.D., 1957; Haute Distinction Honoris Causa, Ecole Central de Lyon, France, 1987. Postdoctoral fellow Johns Hopkins, 1957-59; mem. faculty Pa. State U., 1959-77, prof. aerospace engring., 1963-74, Evan Pugh prof. aerospace engring., 1974-77; Willis H. Carrier prof. engring. Cornell U., 1977—; prof. d'echange U. d'Aix-Marseille, France, 1966-67; Fulbright sr. lectr. U. Liege; vis. prof. U. Louvain-La-Neuve, Belgium; Guggenheim fellow U. Provence and Ecole Centrale de Lyon, France, 1973-74. Author: (with H.A. Panofsky) Structure of Atmospheric Turbulence, 1964, Stochastic Tools for Turbulence, 1970, (with H. Tennekes) A First Course in Turbulence, 1971; also articles.; tech. editor: Statistical Fluid Mechanics, 1971, 75, Variability of the Oceans, 1977; assoc. editor: Physics of Fluids, 1971-73; assoc. editor Ann. Rev. of Fluid Mechanics, 1976-85, co-editor, 1986—; chmn. tech. editorial bd.: Izvestiya: Atmospheric and Oceanic Physics, 1971—; editorial bd.: Fluid Mechanics: Soviet Research, 1972—; editor Theoretical and Computational Fluid Dynamics, 1989—; prin.: films Deformation of Continuous Media, 1963, Eulerian and Lagrangian Frames in Fluid Mechanics, 1968. Recipient medallion U. Liege, Belgium, 1971. Fellow Am. Acad. Arts and Scis., Am. Acad. Mechanics, Am. Phys. Soc. (exec. com. div. fluid dynamics 1972-75, 81-84, chmn. exec. com. div. fluid dynamics 1982, 87-89, fluid dynamics prize 1990), AIAA (assoc., fluid and plasma dynamics award 1982); mem. ASME (Timoshenko medal 1993), NAE, N.Y. Acad. Sci., Soc. Natural Philosophy, AAAS, Am. Geophys. Union, Johns Hopkins Soc. Scholars (charter), Sigma Xi. Home: 743 Snyder Hill Rd Ithaca NY 14850-8708 Office: Cornell U 238 Upson Hall Ithaca NY 14853

LUMPKIN, ALVA MOORE, III, electrical engineer, marketing professional; b. Columbia, S.C., Dec. 4, 1948; s. Alva Moore Jr. and Willodene Evelyn (Rion) L. Student, Washington and Lee U., 1967-68; BS in Applied Math., U. S.C., 1976; MSEE, Ga. Inst. Tech., 1980; student, Rice U. Lic. real estate broker, Ga. Teaching asst. U. S.C., Columbia, 1977, Ga. Inst. Tech., Atlanta, 1978-80; rsch. engr. Ga. Tech. Rsch. Inst., Atlanta, 1980-82; microwave engr., project mgr. Electromagnetic Scis., Inc., Norcross, Ga., 1982-85; pres. A.M. Lumpkin, Inc., Atlanta, 1985—; sales engr. Gentry Assocs., Inc., Orlando, Fla., 1988-90; corp. sec. Gentry Electronics and Instrumentation, Inc., Orlando, 1990-93, also bd. dirs. Mem. IEEE, Antennas and Propagation Soc., Microwave Theory and Techniques Soc., Aerospace and Electronics Systems Soc., Am. Inst. Physics, Assn. Old Crows, Ga. Tech. Alumni Assn., Sigma Pi Sigma, Sigma Alpha Epsilon. Episcopalian. Avocations: open water diving, collecting books, computers. Office: Gentry Electronics and Instrumentation 2793A Clairmont Rd Ste 211 Atlanta GA 30329

LUMSDAINE, EDWARD, mechanical engineering educator, university dean; b. Hong Kong, China, Sept. 30, 1937; came to U.S., 1953; s. Clifford Vere and Miao Ying Lumsdaine; m. Monika Amsler, Sept. 8, 1959; children: Andrew, Anne Josephine, Alfred, Arnold. BS in Mech. Engring., N.Mex. State U., Las Cruces, 1963, MS in Mech. Engring., 1964, ScD, 1966. Research engr. Boeing Co., Seattle, 1966-67, 68; asst. prof. to assoc. prof. S.D. State U., Brookings, 1967-72; assoc. prof. to prof. U. Tenn., Knoxville, 1972-77; prof., sr. research engr. phys. sci. lab., dir. N.Mex. solar energy inst. N.Mex. State U., Las Cruces, 1977-81; prof., dir. energy, environ. and resources ctr. U. Tenn., Knoxville, 1981-83; dean engring., prof. U. Mich., Dearborn, 1982-88, U. Toledo, 1988-93; dean of engring. Mich. Technol. U., Houghton, 1993—; vis. prof. Cairo U., Egypt, 1974, Tatung Inst. Tech., Taipei, China, 1978, Qatar U., Doha, 1983; UNESCO expert cons. to Egypt, 1979-80; dir., cons. E&M Lumsdaine Solar Cons., Toledo, 1979—; cons. Oak Ridge Nat. Lab., Tenn., 1979-82, BDM Corp., Albuquerque, 1984, Ford Motor Co., Dearborn, Mich., 1984—, Am. Supplier Inst., Dearborn, 1986—. Author: Industrial Energy Conservation for Developing Countries, 1984, (with Monika Lumsdaine) Creative Problem Solving: An Introductory Course for Engineering Students, 1990, Creative Problem Solving: Thinking Skills for a Changing World, 1993; contbr. software packages articles to profl. jours. Served with USAF, 1954-58. NASA faculty fellow, 1969, 70; grantee NSF, NASA, U.S. Dept. Energy, Navy, ASHRAE, AID, Ford Motor Co. Fellow AIAA (assoc. tech. com. terrestrial energy systems), ASME; mem. NSPE, Am. Soc. Engring. Edn., ASTM (coms. photovoltaics, heating and cooling, and passive solar), Internat. Solar Energy Soc., Am. Solar Energy Soc., Am. Assn. Higher Edn. Presbyterian. Office: Mich Tech U Coll Engring Houghton MI 49931

LUMSDEN, JAMES GERARD, computer scientist; b. Slough, Berkshire, Eng., Nov. 25, 1940; s. James Scobell and Anne Frances (Mahon) L.; m. Ann Gabrielle Hanley, Mar. 12, 1983 (dec. June 1985); m. Sarah Elizabeth Clive-Powell, Feb. 8, 1989; 1 child, Richard. BSc in Physics, Bedford Coll., London, 1971; diploma in computer sci., Brunel U., London, 1974. Lectr. Internat. Computers Ltd., London, 1975-78, operating systems cons., 1978-83, networking cons., 1983-85; project mgr. Reuters, London, 1985-88; European tech. mgr. Instinet UK Ltd., London, 1988-92; ptnr. Experientia Systems, West Clandon, Surrey, Eng., 1993—. Mem. Inst. of Physics, Assn. Project Mgrs. Roman Catholic. Avocations: music, gardening, travel, photography. Home: Surrey, 11 Bennett Way, West Clandon GU4 7TN, England Office: Experientia Systems, Surrey, Paddock House Bennett Way, West Clandon GU4 7TN, England

LUND, DARYL B., college dean; b. San Bernardino, Calif., Nov. 4, 1941; s. Bert Harry and Edna Susan Agnes (McFarlane) L.; m. Dawn Kreft, June 15, 1963; children: Kristine Kay, Eric Scott. BS in Math., U. Wis., 1963, MS in Food Sci., 1965, PhD in Food Sci. and Chem. Engring., 1968. Rsch. asst. U. Wis., Madison, 1963-67, from instr. to assoc. prof. food sci., 1967-77, prof. food sci., 1977-87, chair dept. food sci., 1984-87; chair dept. food sci., assoc. dir. Agrl. Experiment Sta. Rutgers State U. N.J., New Brunswick, 1988-89, interim exec. dean agriculture and natural resouces, dean Cook Coll., 1989-91, exec. dean agriculture and natural resources, exec. dir. N.J. Agrl. Experiment Sta., dean Cook Coll., 1991—; mem. tech. advisors coun. Grand Met., Mpls., 1990—; mem. nutrition adv. bd. RJR Nabisco, East Hanover, N.J., 1991—. Co-author: Handbook of Food Engineering, 1992; author 20 chpts. to books including Advances in Food Engineering, 1992, Aseptic Processing and Packaging, 1993; co-editor 5 books; contbr. 200 tech. papers to profl. jours. Fellow Inst. Food Technologists (pres. 1991-92); mem. Am. Soc. Agrl. Engring. (divsn. chair 1978-79, Food Engring. award 1987), Am. Inst. Chem. Engring., Am. Inst. Nutrition. Avocations: golf, travel, wood working. Home: 18 Red Coat Dr East Brunswick NJ 08816 Office: Rutgers U Cooke Coll 104 Martin Hall PO Box 231 East Brunswick NJ 08816

LUND, GEORGE EDWARD, retired electrical engineer; b. Phila., Feb. 17, 1925; s. Harold White and Hannah (Lawford) L.; m. Shirley Bolton Stevens, Sept. 24, 1960; children: Marsha (Mrs. Donald Barnett), Roger, Sharon Stevens (Mrs. David Bailey), Gretchen (Mrs. Kevin J. Collette). BEE, Drexel U., 1952; MEE, U. Pa., 1959; postgrad. in computer sci., Villanova U., 1981-83. Project engr. Burroughs Corp., Paoli, Pa., 1952-86; project engr. UNISYS Corp., Paoli, 1986-90, ret., 1990. Assoc. editor, contbr.: Digital Applications of Magnetic Devices, 1960; patentee in field. With USN, 1943-46, ETO. Mem. IEEE (sr.), Eta Kappa Nu. Republican. Methodist. Avocations: photography, amateur radio. Home: 923 Pinecroft Rd Berwyn PA 19312-2123

LUND, HAROLD HOWARD, ceramic engineer, civil engineer, consultant; b. Rockford, Ill., May 13, 1928; s. Harold Henry and Ruth (Howard) L.; m. Suzanne Lyle Swarts, Aug. 7, 1948; children: Kurt Alan, Kathryn Elizabeth, Kent Howard. BS in Ceramic Engring., U. Ill., 1949; diploma in Small Bus. Adminstrn., Guilford Coll., 1961. Registered profl. engr., Vt. Plant engr. Marc Bennett Potteries, Chgo., 1949-50, Northwestern Terra Cotta Co., Chgo., 1950-53; rsch. engr. Ill. Inst. Tech. Rsch. Inst., Chgo., 1953-60; dir. rsch. and engring. Pomona Pipe Products Co., Greensboro, N.C., 1960-79; v.p. tech. dir. W.S. Dickey Clay Mfg. Co., Pittsburgh, Kans., 1980-90; engring. cons. Greensboro, 1990—. Co-author: Marble Inst. Handbook; contbr. Sewer Design manual, 1982; contbr. articles to profl. jours. Fellow Am. Ceramic Soc. (emeritus 1991); mem. ASTM (C-4 com., C-15 com., F-17 com.), NSPE, Nat. Inst. Ceramic Engrs., Rotary (Paul Harris fellow 1988). Achievements include patent in consistancy measuring and control system. Home and Office: 7054 Caindale Dr Greensboro NC 27409

LUNDBERG, ALAN DALE, electrical engineer; b. San Antonio, Mar. 11, 1960; s. Paul and Norma Jeanette (Nester) L.; m. Laura Kaye Bunting, Oct. 23, 1982; children: Lauryn Elizabeth, Kaitlyn Michelle, Bryan David. BSEE, Tex. Tech U., 1982; MBA in Fin., West Tex. State U., 1987. Registered profl. engr. Tex. Elec. engr. Southwestern Pub. Svc. Co., Amarillo, Tex., 1982-85, supervisory engr., 1985-92, sr. engr., 1992—. Sunday sch. tchr. 1st Assembly of God, Amarillo, 1989-91. Mem. IEEE, Tex. Soc. Profl. Engrs. (pres. 1988-89, state dir. 1989—, region v.p. 1992—, Young Engr. of Yr. Panhandle chpt. 1990). Republican. Mem. Assembly of God Ch. Office: Southwestern Pub Svc Co PO Box 631 Lubbock TX 79408

LUNDBERG, GEORGE DAVID, II, medical editor, pathologist; b. Pensacola, Fla., Mar. 21, 1933; s. George David and Esther Louise (Johnson) L.; m. Nancy Ware Sharp, Aug. 18, 1956 (div.); children: George David III, Charles William, Carol Jean; m. Patricia Blacklidge Lorimer, Mar. 6, 1983; children: Christopher Leif, Melinda Suzanne. AA, North Park Coll., Chgo., 1950; BS, U. Ala., Tuscaloosa, 1952; MS, Baylor U., Waco, Tex., 1963; MD, Med. Coll. Ala., Birmingham, 1957; ScD (hon.), SUNY, Syracuse, 1988, Thomas Jefferson U., 1993. Intern Tripler Hosp., Hawaii; resident Brooke Hosp., San Antonio; assoc. prof. pathology U. So. Calif., Los Angeles, 1967-72, prof., 1972-77; assoc. dir. labs. Los Angeles County-U. So. Calif. Med. Ctr., 1968-77; prof., chmn. dept. pathology U. Calif.-Davis, Sacramento, 1977-82; v.p. scientific info., editor Jour. AMA, Chgo., 1982-91, editor in chief scientific publ., 1991—; vis. prof. U. London, 1976, Lund U., Sweden, 1976; prof. clin. pathology Northwestern U., Chgo., 1982—; clin. prof. Georgetown U., Washington, 1982-93; adj. prof. health policy Harvard U., Boston, 1993—. Author, editor: Managing the Patient Focused Laboratory in Medical Decision Making, 1983, 51 Landmark Articles in Medicine, 1984, AIDS from the Beginning, 1986, Caring for the Uninsured and Underinsured, 1991, Violence, 1992; contbr. articles to profl. jours. Served to lt. col. M.C., U.S. Army, 1956-67. Fellow Am. Soc. Clin. Pathologists (past pres.), Am. Acad. Forensic Sci.; mem. N.Y. Acad. Scis., Inst. Medicine, Alpha Omega Alpha. Democrat. Episcopalian. Office: JAMA 515 N State St Chicago IL 60610-4320

LUNDELIUS, ERNEST LUTHER, JR., vertebrate paleontologist, educator; b. Austin, Tex., Dec. 2, 1927; s. Ernest Luther and Hazel (Halton) L.; m. Judith Weiser, Sept. 28, 1953; children—Jennifer, Rolf Eric. B.S. in Geology, U. Tex., 1950; Ph.D. in Paleozoology, U. Chgo., 1954. Postdoctoral Fulbright scholar to Western Australia, 1954-55; postdoctoral research fellow Calif. Inst. Tech., 1956-57; mem. faculty U. Tex., Austin, 1957—; prof. vertebrate paleontology U. Tex., 1969; John Andrew Wilson prof. vertebrate paleontology, 1978—. Served with AUS, 1946-47. Fulbright sr. scholar to Australia, 1976. Home: 7310 Running Rope Austin TX 78731-2132 Office: U Tex Dept Geol Scis Austin TX 78712

LUNDQUIST, CHARLES ARTHUR, university official; b. Webster, S.D., Mar. 26, 1928; s. Arthur Reynald and Olive Esther (Parks) L.; m. Patricia Jean Richardson, Nov. 28, 1951; children: Clara Lee, Dawn Elizabeth, Frances Johanna, Eric Arthur, Gary Lars. BS, S.D. State U., 1949, DSc, 1979; PhD, U. Kans., 1953. Asst. prof. engring. research Pa. State U., 1953-54; sect. chief U.S. Army Ballistic Missile Agy., Huntsville, Ala., 1956-60; br. chief (NASA-Marshall Space Flight Center), Huntsville, 1960-62; dir. Space Scis. Lab., 1973-81; asst. dir. sci. Smithsonian Astrophys. Obs., Cambridge, Mass., 1962-73; assoc. Harvard Coll. Obs., 1962-73; dir. rsch. U. Ala., Huntsville, 1982-90, assoc. v.p. for rsch., 1990—. Editor: (with G. Veis) Smithsonian Institution Standard Earth, 1966, The Physics and Astronomy of Space Science, 1966, Skylab's Astronomy and Space Sciences, 1979. Served with U.S. Army, 1954-56. Recipient Exceptional Sci. Achievement medal NASA, 1971; Hermann Oberth award AIAA, 1978. Mem. AAAS, N.Y. Acad. Scis., Am. Grophys. Union, Am. Astron. Soc., Am. Phys. Soc., Nat. Speleological Soc. Home: 214 Jones Valley Dr SW Huntsville AL 35802-1724 Office: U Ala Office of Assoc VP for Rsch Research Instit RM M-65 Huntsville AL 35899

LUNDQUIST, PER BIRGER, physician, medical illustrator, artist, surgeon; b. Stockholm, Jan. 20, 1945; s. Birger Richard Emanuel and Karin Margareta Beata (Bernstone) L.; m. Margareta Ulvas, June 7, 1967 (div. 1975); children: Karolina, Jonas Petter; m. Maria Cecilia Mahlberg, Oct. 21, 1989; children: Sara, Richard. BSc AAM, U. Toronto, 1971; MD, Karolinska Inst., Stockholm, 1978; student, Art Students League, 1967-68; postgrad., Ariz. Heart Inst., 1993—. Gen. vascular surgery residency Karolinska Hosp., Stockholm, 1978-83; handsurgery Sabbatsbergs Hosp., 1982-83; fellow in cardiovascular surgery Baylor Med. Ctr., Dallas, 1983-84; family practice tng. and rsch. Akademiska Hosp., Uppsala, Sweden, 1988-90; family, surg. practice Stureplan Stockholm City, Stockholm, 1991-93; author, illustrator, lectr. Med. Communications PBL, Stockholm, 1991—; med. illustrator dept. Art as Applied to Medicine, Toronto, 1970, Audio Visual Svcs. Regional Hosp., Linkoping, Sweden, 1971-73, 73—. Author, artist: Svenska Dagbladet, 1989, Allmanmedicin, 1990, Cardiovascular Training Program, 1990-91, Human Interferon Production Purpose, 1984; illustrator: The Five Senses, 1972-73, Human Blood Lipids, 1990; exhibits Vastervik, 1987, Stockholm, 1988, Helsingborg, 1989. Mem. N.Y. Acad. Scis., Am. Heart Assn. (coun. arteriosclerosis, epidemiology, vascular surgery), Swedish Med. Assn., Swedish Assn. Family Physicians, Stockholm Med. Assn., Internat. Soc. Endovascular Surgery. Avocations: canoeing, bicycle touring, restoring old farmhouse, pen and ink drawing. Home: 31 Roslagsgatan, Stockholm 113 55, Sweden also: 3500 N Hayden Rd Scottsdale AZ 85251 Office: Faltoverstens Med Offices, Valhallav 146C, 115 24 Stockholm Sweden also: Ariz Heart Inst 2632 N 20th St Phoenix AZ 85006

LUNIN, JESSE, retired soil scientist; b. N.Y.C., May 11, 1918; s. David and Anna (Paley) L.; m. Jean K. Gottlieb, Dec. 26, 1942; children: Alan S., Barry J. (dec.). BS, Okla. A&M U., 1939; MS, Cornell U., 1947, PhD, 1949. Soil scientist USDA-Soil Conservation, 1940-42; rsch. asst. Cornell U., Ithaca, N.Y., 1945-49; head rsch. div. West Indies Sugar Co., Barahona, Dominican Republic, 1949-56; rsch. leader USDA-ARS, Norfolk, Va., 1957-65; br. chief USDA-ARS, Beltsville, Md., 1965-69, staff scientist, nat. program staff, 1969-80; ret.; cons. Waste Mgmt. Co., 1963, others. Contbr. over 40 articles to profl. jours. Lt. USNR, 1942-45. Mem. Am. Soc. Agronomy (chmn. environ. quality div. 1972), Am. Chem. Soc. Home: 711 Hermleigh Rd Silver Spring MD 20902-1646

LUNT, OWEN RAYNAL, biologist, educator; b. El Paso, Tex., Apr. 8, 1921; s. Owen and Velma (Jackson) L.; m. Helen Hickman, Aug. 8, 1953; children: David, Carol, Janet. BA in Chemistry, 1947, PhD in Agronomy, 1951. Mem. faculty UCLA, 1951-93, prof. plant nutrition, 1964-72, prof. biology, 1972—, acting chmn. dept. biophysics, 1965-70, prof. emeritus, 1993; dir. Lab. Biomed. and Environ. Scis., 1968-93; researcher in soil chemistry, fertility, plant physiology; tech. expert Internat. Atomic Energy Agy to Colombia, 1970, Kenya, 1983, Malaysia, 1985, Uruguaay, 1987. Served with USN, 1944-46. Fellow Am. Soc. Agronomy, Soil Sci. Soc. Am.; Internat. Soc. Soil Sci., AAAS, Am. Nuclear Soc. (L.A. chpt.), Sigma Xi. Home: 1200 Roberto Ln Los Angeles CA 90077-2334 Office: UCLA 900 Veteran Ave Los Angeles CA 90024-2703

LUO, HONG YUE, biomedical engineering researcher; b. Jingsu, China, June 21, 1958; came to U.K., 1991; s. Xing Wu and Su Yun (Zhou) L.; m. Yi He Huang, May 13, 1985; 1 child, Lei. MSc, East China U. Chem. Tech., 1985; postgrad., U. Sussex, 1992—. Worker North China Oilfield, 1976-78; lectr. East China U. of Chem. Tech., Shanghai, 1985-91, rsch. asst. prof. Automatic Control Inst., 1986-91; cons. Shanghai Sci. and Tech. Soc., 1989—. Author: (with others) Analytical Instrumentation, China, 1990; contbr. articles to profl. jours. including Jour. of East China U. of Chem. Tech., Analytical Chemistry China, Micro-special Motor, China, Advanced Education, China. Mem. IEEE, Engring. in Medicine and Biology Soc., Acoustical Soc. Am., Analytical Instrument Soc. of China, Automatical Instrumentation Soc., Detection on-line Soc. of Shanghai. Achievements include research on intelligent signal processor for chromatograph signal analysis and processing, microcomputerized infra-red spectrophotometer, infra-red spectrum searching system, infra-red database, microcomputer system for stepping motor variable subdivision, speech signal processing including overlapping speech separation, improving hearing ability for the hearing-impaired people and developing new kind of hearing aid. Home: Rm 301 House 163, Hua Gong Yi Cun, Shanghai 20037, China Office: Biomed Engring U Sussex, Falmer, BN1 9QT Brighton England

LUPIANI, DONALD ANTHONY, psychologist; b. N.Y.C., June 7, 1946; s. Louis and Josephine (Boccia) L.; m. Linda Moyik, June 20, 1970; 1 child, Jennifer. BA, Iona Coll., 1968; MA, Columbia U., 1971, PhD, 1973; post-doctoral, Behavior Therapy Inst., White Plains, N.Y., 1976. Lic. psychologist, N.Y. Clin. assoc. Columbia U., N.Y.C., 1974-85, Fordham U., Bronx, N.Y., 1979-81; dir. psychology and spl. edn. svcs. Riverdale Country Sch., Bronx, 1973-87; chief psychologist Franciscan Order of Priests, N.Y.C., 1983—; pvt. practice Yonkers, N.Y., 1975—; dir. spl. svcs. Riverdale Country Sch., Bronx., 1973-87; bd. dirs. St. Ursula Learning Ctr., Mt. Vernon, N.Y. Contbr. articles to profl. jours. Bd. dirs., mem. The St. Ursula Learning Ctr. Fellow Am. Orthopsychiat. Assn., Am. Coll. Psychology; mem. APA, N.Y. State Psychol. Assn., Westchester County Psychol. Assn. (chmn. ethics com. 1980-87). Roman Catholic. Avocations: woodworking, painting, drawing. Home and Office: 227 Mile Square Rd Yonkers NY 10701-5369

LUPO, MICHAEL VINCENT, aviation engineer; b. St. Louis, May 6, 1952; s. Vincent Joseph and JoAnn Marie (Macke) L.; m. Marilyn Kay Kern, June 5, 1976; children: Maria, Paul. BS in Aero. Engring., U. Mo., Rolla, 1975, MS in Engring., 1982. Registered maintainability engr., Army Materiel Command, Tex. Gen. engr. AVSCOM, St. Louis, 1976-87; product assurance engr. Program Exec. Officer Aviation, St. Louis, 1987—. Mem. AIAA (sec. 1977—), Planetary Soc. (founder), Army Aviation Assn. Am., Soc. Reliability Engrs. (pres.). Roman Catholic. Home: 5619 Gum Tree Ct Saint Louis MO 63129 Office: Utility Helicopters PMO 4300 Goodfellow Saint Louis MO 63120-1798

LUPULESCU, AUREL PETER, medical educator, researcher, physician; b. Manastiur, Banat, Romania, Jan. 1, 1923; came to U.S., 1967, naturalized, 1973; s. Peter Vichentie and Maria Ann (Dragan) L. MD magna cum laude, Sch. Medicine, Bucharest, Romania, 1950; MS in Endocrinology, U. Bucharest, 1965; PhD in Biology, Faculty of Scis., U. Windsor, Ont., Can. Diplomate Am. Bd. Internal Medicine. Chief Lab. Investigations, Inst. Endocrinology, Bucharest, 1950-67; research assoc. SUNY Downstate Med. Ctr., 1968-69; asst. prof. medicine Wayne State U., 1969-72; assoc. prof., 1973—; vis. prof. Inst. Med. Pathology, Rome, 1967; cons. VA Hosp., Allen Park, Mich., 1971-73. Author: Steroid Hormones, 1958, Advances in Endocrinology and Metabolism, 1962, Experimental Pathophysiology of Thyroid Gland, 1963, Ultrastructure of Thyroid Gland, 1968, Hormones and Carcinogenesis, 1983, Hormones and Vitamins in Cancer Treatment, 1990; contbr. chpts., numerous articles to profl. publs.; research on hormones and tumor biology; studies regarding role of hormones and vitamins in carcinogenesis. Fellow Fedn. Am. Socs. for Exptl. Biology; mem. Electron Microscopy Soc. Am., Soc. for Investigative Dermatology, N.Y. Acad. Sci., AMA (physician's recognition award 1983, 86), Am. Soc. Cell Biology, Soc. Exptl. Biology and Medicine, AAAS. Republican. Home: 21480 Mahon Dr Southfield MI 48075-7525 Office: Wayne State U Sch Medicine 540 E Canfield St Detroit MI 48201-1998

LURIX, PAUL LESLIE, JR., chemist; b. Bridgeport, Conn., Apr. 6, 1949; s. Paul Leslie and Shirley Laurel (Ludwig) L.; m. Cynthia Ann Owens, May 30, 1970; children: Paul Christopher, Alexander Tristan, Einar Gabrielson. BA, Drew U., 1971; MS, Purdue U., 1973; postgrad., 1973—. Tech. dir. Analysts, Inc., Linden, N.J., 1976-77; chief chemist Caleb Brett USA, Inc., Linden, 1977-80; v.p. Tex. Labs., Inc., Houston, 1980-82; pres. Lurix Corp., Fulshear, Tex., 1982—; cons. LanData, Inc., Houston, 1980-88, Nat. Cellulose Corp., Houston, 1981-88, Met. Transit Authority, Houston, 1981—, Phillips 66, Houston, 1986—, Conoco, Inc., Houston, 1988—, Caronia Corp., Houston, 1988—, WBC Holdings, Inc., 1989—; dir. research and devel. Stockbridge Software, Inc., Houston, 1986-88; v.p. Diesel King Corp., Houston, 1980-82. Contbr. articles to profl. jour. Patentee distillate fuel additives. Fellow Am. Inst. Chemists; mem. Am. Chem. Soc., ASTM, AAAS, Soc. Applied Spectroscopy, N.Y. Acad. Sci., Phi Kappa Phi, Phi Lambda Upsilon, Sigma Pi Sigma. Republican. Methodist. Lodge: Kiwanis (pres., 1970-71). Current work: Infrared spectroscopy; data base programming for science and industrial applications. Subspecialties: Infrared spectroscopy; Information systems, storage, and retrieval (computer science). Avocations: tennis, golf, piano. Home: 32602 Hepple White Dr Fulshear TX 77441

LUSAS, EDMUND WILLIAM, food processing research executive; b. Woodbury, Conn., Nov. 25, 1931; s. Anton Frank and Damicele Ann (Kasputis) L.; m. Jeannine Marie Muller, Feb. 2, 1957; children—Daniel, Ann, Paul. B.S., U. Conn., 1954; M.S., Iowa State U., 1955; Ph.D., U. Wis., 1958; M.B.A., U. Chgo., 1972. Project leader Quaker Oats Research Labs., Barrington, Ill., 1958-61, mgr. canned pet foods research, 1961-67, mgr. pet foods research, 1967-72, mgr. sci. services, 1972-77; assoc. dir. Food Protein Research and Devel. Ctr., Tex. A&M U., College Station, 1977-78, dir., 1978—. Contbr. articles to jours., chpts. to books. Assoc. editor Jour. Am. Oil Chem. Soc., 1980—. Fund raiser YMCA, Crystal Lake, Ill., 1970-77, chmn. fin. com., 1977. Recipient F.N. Peters research award Quaker Oats Co., 1968. Mem. Am. Oil Chemists Soc., Inst. Food Technologists (Gen. Foods research fellow 1956, 57), Am. Chem. Soc., Am. Assn. Cereal Chemists, Am. Soc. Agrl. Engrs., Nutrition Today Soc., Guayule Soc. Am., Sigma Xi, Phi Tau Sigma. Avocation: fishing. Home: 3604 Old Oaks Dr Bryan TX 77802-4743 Office: Texas A & M Univ Food Protein Rsch & Devel Ctr College Station TX 77843

LUSCH, CHARLES JACK, physician; b. Lehighton, Pa., Feb. 15, 1936; s. Charles Norman and Loretta (Gaumer) L.; m. Carole Faye Eckart, Aug. 17, 1957; children: Marjorie, Susan, Stephen, Robert. AB in Biology magna cum laude, Lafayette Coll., Easton, Pa., 1957; MD, Temple U., 1961. Bd. cert. in med. oncology, Hematology, and internal medicine. Pres. Berks Hematology-Oncology Assocs., Reading, Pa., 1968—; chief sect. of med. oncology & hematology Reading Hosp. & Med. Ctr., Reading, 1970—; dir. Pa. State Hemophilia Ctr., Reading Hosp. & Med. Ctr., 1973—; v.p. Lusch Motor Parts, Lehighton, Pa., 1975—; chief sect. med. oncology & hematology Community Gen. Hosp., Reading, 1980—; asst. chief medicine Reading Hosp. and Med. Ctr., 1986—; med. dir. Pocono Internat. Raceway, 1980-85; chmn. institutional rev. bd. Reading Hosp. and Med. Ctr., 1986—, dir. continuing med. edn., 1987—; med. dir. Berks County Hospice, Berks County Vis. Nurse Assn., Reading, 1987—; dir. oncology svcs. Reading Hosp. and Med. Ctr., 1990—; med. adv. com. Pa. Blue Shield, Camp Hill, Pa., 1987—; bd. dirs. Berks Home Health Car, Reading Cancer Ctr., Reading Hosp.; malpractice cons. Med. Protective Ins. Co., Ft. Wayne, Ind., 1985—; cons. in hematology and oncology Pottsville (Pa.) Hosp. and Good Samaritan Hosp., 1975—; clin. asst. prof. medicine Pa. Med. Sch., 1984—, Pa. State Med. Sch., 1981—, Temple U. Med. Sch., clin. assoc. prof. 1990; sr. clin. instr. Mahnemann U. Med. Sch., 1968—; prin. investigator Ea. Coop Oncology Group, 1975-90, Nat. Surg. Adj. & Breast Project, 1986—. Contbr. articles to profl. jours.; editor The Med. Record (regional med. jour.), 1970-71. Advisor Future Physicians Am., Reading, 1965; bd. dirs. Berks County unit Am. Cancer Soc., Reading, 1968-78, Keystone Community Blood Bank, Reading, 1970-80; adv. com. The Women's Ctr., Reading Hosp., 1987-88. Lt. comdr. USPHS, 1965-67. Fellow ACP; mem. Pa. Soc. Hematology-Oncology (sec.-treas. 1986-87), Am. Soc. Clin. Oncology, Am. Soc. Hematology, Am. Fedn. Clin. Rsch., Acad. Hospice Physicians (publs. com. 1989—), Lafayette Coll. "The Graduates" Choir, U.S. Amateur Ballroom Dance Assn. (pres. Reading chpt.), Sports Car Club Am., Phi Beta Kappa, Alpha Omega Alpha. Republican. Lutheran. Avocations: competition ballroom dancing, tennis, motor racing. Home: 1617 Meadowlark Rd Reading PA 19610-2820 Office: Berks County Oncology Assoc 301 S 7th Ave Reading PA 19611-1410

LUST, ROBERT MAURICE, JR., physiologist, educator, researcher; b. Rantoul, Ill., Aug. 20, 1955; s. Robert Maurice and Margaretta Regina (McLaughlin) L.; m. Carol Ann White, June 24, 1978; children: Catherine Colleen, Jennifer Ann. BA, Hamilton Coll., 1977; PhD, Tex. Tech U., 1981. Rsch. asst. Tex. Tech U. Health Scis. Ctr., Lubbock, 1977-79, NIH predoctoral fellow, 1979-81, post-doctoral rsch. fellow, 1981-84, asst. prof., 1984-86; asst. prof. East Carolina U. Sch. Medicine, Greenville, N.C., 1986-90, assoc. prof., 1990—, dir. cardiac surgery rsch. labs., 1986—; bd. dirs. Heart Ctr. Ea. N.C.; cons. IVAC Inc., San Diego, 1990—, UpJohn Co., Kalamazoo, 1990—. Contbr. chpt. to book, 50 abstracts and 35 articles to profl. jours.; mem. editorial bd. Jour. Investigative Surgery, 1990—, Current Surgery, 1991—. Mem. Am. Heart Assn., 1981—; state coaching staff North Tex. State Soccer Assn., 1979-86; head coach men's soccer Tex. Tech U., 1979-86, East Carolina U., 1988-91. Grantee Am. Heart Assn., 1987, 88, VA, 1988, UpJohn Inc., 1991, and numerous others. Mem. Am. Physiol. Assn., Assn. for Acad. Surgery, Acad. for Surgical Rsch., Shock Soc. (young investigator 1982). Achievements include research on consequences of ventricular hypertrophy secondary to aortic valve disease; retrograde cardioplegia; mammary artery usage on sternal blood flow in surgery; pathophysiology of vein graft disease. Home: 401 Cedarhurst Rd Greenville NC 27834-6954 Office: East Carolina U Medicine Dept Cardiac Surgery & Physiology Greenville NC 27858

LÜTHI, BRUNO, physicist, educator. Prof. U. Frankfurt, Germany. Recipient Robert-Wichard-Pohl-Preis, Deutsche Physikalische Gesellschaft, 1993. Office: Univ Frankfurt Postfach 11932, Sencken Bergancage 31, D-6000 Frankfurt Germany*

LUTHY, RICHARD GODFREY, environmental engineering educator; b. June 11, 1945; s. Robert Godfrey Luthy and Marian Ruth (Ireland) Haines; m. Mary Frances Sullivan, Nov. 22, 1967; children: Matthew Robert, Mara Catherine, Jessica Bethlin. BSChemE, U. Calif., Berkeley, 1967; MS in Ocean Engring., U. Hawaii, 1969; MSCE, U. Calif., Berkeley, 1974, PhDCE, 1976. Registered profl. engr., Pa.; diplomate Am. Acad. Environ. Engrs. Rsch. asst. dept. civil engring. U. Hawaii, Honolulu, 1968-69; rsch. asst. div. san. and hydraulic engring. U. Calif., Berkeley, 1973-75; asst. prof. civil engring. Carnegie Mellon U., Pitts., 1975-80, assoc. prof., 1980-83, prof., 1983—, assoc. dean Carnegie Inst. Tech., 1986-89, head dept. civil engring., 1989—; cons. sci. adv. bd. U.S. EPA, 1983—, Bioremediation Action Com., 1990-92; cons. U.S. Det. Energy, 1978—, various pvt. industries; del. water sci. and tech. bd. NAE, Washington and Beijing, 1988; mem. tech. adv. bd. Remediation Techs. Inc., Concord, Mass., 1989—, Fostin Capital, Pitts., 1991—, Balt. Gas & Elec., 1992—. Contbr. articles to tech. and sci. jours. Chmn. Conf. on Fundamental Rsch. Directions in Environ. Engring, Washington, 1988. Lt. C.E. Corps, USN, 1969-72. Recipient George Tallman Ladd award Carnegie Inst. Tech., 1977. Mem. ASCE (Pitts. sect. Prof. of Yr. award 1987), Assn. Environ. Engring. Profs. (pres. 1987-88, Nalco award 1978, 82, Engring. Sci. award 1988), Water Pollution Control Fedn. (rsch. com. 1982-86, awards com. 1981-84, 89-93, std. methods com. 1977—, groundwater com. 1989-90, editor jour. 1989-92, Eddy medal 1980), Internat. Assn. on Water Pollution Rsch. and Control (Founders award U.S. Nat. Com. 1986, orgnl. com. 16th Biennial Conf. Washington 1992), Am. Chem. Soc. (div. environ. chemistry, mem. editorial adv. bd. Environ. Sci. Tech. 1992—). Presbyterian. Home: 620 S Linden Ave Pittsburgh PA 15208-2813 Office: Carnegie Mellon U Dept Civil Engring Pittsburgh PA 15213-3890

LUTON, JEAN MARIE, space agency administrator; b. Chamalières, Puy-de Dome, France, Aug. 4, 1942; s. Pierre Luton and Marie Piatot; m. Cécile Robine, 1967; children: Grégoire, Augustin, Clément. Lycées Blaise Pascal, France, Faculty of Science, France; Engineer Diploma in Math. and Physics, Ecole Polytechnique, France. Researcher Centre National de la Recherche Scientifique, 1964-70; dir. of programs Ministry for Indsl. and Scientific Devel., 1971-74; with program and indsl. policy directorate Centre National d'Etudes Spatiales, 1974; head of rsch. program divsn. Centre National d Etudes Spatiales, 1974-75, head of planning and projections divsn., 1975-78, dir. programs and planning, 1978-84, dep. dir. gen., 1984-87, dir. gen., 1989-90; dir. space programs Aérospatiale, 1987-89; dir. gen. European Space Agency, 1990—. Decorated officier Order National du Mérite, chevalier Ordre National de la Légion d'Honneur; recipient Astronautics prize French Assn. for Aeronautics and Astronautics, 1985. Mem. Internat. Acad. Astronautics. Office: European Space Agency ESA, 8-10 rue Mario Nikis, Paris F-75738, France

LUTTERBACH, ROGERIO ALVES, aeronautical engineer, consultant, researcher; b. Belo Horizonte, Brazil, Apr. 29, 1958; s. Renato Moraes and Alice Alves (Borges) L. Student, Cath. U., Belo Horizonte, 1977-78; BSME with aero. emphasis, Fed. U., Belo Horizonte, 1982. Trainee Açominas/Eletroprojetos S.A., Belo Horizonte, 1978, Lider Taxi Aereo S.A., air taxi operators, Belo Horizonte, 1982; trainee Embraer S.A., Sao Jose dos Campos, Brazil, 1982, flight performance engr., instr., 1982—. Mem. AIAA, Air Force Technol Inst. Roman Catholic. Achievements include development computer program for calculation of propeller parameters, software for trajectory determination using microwave transmitter and receivers. Home: Apt 42, Ave Cassiano Ricardo 735, 12240540 Sao Jose dos Campos Brazil

LUTTMAN, HORACE CHARLES, retired aeronautical engineer; b. Banbury, Oxford, Eng., May 30, 1908; arrived in Can., 1947; s. Walter Charles and Ruth Mary (Wilkinson) L.; M. Jean I. M. Morrison, Oct. 6, 1934 (dec. 1987); children: 1 child, Rachel Mary. BA in Mech. Scis., Cambridge U., 1930, MA in Mech. Scis., 1947. Apprentice Armstrong Whitworth Aircraft, Coventry, Eng., 1930-33; chief ground engr. London, Scottish & Provincial Airways, London, 1934; examiner Vickers Aircraft Ltd. Aero. Inspection Directorate British Air Ministry, Weybridge, Eng., 1935-36; asst. insp. Fairey Aviation Ltd. Aero. Inspection Directorate British Air Ministry, Stockport, Eng., 1936-38; insp. N. Am. Aviation, Inc. Aero. Inspection Directorate British Air Ministry, Inglewood, Calif., 1938-40; insp. in charge ea. group British Air Commn., N.Y.C., Washington, 1940-41; asst. chief insp. aircraft British Air Commn., Washington, 1941-45; prin. prodn. officer standardization Ministry of Supply (Air), London, 1946-47; patents officer A.V. Roe Can., Ltd., Malton, Ontario, 1947-52, contracts adminstr., 1952-54; sec., treas. Canadian Aeronautics and Space Inst., Ottawa, 1954-73. Editor Canadian Aero. Jour., 1955-62, Canadian Aeronautics and Space Jour., 1962-73, Canadian Aeronautics and Space Inst. Transactions, 1968-73. Recipient Canada Centennial Medal, Canadian Govt., 1967. Fellow AIAA (assoc.), Canadian Aeronautics and Space Inst., Royal Aero. Soc.; mem. Sigma Gamma Tau. Achievements include initiating. in 1947, standardization of single high tensile bolt in place of two types (high and low tensile) then standard in British aero. mfg.; founder (with others) Canadian Aero. Inst.; holds 2 or 3 aircraft related patents. Home: Rose Cottage Seal Chart, Sevenoaks Kent TN15 OEZ, England

LUTZ, ROBERT BRADY, JR., engineering executive, consultant; b. Akron, Ohio, Nov. 17, 1944; s. Robert Brady and Kathryn Mae (Vallen) L.; m. Anna Lee Burns, Aug. 26, 1967. BS in Engring. Mgmt., Rensselaer Polytech., 1970. Reg. profl. engr., Hawaii, Tex., Okla. Resident engr. Goodkind & O'Dea, Inc., Montclair, N.J., 1967-69; project engr. to program control cons. Parsons, Brinckerhoff, Quade & Douglas, Inc., Portland, Oreg., and internat. cities, 1970-81; constrn. mgr. CH 2M Hill, Portland, Oreg., 1981-83; project engr. Bechtel Constrn., Inc., Portland, 1983-84; asst. exec. dir., dir of facilities engring. & constrn. Dallas Area Rapid Transit, 1984-88; v.p., transp. sect. mgr. HDR Engring., Inc., Dallas, 1988—. Bd. dirs. Garland (Tex.) Civic Theater, 1985-87. Mem. Am. Soc. Civil Engrs. (urban transp. com. 1988—) Am. Assn. Cost Engrs. (coun., bd. dirs., 1982-84, v.p. Oreg. sect. 1984), Am. Bus. Clubs (v.p. Duck Creek chpt., Garland Tex., 1986). Democrat. Home: 3705 Oakridge Cir Garland TX 75040-3559 Office: HDR Engring Inc Ste 125 12700 Hillcrest Rd Dallas TX 75230-2096

LUU, JANE, astronomer. With Harvard Smithsonian Ctr., Cambridge, Mass. Recipient Annie Jump Cannon award Am. Astron. Soc., 1991. Office: Harvard Smithsonian Ct 60 Garden St MS 42 Cambridge MA 02138*

LUXEN, ANDRE JULES MARIE, chemist; b. Malmedy, Belgium, Dec. 1, 1954; s. Leon Luxen and Maria Mathonet. MS in Chemistry, U. Liege, Belgium, 1977, DSc in Organic Chemistry, 1983. Vis. asst. prof. UCLA, 1986-87; asst. prof. U. Calif., L.A., 1987-89; co-dir. cyclotron U. Brussels, Belgium, 1989—; cons. Ion Beam Application, Louvain-La-Neuve, Belgium, 1989—. Contbr. articles to profl. jours. Mem. Am. Chem. Soc., Soc. Chimiques de Belgique, European Assn. Nuclear Medicine, Nuclear Medicine Soc. Achievements include development of new radiopharmaceuticals labeled with short half-life isotopes for positron emission tomography, automatisation of chemical synthesis of these compounds and in-vivo study of these new radiopharmaceuticals. Home: Floriheid 1, 4960 Malmedy Belgium Office: U Brussels, Cyclotrow, 808 Route De Lennik, 4070 Brussels Belgium

LUYENDYK, BRUCE PETER, geophysicist, educator, institution administrator; b. Freeport, N.Y., Feb. 23, 1943; s. Pieter Johannes and Frances Marie (Blakeney) L.; m. Linda Kay Taylor, Sept. 7, 1967 (div. 1979); 1 child, Loren Taylor Luyendyk; m. Jaye Ellen UpDeGraff, Oct. 12, 1984 (div. 1987). BS Geophysics, San Diego State Coll., 1965; PhD Marine Geophysics, Scripps Inst. Oceanography, 1969. Registered geophysicist, Calif. Geophysicist Arctic Sci. and Tech. Lab. USN Electronics Lab. Ctr., 1965; lectr. San Diego State Coll., 1967-68; postgrad rsch. geologist Scripps Inst. Oceanography, 1969; postdoctoral fellow Dept. Geology and Geophysics Woods Hole Oceanographic Instn., 1969-70, asst. sci. dept. geology and geophysics, 1970-73; lectr. dept. meteorology and oceanography U. Mich. at Woods Hole Oceanographic Inst., 1971-72; lectr. dept. geol. scis. U. Calif., Santa Barbara, 1973—, acting dir. Inst. Crustal Studies, 1987-88, dir. Inst. Crustal Studies, 1988—; com. marine geophysical formats NASCO, 1971; working group problems mid-Atlantic ridge NAS NRC, 1972; working group Inter-Union commn. Geodynamics; participant,chief scientist oceanographic cruises, geol. expeditions. Editorial bd. Geology, 1975-79, Marine Geophysical Rschs., 1976-92, Jour. Geophysical Rsch., 1982-84, Tectonophysics, 1988-92, Pageoph, 1988—; contbr. articles to profl. jours., chpts. to books, encys. Recipient Newcomb Cleve. prize AAAS, 1980, Antartic Svc. medal U.S. NSF, Dept. Navy, 1990, numerous rsch. grants, 1970-93. Fellow Geol. Soc. Am.; mem. Am. Geophysical Union, Soc. Exploration Geophysics. Home: 4645 Via Vistosa Santa Barbara CA 93110 Office: Univ of Cal Santa Barbara Inst Institute for Crustal Studies Santa Barbara CA 93106-1100

LWOFF, ANDRÉ MICHEL, retired microbiologist, virologist; b. Ainay-le-Chateau, France, May 8, 1902; s. Salomon and Marie (Siminovitch) L.; m. Marguerite Bourdaleix, Dec. 5, 1925. Licence es Scis. Naturelles, Paris, 1921, D Med., 1927, D Scis. Naturelles, 1932; hon. doctorate, U. Glasgow (Scotland), 1963, U. Chgo., Oxford (Eng.) U., Belgium, 1964, U. Brussels, 1969, Harvard U., 1986. Became fellow Pasteur Inst., Paris, 1921, asst., 1925, head lab., 1929, head dept. microbiol. physiology, 1938; prof. microbiology Faculty Scis., U. Sorbonne, Paris, 1959-68; head Cancer Research Inst., Villejuif, 1968-72; pres. French Family Planning Movement, 1970-74; Researcher nature and function of growth factors, physiology of viruses, induction and repression of viruses, phenomenon of lysogenic bacteria, latent bacterial virus, protozoa nutrition, vitamins as microbial growth factors, vitamin function as co-enzyme, thermoresistance of viral devel. and virulenza, role of fever in fight of animals against viral infections. Author: Problems of Morphogenesis in Ciliates: the Kinetosomes in Development, Reproduction and Evolution, 1950; Biological Order, 1962; also articles. Recipient Nobel prize (with François Jacob and Jacques Monod) in medicine and physiology, 1965. Fellow Royal Soc. (fgn.), Nat. Acad. Scis.; mem. N.Y. Acad. Scis., Nat. Acad. Scis., Academie des Sciences, 1976. Home and Office: 69 ave de Suffren, 75007 Paris France

LYBEROPOULOS, ATHANASIOS NIKOLAOS, mathematician, engineer; b. Athens, Greece, Dec. 2, 1959; came to U.S. 1984; s. Nikolaos Athanasios and Evgenia (Paraskevopoulos) L. ScM in Marine and Mech. Engring., Nat. Tech. U. Athens, 1984; ScM in Applied Math., Brown U., Providence, 1985; PhD in Applied Math., Brown U., 1989. Vis. asst. prof. math. U. Okla., Norman, 1989—. Brown U. grad. fellow, 1984-85. Mem. Am. Math. Soc., Soc. for Indsl. and Applied Math. Home: 17 Gennimata St, Athens Greece 11524 Office: U Okla Dept Math 601 Elm Ave Norman OK 73019-0001

LYDING, JOSEPH WILLIAM, electrical and computer engineer; b. Waukegan, Ill., May 12, 1954; s. Joseph William and Helen (Tamerian) L.; m. Barbara Jeanne Cohen, Nov. 7, 1982. BS, Northwestern U., Evanston, Ill., 1976, MS, 1978, PhD, 1983. Asst. prof. U. Ill., Urbana, 1984-88, assoc. prof., 1988-93; prof., 1993—; visiting asst. prof. Northwestern U., Evanston, 1983-84. Contbr. articles to Rev. Sci. Instrum., Jour. Vac. Sci. Technol. IBM Postdoctoral fellow, 1983-84, Beckman fellow U. Ill., 1987-88; Univ. Rsch. Initiative grantee Office Naval Rsch., 1992—; recipient Arthur K. Doolittle award Am. Chem. Soc., 1983, Outstanding Teaching award Tau Beta Pi, 1984. Mem. Am. Phys. Soc., Sigma Xi. Achievements include patent for variable temperature scanning tunneling microscope. Home: 3010 Meadowbrook Ct Champaign IL 61821 Office: Univ Illinois Beckman Inst 405 N Mathews Urbana IL 61801

LYLE, ROBERT EDWARD, chemist; b. Atlanta, Jan. 26, 1926; s. Robert Edward and Adaline (Cason) L.; m. Gloria Gilbert, Aug. 28, 1947. B.A., Emory U., 1945, M.S., 1946; Ph.D., U. Wis.-Madison, 1949. Asst. prof. Oberlin Coll., Ohio, 1949-51; asst. prof. U. N.H., Durham, 1951-53; assoc. prof. U. N.H., 1953-57, prof., 1957-76; prof., chmn. dept. chemistry U. North Tex., Denton, 1977-79; v.p. chemistry, chem. engr. S.W Rsch. Inst., San Antonio, 1979-91; v.p. GRL Cons., San Antonio, 1992—; vis. prof. U. Va., Charlottesville, 1973-74, U. Grenoble, France, 1976; adj. prof. Bowdoin Coll., Brunswick, Maine, 1975-79, U. Tex., San Antonio, 1985—. Mem. editorial bd. Index Chemicus, 1976—. USPHS fellow Oxford U., Eng., 1965; recipient honor scroll award Mass. chpt. Am. Inst. Chemistry, 1971; Harry and Carol Mosher awardee, 1986. Fellow AAAS; mem. AACR, Am. Chem. Soc. (councilor 1965-84, 86-92, medicinal chemistry div.), Royal Soc. Chemistry, Alpha Chi Sigma (editor Hexagon 1992—). Methodist. Home: 12814 Kings Forest St San Antonio TX 78230-1511 Office: GRL Cons 12814 Kings Forest St San Antonio TX 78230

LYLES, ANNA MARIE, zoo curator, ornithologist; b. Santa Maria, Calif., Mar. 3, 1961; d. William Murray and Valera Josephine (Whitford) L. BS, Yale U., 1983; PhD, Princeton U., 1990. Rsch. asst. San Diego Zoo, 1983-84; teaching asst. Princeton (N.J.) U., 1984-88; curatorial intern N.Y. Zool. Soc., Bronx, 1989-91, asst. curator, 1991—; exec. bd. Yale Sci. and Engring. Assn., New Haven, 1985—; mem. grad. com. on sci., tech. and human values Princeton U., 1986-89. Author: N.Am. Regional Stud Book for Scarlet Ibis, 1991; contbr. to Inbreeding in Natural Populations, 1986; editor Procs. Colonial Waterbird Workshop, 1992. Grad. fellow NSF, 1985-88; grantee Marine Biol. Lab., 1988. Fellow Am. Assn. Zool. Parks and Aquariums (chair heron and ibis adv. group 1991—, small population mgmt. adviser 1992—); mem. Soc. Conservation Biology, Colonial Waterbird Soc., Internat. Union Conservation of Nature (captive breeding specialist group), Sigma Xi (grantee 1986). Office: NYZS/Wildlife Consv Soc Internat Wildlife Park Ornithology Dept Bronx NY 10460

LYMAN, BEVERLY ANN, biochemical toxicologist; b. Phila., Aug. 22, 1956; m. Henry M. Laboda, Aug. 22, 1981; 1 child, Alex. MS, U. Pa., 1982, Hahnemann Med. Coll., 1984; PhD, Hahnemann Med. Coll., 1986. Lab. supr. Children's Hosp. of Phila., 1978-81; edn. coord. Cooper Hosp., Rutgers U., Camden, N.J., 1981-82; postdoctoral fellow Chem. Industry Inst. of Toxicology, Research Triangle Park, N.C., 1987-88; asst. prof. U. Tenn., Memphis, 1989-93; assoc. prof., 1993—. Author: Applied Sciences Review-Biochemistry, 1993; contbr. book: Clinical Chemistry-Concepts and Applications, 1992. Mem. Germantown (Tenn.) Dem. Party Orgn., 1991—. Recipient Excellence in Teaching award U. Tenn. Student Govt. Assn., 1992, 93. Mem. AAAS, Am. Soc. Med. Tech. (cons. editor 1990—), Soc. Toxicology, Am. Soc. Biochemistry and Molecular Biology. Office: U Tenn 800 Madison Ave Memphis TN 38163

LYMAN, CHARLES EDSON, materials scientist, educator; b. Willimantic, Conn., Mar. 7, 1946; s. Edson Hunt and Sylvia (Hill) L.; m. Valerie Ann Livingston, Aug. 30, 1984. BS, Cornell U., 1968; PhD, MIT, 1974. Postdoctoral fellow dept. metallurgy Oxford (England) U., 1974-76; asst. prof. Rensselaer Poly. Inst., Troy, N.Y., 1976-80; staff scientist E.I. DuPont de Nemours, Wilmington, Del., 1980-84; assoc. prof. Lehigh U., Bethlehem, Pa., 1984-90, prof., 1990—; electron microscopy steering com. Argonne (Ill.) Nat. Lab., 1984—. Author, editor: Scanning Electron Microscopy, X-Ray Microanalysis, and Analytical Electron Microscopy: A Laboratory Workbook, 1990; contbr. articles to profl. jours. Mem. Electron Microscopy Soc. Am. (pres. 1991), Microbeam Analysis Soc. (bd. dirs. 1993—), Am. Soc. Materials Internat., Am. Chem. Soc., Burnside Plantation Inc. (v.p. 1987-93, pres. 1993—). Home: 44 N New St Bethlehem PA 18018-5814 Office: Lehigh U Whitaker Lab 5 E Packer Ave Bethlehem PA 18015

LYMAN, HOWARD B(URBECK), psychologist; b. Athol, Mass., Feb. 12, 1920; s. Stanley B(urbeck) and Ruth Mary (Gray) L.; A.B., Brown U., 1942; M.A., U. Minn., 1948; Ph.D., U. Ky., 1951; m. Patricia Malone Taylor, May 4, 1966; children—David S., Nancy M., D. Jane Lyman Paraskevopoulos; stepchildren—Richard P. Taylor, Martha C. Kitsinis, Robert M. Taylor,

David P. Taylor. Acting dir. student personnel E. Tex. State Tchrs. Coll., Commerce, 1948-49; counselor, research asst. univ. personnel office U. Ky., Lexington, 1949-51; research psychologist tests and measurements U.S. Naval Exam. Center, Norfolk, Va. and Gt. Lakes, Ill., 1951-52; asst. prof. psychology U. Cin., 1952-62, asso. prof., 1962-85 , prof. emeritus, 1986—; dir. Acad. Edn. and Research in Profl. Psychology Ohio 1975-84. Served with AUS, 1942-46. Licensed psychologist, Ohio. Fellow APA, Am. Psychol. Soc.; mem. Ohio Psychol. Assn. (dir. 1960-84 , Distinguished Service award 1974), Midwestern, Cin. psychol. assns., Assn. Measurement and Evaluation in Guidance, Nat. Council Measurement in Edn., Cheiron Soc., Psi Chi. Author: Single Again, 1971; Test Scores and What They Mean, 5th edit., 1990; editor Ohio Psychologist, 1967-79. Home: 3422 Whitfield Ave Cincinnati OH 45220-1525

LYMBERIS, COSTAS TRIANTAFILLOS, environmental engineer; b. Athens, Greece, Mar. 29, 1944; came to U.S. 1962; s. Triantafillos and Stella (Orologa) L.; m. Gwyneth Johnson, Dec. 20, 1967 (div. 1971); m. Dionisia Helen Theofilopoulos, Sept. 24, 1972; children: Stella C., Cleo C. BS in Physics, UCLA, 1967; MS in Physics, Calif. State U., L.A., 1970; postgrad., U. Calif., Berkeley, 1971. Mech. and corrosion specialist Petrola Internat., Athens, 1977-83; mech. engr. AMAX-USMR, Carteret, N.J., 1983-85; engring. mgr. Mimikos Group of Cos., Pireaus, Greece, 1985-86; environ. mgr. Elsco, N.J. div., South Plainfield, 1986-88; engiron. engr. TRC Environ. Cons., Somerset, N.J., 1988-89; mech. engr. Leonard Engring., Cranford, N.J., 1989-90; corrosion engr. Ebasco Svcs., Inc., N.Y.C., 1991-92; environ. engr. Corps of Engrs., U.S. Army, N.Y.C., 1992—; lectr./instr. U. Laverne, Greece, 1975-82; instr. physics Seton Hall U., South Orange, N.J., 1984, 88, 90; prof. physics, math. Middlesex County Coll., Edison, N.J., 1987—. Mem. ASME, AAAS, Am. Assn. Physics Tchrs., Nat. Assn. Corrosion Engrs. (cert. corrosion specialist), Sigma Pi Sigma. Democrat. Greek Orthodox. Home: 34 Robin Rd Rumson NJ 07760-1829 Office: US Army Corps of Engrs Rm 1811 26 Federal Plaza New York NY 10278-0090

LYNAM, DONALD RAY, environmental health scientist; b. Carlisle, Ky., Oct. 15, 1938; s. T.R. and Clarine S. (Stephenson) L.; m. Judith Galbraith, Sept. 3, 1960; children: Donald R. Jr., Sean Galbraith, Tina Clarissa. BS, U. Ky., 1961, MS, 1963; MS, U. Cin., 1968, PhD, 1972. Registered profl. engr.; cert. indsl. hygienist. Materials engr. Ky. Dept. Hwys., Frankfort, 1966; dir. environ. health Internat. Lead Zinc Rsch. Orgn., Inc., N.Y.C., 1971-81; dir. air conservation indsl. hygiene and safety Ethyl Corp., Baton Rouge, 1981—. Author: (chpt.) Highway Pollution, 1981; editor: Environmental Lead, 1981; contbr. articles to profl. jours. Capt. USAF, 1963-66. Mem. Am. Indsl. Hygiene Assn. (chmn. N.Y. Met. sect. 1976, gen. conf. chmn. 1992—), Air and Waste Mgmt. Assn., N.Y. Acad. Scis., Soc. Occupations and Environ. Health. Home: 15123 Seven Pines Ave Baton Rouge LA 70817 Office: Ethyl Corp 451 Florida Baton Rouge LA 70801

LYNCH, CHARLES ANDREW, chemical company executive; b. Bklyn., Jan. 6, 1935; s. Charles Andrew and Mary Martina (McEvoy) L.; m. Marilyn Anne Monaco, July 30, 1960; children: Nancy Callan, Cara Martina. BS, Manhattan Coll., 1956; PhD, U. Notre Dame, 1960. Rsch. chemist Esso Rsch. & Engring. Co., Linden, N.J., 1960-65; rsch. supr. FMC Corp., Organic Chem. Div., Balt., 1965-72; rsch. mgr. FMC Corp., Indsl. Chem. Div., Princeton, N.J., 1972-74; exec. v.p. Am. Oil & Supply Co., Newark, 1974-80; tech. dir., dir. sales & mktg., dir. rsch. & bus. devel., v.p. tech. Hatco Corp., Fords, N.J., 1981—. Contbr. articles to profl. jours.; patentee in field (U.S. and foreign). Mem. Am. Chem. Soc., Am. Oil Chemists Soc., Soc. Automotive Engrs., Soc. Tribiologists and Lubrication Engrs. (N.Y. sect. chmn. 1980-81), Ind. Lubricant Mfrs. Assn. (bd. dirs. 1985-88), Commercial Devel. Assn., Chem. Mgmt. and Resources Assn. Office: Hatco Corp 1020 King George Post Rd Fords NJ 08863-0601

LYNCH, CHARLES THEODORE, SR., materials science engineering researcher, administrator, educator; b. Lima, Ohio, May 17, 1932; s. John Richard and Helen (Dunn) L.; m. Betty Ann Korkolis, Feb. 3, 1956; children: Karen Elaine Ostdiek, Charles Theodore Jr., Richard Anthony, Thomas Edward. BS, George Washington U., 1955; MS, U. Ill., 1957, PhD in Analytical Chemistry, 1960. Group leader ceramics div. Air Force Materials Lab., Wright-Patterson AFB, Ohio, 1962-66; lectr. in chemistry Wright State U., Dayton, Ohio, 1964-66; chief advanced metall. studies br. Air Force Materials Lab., Wright-Patterson AFB, Ohio, 1966-74, sr. scientist, 1974-81; head materials div. Office of Naval Rsch., Arlington, Va., 1981-85; pvt. practice cons. Washington, 1985-88; sr. engr. space ops. Vitro Corp., Washington, 1988—; USAF liaison mem. NMAB Panels on Solids Processing, Ion Implantation and Environ. Cracking, Washington, 1965-68, 78, 81; U.S. rep. AGARD structures and materials panel NATO, 1983-85. Co-author: Metal Matrix Composites, 1972; editor, author: Practical Handbook of Materials Science, 1989; editor: (series) Handbook of Materials Science, vol. I, 1974, vol. II, 1975, vol. III, 1975; vice chmn. editorial bd. Vitro Corp. Tech. Jour., 1989-92, chmn., 1993—; contbr. articles to profl. jours. including Jour. Am. Ceramics Soc., Analytical Chemistry, Sci., Transactions AIME, Corr. Jour., Jour. Inorganic Chemistry, SAMPE, Jour. Less Common Metals. Mem., soloist George Washington U. Traveling Troubadours, Washington, 1950-55; choir dir. Trinity United Ch. of Christ, Fairborn, Ohio, 1966-81, Univ. Bapt. Ch., Champaign, Ill., 1957-60, Chapel II, Wright-Patterson AFB, Ohio, 1960-64; pres. Pub. Sch. PTO, 1967-69. 1st lt. USAF, 1960-62. Bailey scholar U. Ill., 1958-60; recipient Commendation medal USAF, 1962, Outstanding Achievement cert. NASA, 1992, award Soc. for Tech. Comm. Publ., 1993. Mem. Am. Chem. Soc. (treas. 1966-67, chmn. audit sect. 1967-68), ASM Internat. (sec. oxidation and corrosion com. 1980-81, chmn. 1981-82). Presbyterian. Achievements include patents for new corrosion inhibitors including encapsulated types, and for alkoxides and oxides; co-development of the refractory ceramic Zyttrite, the first high density translucent zirconia made from thermal or hydrolytic decomposition of mixed alkoxides followed by hot pressing; pioneered general approach of organometallic compounds as precursors of high purity, fine particulate, materials. Home: 5629 Kemp Ln Burke VA 22015-2041 Office: Vitro Corp Space Ops 400 Virginia Ave SW Ste 825 Washington DC 20024

LYNCH, HARRY JAMES, biologist; b. Glenfield, Pa., Jan. 18, 1929; s. Harry James and Rachel (McComb) L.; m. Pokum Lee Lynch. BS, Geneva Coll., Beaver Falls, Pa., 1957; PhD, U. Pitts., 1971; postgrad. Bio-Space Tech. Tng. Program, NASA and U. Va., 1970. Clin. chemist West Penn Hosp., Pitts., 1955-56; grad. teaching asst. U. Pitts., 1966-71, sr. teaching fellow, 1971; postdoctoral fellow MIT, Cambridge, 1973-75, rsch. assoc. dept. nutrition, lab. neuroendocrine regulation, 1973-75, lectr., 1976-81, rsch. scientist dept. brain and cognitive sci., 1982-92; cons. Ctr. for Brain Scis. and Metabolism Charitable Trust, 1992—; cons. Ctr. for Brain Scis. and Metabolism Charitable Trust. Contbr. more than 60 articles on the pineal gland to profl. jours. and books; patentee on implantable programmed microinfusion apparatus, 1981. With USN, 1950-54. NIH postdoctoral fellow 1971-73. Mem. Soc. Light Treatment and Biol. Rhythms. Democrat. Avocation: study of animal behavior. Office: MIT E25-615 77 Massachusetts Ave Cambridge MA 02139-4307

LYNCH, JOHN JOSEPH, city and regional planner, consultant; b. N.Y.C., Nov. 7, 1963; s. John Joseph Lynch and Margaret Frances (Woods) Mulligan. BA, NYU, 1985; MS, Pratt Inst., 1990. Cert. planner. Consulting planner Jed S. Marcus, Esq.; Insite, Bklyn., 1986-87; dir. planning Tim Miller Assocs., Inc., Cold Spring, N.Y., 1987—. Democrat. Home: 410 W 24th St # 6F New York NY 10011

LYNCH, JOHN PATRICK, aerospace engineer; b. Morgantown, W.Va., Aug. 29, 1956; s. George Lamont and Marian (Cira) L. BS in Physics, George Mason U., 1979. Computer scientist, mem. tech. staff Computer Scis. Corp., Silver Spring, Md., 1979-89; aerospace engr. NASA Goddard Space Flight Ctr., Greenbelt, Md., 1989—. Contbr. articles to profl. jours. Mem. Am. Astron. Soc., AIAA (sr.), Sigma Pi Sigma, Alpha Chi. Republican. Achievements include responsibility for flight dynamics mission analysis and operations support for both U.S. and internat. cooperative spacelight projects, including Space Transportation System, expendable launch vehicles, long duration balloons and communications, meteorological, planetary, and other scientific research satellites. Home: 4229 Sleepy Lake Dr Fairfax VA 22033-2841 Office: NASA Goddard Space Flight Ctr Flight Dynamics Div 553.3 Greenbelt MD 20771

LYNCH, NANCY ANN, computer scientist, educator; b. Bklyn., Jan. 19, 1948; d. Roland David and Marie Catherine (Adinolfi) Evraets; m. Dennis Christopher Lynch, June 14, 1969; children: Patrick, Kathleen (dec.), Mary. BS, Bklyn. Coll., 1968; PhD, MIT, 1972. Asst. prof. math. Tufts U., Medford, Mass., 1972-73, U. So. Calif., Los Angeles, 1973-76, Fla. Internat. U., Miami, 1976-77; assoc. prof. computer sci. Ga. Tech. U., Atlanta, 1977-82; assoc. prof. computer sci. MIT, Cambridge, 1982-86, prof. computer sci., 1986—; Ellen Swallow Richards chair MIT, 1982-87; cons. Computer Corp. Am., Cambridge, 1984-86, Apollo Computer, Chelmsford, Mass., 1986-89, AT&T Bell Labs, Murray Hill, N.J., 1986-89, Digital Equipment Corp., 1990. Contbr. numerous articles to profl. jours. Mem. Assn. Computing Machinery. Roman Catholic. Office: MIT NE43-525 Cambridge MA 02139

LYNCH, SONIA, data processing consultant; b. N.Y.C., Sept. 17, 1938; d. Espriela and Sadie Beatrice (Scales) Sarreals; m. Waldro Lynch, Sept. 18, 1981 (div. Oct. 1983). BA in Langs. summa cum laude, CCNY, 1960; cert. in French, Sorbonne, 1961. Systems engr. IBM, N.Y.C., 1963-69; cons. Babbage Systems, N.Y.C., 1969-70; project leader Touche Ross, N.Y.C., 1970-73; sr. programmer McGraw-Hill, Inc., Hightstown, N.J., 1973-78; staff data processing cons. Cin. Bell Info. Systems, 1978-89; sr. analyst AT&T, 1989-92; sr. analyst Automated Concepts Inc. Automated Concepts Inc., Arlington, Va., 1992—. Elder St. Andrew Luth. Ch., Silver Spring, 1992—. Downer scholar CUNY, 1960, Dickman Inst. fellow Columbia U., 1960-61. Mem. Assn. for Computing Machinery, Phi Beta Kappa. Democrat. Avocations: needlework, sewing. Home: 13705 Beret Pl Silver Spring MD 20906-3030

LYNCH, THOMAS BRENDAN, pathologist; b. Melbourne, Victoria, Australia, Mar. 27, 1937; s. Francis Michael and Mary Josephine (Fitzgerald) L.; m. Patricia Noreen Behl, Feb. 24, 1962; children: Sophie, Denis, Carla, Bernadette, Rosina, Justin. MBBS, U. Sydney, Australia, 1961; FRCPA, Royal Coll. Pathologists, Australia, 1967. Registrar in pathology Mater Misericordiae Hosp., Brisbane, Queensland, Australia, 1962-67; pathologist in charge Commonwealth Health Lab., Rockhampton, Queensland, 1968-73; cons. pathologist Base Hosp., Rockhampton, 1974—. Discovered Duffy 3; cured amoebic meningitis, 1971. Mem. Royal Coll. Pathologists, Australian Family Assn. Roman Catholic. Office: Anzac House, Rockhampton Queensland 4700, Australia

LYNDEN-BELL, DONALD, astronomer; b. Dover, Kent, Eng., Apr. 5, 1935; s. Lachlan Arthur and Monica Rose (Thring) L.; m. Ruth Marion Truscott, July 1, 1961; children: Marion, Edward. BA, U. Cambridge, Eng., 1956; PhD, U. Cambridge, 1960; DSc (hon.), U Sussex, Eng., 1987. Asst. lectr. math. Clare Coll., Cambridge, 1962-65; prin. sci. officer Royal Greenwich Obs., Sussex, 1965-72; prof. astrophysics U. Cambridge and Clare Coll., 1972—; dir. Inst. Astronomy, U. Cambridge, 1972-77, 82-87, 92—; vis. prof. U. Sussex, 1970-72; vis. Oort prof. Leiden U., Netherlands, 1992. Harkness fellow Calif. Inst. Tech., Pasadena, 1960-62, rsch. fellow Clare Coll., 1960-65; recipient Schwarzsch medal Astronomy Assn., Fed. Republic of Germany, Dirk Brouwer prize Am. Astron. Soc., 1990; Einstein fellow Israeli Acad., 1990. Fellow Royal Soc., Royal Astron. Soc. (pres., Eddington medal 1984, Gold medal 1993), Cambridge Philos. Soc. (pres.); mem. U.S. Nat. Acad. Scis. (fgn. assoc.). Mem. Ch. of Eng. Avocation: hill walking. Home: 9 Storey's Way, Cambridge CB3 0DP, England Office: Inst Astronomy, Obs, Madingley Rd, Cambridge CB3 0HA, England

LYNT, RICHARD KING, microbiologist; b. Washington, Feb. 25, 1917; s. Richard King and Elsie Ackerman (King) L.; m. Elizabeth Mackenzie Cissel, Nov. 17, 1944; children: Richard, Margaret, David. BS in Bacteriology, U. Md., 1939, MS in Bacteriology and Food Technology, 1942. Grad. asst. bacteriology U. Md., College Pk., Md., 1939-40; food inspector, bacteriologist D.C. Health Dept., Washington, 1940-42; lab officer, hosp. corps, med. svc. corps USNR, Bethesda, Md., Oceanside, Calif., 1942-46; bacteriologist E.R. Squibb & Sons, New Brunswick, N.J., 1946-48; rsch. fellow Rutgers U., New Brunswick, 1948-49; bacteriologist NIH Nat. Inst. Allergy and Infectious Disease, Bethesda, 1951-63; microbiologist U.S. FDA Div. Microbiology, Washington, 1963-83; microbiologist cons., retired Silver Spring, Md., 1983—; mem. interagy. botulism rsch. coordinating com., Washington, 1968-83; faculty mem. Am. Soc. for Microbiology Continuing Edn. Com., Dallas, 1981; mem. People to People Enzymology Del. to People's Republic of China, 1985. Author: (with others) Clostridium Botulinum, 1970-84; contbr. chpts to books, articles to profl. jours. Mem. Pinecrest Citizens Assn., Silver Spring, 1960—; adminstrv. bd. Marvin Meml. United Meth. Ch., Silver Spring, 1985—; judge of elections Bd. Suprs. Elections, Montgomery County, Md., 1988-92. Served to lt. Med. Svc. Corps, USNR, 1942-46. Recipient Awards of Merit, U.S. FDA, 1974. Mem. N.Y. Acad. Sci., Inst. Food Technologists, Naval Res. Assn., Am. Soc. for Microbiology, AAAS. Republican. Methodist. Achievements include isolation of type B Influenza Virus in monkey kidney tissue culture; antigenicity and heat resistance of Clostridium Botulinum, grouping proteolytic type A,B, and F strains and nonproteolytic types B,E,F strains in homogeneous gr. Home: 316 Penwood Rd Silver Spring MD 20901-2716

LYON, BETTY CLAYTON, mathematics educator; b. Amarillo, Tex., Dec. 5, 1929; d. John Ratliff and Dovie Belle (Strasburger) Clayton; m. James Keith Lyon; 1 child, Jane Randolph Lyon Hart. BS in Math., Purdue U., 1952; MS in Math., U. Nebr., Omaha, 1969; PhD in Math. EDn., U. Nebr., 1977. Cert. elem. and secondary tchr., Nebr. Tchr. Central High Sch., Omaha, 1952-54; asst. prof., chmn. dept. math. Bellevue (Nebr.) Coll., 1969-74; grad. asst. dept. math. U. Nebr., Lincoln, 1974-78; asst. prof. dept. math. Kans. State U., Manhattan, 1979-83, U. Nebr., Omaha, 1983-85; prof. math. scis. Eastern N.Mex. U., Portales, 1985—; presentor math. topics confs. and schs. Editor: Journal of the New Mexico Council of Teachers of Mathematics; contbr. articles to profl. jours. Mem. chancel choir First Presbyn. Ch., Clovis, N.Mex., 1985—. NSF scholar, 1972; faculty fellow Tex. Christian U., 1973; grantee P.E.O., 1975, State of N.Mex., 1986, 88, 89, Eisenhower, 1990, 91, 92, 93; recipient NUCEA award. Mem. Nat. Coun. Tchrs. Math., Math. Assn. Am., N.Mex. Coun. Tchrs. Math., N.Mex. Acad. Sci., Sch. Sci. and Math., PEO, Phi Kappa Phi. Office: Ea N Mex U Dept Math Scis Sta 18 Portales NM 88130

LYON, RICHARD HAROLD, educator, physicist; b. Evansville, Ind., Aug. 24, 1929; s. Chester Clyde and Gertrude (Schucker) L.; m. Jean Wheaton; children—Katherine Ruth, Geoffrey Cleveland, Suzanne Marie. A.B., Evansville Coll., 1952; Ph.D. in Physics (Owens-Corning fellow), Mass. Inst. Tech., 1955; D.Eng., U. Evansville, 1976. Asst. prof. elec. engring. U. Minn., Mpls., 1956-59; Mem. research staff Mass. Inst. Tech., 1955-56, lectr. mech. engring., 1963-69, prof. mech. engring., 1970—, head mechanics and materials div., 1981-86; NSF postdoctoral fellow U. Manchester, Eng., 1959-60; sr. scientist Bolt Beranek & Newman, Cambridge, 1960-66, v.p., 1966-70; chmn. Cambridge Collaborative, Inc., 1972-90; v.p. Grozier Pub., Inc., 1972; pres. Grozier Tech. Systems, 1976-82, RH Lyon Corp., 1976—. Author: Transportation Noise, 1974, Theory and Applications of Statistical Energy Analysis, 1975, Machinery Noise and Diagnostics, 1987. Bd. dirs. Boston Light Opera, Ltd., 1975; mem. alumni bd. U. Evansville, 1988—. Fellow Acoustical Soc. Am. (assoc. editor Jour. 1967-74, exec. coun. 1976-79, v.p. elect 1988-89, v.p. 1989-90, pres. elect 1992-93, pres. 1993—); mem. Sigma Xi, Sigma Pi Sigma. Research, publs. in fields of nonlinear random oscillations, energy transfer in complex structures, sound transmission in marine and aerospace vehicles, building acoustics, environmental noise, machinery diagnostics, home theater audio systems. Home: 60 Prentiss Ln Belmont MA 02178-2021 Office: MIT Rm 3-366 Dept Mechanical Engring Cambridge MA 02139 also: RH Lyon Corp 691 Concord Ave Cambridge MA 02138

LYON, ROGER WAYNE, information scientist; b. Albany, Oreg., July 1, 1958; s. Ralph Wesley and Claudina Elaine (Hoke) L.; m. Cherie D. Buske, Apr. 8, 1980; children; Mark Allen, James Asher. Student, Rock Valley Coll., 1984. Operator Clinicare, Inc., Loves Park, Ill., 1988-89, systems mgr., 1989-91, supr. info. svcs. opts., 1991—; co-founder Catalyst Cons. Mem. Digital Equip. Corp. Users Soc. Office: CliniCare Inc 7124 Windsor Lake Pkwy Loves Park IL 61111

LYONS, JERRY LEE, mechanical engineer; b. St. Louis, Apr. 2, 1939; s. Ferd H. and Edna T. Lyons. Diploma in Mech. Engring., Okla. Inst. Tech., 1964; M.S.M.E., Southwest U., 1983, Ph.D. in Engring. Mgmt., 1984. Registered profl. engr., Calif.; cert. mfg. engr. in product design. Project engr. Harris Mfg. Co., St. Louis, 1965-70, Essex Cryogenics Industries, St. Louis, 1970-73; mgr. engring. rsch. Chemetron Corp., St. Louis, 1973-77; cons. fluid controls Wis. U., 1977—; pres., chief exec. Yankee Ingenuity, Inc., St. Louis, 1974—; v.p., gen. mgr. engring. rsch. and devel. Essex Fluid Controls div. Essex Industries, Inc., St. Louis, 1977-90; pres. Lyons Pub. Co., St. Louis, 1983—; pres., chief exec. officer Innovative Controls div. Yankee Ingenuity, Inc., Ft. Wayne, Ind., 1991—; chmn. exec. bd. continuing engring. edn. in St. Louis for U. Mo., Columbia, 1980-81; bd. dirs. Intertech., Inc., Houston; cons. fluid power dept. Bradley U., Peoria, 1977-84. Author: Home Study Series Course on Actuators and Accessories, 1977, The Valve Designers Handbook, 1983, The Lyons' Encyclopedia of Valves, 1975, 93, The Designers Handbook of Pressure Sensing Devices, 1980, Special Process Applications, 1980; co-author: Handbook of Product Liability, 1991; contbr. articles to profl. jours. Served with USAF, 1957-62. Recipient Winston Churchill medal, 1988, Dwight D. Eisenhower Achievement award of honor, 1990; named Businessman of Week (KEZK radio), Eminent Churchill fellow Winston Churchill Wisdom Soc. Fellow ASME; Mem. Soc. Mfg. Engrs. (chmn. Mo. registration com. 1975-90, chmn. St. Louis chpt. 1979-80, internat. dir. 1982-84, 85-87, Engr. of Yr. 1984, internat. award of merit 1985), Nat. Soc. Profl. Engrs., Mo. Soc. Profl. Engrs., St. Louis Soc. Mfg. Engrs. (chmn. profl. devel., registration and cert. com. 1975-79), Instrument Soc. Am. (mem. control valve stability com. 1978-84), Computer and Automated Systems Assn. (1st chmn. St. Louis chpt. 1980-81), St. Louis Engrs. Club (award of merit 1977, Wisdom award of Honor 1987, mem. Wisdom Hall of Fame 1987), Am. Security Council (committeeman 1976—), Nat. Fluid Power Assn. (com. on pressure ratings 1975-77), Am. Legion. Lutheran. Home and Office: 2607 Northgate Blvd Fort Wayne IN 46835-2986

LYONS, JOHN ROLLAND, civil engineer; b. Cedar Rapids, Iowa, Apr. 27, 1909; s. Neen T. and Goldie N. (Hill) L.; BS, U. Iowa, 1933; m. Mary Jane Doht, June 10, 1924; children: Marlene R. Sparks, Sharon K. Hutson, Mary Lynn Lyons. Jr. hwy. engr. Works Projects Adminstrn. field engr. Dept. Transp., State Ill., Peoria, 1930-31, civil engr. I-IV Cen. Office, Springfield, 1934-53, civil engr. V, 1953-66, municipal sect. chief, civil engr. VI, 1966-72. Civil Def. radio officer Springfield and Sangamon County (Ill.) Civil Def. Agy., 1952—. Recipient Meritorious Service award Am. Assn. State Hwy. Ofcls., 1968, 25 Yr. Career Service award State Ill., 1966, Cert. Appreciation Ill. Mcpl. League, 1971. Registered profl. engr., Ill.; registered land surveyor, Ill. Mem. Ill. Assn. State Hwy. Engrs., State Ill. Employees Assn., Am. Pub. Works Assn., Am. Assn. State Hwy. Ofcls., Amateur Trapshooters Assn., Sangamon Valley Radio Club, Lakewood Golf and Country Club, KC, Abe Lincoln Gun Club. Address: 3642 Lancaster Rd Springfield IL 62703-5022

LYONS, PATRICK JOSEPH, management educator; b. N.Y.C., Dec. 12, 1943; s. Joseph Raphael and Catherine (Albrecht) L.; m. Georgette Marie Tumasonis, June 27, 1970; children: Michael, Theresa, George. BEE, Manhattan Coll., 1965; MS in Applied Math., Case Western Res. U., 1967; PhD in Applied Math., Adelphi U., 1973. Systems analyst Grumman Aerospace, Bethpage, N.Y., 1967-75, asst. mgr., 1975-76; prof. mgmt. St. John's U., Jamaica, N.Y., 1976—; cons. mgmt. sci., 1978—. Contbr. articles to profl. jours. Adult edn. instr. Sacred Heart Ch., Bayside, N.Y., 1983—. Mem. Am. Assn. for Artificial Intelligence, Inst. Mgmt. Sci. Roman Catholic. Avocations: photography, jogging. Office: St John's U Coll Bus Jamaica NY 11439

LYTLE, JOHN ARDEN, chemical engineer; b. Columbus, Ohio, Oct. 29, 1949; s. Richard Arden and Helen Marie (Hiser) L.; m. Cynthia Anne Kloss, Sept. 5, 1971; children: Peter, Liz. BS, Miami U., Oxford, Ohio, 1971; M of Environ. Scis., Miami U., 1973; B in Chem. Engring., Ohio State U., 1986. Registered profl. engr. Ohio, Mich. Vol. Peace Corps, Belo Horizonte, Brazil, 1974-75; environ. scientist Burgess & Niple, Ltd., Columbus, Ohio, 1976-86; dir. of Air Pollution Svcs. Burgess & Niple, Ltd., Columbus, 1980-85; engr. Burgess & Niple, Ltd., 1986-90, sect. dir., 1990—; tech. counsel Ohio Air Quality Devel. Authority, Columbus, 1992—. Mem. NSPE, Am. Inst. Chem. Engrs., Ohio Soc. Profl. Engrs. Avocations: backpacking, bicycling, tennis, golf, computers. Office: Burgess & Niple Ltd 5085 Reed Rd Columbus OH 43220

LYTTON, ROBERT LEONARD, civil engineer, educator; b. Port Arthur, Tex., Oct. 23, 1937; m. Robert Odell and Nora Mae (Verrett) L.; m. Eleanor Marilyn Anderson, Sept. 9, 1961; children: Lynn Elizabeth, Robert Douglas, John Kirby. BSCE, U. Tex., 1960, MSCE, 1961, PhD, 1967. Registered profl. engr., Tex., La.; registered land surveyor, La. Assoc. Dannenbaum and Assocs., Cons. Engrs., Houston, 1963-65; U.S. NSF fellow U. Tex., Austin, 1965-67, asst. prof., 1967-68; NSF postdoctoral fellow Australian Commonwealth Sci. & Indsl. Rsch. Orgn., Melbourne, Australia, 1969-70; assoc. prof. Tex. A&M U., College Station, 1971-76, prof., 1976-90, Wiley chair prof., 1990—, dir. ctr. for infrastructure engring., 1991—; divsn. head Tex. Transp. Inst., 1982-91; bd. dirs. MLA Labs., Inc., Austin, 1980—; v.p. bd. dirs. MLAW Cons., Inc., 1980—, Austin, ERES Cons., Inc., 1981—, Champaign, Ill. Mem. St. Vincent de Paul Soc., Houston, 1963-65, Redemptorist Lay Mission Soc., Melbourne, Australia, 1969-70. Capt. U.S. Army, 1961-63. Mem. Transp. Rsch. Bd. (chmn. com. A2L06 1987-93), ASCE, Internat. Soc. for Soil Mechanics & Found. Engring. (U.S. rep. tech. com. TC-6 1987-93), Assn. Asphalt Paving Technologists, Post-Tensioning Inst. (adv. bd.), Tex. Soc. Profl. Engrs., NSPE, Internat. Soc. Asphalt Pavements. Roman Catholic. Office: Tex A&M U 508G CE/TTI Bldg College Station TX 77843

LYUBAVINA, OLGA SAMUILOVNA, economist; b. Sverdlovsk, Russia, Feb. 6, 1945; came to U.S., 1988; d. Samuil U. and Ann L. (Gendeleva) L.; m. Alexander Bolonkin, Fb. 5, 1987. MS in Engring. with honors, Moscow Inst. Tech., 1968; PhD in Econs., Leningrad U., 1980; MS in Econs., CUNY, 1991. Researcher Moscow Inst. Rsch. & Devel., 1960-73, sr. researcher Moscow Rsch. Inst. Econs., 1973-88; cons. Delphic Assocs. Inc., Falls Church, Va., 1990, Ctr. Practical Solutions, N.Y.C., 1990—; freelance writer Radio Liberty, N.Y.C., 1990; researcher Coun. Econ. Priorities, N.Y.C., 1989-90, N.Y.C. Bd. Edn., 1992. Author: Effectiveness of New Technology, 1980; contbr. scientific papers and articles to profl. jours. Mem. AAAS, Am. Econ. Assn. Home: 1001 Ave H #9-C Brooklyn NY 11230

LYUBLINSKAYA, IRINA E., physicist, researcher; b. Leningrad, Russia, Jan. 5, 1963; came to U.S., 1991; d. Efim Ya and Esther S. (Stopskaya) L.; m. Vladimir V. Korolev, May 20, 1992; children: Olga V. Korolev, Maria V. Korolev, Samuel J. Korolev. MS, Leningrad State U., 1986, PhD, 1991. Cert. English-Russian Technical Referent-Interpreter. Engr., teaching asst. Leningrad (Russia) Tech. Inst., 1986-88, jr. scientist, asst. prof., 1988-91; vis. lectr. U. Ark., Fayetteville, 1991-92; postdoctoral fellow U. Ark., Little Rock, 1992—. Author: Hydrodynamics and Mass-Exchange in Dispersive Systems "Liquid-Solid", 1987, Hydrodynamcis and Mass-Exchange in Systems "Gas-Liquid", 1990; contbr. articles to profl. jours. Recipient Gold medal Treatise in Biophysics, Moscow, 1980, Julio-Curie scholarship Russian Ministry of Edn., Moscow, 1982-85, Lenin scholarship USSR Ministry of Edn., Moscow, 1985-86, 2d prize for book City Competition for Scientists, Leningrad, 1987, 3d prize for rsch. City Competition for Rschrs., Leningrad, 1989. Jewish. Office: Univ Ark 2801 S University Little Rock AR 72204

MA, CHUENG-SHYANG (ROBERT MA), reproductive physiology educator, geneticist; b. Chenghsien, Chekiang, China, Apr. 5, 1922; s. Yun-Sun and Tung-Chih (Tung) M.; m. Jyu-Fen Shih, Dec. 27, 1942; children: Jim Young, Jean Hsin. BS, Nat. Cen. U., Nanking, People's Republic China, 1948; MS, U. Tenn., 1957; PhD, U. Ill., 1960. Teaching asst. Nat. Taiwan U., Taipei, 1948-54, instr. reproductive physiology farm animals, 1954-55, assoc. prof., 1960-64, prof., 1964-92, prof. emeritus, 1992—; vis. rsch. prof. U. Ill., Champaign, 1973-74. Author: Animal Breeding, 1980; contbr. articles to Poultry Sci., Advances in Neuroendocrinology, Proc. Sabrao Workshop on Animal Genetic Resources, Zeo-Agr. Recipient Disting. Prof. award Taiwan Ministry Edn., 1988, Disting. Agrl. Educator award Taiwan Coun. Agr., 1988. Mem. Chinese Soc. Animal Sci. (exec. com. 1972-83, Disting. Researcher award 1977, Disting. Tchr. award 1980). Home: 2 Alley 3 Ln 30, Chou-Shan Rd, Taipei 10670, Taiwan Office: Nat Taiwan U Dept Animal Sci, 1 Roosevelt Rd Sect 4, Taipei 10764, Taiwan

MA, DAVID I, electronics engineer; b. Taipei, Republic of China, June 14, 1952; came to U.S., 1977; s. Peter P. Y. Ma and Iris (Di) Stoffels; m. Josephine Hsin-Hsin Yung, Oct. 24, 1981; 1 child, Peter Hung-Chi. BS in Physics, Tung Hai U., Tai-Chung, Republic of China, 1975; MS in Physics, U. Md., 1981, PHD in Elec. Engring., 1991. Teaching asst. U. Md., College Park, 1977-81; rsch. physicist Sachs/Freeman Assn., Bowie, Md., 1981-85; rsch. physicist Naval Rsch. Lab., Washington, 1985-89, electronics engr., 1989—. Contbr. articles to profl. publs. 2d lt. Rep. of Taiwan Army, 1975-77. Mem. IEEE, Phi Kappa Phi, Phi Tau Phi. Achievements include patent for planar detector-amplifier arrays and detector-amplifier circuit channels with an insulating layer between the detector and transistor layers. Office: Naval Rsch Lab Code 6804 4555 Overlook Ave SW Washington DC 20375-5347

MA, FENGCHOW CLARENCE, agricultural engineering consultant; b. Kaifeng, Honan, China, Sept. 4, 1919; came to U.S., 1972; s. Chao-Hsiang and Wen-Chieh (Yang) Ma; m. Fanny Luisa Corvera-Achá, Jan. 20, 1963; 1 child, Fernando. BS in Agr., Nat. Chekiang U., Maytan, Kweichow, China, 1942; postgrad., Iowa State U., 1945-46. Cert. profl. agronomist, Republic of China, 1944; registered profl. agrl. engr., Calif. Chief dept. ops. Agrl. Machinery Operation and Mgmt. Office, Shanghai, China, 1946-49; sr. farm machinery specialist Sino-Am. Joint Commn. on Rural Reconstrn., Taipei, Taiwan, Republic of China, 1950-62; agrl. engring. adviser in Bolivia, Peru, Chile, Ecuador, Liberia, Honduras, Grenada, Bangladesh FAO, Rome, 1962-80; consulting agrl. engr. to USAID projects in Guyana & Peru IRI Rsch. Inst., Inc., Stamford, Conn., 1981-82, 83, 85; chief adviser Com. Internat. Tech. Coop., Taipei, 1984-85; prin., cons. agrl. engr. Fengchow C. Ma and Assocs., Inc., Sunnyvale, Calif., 1962—; short consulting missions to Paraguay, Saudi Arabia, Indonesia, Malawi, Swaziland, Barbados, Dominica, Ivory Coast, Vietnam, Philippines and others. Author papers, studies; contbr. articles to profl. publs. Mem. Am. Soc. Agrl. Engrs. Avocations: reading, stamp and coin collecting. Home: 1004 Azalea Dr Sunnyvale CA 94086-6747 Office: PO Box 70096 Sunnyvale CA 94086-0096

MA, JEANETTA PING CHAN, elementary education educator; b. Honolulu, May 21, 1952; d. Kwai and Doris (Yu) Chan; m. John Yum Hung Ma, June 15, 1974; children: Jermaine, Jonathan, Josalyne, Jerrauld. BFA and BEd, U. Manoa, Honolulu, 1974, MEd, 1988. Tchr. adult edn. overseas St. Louis High Sch., Seoul, Korea, 1974-75; tchr. creative cooking Community Sch. for Adults, Aiea and Wapahu, Hawaii, 1977-83; tchr. arts and crafts Waipahu (Hawaii) Intermediate Sch., 1983-86, tchr. maths., 1983-90; instr. Wheeler Intermediate Sch., Wahiawa, 1990—; christian edn. dir. Calvary Assembly of God, Honolulu. Mem. NEA, Hawaii State Tchrs. Assn., Educators for Tech. and Computing in Hawaii, Internat. Soc. Tech. in Edn. Avocations: tennis, swimming, singing, sewing, cooking. Home: 99-598 Hoio St Aiea HI 96701

MA, TONY YONG, chemist, international business consultant; b. Beijing, Mar. 16, 1955; came to U.S., 1985; s. Weixiang and Huiying Ma; m. Susan Wei Wan, Sept. 11, 1982; children: Michael, Lisa. PhD, U. Pacific, 1988. Rsch. asst. U. Pacific, Stockton, Calif., 1985-87; asst. scientist Roy Weston, Inc., Stockton, 1988; rsch. assoc. Johns Hopkins U., Balt., 1989-90; scientist Johnson & Johnson, Skillman, N.J., 1990—; sr. cons. Ribosome, Inc., Princeton, N.J., 1992—; vice dir. scientific info. exch. ctr. SINO-U.S.A., bd. dirs. biochem. & med. soc. Contbr. articles to profl. jours. Postdoctoral fellow Johns Hopkins U., 1989. Mem. Am. Chem. Soc., Phi Kappa Phi. Achievements include research in new methods for antifungal drugs analysis in OTC products, degradation compounds of antifungal drugs, improved purification process of cell adhesion molecules in chicken livers. Office: Johnson & Johnson 199 Grandview Rd Skillman NJ 08558

MA, TSU SHENG, chemist, educator, consultant; b. Guangdong, China, Oct. 15, 1911; came to U.S., 1934, naturalized 1956; s. Shao-ching and Sze (Mai) M.; m. Gioh-Fang Dju, Aug. 27, 1942; children: Chopo, Mei-Mei. BS, Tsinghua U., Peking, 1931; PhD, U. Chgo., 1938. Mem. faculty U. Chgo., 1938-46; prof. Peking U., 1946-49; sr. lectr. U. Otago (N.Z.), 1949-51; mem. faculty NYU, 1951-54; mem. faculty CUNY, 1954—, prof. chemistry, 1958—, prof. emeritus, 1980—; vis. prof. Tsinghua U., 1947, Lingnan U., 1949, NYU, 1954-60, Taiwan U., 1961, Chiangmei U., 1968, Singapore U., 1975; specialist Bur. Ednl. and Cultural Affairs State Dept., 1964, Hongkong, Philippines, Burma, Sri Lanka. Fulbright lectr., 1961-62, 68-69; recipient Benedetti-Pichler award in microchemistry, 1976. Fellow N.Y. Acad. Sci., AAAS, Royal Soc. Chemistry, Am. Inst. Chemists; mem. Am. Chem. Soc., Soc. Applied Spectroscopy, Am. Microchem. Soc., Sigma Xi. Author: Small-Scale Experiments in Chemistry, 1962; Organic Functional Group Analysis, 1964; Microscale Manipulations in Chemistry, 1976; Organic Functional Group Analysis by Gas Chromatography, 1976; Quantitative Analysis of Organic Mixtures, 1979; Modern Organic Elemental Analysis, 1979; Organic Analysis Using Ion-Selective Electrodes, 1982; Trace Element Determination in Organic Materials, 1988; editor: Mikrochimica Acta, 1965-89; contbr. articles to profl. jours., chpts. to 8 books. Achievements include 1 patent; research in trace element analysis, microchemical investigation of medicinal plants, organic analysis and synthesis in the milligram to microgram range, and the use of small-scale, inexpensive equipment to teach chemistry. Home: 7 Banbury Ln Chapel Hill NC 27514-2500 Office: CUNY Dept Chemistry Brooklyn NY 11210

MA, ZUGUANG, optical scientist, quantum electronics educator; b. Beijing, China, Apr. 11, 1928; s. Ke-Cian and Cai-Zhang (Zhu) M.; m. Yue-Zhen Sun, Nov. 7, 1954; children: Tian Zong, Tian Yi. BA, Shang Dong U., Qing-Dao, Shang Dong, 1950; MS, Harbin Inst. Tech., Harbin-Heilongjiong, 1953. Assoc. prof. Harbin Inst. Tech., Harbin, Heilongjig, 1963-82, prof. of physics, 1982—; prof. and dir. Inst. of Optoelectronics, Harbin Inst. Tech., Harbin, Heilongjig, 1988—; sci. and tech. com. mem. State Edn. Comn., China, 1984—, Ministry of Aero-Space, China, 1991—, academic degrees com. mem. State Coun. of China, 1988—. Contbr. articles to profl. jours. Recipient Nat. Natural Sci. prize Nat. Natural Sci. Award Com., China, 1989. Mem. Optical Soc. China (dir. Heilonjiong Provice chpt.), Optical Soc. Am., Soc. Chinese Electronics (sr.). Home: No 15 FUHUA 4th St, 150006 Harbin China Office: Harbin Inst Tech, PO Box 309, 150006 Harbin China

MAASOUMI, ESFANDIAR, economics educator; b. Tehran, Iran, Mar. 5, 1950; came to U.S., 1977; s. Ahmad and Sharifeh (Fakhri) M. BS in Econs., U. London, 1972, MS in Stats., 1973, PhD in Econometrics, 1977. Lectr. London Sch. Econs., 1975-76, U. Birmingham, Eng., 1976-77; asst. prof. U. So. Calif., L.A., 1977-81; assoc. prof. Ind. U., Bloomington, 1982-86, prof., 1986-89; Robert and Nancy Dedman prof. econs., adj. prof. stats. So. Meth. U., Dallas, 1989—; vis. scholar MIT, Boston, 1981; vis. prof. U. Calif., Santa Barbara, 1987-88; presenter over 100 lectures and seminars. Editor Econometric Revs., 1987—; contbr. over 50 books, articles and revs. Rsch. grantee NSF, 1980-82; fellow Jour. Econometrics, others. Fellow Royal Statis. Soc.; mem. Econometric Soc. (conf. organizer 1983), Am. Statis. Assn., Inst. Math. Stats. Avocations: reading, sports, chess, traveling. Office: So Meth U Dallas TX 75275

MAASSEN, JOHANNES ANTONIE, chemistry educator, consultant; b. Deventer, The Netherlands, Oct. 11, 1945; s. Johannes Antonie and Anna (Heckmann) M. M of Chemistry, U. Amsterdam, The Netherlands, 1969, PhD in Biochemistry, 1972. Rsch. asst. U. Amsterdam, 1967-72; asst. prof. biochemistry Leiden (The Netherlands) U., 1972-79; vis. prof. Harvard Med. Sch., Boston, 1979; assoc. prof. Leiden Med. Sch., 1980—; bd. dirs. Leiden U. Med. Sch. Author of 4 books; contbr. over 70 articles to sci. publs. advisor Social Dem. Party environ. com., Amsterdam, 1975-78. Rsch. grantee Dutch Soc. for Advancement of Pure Rsch., 1986, 89. Mem. Am. Diabetes Assn., Royal Dutch Chem. Soc., Dutch Diabetes Assn. (rsch. grantee 1986, 90), Dutch Soc. Anthropogenetics. Mem. Dutch Reformed Ch. Avocations: tennis, European history. Home: Walborg 5, 1082AM Amsterdam The Netherlands Office: Sylvius Lab, Wassenaarseweg 72, 2333AL Leiden The Netherlands

MAAT, BENJAMIN, radiation oncologist; b. Vlaardingen, The Netherlands, Apr. 17, 1947; s. Teunis and Evelien (van de Ridder) M.; m. Suzanne Faase, Dec. 19, 1969; children: Arthur, Menno, Michiel. MD, State U. Leiden, 1973, PhD, 1986. Intern State U. Hosp. Leiden, The Hague, The Netherlands, 1971-72; resident in radiother St. Johannes de Deo Hosp., The Hague, 1976-77; intern Dr. Daniel den Hoed Kliniek Rotterdam Radiotherapeutisch Inst., 1977-80; scientist Radiobiologic Inst. TNO, Dutch Health Orgn., Delft, The Netherlands, 1971-80; cons. radiotherapist Dr. B. Verbeeten Inst., Tilburg, The Netherlands, 1980—; cons. neuro-oncologist St. Elisabeth Hosp., Tilburg, 1980—; cons. radiotherapist St. Nicolaas Hosp. and St. Elizabeth Hosp., Tilburg, 1980—. Author: The Use of Anticoagulants in Antimetastatic Therapy, 1986; contbr. articles to profl. jours. Police doctor Mcpl. Police, Delft, 1976-80. Mem. European Orgn. Rsch. Treatment of Cancer (Brain Tumor Group and Radiotherapy Group), Dutch Soc. Radiotherapy, Dutch Soc. Radiobiology, European Soc. Therapeutic Radiation Oncology. Avocations: piano playing, collecting antique model steam locomotives. Home: Regge 93, N Brabant, 5032 RB Tilburg The Netherlands Office: Dr B Verbeeten Inst, Brugstraat 10, Tilburg 5042 SB, The Netherlands

MABRY, SAMUEL STEWART, petroleum engineer, consultant; b. Kaw, Okla., Mar. 16, 1925; s. Samuel Stewart and Bertha (Hudson) M.; m. Edith Joyce McGill, June 21, 1944; children: Michael Stewart, Paul Martin. BS in Petroleum Engring., U. Okla., 1949. Registered profl. engr., Okla. Petroleum engr. Cities Svc. Co., Bartlesville, Okla., 1949-52; petroleum engr., secondary recovery engr. Cities Svc Co., Madison, Kans., 1952-53; petroleum engr., secondary recovery engr. Cities Svc. Co., Bartlesville, 1953-67, secondary recovery coord., 1967-69, prodn. computing coord., 1969-73; mgr. petroleum engring. Cities Svc. Co., Tulsa, 1973-85; v.p. Challenger Engring., Tulsa, 1986-88; v.p. Tulsa Petroleum Cons., 1986-90, cons., 1990—; ind. cons. SEMCO, Tulsa, 1990—. With U.S. Army, 1943-46, ETO. Mem. NSPE, Soc. Petroleum Engrs. (chmn. Bartlesville chpt. 1964-65, chmn. Tulsa chpt. 1984-85, nat. bd. dirs. 1983-85, visitor/reviewer Accreditation Bd. for Engring. and Tech. 1980-90, Regional Svc. award 1987, Disting. Mem. award 1989), Okla. Soc. Profl. Engrs. Republican. Baptist. Avocations: photography, computers, aviation. Home: 6612 S 77 E Ave Tulsa OK 74133

MAC AN AIRCHINNIGH, MÍCHEÁL, computer science educator; b. Magheralin, Down, Ireland, Mar. 30, 1950; s. Stephen and Mary Rose (Mc Ilduff) Mc Inerney; m. Anne-Isabelle Coureaux, Aug. 16, 1979; children: Eloïse, Thérèse. BSc, U. London, 1978; MSc, U. Dublin (Ireland), 1981, MA, 1985, PhD, 1991. Sr. lectr. U. Dublin Trinity Coll., 1980—; bd. dirs. Generics (Software) Ltd., Dublin, K & M Techs. Ltd., Dublin; cons. Motorola, Arlingtn Heights, Ill., 1989-91. Mem. Assn. for Computer Machinery, IEEE, Am. Math. Soc., Irish Computer Soc., Irish Math. Soc., Eurographics. Avocations: literature, history, mathematics. Office: Trinity Coll, Dept Computer Sci, Dublin Ireland

MACARI, EMIR JOSÉ, civil engineering educator; b. Mexico City, July 22, 1957; came to U.S., 1975; s. Emir and Anna Laura (Pasqualino) M.; m. Jill Kathleen Gregg, Aug. 26, 1989; 1 child, Anna Christina. BSCE, Va. Poly. Inst. and State U., 1979; MSCE, U. Colo., 1982, PhDEE, 1989. Registered engr.-in-tng., Va. Geotech. engr. McClelland Engrs., Inc., Houston, 1983-84, Seafloor Engrs., Houston, 1984-85; chief engr. Kiso-Jiban, L.T.D., Tokyo, Singapore, 1985; rsch. asst. U. Colo., Boulder, 1985-89, rsch. assoc., 1989; asst. prof. civil engring. U. P.R., Mayaguez, 1990-92, assoc. prof., 1992-93; assoc. prof. civil engring. Ga. Tech., Atlanta, 1993—; rsch. fellow NASA Marshall Space Flight Ctr., Huntsville, Ala., summers 1990-91; mem. soil and rock properties com. Transp. Rsch. Bd., 1991—; bd. dirs. Civil Infrastructure Rsch. Ctr. Mayaguez, 1993—. Contbr. articles to Ingenieur Archive, Jour. ASCE. Bd. dirs. Good Shephard Day Cre Ctr., Houston, 1982-85; mem. P.R. Ecol. League, Rincon, 1992. Recipient Disting. Visitor award City of Veracruz, Mex., 1992, scholarly productivity award U. P.R., San Juan, 1992; presdl. faculty fellow White House and NSF, 1992-97. Mem. ASCE (editorial bd. jour. 1993, assoc., faculty advisor 1990—), Internat. Soc. Soil Mechanics and Founds., Sigma Xi, Chi Epsilon. Democrat. Roman Catholic. Office: Ga Tech Dept Civil Engring Atlanta GA 30332

MACAYA, ROMAN FEDERICO, biochemist; b. Fla., Sept. 19, 1966; s. Ernesto and Roberta (Hayes) M. BS, Middlebury Coll., 1988; PhD, UCLA, 1993. Rsch. asst. U. Costa Rica, San Pedro, Costa Rica, 1985; mgmt. asst. RIMAC Agrochems., Cartago, Costa Rica, 1986-87; rsch. asst. UCLA, 1988-93; sr. scientist Pharma Genics, Inc., Allendale, N.J., 1993—. Contbr. articles to profl. jours. Mem. Biophys. Soc. Roman Catholic. Achievements include elucidation of structural features of Triplex and Quadruplex DNA using two-dimensional NMR. Home: 240 Prospect Ave # 509 Hackensack NJ 07601 Office: Pharma Genics Inc 4 Pearl Ct Allendale NJ 07401

MACCALLUM, (EDYTHE) LORENE, pharmacist; b. Monte Vista, Colo., Nov. 29, 1928; d. Francis Whittier and Berniece Viola (Martin) Scott; m. David Robertson MacCallum, June 12, 1952; children: Suzanne Rae MacCallum Barslund and Roxanne Kay MacCallum Batezel (twins), Tracy Scott, Tamara Lee MacCallum Johnson, Shauna Marie MacCallum Bost. BS in Pharmacy U. Colo., 1950. Registered pharmacist, Colo. Pharmacist Presbyn. Hosp., Denver, 1950, Corner Pharmacy, Lamar, Colo., 1950-53; rsch. pharmacist Nat. Chlorophyll Co., Lamar, 1953; relief pharmacist, various stores, Delta, Colo., 1957-59, Farmington, N.Mex., 1960-62, 71-79, Aztec, N.Mex., 1971-79; mgr. Med. Arts Pharmacy, Farmington, 1966-67; cons. pharmacist Navajo Hosp., Brethren in Christ Mission, Farmington, 1967-77; sales agt. Norris Realty, Farmington, 1977-78; pharmacist, owner, mgr. Lorene's Pharmacy, Farmington, 1979-88; tax cons. H&R Block, Farmington, 1968; cons. Pub. Svc. Co., N.Mex. Intermediate Clinic, Planned Parenthood, Farmington. Author numerous poems for mag. Advisor Order Rainbow for Girls, Farmington, 1975-78. Mem. Nat. Assn. Bds. Pharmacy (com. on internship tng., com. edn., sec., treas. dist. 8, mem. impaired pharmacists adv. com., chmn. impaired pharmacists program N.Mex., 1987—, mem. law enforcement legis. com., chmn. nominating com. 1992), Nat. Assn. Retail Druggists, N.Mex. Pharm. Assn. (mem. exec. coun. 1977-81), Order Eastern Star (Farmington). Methodist. Home and Office: 1301 Camino Sol Farmington NM 87401-8075

MACCHESNEY, JOHN BURNETTE, materials scientist, researcher; b. Caldwell, N.J., July 8, 1929; s. Samuel Burnette and Helen Frances (Van Houten) MacC.; m. Janice Hoyt, Mar. 22, 1952; 1 child, John Burnette. A.B., Bowdoin Coll., 1951; Ph.D., Pa. State U., 1959. Materials scientist AT&T Bell Labs., Murray Hill, N.J., 1959—. Contbr. numerous articles on processing glass for use in fiber optics; patentee in field. Served with U.S. Army, 1951-53. Bell Labs. fellow, 1982. Fellow Am. Ceramic Soc. (George M. Murey award 1976, John Jeppson medal 1992); mem. IEEE (N. Libman award 1978), Nat. Acad. Engring., Sigma Xi (Commonwealth award). Home: Box 187 Cratetown Rd Lebanon NJ 08833 Office: AT&T Bell Labs 600 Mountain Ave New Providence NJ 07974-2010

MACCONNELL, GARY SCOTT, environmental engineer; b. Sarasota, Fla., Dec. 5, 1958; s. David Charles and Barbara Ann (Smith) MacC.; m. Chrissa Lynn Sellers, Apr. 22, 1984; children: Monica Lynn, Forrest Ryan, Jessica Dawn, Tyler Scott. BA in Biology, Gettysburg Coll., 1981; M in Environ. Mgmt., Duke U., 1984, MS in Envrion. Engring., 1984. Registered profl. engr., Fla., Va., N.C. Engr. Fla. Dept. Environ. Regulation, Tallahasse, Fla., 1984-85, Camp Dresser & McKee, Inc., Orlando, Fla., 1985-87; supr. engr. James M. Montgomery, Consulting Engring., Inc., Psadena, Calif., 1987-90; sr. engr. Hazand Sawyer Consulting Engrs., Raleigh, N.C., 1990-91, Piedmont Olson Hensley, Inc., Raleigh, N.C., 1991—. Contbr. over 23 tech. publs. to profl. jours. Mem. NSPE, ASCE, Water Environment Fedn., Am. Chem. Soc. Home: 1602 Sunrise Rd Cary NC 27513 Office: Piedmont Olsen Hensley Inc 2301 Rexwoods Dr Raleigh NC 27622

MACCRACKEN, MARY JO, physical education educator; b. Akron, Ohio, Oct. 6, 1943; d. Joel Milton and Mary Ellen (Frame) Weaver; m. Alan Lemuel MacCracken Jr., Aug. 23, 1969; 1 child, Alan Lemuel III. BA, Coll. of Wooster, 1965; MA, U. Akron, 1969; PhD, Kent State U., 1980. Tchr., coach Hudson (Ohio) Pub. Schs., 1965-68; instr. U. Akron, 1968-78, asst. prof., then assoc. prof., 1978-88, prof., 1988—, dir. Motor Behavior Lab., 1986—; collaboration tchr. Ritzman Sch., Akron, 1978-93, Mason Sch., Akron, 1978-93, St. Martha's Sch., Akron, 1993—; presenter at profl. confs. Contbr. articles to refereed jours. Sunday sch. tchr. Christ Ch. Episcopal, Hudson, 1979-92; vol. Liltin' Leaguers, Jr. League Cleve., 1979—; faculty mentor Akron High Sch. Drop-Out program, 1989. Grantee Ohio Bd. Regents, 1987-92. Mem. AAHPERD (v.p. health Midwest dist. 1990-92; meritorious honor award Ohio assn. 1988), Am. Psychol. Assn., N.Am. Soc. for Psychology of Sport and Phys. Activity, Nat. Assn. Phys. Edn. in Higher Edn. (sec.), Delta Kappa Gamma (Annie Webb Blanton award 1980). Republican. Avocations: antiques, genealogy, tennis, golf, sailing. Home: Box 631 431 N Main St Hudson OH 44236 Office: U Akron Motor Behavior Lab MH 81 Akron OH 44325-5103

MACCULLOCH, PATRICK C., oil industry executive. Sr. v.p. B.P. Canada Inc., Calgary. Recipient Inco medal Canadian Inst. Mining and Metallurgy, 1990. Office: BP Canada Inc, BP House/333 5th Ave SW, Calgary, AB Canada T2P 5C3*

MACDIARMID, ALAN GRAHAM, metallurgist, educator; b. Masterton, New Zealand, Apr. 14, 1927; married 1954; 4 children. BSc, U. New Zealand, 1948, MSc, 1950; MS, U. Wis., 1952, PhD in Chemistry, 1953; PhD in Chemistry, Cambridge U., 1955. Asst. lectr. in chemistry St. Andrews U., 1955; from instr. to assoc. prof. U. Pa., Phila., 1955-64, Sloan fellowship, 1959-63, prof. chemistry, 1964—. Recipient Francis J. Clamer medal Franklin Inst., 1993. Mem. Am. Chem. Soc., Royal Soc. Chemistry. Achievements include preparation and characterization of organosilicon compounds, derivatives of sulfur nitrides and quasi one-dimensional semiconducting and metallic covalent polymers such as polyacetylene and its derivatives. Office: U Pa Dept Chemistry 34th & Spruce Sts Philadelphia PA 19104*

MACDONALD, ALEXANDER DANIEL, physics consultant; b. Sydney, N.S., Can., Apr. 8, 1923; came to U.S. 1960; s. Daniel Malcolm and Alexandrina (MacLeod) MacD.; m. Lois Roberta Stevenson, May 1, 1946; children: Muriel Ruth, Susan Lex, Robert Bronson, Daniel Rufus. BSc, Dalhousie U., Halifax, N.S., 1945, MSc, 1947; PhD, MIT, 1949. Prof. physics Dalhousie U., Halifax, 1949-60, head div. applied math., 1962-65; sr. scientist G.T.E. Microwave Physics Lab., Palo Alto, Calif., 1960-62; sr. mem. rsch. lab. Lockheed Missile & Space Ctr., Palo Alto, Calif., 1965-73, dir. comm. and electronics lab., 1973-78; sr. scientist Lockheed Space Systems Div., Sunnyvale, Calif., 1978-81; cons. in physics MacDonald Cons., Palo Alto, 1981—; mem. space rsch. com. NRC-Can., 1958-60; cons. radiation physics V.O. Hosp., Halifax, 1953-60. Author: Microwave Breakdown in Gases, 1966, Russian translation, 1969; contbr. articles to profl. jours. Fellow Am. Phys. Soc. Democrat. Soc. of Friends. Home: 3056 Greer Rd Palo Alto CA 94303

MACDONALD, DAVID RICHARD, industrial psychologist; b. Dowagiac, Mich., May 20, 1953; s. Jerrold Brewster and Shirley Ann (Shaffer) MacD.; m. Mary Elizabeth Olson, Dec. 20, 1975; 1 child, Sarah Ann. AS, Southwestern Mich. Coll., 1973; BBA, Western Mich. U., 1975, MA, 1976, EdS, 1979; PhD, Mich. State U., 1986. Announcer, boardman WDOW AM/FM, Dowagiac, Mich., 1969-72; mgmt. devel. specialist Interstate Motor Freight System, Grand Rapids, Mich., 1977-79; sr. mgmt. tng. instr. GTE Gen. Telephone Co. Mich., Muskegon, 1979-82; cons. human resources devel. Steelcase, Inc., Grand Rapids, 1982-86, mgr. program devel. and tng., 1986—; asst. prof. grad. mgmt. Aquinas Coll., Grand Rapids, 1983—; cons., speaker in field; facilitator, program dir. Devel. Dimensions Internat., Pitts., 1981; facilitator Alamo Learning Systems, Southfield, Mich., 1983, 86, Wilson Learning Corp., Eden Prairie, Minn., 1983; job analysis program mgr. Barry M. Cohen & Assocs., Largo, Fla., 1985. Co-chair United Way Steelcase campaign, Grand Rapids, 1986. Mem. ASTD (sec. W. Mich. chpt. 1977-79), Soc. Indsl.-Orgnl. Psychology, Am. Psychol. Assn., Nat. Soc. for Performance and Instrn., Mensa, Phi Kappa Phi. Republican. Avocations: building harpsichords, stained glass, brewing, gardening, early music. Home: 2306 Prospect Ave SE Grand Rapids MI 49507-3159 Office: Steelcase Inc PO Box 1967 Grand Rapids MI 49501-1967

MACDONALD, DIGBY DONALD, scientist, science administrator; b. Thames, New Zealand, Dec. 7, 1943; came to U.S., 1977; s. Leslie Graham and Francis Helena (Verry) M.; m. Cynthia Lynch, 1969; m. Mirna Urquidi, July 6, 1985; children: Leigh Vanessa, Matthew Digby, Duncan Paul, Nahline. BS in Chemistry, U. Auckland, New Zealand, 1965; MS in Chemistry with honors, U. Auckland, 1966; PhD, U. Calgary, Alta., Can., 1969. Asst. research officer Atomic Energy of Can., Pinawa, Man., Can., 1969-72; lectr. Victoria U., Wellington, New Zealand, 1972-75; sr. research assoc., assoc prof. Alta. Sulfur Research U. Calgary, 1975-77; sr. metallurgist SRI Internat., Menlo Park, Calif., 1977-79; prof. metall. engring. Ohio State U., Columbus, Ohio, 1979-84; lab dir., dep. dir. phys. scis. divsn. SRI Internat., Menlo Park, 1984-91; prof. material sci. engring., dir. Ctr. Adv. Mat. Pa. State U., 1991—; adj. prof. Ohio State U., 1984; cons. in field. Author: Transient Techniques in Electrochemistry, 1977; contbr. numerous articles to profl. jours.; patentee in field. Nat. Research Council scholar, Ottawa, Can., 1967-69; recipient Research award Ohio State U., 1983. Mem. Electrochem. Soc. (divsn. editor 1982-84, C. Wagner Meml. award 1991), Nat. Assn. Corrosion Engrs. (pub. com. 1982-85, Whitney award 1992), Am. Chem. Soc. Avocations: sailing, flying. Home: 1010 Greenbriar Dr State College PA 16801-6935 Office: Pa State U Ctr Advanced Materials 517 Deike Bldg University Park PA 16802

MACDONALD, HUBERT CLARENCE, analytical chemist; b. Detroit, Aug. 3, 1941; s. Hubert Clarence and Edna (Weldon) MacD.; m. Janet Elizabeth Polachek, May 13, 1967; children: Kristin Marie, Michael Weldon, Cindy Anne, Wendy Lynne. BS in Chemistry, Wheeling (W.Va.) U., 1963; MS in Inorganic Chemistry, U. Mich., 1965, PhD in Analytical Chemistry, 1969. Scientist Koppers Co. Inc., Pitts., 1969-78; dir. Senate Environ. Testing Lab., Cheswick, Pa., 1978-82; tech. dir. Westinghouse Electric Corp., Pitts., 1982—. Co-editor: Computers in Chemistry and Instrumentation, 1976; contbr. articles to Jour. Phys. Chemistry. Judge, sponsor, Pa. Jr. Acad. Sci., Pitts., 1975—. Mem. ASTM, Am. Chem. Soc., Soc. Analytical Chemists of Pitts., Spectroscopy Soc. of Pitts. Roman Catholic. Home: 4103 Tartan Ct Murrysville PA 15668 Office: Westinghouse Electric Corp 517 Parkway View Dr Pittsburgh PA 15205-1466

MACDONALD, JAMES ROSS, physicist, educator; b. Savannah, Ga., Feb. 27, 1923; s. John Elwood and Antonina Jones (Hansell) M.; m. Margaret Milward Taylor, Aug. 3, 1946; children: Antonina Hansell, James Ross IV, William Taylor. B.A., Williams Coll., 1944; S.B., Mass. Inst. Tech., 1944, S.M., 1947; D.Phil. (Rhodes scholar), Oxford (Eng.) U., 1950, D.Sc., 1967. Mem. staff Digital Computer Lab., Mass. Inst. Tech., 1946-47; physicist Armour Research Found., Chgo., 1950-52; asso. physicist Argonne Nat. Lab., 1952-53; with Tex. Instruments Inc., Dallas, 1953-74; v.p. corporate research and engring. Tex. Instruments Inc., 1968-73, v.p. corporate research and devel., 1973-74; cons., 1974—; dir. Simmonds Precision Products Inc., 1979-83; William Rand Kenan Jr. prof. physics U. N.C., Chapel Hill, 1974-91, prof. emeritus, 1991—; mem. editorial bd. Jour. Applied Physics, 1984-86; adj. prof. biophysics U. Tex. Med. Sch., Dallas, 1954-74; mem. solid state scis. panel NRC, 1965-73; mem. adv. com. for sci. edn. NSF, 1971-73; mem. vis. com. physics Mass. Inst. Tech., 1971-74; mem. external adv. com. Engring. Expt. Sta., Ga. Inst. Tech., 1976-79. Editor, contbr.: Impedance Spectroscopy-Emphasizing Solid Materials and Systems, 1987; contbr. more than 190 articles to profl. jours. Mem. Dallas Radio Commn., 1967-71; mem. sci. adv. coun. Callier Hearing and Speech Ctr., Dallas, 1974-78; bd. dirs. League for Ednl. Advancement in Dallas, 1965-70; mem. adv. com. Weber Rsch. Inst., 1985-90. Fellow Am. Phys. Soc. (com. on edn. 1973-75, com. on applications of physics 1975-78, George E. Pake prize 1985), IEEE (awards 1962, 74, assoc. editor Transactions of Profl. Group on Audio 1961-66, Transactions on Audio and Electroacoustics 1966-73, recipient Edison Gold medal 1988), AAAS; mem. Nat. Acad. Engring. (exec. com. assembly of engring. 1975-78, council 1971-74), Nat. Acad. Scis. (chmn. numerical data adv. bd. 1970-74, mem. com. on motor vehicle emissions 1971-74, chmn. com. on motor vehicle emissions 1973-74, mem. com. on satellite power systems 1979-81, mem. com. on sci., engring., and pub. policy 1981-83, mem. commn. on phys. scis., math. and resources 1985-88, mem. report review com., 1990-93), Am. Inst. Physics. (governing bd. 1975-78, chmn. com. on profl. concerns 1976-78), Electrochem. Soc., Audio Engring. Soc., Phi Beta Kappa, Sigma Xi, Tau Beta Pi. Achievements include 10 patents in field. Office: Univ NC Dept Physics and Astronomy Chapel Hill NC 27599-3255

MACDONALD, JEROME EDWARD, consultant, school psychologist; b. Newark, Aug. 16, 1925; s. Jerome A. and Olvinia Regina (McKenna) MacD.; m. Nan Elizabeth Kennington, June 2, 1951; children: Jerome C., Mary Jane, Charles, Blanche Kohler, Ruth, Gregory, Paul, Robert, Carol. BS, Niagara U., 1947, MA (grad. fellow), 1950; MA in Ednl.

Psychology (experienced tchr. fellow), profl. diploma in sch. psychology, Jersey City State Coll., 1970; postgrad., Fordham U., 1950-55. Asst. prof. philosophy Seton Hall U., South Orange, N.J., 1948-55; lectr. in philosophy, edn. Seton Hall U., South Orange, 1955-61; tchr. English Newark Pub. Schs., 1955-60, guidance counselor, 1960-62, chmn. dept., 1963-69, psychologist, 1969-71; psychologist Metuchen (N.J.) Pub. Schs., 1971-86; vis. tchr. NDEA Reading Inst. Bowling Green (Ohio) U., 1966-67; extern psychologist N.J. Diagnostic Ctr., Menlo Park, 1969; consulting psychologist Dept. Health and Social Svcs., Province of Prince Edward Island, Can., 1987—. Editor: (with Eli Levinson) The English Curriculum in Secondary Schools: Ninth Grade, 1964. Troop treas. Boy Scouts Am., 1967-69. With inf., AUS, 1943-46. Decorated Bronze Star medal. Mem. Nat. Assn. Sch. Psychologists, Internat. Reading Assn., NEA, Am. Psychol. Assn., N.J. Psychol. Assn., N.J. Assn. Sch. Psychologists, Middlesex County Sch. Psychologists Assn. (pres. 1976-77, 81-82), Psychol. Assn. Prince Edward Island, N.J. Catholic Tchrs. Guild (pres. 1966), VFW, Am. Legion, DAV, Holy Name Soc., Can. Legion, Mensa, Phi Delta Kappa, Lions. Roman Catholic. Home: 1 MacDonald Rd Cavendish, PO Box 71, North Rustico, PE Canada C0A 1X0

MACDONALD, MHAIRI GRAHAM, neonatologist; b. Scotland, Oct. 11, 1945; came to U.S., 1973; s. Donald Francis and Jane Graham (Wild) MacD.; m. Harold Myron Ginzburg, Nov. 30, 1977; children: Rebecca MacDonald, James Stuart. MBChB, Edinburgh (Scotland) U., 1969. Diplomate Am. Bd. Pediatrics, Am. Bd. Neonatal-Perinatal Medicine. Intern pediatric surgery and medicine Sheffield (Eng.) Children's Hosp., 1969-70; pediatric resident Royal Hosp. for Sick Children, Edinburgh, 1970-73; fellow in neonatal-perinatal medicine John's Hopkins U. and Albany (N.Y.) Med. Ctr., Balt., 1974-76; attending neonatologist Albany Med. Ctr. Hosp., 1976-78; sr. attending neonatologist, vice-chair dept. neonatology, dir. Neonatal Intensive Care Unit Children's Nat. Med. Ctr., Washington, 1978—; guest researcher N.Y. State Health Dept., Albany, 1976-78, NIH and FDA, Bethesda, Md., 1978-83; prof. pediatrics George Washington U., Washington, 1988. Editor, contbr.: Atlas of Procedures in Neonatology, 1983, 2nd edit., 1993, Emergency Transport of the Perinatal Patient, 1989, Neonatology - Pathophysiology and Management of the Newborn, 1993; editor, founder Pediatric AIDS and HIV Infection: Fetus to Adolescent, 1989—; contbr. articles to profl. jours. Com. mem., chair multiple AIDS/HIV related coms. for govt. and other agys., Washington, 1985—. Fellow Royal Coll. Physicians Edinburgh, Am. Acad. Pediatrics; mem. Soc. Pediatric Rsch. (sr.). Jewish. Achievements include research on unique toxic side-effects of some substances that are mixed with therapeutic drugs, seen when these drugs are used in newborns; etiology and treatment of apnea in premature infants. Home: 808 Edelblut Dr Silver Spring MD 20901 Office: Childrens Nat Med Ctr 111 Michigan Ave NW Washington DC 20010

MACDONALD, NORVAL (WOODROW), safety engineer; b. Medford, Oreg., Dec. 8, 1913; s. Orion and Edith (Anderson) MacD.; m. Elizabeth Ann Clifford, Dec. 8, 1937; children: Linda (Mrs. Bob Comings), Peggy (Mrs. Don Lake), Kathleen (Mrs. Michael Nissenberg). Student, U. So. Calif., 1932-34. Registered profl. safety engr., Calif. Safety engr. Todd Shipyards, San Pedro, Calif., 1942-44, Pacific Indemnity Ins. Co., San Francisco, 1944-50; area safety engring. chief safety engr. Indsl. Ind., San Francisco, 1950-76; v.p. loss control Beaver Ins. Co., 1982-88; tchr. adult evening classes U. San Francisco, 1960-63, Golden Gate U., 1969—. Contbr. articles to profl. jours.; producer safety training films. Mem. ASME, Am. Soc. Safety Engrs. (pres. 1958, 59), Las Posas Country Club, Masons, Shriners. Methodist. Home: 1710 Shoreline Dr Camarillo CA 93010-6018

MACDONALD, STEWART DIXON, ornithologist, ecologist, biologist. BS, Iowa State U., 1959. Technician Can. Mus. of Nature, Ottawa, Ont., 1952-56, asst. curator of birds and vertebrate ethnology, 1959—; mem. panel, tundra panel, internat. biol. program Smithsonian Inst. expedition to Prince Patrick Island, 1949; mem. expedition to Antarctica, NSF, 1972. Contbr. articles to profl. jours. Recipient Massey medal Royal Can. Geog. Soc., 1992. Achievements include discovery of first North American nesting site of Ross' gulls, only known breeding colony in Canadian arctic of Ivory gulls; development of Seymour Island bird sanctuary, national wildlife area and first inland arctic research station at Polar Bear Pass on Bathurst Island. *

MACDONALD, TIMOTHY LEE, chemistry educator; b. Long Beach, Calif., Mar. 12, 1948; m. Deborah L. Patrick; children: Kate, Alice. BS with honors, UCLA, 1971; PhD, Columbia U., 1975. Asst. prof. chemistry Vanderbilt U., Nashville, 1977-82; assoc. prof. chemistry U. Va., Charlottesville, 1982-89, prof. chemistry, 1989—. Contbr. articles to profl. publs.; patentee in field. Office: U Va Dept Chemistry Mccormick Rd Charlottesville VA 22904-0002

MACDUFFEE, ROBERT COLTON, family physician, pathologist; b. Princeton, N.J., Apr. 23, 1923; s. Cyrus Colton and Mary Augusta (Bean) MacD.; m. Elizabeth Ann Jessup, Aug. 30, 1984; children: Martha, Jennifer, Susan. BS, U. Chgo., 1944, MD, 1946. Diplomate, Am. Bd. Pathology, Am. Bd. Family Practice. Asst. prof. pathology Hahnemann Med. Sch., Phila., 1959-60; chief clin. pathologist Grad. Hosp. U. Pa., Phila., 1960-63; pathologist Lock Haven (Pa.) Hosp., 1963-64; pathologist, chief Altoona (Pa.) Hosp., 1964-71; pathologist, assoc. SmithKline Labs., Tampa, Fla., 1971-84; dir. walk-in clinic Naples (Fla.) Med. Ctr., 1984—. Maj. MC, U.S. Army, 1946-54, Korea. Fellow Coll. Am. Pathologists; mem. AMA, Fla. Med. Assn., Collier County Med. Soc. Presbyterian.

MACE, JOHN WELDON, pediatrician; b. Buena Vista, Va., July 9, 1938; s. John Henry and Gladys Elizabeth (Edwards) M.; m. Janice Mace, Jan. 28, 1962; children—Karin E., John E., James E. B.A., Columbia Union Coll., 1960; M.D., Loma Linda U., 1964. Diplomate: Am. Bd. Pediatrics, Sub-bd. Pediatric Endocrinology. Intern U.S. Naval Hosp., San Diego, 1964-65, resident in pediatrics, 1966-68; fellow in endocrinology and metabolism U. Colo., 1970-72; asst. prof. pediatrics Loma Linda (Calif.) U. Med. Center, 1972-75, prof., chmn. dept., 1975—; med. dir. Loma Linda U. Children's Hosp., 1990-92, physician-in-chief, 1992—. Contbr. articles to profl. jours. Treas. Found. for Med. Care, San Bernardino County, 1979-80, pres., 1980-82; mem. Congl. Adv. Bd., 1984-87; pres. So. Calif. affiliate Am. Diabetes Assn., 1985-86, dir., 1987-89; chmn. adv. bd. State Calif. Children's Svcs., 1986—. With USN, 1962-70. Mem. AAAS, N.Y. Acad. Sci., Calif. Med. Soc. (adv. panel genetic diseases State Calif., 1975—), Western Soc. Pediatric Rsch., Lawson Wilkens Pediatric Endocrine Soc., Assn. Med. Pediatric Dept. Chairmen, Sigma Xi, Alpha Omega Alpha. Office: Loma Linda U Sch Medicine Barton & Anderson Sts Loma Linda CA 92350

MACEK, KAREL, analytical chemistry educator; b. Prague, Czechoslovakia, Oct. 31, 1928; s. Karel Macek and Pavla (Kotaskova) Mackova; m. Olga Haaszova, June 17, 1950; children: Jiri, Jan. Dr. rerum naturalim, Charles U., Prague, 1951; Dr. Sc., Tech. U., Prague, 1983. Head of labs. Rsch. Inst. for Pharmacy and Biochemistry, Prague, 1950-68, head of labs. internal dept., 1970-71; head of labs. 3rd Internal Clinic Charles U., Prague, 1971-77; leading scientist Inst. of Physiology, Czechoslovak Acad. of Scis., Prague, 1977-91; assoc. editor Jour. of Chromatography, Prague, 1968-88, editor, 1977—; chmn. 17th Internat. Symposia on Chromatography, 1961-90. Author, editor 16 books on chromatography; contbr. chpts. to books and 197 articles to profl. jours. Recipient 6 awards for chromatography. Mem. Czechoslovak Chem. Soc. (chmn. chromatography sect. 1961-91), Gesellschaft f. Toxikologie (hon. mem.). Achievements include two patents in field. Home: Lukesova 16, 14200 Prague Czech Republic Office: Jour of Chromatography, PO Box 70, 14200 Prague Czech Republic

MACESIC, NEDELJKO, electrical engineer; b. Karlovac, Croatia, Yugoslavia, Aug. 10, 1956; s. Savo and Milka (Ljepovic) M.; m. Vesna Jutrisa, Sept. 20, 1980; children: Nenad, Lea. BSEE, Electrotech. Fac., Zagreb, Yugoslavia, 1979; MSCE, Electrotech. Fac., 1983. Jr. sw engr. Rade Končar, Zagreb, 1979-82; sr. sw engr. Rade Končar, 1983-85, system analyst, 1985-86, project mgr. gas and elec. distbn., food and steel mfg., fin. and acctg., 1987-89, corp. mgr., 1989-90; software project mgr. mobile telecomm. Telecom Australia, Melbourne, 1991-92; projects dir. noise and flight paths monitoring system Manchester, Amsterdam, Zürich, Sydney, Brisbane, Melbourne, Perth, Adelaide, Madrid airports Lochard Environ Systems, Melbourne, 1993—. Author: Encyclopedia of Computing, 1986. Recipient Braca Ribar award Com. for Edn., Zagreb, 1970, Josip Loncar award for edn., 1979. Avocation: tennis. Office: Lochard Environ Systems, 6/875 Glenhuntly Rd, Caulfield South Victoria 3000, Australia

MACFADDEN, KENNETH ORVILLE, chemist; b. Phila., Sept. 5, 1945; s. Kennth Pennel and Alice Marie (Whitehorn) MacF.; m. Lois Nellie Rierson, June 8, 1968; children: Michelle, Kira. BS, Juniata Coll., 1966; PhD, Georgetown U., 1971. Asst. prof. Stockton State Coll., Pomona, N.J., 1972-75; rsch. chemist Air Products and Chem., Marcus Hook, Pa., 1975-80; rsch. mgr. Air Products and Chem., Allentown, Pa., 1980-84; dir. analytical rsch. W.R. Grace & Co., Columbia, Md., 1984-90, v.p. rsch. in electrochemistry, analytical chemistry, 1990—, v.p. rsch. in bioproducts, 1992—. Contbr. articles to profl. jours. Pres., bd. dirs. Cancerscope, Columbia, Md., 1991, 92. Mem. AAAS, Am. Chem. Soc., Sigma Xi. Home: 6575 River Clyde Dr Highland MD 20777

MACFARLANE, DAVID B., physicist, educator. Prof. physics dept. McGill U., Montreal, Que., Can. Recipient Gerhard Herzberg medal Can. Assn. Physicists, 1991. Office: McGill Univ, Physics Dept, Montreal, PQ Canada H3A 2T8*

MACGINITIE, LAURA ANNE, electrical engineer; b. N.Y.C., Aug. 30, 1958; d. Walter Harold and Ruth (Kilpatrick) MacG. MS, MIT, 1982, PhD, 1988. Asst. prof. dept. Engring. Pacific Lutherna Univ., Tacoma, Wash., 1993—. Author: (with others) Mechanistic Approaches to Interaction of Electrical and EM fields, 1987; contbr. articles to profl. jours. Mem. SPARC, Goshen, N.Y., 1990—, Rails to Trails Conservance, Washington, 1989—; coach Pacific Luth. JV Men's Crew. NSF grantee, 1991, 92; recipient Federation International de Societes de L'Auiron. Gold medal as mem. U.S. Lightweight Rowing Team, 1984. Mem. IEEE, Bioelectromagnetics Soc., Bioelec. Regeneration and Growth Soc. (coun. mem. 1991-93), Biomed. Engring. Soc. (membership com. 1991-91), Sigma Xi. Democrat. Achievements include initial indication that sinusoidal electric current densities 10mA/cm2 alter protein synthesis in cartilage tissue; demonstrated experimentally that small (A200 A) pores produce convective current during mechanical generation of streaming potentials in bone; demonstrated experimentally that conductive current flows in large (5um) pores in mechanical generation of streaming potentials in bone. Office: Pacific Luth Univ Dept Engring Rt 9W Tacoma WA 98447

MACGREGOR, JAMES THOMAS, toxicologist; b. N.Y.C., Jan. 14, 1944; s. James and Phyllis (Bowman) MacG.; m. Judith Anne Anello, July 12, 1969; 1 child, Jennifer Lee. BS in Chemistry, Union Coll., Schenectady, N.Y., 1965; PhD in Toxicology, U. Rochester, 1970. Diplomate Am. Bd. Toxicology. Postdoctoral fellow U. Calif., San Francisco, 1970-72; dir. food safety rsch. USDA, Berkeley, Calif., 1972-88; assoc. prof. U. Calif., Berkeley, 1978-88; pres. Toxicology Consulting Svcs., Danville, Calif., 1988-90; dir. toxicology lab. SRI Internat., Menlo Park, Calif., 1990—. Mem. editorial bd.: Environ. Molecular Mutagenesis, N.Y.C., 1986-88, Mutation Res., Amsterdam, 1989-91, Mutagenesis, Oxford, 1989-93. NIH fellow, 1965-72. Mem. Am. Assn. Cancer Rsch., Soc. Toxicology, Environ. Mutagen Soc. (treas. 1986-89, pres. 1992-93), Genetic Environ. Toxicology Assn. No. Calif. (pres. 1982). Home: 50 Mackenzie Pl Danville CA 94526 Office: SRI Internat 333 Ravenswood Ave Menlo Park CA 94025

MACHADO, ADELIO ALCINO SAMPAIO CASTRO, chemistry educator; b. Fafe, Minho, Portugal, May 10, 1942; s. Henrique C. and Maria Adelia (Castro) M.; m. Maria Helena Sousa, Sept. 17, 1967; children: Claudia, Henrique. BSc, Engring. Faculty, Porto, Portugal, 1965; PhD in Chemistry, Imperial Coll., 1971; Doutor (Quimica) (hon.), Universidade, Portugal, 1971. Asst. lectr. Faculty of Sci., Porto, 1965-67, lectr., 1971-72, assoc. prof., 1972-78, prof., 1978—; head of sci. bd. Chemistry Dept., Porto, 1980, head sci. bd. Faculty of Sci., Porto, 1983-84. Contbr. articles to profl. jours. including Inorganic Chemistry, Inorganic Chemica Acta, Polyhedron, Analyst, Analytical Letters, Analusis, Sci. of Total Environ., others. Mem. Am. Chem. Soc., Sociedade Portuguese de Electroquimica, Internat. Humic Sustances Soc. Achievements include devel. of a gen. procedure for constrn. of ion-selective electrodes based on conductive epoxies. Office: Faculty of Sci, PR Gomes Teixeira, P4100 Porto Portugal

MACHELL, ARTHUR R., retired mechanical engineer, association board member. BS in Mech. Engring., U. N.H., 1948. Various engring. positions Gen. Electric, 1948-62; from mgr. engring. standards program to producability specialist Xerox; ret. Mem. ASME (various couns. codes and standards, v.p. 1972-76, 86-88, exec. com., bd. govs., Codes and Standards medal 1991), Am. nat. Standards Inst. (bd. dirs., mem. coun., internat. standards coun., exec. standards bd., bd. standards review), Computer Bus. Equipment Mfrs. Assn. (standards mgmt. com. 1977-85), Internat. Standards Orgn. (U.S. tech. adv. group), Am. Nat. Metric Coun., Am. Soc. Testing and Materials. Achievements include development of national and international standards including fasteners, paper, sheet steel, synchronous belts and pulleys, sapphire jewel bearing test patterns and operator function symbols. Home: 901 Salt Road Webster NY 14580*

MACHER, JANET MARIE, industrial hygienist; b. St. Louis, Oct. 21, 1950; d. Robert Anthony and Rosalia Ann (Standfast) M. BA summa cum laude, Ottawa U., 1972; MPH, U. Calif., Berkeley, 1978; ScD, Harvard U., 1984. Environ. health and safety asst. USPHS, San Francisco, 1977-78; rsch. asst. U. Calif., Berkeley, 1978-79; biol. safety officer Harvard U., Boston, 1979-81, teaching asst., fellow, 1981-84; guest researcher Nat. Def. Rsch. Inst., Umeå, Sweden, 1984-85; air pollution researcher Calif. Dept. Health Svcs., Berkeley, 1985—. Contbr. articles to profl. publs. Fulbright grantee, 1985; Leslie Silverman fellow, 1982, E.I. duPont de Nemours & Co. fellow, 1981. Mem. Am. Nat. Standards Inst., Am. Conf. Govtl. Indsl. Hygienists (mem. bioaerosols com. 1991—), Am. Biol. Safety Assn., Am. Pub. Health Assn., Calif. Pub. Health Assn., Internat. Assn. Aerobiology (mem. coun. 1986—), No. Calif. Soc. Microbiology, Am. Soc. Microbiology, No. Calif. Assn. Pub. Health Microbiologists, Pan-Am. Aerobiology Assn., Am. Assn. Aerosol Rsch., Am. Indsl. Hygiene Assn. Office: Calif Dept Health Svcs 2151 Berkeley Way Berkeley CA 94704-1011

MACHINIS, PETER ALEXANDER, civil engineer; b. Chgo., Mar. 12, 1912; s. Alexander and Catherine (Lessares) M.; m. Fay Mezilson, Aug. 5, 1945; children: Cathy, Alexander. BS, Ill. Inst. Tech., 1934. Civil engr. Ill. Hwy. Dept., 1935-36 engr., estimator Harvey Co., Chgo., 1937; project engr. PWA, Chgo., 1938-40; supervisory civil engr. C.E., Dept. Army, Chgo., 1941-78; asst. to exec. dir. Chgo. Urban Transp. Dist., 1978-84; sr. civil engr. Parsons Brinckerhoff, 1985—; partner MSL Engring. Cons., Park Ridge, Ill., 1952—. Apptd. by gov. Civil Def. Adv. Coun. Ill., 1967—. Served with USAF, also C.E., U.S. Army, 1943-45; ETO; lt. col. Res. ret. Registered profl. engr., Ill. Recipient Emeritus Club award Ill. Inst. Tech., 1989. Fellow Soc. Am. Mil. Engrs. (past pres. Chgo. chpt.); mem. ASCE (life, presenter tech. papers), Nat. Soc. Profl. Engrs. (life, nat. chmn. sessions program com.), Am. Congress Surveying and Mapping, Assn. U.S. Army, Ill. Engring. Coun., Mil. Order World Wars (life). Greek Orthodox (ch. trustee). Home: 10247 S Oakley Ave Chicago IL 60643-1915

MACHLACHLAN, JULIA BRONWYN, materials scientist, researcher; b. Lancaster, Eng., Aug. 15, 1969; d. Richard Mortimer and Mary Vivien (Evans) Morris; m. Roderick James Maclachlan, Aug. 25, 1990. BA with hons., U. Cambridge, Eng., 1990. Rsch. scientist Pilkington, Lancashire, 1990. Mem. Inst. Materials (assoc.). Anglican. Office: Pilkington Tech Ctr, Hall Lane Lathom, Lancs Ormskirk L40 5UF, England

MACHOTKA, PAVEL, psychology and art educator; b. Prague, Czechoslovakia, Aug. 21, 1936; came to U.S., 1948; s. Otakar Richard and Jarmila Marie (Mohr) M.; m. Hannelore Gothe, Apr. 6, 1963 (div. Dec. 1980); children: Danielle, Julia; m. Nina Jane Hansen, Sept. 10, 1989. AB, U. Chgo., 1956; MA, Harvard U., 1958, PhD, 1962. Instr. Harvard U., Cambridge, Mass., 1962-65; asst. prof. Med. Sch. U. Colo., Denver, 1965-70; from assoc. prof. to prof. U. Calif., Santa Cruz, 1970—, provost coll. V, 1976-79, chair acad. senate, 1992—. Mem. editorial bd. Empirical Studies of the Arts, 1980—; author: The Nude, 1979; co-author: Messages of the Body, 1976. Precinct worker, alt. del. Dem. Party, Denver, 1968. Woodrow Wilson Nat. scholar, 1956; Fulbright fellow, 1958-60. Fellow APA, Am. Psychol. Soc.; mem. Czechoslovak Soc. for Arts and Scis. (sec.-gen. 1992—), Internat. Assn. for Empirical Aesthetics (pres. 1980-88). Office: U Calif Santa Cruz CA 95064

MACHOVER, CARL, computer graphics consultant; b. Bklyn., Mar. 26, 1927; s. John Herman and Rose (Alter) M.; m. Wilma Doris Simon, June 18, 1950; children: Tod, Julie, Linda. BEE, Rensselaer Poly. Inst., 1951; postgrad., NYU, 1953-56. Mgr. applied engring. Norden div. United A/C Corp., 1951-59; mgr. sales Skiatron Electronics & TV, N.Y.C., 1959-60; v.p. mktg., dir. Info. Displays, Inc., Mount Kisco, N.Y., 1960-73; v.p., gen. mgr., Info. Displays, Inc., Mount Kisco, 1973-76; pres. Machover Assocs. Corp., White Plains, N.Y., 1976—; adj. prof. Rensselaer Poly. Inst. Author: Gyro Primer, 1957, Basics of Gyroscopes, 1958; mem. editorial bd. IEEE Computer Graphics and Applications, Computers and Graphics, Spectrum; editor C4 Handbook, 1989; co-editor CAD/CAM Handbook, 1980, Computer Graphics Rev.; contbr. articles to profl. jours. Mem. adv. bd. Pratt Ctr. for Computer Graphics in Design. With USNR, 1945-46. Recipient Frank Oppenheimer award Am. Soc. for Engring. Edn., 1971, Orthagonal award N.C. State U., 1988, Vanguard award Nat. Comp. Graphics Assn., 1993; named to Computer Graphics Hall of Fame Fine Arts Mus. of L.I., Hempstead, N.Y., 1988. Fellow Soc. for Info. Display (pres. 1968-70); mem. IEEE, Assn. for Computing Machinery, Am. Inst. Design and Drafting, Nat. Soc. Profl. Engrs., Nat. Computer Graphics Assn. (bd. dir., pres. 1989-90), Computer Graphics Pioneer, Sigma Xi, Tau Beta Pi, Eta Kappa Nu. Home: 152 Longview Ave White Plains NY 10605-2314 Office: Machover Assocs Corp 199 Main St White Plains NY 10601-3200

MACIONSKI, LAWRENCE EDWARD, electronics engineer; b. Detroit, June 24, 1949; s. Albert Thomas and Olive Ilene (Julian) M.; m. Cheryl McBride (div. Nov. 1989); children: Catherine, Karen, Megan, Christopher; m. Ann Marie Boudreau, Apr. 26, 1990; 1 child, James Edward. AAS, Henry Ford C.C., Dearborn, Mich., 1969; BA in Vocat. Edn., SUNY, Oswego, 1973. Instr. Onondaga County Bd. Coop. Ednl. Svcs., Syracuse, N.Y., 1973-75; alarm supr. Unistate Security, Elbridge, N.Y., 1975-76; field engr. heavy mil. dept. GE, Shemya, Alaska, 1976-78, Ohio Nuclear, Solon, 1978-80; night supr. Detroit Free Press, 1980-88; site mgr. Janus Systems, Inc., Troy, Mich., 1988-92; sr. field engr. Maintech, Chantilly, Va., 1992—. Contbr. articles to various publs. Res. officer Royal Oak (Mich.) Police Dept., 1978-84; mem. amateur aux. FCC, Farmington Hills, Mich., 1986-92; vol. probation officer 44th Dist. Ct., Royal Oak, 1990-92. Recipient citation Royal Oak Police, 1983, Amateur of Yr. award Oak Park (Mich.) Amateur Radio Club, 1987. Mem. Am. Radio Relay League (coord. Newington, Conn. 1985-92), Quarter Century Wireless Assn., Fraternal Order Police. Republican. Achievements include research on understanding frequency agile repeating, adapting cellular antennas to ham radio. Home: Rte 2 Box 255-G Bluemont VA 22012 Office: Maintech 3800 Concorde Pky Ste 500 Chantilly VA 22110

MACIOROWSKI, ANTHONY FRANCIS, ecological toxicologist; b. Detroit, Mar. 9, 1948; s. Edmund Bernard and Phyllis (Chojna) M.; m. Donna Johnston, June 5, 1970. BS, Ea. Mich. U., 1970; PhD, Va. Poly. Inst., 1978. Cert. fisheries scientist. Rsch. assoc. U. N.C., Chapel Hill, 1978-80; lab. mgr. E.A. Engring. Sci. and Tech., Sparks, Md., 1980-83, Tex. Park and Wildlife Dept., Palacios, 1983-87; sr. scientist Battelle Meml. Inst., Columbus, Ohio, 1987-91; br. chief Nat. Fisheries Rsch. Ctr., U.S. Fish and Wildlife Svc., Kearneysville, W.Va., 1991-92; br. chief ecol. effects br. Office Pesticide Programs, U.S. EPA, Washington, 1992—; vis. prof. Tex. A&M U., College Station, 1986-87. Editorial bd. Environ. Toxicology and Chemistry, 1989-91; contbr. articles to profl. jours. Mem. Water Pollution Control Fedn. (rsch. com. 1977-87, ecology com. 1989-91), Am. Fisheries Soc., Am. Microscopical Soc., Soc. Environ. Toxicology and Chemistry. Roman Catholic. Achievements include research in ecological hazard evaluation and assessment of chemicals, spawning and culture of recreationally important marine fishes, ecological effects of pesticides, biological criteria development. Home: 950 Norwood Harpers Ferry WV 25425 Office: US EPA H7507C Office Pesticide Programs 401 M St SW Washington DC 20460

MACIULIS, LINDA S., computer coordinator, consultant; b. Chgo., May 23, 1949; d. Edward J. and Geraldine (Prohaska) Tesarek; divorced; children: Ian, Olivia. BA, U. Ill., 1971; Cert. Advanced Studies, Lewis U., 1977; MA, Nat. Coll. Edn., 1985, EdD, 1987. Tchr. English High Sch. Dist. 230, Palos Hills, Ill., 1976-87, coord., instr. computers, 1987—; owner, cons. Ill. Computing Educators Consortium, Palos Hills, 1987—; presenter Ill. Gifted Edn. Coun., 1984-87; designer computer systems Banco Argentina, Buenos Aires, 1988; mem. strategic planning com. High Sch. Dist. 230, Palos Hills, 1989—; judge Educationis Lumen Award Com., Lewis U., 1991—. Vol. Crisis Ctr. for South Suburbia and High Sch. Dist. 230, 1985—; active Hartigan for Gov. campaign, Chgo., 1989-90, Clinton/Gore Presdl. Campaign, Ill., Carol Moseley Braun for Senator Campaign, Chgo.; del. Dem. Nat. Conv., N.Y.C., 1992. Ill. Dept. Edn. grantee, 1984-85; named Suburban Educator of Yr., Lewis U., 1990. Mem. AAUW (bd. dirs., past pres.), NEA (bd. dirs. 1979—, adv. pub. rels. 1979—, Excellence in Gifted Edn. Design award 1985), Ill. Edn. Assn. (bd. dirs. 1979—). Roman Catholic. Avocations: oil painting, opera, classical music, international travel. Home: 3450 N Lake Shore Dr Chicago IL 60657 Office: Ill Computing Educators Consortium 10705 S Roberts Rd Ste 21 Palos Hills IL 60465

MAC IVER, DOUGLAS YANEY, geologist; b. Lone Pine, Calif., Jan. 24, 1930; s. Joseph Kenneth and Frances Eva (Yaney) MacI.; m. Albertine Dorothy Brooks, Aug. 8, 1953; children; Yaney Lee Ann, Douglas Joseph, Andrew Conray. BS, U. Calif., Berkeley, 1951; MS, U. Calif., 1959. Registered geologist, Calif. Mine geologist Wah Chang Mining Corp., Bishop, Calif., 1955-58; sr. rsch. chemist Southwestern Portland Cement, Victorville, Calif., 1960-68, plant process engr., 1969-78, environ. process specialist, 1978-81, mgr. environ. engring., 1981-89; mgr. environ. engring. Southdown, Inc., Victorville, Calif., 1989-91; cons. Doug Mac Iver Consulting, Big Bear Lake, Calif., 1991—; chmn., mem. San Bernardino County Air Pollution Control Dist. Adv. Bd., Victorville, 1977—; mem. sci. adv. com. So. Calif. Air Pollution Control Dist., El Monte, Calif., 1975-77, Tech. Adv. Com. Victorville Regional Sewage Project, 1971-73. Contbr. articles to profl. jours. Advisor Jr. Achievement, Victorville, 1961-63; chief Indian Guides, Victorville, 1973; elder, deacon Ch. of the Valley, Apple Valley, Calif., 1962-81. With U.S. Army, 1953-55. Mem. Air and Waste Mgmt. Assn., Soc. Mining Engrs., Inland Geol. Soc., Portland Cement Assn.-Environ. Affairs. Republican. Presbyterian. Home and Office: Doug Mac Iver Consulting PO Box 5484 Big Bear Lake CA 92315

MACKAY, ALEXANDER RUSSELL, surgeon; b. Bottineau, N.D., Oct. 8, 1911; s. Alexander Russell and Eleanor (Watson) M.; BS, Northwestern U., 1932, MD, 1936; MS in Surgery, U. Minn., 1940; m. Marjorie Andres, July 16, 1941; children: Andrea, Alexander Russell. Intern, Med. Center, Jersey City, 1935-37; fellow in surgery Mayo Clinic, Rochester, Minn. 1937-41; practiced medicine specializing in gen. surgery, Spokane, Wash., 1941-82, now ret.; former staff Deaconess, Sacred Heart hosps., Spokane. Capt., M.C., AUS, 1942-45. Diplomate Am. Bd. Surgery. Fellow ACS; mem. Spokane Surg. Soc., North Pacific Surg. Assn., Alpha Omega Alpha, Phi Delta Theta, Nu Sigma Nu, Phi Beta Kappa. Home: 540 E Rockwood Blvd Spokane WA 99202-1143

MACKAY, EDWARD, engineer; b. Kilmarnock, Ayrshire, Scotland, Feb. 29, 1936; s. Edward and Gertrude (Black) M.; widowed. Higher Nat. Cert., Glasgow Tech. Coll., Scotland, 1957. Layout engr. Stanley Works, New Britain, Conn., 1957-58; project engr. Grumman Olson, Athens, N.Y., 1961-69, asst. chief engr., 1969-74; chief engr. Grumman Olson, Sturgis, Mich., 1974-78, v.p. engring., 1978—. With U.S. Army, 1959-61. Presbyterian. Avocation: sheep farming. Office: Grumman Olson 1801 S Nottawa St Sturgis MI 49091

MACKAY, JOHN, mechanical engineer; b. Stockport, Eng., Mar. 26, 1914; s. Frederick and Annie MacK.; m. Barbara Hinnell, Jan. 11, 1939; 1 child, Penelope; m. Veronica Hwang, Dec. 2, 1960; 1 child, Teresa. Student, Malvern Coll., 1927-31; BS, U. Manchester, 1936. Registered profl. engr., N.Y. Mng. dir. Industrial Gases (Malaya) Ltd., Singapore, 1947-50; dir. Saturn Oxygen Co. Ltd. & Group, London, 1950-52; supt. Am. Cyanamid, New Orleans, 1952-56; project mgr. M.W. Kellogg Co., N.Y.C., 1956-67, Union

Carbide, N.Y.C., 1967-70; v.p. Procon Internat., Chgo., 1970-75; mgr. sales Davy Powergas, Houston, 1975-78; pres. Davy Corp. (Korea) Ltd., Seoul, 1978-80; v.p. Davy McKee Overseas Corp., Singapore, 1980-82; regional rep. Davy Corp. Ltd., Singapore, 1980-82; cons. Petronas, Kuala Lumpur, Malaysia, 1982-83; asst. dept. head cryogenics Superconducting Supercollider Lab./U. Rsch. Assocs., Waxahachie, Tex., 1989-93; cons. Lotepro Corp., Valhalla, N.Y., 1993—. Lt. col. Brit. Army, 1939-46. Decorated Brit. Terr. decoration, Chevalier l'Ordre Leopold II avec palme, Croix de Guerre avec Palme. Mem. ASME (life), Instn. Mech. Engrs., Sports Car Club Am., Caledonian Club London. Republican. Home: 4822 E Cascalote Dr Cave Creek AZ 85331

MACKAY, KENNETH DONALD, environmental services company executive; b. Detroit, July 18, 1942; s. John and Ina (Finlayson) M.; m. Bonnie Young, Aug. 15, 1964; children: Heather, Laurel. BS, U. Mich., 1964; PhD, U. Minn., 1968. Sr. rsch. chemist Gen. Mills., Mpls., 1968-73, group leader, 1973-77; tech. mgr. Henkel Corp., Mpls., 1977-80, assoc. dir. R&D, 1980-82, v.p., dir. rsch., 1982-86; pres. Henkel Rsch. Corp., Santa Rosa, Calif., 1986-91, Cognis, Inc., Santa Rosa, 1991—. Contbr. articles to profl. jours.; patentee in field. Mayor City of Circle Pines, Minn., 1971-77; mem. Santa Rosa Econ. Devel. Commn., 1987-88; mem. indsl. adv. bd. U. Calif., Berkeley, 1987—, L.A.; bd. dirs. Santa Rosa Symphony, 1987-88. Mem. Am. Chem. Soc., Indsl. Rsch. Inst. Avocations: fishing, tennis, golf. Office: Cognis Inc 2330 Circadian Way Santa Rosa CA 95407-5441

MACKAY, RAYMOND ARTHUR, chemist; b. N.Y.C., Oct. 30, 1939; s. Theodore Henry and Helen Marie (Cusack) M.; m. Mary Dilberian, Aug. 13, 1966; 1 child, Chelsea Christine; children by previous marriage—Brett, Edward. B.S. in Chemistry, Rensselaer Poly. Inst., 1961; Ph.D. in Chemistry, SUNY-Stony Brook, 1966. Research assoc. Brookhaven Nat. Lab., Upton, N.Y., 1966-67; prof. Drexel U., Phila., 1969-83; chief chem. div. Chem. Research and Devel. Ctr., Aberdeen Proving Ground, Md., 1983—. Contbr. articles to profl. jours. Served to capt. U.S. Army, 1967-69. Grantee, U.S. Army, Dept. Energy, Army Research Office, NSF, Acad. Applied Scis., 1972-83, NATO, 1982-86. Mem. Am. Chem. Soc., Am. Oil Chemists Soc. (assoc. editor), Sigma Xi. Office: Chem Div Research Directorate Research and Devel Center Aberdeen Proving Ground MD 21010

MACKEN, DANIEL LOOS, cardiologist, educator; b. Rochester, N.Y., May 7, 1933; s. Daniel Edward and Mary Frances (Loos) M.; m. Elaine Kathryn Audi (div. 1979); children: Elizabeth Redford, Diana Loos; m. Maria Luisa Medina de Palma, Nov. 16, 1979. AB, Holy Cross Coll., Worcester, Mass., 1955; postgrad., Yale U., 1956-57; MD, Boston U., 1960. Resident Roosevelt & Columbia-Presbyn. Hosps., N.Y.C., 1960-63; fellow Am. Heart Assn., 1964-65; asst. physician Presbyn. Hosp., N.Y.C., 1965-67; dir. coronary care unit Walter Reed Gen. Hosp., Washington, 1968; staff rsch. physician Walter Reed Army Inst. of Rsch., Washington, 1970; instr. Columbia U., N.Y.C., 1966-78, asst. clin. prof., 1979—; pres. Medica Found., Inc., N.Y.C., 1971—; bd. dirs. Medica Endowment Fund, N.Y.C., 1975—. Contbr. chpts. in book and articles to profl. jours. Lt. Col. U.S. Army, Med. Corp, 1967-70, Vietnam. Recipient Bronze Star medal U.S.A. 1970; Vietnam Cross 1969. Fellow Am. Coll. Cardiology, Royal Soc. Medicine, N.Y. Acad. Medicine, Harvey Soc.; mem. AMA, Assn. Mil. Surgeons of U.S., Am. Heart Assn., Met. Govs. Island Officers Club. Republican. Roman Catholic. Avocations: music, sports, photography. Home: 570 Park Ave New York NY 10021-7370 Office: Columbia-Presbyn Med Ctr 161 Ft Washington Ave New York NY 10032-3713

MACKEY, GEORGE WHITELAW, educator, mathematician; b. St. Louis, Feb. 1, 1916; s. William Sturges and Dorothy Frances (Allison) M.; m. Alice Willard, Dec. 9, 1960; 1 child, Ann Sturges Mackey. B.A., Rice Inst., 1938; A.M., Harvard U., 1939, Ph.D., 1942; M.A., Oxford, 1966. Instr. math. Ill. Inst. Tech., 1942-43; faculty instr. math. Harvard U., 1943-46, asst. prof., 1946-48, assoc. prof., 1948-56, prof. math., 1956-69, Landon T. Clay prof. math. and theoretical sci., 1969-85, prof. emeritus, 1985—; vis. prof. U. Chgo., summer 1955, UCLA, summer 1959, Tata Inst. Fundamental Research, Bombay, India, 1970-71; Walker Ames vis. prof. U. Wash., summer 1961; Eastman vis. prof. Oxford, 1966-67; assoc. prof. U. Paris, 1978; vis. researcher Math. Sci. Inst., Berkeley, 6 mos. 1983; vis. prof. U. Calif., Berkeley, 1984; lectr. Heedelberg, 1988, CUNY, 1987, U. Iowa, 1988, Kings Coll., 1991. Author: Mathematical Foundations of Quantum Mechanics, 1963, Lectures on the Theory of Functions of a Complex Variable, 1967, Induced Representations and Quantum Mechanics, 1968, The Theory of Unitary Group Representations, 1976, Unitary Group Representations in Physics, Probability and Number Theory, 1978; Contbr. articles math. jours. Served as civilian, operational research sect. 8th Air Force, 1944; applied math. panel NDRC, 1945. Guggenheim fellow, 1949-50, 61-62, 70-71; recipient Humboldt prize Max Planck Inst., Bonn, Fed. Republic of Germany, 1985-86. Mem. Am. Math. Soc. (v.p. 1964-65, Steele prize 1974), Nat. Acad. Scis., Am. Philos. Soc., Am. Acad. Arts and Scis., Phi Beta Kappa, Sigma Xi. Office: Harvard U Dept Math Cambridge MA 02138

MACKEY, ROBERT EUGENE, environmental engineer; b. Jasper, Ind., Nov. 1, 1956; s. Thomas Andrew and Josephine Ann (Mehling) M. BS in Civil Engring., U. Evansville, 1979; MS in Environ. Engring., U. Cin., 1986. Registered profl. engr., Ohio, Fla., N.C., Tenn., Va. Rsch. asst. U. Cin., 1980-86; engr. Post, Buckley, Schuh & Jernigan, Inc., Orlando, Fla., 1986—. Mem. ASTM (D-35 sub-task group leader 1991—), ASCE (assoc.), Fla. Engring. Soc. (math. events chmn. 1987—), Solid Waste Assn. N.Am. Republican. Roman Catholic. Achievements include rsch. in landfill closure using VLDPE, landfill excavation and recycling and evaluation of 30 mil PVC liner from ten yr. old landfill. Home: 1521 Minnesota St Orlando FL 32803 Office: Post Buckley Schuh & Jernigan 1560 Orange Ave Ste 700 Winter Park FL 32789-5542

MACKEY, ROBERT JOSEPH, video publisher; b. Detroit, Apr. 28, 1946; s. Robert and Bridget (Degnan) M.; m. Regina E. Richmond, July 27, 1968; children: Robert, Scott. BS in Indsl. Mgmt., Lawrence Inst., Southfield, Mich., 1971; MBA, Wayne State U., 1979. Dir. bus. affairs Harper Grace Hosp., Detroit, 1975-79; v.p. Nat. Health Corp., Southfield, 1979-83; v.p., founder Health Resources Mgmt., Southfield, 1983-86; pres., founder Medview, Inc., Farmington Hills, Mich., 1986-91, CompPro, Inc., Farmington Hills, Mich., 1986-91, CompPro Calif., Inc., Long Beach, 1986-91; founder, owner Regulation Enterprises, Inc., Southfield, 1991—; founder, bd. dirs. Animal Provider Orgn., Inc., Farmington Hills, 1989—. Bd. dirs. Am. Cancer Soc., Southfield, 1986-91, Dad's Club, Cath. Cen. High Sch., Redford, 1987-93; mem. Gov.'s Task Force on Health Care Cost Containment, Mich., 1989, 90. Mem. Am. Assn. Profl. Providers, Detroit Econ. Club, Marina City Club. Avocations: skiing, racquetball, golf. Home: 33905 Schulte St Farmington MI 48335-4162 Office: Regulation Enterprises Inc 24471 W 10 Mile Rd Southfield MI 48034

MACKIE, JOHN CHARLES, chemistry educator; b. Sydney, Australia, May 1, 1938; s. Charles and Marion Alice (Bradley) M.; m. Marie Elizabeth Cree, Aug. 5, 1961; children; Fiona Elizabeth, Joanna Mary. BSc with honors, U. Sydney, 1959, MSc, 1960, PhD, 1964. Chartered chemist, Australia. Rsch. assoc. Yale U., New Haven, 1963-64, U. Chgo., 1964-65; postdoctoral fellow U. Coll., London, 1965-66; lectr. in chemistry LaTrobe U., Melbourne, 1966-69; lectr. in phys. chemistry U. Sydney, 1969-79; vis. scientist United Tehcnologies Rsch. Ctr., East Hartford, Conn., 1987; prof. U. Sydney, 1980—; sr. vis. fellow U. Southampton, U.K., 1974-75; cons. CSIRO, Australia, 1986—. Editorial bd. mem. SEARCH, 1982-85; contbr. over 70 articles to profl. jours. including Australian Jour. of Chemistry, Jour. Chem. Soc., Jour. of Phys. Chemistry, and Jour. Chem. Physics. Fellow Royal Australian Chem. Inst. (Olle Prize for Chem. Lit. 1992), Combustion Inst. Achievements include rsch. on high temperature chemical kinetics, natural gas conversion studies, pyrolysis and combusion of coals and coal-model compounds. Home: 48 Wigram Rd, Glebe NSW 2037, Australia Office: U Sydney, Dept Phys. and Theoretical, Chemistry, Sydney 2006, Australia

MACKINTOSH, FREDERICK ROY, oncologist; b. Miami, Fla., Oct. 4, 1943; s. John Harris and Mary Carlotta (King) MacK.; m. Judith Jane Parnell, Oct. 2, 1961 (div. Aug. 1977); children: Lisa Lynn, Wendy Sue; m. Claudia Lizanne Flournoy, Jan. 7, 1984; 1 child, Gregory Warren. BS, MIT, 1964, PhD, 1968; MD, U. Miami, 1976. Intern then resident in gen. medicine Stanford (Calif.) U., 1976-78, fellow in oncology, 1978-81; asst. prof. med. U. Nev., Reno, 1981-85, assoc. prof., 1985-92, prof. medicine, 1992—. Contbr. articles to profl. jours. Fellow ACP; mem. Am. Soc. Clin. Oncology, Am. Cancer Soc. (pres. Nev. chpt. 1987-89, Washoe chpt. 1988-90), No. Nev. Cancer Coun. (bd. dirs. 1981—), No. Calif. Cancer Program (bd. dirs. alt. 1983-87, bd. dirs. 1987-91). Avocation: bicycling. Office: Nev Med Group 781 Mill St Reno NV 89502-1320

MACKLEY, ERNEST ALBERT, aeropropulsion engineer, consultant; b. Grand Junction, Colo., Dec. 7, 1925; s. Ellison Albert and Julia Ruth (Patton) M.; m. E. Ruth Sharpe, Apr. 25, 1959; children: Anna Barbara, Jane Ellen. AS, Mesa Jr. Coll., 1946; BS in Aero. Engring., U. Colo., 1949. Rodman Denver and Rio Grande R.R., Grand Junction, 1950-51; jr. engr. Nat. Adv. Com. for Aeronautics, Langley Field, Va., 1951; rsch. scientist Nat. Adv. Com. for Aeronautics, Hampton, Va., 1958, NASA, Langley, 1958-73; project mgr. NASA, Hampton, Va., 1973-75; chief propulsion scientist NASA, Hampton, 1975-85, asst. br. head, 1985-90, sr. scientist Analytical Svcs. and Materials, 1991—. Contbr. articles on hypersonic propulsion topics. Recipient Lifetime Contbn. to High-Speed Airbreathing Propulsion Tech. Space Act award NASA, 1993. Fellow AIAA (assoc.). Home: 11 Fallmeadow Ct Hampton VA 23666 Office: Analytical Svcs Materials NASA Mail Stop 168 NASA Langley RC Hampton VA 23665

MACKNIGHT, WILLIAM JOHN, chemist, educator; b. N.Y.C., May 5, 1936; s. William John and Margaret Ann (Stuart) M.; m. Carol Marie Bernier, Aug. 19, 1967. B.S., Rochester U., N.Y., 1958; M.A., Princeton U., N.J., 1963, Ph.D., 1964. Research assoc. Princeton U., N.J., 1964-65; asst. prof. chemistry U. Mass., Amherst, 1965-69, assoc. prof. chemistry, 1969-74, prof. chemistry, 1974-76, dept. head polymer sci., 1976-85, prof. polymer sci. and engring., 1985-88, head dept. polymer sci. & engring., 1988—; mem. sci. and tech. adv. bd. Alcoa, Pitts., 1984-86, Diversitech Gen., Akron, Ohio, 1985—; mem. panel for materials sci. Nat. Bur. Standards, Washington, 1983-89. Author: Polymeric Sulfur and Related Polymers, 1965; Introduction to Polymer Viscoelasticity, 2d edit., 1983. Served to lt. USN, 1958-61. Recipient Ford prize in high polymer physics Am. Phys. Soc., 1984; Guggenheim fellow, 1985. Fellow AAAS, Am. Phys. Soc. (exec. com. 1975-76); mem. Am. Chem. Soc., Am. Soc. Rheology. Club: Cosmos. Avocations: Music; sports. Home: 127 Sunset Ave Amherst MA 01002-2019 Office: Polymer Sci & Engring Dept U Mass Graduate Rsch Ctr Amherst MA 01003

MACLACHLAN, ALEXANDER, chemical company executive; b. Boston, Jan. 22, 1933; s. Hugh and Catherine (Sullivan) MacL.; m. Elizabeth Pegues, Jan. 25, 1958; children: Katherine Ellen, Amy Elizabeth, Mary Emily, Alexander Hugh. BS in Chemistry, Tufts U., 1954; PhD in Organic Chemistry, MIT, 1957. With E.I. duPont de Nemours & Co., Wilmington, Del., 1957—, mktg. dir., printing product dir., 1976-78, dir. plastics dept., 1978-80, dir. research and devel. div. chemical and pigments div., 1980-82, asst. dir. ctrl. R&D dept., 1982-83, dir. ctrl. R&D dept., 1983-86, sr. v.p. tech., 1986-90, sr. v.p. R&D, 1990—. Mem. nat. sci. adv. bd. Howard U., Washington, 1985—; mem. adv. com. Coun. on Competitiveness, 1989—; trustee Mt. Cuba Astron. Obs., Wilmington, 1986—, Bartol Rsch. Inst.; dir. Indsl. Rsch. Inst.; bd. oversees Fermilab; mem. Nat. Acad. Engring. Mem. Am. Chem. Soc., Nat. Sci. Assn. of U. Pa. Office: E I du Pont de Nemours & Co Experimental Sta #326 PO Box 80-326 Wilmington DE 19880

MACLAY, TIMOTHY DEAN, aerospace engineer; b. Frankfurt, Germany, Jan. 19, 1965; s. Donald Merle and Nancy Margaret (Hixenbaugh) M.; m. Sharon Marie Link, Jan. 3, 1987. BS, Bucknell U., 1986; MS, U. Colo., 1987, PhD, 1993. Electronic tech. Bendix Field Engring. Corp., Columbia, Md., 1984, 85; cryptologic mathematician Nat. Security Agy., Fort Meade, Md., 1986; grad. rsch. asst. U. Colo., Boulder, 1986—; engr. Martin Marietta Def. Space and Communications Co., Denver, 1991—; mem. Satellite Orbital Debris Characterization Impact Test Working Group, 1989—, SMART Catalog Working Group/Hybrid Data Base Modelling Com., 1987-88. Contbr. articles to profl. jours. Recipient 1982 Space Shuttle Student Involvement Project award NASA, 1982. Mem. AIAA, Am. Astron. Soc., U. Colo. Water Polo Club, Pi Mu Epsilon. Office: U Colo Ctr Astrodynamic Rsc Campus Box 431 Boulder CO 80309

MACLEAN, MARY ELISE, chemist; b. Winchester, Mass., July 22, 1963; d. John Richard and Helen Veronica (Hart) MacL. BA in Chemistry, Coll. of Holy Cross, Worcester, Mass., 1985; MA in Sci. Teaching, Tufts U., 1989. Cert. tchr., Mass. Mgr. support svcs. Am. Hosp. Supply Corp., Secaucus, N.J., 1985-86; loss control rep. Chubb Ins. Co., N.Y.C., 1986-88; sales engr. Anacon Corp. (divsn. Moisture Systems Corp.), Hopkinton, Mass., 1989-91; product specialist Anacon Corp./Moisture Systems Corp., Hopkinton, 1992; mgr. process liquids div. Moisture Systems Corp., Hopkinton, 1992—. Vol. Greater Boston Food Bank. Mem. NAFE, Am. Chem. Soc., Soc. Applied Spectroscopy, Instrument Soc. Am., Holy Cross Alumni Admissions Program, Holy Cross Varsity Club. Home: Apt 4 18 Lincoln St Newton MA 02161 Office: Moisture Systems Corp 117 South St Hopkinton MA 01748

MACLENNAN, DAVID HERMAN, scientist, educator; b. Swan River, Man., Can., July 3, 1937; s. Douglas Henry and Sigridur (Sigurdson) MacL.; m. Linda Carol Vass, Aug. 18, 1965; children—Jeremy Douglas, Jonathan David. B.S.A., U. Man., 1959; M.S., Purdue U., 1961, Ph.D., 1963. Postdoctoral fellow Inst. Enzyme Research, U. Wis., Madison, 1963-64; asst. prof. Inst. Enzyme Research, U. Wis., 1964-68; assoc. prof. Banting and Best Dept. Med. Rsch. U. Toronto, Ont., Can., Can., 1969-74; prof. Banting and Best Dept. Med. Rsch. U. Toronto, Ont., 1974-93; John W. Billes prof. med. research U. Toronto, 1987—; univ. prof. Banting and Best Dept. Med. Rsch. U. Toronto, Ont., 1993—; acting chmn. U. Toronto, 1978-80, chmn., 1980-90; mem. med. adv. bd. Muscular Dystrophy Assn., 1976-87; scientists rev. panel MRC, 1988-90. Contbr. articles on muscle membrane biochemistry to profl. jours.; mem. editorial bd. Jour. Biol. Chemistry, 1975-80, 82-87. Recipient Gairdner Found. Internat. award, 1991; Can. Med. Rsch. Coun. scholar, 1969-71, I.W. Killam Meml. scholar, 1977-78. Fellow Royal Soc. Can.; mem. Can. Biochem. Soc. (Ayerst award 1974), Am. Soc. Biol. Chemists, Biophys. Soc. (Internat. Lectr. award 1990). Home: 293 Lytton Blvd, Toronto, ON Canada M5N 1R7 Office: U Toronto—Banting & Best Med Rsch, 112 College St, Toronto, ON Canada M5G 1L6

MACLEOD, JOHN MUNROE, radio astronomer, academic administrator; b. Vermilion, Alta., Can., Sept. 3, 1937; s. Munroe and Ruth Alberta (Williams) MacL.; m. Patricia Irene Nichols, Dec. 30, 1959; children: Carolyn, Audrey, Darryl. BSc, U. Alta., Edmonton, 1959; PhD, U. Ill., 1964. Radio astronomer NRC Can., Ottawa, 1964-86, head James Clerk Maxwell telescope group Herzberg Inst., 1986—. Union Carbide scholar, 1955-59, McKinley Found. scholar, 1963-64. Mem. Am. Astron. Soc., Can. Astron. Soc. (councillor 1977-80, v.p. 1980-84, pres. 1984-86). Mem. United Ch. of Can. Achievements include co-discovery of the radio variability of BL Lacertae, of long-chain interstellar molecules HC5N, HC7N, HC9N. Office: NRC Herzberg Inst Astrophysics, 100 Sussex Dr, Ottawa, ON Canada K1A 0R6

MACLEOD, RICHARD PATRICK, foundation administrator; b. Boston, Apr. 2, 1937; s. Thomas Everett and Margaret Gertrude (Fay) MacL.; m. Sarah Frances Mancari, Sept. 7, 1963; children: Kimberly Margaret Hamelin, Richard Alexander MacLeod. BA in Govt., U. Mass., 1960; MA in Internat. Rels., U. So. Calif., 1968. Commd. 2d lt. USAF, 1960, advanced through grades to col., 1981; sr. rsch. fellow The Nat. Def. U., Washington, 1978-79; chief Space Policy Br., dep. chief Plans USAF Aerospace Def. Command, 1979-80; exec. officer to the comdr. in chief USAF Aerospace Def. Command, NORAD, 1980-81; chief of staff NORAD, 1981-84, USAF Space Command, 1982-84; ret. U.S. Space Found., 1985; exec. dir. U.S. Space Found., Colorado Springs, Colo., 1985-88; pres. U.S. Space Found., Colorado Springs, 1988—; bd. dirs. Analytical Surveys, Inc., Colorado Springs, 1985—. Author: Peoples War in Thailand, Insurgency in the Modern World, 1980. Mem. White House Space Policy Adv. Bd.; bd. dirs. Pike's Peak Coun. Boy Scouts Am., Colorado Springs; past pres. Colorado Springs Symphony Coun.; past dir. World Affairs Coun., Colorado Springs. Named Outstanding Young Man of Am., 1969, Nat. War Coll., 1979; disting. grad. Indsl. Coll. Armed Forces. Fellow Brit. Interplanetary Soc.; mem. AIAA, Air Force Acad. Found. (bd. dirs., trustee), U.S. Space Found. (founding). Office: U S Space Found 2800 S Circle Dr # 2301 Colorado Springs CO 80906

MACMILLAN, ROBERT SMITH, electronics engineer; b. L.A., Aug. 28, 1924; s. Andrew James and Moneta (Smith) M.; BS in Physics, Calif. Inst. Tech., 1948, MS in Elec. Engring., 1949, PhD in Elec. Engring. and Physics cum laude, 1954; m. Barbara Macmillan, Aug. 18, 1962; 1 son, Robert G. Rsch. engr. Jet Propulsion lab. Calif. Inst. Tech., Pasadena, 1951-55, asst. prof. elec. engring., 1955-58; assoc. prof. elec. engring. U. So. Calif., L.A., 1958-70; mem. sr. tech. staff Litton Systems, Inc., Van Nuys, Calif., 1969-79; dir. systems engring. Litton Data Command Systems, Agoura Hills, Calif., 1979-89; pres. The Macmillan Group, Tarzana, Calif., 1989—; treas., v.p. Video Color Corp., Inglewood, 1965-66. Cons. fgn. tech. div. USAF, Wright-Patterson AFB, Ohio, 1957-74, Space Tech. Labs., Inglewood, Calif., 1956-60, Space Gen. Corp., El Monte, Calif., 1960-63. With USAAF, 1943-46. Mem. IEEE, Am. Inst. Physics, Am. Phys. Soc., Sigma Xi, Tau Beta Pi, Eta Kappa Nu. Research in ionospheric, radio-wave, propagation; very low frequency radio-transmitting antennas; optical coherence and statist. optics. Home: 350 Starlight Crest Dr La Canada Flintridge CA 91011-2839 Office: The Macmillan Group 5700 Etiwanda Ave Unit 260 Tarzana CA 91356-2546

MACMURREN, HAROLD HENRY, JR., psychologist, lawyer; b. Jersey City, Sept. 18, 1942; s. Harold Sr. and Evelyn (Almone) MacM.; m. Margaret Bartro, Nov. 21, 1970. BA, William Paterson Coll., Wayne, N.J., 1965; MA, Jersey City Coll., 1973; EdD, St. Johns U., N.Y.C., 1985; JD, Rutgers U., 1989. Cert. secondary tchr., N.J.; Bar: N.J. 1989. Instr. Wanaque (N.J.) Bd. Edn., 1965-66, cons. psychologist, 1983-84; instr. Elmwood Park (N.J.) Bd. Edn., 1967-70; coll. faculty mem., psychologist Assoc. Clinic, Jersey City, 1971-72; cons. psychologist Rockaway (N.J.) Bd. Edn., 1972-83; intern lawyer Environ. Law Clinic, Newark, N.J., 1988-89; cons. psychologist Pequannock (N.J.) Bd. Edn., 1984—; speaker and writer in field. Mem. N.J. Edn. Assn., NEA, N.J. Psychologist Assn., N.J. Bar Assn., ABA, Sierra Club. Avocations: reading, travel, skiing, hiking. Home: 4 Sytsema Pl Sussex NJ 07461 Office: Pequanock Bd Edn Pequannock NJ

MAC NEISH, RICHARD STOCKTON, archaeologist, educator; b. N.Y.C., Apr. 29, 1918; s. Harris Franklin and Elizabeth (Stockton) MacN.; m. Phyllis Diana Walter, Sept. 26, 1963; children: Richard Roderick, Alexander Stockton. B.A., U. Chgo., 1940, M.A., 1944, Ph.D., 1949; LL.D. (hon.), Simon Frazer U., 1980; LL.D. Guggenheim fellow, Harvard U., 1956; LL.D. Aboriginal fellow, U. Mich., 1946. Supr., dir. U. Chgo. (W.P.A.), 1941-46; head dept. archaeology U. Calgary, 1964-68; anthropologist Nat. Mus. Can., 1949-62; dir. R.S. Peabody Found. for Archaeology, Andover, Mass., 1968-83; prof. dept. archeology Boston U., 1982-86; dir. Andover Found. for Archaeol. Research, 1986—. Contbr. numerous articles and revs. to profl. jours. Served with AUS, 1942-43. Recipient Spinden medal for archaeology, 1964, Lucy Wharton Drexel medal for archaeol. research U. Pa. Mus., 1965, Addison Emery Verrill medal Peabody Mus., Yale U., 1966, hon. Disting. Prof. award Universidad Nacional de San Cristobal de Huamanga, Ayacucho, Peru, 1970, Cornplanter medal for Iroquois research Auburn, N.Y., 1977. Mem. AAAS, Soc. Am. Archaeology (exec. council), Nat. Acad. Scis., Brit. Acad., Soc. Am. Archaeology (pres. 1971-72), Am. Anthrop. Assn. (Alfred Vincent Kidder award 1971), Sigma Psi, Alpha Tau Omega. Office: Andover Found Archaeol Research PO Box 83 Andover MA 01810-0002

MACON, IRENE ELIZABETH, interior designer, consultant; b. East St. Louis, Ill., May 11, 1935; d. David and Thelma (Eastlen) Dunn; m. Robert Teco Macon, Feb. 12, 1954; children: Leland Sean, Walter Edwin, Gary Keith, Jill Renee Macon Martin, Robin Jeffrey, Lamont. Student Forest Park Coll., Washington U., St. Louis, 1970, Bailey Tech. Coll., 1975, Lindenwood Coll., 1981. Office mgr. Cardinal Glennon Hosp., St. Louis, 1965-72; interior designer J.C. Penney Co., Jennings, Mo., 1972-73; entrepreneur Irene Designs Unltd., St. Louis, 1974—; vol. liaison Pub. Sch. System, St. Louis, 1980-82; cons. in field. Inventor venetian blinds for autos, 1981, T-blouse and diaper wrap, 1986; Author 26th Word newsletter, 1986, (songs) My God's Child Teach Free Will, God is Hiring Now, 1993. Committeewoman Republican party, St. Louis, 1984; vice chair 4th Senatorial Dist. of Mo., 1984, vol. St. Louis Assn. Community Orgns., 1983; instr. first aid Bi-State chpt. ARC, St. Louis, 1984, mem. speakers bur., 1991; cubmaster pack #80 Keystone dist. Boy Scouts Am.; block capt. Operation Brightside, St. Louis, 1984; co-chair status and role of women Union Meml. United Meth. Ch., 1986—, program resource sec., 1990—; trustee Wofit Found., 1989; spokesperson Minority Affairs Initiative Program Am. Assn. Retired Persons, 1991. Composer religious music. Named One of Top Ladies of Distinction St. Louis, 1983. Mem. NAACP, Am. Soc. Interior Designers (assoc.), Nat. Mus. Women in the Arts (charter), Internat. Platform Assn., Nat. Coun. Negro Women (1st v.p. 1984), Invention Assn. of St. Louis (subcom. head 1985), Coalition of 100 Black Women, St. Louis Assn. Fashion Designers, Pres. Club. Methodist. Achievements include invention of Irene's Autoshade, an accordian type of pleated material designed to adhere to automobile windows for the purpose of protecting it from the sun. Avocations: reading, designing personal wardrobe, modeling, horseback riding, boating. Home and Office: PO Box 20370 Saint Louis MO 63112-0370

MACON, JORGE, fiscal economist; b. C. Suarez, Argentina, Feb. 16, 1924; s. Samuel and Sara (Gutzait) M.; m. Sara Marta Garciandia, Nov. 20, 1959; 1 child, Cecilia Marta. Pub. acct. degree, U. Buenos Aires, 1960, M. in Econs., 1965, D. in Econs., 1973. Cert. economist. Dir. taxes Govt. of Province of Buenos Aires, La Plata, Argentina, 1958-60; chief fiscal policy dept. Fed. Investments Coun., Buenos Aires, 1960-66; prof. pub. fin. U. La Plata, 1966-85, U. Buenos Aires, 1978—; expert UN, Montevideo, Uruguay, 1971-74; dir. econ. rsch. U. La Plata, 1978-81; dir. applied econs. U. Buenos Aires, 1986-87; undersec. tax policy Nat. Govt. of Argentina, Buenos Aires, 1989-90; dir. postgrad. area in econs. U. Buenos Aires, 1988—; cons. in field. Author several books, numerous articles and papers on tax policy and intergovtl. fiscal rels. Mem. Internat. Inst. Pub. Fin. (hon.), Nat. Tax Assn., Am. Econ. Assn., Interant. Fiscal Assn., Argentine Fiscal Assn., Argentine Assn. Polit. Economy, Inst. Econ. and Social Devel. Home: Marcelo T de Alvear 1648, 1060 Buenos Aires Argentina

MACOSKO, PAUL JOHN, II, psychotherapist; b. Erie, Pa., May 15, 1952; s. Paul Sr. and Susan Ann (Miraldi) M.; m. Marsha Gail Blystone, July 1, 1978; children: Paul John III, Benjamin Jamison. BA in Psychology, Mercyhurst Coll., 1976; postgrad., Grand Rapids Bapt. Bible Sem., and Edinboro U., 1976-78; MA in Bibl. Counseling, Grace Theol. Sem., Winona Lake, Ind., 1983; DPhil, Oxford U., 1992. Protective svc. caseworker I Trumbull County Children's Svcs. Bd., Warren, Ohio, 1976-77; counselor Regular Bapt. Children's Agy., St. Louis, 1978-81; coord. devel. dual diagnosis program Dr. Gertrude A. Barber Ctr., Erie, 1987-92; pvt. practice marriage, family and crisis counseling Erie, 1984—; dir. Christian Care Ministry Grace Bapt. Ch., Erie, 1992—. Mem. ACA, Am. Assn. Christian Counselors, Assn. Religious and Value Issues in Counseling, Pa. Counseling Assn., Nat. Disting. Svc. Registry. Baptist. Avocations: reading, biking, hiking, fishing, camping. Office: 1561 W 38th St Ste 10 Erie PA 16505

MACOVSKI, ALBERT, electrical engineering educator; b. N.Y.C., May 2, 1929; s. Philip and Rose (Winogr) M.; m. Adelaide Paris, Aug. 5, 1950; children—Michael, Nancy. B.E.E., City Coll. N.Y., 1950; M.E.E., Poly. Inst. Bklyn., 1953; Ph.D., Stanford U., 1968. Mem. tech. staff RCA Labs., Princeton, N.J., 1950-57; asst. prof., then assoc. prof. Poly. Inst. Bklyn., 1957-60; staff scientist Stanford Research Inst., Menlo Park, Calif., 1960-71; fellow U. Calif. Med. Center San Francisco, 1971-72; prof. elec. engring. and radiology Stanford U., 1972—, endowed chair, Canon USA prof. engring., 1991—; dir. Magnetic Resonance Systems Research Lab.; cons. to industry. Author. Recipient Achievement award RCA Labs., 1952, 54; award for color TV circuits Inst. Radio Engrs., 1958; NIH spl. fellow, 1971. Fellow IEEE (Zworykin award 1973), Am. Inst. Med. Biol. Engring., Optical Soc. Am.; mem. NAE, Inst. of Medicine, Am. Assn. Physicists in Medicine, Soc. Magnetic Resonance in Medicine (trustee), Sigma Xi, Eta Kappa Nu. Jewish. Achievements include patents in field. Home: 2505 Alpine Rd Menlo Park CA 94025-6314 Office: Stanford U Dept Elec Engring Stanford CA 94305

MACPHERSON, ROBERT DUNCAN, mathematician, educator; b. Lakewood, Ohio, May 25, 1944; s. Herbert G. and Jeanette (Wolfenden) MacP. BA, Swarthmore Coll., 1966; MA, PhD, Harvard U., 19770. Instr. Brown U., Providence, 1970-72, asst. prof., 1972-74, assoc. prof., 1974-77, prof., 1977-85, Florence Pirce Grant prof., 1985-87; prof. MIT, Cambridge, Mass., 1987—; mem. Inst. des Hautes Etudes Sci., Paris, France, 1980-81, Steklov Math Inst., Moscow, USSR, 1980; visiting prof. U. Rome, 1985. Co-author: Stratified Morse Theory, 1988, Nilpotent Orbits, 1989; contbr. numerous articles to profl. jours. Recipient Research Grant NSF, 1970—; named Herman Weyl Lectr. Inst. for Advanced Study, 1982. Mem. NAS (Math. award 1992), Am. Acad. Arts and Scis., Am. Math. Soc., Soc. for Applied and Indsl. Math., Phi Beta Kappa. Home: 77 Sea Ave Quincy MA 02169-3127 Office: MIT Dept Math 77 Massachusetts Ave Cambridge MA 02139

MACQUEEN, ROBERT MOFFAT, solar physicist; b. Memphis, Mar. 28, 1938; s. Marion Leigh and Grace (Gilfillan) MacQ.; m. Caroline Gibbs, June 25, 1960; children: Andrew, Marjorie. BS, Rhodes Coll., 1960; PhD, Johns Hopkins U., 1968. Asst. prof. physics Rhodes Coll., 1961-63; instr. physics and astronomy Goucher Coll., Towson, Md., 1964-66; sr. research scientist Nat. Ctr. for Atmospheric Research, Boulder, Colo., 1967-90, dir. High Altitude Obs., 1979-86, asst. dir., 1986-87, assoc. dir., 1987-89; prof. physics Rhodes Coll., Memphis, 1990—; prin. investigator NASA Apollo program, 1971-75, NASA Skylab program, 1970-76, NASA Solar Maximum Mission, 1976-79, NASA/ESA Internat. Solar Polar Mission, 1978-83; lectr. U. Colo., 1968-79, adj. prof., 1979-90; mem. com. on space astronomy Nat. Acad. Scis., 1973-76, mem. com. on space physics, 1977-79; mem. Space Sci. Bd., 1983-86. Recipient Exceptional Sci. Achievement medal NASA, 1974. Fellow Optical Soc. Am.; mem. Am. Astron. Soc. (chmn. solar physics div. 1976-78), Assn. Univ. Research Astronomy (dir.-at-large 1984-93, chmn. bd. 1989-92), Am. Assn. Physics Tchrs., Sigma Xi.

MACROE-WIEGAND, VIOLA LUCILLE (COUNTESS DES ESCHEROLLES), psychiatrist, psychoanalyst; b. Indiana, Pa., May 17, 1920; d. Joseph Cyprian and Lucy E. (Colson) Macro; m. Thomas F. Gordon, Nov. 23, 1977 (div. 1982); m. Count Alexander des Escherolles-Krusper, Dec. 31, 1982. BA, St. Joseph's Coll. for Women, 1941; MA, Columbia U., 1942, PhD, 1958; MD, U. Hamburg (Germany), 1962. Instr. and chief psychologist Manhattan Eye and Ear Hosp., N.Y.U. Med. Sch., N.Y.C., 1952-58; lectr. dept. psychiatry SUNY Downstate Med. Ctr., Bklyn., 1962-63; psychiat. fellow Creedmore State Hosp., Queens, N.Y. 1962-63; intern U. Hamburg, 1962-63; resident St. George's Hosp., Hamburg, 1963-64; research fellow in neurology Mt. Sinai Hosp., 1963-64; resident in psychiatry P.R. Inst. of Psychiatry, 1976-79; practice internal medicine and psychiatry, San Juan, P.R., 1974—; psychologist geriatrics Little Sisters of Poor Hosp., Bklyn., 1965-67; mem. staff dept. neurology Kingsbrook Med. Ctr., Bklyn., 1967-68; asst. prof. psychology Kingsborough Community Coll., N.Y., 1966-67, CCNY, summer, 1968; mem. staff psychiatry Rio Piedras State Hosp., San Juan, 1974-82, P.R. Inst. of Psychiatry, San Juan, 1976-82; psychiatrist dept. mental health Knud Hansen Meml. Hosp., St. Thomas, V.I., 1979-82; neurol. and psychol. research dir., adminstr. Humboldt Med. Arts Bldg., 1987-88. Fellow Am. Assn. Mental Deficiency; mem. Am. Psychol. Assn., Ea. Psychol. Assn., N.Y. State Psychol. Assn., AMA, Am. Psychiat. Assn., P.R. Psychiatric Assn., Associación Hermandad en las Carreteras de P.R. (v.p. 1975—), Pi Lambda Theta, Kappa Delta Pi. Roman Catholic. Avocation: speaking fgn. langs. (German, Spanish, French, Italian, Russian). Contbr. articles on physiol. psychology to profl. jours.; research in visual and auditory perception. Home: 185 Clinton Ave Brooklyn NY 11205-3569 also: Condo El Monte 175 Avenida de Hostos Apt 702A Hato Rey PR 00918 also: PO Box 6252 Loiza Sta Santurce PR 00630 also: Michael Balint Inst für, Psychoanalysis, 62 Hamburg Germany also: Nagyvarad Hungary also: Chateau des Escherolles, Saint Etienne de Lólm France

MADABHUSHI, GOVINDACHARI VENKATA, civil engineer; b. Guntur, India, Oct. 13, 1933; came to U.S., 1974; s. Narasimhachari Vedantam and Venkatamma M. MSCE, W.Va. U., 1966; PhD, Utah State U., 1985. Registered profl. engr., Calif. Asst. engr. Bhubaneswar Dept. Pub. Works, Orissa, India, 1957-61; instr. Coll. Engring. Duke U., Durham, N.C., 1961-63; rsch. scholar Indian Inst. Tech., Bombay, 1967-69; lectr. in civil engring. Victoria Jubilee Tech. Inst., Bombay, 1969-74; water resources engr. Ark. Soil and Water Conservation, Little Rock, 1977-80; rsch. asst. Utah Water Rsch. Lab., Logan, 1980-83, Assn. Western Univs., Salt Lake City, 1983-84; water resources control engr. Regional Water Quality Control Bd., Riverside, Calif., 1985-86; assoc. waste mgmt. engr. Dept. Toxic Substances Ctrl., Calif. EPA (formerly Berkeley Dept. Health Svcs.), 1986—; mem. shallow founds. cod of practice Indian Standards Inst., New Delhi, 1969. Author profl. conf. procs. Mem. ASCE (ground water sect., mem. earthquake lifeline). Research on negative friction on pile foundations, effect of temperature changes on chemical equilibria in groundwater due to groundwater heat pumps, kinetics of water, mineral reactions in groundwater. Home: 2856 Fruitvale Ave Apt 41 Oakland CA 94601-2043 Office: Dept Toxic Substances Ctrl CA EPA 700 Heinz Ave Bldg F Berkeley CA 94710-2721

MADAIO, MICHAEL PETER, medical educator. BS in Biology cum laude, Fairfield U., 1970; MD, Albany Med. Coll., 1974. Diplomate Am. Bd. Internal Medicine, Am. Bd. Nephrology. Intern Med. Coll. Va., Richmond, 1974-75, resident, 1975-77, chief resident, 1977-78; clin. fellow in nephrology Boston U., 1978-79, rsch. fellow in nephrology, 1979-81; rsch. fellow in immunology Tufts U., Boston, 1981-82, instr. in medicine, 1981-82, asst. to assoc. prof., 1982-90; assoc. prof. medicine U. Pa., 1990—, assoc. chief renal electrolye div., 1992—; asst. physician New Eng. Med. Ctr., Boston, 1981-87, physician, 1988-90; reviewer manuscripts an dgrants in field; lectr. Pathology A Study Sect., NIH, 1992-96. Contbr. articles to profl. jours., publs. Vice-pres. Nat. Kidney Found. Mass., 1987-89, pres. 1989-90; active Lupus Found. Am., Phila., 1991—. Mem. AAAS, Internat. Soc. Nephorology, Am. Soc. Nephrology, Am. Fedn. Clin. Rsch., Am. Heart Assn., Fedn. Am. Socs. for Exptl. Biology, Am. Assn. Immunologists, Nat. Kidney Found., Am. Soc. Clin. Investigation, Alpha Epsilon Delta. Office: Univ Pa 700 Clin Rsch Bldg 422 Curie Blvd Philadelphia PA 19104-6144

MADAKSON, PETER BITRUS, physicist, engineer; b. Kagoro, Kaduna, Nigeria, Feb. 3, 1953; came to U.S., 1986; s. Madaki and Kachio (Bitiyak) Shemang; m. Rebecca Gwamna, Apr. 13, 1985; children: Peter Albert, Peter Ramses. BS, King's Coll., London, 1980, PhD, 1983. Rsch. asst. King's Coll., 1982-83, rsch. fellow, 1984-85; vis. scientist T.J. Watson Rsch. Ctr., IBM, Yorktown Heights, N.Y., 1986-87, mem. rsch. staff, 1987—; exec. chmn. Afan Electronics Ltd., Lagos, Nigeria, 1993—. Contbr. over 44 articles to sci. publs., 1983-90. Mem. Materials Rsch. Soc., Bohmische Phys. Soc. Achievements include research in ion implantation, ion channeling, thin film interdiffusion and ultra-low temperature properties of materials. Office: Afan Electronics Ltd, 30 Oyinkam Abayomi Dr, Ikoyi Nigeria

MADAN, SUDHIR YASHPAL, chemical engineer; b. Bombay, India, Dec. 20, 1961; came to U.S., 1984; s. Yashpal and Kamala M. BS, Panjab U., Chandigarh, India, 1984; MS, U. Akron, 1985; MBA, U. Dallas, 1992. Sr. process engr. Novacor, Inc., Akron, Ohio, 1985-88; engring. group leader Avery-Dennison, Painesville, Ohio, 1988-89; mgr. tech. and engring. Scott Paper Co., Fort Forth, 1989—. Contbr. articles to profl. jours. Pres. Lion's Internat., India, 1984, Internat. Students Orgn., U. Dallas, 1991-92. Recipient Disting. Svc. award Novacor, Inc., 1988. Mem. AICE, Am. Chem. Soc., Soc. Plastics Engrs. Achievements include patent (with others) for devolatilization technique at Novacor, Inc. Home: 405 Brazil Dr Hurst TX 76054 Office: Scott Paper Co 3607 N Sylvania Ave Fort Worth TX 76111

MADANAT, SAMER MICHEL, civil engineer, educator; b. Amman, Jordan, May 24, 1963; came to U.S., 1986; s. Michel Issa and Samira Farah (Ammari) M. MSCE, MIT, 1988, PhD, 1991. Registered profl. engr., Jordan. Trainee Kjessler-Mannerstrale, Stockholm, 1985; rsch. asst., then teaching asst. MIT, Cambridge, 1987-90; sr. analyst Cambridge Systematics, 1991; asst. prof. civil engring. Purdue U., West Lafayette, Ind., 1992—; cons. World Bank, Washington, 1992. Contbr. articles to Jour. Transp. Engring., Transp. Rsch. Record, Transp. Rsch. Recipient David Ross grant Purdue Rsch. Found., 1992. Mem. World Transp. Rsch. Soc., Transp. Rsch. Bd., Sigma Xi. Achievements include development of methodology for managing infrastructure systems under condition measurement and forecasting uncertainty using stochastic optimization techniques. Office: Purdue Univ Civil Engring Bldg West Lafayette IN 47907

MADANAYAKE, LALITH PRASANNA, marine engineer; b. Galle, Sri Lanka, May 15, 1965; s. Dharmapala and Madura (Jalath) M. Gen. cert. in engring., Richmond Coll., Sri Lanka, 1983; nat. diploma in tech. in marine engring., U. Moratuwa, Sri Lanka, 1986, BScME, 1989, MSc in Marine Engring, 1991. Cert. class 1 marine engr., U.K. Fitting shop engr. Colombo (Sri Lanka) Dockyards Ltd., 1989-90; marine engr. Sri Lanka Ports Authority, Colombo, 1990-91; marine engr., chief exec. Ceylon Shipping Corp., Colombo, 1991—; marine mgr. Colombo Dockyards Pvt. Ltd., 1993—. Author: (book) Prize Crew; Report on Inplant Training, 1989, Marine Propulsion Systems. Mem. Inst. Engring. Diplomates in Sri Lanka, Internat. Maritime Orgn. Buddhist. Avocations: collecting valuable books, swimming, dancing, karate. Home: Gothatuwa Baddegama, Galle Sri Lanka Office: Colombo Dockyards Pvt Ltd, Port of Colombo, Colombo Sri Lanka

MADDALENA, FREDERICK LOUIS, chemical engineer; b. Wakefield, R.I., Sept. 1, 1947; s. Harold L. and Ursula (Briggs) Strout; m. Linda J. Pelletier, Jan. 25, 1969; children: Cheryl J., Roger L. BSChemE, U. R.I., 1969. With U.S. Steel Clairton (Pa.) Works, 1969--, area mgr. chem. recovery, 1984-89, process mgr. chems. and energy, 1989--. Author: (tng. manual) Continuous Improvement to Environment, 1990. Treas., coach Mon Valley Youth Hockey Assn., Rostraver, Pa., 1982-90; pres., coach Belle Vernon High Sch. Club, Pa., 1988--. Mem. Am. Inst. Chem. Engrs. Achievements include patent for flushing liquor decanter drag modification. Home: 75 Yakubic Ln Belle Vernon PA 15012 Office: US Steel Clairton Works 400 State St Clairton PA 15025

MADDEN, ARTHUR ALLEN, nuclear pharmacist; b. Atlanta, Sept. 19, 1960; s. Arthur Allen and Lillian Brandon (Vaughan) M.; m. Rebecca Kaye Teague, June 25, 1988; 1 child, Kelley Vaughan. BA in English, U. of the South, Sewanee, Tenn., 1982; BS in Pharmacy, U. S.C., 1988, PharmD, 1990. Registered pharmacist. Poison control specialist Palmetto Poison Ctr., Columbia, 1988-90; relief pharmacist Wal-Mart, Columbia, 1989—; dir. S.C. Nuclear Pharmacy, Columbia, 1990-91; nuclear pharmacist Syncor Internat., Columbia, 1991—; faculty U. S.C. Sch. Medicine, Columbia, 1990—; asst. prof. U. S.C. Coll. Pharmacy, 1990—; radiation safety cons. Syncor Internat. Corp., Columbia, 1990—; third party ins. expert Dr. Arthur A. Madden, Columbia, 1983—; mem. ad hoc com. for Infectious Disease Policy. Mem. Am. Pharm. Assn., Am. Soc. Hosp. Pharmacists, S.C. Nuclear Medicine Soc., S.C. Sch. Medicine Hemotology/Oncology Jour. Club, Phi Lambda Sigma (sec. 1987-89), Order of the Thistle, Order of the Highlander. Home: 4626 Reamer Ave Columbia SC 29206-1541 Office: Syncor Internat Corp 5820 Shakespeare Rd Columbia SC 29223-7210

MADDEN, JAMES DESMOND, forensic engineer; b. Jersey City, Mar. 1, 1940; s. Louis A. and Ann (Desmond) M. BSChemE, U. S.C., 1963, ME, 1966. Registered profl. engr., Ohio. Process engr. Monsanto Co., Alvin, Tex., 1966-67; process and project engr. Union Carbide Corp., Houston, 1967-70; systems engr. M.W. Kellogg Co., Houston, 1970-73, prin. systems engr., 1974-77; sr. process engr. Litwin Co., Houston, 1973-74; sr. project engr. Davy Powergas, Houston, 1977-78, supervising project engr., 1978-79; mgr. equipment engring. DM Internat., Houston, 1979-80, project engring. mgr., 1980-83; owner, forensic engr. Madden Forensic Engring., Parma and Parma Heights, Ohio, 1983—. Pres. Houston Young Adult Rep. Club, 1970-73; chmn. Tex. Young Adult Rep. Clubs, 1973. NSF rsch. grantee, 1963; NASA fellow, 1963. Mem. ASTM, ASME, Soc. Automotive Engrs., Nat. Fire Protection Assn., Am. Chem. Soc., Am. Inst. Chem. Engrs., Inst. Transp. Engrs., Transp. Rsch. Bd. (individual assoc.), Sigma Xi, Sigma Pi Sigma, Tau Beta Pi, Omicron Delta Kappa. Office: 6415 Stumph Rd Ste 306 Cleveland OH 44130-2945

MADDEN, LAURENCE VINCENT, plant pathology educator; b. Ashland, Pa., Oct. 10, 1953; s. Lawrence Vincent and Janet Elizabeth (Wewer) M.; m. Susan Elizabeth Heady, July 7, 1984. BS, Pa. State U., 1975, MS, 1977, PhD, 1980. Research scientist Ohio State U., Wooster, 1980-82, asst. prof., 1983-86, assoc. prof., 1986-91, prof., 1991—; invited univ. lectr. on plant disease epidemiology in more than 10 countries. Author: Introduction to Plant Disease Epidemiology; sr. editor Phytopathology, 1988-90, APS Press; editor-in-chief Phytopathology, 1991—; contbr. 90 articles to profl. jours. Recipient CIBA-GEIGY Agronomy award Am. Soc. Agronomy, 1990; U.S. Dept. Rsch. grantee, 1984, 85, 86, 87, 89, 90; Disting. scholar Ohio State U., 1991. Fellow AAAS; mem. Am. Phytopathology Soc. (chmn. com. 1983, 86, coun. 1991—), Biometric Soc., Brit. Soc. Plant Pathology, Sigma Xi (chpt. pres. 1985). Achievements include development of statistical models for understanding and comparing epidemics. Avocations: photography, travel. Home: 677 Greenwood Blvd Wooster OH 44691-4923 Office: Ohio State U OARDC Dept Plant Pathology Wooster OH 44691

MADDEN, ROBERT WILLIAM, physicist; b. Worcester, Mass., Mar. 14, 1927; s. William Francis and Julia Mary (Jurre) M.; m. Inge Renner, June 1, 1967 (dec. May 1978); children: Madelyn Renee, Mary Ann; m. Davida Dare Gordon, Aug. 28, 1991. BSc, Calif. Inst. Tech., 1951; postgrad., MIT, 1957-58. Physicist IBM, Vestal, N.Y., 1951-56; mathematician RCA, Burlington, Mass., 1956-60; staff scientist Inst. for Def. Analyses, Arlington, Va., 1960-67; sci. advisor Nat. Security Agy., Ft. Meade, Md., 1967—. Contbr. articles to profl. publs. With USN, 1945-46. Achievements include patents for navigation, stable platforms, high altitute bombing, shortwave antennas. Home: 106 St Martin's Rd Baltimore MD 21218 Office: NSA K1 Fort Meade MD 20755

MADDOCK, JEROME TORRENCE, information services specialist; b. Darby, Pa., Feb. 7, 1940; s. Richard Cotton and Isobel Louise (Mezger) M.; m. Karen Rheuama Weygand, Oct. 2, 1965. BS in Biology, Muhlenberg Coll., 1961; MS in Info. Sci., Drexel U., 1968. Editorial assoc. Biol. Abstracts, Phila., 1962-63; mgr. rsch. info. Merck & Co., West Point, Pa., 1963-72; sr. cons. Auerbach Assocs., Inc., Phila., 1972-79; mgr. libr. and info. svcs. Solar Energy Rsch. Inst., Golden, Colo., 1979-88; mgr. info. svcs. Transp. Rsch. Bd., Washington, 1988—; del. Gov.;s Conf. on Libr. and Info. Svc., Pa., 1978; mem. blue ribbon panel to select archivist of U.S., Washington, 1979; U.S. del. to ops. com. on transp. rsch. info. Orgn. for Econ. Cooperation and Devel., 1988—. Bd. dirs. Paoli (Pa.) Pub. Libr., 1976-77. With USAFR, 1962-68. Mem. AAAS, Am. Soc. Info. Sci. (chmn. 1974-75), Elks, Beta Phi Mu, Pi Delta Epsilon. Republican. Episcopalian. Achievements include projection of information science operations 10 years into the future. Home: 20517 Aspenwood Ln Gaithersburg MD 20879 Office: Transp Rsch Bd 2101 Constitution Ave NW Washington DC 20879

MADDOCK, THOMAS SMOTHERS, engineering company executive, civil engineer; b. Norfolk, Va., Mar. 23, 1928; s. John Francis and Mary Eileen (Smothers) M.; m. Caroline Diane Street, Nov. 5, 1977; children: Kimberly Anne, Victoria Anne. BCE with honors, Va. Poly. Inst., 1950; MCE, MIT, 1951; MBA, Stanford U., 1957. Registered profl. engr. in 8 states including Va., Calif., Nev., Ariz., Fla., D.C. V.p., project mgr. Boyle Engring. Corp., Newport Beach, Calif., 1957-72, pres., 1972—; presentations in field; project mgr. AID-financed Master Plan of Water Resources, Santo Domingo, Dominican Republic; project dir. for design and constrn. supervision mcpl. waterworks and irrigation facilities, Libya; supervisor preliminary designs for seawater delivery system ARAMCO, Saudi Arabia; dir. preparation of water wastewater facilities master plans (HUD 701 planning studies for cities), San Luis Obispo County, Calif. Contbr. articles to profl. jours. Pres. Seabee Meml. Fund Scholarship Fund, Washington, 1981-83. Lt. (j.g.) C.E., USN, 1952-55; rear adm. C.E., USNR. Decorated Navy Commendation medal, Meritorious Svc. medal, Legion of Merit. Fellow Los Angeles Inst. Advancement Engring., ASCE (v.p. Los Angeles sect. 1979); mem. Cons. Engrs. Assn. Calif., Nat. Soc. Profl. Engrs., Soc. Am. Mil. Engrs. Office: Boyle Engring Corp PO Box 7350 Newport Beach CA 92658-7350

MADDOX, ROBERT ALAN, atmospheric scientist; b. Granite City, Ill., July 12, 1944; s. Robert Alvin and Maxine Madeline (Elledge) M.; m. Rebecca Ann Speer, Dec. 27, 1967; children—Timothy Alan, Jason Robert. Student Purdue U., 1962-63; B.S., Tex. A&M U., 1967; M.S. in Atmospheric Sci., Colo. State U., 1973, Ph.D. in Atmospheric Sci., 1981. Meteorologist, Nat. Weather Service, Hazelwood, Mo., 1967; research meteorologist Geophys. Research & Devel. Corp., Ft. Collins, Colo., 1975-76; research meteorologist Atmospheric Physics and Chemistry Lab., NOAA-Environ. Research Labs., Boulder, Colo., 1976-79, meteorologist Office Weather Research and Modification, 1979-82, program mgr. weather analysis and storm prediction, 1982-83, dir. weather research program, 1983-84, program mgr. mesoscale studies, 1984-86; dir. Nat. Severe Storms Lab., Norman, Okla., 1986—; participant numerous sci. workshops; tchr. weather analysis, forecasting, mesoscale phenomena; presenter workshops on mesoscale analysis and heavy precipitation forecasting Nat. Weather Service Forecast Offices; cons. in field. Served to capt. USAF, 1967-75. Decorated Air Force Commendation medal with oak leaf cluster; recipient Superior Performance award NOAA, 1981, Outstanding Sci. Paper award, 1984. Fellow Am. Meteorol. Soc. (councilor 1989-92, chmn. severe elect. storms com. 1991-92, Clarence Leroy Meisinger award 1983, co-editor, assoc. editor Monthly Weather Rev., assoc. editor Weather & Forecasting); mem. Nat. Weather Assn. (award for outstanding contbns. to operational meteorology 1981, co-editor Nat. Weather Digest, Mem. of Yr. award 1992), Sigma Xi, Phi Kappa Phi, Chi Epsilon Pi. Contbr. articles to profl. publs. Office: NOAA/ERL Nat Severe Storms 1313 Halley Cir Norman OK 73069-8493

MADER, BRYN JOHN, vertebrate paleontologist; b. N.Y.C., July 29, 1959; s. Walter Richard and Audrey Jeanne (Hargest) M. BS, SUNY, Stony Brook, 1982; MS, U. Mass., 1987, PhD, 1991. Curatorial asst. dept. vertebrate paleontology Am. Mus. Natural History, N.Y.C., 1982-83, asst. collection mgr., 1990-93, collections registrar dept. mammalogy, 1993—. Contbr. articles to profl. jours., chpts. to books. Mem. N.Y. Acad. Scis., Soc. Vertebrate Paleontology, Paleontol. Soc., Sigma Xi. Presbyterian. Achievements include publication of first and only significant review of brontotheres in almost 50 years (a major perissodactyl lineage known primarily from the Eocene and Oligocene epochs of North America and Central Asia). Office: Am Mus Natural History Cen Park West at 79th St New York NY 10024

MADER, DOUGLAS PAUL, quality educator; b. Brookings, S.D., May 16, 1963; s. Lawrence Harold Mader and Susan Margaret (Littleton) Burk; m. Darla Susan Hower, Dec. 30, 1991; children: Alyssa, Megan. BS in Engring. Physics, S.D. State U., 1985; MS in Math., Colo. Sch. of Mines, 1990. Cert. quality engr. Quality control engr. Govt. Electronics Group, Motorola, Scottsdale, Ariz., 1985-87; integrated circuit test engr. Semiconductor Products sector, Motorola, Mesa, Ariz., 1987-88; sr. staff engr. Six Sigma Rsch. Inst., Motorola, Schaumburg, Ill., 1990-92, prin. staff scientist, 1992; cons. Rockwell Internat., Cedar Rapids, Iowa, 1992-93; statis. quality mgr. Advanced Energy Industries, Ft. Collins, Colo., 1993—. Editor: The Encyclopedia of Six Sigma Tools, 1992, The Encyclopedia of Six Sigma Applications; author: (videotapes) Concurrent Engineering-The Foundation of Six Sigma Quality, 1992, (software) The Black Box Simulator, 1991. Acad. scholarship Colo. Sch. Mines, 1990-91, S.D. State U., 1981-85. Mem. Am. Statis. Assn., Inst. of Indsl. Engrs., Am. Soc. for Quality Control, Inst. of Mgmt. Scis. Office: Advanced Energy Industries 1600 Prospect Pky Fort Collins CO 80525

MADHANAGOPAL, THIRUVENGADATHAN, environmental engineer; b. Madurai, India, May 28, 1955; s. Krishnan and Lakshmi Thiruvengadathan; m. Vyjayanthi T. Madhanagopal, May 22, 1981; children: Harry T., Hema L. BE, U. Madras, India, 1977; MS, Wayne State U., 1981; MBA, U. Cent. Fla., 1992. Profl. engr., Fla., Mich., Wis. Project engr. Land S.E. A. Corp., Detroit, 1978; jr. engr. Detroit Water and Sewerage Dept., 1979, asst. engr., 1980, sr. asst. engr., 1981-82, assoc. engr., 1983-84; staff engr. Orange County Pub. Utilities, Orlando, Fla., 1985-88, sr. engr., 1989—; mem. water quailty task force Orange County, 1992. Compact mentor Orange County Compact Program, Orlando, 1990-92; pres. Cultural Orgn., Orlando, 1988-89. UCF Found. fellow, 1990. Mem. NSPE (mgmt. fellowship 1992), Am. Water Works Assn., Water Environ. Fedn., Fla. Engring. Soc. (com. mem. mathcounts contests 1990-92, scholarship chmn. 1992, 93, fellowship 1990, 91, environ. concerns com. 1993), Fla. Water Environ. Assn. (pretreatment com. 1993, local chpt. steering com. 1993), Phi Kappa Phi, Beta Gamma Sigma, Omicron Delta Kappa. Hindu. Home: 1513 Southwind Ct Casselberry FL 32707 Office: Orange County Pub Utilities 109 E Church St Orlando FL 32801

MADHOUN, FADI SALAH, civil engineer; b. Beirut, Lebanon, July 7, 1962; came to U.S., 1985; s. Salah Ragheb and Wasila (Fanous) M. B of Engring., Am. U. Beirut, 1985; MS in Civil Engring., Syracuse U., 1987, MBA, 1992. Registered profl. engr., N.Y., Lebanon. Structural engr. Ramco Trading & Contracting Co., Beirut, 1984; project engr. Stearns and Wheeler, Cazenovia, N.Y., 1989-90, sr. project engr., 1990-92, project mgr., 1992—; sec. tech. practices com. Stearns and Wheeler, Cazenovia, 1991—. Editor: (manual) Technical Engineers Standards and Guidelines. Recipient Hariri Found., 1986; scholar Syracuse U., 1986. Mem. ASCE, Am. Concrete Inst., Lebanese Soc. Engrs. Home: # 2 Woodview Ter Fayetteville NY 13066 Office: Stearns and Wheeler 1 Remington Park Dr Cazenovia NY 13035

MADLANG, RODOLFO MOJICA, urologic surgeon; b. Indang, Cavite, The Philippines, Apr. 9, 1918; came to U.S., 1953; s. Simeon Fajardo and Eugenia R. (Mojica) Madlangsacay; m. Lourdes Recto Gregorio, Dec. 8, 1946; children: Cesar, Rodolfo G., Mercy Lynn. AA, U. Philippines, Manila, 1939, MD, 1945. Diplomate Am. Bd. Urology. Resident in gen. surgery Philippine Gen. Hosp., Manila, 1946-49; resident in urology St. Francis Hosp., Peoria, Ill., 1953-55; asst. prof. physiology Far Ea. U. Inst. Medicine, Manila, 1956-58, cons. in urology, 1956-58; attending urologist St. Catherine Hosp., East Chicago, Ind., 1958-81, chief surgery, 1977-79; attending urologist St. Margaret Hosp., Hammond, Ind., 1960-81; chief urology U.S. VA Outpatient Clinic, L.A., 1982—. Fellow ACS; mem. AMA, Am. Urol. Assn., Pan Pacific Surg. Assn., Assn. Mil. Surgeons of the U.S., Ind. State Med. Assn., N.Y. Acad. Scis. Republican. Roman Catholic. Office: VA Outpatient Clinic 351 E Temple St Los Angeles CA 90012

MADORY, JAMES RICHARD, hospital administrator, former air force officer; b. Staten Island, N.Y., June 11, 1940; s. Eugene and Agnes (Gerner) M.; m. Karen James Clifford, Sept. 26, 1964; children: James E., Lynn Anne, Scott J., Elizabeth Anne, Joseph M. BS, Syracuse U., 1964; MHA, Med. Coll. Va., 1971. Enlisted USAF, 1958, commd. 2d lt., 1964, advanced through grades to maj., 1978; adminstr. Charleston (S.C.) Clinic, 1971-74, Beale Hosp., Calif., 1974-77; assoc. adminstr. Shaw Regional Hosp., S.C., 1977-79; ret. USAF, 1979; asst. adminstr. Raleigh Gen. Hosp., Beckley, W.Va., 1979-81; adminstr., dir., sec. bd. Chesterfield Gen. Hosp., Cheraw, S.C., 1981-87; pres., CEO Grand Strand Hosp., Myrtle Beach, S.C., 1987—, trustee, 1987-91; mem. adv. bd. Cheraw Nursing Home, 1984-85. Contbr. articles to profl. publs. Chmn. bd. W.Va. Kidney Found., Charleston, 1980-81; chmn. youth bd. S.C. TB and Respiratory Disease Assn., Charleston, 1972-73; county chmn. Easter Seal Soc., Chesterfield County, S.C., 1984-85; campaign crusade chmn. Am. Cancer Soc., Chesterfield County, 1985-86; chmn. dist. advancement com. Boy Scouts Am., 1987-90; bd. dirs. Horry County United Way, 1989-91, Horry County Access Care, 1989-91; trustee Cheraw Acad., 1982-85, Grand Strand Gen. Hosp., 1987-91, Coastal Acad., 1988-90. Decorated Bronze Star, Vietnamese Cross of Gallantry, Vietnamese Medal of Honor. Fellow Am. Coll. Hosp. Adminstrs., Am. Coll. Health Care Execs.; mem. S.C. Hosp. Assn. (com. on legislation 1984-86, trustee 1989—), Cheraw C. of C. (bd. dirs. 1982-83), Rotary (pres. 1984-85). Republican. Roman Catholic. Office: Grand Strand Hosp 809 82nd Pkwy Myrtle Beach SC 29577-1413

MADRAS, BERTHA KALIFON, neuroscientist, consultant; b. Montreal, Quebec, Canada, Dec. 9, 1942; m. Peter Madras, June 21, 1964; children: Cynthia G., Claudine D. BSc, McGill Univ., 1963, PhD, 1967. Postdoctoral fellow Tufts Univ., Boston, 1966-67; postdoctoral fwllow rsch. assoc. Mass. Inst. Tech., Cambridge, 1967-69, 72-74; asst. prof. Univ. Toronto, 1979—; asst. prof. Harvard Medical Sch., Boston, 1986-90, assoc. prof., 1990—; vist. com. Brookhaven Nat. Lab., Upton, N.Y., 1992—; pub. info. com. Soc. Neuroscience, Washington, 1992—; ad hoc reviewer, cons. Nat. Inst. Drug Abuse, Ont. Mental Health Found. 1984-90, chmn. fellowships and awards com., 1988-90; ad hoc reviewer NIMH; chmn. radiation safety New England Regional Pvt. Rsch. Ctr. 1987—; radiation safety com. Harvard U.; rsch assoc. Mass. Gen. Hosp., 1991—; ctr. faculty Norman E. Zinberg Ctr. for Addiction Studies, 1992—; steering com. divsn. on addictions Harvard Med. Sch., 1992—. Author: (book chpt.) Dopamine, 1984; contbr. articles to profl. jours.; editorial bd. Synapse, 1991—. Sci. fair judge. Recipient rsch. grant Nat. Inst. Drug Abuse, 1992-97. Sci. Edn. Partnership award NIDA, 1992-94, grant. parkinson's Disease Found., 1990-91. Mem. Soc. for Neuroscience, Coll. Probs. Drug Dependence. Achievements include development of a marker for Parkinson's disease and a probe for cocaine binding sites in brain, mapping cocaine binding sites in the brain which are relevant to the behavioral effects of cocaine, developed a PET imaging drug to image cocaine binding sites in living brain, developed a PET imaging drug to monitor Parkinsonism in brain. Office: Harvard Medical Sch 1 Pine Hill Dr Southborough MA 01772-9102

MADSON, PHILIP WARD, engineering executive; b. Atlantic, Iowa, Aug. 27, 1948; s. Philip Ward and Pearl Elaine (Thomson) M.; m. Maria Concepcion Casamitjana, Aug. 11, 1968; children: Peter Wesley, David Philip. BSChemE, Iowa State U., 1969, MSChemE, 1970. Registered profl. engr., Ohio. Process engr. Procter & Gamble, Cin., 1971-74, tech. brand mgr., 1974-76, sect. head, 1976-77, assoc. dir., 1977-79; cons. engr. Raphael Katzen Assocs., Cin., 1979-84, v.p., 1984-90; sr. v.p. tech. and mktg., 1990-92, pres., 1993—. Chmn. Police/Community Rels. Com. Kennedy Heights, 1978-84; chmn. media response team S.W. Ohio Sportsmen's Assn., 1990—; chmn. pub. rels. com. Firearms Facts Com., 1991—; founder, chmn. Kennedy Heights Concerned Citizens, Cin., 1979-85; mem. exec. bd. dirs. Kennedy Heights Community Coun., 1979-84. Recipient Spl. award Kennedy Heights Community Coun., 1984, Recognition Hamilton County Crime Prevention Assn., 1983. Mem. Am. Inst. Chem. Engrs. Avocations: music, hunting, target shooting. Home: 3749 Davenant Ave Cincinnati OH 45213-2218 Office: Raphael Katzen Assocs Internat Inc 2300 Wall St Ste K Cincinnati OH 45212

MADUEME, GODSWILL C., nuclear scientist, international safeguards agency administrator; b. Port Harcourt, Nigeria, July 9, 1943; s. Rueben Onwudiwe and Susana Nwinyinya (Ume) M.; m. Adaora Mercy Onwuagba, Sept. 7, 1974; children: Hans Ifeanyi, Elvira Chika, Chike, Linda Ebele. BSc, U. Ibadan, Nigeria, 1964; Fil.Kand., U. Uppsala, Sweden, 1968; PhD, U. Uppsala (Sweden), 1972, Fil.Dr., 1975. Cert. nuclear physicist. Rsch. assoc. U. Uppsala (Sweden), 1969-71, asst. lectr., 1972-74; asst. prof. U. Tokyo, 1975-76; sr. lectr. U. Ife (Nigeria), 1977-80; second officer Internat. Atomic Energy Agy., Vienna, Austria, 1981-83, first officer and group leader, 1984-90, sr. officer and unit head, 1991—; assoc. ICTP, Trieste, Italy, 1981-85. Contbr. articles to profl. jours. and conf. proc. Uppsala U. grantee, 1970. Fellow Italian Phys. Soc., Japan Soc. for Promotion of Sci.; mem. Swedish Phys. Soc., European Phys. Soc., Am. Phys. Soc., Inst. for Nuclear Materials Mgmt., N.Y. Acad. Sci.

MADURA, JEFFRY DAVID, chemistry educator; b. Greenville, Pa., Dec. 15, 1957; s. Kenneth J. and Bonita (Reinhart) M.; m. Alana Leigh Zimmer; 1 child, Brandon Zimmer. BA, Thiel Coll., 1980; PhD, Purdue U., 1985. Postdoctoral fellow U. Houston, 1986-88, dir. chem. computations, 1988-91; chief scientist U. Houston Inst. for Molecular Design, 1988-91; asst. prof. U. South Ala., Mobile, 1991—; vis. asst. prof., U. Houston, 1988-91. Contbr. articles to profl. jours. Recipient Analytcial Chemists Sr. award, Soc. Analytical Chemists, 1980; grantee Rsch. Corp., 1991, Petroleum Rsch. Fund, 1991. Mem. Am. Chem. Soc. (Outstanding Sophomore Chemist 1978), N.Y. Acad. Scis., Ala. Acad. Sci., Mason. Democrat. Lutheran. Office: Chem Bldg Rm 223 U South Ala Mobile AL 36688

MAEDA, HIROSHI, medical educator; b. Hyogo-ken, Japan, Dec. 22, 1938; s. Muraji and Yoshiko Maeda; m. Norico Soma, Oct. 1, 1967; children: Jun-ichiro, Kei. BS, Tohoku U., Sendai, Japan, 1962; MS, U. Calif., Davis, 1964; PhD, Tohoku U., 1967, MD, 1972. Rsch. assoc. Sidney Farber Cancer Inst., Harvard U., Boston, 1967-71; assoc. prof. Med. Sch. Kumamoto (Japan) U., 1971-80, prof., 1980—; Lichfield lectr. John Radcliffe Med. Sch., Oxford U., Eng., 1990. pioneer in macromolecular anti-cancer agt. discovery, cancer treatment methods, viral disease rsch. Fulbright fellow, 1962; recipient Internat. Career Tech. Transfer award Internat. Union Against Cancer, Geneva, 1974, Princess Takamatsu Cancer Rsch. award, Tokyo, 1982, Sapporo Life Sci. Rsch. award, 1983; named Hon. Mayor of San Antonio, Hon. Citizen of Okla. State. Mem. Japanese Biochem. Soc., Japanese Cancer Assn., AAAS, Am. Assn. Cancer Rsch., Am. Chem. Soc., N.Y. Acad. Sci., Japan Soc. Bacteriology, Am. Soc. Microbiology, Soc. Exptl. Med. Biology, Sigma Xi. Home: Hotakubo Honmachi 631-3, Kumamoto 862, Japan

MAEDA, TOSHIHIDE MUNENOBU, spacecraft system engineer; b. Sagamihara, Kanagawa, Japan, Apr. 11, 1962; s. Mamoru and Yasuko Hiromi (Yamagishi) M. BS, U. Tokyo, 1986; postgrad., Internat. Space U., Cambridge, Mass., 1988. Engr. space systems div. Hitachi, Ltd., Yokohama, Japan, 1986-89, 92—; engr. Nat. Spece Devel. Agy. Japan, Tsukuba, 1990-92. Supr. Newton jour., 1991. Mem. AIAA, Nat. Space Soc., Planetary Soc., Remote Sensing Soc. Japan, Japan Soc. Aero. and Space Sci. Achievements include patents on thermal control technology for spacecraft components, structural test equipment. Home: 2349 Kamitsuruma, Sagamihara Kanagawa 228, Japan

MAEDA, YUKIO, engineering educator; b. Toyohara, Karafuto, Japan, Mar. 29, 1922; s. Tsuruji and Fusa (Sato) M.; m. Junko Minakata, Dec. 8, 1961. B in Engring., Hokkaido U., Sapporo, Japan, 1945, postgrad., 1947; D in Engring., U. Tokyo, 1966. From lectr. to asst. prof. engring. Hokkaido U., Sapporo, 1947-57; tech. dir. Sakurada Iron Works Co. Ltd., Tokyo, 1958-66; from assoc. prof. to prof. emeritus engring. Osaka U., Suita, 1966—; prof. Kinki U., Higashi-Osaka, 1985-89; head Structural Rsch. Ctr., Osaka, 1989—. Contbr. articles to profl. jours. Mem. bridge com. Osaka City, 1976—. Recipient Achievement prize, Min. Labour, Japanese Govt., Tokyo, 1988; grantee Fulbright, U.S. Edn. Commn., Tokyo, 1955, HIROI-OSAMU, Hokkaido U., Sapporo, 1945. Fellow ASCE (pres. Kansai br. Japan sect. 1989-91); mem. Japan Soc. Civil Engrs. (hon., dir. 1974-76, TANAKA prize 1981), Internat. Assn. for Bridge and Structural Engring. (chmn. tech. com. 1987-91, hon.), Fulbright Assn. (life), Japan Welding Soc., Japan-Am. Soc. Osaka. Avocation: photography. Home: 404 Iwazono 10-16, Ashiya 659 Hyogo, Japan

MAEKAWA, MAMORU, computer science educator; b. Kyoto, Japan, Mar. 8, 1942; s. Teiichi and Yoshi Maekawa; m. Akiko Morimoto, Sept. 4, 1970; children: Miko, Maya. BS, Kyoto U., 1965; MS, U. Minn., 1971, PhD, 1973. Systems programmer Toshiba Co. Ltd., Kawasaki, Japan, 1965-70; rsch. asst. U. Minn., Mpls., 1971-73; vis. asst. prof. U. Iowa, Iowa City, 1973-75; prin. researcher Toshiba Co. Ltd., Tokyo, 1975-79; assoc. prof. computer sci. U. Tokyo, 1979-92; prof. info. systems U. Electro-Comm., 1992—; adj. assoc. prof. Keio U., Yokohama, Japan, 1977-80; vis. assoc. prof. U. Tex., Austin, 1981-82. Author: Hardware, 1985, Operating Systems Advanced Concepts, 1987, Operating Systems, 1988, Computer Architecture, 1988, Software Execution/Development Environments, 1992. Mem. IEEE (cert. of recognition 1983), Assn. for Computing Machinery, Info. Processing Soc. Japan. Home: 4-4-13 Masuodai, Kashiwa Chiba 277, Japan Office: U Electro-Comm, 1-5-1 Chofu, Tokyo 182, Japan

MAENO, NORIKAZU, geophysics educator; b. Hokkaido, Japan, Sept. 1, 1940; s. Juzo and Hanano (Komiyama) M.; m. Yuko Higuchi, Feb. 1, 1969. BA, Hokkaido U., Sapporo, 1963, MS, 1965, DSc, 1972. Rsch. fellow Inst. Low Temperature Sci., Hokkaido U., 1965-73, asst. prof. sci., 1973-84, prof. snow and ice physics and earth and planetary glaciology, 1984—. Author: Ice Science, 1981, Nauka o L'te (transl. into Russian 1988), Fundamentals of Glaciology-Structure and Physical Properties of Snow and Ice, 1986. Recipient sci. prize Japanese Soc. Snow and Ice, 1980. Home: Hanakawa Minami 7-2-133, Ishikari-Cho 061-32, Japan Office: Hokkaido U Inst Low Temp, Sci, N-19, Nishi-8, Kita-ku, Sapporo 060, Japan

MAENPAA, PEKKA HEIKKI, biochemistry educator; b. Tampere, Finland, June 6, 1939; s. Heikki Artturi and Kerttu Elisabet (Paasikoski) M.; m. A. Mirjam Piironen, 1959 (div. 1988); children: Petri, Johanna; m. Leena Kirkkomäki, 1989; 1 child, Louhi. MD, U. Helsinki, 1966, DMS, 1968. Rsch. fellow State Alcohol Co., Helsinki, 1962-64; instr. U. Helsinki, 1964-66, lectr., 1970-72, 73-74; chief clin. lab. Children's Hosp., Helsinki, 1966-67; postdoctoral fellow Stanford (Calif.) U., 1968-70, rsch. assoc., 1972-73; prof. biochemistry U. Kuopio, Finland, 1974—, chmn. dept. biochemistry and biotech., 1980—, dean faculty gen. biology, 1984-85; vis. prof. U. Calif., Irvine, 1977. Contbr. over 100 articles to profl. jours. Mem. Assn. Finnish Chemists (Heikki Suomalainen award 1987), Acad. Finland (coun. natural scis. 1983-88). Avocation: horses. Home: Ahkiotie 12A 14, SF-70200 Kuopio Finland Office: U Kuopio Dept Biochemistry, & Biotech, PO Box 1627, SF-70211 Kuopio Finland

MAFI, MOHAMMAD, civil engineer, educator; b. Tehran, Iran, Jan. 12, 1954; came to U.S., 1978; s. Javad and Mansoureh Mafi; m. Simin K. Mafi, July 2, 1975; children: Iman, Mona, Ehsan. MS, Pa. State U., 1980, PhD, 1985. Registered profl. engr. N.Y., Pa. Asst. prof. dept. civil engring. Union Coll., Schenectady, N.Y., 1985-90, assoc. prof., chmn. dept., 1990—; pres. SAFE Cons. Bridge Inspection, Design and Rehab., Schenectady, 1985—. Contbr. articles to sci. and engring. publs. Mem. ASCE, Am. Soc. Engring. Edn. (Dow Outstanding Young Faculty award 1989), Am. Concrete Inst., Concrete Reinforcing Steel Inst., Chi Epsilon, Tau Beta Pi. Office: Union Coll Dept Civil Engring 1 Union St Schenectady NY 12308

MAGA, JOSEPH ANDREW, food science educator; b. New Kensington, Pa., Dec. 25, 1940; s. John and Rose Maga; m. Andrea H. Verporian, June 13, 1964; children: Elizabeth, John. BS, Pa. State U., 1962, MS, 1964; PhD, Kans. State U., 1970. Project leader Borden Foods Co., Syracuse, N.Y., 1964-66; group leader Cen. Soya Co., Chgo., 1966-68; asst. prof. Colo. State U., Ft. Collins, Colo., 1970-72, assoc. prof., 1972-74, prof. food sci., 1974—. Contbr. numerous articles to profl. jours. Mem. Inst. Food Technologists, Am. Chem. Soc., Am. Assn. Cereal Scientists. Office: Colo State U Dept Food Sci & Human Nutrition Fort Collins CO 80523

MAGARILL, SIMON, optical engineer; b. Barnaul, USSR, Apr. 24, 1953; came to U.S., 1990; s. Jacob and Liliya (Krashke) Livshites; m. Marina Magarill, Nov. 2, 1974; children: Nelly, Paul. MS in Optics, Inst. Fine Mechanics & Optics, St. Petersburg, USSR, 1976, PhD in Optics, 1983. Engr. Opto-Mech. Corp., St. Petersburg, 1976-79, devel. engr., 1979-84, prin. engr., 1984-88; sr. researcher Spl. OE Lab., St. Petersburg, 1988-90; devel. engr. U.S. Precision Lens Inc., Cin., 1990—. Contbr. articles to profl. publs. Home: 9501 Linfield Dr Cincinnati OH 45242 Office: US Precision Lens 3997 McMann Rd Cincinnati OH 45245

MAGASANIK, BORIS, microbiology educator; b. Kharkoff, U.S.S.R., Dec. 19, 1919; came to U.S., 1938; s. Naum and Charlotte (Schreiber) M.; m. Adele Karp, Aug. 9, 1949. BS, CCNY, 1941; PhD, Columbia U., 1948; MS (hon.), Harvard U., 1958. Tech. asst. Mt. Sinai Hosp., N.Y.C., 1939-41; rsch. asst. Columbia U., N.Y.C., 1948-49; Ernst fellow Harvard U. Med. Sch., Boston, 1949-51, assoc. to assoc. prof., 1951-59; prof. microbiology MIT, Cambridge, 1960-77, Jacques Monod prof., 1977—, head dept. biology, 1967-77; tutor in biochem. scis. Harvard U., 1951—. Contbr. over 240 sci. articles and revs. to profl. publs. With M.C., U.S. Army, 1942-45, ETO. Guggenheim fellow, 1959; Markle scholar in med. scis., 1951-56. Mem. NAS (Selman A. Waksman microbiology award 1993), Am. Soc. Microbiology, Am. Soc. Biol. Chemists. Home: 120 Collins Rd Newton MA 02168-2234 Office: MIT Dept Biology 77 Massachusetts Ave Cambridge MA 02139-4307

MAGAW, JEFFREY DONALD, civil engineer. BSCE, Norwich U., 1980. Registered profl. engr., Mass., N.H., Mich. Sr. civil engr. Alliance Technologies, Bedford, Mass., 1986-87; town engr. Town of Hudson (N.H.), 1987-89; region project mgr. Waste Mgmt. Inc.-East, Wakefield, Mass., 1989—. With USAF, 1980-85, USAF Res., 1986—. Mem. ASCE, NSPE, Internat. Facility Mgmt. Assn. Home: 11 Bennington Rd Nashua NH 03060-8100 Office: Waste Mgmt Inc East 580 Edgewater Dr Wakefield MA 01880

MAGDA, MARGARETA TATIANA, physicist; b. Bucharest, Romania, July 10, 1936; came to U.S., 1987; d. Alexandru and Marie (Constantinescu) M. MS in Physics, Faculty of Physics, Bucharest, 1958; PhD in Nuclear Physics, Bucharest U., 1969. Researcher Inst. for Atomic Physics, Bucharest, 1958-69, sr. researcher, 1969-87; assoc. sr. researcher chemistry dept. SUNY, Stony Brook, 1988—; prof. Ctr. for Tng. Pers. Working in Nuclear Field, Bucharest, 1971-73. Author: Atomic and Nuclear Physics, 1976; editor: Horia Hulubei, 1987; contbr. articles to Nuclear Physics, Jour. Physics, Yadernaja Fizika. Recipient Dragomir Hurmuzescu award for physics Romanian Acad., Bucharest, 1965. Mem. AAAS, Am. Phys. Soc., Am. Chem. Soc. Achievements include initial research in experimental preequilibrium reactions and breakup processes at low energy heavy ion reactions; model for transfer reactions leading to heavy actinides. Office: SUNY Stony Brook NY 11794-3400

MAGEE, RICHARD STEPHEN, mechanical engineering educator; b. East Orange, N.J., Mar. 26, 1941. BEngring., Stevens Inst. Tech., 1963, MS, 1964, DSc in Applied Mechanics, 1968. Rsch. engr. combustion lab. Stevens Inst. Tech., 1966-68, from asst. to assoc. prof. mech. engring., 1968-77, prof., 1977—; cons. Photochem, Inc., 1970—. Grantee NSF, NASA, 1969-70. Mem. ASME, Am. Soc. Engring. Edn., Combustion Inst. Achievements include research in flammability characteristics of combustible materials, including ignition, flame spread and burning characteristics, pyrolysis characteristics of polymeric materials, basic incinerator design parameters. Office: NJ Inst Tech Hazardous Substance Mgmt 323 High St Nutley NJ 07110-1434*

MAGGARD, BILL NEAL, mechanical engineer; b. Wise, Va., Oct. 18, 1939; s. Edward Grady and Beulah C. (Kennedy) M.; m. Linda Lou Johnson, July 30, 1960; children: Teresa, Latricia, Neal. BSME, Va. Poly. Inst. and State U., 1963. Registered profl. engr., Tenn. Tool and die designer Coop. at Monroe Calculating Co., Bristol, Va., 1960-61; mech. engr. Poly-Sci. Corp., Blacksburg, Va., 1963-64; mech. engr. Tenn Eastman, Kingsport, 1964-70, sr. mech. negr., 1970-80, assoc. engr., 1980-82, dept. supt., 1982-86, TMP mgr., organization devel. cons., 1986—; speaker in field. Author: TPM That Works, 1992, publs. in field. Leader Tenn. Eastman's United Way, Kingsport, 1992. Recipient Best Speaker Recognition award Inst. Internat. Rsch., Can., 1989, 90, 91. Mem. Inst. Indsl. Engrs. Internat. Maintenance Con. (mem. com. 1987-92, track chmn. 1987-92). Republican. Baptist. Achievements include research in total productive maintenance. Home: 495 Barr Rd Blountville TN 37617 Office: Tenn Eastman Bldg 54D Kingsport TN 37662-5054

MAGGAY, ISIDORE, III, engineering executive, food processing engineer; b. San Diego, Calif., Sept. 12, 1952; s. Isidore Jr. and Dolores (Ambay) M.; m. Karen Elizabeth, Dec. 25, 1981; children: Adrienne Leigh, Brittany Elizabeth. BSME, Calif. Maritime Acad., 1973; MBA, Nat. U., 1980. Registered environ. assessor Calif., 1990; cert. hazardous material contractor Calif., 1989. Project engr. Ralston Purina, San Diego, 1976-78; dist. engr. Carnation Co., L.A., 1978-81; dir. engring. Sara Lee Corp., San Francisco, 1981-85; v.p. engring. Alex Foods Inc., Anaheim, Calif., 1986-89; pres. Acad. Engring., Vista, Calif., 1989—; gen. engring. contractor, Contractors State Lic. Bd., Calif., 1989—. Commr. Environ. Quality Commn., Vista, 1991-92. Lt. Commdr. USNR, 1973—. Mem. Am. Inst. Plant Engrs., Environ. Assessment Assn., Nat. Soc. Profl. Engrs. Roman Catholic. Office: Acad Engring 533 Jobe Hill Dr Vista CA 92083

MAGLATY, JOSEPH LOUIS, data processor, consultant; b. Hartford, Conn., Aug. 31, 1955. BS in Chemistry, Providence Coll., 1977; PhD of Analytical Chemistry, Colo. State U., 1983. Sr. applications specialist Merck & Co., Inc., Rahway, N.J., 1982-86; from sr. systems analyst to project specialist Merck & Co., Inc., West Point, Pa., 1986-91; systems cons. Merck & Co., Inc., West Point, 1991—; adv. bd. mem. child care task force North Penn Sch. Dist., Lansdale, Pa., 1991-92, data processing com., 1992—, sci.-by-mail program coord., 1989—, sci. curriculum rev. com., 1993. Mem. Del. Valley Mass-II Users Group (pres. 1989-91), Del. Valley Lab. Automation Group (co-chairperson 1990-91). Republican. Roman Catholic. Office: Merck & Co Inc 760 Sumneytown Pike West Point PA 19486-0004

MAGNANTI, THOMAS L., business management educator; b. Omaha, Oct. 7, 1945; s. Leo A. and Florence L. Magnanti; m. Beverly A. McVinney, June 10, 1967; 1 child, R. Randall. BS in Chem. Engring., Syracuse U., 1967; MS in Stats., Stanford U., 1969, MS in Math., 1971, PhD in Ops. Rsch., 1972. Asst. prof. Alfred P. Sloan Sch. Mgmt. MIT, Cambridge, Mass., 1971-75; rsch. fellow, vis. prof. Ctr. for Ops. Rsch. and Econometrics Univ. Catholique de Louvain, 1976-77, 89; assoc. prof. Alfred P. Sloan Sch. Mgmt. MIT, 1975-79, prof., 1979-85, George Eastman prof. of mgmt. sci., 1985—, head mgmt. sci. area, 1982-88, co-dir. Ops. Rsch. Ctr., 1986—, founding co-dir. Leaders for Mfg. Program, 1988—; vis. scientist Bell Labs., 1977, GTE Labs., 1989; vis. scholar Grad. Sch. Bus. Adminstrn., Harvard U., 1980-81; mem. corp. mfg. staff Digital Equipment Corp., 1990; mem. editorial bd. Wadsworth Publishing Co. Series on Ops. Rsch., Jour. Computational Optimization and Applications; mem. adv. bd. North Holland Handbooks in Ops. Rsch. and Mgmt. Sci.; former mem. adv. bd. mgmt. dept. Worcester Poly. Inst.; chmn. exec. com. mgmt. in the 90's Sloan Sch. Mgmt., 1987-88; co-organizer workshop on ops. rsch. and systems theory NSF, Cambridge, 1983; speaker in field. Author: Applied Mathematical Programming, 1977, Network Flows, 1993; editor: Jour. Ops. Rsch., 1983-87; co-editor: Math. Programming, 1981-83; assoc. editor SIAM Jour. Algebraic and Discrete Methods, 1981-83, Mgmt. Sci., 1978-81, Ops. Rsch., 1978-81, SIAM Jour. Applied Math., 1976-81, Math. Programming, 1988—; adv. editor Transp. Sci., 1985—, Mktg. Sci., Math. of Artificial Intelligence; contbr. numerous articles to profl. jours. Mem. NSF Sci. and Tech. Exchange Delegation to Soviet Union, 1977, NSF Rsch. Initiation Grant panels, 1985, 90; advisor NSF program on decision, risk and mgmt. sci., 1988, 89; mem. mfg. studies bd. Nat. Rsch. Coun., 1993—. Recipient MIT Convocation Program award, 1992, Mgmt. Program Exchange grant IREX, Curriculum Devel. grant Sloan Found., 1990-92. Mem. IEEE (com. on large scale systems 1979-83), TIMS (sec., treas. Boston chpt. 1976), Nat. Acad. Engring, Ops. Rsch. Soc. Am. (pres. 1988-89, chmn. various coms., Lanchester prize com. 1979, coun. mem. computer sci. tech. sect. 1983-87, co-organizer 1st doctoral consortium 1983, plenary speaker conf. on telecom. 1990), Tau Beta Phi, Pi Mu Epsilon, Phi Kappa Phi. Achievements include research in transportation network analysis and decomposition methods, network analysis and optimization, workshop on operations research and systems theory, development of improved analytic methods in metro planning, network design and combinatorial optimization. Home: 33 School St Hopkinton MA 01748 Office: MIT Dept Mgmt 77 Massachusetts Ave Cambridge MA 02139

MAGNAVITA, JEFFREY JOSEPH, psychologist; b. Phila., Oct. 13, 1953; s. Hugo and Elsie (Gross) M.; m. Anne Gardner, Sept. 14, 1991; 1 child, Elizabeth Gardner. BA in Psychology, Temple U., 1975; PhD in Psychology, U. Conn., 1981. Diplomate in counseling Am. Bd. Profl. Psychology. Doctoral asst. dept. counseling psychology Univ. Conn., Storrs, 1978-80; clin. psychology intern Elmcrest Psychiat. Inst., Portland, Conn., 1980-81; unit chief, psychologist Elmcrest Psychiat. Inst., Portland, 1981-84; dir., founder Glastonbury (Conn.) Psychol. Assocs., 1984—; affiliate med. staff Manchester (Conn.) Meml. Hosp., 1984—. Mem. APA, Nat. Register for Health Care Providers in Psychology. Achievements include rsch. publs. in short-term dynamic psychotherapy. Office: Glastonbury Psychol Assocs Glastonbury Med Arts Ctr 300 Hebron Ave Ste 215 Glastonbury CT 06033

MAGNELI, ARNE, chemist; b. Stockholm, Dec. 6, 1914; s. Agge and Valborg (Hultman) M.; m. Barbro Wigh, Aug. 10, 1946; children: Christina, Lars, Peter. PhD, U. Stockholm, 1942; DSc, U. Uppsala, 1950; Dr.Hon.Causa, U. Pierre et Marie Curie, Paris, 1988. Asst. prof. U. Uppsala, 1950-53; assoc. prof. U. Stockholm, 1953-61, prof. chemistry, 1961-80, emeritus prof. chemistry, 1981—; sec. Nobel Com. for Physics, 1966-73, Nobel Com. for Chemistry, 1966-86; bd. dirs. Nobel Found., 1973-85. Contbr. articles to profl. jours. Recipient Georges Chaudron Gold medal French Soc. for High Temperatures and Refractories, 1990. Mem. Royal Swedish Acad. Scis. (Gregori Aminoff prize 1989), French Acad. Scis., Internat. Union Crystallography (pres. 1975-78). Home: Odensgatan 5A, 75315 Uppsala Sweden Office: Arrhenius Lab, Univ of Stockholm, 10691 Stockholm Sweden

MAGNUS, PHILIP DOUGLAS, chemistry educator. BS, Imperial Coll., England, 1965, PhD, 1968. With Ind. U., Bloomington, 1981—, now Disting. prof. chemistry. Office: Ind U Dept of Chemistry Bloomington IN 47405

MAGNUSON, CHARLES EMIL, physicist; b. Rushville, Nebr., Dec. 19, 1939; s. Ivan N. and Lena C. (Ray) M.; m. Denise Therese Maynard, Aug. 20, 1971; children: Ivan Geard, Curt Emil, Todd Maynard. BA, Nebr. Wesleyan U., 1962; MA, SUNY, Buffalo, 1965; PhD, Tex. A&M U., 1974. Cert. pers. protection and safety tng. Asst. prof. physics Carson Newman Coll., Jefferson City, Tenn., 1966-70; grad. teaching asst. Tex. A&M U., College Station, 1970-74, lectr., rsch. assoc. dept. physics, 1975, rsch. scientist indstrl. engring., 1975-80; sr. rsch. scientist Duke U., Durham, N.C., 1981-83; pres. Biosystems Techs. Inc., Durham, 1988—; sci. fellow KWBES, College Station, 1992—; cons. Swike Anderson & Assocs., Bryan, Tex., 1980—. Contbr. over 50 articles to profl. jours. Scoutmaster Boy Scouts Am., College Station, 1989—; precinct chmn. Rep. Party, Tex., 1978, 79, 82. Mem. AAAS, Am. Phys. Soc., Kiwanis (youth svcs. chmn.), Sigma Xi. Presbyterian. Home: PO Box 1955 College Station TX 77841-1955 Office: KWBES 500 Graham Rd College Station TX 77845

MAGUIRE, CHARLOTTE EDWARDS, retired physician; b. Richmond, Ind., Sept. 1, 1918; d. Joel Blaine and Lydia (Betscher) Edwards; m. Raymer Francis Maguire, Sept. 1, 1948 (dec.); children—Barbara, Thomas Clair II. Student, Stetson U., 1936-38, U. Wichita, 1938-39; B.S., Memphis Tchrs. Coll., 1940; M.D., U. Ark., 1944. Intern, resident Orange Meml. Hosp., Orlando, Fla., 1944-46; resident Bellevue Hosp. and Med. Ctr., NYU, N.Y.C., 1954, 55; instr. nurses Orange Meml. Hosp., 1947-57, staff mem., 1946-68; staff mem. Fla. Santarium and Hosp., Orlando, 1946-56, Holiday House and Hosp., Orlando, 1950-62; mem. courtesy and cons. staff West Orange Meml. Hosp., Winter Garden, Fla., 1952-67; active staff, chief dept. pediatrics Mercy Hosp., Orlando, 1965-68; med. dir. med. services and basic care Fla. Dept. Health and Rehab. Services, 1975-84; med. exec. dir., med. services div. worker's compensation Fla. Dept. Labor, Tallahassee, 1984-87; chief of staff physicians and dentists Central Fla. div. Children's Home Soc. of Fla., 1947-56; dir. Orlando Child Health Clinic 1949-58; pvt. practice medicine Orlando, 1946-68; asst. regional dir. HEW, 1970-72; pediatric cons. Fla. Crippled Children's Commn., 1952-70, dir., 1968-70; med. dir. Office Med. Services and Basic Care, sr. physician Office of Asst. Sec. Ops., Fla. Dept. Health and Rehab. Services; clin. prof. dept. pediatrics U. Fla. Coll. Medicine, Gainesville, 1980-87; mem. Fla. Drug Utilization Rev/. 1983-87; real estate salesperson, Investors Realty, 1982—. Mem. profl. adv. com. Fla. Center for Clin. Services at U. Fla., 1952-60; del. to Mid-century White House Conf. on Children and Youth, 1950; U.S. del from Nat. Soc. for Crippled Children to World Congress for Welfare of Cripples, Inc., London, 1957; pres of corp. Eccleston-Callahan Hosp. for Colored Crippled Children, 1956-58; sec. Fla. chpt. Nat. Doctor's Com. for Improved Med. Services, 1951-52; med. adv. com. Gateway Sch. for Mentally Retarded, 1959-62; bd. dirs. Forest Park Sch. for Spl. Edn. Crippled Children, 1949-54, mem. med. adv. com., 1955-68, chmn., 1957-68; mem. Fla. Adv. Council for Mentally Retarded, 1965-70; dir. central Fla. poison control Orange Meml. Hosp.; mem. orgn. com., chmn. com. for admissions and selection policies Camp Challenge; participant 12th session Fed. Exec. Inst., 1971; del. White House Conf. on Aging, 1980. Mem. Nat. Rehab. Assn., Am. Congress Phys. Medicine and Rehab., Fla. Soc. Crippled Children and Adults, Central Fla. Soc. Crippled Children and Adults (dir. 1949-58, pres. 1956-57), Am. Assn. Cleft Palate, Fla. Soc. Crippled Children (trustee 1951-57, v.p. 1956-57, profl. adv. com. 1957-68), Mental Health Assn. Orange County (charter mem.; pres. 1949-50, dir. 1947-52, chmn. exec. com. 1950-52, dir. 1963-65), Fla. Orange County Heart Assn., AMA, Am. Med. Women's Assn., Am. Acad. Med. Dirs., Fla. Med. Assn. (chmn. com. on mental retardation), Orange County Med. Assn., Fla. Pediatric Soc. (pres. 1952-53), Fla. Cleft Palate Assn. (counselor-at-large, sec.), Nat. Inst. Geneal. Rsch., Nat. Geneal. Soc., Assn. Profl. Genealogists, Tallahassee Geneal. Soc. Club: Governors. Home: 4158 Covenant Ln Tallahassee FL 32308-5765

MAGUIRE, MILDRED MAY, chemistry educator, magnetic resonance researcher; b. Leetsdale, Pa., May 7, 1933; d. John and Mildred (Sklarsky) Magura. BS in Chemistry, Carnegie-Mellon U., 1955; MS in Phys. Chemistry, U. Wis., 1960; PhD in Phys. Chemistry, Pa. State U., 1967. Devel. chemist Koppers Co., Monaca, Pa., 1955-58; rsch. chemist Am. Cyanamid Co., Stamford, Conn., 1960-63; asst. prof. chemistry Waynesburg

(Pa.) Coll., 1967-70, assoc. prof., 1970-74, prof., 1974—; Leverhulme vis. prof. U. Leicester, Eng., 1980-81, summer 1989; cons. Pitts. Energy Tech. Ctr., summers 1978-86, Oak Ridge Assoc. univs. faculty rsch. participant, summers 1978-80, 82-85. Contbr. articles to sci. jours., chpt. to book. Sec. Waynesburg Women's Club, 1981-82. Recipient Woman of the Yr. award AAUW, Waynesburg, 1983; Cottrell grantee Rsch. Corp. N.Y., 1970-71; Leverhulme vis. fellow U.K., 1980-81; Curie Internat. fellow AAUW, U.K., 1980-81. Mem. AAUP, Am. Chem. Soc. Avocations: gardening, painting, swimming, classical music, reading. Home: 1550 Crescent Hls Waynesburg PA 15370-1654 Office: Waynesburg Coll College St Waynesburg PA 15370-1318

MAHADEVA, WIJEYARAJ ANANDAKUMAR, information company executive; b. Colombo, Sri Lanka, Aug. 26, 1952; came to U.S., 1976; s. Balakumara and Sundareswari (Tyagaraja) M. BSEE, Cambridge U., 1973, MSEE, 1980; MBA, Harvard Bus. Sch., 1978. Rsch. and devel. engr. British Broadcasting Corp., London, 1973-76; sr. engagement mgr. McKinsey and Co., N.Y.C., 1978-85; corp. dir. AT&T, Bridgewater, N.J., 1985-89, Dun & Bradstreet Corp., N.Y.C., 1989—. Contbr. articles to profl. jours. Home: 201 E 87th St Apt 15-S New York NY 10128-3203

MAHAFFEY, RICHARD ROBERTS, information analyst; b. Norton, Va., Sept. 22, 1950; s. Donald Baker and Helen (Roberts) M.; m. Debra Lynn Carter, Aug. 16, 1977; 1 child, Landis Harmony. BS in Chemistry, Clinch Valley Coll., U. Va., Wise, 1973. Chemist UNIVAC, Bristol, Tenn., 1973-74; chemist Eastman Chem. Co., Kingsport, Tenn., 1974-75, systems analyst, 1975-87, mgr. rsch. computer svcs., 1987-91, info. analyst, 1991—; exec. bd. dirs. LIMS Inst., Pitts., 1989—; customer adv. bd. chem. industry Digital Equip. Corp., Maynard, Mass., 1988-91. Author: LIMS: Applied Information Technology in the Laboratory, 1990; No.Am. editor Lab. Info. Mgmt. Jour., 1990—; contbr. articles to profl. jours. Office: Eastman Chem Co PO Box 1974 Kingsport TN 37662

MAHAN, CLARENCE, federal agency administrator, writer; b. Dayton, Ohio, Jan. 1, 1939; s. Clarence Mahan and Elsie (Crouch) Diltz; m. Suky Mahan, May 27, 1962; children: Sean, Christine Elizabeth. BA, U. Md., 1963; MA, Am. U., 1968; MBA, Syracuse U., 1969. Dep. comtroller U.S. Army, Japan, 1974-76; dep. chief program and budget Defense Commn. Agy., Arlington, Va., 1976; aide Asst. Sec. Army, Washington, 1976-77; chief operating appropriations Dept. AF, Washington, 1979-80; dir. fin. and acctg. Dept. Edn., Washington, 1980-81; dep. comptroller Dept. Energy, Washington, 1981-82; dir. fiscal and contracts mgmt. EPA, Washington, 1982-83, dep. comptroller, 1983-85, dir. rsch. mgmt. program office, 1985—; instr., lectr. in field. Contbr. articles to profl. jours. With U.S. Army, 1959-62, Korea. Mem. Am. Iris Soc. (dir., 2d v.p. 1991—), Hist. Iris Preservation Soc. (pres. 1991-93), Soc. Japanese Irises (pres. 1989-92), Reblooming Iris Soc. (dir. 1986—). Democrat. Home: 7311 Churchill Rd Mc Lean VA 22101 Office: EPA Rsch Program Mgmt Office 401 M St SW Washington DC 20460-0002

MAHER, FRANCIS RANDOLPH, engineer, consultant; b. Memphis, Nov. 10; s. Francis R. and Laura (Walker) M.; m. Anecita Datuin Jabagat, May 16, 1983; children: Francis Rose, Andrew Carnegie, Francis J. Jr. Student, Fordham U., 1940-41, MIT, 1941-42; cert. in engring., U. Fla., 1944; BSCE, Far Eastern U., Manila, 1952; postgrad., Nat. U., Manila, 1953-56; MBE, Ateneo U., Manila, 1971. Mgr. Promenador Pub. Co., Manila, 1945-46, Atom Gas Co., Manila, 1946-48, Am. Wrought Iron & Metal, Manila, 1948-53; mem. constrn. dept. Foster Wheeler Corp., Batangas City, 1953; mgr. machines dept. J.P. Heilbronn Co., Makati, 1953-62; chmn. bd. Am. Steel Windows Inc., Manila, 1962-78, Maher Graphic Machinery, Inc., Manila, 1966-78, Maher Enterprises, Inc., Manila, 1978—; prof. Ateneo U., 1974—, Feati U., Manila, 1950-53; pres. Maher Enterprises, Wilmington, Del., 1988—. Mem. Am. Legion, Elks (trustee 1990—, bd. dirs. Cerebral Palsy Found. 1990—), Rotary Club (Mandaluyong and San Juan), Masons (32 degree). Republican. Episcopalian. Avocations: swimming, bowling, disco dancing, sightseeing, reading. Home and Office: Maher Enterprises, 831 B Balagtas St, Mandaluyong Manila 1501, The Philippines

MAHER, JOHN FRANCIS, physician, educator; b. Hempstead, N.Y., Aug. 3, 1929; s. William Lawrence and Marie Elizabeth (Duffy) M.; m. Margaret Helen Ulincy, Oct. 24, 1953; children: John Michael, Andrew, Mary, George, Paul Duffy. BS, Georgetown U., 1949, MD, 1953. Diplomate Am. Bd. Internal Medicine. Intern Boston City Hosp., 1953-54; resident Georgetown U. Med. Ctr., Washington, 1956-58, fellow in nephrology, 1958-60; instr. medicine Georgetown U., 1960-62, asst. prof., 1962-68, assoc. prof., 1968-69; prof. U. Mo. Med. Ctr., Columbia, 1969-74, U. Conn. Health Sci. Ctr., Farmington, 1974-79, Uniformed Services Univ. Health Scis., Bethesda, Md., 1979—; dir. Georgetown Univ. Med. Ctr. Renal Clinic, 1962-69, Univ. Mo. Med. Ctr. Nephrology Div. and Clin. Research Ctr., 1969-74, Univ. Conn. Health Ctr. Nephrology Div. and Clin. Research Ctr., 1974-79, Uniformed Service Univ. Health Scis. Nephrology Div., 1979—. Author: Uremia, 1961, The Kidney, 1971; editor: Replacement of Renal Function by Dialysis, 1978; contbr. more than 200 articles to profl. jours. Served to capt. USAF, 1954-56. Recipient Karol Marcinkowski award Univ. Poznan, Poland, 1985. Fellow ACP (mem. editorial bd. 1975-78), Royal Coll. Physicians of Ireland (hon.); mem. Am. Soc. Nephrology (audit com. chmn. 1973), Am. Soc. Artificial Internal Organs (pres. 1975-76), Internat. Soc. Nephrology (mem. program com. 1966, nominating com. 1975), Am. Fedn. Clin. Rsch. (editor 1967-69). Republican. Roman Catholic. Home: 8104 Gainsborough Ct E Rockville MD 20854-4271 Office: Uniformed Services U Health Scis 4301 Jones Bridge Rd Bethesda MD 20814-4799 *Died Sept. 13, 1992.*

MAHER, LOUIS JAMES, JR., geologist, educator; b. Iowa City, Iowa, Dec. 18, 1933; s. Louis James and Edith Marie (Ham) M.; m. Elizabeth Jane Crawford, June 7, 1956; children: Louis James, Robert Crawford, Barbara Ruth. B.A., U. Iowa, 1955, M.S., 1959; Ph.D., U. Minn., 1961. Mem. faculty dept. geology and geophysics U. Wis.-Madison, 1962—, prof., 1970—, chmn. dept., 1980-84. Contbr. articles to profl. jours. Served with U.S. Army, 1956-58. Danforth fellow, 1955-61; NSF fellow, 1959-61; NATO fellow, 1961-62. Fellow AAAS, Geol. Soc. Am.; mem. Am. Quaternary Assn., Ecol. Soc. Am., Wis. Acad. Sci., Arts and Letters, Sigma Xi. Episcopalian. Office: U Wis Dept Geology and Geophysics Madison WI 53706

MAHER, ROBERT CRAWFORD, electrical engineer, educator; b. Cambridge, U.K., Jan. 24, 1962; came to U.S., 1962; s. Louis James Jr. and Elizabeth Jane (Crawford) M.; m. Lynn Marie Peterson, Aug. 17, 1985; children: Maxwell Lloyd, Henry Peterson. BSEE, Washington U., St. Louis, 1984; MSEE, U. Wis., 1985; PhD, U. Ill., 1989. Rsch. asst. U. Ill., Urbana, 1985-89; asst. prof. elec. engring. U. Nebr., Lincoln, 1989—. Contbr. articles to profl. jours. Mem. IEEE, Audio Engring. Soc. (grantee 1988), Acoustical Soc. Am., Am. Soc. Engring. Edn. Episcopalian. Achievements include research and development in digital audio signal processing and electronic and computer music. Office: Univ Nebr 209 N WSEC Lincoln NE 68588-0511

MAHESH, MAHADEVAPPA MYSORE, medical physicist, researcher; b. Mysore, Karnataka, India, Oct. 16, 1963; came to U.S., 1987; s. Mahadevappa and Girijamma Mysore. MS, Mysore U., 1986, Marquette U., 1988; PhD, Med. Coll. Wis., 1993. Asst. prof. Marimallappa's Jr. Coll., Mysore, 1986; teaching asst. Marquette U., Milw., 1987-88, rsch. asst., 1987-88; rsch. fellow Med. Coll. Wis., Milw., 1988-93; chief med. physicist Johns Hopkins Hosp., Balt., 1993—. Contbr. articles to profl. jours. Active Circle-K, Milw., 1987-90. Recipient Charles Lescrenier fellowship Gammex Inc., 1989-93. Mem. Am. Assn. Physicist in Medicine, Soc. Physics Students, Am. Assn. Physicians India. Office: Johns Hopkins Hosp Baltimore MD 21287

MAHESH, VIRENDRA BHUSHAN, endocrinologist; b. India, Apr. 25, 1932; came to U.S., 1958, naturalized, 1968; s. Narinjan Prasad and Sobhagyawati; m. Sushila Kumari Aggarwal, June 29, 1955; children: Anita Rani, Vinit Kumar. BSc with honors, Patna U., India, 1951; MSc in Chemistry, Delhi U., India, 1953, PhD, 1955; DPhil in Biol. Sci, Oxford U., 1958. James Hudson Brown Meml. fellow Yale U., 1958-59; asst. research prof. endocrinology Med. Coll. Ga., Augusta, 1959-63; assoc. research prof. endocrinology Med. Coll. Ga., 1963-66, prof., 1966-70, Regents prof. endocrinology, 1970-86, Robert B. Greenblatt prof., 1979—, chmn. endocrinology, 1972-86, chmn., Regents prof. physiology and endocrinology, 1986—; dir. Ctr. for Population Studies, 1971—; mem. reproductive biology study sect. NIH, 1977-81, mem. human embryology and devel. study sect. NIH, 1982-86, 90-93, chmn., 1991-93. Contbr. articles to profl. jours., chpts. to books; editor: The Pituitary, a Current Review, Functional Correlates of Hormone Receptors in Reproduction, Recent Advances in Fertility Research, Hirsutism and Virilism, Regulation of Ovarian and Testicular Function; mem. editorial bd. Steroids, 1963—, Jour. of Clin. Endocrinology and Metabolism, 1976-81, Jour. Steroid Biochemistry and Molecular Biology, 1991—, Assisted Reproductive Tech./Andrology, 1993—; mem. adv. bd. Maturitas, 1977-81. Recipient Rubin award Am. Soc. Study of Sterility, 1963, Billings Silver medal, 1965, Best tchr. award Med. Coll. Ga. Sch. Medicine freshman class of 1969, Outstanding Faculty award Med. Coll. Ga. Sch. Grad. Studies, 1981, Disting. Teaching award Med. Coll. Ga. Sch. Grad. Studies, 1988, Excellence in Rsch. award Grad. Faculty Assembly Med. Coll. Ga., 1987, 88, 89, 90, 91, Disting. Scientist award Assn. Scientists of Indian Origin in Am., 1989, Outstanding Faculty award Med. Coll. Ga. Sch. Medicine, 1992; NIH rsch. grantee, 1960—. Mem. Chem. Soc. (Eng.), Soc. Biochem. & Molecular Biol., Soc. Neurosci., Endocrine Soc., Soc. for Gynecologic Investigation, Internat. Soc. Neuroendocrinology, Soc. for Study Reproduction, Am. Physiol. Soc., Internat. Soc. Reproductive Medicine (pres. 1980-82), Soc. Exptl. Biology and Medicine, Am. Fertility Soc., Am. Assn. Lab. Animal Sci., N.Y. Acad. Scis., AAUP, Sigma Xi. Office: Med Coll of Ga Dept Physiology and Endocrinology Augusta GA 30912-3000

MAHJOUB, ELISABETH MUELLER, health facility administrator; b. Lorain, Ohio, Jan. 13, 1937; d. William Kurt and Elizabeth Sophia (Dietz) Mueller; m. Mohamed Salah Mahjoub, Aug. 15, 1972 (div. Aug. 1981); 1 child, Ramsy Mahjoub. AB in Microbiology, Miami U., Oxford, Ohio, 1959; MPH, Yale U., 1967. Microbiology technologist Univ. Hosps., Cleve., 1959-60; rsch. asst. Harvard Sch. Pub. Health, Boston, 1960-65; trainee internat. family planning population coun. Ministry of Health, Tunis, Tunisia, 1967-69; cons., expert family planning population control Ministry of Edn., Tunis, 1969-71; cons. pub. health edn. Project Hope-Sch. of Pub. Health, Tunis, 1972-78; educator pub. health rural water, project coord. Care Medico, Tunis, 1978-81; house dir. Oberlin (Ohio) Coll., 1982-85; coord. spl. projects Elyria (Ohio) Meml. Hosp., 1985-92, hosp. found. coord., 1993—. Author: (with others) Change in Tunisia, 1976. Sec. softball com. Oberlin Youth Baseball, 1988-90; mem. Oberlin Strategic Plan Recreation Task Force, 1990-91, FME Ch. Fin. Adv. Com., 1990-91; bd. mem. Elyria Sr. Citizen Ctr. Adv., 1990-91; active OBHS Sports Booster Club. Methodist. Avocations: choral singing, downhill skiing, hiking. Home: 614 Beech St Oberlin OH 44074-1417 Office: Elyria Meml Hosp & Med Ctr 630 E River St Elyria OH 44035-5902

MAHL, GEORGE JOHN, III, electrical engineering consultant; b. Memphis, May 14, 1944; s. George John Jr. and Shirley (Parker) M.; m. Yvonne Rita Jones, Mar. 11, 1967; children: Michele Marie, George Brian, Ann Marie. BSEE, La. State U., Baton Rouge, 1967. REgistered profl. engr., La., Tex., Ala., Nev., Calif. Elec. engr. Freeport-McMoRan, Inc., New Orleans, 1967-73; project engr. Waldemar S. Nelson & Co., Inc., New Orleans, 1973-77; owner Mahl & Assocs., Inc., Harahan, La., 1978—. Chmn. Planning and Zoning Bd., Harahan, 1985-89; commr. East Jefferson Levee Dist., Jefferson, La., 1989-92; founder Compact Six-Six Parish Edn. Partnership, SE La., 1986. Named Rotarian of Yr., Harahan Rotary Club, 1992. Mem. IEEE (sr.), ASME (chmn. New Orleans sect. 1989-90, Svc. award 1991), La. Engring. Soc. (pres. New Orleans chpt. 1992-93), Instrument Soc. Am. (sr., pres. New Orleans sect. 1984-85, Svc. award 1991). Roman Catholic. Home: 7217 Westminster Dr Harahan LA 70123

MAHLER, DAVID, chemical company executive; b. San Francisco; s. John and Jennie (Morgan) M.; PhC, U. So. Calif., 1932; children: Darrell, Glenn. Pres., United Drug Co., Glendale, Calif., 1934-37, Blue Cross Labs., Inc., Saugus, Calif., 1937—. Active Fund for Animals, Friends of Animals, Com. for Humane Legislations; patron Huntington Hartford Theatre, Hollywood, Calif. Mem. Packaging and Rsch. Devel. Inst. (hon.), Anti-Defamation League, Skull and Daggar, Rho Pi Phi. Office: 26411 Golden Valley Rd Santa Clarita CA 91350-2988

MAHLMAN, JERRY DAVID, research meteorologist; b. Crawford, Nebr., Feb. 21, 1940; s. Earl Lewis and Ruth Margaret (Callendar) M.; m. Janet Kay Hilgenberg, June 10, 1962; children—Gary Martin, Julie Kay. A.B., Chadron State Coll., Nebr., 1962; M.S., Colo. State U., 1964, Ph.D., 1967. Instr. Colo. State U., Fort Collins, 1964-67; from asst. prof. to assoc. prof. Naval Postgrad. Sch., Monterey, Calif., 1967-70; rsch. meteorologist NOAA Geophys. Fluid Dynamics Lab., Princeton, N.J., 1970-84, lab. dir., 1984—; lectr. with rank of prof. Princeton U., 1980—; lectr. with rank prof. Princeton U., 1980—; chmn. panel on mid-atmosphere program NAS-NRC, 1982-84, mem. climate rsch. com., 1986-89, mem. panel on dynamic extended range forecasting, 1987-90; mem. U.S.-USSR Commn. on Global Ecology, 1989-92; mem. Bd. on Global Change, 1991—; U.S. rep. world climate rsch. programm Joint Sci. Commn., 1991—. Contbr. over 70 articles to profl. jours. Elder Monterey Presbyterian Ch., 1968-70, Lawrence Road Presbyn. Ch., Lawrenceville, N.J., 1972-75; bd. dirs. Lawrence Non-Profit Housing Inc., 1978-88. Recipient Disting. Authorship award Dept. Commerce, 1980, 81, Gold medal, 1986; Disting. Service award Chadron State Coll., 1984. Fellow Am. Geophys. Union (Jule Charney lectr. 1993), Am. Meteorol Soc. (awards com. 1984, chmn. upper atmosphere com. 1979, assoc. editor Jour. Atmospheric Sci. 1979-86, councilor 1991—, Editor's award 1978); mem. Sigma Xi. Democrat. Home: 9 Camelia Ct Lawrenceville NJ 08648-3201 Office: Princeton U Geophys Fluid Dynamics Lab PO Box 308/NOAA Princeton NJ 08542-0308

MAHMUD, SYED MASUD, engineering educator, researcher; b. Faridpur, Bangladesh, Dec. 29, 1955; came to U.S., 1980; s. Syed Mohammad Abdul and Begum (Marium) Ghani; m. Nadira Begum Mahmud, Aug. 22, 1983; children: Neeaz Mahdee, Irene Sarah. BS, Bangladesh U. Engring. & Tech., Dhaka, 1978; PhD, U. Wash., 1984. Teaching & rsch. asst. U. Wash., Seattle, 1980-82, teaching assoc., 1982-84; asst. prof. Oakland U., Rochester, Mich., 1984-88; asst. prof. Wayne State U., Detroit, 1988-91, assoc. prof., 1991—; tchr., rschr. Ford Motor Co., Dunton, Eng., 1991; tech. cons. Dimango Products Corp., Brighton, Mich., 1986-88. Author 1 book; contbr. 45 articles to profl. jours. and conf. proceedings. Recipient 4 grants Ford Motor Corp., 1989-93. Mem. IEEE, Engring. Soc. Detroit, Sigma Xi. Office: Wayne State U Elec Engring Dept 5050 Anthony Wayne Dr Detroit MI 48202

MAHNKE, KURT LUTHER, psychotherapist, clergyman; b. Milw., Feb. 18, 1945; s. Jonathan Henry and Lydia Ann (Pickron) M.; m. Dana Moore, Mar. 19, 1971; children: Rachel Lee, Timothy Kurt, Jonathan Roy. BA, Northwestern Coll., Watertown, Wis., 1967; MDiv, Wis. Luth. Sem., 1971; MA, No. Ariz. U., 1984. Cert. mental health counselor, Wis. Pastor Redeemer/Grace Luth. Chs., Phoenix & Casa Grande, Ariz., 1971-75, St. Philips Luth. Ch., Milw., 1975-78, 1st Luth. Ch., Prescott, Ariz., 1978-82; counselor NAU Counseling/Testing Ctr., Flagstaff, Ariz., 1983-84, Wis. Luth. Child & Family Svc., Wausau, Wis., 1984-86; area adminstr. Wis. Luth. Child & Family Svc., Appleton, Wis., 1986-89; founder, psychotherapist Family Therapy & Anxiety Ctr., Menasha, Wis., 1989—; part-time min. St. Paul Luth. Ch., Appleton, 1993—; presenter 13th Nat. Conf. on Anxiety Disorders, Charleston, S.C., 1993; cons. editor Northwestern Pub. House, Milw., 1990—. Cons. editor Counseling at the Cross, 1990; contbr. articles to profl. publs. Cons. Wis. Evang. Luth. Synod, Milw., 1986—; cons. crisis counselor Fox Valley Luth. High Sch., Appleton, also mem. acad. program com.; crisic counselor Critical Incident Stress Debriefing Team, Fox Cities, 1991—; mem. parent and family com. Fox Valley Unites; mem. exec. com. Fox Cities Community Counsel. Mem. ACA, Am. Mental Health Counselors Assn., Anxiety Disorders Assn. Am.. (charter), Internat. Assn. Marriage and Family Counselors, Assn. Specialists in Group Work, Obsessive Compulsive Found., Wis. Outpatient Mental Health Facilities. Republican. Lutheran. Avocations: fishing, camping, hunting, boating, officiating sports events. Office: Family Therapy/Anxiety Ctr 1477 Kenwood Ctr Menasha WI 54952-1160

MAIDES-KEANE, SHIRLEY ALLEN, psychologist; b. Roanoke Rapids, N.C., Sept. 16, 1951; d. John Thomas and Mary Shirley (Allen) Maides; m. John Thomas Keane, Nov. 8, 1980; children: Josephine Claire Keane, Michael Allen Keane. BA, U. N.C., 1973; MA, Vanderbilt U., 1975, PhD, 1979. Intern U. Calif., Davis, 1977-78; fellow Michael Reese Hosp and Med. Ctr., Chgo., 1978-79; psychologist Chgo., 1980-81, pvt. practice, 1981—; mem. allied profl. staff Linden Oaks Hosp.; adj. faculty Ill. Sch. Profl. Psychology, 1980-82; mgmt. cons. Mgmt. Sci. Assocs., 1982-86. Contbr. articles to profl. jours. Mem. APA, Ill. Psychol. Assn., Assn. DuPage County Psychologists in Profl. Practice (past pres.). Methodist. Office: 1100 Jorie Blvd Ste 255 Oak Brook IL 60521

MAIER, GERALD JAMES, natural gas transmission and marketing company executive; b. Regina, Sask., Can., Sept. 22, 1928; s. John Joseph and Mary (Passler) M. Student, Notre Dame Coll. (Wilcox), U. Man., U. Alta., U. Western Ont. With petroleum and mining industries Can., U.S., Australia, U.K.; responsible for petroleum ops. Africa, United Arab Emirates, S.E. Asia; chmn., CEO TransCan. PipeLines, Toronto, 1985—, also bd. dirs., chmn., pres., ceo; bd. dirs. BCE Inc., Bank of N.S., TransAlta Utilities Corp., Du Pont Can. Inc., Alberta Nat. Gas Co., Ltd., Great Lakes Gas Transmission Co.; chmn. Can. Nat. com. for World Petroleum Congresses, Van Horne Inst. for Internat. Transp.; bd. govs. Bus. Coun. on Nat. Issues; chmn. bd. dirs. Western Gas Mktg. Ltd. Named Hon. Col. King's Own Calgary Regt., Resource Man of Yr. Alta. Chamber of Resources, 1990; recipient Can. Engr.'s Gold medal Can. Coun. Profl. Engrs., 1990, Disting. Alumni award U. Alberta, 1992, Mgmt. award McGill U., 1993. Fellow Can. Acad. Engring.; mem. Am. Gas Assn. (bd. govs.), Can. Petroleum Assn. (bd. govs.), Assn. Profl. Engrs., Geologists and Geophysicists Alta (past pres.), Can. Inst. Mining and Metallurgy (Past President's Meml. medal 1971), Interstate Natural Gas Assn. Am. (bd. govs.). Avocations: golf, downhill skiing, shooting, fishing. Office: TransCan PipeLines Ltd, PO Box 1000 Sta M, Calgary, AB Canada T2P 4K5

MAIER, HENRY B., environmental engineer; b. Yonkers, N.Y., July 11, 1925; s. Henry and Adelaide (Boyce) M.; m. Elizabeth A. Maier, May 4, 1968. BA, Columbia U., 1947; postgrad., Adelphi U., Hofstra U. Prin. Maier Solar Developments, Hempstead, N.Y. Author: Techniques for Seascape Painting. Mem. Am. Chem. Soc., N.Y. Acad. Scis. Achievements include patents for elapsed time indicator, multiple reflecting solar collecting system, electroresponsive coatings, fusion power pellets, and fusion power; design of initial stage of work for aerospace vehicle; development of rapid method for perspective visualizations, for views of engineering and design concepts; definition of geometrics for placement of measuring points by approximation; research on inorganic sulfur and chlorine pollutants from combustion of fossil fuels and from incinerator processes, and their interactive roles in the progressive deterioration of the stratospheric ozone shield previously blocking frequencies in the infrared, far infrared and microwave frequencies, with particular regard to the prediction and pattern formation of major North Atlantic storm systems. Home: 6 Sealey Ave Apt 3K Hempstead NY 11550-1232

MAIESE, KENNETH, neurologist; b. Audubon, N.J., Dec. 5, 1958; s. Charles and Margaret (Fioretti) M. BA summa cum laude, U. Pa., 1981; MD, Cornell U., 1985. Intern N.Y. Hosp., 1985-86, resident in neurology, 1986-89, asst. attending physician, 1989—; asst. prof. Cornell U. Med. Coll., N.Y.C., 1989—; dir. neurolog. diagnosis N.Y. Hosp., 1991—. Author: Neurology and General Medicine, 1989, Neurological and Neurosurgical ICU Medicine, 1988; contbr. articles to Neurology, Jour. Cerebral Blood Flow and Metabolism, Jour. Intensive Care Medicine, Jour. Neuroscience. Joseph Collins scholar, 1981-85, Teagle Found. scholar, 1982, Grupe Found. scholar, 1985; NIH grantee, 1990—; recipient Young Scientist award Jour. Cerebral Blood Flow, 1991, Hoechst Investigator award, 1993. Mem. Am. Acad. Neurology, N.Y. Acad. Scis., Assn. for Rsch. in Nervous and Mental Diseases, Am. Neurol. Assn. Roman Catholic. Achievements include rsch. in imidazole receptors, cerebral ischemia, nitric oxide toxicity, growth factor neuroprotection. Office: Cornell New York Hosp Dept Neurology Rm A579 525 E 68th St New York NY 10021-4873

MAILANDER, MICHAEL PETER, biological and agricultural engineering educator; b. Denver, Nov. 20, 1954; s. Leo R. and Marie J. Mailander. BS, Regis U., 1977; PhD, Purdue U., 1985. Rsch. asst. Purdue U., West Lafayette, Ind., 1977-81, rsch. assoc., 1981-84; asst. prof. La. State U., Baton Rouge, 1985-90, assoc. prof. biol. and agrl. engring., 1991—; cons. So. U., Baton Rouge, 1989. Contbr. articles to profl. jours., chpt. to book. Mem. SAE (Speaker Hall of Fame 1981), ASM, Am. Soc. Agrl. Engrs., Sigma Xi. Office: La State U 167 Doran Baton Rouge LA 70803

MAIN, MYRNA JOAN, mathematics educator; b. Kirksville, Mo., Oct. 31, 1947; d. Stanford H. and Jennie Vee (Nuhn) Morris; m. Carl Donet Main, Feb. 22, 1968; children: D. Christopher, Laura S. BSE, Northeast Mo. State U., 1968, MA, 1970. Instr. math. Callao (Mo.) Sch., 1968-73; tchr., chair dept. math. Macon (Mo.) R-I Schs., 1973—; extension staff Moberly (Mo.) Area C.C., 1983—; adj. faculty mem. Northeast Mo. State U., Kirksville, 1987-93; mentor Mo. Math. Mentoring Project, Moberly, 1989—. Organist, UBS tchr. Crossroads Christian Ch., Macon, 1981—; troop #503 leader Becky Thatcher coun. Girl Scouts U.S. Recipient Presdl. award for Excellence in Math., 1989. Mem. AAUW (chpt. pres. 1980-81), Nat. Coun. Tchrs. Math., Mo. Coun. Tchrs. Math. (treas. 1978-79, v.p. 1976), Mo. Alliance for Sci., Math. and Tech. Edn. (bd. dirs. 1988—), Mo. Math. Coalition (bd. dirs. 1988-93), Math. Assn. Am., Phi Delta Kappa, Delta Kappa Gamma. Democrat.

MAINHARDT, DOUGLAS ROBERT, chemical engineer; b. Huntington, N.Y., Jan. 3, 1961; s. Joseph Henry and Olive (Stewart) M. BS in Biomed. Engring., N.Y. Inst. Tech., 1988. Analyst Pall Corp., Glen Cove, N.Y., 1988, test engr., 1988-90, sr. test engr., 1990-92, field engr., 1992—. Office: Pall Corp 30 Sea Cliff Ave Glen Cove NY 11542

MAIORIELLO, RICHARD PATRICK, otolaryngologist; b. Phila., Mar. 17, 1936; s. Gesumino Theodore and Angelina (Del Rossi) M.; A.B., U. Pa., 1960; M.D., Jefferson Med. Coll., 1964; M.S., Thomas Jefferson U., 1972; m. Susan Hemenway, Mar. 6, 1979; children—Gabriel, Angela, Richard. Commd. 2d lt., U.S. Air Force, 1963, advanced through grades to col., 1977; ret., 1979; intern Keesler Hosp., 1965-67; chief flight medicine USAF Base, Bitburg, W. Ger., 1965-68; resident in otolaryngology Thomas Jefferson Hosp., Phila., 1968-71, 72-73; fellow in physiology Thomas Jefferson U., 1971-72; dir. med. edn. Andrews AFB, 1974-78; assoc. prof. uniformed services Univ. Health Scis., 1978-79; assoc. prof. Northeastern Ohio U. of Medicine, 1983—; mem. staff Aultman Hosp., 1979—; assoc. staff Timken Mercy Med. Ctr., 1981—, Union Hosp., 1988—; cons. otolaryngology to Surgeon Gen., 1977—; pres. Mid-Ohio Dressage Assn. Served with USNR, 1954-58. Decorated Air Force Commendation medal; diplomate Nat. Bd. Med. Examiners, Am. Bd. Otolaryngology. Fellow ACS, Am. Soc. Head and Neck Surgery; mem. Am. Acad. Otolaryngology, Am. Acad. Facial Plastic and Reconstructive Surgery, Am. Assn. Cosmetic Surgery, Vail Cosmetic Surg. Soc., Hanoverian Soc. (exec. v.p.), U.S. Dressage Fedn. (chmn. all-breeds coun.), Centurion Club. Republican. Roman Catholic. Office: 1445 Harrison St NW Canton OH 44708

MAIR, BRUCE LOGAN, interior designer, company executive; b. Chgo., June 5, 1951; s. William Logan and Josephine (Lee) M. BFA, Drake U., 1973; postgrad., Ind. Wesleyan U., 1990—. Mgr., head designer Reifers of Indpls., 1973-79; pres. Interiors Internat., Indpls., 1979-87; sr. designer Kasler Group, Indpls., 1987-89; dir. devel. Tillery Interiors and Imports, Greenwood, Ind., 1990; v.p. Tillery Interiors and Imports, Indpls., 1990-92; owner Mair Interior Design Group, Indpls., 1992—; pres. Tokens Inc., Indpls., 1982-88, Meg-A-Wat Enterprises Inc., Indpls., 1985-87, Luxury Ice Creams Inc., Indpls., 1986-87. Cover designer Indpls. Home and Garden mag., 1978, feature designer 1980; feature designer Builder mag., 1979; co-designer feature Indpls. At Home mag., 1979. Campaigner Anderson for Pres., 1980. Mem. Am. Soc. Interior Designers (treas. Ind. chpt. 1982-83, Pres. awards 1981-82), U.S. Rowing Assn. (master 1987—), St. Joseph Hist. Neighborhood Assn., Columbia Club (rowing crew coxswain 1986—), Highland Model A Club, Alpha Epsilon Pi. Avocations: sculling, historic preservation, model A Ford restoration, fishing, farming. Home: 219 E 10th

St Indianapolis IN 46202-3303 Office: Mair Interior Design Group 219 E 10th St Indianapolis IN 46202

MAIR, WILLIAM AUSTYN, aeronautical engineer; b. Epsom, Surrey, Eng., Feb. 24, 1917; s. William and Catharine Millicent (Fyfe) M.; m. Mary Woodhouse Crofts, Apr. 15, 1944; children: Christopher William, Robert James. BA, Cambridge (Eng.) U., 1939; DSc (hon.), Cranfield Inst. Tech., 1990. Chartered engr., Eng. Dir. fluid motion lab. U. Manchester, Eng., 1946-52; prof. aero. engring. U. Cambridge, 1952-83, head engring. dept., 1973-83; engring. cons., 1983—; mem. Def. Sci. Adv. Coun., Eng.; mem. various coms. Aero. Rsch. Coun., Eng., 1946-80. Author: (with D. L. Birdsall) Aircraft Performance, 1992; contbr. articles to profl. publs. With RAF, 1940-46. Downing Coll. U. Cambridge fellow, 1953-83, hon. fellow, 1983. Fellow AIAA (assoc.) Royal Acad. Engring., Royal Aero. Soc. (Silver medal 1975). Home: 7 The Oast House, Pinehurst Grange Rd, Cambridge CB3 9AP, England

MAIROSE, PAUL TIMOTHY, mechanical engineer, consultant; b. Mitchell, S.D., Aug. 4, 1956; s. Joseph E. and Phyllis R. (Glissendorf) M.; m. Connie L. Nickell, Apr. 1, 1989 (dec. June 8, 1992); m. Donna M. Ward, Sept. 10, 1993. BSME, S.D. Sch. Mines and Tech., 1978; postgrad., Tulane U., 1986. Registered profl. engr., Wash. Mech. engr. UNC Nuclear Industries, Richland, Wash., 1979-80, Wash. Pub. Power Supply System, Richland, 1980-85, 89; cons. La. Power & Light Co., New Orleans, 1985-86, Erin Engring. & Rsch. Inc., Walnut Creek, Calif., 1986-87, Sacramento Mcpl. Utility Dist., 1987-89; mech. engr. GE, Portland, Oreg., 1989-90; sr. cons. Rocky Flats Project Cygna Energy Svcs., 1990-91; v.p. mktg. Data Max, 1991—; pvt. practice cons. engr. Vancouver, Wash., 1991—; project engr. MatTec, Inc., Richland, Wash., 1990-91; pres. Project Tech. Mgmt., 1990—; chief engr. S.W. Air Pollution Control Authority, Vancouver, Wash., 1992—. Co-author: Topical Report on Extreme Erosion at Yucca Mountain, Nevada, 1993. Mem. polit. action com. Sacramento Mcpl. Utility Dist., 1988. Mem. ASME (assoc.), ASHRAE (assoc.), Aircraft Owners and Pilots Assn., Profl. Assn. Diving Instrs., Air & Water Mgmt. Assn., Sierra Club, Bards of Bohemia. Republican. Roman Catholic. Avocations: foreign travel, hiking, bicycling, private piloting, scuba diving. Home: 1610 NW 137th St Vancouver WA 98685-1513

MAJDA, ANDREW J., mathematician, educator; b. East Chicago, Ind., Jan. 30, 1949; m. Gerta Keller. BS, Purdue U., 1970; MS, Stanford U., 1971, PhD, 1973. Instr. Courant Inst. NYU, 1973-76; from asst. prof. to assoc. prof. U. Calif., L.A., 1976-78, prof., 1978; prof. U. Calif., Berkeley, 1979-84; vis. prof. Princeton (N.J.) U., 1984-85, prof., 1985—. Recipient medal of college de France, 1982, John von Neumann award Soc. for Indsl. and Applied Math., 1990, Applied Math. and Numerical Analysis award NAS, 1992; Alfred P. Sloan Found. fellow, 1977-79. Office: Princeton U Dept Math Fine Hall Princeton NJ 08544

MAJDALAWI, FOUAD FAROUK, aeronautical engineer; b. Beirut, Lebanon, May 28, 1969; s. Farouk Said and Hiyam Mohammad M. BS in Aeronautical Engring., Embry-Riddle Aeronaut. U., Daytona Beach, Fla., 1988; MS in Aeronautics, Imperial Coll., London, 1990. Tng. engr. Royal Jordanian Airline, Amman, Jordan, 1987; tng. aerodynamicist Airbus Industrie, Toulouse, France, 1987; aero. engr. Brit. Aerospace Airbus Ltd., Bristol, Eng., 1990—. Mem. AIAA. Muslim. Achievements include patent (with others) on the process of boundary layer transition from laminar to turbulent state and laminar flow control technology. Office: PO Box 1819, Amman Jordan

MAJEED, ABDUL, mathematics educator; b. Tanda, Punjab, Pakistan, Nov. 11, 1937; parents: Noor Muhammad and Fazal Chaundhry; m. Hameeda Majeed, July 25, 1959; children: Kausar, Shahid, Tahir, Athar, Shazia, Saima. BA with honors, Punjab U., Lahore, 1959, MA, 1961; MA, Australian Nat. U., Canberra, 1966; PhD, Carleton U., Ottawa, Ontario, Can., 1974. Lectr., asst. prof., assoc. prof., prof. Punjab U., 1982—, chmn., math. dept., 1979-81, 82—. Author: Theory of Groups, 1988 (1st prize), Elements of Topology and Fuctional Analysis, 1990. Postdoctoral fellow Carleton U., 1978, 82. Mem. Am. Math. Soc. (reviewer 1974—), Pakistan Math. Assn., Punjab Math. Soc. (nat. review com. 1983—), Punjab U. Acad. Staff Assn. Home: 8-B Quaid-e-Azam Campus, Lahore 54590, Pakistan Office: Punjab U Dept Math, Quaid-e-Azam Campus, Lahore 54590, Pakistan

MAJEWSKI, THEODORE EUGENE, chemist; b. Boonton, N.J., July 5, 1925; s. Witold Charles and Felixa (Tkacz) M.; m. Cynthia Ann Davis, Sept. 26, 1953; children: Andrea, Theodore, Steven, Felicia, Cynthia, Melissa. BA, Syracuse U., 1951; MS, U. Del., 1953, PhD, 1960. Chemist Dow Chem. Co., Midland, Mich., 1957-69; rsch. chemist Philip Morris USA, Richmond, Va., 1969-92; ret., 1992; cons. Herald Pharmacal, Richmond, 1979-81. Contbr. articles to profl. jours.; patentee in field. Bd. dirs. Boy Scouts Am., Richmond, 1957-91. With USN, 1943-46, PTO. Recipient Silver Beaver award Boy Scouts Am., 1980. Mem. Am. Chem. Soc., AAAS, Alpha Ci Sigma. Avocations: travel, fishing, reading, camping. Home: PO Box 8117 Duck NC 27949-9999

MAJI, ARUP KANTI, civil engineering educator; b. Calcutta, Bengal, India, Mar. 2, 1962; came to U.S., 1983; s. Paritosh and Lakshmi (Samanta) M.; m. Manidipa Dutta, Dec. 15, 1987. B Tech. with honors, Indian Inst. Tech. Kharagpur, India, 1983; MS in Structural Engring., U. Miami, Coral Gables, Fla., 1984; PhD in Civil Engring., Northwestern U., Evanston, Ill., 1989. Teaching asst. U. Miami, Coral Gables, 1983-84; rsch. asst. Northwestern U., Evanston, 1984-88; asst. prof. civil engring. U. N.Mex., Albuquerque, 1988—; cons. to several local R&D orgns., 1990-92. Contbr. over 70 articles to profl. jours., procs. and reports. Mem. ASCE (sec. subcom. exptl. analysis 1991-92), Soc. Exptl. Mechanics (sec. structural testing 1991-92), Am. Concrete Inst., Am. Soc. Nondestructive Testing, Sigma Xi, Tau Beta Pi, Phi Kappa Phi. Achievements include prin. investigator for over one million dollars of rsch. grants from the NSF, Dept. of Energy, others. Office: U NMex Dept Civil Engring Albuquerque NM 87131

MAJOR, THOMAS D., academic program director; b. South Rim Grand Canyon, Ariz., Aug. 21, 1954; s. Thomas LeGrand and Delna (Farnsworth) M.; m. Vikki Vincent, Aug. 30, 1975; children: Vincent, Madison. BS in Chemistry, So. Utah State U., 1976; MSChemE, Brigham Young U., 1978; MBA, So. Ill. U., 1981. Registered profl. engr. Design engr. Monsanto Co. Corp. Engrs., St. Louis, 1978-81; prodn. engr. Monsanto Oil Co., Denver, 1981-83; asst. v.p. E.F. Hutton, Denver, 1983-85; engring. mgr. MAZE Exploration, Denver, 1985-87; dir. U. Utah Tech. Transfer Office, Salt Lake City, 1987—; bd. dirs. Western Inst. Biomed. Rsch., Salt Lake City. Disting. scholar So. Utah U., 1976. Mem. Assn. Univ. Tech. Mgrs., Lic. Exec. Soc. Office: U Utah Tech Transfer 421 Wakara Way Ste 170 Salt Lake City UT 84108

MAJUMDAR, ARUNAVA, mechanical engineer, educator. BTech, Indian. Inst. Tech., Bombay, 1985; MS in Mech. Engring., U. Calif., Berkeley, 1987, PhD in Mech. Engring., 1989. Rsch. asst. U. Calif., Berkeley, 1985-89; asst. prof. Ariz. State U., 1989-92, U. Calif., Santa Barbara, 1992—; co-chmn. U.S.-Japan seminar on Molecular and Microscale Transport Phenomena, NSF, 1993; co-chmn. 22nd Internat. Thermal Conductivity Conf., Ariz. State U., Tempe, 1993. Reviewer Internat. Jour. Heat and Mass Transfer, Internat. Jour. Wear, NSF, Am. Chem. Soc. Petroleum Rsch. Fund, Solid State Electronics, Am. Inst. Physics, Soc. Photo-Instrumentation Engrs., Biotechnology Progress; contbr. over 20 articles to sci. jours. Scroeder-Scovill-Duncan scholar Indian Inst. Tech., 1982-84, D. K. Merchant scholar, 1984-85; Regents fellow U. Calif., Berkeley, 1985-86; recipient Young Investigator award NSF, 1992—; grantee NSF, 1990—, 91— (two grants), 92— (two grants). Mem. ASME (sec. K-8 com. fundamentals heat transfer heat transfer divsn., reviewer Jour. Heat Transfer, Jour. Tribology, Jour. Applied Mechanics, Melville medal 1992), AAAS, Am. Vacuum Soc., Materials Rsch. Soc. Achievements include research in heat generation and transport in nanometer scale devices and structures, nanomechanics of ductile grinding of brittle materials, contact mechanics of surfaces and application to micro-tribology, thermal and mechanical property measurement of very thin films. Office: Univ of Calif Dept of Mech & Envir Eng Santa Barbara CA 93106*

MAJUMDER, SABIR AHMED, biophysicist; b. Chandpur, Comilla, Bangladesh, July 15, 1957; came to U.S., 1986; s. Quashem Majumder and Momtaz Begum; m. Hamida Khanam, Dec. 15, 1985; children:Faryha, Nabilah. BS in Chemistry with honors, U. Dhaka, Bangladesh, 1981, MS, 1983; MS, Duquesne U., 1988; PhD, U. New Mex., 1993. Corr. The Daily Janapad, Dhaka, Bangladesh, 1973-74; rsch. fellow U. Dhaka, 1983-84, lectr. in Chemistry, 1984-86; teaching asst. Duquesne U., Pitts., 1986-88, U. N.Mex., Albuquerque, 1988-90; Assoc. Western Univs. grad. lab. fellow Sandia Nat. Labs., Albuquerque, 1991—. Contbr. articles to profl. jours. Gen. Sec. Bangladesh Youth Coun., Dhaka, 1982; mem. Nat. Student League Ctrl., Dhaka, 1983. Trainee Youth Leadership Tng. Inst., Singapore, 1981; recipient Link Energy Fellowship Link Found., Rochester, N.Y., 1987. Mem. AAAS, Am. Chem. Soc., Biophys. Soc., Bangladesh Chem. and Biol. Soc. N.Am. (elected gen. sec. 1993-94). Democrat. Muslim. Achievements include patent disclosure: photocatalytic degradation of aromatic compounds by metalloporhyrins adsorbed into alumina using visible light. Home: 941 Buena Vista Dr SE G204 Albuquerque NM 87106 Office: Sandia Nat Labs Fuel Sci Dept 6211 Albuquerque NM 87185

MAK, KOON HOU, cardiologist, researcher; b. Singapore, June 23, 1961; s. Kok Thye and Ayen (Fong) M.; m. Li-Hwei Sng, May 1, 1993. B of Medicine, B of Surgery, Nat. U. of Singapore, 1985, M of Medicine (internal medicine), 1989; diploma, Singapore Bible Coll., 1990. House officer Ministry of Health, Singapore, 1985, med. officer trainee, 1988-89, registrar, 1991-92; intern Nat. Univ. Hosp., Singapore, 1986; med. officer Ministry of Defense, Singapore, 1986-87; head clin. br., med. officer (cardiology) Med. Classification Ctr. Cen. Manpower Base, Singapore, 1987-88; ag registrar Singapore Gen. Hosp., 1990-91, registrar dept. cardiology, 1991—; examining med. officer Civil Aviation Bd., Singapore, 1988-89, 92; clin. tchr. Nat. U. Singapore, 1989—; clin. supr. Singapore Armed Forces, 1990; vis. cardiology specialist Ministry Def., 1990—. Contbr. sci. papers to profl. jours. Speaker Singapore Lion and Lioness Club, 1988; dr. Orissa (India) Follow-Up, 1990; participant Japan Internat. Coop. Agy., 1990; mem. com. Singapore Nat. Heart Week, 1990, 92, 93, Singapore Cancer Soc.(chmn. Stay Fresh club 1992—, Smoke Free Day 1992-93). Capt. Med. Svcs., Singapore Armed Forces, 1986-88, res., 1988—. Recipient Sch. Merit award Raffles Instn., Singapore, 1979, Lim Boon Keng medal, Singapore Med. Assn. medal and First Meckie Book prize Nat. U. Singapore, 1985; local merit scholar Pub. Svc. Commn., Singapore, 1980. Mem. ACP, Royal Coll. Physicians Edinburgh, Royal Coll. Physicians and Surgeons Glasgow, Singapore Cardiac Soc., N.Y. Acad. Scis., Singapore Cancer Soc. Stay Fresh Club (mem. com.), Singapore Nat. Heart Assn. (bd. dirs. 1993). Methodist. Office: Tan Tock Seng Hosp, Dept Cardiology Moulmein Rd, Singapore 1130, Singapore

MAK, TAK WAH, biochemist; b. Canton, Republic of China, Oct. 4, 1946; s. Kent and Linda (Chan) M.; m. Shirley Lau, June 7, 1969; children: Julie Shi-Lan, Jennifer Shi-yan. BSc, U. Wis., 1967, MSc, 1969; PhD, U. Alta., Edmonton, Can., 1972; ScD (hon.), Carlton U., Ottawa, 1989, Laurentian U., 1992. Research asst. U. Wis., Madison, 1967-69, U. Alta., Edmonton, 1969-72; postdoctoral fellow U. Toronto, Ont., Can., 1972-74, asst. prof., 1974-78, assoc. prof., 1978-84, prof., 1984—; vis. prof. U. Wis., Madison, 1980; hon. prof. Beijing Union Med. U., 1986—. Editor: Molecular and Cellular Biology of Hemopolitic Stem Cell Differentiation, 1981, Molecular and Cellular Biology of Neiplasia, 1983, Cancer: Perspective for Control, 1986, The T Receptor, 1987, AIDS: Ten Years Later, 1991; contbr. over 250 sci. articles. Recipient E.W.R. Steacie prize, 1986, Ayerst award Can. Biochem. Soc., 1985, Emil Von Behring prize Marburg (Germany) U., 1986, Gairdner Internat. award Gairdner Found., 1989; hon. mem. Beijing Union Med. U., 1986, rsch. award Can. Found. for AIDS Rsch., 1992; E.W.R. Steacie fellow Nat. Sci. and Engring. Rsch. Coun. Can., 1984. Mem. Royal Soc. Can. (McLaughlin medal 1990), Chinese Acad. Med. Sci. (hon.). Avocations: classical music, tennis. Home: 130 Glen Rd, Toronto, ON Canada M4W 2W3 Office: Ont Cancer Inst, 500 Sherbourne St, Toronto, ON Canada M4X 1K9

MAKAROWSKI, WILLIAM STEPHEN, rheumatologist; b. Elmira, N.Y., Dec. 31, 1948; s. William John and Irene (Obuhanich) M.; m. Barbara Ann Payne; children: Elizabeth, Kathleen, Mary Lou. BS cum laude, Saint Bonaventure U., 1970; MD, Loyola-Stritch Coll., 1974; Rheumatology fellow, Cleve. Clinic, 1977-79. Diplomate Nat. Bd. of Med. Examiners, Am. Bd. of Pain Practice Mgmt. Resident in internal medicine Robert Packer Hosp., Sayre, Pa., 1974-77; pvt. practice Erie, Pa., 1984; chief rheumatology div. St. Vincent Hosp., Erie, Pa., 1979-86; med. dir. Musculio Occupational Rehab. Pain Mgmt. Program Great Lakes Rehab. Hosp., Erie, Pa., 1986-92, pres., 1991—; cons. Shriner's Hosp., Erie, 1980—, Metro Health Ctr., 1980—, VA Hosp., 1981—; mem. Hamot Med. Ctr. med. dept., 1979—; clin. asst. prof. Gannor U. Columnist: The Joint Achievement newsletter, 1981-88. Chmn. med. adv. bd. The Lupus Found., 1987-89; advisor Arthritis Discussion Group, 1981-91; mem. bd. govs. Arthritis Found., 1981-91. Fellow Am. Coll. Rheumatology (founding); mem. AMA, Pa. Med. Soc., Erie County Med. Soc., So. Med. Assn., Am. Pain Soc., Internat. Soc. for Rheumatic Therapy, Can. Pain Soc., Erie Yacht Club, Kahkwa Country Club, Erie Club, Aviation Club. Avocations: music, tennis, hist. novels. Home: 5075 Tramarlac Ln Erie PA 16505 Office: Rheumatology Assocs NW Pa 1781 W 26th St Erie PA 16508-1256

MAKHIJA, MOHAN, nuclear medicine physician; b. Bombay, Oct. 1, 1941; came to U.S., 1969; m. Arlene Zambito, Nov. 11, 1978. MD, Bombay U., 1966. Diplomate Am. Bd. Nuclear Medicine, Am. Bd. Radiology; cert. spl. competence in nuclear radiology. Resident in radiology Morristown (N.J.) Meml. Hosp., 1972-75; resident in nuclear medicine Yale-New Haven Hosp., 1975-76; post-doctoral fellow Yale U. Sch. Medicine, New Haven, 1976-77; jr. attending physician Helene Fuld Med. Ctr., Trenton, N.J., 1977-78; acting dir. dept. nuclear medicine Monmouth Med. Ctr., Long Branch, N.J., 1978, dir. nuclear medicine sect., 1979—, asst. attending radiology, 1978-80, assoc. attending radiology, 1980-83, attending radiologist, 1983—; sr. instr. Hahneman U., Phila., 1978-80, clin. asst. prof., 1980-83, clin. assoc. prof., 1983-91, clin. prof., 1991—. Contbr. articles to profl. jours. Fellow ACP, Am. Coll. Nuclear Physicians (speaker ho. of dels. 1992-93), Am. Coll. Radiology; mem. Monmouth (N.J.) County Med. Soc. (pres. 1991-92), Radiol. Soc. of N.J. (chmn. nuclear medicine 1988—), Indo-Am. Soc. Nuclear Medicine (pres. 1992-93). Home: 5 High Ridge Rd Ocean NJ 07712 Office: Monmouth Med Ctr 300 Second Ave Long Branch NJ 07740

MAKHIJA, SUBHASH, research chemical and polymer engineer; b. Thari, Haryana, India, Apr. 7, 1963; came to U.S., 1986; s. Mukand Lal and Sumitra (Devi) M.; m. Roopa Gandhi, July 5, 1991. PhD, Poly. U., Bklyn., 1992, MS in Mgmt., 1991. Rsch. engr. Hoechst Celanese Corp., Summit, N.J., 1991—; presenter profl. meetings. Contbr. articles to profl. jours. Recipient Dow Jones Indsl. award, 1992. Mem. AICE, Am. Chem. Soc., Am. Phys. Soc. Achievements include patents in field. Home: 922 Ripley Ave Westfield NJ 07090 Office: Hoechst Celanese Corp 86 Morris Ave Summit NJ 07901

MAKI, ATSUSHI, economics educator; b. Kanagawa, Japan, Jan. 14, 1948; s. Sadao and Eiko (Yamaguchi) M.; m. Michie Yabu, Feb. 28, 1975; children: Chiori, Hisashi. BA, Keio U., 1971, MA, 1973, PhD, 1993. Asst. prof. Keio U., Tokyo, 1973-79, assoc. prof., 1979-87, prof., 1987—; vis. scholar Harvard U., Cambridge, Mass., 1982-84; guest rsch. officer Ministry Posts and Telecommunications, 1988-90; vis. prof. Osaka (Japan) U., 1989; vis. fellow Australian Nat. U., Canberra, 1990, Massey U., New Zealand, 1991; vis. prof. Japan Found., 1990. Author: Consumer Preferences & Measurement of Demand, 1983. Grantee Japan Ministry Edn., 1986, Japan Econ. Rsch. Promotion Found., 1988, Nomura Rsch. Promotion Found., 1989, Inamori Rsch. Promotion Found., 1990, Japan Securities Scholarship Found., 1992; recipient award Union of Nat. Econ. Assns. in Japan, 1986, 92, Nomura Travel award, 1991, Japan Found. Travel award, 1991. Mem. Am. Econ. Assn., Econometric Soc., Japan Assn. Econs. and Econometrics, Japan Assn. Stats., Japan Soc. Household Econs. Home: Sunlight Mansion # 1102, 2602-3 Tsuruma, Fujimi-shi Saitama 354, Japan Office: Keio U, 2-15-45 Mita Minato-ku, Tokyo 108, Japan

MAKI, DENNIS G., medical educator, researcher, clinician; b. River Falls, Wis., May 8, 1940; m. Gail Dawson, 1962; children: Kimberly, Sarah, Daniel. BS with high honors in Physics, U. Wis., 1962, MS in Physics, 1964, MD, 1967. Diplomate Am. Bd. Internal Medicine, Am. Bd. Infectious Diseases, Am. Bd. Critical Care Medicine. Physicist, computer programmer Lawrence Radiation Lab., AEC, Livermore, Calif., 1962; intern, asst. resident Harvard Med. unit Boston City Hosp., 1967-69, chief resident, 1972-73; with Hosp. Infections sect. Ctrs. for Disease Control, USPHS, Atlanta, 1969-71; acting chief nat. nosocomial infections study Ctr. for Disease Control, USPHS, Atlanta, 1970-71; sr. resident dept. medicine Mass. Gen. Hosp., 1971-72, clin. and research fellow infectious disease unit, 1973-74; asst. prof. medicine U. Wis., Madison, 1974-78, assoc. prof., 1978-82, prof., 1982—; hosp. epidemiologist, U. Wis. Hosp. and Clinic, Madison, 1974—; Ovid O. Meyer chair in medicine U. Wis., Madison, 1975—, head sec. infectious diseases, 1979—; attending physician Ctr. for Trauma and Life Support U. Wis., 1976—; mem. com. Critical Care Medicine Am. Bd. Internal Medicine, 1989-95; clinician, researcher, educator in field. Contbr. research articles to med. jours.; sr. assoc. editor Infection Control and Hosp. Epidemiology, 1979—; mem. editorial bd. Jour. Lab. and Clin. Investigation, 1980-86, Jour. Critical Care, 1985—, Jour. Infectious Diseases, 1988-90, Critical Care Medicine, 1989—. Mem. program com. Interscience Conf. on Antimicrobial Agents and Chemotherapy, 1987-93. Recipient 1st award for disting. rsch. in Antibiotic Rev., 1980, numerous teaching awards and hon. lectrs. Fellow ACP, ACCP, Infectious Diseases Soc. Am. (coun. 1993-96), Soc. for Critical Care Medicine, Surg. Infection Soc.; mem. AOA (nat. bd. dirs. 1989-91), Soc. Hosp. Epidemiologists Am. (pres. 1990—), Cen. Soc. for Clin. Rsch., Am. Soc. Microbiology, Am. Fedn. for Clin. Rsch., Alpha Omega Alpha (bd. dirs. 1983-89). Office: U Wis Hosp and Clinics H4/574 Madison WI 53792

MAKI, FUMIHIKO, architect, educator; b. Tokyo, Sept. 6, 1928; m. Misao, 1960; 2 children. Ed., U. Tokyo, Cranbrook Sch. Art, Mich. and Harvard U.; MArch (hon.), Washington U. Assoc. prof. Washington U., St. Louis, 1956-62, Harvard U., 1962-66; lectr. dept. urban engring. U. Tokyo, 1964—, prof. architecture, 1979—; prin., ptnr. Maki and Assocs., 1964—; mem. Trilateral Commn., 1975—; vis. prof. U. Calif.-Berkeley, 1970, UCLA, 1977, Colombia U., 1977, Tech. U. Vienna, 1978. Prin. works include Toyoda Meml. Hall, Nagoya U., 1960; Rissho U. Campus, 1966; Nat. Aquarium, Okinawa, 1975; Tsukuba U. Complex, 1976; Hillside Terr. Housing Complex, 1978; The Royal Danish Embassy in Tokyo, 1979, Mitsubishi Bank Hiroo Br. Office, Minato Ward, Tokyo, 1881, Dentsu Advt. Co. Offices, Kita Ward, Osaka, 1983, Nat. Mus. Modern art, Kyoto, 1986; author: Investigations in Collective Form, 1964; Movement Systems in the City, 1965; Metabolism, 1960. Recipient Gold medal Japan Inst. Architects, 1964, 1st prize Low Cost Housing Internat. Competition, Lima, Peru, 1969, Art award Mainichi Press, 1969, Pritzker Architecture prize, 1993. Fellow AIA (hon.); mem. Japan Inst. Architecture. *

MAKI, KAZUMI, physicist, educator; b. Takamatsu, Japan, Jan. 27, 1936; s. Toshio and Hideko M.; m. Masako Tanaka, Sept. 21, 1969. B.S., Kyoto U., 1959, Ph.D., 1964. Research asso. Inst. for Math. Scis., Kyoto U., 1964; research asso. Fermi Inst., U. Chgo., 1964-65; asst. prof. physics U. Calif., San Diego, 1965-67; prof. Tohoku U., Sendai, Japan, 1967-74; vis. prof. Universite Paris-Sud, Orsay, France, 1969-70; prof. physics U. So. Calif., Los Angeles, 1974—; vis. prof. Inst. Laue-Langevin, U. Paris-Sud, France, 1979-80, Max-Planck Inst. für Festkörper Forschung, Stuttgart, Fed. Republic Germany, 1986-87, U. Paris-7, 1990. Assoc. editor Jour. Low Temperature Physics, 1969-91; contbr. articles to profl. jours. Recipient Nishina prize, 1972, Alexander von Humboldt award, 1986-87; Fulbright scholar, 1964-65; Guggenheim fellow, 1979-80. Fellow Am. Phys. Soc.; mem. Phys. Soc. Japan, AAAS. Office: U So Calif Dept Physics Los Angeles CA 90089-0484

MAKIGAMI, YASUJI, transportation engineering educator; b. Shenyan, Liaoning, China, June 3, 1936; arrived in Japan, 1946; s. Shikajiro and Masano (Kaneko) M.; m. Fumiko Nishina, June 9, 1964; children: Takeshi, Hiroshi, Satoshi. B. of Engring., Kyoto (Japan) U., 1960, D. of Engring., 1975; MS, U. Calif., Berkeley, 1970. Asst. engr. Fushimi Constrn. Office Japan Hwy. Pub. Corp., Kyoto, 1960-62; sngr. spl. design sect. Keihin Constrn. Bur. Japan Hwy. Pub. Corp., Kawasaki, Japan, 1962-65; rsch. engr. tech. devel. sect., head office Japan Hwy. Pub. Corp., Tokyo, 1965-69; asst. sect. chief Nagoya Operation Bur. Japan Hwy. Pub. Corp., 1971-73; sect. chief Japan Hwy. Pub. Corp., Shimonoseki, 1973-75; sect. chief Tokyo 1st Operation Bur. Japan Hwy. Pub. Corp., Kawasaki, 1975-77; sect. chief Tokyo 1st Constrn. Bur. Japan Hwy. Pub. Corp., 1978; prof. of transp. engring., councillor Ritsumeikan U., Kyoto, 1978—; bd. dirs. System Sci. Rsch. Inst., Kyoto, 1979—; mem. com. Expressway Tech. Ctr., Tokyo, 1990—; council specialist Ministry of Edn., Tokyo, 1991—. Author: Traffic Engineering, 1984, Civil Engineering Practice, 1988; author, editor: Highway Engineering, 1988, Traffic Engineering, 1990. Fulbright travel grantee, 1968-70. Mem. Japan Rd. Assn., Expressway Rsch. Found., Traffic Engring. Assn., Japan Soc. Civil Engrs. (councillor 1986-87), World Conf. of Transp. Rsch. Avocations: golf, hiking, detective stories. Home: Bamba 2-1-15, Otsu Shiga 520, Japan Office: Ritsumeikan U, 56-1 Tojiinkita-machi, Kita-ku Kyoto 603, Japan

MAKINS, JAMES EDWARD, retired dentist, dental educator, educational administrator; b. Galveston, Tex., Feb. 22, 1923; s. James and Hazel Alberta (Morton) M.; m. Jane Hopkins, Mar. 4, 1943; children: James E. Jr., Michael William, Patrick Clarence, Scott Roger. DDS, U. Tex.-Houston, 1945; postdoctoral, SUNY-Buffalo, 1948-49. Lic. dentist, Tex. Practice dentistry specializing in orthodontics, Lubbock, Tex., 1949-77; dir. clinics Dallas City Dental Health Program, 1977-78; dir. continuing edn. Baylor Coll. Dentistry, Dallas, 1978-92, ret., 1992, prof. emeritus Baylor Coll. Dentistry. Author: (book chpt.) Handbook of Texas, 1986. Chmn. profl. div. United Fund, Lubbock, 1958; pres. Tex. State Bd. Dental Examiners, Austin, Tex., 1968; instl. chmn. United Way, Dallas, 1983. Served to lt. comdr., USNR, 1945-47. Recipient Community Service award W. Tex. C. of C., Abilene, 1968, Clinic award Dallas County Dental Soc., 1981. Fellow Am. Coll. Dentists, Internat. Coll. Dentists; mem. Tex. Dental Assn. (life, v.p. 1954, Goodfellow 1973), West Tex. Dental Assn. (pres. 1955), Am. Assn. Dental Examiners, Park City Club, Rotary, Omicron Kappa Upsilon. Methodist. Avocation: dental history.

MAKIOS, VASILIOS, electronics educator; b. Kavala, Greece, Dec. 31, 1938; s. Thrassivoulos and Sophia M. Dipl.Ing., Tech. U. Munich, 1962; Dr. Ing., Max Planck Inst. for Plasmaphysics and Tech. U. Munich, 1966. Profl. engr., Ger., Ont., Greece. Research assoc. Max Planck Inst., Munich, 1962-67; asst. prof. dept. electronics Carleton U., Ottawa, Ont., 1967-70, assoc. prof., 1970-73, prof., 1973-77; prof. and head Electromagnetics lab. U. Patras, Greece, 1975—; cons. in field; dean engring. U. Patras, 1980-82; hon. adj. prof. Carleton U., 1977—. Contbr. over 100 articles to profl. jours. Patentee in field. Recipient Silver medal German Elec. Engring. Soc., 1984; numerous grants for research in Can., Greece and European community. Mem. IEEE, German Phys. Soc., German Inst. Elec. Engrs., Can. Assn. Physicists & Engrs., Greek Tech. Chamber. Greek Orthodox. Avocations: classical music; swimming; skiing. Home: 2 Lefkosias St, 26441 Patras 41, Greece Office: U Patras, Lab Electromagnetics, Patras Greece

MAKOWSKI, GERD, aerospace engineer; b. Neviges, Fed. Republic Germany, Feb. 5, 1964; s. Ewald and Grete (Massmann) M.; m. Birgit Henke, Apr. 24, 1992; 1 child, Viola. MS, U. Stuttgart, Germany, 1992. Quality assurance engr. Siemens Power Corp., Berlin, 1992—. Cpl. armed forces, 1983-84. Mem. AIAA, Verein Deutscher Ingenieur. Home: Albert Einstein Str 16, 8771 Triefenstein 1, Germany

MAKSYMOWICZ, JOHN, electrical engineer; b. Bklyn., Feb. 3, 1956; s. Theodore John and Helen Mary (Kisinski) M. BEE with highest honors, Pratt Inst., Bklyn., 1983. Elec. engr. RF and digital automated test equipment IBM, Poughkeepsie, N.Y., 1983; elec. engr. AWACS airborne early warning radar Grumman Aerospace Corp., Bethpage, N.Y., 1983-87; elec. engr. radar and spread spectrum comm. Plessey Electronics, Totowa, N.J., 1987-88; sr. mem. tech. staff, radar designer The Aerospace Corp., L.A., 1989—. Recipient Cook-Marsh scholarship Pratt Inst., 1979-83, Samuel Brown scholarship, 1979-83. Mem. IEEE, SPIE, U.S. Space Found., Old Crows Assn., Tau Beta Pi (coll. chpt. pres. 1981-82), Eta Kappa Nu (coll. chpt. pres. 1981-82). Roman Catholic. Avocations: photography, reading, music, running, astronomy. Office: The Aerospace Corp PO Box 92957 Los Angeles CA 90009-2957

MALA, THEODORE ANTHONY, physician, state official; b. Santa Monica, Calif., Feb. 3, 1946; s. Ray and Galina (Liss) M.; children: Theodore S., Galina T. BA in Philosophy, DePaul U., 1972; MD, Autonomous U., Guadalajara, Mex., 1976; MPH, Harvard U., 1980. Spl. asst. for health affairs Alaska Fedn. Natives, Anchorage, 1977-78; chief health svcs. Alaska State Div. of Corrections, Anchorage, 1978-79; assoc. prof., founder, dir. Inst. for Circumpolar Health Studies, U. Alaska, Anchorage, 1982-90; founder Siberian med. rsch. program U. Alaska, Anchorage, 1982, founder Magadan (USSR) med. rsch. program, 1988; commr. Health and Social Svcs. State of Alaska, Juneau, 1990—; mem. Alaska rsch. and publs. com. Indian Health Svc., USPHS, 1987-90; advisor Nordic Coun. Meeting, WHO, Greenland, 1985; mem. Internat. Organizing Com., Circumpolar Health Congress, Iceland, 1992-93; chmn. bd. govs. Alaska Psychiat. Inst., Anchorage, 1990—; cabinet mem. Gov. Walter J. Hickel, Juneau, 1990—; advisor humanitarian aid to Russian Far East U.S. Dept. State, 1992—. Former columnist Tundra Times; contbr. articles to profl. jours. Trustee United Way Anchorage, 1978-79. Recipient Gov.'s award, 1988, Outstanding Svc. award Alaska Commr. Health, 1979, Ministry of Health citiation USSR Govt., 1989; Citation award Alaska State Legislature, 1989, 90, Commendation award State of Alaska, 1990, Honor Kempton Svc. to Humanity award, 1989, citation Med. Community of Magadan region, USSR, 1989; Nat. Indian fellow U.S. Dept. Edn., 1979. Mem. Assn. Am. Indian Physicians, Am. Assn. University Profs., N.Y. Acad. Scis., Internat. Union for Circumpolar Health (permanent sec.-gen. 1987-90, mem. organizing com. 8th Internat. Congress on Circumpolar Health 1987-90), Am. Pub. Health Assn., Alaska Pub. Health Assn. Avocations: hiking, photography. Home: PO Box 232228 Anchorage AK 99523-2228 Office: Alaska Dept Health and Social Svcs Office of Commr PO Box 110601 Juneau AK 99811

MALABANAN, ERNESTO HERELLA, internist; b. Lemery, Batangas, Philippines, Aug. 8, 1919; s. Lazaro Mendoza and Emilia Atienza (Herella) M.; m. Simplicia de Guzman Diego, Sept. 29, 1957; children: Maria Susan, Sheila Rosalia, Ernesto, Emmanuel, Edgar, Edwin. AA, U. Philippines, 1938, MD, 1943. Diplomate and fellow Philippines Coll. Chest Physicians. Intern Philippine Gen. Hosp., Manila, 1942-43; pvt. practice medicine Batangas City, 1946-54; resident in medicine Goldwater Meml. Hosp., N.Y.C., 1954-55; resident in cardio-pulmonary svc. Bellevue Hosp., N.Y.C., 1955-56; resident in pulmonary svc. Cleve. City Hosp., 1956-57; pvt. practice specializing in internal medicine Batangas City, 1957—; cons. St. Patrick Hosp., Batangas City, 1966—, Philippine Heart Ctr., Quezon City, 1975—. Mem. Police Adv. Coun. and Peace and Order Coun., Batangas City, 1975—, Adv. Coun., Batangas City Mayor, 1986—. 1st lt. Philippine Army, 1944-46. Named Most Outstanding Physician, Philippine Med. Assn., 1966, Frederick Stevens Award Masons, 1972, Grand Lodge of Philippines award, 1990-91, Gov.'s Achievement award, Lions, 1970, 73, others. Fellow Philippine Coll. Cardiology, Internat. Coll. Angiology, Am. Coll. Chest Physicians (chpt. pres. 1979-81), Philippine Acad. Family Physicians, Fedn. Pvt. Med. Practitioners (pres. 1963-65), Batangas Med. Soc. (pres. 1965-67), Philippine Heart Assn. (chpt. pres. 1979-81), Lions (pres. 1967-69), Knights of Rizal (comdr. 1986—), Masons (33 deg., named one of top ten outstanding masons in medicine). Roman Catholic. Avocations: sports, basketball, baseball, swimming. Home: Hilltop, Batangas City The Philippines 4201 Office: Malabanan Med Offices, 164 Rizal Ave, Batangas The Philippines 4200

MALAMUD, HERBERT, physicist; b. N.Y.C., June 28, 1925; s. Max and Anna (Mintzer) M.; m. Sylvia Kolkin, Oct. 27, 1951; children: Ronni Sue, Marc David, Kathi Jan Malamud Knill. MS, U. Md., 1951; PhD, NYU, 1958; MS, C.W. Post Coll., 1976. Diplomate Am. Bd. Sci. in Nuclear Medicine. Rsch. asst. U. Md., College Park, 1950-53, NYU, Bronx, 1953-57; adv. rsch. engr. Sylvania Elec. Prodn. Co., Bayside, N.Y., 1957-59; sci. rsch. engr. Republic Aviation Corp., Farmingdale, N.Y., 1959-64; head rsch. sect. Sperry Gyroscope Co., Great Neck, N.Y., 1964-65; dir. physics Radiation RSch. Corp., Westbury, N.Y., 1965-67; v.p. Plasma Physics Corp., Hicksville, N.Y., 1967-70; physicist Queens Hosp. Ctr., Jamaica, N.Y., 1970-79; tech. dir. Nuclear Assocs., Carle Place, N.Y., 1979-89; hosp. physicist Nassau County Med. Ctr., East Meadow, N.Y., 1989—; adj. instr. physics CCNY, 1958-59; adj. prof. physics Hofstra U., 1960-61, Poly. Inst. Bklyn., 1961-64, N.Y. Inst. Tech., 1966-67; adj. prof. mgmt. engring. C.W. Post Coll., 1975—; adj. prof. computer sci. Westchester C.C., 1978; asst. prof. medicine SUNY, Stony Brook, 1975—; cons. North Shore U. Hosp., 1973-78, VA Hosp., Bronx, 1977-78. Contbr. over 20 articles to Physics Jour.; referee Jour. Nuclear Medicine; reviewer Am. Scientist, Clin. Nuclear Medicine. Mem. radioisotope and radiation safety com. L.I. Jewish/Hillside Med. Ctr.; mem. com. human rsch., radiation protection and pubs. North Shore U. Hosp. Cpl. U.S. Army, 1943-46, ETO. Mem. AAAS, Am. Physic Soc., Am. Inst. Physics, Am. Assn. Physicists in Medicine, Am. Inst. Ultrasound in Medicine, Soc.Nuclear Medicine, Health Phys. Soc., Sigma Xi, Sigma Pi Sigma. Achievements include patents for Medical Accessory Equipment. Home: 30 Wedgewood Dr Westbury NY 11590 Office: Nassau County Med Ctr Physics Div 2201 Hempstead Turnpike East Meadow NY 11554

MALCHOW, DOUGLAS BYRON, engineering executive; b. Windom, Minn., Mar. 20, 1938; s. Byron S. and gladys A. (Jacobson) M.; children: Deborah Ruth, Steven Byron, Michael Douglas. BSME, U. Minn., 1961. Design engr. Bemis Co., Mpls., 1961-73; dir. of engring. Warner Mfg. Co., Mpls., 1973-93, v.p., 1993—. Bd. dirs. YMCA, Mpls. Mem. ASM Internat., Am. Soc. for Quality Control, Am. Cutlery Mfg. Assn., Inst. Packaging Profls., Soc. Plastics Engrs. Methodist. Achievements include two patents in the field of hand tools. Office: Warner Mfg Co 13435 Industrial Park Blvd Minneapolis MN 55441

MALDEN, JOAN WILLIAMS, physical therapist; b. Bayshore, N.Y., Apr. 14; d. Sidney S. and Myrtle L. (Williams) Siegel; B.S., N.Y. U., 1957; m. Alan A. Chasnov, Jan. 20, 1951; children—Marc, Robin, Debra and David (twins); m. 2d, Miroslav Mladenovic, Sept. 14, 1967; 1 child, Kristine. Phys. therapist hosps. and orgns. in N.Y.C. area, 1956-57; phys. therapist Brunswick Hosp. Center, Amityville, N.Y., 1968-69; pvt. practice phys. therapy, Wantagh, N.Y., 1968—; licensure examiner, N.Y. State; cons., tchr. in field; lectr. L.I. U., NYU, Citizen Amb. Program to China; clin. coord. Hunter Coll., L.I. Coll., L.I. U., Daemen Coll., NYU, Columbia U., SUNY at Stony Brook, Touro Coll., Springfield Coll.; researcher in field; contbr. articles to profl. jours. Pres. internat. scholarships com. Massapequa chpt. Am. Field Service, 1962-64. Mem. Am. Acad. Cerebral Palsy, Am. Phys. Therapy Assn. (chmn. polit. action com. N.Y. chpt., chmn. L.I. dist.), AAUW (pres. Massapequa chpt. 1962-64), N.Y. State Soc. Continuing Edn. in Phys. Therapy, L.I. Assn. Ind. Phys. Therapists, Airplane Owners and Pilots Assn., Ninety-Nines, Exptl. Aviation Assn., Farmingdale Flyers (officer). Democrat. Unitarian. Home: 35 S Bay Ave Massapequa NY 11758-7847 Office: Wantagh Med Plz 3305 Jerusalem Ave Wantagh NY 11793-2209 also: 161 E Main St Huntington NY 11743

MALECH, HARRY LEWIS, immunologist, researcher; b. Carlstadt, N.J., Nov. 10, 1946; s. Morris and Freda M. (Lipowitz) M.; m. Emily Ann Root, June 4, 1972; children: Sarah Ruth, Dora Rachel, Daniel Lewis. BA, Brandeis U., 1968; MD, Yale U., 1972. Diplomate Am. Bd. Internal Medicine, 1974. Resident in internal medicine Hosp. of U. Pa., Phila., 1972-74; rsch. assoc. Nat. Cancer Inst., Bethesda, Md., 1974-76; fellow in infectious disease Yale U., New Haven, 1976-78, asst. prof. medicine, 1978-83, assoc. prof. medicine, 1983-86; sr. investigator, asst. chief Lab. Host Defenses Nat. Inst. Allergy and Infectious Diseases, Bethesda, 1986—; meeting coord. yearly phagocyte workshop, Washington, 1989—; sci. adv. bd. Cadus Pharms., N.Y.C., 1992—; cons. U.S. FDA, Bethesda, 1992. Assoc. editor Jour. Immunology, 1986-91, Jour. Immunologic Methods, 1991—; contbr. articles to sci. jours., chpts. to books. Fund raiser Yale U. Sch. Medicine Alumni Fund, 1973—, United Jewish Appeal Fedn., Rockville, Md., 1992; cons. Simon Wiesenthal Ctr., L.A., 1991. Surgeon USPHS, 1974-76. Recipient Dir.'s award NIH, 1990. Fellow Infectious Disease Soc. Am.; mem. AAAS, ACP, Am. Soc. Cell Biology, Am. Assn. Immunologists, Am. Soc. Clin. Investigation, Assn. Am. Physicians. Democrat. Jewish. Achievements include determination of genetic basis of two autosomal forms of chronic granulomatous disease, research on blood phagocytic cells, genetic therapy of immunodeficiency disorders. Office: Nat Insts Health Bldg 10 Rm 11 N 113 Bethesda MD 20892

MALENKA, BERTRAM JULIAN, physicist, educator; b. N.Y.C., June 8, 1923; s. Morris and Mollie (Wichtel) M.; m. Ruth D. Stolper, Mar. 28, 1948; children—David Jonathan, Robert Charles. A.B., Columbia, 1947; M.A., Harvard, 1949, Ph.D., 1951. Research fellow Harvard, 1951-54; asst. prof. physics Washington U., St. Louis, 1954-56; asso. prof. Tufts U., Medford, Mass., 1956-60; faculty Northeastern U., Boston, 1960—; prof. physics Northeastern U., 1962-93, prof. emeritus, 1993—; Mem. sci. adv. group Harvard-Mass. Inst. Tech. Cambridge Electron Accelerator, 1956—. Mem. Am., Italian phys. socs., N.Y. Acad. Scis., Phi Beta Kappa, Sigma Xi. Research and publs. on theory of nuclear forces and structure of nucleus, explanation polarization phenomena in high-energy scattering, gamma radiation, electric polarization deuteron, accelerator design. Home: 16 Rutledge Rd Belmont MA 02178-3323 Office: Northeastern Univ Dept of Physics Boston MA 02115

MALEY, WAYNE ALLEN, engineering consultant; b. Stanley, Iowa, Mar. 9, 1927; s. Neil Gordon and Flossie Amelia (Wharram) M.; m. Marianne Nelson, Aug. 2, 1959; children: James G., Mary G., Mark N. BS in Agrl. Engring., Iowa State U., 1949; postgrad., Purdue U., Ga. Tech., IIT. Power use advisor Southwestern Electric, Greenville, Ill., 1949-53; field agt. Am. Zinc Inst., Lafayette, Ind., 1953-59; mktg. devel. specialist U.S. Steel, Des Moines, Iowa, 1959-65; mktg. rep. U.S. Steel, Pitts., 1965-71, bar products rep., 1972-76; assoc. Taylor Equipment, Pitts., 1977-81; mgr. pub. rels. Am. Soc. Agrl. Engrs., St. Joseph, Mich., 1981-84, dir. mem. svcs., 1984-92; cons. Tech. Tours, St. Joseph, 1992—. Author: Iowa Really Isn't Boring, 1992, (textbook) Farm Structures, 1957, (computer program/workbook) Rim Lift Material Handling, 1970 (Blue Ribbon award 1971); editor: Agriculture's Contract with Society, 1991. Pres. Ednl. Concerns for Hunger Orgn., Ft. Myers, Fla., 1979-81; dist. activity dir. Boy Scouts Am., Moon Twp., 1969-70. With USN, 1945-46. Named Hon. Star Farmer, FIA Ill., 1958. Fellow Am. Soc. Agrl. Engrs. (bd. dirs. 1979-81 hon. for forum leadership membership coun. 1991); mem. Agrl. Editors Assn., Coun. Engring. Soc. Execs. (bd. dirs. 1984-85), Sigma Xi (pres./del. Whirlpool chpt. 1993-94). Presbyterian. Achievements include patent for fence building machine, for material landing system; design of cable fences; design and installation of steel beverage can recycling center. Home and Office: Tech Tours 2592 Stratford Dr Saint Joseph MI 49085

MALHERBE, BERNARD, surgeon; b. Rethel, France, Apr. 9, 1930; s. Maurice Eugene and Marguerite Elise (Drapier) M.; m. Camille Redon, May 10, 1952; children—Chantal, Philippe, Magali. MD, U. Paris, 1962. Intern, Paris Hosp.; resident in surgery various hosps.; clinic chief Hosp. Pitie-Salpetaient, Paris, 1962; mem. med. faculty U. Paris, 1962—; surgeon Clinic Leonardo da Vinci, Paris, 1970—; pres. Coll. Med. L'Hospitalisation Privee. Served to lt. French Air Force, 1956-58. Roman Catholic. Lodge: Lions. Home: 12 rue Felicien David, 75016 Paris France Office: Medecins Sans Frontieres, 8 rue St Sabin, 75011 Paris France

MALHOTRA, VIJAY KUMAR, mathematics educator; b. Punjab, India, Sept. 23, 1946; came to U.S., 1969; s. Anand K. and Swarn Kanta (Chadha) M.; m. Madhu Chadha, Aug. 18, 1973; children: Jaishri, Vaishali, Vivek.. BA, Delhi (India) U., 1965; MA, Meerut U., India, 1968, Pepperdine U., 1970. Cert. instr. community colls., Calif. Head math. dept. Le Lycee de L.A., 1971-78; instr. math. L.A. Trade Tech. Coll., 1978-84; prof. El Camino Coll., Torrance, Calif., 1984—. Mem. Am. Fedn. Tchrs. Office: El Camino Coll 16007 Crenshaw Blvd Torrance CA 90506-0001

MALIGAS, MANUEL NICK, metallurgical engineer; b. Thimena, Greece, May 9, 1943; came to U.S., 1950; s. Nick and Jane M.; children: James Paul, John Michael. BE, Youngstown U., 1966; MS, Youngstown State U., 1974. Sr. material engr. Goodyear Aerospace, Akron, Ohio, 1966-76; plant metallurgist Tex. Bolt, Houston, 1976-81; sr. material engr. N.L. Shaffer, Houston, 1981-83; ind. cons. M&M Metall., Houston, 1983; group leader materials FMC Corp., Houston, 1983—; presenter symposia. Contbr. articles to profl. pubis. on welding overlays and life predication of elastomers. Chmn. Jersey Village (Tex.) Parks Com., 1974; tchr. Sunday sch. Greek Orthodox Ch., Houston, 1982. Mem. Am. Soc. Metals (past officer), Am. Petroleum Inst. (chmn. material and welding coms.), Nat. Assn. Corrosion Engrs. Achievements include early design of carbon composite brake for commercial aircraft, use of laser and HUOF process for hard facing. Home: 5907 Winged Foot Dr Houston TX 77069 Office: FMC Corp PO Box 3091 Houston TX 77253-3091

MALIK, ABDUL HAMID, engineering executive; b. Daudkhel, Punjab, Pakistan, Sept. 25, 1948; s. Noor Muhammad and Zehrah (Noor Muhammad) Awan; m. Razia Hamid Malik, Apr. 14, 1974; children: Shahzad Kaleem, Mazhar Hamid. BA, Govt. Coll., Mianwali, Pakistan, 1970; postgrad., U. Punjab, Lahore, Pakistan, 1972—. Pub. and distbn. officer Nat. Book Found., Lahore, 1973-75; asst. mgr. personnel Nat. Fertilizer Corp./USAID Rsch. Project, Lahore, 1976-78; dep. mgr. personnel Nat. Fertilizer Mktg. Ltd., Lahore, 1979-83, mgr. personnel, 1984-87; dep. gen. mgr. personnel State Engring. Corp., Islamabad, Pakistan, 1988-89; gen. mgr. adminstrn. and indsl. rels. Pakistan Engring. Co., Lahore, 1990; gen. mgr. adminstrn. and personnel State Engring. Corp., Islamabad, 1991-92; gen. mgr. adminstn. and pers. Heavy Mech. Complex, Taxila, Pakistan, 1992; cons. Econ. Rsch. Inst., Lahore, 1977; vis. tchr., personnel mgmt. bus. edn. dept. Punjab U., Lahore, 1986-87; team leader, task force Nat. Fertilizer Mktg., Lahore, 1987; vis. instr., personnel mgmt. Nat. Inst. Pub. Adminstrn., Lahore, 1990. Editor: State Engring. Corp. newsletter, 1987. Sec. gen. Nat. Student Fedn. Punjab, Lahore, 1971-72; student del. Pakistan Ednl. Conf., Islamabad, 1972; employer's del. Pakistan Tripartite Labour Conf., Islamabad, 1988. Fellow Pakistan Inst. Personal Adminstrn.; mem. Am. Mgmt. Assn. Avocations: modern concepts of human resource management, teaching personnel management. Home: House No 737 St, 10 G-9/3, Islamabad Pakistan Office: Heavy Mech Complex, Taxila Dist, Rawalpindi Pakistan

MALIK, TARIQ MAHMOOD, scientist, chemical engineer; b. Lahore, Punjab, Pakistan, Nov. 4, 1953; arrived in Can., 1980; came to U.S., 1991; s. Abdul Raheem and Ruzia Begum M.; m. Mehfooza Athar; children: Zunaira, Arsheen, Nida, Sarah. MS, Laval, Quebec, Can., 1981, PhD, 1984. Assoc. prof. Chem. Engring. U. of Montreal, Can., 1985-91; group leader, rsch. and devel. assoc. Tremco Inc., ABF Goodrich Co., Cleve., 1991—. Contbr. over 25 articles to profl. jours. Recipient operating grants Nat. Scis. and Engring. Rsch. Coun. of Can., Montreal, 1990. Home: 32701 S Roundhead Solon OH 44139 Office: Tremco Inc 3777 Green Rd Beachwood OH 44122

MALING, GEORGE CROSWELL, JR., physicist; b. Boston, Mass., Feb. 24, 1931; s. George Croswell and Marjory (Bell) M.; m. Norah J. Horsfield, Dec. 29, 1960; children: Ellen P., Barbara J., Jeffrey C. A.B., Bowdoin Coll., 1954; S.B., S.M., MIT, 1954, Elec. Engr., 1958, Ph.D., 1963. Rsch. asst., postdoctoral fellow MIT, 1957-65; adv. physicist IBM Corp., 1965-71; sr. physicist IBM Corp., Poughkeepsie, N.Y., 1971-92; pres. Empire State Software Systems, Ltd., 1992—; dir. Noise Control Found., Inc., Poughkeepsie, 1975—; chmn. com. Sl-acoustics Am. Nat. Standards Com., 1976-79. Ediotr: Noise/News, 1972-92; mng. editor: Noise/News Internat., 1993—; assoc. editor Jour. Acoustical Soc. Am., 1976-83; editor tech. proc.; contbr. numerous articles to profl. jours. Served with U.S. Army, 1955-57. Fellow IEEE, AAAS, Acoustical Soc. Am. (exec. coun. 1980-83, Silver medal in noise 1992), Audio Engring. Soc.; mem. Inst. Noise Control Engring. (bd. dirs. 1972-77, pres. 1975), Internat. Inst. Noise Control Engring. (bd. dirs. 1980-86, 90—). Office: ESSS PO Box 2880 Poughkeepsie NY 12603

MALIS, BERNARD JAY, pharmacologist; b. Phila., Feb. 25, 1923; s. Louis J. and Jennie L. (Josselowitz) M.; children: David Joel, Alexa Faith, Olga Lee, Kenneth Andrew, Jonathan Martin. BSc in Pharmacy and Chemistry, Phila. Coll. Pharmacy and Sci., 1944, MSc in Pharmacology/Indsl. Chemistry, 1947. Registered pharmacist, Pa., N.J. Pres. Ogontz Manor Pharmacies, Inc., Phila., 1950-56; dir. pharmacy and cosmetics Sav-Fair, Food Fair Stores, Phila. and Miami, 1956-58; rsch. dir. Haelan Labs., Phila., 1958-60; prin BJM Assocs., Phila., 1960-85, ret., 1985; U.S. del. Fedn. Internat. Pharm., The Hague, 1960-80; bd. incorporators The Phila. Coll. of Pharmacy and Sci., 1950-80. Contbr. articles to profl. jours. including The Explorers Jour. Chmn. The Forum, Phila., 1988-90; assoc. fgn. policy rsch. inst. and Middle East coun. divsn. U. Pa., univ. mus. archaeology and anthropology. Recipient Legion of Honor award Chapel of Four Chaplains, 1989; named Col. Gov. Wallace G. Wilkinson, 1988. Fellow Royal Soc. (London), AAAS; mem. Am. Pharm. Assn. (life), C. of C. (ambassador 1985-91), The Pickwick Club, Am. Legion (Benjamin Franklin Post), Navy League, Masons (hon. life mem.), The Explorers Club (chmn. 1990-91), Rotary (dir. club and found. 1988-90). Home and Office: The Warwick Hotel Ste 1502 1701 Locust St Philadelphia PA 19103-6114

MALLA, PRAKASH BABU, research materials chemist; b. Gorkha, Nepal, June 13, 1957; came to U.S., 1982; s. Padma Narayan and Dmarma Kumari (Shrestha) M.; m. Anju Shrestha, July 6, 1988. BSc, Rajendra Agrl. U., Pusa Bihar, India, 1980; PhD, Rutgers U., 1987. Cert. profl. soil scientist. Asst. lectr. Tribhuwan U., Paklihawa, 1984; rsch. asst. Rutgers U., New Brunswick, N.J., 1983-87; rsch. assoc. Pa. State U., University Park, Pa., 1987-92; rsch. materials chemist/group leader Thiele Kaolin Co., Sandersville, Ga., 1992—. Contbr. articles to profl. jours. including Nature, Jour. Matrials Chemistry, Jour. Materials Sci. and others. Mem. Am. Chem. Soc., Am. Ceramic Soc., Materials Rsch. Soc., Clay Minerals Soc. Achievements include patent for dessicant materials for use on gas fired cooling and dehumidification equipment. Office: Thiele Kaolin Co R&D PO Box 1056 Sandersville GA 31082

MALLA, RAMESH BABU, structural and mechanical engineering educator; b. Chhoprak, Gorkha, Nepal, Mar. 11, 1955; came to U.S., 1979; s. Padma Narayan and Dharma Kumari (Shrestha) M.; m. Sun-Kyeong Lee, July 12, 1989. ISc, Tribhuvan U., Kathmandu, Nepal, 1973; BSCE, Indian Inst. Tech., Kanpur, India, 1979; MSCE, U. Del., 1981; PhD, U. Mass., 1986. Rsch. asst. U. Del., Newark, 1979-81; structural engr. United Engrs. Constructors, Inc., Phila., 1981-83; rsch. and teaching asst. U. Mass., Amherst, 1983-85; vis. asst. prof. U. Conn., Storrs, 1985-90, asst. prof., 1990—; campus dir. NASA-Conn. Space Grant Consortium, 1991—. Contbr. articles to profl. jours. Mem. ASCE (com. on spl. structures, space structures and materials, space engring. and constrn., task com. lunar base structures, task com. double later grids aerospace com. synamics and contrils chair, publ. com. Jour. Aerospace Engring.), ASME, AIAA, AAUP, Am. Acad. Mechanics, Internat. Design for Extreme Environ. Assn. (founding mem.), Tau Beta Pi. Avocations: public service, games, sports, travel. Office: U Conn Dept of Civil Engineering U-37 Storrs Mansfield CT 06269

MALLARAPU, RUPA LATHA, transportation engineer, consultant; b. Tirupati, India, Mar. 20, 1965; came to U.S., 1988; d. Narayana Naidu and Vijaya Lakshmi (Peravali) M.; m. Prabhakar Somavarapu, Jan. 26, 1991. B Tech., S.V. U., Tirupati, 1988; MS, U. Nebr., Lincoln, 1990. Grad. asst. dept. civil engring. U. Nebr., Lincoln, 1989-90; transp. engr. JHK and Assocs., Sacramento, Calif., 1990—. Mem. Inst. Transp. Engrs. (assoc.), Soc. Women Engrs., Women's Transp. Seminar. Hindu. Office: JHK and Assocs Ste 100 1001 G St Sacramento CA 95814

MALLET, MICHEL FRANCOIS, numerical analyst; b. Caen, France, May 23, 1960; s. Albert Jean and Josette (Grange) M.; m. Francoise Caroline Moskowitz, Mar. 19, 1988; children: Sophie, Elizabeth. Ingenieur, Ponts et Chaussee, Paris, 1983; PhD, Stanford U., 1985. Rsch. affiliate Stanford U., 1985-86; rsch. engr. Dassault Aviation, St. Cloud, France, 1986—; mem. steering com. GAMNI-SMAI, Paris, 1990—; mem. organizing com. various internat. confs. Contbr. articles to profl. jours. IBM fellow, 1984, 85. Mem. Stanford Club of France. Achievements include contribution to the development of innovative numerical techniques to simulate high speed viscous compressible fluid flows; application of research to design of hypersonic reentry vehicles, most notably the European Space Plane. Home: 21 Avenue du Bel Air, Paris 75012, France Office: Dassault Aviation, 78 quai Dassault, Saint-Cloud 92214, France

MALLEVIALLE, JOËL CHRISTIAN, marine engineer; b. Auch, Gers, France, Dec. 21, 1944; s. Lucien and Albertine (Brouxel) M.; m. Fabienne Bourdon, Sept. 17, 1981; children: Cedric, Rudy. MSCE, Institut National des Sciences Appliquées, Toulouse, France, 1969, PhD, 1974. Rsch. engr. Oceanographic Inst., Paris, 1967, Organic Chem. Inst., Zûrich, Switzerland, 1968-69, Lyonnaise des Eaux, Le Pecq, France, 1969; engr. attaché Environ. Inst. Drexel U., Phila., 1980; mgr. Internat. Ctr. Rsch. for Water and Environ. Lyonnaise des Eaux, Le Pecq, France, 1991; bd. dirs. Am. Water Works Assn., Denver, 1981, IOA, Lille, France, 1980, Am. Chem. Soc., Washington, 1982. Editor books Identification and Treatment of Tastes and Odors in Drinking Water, 1987, Influence and Removal of Organics in Drinking Water, 1991, Treatment of Taste and Odor, 1992. Recipient Chemviron award, Chemviron Carbon br. Calgon Carbon Corp., 1980, Gold medal for Rsch. and Invention, Soc. d'Encouragement pour la Recherche et l'invention Paris, 1992. Office: Cirsee/Lyonnaise des Eaux, 38 rue du Président Wilson, 78230 Le Pecq France

MALLEY, JAMES HENRY MICHAEL, industrial engineer; b. Providence, Oct. 15, 1940; s. Leo Henry and Gladys Elizabeth (Canning) M.; children: James Michael, Julie Michele; m. Joyce Sue Marie Greenwell, Aug. 28, 1993. BS in Engring., U.S. Mil. Acad., 1962; MS in Indsl. Engring., U. R.I., 1977. Commd. U.S. Army, 1962-84, advanced through grades to lt. col., ret., 1984; milt. advisor U.S. Army, Rep. of Vietnam, 1964-65; co. comdr. Army Tng. Ctr., Ft. Benning, Ga., 1965-67; ops. and exec. officer First Air Cavalry Div., Vietnam, 1968-69; asst. prof. U. R.I., Kingston, 1969-73; asst. inspector gen. U.S. Army Criminal Investigation Command, Washington, 1973-76; ops. rsch. analyst and study dir. U.S. Army Concepts Analysis Agy., Bethesda, Md., 1977-80; dir. tng. U.S. 7th Army Combined Arms Tng. Ctr., Vilseck, Ger., 1980-81; chief of ops. rsch. and sys. analysis U.S. Army Europe, Heidelberg, Ger., 1981-84; mgr. engring. svcs. Orion Internat. Tech., Inc., Albuquerque, 1985-90; temp. recall, Ops. Desert Shield/Desert Storm U.S. Army, 1991; army after action report integrator ODCSOPS-HQDA, Washington, 1991; prin. analyst Gen. Rsch. Corp., Washington, 1992; ops. rsch. and analysis exec. Lockheed-Sanders, Merrimack, N.H., 1992—; mgmt. advisor to pres. PC Support, Inc., Albuquerque, 1986—. Decorated Silver Stars (2), Legion of Merit, Bronze Stars (3), Air medals (4), Purple Heart, Vietnamese Cross of Gallantry with Gold Star (1) with Palm (2). Mem. Ops. Rsch. Soc. Am., Assn. of U.S. Army, U.S. Naval Inst., Am. Def. Preparedness Assn., Internat. Test & Evaluation Assn. Avocations: volksmarching, kayaking, skiing, rafting, mathematics. Home: PO Box 746 Merrimack NH 03054-0746

MALLORY, CHARLES WILLIAM, consulting engineer, marketing professional; b. Brewster, Kans., Sept. 17, 1925; s. Wilbur Lloyd and Carolina Metilda (Andregg) M.; m. Anne Veronica Duffy, July 8, 1950; 1 child, Michael Joseph. BSME with honors, U. Colo., 1946; BCE, Rensselaer Poly. Inst., 1950. Registered profl. engr., Md., 1968. Commd. ensign USN, 1944, advanced through grades to comdr., 1962; project officer Schenectady (N.Y.) ops. U.S. AEC, 1952-58; br. mgr. reactor devel. U.S. AEC, Germantown, Md., 1958-62; dir. nuclear power div. NAVFAC, Arlington, Va., 1962-65; ret. USN, 1965; chief engr., v.p. engring. Hittman Assocs., Inc., Columbia, Md., 1965-77; v.p. mktg., v.p. engring. Hittman Nuclear and Devel., Columbia, 1977-82; sr. tech. adviser Westinghouse Hittman Nuclear, Columbia, 1982-88; prin. Engring. and Mgmt. Svcs., Severna Park, Md., 1988—; industry rep. Atomic Industry Forum QA Waste Shipping, Washington, 1980-82, U.S. Dept. Transp. Radwaste Transp., Washington, 1981-85. Author over 40 rsch. reports, papers. Mem. expedition Mallory Point, Anarctica, 1947-48. Mem. ASME (radwaste mgmt. com. 1980-90), Chartwell Golf and Country Club (pres. 1970-72, citation 1974). Republican. Roman Catholic. Achievements include patents for improved extended aeration wastewater treatment system, closure for high-integrity containers, grappling system for high-integrity containers, closure for radioactive shipping casks, tie-downs for shipping casks, others. Home: 536 Heavitree Hl Severna Park MD 21146-1009 Office: Engring and Mgmt Svcs 536 Heavitree Hl Severna Park MD 21146-1009

MALLORY, DAVID STANTON, physiology educator; b. Boston, Feb. 24, 1958; s. Alvah Theophilus and Marilyn Louise (Cox) M.; children: Caitlin Marie, Max Theophilus. BS, Cornell U., 1980; MS, U. Maine, 1983; PhD, W.Va. U., 1987. Postdoctoral fellow U. Conn., Storrs, 1987-89; asst. prof. Marshall U., Huntington, W.Va., 1989—. Contbr. articles to profl. jours. Worker Habitat for Humanity, Huntington, 1989-90. Grantee Huntington Found., 1991, 93; recipient Heatly Green scholarship Cornell U., 1978, Carl

E. Ladd scholarship, 1977, John R. McKernan scholarship, 1976. Mem. Soc. for Study of Reproduction, Sigma Xi. Achievements include discovery that anterior pituitary gland (ovine) secretes LH/FSH after transection of the hypothalamo-hypophyseal stalk. Office: Marshall U Dept of Biology 400 Hal Greer Blvd Huntington WV 25755-2510

MALLORY, MARY EDITH, psychology educator; b. Oakland, Calif., July 17, 1952; d. James Irving and Margaret Gene (Kimball) M.; m. Michael Gene Heaton, Aug. 12, 1978; children: Amanda Mallory Heaton, Amelia Mallory Heaton. BA, U. Calif., Santa Barbara, 1979; MA, U. Calif., Davis, 1983, PhD, 1983; cert. in Russian aural comprehension, Def. Lang. Inst., 1975. Rsch. asst. U. Calif., Davis, 1979-82, lectr., 1982-83; mental health counselor Yolo County Care Continuum, Davis, 1980-83; rsch. assoc. Ctr. for Consumer Rsch., Davis, 1984; lectr. in psychology Calif. State U., Fresno, 1986—; pvt. practice rsch. cons. Fresno, 1986—; cons., speaker Abbott and Assocs., Fresno, 1991—; cons. Calif. Task Force to Promote Self Esteem and Personal and Social Responsibility, Sacramento, 1988-90; co-chair Fresno County Self Esteem Task Force, 1990—. Contbr. articles to profl. publs. With U.S. Army, 1975-76. Fellow Am. Psychol. Soc. (charter); mem. AAUW (pres. 1993-94), Sigma Xi. Democrat. Congregationalist. Office: Calif State U Psychology Dept 5310 N Campus Dr MS.11 Fresno CA 93740-0011

MALLUCHE, HARTMUT HORST, nephrologist, medical educator; b. Breslau, Fed. Republic Germany, Jan. 1, 1943; came to U.S., 1975, naturalized, 1985; s. Harald E. and Renate (Muenzberg) M.; m. Gisela Gleich, Dec. 19, 1975; children: Nadine, Danielle, Tiffany. Abitur, Albertus Magnus Coll., Koenigstein, Germany, 1963; postgrad. Phillips U., Marburg/Lahn, Fed. Republic of Germany, 1963-65, U. Innsbruck, Austria, 1965-66, U. Vienna, Austria, 1966; MD, J. W. Goethe U., Frankfurt, Fed. Republic of Germany, 1969. Diplomate German Bd. Internal Medicine. Intern, County Hosp., Aichach, Fed. Republic of Germany, 1969-70; resident in internal medicine and fellow in nephrology Ctr. Internal Medicine, Univ. Hosp., Frankfurt am. Main, 1970-75, asst. prof. medicine U. So. Calif., Los Angeles, 1975-78, assoc. prof., 1978-81; prof., dir. Div. Nephrology, Bone and Mineral Metabolism U. Ky. Med. Ctr., Lexington, 1981—; cons. NIH, FDA; Va. merit Rev. bd. nephrology. Author (monograph) Atlas of Mineralized Bone Histology, 1986; contbr. articles to profl. jours. and books. Grantee NIH, 1982—, Shriner's Hosp. for Crippled Children, Lexington, 1982—. Fellow ACP; mem. Am. Soc. Nephrology, Am. Soc. Clin. Investigation, Am. Soc. Bone and Mineral Research, Am. Soc. Physiol. Endocrinology, European Dialysis and Transplantation Assn., Am. Fedn. Clin. Research, Internat. Soc. Nephrology, AAAS.

MALMBERG, TORSTEN, human ecologist; b. Helsingborg, Scania, Sweden, July 5, 1923; s. Sigfrid Vilhelm and Lydia Henrietta (Bergsten) M.; m. Aimee Birgit Magnussen, May 15, 1948; children: Ole Vilhelm, Anna Birgitta. BA, U. Lund, Sweden, 1945, MA, 1951, DSc, 1973. From asst. to libr. to mus. keeper dept. zoology U. Lund, 1946-55, from rsch. asst. to asst. prof. dept. geography, 1978-83, asst. prof., chief human ecology div. dept. history, 1984-91, prof., chief human ecology div. dept. history, 1991—; tchr. pub. sch. system Sweden, 1953-77; environment and bird expert Swedish Environ. Protection Bd., Stockholm, Sweden, 1948-68; examiner inst. librs. U. Libr., Lund, 1953; ecol. adviser WHO, Geneva, 1970. Initiator, editor: Popular Sci. Jour., 1982—, Acta Oecologiae Hominis, 1989—; author: Human Territoriality, 1980. Recipient Bronze medal Swedish Ornithological Soc., Silver medal Swedish Nature Protection Soc.; fellow UNEP, Nairobi, 1992, Global 500 award. Fellow Inst. Human Ecology; mMem. Ornithological Soc. Scania (founder, bd. dirs., sec. Lund chpt. 1948-55), Ornithological Soc. Sweden (founder, bd. dirs. Stockholm chpt. 1949-62), Scanian Nature Protection Soc. (bd. dirs. Lund chpt. 1948-63), Nordic Soc. for Human Ecology (bd. dirs., chmn. Lund chpt. 1980-90, disting. leadership award 1986), Internat. Soc. Human Ecology (bd. dirs., vice chmn. Vienna, Austria chpt. 1980-84). Avocations: music, literature, history, politics, sports. Home and Office: U Lund Human Ecology Div, Box 2015, 220 02 Lund 2, Sweden

MALMCRONA-FRIBERG, KARIN ELISABET, microbiologist; b. Göteborg, Sweden, May 2, 1959. PhD, U. Göteborg, 1990. Project coord. Nordic Marine Biotech. Network, Göteborg, 1990-93; sr. microbiologist Swedish Inst. Food Rsch., Göteborg, 1991—. Home: Torsgatan 40, S-431 38 Molndal Sweden Office: SIK, PO Box 5401, S-402 29 Göteborg Sweden

MALMUTH, NORMAN DAVID, program manager; b. Brooklyn, N.Y., Jan. 22, 1931; s. Jacob and Selma Malmuth; m. Constance Nelson, 1970; children: Kenneth, Jill. AE, U. Cin., 1953; MA in Aero. Engring., Polytech. Inst. of N.Y., 1956; PhD in Aeronautics, Calif. Inst. Tech., 1962. Rsch. engr. Grumman Aircraft Engring. Corp., 1953-56; preliminary design engr. N.A. Aviation Div., L.A., 1956-68; teaching asst. Calif. Inst. Tech., L.A., 1961; mem. maths. sci. group Rockwell Internat. Sci. Ctr., 1968-75, project mgr. fluid dynamics rsch., 1975-80, mgr. fluid dynamics group, 1980-82, program mgr. spl. projects, 1982—; cons. Aerojet Gen., 1986-89; lectr. UCLA, 1971-72. Referee AIAA Jour.; bd. editors Jour. Aircraft; contbr. articles to Jour. of Heat Transfer, Internat. Jour. Heat Mass Transfer, and others. Named Calif. Inst. Tech. fellow; recipient Outstanding Alumnus award Univ. Cin., 1990. Assoc. fellow AIAA (editorial adv. bd. AIAA Jour. Aircraft, AIAA Aerodynamics award 1991); mem. Am. Acad. Mechanics, Am. Inst. Physics (fluid dynamics divsn.), Soc. Indsl. and Applied Maths. Achievements include patent in Methods and Apparatus for Controlling Laser Welding; pioneering development of high aerodynamic efficiency of hypersonic delta wing body combinations, hypersonic boundary layer stability, transonic wind tunnel interference web dynamics, combined asymptotic and numerical methods in fluid dynamics and aerodynamics. Home: 182 Maple Rd Newbury Park CA 91320-4718 Office: Rockwell Sci Ctr PO Box 1085 Thousand Oaks CA 91358-0085

MALONE, CHARLES TRESCOTT, photovoltaic specialist; b. Attleboro, Mass., Dec. 13, 1966; s. William Matthew and Barbara Ann (Trescott) M. BSc, Pa. State Univ., 1989, MSc, 1991, postgrad., 1993—. Computer lab. operator dept. engring. sci. Pa. State Univ., University Park, 1987-88; rsch. asst. Pa. State Ctr. Electronic Materials, University Park, 1988-89, 1989-90, 1990—; rsch. engr. Solarex Thin Film Division, Newtown, Pa., 1989, Glasstech Solar Inc., Golden, Colo., 1990. Contbr. articles to profl. jours. Recipient Theodore H. Thomas Jr. scholarship, 1987, scholarship Am. Vacuum Soc., 1990. Mem. IEEE, Materials Rsch. Soc., Pa. State Cycling Team. Home: 148 West Hamilton Ave State College PA 16801 Office: The Pa State Univ 227 Hammond Bldg University Park PA 16802

MALONE, STEPHEN ROBERT, plant physiologist; b. Searcy, Ark., Apr. 19, 1960; s. Bobby J. and Marjorie M.; m. Georgann Tracy, May 16, 1987; 1 child, Bryan J. BS with high honors, U. Ark., 1982, MS, 1983; PhD, Iowa State U., 1989. Postdoctoral rsch. assoc. horticulture dept. Purdue U., West Lafayette, Ind., 1989-91; plant physiologist Agrl. Rsch. Svc.-USDA, Temple, Tex., 1991-93; asst. prof. biology Ga. So. U., Statesboro, 1993—. Contbr. articles to profl. jours. Den leader Cub Scouts, Temple, 1991—; baseball coach Temple Youth Baseball Assn., 1991-92; industry ptnr. Ctrl. Tex. Tech-Prep Consotrium, Temple, 1992. Mem. AAAS, Am. Soc. Plant Physiologists, Bot. Soc. Am., Toastmasters Internat. Republican. Mem. Ch. of Christ. Achievements include development of technique for preparing frozen woody plant stem samples for low-temperature SEM without thawing, which allowed ice to be viewed in situ; demonstration of lack of theorized cell collapse in extremely cold-hardy nonsupercooling woody plant species; current research on plant anatomical and physiological response to increasing atmospheric carbon dioxide. Office: USDA-Agrl Rsch Svc 808 E Blackland Rd Temple TX 76502

MALONE, WINFRED FRANCIS, toxicologist; b. Revere, Mass., Feb. 10, 1935; s. Winfred and Margurite (Meehan) M.; m. Eleanor Malone, Aug. 1974. BS, U. Mass., 1957, MS, 1961; MS, Rutgers U., 1963; PhD, U. Mich., 1970. Scientist Nat. Cancer Inst., Bethesda, Md., 1970-81, chief chemoprevention br., 1981—, acting assoc. dir., 1991—. Contbr. articles to profl. jours. Col. USAR, 1957-88. Mem. AAAS, Am. Coll. Toxicology, N.Y. Acad. Scis. Home: 3209 Wake Dr Kensington MD 20895-3216 Office: Nat Cancer Inst EPN 201 Bethesda MD 20892

MALONEY, FRANCIS PATRICK, physiatrist; b. Pitts., Mar. 4, 1936; s. Francis Barrington and Esther Elizabeth (Kuhn) M.; m. Kathryn Brassell Anderson, June 25, 1960 (dec. June 6, 1987); children: Timothy J., Kevin P., J. Christopher; m. Billie Barbara Galloway, Feb. 14, 1990. BA, St. Vincent Coll., 1958; MD, U. Pitts., 1962; MPH, Johns Hopkins U., 1966. Intern St. Francis Hosp., Pitts., 1962-63; resident gen. preventive medicine Johns Hopkins U. Sch. of Hygiene & Pub. Health, Balt., 1965-67; fellow medicine, med. genetics Johns Hopkins U. Sch. of Medicine, Balt., 1966-68; resident phys. medicine and rehab. U. Minn., Mpls., 1968-70; staff physician Sister Kenny Inst., Mpls., 1970-72; asst. clin. prof. U. Minn., Mpls., 1970-72; asst. prof. phys. medicine and rehab., assoc. prof. U. Colo., Denver, 1972-78, 78-84; prof. head div. of rehab. medicine U. Ark., Little Rock, 1984-91, prof., chmn. dept. phys. medicine and rehab., 1991—; med. dir. Bapt. Rehab. Inst., Little Rock, 1985—; chief rehab. medicine svc. VA Med. Ctr., Little Rock, 1984—. Editor: Interdisciplinary Rehabilitation of Multiple Sclerosis and Neuromuscular Disease, 1984; editor, author: Physical Medicine & Rehabilitation State of the Art Reviews, 1987, Primer on Management, 1987, Rehabilitation of Aging, 1989, Management for Rehabilitation Medicine II, 1993; alt. editor: Archives of Physical Medicine and Rehabilitation, 1989-93. Mem. exec. bd. Greater No. Colo. Chpt. of Muscular Dystrophy Assn. of Am., 1972-82; spl. edn. adv. com. Cherry Creek Sch. Dist., Denver, 1975, vice chmn., 1976, chmn., 1977; med. advisor Denver Commn. on Disabled and Coun. on Aging, Denver, 1980-82, Denver Commn. on Human Svcs., 1982-84; external examiner King Saud U. Med. Sch., Saudi Arabia, 1983; med. adv. bd. Ark. Multiple Sclerosis Soc., Little Rock, 1985-88; chmn. coun. Assn. Acad. Physiatrists, Indpls., 1992—. Fellow Am. Acad. Phys. and Rehab.; mem. AMA, Am. Congress of Rehab. Medicine, Am. Acad. Cerebral Palsy, Am. Pub. Health Assn., Am. Bd. Physical Medicine and Rehbilitation (dir. 1988—), Soc. for Exptl. Biology and Medicine, Assn. Acad. Physiatrists, Ark. Med. Soc., Pulaski County Med. Soc., Soc. for Neuroscis. Office: U of Ark for Med Scis 4301 W Markham Slot 602 Little Rock AR 72205

MALONEY, MILFORD CHARLES, internal medicine educator; b. Buffalo, Mar. 15, 1927; s. John Angelus Maloney and Winifred Hill; m. Dione Ethyl Sheppard. BS, Canisius Coll., 1947, postgrad., 1947-49; MD, U. Buffalo, 1953. Diplomate Am. Bd. Internal Medicine. Rsch. chemist Buffalo Electrochem. Co., 1947-49; intership Mercy Hosp./Georgetown U., 1953-54; med. residency Buffalo VA Hosp., 1954-56; cardiology fellow Buffalo Gen. Hosp., 1956-57; chmn. dept. medicine Mercy Hosp., 1969—; program dir., internal medicine residency Mercy Hosp., Buffalo, 1972-89; with steering com. Assn. Program Dirs. in Internal Medicine, 1976, coun. mem., 1977-80; clin. prof. medicine SUNY, Buffalo, 1981—; trustee Am. Soc. Internal Medicine, 1984—; edn. leader med. seminar Am. Soc. Internal Medicine, Austria, Switzerland, France, 1987, Argentina, Brazil, Paraguay, 1988; bd. dirs. Internal Medicine Ctr. for Advancement and Rsch. Edn.; pres. Heart Assn. Western N.Y., Buffalo, 1969; sr. cancer rsch. physician Roswell Park Meml. Cancer Inst., 1959-62. Editor (newsletter) N.Y. State Soc. Internal Medicine, 1972-78. bd. dirs. Health Systems Agy. Western N.Y., Buffalo, 1981; exec. com., bd. dirs. Blue Cross Western N.Y., Buffalo, 1987; bd. regents Canisius Coll., Buffalo, 1987—; mem. pres. assocs. SUNY, Buffalo. Capt. M.C., U.S. Army, 1957-59. Recipient Award of Merit N.Y. State Soc. Internal Medicine, 1980, Man of Yr. award Heart Assn. Western N.Y., 1982, Ann. Honoree award Trocai Coll., 1986, Disting. Alumni award Canisius Coll., 1991, Berkson Excellence award in Teaching and Art of Medicine SUNY at Buffalo, 1992; named to Sports Hall of Fame, Canisius Coll., 1978. Fellow ACP (Upstate Physician Recognition award 1989), Am. Coll. Cardiology; mem. N.Y. State Soc. Internal Medicine (pres.), Med. Alumni Assn. SUNY (pres. 1975), Med. Soc. County Erie (pres. 1969), Am. Soc. Internal Medicine (trustee 1984—, pres. 1990-91, chmn. long range planning com.), Internal Medicine Purchasing Group of Am. (bd. dirs.), Buffalo Club, N.Y. Athletic Club. Home: 116 Covepoint Ln Williamsburg VA 23185 Office: Mercy Hosp Buffalo 565 Abbott Rd Buffalo NY 14220-2095

MALOOLEY, DAVID JOSEPH, electronics and computer technology educator; b. Terre Haute, Ind., Aug. 20, 1951; s. Edward Joseph and Vula (Starn) M. B.S., Ind. State U., 1975: M.S., Ind. U., 1981, doctoral candidate. Supr., Zenith Radio Corp., Paris, Ill., 1978-79; assoc. prof. electronics and computer tech. Ind. State U., Terre Haute, 1979—; cons. in field. Served to 1st lt. U.S. Army, 1975-78. Mem. ASCD, Soc. Mfg. Engrs., Nat. Assn. Indsl. Tech., Am. Vocat. Assn., Instrument Soc. Am. (sr.), Phi Delta Kappa, Pi Lambda Theta, Epsilon Pi Tau. Democrat. Christian. Home: 11420 Spring Creek Rd Terre Haute IN 47805-9679 Office: Ind State U Terre Haute IN 47809

MALPASS, ROY SOUTHWELL, psychology educator; b. Amsterdam, N.Y., Sept. 23, 1937; s. Clement Southwell and Ursula Caroline (Smith) M.; m. Nancy Jo Smith, Dec. 26, 1960; children: Andrea Lyn, Wendy Jo. BS, Union Coll., 1959; MA, New Sch. for Social Rsch., 1961; PhD, Syracuse U., 1968. Rsch. asst. Rsch. Found Children's Hosp. D.C., 1961-62; lectr. psychology Mt. Allison U., Sackville, N.B., Can., 1962-65; NDEA title IV fellow Syracuse (N.Y.) U., 1965-67; asst. prof. psychology U. Ill., Champaign, 1967-73; assoc. prof., prof. behavioral sci. SUNY, Plattsburgh, 1973-92; prof. psychology, dir. criminal justice program U. Tex., El Paso, 1992—; vis. prof. psychology Nanjing Normal U., China, 1991. Co-editor: Psychology and Culture, 1992, Field Methods in Cross Cultural Rsch., 1986, Eyewitness Testimony: Psychological Perspectives, 1984; editor: Jour. of Cross Cultural Psychology, 1982-86. Recipient Excellence in Teaching award Grad. Student Assn., 1973, Chancellor's award SUNY, 1982. Fellow Am. Psychology Assn., Am. Psychol. Soc.; mem. Soc. for Exptl. Social Psychology, Internat. Assn. Cross Cultural Psychology (pres.), Internat. Assn. Applied Psychology (pres. psychology and law divsn.), Soc. for Cross Cultural Rsch. (pres.). Achievements include initial study of cross racial of differences in face recognition and context reinstatement effect in eyewitness identification. Office: U Tex Dept Psychology El Paso TX 79968-0553

MALWITZ, NELSON EDWARD, chemical engineer; b. Newark, Aug. 14, 1946; s. Walter and Virginia (Doerr) M.; m. Marguerite Ann Schwartz, July 31, 1947; children: Jonathan Edward, David Jared. BS in Chem. Engring., Newark Coll. Engring., 1968; MS in Chem. Engring., Lehigh U., 1970. Process engr. Air Products & Chems., Allentown, Pa., 1969-75; rsch. mgr. Sealed Air Corp., Danbury, Conn., 1975-85, dir. comml. devel., 1985-90, rsch. dir., 1990—; environ. affairs chair Sealed Air Corp., Saddle Brook, N.J., 1986—. Contbr. articles in chemistry kinetics to profl. jours. V.p. Here's Life Danbury, 1975-85; bd. mem. Grace Community Ch., Brookfield, Conn., 1982-92; mem. Creative Edn. Found., Buffalo, 1984—. Mem. ASTM (task chmn. 1968-71), Am. Chem. Soc., Soc. Plastics Engrs., Electric Overstress/Electrostatic Discharge, Nat. Assn. Environ. Mgrs. Republican. Achievements include 7 patents on polyurethane insulation and fire retarding, fire retarding other plastics, polyvinyl alcohol foam; development of polyurethane packaging foams now std. in the industry. Office: Sealed Air Corp 10 Sherman Turnpike Danbury CT 06810

MAMMI, MARIO, chemist, educator; b. Reggio Emilia, Italy, July 2, 1932; s. Aldo and Nerina (Cattani) M.; m. Romea Torreggiani, Oct. 12, 1957; children: Stefano, Isabella, Cristina. Diploma di Maturita, Liceo Scientifico, Reggio Emilia, 1951; Laurea in Chimica, U. Padova, Italy, 1956; Libera Docenza, U. Padua, Italy, 1963. Lectr. U. Padua, 1959-71, asst., 1961-71, prof. chemistry, 1971—; sci. leader Structural Chemistry Group, Padova, 1964—; dir. Inst. Organic Chemistry, Padova, 1972-74, CNR Biopolymer Rsch. Ctr., Padova, 1975—; pres. CNR Rsch. Area, Padova; mem. Univ. Adminstrn. Coun. of U. Padova, 1984—; coun. mem. Univ. Computing Ctr., 1982—. Italy sub-editor: 6th, 7th, and 8th World Directory of Crystallographers, 1981, 86, 90. Recipient Mion award U. Padova, 1971, Musaia Gold medal, 1983. Mem. Italian Crystallographic Assn. (sec. 1971-75, pres. 1985-87), Italian Crystallographic Commn. (pres. 1989—), Consortium Padova Richerche (coun. mem. 1987—). Home: Via Marco Lando 20, Padua Italy 35133 Office: U Padua Dept Organic Chem, Via Marzolo 1, Padua Italy 35131

MAMMONE, RICHARD JAMES, engineering educator; b. N.Y.C., Sept. 3, 1953; s. Americo Anth and Helen (Kowalski) M.; m. Valerie Altman, June 29, 1981 (div. 1988); m. Christine Podilchuk, Aug. 19, 1989; children from previous marriage: Robert, Jason, Richard, James Jr.. BE, CCNY, 1975, M.E., 1977; PhD, CUNY, 1981. Computer systems analyst Picatinny Arsenal, Dover., N.J., 1975-77; research fellow CCNY, 1977-81; asst. prof. Manhattan Coll., Riverdale, N.Y., 1981-82; assoc. prof. engring. Rutgers U., Piscataway, N.J., 1981-93, prof., 1993—; co-founder Computed Anatonomy Inc., N.Y.C., 1982; cons. in field. Co-Author Image Recovery: Theory and Applications, Acad. Press Pubs., 1987, Computational Methods of Signal Recovery and Recognition, 1992; co-editor: Neural Networks: Theory and Applications, 1991; editor: Artificial Neural Networks for Speech and Vision, 1993; editor Pattern Recognition Jour., 1989—; series editor Chapman-Hall on Neural Networks, 1991—; editor artificial neural networks speech and vision Chapman-Hall Pubs.; contbr. articles to profl. jours.; patentee in field. Assoc. Whitaker Found. grantee, 1982, NSF grantee, 1992; Internat. Tel. & Tel. grantee, 1984; CAIP Research Ctr. grantee, 1985; Henry Rutgers fellow, 1985-87; U.S. Nat. Security Agy. grantee, 1986—; USAF grantee, 1986—; Temeplex grantee, 1986—. Mem. IEEE (sr., editor Communications Jour. 1983-89), N.Y. Acad. Scis. Office: Rutgers U Dept Elec Engring Piscataway NJ 08854

MAMON, GARY ALLAN, astrophysicist; b. N.Y.C., Feb. 7, 1958; s. Michel and Enna (Grebelsky) M.; m. Gabrielle Rodan, Aug. 24, 1990. SB in Physics, MIT, 1979; PhD in Astrophys. Scis., Princeton U., 1985. Asst. rsch. scientist dept. physics NYU, N.Y.C., 1985-88; maître de confs. UFR de Physique, U. de Paris 7, 1988-89; computer engr. ANSTEL, SA, Paris, 1989-90; astronome adjoint DAEC, Observatoire de Paris-Meudon, Meudon, France, 1990—. Prin. editor 2nd DAEC Meeting, Distribution of Matter in the Universe, 1992; contbr. articles to Astrophys. Jour., Astronomy & Astrophysics, Astrophys. Jour. Letters, Monthly Notices of the Royal Astron. Soc. Thaw fellow Princeton U. Obs., 1979. Mem. Am. Astron. Soc. (assoc.), Sigma Xi (assoc.), Nu Delta (v.p. 1978-79). Achievements include research in galaxy dynamics in groups and clusters, in galaxy formation, in stellar dynamics in elliptical galaxies, and in circumstellar chemistry; co-investigator DENIS two-micron survey. Office: DAEC Observatoire de Meudon, F 92195 Meudon France

MAMULA, MARK, aerospace engineer; b. Pitts., June 27, 1953; s. Milan and Millicent (Brnlovich) M.; m. Brenda Susan Allen, June 8, 1975; children: Sarah, Matthew, Rebecca. BS in Aerospace Engring., VPI & SU, Blacksburg, Va., 1975; MS in Aeronaut. Engring., U. Dayton, 1979. Indsl. engr. Fairchild Republic Co., Hagerstown, Md., 1975-76; commd. USAF, 1976, advanced through ranks to lt. col., 1976—; C141B ops. officer 7th Airlift Squadron, Travis AFB, Calif.; asst. prof. U.S. Mil. Acad., West Point, N.Y., 1986-89; instructor pilot USAF, McGuire AFB, N.J., 1981-86. V.p. Nativity of Virgin Mary Ch., Madison, Ill., 1992—. Mem. AIAA, Order Daedalions. Republican. Eastern Orthodox. Home: 3345 Wallace Dr Pittsburgh PA 15227 Office: 7th Airlift Squadron Travis AFB CA 94535

MAN, CHI-SING, mathematician, educator; b. Hong Kong, Aug. 23, 1947; s. Yip and Sau-Ying (Leung) M.; m. May Lai-Ming Chan, July 5, 1973; children: Li-Xing, Yi-Heng. BSc, U. Hong Kong, 1968, MPhil, 1976; PhD, Johns Hopkins U., 1980. Tutor of math. and physics Hong Kong Bapt. Coll., 1970-72, asst. lectr. of physics, 1972-76; postdoctoral fellow Johns Hopkins U., Balt., 1980-81; asst. prof. civil engring. U. Manitoba, Winnipeg, 1981-85; rsch. assoc. Inst. for Math. and Its Applications U. Minn., Mpls., 1984-85; asst. prof. of math. U. Ky., Lexington, 1985-88, assoc. prof., 1988—; vis. assoc. prof. U. Minn., 1991, U. Manitoba, 1992. Contbr. articles to profl. jours. Grantee Natural Scis. and Engring. Rsch. Coun. of Can., 1982-86, NSF, 1987—. Mem. Am. Math. Soc., Internat. Soc. Offshore and Polar Engrs., Soc. Natural Philosophy (svc. mem., sec. 1992-93). Achievements include research on acoustoelastic measurement of residual stress, elasticity with initial stress, stress waves in lungs, constitutive equation for creep of ice, foundations of continuum thermodynamics and Gibbsian thermostatics. Home: 348 Melbourne Way Lexington KY 40502 Office: U Ky Dept of Math Lexington KY 40506-0027

MAN, KIN FUNG, physicist, researcher; b. Hong Kong, July 1, 1957; came to U.S., 1987; s. Kam Mun and Hang Mui Man; m. Vivian Kay-Wan, 1986; children: Han Bin, Yan Bin. BSc, U. London, 1980; PhD, U. Sussex, Brighton, England, 1984. Chartered physicist. Rsch. fellow U. Sussex, Brighton, 1984-85; rsch. asst. U. Coll. London, England, 1985-87; rsch. assoc. Ukaea Culham Lab., Oxon, England, 1985-87, Nat. Rsch. Coun., Pasadena, Calif., 1987-89; mem. tech. staff, task leader Calif. Inst. Tech., Pasadena, 1989—. Contbr. articles to Jour. of Physics B, Phys. Rev. Letters, Rev. Sci. Instruments, Phys. Rev. A, Am. Inst. Physics Proceedings, Jour. Chem. Physics. Recipient Rsch. Studentship award Sci. and Engring. Rsch. Coun. England, Brighton, 1980-83, Rsch. Associateship award Nat. Rsch. Coun.-Nat. Aero. and Space Adminstrn., Pasadena, 1987-89. Mem. Inst. Physics (London), European Phys. Soc., Am. Phys. Soc. Achievements include development of a merged electron-ion beams apparatus for the measurement of absolute, cascade-free excitation cross sections of ions and an electrostatic levitation system for microgravity containerless materials processing. Office: Calif Inst Tech Jet Propulsion Lab 183-901 4800 Oak Grove Dr Pasadena CA 91109

MAN, XIUTING CHENG, acoustic and ultrasonic engineer, consultant; b. Liaoning, Peoples Republic China, Oct. 5, 1965. BSc, Beijing U., 1987, MSc, 1990. Rsch. asst. Beijing U., 1987-90; teaching asst. dept. mech. engring. U. Houston, 1990-91, rsch. asst., 1991—. Contbr. rsch. papers, abstracts to Jour. Acoustical Soc. Am. Mem. Acoustical Soc. Am., Acoustical Soc. China. Office: Univ Houston Dept Mech Engring Houston TX 77204-4792

MANABE, SHUNJI, control engineering educator; b. Tokyo, Jan. 16, 1930; s. Yoshio and Katsuko (Nishimura) M.; m. Yasuko Sasaki, Mar. 16, 1958; children: Kanji, Kohji. B. Engring., Tokyo U., 1952; MS, Ohio State U., 1954; PhD, Tokyo U., 1962. Rsch. engr. Mitsubishi Electric Corp. Cen. Rsch. Lab., Amagasaki, Hyogo, Japan, 1952-65; sect. mgr. Mitsubishi Precision Co., Kamakura, Kanagawa, Japan, 1965-71; mgr. control and simulation Mitsubishi Electric Corp. Computer Works, Kamakura, Kanagawa, 1971-78; chief engr. Mitsubishi Electric Corp. Kamakuka Works, Kamakura, Kanagawa, 1978-90; prof. control engring. dept. Tokai U., Hiratsuka, Kanagawa, Japan, 1990—. Author: Space Technology, 1986; contbr. articles to Jour. of Inst. Elec. Engrs. of Japan, Jour. of Soc. Instrument and Control Engrs. Referee Inst. Elec. Engrs. of Japan, Tokyo, 1974-77, bd. dirs., 1977-79; bd. dirs. Soc. Instrument and Control Engrs., Tokyo, 1985-87. Achievements include development of design theory of automatic control systems; research in wind tunnel drive; in radar antenna tracking control; in flight simulators; in spacecraft control. Home: 1-8-12 Kataseyama, Fujisawa Kanagawa 251, Japan Office: Tokai U, 1117 Kitakaname, Hiratsuka Kanagawa 259-12, Japan

MANABE, SYUKURO, climatologist; b. Shingu-Mura, Uma-Gun, Ehimeken, Japan, Sept. 21, 1931; came to U.S., 1958; s. Seiichi and Sueko (Akashi) M.; m. Nobuko Nakamura, Jan. 21, 1962; children: Nagisa M. Bianchini, Yukari C. BS, Tokyo U., 1953, MS, 1955, DS, 1958. Rsch. meteorologist U.S. Weather Bur., Washington, 1958-63; sr. rsch. meteorologist Geophysical Fluid Dynamics Lab., NOAA, Washington, 1963-68; sr. rsch. meteorologist Geophysical Fluid Dynamics Lab., NOAA, Princeton, N.J., 1968—, mem. sr. exec. svc., 1979—; adj. prof. Princeton U., 1968—; mem. joint sci. com. World Climate Rsch. Program, 1981-87; mem. bd. on atmospheric sci. and climate NRC, 1988-91, mem. Commn. on Geoscis., Environment and Resources, 1990-93; mem. panel on climate and global change NOAA, 1988—. Recipient Fujiwara award Japan Meteorol. Soc., 1967, gold medal U.S. Dept. Commerce, 1970, Presdl. Rank Meritorious Exec. award Pres. of U.S., 1989, Acad. award of Blue Planet Prize, Asahi Glass Found., Japan, 1992. Fellow Am. Geophys. Union (Revelle medal 1992), Am. Meteorol. Soc. (Meisinger award 1967, 2nd Half Century award 1977, Carl-Gustav Rossby Rsch. medal 1992); mem. NAS. Home: 8 Princeton Ave Princeton NJ 08540-5236 Office: NOAA Geophys Fluid Dynamics Lab Princeton U PO Box 308 Princeton NJ 08542-0308

MANAKER, ARNOLD MARTIN, mechanical engineer, consultant; b. N.Y.C., Feb. 11, 1947; s. Paul Bernard and Rose Norma (Malakoff) M.; m. Ellen Conant, Nov. 21, 1970; children: Ryan Scott, Heidi Cora, Jana Ashley. BSME, Newark Coll. of Eng., 1968; MS in Mech. Engring., U. Mass., 1970, PhD in Mech. Engring., 1973. Asst. to plant engr. J. Wiss & Sons Co., Newark, 1965-68; rsch. asst. Dept. Mech. and Aerospace Engring., U. Mass., Amherst, 1968-73, teaching asst., 1970-72; pvt. practice cons., 1972—; staff engr., Clinch River Breeder Reaction Project TVA, Oak Ridge,

1976-77; mech. engr. TVA, Chattanooga, 1977-79, project mgr. adv. energy, 1979, project mgr. AFBC R & D, 1979-81, project mgr. AFBC Demonstration Plant, 1984-88, AFBC devel. project engr., 1988-91, mgr. NOx/CEMS projects, 1991—; adminstrv. asst. NASA, 1968-69; rsch. asst. NSF, 1971-73. Contbr. numerous articles to profl. jours. Vol. United Way Leadership Club, 1987-92, Friends Always Indian Guide Program/YMCA, 1988-92. Named Engr. of Yr., TVA, 1989; recipient Product Champion award EPRI, 1992. Fellow ASME (officer coms.). Jewish. Avocations: racquetball, tennis, baseball, sports car racing. Home: 9420 Mountain Shadows Dr Chattanooga TN 37421-3444 Office: TVA 1101 Market St Chattanooga TN 37402-2801

MANASREH, OMAR M., electronics engineer, physicist; b. Hebron, Palestine, July 1, 1952; came to U.S., 1978; s. Omar A. and Sarah Manasreh; m. Ann Rutledge, Aug. 20, 1982; children: Sarah, Hannah. BS in Physics, U. Jordan, Ammon, 1976; MS in Physics, U. Puerto Rico, Rio Piedras, 1980; PhD in Physics, U. Ark., 1984. Instr. dept. physics Yarmouk U., Irbid, Jordan, 1977; rsch. asst. dept. physics U. Puerto Rico, Rio Piedras, 1978-80; elem. lab. technician dept. physics U. Ark., Fayetteville, 1980, teaching asst., 1980-84, grad. asst. dept. elec. engring., 1984-85; rsch. adjoint dept. physics U. Sherbrooke, Can., 1985-86; postdoctoral rsch. assoc. physics dept. Sam Houston State U., Huntsville, Tex., 1986-87; rsch. assoc. Nat. Rsch. Coun. Materials Lab., Wright-Ptterson AFB, Ohio, 1988-89; electronics engr. Solid State Electronics Directorate, Wright Lab., Wright-Ptterson AFB, 1989—, project leader, 1992—; vis. scientist Materials Lab., Wright-Patterson AFB, 1987-88; speaker at numerous tech. meetings., confs., presentations. Editor: (with others) Degradation Mechanisms in III-V Compound Semiconductor Devices and Structures, 1990, Semiconductor Quantum Wells and Superlattices for Long Wave-Length Infrared Detector, 1993, Low Temperature Molecular Beam Epitaxial III-V Materials: Physics and Applications, 1993; contbr. 80 articles to Synthetic Metals, Physics Rev., Applied Physics Letter, numerous others. Mem. IEEE, Am. Phys. Soc., Materials Rsch. Soc., Electrochem. Soc., Sigma Xi (Aubrey E. Harvey award 1985). Home: 1906 Sugar Run Trl Bellbrook OH 45305 Office: Solid State Electronics Wright-Patterson AFB Dayton OH 45433-6543

MANBECK, HARVEY B., agriculturist, educator; b. Reading, Pa., Jan. 11, 1942; m. Glenda Manbeck; children: Eric, Christina. BS, Pa. State U., 1963, MS in Agrl. Engring., 1965; PhD in Engring., Okla. State U., 1970. Rsch. assoc. agrl. engring. dept. Pa. State U., 1965, instr. agrl. engring. dept., 1966, prof. agrl. engring., 1980—; asst. prof. agrl engring. dept. U. Ga., 1970-75, assoc. prof. agrl. engring. dept., 1977-80; assoc. prof., extension agrl. engr. Ohio State U., 1975-77; adminstrv. intern rsch. office Pa. Agrl. Experiment Sta., 1991-92; vis. prof. agrl. engring. U. Manitoba, 1986-87, Shenyang Agrl. U., 1988. Contbr. chpts. to books, articles to profl. jours. Coach Little League Baseball, 1981-84, leader YMCA Indian Princess Longhouse, 1983-85, Webelo's Cub Scouts 1984, com. mem. Troop 31 Boy Scouts of Am. 1985—. Mem. ASCE, Am. Soc. Agrl. Engrs. (mem. structures group, vice chair 1978-79, chair 79-81, Pa. state sect., sec.-treas. 1983-84, chair 1985-86, named Outstanding Tchr. in Agrl. Engring. 1972, various other coms.), Nat. Soc. Profl. Engrs., Nat. Frame Builders Assn. (mem. editorial review com. for the post-frame profl., chair 1988—), Ga. Soc. Profl. Engrs. (state dir. at large 1974-75, named Outstanding Young Engr. of the Yr. 1972, recipient Outstanding Chpt. pres. award, 1974, various other coms.), Ohio Soc. Profl. Engrs., Am. Soc. for Engring. Edn., Forest Products Rsch. Soc., Jaycees, Penn. State Coaly Soc., Gamma Sigma Delta, Alpha Epsilon, Sigma Xi, numerous other mems. Achievements include design and construction of an addition to a controlled environment broiler housing research facility at U. Ga., a transport system for handling live broilers in bulk, a model grain bin for measurement of thermally induced loads. Home: 912 Anna St Boalsburg PA 16827 Office: Penn St U 210 Agr Engring Bldg University Park PA 16802

MANCALL, ELLIOTT LEE, neurologist, educator; b. Hartford, Conn., July 31, 1927; s. Nicholas and Bess Tuch M.; m. Jacqueline Sue Cooper, Dec. 27, 1953; children: Andrew Cooper, Peter Cooper. B.S., Trinity Coll., Hartford, 1948; M.D., U. Pa., 1952. Diplomate: Am. Bd. Psychiatry and Neurology (dir. 1983—). Intern Hartford Hosp., 1952-54; clk. in neurology Nat. Hosp. Nervous Disease, London, 1954-55; asst. resident neurology Neurol. Inst. N.Y., 1955-56; resident neuropathology Mass. Gen. Hosp., 1956-57, also clin. and research fellow, 1957-58; teaching fellow neuropathology Harvard Med. Sch., 1956-57; asst. prof. neurology Jefferson Med. Coll., 1958-64, asso. prof., 1964-65; prof. medicine Hahnemann Med. Coll. and Hosp., 1965-76, prof. neurology, chmn. dept., 1976—; dir. Hahnemann Univ. Hosp. (Huntington's Disease Diagnostic and Referral Center); dir. Hahnemann U. ALS Clinic; chmn. bd. dirs. Phila. Profl. Standards Rev. Orgn., 1981-84; del. Am. Bd. Med. Specialties, 1984—. Author: (with others) The Human Cerebellum: A Topographical Atlas, 1961, (with B.J. Alpers) Clinical Neurology, 1971, Essentials of the Neurological Examination, 1971, 81; contbr. numerous articles to profl. jours. Served with USN, 1945-47. Recipient Christian R. and Mary F. Lindback award, 1969; Oliver Meml. prize ophthalmology U. Pa., 1952. Fellow Am. Acad. Neurology (alt. del. to AMA 1982-86, gen. editor CONTINUUM 1991—); mem. Am. Neurol. Assn., Am. Assn. Neuropathology, Assn. Research in Nervous and Mental Diseases, Soc. Neuroscience, AAUP, Pa. Med. Peer Rev. Orgn. (dir. 1979-84), Phila. Neurol. Soc., Alpers Soc. Clin. Neurology, Coll. Physicians Phila., Neurology, Sydenham Coterie, Phila. County Med. Soc., Pa. State Med. Soc., AMA (sec.-treas. sect. council neurology 1983-86), Am. Med. Soc. on Alcoholism, Neurology Intersoc. Liaison Group, Intersoc. Com. Neurol. Resources, Assn. Univ. Profs. Neurology (pres. 1988-90), Soc. for Exptl. Neurology, Am. Bd. Med. Specialties (exec. bd., chair COSEP, 1992—), Am. Bd. Psychiatry and Neurology (v.p. 1990, del. to Am. Bd. Med. Specialities, emeritus dir. 1991—). Home: 2088 Harts Ln Miquon PA 19452 Office: Broad and Vine Sts Philadelphia PA 19102*

MANCE, ANDREW MARK, chemist; b. Braddock, Pa., Jan. 21, 1952; s. Andrew and Virginia (Borushko) M.; m. Susan Beth Duffey, June 8, 1974; children: William Andrew, Steven Mark. BA, Thiel Coll., Greenville, Pa., 1973; PhD, Carnegie Mellon U., 1979. Sr. rsch. scientist electrochemistry dept. GM Rsch. Labs., Warren, Mich., 1978-87, sr. rsch. scientist elec. and electronics engring. dept., 1987-89, staff rsch. scientist, 1989—. Mem. AAAS, Am. Chem. Soc., Am. Ceramic Soc., Am. Electroplaters and Surface Finishers Soc. Achievements include patents for wire-glass composite and method of making same, UV induced copper-catalyzed electroless deposition onto styrene-derivative polymer surfaces, selective laser pyrolysis of metallorganics as a method of forming patterned thin films, a method of forming tungsten oxide films, a method of applying metal catalyst patterns onto ceramic for electroless copper deposition. Home: 1127 Grove Ave Royal Oak MI 48067 Office: GM Rsch Labs 30500 Mount Rd Box 9055 Warren MI 48090

MANCHESTER, CAROL ANN FRESHWATER, psychologist; b. Coshocton, Ohio, Sept. 30, 1942; d. James M. and Kathleen C. (Call) Freshwater; m. Crosby Manchester, Mar. 16, 1963 (dec. 1973). BS, Ohio State U., 1963, MS, 1973, PhD, 1977. Diplomate Internat. Soc. Psychotherapy and Behavioral Medicine. Elem. counselor Columbus (Ohio) Pub. Schs., 1973-79; counselor Regional Alcoholism & Tng. Ctr., Columbus, 1977-79; therapist Beechwold C.inic, Columbus, 1977-80; counselor Gifted and Talented Program, Columbus, 1979-81; dir. Freshwater Mental Health Clinic, Columbus, 1982—; asst. clin. prof. Coll. Medicine Ohio State U., 1990-92; instr. psychology Urbana Coll., Columbus, 1977-79; dir. Freshwater House Clinic, Columbus, 1983—; bd. dirs. Ecole Francaise, Columbus, 1985—; cons. Columbus Community Hosp., 1988—, Mt. Carmel Med. Ctr., Park Med. Ctr., The Halterman Ctr., Columbus, 1990—; presentor in field. Author: Affective Model The Gifted and Talented Handbook for Columbus Public Schools, 1981. Active Gov.'s Task Force on Child Abuse, Columbus. Recipient Disting. Svc. award Ohio Counselor's Assn., Valley Forge Freedom award. Mem. ACLU, AOA, Am. Acad. Cert. Neurotherapists (v.p. 1993), Nat. Soc. Clin. Hypnosis, Meninger Soc., Internat. Soc. Post Traumatic Stress, Internat. Soc. Multiple Personality Disorder, Assn. Applied Psychophysiology and Biofeedback, Ohio Psychol. Assn., Delta Omicron, Tau Beta Sigma. Office: Freshwater House Clinic 114 Buttles Ave 6065 Glick Rd Ste C Powell OH 43065

MANCIET, LORRAINE HANNA, physiologist, educator; b. Tucson, Ariz., Dec. 19, 1950; d. Hector Juan Benitez Encinas and Lilian Eloisa (Hanna) M.; children: Salvatore Rocco, Michael Hector, Anthony Joseph. BA, U. Ariz., 1972, PhD, 1989. Budget analyst City of Tucson, 1972-75; rsch. assoc. dept. surgery U. Ariz., Tucson, 1989-90, rsch. asst. prof., 1990—; mem. grant peer rev. com. Am. Heart Assn., Ariz., 1992—; interviewer med. sch. applicants U. Ariz., 1992-93; mentor NIH Minority High Sch. Rsch. Apprentice Program, Tucson, 1990, 92, 93. Contbr. articles to Jour. Thoracic Cardiovascular Surgery, Jour. Heart Lung Transplant. Bd. dirs. Acad. Preparation for Excellence, Tucson, 1990—; active Ariz. Assn. for Chicanos in Higher Edn., Tucson, 1991—; mgr. Sahuaro Little League Baseball Team, Tucson, 1990; sec. Western Little League Bd. Dirs., 1992—. Mem. Am. Physiol. Soc. (assoc.), Microcirculatory Soc., AAAS, U. Ariz. Hispanic Alumni Assn. Democrat. Office: U Ariz Health Scis Ctr Dept Surgery 1501 N Campbell Ave Tucson AZ 85724

MANCINI, ERNEST ANTHONY, geologist, educator, researcher; b. Reading, Pa., Feb. 27, 1947; s. Ernest and Marian K. (Filbert) M.; m. Marilyn E. Lee, Dec. 27, 1969; children—Lisa L., Lauren N. B.S., Albright Coll., 1969; M.S., So. Ill. U., 1972; Ph.D., Tex. A&M U., 1974. Petroleum exploration geologist Cities Service Oil Co., Denver, 1974-76; asst. prof. geology U. Ala., Tuscaloosa, 1976-79, assoc. prof., 1979-84, prof., 1984—; state geologist, oil and gas supr. State Ala., Tuscaloosa, 1982—. Recipient Nat. Council Citation, Albright Coll., 1983, Pratt-Haas Disting. Lectr. award, 1987-88. Fellow Cushman Found.; mem. Am. Assn. Petroleum Geologists (A.I. Levorsen petroleum geology Meml. award Gulf Coast assn., geol. socs. sect. 1980), Assn. Am. State Geologists (past pres.), Geol. Soc. Am., Soc. Econ. Paleontologists and Mineralogists Gulf Coast sect. (hon., past pres.), Paleontol. Soc. (past pres. southeast sect.), N.Am. Micropaleontology Soc., Internat. Micropaleontology Soc., Ala. Geol. Soc. (past pres.), Sigma Xi, Phi Kappa Phi (past chpt. pres.), Phi Sigma. Democrat. Presbyterian. Contbr. articles to profl. jours. Home: 15271 Four Winds Loop Northport AL 35476 Office: Geological Survey of Ala PO Box O 420 Hackberry Ln Tuscaloosa AL 35486

MANCINI, MARY CATHERINE, cardiothoracic surgeon, researcher; b. Scranton, Pa., Dec. 15, 1953; d. Peter Louis and Ferminia Teresa (Massi) M. BS in Chemistry, U. Pitts., 1974, MD, 1978. Diplomate Am. Bd. Surgery (speciality cert. critical care medicine), Am. Bd. Thoracic Surgery. Intern in surgery U. Pitts., 1978-79, resident in surgery, 1979-87; fellow pediatric cardiac surgery Mayo Clinic, 1987-88; asst. prof. surgery, dir. cardiothoracic transplantation Med. Coll. Ohio, Toledo, 1988-91; assoc. prof. surgery, dir. cardiothoracic transplantation La. State U. Med. Ctr., Shreveport, 1991—. Author: Operative Techniques for Medical Students, 1983; contbr. articles to profl. jours. Recipient Pres.'s award Internat. Soc. Heart Transplantation, 1983, Charles C. Moore Teaching award U. Pitts., 1985, Internat. Woman of Yr. award 1992-93; rsch. grantee Am. Heart Assn., 1988. Fellow ACS, Am. Coll. Chest Physicians, Internat. Coll. Surgeons (councillor 1991—); mem. Assn. Women Surgeons, Rotary (gift of life program 1991). Roman Catholic. Achievements include first multiple organ transplant, in Louisiana, 1993. Office: La State U Med Ctr 1501 Kings Hwy Shreveport LA 71103-4228

MANCINI, NICHOLAS ANGELO, psychologist; b. Newark, May 30, 1944; s. Joseph Michael and Anne Marie (Caputo) M.; m. Gail Helene Reichman, Mar. 20, 1976; children: Michael, Daniel, Brant. BBA in Bus. Adminstrn., Upsala Coll., 1967; MA in Gen. Exper. Psychology, Fairleigh Dickinson, 1975; postgrad., Profl. Sch. Psychology, 1985. Diplomate Psychotherapy and Behavioral Medicine, Family Counseling; cert. biofeedback therapist; cert. clin. psychology, N.J.; lic. psychologist Pa., Calif. County clin. psychologist Cape May County and Mental Health Svc., Inc., Crest Haven, N.J., 1977-85; clin. psychologist N.J. Div. Devel. Disabilities Vineland Devel. Ctr./Hosp., 1985-86, N.J. Div. Youth/Family Svcs., Vineland Childrens Residential, 1986-87; clin. dir., community re-integration P.L.U.S. Inc., for traumatic brain injured, Bensalem, Pa., 1987-89; dir. psychology and behavioral medicine Rising Sun, Mirman and St. Vincent med. ctrs., Phila., Springfield, 1987—; dir. psychology Pa. Pain Rehab. Ctr., 1989-90; dir. biofeedback rehab. and behavior therapy Univ. Med. Ctr./Cooper Hosp.-U. Medicine and Surgery N.J., Camden, 1990-91; founder, dir. Adaptive Behavior Ctrs., Inc. Mem. Phila. Neuropsychology Group, Soc. Behavioral Medicine, Am. Psychol. Soc. (charter), Assn. of Applied Psychophysiology and Biofeedback, Lions. Avocations: tennis, swimming, boating, classical music, wood working. Home: 4 Chapel Hill RD Huntingdon Valley PA 19006 Office: Rising Sun Med Ctr 6412 Rising Sun Ave Philadelphia PA 19111-5229

MANCINI, ROBERT KARL, computer analyst, consultant; b. Burbank, Calif., May 13, 1954; s. Alfred Robert and Phyllis Elaine (Pflugel) M.; m. Barbara Diane Bacon, Aug. 4, 1979; children: Benjamin Robert, Bonnie Kathryn, Brandon Peter. BA in Econs., UCLA, 1976; cert. in bibl. studies, Multonmah Sch. of the Bible, 1981; MBA, Santa Clara (Calif.) U., 1987. Process clk. Am. Funds Svc. Co., L.A., 1976-77; exec. asst. Sierra Thrift & Loan Co., San Mateo, Calif., 1977-78; sci. programming specialist Lockheed Missiles & Space Co., Sunnyvale, Calif., 1978-90; product mgr. Diversified Software Systems Inc., Morgan Hill, Calif., 1990—; cons. Mancini Computer Svcs., San Jose and Morgan Hill, 1985—; instr. Heald Coll., San Jose, Calif., 1990. Mem. fin. coun. Hillside Ch., 1990-91; mem. blue ribbon budget rev. com. City of Morgan Hill, 1992. Mem. Phi Kappa Sigma (expansion com. 1976-78). Republican. Avocations: tennis, gardening, restoring vintage autos. Home: PO Box 1602 Morgan Hill CA 95038-1602

MANCINI, ROCCO ANTHONY, civil engineer; b. Prezza, Abruzzi, Italy, Aug. 16, 1931; came to U.S., 1940; s. Salvatore and Bambina (Tulliani) M.; m. Eileen Clifford, Apr. 11, 1959; children: Charles V., Ann Marie, Linda F., Donna. BSCE, MIT, 1953; cert. in transp. engring., Yale U., 1958. Registered profl. engr., Mass. Supr. devel. planning Mass. Bay Transp. Authority, Quincy, 1966-68, project mgr. traffic engring., devel., 1976-79, mgr. transp. systems mgmt. projects, 1979-80, acting dir. program devel., 1980-82, rep. to Met. Planning Orgn., 1982-87, project mgr. constrn. office, 1987-89, asst. dir. constrn., 1989-92, asst. dir. design and constrn., 1992—; cons. spl. transp. projects, City of Honolulu, 1968-69, Sao Paulo, Brazil, 1968-69; prin. transp. engr. Wilbur Smith and Assocs., 1969-70; New Eng. regional mgr. Tippetts-Abbett-McCarthy-Stratton, 1970-76. Contbr. articles to profl. publs. Mem. Milton (Mass.) Town Meeting, active various town coms. With U.S. Army, 1954-55. Automotive Safety Found. fellow Yale U., 1957-58. Fellow Inst. Transp. Engrs. (pres. New Eng. sect. 1965-66); mem. ASCE (transp. com. 1953—). Achievements include development, design and supervision of transportation construction projects. Home: 49 Columbine Rd Milton MA 02186 Office: Mass Bay Transp Authority 1515 Hancock St Quincy MA 02169

MAND, RANJIT SINGH, device physicist, educator; b. Nairobi, Kenya, Sept. 28, 1956; came to U.S., 1989; s. Naranjan Singh and Gurdev Kaur (Dhanova) M.; 1 child, Anreet Kaur. BS in Solid State Electronics, U. Bradford, Eng., 1981, PhD, 1985; MS, U. Wales, Bangor, 1983. Cons. AT&T Bell Labs., Murray Hill, N.J., 1982-85; Toshiba fellow Toshiba R&D Centre, Kawasaki, Japan, 1985-86; mem. sci. staff Bell Northern Rsch., Ottawa, Ont., Can., 1987-89; lectr. U. Carlton, Ottowa, 1988-89; mgr. optoelectronics Furukawa Electric Techs., Santa Clara, Calif., 1990—; lectr. U. Santa Clara, Calif., 1990—; bd. dirs. Optical Communications Products, Inc., Chatsworth, Calif. Contbr. articles to Electronic Letters, Applied Physics Letters, Jour. Applied Physics, Electron Device Letters, 1985-89, noew assoc. editor; referee IEE Electronic Letters, 1987—. Toshiba R & D Ctr. fellow, 1985; recipient City & Guilds of London Insigna award, 1989. Mem. IEEE (sr.), Inst. Elec. Engrs. (corp.), Inst. Physics (corp.), Engring. Coun. of Eng. (charterd), Am. Phys. Soc. Achievements include patents for optoelectronic distable apparatus, monolithic integration of electronic and optoelectronic devices; research on device physics of electronic and optoelectronic devices and OEICs. Office: Furukawa Electric Techs 900 Lafayette St Ste 401 Santa Clara CA 95050-4966

MANDAL, KRISHNA PADA, radiation physicist; b. Jessore, Bengal, India, June 30, 1937; came to U.S., 1967; s. Panchanan and Rupasree (Dhar) M.; m. Bharati Das, Feb. 22, 1966; children: Konoy, Binita. BS (with honors), Dacca U., East Pakistan, 1957; MS, Dacca U., 1959, Fla. state U., 1968; PhD, Georgetown U., 1974. Cert. in radiation and med. physics. Postdoctoral trainee in med. physics Howard U. Cancer Ctr., Washington, 1974-75; instr., 1975-76; asst. prof. radiology Albany (N.Y.) Med. Ctr., 1977-84; radiation physicist Radiology Cons., P.A., Winter Haven, Fla., 1984—; radiation safety officer Winter Haven (Fla.) Hosp., 1984—. Author: Phys. Rev., 1977, Internat. Jour. of Radiation, Oncology, Biology, Physics, Vol. 6, 4, 1980, AMPI Med. Physics Bulletin, Vol. 7, 4, 1982, Physics in Canada, Vol. 32, 1976. V.p., pres. Tri-City India Assn., Albany, 1980-84; pres. Bengali Soc. of Fla., Lakeland, Orlando, Tampa, Fla., 1988-92. Mem. Am. Assn. Physicists in Medicine, Assn. Med. Physicists in India. Republican. Hindu. Office: Winter Haven Hosp 200 Ave F NE Radiation Oncology Winter Haven FL 33880

MANDAVILLI, SATYA NARAYANA, chemical engineering educator, researcher; b. Visakha, India, Sept. 27, 1932; came to U.S., 1991; s. Srinivasa Rao and Sreerama (Ratnamma) M.; m. Ranga Lakshmi, May 22, 1959; children: Raja Sekaran, Rathan Shekar, Uma. BS with honors, Andhra (India) U., 1954, MS, 1955; PhD, Indian Inst. Tech., Kharagpur, 1959. Prof. Indian Inst. Tech., Madras, 1960-90, chmn. dept., 1975-78; prof. chem. engring. U. P.R., Mayaguez, 1991—; postdoctoral fellow U. Calif., Davis, 1963-64; vis. prof. Berlin and Karlsruhe Univs., Germany, 1979, U. Hannover, Germany, 1984, 85, 86; Fulbright prof. Drexel U., Phila., 1981; mem. Ctrl. Crisis Group for Chem. Disasters, Delhi, India, 1988-92; mem. Indian Bur. of Standards, Delhi, 1986-91; mem. All India Coun. Tech. Edn., Delhi, 1982-83. Editor: Indian Chemical Engineer, 1984; author 2 books, numerous tech. articles. Mem. Am. Inst. Chem. Engrs., Am. Chem. Soc., Biomed Engring. Soc., Indian Inst. Chem. Engrs. (life; pres. 1982-83, chmn. tech. sessions 1988-90), Sigma Xi. Hindu. Achievements include Indian patents for method for improvment of product compositon for transesterification process, for tubular flow reactor with motionless mixing elements. Office: Post Box 5814 College Sta Mayaguez PR 00681-5814

MANDEL, ELLIOTT DAVID, structural engineer, consultant; b. White Plains, N.Y., May 22, 1961; s. Herbert Maurice and Charlotte (Feldman) M.; m. Marlene Miller, Apr. 12, 1992; 1 child, Jordan Aaron Mandel. BSCE, U. Tex., 1984, MS in Engring., 1989. Lic. profl. engr., Va., Calif. Asst. structural engr. Howard Needles Tammen & Bergendoff, Boston, 1985-86; structural engr. Parsons DeLeuw Inc., Washington, 1989—. Mem. ASCE, Am. Concrete Inst. Achievements include seismic retrofit of bridges, San Francisco, seismic retrofit of I-280 double deck viaduct damaged in 1989 Loma Prieta earthquake, new bridge design schemes for behavior of bridges in seismic regions. Home: 115 N Cleveland St Arlington VA 22201 Office: Parsons Deleuw Inc 1133 15th St NW Washington DC 20005

MANDEL, HERBERT MAURICE, civil engineer; b. Port Chester, N.Y., May 11, 1924; s. Arthur William and Rose (Schmeiser) M.; m. Charlotte Feldman, Aug. 22, 1954; children: Rosanne Mandel Levine, Elliott D., Arthur M. B.S.C.E., Va. Poly. Inst., 1948, M.Engring., Yale U., 1949. Registered profl. engr., N.Y., Conn., Fla., Md., Mich., Minn., Ohio, Pa., R.I., Va., W.Va. Structural engr. Madigan Hyland Co., Long Island City, N.Y., 1949-50; with firm Parsons, Brinckerhoff, Quade & Douglas, Inc., 1950-86; v.p. GAI Cons. Inc., 1986—, project mgr., Atlanta, 1962, N.Y.C., 1963-70, Honolulu, 1970-74, v.p., 1974, sr. v.p., Pitts., 1977-86; faculty Yale U., 1948-49; adj. faculty Bklyn. Poly. Inst., 1956-64, U. Pitts., 1986; gen. chmn. 6th Internat. Bridge Conf., Pitts., 1989. Author tech. papers; prin. works include (prin.-in-charge) Williamstown-Marietta Bridge, W.Va.-Ohio, Dunbar Bridge, W.Va., I-64 Bridge over Big Sandy River, W.Va.-Kentucky, Davis Creek Bridge, Charleston, W.Va., Tygart R. Bridge, W.Va. (project mgr.) Newport Bridge, Narragansett Bay, R.I., (designer/project engr.) Hackensack River Bridge, N.J., Housatonic River Bridge, Conn., Arthur Kill Vertical Lift R.R. Bridge, S.I., N.Y., 62nd St. Bridge, Pitts., Savannah River Cantilever Bridge, Ga., I-84 Bridges, Danbury, Conn., (structural rehab designer) Avondale Bridge, N.J., Lincoln Bridge, N.J., B&O R.R. Bridge, Ind., Hawk St. Viaduct, Albany, N.Y., Congress Ave. Bridge, Austin, Tex., Ohio St. Bridge, Buffalo, Panhandle Bridge, Pitts.; project mgr. Interstate Rt. H-3, Honolulu, 1970-74; project dir. design and constrn. Pitts. Light Rail Transit System, 1977-84; designer Elizabeth R. Tunnel, Norfolk, Va., 1950. Served to 1st lt. U.S. Army, 1943-46, 50-52; ETO. Fellow ASCE, Soc. Am. Mil. Engrs. (pres. Pitts. post 1987-88); mem. Am. Ry. Engring. Assn. (steel structures specifications com., 1974—), Nat. Soc. Profl. Engrs., Profl. Engrs. in Pvt. Practice, Pa. Profl. Engrs. in Pvt. Practice (state vice-chmn. 1992—), Internat. Assn. Bridge and Structural Engring., Assn. for Bridge Constrn. and Design, Tau Beta Pi, Chi Epsilon, Omicron Delta Kappa, Phi Kappa Phi, Pi Delta Epsilon, Scabbard and Blade. Jewish. Club: Engineers (Pitts.). Home: 920 Parkview Dr Pittsburgh PA 15243-1116 Office: GAI Cons Inc 570 Beatty Rd Monroeville PA 15146-1300

MANDEL, MORTON, molecular biologist; b. Bklyn., July 6, 1924; s. Barnet and Rose (Kliner) M.; m. Florence H. Goodman, Apr. 1, 1952; children: Robert, Leslie. BCE, CUNY, 1944; MS, Columbia U., 1949, PhD in Physics, 1957. Scientist Bell Telephone Labs., Murray Hill, N.J., 1956-57; asst. prof. physics dept. Stanford (Calif.) U., 1957-61; scientist Gen. Telephone & Telegraph, Mountain View, Calif., 1961-63; rsch. assoc. dept. genetics Stanford U., 1963-64; rsch. fellow Karolinska Inst., Stockholm, Sweden, 1964-66; assoc. prof. sch. of medicine U. Hawaii, Honolulu, 1966-68, prof., 1968—; cons. Fairchild Semiconductor, Hewlett Packard, Lockheed, Rheem, Palo Alto, Calif., 1957-61. Contbr. articles to profl. jours. Lt. (j.g.) USN, 1944-46. Recipient Am. Cancer Soc. Scholar award Am. Cancer Soc., 1979-80, Eleanor Roosevelt Internat. Cancer fellowship, 1979; named NIH Spl. fellow Karolinska Inst., 1964-66. Fellow Am. Phys. Soc.; mem. Sigma Xi. Achievements include citation classics, optional conditions for mutagenesis by N-methyl-N1-nitro-N-nitrosoquauidine in E. coli K12; calcium dependent bacteriophage DNA infection. Office: Dept of Biochemistry 1960 East-West Rd Honolulu HI 96844

MANDEL, RONALD JAMES, neuroscientist; b. Atlanta, Nov. 22, 1957; s. Benjamin A. and Suzanne (Carr) M.; m. Marie Perna, Aug. 5, 1984; children: Samantha Annika, Grace Emily. BS, Duke U., 1979; PhD, U. So. Calif., 1986. Post doct. fellow U. Calif. San Diego, La Jolla, 1986-88, U. Lund (Sweden), 1988-91, asst. prof. U. Ill., Champaign, 1991-93, sect. head Somatix Therapy Corp., Almeda, Calif., 1992—. Contbr. articles to profl. jours. Recipient Young Investigator award Alzheimer's Disease Assn., Chgo., 1991-94, Vis. Scientist award Swedish Med. Coun., 1988-89; Internat. Brain Rsch. Orgn. fellow, 1990, John E. Fogarty fellow NIH, 1990. Mem. AAAS, Am. Psm. Psychol. Soc., Soc. Neurosci. Office: Somatix Therapy Corp 850 Marina Village Pkwy Alameda CA 94501

MANDELBROT, BENOIT B., mathematician, scientist, educator; b. Warsaw, Poland, Nov. 20, 1924; came to U.S., 1958; s. Charles and Belle (Lurie) M.; m. Aliette Kagan, Nov. 5, 1955; children: Laurent, Didier. Diploma, Ecole Polytechnique, Paris, 1947; MS in Aerospace, Calif. Inst. Tech., 1948; PhD in Math., U. Paris, 1952; D.Sc. (hon.), Syracuse U., 1986, Laurentian U., 1986, Boston U., 1987, SUNY, 1988, U. Bremen, 1968; DSc (hon), Union Coll., 1993; D.Sc. (hon.), U. Dallas, 1992; Union Coll., 1993, D.Sc. (hon.), 1993; DHL (hon.), Pace U., 1989. Jr. mem. and Rockefeller scholar Inst. for Advanced Study, Princeton, N.J., 1953-54; jr. prof. math. U. Geneva, Switzerland, 1955-57, U. Lille and Ecole Polytechnique, Paris, 1957-58; research staff mem. IBM Watson Research Center, Yorktown Heights, N.Y., 1958-74; IBM fellow IBM Watson Rsch. Center, Yorktown Heights, N.Y., 1974-93; vis. prof. econs. Harvard U., 1962-63, vis. prof. applied math., 1963-64, vis. prof. math., 1979-80, prof. practice math., 1984-87; vis. prof. engring. Yale U., 1970, prof. math. scis., 1987—; vis. prof. physiology Einstein Coll. Medicine, 1970; Hitchcock prof. U. Calif., Berkeley, 1992; visitor MIT, 1953; also Inst. lectr.; visitor U. Paris, 1966; visitor Coll. de France, Paris, 1973, and various times, Institut des Hautes Etudes Scientifiques, Bures, 1980, Mittag-Leffler Inst., Sweden, 1984, Max Planck Inst. Math., Bonn, 1988; lectr. Yale U., 1970, Cambridge U., 1990, Oxford U., 1990, Imperial Coll., London, 1991; speaker and organizer profl. confs. Author: Logique, Langage et Théorie de l'Information, 1957, Les Objets Fractals: Forme, Hasard et Dimension, 1975, 3rd edit., 1989 (trans. to Italian, Spanish, Basque, Bulgarian and Portuguese), Fractals: Form, Chance and Dimension, 1977, The Fractal Geometry of Nature, 1982 (trans. to Chinese, German, Japanese, Polish and Spanish), La Geometria della Natura, 1987; contbr. articles to profl. jours. Recipient Franklin medal Franklin Inst., 1986, Alexander von Humboldt Preis, 1987, Caltech Disting. Svc. award, 1988, Moet-Hennessy prize, 1988, Harvey prize, 1989, Nev. prize U. Nev. System, 1991, Wolf prize for physics, 1993; nat. lectr. Sigma Xi, 1980-82; Guggenheim fellow, 1968. Fellow AAAS, IEEE (Charles Proteus Steinmetz medal 1988), Am. Acad. Arts and Scis., European Acad. Arts,

Scis. and Humanities, Am. Phys. Soc., IBM Acad. Tech., Inst. Math. Stats., Econometric Soc., Am. Geophys. Union, Am. Statistic Assn.; mem. NAS U.S.A. (fgn. assoc., Barnard medal 1985), Internat. Statis. Inst. (elected), Am. Math. Soc., French Math. Soc. Achievements include orgination of theory of fractals, an interdisciplinary enterprise concerned with financial data and all other shapes and phenomena that are equally rough, irregular or broken-up at all scales. Office: IBM PO Box 218 Yorktown Heights NY 10598-0218 also: Yale U Math Dept PO Box 2155 New Haven CT 06520

MANDELKERN, LEO, biophysics and chemistry educator; b. N.Y.C., Feb. 23, 1922; s. Israel and Gussie (Krostich) M.; m. Berdie Medvedoff, May, 1946; children: I. Paul, Marshal, David. BA, Cornell U., 1942, PhD, 1949. Postdoctoral rsch. assoc. Cornell U., Ithaca, N.Y., 1949-52; phys. chemist Nat. Bur. Standards, Washington, 1952-62; prof. chemistry and biophysics Fla. State U., Tallahassee, 1962—, R.O. Lawton Disting. prof., 1984—; vis. prof. U. Miami (Fla.) Med. Sch., 1963, U. Calif. Med. Sch., San Francisco, 1964, Cornell U., 1967; mem. biophysics fellowship com. NIH, 1967-70; mem. study panel crystal growth and morphology NRC, 1960; cons. in field. Author: Crystallization of Polymers, 1964, An Introduction to Macromolecules, 1972, 1983; contbr. numerous articles to profl. jours. 1st lt. USAAF, 1942-46, PTO. Recipient Meritorious Svc. award U.S. Dept. Commerce, 1957, Arthur S. Fleming award Washington Jaycees, 1958, Mettler award N.Am. Thermal Analysis Soc., Phila., 1984, Disting. Svc. in Advancement of Polymer Sci. award Soc. Polymer Sci., Japan, 1993. Mem. AAAS, N.Y. Acad. Scis., Am. Inst. Chemists, Am. Chem. Soc. (Polymer Chemistry award 1975, Fla. award 1984, Rubber div. Whitby award 1988, Charles Goodyear medal 1993, Applied Polymer Sci. award 1989, award for disting. svc. in advancement of polymer sci. 1993), Polymer Soc. Japan, Biophys. Soc., Am. Phys. Soc. (Outstanding Educator of Am. 1973, 75), Cosmos Club Washington, Alpha Epsilon Pi. Home: 1503 Old Ft Dr Tallahassee FL 32301-5637 Office: Fla State U Dept Chemistry Tallahassee FL 32306

MANDELL, ROBERT LINDSAY, periodontist, researcher; b. Lowell, Mass., Feb. 2, 1951. BS in Biology, Boston Coll., Newton, Mass., 1972; DMD, Boston U., 1975; CAGS and MMSc, Harvard U., 1980. Diplomate Am. Bd. Peridontology; lic. dentist, Mass., Ga., N.H.; cert. dentist N.E. Regional Dental Bds. Intern in gen. dentistry Eastman Dental Ctr., Rochester, N.Y., 1975-76; gen. practice fellow Beth Israel Hosp., Boston, 1976-77; postdoctoral fellow in periodontology and oral microbiology Harvard Dental Sch. and Forsyth Dental Ctr., Boston, 1977-80; asst. prof. in periodontology Med. Coll. Va., Richmond, 1980-81, Emory Sch. Dentistry, Atlanta, 1982; staff assoc. Forsyth Dental Ctr., Boston, 1983—; pvt. practice in periodontics Tyngsboro, Mass., 1984—. Contbr. articles to Jour. Clin. Microbiology, Jour. Periodontology, Jour. Dental Rsch., Infection Immunology, Jour. Dentistry for Children, Jour. Clin. Periodontology. NIH grantee, 1977-80; recipient Young Investigators award, 1982-85, Hoyt Labs. award, 1982, Whitehall Labs. award, 1991, Omnigene Inc. award, 1992. Mem. Internat. Assn. Dental Rsch., Am. Acad. Periodontology (Orban prize 1980), Am. Dental Soc., Mass. Dental Soc., Met. Dental Soc., Lowell Soc., Boston Coll. Order of Cross and Crown. Achievements include research on juvenile periodontitis, on effectiveness of NaF in controlling suspected periodontopathic organisms, and on smoking, diabetes and periodontal disease. Office: Forsyth Dental Ctr 140 Fenway Boston MA 02115-3799

MANDELSTAM, STANLEY, physicist; b. Johannesburg, South Africa, Dec. 12, 1928; came to U.S., 1963; s. Boris and Beatrice (Liknaitzky) M. BSc, U. Witwatersrand, Johannesburg, 1952; BA, Cambridge U., Eng., 1954; PhD, Birmingham U., Eng., 1956. Boese postdoctoral fellow Columbia U., N.Y.C., 1957-58; asst. rsch. physicist U. Calif., Berkeley, 1958-60, prof. physics, 1963—; prof. of math. physics U. Birmingham, 1960-63; vis. prof. physics Harvard U., Cambridge, Mass., 1965-66, Univ. de Paris, Paris, 1979-80, 84-85. Editorial bd. The Phys. Rev. jour., 1978-81, 85-88; contbr. articles to profl. jours. Recipient Dirac medal and prize Internat. Ctr. for Theoretical Physics, 1991. Fellow AAAS, Royal Soc. London, Am. Phys. Soc. (Dannie N. Heineman Math. Physics prize 1992). Jewish. Office: Univ of Calif Dept of Physics Berkeley CA 94720

MANDERS, KARL LEE, neurosurgeon; b. Rochester, N.Y., Jan. 21, 1927; s. David Bert and Frances Edna (Cohan) Mendelson; m. Ann Laprell, July 28, 1969; children: Karlanna, Maidena; children by previous marriage: Karl, Kerry, Kristine. Student, Cornell U., 1946; MD, U. Buffalo, 1950. Diplomate Am. Bd. Neurol. Surgery, Am. Bd. Clin. Biofeedback, Am. Bd. Hyperbaric Medicine, Nat. Bd. Med. Examiners. Intern U. Va. Hosp., Charlottesville, 1950-51, resident in neurol. surgery, 1951-52; resident in neurol. surgery Henry Ford Hosp., Detroit, 1954-56; pvt. practice Indpls., 1956—; med. dir. Community Hosp. Rehab. Ctr. for Pain, 1973—; chief hosp. med. and surg. neurology Community Hosp., 1983, 93; coroner Marion County, Ind., 1977-85, 92—. Served with USN, 1952-54, Korea. Recipient cert. achievement Dept. Army, 1969. Fellow ACS, Internat. Coll Surgeons, Am. Acad. Neurology; mem. Am. Assn. Neurol. Surgery, Congress Neurol. Surgery, Internat. Assn. Study of Pain, Am. Assn. Study of Headache, N.Y. Acad. Sci., Am. Coll. Angiology, Am. Soc. Contemporary Medicine and Surgery, Am. Holistic Med. Assn. (co-founder), Undersea Med. Soc., Am. Acad. Forensic Sci., Am. Assn. Biofeedback Clinicians, Soc. Cryosurgery, Pan Pacific Surg. Assn., Biofeedback Soc. Am., Acad. Psychosomatic Medicine, Pan Am. Med. Assn., Internat. Back Pain Soc., North Am. Spine Soc., Am. Soc. Stereotaxic and Functional Neurosurgery, Soc. for Computerized Tomography and Neuroimaging, Ind. Coroners Assn. (pres. 1979), Royal Soc. Medicine, Nat. Assn. Med. Examiners, Am. Pain Soc., Midwest Pain Soc. (pres. 1988), Am. Acad. Pain Medicine, Cen. Neurol. Soc., Interurban Neurosurg. Soc., Internat. Soc. Aquatic Medicine, James A. Gibson Anat. Soc., Am. Bd. Med. Psychotherapists (mem. profl. adv. council), James McClure Surg. Soc., Brendonwood Country Club, Highland Country Club. Home: 5845 High Fall Rd Indianapolis IN 46226-1017 Office: 7209 N Shadeland Ave Indianapolis IN 46250-2021

MANDICH, NENAD VOJINOV, chemical industry executive; b. Kos Mitrovica, Serbia, Yugoslavia, Dec. 25, 1944; came to U.S., 1971; s. Vojin P. and Jelica (Zarkovic) M.; m. Olga M. Mamula, 1975; children: Petar, Nikola. Diploma in engring., U. Belgrade, Yugoslavia, 1970; MSc, Roosevelt U., 1990; postgrad., Aston U., Birmingham, U.K. cert. electroplater-finisher. Rsch. chemist Coral Chem. Co., Waukegan, Ill., 1971-72; tech. dir. KCI Chem. Co., LaPorte, Ind., 1972-74; sr. process engr. Sunbeam Corp., Chgo., 1974-78; plant mgr. Fed. Tool & Plastics, Evanston, Ill., 1978-79; sr. corp. project engr. Apollo Metals, Bridgeview, Ill., 1979-80; project mgr. Bunker-Ramo, Brodview, Ill., 1981-82; pres., chief exec. officer HBM Electrochem. & Engring. Co., Lansing, Ill., 1982—. Contbr. more than 20 articles to profl. jours. Fellow Inst. Metal Finishing; mem. Am. Electroplating Soc. (chmn. hard chrome plating com. 1990—), Internat. Soc. Electrochemistry, Electrochem. Soc., N.Y. Acad. Sci., Tesla Meml. Soc. (bd. dirs. 1989—). Mem. Eastern Orthodox Serbian Ch. Achievements include 8 patents pending. Home: 1650 Ridge Rd Homewood IL 60430-1831 Office: HBM Electrochem Engring 2800 Bernice Rd Lansing IL 60438-1271

MANDORINI, VITTORIO, research physicist; b. Taranto, Italy, Apr. 30, 1941; s. Antonio and Rosa (De Bellis) M. D Phys. Scis., U. Bari, Italy, 1966. With Istituto Richerche Breda, Milan, 1966—, head dept. strenght of materials, 1975-84, asst. to pres., 1984-88, rsch. mgr., 1988—. Contbr. articles to profl. publs. Mem. ASME, Assn. Italiana Analisi Sollecitazioni, European Group on Fracture, Associazione Italiana Prove non Distruttive. Office: Istituto Richerche Breda, Viale Sarca 336, 20126 Milan Italy

MANDRI, DANIEL FRANCISCO, psychiatrist; b. Camaguey, Cuba, Apr. 22, 1950; came to U.S., 1962; s. Adalberto Froilan and Estrella (Pereiro) M.; m. Monica A. Ruffing, May 21, 1983; 1 child, Nicholas. MD, U. Cen. Del Este, Dominican Republic, 1977. Diplomate Am. Bd. Psychiatry and Neurology. With internal medicine PGY-1 Christ Hosp., Oak Lawn, Ill., 1979-80; with psychiatry PGY 2 plus 3 U. Miami/Jackson Meml. Hosp., Miami, Fla., 1980-82, chief resident psychiatry, 1982-83; pvt. practice psychiatry Coral Gables, Fla., 1983-86; dir. acute care unit Broward County Mental Health Div., Hollywood, Fla., 1986-87; dir. psychiat. svcs. Douglas Gardens Community Mental Health Ctr., Miami, 1987—, Douglas Gardens Home and Hosp. for the Aged, Miami, 1989-92; asst. instr. psychiatry dept. of psychiatry U. Miami, 1982-83. Mem. N.Y. Acad. Scis., Am. Psychiatry Assn., World Psychiat. Assn., World Fedn. for Mental Helath, Am. Assn. Community Psychiatrists. Office: Douglas Gardens Community Mental Health Ctr 701 Lincoln Rd Miami FL 33139-2879

MANDRIOLI, DINO GIUSTO, computer scientist, educator; b. Roma, Lazio, Italy, Mar. 5, 1949; s. Crisanto and Adriana (Serafini) M.; m. Maria Cristina Malinverni, Nov. 30, 1980; children: Leonardo, Laura. Maturità Classica, Liceo Berchet, Milano, Italy, 1967; Laurea Ing. Elettronica, Politecnico, Milano, 1972; Laurea Matematica, Univ. Statale, Milano, 1976. Profl. engr. Researcher CNR (Italian Rsch. Coun.), Milano, 1975-80; asst. prof. Politecnico Di Milano, 1976-79, assoc. prof., 1979-80, prof., 1984—; prof. Univ. Statale, Udine, Italy, 1981-83; cons. Olivetti, Ivrea, Italy, 1976-80, Enel Italia Energy Agy., Milan, 1986-91, ENEA (Italian Nuclear Power), Rome, 1983-88; vis. scholar UCLA, U. Calif., Santa Barbara, HP Rsch. Labs., L.A., Palo Alto, Calif., 1976, 81, 89. Author: Theoretical Computer Science, 1987, Fundamentals of Software Engineering, 1991; editor/author: Advances in Object Oriented SE, 1991; contbr. over 70 articles to profl. jours. With Italian armed forces, 1973-74. Recipient Univac award, 1983, Cray award, 1992, Phillip Morris award, 1992. Mem. ACM, IEEE, N.Y. Acad. Scis. Avocations: fencing, mountain hiking, music. Office: Poly Milan Dept Electronics, PL da Vinci 32, 20133 Milan Italy

MANEGGIO, LIZETTE, civil engineer; b. N.Y.C., Aug. 17, 1962; d. Edward John and Theresa Rose (Mastroserio) Sukla. BSCE, Manhattan Coll., 1984. Registered profl. engr., Nev. Civil engr. Holmes & Narver, Inc., Las Vegas, 1984-87; civil engr. USAF-Nellis AFB, Las Vegas, 1987-90, chief engr., 1990; chief environ. mgmt. USAF, Nellis AFB, 1990-91; dep. dir. environ. mgmt. USAF-Wright Patterson AFB, Dayton, Ohio, 1992—. Recipient Cash Notable Achievement award, 1987, Superior Performance award, 1990, 91; named Engr. of the Yr., 1989. Mem. NSPE, Soc. Am. Engrs., Federally Employed Women (v.p. 1990). Roman Catholic. Home: 2037 Mill Run Ln Bellbrook OH 45305

MANFREDI, MARIO ERMINIO, neurologist educator; b. Genoa, Liguria, Italy, Aug. 5, 1934; s. Leonardo Carlo and Alessandra (Bianchetti) M.; m. Maddalena Irene Canepa, June 6, 1966; children: Carlo Luigi, Giovanni Walter. MD, U. Genoa, Italy, 1958, Bd. in Neuropsychiatry, 1961; Libera Docenza, Nat. Examination, Rome, Italy, 1966, Idoneita Primariale, 1970. Rsch. fellow Nat. Rsch. Coun., Genoa, Italy, 1962-63; asst. prof. U. Genoa, Italy, 1964-70; fellow NIH Washington U., St. Louis, 1966-67; asst. prof. U. Rome Sapienza, 1970-84; prof. U. Aquila, 1972-75; prof. neurology Nat. U. Somalia, Mogadishu, 1976; prof. U. Rome Sapienza, 1984—; cons. neurologist Santa Corona Hosp., Pietra Ligure, Italy, 1965, Nuovo Regina Margherita Hosp., Rome, 1970-81. Contbr. numerous articles to profl. jours. Grantee Nat. Rsch. Coun., 1969—, Ministry of Health, 1980—. Mem. Italian League Against Epilepsy (councillor 1973-79), Italian Soc. EEG (councillor 1978-80), Italian Soc. Neurology (treas. 1978-81, sec. 1981-83), Internat. Med. Soc. Motor Disturbances (pres. 1986-88), FOREP Epilepsy and Related Syndromes Rsch. Found. (pres. 1989—). Home: Via Montevideo 21, 00198 Rome Italy Office: Dept Neurosciences, Viale dell'Universita 30, 00185 Rome Italy

MANFREDINI, STEFANO, medicinal chemistry educator; b. Vigarano Mainarda, Ferrara, Italy, Dec. 29, 1956; s. Lino and Anna Maria (Govoni) M.; m. Francesca Casotti, May 3, 1987. D., U. Ferrara, 1980. Secondary sch. tchr. Ferrara, 1982-83; postdoctoral fellow Glaxo Co., Verona, Italy, 1983-87; asst. prof. medicinal chemistry U. Ferrara, 1987—; vis. scientist Brigham Young U., Provo, Utah, 1987, 88. Contbr. articles to Jour. Med. Chemistry. Cpl. Italian Army, 1981-82. Mem. Am. Chem. Soc., Internat. Soc. Antiviral Rsch. Achievements include patent in field. Office: Dept Medicinal Scis U Ferrara, Via Fossato Mortara 17-19, I-44100 Ferrara Italy

MANGANIELLO, EUGENE JOSEPH, retired aerospace engineer; b. N.Y.C., June 8, 1914; s. Joseph and Mary (Pascuso) M.; m. Helen M. Gosney, Nov. 1974; children: Eugene J., Gail, Gui, Gwen. BS in Engring., CCNY, 1934, MS in Engring., 1935. Jr. engr. NASA, Langley Field, Va., 1963-42; sect. head NASA, Cleve., 1942-45, rsch. chief, 1945-70, deputy dir., 1970-74. Mem. AIAA, Soc. Aerospace Engrs., Soc. Automotive and Aerospace. Republican. Home: 329 Hillcrest Dr Encinitas CA 92024

MANGASER, AMANTE APOSTOL, computer engineer; b. Quezon City, The Philippines, Mar. 1, 1953; s. Anacleto C. and Paz J. (Apostol) M.; m. Maria Bernardita M. Roman, Dec. 2, 1978; children: Diana Marie, Amanda Bernadine. BSEE magna cum laude, U. Philippines, 1975; MSEE, Philips Internat. Inst., Eindhoven, The Netherlands, 1980; MS in Computer Engring., U. Philippines, 1982; PhD in Elec. and Computer Engring., U. Calif., Santa Barbara, 1990. Asst. prof. U. Philippines, Quezon City, 1980-85; teaching asst. U. Calif., Santa Barbara, 1985-86, researcher, 1986-90; sr. mem. tech. staff Computer Motion Inc., Goleta, Calif., 1989—; cons. to numerous govt. offices, pvt. corps., The Philippines, 1977-85. Contbr. articles to profl. jours. Mem. Tau Beta Pi, Phi Kappa Phi. Roman Catholic. Office: Computer Motion Inc 250 Storke Rd Ste # A Goleta CA 93117

MANGELS, ANN REED, nutrition educator, researcher; b. Jacksonville, Fla., July 31, 1956; d. John Jr. and Ann (Lyon) M.; m. Arnold L. Alper; 1 child, Sarah Ida. BS, Fla. State U., 1977; MS, Case Western Reserve U., 1978; PhD, U. Md., 1989. Registered dietition; lic. dietitian. Clin. dietitian U. Hosps. Cleve., 1979-80, systems dietitian, 1980-84; rsch. asst. U. Wis., Madison, 1984-85; rsch./tng. asst. U. Md., College Park, 1986-89; rsch. assoc. Georgetown U., Washington, 1989-92; instr. U. Md., College Park, 1992; nutrition advisor Vegetarian Resource Group, Balt., 1990—; thesis advisor Hood Coll., Frederick, Md., 1991-92. Reviewer Jour. Am. Dietetic Assn., 1988—, Am. Jour. Clin. Nutrition, 1989—; contbr. articles to profl. jours. Prenatal nutritionist Washington Free Clinic, 1987-89. Recipient fellowship U. Md., 1988, scholarship Mead Johnson, 1988; named Outstanding Female Grad., AAUW, 1989. Mem. Am. Dietetic Assn., Am. Inst. Nutrition, Am. Soc. Clin. Nutrition, Vegetarian Nutrition Dietetic Practice Group (chmn.-elect 1992—), Sigma Xi.

MANGER, WILLIAM MUIR, internist; b. Greenwich, Conn., Aug. 13, 1920; s. Julius and Lilian (Weissinger) M.; m. Lynn Seymour Sheppard, May 30, 1964; children: William Muir, Jr., Lilian Wade, Stewart Sheppard, Charles Seymour. BS, Yale U., 1944; MD, Columbia U., 1946; PhD, Mayo Found., U. Minn., 1958. Intern, Presbyn. Hosp., N.Y.C., 1946-47, resident, 1949-50; fellow internal medicine Mayo Found., 1950-57; asst. physician Presbyn. Hosp., 1957—; dir. Manger Rsch. Found., 1961-77; clin. asst. vis. physician Columbia div. Bellevue Hosp., 1964-68; asst. attending physician NYU Bellevue Hosp. 1969-77; assoc. attending physician, 1977-83, attending physician, 1983—; instr. medicine Columbia U. Coll. Phys. and Surg., 1957-66, assoc. medicine, 1966-70, lectr., 1981—; asst. attending physician Presbyn. Hosp., 1966—; asst. clin. prof. medicine N.Y.U. Med. Ctr., 1968-75, assoc. clin. prof. medicine, 1975-83, prof. clin. medicine, 1983—; mem. Internat. Med. Council on Drug Use, 1977—; mem. devel. com. Mayo Clinic, 1981-87; vice chmn. bd. Manger Hotels, Inc., 1957-73. Mem. bd. govs. St. Albans Sch., Washington, 1958-64, 67-73, 83-89, chmn., 1967-69; trustee Found. Rsch. in Medicine and Biology, 1971-77, Buckley Sch., 1975-85, Found. for Advancement Internat. Rsch. in Microbiology, 1977-82, Thyroid Found., 1980-85; mem. bd. visitors Boston U. Med. Sch., 1992—; trustee Found. for Depression and Manic Depression, 1978-89, pres., 1980-89; elder Presbyn. Ch., 1968-70, 92-93, trustee, 1962-67, 80-84, deacon, 1959-61. Lt. (j.g.) M.C., USNR, 1947-49. Recipient Meritorious Rsch. award Mayo Found., 1955, Disting. Alumnus award, 1992. Diplomate Nat. Bd. Med. Examiners, Am. Bd. Internal Medicine. Fellow ACP, Acad. Psychosomatic Medicine, Am. Geriatric Soc., Coun. on Geriatric Cardiology, N.Y. Acad. Medicine (admission com. 1976-78, edn. com. 1979-92), Am. Coll. Cardiology, Am. Coll. Clin. Pharmacology, Royal Soc. Health, Am. Inst. Chemists; trustee Nat. Hypertension Assn. (chmn. 1977—), AMA, Am. Soc. Internal Medicine, N.Y. State Med. Soc., N.Y. County Med. Soc., Am. Heart Assn. (fellow council on circulation and council for high blood pressure rsch.), Nat. High Blood Pressure Edn. Program (mem. Coord. Com.), Inter-Am. Soc. Hypertension, Internat. Soc. Hypertension, Am. Thoracic Soc., N.Y. Acad. Sci., AAAS, Am. Physiol. Soc., Am. Chem. Soc., Am. Soc. Pharmacology and Exptl. Therapeutics, Am. Soc. for Clin. Pharmacology and Therapeutics, Clin. Autonomic Rsch. Soc., Med. Strollers, N.Y.C., Endocrine Soc., Pan Am. Med. Assn., Harvey Soc., Soc. Exptl. Biology and Medicine, Rsch. Discussion Group (founding mem., sec.-treas. 1958-80), Am. Fedn. Clin. Rsch., Am. Soc. Nephrology, Royal Soc. Medicine (affiliate), Fellows Assn. Mayo Found. (v.p., pres. 1953), Mayo Alumni Assn. (v.p. 1981-82, exec. com. 1981-89, pres. elect 1982-85, pres. 1985-87), Catecholamine Club (founder, sec.-treas. 1967-80, pres. 1981-82), Doctors Mayo Soc., Albert Gallatin Assos., New Eng. Soc., S.R. (chmn. admissions com. 1959-67, bd. mgrs. 1959-67, 69-70), Soc. Colonial Wars, Sigma Xi, Nu Sigma Nu, Phi Delta Theta, Explorers, Meadow (L.I., N.Y.); Univ.; Yale; N.Y. Athletic (N.Y.C.); Devon Yacht; Southampton Bathing Corp.; Jupiter Island. Co-author: Chemical Quantitation of Epinephrine and Norepinephrine in Plasma, 1959, Pheochromocytoma, 1977; author: Catecholamines in Normal and Abnormal Cardiac Function; editor, contbr. Hormones and Hypertension, 1966; editor: Am. Lecture Series in Endocrinology, 1962-75; guest editor First Irvine H. Page Internat. Hypertension Rsch. Symposium; contbr. articles to profl. and lay jours. Achievements include research on the mechanism of salt-induced hypertension, and on pheochromocytoma. Home: 8 E 81st St New York NY 10028-0201

MANGOLD, VERNON LEE, physicist; b. Dayton, Ohio, Aug. 15, 1935; s. Herman Leo and Grace Emma (Galvin) M.; m. Henrietta C. Heeter, Aug. 10, 1956 (div. 1977); children: Vernon Jr., Vincent, Cathy, Leona, Anthony, Edward; m. Laura Marie Snow, Aug. 15, 1978; children: Michael, Lee. BS in Math. and Physics, Cen. State U., Wilboforce, Ohio, 1961; MS in Physics, Southeastern U., 1981, PhD in Physics, 1983. Cert. environ. profl., environ. assessor. Physicist Flight Dynamics Lab., Dayton, Ohio, 1961-79, AMTC Lasers Inc., Orlando, Fla., 1979-92; physicist radiation control State of Fla., Orlando, 1990—; cons. Electromagnet and Environ. Protection Cons., Orlando, 1989—. Contbr. articles to profl. jours. With USMC, 1954-56. Mem. Health Physicist Soc., Fla. Assn. Environ. Profs. Republican. Roman Catholic. Achievements include establishment the Electromagnetic hazards Group; developed a fiber optics DOT/script laser marker. Office: AMTC Lasers Inc 8121 Forest City Rd Orlando FL 32810

MANGONE, RALPH JOSEPH, metallurgical engineer; b. Denver, July 6, 1923; s. Vincent M. and Josephine Carmella (Labriola) M.; m. Evangeline Nau, Sept. 28, 1950; children: Vincent, Jeannine, Carl, Mary Jo. Degree in metall. engring., Colo. Sch. Mines, 1948; MS in Metall. Engring., Ohio State U., 1958. Registered profl. engr., Colo. Metall. engr. U.S. Bur. Mines, Salt Lake City, 1950-52; project engr. Battelle Meml. Inst., Columbus, Ohio, 1952-59, Colo. Sch. Mines Rsch. Found., Golden, Colo., 1964-67; head materials and processing engring. Rohr Aircraft Corp., Riverside, Calif., 1959-61; sr. lead engr. Lockheed Missiles and Space Co., Sunnyvale, Calif., 1961-64; pres. Mangone Lab., Inc., Golden, 1967—. With USN, 1944-46. Mem. ASTM, Am. Soc. Metals (chmn. Rocky Mountain chpt. 1972-73), Internat. Metall. Soc., Internat. Soc. Air Safety Investigators (assoc.). Achievements include alloy development, failure analysis. Office: Mangone Lab Inc 14335 W 44th Ave Golden CO 80403

MANGUS, CARL WILLIAM, technical safety and standards consultant, engineer; b. Broken Bow, Okla., Aug. 20, 1930; s. Nathaniel M. and Eva Tennessee (Johnson) M.; m. Dorotha Marie Wood; children: Steven Neal, Roy Gene, Carla Anne. BSME, Okla. State U., 1958. Registered profl. engr., La. Various positions, 1948-63; project mgr. Chalkley Gas Processing Plant, 1964; project devel. Seven Natural Gas Plants, 1965; project mgr. N. Terrebonne Plant Expansion & Dual 36 Pipeline Loop, 1966-67; with Project Devel.-Two Natural Gas Plants, 1967-68; project mgr. Calumet Gas Processing Plant, 1968, offshore engring. sect. leader facilities, 1969; offshore prodn. supt. Maintenance and Operating Standards, 1970; with tech. safety rev. & approval engring. procedures Plus Regulations and Industry Standards, 1971; mgr. regulatory affairs Shell Offshore Inc., 1982, sr. staff tech. safety specialist, 1985; pvt. practice tech. safety and standards Lacombe, La., 1986—; com. mem. Am. Bur. Shipping, N.Y.C., 1974-88; Dept. State cons. Internat. Maritime Orgn., London, 1975-79; Am. Petroleum Inst. rep. to Exploration/Prodn. Forum, London, 1979-83; com. mem. NAS, Washington, 1979-84; mem. spl. adv. ad hoc com. Internat. Assn. Drilling Contractors, Houston, 1978, human resources com., 1978-80; past mem. offshore operators com. Am. Petroleum Inst., New Orleans; past U.S. industry rep. safety code for constrn. offshore structures, ILO, Geneva. Staff sgt. USAF, 1951-55. Recipient Am. Petroleum citation for svc. Am. Petroleum Inst., 1987. Mem. La. Engring. Soc., Am. Soc. Safety Engrs., Gulf Coast Safety & Tng. Group, Soc. Petroleum Engrs., Pine Island Club. Republican. Avocations: hunting, fishing, woodworking, swimming, boating, traveling. Home and Office: PO Box 250 Lacombe LA 70445-0250

MANIER, AUGUST EDWARD, philosophy of biology educator; b. Versailles, Ohio, Apr. 7, 1931; s. Francis Vitalis and Whilma Anna (Grillot) M.; m. Dorothy Frances Hickey, June 18, 1955 (div. June 1983); children: Michael, David, Maureen, August Edward Jr., John, Daniel, Jeremy; m. Jenny Ann Pitts, Oct. 18, 1985. BS with high honors, U. Notre Dame, 1953; PhD, St. Louis U., 1961. Lectr. Webster Coll., Webster Groves, Mo., 1956-59; instr. philosophy of biology U. Notre Dame, Ind., 1959-61, asst. prof., 1961-67, assoc. prof., 1967-88, prof., 1988—; NEH vis. fellow Cambridge (Eng.) U., 1971-72. Author: The Young Darwin and His Cultural Circle, 1978; author, editor: Academic Freedom and The Catholic University, 1968, Abortion: New Directions for Policy Studies, 1977. Precinct committeeman South Bend (Ind.) Dem. Com., 1970-80; pres. Woods and Wedge Neighborhood Assn., South Bend, 1975-77, East Side Little League Baseball, 1981-83. Fellow Nat. Humanities Inst., U. Chgo., 1977-78; grantee NSF, 1981-82, 87-88. Mem. AAAS (sec. history and philosophy of sci. 1989—), Am. Philos. Assn., Philosophy of Sci. Assn., History of Sci. Soc., Soc. for Social Studies of Sci. Home: 1444 Sunnymede South Bend IN 46615 Office: U Notre Dame 314 Decio Hall Notre Dame IN 46556

MANKIN, CHARLES JOHN, geology educator; b. Dallas, Jan. 15, 1932; s. Green and Myla Carolyn (Bohmert) M.; m. Mildred Helen Hahn, Sept. 6, 1953; children: Sally Carol, Helen Francis, Laura Kay. Student, U. N.Mex., 1949-50; B.S., U. Tex. at Austin, 1954, M.A., 1955, Ph.D., 1958. Asst. prof. geology Calif. Inst. Tech., 1958-59; asst. prof. geology U. Okla., 1959-63, asso. prof., 1963-64, prof., 1964—; dir. Sch. Geology and Geophysics, 1963-77, Energy Resources Inst., 1978-87; mem. U.S. Nat. Commn. on Geology, 1977-80; dir. Okla. Geol. Survey, 1967—; former chmn. bd. mineral and energy resources, former mem. commn. on phys. sci., math. and resources Nat. Acad. Scis.; former commr. Commn. Fiscal Accountability of Nation's Energy Resources; former chmn. Royalty Mgmt. Adv. Com. Dept. Interior; bd. dirs. Environ. Inst. for Waste Mgmt. Studies, U. Ala. Contbr. articles profl. jours. Recipient Conservation Service award Dept. Interior, 1983. Fellow Geol. Soc. Am. (co-project leader Decade N.Am. Geology, former councillor, chmn. found.), Mineral. Soc. Am.; mem. Am. Assn. Petroleum Geologists (Pub. Service award 1988), Am. Inst. Profl. Geologists (v.p., past pres., Martin Van Couvering Meml. award 1988, mem. found.), Clay Minerals Soc., Geochem. Soc., AAAS, Assn. Am. State Geologists (past pres.), Am. Geol. Inst. (past pres., Ian Campbell medal 1987), Soc. Econ. Paleontologists and Mineralogists (pres. Mid-Continent sect.), Sigma Gamma Epsilon (nat. sec.-treas.). Home: 2220 Forister Ct Norman OK 73069-5120 Office: Okla Geol Survey Energy Ctr 100 E Boyd St Rm 131N Norman OK 73019-0001

MANLEY, LANCE FILSON, data processing consultant; b. Atlanta, Dec. 8, 1945; s. Vern Paul Manley and Beth (Filson) Morgan; m. Sandra Faye Parris, Oct. 31, 1964 (div. 1967); 1 child, Lance Filson Jr.; m. Elizabeth Jane Wallace, Oct. 31, 1968; children: Jeffrey Lance, Heather Leigh. Student, John Marshall Law Sch., 1964-66, Shorter Coll., Rome, Ga., 1967-68, Brevard Coll., Cocoa, Fla., 1968-69, U. Mid Fla., 1972-74; tech. cert., Programming Systems Inst., Atlanta, 1967, RCA Edn. Ctr., L.A., 1968, Burroughs Edn. Ctr., Detroit, 1973, Honeywell Edn. Ctr., Atlanta, 1976, Info. Sci., San Antonio, 1977, IBM Edn. Ctr., Atlanta, 1978, Emory U., 1988, Platinum Tech. Inc., Atlanta, 1993. Sr. computer operator Fed. Elec. Corp., Cape Kennedy, Fla., 1966-68, Universal Studios, Universal City, Calif., 1968-70; program analyst State of Fla., Jacksonville, 1970-73, Fla. Nat. Bank, Jacksonville, 1973-76; sr. program analyst Fulton Nat. Bank, Atlanta, 1977-78, Ist Nat. Bank Atlanta, 1978-80; cons. Ins. Systems Am., Atlanta, 1980, Cotton States Ins., Atlanta, 1981, State of Ga., Atlanta, 1981-83, Decatur Fed., Atlanta, 1981, C&S Bank, Atlanta, 1982, Cox Communication, Atlanta, 1983, Emory U., Atlanta, 1984-88; staff cons. So. Co. Svcs., Atlanta, 1988—. Avocations: skiing, camping, fishing, music. Home: 4777 Alpine Dr SW Lilburn GA 30247-4605 Office: So Co Svcs 64 Perimeter Center E Atlanta GA 30346

MANLY, CAROL ANN, speech pathologist; b. Canton, Ohio, Nov. 21, 1947; d. William George and Florence L. (Parrish) M.; m. William Merget, Sept. 19, 1992; 1 child, John Michael. MA, U. Cin., 1970; PhD, NYU, 1988. Instr. U. Cin. Med. Ctr., 1970-72; asst. dir. Goldwater Hosp. NYU, 1972-83; pvt. practice N.Y.C., 1983—; cons. Mary Manning Walsh Home, N.Y.C., 1974-85, Beth Israel Hosp. North, N.Y.C., 1983—; adj. asst. prof. NYU, 1989-90, L.I. U., C.W. Post Campus, Brookville, N.Y., 1990-91. Author: (with others) Current Therapy in Physiatry, 1984, Understanding Communication Disorders of the Older Adult: A Practical Handbook for Health Care Professionals, 1993; contbr. articles to profl. jours. Mem. NOW, Am. Speech, Lang., and Hearing Assn., Nat. Action Rights Abortion League, N.Y. Acad. Scis., N.Y. Neuropsychology Group. Achievements include development of new diagnostic and treatment procedures for oral-pharyngeal dysphagia in neurologically impaired adults. Office: 360 E 65th St Apt 21D New York NY 10021-6726

MANLY, WILLIAM DONALD, metallurgist; b. Malta, Ohio, Jan. 13, 1923; s. Edward James and Thelma (Campbell) M.; m. Jane Wilden, Feb. 9, 1949; children—Hugh, Ann, Marc, David. Student, Antioch Coll., 1941-42; B.S., U. Notre Dame, 1947, M.S., 1949; postgrad., U. Tenn., 1950-55. Metallurgist Oak Ridge Nat. Lab., 1949-60, mgr. gas cooled reactor program, 1960-64; mgr. materials research Union Carbide Corp., N.Y.C., 1964-65; gen. mgr. Union Carbide Corp. (Stellite div.), N.Y.C., 1967-69; v.p. Union Carbide Corp. (Stellite div.), Kokomo, Ind., 1969-70; sr. v.p. Cabot Corp., Boston, 1970-83; exec. v.p. Cabot Corp., 1983-86; ret., 1986; also dir. chmn. adv. com. for reactor safety AEC, 1964-65. Served with USMC, 1943-46. Recipient Honor award U. Notre Dame, 1974, Nat. Medal of Tech., Nat. Sci. Found., 1993. Fellow Am. Soc. Metals (pres. 1972-73), AIME, Am. Nuclear Soc. (Merit award 1966); mem. Nat. Acad. Engring., Nat. Assn. Corrosion Engrs., Metall. Soc. Presbyterian. Clubs: Cosmos, Masons. Home: RR 1 Box 197A Kingston TN 37763-9568

MANN, DAVID MARK, researcher; b. Pasadena, Calif., Apr. 1, 1943; s. Marvin Matthew and Mary Belle (Coleman) M.; m. Sherrill Robin Cartt, Apr. 23, 1965; 1 child, Carolyn Marie. BA in Math., Whittier Coll., 1964; MA in Physics, U. Calif., Santa Barbara, 1967, PhD in Physics, 1970. Rsch. physicist Phillips Rsch. Labs., Eindhoven, Netherlands, 1970-72; prin. rsch. scientist Aerodyne Rsch., Inc., Burlington, Mass., 1973-76; rsch. phys. scientist Air Force Rocket Propulsion Lab., Edwards AFB, Calif., 1976-82; assoc. dir. engring. scis. div. U.S. Army Rsch. Office, Research Triangle Park, N.C., 1982—; spl. asst. to dir. rsch. and lab. mgmt. Office Dep. Undersec. of Def. for Rsch. and Advanced Tech., 1985-86; exec. intern Dep. for Rsch. and Tech., Office of Asst. Sec. of Army for Rsch. Devel. and Acquisition, 1989; combustion subcom. Tech. Steering Group of Joint Army, Navy, NASA, Air Force, 1982—, chmn. exhaust plume tech. subcom., co-chmn. panel study of advanced diagnostic; mem. plume sci. adv. group Strategic Defense Initiative Orgn., 1986-91. Contbr. articles to profl. jours. Chmn. adminstrv. bd. Aldersgate Meth. Ch., Durham, N.C., 1988—. NSF grad. fellow, 1965. Mem. AIAA (sect. chmn. local chpt., propellants and combustion tech. com.), ASME (session chmn.), Soc. Automotive Engrs., Am. Phys. Soc., Combustion Inst. (session chmn., program com.), Soc. Photo-Optic Instrumentation Engrs. (session chmn.). Achievements include research in propellant ignition micromechanics, shock-layer induced ultraviolet emissions measured by rocket payloads, ultraviolet emissions from in-flight plume and hardbody flowfields, propellants and combustion. Office: US Army Rsch Office PO Box 12211 Research Triangle Park NC 27709-2211

MANN, JAMES DARWIN, mathematics educator; b. Lambric, Ky., Feb. 27, 1936; s. Glinn W. and Wanda (Collins) M.; 1 child, Terry Brian. BS, Morehead State U., 1962; M in Math., U. S.C., 1965; postgrad., Ind. U., 1968-69, Obelin Coll., 1968. High sch. tchr. math., 1962-64; instr. math. Presbyn. Coll., Clinton, S.C., 1965-66; assoc. prof. math. Morehead (Ky.) State U., 1966—. Fundraiser United Way, 1977-78; vol. coach Little League Baseball, 1973-76; coach Babe Ruth Baseball, 1977; chmn. N.E. Ky. Sci. Fair Rules, 1969-76; judge N.E. Ky. Sci. Fair, 1967, 68, 77, 79, 80, 81. NSF grantee U. S.C., 1964-65, Vanderbilt U., 1967, Oberlin Coll., 1968, N.C. State U., 1972; recipient Outstanding Alumni award Ky. Zeta chpt. Sigma Phi Epsilon, 1977. Mem. Math. Assn. Am., Nat. Coun. Tchrs. Math. Baptist. Home: 4200 Christy Creek Rt 6 Morehead KY 40351-9806 Office: Morehead State U PO Box 1231 Morehead KY 40351-5231

MANN, KENNETH GERARD, biochemist, educator; b. Floral Park, N.Y., Jan. 1, 1941; s. Arthur Welsley and Helen Francis (Reiger) M.; m. Jeanette Marie Doner, Aug. 22, 1964; children: Kevin, Philip, David, Stephen. BS in Chemistry, Manhattan Coll., 1963; PhD in Biochemistry, U. Iowa, 1967, postdoctoral, 1967-68; postdoctoral, Duke U., 1968-70. Asst. prof. dept. biochemistry U. Minn., St. Paul, 1970-75, assoc. prof., 1975-80, prof., 1980-84; cons. hematology rsch. Mayo Clinic, Mayo Found., Minn., 1972-84; asst. prof. dept. biochemistry Mayo Med. Sch., 1972-74, assoc. prof., 1974-78, prof., 1978-84; prof., chmn. dept. biochemistry U. Vt., Burlington, 1984—; cons. in hematology research Mayo Clinic and Mayo Found., Minn., 1972-84; chmn. sci. subcom. on hemostasis Am. Soc. Hematology, 1987; mem. blood services sci. council ARC, 1985—, chmn. platelets/coagulation initial rev. group, 1985—; chmn. ISTH, 1990-92; assoc. editor Blood, 1989-92; sec., assoc. Med. Sch. Depts. Biochemistry, 1993; Camille & Henry Dreyfus Tchr. Scholar, 1971-76. Inventor composition and method to control hemophilia. Co-founder, sec. Citizens for a Clean Miss.; cubmaster Cub Scout pack 101; past pres. Rochester Aero Model Soc. Recipient Disting. Career award for Contbn. to Hemostasis, 1987, Scholar award U. Vt., 1988; named to Presdl. symposium, Am. Soc. Hematology, 1987. Mem. NHLBI (hematology study sect. 1978-88), Am. Heart Assn. (chmn. council on thrombosis 1989-91, sci. adv. com. 1991—, rsch. com. 1993—, est. investigator 1974-79, Sol Sherry award 1992), Am. Soc. Hematology (sci. affairs com. 1989, chmn. sci. subcom. on hemostasis 1987-92, presdl. symp. 1987, Henry Stratton medal 1992), Am. Soc. Biol. Chemists (membership com. 1982-84), Am. Soc. Biol. Scientists, Am. Chem. Soc., Internat. Soc. Thrombosis and Haemostasis. Roman Catholic. Clubs: Green Mountain Recreational, Austin Healey of Am. Avocations: model airplanes, fishing, antique cars, sailing. Office: U of Vt Coll of Medicine Dept of Biochemistry Burlington VT 05405

MANN, LAURA SUSAN, aerospace engineer; b. Houston, Sept. 20, 1958; d. Manfred Walter and Sally Mae (Hennels) Schaefer; m. Richard Drew Mann, Aug. 1, 1987; 1 child, W. Cole. BS in Physics cum laude, U. Houston, 1986. Mktg. sec. Vector Cable/Schlumberger, Sugar Land, Tex., 1981-83; adminstrv. asst. Bekaert Internat. Trade, Inc., Houston, 1983-84; polit. pollster, rsch. and teaching asst. U. Houston, 1984-86; flight contr. Johnson Space Ctr., NASA, Houston, 1986-91, mgr. grapple fixture subsystem, 1991-92, mgr. space sta. engring. configuration, 1992-93; part time beauty cons. Mary Kay cosmetics, 1992—; mem. tech. adv. com. flight telerobotic servicer Goddard Space Flight Ctr., NASA, Greenbelt, Md., 1989; mission ops. directorate rep. hand contr. commonality study, leader space shuttle payload and deployment system tech. team/space sta. flight compatability rev. Johnson Space Ctr., NASA, 1990; rsch. asst. medium energy physics expt. U. Houston at Brookhaven Nat. Lab., Upton, N.Y., 1985, 86; mem. configuration mgmt. process improvement team for Space Sta. Freedom Program, 1992-93. Pres. Durham Pk. Homeowners Assn., Houston, 1990-92. Mem. Am. Horse Shows Assn., Tex. Hunter Jumper Assn., Greater Houston Hunter Jumper Assn. (bd. dirs. 1991-92, contbg. newsletter columnist 1991-92, Jr./Adult Jumper Champion 1990, 4th in open jumper ann. awards 1992), U.S. Equestrian Team, Inc., U. Houston Alumni Orgn.., Nat. Arbor Day Found., MENSA, Phi Theta Kappa. Avocations: horse showing, camping, walking, aerobics.

MANN, LESTER PERRY, mathematics educator; b. Milford, Mass., May 30, 1921; s. Lester P. and Viola E. (Tracy) M.; m. Dorothy M. Davis, Oct. 11, 1947; children: Kelly P., Leslie P. BS with high honors, U. Md., 1964; MEd, U. Alaska, Anchorage, 1974; EdD, Boston U., 1983. Cert. elem. tchr., reading specialist and supr., Mass.; cert. elem. tchr., reading specialist, Alaska. Commd. 2nd lt. USAAF, 1941; advanced through grades to maj. USAF, 1954, navigator, weather officer, 1941-64; ret., 1964; resident counselor OEO-Job Corps, 1965-66; flight navigator Südflug, Braniff, Capitol and Japan Air Lines, 1966-73; instr. math., adminstr., curriculum developer U. Alaska, 1974-86, adj. instr., 1987—; instrnl. assoc. Mann Assocs., Applied Lifelong Learning, Anchorage, 1983—; instr. Anchorage Community Coll., 1974-86; asst. prof. Embry-Riddle Aero. U., Anchorage, 1987—, acad. advisor, 1987-90; mem. for remedial reading Alaska Talent Bank; vis. adult educator German Adult Edn. Assn., 1984. Mem. Math. Assn. Am., Nat. Coun. Tchrs. Math., Internat. Reading Assn., Am. Assn. Adult and Continuing Edn. (profl., past mem. nomination and election com.), Am. Meteorol. Soc., Phi Alpha Theta, Phi Kappa Phi. Avocations: fishing, sport flying, classical guitar. Home and Office: 2304 Turnagain Pky Anchorage AK 99517-1124

MANN, MICHAEL MARTIN, electronics company executive; b. N.Y.C., Nov. 28, 1939; s. Herbert and Rosalind (Kaplan) M.; m. Mariel Joy Steinberg, Apr. 25, 1965. BSEE, Calif. Inst. Tech., 1960, MSEE, 1961; PhD in Elec. Engring. and Physics, U. So. Calif., 1969; MBA, UCLA, 1984. Cert. bus. appraiser, profl. cons., mgmt. cons., lic. real estate broker, Calif. Mgr. high power laser programs office Northrop Corp., Hawthorne, Calif., 1969-76; mgr. high energy laser systems lab. Hughes Aircraft Co., El Segundo, Calif., 1976-78; mgr. E-0 control systems labs. Hughes Aircraft Co., El Segundo, 1978-83, asst. to v.p., space & strategic, 1983-84; exec. v.p. Helionetics Inc., Irvine, Calif., 1984-85, pres., chief exec. officer, 1985-86, also bd. dirs.; ptnr. Mann Kavanaugh Chernove, 1986-87; sr. cons. Arthur D. Little, Inc., 1987-88; chmn. bd., pres., chief exec. officer Blue Marble Devel. Group, Inc., 1988—; exec. assoc. Ctr. Internat. Cooperation and Trade, 1989—; sr. assoc. Corp. Fin. Assocs., 1990—; exec. assoc. Reece and Assocs., 1991—; dir. Reece & Assocs., 1991—; mng. dir. Blue Marble Ptnrs. Ltd, 1991—; chmn. bd. dirs., CEO Blue Marble Ptnrs., 1992—; mem. Army Sci. Bd., Dept. Army, Washington, 1986-91; chmn. Ballistic Missile Def. Panel, Directed Energy Weapon Panel, Rsch. and New Initiatives Panel; cons. Office of Sec. of Army, Washington, 1986—, Inst. of Def. Analysis, Washington, 1978—, Dept. Energy, 1988—, Nat. Riverside Rsch. Inst., 1990—; bd. dirs. Datum, Inc.,1988—, Fail-Safe Tech., Corp., 1989-90, Safeguard Health Enterprises, Inc., 1988—, Am. Video Communications, Inc., Meck Industries, Inc., 1987-88, Decade Optical Systems, Inc., 1990—, Forum Mil. Application Directed Energy, 1992—, Am. Bus. Consultants, Inc., 1993—; chmn. bd. Mgmt. Tech., Inc. 1991—; bd. dirs., mem. adv. bd. Micro-Frame, Inc., 1988-91; chmn. bd. HLX Laser, Inc., 1984-86; bd. dirs. Cons's. Roundtable, 1992—, Am. Bus. Cons., Inc., 1993—; rsch. assoc., mem. extension teaching staff U. So. Calif., L.A., 1964-70; chmn. Ballistic Missile Def. Subgroup, 1989-90, Tactical Directed Energy Weapons Subgroup, 1988-90; chmn., chief exec. officer Mgmt. Tech., Inc., 1991—; dir. Am. Bus. Cons., Inc., 1993—. Contbg. editor, mem. adv. bd. Calif. High-Tech Funding Jour., 1989-90; contbr. over 50 tech. articles to profl. jours.; patentee in field. Adv. com. to Engring. Sch., Calif. State U., Long Beach, 1985—; chmn. polit. affairs Am. Electronics Assn., Orange County Coun., 1986-87, mem. exec. com., 1986-88; adv. com. several Calif. congressmen, 1985—; mem. dean's coun. UCLA Grad. Sch. Mgmt., 1984-85; bd. dirs. Archimedes Circle U. So. Calif., 1983-85, Ctr. for Innovation and Entrepreneurship, 1986-90, Caltech/MIT Venture Forum, 1987-91. Hicks fellow in Indsl. Rels. Calif. Inst. Tech., 1961, Hewlett Packard fellow. Mem. So. Calif. Tech. Execs. Network, IEEE (sr.), Orange County CEO's Roundtable, Pres.' Roundtable, Nat. Assn. Corp. Dirs., Aerospace/Def. CEO's Roundtable, Am. Def. Preparedness Assn., Security Affairs Support Assn., Acad. Profl. Cons. and Advisors, Internat. Platform Assn., Inst. of Mgmt. Cons's., Pres. Assn., Nat. Assn. Corp. Dirs., Cons's. Roundtable, Pres. Assn., King Harbor Yacht Club. Republican. Avocations: sailing, photography, writing. Home: 4248 Via Alondra Palos Verdes Peninsula CA 90274-1545 Office: Blue Marble Partners 406 Amapola Ave Ste 200 Torrance CA 90501-1475

MANN, OSCAR, physician, internist, educator; b. Paris, Oct. 13, 1934; came to U.S., 1953; s. Aron and Helen (Biegun) M.; m. Amy S. Mann, July 19, 1964; children: Adriana, Karen. AA with distinction, George Washington U., 1958; MD cum laude, Georgetown U., 1962. Diplomate Am. Bd. Med. Examiners, Am. Bd. Internal Medicine, Am. Bd. Internal Medicine subspecialty Cardiovascular Disease; cert. advanced achievement in internal medicine; re-cert. in internal medicine. Intern Georgetown U. Med. Ctr., Washington, 1962-63, jr. asst. med. resident, 1963-64, clin. fellow in cardiology with Proctor Harvey program, 1965-66; sr. asst. resident in medicine Georgetown svc. D.C. Gen. Hosp., Washington, 1964-65; clin. prof. medicine Georgetown U. Sch. Medicine, 1985—, also regional chmn. Med. Alumni Fund; pvt. practice internal medicine and cardiology, Washington, 1966—; mem. Med.-Nursing Audit Com., CME adv. com., teaching. adv. com., Opthamology dept. rev. com., surgery dept. rev. com., faculty com., search com. for a new dean for acad. affairs Georgetown U. Med. Ctr.; appointed coun. to the dean Georgetown U. Sch. Medicine, 1977—; mem. Instnl. Self Study Task Force. Contbr. articles to profl. jours. Served with the U.S. Army, 1953-55. Recipient Mead Johnson Postgrad. Scholar ACP, 1964-65, Physicians Recognition award AMA, 1987—, Advanced Achievement in Internal Medicine, 1987. Fellow ACP, Am. Coll. Cardiology, Am. Coll. Chest Physicians; mem. AMA, Am. Soc. Internal Med., Am. Heart Assn. (coun. clin. cardiology), Med. Soc. D.C., Cosmos Club, Georgetown U. Alumni Assn. (bd. govs., nat. chmn. of med. alumni fund 1993), Alpha Omega Alpha, Phi Delta Epsilon. Avocations: reading, swimming, biking, fitness, racquetball. Home: 4925 Weaver Ter NW Washington DC 20016-2660 Office: Foxhall Internists PC 3301 New Mexico Ave NW Washington DC 20016-3622

MANN, ROBERT WELLESLEY, biomedical engineer, educator; b. Bklyn., Oct. 6, 1924; s. Arthur Wellesley and Helen (Rieger) M.; m. Margaret Ida Florencourt, Sept. 4, 1950; children: Robert Wellesley, Catherine Louise. SB, MIT, 1950, SM, 1951, ScD, 1957. With Bell Telephone Labs., N.Y.C., 1942-43, 46-47; research engr. MIT, 1951-52, rsch. supr., 1952, mem. faculty, 1953—, prof. mech. engring., 1963-70, Germeshausen prof., 1970-72, prof. engring., 1972-74, Whitaker prof. biomed. engring., 1974-92, Whitaker prof. emeritus, sr. lectr., 1992—, head systems and design div., mech. engring. dept., 1957-68, 82-83, founder, dir. engring. projects lab., 1959-62; founder, chmn. steering com. Center Sensory Aids Evaluation and Devel., 1964-86, chmn. div. health scis., tech., planning and mgmt., 1972-74, founder, dir. Newman biomechanics and human rehab. lab., 1975-92; dir. bioengring. programs Whitaker Coll. MIT, 1986-89; dir. Harvard-MIT Rehab. Engring. Ctr., 1988-93; exec. com. Div. Health Scis. and Tech. Harvard U. MIT joint program, 1972-85; prof., 1979—, mem. Com. on Use of Humans as Exptl. Subjects MIT, 1984-93, co-chair Pub. Svc. Ctr., 1988-92; lectr. engring. Faculty of Medicine, Harvard U., 1973-79; research assoc. in orthopedic surgery Children's Hosp. Med. Center, 1973—; cons. in engring. sci. Mass. Gen. Hosp., 1969—; cons. in field, 1953—; mem. Nat. Commn. Engring. Edn., 1962-69; com. prosthetics research and devel. NRC, 1963-69, chmn. sensory aids subcom., 1965-68, com. skeletal system, 1969-71; mem. com. interplay engring. with biology and medicine Nat. Acad. Engring., 1969-73; mem. bd. health scis. policy Inst. Medicine, 1973-74, 82-86; mem. com. on nat. needs for rehab. physically handicapped Nat. Acad. Scis., 1975-76; mem.-at-large confs. com. Engring. Found., 1975-81; Mem. Commn. on Life Scis., NRC, 1984-88, Commn. on Strategic Tech. for U.S. Army, 1989-93; NRC, commn. on Space Biology and Medicine, 1992-95. Cons. editor: Ency. Sci. and Tech.; assoc. editor: IEEE Trans. in Biomed. Engrin., 1969-78, ASME Jour. Biomed. Engring., 1976-82; mem. editorial bd. Jour. Visual Impairment and Blindness, 1976-80, SOMA, 1986-92, chmn., 1993—; mem. editorial adv. bd. new liberal arts program Alfred P. Sloan Found., 1986-92; contbr. numerous articles to profl. jours. Pres., trustee Amanda Caroline Payson Scholarship Fund, 1965-86; bd. dirs. Carroll Ctr. for Blind, 1967-74, pres., 1968-74; mem. corp. Perkins Sch. for Blind, 1970—, Mt. Auburn Hosp., 1972—; trustee Nat. Braille Press, 1982—, pres., 1990—; mem. Cardinal's adv. com. on social justice Archdiocese Boston, 1993—. With AUS, 1943-46. Recipient Sloan award for outstanding performance, 1957; Talbert Abrams photogrammetry award, 1962; award Assn. Blind of Mass., 1969; IR-100 award for Braillemboss, 1972; Bronze Beaver award Mass. Inst. Tech., 1975; UCP Goldenson Research for Handicapped award, 1976; H.R. Lissner award, 1977; New Eng. award, 1979; J.R. Killian Faculty achievement award MIT, 1983. Fellow Am. Acad. Arts and Scis., Am. Inst. Med. and Biol. Engring., IEEE (editorial bd. spectrum 1984-86), AAAS, ASME (Gold medal 1977); mem. NAS, Inst. Medicine NAS, NAE, Biomed. Engring. Soc. (bd. dirs. 1981-84), Orthopedic Rsch. Soc., Rehab. Soc. N.Am., MIT Alumni/ae Assn. (pres. 1983-84, Alumni Fund Bd. 1978-80, bd. dirs. 1980-86, corp. joint adv. com. 1983-84, chair nat. selector com. 1985-88, awards com. 1992—, chmn., bd. Tech. Rev. 1986—, chmn., 1993—), Sigma Xi (nat. lectr. 1979-81), Tau Beta Pi, Pi Tau Sigma, Sigma Xi. Roman Catholic. Achievements include patents in field. Home: 5 Pelham Rd Lexington MA 02173-5707 Office: Mass Institute of Technology 77 Massachusetts Ave Rm 3144 Cambridge MA 02139-4307

MANN, ULRICH, petroleum geologist; b. Ludwigshafen, Rheinland-Pfalz, Fed. Republic Germany, May 7, 1952; s. Heinz and Ruth (Trautmann) M.; m. Petra Ruth Metz, Mar. 28, 1980; 1 child, Tatjana. Diplom, U. Heidelberg, Fed. Republic Germany, 1977, Dr.rer.nat., 1980. Cert. petroleum geologist. Sedimentologist U. Heidelberg, 1978-80; petroleum geologist/geochemistry Rsch. Centre Juelich, Nordrhein-Westfalen, 1980—; sedimentologist, physic properties specialist DSDP Leg 68, Carribean, 1979; field geologist, Greenland Geol. Survey, 1983; co-chmn. 78 Ann. Meeting of the Geol., 1988; lectr. in organic geochemistry U. Nuernberg-Erlangen, Bavaria, 1991—. Co-author: Sediments and Environmental Geochemistry, 1990; contbr. articles to profl. jours. Mem. Cactus Club, 1981; founding mem. Spielwiese e.V., Aachen, 1990. Mem. Am. Assn. of Petroleum Geologists, Soc. of Sedimentary Geology, European Assn.s of Petroleum Geoscientists (founder), Am. Chem. Soc. (div. of geochemistry). Achievements include rsch. petrophys. approach to primary migration in petroleum source rocks, relation between petrophys. properties and residual oil in petroleum source rocks. Home: Am Gut Bau 27, Aachen D-5100, Germany Office: Rsch Centre Juelich, PO Box 1913, Jülich D-5170, Germany

MANNHEIM, WALTER, medical microbiologist; b. Kaiserslautern, Palatinate, Germany, Nov. 27, 1930; s. Karl and Charlotte (Gork) M.; m. Anna Barbara Hager, Sept. 27, 1959; children: Korinna, Rolf, Karl, Berthold. MD, U. Bonn, Federal Republic of Germany, 1959; PhD, U. Marburg, Federal Republic of Germany, 1967. Med. asst. U. Bonn, 1956-57; med. asst. surg. and obstet. dept. Hosp. Remscheid-Lennep, 1957-58; sci. asst. Inst. Hygiene, U. Heidelberg, 1959-63; sci. asst. Inst. Hygiene, U. Marburg, 1963-67, from lectr. to prof., 1967-71, dir. dept. med. microbiology, 1980. Contbr. articles to profl. jours. Mem. Internat. Orgn. for Mycoplasmology, Am. Soc. Microbiology, Internat. Com. on Systematic Bacteriology (subcom. on Pasteurellaceae and related organisms). Office: Philipps U Inst Med Microbiology, Pilgrimstein 2 PO Box2360, D 35011 Marburg Germany

MANNICK, JOHN ANTHONY, surgeon; b. Deadwood, S.D., Mar. 24, 1928; s. Alfred and Catherine Elizabeth (Schuster) M.; m. Alice Virginia Gossard, June 9, 1952; children—Catherine Virginia, Elizabeth Eleanor, Joan Barbara. B.A., Harvard U., 1949, M.D., 1953. Diplomate: Am. Bd. Surgery (dir. 1971-77). Intern Mass. Gen. Hosp., 1953-54, resident in surgery, 1956-60; instr. in surgery to asst. prof. Med. Coll. Va., 1960-64; asso. prof. to prof. surgery Boston U., 1964-76, chmn. div. surgery, 1973-76; Moseley prof. surgery Harvard U., 1976; chmn. dept. surgery Peter Bent Brigham Hosp. and Brigham and Women's Hosp., Boston, 1976—; mem. surgery, anesthesiology and trauma study sect. NIH, 1978-82, mem. medicine study sect., 1967-70; research com. Med. Found., Inc., 1970-76. Author: (with others) Modern Surgery, 1970, Core Textbook of Surgery, 1972, Surgery of Ischemic Limbs, 1972, The Cause and Management of Aneurysms, 1990; mem. editorial bd. AMA Archives of Surgery, 1973-84, Clin. Immunology and Immunopathology, 1972-84, Surgery, 1982—, Brit. Jour. Surgery, 1982-92, European Jour. Vascular Surgery, 1988—; mem. editorial bd. Advances in Surgery, 1979—, editor, 1984-86; mem. editorial bd. Jour. Vascular Surgery, 1984—, assoc. editor, 1990—; also articles. Served to capt. M.C. USAF, 1954-56. Markle scholar in acad. medicine, 1961-66. Fellow ACS (gov.), Royal Coll. Surgeons (England; hon.); mem. Am. Fedn. Clin. Research, Am. Assn. Immunologists, Am. Soc. Exptl. Pathology, Soc. Clin. Investigation, Soc. Clin. Surgery, Soc. Univ. Surgeons, Soc. Surg. Chmn. (sec. 1985-87, pres. 1987-88), Am. Surg. Assn. (pres. 1989-90), Internat. Cardiovascular Soc. (recorder N.Am. chpt. 1973-76, pres. 1991-92), Soc. Vascular Surgery (pres. 1981), N.E. Surg. Soc., New Eng. Soc. for Vascular Surgery, So. Surg. Assn., So. Soc. for Vascular Surgery, Surg. Infection Soc., Halstead Soc., Phi Beta Kappa. Home: 81 Bogle St Weston MA 02193-1056 Office: 75 Francis St Boston MA 02115-6195

MANNINO, J. DAVIS, psychotherapist; b. Patchoque, N.Y., Sept. 27, 1949; s. Joseph I. and Adrienne Adele (Davis) M. BA magna cum laude, SUNY, Stony Brook, 1971; MSW summa cum laude, San Francisco State U., 1974; EdD in Counseling and Ednl. Psychology, U. San Francisco, 1989. Lic. psychotherapist, Calif.; lic. clin. social worker, Calif., marriage, family and child counselor. Instr. U. Malaysia, 1974-76; dir. refugee programs City San Francisco, 1979-82; instr. U. San Francisco, 1979-85; pvt. practice specializing in psychology San Francisc, Sonoma Counties, 1979—; cons. foster care Calif. State Legis., 1980, community rels., San Francisco Police Dept., 1982-87, Hospice Sonoma County, 1990, Sonoma County Mental Health; forensic task force on A.I.D.S., San Francisco Pub. Health Dept., 1984-85; child abuse investigation supr. City of San Francisco, 1985-88; supr. Reasonable Efforts to Families Unit, 1988-90; project coord. Edna McConnell Clark Found. Family Mediation Demonstration Grant; instr. child growth and devel., death and dying, Intro. to Psychology Santa Rosa Jr. Coll., 1990—; commr. Calif. Bd. Behavioral Sci. Examiners, 1990. Contbr. articles to profl. jours.; local psychology columnist, 1986—. Mem. Am. Psychol. Assn., Nat. Assn. Social Workers (diplomate clin. social work), Orthopsychiat. Assn., Am. Assn. Counseling and Devel., Calif. Assn. Marriage Family and Child Therapists, Golden Gate Bus. Assn. (ethics com. 1986, Disting. Svc. award, 1985), Am. Assn. Marriage and Family Therapists, Nat. Register Clin. Social Workers, Lions (Helen Keller Humanitarian award, bd. dirs. San Francisco chpt. 1986). Avocations: running, weight lifting, writing, gardening. Office: PO Box 14031 San Francisco CA 94114-0031

MANNS, ROY LOKUMAL, polymer technologist, researcher, executive; b. Sind, India, July 1, 1937; came to U.S., 1978; s. Lokumal H. and Parpati Mansukhani; m. Joy Jupe, July 4, 1964 (div. 1985); children: William John, Robert David, Julie Helen; m. Maureen Young, Sept. 12, 1987. Diploma in Design Engring., St. Peters, Panchgani, India, 1954. Chief engr. Kenwood Appliances, Havant, Eng., 1967-70; sr. ptnr. Roy Manns, Salmon & Assoc., Brighton, Eng., 1970-73; mng. dir. IPEC Polymers Ltd., Brighton, 1974-78; group mgr. Baxter Travenol Inc., Deerfield, Ill., 1978-81; dir. R & D Costar Corp., Cambridge, Mass., 1981-82; chmn. v.p. R & D Polyfiltronics Group Inc., Rockland, Mass., 1984—. Editor PlasMed News, 1986-87; cons. editor Plastics and Rubber Weekly, 1966-76. Recipient HRH Prince Philip award Coun. Indsl. Design, London, 1976. Fellow Plastics and Rubber Inst.; mem. Soc. Plastics Industry (exec. mem. med. products chpt.), Brit. Officers Club (life mem. Boston chpt.). Mem. United Ch. of Christ. Achievements include patents in laboratory disposables, and bonding of filters. Home: 811 S Riv Box # 358 Marshfield Hills MA 02051-9999 Office: Polyfiltronics Group Inc 100 Weymouth St Rockland MA 02370-1147

MANOHARAN, RAMASAMY, biomedical scientist; b. Viralippatty, Madurai, India, Apr. 19, 1962; came to U.S., 1989; s. Gurusamy and Alagammal (Perumal) Ramasamy; m. Lakshmi Manoharan, Nov. 13, 1991; 1 child, Rahul. MSc, Madurai U., Tamilnadu, India, 1984; PhD, Indian Inst. Tech., Kanpur, India, 1988. Postdoctoral assoc. U. R.I., Kingston, 1989-90; rsch. staff MIT, Cambridge, Mass., 1990-92, rsch. scientist, 1993—. Contbr. articles to profl. jours. Rsch. fellowship Coun. of Scientific and Indsl. Rsch., 1988; recipient Rsch. award Humboldt Found., 1990. Mem. AAAS, Am. Chem. Soc., Coblentz Soc. Achievements include development of spectroscopic method for detection of bacteria, spores, laser spectroscopic method for disease diagnosis, quantitative model for Raman spectral analysis of human tissue. Office: MIT Spectroscopy Lab 77 Massachusetts Ave Cambridge MA 02139

MANSFIELD, LOIS EDNA, mathematics educator, researcher; b. Portland, Maine, Jan. 2, 1941; d. R. Carleton and Mary (Bowdish) M. BS, U. Mich., 1962; MS, U. Utah, 1966, PhD, 1969. Vis. asst. prof. computer sci. Purdue U., 1969-70; asst. prof. computer Sci. U. Kans., Lawrence, 1970-74, assoc. prof., 1974-78; assoc. prof. math. N.C. State U., Raleigh, 1978-79; assoc. prof. applied math. U. Va., Charlottesville, 1979-83, prof., 1983—; mem. adv. panel computer sci. NSF, 1975-78; cons., vis. scientist Inst. Computer Applications in Sci. and Engring., Hampton, Va., 1976-78. Mem. editorial bd. Jour. Sci. Statis. Computing, 1979-88; contbr. articles to profl. jours. Grantee NSF and DOE. Mem. Am. Math. Soc., Soc. Indsl. and Applied Math., Assn. Computing Machinery (bd. dirs. SIGNUM 1980-83). Office: U Va Dept Applied Math Thornton Hall Charlottesville VA 22903

MANSKY, ARTHUR WILLIAM, computer engineer; b. Chester, Pa., Feb. 29, 1956; s. Joseph and Violet (Levin) M.; m. Shelley Anne Raymond, May 18, 1986; children: William, Joseph, Benjamin. BS in Math., West Chester U., 1978; MS in Computer Sci., U. Del., 1980. Tech. staff mem. Bell Labs., Holmdel, N.J., 1980-82; sr. software engr. Gen. Instrument, Hatboro, Pa., 1982-87; supr. Vitro Corp., Silver Spring, Md., 1987—. Mem. IEEE, Assn. for Computing Machinery, Soc. for Info. Display. Office: Vitro Corp 14000 Georgia Ave Silver Spring MD 20906

MANSOUR, AWAD RASHEED, chemical engineering educator; b. Jenin, Jordan, Mar. 13, 1951; s. Rasheed Suleiman and Amenah Ahmad (Saeed) M.; m. Aisheh Moh'd Hussain, Oct. 16, 1976; children: Fatimah, Mariam, Ahmad, Omar, Ammar. BSChemE, Baghdad U., 1975; MSChemE, U. Tulsa, 1979, PhDChemE, 1980. Lab. engr. Ministry Pub. Works, Amman, Jordan, 1975-76; process engr. Cement Factory, Amman, 1976-77; asst. prof. Yarmouk U., Irbid, Jordan, 1980-86, assoc. prof., 1986-90; founder dept. chem. engring., prof. Jordan U. Sci. and Tech., Irbid, 1991—, past chmn. dept. Author 40 books on computer and engring. applications; contbr. over 70 articles to profl. jours.; patentee chem. toilet, chem. solvent to solve oil spills and to clean motors and engines. Scholar Ministry Edn., 1970, Yarmouk U., 1977, 78. Office: Yarmouk U, PO Box 4455, Irbid Jordan

MANSOUR, FARID FAM, civil engineer; b. Menovfeya, Egypt, June 27, 1942; came to U.S., 1971; s. Fam F. Mansour and Elaine (Fam) Ibrahim; m. Samiha Helmy Farage, Aug. 25, 1968; children: Amany F., Michael G. BS with honors, Alexandria (Egypt) U., 1964, MS with honors, 1970; MS in Engring., Lehigh U., 1973; DEng, U. Manitoba, 1974. Registered profl. engr., Pa., Ky.; cert. cost engr., cert. value specialist. Lectr. asst. Alexandria U., 1964-70; teaching asst. U. Manitoba, Winnipeg, Can., 1970-71; grad. asst. Lehigh U., Bethlehem, Pa., 1971-73; project engr. GFCC, Camp Hill, Pa., 1973-80; prin. engr. United Engrs. and Constructors, Phila., 1980-84, supervising cost engr., 1984-87, mgr. project controls and value engring., 1987—; mem. U.S. Com. on Large Dams, 1980-85. V.p. St. George Orthodox Ch., Conshohocken, Pa., 1985-89. Mem. ASCE (pres. sect. hydraulics 1987-88, author papers on hydraulics), Am. Assn. Cost Engrs. (author papers), Soc. Am. Value Engrs. (author papers). Achievements include rsch. in value engring., hydraulics project mgmt. and cost engring. Home: 520 Country Club Dr Cherry Hill NJ 08003 Office: Raytheon Engrs and Constructors 30 S 17th St Philadelphia PA 19101

MANTAS, JOHN, health informatics educator; b. Athens, Greece, Oct. 1, 1954; s. Gerassimos and Despina (Deliolanis) M. BSc with honors, U. Manchester, Eng., 1979, MSc, 1980, PhD, 1983. Teaching asst. U. Manchester, 1979-81, rsch. fellow, 1981-83; researcher Greek Armed Forces, Athens, 1983-85; lectr. U. Athens, 1985-87, asst. prof., 1987-92, assoc. prof., 1992—, coord. prof. Erasmus course, 1989—; mem. tech. panel Aim Office, European Econ. Communities, Brussels, 1990, evaluator telematics, 1991, project ptnr., 1989-91, coord. Erasmus Bur., 1991; head delegation European Standardization Com., Brussels, 1990-91. Author: Introduction to Informatics, 1989, MSc Course in Health Informatics, 1990, Health Informatics, 1990. European Econ. Communities fellow 1989-90, grantee, 1989-91. Mem. IEEE, Inst. Elec. Engrs., Assn. Computer Machinery, Internat. Fuzzy Sets Assn., Pattern Recognition Soc. Christian Orthodox. Home: PO Box 77313, P Faliro, GR-17510 Athens Greece Office: U Athens, 75 Mikras Asias St, GR-11527 Athens Greece

MANTHEY, FRANK ANTHONY, physician, director; b. N.Y.C., Dec. 2, 1933; s. Frank A.J. and Josephine (Roth) M.; m. Douglas Susan Falvey, Sept. 14 1958 (div. 1979, dec. 1989); children: Michael P., Susan M., Peter J.; m. Doris Jean Pulley, Oct. 11, 1979. BS, Fordham U., 1954; MD, SUNY, Syracuse, 1958. Diplomate Am. Bd. Anesthesiology, Am. Bd. Med. Examiners. Intern Upstate Med. Ctr., Syracuse, 1958-59; resident in anesthesiology Yale-New Haven Med. Ctr., 1962-64; physician Yale-New Haven Hosp., 1964-75; pvt. practice medicine Illmo, Mo., 1975-79; dir. Manthey Med. Clinic, Elkton, Ky., 1979—; clin. instr. anesthesiology Yale U. Med. Sch., New Haven, 1964-69, asst. clin. prof. anesthesiology, 1969-75; cons. Conn. Dept. Aeros., Hartford, 1969-70; sr. med. examiner Fed. Aviation Adminstrn., Illmo, 1975-79. Contbr. articles to profl. jours. Chmn. gen. works Little Folks Fair, Guilford, Conn., 1967-71; mem. Rep. Town Com., Guilford, 1969-75; chmn. Guilford Sch. Bldg. Com., 1973-75. Capt. USAF (M.C.), 1956-62. Mem. Ky. Med. Assn., Aerospace Med. Assn. (assoc. fellow 1973-75), Flying Physicians Assn. (v.p. NE chpt. 1973-75, v.p. nat. 1974-75, 79-80, bd. dirs. 1970-73, 75-78, bd. dirs. nat. 1975-78), Aircraft Owners and Pilots Assn., Mercedes Benz Club Am., Alpha Kappa Kappa. Avocations: philately, aviation, skiing, automobile mechanics and restoration, photography. Home: 105 Sunset Dr Elkton KY 42220-9257 Office: Manthey Family Practice Clinic 203 Allensville St PO Box 368 Elkton KY 42220

MANTHORPE, ROLF, medical educator; b. Copenhagen, July 19, 1942; s. Christian and Ruth (Kjaer) M.; m. Tove Bunken, Dec. 4, 1965; 1 child, Martin. Student, U. Frederiksberg, Denmark, 1961; MD, U. Copenhagen, 1969, D in Medicine, 1983. Resident, registrar U. Hosps. in Copenhagen, 1969-78; sr. registrar State U. Hosp., Denmark, 1978-84; assoc. prof. Lund U., Malmoe, Sweden. Editor: 1st International Seminar on Sjogren's Syndrome, 1986. Home: Vejlesoeparken 11, DK-2840 Holte Denmark Office: Malmoe Acad Hosp Sjoegren Ctr, Rheumatology Unit, S-214 01 Malmö Sweden

MANTIL, JOSEPH CHACKO, nuclear medicine physician, researcher; b. Kottayam, Kerala, India, Apr. 22, 1937; came to U.S., 1958; s. Chacko C. and Mary C. Manthuruthil; m. Joan J. Cunningham, June 18, 1966; children: Ann Marie, Lisa Susan. BS in Physics, Chemistry and Math. with distinction, Poona U., 1956; MS, U. Detroit, 1960; PhD, Ind. U., 1965; MS in Biological Scis., Wright State U., 1975; MD, U. Autonoma de Ciudad Juarez, Mex., 1977. Diplomate Am. Bd. Internal Medicine, Am. Bd. Nuclear Medicine; lic. physician, Ohio, Ind., Ky. Rsch. physicist Aerospace Rsch. Lab, Wright Patterson AFB, Ohio, 1964-75; chief resident, resident in internal medicine Good Samaritan Hosp., Dayton, Ohio, 1977-80; chief resident, resident in nuclear medicine Cin. Med. Ctr., 1980-82; assoc. dir., divsn. nuclear med. Kettering (Ohio) Med. Ctr., 1982-86, dir. dept. nuclear medicine/PET, 1986—; dir. Kettering-Scott Magnetic Resonance Lab., Kettering Med. Ctr. Wright State U. Sch. Medicine, Kettering, 1985—, clin. prof. medicine, chief divsn. nuclear medicine, dept. medicine, 1988—; served as session chmn., speaker, amd co-organizer for five internat. confs. Author: Radioactivity in Nuclear Spectroscopy Vol. I and II, 1972; contbr. 38 articles to profl. jours. Mem. ACP, Am. Physical Soc., Soc. Nuclear Medicine, Soc. Magnetic Resonance in Medicine, Soc. Magnetic Resonance Imaging. Achievements include research in proton and phosphorous NMR spectroscopy and glucose metabolism (using PET) in various types of dementia; use of carbon-13 NMR spectroscopy in the study of a saturated and unsaturated fatty acid composition of body fat in patients with heart disease, diabetes and other diseases; normal pressure hydrocephalus with NMR spectroscopy and radionuclide cisternography; general muscle metabolism study with phosphorous NMR spectroscopy and PET and in atrophy, steroid myopathy, and various muscle disorders; use of NMR spectroscopy (both proton and phosphorous) and positron emission tomography (measurement of glucose and protein metabolism) in the study of tumors and assessment of thier reponse to chemotherapy and radiation therapy; positron emission tomography in the study of myocardial viabilty; PET in the diagnosis of coronary artery disease; PET in seizire disorders; use of monoclonal antibody in the diagnosis of maglignant tumors. Home: 6040 Mad River Rd Dayton OH 45459 Office: Kettering Med Ctr 3535 Southern Blvd Kettering OH 45429

MANUEL, PHILLIP EARNEST, meteorologist; b. Martinsville, Va., Oct. 18, 1966; s. James E. and Delorese (Overstreet) M.; m. Robin E. Kellerman, Apr. 21, 1990. BS in Meteorology, Penn State U., 1989. Student trainee Nat. Weather Svc., Charleston, W.Va., 1988; meteorologist Nat. Weather Svc., Cin., 1989-92, Pitts., 1993—; radar focal point Nat. Weather Svc., Cin., 1990-92, storm data focal point, 1993—; asst. focal point WSR-88D, 1993—; speaker in field. Author: Eastern Region Technical Attachment, 1992, Multiscale Evaluation of the 2 June 1990 Tornado Outbreak. Cameraman Cable One, Cin., 1990-92. Recipient Spl. Act reward Nat. Weather Svc., 1990, scholarship Accu-Weather, 1989, First Pl. Forecasting award Penn State U., 1987-88, Scout of the Yr. award Boy Scouts of Am., 1981, 82, Svc. award Snow Creek Rescue Squad, 1985. Mem. Am. Meteorol. Soc. (local chpt. v.p. 1992), Nat. Chpt. Am. Meteorol. Soc., Penn State U. Alumni Assn. Office: Nat Weather Svc 192 Shafer Rd Coraopolis PA 15108

MANUELIDIS, LAURA, pathologist, neuropathologist, experimentalist; b. Bklyn., Sept. 8, 1942; d. Milton Fredrick and Rose (Epstein) Kirchman; m. Elias E. Manuelidis, Nov. 19, 1966; children: Emmanuel Elias, Laertes Alexis. BA, Sarah Lawrence Coll., Bronxville, N.Y., 1963; MD, Yale U, 1967. Asst. Finlays, Balisera, Bangladesh, 1966; intern, resident dept. pathology Yale Med. Sch., New Haven, 1967-69, postdoctoral fellow, instr. neuropathology, 1969-72, asst. prof., 1972-78, assoc. prof., chief neuropathology, 1978-89, prof., head, 1989—; cons. DHHS adv. panel on Alzheimers' Disease, 1993—, NIH, Washington, 1982—, NSF, VA, March of Dimes, U.S.A., Can. Rsch. Coun., MRC, ARC, Eng., 1985; lectr. nat. and internat. meetings. Mem. editorial bd. Acta Neuropathology, 1986-91, Chromosoma, 1990—; author 100 sci. publs., 1969—. Recipient Rsch. Career Devel. award NIH, 1974-79, The Chromosome Prize, 1989; NIH grantee, 1970—. Mem. AAAS, Am. Assn. Neuropathology, Fedn. Socs. for Exptl. Biology. Achievements include patent on in-situ hybridization; rsch. on chromosomes, genetics, dementias. Office: Yale Med Sch 310 Cedar St New Haven CT 06510

MANZI, JOSEPH EDWARD, construction executive; b. N.Y.C., Nov. 16, 1945; s. Frederick Albert and Connie Manzi. BS, Polytech. Inst. Bklyn., 1967; MS, NYU, 1969. Registered profl. engr., N.Y., N.J., Ky., Colo., Ill. Mgr. Bd. Water Supply City of N.Y., 1971-79; group mgr. Kellogg Corp., Littleton, Colo., 1979-85; mng. prin. J.E. Manzi & Assocs., Inc., Park Ridge, Ill., 1985—. Contbr. more than 50 articles on constrn. contract disputes various mags. Testified as expert witness. Mem. NSPE, Def. Rsch. Inst., Am. Soc. Civil Engrs., Constrn. Specifications Inst. Office: JE Manzi & Assocs Inc 826 Busse Hwy Park Ridge IL 60068-2302

MAO, HO-KWANG, geophysicist, educator; b. Shanghai, China, June 18, 1941; came to U.S., 1964; s. Sen and Tak-chun (Hu) M.; m. Agnes Liu, Feb. 10, 1968; children: Cynthia, Linda, Wendy. BS, Nat. Taiwan U., Taepei, 1963; MS, U. Rochester, 1966, PhD, 1968. Rsch. asst., teaching asst. U. Rochester, N.Y., 1964-67, rsch. assoc., 1967-68; postdoctoral fellow Geophys. Lab., Carnegie Instn., Washington, 1968-70, rsch. assoc., 1970-72, geophysicist, 1972—. Adv. editor Physics of Earth and Planetary Interior, 1986—; mem. editorial bd. Jour. of High Pressure Rsch., 1988—. Recipient Bridgman Gold medal Internat. Assn. for Advancement of High Pressure Sci. and Tech., 1989. Fellow Am. Geophys. Union, Mineral Soc. Am. (award 1979); mem. AAAS, NAS (Arthur L. Day Sprice award 1990, Arthur L. Day prize and lectureship 1990), Geol. Soc. Washington, Sigma Xi. Home: 11322 Edenderry Dr Fairfax VA 22030-5441 Office: Carnegie Inst Geophysics Lab 5251 Broad Branch Rd NW Washington DC 20015-1305

MAO, KENT KEQIANG, consulting civil engineer; b. Beijing, People's Republic of China, July 11, 1956; came to U.S., 1984; s. Zhicheng and Shuqing (Dai) M.; m. Yue Zhang, Aug. 20, 1983; 1 child, Jennifer May. BCE, Tsinghua U., Beijing, 1982; MS, Colo. State U., 1985, PhD, 1990. Cert. profl. engr., Colo., Wash. High sch. physics tchr. Beijing #32 High Sch., 1975-78; rsch. engr. Inst. Water Conservancy and Hydroelectric Power Rsch., Beijing, 1982-84; water resources engr. Water and Wastewater Utilities, Ft. Collins, Colo., 1986-91; sr. project water resources engr. HDR Engring., Inc., Bellevue, Wash., 1991-93; sr. water resources engr. KCM, Inc., Seattle, 1993—. Co-author: Environmental Engineers' Handbook, 2d edit., 1994; contbr. articles to profl. jours. Pres. Chinese Student Assn., Colo. State U., 1985-86; pres. Tsinghua Alumni Assn. in Am., 1991. Named Outstanding Tchr. of Beijing Dept. Edn. of Beijing, 1976; recipient Rsch. Scholarship award Inst. Water Conservancy and Hydroelectric Power Rsch., Beijing, 1984. Mem. ASCE (co-chmn. internat. Asian affairs com. 1992—), Am. Water Resources Assn. (vice chair internat. affairs com.). Achievements include development of probablistic and stochastic approaches to forecasting peak water requirements for urban water supply systems, concept and methodology of using optimally combined water treatment and storage capacities for meeting future peak water demands. Home: 17014 NE 38th Pl Bellevue WA 98008-6120 Office: KCM Inc 1917 First Ave Seattle WA 98101-1027

MAO, XIAOPING, fiber optics engineer, researcher; b. Nantong, Jiangsu, China, Sept. 27, 1958; came to U.S., 1988; s. Jiashou Chu and Meifang Mao; m. Lifen Liang, Jan. 24, 1984; 1 child, Lizabeth. PhD, Poly. Inst., Bklyn., 1992. Elec. engr. Wuhan (China) Rsch. Inst. Posts and Telecom., 1984-88; cons. AT&T Bell Labs., Holmdel, N.J., 1989-92; chief fiber optic engr. C-CoR Electronics, State College, Pa., 1992—. Contbr. articles to IEEE Photonic Tech. Letters. Mem. IEEE, Soc. Cable TV Engrs. Achievements include patent pending for Suppression of Brillouin Scattering in Lightwave Transmission System. Home: 31 Abby Pl State College PA 16803 Office: C-COR Electronics 60 Decibel Rd State College PA 16801

MAO, YU-SHI, economist, engineer, educator; b. Nanjing, Jiangsu, China, Jan. 14, 1929; s. Yi-xin Mao and Jing-xiang Chen; m. Yan-ling Zhao, June 20, 1955; children: Wei-xin, Yan-xin. B in Mech. Engring., Jiaotong U., Shanghai, China, 1950. Steam locomotive driver Qiqihar (Heilongjiang) Railway Adminstrn., China, 1950-52; railway mech. engr. Quqihar (Heilongjiang) Railway Adminstrn., China, 1952-55; railway rsch. engr. Inst. Railway Rsch., Beijing, 1955-68; designing engr. Datong Locomotive Factory, Datong, Shanxi, China, 1968-78; rsch. economist Inst. Railway Rsch., 1978-84; sr. rsch. fellow Inst. Am. Studies Chinese Acad. Social Scis., Beijing, 1984—; prof. China Inst. Mining and Tech., Beijing, 1987-91; chmn. Unirule Inst. Econs., Beijing, 1993—; prof. Beijing Inst. Econs., 1982; vis. prof. U. Queensland, Australia, 1990; cons. Asia Devel. Bank, Manila, 1990—, UN Devel. Programme, 1990; vis. scholar Harvard U., Cambridge, Mass., 1986; resource person Africa Energy Policy Rsch. Network, Gaborone, Botswana, 1987—. Author: Principal of Optimal Allocation: Economics and Its Mathematical Foundations, 1985, Economics in Dialy Life: A Survey of U.S. Market, 1993; editor China Econ. Rev. jour., 1988—; cons. editor China Econ. Rev. (domestic and internat. issues), 1988—, Reform and Strategy jour., 1985—; editorial bd. Sci. and Tech. Rev. jour., 1988—; contbr. articles to prof. jours. Participant InterAction Coun., Amsterdam, 1990; Global Challenge Network, Munich, 1988; Internat. Found. Survival and Devel., Leningrad, 1988; speaker 1st, 2d Internat. Conf. on Ethics and Environ. Policies, Venice, 1990, Athens, U.S., 1992. Recipient Ford Found. fellowship, 1986, Silver Ring prize State Sci. and Tech. Commn., 1988. Fellow Chinese Economists Soc.; mem. China Energy Rsch. Soc. (bd. dirs. 1980—), China Soc. Quantitative Econs. (bd. dirs. 1990—), Chinese Economists Soc. Avocations: classical music, Tai-chi fighting. Home: 2-1-1 Nan Sha Gou, San Li He, Beijing 100045, China Office: Inst Am. Studies Chinese Acad Social Scis, 5 Jian Guo Men Nei, Beijing 100732, China

MAPLES, WILLIAM ROSS, anthropology educator, consultant; b. Dallas, Tex., Aug. 7, 1937; s. William Hunter and Agnes Ross (Bliss) M.; m. Margaret Jane Kelley, Dec. 20, 1958; children: Lisa Linda, Cynthia Lynn. BA, U. Tex., 1959, MA, 1962, PhD, 1967. Diplomate Am. Bd. Forensic Anthropology. With Darajani (Kenya East Africa) Primate Rsch. Sta., 1962-63; teaching asst. U. Tex., Austin, 1963-64; mgr. S.W. Primate Rsch. Ctr., Nairobi, Kenya, East Africa, 1964-66; asst. prof. Western Mich. U., Kalamazoo, 1966-68; from asst. prof. to prof. anthropology U. Fla., Gainesville, 1968—; with Fla. Mus. Natural History, Gainesville, 1972-78, chmn. dept. social scis., 1973-87, curator in charge human indentificaion lab., 1986—; cons. Armed Forces Graves Registration Office, Honolulu, 1986—, Armed Forces Inst. Pathology, Washington, 1989—, N.Y. State Police Forensic Unit, Albany, 1987—; cons. in residence U.S. Army Cen. Identification Lab., Honolulu, 1986-87. Contbr. articles to profl. jours. Recipient Disting. Tchr. Cert., U. Fla., 1973, Cert. of Honor, City of Lima, Peru, 1985. Fellow Am. Anthropol. Assn., Am. Assn. Phys. Anthropologists; mem. Am. Acad. Forensic Scis. (v.p. 1986-87, bd. dirs. 1989—), Forensic Scis. Found. (trustee, treas. 1988—), Am. Bd. Forensic Anthropology (treas. 1984-87, pres. 1987-89). Avocations: photography, sailing. Office: U Florida Florida Museum Natural History Gainesville FL 32611

MARAMAN, WILLIAM JOSEPH, nuclear engineering company executive; b. El Paso, Tex., May 19, 1923; s. William Minor and Katherine (Hawkins) M.; m. Katherine Ann Thorpe, Oct. 12, 1948; children: Katherine Ann, Linda Susan. BS in Chem. Engring., U. Tex., Austin, 1944; MS, U. N.Mex., 1960. Registered profl. engr., N.Mex. Staff mem. Los Alamos (N.Mex.) Nat. Lab., 1946-56, group leader, 1956-79, div. leader, 1979-83; dir. TRU Engring. Co., Inc., Santa Fe, 1984—; tech. mgr. LANL Plutonium Facility, Los Alamos, 1969-79; cons. in field. Patentee in field. Income tax preparer Am. Assn. Ret. Persons, Los Alamos, 1985-92. Lt. (j.g.) USN, 1944-46, PTO. Home: 288 Connie Ave Los Alamos NM 87544-3613

MARAN, STEPHEN PAUL, astronomer; b. Bklyn., Dec. 25, 1938; s. Alexander P. and Clara F. (Schoenfeld) M.; m. Sally Ann Scott, Feb. 14, 1971; children: Michael Scott, Enid Rebecca, Elissa Jean. B.S., Bklyn. Coll., 1959; M.A., U. Mich., 1961, Ph.D., 1964. Astronomer Kitt Peak Nat. Obs., Tucson, 1964-69; project scientist for orbiting solar observatories NASA-Goddard Space Flight Center, Greenbelt, Md., 1969-75; head advanced systems and ground observations br. NASA-Goddard Space Flight Center, 1970-77; mgr. operation Kohoutek, 1973-74, sr. staff scientist, 1977—; Cons. Westinghouse Rsch. Labs., 1966; vis. lectr. U. Md., College Park, 1969-70; sr. lectr. UCLA, 1976; A. Dixon Johnson lectr. in sci. communication, Pa. State U., 1990; lectr. on astronomy cruises and eclipse tours. Author: (with John C. Brandt) New Horizons in Astronomy, 1972, 2d edit., 1979, Arabic edit., 1979; Editor: Physics of Nonthermal Radio Sources, 1964, The Gum Nebula and Related Problems, 1971, Possible Relations Between Solar Activity and Meteorological Phenomena, 1975, New Astronomy and Space Science Reader, 1977, A Meeting with the Universe, 1981, Astrophysics of Brown Dwarfs, 1986, The Astronomy and Astrophysics Encyclopedia, 1991; assoc. editor: Earth, Extraterrestrial Scis, 1969-79; editor: Astrophys. Letters, 1974-77, assoc. editor, 1977-85; contbg. editor Air & Space/Smithsonian, 1990—; contbr. articles on astronomy, space to popular mags. Named Disting. Visitor Boston U., 1970; recipient Group Achievement awards NASA, 1969, 74, Exceptional Achievement medal, 1991. Mem. AAAS, Internat. Astron. Union (editor daily newspaper 1988), Am. Astron. Soc. (Harlow Shapley vis. lectr. 1981—), press. officer 1985—), Royal Astron. Soc., Am. Phys. Soc., Am. Geophys. Union. Office: NASA Goddard Space Flight Ctr Code 680 Greenbelt MD 20771

MARANZANO, MIGUEL FRANSCISCO, engineer; b. Buenos Aires, Argentina, Aug. 14, 1941; came to U.S., 1977; s. Miguel and Nelida (Pizzini) M.; m. Ana Isabel Carranza, Mar. 25, 1978; children: Cynthia, Rossana, Giancarlo, Gabriel. BS, SUNY, 1982; MS, Rensselaer Poly., 1985. Field engr. Johnson Controls, Phila., 1978-82; sr. rsch. engr. Hi-G Rsch., Bloomfield, Conn., 1979-82; engr. mgr. Locknetics, Hamden, Conn., 1982-85; group leader Superior Elec., Bristol, Conn., 1985-86; mgr. elec. testing Springborn Labs., Enfield, Conn., 1986-88; sr. product engr. Skinner Valve, New Britain, Conn., 1988-89, Kip, Inc., Farmington, Conn., 1989—; mem. adj. faculty dept. math. and sci. Manchester C.C., 1986-88; presenter 39th internat. conf. Nat. Assn. Relay Mfrs. Contbr. articles to profl. jours. Mem. Hispanic-Am. Cultural Coun., New Britain, Conn., 1990. Achievements include patent on electronic game and solenoid valve. Home: 110 Pheasant Run Bristol CT 06010 Office: Kip Inc 72 Spring Ln Farmington CT 06032

MARAVELIAS, PETER, systems engineer; b. Lynn, Mass., Sept. 6, 1949; s. James Angelo and Dora (Nicholopoulos) M.; m. Naomi Thomaides, Apr. 23, 1972; children: Sophia, James Peter. BS in Math., Lowell Tech. Inst., 1971; MS in Systems Mgmt., U. So. Calif., 1976; grad., Air Command and Staff Coll., 1985. Commd. 2nd lt. USAF, 1971, advanced through grades to lt. col., 1987, mgr. info. systems, 1971-78, chief program control, 1978-80; compt. Iraklion Air Sta. USAF, Crete, Greece, 1982-84; exec. fin. advisor USAF, 1985-86, program dir. NORAD Granite Sentry Program, 1986-89; program dir. Software Engring. Inst. Carnegie Mellon U., Pitts., 1989-90; spl. project dir. Operation Desert Shield/Storm USAF, 1990—; v.p. Mosaic Data Systems, Bedford, Mass.; tech. mgmt. instr. U. Md.; cons. Wells Fargo Bank, San Francisco, 1981-82. Decorated Meritorious Svc. medal with four oak leaf clusters. Achievements include design of a command ctr. generic info. processing system software architecture utilizing open system stds. and reusable software components to achieve drastic reductions in time and costs of software devel.; led senior engineering team identifying improvements for the Tactical Air Control System deployed to the Persian Gulf in 1990-91 in support of Operation Desert Storm; deployed a modern automated air situation display system to Saudi Arabia that significantly increased command, control and battle management of the Persian Gulf Air War; provided the first major upgrade in 20 years to the North American Air Defense (NORAD) Cheyenne Mountain Complex by replacing the Air Defense Operations Center and upgrading the NORAD Command Center which provided a major deterrent to the then Soviet Union threat. Office: Mosaic Data Systems 19 Crosby Dr Bedford MA 01730

MARAZZO, JOSEPH JOHN, civil engineer; b. Staten Island, N.Y., Sept. 17, 1958; s. Alphonse V. and Josephine (Nuzzo) M.; m. Jennifer Lin Bates, Oct. 8, 1988; children: Christopher J., Alyssa C. BS in Civil Engring., N.J. Inst. Tech., 1980; MBA in Mktg., St. Johns U., Jamaica, Queens, N.Y., 1987. Jr. engr. Bklyn. Union Gas, 1980-82, field engr., 1982-83, sr. rsch. engr., 1983-85, sr. field engr., 1986-88, mgr. leak surveillance, 1988-91, mgr. distbn. ops., 1991—; advisor Gas Rsch. Inst., Chgo., 1984-85. V.p. Pocono County Pl. Property Owners Assn., Tobyhanna, Pa., 1989-92, dir., 1989-90. Mem. ASCE (assoc.), Am. Gas Assn. Roman Catholic. Achievements include patent for steel pipe splitting/expanding for replacement of piping systems. Home: 564 Sterling Pl Scotch Plains NJ 07076 Office: Bklyn Union Gas Co 287 Maspeth Ave Brooklyn NY 11211

MARCALI, JEAN GREGORY, chemist; b. Jermyn, Pa., May 29, 1926; d. John Robert and Anna Marie Gregory; student U. Pa., 1948-52, U. Del., 1971-72; m. Kalman Marcali, Oct. 6, 1956; children—Coleman, Frederick. Microanalyst E. I. du Pont de Nemours & Co., Deepwater, N.J., 1943-60, tech. info. analyst, Jackson Lab., Deepwater, N.J. also Wilmington, Del., 1960-67, sr. adviser tech. info., Wilmington, 1967-70, supr. tech. info., 1970-82, 85-89, supr. adminstrv. svcs., 1982-85, cons., 1989-92, ret., 1992. Sec., Alfred I. DuPont Elem. PTA, 1971, pres., 1972; pres. PTA of Brandywine Sch. Dist., 1973; mem. Wilmington Dist. Republican Com., 1976—. Mem. Am. Chem. Soc. (treas. div. chem. info. 1976-81, chmn.-elect 1981, chmn. 1982, 83, div. councilor 1983-90), Am. Chem. Soc. (com. on chem. abstracts service 1983-85, 87-93). Lutheran. Clubs: Order Eastern Star, Du Pont Country. Home: 312 Waycross Rd Wilmington DE 19803-2950

MARCELLAS, THOMAS WILSON, electronics company executive; b. Owings, Md., June 22, 1937; s. Carroll Wilson and Mabel Ellice (Hardesty) M.; B. Engring. Sci., Johns Hopkins U., 1960; M.Engring. Adminstrn., George Washington U., 1980; m. Janet Fay Hardesty, June 20, 1964; children: David Carroll, Diane Elizabeth. Project engr. Bendix Radio Corp., Towson, Md., 1960-61, 63-66; electronic systems engr. Electronic Modules Corp., Hunt Valley, Md., 1966-78, product devel. mgr., 1978-81; dir. research and devel. EMC Controls Inc., Hunt Valley, 1981-82, v.p. research and devel., 1982-84; v.p. research and devel. EMC Ops. Div., Hunt Valley, 1984-85; v.p. research and devel. Rexnord Automation Inc., Hunt Valley Md., 1986-87, v.p. mfg. and engring. Electronic Ops. Group, 1987-89; prodn. program mgr. Bendix Communications div. Allied Signal Aerospace Co, Towson, 1989—. Mem. adv. bd. Dundalk Community Coll. Served to 1st lt. U.S. Army, 1961-63. Mem. IEEE, Am. Mgmt. Assn., Surface Mount Tech. Assn., Assn. Mfg. Excellence. Democrat. Methodist. Home: 13804 Princess Anne Way Phoenix MD 21131-1522 Office: Allied Signal Aerospace Co Bendix Communications Div 1300 E Joppa Rd Baltimore MD 21204-5917

MARCELLUS, MANLEY CLARK, JR., chemical engineer; b. Duluth, Minn., Jan. 1, 1921; s. Manley Clark and Eva (Scheideker) M.; m. Nadine Augusta Webb, Nov. 16, 1945; children: John Robert, Charles Alan, Lynn Elaine. BChE, U. Minn., 1943. Chem. engr. Westvaco Chi Prop FMC, Newark, Calif., 1943-50; group leader FMC, Newark, 1950-57; chem. engr. Minn. Mining and Mfg. Co., Maplewood, Minn., 1957-70; specialist then sr. specialist Minn. Mining and Mfg. Co., St. Paul, 1970-89; cons. advance systems, indsl. heat Minn. Mining and Mfg. Co., various locations, 1989-93; ret., 1993; chair Engring. Specialists and Prin. Engrs., 1985. Contbr. articles to profl. publs. Mem. AIChE. Presbyterian. Achievements include patents for Pelletization of phosphate shale, Barium silicate process, Periclase ramming mixture, Auto exhause muffler. Home: 4611 Birchbark Tr Lake Elmo MN 55042 Office: 3M Co 900 Bush Ave Saint Paul MN 55144

MARCH, JACQUELINE FRONT, chemist; b. Wheeling, W.Va.; m. A.W. March (dec.); children: Wayne Front, Gail March Cohen. BS, Case Western Res. U., 1937, MA, 1939; postgrad. U. Chgo., U. Pitts., Ohio State U. Clin. chemist, Mt. Sinai Hosp., Cleve.; med. rsch. chemist U. Chgo.; rsch. analyst Koppers Co., also info. scientist Union Carbide Corp., Mellon Inst., Pitts.; propr. March Med. Rsch. Lab., etiology of diabetes, Dayton, Ohio; guest scientist Kettering Found., Yellow Springs, Ohio; Dayton Found. fellow Miami Valley Hosp. Rsch. Inst.; mem. chemistry faculty U. Dayton, info. scientist Rsch. Inst. U. Dayton; on-base supr. Air Force Info. Ctr. Wright-Patterson AFB, 1969-79; chem. info. specialist Nat. Inst. Occupational Safety and Health,PHS, HHS, CDC, Cin., 1979-90; propr. JFM Cons., Ft. Myers, Fla., 1990—; designer info. systems, speaker in field. Contbr. articles to profl. publs. Wyeth fellow med. rsch. U. Chgo., 1940-42. Mem. AAUP (exec. bd. 1978-79), Am. Soc. Info. Sci. (treas. South Ohio chpt. 1973-75), Am. Chem. Soc. (pres. Dayton 1977), Dayton Engring. Soc. (hon.), Soc. Advancement Materials & Process Engring. (pres. Midwest chpt. 1977-78), Affiliated Tech. Socs. (Outstanding Scientist and Engr. award 1978), Sigma Xi, the Scientific Rsch. Soc. (pres. Cin. fed. environ. chpt. 1986-87). Home and Office: # 5 13201 Oakmont Dr Fort Myers FL 33907

MARCHALONIS, JOHN JACOB, immunologist, educator; b. Scranton, Pa., July 22, 1940; s. John Louis and Anna Irene (Stadner) M.; m. Sally Ann Sevy, May 5, 1978; children: Lee, Elizabeth, Emily. A.B. summa cum Laude, Lafayette Coll., 1962; Ph.D., Rockefeller U., 1967. Grad. fellow Rockefeller U., 1962-67; fellow Am. Cancer Soc. Walter and Eliza Hall Inst. Med. Research, 1967-68; asst. prof. biomed. scis. Brown U., 1969-70; head molecular immunology lab. Walter and Eliza Hall Inst. Med. Research, Melbourne, Australia, 1970-76; head cell biology and biochemistry sect. Frederick Cancer Research Ctr., 1977-80; prof. adj. faculty dept. pathology U. Pa., 1977-83; prof., chmn. dept. biochemistry and molecular biology Med. U. S.C., Charleston, 1980-88; prof., chmn. dept. microbiology and immunology U. Ariz., Tucson, 1988—, prof. pathology, 1991—, prof. medicine, 1992—; bd. dirs. Am. Type Tissue Culture Collection. Author: Immunity in Evolution, 1977; editor: Comparative Immunology, 1976, The Lymphocyte: Structure and Function, 1977, (with N. Cohen) Self/Non-Self Discrimination, 1980, (with G.W. Warr) Antibody as a Tool, 1982, The Immunobiology and Molecular Biology of Parasitic Infections, 1983, Antigen-Specific T Cell Receptors and Factors, 1987, The Lymphocyte: Structure and Function, 2d edit., 1987, (with Carol Reinisch) Defense Molecules, 1989; contbr. articles to profl. jours. Active Nat. Commn. Damon Runyon-Walter Winchel Cancer Fund. Named among 1,000 most highly cited sci. authors Inst. for Sci. Info.; Frank R. Lillie fellow, 1974; grantee in field. Fellow Am. Inst. Chemists, Am. Acad. Microbiology; mem. AAAS, Am. Assn. Immunology, Am. Soc. Biol. Chemists, Sigma Xi, Phi Beta Kappa. Episcopalian. Achievements include development of microchemical (radioimmunochemical) approaches for proteins and surface receptors of living cells; characterization of immunoglobulin-like antigen receptors of thymus-derived lymphocytes; application of synthetic peptide technology to antibodies, T cell receptors and autoimmunity. Home: 5661 N Camino Arturo Tucson AZ 85718-3933 Office: U Ariz Health Scis Ctr Tucson AZ 85724

MARCHESANO, JOHN EDWARD, electro-optical engineer; b. N.Y.C., Aug. 20, 1927; s. John R. and Maria J. (Mollino) M.; divorced; children: Pamela, Debra, Scott, Neal. BE, CCNY; postgrad., U. Pa. Project engr. Philco Rsch., Pa., 1951-56; sr. engr. Am. Bosch Arma, Garden City, N.Y., 1956-60; pres. Automation Labs. Inc., Mineola, N.Y., 1960-66; now pres., CEO Decilog, Inc., Melville, N.Y. Achievements include development and design of wide angle lens for aircraft collision avoidance, of modulation transfer system for low light level TV evaluation, long wavelength infared missile research. Office: Decilog Inc 555 Broadhollow Rd Melville NY 11747-5001

MARCHESE, ANTHONY JOHN, research engineer; b. Livingston, N.J., Sept. 4, 1967; s. Lawrence Nicholas and Elaine Marie (Andre) M.; m. Debra Ann Cooper, July 28, 1990. BSME, Resselaer Polytech, 1989, MSME, 1992. Student trainee rsch. engr. NASA Lewis Rsch. Ctr., Cleve., 1987-89; asst. rsch. engr. United Techs. Rsch. Ctr., East Hartford, Conn., 1989—; rsch. assoc. Princeton (N.J.) U., 1992—. Contbr. articles to profl. jours. Mem. ASME, AIAA, Tau Beta Pi, Pi Tau Sigma. Independent. Roman Catholic. Achievements include 3 patent disclosures pertaining to scroll compressors. Office: Princeton Univ Dept Mech Dept Mech and Aerospace Eng Princeton NJ 08544-5263

MARCHESI, GIAN FRANCO, psychiatry educator; b. Perugia, Umbria, Italy, Mar. 11, 1940; s. Mario and Pina (De Notaris) M; m. Maria Ida Catagna, Jan. 20, 1973; 1 child, Piergiorgio. MA, U. Perugia, 1958, MD, 1964; PhD in Neuro-Psychiatry, U. Ancona (Italy), 1975. Asst. prof. U. Perugia, 1971-73; vis. prof. U. Glasgow (Scotland), 1974; assoc. prof. U. Ancona, 1975-79, prof., 1981—, counselor Univ. Office Data Processing, 1978-82; vis. prof. SUNY, N.Y.C., 1980, Nat. Acad. Scis., Prague, Czechoslovakia, 1982; dir. Inst. Psychiatry, Ancona, 1984—; dir. Clin. Neurophysiology Specialization, Ancona, 1984-87, Pyschiatry Specialization Sch., Ancona, 1984—, Psychiat. Rehab. Specialization, Ancona, 1989—, Psychiat. Social Worker Specialization, Ancona, 1990—. Author, co-editor: Manual of Neurology, 1979, Principles of Neurology and Psychiatry, 1985; author, editor: Manual of Psychiatry, 1992; contbr. articles to profl. jours. Counselor Adminstrv. Univ. Coun., Ancona, 1979-82, Region Office Univ. Student Svcs., Ancona, 1980-88; pres. Assn. Against Psychol. Suffering, Ancona, 1986—. Grantee Ministry Edn., Rome, 1971-88, NRC, Rome, 1978-86, Dept. Health, Marche Region, Ancona, 1980—, Ministry of Univs., Rome, 1989—. Mem. AAAS, Am. Med. Electroencephalographic Assn., Am. Acad. Neurology, World Fedn. Soc. Biol. Psychiatry, Collegium Internat. Neuro-Psychopharmacologicum, Am. Acad. Clin. Neurophysiology, World Fedn. Sleep Rsch. Socs. Office: Inst Psychiatry, Policlinico Umberto I, I-60121 Ancona Italy

MARCHESSAULT, ROBERT H., chemical engineer. Recipient Montreal medal Chem. Inst. Can., 1991. *

MARCHIONE, SHARYN LEE, computer scientist; b. Schenectady, Oct. 1, 1947; d. Albert Jr. and Estelle Mabelle (Christiansen) O'Brien; m. Joseph Michael Marchione, May 4, 1972; 1 child, Heather E. AS in Engring., Hudson Valley Community Coll., Troy, N.Y., 1967; BS in Computer Sci., Skidmore Coll., 1987; MBA, Coll. of St. Rose, Albany, N.Y., 1993. Computer programmer info. systems GE, Schenectady, 1967-72, 78-81, shift leader CAD-CAM systems, 1981-84, advanced techniques specialist, 1984-88, mgr. end. user computing decision support ops., 1988—, chmn. windows spl. interest group, 1990—; mem. adv. coun. Software Pub. Co., 1991. Vol. Rep. Town Supr. Campaign, Halfmoon, N.Y., 1987-91, Concerned for Hungry, Schenectady, 1988—; cons. Schenectady Econ. Devel. Coun., 1991—, Cobleskill Coll., SUNY, 1991—. Avocations: mountain hiking, camping, bicycling, reading, tennis. Home: 10 Herlihy Rd Round Lake NY 12151 Office: GE Decision Support Ops # 4-305 1 River Rd Schenectady NY 12345-6001

MARCHISOTTO, ROBERT, pharmacologist; b. N.Y.C., Nov. 9, 1929; s. G.J. and Annetta (Franco) M; m. Jo Guaragna, Aug. 23, 1952 (div. 1979); children: Laura Anne, Denise Frances, Robert Andrew. BS, L.I. U., 1952; MS, Purdue U., 1954, PhD, 1956. Cert. pharmacist N.Y., N.J. Ind. Group leader rsch. div. Johnson and Johnson, North Brunswick, N.J., 1956-61; dir. internat. labs. Bristol-Myers, Hillside, N.J., 1961-65; dir. scientific info. Richardson-Merrell, Vick divsn., Mount Vernon, N.Y., 1965-68; dir. R & D Schering Corp. Pharmaco divsn., Kenilworth, N.J., 1968-71; mng. dir. Biosis U.K. ltd., York, Eng., 1980-81; dir. scientific affairs Bioscis. Info. Svcs., Phila., 1976-85; exec. dir. Purdue Assocs., Princeton Junction, N.J., 1971—; bd. dirs. AIHA, Staten Island, N.Y.; sec. Nat. Fedn. Abstrating Inf. Svcs., Phila., 1979-84. Contbr. articles to profl. jours. mem. Bd. Health, West Windsor Twp., N.J., 1988, Bd. Edn., East Brunswick, 1961-63; sci. adv. com. Bd. Edn. East Brunswick, 1965-68. Mem. Am. Pharm. Assn., Nat. Coalition Ind. Scholars, Princeton Rsch. Forum, Phi Lambda Upsilon, Rho Chi, Sigma Xi. Achievements include patents for Novel First Aid Products, Denture Adhesive Composition. Home and Office: 4 Canoe Brook Dr Princeton Junction NJ 08550-1602

MARCOTTE, MICHEL CLAUDE, geotechnical engineer, consultant; b. Montreal, Quebec, Can., Sept. 17, 1955; s. Andre and Therese (LaFond) M.; m. Suzanne Mordente; 1 child, Miriam. B of Engring., Polytech., Montreal, 1978, Maitrise en Scis. Appliquées, 1982. Rsch. assoc. Ecole Poly., Montreal, 1981-84; tech. dir. Queformat Ltd., Longueuil, Quebec, 1984-88; exec. dir. Solmers Internat., Boucherville, Quebec, 1988-91, pres., 1991—; mem. std. bd. Can. Gen., Ottawa, 1988. Mem. N.Am. Geosynthetics Soc. (Award of Excellence 1991). Office: Solmers Internat, 25 ch du Tremblay Ste 200, Boucherville, PQ Canada J4B 7L6

MARCOTTE, VINCENT CHARLES, metallurgist; b. Langdon, N.D., June 4, 1932; s. Charles Arthur and Lucille Mary (McKowen) M.; m. Candee Frances Kinneer, Jan. 19, 1963; 1 child, Gregory. BA in Chemistry, St. Johns U., Collegeville, Minn., 1957; MS in Metallurgy, Iowa State U., 1963. Chemist AEC, Ames, Iowa, 1957-65; metallurgist IBM, Hopewell Junction, N.Y., 1965—. Author: (with others) Solder Mechanics, 1990; contbr. articles to Jour. of Less-Common Metals, Scripta Metallurgica. Cpl. U.S. Army, 1952-54. Fellow Am. Soc. Material Internat.; mem. Minerals, Metals and Materials Soc., Sigma Xi. Achievements include patents in the process for in-site modification of solder composition, the process of brazing using low temperature braze alloy of solid indium tin. Home: 39 Kuchler Dr LaGrangeville NY 12540 Office: IBM Hopewell Junction NY 12533

MARCUM, JAMES ARTHUR, physiology educator; b. Hamilton, Ohio, June 11, 1951; s. Richard C. and Madonna M. (Rohrkemper) M.; m. Sarah Hite Johnson, June 20, 1992. BSEd, Miami U., Oxford, Ohio, 1972, MS, 1974; PhD, U. Cin., 1978; MATS, Gordon-Conwell Sem., 1982; MA, Boston Coll., 1987. Rsch. assoc. MIT, Cambridge, Mass., 1982-84; instr. Harvard U. Med. Sch., Boston, 1984-85, asst. prof., 1985—. Contbr. articles to profl. jours. including Jour. Clin. Investigation, Jour. Biol. Chemistry, Biol. Bull., Annals N.Y. Acad. Scis., Perspectives in Biology and Medicine. Predoctoral fellow U. Cin., 1974, postdoctoral fellow NIH, 1983, Frederik B. Bang fellow Marine Biol. Lab., Woods Hole, Mass., 1985; recipient Young Investigator award NIH, 1985; grantee-in-aid Am. Heart Assn., 1990. Achievements include demonstration that the non-thrombogenic property of the vascular endothelium is due in part to anticoagulantly active proteoheparan sulfate synthesized by endothelial cells. Office: Beth Israel Hosp Pathology Dept 330 Brookline Ave Boston MA 02215

MARCUS, DAVID ALAN, physicist; b. Newport News, Va., July 17, 1958; s. Allen C. and Lenore (Berger) Mills; m. Barbara A. Higson, July 28, 1990; 1 child, Michael. BS, Tufts U., 1980; MS, Trinity U., 1982; postgrad., Ariz. State U., 1989—. Rsch. engr. Inst. Desert Rsch. Ben-Gurion U., Beer Sheva, Israel, 1983-85; optical thin films scientist Omega Optical, Inc., Brattleboro, Vt., 1985-89; founder, consulting physicist Custom Sci., Phoenix, 1990—; rsch. asst. physics dept. Ariz. State U., Tempe, 1989—. Office: Ariz State U Physics Dept Tempe AZ 85287-1504

MARCUS, DONALD MARTIN, internist, educator; b. N.Y.C., Dec. 8, 1930; s. James Louis and Sophia (Horne) M.; m. Marianne Taft, Feb. 15, 1958; children: Laura Ruth, Susan Marcus Mendoza, James William. BA, Princeton U., 1951; MD, Columbia U., 1955. Diplomate Am. Bd. Internal Medicine. Med. resident in internal medicine Columbia-Presbyn. Med. Ctr., N.Y.C., 1955-57; resident in internal medicine Stroug Meml. Hosp., Rochester, N.Y., 1959-60; Helen Hay Whitney Found. fellow dept. microbiology and immunology Coll. Physicians and Surgeons Columbia U., 1960-63; mem. faculty Albert Einstein Coll. Medicine, Bronx, N.Y., 1963-80; prof. medicine, microbiology and immunology Baylor Coll. Medicine, Houston, 1980—; mem. study sect. NIH, Nat. Cancer Inst., Bethesda, Md., 1973-77, NIAID, Bethesda, 1981-85; rsch. com. Arthritis Found., Atlanta, 1987-90; sci. adv. coun. ARC, Rockville, Md., 1988—. Contbr. over 100 articles to sci. publs. Capt. U.S. Army, 1957-59. Recipient Karl Landsteiner Meml. award Am. Assn. Blood Banks, 1980, Philip Levine award Am. Soc. Clin. Pathologists, 1985. Mem. Am. Assn. Immunologists, Am. Coll. Rheumatology, Am. Soc. Clin. Investigation, Am. Assn. Physicians, Am. Soc. Biochemistry and Molecular Biology. Achievements include study of biochemical basis of human blood group P system; fundamental investigations into immunochemistry of carbohydrates. Office: Baylor Coll Medicine 1 Baylor Plz Houston TX 77030

MARCUS, ERIC ROBERT, psychiatrist; b. N.Y.C., Feb. 16, 1944; s. Victor and Pearl (Maddow) M.; m Eslee Samberg, Nov. 24, 1985; children: Max Thomas, Pia. AB, Columbia U., 1965; MD, U. Wis., 1969. Diplomate Am. Bd. Psychiatry and Neurology. Intern NYU Med. Ctr. Bellevue Hosp., 1969-70; resident Columbia Presbyn. Med. Ctr.-N.Y. State Psychiatric Inst., 1972-75; dir. St. Marks Free Clinic, N.Y.C., 1971-75; from co-dir. to dir. neuropsychiatric/diagnostic treatment unit Columbia-Presbyn. Med. Ctr., N.Y.C., 1975-84; dir. med. student edn. in psychiatry Columbia U. Coll. Physicians and Surgeons, N.Y.C., 1981—; assoc. clin. prof. psychiatry and social medicine Columbia U. Coll. Physicians and Surgeons, 1981—; mem. faculty Columbia U. Ctr. for Psychoanalytic Tng. and Rsch., 1987—; bd. govs. student health Columbia U., 1986. Author: Psychosis and Near Psychosis, 1992; mem. editorial bd. The Psychoanalytic Study of Society, 1989—; contbr. articles to profl. jours. Grants and edn. com. Am. Cancer Soc., N.Y.C., 1988—. Recipient Weber award Columbia U. Psychoanalytic Ctr., 1991. Fellow Am. Psychiat. Assn. (Roeske award 1991), Am. Psychoanalytic Assn., N.Y. Acad. Medicine. Avocations: classical music, photography, swimming, reading. Office: Columbia U Dept Psychiatry 722 W 168th St New York NY 10032-2603

MARCUS, HARRIS LEON, mechanical engineering and materials science educator; b. Ellenville, N.Y., July 5, 1931; s. David and Bertha (Messite) M.; m. Leona Gorker, Aug. 28, 1962; children: Leland, M'Risa. BS, Purdue U., 1963; PhD, Northwestern U., 1966. Registered profl. engr., Tex. Tech. staff Tex. Instruments, Dallas, 1966-68; tech. staff Rockwell Sci. Ctr., 1968-70, group leader, 1971-75; prof. mech. engring. U. Tex., Austin, 1975-79, Harry L. Kent Jr. prof. mech. engring., 1979-90, Cullen Found. prof., 1990—, dir. ctr. for Materials Sci. and Engring., dir. program, 1979—; cons. numerous orgns. Contbr. numerous articles to profl. publs. Recipient U. Tex. faculty U. Tex. Engring. Found., 1983; Krengel lectr. Technion, Israel, 1983; Alumni Merit medal Northwestern U., 1988. Fellow Am. Soc. Metals; mem. ASME, ASTM, AIME (bd. dirs. Metall. Soc. 1976-78, 84-86), Am. Phys. Soc., Materials Rsch. Soc., Univ. Materials Coun. Achievements include 10 patents. Home: 4102 Hyridge Dr Austin TX 78759-8022 Office: U Tex Ctr Rsch Mat Sci & Engring ETC 9 104 Austin TX 78712-1063

MARCUS, JOY JOHN, pharmacist, educator, consultant; b. Charleston, S.C., Aug. 13, 1951; d. John Basil and Penelope (Polizos) M. AS, Anderson (S.C.) Jr. Coll., 1971; BS in Health and Phys. Edn., U. S.C., 1976; MS in Sports Adminstrn., St. Thomas U., Miami, 1987; BS in Pharmacy, Southeastern U., Miami, 1992. Assoc. prof. Miami Dade C.C. N, 1987—; pharmacy technician/intern Bay Rexall Drug Store, Miami, 1974-91; intern Eckerd's Pharmacy, Miami, 1991-92, extern and intern, 1991-92; pharmacist Eckerd's Pharmacy, 1992—; lab. instr. Southeastern Coll. of Pharmacy, 1991, asst. prof. pharm. sci., pharmacy practice, 1992—; adj. prof. St. Thomas U., 1988. Treas., Miami Shore Bus. Assn., 1985. Recipient award for Campuses Addressing Substance Abuse, 1991; recipient several grants. Mem. Am. Pharm. Assn., Nat. Assn. Retail Druggists, Fla. Pharmacy Assn. (exec. com. 1991-92, pub. affairs com. 1991-92, futuristic com. 1992-93), Dade County Pharmacy Assn. (resolution com. 1992, exec. com. 1992, pres. elect 1993-94, pres. 1994-95), Phi Lambda Sigma (sec., treas. 1991-92), Alpha Zeta Omega (v.p. 1989-90, sec. 1990-91). Republican. Greek Orthodox. Avocations: martial arts, tennis, running,. scuba diving, reading. Home: 13105 Ixora Ct Apt 317 Miami FL 33181-2322

MARCUS, JULES ALEXANDER, physicist, educator; b. Coytesville, N.J., May 10, 1919; s. Alexander and Julia Hollister (Parks) M.; m. Ruth Charlotte Barcan, Aug. 28, 1942; children: James S., Peter W., Katherine H., Elizabeth P. BS, Yale U., 1940, PhD, 1947. Physicist Johns Hopkins U., Balt., 1944-46; rsch. asst. Yale U., New Haven, 1946-47; postdoctoral fellow U. Chgo., 1947-49; asst. prof. physics Northwestern U., Evanston, Ill., 1949-61, prof. physics, 1961-89, prof. emeritus physics, 1989—. Contbr. articles to profl. jours. Union Carbon and Carbide fellow, U. Chgo., 1947-49; NSF rsch. grantee, 1952-75. Fellow Am. Phys. Soc.; mem. Am. Assn. Physics Tchrs., Sigma Xi. Achievements include first observation of quantum oscillations in magnetic properties of ordinary metals; spin-density wave modification of electronic structure of antiferromagnetic chromium; research in low-temperature solid-state physics. Home: 142 Sima Rd Higganum CT 06441-4354

MARCUS, ROBERT BORIS, physicist; b. Chgo., Nov. 26, 1934; s. H.L. and Ethel (Leaf) M.; m. Johanna E. Marcus, Sept. 1, 1957; children: Karen T., Suzanne E. MS, U. Chgo., 1958; PhD, U. Mich., 1962. Mgr. tech. svc. Bell Labs., Murray Hill, N.J., 1963-67, supr., 1967-84; mgr. Bellcore, Red Bank, N.J., 1984-92; prof. physics and chemistry N.J. Inst. Tech., Newark, 1992—. Author: Transmission Electron Microscopy of VLSI Circuits and Devices, 1983; editor: Measurement of High Speed Signals, 1990; contbr. over 100 articles to profl. jours. Achievements include patents in field in microstructures and emitters. Office: NJ Inst Tech Dept Physics University Heights Newark NJ 07102

MARCUS, RUDOLPH ARTHUR, chemist, educator; b. Montreal, Que., Can., July 21, 1923; came to U.S., 1949, naturalized, 1958; s. Myer and Esther (Cohen) M.; m. Laura Hearne, Aug. 27, 1949; children: Alan Rudolph, Kenneth Hearne, Raymond Arthur. BSc in Chemistry, McGill U., 1943, PhD in Chemistry, 1946, DSc (hon.), 1988; DSc (hon.), U. Chgo., 1983, Poly. U., 1986, U. Göteborg, Sweden, 1987, U. N.B., Can., 1993, Queens U., Can., 1993. Rsch. staff mem. RDX Project, Montreal, 1944-46; postdoctoral rsch. assoc. NRC of Can., Ottawa, Ont., 1946-49, U. N.C., 1949-51; asst. prof. Poly. Inst. Bklyn., 1951-54, assoc. prof., 1954-58, prof., 1958-64; prof. U. Ill., Urbana, 1964-78; Arthur Amos Noyes prof. chemistry Calif. Inst. Tech., Pasadena, 1978—; temp. mem. Courant Inst. Math. Scis., NYU, 1960-61; trustee Gordon Rsch. Confs., 1966-69, chmn. bd., 1968-69, mem. coun., 1965-68; mem. rev. panel Argonne Nat. Lab., 1966-72, chmn., 1967-68; mem. rev. panel Brookhaven Nat. Lab., 1971-74; mem. rev. com. Radiation Lab., U. Notre Dame, 1975-80; mem. panel on atmospheric chemistry climatic impact com. NAS-NRC, 1975-78, mem. com. kinetics of chem. reactions, 1973-77, chmn., 1975-77, mem. com. chem. scis., 1977-79, mem. com. to survey opportunities in chem. scis., 1982-86; adv. com. for chemistry NSF, 1977-80; vis. prof. theoretical chemistry U. Oxford, Eng., IBM, 1975-76; also professorial fellow Univ. Coll. Former mem. editorial bd. Jour. Chem. Physics, Ann. Rev. Phys. Chemistry, Jour. Phys. Chemistry, Accounts Chem. Rsch., Internat. Jour. Chem. Kinetics Molecular Physics, Theoretica Chimica Acta, Chem. Physics Letters; mem. editorial bd. Laser Chemistry, 1982—, Advances in Chem. Physics, 1984—, World Sci. Pub., 1987—, Internat. Revs. in Phys. Chemistry, 1988—, Faraday Trans., Jour. Chem. Soc., 1990—, Progress in Physics, Chemistry and Mechanics (China), 1989—, Perkins Transactions 2, Jour. Chem. Soc., 1992—, Chem. Physics Rsch. (India), 1992—, Trends in Chem. Physics Rsch. (India), 1992—. Recipient Sr. U.S. Scientist award Alexander von Humboldt-Stiftung, 1976, Electrochem. Soc. Lecture award Electrochem. Soc., 1979, Robinson medal Faraday div. Royal Soc. Chemistry, 1982, Centenary medal Faraday div., 1988, Chandler medal Columbia U., 1983, Wolf Prize in Chemistry, 1985, Nat. Medal of Sci., 1989, Evans award Ohio State U., 1990, Nobel prize in Chem., 1992, Hirschfelder prize in Theoretical Chem. U. Wis., 1993, Golden Plate award Am. Acad. Achievement, 1993; Alfred P. Sloan fellow, 1960-61; NSF sr. postdoctoral fellow, 1960-61; sr. Fulbright-Hays scholar, 1972. Fellow AAAS (exec. com. western sect., co-chmn. 1981-84, rsch. and planning com. 1989-91), Royal Soc. Chemistry (hon.), Royal Soc. Can. (fgn.); mem. NAS, Am. Philos. Soc., Am. Chem. Soc. (past div. chmn., mem. exec. com., mem. adv. bd. petroleum rsch. fund, Irving Langmuir award Chem. Physics 1978, Peter Debye award Phys. Chemistry 1988, Willard Gibbs medal Chgo. sect. 1988, S.C. Lind Lecture, East Tenn. sect. 1988, Theodore William Richards medal Northeastern sect. 1990, Edgar Fahs Smith award Phila. sect. 1991, Ira Remsen Meml. award Md. sect. 1991, Pauling medal Portland, Oreg. and Puget Sound sect. 1991), Royal Soc. London (fgn. mem.), Internat. Acad. Quantum Molecular Sci. Achievements include responsiblity for the Marcus Theory of electron transfer reactions in chemical systems. Home: 331 S Hill Ave Pasadena CA 91106-3405

MARCUS, STEVEN IRL, electrical engineering educator; b. St. Louis, Apr. 2, 1949; s. Herbert A. and Peggy L. (Polishuk) M.; m. Jeanne M. Wilde, June 4, 1978; children: Jeremy A., Tobin L. BA, Rice U., 1971; SM, MIT, 1972, PhD, 1975. Research engr. The Analytic Scis. Corp., Reading, Mass., 1973; asst. prof. U. Tex., Austin, 1975-80, assoc. prof., 1980-84, prof., 1984-91, assoc. chmn., dept. elec. and computer engring., 1984-89, L.B. Meaders prof. engring., 1987-91; prof. elec. engring., dir. Inst. for Systems Rsch. U. Md., College Park, 1991—; cons. Tracor Inc., Austin, 1977, 90. Assoc. editor Math. of Control Signals and Systems, 1987—, Jour. on Discrete Event Dynamic Systems, 1990—. NSF fellow, 1971-74; Werner W. Dornberger Centennial Teaching fellowship in engring., U. Tex., Austin, 1982-84. Fellow IEEE (prize paper awards com. 1985-88, field awards com. 1989-90, assoc. editor Transactions Info. Theory 1990-92), IEEE Control Systems Soc. (bd. govs. 1985-90, chmn. conf. on decision and control program com. 1983, chmn. working group on stochastic control and estimation 1984-87, assoc. editor Transactions Automatic Control 1980-81); mem. Am. Math. Soc., Soc. Indsl. and Applied Math. (editor Jour. Control and Optimalization 1990—), Eta Kappa Nu, Tau Beta Pi. Home: 9516 Thornhill Rd Silver Spring MD 20901-4836 Office: U Md Inst for Systems Rsch AV Williams Bldg 2167 College Park MD 20742

MARCUS, YIZHAK, chemistry educator; b. Kolberg, Germany, Mar. 17, 1931; arrived in Israel, 1936; s. Fritz and Rosa (Nelken) M.; m. Tova Semel, Oct. 10, 1954; children: Tamar, Ruth. MSc, Hebrew U., Jerusalem, 1952, PhD, 1956. Researcher Soreq Nuclear Rsch. Ctr., Yaveh, Israel, 1952-63; dir. chemistry Soreg Nuclear Rsch. Ctr., Yaveh, Israel, 1963-65; prof. chemistry Hebrew U., Jerusalem, 1965—; vis. prof. Royal Soc., 1993. Author: Introduction to Liquid State Chemistry, 1977, Ion Solvation, 1985; co-author: Ion Exchange and Solvent Extraction, 1969. Sr. fellowship A. v. Humboldt Found., 1971. Mem. Internat. Union Pure Applied Chemistry, Israel Chem. Soc., Am. Chem. Soc. Jewish. Office: Hebrew U, Dept Inorganic Chemistry, Jerusalem 91904, Israel

MARCY, WILLARD, chemist, chemical engineer, retired; b. Newton, Mass., Sept. 27, 1916; s. Willard Adna and Jane (Locke) M.; m. Helen Butler, Oct. 8, 1938; children: Martha Ann Marcy Simoneau, Ellen Louise. BSChemE, MIT, 1937, PhD in Organic Chemistry, 1949. Refinery asst. supt. Am. Sugar Co., Bklyn., 1937-42; rsch. asst. chemistry MIT, Cambridge, 1946-49; head dept. refinery process devel. Am. Sugar Co., Bklyn., 1946-64; v.p. Rsch. Corp., N.Y.C., 1964-82; pres. ARDUS, Inc., Charleston, S.C., 1983-85; mem. com. on patent matters Am. Chem. Soc., Washington, 1969-83, chmn., 1977-83; chmn. bd. editors Indsl. Rsch. Inst., N.Y.C., 1973-78. Editor: Patent Policy, 1978; contbr. articles to profl. jours. Maj. U.S. Chem. Corps, 1942-46. Recipient grant NSF, 1980. Fellow AAAS, Am. Inst. Chemist (pres. 1984-85), N.Y. Acad. Sci.; mem. AICE, N.Mex. Acad. Scis. , Newcomen Soc., Chemists' Club N.Y. (trustee 1981-83, chmn. libr. com. 1982-86), Sigma Xi, Alpha Chi Sigma. Republican. Achievements include 2 patents; development of continuous process for absorption of impurities using bonechar; developed procedures for patenting and licensing inventions from academic and non-profit insts.; development of govt. patent policies. Home: 621 Caminito del Sol Santa Fe NM 87505

MARDER, WILLIAM ZEV, technological consultant; b. Phila., Nov. 4, 1947; s. Isadore Myron and Nancy Annette (Segall) M. B.S. in Mech. Engring., U. Pa., 1970, B.A., 1970; m. Mona Marlene Kaufman, June 28, 1970. Div. mgr. Kulicke and Soffa Industries, Horsham, Pa., 1972-74; pres. Zevco Enterprises, Inc., Penllyn, Pa., 1974-77; sr. devel. engr. Air Shields, Inc., Hatboro, Pa., 1977-78; mem. tech. staff RCA, Princeton, N.J., 1978-81; v.p. engring. Kulicke Design, Inc., Ivyland, Pa., 1981-83; tech. mgr., cons. PA Cons. Group, Hightstown, N.J., 1984-90; proprietor Strategic Initiatives, Pennington, N.J., 1990—. Patentee self-priming centrifugal pump, knife sharpener, apparatus for handling deformable components supported in a matrix, numerically controlled method of machining cams and other parts, variable permeability steering torque sensor. Home and Office: 147 S Main St Pennington NJ 08534-2824

MARDIS, HAL KENNEDY, urological surgeon, educator, researcher; b. Lincoln, Nebr., Apr. 4, 1934; s. Harold Corson and Marie (Swaim) M.; m. Janet Reimers Schenken, June 22, 1956; children: Michael Corson, Anne Lucille, Jeanne Marie. BS, U. Nebr., Lincoln, 1955; MD, U. Nebr., Omaha, 1958. Diplomate Am. Bd. Urology. Intern Nebr. Meth. Hosp., Omaha, 1958-59, med. dir. The Stone Ctr., 1966—; resident in urology Charity Hosp. La., New Orleans, 1959-62, chief resident in urology, 1962-63; pvt.

practice Omaha, 1965—; instr., asst. prof. La. State U. Sch. Medicine, New Orleans, 1963-65; asst. prof., assoc. prof. surgery U. Nebr. Med. Ctr., 1965-85, prof., 1985—; investigator North Cen. Cancer Treatment Group, Rochester, Minn., 1988—, Technomed Internat., Inc., Danvers, Mass., 1988—; cons. Boston Sci. Corp., Watertown, Mass., 1988—. Assoc. editor Jour. Stone Disease; contbr. articles to Jour. AMA, So. Med. Jour., Jour. Urology, Urology, Urol. Clinics N.Am., Seminars in Interventional Radiology. Sec., pres. Omaha Symphony Assn., 1973-76; advisor United Arts Omaha, 1983-88. Recipient Outstanding Contbn. award dept. surgery U. Nebr. Med. Ctr., 1990. Fellow ACS; mem. AMa (del. med. staff sect. 1983-86), Am. Urol. Assn. (pres. South Cen. chpt. 1990-91, 1st prize 1976, best clin. exhibit award 1977), Am. Lithotripsy Soc. (pres. 1989-90), Alpha Omega Alpha (pres. 1991-92). Republican. Achievements include development of guidewire techniques for angiography and endourology, thermoplastic internal ureteral stent; description of benefits of hydrophilic polymers for endourologic devices. Office: The Urology Ctr 111 S 90th St Omaha NE 68114-3907

MARES, JOSEPH THOMAS, entomologist; b. Salt Lake City, Aug. 18, 1960; s. Thomas Philip and Mary Patricia (Peake) M.; m. Melanie Jean Pickens, July 2, 1983; children: Patricia Grace, Daniel Joseph, Jillian Marie. BS, James Madison U., 1982; MS, Va. Poly. Inst. and State U., 1984. Biologist Mobay Corp., Kansas City, Mo., 1984-86; project leader Boyle-Midway, Cranford, N.J., 1986-89; rsch. entomologist Shulton Inc., Clifton, N.J., 1989; entomologist Griffin Corp., Valdosta, Ga., 1990—. Instr. ARC, Valdosta, 1991—; ambassador So. Agrl. Chems. Assn., Valdosta, 1991—. Mem. Entomol. Soc. Am., Sigma Xi, Pi Chi Omega, Gamma Sigma Delta. Republican. Roman Catholic. Achievements include patent for insecticidal bait station. Office: Griffin Corp PO Box 1847 Rocky Ford Rd Valdosta GA 31603-1847

MARESCA, LOUIS M., chemicals executive. BS in Chemistry, CUNY, 1972; PhD in Organic Chemistry, Columbia U., 1976. Scientist high performance engrng. plastics divsn. Union Carbide; with G.E. Plastics, gen. mgr. Am.'s product tech.; with BFGoodrich; v.p. rsch. and devel. Geon Co. (formerly a divsn. of BFGoodrich Co.), Avon Lake, Ohio. Contbr. chpts. to books and tech. papers to profl. jours. Achievements include over 50 patents in field. Office: Geon Co Tech Ctr PO Box 122 Avon Lake OH 44012-0122

MARESCHI, JEAN PIERRE, biochemical engineer; b. Pinzano, Italy, Apr. 11, 1937; s. Thomas and Marie Mareschi; m. Françoise Comte, Feb. 11, 1966. B in Math., Paris, 1957; degree in biochem. engring., Inst. Nat. Sci. Appl. Lyon, Lyon, France, 1962; cert. adminstrn. entre, Bus. Sch., Dijon, France, 1966. Cert. biochem. engr., Nutrition Toxicology Regulatory. Researcher French Army Lab., Lyon, 1962-64; head lab. Rsch. Ctr., Sterling Winthrop, Dijon, 1964-70; head lab. fermentation rsch. and mktg. Air Liquide, Grenoble and Paris, France, 1970-75; dir. sci. asst. Hoffman La Roche, Paris, 1975-82; dir. relation sci. regulatory BSN Group Paris, 1982-89; dir. internat. affairs BSN, Paris, 1989—; tchr. Faculty of Medicine, Nancy, France, 1985—; mem. mng. com. sci. schs., Strasbourg, France, 1988, engring. schs., Lyon, 1991. Contbr. articles to sci. nutrition publs.; patentee in field of fermentation. Chevalier Tastevin. Clos Vougeot, Cote d'Or, France, 1991, Agrl. Merit award Ministry of Agr., France, 1992. Mem. Internat. Life Sci. Inst. Europe (bd. dirs., mem. exec. com., vice chmn. 1991). Avocations: sports, cinema, theatre, concerts. Home: 16 BD du Parc, 92200 Neuilly-Seine France Office: BSN, 7 Rue de Teheran, F75008 Paris France

MARGARÉTHA, HERBERT MORIZ PAUL MARIA, chemical consultant; b. Vienna, Austria, Oct. 29, 1911; s. Eugen and Katharina (Seidel) M.; m. Christine Veronica Mauthner, Sept. 14, 1940; children: Elisabeth, Pavl. LLD, U. Vienna. Dir. Magyar Textil Festogyar, Budapest, 1938-40; supt. Magyar Poszto Gyar, Budapest, 1940-44; tech. svc. rep. Ciba Ag, Rio de Janeiro, 1948-58; mem. bd. Danubia Petrochemie, Schwechat, 1958-63; dir. Gebrueder Schoeller, Vienna, 1963-65; mem. bd. Montana Ag, Vienna, 1965-75, Jungbunzlauer Ag, Vienna, 1967-70; ret. Jung Bunzlauer Ag, Vienna, 1970. Contbr. articles to profl. jours. Mem. Am. Chem. Soc., Alt-Hietzinger (v.p. 1965—), Techniker Cercle (pres. 1982-88, honor pres. 1988). Home: 97 Hietzinger Hauptstr, A-1130 Vienna Austria Office: Montana Ag, Schwarzenberg Platz 16, A-1010 Vienna Austria

MARGERUM, DALE WILLIAM, chemistry educator; b. St. Louis, Oct. 20, 1929; s. Donald C. and Ida Lee (Nunley) M.; m. Sonya Lora Pedersen, May 16, 1953; children: Lawrence Donald, Eric William, Richard Dale. B.A., S.E. Mo. State U., 1950; Ph.D., Iowa State U., 1955. Research chemist Ames Lab., AEC, Iowa, 1952-53; instr. Purdue U., West Lafayette, Ind., 1954-57; asst. prof. Purdue U., 1957-61, asso. prof., 1961-65, prof., 1965—, head dept. chemistry, 1978-83; inorganic-analytical chemist, vis. scientist Max Planck Inst., 1963, 70; vis. prof. U. Kent, Canterbury, Eng., 1970; mem. med. chem. study sect. NIH, 1965-69; mem. adv. com. Research Corp., 1973-78; mem. chemistry evaluation panel Air Force Office Sci. Research, 1978-82. Cons. editor McGraw Hill, 1962-72; mem. editorial bd. Jour. Coordination Chemistry, 1971-81, Analytical Chemistry, 1967-69, Inorganic Chemistry, 1985-88. Recipient Grad. Rsch. award Phi Lambda Upsilon, 1954, Alumni Merit award S.E. Mo. State U., 1991; NSF sr. postdoctoral fellow, 1963-64. Fellow AAAS; mem. Am. Chem. Soc. (chmn. Purdue sect. 1965-66), AAUP, Sigma Xi, Phi Lambda Upsilon. Office: Dept Chemistry Purdue U West Lafayette IN 47907

MARGEVICH, DOUGLAS EDWARD, spectroscopist; b. Lockport, N.Y., Jan. 22, 1964; s. Edward S. and Rosa (Sutton) M.; m. Maureen Collins, Aug. 8, 1992. AAS, Niagara Coll., 1985. Analytical technician Eastman Kodak Co., Rochester, N.Y., 1985-89, rsch. technician, 1989—. Contbr. articles to profl. publs. Mem. Am. Chem. Soc., Soc. Applied Spectroscopy. Office: Eastman Kodak Co Kodak Rsch Labs Rochester NY 14650-2132

MARGIOTTA, MARY-LOU ANN, software engineer; b. Waterbury, Conn., June 14, 1956; d. Rocco Donato and Louise Antoinette (Carosella) M. AS in Gen. Edn., Mattatuck Community Coll., Waterbury, 1982; BS in Bus. Mgmt., Teikyo Post U., 1983; MS in Computer Sci., Rensselaer Polytech. Inst., 1989. Programmer analyst Travelers Ins. Co., Hartford, Conn., 1985-87; sr. programmer analyst Conn. Bank and Trust Co., East Hartford, Conn., 1987-88; programmer analyst The Torrrington Co., Torrington, Conn., 1990-91; sr. programmer analyst Orion Capital Cos. Inc., Farmington, Conn., 1991-92, A.M. Cons., New Britain, Conn., 1993—; César Ritz MIS dir. Swiss Hospitality Inst., Washington, Conn., 1993—. Mem. social action com. St. Helena's Parish, West Hartford, Conn., 1988—; advisor Jr. Achievement, Waterbury, 1981-83; tutor Traveler's Ins. Co. Tutorial Program, West Hartford, 1986-87; trainer CPR, ARC, Hartford, 1986-87. Clayborn Pell grantee Post Coll., 1982-83, State of Conn. grantee, 1982-83; recipient Citation, Jr. Achievement, 1982. Mem. IEEE, Assn. for Systems Mgmt., Am. Mktg. Assn., Women in Am. Bus., Toastmasters Internat., Tau Alpha, Beta Gamma. Roman Catholic. Avocations: European travel, gourmet cooking, reading, tennis, golf. Home: 210E Brittany Farms Rd New Britain CT 06053-1161

MARGOLIN, HAROLD, metallurgical educator; b. Hartford, Conn., July 12, 1922; s. Aaron David and Sonia (Krupnikoff) M.; m. Elaine Marjorie Rose, July 4, 1946; children—Shelley, Deborah, Amy. B.Engring., Yale U., 1943, M.Engring., 1947, D.Eng. (Internat. Nickel fellow) 1947-49), 1950. Research assoc., research scientist, research div. N.Y. U., N.Y.C., 1949-56; assoc. prof. metall. engring., 1956-62, prof., 1962-73; prof. phys. metallurgy Poly. U. N.Y., Bklyn., 1973-93; cons. in field. Contbg. author books; contbr. articles to profl. publs. Served with USNR, 1944-46. Theodore W. Krengel vis. prof. Technion, Haifa, Israel, 1983; named Disting. Rsch. Prof., 1993. Fellow Am. Soc. Metals (edn. award N.Y. chpt. 1967); mem. AAAS, The Metall. Soc., Am. Soc. Materials, Sigma Xi. Democrat. Jewish. Patentee in field. Home: 19 Crescent Rd Larchmont NY 10538-1733 Office: 333 Jay St Brooklyn NY 11201-2990

MARGOLIN, SOLOMON BEGELFOR, pharmacologist; b. Phila., May 16, 1920; s. Nathan and Fannie (Begelfor) M.; m. Gerda Levy, Jan. 17, 1947 (div. Feb. 1985); children: David, Bernard, Daniel; m. Nancy A. Cox, Apr. 30, 1987. BSc, Rutgers U., 1941, MSc, 1943, PhD, 1945. Asst. Rutgers U., New Brunswick, N.J., 1943-45; rsch. biologist Silmo Chem. Co., Vineland, N.J., 1947-48; rsch. pharmacologist Schering Corp., Bloomfield, N.J., 1948-52, dir. pharmacology dept., 1952-54; chief pharmacologist Maltbie Labs., Belleville, N.J., 1954-56; chief pharmacologist Wallace Labs, Carter-Wallace, Inc., Cranbury, N.J., 1956-60, dir. pharmacology dept., 1960-64, v.p. biol. rsch., 1964-68; pres. AMR Biol. Rsch., Inc., Princeton, N.J., 1968-78; from prof., chmn. pharmacology dept. to emeritus prof. St. George's (Grenada) U. Sch. Medicine, 1978—; pres. MARNAC, Inc., Dallas, 1990—. Author: Harper's Handbook Therapeutic Pharmacology, 1981; author: (with others) Physiological Pharmacology, 1963, World Review, Nutrition & Dietetics, 1980; contbr. over 50 articles to profl. jours. including Annals of Allergy, Proc. Soc. Exptl. Biol. & Med., Nature. Mem. AAAS, Endocrino Soc., Am. Chem. Soc., Soc. Exptl. Biology and Medicine, Am. Soc. Pharmacology and Exptl. Therapeutics, N.Y. Acad. Scis., Drug Information Assn. Achievements include U.S., European, and Japanese patents for Prevention and Treatment of Fibrotic Lesions; research in anti-histamines anti-cholinergics, endorphins, sedative-hypnotics, tranquilizers, muscle relaxants, glucocorticoids, cardiovascular agents, anti-inflammatory drugs, anti-fibrotic agents. Home: 6723 Desco Dr Dallas TX 75225-2704 Office: Marnac Inc 6723 Desco Dr Dallas TX 75225

MARGOLIS, JAMES MARK, international industrial developer; b. Boston, May 4, 1930; s. Herbert and Annette Florence (Marquette) Margolis; m. Erini Theodoridou, July 8, 1977. SB, MIT, 1952. lectr., seminar presenter, 1958—. Contbr. numerous articles to various publs. Achievements include management, marketing and research planning for petroleum refinery, petrochemical resin polymer industries. Home and Office: 40 Twilight Ave Keansburg NJ 07734

MARGOLIUS, HARRY STEPHEN, pharmacologist, physician; b. Albany, N.Y., Jan. 29, 1938; s. Irving Robert and Betty (Zweig) M.; m. Francine Rockwood, May 22, 1964; children: Elizabeth Anne, Craig Matthew. BS, Union U., 1959, PhD, 1963; MD, U. Cin., 1968. Diplomate Nat. Bd. Med. Examiners, 1969, chmn. pharmacology test com., 1990—. Intern, resident Harvard Med. Svc. Boston City Hosp., 1968-70, pharmcology rsch. assoc., 1970-72; sr. clin. investigator NHLBI NIH, Bethesda, 1972-74; assoc. prof. pharmacology, asst. prof. medicine Med. U. S.C., Charleston, 1974-77, prof. pharmacology, assoc. prof. medicine, 1977-80, prof. pharmacology, prof. medicine, 1980—, chmn. pharmacology, 1989—; cons. NIH, FDA, VA, NSF, Washington, Bethesda, 1975—; mem. editorial bd. Am. Heart Assn., Dallas, 1980—. Editor: Kinins IV, 1986, Renal Function, Hypertension and Kallikrein-Kinin System, 1988; contbr. numerous articles to profl. jours. Commdr. USPHS, 1967-74. Recipient S.C. Gov.'s award for sci. S.C. Acad. Scis., 1988; Burroughs-Wellcome scholar, 1976; vis. scholar U. Cambridge, Eng., 1980-81; NIH grantee, 1975—. Fellow Coun. for High Blood Pressure Rsch., Am. Heart Assn.; mem. Am. Soc. for Pharmacology and Exptl. Therapeutics, Am. Soc. for Clin. Investigation and 10 additional med., sci. socs. Jewish. Achievements include studies of the role of kallikreins and kinins in human and animal forms of hypertension; discovery of abnormalities which signify possible roles in causing high blood pressure. Office: Medical Univ of SC College of Medicine 171 Ashley Ave Charleston SC 29425-0001

MARGOSHES, MARVIN, chemist, consultant; b. N.Y.C., May 23, 1925; s. Israel and Lillian (Lenorowitz) M.; m. Miriam Kagan, Aug. 7, 1955; children: Bethia, Sara, Jessa, Dan. BS in Chemistry, Poly. Inst. Bklyn., 1951; PhD in Phys. Chemistry, Iowa State Coll., 1953. Rsch. fellow Harvard Med. Sch., Boston, 1954-57; chemist Nat. Bur. Standards, Gaithersburg, Md., 1957-69; project mgr. Digilab, Inc., Cambridge, Mass., 1969-70; tech. dir. Technicon Instruments, Tarrytown, N.Y., 1971-89; pres. Tech. Transfer Svcs., Tarrytown, 1989—; instrumentation adv. bd. Analytical Chemistry, Washington, 1970-75, editorial adv. bd., 1978-80. Co-editor: Procs Xth Colloquium Spectrochem. Internat., 1963; editor Spectrochimica Acta, 1966-72; contbr. articles to sci. publs. Mem. Sch. Bd. of Tarrytown, 1981-90, v.p., 1985-87, pres., 1988-90. With U.S. Army, 1943-46, PTO. Mem. Soc. Applied Spectroscopy (pres. 1974, Gold medal 1976), Am. Chem. Soc., N.Y. Acad. Scis., Sigma Xi. Achievements include 2 patents, original isolation of metallothionein; discovery of high potassium level of bananas; invention of D.C. plasma source, other instruments. Home: 69 Midland Ave Tarrytown NY 10591 Office: Tech Transfer Svcs 69 Midland Ave Tarrytown NY 10591

MARGRAVE, JOHN LEE, chemist, educator, university administrator; b. Kansas City, Kans., Apr. 13, 1924; s. Orville Frank and Bernice J. (Hamilton) M.; m. Mary Lou Davis, June 11, 1950; children: David Russell, Karen Sue. B.S. in Engring. Physics, U. Kans., 1948, Ph.D. in Chemistry, 1950. AEC postdoctoral fellow U. Calif. at Berkeley, 1951-52; from instr. to prof. chemistry U. Wis., Madison, 1952-63; prof. chemistry Rice U., 1963—, E.D. Butcher chair, 1986—, chmn. dept., 1967-72, dean advanced studies and research, 1972-80, v.p., 1980-86; v.p. for research Houston Advanced Research Ctr., 1986-89, chief sci. officer, 1989—; dir. Materials Sci. Ctr., 1986—; vis. prof. chem. Tex. So. U., 1993—; dir. Council for Chem. Research, 1985-88; Reilly lectr. Notre Dame, 1968; vis. distinguished prof. U. Wis., 1968, U. Iowa, 1969, U. Colo., 1975, Ga. Inst. Tech., 1978, U. Tex. at Austin, 1978, U. Utah, 1982; Seydel-Wooley lectr. Ga. Inst. Tech., 1970; Dupont lectr. U. S.C., 1971; Abbott lectr. U. N.D., 1972; Cyanamid lectr. U. Conn., 1973; Sandia lectr. U. N.Mex., 1981; R.A. Welch lectr., 1985; NSF-Japan Joint Thermophys. Properties Symposium, 1983; chmn. com. on chem. processes in severe nuclear accidents NRC, 1987-88; mem. Wilhelm und Else Heraeus Stiffung Found. Symposium on Alkali Metal Reactions, Fed. Republic Germany, 1988; various nat. and internat. confs. on chem. vapor deposition of thin diamond films, 1989, 90, 91,92; orgnl. com. First, Second and Third World Superconductivity Congresses, 1989, 90, 92; mem. adv. coms. chem., materials sci., rsch. U. Tenn. Knoxville, Ohio State U., Tex. So. U., La. Bd. Regents, sci. adv. bd. SI Diamond Tech., 1992—, BioNumerik, 1993—; cons. to govt. and industry, 1954—; pres. Mar Chem., Inc., 1970—, High Temperature Sci., Inc., 1976—; dir. Rice Design Center, Houston Area Research Ctr.; U. Kans. Research Found., Gulf Univs. Research Consortium, Energy Research and Edn. Found., Spectroscopic Assocs., World Congress on Superconductivity. Editor: Modern High Temperature Sci., 1984; contbg. editor Characterization of High Temperature Vapors, 1967, Mass. Spectrometry in Inorganic Chemistry, 1968; editor High Temperature Sci., 1969—, Procs. XXIII and XXIV Confs. on Mass Spectrometry, 1975, 76; author: (with others) Bibliography of Matrix Isolation Spectroscopy, 1950-85, 87; contbr. articles to profl. jours. Served with AUS, 1943-46; capt. Res. ret. Sloan research fellow, 1957-58; Guggenheim fellow, 1960; recipient Kiekhofer Teaching award U. Wis., 1957; IR-100 award for CFX lubricant powder, 1970, IR-100 award for Cryolink, 1986; Tex. Honor Scroll award, 1978; Disting. Alumni citation U. Kans., 1981. Fellow Am. Inst. Chemists, AAAS, Tex. Acad. Sci., Am. Phys. Soc.; mem. AAUP, NAS, Am. Chem. Soc. (Inorganic Chemistry award 1967, S.W. Regional award 1978, Fluorine Chemistry award 1980, S.E. Tex. Sect. award 1993), Am. Ceramic Soc., Electrochem. Soc., Chem. Soc. (London), Tex. Philos. Soc., Am. Soc. Mass Spectrometry (dir.), Am. Soc. Metals, Materials Rsch. Soc., Sigma Xi, Omicron Delta Kappa, Sigma Tau, Tau Beta Pi, Alpha Chi Sigma. Methodist. Patentee in field. Home: 5012 Tangle Ln Houston TX 77056-2114 Office: Rice University PO Box 1892 6100 South Main Houston TX 77251

MARGRON, FREDERICK JOSEPH, environmental engineer; b. N.Y.C., May 3, 1963; s. Gaston G. and Michael M. (Salgado) M.; m. Lola T. Velez, June 25, 1988. B Engring., Hofstra U., 1986. Registered environ. mgr.; cert. environ. auditor. Resident rep. Charles A. Manganaro Cons. Engrs., Hackensack, N.J., 1983-84, project engr., 1985-90, project mgr., 1990—. Bd. dirs. Penn Estates, East Stroudsburg, 1992. Mem. ASCE (assoc.), Nat. Registry Environ. Profls. Republican. Roman Catholic. Office: Charles A. Manganaro Cons Engrs 25 Main St Hackensack NJ 07601

MARGULIES, ANDREW MICHAEL, chiropractor; b. Bklyn.; s. Irving R. and Marion (Steiner) M.; m. Lorraine Raffa, Dec. 23, 1990; 1 child, Samantha Cara. D. Chiropractic, Palmer Coll. Chiropractic, Davenport, Iowa, 1981. Diplomate Nat. Bd. Chiropractic, Am. Acad. Pain Mgmt.; cert. chiropractic sports physician. Dir., chiropractic physician Margulies Chiropractic and Sports Injuries Ctr., Massapequa, N.Y., 1981—; chiropractor, mem. med. team N.Y. Long Island Marathon, 1986—, USA/Mobil Track and Field Nat. Championships, 1991—; cons. Massapequa Rd. Runners, Long Island, 1985—. Recipient Silver Star award Markson/Svc. to Community, Flushing, N.Y., 1984, Markson Mgmt. Annual award, 1984, Community Svc. and Profl. award Success Systems, 1993. Fellow Am. Acad. Applied Spinal Biochem. Engring.; mem. APHA, AAAS, N.Y. Acad. Scis., Am. Chiropractic Assn. (coun. on sports injuries and phys. fitness, coun. on diagnostic imaging), Found. for Chiropractic Edn. and Rsch., N.Y. State Chiropractic Assn., N.Y. Chiropractic Coun. Office: Margulies Chiropractic and Sports Injury Ctr 1148 Hicksville Rd Massapequa NY 11758-1222

MARGULIES, ROBERT ALLAN, physician, educator, emergency medical service director; b. Jersey City, N.J., May 29, 1942; s. Max and Edith (Cromnick) M.; m. Edythe Rhoda Levy, Dec. 25, 1966 (div. Sept. 1982); children: Steven Brian, Marcy Lee, Shari Lynn; m. Victoria Anne Cassano, May 20, 1984; 1 child, Johnathan Maxwell Cassano Margulies. AA, Union Coll., 1963; BA, Rutgers U., 1965; MD, N.J. Coll. Medicine/Dentistry, 1969; MPH, Johns Hopkins U., 1978. Diplomate Am. Bd. Preventive Medicine, Am. Bd. Emergency Medicine. Commd. ensign USN, 1968, advanced through grades to capt., 1980; commdg. officer Navy Submarine Rsch. and Devel. Lab., Groton, Conn., 1978-81; chair dept. operational and emergency medicine U. Health Sci., Bethesda, Md., 1981-84; exec. officer Navy Hosp., Beaufort, S.C., 1984-86; commdg. officer Navy Hosp. Camp LeJeune, Jacksonville, N.C., 1986-88; ret. USN, 1989; mem. staff St. Mary's Med. Ctr., Saginaw, Mich., 1989-92, med. dir., flight care, 1991-92; dir. ground/prehosp. emergency med. svc., sr. staff mem. Hartford (Conn.) Hosp., 1992—; clin. asst. prof. surgery U. Conn., Farmington, 1993—; clin. assoc. prof. preventive medicine U. S.C., Columbia, 1984—, clin. asst. prof. preventive medicine and biometrics USPHS, Bethesda, Md., 1984—; mem. med. and physiol. rev. panels NASA, Washington, 1977-84; reviewer aerospace medicine Residency Rev. Com., 1980; nat. evaluator Nat. Sci. Tchrs. Assn./NASA, Washington, 1982-85. Author: (with others) Trauma, Emergency Surgery and Critical Care, 1987, Emergency Transport of the Perinatal Patient, 1989; mem. editorial bd. Aviation, Space and Environ. Medicine, 1987-92, Jour. Emergency Medicine, 1984—; contbr. articles and papers to profl. jours. Fellow Am. Coll. Emergency Physicians, Am. Coll. Preventive Medicine, Aerospace Med. Assn.; mem. AMA, AIAA, Soc. for Acad. Emergency Medicine, Soc. of Tchrs. Preventive Medicine, Internat. Soc. Air Safety Investigators, Undersea and Hyperbaric Med. Soc., Wash. Soc. for History of Medicine. Home: 24 Park Pl Apt 23C Hartford CT 06106 Office: Hartford Hosp 80 Seymour St Hartford CT 06115

MARINAS, MANUEL GUILLERMO, JR., psychiatrist; b. Habana, Cuba, Oct. 23, 1954; came to U.S., 1962; s. Manuel G. and Isabel (Ulloa) M. BS, U. Miami, 1977, MD, 1981. Intern St. Luke's-Roosevelt Hosp., 1981-82; resident Jackson Meml. Hosp., 1982-85; fellow Bellevue Hosp., N.Y.C.; attending psychiatrist Gouverneur Hosp., N.Y.C., 1987—; cons. psychiatrist North Side Ctr. for Child Devel., N.Y.C., 1989—. Mem. AMA, Am. Psychiatric Assn., Acad. Child and Adolescent Psychiatry, Orthopsyciatry Soc. Roman Catholic.

MARINENKO, GEORGE, chemist; b. Yoronezh, USSR, Sept. 16, 1935; came to U.S., 1951; s. Vladimir Filipovich and Angelina Mikhilovna (Isayeva) M.; m. Joan McCormack, 1959 (div. 1973); m. Ryna Beth Joseph, Feb. 23, 1974; children: George L., Natalia, Tatyana, Andrei M., Aleksei. MS in Analytical Chemistry, Am. U., Washington, 1961, PhD in Phys. Chemistry, 1972. Russian sci. lexicographer Libr. of Congress, Washington, 1958-60; rsch. chemist Nat. Bur. of Standards, Gaithersburg, Md., 1960-88; systems scientist Mitre Corp., McLean, Va., 1980—. Contbr. 80 articles to profl. jours. Recipient Silver medal U.S. Dept. Commerce, 1971. Mem. ASTM, Am. Chem. Soc., Electrochem. Soc., Sigma Xi. Russian Orthodox. Achievements include research in atomic weight of zinc; patent for Electrochemical Chlorine Flox-Monitor; development of high presiciosn coulometric methods of analyses. Home: 17401 Siever Ct Germantown MD 20874 Office: Mitre Corp 7525 Colshire Dr Mc Lean VA 22102

MARINI BETTOLO, GIOVANNI BATTISTA, chemistry educator, researcher; b. Rome, June 27, 1915; s. Rinaldo and Evelina (Bettolo) Marini; m. Luisa Piva, Sept. 12, 1945; children: Rinaldo, Priscilla, Umberto, Maria Vittoria. D. in Chemistry, U. Rome, 1937; D honoris causa, U. Nancy (France), 1968, U. Complutense, Madrid, 1989. Asst. U. Rome, 1938-46; prof. chemistry Cath. U., Santiago, Chile, 1947-48, U. Montevideo (Uruguay), 1949; rsch. prof. Inst. Superiore Sanità, Rome, 1950-64, dir., 1964-71; prof. chemistry U. Rome La Sapienza, Rome, 1971-90; pres. European Pharmacopoeia Commn., Strasbourg, France, 1964-68; dir. Consiglio Nazionale Ricerche Centro Chimica Recettori, Rome, 1963-89. Contbr. over 250 papers on natural products and pharm. chemistry. Fellow Nat. Acad. Sci. XL (pres. 1981-89), Nat. Acad. Lincei, Acad. Pontificia Scienze (pres. 1988-93), Acad. Naz. delle Scienze (pres. 1981-89). Roman Catholic. Home: Via Principessa Clotilde, 00196 Rome Italy Office: U Rome La Sapienza, Piazzale Aldo Moro 5, 00185 Rome Italy

MARINO, IGNAZIO ROBERTO, transplant surgeon, researcher; b. Genova, Italy, Mar. 10, 1955; s. Pietro Rosario and Valeria (Mazzanti) M.; m. Rossana Parisen-Toldin, Sept. 15, 1990. Maturità-Classica, Coll. of Merode, Rome, 1973; MD, Cath. U., Rome, 1979. Diplomate Nat. Bd. Gen. Surgery. Intern, then resident Gemelli U. Hosp., Rome, 1979-84; temp. asst. dept. surgery Cath. U., Rome, 1981, asst. prof. surgery, 1983-92; asst. prof. surgery Transplantation Inst., U. Pitts., 1991—; prof. surgery Post-grad. Sch. Thoracic Surgery, U. Milan, 1991—; attending surgeon Presbyn. Hosp., Pitts., 1991—, VA Med. Ctr., Pitts., 1992—, Children's Hosp. Pitts., 1993—; sci. journalist Agenzia Nazionale Stampa Associata, 1992—. Author: New Technique in Extracorporeal Circulation, 1985 (Annual Prize of the Italian Soc. of Surgery 1986), New Technique in Liver Transplantation, 1986 (De Angelis award 1986). Recipient grant Italian Nat. Coun. Rsch., 1979, 86, 90, Gastroenterology Soc., 1988, award Istituto Nazionale Previdenza Dirigenti Aziende Industriali, 1982. Mem. Am. Soc. Transplant Surgeons, Italian Soc. Surgery, Transplantation Soc. (grant 1988), European Soc. for Organ TX, Italian Soc. Hepatobiliary Surgery, Italian Soc. Surgeons Under 40 (Annual prize 1986), Xenotransplantation Club (founding mem.), Cell Transplant Soc. (founding mem.). Avocations: reading (history books), sailing. Home: Corso Italia 29, Rome 00198, Italy Office: Univ Pitts Transplant Inst 5C Falk Clinic 3601 Fifth Ave Pittsburgh PA 15213

MARINO, PAUL LAWRENCE, physician, researcher; b. Everett, Mass., Feb. 10, 1946; s. Charles Joseph and Jean Marie (Casale) M.; m. Jeanne Marie Burns, Feb. 15, 1987; 1 child, Daniel Joseph. BA, Merrimack Coll., 1967; MD, PhD, U. Va., 1974. Diplomate Am. Bd. Internal Med. Resident physician U. Mich., Ann Arbor, 1974-77; fellow in pulmonary medicine U. Pa., Phila., 1977-80; med. faculty U. Pa. Sch. Medicine, Phila., 1980-90; assoc. prof. Mt. Sinai Sch. Medicine, N.Y.C., 1990-92; dir. critical care Presbyn. Med. Ctr., Phila., 1992—. Author: The ICU Book, 1991; editorial bd. Jour. Internal Medicine; adv. bd. Medical Tribune; software designer The Expert Series. Recipient Nat. Rsch. Svc. award NIH, 1969-74. Fellow Am. Coll. Critical Care Medicine, Phila. Coll. Physicians (hon. 1990); mem. Found. Med. Edn. (founder, chmn.), Pa. Thoracic Soc. (critical care chmn. 1989-91). Home: 1830 Spruce St Philadelphia PA 19103 Office: Presbyn Med Hosp 19th and Market Sts Philadelphia PA 19104

MARINUZZI, FRANCESCO, computer science educator. M of Computer Sci., U. Rome, 1987, PhD in Computer Sci., 1992. Researcher Engring. S.P.A., Rome; cons. Softlab S.P.A., Rome, 1991, dir. R&D, 1992; prof. electronic calculator U. Rome, 1990—; bd. dirs. Advances Bus. Systems, Rome. Author: Advanced Business System, 1990, Geos: Generation and Evolution of Software Organisms, 1991, Dynamic Information Systems, 1992, Philosophy Versus Informatic; contbr. articles to profl. publs. Fellow Euro Mgr. Soc.; mem. IEEE, Assn. Computing Machinery, Associazione Italiana Intelligenza Artificiale, Mensa. Avocations: chess, golf, philosophy. Home: Via Casilina 1204, 00133 Rome Italy Office: AIS, Via Casiline 1204, 00133 Rome Italy

MARIOTTE, MICHAEL LEE, environmental activist, environmental publication director; b. Indpls., Dec. 9, 1952; s. Richard H. and Rozetta Mae (Dorton) M.; m. Lynn W. Thorp, Mar. 3, 1984; children: Nicole Lynn, Richard Matthew. BA, Antioch Coll., 1978. Editorial asst. ABA, Washington, 1979-81; mng. editor, gen. mgr. City Paper, Washington, 1981-84; editor Nuclear Info. & Resource Svc., Washington, 1985-86, exec. dir., 1986—; dir. Safe Energy Comm. Coun., Washington, 1990—; mem. adv. bd. GE Stockholder Alliance, 1991—; advisor Coun. on Econ. Priorities, 1990—.

Editor (newsletter) Nuclear Monitor, 1985—. Office: Nuclear Info & Resource Svc 1424 16th St NW # 601 Washington DC 20036-2211

MARIOTTO, MARCO JEROME, psychology educator, researcher; b. Ill., Oct. 21, 1946; s. Marco Anibele and Sally (Hughes) M.; m. Danita Irene Czyzewski, May 4, 1985; 1 child, Ana-Sofia Antonia. BS, U. Ill., 1968, PhD, 1974. Diplomate Am. Bd. Sexology; lic. psychologist; cert. sex therapist, cert. health svcs. provider. Asst. rsch. dir. Adolf Meyer Ctr. Rsch. Units, Decatur, Ill., 1972-74; psychologist U.S. Army Acad. Health Scis., San Antonio, 1974; asst. prof. Purdue U., West Lafayette, Ind., 1975-79; assoc. prof. U. Houston, 1979-90, supervisory psychologist, 1979—, prof., 1990—; cons. NIMH, Bethesda, Md., 1977—, NSF, Washington, 1980-84, Nat. Inst. Drug Abuse, Bethesda, 1986-89; adj. prof. U. Tex. Health Scis., Houston, 1980—. Contbr. 16 chpts. to books, 27 articles to jours., also rsch. monographs and tech. reports. Forensic cons. Harris County Dist. Atty.'s Office, Houston, 1988—, ABA, 1989—; founding mem. Gulf Coast Consortium on Mental Health, Houston and Galveston, Tex., 1989. Capt. U.S. Army, 1968-74. Named one of top 35 Young Scientist Profls. Jour. Cons. and Clin. Psychology, 1988; David Ross fellow Purdue U., 1977. Mem. APA, Am. Psychol. Soc.; mem. AAAS, Midwestern Psychol. Assn. (local rep. 1979—), Sigma Xi. Achievements include co-devel. of TSBC/SRIC planned access infosystem for rsch. and svc. for patients in residential treatment settings; rsch. in observational measurement in mental health, schizophrenia, chronic mental patients. Office: U Houston Dept Psychology Houston TX 77004

MARIOTTO, PAULO ANTONIO, electronics educator; b. Sao Paulo, Oct. 5, 1937; s. Mario and Orietta (Prandini) M.; m. Maria Auxiliadora Frugis, July 16, 1965; children: Ana Paula, Maria Teresa, Alberto. BSEE, U. Sao Paulo, 1960, D in Engring., 1966; MSEE, Stanford U., 1968. Instr. U. Sao Paulo, 1961-66, asst. prof., 1969-73, assoc. prof., 1974-90, prof., 1990—, head dept. elec. engring., 1990-91, head dept. electronic engring., 1991—; rsch. asst. Stanford (Calif.) U., 1966-68; mem. Brazilian Union Radio-Scientifique Internat. Com., 1969-71. Author: Waves and Lines, 1981; contbr. articles to profl. publs. 2d lt. Brazilian armed forces, 1958-61. Mem. Regional Coun. of Engring., Architecture and Agronomy, Brazilian Computer Soc. (founder). Home: Av Bagiru 661, 05469-020 São Paulo SP, Brazil Office: U São Paulo, Caixa Postal 8174, 01065-970 São Paulo Brazil

MARK, HANS MICHAEL, aerospace engineering educator, physicist; b. Mannheim, Germany, June 17, 1929; came to U.S., 1940, naturalized, 1945; s. Herman Francis and Maria (Schramek) M.; m. Marion G. Thorpe, Jan. 28, 1951; children: Jane H., James P. A.B. in Physics, U. Calif. at Berkeley, 1951; Ph.D., MIT, 1954; Sc.D. (hon.), Fla. Inst. Tech., 1978; D. Eng. (hon.), Poly. U. N.Y., 1982; DEng (hon.), Milw. Sch. Engring., 1991; LHD (hon.), St. Edward's U., 1993. Research assoc. MIT, 1954-55, asst. prof., 1958-60; research physicist Lawrence Radiation Lab., U. Calif. at Livermore, 1955-58, 60-69, exptl. physics div. leader, 1960-64; assoc. prof. nuclear engring. U. Calif. at Berkeley, 1960-66, prof., 1966-69, chmn. dept. nuclear engring., 1964-69; lectr. dept. applied sci. U. Calif. at Davis, 1969-73; cons. prof. engring. Stanford, 1973-84; dir. NASA-Ames Research Center, 1969-77; undersec. Air Force, Washington, 1977-79; sec. Air Force, 1979-81; dep. adminstr. NASA, Washington, 1981-84; chancellor U. Tex. System, Austin, 1984-92; prof. aerospace engring. and engring. mechanics U. Tex., Austin, 1988—; mem. Pres.'s Adv. Group Sci. and Tech., 1975-76; bd. dirs. BDM Internat. Corp., Astronautics Corp. Am., MAC Equipment Co., Tex. Biotech. Corp.; trustee Poly. U., 1984—, MITRE Corp., 1985-88. Author: (with N.T. Olson) Experiments in Modern Physics, 1966 (with E. Teller and J.S. Foster, Jr.) Power and Security, 1976, (with A. Levine) The Management of Research Institutions, 1983, The Space Station-A Personal Journey, 1987; also numerous articles.; Editor: (with S. Fernbach) Properties of Matter Under Unusual Conditions, 1969, (with Lowell Wood) Energy in Physics, War and Peace, 1988. Recipient Disting. Svc. medal NASA, 1972, 77, medal for Exceptional Sci. Achievement, NASA, 1984, Exceptional Civilian Svc. award USAF, 1979, Disting. Pub. Svc. medal Dept. Def., 1981. Fellow AIAA (Von Karmen Lectr. Astronautics 1992), Am. Phys. Soc.; mem. Nat. Acad. Engring., Am. Nuclear Soc., Am. Geophys. Union, Coun. Fgn. Rels., Cosmos Club. Achievements include research on nuclear energy levels, nuclear reactions, applications, nuclear energy for practical purposes, atomic flourescence yields, measurement X-rays above atmosphere, spacecraft and experimental aircraft design. Office: U Tex Dept Aerospace Engring and Engring Mechanics Austin TX 78712-1085

MARK, KATHLEEN ABBOTT, writer; b. Toronto, Ont., Can., Oct. 4, 1911; came to U.S., 1945; d. Arthur and Clara Barker (Foulkes) Abbott; m. Jordan Carson Mark, Nov. 22, 1935; children: Joan Neary, Thomas, Elizabeth Smith, Graham, Christopher, Mary Deyhle. Student, U. N.Mex., 1971-75. Free lance writer Los Alamos, N. Mex., 1964—. Author: Meteorite Craters, 1987; contbr. articles to Sea Frontiers. Co-recipient Nininger award U. Ariz., 1974-75. Mem. AAAS, Geol. Soc. Am., Meteoretical Soc., History of Earth Scis. Soc., Soc. of Meteoritophiles (v.p.). Home and Office: 4900 Sandia Dr Los Alamos NM 87544

MARKANDA, RAJ KUMAR, mathematics educator; b. Amritsar, Panjab, India, Nov. 15, 1940; came to U.S. 1983; s. Guranditta Mall and Maya Devi (Sharma) Markanda; m. Manjula Shukla, Oct. 16, 1974; children: Neha, Sonal. BA with honors, Panjab U., 1959, MA, 1961; PhD, U. Colo., 1973. Sr. rsch. asst. Irrigation & Power Rsch. Inst., Amritsar, 1961-63; lectr. Govt. Coll., Kapurthala, Panjab, India, 1963-64; jr. rsch. fellow Panjab U., 1964-68; teaching asst. U. Colo., Boulder, 1968-69, 70-71, fellow, 1969-70, 71-72; fellow Universite' de Paris X, Paris, 1972-73; vis. mem. Tata Inst., Bombay, India, 1973-74; lectr. Panjab U., Chandigarh, India, 1974-75; assoc. prof. Nat. U. Colombia, Bogota, 1975-77; assoc. prof. math. No. State U., Aberdeen, S.D., 1986-91, prof., 1991—; assoc. prof. Universidad de Los Andes, Merida, Venezuela, 1977-83; vis. assoc. prof. U. Iowa, 1983-86. Co-author: College Algebra Exam File, 1990; contbr. articles to profl. jours. S.D. Bd. Regents grantee, 1990, 92, 93. Mem. Am. Math. Soc., Indian Math. Soc. (life mem.), S.D. Acad. Sci. Achievements include discovery of a new infinite class of arithmetic Euclidean rings whose existence had not been suspected. Office: No State Univ Dept Math Aberdeen SD 57401

MARKATOS, NICOLAS-CHRIS GREGORY, chemical engineering educator; b. Athens, Greece, Mar. 1, 1944; s. Gregory N. and Kathreen J. Markatos; m. Marina Verykios, Aug. 2, 1989. Dipl. Chem. Engring., Nat. Tech. U., Athens, 1967; MA in Bus. Adminstrn., Athens Sch. Econs., 1969; diploma, U. London, 1973, PhD in Engring., 1974. Process mgr. Procter & Gamble, Athens, 1969-70; rsch. fellow Imperial Coll., London, 1973-75; group leader Cham Ltd., London, 1975-78, tech. mgr., 1978-82; reader, dir. U. Greenwich, London, 1982-86; prof. Nat. Tech. U., Athens, 1985—, head. chem. engring. dept., 1990—, rector, 1991—; cons. NASA Langley Rsch. Ctr., Nat. Maritime Inst., Comb, Eng., Boeing, Brit. Leyland, others; vis. lectr. Computational Fluid Dynamics Unit Imperial Coll., U. London, 1983-86; referee Internat. Jour. Heat and Mass Transfer, Transactions of AICE, Chem. Engring. Process, others. Editor: Computational Fluid Mechanics, 1987, Internal Combustion Engines, 1990; editor jour. Applied Math. Modeling, 1985—; author 2 books, over 100 sci. papers in field; contbr. articles to sci. publs. Served with Hellenic Army, 1967-69. Recipient cert. Inventions Coun. NASA, 1980. Fellow Inst. Chem. Engrs.; mem. AIAA, AICE, Tech. Chambers Greece, U. London Convocation, Engr. Coun. Gt. Britain (chartered engr.), Brit. Coun. Computer Simulation, Internat. Soc. Computational Methods in Engring., Soc. Computer Simulation. Home: 27-29 Hippodamou St, 116 35 Athens Greece Office: Nat Tech U, Zografou Campus, 157 73 Athens Greece

MARKEL, DORENE SAMUELS, geneticist; b. Detroit, Mar. 5, 1959; d. Thomas and Concetta (Leo) Samuels; m. Timothy Joseph Markel, July 3, 1981; children: Aaron Samuels, Arielle Sofia. BS, Mich. Tech. U., 1981; MS, U. Mich., 1983, MHSA, 1990. Rsch. asst. human genetics U. Mich., Ann Arbor, 1980-83, rsch. assoc. neurology, 1984-86, genetic counselor, 1986-90, dir. family studies human genome ctr., 1990—, genetic counselor div. med. genetics, 1990—; mem. exec. com. NIH Human Genome Ctr. U. Mich., 1990—. Contbr. chpts.: Contributions to Contemporary Neurology, 1988, Strategies in Genetic Counseling: The Challenge of the Future, 1988; contbr. articles to Am. Jour. Med. Genetics, Annals of Neurology, Archives of Neurology, New England Jour. Medicine. Mem. adv. bd. U. Mich. Hosp. Child Care Ctr., Ann Arbor, 1991—; mem. PTO bd. Thurston Elem. Sch., Ann Arbor, 1992—. Mem. Nat. Soc. Genetic Counselors (human genome project liaison 1992—, chairperson subcom. on genetic rsch. issues 1992—), Am. Soc. Human Genetics, AAAS, Alliance Genetic Support Groups. Roman Catholic. Achievements include research in Huntington's Disease genetic counseling and testing and the involvement of individuals and families in genetic research. Office: U Mich Human Genome Ctr 2570A MSRB II PO Box 0674 Ann Arbor MI 48109

MARKESBERY, WILLIAM R., neurology and pathology educator, physician; b. Florence, Ky., Sept. 30, 1932; s. William M. and Sarah E. (Tanner) M.; m. Barbara A. Abram, Sept. 5, 1958; children—Susanne Hartley, Catherine Kendall, Elizabeth Allison. B.A., U. Ky., 1960, M.D. with distinction, 1964. Diplomate Am. Bd. Neurology and Psychiatry Diplomate Am. Bd. Pathology. Resident neurology, neuropathology Coll. Physicians and Surgeons, Columbia U., N.Y.C., 1965-69; asst. prof. pathology, neurology U. Rochester, N.Y., 1969-72; assoc. prof. pathology, neurology U. Ky., Lexington, 1972-77, prof. neurology, pathology, anatomy, 1977—, dir. Ctr. on Aging, 1979—, prof. neurology, pathology, dir., 1977—; mem. pathology study sect. NIH, Washington, 1982-85, chmn. nat. adv. coun. NIH, 1990—, Med. Sci. Adv. Bd., Chgo., 1989—, Nat. Alzheimer's Assn., Chgo., 1985-86, adv. panel on dementia U.S. Congress of Tech., Washington, 1985-86; dir. Alzheimer's Disease Research Ctr., 1985—, Alzheimer's Diseases Program Project Grant, 1984—. Mem. editorial bd. Jour. Neuropathology and Exptl. Neurology, 1983-86, 89—, Neurobiology of Aging, 1986—, Ann. Neurology, 1990—; contbr. numerous articles to profl. jours. With U.S. Army, 1954-56. Recipient Disting. Achievement award Ky. Research Found., Lexington, 1978; named U. Ky. Disting. Alumni prof., 1985, Disting. Research prof., U. Ky., 1977,; inductee U. Ky. Disting Alumni, 1989; prin. investigator NIH, Washington, 1977—. Mem. Am. Acad. Neurology, Am. Assn. Neuropathologists (exec. com. 1984-86, pres1991—), Soc. Neurosci., Am. Neurol. Assn., Alpha Omega Alpha. Home: 1555 Tates Creek Rd Lexington KY 40502-2229 Office: U Ky Sanders Brown Ctr of Aging 101 Sanders-Brown Bldg Lexington KY 40536

MARKHAM, SISTER M(ARIA) CLARE, chemistry educator, college administrator; b. New Haven, Aug. 12, 1919; d. James J. and Agnes V. (Manning) M. BA in Chemistry, St. Joseph Coll., West Hartford, Conn., 1940, LHD (hon.), 1989; PhD, Cath. U. Am., 1952. Asst. prof. chemistry St. Joseph Coll., West Hartford, Conn., 1952-59, assoc. prof., 1960-67, prof., 1968—, dean grad. div., 1979-87, asst. pres. acad. affairs, 1987-91, dir. inst. rsch., 1991—; chmn. chemistry dept., 1960-70, mem. pres.'s council, 1980—; faculty fellow chem. biodynamics U. Calif., Berkeley, 1967-68, Inst. Tech., Trondheim, Norway, 1967; experimentation chair Sisters of Mercy, West Hartford, 1969-73, councilor, 1969-77; cons. ITT, Madras, India, 1974-77; mem. planning com. Grad. Consortium, Hartford, Conn., 1976-87; under sec. energy State of Conn. Office Planning and Mgmt., 1977-79. Editor: Basic Science Series, 1962-67. Contbr. articles to profl. jours. Mem. White House Conf. on Energy Prodn., Washington, 1978; chmn. adv. bd. Conn. Environ. Mediation Ctr., Hartford, 1982-85; bd. dirs. Conn. Energy Council for Tchrs., Hartford, 1982-87; mem. adv. bd. Conn. Energy Round Table, Hartford, 1982. Research grantee Am. Chem. Soc., NSF, Hartford, 1963-73; faculty fellow NSF, Norway, Berkeley, 1967-78, travel grantee NSF, Madras, India, 1974, 76, 77; sci. edn. grantee NSF, U.S. Dept. Energy, Hartford, 1959-80; recipient Excellence in Equity award AAUW, 1992. Mem. AAAS, Am. Chem. Soc. (councilor 1968-70, 1974-88, chmn. Connecticut Valley sect. 1971-73), Conn. Acad. Sci. and Engring. (chmn. membership com. 1978-79), Sigma Xi (pres. Hartford chpt. 1992—). Office: St Joseph Coll 1678 Asylum Ave West Hartford CT 06117-2700

MARKIEWICZ, LESZEK, research biochemist; b. Sowliny, Krakow, Poland, Sept. 10, 1928; came to U.S., 1965; s. Jan and Maria (Drozdowicz) M. MS, Coll. Agr., Krakow, 1956; PhD, Coll. Agr., 1963. Rsch. assoc. Polish Acad. Sci., Krakow, 1955-63; adj. Coll. Agr., 1964-65; rsch. assoc. Cornell U. Coll. Agr., Ithaca, N.Y., 1965-67, Inst. for Muscle Disease, Inc., N.Y.C., 1968-74; rsch. biochemist dept. neurology U. Calif. Sch. Medicine, San Francisco, 1974-79; rsch. asst. prof. dept ob-gyn and reproductive sci. Mt. Sinai Sch. Medicine, N.Y.C., 1982—. Contbr. over 30 articles to Jour. Steroid Biochemistry, Endocrinology, Jour. Clin. Endocrinology and Metabolism, also chpts. to books. Mem. Endocrine Soc. Office: Mt Sinai Med Ctr One Gustave L Levy Pl New York NY 10029

MARKLE, DOUGLAS FRANK, ichthyologist, educator; b. Terre Haute, Ind., Aug. 29, 1947; s. Earl Frederick and Roena Jane (Wegrich) M.; 1 child, Bradley Ross. BS, Cornell U., 1969; PhD, William & Mary Coll., 1976. Rsch. scientist Huntsman Marine Lab., St. Andrews, Can., 1977-85; prof. Oreg. State U., Corvallis, 1985—.

MARKO, JOHN FREDERICK, theoretical physicist; b. Kingston-Upon-Thames, England, Oct. 25, 1962; came to U.S., 1984; s. John and Peggy Marie (Kubik) M. BSc in Physics, U. Alberta, Edmonton, Can., 1984; PhD, MIT, 1989. Rsch. asst. dept. physics MIT, Cambridge, 1984-89; postdoctoral fellow James Franck Inst., U. Chgo., 1989-91; postdoctoral fellow Lab. Atomic and Solid State Physics Cornell U., Ithaca, N.Y., 1991—. Contbr. articles to Phys. Rev. Letters. Postdoctoral fellow NSERC, 1989-91, postgrad. fellow, 1984-88. Mem. Am. Phys. Soc., Materials Rsch. Soc. Achievements include research in correlation structure of grafted polymers and microphase formation in surfactants. Office: Lab Atomic & Solid State Physics Clark Hall Cornell Univ Ithaca NY 14853-2501

MARKÓ, LÁSZLÓ, chemist; b. Oct. 16, 1928. Student, Budapest Tech. U. Rschr. Hungarian Mineral Oil and Gas Exptl. Inst., 1951-65; asst. prof. dept. organic chemistry U. Chem. Industry, Veszprém, 1965-69, dept. head, prof., 1969—; corr. mem. Hungarian Acad. Scis., 1976—; dir. rsch. group for petrochemistry, Veszprém. Author: Az oxo szintézis, 1952; co-author: Mélyfúrási geofizika, 1970. Achievements include research in homogenous catalytic reactions of hydrocarbons and their derivatives, in asymmetrical homogenous catalysis; production and complex chemical examination of transitional fero-organic homogenous catalysts. Office: Magyar Tudományos Akadémia, roosevelt-tér 9, 1051 Budapest Hungary*

MARKOE, ARNOLD MICHAEL, radiation oncologist; b. N.Y.C., Apr. 15, 1942; s. Joseph Markoe and Claire (Hershkowitz) Markoe Berger; m. Tana Kates, Sept. 3, 1967; 1 child, Zaharah. BA, Adelphi U., 1963; MS, U. Rochester, 1966; ScD, U. Pitts., 1972; MD, Hahnemann U., 1977. Diplomate, Am. Bd. Radiology (Therapeutic Radiology). Rsch. assoc. Albert Einstein Coll. Medicine, Bronx, N.Y., 1966-69; USPHS postdoctoral fellow Allegheny Gen. Hosp., Pitts., 1972-73; Am. Cancer Soc. spl. postdoctoral fellow Hahnemann Med. Coll., Phila., 1975-77; from sr. instr. to assoc. prof. radiation oncology Hahnemann U., Phila., 1977-89; staff physician Jackson Meml. Hosp., Miami, Fla., 1990—; mem. Sylvester Comprehensive Cancer Ctr., Miami, 1990—; assoc. prof. Sch. Medicine, U. Miami, 1989-92, prof., 1992—; staff physician U. Miami Health Ctr., 1990—; cons. Anna Bates Leach Hosp. of Bascom-Palmer Eye Inst., 1990—, Cancergrams Info. Ventures, Inc., Phila., 1989-92; mem. adv. bd. radiation therapy tech. tng. program Gwynedd-Mercy Coll., Gwynedd Valley, Pa., 1988-89, Miami Dade C.C./Jackson Meml. Hosp. Consortium, 1989—; mem. adv. panel Radiation Oncology Self-Assessment Program, 1992—. Mem. editorial bd. Am. Jour. Clin. Oncology, 1991—, Radiation Oncology Investigations, 1992—; contbr. numerous articles to med. publs., chpts. to med. textbooks. Grantee, Soc. Nuclear Medicine, 1976. Mem., Am. Radium Soc., Am. Soc. Clin. Oncology, Am. Coll. Radiology, Am. Coll. Radiation Oncology, Am. Soc. Therapeutic Radiation Oncology, So. Med. Soc., Fla. Med. Soc., Dade County Med. Soc., Fla. Soc. Clin. Oncology, Alpha Omega Alpha, Beta Beta Beta. Avocations: reading, music, fishing.

MARKOV, VLADIMIR VASILIEVICH, radio engineer; b. St. Petersburg, Russia, June 9, 1954; m. Irina E. Lukina, Apr. 10, 1976; 1 child, Igor V. Radio Engr., Radio Engring. Inst., Taganrog, 1979; Cand. Tech. Sci., Radio Engring. Inst., 1985. Radio engr. Radio Engring. Inst., Taganrog, 1979-84, asst. lectr., 1984-86, assoc. prof., 1986—. Reviewer and author books and papers. Achievements include research in CAD systems field. Home: 2 Garibaldi Str Apt 25, Taganrog Russia 347922 Office: Radio Engring Inst, 22 Chekhov Str, Taganrog Russia 347915

MARKOVICH-TREECE, PATRICIA, economist; b. Oakland, Calif.; d. Patrick Joseph and Helen Emily (Prydz) Markovich; BA in Econs., MS in Econs., U. Calif.-Berkeley; postgrad. (Lilly Found. grantee) Stanford U., (NSF grantee) Oreg. Grad. Rsch. Ctr.; children: Michael Sean, Bryan Jeffry, Tiffany Helene. With pub. rels. dept. Pettler Advt., Inc.; pvt. practice polit. and econs. cons.; aide to majority whip Oreg. Ho. of Reps.; lectr., instr., various Calif. instns., Chemeketa (Oreg.) Coll., Portland (Oreg.) State U.; commr. City of Oakland (Calif.), 1970-74; chairperson, bd. dirs. Cable Sta. KCOM, Piedmont; coord. City of Piedmont, Calif. Gen. Planning Commn.; mem. Piedmont Civic Assn., Oakland Mus. Archives of Calif. Artists.; commr. Core Adv. Com. City of Oakland, Calif. Mem. Internat. Soc. Philos. Enquiry, Mensa (officer San Francisco region), Bay Area Artists Assn. (coord., founding mem.), Berkeley Art Ctr. Assn., San Francisco Arts Commn. File, Calif. Index for Contemporary Arts, Pro Arts, No. Calif. Pub. Ednl. and Govt. Access Cable TV Com. (founding), Triple Nine Soc.

MARKOWSKA, ALICJA LIDIA, neuroscientist, researcher; b. Warsaw, Poland, Aug. 22, 1948; came to U.S., 1986; d. Marian Boleslaw and Eugenia Krystyna (Wodzynska) Pawlak; m. Janusz Jozef Markowski, Oct. 23, 1971; children: Marta Agnieszka, Michal Jacek. BA, MSc, Warsaw U., 1971; PhD, Nencki Inst., Warsaw, 1979. Postdoctoral fellow Nencki Inst., 1979-81, asst. prof., 1981-86; assoc. rschr. Johns Hopkins U., Balt., 1987-91, rsch. scientist, 1991-92, prin. rsch. scientist, 1992—; vis. fellow Czechoslovak Acad. Sci., Prague, 1981; rschr., lectr. U. Bergen, Norway, 1983; vis. faculty Johns Hopkins U., 1986-87; cons. Sigma Tau & Otsuka Co., Italy, Japan, 1990-92. Reviewer Neurobiology of Aging, 1992—, Behavioral Brain Rsch., 1992—' contbr. chpts. to Preoperative Events, 1989, Prospective on Cognitive Neuroscience, 1990, Encyclopedia of Memory, 1992, Neuropsychology of Memory, 1992, Methods in Behavioral Pharmacology, 1993. Grantee Nat. Inst. Age, 1989—, NSF, 1990-93, NIH, 1992—. Mem. AAAS, Soc. for Neuroscience, Internat. Brain Rsch., N.Y. Acad. Sci. Achievements include first evidence that pharmacological interventions with cholinergic against, oxotremorine, through intracranial stimulation of the septohippocampal system can alleviate age-related mnemonic impairments, research has focused on an importance of the septohippocampal cholinergic system in memory function, brain mechanisms involved in different kinds of memory and sensorimotor skills and their relations to aging, amnesia, and dementia, animal models to examine the effect of nerve growth factor treatment. Home: 2 Oberlin Ct Baltimore MD 21204 Office: Johns Hopkins U 34th & Charles St Baltimore MD 21218

MARKOWSKI, ROBERTA JEAN, electrical engineer; b. West Allis, Wis., Apr. 10, 1967; d. Robert Alois Sadowski and Donna Jean (Sachi) Hamburgur; m. Thomas Lee Markowski, Sept. 8, 1990. BSEE, Milw. Sch. of Engring., 1989. Applications engr. Pillar Industries, Menomonee Falls, Wis., 1989—. Mem. IEEE, NSPE. Avocations: crafts, aerobics, camping, fishing, landscaping. Home: 1306 S 86th St West Allis WI 53214 Office: Pillar Industries N92 W15800 Megal Dr Menomonee Falls WI 53051

MARKS, ANDREW ROBERT, molecular biologist; b. N.Y.C., Feb. 22, 1955; s. Paul Alan and Joan Harriet (Rosen) M.; m. Margaret Foster, Jan. 14, 1984; children: Joshua, Daniel, Sarah. BA (magna cum laude), Amherst Coll., 1976; MD, Harvard Med. Sch., 1980. Diplomate Am. Bd. Internal Medicine; cert. in cardiovascular diseases. Intern, resident Mass. Gen. Hosp., Boston, 1980-83, cardiology fellow, 1985-87; fellow genetics Harvard Med. Sch., Boston, 1983-85, instr. medicine, 1987-90; asst. prof. molecular biology Mt. Sinai Sch. Medicine, N.Y.C., 1990-93; assoc. prof. molecular biology and medicine, 1993—; Contbr. articles to profl. jours. Mem. Sierra Club, 1986—. Established investigatorship, Am. Heart Assn., 1993. Recipient Clinician-Scientist award Am. Heart Assn., 1986, Excellence in Rsch. award Am. Fedn. for Clin. Rsch., 1990, Syntex Scholars award, 1991. Mem. Am. Assn. Biol. Chemistry and Molecular Biology, Am. Assn. Cell Biology, Biophys. Soc., Harvey Soc., Sierra Club. Achievements include research in cloned DNA encoding the calcium release channel in skeletal muscle; characterization of smooth muscle calcium release channel. Office: Mount Sinai Med Ctr 1 Gustave L Levy Pl New York NY 10029-6504

MARKS, DAVID HUNTER, civil engineering educator; b. White Plains, N.Y., Feb. 22, 1939; s. Sidney M. and Jean (Berger) M.; div.; 1 child, Joanna. BCE, Cornell U., 1962, MS in Environ. Engring., 1964; PhD, Johns Hopkins U., 1969. Registered profl. engr., N.Y., Mass.; registered hydrologist, Am. Inst. Hydrology. Sr. sanitary engr. USPHS, Phila., 1964-66; asst. prof. civil engring. MIT, Cambridge, Mass., 1969-72, assoc. prof., 1972-75, prof., 1975—, head dept., 1985-92, dir. program in environ. engring. edn. and rsch., 1991—, James Mason Crafts Prof., 1992—; bd. dirs. Camp Dresser and McKee, Environ. Engrs., Boston. Office: MIT Rm 48-305 Dept Civil & Environ Engring Cambridge MA 02139

MARKS, ERNEST E., maintenance engineer; b. Glasgow, Mont., Mar. 30, 1950; s. William Ernest and Christina Ellen (Mix) M.; m. Maris Ellen Dowrie, Aug. 15, 1972 (div. 1987); children: Anne, Tina, Kellie; m. Cindy Kline Hague, Nov. 27, 1991. Grad. high sch. Technician EMI Therapy Systems, Sunnyvale, Calif., 1978-79; mfg. engr. Rapicom, Sunnyvale, Calif., 1979-80; mfg. engr. Eaton Vacuum Systems, Santa Clara, Calif., 1980-81, field svc. mgr., 1981-83; technician Micron Tech., Boise, Idaho, 1983-84, equip. maint. mgr., 1984—. With USN, 1969-78. Achievements include research vacuum pneumatic wafer transfer machine, rinser dryer resistance probe fixture, rinser dryer improvements. Office: Micron Semiconductor 2805 E Columbia Rd Boise ID 83706

MARKS, JAMES FREDERIC, pediatric endocrinologist, educator; b. Pitts., Dec. 18, 1928; s. Alfred Mozelle and Cecil (Cuff) M.; m. Mary Fay Clement, Jan. 29, 1959; 1 child, Roland Phillip. BA, Princeton U., 1950; MD, Harvard U., 1954; MPH, U. Pitts., 1984. Intern Montefiore Hosp., Pitts., 1954-55; resident in pediatrics Children's Hosp. of Pitts., 1955-57; rsch. fellow in pediatric endocrinology U. Pitts., 1959-61; asst. prof. dept. pediatrics U. Tex. Southwestern Med., Dallas, 1961-68, assoc. prof. pediatrics, 1968—; bd. dirs. State Newborn Screening Program, Tex., 1980—. Contbr. articles to profl. publs., chpts. to med. textbooks. Capt. U.S. Army, 1957-59. Sr. rsch. fellow USPHS, 1983-84. Mem. Am. Diabetes Assn. (bd. dirs. Tex. affiliate 1992—), Am. Acad. Pediatrics, Endocrine Soc., Soc. for Pediatric Rsch. Achievements include research in thyroid function in infancy, delineation of early clinical course in Lesch-Nyhan disease, observations on the possible genetic factors in diabetic microvascular disease. Office: U Tex Southwestern Med 5323 Harry Hines Blvd Dallas TX 75235-9063

MARKS, PAUL ALAN, oncologist, cell biologist; b. N.Y.C., Aug. 16, 1926; s. Robert R. and Sarah (Bohorad) M.; m. Joan Harriet Rosen, Nov. 28, 1953; children: Andrew Robert, Elizabeth Susan Marks Ostrer, Matthew Stuart. AB with gen. honors, Columbia U., 1945, MD, 1949; D in Biol. Sci. (hon.), U. Urbino, Italy, 1982; PhD (hon.), Hebrew U., Jerusalem, Israel, 1987, U. Tel Aviv, 1992. Fellow Columbia U. Coll. Physicians and Surgeons, 1952-53; assoc. Columbia Coll. Physicians and Surgeons, 1955-56, mem. faculty, 1956-82, dir. hematology tng., 1961-74, prof. medicine, 1967-82, dean faculty of medicine, v.p. med. affairs, 1970-73, dir. Comprehensive Cancer Ctr., 1972-80, v.p. health scis., 1973-80, prof. human genetics and devel., 1969-82, Frode Jensen prof. medicine, 1974-80; prof. medicine and genetics Cornell U. Coll. Medicine, N.Y.C., 1982—; prof. medicine Grad. Sch. Med. Scis. Cornell U., N.Y.C., 1983—; instr. Sch. Medicine, George Washington U., 1954-55; cons. VA Hosp., N.Y.C., 1962-66; attending physician Presbyn. Hosp., N.Y.C., 1967-82; pres., chief exec. officer Meml. Sloan-Kettering Cancer Ctr., 1980—; attending physician Meml. Hosp. for Cancer and Allied Diseases, 1980—; mem. Sloan-Kettering Inst. for Cancer Rsch., 1980—; adj. prof. Rockefeller U., 1980—; vis. physician Rockefeller U. Hosp., 1980—; hon. staff N.Y. Hosp., 1981—; bd. sci. counselors div. cancer treatment Nat. Cancer Inst., 1980-83; mem. steering com. Frederick Cancer Rsch. Facility Nat. Cancer Inst., 1982-86; chmn. program adv. com. Robert Wood Johnson Found., 1983-89; mem. Gov.'s Commn. on Shoreham Nuclear Plant, 1983; mem. Mayor's Commn. Sci. and Tech. City of N.Y., 1984-87; mem. adv. com. on NIH to Sec. HHS, 1989-90, 93—; mem. coun. biol. scis. Pritzker Sch. Medicine U. Chgo., 1977-88; first lectr. Nakasone Program for Cancer Control U. Tokyo, 1984; Ayrey fellow, vis. prof. Royal Postgrad. Med. Sch. U. London, 1985; William Dameshek vis. prof. hematology Mt. Sinai Med. Ctr., 1985; nat. vis. com. CUNY Med. Sch., 1986-89; trustee Feinberg Grad. Sch. Weizmann Inst. Sci., Rehovot, Israel, 1986—; William H. Resnick lectr. in medicine Stamford Hosp., 1986; disting. faculty lectr. M.D. Anderson Hosp. U. Tex., 1986; Maurice C. Pincoffs lectr. U. Md., Balt., 1987; Japan Soc. Hematology Disting. lectr., 1989; vis. prof.

Coll. de France, 1988; Alpha Omega Alpha vis. prof. N.Y. Med. Coll., 1990; Mario A. Baldini vis. prof. Harvard Med. Sch., 1991; mem. sci. adv. bd. City of Hope Nat. Med. Ctr., Duarte, Calif., 1987-92, Raymond and Beverly Sackler Found., Inc., 1989, Jefferson Cancer Inst., Phila., 1989; mem. Found. Biomed. Rsch., 1989—; advisor Third World Acad. Sci. Editor: Monographs in Human Biology, 1963; author 9 books; contbr. over 350 articles to profl. jours.; mem. editorial bd. Blood, 1964-71, assoc. editor, 1976-77, editor-in-chief, 1978-82; editor-in-chief Jour. Clin. Investigation, 1967-71; mem. editorial bd. Cancer Treatment Revs., 1981—, Cancer Preventions, 1989; guest editorial bd. Japanese Jour. Cancer Rsch., 1985—; assoc. editor Molecular Reprodn. and Devel., 1988—; expert analyst Chemistry and Molecular Biology edit. of Chemtracts, 1990-92; mem. adv. bd. Internat. Jour. Hematology, 1992; mem. editorial bd. Sci., 1990. Trustee St. Luke's Hosp., 1970-80, Roosevelt Hosp., 1970-80, Presbyn. Hosp., 1972-80, Metpath Inst. Med. Edn., 1977-79; mem. jury Albert Lasker Awards, 1974-82; bd. govs. Wiezmann Inst., 1976—; bd. dirs. Revson Found., 1976-91, Am. Found. for Basic Res. Israel, Israel Acad. Scis., 1991; mem. tech. bd. Milbank Meml. Fund, 1978-85. Recipient Charles Janeway prize Columbia, 1949, Joseph Mather Smith prize, 1959, Stevens Triennial prize, 1960, Swiss-Am. Found. award in med. rsch., 1965, Columbia U. Coll. Physicians and Surgeons Disting. Achievement medal, 1980, Centenary Inst. Pasteur medal, 1987, Disting. Oncologist award Hipple Cancer Ctr. and Kettering Ctr., 1987, Found. for Promotion of Cancer Rsch. medal, 1984 (Japan), Disting. Svc. od Johnson Found., 1989, award Outstanding Achievement in Hematopoiesis U. Innsbruck, 1991, Pres'. Nat. medal Sci., 1991; Commonwealth Fund fellow Pasteur Inst., 1961-62. Master ACP; fellow Royal Soc. Medicine, Am. Acad. Arts and Scis., AAAS, Pasteur Inst. Paris; mem. NAS (chmn. sect. med. genetics, hematology and oncology 1980-83, chmn. Acad. Forum Adv. Com. 1980-81, mem. council 1984-87, mem. coun. com. internat. affairs 1985-87, del. biol. warfare com. Internat. Security and Arms Control 1986-89), Royal Soc Med. (London), Inst. Medicine (mem. coun. 1973-76, chmn. com. study resources clin. investigation with NAS 1988), N.Y. Acad. Medicine (awards com. 1988—), Red Cell Club (past chmn.), Am. Fedn. Clin. Rsch. (past councillor Eastern dist.), Am. Soc. Clin. Investigation (pres. 1972-73), Am. Soc. Biol. Chemists, Am. Soc. Human Genetics (past mem. program com.), Am. Assn. Cancer Rsch., Assn. Am. Cancer Insts. (bd. dirs. 1983-88), Soc. Cell Biology, Am. Soc. Hematology (pres.-elect 1983, pres. 1984, chmn. adv. bd. 1985), Assn. Am. Physicians, Econ. Club (N.Y.C.), Harvey Soc. (pres. 1973-74), Internat. Soc. Devel. Biologists, Italian Assn. Cell Biology and Differentiation (hon.), Chinese Anti-Cancer Assn. (hon.), Soc. for Devel. Biology, Japanese Cancer Assn. (hon.), Japan Soc. Hematology (Disting. lectr. 1989), Internat. Leadership Ctr. on Longevity and Soc., Interurban Clin. Club, Soc. for Study Devel. and Growth, Third World Acad. Scis., Century Assn., Econ. CLub, N.Y.C., University Club N.Y.C., Alpha Omega Alpha. Home: PO Box 1485 Washington CT 06793-0485 Office: Meml Sloan-Kettering Cancer Ctr 1275 York Ave New York NY 10021

MARKS, RICHARD HENRY LEE, biochemist, educator; b. Richmond, Va., Nov. 23, 1943; s. Henry Lee and Helen Campbell (Hutchison) M.; m. Lynne Evelyn Griffith, Aug. 21, 1966; children: Christopher Scott, Brian Stuart. BS, U. Richmond, 1965; PhD, Ind. U., 1969. Postdoctoral fellow U. Calif., Santa Barbara, 1969-72; asst. prof. U. of Medicine and Dentistry of N.J., Newark, 1972-76; asst. prof. East Carolina U. Sch. of Medicine, Greenville, N.C., 1976-77, assoc. prof., 1977-92, prof., 1992—. Mem. Am. Chem. Soc., Am. Soc. for Biochemistry and Molecular Biology, Arteriosclerosis Coun., Am. Heart Assn., Sigma Xi. Home: 106 Worthington Ln Greenville NC 27858 Office: Sch of Medicine Dept of Biochemistry East Carolina U Greenville NC 27858

MARKS, TOBIN JAY, chemistry educator; b. Washington, Nov. 25, 1944; s. Eli Sidney and Miriam (Heller) M.; m. Indrani Mukharji, May 19, 1985. B.S., U. Md., 1966; Ph.D., MIT, 1970. Asst. prof. Northwestern U., Evanston, Ill., 1970-74, assoc. prof., 1974-78, prof. chemistry, 1978—, Morrison prof. chemistry, 1986—, prof. materials sci. and engring., 1987—. Editor: Organometallics of the F-Elements, 1979; assoc. editor Organometallics, 1986—; editor: Fundamental and Technological Aspects of Organo-F-Element Chemistry, 1985. Sloan Found. fellow, 1974, Guggenheim fellow, 1989—; Dreyfus Found. scholar, 1975; recipient Nat. Fesenius award Phi Lambda Upsilon, 1979, Mack award Ohio State U., 1987. Fellow Am. Aacd. Arts and Scis.; mem. NAS, Am. Chem. Soc. (A.K. Doolittle award 1984, award in Organometallic Chemistry 1989), Soc. for Applied Spectroscopy, Materials Rsch. Soc., Phi Beta Kappa. Home: 2300 Central Park Ave Evanston IL 60201-1810

MARKTUKANITZ, RICHARD PETER, metallurgical engineer; b. McKeesport, Pa., Feb. 9, 1953; s. Peter John and Frances Mary (Greco) M.; m. Karen Rae Caldwell, Jan. 12, 1974; children: Mark Joseph, Pamela Marie. BS in Metallurgy, Pa. State U., 1978, postgrad.; MS in Metall. Engring., U. Pitts., 1985. Staff engr. Alcoa Labs., Alcoa Center, Pa., 1978-89; mgr. joining and cutting techs. Nat. Ctr. for Excellence in Metalworking Tech., Johnstown, Pa., 1989-90; rsch. assoc. applied rsch. lab. Pa. State U., State College, 1990—; corp. metall. cons. Cannondale Corp., Georgetown, Conn., 1992—. Contbr. articles to Jour. of Metals, Aluminium. With USNR, 1971-74. Mem. Am. Soc. for Metals, Am. Welding Soc. (prin. reviewer 1989—), Metall. Soc. Am. Inst. Mining, Metall., and Petroleum Engrs., Sigma Xi. Democrat. Roman Catholic. Achievements include patents for method for welding aluminum alloys, filler alloy for welding aluminum-lithium alloys; discovery of formation of weld porosity model for aluminum welding based on available sources of hydrogen. Home: 610 Old Farm Ln State College PA 16803 Office: Applied Rsch Lab PO Box 30 State College PA 16804

MARKUSSEN, JOANNE MARIE, chemical engineer; b. Rolla, Mo., Apr. 29, 1956; d. Kenwood Markus and Lillian (Heaton) M. BS in Chemistry and Math., Mary Washington Coll., 1978; MS in Chem. Engring., Carnegie-Mellon U., 1980, PhD in Chem. Engring., 1991. R&D project leader U.S. Dept. Energy, PETC, Pitts., 1984—; lectr. in field. Contbr. articles to profl. jours. Mem. Air and Waste Mgmt. Assn. (AE-1 com. 1989—), Am. Chem. Soc. (nat. chemistry wk. com. mem. 1992, mem. coal tech. group), Am. Inst. Chem. Engrs., Mortar Bd., Sigma Xi, Phi Beta Kappa, Alpha Phi Sigma, Chi Beta Phi. Achievements include research on combined S02 and NOx removal in spray drying flue gas desulfurization systems; regeneration kinetics of copper oxide sorbent in the copper oxide process; others. Home: 981 Crest Ln Carnegie PA 15106 Office: US Dept Energy Pitts Energy Tech Ctr PO Box 10940 Pittsburgh PA 15236

MARKVART, TOMAS, physicist; b. Praque, Czechoslovakia, June 24, 1950; came to U.K. 1968; s. Jaroslav and Edith M. BSc, Birmingham U., U.K., 1971, PhD, 1975. Demonstrator U. Keele, U.K., 1974-77; sr. rsch. fellow U. Southampton, U.K., 1977-91; assoc. prof. Universidad Politecnica de Madrid, Spain, 1991-92; head solar energy ctr. U. Southampton, 1992—. Editor: Solar Electricity, 1992; contbr. book: Recombination in Semiconductors; contbr. articles to profl. jours. Recipient Allocation de Sejour Scientifique de Haut Niveau, French Govt., 1986, Corbett prize, U. Birmingham, 1971, Nora Calderwood prize, 1970. Achievements include 1 patent on radiation resistant silicon solar cell; research includes solar cells, photovoltaic systems, defects and radiation damage in semiconductors. Office: U Southampton, Highfield, Southampton England SO95NH

MARLAND, ALKIS JOSEPH, leasing company executive, computer science educator, financial planner; b. Athens, Greece, Mar. 8, 1943; came to U.S., 1961, naturalized, 1974; s. Basil and Maria (Pervanides) Mouradoglu; m. Anita Louise Malone, Dec. 19, 1970; children: Andrea, Alyssa. BS, Southwestern U., 1963; MA, U. Tex., Austin, 1967; MS in Engring. Adminstrn., So. Meth. U., 1971. Cert. in data processing, enrolled agt., fund specialist, ChFC., CLU, CFP, RFP, CTP. With Sun Co., Richardson, Tex., 1968-71, Phila., 1971-76; mgr. planning and acquisitions Sun Info. Svcs. subs. Sun Co., Dallas, 1976-78; v.p. Helios Capital Corp. subs. Sun Co., Radnor, Pa., 1978-83; pres. ALKAN Leasing Corp., Wayne, Pa., 1983—; prof. dept. computer scis. and bus. adminstrn. Eastern Coll., St. David's, Pa., 1985-87; prof. math. Villanova (Pa.) U., 1987-89; bd. dirs. Alkan Leasing Corp., 1983—. Bd. dirs. Radnor Twp. Sch. Dist., 1987-91, Delaware County Intermediate Unit, 1988-91, Phila. Fin. Assn., 1989-92. Mem. IEEE, Assn. Computing Machinery, Data Processing Mgmt. Assn., Internat. Assn. Fin. Planners, Am. Soc. CLUs and ChFC, Am. Assn. Equipment Lessors, Inst. Cert. Fin. Planners, Nat. Assn. Enrolled Agts., Nat. Assn. Tax Practitioners, Nat. Assn. Pub. Accts., Fin. Analysts Phila., Phila. Fin. Assn. (sec. 1989-92, mem. award 1988), Fgn. Policy Rsch. Inst., World Affairs Coun. Phila., Phila. Union League, Main Line C. of C., Assn. Investment Mgmt. and Rsch., Rotary (pres. Wayne club 1989-90, gov.'s rep. dist. 7450, 1990-91, 93—), Masons (32 degree). Republican. Home: 736 Brooke Rd Wayne PA 19087-4709 Office: PO Box 8301 Radnor PA 19087-8301

MARLAY, ROBERT CHARLES, physicist, engineer; b. Seaside, Oreg., Dec. 19, 1946; s. Myron George Jr. and Margaret Alice (Bump) M.; m. Nancy Evelyn Tate, Oct. 18, 1980; children: Jennifer Lynn, Sarah Elizabeth. BS in Engring., Duke U., 1969; MS, MIT, 1971, M City Planning, 1971, PhD, 1983. Profl. engr., D.C. Engr. Office of Policy U.S. Dept. Energy, Washington, 1977-83, engr. office of energy conservation, 1983-85, scientist office of energy rsch., 1985-89, coord. nat. energy strategy devel. office, 1989-90, dir. office of program rev. & analysis, office of policy, 1990-92, dir. Office Tech. Policy, 1993—. Contbr., editor (govt. publ.) Nat. Energy Strategy, 1991. Lt. Civil Engring. Corps, USNR, 1971-74. Mem. AAAS, Internat. Assn. for Energy Econs., Sigma Xi. Episcopalian. Achievements include development of quantitative methods for characterizing structural change in industrial economies; development of a national energy strategy; use of technology assessments to guide federal funding priorities for energy technology research and development. Office: US Dept Energy 1000 Independence Ave SW Washington DC 20585

MARLIN, DONNELL CHARLES, dental educator; b. Rensselaer, Ind., Nov. 26, 1930; s. Charles Franklin and Helen (Batterton) M. AB, Ind. U., 1952; DDS, Ind. U., Indpls., 1956. Diplomate Am. Bd. Forensic Odontology. Pvt. practice Noblesville, Ind., 1962-74; dental educator Ind. U. Sch. Dentistry, Indpls., 1974-81, asst. prof., 1981—. Contbr. article to Jour. Forensic Scis. Lt. USN, 1957-61. Named Ky. Col. Commonwealth of Ky., 1967. Mem. Am. Acad. Forensic Scis. Office: Ind U Sch Dentistry 1121 W Michigan Indianapolis IN 46202

MARLIN, JOHN TEPPER, economist, writer, consultant; b. Washington, Mar. 1, 1942; s. Ervin Ross and Hilda (van Stockum) M.; AB cum laude, Harvard U., 1962; BA, Oxford (Eng.) U., 1965, MA, 1969; PhD in Econs., George Washington U., 1968; m. Alice Rose Tepper, Sept. 25, 1971; children: John Joseph Tepper, Caroline Alice Tepper. Fin. economist Fed. Res. Bd., FDIC and SBA, Washington, 1964-69; asst. prof. Baruch Coll., City U. N.Y., 1969-73; founder, pres. Council Mcpl. Performance, N.Y.C., 1973-88; pres. JTM Reports, Inc., 1989-92; social auditor Ben and Jerry's Homemade, 1989; dir. Conversion Info. Ctr., Coun. on Econ. Priorities, 1991-92; chmn., bd. advisors CIC, CEP, 1992—; cons. J.M. Kaplan Fund, 1991; chief economist for Office of Comptr., City of N.Y., 1992—. Mem. Am. Econ. Assn. (life), Fin. Mgmt. Assn. (life), Economists Allied for Arms Reduction (bd. dirs.), City Club (N.Y.C.), Harvard Club (N.Y.C.), Devon Yacht Club, Trinity (Oxford) Soc. U.S.A. (pres.), Oxford Soc., N.Y. (branch sec., pres. oxford-cambridge dinner com.). Author: The Wealth of Cities, 1974, Cities of Opportunity, 1988, Catalogue of Healthy Food, 1990, The Livable Cities Almanac, 1992, (with others) Book of American City Rankings, 1983, Contracting Municipal Services, 1984, Book of World City Rankings, 1986, Soviet Conversion, 1991, Building a Peace Economy, 1992 ; founding editor Jour. Fin. Edn., 1972-73; editor Nat. Civic Rev., 1987-88, Privatization Report, 1986-88. Home: 360 W 22d St New York NY 10011 Office: Office of Comptr City of New York # 510 1 Centre St New York NY 10007

MARLOW, JOSEPH, automation engineer; b. Timisoara, Banat, Romania, Oct. 9, 1939; came to U.S. 1970; s. Ian Marcovici and Rety Marlow; m. Lee Dascalu, Aug. 20, 1960; children: Janik, Alvin. MS, Poly. U., N.Y., 1975, PhD in Engring., 1979. Design engr. Treadwell, N.Y.C., 1970-76; system engr. Litton Ind., Totowa, N.J., 1976-81; specialist engr. Breeze Corp., Union, N.J., 1981-84; group leader Machine Tech. Inc., Whippany, N.J., 1984-85; v.p. Tech., Maplewood, N.J., 1985—; cons. Cerro, N.Y.C., 1975-79, Bell Communication Rsch., Morristown, N.J., 1984-87, Slaby Engring., Morris Plains, N.J., 1987-90, Applied Energy, Bloomfield, N.J., 1991—. Author: Industrial Manufacturing, 1969, Welding Society and Industry, 1971, Aerial Stations, 1989; contbr. articles to profl. jours. Mem. IEEE, Lions. Achievements include research in automation, robotics, TV control, ultrasonic processes, magnetic field handling. Home: 20 Colonial Ter Maplewood NJ 07040-1021

MARMANN, SIGRID, software development company executive; b. Voelklingen, Saarland, Fed. Republic Germany, Feb. 8, 1938; s. Leo and Karoline Anna (Weidenhof) M. Postgrad., Norwood Coll., London, 1962; BS in Acctg., Ind. & Handelskammer, Saarbruecken, Fed. Republic Germany, 1956; postgrad., Golden Gate U., 1970-85; BA in Mgmt., St. Mary's Coll., Moraga, Calif., 1984. Controller M.O.M., Paris, 1965-69; bookkeeper Chrissa Imports, Brisbane, Calif., 1970-78; acctg. mgr. Highcity Internat., San Anselmo, Calif., 1978-80; acctg. mgr., system analyst Kukje Korean Trading Co., Rutherford, N.J., 1980-81; asst. treas. Am. Mercantile Co., Brisbane, 1981-84; controller Provident Credit Union, Burlingame, Calif., 1984; owner Datatech EDI Systems, San Rafael, Calif., 1984—; pres., chief owner Datatech EDI Systems, San Rafael, 1989—; pres. Telepay Express, Inc., 1989. Founder No. Calif. Electronic Data Interchange Users Group, San Francisco, 1990. Mem. ANSI ASC X12 Electronic Data Interchange (fin. subcom. Alexandria, Va. chpt., nominee Membership award 1990) Great Plains Software (qualified installer), Computer Assocs. Internat. (installer). Avocations: travelling, skiing, swimming, sailing, fishing, baking. Home: 30 Newport Way San Rafael CA 94901-4411

MARMOR, THEODORE RICHARD, political science and public management educator; b. Bklyn., Feb. 24, 1939; s. James and Mira Bernice (Karpf) M.; m. Jan Schmidt, Oct. 20, 1961; children—Laura Carleton, Sarah Rogers. B.A., Harvard U., Cambridge, Mass., 1960, Ph.D., 1966; postgrad., Wadham Coll., Oxford U., Eng., 1961-62. Asst. and assoc. prof. polit. sci. U. Wis.-Madison, 1967-69; assoc. prof. pub. affairs U. Minn.-Mpls., 1970-73; prof. U. Chgo., 1973-79; prof. polit. sci. Yale U., New Haven, 1979—, chmn. Ctr. Health Studies, 1979—; prof. pub. mgmt. Yale U. Sch. Orgn. & Mgmt., New Haven, 1983—; vis. fellow Russell Sage Found., 1987-88; cons., lectr. in field. Author: The Politics of Medicare, 1970, 73, Political Analysis & American Medical Care, 1983; co-author: Health Care Policy, 1982; editor: Poverty Policy, 1971, National Health Insurance, 1980, Social Security: Beyond the Rhetoric of Crisis, 1988; editor Jour. Health Politics Policy and Law, 1980-84; contbr. articles to profl. jours. Mem. Council on Fgn. Relations, N.Y.C., 1979-80, Pres.' Commn. on 1980s, 1980; social policy adviser Walter Mondale Presdl. Campaign, 1984. Can. Inst. Advanced Research fellow, 1987-91. Fellow Adlai Stevenson Inst., JFK Inst. of Politics; mem. U.S. Squash Racquets Assn. (bd. dirs. 1983—). Democrat. Jewish. Clubs: United Oxford and Cambridge (London); Lawn (New Haven); Yale (N.Y.C.). Home: 139 Armory St Hamden CT 06517-4005 Office: Yale Univ Sch Orgn & Mgmt 111 Prospect St New Haven CT 06511-3729

MAROIS, JIM, plant pathologist, educator. Prof. dept. plant pathology U. Calif., Davis. Recipient CIBA-GEIGY award Am. Phytopathological Soc., 1991. Office: Univ of California Hutchinson Hall Dept of Plant Pathology Davis CA 95616*

MARON, MICHAEL BRENT, physiologist; b. Long Beach, Calif., Oct. 8, 1949; s. Victor and Florence Kathryn (Perlman) M.; m. Dawn Nelson Wallace, Aug. 19, 1983. AA, Long Beach City Coll., 1969; BA, U. Calif., Santa Barbara, 1972, PhD, 1976. Postdoctoral fellow Med. Coll. Wis., Milw., 1977-79; asst. prof. physiology N.E. Ohio U. Coll. Medicine, Rootstown, 1979-84, assoc. prof. physiology, 1984-89, interim chmn. physiology, 1985-86, chmn., 1986—, prof. physiology, 1989—. Contbr. articles to profl. jours. NIH grantee, 1983—. Office: NE Ohio U Coll Medicine PO Box 95 Rootstown OH 44272

MARONDE, ROBERT FRANCIS, internist, clinical pharmacologist, educator; b. Monterey Park, Calif., Jan. 13, 1920; s. John August and Emma Florence (Palmer) M.; m. Yolanda Cerda, Apr. 15, 1970; children—Robert George, Donna F. Maronde Varnau, James Augustus, Craig DeWald. B.A., U. So. Calif., 1941, M.D., 1944. Diplomate: Am. Bd. Internal Medicine. Intern L.A. County-U. So. Calif. Med. Ctr., 1943-44, resident, 1944-45, 47-48; asst. prof. physiology U. So. Calif., L.A., 1948-49, asst. clin. prof. medicine, 1949-60, assoc. clin. prof. medicine, 1960-65, assoc. prof. medicine and pharmacology, 1965-67, prof. medicine and pharmacology, 1968-90, emeritus, 1990—, prof. emeritus, 1990—; spl. asst. v.p. for health affairs, 1990—; cons. FDA, 1973, Med. Co. Containment, Inc., 1991—, State of Calif., Dept. Health Svcs., 1993. Served to lt. (j.g.) USNR, 1945-47. Fellow ACP; mem. Am. Soc. Clin. Pharmacology and Therapeutics, Alpha Omega Alpha. Home: 785 Ridgecrest St Monterey Park CA 91754-3759 Office: U So Calif 2025 Zonal Ave Los Angeles CA 90033-4526

MARONEY, MICHAEL JAMES, chemistry educator; b. Ames, Iowa, Dec. 16, 1954; s. Donald James and Dorothy (Queal) M. BS, Iowa State U., 1977; PhD, U. Wash., 1981. Rsch. chemist Chevron Rsch. Co., Richmond, Calif., 1981-83; postdoctoral fellow Northwestern U., Evanston, Ill., 1983; postdoctoral assoc. U. Minn., Mpls., 1983-85; asst. prof. U. Mass., Amherst, 1985-90, assoc. prof., 1990—; cons. Shriner's Hosp., Boston, 1992—, Abbott Labs., Abbott Park, Ill., 1987-90; spl. study sect. mem. NIH, Bethesda, Md., 1990-91; mem. faculty program in molecular and cellular biology U. Mass., 1987—. Contbr. articles to prof. jours. Recipient NIH First award, 1987-92; named Lilly Endowment Faculty Teaching fellow U. Mass., 1989-90, Nat. Acad. Scis. Interacad. Exch. fellow, 1990, Chevron fellow U. Wash., 1980, Ga. Pacific fellow U. Wash., 1979-80. Mem. AAAS, Am. Chem. Soc., Sigma Xi. Achievements include research in the structure and function of biol. metal clusters, and for the design and synthesis of structural, functional, and spectroscopic models of biol. metal sites (particularly nickel sites in hydrogenases). Office: Univ of Mass Chemistry Dept Amherst MA 01003

MARONI, GUSTAVO PRIMO, geneticist, educator; b. Merlo, B.A., Argentina, Nov. 20, 1941; came to U.S, 1968; s. Victor and Iole (Brighi) M.; m. Donna Farolino, Dec. 16, 1974. Licenciado, U. Buenos Aires, Argentina, 1967; PhD, U. Wis., 1972. Asst. prof. U. N.C., Chapel Hill, 1975-81, assoc. prof., 1981—. Author: An Atlas of Drosophilia Genes, 1993; contbr. numerous articles to profl. jours. Grantee NSF, 1975-78; NIH, 1978—. Mem. AAAS, Genetics Soc. Am. Office: Univ NC Dept Biology Fordham Hall Chapel Hill NC 27599-3280

MAROPIS, NICHOLAS, engineering executive; b. Slovan, Pa., May 14, 1923; s. Speros N. and Argero (Skinakis) M.; widowed; children: Samuel, Colin, Janice, Michelle. BA, Washington and Jefferson U., 1949; MS, Pa. State U., 1967. Physicist Naval Ordinance Lab., White Oak, Md., 1950-53; sr. project engr., physicist RM Parsons Inc., Frederick, Md., 1953-55; v.p. engring. Aeroprojects Inc., Westchester, Pa., 1955-71, UTI Corp., Collegville, Pa., 1972-91; prin. Maropis Tech. Enterprises, Inc. (M-TEI), Baden, Pa., 1991—. Mem. allocations Com. United Way Chester County, Exton, Pa., 1989-90; pres. St. Sophia Greek Orthodox Ch., 1985-87. Sgt. USAAF, 1942-45. Recipient Commendation Atomic Energy Commn., Oak Ridge, Tenn., 1964, NASA, 1968. Mem. Hellenic Ednl. Progressive Assn. (officer 1979-91). Republican. Achievements include patents on High Powered Ultrasonic Systems and their applications to metal deformations.

MAROTTA, SABATH FRED, physiology educator; b. Chgo., Aug. 26, 1929. BS, Loyola U., 1951; MS, U. Ill., 1953, PhD in physiology, 1957. Rsch. assoc. animal sci. U. Ill., Chgo., 1957-58, from instr. to assoc. prof. physiology, 1958-70; rsch. assoc. Aeromed Lab. Med. Ctr., U. Ill., Chgo., 1958-60, asst. dir., 1960-64; prof. physiology U. Ill., Chgo., 1970—, assoc. dean grad. coll., 1975; assoc. dir. Rsch. Resources Ctr., U. Ill., Chgo., 1975-80, dir., 1980—; adv. U. Ill Chiengmai Project, Thailand, 1964-66. Mem. Aerospace Med. Assn., Am. Physiology Soc., Soc. Exptl. Biology and Medicine. Achievements include research in neuroendocrinology; biologic rhythms; role of the adrenal cortex in the adaption to environmental stresses. office: Univ of Illinois at Chicago UIC Eye Ctr 1855 W Taylor Chicago IL 60612*

MAROTTI, KEITH RICHARD, molecular biologist, researcher; b. Beaver Falls, Pa., May 26, 1952; s. Anthony James and Marie Helen (Pavkovich) M.; m. Hope Marie Gregorius, Oct. 28, 1978; children: Richard Francis, Rosemarie Francis, Francis Anthony, Marie Helen Francis. BS, Pa. State U., 1974; PhD, U. Pitts., 1979. Rsch. assoc. U. Ill. Med. Sch., Chgo., 1974-75; postdoctoral fellow Rockefeller U., N.Y.C., 1979-82; rsch. scientist Upjohn Labs., Kalamazoo, Mich., 1982-87, sr. rsch. scientist, 1987-91, sr. scientist, 1991—. Contbr. articles to profl. jours. Nat. Family Planning instr. Diocese of Kalamazoo, 1988—; marriage preparation St. Monica Parish, Kalamazoo, 1989—. Recipient Postdoctoral fellowship Damon-Runyon Walter Winchell Cancer Fund, 1980-82, Kagan award Upjohn Labs., 1987. Fellow Am. Heart Assn.; mem. N.Y. Acad. Sci., Scientists for Life. Roman Catholic. Achievements include 1 patents in tPA analogs. Office: Upjohn Labs 301 Henrietta St Kalamazoo MI 49001

MARR, KATHLEEN MARY, biologist, educator; b. Sheboygan, Wis., Sept. 20, 1954; d. David William Rath and Gloria Agnes (Carus) Otto; m. Philip Dean Marr, Jan. 3, 1976; children: Amanda, Samantha, Cornelius, Emerson. BS, Lakeland Coll., 1976; MS, Marquette U., 1986, postgrad. Instr. U. Wis., Sheboygan, 1978-82, Manitowoc, 1982; teaching asst. Marquette U., Milw., 1982-85; asst. prof. Divine Word Coll., Epworth, Iowa, 1985-87; asst. prof. Lakeland Coll., Sheboygan, 1987—, chair dept. biology, dir. pre-med. program. Author Lab Studies in Intro Biology, Lab Studies in Human Anatomy & Physiology. Educator Elderhostels, Lakeland Coll., 1989-91; accordianist Cedar Grove (Wis.) Klompen Dancers, 1982-91; ethicist Speakers Bur., Union of Concerned Scientists, Washington, 1990-91. Mem. AAAS, Am. Soc. Zoologists, Midwest Coll. Biology Tchrs. Roman Catholic. Office: Lakeland Coll County Trunk M Sheboygan WI 53082

MARRONE, DANIEL SCOTT, business, production and quality management educator; b. Bklyn., July 23, 1950; s. Daniel and Esther (Goodman) M.; m. Portia Terrone, Sept. 1, 1979; children: Jamie Ann. BA, Queens Coll., 1972, MLS, 1973; MBA, N.Y. Inst. Tech., 1975; PhD, NYU, 1988; diploma in Quality Engring., Mfg. Engring., Quality Inst. L.I., 1992, diploma in Mfg. Engring., 1993. Cert. fellow prodn. and inventory mgmt., quality auditor, quality engr., sr. indsl. technologist, purchasing mgr., quality systems provisional auditor. Auditor/investigator N.Y. State Spl. Pros., N.Y.C., 1977-78; asst. prof. Delehanty Inst., N.Y.C., 1978-79, Ladycliff Coll., Highland Falls, N.Y., 1979-80, Am. Bus. Inst., Bklyn., 1980-82; asst. dir. Adelphi Inst., Bklyn., 1982-85; asst. prof. Coll. St. Elizabeth, Convent Station, N.J., 1986-88; adj. asst. prof. NYU, 1986-88, Dowling Coll., Oakdale, N.Y., 1989-90; assoc. prof., co-dir. computer integrated mfg. ctr., dir. mgmt. tech. SUNY Coll. of Tech., Farmingdale, 1987—. Editor: Research Techniques in Business Education, NYU Business Education Doctoral Abstracts, 1981—, Agnew lecture by P.M. Sapre, 1989, NYU Symposium, 1989. Recipient Paul S. Lomax award, NYU, 1989. Mem. Nat. Assn. Indsl. Tech., Inst. Mgmt. Accts., Am. Prodn. and Inventory Control Soc., Nat. Assn. Purchasing Mgmt., Am. Soc. for Quality Control, Delta Pi Epsilon (Cert. of Merit 1988). Republican. Home: 493 Lariat Ln Bethpage NY 11714-4017

MARSCHER, WILLIAM DONNELLY, engineering company executive; b. Utica, N.Y., Feb. 18, 1948; s. William Ransford and Margaret Elizabeth (Donnelly) M.; m. Deborah Lynn Schmidt, May 27, 1972; children: Michael, Colleen, Megan. BS, Cornell U., 1970, MS in Engring., 1972; MS in Applied Mechanics, Rensselaer Poly. Inst., 1976. Sr. engr. Bendix EFI Div., Troy, Mich., 1970-73, Pratt & Whitney Aircraft, East Hartford, Conn., 1973-78; project supr. Creare Inc., Hanover, N.H., 1978-82; engring. mgr. Dresser Pump Divsn., Liberty Corner, N.J., 1982-92; v.p. Concepts ETI, Inc., Norwich, Vt., 1992—; short course lectr. U. Va., Tex. A&M U., Concepts ETI, Brit. Pump Mfrs. Assn. Author: Predictive Maintenance, 1993; editor: Tribology Transaction Mag., 1988—; contbr. chpts. to books, articles to profl. jours. Edn. dir. St. Denis Ch., Hanover, 1980-82; tchr. St. Margaret's Ch., Morristown, N.J., 1982—; head coach Morris County 12/14 yr. old Soccer, Morristown, 1985-90. NASA fellow, 1971-72; recipient Creativity Gold medal Dresser Industries, 1986, Hodson Best Paper award, 1983. Mem. ASME (vice chmn. tribol. conf. organizing com. 1993, co-chmn. rotating machinery conf. 1993), ASTM (fatigue standards com. 1980—), Soc. Tribologists and Lubrication Engrs. (chmn. wear com. 1988-90, bd. dirs. 1989—, chmn. awards com. 1992, chmn. ann. meeting 1989). Republican. Roman Catholic. Achievements include patents in field. Office: Concepts ETI Ste 360 9 Sylvan Way Parsippany NJ 07054

MARSDEN, BRIAN GEOFFREY, astronomer; b. Cambridge, Eng., Aug. 5, 1937; came to U.S., 1959; s. Thomas and Eileen (West) M.; m. Nancy Lou

Zissell, Dec. 26, 1964; children: Cynthia Louise, Jonathan Brian. BA, Oxford U., U.K., 1959, MA, 1963; PhD, Yale U., 1965. Rsch. asst. Yale U., New Haven, 1959-65; lectr. astronomy Harvard U., Cambridge, Mass., 1966-83; astronomer Smithsonian Astrophys. Obs., Cambridge, 1965-86; assoc. dir. planetary scis. Harvard-Smithsonian Ctr. for Astrophysics, Cambridge, 1987—; dir. Cen. Bur. Astron. Telegrams, 1968—, Minor Planet Ctr. Internat. Astron. Union, 1978—. Editor: The Earth-Moon System, 1966, The Motion, Evolution of Orbits and Origin of Comets, 1972, Catalogue of Cometary Orbits, 1992, Catalogue of Orbits of Unnumbered Minor Planets, 1992. Recipient Merlin medal Brit. Astron. Assn., 1965, Goodacre medal, 1979; Van Biesbroeck award U. Ariz., 1989, Camus-Waitz prize Societé astronomique de France, 1993. Fellow Royal Astron. Soc.; mem. Am. Astron. Soc. (chmn. div. on dynamical astronomy 1976-78), Internat. Astron. Union (pres. commn. 1976-79), Astron. Soc. Pacific, Sigma Xi. Office: Harvard-Smithsonian Ctr for Astrophysics 60 Garden St Cambridge MA 02138-1596

MARSDEN, JERROLD ELDON, mathematician, educator; b. Ocean Falls, British Colombia, Aug. 17, 1942; married 1965; 1 child. BSc, U. Toronto, Canada, 1965; PhD in Math., Princeton U., 1968. Instr. math. Princeton U., N.J., 1968; lectr. U. Calif., Berkeley, 1968-69, asst. prof., 1969-72, assoc. prof., 1972-77, prof. math., 1977—; asst. prof. U. Toronto, Canada, 1970-71. Recipient Norbert Weiner Applied Math. prize Am. Math. Soc., 1990. Mem. Am. Phys. Soc. Achievements include research in mathematical physics, global analysis, hydrodynamics, quantum mechanics, nonlinear Hamiltonian systems. Office: Univ of Calif Evans Hall 2120 Oxford St Berkeley CA 94720*

MARSEE, DEWEY ROBERT, chemical engineer; b. Norton, Va., Sept. 2, 1932; s. William Robert and Retha Jane (Meade) M.; m. Jean Catherine Donnelly, Apr. 23, 1955; children: Sharon Elizabeth, Dale Robert, Brenda Carol. BSChemE, U. Louisville, 1961. Mfg. specialist Olin Corp., Stamford, Conn., 1961-86; plant mgr., v.p. Ky. Agrl. Energy Corp., Franklin, 1986-89; plant mgr. Wako USA, Richmond, Va., 1989-91; mgr. terminal Regional Enterprise, Hopewell, Va., 1991-92; project and design engr. Delta-T Corp., Williamsburg, Va., 1992—. With U.S. Army, 1952-54. Achievements include contributions to fuel alcohol industry; international introductions of new manufacturing technologies. Home: 4748 Cochise Trail Richmond VA 23237

MARSH, HERBERT RHEA, JR., dentist; b. Houston, Dec. 30, 1957; s. Herbert Rhea Sr. and Lois Louise (Carby) M.; m. Cheri Lynn Smith, Nov. 2, 1989. BS, Abilene Christian U., 1980; DDS, Baylor U., Dallas, 1985. Pvt. practice Richardson, Tex., 1985-89, DeSoto, Tex., 1989—. Exhibitor High-land Oaks Ch. Health Fair, Garland, Tex., 1989; judge Garland High Sch. Sci. Fair, 1981—. Mem. ADA, Acad. Gen. Dentistry, Am. Acad. Oral Medicine, Am. Soc. Dentistry for Children, Tex. Dental Assn., Dallas County Dental Soc., DeSoto C. of C., Beta Beta Beta. Mem. Ch. of Christ. Avocations: music, travel, scuba, shooting. Home and Office: 216 Dalton Dr De Soto TX 75115-4414

MARSH, ROBERT HARRY, chemical company executive; b. Camden, N.J., Sept. 6, 1946; s. Harry Louis and Margaret Charlotte (Starke) M.; B.A., B.S. in Mech. Engring., Rutgers U., 1969; M.B.A. in Mgmt. and Fin., Temple U., 1980; m. Margaret Sammartino, Mar. 21, 1970. From mech. engr. to mech. specialist and project engr. Rohm & Haas Engring., Bristol, Pa., 1969-76; from staff engr. to sr. engring. specialist Hercules, Inc., Wilmington, Del., 1976-80, sr. fin. analyst for corp. strategic planning, 1980-81, sr. bus. analyst bus. group, 1982-83; mgr. bus. analysis Himont, Inc., 1983-86, dir. strategy and planning, 1988, dir., bus. mgmt., 1988-91, mng. dir. China, 1991—, dir. strategy 1991—. Bd. dirs. Indelpro. Active Moorestown civic affairs. Mem. ASME (nat. power com. 1977-84, vice chmn. awards com. 1980, membership chmn. 1982), Nat. Soc. Profl. Engrs., Beta Gamma Sigma, Engrs. Club Phila. Club: Hercules Country. Contbr. articles to profl. jours. Home: 3 Bartram Ct Moorestown NJ 08057-1832 Office: 3 Little Falls Ctr Wilmington DE 19807-2421

MARSHAK, MARVIN LLOYD, physicist, educator; b. Buffalo, Mar. 11, 1946; s. Kalman and Goldie (Hait) M.; m. Anita Sue Kolman, Sept. 24, 1972; children: Rachel Kolman, Adam Kolman. AB in Physics, Cornell U., 1967; MS in Physics, U. Mich., PhD in Physics, 1970. Rsch. assoc. U. Minn., Mpls., 1970-74, asst. prof., 1974-78, assoc. prof., 1978-83, prof. physics, 1983—, dir. grad. studies in physics, 1983-86, prin. investigator high energy physics, 1982-86, head sch. of physics and astronomy, 1986—. Contbr. articles to profl. jours. Trustee Children's Theater Co., 1989—. Mem. Am. Phys. Soc. Home: 2855 Ottawa Ave S Minneapolis MN 55416-1946 Office: U Minn Dept Physics 116 Church St SE Minneapolis MN 55455

MARSHALL, ALAN GEORGE, chemistry and biochemistry educator; b. Bluffton, Ohio, May 26, 1944; s. Herbert Boyer Marshall Jr. and Cecile (Mogil) Rosser; m. Marilyn Gard, June 13, 1965; children: Gwendolyn Scott, Brian George. BA in Chemistry with honors, Northwestern U., 1965; PhD in Phys. Chemistry, Stanford U., 1970. Instr. II U. B.C., Vancouver, Can., 1969-71, asst. prof., 1971-76, assoc. prof., 1976-80; prof. chemistry and biochemistry Ohio State U., Columbus, 1980—; cons. Extrel FTMS, Madison, Wis., 1989—, Oak Ridge (Tenn.) Nat. Lab., 1990—. Author: Biophysical Chemistry, 1978, Fourier Transforms in Spectroscopy, 1990; also over 140 articles. Recipient Disting. Scholar award Ohio State U. Fellow AAAS, Am. Phys. Soc.; mem. Am. Chem. Soc. (award in chem. instrumentation, Akron sect. award, award in analytical chemistry Eastern Analytical Symposium 1991), Soc. for Applied Spectroscopy (chmn. local sect. 1990-91), Am. Soc. for Mass Spectroscopy (bd. dirs. 1991-92). Office: Ohio State U Chemistry Dept 120 W 18th Ave Columbus OH 43210 1153

MARSHALL, ARTHUR HAROLD, architectural engineer; b. Auckland, New Zealand, Sept. 15, 1931; s. Arthur Cecil and Flossie May (Woodward) M.; m. Shirley Anne Lindsey, Mar. 17, 1956; children: Simon Lynn, Paul David, John Harold, Lindsey Richard. BArch with hons., U. New Zealand, 1956; BS, U. Auckland, 1956; PhD, U. Southampton, 1967. Pvt. practice Auckland, 1959-61; sr. lectr. U. Auckland, 1961-66, prof. architecture, 1973-87, personal chair, 1987—, head acoustics rsch., 1987—; sr. lectr. U. Western Australia, Perth, 1967-70, assoc. prof. bldg. sci., 1970-73; cons. A. Harold Marshall and Assts., 1973-81; sr. ptnr. Marshall Day Assocs., Auckland, New Zealand, 1981-91; group cons. Marshall Day Group, New Zealand, Australia, Malaysia, 1991—. Author two books, six poems; contbr. about 80 articles to profl. jours. Bass soloist Auckland Bach Cantata Soc., 1986-92. Grantee Australian Rsch. Grants Com., 1968-72, New Zealand U. Grants Commn., 1973-91; recipient Fowlds Meml. prize, New Zealand U. Grants Commn., 1956, travel scholarship New Zealand U., 1957; DAAD fellow Fed. German Govt., 1957, 66, 77. Fellow Acoustical Soc. Am., New Zealand Inst. of Architects, Royal Australian Inst. of Architects. Achievements include acoustical design of Perth Concert Hall, Christ Ch. Town Hall, Segerstrom Hall, Orange County, Calif., Hong Kong Cultural Centre, renovation Acad. Music of Phila., others; discovered significance of early lateral reflections of sound, 1967; first measured the necessary and sufficient acoustical conditions for ensemble for singers and instrumentalists 1978-87; authority on acoustical design process, joint developer of Midas acoustical modeling system. Office: Marshall Day Assocs, 56 Vincent St, Auckland New Zealand

MARSHALL, CAROL JOYCE, clinical research data coordinator; b. Mt. Holly, N.J., July 29, 1967; d. Oliver Jr. and Ruby Jean (Bennefield-Smith) M. BA in Biol. Scis., Rutgers U., 1985-89. Transplant-procurement coord. Nat. Disease Rsch. Interchange, Phila., 1989-90, supr. procurement dept., 1990-91, rsch. mgr., 1991-92; clin. rsch. data coord. U.S. Biosci., West Conshohocken, Pa., 1992-93, G.H. Besselaar Assocs., Princeton, N.J., 1993—. Avocations: piano, flute, sewing, calligraphy, swimming. Home: 603 Flynn Ave Moorestown NJ 08057-1731 Office: GH Besselaar Assocs Princeton Forrestal Ctr 103 College Rd E Princeton NJ 08540-6681

MARSHALL, CHARLES FRANCIS, aerospace engineer; b. St. Louis, Sept. 24, 1943; s. Charles Henry and Lucille Frances (Biersmith) M.; m. Mary Ellen Heep, May 14, 1966; children: Charles John, Amy Courtenay. BSME, U. Mo., Rolla, 1972. Enlisted USN, 1961, advanced to chief petty officer, 1967; missile officer USN, Charlestown, S.C., 1962-69; resigned USN, 1969; asst. ship missile systems, cruise missile systems McDonnell Douglas Co., St. Louis, 1971-81; with Martin Marietta Co., Denver, 1981—, mgr. advanced Titan space launcher systems, 1990—. Editor: Soviet ICBM Silos, 1989; contbr. to profl. publs. Bd. dirs. Paddock Home Owners Orgn., Florissant, Mo., 1978-79. Mem. AIAA (sr.). Achievements include development of space transportation architecture, design of next generation high energy upper stage. Office: Martin Marietta Astronautic Denver CO 80127

MARSHALL, DAVID, orthodontist; b. Syracuse, N.Y., Feb. 4, 1914; s. Moses and Fanny (Bagelman) Salutsky; B.S., Syracuse U., 1932-35; D.D.S., U. Md., 1938-42; postgrad. Columbia, 1943-45, Tufts Coll., Northwestern U.; children from previous marriage: Robert Andrew, Howard Randy, Douglas S. (dec.), Susan Beth, Robin (dec.); m. Marjorie Kaufman, Sept. 7, 1973. Practice dentistry specializing in orthodontics, Syracuse, mem. staff St. Joseph's Hosp., Crouse-Irving Hosp., University Hosp., Meml. Hosp.; mem. cons. School Speech, Syracuse U.; orthodontic cons. N.Y. State Health Dept.; lectr. in field, producer sci. exbns., Anat. Mus. Recipient Hektoen medal AMA, 1970. Diplomate Am. Bd. Orthodontists. Mem. Royal Soc. Medicine, ADA, N.Y. Dental Soc., Syracuse Dental Soc., 5th Dist. Dental Soc., Syracuse C. of C., Northeastern (qualifying com.), Am. orthodontists socs., Pierre Fauchard Acad. Contbg. author textbooks dentistry and orthodontics; contbr. articles to dental publs. Home: 5231 Brockway Ln Fayetteville NY 13066-1705 Office: 1124 E Genesee St Syracuse NY 13210

MARSHALL, FREDERICK JOSEPH, retired research chemist; b. Detroit, Aug. 14, 1920; s. Frederick Joseph and Nora Louise (Orleman) M.; m. Marcella Edith Campbell, Dec. 28, 1946; children: Mary Margaret, Suzanne, Frederick III, Rita, Jane, Timothy, Maureen. BS, U. Detroit, 1941, MS, 1943; PhD, Iowa State U., 1948. Sr. rsch. chemist Eli Lilly & Co., Indpls., 1948-75, rsch. scientist, 1976-83; ret., 1983. Author rsch. papers, jour. articles. Vice pres. Jesuit Alumni Assn., Indpls., 1964-66; treas. Indpls. Cath. Interracial Coun., 1968-74; Dem. precinct vice-committeeman, Indpls., 1975-79. Mem. Am. Chem. Soc., N.Y. Acad. Scis., Alpha Chi Sigma. Achievements include patent on marketed hypoglycemic agent, others. Home: 3120 Shady Grove Ct Indianapolis IN 46222

MARSHALL, GARY SCOTT, pediatrician, educator; b. Lincoln, Nebr., Oct. 15, 1957; s. Leon and Sylvia (Silverman) M.; m. Cherie McKinley, May 19, 1984; children: Emily Corinne, Cullen Spencer. BA, U. Pa., 1979; MD, Vanderbilt U., 1983. Diplomate Nat. Bd. Med. Examiners, Am. Bd. Pediatrics. Resident pediatrics Vanderbilt Children's Hosp., Nashville, 1983-86; fellow infectious diseases Children's Hosp. Phila., 1986-89; asst. prof. pediatrics U. Louisville, 1989—; attending physician Kosair Children's Hosp., Louisville, 1989—. Contbr. articles to profl. jours. Recipient David T. Karzon award Vanderbilt U., 1986, Outstanding Clin. Prof. award U. Louisville, 1992, Young Investigator award U. Louisville, 1992. Fellow Am. Acad. Pediatrics; mem. So. Soc. for Pediatric Rsch., Am. Soc. for Microbiology, Pediatric Infectious Diseases Soc. Achievements include characterization of chronic fatigue in children; description of new clin. syndrome of periodic fever; construction of recombinant adenovirus that expresses a cytomegalovirus protein and characterization of immune response to cytomegalovirus. Office: Dept of Pediatrics U of Louisville Louisville KY 40292

MARSHALL, JEFFREY SCOTT, mechanical engineer, educator; b. Cin., Feb. 10, 1961; s. James C. and Norma E. (Everett) M.; m. Marilyn Jane Patterson, July 16, 1983; children: Judith K., Eric G., Emily J. BS summa cum laude, UCLA, 1983, MS, 1984; PhD, U. Calif., Berkeley, 1987. Asst. rsch. engr. U. Calif., Berkeley, 1988; engr. Creare, Inc., Hanover, N.H., 1988-89; from asst. to assoc. prof. dept. ocean engring. Fla. Atlantic U., 1989-93; assoc. prof. dept. mech. engring., rsch. scientist Iowa Inst. Hydraulic Rsch. U. Iowa, Iowa City, 1993—; dir. hydrodynamics lab. dept. ocean engring. Fla. Atlantic U., 1989—, mem. univ. senate, 1990—. Reviewer Jour. Fluid Mechanics, Physics Fluids A, Acta Mechanica, ASCE Jour. Engring. Mechanics, NSF; contbr. articles to profl. jours. Rsch. grantee ASEE/USN Summer Faculty Rsch. Programs, 1991, 92, ARO Young Investigator Program, 1992, NASA, 1992. Mem. ASME (assoc., Henry Hess award 1992), Am. Phys. Soc. (chmn. tech. session on three-dimensional vortex dynamics fluid dynamics divsn. 1991), Tau Beta Pi. Achievements include research in fluid mechanics, three-dimensional vortex dynamics, coherent structure in turbulent flows, geophysical flows and environmental fluid mechanics. Office: U Iowa Iowa Inst Hydraulic Rsch Dept Mech Engring Iowa City IA 52242

MARSHALL, JOHN, association administrator; b. Sandwich, Mass., June 30, 1917; s. Walton H. and Vira F. (Stowe) M.; m. Anna Silk, May 11, 1961. BA cum laude, Williams Coll., 1939. Trainee Am. Tobacco C., Va., 1939-41; asst. to sr. v.p. European dept. Sugar Serving Machine Co., Europe, Near East and Africa, 1946-56; past pres. Amateur Astronomers Assn., CEO; lectr. in field. Contbr. articles to profl. jours. Ensign USN, 1941-45, ETO. Recipient Excellence in French Studies award Govt. of France, 1939, Amateur Astronomers medal, 1985. Mem. N.Y. Acad. Scis. Home: 1 Gracie Ter New York NY 10028

MARSHALL, JOHN DAVID, forest biologist; b. Fremont, Ohio, Oct. 27, 1956; s. John Vernon and Helene Ann (Wilhelm) M.; m. Elizabeth Ann Aulph, Sept. 20, 1980; children: Lisa Marie, Katharine Theresa, Jacqueline Claire. BS, Mich. State U., 1978, MS, 1980; PhD, Oreg. State U., 1985. Rsch. assoc. Oreg. State U., Corvallis, 1984-85; sr. rsch. scientist GM Rsch. Lab., Warren, Mich., 1985-88; postdoctoral fellow U. Utah, Salt Lake City, 1988-89; instr. Oakland U., Auburn Hills, Mich., 1989; asst. prof. forest resources U. Idaho, Moscow, 1990—; cons. U. Utah, 1990; adj. asst. prof. U. Mich., Ann Arbor, 1987-88; panel mem. USDA Competitive Grants Program, Washington, 1991. Contbr. articles to profl. jours. Grantee NSF, 1985, USDA, 1990, 91, McIntyre Stennis Program, CSRS, USDA, 1990, 92, Stillinger Found., 1990, 91, 92, U. Idaho Seed Grant, 1990, 91. Mem. Soc. Am. Foresters, Ecol. Soc. Am. Achievements include description of relationships between root food reserves and root longevity; definition of upper limits to leaf area index in northwestern U.S. forests; having designed and built pollutant exposure facility for analysis of respiration; quantification of proportional heterotrophy in a xylem-tapping mistletoe; analysis of genetic and environmental variation in water-use efficiency of conifers. Office: Univ of Idaho Dept of Forest Res Moscow ID 83844-1133

MARSHALL, JOHN HARRIS, JR., geologist, oil company executive; b. Dallas, Mar. 12, 1924; s. John Harris and Jessie Elizabeth (Mosley) M.; BA in Geology, U. Mo., 1949, MA in Geology, 1950; m. Betty Eugenia Zarecor, Aug. 9, 1947; children: John Harris III, George Z., Jacqueline Anne Marshall Leibach. Geologist, Magnolia Oil Co., Jackson, Miss., 1950-59, assoc. geologist Magnolia/Mobil Oil, Oklahoma City, 1959-63, dist. and div. geologist Mobil Oil Corp., Los Angeles and Santa Fe Springs, Calif., 1963-69, div. geologist, Los Angeles and Anchorage, 1969-71, exploration supt., Anchorage, 1971-72, western region geologist, Denver, 1972-76, geol. mgr., Dallas, 1976-78, chief geologist Mobil Oil Corp., N.Y.C., 1978-81, gen. mgr. exploration for Western Hemisphere, 1981-82; chief exec. officer Marshall Energetics, Inc., Dallas, 1982—; dir. exploration Anschutz, 1985-91; pres. Madera Prodn. Co., 1992—; mem. Geology Devel. Bd. U. Mo., 1982— . Councilman, City of Warr Acres (Okla.), 1962-63; various positions Meth. Ch., 1951—, Boy Scouts Am., 1960-68; Manhattan adv. bd. Salvation Army, 1980-82. Served with U.S. Army, 1943-46. Recipient U. Mo. Bd. Curators medal, ROTC Most Outstanding Student, 1949; registered geologist, Calif., Wyo. Mem. Am. Assn. Petroleum Geologists (Pacific sect.), Am. Geol. Inst., Petroleum Exploration Soc. N.Y., Dallas Geol. Soc., Rocky Mountain Assn. Geologists, Alaska Geol. Soc., Oklahoma City Geol. Soc., N.Y. Acad. Sci., Los Angeles Basin Geol. Soc. (pres. 1969-70)., Am. Sci. Affiliation, Assn. Christian Geologists, Meth. Men Club, Denver Petroleum Club, Sigma Xi. Democrat. Home: 9526 Moss Haven Dr Dallas TX 75231-2608 Office: Marshall Energetics Inc 8001 LBJ Fwy Ste 400 Dallas TX 75251-1300

MARSHALL, JOSEPH FRANK, electronic engineer; b. Wyoming, Pa., Mar. 2, 1917; s. Anthony Marchel and Mary (Moosic) M.; m. Margaret Mary Kennedy, June 17, 1961. BSEE, Pa. State Coll., 1941; MSEE, Harvard U., 1951. Registered profl. engr., Mass., N.J. Devel. engr., project mgr. Stromberg Carlson Co., Rochester, N.Y., 1941-49; design engr., staff engr. Bell Aircraft Corp., Buffalo, 1952-60; fellow engr. Electronics div. Westinghouse, Balt., 1961-62; sr. staff engr. Avco Corp., Wilmington, Mass., 1962-64; rsch. electrical engr. Cornell Lab., Buffalo, 1964-65; systems engr. Radio div. Bendix, Balt., 1966-67, Raytheon Corp., Sudbury, Mass., 1967-69, Astro Electronics div. RCA, Princeton, N.J., 1969-72; broadcast engr. N.J. Pub. TV, Princeton, 1972-74; sr. staff engr. Office of Engring. Tech. FCC, Washington, 1974-92; with Luthier Acoustic Rsch., Pittsford, N.Y.; IRE subcom. mem. Industry Audio Amplifier Standards, 1944-46; served on EIA TR-8 ad hoc com. Nationwide Cellular Mobile Radio Standards, 1979-80. Violinist Pa. State Coll. Symphony Orch., 1937-40. Fellow Radio Club of Am., Inc.; mem. Inst. Elec./Electronics Inc. (life), Radio Club of Washington, Violin Soc. Am., Catgut Acoustical Soc. Inc., Harvard Club of Rochester, Eta Kappa Nu, Sigma Tau, Tau Beta Pi, Pi Mu Epsilon. Democrat. Roman Catholic. Achievements include patents for selective tuning and damping of partials of rods and method of clamping tunable rods for electronic carillons, for critical components employed in a Navy secure missile command guidance system. Home: 9 Kimberley Rd Pittsford NY 14534

MARSHALL, LARRY RONALD, laser physicist; b. Sydney, Australia, Feb. 24, 1962; came to U.S., 1990; s. Harold Leslie and Kaye (Gelmi) M. BSc, Macquarie U., N.S.W., Australia, 1984, BSc with honors, 1985; PhD, Ctr. Lasers and Applications, N.S.W., Australia, 1989. Physicist Commonwealth Sci. and Indsl. Rsch. Orgn., Sydney, 1984; laser researcher Ctr. Lasers and Applications, Sydney, 1984-89; rsch. scientist Def. Sci. and Tech. Orgn., Adelaide, Australia, 1985-89; laser physicist Fibertek, Inc., Herndon, Va., 1990, mgr. tech. program, 1991—. Author: Life by Misadventure, 1992; contbr. articles to profl. jours. Mem. Australian Embassy Club. Achievements include development of optical parametric oscillators; patents for eyesafe lasers, OPOs, green lasers, CW lasers; high-power lasers for underwater imagery; lasers for submarine communication; power scaling CW diode pumped lasers. Office: Fibertek Inc 510 Herndon Pky Herndon VA 22070

MARSHALL, L(UTHER) GERALD, acoustic consultant; b. Oklahoma City, Sept. 15, 1931; s. Luther N. and Agnes Irene (Gorgas) M.; m. Allie Elizabeth Williams, Oct. 1, 1956; children: Mary Beth Parrott, Rian Luther, Darla Sue Stock, Jerry William. B in Mus. Edn., U. Okla., 1953, M in Mus. Edn., 1954; B in Archtl. Engring., U. Colo., 1965. Musician Okla. City Symphony, 1952-54, U.S. Mil. Acad. Band, West Point, N.Y., 1955-57 Buffalo (N.Y.) Philharmonic, 1957-58, Denver Symphony Orch., 1958-63 structural engr. Ketchum & Konkel, Denver, 1964; acoustic cons. Bolt Beranek & Newman, N.Y.C. and Cambridge, Mass., 1965-71; acoustic cons. Klepper Marshall King Assocs., White Plains, N.Y., 1971—, v.p., sec., 1971—; mem. adv. bd. Concert Hall Rsch. Group, Cambridge, 1991—. Contbr. articles to Symphony News, Theatre Design and Tech., Absolute Sound, Instrumentalist. CEO Empire State Pops Orch., Scarsdale, N.Y.; chmn. Zoning Bd. Appeals, Town of Cortlandt, N.Y., 1975-85, vice chmn. Master Plan Com., 1988—. With U.S. Army, 1954-57. Mem. Acoustical Soc. Am., Am. Fedn. Musicians, Friends of Mozart, Inc., USMA Band Retirees/Alumni Assn. Achievements include development of automated acoustics measurement program based on the early/late sound energy ratio for evaluating auditoriums and related spaces. Home: 17 Hillcrest Dr Cortlandt Manor NY 10566 Office: Klepper Marshall King Assoc 7 Holland Ave White Plains NY 10603

MARSHALL, RICHARD DALE, structural engineer; b. Mandan, N.D., Dec. 12, 1934; s. Vernard Dale and Gertrude (Gordon) M.; m. Virginia Rae Cameron, June 6, 1959; children: Adrian Stanton, Bruce Cameron, Leslie Anne. BSCE, N.D. State U., 1956; MSCE, U. Colo., 1959; PhD, Colo. State U., 1968. Instr. U. Ariz., Tucson, 1959-61; sr. engr. Engring. Cons., Inc., Denver, 1961-63; rsch. asst. Fluid Dynamics & Diffusion Lab., Ft. Collins, Colo., 1963-68; structural engr. Nat. Bur. Standards, Gaithersburg, Md., 1968-78, group leader, 1979-83; rsch. structural engr. Nat. Inst. Standards and Tech. (NIST), Gaithersburg, 1984—; cons. Govt. of Australia, Darwin, 1975; guest researcher Ministry of Constrn., Tsukuba, Japan, 1982; lectr. Annual Engrs. Week, Mont. Co. High Schs., 1989—. Author: (investigative reports) Hyatt Regency Walkways Collapse, 1982, L'Ambiance Plaza Collapse, 1987; mem. editorial bd. Jour. Indsl. Aerodynamics, 1975—; contbr. over 100 articles to profl. jours. Recipient Silver medal U.S. Dept. Commerce, 1975, Gold medal, 1982, Rsch. award Sci. and Tech. Agy., 1982, Excellence award Soc. for Tech. Communication, 1983, Engring. award Nat. Hurricane Conf., 1991. Mem. ASCE (chmn. com. on structural performance 1990—), Wind Engring. Rsch. Coun. (v.p. 1985-90), Sigma Xi. Achievements include patent for Ambient Pressure Probe; development of design loads for U.S. building codes and structural design standards; research into typhoon effects on buildings in the Philippines, on hurricane surface wind speeds in the Caribbean, Atlantic and Gulf coasts. Office: Nat Inst Standards and Tech Gaithersburg MD 20899

MARSHALL, WAYNE KEITH, anesthesiology educator; b. Richmond, Va., Feb. 9, 1948; s. Chester Truman and Lois Ann (Tiller) M.; m. Dale Claire Reynolds, June 18, 1977; children: Meredith Reynolds, Catherine Truman, Whitney Wood. BS in Biology, Va. Poly. Inst. and State U., 1970; MD, Va. Commonwealth U., 1974. Diplomate Am. Bd. Anesthesiology, Nat. Bd. Med. Examiners. Surg. intern U. Cin., 1974-75, resident in surgery, 1975-77; resident in anesthesiology U. Va. Coll. Medicine, Charlottesville, 1977-79, rsch. fellow, 1979-80; asst. prof. anesthesia Pa. State U. Coll. Medicine, Hershey, 1980-86, assoc. prof., 1986—, assoc. clin. dir. oper. rm., 1982—, dir. pain mgmt. svc., 1984—, chief divsn. pain mgmt., 1992—; moderator nat. meetings. Mem. editorial bd. Anesthesiology Rev., 1987—, Jour. Neurosurg. Anesthesiology, 1988—; contbr. articles and abstracts to med. jours. Recipient Antarctic Svc. medal NSF, 1980. Mem. AMA, AAAS, Soc. Neurosurg. Anesthesia and Critical Care (sec.-treas. 1985-87, v.p. 1987-88, pres. 1989-90, bd. dirs. 1985-91), Assn. Univ. Anesthetists, Am. Soc. Anesthesiologists (del. ASA ho. of dels. 1990-92), Internat. Anesthesia Rsch. Soc., Pa. Soc. Anesthesiology, N.Y. Acad. Scis. Republican. Baptist. Avocations: physical fitness, American history, cycling. Office: Pa State U Dept Anesthesiology PO Box 850 Hershey PA 17033-0850

MARSHALL, WILLIAM LEITCH, chemist; b. Columbia, S.C., Dec. 3, 1925; s. William Leitch and Georgia (Kittrell) M.; m. Joanne Fox, Apr. 16, 1949; children: Nancy Diane, William Fox. B.S., Clemson U., 1945; Ph.D., Ohio State U., 1949. Teaching asst. Clemson U., 1944-45, Ohio State U., 1945-46; Naval research fellow Ohio State U., 1947-49; mem. sr. research staff (chemistry) Oak Ridge Nat. Lab., 1949—, research group leader, 1957-75; Plenary lectr. internat. congresses on oceanography, electrochemistry, geochemistry, high temperature water chemistry, high pressure fluids; mem. orgn. coms. internat. sci. congresses. Guggenheim fellow van der Waals Lab., U. Amsterdam, 1956-57. Contbr. articles to profl. jours. Patentee in field. Mem. Am. Chem. Soc. (nat. council 1968-83, nat. membership affairs com. 1980-82, nat. council com. chem. edn. 1970-80, nat. com. chem. edn. 1980-82, chmn. nat. subcom. on high sch. chem. edn. 1970-75, mem. nat. high sch. chemistry com. 1978-81, nat. congl. sci. counselor 1974-83, nat. com. tchr. tng. guidelines 1975-77, Charles Holmes Herty Gold medal 1977), AAAS, Geochem. Soc., Am. Geophys. Union, Sigma Xi (v.p. chpt. 1974-75), N.Y. Acad. Sci., Tenn. Acad. Sci. (vis. scientist program 1975), Internat. Assn. Properties of Water and Steam (working groups 1975—), Internat. Platform Assn., Phi Kappa Phi. Achievements include research on water solutions over wide ranges of temperature and pressure of basic and applied interest. Home: 101 Oak Ln Oak Ridge TN 37830-4046 Office: Oak Ridge Nat Lab Chemistry Div MS 6110 PO Box 2008 Oak Ridge TN 37831-6110

MARSHALL, WILLIS HENRY, psychiatrist; s. Willis Henry Sr. and Pauline Elizabeth (Murphy) M.; m. Carolyn Mae Kowalski; children: Louann Lorinda Marshall Johnson, John Willis. AB cum laude, U. Evansville, 1957; MD, Ind. U., 1961. Intern Detroit Meml. Hosp., 1961-62; resident psychiatry Mental Health Inst., Cherokee, Iowa, 1965-67, 69-70, staff psychiatrist, 1967-69; staff psychiatrist Mental Health Ctr., Muskegon, Mich., 1970-71, Grand Haven, Mich., 1971-73; pvt. practice psychiatry Madison, Tenn., 1974-85; staff psychiatrist evaluation unit forensic svcs. div. Mid. Tenn. Mental Health Inst., Nashville, 1985-87, chief of staff, 1986-87; forensic psychiatrist State of Tenn., 1985-87; pvt. practice psychiatry Bowling Green, Ky., 1987—; psychiat. cons. Allegan County Mental Health Ctr., Allegen, Mich., 1973; med. svcs.. cons. dept. of forensic svcs. Mid. Tenn. Mental health Inst., Nashville, 1983-84; pvt. practice psychiatry, Muskegon, Mich., 1970-74, Madison, 1986-87; assoc. clin. dir. mental health unit Med. Ctr., Bowling Green, Ky., 1987-91; staff psychiatrist Lifeskills, Inc., Glasgow,

Ky., Franklin, Ky., 1987-89; acting med. dir. Rivendell Children's Psychiat. Hosp., Bowling Green, 1989. Commd. officer, surgeon USPHS, 1962-65. Recipient AMA Physicians Recognition award, 1969, 79, 83, 86, 89. Mem. Am. Psychiat. Assn. (art assn. 1976—), Ky. Med. Assn., Warren County Med. Soc., Am. Profl. Practice Assn., Am. Acad. Clin. Psychiatrists, Am. Physicians Art Assn., NRA, Nat. Geog. Soc., AAA Automobile Club, Gallatin Gun Club, Alpha Omega Alpha. Avocations: sculpture, photography, painting, hunting, fixing old guns. Home: 315 Matlock Pike Bowling Green KY 42104-7431 Office: Commonwealth Medical Plaza 720 2d St Ste 207 Bowling Green KY 42101-1706

MARSIK, FREDERIC JOHN, microbiologist; b. Camden, N.J., June 22, 1943; s. Ferdinand Vincent and Helen (Reidl) M.; children: Teri Jean, Kristi Ann Marsik McCann. BA, Lebanon Valley Coll., 1965; MS, U. Mo., 1970, PhD, 1973. Diplomate Am. Bd. Med. Microbiology. Asst. prof. Sch. Medicine, U. Va., Charlottesville, 1976-80; tech. dir. microbiology and serology Children's Hosp. Wis., Milw., 1980-84; assoc. prof. microbiology and internal medicine Sch. Medicine, Oral Roberts U., Tulsa, 1984-87; dir. microbiology Crozer-Chester Med. Ctr., Upland, Pa., 1987-88; mgr. media systems Becton Dickinson Microbiology Systems, Cockeysville, Md., 1988—; mem. adv. com. Milw. Area Tech. Coll., Milw., 1983-84, Tulsa Jr. Coll., 1985-87; mem. rev. bd. Clin. Lab. Sci. Publ., Washington, 1990—. Contbr. chpts. to textbooks. Treas. Rose Fire Co. and Ambulance Svc., New Freedom, Pa., 1989—; bd. govs. New Freedom Community Ctr., 1989—; mem. adult edn. com. So. York County Sch. Dist., Glen Rock, Pa., 1989—. Lt. col. USAR. Recipient Best Rsch. Project award S.W. Assn. for Clin. Microbiology, 1984. Mem. Am. Soc. Microbiology (mem. lab. practices com. 1990—), Am. Soc. Med. Tech., N.Y. Acad. Scis. Congregationalist. Avocations: fishing, camping, basketball. Home: 6 Keesey Rd New Freedom PA 17349-9638 Office: Becton Dickinson 250 Schilling Cir Cockeysville Hunt Valley MD 21030

MARTEGANI, ENZO, molecular biology educator; b. Legnano, Milan, Italy, Nov. 14, 1950; s. Ferruccio and Libera (Macchi) M.; m. Giovanna Berruti, Dec. 6, 1975; 1 child, Alessandro. Laurea Scienze Biologiche, U. Milan, 1973, diploma perfezionamento rsch. biology, 1975. Postdoctoral fellow plant molecular biology Consiglio Nationale delle Ricerche, Milan, 1975-77; asst. in plant physiology U. Milan, 1978-79, asst. in biochemistry, 1979-83, assoc. prof. molecular biology, 1983—; cons. biologist Istituto Farmacobiologico Giustini, Orago, 1976-81. Contbr. articles to internat. jours. and books. Mem. Soc. Math. Biology, Soc. Italiana Biofisica e Biologia Molecolare, Gruppo Italiano di Citometria, Soc. Italiana di Biochimica, N.Y. Acad. Sci. Avocations: computer graphics, astronomy, photography, music. Office: U Milan Dept Fisologia e, Biochimica Generali, Via Celoria 26, 20133 Milan Italy

MARTENS, FREDERICK HILBERT, nuclear engineer; b. Peoria, Ill., Dec. 15, 1921; s. Hilbert Christian and Anna Danelia (Seebergen) M.; m. Carolyn Lee Arnold, Aug. 23, 1947; children: Phillip A., Christine L., Susan E. (dec.). BS, U. Chgo.; MS, U. N.Mex. Cert. quality engr. Staff physicist U. N.Mex., Albuquerque, 1948-50; various poisitions Argonne (Ill.) Nat. Lab., 1950-83; cons. Nat. Bur. Stds., Gaithersburg, Md., 1966—. Contbr. articles to profl. jours. 2d lt. U.S. Air Corps, 1942-45. Mem. Am. Interprofl. Inst., Am. Nuclear Soc., Sigma Xi. Lutheran. Achievements include patent on fuel assay reactor. Home: 502 Union St Plainfield IL 60544

MARTIC, PETER ANTE, research chemist; b. Split, Croatia, July 6, 1938; came to U.S., 1968; s. Ante P. and Antonietta (Hanzalek) M.; children: Bosiljka, Karla, Katarina. MS, Okla. State U., 1966. Rsch. chemist Eastman Kodak Co., Rochester, N.Y., 1968—. Contbr. articles to profl. jours. Roman Catholic. Achievements include patents in radiographic elements and binders; in photographic element with optical brighteners having reduced migration. Home: 40 Elwell Dr Rochester NY 14618

MARTICORENA, ERNESTO JESUS, mechanical engineer, educator; b. Havana, Cuba, Mar. 14, 1941; came to U.S., 1962; s. Ernesto and Rosa (Rodil) M.; m. Otilia Maria Mejias, Nov. 7, 1964; children: Ernesto J., Otilia. B Engring., Stevens Inst. Tech., 1968, MME, 1970. Registered profl. engr., mcpl. engr., profl. planner, N.J. Plant engring. supt. N.L. Industries, Inc., Sayreville, N.J., 1968-75; mgr. mech. engring. C.F. Braun and Co., Murray Hill, N.J., 1975-82; dir. engring. White Storage and Retrieval Systems, Kenilworth, N.J., 1982-87; mcpl. engr. City of Elizabeth, N.J., 1987—; pres. Mero Engring., Inc., Elizabeth, 1987—; adv. com. dept. engring. Union County Coll., Elizabeth, 1988—. Mem. ASME, NSPE, N.J. Soc. Profl. Engrs., N.J. Soc. Mcpl. Engrs. Office: City of Elizabeth 50 Winfield Scott Plz Elizabeth NJ 07201

MARTIN, ARCHER JOHN PORTER, retired chemistry educator; b. London, Mar. 1, 1910; s. William Archer Porter and Lilian Kate (Brown) M.; m. Judith Bagenal, Jan. 9, 1943; 5 children. Student, Peterhouse, Cambridge, Eng., 1929-32, MA, 1936; PhD, DSc, Leeds U., 1968; LLD (hon.), U. Glasgow, Scotland, 1973; laurea honoris causa, U. Urbino, Italy, 1985. Chemist Nutritional Lab., Cambridge, 1933-38, Wool Industries Rsch. Assn., Leeds, 1938-46; mem. rsch. dept. Boots Pure Drug Co., Nottingham, Eng., 1946-48; mem. staff Med. Rsch. Coun., 1948-52; head phys. chemistry div. Nat. Inst. Med. Rsch., Mill Hill, 1952-56; chem. cons., 1956-59; dir. Abbotsbury (Eng.) Labs. Ltd., 1959-73; profl. fellow U. Sussex, Eng., 1973-77; Robert A. Welch prof. chemistry U. Houston, 1974-78; guest prof. Ecole Polytechnique Fed. de Lausanne, Switzerland, 1980-85; cons. Wellcome Rsch. Labs., Beckenham, Eng., 1970-73; Extraordinary prof. Tech. U., Eindhoven, The Netherlands, 1965-74. Decorated comdr. Brit. Empire; Order of Rising Sun Japan; recipient Berzelius Gold medal Swedish Med. Soc., 1951, Nobel prize chemistry (with R.L.M. Synge) for invention of partition chromatography, 1952, John Scott award, 1958, John Price Wetherill medal, 1959, Franklin Inst. medal, 1959, Koltoff medal Acad. Pharm. Sci., 1969, Callendar medal Inst. Measurement & Control, 1971, Fritz-Pregl medal Austrian Chemical Soc., 1985. Fellow Royal Soc. (Leverhulme medal 1964). Club: Chemist's (N.Y.C.) (hon.). Home: 47 Roseford Rd, Cambridge CB4 2HA, England

MARTIN, CARL NIGEL, biotechnologist; b. Leeds, Yorkshire, Eng., Oct. 23, 1949; s. Charles Musgrave and Muriel (Barrand) M.; m. Kathleen Dianne Smith, Aug. 21, 1976; children: Fraser Charles, Robyn Alexandra. BSc with honors, Leeds (Eng.) U., 1972; PhD, U. London, 1975. Cert. toxicologist. Rsch. scientist York (Eng.) U., 1975-85; project leader Nat. Ctr. Toxicology Rsch., Ark., 1981; head biotech. svcs. Hazleton, Harrogate, Eng., 1985—. Author: Biochemical Toxicology, 1987; author: (with others) Chemical Carcinogens & DNA, 1979, Chemical Carcinogens, 1985; editor Sci. Jour., 1980-89. Councillor Thorganby (Eng.) Parish Coun., 1989—. Mem. Royal Coll. Pathologists. Achievements include patents (with others) for novel immunological method of product protection against counterfeiting; discovery of structure of benzidine-DNA adduct, of structure of MOCA-DNA adduct; development of viral screening assays for biopharmaceuticals. Office: Hazleton Europe, Otley Rd, Harrogate North Yorkshire HG3 1PY, England

MARTIN, CHARLES RAYMOND, chemist, educator; b. Cin., Nov. 20, 1953; s. Samuel James and Flora Joy (Clifton) M.; m. Deborah Susan Greco, May 2, 1987. BS, Centre Coll., Danville, Ky., 1975; PhD, U. Ariz., 1980. Asst. prof. chemistry Tex. A&M U., College Station, 1981-85, assoc. prof., 1985-87, prof., 1987-90; prof. chemistry Colo. State U., Ft. Collins, 1990—; cons. Dow Chem. Co., Midland, Mich., 1983—. Editor: Functional Polymers, 1990; contbr. peer-reviewed rsch. articles to sci. publs. Mem. AAAS, Am. Chem. Soc., Electrochem. Soc., Materials Rsch. Soc. Achievements include fundamental research in membrane science, electronically conductive polymers, electrochemistry.

MARTIN, CHRISTINE KALER, chemist; b. Detroit, Nov. 15, 1958; d. John Bailey and Marion Elizabeth (Phillips) Kaler; m. Michael Kent Martin, Sept. 6, 1980; children: Kent Alexander, Brock Andrew Lee. BA, Va. Tech., 1980. Rsch. asst. Exxon Chem., Houston, 1981-82; analytical chemist Witco Chem., Houston, 1982-83; chromatography tech. support Varian, Houston, 1983-84; supr. analytical labs. 3M, St. Paul, 1984-89; mgr. color tech. B.F. Goodrich, Avon Lake, Ohio, 1989—. Mem. Ohio Acad. Sci. Exemplar, Iota Sigma Pi. Office: B F Goodrich PO Box 122 Avon Lake OH 44012

MARTIN, CLYDE VERNE, psychiatrist; b. Coffeyville, Kans., Apr. 7, 1933; s. Howard Verne and Elfrieda Louise (Moehn) M.; m. Barbara Jean McNeilly, June 24, 1956; children: Kent Clyde, Kristin Claire, Kerry Constance, Kyle Curtis. Student Coffeyville Coll., 1951-52; AB, U. Kans., 1955; MD, 1958; MA, Webster Coll., St. Louis, 1977; JD, Thomas Jefferson Coll. Law, Los Angeles, 1985. Diplomate Am. Bd. Psychiatry and Neurology. Intern, Lewis Gale Hosp., Roanoke, Va., 1958-59; resident in psychiatry U. Kans. Med. Ctr., Kansas City, 1959-62, Fresno br. U. Calif.-San Francisco, 1978; staff psychiatrist Neurol. Hosp., Kansas City, 1962; practice medicine specializing in psychiatry, Kansas City, Mo., 1964-84; founder, med. dir., pres. bd. dirs. Mid-Continent Psychiat. Hosp., Olathe, Kans., 1972-84; adj. prof. psychology Baker U., Baldwin City, Kans., 1969-84; staff psychiatrist Atascadero State Hosp., Calif., 1984-85; clin. prof. psychiatry U. Calif., San Francisco, 1985—; chief psychiatrist Calif. Med. Facility, Vacaville, 1985-87; pres., editor Corrective and Social Psychiatry, Olathe, 1970-84, Atascadero, 1984-85, Fairfield, 1985—. Contbr. articles to profl. jours. Bd. dirs. Meth. Youthville, Newton, Kans. 1965-75, Spofford Home, Kansas City, 1974-78. Served to capt. USAF, 1962-64, ret. col. USAFR. Oxford Law & Soc. scholar, 1993. Fellow Am. Psychiat. Assn., Royal Soc. Health, Am. Assn. Mental Health Profls. in Corrections, World Assn. Social Psychiatry, Am. Orthopsychiat. Assn.; mem. AMA, Assn. for Advancement Psychotherapy, Am. Assn. Sex Educators, Counselors and Therapists (cert.), Assn. Mental Health Adminstrs. (cert.), Kansas City Club, Masons, Phi Beta Pi, Pi Kappa Alpha. Methodist (del. Kans. East Conf. 1972-80, bd. global ministries 1974-80). Office: PO Box 3365 Fairfield CA 94533-0587

MARTIN, DAVID CHARLES, materials science engineering educator; b. Kalamazoo, Sept. 24, 1961; s. Ernest Charles and Shirley Jean (Calkins) M.; m. Kim Louise, May 12, 1990; children: Nathaniel Ernest, Timothy Walker. MS, U. Mich., 1985; PhD, U. Mass., 1989. Process devel. staff IBM, Burlington, Vt., 1983; rsch. engr. GM Corp., Warren, Mich., 1983-85; vis. scientist E.I. duPont De Nemours & Co., Inc., Wilmington, Del., 1989-90; asst. prof. materials sci. U. Mich., Ann Arbor, 1990—. Named Nat. Young Investigator NSF, 1992. Mem. Am. Soc. Metals Internat. (chpt. ops. com. 1992), TMS (exec. com. 1992, pres. Detroit chpt.), Am. Phys. Soc., Am. Chem. Soc., Materials Rsch. Soc., Microscopy Soc. Am. Home: 1660 Fulmer St Ann Arbor MI 48103-2453 Office: Univ of Mich 2022 H H Dow Bldg Ann Arbor MI 48109-2136

MARTIN, DENNIS CHARLES, dentist; b. Osage, Iowa, Sept. 30, 1960; s. Charles H. and Sandra L. (Skou) M.; m. Lori Jane Reisel, July 10, 1982; children: Andrew, Michael. DDS, U. Nebr., 1986. Dentist Ridgeview Dental Office, Clinton, Iowa, 1986-; owner, mgr. Toy Cellar, Clinton, 1991—. Recipient Progress award River City C of C., Clinton, 1986. ADA, Iowa Dental Assn., Am. Acad. Gen. Dentistry, Chgo. Dental. Soc., River City C. of C., Lions. Republican. Lutheran. Avocations: computer programming, radio control airplane modeling. Home: RR 2 Box 163 Clinton IA 52732-9636 Office: Ridgeview Dental Office 4023 Hwy 136 Clinton IA 52732-9636

MARTIN, FRANK SCOTT, chemist; b. Shreveport, La., Dec. 30, 1960; s. Dorothy (Martin) Rice. BA, Rutgers Univ., 1984. Tech. Permabond Internat. Nat. Starch, Plainfield, N.J., 1985-88; sr. tech., jr. chemist Microporous Materials Nat. Starch, Plainfield, N.J., 1988-90; chemist, tech. Norland Products Inc, New Brunswick, N.J., 1990—. Contbr. articles to profl. jours. Vice-chmn. Plainfield Action Svcs., 1986; steward Mt. Zion Ch., 1987; vice-chmn. Plainfield Bd. Edn., 1988. Named Outstanding Young Man of Am., OYM, 1988; recipient Commendation award City of Plainfield, 1988, Plainfield Bd. Edn., 1989. Mem. Am. Chem. Soc. (exec. com. N.J. sect.), Am. Mgmt. Assn., Young Chemist Com. (chmn. 1992-93). Democrat. Achievements include patent for form in place gasket, white cyamocarylate, synthesign polyhipe, development of 5 proprietary adhesive formulations. Office: Norland Products Inc 695 Joyce Kilmer Ave New Brunswick NJ 08901-3307

MARTIN, FREDDIE ANTHONY, agronomist, educator; b. Raceland, La., Nov. 17, 1945; s. of Abraham and Flossie Margarette (Foret) M.; m. Rose Ann Hill, Aug. 23, 1969; children: Samson, Jonathan, Robert. BS, Francis T. Nicholls State Coll, Thibodaux, La., 1966; MS, Cornell U., 1968, PhD, 1970. Asst. prof. Plant Pathology Dept. La. State U., Baton Rouge, 1971-76; assoc. prof. Plant Pathology Dept. La. State U., 1976-80, prof. Plant Pathology Dept., 1980, prof. Agronomy Dept., 1980—, head Sugar Sta./ Audubon Sugar, 1988—. Editor: profl. jour. Am. Soc. Sugar Cane Technologists, 1980—; author: profl. manuscripts, jours. Recipient Rsch. Excellence Award, La. Agricultural Experimental Sta., 1984, Svc. Award, St. James Sugar Growers, 1989. Mem. Am. Soc. Agronomy, Crop Sci. Soc. Am., Am. Soc. Sugarcane Tech., Gamma Sigma Delta. Office: Louisiana State Univ Sugar Station Sugar Sta/Audobon Sugar Inst Baton Rouge LA 70803

MARTIN, GEORGE M., pathologist, gerontologist; b. N.Y.C., June 30, 1927; s. Barnett J. and Estelle (Weiss) M.; m. Julaine Ruth Miller, Dec. 2, 1952; children: Peter C., Kelsey C., Thomas M., Andrew C. BS, U. Wash., 1949, MD, 1953. Diplomate Am. Bd. Pathology, Am. Bd. Med. Genetics. Intern Montreal Gen. Hosp., Quebec, Can., 1953-54; resident-instr. U. Chgo., 1954-57; instr.-prof. U. Wash., Seattle, 1957—; vis. scientist Dept. Genetics Albert Einstein Coll., N.Y.C., 1964; chmn. Gordon Confs. Molecular Pathology, Biology of Aging, 1974-79; chmn., nat. res. Plan on Aging Nat. Inst. on Aging, Bethesda, Md., 1985-89; dir. Alzheimer's Disease Rsch. Ctr. U. Wash., 1985—. Editor Werner's Syndrome and Human Aging, 1985; contbr. articles in field to profl jours. Active Fedn. Am. Scientists. With USN, 1945-46. Recipient Allied Signal award in Aging, 1991; named Disting. alumnus U. Wash. Sch. Medicine, 1987; USPHS rsch. fellow dept. genetics Glasgow U., 1961-62; Eleanor Roosevelt Inst. Cancer Rsch. fellow Inst. de Biologie, Physiologie, Chimie, Paris, 1968-69; Josiah Macy faculty scholar Sir William Din Sch. Pathology, Oxford (Eng.) U., 1978-79; Humboldt Disting. scientist dept. genetics U. Wurzburg, Germany, 1991. Fellow AAAS, Gerontological Soc. Am. (chmn. Biol. Sci. 1979, Brookdale award 1981), Tissue Culture Assn. (pres. 1986-88); mem. Inst. Medicine, Am. Assn. Univ. Pathologists (emeritus), Am. Soc. Human Genetics, Am. Assn. Pathologists. Democrat. Avocations: internat. travel, jazz music, biography. Home: 2223 E Howe Seattle WA 98112 Office: U Wash Sch Medicine Dept Pathology SM 30 Seattle WA 98195

MARTIN, GEORGE REILLY, federal agency administrator; b. Boston, Jan. 20, 1933. BS, Colgate U., 1955; PhD, U. Rochester, 1958. Rsch. asst. U. Rochester, N.Y., 1955-58; guest researcher LCP, Nat. Heart Inst., NIH, Bethesda, Md., 1958-59; pharmacologist Nat. Inst. Dental Rsch., NIH, Bethesda, 1959-66, Weizmann Inst. Sci., Rehovet, Israel, 1966-67; sect. chief CTS/LBC Nat. Inst. Dental Rsch., NIH, Bethesda, 1967-74, acting chief LBC/LDBA, 1972-74; with Max Planck Institut fur Biochemie, Munich, 1975-76; lab. chief Lab. Devel. Biol. and Anomalies, NIDR, NIH, Bethesda, 1974-88; sci. dir. Nat. Inst. on Aging, NIH, Balt., 1988—. Office: NIH Gerontology Rsch Ctr NIA Bldg 31 Rm 2C04 4940 Eastern Ave Baltimore MD 21224

MARTIN, H(ARRY) LEE, mechanical engineer, robotics company executive; b. Nashville, Mar. 10, 1956; s. Paul Walter and Kathleen May (Lee) M.; 1 child, Nina Marie. BSME, U. Tenn., 1978, PhD in Engring., 1986; MSME, Purdue U., 1979. Registered profl. engr., Tenn. Devel. engr. Oak Ridge (Tenn.) Nat. Lab., 1980-86; pres., founder TeleRobotics Internat., Inc., Knoxville, Tenn., 1986—; cons. Remotec, Oak Ridge, 1985-86. Editor: Teleoperated Robotics in Hostile Environments, 1985; contbr. over 20 papers on controls for remote robotic systems, 1983-90. Trombonist Oak Ridge Brass Quintet, 1983—, Knoxville Symphony Orch., 1978. Recipient Indsl. Rsch. 100 award Indsl. Rsch. Mag., 1984. Mem. NSPE (named Nat. Young Engr. of Yr. 1989, mem. adv. rev. bd. 1989-90), Soc. Mfg. Engrs. (named Outstanding Mfg. Engr. 1986), Am. Nuclear Soc. Methodist. Achievements include patents for State of Charge Indicator for Electric Vehicles, Advanced Servomanipulator System, Multimorphic Kinematics for Manipulator, and Bilateral Simulated End-Effector System. Office: TeleRobotics Internat Inc 7325 Oak Ridge Hwy Knoxville TN 37931-3476

MARTIN, HORACE FELECIANO, pathologist, law educator; b. St. Miquel, Azores, Portugal, Jan. 11, 1931; came to U.S., 1941; s. Manuel Feleciano and Maria (Rapozo) M.; m. Florence Jadach, Nov. 25, 1954; children: Michael, Paul, Kathleen, Peter, Susan, John, Mary. BS in Biology, Providence Coll., 1953; MS, U. R.I., 1955; PhD, Boston U., 1963; MS (hon.), Brown U., 1965, MD, 1975; JD, New England Sch. Law, 1991. Diplomate Am. Bd. Med. Examiners, Am. Bd. Clin. Chemistry. Rsch. group leader Monsanto Rsch., Boston, 1957-59; dir. clin. chemistry R.I. Hosp., Providence, 1963—; prof. pathology Brown U., Providence, 1963—; adj. prof. law Providence Coll., 1991—; mem. device panel FDA, Washington, 1989—. Author: Determination of Normal Values, 1975. Named Man of Yr., Commn. of Volunteerism. Fellow Coll. Legal Medicine; mem. Maine Toxicological Inst. (bd. dirs. 1992), Brown U. Med. Alumni (pres. 1991), Sigma Xi. Home: 57 Spring St Pawtucket RI 02860 Office: RI Hosp APC-1142-A Providence RI 02903

MARTIN, JAMES ARTHUR, aerospace educator; b. Ft. Benning, Ga., Aug. 9, 1944; s. Paul Arthur and Mildred Ruby (McDowell) M.; m. Carol Sue Feather, Dec. 31, 1966; children: Fredrick, Heather, Andrew. MS in Aeronautics and Astronautics, MIT, 1967; DSc, George Washington U., 1982. Aerospace engr. NASA Langley Rsch. Ctr., Hampton, Va., 1967-90; assoc. prof. U. Ala., Tuscaloosa, 1991—. Assoc. editor AIAA Jour. Spacecraft and Rockets, Washington, 1985—. Achievements include 4 patents on earth-to-orbit vehicles and rocket propulsion. Home: 11571 Country Club Dr Northport AL 35476 Office: Aerospace Engineering U Ala Tuscaloosa AL 35487-0280

MARTIN, JAMES CULLEN, chemistry educator; b. Dover, Tenn., Jan. 14, 1928; s. Joseph Emmett and Myrtle (Futrell) M.; m. June Echols, Aug. 25, 1951; children: Joseph Vinson, Bruce Thomas, Kendall James, Steven Echols, Christopher Cullen. BA, Vanderbilt U., 1951, MS, 1952; PhD, Harvard U., 1956. Instr. organic chemistry U. Ill., Urbana, 1956-59, asst. prof., 1959-62, assoc. prof., 1962-65, prof. chemistry, 1965-85; Disting. prof. chemistry Vanderbilt U., Nashville, 1985—; mem. adv. com. on chemistry NSF, Washington, 1986-89. Served with U.S. Army, 1946-48. Recipient Alexander von Humboldt U.S. Sr. Scientist award Humboldt Stiftung, Bonn, Fed. Republic of Germany, 1979, 88; Alfred P. Sloan fellow, 1962, Guggenheim fellow, 1966. Fellow AAAS, Japan Soc. for Promotion Sci.; mem. Am. Chem. Soc. (com. publs. 1978-87, chem. abstracts svc. com. 1988-90, chmn. organic div. 1981-82, editorial adv. bd. 1986-87, Buck-Whitney medal Ea. N.Y. chpt. 1979), Internat. Union Pure and Applied Chemistry (U.S. nat. com. Washington, 1985-88, observer Commn. on Nomenclature of Organic Chemistry 1991—). Avocations: camping, singing. Home: 5031 Lakeview Dr Nashville TN 37220-1407

MARTIN, JAMES WALTER, chemist, technology executive; b. Easton, Pa., May 14, 1927; s. William F. and Lucinda (Schwar) M.; m. Elizabeth Donovan, Sept. 15, 1951; children: Teresa, Carol, Rosemary, James Jr., Melissa. BS, Mt. St. Mary's Coll., 1951. Chemist William Zinsser Co. Inc., N.Y.C., 1952-59, sr. chemist, 1959-65; tech. dir. William Zinsser Co. Inc., Somerset, N.J., 1965-70, dir. rsch. and devel., 1970-81; v.p. Bradshaw Praeger Co., Chgo., 1981—; v.p. tech. William Zinsser Co. Inc., Somerset, 1981—. Contbr. chpts. to books, articles to profl. jours. With USN, 1944-46. Fellow AICs; mem. ACS, Soc. Plant Tech., ASTME (chmn. 1965—). Roman Catholic. Home: 158 Hope Rd Blairstown NJ 07825

MARTIN, JEFFREY ALAN, chemical company executive; b. Hanover, Pa., Aug. 14, 1953; s. Wilbur James and Agnes (McGinnis) M.; m. Suzanne Helscher, May 19, 1979; children: Laura Elizabeth, Adam Charles, Jeffrey McGinnis. BS in Mktg., U. Del., 1976. Sales coord. I.C.I. Americas, Wilmington, Del., 1977-80; sales rep. CYRO Industries, Wallingford, Conn., 1980-81, Mobil Chem. Corp., Paramus, N.J., 1981-84; regional sales mgr. Huntsman Chem. Corp., Reston, Va., 1984-87, Berwyn, Pa., 1984-87; gen. sales mgr. Huntsman Chem. Corp., Chesapeake, Va., 1987-89, dir. internat. accts., 1989; v.p., Asia Pacific region Huntsman Chem. Corp., Hong Kong, 1989-91; regional sales mgr. Huntsman Chem. Corp., Hinsdale, Ill., 1991-92, dir. of sales, 1993—. Republican. Methodist. Home: 444 S Clay St Hinsdale IL 60521-4036 Office: Huntsman Chem Corp 15 Salt Creek Ln Ste 410 Hinsdale IL 60521-2965

MARTIN, JOANNE DIODATO, consulting engineer; b. Harrisburg, Pa., May 6, 1959; d. Joseph Louis and Helen (Halapy) Diodato; 1 child, James. BS, Pa. State U., 1981. Nuclear analyst Pa. State U., University Park, 1979-81; nuclear engr. Gen. Physics Corp., Limerick, Pa., 1981-83; tng. specialist Pub. Svc. Electric and Gas, Salem, N.J., 1983-84; simulator test dir. Modification Systems, Inc., Columbia, Md., 1984-85; radiation cons. CDS Techs., Inc., Allentown, Pa., 1984-86; lab. dir. DMA Radtech, Inc., Allentown, 1986-91; ind. cons. engr. Hanover, Pa., 1984—. Author: BWR Reactor Theory, 1983; contbr. articles to profl. jours. Mem. ASTM, Pa. State Engring. Soc. (bd. dirs. 1991-92), Am. Nuclear Soc. (sect. v.p. 1980-81), Health Physics Soc. (legis. com. 1990-92), Am. Assn. Radon Scientists and Technologists (chair ethics com. 1988-91). Republican. Roman Catholic. Achievements include being the first U.S. woman certified as BWR senior reactor operator instructor.

MARTIN, JOANNE LEA, computer scientist, researcher; b. Balt., Aug. 17, 1954; d. Gruver Howard and Lillian (Jackson) M.; m. Brian David Crawford, Mar. 27, 1976; children: Jonathan David Martin-Crawford, Kelly Michelle Martin-Crawford, Courtney Elise Martin-Crawford. BA summa cum laude, Appalachian State U., 1975; MA in Math., Johns Hopkins U., 1977, PhD in Math., 1981. Grad. teaching fellow dept. math. Johns Hopkins U., Balt., 1976-81; instr. dept. of math Loyola Coll., Balt., 1978-81; mem. rsch. staff Los Alamos (N.Mex.) Nat. Lab., 1981-84; mem. rsch. staff T.J. Watson Rsch. Ctr. IBM, Yorktown Heights, N.Y., 1984-87; staff asst. to v.p. for Eng./Sci. Processing, Data Systems Div. IBM, White Plains, N.Y., 1987-88; sr. systems analyst Data Systems Div. IBM, Kingston, N.Y., 1987—; project mgr. high performance systems studies IBM, Kingston, 1988-93; sr. mem. tech. staff IBM, Kingston, N.Y., 1993—; Mem. NSF Program Adv. Com. Div. of Advanced Sci. Computing, 1989— (chmn. 1990—), NAS Com. on Supercomputer Performance and Devel., 1986, IEEE/Assn. Computer Machinery Supercomputing 'XY Confs., 1988— (mem. steering com. 1988—, dep. chair 1989, gen. chair 1990), Nat. Rsch. Coun. panel Computer and Applied Math. for Nat. Inst. Sci. and Tech., 1988—; lectr. on supercomputign various confs. and workshops, 1985—. Founding editor, editor-in-chief Internat. Jour. Supercomputer Applications, 1987—. Mem. IEEE Computer Soc. (TC Chair 1988-90), Assn. Computer Machinery (Recognition of Service award, 1989). Democratic. Presbyterian. Office: IBM Corp PO Box 218 Mail Stop 601 Neighborhood Rd Kingston NY 12401

MARTIN, JOEL JEROME, physics educator; b. Jamestown, N.D., Mar. 27, 1939; s. Clarence and Marian (Stelter) M. BS in Physics, S.D. State U., 1961, MS in Physics, 1962; PhD in Physics, Iowa State U., 1967. Postdoctoral fellow Ames Lab. Iowa State U., Ames, 1967-69; asst. prof. Okla. State U., Stillwater, 1969-73, assoc. prof., 1973-79, prof., 1979—. Contbr. articles to profl. jours. Mem. Am. Phys. Soc., Am. Assn. Physics Tchrs., Am. Assn. for Crystal Growth. Office: Okla State U Dept Physics Stillwater OK 74078

MARTIN, JOHN BRAND, engineering educator, researcher; b. Durban, South Africa, Apr. 20, 1937; s. Edwin and Mavis Elizabeth (Wilkinson) M.; m. Nancy Jill Hickman, Sept. 5, 1961; children: Neil, Andrew, Peter. BSc in Engring., Natal U., South Africa, 1957; PhD, Cambridge U., U.K., 1962; DSc, Natal U., 1978. Profl. engr., South Africa. Asst. prof., assoc. prof. & prof. engring. Brown U., Providence, 1962-72; Corp. prof. civil engring. U. Cape Town, South Africa, 1973-82, dean faculty of engring., 1983—; dir. info. tech. U. Cape Town, 1984-91; chmn. Coun. for Nuclear Safety, South Africa, 1986—, Finite Element Analysis Svcs., Cape Town, 1988—. Author: Plasticity, 1976; contbr. over 120 articles to profl. jours. Fellow U. Cape Town, 1981; recipient Percy Fox Found. ann. award, 1992. Fellow ASME, South Africa Inst. Civil Engrs., Acad. of Engrs., South Africa Soc. Profl. Engrs., Royal Soc. South Africa (John FW Herschel Medal 1990).

MARTIN, JOHN RICHARD, pharmacology educator, researcher; b. Mpls., Dec. 26, 1951; s. Roderick John and Donna Rae (Crawbuck) M.; m. Moira Jean Johnson, Mar. 31, 1979; children: Sean Charles, Sarah Elizabeth, Robert John. BS, U. Calif., Davis, 1975; MS, U. Pacific, Stockton, Calif., 1979; PhD, U. Minn., 1985. Postdoctoral fellow St. Louis U. Sch. Medicine, 1985-88; asst. prof. pharmacology Kirksville (Mo.) Coll. Osteo. Medicine,

1988-93, assoc. prof. pharmacology, 1993—. Co-author book chpt., 1991; contbr. articles to profl. jours. Deacon First Presbyn. Ch., Kirksville, 1991—; cubmaster Cub Scout Pack 22, Kirksville, 1992; fundraiser, coach Kirksville Soccer Club, 1992. Recipient Scholarship award N. Calif. chpt. Achievement Rewards for Coll. Scientists Found., 1989, Mo. Affiliate Am. Heart Assn. grant, 1989-90, 90-91, Acad. Rsch. Enhancement award NIH, 1991-94, Rsch. Experience for Undergrads., Nat. Sci. Found., 1992-93, 93-95. Mem. Am. Soc. Pharmacology and Exptl. Therapeutics, Soc. Neurosci., Am. Heart Assn. (basic. sci. coun.), Sigma Xi. Achievements include research of presence of supersensitive dopamine receptors in hypothalamus of spontaneously hypertensive rats; of a single admininstration of opiate can cause development of supersensitive dopamine receptors in mice brain; in microinjection into posterior hypothalamus of neuropeptide Y increases blood pressure of rat; of muscarinic M3 receptor mediates increase in blood pressure by microinjection of carbachol into posterior hypothalamus of rat. Office: Kirksville Coll Osteo Medicine Dept Pharmacology 800 W Jefferson St Kirksville MO 63501

MARTÍN, JOSÉ GINORIS, nuclear and solar energy engineer, educator; b. Havana, Cuba, Feb. 4, 1941; came to U.S., 1961, naturalized, 1967; s. José and María Ginoris; m. Dagma Faria Neto, Sept. 2, 1976, B.S. with honors in Nuclear Engring., Miss. State U., 1964; M.S., U. Wis., 1966, Ph.D., 1971. OAS prof. Instituto Politécnico Nacional del Méx., 1971, prof., 1972-75; vis. prof. Instituto Militar de Engenharia, Rio de Janeiro, Brazil, 1973; mem. faculty U. Lowell (Mass.), 1975—, prof., grad. chmn., coord. for energy engring. program, 1980—; chmn. dept. chem. and nuclear engring, 1990—; chief evaluator Internat. Energy Agy. Small Solar Power Systems, Spain, 1983-85; energy cons. Universal Expo '92, Seville, Spain; vis. prof. U. Mex., Ariz. State U. Coll. Architecture; instr. Internat. Solar Sch., Igls, Austria, 1985; Prof. alternative energy, Urbino, Italy, 1985. Contbr. articles to profl. jours.; co-editor: Procs. Internat. Workshop on Distributed Solar Collectors, 1983. Mem. U.S. del. Internat. Energy Agy. Solar Project. Fellow Wis. Alumni Research Found., 1974-75; U.S. AEC fellow, 1975-78; NSF grantee, 1979-80; prin. investigator NRC, summer 1976-80. Mem. Am. Nuclear Soc., Am. Solar Energy Soc., Sigma Xi, Tau Beta Pi, Phi Kappa Phi. Home: 85 Mansur St Lowell MA 01852-2817 Office: U Lowell Energy Ctr 1 University Ave Lowell MA 01854-2881

MARTIN, KEVIN PAUL, physicist; b. Belvidere, Ill., Sept. 28, 1954; s. Dennis John and Joan (Wolf) M. MS in Physics, Ohio State U., 1978, PhD in Physics, 1982. Rsch. assoc. U. Oreg., Eugene, 1985-87; sr. rsch. scientist Ga. Inst. Tech., Atlanta, 1987—; vis. scientist MIT, Cambridge, 1982-85; pres. Nano Scribe, Inc., Atlanta, 1991—. Vol. Amnesty Internat., Atlanta, 1992. Mem. IEEE, Am. Phys. Soc., Sigma Pi Sigma. Office: Ga Inst Tech 791 Atlantic Dr Atlanta GA 30332

MARTIN, LARRY J., health educator, counselor, naturalist; b. Martins Ferry, Ohio, Feb. 8, 1952; s. Irving and Olva M.; m. Joy L. Cooper, Nov. 22, 1984; children: Tom, Beth. BS in Edn. with honors, Mich. State U., 1975, MA in Edn., 1984; MEd in Edn., Xavier U., 1987; PhD in Human Svcs., Walden U., 1992. Cert. tchr. non-tax supported schs. Ohio; cert. credited tchr. elem., secondary Mich.; lic. ltd. psychologist Mich., lic. profl. counselor Mich., registered social worker Mich. Youth dir. First Presbyn. Ch., Lansing, Mich., 1973-75; naturalist tchr. and program coord. Traveling Naturalist Ctr., Lansing, 1974-78; naturalist instr. and developer Eaton County Naturalist Svcs., Grand Lodge, Mich., 1979-81; tchr. Tree of Life Christian Schs., Columbus, Ohio, 1984-86; counselor Christian Pschol. Assocs., Columbus, Ohio, 1986-88; social worker, ltd. lic. psychologist Southwestern Med. Clinic, Berrien Ctr., Mich., 1988—; instr. Lifetime Studies, Lansing C.C., 1975-78; nature instr. Camp Discovery, Woldumar Nature Ctr., 1974-77, acting dir., 1973 (summers); speaker on nature for numerous orgns. in Mich., conducted workshops and nature seminars. Author: A Week in the Wilderness, 1983, Broken,Tattered, Torn and Mendable, 1991; contbr, chpts. to books and numerous pubs. on nature trails, etc. for parks. Liason between State Dept. Recreation Svcs. and Mich. State U. Recreation and Youth Leadership Dept., 1974-75; merit badge counselor Boy Souts Am., 1976-77. Mem. Am. Psychol. Assn., Ledge Craft Lane Artists (Outstanding Young Man Am. 1982). Home: PO Box 182 12988 Peck Ave Sawyer MI 49125 Office: Southwestern Med Clinic 5675 Fairview Ave Stevensville MI 49127

MARTIN, LENORE MARIE, bioorganic researcher, educator; b. Ann Arbor, Mich., June 26, 1963; d. Donald Leigh and Alice Fay (Opaskar) M. BA, Northwestern Univ., 1983; PhD, UCLA, 1988. Rsch. assoc. The Rockefeller Univ., N.Y.C., 1988—; safety com. UCLA Chem. dept., 1985-87; orgn. symposium Am. Chem. Soc., L.A., 1987. Contbr. articles to profl. jours. Recipient Best Paper in Organic Chem. award Am. Chem. Soc., 1983. Mem. Am. Chem. Soc., Am. Peptide Soc., Am. Assn. Advancement of Sci., Sigma Xi., Phi Lambda Upsilon, Sigma Delta Epsilon. Achievements include first synthesis of an oligonucleotide on solid polystyreme support; pioneer in the use of capillary electrophoresis on antibodies. Home: 1161 York Ave 1D New York NY 10021 Office: The Rockefeller U 1230 York Ave New York NY 10021

MARTIN, LEONARD LOUIS, social psychologist, educator; b. New Orleans, Feb. 13, 1952; s. George Nicholas and Celeste (Fossier) M. MA, U. N.C., Greensboro, 1980, PhD, 1983. Vis. asst. prof. U. Ill., Champaign, 1984, NIMH post doctoral fellow, 1984-85; asst. prof. psychology U. Ga., 1985-91, assoc. prof., 1991—; rsch. fellow Inst. for Behavioral Rsch., Athens, Ga., 1989—; mem. editorial bd. Jour. of Personality and Social Psychology, Athens, 1990—. Author: Set, Reset, Comparison, 1985; co-editor The Construction of Social Judgments, 1992. Recipient Rsch. fellowship Humboldt Found., Munich, Germany, 1990; grantee NSF, Washington, 1992. Fellow Inst. for Behavioral Rsch., Am. Psychol. Soc.; mem. Soc. Exptl. Social Psychology (Dissertation of Yr. 1984), Midwestern Person Psychol. Assn. Achievements include devel. theory of context effects in person perception, theory of ruminative thought, theory of happiness, theory of the role of mood in motivation. Office: U Ga Psychology Dept Athens GA 30602

MARTIN, LOUIS FRANK, surgery and physiology educator; b. Troy, N.Y., Nov. 7, 1951; s. Eugene Lavern and Lois Jane (Perkins) Martin; m. Deborah Lynn Tjarnberg, Mar. 12, 1977; children: Jesse Tjarnberg, James Casey, Tyler Gene. BA, Brown U., 1973, MD, 1976; MS in Health Adminstrn., U. Louisville, 1993. Diplomate Am. Bd. Surgery. Resident in gen. surgery U. Wash. Affiliated Hosps., Seattle, 1977-78; resident in gen. surgery U. Louisville, 1978-83, rsch. fellow trauma rsch. and health care edn. adminstrn., 1980-82; asst. prof. surgery Pa. State U., Hershey, 1983-88, asst. prof. physiology, 1986-88, assoc. prof. surgery and cellular and molecular physiology, 1988-92; prof. surgery, assoc. chmn. dept. La. State U., New Orleans, 1992—; vis. scientist INSERM, Poste Orange, France, 1990-91. Contbr. articles to newspapers and profl. jours. Recipient Loyal Davis Traveling Surg. scholar ACS, 1990, Clin. Investigator award NIH, 1985-90. Mem. ACS, Am. Coll. Critical Care Medicine, Am. Physiol. Soc., Assn. for Acad. Surgery (councilman 1988-90), Collegium Internat. Chirurgiae Digestivae, Soc. Internat. Chirurgie, Soc. Univ. Surgeons. Home: 3005 Palm Vista Dr Kenner LA 70065 Office: La State U Surgery Dept 1542 Tulane Ave New Orleans LA 70112-2822

MARTIN, MICHAEL RAY, transportation engineer; b. Knoxville, Tenn., Feb. 28, 1952; s. Ulyess Ray and Sarah Ruth (Grooms) M.; m. Kathryn Frances Parker, Sept. 6, 1975. BSCE, Va. Poly. Inst. and State U., 1974; MSCE, U. Va., 1980. Registered profl. engr., Va., Md., Del.; registered Am. Inst. Cert. Planners. Design engr. Baldwin & Gregg, Ltd., Norfolk, Va., 1974-75; transp. engr. Wilbur Smith and Assocs., Richmond, Va., 1975-77; asst. traffic engr. County of Henrico, Va., Richmond, Va., 1977-78; rsch. engr. Va. Hwy and Transp. Rsch. Coun., Charlottesville, Va., 1978-79; mgr. traffic ops. Kelerco, McLean, Va., 1979-85; v.p. Patton Harris Rust and Assocs., Fairfax, Va., 1985-93; pres. Martin Enterprises & Assocs., 1993—; mem. Dulles Area Transp. Assn. Adv. Bd., Fairfax, 1992—, transp. com. Fairfax County C. of C., 1986-91; mem. citizens adv. com. Met. Washington Coun. Govts., 1993—. Author: (reports) Vehicle Accidents in Highway Work Zones, 1980, Evaluation of Motor Vehicle Accidents Within Highway Construction Work Zones, 1979, Using Geographic Information Systems for Highway Maintenance, 1992. Chmn. Vienna (Va.) Transp. Safety Commn., 1987-89; mem. Va. Railway Express-Land Use Task Force, 1992—; dir. Wesley G. Harris Jr. Meml. Scholarship Fund, Fairfax, 1987-93. Mem. ASCE, Am. Soc. Hwy. Engrs. (nat. dir. 1989—, Potomac sect. pres. 1990-91), Inst. Transp. Engrs. (Va. sect. dir. 1984-86, 87-89, Washington sect. dir. 1992), Am. Planning Assn.(Northern Va. chpt.), Nat. Assn. of Indsl. and Office Pks. (legis. review com. 1988—), Northern Va. Builders Assn. (transp. com. 1986—).

MARTIN, OSVALDO JOSE, investment consultant, entrepreneur; b. Buenos Aires, Argentina, Mar. 8, 1952; s. Osvaldo Jose and Lidia (Paz Videla) M.; m. Elena Isabel Starzenski, Jan. 24, 1982; children: Tomas, Sebastian, Sofia. BS, Manuel Belgrano U., 1969; Navy Pilot Officer, Naval Acad., La Plata, Argentina, 1970-74; Systems Specialist, Tech. Devel., Buenos Aires, Argentina, 1980-85; Airline Capt., Aerolineas Argentinas, 1989. Officer Argentine Navy, La Plata, 1969-78; advanced through grades to lt. sr. Argentine Navy; airline capt. Aerolineas Argentinas, Buenos Aires, 1979-91; pres., chief exec. officer Startin, S.A., Buenos Aires, 1979-91; comml. rep. Camenex Corp., USA, Buenos Aires, 1979-80, Sensor Engring. Co., USA, Buenos Aires, 1980-83; pres., chmn. of bd. Datasystem, S.A., Buenos Aires, 1984-87; cons., advisor Ministry of Def., Buenos Aires, 1984-87; ptnr. Semeria, Martin & Assocs., Buenos Aires, 1987-91; owner Osvaldo Martin & Assocs., Buenos Aires, 1991—; cons. proyectos de inversion para la representacion Argentina Chartered West Bank, 1991—; project leader Datasystem/Startin, S.A., Buenos Aires, 1979-91; cons. Argentine Govt. Projects, Buenos Aires., 1985-91; chmn. bd. dirs. Startin, S.A., Buenos Aires, 1985-91; European Roux, Seguela, Coursad, Gayzac Sopha Design rep., Argentina, 1992; internat. rep. Societe Generale Paris; owner, chmn. bd. dirs. Inteltech Field Svcs S.A. Patentee in field; contbr. articles to profl. jours. Recipient Svcs. recognition Navy Command Security Corp., 1977, Disting. Svcs. recognition award Def. Rsch. Inst., 1984. Mem. IEEE, Assn. for Computing Machinery, Soc. for Computer Simulation, N.Y. Acad. Scis., Soc. Gen. Paris (ARG pres. 1993), Planetary Soc., Airlines Pilots Assn., Navy Club, Yacht Club. Roman Catholic. Avocations: aircraft pilot, snow skiing, tennis, sailing. Home: Soldado de la Independencia 1267, 1426 Buenos Aires Argentina also: Osvaldo Martin & Assocs, Avd del Libertador 4854 4B, 1426 Buenos Aires Argentina

MARTIN, PAUL CECIL, physicist, educator; b. Brooklyn, N.Y., Jan. 31, 1931; s. Harry and Helen (Salzberger) M.; m. Ann Wallace Bradley, Aug. 7, 1957; children: Peter, Stephanie, Glennon, Daniel. A.B., Harvard U., 1952, Ph.D., 1954. Faculty Harvard U., 1957—, prof. physics, 1964-82; J. H. VanVleck prof. pure and applied physics; chmn. dept. physics Harvard U., 1972-75, dean div. applied scis., 1977—, assoc. dean Faculty Arts and Scis., 1981—; vis. prof. Ecole Normale Superieure, Paris, 1963, 66, U. Paris (Orsay), 1971; mem. materials rsch. adv. coun. NSF, 1986-89; bd. dirs. Mass. Tech. Pk. Corp., 1990—, exec. com. 1992-93. Bd. editors: Jour. Math Physics, 1965-68, Annals of Physics, 1968-82, Jour. Statis. Physics, 1975-80. Bd. dirs. Assoc. Univs. for Rsch. in Astronomy, 1979-85; trustee Assoc. Univs., Inc., 1981—, exec. com., 1986-90, 92—. NSF postdoctoral fellow, 1955; Sloan Found. fellow, 1959; Guggenheim fellow, 1966, 71. Fellow AAAS (chair physics sect. 1986), Nat. Acad. Scis., Am. Acad. Arts and Scis., Am. Phys. Soc. (councillor-at-large 1982-84, panel on pub. affairs 1983-86, chmn.-elect nominating com. 1993). Office: Harvard U Lyman Lab Physics and Pierce Hall Cambridge MA 02138

MARTIN, RICHARD DOUGLAS, nuclear engineer, consultant; b. Englewood, N.J., May 17, 1959; s. Douglas Harry and Christine (Jacob) M. B.S. in Nuclear Engring., Pa. State U., 1981; postgrad. Poly. Inst. N.Y., White Plains, 1984. Cert. sr. reactor operator. Staff nuclear engr. Gen. Physics, Linfield, Pa., 1981-83; tng. coord. PSE&G, Salem, N.J., 1983; tech. dir. LOM-TECH, Inc., Elmsford, N.Y., 1983-85; v.p., dir. CDS Techs., Inc., Allentown, Pa., 1985-87; pres. Douglas Martin & Assocs. Inc., Allentown, 1987-90; pres. DMA Radtech, Inc., 1990—, chmn. bd. dirs., 1991—; bd. dirs. Electro-Kinetic Systems, Inc., Trainer, Pa., 1990; adj. faculty mem. Rutgers U., N.J., 1990—; del. Gov.'s Conf. on Small Bus., Pa., 1988; del. Citizen Amb. Program to Commonwealth Ind. States in radiation matters, 1992. Co-author: Mitigating Core Damage, 1982, Basic Chemistry Principles, 1986, Waterborne Radon and Analysis, 1991; contbr. articles to profl. publs. Mem. alumni admissions bd. Pa. State U., 1983. Mem. Am. Nuclear Soc. (John and Muriel Landis award 1980), Am. Assn. Radon Scientists and Technologists (reg., nat. dir.), IEEE, AAAS, EPC-AARST (past pres.), Allentown/Lehigh County C. of C. (chmn. environ. com., v.p. small bus. coun., pres. 1990), Am. Waterworks Assn. Republican. Presbyterian. Current work: Operates full service radiation analytical laboratory.Expert in waterborne radon analytics using liquid scintillation techniques. Cons. on indoor radon remedial techniques. Operator gamma spectroscopy lab. and liquid scintillaton. Subspecialties: Nuclear power plant operations and control, training, emergency response and condition evaluation. Office: DMA-Radtech Inc 701 Chestnut St Trainer PA 19061

MARTIN, ROBERT ANTHONY, mechanical engineer; b. Cedar Rapids, Iowa, Aug. 23, 1960; s. Clair Henry and Kathleen Francis (Wachter) M.; m. Sandra Kaye Jacobs, Dec. 29, 1984; children: Adam Michael, Alex Jacob, Annie Elizabeth. Diploma in diesel truck mechanics, Hawkeye Inst. Tech., 1980; BA in Indsl. Arts Edn., U. No. Iowa, 1983. Sr. engring. aide Wyle Labs., Huntsville, Ala., 1984-87; maintenance planner Iowa Electric Light and Power, Cedar Rapids, 1987-88, predictive maintenance specialist in machinery vibration analysis, 1988—. Mem. Vibration Inst. Roman Catholic. Home: 3102 Coral Ln SW Cedar Rapids IA 52404 Office: Iowa Electric Light & Power 3277 Daec Rd Palo IA 52327

MARTIN, ROBERT BRUCE, chemistry educator; b. Chgo., Apr. 29, 1929; s. Robert Frank and Helen (Woelffer) M.; m. Frances May Young, June 7, 1953. B.S., Northwestern U., 1950; Ph.D., U. Rochester, 1953. Asst. prof. chemistry Am. U., Beirut, Lebanon, 1953-56; research fellow Calif. Inst. Tech., 1956-57, Harvard U., 1957-59; asst. prof. chemistry U. Va., Charlottesville, 1959-61, assoc. prof., 1961-65, prof., 1965—, chmn. dept., 1968-71; spl. fellow Oxford U., 1961-62; Program dir. Molecular Biology Sect., NSF, 1965-66. Author: Introduction to Biophysical Chemistry, 1964. Fellow AAAS; mem. Am. Chem. Soc. Office: Univ of Va Dept Chemistry Charlottesville VA 22901

MARTIN, RODERICK HARRY, aeronautics research scientist; b. Rustington, Eng., Nov. 28, 1961; came to U.S., 1987; s. Dennis Harry and Patricia Dorothy (Clark) M.; m. Joanne Patricia Smith, May 9, 1987: children: Ryan Dennis, Tobias George, Luke Roderick. BSc in Aero. Engring., City U., London, 1983, PhD in Aero. Engring., 1987. Rsch. asst. City U., London, 1983-87; rsch. assoc. Nat. Rsch. Coun.-NASA Langley Rsch. Ctr., Hampton, Va., 1987-89; sr. rsch. scientist Analytical Svcs. and Materials, Inc., Hampton, 1989—. Editor tech. publs.; contbr. articles to profl. publs. Mem. ASTM (chmn. task group 1987—, chmn. subcom. 1993—), Am. Soc. Composites, Brit. Soc. Composites. Office: Analytical Svcs & Materials 107 Research Dr Hampton VA 23666

MARTIN, ROY ERIK, marine biologist, consultant; b. Jacksonville, Fla., May 24, 1949; s. Edgar Arthur and Hedvig (Abarick) M.; m. Elizabeth Gage Smith, Sept. 19, 1970; 1 child, Shannon Elizabeth. BS, Fla. State U., 1972. Lab. technologist dept. oceanography Fla. State U., Tallahassee, 1972-74; marine biologist Marine Rsch. Lab., Fla. Dept. Natural Resources, St. Petersburg, 1974-76; sr. scientist Applied Biology, Inc., Jensen Beach, Fla., 1976—; mem. marine turtle specialist group Internat. Union Conservation of Nature, Natural Resources, Gland, Switzerland, 1989—. Contbr. articles to sci. jours. Treas. Treasure Coast Environ. Coalition, Ft. Pierce, Fla., 1987; v.p. Conservation Alliance St. Lucie County, Ft. Pierce, 1990; mem. St. Lucie County environ. adv. com., Ft.Pierce, 1993. Mem. Am. Inst. Biol. Scis., Am. Soc. Ichthyologists and Herpetologists, Nat. Assn. Environ. Profls., Soc. for Conservation Biology. Achievements include research on clam mariculture, marine ecology, sea turtle biology. Office: Applied Biology Inc PO Box 974 Jensen Beach FL 34958-0974

MARTIN, SCOTT LAWRENCE, psychologist; b. Camden, N.J., Mar. 3, 1960; s. Donald James and Barbara Ann (Martel) M. MA, U. New Haven, 1984; PhD, Ohio State U., 1987. Founder Applied Personnel Strategies, Northbrook, Ill., 1987-89; dir. London House, Rosemont, Ill., 1989—; cons. IBM, Armonk, Ill., 1985, Gen. Electric, Fairfield, Conn., 1987-88, DuPont, Wilmington, Del., 1988. Contbr. articles to profl. jours. Mem. Toastmasters (CTM award 1989). Office: London House/Sci Rsch Assoc 9701 W Higgins Rd Rosemont IL 60018

MARTIN, WESLEY DAVIS, JR., entomologist, research biologist; b. Halifax, Va., Dec. 18, 1956; s. Wesley Davis and Marian (Blankenship) M.; m. Debra Ann Laska, Nov. 10, 1983; children: Toni Michele, Wesley Davis III, Debra Ann. MS, U. Ga., 1985; PhD, Clemson U., 1989. Rsch. entomologist U. Ga., Tifton, 1983-85, rsch. asst., 1985-87; rsch. entomologist Clemson (S.C.) U., 1987-89, ARS/USDA, Byron, Ga., 1989-90; rsch. biologist FMC Corp., Halifax, 1990—. Mem. Entomol. Soc. Am., Entomol. Soc. S.C., Northeastern Weed Sci. Soc., Sigma Xi. Democrat. Episcopalian. Home and Office: Rte 3 Box 785 Halifax VA 24558

MARTIN, WILLIAM ROYALL, JR., association executive; b. Raleigh, N.C., Sept. 3, 1926; s. William Royall and Edith Ruth (Crocker) M.; m. Betty Anne Rader, June 14, 1952; children: Sallie Rader Martin Busby, Amy Kemp Martin Bass. A.B., U. N.C., 1948, M.B.A., 1964; B.S., N.C. State U., 1952. Chemist Stamford (Conn.) research labs. Am. Cyanamid Co., 1952-54, Dan River Mills, Danville, Va., 1954-56, Union Carbide Corp., South Charleston, W.Va., 1956-59; research asso. Sch. Textiles, N.C. State U., 1959-63; tech. dir. Am. Assn. Textile Chemists and Colorists, Research Triangle Park, N.C., 1963-73, exec. dir., 1974—; adj. asst. prof. Coll. Textiles, N.C. State U., 1966-88, adj. assoc. prof., 1989—, spl. lectr. dept. econs. and bus., 1966—; del. Internat. Orgn. Standardization, Pan Am. Standards Commn. Served with USNR, 1944-46. Fellow Am. Inst. Chemists, Soc. Dyers and Colourists, Textile Inst.; mem. Am. Chem. Soc., Coun. Engring. and Sci. Soc. Execs. (past pres. 1992-93), Fiber Soc., Am. Assn. Textile Chemists and Colorists, Masons, Rotary, Phi Kappa Phi, Phi Gamma Delta. Methodist. Home: 224 Briarcliff Ln Cary NC 27511-3901 Office: AATCC 1 Davis Dr PO Box 12215 Research Triangle Park NC 27709-2215

MARTIN DE AGAR, PILAR MARIA, ecology educator; b. Cordoba, Andalucia, Spain, Sept. 13, 1956; d. Federico-Carlos and Carmen (Valverde) Martin de Agar; m. Carlos Montes, July 27, 1984; children: Carlos, Rocio. Licenciada, U. Seville, Spain, 1978; D Biology, U. Murcia, Spain, 1983. Becaria U. Murcia, 1979-81, asst. prof. ecology, 1982-84; asst. prof. ecology U. Complutense, Madrid, 1984-86; prof. titular in ecology U. Complutense, 1987—. Author: Ecologia y Ordenacion del Territorio, 1984; contbr. articles to profl. publs. Mem. Ecol. Soc. Am., Asociacion Española de Ecologia Terrestre, Soc. Ecol. Restoration. Office: U Complutense Dept Ecologia, Facultad de Biologia, 28040 Madrid Spain

MARTINEK, FRANK JOSEPH, chemical company executive; b. Cleve., Apr. 10, 1921; s. Michael Joseph and Sophie Anna (Fortuna) M.; B.S. in Chemistry, Ohio State U., 1950; m. Julianne Christine Radics, Oct. 9, 1954; children—Frank Joseph, Janet Arlene. Chemist, Diamond Alkali Co., Painesville, Ohio, 1950, Sherwin Williams Co., Cleve., 1950-52; group supr. Sherwin Williams, Cleve., 1953-63, tech. mgr., 1964-72; pres. Mar-Bal, Inc., Cleve., 1972-78; pres. Mid-Am. Chem. Corp., Cleve., 1978—. Served with AUS, 1942-45. Decorated Bronze Star (2). Mem. Am. Chem. Soc., Soc. Plastics Industries, Cleve. Soc. Coatings Tech., Cleve. Chem. Assn. Roman Catholic. Home: 1745 Jonathans Trce Cleveland OH 44147-3288 Office: 4701 Spring Rd Cleveland OH 44131-1025

MARTINEZ, LUIS ENRIQUE, JR., chemist researcher; b. El Paso, Feb. 20, 1969; s. Luis Enrique Sr. and Consuelo (Esparza) M. BS, Trinity U., 1991. Rsch. asst. Trinity U., San Antonio, 1988-91; rsch. chemist Exxon R&D Labs., Baton Rouge, 1991; fellow U. Ill., Urbana, 1991-93, Harvard U., 1993—; cons. Exxon R&D Labs., Baton Rouge, 1991-92. Fellow Ford Found. 1992-95. Mem. Am. Chem. Soc., Am. Inst. Chemists, N.Y. Acad. Sci., Sigma Xi. Achievements include co-discovery and investigation of novel refractory bond cleavage for use in heavy petroleum and coal sources. Further modification of the catalytic asymmetric epoxidation of unfunctionalized olefins by chiral, C-2 symmetric Mn(salen) complexes. Home: 1025 Hancock # 5E Quincy MA 02169 Office: Harvard U Box 196 12 Oxford St Cambridge MA 02138

MARTINEZ, MARIO ANTONIO, chemical engineer; b. Reynosa, Mexico, June 13, 1965; came to the U.S., 1978; s. J. Guadalupe and Severa (Trevino) M. BSChE, Tex. A&I U., 1987, BS in Natural Gas Engring., 1989, postgrad. Process chem. engr. ethylene glycol/air separation plant Formosa Plastics Corp., Point Comfort, Tex., 1990—. Roman Catholic. Home: 2624 Quebec McAllen TX 78503 Office: Formosa Plastics Corp 201 Formosa Ave Point Comfort TX 77978

MARTINEZ, MIGUEL ACEVEDO, urologist, consultant, lecturer; b. Chihuahua, Mex., Aug. 18, 1953; came to U.S., 1956; s. Miguel Nuñez and Velia (Acevedo) M.; m. Lilly Mary Barba, June 18, 1983. AB, Stanford U., 1976; MD, Yale U., 1983. Intern U. S.C. Med. Ctr., 1983-84; resident in urology White Meml. Med. Ctr., L.A., 1984-89, urologist, 1989—; cons., lectr. physician asst. program U. So. Calif., L.A., 1990—; patient edn. cons. ICI Pharm., Del., 1991—; patient edn., med. cons., lectr. Abbott Lab., 1991—; med. edn. cons. several radio/TV stas., 1991—; mem. subcom. for diseases on kidney and transplantation NIH, Washington, 1991. Author: Intercellular Pathways, 1981. Polit. cons. Xavier Becerra, U.S. Congress, 1992, Luis Caldera, State Assembly, Calif., 1992. Named one of Outstanding Young Men of Am., 1981. Mem. AMA, Am. Urological Assn., Calif. Med. Assn., L.A. Med. Assn. (polit. action com. 1992—), L.A. County Med. Assn., Yale Alumni Assn., Stanford Alumni Assn., L.A. Athletic Club. Office: White Meml Med Ctr 1700 Brooklyn Ave Ste 15 Los Angeles CA 90033

MARTINEZ, SALVADOR, electronic technology educator; b. Cintruenigo, Navarra, Spain, Mar. 5, 1942; s. Guillermo and Josefina (Garcia) M.; m. Maria Dolores Almenara, June 27, 1969; children: Ana, Sofia. Degree in Indsl. Engring., High Tech. Sch. Indsl. Engrs., Madrid, Spain, 1966, DEng, 1969. Prof. electronics Politechnic U., Madrid, Spain, 1967-79; prof. electronics U. a Distancia, Madrid, Spain, 1978-82, full prof., 1982—; electronic design engr. Isodel Sprecher, S.A., Madrid, Spain, 1971-72; design engr. dept. head I & D Isolux S.A., Madrid, Spain, 1972-73, I & D C.D.E. Electronica S.A., Madrid, Spain, 1973-77; electronic design mgr. Construcciones Aeronauticas, Getafe, Spain, 1977-80; cons. engr. in field. Co-author 3 books on electronics; author: (handbook) Prontuario para el Diséno Electrico y Electronico, 1989, Alimentacion de Cargas Criticas, Electrotecnologias Avanzadas Series, 1992; patentee on static converters and educative simulators. Grantee Ministerio de Edn., 1969-71, Fundacion Juan March, 1966; recipient Best Profl. Electronics Paper, 1981, 82. Mem. IEEE (sr.), Mundo Electronico. Avocations: poetry, gardening, piano, philosophy.

MARTINEZ-GALARCE, DENNIS STANLEY, physicist; b. Guayaquil, Guayas, Ecuador, Aug. 18, 1961; came to U.S., 1965; s. Marco Antonio and Alicia Elena (Galarce) M. AB in Physics, Cornell U., 1983; MS in Applied Physics, Stanford U., 1986. Devel. engr. Intel Corp., Santa Clara, Calif., 1983-84; adj. prof. Nat. U., San Diego, 1987-88; rsch. engr. Lockheed Missiles and Space Co., Sunnyvale, Calif., 1989—; grad. student elec. engring. dept. Stanford (Calif.) U., 1990—; bd. dirs. MARGAL Prodns. Inc., N.Y.C., 1988—. Contbr. articles to profl. jours. Recipient NASA Grad. Rsch. fellowship, 1992, Stanford Honors Coop. fellowship Lockheed Missiles and Space Co., 1990—, 83-84. Mem. Soc. Hispanic Profl. Engrs. Democrat. Office: Lockheed Missiles and Space 0/6H-34 B/177 1111 Lockheed Way Sunnyvale CA 94089

MARTINEZ SANZ, ANTONIO F., chemist; b. Madrid, Feb. 18, 1950; s. Felix Martinez Vallejo and Soledad (Sanz) Velasco; m. Rosario Poles Lopez, Dec. 19, 1981; children: Alejandro, Javier. Degree in Chemistry, Complutense U., Madrid, 1972, PhD, 1978. Prof. Complutense U., Madrid, 1973-79, Ciudad Real, Spain, 1979-82; rschr. Laboratorio Lafarquim, Madrid, 1982-84, Laboratorio Juste Saqf, Madrid, 1984-90, Centro Inves Justesa Imagen SA, Madrid, 1990—. Contbr. articles to profl. jours. Achievements include patents in field. Home: Valmojado 83 5C, 28047 Madrid Spain Office: Centro Inves Justesa Imagen, Roma 19, 28028 Madrid Spain

MARTINO, PETER DOMINIC, software company executive, military officer; b. N.Y.C., Sept. 21, 1963; s. Rocco Leonard and Barbara Italia (D'Iorio) M.; m. Martha Dorothy Laffey, Sept. 9, 1989; 1 child, Elizabeth Marie. BS, U.S. Naval Acad., 1985. Cert. cash manager, 1992. Commd. ensign USN, 1985, advanced through grades to lt., 1989, resigned, 1990; with

USNR, 1990—; v.p. mktg. XRT, Inc., Wayne, Pa., 1990, exec. v.p., COO, 1990-93, pres., 1993—, also bd. dirs.; co-owner, co-founder Medical Check, Inc., Wayne, 1991. Sustaining mem. Rep. Nat. Com., 1981—; mem. Chester County Rep. Party, 1993—; coun. mem. Phoenixville Borough Coun., 1993—. Mem. Treasury Mgmt. Assn. (assoc. mem. coun., bd. dirs.), Naval Acad. Alumni Assn., Naval Acad. Athletic Assn., Army Navy Club, Army Navy Country Club, Naval Submarine League, Pyramid Club (Phila.), Jaycees (Phoenixville). Roman Catholic. Avocations: sailing, boating, computers, art collector, rare coins. Office: XRT Inc 989 Old Eagle School Rd Wayne PA 19087-1704

MARTINS, ANA PAULA, economics educator; b. Lisbon, Portugal, Feb. 20, 1960; d. Adelino and Maria de Lourdes (Correia) M.; children: Marta C.N. Nascimento, David C.M. Nascimento, Madalena C.M. Nascimento. BA, U. Catolica Portuguesa, Lisbon, 1982; MA, Columbia U., 1985, PhD, 1987. Instr. Columbia U., N.Y.C., 1986-87; asst. prof. U. Católica Portuguesa, Lisbon, 1987—; invited prof. U. Nova de Lisboa, Lisbon, 1988—; cons. Comissão Interministerial para O Emprego, Lisbon, 1988-89; adj. cons. Conselho Nacional do Plano, 1990-92. Contbr. articles to profl. jours. Pres.'s fellow Columbia U., 1984-85, 85-86. Mem. Am. Econ. Assn., Econometric Soc., European Assn. for Rsch. in Indsl. Econs., European Assn. Labor Econs. Roman Catholic. Office: U Católica Portuguesa, Cam Palma de Cima, 1600 Lisbon Portugal

MARTON, JOSEPH, paper chemistry consultant, educator; b. Budapest, Hungary, Mar. 5, 1919; came to U.S., 1960; s. Stephan and Roza (Prober) Mechner; m. Terezia Kellner Marton, Jan. 9, 1949; 1 child, Marianne Evelyn. PhD, Pazmany Peter U., Budapest, 1943. Dept. head Hungarian Inst. for Applied Organic Chemistry, Budapest, 1947-57; fellow Chalmers Tech. U., Goteburg, Sweden, 1957-60; rsch. assoc. Westvaco, Charleston, S.C., 1960-66; sr. rsch. assoc. Westvaco, Laurel, Md., 1966-87; cons. J&T Assocs., Silver Spring, Md., 1987—; adj. prof. Coll. Environ. Sci. and Forestry SUNY, Syracuse, 1983—. Author: (with others) Reactions in Alkaline Pulping, 1971, Printing Fundamentals, 1985, Paper Chemistry, 1991; editor: Lignin Chemistry; contbr. over 100 articles to profl. jours. Fellow Tech. Assn. Pulp and Paper Industry (Paper divsn. award 1986, Harris O. Ware prize 1986), Am. Inst. Chemistry; mem. AAAS, Am. Chem. Soc., Can. Pulp and Paper Assn., Cosmos Club (Washington). Achievements include 7 patents in field of paper chemistry. Home and Office: 8 Northrup Ct Newtown PA 18940

MARTS, KENNETH PATRICK, materials engineer; b. Denver, May 31, 1956; s. Donald LaVern and JoAnn Margurite (Deluzio) M.; m. Cheryl Lynn White, Aug. 18, 1979; children: Jonathan, Justin, Jason. BSMetE, Colo. Sch. of Mines, 1978, MSMetE, 1987; MEd in Engring. Mgmt., U. Colo., Boulder, 1990. Engr. Martin Marietta Corp., Denver, 1978—. Achievements include patent pending on precision cleaning device to eliminate use of CFC113. Home: 4846 S Zang Way Morrison CO 80465-1629 Office: Martin Marietta PO Boxc 179 MS 8048 Denver CO 80201

MARTYNYUK, ANATOLY ANDREEVICH, mathematician; b. Ukraine, Mar. 6, 1941; s. Andrei Gerasimovich and Tatyana Fomovna (Kotik) M.; m. Alevtina Grigorievna, Aug. 1, 1942; children: Vladislav Anatolievich and Julya Anatoljevna. Cand. in Physics and Math, 1967, Dr.Sci. in Physics and Math, 1973. Rschr. inst. math. Acad. Sci., Kiev, 1968-78, head dept. Stability of Processes, 1979—, prof., 1985—. Author: Techinical Stability in Dynamics, 1973, Motion Stability of Composite Systems, 1975, Practical Stability of Motion, 1983. Mem. Acad. Sci. Kiev (corr., N.M. Krylov prize, 1981). Avocations: history, ecology. Home: Vernadsky Prospect 85 Apt 78, 252142 Kiev Ukraine Office: Inst Mech Acad Sci, Nestvor Str 3, 252057 Kiev Ukraine

MARUSIAK, RONALD JOHN, quality engineer, electronics executive; b. Washington, Jan. 6, 1948; s. John Jr. and Angeline Francis (Gadinski) M.; m. Barbara Audrey Stegemann, Nov. 18, 1978; children: Lisa Marie, Erika Coleen. BS in History, USAF Acad., 1971; MS in Mgmt., LaVerne U., 1981. Commd. 2d lt. USAF, 1971, advanced through grades to capt., 1975, served in various locations, 1971-81, resigned, 1981; v.p. Micro-Tronics, Inc., Phoenix, 1981-85; pres. Micro-Tronics, Inc., Tempe, Ariz., 1985—. Dir. logistics Ariz. Air N.G., Phoenix, 1989—. Lt. col. USAF. Mem. N.G. Assn., Soaring Assn. (life), Daedalian Soc., Assn. Grads. USAF Acad. (v.p. 1990—). Republican. Roman Catholic. Office: Micro Tronics Inc 2905 S Potter Dr Tempe AZ 85282

MARUYAMA, KOSHI, pathologist, educator; b. Sapporo, Hokkaido, Japan, Feb. 19, 1932; s. Kotaro and Eiko (Nakamura) M.; m. Rumy Misawa, May 6, 1961; children: Nariyuki, Narihiro, Yumie. MD, U. Hokkaido Sch. Medicine, 1957, PhD, 1962. Diplomate Japanese Bd. Pathology. Staff pathologist Nat. Inst. Leprosy Research, Tokyo, 1962-65, Nat. Cancer Ctr. Research Inst., Tokyo, 1965-67; assoc. prof., assoc. virologist U. Tex. M.D. Anderson Hosp. and Tumor Inst., Houston, 1967-75; dir. dept. pathology Chiba Cancer Ctr. Research Inst., Japan, 1975—. Contbr. articles to profl. jours. Assoc. editor: Japanese Jour. Cancer Clinic, 1978—, The Cancer Bull., 1978-89, The Year Book of Cancer, 1978. Trustee Tex. Gulf Coast chpt. Leukemia Soc. Am., Houston, 1973-75. Leukemia Soc. Am. scholar, 1968; hon. prof. Liaoning Cancer Hosp. and Inst., Shenyang, 1992; mem. world com. The Internat. Assn. Comparative Rsch. on Leukemia and Related Diseases, 1993. Fellow N.Y. Acad. Sci., Japanese Pathol. Soc., Japanese Cancer Assn., Japan Soc. Reticuloendothelial System; mem. Am. Assn. Cancer Research, AAAS, Am. Soc. Microbiology, Am. Assn. Investigative Pathology, Japan Assn. Hosp. Pathologists, Internat. Assn. Comparative Research on Leukemia and Related Diseases. Office: Chiba Cancer Ctr Rsch Inst, Dept Pathology 666-2 Nitona-cho, Chuo-ku Chiba 260, Japan

MARUYAMA, YOSH, physician, educator; b. Pasadena, Calif., Apr. 30, 1930; s. Edward Yasaki and Chiyo (Sakai) M.; m. Fudeko Tsuji, July 18, 1954; children: Warren H., Nancy C., Marian M., Karen A. AB, U. Calif., Berkeley, 1951; MD, U. Calif., San Francisco, 1955. Diplomate Am. Bd. Radiology. Intern San Francisco Hosp., 1955-56; resident Mass. Gen. Hosp., Boston, 1958-61; James Picker advanced acad. fellow Stanford U., 1962-64; asst. prof. radiology Coll. Med. Scis., U. Minn., Mpls., 1964-67, assoc. prof., 1967-70, dir. div. radiotherapy, 1968-70; prof., chmn. dept. radiation medicine Coll. Medicine, U. Ky., Lexington, 1970-92, dir. Radiation Cancer Ctr., 1975-92; prof. radiation oncology Wayne State U.; dir. clin. neutron therapy Gershenson Oncology Ctr., Detroit, Mich., 1993—; bd. dirs. Markey Cancer Ctr.; cons. VA Hosp., Lexington; examiner Am. Bd. Radiology; mem. spl. study sect. Nat. Cancer Inst., Bethesda, Md.; mem. Panel Transplutonium Rsch., Nat. Rsch. Coun.; convener Internat. Cf-252 Neutron Therapy Workshop, 1985, 90; disting. oncology lectr. Wayne State U., 1993. Author: CF-252 Neutron Brachytherapy: Advance for Bulky Localized Cancer Therapy, 1984; assoc. editor: Applied Radiology, Endocuriether. Hypertherm. Oncology; editor: New Methods in Tumor Localization, 1977, (with others) CF-252 Brachytherapy and Fast Neutron Beam Therapy: Proceedings of the Workshop, 1986, Internat. Neutron Therapy Workshop, 1991; contbr. articles to profl. jours. Served with M.C. AUS, 1956-58. Am. Cancer Soc. fellow, 1960-61; recipient Nat. award Ky. div. Am. Cancer Soc., 1988. Fellow Am. Coll. Radiology (commn. on radiation therapy and patterns of care study), Royal Soc. Medicine; mem. AAAS, Am. Radium Soc., Am. Cancer Soc. (pres. Ky. div.), Am. Endocurietherapy Soc., Cell Kinetics Soc., Soc. Exptl. Biol. Medicine, Am. Assn. Cancer Rsch., European Soc. Therapeutic Radiology and Oncology, N.Am. Hyperthermia Group, Radiol. Soc. N.Am., Am. Soc. Therapeutic Radiology and Oncology, Soc. Chmn. Acad. Radiation Oncology Programs, Soc. Radiology Japan, Am. Assn. Immunologists, Southeastern Cancer Rsch. Assn. (bd. dirs.), Southeastern Cancer Group, Southwestern Oncology Group, Ky. Med. Assn., Order Ky. Cols., Minn. Acad. Scis., N.Y. Acad. Scis., Japan Soc. N.Y., Japan Club Bluegrass (pres.), Shodan Judoka, Kodokan Inst. (Tokyo), Spindletop Hall Club (Lexington), Sigma Xi (chpt. pres.), Phi Beta Kappa, Alpha Omega Alpha. Home: 1739 Lakewood Dr N Lexington KY 40502-2887

MARVEL, JOHN THOMAS, chemical company executive; b. Champaign, Ill., Sept. 14, 1938; s. Carl Shipp and Alberta (Hughes) M.; m. Joyce Elizabeth Strand, June 30, 1961 (div. Feb. 1981); children: Scott T., Chris A., Carl R.; m. Mary Anne Hamilton, July 24, 1982. Student, DePauw U., 1955-57; AB, U. Ill., 1959; PhD, MIT, 1964; postgrad., Stanford U., 1977. Instr. U. Ariz., Tucson, 1964-65, asst. prof., 1965-68; sr. rsch. chemist Monsanto, St. Louis, 1968-70, rsch. specialist, 1970-72, from sr. rsch. group leader to mgr., 1972-78, from assoc. dir. rsch. to dir. rsch., 1978-81, gen. mgr., 1981-85; v.p. Hybritech Internat., St. Louis, 1983-85; gen. mgr. Monsanto, Brussels, 1985-87; v.p. R&D Ethyl Corp., Baton Rouge, 1988—; mem. USDA Nat. Agrl. Rsch. and Extension Users Adv. Bd., Washington, 1988-91, chmn., 1988-89. Recipient USDA Commendation, 1991, Coll. of Basic Scis. Outstanding Svc. award La. State U., 1992. Mem. AAAS, Indsl. Rsch. Inst. (ethyl rep. 1988—), Am. Chem. Soc., N.Y. Acad. Scis., Chem. Soc., Soc. Chem. Industry, Internat. Union Pure and Applied Chemistry, Rotary Club of Baton Rouge, Sigma Xi, Lambda Upsilon. Office: Ethyl Corp 8000 GSRI Rd Baton Rouge LA 70820

MARWILL, ROBERT DOUGLAS, aircraft design engineer; b. Clarkton, Mo., Feb. 4, 1945; s. Robert Louis and Lorena Cleo (Holt) M.; m. Paula Roselind Reffel, Aug. 17, 1968; children: Tim, Jeff. BSME, Wichita State U., 1977. Registered profl. engr., Tex. Engr. draftsman Lockheed Missiles and Space, Sunnyvale, Calif., 1967-68; design engr. Beech Aircraft Corp., Wichita, Kans., 1969-78; supr. piston propulsion systems Beech Aircraft Corp., Wichita, 1978-82; chief of propulsion systems Fairchild Aircraft Corp., San Antonio, 1982-84; chief engr. R. D. Marwill & Assocs., San Antonio, 1984—; lectr. various seminars. Mem. NSPE, Tex. Soc. Profl. Engrs., Soc. Automotive Engrs.

MARX, BRIAN, statistician; b. Detroit, Mar. 29, 1960. BS in Physiology, Mich. State U., 1982; MA in Statistics, Pa. State U., 1984; PhD, Va. Poly. Inst. and State U., 1988. With Ford Motor Co., Dearborn, Mich., summer 1978, Ford Motor Credit Co., Dearborn, Mich., summer 1985; process engr. Corning (N.Y.) Glass Works, summer 1986; statis. cons. Gen. Foods Corp., Tarrytown, N.Y., summer 1987; lectr. Va. Poly. Inst. and State U., Blacksburg, 1985-88, grad. asst. in statistics, 1984-88; asst. prof. statistics Dept. Exptl. Statistics, La. State U., Baton Rouge, 1988—; lectr. in field; invited mem. sci. com. GLIM 92 and the 7th Internat. Workshop on Statis. Modelling, Munich, 1992, 6th Internat. Workshop on Statis. Modelling, Utrecht, Netherlands, 1991; statis. cons. Contbr. articles to profl. jours.; book reviewer for statis. jours. Coun. on Rsch. Summer Faculty Rsch. awardee, 1989, 91; Miller-Perkins Funding Summer rsch. grantee, 1989. Mem. Am. Statis. Assn. (La. chpt. pres. 1992—, coun. rep. 1990—), Inst. Math. Statictics, Mu Sigma Rho. Office: La State U Dept Exptl Statistics Baton Rouge LA 70803-5606

MARX, JAMES JOHN, immunologist; b. Paris, Tex., Dec. 17, 1944; s. James J. and Grace (Beckfeld) M.; m. Mary Alice Kettrick, Aug. 28, 1973; children: Jonathyn, Christopher, Scott. MS, W.Va. U., 1969, PhD, 1972. Rschr. Marshfield (Wis.) Med. Rsch. Found., 1973-75, sr. rsch. scientist, 1975—. Contbr. articles to profl. jours. Active Boy Scouts Am., 1973—, Nat. Ski Patrol; chair edn. com. OLP, Marshfield, 1988-92; chair Med. Ctr. Credit Union, 1990-91. Mem. Am. Acad. Allergy Immunology, Am. Soc. Microbiology, Am. Med. Lab. Immunology, Am. Thorasic Soc., Am. Assn., Immunology, Clin. Immunology Soc. Achievements include rsch. in hypersensitivity pneumonitis, lyme disease in dairy cattle, specialized tests. Home: M204 Marsh Ln Marshfield WI 54449 Office: Marshfield Med Rsch Fedn 1000 N Oak Marshfield WI 54449

MARX, RICHARD BRIAN, forensic scientist; b. Huntsville, Ala., Mar. 1, 1968; s. Adolph Richard Jr. and Ardelle (Coble) M. BS in Chemistry, U. Ala., 1991, postgrad., 1992—. Lab. analyst Ala. Dept. of Forensic Scis., Huntsville, 1985-93, forensic scientist, 1985-93; cons. Analytical and Forensic Assocs., Huntsville, 1990—. Mem. Circle K Internat., U. Ala., 1986-91. Named Vol. of Yr. State of Ala., 1986. Mem. Am. Chem. Soc., Ala. Assn. of Forensic Scis., Internat. Assn. of Identification, British Fingerprint Soc. Office: Ala Dept Forensic Scis 716 Arcadia Circle Huntsville AL 35801

MARZ, LOREN CARL, environmental engineer and scientist, chemist; b. Jamestown, N.Y., June 11, 1951; s. Maurice Carl and Dorothy May (Anderson) M.; m. Sharon Lee Mekus, June 2, 1979; children: Brandon, Stephen. BS, Gannon U., Erie, Pa., 1975; MS, SUNY, Fredonia, 1990. Registered environ. profl. Profl. baseball player Milw. Brewers Class A Team, Newark, N.Y., 1973; analyst chem. lab. Dunkirk (N.Y.) Ice Cream Co., 1975-80, Ralston-Purina Co., Inc., Dunkirk, 1980-84; environ. engr. CPS, Dunkirk, 1985-89; environ. scientist U.S. Army Med. Rsch. Inst. Chem. Def., Aberdeen Proving Ground, Md., 1989-91; environ. engr. U.S. Dept. Energy, Oak Ridge, Tenn., 1991—; site operator Nat. Atmospheric Deposition Program-Nat. Trends Network, Ft. Collins, Co., 1987-89. Mem. Mayville (N.Y.) Emergency Planning Com., 1988-89. Fellow Am. Inst. Chemists; mem. AAAS, Am. Chem. Soc. Baptist. Home: 410 Jefferson Ave Clinton TN 37716-9558 Office: Dept Energy PO Box 2001 Oak Ridge TN 37831-8723

MASAI, MITSUO, chemical engineer, educator; b. Kobe, Hyogo, Japan, Sept. 30, 1932; s. Ei-ichi and Fumiko (Kimoto) M.; m. Rei Yamamura, May 1960; 1 child, Yohsuke. BS, Osaka U., 1956; PhD, Tokyo Inst. Tech., 1969. Researcher Showa Oil Co., Ltd., Tokyo, 1956-62; rsch. assoc. Tokyo Inst. Tech., 1962-69; from assoc. prof. to prof. catalysis Kobe U., 1969—. Contbr. articles to profl. jours. Mem. Chem. Soc. Japan, Catalysis Soc. Japan (achievement award 1991), Japan Petroleum Inst., Soc. Chem. Engrs., Surface Sci. Soc. Japan, Camerata Muti Club, Nippon Gi-in Club. Avocations: classical music, audio, photography, visiting art museums. Home: Minami-Showa 7-28-204, Nishinomiya 662, Japan Office: Kobe U/Faculty Engring, Rokkodai Nada, Kobe 657, Japan

MASAMOTO, JUNZO, chemist, researcher; b. Shizuki, Japan, May 14, 1937; s. Kazuo and Kiyoko Masamoto; m. Noriko Masamoto, Oct. 1, 1967; children: Kyoko, Kenichi. BS, Kyoto U., 1961, MS, 1963, PhD, 1969. Rsch. chemist Asahi Chem. Industry, Japan, 1963—, mgr., 1975—; asst. gen. mgr. Asochi Chem. Industry, Japan, 1983—, gen. mgr., 1987—, rsch. fellow, 1992—. Author: Poly-B-alanine Fiber, 1976; contbr. articles to profl. jours. Recipient Atsugi award Fiber Sci. and Tech. Japan, 1971, award Soc. Chem. Engrs., 1988-89, Okochi Meml. prize, 1989, Ichimura prize, 1989, Mainichi Newspaper Tech. award 1990. Mem. Japan Chem. Assn. (tech. award 1990), Chem. Soc. Japan (award 1990), Soc. Polymer Sci. (award 1990), Sci. and Tech. Agy. (dir. gen., award 1991), Soc. Polymer Sci. (v.p. Chugoku and Shikoku regional br. 1992—). Achievements include pioneering in development of synthetic fiber-nylon 3, new process for polyacetal resins, new formaldehyde process methylal oxidation, new acetal homo-polymer process, new acetal-copolymer process. Home: 398-8 Nishinakashinden, Kurashiki 710, Japan Office: Asahi Chem, 3-13 Ushiodori, Kurashiki 712, Japan

MASCARENHAS, JOSEPH PETER, biologist, educator, researcher, consultant; b. Nairobi, Kenya, Nov. 19, 1929; came to U.S., 1957.; s. Theotonio C. and Philomena Olive (D'Sousa) M.; m. Patricia Schneider, Jan. 23, 1960; children: Nonika, Shaun. BSc in Agriculture, U. Poona, India, 1952, MSc in Agriculture, 1954; PhD, U. Calif., Berkeley, 1962. Rsch. officer Parry & Co., Thiruvalla, India, 1953-56; instr. biol. Amherst Coll., 1962-63; instr. biol. scis. Wellesley Coll., 1963-64, asst. prof. biol. scis., 1964-67; vis. asst. prof., rsch. assoc. dept. biol. MIT, 1966-68; assoc. prof. biol. scis. SUNY, Albany, 1968-74, prof., 1974—, chair biol. scis., 1991—; program dir. devel. biol. NSF, Washington, 1985-86, 1986-87; cons. in field. Contbr. numerous articles to sci. jours. Grantee Brown Hazen Fund of Rsch. Corp., 1963-64; NSF, 1965—, Rsch. Found. SUNY, 1969-78, Am. Soybean Assn., 1980-81, USDA, 1983, 92—0-81. Fellow AAAS; mem. Am. Soc. Plant Physiologists (mng. editor Sexual Plant Reprodn. 1993—), Bot. Soc. Am., Soc. Devel. Biol., Am. Soc. Cell Biol., Plant Molecular Biol. Assn., Internat. Soc. Devel. Biol., Internat. Plant Tissue Culture Assn., Internat. Assn. Sexual Plant Reprodn. Rsch. Achievements include molecular regulation of plant devel. with emphasis on pollen, engring. of plants for heat stress tolerance. Subspecialties: genetics, genetic engring.(agriculture), devel. biol. Office: U Albany-SUNY Dept Biol Scis 1400 Washington Ave Albany NY 12222

MASHBURN, JOHN WALTER, quality control engineer; b. Athens, Tenn., Oct. 17, 1945; s. Edgar Newton and Grace Victoria (Ellis) M.; m. Lillian Loyall Tauxe, Sept. 17, 1965; children: Samuel Louis, Laura Jean. BS in Engring. Physics, U. Tenn., 1966, MS in Physics, 1970. Registered profl. engr., Tenn.; cert. quality engr., profl. mgr. Sr. prodn. engr. EG&G ORTEC, Oak Ridge, Tenn., 1966-75; nuclear engr. TVA, Knoxville, Tenn., 1975-86; v.p. mfg. Delta M Corp., Oak Ridge, 1986-89; cons. engr. LTM Cons., Knoxville, 1982-89; quality engr. Martin Marietta Energy Systems, Oak Ridge, 1989—. Chmn. Gulf Park Civic Assn., Knoxville, 1978. Mem. IEEE (sr., chair East Tenn. sect. 1990-91), Am. Nuclear Soc. (sr., sect. chair 1986-87), Am. Soc. Quality Control, Tau Beta Pi. Democrat. Presbyterian. Home: 920 Venice Rd Knoxville TN 37923-2098

MASI, JAMES VINCENT, electrical engineering educator; b. Norwalk, Conn., Sept. 21, 1938; s. James V. and Theresa G. (Nardi) M.; m. Sara C. Natale, Oct. 24, 1964 (dec. Feb. 1981); James V., Louis C., Edmund J., Terese L., Stephen F., Catherine M.; m. Patricia Begley, June 5, 1983. BS in Physics, Fairfield (Conn.) U., 1960; MS in Physics, L.I. U., 1970; PhD in Applied Sci., U. Del., 1980. Materials engr. Transitron Electronics Corp., Wakefield, Mass., 1960-62; sr. engr. Space Age Materials/Pfizer, Woodside, N.Y., 1962-65; sr. staff scientist Hartman Systems/ATO, Huntington, N.Y., 1965-69, Bunker Rame Corp., Trumbull, Conn., 1969-73; v.p. R&D U.C.E., Inc./Innotech, Norwalk, 1973-75; program devel. mgr. U. Del., Inst. of Energy Conservation, Newark, 1975-80; prof. elec. engring. Western New England Coll., Springfield, 1980—; dir. of rsch. Shriners Hosp., Springfield, 1989—. Author: Some Properties of (Znx Cd(1-x)3 P2, 1982, Electrical Materials and Devices, 1992; co-author: Laboratory Book of Power, 1988; contbr. over 100 articles to profl. jours.; more than 70 patents in field. Planning bd. Town of Wilbraham, Mass., 1986-88, bd. appeals, 1988-91. Fellowship NSF, 1961-62; recipient Rsch. Teaching Excellence award AT&T, 1987. Mem. IEEE (pres. local chpt. 1984-86), ASM (sec., treas., v.p., pres. local chpt. 1985-89), N.Y. Acad. Scis., Electrochem. Soc., Am. Soc. Engring. Edn., Am. Assn. Physics Tchrs., Soc. Photographic Instrumentation Engrs., Am. Vacuum Soc., Materials Rsch. Soc., Engring. Soc. Western Mass. (v.p., pres., bd. dirs. 1989-90), Engring. in Medicne and Biology Soc., ASM Internat. (sec., pres. 1986-89). Achievements include over 70 patents and inventions. Home: 40 W Colonial Rd Wilbraham MA 01095 Office: Dept Elect Engring Western New Eng Coll Springfield MA 01119

MASKELL, DONALD ANDREW, contracts administrator; b. San Bernadino, Calif., June 22, 1963; s. Howard Andrew Maskell and Gloria Evelyn (Iglesias) White. BA, U. Puget Sound, 1985. Adminstrv. asst. State of Wash., Kent, 1986-87; data analyst Boeing Co., Seattle, 1987-93, engring. contract requirements coord., 1993—. Mem. Elks. Republican. Presbyterian. Avocations: travel, computers, golf, theater, history.

MASLANSKY, CAROL JEANNE, toxicologist; b. N.Y.C., Mar. 3, 1949; d. Paul Jeremiah and Jeanne Marie (Filiatrault) Lane; m. Steven Paul Maslansky, May 28, 1973. BA, SUNY, 1971; PhD, N.Y. Med. Coll., 1983. Diplomate Am. Bd. Toxicology; cert. gen. toxicology. Asst. entomologist N.Y. State Dept. Health, White Plains, 1973-74; sr. biologist Am. Health Found., Valhalla, N.Y., 1974-76; rsch. fellow N.Y. Med. Coll., Valhalla, 1977-83, Albert Einstein Coll. Medicine, Bronx, N.Y., 1983; copr. toxicologist Texaco, Inc., Beacon, N.Y., 1984-85; prin. GeoEnviron. Cons., Inc., White Plains, N.Y., 1982—; lectr. in entomology Westchester County Parks and Preserves, 1973—, lectr. toxicology and hazardous materials, 1985—. Author: Air Monitoring Instrumentation, 1993, (with others) Training for Hazardous Materials Team Members, 1991 (manual, video) The Poison Control Response to Chemical Emergencies, 1993. Mem. Harrison (N.Y.) Vol. Ambulance Corps., 1986-91, Westchester County (N.Y.) Hazardous Materials Response Team, 1987—. Monsanto Fund Fellowship in Toxicology, 1988-90; grad. fellowship N.Y. Med. Coll., 1977-83. Mem. AAAS, N.Y. Acad. Sci., Am. Coll. Toxicology, Am. Indsl. Hygiene Assn. (NYCON chpt.). Achievements include participation in development of genetic toxicity assays to identify potential carcinogens; rsch. on air monitoring instrumentation at hazardous materials sites, health and safety for hazardous waste site workers, environmental and chemical toxicology, genetic toxicology. Home: 122 Saxon Woods Rd White Plains NY 10605

MASLING, JOSEPH MELVIN, psychology educator; b. Rochester, N.Y., Dec. 18, 1923; s. Samuel Harold and Helen (Schoffman) M.; m. Annette Chernoff, Sept. 25, 1952; children: Mark, Susan Masling Rubel. BA, Syracuse U., 1947; PhD, Ohio State U., 1952. Diplomate Am. Bd. Examiners in Profl. Psychology. Asst. prof. to prof. Syracuse (N.Y.) U., 1953-65; prof. SUNY, Buffalo, 1965—. Editor: Empirical Studies of Psychoanalytic Theories, Vol. 1, 1983, Vol. 2, 1986, Vol. 3, 1990. Sgt. U.S. Army Air Force, 1943-46. Recipient fellowships Fulbright Commn., 1964-65, Founds.' Fund for Rsch. in Psychiatry, 1972-73. Mem. Phi Beta Kappa. Office: SUNY Buffalo Psychology Dept Buffalo NY 14265

MASLOV, VIKTOR PAVLOVICH, mathematician, educator; b. Moscow, June 15, 1930; m. Le Vu Ann, 1977; 3 children. Student, Moscow U., postgrad., 1953-56. Asst. prof. Moscow U., 1956-64, dean, 1964-67, sr. rsch. asst., 1967-73; tchr. Moscow Inst. Electronic Machine-Constrn., 1968-73, head dept., 1973—. Editor in Chief Mathematicheskiye Zametki, 1988—. Recipient State prize USSR, 1978, Lenin prize, 1986. Mem. Russian Acad. Scis. (Lyapunov Gold medal 1983). Office: Scientific Council of Appl Math, Ulitsa Vavilova 44 Korpus 2, 117333 Moscow Russia*

MASON, DEAN TOWLE, cardiologist; b. Berkeley, Calif., Sept. 20, 1932; s. Ira Jenckes and Florence Mabel (Towle) M.; m. Maureen O'Brien, June 22, 1957; children: Kathleen, Alison. B.A. in Chemistry, Duke U., 1954, M.D., 1958. Diplomate: Nat. Bd. Med. Examiners, Am. Bd. Internal Medicine (cardiovascular diseases). Intern, then resident in medicine Johns Hopkins Hosp., 1958-61; clin. assoc. cardiology br., sr. asst. surgeon USPHS, Nat. Heart Inst., NIH, 1961-63, asst. sect. dir. cardiovascular diagnosis, attending physician, sr. investigator cardiology br., 1963-68; prof. medicine, prof. physiology, chief cardiovascular medicine U. Calif. Med. Sch., Davis-Sacramento Med. Center, 1968-82; dir. cardiac ctr. Cedars Med. Ctr., Miami, Fla., 1982-83; physician-in chief Western Heart Inst., San Francisco, 1983—; chmn. dept. cardiovascular medicine St. Mary's Med. Ctr., San Francisco, 1986—; co-chmn. cardiovascular-renal drugs U.S. Pharmacopeia Com. Revision, 1970-75; mem. life scis. com. NASA; med. rsch. rev. bd. VA, NIH; vis. prof. numerous univs., cons. in field; mem. Am. Cardiovascular Splty. Cert. Bd., 1970-78. Contbr. numerous articles to profl. publs. Recipient Research award Am. Therapeutic Soc., 1965; Theodore and Susan B. Cummings Humanitarian award State Dept.-Am. Coll. Cardiology, 1972, 73, 75, 78; Skylab Achievement award NASA, 1974; U. Calif. Faculty Research award, 1978; named Outstanding Prof. U. Calif. Med. Sch., Davis, 1972. Fellow Am. Coll. Cardiology (pres. 1977-78), A.C.P., Am. Heart Assn., Am. Coll. Chest Physicians, Royal Soc. Medicine; mem. Am. Soc. Clin. Investigation, Am. Physiol. Soc., Am. Soc. Pharmacology and Exptl. Therapeutics (Exptl. Therapeutics award 1973), Am. Fedn. Clin. Research, N.Y. Acad. Scis., Am. Assn. U. Cardiologists, Am. Soc. Clin. Pharmacology and Therapeutics, Western Assn. Physicians, AAUP, Western Soc. Clin. Research (past pres.), Phi Beta Kappa, Alpha Omega Alpha. Republican. Methodist. Club: El Marcero Country. Home: 44725 Country Club Dr El Macero CA 95618-1047 Office: Western Heart Inst St Mary's Med Ctr 450 Stanyan St San Francisco CA 94117-1079

MASON, JAMES MICHAEL, biomedical laboratories executive; b. Kingsport, Tenn., Mar. 19, 1943; s. William Tilson and Mary Thelma (Epperson) M; m. Linda Kaye Hussung, June 14, 1969. BS, Memphis State U., 1966; PhD, U. Tenn., 1972. Instr. Health Sci. Ctr., U. Tenn., Memphis, 1972-74, asst. prof. pathology, 1974-84, assoc. prof. pathology, 1984-89; dir. paternity evaluation and histocompatibility testing Roche Biomed. Labs., Inc., Burlington, N.C., 1989—, asst. v.p., 1992—; HLA cons. Roche Biomed. Labs., Burlington, 1986-89; blood bank dir. Regional Med. Ctr. at Memphis, 1977-89. Contbr. articles to profl. jours. Rsch. grantee NIH, U. Tenn., 1973-76; recipient Astra Pharm. award Soc. of Perinatal Obstetrics. Fellow Assn. of Clin. Scientists; mem. Am. Assn. of Blood Banks (chmn. transp. immunology com. 1989-92), Internat. Soc. Forensic Hemogenetics, Nat. Child Support Enforcement Assn., Am. Soc. for Histocompatibility and Immunogenetics, Kiwanis. Presbyterian. Achievements include patent for cryopreservation of platelets; research in T-lymphocyte counting in neoplastic disease, pathogenesis of SIDS, automated reading of HLA reactions for paternity testing, innovative approaches to DNA analysis in parentage and identity testing. Home: 50 Driftwood Ct Gibsonville NC 27249 Office: Roche Biomedical Labs Inc 1447 York Ct Burlington NC 27215

MASON, JAMES OSTERMANN, public health administrator; b. Salt Lake City, June 19, 1930; s. Ambrose Stanton and Neoma (Thorup) M.; m. Lydia Maria Smith, Dec. 29, 1952; children: James, Susan, Bruce, Ralph, Samuel, Sara, Benjamin. BA, U. Utah, 1954, MD, 1958; MPH, Harvard U.,

1963, DPH, 1967. Diplomate Am. Bd. Preventive Medicine. Intern Johns Hopkins Hosp., Balt., 1958-59; resident in internal medicine Peter Bent Brigham Hosp.-Harvard Med. Service, Boston, 1961-62; chief infectious diseases Latter-day Saints Hosp., Salt Lake City, 1968-69; commr. Health Services Corp., Ch. of Jesus Christ of Latter-day Saints, 1970-76; dep. dir. health Utah Div. Health, 1976-78, exec. dir., 1979-83; chief epidemic intelligence service Ctr. Disease Control, Atlanta, 1959, chief hepatitis surveillance unit epidemiology br., 1960, chief surveillance sect. epidemiology br., 1961, dep. dir. bur. labs., 1964-68, dep. dir. of Ctr., 1969-70; dir. Ctrs. for Disease Control, Atlanta; adminstr. Agy. for Toxic Substances and Disease Registry, 1983-89; acting asst. sec. health HHS, Washington, 1985, asst. sec. for health, acting surgeon gen., 1989-90, asst. sec. for health, 1990-93; asst. prof. dept. medicine and preventive medicine U. Utah, Salt Lake City, 1968-69; assoc. prof., chmn. div. community medicine, dept. family and community medicine U. Utah, 1978-79; v.p. planning, devel., prof. preventive medicine and biometrics Uniformed Svcs. U. Health Scis., 1993—; physician, cons. to med. services Salt Lake VA Hosp., 1977-83; clin. prof. dept. family and community medicine, U. Utah. Coll. Medicine, 1979-83, clin. prof. dept. pathology, 1980-83; clin. prof. community health Emory U. Sch. Medicine, 1984-86; chmn. joint residency com. in preventive medicine and pub. health Utah Coll. Medicine, 1975-80; mem. Utah Cancer Registry Research Adv. Com., 1976-83; mem. adv. com. Utah Ctr. Health Stats., 1977-79; chmn. bd. Hosp. Coop. Utah, 1977-79; chmn. exec. com. Utah Health Planning and Resource Devel. Adv. Group, 1977-79; chmn. Utah Gov.'s Adv. Com. for Comprehensive Health Planning, 1975-77; mem. recombinant DNA adv. com. NIH, 1979-83; mem. Gov.'s Nuclear Waste Repository Task Force, 1980-83, chmn., 1980-82; bd. dirs. Utah Health Cost Mgmt. Found., 1980-83; mem. adv. com. for programs and policies Ctrs. for Disease Control, 1980; mem. com. on future of local health depts., Inst. Medicine, 1980-82; mem. exec. com., chmn. tech. adv. com. Thrasher Research Found., 1980-89; mem. Robert Wood Johnson Found. Program for Hosp. Initiatives in Long-Term Care, 1982-84; mem. sci. and tech. adv. com. UNDP-World Bank-WHO Spl. Programme for Research and Tng. in Tropical Diseases, 1984-89; mem. Utah Resource for Genetic and Epidemiologic Research, 1982-85, chmn. bd., 1982-83. Author: (with H.L. Bodily and E.L. Updyke) Diagnostic Procedures for Bacterial, Mycotic and Parasitic Infections, 5th edit., 1970; (with M.H. Maxell, K.H. Bousfield and D.A. Ostler) Funding Water Quality Control in Utah, Procs. for Lincoln Inst., 1982; contbr. articles to profl. jours. Mem. nat. scouting com. Boy Scouts Am., 1974-78. Recipient Roche award U. Utah, 1957, Wintrobe award U. Utah, 1958, Disting. Alumni award U. Utah, 1973, Adminstr. of Yr. award Brigham U., 1980, spl. award for outstanding pub. svc. Am. Soc. Pub. Adminstrn. 1984, Disting. Svc. medal USPHS, 1988, LDS Hosp. Deseret Found. Legacy of Life award, 1992. Mem. Inst. Medicine of NAS, AMA, Am. Pub. Health Assn. (task force for credentialing of lab. personnel 1976-78, program devel. bd. 1979-81), Utah State Med. Assn. (trustee 1979-83), Utah Acad. Preventive Medicine (pres. 1982-83), Utah Pub. Health Assn. (pres. 1980-82, Beatty award 1979), Sigma Xi, Alpha Epsilon Delta, Phi Kappa Phi, Alpha Omega Alpha, Delta Omega. Mem. LDS Ch. Lodge: Rotary. Home: 2 West Dr Bethesda MD 20814-1508 Office: USUHS Rm C2117 4301 Jones Bridge Rd Bethesda MD 20814-4799

MASON, JOHN THOMAS, III, chemical engineering educator, consultant; b. McMinnville, Tenn., Jan. 12, 1938; s. John Thomas Jr. and Elizabeth (Smith) M.; m. Linda Sain, Dec. 20, 1959; children: Malissa Leanne, John Thomas IV. BS, Tenn. Technol. U., 1960; MS, U. Mo., Rolla, 1966, PhD, 1971. Registered profl. engr., Tenn.; lic. residential appraiser, Tenn. Commd. 2d lt. U.S. Army, 1960, advanced through grades to col., 1980; ret., 1982; assoc. prof. chem. engring. Tenn. Technol. U., Cookeville, 1982-86, prof., asst. dean, 1986—; bd. dirs. Security Fed. Savs. & Loan, McMinnville. Elder Collegeside Ch. of Christ, Cookeville, 1983-90. Named to ROTC Hall of Fame, Tenn. Technol. U., 1991. Mem. AICE, Tenn. Acad. Sci. Republican. Home: 1241 Woodlake Trace Cookeville TN 38501 Office: Tenn Technol U Coll Engring Box 5005 Cookeville TN 38505

MASON, MICHAEL EDWARD, mechanical engineer; b. Charleston, W.Va., Feb. 26, 1951; s. Weldon W. and Helen (Coleman) M.; m. Brenda McWilliams, Mar. 29, 1980; children: John, Matt, Stephen. BSME, W.Va. Inst. Tech., 1973. Registered profl. engr., Tenn. Mech. engr. TVA, Chattanooga, 1974—. Presbyterian. Office: TVA 1101 Market St Chattanooga TN 37402

MASON, RICHARD RANDOLPH, entomologist; b. St. Louis, Oct. 3, 1930; s. Max Marion and Lucy (Smith) M.; m. Lois Oberdorfer, July 25, 1956; children: Richard R. Jr., John N., Philip M., Anne D. BS, U. Mich., 1952, MF, 1956, PhD, 1966. Rsch. forester Bowaters Southern Paper Corp., Calhoun, Tenn., 1956-65; forest entomologist USDA Forest SVc., Corvallis, Oreg., 1965-82; research entomologist Pacific Northwest Rsch. Sta., La Grande, Oreg., 1982—. Contbr. articles to profl. jours. With U.S. Army, 1952-54. Mem. AAAS, Soc. Am. Foresters, Entomological Soc. Am., Ecol. Soc. Am. Republican. Roman Catholic. Achievements include developed first statistical sampling methods for Douglas-fir tussock moth; determined and described the long-term dynamics of forest pests like the Douglas-fir tussock moth, western spruce budworm and lodgepole needle miner. Office: Pacific Northwest Rsch Sta 1401 Gekeler Ln La Grande OR 97850

MASON, ROBERT THOMAS, zoologist; b. Hartford, Conn., Dec. 18, 1959; s. Millard Howard and Lorraine (Hansen) M. BA, Coll. of Holy Cross, 1982; PhD, U. Tex., 1987. Sr. staff fellow Nat. Heart, Lung, Blood Inst. NIH, Bethesda, Md., 1987-91; asst. prof. zoology Oreg. State U., Corvallis, 1991—. Contbr. revs. to Biology of the Reptilia, Science, Nature. Recipient NSF Young Investigator award 1993-98. Mem. AAAS, Am. Soc. Zoologists, Sigma Xi, Phi Kappa Phi. Achievements include identification of 1st pheromone in a reptile; 1 patent in field. Office: Oreg State U Dept Zoology Corvallis OR 97331

MASON, SCOTT MACGREGOR, entrepreneur, inventor, consultant; b. N.Y.C., Feb. 11, 1923; s. Gregory Mason and Mary Louise Turner; m. Mildred Davidson, Mar. 13, 1949 (div. 1970); children: Alan Gregory, Phoebe Louise, Caleb; m. Virginia Frances Perkins, May 5, 1970 (dec. 1990). AB, Princeton U., 1943; MS, NYU, 1947. Control chemist Firestone Tire & Rubber Co., Akron, Ohio, 1943-44; R & D chemist Am. Cyanamid Co. Rsch. Labs., Stamford, Conn., 1948-52; mgr. stearate dept. Warwick Chem. div. Sun Chem. Corp., Wood River Junction, R.I., 1952-58; cons., Stonington, Conn., 1958-59; instr. Williams Meml. Inst., New London, Conn., 1959-63; NSF fellow Brown U., Providence, 1963-64; tchr. Moses Brown Sch., Providence, 1964-70; owner, mgr. Innoventures, Wakefield, R.I., 1970—; cons. Greene Plastics Corp., Canonchet, R.I., 1972-80, Dorette Inc., Pawtucket, R.I., 1982-83. Patentee in field. Trustee Pine Point Sch., Stonington, 1956-62, pres. bd., 1959-61. With AUS, 1944-46, ETO. Named Tchr. of Week, Sta. WPRO, Providence, 1968; summer rsch. fellow NSF, U. R.I., 1960. Mem. AAAS, N.Y. Acad. Scis. Avocations: tennis, fishing, snorkeling, photography, music. Office: Innoventures PO Box 369 Wakefield RI 02880-0369

MASON, STEPHEN FINNEY, chemistry educator; b. Leicester, Eng., July 6, 1923; s. Leonard Stephen and Christine (Finney) M.; m. Joan Banus, June 3, 1955; children: Oliver, Andrew, Lionel. BA, U. Oxford, U.K., 1945; MA, D.Phil., U. Oxford, 1947, ScD, 1967. Chemistry tutor Wadham Coll., Oxford, 1947-53; rsch. fellow Australian Nat. U., 1953-56; reader Exeter U., U.K., 1956-64; prof. chemistry U. E. Anglia, U.K., 1964-70; prof. chemistry King's Coll., London, 1970-88, emeritus prof. chemistry, 1988—; rsch. assoc. U. Cambridge, 1988—. Author: History of the Sciences, 1953, 2d edit. 1962, Molecular Optical Activity, 1982, Chemical Evolution, 1991. Recipient Gold medal, Italian Chem. Soc., 1991; Wilkins lectr. Royal Soc., 1991. Fellow Royal Soc. London. Avocations: history and philosophy of science. Home: 12 Hills Ave, Cambridge England CB1 4XA Office: King's Coll London, Strand, London England WC1R2LS

MASOVER, GERALD KENNETH, microbiologist; b. Chgo., May 12, 1935; s. Morris H. and Lillian (Perelgut) M.; m. Bonnie Blumenthal, Mar. 30, 1958 (dec. 1992); children: Steven, Laurie, David. BS, U. Ill., Chgo., 1957, MS, 1970; PhD, Stanford U., 1973. Registered pharmacist, Calif., Ill. Owner, operator Ropert Pharmacy, Chgo., 1960-68; rsch. assoc. Stanford U. Med. Sch., Palo Alto, Calif., 1974-80; assoc. rsch. cell biologist Children's Hosp., Oakland, Calif., 1980-83; rsch. microbiologist Hana Biologics, Berkeley, Calif., 1983-86; pharmacist various locations, 1970—; quality control sect. head Genentech, Inc., South San Francisco, 1986-90, quality control sr. microbiologist, 1990—. Contbr. articles to profl. jours., chpts. to books. 1st Lt. USAR, 1957-66. NSF predoctoral fellow, 1970-73; NIH rsch. grantee, 1974-78. Mem. Tissue Culture Assn., Internat. Orgn. for Mycoplasmology, Parenteral Drug Assn., Sigma Xi. Republican. Jewish. Achievements include patents on triphasic mycoplasmatales detection method; triphasic mycoplasmatales detection device. Home: 2091 Camino De Los Robles Menlo Park CA 94025 Office: Genentech Inc 460 Point San Bruno Blvd South San Francisco CA 94080

MASRI, MERLE SID, biochemist, consultant; b. Jerusalem, Palestine, Sept. 12, 1927; came to U.S., 1947; s. Said Rajab and Fatima (Muneimné) M.; m. Maryjean Loretta Anderson, June 28, 1952 (div. 1974); children: Kristin Corinne, Allan Eric, Wendy Joan, Heather Anderson. BA in Physiology, U. Calif., Berkeley, 1950; PhD in Mammalian Physiology and Biochemistry, U. Calif. Berkeley, 1953. Rsch. asst. Dept. Physiology, Univ. Calif., Berkeley, 1950-53; predoctoral fellow Baxter Labs., Berkeley, 1952-53; rsch. assoc. hematology Med. Rsch. Inst., Michael Reese Hosp., Chgo., 1954-56; sr. rsch. biochemist Agrl. Rsch. Svc., USDA, Berkeley, 1956-87; supervisory rsch. scientist Agrl. Rsch. Svc., USDA, N.D. State U. Sta., Fargo, N.D., 1987-89; pvt. practice as cons. Emeryville, Calif., 1989—; lectr. numerous confs. Contbr. more than 120 articles to profl. jours. Fellow Am. Inst. Chemists; mem. AAAS, Am. Chem. Soc., Am. Oil Chemists Soc., Am. Assn. Cereal Chemists, N.Y. Acad. Scis., Sigma Xi. Achievements include patents for detoxification of aflatoxin in agrl. crops, new closed-circuit raw wool scouring tech. to conserve water and energy and control pollution, synthesis and use of polymers for waste treatment and for enzyme immobilization, textile finishing treatment; discovered new metabolic pathways; patent for inactivation of aflatoxins in crops; patents for non-polluting new technology for scouring raw wool in a closed circuit with water recycling and re-use and waste effluent control; developed other non-polluting textile finishing treatments. Home: 9 Commodore Dr Emeryville CA 94608

MASSARO, EDWARD JOSEPH, cell biology, biochemistry research scientist, experimental pathology educator; b. Passaic, N.J., June 7, 1933; s. Anthony and Sarah Leah (Topchik) M.; m. Arlene Margaret Mahood, May 31, 1978; children: David Alan, Anita Diane Massaro Umek, Paul Anthony, Steven Joseph. AB, Rutgers U., 1955; MA, U. Tex., 1958; PhD, U. Tex. Med. Br., Galveston, 1962. USPHS postdoctoral fellow dept. physiol. chemistry Johns Hopkins Med. Sch., 1962-65; rsch. assoc. dept. biology Johns Hopkins U., Balt., 1965-66; instr. dept. biology Yale U., New Haven, 1966-68; prof. biochemistry SUNY Sch. Med. and Biomed. Scis., Buffalo, 1968-78; fellow Rachel Carson Coll., SUNY, Buffalo, 1968-79; dir. toxicology and chem. carcinogenesis Mason Rsch. Inst., Worcester, Mass., 1978; prof., dir. Ctr. for Air Environment Studies, Pa. State U., University Park, 1978-86; dir. inhalation toxicology div. U.S. EPA Health Effects Rsch. Lab., Research Triangle Park, N.C., 1983-85, sr. rsch. scientists, 1985—; rsch. prof. Ctr. for Biochem. Engring., Duke U., Durham, N.C., 1990—; cons. State College (Pa.) Bd. Health, 1979-84. Co-editor (with J.A. Crapo and D.E. Gardner): Toxicology of the Lung, 1988; contbr. over 100 articles to profl. publs. Founder, bd. dirs. Environ. Clearing House, Inc., Buffalo, 1973-76; v.p. Cen. Pa. Lung and Health Assn., State College, 1981-84. Recipient Achievement award Neurotoxicology Jour., 1989, Scientific and Technological Achievement award, U.S. EPA, 1992. Fellow AAAS; mem. Am. Soc. for Biol. Chemistry and Molecular Biology, Am. Soc. for Cell Biology, Am. Soc. Investigative Pathology, Am. Soc. for Pharmacology and Exptl. Therapeutics, Soc. of Toxicology (councilor metals specialty sect. 1987-88, founding pres. 1980-81), Am. Coll. Toxicology (councilor 1984-87), Sigma Xi(resch. award Univ. Tex. medical branch chpt. 1962. Office: Duke U Ctr for Biochem Engring 142 Engineering Bldg Durham NC 27706

MASSEY, CAROLYN SHAVERS, mathematics educator; b. Phenix City, Ala., Dec. 11, 1941; d. James William Sr. and Edna Lurline (Sanders) Shavers; m. Carlton Linwood Massey, June 11, 1966 (dec. 1967). BS, Auburn (Ala.) U., 1963, MEd, 1971, cert. edn. specialist, 1977. Math. tchr. Muscogee County Schs., Columbus, Ga., 1963—, Cattahoochee Valley Community Coll., Phenix City, 1975; star tchr. Jordan High Sch., 1991; mem. tchr. adv. bd. Columbus Regional Math. Collaborative, 1989-92; sponsor Jordan High Math. Team, Columbus, 1975—; curriculum revision com. Muscogee County Sch. Dist., Columbus, 1986—. Pianist, organist Ladonia Bapt. Ch., Phenix City, 1963-87, 92—; fundraiser March of Dimes Walkathon, Columbus. Mem. Cattahoochee Coun. for Tchrs. of Math., Ga. Coun. for Tchrs. of Math., Nat. Coun. for Tchrs. of Math., Profl. Assn. Ga. Educators, Math. Assn. Am. Avocations: playing piano and organ, needlework, reading, traveling. Office: Jordan High Sch 3200 Howard Ave Columbus GA 31995-3699

MASSEY, L. EDWARD, chemical marketing executive; b. Greeneville, Tenn., July 23, 1949; s. Claude Hobart and Mary Agnes (Buck) M.; m. Sandra Marie Gossen, Aug. 8, 1970; children: Eric Christopher, Alexis Christine. BA, U. Ill., 1971, MA, 1973; MBA, U. Wyo., 1978. Lead control rm. supr. Merck, Sharp & Dohme, West Point, Pa., 1977-80; prin. instr. Honeywell process mgmt. system divsn., Ft. Washington, Pa., 1980-82; product mgr. Fisher Controls Internat., Inc., Marshalltown, Iowa, 1982-84; systems cons. Taylor Instrument Co., Rochester, N.Y., 1984-85; mktg. specialist Yokogawa Corp. Am., Peach Tree City, Ga., 1985-88; batch products mgr. Honeywell Indsl. Automation and Control Divsn., Phoenix, 1988-90, dir. chem. mktg., 1990—. Contbr. articles to profl. jours. Capt. USAR, 1973-77. Mem. Instrument Soc. Am. (co-chair subcom. 1990-91), Phi Beta Kappa, Phi Alpha Theta. Achievements include application of automated control techniques (microprocessors) to the automation of complex, batch processes, including both exothermic and endothermic reactions. Office: Honeywell IAC 16404 N Black Canyon Hwy Phoenix AZ 85023

MASSEY, LAWRENCE JEREMIAH, broadband engineer. Student, Nat. Cable TV Inst., 1981, 92. Lic. commercial 1st class, amateur extra class Fed. Communication Commn; cert. engr. Soc. Cable TV Engrs Broadband Communications Tech./Engr. CATV tech. Liberty Cable TV, Portland, Oreg., 1979-82; field svc. rep. Reuters U.S., Inc., San francisco, 1986; CATV tech. Tucson Cablevision, 1982-86, 87-91, CATV engr., 1991—. Home: PO Box 732 Sahuarita AZ 85629 Office: Tucson Cablevision 1511 E 16th St Tucson AZ 85719-6413

MASSIMINI, JOSEPH NICHOLAS, software engineer; b. Pitts., Dec. 27, 1959; s. Ralph J. Massimini and Geraldine B. (Hickly) Brandon; m. Amy L. French, July 10, 1982; children: Gerianna, William. BS, U. Pitts., 1981; MS, Bowie (Md.) State U., 1991. Engr. HRB Singer, State College, Pa., 1981; engr. HRB Singer, Lanham, Md., 1981-83, advanced engr., 1983-86, sr. engr., 1986-87; sr. engr. HRB Systems, Linthicum, Md., 1987-90, prin. engr., 1990—. Usher St. Elizabeth Ann Seton Parish, Crofton, Md., 1987-91; v.p. Holmehurst South Civic Assn., Bowie, Md., 1989, membership chmn., 1990. Recipient Letter of Appreciation Nat. Security Agy., Ft. Meade, Md., 1987. Mem. Assn. Computing Machinery, Spl. Interest Group on Programming Langs., Spl. Interest Group on Ada, Alpha Epsilon Pi. Democrat. Roman Catholic. Home: 4409 Holmehurst Way West Bowie MD 20720 Office: HRB Systems 800 International Dr Linthicum MD 21090

MASSOUD, HISHAM ZAKARIA, electrical engineering educator; b. Cairo, July 31, 1949; came to U.S., 1975; s. Zakaria M. and Zeinab (Salama) M.; m. Kelly L. Watkins, Oct. 23, 1983; 1 child, Nadia. MS, Stanford U., 1976, PhD, 1983. Rsch. asst. Stanford (Calif.) U., 1975-82, rsch. assoc., 1983; rsch. staff mem. IBM T. J. Watson Rsch. Ctr., Yorktown Heights, N.Y., 1977, 80-81; assoc. prof. Duke U., Durham, N.C., 1983—; mem. adv. bd. Microelectronics Ctr., Research Triangle Park, N.C., 1986—; presenter in field. Author over 50 rsch. papers in field, book chpt. Bd. dirs. Montessori Community Sch., Durham, N.C., 1991-92. Rotary Internat. grad. fellow, 1975. Mem. IEEE-IEDM (solid state devices com. 1990-92), IEEE-SISC (tech. program chair 1993), Electrochem. Soc. (program planning com. 1988—, elec. div. exec. com. 1990—), Am. Phys. Soc. Achievements include patent for in situ process monitoring and control using ellipsometry, applications in rapid thermal processing. Office: Duke U Dept Elec Engring Box 90291 Durham NC 27708-0291

MASTERS, BETTIE SUE SILER, biochemist, educator; b. Lexington, Va., June 13, 1937; d. Wendell Hamilton and Mildred Virginia (Cromer) Siler; m. Robert Sherman Masters, Aug. 6, 1960; children: Diane Elizabeth, Deborah Ann. B.S. in Chemistry, Roanoke Coll., 1959, D.Sc. (hon.), 1983; Ph.D. in Biochemistry, Duke U., 1963. Postdoctoral fellow Duke U., 1963-66, advanced research fellow, 1966-68, assoc. on faculty, 1967-68; mem. faculty U. Tex. Health Sci. Ctr. (Southwestern Med. Sch.), Dallas, 1968-82; assoc. prof. biochemistry U. Tex. Health Sci. Ctr. (Southwestern Med. Sch.), 1972-76, prof., 1976-82, research prof. surgery, dir. biochem. burn research, 1979-82; prof. biochemistry, chmn. dept. Med. Coll. Wis., Milw., 1982-90; Robert A. Welch prof. chemistry, dept. biochemistry U. Tex. Health Sci. Ctr., San Antonio, 1990—; mem. pharmacology-toxicology rsch. rev. com. Nat. Inst. Gen. Med. Scis., NIH, 1975-79; mem. bd. sci. counselors Nat. Inst. Environ. Health Scis., 1982-86, chmn., 1984-86; mem. adv. com. on biochemistry and endocrinology Am. Cancer Soc., 1989-92, chmn., 1991-92; mem. phys. biochemistry study sect. NIH, 1989-90; vis. scientist Japan Soc. for Promotion Sci., 1978. Author papers, chpts., revs. and abstracts; editorial bd. Jour. Biol. Chemistry, 1976-81, Archives Biochemistry and Biophysics, 1991—. Recipient Merit award Nat Heart, Lung, and Blood Inst., 1988-98, Women's Excellence in Sci. award Fedn. Am. Socs. for Exptl. Biology, 1992, Am. Cancer Soc. postdoctoral fellow, 1963-65, Advanced Rsch. fellow Am. Heart Assn., 1966-68; established investigator, 1968-73; NIH rsch. grantee, 1970—; Robert A. Welch Found. rsch. grantee, 1971-82, 90—, grantee Nat. Inst. Gen. Med. Scis., 1980—, Nat. Heart, Lung and Blood Inst., 1970—. Mem. AAAS, Am. Soc. Biochemistry and Molecular Biology (nominating com. 1983, coun. 1985-86, awards com. 1992—, fin. com. 1993—), Am. Soc. Pharmacology and Exptl. Therapeutics (exec. com. of drug metabolism div. 1979-81, chmn. exec. com. of drug metabolism div. 1993—, bd. publs. trustees 1982-87), Am. Soc. Cell Biology, Am. Chem. Soc., Internat. Union of Biochemistry (U.S. nat. com.), Sigma Xi, Alpha Omega Alpha. Office: U Tex Health Sci Ctr Dept Biochemistry 7703 Floyd Curl Dr San Antonio TX 78284-7760

MASTERS, EUGENE RICHARD, environmental engineer; b. N.Y.C., May 9, 1939; s. Gene Robert and Phyllis Ann (Collaro) M.; m. Ruth Talley, May 4, 1968; children: Giannine Marie Morris, Ann Elaine. BS, U. Notre Dame, 1960; MBA, Columbia U., 1966; MS, U. South Fla., 1991. Registered profl. engr., Fla. Planning engr. Gen. Dynamics, Groton, Conn., 1963-64; plant engr. Monsanto Corp., Yardville, N.J., 1966-71; v.p. tech. svcs. Briggs div. Celotex, Tampa, Fla., 1971-76; plant mgr. Celotex, Elizabethtown, Ky.., 1976-77; sr. engr. Gardinier Inc., Tampa, 1977-89; sr. project engr. ERM-South, Inc., Tampa, 1989-91; dir. engring. Blasland, Bouck & Lee, Tampa, 1991—. Contbr. articles to profl. jours. Lt. USNR, 1960-63. Mem. ASME, Air and Waste Mgmt. Assn., Tau Beta Pi, Phi Kappa Phi. Republican. Roman Catholic. Office: Blasland Bouck & Lee Ste 100 3350 W Busch Blvd Tampa FL 33618

MASTERS, LARRY WILLIAM, physical science administrator; b. Martinsburg, W.Va., Nov. 18, 1941; s. William Ercel and Pauline Frances (Armbrester) M.; m. Rebecca Jo Riner, Aug. 2, 1963; children: Cynthia R., Mark W. BS, Shepherd Coll., 1963; MS, Am. U., 1968. Chemist Hazleton Labs., Falls Church, Va., 1963-64; rsch. chemist Nat. Inst. Standards and Tech., Gaithersburg, Md., 1964-77, group leader bldg. materials div., 1977-92, dep. chief factory automation systems div., 1992—; chair svc. life com. Reunion Imtermat. de Labs. d'Essais et Recherches sur les Materiaux et les Constructions, Paris, 1981-87, chair NATO Advanced Rsch. Workshop, Paris, 1984; com. chair ASTM E44, Phila., 1984-89. Editor: Problems in Service Life Prediction, 1985. Chair adminstrv. coun. Hedgesville (W.Va.) United Meth. Ch., 1985—, lay leader, 1982-90. Recipient Bronze medal Dept. Commerce, Washington, 1980. Republican. Methodist. Achievements include ISO standard on service life prediction. Office: Nat Inst Stds and Tech Mfg Engring Lab A127 Bldg 220 Gaithersburg MD 20899

MASTERS, ROBERT EDWARD LEE, neural re-education researcher, psychotherapist, human potential educator; b. St. Joseph, Mo., Jan. 4, 1927; s. Robert and Katherine (Leeper) M.; m. Jean Houston, May 8, 1965. BA in Philosophy, U. Mo., 1951; PhD in Clin. Psychology, Humanistic Psychology Inst., 1974. Dir. Library of Sex Research, N.Y.C., 1962-66, Sensory Imagery Program, 1965-68; dir. research Found. for Mind Research. N.Y.C. and Pomona, N.Y., 1965—; dir. Zarathustra Project, Pomona, 1980—; co-dir. Human Capacities Tng. Program, Ramapo, N.J., 1982—; pres. Kontrakundabuffer Corp.; Pomona, 1983—; pvt. practice psychotherapy, neural re-edn.; prin. tchr. Psychophys. Method Tchr. Tng. Programs, 1980—; pres. Human Capacities Corp.; Pomona, 1982—; prin. tchr. Hypnotherapist Tng., Pomona, 1982—. Author: books (25) including Eros and Evil, 1962, Mind Games, 1972, Listening to the Body, 1978, Psychophysical Method Exercises Vols. I-VI, 1983, The Goddess Sekhmet, 1987, The Masters Technique, 1987; (with J. Houston) Varieties of Psychedelic Experience, 1966; also sci. publs., poetry, fiction, essays, lit. and art criticism, book revs. Served with USN, 1945-46, PTO. Grantee Erickson Found, 1966, Kleiner Found., 1968, Babcock Found, 1970, Doris Duke Found., 1972. Fellow Am. Acad. Clin. Sexologists (founder), Mem. Am. Psychol. Assn., Can. Psychol. Assn., N.Y. Acad. Scis., Am. Anthrop. Assn., Am. Bd. Sexology (clin. supr.), Am. Assn. Sex Educators, Counselors and Therapists, AAAS, Assn. Humanistic Psychology. Office: Found Mind Rsch PO Box 3300 Pomona NY 10970-0523

MASTERS, RON ANTHONY, research chemist; b. Reno, July 16, 1961; s. Donald Eugene and Jane Elizabeth (Clark) M.; m. Marla Joan Flynn, May 26, 1984; children: Daniel, Aaron, Joseph. BS with honors, Oreg. State U., 1983; PhD, Purdue U., 1988. Analytical staff Procter and Gamble Co., Cin., 1988-92, rsch. chemist, technologist, 1992-93, sr. scientist, 1993—. Contbr. articles to Talanta, Applied Optics, Spectrochimica Acta. Recipient Chemists award Am. Inst. Chemists, 1983, W.R. Grace fellow, 1986. Mem. Am. Chem. Soc. (mem. Cin. local sect., Undergrad. award 1982), Mason Area Amateur Radio Repeater Club. Achievements include rsch. in coupling spectroscopic imaging detectors to echelle spectrographs, high angle optical effects associated with echelle grating. Home: 2300 Catalpa Dr Loveland OH 45140

MASTERSON, LINDA HISTEN, medical company executive; b. N.Y.C., May 21, 1951; d. George and Dorothy (Postler) Riddell; m. Robert P. Masterson, March 6, 1982; m. William J. Histen, May 24, 1971 (div. 1979). BS in med. tech., U. R.I., 1973; MS in microbiology, U. Md., 1977; student, Wharton U. Pa., Phila., 1988. Med. technologist various hosps., 1972-78; microbiology specialist Gen. Diagnostics, Warner-Lambert, Morris Plains, N.J., 1978-80; from tech. sales rep. to dir. internat. mktg. Micro-Scan, Baxter Internat., Sacramento, 1980-87; dir. mktg. Ortho Diagnostics, Johnson & Johnson, Raritan, N.J., 1987-89; sr. v.p. mktg/sales GenProbe, San Diego, 1989-92; v.p. mktg./sales Bio Star, Boulder, Colo., 1992—; bd. dirs. Ethicon Employee Fed. Credit Union, Sommerville, N.J., 1988-89. Tribute to women in industry Young Women's Christian Assn., N.J., 1989. Mem. Biomedical Mktg. Assn., Med. Mktg. Assn., Phi Kappa Phi. Avocations: skiing, kayaking, racketball. Office: Bio Star 6655 Lookout Rd Boulder CO 80301

MASTRIANA, ROBERT ALAN, architect; b. Youngstown, Ohio, July 6, 1949; s. Fred Paul and Rose L. (Fusco) M.; m. Kathy Ann Peloe, June 26, 1971; children: Robert Byron, Kathryn Olivia. BArch, Kent State U., 1972, U. D'Florence (Italy), 1972. Registered profl. architect, Fla., N.J., Ohio, Pa., Tenn., Tex., W.Va., Va.; cert. gen. contractor, Fla. Project architect Allan M. Sveda, Architect, Cuyahoga Falls, Ohio, 1972-75; pvt. practice architect Poland, Ohio, 1976-78; ptnr., architect 4M Co., Youngstown, 1978—; mem., past chmn. City of Youngstown Design Rev. Bd.; past chmn. Poland Village Design Rev. Bd.; past asst. chmn. Poland Bd. Appeals; tchr. archtl. awareness class to realtors for continuing edn. Trustee Circle of Butler Art; bd. trustees First Covenant Ch.; trustee Poland Twp. Historic Soc.; patron Youngstown Symphony Soc.; mem. Boardman Civic Assn.; mem. community adv. bd. YWCA. Recipient Builder award Poland Masonic Lodge, 1988, Charles Marr award Archtl. Soc. Ohio Found., 1990; Paul Harris fellow Boardman Rotary, 1988. Mem. AIA (pres. Ohio chpt., past v.p., bd. dirs., past pres. ea. Ohio chpt., chmn. design awards), AIA Ohio (bd. dirs., design awards chmn., state conv. chmn., past treas., pres.), Ohio Archtl. Found. (bd. dirs.), Youngstown Warren Region C. of C., Dover Club, Rotary, Phi Gamma Delta. Avocations: running, golf, tennis, antique cars. Home: 34 Botsford St Youngstown OH 44514-1755 Office: 4M Company 4251 Glenwood Ave Ste 3 Youngstown OH 44512-1062

MASTRO, VICTOR JOHN, mathematician, educator; b. N.Y.C., Sept. 15, 1948; s. Felix and Marie (Scardino) M.; m. Miriam Elena Bonano, Aug. 19, 1973; 1 child, Christopher. BS in Secondary Edn., Fordham U., 1970, MS in Secondary Edn., Math., 1973. Edn. cons., curriculum cons. Ind. Sch. 138-184, Bronx, N.Y., 1974-79; art sales cons. Intercraft Industries, Inc., Chgo., 1976-78; prof., coord. Hudson County C.C., Jersey City, N.J., 1979—; asst. prof. math. Fordham U., Bronx, 1984—; Tutor vol. Hudson County C.C., 1979-92, faculty adviser, 1989-91; mentor, counselor Forum Italian Am. Educators, Bronx, 1982-87; speaker local radio programs; presenter workshops. Author multicultural math. program, math. poem; contbr. articles to profl. publs. Mem. AAAS (text and video evaluator 1984-88), Am. Math. Assn., Nat. Coun. Tchrs. Math. (text evaluator). Roman Catholic. Achievements include development of multicultural mathematics programs for Afro-Americans, Native Americans, Hispanics, women, Italian, near- and far-East cultures. Home: 1907 Narragansett Ave New York NY 10461 Office: Hudson County C C 168 Sip Ave Jersey City NJ 07306

MASTROIANNI, LUIGI, JR., physician, educator; b. New Haven, Nov. 8, 1925; s. Marion (Dallas) M.; m. Elaine Catherine Pierson, Nov. 4, 1957; children: John James, Anna Catherine, Robert Luigi. AB, Yale U., 1946; MD, Boston U., 1950, DSc (hon.), 1973; MA (hon.), U. Pa., 1970. Diplomate Am. Bd. Ob-gyn. and Reproductive Endocrinology and Infertility. Intern, then resident ob.-gyn. Met. Hosp. N.Y., 1950-54; fellow rsch. Harvard Med. Sch. and Free Hosp. for Women, Boston, 1954-55; instr. dept. ob-gyn. Sch. of Medicine Yale U., New Haven, 1955-56, asst. prof. ob.-gyn. dept., 1956-61; prof. U. Calif., L.A., 1961-65; chief ob-gyn Harbor Gen. Hosp., L.A., 1961-65; William Goodell prof. ob.-gyn., chmn. dept. U. Pa. Sch. of Medicine, Phila., 1965-87, William Goodell prof. ob.-gyn. dept., dir. human reproduction div., 1987—. Contbr. numerous articles to profl. jours. Recipient Squibb prize Pacific Coast Fertility Soc., 1965, Christian R. and Mary Lindback award, 1969, Gold medal Barren Found., 1977, King Faisal prize in medicine, 1989, Pub. Recognition award Ass. Profls. of Gynecology and Obstets., 1990, Disting. Svc. award Soc. Study of Reproduction, 1992. Mem. ACS, Am. Gyneol. and Ostet. Soc., Am. Gynecol. Club, Am. Fertility Soc., Am. Physiol. Soc., Am. Coll. Obs.-Gyns., Inst. Medicine of NAS, Soc. Gynecology Investigation, Soc. for Exptl. Biology and Medicine, Endocrine Soc., Soc. for Study Reproduction (Disting. Svc. award 1992), Pacific Coast Fertility Soc. (hon.), Cen. Assn. Ob.-Gyns. (hon.), Tex. Assn. Ob.-Gyns. (hon.), N.C. Gynecol. Soc. (hon.), Assn. Profs. Ob.-Gyns., Brazilian Fertility Soc. (hon.), Italian Soc. Ob-Gyns. (hon.), Argentina Fertility Soc. (hon.), Peruvian Fertility Soc. (hon.), Sociedad Espanola de Fertilidad (hon.), Israel Soc. Ob-Gyn. (hon.), Uruguan Soc. Sterility and Fertility (hon.), Inst. Medicine, Sigma Xi, Alpha Omega Alpha. Home: 561 Ferndale Ln Haverford PA 19041-1614 Office: Hosp U Pa 3400 Spruce St Philadelphia PA 19104-4378

MASUDA, GOHTA, physician, educator; b. Tokyo, Japan, Nov. 21, 1940; s. Ryota and Chiyo (Ikeuchi) M.; m. Mitsuko Taguchi, May 14, 1983. MD, Keio U., 1966, PhD, 1977. Intern, Keio Univ. Hosp., Tokyo, 1966-67; instr. Keio U., Tokyo, 1967-74, 76-78; asst. prof. Kitasato Med. Coll., Kanagawa-Ken, Japan, 1974-76; chief dept. infectious diseases Tokyo Met. Komagome Hosp., 1978—; asst. prof. Toho U., Tokyo, 1985—, Keio U., 1986—, Gunma U., 1989—. Contbr. articles to profl. jours. Mem. Japanese Assn. for Infectious Diseases, Japan Soc. Chemotherapy, Am. Soc. Microbiology, Brit. Soc. for Antimicrobial Chemotherapy, N.Y. Acad. Scis. Buddhist. Home: 1-14-12-305 Komagome, Toshima-ku, 170 Tokyo Japan Office: Tokyo Met Komagome Hosp, Dept Infectious Diseases, 3-18-22 Honkomagome, Bunkyo-ku 113 Tokyo Japan

MASUHARA, HIROSHI, chemist, educator; b. Tokyo, Mar. 29, 1944; s. Ryoji and Midori M.; m. Nobuyo Yoshimura, Aug. 25, 1968; children: Akito, Aya. BSc, Tohoku U., Sendai, Japan, 1966, MSc, 1968; PhD, Osaka (Japan) U., 1971. Rsch. assoc. Osaka (Japan) U., 1972-84; prof. Kyoto (Japan) Inst. Tech., 1984-91, Osaka U., 1991—; vis. lectr. Cath. Univ., Leuven, Belgium, 1981; vis. prof. Tokyo Inst. Tech., Yokohama, 1985-86, Tohoku Univ., Sendai, 1991-92; fellow IBM Almaden Rsch. Ctr., San Jose, Calif., 1986; project leader Microphotoconversion Project, Kyoto, Japan, 1988-93. Co-author: Lasers in Polymer Science and Technology, 1990; editor: Photochemical Processes in Organized Molecular Systems, 1991, Dynamics and Mechanisms of PHotoinduced Electron Transfer and Related Phenomena, 1992. Recipient award Japanese Photochemistry Assn. 1989, Sci. for Art prize, 1993. Home: Minami-Konoike, Higashi-Osaka 578, Japan Office: Osaka U, Dept Applied Physics, Yamada-oka, Suita Osaka 565, Japan

MASUI, YOSHIO, zoology educator; b. Kyoto, Japan, June 10, 1931; arrived in Can., 1969; s. Fusa and Toyo Masui; m. Yuriko Masui, Sept. 5, 1959; children: Sayuri, Hitoshi. BSc, Kyoto U., 1953, MSc, 1955, PhD, 1961. Asst. prof. Konan U., Kobe, Japan, 1965; rsch. staff biologist Yale U., New Haven, 1966-69, lectr., 1969; assoc. prof. U. Toronto, Ont., 1969-78, prof., 1978—. Recipient Manning award Manning Found., Calgary, Alta., 1991, Gairdner Internat. award Gairdner Found., Toronto, 1992. Achievements include discovery of Maturation Promoting Factor (MPF) and Cytostatic Factor (CSF) and their roles in cell divison control. Office: Univ of Toronto, Dept of Zoology, Toronto, ON Canada M5S 1A1

MASUO, RYUICHI, mechanical engineering educator; b. Kyoto, Japan, Mar. 24, 1928; s. Takejiro and Mitsu (Masui) M.; m. Takeko Okamoto, Mar. 28, 1956. B of Engring., Kyoto U., 1951, PhD in Engring., 1961. Rsch. asst. Kyoto U., Sakyo-Ku, 1951-56; asst. Osaka Inst. Tech., Asahi-Ku, 1954-56, lectr., 1956-60; assoc. prof. mech. engring. Osaka Inst. Tech., Japan, 1960-63, prof., 1963—, dean, 1972-73, chief libr., 1991-92; cons. Kubota Ltd., Osaka, other orgns., 1963—. Patentee weighing scales; contbr. articles on mech. engring. to profl. publs. Mem. Japan Soc. Mech. Engring., Soc. Instrument and Control Engrs., Kisaichi Country Club. Home: 7-11 Korigaoka 7-Chome, Hirakata, Osaka Japan 573 Office: Osaka Inst Tech, 16-1 Omiya 5-Chome Asahi-ku, Osaka Japan 535

MATA, ZOILA, chemist; b. Galveston, Tex., Aug. 8, 1937; d. Francisco Zuniga and Leonarda (Sustaita) M. BS in Biology, Chemistry, Tex. A&I U., 1975. Office asst. Galveston Pub. Health Nursing Service, 1959-63; draftswoman Wilson Real Estate Index and Pub, Houston, 1964-65; bookkeeper City Products Corp, Galveston, 1966-67; research asst. U. Tex. Med. Br., Galveston, 1967-70; clk. State Dept. Pub. Welfare, Houston, 1971-72, Quinby Temporary, Houston, 1972-76; research technician Baylor Coll. Medicine, Houston, 1976; sr. chemist Nalco Chem. Co., Sugarland, Tex., 1976—. Active Rep. Nat. Hispanic Assembly of Tex. (chair membership credentials 1986—), Rep. Nat. Hispanic Assembly of Harris (vice chair 1986, treas. 1991—), Iota Sigma Pi. Named one of Notable Woman of Tex., 1984-85. Mem. Am. Chem. Soc. (rubber div.), Amigas de las Americas, Nat. Chicano Health Orgn. Avocations: needlepoint, gardening, reading, travel. Home: 7733 Dixie Dr Houston TX 77087-5507 Office: Nalco Chemical Co PO Box 87 Sugar Land TX 77487-0087

MATAGA, NOBORU, scientist; b. Masuda, Japan, May 20, 1927; s. Saburo and Fusae Mataga; m. Shizuyo Tsuno, May 29, 1958; 1 child, Jun. BS, U. Tokyo, 1951, DSc, 1959. Rsch. assoc. chemistry Osaka (Japan) City U., 1951-58, lectr. chemistry, 1958-62, assoc. prof. chemistry, 1962-64; prof. chemistry Osaka (Japan) U., 1964-91, prof. emeritus, 1991—, mem. spl. rsch. Inst. for Laser Tech., 1992—; steering com. Inst. Molecular Sci., 1985-87, prof., 1986-88; mem. editorial bd. Chem. Physics, Zeitschrift fur Physikalische Chemie, Progress in Reaction Kinetics. Author: Molecular Interactions and Electronic Spectra, 1970; contbr. articles to profl. jours.; editor, assoc. editor various publs.; reviewer articles in field. Sci. and tech. grantee Toray Sci. Found., 1976, sci. rsch. grantee Mitsubishi Found., 1977, Yamada Sci. Found., 1982. Mem. Chem. Soc. Japan (award 1986), Am. Chem. Soc., Am. Soc. for Photobiology, AAAS, Phys. Soc. Japan, Laser Soc. Japan. Avocations: hiking, playing with cats. Office: Inst Laser Tech, Utsubo-Hommachi, Nishiku Osaka, Japan 550

MATALON, MOSHE, applied mathematician, educator; b. Cairo, Feb. 25, 1949; came to U.S., 1978; s. Ernest and Elise (Saadia) M.; m. Hedva Waizmann, Nov. 1, 1978; children: Leehe, Omer. BSc, MSc, Tel Aviv (Israel) U., 1973; PhD, Cornell U., 1977. Asst. prof. Poly. Inst. N.Y., Farmingdale, 1978-80; asst. prof. Northwestern U., Evanston, Ill., 1980-85, assoc. prof., 1985-91, prof., applied mathematician, 1991—; vis. prof. Tel Aviv U., 1985, Technion, Haifa, Israel, 1988, Ecole Polytechnique Federal Lausanne, Switzerland, 1989; cons. NASA Langley, Hampton, Va., 1991. Mem. AIAA, Am. Phys. Soc., Soc. Indsl. and Applied Math., Combustion Inst. Office: Northwestern U McCormick Sch Engring 2145 Sheridan Rd Evanston IL 60208-3125

MATEKER, EMIL JOSEPH, JR., geophysicist; b. St. Louis, Apr. 25, 1931; s. Emil Joseph and Lillian (Broz) M.; m. Lolita Ann Winter, Nov. 25, 1954; children: Mark Steven, Anne Marie, John David. BS in Geophys. Engring., St. Louis U., 1956, MS in Rsch. Geophysics, 1959, PhD in Seismology, 1964. Registered geologist and geophysicist, Calif. Assoc. prof. geophysics Washington U., St. Louis, 1966-69; mgr. geophys. rsch. Western Geophys. Co. of Am., Houston, 1969-70; v.p. R & D Western Geophys. Co. of Am., 1970-74; pres. Litton Resources Sys., Houston, 1977, Litton Westrex, Houston, 1974-79; pres. Aero Svc. div. Western Geophys. Co. of Am., 1974-87, v.p., 1974-87; v.p. Western Atlas Internat., Inc., Houston, 1987—; pres. Aero Svc. div. Western Atlas Internat., Inc., Houston, 1987-90; v.p. tech. Western Geophys. div. Western Atlas Internat., Inc., Houston, 1990—; pres. Western Atlas Software Divsn. Western Atlas Internat. Inc., Houston, 1993—; mem. State of Calif. Bd. Registration for Geophysicists, Sacramento, 1974—, State of Calif. Bd. Registration for Geologists, Sacramento, 1974—. Author: A Treatise on Modern Exploration Seismology, 2 vols., 1965; contbr. articles to profl. jours.; asst. editor Geophysica, 1969-70. Baseball mgr. Westchester High Sch., 1969-74; soccer coach Spring Forest Jr. High Sch., Houston, 1974; bd. dirs. St. Agnes Acad., Houston, 1977-82; pres. Strake Jesuits Booster Club, Houston, 1977-78. 2nd lt. U.S. Army, 1951-54. Recipient St. Louis U. Alumni award, 1976. Mem. AAAS, Am. Geophys. Union, Seismological Soc. Am., Geophys. Soc. Houston, European Assn. Exploration Geophysicists, Soc. Exploration Geophysicists (chmn. 1974). Roman Catholic. Avocations: racquetball, golf, literature, running, fishing. Home: 419 Hickory Post Ln Houston TX 77079-7430 Office: Western Atlas Software Div Western Atlas Internat 10001 Richmond Ave Houston TX 77042-9999

MATERNA, THOMAS WALTER, ophthalmologist; b. Passaic, N.J., Oct. 24, 1944; s. Anthony and Ann (Popowich) M.; m. Jorunn Pauline Aronsen, Aug. 17, 1973; children: Richard C., Barbara L. BA, Coll. Holy Cross, Worcester, Mass., 1966; MD, SUNY, N.Y.C., 1971; postgrad., Rutgers U., Newark, 1990. Diplomate Am. Bd. Ophthalmology. Intern N.Y. Hosp.-Cornell U. Med. Ctr., N.Y.C., 1971-72; resident N.Y. Eye and Ear Infirmary, N.Y.C., 1975-78; pvt. practice ophthalmology San Francisco, 1986; ophthalmologist N.J. Eye Physicians & Surgeons, Newark; pres., chief exec. officer US Try Zub Ent., Inc., Newark. Com. mem. N.J. Sch. for the Arts, Montclair, 1991—. Lt. USN, 1972-74, comdr. USNR, 1974—. Fellow ACS, Am. Acad. Ophthalmology; mem. Rotary, Army-Navy Club. Democrat. Roman Catholic. Avocations: coin collecting, rare document collecting, tennis, art history. Home: 87 Lorraine Ave Montclair NJ 07043-2304 Office: NJ Eye Physicians and Surgeons 16 Ferry St Newark NJ 07105-1420

MATERSON, RICHARD STEPHEN, physician, educator; b. Phila., Feb. 11, 1941; s. Alfred Lawrence and June Eileen (Slakoff) M.; m. Rosa Maria Navarro, Aug. 22, 1964; children: Lisa Gail, Lawrence Mark. MD, U. Miami, Coral Gables, Fla., 1965. Diplomate Am. Bd. Phys. Medicine and Rehab. Intern Walter Reed Gen. Hosp., Washington, 1965-66; resident Letterman Gen. Hosp., San Francisco, 1966-68; chief phys. medicine and rehab. Tripler Gen. Hosp., Honolulu, 1968-72; asst. prof. phys. medicine and rehab. Ohio State U., Columbus, 1972-76; assoc. clin. prof. phys. medicine and rehab. Baylor Coll. Medicine, Houston, 1976—; pres. Materson MD, PA, Houston, 1976-90; sr. v.p. for med. affairs, med. dir. Nat. Rehab. Hosp., Washington, 1990—; med. dir. Dept. Phys. Medicine and Rehab., Meml. Hosp. SE, Houston, 1978-90, Ctr. for Sports Medicine and Rehab., 1987-90, Electromyography Lab., 1978-90. Co-author: Physical Medicine and Rehabilitation, 1977, 2nd rev. edit., 1980, The Practice of Rehabilitation Medicine, 1982; contbg. author: Practice of Medicine, 1978. Trustee Meml. Hosp. System, Houston, 1986-90, Nat. Rehab. Hosp., Washington, 1990—; host family Experiment in Internat. Living, 1985, 86, 87. Served to maj. U.S. Army, 1965-72. Fellow Am. Acad. Phys. Medicine and Rehab. (pres. 1986-87), Am. Assn. Electrodiagnostic Medicine; mem. AMA (del. 1978—), Phys. Medicine and Rehab. Edn. and Rsch. Found. (founder, pres. 1982-90, bd. dirs. 1982—), Houston Acad. Phys. Medicine and Rehab. (pres. 1979-80), Am. Acad. Pain Mgmt. (chmn. bd. advisors 1989—), Internat. Wine and Food Soc., Knights of Vine (master comdr. 1982—), Confrerie des Chevaliers du Tastevin, Chaine des Rotisseurs. Jewish.

MATHAUDHU, SUKHDEV SINGH, mechanical engineer; b. Dhamtan Sahib, Haryana, India, Sept. 11, 1946; came to U.S., 1965; s. Kesho Ram and Channo Devi (Dhiman) M.; m. Veena Chand, Aug. 20, 1972; children: Suveen Nigel, Suneel Adrian. BSME, Walla Walla (Wash.) Coll., 1970. Registered profl. engr., Calif., Pa. Mech. engr. McGinnis Engring., Inc., Portland, Oreg., 1970-71, Can. Union Coll., LaCombe, Alta., Can., 1971-72, H.D. Nottingham & Assocs., McLean, Va., 1972; project engr. Shefferman & Bigelson Co., Silver Spring, Md., 1973-77; mech. engr. Buchart Assocs., York, Pa., 1977-78; sr. mech. engr. Gannett Fleming, Harrisburg, Pa., 1978-80; chief mech. engr. Popov Engrs., Newport Beach, Calif., 1981-83; pres. Mathaudhu Engring., Inc., Riverside, Calif., 1983—. Vice chmn. LaSierra Acad. of SDA, Riverside, 1988-92; law adv. counselor SE Conf. SDA, 1987-92. Mem. ASHRAE (chpt. pres. 1988-90, regional vice-chmn. 1990-92, jour. com. 1992-93, bd. dirs., region chmn. 1993—), NSPE, Am. Soc. Plumbing Engrs., Soc. Am. Mil. Engrs., Am. Cons. Engrs. Coun. Calif. Soc. Profl. Egnrs. (pres. 1985-86, state dir. 1986-87), Cons. Engrs. Assn. Calif. Republican. Seventh-Day Adventist. Home: 5394 College Ave Riverside CA 92505-3123 Office: 3903 Brockton Ave Ste 5 Riverside CA 92501-3212

MATHAVAN, SUDERSHAN KUMAR, nuclear power engineer; b. Muzfrabad, Kashmir, India, Aug. 18, 1945; came to U.S., 1968; s. Kartar Chand and Ram Rakhi (Makoli) M.; m. Alka Rani Ajrawat, Oct. 23, 1979; children: Erik, Sarita, Manika, Ketan. BS, Kasmir U., Srinagar, India, 1967; MS, U. Miami, 1970, PhD, 1977. Registered profl. engr., Fla. Engr. Ground Support Engring., Miami, Fla., 1969-73, Smith, Korach A/E, Miami, 1973-75; cons. engr. U. Miami, Coral Gables, Fla., 1975-77; sr. engr. Duke Power, Charlotte, N.C., 1977-79; prin. engr. Fla. Power & Light Co., West Palm Beach, 1979—; mem. analysis subcom. Westinghouse Owners Group, Pitts., 1983—. Contbr. articles to profl. jours. Pres. India Soc., Miami, 1984-86; sec. Hindu Temple, Ft. Lauderdale, Fla., 1986—. Recipient Outstanding Archtl. Designs award Am. Inst. Architecture, 1973. Mem. Am. Nuclear Soc. Democrat. Achievements include safety analyses of St. Lucie and Turkey Point nuclear power plants. Home: 1130 Fairdale Way West Palm Beach FL 33414 Office: Fla Power & Light Co Universe Blvd North Palm Beach FL 33408-2657

MATHENY, ADAM PENCE, JR., child psychologist, educator, consultant, researcher; b. Stanford, Ky., Sept. 6, 1932; s. Adam Pence and Dorotha (Steele) M.; m. Ute I. Debus, July 10, 1962 (div.); m. Mary P. Tolbert, June 24, 1967; children—Laura Steele, Jason Gaverick. B.S., Columbia U., 1958; Ph.D., Vanderbilt U., 1962. Sr. human factors engr. Martin Aerospace div., Balt., 1962-63; instr. Johns Hopkins U. Med. Sch., 1963-65; staff fellow Nat. Inst. Child Health and Human Devel., 1965-67; from asst. prof. to prof. pediatrics U. Louisville Med. Sch., 1967-86, assoc. dir. Louisville Twin Study, 1986—, dir. Served with USN, 1951-55. Fellow Internat. Soc. Twin Research, Am. Psychol. Assn., Am. Psychol. Soc.; mem. Soc. Research Child Devel., AAAS, Behavior Genetics Soc., Internat. Soc. Behavior Devel., Internat. Soc. Infant Study, Phi Beta Kappa, Sigma Xi. Co-author: Genetics and Counseling in Medical Practice, 1969; contbr. articles to profl. jours.

MATHER, ELIZABETH VIVIAN, health care executive; b. Richmond, Ind., Sept. 19, 1941; d. Willie Samuel and Lillie Mae (Harper) Fuqua; m. Roland Donald Mather, Dec. 26, 1966. BS, Maryville (Tenn.) Coll., 1963; postgrad., Columbia U., 1965-66. Tchr. Richmond Community Schs., 1963-67, Indpls. Pub. Schs., 1967-68; systems analyst Ind. Blue Cross Blue Shield, Indpls., 1968-71, Ind. Nat. Bank, Indpls., 1971; med. cons. Ind. State Dept. Pub. Welfare, Indpls., 1971-78, cons. supr., 1978-86; systems analyst Ky. Blue Cross Blue Shield, Louisville, 1988-89; contracts specialist Humana Corp., Louisville, 1989—. Active Rep. Cen. Com. Montgomery County, Crawfordsville, 1976-86, Centenary Meth. Ch., adminstrv. bd., 1990. Mem. DAR (treas. 1963-66, sec. 1978-86). Avocations: designing and sewing clothes. Home: 6106 Partridge Pl Floyd Knobs IN 47119 Office: Humana Corp 500 W Main St Louisville KY 40202

MATHER, JOHN CROMWELL, astrophysicist; b. Roanoke, Va., Aug. 7, 1946; s. Robert Eugene and Martha Belle (Cromwell) M.; m. Jane Anne Hauser, Nov. 22, 1980. BA, Swarthmore (Pa.) Coll., 1968; PhD, U. Calif., Berkeley, 1974. NAS/NRC rsch. assoc. NASA/Goddard Inst. for Space Studies, N.Y.C., 1974-76; lectr. in astronomy Columbia U., N.Y.C., 1975-76; astrophysicist NASA/Goddard Space Flight Ctr., Greenbelt, Md., 1976—, head infrared astrophysics br., 1988-89, 90-93; sr. scientist NASA/Goddard Space Fligh Ctr., Greenbelt, Md., 1989-90, 93—, study scientist, Cosmic Background Explorer Satellite, 1976-82, project scientist COBE, 1982—, prin. investigator FIRAS on COBE, 1976—; chmn. external adv. bd. Ctr. for Astrophys. Rsch. in the Antarctic, U. Chgo., 1992—; mem. lunar astrophysics mgmt. ops. working group NASA Hdqrs., Washington, 1992. Contbr. over 50 articles to profl. jours. Recipient Space Sci. award AIAA, 1993, Nat. Space Achievement award Rotary, 1991, Laurels awd., Aviation Week & Space Technology, 1993. Mem. Am. Astron. Soc. (Dannie Heineman prize astrophysics 1993), Am. Phys. Soc., Sigma Xi. Democrat. Unitarian. Achievements include proposed Cosmic Background Explorer Satellite, led team to successful launch in 1989, measured spectrum of cosmic microwave background radiation to unprecedented accuracy. Office: NASA/Goddard Space Flight Code 685 Code G85 Greenbelt MD 20771

MATHEWS, BARBARA EDITH, gynecologist; b. Santa Barbara, Calif., Oct. 5, 1946; d. Joseph Chesley and Pearl (Cieri) Mathews; AB, U. Calif., 1969; MD, Tufts U., 1972. Intern, Cottage Hosp., Santa Barbara, 1972-73, Santa Barbara Gen. Hosp., 1972-73; resident in ob-gyn Beth Israel Hosp., Boston, 1973-77; clin. fellow in ob-gyn Harvard U., 1973-76, instr., 1976-77; gynecologist Sansum Med. Clinic, Santa Barbara, 1977—. faculty mem. ann. postgrad. course Harvard Med. Sch.; bd. dirs. Sansum Med. Clinic; dir. ann. postgrad course UCLA Med. Sch. Bd. dirs. Meml. Rehab. Found., Santa Barbara, Channel City Club, Santa Barbara, Music Acad. of the West, Santa Barbara; mem. citizen's continuing edn. adv. council Santa Barbara Community Coll. Diplomate Am. Bd. Ob-Gyn. Fellow ACS, Am. Coll. Obstetricians and Gynecologists; mem. AMA, Am. Soc. Colposcopy and Cervical Pathology (dir. 1982-84), Harvard U. Alumni Assn., Tri-counties Obstet. and Gynecol. Soc. (pres. 1981-82), Phi Beta Kappa. Clubs: Birnam Wood Golf (Santa Barbara). Author: (with L. Burke) Colposcopy in Clinical Practice, 1977; contbg. author Manual of Ambulatory Surgery, 1982. Home: 2105 Anacapa St Santa Barbara CA 93105-3503 Office: 317 W Pueblo St Santa Barbara CA 93105-4365

MATHEWS, DAVID, foundation executive; b. Grove Hill, Ala., Dec. 6, 1935; s. Forrest Lee and Doris (Pearson) M.; m. Mary Chapman, Jan. 24, 1960; children: Lee Ann Mathews Hester, Lucy Mathews Heegaard. A.B., U. Ala., 1958; Ph.D., Columbia U., 1965; LL.D., U. Ala., 1969, Mercer U., 1976; L.H.D., William and Mary Coll., 1976, Med. U. S.C., 1976, Samford U., 1978, Transylvania U., 1978, Stillman Coll., 1980, Miami U., 1982; H.H.D., Birmingham-So. Coll., 1976, Wash. U., St. Louis, 1984; L.H.D., Ctr. Coll., 1985; L.L.D., Ohio Wesleyan U., 1987, Lynchburg Coll., 1987; L.H.D., U. New Eng., 1988. Exec. v.p. U. Ala., 1968-69, pres., 1969-80, prof. history, 1977-81; pres., chief exec. officer Charles F. Kettering Found., Dayton, Ohio, 1981—; sec. HEW, Washington, 1975-77; dir. Birmingham br. Fed. Res. Bank of Atlanta, 1970-72, chmn., 1973-75; mem. council SRI Internat., 1978-85; chmn. Council Public Policy Edn., 1980—. Contbr. articles to profl. jours. Trustee Judson Coll., 1968-75, Am. Univs. Field Staff, 1969-80; bd. dirs. Birmingham Festival of Arts Assn., Inc., 1969-75; mem. Nat. Programming Council for Public TV, 1970-73, So. Regional Edn. Bd., 1969-75, Ala. Council on Humanities, 1973-75; vice chmn. Commn. on Future of South, 1974; mem. So. Growth Policies Bd., 1974-75; mem. nat. adv. council Am. Revolution Bicentennial Adminstrn., 1975; mem. Ala. State Oil and Gas Bd., 1975, 77-79; bd. dirs. Acad. Ednl. Devel., 1975—, Ind. Sector, 1982-88, ; chmn. Pres.'s Com. on Mental Retardation, 1975-77; chmn. income security com. aging com. Health Ins. Com. of Domestic Council, 1975-77; bd. govs. nat. ARC, 1975-77; bd. govs., bd. visitors Washington Coll., 1982-86 ; trustee John F. Kennedy Center for Performing Arts, 1975-77, Woodrow Wilson Internat. Center for Scholars, 1975-77; fed. trustee Fed. City Council, 1975-77; bd. dirs. A Presdl. Classroom for Young Americans, Inc., 1975-76; trustee Tchrs. Coll., Columbia U., 1977—, Nat. Found. March of Dimes, 1977-83, Coun. om Learning, 1977-84, Miles Coll., 1978—; mem. nat. adv. bd. Nat. Inst. on Mgmt. Lifelong Edn., 1979-84; mem. Ala. 2000, 1980—; spl. adviser Aspen Inst., 1980-84; mem. bd. trustees Gerald R. Ford Found., 1988—, bd. visitors Mershon Ctr. Ohio State U., 1988-91. Served with U.S. Army, 1959-60. Recipient Nicholas Murray Butler medal Columbia U., 1976, Ala. Adminstr. of Year award Am. Assn. Univ. Adminstrs., 1976, Educator of Year award Ala. Conf. Black Mayors, 1977, Brotherhood award NCCJ, 1979. Mem. Newcomen Soc. Am., Phi Beta Kappa, Phi Alpha Theta, Omicron Delta Kappa, Delta Theta Phi. Home: 6050 Mad River Rd Dayton OH 45459-1508 Office: Charles F Kettering Found 200 Commons Rd Dayton OH 45459-2799

MATHEWS, JOHN DAVID, electrical engineering educator, consultant; b. Kenton, Ohio, Apr. 3, 1947; s. John Joseph and Mary (Long) M.; children: John Todd, Debra Juanita, Alex David. BS in Physics with honors, Case Western Res. U., 1969, MS in Elec. Engring. and Applied Physics, 1972, PhD in Elec. Engring. and Applied Physics, 1972. Lectr. dept. elec. engring. and applied physics Case Western Res. U., Cleve., 1969-72, asst. prof., 1975-79, assoc. prof., 1979-85, prof., 1985-87; prof. dept. elec. engring. Pa. State U., University Park, 1987—, dir. Communications and Space Scis. Lab. Coll. Engring., 1988—; vis. scientist Nat. Astronomy and Ionosphere Ctr., Case Western Res. U., Arecibo, P.R., 1972-75; adj. asst. prof. dept. elec. engring. Pa. State U., 1978-86, dir. artist-in-residence program Coll. Engring., 1990-92; mem. Arecibo adv. bd. and visiting com. Cornell U., 1989-92, chmn., 1991-92; cons. to engring. svc. group and laser engring. group Lawrence Livermore Nat. Labs., 1979-87; cons. Nat. Astronomy and Ionosphere Ctr. Arecibo Obs., 1980—; cons. engr., 1975—; presenter, speaker numerous profl. meetings. Contbr. articles to profl. jours. including Jour. Atmos. Terr. Physics, Jour. Geophys. Rsch., Radio Sci., numerous others. Achievements include research on observations of narrow sodium and narrow ionization layers using both lidar and incoherent scatter radar techniques, on tides and acoustic-gravity waves as observed in the motions of ionospheric E region meteoric ion layers, on electron concentration configurations during geomagnetic storms, and on detection and correction of coherent interference in incoherent scatter radar data processing. Office: Pa State U Communications & Space Scis Lab 316 Electrical Engineering E University Park PA 16802-2707

MATHEWS, NANCY ELLEN, wildlife ecologist, educator; b. Kettering, Ohio, Mar. 9, 1958; d. Robert William and Sara (Schmidt) M. BS, Pa. State U., 1980; MS, SUNY, Syracuse, 1982, PhD, 1989. Cert wildlife biologist. Rsch. asst. SUNY Coll. of Environ. Sci. and Forestry, Syracuse, 1980-82, 85-89; endangered species biologist EG&G, Inc., Tupman, Calif., 1982-85; asst. rsch. ecologist U. Ga.-Savannah River Ecol. Lab., AIken, S.C., 1990; asst. unit leader, asst. prof. U.S. Fish and Wildlife Svc., Tex. Coop. Rsch. Unit, Tex. Tech. U., Lubbock, 1990—. Contbr. articles to profl. jours. Alumni Meml. scholar SUNY Coll. Environ. Sci. and Forestry Alumni Assn., 1982; recipient Wilford A. Dence award Wilford A. Dence Found., 1989; named Outstanding Young Woman of Am., Outstanding Young Women of Am., 1988. Mem. AAAS, Am. Soc. Naturalists, Am. Soc. Mammalogists, Ecol. Soc. Am. (membership com. 1990-93), The Wildlife Soc. Achievements include research on the integration of behavioral ecology with wildlife management, predator and prey ecology, conservation biology, long-term changes in wildlife populations. Office: Tex Coop Unit Tex Tech Univ Rm 9 Goddard Bldg Lubbock TX 79409

MATHEWS, WILLIAM HENRY, geologist, educator; b. Vancouver, B.C., Can., Feb. 2, 1919; s. Thomas Mathews and Jean Wilson; m. Laura Lu Tolsted, June 17, 1948; children: Thomas William, Margaret Jean, Janet Marian. BASc, U. B.C., Vancouver, 1940, MASc, 1941; PhD, U. Calif., Berkeley, 1948. Registered profl. engr., B.C. Assoc. mining engr. B.C. Dept. Mines, Victoria, 1942-49; asst. prof. U. Calif., Berkeley, 1949-51; assoc. prof. geology U. B.C., 1951-59, prof., 1959-84, head dept., 1964-71, prof. emeritus, 1984—; cons. in field. Contbr. over 150 articles to profl. jours. Fellow Geol. Soc. Am. (vice chmn., chmn. cordilleran sect. 1980-82), Royal Soc. Can. (Willet G. Miller medal 1989); mem. Geol. Assn. Can. (hon., v.p., pres.

cordilleran sect. 1978-80), Can. Assn. for Quaternary (Johnston medal 1991), Alpine Club Can. (chmn. Vancouver sect. 1942). Office: U BC, Dept Geol Sci, Vancouver, BC Canada V6T 2B4

MATHEWSON, CHRISTOPHER COLVILLE, engineering geologist, educator; b. Plainfield, N.J., Aug. 12, 1941; s. George Anderson and Elsa Rae (Shrimpton) M.; m. Janet Marie Olmsted, Nov. 2, 1968; children: Heather Alexis, Glenn George Anderson. BSCE, Case Inst. Tech., 1963; MS in Geol. Engring., U. Ariz., 1965, PhD in Geol. Engring., 1971. Registered profl. engr., Tex., Ariz., geologist, Oreg., Alaska. Officer, lt. Nat. Ocean Survey, 1965-71; prof., dir. Tex. A&M U., College Station, 1981—; speaker, cons. in field. Author: Engineering Geology, 1981 (C.P. Holdredge award); contbr. articles to profl. publs. Chmn. College Station Planning and Zoning Commn., 1973-81. Fellow Geol. Soc. Am. (chmn. engring. geology div. 1986-87, Meritorious Svc. award 1991); mem. Assn. Engring. Geologists (editor bull. 1981-88, pres. 1988-89), Am. Geol. Inst. (pres. 1991-92), Nat. Coal Coun. Office: Tex A&M U Dept Geology College Station TX 77843-3115

MATHEWSON, HUGH SPALDING, anesthesiologist, educator; b. Washington, Sept. 20, 1921; s. Walter Eldridge and Jennie Lind (Jones) M.; m. Dorothy Ann Gordon, 1943 (div. 1952); 1 child, Jane Mathewson Holcombe; m. Hazel M. Jones, 1953 (div. 1978); children: Geoffrey K., Brian E., Catherine E. Brock, Jennifer A. Jehle; m. Judith Ann Mahoney, 1979 (div. 1990). Student, Washburn U., 1938-39; A.B., U. Kans., 1942, M.D., 1944. Intern Wesley Hosp., Wichita, Kans., 1944-45; resident anesthesiology U. Kans. Med. Ctr., Kansas City, 1946-48; pvt. practice specializing in anesthesiology Kansas City, Mo., 1948-69; chief anesthesiologist St. Luke's Hosp., Kansas City, 1953-69; med. dir., sect. respiratory therapy U. Kans. Med. Ctr., 1969-92, assoc. prof., 1969-92, prof., 1975-92, prof. anesthesiology emeritus, respiratory care edn., 1992—; examiner schs. respiratory therapy, 1975—; oral examiner Nat. Bd. Respiratory Therapy; mem. Coun. Nurse Anesthesia Practice, 1974-78; prof. phys. therapy edn., 1993—. Author: Structural Forms of Anesthetic Compounds, 1961, Respiratory Therapy in Critical Care, 1976, Pharmacology for Respiratory Therapists, 1977; contbr. articles to profl. publs.; mem. editorial bd. Anesthesia Staff News, 1975-84; assoc. editor: Respiratory Care, 1980—, cons. editor, 1980—, editor-in-chief Respiratory Mgmt., 1989-92. Trustee Kansas City Mus., Kansas City Conservatory of Music, 1993—. Served to lt. comdr. USNR, 1956. Recipient Bird Lit. prize Am. Assn. Respiratory Therapists, 1976. Mem. Mo. Soc. Anesthesiologists (pres. 1963), Kans. Soc. Anesthesiologists (pres. 1974-77), Kans. Med. Soc. (council), Phi Beta Kappa, Sigma Xi, Lambda Beta (hon.). Office: Kans Med Ctr 39th and Rainbow Sts Kansas City KS 66103

MATHIAS, ALICE IRENE, health plan company executive; b. N.Y.C., Mar 2, 1949; d. Murray and Charlotte (Kottle) M. B.S. in Math., Western New Eng. Coll., 1972. Programmer, Carnation Co., Los Angeles, 1973-78; programmer/analyst Cedars-Sinai Med. Ctr., Los Angeles, 1978-79, Union Bank, Los Angeles, 1979-81; group leader Kaiser Found. Health Plan, Pasadena, Calif., 1981—. Mem. Nat. Assn. Female Execs., Am. Mgmt. Assn., Kaiser Mgmt. Assn., Kaiser Women in Mgmt., Los Angeles County Mus. Art (sponsor), Smithsonian Inst., KCET Pub. TV, Choice In Dying, U.S. Holocaust Meml. Mus. (charter mem.), Caithness Collectors Club, Statue of Liberty Ellis Island Found. Home: 4210 Via Arbolada Apt 311 Los Angeles CA 90042-5124 Office: Kaiser Found Health Plan Info Svcs Dept 393 E Walnut St Pasadena CA 91188-0001

MATHIAS, EDWARD CHARLES, aerospace engineer; b. Austin, Tex., Aug. 14, 1959; s. Duane Merle and Helen Cathine (Ehlers) M.; m. Susan Hansen, May 12, 1990; 1 child, Spencer Duane. BSME, U. Utah, 1985; MS in Aerospace Engring., Ga. Inst. Tech., 1986. Engr. Thiokol, Brigham City, Utah, 1987-88, sr. engr. assoc., 1988-90, sr. engr., 1990-92, prin. engr., 1992—. Contbr. articles to profl. jours. Republican. Mem. LDS Ch. Achievements include research on cooling system for the statically tested Space Shuttle full scale solid rocket motors, thermal modeling of the shuttle solid rocket motors on the launch pad, gaseous oxygen cooling of the space transportation system launch pad environment, negatively bouyant flow along verticle cylinders at high Rayleigh numbers. Home: 297 East 1240 North Logan UT 84321 Office: Thiokol PO Box 707 Brigham City UT 84302-0707

MATHIAS, MILDRED ESTHER, botany educator; b. Sappington, Mo., Sept. 19, 1906; d. John Oliver and Julia Hannah (Fawcett) M.; m. Gerald L. Hassler, Aug. 30, 1930 (dec.); children: Frances, John, Julia, James (dec.). A.B., Washington U., 1926, M.S., 1927, Ph.D. in Systematic Botany, 1929. Asst. Mo. Bot. Garden, 1929-30; research assoc. N.Y. Bot. Garden, 1932-36, U. Calif., 1937-42; herbarium botanist UCLA, 1947-51, lectr. botany, 1951-55, from asst. prof. to prof., 1955-74, vice chmn. dept., 1955-62, dir. bot. garden, 1956-74, prof. emeritus, 1974—; asst. specialist UCLA Exptl. Sta., 1951-55, asst. plant systematist, 1955-57, assoc. plant systematist, 1957-62; pres. Orgn. Tropical Studies, 1968-70; sec. bd. trustees Inst. Ecology, 1975-77; exec. dir. Am. Assn. Bot. Gardens and Arboreta, 1977-81. Hon. trustee Mo. Botanical Garden, St. Louis, 1989-93. Recipient Medal of Honor Garden Club Am., 1982, Charles Laurence Hutchinson medal Chgo. Hort. Soc., 1988; Mildred E. Mathias Bot. Garden named in her honor, UCLA, 1979; B.Y. Morrison lectr. USDA, 1989. Fellow Calif. Acad. Scis.; mem. AAAS (pres. Pacific div. 1977) Am. Soc. Plant Taxonomy (pres. 1964), Bot. Soc. Am. (Merit award 1973, pres. 1984), Soc. Study of Evolution, Am. Soc. Naturalist, Am. Horticulture Soc. (Sci. citation 1974, Liberty Hyde Bailey medal 1980). Office: UCLA Dept Biology Los Angeles CA 90024-1606

MATHIS, JAMES FORREST, retired petroleum company executive; b. Dallas, Sept. 28, 1925; s. Forrest and Martha (Godbold) M.; m. Frances Ellisor, Sept. 4, 1948; children: Alan Forrest (dec.), Lisa Lynn Lambeth. BS in Chem. Engring., Tex. A&M U., 1946; MS, U. Wis., 1951, PhD, 1953. Rsch. engr. Humble Oil & Refining Co., Baytown, Tex., 1946-49, 53-61, mgr. rsch. and devel., 1961-63, mgr. Splty. products planning, 1963-65; v.p. Exxon Rsch. & Engring. Co., Linden, N.J., 1966-68; sr. v.p., dir. Imperial Oil Ltd., Toronto, Ont., Can., 1968-71; v.p. tech. Exxon Chem. Co., Florham Park, N.J., 1971-80; v.p. sci. and tech. Exxon Corp., N.Y.C., 1980-84; ret., 1984; cons. Arthur D. Little, 1985-92, ChemShare Corp., 1989-92; chmn. N.J. Commn. Sci and Tech., 1988—; dir. Laser Recording Systems, Inc., 1989—, N L Industries, 1985-86, Hanlin Corp., 1989—. Bd. dirs. Chem. Industry Inst. Toxicology, 1975-83, treas., 1977-80, chmn., 1980-83; trustee Wis. Alumni Rsch. Found., 1984—, pres. 1993—; bd. chem. sci. and tech. of Nat. Rsch. Coun., 1987-89. Served with AC, USNR, 1944-45. Recipient Disting. Alumni award Coll. Engring. Tex. A&M U., 1982, Disting. Svc. citation Coll. Enring. U. Wis., 1969. Fellow Am. Inst. Chem. Engrs. (interim exec. dir., sec., 1987-88, Robert L. Jacks award in Mgmt. 1985, Van Antwerpen award for Svc. to Inst. 1989); mem. AAAS, NAE, Am. Chem. Soc. (Earle B. Barnes award for Chem. Rsch. Mgmt. 1984), Sigma Xi, Phi Lambda Upsilon, Tau Beta Pi. Presbyterian. Achievements include 2 patents in field. Home: 96 Colt Rd Summit NJ 07901-3042 Office: PO Box 3 Summit NJ 07902-0003

MATHLOUTHI, MOHAMED, chemistry educator; b. Kalaa-Kebira, Tunisia, Nov. 14, 1940; s. Abderrahman and Khadija (Benameur) M.; m. Zahia Bentayeb, July 26, 1986; children: Jazia, Ridha. Engr., Ensia, Massy, France, 1966; Dr. Engring., U. Dijon, France, 1973, Dr. Sci., 1980. Head lab. Tunisian Sugar Co., Beja, 1967-69; asst. prof. chemistry U. Dijon, 1970-79, maitre asst., 1980-84, maitre de conf., 1985-86; prof. chemistry U. Reims, France, 1987—. Editor: Food Packaging, 1986, Sweet Taste Chemoreception, 1993, Food Packaging and Preservation, 1993; author over 80 sci. and tech. papers, book chpts. Recipient B. Delessert medal, C.E.D.U.S., Paris, 1981. Mem. Am. Chem. Soc., French CArbohydrate Group. Avocations: poetry, music, psychology, food. Office: Faculte des Sciences, BP 347, Reims France 51062

MATHUR, RADHEY MOHAN, electrical engineering educator, dean; b. Alwar, India, Feb. 2, 1936; married, 1965; 2 children. BSc, U. Rajasthan, India, 1956; BTech, Indian Inst. Tech., India, 1960; PhD in Elec. Engring., U. Leeds, 1969. Rsch. fellow Nat. Rsch. Can. Coun., U. Man, 1969-70; from asst. prof. to prof. U. Man, 1970-87; rsch. fellow Nat. Rsch. Can. Coun., U. Man, 1971-72; head elec. engring. U. Man, 1980-87; prof., dean Sch. Engring. Sci. U. Western Ont., London, Ont., Can., 1987—; lectr. elec. engr. U. Jodhpur, 1960-64, Malaviya Regional Engring. Coll., 1964-65, Indian Inst. Tech., 1965-66. Recipient Indian Inst. Engring. prize, 1964; Centennial award Inst. Elec. and Electronics Engrs., 1984. Mem. Can. Elec. Assn., Inst. Elec. Electronics Engrs. Achievements include research in rotating machines; reluctance and stepper motors, transient and steady state performance and design optimization; power systems modeling; HVDC systems; static compensators; power systems. Office: University of Western Ontario, Faculty of Engineering, London, ON Canada N6A 3K7*

MATHUR, VEERENDRA KUMAR, physicist, researcher, project manager; b. Jobner Road, Rajasthan, India, July 11, 1935; came to U.S., 1979; s. Niranjan Lal and Sarla (Rani) M.; m. Sushma Mathur, Dec. 9, 1961; children: Himanshu, Sudhanshu. MS, Agra (India) U., 1955; PhD, Saugor U., Sagar, India, 1963. Lectr. S.B.R. Coll., Bilaspur, India, 1955-59; asst. prof. Saugor U., 1959-63; from asst. prof. to assoc. prof. Kurukshetra (India) U., 1963-82; commonwealth rsch. fellow Heriot-Watt U./AERE, Edinburgh and Harwell, Eng., 1975-76; rsch. assoc. U. Md., College Park, 1979-82; project engr. Bendix Field Engring., Laurel, Md., 1982; sr. rsch. physicist, rsch. physicist Naval Surface Warfare Ctr., Silver Spring, Md., 1982—; sessional chmn. Nat. Symposium on Thermoluminescence, Kalpakkam, Madras, India, 1975; sessional co-chmn. 10th Internat. Conf. on Solid State Dosimetry, Washington, 1992. Author: (chpt.) Optical Properties of CAF, 1991. Director Ardmore Enterprises for Handicapped, Mitchelville, Md., 1986—. Recipient award of merit Fed. Lab. Consortium for Tech. Transfer, 1991. Mem. Indian Physics Assn. (vice chmn. Kurukshetra, 1975-76), Optical Soc. Am. Achievements include patents for ceramic arc-discharge source for an ion implanter (India), for method of laser discrimination using stimulated luminescence, for laser detection and discrimination system; patent pending for real-time infra-red imaging using a graded band gap semiconductor. Home: 3680 Sellman Rd Beltsville MD 20705-2840 Office: Naval Surface Warfare Ctr 10901 New Hampshire Ave Silver Spring MD 20903-5000

MATIENZO, RAFAEL ANTONIO, computer engineer; b. Maracaibo, Zulia, Venezuela, Dec. 2, 1956; s. Manuel and Eileen (Pardo) M. Grad. systems engr. cum laude, U. Met., Caracas, Venezuela, 1978. Advisor to the minister Ministry Fin., Caracas, Venezuela, 1979-83; cons. data processing pvt. practice, Caracas, 1984—; adj. instr. Univ. Met., Caracas, 1978—, Armed Forces Polytech. Inst., 1988-91. Author: Simulation of a Cement Plant, 1978; inventor: software programs for acad. use, 1986-92; composer: Venezuelan Folk Songs, 1971—. Mem. IEEE, Venezuelan Engring. Assn. (M.C. Perez prize 1978), Data Processing Mgmt. Assn., Assn. for Computing Machinery, Nat. Drum Assn., Caracas Country Club. Avocation: music. Home: Edf Zoila # 3A Calle F, Sta Rosa Lima, Caracas 1061, Venezuela Office: Apartado 66505, Caracas 1061-A, Venezuela

MATIGAN, ROBERT, electrical engineer; b. Forest Hills, N.Y., Sept. 4, 1961; s. Armen and Christin (Haroutunian) M. BSEE, N.Y. Inst. Tech., 1985; MSEE, SUNY, Stony Brook, 1987. Radar systems engr. Grumman Corp., Bethpage, N.Y., 1987-89, radar flight test engr., 1989-91; radar flight test engr. Grumman Corp., Point Mugu, Calif., 1991—. Home: 555 Laurie Ln Thousand Oaks CA 91360 Office: Grumman Corp PO Box 42232 Port Hueneme CA 93044-4532

MATIJEVIC, EGON, chemistry educator, researcher, consultant; b. Otocac, Croatia, Apr. 27, 1922; came to U.S., 1957; s. Grgur and Stefica (Spiegel) M.; m. Bozica Biscan, Feb. 27, 1947. Diploma in chem. engring., U. Zagreb, 1944, PhD in Chemistry, 1948, Dr. Habil. in Phys. Chemistry, 1952; DSc (hon.), Lehigh U., 1977, M. Curie-Sklodowska U., Lublin, Poland, 1990; DSc. (hon.), V. Clarkson U., 1992. Instr. chemistry U. Zagreb, Yugoslavia, 1944-47; sr. instr. phys. chemistry U. Zagreb, 1949-52, privat dozent in colloid chemistry, 1952-54, dozent in phys. and colloid chemistry, 1955-56, on leave, 1956-59; rsch. assoc. Inst. Cinematography, Zagreb, 1948; rsch. fellow dept. colloid sci. U. Cambridge, Eng., 1956-57; vis. prof. Clarkson Coll. Tech., Potsdam, N.Y., 1957-59; assoc. prof. chemistry Clarkson Coll. Tech., Postdam, N.Y., 1960-62; prof. Clarkson U., Postdam, 1962-86, disting. univ. prof., 1986—; assoc. dir. Inst. Colloid and Surface Sci. Clarkson Coll. Tech., 1966-68; dir. inst., 1968-81, chmn. dept. chemistry, 1981-87; vis. prof. Japan Soc. for Promotion Sci., 1973, U. Melbourne, Australia, 1976, Sci. U. Tokyo, 1979, 84; vis. scientist U. Leningrad, USSR, 1977; Internat. Atomic Energy Agy. adviser Buenos Aires, Argentina, 1978, 80; fgn. guest Inst. Colloid and Interface Sci. Sci. U. Tokyo, 1982; lectr. in field; mem. adv. com. Univs. and Space Research Assn.; referee NATO Advanced Study Inst. Author: (with M. Kesler) General and INorganic Chemistry for Senior High Schools, 11 edits., including Croatian, Macedonian, Hungarian, Italian, 1943-63; translator: Einfuhrung in die Stochiometrie (Nylen and Wigren), 1948; editor: (with Walter J. Weber) Adsorption from Aqueous Solution, 1968, Surface and Colloid Science, vols. 1-15, 1969-92; regional editor Colloid and Polymer Sci., 1966—; contbr. numerous articles to profl. publs. Recipient Gold medal Am. Electroplaters Soc., 1976; guest of honor 56th and 63rd Colloid and Surface Sci. Symposiums, Blacksburg, Va., 1982, Seattle, 1989. Mem. Am. Chem. Soc. (councilor div. colloid and surface chemistry 1982-87, chmn. 1969-70, Kendall award 1972, Langmuir Disting. Lectureship award 1985, Ralph K. Iler award 1993), Kolloid Gesellschaft (Thomas Graham award 1985), Internat. Assn. Colloid Interface Sci. (pres. 1985-87), Chem. Soc. Japan, Inst. Colloid and Interface Sci. of Sci. of Tokyo (hon.), Phalanx Soc., Croatian Acad. Scis. and Arts (fgn.), Am. Ceramic Soc. (hon.), Materials Rsch. Soc. Japan (hon.), Acad. Ceramics (Italy), Croatian Chem. Soc. (Bozo Tezak medal 1991), Sigma Xi (Clarkson Coll. Tech. chpt. award 1972, nat. lectr. 1987-89). Roman Catholic. Office: Clarkson U Dept Chemistry Potsdam NY 13699-5814

MATILAINEN, RIITTA MARJA, pediatrician, pediatric neurologist; b. Hankasalmi, Finland, Dec. 10, 1948; s. Paavo Eskil and Lyyli Dagmar M. Licentiate Medicine, U. Turku, Finland, 1974; student pediatrics, U. Kuopio, Finland, 1982, student ped. neurology, 1988, MD, 1988. Gen. practice medicine Communal Health Svc., Heinola, Finland, 1974-77; asst. surgeon Childrens Hosp. U. Kuopio, Finland, 1977-82, asst. surgeon pediatric neurology, 1983-84; asst. surgeon neurology Vaajasalo Hosp., Kuopio, Finland, 1982-83; sub-chief med. officer Vaajasalo Hosp., Kuopio, 1984-86, chief physician pediatric neurology, 1987-93. Author: (with others) A Multidisciplinary Case Control Study of Mental Retardation in Children of Four Birth Cohorts, 1985; The Significance of Intrauterine Growth Retardation for the Prognosis of Preterm Children, 1988, MBD and Rehabilitation Report on a Group Therapy; Effect of Vigabatrium on Epilepsy in Mentally Retarded Children, 1988; contbr. articles to profl. jours. Mem. Jr. C. of C. (exec. v.p. youth sect.). Achievements include research in the high prevalence of aspartylglucosaminuna among school-age children in eastern Finland, the prevalence of the fragile X syndrome in four birth cohorts of children of school-age, population cytogenetics of folate sensitive fragile sites II autosomal rare fragile sites. Home: Kasarmik 10 C 39, SF-70110 Kuopio Finland Office: Vaajasalo Hosp, Kortejoki, SF7110 Kuopio Finland

MATIN, ABDUL, microbiology educator, consultant; b. Delhi, India, May 8, 1941; came to U.S., 1964, naturalized, 1983; s. Mohammed and Zohra (Begum) Said; m. Mimi Keyhan, June 21, 1968. BS, U. Karachi, Pakistan, 1960, MS, 1962; PhD, UCLA, 1969. Lectr. St. Joseph's Coll., Karachi, 1962-64; research assoc. UCLA, 1964-71; sci. officer U. Groningen, Kerklaan, The Netherlands, 1971-75; from asst. to assoc. prof. microbiology Stanford U., Calif., 1975—; cons. Engenics, 1982-84, Monsanto, 1984—, Chlorox, 1992—; chmn. Stanford Recombinant DNA panel; lectr. ASM Found.; mem. Accreditation Bd. for Engring. and Tech.; convener of microbiol. workshop and confs. Mem. editorial bd. Jour. of Bacteriology; bd. dirs. Ann. Rev. Microbiol., Rev. of NSF and other Grants; contbr. numerous publs. to sci. jours. Fellow Fulbright Found., 1964, NSF, 1981—, Ctr. for Biotech. Research, 1981-85, EPA, 1981-84, NIH, Coll. Biotech., U.N. Tokten, 1987. Mem. AAAS, AAUP, Am Soc. for Microbiology (Found. lectr. 1991-92), Soc. Gen. Microbiology, Soc. Indsl. Microbiology, No. Soc. Indsl. Microbiology (bd. dirs.), Biophys. Soc. Avocations: reading, music, walking. Home: 690 Coronado Ave Palo Alto CA 94305-1039 Office: Stanford U Dept Microbiology and Immunology Fairchild Sci Bldg Stanford CA 94305-5402

MATKOWSKY, BERNARD JUDAH, applied mathematician, educator; b. N.Y.C., Aug. 19, 1939; s. Morris N. and Ethel H. M.; m. Florence Knobel, Apr. 11, 1965; children: David, Daniel, Devorah. B.S., CCNY, 1960; M.E.E., NYU, 1961, M.S., 1963, Ph.D., 1966. Fellow Courant Inst. Math. Scis., NYU, 1961-66; mem. faculty dept. math. Rensselaer Poly. Inst., 1966-77; John Evans prof. applied math. and mech. engring. Northwestern U., Evanston, Ill., 1977—; vis. prof. Tel Aviv U., 1972-73; vis. scientist Weizmann Inst. Sci., Israel, summer 1976, summer 1980, Tel Aviv U., summer 1980; cons. Argonne Nat. Lab., Sandia Labs., Lawrence Livermore Nat. Lab., Exxon Research and Engring. Co. Editor Wave Motion-An Internat. Jour., 1979—, Applied Math. Letters, 1987—, SIAM Jour. Applied Math., 1976—, European Jour. Applied Math., 1989—, Random and Computational Dynamics, 1991—, Internat. Jour. SHS, 1992—, Jour. Materials Synthesis and Processing, 1992—; mem. editorial adv. bd. Springer Verlag Applied Math. Scis. Series; contbr. chpts. to books, articles to profl. jours. Fulbright grantee, 1972-73; Guggenheim fellow, 1982-83. Fellow Am. Acad. Mechanics; mem. AAAS, Soc. Indsl. and Applied Math., Am. Math. Soc., Combustion Inst., Am. Phys. Soc., Am. Assn. Combustion Synthesis, Conf. Bd. Math. Scis. (coun., com. human rights of math. scientists), Com. Concerned Scientists, Soc. Natural Philosophy, Sigma Xi, Eta Kappa Nu. Home: 3704 Davis St Skokie IL 60076-1745 Office: Northwestern U Technological Inst Evanston IL 60208

MATNEY, WILLIAM BROOKS, VII, electrical engineer, marine engineer; b. Detroit, June 14, 1935; s. William Brooks VI and Maurine (Huff) M.; m. Carolyn Weaver, Dec. 29, 1959; children: C. Melinda Matney Levin, William Brooks VIII, James Richard, Robert Weaver. BSEE, Marine Engring., U.S. Naval Acad., 1957. Commd. ensign USN, 1957, advanced through grades to lt. j.g., 1962; div. officer aircraft carrier USS Bon Homme Richard, 1957-59; combat, info. officer attack transport USS Paul Revere, 1959-62; design engr. gen. svcs. dept. Exxon Co. USA, Houston, 1962-70; constrn. engr. mktg. dept. Exxon Co. USA, Dallas, 1970-80; oil prodn. engr. prodn. dept. Exxon Co. USA, Ventura, Calif., 1980-85; underground tank engr. mktg. dept. Exxon Co. USA, Houston, 1985-89, coord. regulatory compliance mktg. dept., 1989-92; mgr. aboveground tank svcs. Tanknology Corp. Internat., Houston, 1992—; work group leader underground tank certification Am. Petroleum Inst., Washington, 1991-92; committeeman leak detector group U.S. EPA, Washington, 1991-92. Speaker OSHA safety rules Nat. Safety Coun., 1990; contbr. articles to profl. jours. Trustee Ind. Sch. Dist., Richardson, Tex., 1979-80; hon. life mem. Tex. PTA, 1974—, Calif. PTA, 1982—. Guest of Astronauts, Nat. Space and Aeronautics Agy., 1970. Mem. Nat. Assn. Corrosion Engrs., Am. Contract Bridge League (cert. dir., Regional Master award), Environ. Info. Assn. (underground tank task force leader 1992—, sec. Tex. chpt. 1990-92). Southern Baptist. Achievements include invention and development of silent ship tracking method (at sea), flood control device, cathodic protection applications. Home: 1503 Anvil Houston TX 77090

MATOS, JOSE GILVOMAR ROCHA, civil engineer; b. Fortaleza, Ceara, Brazil, Nov. 25, 1944; s. Gilvan Souza and Vicencia Rocha (Fernandes) M.; m. Rosa Maria Barbosa Carneiro, Oct. 11, 1969; children: Ana Gabriela, Carolina, Bernardo. Student, Brazilian Air Force, 1966-69, Cath. U., Rio de Janeiro, 1970-75. Cert. civil engring. Fin. mgr. Furnas Centrais Eletricas S.A., Rio de Janeiro, 1976—; dir. SBS Biotecnologia Agricola Ltda., Rio de Janeiro, 1985, dir., pres., 1986—; deliberative bd. mem. Rio de Janeiro Sci. Park for Biotech., Rio de Janeiro, 1987—. Author: (with others) BID/Colciencias, 1988; contbr. articles to jours. and newspapers. Mem. PMDB polit. party, Rio de Janeiro, 1991. Lt. Brazilian Air Force, 1969-76. Decorated Ten Yr. Hon. Svc. medal Brazilian Air Force, 1975. Mem. Brazilian Biotech. Enterprises Assn.. Roman Catholic. Home: Rua Guimaraes Rosa 143, Apt 301, Rio de Janeiro 22.793, Brazil Office: SBS Estrada dos Tres Rios, 90 Sala 227, Rio de Janeiro 22.755, Brazil

MATOSSIAN, JESSE NERSES, physicist; b. L.A., Feb. 2, 1952; s. Hagop Sarkis and Alice Elizabeth (Barsoomian) M. BS in Physics, U. So. Calif., L.A., 1975; MS in Physics, Stevens Inst. Tech., Hoboken, N.J., 1976; PhD in Physics, Stevens Inst. Tech., 1983. Mem. tech. staff Hughes Rsch. Labs., Plasma Physics Lab., Malibu, Calif., 1983-91, sr. mem. tech. staff, sr. staff physicist, 1992—. Reviewer Jour. Propulsion and Power, 1987-91; contbr. articles to profl. jours.; 6 patents in field. Patron mem. Los Angeles County Mus. of Art, sustaining mem. graphic arts coun. Recipient Superior Performance award Hughes Rsch. Labs., 1991, also 33 div. invention awards. Mem. AIAA, IEEE, Am. Phys. Soc. (life), N.Y. Acad. Scis., Sigma Xi. Avocations: art history, collecting 19th and 20th century European paintings and 16th century engravings, classical music, travel.

MATSEN, JOHN MARTIN, pathology educator, microbiologist; b. Salt Lake City, Feb. 7, 1933; s. John M. and Bessie (Jackson) M.; m. Joneen Johnson, June 6, 1959; children: Marilee, Sharon, Coleen, Sally, John H., Martin K., Maureen, Catherine, Carl, Jeri. BA, Brigham Young U., 1958; MD, UCLA, 1963. Diplomate Am. Bd. Pediatrics, Am. Bd. Pathology, Spl. Competence in Med. Microbiology. Intern UCLA, L.A., 1963-64; resident L.A. County Harbor/UCLA, Torrance, Calif., 1964-66; USPHS fellow U. Minn., Mpls., 1966-68, asst. prof., 1968-70, assoc. prof., 1971-74, prof., 1974; prof. U. Utah, Salt Lake City, 1974—, assoc. dean, 1979-81, chmn. Dept. of Pathology, 1981-93, v.p. health scis., 1993—; pres. Associated Regional and Univ. Pathologists, Inc., Salt Lake City, 1983-93, chmn. bd. dirs., 1993—. Author over 200 publs. in field. Recipient Sonnenwirth Meml. award Am. Soc. Microbiology, 1993. Mem. Acad. Clin. Lab. Physicians and Scientists (pres. 1978-79), Assn. of Pathology Chmn. (pres. 1990-92). Mem. LDS Church. Home: 2845 St Marys Way Salt Lake City UT 84108-2041 Office: U Utah Health Scis Ctr 50 N Medical Dr Salt Lake City UT 84132

MATSEN, JOHN MORRIS, engineer; b. Neenah, Wis., May 30, 1936; s. Morris and Bertha Rowena (Witt) M.; m. Sandra Louise Schwartz, May 8, 1971. BS in Engring., Princeton (N.J.) U., 1957; MS, Columbia U., 1959, PhD, 1963. Instr. Columbia U., N.Y.C., 1959-61; engr. Exxon Rsch. & Engring. Co., Florham Park, N.J., 1961-66, sr. engr., 1966-73, engring. assoc., 1973—; mem. tech. com. Particulate Solids Rsch. Inc., Chgo., 1976—. Editor: Fluidization Technology, 1976, Fluidization, 1980; patentee in field; contbr. articles to profl. jours. Member Clinton (N.J.) Twp. Planning Bd., 1976—, chmn., 1978-82; mem. Hunterdon County Agrl. Devel. Bd., N.J., 1982-87; councilman Clinton Twp., 1983—, mayor, 1986; trustee N.J. Symphony Orch. League, N.J., 1966-82, pres., 1973-79. NSF fellow, 1959-60. Mem. Am. Inst. Chem. Engrs., Am. Chem. Soc., Raritan Yacht Club, Princeton Club of N.Y., Rolls Royce Owners Club, Phi Lambda Upsilon. Republican. Avocations: organology, yachting, historic preservation. Home: 39 Sand Hill Rd Annandale NJ 08801-3111 Office: Exxon Rsch & Engring Co PO Box 101 Florham Park NJ 07932-0101

MATSUBARA, TOMOO, software scientist; b. Tokyo, June 24, 1929; s. Goro and Fumiko (Miyako) M.; m. Mariko Matsubara, May 17, 1960; children: Yoko, Osamu. BS in Mech. Engring., Waseda U., Tokyo, 1950. System engr. USAF, Tachikawa (Japan) Base, 1947-56; machine designer Kameari Works Hitachi Ltd., Tokyo, 1956-65, sect. mgr. computer div., 1965-67; project mgr. Kanagawa Works Hitachi Ltd., Hadano, 1967-69; sect. mgr. software works Hitachi Ltd., Yokohama, Japan, 1969-70; dept. mgr. Hitachi Software Engring., Yokohama, 1970-83, chief engr., 1983-91; chmn. standardization subcom. Japan Info. Svc. Industry Assn., Tokyo, 1980-91; mem. program com. 10th, 11th Internat. Conf. on Software Engring., Singapore, Pitts, 1988-89. Mem. editorial bd. Am. Programmer, 1993—, Internat. and Software Tech., 1990—; contbr. tech. articles to profl. jours. Recipient awards Software Industry Assn., Tokyo, 1984, Japan Info. Svc. Industry Assn., 1989. Mem. IEEE (industry adv. bd. software mag.), Japan Software Engrs. Assn. (bd. mem. 1986—). Avocations: folk music collection, music composition. Home and office: 1-9-6 Fujimigaoka Ninomiya, Nakagun, Kanagawa 259-01, Japan

MATSUDA, YASUHIRO, computer scientist; b. Ako, Hyogo, Japan, Jan. 29, 1947; s. Eiichi and Kiyoko M.; m. Eiko Miura, Sept. 19, 1981. BS, Osaka U., Suita, Osaka, Japan, 1969, M in Engring., 1971, D in Engring., 1980; MS, Yale U., 1982. Systems engr. IBM Japan Ltd., Osaka, 1971-80, adv. researcher, Tokyo, 1983-85, mgr. rsch. group, 1985-87; prof., Shinshu U., Nagano, 1987—; postdoctoral assoc. Yale U., New Haven, 1981-82; lectr., author corr. course Japan Daily Indsl. Newspaper Co., Ltd., Tokyo, 1977—. Recipient Ann. award for best paper Textile Machinery Soc. Japan, 1975; IBM Japan Ltd. grantee, 1980-82. Mem. Japan Soc. Mech. Engrs., The Soc. for Computer Simulation, Japan Info. Processing Soc. Home: N 201, 53

Wakasato, 380 Nagano Japan Office: Shinshu U, Dept Mech Systems Engring, 500 Wakasato, 380 Nagano Japan

MATSUGO, SEIICHI, chemistry educator, chemistry researcher; b. Takaoka, Toyama, Japan, Nov. 18, 1952; s. Nosaku Noto and Mitsu Matsugo; m. Michiko Kawami, Oct. 7, 1984; children: Motohisa, Masanori, Hiromichi. Bachelor, Kyoto (Japan) U., 1975, Master, 1978, PhD, 1981. Rsch. assoc. Niigata (Japan) Coll. Pharmacy, 1981-86; assoc. prof. Kobe (Japan) U. Mercantile Marine, 1986-92; assoc. prof. dept. che, and biochem. engring Toyama U., 1992—. Author: Organic Peroxides, 1991, Active Oxygen and Luminescence, 1989, Active Oxygen, 1988, Free Radicals in Clinical Medicine, 1991. Office: Toyama U, Gofuku 3190, Toyama 930, Japan

MATSUI, EIICHI, economics and sociology educator; b. Osaka-shi, Osaka-fu, Japan, May 12, 1925; s. Kyoichi and Uta M.; m. Hiroko Masukura, May 23, 1959; children: Hina, Kyona. B in Econs., Kyoto U., Kyoto-shi, Kyoto-fu, 1951. Asst. of econs. Kochi U., Kochi-shi, Kochi-ken, 1952-60; lectr. econs. Kochi U., Kochi-Shi, Kochi-Ken, 1960-62, assoc. prof. econs., 1962-68, prof. econs., 1968-89, prof. emeritus, 1989—; prof. sociology Kibi Internat. U. of Takahashi Gaukuen, Takahashi-shi, Okayama-Ken, 1989—; mem. coun. Kochi U., 1979-85; mem. com. Minimum Wage Coun., Kochi-Shi, Kochi-Ken, 1969-78, Comml. Activities Coordination Coun., Kochi-Shi, 1976-90; chmn. Employment Security Coun., Kochi-Shi, 1984—; bd. dirs. Shikoku Productivity Ctr., Takamatsu-Shi, Kagawa-Ken, 1985—; lectr. in field; cons. Kuroshio Region Rsch. Inst., Kochi-Shi, 1989—; chmn. Kochi-Ken Local Employment Devel. Coun., 1990—. Editor: Theories on Industrial Relations, 1975, Welfare State and Labor, 1981, Program of Commerce in Aki Area, 1987, Employment in Kochi-Ken, 1988. Mem. Kyoto U. Econ. Soc. (writer 1960, 70, 85), Social Policy Soc. (mgr. 1978-79), Inst. Laborer's Welfare (counselor 1977), Econ. Soc. Kochi U. (editor 1971-79), Chu-Shikoku Comml. and Econ. Soc. (bd. dirs. 1972-76). Avocation: horticulture. Home: 1504-6 Kamobe, 780 Kochi-Shi Kochi-Ken Japan Office: Kibi Internat U Takahashi Gaukuen, Iga-Cho 8, Takahashi 716 Okayama, Japan

MATSUI, IWAO, chemical engineer; b. Tokyo, July 10, 1936; s. Kiichi and Mie (Kobayashi) M.;m. Rieko Hata, Sept. 12, 1967; 1 child, Ryoko. PhD in Chem. Engring., Lehigh U., 1973; 5th yr. cert. pulp and paper mgmt., U. Maine, 1978. Mgr. Nalco-Hakuto Chem. Co., Tokyo, 1973-79; dep. gen. mgr. United Petroleum Devel. Co., Tokyo, 1980-84; chief engr. Hakuto Chem. Co., Tokyo, 1984-86; pres. Assoc. Coating Inspectors, Tokyo, 1986—; chief cons. Tokai Toso Co., Tokyo, 1988-92; rsch. asst. NSF, Lehigh U., 1967. Recipient Fellowship William Gotshall, Lehigh U., 1971, Mobil Oil Co., Lehigh U., 1972. Mem. Nat. Assn. Corrosion Engrs. (lic. internat. coating insp.), Steel Structures Painting Coun., Soc. Am. Mil. Engrs., Fed. Soc. Coating Tech. Achievements include rsch. in fluorinated polyurethanes and silicones as applied to decontaminable and foulant-release paints and development of computerized program for relative humidity, absolute humidity and dew point calculation. Home and Office: 3-5-5 Ebisu Shibuya-Ku, Tokyo 150, Japan

MATSUMOTO, HIROSHI, mechanical engineer; b. Akashi, Hyogo, Japan, Mar. 11, 1948; s. Yutaka and Sumie Matsumoto; m. Fuchiko Takahashi; children: Asuka, Takumi. B. Engring., Akashi Tech. Coll., 1968. Devel. engr. Nippon Electric Co., Ltd., Tokyo, 1968-80, Olympus Optical Industry Co., Ltd., Tokyo, 1980-81; chief engr. Seiko-Sha Corp., Tokyo, 1981-83, Digital Equipment Corp. Japan, Tokyo, 1983-91; dir planning div. Systems Intelligence Products, 1991-92; sr. quality control engr. Seagate Japan, 1992-93; chief engr. Advanced Mobile System, 1993—; cons. Nippon Denki Seiki Co., Ltd., Tokyo, 1980, Sekisui Chem. Co., Ltd., Tokyo, 1987-88; lectr. Kao Corp., Tokyo, 1987. Recipient Commendation Mgmt. Course, Japan Exec. Confabulation Ctr., Tokyo, 1986. Mem. Laser Soc. Japan, So. Pacific Club. Home: 4-15-5-402 Motogou, Kawaguchi 332, Japan

MATSUMOTO, KAZUKO, chemistry educator; b. Tokyo, Oct. 27, 1949; d. Sakae and Tsuneko (Yokoyama) Yamamura; m. Akira Matsumoto, Oct. 30, 1974; children: Masahiko, Hidehiko. BS, U. Tokyo, 1972, MS, 1974, PhD, 1977. Rsch. assoc. dept. chemistry U. Tokyo, 1977-84; assoc. prof. dept. chemistry Waseda U., Tokyo, 1984-89, prof., 1989—. Recipient award Japan Chem. Soc., Tokyo, 1990. Mem. Japan Soc. for Analytical Chemistry (bd. dirs. 1990-91, award for younger scientists 1984). Achievements include patent for antitumor-active tetranuclear platinum complexes. Home: Noborito Tama-ku, 2578-1-708, 214 Kawasaki Japan Office: Waseda Univ, Dept Chemistry, Shinjuku-ku, 169 Tokyo Japan

MATSUMURA, FUMITAKE, economics educator; b. Tokyo, Sept. 5, 1942; s. Kyuuya and Tamae Matsumura; m. Mariko Morishita, Oct. 29, 1970; children: Naoto, Kana. B in Commerce, Waseda U., 1965, M in commerce, 1968; D in Econs., Kyoto U., 1987. Asst. lectr. Kinki U., Higashiosaka City, Japan, 1968-70, lectr., 1971-74, assoc. prof., 1974-76; assoc. prof. Osaka Keizai U., Osaka City, Japan, 1976-82; prof. econs. Osaka Keizai U., Osaka City, 1982—, dean Sch. Econs., 1989-91, chmn. publicity com., 1991-93. Author: The Structure of American Economy as a Debtor Nation, 1988. Mem. Japan Soc. Internat. Econs. Avocations: baseball, rugby, jazz. Home: 2 4 12 Hiyoshidai, Takatsuki 569, Japan Office: Osaka Keizai U, 2 2 8 Osumi, Higashiyodogawa-Ku, Osaka 533, Japan

MATSUNO, KOICHIRO, biophysics educator; b. Asahikawa, Hokkaido, Japan, Mar. 14, 1940; s. Kunie and Misue (Miyatake) M.; m. Yukiko Kondoh, Mar. 23, 1964; children: Kyoko, Nahoko. BSc, U. Tokyo, 1963, MSc, 1965; PhD, MIT, 1971. Mem. tech. staff NEC Corp., Tokyo, 1965-68, 72-77; assoc. prof. Toyo U., Tokyo, 1978-79; assoc. prof. Nagaoka (Japan) U. Tech., 1980-84, prof., 1985—; vis. prof. U. Miami, Fla., 1982-84. Author: Molecular Evolution and Protobiology, 1984, Protobiology: Physical Basis of Biology, 1989, The Origin and Evolution of the Cell, 1992; editor Jour. BioSystems, 1988, Jour. Uroboros, 1991. Nat. Found. for Cancer Rsch. grantee, 1984. Mem. Sigma Xi. Home: 606-55 Mizuno, Sayama 350-13, Japan Office: Nagaoka U Tech, 1603-1 Kamitomioka, Nagaoka 940-21, Japan

MATSUSHITA, KEIICHIRO, sociology educator; b. Kobe, Japan, Oct. 8, 1953; s. Haruaki and Shoko (Minami) M.; m. Uyen Ton-Nu-Le, June 27, 1981; children: Kenichi, Kaoru, Megumi. BA, Kyoto (Japan) U., 1976, MA, 1978; PhD in Econs., U. Mich., 1986. Rsch. assoc. Ctr. Southeast Asian Studies Kyoto U., 1980-85; rsch. scientist Inst. Population Problems Ministry of Health and Welfare, Tokyo, 1985-88; demographer Japan Internat. Cooperation Agy., Dept. Census and Stats., Colombo, Sri Lanka, 1988-89; assoc. prof. sociology, coord. Ctr. Regional Studies Ryukoku U., Otsu, Shiga, Japan, 1989—. Mem. Population Assn. Am., Am. Econ. Assn., European Soc. Population Econs. Office: Ryukoku U, Seta, Otsu 520-21, Japan

MATSUURA, TERUO, chemistry educator; b. Osaka, Japan, Dec. 19, 1925; s. Takeo and Fumi (Nakata) M.; m. Fumi Ohshiro, Oct. 28, 1956; children: Masato, Jiro. BSc, Osaka U., 1949, DSc, 1956. Rsch. asst. dept. chemistry Osaka City U., 1949-57; vis. scientist NIH, Bethesda, Md., 1957-58; rsch. assoc. dept. chemistry MIT, Cambridge, Mass., 1958-59; asst. prof. dept. chemistry Osaka City U., 1959-63; assoc. prof. dept. synthetic chemistry Kyoto (Japan) U., 1963, prof. dept. synthetic chemistry, 1963-89; prof. dept. materials chemistry Ryukoku U., Otsu, Japan, 1990—; prof. emeritus Kyoto U., 1989—. Author: Organic Photochemistry, 1970, Oxygenation Reactions, 1977; contbr. about 400 articles to profl. jours. including Organic Photochemistry, Reactive Oxygen Species, Bioorganic Chemistry; mem. editorial bd. Jour. Photochemistry and Photobiology, 1984—, Bioorganic Chemistry, 1983—. Pres. Photobiology Assn. Japan, 1989-90, Japanese Photochemistry Assn., Japan, 1989-91. Fogarty scholar NIH, 1985; recipient Chem. Soc. Japan prize, 1987, Soc. prize Japan Soc. Synthetic Organic Chemistry, 1974. Mem. Am. Chem. Soc., Royal Soc. Chemistry (London). Home: 21-26 Kawashima Gondencho, Saikyoku Kyoto 615, Japan Office: Ryukoku U Faculty Sci Tech, Seta, Otsu 520-21, Japan

MATSUZAKI, TAKAO, chemist; b. Iwate, Japan, Aug. 1, 1945; s. Shigeru and Taki (Ueda) M.; m. Eriko Miyaoka, Dec. 6,1978; children: Yuko, Mika. BS in Pharm. Scis., U. Tokyo, 1968, MS in Pharm. Scis., 1970, D in Pharm. Scis., 1988. Rsch. scientist Mitsubishi Kasei Corp., Yokohama, Japan, 1970-73, chief rsch. scientist, 1975—; cancer rsch. scientist Roswell Park Meml. Inst., Buffalo, 1973-75. Co-author: X-ray Diffraction, 1988, Molecular Science of Crystals, 1989; contbr. articles to profl. jours. Mem. Chem. Soc. Japan, Pharm. Soc. Japan (Incentive award 1991), Am. Crystallographic Assn., Crystallographic Soc. Japan (editorial com. 1992—, councilor 1993—), Japan Soc. Analytical Chemistry (mem. subcom. 1989—, councilor 1993—). Home: 3-34-1 Kitanodai, Hachioji 192, Japan Office: Mitsubishi Kasei Corp, Midoriku, Yokohama 227, Japan

MATSUZAKI, YUJI, aerospace and biomechanics educator, researcher; b. Tokyo, Aug. 23, 1939; s. Nobuo and Fujiko (So) M.; m. Ayako Matsumoto, Mar. 21, 1969; 1 child, Kenji. BS, U. Tokyo, 1964, MS, 1966, PhD, 1969. Researcher Nat. Aerospace Lab., Tokyo, 1969-73, sect. chief, prin. researcher, 1973-84; prof. Nagoya (Japan) U., 1984—; vis. researcher U. Calif. San Diego, 1973-75; mem.liason com. for aerospace engring. rsch,, Japan Acad. Sci., Tokyo, 1991—. Co-author: Cardiovascular System and Basic Management (Japanese), 1985, Encyclopedia of Fluid Mechanics Vol. 1, 1986, Biomechanics (Japanese) 1991, Introduction to Biomechanics, (Japanese), 1992; editor: (proceedings) 2nd Joint Japan-U.S. conf. on Adaptive Structures, 1992. Grantee Artificial Intelligence Rsch. Promotion Found., Higahi-ku, Nagoya, Japan, 1992. Mem. AIAA, ASME, Japan Soc. Aeronautics and Astronautics, Japan Soc. Mech. Engrs. Achievements include pending patent on shape variable truss-type space structure with docking mechanism; proposed a method to predict airplanes flutter boundary; rsch. into structural dynamics, smart structures and aeroelasticity in Aerospace Engineering and Hydrodynamics in Bioengineering. Office: Nagoya U Dept Aerospace Engring, Furocho Chikusa, Nagoya 464-01, Japan

MATTAUCH, ROBERT JOSEPH, electrical engineering educator; b. Rochester, Pa., May 30, 1940; s. Henry Paul and Anna Marie (Mlinarcik) M.; m. Frances Sabo, Dec. 29, 1962; children: Lori Ann, Thomas J. BS, Carnegie Inst. Tech., Pitts., 1962; MEE, N.C. State U., Raleigh, 1963, PhD, 1967. Asst. prof. elec. engring. U. Va., Charlottesville, 1966-70, assoc. prof. elec. engring., 1970-76, prof. elec. engring., 1976-83, Wilson prof. elec. engring., 1983-86, Standard Oil Co. prof. sci. and tech., 1986-89, chmn. dept. elect. engring., 1987—, BP Am. prof. sci. and tech., 1989—; cons. The Rochester Corp., Culpeper, Va., 1983-88, Milltech Corp., Deerfield, Mass., 1985. Patentee: infrared detector; solid state switching capacitor; thin wire pointing method, whiskerless Schottby diode, controlled in-situ etch back growth technique. Bd. dirs. Children's Home Soc., Richmond, Va. Recipient Excellence in Instruction of Engring. Students award Western Electric, 1980. Fellow IEEE (Centennial medal 1984); mem. Eta Kappa Nu (recipient Oustanding Prof. in Elec. Engring. 1975), Sigma Xi, Tau Beta Pi. Office: U Va Dept Elec Engring Thornton Hall Charlottesville VA 22903-2442

MATTERN, DOUGLAS JAMES, electronics reliability engineer; b. Creede, Colo., May 19, 1933; s. John A. and Ethel (Franklin) M.; student San Jose (Calif.) City Coll., San Jose State U., 1956-58; m. Noemi E. Del Cippo, May 4, 1963. Reliability engr. Intersil, Sunnyvale, Calif., 1973-80; sr. engr. Data Gen. Corp., Sunnyvale, 1981-87; staff engr. Apple Computer, 1987—. Sec. Gen. World Citizens Assembly, San Francisco, 1975-86; dir. World Citizens Internat. Registry, U.S. Ctr., San Francisco, 1976—, World Citizen Diplomats, Palo Alto, Calif., 1988—; del. Peoples Congress, Paris, 1980—; pres. Assn. World Citizens, San Francisco, 1989—; pres. World Citizens Found., 1991—. Served with USN, 1951-55. Contbg. author: Building a More Democratic United Nations, 1991; editor World Citizen Newspaper, 1973—; contbr. 37 articles to profl. jours. Mem. Electron Microbean Analysis Soc., Union of Concerned Scientists Promoting Enduring Peace. Home: 2671 South Court St Palo Alto CA 94306-2462 Office: PO Box 51867 Palo Alto CA 94303

MATTERN, GERRY A., engineering consultant; b. Attica, Ind., June 16, 1935; s. George Edward and Wanda Mae (McCann) M.; p. Jane Ann Snell, Dec. 27, 1956; children: Kimberly Kaye, Geoffrey Kurtis, Kamala Anne, Kristin Annette. BSEE, Rose Polytech. Instit., 1958. Registered profl. engr., Penn., W.Va., Ind. With Mattern Electric Co., Attica, 1958; draftsman Yeager Architects, Terre Haute, Ind., 1956-58; application engr. W. Penn. Power Co., Greensburg, 1958-60; indsl. power engr. W. Penn. Power Co., Jeannette, 1960-62; product mgr. Pitts. Reflector Co., Irwin, Penn, 1962-63; owner, operator G.A. Mattern & Assocs., Ligonier, Penn., 1963—; ptnr. Palco, Inc., Greensburg, Penn., 1965—; owner, operator Gay 90's Dairy Queen, Ligonier, Penn., 1972—; instr. Profl. Engr's Review, Penn State U., Am. Instit. of Architects Review Class; adj. prof. Carnegie-Mellon U., Pitts., 1982—; design cons. Pitts. Reflector Co. Inventor infra-red electric furnace; designer electric heating equipment, emergency lighting equipment. Bd. dirs. Ligonier Twp. Planning Commn., 1982, Ligonier Twp. Sewerage Authority, 1991, Westmoreland County Coun. Boy Scouts Am., Heritage United Meth. Ch., Ligonier YMCA; pres. Ligonier C. of C., 1974-78. Recipient of Power-up award Westinghouse Electric Co., 1961. Mem. (life) Ligonier Booster's Club, (life) Fire Co. number 1, (arbitrator) Am. Arbitration Assn. Independent. Club: Tall Cedars (Westmoreland County), Lodge: Masons (Ligonier). Avocations: devel. of local YMCA, coaching basketball, orchard farming, antiques, house restoration. Home: RR 1 Box 230 Ligonier PA 15658-9728 Office: GA Mattern & Assocs 205 N Market St Ligonier PA 15658-1258

MATTERN, JAMES MICHAEL, physicist; b. Biloxi, Miss., Aug. 9, 1962; s. James L. and Ann Elizabeth (Midwood) M. BS in Nuclear Engring., Tex. A&M, 1986; MSc in Health Physics, Ga. Inst. Tech., 1988. Lic. profl. med. physicist, Tex. Health physicist Neely Nuclear Rsch. Ctr., Atlanta, 1987-88; radiation physicist Hendrick Med. Ctr., Abilene, Tex., 1989-90, St. Agnes Hosp., Fond Du Lac, Wis., 1990-92; radiol. physicist East Tex. Cancer Ctr., Tyler, Tex., 1992—; instr. Hendrick Med. Ctr., Abilene, 1989-90; cons., Tex., 1989—, Wis. Power and Light, Fond Du Lac, 1991. Mem. NSPE, Am. Assn. Physicists in Medicine, Health Physics Soc. Achievements include development of stereotactic radiosurgery program, design and implementation of self-powered neutron detector. Office: East Tex Cancer Ctr 721 Clinic Dr Tyler TX 75701

MATTESON, THOMAS DICKENS, aeronautical engineer, consultant; b. Mpls., Oct. 16, 1920; s. Herbert Sumner and Edna Gertrude (Dickens) M.; m. Rosemary Ann Hamilton, Jan. 11, 1947; children: Ann Claire, John Thomas. B Aero. Engring., U. Minn., 1942; MBA in Mgmt., NYU, 1956. Various managerial positions Pan Am. Airways, N.Y.C., San Francisco, 1946-59; asst. to exec. v.p. Pacific div. Pan Am. Airways, San Francisco, 1959-60; various managerial positions United Airlines, San Francisco, 1960-70, dir. maintenance planning, 1970-75, v.p. maintenance adminstrn., 1975-78; sr. cons. Am. Mgmt. Systems, Arlington, Va., 1978-85; cons. engr., Flat Rock, N.C., 1978—; guest lectr. U. Calif. Ext., Berkeley, L.A., 1966, 70; IBM vis. disting. scholar Northeastern U., Boston, 1983; chmn., mem. maintenance program planning coms. for B727, B737, B747; cons. on nuclear reactor maintenance programs EPRI/NRC; mem. sr. rev. panel Savannah River PRA, 1985-93. Author: NAVSEA Reliability-Centered Maintenance Handbook, 1980; contbr. numerous articles on aircraft, naval ship and nuclear systems maintenance mgnt. to profl. publs. Team leader Grace Commn., Washington, 1982. Lt. USNR, 1942-46, PTO. Assoc. fellow AIAA (chmn. tech. com. 1964-66, Systems Effectiveness and Safety medal 1976), Trout Unlimited. Republican. Christian Scientist. Avocations: cabinet maker, carver, fly fisherman, gourmet. Home and Office: 1933 Little River Rd Flat Rock NC 28731-9766

MATTHAEI, GAY HUMPHREY, interior designer; b. N.Y.C., Mar. 13, 1931; d. Robert Louis and Ethel Gladys Humphrey; m. Konrad Henry Matthaei, Nov. 16, 1956; children: Marcella, Leslie, Konrad. BA, Mt. Holyoke Coll., 1952; MIA, Columbia U., 1954; MA, cert. Russian Inst. Columbia U., 1954; grad. Parsons Sch. Design, 1970. Lectr., cons. NBC, 1956; dir. Radrick Prodns., Where Time Is a River, 1966-67; cons. N.Y.C. Parks Recreation and Cultural Adminstrn., 1970-72; assoc. Pearl R. Mitchell A.S.I.D., 1972-74, owner, 1974-91; owner, mgr. Gay Matthaei Interiors, N.Y.C., 1976-86. Trustee Mt. Holyoke Coll.; mem. Commn. on State Capital Preservation and Restoration, Conn., 1977-82; mem. Nat. Trust for Hist. Preservation. Mem. Asia Soc., River Club, Phi Beta Kappa. Restorations include Town Farms Inn, 1978, State Capital of Conn., 1977-78, Pres.'s House, Mt. Holyoke Coll., 1982, Samuel Russell House, Wesleyan Coll., 1984, Courtly Manor, Greenwich, Ct., 1987, Buhl Family Found., 1991. Home: 51 E 90th St New York NY 10028 Office: 505 Park Ave New York NY 10022

MATTHES, HOWARD KURT, computer consultant and researcher; b. Chgo., June 18, 1929; s. Otto Kurt and Dora Ella (Fleischer) M.; m. Theressa Burton, Aug. 31, 1952; children: Patrice, Marcia, Linda, David Kurt, Ruth. BS, U. Utah, 1954; MS, Utah State U., 1967; PhD, Rutgers U., 1973. Communs. staff cons. Sperry Univac, Salt Lake City, 1972-74; mgr. network systems tech. support Americas div. Sperry Univac, Blue Bell, Pa., 1974-77, mgr. communs. design and devel. Devel. and Mfg. div., 1977-78; dir. network systems Billings Computer Corp., Provo, Utah, 1978-79; mgr. data communs. devel. Gould/Modicon, Andover, Mass., 1980-82; mgr. software devel. ITT Courier Terminal Systems, Tempe, Ariz., 1982-85; founder, pres. Computer Rsch. Inc., Tempe, 1985-89; dir. tng./computer Tempe Tech. Inst., Phoenix, 1989-90; pres. Computer Rsch. Inc., Salt Lake City, 1991—; adj. prof. computer sci. Salt Lake C.C., 1991-92. Author: Introduction to DOS, 1991, Introduction to Word Processing, 1991, Increasing Your Productivity with Lotus1-2-3 Release 2.2, 1992. Chair dist. Rep. Party, Salt Lake City, 1958. With U.S. Army, 1954-56, Germany. Mormon.

MATTHEWS, CHARLES SEDWICK, petroleum engineering consultant, research advisor; b. Houston, Mar. 27, 1920; s. Charles James and Zadoc Coleman (Sedwick) M.; m. Miriam Loraine Ormerod, June 2, 1945; children—Joan Gail, Wendy Loraine. B.S. in Chem. Engring., Rice U., 1941, M.S. in Chem. Engring., 1943, Ph.D. in Chemistry, 1944. Registered profl. engr., Tex. Engr. Shell Devel. Co., San Francisco, 1944-48; research engr. Shell Devel. Co., Houston, 1948-56; dir. research Shell Devel. Co., 1967-72; chief reservoir engr. Shell Oil Co., Houston, 1965; mgr. engring. Shell Oil Co., 1972-73, sr. petroleum engring. cons., 1973-89; mem. engring. adv. com. Rice U., Houston, 1973-77; cons. Dept. Energy, Washington, 1974-78, mem. adv. com., 1975-79; spl. asst. Nat. Petroleum Council, Washington, 1981-83; mem. reserves com. Am. Petroleum Inst. Author: Pressure Buildup and Flow Tests in Wells, 1967; contbr. articles to profl. jours.; patentee in field. Chmn. Tex. Engrs. for Conservation, Houston, 1973. Mem. Soc. Petroleum Engrs. (Lester Uren award 1975, Disting. author, Disting. lectr. 1968, Disting. lectr. emeritus 1986), Nat. Acad. Engring., Phi Beta Kappa, Sigma Xi, Tau Beta Pi, Phi Lambda Upsilon. Republican. Methodist. Clubs: Houston, Meyerland (treas. 1982-85). Avocations: swimming; fishing. Home: 5307 S Braeswood Blvd Houston TX 77096-4149

MATTHEWS, DALE SAMUEL, information scientist; b. Van Nuys, Calif., Oct. 30, 1954. AA cum laude, Coll. of Redwoods, 1974; BA magna cum laude, San Francisco State U., 1976. With physics dept. Georgetown U., Washington; founder Exec. Briefing Technology Alliance; speaker in field. Author: Executive Briefing System. Office: 23 Observation Ct 102 Germantown MD 20876

MATTHEWS, DREXEL GENE, quality control executive; b. Vanzant, Ky., Feb. 1, 1952; s. Marcus Ivan and Lillia Mae (Lake) M.; m. Roberta June Eby, Oct. 16, 1971; children: Tracie Marie, Marcia Nichole. Student, Brescia Coll., Owensboro, Ky., 1976-79, Morehead State U., 1969-71. With Nat. Aluminum div., Nat. Steel Corp., Hawesville, Ky., 1971-78; customer service mgr. Nat. Aluminum div., Nat. Steel Corp., Hawesville, 1977-78; quality control mgr. Hunter Douglas Bldg. Products div., Roxboro, N.C., 1979-81; process engring. mgr., mgr. quality control and specification engring. MEPCO-ELECTRA Co., Roxboro, N.C., 1981-84; quality assurance sr., mng. engr. Sumitomo Electric Co., Research Triangle Park, N.C., 1984-87; quality assurance supplier, quality engring. resource Consol. Diesel Co., Whitakers, N.C., 1987—; quality assurance mgr. Fuel Systems Bus., Whitakers, N.C., 1987—. Mem. ASM, SAE (diesel stds. com. 1993—), Am. Soc. Quality Control (guest speaker 1987, 91), Am. Statis. Assn., Am. Nat. Stds. Inst. (fiber optics com. 1986-90), Durham (N.C.) Kennel Club, Cen. N.C. Siberian Husky Club. Republican. Baptist. Avocation: training, breeding and showing Siberian Huskies. Home: 5057 Netherwood Rd Rocky Mount NC 27803-1422

MATTHEWS, LARRYL KENT, mechanical engineering educator; b. Lubbock, Tex., Sept. 18, 1951; s. Morrison Arliss and Juanita Ruby (Parr) M.; m. Marie Elizabeth Twist, May 15, 1972. MS, N.Mex. State U., 1975; PhD, Purdue U., 1982. Test engr. Sandia Nat. Labs., Albuquerque, 1976-81; rsch. dir., educator N.Mex. State U., Las Cruces, 1982—; cons. Sandia Nat. Labs., 1985-89, ISOTEC, Santa Fe, N.Mex., 1986-88; bd. dirs. Waste-Mgmt. Edn. Rsch. Consortium, Las Cruces, 1990—; mem. mgmt. bd. Ctr. for Space Power. Contbr. articles to Jour. of Solar Energy, Internat. Jour. Exptl. Heat Transfer, ASME Jour. of Solar Engring., Internat. Jour. Heat and Mass Transfer, Intech Mag. Mem. AAAS, ASME, Am. Astron. Soc., Soc. de Ingenieros (founder), Pi Tau Sigma. Democrat. Methodist. Achievements include development of multiple-head radiometer for large pool-fire environments, of CSMP (Circum Solar Measurement Package) for concentrating solar applications. Office: NMex State U Engring Rsch Ctr Box 3449 Box 30001 Las Cruces NM 88003

MATTHEWS, MICHAEL ROLAND, environmental engineer; b. Frankfort, Mich., Apr. 26, 1948; s. Roland Elmer and Marcella Mae (Hyrns) M.; m. Susie Jo Kelly, Aug. 29, 1970. BA, Grand Valley State U., 1970; MS, U. Tenn., 1975. Profl. engr., Tenn., Ga., Ala., N.C.; profl. geologist, Tenn. Environ. engr. TVA, Knoxville, 1974-77, Chattanooga, 1977-86; sole proprietor, cons. Chattanooga, Tenn., 1986-87; environ. engr. Hensley-Schmidt, Inc., Chattanooga, 1987-89, Tricil Environ. Mgmt., Inc., Chattanooga, 1989-90, Laidlaw Environ. Svcs., Inc., Chattanooga, 1990-91; prin. Signal Environ. Svcs., Inc., Chattanooga, 1992—; mem. adv. com. Chattanooga State Hazardous Material Tech. Program, 1991--, Chattanooga State Civil Engring. Tech. Program, 1991--. Co-author: Groundwater Situation Assessment of the Tenn. Valley Region, 1986; contbr. articles to profl. jours. Judge Chattanooga Regional Sci. Fair, 1989-92; pres. Hamilton County Wastewater Regulations Bd., Chattanooga, 1985; fin. chmn. Chattanooga Regional Sci. and Engring. Fair., 1984. Mem. Am. Soc. Civil Engrs. (br. pres. 1986), Chattanooga Engrs. Club. Office: Signal Environ Svcs Inc 419 N Market Ste 200 Chattanooga TN 37405

MATTHEWS, RICHARD J., pharmaceutical research company executive; b. Scranton, Pa., Apr. 11, 1927; s. Richard John and Ruth (Duffy) M.; m. Sally Griffith (Aug. 8, 1953); children: Todd (dec.), Nanette Matthews Field, Pamela Matthews Otto, Wesley, Richard. BS, Phila. Coll. Pharmacy & Sci., 1951; MS, PhD, Jefferson Med. Coll., 1955. Head neuropharmacology rsch. The Upjohn Co., Kalamazoo, 1956-62; owner Pharmakon Labs., Scranton, 1962-66; dir. pharmacology Union Carbide Corp., Tuxedo, N.Y., 1966-69; pres. Pharmakon Rsch. Internat., Waverly, Pa., 1969—. Bd. dirs. Pa. State U., Scranton, Community Med. Ctr., Scranton, 1974-89. Served with U.S. Army, 1945-46. Mem. Am. Soc. Pharmacology and Exptl. Therapeutics. Republican. Presbyterian. Avocations: golf, gun collecting. Office: Pharmakon Rsch Internat Inc PO Box 609 Waverly PA 18471-0609

MATTHEWS, SHAW HALL, III, reliability engineer; b. Washington, May 29, 1942; s. Shaw Hall Matthews Jr. and Helen Louise (Evans) Floyd; m. Judith Arlene Jones, Aug. 2, 1976; children: Louise Anna, Alyson Ross. BS in Math., U. Ill., Chgo., 1972; MS in Ops. Rsch., Ill. Inst. Tech., 1979. Reliability engr. Zenith Corp., Chgo., 1967-73, reliability engring. mgr., 1973-76; component engring. mgr. Zenith Corp., Glenview, Ill., 1976-79; reliability and quality assurance mgr. Burr-Brown Corp., Tucson, 1979-82; systms reliability mgr. Storage Tech. Corp., Louisville, 1982—; mem. Joint Electron Devices Engring. Coun., 1980-82; chmn., mem. Electronics Adv. Group, State Bd. Community Colls. and Occupational Edn., Colo., 1984-86. Contbr. articles to profl. jours. Mem. Longmont (Colo.) Symphony Orch., 1988—, Mahler Fest Orch., Boulder, Colo., 1988—. Sgt. USAF, 1963-67. Mem. Soc. Applied and Indsl. Math., IEEE (treas. 1974-75). Office: Storage Tech Corp 2270 S 88th St Louisville CO 80028-5207

MATTINGLY, THOMAS K., astronaut; b. Chgo., Mar. 17, 1936; s. Thomas K. Mattingly; separated; 1 child, Thomas K. III. BS in Aero. Engring., Auburn U., 1958; attended, Air Force Aerospace Rsch. Pilot Sch. Commd. officer USN, 1958; comdr. assigned to space shuttle ops. NASA; astronaut NASA Manned Spacecraft Ctr., Houston; crew mem. Apollo 16, 1972, 4th Test Mission, Columbia, 1982, Shuttle Mission 51-C, 1985; rear admiral in charge Space Sensor Systems, Dept. Navy, Washington, from 1986. Decorated NASA Disting. Svc. Medals (2), JSC Cert. Commendation,

1970, JSC Group Achievement award, 1972, Navy Disting. Svc. Medal, Navy Astronaut Wings; recipient SETP Ivan C. Kincheloe award, 1972, Delta Tau Delta Achievement award, 1972, Outstanding Achievement award Auburn Alumni Engrs. Coun., 1972, AAS Flight Achievement award, 1972, Fedn. Aeronautique Internationale V.M. Komarov Diploma, 1973, Disting. Svc. Medal Dept. Def., 1982. Fellow AIAA (assoc., Haley Astronautics award 1973), Am. Astronautical Soc.; mem. Socl Exptl. Test Pilots, U.S. Naval Inst. Office: Space Sensor Systems Dept Navy PD-40 care Linda Campus Washington DC 20363-5100

MATTISON, GEORGE CHESTER, JR., chemical company executive, consultant; b. Eutaw, Ala., May 9, 1940; s. George Chester and Martha Pauline (Chilton) M; m. Barbara Peppenhorst, Aug. 20, 1963 (div. 1979); children: Mary Martha, George Chester III, William Grant; m. Linda Morris, May 23, 1987; step-daughter, Lisa Anne. MBA, Winthrop U., 1989. Shift chemist Gulf States Paper Corp., Demopolis, Ala., 1960-66; lab. technician Ala. Kraft Co., Mahrt, 1966-68; sales rep. Drew Chem. Co., Boonton, N.J., 1968-70, Betz Labs., Trevose, Pa., 1970-73; dist. mgr. Nalco Chem. Co., Naperville, Ill., 1976-85; regional mgr. Sandoz Chem. Co., Charlotte, N.C., 1985-89, Procomp (DuPont/Eka Nobel), Marietta, Ga., 1989-90; pres. OmniKem, Rock Hill, S.C., 1990—; bd. dirs. OmniKem; adv. bd. Haas Corp., Phila., 1990—, Horizon Industries. Mem. adv. bd. Rock Hill High, 1985; mem. Rep. Re-election com., Rock Hill, 1990; coach YMCA Pee-Wee Football, Rock Hill, 1973-78. Mem. TAPPI, Paper Industry Mgmt. Assn., Nat. Eagle Scout Assn., Nat. MBA Assn., Shriners, Masons, Rock Hill C. of C. Baptist. Avocations: golf, flying, computers. Home: 441 Lakeside Dr Rock Hill SC 29730-6105 Office: OmniKem 454 S Anderson Rd Ste 210 Rock Hill SC 29730-3392

MATTISON, RICHARD, psychiatry educator; b. Bradford, Pa., Nov. 14, 1946. BA in Biology magna cum laude, Lafayette Coll., Easton, Pa., 1968; MD, Cornell U., 1970. Diplomate Am. Bd. Psychiatry and Neurology (examiner child psychiatry); cert. Nat. Bd. Med. Examiners; bd. cert. psychiatry, 1980, child psychiatry, 1981; lic. physician, Mo., registered DEA. Intern Buffalo Children's Hosp., 1972-73; residency in adult psychiatry Cornell U. Medical Coll., White Plains, N.Y., 1975-77; rsch. fellow UCLA Ctr. for the Health Scis. Neuropsychiatric Inst., 1977-79, chief psychiatrist, 1979-80; from asst. prof. to assoc. prof. dept. psychiatry Pa. State U., Hershey, 1980-90, dir. children's diagnostic clinic, 1980-90, dir. child psychiatry inpatient unit, 1983-84, dir. sch. consulation, 1988-90; dir. divsn. child psychiatry Washington U. Sch. Medicine, St. Louis, 1990—, Blanche F. Ittleson assoc. prof., 1990—; cons. Hyperkinetic Children Clinic Gateway Hosp., L.A., 1976-78, Mental Health/Mental Retardation Program, Harrisburg (Pa.) Hosp., 1980-89; sch. psychiatric cons. Capital Area Intermediate Unit, Harrisburg, 1980-90, Lincoln Intermediate Unit, York, Pa., 1986-90, Lancaster-Lebanon Intermediate Unit, Lancaster, Pa., 1988-90, Spl. Sch. Dist., St. Louis, 1990—; psychiatrist-in-chief St. Louis Children's Hosp., 1990—, med. exec. com., 1990—, child abuse subcom., 1991-92; dir. child psychiatry fellowship training program, Washington U. Sch. Medicine, 1990—; staff Barnes Hosp., St. Louis, 1990—, exec. com. dept. psychiatry, 1990—, clin. affairs com., 1992—; vis. cons., Brown U. Sch. Medicine, Providence, 1986; vis. prof. U. Rochester (N.Y.) Sch. Medicine, 1986; cons., speaker in field. Co-author: Child and Adolescent Mental Health Consultation in Hospitals, Schools, and Courts, 1993; editor: Child Psychopathology; Diagnostic Criteria and Assessment, 1992, Developmental Disorders: Diagnostic Criteria and Clinical Assessment, 1992; editor Jour. Abnormal Child Psychology, Contemporary Child; contbr. numerous articles to profl. jours., chpts. to books. Fellow UCLA Ctr. for the Health Scis., 1975-77; recipient Faculty Devel. award ALCOA Found., 1981-82. Mem. Am. Psychiatric Assn., Am. Acad. Child and Adolescent Psychiatry (program com. 1987-89, editor jour.), Nat. Assn. Sch. Psychologists, Coun. for Exceptional Children, Soc. for Rsch. in Child and Adolescent Psychopathology (mem. com. 1992—), Soc. Profs. Child and Adolescent Psychiatry (Peter Henderson Mem. award com. 1993), Phi Beta Kappa. Home: 14725 Westerly Pl Chesterfield MO 63017 Office: Washington U Sch Medicine Dept Psychiatry 4940 Children's Pl Saint Louis MO 63110*

MATTOUSSI, HEDI MOHAMED, physicist; b. Medjez, Tunisia, May 11, 1959; s. Amor Mattoussi and Bariza Trabelsi. MS in Physics, U. Paris, 1984, PhD in Physics, 1987. Postdoctoral assoc. U. Mass., Amherst, 1987-88; postdoctoral assoc. Carnegie Mellon U., Pitts., 1989-90, rsch. scientist, 1990-92; asst. in physics U. Fla., Gainesville, 1992—. Contbr. articles to profl. jours. Mem. Am. Phys. Soc., Materials Rsch. Soc. Achievements include first x-ray scattering on side-chain polymer liquid crystals, proved the anisotropy of conformation of these materials; measured the order parameter in lyotropic liquid crystal polymers using birefringince technique; nonlinear optical properties of organic materials; polyelectrolyte materials, stability of quantum dots.

MATTSON, CLARENCE RUSSELL, safety engineer; b. Norwood, Mass., Nov. 3, 1924; s. Clarence R. and Jane P. (Dawson) M.; m. Constance W. Towne, June 7, 1953; children: Jennifer Lynn, Sue Ann. AA in Transp., Northeastern U., 1953, BBA, 1956. Cert. safety profl.; registered profl. engr., Calif. Ins. industry safety engr., 1953-62; mgr. accident prevention Dravo Corp., Pitts., 1962-72; corp. mgr. safety and environ. affairs Perini Corp., Framingham, Mass., 1972-84; dir. safety and tng. The Marr Co., South Boston, Mass., 1984; mng. dir. Long Beach-L.A. rail project Transit Ins. Adminstrs.-L.A. County Transp. Commn., 1984-86; v.p. tech. svcs. Fred S. James & Co., Short Hills, N.J., 1987-89; pres. Athena Assocs. Ltd., Safety Mgmt. Cons., 1990—. Deacon Scituate (Mass.) Congl. Ch. Recipient Disting. Svc. award Nat. Safety Coun., 1988. Mem. Am. Soc. Safety Engrs., Nat. Safety Coun. (past gen. chmn. constrn. exec. com., disting. svc. award 1988), Assn. Gen. Contractors Am. (past chmn. safety and health com., safety engrs. adv. com.), Nat. Constructors Assn., Vets. of Safety, Mass. Safety Coun. (bd. dirs.), Elks. Republican. Home and Office: Ll Abigails Way Sandwich MA 02563

MATULA, RICHARD A(LLAN), academic administrator; b. Chgo., Aug. 22, 1939; s. Ludvig A. and Leone O. (Dufeck) M.; m. Brenda C. Mather, Sept. 5, 1959; children: Scott, Kristopher, Daniel, Tiffiny. BS, Purdue U., 1961, MS, 1962, PhD, 1964. Instr. Purdue U., 1963-64; asst. prof. mech. engring. U. Calif., Santa Barbara, 1964-66, U. Mich., 1966-68; assoc. prof. mech. engring. Drexel U., Phila., 1968-70; prof. Drexel U., 1970-76, chmn. thermal and fluid sci. advanced study group, 1969-72; chmn. Drexel U. (Environ. Studies Inst.), 1972-73, chmn. dept. mech. engring. and mechanics, 1973-76; dean Coll. Engring.; prof. mech. engring. La. State U., Baton Rouge, after 1976; pres. Inst. Paper Chemistry, Appleton, Wis., 1986-89, Inst. Paper Sci. and Tech., Atlanta, 1989—. Contbr. articles to profl. jours. Treas., bd. dirs. Wexford Leas Swim and Racquet Club, Inc., 1968-73; v.p. Wexford Leas Civic Assn., 1969-71. Mem. Air Pollution Control Assn., Am. Soc. Engring. Edn., ASME, AAAS, Combustion Inst., Soc. Automotive Engrs., Sigma Xi, Pi Kappa Phi, Pi Tau Sigma, Sigma Pi Sigma, Tau Beta Pi. Roman Catholic. Home: 3143 St Ives Country Club Pky Duluth GA 30136-2001 Office: Inst Paper Sci and Tech 500 10th St NW Atlanta GA 30318-5454

MATUSZKO, ANTHONY JOSEPH, research chemist, administrator; b. Hadley, Mass., Jan. 31, 1926; s. Joseph Anthony and Katherine (Narog) M.; m. Anita Colley, Oct. 26, 1956; children—Martha, Mary, Stephen, Richard. BA, Amherst Coll., 1946; MS in Chemistry, U. Mass., 1951; PhD in Chemistry, McGill U., 1953. Demonstrator in chemistry McGill U., Montreal, Que., Can., 1950-52; from instr. to assoc. prof. chemistry Lafayette Coll., Easton, Pa., 1952-58; head fundamental process div. Naval Propellant Lab., Indian Head, Md., 1958-62; program mgr. in chemistry Air Force Office Sci. Research, Washington, 1962-89; cons., Annandale, Va., 1989—. Contbr. articles to tech. jours. Patentee in field. Pres. Forest Heights PTA, Md., 1967. Served with U.S. Army, 1946-48. Named Hon. Fellow in Chemistry, U. Wis.-Madison, 1967-68, recipient Superior Performance award USAF, Outstanding Career Svc. award U.S. Govt. Fellow AAAS, Am. Inst. Chemists (life); mem. Am. Chem. Soc., Cosmos Club, Sigma Xi. Home: 4210 Elizabeth Ln Annandale VA 22003-3654

MATWAY, ROY JOSEPH, material scientist; b. Pitts., Oct. 18, 1956; s. Donald Paul and Ruth Agnes (Fleckenstein) M. BS, Carnegie Mellon U., 1979, MS, 1979, PhD, 1986. Product metallurgist Cameron Iron Works, Houston, 1979-82; vis. scientist Max Planck Inst., Düsseldorf, Germany, 1986-87; mgr. process rsch. J & L Specialty Products, Pitts., 1987—; mem. adv. bd. Ctr. for Iron and Steel Rsch., Pitts., 1988—. Contbr. articles to profl. jours. Nat. Steel fellow, 1982, Max Planck Soc. fellow, 1986. Mem. Am. Soc. Materials, Pitts. Oratorio Soc., Iron and Steel Soc. (steering com. process tech. div.), Sigma Xi., Delta Alpha (bd. dirs. 1988—). Achievements include development of refinement of liquid steel processing practices for improved product quality, physical modeling technique for liquid, liquid mass transfer studies, elucidation of common design variable changes on mass transfer rates. Home: 5456 Bartlett St Pittsburgh PA 15217

MATZAT, GREGORY MARK, naval architect; b. St. Louis, Nov. 18, 1967; s. George Richard and Rowena Lee (Moeller) M. BS Naval Architecture and Marine Engring, Webb Inst., 1989. Registered profl. engr., N.Y. Naval architect Sparkman & Stephens, Inc., N.Y.C., 1989—. Mem. Soc. Naval Architects and Marine Engrs. (assoc.), Inst. Marine Engrs. Office: Sparkman & Stephens Inc 79 Madison Ave New York NY 10016

MATZIORINIS, KENNETH N., economist; b. N.Y.C., May 4, 1954; s. Neocles N. and Penelope (Gregoratos) M.; m. Catherine Marina Astrakianakis, July 27, 1985; children: Anna Maria, Angela Ellen Rose. BA, McGill U., 1976, MA, 1979, PhD, 1988. Cert. mgmt. cons. Asst. economist Nat. Bank Greece (Can.), Montreal, 1978-81; lectr. econs. McGill U., Montreal, 1977—; prof. econs. John Abbott Coll., Montreal, 1981—; pres. Canbek Econ. Cons., Inc., Montreal, 1983—. Econs. adviser to bd. dirs. Internat. Orgn. Psychophysiology, 1982-89; bd. dirs. Nat. Bank of Greece, Can., 1991—. Author: Introduction to Macro Economics: An Applied Approach,, 1988; editor: Vital Graphs of Canadian Economy, 1984; contbr. articles to profl. jours. V.p. Westmount Liberal Riding Assn., Montreal, 1975-77; bd. govs. McGill U., 1978-81; bd. govs. John Abbott Coll., 1988-91; chmn. bd. dirs. Community Service Ctr. St. Louis, Montreal, 1978-80. Mem. Am. Econ. Assn., Am. Hellenic Ednl. and Progressive Assn., Can. Econ. Assn., Que. Inst. Cert. Mgmt. Cons., Nat. Assn. Bus. Economists, Grad. Club Montreal. Greek Orthodox. Home: 615 67th Ave, Laval, Montreal, PQ Canada H7V 3N9

MAUDLIN, ROBERT V., economics and government affairs consultant; b. Washington, June 8, 1927; s. Cecil V. and Eva Jane (Wright) M.; m. Carole M. Jackson, Sept. 3, 1949; children: Lynda C., David V., Tim W.E. Student, MIT, 1945; BS, Am. U., 1951. Ptnr. C.V. & R.V. Maudlin, Washington, 1952-72, owner, 1972—; exec. dir. Joint Govt. Liaison Com., 1973-81; mem. Industry Sector Adv. Com. U.S. Dept. Commerce, Washington, 1975—; sec. Nat. Assn. Scissors and Shears Mfrs., 1970—; mng. dir. Bur. Applied Econs., Washington, 1960—. Author statis. reports. Pres. Forest Hills Citizens Assn., Washington, 1964; chmn. Boy Scouts Am., Washington, 1972. 2nd lt., Corps of Engrs., AUS, 1945-47. Republican. Home: 2906 Ellicott Ter NW Washington DC 20008-1023 Office: CV & RV Maudlin 1511 K St NW Washington DC 20005-1401

MAUER, ALVIN MARX, physician, medical educator; b. LeMars, Iowa, Jan. 10, 1928; s. Alvin Milton and Bertha Elizabeth (Marx) M.; m. Theresa Ann McGivern, Dec. 2, 1950; children: Stephen James, Timothy John, Daria Maureen, Elizabeth Claire. B.A., State U. Iowa, 1950, M.D., 1953. Intern Cin. Gen. Hosp., 1953-54; resident in pediatrics Children's Hosp. Cin., 1954-56; fellow in hematology dept. medicine U. Utah, Salt Lake City, 1956-59; dir. div. hematology Children's Hosp. Cin., prof. dept. hematology, 1959-73; prof. dept. pediatrics U. Cin. Coll. Medicine, 1959-73; prof. pediatrics U. Tenn. Coll. Medicine, Memphis, 1973—, prof. medicine, 1983—, chief med. oncology/hematology; dir. cancer program U. Tenn. Coll. Health Scis.; dir. St. Jude Children's Research Hosp., Memphis, 1973-83; mem. hematology study sect. NIH; mem. clin. cancer investigation rev. com. Nat. Cancer Inst.; mem. com. on maternal and infant nutrition NRC. Author: Pediatric Hematology, 1969; editor: The Biology of Human Leukemia, 1990. Served with U.S. Army, 1946. Mem. Am. Soc. Hematology (pres. 1980-81), Assn. Am. Cancer Insts. (pres. 1980), am. Acad. Pediatrics (com. on nutrition), Am. Assn. Cancer Edn., Am. Soc. Clin. Investigation, Am. Fedn. Clin. Rsch., Assn. Am. Physicians, Am. Pediatric Soc., Cen. Soc. Clin. Investigation, Cen. Soc. Clin. Rsch., Internat. Soc. Hematology (pres. 1988-90, chmn. 1992—, bd. councilors 1992—), Am. Cancer Soc. (pres. Tenn. divsn. 1992-93), Midwest Soc. Pediat. Rsch., N.Y. Acad. Scis., Soc. Pediat. Rsch., Am. Assn. Cancer Rsch., Phi Beta Kappa, Sigma Xi, Alpha Omega Alpha. Democrat. Roman Catholic. Office: U Tenn Memphis Cancer Ctr N327 Van Vleet Bldg 3 N Dunlap St Memphis TN 38163

MAULDIN, CHARLES ROBERT, aerospace engineer; b. Birmingham, Ala., Feb. 15, 1938; s. Roy Leon and Edna (Vickers) M.; m. Susan Rowley Brady, Mar. 21, 1964; children: Lara, Charles, Alison. BSEE, Auburn U., 1960. Project engr. NASA Marshall Space Flight Ctr., Marshall Space Flight Ctr., Ala., 1961, 63-76, chief solid rocket booster integration br., 1976-82, chief Space Shuttle systems br., 1982-86, chief projects systems div., 1986-87, dir. Systems Safety and Reliability Office, 1987—. 1st lt. USAR, 1961-63. Recipient honor award Armed Forces Comm. and Electronics Assn., 1961; Silver Snoopy award NASA Apollo Astronauts, 1972, NASA Space Shuttle Astronauts, 1981; Exceptional Svc. medal NASA, 1981, Outstanding Leadership medal, 1988. Fellow AIAA (assoc.); mem. Sr. Exec. Assn., NRA (life), Auburn U. Alumni Assn. (life). Baptist. Home: 4009 Centaur Blvd SW Huntsville AL 35805 Office: NASA Marshall Space Flight Ctr CTO1 Huntsville AL 35812

MAUMENEE, IRENE HUSSELS, opthalmology educator; b. Bad Pyrmont, West Germany, Apr. 30, 1940; married, 1972; 2 children. MD, U. Gottingen, 1964. Cert. Am. Bd. Opthalmology, Am. Bd. Med. Genetics. Rsch. asst. U. Hawaii, 1968; vis. genetics Population Genetics Lab., 1968-69; fellow dept. medicine Johns Hopkins U., 1969-71; ophthalmology preceptorship Wilmer Inst., Johns Hopkins Hosp., 1969-73; from asst. prof. to assoc. prof. Wilmer Ophthalmology Inst. Johns Hopkins Hosp., 1972-87; prof. ophthalmology and medicine, divsn. med. genetics Wilmer Ophthalmology Inst. Johns Hopkins Hosp., Moore Clinic, 1987—; dir. Johns Hopkins Ctr. Hereditary Eye Disease, Wilmer Inst., 1979—; cons. John F. kennedy Inst. Visually & Mentally Handicapped Children, 1974—; dir. Low Vision Clinic, Wilmer Inst., 1977—; vis. prof. British Royal Soc. Med., 1987-88, French Opthalmology Soc., Paris & French Acad. Medicine, 1988; adv. Nat. Eye Inst. Task Forces, 1976, 81. Mem. AMA, Am. Soc. Human Genetics, Am. Acad. Ophthalmology, Assn. Rsch. Vision & Opthalmology, Internat. Soc. Genetic Eye Disease, Am. Ophthalmology Soc., Pan Am. Assn. Ophthalmology. Achievements include research in nosology and management in ophthalmic and general medical genetics; population genetics; computer application to genetic analysis; molecular genetics; over 120 publications on human genetics and eye diseases. Office: Johns Hopkins Univ Wilmer Ophthal Inst 601 N Broadway Baltimore MD 21205*

MAURATH, GARRY CALDWELL, hydrogeologist; b. Cleve., July 24, 1952; s. George Anthony and Iona May (Caldwell) M.; m. Lesa Maria, Aug. 28, 1977. BS, Lehigh U., 1974; MS, Kent State U., 1980, PhD, 1989. Lic. geologist, N.C.; registered environ. assessor, Calif. Sr. geologist O'Brien Resources, Grass Valley, Calif., 1980-82; prin. hydrogeologist Ebasco Svcs., Sacramento, Calif., 1979—. Contbr. over 30 articles to profl. jours. Staff sgt. U.S. Army, 1974-77, USAR, 1977-80. Mem. AAAS, AGU, Geol. Soc. Am., Groundwater Resources Assn. Calif., Assn. Ground Water Scientists and Engrs., Sigma Xi. Roman Catholic. Office: Ebasco Svcs Ste 250 2525 Natomas Park Dr Sacramento CA 95833

MAURER, GERNANT ELMER, metallurgical executive, consultant; b. Sayre, Pa., May 5, 1949; s. Elmer L. and Joyce F. (Fox) M.; m. Suzanne Walker Berry, Aug. 19, 1972. BES, Johns Hopkins U., 1971; PhD, Rensselaer Poly. Inst., 1976. Materials engr. Spl. Metals Corp., New Hartford, N.Y., 1976-80, mgr. R & D, 1981-84, dir. R & D, 1985-87, v.p. tech., 1987—; founding v.p. Specialty Metals Processing Consortium, pres. 1992-93; chmn. Internat. Symposium on Superalloys, 1984-88. Co-editor: Superalloys, 1980, 2d edit., 1988; contbr. tech. papers to profl. jours.; inventor various superalloys. Dir. devel. com. Munson William Proctor Inst., Utica, N.Y., 1988-89. Fellow Am. Soc. for Metals (trustee 1992-95); mem. The Metall. Soc., Am. Vacuum Soc. (bd. dirs. metall. div. 1988), Utica Area C. of C. (bd. dirs. 1986-89), Yahnundasis Club. Avocations: golf, fly fishing, art, photography, metal working. Home: 23 Sherman Cir Utica NY 13501-5808 Office: Spl Metals Corp Middle Settlement Rd New Hartford NY 13413-9558

MAURER, HANS HILARIUS, pharmacology educator, researcher; b. Homburg, Germany, Nov. 25, 1950; s. Hilarius Andreas and Elisabeth (Greff) M.; m. Claudia Regina Kessler, Oct. 9, 1981; children: Christine, Johannes. Lic. in pharmacy, U. Saarland, Saarbrücken, Germany, 1977, PhD, 1983, Habilitation, 1988. Asst. prof. pharmacology U. Saarland, 1983-88, Univ. lectr., 1988-92; prof. pharmacology/toxicology, head dept. clin. toxicology U. Saarland, Homburg, Germany, 1992—. Author: (with Pfleger and Weber) Mass Spectral and GC Data, 1985, 92, Mass Spectral Library, 1987, 2d edit., 1993, (with De Zeeuw, Franke, and Pfleger) Gas Chromatographic Retention Indices, 1992, (with Schaefer) Diagnosis and Therapy of Intoxications, 1993. Capt. M.C., German Army, 1977-78. Mem. German Pharm. Soc. (chpt. pres. 1990-92), Soc. Toxicology and Forensic Chemistry (treas. 1987—), Internat. Assn. Forensic Toxicologists, German Soc. Pharmacology and Toxicology, Lions Club Internat. Roman Catholic. Avocations: music, art, theater, history of toxicology. Office: U Saarland, Univ Clinics, D-66421 Hamburg Germany

MAURER, RENÉ, engineer, economist; b. Switzerland, Mar. 17, 1946; s. Alfred and Gerda (Naef) M. Degree in engring., Lucerne State Coll. Tech., Switzerland, 1971; lic. in econs., Sch. Econs. & Bus. Adminstrn., Lucerne, Switzerland, 1975. Devel. engr. AG. für Zweckbauten, Lucerne, 1975-77; pres. René Maurer AG, Lucerne, 1977-87, M.I. Maurer Instruments AG, Baar, 1987—. Inventor manual operating and programming system (MOPS). Recipient Innovation prize Swiss C. of C., 1988, Technology Made in Switzerland prize, 1992. Office: MI Maurer Instruments AG, Lättichstrasse 1, CH-6340 Baar Switzerland

MAURER, RICHARD HORNSBY, physicist; b. Reading, Pa., Apr. 27, 1942; s. Samuel Forest and Marian E. (Hornsby) M.; m. Marian Ross Harvey, May 3, 1975; children: Jonathan, Andrew. BS, L.I. U., 1964; PhD, U. Pitts., 1970. Postdoctoral fellow Bartol Rsch. Found., Swarthmore, Pa., 1970-73; environ. engr. AMP Inc., Harrisburg, Pa., 1973-81; physicist Applied Physics Lab. Johns Hopkins U., Laurel, Md., 1981—, reliability group supr. test and evaluation sect., 1986—; educator Whiting Sch. Engring., Johns Hopkins U., 1989-93. Contbr. chpt. to: Space Systems Reliability and Survivability, 1994; contbr. articles to Jour. IEEE Transactions, Transactions Nuclear Sci., Jour. Spacecraft and Rockets, Internat. Reliability Physics Symposium. Baseball mgr. Howard County Youth Program, Ellicott City, Md., 1985-93; mem. choir Bethany United Meth. Ch., Ellicott City, 1987-92. Mem. IEEE, Am. Phys. Soc., Sigma Xi. Achievements include patent for fabrication of thermal batteries by multi-layer ceramic of organic printed circuit board methods; research on effects of radiation on electronic devices, on reliability of electronic packaging designs and gallium arsenide devices. Office: Johns Hopkins U Applied Physics Lab Johns Hopkins Rd Laurel MD 20723

MAURER, ROBERT (STANLEY), osteopathic physician; b. Bklyn., Feb. 10, 1933; s. Gustav and Hilda Maurer; A.B. in Chemistry, U. Pa., 1953; D.O., Phila. Coll. Osteo. Medicine, 1962; m. Beverly Greenberg, Sept. 4, 1960; children: Ellen Jo, David, Andrew. Cert. Am. Bd. Quality Assurance and Utilization Rev. Physicians, 1989, Am. Coll. Gen. Practice, 1977. Intern, Phila. Coll. Osteo. Medicine, 1962-63; gen. practice osteo. medicine, Woodbridge Twp., N.J., 1963-93; mem. med. staff Meml. Gen. Hosp., Union, N.J.; mem. med. staff JFK Med. Ctr., Edison, N.J., chmn. utilization rev., 1985-93; dir. Med. Inter-Ins. Exch. N.J., 1977-93; police and fire surgeon City Iselin; clin. asst. prof. N.J. Sch. Osteo. Medicine; clin. assoc. prof. Robert Wood Johnson Med. Sch.; dir. alumni affairs UMDNJ Sch. Osteo. Medicine, 1986-93. Mem. N.J. State Gov.'s Task Force on Med. Malpractice, 1985-89; chmn. legis. com. N.J. State Med. Underwriters; candidate for N.J. State Senate, 1983, for N.J. State Assembly, 1987. Lt. USNR, 1953-58; Korea. Named Physician of Yr., 1990. Fellow Am. Osteo. Coll. Rheumatology; mem. VFW, N.J. Council Ambulatory Physicians (v.p. 1977-80), N.J. Assn. Osteo. Physicians and Surgeons (pres. 1976-77), Middlesex County Osteo. Soc. (pres. 1973-74), Am. Osteo. Assn., N.J. Osteo. Edn. Found. (pres. 1985-89), Physicians Rev. Orgn. (bd. dirs. N.J. 1987-93), Am. Coll. Utilization Rev. Physicians, Phila. Coll. Osteo. Medicine Alumni Assn. (pres. 1981, sec. 1991-93).

MAURITZ, KARL HEINZ, neurology educator; b. Krumlov, Czechoslovakia, Aug. 7, 1944; s. Karl and Hanna (Lepschy) M.; m. Angelika Gisela Riedel, Nov. 8, 1974. MD, Med. Sch., Saarland, Fed. Republic Germany, 1970, Habilitation, 1979. Asst. prof. physiology Med. Sch., Saarland, 1970-73; asst. prof. psychology U. Konstanz (Fed. Republic Germany), 1973-74; asst. prof. neurology med. sch. U. Freiburg, 1974-79, assoc. prof. neurology med. sch., 1979-81; rsch. fellow NIH, Bethesda, Md., 1981-84; rsch. fellow neurology U. Düsseldorf (Fed. Republic Germany), 1985-86; prof., dir. Rehab. Ctr., U. Cologne (Fed. Republic Germany), 1986-89, Klinik Berlin, 1989—. Mem. German Soc. for Neurologic Rehab. (pres. 1989—). Office: Klinik Berlin, Kladower Damm 223, W 1000 Berlin 20, Germany

MAURO, ARTHUR, financial executive, university chancellor; b. Thunder Bay, Ont., Can., Feb. 15, 1927; s. Arthur George and Maria (Fortezza) M.; m. Nancie June Tooley, Sept. 1, 1951; children: Barbara, Christine, Jennifer, Gregory. BA, St. Paul's, Winnipeg, 1949; LLB, Univ. Man., 1953, LLM, 1956. Bar. Manitoba Can. 1953. Spl. counsel Province of Manitoba, 1958-69, chmn. royal commn. on northern transp., 1967-69; lect. transp. and comm. law Univ. Manitoba, Can., 1967-69; chmn., dir. Investors Group Inc., 1985—; also chancellor Univ. Manitoba; bd. dirs. Investors Syndicate Ltd., Investors Group Trust Co. Ltd., Investors Syndicate Property Corp., I.G. Investment Mgnt. Ltd., Power Fin. Corp., PWA Corp., Fed. Industries Ltd., Can. Pacific Hotels Corp. Named knight of St. Gregory, 1967. Mem. Manitoba Bar Assn. Roman Catholic. Office: University of Manitoba, Winnipeg, MB Canada R3T 2N2*

MAUZY, MICHAEL PHILIP, environmental consultant, chemical engineer; b. Keyser, W.Va., Nov. 14, 1928; s. Frank and Margery Ola (Nelson) M.; m. Nancy Shepherd Watson, Mar. 27, 1949; children: Michael P. Jr., Jeffrey A., Rebecca A. BSChemE, Va. Poly. Inst., 1950; MSChemE, U. Tenn., 1951. Registered profl. engr., Va., Ill. With Monsanto Co., St. Louis, 1951-71, dir. engring. and mfg., 1968-71; mgr. comml. devel. Kummer Corp., Creve Coeur, Mo., 1971-72; mgr. labs. Ill. EPA, Springfield, 1972-73, mgr. water pollution control, 1973-74, mgr. environ. programs, 1974-77, dir., 1977-81; v.p. Roy F. Weston, Inc., West Chester, Pa., 1981-88, Vernon Hills, Ill., 1988-93, Albuquerque, 1993—; also bd. dirs. Roy F. Weston, Inc., West Chester, Pa.; bd. dirs. DeTox Internat. Corp., St. Charles, Ill.; provider Congl. testimony, 1974-81; presenter various workshops, symposia and seminars, 1974—. Contbr. articles on environ. mgmt. to profl. publs., 1974—. Mem. Ohio River Valley Water Sanitary Commn., Cin., 1976-81. 1st lt. U.S. Army, 1951-53. Recipient Environ. Quality award Region V, U.S. EPA, Chgo., 1976, Disting. Svc. award Cons. Engrs. Coun. of Ill., 1978, Ill. award Ill. Assn. Sanitary Dists., 1979, Clarence W. Klassen award Ill. Assn. Water Pollution Control Ops., 1984. Mem. Am. Pub. Works Assn., Am. Inst. Chem. Engring., Water Pollution Control Assn., Am. Mgmt. Assn. Avocations: reading, travel, home improvements. Home: 2032 Calle Pajaro Azol NW Albuquerque NM 97120-3102

MAVINIC, DONALD STEPHEN, civil engineering educator; b. Detroit, Apr. 3, 1946. BS in Civil Engring., U. Windsor, 1969, MS in Sanitary Engring., 1970, PhD in Sanitary Engring., 1973. From asst. to assoc. prof. civil engring. U. B.C., Vancouver, 1973-83, prof., 1983—, dir. environ. engring. program, 1984—; mem. NSERC site-visit team ind. rsch. chair McMaster U., 1986-92; keynote presenter EPA Specialty Conf. Nutrient Removal, Charleston, 1987; keynote speaker Leachate Treatment Symposium, Toronto, 1991; appointee Com. Re-write Mcpl. Waste Discharge Guidelines, B.C., 1992-93; cons. in field. Regional editor Water Pollution Rsch. Jour. Can., 1984-89, mem. editorial bd., 1989—; mem. editorial bd. Environ. Tech. Letters, 1986—; contbr. over 100 papers and tech. reports to profl. jours. Recipient Keefer Gold medal, 1977, 86. Mem. APEBC (Eng., bd. examiners 1981-83), ASCE (mem. conf. steering com. environ. conf. 1991, mem. com. sludge treatment, utilization, reclamation and disposal 1993—), GVRD (mem. adv. com. waste residuals mgmt. 1992—), Can. Soc. Civil Engring. (bd. dirs. 1977-80, 87-91, chmn. environ. divsn. 1987-91, chmn. joint CSCE-ASCE nat. conf. environ. engring. 1988, co-editor procs. 1988, mem. tech. com. environ. specialty conf. 1990, vice-chmn. ann. conf. 1991, Albert E. Berry award 1992), Internat. Assn. Water Pollution Rsch. and Control (mem. internat. rev. bd. biennial conf. 1986, 88,

90, 92, mem. steering com. conf. 1988). Achievements include research in aerobic sludge digestion, Leachate Treatment, high-ammonia leachate treatment, (with others) metal toxicity problems in treatment of high-ammonia leachates, problems of metal corrosion in soft water distribution systems. Office: U British Columbia, Dept Civil Engineering, Vancouver, BC Canada V6T 1W5*

MAVRIKAKIS, MANOS, chemical engineer; b. Heraklion, Greece, Nov. 24, 1963; came to U.S., 1988; s. Stereos and Fitini (Papyrakis) M. Diploma in chem. engring., Nat. Tech. U. Athens, 1987; MS in Chem. Engring., U. Mich., 1989, MS in Applied Math., 1993, PhD in Chem. Engring./Sci. Computing, 1993. Rsch. asst. in chem. engring. U. Mich., Ann Arbor, 1988—. Presenter in field. Rackham predoctoral fellow, 1992-93, Korgialenion Found. fellow, 1988-92. Mem. Sigma Xi (affiliate), Tau Beta Pi. Office: U Mich Chem Engring Dept Ann Arbor MI 48109

MAVROS, GEORGE S., clinical laboratory director; b. Adelaide, Australia, Oct. 14, 1957; came to U.S., 1970; s. Sotirios George and Angeliki (Korogiannis) M.; m. Renee Ann Cuddeback, June 24, 1979. BA in Microbiology, U. South Fla., 1979, MS in Microbiology, 1987; MBA, Nova U., 1991. Cert. lab. dir. Nat. Certifying Agy. for Clin. Lab. Pers. Med. technologist Jackson Meml. Hosp., Dade City, Fla., 1979-81; microbiology supr. HCA Bayonet Point-Hudson Med. Ctr., Hudson, Fla., 1981-82, dir. labs., 1982-88; lab. mgr., adminstrv. and tch. dir. Citrus Meml. Hosp., Inverness, Fla., 1988—; lab. cons. HCA Oak Hill Hosp., Spring Hill, Fla., 1983-84; cons. lab. info. systems Citation Computer Systems, St. Louis, 1983—, Hosp. Corp. of Am., Nashville, 1986; instr. Microbiology Pasco Hernando Com. Coll., New Port Richey, Fla., 1986-88, Inst. Biolog. Scis. Cen. Fla. Community Coll., Lecanto, 1989—; bd. dirs. Gulf Coast chpt. Clin. Lab. Mgrs. Assn., Tampa, Fla., 1987, pres., 1987-89. Parish pres. Greek Orthodox Ch. of West Cen., Inverness, Fla.; chmn. Bayonet Point Hosp. Good Govt. Group, Hudson, 1986-88. Mem. APHA, Am. Mgmt. Assn., Am. Soc. Microbiology, Am. Soc. Clin. Pathologists (cert. in lab. mgmt.), Am. Soc. Med. Technologists (cert.), Fla. Soc. Med. Technologists, Clin. Lab. Mgmt. Assn. (pres. Gulf Coast chpt. 1988-90), Am. Assn. Clin. Chemists, Am. Acad. Microbiology (cert.), Am. Coll. Health Care Execs., Fla. State Bd. Clin. Lab. Pers. Democrat. Clubs: Greek Orthodox Youth Am. (Clearwater, Fla.). Lodges: Order of DeMolay, Sons of Pericles (sec.). Home: 6 Byrsonima Ct W Homosassa FL 34446-9111 Office: Citrus Meml Hosp 502 W Highland Blvd Inverness FL 34452-4720

MAVROVIC, IVO, chemical engineer; b. Fiume, Italy, Dec. 5, 1927; came to U.S., 1959; s. Janko and Milica (Gregorina) M.; m. Erna Gallian, oct. 14, 1955; 1 child, Paul. BSChemE, U. of Zagres, Yugoslavia, 1952, MSChemE, 1955. Registered profl. engr, N.Y. Chem. engr. Dorr-Oliver, Milan, Italy, 1956-59, Chemico, N.Y.C., 1960-65; cons. N.Y.C., 1965-77; pres. UTI/UTI Constrn. Inc., Hackensack, N.J., 1977—. Patentee in field; contbr. articles to profl. jours. Mem. AICE. Roman Catholic. Office: UTI/UTI Constrn Inc 2 University Pla Hackensack NJ 07601

MAXIMOVICH, MICHAEL JOSEPH, chemist, consultant; b. Akron, Ohio, Mar. 3, 1932; s. Michael Stephen and Josephine Anna (Salzwimmer) M. BS, Kent State U., 1958, MS, 1964. Rsch. chemist PPG Industries, Barberton, Ohio, 1964-69; sr. devel. engr. Goodyear Chems., Akron, 1969-80; sr. rsch. chemist Polymer Industries, Stanford, Conn., 1980-82, Mooney Chems., Cleve., 1982-84; cons. All-Chem Techs., Lakewood, Ohio, 1984—. Pres. St. Mary Holy Name Soc., Akron, 1978-80, Akron Deanery Holy Name Soc., 1978-80; del. Cleve. Holy Name Soc., 1978-80. With USN, 1950-54. Mem. Am. Chem. Soc., Assn. Cons. Chemists and Chem. Engrs. Republican. Roman Catholic. Achievements include patents for liquid phase hydrofluorination of alkynes, for polyglycol carbonates using CO2 and catalysts, for water borne styrene-acrylates; first proof of the mechanism of the Schmidt Reaction of Substituted Acetophenones. Home and Office: 1224 Elbur Ave Cleveland OH 44107-2716

MAXON, HARRY RUSSELL, III, nuclear medicine physician; b. Muncie, Ind., Aug. 28, 1941; m. Mary Isabelle Moss, June 17, 1967; children: Russell, Mary Evelyn, Ashley. BA, Stanford U., 1963; MD with honors, Tulane U., 1967. Diplomate Am. Bd. Nuclear Medicine, Am. Bd. Internal Medicine. Intern U.S. Naval Hosp., Portsmouth, Va., 1967-68, resident in internal medicine, 1968-71; dir. nuclear medicine Portsmouth (Va.) Naval Hosp., 1971-73; fellow Univ. Cin., 1973-74; dir. nuclear medicine tech. Nuclear Medicine, Univ. Cin., 1974-82; clin. dir. Radiobiology Lab., Univ. Cin., 1977-79, assoc. dir., 1979-87; dir. Nuclear Medicine, Univ. Cin., 1987—; assoc. prof. Univ. Cin., 1979-84, prof., 1984—. Author, co-author 9 monographs, 21 books, book chpts., 62 articles in med. jours. and 124 sci. abstracts. Instr. Recreation Program, Terrace Park, Ohio, 1975-84; mem. adv. group for sch. bd., Mariemont, Ohio, 1981-82; co-chmn. tax levy Cin. Gen. Children's Hosp., Cin., 1976, 81. With USN, 1963-73. Recipient Walter Reed Meml. award in Tropical Medicine, 1967, Best Sci. Contbn. award Jour. Nuclear Medicine, L.A., 1992. Fellow ACP; mem. AMA, Am. Soc. Nuclear Medicine, Am. Thyroid Assn., Nat. Coun. on Radiation and Measurements, Internat. Coun. Control Iodine Deficiencies. Achievements include patents for Radiolabeled Dihematoporphyrin Ether and its use in detecting and treating neoplastic tissue and a procedure for isolating and purifying radioactive ligated rhenium pharms. and use thereof and kit. Home: 12 Denison Ln Terrace Park OH 45174 Office: U Cin ML 577 234 Goodman St Cincinnati OH 45267-0577

MAXTON, ROBERT CONNELL, metallurgical engineer; b. McKeesport, Pa., Apr. 14, 1930; s. John Edward and Helen Gertrude (Connell) M.; m. Mary R. Maxton, Dec. 19, 1953; children: Karen Ann, George Edward. BS, Carnegie Mellon U., 1952. Lab. mgr. Colonial div. Vasco A Teledyne Co., Monaca, Pa., 1956-58, quality control met., 1958-61, chief metallurgist, 1962-69; asst. dir. rsch. metallurgy Mpls. Electric Steel, 1969-72, tech. dir., 1972-87; tech. dir. M E Internat., Mpls., 1988—; vice chmn. Steel Founders Soc. Am. Tech. and Oper. Conf., Cleve., 1976, chmn. Steel Founders Soc. Am. Carbon and Low Alloy Rsch. Com., 1980-84. Contbr. articles to profl. jours. 1st lt. U.S. Army, 1953-55. Mem. Am. Soc. for Metals (chmn. Beaver Valley chpt. 1964-65, Twin Cities chpt. 1977-78), Am. Foundrymen's Soc. (chmn. Twin Cities chpt. 1982-83), Beaver Valley Caravan (pres. 1967), Masquers (v.p. 1993). Achievements include co-patent for multiple direction forging process-high alloy steels. Home: 2125 Argonne Dr Minneapolis MN 55421

MAXWELL, ARTHUR EUGENE, oceanographer, marine geophysicist, educator; b. Maywood, Calif., Apr. 11, 1925; s. John Henry and Nelle Irene (Arnold) M.; m. Colleen Oleary, July 1, 1988; children: Delle, Eric, Lynn, Brett, Gregory, Sam Wade, Henry Wade. BS in Physics with honors, N.Mex. State U., 1949; MS in Oceanography, Scripps Instn. Oceanography, 1952, PhD in Oceanography, 1959. Jr. rsch. geophysicist Scripps Instn. Oceanography, La Jolla, Calif., 1950-55; head oceanographer Office Naval Rsch., Washington, 1955-59, head br. geophysics, 1959-65; assoc. dir. Woods Hole (Mass.) Oceanographic Instn., 1965-69, dir. rsch., 1969-71, provost, 1971-81; prof. dept. geol. scis., dir. Inst. Geophysics U. Tex., Austin, 1982—; chmn. bd. govs. Joint Oceanographic Instns., 1985-86, chmn. planning com. deep earth sampling, 1968-70, chmn. exec. com. deep earth sampling, 1971-72, 78-79, 91-92; mem. joint U.S./USSR com. for coop. studies of the world ocean NAS/NRC, 1973-80, chmn. U.S. nat. com. to Internat. Union Geodesy and Geophysics, 1976-80, vice chmn. outer continental shelf/environ. studies rev. com., 1986-93; chmn. U.S. nat. com. on geology NAS, 1979-83, chmn. geophysics rsch. bd. geophysics study com., 1982-87; mem. vis. com. Rosenstiel Sch. Marine and Atmospheric Studies U. Miami, 1982-86, dept. physics N.Mex. State U., 1986—. Editor: The Sea, Vol. 4, Parts I and II, 1970; editorial adv. bd. Oceanus, 1981-92; contbr. articles to profl. jours. Chmn. tech. adv. com. Navy Thresher Search, 1963; mem. Mass. Gov's. Adv. Com. on Sci. and Tech., 1965-71. With USN, 1942-46, PTO. Recipient Meritorious Civilian Svc. award Chief Naval Rsch., 1958, Superior Civilian Svc. award Asst. Sec. of Navy, 1963, Disting. Civilian Svc. award Sec. of Navy, 1964, Disting. Alumni award N.Mex. State U., 1965, Bruun Meml. Lecture award Intergovtl. Oceanographic Commn., 1969, Outstanding Centennial Alumnus award N.Mex. State U., 1988. Fellow Am. Geophys. Union (pres. 1976-78, pres. oceanography sect. 1970-72); mem. Marine Tech. Soc. (charter, pres. 1981-82), Cosmos Club. Achievements include research in heat flow through the ocean floor, in structure and tectonics of the sea floor. Home: # 20 4408 Long Champ Dr Austin TX 78746-1172 Office: U Tex Inst for Geophysics 8701 Mopac Blvd Austin TX 78759

MAXWELL, BRUCE DALE, plant ecologist, educator; b. Blackfoot, Idaho, Feb. 24, 1954; s. Bruce A. and Margurite (Daily) M.; m. Anne S. Sokoloski, Oct. 28, 1977; children: Tyler, Peter. MS, Mont. State U., 1984; PhD, Oreg. State U., 1990. Asst. prof. U. Minn., St. Paul, 1990-92, Mont. State U., Bozeman, 1992—. Contbr. articles to profl. jours. Mem. Am. Inst. Biol. Sci., Ecol. Soc. Am., Weed Sci. Soc. Am. (Grad. Student of Yr. 1989, best paper 1991). Office: Mont State U Dept Plant and Soil Sci Bozeman MT 59717

MAXWELL, ROBERT HAWORTH, college dean; b. Earlham, Iowa, Oct. 8, 1927; s. Charles Erich and Mildred Grace M.; m. Betty Ruth Michener, Dec. 24, 1950; children: Robert Steven, Daniel Guy, Timothy Charles, Kristen Kimuli. Student, Earlham Coll., 1946-48; BS in Farm Ops., Iowa State U., 1950, MS in Agrl. Edn., 1964; PhD in Agrl. Edn., Cornell U., 1970. Cert. tchr., Iowa. Farm operator Iowa, 1952-60; instr. Earlham Coll., Richmond, Ind., 1960-62; tchr. vocat. agr. Earlham (Iowa) Community Sch., 1963-64; asst. prof. agrl. edn. W.Va. U., Morgantown, 1964-68, assoc. prof. agrl. edn., 1970-75, prof. agrl. edn., 1975-79, asst. dean coll. agr. & forestry, acting chmn. divsn. animal & vet. scis., 1980, assoc. dean coll. agr. & forestry, chmn. divsn. internat. agr. & forestry, 1980-84, interim dean coll. agr. & forestry, interim dir. W.Va. agrl. & forestry experiment sta., 1984-85, dean coll. agr. & forestry, dir. W.Va. agrl. & forestry experiment sta., 1985—; vocat. agr. tchr. Dexfield Community Schs., Redfield, Iowa, 1958-60; grad. asst. dept. edn. Cornell U., 1969-70; contract AID agrl. edn. advisor Kenya Ministry Edn., 1960-62, 64-68, spl. asst. dir. manpower devel. divsn. Tanzania Ministry Agr., 1975-79; dir. Allegheny Highlands Project, 1970-75; bd. dirs. Northeast Regional Ctr. Rural Devel., W.Va. U. Rsch. Corp.; lectr. in field. Author: (with others) Agriculture for Primary School series, 1979, 82, 85; editor Empire State Vo-Ag Teacher Jour., 1969-70; pub. papers and reports; contbr. articles to profl. jours. Unit leader United Way, Elkins, W.Va., 1970-75, Morgantown, 1979—; with U.S. Army, 1950-52. Named Disting. West Virginian, Gov. W.Va., 1993; recipient Commemorative medal U. Agr., Nitra, Slovakia, 1993. Mem. AAAS, Am. Assn. Agrl. Edn., Am. Farmland Trust (life), Am. Soc. Agrl. Conss. (cert., bd. dirs.), Am. Soc. Agrl. Conss. Internat. (cert., bd. dirs.), Nat. Assn. Colls. and Tchrs. Agr., Nat. Peace Inst., World Future Soc., Soc. Internat. Devel., Assn. Internat. Agrl. and Extension Educators (pres.-elect, Outstanding Svc. 1991), Assn. Internat. Agr. and Rural Devel. (Outstanding Svc. 1992), Coun. Agrl. Sci. and Tech., Northeast Regional Assn. Agrl. Experiment Sta. Dirs. (Svc. Commendation 1993), UN Assn. of the USA, Soil and Water Conservation Soc., Kenya Assn. Tchrs. Agr., Tanzanian Soc. Agr. Edn. and Extension, W.Va. Shepherd's Fedn., W.Va. Poultry Assn., W.Va. Cattlemen's Assn., W.Va. Farm Bur., Iowa Farm Bur., W.Va. Grassland Coun., W.Va. Horticultural Soc., Upshur Livestock Assn. (life), Morgantown USDA Club, Morgantown Rotary Club, Iowa State U. Alumni Assn. (life), Cornell U. Alumni Assn. (life), W.Va. U. Alumni Assn. (life), FFA Alumni Assn. (life), Phi Delta Kappa, Gamma Sigma Delta, Alpha Zeta, Phi Mu Alpha (life). Mem. Soc. of Friends. Home: 1295-B Stewartstown Rd Morgantown WV 26505 Office: W Va University Coll Agriculture & Forestry Evansdale Campus PO Box 6108 Morgantown WV 26506-6108

MAXWELL, STEVE A., molecular biologist, researcher; b. Lubbock, Tex., Sept. 19, 1956; s. Fowden Gene and Katherine (Gant) M.; m. Elsa Leticia Resendez, Jul. 16, 1983; children: Kathy, Michael. BS, BA, Abilene Christian Univ., 1980; MS, PhD, Univ. Tex., 1985. Post-doctoral Univ. Tex., Houston, 1985-87, Baylor Coll. Medicine, Houston, 1988-90; asst. prof. Univ. Tex., 1991—. Author: (book chpt.) SV40 T-antigen as a Dual Oncogene, 1989; contbr. articles to profl. jours. Recipient NIH First Investigator award Nat. Cancer Inst., 1993, Rsch. grant Tex. Higher Edn. Authority, 1991. Mem. N.Y. Acad. Scis., Am. Assn. Cancer Rsch., Am. Assn. Advancement of Sci. Achievements include finding protein serine kinase activity intrinsic to an oncogene protein. This was first example of an encogene encoding a serine kinuse; discovered several novel proteins that bind to the p53 tumor suppressor gene products. Home: 27 Silver Elm The Woodlands TX 77381

MAXWELL, WILLIAM LAUGHLIN, industrial engineering educator; b. Phila., July 11, 1934; s. William Henry and Elizabeth (Laughlin) M.; m. Judith Behrens, July 5, 1969; children: Deborah, William, Judith, Keely. BMechE, Cornell U., 1957, PhD, 1961. Andrew Schultz Jr. prof. dept. indsl. engring. Cornell U., Ithaca, N.Y., 1961—. Author: Theory of Scheduling, 1967. Recipient Disting. Teaching award Cornell Soc. Engrs., 1968. Fellow Inst. Indsl. Engrs.; mem. Ops. Rsch. Soc., Soc. Mfg. Engrs. Home: 106 Lake Ave Ithaca NY 14850-3537 Office: Cornell U Dept Indsl Engring Ithaca NY 14853

MAXWORTHY, TONY, mechanical and aerospace engineering educator; b. London, May 21, 1933; came to U.S., 1954, naturalized, 1961; s. Ernest Charles and Gladys May (Butson) M.; m. Emily Jean Parkinson, June 20, 1956 (div. 1974); children: Kirsten, Kara; m. Anna Barbara Parks, May 21, 1979. BS in Engring., U. London, 1954; MSE, Princeton U., 1955; PhD, Harvard U., 1959. Research asst. Harvard U., Cambridge, Mass., 1955-59; sr. scientist, group supr. Jet Propulsion Lab., Pasadena, Calif., 1960-67, cons., 1968—; assoc. prof. U. So. Calif., Los Angeles, 1967-70, prof., 1970—, Smith Internat. prof. mech. and aero. engring., 1988—, chmn. dept. mech. engring., 1979-89; cons. BBC Rsch. Ctr., Baden, Switzerland, 1972—, J.P.L., Pasadena, Calif., 1968—; lectr. Woods Hole Oceanographic Inst., Mass., summers 1965, 70, 72, 83; Forman vis. prof. in aeronautics Tech. Haifa, 1986; vis. prof. U. Poly, Madrid, 1988, Inst. Sop. Tech., Lisbon, 1988, E.T.H., Zürich, 1989, E.P.F., Lausanne, 1989—. Mem. editorial bd. Geophys. Fluid Dynamics, 1973-79, 88—, Dynamic Atmospheric Oceans, 1976-83, Phys. Fluids, 1978-81, Zeitschrift fuer Angewandte Mathematik und Physik, 1987—; contbr. articles to profl. jours. Recipient Humboldt Sr. Scientist award, 1981-83; fellow Cambridge U., 1974, Australian Nat. U., 1978, Nat. Ctr. Atmospheric Research, 1976, Glennon fellow U. Western Australia, 1990, Sr. Queen's fellow in Marine Scis., Commonwealth of Australia, 1984; recipient Halliburton award U. So. Calif., 1980, Otto Laporte award Am. Physics Soc., 1990. Fellow Am. Phys. Soc. (chmn. exec. com. fluid dynamics div. 1974-79); mem. NAE, Am. Meteorol. Soc., Am. Geophys. Union, ASME (fluid mechs. com.), European Geophys. Soc., Acad. Applied Mechanics. Office: U So Calif Dept Mech Engring Exposition Park Los Angeles CA 90089-1453

MAY, EVERETTE LEE, JR., pharmacologist, educator; b. Timberville, Va., Aug. 1, 1914; married, 1940; 4 children. AB, Bridgewater Coll., 1935; PhD in Organic Chemistry, U. Va., 1939. Rsch. chemist Nat. Oil Products Co., 1939-41; from assoc. chemist to sr. chemist NIH, 1941-53; from scientist to sr. scientist Commd. Corps., 1953-58, scientist dir., 1959; chief sect. med. chemistry Nat. Inst. Arthritis and Metabolic Disorders, USPHS,, 1960-78; adj. prof. pharmacology Med. Coll. Va., 1974-77, prof. pharmacology, 1977—; mem. expert adv. panel, 1958-78; chem. adv. panel mem. Walter Reed Army Inst., 1965—; ad hoc review com. Nat. Cancer Inst., 1978. Recipient E. E. Smissman award Am. Chem. Soc., 1979, Nathan B. Eddy Meml. award, 1981, Alfred Burger Medicinal Chemistry award Am. Chem. Soc., 1992. Mem. Am. Chem. Soc. Achievements include research in surface active agents, vitamins of the B complex, antimalarial agents, analgesic drugs, antitubetcular compounds, carcinolytic agents, chemical and pharmacological investigations on central nervous system and anti-inflammatory agents. Home: 606 Irene Ave Salisbury MD 21801*

MAY, KENNETH MYRON, information services professional, consultant; b. Indpls., July 25, 1949; s. William John and Clara June (Powell) M.; m. Laura V. Loumeau, Apr. 24, 1988; children: Kendell Morgan, Richard William. BA, U. Colo., 1971; MA in Teaching, U. Wis., Eau Claire, 1974. CPCU; CLU; cert. data processor Inst. for Cert. Computer Profls.; assoc. in risk mgmt. Ins. Inst. Am. Program asst. Fresno County Dept. Health, Fresno, Calif., 1975-77; br. mgr., sales rep. Proteus, Fresno, 1978-81; ins. underwriter Hartford Steam Boiler Inspection and Ins. Co., Basking Ridge, N.J., 1982-83; systems analyst Crum & Forster, Basking Ridge, 1983-85; sr. systems analyst Continental Ins. Co., Piscataway, N.J., 1985-88; software designer ISI Systems, inc., Andover, Mass., 1989-92; mgr. info. svcs. Mchts. Ins. Group, Buffalo, 1992-93; project mgr. Cover-All Systems, Fair Lawn, N.J., 1993—; speaker, conf. chmn. 16 computer and ins. trade confs. and meetings. Author: AI and Insurance: Rewards from Technology, 1991; editor: Electronic Underwriter: AI and Ins., 1991; contbr. articles to CASE Trends, Nat. Underwriter, Chief Info. Officer Jour., Best's Rev., Data Pro, Sytesms AI. Mem. Soc. CPCU's (rsch. grantee Harry J. Loman Found.), Am. Assn. Artificial Intelligence, Internat. Assn. Knowledge Engrs., Data Processing Mgmt. Assn. (pres. spl. interest group on artificial intelligence 1990-92). Home: 330 Sherbrooke Ave Williamsville NY 14221

MAY, MICHAEL LEE, magazine editor; b. Dayton, Ohio, Oct. 13, 1959; s. John William and Sarah (Allen) M. BA, Earlham Coll., 1982; MS, U. Conn., 1984; PhD, Cornell U., 1990. Assoc. editor Am. Scientist, Rsch. Triangle Park, N.C., 1991—. Contbr. articles to profl. jours. Mem. Nat. Assn. Sci. Writers, Sigma Xi. Avocations: bicycling, bird watching. Office: Am Scientist PO Box 13975 Research Triangle Park NC 27709

MAY, RICHARD PAUL, data processing professional; b. Milw., Oct. 19, 1946; s. Gorden Elliot and Marie Karen (Leidgen) M.; m. Amy Yamashiro, Mar. 5, 1982. BBA, U. Hawaii, 1972; Cert. in Owners/Pres's. Mgmt., Harvard U., 1989; MBA, U. Hawaii, 1990. Customer engr. IBM, Honolulu, 1968; co-founder, chief exec. officer Aloha Tax Svc., Honolulu, 1972-88; founder, pres., chief exec. officer Honolulu Bar Supply Ltd./May Foodsvc., 1972-90; co-founder Indtl. Distbrs., Kahului, Hawaii, 1976-88; mng. ptnr. Data Capture Systems, Kaneohe, Hawaii, 1990—; pres. Rick May, Inc., Honolulu, 1989—. With U.S. Army, 1968-70, Vietnam. Decorated Bronze Star medal, Air medal, 2 Purple Heart medals. Mem. Inst. Indsl. Engrs., Harvard Bus. Sch. Club, C. of C., Hawaii Visitors Bur., Beta Gamma Sigma. Republican. Roman Catholic. Office: Data Capture Systems 46-001 Kamehameha Hwy Ste 317 Kaneohe HI 96744

MAYBERRY, WILLIAM EUGENE, retired physician; b. Cookeville, Tenn., Aug. 22, 1929; s. Henry Eugene and Beatrice Lucille (Maynard) M.; m. Jane G. Foster, Dec. 29, 1953; children: Ann Graves, Paul Foster. Student, Tenn. Tech. U., 1947-49; M.D., U. Tenn., 1953; M.S. in Medicine, U. Minn., 1959; D.H.L. (hon.), Jacksonville U., 1983. Diplomate Am. Bd. Internal Medicine. Intern U.S. Naval Hosp., Phila., 1953-54; resident Mayo Grad. Sch. Medicine, Rochester, Minn., 1956-59; mem. staff New Eng. Med. Ctr., Boston, 1959-60, Nat. Inst. Arthritis and Metabolic Diseases, 1962-64; cons. internal medicine, endocrine research and lab. medicine, chmn. dept. lab. medicine Mayo Clinic, Rochester, 1971-75, bd. govs., 1971-87, vice chmn., 1974-75, chmn., chief exec. officer, 1976-87, prof. lab. medicine, 1971—, prof. medicine, 1983-92; asst. in medicine Tufts U. Med. Sch., 1959-60; mem. faculty Mayo Grad. Sch. Medicine and Mayo Med. Sch., 1960-92; trustee Mayo Found., 1971-87, vice chmn., 1974-85, pres. 1986-87, chmn. bd. devel. 1988—; trustee Minn. Mut. Life Ins., 1983-92; bd. dirs. George A. Hormel & Co., 1986-92. Mem. editorial bd. (Jour. of Clin. Endocrinology and Metabolism), 1971-73; contbr. articles to profl. jours. Trustee Mpls. Soc. Fine Arts, 1983-91, Cumberland U., 1984-86, Twin Cities Pub. TV, Inc., 1991-92, trustee, 1991-92; bd. overseers Mpls. Coll. Art and Design, 1983-86, U. Minn. Sch. Mgmt., 1985-88; bd. dirs. Greater Rochester Area Univ. Ctr., 1986-87, Minn. Acad. Excellence Found., 1986-87, U.S. West-Minn. Exec. Bd., 1988-92; rep. Congl. Dist. I State of Minn. Compensation Council, 1986; chmn. Presdl. Commn. on Human Immunodeficiency Virus Epidemic, 1987. Recipient Disting. Alumni award Tenn. Technol. U., 1976, chair of excellence in bus. adminstrn. named in his honor, 1989; recipient Outstanding Alumni award U. Tenn., 1982, Med. Exec. Award Am. Coll. Med. Group Adminstrs., 1986; rsch. fellow NIH, 1959-60, Am. Cancer Soc., 1962-64; NIH research grantee, 1965-71. Fellow ACP; mem. Inst. Medicine of NAS, Am. Thyroid Assn., Am. Clin. and Climatological Soc., Endocrine Soc., Soc. Med. Adminstrs., Am. Acad. Med. Dirs., Am. Coll. Physician Execs. (bd. regents 1983, vice chmn. 1985-86), Sigma Xi. Clubs: Mpls. Club, Rochester Golf & Country, The Club at Pelican Bay (Naples, Fla.). Home: 826 Rue de Ville Naples FL 33963-8531 Office: Emeritus Siebens 9 200 First St SW Rochester MN 55905

MAYER, ALEJANDRO MIGUEL, biologist, educator; b. Buenos Aires, Apr. 13, 1950; came to U.S., 1990; s. Bernardo A.C.T. and Lise Lotte C. (Magnus) M.; m. Marta Beatriz Borgese, Sept. 19, 1974; children: Veronica Alejandra, Leonardo Miguel. PhD, U. Buenos Aires, 1980; postdoctoral student, U. Calif., Santa Barbara, 1988. Ind. investigator Nat. Sci. Rsch. Coun., Buenos Aires, 1981-85, head rsch. group, 1988-90; postdoctoral fellow, then adj. asst. prof. U. Calif., Santa Barbara, 1985-88; asst. prof. rsch. La. State U. Med. Ctr., New Orleans, 1990-93; asst. prof. pharmacology Midwestern U., Downers Grove, Ill., 1993—; head applied rsch. dept. Patagonia Comercial SRL, Buenos Aires, 1974-85. Contbr. over 26 articles to profl. jours. Recipient Federico Schlottman prize Nat. Acad. Medicine, Argentina, 1978. Mem. AAAS, Soc. Exptl. Biology and Medicine, Internat. Phycol. Soc. Roman Catholic. Achievements include research on mechanisms of signal transduction focusing on use of marine natural products as molecular probes. Home: 1600 Harvard Ave Naperville IL 60565 Office: Midwestern U Dept of Pharmacology 555 31st St Downers Grove IL 60515

MAYER, PETER CONRAD, economics educator; b. Balt., Mar. 16, 1938; s. Joseph E. and Maria (Goeppert) M.; m. Mary Fay Marmon, Aug. 23, 1974. BS in Physics, Calif. Inst. Tech., 1961; MA in Econs., U. Calif., Berkeley, 1967, PhD in Econs., 1969. Asst. prof. econs. Miami U., Oxford, Ohio, 1968-71; asst. prof. econs. U. Guam, Mangilao, 1971-74, assoc. prof., 1982-84; sr. economist Guam Dept. Commerce, Tamuning, 1974-80, chief economist, 1981-82; asst. prof. U. Petroleum and Minerals, Dhahran, Saudi Arabia, 1980-81; cons. on stats. Govt. of Guam, Agana, 1984-85; fgn. expert Shanghai U. Fin. and Econs., People's Republic China, 1985-86; fgn. tchr. econs. Kobe (Japan) U., 1986-89; sr. economist Guam Dept. of Commerce, Tamuning, 1989—; cons., expert witness U.S. Dept. Justice, Agana, 1983. Author: (monographs) Guam's Tax System, 1976, Guam's Generating Option, 1982, Notes on Tax Reform No. 1: How to Encourage Home Ownership, 1990, Notes on Tax Reform No. 2: Liquor Taxation and Alcohol Abuse Among Youth, 1991; columnist Pacific Sun. News, 1973-74; contbr. articles to profl. jours. Mem. Am. Econ. Assn., Western Econ. Assn. Democrat. Avocations: writing fantasy, conservation, swimming, scuba diving, hiking. Office: Dept Commerce 590 S Marine Dr Tamuning GU 96911-3507

MAYER, RICHARD THOMAS, laboratory director, entomologist; b. Pensacola, Fla., May 11, 1945; s. Richard Thomas and Lucy (Daniel) M.; divorced; children: Christopher Burton, Stefany Paige; m. Gabriele Eve Klitschka, June 19, 1986. BS in Chemistry, U. Ga., 1967, PhD in Entomology, 1970. Instr. U. Ga., Athens, 1970-71; rsch. entomologist, rsch. leader Dept. Agr. Agrl. Rsch. Svc., College Station, Tex., 1971-84; lab. dir. Dept. Agr. Agrl. Rsch. Svc., Orlando, Fla., 1984—. Editor, founder Archives Insect Biochemistry and Physiology, 1980-92; contbr. articles to profl. jours.; patentee in field. Fellow Royal Entomol. Soc. London; mem. Am. Chem. Soc., Am. Registery Profl. Entomologists (examiner), Entomol. Soc. Am., Ga. Entomol. Soc., AAAS. Episcopalian. Avocations: fishing, woodworking. Office: USDA Agrl Rsch Svc USHRL 2120 Camden Rd Orlando FL 32803-1498

MAYES, MARK EDWARD, molecular toxicologist; b. New Orleans, Apr. 11, 1962; s. Jimmie Ray and Judith Davis (Fergusson) M.; m. Susan Doyle Baldridge, June 9, 1984; children: Lauren Ray, Alec Christopher. BS in Zoology, La. Tech. U., 1985; MS in Toxicology, N.C. State U., 1988, PhD in Toxicology, 1993. Author: (with others) Pesticides and the Future, 1992; contbr. articles to Fundamental and Applied Toxicology and Pesticide Biochemistry and Physiology. Mem. Am. Soc. for Microbiology, Sigma Xi. Home: 4350 Hunters Club Dr Raleigh NC 27606 Office: US EPA Herl MD-67 Research Triangle Park NC 27711

MAYES, MAUREEN DAVIDICA, physician, educator; b. Phila., Oct. 16, 1945; d. David M. and Marguerite Cecilia (Fineran) M.; m. Charles William Houser, Dec. 18, 1976; children: David Steven, Edward Charles. BA, Coll. Notre Dame, 1967; MD, Ea. Va. Med. Sch., 1976. Resident in internal medicine Cleve. Clinic Found., 1977-79, fellow in rheumatology, 1979-81; asst. prof. medicine W.Va. U., Morgantown, 1981-85; asst. prof. medicine Wayne State U., Detroit, 1985-90, assoc. prof. medicine, 1990—; bd. dirs. United Scleroderma Found., Watsonville, Calif., 1986—; dir. scleroderma unit Wayne State U., Detroit, 1991—. Contbr. articles to profl. jours. Pres. bd. United Scleroderma Found., 1988-89. Fellow Am. Coll. Rheumatology, Am. Coll. Physicians; mem. Am. Fedn. Clin. Rsch., Mich. Rheumatism Soc.

Office: Wayne State U Hutzel Hsp 4707 St Antoine St Detroit MI 48201-1498

MAYHEW, ERIC GEORGE, cancer researcher, educator; b. London, Eng., June 22, 1938; came to U.S., 1964; s. George James and Doris Ivy (Tipping) M.; m. Barbara Doe, Sept. 28, 1966 (div. 1976); 1 child, Miles; m. Karen Ann Caruana, Apr. 1, 1978; children: Ian, Andrea. BS, U. London, 1960, MS, 1963, DSc, 1993. Rsch. asst. Chester Beatty Rsch. Inst., London, 1960-64; cancer rsch. scientist Roswell Pk. Meml. Inst., Buffalo, 1964-68, sr. cancer rsch. scientist, 1968-72, assoc. cancer rsch. scientist, 1979—, dep. dir. exptl. pathology, 1988—; assoc. rsch. prof. SUNY, Buffalo, 1979—; ad-hoc mem. NIH study sects., 1982—. Editor jour. Selective Cancer Therapeutics, 1989—; contbr. articles to Jour. Nat. Cancer Inst., Cancer Rsch., other profl. jours. Grantee NIH, Am. Heart Assn., and pvt. industry, 1972—. Mem. Am. Assn. Cancer Rsch., Am. Soc. Cell Biology, N.Y. Acad. Scis. Achievements include development of liposomes for drug delivery and patent for liposome delivery. Office: Roswell Park Cancer Inst Elm and Carlton Sts Buffalo NY 14263

MAYO, CLYDE CALVIN, organizational psychologist; b. Robstown, Tex., Feb. 2, 1940; s. Clyde Culberson and Velma (Oxford) M.; m. Jeanne Lynn McCain, Aug. 24, 1963; children: Brady Scott, Amber Camille. BA, Rice U., 1961; BS, U. Houston, 1964, PhD, 1972; MS, Trinity U., 1966. Lic. psychologist, Tex. Mgmt. engr. LWFW, Inc., Houston, 1966-72, sr. cons., 1972-78, prin., 1978-81; ptnr. Mayo, Thompson, Bigby, Houston, 1981-83, founder Mgmt. and Personnel Systems, Houston, 1983—; counselor Interface Counseling Ctr., Houston, 1976-79; dir. Mental Health HMO Group, 1985-87; instr. St. Thomas U., Houston, 1979—, U. Houston Downtown Sch., 1972, U. Houston-Clear Lake, 1983-88, U. Houston-Central Campus, 1984—; dir. mgmt. devel. insts. U. Houston Woodlands and West Houston, 1986-1991, adj. prof. U. Houston, 1991—. Author: Bi/Polar Inventory of Strengths, 1978, LWFW Annual Survey of Manufacturers, 1966-1981. Coach, mgr. Meyerland Little League, 1974-78, So. Belles Softball, 1979-80, S.W. Colt Baseball, 1982-83, Friends of Fondren Library of Rice U., 1988—. Mem. Soc. Indsl. Organizational Psychologists, Houston Psychol. Assn. (membership dir. 1978, sec. 1984), Houston Area Quality Circle Soc., Tex. Psychol. Assn., Am. Psychol. Assn., Bus. Execs. for Nat. Security, Houston Area Indsl. Orgnl. Psychologists (bd. dirs. 1989-92), Forum Club. Methodist. Club: Meyerland (bd. dirs. 1988-92, pres. 1991). Home: 8723 Ferris Dr Houston TX 77096-1409 Office: Mgmt and Personnel Systems 4545 Bissonnet St Bellaire TX 77401-3114

MAYO, JOAN BRADLEY, microbiologist, epidemiologist; b. Ada, Okla., Oct. 24, 1942; d. Samuel S. and Norene (Parker) Bradley; m. Harry D. Mayo III, Sept. 30, 1967. BA, Drake U., 1964; MS in Microbiology, NYU, 1978; MBA in Mgmt., Fairleigh Dickinson U., 1989. Technologist clin. labs. St. John's Episc. Hosp., Bklyn., 1964-66; supr. Med. Tech. Sch. Bklyn.-Cumberland Med. Ctr., 1966-71; clin. instr., technologist SUNY Downstate Med. Ctr., Bklyn., 1970-73; supr. bacteriology lab. Meml. Sloan-Kettering Cancer Ctr., N.Y.C., 1973-82, mgr. microbiology labs., 1982-87; dir. infection control svc. N.Y.C. Health and Hosp. Corp./Harlem Hosp. Ctr., 1987—; mem. com. for prevention of bloodborne diseases N.Y.C. Health and Hosp. Corp., 1990—. Contbr. articles to profl. publs. Active Friends of Harlem Hosp., 1988—, North Bergen (N.J.) Action Group, 1987—. Mem. Am. Soc. Microbiology, Am. Pub. Health Assn., Assn. Practitioners in Infection Control, Delta Mu Delta. Avocations: writing, travel, reading. Home: 7855 Boulevard East North Bergen NJ 07047 Office: NYC Health/Hosp Corp Harlem Hosp Ctr 506 Lenox Ave New York NY 10037

MAYO, JOHN SULLIVAN, telecommunications company executive; b. Greenville, N.C., Feb. 26, 1930; s. William Louis and Mattie (Harris) M.; m. Lucille Dodgson, Apr. 1957; children: Mark Dodgson, David Thomas, Nancy Ann, Lynn Marie. BS, N.C. State U., 1952, MS, 1953, PhD, 1955. With AT&T Bell Labs., Murray Hill, N.J., 1955—, exec. dir. toll electronic switching div., 1973-75, v.p. electronics tech., 1975-79, sr. v.p. network systems and network svcs., 1979-91, pres., 1991—; mem. N.Y.C. Partnership's High Tech. Com., adv. bd. Coll. Engring., U. Calif., Berkeley, com. on engring. utilization Am. Assn. Engring. Socs., sci. adv. group House Subcom. on Renewing U.S. Sci. and Tech. Policy; bd. dirs. Johnson and Johnson, Sandia Corp., Found. for Nat. Medals of Sci. and Tech. Contbr. articles to profl. jours.; patentee in field. Trustee Polytech U., The Kenan Inst. for Engring., Tech. and Sci at N.C. State U., Liberty Sci. Ctr.; mem. bd. overseers N.J. Inst. Tech. Named Outstanding Engring. Alumnus N.C. State U., 1977, Nat. medal Tech., 1990, Indsl. Rsch. Inst. medal, 1992. Fellow IEEE (Alexander Graham Bell award 1978, Simon Ramo medal 1988, C&C prize 1988, Nat. Medal Tech. 1990); mem. NAE, Sigma Xi, Phi Kappa Phi. Baptist. Avocations: fishing, gardening, bicycling, jogging. Office: AT&T Bell Labs 600 Mountain Ave Murray Hill NJ 07974

MAYO, OLIVER, biology researcher; b. Adelaide, Australia, May 29, 1942; s. Eric Elton and Edith Janet Allen (Simpson) M.; m. Margaret Gwendoline Burton, Aug. 11, 1967; children: Charles, Rebecca, Rupert. BSc, U. Adelaide, 1964, BSc with honors, 1965, PhD, 1968, diploma in bus. mgmt., 1976, DSc, 1989. Chartered Statistician, RSS. Head. biometry sect. Waite Agrl. Rsch. Inst. U. Adelaide, 1971-89, dean Faculty Agrl. Sci., 1986-88; chief div. animal prodn. Commonwealth Sci. and Indsl. Rsch. Orgn., Sydney, NSW, Australia, 1989—. Author: The Theory of Plant Breeding, 1980, R. Austin Freeman, 1980, Natural Selection and its Constraints, 1983, The Wines of Australia, 1986. Mem. Russian Acad. Agrl. Sci. (fgn.). Avocations: skiing, detective stories, wine. Home: 146 Hereford St, New South Wales, Forest Lodge 2037, Australia Office: CSIRO Div Animal Prodn, Clunies Ross St New South Wales, Prospect 2149, Australia

MAYO, ROBERT MICHAEL, nuclear engineering educator, physicist; b. Phila., Sept. 23, 1962; s. Robert Norman and Joan Ann (Oeschner) M.; m. Judith Ann Hopkins, June 24, 1989. BS, Pa. State U., 1984; MS, Purdue U., 1987, PhD, 1989. Rsch. asst. Argonne (Ill.) Nat. Lab., 1984; rsch. asst. Los Alamos (N.Mex.) Nat. Lab., 1985, physicist, 1989-91, instr. Purdue U., W. Lafayette, Ind., 1986; rsch. asst. Purdue U., W. Lafayette, 1984-86, Princeton (N.J.) U., 1986-88; asst. prof. N.C. State U., Raleigh, 1991—; faculty advisor ANS, N.C. State U., 1991—. Contbr. articles to profl. jours. Recipient David Ross fellowship Purdue U., 1986-88. Mem. Am. Phys. Soc., Am. Nuclear Soc. (R.A. Daniels Meml. scholar Purdue U. 1988), Fusion Power Assocs. Achievements include research on transport properties in magnetic confinement plasma devices and space nuclear propulsion. Office: NC State U Dept Nuclear Engring Box 7909 2101 Burlington Labs Raleigh NC 27695-7909

MAYOCK, ROBERT LEE, internist; b. Wilkes-Barre, Pa., Jan. 19, 1917; s. John F. and Mathilde M.; m. Constance M. Peruzzi, July 2, 1949; children: Robert Lee, Stephen Philip, Holly Peruzzi. B.S., Bucknell U., 1938; M.D., U. Pa., 1942. Diplomate Am. Bd. Internal Medicine. Intern Hosp. U. Pa., Phila., 1943-44; resident Hosp. U. Pa., 1944-45, chief med. resident, 1945-46, attending physician, 1946—; chief pulmonary disease Phila. Gen. Hosp., 1955-72, chief pulmonary disease sect., 1959-72, sr. cons. pulmonary disease sect., 1972—; asst. prof. clin. medicine U. Pa., 1949-59, assoc. prof., 1959-70, prof. medicine, 1970-87, prof. emeritus, 1987—; mem. med. adv. com. for Tb Commonwealth of Pa., 1965-74, mem. med. adv. com. on chronic respiratory disease, 1974-92, chmn. adv. com., 1981-90; mem. subsplty bd. pulmonary disease Am. Bd. Internal Medicine, 1965-76; nat. bd. dirs. Am. Lung Assn., 1983-92, local bd. dirs., 1961, local pres., 1966-69, dir. at large, 1983—. Contbr. articles in field to med. jours. Served to capt. U.S. Army, 1952-54. Fellow ACP, Am. Coll. Chest Physicians (regent 1972-79); mem. AMA, Am. Thoracic Soc., Am. Fedn. Clin. Rsch., Am. Heart Assn., Pa. Lung Assn. (dir. 1976—), N.Y. Acad. Scis., Pa. Med. Soc., Phila. County Med. Soc., Physiology Soc. Phila., Laennec Soc. Phila., Merion Cricket Club, Westmoreland Club, Swiftwater Res., Sigma Xi, Alpha Omega Alpha. Home: 244 Gypsy Ln Wynnewood PA 19096-1113 Office: U Penn 3rd Fl Ste 1 Ravdin Bldg Philadelphia PA 19104

MAYR, ERNST, emeritus zoology educator, author; b. Kempten, Germany, July 5, 1904; came to U.S., 1931; s. Otto and Helene (Pusinelli) M.; m. Margarete Simon, May 4, 1935; children: Christa E., Susanne. Cand. med., U. Greifswald, 1925; Ph.D., U. Berlin, 1926; Ph.D. (hon.), Uppsala U., Sweden, 1957; D.Sc. (hon.), Yale U., 1959, U. Melbourne, 1959, Oxford U., 1966, U. Munich, 1968, U. Paris, 1974, Harvard U., 1980, Guelph U., U. Cambridge, 1982, U. Vt., 1984; DSc (hon.), U. Mass., 1993. Asst. curator zool. mus. U. Berlin, 1926-32; mem. Rothschild expdn. to Dutch New Guinea, 1928, expdn. to Mandated Ty. of New Guinea, 1928-29, Whitney Expdn., 1929-30; research asso. Am. Mus. Natural History, N.Y.C., 1931-32; asso. curator Am. Mus. Natural History, 1932-44, curator, 1944-53; Jesup lectr. Columbia U., 1941; Alexander Agassiz prof. zoology Harvard U., 1953-75, emeritus, 1975—; dir. Mus. Comparative Zoology, Harvard U., 1961-70; Messenger lectr. Cornell U., 1985; Hitchcock prof. U. Calif., 1987. Author: List of New Guinea Birds, 1941, Systematics and the Origin of Species, 1942, Birds of the Southwest Pacific, 1945, Birds of the Philippines, (with Jean Delacour), 1946, Methods and Principles of Systematic Zoology, (with E. G. Linsley and R. L. Usinger), 1953, Animal Species and Evolution, 1963, Principles of Systematic Zoology, 1969, Populations, Species and Evolution, 1970, Evolution and the Diversity of Life, 1976, (with W. Provine) Evolutionary Synthesis, 1980, Biologie de l'Evolution, 1981, The Growth of Biological Thought, 1982, Toward a New Philosophy of Biology, 1988, One Long Argument, 1991; editor: Evolution, 1947-49. Pres. XIII Internat. Ornith. Congress, 1962. Recipient Leidy medal, 1946, Wallace Darwin medal, 1958, Brewster medal Am. Ornithologists Union, 1965, Daniel Giraud Elliot medal, 1967, Nat. Medal of Sci., 1970, Molina prize Accademia delle Scienze, Bologna, Italy, 1972, Linnean medal, 1977, Gregor Mendel medal, 1980, Balzan prize, 1983, Darwin medal Royal Soc., 1987. Fellow Linnaean Soc. N.Y. (past sec. editor), Am. Ornithol. Union (pres. 1956-59), New York Zool. Soc.; mem. NAS, Am. Philos. Soc., Am. Acad. Arts and Scis., Am. Soc. Zoologists, Soc. Systematic Zoology (pres. 1966), Soc. Study Evolution (sec. 1946, pres. 1950); hon. or corr. mem. Royal Soc., Royal Australian, Brit. ornithol. unions, Zool. Soc. London, Soc. Ornithol. France, Royal Soc. New Zealand, Bot. Gardens Indonesia, S. Africa Ornithol. Soc., Linnean Soc. London, Deutsche Akademie der Naturforsch Leopoldina., Accad. Naz. dei Lincei, Royal Soc., Académie des Sci. Office: Mus Comparative Zoology Harvard U Cambridge MA 02138

MAYS, DAVID ARTHUR, agronomy educator, small business owner; b. Waynesburg, Pa., Apr. 17, 1929; s. Arthur Lynn and Edith N. (Breakey) M.; m. Betty Ann Sellers, Aug. 7, 1954; children: Gregory D., Laurie Ann. MS in Agronomy, Pa. State U., 1959, PhD in Agronomy, 1961. Cert. profl. agronomist. Asst. county agrl. agt. Pa. State, Washington, Pa., 1954-57; grad. rsch. asst. Pa. State U., University Park, Pa., 1957-61; asst. agronomist Va. Polytech. Inst. & State U., Middleburg, 1961-63; rsch. agronomist TVA, Muscle Shoals, Ala., 1963-88; prof. agronomy Ala. A&M Univ., Normal, Ala., 1989—. Editor: Forage Fertilization, 1974; Contbr. articles to profl. jours. and chpts. to books, 1959—. 1st lt. U.S. Army, 1951-54, Korea. Mem. Am. Soc. Agronomy, Am. Forage and Grassland Coun. (Merit Cert. award 1976), Ala. Turfgrass Assn. Presbyterian. Home: 114 Kathy St Florence AL 35633-1428 Office: Ala A&M U Dept Plant and Soil Sci PO Box 1208 Normal AL 35762-1208

MAYSTRICK, DAVID PAUL, engineering executive; b. Omaha, Sept. 23, 1951; s. Robert Fred and Marcella Marie (Pietzyk) M.; m. Mary Theresa Ambrose, June 8, 1985; children: Matthew, Joseph. BS in Civil Engring., U. Nebr., Omaha, 1973; MS in Civil Engring., U. Nebr., Lincoln, 1976. Diplomate Am. Acad. Environ. Engrs.; registered profl. engr., Nebr., Del., Calif., Va. Engr. technician Nebr. Testing Lab., Omaha, 1972-73; structural engr. Gibbs & Hill Inc., Omaha, 1973-77; project engr. Bell, Galyardt and Wells Inc., Omaha, 1977-78; sr. project mgr. HDR Engring., Inc., Omaha, 1978—. Mem. steering com. U. Nebr., Omaha, 1972, St. Vincent de Paul, Omaha, 1991—, mem. building com. Mem. ASCE (student coord. 1990-92), ASME Solid Waste Assn., Solid Waste Assn. N.Am., Nat. Recycling Coalition, Tau Beta Pi, Chi Epsilon, Phi Kappa Phi. Democrat. Roman Catholic. Achievements include project design engineer for first city owned refuse derive fuel facility in U.S.; project mgr. of first and largest full svc. mass burn facility in U.S.; devel. of concepts to optimize recovery of recyclables and conversion of remaining solid waste to energy. Office: HDR Engring Inc 8404 Indian Hills Dr Omaha NE 68114-4049

MAZARAKIS, MICHAEL GERASSIMOS, physicist, researcher; b. Volos, Greece; came to U.S., 1966, naturalized, 1980; s. Gerassimos Nikolaos and Anthie Gerassimos (Kappatos) M.; m. Carolyn Seidel, June 30, 1990. BS in Physics, U. Athens, Greece, 1960; MS in Physics, U. Sorbonne, Paris, 1963, PhD in Physics, 1965; PhD in Physics, Princeton U. and U. Pa., 1971; cert. in mgmt., MIT, 1976. Mem. faculty Rutgers U., New Brunswick, N.J., 1971-74; v.p. and dir. exptl. program Fusion Energy Corp., Princeton, N.J., 1974-77, also exec. v.p., 1975-77; research physicist Argonne Nat. Lab., U. Chgo., 1978-81; research physicist Sandia Nat. Lab. Div. 1242, Albuquerque, 1981—. Contbr. articles to profl. jours. Patentee in field. Bd. dirs. Orthodox Ch., Albuquerque, 1981—; Served to maj. Greek Army, 1960-62. Recipient award Italian. Govt., 1956, Greek Govt., 1956-60, French Govt., 1962-65; Yale U. grantee, 1966. Mem. Am. Phys. Soc., IEEE, Alliance Francaise, N. Mex. Mountain Club, N.Y. Acad. Sci., Sigma Xi. Current work: Particle beam physics, accelerator research and development, inertial fusion, pulse power technology, plasma physics. Subspecialty: Nuclear fusion, particle beam physics.

MAZENKO, GENE FRANCIS, physics educator; b. July 5, 1945; m. Judy Oakley, Aug. 9, 1969. BS in Physics, Stanford U., 1967; PhD in Phsyics, MIT, 1971. Research assoc. Dept. Physics Brandeis U., Waltham, Mass., 1971-72, Dept. Nuclear Engring. MIT and Dept. Physics Harvard U., Cambridge, 1972-73, Dept. Applied Physics Stanford (Calif.) U., 1973-74; asst. prof. James Franck Inst. and Dept. Physics, U. Chgo., 1975-80, assoc. prof., 1980-85, prof. physics, inst. dir., 1985—; vis. asst. prof. Dept. Physics U. Calif., San Diego, 1980; vis. prof. Dept. Theoretical physics Oxford U., 1985; cons. Argonne (Ill.) Nat. Lab., 1976-80; vis. scientist Service de Physique Theorique Ctr. d' Etudes Nucleaires de Saclay, France, 1979. Contbr. over 90 articles to profl. jours. Alfred P. Sloan Found. fellow, 1977-81. Office: James Franck Inst 5640 S Ellis Ave Chicago IL 60637-1467

MAZUMDER, JYOTIRMOY, mechanical and industrial engineering educator; b. Calcutta, India, July 9, 1951; came to U.S., 1978; s. Jitendra Mohan and Gouri (Sen) M.; m. Aparajita, June 17, 1982; children: Debashis, Debayan. B in Engring., Calcutta U., 1973; diploma, PhD, Imperial Coll., London U., 1978. Rsch. scientist U. So. Calif., L.A., 1978-80; asst. prof. mechanical and indsl. engring. U. Ill., Urbana, 1980-84, assoc. prof., 1984-88, prof., 1988—; co-dir. ctr. laser aided material processing U. Ill., 1990—; dir. Quantum Laser Corp., Edison, N.J., 1982-89; pres. Laser Scis., Inc., Urbana, 1988—; vis. scholar physics dept. Stanford (Calif.) U., 1990. Author: (with others) Laser Welding; co-editor: Laser Materials Processing, 1984, 88; contbr. over 150 articles to profl. jours. Fellow Am. Soc. of Metals and Laser Inst. of Am. (life, sr. editor Jour. Laser Application); mem. Am. Inst. Metallurgical Engrs. (phys. mets. com. 1980—), Optical Soc. Am. Achievements include non-equilibrium synthesis of Ni-Cr-Al-Hf alloy by laser; patent applied for by USAF Office Scientific Research; weld pool visualization system for measurement of free surface deformation. Office: U Ill Dept Mech and Ind Engring 1206 W Green St Urbana IL 61801

MAZURKIEWICZ, JOHN ANTHONY, plant process engineer; b. Elizabeth, N.J., Apr. 11, 1966; s. John Paul and Constance Stephanie (Samsel) M. BS in Indsl. Engring., Rutgers U., 1984-88. Process engr., mfg. supr. Ecolab, Avenel, N.J., 1988-89, project engr., 1989-90, plant engr., 1990—; career adv. Woodbridge (N.J.) Sch. System, 1985—. Mem. Soc. Mfg. Engrs. (assoc.), Woodbridge C. of C. Office: Ecolab Inc 255 Blair Rd Avenel NJ 07001

MAZZA, A. J., electrical engineer, consultant; b. Scranton, Pa., Apr. 24, 1963; s. Albert Philip and Josie (Simsick) M.; m. Michelle Woldanski, Aug. 5, 1989. A of Engring., Pa. State U., Scranton, 1983; B of Engring., Gannon U., 1987. Elec. project engr. Denk Assocs., Inc., Cleve., 1988—. Mem. Illuminating Engrs. Soc. (Cleve. sect. pres. 1993-94). Home: 983 E 179th St Cleveland OH 44119 Office: Denk Assocs Inc 503 E 200th St Cleveland OH 44119

MAZZO, DAVID JOSEPH, analytical chemist; b. Hackensack, N.J., Dec. 22, 1956; s. Rocco John and Frances Catherine (Morone) M.; m. Cheryl Ann Sabarese, June 9, 1979; children: Joseph Rocco, Michael Louis, John Francis. BA in Interdisciplinary Humanities in honors program, Villanova U., 1979, BS in Chemistry, 1979; MS in Chemistry, U. Mass., 1981, PhD in Analytical Chemistry, 1983. Sr. rsch. chemist Merck, Sharp and Dohme Rsch. Labs., West Point, Pa., 1983-86; mgr. R&D Baxter Healthcare Corp., Round Lake, Ill., 1986-89; dir. Rhone-Poulenc Rorer, Fort Washington, Pa., 1989-92; sr. dir. Rhone-Poulenc Rorer, Antony, France, 1992—; rsch. fellow Ecole Poly. Fed. de Lausanne, Switzerland, 1981; vis. instr. U. Mass., 1987, 88, 89, 91, U. Tex., 1989; mem. referee com., mem. exec. bd. Analusis, 1991—; presenter various sci. symposiums, profl. seminars and workshops. Mem. editorial bd. Analytical Profiles of Drug Substances and Excipients, 1989—; contbr. articles to profl. jours., including Jour. Pharm. and Biomed. Analysis, Jour. Chromatography, Analytical Profiles of Drug Substances. Recipient UNICO nat. scholarship, 1975-79, 3 student awards Am. Chem. Soc., 1978,'79, Merck Summer Intern fellowship, 1979; named rsch. assoc. U. Mass., 1982, '83. Mem. Am. Assn. Pharm. Scientists (charter, nat. program com. 1988-90), Am. Chem. Soc., Assn. Official Analytical Chemists, Syndicat Nat. Industrie Pharm. (quality task force, stability working group leader 1992—), Groupe pour l'Avancement des Scis. Analytiques. Office: Rhône-Poulenc Rorer, 20 Ave Raymond Aron, 92165 Antony Cedex, France

MAZZOLA, CLAUDE JOSEPH, physicist, small business owner; b. Newton, Mass., May 24, 1936; s. Gradinola and Anne (Cicconi) M.; m. Helen Alamanos, July 25, 1965; children: Peppina, Jean-Claude. BS in Physics, Boston Coll., 1959; postgrad. in Physics, MIT, 1961-62. Jr. engr. Lab. for Electronics, Boston, 1959-61; scientist AVCO R&D, Wilmington, Mass., 1961-62; engr. Space Scis., Waltham, Mass., 1962-63, BBN, Cambridge, Mass., 1963-72; staff engr. Edo, College Point, N.Y., 1972-82; engr. Sperry, Great Neck, N.Y., 1982-89; sole proprietor Namlak, Mamaroneck, N.Y., 1989—. Author: Active Sound Absorption, 1993; patentee Magnetic Storage Device (floppy disc), 1963; contbr. articles to profl. jours., papers to sci. confs. and meetings. Mem. IEEE. Republican. Roman Catholic. Avocations: jogging, skiing, sailing, foreign travel. Home: 106 Lawn Ter Mamaroneck NY 10543 Office: Namlak PO Box 804 Mamaroneck NY 10543-8040

MCADAMS, STEPHEN EDWARD, experimental psychologist, educator; b. Lubbock, Tex., May 5, 1953; s. Edward Edwards McAdams Jr. and Don Carolyn (Hardin) Whitworth; m. Cécile Marie-Hélène Marin, July 8, 1988; 1 child, Minnetta Thérèse-Wilma. BSc, McGill U., Montreal, Quebec, Can., 1977; PhD, Stanford U., 1984. Rsch. asst. dept. psychology McGill U., Montreal, 1976-77; rsch. asst. Psycoacoustics Lab. Northwestern U., Evanston, Ill., 1977-78; rsch. asst. med. sch. Stanford (Calif.) U., 1979-81; rsch. scientist Inst. Rsch. and Coordination Acoustique/Musique, Paris, 1981-86; dir. edn. IRCAM, Paris, 1986-89; sr. rsch. scientist Nat. Ctr. for Sci. Rsch., Paris, 1989—. Editor: Music and Psychology, 1987; co-editor: Music and the Cognitive Scis., 1989, Thinking in Sound, 1992; regional editor (France) Contemporary Music Review, 1990—; contbr. articles to Jour. Acoustical Soc. Am., Brain Lang. and Music Perception. Bd. dirs. Computer Music Assn., U.S., 1986-89. Rsch. fellow French Ministry Fgn. Affairs, Paris, 1982, Fyssen Found., Paris, 1988; recipient Nat. Ctr. for Sci. Rsch. Bronze medal, Paris 1991. Mem. Acoustical Soc. Am., Behavioral and Brain Scis. Assn., French Acoustical Soc. (hearing tech. com. 1988—, Philips prize 1992). Achievements include research in psychological processes underlying perceptual organization of the sound environment; research in perceptual and cognitive processes underlying musical experience. Office: U R Descartes Lab Psychol, Expérimentale, 28 rue Serpente, F 75006 Paris France

MCALEXANDER, THOMAS VICTOR, government executive; b. Queen City, Tex., July 24, 1939; s. Roy Victor and Bernice Corinne (Brown) McA.; m. Emma Elizabeth Coppinger, Feb. 2, 1962; 1 child, Cathi Lynne Rydell. BS, Am. U., 1971. Diplomate Am. Bd. Forensic Document Examiners (dir. 1988—). Document analyst FBI, Washington, 1960-67; document examiner CIA, Washington, 1968-71, Metro. Police Dept., Washington, 1971-73; sr. document examiner U.S. Secret Svc., Washington, 1973—; del. to china through Citizen Ambassador Program, Spokane, Wash., 1992. Contbr. articles to profl. jours. Fellow Am. Acad. Forensic Sci. (com. chmn. 1988-91, Ordway Hilton award 1992). Lutheran. Achievements include research on standardization of handwriting opinion terminology. Office: US Secret Svc 1800 G St NW # 929 Washington DC 20223

MCALLISTER, DAVID FRANKLIN, computer science educator; b. Richmond, Va., July 2, 1941; s. John Thompson and Dorothy (Waits) McA.; m. Beth Clinkscales (div. Dec. 1983); 1 child, Timothy Walt. BS, U. N.C., 1963; MS, Purdue U., 1967; PhD, U. N.C., 1972. Instr. computer sci. U. N.C., Greensboro, 1967-72; asst. prof. N.C. State U., Raleigh, 1972-76, assoc. prof., 1976-83, prof., 1983—; grad. adminstr. dept. computer sci. N.C. State U., 1984-86. Author: Discrete Mathematics in Computer Science, 1977. Lt. (j.g.) USNR, 1963-65. Mem. IEEE, AAUP, Assn. Computing Machinery, Soc. Photogrammetric Instrumentation Engring., Soc. for Info. Displays. Avocation: piano playing. Home: 813 Roanoke Dr Cary NC 27513-3913 Office: NC State U Computer Sci Dept Raleigh NC 27695-8206

MCALLISTER, RUSSELL BENTON, biologist, researcher; b. Paragould, Ark., Apr. 18, 1958; s. Ray Gene and Wanda Lee (Grasher) McA.; m. Pamela Sue Coleman, July 15, 1989. BS, Ark. State U., 1988, MS, 1991. Grad. asst. dept. biology Ark. State U., Jonesboro, 1986-91; foreman, rsch. technician Inst. Mosquito Control, Jonesboro, 1988-92; program coord., sect. mgr. Ark. Dept. Pollution Control and Ecology, Little Rock, 1992—, vice chmn. computer users group, 1992—. Contbr. articles to Texx. Jour. Herpetology, Ark. Acad. Sci. Mem. Ark. Acad. Sci. Methodist. Home: 5805 Green Valley Dr North Little Rock AR 72118 Office: Ark Dept Pollution Control and Ecology 8001 National Dr Little Rock AR 72219

MCALOON, TODD RICHARD, food microbiologist; b. Pierre, S.D., Mar. 23, 1962; s. John Richard and Grace Isadore (Honrath) McA.; m. Shelly Ann Hougland, June 24, 1983; children: Crystal Ann, Michael John, Laura Beth. BS in Microbiology, S.D. State U., 1985, MS in Microbiology, 1987. Genetic rsch. asst. USDA Grain and Insect Rsch. Lab., Brookings, S.D., 1982-84; City health inspector Brookings (S.D.) City Health Dept., 1984-85; grad. asst. S.D. State U., Brookings, 1985-87; lab. mgr. Cargill Inc./Excel Div., Plainview, Tex., 1987-90; corp. lab. coord. Cargill Inc./Excel Div., Wichita, Kans., 1990-91; corp. food safety coord. Cargill Inc., Mpls., 1991—. Mem. Inst. Food Tech., Gamma Sigma Delta, Sigma Xi. Independent. Roman Catholic. Office: Cargill Inc PO Box 9300 Minneapolis MN 55440

MCALPINE, KENNETH DONALD, systems engineer, researcher; b. Morenci, Ariz., Oct. 31, 1953; s. Donald Sinclair and Beverly (Patton) McA.; m. Susan Stephanie Piesik, Oct. 15, 1982; 1 child, Christina Rosetta. BSEE, U. Ariz., 1975, MSEE, 1977. Rsch. scientist Engring. Lab., U. Ariz., Tucson, 1975-77; engr. Goodyear Aerospace, Litchfield Park, Ariz., 1977-79, Boeing Aerospace, Wichita, Kans., 1979-81; threat specialist Bell Tech. Ops., Eglin AFB, Fla., 1981-85; staff electrooptics and radar engr. Martin Marietta Aerospace, Orlando, Fla., 1986-90, 93; sr. engr. Pine Cap Assocs., Winter Park, Fla., 1972—; staff sr. system analyst Computer Sci. Raytheon Co., Patrick AFB, Fla., 1990-93; cons. Dept. Def., Orlando, 1986-90. With USAF 1972-73. Mem. Nat. Assn. Rocketry, Soc. for Advancement of Material and Process Engring. Democrat. Presbyterian. Achievements include design of space solar photovoltaic for orbit of Venus, 3-dimensional color sensitive 3/4" photovoltaic, high voltage, stacked junction solar cell, first 4" concentration solar cell at University of Arizona; worked on calibration of UV sensor on the Voyager 2 space probe; worked on titanium alloy at Boeing; at Martin, worked on special tungsten alloy for Last Glance II mirror used by NASA for tracking reentry of shuttle. Office: Pine Cap Assocs PO Box 5279 Winter Park FL 32793

MC ANINCH, ROBERT DANFORD, philosophy and government affairs educator; b. Wheeling, W.Va., May 21, 1942; s. Robert Danford and Dorothy Elizabeth (Goudy) McA.; 1 child, Robert Michael; m. Helen M. Perry, June 5, 1993. AB, West Liberty State Coll., 1969; MA, W.Va. U., 1970; MA, Morehead State U., 1975; postgrad. U. Hawaii, U. Ky. Engring. technician Hydro-Space Rsch., Inc., Rockville, Md., 1965-66; prof. govt., philosophy Prestonsburg (Ky.) Community Coll., 1970—; v.p. Calico Corner, Inc.; dir. Chase-Options, Inc., Medisin, Inc. Bd. dirs. Big Sandy Area Community Action Program, Inc., 1973-76; chmn. Floyd County Solid Waste, Inc.; mem. War on Drug Task Force. Served with AUS, 1962-65. Recipient Great Tchr. award Prestonsburg Community Coll., 1971; named Ky. col., 1977. Mem. Am. Polit. Sci. Assn., Am. Philos. Soc., Ky. Philosophy Assn., Ky. Assn. Colls. and Jr. Colls. Achievements include designed Cosmic ray chamber, artificial human circulatory system, Wilson type cloud chamber, TOTO 1, 2. Home: Bert Combs Dr Prestonsburg KY 41653-9500

MCARDLE, JOAN TERRUSO, parochial school mathematics and science educator; b. Phila., July 1, 1947; d. Paul Robert and Alexandria (Cicchitti) Terruso; m. Thomas William McArdle, June 20, 1970; children: Gia, Paula. BS cum laude, Cabrini Coll., Radnor, Pa., 1969; postgrad., Villanova U., 1969, 87. Tchr. math. Sharon Hill (Pa.) Jr.-Sr. High Sch., 1970-71; tchr. math. St. Katharine of Siena Sch., Wayne, Pa., 1980—, supr. math., 1985—; mathletes coach St. Katharine of Siena Sch., Wayne, 1982—; contest sponsor Pa. Math. League, 1983—. Mem. Upper Merion Park and Historic Found., Wayne, 1978—; mem., orator OSIA Salvatore Terruse Lodge 205, Phila., 1990; sch. coord. Marathon Fundraising for St. Jude Children's Rsch. Hosp., 1992-93. Recipient Disting. Educator award Archdiocese of Phila., 1990. Mem. ASCD, Nat. Coun. Tchrs. Math., Nat. Cath. Educators Assn., Assn. Tchrs. Math. Phila. and Vicinity, Soroptimist Internat. of the Main Line, Cabrini Coll. Alumni Assn. (class agt. 1975—). Roman Catholic. Avocations: antiques and collectables, cooking, swimming, travel. Office: St Katharine of Siena Sch Midland & Aberdeen Aves Wayne PA 19087

MCARTHUR, ELDON DURANT, geneticist, researcher; b. Hurricane, Utah, Mar. 12, 1941; s. Eldon and Denise (Dalton) McA.; m. Virginia Johnson, Dec. 20, 1963; children: Curtis D., Monica McArthur Bennion, Denise, Ted O. AS with high honors, Dixie Coll., 1963; BS cum laude, U. Utah, 1965, MS, 1967, PhD, 1970. Postdoctoral rsch. fellow, dept. demonstrator Agrl. Rsch. Coun. Gt. Britain, Leeds, Eng., 1970-71; rsch. geneticist Intermountain Rsch. Sta. USDA Forest Svc., Ephraim, Utah, 1972-75; rsch. geneticist Shrub Scis. Lab., Intermountain Rsch. Sta. USDA Forest Svc., Provo, Utah, 1975-83, project leader, chief rsch. geneticist, 1983—; adj. prof. dept. botany and range sci. Brigham Young U., Provo, 1976—. Author over 220 rsch. papers; contbr. chpts. to books; editor symposium procs. Grantee Sigma Xi, 1970, NSF, 1981, 85, Coop. State Rsch. Svc., 1986, 91. Mem. Soc. Range Mgmt. (pres. Utah sect. 1987), Botan. Soc. Am., Soc. Study Evolution, Am. Genetic Assn., Shrub Rsch. Consortium (chmn. 1983—), Intermountain Consortium for Aridlands Rsch. (pres. 1991—). Mormon. Avocations: hiking, cycling, basketball. Home: 555 N 1200 E Orem UT 84057-4350 Office: USDA Forest Svc Shrub Scis Lab 735 N 500 E Provo UT 84606-1899

MCARTHUR, GREGORY ROBERT, biomedical engineer; b. Salt Lake City, Oct. 13, 1952; s. Robert L. and Julia G. (Morley) McA.; m. Tracy J. Chamberlain, Aug. 14, 1974; children: Melinda, Christopher, Lindsey. BS, U. Utah, 1976, MS, 1980, MPhil, 1983. Biomed. engr., designer HGM Laser Systems, Salt Lake City, 1984-87, sr. product mgr., 1987-90, clin. trials monitor, 1990-91, dir. clin. affairs, 1991-92; biomed. engr. Merit Med., Salt Lake City, 1992—; industry rep. Gynecol. Laser Soc., Balt., 1990-92. Mem. Am. Fertility Soc., Biomed. Engring. Soc., Sigma Xi. Achievements include research in advanced endoscopic gen. surgery, lasers in obstetrics and gynecology. Office: Merit Med Systems Inc 79 W 4500 S Ste 9 Salt Lake City UT 84107

MCAULEY, VAN ALFON, aerospace mathematician; b. Travelers Rest., S.C., Aug. 28, 1926; s. Stephen Floyd and Emily Floree (Cox) McA. BA, U. N.C., Chapel Hill, 1951; postgrad. U. Ala., Huntsville, 1956-57, 60-63. Mathematician Army Ballistic Missile Agy., Huntsville, Ala., 1956-59; physicist NASA, Marshall Center, Huntsville, 1960-61, research mathematician, 1962-70, mathematician, 1970-81. Contbr. articles to profl. jours. Served with U.S. Army, 1944-46. Recipient Apollo achievement award NASA, 1969, cost savs. award, 1973, Skylab achievement award, 1974, Outstanding Performance award, 1976. Mem. AAAS, Am. Math. Soc., N.Y. Acad. Scis., Phi Beta Kappa. Patentee for aerospace control invention; publ. method for solution of polynomial equations; devised methods for numerical solution of heat flow partial differential equations. Home: 3529 Rosedale Dr NW Huntsville AL 35810-2573

MCAVOY, GILBERT PAUL, retired engineer; b. Fall River, Mass., Feb. 28, 1929; s. William Lester and Rose Marie (Heon) McA.; m. Evelyn Marie Berard, Apr. 11, 1953; children: David, Margaret, Steve. Student, Mass. Maritime Acad., 1950. Lic. marine engr. Systems engr. Martin Marietta Aircraft Co., Balt., 1952-57; with Martin Marietta Aerospace, Orlando, 1957-84, adv. program mgr., 1984-86, strategic defense initiative, 1986-89; strategic defense initiative Martin Marietta Aerospace, Denver, 1986-89; test program mgr. joint venture longbow system Martin Marietta Aerospace, Orlando, 1988-92; retired, 1992—; cons. in field. Lt. USNR, 1950-63. Fellow AIAA (assoc.). Achievements include patent for cockpit control mechanism for engines having thrust reversing control means. Home: 203 Sweet Gum Way Longwood FL 32779

MCBATH, DONALD LINUS, osteopathic physician; b. Chgo., May 19, 1935; s. Earl and Phyllis (Michalski) McB.; m. Ruth Southerland, Jan. 18, 1956; children: Donald L. Jr., Donna Ruth McBath Bassett, Daniel P. BA in Polit. Sci., U. Fla., 1957, BS in Pre Med., 1962; DO, Kansas City (Mo.) Coll. Osteopathy and Surgery, 1969; MA, St. Leo Coll., 1981. Diplomate Nat. Bd. of Examiners; Cert. Gen. Practice Am. Osteo. Bd. Gen. Practice, Correctional Health Profl. Med. dir. various orgns., Dade City, Fla., 1971; chief of staff Jackson Meml. Hosp., Dade City, Fla., 1969—; past chief of staff, med. dir. East Pasco Med. Ctr., Zephyrills, Fla.; med. dir. Pasco County, Hernando County Prison Systems, Fla.; bd. dirs. East Pasco Med. Ctr., Zephyrhills, AV-MED Health Plan., Tampa, Fla.; trustee, pres., exec. com. Fla. Osteo. Med. Assn.; sports physician Pasco Comprehensive High Sch., Dade City, grand marshall, past chmn. adv. coun.; trustee East Pasco Med. Ctr. Found.; assoc. prof. clin. sci. Southeastern U. Health Scis., Miami, Fla. Trustee St. Leo (Fla.) Coll.; adv. dir. First Union Nat. Bank Fla., Dade City; com. chmn. Hall of Fame Bowl, Pasco County, Fla., 1987. Recipient Pump Handle award Pasco County Health Authority, 1988, Outstanding Contbn. award H.R.S. Pasco County Pub. Health Unit, 1988; named Gen. Practitioner of Yr., Fla. Acad. Gen. Practice, 1990-91. Mem. Am. Osteo. Assn. (nat. program convention chmn., convention com., exhibit adv. com.), Fla. Soc. of Am. Coll. of Gen. Practitioners (nat. conv. com., bd. mem. 1989-90, pres. 1991-92). Roman Catholic. Avocation: sports. Home and Office: McBath Med Ctr 519 S 17th St Dade City FL 33525-4699

MCBAY, ARTHUR JOHN, toxicologist, consultant; b. Medford, Mass., Jan. 6, 1919; s. Arthur and Virginia (Davito) McB.; m. Avis Louise Botsford, Aug. 24, 1946; children: John, Robert. B.S., Mass. Coll. Pharmacy, 1940, M.S., 1942; Ph.D., Purdue U., 1948. Diplomate Am. Bd. Forensic Toxicology; cert. toxicol. chemist Am. Bd. Clin. Chemistry; registered pharmacist, Mass. Asst. prof. chemistry Mass. Coll. Pharmacy, Boston, 1948-53, asst. in legal medicine, dept. legal medicine, 1953-63; lectr. legal medicine Harvard U.; toxicologist, criminalist, cons. Mass. State Police Chemistry Lab., 1955-63; instr. Northeastern U., 1962-63; assoc. prof. toxicology Law-Medicine Inst. Boston U., 1963-69, assoc. prof. pharmacology Med. Sch., 1963-69; supr. lab. Mass. Dept. Pub. Safety, Boston, 1963-69; assoc. prof. pathology and toxicology U. N.C., Chapel Hill, 1969-73, prof., 1973-89, prof. emeritus Sch. Pharmacy, adj. prof. pathology, 1989—; chief toxicologist Office Chief Med. Examiner, Chapel Hill, 1969-89; mem. task force on alcohol, other drugs and transp. NRC; cons. toxicology resource com. Coll. Am. Pathologists, 1975—, Bur. Med. Devices and Diagnostic Products, FDA, 1975-91, N.C. Drug Authority, 1971-75; dir. Mass. Alcohol Project, 1968-69. Mem. editorial bd. Jour. Forensic Scis., 1981—; bd. editors Yearbook of Pathology, 1981-91; contbr. numerous articles on toxicology to profl. jours. Served to capt. USAAF, 1943-45. Fellow Am. Acad. Forensic Scis.; mem. Internat. Assn. Forensic Toxicologists, Nat. Safety Coun. (exec. bd. com. on alcohol and drugs 1981-91), Am. Pharm. Assn. (sec., treas. sci. sect. 1954-57), Soc. Forensic Toxicologists (dir. 1978), Am. Chem. Soc., Sigma Xi, Rho Chi, Phi Lambda Upsilon. Democrat. Roman Catholic. Home: V-306 Carolina Meadows Chapel Hill NC 27514

MCBEE, GARY L., chemist; b. June 13, 1965. BS in Chemistry, Hope Coll., 1987. Phys. chemist/microscopist Flint Ink Corp., Ann Arbor, Mich., 1988—. Mem. Electron Microscopy Soc. Am. Office: Flint Ink Corp 4600 Arrowhead Dr PO Box 8609 Ann Arbor MI 48107-8609

MCBRAYER, H. EUGENE, retired petroleum industry executive. Former v.p. Exxon Corp., Irving, Tex. Recipient Chem. Industry medal Am. sect. Soc. Chem. Industry, 1993. Office: Exxon Corp 225 E John W Carpenter Fwy Irving TX 75062*

MCBREARTY, SALLY ANN, archaeologist; b. L.A., May 13, 1949; d. Jerome Fredrick and Margrette Mary (Conlan) McB. AB, U. Calif., 1972; MA, U. Ill., 1978, PhD, 1986. Vis. asst. prof. Coll. of William and Mary, Williamsburg, Va., 1987-88; teaching fellow Harvard U., Cambridge, Mass., 1983; vis. lectr. Yale U., New Haven, Conn., 1986-87; postdoctoral fellow Yale U., New Haven, 1985-86, rsch. assoc., 1988—; asst. prof. Brandeis U., Waltham, Mass., 1988—; dir. Archaeology of Muguruk Site, Kenya, 1978-81, Paleoanthropology of Simbi Site, Kenya, 1984-90, Archaeology of Kapthurin Formation, Kenya, 1992—; mem. Geology of Baynunah Formation, Abu Dhabi, 1989—. Contbr. articles to profl. jours. Rsch. grantee NSF, 1992, Nat. Geographic Soc., 1991, Wenner-Gren Fedn., 1990, LSB Leakey Fedn., 1988, 90. Mem. Am. Anthropol. Assn., Am. Assn. of Phys. Anthropologists, Soc. for Am. Archaeology, British Inst. in East Africa, Paleoanthropology Soc. Democrat. Achievements include research on documenting age and context of transition from homo erectus to homo sapiens in East Africa clarifying nature of early human technology and adaptation. Office: Dept Anthropology Brandeis U Waltham MA 02254

MCBRIDE, ANGELA BARRON, nursing educator; b. Balt., Jan. 16, 1941; d. John Stanley and Mary C. (Szczepanska) Barron; m. William Leon McBride, June 12, 1965; children: Catherine, Kara. BS in Nursing, Georgetown U., 1962; LHD (hon.), %, 1993; MS in Nursing, Yale U., 1964; PhD, Purdue U., 1978; D of Pub. Svc. (hon.), U. Cin., 1983; LLD (hon.), Ea. Ky. U., 1991; HHD, Georgetown U., 193. Asst. prof., rsch. asst. inst. Yale U., New Haven, 1964-73; assoc. prof., chairperson Ind. U. Sch. Nursing, Indpls., 1978-81, 80-84, prof., 1981-92, disting. prof., 1992—; assoc. dean rsch. Ind. U. Sch. Nursing, 1985-91, interim dean, 1991-92, univ. dean, 1992—. Author: The Growth and Development of Mothers, 1973, Living with Contradictions. A Married Feminist, 1976, How to Enjoy A Good Life With Your Teenager, 1987. Recipient Disting. Alumna award Yale U., Disting. Alumna award Purdue U., Univ. Medallion, U. San Francisco, 1993; Kellog nat. fellow; Am. Nurses Found. scholar. Fellow Am. Acad. Nursing (pres. elect), Am. Psychol. Assn., Nat. Acads. Practice; mem. Midwest Nursing Rsch. Soc., Soc. for Rsch. in Child Devel., Sigma Theta Tau Internat. (past pres.). Office: Ind U Sch Nursing 1111 Middle Dr Indianapolis IN 46202-5107

MCBRIDE, GEORGE GUSTAVE, electronics engineer; b. Red Bank, N.J., Mar. 27, 1968; s. George Francis McBride and Barbara Ann (Dirner) Nolan. BSEE, Monmouth Coll., West Long Branch, N.J., 1991. Field engr. Gen. Instruments, West Long Branch, 1988-91; electronics engr. U.S. Army Night Vision and Electronic Sensors Directorate, Ft. Monmouth, 1991—. Mem. IEEE (chmn.m, vice chair, sec. 1986—, Appreciation/Svc. awards 1988-93), Assn. Old Crows (assoc.). Roman Catholic. Office: AMSEL-RD-NV-SE-RFCM Fort Monmouth NJ 07703-5206

MC BRIDE, JOHN ALEXANDER, retired chemical engineer; b. Altoona, Pa., Mar. 29, 1918; s. Raymond E. and Carolyn (Tinker) McB.; m. Elizabeth Anne Vogel, Aug. 28, 1942; children: Katherine M. Harris, Susan McBride Malick, Carolyn McBride Nafziger. A.B., Miami U., Oxford, Ohio, 1940; M.Sc., Ohio State U., 1941; Ph.D., U. Ill., 1944. Registered profl. engr., Calif. Various positions in research and devel. dept. Phillips Petroleum Co., 1944-58, 59-65; dir. chem. tech. Phillips Petroleum Co. (Atomic energy div.), 1963-65; chief applications engring. Astrodyne, Inc., 1958-59; dir. div. materials licensing AEC, 1965-70; v.p. E.R. Johnson Assocs., Inc., Fairfax, Va., 1970-92; asst. gen. mgr. Nuclear Chems. & Metals Corp., 1970-71; Adviser U.S. del. 3d Internat. Conf. Peaceful Uses Atomic Energy, 1964. Mem. Am. Chem. Soc., Am. Inst. Chem. Engrs. (chmn. nuclear engring. div. 1966, Robert E. Wilson award 1991), Am. Nuclear Soc., Alpha Chi Sigma, Phi Kappa Tau. Achievements include publications and patents on petrochemical products and processes, solid propellant rockets, irradiated fuel reprocessing, radioactive waste fixation and disposal. Home: 1727 Sherman Ave Canon City CO 81212-4354 Address: PO Box 1482 Canon City CO 81215-1482

MCBRIDE, JON ANDREW, astronaut, aerospace engineer; b. Charleston, W.Va., Aug. 14, 1943; s. William Lester and Catherine (Byus) McB.; m. Sharon White; children: Richard, Melissa, Jon II, Michael. BS, Aero. Engring. Naval Postgrad. Sch., Monterey, Calif., 1971; PhD (hon.), Salem Coll., W.Va., 1984, U. Charleston, W.Va., 1987. Commd. ensign USN, 1965, advanced through grades to capt., 1985, served as naval officer/aviator, 1965-78; astronaut, asst. administrator NASA, Houston, 1978—. Avocations: athletics, carpentry, cooking, numismatics. Office: NASA JSC Houston TX 77058

MC BRIDE, RAYMOND ANDREW, pathologist, physician, educator; b. Houston, Dec. 27, 1927; s. Raymond Andrew and Rita (Mullane) McB.; m. Isabelle Shepherd Davis, May 10, 1958 (div. 1978); children—James Bradley, Elizabeth Conway, Christopher Ramsey, Andrew Gore. B.S., Tulane U., 1952, M.D., 1956. Diplomate: Am. Bd. Pathology. Surg. intern Jefferson Davis Hosp., Baylor U. Coll. Medicine, Houston, 1956-57; asst. in pathology Peter Bent Brigham Hosp., Boston, 1957-60; sr. resident pathologist Peter Bent Brigham Hosp., 1960-61; resident pathologist Free Hosp. for Women, Brookline, Mass., 1959; asst. resident pathologist Children's Hosp. Med. Center, Boston, 1960; teaching fellow pathology Harvard Med. Sch., Boston, 1958-61; research trainee Nat. Heart Inst., NIH, HEW, 1958-61; spl. postdoctoral fellow Nat. Cancer Inst., HEW, McIndoe Meml. Research unit Blond Labs., East Grimstead, Sussex, Eng., 1961-63; asst. attending pathologist Presbyn. Hosp., N.Y.C., 1963-65; asst. prof. pathology Coll. Physicians and Surgeons, Columbia U., 1963-65; research asso. Mt. Sinai Hosp., N.Y.C., 1965-68; assoc. prof. surgery and immunogenetics Mt. Sinai Sch. Medicine, N.Y.C., 1965-68; career scientist Health Research Council City N.Y., 1967-73; attending pathologist Flower and Fifth Ave. Hosps., N.Y.C., 1968-78, Met. Hosp. Center, N.Y.C., 1968-78; prof. pathology N.Y. Med. Coll., 1968-78, Baylor Coll. Medicine, Houston, 1978—; attending pathologist Harris County Hosp. Dist., Ben Taub Gen. Hosp., Houston, 1978—; chief pathology svcs. Harris County Hosp. Dist., 1988—; assoc. staff Meth. Hosp., Houston, 1978-81, active staff, 1981—; vis. grad. faculty Tex. A & M U., College Station, 1979—; clin. prof. pathology U. Tex. Grad. Sch. Biomed. Scis., Galveston, 1982—, U. Tex. Med. Br., Galveston, 1982—; prof., chmn. dept. pathology Libero Instituto Universitario Campus Bio-Medico, Rome, 1993—; adj. prof. dept. stats. Rice U., Houston; mem. sci. com. Libero Instituto Universitario Campus Bio-Medico, Rome, 1991—; exec. dean N.Y. Med. Coll., Valhalla, 1973-75; exec. dir., COO, bd. dirs. Westchester Med. Ctr. Devel. Bd., Valhalla, 1974-76. Mem. editorial bd. Jour. Immunogenetics, Exptl. and Clin. Immunogenetics, European Jour. Immunogenetics; contbr. articles to profl. jours. Bd. dirs. Westchester Artificial Kidney Found., Inc., 1974-78, Westchester Med. Center Library, 1974-78, Westchester div. Am. Cancer Soc., 1973-78, Magnificat House, 1989—, Foundation for Life, 1989—, Tuxedo Library, 1976-78; co-chmn. Westchester Burn Center Task Force, 1975-76. Grantee Health Research Council, N.Y.C., 1963-73; Grantee Am. Cancer Soc., 1971-72; Grantee NIH, USPHS, 1964—; Grantee NSF, 1965-68. Fellow Royal Soc. Medicine; mem. Transplantation Soc., Am. Soc. Exptl. Pathology, Reticuloendothelial Soc., AAAS, Am. Assn. Pathologists and Bacteriologists, Am. Assn. Immunologists, AAUP, AMA, Tex. Med. Assn., Harris County Med. Soc., Tex. Soc. Pathologists, Coll. Am. Pathologists, Am. Assn. Clin. Pathologists, Houston Acad. Medicine, Houston Soc. Clin. Pathologists, Assn. Am. Med. Colls., Fedn. Am. Scientists, Am., N.Y. cancer socs., Soc. Health and Human Values, Am. Acad. Med. Ethics, Alpha Omega Alpha (hon. med. soc.). Republican. Roman Catholic. Club: Tuxedo (Tuxedo Park, N.Y.). Achievements include research on relationship between genes of the major histocompatibility locus and the ability to regress tumors induced by different groups of avian sarcoma retroviruses; haptencarrier relationship between erythrocyte isoantigens providing a strategy for the production of antibodies to weakly immunogenic differentiation antigens; complementation of MHC and non-MHC genes in the ability to regress avian sarcoma retrovirus induced tumors; demonstration that the induction of skin graft tolerance in adult inbred mouse strains by means of parabiosis is accompanied by lymphoid cell chimerism; development of an assay for quantitation of isoimmune plaque forming cells in a non-hemolytic system. Home: 12431 Woodthorpe Ln Houston TX 77024-4109 Office: Baylor Coll Medicine Dept Pathology Tex Med Ctr One Baylor Plaza Houston TX 77030 also: via Lancellotti, 18, Piazzo Lancellotti, 18, 00186 Rome Italy

MCBRIDE, THOMAS CRAIG, physician, medical director; b. Chgo., Feb. 26, 1932; s. Lawrence H. and Marion (Hickey) McB.; m. Alice M. Aho, June 27, 1959; children: Thomas, Peter, Susan. BA, Dartmouth, 1953; MD, Univ. Vt., 1957. Diplomate Am. Bd. Family Practice. Intern, asst. resident Univ. Rochester Medical Ctr. Strong Meml. Hosp., N.Y., 1957-59; dir. heart study Honululu (Hawaii) Heart Disease Control Svc., 1959-61; staff physician Univ. Health Svcs. Univ. Mass., Amherst, 1961-69, asst. dir., 1969-71; dir. health svcs. Hampshire Coll., Amherst, 1970-78; medical dir. Univ. Health Svcs. Univ. Mass., 1971—, assoc. exec. dir., 1983—; medical dir. Kaiser Permanente Health Plan, Amherst, 1976—; prof. family and community medicine U. Mass. Med. Sch., 1972—. Contbr. articles to profl. jours. Mem. Am. Coll. Health Assn., Mass. Medical Assn., Mass. Medical Soc. (counselor), Acad. Family Physicians (sec.). Office: Univ Health Svcs Amherst MA 01003

MCBRIDE, WILLIAM GRIFFITH, research gynecologist; b. Sydney, Australia, May 25, 1927; s. John and Myrine (Griffith) McB.; m. Patricia Mary Glover, Feb. 16, 1959; children: Louise, Catherine, John R., David W. MBBS, U. Sydney, 1950; MD, 1961. Resident med. officer St. George Hosp., 1950, Launceston Hosp., 1951, Women's Hosp., Sydney, 1952-53; med. supt. Women's Hosp., 1955-57, gynecologist, 1958-82; lectr. Northwestern U., Case Western Res. U., U. Calif., Davis, Columbia U., Oxford U., Free U. Berlin, Bangkok U., Bombay U., Singapore U.; cons. St. George Hosp., 1958—, West Mead Hosp., 1983—; founder, dir. Found. 41, birth defects rsch., Sydney, 1972—; cons. ob-gyn Royal Hosp. for Women, 1983-88, WHO study com. safety oral contraceptives, 1971—; mem. expert com. EPA. examiner in Ob-Gyn, U. Sydney, U. New South Wales; mem. NSW Maternal and Perinatal Com., 1977-83; dir. Barrington Pastoral Co. Pty. Ltd. Bd. dirs. Australian Opera, 1979—, Women's Hosp. of Sydney, 1973-79; internat. judge Hereford Cattle, UK. Recipient Gold medal Brit. Petroleum/Inst. de la Vie for discovering teratological effects of thalidomide, 1971; decorated comdr. Order British Empire, Order Australia. Fellow Royal Coll. Ob-Gyn, Royal Soc. Medicine, Senate U. of Sydney, Royal Australian Coll. Obstetricians and Gynecologists, Royal Soc. Medicine; mem. AAAS, Teratology Soc., Reproductive Biology Soc., Endocrine Soc., Australian Med. Assn. (pres. Ob-Gyn sect. 1966-76), Soc. Risk Analysis, Am. Coll. Toxicology, Royal Agrl. Soc. New South Wales (council), N.Y. Acad. Sci., Royal Soc. Medicine, Behavioural Toxicology Soc., Neurobehavioral Teratology Soc. Clubs: Union, Australian Jockey, Royal Sydney, Palm Beach Surf. Edirotial bd. Teratogensis, Carcinogenesis and Mutagenesis; contbr. numerous articles to books, profl. jours. Home: 14/1 Greenknowe Ave, Potts Point 2011, Australia Office: Foundation 41, 111 Bourke St, Sydney 2010, Australia

MCBROOM, THOMAS WILLIAM, mechanical engineer; b. Atlanta, Mar. 29, 1963; s. William Ralph and Ethel Irene (Bradley) McB. B in Mech. Engring., Ga. Tech., 1985, MS in Mech. Engring., 1987; JD, Ga. State U., 1992, MBA, 1992. Bar: Ga., D.C.; registered profl. engr., Ga.; lic. comml. pilot and flight instr. Mfg. engr. AT & T Techs., Norcross, Ga., 1985-86; energy systems engr. Atlanta Gas Light Co., 1987-89, sales engr., 1989-90, dir. power systems markets, 1991—. Mem. ABA, Am. Soc. Heating Refrigerating & Air Conditioning Engrs., Ga. Bar Assn., Ga. Soc. Profl. Engrs. (treas. 1990-91, sec. 1991-92, pres. 1992-93, young engr. of yr. 1991), Phi Delta Phi (exchequer 1991). Avocations: flying, basketball, snow skiing, golf, fishing. Home: 190 Grandchester Way Fayetteville GA 30214-2637 Office: Atlanta Gas Light Co PO Box 4569 Atlanta GA 30302-4569

MC BRYDE, FELIX WEBSTER, geographer, ecologist, consultant; b. Lynchburg, Va., Apr. 23, 1908; s. John McLaren and Flora O'N. (Webster) McB. B.A., Tulane U., 1930, LL.D. (hon.), 1967; Ph.D., U. Calif., Berkeley, 1940; postgrad. (rsch. fellow) U. Colo., 1930-31, Clark U., 1931-32; m. Frances Van Winkle, July 23, 1934; children: Richard Webster, Sarah Elva, John McLaren. Geographer-photographer 4th Tulane Expdn. across Cen. Am. Maya Area, 1927-28; geology teaching asst. Tulane U., 1929-30, U. of Utah-Smithsonian Uinta Ute Expdn., No. Utah, 1931; field fellow Clark U.-Carnegie Inst., Washington, Guatemala, 1932; rsch. fellow Middle Am. Rsch. Inst., Tulane U., 1932-33; teaching asst. geography U. Calif., Berkeley, 1933-35, 37; predoctoral field fellow social sci. Social Sci. Rsch. Coun. N.Y., Guatemala and El Salvador, 1935-36; instr. geography Ohio State U., 1937-42, UCLA, 1940; field fellow in natural scis.NRC, Washington, also Berkeley, Guatemala, Mexico, 1940-41; expert cons., sr. geographer M.I., War Dept., Washington, 1942-45; lectr. geography Western Res. U., 1944; dir. Peruvian office Inst. Social Anthropology, Smithsonian Instn., Washington, Lima, 1945-47; dir., organizer and writer of curriculum Inst. Geography, U. San Marcos, Lima, 1945-47; spl. rep. Inst. Andean Rsch., Lima, 1947-48; lectr. Fgn. Svc. Inst., Dept. State, 1949-53; prof. geography U. Md., 1948-59, cons. prof., 1959-63; chief geographer office of coord. Internat. Stats. U.S. Bur. Census, Washington and Latin Am., 1948-56; geographer com. on 1950 Census of Ams., Inter-Am. Statis. Inst., cons. all Am. nations, 1948-51; chief U.S. Census Mission, tech. advisor 1st Nat. Census of Ecuador, Quito, 1949-51; dir. regional planning Gordon A. Friesen Assos., Inc., Washington and San Jose, Costa Rica nat. master hosp. plan, 1956-58; on contract to Exec. Rsch. Inc., N.Y.C., presdl. campaign advisor Pres. Villeda Morales, Honduras, 1957; U.S. rep., electoral advisor to Pres. Ydigoras Fuentes, Guatemala, 1958-64; pres. F.W. McBryde Assocs., Inc., Washington and Guatemala, 1958-64; founder-pres. Inter-Am. Inst. Modern Langs., Guatemala, 1962-66; Latin Am. cons. Inst. Modern Langs., Washington, 1962-66; chief phys. and cultural geography br., natural resources div. Inter-Am. Geodetic Survey, U.S. Army, Fort Clayton, C.Z., 1964-65; field dir. Bioenviron. Program, Atlantic-Pacific Interoceanic Sea-Level Canal Studies in Panama and Colombia (AEC contract), 1965-70, field dir. U.S. Army, Natick, Mass., Andean ecology project, S.Am., 1967-69, dir. project devel. program, Cen. Am. and Mex., 1968-69; cons. in ecology Battelle Meml. Inst., Columbus, Ohio, 1970—; founder-dir. McBryde Ctr. for Human Ecology, 1969—; cons. in human ecology and Latin Am., Transemantics, Inc., Washington, 1970—; with UN Devel. Program, ecologist (tourism), expert Jamaica, W.I., 1971; hydrology ecologist, expert Parana River Nav. Improvement Project, Argentina, 1972; ecol. cons. Battelle Meml. Inst., Panama and Brazil, 1972; U.S. Bur. Census geography adviser to Govt. of Honduras on cartography for 1973 population census, 1972; Battelle cons., procedural analysis in internat. project devel., 1972; ecologist World Bank environ. impact analysis Bayano River Hydroelectric Project, one-man mission to Panama; Battelle cons., ecologist, prin. investigator and field coord., environ. impact study Darien Gap Hwy., Panama-Colombia for U.S. Dept. Transp., 1973; cons. Enviro Plan. Chesapeake Bay ecology; expert ecologist (biology) Engr. Agy. for Resources Inventories, C.E., U.S. Army, Washington, 1974; dir. recruitment, dir. internat. bus. intelligence, 1975-80, dir. Geog. Rsch. div. Transemantics, Inc., Washington, 1975—; cons. geographer Census Office, Govt. of Honduras, 1981; cons. ecologist UN Tech. Cooperation and Devel., Cerro Colorado Copper Mine, Panama, 1981. Mem. nat. adv. bd. Am. Security Coun.; state advisor U.S. Congl. Adv. Bd., Am. Security Coun. Found.; charter founder Ronald Reagan Rep. Ctr., Washington, 1988, founder PrTrust Rep. Nat. Com., 1988, mem. Chmns. Coun., 1993; charter mem. Pres. Bush's Rep. Presdl. Task Force, 1989; charter mem. Rep. Presdl. Trust, 1992; mem. Rep. Senatorial Inner Circle, 1992. Fellow Explorers Club (life); mem. Am. Anthrop. Assn., AAAS, Am. Cartographic Assn., Am. Congress on Surveying and Mapping, Am. Geog. Soc., Assn. Am. Geographers (formerly Am. Soc. for Profl. Geographers founding pres., sec.(1 yr.), treas., meetings coord., editor publs. 1943-45, creator 8 regional divs. in U.S., now 9 quasi-socs. in U.S. and Can.), Am. Geophys. Union (life), Am. Inst. Biol. Scis., Conf. Latin Americanist Geographers, Arctic Inst. N.A., Assn. Tropical Biology, Chesapeake Bay Found., Am. Soc. Photogrammetry and Remote Sensing, Ecuadorian Inst. Anthropology and Geography (founder dir. 1950-52, hon. dir. 1952—), Inter-Am. Coun. (organizing sec. 1953-59, pres. 1959-62), N.Y. Acad. Scis., Washington Acad. (adv. bd. 1981—), Lima Geog. Soc., Oceanography Soc. (charter mem.), Soc. Am. Archeology, Soc. Am. Mil. Engrs., Soc. for Med. Anthropology, N.Am. Cartographic Info. Soc., Guatemalan Soc. Geography and History, Internat. Oceanographic Found., Nature Conservancy Internat. Program, Mexican Soc. Geography and Stats., U.S. Naval Inst., World Wildlife Fund, Nat. Wildlife Fedn. (world assoc.), Phi Beta Kappa, Sigma Nu, others. Episcopalian. Author: Solola, 1933; Cultural and Historical Geography of Southwest Guatemala, 1947, Spanish , 1969, 2 vols. transl. by Francis Gall, Guatemala, 1969, Greenwood reprint (English), 1971; (with P.

Thomas) Equal-Area Projections for World Statistical Maps, 1949; founding editor Profl. Geographer; contbr. numerous articles to profl. jours. Achievements include patent for equal-area designs and methods of constructing original projections for world maps, wherein median representations between true overall global linear scale (equidistance), true azimuth (indicated by directional bearings of intersecting graticule lines, hence shape of terrestrial features), and equivalence (or true relative size of land and water bodies on the map) are plotted to attain the closest similarity to earth features in all dimensions, which are completely true only on the spherical surface of the terrestrial globe; discovery of origin of beans phaseolus vulgaris and p. lunatus in western Guatemala and Chiapas; of maize varieties; development of new system of biological/ecological classification keyed to environmental factors, of new micro-geodemographic planning techniques employing data graphics plotted on thematic maps, especially useful in hospital, health, and economic surveys; research in Mayan and Andean archeology, ecology, on tectonic and seismic determinants and geomorphology. Home: 10100 Falls Rd Potomac MD 20854

MCCABE, DOUGLAS RAYMOND, analytical chemist; b. Ticonderoga, N.Y., Aug. 28, 1964; s. Raymond Francis and Beverly Jane (Nels) McC.; m. Karen Jeanne Chaput, Sept. 19, 1992. BS in Chemistry, SUNY, Plattsburgh, 1986; MS in Chemistry, U. N.H., 1993. Chemist Ayerst Labs., Rouses Point, N.Y., 1986-87; rsch. asst. U. N.H., Durham, 1987-89; sr. analytical chemist Ionpure Tech. Corp., Lowell, Mass., 1989—. Mem. Am. Chem. Soc. Democrat. Roman Catholic. Office: Ionpure Techs Corp 10 Technology Dr Lowell MA 01851

MCCABE, JOHN CORDELL, surgeon; b. Urbana, Ill., Jan. 29, 1941; s. Louis Cordell and Catherine (Hesselschwerdt) McC.; m. Monique Thomas, July 14, 1977 (div. 1987); children: Timothy, Valerie. AB, U. N.C., 1963; MD, George Washington U., 1967. Diplomate Am. Bd. Surgery, Am. Bd. Thoracic Surgery. Intern and resident N.Y. Hosp.-Cornell Med. Ctr.; asst. prof. surgery Cornell Medica Ctr., N.Y.C., 1976-82, assoc. prof. surgery, 1982—; attending surgeon The N.Y. Hosp., N.Y.C., 1976—; asst. chief, div. thoracic surgery The Lenox Hill Hosp., N.Y.C., 1987—. Contbg. author: Principles of Trauma Care, 1982, Patent Ductus; contbr. articles to profl. jours. With USNR, Vietnam. Decorated Legion of Merit with Combat "V". Mem. numerous profl. orgns. in field. Home: 140 Riverside Dr New York NY 10024 Office: 130 E 77th St New York NY 10021

MCCABE, JOHN LEE, engineer, educator; b. Fond du Lac, Wis., Mar. 26, 1923; s. Arthur Lee and Florence Gertrude (Molleson) McC.; m. M. Leora Harvey, Mar. 17, 1946; 1 child, Steven Lee. Student, Western Mich. U., 1941-42, U. Colo., 1946-47; C.C. Aurora, 1984-85. Designer project assignments, Denver, 1947-50; archtl. engr. The Austin Co., Denver, 1950-52; resident engr. Peter Kiewit Sons Co., Portsmouth, Ohio, 1953; dist. mgr. Hugh J. Baker Co., Evansville, Ind., 1953-56; engr. Lauren Burt Inc., Denver, 1956-58; project mgr. Denver Steel Products Co., Commerce City, Colo., 1958-66; pres. corp. McCabe and Co., Aurora, Colo., 1966-75; master tchr. high sch. Sch. Dist. 50, Westminster, Colo., 1975-83; tchr. Aurora pub. schs., 1983-92, continuing edn. dept. Lawrence Pub. Schs., 1992—. Author: Word Problems Simplified, 1986, Everyday Algebra, Everyday Geometry, 1987, Everyday Mathematics-A Study Guide , 1988, Mathematics Workbook series for Technical Schools, Drafting, Machine Shop, Auto Body, Welding, Horticulture, 1987-88, Mathematics Workbook series for Middle-Schools, Problem Solving, 1987, Whole Numbers, Fractions-Decimals-Percents, 1989, Measurements, Metrics, 1990, Basic Algebra, Basic Geometry, 1991, The Consumer's Handbook, 1991, Everyday Metrics, 1992, How to Get Started As a Contractor 2d edit., 1992, Essentials of Algebra, 1993, (novel) The Survivor's, 1991; PTO. Mem. U.S. Metric Assn., Colo. Soc. Engrs. (life). Roman Catholic. Club: Nat. Writers'. Home: 811 Moundridge Dr Lawrence KS 66049 Office: Lawrence Pub Schs Continuing Edn Dept 2017 Louisiana St Lawrence KS 66046

MCCABE, ROBERT JAMES, engineering executive; b. Long Branch, N.J., June 9, 1953; s. Wilfred James and Angeline (Delferraro) McC. BEE, U. Del., 1975; MSEE, Drexel U., 1983; postgrad., Johns Hopkins U., 1977-80. Profl. engr., N.J. Elec. engr. Westinghouse Electric Corp., Balt., 1975-80; prin. engr. Honeywell, Inc., Fort Washington, Pa., 1980-83; mgr. engring. Gen. Electric Co., Camden, N.J., 1983—; cons. McCabe, D'Ascenzo & Assocs., Cherry Hill, N.J., 1984—; co-chmn. High Speed A/D Conversion Workshop, nationwide/internat., 1978-80. Editor/contbg. author: Worst Case Analysis Handbook, 1986—; patentee in field; contbr. articles to profl. jours. Mem. Elfun, Phila., 1989—; bd. dirs. Cath. Alumni Club of Phila., 1980-83, Search for Christian Maturity, Balt., 1977-80. Recipient 2nd place Table Topics award Toastmasters Internat., Camden, 1985. Mem. Eta Kappa Nu (treas. 1974-75), Tau Beta Pi. Republican. Roman Catholic. Avocations: commercial pilot, instrumentalist, tennis, bicycling, sailing. Home: 112 Old Carriage Rd Cherry Hill NJ 08034-3330

MCCABE, STEVEN LEE, structural engineer; b. Denver, July 11, 1950; s. John L. and M. Leora (Shaw) McC.; m. Ann McCabe, Aug. 10, 1974; 1 child, Stephanie A. BSME, Colo. State U., 1972, MSME, 1974; PhDCE, U. Ill., 1987. Registered profl. engr., Colo., Kans., Okla. Engr. Pub. Svc. Co. of Colo., Denver, 1974-77; sr. engr. R.W. Beck and Assocs., Denver, 1977-78; engr., project engr. Black & Veatch Cons. Engrs., Kansas City, Mo., 1978-81; asst. prof. civil engring. U. Kans., Lawrence, 1985-91, assoc. prof., 1991—. Contbr. articles to profl. jours. Grantee Am. Inst. Steel Constrn., 1990-91, NSF, 1989-91, 91—, Civil Engring. Rsch. Found., 1991—; Ill. fellow, 1981-82; recipient Mech. Coupler Industry Testing Consortium award, 1992—. Mem. ASME (pressure vessels and piping div. honor paper award 1989, cert. of recognition for svc. 1993), ASTM, ASCE (assoc. editor Jour. Structural Engring. 1992—), ACI (pres. Kans. chpt. 1992), IEEE Computer Soc., Am. Soc. Engring. Edn., Earthquake Engring. Rsch. Inst., Sigma Xi, Sigma Tau, Pi Tau Sigma, Phi Kappa Phi, Chi Epsilon. Republican. Roman Catholic. Achievements include development of improved damage mechanics techniques for prediction of earthquake effects on structures, seismic design criteria for power plants, research on inelastic cyclic behavior of mechanical reinforcing bars and couplers, on structural dynamics and earthquake engineering as well as computational mechanics, on the evaluation of response and damage and predictions of reserve capacity in structures and members subjected to earthquake strong ground motion, on use of finite element analysis for the response of structures and machines to various types of loading. Office: U Kans 2015 Learned Hall Lawrence KS 66045-0001

MCCAHILL, THOMAS DAY, physician; b. Moncure, N.C., Jan. 3, 1918; s. Edward Joseph and Mary Fuller (Day) McC.; m. Annelyse Bochinger, July 12, 1967. BS, Coll. William and Mary, 1939; MD, Med. Coll. Va., 1951. Tchr. Pittsylvania County Pub. Schs., Whitmell, Va., 1939-40, Norfolk County Pub. Schs., Cradock, Va., 1946-71; intern Worcester (Mass.) Meml. Hosp., 1951-52; pvt. practice as physician Richmond, Va., 1952—. 1st lt. U.S. Army, 1941-46. Mem. AMA, Va. Heart Inst. (staff mem., bd. dirs. IRB), Med. Soc. Va., Richmond Acad. Medicine. Office: 6108 Lee Ave Mechanicsville VA 23111

MCCALL, DARYL LYNN, avionics engineer; b. Covington, Ky., Apr. 24, 1956; s. Harry Clarkson McCall and Gladys Edna (Wix) Spencer; m. Marilyn Kay Sukke, June 23, 1990; 1 child, Sean William. AAS, Cin. Tech. Coll., 1978; BSEE, Ohio U., 1980, MSEE, 1985. Intern Avionics Engring. Ctr. Ohio U., Athens, 1978-83, rsch. engr. Avionics Engring. Ctr., 1983-86; avionics engr. Rockwell Internat./Collins Comml. Avionics adv. tech. div., Cedar Rapids, Iowa, 1986—. Mem. sci. sta. bd. dirs., Cedar Rapids, 1993—. Mem. IEEE (chmn. and others exec. positions, Ted A. Hunter award), AIAA, Inst. of Navigation, Sigma Xi. Democrat. Home: 3185 Springville Rd Springville IA 52336 Office: Rockwell Internat/Collins Radio 400 Collins Rd MS 124-300 Cedar Rapids IA 52498

MCCALL, DAVID W., chemist, administrator, materials consultant; b. Omaha, Dec. 1, 1928; s. H. Bryron and Grace (Cox) McC.; m. Charlotte Marion Dunham, July 30, 1955; children—William Christopher, John Dunham. B.S., U. Wichita, 1950; M.S., U. Ill., 1951, Ph.D., 1953. Mem. tech. staff AT&T Bell Labs, Murray Hill, N.J., 1953-62; head dept. phys. chemistry AT&T Bell Labs, Murray Hill, NJ, 1962-69, asst. chem. dir., 1969-73, chem. dir., 1973-91; dir. environ. chemistry rsch. AT&T Bell Labs, Murray Hill, N.J., 1991-92; chmn. bd. trustees Gordon Rsch. Confs.; mem. adv. bd. Chem. Abstract Svcs.; chmn. Nat. Commn. on Super-conductivity; chmn. panels on advanced composites and electronic packaging NRC; mem. Naval Studies Bd., NRC; bd. dirs., v.p. Randolph (Vt.) Corp. Trustee Matheny Sch. Hosp. Fellow AAAS, Am. Phys. Soc., Royal Soc. Chemistry London; mem. NAE, AICE. Am. Chem Soc. (Barnes award 1992), Soc. Chem. Industry, Coun. Chem. Rsch. Home: 160 S Finley Ave Basking Ridge NJ 07920-1428

MCCALL, RICHARD POWELL, physics educator; b. Magnolia, Ark., Dec. 29, 1955; s. Powell Crimm and Eula P. (Bauman) McC.; m. Melanie Elise Youngblood, Dec. 27, 1977; children: Tyler Powell, Colin Wayne. BS, Northeast La. U., 1977; PhD, Ohio State U., 1984. Grad. teaching assoc. Ohio State U., Columbus, 1977-79, grad. rsch. assoc., 1979-82; adj. asst. in physics U. Fla., Gainesville, 1982-84; postdoctoral rsch. scientist U. Calif., Santa Barbara, 1984-85; asst. prof. U. So. Miss., Hattiesburg, 1985-87; sr. rsch. assoc. Ohio State U., Columbus, 1987-92; asst. prof. La. Tech U., Ruston, 1992—; reviewer Synthetic Metals Jour., 1992. Contbr. articles to profl. jours. Tchr. Lane Ave. Bapt. Ch., Columbus, 1987-92, deacon, 1991-92. Mem. Am. Phys. Soc., Am. Assn. Physics Tchrs., Materials Rsch. Soc., Optical Soc. of Am. Baptist. Achievements include patent in Erasable Optical Information Storage System. Office: La Tech U Dept Physics Ruston LA 71272

MCCALLUM, CHARLES JOHN, JR., communications company executive; b. Tacoma, Apr. 26, 1943; s. Charles John and Phyllis Elizabeth (Hollaway) McC.; m. Pennie J. Firestone, Sept. 16, 1967; children: Christine Lynn, Charles John III. SB in Math., MIT, 1965; MS in Stats., Stanford (Calif.) U., 1967, PhD in ops. rsch., 1970. Mem. tech. staff AT&T Bell Labs., Holmdel, N.J., 1970-75, supr., 1976-80, head ops. rsch. dept., 1981—; coadjutor in ops. rsch. Rutgers U., New Brunswick, 1975. Contbr. articles to profl. jours. Mem. IEEE, Ops. Rsch. Soc. Am. (treas. 1986-89, pres. 1991-92), Inst. of Mgmt. Scis. (hon. mention CPMS prize competition 1977), Math. Programming Soc., Mgmt. Sci. Roundtable (exec. com. 1989-90), Sigma Xi. Presbyterian. Office: AT&T Bell Labs 101 Crawfords Corner Rd Rm 3L-323 Holmdel NJ 07733-3030

MCCAMBRIDGE, JOHN JAMES, civil engineer; b. Bklyn., Oct. 27, 1933; s. John Joseph and Florence Josita (McDonnell) McC.; m. Dorothy Antoinette Cook, Mar. 17, 1962; children: Sharon J., John S., Patrick J., Kathleen C. BCE, Manhattan Coll., 1955; MS, Vanderbilt U., 1958; postgrad., UCLA, 1963-66. Civil engr. Raymond Concrete Pile Co., N.Y.C., 1955; commd. 2d lt. USAF, 1955, advanced through grades to col., 1972; exec. sec. Defense Com. On Rsch., Washington, 1971-73, DOD-NASA Supportive Rsch. Tech. Panel, Washington, 1972-74; asst. dir. Def. Rsch. and Engring. (for Life Scis.) Office Sec. Def., Washington, 1974-75; dir. Air Force Life Support Systems Program Office, Wright Patterson AFB, Ohio, 1975-79; retired USAF, 1979; prin. Booz, Allen & Hamilton, Inc., Bethesda, Md., 1979-86; v.p. Espey, Huston & Assoc., Inc., Falls Church, Va., 1986-90; mng. prin. JMC Cons. Group, McLean, Va., 1990—; chmn. air panel on NBC Def., NATO, Evere, Belgium, 1970-71; exec. sec. Def. Com. on Rsch., Washington, 1971-73; def. dept. rep. to physics survey com., Nat. Acad. Scis., Washington, 1971. Contbr. articles to profl. jours. Decorated Legion of Merit with oak leaf cluster. Fellow Aerospace Med. Assn. (exec. coun. 1972-73); mem. Inst. Hazardous Materials Mgmt. (chmn. 1988—), Acad. Cert. Hazardous Materials Mgrs. (pres. 1984-86), Survival & Flight Equipment Assn. (nat. sec. 1977-78), River Bend Golf and Country Club, Fairfax Hunt Club, Tower Club, K.C. Republican. Roman Catholic. Office: JMC Cons Group 9200 Falls Run Rd Mc Lean VA 22102-1028

MCCAMMON, DONALD LEE, civil engineer; b. Missoula, Mont., Apr. 15, 1955; s. George Eli and Lillian Agnes (Parkin) McC.; m. Donna Rae Dige, Mar. 1, 1980. Student, U. Mont., 1973-75; BS in Civil Engring., Mont. State U., 1982. Registered profl. engr., Kans., Mont., Colo., Wyo., Nebr., N.Mex., S.D., Wash. Engring. technician II Mont. Dept. Hwys., Missoula and Bozeman, 1975-82; teaching asst. Mont. State U., Bozeman, 1978-82; various engring. positions Burlington No. R.R. Co., Billings, Mont., 1982-84; asst. structural engr. Burlington No. R.R. Co., Overland Park, Kans., 1985, mgr. info. systems, 1985-88; mgr. structure maintenance Burlington No. R.R. Co., Denver, 1988-92, design engr., 1993; dir. ops. Howard, Needles, Tammen & Bergendoff, Denver, Colo., 1993—; presenter, author confs. Tng. grantee Fed. Hwy. Adminstrn., 1977. Mem. ASCE, NSPE, Am. R.R. Engring. Assn. (vice-chair 1991—, lectr. symposium 1993), Am. Rlwy. Bridge and Bldg. Assn. (bd. dirs. 1990-93, tchr., coord. inspection sch., co-developer, co-instr. bridge inspection seminars 1992-93), Clan Buchanan Soc. Am. (regent 1990-92, Red Seal award 1992, 2d v.p. 1993—), Denver Yacht Club (vice-commodore 1991-93, fleet capt. Capri 22 Fleet 12, Colo. Front Range Area, 1990—). United Methodist. Achievements include devel. of lang. for sect. of Am. R.R. Engring. Assn. manual of recommended practice; rsch. in computerized r.r. structure mgmt. systems, tunnel rehab. project, ways to upgrade and extend life of timber bridges and structures. Home: 18169 E Asbury Dr Aurora CO 80013 Office: Howard Needles Tammen & Bergendoff Ste 220 3609 South Eadsworth Blvd Denver CO 80235-2103

MCCAMY, CALVIN SAMUEL, optics scientist; b. St. Joseph, Mo., Sept. 22, 1924; s. Benjamin Samuel and Della Emma (Cervenka) McC.; m. Mabel Alice Bellerud, Nov. 4, 1945; children: Susan, Nicholas, Carter. BSChemE, U. Minn., 1945, M in Physics, 1950. Instr. math. U. Minn., Mpls., 1947-50; instr. physics Clemson (S.C.) U., 1950-52; chief image optics and photography Nat. Bur. Standards, Gaithersburg, Md., 1952-70; v.p. for rsch. Macbeth, Newburgh, N.Y., 1970-89; pvt. practice cons. in color sci. Wappingers Falls, N.Y., 1990—; adj. prof. chemistry Rensselaer Poly. Inst., Troy, N.Y., 1980-85, adv. bd. Munsell Color Sci. Lab. Rochester (N.Y.) Inst. Tech., 1985—; pres. Kollmorgen Found., Hartford, Conn., 1980-92; trustee Munsell Found., Balt., 1979-83; photographic analyst Ho. of Reps. investigation of shooting of Pres. John F. Kennedy, Washington, 1978. Editor: Papers on Image Optics from National Bureau of Standards, 1973; author: Information in a Microphotograph, 1965; contbr. articles to profl. jours. and chpt. to History of Sensitometry and Densitometry, 1986. Lt. (j.g.) USN, 1943-47. Fellow Optical Soc. Am. (chmn. color com. 1978), Soc. Photographic Scientists and Engrs. (v.p. sci. 1968-72, vis. lectr. 1986), Royal Photographic Soc. Great Britain, Soc. Motion Picture and TV Engrs., Washington Acad. Scis. Unitarian. Achievements include development of new principle of absolute radiometry, the compensated variable aperture; discovery of cause of redox blemishes threatening federal microfilm records. Home: 54 All Angels Hill Rd Wappingers Falls NY 12590-1804

MC CANDLESS, BRUCE, II, engineer, former astronaut; b. Boston, Mass., June 8, 1937; s. Bruce and Sue (Bradley) McC.; m. Alfreda Bernice Doyle, Aug. 6, 1960; children: Bruce III, Tracy. BS, U.S. Naval Acad., 1958; MS in Elec. Engring., Stanford U., 1965; MBA, U. Houston, Clear Lake, 1987. Commd. ensign USN, 1958, advanced through grades to capt., 1979, naval aviator, 1960, with Fighter Squadron 102, 1960-64; astronaut Johnson Space Ctr., NASA, Houston, 1966-90; mem. Skylab 1 backup crew Johnson Space Center, NASA, Houston, mem. STS-11 shuttle crew, mem. STS-31 Hubble Space Telescope deployment crew; ret. USN, 1990; prin. staff engr. Martin Marietta Corp., Denver, 1990—. Decorated Legion of Merit; recipient Def. Superior Service medal, NASA Exceptional Service medal, NASA Spaceflight medal, NASA Exceptional Engring. Achievement medal, Collier Trophy, 1985, Haley Space Flight award AIAA, 1991. Fellow Am. Astron. Soc.; mem. IEEE, U.S. Naval Inst., Nat. Audubon Soc., Houston Audubon Soc. (past pres.). Episcopalian. Achievements include executing 1st untethered free flight in space using Manned Maneuvering Unit. Office: Martin Marietta Astronautics Co PO Box 179 Group Mail Stop S5060 Denver CO 80201

MCCANDLESS, DAVID WAYNE, neuroscientist, anatomy educator; b. Dayton, Ohio, Dec. 16, 1941; s. Wayne Hunter and Janet (Bales) McC.; m. Sue Post, Dec. 21, 1963; children: Jeffrey Wayne, Steven Francis. BS, U. Cin., 1967; MS, George Washington U., 1969, MPhil, 1972, PhD, 1973. Asst. prof. U. Vt., Burlington, 1973-76; staff fellow NIH, Bethesda, Md., 1976-78; assoc. prof. U. Tex., Houston, 1978-86; John J. Sheinin prof. of anatomy, interim chmn. Chgo. Med. Sch., North Chicago, Ill., 1986—. Editor-in-chief Metabolic Brain Disease, 1985—; editor: Cerebral Energy Metabolism/Metabolic Encephalopathies, 1985, Developmental Neurochemistry, 1985; contbr. over 70 articles to profl. jours. Sft. USAFR, 1961-67. Grantee NIH/NIAAA, 1975, NIH/NINCDS, 1981, 83, Epilepsy Found., 1980. Mem. Am. Soc. for Neurochemistry, Am. Inst. Nutrition, Fedn. Am. Socs. for Exptl. Biology, Teratology Soc. Presbyterian. Achievements include research in delineating basic mechanisms of experimentally induced metabolic encephalopathies. Office: Chgo Med Sch 3333 Green Bay Rd North Chicago IL 60064-3095

MCCANN, MICHAEL JOHN, industrial engineer, educator, consultant; b. Iowa City, June 21, 1946; s. John Philip and Clara Louise (Lounsberry) McC. BArch, Iowa State U., Ames, 1969, MS, 1972, PhD, 1986. Designer Brooks, Borg and Skiles, Des Moines, Iowa, 1969-72; indsl. engr. Eastman Kodak, Rochester, N.Y., 1972-79; engr. Peter Kiewit Sons, Omaha, 1979-82; prof. Drake U., Des Moines, 1989-90, Iowa State U., Ames, 1990-93; treas. McCann Farms Inc., Marshalltown, Iowa, 1970—. Capt. USAR, 1969-76. Mem. NSPE (hon.), Am. Inst. Contractors (assoc.). Home and Office: McCann Farms Inc 2711 160th St Marshalltown IA 50158

MCCANTS, MALCOLM THOMAS, chemical engineer; b. Houston, Jan. 17, 1917; s. John Thomas and Julia Adele (Yeatman) McC.; m. Flora Olivia Jackson, Feb. 25, 1941; children: Michael, Julianne. BA, Rice Inst., 1937, BSChemE, 1939; MSChemE, MIT, 1940. Registered profl. engr., Tex., Calif. Chem. engr. Humble Oil & Refining Co., Baytown, Tex., 1940-43; asst. chemist Great. So. Corp., Corpus Christi, Tex., 1943-47, gen. supt., 1950-53; process engr. Fluor Corp., L.A., 1947-50; plant mgr. Great. No. Oil Co., St. Paul, 1954-56; cons. engr. Petroleum Technologists, Houston, 1957—. Mem. Sigma Tau. Achievements include patents (3), pending (4); responsible for building the first BTX plant, refinery in Minn., Crude Unit in Bolivia, Crude Unit in Chile. Home: 5481 Cedar Creek Houston TX 77056 Office: Petroleum Technologists Inc Ste 260 2400 Augusta Houston TX 77057

MCCARD, HAROLD KENNETH, aerospace company executive; b. Corinth, Maine, Dec. 18, 1931; s. Fred Leslie and Ada (Drake) McC.; m. Charlotte Marie Despres, June 29, 1957; children: Robert Fred, Renee Glen. BEE, U. Maine, Orono, 1959; MEE, Northeastern U., 1963; MS in Mgmt., MIT, 1977. Engr. Avco Systems div. Textron, Wilmington, Mass., 1959-60, group leader, 1960-62, section chief, 1962-65, dept. mgr., 1965-72, dir./staff dir., 1972-77, chief engr., 1977-79, v.p. ops., 1979-82, v.p., gen. mgr., 1982-85; pres. Textron Def. Systems (formerly Avco Systems div./Avco Systems Textron), Wilmington, Mass., 1986—. Mem. Nat. Indsl. Security Assn. (bd. dirs. 1986—), AIAA, Armament Research and Devel. Council, Am. Def. Preparedness Assn. (exec. com.), North Shore C. of C. (bd. dirs. 1986—). Avocations: golf, tennis, skiing. Home: 6 Lantern Ln Lynnfield MA 01940-1347 Office: Textron Def Systems 201 Lowell St Wilmington MA 01887-2969

MCCARL, HENRY N., economics and geology educator; b. Balt., Jan. 24, 1941; s. Fred Henderson and Mary Bertha (Yaeger) McC.; m. Louise Becker Rys, June 8, 1963 (div. 1986); children: Katherine Lynne, Patricia Louise, Fredrick James; m. Mary Frederica Rhinelander, Jan. 31, 1987; 1 stepchild, Francesca C. Morgan. BS in Earth Sci., MIT, 1962; MS in Geology, Pa. State, 1964, PhD in Mineral Econ., 1969. Cert. profl. geologist. Market rsch. analyst Vulcan Materials Co., 1966-69; asst. prof. econs., asst. prof. geology U. Ala., Birmingham, 1969-72, assoc. prof. econs., 1973-77, assoc. prof. econs. and geology, 1978-91; prof. econs. and geology U. Ala., 1991—; dir. Ctr. for Econ. Edn., Sch. Bus. U. Ala., Birmingham, 1987—; chief econs. div. Ala. Energy Mgmt. Bd., Montgomery, 1973-74; sr. lectr. in energy econs. Fulbright-Hays Program, Bucharest, Romania, 1977-78; mng. dir. McCarl & Assocs., Birmingham, 1969—; vis. fellow Grad. Sch. Arts and Scis., Harvard U., Cambridge, Mass., 1987. Co-author: (book) Energy Conservation Economics, 1986; Introduction to Energy Conservation, 1987; contbr. articles to profl. jours. Mem. Zoning Bd. of Adjustments, Birmingham, 1974-79, Birmingham Planning Commn., 1974-86, chmn., 1980-86; dist. commr. Boy Scouts Am., Birmingham, 1988—. Mem. Soc. Mining Engrs. of AIME (bd. dirs. 1978-80), Am. Inst. Profl. Geologists (sect. pres. 1981-83), Mineral Econs. and Mgmt. Soc. (pres. 1992-93), Ala. Geol. Soc., Nat. Assn. Econ. Educators, St. Andrews Soc., SAR. Democrat. Episcopal. Avocations: hunting, woodworking, collections, art collections. Home: 1828 Mission Rd Birmingham AL 35216-2229 Office: U Ala Sch Bus Dept Econs Birmingham AL 35294-4460

MCCARROLL, KATHLEEN ANN, radiologist, educator; b. Lincoln, Nebr., July 7, 1948; d. James Richard and Ruth B. (Wagenknecht) McC.; m. Steven Mark Beerbohm, July 10, 1977 (div. 1991); 1 child, Palmer Brooke. BS, Wayne State U., 1974; MD, Mich. State U., 1978. Diplomate Am. Bd. Radiology. Intern/resident in diagnostic radiology William Beaumont Hosp., Royal Oak, Mich., 1982, fellow in computed tomography and ultrasound, 1983; radiologist, dir. radiologic edn. Detroit Receiving Hosp., 1984—, vice-chief dept. radiology, 1988—; pres.-elect med. staff Detroit Receiving Hosp., 1992-94, pres. 1994-96; mem. admissions com. Wayne State U. Coll. Medicine, Detroit, 1991—; officer bd. dirs. Dr. L. Reynolds, Assoc., P.C., Detroit; presenter at profl. confs. Editor: Critical Care Clinics, 1992; mem. editorial bd. Emergency Radiology; contbr. articles to profl. publs. Mem. AMA, Radiol. Soc. N.Am., Assn. Univ. Radiologists, Am. Roentgen Ray Soc., Am. Soc. Emergency Radiologists, Mich. State Med. Soc., Wayne/Oakland County Med. Soc., Phi Beta Kappa. Avocations: swimming, skiing, reading. Office: Detroit Receiving Hosp 3L-8 4201 Saint Antoine St Detroit MI 48201

MCCARRON, ROBERT FREDERICK, II, orthopedic surgeon; b. Hot Springs, Ark., Oct. 31, 1952; s. Robert Frederick and Irene (Shanks) McC.; m. Vicki Lynn Nichols, June 10, 1977; children: Elizabeth, Jennifer. BS, La. Tech. U., 1974; MD, U. Ark., 1977. Diplomate Am. Bd. Orthopedic Surgery. Intern U. Ark., Little Rock, 1977-78; resident Tex. Tech U., Lubbock, 1978-82, instr. dept. orthopedics, 1983-84, asst. prof., 1984-88; trauma fellow Kantonspittal, Basel, Switzerland, 1982; spine fellow St. Vincent's Hosp., Melbourne, Australia, 1983; pvt. practice orthopedic surgery Conway (Ark.) Orthopaedic Clinic, P.A., 1988—; cons. physician U. Cen. Ark., Conway, 1989; chief of surgery Conway Regional Hosp., 1991—; presenter, exhibitor in field. Contbr. articles to profl. publs. Mem. Arkansas for Gifted and Talented Edn., Conway, 1990; bd. dirs. Clifton Day Care Ctr.; mem. task force Suspected Child Abuse and Neglect. Fellow Am. Acad. Orthopedic Surgeons; mem. Ark. Med. Soc., Faulkner County Med. Soc., Ark. Orthopedic Soc., Conway Area C. of C., Sigma Nu. Republican. Mem. Christian Ch. (Disciples of Christ). Avocations: reading, gardening, basketball, baseball cards, trumpet. Office: Conway Orthopaedic Clinic 525 Western Ste 202 Conway AR 72032

MCCARTHY, DAVID EDWARD, energy conservation engineer; b. Houston, Mar. 20, 1960; s. Thomas William and Marian (Castner) McC. BS in Aero. and Astro. Engring., U. Ill., 1987; MS in Mech. Engring., Ariz. State U., 1990. Cert. energy mgr. Faculty assoc. Ariz. State U., Tempe, 1991—; mem. Aso Demand Side Mgmt. Com., Tempe, 1992—. Mem. ASME (assoc. mem., energy mgr.). Achievements include research in effects of various load profiles on solar storage tank stratification parameters. Office: Ariz State U Energy Energy Analysis and Diagnostic Ctr Tempe AZ 85287-6806

MCCARTHY, JAMES JOSEPH, oceanography educator, museum director; b. Ashland, Oreg., Jan. 25, 1944; m. 1969, 2 children. B.S., Gonzaga U., 1966; Ph.D., Scripps Inst. Oceanography, U. Calif.-San Diego, 1971. Research assoc. biol. oceanography Chesapeake Bay Inst., Johns Hopkins U., 1971-72, assoc. research scientist, 1972-74; asst. prof. Harvard U., 1974-77, assoc. prof., 1977-80, prof. biol. oceanography, 1980—, assoc. dean faculty Arts and Scis., 1986-90; dir. Agassiz Mus. Comparative Zoology, 1982—. Editor: Global Biogeochemical Cycles, 1986-90. Chmn. Internat. Coun. of Scientific Unions, Scientific Com. Internat. Geosphere-Biosphere Program, 1987—. Fellow AAAS, Am. Acad. Arts and Scis.; mem. Phycol. Soc. Am., Am. Soc. Limnology and Oceanography, Am. Geophys. Union. Office: Mus Comparative Zoology 26 Oxford St Cambridge MA 02138-2902

MCCARTHY, JAMES RAY, organic chemist; b. N.Y.C., Dec. 16, 1943; s. James R. Sr. and Florence (Moses) McC.; m. Sharon Foulds, June 25, 1977. BS, Ariz. State U., 1965; PhD, U. Utah, 1969. Rsch. chemist Dow Chem., Midland, Mich., 1968-72, rsch. specialist Pharm. div., 1973-76; rsch. specialist Pharm. div. Dow Chem., Indpls., 1977-79; adj. assoc. prof. dept.

chemistry Ind.-Purdue U., Indpls., 1979-86; group leader Merrell Dow, Indpls., 1986, Cin., 1986-90; dir. chemistry Marion Merrell Dow, Cin., 1990—. Contbr. articles to profl. publs. Mem. AAAS, Am. Chem. Soc., N.Y. Acad. Scis. Achievements include discovery of new organic reaction, fluoro-pummerer reaction, 28 patents. Office: Marion Merrell Dow 2110 E Galbraith Rd Cincinnati OH 45215

MCCARTHY, JEREMIAH JUSTIN, environmental engineer; b. Pitts., July 2, 1943; s. Justin Jeremiah and Mary Ann (Sinatra) McC.; m. Dorris Marie Wilcox, Dec. 29, 1965; children: Shelley M., Justin J. BS in Civil Engring., So. Meth. U., 1966, MS in Engring. Adminstrn., 1967, PhD in Civil Engring., 1974. Diplomate Am. Acad. Environ. Engrs.; Inst. Hazardous Materials Mgmt.; cert. hazardous material mgr.; registered profl. engr. Planning engr. Ling Tempco Vought Aerospace, Arlington, Tex., 1967-68; officer, rsch. engr. USPHS, Cin., 1980-81; mgr., chem. and process engr. EG&G Idaho, Inc., Idaho Falls, 1990—; vis. indsl. prof. Hood Coll., Frederick, Md., 1975-80, 88-90. Contbr. articles to profl. jours. Asst. scoutmaster Boy Scouts Am., Woodbridge, Va., 1984-88. Lt. col. U.S. Army, 1970-80, 82-90. Recipient Water Quality Div. award Am. Water Works Assn., 1973, Medal of Excellence Internat. Ozone Inst., 1976; decorated Def. Superior Svc. medal, 1990. Mem. ASCE, Sigma Xi. Roman Catholic. Achievements include research in ozonation, water reuse, waste minimization, environ. investigation, unit process evaluation. Home: 915 S Karey Ln Idaho Falls ID 83402 Office: EG&G Idaho Inc PO Box 1625 1955 Fremont St Idaho Falls ID 83415-3625

MCCARTHY, JOHN, computer scientist, educator; b. Boston, Sept. 4, 1927; s. Patrick Joseph and Ida McC.; children: Susan Joanne, Sarah Kathleen, Timothy Talcott. B.S., Calif. Inst. Tech., 1948; Ph.D., Princeton U., 1951. Instr. Princeton U., 1951-53; acting asst. prof. math. Stanford U., 1953-55; asst. prof. Dartmouth Coll., 1955-58; asst. and asso. prof. communications scis. M.I.T., Cambridge, 1958-62; prof. computer sci. Stanford U., 1962—, Charles M. Pigott prof. Sch. Engring., 1987—; bd. dirs. Info. Internat. Inc. Served with AUS, 1945-46. Recipient Kyoto prize, 1988, Nat. medal of Sci. NSF, 1990. Mem. NAS, NAE, Assn. for Computing Machinery (A.M. Turing award 1971), Am. Math. Soc., Am. Assn. Artificial Intelligence (pres. 1983-84). Home: 885 Allardice Way Stanford CA 94305-1050 Office: Stanford U Dept Computer Sci Stanford CA 94305

MCCARTHY, JOSEPH GERALD, plastic surgeon, educator; b. Lowell, Mass., Nov. 28, 1938; s. Joseph H. and Eva (Murphy) McC.; m. Karlan von L. Sloan, June 6, 1964; children: Cara, Stephen. A.B., Harvard U., 1960; M.D., Columbia U., 1964. Diplomate: Am. Bd. Surgery, Am. Bd. Plastic Surgery. Surg. intern and resident Columbia-Presbyn. Med. Ctr.; resident in plastic surgery NYU Med. Ctr., N.Y.C., 1964-73; Lawrence D. Bell prof. plastic surgery NYU Sch. Medicine, N.Y.C., 1981—; dir. NYU Med. Ctr. Inst. Reconstructive Plastic Surgery; attending physician Univ. Hosp.; vis. plastic surgeon Bellevue Hosp.; attending surgeon Manhattan Eye, Ear and Throat Hosp., N.Y.C. VA Hosp. Editor: Symposium on Diagnosis and Treatment of Craniofacial Anomalies, 1979, Plastic Surgery, 1990; assoc. editor Reconstructive Plastic Surgery, 1977, Jour. Plastic and Reconstructive Surgery, Jour. Craniofacial, Genetics and Developmental Biology. Served to lt. comdr. USPHS, 1965-67. Recipient Columbia U. Joseph Garrison Parker prize, 1964, 1st prize Plastic Surgery Edn. Found., 1980, 1st prize Am. Soc. Maxillofoc Surg., 1991, 93; Am. Cancer Soc. fellow Presbyn. Hosp., N.Y.C., 1969-70; prin. investigator NIH, 1974. Fellow ACS; mem. Am. Soc. Plastic and Reconstructive Surgeons, Assn. Acad. Chmns. Plastic Surgery (pres. 1988-89), N.Y. Regional Soc. Plastic and Reconstructive Surgeons (pres. 1984-85), Am. Assn. Plastic Surgeons (historian 1990—), Internat. Soc. Craniomaillofacial Surgeons (pres. 1990—). Office: NYU Med Ctr Univ Hosp 550 1st Ave New York NY 10016-6402

MCCARTHY, JUSTIN HUNTLY, naval architect; b. N.Y.C., Dec. 27, 1933; s. Justin H.T. and Edith M. (Harrell) McC.; m. Sarah Frances Luchars, Nov. 15, 1958; children: Katherine Sarah, Brian Justin, Elizabeth Hayes. BS, Webb Inst., 1955; postgrad., Columbia U., 1955-56, Kings Coll., Durham U., U.K., 1957-58, Johns Hopkins U., 1965-68. Naval architect Elec. Boat divsn. Gen. Dynamics, Groton, Conn., 1956-60; naval architect David Taylor Model Basin, Bethesda, Md., 1960-70, head hydrodynamics br., 1970-80, head naval hydromechanics div., 1980—; liaison scientist Office of Naval Rsch., Tokyo, 1986-87; mem. and com. chmn. Internat. Towing Tank Conf., 1975-87. Editor Jour. Ship Rsch., 1989—; contbr. articles to profl. jours.; editor conf. proceedings Numerical Ship Hydrodynamics, 1985. Pres. Woodburn Forest Citizens Assn., Bethesda, 1967-69; v.p. Potomac Valley League, Bethesda, 1969-71. Webb Inst. scholar, 1951-55, recipient Lewis Nixon prize, 1955; recipient Meritorious Civilian Svc. award USN, 1974, Superior Civilian Svc. award, 1982. Fellow Soc. Naval Architects and Marine Engrs.; mem. AAAS, Am. Soc. Naval Engrs., Soc. Naval Architects of Korea, Soc. Naval Architects of Japan, Sigma Xi. Achievements include development of widely-used computational methods for propeller unsteady forces and stresses; polymer drag reduction; identification of submarine drag-reduction features and Reynolds number scaling procedures; research in fundamental hydromechanics relating to viscous and free-surface flows, including cavitation. Home: 6408 Winston Dr Bethesda MD 20817-5822 Office: David Taylor Model Basin Naval Hydromechanics Div Bethesda MD 20084

MCCARTHY, LAURENCE JAMES, physician, pathologist; b. Boston, Aug. 11, 1934; s. Theodore Clifford and Mary Barrett (Moran) McC.; m. Cynthia Marion DeRoch, Aug. 28, 1978; children: Laurence J. Jr., Jeffrey A., Karen E., Patrick K., Ryan N. BA, Yale U., 1956; student, Georgetown U. Sch. Med., 1956-58; MD, Harvard U., 1960; MS, U. Minn., 1965. Cert. Am. Bd. Pathology, 1965. Intern Boston City Hosp., 1960-61; resident in pathology Mayo Clinic, Rochester, Minn., 1961-65; pathologist Honolulu Heart Program, 1965-67; chief pathology Kelsey-Seybold Clinic, Houston, 1967-68; clin. asst. pathologist M.D. Anderson Hosp., Houston, 1967-68; chief pathology Straub Clinic, Honolulu, 1968-72; assoc. pathologist Wilcox Hosp., Lihue, Hawaii, 1972-74; chief pathology A.R. Gould Hosp., Presque Isle, Maine, 1975-78; assoc. pathologist Kuakini Med. Ctr., Honolulu, 1978—. Med. dir. USPHS, 1965-67. Fellow Coll. Am. Pathologists, Am. Soc. Clin. Pathologists; mem. AMA, Hawaii Soc. Pathologists (pres. 1970), Am. Acad. Forensic Sci., Hawaii Med. Assn., Honolulu County Med. Soc. (del. 1982-83). Roman Catholic. Home: 249 Kaelepulu Dr Kailua HI 96734-3311 Office: Kuakini Med Ctr 347 N Kuakini St Honolulu HI 96817-2372

MCCARTHY, MARY ANN BARTLEY, electrical engineer; b. Drummond, Okla., Nov. 27, 1923; d. William Clifford and Estella Florence (Williams) Bartley; m. Joseph Manderfield McCarthy, Aug. 23, 1946 (dec. 1983); 1 child, Mary Ann McCarthy Morales. BEE, B of Material Sci., U. Calif., Berkeley, 1976. Aircraft radio technician U.S. Civil Svc., San Antonio and Honolulu, 1942-46; salesperson Sears Roebuck & Co., Enid, Okla., 1954-56; specialist reliability engring. Lockheed Corp., Sunnyvale, Calif., 1977-82, program responsible parts engr., 1986—; rsch. engr. Lockheed Corp., Austin, Tex., 1982-86; presenter 9th Internat. Conf. Women Engrs. and Scientists U. Warwick, Eng., 1991. Contbr. articles to profl. jours. Fellow Soc. Women Engrs. (sr. life mem., pres. S.W. Tex. chpt. 1984, counsel reps. sec. 1985, pres. Santa Clara Valley chpt. 1986-87, nat. v.p. 1987-88, 88-89, chmn. nat. career guidance 1988-89, coord. 1990-91, counsel sect. rep 1991-92, coord. Resnik Challenger medal 1990-91, 91-92, 92-93); mem. AAUW (life, com. chmn. 1984, co-chmn. literacy com. 1984), Toastmasters (Vanderhoof award 1992, vol. coord. 4-H series, Excel program 1992-94), Advanced Toastmaster Silver. Republican. Roman Catholic. Home: 6103 Edenhall Dr San Jose CA 95129-3006

MCCARTHY, WILBERT ALAN, mechanical engineer; b. Wadesboro, N.C., Mar. 22, 1945; s. Wilbert Albert and Myrtle (Greene) McC.; m. Judy Karen Prince, Aug. 19, 1972. Student, Va. Poly. Inst., 1963-66, U. Va. Extension, Falls Church, 1967-70, SUNY, Albany, 1990-91; BS in Mech. Engring., Pacific Western U., 1991, MS in Engring. and Constrn. Mgmt., 1993. Registered profl. engr., Wis., Vt., D.C., N.H., Maine, W.Va. Mech. engring. designer Edward L. Middleton and Assocs., Marlow Heights, Md., 1967-72; mech. engr., project mgr. H.D. Nottingham & Assocs., McLean, Va., 1972-79, sr. mech. engr., project mgr., profl. engr., 1982-83; mech. engr., project mgr., profl. engr. E/A Design Group, Washington, 1979-82; firm assoc., mech. dept. head, sr. mech. engr., profl. engr. Engring. Applications Cons., Burke, Va., 1983-87; firm assoc., dep. ops. mgr., mktg. coord., dir. engring. SAIC Architects, Inc., McLean, 1987-90; pvt. practice Dumfries, Va., 1990-91; project dir., project mgr., mktg. coord., sr. mech. engr. Advanced Cons. Engring., Arlington, Va., 1991-92; program dir., project mgr., mktg. field rep., sr. mech. engr. Engring. Design Group, Washington, 1992—; mem. panel arbitrators Am. Arbitration Assn. for disputes in govt., commercial and private constrn. industry. lectr., speaker Inst. for Internat. Rsch., N.Y.C., 1988—, Nat. Energy Mgmt. Inst., N.Y.C., 1988—. Sgt. USAFR, 1966-72. Recipient Award of Excellence for Design of Hynson Bldg. Rehab. Va. Downtown Devel. Assn., 1988, Commendation for Engring. and Mgmt. Proficiency U.S. Dept. Navy, 1987, 88. Mem. Assn. Energy Engrs. (sr.), NSPE, ASHRAE, Am. Soc. Plumbing Engrs., Nat. Coun. Examiners for Engring. and Surveying (cert. profl. engr.). Methodist. Achievements include development of HVAC design projects for American embassies and consulates throughout the world, complete redesign office space expansion Joint Data Systems Support Ctr. Pentagon, mechanical systems design National Fire Academy. Home: 15437 Silvan Glen Dr Dumfries VA 22026 Office: Advanced Cons Engring Inc 1001 N Highland St Arlington VA 22201

MCCARTNEY, BRUCE LLOYD, hydraulic engineer, consultant; b. Portland, Oreg., Dec. 4, 1939; s. Lloyd Thomas and Gwen Elizabeth (Muschamp) McC.; m. Janice May Herring, June 29, 1963; children: Heather, Shannon. BS in Civil Engring., Oreg. State U., 1962. Profl. Engr. Civil and Hydraulic. Lt. jr. grade U.S. Coast & Geodetic Survey, Seattle, 1962-64; hydraulic engr. U.S. Army Corps of Engrs., Seattle, 1964-73; rsch. engr. Coastal Engring. Rsch. Ctr., Ft. Belvoir, Va., 1973-75; hydraulic engr. U.S. Army Corps. of Engrs., Washington, 1975-87, Portland, Oreg., 1987—; cons. engr. U.S. State Dept., Brazil, Korea, 1979-80, 1980-89. Contbr. articles to profl. jours. Mem. ASCE. Home: 12581 SE 127th Ct Clackamas OR 97015 Office: Army Corps Engineers N Pacific Division 220 NW 8th Ave Portland OR 97208-2870

MCCARTNEY, LOUIS NEIL, research scientist; b. Barrow-in-Furness, Cumbria, Eng., May 11, 1943; s. Louis McEwan and Ruth Marjorie (Cooper) McC.; m. Irene Bell, Sept. 14, 1968; children: Gavin Andrew, Janette Louise. BSc in Math. with honors, U. Manchester (Eng.), 1965, MSc in Math., 1966, PhD in Math., 1968. Chartered mathematician. Jr. rsch. fellow Nat. Phys. Lab., Teddington, Middlesex, Eng., sr. scientific officer, prin. scientific officer, individual merit (G6) officer, 1990—; rsch. in theories of fracture in materials and damage processes in composites. Fellow Inst. Math. and its Applications: mem. Inst. Materials (composites div.). Avocations: fell walking, tennis, piano playing. Office: National Physical Lab, Queens Road, Middlesex, Teddington TW11 0LW, England

MCCARTY, MACLYN, medical scientist; b. South Bend, Ind., June 9, 1911; s. Earl Hauser and Hazel Dell (Beagle) McC.; m. Anita Alleyne Davies, June 20, 1934 (div. 1966); children: Maclyn, Richard E., Dale, Colin; m. Marjorie Steiner, Sept. 3, 1966. AB, Stanford U., 1933; MD, Johns Hopkins U., 1937; ScD (hon.), Columbia U., 1976, U. Fla., 1977, Rockefeller U., 1982, Med. Coll. Ohio, 1985, Emory U., 1987, Wittenberg U., 1989; MD (hon.), U. Cologne, Fed. Republic Germany, 1988. House officer, asst. resident physician Johns Hopkins Hosp., 1937-40; assoc. Rockefeller Inst., 1946-48, assoc. mem., 1948-50, mem., 1950—, prof., 1957—, v.p., 1965-78, physician in chief to hosp., 1961-74; research in streptococcal disease and rheumatic fever; Cons. USPHS, NIH. Author: The Transforming Principle: Discovering That Genes are Made of DNA, 1985. Mem. distbn. com. N.Y. Community Trust, 1966-74; chmn. Health Research Council City N.Y., 1972-75; Mem. bd. trustees Helen Hay Whitney Found; chmn. bd. dirs. Pub. Health Research Inst. of N.Y., 1985-92. Served with Naval Med. Research Unit, Rockefeller Hosp. USNR, 1942-46. Fellow medicine N.Y. U. Coll. Medicine, 1940-41; NRC fellow med. scis. Rockefeller Inst., 1941-42; Recipient Eli Lilly award in bacteriology and immunology, 1946, 1st Waterford Biomed. Rsch. award, 1977, Wolf Found. prize in medicine, Israel, 1990. Mem. Am. Soc. for Clin. Investigation, Am. Assn. Immunologists, Soc. Am. Bacteriologists, Soc. for Exptl. Biology and Medicine (pres. 1973-75), Harvey Soc. (sec. 1947-50, pres. 1971-72), N.Y. Acad. Medicine, Assn. Am. Physicians (Kober medal 1989), Nat. Acad. Scis. (Kovalenko medal 1988), Am. Acad. Arts and Scis., N.Y. Heart Assn. (1st v.p. 1967, pres. 1969-71), Am. Philos. Soc. Home: 500 E 63d St New York NY 10021 Office: Rockefeller U 66th St and York Ave New York NY 10021

MCCARTY, RICHARD CHARLES, psychology educator; b. Portsmouth, Va., July 12, 1947; s. Constantine Ambrose and Helen Marie (Householder) McC.; m. Sheila Adair Miltier, July 12, 1965; children: Christopher Charles, Lorraine Marie, Ryan Lester, Patrick James. BS in Biology, Old Dominion U., 1970, MS in Zoology, 1972; PhD in Pathobiology, Johns Hopkins U., 1976. Rsch. assoc. NIMH, Bethesda, Md., 1976-78; asst. prof. U. Va., Charlottesville, 1978-84, assoc. prof., 1984-88, prof., 1988—, chair psychology, 1990—. Co-editor: Stress: Neuroendocrine and Molecular Approaches, 1992; mem. editorial bd. Behavioral and Neural Biology, 1985-90, Physiology and Behavior, 1989—; exec. sec. 5th Symposium on Stress, 1991; contbr. articles to profl. jours. Lt. comdr. USPHS, 1976-78. Recipient Rsch. Scientist Devel. award NIMH, 1985-90; sr. fellow Nat. Heart Lung Blood Inst., NIH, 1984-85. Fellow Soc. Behavioral Medicine, Acad. Behavioral Med. Rsch., Am. Psychol. Soc., Am. Inst. Stress. Roman Catholic. Avocations: sports, gardening. Office: U Va Dept Psychology 102 Gilmer Hall Charlottesville VA 22903-2477

MCCARTY, RICHARD EARL, biochemist, biochemistry educator; b. Balt., May 3, 1938; s. Maclyn and Anita (Davies) McC.; m. Kathleen Connolly, June 17, 1961; children—Jennifer A., Richard E., Jr., Gregory P. A.B., Johns Hopkins U., 1960, Ph.D., 1964. Postdoctoral assoc. Pub. Health Research Inst., N.Y.C., 1964-66; asst. prof. Cornell U., Ithaca, N.Y., 1966-72, assoc. prof., 1972-77, prof., 1977-90, prof., chmn., 1981-85; dir. biotech. program Cornell U., Ithaca, 1988-90; prof., chmn. Johns Hopkins U., Balt., 1990—; mem. editorial bd. Jour. Biol. Chemistry, Bethesda, Md., 1978-88, assoc. editor, 1981-82; mem. panel NSF, Washington, 1985—. Author: (with D. Wharton) Experiments and Methods in Biochemistry, 1972; contbr. articles to profl. jours. Career Devel. award NIH, 1968-73. Mem. AAAS, Am. Soc. Biochemistry and Molecular Biology, Am. Soc. Plant Physiologists. Home: 2204 Dalewood Rd Lutherville Timonium MD 21093-2701 Office: Johns Hopkins U Dept Biology 3400 N Charles St Baltimore MD 21218

MCCARTY, WILLIAM BRITT, natural resource company executive, educator; b. Shawnee, Okla., Apr. 6, 1953; s. William B. and Georgia M. (Lindsay) McC.; m. Jennifer Serio, Apr. 10, 1976; children: Patrick, Sara. BS in Computer Sci., Calif. State U., Fullerton, 1979; MBA in Fin., Claremont Grad. Sch., 1983, postgrad., 1990—. Chief fin. officer Republic Geothermal, Inc., Santa Fe Springs, Calif., 1980-88; v.p. CRM, Fullerton, Calif., 1988-89; mgr. corp. planning Republic Cos., Santa Fe Springs, 1989—; asst. prof. computer sci. Azusa (Calif.) Pacific U., 1990—. Bd. dirs. Eternal Truth Ministry, Huntington Beach, Calif., 1986-90. Office: Republic Cos 11823 Slauson Ave Santa Fe Springs CA 90670-2236

MCCASLIN, BOBBY D., soil scientist, educator; b. Wichita, Kans., July 26, 1943; s. Floyd and Alice (Rosendale) McC.; m. Betty L. Shinn, Dec. 19, 1964; children: Michael D., Nancy L. MS, Colo. State U., 1969; PhD, U. Minn., 1974. Rsch. asst. Colo. State U., Ft. Collins, 1966-69; rsch. fellow in soil sci. U. Minn., St. Paul, 1969-74; asst. prof. agronomy N.Mex. State U., Las Cruces, 1974-80, assoc. prof., 1980-87, prof. agronomy, 1987—, asst. dept. head, 1990—; program bd. adminstr. Great Plains Soil Fertilizer Workshop, Soil, Water, Air Testing Lab. Contbr. numerous articles to profl. jours. Recipient Svc. award, Coll. Agr., Las Cruces, 1990. Mem. Am. Soc. Agronomy, Soil Sci. Soc. Am., Western Soil Sci. Soc., Sigma Xi. Achievements include research on utilization of wastes for plant nutrients and soil amendments; plant mineral nutrition and soil management, genetic selection of plants to better fit problem soils. Home: 2030 Tyre Cir Las Cruces NM 88001-5816 Office: NM State U Agl and Hort Dept 3Q Las Cruces NM 88003

MCCAULEY, CHARLES IRVIN, manufacturing engineer; b. West Chester, Pa., Mar. 8, 1959; s. Charles Rowland and Mary Evelyn (Ricketts) McC. SB in Materials Sci./Engring., MIT, 1982, MS in Materials Engring., 1982. Process engr. AT&T Tech., Reading, Pa., 1982-87; sr. mfg. engr. Itek Optical Systems, Lexington, Mass., 1987-89; prin./mfg. engr. Cimbiosis Inc., Nashua, N.H., 1989-90; sr. mfg. engr. Vicor Corp., Andover, Mass., 1990—. Mem. Am. Assn. Crystal Growth, Am. Ceramic Soc., Soc. Mfg. Engrs. Office: Vicor Corp 23 Frontage Rd Andover MA 01810

MCCAULEY, HUGH WAYNE, human factors engineer, industrial designer; b. Flagstaff, Ariz., Apr. 24, 1935; s. Shelby Samuel and Agnes Ione (Anderson) McC.; m. Lois Cook, Oct. 20, 1962; 1 child, Meghan Cybele McCauley McCormick. BS in Naval Engring., U.S. Naval Acad., 1957; MS in Product Design, Ill. Inst. Tech., 1974. Commd. ensign USN, 1957, advanced through grades to capt., 1979; commdg. officer flag adminstrv. unit COMFAIR, Alameda, Calif., 1966-68; asst. prof. U. Ill., Chgo., 1971-81; prin. engr., scientist Douglas Aircraft Co., Long Beach, Calif., 1981-84; sr. tech. specialist Northrop Aircraft Div., Hawthorne, Calif., 1984-86; sr. staff engr. McDonnell Douglas Aerospace, Long Beach, 1986—; pres. Designtext, Chgo., 1971-81, Long Beach, 1981—; mem. aircrew systems standardization panel for industry and Dept. Def., Washington, 1982—. Pres. Parkview Plaza Owners Assn., 1990—. Mem. Human Factors Soc. (pres. Orange County chpt. 1988, bd. dirs.), McDonnell Douglas Apple User Group (pres. 1986—), Odd Fellows. Achievements include research on flight deck design methodology using computerized anthropometric models, retrofit of data link, assessment of cockpit interface concepts for data link retrofit. Office: 3855 Lakewood Blvd 78-83 Long Beach CA 90846-0001

MCCAULEY, JAMES WEYMANN, ceramics engineer, educator; b. Phila., Mar. 21, 1940; s. Edward Joseph and Emily Marie (Weymann) McC.; m. Mary Ann Malone, June 6, 1964; children: Patrick, Kathleen, Daniel. BS, St. Joseph's Coll., 1961; MS, Pa. State U., 1965, PhD, 1968. Rsch. asst. solid state sci. Pa. State U., University Park, 1966-68; rsch. sicentist Tech. Lab. U.S. Army, Watertown, Mass., 1968-74, group leader, 1974-81, div. chief, 1981-90; liaison scientist Far East Rsch. Office U. S. Army, Tokyo, 1988; dean N.Y. State (SUNY) Coll. Ceramics Alfred (N.Y.) U., 1990—; adj. prof. Boston U., 1984-86; bd. dirs. Alfred Tech. Resources Inc., Unipeg. Editor books; contbr. articles to sci. and profl. jours. Vice chmn. Wakefield Cable TV Com., Mass., 1970-80; chmn. Wakefield Conservatin Com., 1980-90. Felow Am. Ceramic Soc. (trustee, v.p. 1987-91, F.H. Norton award 1987, J.I. Mueller award 1988). Home: HC 64 Box 64L Hillcrest Ext Wellsville NY 14895 Office: Alfred U Coll of Ceramics Office of Dean Alfred NY 14802

MCCAW, VALERIE SUE, civil engineer; b. Kansas City, Kans., Aug. 8, 1960; m. Loren McCaw. BS, Okla. State U., 1982. Profl. engr. Okla. Suvey lab instr. Okla. State U., Stillwater, 1981-82; civil engr. U.S. Corps. Engrs., Tulsa, Okla., 1982-83; asst. city engr. City of Broken Arrow, Okla., 1983-87; civil engr. The Benham Group, Tulsa, 1987-91, project mgr., 1991-93; project mgr. City of Tulsa, 1993—; grad. Leadership Broken Arrow, 1985-86; participant Discover "E", Tulsa, 1991-92. Dir. Pride in Tulsa, 1990-91. Okla. State U. scholar, 1978-82. Mem. Okla. Soc. Profl. Engrs. (state v.p., named Young Engr. of Yr. 1992), Am. Soc. Civil Engrs., Assn. State Dam Safety Officials. Office: The City of Tulsa Rm 312 2317 S Jackson Tulsa OK 74107

MCCHESNEY, JAMES DEWEY, pharmaceutical scientist; b. Hatfield, Mo., Aug. 27, 1939; s. Ernest Perry and Florence Iona (Cummins) McC.; m. Sally Ann Cassidy, June 13, 1959; children: Daniel, Lisa, Thomas, Elizabeth, Margaret, Amy, Matthew. BS, Iowa State U., 1961; MA, Ind. U., 1964, PhD, 1965. Prof. U. Kans., Lawrence, 1965-78; prof., chmn. U. Miss., University, 1978-86; dir. Rsch. Inst. Pharm. Scis., U. Miss., University, 1987—. Contbr. more than 125 rsch. articles to profl. jours. Mem. Sch. Bd., Lafayette County, Miss., 1987—. Recipient Outstanding Teaching award U. Kans. Sch. Pharmacy, 1968; named Fulbright Lectr., 1985, Eli Lilly Rsch. fellow, 1967-68, NSF fellow 1962-65, Ford Motor Co. Fund scholar, 1957-61. Democrat. Roman Catholic. Achievements include 5 patents in field. Office: U Miss Rsch Instit Pharmaceutical Sch Pharmacy University MS 38677

MCCLANAHAN, LARRY DUNCAN, civil engineer, consultant; b. Franklin, Ky., July 30, 1938; s. Ernest William McClanahan and Anne Isabel (Henderson) Hodges; m. Betty Jane Marquess, Mar. 16, 1963; children: Michael Curtis, Marta Suzette, Meredith Angela. BS in Civil Engring., Tenn. Poly. Inst., 1965; MS in Civil Engring., Tenn. Tech. U., 1974. Registered profl. engr., Tenn., Ky., Ala.; registered land surveyor, Tenn. Sr. civil engr. Tenn. Bur. Aero., Nashville, 1965-67; airport planning engr. FAA, Memphis, 1967-68; assoc. R. Dixon Speas & Assocs., Atlanta, 1968-70; prin. Larry D. McClanahan & Assocs., Nashville, 1970—; pres. Agtech-Ky., Inc., Franklin, Ky., 1984-87. Pres. Nannie Perry Sch. PTA, Hendersonville, Tenn., 1975; v.p. Jackson Park Community Assn., Nashville, 1985-86, pres. 1986-87, 90-93; chmn. bd. dirs. Hendersonville Mcpl. Airport Authority, 1973-75, Nashville Met. Pub. Records Commn., 1992—, chmn. bd. dirs., 1992—, geneal. mem., 1992—; mem. Subarea 4 citizens adv. com. Metro Planning Commn.; pres. Woodlands Homeowners Assn., 1992-93. With USAF, 1956-59. Recipient Sustained Superior Service award FAA, 1968. Mem. SAR (Tenn. soc. state sec., registrar 1985-90, state v.p. 1990-91, pres.-elect 1992, pres. 1992-93, Sumner County chpt. 1990-91, alt. nat. trustee 1993-94, cert. disting. svc., 1986, meritorious svc. medal 1987, mem. Nat. Soc. Sons Am. Revolution nat. coms., Florence Kendall medal/award 1991, Liberty medal with 3 oakleaf clusters 1991, Patriot medal 1993), ASCE (chmn. landing area subcom. 1969-71), Soc. Boonesborough, McClanahan Family Assn., Order of Engrs., No. Neck of Va. Hist. Soc., Gen. Soc. Colonial Wars (dep. gov. Tenn. 1993-94). Republican. Baptist. Lodge: Masons. Avocations: geneology, history, travel, basketball and baseball (amateur coach). Home: 1119 Winding Way Rd Nashville TN 37216-2213 Office: 203A Point East Dr Nashville TN 37216-1403

MCCLANAHAN, MICHAEL NELSON, digital systems analyst; b. Cin., Oct. 28, 1953; s. Roland Nelson and Jeanne Ann (Stevens) McC.; m. Tina Rosanne Swiecki, Mar. 8, 1986; 1 child, Sean Gabriel. Student, U. Cin., 1972-73, Goldenwest Coll., 1979-80, Riverside Community Coll., 1980-83, 90-92. Pres. Riverside (Calif.) Mktg., 1983-88; digital systems analyst Wyle Labs., Norco, Calif., 1988—. Author: (software) SDAS, 1989, HCSS DAS System, 1990, (book) HCSS Systems Operation, 1990, (manual) Software Quality Assurance, 1991. Recipient Svc. award Wyle Labs., 1991. Mem. IEEE, Instrument Soc. of Am. Achievements include design and writing of numerous software systems and integration of these with data acquisition hardware systems for the purposes of acquiring rsch. data from unique test systems in aerospace, nuclear and defense industries. Address: Wyle Labs 985 4th St Norco CA 91760 Office: 1841 Hillside Ave Norco CA 91760

MCCLANAHAN, WALTER VAL, psychologist; b. Oklahoma City, Oct. 3, 1946; s. Homer Highley and June Lenora (Inscho) McC.; m. Cheryl Kathryn Rameé, July 28, 1972; children: Valerie Kathryn, Natalie Joanne. BA, U. Okla., 1968, MEd, 1973, PhD, 1980. Lic. psychologist, health svc. psychologist, Okla.; cert. sch. psychologist, Okla. Psychol. technician Children's Ctr., Cen. State Hosp., Norman, 1971-74; psychology trainee Phil Smalley Children's Ctr., Norman, 1974-80, sch. psychologist, 1980-82; sch. psychology intern Okla. State U., Stillwater, 1982-83; dir. sch. psychology svcs. Okla. Youth Ctr., Norman, 1983—; pvt. practice psychology Norman Counseling Clinic, Norman, 1991—; adj. asst. prof. dept. ednl. psychology U. Okla., Norman, 1986-91. Mem. APA, Okla. Psychol. Assn. (div. pres. 1984-85), Nat. Assn. Sch. Psychologists, Okla. Sch. Psychol. Assn. (exec. bd. 1982-83, v.p. 1983-84, pres. 1991-92, Faye Catlett award 1989). Office: Norman Counseling Clinic 2416 Tee Cir Norman OK 73069

MCCLAY, HARVEY CURTIS, data processing executive; b. Houston, Jan. 2, 1939; s. Clarence and Agnes E. McC.; m. Patricia Lott, Jan. 8, 1961; children: James, John, Susan, Robert. BA in Math., Rice U., 1960. Field engr. Western Electric Co., Marysville, Calif., 1960-62; analyst math. Litton Data Systems Co., Canoga Park, Calif., 1962-63; mgr. programming Lockheed Electronics Co., Houston, 1967-75; systems mgr. City of Houston, 1975-77; project mgr. fin. systems devel. Brown & Root, Houston, 1977-81; mgr. data processing Nat. Supply, Houston, 1981-84; project mgr. Computer Scis. Corp., Houston, 1984-92, Grumman Tech. Svcs., Houston, 1992—; part-time instr. data processing and mgmt. Houston Community Coll. Home: 2911 Huckleberry St Pasadena TX 77502 Office: Grumman Technical Svcs 16511 Space Ctr Blvd Houston TX 77058

MCCLELLAN, DAVID LAWRENCE, physician, medical facility administrator; b. Burlington, Iowa, Feb. 13, 1930; s. Harold L. and LaVon H. McC.; children: David, Steven, Mark, Jeffrey. BA, U. Iowa, 1952, MD, 1955. Intern U.S. Naval Hosp., San Diego, 1955-56; med. officer USS Nereus, 1956-58; pvt. practice Garland Med. Office, Spokane, Wash., 1958-64; pres. DeRe Medica Med. Clinic, Spokane, 1964—; pres., chmn. bd. North Spokane Profl. Bldg., Inc., 1966-86; pres. Inland Health Assocs., Inc., Spokane, 1984-92; pres., bd. dirs. Bio-Chem Environ. Svcs., Seattle, 1984; v.p. Forest Resources, 1978-86; pres. Omnex, 1980-86; chief exec. officer Double N Orchards, 1980-87; gen. ptnr. Double N Investments, 1987—; ptnr. Recovre, Inc.; mem. staff Holy Family Hosp. Contbr. articles and papers to med. jours. Lt. USN, 1955-58. Fellow Am. Acad. Family Practice; mem. AMA, Am. Profl. Practice Assn., Wash. State Med. Soc., Spokane County Med. Soc., Inland Empire Acad. Family Practice, Med. Group Mgmt. Assn., Internat. Platform Assn. Republican. Roman Catholic. Avocations: art collecting, antique collecting, boating, miniature sculpting, landscaping. Home: 7151 N Audubon Dr Spokane WA 99208-4516 Office: DeRe Medica 5901 N Lidgerwood St Spokane WA 99207-1055

MCCLELLAN, ROGER ORVILLE, toxicologist; b. Tracy, Minn., Jan. 5, 1937; s. Orville and Gladys (Paulson) McC.; m. Kathleen Mary Dunagan, June 23, 1962; children: Eric John, Elizabeth Christine, Katherine Ruth. DVM with highest honors, Wash. State U., 1960; M of Mgmt, U. N.Mex., 1980. diplomate Am. Bd. Vet. Toxicology, cert. Am. Bd. Toxicology. From biol. scientist to sr. scientist Gen. Electric Co., Richland, Wash., 1957-64; sr. scientist biology dept. Pacific N.W. Labs., Richland, Wash., 1965; scientist med. research br. div. biology and medicine AEC, Washington, 1965-66; asst. dir. research, dir. fission product inhalation program Lovelace Found. Med. Edn. and Research, Albuquerque, 1966-73; v.p., dir. research adminstrn., dir. Lovelace Inhalation Toxicology Research Inst., Albuquerque, 1973-76, pres., dir., 1976-88; chmn. bd. dirs. Lovelace Biomedical and Environ. Research Inst., Albuquerque, 1988—; pres. Chem. Industry Inst. Toxicology Research, Triangle Park, N.C., 1988—; mem. research com. Health Effects Inst., 1981-92; bd. dirs. Toxicology Lab. Accreditation Bd., 1982-90, treas., 1984-90; adj. prof. Wash. State U., 1980—, U. Ark., 1970-88; clin. assoc. U. N.Mex., 1971—, adj. prof. toxicology, 1985—; adj. prof. toxicology Duke U., 1988—; adj. prof. toxicology U. N.C. Chapel Hill, 1989—; adj. prof. toxicology N.C. State Univ., 1991—; mem. dose assessment adv. group U.S. Dept. Energy, 1980-87, mem. health and environ. research adv. com., 1984-85; mem. exec. com. sci. adv. bd. EPA, 1974—, mem. environ. health com., 1980-83, chmn., 1982-83, chmn. radionuclide emissions rev. com., 1984-85, chmn. Clean Air Sci. Adv. Com., 1987-92, chmn. rsch. strategies adv. com., 1992—; mem. com. on toxicology NAS-NRC, 1979-87, chmn., 1980-87; bd. dirs. Lovelace Anderson Endowment Found.; mem. com. risk assessment methodology for hazardous air pollution NAS-NRC, 1991—; pres. Am. Bd. Vet. Toxicology, 1970-73; mem. adv. council Ctr. for Risk Mgmt., Resources for the Future, 1987—; council mem. Nat. Council for Radiation Protection, 1970—; bd. dirs. N.C. Assn. Biomedical Rsch., N.C. Vet. Medical Found. Contbr. articles to profl. jours. Editorial bd. Jour. Toxicology and Environ. Health, 1980—, assoc. editor, 1982—; editorial bd. Fundamental and Applied Toxicology, 1984-89, assoc. editor, 1987-89; editorial bd. Toxicology and Indsl. Health, 1984—; editor CRC Critical Revs. in Toxicology, 1987—; assoc. editor Inhalation Toxicology Jour., 1987—; mem. edit. bd. Regulatory Toxicology and Pharmacology, 1993—. Recipient Herbert E. Stokinger award Am. Conf. Govtl. Indsl. Hygienists, 1985, Alumni Achievement award Wash. State U., 1987, Disting. Assoc. award Dept. Energy, 1987, 88, Arnold Lehman award Soc. Toxicology, 1992; co-recipient Frank R. Blood award Soc. Toxicology, 1989. Fellow AAAS, Am. Acad. Vet. and Comparative Toxicology; mem. Am. Chem. Soc., Inst. Medicine, NAS, Radiation Research Soc. (sec.-treas. 1982-84, chmn. fin. com. 1979-82), Health Physics Soc. (chmn. program com. 1972, Elda E. Anderson award 1974), Soc. Toxicology (v.p.-elect to pres. 1987-90; inhalation specialty sect. v.p. to pres. 1983-86; bd. publs. 1983-86, chmn. 1983-85), Am. Assn. Aerosol Research (bd. dirs. 1982—, treas. 1986-90, v.p., pres. 1990-93), Soc. Risk Analysis, Am. Vet. Med. Assn., Gesellschaft fur Aerosolforschung, Sigma Xi, Phi Kappa Phi, Phi Zeta. Republican. Lutheran. Home: 1111 Cuatro Cerros Trl SE Albuquerque NM 87123-4149 also: 2903-Q Bainbridge Dr Durham NC 27713-1448 Office: Chem Industry Inst Toxicology PO Box 12137 Durham NC 27709-2137

MCCLELLAND, ALAN, molecular biologist, laboratory administrator; b. Glasgow, Scotland, Dec. 7, 1955; came to U.S., 1981; s. James Duncan and Elizabeth Innes (Smith) McC.; m. Monica Anne Lyons, June 6, 1981; children: Lindsay, Stephanie. BSc with honors, U. Edinburgh, Scotland, 1977; PhD, Imperial Coll., London, Eng., 1981. Postdoctoral fellow biology Yale U., New Haven, Conn., 1981-85; sr. scientist Molecular Therapeutics Inc., West Haven, Conn., 1985-90, prin. scientist, 1990-92; lab. dir. Genetic Therapy, Inc., Gaithersburg, Md., 1992—; vis. fellow Yale U., New Haven, 1985-92. Contbr. articles to Cell, PNAS. Recipient Corp. Sci. award Miles, Inc., 1989; Sci. and Engring. Rsch. Coun. fellow, U.K., 1981. Achievements include key paper in 1989 describing the identification and isolation of the receptor for the common cold virus; postdoctoral research led to the cloning of the transferrin receptor gene. Office: Genetic Therapy Inc 19 First Field Rd Gaithersburg MD 20873

MCCLELLAND, SHEARWOOD JUNIOR, orthopaedic surgeon; b. Gary, Ind., Aug. 1, 1947; s. Shearwood and Zenobia Pearl (Pruitt) McC.; m. Yvonne Shirley Thornton, June 8, 1974; children: Shearwood III, Kimberly Itaska. AB, Princeton U., 1969; MD, Columbia U., 1974. Diplomate Am. Bd. Orthopaedic Surgery. Intern, St. Luke's Hosp., N.Y.C., 1974-75, resident, 1975-76; asst. resident in orthopaedic surgery N.Y. Orthopaedic Hosp., 1976-79; commd. lt. USNR, 1979-82, advanced through grades to lt. comdr., 1980; staff orthopaedic surgeon Nat. Naval Med. Center, Bethesda, Md., 1979-82; asst. prof. surgery Uniformed Services U. of Health Scis., 1980-82; acting chief orthopaedic surgery Harlem Hosp. Center, 1983-84, assoc. dir. orthopaedic surgery, 1985-92; acting dir. orthopaedic surgery, 1992—; asst. prof. clin. orthopaedic surgery Columbia U., 1983—; oral examiner Am. Bd. Othopaedic Surgery, mem. N.Y. State Office of Profl. Med. Conduct. Annie C. Kane fellow in orthopaedic surgery, 1978-79; fellow in total joint implant surgery Ohio State U., 1982 . Fellow Internat. Coll. Surgeons, ACS, Am. Acad. Orthopaedic Surgeons, N.Y. Acad. Medicine; mem. Assn. Mil. Surgeons of U.S., Eastern Orthopaedic Assn., N.Y. Orthopaedic Hosp. Alumni Assn., Mensa, No. N.J. Princeton Alumni Assn. Office: Harlem Hosp Ctr KP-9101 506 Lenox Ave New York NY 10037-1802

MCCLELLAND, THOMAS MELVILLE, meteorologist, researcher; b. Berkeley, Calif., Oct. 14, 1963; s. Wilson Melville and Margaret Gordon Mahy (Lloyd) McC; m. Amy Maureen Kennard, Oct. 23, 1993. BS in Atmospheric Sci., U. Calif., Davis, 1985; MS in Atmospheric Sci., Purdue U., 1988, postgrad., 1988—. Rsch. asst. U. Calif., Davis, 1985-86; grad. teaching asst. Purdue U., West Lafayette, Ind., 1986-87, grad. rsch. asst., 1988—; cons. GCV Cons., West Lafayette, 1992. Contbr. articles to profl. jours. Mem. Am. Meteorol. Soc. (travel grantee 1990), Sigma Xi. Office: Purdue U 1397 Civil Engring Bldg West Lafayette IN 47907

MCCLENDON, IRVIN LEE, SR., technical writer and editor; b. Waco, Tex., June 12, 1945; s. Irvin Nicholas and Evelyn Lucile (Maycumber) McC.; divorced; children: Michael Boyd, Irvin Lee Jr., Laura Ann, Paul Nicholas, Richard Lester. Student El Camino Coll., 1961-63, U. So. Calif., 1962-66; BA in Math., Calif. State U.-Fullerton, 1970, postgrad. in bus. adminstrn., 1971-76; cert. nat. security mgmt. Indsl. Coll. Armed Forces, 1974; postgrad. in religion Summit Sch. Theology, 1982-84. Engring. lab. asst. Rockwell Internat. Corp., Anaheim, Calif., 1967-68, test data analyst, 1968, assoc. computer programmer, 1968-70, mem. tech. staff, 1970-82; systems programmer A-Auto-trol Tech. Corp., Denver, 1982-84, sr. tech. writer, 1984-86; sr. tech. writer, editor Colo. Data Systems, Inc., Englewood, Colo., 1986-87; engring. writer III CalComp subs. Lockheed Co., Hudson, N.H., 1987; sr. tech. writer CDI Corp., Arvada, Colo., 1987-88; staff cons. CAP GEMINI AM., Englewood, 1989; sr. tech./instrnl. writer & editor Tech. Tng. Systems, Inc., Aurora, Colo., 1990—. Sec. of governing bd. Yorba Linda Libr. Dist., 1972-77; trustee Ch. of God (Seventh Day), Bloomington, Calif., 1979-81, treas., 1980-81, mem. Calif. State U. and Coll. Statewide Alumni Coun., 1976-77; 2d v.p. Orange County chpt. Calif. Spl. Dists. Assn., 1976, pres., 1977; mem. Adams County Rep. Cen. Com., 1984-90, mem. Denver County Republican Ctrl. Com., 1992—; charter mem. Harmony: A Colo. Chorale, 1991— (treas., bd. dirs. 1992—). With USAFR, 1967-71. USAF Nat. Merit scholar, 1963-67. Mem. Calif. Assn. Libr. Trustees and Commrs. (exec. bd., So. Calif. rep. 1976-77), Nat. Eagle Scout Assn. (life), Scottish-Am. Mil. Soc., St. Andrew Soc. Colo., Am. Coll. Heraldry, Calif. State U.-Fullerton Alumni Assn. (dir. 1975-77). Republican. Home: 13870 Albrook Dr Apt C-106 Denver CO 80239-4736 Office: 3131 S Vaughn Way Ste 300 Aurora CO 80014

MCCLINTIC, FRED FRAZIER, simulation engineer; b. Chester, Pa., Aug. 15, 1948; s. Fred F. and Maxene Mary (Felter) McC.; m. Janet Mary DeVitis, May 23, 1970; children: Shanon Janet, Sharron Marie. BS in Engring., Widener U., Chester, Pa., 1970; MS in Indsl. Engring., Tex. A&M U., 1972. Ops. rsch. analyst U.S. Army Civil Svc., Ft. Knox, Ky., 1970-78; prof. wargaming U.S. Army War Coll., Carlisle, Pa., 1978-83, tech. dir. dept. wargaming, 1978-83; pres. McClintic Wargaming Inc., Media, Pa., 1983-85; sr. assoc. CACI, Mechanicsburg, Pa., 1985-86; lead scientist Computer Scis. Corp., Moorestown, N.J., 1986-89; mgr. rsch. and devel. GE Aerospace-Strategic Systems, Blue Bell, Pa., 1989—. Author: The Armor Development Plan, 1976-78; (wargames) McClintic Theater Model, 1979, VII Corps Model, 1981, Warrior Preparation Model, 1983, TACAIR, 1975. Mem. TAu Beta Pi, Sigma Pi Sigma, Alpha Pi Mu, Phi Kappa Phi, Alpha Chi. Roman Catholic. Achievements include tech. leadership in Army, Navy and Air Force interactive simulations, models and wargames; designing and setup of USAF Warrior Preprations Ctr. Models; lead scientist for Enhanced Naval Wargaming System. Home: 1960 Colt Rd Media PA 19063

MCCLINTIC, GEORGE VANCE, III, petroleum engineer, real estate broker; b. Sayre, Okla., Jan. 27, 1925; s. George Vance and Myrtle Jane (Rogers) M.; m. Margaret Ruth, 1945; m. Betsy Ross, 1969 (dec. 1977); children—Kathern, Michael; m. Judy Prince, 1978 (dec. 1985); m. Caroline Hall, 1986. B.S., Okla. U., 1950. Warehouseman Mid Continent Supply Co., 1950-52, dist. machinery sales mgr., 1952-56; with Sabre Drilling Co., also v.p. 3 Rig Co., 1956-58; Okla. dist. mgr. Republic Supply Co., 1958-59; v.p. Chief Oil Tool Co., Oklahoma City, 1959-62; in real estate, Oklahoma City, 1962—; owner, mgr. McClintic Realty, and Engring. Cos., 1966—; with Jack Callaway Co., 1968-70; pres., engring. officer Geothermal Engring. and Operating Co., Oklahoma City and Carson City, Nev., 1972-76; petroleum engr. SW Mineral Energy Co., Oklahoma City, 1976-79; pres. Flat Tire Caddie Co., Oklahoma City, 1979—. Republican candidate for 5th Dist. Congress, 1976; mayoral candidate for Oklahoma City, 1983. Served with USAAF, 1943-45; ETO. Decorated D.F.C., Air medal with 11 clusters, ETO medal with 5 Campaign bronze stars. Clubs: Oklahoma, Petroleum. Lodge: Masons. Inventor flat tire caddie. Office: McClintic Realty Engring Co 2237 NW 26th St Oklahoma City OK 73107-2509

MCCLINTOCK, SAMUEL ALAN, engineering educator; b. Richlands, Va., Mar. 9, 1957; s. John W. Jr. and Jayne (Christie) McC.; m. Victoria McClintock, Mar. 17, 1979; children: Erica, Amy. MS in Environ. Engring., Va. Poly. Inst. and State U., 1986, PhD in Civil Engring., 1990. Registered profl. engr., Pa., Va. Asst. prof. engring. Pa. State U., Middletown, 1990—. Recipient Lindbergh award Charles Lindbergh Fund, 1992. Office: Pa State U Sci Engring and Tech Bldg Rm 105 Middletown PA 17057

MCCLINTOCK, WILLIAM THOMAS, health care administrator; b. Pittsfield, Mass., Oct. 23, 1934; s. Ernest William and Helen Elizabeth (Clum) M.; m. Wendolyn Hope Eckerman, June 22, 1963; children: Anne Elizabeth, Carol Jean, Thomas Daniel. BA, St. Lawrence U., Canton, N.Y., 1956; MBA, U. Chgo., 1959, MHA, 1962. Prodn. planner Corning Glass Works, Corning, N.Y., 1959-60; adminstrv. asst. Univ. Hosps. of Cleve., 1962-65; asst. adminstr. Presbyn. Hosp., Whittier, Calif., 1965-68; regional asst. Kaiser Found. Hosps., Oakland, Calif., 1968-70; assoc. dir., exec. dir. Conn. Hosp. Planning Commn., New Haven, 1970-75; project dir., lectr. sch. health studies U. N.H., Durham, 1975-77; regional mgr. Tex. Med. Found., Austin, 1977-81; adminstr. Schick Shadel Hosp., Fort Worth, 1981-87; mgmt. cons. George S. May Internat. Co., Park Ridge, Ill., 1987-88; mgr. Nat. Ctr. Rsch. Programs Am. Heart Assn., Dallas, 1988-89; adminstr. Ambulatory Svcs. Health Care of Tex., Ft. Worth, 1990-92; CEO Boundary Community Hosp. & Nursing Home, Bonners Ferry, Idaho, 1992—. 1st lt. U.S. Army, 1957. Fellow Am. Coll. Health Care Execs., Am. Coll. Addiction Treatment Adminstrs.; mem. Am. Hosp. Assn. (life), Unity Lodge No. 9 , F&AM, N.Y. Republican. Presbyterian. Avocations: reading, gardening, photography. Home: PO Box 1226 128 Mohawk Bonners Ferry ID 83805-1226 Office: 551 Kaniksu St Bonners Ferry ID 83805

MCCLOSKEY, THOMAS HENRY, mechanical engineer, consultant; b. Phila., Dec. 11, 1946; s. Thomas H. McCloskey; m. Rosemary Loscalzo, July 11, 1970. BSME, Drexel U., 1969. Rsch. engr. Westinghouse Elec. Corp., Phila., 1969-80; mgr. turbo-machinery Elec. Power Rsch. Inst., Palo Alto, Calif., 1980-92, cons. turbo-machinery, 1992—; mem. adv. bd. Internat. Pump Symposium, Houston, 1987—. Author: ASME Specification Guidelines for Large Steam Turbines, 1987; contbr. articles to profl. jours. Fellow ASME (dir. turbine design course 1988—, mem. rsch. bd. 1989—, Edison Elec. Prime Mover award 1984). Achievements include 6 patents in turbomachinery design; development and field application of finite element/fracture mechanic techniques and erosion/corrosion resistant materials for life assessment/optimization of large turbine generators and pumps. Office: Elec Power Rsch Inst 3412 Hillview Ave Palo Alto CA 94303

MCCLOSKEY, THOMAS WARREN, flow cytometrist, immunologist; b. N.Y.C., Sept. 10, 1964; s. Edward Ireland and Margaret Bayne (McPherson) McC.; m. Pamela Sue Miller, Aug. 3, 1991. BS with highest honors, Hofstra U., 1986; PhD, Rutgers U., 1991. Rsch. assoc. North Shore U. Hosp., Manhasset, N.Y., 1991-92, flow cytometry specialist, 1992—. Mem. Phi Beta Kappa, Beta Beta Beta. Achievements include operator during beta site testing of Epics Elite. Office: North Shore U Hosp 350 Community Dr Manhasset NY 11030

MCCLUNEY, (WILLIAM) ROSS, physics researcher, technical consultant; b. Cape Girardeau, Mo., Dec. 26, 1940; s. William Jones and Luella (Benjamin) McC.; m. Judith Lane, June 1969 (div. 1983); children: Alan M., Kevin E. BA in Physics, Rhodes Coll., Memphis, 1963; MS in Physics, U. Tenn., 1966; PhD in Physics, U. Miami, 1973. Devel. engr. Eastman Kodak Co., Rochester, N.Y., 1966-67; rsch. scientist NASA Goddard Space Flight Ctr., Greenbelt, Md., 1973-76; prin. rsch. scientist Fla. Solar Energy Ctr., Cape Canaveral, 1976—; cons. 3M co., Mpls., 1983, Holder Constrn. Co., Atlanta, 1990 Office Energy-Related Inventions, Nat. Inst. Stds. Tech., Gaithersburg, Md., 1987—, Queen's U., Kingston, Ont., Can., 1991—; tech. cons. for world's largest sundial Disney World, 1990; mem. Fla.-NASA Coun. on Red Tide Rsch., 1974-76; numerous presentations at tech. meetings. Author: Introduction to Radiometry and Photometry, 1993; editor: The Environmental Destruction of South Florida, 1971; also over 50 articles to tech. jours. and mags. Recipient cert. of recognition NASA, 1976, Rschr. of Yr. award Fla. Solar Energy Ctr., 1986. Mem. AAAS, ASHRAE, Illuminating Engring. Soc. N.Am., Optical Soc. Am., Soc. Photo-Optical Instrumentation Engrs., Earth Ethics Rsch. Group (v.p. 1988—), Fla. Audubon Soc. (bd. dirs. 1977-81), Sigma Xi, Sigma Pi Sigma. Achievements include patent for scattering meter; collaborated with J.Y. Cousteau on series of oceanographic studies, 1974-75. Office: Fla Solar Energy Ctr 300 State Rd 401 Cape Canaveral FL 32920

MCCLUNG, JOHN ARTHUR, cardiologist; b. Oneonta, N.Y., Mar. 18, 1949; s. Charles Harvey and Ruth Steiner (Voegtly) McC.; m. Jane Giles, June 29, 1985; children: Daniel James, Timothy John. AB, Johns Hopkins U., 1971; MD, N.Y. Med. Coll., 1975. Diplomate Am. Bd. Internal Medicine with specialties in cardiovascular disease, critcal care medicine. Instr. in clin. medicine N.Y. Med. Coll., Valhalla, 1979-82, asst. prof. medicine, 1982-89, assoc. prof. medicine, 1989—; asst. dir. Cardiac Catheterization Lab./Westchester Med. Ctr., Valhalla, 1987—; assoc. dir. Inst. for Human Values in Med. Ethics/N.Y. Med. Coll., 1989—; chief, critical care sect. Westchester Med. Ctr., Valhalla, 1982—, chief, div. clin. ethics, 1990—; adv. bd. Met. N.Y. Ethics Com. Network, N.Y.C., 1992—; regional corrs. Soc. Bioethics Cons., Cleve., 1992—. Contbr. articles to profl. jours. Dir. EMT defibrillation project, Community Gen. Hosp. of Sullivan County, Harris, N.Y., 1987-91. Fellow Am. Coll. Cardiology, Am. Coll. Physicians, Am. Heart Assn. (coun. clin. cardiology), N.Y. Cardiol. Soc., Soc. Cardiac Angiography and Interventions. Episcopalian. Office: Westchester Med Ctr Divsn Cardiology Valhalla NY 10595

MCCLURE, ALLAN HOWARD, materials engineer, space contamination specialist, space materials consultant; b. Phila., Mar. 29, 1925; s. C. Howard and Edda Cherry (Speirs) McC.; m. Jean Florence Hall, May 31, 1947; children: Joyce Ann, Allan Hall. BS, Widener U., 1949; postgrad., Command & Gen. Staff Coll., 1972. Chemist Am. Cyanamid, Pitts., 1950-52; materials engr. Piasecki/Vertol Helicopter Co., Morton, Pa., 1952-59; lead engr. Boeing Aerospace Co., Seattle, 1959-71; sr. specialist engr. Boeing Aerospace Co., Kent, Wash., 1974-85; tech. cons. Adhesive Engring. Co., San Carlos, Calif., 1971-74. Author, investigator spacecraft contamination control documents and govt. reports. Pres. Seattle Crime Prevention League, 1974-84. Served to maj. U.S. Army, 1943-46, ETO, PTO; sec. Boeing Employees Amateur Radio Soc., 1984; membership chmn. Amateur Radio Emergency Services, 1984-85. Recipient Silver Beaver award and William H. Spurgeon III award Boy Scouts Am., Seattle, 1964. Mem. Am. Chem. Soc., Soc. for Advancement of Material and Process Engring. (nat. dir., pres. Seattle chpt.), Rainier C. of C., Res. Officers Assn. (life). Republican. Avocations: amateur radio, hiking, coin collecting, canoeing, photography. Home: 12026 SE 216th St Kent WA 98031-2272

MCCLURE, DAVID WOODARD, electrical engineer; b. Springfield, Mass., Aug. 31, 1965; s. Frederick Dodds and Diane Evelyn (Woodard) McC.; m. Kim Elizabeth Wilgo, June 22, 1991. AS in Electrical Engring., Wentworth Inst. Tech., 1985, BS in Electronic Engring. Tech., 1989, BSEE, 1991. Repair technician ESL, Rockland, Mass., 1983-85; rsch. and devel. technician MIT Lincoln Lab., Lexington, Mass., 1985-90, electrical engr. rsch. and devel., 1990–; part time instr. Wentworth Inst. Tech., Boston, 1991-93. Office: MIT Lincoln Lab 244 Wood St M/S J-108 Lexington MA 02173

MC CLURE, DONALD STUART, physical chemist, educator; b. Yonkers, N.Y., Aug. 27, 1920; s. Robert Hirt and Helen (Campbell) McC.; m. Laura Lee Thompson, July 9, 1949; children: Edward, Katherine, Kevin. B.Chemistry, U. Minn., 1942; Ph.D., U. Calif., Berkeley, 1948. With war research div. Columbia U., 1942-46; mem. faculty U. Calif., Berkeley, 1948-55; group leader, mem. profl. staff RCA Labs., 1955-62; prof. chemistry U. Chgo., 1962-67; prof. chemistry Princeton (N.J.) U., 1967-91, prof. emeritus, 1991—; vis. lectr. various univs.; cons. to govt. and industry. Author: Electronic Spectra of Molecules and Ions in Crystals, 1959, Some Aspects of Crystal Field Theory, 1964; also articles. Guggenheim fellow Oxford (Eng.) U., 1972-73; Humboldt fellow, 1980; recipient Irving Langmuir prize, 1979. Fellow Am. Acad. Arts and Scis., Nat. Acad. Scis.; mem. Am. Chem. Soc., Am. Phys. Soc. Home: 23 Hemlock Cir Princeton NJ 08540-5405

MCCLURE, HAROLD MONROE, veterinary pathologist; b. Hayesville, N.C., Oct. 2, 1937; s. Elmer S. and Willie S. (Davenport) McC.; m. Joan V. Cunningham, June 7, 1958; children: Michael, Michelle, Barry. BS, N.C. State U., 1963; DVM, U. Ga., 1963. Postdoctoral fellow in pathology U. Wis., Madison, 1963-66; vet. pathologist Yerkes Primate Ctr., Emory U., Atlanta, 1966—, chief div. pathology, 1979—, assoc. dir., 1982—, asst. prof. pathology, 1981—; mem. rev. com.s NIH, bethesda, Md.; cons. Am. Assn. Accreditation of Lab. Animal Care, Rockville, Md., 1989—; mem. AIDS adv. com. NIAID, Bethesda, 1989-91. Contbr. articles to sci. publs. Recipient numerous grants NIH, USDA, NASA. Mem. Am. Soc. Microbiology, Internat. Acad. Pathology, Phi Kappa Phi, Phi Zeta, Gamma Sigma Delta. Republican. Baptist. Achievements include patent on acutely lethal simian immunodeficiency virus. Home: 2301 Poplar Springs Dr Atlanta GA 30319 Office: Emory U Yerkes Primate Ctr Atlanta GA 30322

MCCLURE, JOHN CASPER, materials scientist, educator; b. Weymouth, Mass., Mar. 1, 1944; s. John Casper and Faith (Sullivan) McC. BS in Physics, U. Ill., 1966; MS, PhD in Materials Sci., Syracuse U., 1975. Scientist IBM, Poughkeepsie, N.Y., 1966-68; tchr. U.S. Peace Corps, Ghana, 1968-69, L'Ecole Anglaise, Paris, 1969-70; engr. NASA, Huntsville, Ala., 1976-82; rsch. engr. Nat. Steel, Wierton, W.Va., 1982-85; asst. prof. U. Tex., El Paso, 1985-91, assoc. prof., 1991—. Contbr. articles to profl. jours. Grantee NASA, NSF, 1987—. Mem. ASTM, Am. Welding Soc., Sigma Xi. Achievements include patents for method for measurement of delta noise in ferrite memory cores, method for control of microstructure in monotectic alloys. Home: 3300 Piedmont Ave El Paso TX 79902 Office: U Tex Dept Metallurgy and Materials Engring El Paso TX 79968

MCCLURE, RICHARD BRUCE, satellite communications engineer; b. Yonkers, N.Y., Nov. 22, 1935; s. Robert Hirt and Helen (Campbell) McC.; m. Patricia Lauretta McCarthy, Aug. 16, 1964 (separated); 1 child, Carol Ann. BEE, NYU, 1959, MEE, 1968. From engr. to sr. engr. ITT Fed. Labs., Nutley, N.J., 1959-65; sr. engr., br. mgr., dept. mgr. Modulation Tech. Dept., Comsat Labs., Clarksburg, Md., 1965-76; v.p. engring. Telecomm. Techniques Corp., Germantown, Md., 1976-83; dir. systems engr. Dama Telecomms. Corp., Rockville, Md., 1983-85; dir. Washington office Satellite Transmission Systems, Hauppage, N.Y., 1985-87; v.p. engring. Govt. Commn's Divsn. Calif. Microwave Inc., Annapolis Junction, Md., 1987—; cons. in field. Contbr. papers to confs. Active Balt. Symphony Chorus. Mem. IEEE (sr.). Achievements include patent (with others) for SPADE satellite communications system; research on effect of earth station and satellite parameters on the SPADE system; maritime satellite communications terminal implementation. Office: Calif Microwave Inc 10820 Guilford Rd Annapolis Junction MD 20701

MCCLURE, WILLIAM OWEN, biologist; b. Yakima, Wash., Sept. 29, 1937; s. Rexford Delmont and Ruth Josephine (Owen) McC.; m. Pamela Preston Harris, Mar. 9, 1968 (div. 1979); children: Heather Harris, Rexford Owen; m. Sara Joan Rorke, July 27, 1980. BSc, Calif. Inst. Tech., 1959; PhD, U. Wash., 1964. Postdoctoral fellow Rockefeller U., N.Y.C., 1964-65; rsch. assoc. Rockefeller U., 1965-68, asst. prof. U. Ill., Urbana, 1968-75; assoc. prof. U. So. Calif., L.A., 1975-79; prof. biology, prof. neurology U. So. Calif., 1979—; v.p. sci. affairs Nelson Rsch. & Devel. Co., Irvine, Calif., 1981-82; acting v.p. rsch. & devel. Nelson Rsch. & Devel. Co., 1985-86; dir. programs info. scis. U. So. Calif., 1982—; dir. cellular biology U. So. Calif., 1979-81, dir. neurobiology, 1982-88; cons. in field; dir. Marine & Freshwater Biomed. Ctr., U. So. Calif., 1982-83; co-dir. Baja Calif. Expedition of the R/V Alpha Helix, 1974, others; chmn. Winter Conf. on Brain Rsch., 1979, 80, others; lectr. in field; sci. adv. bd. Nelson R & D, 1972-91; mem. bd. commentators Brain and Behavioral Scis., 1978—. Editor or author 3 books; co-editor: Wednesday Night at the Lab; patentee in field; mem. editorial bd. Neurochem. Rsch., 1975-81, Jour. Neurochemistry, 1977-84, Jour. Neurosci. Rsch., 1980-86; contbr. over 100 articles to profl. jours. Bd. dirs. San Pedro & Peninsula Hosp. Found., 1989—, Faculty Ctr., U. So. Calif., 1991—, San Pedro Health Svcs., 1992—. Scripps Inst. fellow, 1958, NIH fellow, 1959-64, 64-65, Alfred P. Sloan fellow, 1972-76, others; recipient rsch. grants, various sources, 1968—; Intersci. Rsch. Inst. fellow, 1989. Mem. AAAS, Am. Soc. Neurochemistry, Soc. for Neurosci., Am. Soc. Biol. Chemistry and Molecular Biology, Internat. Soc. Neurochemistry, Assn. Neurosci. Depts. and Programs, Univ. Park Investment Group, N.Y. Acad. Scis., Univ. Club L.A. Republican. Presbyterian. Avocations: computing, travel. Home: 30533 Rhone Dr Palos Verdes Peninsula CA 90274-5742 Office: U So Calif Dept Biol Scis Los Angeles CA 90089-2520

MCCLUSKEY, EDWARD JOSEPH, engineering educator; b. N.Y.C., Oct. 16, 1929; s. Edward Joseph and Rose (Slavin) McC.; m. Lois Thornhill, Feb. 14, 1981; children by previous marriage—Edward Robert, Rosemary, Therese, Joseph, Kevin, David. A.B. in Math. and Physics, Bowdoin Coll., Brunswick, Maine, 1953, B.S., M.S. in Elec. Engring., 1953, Sc.D., MIT, 1956. With Bell Telephone Labs., Whippany, N.J., 1955-59; assoc. prof. elec. engring. Princeton, 1959-63, prof., 1963-66, dir. Computer Center, 1961-66; prof. elec. engring. and computer sci. Stanford (Calif.) U., 1967—, dir. Digital Systems Lab., 1969-78; dir. Center for Reliable Computing, 1976—; tech. advisor High Performance Systems, 1987-90. Author: A Survey of Switching Circuit Theory, 1962; Introduction to the Theory of Switching Circuits, 1965; Design of Digital Computers, 1975; Logic Design Principles with Emphasis on Testable Semicustom Circuits, 1986. Editor: Prentice-Hall Computer Engineering Series, 1988-90; assoc. editor: IRE Transactions on Computers, 1959-65, ACM Jour., 1963-69; editorial bd. IEEE Design and Test, 1984-86; assoc. editor, IEEE Trans. Computer Aided Design, 1986-87. Patentee in field. Fellow AAAS, IEEE (pres. computer soc. 1970-71, Centennial medal 1984); mem. IEEE Computer Soc. (Tech. Achievement award 1984, Taylor L. Booth Edn. award 1991), Am. Fedn. Info. Processing Socs. (dir., exec. com.), Internat. Fedn. Info. Processing (charter), Japan Soc. Promotion of Sci. Office: Stanford University Ctr for Reliable Computing ERL-460 Stanford CA 94305-4055

MCCLUSKEY, KEVIN, fungal molecular geneticist; b. Princeton, N.J., Aug. 22, 1961; s. Edward Joseph and Roberta Jean (Erickson) McC.; m. Ann Sherman Daggett, Oct. 26, 1985; 1 child, Morgan Daggett. BS, MS, Stanford (Calif.) U., 1985; PhD, Oreg. State U., 1991. Rsch. technologist physics rsch. Mass. Gen. Hosp., Boston, 1986-87; rsch. asst. botany and plant pathology Oreg. State U., Corvallis, 1987-91; rsch. assoc. dept. of plant pathology U. Ariz., Tucson, 1991—. Contbr. articles to profl. jours. Travel grant NSF, 1990, Novo/Nordisk, 1990; Arco Found. fellow Oreg. State, 1987-90. Mem. AAAS, Internat. Soc. for Molecular Plant-Microbe Interactions, Am. Phytopathol. Soc., Sigma Xi, Gamma Sigma Delta (Achievement award), Phi Eta Sigma. Home: 10700 N La Reserve Dr # 10205 Oro Valley AZ 85737 Office: U Ariz Plant Pathology Forbes Bldg # 104 Tucson AZ 85721

MCCOLLOM, JEAN MARGARET, ecologist; b. Milw.; d. Russell L. Jr. and Violet (Anderson) McC.; m. Michael J. Duever, Dec. 1981. BA cum laude, Western Mich. U., 1981; MS in Watershed Mgmt., U. Ariz., 1990. Rsch. ecologist ecosystem studies program Nat. Audubon Soc., Naples, Fla., 1982—. Contbr. articles to profl. publs. Mem. Ecol. Soc. Am., Natural Area Assn. Office: Nat Audubon Soc Ecosystem Studies Program 479 Sanctuary Rd W Naples FL 33964

MCCOLLOUGH, MICHAEL LEON, astronomer; b. Sylva, N.C., Nov. 3, 1953; s. Stribling Mancell and Vivian Hazel (Bradley) McC. B.S., Auburn U., 1975, M.S., 1981; PhD, Ind. U., 1989. Lab. instr. Auburn (Ala.) U., 1974-75, grad. asst., 1975-77, lab. technician, 1977-78; assoc. instr. Ind. U., Bloomington, 1978-86; ops. astronomer Computer Scis. Corp., Balt., 1988-90, sci. planning and scheduling system dep. br. chief, 1990-92; data processing and distbn. mgr. U.S. ROSAT Sci. Data Ctr., 1992-93; asst. system mgr. BATSE Data Analysis System, 1993—; vis. lectr. Okla. State U., 1986-87; vis. asst. prof. U. Okla., 1987-88. Recipient Achievement award Space Telescope Sci. Inst., 1990, 91, Pub. Svc. Group Achievement award NASA, 1991, Cert. Recognition, 1991. Mem. Am. Astron. Soc., Royal Astron. Soc., Astron. Soc. Pacific, Am. Phys. Soc., Soc. Physics Students, Sigma Xi (assoc.), Sigma Pi Sigma. Baptist. Home: 201 Water Hill Rd Apt # G13 Madison AL 35758 Office: NASA/MSFC Code ES66 Huntsville AL 35812

MCCOMBS, JEROME LESTER, clinical cytogenetics laboratory director; b. Birmingham, Ala., Dec. 16, 1951; s. J.L. and Etta Mae (Hallmark) McC.; m. Glenda Kay Adkins, Aug. 24, 1974; children: Adam, Mark. BS, Auburn Univ., 1974; PhD, Univ. Ala., 1982. Diplomate Am. Bd. Medical Genetics. Asst. prof. health sci. ctr. Univ. Tex., San Antonio, 1984-87; asst. prof. medical br. Univ. Tex., Galveston, 1988—, lab. dir. medical br., 1988—; cons. Lab. Genetic Sci., Houston, 1992—; bd. dirs. Tex. Genetics Soc., 1993—. Author: (chpt. in book) Clinical Identification of Chromosomes Syndrome, 1991, In Situ Lydridization, 1984; contbr. articles to profl. jours. Cons. LaMarque Ind. Sch. Dist., 1989, March of Dimes, 1988-90. Recipient grant Lalor Found., Sealy Smith Found. Fellow Am. Coll. Medical Genetics (founding); mem. Am. Soc. Human Genetics. Democrat. Methodist. Achievements include use of DNA probes to identify marker chromosomes. Office: Univ Tex Med Br C-59 Dept Pediatrics Galveston TX 77555-0359

MCCONATHY, DONALD REED, JR., meteorologist, remote sensing program manager, systems engineer; b. New Haven, Conn., Jan. 19, 1942; s. Donald Reed and Ruth Mable (Hallmark) McC.; m. Evelyn Dorothea Harner, Mar. 26, 1966; children: Michael David, Jeffrey Thomas, Christopher Sean. BS in Physics, Bucknell U., 1963; BS in Meteorology, Pa. State U., 1965; MS in Meteorology, Naval Post Grad. Sch., 1972. Commd. ensign USN, 1966, advanced through grades to comdr., 1986; forecaster Fleet Weather Ctrl., Pearl Harbor, Hawaii, 1966-70; project officer Fleet Numerical Oceanography Ctr., Monterey, Calif., 1972-75; def. meteorol. satellite program project mgr. Navy Space Systems Activity, El Segundo, Calif., 1975-78; meteorologist USS Enterprise, Bremerton, Wash., 1978-80; project mgr. satellite programs Space and Naval Warfare Systems Command, Arlington, Va., 1981-86; earth observation system data and info. system sr. engr. CTA Inc., McLean, Va., 1986-89; sr. assoc. Booz, Allen & Hamilton Inc., McLean, Va., 1989—. Contbr. articles to profl. jours. Coach Vienna (Va.) Youth Soccer, 1981-91; com. chmn., scoutmaster Boy Scouts Am., Va., Calif., Wash., 1975-92. Mem. AIAA (sr.), Am. Astron. Soc. (sec. D.C. sect. 1991-93, chmn. planning com. Goddard Meml. Symposium 1993-94), Sigma Xi. Republican. Office: Booz Allen and Hamilton Inc 8283 Greensboro Dr Mc Lean VA 22102

MCCONNEL, RICHARD APPLETON, aerospace company official; b. Rochester, Pa., May 29, 1933; s. Richard Appleton Sr. and Dorothy (Merriman) McC.; m. Mary Francis McInnis, 1964 (div. 1984); children: Amy Ellen, Sarah Catherine; m. Penny Kendzie, 1993. BS in Naval Engring., U.S. Naval Acad., 1957; MS in Aerospace Engring., USN Postgrad. Sch., 1966. Commd. ensign USN, 1957; naval aviator Operation ASW, 1959-63, 68-71, 75-79; asst. prof. math. U.S. Naval Acad., 1966-68; program mgr. P3C update Naval Air Devel. Ctr., 1971-75; range program mgr. Pacific Missile Test Ctr., 1979-82; ret. USN, 1982; program mgr. Electromagnetic Systems div. Raytheon Co., Goleta, Calif., 1982-87; sr. engr. SRS Techs., Inc., Camarillo, Calif., 1987-92, High Tech. Solutions, Inc., Camarillo, Calif., 1992—. Mem. Internat. Test and Evaluation Assn., Assn. Old Crows. Republican. Office: High Tech Solutions 80 Wood Rd Camarillo CA 93010

MCCONNELL, ANTHONY, polymer scientist, consultant; b. Manchester, Eng., June 27, 1959; came to U.S., 1988; s. Anthony Barry McConnell and Ann (Barber) Strickland; m. Julia Hadfield, Feb. 12, 1983; children: David Anthony, Tyler James, Kevin Michael. BS in Polymer Sci. and Tech., Manchester Poly. U., 1986. Analytical chemist Hawker Siddely Aviation, Manchester, 1975-81; mgr. plastics lab. Brit. Aerospace, Manchester, 1981-84, composites specialist, 1984-86; mgr. devel. John Cotton Ltd., Lancashire, Eng., 1986-88; mgr. advanced product Prince Corp., Holland, Mich., 1988-93; with Haworth Inc., Holland, 1993—; ind. polymers cons., Manchester, 1988. Mem. AAAS, Soc. Plastics Engrs. Achievements include patents in polyurethane composites for automotive interior trim, contribution to aerospace development programs on advanced composite structures for satellite, military and civil aviation applications. Office: Haworth Inc 1 Haworth Ctr Holland MI 49423

MCCONNELL, HARDEN MARSDEN, biophysical chemistry researcher, chemistry educator; b. Richmond, Va., July 18, 1927; s. Harry Raymond and Frances (Coffee) McC.; m. Sophia Milo Glogovac, Oct. 6, 1956; children: Hunter, Trevor, Jane. BS, George Washington U., 1947; PhD, Calif. Inst. Tech., 1951; DSc (hon.), U. Chgo., 1991, George Washington U., 1993. NRC fellow dept. physics U. Chgo., 1950-52; research chemist Shell Devel. Co., Emeryville, Calif., 1952-56; asst. prof. chemistry Calif. Inst. Tech., 1956-58, prof. chemistry and physics, 1963-64; prof. chemistry Stanford U., Calif., 1964-79, Robert Eckles prof. chemistry, 1979—, chmn. dept., 1989—; founder Molecular Devices Corp., 1983—; cons. in field. Contbr. numerous articles to profl. publs.; patentee (in field). Pres. Found. for Basic Rsch. in Chemistry, 1990—; hon. assoc. Neurosci. Rsch. Program. Recipient Calif. sect. award Am. Chem. Soc., 1961, award in pure chemistry Am. Chem. Soc., 1962, Harrison Howe award, 1968, Irving Langmuir award in chem. physics, 1971, Pauling medal Puget Sound and Oreg. sects., 1987, Peter Debye award in phys. chemistry, 1990; Alumni Achievement award George Washington U., 1971; Disting. Alumni award Calif. Inst. Tech., 1982, Sherman Fairchild Disting. scholar, 1988; Dickson prize for sci. Carnegie-Mellon U., 1982, Wolf prize in chemistry, 1984, ISCO award, 1984; Wheland medal U. Chgo., 1988; Nat. Medal Sci., 1989. Fellow AAAS, Am. Phys. Soc.; mem. Nat. Acad. Scis. (award in chem. scis. 1988), Am. Acad. Arts and Scis., Am. Soc. Biol. Chemists, Swedish Biophysical Soc., Internat. Acad. Quantum Molecular Scis., Am. Chem. Soc., Biophysical Soc., Internat. Coun. on Magnetic Resource in Biol. Systems, Brit. Biophysical Soc. Office: Stanford U Dept Chemistry Stanford CA 94305

MCCONNELL, JAMES JOSEPH, internist; b. Lynchburg, Va., Sept. 4, 1946; s. Willis Samson and Hope (Lewis) McC.; m. Pamela Marie Sabatino, Apr. 7, 1979. BS, Lynchburg Coll., 1968; MD, Med. Coll. Va., 1972. Diplomate Am. Bd. Internal Medicine. Intern USN, Portsmouth, Va., 1972-73; dir. occupational medicine Norfolk Naval Shipyard, Portsmouth, Va., 1973-75; resident internal medicine USN, Portsmouth, Va., 1975-78; dir. internal medicine clinic Naval Amphibian Base, Virginia Beach, Va., 1978-79; physician internal medicine Wythe Med. Assn., Wytheville, Va., 1979-80; private practice Wytheville, Va., 1981—; bd. dirs. Spl. Care Svcs., Wytheville, Echo Vascular Lab., Advanced Cardiac Life Support. Recipient Outstanding Citizenship award M.T. Rodgers, 1982, Woodsmen of World, 1984. Mem. AMA, Am. Coll. Physicians, Am. Soc. Internal Medicine. Avocations: model railroading, stamp collecting. Home: 625 S 9th St Wytheville VA 24382-3213 Office: James J McConnell Md PC 365 W Ridge Rd Wytheville VA 24382-1008

MCCONNELL, JOHN EDWARD, electrical engineering company executive; b. Minot, N.D., July 28, 1931; s. Lloyd Waldorf and Sarah Gladys (Mathis) McC.; m. Carol Claire Myers, July 4, 1952 (dec. Feb. 1989); children: Kathleen Anne, James Mathis, Amy Lynn; m. Heidi Banziger, Sept. 29, 1990. Registered profl. engr., Pa. B.S. in Mech. Engring., U. Pitts., 1952; M.S., Drexel Inst. Tech., 1958. With mktg. and design depts. for turbomachinery Westinghouse Electric Corp., Lester, Pa., 1954-60, 63-67, Pitts., 1960-63; mgr. power generation equipment activities in U.S., ASEA Inc., White Plains, N.Y., 1967-79; regional mgr. power equipment activities Middle Atlantic and Southeastern U.S. regions, 1967-79, mgr. turbine generator dept., 1980-83, mgr. internat. ops. Power Systems div., 1983-84, mgr. transmission substas. dept., 1984-85; mgr. Eastern U.S. ops. ASEA Power Systems Inc., 1985-86, mgr. eastern ops. measurements div. GEC, 1986-91; mgr. eastern region Protection and Control div. GEC Alsthom T&D Inc., 1991—; adviser on energy matters to U.S. congressman 1968-74; speaker and author on energy and electric power topics. Served to 1st lt., C.E., U.S. Army, 1952-54. Mem. IEEE (sr.; energy com., past chmn. subcom. cogeneration, chmn. membership com., power sys. relay com.), IEEE Power Engring. Soc. (sr.; past chmn. chpts. public affairs subcom.), ASME. Republican. Contbr. numerous articles on energy and electric power to industry publs.; developer analytical techniques for power systems performance characteristics and econs. of cogeneration systems. Home: 173 Remington Rd Ridgefield CT 06877-4324 Office: GEC Alsthom T&D Inc 4 Skyline Dr Hawthorne NY 10532

MCCONNELL, PATRICIA ANN, health facility administrator; b. Bklyn., Feb. 28, 1935; d. Philip P. and Dagney C. (Petersen) Powers; m. Alexander McConnell, Jan . 15, 1955; children: Francis X., Robert M., Bonnie J., Douglas P. AAS in Nursing, Milw. Area Tech. Coll., Milw., 1978; student, U. Wis., 1980; BA, Nat. Lewis U., 1989. RN, Ill., Iowa, Wis., Inc.; registered profl. nurse; cert. case mgr.; cert. pain mgr.; cert. ins. rehab. specialist; cert. occupational hearing conservationist. Nursing asst., RN oncology dept. St. Luke's Hosp., Milw., 1976-79; supr. employee health dept. Harnisfeger P&H, Cudahy, Wis., 1979-82; staff nurse employee health dept. 1st Wis. Nat. Bank, Milw., 1982-83; med. svcs. cons. Crawford Risk Mgmt. Svcs., Schaumburg, Ill., 1983-86; case mgmt. specialist Nat. Rehab. Cons., Westmont, Ill., 1986-87; pres., dir. Mid-State Health and Rehab., Westmont, 1987—; bd. dirs. Women in Workers Compensation of Ill. Founding mem. Rape Recovery Project Hot Line, vol. support group, Chgo., 1989—, bd. dirs. 1990—; active Ill. Coalition Against Sexual Assault, 1990; vol. literacy tutor World Relief Orgn. & Literacy Vols. Am., 1990—; den mother Cub Scouts of Am., 1966-68; first aid instr. ARC, 1981—; BLS instr. Am. Heart Assn., 1981-84; religion instr. St. Helena Cath. Ch., Greendale, Wis., 1973-75. Mem. LWV, Am. Assn. Occupational Health Nurses (local treas. 1986-87), Rehab. Ins. Nurses Group. (pres. 1992—, treas. 1990-91), Assn. Rehab. Nurses, Am. Acad. Pain Mgmt. (clin. assoc. 1991—), Assn. Vocat. Rehabilitationist in Ill., Oak Brook Assn. Commerce and Industry (small bus. com. 1989-90), Dolton Regional Hosp. Aux. (charter, nominating com., publicity chair 1965), Women in Mgmt. (hospitality chair). Roman Catholic. Avocations: reading, cross country skiing, writing, walking, theater and opera. Home: 821 Oakwood Dr Westmont IL 60559-1035 Office: 504A E Ogden Ave Ste 249 Westmont IL 60559-1277

MCCORD, JAMES RICHARD, III, chemical engineer, mathematician; b. Norristown, Ga., Sept. 2, 1932; s. Zachariah Thigpen Houser Jr. and Neilie Mae (Sumner) McC.; m. Louise France Manning, Oct. 1956 (div. 1974); children: Neil Alexander, Stuart James, Valerie France, Kent Richard. Student, Abraham Baldwin Agrl. Coll., Tifton, Ga., 1949-50; BChE with honors, Ga. Inst. Tech., 1955; postgrad., U. Pitts., 1955-56, Carnegie Inst. Tech., 1956-57; MS, MIT, 1959, PhD, 1961. Asst. chem. engr. TVA, Wilson Dam, Ala., 1951-54; assoc. engr. Westinghouse Electric Corp., Pitts., 1955-57; research asst. ops. research MIT Dept. Math., Cambridge, Mass., 1957-59, teaching asst., 1959-61; research assoc. MIT Dept. Chem. Engring., Cambridge, Mass., 1957-62, asst. prof. engring., postdoctoral fellow, 1962-64; sr. engr., project analyst Esso Research and Engring. Co., Florham Park, N.J., 1964-68; asst. prof. Emory U., Atlanta, 1968-71; pvt. practice math. cons. Atlanta, 1971-80; instr. in math. Ga. So. Coll., Statesboro, 1980-81; inventory control Lovett & Tharpe, Inc., Dublin, Ga., 1981-84; farmer Norristown, 1984—. Contbr. numerous articles to sci. and math. jours. WEBELOS den leader Boy Scouts Am., Dunwoody, Ga., 1969-70; mem., vol. worker Key Meml. Found., Adrian-Norristown, Ga., 1965—. Mem. Am. Inst. Chem. Engrs., Ga. Tech. Alumni Assn., MIT Alumni Assn., Sigma Xi. Republican. Methodist. Avocations: music, fishing, gardening, mathematical puzzles. Home and Office: PO Box 61 Norristown GA 30447-0061

MCCORKLE, RICHARD ANTHONY, physicist; b. Gastonia, N.C., Aug. 6, 1940; s. Grier Randolph and Mary Katherine (Moore) McC.; m. Betty Albert Sodeman, June 9, 1964 (div. Mar. 1991); 1 child, Laura Katherine. BS, N.C. State U., Raleigh, 1962, PhD, 1970. Teaching asst. N.C. State U., 1967-68; asst. prof. East Carolina U., Greenville, N.C., 1968-73; rsch. scientist IBM Watson Rsch. Ctr., Yorktown Heights, N.Y., 1973-91; prof. U. R.I., Kingston, 1992—; faculty fellow NASA, Hampton-Langley, Va., 1970; instr. NSF Inst., East Carolina U., Greenville, N.C., 1971. Contbr. articles to profl. jours. Bd. dirs. Twin Lakes Water Works, South Salem, N.Y., 1973-76. Mem. Am. Phys. Soc. Democrat. Achievements include patents for X-Ray Generators, Ion Sources, X-Ray Optics and Fusion Device; research in unified field theory including gravity as well as electroweak forces. Home: 62 W Bay Dr Narragansett RI 02882 Office: U RI Physics Dept Kingston RI 02881

MCCORMACK, GRACE, retired microbiology educator; b. Rochester, N.Y., Feb. 16, 1908; d. Walter and Maud (Brimacomb) McC. AB, U. Rochester, 1941; MS, U. Md., 1951. Technician U. Rochester Sch. Medicine and Dentistry and Atomic Energy, 1942-48; bacteriologist Dept. Interior U.S. Fish and Wildlife Svc., Coll. Park, East Boston, 1948-53, Md. State Dept. Health, Balt., 1953-55, VA Hosp., Canandaigua, N.Y., 1955-66; asst. to assoc. to microbiology prof. Monroe Community Coll., Rochester, N.Y., 1966-77; prof. Community Coll. of the Finger Lakes, Canandaigua, 1982-86, 88. Fellow Am. Inst. of Chems., Am. Biog. Inst. (hon. mem. rsch. bd. of advisors 1987—, Outstanding Educator of Yr. 1987), Royal Soc. of Health (Eng.), Intercontinental Biog. Assn. (Eng.); mem. Am. Soc. Microbiologists, N.Y. State Pub. Health Assn., Am. Inst. of Food Technologists, Nat. Found. of Infectious Diseases. Avocations: travel, reading. Home: 162 Raleigh St Rochester NY 14620-4148

MCCORMACK, GRACE LYNETTE, engineering technician; b. Dallas, Nov. 2; d. Audley and Janice Meredith (Metcalf) McC. Tech. degree, Durham's Coll., 1958; grad. in civil engring., El Centro Coll., 1972; grad. in advanced surveying, Eastfield, 1975. Cert. sr. engr. technician. Contract design technician various engring firms, Dallas, 1958-70; sr. design engr. technician City of Dallas Survey Div., 1970-80, street light div., 1980—. Mem. NAFE, Women's Forum of Am. Mem. Unity Ch. Avocations: numerology, astrology, metaphysics, Egyptian-Arabian horses, lighting and designing black and white portrait photography. Home: 1428 Meadowbrook Ln Irving TX 75061-4435

MCCORMACK, ROBERT PAUL, environmental engineer; b. Carbondale, Ill., May 31, 1950; s. Robert Leroy and Dorothy Lucille (Jones) McC.; m. Barbara Stacine Gurley, June 5, 1982; children: Stacy, Tara, Robert B. BS in Engring., So. Ill. U., 1973. Registered profl. engr., Ill., Mo., Ky. Design engr. Watkins and Assocs., Lexington, Ky., 1973-76, J.T. Blankinship and Assocs., Murphysboro, Ill., 1976-85; environ. engr. Freeman Coal Co., West Frankfort, Ill., 1985; constrn. engr. Knowles Constrn. Co., Chester, Ill., 1986-88; constrn. mgr. Evansville (Ill.) Cement Finishers, 1989-90; owner, chief engr. Engart, Inc., Murphysboro, 1990—; pres. Midwest Environ. Engring., Inc., 1993; ind. cons. for various industries, contractors, engring. firms, 1985—. Mem. Elks, Sons Am. Legion. Achievements include design of over 100 pollution control projects, landfills, water treatment facilities, work in hazardous and oily waste remediation. Office: Engart Inc PO Box 851 Murphysboro IL 62966

MCCORMICK, BARNES WARNOCK, aerospace engineering educator; b. Waycross, Ga., July 15, 1926; s. Barnes Warnock and Edwina (Brogdon) McC.; m. Emily Joan Hess, July 18, 1946; 1 dau., Cynthia Joan. B.S. in Aero. Engring, Pa. State U., 1948, M.S., 1949, Ph.D., 1954. Research assoc. Pa. State U., University Park, 1949-54, assoc. prof., 1954-55, prof. aero. engring., 1959-92, Boeing prof. aero. engring., 1985-92, prof. emeritus, cons., 1992—, head dept. aerospace engring., 1969-85; assoc. prof., chmn. aero. dept. Wichita U., 1958-59; chief aerodynamics Vertol Helicopter Co., 1955-58; mem. Congl. Adv. Com. Aeros., 1984-86; U.S. coord. flight mechanics panel Adv. Group for Aerospace R&D, 1988—; cons. to industry. Author: Aerodynamics of V/Stol Flight, 1967, Aerodynamics, Aeronautics and Flight Mechanics, 1979; Contbr. articles to profl. jours. Served with USNR, 1944-46. Recipient joint award for achievement in aerospace edn. Am. Soc. Engring. Edn.-Am. Inst. Aeros. and Astronautics, 1976. Fellow Am. Inst. Aeros. and Astronautics (assoc.); mem. ASEE, Am. Helicopter Soc. (tech. council, hon. fellow), Sigma Xi, Sigma Gamma Tau, Tau Beta Pi. Club: Masons. Patentee in field. Home: 611 Glenn Rd State College PA 16803-3475 Office: Pa State U Coll Engring University Park PA 16802

MCCORMICK, DOUGLAS KRAMER, editor; b. Bklyn., Jan. 20, 1952; s. William Tanner and Ann Kellog (Kramer) McC.; m. Susan J. Hodesblatt, June 5, 1983; children: Rachel Joanna, Emma Elizabeth. BA, Yale U., 1979; postgrad., Stevens Inst. Tech., 1981-82. Sr. editor Design Engring. Mag., N.Y.C., 1980-82; computer sci. editor Hayden Pub. Co., Hasbrouck Heights, N.J., 1982-84; editor Bio/Tech. Mag., N.Y.C., 1984—; v.p., editorial dir. Nature Pub. Co., N.Y.C., 1990—; adj. assoc. prof. Med. Sch., U. Miami, Fla., 1988—; mem. editorial com. Am. Bus. Press, N.Y.C., 1988-91; mem. biotech. forum Keystone (Colo.) Ctr. Author: (with others) Cooperation and Competition in the Global Economy, 1988, Biotechnology: Science, Education and Commercialization, 1990, Biotechnology: The Science and the Business, 1991; acquisitions editor of more than 75 books on computer sci. Office: Nature Pub Co 65 Bleecker St New York NY 10012

MCCORMICK, J. PHILIP, natural gas company executive; b. San Antonio, Feb. 21, 1942; s. Eugene Hay and Beulah (Barber) McC.; m. Jo Ann Wendland, July 17, 1965; children; J. Philip Jr., Scott Daniel. BBA, Tex. A&I U., 1964, MS in Bus. and Econs., 1965. CPA, Tex. Sr. acct. Price Waterhouse, Houston, 1965-70; mgr. KPMG Main Hurdman, Houston, N.Y.C., 1970-73; ptnr. KPMG Main Hurdman, El Paso, Houston, 1973-83; mng. ptnr. KPMG Main Hurdman, Houston, 1983-85, mng. ptnr. So. region, 1985-87; mng. ptnr. KPMG Peat Marwick, Austin, Tex., 1987-91; sr. v.p. fin. Lone Star Gas Co., Dallas, 1991-93, also bd. dirs., sr. v.p. transmission, 1993—; pres. Enserch Gas Co., Dallas, 1993—; bd. dirs. The Dallas Opera, 1993—. Bd. dirs. Austin Symphony, 1988-92, Tex. Bus. Hall of Fame, Houston, 1990—, Tex. A&I Found., 1989— (chmn. 1992—); fin. coun. Seton Hosp., Austin, 1990-91; search com. bus. sch. dean Tex. A&I U., 1990, Tex. A&I alumni assn. (pres. 1977-78, nat. chmn. ann. alumni fund drive 1978-80), chmn. industry adv. coun. to dean, 1968-91; sec. bd. regents Univ. System South Tex., Corpus Christi, 1988-89; steering com. St. Edward's U., Austin, 1990-91; founding dir. Escape Ctr. for Prevention Child Abuse, Houston, 1980-89, others. Named Disting. Alumnus Tex. A&I U., 1991, one of Outstanding Young Men Am., 1971. Mem. Am. Gas Assn. (fin. and adminstrv. sect., acctg. adv. coun., mng. com., corp. planning com. 1991), AICPA, Tex. Soc. CPAs, Fin. Execs. Inst. (assoc.), Tower Club Dallas, Austin Country Club, Petroleum Club Houston, Houston Club, Headliners Club Austin, Delta Sigma Pi (life.). Office: Lone Star Gas Co 301 S Harwood St Dallas TX 75201-5619

MCCORMICK, JERRY ROBERT DANIEL, chemistry consultant; b. St. Albans, W.Va., Feb. 24, 1921; s. James Moody and Georgia Alma (Fowler) McC.; m. Catherine Mary Rello, Oct. 24, 1953; 1 child, Joshua F.A. BS in Chemistry, Rensselaer Poly. Inst., 1943; PhD in Organic Chemistry, UCLA, 1949. Rsch. chemist Winthrop Chem. Co., Rensselaer, N.Y., 1942-46; Rsch. chemist Lederle Labs., Am. Cyanamid Co., Pearl River, N.Y., 1949-61, rsch. assoc., 1961-65; rsch. fellow Am. Cyanamid Co., Pearl River, N.Y., 1965-83; cons. Teltech, 1988—. Contbr. articles to profl. jours. Fellow AAAS, Am. Inst. Chemists; mem. Am. Chem. Soc., N.Y. Acad. Scis. Achievements include 44 patents in fields of tetracyclines and fermentation. Home and Office: 19 Pomona Ln Spring Valley NY 10977

MCCORMICK, KATHLEEN ANN KRYM, geriatrics nurse, federal agency administrator; b. Manchester, N.H., June 27, 1947. BSN, Barry Coll., 1969; MSN, Boston U., 1971; MS, U. Wis., 1975, PhD, 1978. Commd. med. officer USPHS; COSTEP nurse USPHS, Staten Island, N.Y., 1968; staff nurse, instr. USPHS, Brighton, Mass., 1970; staff nurse Mercy Hosp., Miami, 1969, St. Elizabeth's Hosp., Brighton, 1970-71; clin. nurse specialist Boston U. Hosp., 1970-71; clin. nurse specialist, instr. U. Wis., Madison, 1971-72; asst. rsch. to chief nursing clin. ctr., dept. Nursing NIH, Bethesda, 1978-83; rsch. nurse, co-dir. inpatient geriatric continence project, Lab. Behavioral Scis., Gerontology Rsch. Ctr. Nat. Inst. Aging, Balt., 1983-88, dir. nursing rsch., 1988-91; dir. office forum quality and effectiveness health care Agy. Helath Care Policy and Rsch., Rockville, Md., 1991—; adj. asst. prof. Cath. U., Washington, 1979, 82; faculty assoc. U. Md., Balt., 1979-91; ad hoc reviewer biomed. rsch. grants NIH, 1979-80, divsn. nursing rsch. and tng. HRA, 1979-82; instr. Found. Advanced Edn. in Scis., 1981-82; exec. com. Nat. Inst. Aging Liaison Ctr. Nursing Rsch., 1986-91; Surgeon Gen.'s rep. Sec. Alzheimer's Task Force, 1989—; Surgeon Gen. alternate to Bd. Regents Nat. Libr. Medicine, 1989—; co-chair panel guidelines for urinary incontinence in the adult, 1990—; speaker numerous confs. Editor Nursing Outlook, 1988-90; mem. editorial staff Military Medicine, 1985—; assoc. editor International Jour. Technology and Aging, 1985—. Recipient award Jour. Acad. Sci., 1965, travel award NSF, 1973, J.D. Lane Jr. Investigator award USPHS Profl. Assn., 1979, Excellence in Writing award Nat. League Nursing/Humana, 1983, award Spl. Recognition Rsch., U. Pa. Sch. Nursing, 1986, Federal Svc. Nursing award, 1987, Surgeon Gen.'s medallion, 1989,; grantee U. Wis. Grad. Sch., 1977, Upjohn Co., 1977; Queen's Vis. scholar Royal Adelaide (Austrailia) Hosp., 1990. Fellow Am. Acad. Nursing, Royal Coll. Nursing, Coll. Am. Med. Informatics, Gerontology Soc. Am. (clin. med. sect., computer program coord. 1985-87, clin. medicine rep. publs. com., Nurse of Yr. 1992), Nat. Acad. Scis. Inst. Medicine; mem. ANA (sec., editor newsletter coun. nurse rschrs., 1980-85, vice chairperson exec. com. coun. computer applications in nursing, 1984-86), Am. Lung Assn./Am. Thoracic Soc. (cert. appreciation mid.-Md. chpt. 1982, 83, 84, chairperson 1983-84, nominating com. 1989, nat. rsch. review com.), Am. Assn. Critical Care Nursing (strategic planning com. 1987-89, Disting. Rsch. award 1986), Am. News Women's Club, Internat. Medical Informatics Assn. (working group 8, program com. 1984—), Assn. Mil. Surgeons (sustaining mem. award 1982), Commd. Officers Assn., Met. Area Nursing Rsch. Consortium, Md. Lung Assn. (awards and grants subcom. 1978,92, v.p. exec. bd. bd. dirs. 1980-87), Captial Speakers Club, Lambda Sigma, Sigma Delta Epsilon (Eloise Gerry grant-in-aid fellow 1979-80), Sigma Theta Tau (grantee 1976). Office: US Agency for Health Care Quality & Effectiveness in Health Care 2101 E Jefferson St Rockville MD 20852

MCCORMICK, NORMAN JOSEPH, mechanical engineer, nuclear engineer, educator; b. Hays, Kans., Dec. 9, 1938; s. Clyde Truman and Vera Mae (Miller) McC.; m. Mildred Mirring, Aug. 20, 1961; children: Kenneth John, Nancy Lynn. BSME, U. Ill., 1960, MS in Nuclear Engring., 1961; PhD in Nuclear Engring., U. Mich., 1965. Postdoctoral researcher NSF, Ljubljana, Yugoslavia, 1965-66; asst. prof. Univ. Wash., Seattle, 1966-70; assoc. prof. Univ. Wash., 1970-75, prof., 1975—; scientist Sci. Applications, Inc., Palo Alto, Calif., 1974-75; mem. editorial bd. Transport Theory and Statis. Physics, 1982—. Author: Reliability and Risk Analysis, 1981; editor Jour. Progress in Nuclear Energy, 1980-85; contbr. more than 100 articles to profl. jours. Named Outstanding Alumnus in Nuclear Engring., Univ. Mich., Ann Arbor, 1986, Disting. Alumnus in Mech. Engring., Univ. Ill., Urbana, 1991. Fellow Am. Nuclear Soc.; mem. Biomed. Optics Soc., Am. Nuclear Soc., Optical Soc. Am. Achievements include patent in field, Method of Preparing Gas Tags for Identification of Single and Multiple Failures of Nuclear Reactor Fuel Assemblies. Office: U Washington Mechanical Engring, FU-10 Seattle WA 98195

MCCORMICK, ROBERT JEFFERSON, chemical engineer; b. Kettering, Ohio, May 18, 1955; s. Robert E. and Beatrice (Frazee) McC.; m. Lydia S. Sloan, Sept. 21, 1985. BSChemE, Washington U., St. Louis, 1977. Chem. engr. Monsanta Rsch. Corp., Dayton, Ohio, 1977-81; project engr. Acurex Corp., Mountain View, Calif., 1981-83; mgr. ops. Acurex Waste Techs., Cin., 1983-84; dir. mktg. Pedco Tech. Corp., Cin., 1984-85; project engr. ENSCO, Franklin, Tenn., 1985-87; exec. v.p. DRE Techs. Inc., Franklin, Tenn., 1987—; guest lectr. Vanderbilt U., Nashville, 1989—; dir. Federated Techs., Nashville, 1989—, DRE, Inc., 1987—. Author: Hazardous Waste Incineration Engineering, 1981, Costs for Hazardous Waste Incineration, 1985. Mem. Air & Waste Mgmt. Assn. Office: DRE Techs Inc PO Box 987 Brentwood TN 37024-0987

MCCORMICK, THOMAS JAY, infosystems engineer; b. Pitts., Nov. 23, 1946; s. Thomas Jay and Marion (Smith) McC.; m. Patricia Michelle McCormick, Dec. 1, 1990; 1 child, Randall James. BA, Dickinson Coll., 1968; MS, Troy State U., 1976; MA, Boston U., 1985. Tech. officer U.S. Army Units, various cities, 1968-78; army research & devel. coord. BETA Joint Program Office, Washington, 1978-80; student officer Armed Forces Staff Coll., Norfolk, Va., 1981; branch exec. officer Defense Intelligence Agy., Washington, 1981-83; ground forces branch chief Defense Liaison Detachment, Bonn, Germany, 1983-86; army operational test officer Army Operational Test and Evaluation Agy., Washington, 1986-87; chief, intelligence systems branch Army Operational Test and Evaluation AGy., Washington, 1987-88; intelligence systems specialist GTE Govt. Systems Corp., Chantilly, Va., 1988-92; spl. programs bus. devel. mgr., 1992—. co-author: Notes and Cases in Military Management, 1976. Council pres. St. Thomas More Parish, Bonn, Germany, 1985-86. Recipient Disting. Svc. award Crofron Civic Assn., 1983. Mem. Internat. Test and Evaluation Assn., Armed Forces Communications Electronics Assn., Nat. Mil. Intelligence Assn., Kiwanis (sec. 1982-83), Theta Chi. Republican. Methodist. Avocations: skiing, golf. Home: 13532 Gray Bill Ct Clifton VA 22024 Office: GTE Govt Systems Corp 15000 Conference Center Dr Chantilly VA 22021-3800

MCCOTTER, MICHAEL WAYNE, civil engineer; b. Hackensack, N.J., Oct. 26, 1967; s. Wayne Eugene and Kathleen Ann (Chirichella) McC. B Engring., Stevens Inst. Tech., 1989. Cert. engr.-in-tng., N.Y. Jr. engr. N.Y. State Dept. Transp., N.Y.C., 1989-90, civil engr. I, 1990-91, civil engr., project mgr., 1991—. Mem. ASCE. Roman Catholic. Office: NY State Dept Transp 47-40 21st St 9th Fl Long Island City NY 11101

MCCOWN, SHAUN MICHAEL PATRICK, chemist, consultant; b. Canton, Ohio, Sept. 10, 1949; s. Richard Eugene and Jessie Mae (Jarrard) McC.; m. Sandra Kaye Adkins, Oct. 19, 1972 (div. 1988); children: Alexandra Silveria, Adrian Christopher Patrick, Julia Roxanne; m. Laurie Gail Rutherford, Oct. 28, 1989. BS in Chemistry summa cum laude, U. Carabobo, Valencia, Venezuela, 1968, BS in Physics magna cum laude, 1969; MS in Analytical Chemistry, Northeast La. U., 1978; postgrad., S. Tex. Coll. Law, 1990—. Registered environ. assessor, Calif. Asst. chemist Vinings (Ga.) Chem. Co., 1969-73; group leader, chemist Woodson-Tennent Labs., Memphis, 1973-76; sr. chemist Water and Air Rsch., Inc., Gainesville, Fla., 1978-80, Beckman Instrumenta/Altex Div., Berkeley, Calif., 1980-85; sr. product specialist Perkin-Elmer Corp., Houston, 1985-87; propr. Analytical Chromatography Support and LegalChem. Co., Houston, 1987—; assoc. referee chemist Assn. Ofcl. Analytical Chemists, Arlington, Va., 1978-83, gen. referee, 1987-90; rschr., writer ABA Ctrl. and East European Law Initiative, 1992-93; mediator, coord. Victim Offender Restitution Program Mediation Clinic, South Tex. Coll. Law. Author: Quantitative Chromatography, 1990, Handbook of Regulated Chemicals, 1992; contbr. articles to tech. jours.; composer Introit for Advent (A Voices), Mass in A minor for Guitar and 4 Voices; arranger Suite in E flat for Mil. Band, Sea Songs March. Baritone Houston Oratorio Soc., 1987-90; soloist, St. Mark's Episc. Ch., Houston, 1989—; counter tenor 1st Presbyn. Ch., 1987-89, St. Andrews Presbyn. Ch., 1990-92; prin. voice Houston Gilbert and Sullivan Soc., 1991—. Recipient hon. mention Roscoe Pound Environ. Law Writing Competition, 1991. Fellow Am. Inst. Chemists; mem. ABA, State Bar Tex., Royal Soc. Chemistry (London) (mem. and chartered chemist), Am. Chem. Soc., Phi Delta Phi. Office: Legal Chem Co 11411 Cliffgate Dr Houston TX 77072-4215

MCCOY, JOHN GREENE, computer applications consultant; b. Little Rock, Aug. 5, 1946; s. Travis Walton and Evelyn Lois (Greene) McC. Grad. in Computer Programmimg, Ark. Coll. Tech., 1981. Commodity broker Keale, Inc., Little Rock, 1971-79; pvt. cons., programmer Little Rock, 1985—; vol. cons. Ark. Arts. Ctr., Little Rock, 1984—; presenter in field. Cpl. USMC, 1967-69. Home: 9201 Kanis Rd Apt 11F Little Rock AR 72205-6444 Office: John McCoy Co PO Box 7713 Little Rock AR 72205

MCCOY, R. WESLEY, biology educator; b. Augusta, Ga., Sept. 20, 1954; s. Roger and Frances (Amick) McC.; m. Deborah Stringer, June 16, 1984. BS in Biology, Ga. State U., 1975; MEd in Sci. Edn., U. Ga., 1977. Tchr. North Cobb High Sch., Kennesaw, Ga., 1978-83; edn. specialist NASA, Kennedy Space Ctr. (Fla.), 1983-87; tchr. Ga. Govs. Honors Program, Dahlonega, 1981-82, sci. dept. chmn. North Cobb High Sch., 1987—; adj. asst. prof. Okla. State U., Stillwater, 1983-87. Christa McAuliffe fellow, 1992, SCI-MAT fellow, NSF, 1992, Ga. Sci. Tchr. of Yr., GTE G.I.F.T. fellow, 1993. Mem. NSTA, Ga. Sci. Tchrs. Assn., Fulbright Assn., Phi Delta Kappa. Lutheran. Achievements include NSF DNA literacy program, Cold Spring Harbor Lab.; NOAA Nat. Undersea Research Program Marine Biology Workshop; Delegation leader People toPeople Youth Science Exchange to Soviet Union, Fulbright tchr. exch. to U.K. Home: 350 Ridgeland Terrace Marietta GA 30062 Office: North Cobb High Sch 3400 Old Hwy 41 Kennesaw GA 30144

MCCRACKEN, ALEXANDER WALKER, pathologist; b. Motherwell, Lanarkshire, Scotland, Nov. 24, 1931; came to U.S., 1968; s. William and Mary Snedden (Walker) McC.; m. Theresa Credgington, June 4, 1960; children: Fiona Jane, Claire Louise. MD, U. Glasgow, Scotland, 1956. Resident in surgery Glasgow Royal Inf., 1956-57; resident in pathology Royal Air Force, U.K., 1957-61, pathologist, 1962-68; fellow in pathology Royal Postgrad. Med. Sch., London, 1961-62; assoc. prof. Med. Sch., U. Tex., San Antonio, 1968-72; prof. Med. Sch., U. Tex., Houston, 1972-73; dir. of microbiology Baylor U. Med. Ctr., Dallas, 1973-81; dir. of labs. Meth. Hosps. of Dallas, 1982—; adj. prof. pathology Baylor U. Coll. of Dentistry, Dallas, 1982—; clin. prof. Southwestern Med. Sch., U. Tex., Dallas, 1986—. Author: (textbook) Pathologic Mechanisms of Human Disease, 1985, Oral & Clinical Microbiology, 1986, Pathology, 1990; author (play) Mister Gilbert, 1985, 89. With Royal Air Force, 1957-68. Decorated G.S.M. Fellow Royal Coll. Pathologists, Coll. Am. Pathologists; mem. AMA, Am. Soc. Microbiology, Tex. Med. Assn., Dallas County Med. Soc., Tex. Soc. Infectious Diseases, Tex. Med. Found., Chapparal club Dallas, Masons. Republican. Anglican. Avocations: theater, music, gardening. Home: 607 Kessler Lake Dr Dallas TX 75208-3943 Office: Lab Physicians Assn 221 W Colorado Blvd Dallas TX 75208-2359

MCCRACKEN, GEORGE H., microbiologist. With SW Med. Ctr., U. Texas, Dallas. Recipient Hoechst Roussel award Am. Soc. Microbiology, Washington, 1991. Office: U Texas TT SW Med Ctr Pediatrics F3202 5323 Harry Hines Blvd Dallas TX 75235*

MCCRAY, RICHARD ALAN, astrophysicist, educator; b. Los Angeles, Nov. 24, 1937; s. Alan Archer and Ruth Elizabeth (Woodworth) McC.; m. Sandra Broomfield; children—Julia, Carla. B.S., Stanford U., 1959; Ph.D., UCLA, 1967. Research fellow Calif. Inst. Tech, Pasadena, 1967-68; asst. prof. astronomy Harvard U., Cambridge, Mass., 1968-71; assoc. prof. astrophysics U. Colo., Boulder, 1971-75, prof., 1975—, chmn. Joint Inst. Lab. Astrophysics, 1981-82, chmn. Ctr. for Astrophysics and Space Astromony, 1985-86. Contbr. articles to profl. jours. Guggenheim fellow, 1975-76. Mem. NAS, Am. Astron. Soc. (councilor 1980-83, chmn. high energy astrophysics div. 1986-87, Heineman Prize for Astrophysics, 1990), Internat. Astron. Union. Office: U Colo Joint Inst Lab Astrophysics Boulder CO 80309-0440

MCCREA, RUSSELL JAMES, mechanical engineer, consultant; b. Chgo., Apr. 11, 1917; s. Arthur James and Marguerite (Murray) McC.; divorced. BSME, Stanford U., 1947. Profl. engr., Calif. Sr. aerodynamicist Convair div. Gen. Dynamics, San Diego, 1940-49; exec. dir. aerospace facilities Office of the Asst. Sec. Def., Washington, 1949-54; asst. econ. commr. U.S. Embassy, Rome, 1954-60; dir. internat. mktg. Hughes Aircraft Corp., Culver City, Calif., 1960-75; dir. engring., contracts, fin. Holmes & Narver Inc., Orange, Calif., 1975-77; pres. McCrea Engring. Assocs., San Diego, 1977—; mgr. constrn. quality control Am. Engring. div. Profl. Svc. Industries, Corona, Calif., 1984-93. Commr. archtl. rev. comm. City of Palm Desert, Calif., 1985—. Maj. USMC, ret. Mem. AIAA (assoc. fellow, sect. chmn.), Mensa. Republican. Achievements include invention of auto turn signal and automatic headlight beam change for approaching autos. Home: PO Box 60469 San Diego CA 92166 Office: McCrea Engring Assocs 89 Santa Anita Rancho Mirage CA 92270

MCCREADIE, ALLAN ROBERT, chemical engineer, consultant; b. Dunedin, Otago, New Zealand, Sept. 3, 1951; s. Morris Sinclair and Marie Ethel McCreadie; m. Linda Mavis Grubb, Mar. 10, 1973; children: Ross Sinclair, Merrin Joy. BE in Chem. and Materials Engring., Auckland (New Zealand) U., 1973. Trainee cons. Stanzol Chems. Ltd., Auckland, 1973-76; indsl. contracts engr. Fisher & Paykel Engrs. Ltd., Auckland, 1975-79; sr. refrigeration engr. Ralph Engle Systems Ltd., Auckland, 1980-82; governing dir. Armadillo Engring. Ltd., Manurewa, New Zealand, 1983—. Mem. AIAA, Exptl. Aircraft Assn. (chpt. pres. 1990—), Refrigeration and Air Conditioning Cos. Assn. (contract arbitrator 1989—), New Zealand Amateur Aircraft Constructors Assn. (pres. Auckland chpt.). Achievements include development of low noise ducted fan systems for hovercraft and airboats, etc., ammonia evaporators and evaporative condensers plus associated production methods and machinery, continuous fluidized bed driers, resin reactor kettles, project management of refrigeration/chemical plants. Office: Armadillo Engring Ltd, PO Box 97, Manurewa New Zealand

MCCREARY, TERRY WADE, chemist, educator; b. Roaring Spring, Pa., Oct. 13, 1955; s. Harry Eugene and Charlotta Joan (Dodson) McC.; m. Geniece Ann Parker, Aug. 9, 1981; children: Corinne, Yvette. BS, St. Francis Coll., Loretto, Pa., 1977; PhD, Va. Poly. Inst. and State U., 1988. Instr. Cumberland Coll., Williamsburg, Ky., 1979-85; project asst. Va. Poly. Inst. and State U., Blacksburg, 1985-88; asst. prof. chemistry Murray (Ky.) State U., 1988—. Author: Laboratory Techniques and Experiments in Chemistry, 1991; contbr. articles to profl. publs. Mem. area recycling com. ARC, Murray, 1992. Grantee NSF, 1992—. Mem. Am. Chem. Soc., Soc. Applied Spectroscopy, Ky. Acad. Sci. Office: Murray State U Dept Chemistry Murray KY 42071

MCCREATH, PETER S., Canadian government official, civil engineer; b. Halifax, Nova Scotia, Can., July 5, 1944; m. Judy Duncan, July, 5, 1982. MCSCE. Former tchr., journalist, parliamentary sec. to minister of state fin., minister vets. affairs, 1993—. Recipient T.C. Keffer medal Canadian Soc. Civil Engring., 1990. Home: 38 Mont St, Guelph, ON Canada N1H 2A4 Office: Vets Affairs Can, PO Box 7700 161 Grafton St, Charlottetown, PE Canada C1A 8M9

MCCRONE, WALTER COX, research institute executive; b. Wilmington, Del., June 9, 1916; s. Walter Cox and Bessie Lillian (Cook) McC.; m. Lucy Morris Beman, July 13, 1957. B Chemistry, Cornell U., 1938, PhD, 1942. Microscopist, sr. scientist Armour Research Found., Ill. Inst. Tech., Chgo., 1944-56; chmn. bd. Walter C. McCrone Assos. Inc., Chgo., 1956-81; sr. research advisor Walter C. McCrone Assos. Inc., 1956-78; pres. McCrone Research Inst., Inc., 1961—; vis. prof. chem. microscopy Cornell U., 1984—; adj. prof. IIT, 1950—, NYU, 1974—, U. Ill., 1989—. Author: Fusion Methods in Chemical Microscopy, 1957, The Particle Atlas, 1967, Particle Atlas Two, 6 vols., 1973-78, Particle Atlas Three Electronic CD-Rom edit., 1992, Polarized Light Microscopy, 1978, Asbestos Particle Atlas, 1980, Asbestos Identification, 1987; editor Microscope Jour., 1962—; contbr. over 300 tech. articles, chpts to books; rschr. on authentification of art and archeology (found Vinland map to be 20th century hoax, 1974 and Holy Shroud of Turin to be a 14th century painting, 1980). Chmn. bd. trustees Ada S. McKinley Community Svcs., 1962—; pres. bd. dirs. VanderCook Coll. Music; trustee Campbell Ctr., 1990—. Recipient Benedetti-Pichler award Am. Microchem. Soc., 1970, Anachem award Assn. Analytical Chemists, 1981, cert. of merit Franklin Inst., 1982, Forensic Sci. Found., 1983, Madden Disting. Svc. award VanderCook Coll. Music, 1988, Fortissimo award, 1991, Irving Selikoff award Nat. Asbestos Coun., 1990, Founder's Day award Calif. Assn. Criminalists, 1990, Roger Green award, 1991, Pub. Affairs award Chgo. Pub. Schs., 1993. Mem. Am. Chem. Soc. (Pub. Affairs award 1993), Am. Phys. Soc., Am. Acad. Forensic Sci. (Disting. Svc. award criminalistics sect. 1984), Am. Inst. Conservators Art (hon.), Internat. Inst. Conservators Art, Australian Micros. Soc., N.Y. Micros. Soc. (Ernst Abbe award 1977), La. Micros. Soc., Can. Micros. Soc., Midwest Micros. Soc., Royal Micros Soc. (hon.), Ill. Micros. Soc., Quekett Micros. Club, Sigma Xi, Phi Kappa Phi, Phi Lambda Upsilon, Alpha Chi Sigma. Achievements include 3 patents in field of explosives technology; proved the Vinland map to be a modern forgery; proved the Turin Shroud to be a medieval painting; developed the analytical methods for detection and identification of asbestos. Home: 501 E 32d St Chicago IL 60616 Office: 2820 S Michigan Ave Chicago IL 60616-3292

MCCRORY, ROBERT LEE, physicist, mechanical engineering educator; b. Lawton, Okla., Apr. 30, 1946; s. Robert Lee Sr. and Marjorie Marie (Garrett) McC.; m. Betsey Christine Wahl, June 14, 1969; children: Katherine Anne, John Damon, George Garrett. BSc, MIT, 1968, PhD, 1973. Physicist Los Alamos Nat. Lab., Albuquerque, 1973-76; scientist, coleader Lab. for Laser Energetics, U. Rochester, N.Y., 1976-77, sr. scientist, 1977—, dir. theoretical dir., 1979-90; dir. Lab. for Laser Energetics, U. Rochester, 1983—; assoc. prof. of physics and astronomy U. Rochester, 1980, prof. of mech. engring., 1983—. Author: Laser Plasma Interactions, 1989, Computer Applications in Plasma Science and Engineering, 1991; contbr. articles to profl. publs. Alfred P. Simon scholar, 1964-67; AEC fellow, 1985. Fellow Am. Phys. Soc. (mem. fellowship com., mem. exec. com. div. plasma physics, Excellence in Plasma Physics award). Office: U Rochester Lab Laser Energetics 250 E River Rd Rochester NY 14623-1299

MCCULLAGH, PETER, statistician, mathematician; b. Plumbridge, No. Ireland, Jan. 8, 1952; s. John A. and Margaret B. (Devlin) McC.; m. Rosa Bogues, Dec. 26, 1977; children: Emma, Martin, Nuala, Ellen. BS, Birmingham U., 1974; PhD, Imperial Coll., 1977. Vis. asst. prof. U. Chgo., 1977-79; lectr. Imperial Coll., London, 1979-85; vis. assoc. prof. U. B.C., Vancouver, Can., 1981-82; mem. tech. staff AT&T, Murray Hill, N.J., 1984-85; prof. U. Chgo., 1985—; assoc. editor Biometrika, London, 1984-92, Ann. Inst. Statis. Math., Tokyo, 1988-92. Author: Generalized Linear Models, 1983, 2d edit. 1989, Tensor Methods in Statistics, 1987. Recipient Guy medal Royal Statistical Soc., 1983, Copss medal N.Am. Statistical Socs., 1990. Fellow AAAS, Inst. Math. Stats. (bd. dirs.); mem. Internat. Stats. Inst. Office: U Chgo Dept Statistics Dept Statistics 5734 University Ave Chicago IL 60637

MC CULLOCH, ERNEST ARMSTRONG, physician, educator; b. Toronto, Ont., Can., Apr. 27, 1926; s. Albert E. and Letitia (Riddell) McC.; m. Ona Mary Morganty, 1953; children: James A., Michael E., Robert E., Cecelia E., Paul A. M.D. with honors, U. Toronto, 1948. Intern Toronto Gen. Hosp., 1949-50, sr. intern, 1951-52; NRC fellow dept. pathology U. Toronto, 1950-51; asst. resident Sunnybrook Hosp., Toronto, 1952-53; pvt. practice specializing in internal medicine Toronto, 1954-67; clin. tchr. dept. medicine U. Toronto, 1954-60, asst. prof. dept. med. biophysics, 1959-64, assoc. prof., 1964-66, prof., 1966, asst. prof. dept. medicine, 1967-68, assoc. prof., 1968-70, prof., 1970—, univ. prof., 1982-91, univ. prof. emeritus, 1991—; mem. grad. faculty U. Toronto (Inst. Med. Sci.), 1968—; dir. Inst. Med. Sci. U. Toronto, 1975-79, asst. dean Sch. Grad. Studies, 1979-82; physician Toronto Gen. Hosp., 1960-67; sr. scientist, sr. physician Ont. Cancer Inst., 1957-91, head div. biol. rsch., 1982-89, head div. cell and molecular biology, 1989-91, sr. scientist emeritus, 1991—; vis. prof. U. Tex. Med. Ctr. Anderson Cancer Ctr., Houston, 1992-93; cons. Nat. Cancer Plan, 1972—; mem. standing com. on health rsch. and devel. Ont. Coun. Health, 1974-82. Author numerous articles on research in hematology; editorial bd.: Blood, 1969-80, Biomedicine, 1973, Clin. Immunology and Immunopathology, 1972-76; assoc. editor: Jour. Cellular Physiology, 1966-68; editor, 1968-91. Trustee Banting Research Found., 1975—, hon. sec. treas., 1958-74, v.p., 1977-79. Decorated officer Order of Can., 1988; recipient William Goldie prize U. Toronto, 1964, Ann. Gairdner award Internat. Gairdner Found., 1969, Starr Medallist award Dept. Anatomy U. Toronto, 1957; Nat. Cancer Inst. Can. fellow, 1954-57. Fellow Royal Soc. Can. (pres. Acad. Sci. 1987-90, Thomas W. Eadie Medal 1991), Royal Coll. Physicians and Surgeons Can.; mem. Can. Acad. Sci., Am. Soc. Exptl. Pathology, Am. Assn. Cancer Rsch., Can. Soc. Cell Biology, Can. Soc. Clin. Investigation, Am., Internat. socs. hematology, Internat. Soc. Exptl. Hematology, Inst. Acad. Medicine (charter mem.). Clubs: Badminton, Racquet. Home: 480 Summerhill Ave, Toronto, ON Canada M4W 2E4 Office: Ont Cancer Inst, 500 Sherborne St, Toronto, ON Canada M4X 1K9

MCCULLOUGH, BENJAMIN FRANKLIN, transportation researcher, educator; b. Austin, Mar. 25, 1934; s. Benjamin Franklin and Mabel Comelia (Kitteridge) McC.; m. Norma Jean Walsh, Sept. 1, 1956; children: Michael Wayne, Bryan Scott, Steven Todd, Franklin Norman, Melanie Jean. MSCE, U. Tex., 1962; PhD of Civil Engring., U. Calif., Berkeley, 1969. Registered profl. engr., Tex. Testing engr. Covair Aircraft Co., Ft. Worth, 1957; design and rsch. engr. Tex. Hwy. Dept., Austin, 1957-66; rsch. engr. Materials R&D, Inc., Oakland, Calif., 1966-68; from asst. to prof. U. Tex., Austin, 1969—, dir. transp. rsch., 1980—. Contbr. articles to profl. jours. Mem. ASCE (Outstanding Paper award 1987), Transp. Rsch. Bd., Coun. Univ. Transp. Ctrs., Univ. Transp. Ctrs. Program, Am. Concrete Inst. Mem. LDS Ch. Avocations: coaching, sports, golfing, U.S. and Tex. history. Office: U Tex Transp Rsch Ctr 3208 Red River Ste 200 Austin TX 78705

MCCULLOUGH, GARY WILLIAM, psychology educator, researcher; b. Rockford, Ill., July 19, 1951; s. William Kennedy and Gloria Maria (Damon) McC.; m. Debra Lynn Rusk, Apr. 14, 1992. BA in Psychology, So. Calif. Coll., 1978; PhD in Psychology, U. Kans., Lawrence, 1991. Rsch. asst. Gerontology Ctr. Kans. U., Lawrence, 1989-91; asst. prof. U. Tex.-Permian Basin, Odessa, 1991—. Contbr. chpt. to book. Tchrs. Andrews (Tex.) Adult Literacy Program, 1992. Sgt. USMC, 1970-74. Mem. Am. Phys. Soc. Democrat. Home: 609 NW 14th St Andrews TX 79714 Office: U Tex Permian Basin Bitnet Garymac 4901 E University Odessa TX 79762

MCCULLOUGH, HENRY G(LENN) L(UTHER), nuclear engineer; b. Waukegan, Ill., Aug. 5, 1939; s. Fredrick Douglas and Octavia Idelphi (Anderson) McC.; m. Suman 'E D'uan Naro, Nov. 6, 1962; children: Barbra, Jeanine, Michelle, Charles, Edmunds, Caroline, Richard, Larry, Robert, Seng, Lynda. BA in Physics, Grinnell Coll., 1961; postgrad. in Nuclear Engring., U. Ariz., 1962; postgrad., No. Ill. U., 1972. Registered profl. engr., Wis., Ill.; cert. quality and reliability engr.; lic. wildlife rehab., Ill. Nuclear safeguards project engr. Sargent & Lundy Engrs., Chgo., 1976-81, project mgmt. engr., 1981-84, cons. engr., 1984—; Lectr. various orgns. Coauthor: Nuclear Glossary, 1986. Bd. dirs. Coun. on Energy Independence, 1979-82; mem. Immaculate Conception Parish Sch. Bd., Waukegan, Ill., 1980-81; mem. Holy Names Soc. Immaculate Conception Ch., Waukegan, 1976—; commr. Northeast Ill. Coun. Boy Scouts Am., 1979-83; min. of the Eucharist St. Joseph Ch., St. Bartholomew Ch., Waukegan, 1992—. Recipient Black Achiever award Chgo. Met. YMCA, 1981; grantee NSF, 1972. Mem. Am. Nuclear Soc. (chmn. stds. working group 1980-86, activist), Am. Soc. Quality Control (sr. mem. project 28 stds. subcom. svc. quality 1986—, task group to internat. stds. orgn. 1989—, apptd. to U.S. tech. adv. group Internat. Orgn. on Standardization TC/176 on Quality Assurance & Quality Mgmt.), Soc. Am. Mil. Engrs. Republican. Roman Catholic. Office: Sargent & Lundy Engrs 55 E Monroe St Chicago IL 60603-5702

MCCULLOUGH, JOHN JAMES, III, civil engineer; b. L.A., Nov. 30, 1959; s. John James Jr. and Marjorie Gail (Flanagan) McC. BS, U. Calif., Davis, 1982. Registered profl. engr., Calif. Resident engr. FAA, L.A., 1982-83; assoc. engr. Oakbrook Assocs., Merced, Calif., 1983-85; project engr. Bedesen Cardoza Andrews, Inc., Merced, 1985—. Mem. citizens adv. com. City of Merced Gen. Plan Undate, 1992. Mem. ASCE, Water Environ. Fedn., Am. Pub. Works Assn., Nat. Ground Water Assn., Am. Water Works Assn., Calif. Aggie Alumni Assn. Home: 2663 9th Ave Merced CA 95340 Office: Bedesen Cardoza Andrews Inc Ste A 777 W 22nd St Merced CA 95340

MC CULLOUGH, JOHN PHILLIP, management consultant, educator; b. Lincoln, Ill., Feb. 2, 1945; s. Phillip and Lucile Ethel (Ornellas) McC.; B.S., Ill. State U., 1967, M.S., 1968; Ph.D., U. N.D., 1971; m. Barbara Elaine Carley, Nov. 29, 1968; children—Carley Jo, Ryan Phillip. Adminstrv. mgr. McCullough Ins. Agy., Atlanta, Ill., 1963-68; ops. supr. Stetson China Co., Lincoln, 1967; asst. mgr. Brandtville Service, Inc., Bloomington, Ill., 1968; instr. in bus. Ill. Cen. Coll., 1968-69; research asst. U. N.D., Grand Forks, 1969-71; assoc. prof. mgmt. West Liberty State Coll., 1971-74, prof., 1974—, chmn. dept. mgmt., 1974-82, dir. Sch. Bus., 1982-86, dean Sch. Bus., 1986—, dir. Small Bus. Inst., 1978—; mgmt. cons., Triadelphia, W.Va., 1971—; instr. Am. Inst. Banking, 1971—; lectr. W.Va. U., 1971—; adj. prof. MBA program Wheeling Coll., 1972—, U. Steubenville, 1982—; lectr. Ohio U., 1982—; profl. asso. Inst. Mgmt. and Human Behavior, 1975—; v.p. West Liberty State Coll. Fed. Credit Union, 1976—; rep. W.Va. Bd. Regents Adv. Council of Faculty. Team leader Wheeling div. Am. Cancer Soc.; coordinator Upper Ohio Valley United Fund, 1972-74; instr. AFL-CIO Community Services Program, Wheeling; project dir. Ctr. for Edn. and Research with Industry; bd. dirs. Ohio Valley Indsl. and Bus. Devel. Corp., Inc., Labor Mgmt. Inst. Wheeling Salvation Army, Progress, Inc., Recipient Service award Bank Adminstrn. Inst., 1974, United Fund, 1973; Acad. Achievement award Harris-Casals Found., 1971. Mem. Soc. Humanistic Mgmt. (nat. chmn.), Orgn. Planning Mgmt. Assn. (exec. com.), Spl. Interest Group for Cert. Bus. Educators (nat. dir.), Soc. Advancement Mgmt. (chpt. adv.), Acad. Mgmt., Adminstrv. Mgmt. Soc. (cert.), Am. Soc. Personnel Adminstrn. (cert.), Nat. Bus. Honor Soc. (Excellence in Teaching award 1976, dir. 1974—), Alpha Kappa Psi (Dist. Service award 1973, Civic award 1977, chpt. adv. 1971—), Merit found. W.Va. Ednl. Excellence award, Delta Mu Delta, Delta Pi Epsilon, Delta Tau Kappa, Phi Gamma Nu, Phi Theta Pi, Pi Gamma Mu, Pi Omega Pi, Omicron Delta Epsilon. Author: (with Howard Fryette) Primer in Supervisory Management, 1973; contbr. articles to profl. jours. Home: 68 Elm Dr Triadelphia WV 26059-9620

MCCULLOUGH, KATHRYN T. BAKER, utilities executive; b. Trenton, Tenn., Jan. 5, 1925; d. John Andrew and Alma Lou (Wharey) Taylor; m. John R. Baker, Sept. 30, 1972 (dec. Oct. 1981); m. T.C. McCullough, May 14, 1988. Student, U. Chgo., 1950, Vanderbilt U., 1950-51; BS, U. Tenn., 1945, MSW, 1954. Lic. social worker, Tenn.; diplomate in clin. social work Am. Bd. Examiners. Home demonstration agt. agrl. extension svc. U. Tenn., Hardeman County, 1946-49; Dyer County, 1949-50; dir. med. social work dept. Le Bonheur Children's Hosp., Memphis, 1954-57; chief clin. social worker clinic mentally retarded children U. Tenn. Dept. Pediatrics, Memphis, 1957-59; clin. social worker Children's Med. Ctr., Tulsa, 1959-60; dir. med. social work dept. Coll. of Medicine U. Tenn., Memphis, 1960-69; dir. community svcs. regional med. program Coll. of Medicine, 1969-76; dir. regional clinic program Child Devel. Ctr. Coll. of Medicine, 1976-85; mem. faculty Coll. of Medicine, Coll. of Social Work U. Tenn., Memphis, 1960-85; admissions rev. bd. Arlington Devel. Ctr., Memphis, 1976—. Author 14 books. Active Gibson County Fedn. Dem. Women, 1987—; commr. Dist. V Gibson Utility Dist., 1990—. Fellow Am. Assn. Mental Retardation; mem. NASW, Acad. Cert. Social Workers, Tenn. Conf. on Social Welfare, Trenton Music Club (pres. 1992—). Mem. Ch. of Christ. Avocations: piano, organ. Home: 627 Riverside Yorkville Rd Trenton TN 38382-9513

MCCULLOUGH, ROBERT DALE, II, osteopath; b. Tulsa, June 2, 1937; s. Robert Dale and Roberta Maud (Purdy) McC.; m. Lindell Arlene Wilcox, Sept. 28, 1963; children: Robert Mark, Lori Lindell. Student, Wheaton (Ill.) Coll., 1955-57; BS, N.E. Mo. State U., 1958; DO, Kansas City (Mo.) Coll. Osteopathy, 1958-62. Cert. Am. Osteo. Bd. Internal Medicine, Internal

Medicine and Med. Oncology. Gen. practice McCullough Clinic, Tulsa, 1963-68; internal medicine resident Detroit Osteo. Hosp., 1968-71; internal medicine Baker-Todd-McCullough-Sutton, Tulsa, 1971-74; fellow med. oncology M.D. Anderson Hosp., Houston, 1974-75; internal medicine-med. oncology Baker-Todd-McCullough-Sutton, Tulsa, 1975-90; pvt. practice Tulsa, 1990—; Trustee Tulsa Regional Med. Ctr., 1983-88, 1990—, Okla. Blue Cross Blue Shield, Tulsa, 1983-92 (vice chmn. 1991-92); adv. coun. Okla. State U. Coll. of Osteo. Medicine, 1988— (chmn. 1988-90). Mem. bd. of editors Patient Care Magazine, Montvale, N.J., 1988-93. Mem. Okla. State Bd. Health, Oklahoma City, 1983-87, Tulsa City/County Bd. Health, 1988—, chmn., 1993. Mem. Am. Osteo. Assn. (vice speaker Ho. of Dels. 1986-92, trustee 1993—), Am. Coll. Osteo. Internists, Am. Soc. Clin. Oncology, Okla. Osteo. Assn. (pres. 1982-83), Tulsa Downtown Lions Club, Soc. for Preservation and Encouragement of Barbershop Quartet Singing in Am. Republican. Southern Baptist. Avocation: barbershop quartet singing. Home: 3951 S Delaware Tulsa OK 74105-3721 Office: 720 W 7th St Tulsa OK 74127-8902

MCCULLOUGH, ROY LYNN, chemical engineering educator; b. Hilsboro, Tex., Mar. 20, 1934; s. Roy Lee and Rubye Maye (Ingram) McC.; m. Jamis Carol Petersen, Sept. 5, 1958; children: Catherine Lynne, Amanda Kaye, Roy Lawrence. BS, Baylor U., 1955; PhD, U. N.Mex., 1960. Mem. staff Los Alamos (N.Mex.) Sci. Lab., 1955-60; group leader, sect. head Monsanto Co., Durham, N.C., 1960-69; sr. scientist Boeing Co., Seattle, 1969-71; prof. chem. engring. U. Del., Newark, 1971—, dir.Ctr. for Composite Material, 1990—. Author: Concepts of Fiber-Resin Composites, 1971; mem. editorial bd. Jour. Composite Sci. and Tech., 1984—; contbr. articles to profl. jours. Mem. Am. Inst. Chem. Engrs., Am. Chem. Soc., Am. Phys. Soc., Am. Soc. Composites. Home: 107 Reynard Dr Landenberg PA 19350 Office: U Del Ctr for Composite Material Newark DE 19716

MC CUNE, WILLIAM JAMES, JR., manufacturing company executive; b. Glens Falls, N.Y., June 2, 1915; s. William James and Brunnhilde (Decker) McC.; m. Janet Waters, Apr. 19, 1940; 1 dau., Constance (Mrs. Leslie Sheppard); m. Elisabeth Johnson, Aug. 8, 1946; children: William Joseph, Heather H.D. SB, MIT, 1937. With Polaroid Corp., Cambridge, Mass., 1939-91, v.p. engring., 1954-69, v.p., asst. gen. mgr., 1963-69, exec. v.p., 1969-75, dir., pres., 1975-83, CEO, 1975-86, chmn. bd., 1982-91, ret., 1991—. Chmn. bd., trustee Mitre Corp.; Trustee Boston Mus. Sci., Mass. Gen. Hosp. Fellow Am. Acad. Arts and Scis.; mem. Nat. Acad. Engring. Office: Polaroid Corp 549 Technology Sq Cambridge MA 02139-3589*

MCCUSKER, CHARLES FREDERICK, psychologist, consultant. BS in Psychology magna cum laude, U. Utah, 1977, MS in Ednl. Psychology, 1980, PhD in Ednl. Psychology, 1988. Lic. psychologist Utah, Ariz.; cert. sch. psychologist, Utah. Mental health specialist psychiatric assessment unit VA Hosp., Salt Lake City, 1975-77; psychiatric specialist dept. child psychiatry Primary Children's Med. Ctr., Salt Lake City, 1978-80; sch. psychologist Granite Sch. Dist., Salt Lake City, 1979-83; guidance specialist Jordan Sch. Dist., Sandy, Utah, 1984-85; psychology svcs. Rivendell of Utah, West Jordan, 1986-88; psychology cons. Utah Dept. Health, Salt Lake City, 1988-90; pvt. practice Salt Lake City, 1990—; cons. Tooele County (Utah) Sch. Dist., 1989-90, Benchmnark Reg. Hosp., Woods Cross, Utah, 1990-91; bd. advisors Med. Equipment Cons., Salt Lake City, 1992—, Progressive Awareness Rsch., Henderson, Nev., 1991—. Presenter in field; contbr. articles to profl. jours. Mem. Am. Psychol. Assn., Nat. Assn. Sch. Psychologists, Utah Psychol. Assn., Utah Assn. Sch. Psychologists. Address: 4700 S 900 E Ste 30-221 Salt Lake City UT 84117

MCCUTCHAN, MARCUS GENE, water utility executive; b. Grand River, Iowa, June 3, 1930; s. Emmett Dalton and Leona May (McIntosh) McC.; m. Sarah Aleene Rankin, Nov. 20, 1949; children: Gregori Shane, Kimberlea Shawn. Div. mgr. Ariz. Water Co., Sedona, 1962—; bd. dirs. Am. Water Works Assn. Ariz. sect. 1992-95; pres. Ariz. Water and Pollution Control Assn., 1982, 83; chmn. Am. Water Works Assn. Ariz. sect., 1982-83, Water Pollution Control Fedn., 1982, 83. Pres. Sedona Rotary, 1968, 69, 91, 92, Sedona C. of C., 1969, 70. Recipient Fuller award Am. WAter Works Assn., 1985; named Operator of Yr. Ariz. Water and Pollution Control Assn. 1981.

MCCUTCHEON, STEVEN CLIFTON, environmental engineer; b. Decatur, Ala., Oct. 29, 1952; s. Bernard Clifton and Rosa May (Askenburg) McC.; m. Sherry Lynn Sharp; children: Michael Ian, Alexander Tavis. BS, Auburn U., 1975; MS, Vanderbilt U., 1977, PhD, 1979. Registered profl. engr., La. Hydrologist U.S. Geol. Survey, Bay St. Louis, Miss., 1977-86; environ. engr. U.S. EPA, Athens, Ga., 1986—; adj. prof. Tulane U., New Orleans, 1984-85; panel mem. Nat. Rsch. Coun., Washington, 1989-92; adj. prof. Inst. Ecology U. Ga., Athens, 1989—; adj. asst.prof. Clemson (S.C.) U., 1990-94. Author: Water Quality Modeling, vol. 1, 1989; (with others) Fate and Transport of Sediment-Associated Contaminants, 1989, Water Quality, Handbook of Hydrology, 1993; editor and author: Manual for Performing Estuarine Waste Load Allocations, 1990; editor Jour. Environ. Engring., 1992-94. Mem. Zoning Commn., St. Tammany Parish, La., 1984-85; vice=chmn. Planning Adv. Bd., St. Tammany Parish, 1985; asst. den leader pack 83 Cub Scouts Am., Athens, 1991-92. Recipient medal and plaque Korea Soc. Water Pollution Rsch. and Control, Seoul, 1986; Engr. of Yr. award in EPA, NSPE, 1992; Whirlpool Corp. fellow, 1976-77. Mem. ASCE (br. pres. 1983-84), Young Civil Engr. of Yr. award 1984), Am. Geophysical Union, Internat. Soc. Environ. Ethics, Phi Kappa Phi, Phi Theta Kappa, Sigma Xi. Achievements include devel. of waste load allocation guidance and response to Exxon Valdez oil spill. Home: 147 Spalding Ct Athens GA 30605 Office: US EPA Envrion Rsch Lab 960 College Station Rd Athens GA 30605-2720

MCDADE, JOSEPH JOHN, microbiologist; b. Taylor, Pa., Nov. 1, 1930; s. Patrick Joseph and Elizabeth (Masters) McD.; m. Sonja Ann DeMaree; children: Joan Elizabeth, Joseph William. BS, U. Scranton, 1952; MS, Syracuse U., 1954; PhD, U. Ky., 1957. Grad. asst. dept. microbiology U. Ky., Lexington, 1954-57; assoc. dept. microbiology and pub. health Chgo. Med. Sch., 1957-59; assoc. dir. microbiology Luth. Gen. Hosp., Park Ridge, Ill., 1959-61; project mgr. for water issues Chems. and Metals, Midland, Mich., 1984—; rsch. microbiologist USDHEW, PHS, DCD, Savannah, Ga., 1961-64; sr. scientist Calif. Inst. Tech., Jet Propulsion Lab., Pasadena, 1964-67; sect. mgr. environ. bio-engring. The Dow Chem. Co., Indpls., 1967-69; dept. head environ. bio-engring. The Dow Chem. Co., Midland, Mich., 1969-74; mgr., site environ. mgr., tech. svc. The Dow Chem. Co., various cities, 1974-81; project mgr. for water issues, chems. and metals The Dow Chem. Co., Midland, 1984—. Editor: NASA Planetary Quarantine Manual; contbr. over 70 publs. to profl. jours. Recipient NASA Planetary Quarantine award. Mem. AAAS, ASTM, Am. Assn. for Contamination Control, Am. Pub. Health Assn., Am. Soc. for Microbiology, Am. water Works Assn., Chem. Mfts. Assn., Chlorine Inst., Inc., Chem. Specialties Mfrs. Assn., EPA Task Force, Internat. Life Scis. Inst., Soc. for Gen. Microbiology, Sigma Xi. Republican. Roman Catholic. Home: 3300 Thornbrook Ct Midland MI 48640 Office: Dow Chem Co 2020 Dow Ctr Midland MI 48674

MCDANIEL, WILLIAM HOWARD TAFT, JR., computer information systems educator; b. Ballinger, Tex., Sept. 28, 1941; s. William Howard Taft and Alta Mae (Broadstreet) McD.; m. Betty Jane Johnson, Aug. 16, 1958; children: William H.T. III, Barry Glenn, Bryan Keith. BS summa cum laude, Miss. State U., 1968, MBA, 1968. Commd. 2d lt. U.S. Army, 1962, advanced through grades to maj., 1981, computer specialist, 1962-72; systems mgr. U.S. Army, Alexandria, Va., 1973-83; ret. U.S. Army, 1983; planning dir. Electronic Data Systems, Balt., 1983; assoc. prof. computer infosystems No. Va. C.C., Alexandria, 1983-88, head computer infosystems, 1988—; instr. Angelo State U., San Angelo, Tex., 1972-73; computer cons. various orgns., 1983—; adj. faculty George Mason U., Annandale, Va., 1990—. Del. Tex. Rep. Conv., San Angelo, 1973; election judge, Prince Georges County, Md., 1984-92; v.p. Skyline Citizens Assn., Prince Georges County, 1985. Decorated Legion of Merit. Mem. SAR, Data Processing Mgmt. Assn. (pres. 1978-79), Mensa, Scottish Rite, Phi Kappa Phi. Republican. Methodist. Achievements include development of computer information system curriculum materials. Home: 6100 Skyline Terr Suitland MD 20746 Office: No Va CC 3001 N Beauregard St Alexandria VA 20023

MCDANIELS, JOHN LOUIS, mathematics educator; b. Alton, Ill., Oct. 3, 1933; s. John Clarence and Carrie Elizabeth (Kortkamp) McD.; m. Betty Lou Verble, June 20, 1964. BS, U. Mo., Rolla, 1960; MS, So. Ill. U., 1977. Registered profl. engr., Ill., Mo. Engr. McDonnell Douglas Corp., St. Louis, 1960-74; prof. Lewis and Clark Community Coll., Godfrey, Ill., 1975—; dist. TEAMS competition coord. Ill. Jr. Engring. Tech. Soc., Lewis and Clark C.C., 1987—, pre-engring. coord., 1975—, water tech. coord., 1975-92. Bd. dirs. Alton (Ill.) Mus. History and Art, 1984-86. With U.S. Army, 1954-56. Mem. Am. Phys. Soc., Am. Math. Assn. Two-Yr. Colls., Am. Soc. Engring. Edn., Ill. Math. Assn. Community Colls., Kiwanis (Alton-Godfrey pres. 1989-90, Disting. Pres. award 1990), Sigma Pi Sigma, Tau Beta Pi, Kappa Delta Pi. Presbyterian. Home: 3208 Greenwood Ln Godfrey IL 62035 Office: Lewis and Clark C C 5800 Godfrey Rd Godfrey IL 62035

MCDERMOTT, JOHN FRANCIS, IV, civil engineer, consultant; b. St. Louis, Mar. 10, 1926; s. John Francis III and Mary Stephanie (Kendrick) McD.; m. Virginia Marie Vogt, Aug. 14, 1948; children: John Francis V, Anne Marie, Martin Kendrick. BSCE, Carnegie Inst. Tech., 1947, MSCE, 1953; PhD in Civil Engring., U. Pitts., 1972. Registered profl. engr., Pa., Mich. Design engr., project engr. Richardson, Gordon and Assocs., Cons. Engrs., Pitts., 1947-57; chief structural engr. Mich. Assocs., Cons. Engrs., Lansing, 1957-60; with U.S. Steel Rsch. Lab., Monroeville, Pa., 1960, now sr. rsch. cons. Contbr. chpts. to textbooks, articles to tech. publs. Mem., vice-chair, chair Monroeville Zoning Hearing Bd., 1970-74. Cpl. U.S. Army, 1944-46. Co-recipient Elbert K. Gary award Am. Iron and Steel Inst., 1984, Kelly award Assn. Iron and Steel Engrs., 1984. Fellow ASCE (life mem., vice-chair structural plastics rsch. coun. 1980-90, subcom. chair reinforced concrete rsch. coun. 1970—), Am. Concrete Inst. (com. chair 1968-74, 86-92). Republican. Achievements include 10 U.S. patents, applications of mathematical modeling to plant processes and to plant equipment and structures, initiating and monitoring significant reinforced concrete research. Home: 1307 Knollwood Dr Monroeville PA 15146 Office: USX Corp US Steel Tech Ctr 4000 Tech Center Dr Monroeville PA 15146

MCDERMOTT, KEVIN J., engineering educator, consultant; b. Teaneck, N.J., Nov. 21, 1935; s. Francis X. and Elizabeth (Casey) McD.; m. Ann McDermott, Aug. 3, 1959; children: Kathleen, Kevin, Donna, Michael. BSEE, N.J. Inst. Tech., 1965; MS Indsl. Engring., Columbia U., 1970; EdD, Fairleigh Dickinson U., 1975. Registered profl. engr., N.J. With Bell Telephone Labs., Murray Hill, N.J., 1960-65, Westinghouse Elecric, Newark, 1965-67, Columbia U., NASA, N.Y.C., 1967-70, RCA Corp., N.Y.C., 1970-76, Ramapo (N.J.) Coll., 1976-80; prof. N.J. Inst. Tech., Newark, 1980—; chmn. engring. dept. N.J. Inst. Tech., 1983—; bd. dirs. Computer Aided Design/Computer Aided Manufacture Robotics Consortium. Contbr. over 30 tech. papers to profl. jours. IBM fellow, 1987. Fellow IEEE, Soc. Mech. Engrs.; mem. Inst. Indsl. Engrs. Achievements include research in industrial robot work cells, manufacturing systems, expert systems, analysis of industrial of Robotics, Flexible Manufacturing Systems, expert and vision systems in computer aided design and manufacturing.

MCDEVITT, HUGH O'NEILL, immunology educator, physician; b. Cin., 1930. M.D., Harvard U., 1955. Diplomate: Am. Bd. Internal Medicine. Intern Peter Bent Brigham Hosp., Boston, 1955-56, sr. asst. resident in medicine, 1961-62; asst. resident Bell Hosp., 1956-57; research fellow dept. bacteriology and immunology Harvard U., 1959-61; USPHS spl. fellow Nat. Inst. Med. Research, Mill Hill, London, 1962-64; physician Stanford U. Hosp., Calif., 1966—; assoc. prof. Stanford U. Sch. Medicine, Calif., 1969-72, prof. med. immunology, 1972—, prof. med. microbiology, 1980—, Burt and Marian Avery Prof. Immunology, 1990—; cons. physician VA Hosp., Palo Alto, Calif., 1968—. Served as capt. M.C., AUS, 1957-59. Mem. NAS, AAAS, Am. Fedn. Clin. Research, Am. Soc. Clin. Investigation, Am. Assn. Immunologists, Transplantation Soc., Inst. Medicine. Office: Stanford U Dept Microbiology and Immunology D-345 Fairchild Bldg Stanford CA 94305

MCDIVITT, JAMES ALTON, defense and aerospace executive, astronaut; b. Chgo., June 10, 1929; s. James A. and Margaret M. (Maxwell) McD.; m. Patricia A. Haas, June 16, 1956 (div. 1981); m. 2d, Judith Odell, June 22, 1985; children: Michael A., Ann L., Patrick W., Kathleen M., Joseph Bagby, Jeffrey Bagby. BS in Aero. Engring., U. Mich., 1959, D Astronautical Sci. (hon.), 1965; DSc (hon.), Seton Hall U., 1969, Miami U., 1970; LLD (hon.), Ea. Mich. U., 1975. Commd. USAF, 1952, grad. exptl. test pilot sch., 1960, grad. aerospace rsch. pilot sch., 1961, advanced through grades to brig. gen., 1972, assigned as astronaut to NASA, 1962, program mgr. Apollo Spacecraft, NASA, 1969-72, ret., 1972; exec. v.p. Consumers Power Co., Jackson, Mich., 1972-75; exec. v.p., dir. Pullman Inc., Chgo., 1975-81; sr. v.p., gov. and internat. ops. Rockwell Internat., Arlington, Va., 1981, sr. v.p. strategic mgmt., sr. v.p. sci. and tech., exec. v.p. def. electronics, exec. v.p. electronics ops., 1981—. Mem. adv. coun. U. Mich. Coll. Engring., 1988—, U. Notre Dame Coll. Engring., 1975-88. Fellow Soc. Exptl. Test Pilots (Kinchloe award 1969). Avocations: hunting, fishing, outdoor activities. Office: Rockwell Internat Corp 1745 Jefferson Davis Hwy Arlington VA 22202-3402

MCDONALD, BERNARD ROBERT, federal agency administrator; b. Kansas City, Kans., Nov. 17, 1940; s. Bernard Luther and Mabel McD.; m. Jean Graves, June 7, 1963; children: Aaron Michael, Elizabeth Kathleen. BA, Park Coll., Parkville, Mo., 1962; MA, Kans. State U., 1964; PhD, Mich. State U., 1968. Prof. math. U. Okla., Norman, 1968-83, chmn. math. dept., 1981-83; program dir. div. math. scis. NSF, Washington, 1983-86, program dir. spl. projects, 1986-88, dep. dir. div. math. scis., head office of spl. projects, 1988—. Author: R-linear Endomorphism, 1983, Geometric Algebra, 1976, Finite Rings, 1974, Ring Theory III, 1980. Mem. Am. Math. Soc., Math. Assn. Am., Soc. Ind. and Applied Math., Assn. Women Math., Sigma Xi. Home: 4001 N 9th St # 721 Arlington VA 22203 Office: NSF Div Math Scis Rm 339 Washington DC 20550

MCDONALD, JAMES HAROLD, chemical engineer, energy specialist; b. Stockbridge, Mich., Mar. 27, 1951; s. Harold Wayne and Bernice Marie (Leach) McD.; m. Dottie Sue Joseph (div. 1992). BS in Chem. Engring., Mich. Tech. U., 1973; MBA, Ea. Mich. U., 1983. Registered profl. engr., Mich. Product engr. Walker Mfg. Co., Grass Lake, Mich., 1974-81; facilities engr. Dart Container Co., Mason, Mich., 1984—. Mem. NSPE, Assn. Energy Engrs., Indsl. Energy Users Group. Home: 42 Fir Dr Mason MI 48854 Office: Dart Container Co 500 Hogsback Rd Mason MI 48854

MCDONALD, JOHN FRANCIS PATRICK, electrical engineering educator; b. Narberth, Pa., Jan. 14, 1942; s. Frank Patrick and Lulu Ann (Hegedus) McD.; m. Karen Marie Knapp, May 26, 1979. B.S.E.E., MIT, 1963; M.S. in Engring., Yale U., 1965, Ph.D., 1969. Instr. Yale U., New Haven, 1968-69, asst. prof., 1969-74; assoc. prof. Rensselaer Poly. Inst., Troy, N.Y.,1974-86, prof., 1986—; founder Rensselaer Ctr. for Integrated Electronics, 1980—. Contbr. articles to 160 profl. publs. Patentee in field. Recipient numerous grants, 1974—. Mem. ACM, IEEE, Optical Soc., Acoustical Soc., Vacuum Soc., Materials Rsch. Soc. Office: Rensselaer Poly Inst Ctr for Integrated Electronics Troy NY 12181

MCDONALD, JULIAN LEROY, JR., manufacturing director; b. Birmingham, Ala., Oct. 9, 1941; s. Julian LeRoy Sr. and Mary Louise (Duncan) McD.; m. Nina Frances Sciara, Feb. 19, 1965; children: Tracie McDonald Houston, Michael A. BSCE, U. Ala., Tuscaloosa, 1965. Supr. wire mill U.S. Steel Corp., Fairfield, Ala., 1965-73, mgr. wire products, 1973-74, staff asst., gen. supt., 1974-76, supt. cold reduction, 1976-79; plant mgr. Sivaco Steel Co.-Can., Quincy, Fla., 1979-86, Insteel Industries, Winston-Salem, N.C., 1986-89; dir. mfg. Insteel Constrn. System, Brunswick, Ga., 1989—; engring. design con. C.A.M.M., Birmingham, 1973-76; design, constrn. cons. Habitat For Humanity, Miami, Fla., 1991. Pres. C. of C., Fairfield, Ala., 1974; v.p. Rotary, Quincy, 1985. Mem. ASTM (com.). Methodist. Achievements include research in technical aspects of concrete truss panel construction. Office: Insteel Constrn Systems 2610 Sidney Lanier Dr Brunswick GA 31525

MCDONALD, MARK DOUGLAS, electrical engineer; b. Princeton, N.J., Aug. 3, 1958; s. James Douglas and Jacquelyn (Milligan) McD.; m. Patricia Joann Watson, Sept. 12, 1980. BSE, Duke U.; MS, N.C. State U. Product engr. Exide Electronics, Raleigh, N.C., 1981-84; rsch. asst. N.C. State U., Raleigh, 1985-86; mem. tech. staff Avantek (Hewlett Packard), Newark, Calif., 1987-90; prin. engr. Nat. Semiconductor, Santa Clara, Calif., 1990-92, engring. project mgr., 1992—; panel session IEEE Solid-State Circuits Conf., San Francisco, 1993; session chmn. Wireless Symposium, Santa Clara, 1993, RF & Microwave Applications Conf., Santa Clara, 1992. Contbr. articles to profl. jours. Precinct capt. various polit. campaigns, Fremont, Calif., 1988. Mem. IEEE. Achievements include U.S. and foreign patents in area of high-speed analog circuits; designed front-end integrated circuits in first digital European cordless telecomm. transceiver (DECT) for voice comm.; design of first selective frequency trip circuit for parallel uninterruptible power supplies. Office: Nat Semiconductor 2900 Semiconductor Dr Santa Clara CA 95052-8090

MCDONALD, MICHAEL SHAWN, nuclear engineer, consultant; b. Palmer, Mass., Mar. 25, 1957; s. Neil Michael and Helen Olive (Plante) McD. BSCE, Worcester Poly. Inst., 1979. Staff engr. Combustion Engring, Windsor, Conn., 1981-85; lead engr. Combustion Engring, Windsor, 1985-87, engring. supr., 1987-90; engring. mgr. ABB-Combustion Engring., Windsor, 1990—. Mem. ASME (mem. working group on erosion and corrosion, mem. working group on responsibilities and program requirements, mem. subgroup on repairs, replacement and modifications), ASCE, Am. Welding Soc., Am. Concrete Inst. (mem. subcom. on radioactive waste storage). Democrat. Roman Catholic. Achievements include patents for permanently installed refueling pool seal and neutrol shielding. Office: ABB-Combustion Engring 1000 Prospect Hill Rd Windsor CT 06095-1564

MCDONALD, THOMAS EDWIN, JR., electrical engineer; b. Wapanucka, Okla., June 19, 1939; s. Thomas Edwin and Rosamond Bell (Enoch) McD.; m. Myrna Kay Booth, Sept. 10, 1961; children: Stephen Thomas, Jennifer Kay, Sarah Lynn. BSEE, U. Okla., 1962, MSEE, 1963; PhDEE, U. Colo., 1969. Asst. prof. elec. engring. U. Okla., Norman, 1969-70; planning engr. Okla. Gas and Electric Co., Oklahoma City, 1970-72; staff mem. Los Alamos (N.Mex.) Nat. Lab., 1972—, group leader, 1974-80, program mgr., 1980—; program mgr. Centurion program Los Alamos (N.Mex.) Nat. Lab., Los Alamos, 1986-90; dep. program dir. inertial confinement fusion program Los Alamos (N.Mex.) Nat. Lab., 1990-92, program coord. mine detection and laser tech., 1992—; adj. prof. elec. engring. U. Okla., 1970-72; cons. Los Alamos Tech. Assocs., 1980—, mgr. design sect., 1980-81. Researcher: Inertial Confinement Fusion; Contbr. articles to profl. jours. Bd. dirs., mem. United Ch. Los Alamos, 1987—, chmn. bd. elders, 1992. Served to capt. U.S. Army, 1963-67. Mem. IEEE (chmn. Los Alamos sect.), AAAS, Los Alamos Gymnastics Club (treas., bd. dirs. 1980-88), Rotary (sec. Los Alamos club, v.p.), Sigma Xi, Etta Kappa Nu. Republican. Avocation: computer science. Home: 4200 Ridgeway Dr Los Alamos NM 87544-1956 Office: Los Alamos Nat Lab PO Box 1663 Los Alamos NM 87544-0010

MCDONNELL, ARCHIE JOSEPH, environmental engineer; b. N.Y.C., June 3, 1936; s. Patrick and Margaret (O'Reilly) McD.; m. Nancy Carol Schaeffer, June 18, 1966; children: Patrick, Sean. BS in Civil Engring., Manhattan Coll., 1958; MS in Civil Engring., Pa. State U., 1960, PhD in Civil Engring., 1963. Prof. Pa. State U., University Park, 1963-93; asst. dir. Water Resources Rsch. Ctr., Pa. State U., 1969-82; dir. Inst. for Rsch. on Land and Water Resources, Pa. State U., 1982-86, Environ. Resources Rsch. Inst., Pa. State U., 1986—; bd. dirs. Pa. Environ. Coun., 1989—, Nat. Assn. State Univs. & Land Grant Colls., 1990-92, chmn. water resources com., 1985-91; mem. rsch. & modeling subcom. EPA Chesapeake Bay Program, 1984-86, sci. & tech. adv. com., 1984—, exec. com., 1988-92; U.S. rep. Internat. Joint Commn., 1976-79, 87-89; mem. Pa. State Conservation Com., 1988-89, water resources policy adv. com. Pa. Dept. Environ. Resources, 1979-82, air & water quality tech. adv. com., 1983—, chmn. water quality subcom., 1986-88; chmn. Northeast Assn. Water Inst. Dirs., 1973-74; mem. exec. com. Nat. Assn. Water Inst. Dirs., 1975-78. Contbr. articles to profl. jours. Fellow U.S. Pub. Health Svc., 1961-62; recipient Commendation cert. Internat. Joint Commn., Conservationist of Yr. award Chesapeake Bay Found., Washington, 1986, Outstanding Rsch. award Pa. State U. Engring. Soc., 1988, Outstanding Profl. Rsch. award Water Pollution Control Assn. Pa., 1990, Karl M. Mason medal Pa. Assn. Environ. Profls., 1991, Gabriel Narutowicz medal Ministry Environ. Protection and Natural Resources, Poland, 1991. Mem. ASCE (chmn. 1972-73, exec. com. 1976-80, J. James R. Croes Rsch. medal 1976, Outstanding Svc. award 1981), Water Environ. Fedn. (co-chmn. 1991—), Fed. Water Pollution Control Fedn., Internat. Assn. Water Pollution Rsch., Am. Soc. Limnology and Oceanography, Chi Epsilon, Sigma Xi, Phi Kappa Phi. Achievements include demonstration of low cost treatment method for renovation of acidmine waters. Office: Pa State U 103 Land & Water Rsch Bldg University Park PA 16802

MCDONNELL, JOHN PATRICK, military officer; b. Mount View, Calif., June 16, 1965; s. James Eugene and Arlene Marie (Dolan) McD. BS in Astronautical Engring., USAF Acad., 1987. Commd. USAF, 1987, advanced through grades to capt.; electronic warfare officer 524th Bombardment Squadron USAF, Wurtsmith AFB, Mich., 1988-91, tactics instr. 379th Ops. Support Squadron, 1991, standardization, evaluation officer 379th Bomb Wing, 1992; instr. electronic warfare officer 325th Bomb Squadron USAF, Fairchild AFB, Wash., 1992—; electronic combat coord. 379th Bomb Wing., Wurtsmith AFB, 1991-92. Editor: (mag.) Talon, 1984-86. Mem. Am. Inst. Aeronautics and Astronautics, Assn. Old Crows, U.S. Parachute Assn. Republican. Roman Catholic. Office: 325th Bomb Squadron Fairchild AFB WA 99011

MCDONOUGH, JOHN GLENNON, project engineer; b. Newark, June 11, 1960; s. Edward Patrick and Susan Mary (Toal) McD.; m. Joanne DiPietro, May 12, 1984; children: Laura, Susan. B of Engring., Stevens Inst. Tech., 1982. Registered profl. engr., Ohio. Mfg. mgmt. program GE, Cin., 1982-84, mfg. systems analyst, 1984-87, sr. systems analyst, 1987-89, reliability engr., 1989-90, staff engr., 1990—. Recipient scholarship Potlatch Found. for Higher Edn., 1978, Outstanding Engring. Achievement award GE, 1990, Managerial award GE, 1989. Mem. NSPE, KC.

MCDOW, RUSSELL EDWARD, JR., surgeon; b. Waynesboro, Va., Mar. 14, 1950; s. Russell Edward and Estelle (Vereen) McD.; m. Anne Stewart, June 24, 1972 (div. 1982); children: Mary Stewart, Sinclair Russell; m. Linda Stack, May 27, 1984; 1 child, Erin Laney. BS, Duke U., 1972; MD, U. Va., 1976. Diplomate Am. Bd. Surgery. Chmn. dept. surgery Loudoun Hosp. Ctr., Leesburg, Va., 1985-89, chief staff, 1989—. Fellow ACS. Home: RR 1 Box 89 Waterford VA 22190-9703 Office: Loudoun Gen Surgery Ctr 211 Gibson St NW # 207 Leesburg VA 22075-2115

MCDOWELL, JENNIFER, sociologist, composer, playwright, publisher; b. Albuquerque, May 19, 1936; d. Willard A. and Margaret Frances (Garrison) McD.; m. Milton Loventhal, July 2, 1973. BA, U. Calif., 1957; MA, San Diego State U., 1958; postgrad., Sorbonne, Paris, 1959; MLS, U. Calif., 1963; PhD, U. Oreg., 1973. Tchr. English Abraham Lincoln High Sch., San Jose, Calif., 1960-61; free-lance editor Soviet field, Berkeley, Calif., 1961-63; rsch. asst. sociology U. Oreg., Eugene, 1964-66; editor, pub. Merlin Papers, San Jose, 1969—, Merlin Press, San Jose, 1973—; rsch. cons. sociology San Jose, 1973—; music pub. Lipstick and Toy Balloons Pub. Co., San Jose, 1978—; composer Paramount Pictures, 1982-88; tchr. writing workshops; poetry readings, 1969-73; co-producer radio show lit. and culture Sta. KALX, Berkeley, 1971-72. Author: (with Milton Loventhal) Black Politics: A Study and Annotated Bibliography of the Mississippi Freedom Democratic Party, 1971 (featured at Smithsonian Instn. spl. event 1992), Contemporary Women Poets: An Anthology of California Poets, 1977, Ronnie Goose Rhymes for Grown-ups, 1984; co-author: (plays off-off Broadway) Betsy and Phyllis, 1986, Mack the Knife Your Friendly Dentist, 1986, The Estrogen Party to End War, 1986, The Oatmeal Party Comes to Order, 1986, (play Burgess Theatre) Betsy Meets the Wacky Iraqi, 1991; contbr. poems, plays, essays, articles, short stories, book revs. to lit. mags., news mags. and anthologies; researcher women's autobiog. writings, contemporary writing in poetry, Soviet studies, civil rights movement and George Orwell, 1962—; writer: (songs) Money Makes a Woman Free, 1976, 3 songs featured in Parade of Am. Music; co-creator: (mus. comedy) Russia's Secret Plot to Take Back Alaska, 1988. Recipient 8 awards Am. Song Festival, 1976-79, Bill Casey award in Letters, 1980; AAUW doctoral fellow, 1971-73; grantee Calif. Arts Council, 1976-77. Mem. Am. Sociol. Assn., Soc. Sci. Study of Religion, Poetry Orgn. for Women, Dramatists Guild, Phi Beta

Kappa, Sigma Alpha Iota, Beta Phi Mu, Kappa Kappa Gamma. Democrat. Office: care Merlin Press PO Box 5602 San Jose CA 95150-5602

MCEACHERN, WILLIAM ARCHIBALD, economics educator; b. Portsmouth, N.H., Jan. 4, 1945; s. Archibald Duncan and Ann Teresa (Regan) McE.; m. Patricia Leonardo, Aug. 18, 1973. AB in Econs., Holy Cross Coll., 1967; MA in Econs., U. Va., 1969, PhD in Econs., 1975. Asst. prof. U. Conn., Storrs, 1973-78, assoc. prof., 1978-84, prof. econs., 1984—, dir. grad. studies, 1981-87; econ. cons. U.S. Dept. Labor, 1977-79, FTC, 1979-82, Conn. Conf. on Municipalities, New Haven, 1975-76, 87-88; dir. Bipartisan Commn. on Conn. Finances, Hartford, 1982-83. Author: Managerial Control and Performance, 1975, Economics: A Contemporary Introduction, 3d edit., 1994; editor Quarterly Rev. on Conn. Economy; founding editor The Teaching Economist; contbr. articles to profl. jours. 1st lt. U.S. Army, 1969-71. Nat. Def. fellow U. Va., 1967-69, 72-73. Mem. Nat. Tax Assn., Am. Econ. Assn., Northeast Bus. and Econs. Assn. (founder, assoc. editor 1978-81), Pub. Choice Soc., So. Econ. Assn., Western Econ. Assn. Office: U Conn Dept Econs U 63 341 Mansfield Rd Storrs Mansfield CT 06269-1063

MCEACHRON, DONALD LYNN, biology educator, researcher; b. Erie, Pa., Nov. 8, 1953; s. Karl Boyer and Marjorie (Blalock) McE.; m. Barbara Anne O'Donnell, Aug. 14, 1987. BA with highest honors, U. Calif., 1977, PhD, 1984. Lab. technician psychiatry VA Med. Ctr., La Jolla, Calif., 1978-82; rsch. technician cell biology U. Tex. Health Sci. Ctr., Dallas, 1983-84; sci. dir. Imaging and Computer Vision Ctr. Drexel U., Phila., 1984-88, vis. asst. prof. dept. biosci., 1986-88, rsch. asst. prof. Biomed. Engring. and Sci. Inst., 1989-92, rsch. assoc. prof., 1992—; vis. asst. prof. Dept. Psychiatry U. Pa., Dept. Psychology Haverford (Pa.) Coll., 1987; lectr. U. Pa., Phila., 1986—; adj. asst. prof. Thomas Jefferson U., Phila., 1989—; cons. Hoffman-La Roche, Nutley, N.J., 1984-86; mem. adv. bd. BioAutomation, Inc., Bridgeport, Pa., 1988—. Editor: Functional Mapping in Biology and Medicine, 1986; contbg. editor Diversity in Biomed. Imaging, 1989, Progress in Imaging in the Neurosciences using Microcomputers and Workstations, 1990; mem. editorial bd. Computerized Med. Imaging and Graphics, 1988—, NeuroImage, 1990. Mem. Fishtown Civic Assn., Phila., 1989. 1st lt. USAR, 1993, capt., 1993—. Regent's fellow U. Calif., 1979. Mem. AAAS, Internat. Soc. Chronobiology, Soc. for Study of Evolution, N.Y. Aca. Sci., Animal Behavior Soc. Republican. Avocations: Am. history, photography. Office: Drexel U Rm 128 Imaging and Computer Vision Ctr Philadelphia PA 19104

MCELGUNN, JAMES DOUGLAS, agriculturist, researcher; b. Vancouver, B.C., Can., Mar. 22, 1939; s. Douglas McElgunn and Margret (Gogan) Lastuka; m. Doris Roseann Johnson; children: Kim, Greg, Colleen. BS, Mont. State U., 1962, MS, 1964; PhD, Mich. State U., 1967. Rsch. scientist Rsch. Sta. Agr. Can., Swift Current, Sask., 1967-80; dir. Rsch. Sta. Agr. Can., Kamloops, B.C., 1980-85, Beaverlodge, Alta., 1985—. Chmn. Saskatoon Mountain Econ. Devel. Authority, Beaverlodge, 1987-89, Amisk Ct., Beaverlodge, 1989—. Home: Box 1204, Beaverlodge, AB Canada T0H 0C0 Office: Agr Can Rsch Sta, Box 29, Beaverlodge, AB Canada T0H 0C0

MCELROY, MICHAEL, physicist, researcher; b. Shercock, County Cavan, Ireland, May 18, 1939; married, 1963. B.A., Queen's U., Belfast, Ireland, 1960; Ph.D. in Math., Belfast, Ireland, 1962. Project assoc. Theoretical Chemistry Inst., U. Wis., 1962-63; from asst. physicist to physicist Kitt Peak Nat. Obs., 1963-71; physicist Ctr. Earth and Planetary Physics Harvard U., Cambridge, Mass., 1971—, now Abbott Lawrence Rotch prof. atmospheric sci.; chmn. Dept. Earth and Planetary Scis. Harvard U., 1986—; mem. Mars panel Lunar and Planetary Missions Bd., NASA, 1968-69, Stratospheric Research Adv. Com., Space and Terrestrial Applied Adv. Com., Com. Atmospheric Sci., Nat. Acad. Sci. Space Sci. Bd.; chmn. Com. Planetary and Lunar Exploration. Recipient James B. Macelwane award Am. Geophys. Union, 1968; recipient Newcomb Cleve. prize AAAS, 1977, Pub. Service medal NASA, 1978. Office: Harvard U Divsn Applied Scis Pierce Hall 100E 20 Oxford St Cambridge MA 02138

MCELWEE, DENNIS JOHN, pharmaceutical company executive; b. New Orleans, July 30, 1947; s. John Joseph and Audrey (Nunez) McE.; m. Nancy Lu Travis, Sept. 3, 1976. BS, Tulane U., 1970; JD., U. Denver, 1992, Hague Acad. Internat. Law, 1990-91. Clean room and quality control analyst Sci. Enterprises Inc., Broomfield, Colo., 1975-76; analytical chemist in toxicology Poisonlab. Inc., Denver, 1977; analytical chemist, then dir. analytical quality control program Colo. Sch. Mines Rsch. Inst., 1977-79; dir. quality control, then dir. compliance Benedict Nuclear Pharms. Co., Golden, Colo., 1979-84; pres. MC Projections, Inc., Morrison, Colo., 1985-86, dir. regulatory affairs, Electromedics Inc., Englewood, Colo., 1986-91. Author: Mineral Research Chemicals, Toxic Properties and Proper Handling, 2d edit., 1979; contbr. articles to profl. jours. Mem. Colo. Bar Assn., Internat. Legal Soc., Regulatory Affairs Profls. Soc., Assn. for the Advancement of Med. Instrumentation. Home: PO Box 56 Morrison CO 80465-0056 Office: 2009 Wadsworth Blvd Lakewood CO 80215

MCENNAN, JAMES JUDD, physicist; b. Ypsilanti, Mich., Feb. 26, 1944. BS, U. Fla., 1965, PhD, 1970. Sr. MTS Computer Sci. Corp., Silver Spring, Md., 1977-79; sr. staff/MTS Hughes Aircraft Co., El Segundo, Calif., 1979-83, sr. scientist, 1983-86, lab. scientist, 1986—; pres., CEO JackPlot Software, Agoura Hills, Calif., 1990—. Mem. Am. Phys. Soc., Phi Beta Kappa. Achievements include patent for method of estimating reference speed for anti-lock braking systems; first calculation of radiative corrections to photoeffect.

MCENROE, CAROLINE ANN, legal assistant; b. Bronx, N.Y., Jan. 21, 1935; d. Patrick David and Annie Marie (Donohue) Fay; m. William T. McEnroe, Apr. 16, 1955; children: Colleen, Patrick, Kieran, Peggyanne. Grad., Krissler Bus. Inst., Poughkeepsie, N.Y., 1953. Cert. legal asst., N.Y. Superior ct. clk. County Ct. Dutchess County, Poughkeepsie, N.Y., 1978-79; legal adminstrv. asst. to pub. defender County of Dutchess, Poughkeepsie, 1979—; bd. dirs. N.Y. State Magistrates Assn., 1974; pres. Dutchess County Magistrates Assn., 1973. Town justice Town of Amenia, N.Y., 1968-76, coun. woman, 1981-85; vol. March of Dimes Dutchess County, Heart Fund Dutchess County; active Ch. Bldg. Fund Campaign, Amenia; tchr. Confraternity of Christian Doctrine classes, 1976-88; eucharistic minister Sharon (Conn.) Hosp. Mem. State Magistrates' Assn., Dutchess County Magistrates' Assn. Republican. Roman Catholic. Home: PO Box 486 Amenia NY 12501 Office: Pub Defender Dutchess County 22 Market St Poughkeepsie NY 12601

MCENTIRE, B. JOSEPH, mechanical engineer; b. Columbia, S.C., Aug. 4, 1962; s. John Thomas Sr. and Loretta (Monts) McE.; m. Karen Carr, July 5, 1986; 1 child, Ellise McKinley. Student, Anderson (S.C.) Coll., 1980-81; BS, Clemson (S.C.) U., 1985; postgrad., Pa. State U., 1988-89; MS, Ga. Inst. Tech., 1993. Design engr. Gentex Corp., Carbondale, Pa., 1986-87; staff engr. VEDA Corp., Warminster, Pa., 1987-88; sr. project engr. Naval Air Devel. Ctr., Warminster, 1988-90; mech. engr. U.S. Army Aeromed. Rsch. Lab., Ft. Rucker, Ala., 1990—; developer, instr. course on crashworthiness Ga. Inst. Tech., 1993. Patentee quick release/positive lock buckle design. Mem. AIAA, Am. Helicopter Soc. (crashworthiness design subcom. 1991), Survival and Flight Equipment Assn. Avocations: camping, boating, reading.

MCEWAN, ROBERT NEAL, molecular biologist; b. Washington, Sept. 6, 1949; s. Thomas Cornealius and Esther (Johnson) McE.; m. Elizabeth Mary Ross, Aug. 24, 1973; 1 child, Amy Elizabeth; stepchildren: Gary W. Tizard, Jacqueline A. Tizard. BS, Va. Mil. Inst., 1971. Med. technician USPHS, Washington, 1971-72; biologist McGuire Va. Hosp., Richmond, 1972-75; rsch. assoc. Frederick (Md.) Cancer Rsch. Facility, 1975-84, The Upjohn Co., Kalamazoo, Mich., 1984-93; hosp. sales specialist The Upjohn Co., Balt., 1993—; adj. faculty Med. Sch. Northwestern U., Chgo., 1984-87; mem. indsl. adv. bd. Cen. Va. Gov.'s Sch., 1992—, Upjohn Acad., 1992. Patentee in field. Leader 4-H, Kalamazoo, Mich., 1988-90; founder, mem. Adventures in Spaces, pres. 1988-90; mem. Gull Lake Middle Sch. PTA, pres. 1987-89, Gull Lake Area Community Vols., pres. 1988. Recipient Outstanding Svc. award Middle Sch. Educators Assn., 1988, STAR award Kalamazoo Gazette, Vol. Activity Ctr., 1989, Svc. Leadership award Gull Lake Area Community Vols., 1990. Mem. Am. Soc. Microbiology, Nat. 4-H Sci. Tech. Design Team. Avocations: golf, skiing, sailing, horseback riding, teaching science. Home: 402 Cypress Ct Bel Air MD 21015 Office: The Upjohn Co 3110 Fairview Park Dr # 600 Falls Church VA 22042

MCFADDEN, JAMES FREDERICK, JR., surgeon; b. St. Louis, Dec. 5, 1920; s. James Frederick and Olivia Genevieve (Imbs) McF.; m. Mary Cella Switzer, Sept. 15, 1956 (div. Sept. 1969); children: James Frederick, Kenneth Michael, John Switzer, Mary Cella, Joseph Robert; m. Deanne Nemec Puls, Apr. 29, 1989. AB, St. Louis U., 1941, MD, 1944. Intern Boston City Hosp., 1944-45; ward surgeon neorsurg. and orthopedics McGuire Gen. Hosp., Richmond, Va., 1945; ward surgeon in internal medicine Regional Hosp., Fort Knox, Ky., 1946; ward surgeon plastic surgery Valley Forge Gen. Hosp., Phoenixville, Pa., 1946-47; intern St. Louis City Hosp., 1947-48; resident in surgery VA Hosp., St. Louis, 1948-52; clin. instr. surgery St. Louis U., 1952-62; gen. practice medicine specializing in surgery St. Louis, 1952—; mem. staff St. Mary's Hosp., 1952-77, St. John's Mercy Hosp., 1952-74, Desloge Hosp., 1952-62, Frisco RR Hosp., 1953-64, DePaul Hosp., 1954—, Christian Hosp., 1955-66, 83—. Mem. St. Louis Ambassadors, 1979-81; officer St. Louis County Aux. Police , 1973-75. Served to capt. AUS, 1945-47. Recipient Eagle Scout award, Order of the Arrow Honor award Boy Scouts Am. Fellow Royal Soc. Medicine; mem. ACS, Internat. Coll. Surgeons, St. Louis Med Soc., Am. Coll. Occupational and Environ. Medicine, Am. Soc. Clin. Hypnosis, Internat. Soc. Hypnosis, Am. Assn. RR Surgeons, St. Louis U. Student Conclave, Alpha Sigma Nu. Roman Catholic. Avocations: hypnosis, photography. Home: PO Box 411933 Saint Louis MO 63141-1933 Office: 11500 Olive Blvd Saint Louis MO 63141-7143

MCFADDEN, PAMELA ANN, architect; b. Detroit, Oct. 25, 1950; d. Donald Arthur and Juanita Mary (Albright) McF. BS, No. Mich. U., 1972; BArch, U. Ill., Chgo., 1979. Registered architect., Colo. Pres. Elements Design Group, Boulder, Colo., 1979—; bd. dirs. Boulder Energy Conservation Ctr., 1981-90. Contbr. articles, photographs to profl. jours., mags., newspapers. Bd. dirs. United Way, Boulder, 1983-86; bd. dirs., mem. vol. task force Boulder County Dept. Social Svcs., 1984-86. Named Vol. of Yr., County Commrs., Boulder, 1983; recipient award Gov. of Colo., 1990. Mem. Am. Solar Energy Soc. (bd. dirs. 1986-91, vice chair 1989-90), Internat. Solar Energy Soc. (steering com. 1988, 89), Denver Solar Energy Soc. (bd. dirs. 1982—). Home and Office: 2888 Bluff St Boulder CO 80301

MCFALL, GREGORY BRENNON, marine biologist; b. Metalline Falls, Wash., Nov. 29, 1959; s. Garry William and Thetta Marie (Stanley) McF.; m. Claire Anne Claude, Jan. 1, 1988. BS, W.Va. U., 1990; MS, U. N.C., 1992. Cinematographer W.Va. U., Morgantown, 1982-83; rsch. diver Sun Marine Inc., Norfolk, Va., 1983-85; rsch. technologist Atlantic Undersea Test and Evaluation Ctr., Andros, Bahamas, 1985-87; marine biologist Nat. Undersea Rsch. Ctr., Wilmington, N.C., 1990—, mem. diving adv. bd., 1990—. With USN, 1978-82. Mem. Am. Acad. Underwater Scis., Golden Key Nat. Honor Soc., Sigma Xi, Phi Kappa Phi. Achievements include invention of underwater laser measuring device, development and application of low-cost paired-laser underwater measuring device for performing accurate assessment of fish size. Home: 541 Hidden Valley Rd Wilmington NC 28409 Office: Nat Undersea Rsch Ctr 7205 Wrightsville Ave Wilmington NC 28403

MCFARLANE, JAMES ROSS, mechanical engineer, educator; b. Winnipeg, Can., June 20, 1934; s. John Ross and Francis Opal (Angus) McF.; m. Noreen Wood, Aug. 31, 1957; 1 child, James Arthur Ross. BSc in Mech. Engring., U. N.B., 1960; MSc, Naval Engring., MIT, 1965; DEng, Royal Mil. Coll., 1988, U. Victoria, 1988; D in Mil. Sci., Royal Road's, 1991; D in Sci., U. N.B., 1992. Commd. ordinary seaman Royal Can. Navy, 1952, advanced through grades to lt. commdr., staff officer constrn. Can.'s Oberon submarines, 1966-68, ret., 1971; v.p. Internat. Hydrodynamics, 1971-74; founder, pres. Internat. Submarine Engring. Ltd., 1974—; adj. prof. Simon Fraser U.; Robert Bruce Wallace speaker MIT, 1986. Recipient Disting. Tech. Achievement award IEEE Oceanic Engring. Soc., 1987, Ernest C. Manning award, 1987, Profl. Engrs. Meritorious Achievement award, 1987, B.C. Sci. & Engring. Gold medal, 1989, Julian C. Smith medal Engring. Inst. Can., 1990; named Officer Order of Can., 1989. Mem. Can. Acad. Engring., Soc. Profl. Engrs. B.C., Sigma Xi. Office: care Engr Inst of Canada, #202, 280 Albert St, Ottawa, ON Canada K1P 5G8*

MCFARLIN, RICHARD FRANCIS, industrial chemist, researcher; b. Oklahoma City, Oct. 12, 1929; s. Loy Lester and Julie Mae (Collins) McF.; m. Clare Jane Burroughs, Apr. 4, 1953; children: Robin Sue McFarlin Godwin, Richard Prescott, Rebecca Lynn McFarlin Bray, Roger Whitsitt. BS, Va. Mil. Inst., 1951; MS, Purdue U., 1953, PhD, 1956. Rsch. chemist Monsanto Chem. Co., St. Louis, 1956-60; supr. inorganic rsch. Internat. Minerals and Chems., Mulberry, Fla., 1961; mgr. Agr. Rsch. Ctr. Armour Agrl. Chem. Co., Atlanta, 1962; v.p. rsch., ops., devel. & adminstrn. div. agri-chems. U.S. Steel, Atlanta, 1986; tech. dir. Lester Labs. Inc., Atlanta, 1986-88; exec. dir. Fla. Inst. Phosphate Rsch., Bartow, 1988—; mem. bd. advisors engring. coun. U. South Fla., Lakeland, 1990—, U. Fla., Gainesville, 1991—; mem. bd. advisors Inst. Recyclable Materials La. State U., Baton Rouge, 1990—. Capt. USAR, 1951-61. M. M. Cohn Found. scholar, 1947, L. D. Wall scholar, 1949, O. M. Baldinger scholar, 1950. Presbyterian. Achievements include eight U.S. and foreign patents for selective organic reducing agents, fertilizer processes and selective biocides. Home: 6611 Sweetbriar Ln Lakeland FL 33813-3598 Office: Fla Inst Phosphate Rsch 1855 W Main St Bartow FL 33830-7718

MCFATE, KENNETH LEVERNE, association administrator; b. LeClaire, Iowa, Feb. 5, 1924; s. Samuel Albert and Margaret (Spear) McF.; m. Imogene Grace Kness, Jan. 27, 1951; children: Daniel Elliott, Kathryn Margaret, Sharon Ann. BS in Agrl. Engring., Iowa State U., 1950; MS in Agrl. Engring., U. Mo., 1959. Registered profl. engr., Iowa, Mo. Agrl. sales engr. Ill. No. Utility Co., Aledo, 1950-51; extension agrl. engr. Iowa State U., Ames, 1951-53, rsch. agrl. engr., 1953-56; prof. agrl. engr. U. Mo., Columbia, 1956-86, prof. emeritus, 1986; dir. Mo. Farm Electric Coun., Columbia, 1956-75; exec. mgr. Nat. Farm Electric Coun., Columbia, 1975-86; pres., exec. mgr. Nat. Food & Energy Coun., Columbia, 1986-91; pres. emeritus, 1991; mgr. Electrotechnology Rsch., 1991-93; bd. dirs. Internat. Com. Agrl. Engrs., Brussels, 1989-94. Editor, author: (with others) Handbook for Elsevier Science, 1984-89; editor: Electrical Energy in Agriculture, 1989; mem. editorial bd. Energy in Agriculture for Elsevier Sci., Amsterdam, The Netherlands, 1981-88. Mem. Kiwanis Club Aledo, 1951, Lions Club, Gilbert, Iowa, 1955. 2d lt. USAAF, 1943-45. Fellow Am. Soc. Agrl. Engrs. (George Kable Elec. award 1974); mem. Future Farmers of Am. (hon. Am. Farmer degree 1991), Alpha Epsilon, Gamma Sigma Delta. Republican. Presbyterian. Avocations: technical writing, gardening, old car restoration, woodworking. Home: 9450 Hwy HH Hallsville MO 65255

MCGARVEY, FRANCIS XAVIER, chemical engineer; b. Kingston, N.Y., Mar. 16, 1919; s. Francis S. and Mabel (Post) McG.; m. Grace J. Leimkuhler, Sept. 20, 1941; children: Francis, Patrica, Ellen, Debra. BS in Chem. Engring., U. Pa., Phila., 1941, MS in Chem. Engring., 1944. Phys. chemist Manhattan Dist. Engrs., Los Alamos, N.Mex., 1944-45; rsch. scientist Rohm & Haas Corp., Phila., 1945-55; tech. rep. Rohm & Haas Corp., Tokyo and Paris, 1955-65; pres. Puricons Inc., Malvern, Pa., 1966-76; tech. dir. Sybron Chem. Corp., Birmingham, N.J., 1976—; cons. various waste and water projects, 1966-76. Contbr. articles to profl. jours., chpts. to books. Sgt. U.S. Army, 1944-45. Mem. Engring. Soc. Western Pa. (adv. group 1955—, award of merit 1987), Am. Inst. Chem. Engrs., Am. Chem. Soc., Am. Soc. Artificial Internal Organs, Am. Waterworks Soc. Achievements include patent on ion exch. mixed bed which has worldwide application in pure water, nuclear energy and sugar industries. Office: Sybron Chem Corp Birmingham Rd Birmingham NJ 08011

MC GAUGH, JAMES LAFAYETTE, psychobiologist; b. Long Beach, Calif., Dec. 17, 1931; s. William Rufus and Daphne (Hermes) McG.; m. Carol J. Becker, Mar. 15, 1952; children: Douglas, Janice, Linda. BA, San Jose State U., 1953; PhD (Abraham Rosenberg fellow), U. Calif. - Berkeley, 1959; sr. postdoctoral fellow, NAS-NRC, Istituto Superiore di Sanità, Rome, 1961-62; DSc (hon.), So. Ill. U., 1991. Asst. prof., assoc. prof. psychology San Jose State U., 1957-61; assoc. prof. psychology U. Oreg., 1961-64; assoc. prof. U. Calif., Irvine, 1964-66, founding chmn. dept. psychobiology, 1964-67, 71-74, 86-89, prof., 1966—, dean Sch. Biol. Sci., 1967-70, vice chancellor acad. affairs, 1975-77, exec. vice chancellor, 1978-82, founding dir. Ctr. Neurobiology of Learning and Memory, 1983—; Mem. adv. coms. NIMH, 1965-78, Mental Health Coun. NIMH, 1992—. Author: (with J.B. Cooper) Integrating Principles of Social Psychology, 1963, (with H.F. Harlow, R.F. Thompson) Psychology, 1971, (with M.J. Herz) Memory Consolidation, 1972, Learning and Memory: An Introduction, 1973, (with R.F. Thompson and T. Nelson) Psychology I, 1977, (with C. Cotman) Behavioral Neuroscience, 1980; editor: (with N.M. Weinberger, R.E. Whalen) Psychobiology, 1966, Psychobiology-Behavior from a Biological Perspective, 1971, The Chemistry of Mood, Motivation and Memory, 1972, (with M. Fink, S.S. Kety, T.A. Williams) Psychobiology of Convulsive Therapy, 1974, (with L.F. Petrinovich) Knowing, Thinking, and Believing, 1976, (with R.R. Drucker-Colín) Neurobiology of Sleep and Memory, 1977, (with S.B. Kiesler) Aging, Biology and Behavior, 1981, (with G. Lynch and N. M. Weinberger) Neurobiology of Learning and Memory, 1984, (with N.M. Weinberger and G. Lynch) Memory Systems of the Brain, 1985, Contemporary Psychology, 1985, (with C.D. Woody and D.L. Alkon) Cellular Mechanisms of Conditioning and Behavioral Plasticity, 1988, (with N.M. Weinberger and G. Lynch) Brain Organization and Memory: Cells, Systems and Circuits, 1990, (with R.C.A. Frederickson and D.L. Felten) Peripheral Signaling of the Brain, 1991, (with L. Squire, G. Lynch and N.M. Weinberger) Memory: Organization and Locus of Change, 1991; author over 300 sci. papers; founding editor Behavioral Biology, 1972-78, Behavioral and Neural Biology, 1979—. Recipient medal U. Calif., Irvine, 1992. Fellow AAAS, Am. Acad. Arts and Scis., Soc. Exptl. Psychologists, APA (chief sci. advisor 1986-88, APA Sci. Contbn. award 1981); mem. NAS (chmn. Psychol. sect. 1992-95), Am. Psychol. Soc. (William James fellow 1989, pres. 1989-91), Western Psychol. Assn. (pres. 1992-93), Internat. Brain Rsch. Orgn., Soc. Neurosci., Am. Coll. Neuropsychopharmacology, Psychonomic Soc., European Behavioral Pharmacology Soc., Phi Beta Kappa, Sigma Xi. Office: U Calif Ctr Neurobiol Learning & Memory Irvine CA 92717

MCGAVIN, JOCK CAMPBELL, airframe design engineer; b. L.A., Sept. 14, 1917; s. Campbell and Irene (LeMarr) McG.; m. Catherine Marcelle Glew, Jan. 12, 1952; 1 child, James Campbell. AA, L.A. City Coll., 1950; AB, U. So. Calif., 1970, MS, 1975; PhD, Calif. Coast U., 1989. Airframe design engr. Rockwell Internat. Corp., L.A., 1946-82; ret., 1982; sr. design engr. X-15 airplane, Apollo Command Module, space shuttle, others. Vol. mem. pub. involvement subcom. Puget Sound Water Quality Authority, Seattle, 1987-89. Capt. C.E., U.S. Army, 1940-46, ETO. Recipient Apollo Achievement award NASA, 1969. Mem. Soc. for History Astronomy, Izaak Walton League Am. (pres. Greater Seattle chpt. 1991-93, vol. worker environ. projects 1985—), U. So. Calif. N.W. Alumni Club (pres. 1987-89). Avocations: world travel, history of astronomy, radio-controlled model airplanes. Home: 12939 NE 146th Pl Woodinville WA 98072

MCGEE, HENRY ALEXANDER, JR., chemical engineering educator; b. Atlanta, Sept. 12, 1929; s. Henry Alexander and Arrie Mae (Mallory) McG.; m. Betty Rose Herndon, July 29, 1951; children: Henry Alexander, Charles Nelson, Kathy Nan. B.Chem. Engring., Ga. Inst. Tech., 1951, Ph.D., 1955; postgrad., U. Wis., 1955-56. Research scientist Army Rocket and Guided Missile Agy. and NASA, Huntsville, Ala., 1956-59; from assoc. prof. to prof. chem. engring. Ga. Inst. Tech., Atlanta, 1959-71; prof. Va. Poly. Inst. and State U., Blacksburg, 1971—, head dept. chem. engring., 1971-82; vis. prof. Calif. Inst. Tech., 1964; dir. chem. and thermal systems div. NSF, Washington 1990—; cons. in field. Author: Molecular Engineering, 1991; editorial adv. bd.: Chemical Abstracts; contbr. numerous articles to profl. publs. Danforth assoc.; recipient various research grants NSF, various research grants NASA, various research grants Air Force Office Sci. Research; named one of five outstanding young men of year Atlanta, 1964. Fellow Am. Inst. Chem. Engrs. (chmn. nat. program com., editorial bd. Jour.), Am. Chem. Soc., AAAS (chmn. sect. on engring. 1985-86); mem. Sigma Xi. Republican. Home: 605 Rainbow Ridge Dr Blacksburg VA 24060 Office: Va Poly Inst and State U Blacksburg VA 24061 also: NSF Washington DC 20550

MCGEE, JAMES PATRICK, information industry executive; b. N.Y.C., Dec. 27, 1941; s. James Edward and Norah Elizabeth (Russell) McG.; m. Roseann Marie Carmody, Feb. 20, 1966; children: James Patrick, Richard Carmody. BBA, CUNY, 1965; MBA, Baruch Coll. N.Y., 1972; PMC, Iona Coll. of New Rochelle, 1974; MS in Computer Sci., Poly. Inst. N.Y., 1978, PhD in computer sci. Southwestern U. Ariz., 1982, PhD in Info. Sci., Walden U. Minn., 1984. Cert. quality engr., records mgr., data processor, computer programmer, info. systems mgr., data mgr., systems profl. Programmer N.Y. Med. Coll., N.Y.C., 1964-65; sr. programmer Shell Oil Co., N.Y.C., 1968-70, sr. systems analyst/rsch. staff mem. IBM T.J. Watson Rsch. Ctr., Yorktown, 1970—; info. systems mgr. IBM, Somers, N.Y., 1990—; program mgr. IBM Worldwide Mgmt. Cons. Practices Tech., White Plains, N.Y., 1991-93; chmn., CEO, pres. Internat. Smart Enterprise Systems Consulting and Svcs., 1993—; lectr. Baruch Coll. of CUNY, 1973-75; ednl. coordinator Assn. Computing Machinery, N.Y.C., 1975-76; lectr. in field. Contbr. articles to profl. jours. Bd. dirs. White Pond Community Ctr., Stormville, N.Y., 1975, Internat. Tech. Inst., 1985-88, Tech. Transfer Soc., 1986-88; leader, instr. Explorer Scout Troop, White Plains, 1973-74. Served with U.S. Army, 1965-67. Decorated Bronze Star. Mem. Am. Assn. Artificial Intelligence, Am. Nat. Standards Inst. (Data Base Systems Study Group adv. group), Assn. Computational Linguistics, Assn. for Computing Machinery, Spl. Interest Group for Mgmt. Data, Assn. of Inst. Certification of Computer Profls., Assoc. Info. Mgrs., Am. Mgmt. Assn., Assn. Records Mgrs. Adminstrs., Assn. for Systems Mgmt., Am. Soc. Quality Control, Data Adminstrn. Mgmt. Assn., Data Entry Mgmt. Assn., Data Processing Mgmt. Assn., Electronic Data Processing Auditors Assn., IEEE, Internat. Tech. Inst. (bd. dirs.), Vietnam Vets. of Am. (life), VFW (life), Omicron Delta Epsilon. Republican. Roman Catholic. Home: Mey Crescent Rd Stormville NY 12582-5624

MCGEE, SAM, laser scientist; b. Louisville, Mar. 4, 1943; s. Walter R. and Sue (Burchett) McG. BA, Vanderbilt U., 1965. Mktg. dir. Brown-Forman Corp., Louisville, 1966-73; pres. FYI Corp., L.A., 1973-75; sr. v.p., gen. mgr Brady Enterprises, East Weymouth, Mass., 1979-82; pres. Laser Images, Inc., L.A., 1982-85; pres., owner Starlasers, L.A., 1985—; mgmt. cons. L.A., 1976-78; cons. in field; bd. dirs. various cos. Mem. Hon. Order Ky. Cols. Achievements include patent for PCM digital recording; development of precision xy mirror mount, laser projector technology. Office: Starlasers 13156 Leadwell St North Hollywood CA 91605-4117

MCGEE, THOMAS JOSEPH, atmospheric scientist; b. Phila., Nov. 16, 1943; s. Thomas J. and Marie K. (Fox) McG.; m. Cheryl A. Simons, Dec. 19, 1970 (div. July 1990); children: Erin N., Kevin T., Meaghan M.; m. Lee Ann Leyo, Aug. 25, 1991; 1 child, Candace. BS, St. Joseph's Coll., 1965; PhD, U. Notre Dame, 1970. Asst. prof. St. Joseph's Coll., Phila., 1969-70; mgmt. cons. Ernst and Ernst, Detroit, 1970-72; rsch. assoc. dept. chemistry U. Md., College Park, 1972-75, vis. rsch. prof. Inst. for Phys. Sci. and Tech., 1975-80; atmospheric scientist Goddard Space Fligh Ctr., NASA, Greenbelt, Md., 1980—. Co-editor: The Stratosphere 1981, 1981; contbr. articles to profl. jours. Recipient Grad. fellowship NSF, 1965-69, Outstanding Performance award NASA, 1989, 91, Tech. Achievement award NASA, 1992. Mem. Am. Meteorol. Soc. (com. on laser atmospheric studies 1992—), Optical Soc. Am. Roman Catholic. Achievements include development of instrumentation for the remote measurement of atmospheric species; of technique for measurement of ozone in the presence of volcanic aerosol results showed ozone depletion related aerosols. Home: 16 Hopton Ct Sterling VA 20165 Office: NASA GSFC Code 916 Greenbelt MD 20771

MCGEEHAN, STEVEN LEWIS, soil scientist; b. Camp Lejune, N.C., July 21, 1955; s. Dudley Foster and Geraldine Pauleta (Barberro) McG.; m. Kathy Alison Beerman, July 28, 1984; children: Anna Rose, Michael Aaron. MS, Oreg. State U., 1985; PhD, U. Idaho, 1992. Cert. prof. soil scientist. Contract inspector/fire suppression U.S. Forest Svc., Somes Bar, Calif., 1977-80; grad. rsch. asst. Oreg. State U., Corvallis, 1981-84, rsch. asst./instr., 1984-85; rsch. and instructional assoc. U. Idaho, Moscow, 1985—; soils cons. Moscow, 1991-92. Contbd. articles to profl. jours.; 1st author manual: Laboratory Manual for General Soil Science, 1992. Tchr., workshop organizer Idaho 4-H Teen Conf. Sci. and Tech. Day, 1988; mentor Young Scholars Program, NSF, 1989. Ursula Kraus Acad. scholar, 1980, Oreg. Lime Assn. scholar, 1980. Mem. Soil Sci. Soc. Am., Gamma Sigma Delta, Alpha Zeta. Democrat. Unitarian. Office: U Idaho Divsn Soil Sci Moscow ID 83843

MCGEER, JAMES PETER, research executive, consultant; b. Vancouver, B.C., Can., May 14, 1922; s. James Arthur and Ada Alice (Schwenger) McG.; m. Catherine Pearson Deas, June 22, 1948; children: Mary, Allison, James, Thomas. BA, U. B.C., 1944, MA, 1946; MA, Princeton U., 1948, PhD, 1949. Researcher Alcan R & D, Arvida, Que., Can., 1949-52, group leader, 1952-59, pilot plant dir., 1960-67; dept. head Alcan Smelters, Arvida, Que., Can., 1968-71, asst. div. head, 1972-73; mgr. tech. transfer Alcan Smelter Svcs., Montreal, Que., 1973-77; dir. rsch. Alcan Internat. Ltd., Kingston, Ont., Can., 1978-82, dir. lab., 1983-87; mng. dir. Ont. Ctr. Materials, Kingston, 1988—; chmn. bd. Can. Rsch. Mgmt. Assn., Toronto, Ont., 1990-91, Welding Inst. Can., Mississauga, Ont., 1988-90, Can. U. Ind. Coun. Advanced Ceramics, Ottawa, Ont., 1986-88; dir. Metall. Soc., Pitts., 1987-89; Can. Coun. lectr. Am. Soc. Metals, 1985-86; disting. lectr. Can. Inst. Mining Metallurgy, 1987. Contbr. articles to profl. jours. Chmn. bd. Que. Assn. Protestant Sch. Bds., Montreal, 1968-70. Mem. Anglican Ch. Office: Ontario Ctr Materials Rsch, PO Box 1146, Kingston, ON Canada K7L 4Y5

MCGERVEY, TERESA ANN, cartographer; b. Pitts., Sept. 27, 1964; d. Walter James and Janet Sarah (Donehue) McG. BS in Geology, Calif. U. Pa., 1986, MS in Earth Sci., 1988. Phys. sci. technician U.S. Geol. Survey, Reston, Va., 1989-90; editor, indexer Am. Geol. Inst., Alexandria, Va., 1990-91; cartographer Def. Mapping Agy., Reston, 1991-93; tech. info. specialist Nat. Tech. Info. Svc., Springfield, Va., 1993—; intern Dept. Mineral Scis., Smithsonian Instn., summers 1985, 1986.

MCGILBERRY, JOE H., food service executive; b. Mobile, Ala., Aug. 6, 1943; s. Thomas Henry and Yvonne (Jorden) McG.; m. Betty Sue Reynolds, Dec. 20, 1964; children: Joe H. Jr., Michele, Brent. BS, Auburn U., 1965; MS, U. Tenn., 1972; PhD, TEx. A&M U., 1978. Gen. plant asst. Gen. Telephone Fla., Tampa, 1966; systems analyst E.I. DuPont de Nemours & Co., Old Hickory, Tenn., 1966-68; research asst. U. Tenn., Knoxville, 1968-69; advanced materials engr. Union Carbide Corp., Texas City, TX, 1969-70; asst. prof. Tenn. Tech. U., Cookeville, 1970-72, 73-75, 76-77; mgr. prodn. support Fleetguard div. Cummins Engine, Cookeville, 1972-73; assoc. research engr. Tex. A&M U., College Station, 1975-76; mgr. food and fiber ctr. Miss. State Coop. Extension Program, 1978—; cons. Dunlap Industries, U. Tenn. at Nashville, Norwalk Furniture Corp., Teledyne-Stillman, Genco Stamping, Inc., Riley Enterprises; lectr. various seminars and presentations, 1978—. Author numerous articles in field. Pres. Starkville (Miss.) High Sch. Athletic Booster Club; coach Starkville Youth Soccer Orgn., Starkville Boys Baseball League; co-organizer Starkville Area Youth Basketball (Leadership award 1969). Mem. Inst. Indsl. Engrs. (sr.), Miss. Indsl. Devel. Council, Order Engrs., Nat. Assn. County Agrl. Agts., Miss. Assn. County Agrl. Agts., Starkville C. of C., Alpha Phi Mu. Office: Food and Fiber Ctr 307 Bost Ctr PO Box 5446 Mississippi State MS 39762-5446

MCGILL, THOMAS CONLEY, physics educator; b. Port Arthur, Tex., Mar. 20, 1942; s. Thomas Conley and Susie Elizabeth (Collins) McG.; m. Toby Elizabeth Cone, Dec. 27, 1966; children: Angela Elizabeth, Sara Elizabeth. BS in Math., Lamar State Coll., 1963, BEE, 1964; MEE, Calif. Inst. Tech., 1965, PHD, 1969. NATO postdoctoral fellow U. Bristol, Eng., 1969-70; NRC postdoctoral fellow Princeton (N.J.) U., 1970-71; from asst. to assoc. prof. applied physics Calif. Inst. Tech., Pasadena, 1971-77, prof., 1977-85, Fletcher Jones prof. applied physics, 1985—; cons. Hughes Research Labs., Malibu, Calif., 1967—, Def. Advance Projects Agy./Materials Research Council, Arlington, Va., 1979—, Xerox Palo Alto (Calif.) Research Corp., 1984—. Alfred P. Sloan Found. fellow, 1974. Fellow Am. Physical Soc.; mem. AAAS, IEEE, Am. Vacuum Soc., Sigma Xi. Office: Calif Inst of Tech Mail Code 128-95 Pasadena CA 91125

MCGILLICUDDY, JOAN MARIE, psychotherapist, consultant; b. Chgo., June 23, 1952; d. James Neal and Muriel (Joy) McG. BA, U. Ariz., 1974, MS, 1976. Cert. nat. counselor. Counselor ACTION, Tucson, 1976; counselor, clin. supr. Behavioral Health Agy. Cen. Ariz., Casa Grande, 1976-81; instr. psychology Cen. Ariz. Coll., Casa Grande, 1978-83; therapist, co-dir. Helping Assocs., Inc., Casa Grande, 1982—, v.p., sec., 1982—; cert. instr. Silva Method Mind Devel., Tucson, 1986—; presenter Silver Mind Control Internat., 1988-91. Mem. Mayor's Com. for Handicapped, Casa Grande, 1989-90, Human Svcs. Planning, Casa Grande, 1985-90. Named Outstanding Am. Lectr. Silva Mind Internat., 1988-91. Mem. AACD. Avocations: jogging, singing. Office: Helping Assocs Inc 1901 N Trekell Rd Casa Grande AZ 85222-4119

MCGLONE, JOHN JAMES, biologist; b. New Rochelle, N.Y., Oct. 27, 1955; s. John Thomas and Maura (Scanlon) McG.; m. Barbara Rose Rattie McGlone, June 12, 1976; children: Molly, Kerry Rose. BS, Wash. State U., 1977, MS, 1979; PhD, U. Ill., 1981. Asst. prof. U. Wyo., Laramie, 1981-84, Tex. Tech. U., Lubbock, 1984-89; assoc. prof. Tex. Tech. Health Sci. Ctr., Lubbock, 1990—, Tex. Tech. U., Lubbock, 1989—; editorial bd. Jour. Animal Sci., ASAS, Champaign, Ill., 1985-90. Author: Guidelines for Swine Husbandry, 1988, Animal Health, 1990; contbr. articles to profl. jours. Advisor Econ. Devel. for Pork Industry, Tex., 1989—; tchr. St. Elizabeth's Cath. Ch., Lubbock, Tex., 1986—. Recipient Undergraduate Rsch. award Washington St. U., Pullman, Wash., 1976, Presdl. Acad. Achievement award Tex. Tech. U., Lubbock, 1990, Harry Frank Guggenheim Rsch. award N.Y.C., 1983, Outstanding Researcher, Coll. Agrl. Sci., Lubbock, Tex., 1992. Mem. Am. SOc. Animal Scis., Soc. For Neuroscience, Animal Behavior Soc. Democrat. Roman Catholic. Achievements include discovery of natural pheromones that modulate pig aggression, odor compound (androstenone) that substantially reduces pig aggression; design of new hide pen that allows pigs to escape from attack. Home: 2225 86th Lubbock TX 79423 Office: Texax Tech University 15th and Detroit Lubbock TX 79409-2141

MCGOVERN, JOHN HUGH, urologist, educator; b. Bayonne, N.J., Dec. 18, 1924; s. Patrick and Mary (McGovern) McG.; m. Mary Alice Cavazos, Aug. 2, 1980; children by previous marriage: John Hugh, Robert, Ward, Raymond. BS, Columbia U., 1947; MD, SUNY, Bklyn., 1952. Diplomate Am. Bd. Urology. Rotating intern Bklyn. Hosp., 1952-53; asst. resident in surgery Bklyn. VA Hosp., 1953-54; with urology N.Y. Hosp., 1954-56; exchange surg. registrar West London Hosp., Eng., 1956-57; resident in urol. surgery N.Y. Hosp., 1957-58, rsch. asst. pediatric urology, 1958-59, asst. attending surgeon James Buchanan Brady Found., 1959-61, assoc. attending surgeon, 1961-66, attending surgeon, 1966—; asst. in surgery Med. Coll. Cornell U., 1957-59, asst. prof. clin. surgery, 1959-64, assoc. prof., 1964-72, prof., 1972—; attending staff in urology Lenox Hill Hosp., 1969—, in-charge urology, 1969-83; cons. urology Rockefeller Inst., St. Vincent's Hosp., Mercy Hosp., Phelps Meml. Hosp.; chmn. coun. on urology Nat. Kidney Found., 1982. Contbr. articles to profl. jours., chpts. to books. Lt. M.C., U.S. Army, 1942-45. Recipient Conatvoy mos medal Chile, 1975, Tree of Life award Nat. Kidney Found., 1990; named Huesped de Honor, Mimunicipalidad de Guayaquil (Ecuador), 1976; award in urology Kidney Found. N.Y., 1977, Sir Peter Freyer medal, Galway, Ireland, 1980. Fellow N.Y. Acad. Medicine (exec. com. urol. sect. 1968-72, chmn. 1972), ACS (credentials com. 1991-92), Am. Acad. Pediatrics (urological); mem. AMA (diagnostic and therapeutic tech. assessment bd. 1991—, diagnostic and therepautic tech. assessment program panel 1991, DATTA panel 1991-92), N.Y. State Med. Soc. (chmn. urol. sect. 1975), Med. Soc. County N.Y., Am. Urol. Assn. (pres.-elect 1988-89, pres. 1989-90, pres. N.Y. sect. 1979-80, N.Y. rep. exec. com. 1982-87, socioecons. com. 1987, chmn. fiscal affairs rev. com. 1987, chmn. awards com. 1990, time and place com. 1989-90), N.Y. State Urol. Soc. (exec. com. 1982—), Pan Pacific Surg. Assn., Am. Assn. Clin. Urologists (pres.-elect 1987-88, pres. 1988-89, bd. dirs. 1984—, mem. interpersonal rels. com. 1975—, govt. rels. com. 1989-90, program com. 1989-90, nominating com. 1989-90), Assn. Am. Physicians and Surgeons, Pan Am. Med. Assn. (diplomate 1981—), Urol. Investigators Forum, Soc. Pediatric Urology (pres.-elect 1979-80, pres. 1980-81), Am. Trauma Soc., Kidney Found. (med. adv. bd. N.Y. sect., trustee, 1979) Société Internationale d'Urologie (exec. com. U.S. sect.); hon. mem. Sociedad Peruana de Urología, Sociedad Guatemale de Urología, Sociedad Ecuadorians de Urología, Royal Coll. Surgeons (London). Home and Office: 53 E 70th St New York NY 10021

MCGOWAN, GEORGE VINCENT, public utility executive; b. Balt., Jan. 30, 1928; s. Joseph H. and Ethna M. (Prahl) McG.; m. Carol Murray, Aug. 6, 1977; children by a previous marriage: Gregg Blair, Bradford Kirby. BS in M.E., U. Md., 1951; LHD (hon.), Villa Julie Coll., 1991, Loyola Coll., Md., 1992. Registered profl. engr., Md. Project engr. nuclear power plant Balt. Gas & Electric Co., 1967-72, chief nuclear engr., 1972-74, pres., chief operating officer, 1980-87, chmn. bd. dirs., CEO, 1988-92, chmn. exec. com., 1993—, mgr. corp. staff services, 1974-78, v.p. mgmt. and staff services, 1978-79; bd. dirs. Balt. Life Ins. co., McCormick & Co., Life of Md. Inc., UNC Inc., Orgn. Resources Counselors, Inc., MNC Fin., Inc. Bd. dirs. U. Md. Med. System, United Way Cen. Md., Pride of Balt.; chmn. bd. regents U. Md. System; chmn. Gov.'s Vol. Coun. State of Md.; chmn. bd. dirs. Balt. Symphony Orch., CollegeBound Found. Recipient Disting. Alumnus award U. Md. Coll. Engring., 1980, U. Md., 1987, Disting. Marylander award Advt. and Profl. Club Balt., 1992, Disting. Citizen award U. Md., 1991, Disting. Citizen of Yr. award Boy Scouts Am. Balt. coun., 1991, Disting. Alumnus award Balt. Poly. Inst., 1992, Nat. Multiple Sclerosis Soc. Corp. Honoree, Md. chpt., 1993, Nat. Soc. of Fund Raising Execs. Outstanding Vol. Fund Raiser, 1993. Mem. ASME (James N. Landis medal 1992), Am. Nuclear Soc., U.S. Energy Assn. of the World Energy Conf., Engring. Soc. Balt. (Founders Day award 1988), Caves Valley Golf Club, The Ctr. Club (pres. bd. govs.), U. Md. M Club, Talbot Country Club, Annapolis Yacht Club, Md. Club. Presbyterian. Office: Balt Gas & Electric Co PO Box 1475 Baltimore MD 21203-1475

MCGOWAN, JOHN JOSEPH, energy manager; b. Phila., Nov. 28, 1950; s. Daniel Joseph and Catherine Theresa (Durkin) McG.; m. Judy Eileen Reed, June 3, 1978; children: Dustin, Kendall. BS, Temple U., 1975; MA, U. N.Mex., 1980. Cert. energy mgr.; cert. corgenation profl.; cert. lighting efficiency profl. Tchr. Zuni Indian Reservation, 1975-79; energy mgr. Haufler Inc., Phila., 1981-83; asst. dir. of energy Svc. Mdse. Corp., Nashville, 1983-86; v.p. Automation Mgmt. Systems, Kansas City, 1986-87; mgr. systems Honeywell, Albuquerque, 1987-91; dir. N.Mex. Energy Conservation Div., Santa Fe, 1991-92; mgr. market develop. Honeywell, Albuquerque, 1992—; chmn. tech. sessions World Energy Engring. Congress, 1985, 89, 90-92; tech. adv. bd. Energy User News Mag., Radnor, Pa., 1992—; prof. U. N.Mex., U. Phoenix; speaker in field and seminar presenter. Author: Networking for Building Automation Systems, 1991, Energy Management for Buildings, 1989, Energy Managment and Control Systems, 1988; contbg. author: New Mexico State Energy Policy, 1992; contbr. articles to profl. jours.; editorial bd. Strategic Planning for Enrgy and Environ., Atlanta, 1990—. Managed pub. agy. N.Mex. State Energy Office, 1991. Mem. ASHRAE, Assn. of Energy Engrs. (v.p. 1992, Energy Profl. Yr. 1991, nat. v.p. 1993—), Assn. of Profl. Energy Mgrs. Office: Honeywell 8500 Blue Water Rd NW Albuquerque NM 87121

MCGRATH, KENNETH JAMES, chemist; b. N.Y.C., Mar. 5, 1953; s. William Francis and Lucille Theresa (Bischoff) McG.; m. Linda Lou Frelier, Aug. 28, 1976; 1 child, Kathleen. BA in Chemistry, SUNY, Geneseo, 1976; PhD in Chemistry, Georgetown U., 1993. Rsch. chemist Eastman Kodak Co., Rochester, N.Y., 1976-80, Andrulis Rsch. Corp., Washington, 1983-86, Naval Rsch. Lab., Washington, 1986—. Contbr. articles to profl. jours. Mem. Am. Chem. Soc. Achievements include patent for Stable Soluble 1,2-Diaminocyclohexane Platinum Complexes, 1987. Office: Naval Rsch Lab Code 6120 Washington DC 20375-5000

MCGRATH, RICHARD WILLIAM, osteopathic physician; b. Hartford, Conn., Nov. 17, 1943; s. William Paul and Stephanie Gertrude (Romash) McG.; B.S., St. Ambrose Coll., 1965; D.Osteo. Medicine and Surgery, Coll. Osteo. Medicine and Surgery, Des Moines, 1971; m. Mariette VanLancker, June 24, 1967; children—Shaun, Megan, Kelley. Osteo. physician Weld County Gen. Hosp., Greeley, Colo., 1971-72, Granby (Colo.) Clinic, 1972-75, Timberline Med. Ctr., P.C., Granby, 1975—; pres. Timberline Med. Center, 1976—, Bighorn Properties Inc., 1978—, Thia of Am. Corp., 1980—; med. coordinator/dir. regional emergency systems Colo. State Health Dept., 1978-79; mem. Colo. Comprehensive Health Planning Agy., 1975-77; assoc. prof. clin. medicine Tex. Coll. Osteo. Medicine; med. advisor Grand County Ambulance System, 1977—; vice chief staff Kremmling Meml. Hosp.; bd. dirs. M&L Bus. Machine Co., Denver, Sun-Flo Internat., Inc., Silver Creek Devel. Co. and Ski Area. Mem. steering com. to develop Colo. Western Slope Health System Agy., 1975-76, bd. dirs., 1977—; bd. dirs. St. Anthony Hosp. Systems Emergency Rooms, 1984—; med. dir. Community Hosp. and Emergency Ctr., Granby, Presbyn. St. Lukes Hosp.; officer, police surgeon Grand Lake and Granby, 1977—; mem. parent adv. bd. Granby Sch. System, 1975-76; chmn. East Grand County Safety Council, 1974-76; dep. coroner Grand County, 1973-75; med. advisor Grand County Rescue Team, 1974-78. Recipient award Ohio State U. Coll. Medicine, 1977. Mem. AMA, ACS (com. on trauma), Am. Coll. Emergency Physicians, Western Slope Physicians Alliance Assn., Colo. State Emergency Med. Technicians (med. chmn. 1982-84), Colo. Union of Physicians (dir.), C. of C. of Granby, Grand Lake and Fraser Valley. Republican. Roman Catholic. Home: PO Box 706 Granby CO 80446-0706 Office: PO Box 857 Granby CO 80446

MCGRAW, KATHERINE ANNETTE, fisheries biologist, environmental consultant; b. Birmingham, Ala., May 18, 1943; d. Robert Lacey and Katherine Elizabeth (Kulp) McG. BS, U. Montevallo, 1965; MS, Auburn U., 1970; PhD, U. Wash., 1980. Tchr. John Carroll High Sch., Birmingham, Ala., 1965-67; marine biologist Gulf Coast Rsch. Lab., Ocean Springs, Miss., 1970-74; rsch. asst. Sch. of Fisheries, U. Wash., Seattle, 1975-80; pvt. cons., fisheries biologist McGraw Environ. Consulting, Inc., Seattle, 1980-83; adj. faculty Seattle U., 1983-84; fisheries biologist U. Wash., Seattle, 1984-92; pvt. cons., fisheries biologist McGraw Environ. Consulting, Radford, Va., 1992—; pk. naturalist Nat. Pk. Svc., Gulf Islands Nat. Seashore, Ocean Springs, Miss., summer 1975; vol. cons. Nat. Com. for New River, Radford, 1992—; adj. faculty Coll. William and Mary, 1993—, Radford U., 1993—. Author: Pacific Coast Clam Fisheries, 1983; contbg. author: Dredging Impacts on Fish, 1990. Vol. instr. Feminist Karate Union, Seattle, 1981-92 (1st degree black belt); mem. exec. bd. Fremont Community Coun., Seattle, 1980-84; precinct officer 32d Dist. Dems., Seattle, 1986-92; founder, chief instr. Women's Karate Union, Radford, Va. Recipient Honors scholarship U. Montevallo, 1961-65, W.M. Chapman Meml. scholarship U. Wash., 1980. Mem. Nat. Shellfisheries Assn. (exec. com. 1980-83), Am. Inst. Fisheries Rsch. Biologists, Estuarine Rsch. Fedn., Sigma Xi. Achievements include development of a mitigation plan for Dungeness crab; research in dredging impacts on Dungeness crab and fish in Grays Harbor, Wash.; in the basic biology, life history and population dynamics of arkshell (blood) clams on Eastern shore of Va. Home and Office: McGraw Environ Consulting 7082 Hickman Cemetery Rd Radford VA 24141

MCGREEVY, MARY, retired psychology educator; b. Kansas City, Kans., Nov. 10, 1935; d. Donald and Emmy Lou (Neubert) McG.; m. Phillip Rosenbaum (dec.); children: David, Steve, Mariya, Chay, Allyn, Jacob, Dora. BA in English with honors, Vassar Coll., 1957; postgrad., New Sch. for Social Rsch., NYU, 1958-59, Columbia U., 1959-60, U. P.R., 1963-65, U. Mo., 1965-68, U. Kans.; PhD, U. Calif., Berkeley, 1969. Formerly exec. Doubleday & Co., N.Y.C., 1957-60; chief libr. San Juan Sch., P.R., 1962-63; NIMH drug researcher Russell Sage Found., Clinico de los Addictos, Rio Piedras, P.R., 1963-65; psychiat. researcher U. P.R. Med. Sch., 1966-68, U. Kans., Lawrence, 1966-68; rsch. assoc. Ednl. Rsch., 1968-69; prof. U. Calif., Berkeley, 1968-69, disting. prof., ret., 1969; yacht owner Encore. Author: (poetry) To a Sailor, 1989, Dreams & Illusions, 1993, also articles, poems. Founder, exec. dir. Dora Achenbach McGreevy Poetry Found. Inc. Sproul fellow; fellow Bancroft Libr., Russell Sage Found.; postdoctoral grantee U. Calif. Mem. AAUW (corr. sec. 1991—, bd. dirs. 1991, honoree Ednl. Found. Fund 1993), Am. Cancer Soc., Pres.'s Coun., Broward Women's History Coalition (archivist), Broward County Hist. Com. (vol. archivist), Women in Psych., Union of Concerned Scientists Broward, Inst. for Applid Philosophy, South Fla. Poetry Inst., Fla. Philos. Assn., Ft. Lauderdale Philharm. Soc., Vassar Alumni Assn. (class historian), Pem-Hill Alumni Assn., Secular Humanists South Fla., Vassar Club Kansas City (bd. dirs. 1993), St. Anthony's Cath. Women's Club, Sierra Club (newspaper reporter, environ. com., archivist 1993—). Home: PO Box 900 Fort Lauderdale FL 33302

MCGREGOR, THEODORE ANTHONY, chemical company executive; b. Detroit, Mar. 28, 1944; s. Lorraine Guyeveve Guyette; m. Bonny-Joan Beach, Sept. 14, 1963; children: Todd, Timothy, Amy. Student, Henry Ford Coll., 1961-63. Mem. sales staff Gen. Binding Corp., GBC Sales and Service, Oak Brook, Ill., 1965-69; with indsl. chem. divsn. Diversey Chem. Corp., Chgo., 1967-69; with Detrex Corp., Detroit, 1969-75, regional mgr. indsl. chem. specialties divsn., 1975-77, asst. gen. mgr., 1977-81, gen. mgr. indsl. chem. specialties divsn., 1981-83, 1983-85, corp. v.p., gen. mgr., 1985-89; group v.p. indsl. chem. specialties divsn. Wayne Chem., RTI, Seibert-Oxidermo, Detroit, 1987-93; exec. v.p., also bd. dirs. Wayne Chem., Detroit; pres. TAM Consulting Svcs., Redford, Mich., 1993—; exec. v.p. Asian Rim and Internat. Mktg. Seibert Oxiderno, 1989-90; pres. TAM, 1993—; exec. v.p. Harbor Group, 1993—; bd. dirs. Viking Chem. Co. Mem. Wire Inst., Am. Electroplaters Soc., Porcelain Enameling Inst., Detroit Socs. Coating Tech. Avocations: reading, chess, boating. Home: 26111 Harbour Point Dr Mount Clemens MI 48045-2620

MCGREGOR, WALTER, medical products company designer, inventor, consultant, educator; b. Kiew, Ukraine, Nov. 2, 1937; came to U.S., 1957; s. William and Lydia (Aplass) McG.; m. Helen McGregor, July 18, 1965; children: Roxanne, Walter. BS, Fairleigh Dickinson U., 1973, MBA, 1975. Sect. leader Ethicon Inc., Somerville, N.J., 1965-68, supr., 1968-76, mgr., 1976-83, dir. surg. products devel. and materials engring., 1983-92, dir. of tech., 1992—; guest cons. Wilmer Inst., Johns Hopkins Hosp. Rsch. Lab., Balt., 1965-70; guest lectr. dept. plastic surgery U. Va. Med. Sch., Charlottesville, 1986—. Contbr. articles to profl. jours. Patentee surg. instruments. Life mem. Rep. Presdl. Task Force, Washington, 1984—; mem. Rep. Nat. Com., 1991. Fellow Soc. for Advancement of Med. Instrumentation; mem. Am. Med. Informatics Assn. (founding), Am. Mktg. Assn. (profl.), Med. Mktg. Assn. Avocations: stamp collecting, fishing, reading in surgical developments. Home: 104 Hoffman Rd Flemington NJ 08822 Office: Ethicon Inc US Rt 22 Somerville NJ 08876

MCGREW, STEPHEN PAUL, physicist, entrepreneur; b. Chgo., Sept. 6, 1945; s. Paul Orman and Winifred (Cook) McG.; m. Sharon Beth (Ganansky), May 28, 1989; children: Paula Sue, Jesse Marx, Benjamin Paul. BA in Physics, Math., Ea. Wash. State Coll., 1969; MS in Physics, U. Wash., 1970. Chief physicist Europlex Holographics, B.V., Amsterdam, The Netherlands, 1975-77; chief physicist Holex Corp., Morristown, Pa., 1977-78; mgr. R&D Holotron Corp., Richland, Wash., 1978-79; chmn./CEO, founder Light Impressions Inc., Santa Cruz, Calif., 1980-91; pres., founder New Light Industries, Ltd., Spokane, Wash., 1991—; bd. dirs. Light Impressions Europe, Ltd., Leatherhead, Eng., 1982—. Contbr. articles to profl. jours. Vol. tchr. with sci. demonstrations at local lower, mid. and upper schs., colls. Mem. IEEE, Soc. Photog. Scientists and Engrs., Optical Soc. Am. Achievements include 17 patents in holography and optics, devel. embossing process that is basis for today's mass produced holograms. Office: New Light Industries Ltd 3610 S Harrison Rd Spokane WA 99204

MC GRODDY, JAMES CLEARY, computer company executive; b. N.Y.C., Apr. 6, 1937; s. Charles B. and Helen F. (Cleary) McG.; children: Kathleen, Sheila, Christine, James. B.S., St. Joseph's U., 1958; Ph.D., U. Md., 1964. With IBM, 1965—; v.p. devel. and mfg. gen. tech. div. IBM, White Plains, N.Y., 1977—; prof. Tech. U. Denmark, 1970-71. Contbr. articles to profl. jours.; patentee in field. Fellow IEEE, Am. Phys. Soc. Office: IBM Science and Tech Old Orchard Rd Armonk NY 10504-1709*

MCGRORY, JOSEPH BENNETT, physicist; b. Phila., Feb. 23, 1934; s. John Reardon and Eleanor Custis (Bennett) McG.; m. Thelma Ruth Houk, July 20, 1957; children: Eleanor Ruth, William Dandridge. BS in Math., U. of the South, 1955; PhD in Physics, Vanderbilt U., 1963. Health physicist USPHS, Washington, 1957-60; nuclear theorist Oak Ridge Nat. Lab., 1963-76, sect. head for nuclear theory, 1976-90, asst. dir. physics div., 1990-91; program mgr. for nuclear theory U.S. Dept. Energy, Washington, 1991—. Contbr. articles to profl. jours. Pres. Oak Ridge Civil Music Assn., 1980; bd. dirs. numerous arts orgns. Fellow AAAS, Am. Phys. Soc.; mem. Phi Beta Kappa, Sigma Xi, Omicron Delta Kappa. Episcopalian. Home: 9805 Meadowcroft Ln Gaithersburg MD 20879 Office: US Dept Energy ER-23 G312/GTN Washington DC 20585

MCGUIRE, JAMES HORTON, physics educator; b. N.Y., June 7, 1942; s. Horton E. and Karolyn W. (Wright) McG.; m. V. Jane Rasmussen, Oct. 10, 1981; children: Carrie Marti, Bruce, Brooke. BS, Rensselaer U., 1964; PhD, Northeastern U., 1969. Asst. prof. Tex. A&M U., College Station, 1969-72; prof. Kans. State U., Manhattan, 1972-91; Murchison-Mallory prof. physics Tulane U., New Orleans, 1991—. Assoc. editor: Encyclopedia Physics; contbr. articles to profl. jours. Grantee NSF, 1992, DOE, 1992. Fellow Am. Phys. Soc. (sec.-treas. div. atomic molecular optical physics 1989-92, sec.-elect Internat. Conf. on Physics of Atomic and Elec. Collisions); mem. Am. Chem. Soc., Am. Assn. Physics Tchrs. Democrat. Office: Tulane U Dept Physics New Orleans LA 70118

MCGUIRE, JOHN ALBERT, dentist; b. Warren, Ohio, June 20, 1950; s. Bernard Leo and Lucille Ann (Guarnieri) McG.; m. Pamela Kay Muter, May 30, 1969; children: John, Jessica. BS, Ohio State U., 1972, DDS, 1975. Dentist, capt. USAF, Bellevue, Nebr., 1975-77; dentist pvt. practice Dayton, Tenn., 1977-83, Knoxville, Tenn., 1983—. Author: (short story) Stirs, 1990. Mem. Sertoma Club, Knoxville, 1983-86, Jaycees, Dayton, 1978-81; vol. United Meth. Ch., Tilaran, Costa Rica, 1985. Recipient Scholarship, Fred M. Roddy Found., 1990. Mem. Phi Kappa Phi. Avocations: fly fishing, photography, mtn. biking, music. Home: 301 Grandeur Dr Knoxville TN 37920-6325 Office: 6017 Chapman Hwy Knoxville TN 37920-5932

MCGUIRE, JOHN LAWRENCE, pharmaceuticals research executive; b. Kittanning, Pa., Nov. 3, 1942; s. Lawrence F. and Florence G. (Jones) McG.; m. Pamela Hale, Aug. 2, 1969; children—Megan L., Christa H. B.S., Butler U., 1965; M.A., Princeton U., 1968, Ph.D., 1969; postgrad., Columbia Sch. Bus., 1981. Asst. in instrn. Princeton U., N.J., 1967-69; pharmacologist Ortho Pharm. Corp., Raritan, N.J., 1969-72, sect. head molecular biology, 1972-75, exec. dir. research, 1975-80, v.p. preclinical research and devel., 1980-88, bd. dirs., 1988-93; sr. v.p. rsch. and devel. worldwide, bd. dirs. R.W. Johnson Pharm. Rsch. Inst., Raritan, N.J., 1988-92; sr. v.p. bus. devel., pharmaceutical/diagnostics sector Johnson & Johnson, New Brunswick, N.J., 1992—; adj. assoc. prof. dept. medicine M.S. Hershey Sch. Medicine Pa. State U., 1978—; adj. prof. dept. animal sci. Rutgers U., 1983-92; adj. prof. ob-gyn East Va. Med. Sch., 1987—; cons. NASA, 1985-87. Mem. editorial bd. Ullman's Ency. Indsl. Chemistry, 1987—; contbr. articles to profl. jours.; patentee in field. Mem. exec. bd. Keystone Area council Boy Scouts Am., Harrisburg, Pa., 1975-92 , George Washington council, Trenton, N.J., 1980-86; trustee Raritan Valley Community Coll., N.J., 1986—, vice-chmn., 1990—, United Way of Hunterdon County, N.J., 1983—, pres., 1985-87; trustee Hunterdon Med. Ctr., Flemington, N.J., 1978—, vice chmn., 1984-88, chmn., 1988—; trustee Atlantic Health Systems Morristown, N.J., 1991-93, vice-chmn., 1992—; trustee Tri State United Way, N.Y., 1987-93; bd. dirs. Hunterdon County YMCA, N.J., 1982-87; chmn. bd. dirs. Mid Jersey Health Corp., 1986-88. Recipient Silver Beaver award Boy Scouts Am., 1984; Population Council fellow, 1969. Mem. Am. Soc. Pharmacology and Exptl. Therapeutics, Soc. Exptl. Biology and Medicine, Am. Physiol. Soc., Endocrine Soc., Am. Coll. Ob-Gyn, Am. Soc. Clin. Pharmacology and Therapetics, Licensing Execs. Soc., Biochemistry Soc. Great Britain, Royal Soc. Medicine (U.K.), Am. Chem. Soc. Club: Princeton (N.Y.C.). Home: 9 Sunnyfield Dr Whitehouse Station NJ 08889 Office: Johnson & Johnson 1 Johnson & Johnson Plz New Brunswick NJ 08933

MCGUIRE, MARK WILLIAM, electrical engineer; b. Franklin Park, Ill., June 7, 1962; s. Claude and Carol (Bognar) McG.; m. Deborah Ann Rayas, Sept. 29, 1990; children: Shannon, Connor. BSEE, U. Ill., 1983; M in Mgmt., Northwestern U., 1990. Engr. Motorola, Inc., Arlington Heights, Ill., 1984-91, product mgr., 1991—. Achievements include patent in field. Office: Motorola Inc 1501 W Shure Dr Arlington Heights IL 60004

MCHALE, MAGDA CORDELL, academic administrator, trend analyst; b. June 24, 1921; widow. Sr. research assoc. Ctr. for Integrative Studies, SUNY-Binghamton, 1968-76; sr. research assoc. Ctr. for Integrative Studies, U. Houston, 1977-79; dir. Ctr. for Integrative Studies, SUNY-Buffalo, 1980—. Contbr. articles, papers, reports in field; mem. editorial bd. profl. publications. Grantee UN Environ. Programme, Ctr. for Econ. and Social Studies of Third World, Intergovernmental Bur. for Informatics, Hubert H. Humphrey Inst. Pub. Affairs, Aspen Inst., Population Reference Bur. Fellow World Acad. of Art and Sci., Royal Soc. Arts; mem. N.Y. Acad. Scis., U.S. Assn. for Club of Rome, World Future Soc., Assn. Internationale Futuribles (hon.), World Futures Studies Fedn. (v.p.). Office: SUNY-Ctr for Integrative Studies Sch of Architecture and Planning 333 Hayes Hall Buffalo NY 14214

MC HARGUE, CARL JACK, research laboratory administrator; b. Corbin, Ky., Jan. 30, 1926; s. John David and Virginia (Thomas) McH. B.S. in Metall. Engring., U. Ky., 1949, M.S., 1951, Ph.D., 1953; m. Edith Trovillion, Aug. 28, 1948; children: Anne Odell McHargue Diegel, Carol Virginia, Margaret Katherine McHargue Behrendt; m. Betty Ford, Sept. 30, 1960. Instr. U. Ky., Lexington, 1949-53; with Oak Ridge Nat. Lab., 1953-90, sect. head, 1960-80, program mgr. for materials scis., 1961-88, sr. rsch. staff 1980-90; prof. metall. engring. U. Tenn., Knoxville, 1963—, dir. ctr. materials processing, 1991—; vis. prof. U. Newcastle upon Tyne, Eng., 1987; adj. prof. Vanderfilt U., 1988—. With AUS, 1944-46. Fellow Metall. Soc. AIME, Am. Soc. for Metals; mem. Am. Nuclear Soc., Materials Rsch. Soc., Sigma Xi, Tau Beta Pi. Republican. Presbyterian. Contbr. numerous articles in field to profl. jours. Home: 7201 Sheffield Dr Knoxville TN 37909-2414 Office: U Tenn 121 Perkins Hall Knoxville TN 37996-2000

MCHENRY, DOUGLAS BRUCE, naturalist; b. Stillwater, Okla., Jan. 6, 1932; s. Donald Edward and Bona May (Ford) McH.; m. Martha Margaret Phelon, Apr. 14, 1957 (dec. May 1985); children: Jonathan Keith, Kenneth Bruce, Martha Margaret; m. Martha Locke Hazen, Apr. 6, 1991. BS, U. Wyo., 1954; MS, Utah State U., 1960. Park ranger Colonial NHP Nat. Park Svc., Yorktown, Va., 1960-62; park naturalist Rock Creek Park Nat. Park Svc., Washington, 1962-64; park naturalist Nat. Park Svc., Grand Canyon, Ariz., 1964-66; asst. chief park naturalist Big Bend (Tex.) Nat. Park Nat. Park Svc., 1966-68; asst. chief park naturalist Shenandoah Nat. Park Nat. Park Svc., Luray, Va., 1968-72; asst. chief interpretation Everglades Nat. Park Nat. Park Svc., Homestead, Fla., 1972-74; regional chief interpretation North Atlantic Region Nat. Park Svc., Boston, 1974-86. Asst. scoutmaster troop 100 Boy Scouts Am., Luray, 1968-72; chmn. Barnstable (Mass.) Conservation Commn., 1978-91. Mem. Assn. Interpretative Naturalists (regional dir. 1976-78, v.p. 1987-89), Nat. Assn. Interpretation (founder), Centerville Osteville Lions (pres. 1987, 90), Sigma Xi. Office: Team Interpretation 15 Chilton St Belmont MA 02178

MCHENRY, KEITH WELLES, JR., oil company executive; b. Champaign, Ill., Apr. 6, 1928; s. Keith Welles and Jayne (Hinton) McH.; m. Lou Petry, Aug. 23, 1952 (dec. Oct. 1990); children: John, William; m. Dolores Leo, Mar. 21, 1992. B.S. in Chem. Engring, U. Ill., 1951; Ph.D. in Chem. Engring, Princeton U., 1958. With Amoco Corp. (and affiliates), 1955-93; various positions in R & D Amoco Corp., Whiting, Ind., 1955-74; mgr. process research Amoco Oil Co., Naperville, Ill., 1974-75, v.p. research and devel., 1975-89; sr. v.p. tech. Amoco Corp., Chgo., 1989-93; retired, 1993; Hurd lectr. Northwestern U., 1981; Thiele lectr. in fuels engring. U. Utah, 1983, Gerster Meml. lectr. U. Del., 1987; mem. adv. council Catalysis Center, U. Del., Newark, 1978-83, chmn., 1981-82; mem. adv. council Sch. Engring. and Applied Sci., Princeton U., 1976-82; mem. indsl. adv. bd. Coll. Engring., U. Ill., Chgo., 1979-89, chmn., 1984-86; bd. overseers Sch. Bus. Adminstrn., Ill. Inst. Tech., 1983-86; bd. dirs. Indsl. Research Inst., 1982-90, pres., 1988-89; mem. U.S. Nat. Com. World Petroleum Congress, 1975-83. Trustee North Central Coll., Naperville, 1978—; chmn. area com. Jr. Achievement, 1981-83, 86-88; ordained elder Presbyn. Ch., 1964. Served with U.S. Army, 1946-47. Recipient award Am. Inst. Chem. Engrs., 1988, Univ. Ill., 1989; Gen. Electric fellow, DuPont fellow, 1952-54. Fellow Am. Inst. Chem. Engrs. (editorial bd. jour. 1974-78); mem. Nat. Acad. Engring., Am. Chem. Soc., AAAS, Sigma Xi, Tau Beta Pi. Home: 1853 N Orchard Chicago IL 60614*

MC HENRY, MARTIN CHRISTOPHER, physician; b. San Francisco, Feb. 9, 1932; s. Merl and Marcella (Bricca) McH.; m. Patricia Grace Hughes, Apr. 27, 1957; children: Michael, Christopher, Timothy, Mary Ann, Jeffrey, Paul, Kevin, William, Monica, Martin Christopher. Student, U. Santa Clara, 1950-53; MD, U. Cin., 1957; MS in Medicine, U. Minn., 1966. Intern, Highland Alameda County (Calif.) Hosp., Oakland, 1957-58; resident, internal medicine fellow Mayo Clinic, Rochester, Minn., 1958-61, spl. appointee in infectious diseases, 1963-64; staff physician infectious diseases Henry Ford Hosp., Detroit, 1964-67; staff physician Cleve. Clinic, 1967-72, chmn. dept. infectious diseases, 1972-92, sr. physician infectious diseases, 1992—. Asst. clin. prof. Case Western Res. U., 1970-77, assoc. clin. prof. medicine, 1977-91, clin. prof. medicine, 1991—; assoc. vis. physician Cleve. Met. Gen. Hosp., 1970—; cons. VA Hosp., Cleve., 1973—. Chmn. manpower com. Swine Influenza Program, Cleve., 1976. Served with USNR, 1961-63. Named Disting. Tchr. in Medicine Cleve. Clinic, 1972, 90; recipient 1st ann. Bruce Hubbard Stewart award Cleve. Clinic Found. for Humanities in Medicine, 1985. Diplomate Am. Bd. Internal Medicine. Fellow ACP, Infectious Diseases Soc. Am., Am. Coll. Chest Physicians (chmn. com. cardiopulmonary infections 1975-77, 81-83), Royal Soc. Medicine of Great Britain; mem. Am. Soc. Clin. Pharmacology and Therapeutics (chmn. sect. infectious diseases and antimicrobial agts., 1970-77, 80-85, dir.), Am. Thoracic Soc., Am. Soc. Clin. Pathologists, Am. Fedn. Clin. Rsch., Am. Soc. Tropical Medicine and Hygiene, Am. Soc. Microbiology, N.Y. Acad. Scis. Contbr. numerous articles to profl. jours., also chpts. to books. Home: 2779 Belgrave Rd Cleveland OH 44124-4601 Office: 9500 Euclid Ave Cleveland OH 44195

MCHUGH, EARL STEPHEN, dentist; b. Colorado Springs, Colo., Feb. 27, 1936; s. Earl Clifton and Margaret Mary (Higgins) M.; m. Joan Bleckwell, Aug. 24, 1957; children: Kevin, Stacey, Julie. BA, Cornell U., 1958; DDS, U. Mo., 1962. Pvt. practice, Kansas City, Mo., 1964—; lectr. U. Mo. Dental Sch., Kansas City, 1988, clin. staff, 1989, 90, 91, 92, 93; cons. Hallmark, Inc., Kansas City, 1988. Contbr. articles to profl. jours. Deacon Presbyn. Ch. Prairie Village, Kans., 1982-84; vol. Shawnee Mission Hosp. Kans., 1985-88; lectr. Drug Recovery Program, Kansas City, Kans., 1988-89, 92, 93. Capt. Dental Corp, U.S. Army, 1962-64. Mem. Valley Hosp. Assn. (bd. dirs. 1989-90), Audubon Soc. Ornithologist of Yr. Kansas City chpt. 1990), Kans. Onrithol. Soc. (v.p. 1989-90, pres. 1990-91), Internat. Coun. Bird Preservation (Kans. del. 1990, coord. Kans. Breeding Bird Atlas 1992, 93), Omicron Kappa Upsilon, Chi Psi.

MCHUGH, HELEN FRANCES, university dean, home economist; b. Tucson, Aug. 19, 1931; d. James Patrick and Mary Catherine (Hochstatter) McH.; m. Herbert J. Brauer, Mar. 26, 1982. B.S. with distinction, U. Mo.-, Columbia, 1958, M.S., 1959; Ph.D., Iowa State U., 1965. Instr. U. Tex., Austin, 1961-63; asst. prof. U. Tex., 1963-66, asso. prof., 1966-67; asso. prof. Ind. State U., Terre Haute, 1967-69; asso. prof., dept. head Oreg. State U., Corvallis, 1969-73; prof., dean Coll. Home Econs., U. Del., Newark, 1973-75; prof. consumer econs., dean Coll. Human Resource Scis. assoc. dir. Colo. Experiment Sta., Colo. State U., Ft. Collins, 1976—; cons. in field. Chmn. policy bd.: Jour. Consumer Research, 1978-80. Recipient Disting. Service award U. Mo. Alumni Assn., 1975. Mem. Am. Econ. Assn., Am. Agrl. Econs. Assn., Am. Home Econs. Assn. (bd. dirs. 1973-75), Assn. Adminstrs. Home Econs. (pres. 1978-79), Sigma Xi, Gamma Sigma Delta, Phi Kappa Phi. Roman Catholic. Office: Colorado State U Agricultural Experiment Station 16 Administration Bldg Fort Collins CO 80523

MC HUGH, PAUL R., psychiatrist, neurologist, educator; b. Lawrence, Mass., May 21, 1931; s. Francis Paul and Mary Dorothea (Herlihy) McH.; m. Jean Barlow, Dec. 27, 1959; children: Clare Mary, Patrick Daniel, Denis Timothy. AB, Harvard U., 1952, MD, 1956. Diplomate: Am. Bd. Psychiatry and Neurology. Intern Peter Bent Brigham Hosp., Boston, 1956-57; resident in neurology Mass. Gen. Hosp., 1957-60, fellow in neuropathology, 1958-59; teaching fellow in neurology and neuropathology Harvard, 1957-60; clin. asst. psychiatry Maudsley Hosp., London, Eng., 1960-61; mem. neuropsychiatry div. Walter Reed Army Inst. Research, Washington, 1961-64; asst. prof. psychiatry and neurology Cornell U., N.Y.C., 1964-68; assoc. prof. Cornell U., 1968-71, prof., 1971; dir. electroencephalography N.Y. Hosp., 1964-68; founder, dir. Westchester div. dept. psychiatry N.Y. Hosp. Bourne Behavioral Resch. Lab., 1967-68, clin. dir., supr. psychiat. edn., 1968-73; prof., chmn. dept. psychiatry U. Oreg. Health Sci. Center, Portland, 1973-75; Henry Phipps prof. psychiatry Johns Hopkins, Balt., 1975—; chmn. dept. psychiatry Johns Hopkins, 1975—, prof. dept. mental hygiene, 1976—; psychiatrist-in-chief Johns Hopkins Hosp., 1975—; dir. Blades Ctr. for Clin. Practice and Rsch. in Alcoholism Johns Hopkins Med. Inst., 1992—; chmn. med. staff Johns Hopkins Hosp., 1983-89, trustee, 1983—; vis. prof. Guys Hosp., London, Eng., 1976; chmn. bio-psychology Study sect. NIH, 1986-89. Author: The Perspectives of Psychiatry, 1983; (with Phillip R. Slavney) Psychiatric Polarities, 1987; contbg. author: Cecil-Loeb Textbook of Medicine, 1974; mem. editorial bd. Jour. Nervous and Mental Disease, Comprehensive Psychiatry, Medicine, and Psychol. Medicine, 1976—; editor (with Victor A. McKusick) Genes, Brain and Behavior, 1990; assoc. editor sects. Am. Jour. Physiology contbr. articles to med. jours. Mem. Md. Gov.'s Adv. Com., 1977-80. Grantee NIH, 1964-68, 67-70, 70-74, 75—; recipient William C. Menninger award ACP, 1987. Fellow Royal Coll. Psychiatry, Am. Psychiat. Assn.; mem. Am. Neurol. Assn., Am. Physiol. Soc., Harvey Soc., Am. Psychopath. Assn., Pavlovian Soc., Inst. Medicine, NAS. Club: W Hamilton St. Home: 3707 St Paul St Baltimore MD 21218-2403 Office: Johns Hopkins Med Insts 600 N Wolfe St Baltimore MD 21205-2104

MCHUGH, WILLIAM DENNIS, dental educator, researcher; b. Berwick, Northumberland, U.K., May 8, 1929; came to U.S., 1970, naturalized, 1977; s. Gerald D. and Jeanne (Stewart) McH.; m. Susan Mary Cort, Feb. 22, 1958; children: Moira A. McHugh Douglas, Alan N.C. B of Dental Surgery cum laude, St. Andrews (U.K.) Univ., 1954, DDS, 1959. House surgeon in oral surgery Dundee (Scotland) Dental Sch., 1950, sr. house surgeon in orthodontics, 1952-53; instr. dept. operative dentistry Royal Dental Sch., Malmö, Sweden, 1954-55; lectr. in periodontology and pathology Royal Dental Hosp., London, 1956-58; lectr. periodontology and preventive dentistry St. Andrews U. Dental Sch., 1959-62, sr. lectr., head dept., 1962-65, prof., chmn. dept., 1965-70; prof. clin. dentistry and dental rsch., assoc. dean for dental affairs Sch. Medicine and Dentistry, U. Rochester, N.Y., 1970—; dir. Eastman Dental Ctr., Rochester, 1970—; cons. Scottish Ea. Regional Hosp. Bd., 1962-65, Genesee (N.Y.) Hosp., 1971—; cons. planning and projects various schs. dentistry including U. Conn., 1984, U. Riyadh, Saudi Arabia, 1984, U. Pa., 1985-90, Columbia U., 1986—, W.Va. U., 1986-87; cons. Nat. Inst. Dental Health, NIH, Bethesda, Md., 1975—; mem. Nat. Adv. Dental Rsch. Coun., 1987-91; dentist-in-chief Strong Meml. Hosp., 1971-75, dentist, 1975—; mem. ambulatory care com. Finger Lakes Health Systems Agy., 1976-78; med. adv. com. Monroe Community Hosp., 1971-83. Editor: Dental Plaque, 1970; co-editor: Systematized Prevention of Oral Disease, 1987; assoc. editor Jour. Periodontal Rsch., 1966-71, Jour. Am. Coll. Dentists, 1987—; mem. editorial bd. Community Dentistry and Oral Epidemiology, 1972-74, Preventive Dentistry, 1973-81, J. Clin. Periodontology, 1989—; editor Proc. Brit. Soc. Periodontology, 1960-62; contbr. numerous articles to profl. jours. Mem. archtl. rev. bd. Town of Brighton, 1982-89. Flight lt. RAF, 1951-53. Rsch. fellow and Nuffield Found. fellow Royal Dental Coll., Sweden, 1954-56, rshc. fellow oral pathology U. Birmingham, 1958-59; NIH grantee, 1970-87, 78-81, 80—, Eastman Kodak Co., grantee 1982-88. Fellow Am. Coll. Dentists, Internat. Coll. Dentists, AAAS; mem. Am. Assn. Dental Rsch. (pres. 1983-84), Internat. Assn. Dental Rsch. (pres. 1988-89), Am. Assn. Dental Schs. (v.p. advanced edn. and fed. dental svcs. 1980-83), ADA (cons. Coun. Dental Therapeutics 1985), Brit. Soc. Periodontology (pres. 1969-70), Brit. Dental Assn. (pres. N. of Scotland br. 1967-68), Monroe County Dental Soc. (community svcs., hosps. profl. conduct com. 1985-86), Dental Soc. State N.Y., Rochester Dental Study Club, Royal Coll. Surgeons of Edinburgh, Internat. Coll. Dentists, Genesee Valley Club (Rochester) (gov. 1987-93), Rotary. Avocations: gardening, tennis, squash. Office: Eastman Dental Ctr 625 Elmwood Ave Rochester NY 14620-2989

MCILHERAN, MARK LANE, mechanical engineer; b. Bryan, Tex., Mar. 9, 1961; s. Louis Colyar and Beverly Ann (Gilliam) McI.; m. Cherri Jill Clifton, July 16, 1988; 1 child, Elizabeth Grace. BSME, U. Tex., Arlington, 1988. Design engr. Purotek, Irving, Tex., 1987-88; div. mgr. EDC Ozone Systems, Irving, 1988-91; engring. cons. Heron Tech., Colleyville, Tex., 1991-92; dir. engring. Greentech Inc., Grapevine, Tex., 1992—. Mem. Internat. Ozone Assn. Republican. Office: Heron Tech 1201 A Minters Chapel Grapevine TX 76051

MCILVAIN, JESS HALL, architect, consultant; b. Denton, Tex., Mar. 29, 1933; s. Charles L. and Edith (Hall) McI.; m. Joni Wimberley, Aug. 23, 1959; children—James Sean, Sheila Maria. B.Arch., Tex. Tech. U., 1959. Designer, Nesmith & Lane, architects, El Paso, 1959, Garland & Hilles, architects, El Paso, 1960-63, William Metcalf, Architect, Washington, 1963; architect, designer Cooper & Auerbach, Washington, 1963-65, Bucher-Meyers, 1966-67; project mgr. Weihe, Black, Kerr, Washington, 1967-68; designer Callmer & Milstead, Washington, 1968-69; dir. archtl. services Tile Council Am., Inc., Washington, 1969-80; lectr. on ceramic tile, 1970—; pres. Jess McIlvain & Assocs., Washington, 1980—; Third v.p. Woodacres PTA, 1975-76, treas., 1978-79; treas. Woodacres Citizens Assn., 1973-75, pres., 1975-76. Served with AUS, 1953-55. Recipient Horizon Homes Regional award, 1962. Mem. AIA (bd. dirs. Washington chpt. 1968, mem. nat. codes and standards com. 1974-77), Constrn. Specifications Inst. (program chmn. Washington chpt. 1972-75, 3d v.p., 1974-76, chmn. CCS Review Com., 1987—, bd. dirs. 1988-90, excellent service award Washington chpt. 1973, 75, 91, 92, award region 2, 1976, tech. commendation 1977, author specification series 1971, 78, cert. constrn. specifier 1984—), ASTM, Bldg. Ofcls. and Code Adminstrn. Internat., Internat. Conf. Bldg. Ofcls., So. Bldg. Code Conf., Bldg. Industry Assn. Reps. (pres. 1976-77), Tex. Tech U. Alumni Assn. (pres. Washington chpt. 1971-73, dist. rep. 1975-77), Capital PC Users Group (chmn. Word Processing Spl. Interest Group 1983-85, v.p. 1986-87), Nat. Tile Contractors Assn. (tech. com. 1986—), Tile Council Am. (handbook com. 1985—, coordinator with CTDA archtl. student tile competition 1979-87, speaker 7th Internat. Symposium on Ceramics, Bologna, Italy, 1988, speaker Cersaie 1989, 8th internat. Rimini, 1992, Internat. Tile & Stone Exhibition, 1987-93), Sigma Chi. Republican. Methodist. Clubs: Kenwood Country, Tex. Tech, Century. Lodge: Lions (pres. 1974-76). Contbr. articles to profl. jours. Address: 6012 Woodacres Dr Bethesda MD 20816

MC ILVEEN, WALTER, mechanical engineer; b. Belfast, Ireland, Aug. 12, 1927; s. Walter and Amelia (Thompson) McI.; came to U.S., 1958, naturalized, 1963; M.E., Queens U., Belfast, 1940, H.V.A.C., Borough Polytechnic, London, 1951; m. Margaret Teresa Ruane, Apr. 17, 1949; children: Walter, Adrian, Peter, Anita, Alan. Mech. engr. Davidson & Co., Belfast, 1943-48; sr. contract engr. Keith Blackman Ltd., London, 1948-58; mech. engr. Fred S. Dubin Assos., Hartford, Conn., 1959-64; chief mech. engr. Koton & Donovan, West Hartfor, Conn., 1964-66; prin., engr. Walter McIlveen Assos., Avon, Conn., 1966—. Mem. IEEE, ASME, ASHRAE, Illuminating Engring. Soc., Hartford Engring. Club, Conn. Engrs. in Pvt. Practice. Mem. Ch. of Ireland. Home: 3 Valley View Dr Weatogue CT 06089-9714 Office: 195 W Main St Avon CT 06001

MCINTEE, GILBERT GEORGE, materials testing engineer; b. Cornwall, Ont., Can., Mar. 15, 1943; s. Ralph Edmond McIntee and Cecile (Lottie) McIntee Hart; m. Marsha Widrick, Oct. 12, 1985 (div. Apr. 1989); m. Madeleine Arseneau, June 3, 1989. BSc, Queen's U., Kingston, Ont., 1967; MBA, U. Toronto (Ont., Can.), 1972; LLB, U. Ottawa (Ont., Can.), 1985. Testing engr. St. Lawrence Seaway, St. Catharines, Ont., Can., 1967-68, constrn. engr., 1968-70; constrn. supt. Inducon Developments, Toronto, 1971, Laing Constrn., Toronto, 1972; testing engr. Warnock Hersey, Toronto, 1972-73, mgr. ops., 1973-75; pres. St. Lawrence Testing, Cornwall, Ont., Can., 1975—; cons. for slate roofs on Heritage bldgs. including the Ont. Legis. Bldg. and Osgoode Hall Law Bldg., Toronto. Mem., chmn. Cornwall Suburban Rds. Commn., 1978—; bd. dirs. Cornwall C. of C., 1976-78, Ont. C. of C., Toronto, 1979-82. Recipient Award of Appreciation ASTM, 1986, Award of Recognition (w), Can. Testing Assn., 1987-90, Award of Honour, Cornwall C. of C., 1979. Mem. Law Soc. Upper Can. (barrister), Engring. Inst. Can., Can. Soc. for Civil Engring., Can. Geotech. Soc., Union Internat. des Laboratoires Independent (bd. dirs. 1989—), Can. Testing Soc. (bd. dirs. 1978—, pres. 1983-85, v.p. 1982-83, sec.-treas. 1978-82), ASTM (chmn. com. C18 on dimension stone 1979-85, 92—, vice-chmn. com. C18 1986-89, chmn. subcom. C18.01 test methods 1975-80, 85—, chmn. subcom. C18.01 task group on flexure, chmn. subcom. C18.01 task group on abrasion, com. on tech. com. ops.), Assn. Profl. Engrs. Ont. (commendation Toronto 1981, registered profl. engr.). Home: 606 Riverdale Ave, Cornwall, ON Canada K6J 2K6 Office: St Lawrence Testing and Inspection Co Ltd, 814 2d St W, Cornwall, ON Canada K6J 1H6

MCINTIRE, MATILDA STEWART, pediatrician, educator, retired; b. Bklyn., July 15, 1920; d. David Horatio and Lucy Ethel (Warner) Stewart; m. Waldean Chester McIntire, July 12, 1947; 1 child, David Stewart. BA, Mt. Holyoke Coll., 1942, DSc, 1989; MD, Albany Med. Coll., 1946. Technician Rockefeller Found. Med. Rsch., N.Y.C., 1942-43; med. dir. Nebr. Regional Poison Control Ctr., Children's Hosp., Omaha, 1955-61, 79—; dir. div. maternal and child health Douglas County Health Dept., Omaha, 1966-73; dir. ambulatory pediatrics Creighton U. Sch. Medicine, Omaha, 1973-89, prof. pediatrics, 1973-90, prof. emeritus pediatrics, 1991—; sr. cons. pediatrics U. Nebr. Coll. Medicine, Omaha, 1991—; Co-author: Patient Care Emergency Handbook, 1988; editor: Accident Prevention, 1987, Injury Control for Children and Youth, 1987, Suicide in Children and Youth, 1980; co-editor: Handbook of Common Poisons in Children, 1976; contbr. articles to profl. jours. Bd. dirs. Health Planning Coun. of the Midlands, Omaha, 1975. With med. corps. U.S. Army, 1948-49. Named Woman of Yr. Omaha Woman's Polit. Caucus, 1974; recipient Recognition award Head Start, 1978, 81, 3rd Annual Community Svc. award Mid Am. coun. Boy Scouts Am., 1985. Fellow Am. Acad. Pediatrics (citation 1980), Am. Assn. Poison Control Ctrs. (Recognition award 1975), Am. Acad. Clin. Toxicology; mem. Ambulatory Pediatric Assn. Episcopalian. Home: 1510 S 80th St Omaha NE 68124 Office: Regional Poison Control Ctr Childrens Meml Hosp 8301 Dodge St Omaha NE 68114

MCINTOSH, DON LESLIE, electrical engineer; b. Ft. Worth, July 22, 1959; s. Robert Leroy and Kala Janice (Miller) McI. BSEE, U. Mo., Rolla, 1982; MS in MIS, No. Ill. U., 1991. Assoc. engr. Ark. Power and Light, Little Rock, 1983-84; engr. Emerson Electric, St. Louis, 1984-85, Ralston Purina, St. Louis, 1985-87, M&M Mars, Chgo., 1987; pres. CIMware, Inc., Westmont, Ill., 1987—; advisor, cons. Met. Water Reclamation Dist., Chgo., 1991—. Vol. Ronald McDonald House, St. Louis, 1992—; big brother Big Bros./Big Sisters, DuPage County, Ill., 1990—. Mem. NSPE, IEEE, Instrument Soc. Am. Achievements include development of direct-drive system for major appliances (washers, air conditioners, etc.). Home and Office: CIMware Inc 5810 Raintree Ct Westmont IL 60559-2122

MCINTOSH, DONALD HARRY, electrical engineering consultant; b. Villisca, Iowa, Dec. 6, 1919; s. Harry Alexander and Maud Ethel (Weimer) McI.; m. Mary Louise Johnson, Sept. 12, 1943; children: Fredric Donald, Ann Louise, William Harry, Jill Marie, Gregory Johnson. BSEE, Iowa State U., 1942; student, Princeton U., 1944, MIT, 1945. Engr. elec. applications Allis-Chalmers Mfg. Co., West Allis, Wis., 1942-56; elec. specialist engr. E.I. du Pont de Nemours & Co., Inc., Wilmington, Del., 1956-72, sr. elec. engr., 1972-74, sr. design cons., 1974-78, prin. design cons., 1978-82; pvt. practice Newark, Del., 1982—; chief elec. engr. Fleetridge, Inc. Engring. Corp., Wilmington, 1982-83. Contbr. articles to profl. jours. Troop com. chmn. Boy Scouts Am., Newark, Del., 1963-78; bd. elders, trustee Sunday sch. tchr. 1st Presbyn. Ch., Newark, 1965-75, Sunday Sch. tchr.; treas. Nottingham Civic Assn., 1992-93. Lt. USNR, 1944-45. Fellow IEEE (various offices, Industry Applications Soc. and PCIC divs. 1st Paper prize 1981, Best Paper award 1971), Nat. Elec. Code (mem. code-making panel #5, tech. coms. 1975-92), Nat. Fire Protection Assn. (panelist on safety-related work practices NFPA ann. meeting 1990), Del. Assn. Profl. Engrs. (mem. law enforcement and ethics com. 1983—), Del. Pony Club (past dist. commr.), Delmarra Morgan Horse Club (pres.), DuPont Country Club. Republican. Presbyterian. Avocations: raising and training Morgan horses, combined driving competitions. Home and Office: Donald H McIntosh Inc 107 Radcliffe Dr Newark DE 19711-3146

MCINTOSH, DONALD WALDRON, retired mechanical engineer; b. Portland, Maine, June 21, 1927; s. Charles Henry and Annie Louise (Waldron) McI.; m. Margaret Mae Mollison, July 20, 1963; 1 child, Meredith. BS in Engring. Physics, U. Maine, 1950. Registered profl. engr., Maine, N.H. Nuclear engr. Portsmouth Naval Shipyard, Kittery, Maine, 1950-83, retired, 1983; cons. engr. York Harbor, Maine, 1983—. Mem. York Recycling Commn. With USN, 1945-46. Recipient Sci. and Math. medal Rensselaer Poly. Inst., 1945. Mem. NSPE, ASME, Nat. Assn. Naval Tech. Suprs. (chpt. pres. 1970), Sigma Pi Sigma. Republican. Congregationalist. Home: 36 Norwood Farms Rd York Harbor ME 03911

MCINTOSH, GREGORY CECIL, manufacturing engineer; b. Ft. Hood, Tex., Dec. 19, 1949; s. Horace Samuel and Phyllis Mary (Mountford) McI.; m. Carol Ann Hackett, 1978 (div. 1986); m. Sandra Lee Lauless, Aug. 18, 1990; children: Kyle Reagan, Sean Michael. BA, Calif. State U., 1981. Corp. pres. H.M.S. Engring., Inc., Gardena, Calif., 1979-80; foreman Essick-Hadco Mfg. Co., L.A., 1980-84; mfg. engr. Hughes Aircraft Co., El Segundo, Calif., 1984-89; sr. mfg. engr. McDonnell Douglas Corp., Long Beach, Calif., 1989—. Contbr. articles to profl. jours. including Terrae Incognitae, Portolan, Vikingship., American Neptune. Mem. Soc. for the History of Discoveries, Internat. Soc. for the History of Cartography. Home: 19615 Donna Ave Cerritos CA 90701 Office: McDonnell Douglas Corp 3855 Lakewood Blvd Long Beach CA 90808

MCINTOSH, HELEN HORTON, research scientist; b. Artesia, N.Mex.; d. Nolan and Laura Virginia (Featherston) Horton; m. John Wallace McIntosh, Aug. 28, 1965; children: Michele Marie, Katherine Lynn, John Douglas. BS in Natural Sci., Okla. State U., 1962; MS in Chemistry, St. Louis U., 1970, PhD in Pharmacology, 1985. Rsch. technician U. Tex. Southwestern Med. Sch., Dallas, 1962-64; tchr. chemistry Dallas Ind. Sch. Dist., 1964-65, St. Louis C.C., 1976-77; rsch. technician Mo. Inst. Psychiat., St. Louis, 1978-80; postdoctoral fellow Wash. U. Sch. Medicine, St. Louis, 1985-89, rsch. assoc., 1990—. Contbr. articles to profl. publs. Bd. dirs. LWV, Met. Coun. St. Louis and St. Louis County, 1970-72; classroom and office vol. Henry Hough Elem. Sch., Kirkwood, Mo., 1977-80; mem. youth adv. com. Eliot Unitarian Chapel, Kirkwood, 1984-86. Grantee Nat. Inst. Aging, St. Louis U., 1982, Nat. Soc. to Prevent Blindness, Washington U., 1991. Mem. AAAS, Soc. Neurosci., Assn. Rsch. in Vision and Ophthalmology. Achievements include discovery that aspects of norepinephrine function in hypothalmus of F-344 rats decrease with increasing age; age-related decline in growth-associated protein (gap-43) levels is comparable to decline in plasticity reported for cat visual cortex. Office: Washington U Sch Medicine 660 S Euclid St Saint Louis MO 63110

MCINTYRE, JOHN PHILIP, JR., physics educator; b. Lancaster, Pa., Sept. 26, 1949; s. John Philip and Jennette Russell (McBurney) McI. BS in Phys. Sci., Mich. State U., 1970, MA in English, 1972. Cert. secondary tchr., Mich. Lab. aide Hort. Dept., Mich. State U., East Lansing, 1971-74; sci. aide East Lansing Pub. Schs., 1974-79; sec. St. Johns (Mich.) Pub. Schs., 1979-84; ednl. equip./supplies technologist Physics/Astronomy dept. Mich. State U., 1984—. Musician casette tape: Totentanz "On Beyond Music", 1992, Mutually Assured Destruction "Ambient Genocide", 1992; contbr. articles to profl. jours. Mem. Am. Assn. Physics Tchrs., Physics Instructional Resource Assn., Sci. Theatre (Appreciation award 1992). Office: Mich State U Physics/Astronomy Dept East Lansing MI 48824

MCINTYRE, PETER MASTIN, physicist, educator; b. Clewiston, Fla., Sept. 26, 1947; s. Peter Mastin and Ruby Eugenia (Richaud) McI.; m. Rebecca Biek, June 29, 1968; children: Peter B., Colin H., Jana M., Robert J. BS, U. Chgo., 1967, MS, 1968, PhD, 1973. Asst. prof. Harvard U., Cambridge, Mass., 1975-80; group leader Fermilab, Batavia, Ill., 1978-80; assoc. prof. Tex. A&M U., College Station, 1980-84, prof. physics, 1985—, assoc. dean Coll. of Sci., 1990-92; pres. Accelerator Tech. Corp., College Station, 1988-92; dir. Tex. Accelerator Ctr., The Woodlands, 1991—. Editorial advisor Plenum Press, Saunders Pub. Co., 1989—; prin. author Tex. SSC Site Proposal, 1988. Sloan Found. fellow, 1976-78; recipient IR-100 award Indsl. Rsch. Mag., 1980. Mem. AAAS, Am. Phys. Soc. (pres. Tex. sect. 1990-91). Achievements include patents for Proton-Antiproton Colliding Beams, Continuous Unitized Tunneling System, Gigatron High Power Microwave Amplifier, Knife-edge Chamber for X-rays and Changed Particles, and X-ray Disinfestation of Foods. Home: 611 Montclair Ave College Station TX 77840-2868 Office: Tex A&M U Dept Physics College Station TX 77843

MCKAIN, THEODORE F., mechanical engineer; b. Gary, Ind., Dec. 11, 1946; s. Robert Frank and Patricia Jean (Grange) McK.; m. Genevieve A. Miller, Aug. 22, 1970; children: Robert P., Brian T., Kurt W. BSME, Rose Poly. Inst., 1968; MSE, Purdue U., 1975. Registered profl. engr., Ind. Compressor aero. staff Allison Gas Turbines, Indpls., 1968-80, compressor aerodynamics supr., 1980-85, performance chief, 1985-87, preliminary design chief, 1987-90, chief engr. turbofans, 1990—. Mem. AIAA (air breathing propulsion com. 1989-93). Home: 9109 Log Run Dr S Indianapolis IN 46234

MCKAY, COLIN BERNARD, biomedical engineer; b. London, Mar. 5, 1949; came to U.S., 1980; s. Bernard John and Hertha (Zimmermann) McK. BSc, U. Strathclyde, Glasgow, Scotland, 1973; PhD, U. Tech., Compiegne, France, 1980. Scholar Centre Nat. de Recherche Scientifique, Marseille, France, 1976; postdoctoral fellow U. So. Calif., L.A., 1980-83; rsch. assoc. Columbia U., N.Y.C., 1983-88; rsch. scientist Rice U., Houston, 1988—. French Govt. scholar, Marseille, 1976; recipient Travel award European Soc. Microcirculation, Oxford, 1983, Nat. Rsch. Svc. award NIH, 1987. Mem. AAAS, Internat. Soc. Biorheology, Biomed. Engring. Soc., N.Y. Acad. Scis., Sigma Xi. Achievements include characterization of temporal and spatial distbn. of blood flow in the microcirculation, viscosity and hematocrit reduction for suspensions of red blood cells with altered cellular deformability in narrow tubes, the determination of oxygen transport to and from blood in artificial capillaries. Office: Rice U Biomed Engring Lab PO Box 1892 Houston TX 77251-1892

MCKAY, DONALD ARTHUR, mechanical contractor; b. Providence, June 10, 1931; s. Benjamin Arthur and Florence (Heeney) McK.; m. Janette Capellaro, Dec. 30, 1978; children by previous marriage: Susan Kelly, Barbara Albury, Laura Lower, Douglas. AB, Harvard U., 1952. Registered profl. engr., Mass. Sales engr. C.P. Blouin, Cambridge, Mass., 1955-60; contract mgr. to v.p. Limbach Co., Boston, 1960-68; exec. v.p. Tougher Heating & Plumbing Co., Albany, N.Y., 1968-74; chmn., chief exec. officer Tougher Industries, Albany, 1986—, pres., 1974-86; v.p. Spunduct Inc. Pres. Fifty Group (Albany) 1991-92; mem. corp. gifts com. Albany Med. Ctr., 1978-84; chmn. 25th reunion fund raising com. of upstate N.Y., Harvard Class '52; mem. curriculum adv. bd. Hudson Valley Community Coll.; trustee Coll. St. Rose; bd. dirs. Empire State Aeroscis. Mus., Ctr. for Econ. Growth, Albany Meml. Hosp. (chmn. 1988-91). With USN, 1951-54; active Albany Airport Citizens Adv. Bd., U. Albany Found. Mem. ASHRAE, NSPE, Am. Soc. Sanitary Engrs., Am. Soc. Plumbing Engrs., Mech. Contractors Assn. Am. (pres. capital dist. 1981-82, pres.-elect 1988, pres. 1989-90), Mech. Contractors Assn. N.Y. State (v.p. 1981-82, pres. 1981-82), Subcontrators Assn. N.Y. (bd. dirs.), AiN.Y. (bd. dirs.), Aircraft Owners and Pilots Assn., Exptl. Aircraft Assn., U.S. C. of C. (small bus. coun.), Albany-Colonie C. of C. (bd. dirs., James Michael's Envoy salute, 1993), Clan Mackay Soc. N.Am. (nat. v.p.). Congregationalist. Clubs: Harvard (pres. N.E. N.Y. chpt. 1987-88), Pvt. Industry Coun., Ft. Orange (Albany), Masons (Dorchester, Mass.), Wolferts Roost Country, Schuyler Meadows. Home: 6 Park Ridge Albany NY 12204-2233 Office: Tougher Industries PO Box 4067 175 Broadway Albany NY 12204-0067

MCKEAN, HENRY P., mathematics institute administrator; b. Wenham, Mass., Dec. 14, 1930. AB, Dartmouth Coll., 1952; PhD, Princeton U., 1955. Instr. Princeton (N.J.) U., 1955-57; instr. MIT, Cambridge, 1958-63, prof., 1964-66; prof. Rockefeller U., N.Y.C., 1966-70, dep. dir., chmn. dept. math., 1984-88; prof. Courant Inst. Math. Scis. NYU, 1970—, dir., 1988—; vis. prof. Kyoto U., 1957-58, Rockefeller U., 1963-64; George Eastman prof. Balliol Coll., Oxford U., 1979-80. Contbr. numerous articles to profl. jours. Mem. NAS, Am. Acad. Arts and Scis. Office: NYU Courant Inst Math Scis 251 Mercer St New York NY 10012-1185*

MCKEARN, THOMAS JOSEPH, immunology and pathology educator, scientist; b. Rockford, Ill., Oct. 4, 1948; s. Gerald Andrew and Dorothy (Green) McK.; m. Trina Marie Berg (div. 1982); 1 child, Joseph; m. Patricia Joan Hickey; children: Marissa, David. BA in Chemistry, Ind. U., 1971; PhD in Immunology, U. Chgo., 1974, MD, 1976. Rsch. assoc., instr. dept. pathology, spl. immunopathology fellow, resident U. Chgo., 1977-78; head immunoprotein lab. Hosp. of U. Pa., Phila., 1978-81; v.p. R&D Cytogen Corp., Princeton, N.J., 1981-87, sr. v.p. sci. affairs, 1987-90, exec. v.p., 1990-91, pres., 1991—, also bd. dirs.; asst. prof. dept. pathology, U. Pa., Phila., 1978-81, adj. asst. prof., 1981-87, adj. assoc. prof., 1987—. Editor: Antibody-Mediated Delivery Systems, 1988, Monoclonal Antibodies Hybridomas: Dimension in Biological Analyses, 1980; patentee in antibody developments. Bd. dirs. New Hope/Solebury (Pa.) Sch. Dist., 1987-91. Arthur Metz scholar Ind. U., 1966-68. Mem. Am. Assn. Immunologists, Sigma Xi, Phi Beta Kappa, Alpha Omega Alpha. Republican. Office: Cytogen Corp CN-5308 600 Coll Rd E CN-5308 Princeton NJ 08540-5308

MCKECHNIE, JOHN CHARLES, gastroenterologist, educator; b. Louisville, Feb. 1, 1935; s. Albert Hay and Edna Scott (Johnson) M.; children: Steven Keith, Kevin Stuart. BA, U. Louisville, 1955; MD, Baylor Coll. Medicine, 1959. Diplomate Am. Bd. Internal Medicine, Am. Bd. Gastroenterology. Intern Jefferson Davis Hosp., Houston, 1959-60; resident in internal medicine Baylor Affiliated Program, Houston, 1960-61, 65-66; gen. practice medicine, Benham, Ky., 1964; practice medicine specializing in gastroenterology, Houston, 1966—; clin. instr. Baylor Coll. Medicine, Houston, 1966-69, asst. prof., 1969-72, assoc. prof., 1972-77, prof., 1977—; mem. staff Methodist Hosp.; cons. Ben Taub Hosp., St. Luke's Episcopal Hosp. Served to capt. USMC, 1962-64. Fellow Am. Coll. Gastroenterology (gov. Tex. 1979-80, trustee 1981-84), ACP; mem. AMA, So. Med. Assn., Tex. Med. Assn., Am. Gastroent. Assn., Digestive Disease Found., Am. Soc. Gastrointestinal Endoscopy, Tex. Soc. Gastrointestinal Endoscopy, Houston Gastroent. Soc. (pres. 1983), Alpha Omega Alpha. Republican. Baptist. Contbr. numerous articles to profl. jours. Office: 6560 Fannin St Ste 1630 Houston TX 77030-2762

MCKEE, DAVID LANNEN, economics educator; b. St. John, N.B., Can., Apr. 18, 1936; came to U.S., 1961; s. Horace George and Mary K. (Lannen) McK. BA magna cum laude, St. Francis Xavier, 1958; MA, U. N.B., Fredericton, 1959; PhD, U. Notre Dame, 1966. Lectr. Ind. U., Ft. Wayne, 1965-66, asst. prof., 1966-67; asst. prof. Kent (Ohio) State U., 1967-69, assoc. prof., 1969-74, prof. econs., 1974—. Author: Growth, Development, and the Service Economy in the Third World, 1988, Schumpeter and the Political Economy of Change, 1991; editor: Canadian American Economic Relations: Conflict and Cooperation on a Continental Scale, 1988, Hostile Takeovers: Issues in Public and Corporate Policy, 1989, Energy, the Environment and Public Policy, 1991, External Linkages and Growth in Small Economies, 1993; co-author: Developmental Issues in Small Island Economics, 1990, Accounting Services, the International Economy and Third World Development, 1992, others; contbr. articles to profl. jours. Mem. Am. Econ. Assn., Can. Econ. Assn., Western Econ. Assn., Internat. Assn. for Impact Assessment. Roman Catholic. Home: 1997 Hastings Dr Kent OH 44240-4613 Office: Kent State U Dept Econs Kent OH 44242

MCKEE, FRANCIS JOHN, medical association executive, lawyer; b. Bklyn., Aug. 31, 1943; s. Francis Joseph and Catherine (Giles) McK.; m. Antoinette Mary Sancis; children: Lisa Ann, Francis Dominic, Michael Christopher, Thomas Joseph. AB, Stonehill Coll., 1965; JD, St. John's U., 1970. Bar: N.Y. 1971. Assoc. Samuel Weinberg, Esquire, Bklyn., 1970-71, Finch & Finch, Esquire, Long Island City, N.Y., 1971-72; staff atty. Med. Soc. of State of N.Y., Lake Success, N.Y., 1972-77; prin. Francis J. McKee Assocs., Clinton, N.Y., 1984—; exec. dir. Suffolk Physicians Rev. Orgn., East Islip, N.Y., 1977-81, N.Y. State Soc. Surgeons, Inc., Clinton, N.Y., 1981—, N.Y. State Soc. Orthopaedic Surgeons, Inc., Clinton, 1981—, Upstate N.Y. chpt. ACS, Inc., Clinton, 1981—, N.Y. State Ophthalmol. Soc., 1984-92, N.Y. State Soc. Obstetricians and Gynecologists, 1985—, Orthopac of N.Y., 1986—, Nat. Com. for the Preservation Orthopaedic Practice, Clinton, 1989—. With U.S. Army, 1966-68. Mem. N.Y. State Bar Assn., Oneida County Bar Assn., Am. Soc. Assn. Execs., Am. Assn. Med. Soc. Execs., Nat. Health Lawyers Assn., Skenandoa Club. Republican. Roman Catholic. Home: 19 Mulberry St Clinton NY 13323-1532 Office: Box 308 40 Chenango Ave Clinton NY 13323-0308

MC KEE, GEORGE MOFFITT, JR., civil engineer, consultant; b. Valparaiso, Nebr., Mar. 27, 1924; s. George Moffitt and Iva (Santrock) McK.; student Kans. State Coll. Agr. and Applied Sci., 1942-43, Bowling Green State U., 1943; B.S. in Civil Engring., U. Mich., 1947; m. Mary Lee Taylor, Aug. 11, 1945; children—Michael Craig, Thomas Lee, Mary Kathleen, Marsha Coleen, Charlotte Anne. Draftsman, Jackson Constrn. Co., Colby, Kans., 1945-46; asst. engr. Thomas County, Colby, 1946; engr. Sherman County, Goodland, Kans., 1947-51; salesman Oehlert Tractor & Equipment Co., Colby, 1951-52; owner, operator George M. McKee, Jr., cons. engrs., Colby, 1952-72; sr. v.p. engring. Contract Surety Consultants, Wichita, Kans., 1974—. Adv. rep. Kans. State U., Manhattan, 1957-62; mem. adv. com. N.W. Kans. Area Vocat. Tech. Sch., Goodland, 1967-71. Served with USMCR, 1942-45. Registered profl. civil engr., Kans., Okla., registered land Surveyor, Kans. Mem. Kans. Engring. Soc. (pres. N.W. profl. engrs. chpt. 1962-63, treas. cons. engrs. sect. 1961-63), Kansas County Engr's. Assn. (dist. v.p. 1950-51), N.W. Kans. Hwy. Ofcls. Assn. (sec. 1948-49), Nat. Soc. Profl. Engrs., Kans. State U. Alumni Assn. (pres. Thomas County 1956-57), Am. Legion (Goodland 1st vice comdr. 1948-49), The Alumni Assn. U. Mich. (life), Colby C. of C. (v.p. 1963-64), Goodland Jr. C. of C. (pres. 1951-52). Methodist (chmn. ofcl. bd. 1966-67). Mason (32 deg., Shriner); Order Eastern Star. Home: 8930 Suncrest St # 502 Wichita KS 67212-4069 Office: 6500 W Kellogg Dr Wichita KS 67209-2298

MCKEE, JAMES STANLEY COLTON, physics educator; b. Belfast, Northern Ireland; m. Christine McKee; children: Conor, Siobhan. BS in Physics with honors, Queen's U., Belfast, 1952, PhD in Theoretical Physics, 1956; DSc in Physics, Birmingham U., Eng., 1968. Asst. lectr. dept. physics Queen's U., Belfast, 1954-56; lectr. dept. physics U. Birmingham, 1956-64, sr. lectr. dept. physics, 1964-74; prof. physics U. Manitoba, Winnipeg, Can., 1974—, acting dir. Cyclotron Lab., 1975-75, dir. Cyclotron Lab./Accelerator Ctr., 1975—; mem. senate U. Manitoba, 1975, mem. senate exec. com., 1982-86, corp. rep. to Can. Nuclear Assn., 1979—, mem. senate planning and priorities com., 1981-86, chmn. senate planning and priorities com., 1983-85, mem. bd. govs., 1984—, exec. com. mem., 1985-88, fin. com., 1988—; mem. evaluation com. Sci. Culture Can., 1988—; hon. mem. adv. com. on sci. and tech. CBC, 1985—; mem. Manitoba Fusion Power Com., 1980—; mem. NSERC Nuclear Physics Grant Selection Com., 1978-81; corr. Info. Radio, Manitoba, 1980-88; adv. bd. Energy Sources, 1988—; presenter 4 papers at internat. confs. Editor: Physics in Can., 1990—; sci. columnist CBC Newsworld TV, 1988—; referee Phys. Rev., Phys. Rev. Letters, Jour. Physics G.; contbr. to C.B.C. Morningside, 1983—; mem. editorial bd. MOSAIC, U. Manitoba, 1988—; contbr. over 200 articles and 19 papers to profl. jours. and confs. Dep. chmn. United Way Campaign, 1992-93; univ. divsn. chmn. United Way of Winnipeg, 1990-92. Recipient Actra award, 1972, Outreach award U. Manitoba, 1986; Kitchener scholar Queen's U., 1948-52, Univ. grad. scholar Queen's U., 1952-55, Fulbright scholar, 1965-66. Mem. Solar Thermal Test Facility Users Assn. (charter assoc. 1977—), Can. Assn. Physicists (v.p.-elect 1985-86, pres. 1986-87, past pres. 1987-88), ACSTP (chair action com. of sci. and tech. pres. 1987), Rotary Club of Winnipeg West (bd. dirs. 1992-93). Achievements include research in high energy proton PIXE; study of damage to materials through particle emissions in fusion and fission reactors, study of proton induced damage to refractory materials such as TiC. Office: U Manitoba, Dept Physics, 301 Allen Bldg, Winnipeg, MB Canada R3T 2N2

MCKEE, KEITH EARL, manufacturing technology executive; b. Chgo., Sept. 9, 1928; s. Charles Richard and Maude Alice (Hamlin) McK.; m. Lorraine Marie Celichowski, Oct. 26, 1951; children: Pamela Ann Houser, Paul Earl. BS, Ill. Inst. Tech., 1950, MS, 1956, PhD, 1962. Engr. Swift & Co., Chgo., 1953-54; rsch. engr. Armour Rsch. Found., Chgo., 1954-62; dir. design and product assurance Andrew Corp., Orland Park, Ill., 1962-67; dir. engring. Rsch. Ctr. Ill. Inst. Tech., Chgo., 1967-80, dir. mfg. prodn. ctr., 1977—; adj. prof. Ill. Inst. Tech., Chgo., 1979—; coord. Nat. Conf. on Fluid Power, Chgo., 1983-88; mem. com. on materials and processing Dept. Def., Washington, 1986—. Author: Productivity and Technology, 1988; editor: Automated Inspection and Process Control, 1987; co-editor: Manufacturing High Technology Handbook, 1987; mng. editor: Manufacturing Competitiveness Frontier, 1977—. Capt. USMC, 1950-54. Recipient outstanding presentation award Am. Soc. of Quality Control, Milw., 1983. Fellow World Acad. Productivity Scis.; mem. ASCE, Am. Def. Preparedness Assn. (pres. Chgo. chpt. 1972—), Am. Assn. Engring. Soc. (Washington) (coor. com. on productivity 1978-88), Inst. of Indsl. Engrs., Soc. Mfg. Engrs. (Gold medal 1991), Am. Assn. for Artificial Intelligence, Robotic Industry Assn. (bd. dir. 1978-81), Assn. for Mfg. Excellence, Soc. for Computer Simulation. Democrat. Roman Catholic. Home: 18519 Clyde Rd Homewood IL 60430-3015 Office: Mfg Productivity Ctr 10 W 35th St Chicago IL 60616-3799

MCKEE, MARGARET CRILE, pulmonary medicine and critical care physician; b. Cleve., Jan. 12, 1945; d. Richard List and Florence Mae (Johnson) McK. BA, Coll. Wooster, 1967; MRP, Cornell U., 1971; MD, SUNY, Stony Brook, 1976. Diplomate Am. Bd. Internal Medicine, Pulmonary Medicine and Critical Care. Social planner Model Cities, Binghamton, N.Y., 1970-71; resident internal medicine Harlem Hosp., N.Y.C., 1976-79; physician Health Ins. Plan, Bedford-Williamsburg, N.Y., 1979-80; pulmonary fellow Columbia Presbyn. Med. Ctr., N.Y.C., 1980-82; chief of medicine Phoenix Indian Med. Ctr., 1983-92; pvt. practice Ariz. Med. Clinic, Sun City, Ariz., 1992—. Mem. Am. Thoracic Soc., Union of Concerned Scientists, Sierra Club. Methodist. Avocations: travel, music, skiing. Office: Ariz Med Clinic 13640 N Plaza del Rio Blvd Peoria AZ 85381

MCKEEVER, SHEILA A., utilities executive; b. Bethesda, Md., May 12, 1948; d. Kenneth and Jean (Harmon) McK.; m. Nathan Samuels, Feb. 14, 1979; children: Kirsten Mary, Lorna Alice. BSEE, Calif. Inst. Tech., 1968, MSEE, 1971. Engr. Calif. Power Co., L.A., 1972-75, rschr., 1975-82; mgr. Werik Gas & Electric, Palo Alto, Calif., 1982-92, v.p., 1992—. Vol. Literacy Action, 1987-90, Make-a-Wish Found., 1988—. Mem. IEEE, IEEE Industry Applications Soc., LWV, Mensa, Edgeview Country Club. Office: Werik Gas & Electric 260 Sheridan Ave Ste #216 Palo Alto CA 94306-2009

MC KELVEY, JOHN CLIFFORD, research institute executive; b. Decatur, Ill., Jan. 25, 1934; s. Clifford Venice and Pauline Lytton (Runkel) McK.; m. Carolyn Tenney, May 23, 1980; children: Sean, Kerry, Tara, Evelyn, Aaron. B.A., Stanford U., 1956, M.B.A., 1958. Research analyst Stanford Research Inst., Palo Alto, Calif., 1959-60; indsl. economist Stanford Research Inst., 1960-64; with Midwest Research Inst., Kansas City, Mo., 1964—; v.p. econs. and mgmt. sci. Midwest Research Inst., 1970-73, exec. v.p., 1973-75, pres., chief exec. officer, 1975—; bd. dirs. Yellow Freight System, Inc.; chmn. bd. Menninger Clinic, 1988. Trustee Rockhurst Coll., 1993; mem. Civic Coun. of Greater Kansas City; bd. regents Rockhurst Coll., Kansas City, Mo., 1973—; bd. dirs. Oxford Park Acad., North Star Found., 1981, Mid-Am. Mfg. Tech. Ctr., 1991; trustee The Menninger Found., 1975. Clubs: Carriage, Mission Hills, Hallbrook Country. Home: 912 W 121st Ter Kansas City MO 64145-1015 Office: Midwest Rsch Inst 425 Volker Blvd Kansas City MO 64110-2299

MCKENDALL, ROBERT ROLAND, neurologist, virologist, educator; b. Providence, Feb. 18, 1944; s. Benjamin Salvatore and Pauline (Nardelli) McK.; m. Joyce Marie Podlesak, Oct. 12, 1973; children: Lauren Patricia, Alexis Victoria. BA, Columbia Coll., N.Y.C., 1965; MD, Tufts U., 1969. Diplomate Am. Bd. Neurology & Psychiatry. Resident in neurology Rush-Presbyn.-St. Lukes Hosp., Chgo., 1970-74; neurovirology fellow U. Calif., San Francisco, 1974-78, asst. prof. neurology, 1981-84; asst. prof. neurology U. Tex. Med. Br., Galveston, 1984-88, assoc. prof. neurology & microbiology, 1988—. Editor: Clinical and Basic Neurovirology, 1993, (textbook) Viral Diseases-Handbook of Clinical Neurology, 1989; contbr. over 40 articles to profl. jours.; editorial reviewer Neurology, 1990—, Annals of Neurology, 1990—. Mem. Multiple Sclerosis Med. Adv. Bd., Houston, 1988—; mem. Sch. Dist. Parent Adv. Bd., Dickinson, Tex., 1992—. Recipient Young Investigator award NIH, 1978-81, Career Devel. award VA Rsch. Svc., 1981-84, Herpes Simplex Infection grant, 1984-86, AIDS Clin. Trials grant NIH, 1992—. Mem. Am. Soc. Microbiology, Am. Fedn. Clin. Rsch., Am. Assn. Immunologists, Soc. for Neurosci. Office: U Tex Med Br Dept Neurology E-39 601 University Ave Galveston TX 77550

MCKENNA, JAMES EMMET, chemist; b. Bayonne, N.J., Mar. 7, 1947; s. James Emmet Jr. and Mildred Elizabeth (Jones) McK.; m. Diane Louise Ashcraft, Aug. 31, 1968; children: James S., Peter, Erin. BS, St. Peter's Coll., 1968; PhD, Fordham U., 1973. Rsch. chemist Rsch. Organic, Belleville, N.J., 1975-76; sr. rsch. chemist Chem-Fleur Inc., Newark, 1977-85; process mgr. Firmenich, Inc., Newark, 1985—. Com. chmn. Boy Scouts Am., Short Hills, N.J., 1980-86. Capt. U.S. Army, 1973-75. Mem. Am. Chem. Soc., Instrument Soc. Am. (sr.), Sigma Xi, Phi Lambda Upsilon (chpt. pres. 1971-72). Republican. Achievements include patents for phenethyl alcohol from styrene oxide, selective decarbonylation. Home: 112 Greenwood Dr Millburn NJ 07041 Office: Firmenich 928 Doremus Ave Port Newark NJ 07114

MCKENNA, JOHN DENNIS, environmental testing engineer; b. N.Y.C., Apr. 1, 1940; s. Hubert Guy and Elizabeth Ann (Record) McK.; BSChemE, Manhattan Coll., 1961; MSChemE, Newark Coll. Engring., 1968; MBA, Rider Coll., 1974, PhD, Walden U., 1991; m. Christel Klages, Dec. 26, 1964; children: Marc, Michelle. Tech. asst. to pres. Eldib Engring. & Rsch. Co., Newark, 1964-67; project mgr. Princeton Chem. Rsch., Inc. (N.J.), 1967-68; projects dir. Rsch. Cottrell Environ. Systems, Bound Brook, N.J., 1968-72; v.p., then pres. Enviro-Systems & Rsch., Inc., Roanoke, Va., 1973-79; pres. ETS, Inc., Roanoke, 1979-91, ETS Internat. Inc., 1991—; bd. dirs. air pollution adv. bd. State of Va. Mem. Air Pollution Control Assn., Am. Inst. Chem. Engrs (treas. Cen. Va. chpt.); div. chmn. tech. coun. Air and Waste Mgmt. Assn., Pitts.; workshop lectr. and sci. reviewer publs. for EPA, 1978-79. Recipient Outstanding Engring. Grad. Manhattan Coll. Centennial award, 1992. Contbr. chpts. to books, articles to profl. jours. Roman Catholic. Home: RR 1 PO Box 1925 Rocky Mount VA 24151-9607 Office: ETS Inc 1401 Municipal Rd NW Roanoke VA 24012-1309

MC KENNA, MALCOLM CARNEGIE, vertebrate paleontologist, curator, educator; b. Pomona, Calif., July 21, 1930; s. Donald Carnegie and Bernice Caroline (Waller) McK.; m. Priscilla Coffey, June 17, 1952; children—Douglas M., Katharine L., Andrew M., Bruce C. B.A., U. Calif., Berkeley, 1954, Ph.D., 1958. Instr. dept. paleontology U. Calif., Berkeley, 1958-59; asst. curator dept. vertebrate paleontology Am. Mus. Natural History, N.Y.C., 1960-64; assoc. curator Am. Mus. Natural History, 1964-65; Frick assoc. curator, chmn. Frick Lab., 1965-68, Frick curator, 1968—; asst. prof. geology Columbia U., N.Y.C., 1960-64, assoc. prof., 1964-72, prof. geol. scis., 1972—; research assoc. U. Colo. Mus., Boulder, 1962—. Contbr. articles on fossil mammals and their evolution, the dating of Mesozoic and Tertiary sedimentary rocks, and paleogeography and plate tectonics to profl. jours. Bd. dirs. Bergen Community (N.J.) Mus., 1964-67, pres., 1965-66; trustee Flat Rock Brook Nature Assn., N.J., 1979—, Raymond Alf Mus., Webb Sch. of Calif., 1980—, Dwight-Englewood Sch., Englewood, N.J., 1968-80; bd. dirs. Flat Rock Brook Nature Assn., N.J., 1979-84; trustee Claremont McKenna Coll., Calif., 1983-91; Planned Parenthood Bergen County, N.J., 1979-88, Mus. No. Ariz., 1978-85, 87—. Nat. Acad. Scis. exchange fellow USSR, 1965. Fellow Explorers Club, Geol. Soc. Am.; mem. Grand Canyon Natural History Assn. (dir. 1972-76), AAAS, Soc. Systematic Zoology (council 1974-77), Soc. Vertebrate Paleontology (v.p. 1975, pres. 1976), Am. Geophys. Union, Am. Soc. Mammalogists, Paleontol. Soc. (award 1992), Soc. for Study Evolution, Polish Acad. of Scis. (fgn. mem. 1991—), Sigma Xi. Office: Am Mus Nat History Vertebrate Paleontology Central Park St W New York NY 10026-4355 Office: Columbia Univ 420 Lewisohn Hall Broadway & 116th St New York NY 10027

MCKENNA, MARK JOSEPH, physicist, educator, researcher; b. Brockton, Mass., Feb. 18, 1962; s. Martin Joseph and Mary Josephine (Doran) McK. BS, Georgetown U., 1984; MSc, Brown U., 1986, PhD, 1989. Teaching asst. Brown U., Providence, R.I., 1984-85; rsch. asst. Brown U., Providence, 1985-89; rsch. assoc. Pa. State U., University Park, Pa., 1989—. Author: (with others) Physical Acoustics XX, 1992. Mem. Am. Phys. Soc., Acoustical Soc. Am., Am. Soc. of Physics Tchrs., Phi Beta Kappa, Sigma Xi. Roman Catholic. Achievements include study of tunneling systems in high Tc superconductors; discovery of a second sound-like mode in superfluid filled aerogels; study of nonlinear effects on Anderson localization. Home: H17 445 Waupelani Dr State College PA 16801 Office: Pa State U 104 Davey Lab University Park PA 16802

MCKENNA, THOMAS EDWARD, hydrogeologist; b. Camden, N.J., Jan. 10, 1962; s. Francis William and Patricia Ann (McGrory) McK.; m. Kimberly Kemble, June 27, 1987. BS, Stockton State Coll., 1984; MS, U. S.C., 1987; postgrad., U. Tex. Teaching asst. U. S.C., Columbia, 1985-87, rsch. asst., 1986; hydrogeologist Geraghty and Miller, Inc., Tampa, Fla., 1987-90; rsch. asst. U. Tex., Austin, 1990—; convenor U. Tex. Hydrogeology Seminar, 1990. Contbr. articles to Jour. Petroleum Geology, Soc. Econ. Paleontologists and Mineralogists Transactions, Palaeogeography, Palaeoclimatology, Palaeoecology; manuscript reviewer Jour. Groundwater, 1990—. Fellow Nat. Assn. Geology Tchrs., 1985; grantee Geol. Soc. Am., 1992, Am. Assn. Petroleum Geologists, 1992. Mem. Am. Geophys. Union, Geol. Soc. Am., Am. Assn. Petroleum Geologists, Assn. Ground Water Scientists and Engrs., Austin Geol. Soc. (student liaison 1990-92), Phi Kappa Phi. Democrat. Office: U Tex Austin Dept Geol Scis 61140 Austin TX 78712

MCKENZIE, RITA LYNN, psychologist; b. Boston, Nov. 25, 1952; d. Wallace Andrew and Angelina Rita (Bagnoli) McK. BA, Framingham State Coll., 1974; MEd, Northeastern U., 1975; PhD, Temple U., 1983. Lic. psychologist, Mass., Conn., Pa. Pvt. practice Fairfield, Conn., 1984-86; psychologist Johnson Life Ctr., Springfield, Mass., 1986-87, dir. outpatient therapy, 1987-88; pvt. practice Springfield, 1988—; adj. faculty Holyoke (Mass.) Community Coll., 1989-90, Springfield Tech. Community Coll., 1989-90; dir. day treatment DuBois Day Treatment Ctr., Stamford, Conn., 1982-86; cons. psychologist Community Care Mental Health Ctr., Springfield, 1989—, Spofford Hall Treatment Ctr., Ludlow, Mass., 1991-92. Trustee Northampton (Mass.) State Hosp.; mem. organizing com. Week of Young Child, Springfield. Mem. Women Bus. Owners alliance, Zonta Internat. Office: 380 Union St Ste 14 West Springfield MA 01089

MCKEREGHAN, PETER FLEMING, hydrogeologist, consultant; b. Denver, Mar. 15, 1960; s. Charles Denn and Suzanne (Clark) McK.; m. Barbara Karin Wach, Aug. 31, 1991. BA in Geology, U. Calif., Berkeley, 1983; MS in Geology, U. Wis., Milw., 1988. Certified engring. geologist, Calif. Geologist U.S. Geol. Survey, Menlo Park, Calif., 1984-86; teaching asst. U. Wis., Milw., 1986-87, rsch. asst., 1987-88; project hydrogeologist Weiss Assocs., Emeryville, Calif., 1988—. Recipient Rsch. grant Wis. Dept. Nat. Resources, 1987-88. Mem. Nat. Ground Water Assn., Assn. Engring. Geologists (assoc.). Democrat. Office: Weiss Assocs 5500 Shellmound St Emeryville CA 94608

MC KETTA, JOHN J., JR., chemical engineering educator; b. Wyano, Pa., Oct. 17, 1915; s. John J. and Mary (Gelet) McK.; m. Helen Elisabeth Smith, Oct. 17, 1943; children: Charles William, John J. III, Robert Andrew, Mary Anne. B.S., Tri-State Coll., Angola, Ind., 1937; B.S.E., U. Mich., 1943, M.S., 1944, Ph.D., 1946; D.Eng. (hon.), Tri-State Coll., 1965, Drexel U., 1977; Sc.D., U. Toledo, 1973. Diplomate: registered profl. engr., Tex., Mich. Group leader tech. dept. Wyandotte Chem. Corp., Mich., 1937-40; asst. supt. caustic soda div. Wyandotte Chem. Corp., 1940-41; teaching fellow U. Mich., 1942-44, instr. chem. engring., 1944-45; faculty U. Tex., Austin, 1946—; successively asst. prof. chem. engring., assoc. prof., then prof. chem. engring. U. Tex., 1951-52, 54—, E.P. Schoch prof. chem. engring., 1970-81, Joe C. Walter chair, 1981—; asst. dir. Tex. petroleum research com., 1951-52, 54-56, chmn. chem. engring. dept., 1950-52, 55-63, dean Coll. Engring., 1963-69; exec. vice chancellor acad. affairs U. Tex. System, 1969-70; editorial dir. Petroleum Refiner, 1952-54; pres. Chemoil Cons., Inc., 1957-73; dir. Gulf Pub. Co., Howell Corp., Houston, Tesoro Petroleum Co., San Antonio; Chmn. Tex. AEC, So. Interstate Nuclear Bd., 1963-70; mem. Tex. Radiation Adv. Bd., 1978-84; chmn. Nat. Energy Policy Com., 1970-72, Nat. Air Quality Control Com., 1972-85; mem. adv. bd. Carnegie-Mellon Inst. Research, 1978-84; pres. Reagans's rep. on U.S. Acid Precipitation Task Force, 1982-88; apptd. mem. Nuclear Waste Tech. Rev. Bd., 1992—. Author: series Advances in Petroleum Chemistry and Refining; Chmn. editorial com.: series Petroleum Refiner; mem. adv. bd.: series Internat. Chem. Engring. mag; editorial bd.: series Ency. of Chem. Tech; exec. editor: series Ency. of Chem. Processing and Design (65 vols.). Bd. regents Tri-State U., 1957—. Recipient Bronze plaque Am. Inst. Chem. Engrs., 1952, Charles Schwab award Am. Steel Inst., 1973, Lamme award as outstanding U.S. educator, 1976, Joe J. King Profl. Engring. Achievement award U. Tex., 1976, Gen. Dynamics Teaching Excellence award, 1979, Triple E award for contbns. to nat. issues on energy, environment and econs. Nat. Environ. Devel. Assn., 1976, Boris Pregal Sci. and Tech. award NAS, 1978, Internat. Chem. Engring. award, Italy, 1984, Pres. Herbert Hoover award for advancing well-being of humanity and developing richer and more enduring civilization Joint Engring. Socs., 1989, Centennial award exceptional contbn. Am. Soc. Engring. Edn., 1993; named Disting. Alumnus U. Mich Coll.

Engring., 1953, Tri-State Coll., 1956; fellow Allied Chem. & Dye, 1945-46; named Disting. fellow Carnegie-Mellon U., 1978. Mem. Am. Inst. Mining Engrs., Am. Chem. Soc. (chmn. Central Tex. sect. 1950), Am. Inst. Chem. Engrs. (chmn. nat. membership com. 1955, regional exec. com., nat. dir., nat. v.p. 1961, pres. 1962, service to soc. award 1975), Am. Soc. Engring. Edn., Chem. Markets Research Assn., Am. Gas Assn. (adv. bd. chems. from gas 1954), Houston C. of C. (chmn. refining div. 1954, vice chmn. research and statistics com. 1954), Engrs. Joint Council (dir.), Engrs. Joint Countil Profl. Devel. (dir. 1963-85), Nat. Acad. Engring., Sigma Xi, Chi Epsilon, Alpha Psi Omega, Tau Omega, Phi Lambda Upsilon, Phi Kappa Phi, Iota Alpha, Omega Chi Epsilon, Tau Beta Pi, Omicron Delta Kappa. Home: 5227 Tortuga Trl Austin TX 78731-4501

MCKINLEY, JOHN MCKEEN, retired physics educator; b. Wichita, Kans., Feb. 1, 1930; s. Lloyd and Ruth Muriel (McKeen) McK.; m. Martha Ann Dicker, Feb. 7, 1953; children: Susan, Kathi, Kevin Michael. BS, U. Kans., 1951; PhD, U. Ill., 1962. Asst. prof. Kans. State U., Manhattan, 1960-66; assoc. prof. Oakland U., Rochester, Mich., 1966-71, prof., 1971-92, ret., 1992; rsch. assoc. Goddard Space Flight Ctr. NASA, Greenbelt, Md., 1980-81, 82-83. Author: Solutions Manual to Accompany Tipler's Modern Physics, 1978; assoc. editor Am. Jour. Physics, 1978-81. 1st lt. U.S. Army, 1951-55, Korea. Mem. AAAS, Am. Phys. Soc., Am. Assn. Physics Tchrs., Sigma Xi. Avocations: family history and genealogy.

MCKINNELL, ROBERT GILMORE, zoology, genetics and cell biology educator; b. Springfield, Mo., Aug. 9, 1926; s. William Parks and Mary Catherine (Gilmore) McK.; m. Beverly Walton Kerr, Jan. 24, 1964; children: Nancy Elizabeth, Robert Gilmore, Susan Kerr. A.B., U. Mo., 1948; B.S., Drury Coll., 1949; Ph.D., U. Minn., 1959; D.Sc. (hon.), Drury Coll., 1993. Research asso. Inst. Cancer Research, Phila., 1958-61; asst. prof. biology Tulane U., New Orleans, 1961-65; asso. prof. Tulane U., 1965-69, prof., 1969-70; prof. zoology U. Minn., Mpls., 1970—; prof. genetics and cell biology U. Minn., St. Paul, 1976—; vis. scientist Dow Chem. Co., Freeport, Tex., 1976; guest dept. zoology U. Calif., Berkeley, 1979; Royal Soc. guest rsch. fellow Nuffield dept. pathology John Radcliffe Hosp., Oxford U., 1981-82; NATO vis. scientist Akademisch Ziekenhuis, Ghent, Belgium, 1984; faculty rsch. assoc. Naval Med. Rsch. Inst., Bethesda, Md., 1988; secretariat Third Internat. Conf. Differentiation, 1978; mem. amphibian com. Inst. Lab. Animal Resources, NRC, 1970-73, mem. adv. coun., 1974; mem. panel genetic and cellular resources program NIH, 1981-82, spl. study sect., Bethesda, 1990. Author: Cloning: Amphibian Nuclear Transplantation, 1978, Cloning, A Biologist Reports, 1979; sr. editor: Differentiation and Neoplasia, 1980, Cloning: Leben aus der Retorte, 1981, Cloning, of Frogs, Mice, and other Animals, 1985; mem. editorial bd. Differentiation, 1973—; assoc. editor: Gamete Research, 1980-86; contbr. articles to profl. jours. Served to lt. USNR, 1944-47, 51-53. Recipient Outstanding Teaching award Newcomb Coll., Tulane U., 1970; Disting. Alumni award Drury Coll., 1979, Morse Alumni Teaching award U. Minn., 1992; Research fellow Nat. Cancer Inst., 1957-58; Sr. Sci. fellow NATO, 1974. Fellow AAAS, Linnean Soc. (London); mem. Am. Assn. Cancer Rsch., Am. Assn. for Cancer Edn., Metastasis Soc., Am. Inst. Biol. Scis., Soc. for Devel. Biology, Internat. Soc. for the Study of Comparative Oncology, Inc., Internat. Soc. Differentiation (exec. com., sec.-treas. 1975-92, pres.-elect 1992—), Gown-In-Town Club, Sigma Xi. Home: 2124 Hoyt Ave W Saint Paul MN 55108-1315 Office: U Minn Dept Genetics & Cell Biology Saint Paul MN 55108-1095

MCKINNEY, BRYAN LEE, chemist; b. San Antonio, Jan. 2, 1946; s. Oscar Bryan and Mamye Maxine (Faubion) McK.; m. Susan Northcutt, June 8, 1968 (div. May 1980); children: Samuel, Jennifer, Joel; m. Joan Diane Miller, June 6, 1981. BS in Chemistry, U. Tex., Arlington, 1968; PhD in Chemistry, U. Oreg., 1972. Rsch. scientist Battelle Meml. Inst., Columbus, Ohio; group leader Gould Inc., Rolling Meadows, Ill.; staff electrochemist Kerr-McGee, Oklahoma City, 1981-82; sr. electrochemist Johnson Controls, Inc., Milw., 1982-84, mgr. materials rsch., 1984-91, dir. cen. rsch. labs., 1991—; mem. adv. bd. Great Lakes Composites Consortium, Kenosha, Wis., 1992—. Contbr. articles to profl. publs. Mem. Am. Chem. Soc., Materials Rsch. Soc., Electrochem. Soc. (chmn. Chgo. sect. 1980, Wood award). Achievements include patents in evaluation of stress-corrosion cracking in bottles, bottle designs, battery explosion protection. Home: 7424 Willowbrook Ct Mequon WI 53092 Office: Johnson Controls Inc 1701 Civic Dr Milwaukee WI 53209

MCKINNEY, COLLIN JO, electrical engineer; b. Colorado Springs, Jan. 24, 1958; s. Collin Jo McKinney and Joy Darlene (Upham) Melton. BSEE, Tex. A&M U., 1981. Rsch. technician Biosystems Rsch. Div., Tex. A&M U., College Station, Tex., 1977-81; cons. Sector Rsch., Inc., Huntsville, Ala., 1982; elec. engr. Tex. Instruments Digital Systems Div., College Station, 1982-85; cons. Coherent Technologies, Inc., Missouri City, Tex., 1990—, Duke U. Med. Ctr. Nuclear Medicine, Durham, N.C., 1988—; vis. sr. engr. dept. botany Duke U., Durham, N.C., 1988—; v.p., dir. engr. Biosystems Technologies, Inc., Durham, N.C., 1985—. Contbr. articles to profl. jours. Recipient Quality award for Excllence, Tex. Instruments, 1983. Mem. IEEE (sr.), Com. for Entrepreneurial Devel. Achievements include patent in Magnetic Tire Monitor System. Office: Biosystems Technologies Inc PO Box 61742 Durham NC 27715-1742

MCKINNEY, CYNTHIA EILEEN, molecular biologist; b. Pitts., Apr. 14, 1949; d. Joel Drexler and Miriam Carolyn (Dillon) McK. BS in Biology, Juniata Coll., 1971; MS, PhD in Zoology, U. Md., 1989. Technician Microbiol. Assocs., Bethesda, Md., 1972-87; teaching asst. U. Md., College Park, 1984-89; postdoctoral fellow Clin. Neurosci. Br., NIMH, NIH, Bethesda, 1990-92, staff fellow, 1992—. Contbr. articles to profl. jours. Recipient Grant-in-Aid Rsch., Sigma Xi., 1987, 88. Mem. AAAS,Am. Assn. Human Genetics, Sigma Xi. Office: Clin Neuroscience Br 49/BIEE16 Bethesda MD 20892

MCKINNEY, DAENE CLAUDE, environmental engineering educator; b. Merced, Calif., Nov. 30, 1954; s. Brian C. and Patricia Anne (Johnson) McK.; m. Carie Goodman, June 16, 1984; children: Caryn Johnson McKinney, James Carey McKinney. BS, Humboldt State U., 1983; MS, Cornell U., 1986, PhD, 1990. Environ. engr. U.S. EPA, San Francisco, 1989-90; asst. prof. U. Tex., Austin, 1990—; cons. Sierra Club, Austin, 1992—. Mem. ASCE, Am. Geophys. Union. Republican. Baptist. Home: 10610 Sierra Oaks Austin TX 78759 Office: U Tex Dept Civil Engring Austin TX 78712

MC KINNEY, ROSS ERWIN, civil engineering educator; b. San Antonio, Aug. 2, 1926; s. Roy Earl and Beatrice (Saylor) McK.; m. Margaret McKinney Curtis, June 21, 1952; children: Ross Erwin, Margaret E., William S., Susanne C. B.A., So. Meth. U., 1948, B.S. in Civil Engring, 1948; S.M., MIT, 1949, Sc.D., 1951. San. scientist S.W. Found. for Research and Edn., San Antonio, 1951-53; asst. prof. MIT, 1953-58, assoc. prof., 1958-60; prof. U. Kans., 1960-63, chmn. dept. civil engring., 1963-66, Parker prof. civil engring., 1966-76, N.T. Veatch prof. environ. engring., 1976-93, prof. emeritus, 1993—; adv. prof. Tongji U., Shanghai, Peoples Rep. China, 1985; v.p. Rolf Eliassen Assocs., Winchester, Mass., 1954-60; pres. Environ. Pollution Control Services, Lawrence, Kans., 1969-73. Author: Microbiology for Sanitary Engineers, 1962; Editor: Nat. Conf. on Solid Waste Research, 1964, 2d Internat. Symposium for Waste Treatment Lagoons, 1970. Mem. Cambridge (Mass.) Water Bd., 1953-59, Lawrence-Douglas County Health Bd., 1969-76, Kans. Water Quality Adv. Council, 1965-76, Kans. Solid Waste Adv. Council, 1970-76, Kans. Environ. Adv. Bd., 1976-85. Served with USNR, 1943-46. Recipient Harrison P. Eddy award, 1962, Water Pollution Control Fedn. Rudolph Hering award, 1964, U.S. Presdl. Commendation, 1971, Environ. Quality award EPA Region VII, 1979, Chancellors Teaching award, U. Kans., 1986. Mem. ASCE (hon.), Am. Water Works Assn., Water Pollution Control Fedn. (Thomas R. Camp medal 1982), Am. Pub. Works Assn., Am. Chem. Soc., Am. Soc. Microbiologists, AAAS, Am. Soc. Engring. Edn., Internat. Assn. Water Pollution Rsch., Am. Acad. Environ. Engrs., Kans. Water Pollution Control Assn. (hon. mem., Gordon M. Fair medal 1991), NAE, N.Y. Acad. Sci., AAUP, Sigma Xi, Sigma Tau., Kappa Mu Epsilon, Chi Epsilon, Tau Beta Pi. Achievements include patent for water treatment process. Home: 2617 Oxford Rd Lawrence KS 66049-2822

MCKINNEY, WAYNE RICHARD, physicist; b. Elkton, Md., Mar. 28, 1947; s. Arthur Richard Jr. and Helen Opal (Fields) McK.; m. Judy Lynn Bowen, July 20, 1968; children: Darren Wayne, David Richard. BA, Johns Hopkins U., 1969, MA, 1971, PhD, 1974. Postdoctoral fellow Brookhaven Nat. Lab., Upton, N.Y., 1974-77, asst. physicist, 1977-79; staff scientist Bausch & Lomb Co., Rochester, N.Y., 1979-81, mgr. rsch. and devel., 1981-85; mgr. rsch. and devel. Milton Roy Co., Rochester, 1985-87; staff scientist Lawrence Berkeley Lab., Berkeley, Calif., 1987—. Leader Weblos Boy Scouts Am., Fairport, N.Y., 1980-81, asst. scoutmaster, 1981-85, scoutmaster, 1985-87, asst. scoutmaster Boy Scouts Am., 1987-89. Mem. Optical Soc. Am. (No. Calif. sect., Rochester sect.), Soc. Photo-Optical Instrumation Engrs., Sigma Xi. Democrat. Achievements include definitive article on fluorescence in scanning electron microscopy applied to biology, toroidal grating monochromator design; devel. of one tenth arc second metrology for X-ray optics constrn. of synchrotron, radiation beamline. Office: Lawrence Berkeley Lab Mail Stop 2-400 1 Cyclotron Rd Berkeley CA 94720

MCKINZIE, HOWARD LEE, petroleum engineer; b. Olustee, Okla., Apr. 16, 1941; s. George Howard and Edna Ruth (Whitten) McK.; m. Bonnie Lea Reid, June 3, 1960; children: Gwen Denise, LaVonna Lea, Scott Howard. BS in Math. & Chemistry, Ctrl. Okla. U., 1963; PhD in Physical Chemistry, Ariz. State U., 1967. Asst. prof. engring. Brown U., Providence, 1969-73; mem. tech. staff GTE Labs., Waltham, Mass., 1973-78; supr. Getty Oil Co., Houston, 1978-84; sect. mgr. Texaco Inc., Bellaire, Tex., 1984—; vis. rsch. scientist Ctr. Nat. Rsch. Sci., Bordeaux, France, 1970, Paris, 1972, 73. Postdoctoral fellow Brown U., 1967-69. Mem. Am. Chem. Soc., Soc. Petroleum Engrs., Soc. Automotive Engrs. (Ralph Tector award 1973), Completion Engring. Assn. (vice chmn. 1989-90, chmn. 1991-92). Achievements include patents in photo-electro catalysis sulfur dioxide removal from stack gases, well completion technology and formation evaluation during drilling. Home: 3410 Crystal Creek Ct Sugarland TX 77478 Office: Texaco Exploration & Prodn 5901 S Rice Bellaire TX 77401

MCKITTRICK, PHILIP THOMAS, JR., analytical chemist; b. Perth Amboy, N.J., June 15, 1964; s. Philip Thomas and Carol Eve (Vollmann) McK. BS, Bradley U., 1986; PhD, Miami U., Oxford, Ohio, 1990. Phys. scientist USDA, Peoria, Ill., 1983-86; vis. instr. Wilmington (Ohio) Coll., 1991; analytical chemist Morton Internat., Woodstock, Ill., 1991—. Mem. Am. Chem. Soc., Soc. Applied Spectroscopy (treas. Chgo. sect. 1992—), Coblentz Soc. Presbyterian. Office: Morton Internat 1275 Lake Ave Woodstock IL 60098

MCKNIGHT, JENNIFER LEE COWLES, molecular biologist; b. Neenah, Wis., May 31, 1952; d. Emil Robert Fischer Jr. and Sallyann (Pratt) Cowles; m. Robert Garland McKnight (div.). AB in Biology, Washington U., St. Louis, 1974; PhD in Biochemistry, U. Pitts., 1983. Postdoctoral fellow U. Chgo., 1983-87; asst. prof. U. Pitts., 1987-92, assoc. prof., 1992—. Mem. AAAS, Am. Soc. for Microbiology, Am. Soc. for Biochemistry and Molecular Biology, Pitts. Cancer Inst. Achievements include research on characterization of viral encoded genes which regulate herpes simplex virus gene expression; Epstein-Barr virus gene expression in post-transplant lymphomas. Office: U Pitts Dept Infectious Diseases 130 DeSoto St Pittsburgh PA 15261

MCKNIGHT, LENORE RAVIN, child psychiatrist; b. Denver, May 15, 1943; d. Abe and Rose (Steed) Ravin; m. Robert Lee McKnight, July 22, 1967; children: Richard Rex, Janet Rose. Student, Occidental Coll., 1961-63; BA, U. Colo., 1965, postgrad. in medicine, 1965-67; MD, U. Calif., San Francisco, 1969. Diplomate Am. Bd. Psychiatry and Neurology. Cert. adult and child psychiatrist Am. Bd. Psychiatry. Intern pediatrics Children's Hosp., San Francisco, 1969-70; resident in gen. psychiatry Langley Porter Neuropsychiat. Inst., 1970-73, fellow child psychiatry, 1972-74; child psychiatrist Youth Guidance Center, San Francisco, 1974-74; pvt. practice medicine specializing in child psychiatry, Walnut Creek, Calif., 1974—; asst. clin. prof. Langley Porter Neuropsychiat. Inst., 1974—; clin. assoc. in psychiatry, U. Calif. at Davis Med. Sch; asst. clin. prof. psychiatry U. Calif. San Francisco Med. Ctr. Internat.; med. dir. CPC Walnut Creek (Calif.) Hosp., 1990—. Insts. Edn. fellow U. Edinburgh, 1964; NIH grantee to study childhood nutrition, 1966. Mem. Am. Acad. Child Psychiatry, Am. Psychiat. Assn., Am. Coll. Physician Execs., Psychiat. Assn. No. Calif., Am. Med. Women's Assn., Internat. Arabian Horse Assn., Diablo Arabian Horse Assn. Avocation: breeding Arabian Horses. Office: 130 La Casa Via Walnut Creek CA 94598-3008

MCKNIGHT, STEVEN LANIER, molecular biologist; b. El Paso, Tex., Aug. 27, 1949; s. Frank Gillespie and Sara Elise (Stevens) McK.; m. Jacquelynn Ann Zimmer, Sept. 16, 1978; children: Nell, Grace, Frances, John Stevens. BA, U. Tex., 1974; PhD, U.Va., 1977. Postdoctoral fellow Carnegie Instn. Washington, Balt., 1977-79, staff assoc., 1979-81, mem. staff, 1984-92; co-founder, dir., dir. rsch. Tularik Inc., 1991—. Contbr. articles to jours. in field. With U.S. Army, 1969-71, Vietnam. Decorated ARCOM medal; recipient Eli Lilly prize Am. Soc. Microbiology, 1987, Newcomb-Cleveland prize Sci. mag., 1989, NAS Molecular Biology award Nat. Acad. Sci., 1991. Fellow Carnegie Inst. (Washington) (hon.); mem. Nat. Acad. Scis., Am. Acad. Arts and Scis., Am. Soc. for Biochemistry and Molecular Biology, Am. Soc. for Cell Biology, Japanese Biochem. Soc. (hon.). Democrat. Home: 530 Roehampton Rd Hillsborough CA 94010 Office: Tularik Inc 270 E Grand Ave S San Francisco CA 94080

MCLAFFERTY, FRED WARREN, chemist, educator; b. Evanston, Ill., May 11, 1923; s. Joel E. and Margaret E. (Keifer) McL.; m. Elizabeth E. Curley, Feb. 5, 1948; children: Sara L., Joel E., Martha A., Samuel A., Ann E. B.S., U. Nebr., 1943, D.Sc. (hon.), 1983, M.S., 1947; Ph.D., Cornell U., 1950; D.Sc. (hon.), U. Liege, Belgium, 1987. Postdoctoral fellow U. Iowa, 1949-50; research chemist, div. leader Dow Chem. Co., 1950-56; dir. Eastern Research Lab., 1956-64; prof. chemistry Purdue U., 1964-68, Cornell U., 1968—; mem. chem. sci. and tech. bd., numerical data adv. bd., bd. Army sci. tech., bd. radioactive waste mgmt. NRC; chem. co-chmn. World Bank's Chinese Univ. Devel. Project. Author: Mass Spectrometry of Organic Ions, 1963, Mass Spectral Correlations, 2d edit., 1981, Interpretation of Mass Spectra, 4th edit., 1993, Tandem Mass Spectrometry, 1983, Advances in Analytical Chemistry and Instrumentation, (with C.N. Reilley), Vols. 4-7, 1967-70, Index and Bibliography of Mass Spectrometry, (with J. Pinzelik), 1967, Atlas of Mass Spectral Data; (with E. Stenhagen and S. Abrahamsson), 1969, Registry of Mass Spectral Data, 1974; (with D.B. Stauffer) Wiley/NBS Registry of Mass Spectral Data, 1989, Important Peak Index of Mass Spectral Data, 1991; editor: Accounts of Chemical Research, 1986—; co-editor: (with E. Stenhagen and S. Abrahamsson) Archives of Mass Spectral Data, 1969-72. Served with AUS, 1943-45, ETO. Decorated Purple Heart, Combat Inf. badge, Bronze Star with 4 oak leaf clusters; recipient Pitts. Spectroscopy award Spectroscopy Soc. Pitts., 1975, Gold medal U. Naples, 1989; John Simon Guggenheim fellow, 1972, Overseas fellow Churchill Coll., Cambridge U., Eng., 1979; W.L. Evans award Ohio State U., 1987. Fellow NAS, AAAS, N.Y. Acad. Scis., Am. Acad. Arts and Scis.; mem. Soc. Analytical Chemists (Pitts. Analytical Chemist award 1987, Pioneer Analytical Instrumentation award 1994), Am. Chem. Soc. (chmn. analytical chem. div. 1969, chmn. Midland sect. 1956, Northeastern sect. 1964, award chem. instrumentation 1971, award analytical chemistry 1981, Nichols medal N.Y. sect. 1984, Oesper award Cin. sect. 1985, Lind award Ea. Tenn. sect. 1986, award mass spectrometry 1989), Internat. Spectrometry Orgn. (Sir J.J. Thomson medal 1985), Assn. Analytical Chemists (Anachem award 1985), Am. Soc. Mass Spectrometry (founder, sec. 1957-58), Royal Soc. Chemistry London (Robert Boyle medal 1992), Sigma Xi, Phi Lambda Upsilon, Alpha Chi Sigma. Home: 103 Needham Pl Ithaca NY 14850-2120

MCLAREN, DIGBY JOHNS, geologist, educator; b. Carrickfergus, Northern Ireland, Dec. 11, 1919; m. Phyllis Mary Matkin, Mar. 25, 1942; children: Ian, Patrick, Alison. Student, Queen's Coll., Cambridge U., 1938-40; BA, Cambridge U., 1941, MA (Harkness scholar), 1948; PhD, Mich. U., 1951; DSc (hon.), U. Ottawa, 1980, Carleton U., 1993. Geologist Geol. Survey Can., Ottawa, Ont., 1948-80; dir. gen. Geol. Survey Can., 1973-80; sr. sci. advisor Can. Dept. Energy, Mines and Resources, Ottawa, 1981-84; vis. prof. U. Ottawa, 1981—; 1st dir. Inst. Sedimentary and Petroleum Geology, Calgary, Alta., Can., 1967-73; pres. Commn. on Stratigraphy, Internat. Union Geol. Scis., 1972-76; apptd. 14th dir. Geol. Survey Can., 1973; chmn. bd. Internat. Geol. Correlation Program, UNESCO, 1976-80. Contbr. memoirs, bulls., papers, geol. maps, sci. articles in field of Devonian geology and paleontology of Western and Arctic Canada, internat. correlation and boundary definition, global extinctions and asteroid impacts, and global change. Served to capt. Royal Arty. Brit. Army, 1940-46. Gold medalist (sci.) Profl. Inst. Pub. Service of Can., 1979; Officer Order of Can., 1987. Fellow Royal Soc. Can. (pres. 1987-90), Royal Soc. London, European Union of Geoscis. (hon.); U.S. Nat. Acad. Scis (fgn. assoc.), Geol. Soc. France (hon.); mem. Geol. Soc. London (hon., Coke medal 1986), Geol. Soc. Germany (hon., Leopold von Buch medal 1983), Geol. Soc. Am. (pres. 1982), Paleontol. Soc. (pres. 1969), Geol. Assn. Can. (Logan medal 1987), Can. Soc. Petroleum Geologists (pres. 1971, hon.). Home: 248 Marilyn Ave, Ottawa, ON Canada K1V 7E5 Office: Royal Soc Can, PO Box 9734, Ottawa, ON Canada K1G 5J4

MCLAREN, JOHN PATERSON, JR., civil engineer; b. Savannah, Ga., Oct. 13, 1952; s. John Paterson and Frances (Marcino) McL.; m. Diane Lynn Hunter, July 26, 1975; children: John, Sara. BSCE, Va. Mil. Inst., 1974. Registered profl. engr., Va. Constrn. area mgr. Atlantic div. Naval Facilities Engring. Command, Norfolk, Va., 1978-83, resident engr., 1983-85, 86-91, asst. to dir. constrn. div., 1985-86, head stateside br. constrn. div., 1991—. Asst. coach Lakrspur Basketball, Virginia Beach, Va., 1990, Kempsville Boys Baseball, Virginia Beach, 1992; coach Larkspur Flag Football, Virginia Beach, 1990. 1st lt. U.S. Army, 1974-78. Major U.S. Army Reserve, 1978—. Mem. ASCE, NSPE, Res. Officers Assn., Assn. U.S. Army, Kappa Alpha. Roman Catholic. Office: Atlantic Divsn Naval Facilities Engring Command Gilbert St Norfolk Naval Base Norfolk VA 23459

MCLAUGHLIN, EDWARD DAVID, surgeon; b. Ridley Park, Pa., Jan. 8, 1931; s. Edward D. and Catherine J. (Hilbert) McL.; m. Mary Louise Hanlon, June 20, 1959; children: Catherine, Louise, Edward, Patricia. BS magna cum laude Georgetown U., 1952; MD, Jefferson Med. Coll., 1956. Intern, Jefferson Med. Coll., Phila., 1956-57, resident in surgery, 1957-59; resident in surgery Jefferson Med. Coll. Hosp., Phila., 1962-64; practice medicine specializing in surgery; surg. asso. Nat. Cancer Inst., NIH, 1959-61, surgeon, 1961-62; teaching fellow Harvard Med. Sch., Boston and clin. research fellow Mass. Gen. Hosp., Boston, 1964-66; sr. surg. registrar Hawkmoor Chest Hosp., Devon, Eng., 1966-67; sr. surgeon Chestnut Hill Hosp., 1967-71; asst. prof. surgery Jefferson Med. Coll., 1968-72, assoc. prof., 1972—, lectr. Jefferson continuing med. edn. program, 1976-77; assoc. chmn. of surgery Mercy Cath. Med. Center, Phila., 1972-88; pres., treas. Cedar Mgmt. Corp., 1981-86; pres., treas. Garnet Moor Ltd., 1981-91, Garnetmoor Pub. Svc. Ltd., 1986-91, Garnet Valley Acad. Alliance, 1991—; treas. Physicians and Surgeons Ltd., 1983-86. Chmn., Bethel Twp. Planning Study Group, 1971-72, Bethel Twp. Sewer Authority, 1972-78, Bethel Twp. Planning Commn., 1989-91; mem. bd. of sch. dirs. Garnet Vally Sch. Dist., 1990—; chmn. curriculum com. Garnet Valley Sch. Bd., 1992—. With USPHS, 1959-62. Recipient Mead Johnson award for research, 1962, Americus award KC, 1963, Lindback award Jefferson Med. Coll., 1974; named Outstanding Prof. of 1976-77, Phi Alpha Sigma. Diplomate Am. Bd. Surgery. Fellow ACS; mem. Phila. Acad. Surgery, N.Y. Acad. Scis., Med. Soc. State Pa., AAAS, Am. Soc. Artificial Internal Organs, AMA, Pa. Thoracic Soc., Georgetown U. Alumni (dir. 1970-72, senator 1972—), Nu Sigma Nu, Alpha Kappa Kappa. Contbr. articles on research in cancer to med. jours. and articles on edn. to ednl. jours. Home and Office: 3112 Garnet Mine Rd Boothwyn PA 19061-1718

MCLAUGHLIN, GERALD LEE, parasitology educator; b. Westwood, Calif., Aug. 9, 1949; s. Alvis Euland and Patricia (Hanlan) McL.; m. Robin Ann Spicher, June 9, 1979; children: Jason Lee, Melissa Kay, Amanda Marie. AB in Biochemistry, U. Calif., Berkeley, 1972; MA in Biology, Chico State U., 1974; PhD in Biology, U. Iowa, 1981. Instr. microbiology Butte Community Coll., Chico, Calif., 1974; instr. parasitology U. Iowa, Iowa City, 1979, 81; NIH postdoctoral trainee U. Notre Dame, Ind., 1981-83; postdoctoral rsch. assoc. Texas A&M U., College Station, 1983-85; NRC fellow, guest researcher Ctrs. for Disease Control, Atlanta, 1985-87; asst. prof. U. Ill., Urbana, 1987-91; assoc. prof. Purdue U., West Lafayette, Ind., 1991—; advisor WHO, Geneva, 1987; cons. Molecular Biosystems, Inc., San Diego, Calif., 1987-90, Ecuador Cholera Project, 1992—; plenary speaker VIII Internat. Conf. on Malaria and Babesiosis, Brazil, 1991. Reviewer jours. and grants WHO, NIH, NSF, USDA, AID; contbr. articles to profl. jours. Cub scout leader Boy Scouts Am., Urbana, 1989, 90; active Champaign County (Ill.) Humane Soc., 1991—. Rsch. grantee Nat. Eye Inst. of NIH, Md., AID, Washington, AID DiaTech PATH, Seattle, WHO TDR, Geneva. Mem. AAAS, Midwest Tropical Disease Rsch. (program officer), Ill. State Vet. Medicine Assn. (assoc.), Sigma Xi. Republican. Achievements include patents pending; development of prototype molecular assays for detection of malaria, primary amoebic meningitis, amoebic keratitis, babesiosis, anaplasmosis, and viral and bacterial encephalitis. Home: 2800 Wilshire Ave Urbana IL 61801 Office: Purdue U 1027 Lynn Hall West Lafayette IN 47907-1027

MCLAUGHLIN, JOHN, production company technical director. BS, MS, PhD in Math.; vis. scholar, Stanford U. Formerly cons. software specialist & mathematician, dir. engring., tech. dir., tech. cons., film editor, prodn. coord. various cos. including 4C Tech., DemoGraFX, deGraf/Wahrman, Whitney/Demos Prodns.; now tech. dir. MetroLight Studios, L.A. Office: MetroLight Studios Ste 400 5724 W 3rd St Los Angeles CA 90036-3078

MC LAUGHLIN, JOHN FRANCIS, civil engineer, educator; b. N.Y.C., Sept. 21, 1927; s. William Francis and Anna (Goodwin) McL.; m. Eleanor Thomas Trethewey, Nov. 22, 1950; children: Susan, Donald, Cynthia, Kevin. B.C.E., Syracuse U., 1950; M.S. in Civil Engring., Purdue U., 1953, Ph.D., 1957. Mem. faculty Purdue U., 1950—, prof. civil engring., 1963—, head Sch. Civil Engring., 1968-78, asst. dean engring. Sch. Civil Engring., 1977-80, assoc. dean engring., 1980—; cons. in field. Served with USAAF, 1945-47. Fellow ASCE, Hwy. Research Bd.; mem. ASTM (bd. dirs. 1984-86), Am. Concrete Inst. (hon. mem., bd. dirs., v.p. 1977-79, pres. 1979), Am. Nat. Studies Inst. (bd. dirs. 1992—), Sigma Xi, Tau Beta Pi, Chi Epsilon, Theta Tau. Home: 112 Sumac Dr West Lafayette IN 47906-2157 Office: Civil Engring Dept Purdue Univ Lafayette IN 47907

MCLAUGHLIN, PHILIP VANDOREN, JR., mechanical engineering educator, researcher, consultant; b. Elizabeth, N.J., Nov. 10, 1939; s. Philip VanDoren and Ruth Evans (Landis) McL.; m. Phoebe Ann Feeney, Aug. 19, 1961; children: Philip VanDoren III, Patrick Evans, Christi Duff. BSCE, U. Pa., 1961, MS in Engring. Mechanics, 1964, PhD in Engring. Mechanics, 1969. Assoc. engr. Boeing-Vertol, Morton, Pa., 1962-63, engr. II, 1963; rsch. engr. Scott Paper Co., Phila., 1963-65, rsch. project engr., 1965-69, sr. rsch. project engr., 1969; asst. prof. theoretical and applied mechanics U. Ill., Urbana, 1969-73, asst. dean engring., 1971-72; project mgr. Materials Scis. Corp., Blue Bell, Pa., 1973-76; assoc. prof. mech. engring. Villanova (Pa.) U., 1976-81, prof., 1981—; judge Cons. Engrs.'s Coun. Ill. 1st Ann. Engring. Excellence Awards Competition, 1972; cons. Naval Air Engring. Ctr., Lakehurst, N.J., 1977-79, U.S. Steel Corp., Trenton, N.J., 1980-82, RCA Corp., Moorestown, N.J., 1986, Coal Tech Corp., Merion Station, Pa., 1986, Air Products and Chems., Inc., Allentown, Pa., 1986; vis. prof. dept. engring. U. Cambridge, Eng., 1990-91. Reviewer Prentice Hall, 1980—, Jour. Engring. Mechanics, 1973-83, AIAA Jour., 1970-87, Materials Evaluation, 1988, Jour. Composite Materials, 1988, Composites Sci. and Tech., 1990—, others; contbr. articles to Jour. Applied Mechanics, Internat. Jour. Solids and Structures, Jour. Engring. Materials and Tech., NDT Internat., others. Rsch. grantee NSF, 1970-72, Naval Air Engring. Ctr., 1978-84, Lawrence Livermore Nat. Lab., 1979-81, Naval Air Warfare Ctr., 1985-86, RCA Corp., 1986-87; sr. rsch. assoc. Nat. Rsch. Coun., Washington, 1983-84. Mem. ASCE (chmn. engring. mechanics div. com. on inelastic behavior 1977-79, assoc. editor Jour. Engring. Mechanics Div. 1977-79, mem. aerospace div. com. on structures and materials 1986—), ASME (chmn. aplied mechanics div. Phila. sect. 1981-83, mem. materials div. com. on composites 1992—), Am. Acad. Mechanics, Am. Soc. for Engring. Edn., Am. Soc. Composites, Sigma Xi. Achievements include research and consulting on composite materials and structures, structural analysis and design and inelastic behavior. Office: Villanova U Dept Mech Engring Villanova PA 19085

MCLAUGHLIN, THOMAS DANIEL, aerospace engineer; b. Odessa, Tex., July 5, 1962; s. James Dozier and Margaret Anne (Tucker) McL. BS, U. Tex., 1984. Assoc. engr. McDonnell-Douglas, Houston, 1984-87; assoc.

engr. sr. Lockheed Engring. and Scis. Co., Houston, 1987-88, engr., 1988-89, sr. engr., 1990—; task leader advanced robotics sect. Lockheed Engring. & Scis. Co., Houston, 1990—, supr., 1993—. Mem. Houston Ballet Guild. Recipient Group Achievement award NASA Program Support, 1991, Superior Assistance award JSC Automation and Robotics, 1992. Mem. AIAA. Office: Lockheed ESC 2400 NASA Rd 1 Houston TX 77058

MCLAURY, RALPH LEON, physician; b. Ponca City, Okla., July 19, 1942; s. Ralph L. and Nina A. (Clark) McL.; m. Pat Owen, Jan. 21, 1967; children: Ralph L. III, Clara Courtney, Molly Reagan. BS, U. Okla., 1964; MD, U. Okla., Oklahoma City, 1967; MPH, Med. Coll. of Wis., 1993. Cert. Am. Bd. Internal Medicine, Am. Bd. Preventive Medicine. Pvt. practice medicine Bartlesville, Okla., 1974-82, 86-88; asst. med. dir. Phillips Petroleum Co., Bartlesville, 1982-86, Exxon Chem. Americas, Houston, 1988-89; regional med. dir. BASF Corp., Geismar, La., 1989-93; assoc. clin. prof. medicine Tulane U. Med. Ctr., New Orleans, 1991—; asst. med. dir. Ethyl Corp., Baton Rouge, La., 1993—; mem. LCA Sci. Adv. Coun., Baton Rouge, 1990—, pub. health com. La. State Med. Soc., 1992—. Contbr. articles to profl. jours. Maj. U.S. Army, 1966-71. H.L. Doherty scholar Cities Svc. Co., 1960-64. Fellow ACP; mem. AAAS, Am. Coll. Occupational and Environ. Medicine, La. State Med. Soc., Ascension Parish Med. Soc. (v.p. 1993), La. Chem. Assn., Chm. Mfrs. Assn. (chmn. 1992-93). Mem. Soc. of Friends. Office: Ethyl Corp 451 Florida Blvd Baton Rouge LA 70801

MCLAWHON, RONALD WILLIAM, pathology educator, biochemist; b. Chgo., Sept. 10, 1957; s. William Columbus and Esther Shirley (Bukowski) McL. AB in Biol. Scis., U. Chgo., 1979, MS in Biochemistry, 1980, PhD in Biochemistry, 1982; MD, Rush Med. Coll., 1986. Diplomate Am. Bd. Pathology. Rsch. assoc. pediatrics Joseph P. Kennedy Jr. Mental Retardation Rsch. Ctr., Chgo., 1982-83, U. Chgo. Pritzker Sch. Medicine, 1982-83; resident in pathology Rush-Presbyn.-St. Luke's Med. Ctr., Chgo., 1986-87, pathologist, 1987-88; instr. Rush Med. Coll., Chgo., 1986-87, asst. prof., 1987-88; resident in pathology U. Chgo. Med. Ctr., 1988-90; asst. prof. U. Chgo. Pritzker Sch. Medicine, 1990—; dir. clin. chemistry, attending physician U. Chgo. Med. Ctr., 1990—. Contbr. articles to Jour. Biol. Chemistry, Molecular Pharmacology, Jour. Neurochemistry, Jour. Membrane Biology, Procs. of NAS. U.S. Pub. Health Predoctoral fellow NIH, 1981-82; James B. Herrick scholar Rush Med. Coll., 1986-87; recipient Young Investigator award Acad. Clin. Lab. Physicians and Scientists, 1990. Mem. AAAS, Am. Soc. for Biochemistry and Molecular Biology, Am. Soc. for Cell Biology, Am. Assn. Pathologists, Coll. Am. Pathologists, Am. Soc. Clin. Pathologists, Sigma Xi. Achievements include research in biochemistry of cell membrane receptors and signal transduction in the nervous system, molecular pharmacology of opiates and opioid peptides, regulation of complex carbohydrate and lipid metabolism. Office: U Chgo Pritzker Sch Medicine Dept Pathology 5841 S Maryland Ave MC 0004 Chicago IL 60637-1470

MCLAWHORN, REBECCA LAWRENCE, mathematics educator; b. Newport News, Va., July 13, 1949; d. Marion Watson and Hazel Estelle (Babb) Lawrence; m. James Richard McLawhorn, June 23, 1973 (dec. 1980); 1 child, Susan Annette. BS, East Carolina U., 1971, MEd, 1974. Tchr., coach Greene Cen. High Sch., Snow Hill, N.C., 1972-76, Ridgecroft Sch., Ahoskie, N.C., 1976-78, Gates County High Sch., Gatesville, N.C., 1978-86; prof. Chowan Coll., Murfreesboro, N.C., 1986—. Pianist Gatesville Bapt. Ch., 1977—; active Athletic Boosters Club, Gatesville, 1984-91. Mem. ASCD, Math. Assn. Am., Nat. Coun. Tchrs. of Math., N.C. Coun. Tchrs. of Math., Parents for Advancement of Gifted Edn. (sec. 1986-90). Democrat. Avocations: piano, volleyball, softball, basketball, crocheting. Office: Chowan Coll Murfreesboro NC 27855

MCLEAN, IAN SMALL, astronomer, physics educator; b. Johnstone, Scotland, U.K., Aug. 21, 1949; s. Ian and Mary (Small) McL.; (div.); 1 child, Jennifer Ann; m. Janet Wheelans Yourston, Mar. 4, 1983; children: Joanna, David Richard, Graham Robert. BS with hons., U. Glasgow, 1971, PhD, 1974. Rsch. fellow Dept. Astronomy U. Glasgow, Scotland, 1974-78; rsch. assoc. Steward Observatory U. Ariz., Tucson, 1978-80; sr. rsch. fellow Royal Observatory U. Edinburgh, Scotland, 1980-81, sr. scientific officer Royal Observatory, 1981-86; prin. scientific officer Joint Astronomy Ctr., Hilo, Hawaii, 1986-89; prof. Dept. Physics and Astronomy UCLA, 1989—. Author: Electronic and Computer-Aided Astronomy: From Eyes To Electronic Sensors, 1989; contbr. articles to profl. jours. Recipient Exceptional Merit award U.K. Serc, Edinburgh, 1989; NSF grantee, 1991. Fellow Royal Astron. Soc.; mem. Internat. Astron. Union (pres. com. Paris chpt. 1988-91, v.p. 1985-88), Inst. Physics, Am. Astron. Soc. Achievements include discovery of relationship between polarization of light and orbital inclination of close binary stars; development of first CCD spectropolarimeter, first fully automated infrared camera for astronomy used to achieve images of faintest high redshift galaxies; research in polarization measurements of radiation from astronomical sources, use of CCDs and infrared array detectors. Office: UCLA Dept Astronomy 405 Hilgard Ave Los Angeles CA 90024-1301

MCLEAN, JOHN, architect, industrial designer; b. N.Y.C., Nov. 14, 1944. B Indsl. Design, Pratt Inst., Bklyn., 1969, BArch, 1975. Licensed/registered architect, N.Y., Ill., N.J., La., Pa., Conn., Israel; cert. Nat. Coun. Archtl. Registration Bds.; lic. tchr. fine arts, N.Y.C. Tchr. fine art and math. Bd. Edn., N.Y.C., 1969-76; staff architect, assoc., prin. various archtl. and indsl. design offices, N.Y.C., 1976-84; prin. John Mclean Architecture and Indsl. Design, White Plains, N.Y., 1985—; mem. N.Y. State Westchester County Rent Guidelines Bd., White Plains, 1979-85, Design Rev. Bd. City of White Plains, 1985-92, Westchester County Planning Bd., White Plains, 1985-92. Author: Architectural High Pressure Laminate Applications, 1984; exhibitor archtl. drawings Art in Engring., 1978. Mem. City of White Plains Citizen Resource Com. for Zoning, 1978-79; bd. dirs. Search for Change, Inc., White Plains, 1976-84. Recipient Gold award NOMMA, Atlanta, 1986, Bronze I, Interiors mag., N.Y.C., 1986. Mem. AIA. Achievements include reseach in fluid column and zero-stressing structural systems and design, expanded polystyrene cold billet molding technology and design. Office: 202 Mamaroneck Ave White Plains NY 10601-5312

MCLEAN, MALCOLM, materials scientist, educator, researcher; b. Ayr, Scotland, Dec. 12, 1939; s. Andrew Bell and Jane Pattison (Kilmartin) McL.; m. Malinda Ruth Conner; children: Andrew Lister, Calum Conner. BSc, U. Glasgow, Scotland, 1962, PhD, 1965. Chartered engr. Rsch. assoc. Ohio State U., Columbus, 1965-67; from sr. sci. officer to sr. prin. sci. officer Nat. Phys. Lab., Teddington, U.K., 1967-90; prof., head dept. materials Imperial Coll. Sci., Tech. & Medicine, London, 1990—. Author: Directionally Solidified Materials for High Temp Service, 1983; contbr. articles to profl. jours. Gov. Teddington (U.K.) Sch., 1989-93. Acta/Scripta Netallurgica lectr. Acta Metallurgica, 1992. Fellow Inst. Materials (Rosenhain medal 1986), ASM Internat.; mem. Inst. Physics. Office: U London Imperial Coll Science, Prince Consort Road, London SW7 2BP, England*

MCLEAN, RYAN JOHN, technical service professional; b. Ashland, Wis., Aug. 29, 1959; s. John Wallace and Dorothy Marie (Johnson) McL. AAS, Vermilion Coll., 1980. Refuge mgr.'s aid U.S. Fish & Wildlife Svc., Trempealeu, Wis., 1979; wildlife technician U.S. Forest Svc., Ely, Minn., 1980; forest inventory specialist Lake County Land & Timber, Two Harbors, Minn., 1980-82; hydrological field asst. U.S. Geol. Survey, Vancouver, Wash., 1982-83; forester Minn. Dept. Natural Resources, Grand Rapids, Minn., 1983-85; pvt. woodlands forester Carlton County Soil & Water Conservation, Barnum, Minn., 1985; tech. dir. La. Pacific Corp., Hayward, Wis., 1986-88; tech. svc. person Dyno Overlay's, Inc., Hayward, Wis., 1988—.

MCLEAN, WILLIAM GEORGE, engineering education consultant; b. Scranton, Pa., Mar. 15, 1910; s. Michael and Matilda Marie (Geueke) McL.; B.S. in Elec. Engring., Lafayette Coll., 1932; M.S., Brown U., 1933. Head math. dept. West Scranton (Pa.) High Sch., 1934-37; asst. prof. mech. engring. Lafayette Coll., Easton, Pa., 1937-44; asst. to supr. spl. products div. Eastman Kodak Co., Rochester, N.Y., 1944-46; prof., head engring. sci. Lafayette Coll., 1946-75, dir. engring., 1962-75; cons. in field, 1950—; mem. Pa. Registration Bd. Profl. Engrs., 1981-87. Chmn. Hugh Moore Park Commn., 1969—; v.p. United Neighborhood Ctrs., Lackawanna County, Pa; mem. Am. Nat. Metric Practice Group, 1974—. Fellow ASME (nat. v.p. 1953-55, 70-72, mem. various bds., Codes and Standards medal 1977, Performance Test Codes medal 1984); mem. Nat. Soc. Profl. Engrs. (pres. Pa. 1965-66), Am. Soc. Engring. Edn., Sigma Xi, Phi Beta Kappa, Tau Beta Pi, Eta Kappa Nu, Pi Tau Sigma, Kappa Delta Rho. Democrat. Roman Catholic. Author: (with E.W. Nelson) Engineering Mechanics, 1952, 4th edito., 1988; (with C.L. Best) Engring. Mechanics, 1965. Home and Office: 333 5th Ave Scranton PA 18505-1022

MCLEAN, WILLIAM RONALD, electrical engineer, consultant b. Bklyn., Mar. 26, 1921; s. Harold W. and Helena Winifred (Farrell) McL.; m. Cecile L. Mills, Aug. 17, 1946 (div.); m. 2d, Evelyn Hupfer, Nov. 29, 1968. BA, Bklyn. Coll., 1980, BS, 1981. Chief electrician U.S. Mcht. Marine, 1942-64; elect. designer, engr., 65-76; sr. elect. engr. Rosenblatt & Son, Inc., N.Y.C., 1976-86; cons., 1986—. Mem. Soc. Naval Architects and Marine Engineers, IEEE, Am. Soc. Naval Engrs. Home and Office: 57 Montague St Brooklyn NY 11201

MCLEAN-WAINWRIGHT, PAMELA LYNNE, educational consultant, college educator, counselor, program developer; b. Rockville Centre, N.Y., Oct. 25, 1948; d. George Clifford Sr. and Violet Maude (Jones) McLean; m. Joseph Charles Everest Wainwright Jr., Jan. 20, 1982; children: Joseph Charles Everest III, Evan Clifford Jerome. BS, NYU, 1973; MEd, Fordham U., 1974; MSW, Adelphi U., 1986. Qualified clin. social worker. Tchr. Martin Deporres Day Care Ctr., Bklyn., 1973-77; dir. student personnel services Ujamaa Acad., Hempstead, N.Y., 1977-78; coordinator Youth Employment and Tng. Program Hempstead, 1978; edni. opportunity counselor SUNY, Farmingdale, 1978-79; assoc. prof. student pers. svcs. Nassau Community Coll., Garden City, N.Y., 1979-93; founder, program dir. Adult Individualized Multi-Service Program, Garden City, 1985—; with counseling and advisement for health occupations program Cen. Fla. Community Coll., Ocala, 1991—. Mem. L.I. Coalition for Full Employment; mem. citizens adv. coun. Nassau Tech. Ctr., Women-on-Job Task Force, Port Washington, N.Y.; mem. adv. bd. Region 2 Displaced Homemakers Network; bd. dirs. Children's Greenhouse Inc., 1987-89, mem. founding com., 1980-81; civil rights adv., 1963—; mem. adv. bd. L.I. Cares, Hempstead, 1986-90. Recipient Women's History month citation Nassau County, N.Y., 1988, honoree in edn. Women-on-Job Task Force, 1989. Mem. Assn. Black Psychologists, Assn. Black Women in Higher Edn. (bd. dirs.), Nat. Assn. Black Coll. Alumni, Nat. Assn. Female Execs., Women's Faculty Assn. Nassau Community Coll. (pres. 1986-88), L.I. Women's Council for Equal Edn. Employment and Tng. Avocations: music collecting, writing poetry, travel, photography, sewing. Office: Cen Fla Community Coll Citrus County Campus 3820 W Educational Path Lecanto FL 34461-8054

MCLELLAN, KATHARINE ESTHER, health physicist, consultant; b. Boston, Apr. 14, 1963; d. Paul Edward and Esther Charlotte (Smith) McL. BA in Chemistry and Biology, Regis Coll., 1985; MS in Health Physics, Ga. Inst. Tech., 1986. Rsch. asst. dept. nuclear engring. Ga. Inst. Tech., Atlanta, 1985-86; asst. analyst NUS Corp., Gaithersburg, Md., 1986-88; health physicist NIH, Bethesda, Md., 1988—; lectr. Radiation Safety Tng. Program, 1988; ex officio mem. Animal Care and Use com. Nat. Inst. Child Health and Devel./NIMH, Bethesda, 1990—. Activities dir. Chase Knoll Apts., Germantown, Md., 1988-90; coord. St. Rose of Lima Young Adults, Gaithersburg, 1990—; mem. St. Rose Coun. Ministries, Gaithersburg, 1990—, chair youth ministry com., 1992—, co-chair fiesta, 1992—. Mem. Nat. Health Physics Soc. (Balt., Washington chpt.), Appalachian Compact Users of Radioactive Isotopes. Roman Catholic. Office: NIH 9000 Rockville Pike Bldg 21 Bethesda MD 20892-0001

MCLELLAND, SLATEN ANTHONY, electrical engineer; b. Prescott, Ark., Aug. 28, 1962; s. Slaten Everett and Betty Jean (Milam) McL.; m. Paula Jean Crenshaw, June 16, 1984; children: Chrystin Ann, Elizabeth Marie, Brittany Nicole. BSEE, Tex. A&M U., 1984. Assoc. engr. Ark. Power and Light, El Dorado, 1984-87; field maintenance supr. Star Enterprise, Charlotte, N.C., 1987-90; constrn. engr. Star Enterprise, Richmond, Va., 1990—. Usher Hugenot Rd. Bapt. Ch., Richmond, 1992. Named Star Employee of Month, 1989, Engr. Annual award, 1991. Mem. IEEE, NSPE, Nat. Soc. Corrosion Engrs. Republican. Baptist. Achievements include teaching quality process classes to Star Enterprise employees as well and outside contractors and vendors. Home: 9804 Aldersmead Pl Richmond VA 23236 Office: Star Enterprise PO Box 34707 Richmond VA 23234

MCLELLON, RICHARD STEVEN, aerospace engineer, consultant; b. Lawton, Okla., May 28, 1952; s. Robert Nelson and Jane (Warriner) McL. BSME, Old Dominion U., 1979. Aerospace engr. Naval Engring. Support Office, Norfolk, Va., 1979-82, U.S. Army Aviation Systems Commd., Ft. Eustis, Va., 1982-86; lead dynamicist Martin Marietta Astronautics Group Launch Systems, Denver, 1986—; cons. Aircraft Devel. Inc., Englewood, Colo., 1991—. Mem. Soaring Soc. Am., Rocky Mountain Aerobatic Club. Office: Aircraft Devel Inc PO Box 814 Englewood CO 80151-0814

MCLENNAN, DONALD ELMORE, physicist; b. London, Ont., Can., Dec. 5, 1919; came to U.S., 1959; s. John Nelson and Viola Winnifred (Hankinson) McL.; m. Winona Beckstead, Aug. 21, 1943 (div. June 1966); children: Peter, Carol, Christine, Douglas, Robert; m. Louise Wylie, June 5, 1966; adopted children: Frank, Patricia. BA with honors, U. Western Ont., 1941; PhD, U. Toronto, 1941. Rsch. scientist Can. Armaments Rsch. & Devel., Quebec, 1950-59; prof. physics Coll. William & Mary, Williamsburg, Va., 1959-67; prof. physics Youngstown (Ohio) State U., 1967-90, prof. emeritus, 1990—; cons. physicist Dow Chem. Co., Williamsburg. Contbr. articles to profl. jours. Achievements include Can. patents on ammunition sleeve for cooling gun barrels; research in electrodynamics and unified field theory. Office: Youngstown State U 410 Wick Ave Youngstown OH 44555

MCLENNAN, SCOTT MELLIN, geochemist, educator; b. London, Ont., Can., Apr. 20, 1952; came to U.S., 1987; s. Norman Marr and June Lyvonne (Clayton) McL.; m. Fiona Elspeth McPherson, Jan. 19, 1985; 1 child, Katherine Louise. BSc, U. Western Ont., London, 1975, MS, 1977; PhD, Australian Nat. U., Canberra, 1981. Rsch. fellow Australian Nat. U., Canberra, 1981-86; asst. prof. SUNY, Stony Brook, 1987-89, assoc. prof., 1989—. Co-author: The Continental Crust: Its Composition and Evolution, 1985; assoc. editor jour. Geochimica et Cosmochimica Acta, 1990—; editor Geochem. Soc. Spl. Publs., 1993—. Recipient Presdl. Young Investigator award NSF, 1989. Mem. Am. Geophys. Union, Geochem. Soc. Achievements include development of models for evolution of continental crust using sedimentary rock compositions. Office: SUNY Dept Earth and Space Scis Stony Brook NY 11794-2100

MCLEOD, JOHN ARTHUR SR., mechanical engineer; b. Bay Shore, N.Y., Dec. 12, 1952; s. John George and June (Nordby) Mc.; m. Linda Alice Sterry Aug. 9, 1975 (dec. Apr. 1988); 1 child, John (Jay) Arthur Jr.; m. Lisa Anne Hess; 1 child, Michael Jay. AA in Math/Sci., Hudson Valley C.C., Troy, N.Y., 1975; ASME, Hartford State Tech., 1987. With USN, Middletown, Conn., 1975-76; sr. lab. technician Olin Ski Co., Middletown, Conn., 1976-86; sr. quality ctrl. technician Electro Mechanics, New Britain, Conn., 1986-88; jr. mfg. engr. Kaman Aerospace, Bloomfield, Conn., 1986-89; Lab. supr. Walbro Automotive Corp., Meriden, Conn., 1989—. Mem. Am. Legion, Powder Ridge Ski Team (head coach 1982-90), Profl. Ski Instrs. Am. (cert.), U.S. Ski Coaches Assn. (cert.). Methodist. Office: Walbro Corp 45 Gracey Ave Meriden CT 06450-2202

MCLIN, WILLIAM MERRIMAN, foundation administrator; b. Attleboro, Mass., Feb. 22, 1945; s. William Hellen and Helen Francis (Orr) McL. B.A., George Washington U., 1967; M.Ed., U. Md., 1974. Exec. dir. Prince Georges County (Md.) chpt. Am. Cancer Soc., 1968-70; dir. programs Am. Cancer Soc. (Md. div.), 1970-72; dir. youth programs nat. office Am. Cancer Soc., N.Y.C., 1972-76; dep. exec. dir. Epilepsy Found. Am., Washington, 1976-79; exec. v.p. Epilepsy Found. Am., 1979—; pres. Nat. Health Coun., N.Y.C., 1989-91; pres. Internat. Bur. for Epilepsy, 1989-93. Mem. Am. Public Health Assn., Am. Soc. Assn. Execs. Office: Epilepsy Foundation of America 4351 Garden City Dr Hyattsville MD 20785-2223*

MCLINDEN, JAMES HUGH, molecular biologist; b. Marion, Kans., July 29, 1949; s. James Edward and Lenora Ann (Waner) McL. BA with honors, Emporia State U., 1971; PhD, U. Kans., 1983. Postdoctoral rsch. asst. biology Ohio State U., Columbus, 1983-87; sr. scientist Am. Biogentic Scis., Inc., Notre Dame, Ind., 1987-89, dir. molecular biology, 1989-91, v.p. molecular biology, 1991—. Author: (with others) Viral Hepatitis, 1990; contbr. articles to Jour. Virology, CRC Critical Revs. in Biotech., Biochem.-Biophysica ACTA, Applied and Environ. Microbiology. Mem. AAAS, Am. Soc. Microbiology, Am. Soc. Virology, N.Y. Acad. Sci., Soc. Indsl. Microbiology, Beta Beta Beta. Achievements include research in methods and material for expression of human plasminogen in eukaryodic cell system lacking a site spectic plasminogen activator; patent for Recombinant Hepatitis A Virus Vaccine. Home: 4232 Hickory Rd Apt 3A Mishawaka IN 46545 Office: Am Biogenetic Scis Inc Douglas Rd Reyniers Bldg Notre Dame IN 46556

MCLUCKEY, JOHN ALEXANDER, JR., electronics company executive; b. Star Junction, Pa., May 28, 1940; s. John Alexander and Louise Ann (Sayre) McL.; m. Sharon Grace Teemer, June 23, 1979; children Derek Alan, John Thomas, Jeffrey Scott, Kristin Dawn Galdys. AA, Cerritos Jr. Coll., 1964; BA, Calif. State U., 1967. Various adminstrv. and managerial positions autonetics div. Rockwell Internat. Corp., Anaheim, Calif., 1959-73; mgr. performance analysis Rockwell Internat. Corp., El Segundo, Calif., 1973-75, dir. western region fin. planning and analysis, corp. ops. analysis, 1973-75, exec. for aerospace and electronics, 1976-78; v.p., contr. electronic systems group Rockwell Internat. Corp., Anaheim, 1978-80, v.p., contr. def. electronics ops., 1980-83, v.p., gen. mgr. autonetics strategic systems div., 1983-84, pres. autonetics strategic systems div., 1984-87, pres. autonetics electronics systems, 1987-90, pres. def. electronics group, 1990—; chmn. pres.'s exec. forum Calif. State U., Fullerton, 1988. Mem. chief execs. roundtable U. Calif., Irvine, 1988, Orange County (Calif.) Bus. Com. for the Arts, 1988; bd. dirs. Orange County Boy Scouts Am., 1990. Named Silver Knight of Mgmt. Nat. Mgmt. Assn., 1989. Mem. Am. Def. Preparedness Assn. (bd. dirs. 1988—), AIAA, Assn. U.S. Army, Am. Electronics Assn., Armed Forces Communications and Electronics Assn., Balboa Bay Club (Newport Beach, Calif.), Center Club (Newport Beach), Yorba Linda Country Club (Calif.). Republican. Methodist. Office: Rockwell Internat Corp Def Electronics PO Box 3105 3370 E Miraloma Ave Anaheim CA 92806-1911*

MCMAHON, JAMES FRANCIS, quality control engineer; b. Buffalo, Apr. 1, 1942; s. Francis James and Rose Mary (Gavin) McM.; m. Carol Ann McCarthy, Sept. 20, 1969. BSEE, SUNY, Buffalo, 1965, MBA, 1976. Cert. quality engr. Sales engr. Worthington, Buffalo, N.Y., 1966-74, sr. sales engr., 1974-76, product mgr., 1976-78, mgr. customer svc., 1978-79, mgr. quality, 1979-85, mgr. compressor parts ops., 1985-88; mgr. quality Carborundum, Niagara Falls, N.Y., 1988—. Sgt. U.S. Army, 1966-67. Mem. Am. Soc. Quality Control, Am. Ceramics Soc., Carburundum Bus. Club. Democrat. Roman Catholic. Home: 45 Creekview Dr West Seneca NY 14224

MCMANIS, KENNETH LOUIS, civil engineer, educator; b. Lake Charles, La., Oct. 20, 1941; s. Louis Barber McManis; m. Josephine Agnes Agnew (div. 1975); 1 child, Patrick James; m. Julie Ann Sander, Aug. 5, 1978; 1 child, Kelly Lynn. BS, U. Southwestern La., 1963; MS, La. State U., 1966, PhD, 1975. Registered profl. engr., profl. land surveyor, La. Prodn. engr. Mobil Oil Co., Morgan City, La., 1963-64; with engring. design dept. W.S. Engrs. and Architects, Inc., New Orleans, 1966-68; prof., dean engring. tech. Delgado Jr. Coll., New Orleans, 1968-78; prof., dept. chmn. dept. civil engring. U. New Orleans, 1978—, dir. urban waste mgmt. and rsch. ctr., 1990—; adv. bd. Inst. for Recyclable Materials La. State U., Baton Rouge, La., 1991—, La. Transp. Rsch. Ctr., Baton Rouge, 1987-89; tech. com. Transp. Rsch. Bd., Washington, 1990—. Contbr. articles to ASTM Geotech. Testing Jour., Transp. Rsch. Record. Chmn. Landfill Siting Pub. Adv. Com., New Orleans, 1991-92; mem. Gov.'s Transition Com. on Solid Waste, La., 1992, Metrovision Subcom. on Solid Waste, New Orleans, 1992. Lt. cpl. USMCR, 1960-66. Named Michael Claus grad. La. State U., 1975; recipient Professionalism award La. Engring. Found., 1992. Mem. ASCR, La. Engring. Soc. (edn. com. 1980-83). Achievements include research in areas of soil sampling, soil stabilization, pile foundations, use of waste by-products in construction and waste management. Office: U New Orleans Dept Civil Engring New Orleans LA 71022

MCMANUS, JAMES WILLIAM, chemist, researcher; b. Atlanta, Oct. 7, 1944; s. Claude William and Sara Louise (Cook) McM.; m. Ruth Krieger, Apr. 10, 1971; children: Angela Ruth, Meagan Joy. BS in Chemistry, Auburn U., 1971. Mgr. Cook's Grocery Co., Atlanta, 1970-73; analytical chemist North Chem. Co., Atlanta, 1973-74; analytical chemist Merck & Co., Inc., Albany, Ga., 1974-75, staff chemist, 1975-76, sr. staff chemist, 1976-78, sr. chemist, 1978-80, 1980-89, rsch. fellow, 1989—; bd. dirs. M. Taylor, Inc., Albany, 1988—. Mem. editorial bd. Process Control and Quality, 1990—; inventor, patentee in field. Mem. Am. Chem. Soc. (cert.). Republican. Baptist. Office: Merck and Co Inc 3517 Radium Springs Rd Albany GA 31708-8301

MCMANUS, JOHN GERARD, software engineer; b. Quantico, Va., Feb. 27, 1960; s. John Francis and Mary Helen (O'Rielly) McM.; m. Linda Jean Prisco, Aug. 14, 1981; children: Christopher John, Allison Marie. BS in Computer Sci., U. N.H., 1982; MS in Ops. Rsch., U. Kans., 1986. Mem. tech. staff Triad Microsystems, Huntsville, Ala., 1986-87; sr. software engr. DSD Labs., Inc., Sudbury, Mass., 1987—. Author: Enhancements to Linear Simplex Algorithm, 1986. Capt. U.S. Army, 1982-86, Res., 1986—. Mem. Armed Forces Comm. Elec. Assn., Common Oper. Environ., John Birch Soc. (life). Roman Catholic. Office: DSD Labs Inc 75 Union Ave Sudbury MA 01776

MCMARTIN, KENNETH ESLER, toxicology educator; b. Madelia, Minn., Aug. 1, 1951; s. Finlay and Barbara (Bechtel) McM.; m. Deborah Ann King, July 1, 1989; children: Eric, F. Jackson. BA, Coe Coll., 1973; PhD, U. Iowa, 1977. Postdoctoral fellow Karolinska Inst., Stockholm, Sweden, 1977-78; asst. rsch. scientist U. Iowa, Iowa City, 1979-80; asst. prof. La. State U. Med. Ctr., Shreveport, 1980-84, assoc. prof., 1984-90, prof., 1990—; vis. assoc. prof. Med. U. S.C., Charleston, 1989-90. Author: (with others) Biochemistry and Pharmacology of Ethanol, 1979, Aspartame: Advances in Biochemistry and Physiology, 1984; contbr. articles to Jour. Pharmacology and Exptl. Therapeutics, Am. Jour. Physiology. Recipient Kenneth Morgareidge award Internat. Life Scis. Inst., 1988; Nat. Presbyn. Coll. scholar, 1969-73; Proctor & Gamble fellow, 1975; Swedish Med. Rsch. Coun., 1977. Mem. Am. Inst. Nutrition, Am. Soc. Pharm. and Exptl. Therapeutics, Am. Acad. Clin. Toxicology, Rsch. Soc. Alcoholism, Soc. Toxicology, Phi Beta Kappa, Phi Kappa Phi. Achievements include research in orphan product designation for 4-methylpyrazole.

MCMILLAN, CHARLES FREDERICK, physicist; b. Fayetteville, Ark., Oct. 25, 1954; s. Robert C. and Betty Jo (Boynton) McM.; m. Janet Faye Robb, June 26, 1977; children: Paul, Katherine, Caroline. BA, Columbia Union Coll., Takoma Park, Md., 1977; PhD, MIT, 1983. Exptl. physicist Lawrence Livermore (Calif.) Lab., 1983-92, computational physics group leader, 1992—; cons. Phoenix Laser Systems, San Francisco, 1988-92. Contbr. articles to profl. jours. Pres. Valley Choral Soc., Livermore, 1984; bd. dirs. San Francisco Bay Revels, Oakland, 1992—. Mem. Optical Soc. Am., Soc. Photo Instrumentation Engrs. Achievements include patents in medical imaging. Office: Lawrence Livermore Nat Lab PO Box 808 L-35 Livermore CA 94550

MCMILLAN, JAMES ALBERT, electronics engineer, educator; b. Lewellen, Nebr., Feb. 6, 1926; s. William H. and Mina H. (Taylor) McM.; m. Mary Virginia Garrett, Aug. 12, 1950 (dec. Feb. 1990); children: Michael, James, Yvette, Ramelle, Robert. BSEE, U. Wash., 1951; MS in Mgmt., Rensselaer Poly. Inst., 1965. Commd. 2d lt. U.S. Air Force, 1950, advanced through grades to lt. col., 1970; jet fighter pilot Columbus AFB, Miss., Webb AFB, Tex., 1951-52, Nellis AFB, Nev., 1953, McChord AFB, Wash., 1953-54; electronic maintenance supr. Lowry AFB, Colo., 1954, Forbes AFB, Kans., 1954-56, also in U.K. and Morocco, 1956-59; electronic engr., program dir. Wright-Patterson AFB, Ohio, 1959-64; facilities dir. Air Force Aero Propulsion Lab., Wright-Patterson AFB, 1965-70, ret., 1970; instr., div. chmn. Chesterfield-Marlboro Tech. Coll., S.C., 1971-75; instr., chmn. indsl. div., Maysville (Ky.) Community Coll., 1976—, asst. prof., 1977, assoc. prof., 1980, prof. 1986-93, prof. emeritus 1993—; chmn. indsl. tech. program, 1976-93; cons. mgmt. and electronic maintenance, 1970—. Served with U.S.

Army, 1943-45. Named to Hon. Order Ky. Cols., 1984. Mem. IEEE (sr., life), Soc. Mfg. Engrs. (sr.), Nat. Rifle Assn. (life), Sigma Xi (life). Republican. Presbyterian (elder). Clubs: Rotary (Maysville, Ky., pres. 1989-90), Masons (32 deg.), Shriners. Author: A Management Survey, 1965. Home: 6945 Scoffield Rd Ripley OH 45167-9682

MCMILLAN, ROBERT WALKER, physicist, consultant; b. Sylacauga, Ala., Apr. 18, 1935; s. Robert Thomas and Alma (Bush) McM.; m. Ann Simmons, Sept. 11, 1955; children: Marisa Ann, Robert Murray, Natalie June. BS, Auburn U., 1957; MS, Rollins Coll., 1966; PhD, U. Fla., 1974. Engr. Westinghouse Electric, Balt., 1960-61; staff engr. Martin Marietta Aerospace, Orlando, Fla., 1961-76; prin. rsch. scientist Ga. Tech. Rsch. Inst., Atlanta, 1976—; cons. Univs. Space Rsch. Assn., Huntsville, Ala., 1977, U.S. Army, 1976-92, numerous cos., 1976-92; program co-chmn. Internat. Conf. Millimeter Wave and Infrared Tech., Beijing, Peoples Republic of China, 1990, 92; program chmn. Internat. Conf. on Lasers, Orlando, Fla., 1986. Contbr. over 75 articles to profl. jours., conference digests. 1st lt. USAF, 1958-60. Mem. IEEE (sr.), Optical Soc. Am. Democrat. Baptist. Achievements include patents in field. Office: Ga Tech Rsch Inst Ga Inst Tech Atlanta GA 30332

MCMILLAN, RONALD THEROW, optician; b. Richmond, Va., Sept. 4, 1951; s. John Johnston and Page Valeria (Rankin) McM.; m. Rohini Dhanda, June 18, 1977; children: Sean, Aaron, Rani Page. BA, George Washington U., 1978. Prodn. supr. Isomet Corp., Springfield, Va., 1976-84, Spectra-Physics, Mountain View, Calif., 1984-87; prodn. mgr. Isomet Corp., Springfield, 1987—; graphic artist in field. Mem. Optical Soc. Am. Achievements include work in perfecting laser quality surfaces on crystalline and glass materials for visible and I.R. laser systems. Home: 13217 Shady Ridge Ln Fairfax VA 22033 Office: Isomet Corp 5263 Port Royal Rd Springfield VA 22151

MCMILLEN, DAVID L., psychology educator; b. Columbus, Ohio, Sept. 3, 1941; s. Luen R. and Lela (Miller) McM.; m. Edith C. McMillen, Dec. 28, 1963 (div. 1977); children: Robert, Eleanor, Randall; m. Dixie T. McMillen, May 23, 1988. BS, Memphis State U., 1963; PhD, U. Tex., 1968. Asst. prof. Miss. State U., 1968-72, assoc. prof., 1972-78, 80-85, prof., 1985—; sr. rsch. scientist U. Mich., Ann Arbor, 1978-81, assoc. dir. Ctr. for Rsch. on Learning and Teaching, 1978-81. Contbr. articles to profl. jours. including Addictive Behaviors, Internat. Jour. of the Addictions, and Jour. of Alcohol Studies. Rsch. grant Alcohol Beverage Med. Rsch. Found., Miss. State U., 1991-92, Miss. Alcohol Safety Edn. Program, 1984-87, U. Mich., 1979-81. Mem. AAUP, APA, Southeastern Psychol. Assn., Southeastern Soc. Psychologists. Achievements include rsch. on personality, behavioral and situational factors associated with alcohol impaired driving. Home: PO Box 130 Starkville MS 39759 Office: Miss State U Dept Psychology PO Drawer 6161 Mississippi State MS 39762

MCMURPHY, MICHAEL ALLEN, energy company executive, lawyer; b. Dothan, Ala., Oct. 1, 1947; s. Allen L. and Mary Emily (Jacobs) McM.; m. Maureen Daly, Aug. 8, 1970; children: Matthew, Kevin, Patrick. BS, USAF Acad., 1969; MA, St. Mary's U., San Antonio, 1972; JD, U. Tex., 1975. Bar: Tex. 1975, U.S. Supreme Ct. 1977, U.S. Ct. Appeals (fed. cir.), D.C. 1978. Commd. 2d lt. USAF, 1969, advanced through grades to capt.; instr. Air U., Ala., 1975-79; resigned USAF, 1979; atty., advisor Oak Ridge (Tenn.) ops. U.S. Dept. Energy, 1979-83; gen. counsel COGEMA, Inc., Washington, 1983-87, v.p., 1987-88; pres., chief exec. officer COGEMA, Inc., Bethesda, Md., 1988—; pres., CEO Va. Fuels, Inc., Lynchburg, 1987-92; bd. dirs. Pathfinder Gold Corp., Bethesda, B&W Fuel Co., Lynchburg, Pathfinder Mines Corp., Bethesda, U.G. USA, Atlanta; chmn. Numatec, Inc., Bethesda, 1989—; bd. dirs. U.S. Com. for Energy Awareness, 1993—; pres. Uranium Producers Am., 1991-92. Mem. editorial bd. Air Force Law Rev., 1977-79. Recipient Nat. Order Merit, Republic of France, 1993. Mem. ABA, Fed. Bar Assn. (pres. East Tenn. chpt. 1982-83), Mensa. Republican. Roman Catholic. Avocation: skiing. Office: COGEMA Inc 7401 Wisconsin Ave Bethesda MD 20814

MC MURTRY, JAMES GILMER, III, neurosurgeon; b. Houston, June 11, 1932; s. James Gilmer and Alberta Elizabeth (Matteson) McM.; student Rice U., Houston, 1950-53; M.D. cum laude, Baylor U., Houston, 1957. Intern, Hosp. U. Pa., Phila., 1957-58; resident gen. surgery Baylor U. Affiliated Hosps., Houston, 1958-59; asst. neurol. surgery Coll. Physicians and Surgeons, Columbia U., N.Y.C., 1959-60; asst. resident neurol. surgery and neurology Neurol. Inst. N.Y., Columbia Presbyn. Med. Center, N.Y.C., 1960-62, chief resident neurol. surgery, 1962-63; Nat. Inst. Neurol. Disease and Blindness spl. fellow neurol. surgery Coll. Physicians and Surgeons, Columbia U., N.Y.C., 1963-64, instr. neurol. surgery, 1963-65, assoc., 1965-68, asst. prof. clin. neurol. surgery, 1968-73, assoc. prof., 1973-89, prof., 1989—; asst. attending neurol. surgeon Neurol. Inst. N.Y., N.Y.C., 1964-73, assoc. attending neurol. surgeon, 1973-89, attending neurol. surgeon, 1989—; chief neurol. surgery clinic Vanderbilt Clinic, Columbia Presbyn. Med. Center, N.Y.C., 1964-68; attending-in-charge neurosurgery Lenox Hill Hosp., N.Y.C., 1970-91; assoc. cons. neurol. surgery Englewood (N.J.) Hosp., 1964—; asst. cons. neurol. surgery Harlem Hosp., N.Y.C., 1964—; cons. neurol. surgery Bronx (N.Y.) VA Hosp., 1964-65; mem. NIH Parkinson Research Group, Columbia U., 1965—; mem. med. adv. bd. N.Y. State Athletic Commn. Jesse H. Jones scholar Baylor U. Coll. Medicine, 1953-57, Allen fellow dept. neurol. surgery Columbia U., 1964-65. Diplomate Am. Bd. Neurol. Surgery. Fellow ACS, Linnean Soc. (London); mem. AAUP, AAAS, AMA, Am. Assn. Neurol. Surgeons, European Congress Pediatric Neurosurgery, Am. Soc. Stereotaxic Surgeons, Pan Am. Med. Assn., N.Y. State Soc. Surgeons, N.Y. State Neurosurgery Soc., N.Y. Acad. Sci., N.Y. Neurosurg. Soc., Med. Soc. State N.Y., N.Y. County Med. Soc., Osler Soc., Baylor U. Coll. Medicine Alumni Assn., Med. Strollers, The Med. Soc. of London, The Harveian Soc., Alpha Omega Alpha. Presbyn. Clubs: The Union (N.Y.C.), The Garrick (London), The Atheneum (London), The Met. Opera (N.Y.C.), The Norfolk Yacht and Country. Author: Medical Examination Review Book-Neurological Surgery, 1970, rev. edit., 1975; Neurological Surgery Case Histories, 1975; contbr. articles to profl. jours. Home: 1 Cobb Ln Tarrytown NY 10591-3003 Office: 710 W 168th St New York NY 10032-2699

MCNAIR, DENNIS MICHAEL, biology educator; b. Dayton, Wash., June 7, 1945; s. Clarence William and Lula Miriam (Williams) McN.; m. Koren Lee Jacobson, June 8, 1968. AB, Whitman Coll., 1967; PhD, So. Ill. U., 1980. Asst. prof. Western Ill. U., Macomb, Ill., 1978-79, Wright State U., Dayton, Ohio, 1979-80; asst., assoc. prof. U. Pitts., Johnstown, Pa., 1980—; cons. Pa. Elec. Co., Johnstown, 1982—; exec. dir. Cambria-Somerset Coun., Johnstown, 1987—. Bd. dirs. Johnstown Symphony Orch., 1989—, Allegheny Plateau Audubon Chpt., Johnstown, 1984—. Recipient Program grant NEH. Mem. AAAS, Am. Soc. Parasitologists, Entomol. Soc. of Am., Sigma Xi. Democrat. Home: 517 Cypress Ave Johnstown PA 15902 Office: U of Pittsburgh Schoolhouse Rd Johnstown PA 15904

MCNALLY, HARRY JOHN, JR., engineer, construction and real estate executive, consultant, researcher, accountant; b. Phila., Nov. 12, 1938; s. Harry John and Jane Sabina (Hub)McN.; m. Lynnette Anne Burley, May 29, 1966 (div. Dec. 1981); children: Megan Kathleen, Harry John III. BA, Lehigh U., 1960; MBA, Columbia U., 1966. Registered profl engr., Pa. Project engr. Turner Constrn. Co., N.Y., 1960-65; devel. engr. Dravo Corp., Pitts., 1966-69; mgr. The Austin Co., Roselle, N.J., 1969-79; sr. mgr. The Lummus Co., Bloomfield, N.J., 1979-82; v.p. The Eagle Group, N.Y., 1983-89; contractor Resolution Trust Corp., Somerset, N.J., 1991—; cons. in field, N.Y., 1965—. Sec., newsletter editor Cook Sch. Adv. Coun., Plainfield, N.J., 1985-86; mem. Nat. Trust Hist. Preservation, Washington, 1985—, Nat. Bldg. Mus., 1986—; treas. Shadyside Young Rep. Club, Pitts., 1966-69; stewardship chmn. Grace Episcopal Ch., Plainfield, 1975-77. Mem. Nat. Soc. Profl. Engrs. (minuteman 1977—), N.J. Soc. Profl. Engrs. (legis. com. 1977—), Lehigh U. Alumni Assn., Columbia U. Alumni Assn., Omicron Delta Kappa, Alpha Kappa Psi, Chi Psi. Republican. Home and Office: 44 2D St Fanwood NJ 07023

MCNALLY, MARK MATTHEW, control systems engineer; b. Paterson, N.J., June 19, 1958; s. Patrick Thomas and Catherine B. (Aylward) McN.; m. Dinah Lynn Etter, Jan. 7, 1984; 1 child, Brenna E. BSEE, U. Notre Dame, 1980; posrgrad., Pa. State U., 1991—. Field engr. GE Co., King of Prussia, Pa., 1980-89; sr. control systems engr. Lukens Steel, Coatesville, Pa., 1989-91, lead control systems engr., 1991—. Mem. IEEE, Assn. Iron and Steel Engrs. Independent. Roman Catholic. Office: Lukens Steel Modena Rd A-100 Coatesville PA 19320

MCNALLY, PATRICK JOSEPH, aerospace engineer, technical program manager; b. Grosse Pointe, Mich., May 19, 1958; s. James Nelson McNally and Mary Joan (Codd) Mannino; m. Gail Marie Barbaza, May 9, 1981; children: Brian Patrick, Jacqueline Marie, Colin James. BSE in Aerospace magna cum laude, U. Mich., 1980, MS in Aerospace, 1981; postgrad., 1991—. Registered profl. engr., Calif. Rsch. asst. U. Mich. Gas Dynamics Lab., Ann Arbor, 1978-80; intern engr. McDonnell Aircraft Co., St. Louis, 1980; mem. tech. staff Caltech Jet Propulsion Lab NASA, Pasadena, Calif., 1981-86; rsch. engr. KMS Fusion, Inc., Ann Arbor, 1986-88, tech. program mgr., 1988-91; dir. programs Canopus Systems, Inc., Ann Arbor, 1991—; cont. edn. instr. Washtenaw Community Coll., Ann Arbor, 1988-89. Rockwell fellow U. Mich. Coll. Engring., 1980-81, Milo E. Oliphant fellow, 1980-81. Mem. AIAA, IEEE (control systems soc.), Tau Beta Pi. Achievements include program management: orbital acceleration research experment, flew on shuttle Columbia STS 40 and STS 50; rarefield-flow acceleration measurement experment developed 1989-91. Office: Canopus Systems Inc 2010 Hogback Rd Ann Arbor MI 48105

MCNAMEE, WILLIAM LAWRENCE, industrial engineer; b. Pitts., Nov. 10, 1931; s. William Lawrence and Mary Cathrine (Kunkel) McN.; m. Evelyn McSwiggen, Nov. 29, 1952; 1 child, Leisa Jean. BSEE, U. Pitts., 1963. Resident engr. Union Switch & Signal, Swissvale, Pa., 1954-58; elec. technician Mason-Shaver Rhodes, East McKeesport, Pa., 1954-58; shop supr. Westinghouse Atomic, Waltz Mills, Pa., 1958-60; field engr. Martin-Orlando, Oakdale, Pa., 1960-61; sr. elec. engr. U. Pitts. Physics & Astronomy Dept., 1961—; advisor Electronics Inst., N.Y.C., 1975-81; cons. Extranuclear Labs., Blawnox, Pa., 1980-86, NMR Inst., Pitts., 1987-89. Co-author: Modern Industrial Electronics, 1993; contbr. articles to profl. jours. Served in USCG, 1950-53. Home: 4940 Brightwood Rd A512 Bethel Park PA 15102 Office: U Pitts Physics and Astronomy Dept Pittsburgh PA 15260

MCNAUGHTON, MICHAEL WALFORD, physicist, educator; b. Durban, South Africa, Mar. 2, 1943; came to U.S., 1972; s. Lionel W. and Margaret P. (Messenger) McN.; m. Kok-Heong Ng, May 9, 1969; children: Jennifer, Elizabeth. BS, U. London, 1962, PhD, 1972; MA, Oxford (Eng.) U., 1965. Physicist Agrl. Rsch. Coun., Zimbabwe, 1963; sci. tchr. King George V. Sch., Seremban, Malaysia, 1966-67; sr. sci. tchr. Hong Kong Internat. Sch., 1968-69; postdoctoral fellow U. Calif., Davis, 1972-75; sr. rsch. assoc. Case Western Res. U., Cleve., 1975-78; mem. staff Los Alamos (N.Mex.) Nat. Lab., 1978—; prof. physics U. N.Mex., Los Alamos, 1987—; liaison program adv. com. Los Alamos Meson Physics Facility, 1980—, mem. scheduling com., 1980—, chair users group, 1990. Editor: Polarization Phenomena in Nuclear Physics, 1980; contbr. articles to profl. publs. Fellow Am. Phys. Soc. Achievements include development of first polarized target, world's most accurate analyzing power measurements for PP and NP, first polarimeters, complete determination of nucleon-nucleon amplitudes, world's most accurate proton-proton cross section measurement. Home: 803 Quartz Los Alamos NM 87544 Office: Los Alamos Nat Lab MP10 Los Alamos NM 87545

MCNEILL, WILLIAM, environmental scientist; b. Evanston, Ill., Jan. 1, 1930; s. John and Ebba Kratina (Hansen) McN.; m. Caryl Mook, June 15, 1951 (dec. 1969); children: Elizabeth Marie, Charles Craig, Margaret Ruth; m. Caecilia Cinquanto, Oct. 10, 1970. BA, Colgate U., 1951; MA, Temple U., 1955, PhD, 1961. Chief phys. chemistry br. Frankford Arsenal U.S. Army, Phila., 1955-70, dir. applied sci., 1970-75; chief scientist, environ. mgr. Rocky Mountain Arsenal U.S. Army, Denver, 1975-80, dir. tech. ops., 1980-85; gen. mgr. Battelle Denver Ops., 1985-88; sr. tech. adviser Sci. Applications Internat. Corp., Golden, Colo., 1989-92; dir. tech. devel. Sci. Applications Internat. Corp., Oak Ridge, Tenn., 1992—; mem. materials adv. bd. ceramics Nat. Acad. Sci./Nat. Rsch. Coun., Washington, 1966; mem. Gov.'s Task Group on Rocky Mountain Arsenal, 1976, Colo. Pollution Prevention Adv. Bd., Denver, 1991—. Contbr. articles to Jour. Che. Physics, Applied Physics Letters, other profl. publs. Mem. Am. Chem. Soc.,Hazardous Material Control Rsch. Inst., Air and Waste Mgmt. Assn. Achievements include 10 patents for electrochemical processes, inorganic materials sythesis, electro-optical devices; demonstration and use of narrow-band optical absorbers for laser protection; leader in development of Army environmental programs. Home: 11910 W 76th Dr Arvada CO 80005 Office: Sci Applications Internat Ste 250 Bldg 52 14062 Denver W Pky Golden CO 80401-3121

MCNOWN, JOHN STEPHENSON, hydraulic engineer, educator; b. Kansas City, Kans., Jan. 15, 1916; s. William Coleman and Florence Marie (Klahr) Mc.N.; m. Miriam Leigh Ellis, Sept. 6, 1938 (div. Nov. 1971); children: Stephen Ellis, Robert Neville, Cynthia Leigh, Mark William. BS, U. Kans., 1936; MS, U. Iowa, 1937; PhD, U. Minn., 1942; DSc, U. Grenoble, France, 1951. Registered profl. engr., Kans. Instr. Math. and mechanics U. Minn., 1937-42; research assoc. div. war research U. Calif., San Diego, 1942-43; from asst. prof. to prof. mechanics and hydraulics Coll. Engring. U. Iowa, Iowa City, 1943-54; research engr. Iowa Inst. Hydraulic Research, 1943-51, assoc. dir., 1951-54; Fulbright research scholar Grenoble, France, 1950-51; prof. engring. mechanics U. Mich., Ann Arbor, 1954-57; prof. engring. mechanics, dean Sch. Engring. and Architecture U. Kans., 1957-65, Albert P. Learned prof. civil engring., 1965-86, prof. emeritus, 1986—; exec. dir. Ctr. Research in Engring. Sci., 1959-62, dir. engring. div, 1962-65; dir. overseas liason com. Am. Council Edn., 1967-69; tech. edn. specialist IBRD, 1972-73; hydraulic engr. Ministry of Agr., Govt. of Swaziland, 1981-82; cons. Ministry of Higher Edn./Govt. of Tunisia, 1982-83; vis. prof. Chalmers U., Goteborg, Sweden, 1977, Royal Inst. Tech., Stockholm, 1978, 83—, U. Karlsruhe, 1987—, Nanyang Tech. Inst., Singapore, 1989; mem. numerous internat. study coms. Author: Technical Education in Africa; contbr. articles to profl. jours. Fellow ASCE (exec. com. engring. mech. divsn. 1956-60, J.C. Stevens award 1946, Rsch. Program prize 1949, J. Jas. R. Croes medal 1955), Am. Acad. Mechanics; mem. NAE, Nat. Conf. Engring. Edn. (coord. 1961), Internat. Assn. Hydraulic Rsch. (hon., coun. 1955-59), Am. Soc. Engring. Edn. (exec. com., vice chmn. internat. div. 1975-77), Com. Engring. Edn. (Mid. Africa), AAAS, Permanent Internat. Assn. for Nav. Congresses, Sigma Xi, Theta Tau, Phi Delta Theta, Tau Beta Pi, Phi Kappa Phi, Chi Epsilon. Congregationalist. Office: Royal Inst Tech, Hydraulics Lab, 100 44 Stockholm Sweden

MCNULTY, FRANK JOHN, laboratory coordinator; b. Phila., July 8, 1923; s. Clarence and Emily (Doyle) McN.; m. Elizabeth Ann Bertram, June 20, 1953; children: MaryAnn, Charles, Jane. BS, LaSalle U., 1948; MS, Villanova U., 1955. Chemist Rohm & Haas Co., Phila., 1948-61; lab. coord. Nopco Chem. Co. div. Diamond Alkali Shamrock Corp., Newark, N.J., 1961-75; editor, abstractor Chem. Abstracts Svc., Columbus, Ohio, 1975-78; lab. coord. Dept. health labs. State of Ohio, Columbus, 1978—; cons. Tech. Adv. Svc. to Attys., Blue Bell, Pa., 1978—. Editor publs. Polymer Sci., 1975-78, Chem. Abstracts, 1975-78. Grantee Ctrl. State U., 1971, SUNY, 1972, Am. U., 1973, Hope Coll., 1974, McCrone Rsch. Inst., 1978. Mem. APHA, Am. Chem. Soc. (sr., treas. Passaic sect. 1961-62), Am. Indsl. Hygiene Assn., Ohio Civil Svc. Employees Assn. (pres. Franklin chpt. 1978-80, bd. dirs.). Achievements include patent for Dispersant Detergent Lubricating Oil Additive; research in polymerization of N-vinyl pyrrolidone. Home: 4500 Midvale Rd Columbus OH 43224 Office: State of Ohio Dept Health Labs PO Box 2568 1571 Perry St Columbus OH 43216-0068

MCNULTY, JOHN ALEXANDER, anatomy educator; b. Bogota, Colombia, July 14, 1946; came to U.S., 1957; s. Alex and Hilda Margaret (Crellin) McN.; children: Margaret Anne, Patrick Alexander. BA, U. of the Pacific, 1968; PhD, U. So. Calif., 1976. Rsch. asst. U. So. Calif., L.A., 1970-71, teaching asst., 1971-76; asst. prof. Loyola U. Stritch Sch. Medicine, Maywood, Ill., 1976-81; assoc. prof. Loyola U. Stritch Sch. Medicine, Maywood, 1981-88, prof., 1988—; bd. dirs. Anatomical Gifts Assn. Author: Pineal Research Reviews, 1984; editorial bd.: Jour. Pineal Rsch., 1992—; contbr. articles to Cell and Tissue Rsch., Microscopy Rsch. Techniques, Jour. Neural Transmission. With U.S. Army, 1968-70. Recipient Rsch. awards NSF, Washington, 1979, 1988, first prize electron microscopy Am. Soc. Clin. Pathologists, Chgo., 1985, 86. Mem. Am. Assn. Anatomists, Am. Soc. for Cell Biology, Am. Soc. Zoologists, Am. Soc. Ichthyologists and Herpetologists. Achievements include research on the comparative neuroendocrinology of the structure and function of the pineal complex in vertebrate systems. Office: Loyola U Dept Cell Biol/Neurobiol 2160 S First Ave Maywood IL 60153

MCNULTY, MATTHEW FRANCIS, JR., health care administration educator, university administrator, consultant, horse and cattle breeder; b. Elizabeth, N.J., Nov. 26, 1914; s. Matthew Francis and Abby Helen (Dwyer) McN.; m. Mary Nell Johnson, May 4, 1946; children: Matthew Francis III, Mary Lauren. BS, St. Peter's Coll., 1938, DHL (hon.), 1978; postgrad., Rutgers U. Law Sch., 1939-41; grad., Officer Candidate Sch., U.S. Army, 1941, U.S. Army Staff and Command Sch., Ft. Leavenworth, 1945; MHA, Northwestern U., 1949; MPH, U. N.C., 1952; ScD (hon.), U. Ala., 1969, Georgetown U., 1986. Contract writer, mgmt. trainee acturial div. Prudential Life Ins. Co., Newark, N.J., 1938-46; dir. med. adminstrn. VA, Chgo. and Washington, 1946-49; project officer to take over and operate new VA Teaching Hosps. VA, Little Rock, Birmingham, Ala. and Chgo., 1949-54; adminstr. U. Ala. Jefferson-Hillman Hosp., Birmingham, 1954-60; founding gen. dir. U. Ala. Hosps. and Clinics, 1960-66; founding prof. hosp. adminstrn. U. Ala. Grad. Sch., 1954-69, vis. prof., 1969—, founding dir. grad. program health adminstrn., 1964-69; prof. epidemiology and preventive medicine Sch. Medicine U. Ala., 1960-69; founding dean Sch. Health Adminstrn. (now Sch. Health Related Profls.), 1965-69; pres. Matthew F. McNulty, Jr. & Assocs., Inc., 1954-91; founding dir. Coun. Teaching Hosps. and assoc. dir. Assn. Am. Med. Colls., 1966-69; prof. community medicine and internat. health Georgetown U., 1969-89, prof. emeritus 1989—, v.p. med. ctr. affairs, 1969-72, exec. v.p., med. ctr. affairs, 1972-74; chancellor, dir. Georgetown U. Med. Ctr., 1974-86; chancellor emeritus Georgetown U., 1986—; chmn. acad. affairs com., trustee Hahnemann U., Phila, 1987—; trustee Fla. Found. for Active Aging, 1989—; cons. VA Adv. Com. on Geriatrics & Gerontology, 1991—; founding chmn. bd. Univ. D.C. Affiliated Health Plan, Inc., 1974-78; founding chmn. bd. trustees Georgetown U. Community Health Plan, Inc., 1972-80; vis. prof. Cen. U., Caracas, Venezuela, 1957-61; hosp. cons., 1953—; bd. dirs. Kaiser-Georgetown Community Health Plan, Inc., Washington, 1980-85, bd. dirs. Kaiser Health Plans and Hosps., Oakland, Calif., 1980-85, emeritus, 1985—; mem. Statuatory VA Spl. Med. Adv. Group, 1978-89, Higher Edn. Com. on Dental Schs. Curriculum, 1978-79; preceptor hosp. adminstrn. Northwestern U., Washington U., U. Iowa, U. Minn., 1953-69; mem. nat. adv. com. health research projects Ga. Inst. Tech., 1959-65, 73-85; nat. adv. com. health rsch. projects U. Pitts., 1956-60; adv. com. W.K. Kellogg Found., 1960-65; vis. cons., lectr. Venezuelan Ministry Health and Social Welfare, 1967-69; dir. Blue Cross-Blue Shield Ala., 1960-61, 65-68; trustee, mem. exec. com. Blue Cross and Blue Shield Nat. Capital Area, 1973-89, Washington Bd. Trade, 1972-86. Bd. dirs. Greater Birmingham United Appeal, 1960-66; trustee, chmn. Jefferson County (Ala.) Tb Sanatorium, 1958-64; mem. health services research study sect. NIH, 1963-67; cons. USPHS, 1959-63; mem. White House Conf. on Health, 1965, on Medicare Implementation, 1966, NIH, USPHS and DHEW Commns., 1967-86, others; trustee Nat. Council Internat. Health, 1975-86; pres. Nat. League Nursing, 1979-81. Served to maj. USAAF, 1941-46, lt. col. USAFR, 1946-55. Recipient Disting. Alumnus award Northwestern U., 1973, Disting. Alumnus award U. N.C., John Benjamin Nichol award Med. Soc. D.C., Mayor and D.C. Coun., Matthew F. McNulty, Jr. Unanimous Recognition Resolution of 1986, Centennial award Georgetown U. Alumni Assn. award, 1982, Patrick Healy Disting. Svc. award, 1985, Alumni Life Senator Election award, 1986; named to Hon. Order Ky Cols., 1984. Fellow Am. Pub. Health Assn., Am. Coll. Healthcare Execs. (life) (bd. regents and council of regents 1961-67, Disting. Health Sci. Exec. award 1976); mem AAAS, Am. Hosp. Assn. (life) (Disting. Service award 1984), Ala. Hosp. Assn. (past pres.), Nat. League for Nursing (past pres.), D.C. League Nursing (past dir.), Nat. Forum Health Planning (past pres., Distin, 1987), Council Med. Adminstrn., Internat. Hosp. Fedn., Jefferson County Ala. Vis. Nursing Assn. (past pres.; Disting. Service award), Ala. Pub. Health Assn. (past chmn. med. care sect.), Southeastern Hosp. Conf. (past dir.), Birmingham Hosp. Council (past pres.), Hosp. Council Nat. Capital Area (pres. 1985-89, exec. com. 1989—, past pres. 1989-93), Assn. Univ. Programs in Hosp. Adminstrn. (Disting. award 1971), Greater Birmingham Area C. of C. (Merit award), Washington Acad. of Medicine, Am. Assn. Med. Colls. (founding chmn. teaching hosp. council 1964-69; Disting. Service Mem.), Royal Soc. Health, Am. Systems Mgmt. Soc. (Disting. award), Orgn. Univ. Health Ctr. Adminstrs., Santa Gertrudis Breeders Internat., Bashkir Curley Horse Breeders Assn., Med. Soc. of D.C. (John Benjamin Nichols award 1982), Univ. Club Ala., Cosmos Club, City Tavern Club, KC (3d degree, coun. 10499 Ocean Springs, 4th degree Francis Deignan Assembly), Knights of Malta, Omicron Kappa Upsilon. Home and Office: Teoc Pentref 3100 Phil Davis Dr Ocean Springs MS 39564-9076

MCNULTY, PETER J., physics educator; b. N.Y.C., Aug. 2, 1941; s. Peter James and Winifred (Bones) McN.; m. Patricia Ann Arnold, Nov. 5,1966; children: Patricia Ann, Peter James. BS, Fordham U., 1962; PhD, SUNY, Buffalo, 1965. Postdoctoral fellow SUNY, Buffalo, 1965-66; asst. prof. Clarkson U., Potsdam, N.Y., 1966-71, assoc. prof., 1971-77, prof., 1977-88; prof., head dept. Clemson (S.C.) U., 1988—; mem. Air Force Spacerad Sci. Team, 1982-92, Combined Release Radiation Effects Satelite Microelectronics Working Goup, 1982-92, Navy's Microelectronics and Photonics Test Bed Working Group, 1993. Contbr. over 80 articles to profl. jours. Recipient Natural Rsch. Coun. associateship, 1971, 77. Mem. IEEE (sr., organizing com. 1990, 92, 93), Am. Phys. Soc., Radiation Rsch. Soc. Democrat. Roman Catholic. Achievements include patent pending for Solid State Microdosimeter; research in computer model of proton induces circuit upsets in space, charge collection spectroscopy of microelectronic circuits, analysis of muon-nucleus interactions. Office: Clemson U 117 Kinard Lab Physics Clemson SC 29634-1911

MCNULTY, RICHARD PAUL, meteorologist; b. Scranton, Pa., Apr. 23, 1946; s. Paul J. and Margaret A. (Dunleavy) McN.; m. Linda L. Glotzbach; children: Daniel, Michael. BS, NYU, 1968, MS, 1972, PhD, 1974. Rsch. meteorologist Nat. Severe Storms Forecast Ctr., Kansas City, Mo., 1976-80; dep. meteorologist in charge Nat. Weather Svc. Forecast Office, Topeka, 1980-90; chief hydrometeorology and mgmt. div. Nat. Weather Svc. Tng. Ctr., Kansas City, 1991—. Assoc. editor Jour. Nat. Weather Digest, 1982—; contbr. articles to Jour. Am. Meteorol. Soc. and Nat. Weather Digest. With USNR, 1968-91. Mem. Am. Meteorol. Soc., Nat. Weather Assn., Am. Geophys. Union, U.S. Naval Inst. Office: Nat Weather Svc Tng Ctr 617 Hardesty Kansas City MO 64124

MCNUTT, R. H., geologist, geochemist, educator; b. Moncton, N.B., Can., July 4, 1937; s. Harold Ashfield and Mary Edith (Biddiscombe) McN.; m. Paula McNutt, 1964; children: Suzanne, Christopher, Amy. BSc, U. N.B., 1959; PhD in Geology, MIT, 1964. Prof. geology McMaster U., Hamilton, Ont., 1965—, dean faculty sci., 1989—. Fellow Geol. Assn. Can.; mem. Geochem. Soc., Am. Geophys. Union. Achievements include research in geochemistry and strontium neodymium lead and osmiumisotopic studies of Archean and Grenville gneissic terrains;isotoic geochemistry of Precambrian shield and sedimentary basin brines. Office: McMaster U, Dean of Faculty of Science, Hamilton, ON Canada L8S 4L8

MCPHERSON, RONALD P., federal agency administrator. BS in Meteorology, U. Tex., MS in Environ. Engring., PhD in Atmospheric Scis. Trainee U.S. Weather Bur., 1959; observer/briefer, asst. aviation forecaster, rsch. meteorologist Dept. Commerce, Nat. Weather Svcs, 1968-80, branch chief devel. divsn., 1980-87, chief meteorol. ops. divsn., 1987-88, dep. asst. adminstr., 1988-90; dir. Nat. Meteorol. Ctr., Nat. Weather Svc., 1990—. Contbr. to profl jours. Fellow Am. Meteorological Soc. Office: Dept of Commerce-Nat Weather Svc National Meteorological Center 5200 Auth Rd Washington DC 20233*

MCQUARRIE, TERRY SCOTT, technical executive; b. Springville, Utah, Dec. 27, 1942; s. Evan Dain and Fay (Torkeldsen) McQ.; m. Judith Lynn Lewellen, June 20, 1970; children: Devin Daniel, Melanie Fay. BA, U. Oreg., 1966; MA, San Jose State U., 1977. Production mgr. Lunastran Co., San Jose, Calif., 1974-76; group leader Koppers Co., Inc., Pitts., 1978-79, industry mgr., 1980-87; v.p., tech. dir. Glasforms, Inc., San Jose, Calif., 1987—; Chmn. Pultrusion Industry Coun. of SPI, N.Y.C., 1988-90; vice-chmn. Panel Coun. of SPI, 1986-87. Contbr. articles to profl. publs. Mem. ASTM, Composites Inst. of Soc. of Plastics Industry (bd. dirs. 1991—), Nat.

Assn. Corrosion Engrs. Republican. Mem. LDS Ch. Achievements include patent for pultrusion polyester resins and process.

MCQUEEN, REBECCA HODGES, health care executive, consultant; b. Dothan, Ala., July 20, 1954; d. Edward Grey and Shirley Louise (Varner) Hodges; m. David Raymond McQueen, Mar. 5, 1982; children: Matthew David, Owen Grey. BS, Emory U., 1976, MPH, 1979. Research assoc. North Cen. Ga. Health Systems Agy., Inc., Atlanta, 1979-80; assoc. dir. Health Services Analysis, Inc., Atlanta, 1980-82; med. group adminstr. Southeastern Health Services, Inc./Prucare, Atlanta, 1982-84; sr. v.p., COO SouthCare Med. Alliance, Atlanta, 1985—; cons. North Cen. Ga. Health Systems Agy., 1980-81, Region 4 HHS, Atlanta, 1980-82, instr. Applied Stats., Washington, 1980-82; mem. Health Data com. and Health Cost subcom. Atlanta Healthcare Alliance, 1985—; cons. Atlanta Com. for the Olympic Games, 1992. Contbr. articles to profl. jours. Adviser to med. support panel Atlanta Com. for Olympic Games; mem. Morningside/Lenox Park Civic Assn., Friends of Atlanta-Fulton Pub. Libr., Atlanta Bot. Garden, Planned Parenthood-Atlanta, Ga. Coun. on Child Abuse, Atlanta Wellness Coun. Recipient research award Nat. Conf. on High Blood Pressure Control, 1981. Mem. ACLU, NOW, Am. Pub. Health Assn. (women's caucus com., presenter 1980, 81), Am. Coll. Healthcare Execs., Women Healthcare Execs., Am. Managed Care and Rev. Orgn. (presenter at nat. conf. 1989), Am. Assn. Preferred Provider Orgns., Delta Delta Delta. Democrat. Baptist.

MCQUENEY, PATRICIA ANN, biologist, researcher; b. Lancaster, Pa., Aug. 7, 1966; d. John Robert Sr. and Helen Patricia (Boyles) McQ. BS in Biology magna cum laude, Millersville (Pa.) U., 1988; MS in Molecular Biology, Lehigh U., 1992. Staff biologist Merck Sharp and Dohme, West Point, Pa., 1992—. Mem. AAAS, Am. Soc. Microbiology, Sigma Xi. Republican. Roman Catholic.

MCRAE, JOHN LEONIDAS, civil engineer; b. Lexington, Miss., Sept. 16, 1917; s. James Wright and Lota (O'Bryant) McR.; m. Thelma Lucile Nabors, Mar. 23, 1940; children—John Malcolm, Virginia Margaret McRae Pugh. B.S. in Civil Engring. and Geotech. Engring., Northwestern U., 1948. Chief bituminous and chem. lab. U.S. Army Engring. Waterways Exptl. Sta., Vicksburg, Miss., 1950-61, research engr. mobility and environ. div., 1961-72; CEO EDCO Inc., Vicksburg, 1960—; cons. on soil mechanics and bituminous pavements. Fellow ASCE; mem. Nat. Soc. Profl. Engr., ASTM, Assn. Asphalt Paving Technologists. Baptist (deacon). Contbr. numerous tech. papers to profl. lit. Patentee in field. Home: 416 Groome Dr Vicksburg MS 39180-5108 Office: PO Box 1109 Vicksburg MS 39181-1109

MCREE, JOHN BROWNING, JR., physician; b. Anderson, S.C., Dec. 9, 1950; s. John Browning and Melinda Bratton (Beaty) McR.; m. Melody Lynnn Jennings, May 29, 1976; children: Ansley, Sarabeth. BS, Presbyn. Coll., Clinton, S.C., 1973; MD, Med. U. of S.C., 1977. Diplomate Am. Bd. Family Physicians. Resident Anderson (S.C.) Meml. Hosp., 1977-80; physician Family Practice Assocs., North Augusta, S.C., 1980—; asst. clin. prof. family medicine Med. Coll. Ga., 1982—. Fellow Am. Acad. Family Physicians. Presbyterian. Home: 201 Oakhurst Dr North Augusta SC 29841-9719 Office: Family Practice Assocs 509 W Martintown Rd North Augusta SC 29841-3108

MCSHANE, EUGENE MAC, psychologist; b. Westerly, R.I., Mar. 25, 1950; s. Eugene H. McShane and Phyl Desimone Pietrahllo; m. Cynthia Calvin, Oct. 7, 1990; 1 child, Benjamin D. BSE, U. Mich., 1972; MEd, U. Mass., 1975; PsyD, U. Denver, 1980. Lic. clin. psychologist, Colo.; cert. tchr. Mass., Colo. Naval architect Hovermarine Corp., Pawcatuck, Conn., 1972-74; sci. tchr. Falmouth (Mass.) High Sch., 1974-75, Bear Creek High Sch., Lakewood, Colo., 1975-78; instr. math., psychology Colo. Women's Coll., Community Coll. Denver, 1976-78; sch. psychologist Cherry Creek Sch. System, Englewood, Colo., 1980-83; clin. psychologist in pvt. practice Denver, 1980—; expdn. leader Infinite Odessy, Boston, 1975, 76, 77; mountain ski guide Copper Mtn., Colo., 1976, 77, 78. Contbr. articles to profl. jours. Mem. APA, Colo. Psychol. Assn. (chmn. membership com.), InterDisciplinary Com. on Child Custody, Profl. Ski Instrs. of Am. Democrat. Office: 950 S Cherry St # 420 Denver CO 80222

MCSHERRY, FRANK D(AVID), JR., writer, editor; b. McAlester, Okla., Dec. 18, 1927; s. Frank D. and Mary (Clinton) McS. BFA, U. Okla., 1953. freelance artist, 1953-68. Co-editor of 32 published anthologies; contbr. numerous articles to profl. jours. Pvt. 1st class USAF, 1945-47. Democrat. Home and Office: 314 W Jackson Mcalester OK 74501

MCSWAIN, RICHARD HORACE, materials engineer, consultant; b. Greenville, Ala., Sept. 27, 1949; s. Howard Horace and La Belle (Henderson) McS.; m. Wanda Lynn Hare, June 9, 1972; children: Rachel Lynn, John Angus, Daniel Richard. BS in Materials Engring., Auburn U., 1972, MS in Materials Engring., 1974; PhD in Materials Engring., U. Fla., 1985. Teaching and rsch. asst. Auburn (Ala.) U., 1972-73; metallurgist So. Rsch. Inst., Birmingham, Ala., 1973-76; materials engr. Naval Aviation Depot, Pensacola, Fla., 1977-88, head metallic materials engring., 1988-90; pres. McSwain Engring., Inc., 1991—; cons. materials engr., Pensacola, 1982-90; presenter in field. Contbr. articles to tech. jours. Mem. ASTM, SAE Internat., ASM Internat. (chpt. edn. chmn. 1975-76), Am. Welding Soc., Nat. Assn. Corrosion Engrs., Electron Microscopy Soc., Internat. Soc. Air Safety Investigators. Presbyterian. Avocations: boating, fishing, running. Home: 1405 Kings Rd Cantonment FL 32533 Office: McSwain Engring Inc PO Box 10847 Pensacola FL 32524-0847

MCSWEENEY, AUSTIN JOHN, psychology educator, researcher; b. Berwyn, Ill., May 2, 1946; s. Austin John and Erna Elanor (DeSollar) McS.; m. June Marilee Erickson; Sept. 28, 1968; children: Andrew John, Patrick Michael. BA, U. Wis., 1969; PhD, N. Ill. U., 1975. Diplomate Am. Bd. Profl. Psychology. Post-doctoral fellow Northwestern U., Evanston, Ill., 1975-77; lectr. Northwestern U., Evanston, 1977-78; asst. prof. W. Va. U., Morgantown, 1978-81; asst. prof. Med. Coll. Ohio, Toledo, 1981-84, assoc. prof., 1985—; cons. VA Outpatient Med Ctr. Toledo, 1981-87, Parkview Hosp., Ft. Wayne, Ind., 1988-89. Editor: Practical Program Evaluation, 1982, COPD: A Behavioral Perspective, 1988; contbr. 25 articles to profl. jours., chpts. to books., 1977—. Bd. dirs. Am. Lung Assn. W. Va., 1978-81, Apple Tree Nursery Sch., Toledo, 1985-86, Teldo Hearing and Speech, !987-88. Mem. Am. Psychol. Assn., Internat. Neuropsychol. Assn., N.W. Ohio Soc. Profl. Psychologists (pres. 1986-87, Disting. Mem. award 1990), N.W. Ohio Psychol. Assn. (pres. 1986-87). Home: 4146 Northmoor Rd Toledo OH 43606 Office: Med Coll Ohio Psychiatry 3000 Arlington Ave Toledo OH 43699-0008

MCSWEENY, PAUL EDWARD, research technologist; b. Niagara Falls, N.Y., Feb. 4, 1942; s. Joseph A. Jr. and Leonora B. (Rozan) McS.; m. Catherine L. Fontana, July 12, 1966; children: Patrick, Michael, David, Theresa, Matthew. BS in Chemistry, SUNY, Buffalo, 1964; MBA, Bowling Green State U., 1973. Various positions Occidental Chem. Co., Houston, 1964-87; mgr. comml. planning W. R. Grace & Co., Columbia, Md., 1981-87; assoc. dir. strategic R & D Borden, Inc., Columbus, Ohio, 1987—; vice-chmn. Bus. Tech. Ctr., Columbus, 1990—. Mem. Inst. Food Technologists, Comml. Devel. Assn. (pres. 1985-86). Licensing Execs. Soc. Office: Borden Inc 1105 Schrock Rd Columbus OH 43229

MCTIGUE, TERESA ANN, biologist, researcher, educator; b. Washington, July 9, 1962; d. William Edward and Bernice Ann (Bakajza) McT. BS in Zoology, U. Md., 1984; MS in Marine Sci., U. S.C., 1986; PhD in Wildlife and Fisheries Scis., Tex. A&M U., 1993. Lab. asst. U. Md., College Park, 1981-83; rsch. asst. U. S.C., Columbia, 1984-86; staff biologist Sea Camp, Galveston, Tex., 1988-93; fisheries biologist Nat. Marine Fisheries Svc., Galveston, 1987—. Contbr. articles to profl. jours. Recipient Grad. Rsch. award Sea Grant, 1986. Mem. Estuarine Rsch. Fedn., Am. Fisheries Soc., Sigma Xi. Democrat. Roman Catholic. Achievements include rsch. on the dietary habits of juvenile penaeid shrimp in salt marshes, role of infauna in shrimps' diets, regulatory effects of predators on the prey's abundances, settlement patterns of brachyuran larvae in Gulf of Mexico. Office: Nat Marine Fisheries Svc 4700 Ave U Galveston TX 77551

MCVAY, BARBARA CHAVES, mathematics educator; b. Dallas, July 6, 1950; d. Joe M. and Dorothy May (Nock) Chaves; m. David Clyde McVay, Dec. 23, 1968; 1 child, Kathryn McVay Hearn. BS in Math., U. Tex., Arlington, 1971. Cert. secondary tchr. math., English, Tex. Tchr. math. C.W. Nimitz High Sch. Irving (Tex.) Ind. Sch. Dist., 1972—; bldg. rep. Dallas Tchrs. Credit Union, 1982—; part time lab. instr. Northlake/Dallas County Community Coll., Irving, 1988—. Tchr. Sunday sch. North Dallas Bapt. Ch., 1971-80; ch. tng. leader 1st Bapt. Ch., Irving, 1981-85. Mem. NEA (rep. 1980—), TSTA, Irving Edn. Assn., Nat. Coun. Tchrs. Math., Tex. Coun. Tchrs. Math., Greater Dallas Coun. Tchrs. Math., Math. Assn. Am., Delta Kappa Gamma. Republican. Avocations: crafts, sewing, needlework. Office: CW Nimitz High Sch 100 W Oakdale Rd Irving TX 75060-6899

MCVICAR, ROBERT WILLIAM, JR., industrial engineer; b. L.A., Mar. 2, 1944. BS, Calif. Poly., San Luis Obispo, 1965; MS, Ga. Inst. Tech., 1967; PhD, Oreg. State U., 1978. Dir. corp. R&D Anheuser Busch Cos., St. Louis, 1980-85; v.p., gen. mgr. Anheuser Busch Wines Inc., St. Louis, 1983-85; sr. v.p. GE Venture Capital Corp., Menlo Park, Calif., 1985-89; v.p. rsch. and tech. Pizza Hut (Pepsi Co.), Wichita, Kans., 1989—; owner SF Brewpartners Inc., San Francisco, 1987-89. Capt. U.S. Army, 1967-69. Office: Pizza Hut Inc Rsch & Tech 9111 E Douglas Wichita KS 67207

MCWATERS, THOMAS DAVID, mining engineer, consultant; b. Hibbing, Minn., June 23, 1942; s. David Thomas and Ida Marie (Morzenti) McW.; m. Elaine Delores Rodenbo, Aug. 25, 1962; children: Katherine Ann, Christopher Aldo. BS in Mining Engring., Mich. Coll. Mining & Tech., 1963. Registered profl. engr., Va. Shift foreman Andes Copper Mining Co., El Salvador, Chile, 1963-65; chief engr. Reynolds Haitian Mines Inc., Miragoane, Haiti, 1965-67; plant engr. Reynolds Guyana Mines Ltd., Kwakwani, Guyana, 1967-70; hwy. engr. Calif. Div. of Hwys., San Luis Obispo, Calif., 1970-71; maint. supt. Reynolds Haitian Mines, Inc., Miragoane, 1971-73; sr. staff engr. Magma Copper Co., Miami, Ariz., 1973—. Author: Hydraulic Mining Manual, 1991; contbr. articles to profl. jours., chpts. to books. Recipient Jon S. Mayer Meml. award Mich. Coll. Mining and Tech., 1963. Mem. Soc. for Mining, Metallurgy and Exploration Inc., Lega Fratellanza. Home: 113 Miami Gardens Dr Miami AZ 85539 Office: Magma Copper Co PO Box 100 Miami AZ 85539-0100

MCWHAN, DENIS BAYMAN, physicist; b. N.Y.C., Dec. 10, 1935; s. Bayman and Evelyn (Inch) McW.; m. Carolyn Quick, June 20, 1959; children: Susan, Jeanette, David. BS, Yale U., 1957; PhD, U. Calif., Berkeley, 1961. Disting. mem. tech. staff AT&T Bell Labs., Murray Hill, N.J., 1962-1990; chmn. Nat. Synchrotron Light Source, Brookhaven Nat. Lab., Upton, N.Y., 1990—. Contbr. articles to sci. jours. Fellow AAAS, Am. Phys. Soc. Achievements include rsch. in condensed matter physics. Office: Nat Synchrotron Light Source Brookhaven Nat Lab Bldg 725B Upton NY 11973

MCWHIRTER, JOAN BRIGHTON, psychologist; b. Urbana, Ill., July 10, 1954; d. Gerald David and Lois (Robbins) Brighton; m. Richard Eugene McWhirter, Aug. 13, 1976. BA in Sociology, So. Ill. U., 1974; MA in Psychology, U. Nev., 1982. VISTA vol. Oper. Life, Las Vegas, 1974-75; grad. asst. U. NEv., Las Vegas, 1978-80; psychologist So. Nev. Adult Mental Health Svcs., Las Vegas, 1984-92; vocational rehab. counselor State Indsl. Ins. System, Las Vegas, 1992—. Treas. Spiritual Assembly of Baha'is, North Las Vegas, 1977-92; vice chmn. Baha'is, Sunrise Manor, 1993—. Pres. scholar So. Ill. U., 1971-74. Mem. Nat. Alliance for Mentally Ill, Nat. Wildlife Fedn., Nat. Wildlife Fund, Nev. Alliance for Mentally Ill (liaison 1986-89), State of Nev. Employee's Assn., Amnesty Internat., Sierra Club (Toiyabe chpt.), Habitat for Humanity, Phi Kappa Phi, Psi Chi. Avocations: pottery, aerobics, hiking, running, photography. Home: 3648 Rochester Ave Las Vegas NV 89115 Office: State Indsl Ins System 1700 W Charleston Blvd Las Vegas NV 89126

MCWHORTER, KATHLEEN, orthodontist; b. Houston, May 29, 1953; d. Archer and Lucile (Taft) McW. BA summa cum laude, U. Houston, 1986; DDS with honors, Baylor Coll., 1990. Mgr. Am. Internat. Rent-A-Car, Houston, 1974-79; mktg. researcher Concoco Oil Co., Houston, 1979-83; orthodontist Baylor Coll. Dentistry, Dallas, 1990—; presenter Am. Assn. Dental Rsch., Montreal, Que., Can., 1988, Cin., 1990; rsch. fellow Baylor Coll. Dentistry, Dallas, 1987, 88, 89. Contbr. articles to profl. jours. Mem. ADA, Am. Assn. Orthodontists, Am. Assn. Women Dentists, Am. Assn. Dentistry for Children, Internat. Assn. Dental Rsch., Am. Assn. Dental Rsch., Tex. Dental Assn., Dallas County Dental Soc., The Crescent Club. Avocations: tennis, walking, music, water skiing. Office: Baylor U Coll Dentistry Dept Orthodontics 3302 Gaston Ave Dallas TX 75246-2098

MCWILLIAMS, C. PAUL, JR., engineering executive; b. Louisville, June 4, 1931; s. Cleo Paul and Audrey Dora (Hale) McW.; m. Barbara Ann Sparks, Feb. 22, 1950 (div. 1962); children: Bruce Kevin, Craig Tinsley; m. Barbara Ann Heintz, Apr. 25, 1980; 1 stepchild, Kimberly Jean Moorhouse Swigert. B Chem. Engring., U. Louisville, 1954, M Engring., 1972. Lic. profl. engr., N.Y., N.C. Sr. process devel. engr. Olin Mathieson Chem. Corp., Brandenburg, Ky., 1958-66, Rochester, N.Y., 1958-66; sr. chem. engr. GTE Sylvania, Seneca Falls, N.Y., 1966-74, Eastman Kodak Co., Rochester, 1974-81; prin., treas. Flint & Sherburne Assocs., P.C., Rochester, 1981-89; project engr. Roy F. Weston, Inc., Rochester, 1989-92; engring. mgr. ECCO, Inc. (Environ. Cons. Co., Inc.), Buffalo, 1992-93; pres. ECCO Engring., Buffalo, 1993—; cons. water tech. Water Tech. Corp., Tonawanda, N.Y., 1973-76; product rsch. panel Chem. Engring. Mag., 1982-83. Author: Waste Disposal Manual, 1976. Life mem. Rep. Presdl. Task Force, Webster, N.Y., 1986—; mem. Rep. Nat. Com., Webster, 1991-92. 1st lt. USAF, 1954-58, ret. lt. col. USAF Res., 1982. Decorated Meritorious Svc. medal. Mem. NSPE, Am. Inst. Chem. Engrs., Soc. Am. Mil. Engrs., Res. Officers Assn. (life), Monroe Profl. Engrs. Soc. (environ. com. 1972-75, chmn. 1973-75, bd. dirs. 1982-84, program chmn. 1984), Cons. Engrs. Coun. N.Y. State (program chmn. Rochester chpt. 1986-87, sec. 1987-88, treas. 1989—). Episcopalian. Achievements include replacing boiler feedwater regulators, related instrumentation and control systems and blowdown at a N.Y. State U. facility; system design for dry fabric dust collectors to remove fly ash from coal-fired boilers' flue gas. Home: 1132 Woodbridge Ln Webster NY 14580 Office: ECCO Engring Brisbane Bldg Ste 515 403 Main St Buffalo NY 14203

MEAD, FRANK WALDRETH, taxonomic entomologist; b. Columbus, Ohio, June 11, 1922; s. Arlington Alfred and Edith May (Harrison) M.; widowed; children: David Harrison, Gregory Scott. BS, Ohio State U., 1947, MS, 1949; PhD, N.C. State U., 1968. Rsch. asst. dept. physiology Ohio State U., Woods Hole, Mass., summer 1941; rsch. asst. dept. entomology Ohio State U., Columbus, 1948-50; Japanese beetle scout bur. entomology and plant quar. USDA, Columbus, summer 1948, biol. aid bur. entomology and plant quar., 1950-53; entomologist div. plant industry Fla. Dept. Agr., Gainesville, 1953-58, 60, biologist IV, 1983—; rsch. asst. N.C. State U., Raleigh, 1958-60; state survey entomologist Fed.-State Coop. Survey, Gainesville, 1969-80; courtesy assoc. prof. dept. entomology U. Fla., Gainesville, 1973—, Fla. A&M U., Tallahassee, 1977—. Co-editor Tri-ology Technical Report; contbr. articles to profl. jours. Bd. dirs., treas. Alachua Audubon Soc., Gainesville, 1968-75, 77-82; bd. dirs. Alachua County Hist. Soc., Gainesville, 1980-82; mem. steering com. Civitan Regional Blood Bank, Gainesville, 1977-79; vol. photographer P.K. Yonge Lab. Sch. U. Fla., Gainesville, 1978—. With U.S. Army, 1943-46, PTO. Ohio Acad. Sci. fellow, 1966. Mem. Entomol. Soc. Am. (bd. dirs. S.E. br. 1978-79), Ga. Entomol. Soc., Fla. Entomol. Soc. (sec. 1966-82, Cert. of Appreciation 1975, 82, 91, Cert. of Merit 1986), Fla. Mosquito Control Assn., Entomol. Soc. Washington, Soc. Systematic Biologists, SAR (Benjamin Franklin chpt. Columbus, Ohio), Fla. Track Club. Avocations: photography, jogging, birding. Home: 2035 NE 6th Terr Gainesville FL 32609-3758 Office: Fla Dept Agr and Cons Svcs Div Plant Industry PO Box 147100 Gainesville FL 32614-7100

MEAD, FRANKLIN BRAIDWOOD, JR., aerospace engineer; b. Chgo., Mar. 29, 1938; s. Franklin Braidwood and Elisa Francis (Ruiz) M. BSME, U. Mich., 1963; MSME, Purdue U., 1969; PhD, Penn State U., 1986. Aerospace engr. AF Rocket Lab. Propulsion Directorate, Edwards AFB, Calif., 1964-76; grad. student Princeton U., 1976-78; grad. student, rsch. technician The Penn State U., State Coll., Pa., 1978-83; aerospace engr. AF Rocket Propulsion Lab., Edwards AFB, 1983-88, sect. chief, 1988-91; sr. scientist Phillips Lab. Propulsion Divsn., Edwards AFB, 1991—. Editor: Advanced Propulsion Concepts, 1973. Recipient Outstanding Tech. Achievement award USAF, 1973. Mem. AIAA, IEEE, Am. Phys. Soc. Office: Phillips Lab/Propulsion Div OLAC PL/RKFE Edwards AFB CA 93524

MEAD, KATHRYN NADIA, astrophysicist, educator; b. Jacksonville, Fla., Aug. 6, 1959; d. Charles A. Mead and Nadia L. Mead. BS in Physics, Rensselaer Poly. Inst., 1981, MS in Physics, 1983, PhD in Physics, 1986. Cooperative rsch. assoc. Naval Rsch. Lab., Washington, 1986-88; adj. asst. prof. Union Coll., Schenectady, N.Y., 1988-90, vis. asst. prof., 1990—. Mem. bd. visitors Bolles Sch., Jacksonville Fla.; mem. Coun. Undergrad. Rsch. Recipient Career Devel. award Dudley Observatory, 1990, Faculty Rsch. Fund award Union Coll., 1990, 92, Fund award for Astrophysical Rsch., 1992. Mem. AAUW, Am. Astron. Soc. (Gaposchkins Rsch. Fund award 1991), Assn. for Women in Sci., Sigma Xi, Sigma Pi Sigma. Achievements include discovery of the existence of molecular clouds and star formation much farther from the center of the Milky Way than previously known; research on molecular clouds and star formations outside the solar circle in our Galaxy, broad CO line wings near T-Tauri stars, the origin and structure of isolated dark globules, high resolution studies of the HII region/molecular cloud interface in NGC1977. Office: Union Coll Physics Dept Schenectady NY 12308

MEAD, SEAN MICHAEL, anthropological researcher, consultant; b. Salem, Ind., Aug. 9, 1966; s. John Walter and Mary Elizabeth (Boling) M. BA in Anthropology, Ind. U., 1991, MLS, 1992. Rschr. Uralic & Altaic studies Ind. U., Bloomington, 1991—; mem. Bloomington Faculty Libr. Com., 1992, Univ. Faculty Libr. Com., 1992. SLIS rep. Ind. U. Acad. Assembly, 1992. Burns-Marshall Meml. scholar Burns-Marshall Meml. Trust, 1988-89; honors divsn. scholar Ind. U., 1986. Mem. ALA, Spl. Librs. Assn., Am. Soc. for Info. Sci., Uralic & Altaic Studies Students Orgn. (pres. 1991). Republican. Roman Catholic. Office: Ind U Uralic and Altaic Studies Goodbody Hall Rm 157 Bloomington IN 47405

MEADER, JOHN LEON, environmental engineer, consultant; b. Boston, Dec. 1, 1956; s. John W. and Mildred Meader. BS with distinction, Worcester Poly. Inst., 1979, MSCE, 1986. Registered profl. engr. Conn., Mass., N.H., R.I. Civil engr. Dewberry and Davis, Fairfax, Va., 1979-83; grad. teaching asst. dept. civil engring. Worcester (Mass.) Poly. Inst., 1983-87; sr. project engr. Weston and Sampson Engrs., Inc., Peabody, Mass., 1987—; lectr. Northeastern U., Dedham, Mass., 1992—. Co-author tech. papers. Mem. Conservation Commn., Northborough, Mass., 1991—, vice-chmn. 1993—, Water and Sewer Commn., Northborough, 1992—. Named Young Engr. of Yr., Va. Soc. Profl. Engrs. (George Washington chpt.), 1983. Mem. ASCE, NSPE (chpt. dir. 1982-83, newsletter editor 1982-83, Outstanding Svc. award 1983), Am. Water Works Assn., Chi Epsilon. Office: Weston and Sampson Engrs 5 Centennial Dr Peabody MA 01960

MEADOR, CHARLES LAWRENCE, management and systems consultant, educator; b. Dallas, Oct. 7, 1946; s. Charles Leon and Dorothy Margaret (Brown), m. Diane E. Collins, May 18, 1985. BSME with honors, U. Tex., 1970; MSME, MS in Mgmt., MIT, 1972. Mem. engring. staff Union Carbide Corp., Houston, 1967-68; instr. Alfred P. Sloan Sch. Mgmt. MIT, Cambridge, 1972-75, asst. dir. Ctr. Info. Systems Rsch., 1976-78, lectr. Sch. Engring., co-dir. Macro-Engring. Rsch. Group, 1978—; founder, pres. Decision Support Tech., Inc., 1974-92; co-founder, vice-chmn., dir. Software Productivity Rsch., Inc., 1985-87; pres., dir. The Softbridge Group, 1989-92, Mgmt. Support Tech. Corp., 1992—. Editor: How Big and Still Beautiful? Macro-Engineering Revisited, 1980, Macro-Engineering: The Rich Potential, 1981, Macro-Engineering and the Future: A Management Perspective, 1982, Macro-Engineering: Global Infrastructure Solutions, 1992; mem. editorial adv. bd. Computer Communication, 1979-91; mem. editorial bd. Communicacion e Informatica, 1980—; contbr. papers in field. Wilfred Lewis fellow, 1971; Draper Lab. fellow, 1974; NSF trainee, 1970. Mem. Computer Soc. IEEE (vice-chmn. Ea. Hemisphere and Latin Am. area com. 1977-83), Am. Soc. for Macro-Engring. (bd. dirs. 1992—), Cosmos Club, St. Botolph's Club, Sigma Xi, Tau Beta Pi, Pi Tau Sigma. Home: 3 Windy Hill Ln Wayland MA 01778-2612 Office: MIT Rm 3-282 Cambridge MA 02139

MEAKIN, JOHN DAVID, university research executive, educator; b. Nottingham, Eng., Feb. 11, 1934; came to U.S., 1958, naturalized, 1972; s. Claude Edwin and Hilda May (Storer) M.; m. Katharine Sadie Glover, July 21, 1956; children: Robert Nicholas, David Harry, Ian James, William Edwin, Andrew John. B.Sc. in Metallurgy, Leeds (Eng.) U., 1955, Ph.D., 1957. Vis. asso. Franklin Inst., Phila., 1958-59; research fellow U. Durham, Eng., 1960-62; sr. rsch. scientist Franklin Inst., 1962-65, prin. scientist, 1965-70, mgr. lab., 1970-74; prof. mech. engring. U. Del., 1974—, chmn. mech. engring. dept., 1987—, sr. scientist Inst. Energy Conversion, 1974—; vis. prof. U. Del., 1967, U. Murdoch, 1983. Contbr. articles to profl. jours. Yorkshire Copper Works (Eng.) Research scholar, 1955; Dept. Sci. and Indsl. Research scholar, 1956-57; Imperial Chem. Industries sr. research fellow, 1961, 62. Mem. IEEE (sr.). Home: 905 Baylor Dr Newark DE 19711-3127

MEAL, LARIE, chemistry educator, consultant; b. Cin., June 15, 1939; d. George Lawrence Meal and Dorothy Louise (Heileman) Fitzpatrick. BS in Chemistry, U. Cin., 1961, PhD in Chemistry, 1966. Rsch. chemist U.S. Indsl. Chems., Cin., 1966-67; instr. chemistry U. Cin., 1968-69, asst. prof., 1969-75, assoc. prof., 1975-90, prof., 1990—, researcher, 1980—; cons. in field. Contbr. articles to sci. jours. Mem. AAAS, N.Y. Acad. Scis., Am. Chem. Soc., Internat. Assn. Arson Investigators, NOW, Planned Parenthood, Iota Sigma Pi. Democrat. Avocations: gardening, yard work. Home: 2231 Slane Ave Norwood OH 45212-3615 Office: U Cin 2220 Victory Pky Cincinnati OH 45206-2822

MEANEY, THOMAS FRANCIS, radiologist; b. Washington, Dec. 4, 1927; s. Thomas James and Alice Lorraine (Andrews) M.; m. Mary R. McCallum, Aug. 25, 1951; children: Michael, Patricia, Thomas, Sean, Matthew, Daniel, Maura, Bridget. BS, Georgetown U., 1949; MD, George Washington U., 1953. Diplomate Am. Bd. Radiology (trustee). Intern Providence Hosp., Washington, 1953-54; resident in radiology Cleve. Clinic, 1954-57; chmn. div. radiology Cleve. Clinic Found., 1966—. Contbr. articles on radiology to profl. jours. With USN, 1946-48. Fellow Am. Coll. Radiology (Gold medal 1991), Am. Coll. Chest Physicians; mem. AMA, Ohio Med. Assn., Ohio. Radiol. Soc., Ea. Radiol. Soc., Cleve. Acad. Medicine, Radiol. Soc. N.Am., Am. Roentgen Ray Soc., Am. Soc. Clinic Radiologists, Am. Heart Assn., Soc. Cardiovascular Radiology, Cleve. Radiol. Soc., Alpha Omega Alpha. Office: Cleve Clinic 1 Clinic Ctr 9500 Euclid Ave Cleveland OH 44195-0002

MEANS, DONALD BRUCE, environmental educator, research ecologist; b. L.A., Mar. 9, 1941; s. Grant Hugh and Margaret Louise (Nolte) M.; divorced; children: Guy Harlan, Ryan Cameron; m. Katherine Ann Steinheimer, Jan. 17, 1993. BS, Fla. State U., 1968, MS, 1972, PhD, 1975. Rsch. biologist Tall Timbers Rsch. Sta., Tallahassee, 1970-77, dir., 1978-84; pres., exec. dir. Coastal Plains Inst., Tallahassee, 1984—; vis. faculty Orgn. for Tropical Studies, Costa Rica, 1977; adj. asst. prof. dept. biol. sci. Fla. State U., Tallahassee, 1976-82, adj. assoc. prof., 1982-89, adj. prof., 1989—; adj. curator herpetology Fla. State Mus., Gainesville, 1984—; rsch. assoc. in zoology Smithsonian Instn., Washington, 1989-92; proposal reviewer NSF, Nat. Geog. Soc., Fla. Game and Fish Commn., 1975—; environ. cons. pub. and pvt. agys., 1972—; organizer, leader Natural History Adventure Tours, 1983—. Contbr. editor S.Am. Explorer, 1990—; contbr. articles on ecology to sci. jours., on environment to popular publs. Named Outdoorsman of Yr., Fla. Wildlife Fedn., 1987; fellow NIH, Panama, 1967, Tall Timbers Rsch. Sta., 1972-77. Mem. AAAS, Soc. for Study Evolution, Ecol. Soc. Am., Am. Soc. Ichthyologists and Herpetologists, Soc. for Study Amphibians and Reptiles, Herpetologists League, Sigma Xi (pres. Fla. State U. chpt. 1980-81). Avocations: scuba diving, kayaking, long distance hiking, Amazon exploration. Home and Office: 1313 N Duval St Tallahassee FL 32303-5512

MEAUX, ALAN DOUGLAS, facilities technician, sculptor; b. Joliet, Ill., Sept. 10, 1951; s. Berry Lee and Luella Ann (Ferguson) M.; m. Letta Sue Nygaard, Sept. 15, 1984; children: Ashley Nicole, Lacey Marie. Student, Joliet Jr. Coll., 1969-71, Bradley U., 1971-72, U.S. Dept. Agr. Grad. Sch.,

1972, Skagit Valley Coll., 1983-85. Photographer J.J.C. Blazer, Joliet Herald News, Joliet, 1969-71; auto mechanic Pohanka Olds and Fiat, Hillcrest Heights, Md., 1972-74, Hoffman Olds and Rolls Royce, Hartford, Conn., 1974-75; carpenter Klappenbach Constrn. Co., Moscow, Idaho, 1975-79; property mgr. Olympic Builders, Oak Harbor, Wash., 1979-86; maintenance technician Troubleshooters Inc., Oak Harbor, 1986-87; facilities technician Island County Govt., Coupeville, Wash., 1987—; bronze sculptor Ronin Art Prodns., Oak Harbor, 1979—; appraiser class A Mid-Am. Appraisers Assn., Springfield, Mo., 1986—; bd. dirs. North West Token Kai, U. Wash., Seattle, 1989—, lectr., 1985; contbr. Nanka Token Kai, L.A., 1985—. Author: Japanese Samurai Weapons, 1989; prin. works exhibited at Mini Guild Children's Orthopedic Show, Ballard, Wash., 1986, Worldfest/Ethnic Heritage Coun., Seattle, 1988, 89, 90, Stanwood (Wash.) Invitational Art Show, 1988. Mem. Japanese Sword Soc. U.S. (life), N.W. Token Kai (charter, bd. dirs. 1989-91), Western Mus. Conf., Wash. Mus. Assn., Ethnic Heritage Coun., Nanka Token Kai, Japan Soc. Inc., Nat. Rifle Assn., Wash. Arms Collectors Assn, North Whidbey Sportmen's Assn., Cen. Whidbey Sportmen's Club. Avocations: hunting, fishing, woodworking, reading, collecting Japanese antiques. Office: Ronin Art Prodns PO Box 1271 Oak Harbor WA 98277-1271

MECH, LUCYAN DAVID, research biologist, conservationist; b. Auburn, N.Y., Jan. 18, 1937; divorced; 4 children. BS, Cornell U., 1958; PhD in Vertebrate Ecology, Purdue U., 1962. NIH fellowship U. Minn., Mlps., 1963-64, rsch. assoc., 1964-66; rsch. assoc. biologist Macalester Coll., 1966-69; wildlife rsch. biologist U.S. Fish & Wildlife Svc., 1969—; adj. prof. U. Minn., 1979—. Recipient Terrestrial Wildlife Publication award Wildlife Soc., 1972, Aldo Leopold Meml. award Wildlife Soc., 1993. Mem. Am. Soc. Mammal, Ecological Soc. Am., Wildlife Soc., Sigma Xi. Achievements include research in predator-prey relations, mammal behavior and natural history, animal movements and factors affecting them, ecology behavior and sociology of wolves, spatial organization of mammals, telemetry and radio-tracking. Office: 1992 Folwell Ave Saint Paul MN 55108*

MEDDIN, JEFFREY DEAN, safety executive; b. Dec. 29, 1946; married; 3 children. BBA, Armstrong State Coll., 1972. Cert. hazard chem. mgr.; cert. safety exec.; cert. safety profl. Prodn. mgr. Home Builders Savannah, Ga., 1972-74; project supt. A&W Constrn. Co., Newport News, Va., 1974; project safety engr. B.F. Diamond Constrn. Co., Savannah, 1974-76; safety/pers. supr. Metric Constructors, Inc. (subs. J.A. Jones Constrn. Co.), Charlotte, N.C., 1976-77; mill and constrn. safety dir. Continental Forest Industries, 1977-80; corp. dir. safety Zurn Industries, Inc., Erie, Pa. and Tampa, Fla., 1980—. Mem. ANSI (mem. A10 com., chmn. A10.22 sub-com.), Am. Soc. Safety Engrs. (past pres. West Fla. chpt., mem. nat. member edn. com.), Am. Indsl. Hygiene Assn. Vet. Safety, Nat. Constrn. Safety Execs., World Safety Orgn., Nat. Safety Coun. (constrn. divsn. & metals sect. exec. com.). Avocations: scuba diving, flying, boating, photography, electronics. Home: 3311 San Jose St Clearwater FL 34619 Office: Zurn Industries Inc Corp Safety Dept 405 N Reo St Ste 110 Tampa FL 33609-1004

MEDZIHRADSKY, FEDOR, biochemist, educator; b. Kikinda, Yugoslavia, Feb. 4, 1932; came to U.S., 1966; s. Miklos and Melanie (Gettmann) M.; m. Mechthild Westmeyer, Sept. 13, 1967; children: Sofia, Oliver. MS in Chemistry, Technische Hochschule Munich, Fed. Republic Germany, 1961. Instr. biochemistry U. Munich, 1965-66; postdoctoral assoc. U. Wis.-Madison and Washington U., St. Louis, 1966-69; asst. prof. biochemistry U. Mich., Ann Arbor, 1969-73, assoc. prof. biochemistry, 1973-81, assoc. prof. pharmacology, 1975-81, prof. biochemistry and pharmacology, 1981—, dir. biochemistry lab., Upjohn Ctr. Clin. Pharmacology, 1969-76; dir. and co-dir. biochemistry core lab. Mich. Diabetes Rsch. and Tng. Ctr., U. Mich. Med. Ctr. 1977-91; vis. assoc. prof. Stanford U., 1975-76; vis. prof. U. Calif., San Diego, 1983-84, U. Utrecht and U. Wageningen, The Netherlands, 1992-93; invited lectr. nat. and internat. meetings. Contbr. articles to profl. jours. and chpts. to books; mem. editorial bd. Drug Metabolism and Disposition, 1982—. Recipient Nat. Rsch. Svc. award USPHS, 1975-76, Kaiser Permanente award for excellence in teaching U. Mich., 1985; USPHS grantee, 1975—; Fogarty Senior Internat. fellowship, 1992-93. Mem. AAAS, Am. Soc. Neurochemistry, Am. Soc. Biochemistry and Molecular Biology, Am. Soc. Pharmacology and Exptl. Therapeutics, Soc. Biol. Chemistry (Germany), Sigma Xi. Achievements include research in neurochemistry, molecular pharmacology, membrane biochemistry.

MEECH, KAREN JEAN, astronomer; b. Denver, July 9, 1959; d. Lloyd Augustus and Patricia Ann (Marshall) M. BA cum laude in Physics, Rice U., 1981; PhD in Planetary Astronomy, MIT, 1987. Rsch. asst. Maria Mitchell Obs., Nantucket, Mass., 1978; rsch. asst. archaeoastronomy EARTHWATCH, Cusco, Peru, 1980; univ. lab. asst. molecular physics Rice U., Houston, 1980-81, quantum physics grader, 1980-81; rsch. specialist MIT, Cambridge, 1981-82, grad. teaching asst., 1982-86, grad. rsch. asst., 1986-87; asst. astronomer, tenure track Inst. for Astronomy, Honolulu, 1987-90, asst. astronomer, 1990-91, assoc. astronomer, 1992—; mem. IFA Computer Adv. Com., 1991—, IFA Endowment Com., 1991—, U. Rsch. Coun., 1990-93, NASA Planetary Astronomy Com. II, 1993-94, NASA Planetary Astronomy Rev. Panel, 1990-91, Cerro Tololo Interamerican Obs. User's Com., 1991-94, USIA Internat. Teleconf., 1991; chair IFA Scholarship Com., 1991—; interviewer Rice U. Alumni, 1989—; reviewer. Contbr. articles to Astronomy Jour., Science, Icarus, Nature, Astrophysics Jour., Bull. Am. Astronomy Soc., Info. Bull. Various Stars, Minor Planets Circulars, IAU Circular. Safety diver U. Hawaii Scuba class, 1988-90; vol. Honolulu Zool. Soc. Zoo Fun Run, 1991-92; active Dept. Edn. High Sch. Student Career Program, Honolulu, 1988. scholar Bd. of Govs., 1980, Grad. Student Rschrs. fellow NASA, 1986-87; recipient Annie Jump Cannon award in Astronomy, 1988. Mem. Am. Astron. Soc. (divsn. planetary scis.), Internat. Astron. Union-Commn. 15, Am. Assn. Variable Star Observers. Achievements include co-discovery of the outburst of Halley's comet at the longest distance from the sun for a recorded outburst; discovery of cometary activity on object 2060 Chiron. Office: Inst for Astronomy 2680 Woodlawn Dr Honolulu HI 96822

MEEHAN, ROBERT HENRY, utilities executive, human resources executive, business educator; b. Hackensack, N.J., June 19, 1946; s. Horace Miles and Pauline Jeannette (Pente) M.; m. Ruth Ann Auletta, Sept. 28, 1969; children: Robert Michael, Brian John. BA, Montclair State Coll., 1968; MA magna cum laude, Fairleigh Dickinson U., 1972; postgrad., Pace U., 1985—. Cert. secondary sch. tchr. of social studies, N.J., compensation profl. Job analyst Citicorp, N.Y.C., 1969-70, sr. job analyst, 1970-72, ofcl. asst., 1972, project specialist pers. practices/policy review, 1973, project specialist attitude surveys, 1973-75, pers. officer nat. banking group, 1975-76; asst. dir. pers. N.Y. Power Authority, White Plains, N.Y., 1976-84; dir. compensation N.Y. Power Authority, White Plains, 1984—; instr. Am. Compensation Assn., Scottsdale, Ariz., 1986—; mem. N.Y. Power Pool Salary com., 1990—. Sr. author: Managing a Direct Pay Program, Cert. Course 4A, 1991, Determining Compensation Costs: An Approach to Estimating and Analyzing Expense, 1991; mem. exec. adv. panel Acad. Mgmt. Exec., 1993—; contbr. articles to profl. jours. Scoutmaster, Boy Scouts Am., Ridgefield Park, N.J., 1968; also scouting coordinator, Maywood, N.J., 1982-83; vestryman, sr. warden St. Martin's Episcopal Ch., Maywood, 1977-84. Mem. Am. Compensation Assn. (cert. instr. 1986—, course coord. 1993—, mem. cert. and currency com. 1988-89, mem. direct compensation com. 1990-91, chmn. direct compensation com. 1992-93, bd. dirs. 1993), Soc. for Human Resource Mgmt., Doctoral Students Assn. Pace U., Acad. Mgmt. (exec. adv. panel jour. The Exec.), Order DeMoley (master councilro 1962, 65, scribe, adv. bd. 1965-68, Meritorious Svcs. award 1965), Psi Chi, Delta Mu Delta. Episcopalian. Avocations: sailing, furniture making, golf. Office: NY Power Authority 123 Main St White Plains NY 10601

MEEHL, GERALD ALLEN, research climatologist; b. Denver, May 21, 1951; s. Paul Edwin and Eileen Marcella (Wall) M.; m. Marla Jeanine Sparn, Dec. 14, 1986. BA, U. Colo., 1974, MA, 1978, PhD, 1987. Field project team Nat. Ctr. for Atmospheric Rsch., Pago Pago, American Samoa, 1975, Christchurch, New Zealand, 1975-76; team scientist Nat. Ctr. for Atmospheric Rsch., Bintulu, Malaysia, 1978, Kathmandu, Nepal, 1979; support scientist II Nat. Ctr. for Atmospheric Rsch., Boulder, Colo., 1979-82, assoc. scientist III, 1982-87, assoc. scientist IV, 1987-90, scientist II, 1990-93, scientist III, 1993—; mem. working group one Intergovtl. Panel on Climatic Change World Meteorol. Orgn., Geneva, 1989—; mem. Steering Group on Global Climate Modeling, 190—. Contbr. articles to Jour. Atmospheric Sci., Jour. Climate, Climate Dynamics; contbr. chpt.: Climate Systems Modeling, 1992. Mem. Am. Metereol. Soc., Phi Beta Kappa. Achievements include documentation of mechanism of ocean-atmosphere biennial oscillation in tropical Indian and Pacific Oceans; documented El Nino-Southern Oscillation in coarse-grid coupled ocean atmosphere model; explained differences of soil moisture change in models with increased carbon dioxide; showed possible changes of El Nino-Southern oscillation effects in model of greenhouse effect; showed role of climate variability in detecting climate change signals from slowly increasing carbon dioxide in a coupled model. Office: Nat Ctr Atmospheric Rsch 1850 Table Mesa Dr Boulder CO 80307

MEEKER, LAWRENCE EDWIN, civil engineer; b. Uniontown, Pa., Mar. 7, 1959; s. Harold E. and Lorraine J. (Bullinger) M. BS in Geology, U. Calif., Davis, 1984; MSCE, U. Nev., 1990. Engr. in tng., Nev. Project mgr. coll. engring. U. Nev., Reno, 1991—; design cons. So. Pacific Lines, San Francisco, 1991—; tech. editor Nev. T2 Ctr., Reno, 1993—. Mem. Am. Ry. Engring. Assn., Transp. Rsch. Bd. Republican. Roman Catholic. Home: PO Box 9878 Reno NV 89507 Office: U Nev Coll Engring # 256 Reno NV 89557

MEEKS, CRAWFORD RUSSELL, JR., mechanical engineer; b. Winston-Salem, N.C., Oct. 17, 1931; s. Crawford Russell and Grace Alice (Sheppard) M.; m. Rebecca Ann Weavil, May 12, 1954 (div. Aug. 1972); 1 child, Clark Eugene; m. Shirlyne Myrtle Parsons, Jan. 31, 1988. BSME with high honors, N.C. State U., 1959. Design engr. Marquardt Aircraft, Van Nuys, Calif., 1959-61; rsch. engr. Atomics Internat., Canoga Park, Calif., 1961-66; chief scientist Hughes Aircraft Co., El Segundo, Calif., 1966-88; pres. AVCON-Advanced Controls Tech., Inc., Northridge, Calif., 1988—. Contbr. over 30 articles to profl. jours. Vol. psychol. counselor to area low-income patients, 1973-79. With U.S. Army, 1953-55. Mem. AIAA, ASME (mem. wear com.), Soc Tribologists and Lubrication Engrs., Phi Kappa Phi. Achievements include patents in magnetic levitation bearings, anemometers, pumps; development of most efficient magnetic levitation bearing in the world, advanced computer program for analysis of high-speed rotating machinery; advanced research on magnetic levitation and tribology. Home: 5540 Mason Ave Woodland Hills CA 91367-6841 Office: AVCOM Advanced Controls Tech Inc 5210 Lewis Rd No 14 Agoura Hills CA 91301

MEEKS, LISA KAYE, hydrogeologist, researcher; b. Pasedena, Tex., Mar. 16, 1965; d. Gerald Eugene Meeks and Dorothy Louise (Perkins) Gable. BS, Univ. Ark., 1987, MS, 1990. Cert. profl. geologist. Teaching asst. geology dept. Univ. Ark., Fayetteville, 1987-90, rsch. asst. geology dept., 1987-90; hydrologist Keer-McGee Corp., Oklahoma City, Okla., 1990-91, MidContinent Geological Cons., Bentonville, Ark., 1991—; rsch. scientist Ark. Mining Inst., Russellville, Ark., 1991-93; pres. interest group Hydrogeology of Bottle Mineral Waters and Spas, Russellville, 1992—. Contbr. articles to profl. jours.; editor Rock Review, 1988. Recipient Grant in Aid Am. Assn. Petroleum Geologist, 1989, faculty grantee Assn. of Ground Water Scientist and Engrs., 1993; scholarship Univ. Ark., 1988, W.A. Tarr award, 1987. Mem. Nat. Ground Water Assn., Geological Soc. Am., Ark. Ground Water Assn., Sigma Gamma Epsilon (v.p.). Democrat. Office: Ark Mining Inst Ark Tech Univ Russellville AR 72801

MEENAKSHI SUNDARAM, KANDASAMY, chemical engineer; b. Tirumangalam, Tamilnadu, Madurai, India, Dec. 24, 1949; came to U.S., 1980; s. Kandasamy and Mahamayee; m. Selvamani Muthusamy, Mar. 26, 1979; 1 child, Varuna. B Tech., U. Madras, India, 1972; ME, Indian Inst. Sci., Bangalore, 1974; PhD, State U. Ghent, Belgium, 1977. Rsch. asst. U. Ghent, 1977-79; postdoctoral fellow U. Del., 1980-81; sr. engr. ABB Lummus Crest, Bloomfield, N.J., 1981-82, prin. engr., 1982-84, sr. prin. engr., 1984-87, process design mgr., 1987-90, tech. devel. mgr., 1990—. Contbr. over 30 articles to profl. jours. Mem. Am. Inst. Chem. Engrs., Am. Chem. Soc. Achievements include development of SRT V heater for producing olefins. Home: 101 Throckmorton Ln Old Bridge NJ 08857 Office: ABB Lummus Crest Inc 1515 Broad St Bloomfield NJ 07003

MEESE, ERNEST HAROLD, thoracic and cardiovascular surgeon; b. Bradford, Pa., June 23, 1929; s. Ernest D. and Blanche (Raub) M.; m. Rockell D. Dombar, Aug. 30, 1985; children: Donyel Hindee, Nathan Samuel; children from previous marriage: Constance Ann, Roderick Bryan, Gregory James. BA, U. Buffalo, 1950, MD, 1954. Diplomate Am. Bd. Surgery, Am. Bd. Thoracic Surgery. Resident in gen. surgery Millard Fillmore Hosp., Buffalo, 1955-59; resident in thoracic surgery U.S. Naval Hosp., St. Albans L.I., N.Y., 1961-63; group practice thoracic and cardiovascular surgery, Cin., 1965-88; pvt. practice, Cin., 1988—; asst. clin. prof. surgery Cin. Med. Ctr., 1972—; head sect. thoracic and cardiovascular surgery St. Francis-St. George Hosp., Deaconess Hosp.; mem. staff Good Samaritan Hosp., Bethesda Hosp., Christ Hosp., Providence Hosp., Childrens Hosp., Jewish Hosp. Kenwood, Jewish Hosp., Cin. Contbr. articles to profl. jours. and textbooks. Pres. bd. dirs., chmn. service com. Cin.-Hamilton County unit Am. Cancer Soc., v.p., trustee, mem. exec. bd., chmn. service com.; pres., bd. trustees, exec. bd. Ohio div. Am. Cancer Soc.; trustee Southwestern Ohio chpt. Am. Heart Assn. Comdr. M.C., USN, 1959-65. Fellow A.C.S., Internat. Coll. Surgeons; mem. Soc. Thoracic Surgeons, Am. Coll. Chest Physicians, Am. Coll. Angiology, Cin. Surg. Soc., Am. Coll. Cardiology; mem. Gibson Anat. Hon. Soc., AMA, Am. Thoracic Soc., Assn. Mil. Surgeons U.S., Acad. Medicine Cin., Assn. Advancement Med. Instrumentation, N.Am. Soc. Pacing and Electrophysiology, Am. Soc. Laser Medicine and Surgery, Phi Beta Kappa, Phi Chi (treas. 1952-54). Clubs: Western Hills Country, Queen City, Mediclub (pres. 1983-85) (Cin.). Lodge: Masons. Home: 174 Pedretti Ave Cincinnati OH 45238-6025 Office: 5049 Crookshank Rd Cincinnati OH 45238-3349

MEETZ, GERALD DAVID, anatomist; b. Aurora, Ill., Aug. 22, 1937; s. Quinlen Blackmer and Beatrice Clara (Friebele) M.; m. Jane Marie Klosiewski, Dec. 12, 1981; 1 child, Johanna Kathleen. AB, North Cen. Coll., 1959; MS, U. Ill. Med. Ctr., Chgo., 1967, PhD, 1969. Asst. prof. Med. Coll. Va., Richmond, 1972-75, Marquette U. Coll. Dentistry, Milw., 1975-79, So. Calif. Coll. Optometry, Fullerton, 1979-85; assoc. prof. U. Osteo. Medicine, Kansas City, Mo., 1985-87; prof. U. Osteo. Med. and Health Sci., Des Moines, 1987—, asst. dean, 1992—. Contbr. articles to profl. jours. With U.S. Army, 1961-63. Office: U Osteo Medicine Health Sci 3200 Grand Des Moines IA 50312

MEGARIDIS, CONSTANTINE MICHAEL, mechanical engineering educator, researcher; b. Athens, Greece, Nov. 30, 1959; came to U.S., 1983; s. Michael K. and Panagiota (Adamopoulos) M.; m. Crystal Anne Sewell, July 14, 1990. MS in Applied Math., Brown U., 1986, PhD in Mech. Engring., 1987. Postdoctoral rschr. Brown U., Providence, 1987-88; asst. specialist U. Calif., Irvine, 1988-90; asst. prof. U. Ill., Chgo., 1990—. Contbr. articles to profl. jours. Recipient Tech. Chamber of Greece award, 1978-82, Bodosakis Found. High Distinction award, Athens, 1982; Brown U. fellow, 1983. Mem. AIAA, ASME, Am. Assn. for Aerosol Rsch., Combustion Inst. Office: U Ill Chgo Dept Mech Engring 842 W Taylor St Chicago IL 60607-7022

MEGHERBI, DALILA, electrical and computer engineer, researcher; b. Algiers, Algeria, May 29, 1957; came to U.S., 1983; d. Med and Astite Megherbi. Diploma in Elec. Engring. with distinction, Ecole Nationale Polytechnique, Algeria, 1983; MSc in Elec. Engring., Brown U., 1986, MSc in Applied Math., 1987, PhD in Elec. Engring., 1993. Rschr. Nat. Ctr. New Energies, Algiers, 1982-83; sr. rsch. engr. Nat. Railway Co., Algiers, 1983; rschr. LEMS lab. divsn. engring. Brown U., Providence, 1986-92, postdoctoral rsch. assoc. LEMS lab. divsn. engring., 1993—; tech. cons. IDS Co., East Arlington, Mass., 1993; guest lectr. Northeastern U., 1988-91. Reviewer tech. articles; contbr. articles to profl. jours. Mem. IEEE, Computer Soc. of IEEE, Robotics Soc. of IEEE, Sigma Xi. Achievements include patent for Method and Apparatus for Robot Motion Near Singularities and for Robot Mechanical Design; research in computer-controlled and sensor-based robotics and automation, expert task planning and machine intelligence, robotics applications in space, computer languages and graphics, data processing and pattern recognition, mathematical modeling and system analysis. Home: 333 E 53rd St Apt 3N New York NY 10022

MEHALIC, MARK ANDREW, electrical engineer, air force officer; b. Latrobe, Pa., May 11, 1958; s. George Andrew and Carrie Darlene (Baird) M.; m. Linda Kay Greene, July 1987; children: Nicholas, Stephanie, Christina. Student, Pa. State U., 1976-80; MSEE, Air Force Inst. Tech., 1983; PhD, U. Ill., 1989. Commd. 2d lt. USAF, 1980, advanced through grades to maj., 1992; project engr. Flight Dynamics Lab., Wright-Patterson AFB, Ohio, 1980-82; test engr. 3246th Test Wing, Eglin AFB, Fla., 1983-86; asst. prof. Air Force Inst. Tech., WPAFB, Ohio, 1989—. Mem. IEEE. Office: Air Force Inst Tech AFIT/ENG Wright Patterson AFB OH 45433

MEHLENBACHER, DOHN HARLOW, engineering executive; b. Huntington Park, Calif., Nov. 18, 1931; s. Virgil Claude and Helga (Sigfridson) M.; m. Nancy Mehlenbacher; children: Dohn Scott, Kimberly Ruth, Mark James, Matthew Lincoln. BS in Civil Engring., U. Ill., 1953; MS in City and Regional Planning, Ill. Inst. Tech., 1961; MBA, U. Chgo., 1972. Structural engr., draftsman Swift & Co., Chgo., 1953-54, 56-57, DeLeuw-Cather Co., Chgo., 1957-59; project engr. Quaker Oats Co., Chgo., 1959-61, mgr. constrn., 1964-70, mgr. real property, 1970-71, mgr. engring. and maintenance, Los Angeles, 1961-64; chief facilities engr. Bell & Howell Co., Chgo., 1972-73; v.p. design Globe Engring. Co., Chgo., 1973-76; project mgr. I.C. Harbour Constrn. Co., Oak Brook, Ill., 1976-78; dir. estimating George A. Fuller Co., Chgo., 1978; pres. Food-Tech Co., Willowbrook, Ill., 1979-80; dir. phys. resources, adj. prof. dept. civil engring. Ill. Inst. Tech., Chgo., 1980-92; pvt. practice facility cons., Chgo., 1993—. Served with USAF, 1954-56. Registered profl. engr. and structural engr., Ill. Fellow ASCE; mem. Assn. Physical Plant Adminstrs., Am. Arbitrators Assn. Office: 3100 S Michigan Ave 901 Chicago IL 60616

MEHLMAN, EDWIN STEPHEN, endodontist; b. Hartford, Conn., Nov. 30, 1935; s. Sol Abraham and Rose (Slitt) M.; m. Lesley Judith Lunin, June 13, 1959; children: Jeffrey Cole, Brian Scott, Erik Van. BA, Wesleyan U., 1957; DDS, U. Pa., 1961; cert. endodontics, Boston U., 1965. Diplomate Am. Bd. Endodontists. Instr. oral medicine Sch. Dental Medicine Harvard U., Boston, 1965-67; clin. instr. endodontics Sch. Dental Medicine Tufts U., Boston, 1968-70; lectr. endodontics Sch. Dental Medicine Harvard U., Boston, 1970-72, asst. clin. prof. endodontics Sch. Dental Medicine, 1972—; staff assoc. Forsyth Dental Ctr., Boston, 1965—; pvt. practice Providence, 1965—; vis. lectr. dental hygiene U. R.I., Kingston, 1965-71, Community Coll. R.I., Lincoln, 1990—; cons. com. on accreditation of Dentists and Dental Aux. Edn. Programs, 1974-78. Contbr. articles to profl. jours. Pres. Temple Habonim, Barrington, R.I., 1968-70, Bur. Jewish Edn. of R.I., 1980-84; area v.p. Jewish Fedn. R.I., 1975-78; mem. R.I. Legis. Commn. to Study Malpractice Crisis, 1985-86; chmn. R.I. Dental Polit. Action Com., 1987-90. Capt. USAF, 1961-63. Fellow Am. Coll. Dentists, Internat. Coll. Dentists, Pierre Fauchard Acad. (Award of Merit); mem. ADA (coun. on govt. affairs and fed. dental svcs. 1988-92, vice chmn. 1991-92), Am. Assn. Endodontists (dir. 1988-91), R.I. Dental Assn. (pres. 1986-87), Alpha Omega. Jewish. Avocations: tennis, reading, civic activities. Home: 6 Ridgeland Rd Barrington RI 02806-4028 Office: 130 Waterman St Providence RI 02906-2010 also: 1090 New London Ave Cranston RI 02920

MEHLMAN, LON DOUGLAS, information systems specialist; b. Los Angeles, Apr. 29, 1959; s. Anton and Diane Mehlman. BA, UCLA, 1981; MBA, Pepperdine U., 1983. Systems programmer Ticom Systems Inc., Century City, Calif., 1978-81; systems analyst NCR Corp., Century City, 1981-83; sr. systems analyst Tandem Computers Inc., L.A., 1983-91; info. systems specialist Computer Scis. Corp., El Segundo, Calif., 1991—. Mem. Am. Mgmt. Assn., Assn. for Info. and Image Mgmt., Armed Forces Communications and Electronics Assn., Sierra Club, Phi Delta Theta. Avocations: sailing, skiing, world travel. Office: Computer Scis Corp 2100 E Grand Ave El Segundo CA 90245

MEHLMAN, MYRON A., environmental and occupational medicine educator, environmental toxicologist; b. Poland, Dec. 21, 1934; m. Sept. 4, 1960; children: Mara, Hope, Alison, Constance Lloyd. BS, CCNY, 1957; PhD, MIT, 1964. Prof. biochemistry Rutgers U., Newark, 1965-69; prof. biochemistry Coll. of Medicine U. Nebr., Omaha, 1967-71; chief biochem. toxicology FDA, Washington, 1972-73; spl. asst. toxicology dept. HEW, Washington, 1973-75; interagy. liaison officer NIH, Bethesda, Md., 1975-77; dir. toxicology Mobil Oil, Princeton, N.J., 1977-89; prof. U. Medicine and Dentistry of N.J., Piscataway, 1990—. Editor Jour. Environ. Pathology & Toxicology, 1977-81, Jour. Toxicology and Indsl. Health, 1975-78, Jour. Clean Tech. and Environ. Sci., 1989—; contbr. over 100 articles to profl. jours.; edited over 60 books. 1st lt. U.S. Army, 1958-60. Fellow Acad. Toxol. Soc., Am. Coll. Toxicology, Collequim Ramazzinic (bd. dirs.). Achievements include research in toxicology, environmental health, and nutritional and biomedical science. Home: 7 Bouvant Dr Princeton NJ 08540-1208 Office: U Medicine and Dentistry NJ 675 Hoes Ln Piscataway NJ 08854-5635

MEHNE, PAUL HERBERT, chemical engineer, engineering consultant; b. Ossining, N.Y., Mar. 10, 1924; s. Carl Albert and Janet Murray (Southard) M.; m. Doris Ruth Longfritz, June 17, 1944; children: Paul Randolph, Robert Arthur. BS in Chem. Engring., Clarkson U., Potsdam, N.Y., 1947. Prodn. control supr. I.F. Laucks div. Monsanto Chem. Co., Lockport, N.Y., 1943-44; SemiWorks supr. fluorocarbon R&D Jackson Lab., E.I. DuPont Co., Deepwater, N.J., 1947-49; supervising engr. pressure rsch. lab. DuPont Exptl. Sta., Wilmington, Del., 1949-68, supr. pressure rsch. lab., ctrl. R&D dept., 1968-85; engring. cons. in pvt. practice Mendenhall, Pa., 1985—. Sgt., Chem. Corps, U.S. Army, 1944-46, ETO. Mem. Am. Inst. Chem. Engrs., Am. Chem. Soc., Am. Soc. Metals Internat., Sigma Xi. Home and Office: PO Box 124 Mendenhall PA 19357

MEHRA, JAGDISH, physicist; b. Meerut, India, Apr. 8, 1937; came to U.S., 1957; s. Bhagwan Das and Shanti Devi (Kakkar) M.; m. Marlis Helene Lehn, Apr. 27, 1959; 1 child Anil. MS, UCLA, 1960; PhD, U. Neuchatel, 1963. Sr. lectr. U. Neuchatel (Switzerland), 1963-64; asst. prof. physics Purdue U., Hammond, Ind., 1964-65; assoc. prof. U. Mass., North Dartmouth, 1965-67; program dir. Sci. Rsch. Assocs. (IBM), Chgo., 1967-69; spl. rsch. assoc. U. Tex., Austin, Tex., 1969-73; prof. Inst. Solvay Inst., Brussels, 1973-88; Sir Julian Huxley prof. UNESCO, Paris, 1989-93; disting. prof. physics The Citadel, Charleston, S.C., 1993—; mem. edit. bd. Founds. of Physics, Denver, 1988—. Author: The Quantum Principle, 1974, Einstein, Hilbert and Theory of Gravitation, 1974, The Solvay Conferences on Physics, 1975, The Historical Development on Theory, 1982, 87, The Beat of a Different Drum: The Life and Science of Richard Geynmen, 1993; editor: The Physicist's COnception of Nature, 1973. Rsch. grantee Krupp Found., 1978-80, J.D. and C.T. MacArthur Found., 1982-85, Minna-James-Heineman Found., 1985-87; recipient Humboldt prize, 1976. Mem. Am. Phys. Soc., Swiss Phys. Soc., History Sci. Soc., Sherlock Holmes Soc. London. Home: 7830 Candle Ln Houston TX 77071

MEHRA, RAMAN KUMAR, data processing executive, automation and control engineering researcher; b. Lahore, Punjab, India, Feb. 10, 1943; came to U.S., 1964; s. Madan Mohan and Vidya Vati (Khanna) M.; m. Anjoo Talwar; children: Archana, Mandira, Kunal. BEE, Punjab Engring. Coll., 1964; MS in Engring., Harvard U., 1965, PhD, 1968. Assoc. prof. Harvard U., Cambridge, Mass., 1972-76; pres., chief exec. officer Sci. Systems, Co., Inc., Woburn, Mass., 1976—. Author: System Identification, 1976; also tech. papers on model algorithmic control (Best Paper award Internat. Fedn. Automatic Control, 1983). Recipient Eckman award Am. Automatic Control Coun., St. Louis, 1971. Fellow IEEE. Avocations: hiking, skiing, tennis. Home: 5 Angier Rd Lexington MA 02173-1608 Office: Sci Systems Co Inc 500 W Cummings Park Woburn MA 01801-6506

MEHRABI, M. REZA, chemical engineer; b. Tehran, Iran, May 27, 1964; came to U.S. 1982; s. Hooshang and Mahin (Hamidi) M. BSE, Princeton U., 1988; MS, MIT, 1991, postgrad, 1991—. Rschr. Princeton (N.J.) U., 1986-88, GE, Albany, N.Y., 1989, Chevron, Richmond, Calif., 1989. David Koch fellow, MIT, 1989-90; recipient Xerox prize Princeton U., 1988. Mem. Am. Inst. Chem. Engrs. (affiliate, Disting. Svc. award 1988), Soc. Indsl. Applied Math., Am. Assn. Crystal Growth, Sigma Xi, Phi Beta Kappa, Tau Beta Pi. Achievements include using parallel computers for modelling complex material processing systems; theoredical studies of growth of synthetic crystals for microelectronics and optoelectronics applications; rsch. in directional solidification. Home: 26 Myrtle St # 5 Boston MA 02114 Office: MIT Rm 66-256 Cambridge MA 02139

MEHRABIAN, ROBERT, academic administrator; b. Tehran, Iran. Former prof. MIT, U. Ill., Urbana; dean Coll. of Engring. U. Calif., Santa Barbara, until 1990; past dir. Ctr. Materials Sci. Nat. Bur. of Standards; pres. Carnegie-Mellon U., Pitts., 1990—. Office: Carnegie-Mellon U Office of the Pres 5000 Forbes Ave Pittsburgh PA 15213*

MEHRING, JAMES WARREN, electrical engineer; b. Denver, June 15, 1950; s. Clinton Warren and Carol Jane (Adams) M. BS, Colo. State U., 1972; MS, UCLA, 1974; PhD in Engr., Stanford U., 1983. Registered profl. engr., Colo. Mem. tech. staff Hughes Aircraft, L.A., 1972-77; sr. project engr. Hughes Aircraft, Sunnyvale, Calif., 1977-83; project mgr. Hughes Aircraft, Arlington, Va., 1983-89; lab. mgr. Hughes Aircraft, Reston, Va., 1989—; prin. mem. new tech. panel Nat. Security Telecom. Adv. Com., Washington, 1990—. Contbr. articles to profl. jours. V.p. Coll. Ter. Homeowner's Assn., Los Gatos, Calif., 1980-83; sec. Lancers, Ft. Collins, Colo., 1970-71. Recipient excellence award USAF Space Div., 1983. Mem. IEEE, AIAA, Armed Forces Communications and Electronics Assn., Sigma Tau, Tau Beta Pi, Phi Kappa Phi. Home: 4005 Old Mill Rd Alexandria VA 22309 Office: Hughes Aircraft 1768 Business Center Dr Herndon VA 22090

MEHROTRA, SUBHASH CHANDRA, hydrologist, consultant; b. Barabanki, India, July 5, 1940; came to U.S., 1965; s. Keshav Chandra and Shanti (Tandon) M.; m. Anju Kapoor, Aug. 13, 1974; children: Anupum, Apurva. PhD, UCLA, 1973. Rsch. asst. U. Iowa, Iowa City, 1965-67; postgrad. rsch. engr. U. Calif., L.A., 1968-73; sr. engr. Bechtel Corp., San Francisco, 1974-77; sr. engring. specialist Bechtel Corp., Oak Ridge, Tenn., 1979—; assoc. prof. Howard U., Washington, 1977-79; cons. O'Brien & Gere, Syracuse, N.Y., 1978-79, Waste Policy Inst., Blacksburg, Va., 1991—. Contbr. articles to profl. jours. Sec. Indian Student Assn., U. Iowa, 1966. Recipient Internat. scholarship U. Iowa, 1966-67, Deutsch Co. fellowship UCLA, 1969. Mem. Am. Geophys. Union, N.Y. Acad. Sci. Home: 260 Gum Hollow Rd Oak Ridge TN 37830 Office: Bechtel Corp 151 Lafayette Dr Oak Ridge TN 37830

MEHROTRA, VIVEK, materials scientist; b. Ranikhet, India, Sept. 7, 1965; came to U.S., 1987; s. Brahma Narain and Sushila (Kapoor) M.; m. Sunita Pandit, Dec. 28, 1992. B of Tech., IIT-Kanpur, India, 1987; MS, Cornell U., 1989, PhD, 1992. Rsch. asst. Cornell U., Ithaca, N.Y., 1987-92; sr. mem. rsch. staff Philips Labs., Briarcliff Manor, N.Y., 1992—. Contbr. articles to profl. jours. Recipient Best B-Tech. Project, IIT-Kanpur, 1987, Cert. of Merit, IIT-Kanpur, 1983-87. Mem. Am. Phys. Soc. Achievements include patents in field. Office: Philips Labs 345 Scarborough Rd Briarcliff Manor NY 10510

MEHTA, JAGJIVAN RAM, research scientist; b. Bilaspur, Himachal, India, June 23, 1951; came to U.S., 1986; s. Nand Lal and Mathura (Kapoor) M.; m. Meena Lumba, July 3, 1983; children: Mohit, Neha. MS with honors, Panjab U., Chandigarh, India, 1977; PhD in Neurosci., Pgimer U., Chandigarh, India, 1984. Rsch. officer Postgrad. med. Inst., Chandigarh, 1984-86; postdoctoral fellow Auburn (Ala.) U., 1986-89, rsch. scientist Scott-Ritchey Rsch. Ctr., 1989—. contbr. over 36 articles to Am. Jour. Vet. Rsch., Neurochem. Rsch., Rsch. in Vet. Sci., Toxicology Letter. Recipient Gold medal E. Merck, India, 1983, Young Scientist award Indian Coun. Med. Rsch., India, 1984. Mem. Am. Assn. Neuropathologists, Soc. Neurosci., Internat. Soc. Neurochemistry, Sigam Xi. Home: 111 S Dean Rd Auburn AL 36830 Office: Auburn U Scott Ritchey Rsch Ctr Auburn AL 36849-5525

MEHTA, RAJENDRA, chemist, researcher; b. Jodhpur, India, May 15, 1955; came to U.S., 1981; s. Moti Mal and Puspha (Bhandari) M.; m. Kamala Parkah, Nov. 28, 1981; children: Nidhi, Neeray. BSc, U. Jodhpur, 1974, MSc in Organic Chemistry, 1976, PhD in Chemistry, 1981. Postdoctoral rsch. fellow Gaylord Rsch. Inst., Whippany, N.J., 1981-83, sr. rsch. chemist, 1983-86; mgr. R & D Epolin Inc., Paterson, N.J., 1987-89; project scientist Standard Register Co., Dayton, Ohio, 1989-90, sr. project scientist, 1990—. Patentee for protective coating on thermal paper, 1990, radiation curable receptive coatings for thermal transfer imaging, 1993. Mem. ASTM, Am. Chem. Soc., Red-Tech Orgn. Avocations: golf, tennis. Home: 220 Estates Dr Dayton OH 45459-2838 Office: Standard Register Co 120 Campbell St Dayton OH 45408-1973

MEHTA, RAKESH KUMAR, physician, consultant; b. Gidderbaha, Punjab, India, Aug. 18, 1952; came to U.S., 1985; s. Parkash Chander and Sheela (Thukral) M.; m. Anita Gupta. MB BS, Med. Coll., Amritsar, Punjab, 1975; MD, Postgrad. Inst. Med. Edn., Chandigarh, 1979. Diplomate Am. Bd. Internal Medicine. Intern Victoria Jubilee, Amritsar, 1975; resident in medicine Hosp. Postgrad. Inst., Chandigarh, India, 1976-78; sr. resident in medicine Hosp. Postgrad. Inst., Chandgarh, India, 1978-79, All India Inst. of Med. Scis., New Delhi, 1979-81, U. Alta., Edmonton, Can., 1981-83; clin. fellow in oncology Cross Cancer Inst., Edmonton, 1984-85, N.Y. Med. Coll., N.Y.C., 1985-86; cons. med. oncologist Vets. Affairs, Castle Point, N.Y., 1986-92; clin. asst. prof. medicine N.Y. Med. Coll., Valhalla, 1992—; chief oncology program Vets. Affairs Med. Ctr., Castle Point, 1986—. Contbr. articles to professional jours. Fellow ACP, Royal Coll. Physicians and Surgeons of Can., Am. Coll. Internat. Physicians; mem. Am. Soc. of Clin. Oncologists, Can. Assn. Med. Oncologists. Avocations: travel, hiking, photography. Office: Vets Affairs VA Med Ctr Castle Point NY 12511

MEHTA MALANI, HINA, biostatistician, educator; b. Songad, Gujrat, India, June 24, 1958; came to U.S. 1978.; d. Jashwantlal and Usha (Modi) Mehta; m. Narendra Malani, Jan. 17, 1982; 1 child, Neil-Kanth. MA, SUNY, Buffalo, 1979; PhD, Columbia U., 1986. Statistician Merck and Co., Rahway, N.J., 1980-82, sr. statistician, 1982-86, biometrician, 1986-87; asst. prof. biostatistics U. Calif., Berkeley, 1987—. Contbr. articles to profl. jours. Mem. Am. Statis. Assn., Biometrics Soc., Sigma Xi.

MEIER, MARK F., research scientist, glaciologist, educator; b. Iowa City, Dec. 19, 1925; s. Norman C. and Clea (Grimes) M.; m. Barbara McKinley, Sept. 16, 1955; children: Lauren G., Mark S., Gretchen A. BSEE, U. Iowa, 1949, MS in Geology, 1951; PhD in Geology and Applied Mechanics, Calif. Inst. Tech., 1957. Instr. Occidental Coll., L.A., 1952-55; chief glaciology project office U.S. Geol. Survey, Tacoma, 1956-85; dir. Inst. Arctic & Alpine Rsch. U. Colo., Boulder, 1985—; vis. prof. Dartmouth Coll., Hanover, N.H., 1964; rsch. prof. U. Wash., Seattle, 1964-86; profl. geol. scis. U. Colo., 1985—; pres. Internat. Comn. on Snow and Ice, 1967-71; pres. Internat. Assn. Hydrol. Scis., 1979-83; Mendenhall lectr. U.S. Geol. Survey, 1982. Contbr. articles to profl. jours. With USN, 1945-46. Recipient 3 medals Acad. Scis., Moscow, USSR, 1970-85, Disting. Svc. award (Gold medal) U.S. Dept. of the Interior, 1968; Meier Valley, Antarctica named in his honor U.S. and U.K. Bd. Geographic Names. Fellow AAAS, Am. Geophys. Union (com. chmn.), Geol. Soc. Am., (com. mem.), Internat. Glaciological Soc. (v.p., coun., Seligman Crystal 1985), Arctic Inst. N.Am. (gov. 1987—). Office: U Colo Inst Inst Arctic and Alpine Rsch 1560 30th St Boulder CO 80309-0450

MEIER, PAUL, statistician, mathematics educator; b. N.Y.C., July 24, 1924. B.S. in Physics, Math., Oberlin Coll., 1945; postgrad. program in applied math., Brown U., 1945; M.A. in Math., Princeton U., 1947, Ph.D., 1951. Asst. prof. math. Lehigh U., 1948-49; research sec. Phila. Tb and Health Assn., 1949-51; research assoc. analytical research group, Forrestal Research Ctr. Princeton U., 1951-52; research assoc. dept. biostats. Sch. of Hygiene and Pub. Health, Johns Hopkins U., 1952-57, asst. prof., 1953-55, assoc. prof., 1955-57; assoc. prof. dept. stats. div. biol. scis. U. Chgo., 1957-62, prof. stats., 1962—, chmn. dept. stats., 1960-66, dir. Biomed. Computation Facilities, 1962-69, prof. theoretical biology, 1968-74, acting chmn. dept. stats., 1970-71, chmn. dept. stats., 1973-74, 83—, Ralph and Mary Otis Isham Disting. Service Prof. Stats. and the Pharm. and Physiol. Scis., 1975-84; spl. fellow NIH U. London Sch. of Hygiene and Tropical Medicine and Imperial Coll., 1966-67; vis. prof. dept. stats., Ctr. for Analysis of Health Practices Harvard Sch. Pub. Health, 1975-76; mem. com. on lung cancer Am. Cancer Soc., 1959-62; mem. spl. study sect. biomath. and stats. Nat. Inst. Gen. Med. Scis. NIH, 1965-70, mem. therapeutic evaluation com., Nat. Heart Inst., 1967-71; mem. adv. bd. Environ. Health Resource Ctr. State of Ill. Inst. Environ. Quality, 1971—; mem. Task Force on Health Considerations of Nat. Energy Policy Am. Pub. Health Assn., 1972-74; mem. adv. bd. VA Coop. Study of the Pathogenic Effects of Sickle Cell Trait, 1973-78; mem. adv. council dept. stats. Princeton U., 1974-76; mem. Computer and Biomath. Scis. Study Sect. NIH, 1974-77; dir. Data Quality Control Ctr. for Persantine-Aspirin Reinfarction Study, 1975-81; mem. Policy Adv. Bd. Multiple Risk Factor Intervention Trial, 1976—; mem. Com. on Nat. Stats., 1978—; mem. Clin. Trials Rev. Com., Nat. Heart, Lung and Blood Inst., NIH, 1978—; dir. Data Audit Ctr. for Internat. Mexiletine and Placebo Antiarrhythmic Coronary Trial (IMPACT), 1979—. Am. Acad. Arts and Scis. fellow; John Simon Guggenheim Meml. Found. fellow; Ctr. for Advanced Study in Behavioral Scis. fellow, Stanford U., 1982-83; Sigma Xi lectr., 1974-76. Fellow Am. Statis. Assn. (bd. dirs., v.p. 1965-67, chmn. com. on computers in stats. 1967, chmn. sect. on tng. 1974), Royal Statis. Soc., Am. Pub. Health Assn., Am. Thoracic Soc., Inst. Math Stats. (chmn. com. on fellows 1979), AAAS, Am. Heart Assn. (council on epidemiology, com. on fellows 1978); mem. Biometric Soc. (exec. com. Ea. N.Am. Region, pres. Ea. N.Am. Region, 1967), Am. Math. Soc., Math Assn. Am., Assn. for Symbolic Logic, Soc. for Indsl. and Applied Math. Office: U Chgo Dept Stats 5734 S University Ave Chicago IL 60637-1546

MEIER, WILBUR LEROY, JR., industrial engineer, educator, former university chancellor; b. Elgin, Tex., Jan. 3, 1939; s. Wilbur Leroy and Ruby (Hall) M.; m. Judy Lee Longbotham, Aug. 30, 1958; children: Melynn, Marla, Melissa. BS, U. Tex., 1962, MS, 1964, PhD, 1967. Planning engr. Tex. Water Devel. Bd., Austin, 1962-66, cons., 1967-72; research engr. U. Tex., Austin, 1966; asst. prof. indsl. engring. Tex. A&M U., College Station, 1967-68; assoc prof. Tex. A&M U., 1968-70, prof., 1970-73, asst. head dept. indsl. engring., 1972-73; prof., chmn. dept. indsl. engring. Iowa State U., Ames, 1973-74; prof., head sch. of indsl. engring. Purdue U., West Lafayette, Ind., 1974-81; dean Coll. Engring., Pa. State U., University Park, 1981-87; chancellor U. Houston System, 1987-89; prof. indsl. engring. Pa. State U., University Park, 1989-91; dir. div. engring. infrastructure devel. NSF, Washington, 1989-91; dean Coll. Engring. N.C. State U., 1991—; mem. bd. visitors Air Force Inst. Technology; cons. Ohio Bd. Regents, 1990, U. Arizona, 1989, Indsl. Rsch. Inst., St. Louis, 1979, Environments for Tomorrow, Inc., Washington, 1970-81, Water Resources Engrs., Inc., Walnut Creek, Calif., 1969-70, Computer Graphics, Inc., Bryan, Tex., 1969-70, Kaiser Engrs., Oakland, Calif., 1971, Tracor, Inc., Austin, 1966-68, div. planning coordination Tex. Gov.'s Office, 1969, Office of Tech. Assessment, 1982-86, Southeast Ctr. for Elec. Engring. Edn., 1978—; mem. rev. team Naval Rsch. Adv. Com. Editor: Marcel Dekker Pub. Co., 1978—; Contbr. articles to profl. jours. Recipient Bliss medal Soc. Am. Mil. Engrs., 1986, Am. Spirit award USAF, 1984; named Outstanding Young Engr. of Yr. Tex. Soc. Profl. Engrs., 1966, Disting. Grad. Coll. Engring., U. Tex. at Austin, 1987; USPHS fellow, 1966. Fellow AAAS, Am. Soc. Engring. Edn. (chmn. indsl. engring. divsn. 1978-83), Inst. Indsl. Engrs. (dir. ops. research div. 1975, pres. Ind. chpt. 1976, program chmn. 1973-75, editorial bd. Trans., publ. chmn., newsletter editor engring. economy div. 1972-73, v.p. region VIII 1977-79, exec. v.p. chpt. ops. 1981-83, pres. 1985-86); mem. Ops. Research Soc. Am., Inst. Mgmt. Scis. (v.p. S.W. chpt. 1971-72), ASCE (sec.-treas. Austin br. 1965-66, chmn. research com., tech. council water resources planning and mgmt. 1972-74), Am. Assn. Engring. Socs. (bd. govs. 1984-86), Nat. Assn. State Univ. and Land Grant Colls. (mem. engring. legis. task force 1983-87), Assn. Engring. Colls. Pa. (pres. 1985-86, treas. 1981-87), Air Force Assn. (advisor sci. and tech. com. 1984-87), Nat. Soc. Profl. Engrs., Profl. Engrs. in Edn. (vice chmn. N.E. region 1985-87, bd. govs. 1983-85), Sigma Xi, Tau Beta Pi, Alpha Pi Mu (asso. editor Cogwheel 1970-75, regional dir. 1976-77, exec. v.p. 1977-80, pres. 1980-82), Phi Kappa Phi, Chi Epsilon. Lodge: Rotary. Home: 7504 Grist Mill Rd Raleigh NC 27615

MEIJS, GORDON FRANCIS, chemist, research scientist; b. Adelaide, South Australia, Australia, Nov. 4, 1956; s. Andre Robert Marie and Gloria June (Mauger) M.; m. Carole Ann Schneider, Jan. 12, 1985; 1 child, David Joseph. BSc with first class hons., U. Adelaide, S. Australia, 1977, PhD in Organic Chemistry, 1981. Postdoctoral fellow U. Calif., Santa Cruz, 1982; sr. teaching fellow U. Adelaide, Australia, 1982-85; rsch. scientist CSIRO Div. of Applied Organic Chemistry, Melbourne, Victoria, Australia, 1988; sr. rsch. scientist CSIRO Div. of Chemicals and Polymers, Melbourne, 1988-92, prin. rsch. scientist, 1992—; mem. adv. panel on silicone implants Therapeutic Device Evaluation Com., Australia, 1992; key researcher Coop. Rsch. Ctr. for Cardiac Tech., Australia, 1992—; researcher Coop. Rsch. Ctr. for Eye Rsch. and Tech. Contbr. articles to Jour. of Macromolecular Sci., Macromolecules, Polymer Internat., Jour Applied Polymer Sci., Makromoleculare Chemie. Mem. Soc. for Biomaterials (US), Australian Soc. for Biomaterials, Am. Chem. Soc., Royal Australian Chem. Inst. (Victorian br. com. 1988-90, sec. polymer group 1991). Achievements include 3 patents in area of molecular weight control of polymers with controlled end group functionality; patent for polyurethanes for med. implant. Office: CSIRO Divsn Chems and Polymers, Bayview Ave, Clayton Victoria 3168, Australia

MEIKLE, PHILIP G., government agency executive; b. Glendale, W.Va., Dec. 5, 1937; s. Philip and Caroline Elizabeth (Stephens) M.; m. Linda Kay Price, July 14, 1961 (div. Aug. 1976); children—Philip Kevin, Melissa Kay. B.S. in Mining Engring., W.Va. U., 1961, M.S. in Mining Engring., 1965; M.Engring. Adminstrn., George Washington U., 1980. Registered profl. engr. Mining engr. Duquesne Light Co., Pitts., 1961-63; research engr. W.Va. U., Morgantown, 1963-66; materials engr. Mobay Chem. Co., New Martinsville, W.Va., 1966-68; asst. dir. Nat. Ash Assn., Washington, 1968-72; staff mining engr. U.S. Bur. Mines, Washington, 1972-82, div. chief, 1982—; mem. U.S. Nat. Com. for Tunneling Tech., Nat. Acad. Scis., Washington, 1985-90, chmn., 1988-89; lectr. in field. Contbr. articles to profl. jours., chpts. to books. Recipient Superior Svc. award Dept. Interior, 1980, Meritorious Svc. award, 1986, Disting. Svc. award, 1991, Presdl. Rank award, 1991. Mem. AIME, Fed. Exec. Inst. Alumni Assn., Sigma Xi (life), Tau Beta Pi (life). Republican. Baptist. Lodge: Masons. Avocations: racquetball; tennis; golf. Office: Bur of Mines 810 7th St NW Washington DC 20241

MEIKLEJOHN, WILLIAM HENRY, physicist; b. Virden, Ill., Jan. 7, 1917; s. John Backus and Ethel (Bone) M.; m. Ella Mae Moore, Nov. 10, 1936 (dec. 1955); children: Jo Ann, William Henry, Anita Louise, Joyce Ellen; m. Arlene Mae Loucks, Mar. 16, 1974. BSEE, U. Ill., 1940; MS in Physics, Union Coll., 1954. Registered profl. engr., Mass. Mgr. GE, Lynn, Mass., 1948-51; scientist GE, Schenectady, N.Y., 1951-84; vis. prof. Carnegie Mellon U., Pitts., 1984-87; with MOVID Info. Tech., Schenectady, N.Y., 1987—; mem. minerals and metals adv. bd. NAS, Washington, 1952-62. Author publs. in field. Scoutmaster Boy Scouts Am., Lynnfield, Mass., 1945-51, mem. coun., 1948-51; chmn. Redfeather Community Chest, Lynnfield, 1950-51. Mem. IEEE (basic sci. com.), Am. Inst. Physics. Achievements include 14 patents in field. Home: 100 Van Buren Rd Scotia NY 12302 Office: MOVID Info Tech 884 Morgan Ave Schenectady NY 12309

MEILGAARD, MORTEN CHRISTIAN, food products executive, international consultant; b. Vigerslev, Denmark, Nov. 11, 1928; s. Anton Christian Meilgaard and Ane Maria Elisa Larsen; m. Manon Meadows, Oct. 29, 1962; children: Stephen Paul, Justin Christian. MSChemE, Tech. U. Denmark, 1952, DS in Food Sci., 1982. Rsch. chemist Carlsberg Breweries, Copenhagen, Denmark, 1947-57; dir. and co-owner Alfred Jorgensen Lab. for Fermentation, Copenhagen, 1957-67; dir. rsch. and devel. Cerveceria Cuauhtemoc, Monterrey, Mex., 1967-73; v.p. rsch. Stroh Brewery Co., Detroit, 1973-89, pres. Strohtech Inc. div., 1986-91; cons., 1991—. Author: Sensory Evaluation Techniques, 1987, 2d edit., 1991; contbr. articles to profl. jours. Recipient Schwarz award, 1974. Fellow Inst. Brewing; mem. Internat. Med. Advisory Group, European Chemoreception Rsch. Orgn., Assn. Chemoreception Scis., Inst. Food Technologists, Am. Chem. Soc., Dansk Ingeniorforening, Am. Wine Soc., Air Pollution Control Assn., Master Brewers Assn. Am. (chmn. various coms., award of merit 1990), Am. Soc. Brewing Chemists (chmn. various coms.), ASTM (chmn. various coms., award of merit 1992), U.S. Hop Rsch. Coun. (pres. 1978-80, 1982-84, founder). Avocations: theatre, music, sailing, skiing. Home: 2938 Moon Lake Dr West Bloomfield MI 48323-1841 Office: Strohtech Inc 100 River Place Dr Detroit MI 48207-4291

MEILINGER, PETER MARTIN, quality assurance manager, analytical chemist; b. Grand Haven, Mich., June 3, 1952; s. Herman Joseph and Elizabeth Marie (Maurer) M.; m. Linda Joe Burke, June 25, 1982. BS, Mich. State U., 1976, MS, 1992. Assoc. chemist Baxter, Burdick & Jackson, Muskegon, Mich., 1976-80, chemist, 1980-83, lab. supr., 1983-89, quality assurance mgr., 1989—. Mem. Am. Chem. Soc. Achievements include development of GC2 line of solvents. Office: Baxter Burdick & Jackson 1953 S Harvey St Muskegon MI 49442

MEINEL, ADEN BAKER, scientist; b. Pasadena, Calif., Nov. 25, 1922; s. John G. and Gertrude (Baker) M.; m. Marjorie Steele Pettit, Sept. 5, 1944; children: Carolyn, Walter, Barbara, Elaine, Edward, Mary, David. AB, U. Calif., Berkeley, 1947, PhD, 1949; DSc (hon.), U. Ariz., 1990, U. Ariz., 1990. Assoc. prof. Yerkes Obs., U. Chgo., Williams Bay, Wis., 1950-57; dir. Kitt Peak Nat. Obs., Tucson, 1958-60; prof. U. Ariz., Tucson, 1961-85; dir. Steward Obs., Tucson, 1962-67, Optical Scis. Ctr., Tucson, 1966-73; Disting. scientist Jet Propulsion Lab., Pasadena, 1985-93; ret., 1993; regent Calif. Luth. Coll., 1961-71; cons. USAF Spl. Projects Office, 1965-80. Co-author: Applied Solar Energy, 1976, Sunsets, Twilights and Evening Skies, 1983. Recipient Warner prize Am. Astron. Soc., 1954, Van Blesbroeck award Astron. Soc. Pacific, 1990, NASA Exceptional Scientific Achievement medal, 1993; Aden B. Meinel bldg. U. Ariz., dedicated 1993. Fellow Am. Acad Arts and Scis., Optical Soc. Am. (pres. 1972-73, Adolph Lomb medal 1952, Ives medal 1980), Internat. Optical Engring. Soc. (Goddard award 1984, Kingslake medal and prize, 1993). Home: 1600 Shoreline Dr Santa Barbara CA 93109

MEINERTZ, JEFFERY ROBERT, physiologist; b. La Crosse, Wis., Aug. 20, 1962. BS in Biology, U. Wis., La Crosse, 1985, MS in Biology, 1989. Rsch. asst. Nat. Fisheries Rsch. Ctr., La Crosse, 1985-87, phys. sci. tech., 1987, rsch. physiologist, 1990—; quality control asst. mgr. Seafreeze, Seattle, 1988. Contbr. articles to profl. publs. Mem. Assn. Ofcl. Analytical Chemists, Am. Fisheries Soc., Sigma Xi (assoc.). Office: Nat Fisheries Rsch Ctr 2630 Fanta Reed Rd La Crosse WI 54603

MEINHART, ROBERT DAVID, analytical toxicologist; b. Columbus, Ohio, July 7, 1954; s. Charles J. and Barbara J. (Parrish) M.; m. Gloria J. Adkins, May 28, 1980; 1 child, Jessica G. BS in Biology, U. Tulsa, 1975. Cert. analytical toxicologist. Tech. in toxicology St. Luke's Hosp., Phoenix, Ariz., 1981-88; tech. specialist Good Samaritan Regional Med. Ctr., Phoenix, 1988—; cons. lawyers and physicians, Phoenix, 1988—. Mem. Am. Assn. Clin. Chemistry. Presbyterian. Office: Good Samaritan Regional Med Ctr 1111 E McDowell Rd Phoenix AZ 85006

MEINWALD, JERROLD, chemist, educator; b. Bklyn., Jan. 16, 1927; s. Herman and Sophie (Baskind) M.; m. Yvonne Chu, June 25, 1955 (div. 1979); children: Constance Chu, Pamela Joan; m. Charlotte Greenspan, Sept. 7, 1980; 1 child, Julia Eve. Ph.B., U. Chgo., 1947, B.S., 1948; M.A., Harvard, 1950, Ph.D., 1952; Ph.D., U. Göteborg, 1989. Mem. faculty Cornell U., 1952-72, 73—, Goldwin Smith prof. chemistry, 1980—, mem. sci. directing group Cornell Inst. Rsch. chem. ecology, 1992—; research dir. Internat. Centre Insect Physiology and Ecology, Nairobi, 1970-77; A. Mellon Term prof., 1992—; prof. chemistry U. Calif. at San Diego, 1972-73; chem. cons. Schering-Plough Rsch. Inst., 1957—, Procter & Gamble Pharms., 1958—, Cambridge Neurosci. Rsch., 1988-92; vis. prof. Rockefeller U., 1970; Camille and Henry Dreyfus Disting. scholar Mt. Holyoke Coll., 1981, Bryn Mawr Coll., 1983; Kolthoff lectr. U. Minn., 1985; Beckman lectr. Calif. Inst. Tech., 1986; Swiss "Troisième Cycle" Lectr., 1986; Russell Marker lectr. Pa. State U., 1987; mem. vis. com. chemistry Brookhaven Nat. Lab., 1969-72, chmn., 1972; mem. med. A chemistry study sect. NIH, 1963-67, chmn., 1965-67; mem. adv. bd. Petroleum Rsch. Found., 1971-73; mem. adv. coun. chemistry dept. Princeton U., 1978-83 ; mem. adv. bd. Rsch. Corp., 1978-83; mem. adv. bd. chemistry div. NSF, 1979-83; organizing chmn. Sino-Am. Symposium on Chemistry of Natural Products, Shanghai, 1980; mem. adv. bd. A.P. Sloan Found., 1985-91; Frontiers of Rsch. lectr. Coun. Chem. Rsch., 1987; mem. sci. adv. bd. Agridyne Corp., 1989-93; Carlton Coll. Convocation, 1993; Mary Aldrich lectr., American U., 1993; K. Pfister lectr. MIT, 1992, Hilldale lectr. U. Wis., 1991, Nat. Undergrad. Rsch. Symposium, Plenary lectr., Mpls., 1992; UNOCAL lectr. Calif. State U., Long Beach, 1992,. Bd. editors Jour. Organic Chemistry, 1962-66, Organic Reactions, 1968-78, Organic Synthesis, 1968-72, Jour. Chem. Ecology, 1974—, Insect Sci., 1979-90; contbr. articles to profl. jours. Sloan fellow, 1958-62, Guggenheim fellow, 1960-61, 76-77, NIH spl. postdoctoral fellow, 1967-68, Japan Soc. Promotion of Sci. fellow, 1983, Ctr. for Advanced Study in Behavioral Sci, fellow, 1990-91; NIH Fogarty internat. scholar, 1983-85, Czechoslovakia Exchange scholar Nat. Acad. Sci., 1987; recipient Tyler Environ. Achievement prize U. So. Calif., 1990, Gustavus J. Esselen award for Chemistry in Pub. Interest, 1991. Mem. NAS, AAAS, Am. Acad. Arts and Scis., Am. Philos. Soc., Am. Chem. Soc. (chmn. organic div. 1969, E. Guenther award 1985, Disting. Scientist award Kalamazoo sect. 1985, A.C. Cope Scholar award 1989), Internat. Soc. Chem. Ecology (pres. 1988, Silver medal 1991), Phi Beta Kappa, Sigma Xi (nat. lectr., 1992—). Office: Cornell U Dept Chemistry Baker Lab Ithaca NY 14853-1301

MEINZER, RICHARD A., research scientist; b. Newark. PhD, U. Ill. Sr. rsch. scientist United Tech. Rsch. Ctr., East Hartford, Conn., 1966—. Author: (with others) HFINF Chemical Lasers, 1987. Mem. ARS, Am. Chem. Soc., Am. Phys. Soc. Achievements include 10 patents in field. Office: United Tech Rsch Ctr MS 92 Silver Ln East Hartford CT 06118

MEISEL, JOHN, political scientist; b. Vienna, Austria, Oct. 23, 1923; s. Fryda and Ann M. BA, U. Toronto, 1948, MA, 1950; PhD in Polit. Sci., London Sch. Econs., 1959; LLD (hon.), Brock U., 1983, U. Guelph, 1985, Carleton U., 1990, U. Toronto, 1993; DU (hon.), U. Ottawa, 1983; D of Scis. Sociales, Laval U., 1988; postgrad., U. Toronto, 1993. Head dept. polit. studies Queen's U., Kingston, Ont., Can., 1963-67, Hardy prof. polit. sci., 1963-80; former chmn. Can. Radio-TV and Telecommunications Commn.; Sir Edward Peacock prof. polit. sci. Queen's U., 1983—. Author: The Canadian General Election of 1957, 1962, Papers on the 1962 Election, 1964, Ethnic Relations in Canadian Voluntary Assocations, 1972, Working Papers on Canadian Politics, 1975; editor: Internat. Polit. Sci. Rev., 1979—; contbr. articles to profl. jours. Decorated officer Order of Can.; recipient Killam award Can. Council, 1968, 73, Northern Telecom Internat. Can. Studies award of excellence, 1991. Fellow Royal Soc. Can. (pres. 1992—). Home: Colimaison, Tichborne, ON Canada K0H 2V0 Office: Queen's U, Kingston, ON Canada K7L 3N6

MEISEL, WERNER PAUL ERNST, physicist; b. Gleiwitz, Silesia, Germany, Apr. 10, 1933; s. Josef and Johanna (Smolka) M. Diploma in physics, U. Leipzig, Germany, 1952, Dr.rer.nat., 1961; Dr.sc.nat, Acad. Sci., Germany, 1972. Mem. staff Acad. Scis., Berlin, 1958-62, head lab., 1962-68, head rsch. group, 1968-79; sr. scientist U. Mainz, Fed. Republic Germany, 1979—; guest prof. U. Nijmegen (The Netherlands), 1983-84. Author: (with others) Physics in Chemistry, 1978; editor: Mössbauer Spectroscopy, 1980; contbr. articles to profl. jours. Mem. German. Phys. Soc., German Chem. Soc. Office: U Inst Inorganic Chemistry, Staudinger Weg 9, D-55099 Mainz Germany

MEISSNER, RUDOLF OTTO, geophysicist, educator; b. Dortmund, Germany, June 15, 1925; s. Alfred Richard and Gertrude Maria (Auffermann) M.; m. Sofia Theresa Hitzegrad, Oct. 15, 1954; children: Monika, Gunter. Diploma, U. Frankfurt, 1953, doctor, 1955, habilitation, 1966. Party leader Praula, Europe, Africa, 1955-59; supr. Shell Oil Co., Tripolis, Libya, 1959-61; asst. prof. Univ. Frankfurt, Hesse, Germany, 1961-69; lectr. Univ. Frankfurt, Hesse, 1966-69, Univ. Mainz, Rhineland, Germany, 1966-69; vis. prof. Hawaiian Inst. Geophysics, 1969-70; prof., dir. Inst. Geophysics, Kiel Univ., Schleswig-Holstein, 1971—; dean of math.-sci. faculty Univ. Kiel, 1981-83; chmn. working group Internat. Lithosphere Project, 1987-90. Author: Der Mond, 1969, Seismische Messungen, 1976, The Continental Crust, 1986; co-author: The Dekorp Atlas, 1990; chief editor: Continental Lithosphere, 1992; contbr. articles to profl. jours; mem. editorial bd. 4 sci. jours. Mem. steering com., bd. dirs. Dekorp, 1985—. Mem. European Geophys. Soc. (pres. 1985-87), Europe Acad., N.Y. Acad., Assn. Royal Astron. Soc., Internat. Geologic Correlation Program. Avocations: music, piano. Home: Struckbrook 2, D-24161 Altenholz Germany Office: Inst Geophysics, Olshausenstr 40, D-2300 Kiel 1, Germany

MEISTER, BERNARD JOHN, chemical engineer; b. Maynard, Mass., Feb. 27, 1941; s. Benjamin C. M. and Gertrude M. (Meister); m. Janet M. White, Dec. 31, 1971; children: Mark, Martin, Kay Ellen. B.S. in Chem. Engring., Worcester Poly. Inst., 1962; Ph.D. in Chem. Engring., Cornell U., 1966. Engring. researcher Dow Chem. Co., Midland, Mich., 1966—, sr. rsch. specialist, 1978-81, assoc. scientist, 1981-85, sr. assoc. scientist, 1985-92, rsch. scientist, 1992—. Contbr. articles to profl. jours. Mem. Am. Inst. Chem. Engrs., Am. Chem. Soc., Soc. Plastics Engrs., Soc. Rheology, Sigma Xi. Mem. Ch. of Nazarene. Home: 2925 Chippewa Ln Midland MI 48640-4181 Office: Dow Chem Co 438 Bldg Midland MI 48640

MEITES, SAMUEL, clinical chemist, educator; b. St. Joseph, Mo., Jan. 3, 1921; s. Benjamin and Frieda (Kaminsky) M.; m. Lois Pauline Maranville, Mar. 11, 1945; 1 child, David Russell. A.S., St. Joseph Jr. Coll., 1940; A.B., U. Mo., 1942; Ph.D., Ohio State U., 1950. Diplomate Am. Bd. Clin. Chemistry. Clin. biochemist VA, Poplar Bluff, Mo., 1950-52, Toledo Hosp., 1953-54, Children's Hosp., Columbus, Ohio, 1954-91; prof. dept. pediatrics Ohio State U. Coll. Medicine, Columbus, 1972-91; prof. dept. pathology, 1974-91; cons. Brown Labs., Columbus, 1968-83, VA, Chillicothe, Ohio, 1980-84. Co-author: Manual of Practical Micro and General Procedures in Clinical Chemistry, 1962. Editor: Standard Methods of Clinical Chemistry, Vol. 5, 1965; Pediatric Clinical Chemistry, 1st edit., 1977, 2d edit., 1981, 3rd edit., 1989; co-editor: Selected Methods for the Small Clinical Chemistry Laboratory, 1982, Biography of Otto Folin, 1989. Contbr. articles to profl. jours. Served to 1st lt. U.S. Army, 1942-46. Fellow AAAS; mem. Am. Chem. Soc., Am. Assn. Clin. Chemistry (sec. 1975-77, Bernard Katchman award Ohio Valley sect. 1971, Fisher award 1981, Miles-Ames award, 1990. chmn. com. on archives, 1982-86, history divsn., 1992—). Democrat. Jewish. Avocations: gardening, history of clinical chemistry. Office: Childrens Hosp 700 Childrens Dr Columbus OH 43205-2696

MEIXNER, JOSEF, emeritus physics educator; b. Percha, Bavaria, Fed. Republic of Germany, Apr. 24, 1908; s. Karl and Maria (Bartl) M.; m. Hildegarde Luise Diemke, Jan. 5, 1933; children: Michael J., Georg H., Reinhard Olaf. PhD, U. Munich, Germany, 1931; Dr. rer. nat. (hon.), U. Cologne, Germany, 1968. Asst. U. Giessen, Fed. Republic Germany, 1934-37, lectr., 1937-39; lectr. U. Berlin, 1939-42; extraordinary prof. Technische Hochschule, Aachen, Germany, 1942-51; prof. Technische Hochschule, Aachen, 1951-74, prof. emeritus, 1974—; vis. prof. Seven U.S. Univs., 1954-70. Hokkaido U., Sapporo, Japan, 1975. Co-author: (with F. W. Schäfke) MathieuscheFunktionen und Sphëroidfunktionen, 1954; contbr. numerous articles to profl. jours. Fellow Am. Physical Soc.; mem. Deutsch Physikal Gesellschaft, Rhein-Westf. Acad.der Wissenschaften, GaMM. Home: Am Blockhaus 31, 52 074 Aachen Germany

MEKALANOS, JOHN J., microbiology educator. Prof. dept. microbiology and molecular genetics Harvard Med. Sch., Boston. Recipient Eli Lilly & Co. Microbiology and Immunology Rsch. award Am. Soc. Microbiology, Washington, 1991. Office: Harvard Med Sch Dept Microbiol Molecular Gen 25 Shattuck St Boston MA 02115-6093*

MELAMED, BENJAMIN, computer scientist; b. Linz, Austria, Feb. 22, 1948; came to U.S., 1972; s. Josef and Felicia (Drewniak) M.; m. Anna S. Zalecka, July 23, 1973; children: Yael, Eitan J.S. MS, U. Mich., 1973, PhD, 1976. Postdoctoral fellow U. Mich., Ann Arbor, 1976-77; asst. prof. Northwestern U., Evanston, Ill., 1977-81; mem. tech. staff AT&T Bell Labs., Holmdel, N.J., 1981-89, fellow, 1988; dept. head NEC USA, Inc., Princeton, N.J., 1989—; chair profl. confs. Contbr. to profl. publs. Mem. IEEE (sr.). Jewish. Achievements include development of visual interactive simulation package for queueing networks, development of method for accurate modeling of stochastic sequences used, e.g. to model video traffic in high-speed communications networks. Home: 5 Miller Ln Warren NJ 07059 Office: NEC USA Inc 4 Independence Way Princeton NJ 08540

MELAMID, ALEXANDER, economics educator, consultant; b. Freiburg, Fed. Republic Germany, Mar. 28, 1914; s. Michael and Zinaida (Gruenholz) M.; m. Ruth Caro, Nov. 28, 1940 (dec. July 1968); m. Ilse Hoenigsberg, Apr. 15, 1975; children: Diane H.C. Melamid Michaels, Suzanna C. Melamid Portnoy. BSc in Econs., London Sch. Econs., 1939; PhD, New Sch. for Social Research, 1952. Asst. prof. New Sch. for Social Research, N.Y.C., 1953-57; assoc. prof. NYU, 1957-65, prof., 1965-87, prof. emeritus, 1987—; v.p. Am. Geog. Soc., N.Y.C., 1975. Author: Turkey, 1956, History of Iran, 1968, New York City Region, 1985. Fellow Am. Geog. Soc. (hon., mem. council, Morse medal 1991); mem. Middle East Studies Assn. Am. Avocations: skiing. Office: NYU Dept Econs Washington Sq N New York NY 10003-6635

MELBY, EDWARD CARLOS, JR., veterinarian; b. Burlington, Vt., Aug. 10, 1929; s. Edward C. and Dorothy H. (Folsom) M.; m. Jean Day File, Aug. 15, 1953; children: Scott E., Susan J., Jeffrey T., Richard A. Student, U. Pa., 1948-50; D.V.M., Cornell U., 1954. Diplomate: Am. Coll. Lab. Animal Medicine. Practice veterinary medicine Middlebury, Vt., 1954-62; instr. lab. animal medicine Johns Hopkins U. Sch. Medicine, Balt., 1962-64; asst. prof. Johns Hopkins U. Sch. Medicine, 1964-66, assoc. prof., 1966-71, prof., dir. div. comparative medicine, 1971-74; prof. medicine, dean Coll. Vet. Medicine, Cornell U., Ithaca, N.Y., 1974-84; v.p. R & D SmithKline Beecham Animal Health, 1985-90, v.p. sci. and tech. assessment, 1990-91; ind. cons., 1992—; cons. VA, Nat. Research Council, NIH. Author: Handbook of Laboratory Animal Science, Vols. I, II, III, 1974-76. Served with USMC, 1946-48. Mem. Am., N.Y. State, Md., Pa. veterinary med. assns., Am. Assn. Lab. Animal Sci., Am. Coll. Lab. Animal Medicine, AAAS, Phi Zeta. Home: 770 Newtown Rd Villanova PA 19085-1199 Office: 770 Newtown Rd Villanova PA 19085

MELCHER, JERRY WILLIAM COOPER, clinical psychologist, army officer; b. Bloomington, Ill., Oct. 17, 1948; m. Margaret Frances Orban; children: Heather, Shawna, Jerry. BS, Lincoln U., Jefferson City, Mo., 1975; MS, Tex. A&I U., 1976; PhD, Tex. A&M U., 1980. Psychometrist Lamar U., Beaumont, Tex., 1978-79, psychologist, 1979-81; commd. 1st lt. U.S. Army, 1981, advanced through grades to maj., 1987; clin. intern William Beaumont Army Med. Ctr., 1981-82; psychologist 1st Cav. Divsn., Fort Hood, Tex., 1982-84; chief psychology svc. Darnall Army Community Hosp., Fort Hood, 1984-85, Blanchfield Army Community Hosp., Fort Campbell, Ky., 1986-87; clin. psychologist Area Counseling Assocs., Millington, Tenn., 1987—, U.S. Army Res. 330th Gen. Hosp., Memphis, 1988—; clin. dir. Genesis Treatment Ctr., Memphis, 1990—; tng. coord. CETA, Beaumont, 1980-81; res. psychologist Operation Desert Storm, Fort Stewart, Ga., 1991. Bd. dirs. Family Aid Network, Killeen, Tex., 1984-85; vol. Rape Crisis Ctr., Beaumont, 1979. Decorated Bronze Star with valor device, Meritorious Svc. medal; Cross of Gallantry with palm (Vietnam). Mem. APA. Avocations: travel, antique collecting, gardening, swimming. Office: Area Counseling Assocs 8222 US Hwy 51 N Millington TN 38053-1708

MELCONIAN, JERRY OHANES, engineering executive; b. Cairo, Egypt, Jan. 22, 1934; came to U.S., 1967; s. Melik Melconian and Zarouca Papazian; m. Kathleen F. Dire, May 8, 1976; 1 child, Terran Kirk. BSc, U. London, 1957. Section leader Otis Elevator Co., London, Eng., 1957-61, Rolls Royce Ltd., Derby, Eng., 1961-66; program coordinator Textron Lycoming, Stratford, Conn., 1967-74; mgr., TF34 Design to Cost Gen. Electric Co., Lynn, Mass., 1974-77; mgr. mktg. No. Rsch. and Engring. Co., Woburn, Mass., 1977-82; pres. SOL-3 Resources Inc., Reading, Mass., 1982—. Editor: Design and Development of Gas Turbine Combustors, 1980; patentee in field. Mem. Am. Inst. Aeronautics and Astronautics. Office: SOL-3 Resources Inc 76 Beaver Rd Reading MA 01867-1310

MELDRUM, DANIEL RICHARD, general surgeon, physician; b. Flint, Mich., Sept. 27, 1965; s. Richard Terry and Patricia Ellen (Klug) M. BS, U. Mich., 1987; MD, Mich. State U., 1992. Teaching asst. dept. biochemistry U. Mich., Ann Arbor, 1986, rsch. asst., 1987-88; rsch. student fellow Mich. State U., East Lansing, 1989-92; resident surgery U. Colo., 1992—; capt. USAR, 1993—; advisor Mich. State U. Adv. Com., 1990-92; supr., adv. Biochemistry Teaching Assts., Ann Arbor, 1986; guest speaker 18 internat. and nat. confs. Contbr. more than 20 articles to profl. jours. Del. Am. Med. Student Assn.; coord. Mich. State U. Red Cross Med. Sch. Blood Drives, 1989-90. Recipient Young Investigator award The Shock Soc., 1992, Student Rsch. award Assn. Acad. Surgery, 1992, Moorhead Rsch. award Gramec Found., 1991, NIH Biomed. Rsch. grant 1989, 90, Excellence award Mich. State U., 1992. Mem. AMA, AAAS, N.Y. Acad. Scis., Shock Soc., Assn. Acad. Surgery, Am. Med. Student Assn. (pres. 1989-90), Am. Assn. Med. Colls. (rep.), Alpha Omega Alpha. Achievements include work in ATP-MgCl2 restores macrophage and lymphocyte energetics and functions after shock; immunoprotective effects of calcium channel blockers after hemorrhage; the energetics of defective macrophage antigen presentation after hemorrhage. Home: 870 Dexter # 301 Denver CO 80220 Office: U Colo Dept Surgery 4200 E 9th Ave Denver CO 80262

MELEY, ROBERT WAYNE, structural engineer; b. Warren, Pa., June 10, 1952; s. Nathan Arnold and Audrey Elizabeth (Dunn) M.; m. Kathy Dorien McDonald, Oct. 20, 1973; children: Kathy Dorien, Helen Renee, January Lynn. BArch, Pa. State U., 1979, M of Engring., 1979; PhD, Columbia Pacific U., 1989. Registered engr. Scientist Douglas Aircraft Co., Long Beach, Calif., 1979-80; sr. devel. engr. Rexnord, Inc., Warren, 1980-85; pres. Meley and Meley Assocs., Cooksburg, Pa., 1979—; lectr. Pa. State U., Beaver, 1986-87; chief engr. Brown Boiler and Tank Works, Ltd., Franklin, Pa., 1987-90; mem. adv. bd. Pa. Dept. Environ. Resources, Harrisburg, Pa., 1989—. Author: Inspecting Above Ground Storage Tanks, 1990, Probabilistic Analysis of Storage Tanks, 1992, Evaluating Storage Tanks, 1992. Mem. Am. Soc. Civil Engrs., Nat. Soc. Architectural Engrs. (founder). Achievements include a new design of leak detection system for storage tanks. Office: Meley & Meley Assocs HC1 Box 10 Cooksburg PA 16217-9704

MELEZINEK, ADOLF, engineering educator; b. Vienna, Austria, Oct. 3, 1932; s. Rudolf and Franziska Melezinek; m. Vera Vysansky, July 21, 1952; children: Adolf, Vera. MEE, Tech. U. Prague, 1957; PhD, U. Prague, 1969. Rsch. asst. Tech. Rsch. Inst., Prague, 1952, chief engr., 1957-61, assoc. prof., 1961-69; prof. HTBLVA, Vienna, 1969-71; full prof.; prof. U. Klagenfurt, Austria, 1971—; head Inst. Ednl. Tech. and Engring. Edn., 1978—; guest prof. U. Karlsruhe, 1971, Tech. U. Vienna, 1972—, Tech. U. Graz, 1977—, Tech. U. Zurich, 1974—, Tech. U. Budapest (Hungary), 1985—. Contbr. numerous articles to profl. publs., books; contbr. to films, videotapes, slide series in field. Recipient Golden Felber medal Czech Tech. U., Prague, 1991, Great Golden medal Govt. Carinthia, 1992. Mem. Internat. Soc. Engring. Edn. (pres. 1972—, Golden Ring award 1992), Soc. Motion Picture and TV Engrs., Czechoslovakian Cybernetics Soc., Soc. for Programmed Instrn. (bd. dirs. 1976-79). Avocations: flying, music. Home: Akazienhofstr 79, A 9020 Klagenfurt Austria Office: U Klagenfurt, Universitatsstr, A 9020 Klagenfurt Austria

MELIA, ANGELA THERESE, biologist, pharmacokineticist; b. Newark, Apr. 12, 1964; d. Arcangelo Italo and Rose Carol (Lardieri) M. BS in Biology, Seton Hall U., 1986. Analytical chemist Shulton, Inc., Clifton, N.J., 1984-85; asst. scientist Hoffmann-LaRoche, Inc., Nutley, N.J., 1986-89, assoc. scientist, 1989-92, clin. rsch. assoc., 1992, sr. clin. rsch. assoc., 1992—. Roman Catholic. Achievements include raising hybridoma cells needed to aid in development of a method to purify phospolipase-D, conducted clinical studies evaluating anti-AIDS, antibacterial, anti-hypertensive, and anti-obesity investigational drugs.

MELICHAR, JOHN ANCELL, aerospace engineer; b. Denver, Apr. 6, 1966; s. Ancell Gene and Margaret (Zuerl) M.; m. Dawn Alison McNeill, July 8, 1989. BS in Aerospace Engring., U. Ariz., 1989; M in Engring. Mgmt., George Washington U., 1993. Aerospace engr. Naval Air Systems Command, Washington, 1989-92, logistics engr., 1992—. Mem. Am. Inst. Aeronautics and Astron. Office: Naval Air Systems Command 1421 Jefferson Davis Hwy Arlington VA 22202

MELICK, GEORGE FLEURY, mechanical engineer, educator; b. Morristown, N.J., Sept. 7, 1924; s. George Fleury and Esther Purdy (Udall) M.; m. Florence Miriam Bevins, Dec. 28, 1946; children: Robert A., Linda S., Judith E., Karen L. BSE, Princeton U., 1944; MS, Stevens Inst. Tech., 1955; ME, Columbia U., 1963; MA, NYU, 1970. Registered profl. engr., N.J., D.C. Asst. chief engr. Worthington Corp., Harrison, N.J., 1946-55; asst. prof. Stevens Inst. Tech., Hoboken, N.J., 1955-58; assoc. in mech. engring. Columbia U., N.Y.C., 1958-61; assoc. prof. mech. engring., dean Rutgers U., New Brunswick, N.J., 1961-77; cons. engr. Stone & Webster Engring. Corp., Cherry Hill, N.J., 1977-87; dir. engring. mgmt. program Drexel U., Phila., 1987-91; chmn. bd. Anastasio & Melick Assocs., Mt. Laurel, N.J., 1987—; cons. Worthington Corp., Harrison, 1956-65, Pub. Svc. Elec. & Gas, Newark, 1966-76. Author: John Mark and the Origin of the Gospels, 1979. Mem. countycom. Dem. Party, Franklin Twp., N.J., 1976. 1st lt. U.S. Army, 1945-52. Decorated Bronze Star medal. Mem. ASME, Am. Soc. Engring. Mgmt. (life), Am. Soc. Engring. Edn., Soc. Biblical Lit., Am. Acad. Religion (charter), Sigma Xi, Pi Tau Sigma, Tau Beta Pi. Presbyterian. Achievements include radical new theory of the origin of the gospels. Home: 6 Raven Ct Mount Laurel NJ 08054 Office: Anastasio & Melick Assocs Ste 13A 4201 Church Rd Mount Laurel NJ 08054

MELIGNANO, CARMINE (EMANUEL MELIGNANO), video engineer; b. N.Y.C., Dec. 19, 1936; s. Salvatore and Lita (Poggialli) M.; m. Eileen Kinzie; children: Lori Ann, Robert, Michael. BS in Elec. Engring., Stevens Inst. Tech., Hoboken, N.J., 1959; postgrad., William Paterson Coll., 1978, Pace U., 1979. Registered profl. engr., N.J. Quality contr. Isomet Corp., Palisades Park, N.J., 1959-63; sales engr. RCA Service Corp., Camden, N.J., 1963-73; video engr. N.J. Sports and Expn. Authority, East Rutherford, 1974-77; chief engr. Price Waterhouse, N.Y.C., 1978—; engring. cons. Passaic County Vocat. Edn. High Sch., Wayne, N.J., 1971-77, Meadowlands Racetrack, East Rutherford, 1973-77, Royal Sound, Eatontown, N.J., 1984-86. Bd. trustees N.Y.C. chpt. Leukemia Soc., 1981-86, pres., 1987-90, nat. bd. trustees, 1990—. Recipient Emmy award NATAS, 1985, Outstanding Svc. award Leukemia Soc. Nat. Bd., 1986, Vincent T. Lombardi Humanitarian award, 1990, Pres.' award Leukemia Soc. Am., 1991, People's award, 1991. Mem. Soc. Motion Picture and TV Engrs. (sec., treas. elect N.Y.C. chpt. 1984-86), Nat. Sports Com., Nat. Performing Arts Com. (vice chmn. 1985—), Friar's Club Internat. (N.Y.C., profl. mem., mem.-elect). Republican. Roman Catholic. Lodge: KC. Avocations: sports, chess, organ and piano. Home: PO Box 3 Lodi NJ 07644-0003 Office: CarMel Prodns 10 Dell Glen Ave Lodi NJ 07644-1740 also: Carmel Home Entertainment Ltd 641 Lexington Ave 21st Fl New York NY 10022

MELISSINOS, ADRIAN CONSTANTIN, physicist, educator; b. Thessaloniki, Greece, July 28, 1929; came to U.S., 1955, naturalized, 1970; s. Constantin John and Olympia (Abbott) M.; m. Mary Joyce Mitchell, June 7, 1960; children: Constantin John, Andrew William. Student, Royal Naval Acad., Greece, 1945-48; M.S., Mass. Inst. Tech., 1956, Ph.D., 1958. Naval cadet Greek Navy, 1945-48, commd. ensign, 1948, advanced through grades to lt., 1951; ret., 1954; teaching and research asst. Mass. Inst. Tech., 1955-58; instr. U. Rochester, N.Y., 1958-60; asst. prof. physics U. Rochester, 1960-63, assoc. prof., 1963-67, prof., 1967—, chmn. dept. physics and astronomy, 1974-77; vis. scientist CERN European Center for Nuclear Research, 1968-69, 77-78, 89-90; cons. Brookhaven Nat. Lab., 1970-72, 75-79. Author: Experiments in Modern Physics, 1966, (with F. Lobkowicz) Physics for Scientists and Engineers, 1975; (with A. Das) Quantum Mechanics, 1985, Principles of Modern Technology, 1990. Decorated Swedish Order of Sword. Fellow Am. Phys. Soc.; mem. Greek Nat. Acad. (corr.). Achievements include experimentation with elementary particles at most major high energy accelerators in the U.S. and Europe, experimentation with high power lasers. Home: 177 Whitewood Ln Rochester NY 14618-3223 Office: U Rochester Dept Physics Rochester NY 14627

MELLBERG, LEONARD EVERT, physicist; b. Springfield, Mass., Dec. 18, 1935; s. Evert and Dorothy (Baker) M.; m. Margaret Ann Baker, Oct. 2, 1987. BS in Physics, U. Mass., 1961; MS in Physics, Trinity Coll., Hartford, Conn., 1968. Rsch. physicist Navy Underwater Sound Lab., New London, Conn., 1961-68, SACLANT Undersea Rsch. Centre, LaSpezia, Italy, 1968-72, Naval Underwater Systems Ctr., Newport, R.I., 1972-91; sr. scientist Marine Acoustics Inc., Newport, 1991—; mem. numerous govt. and profl. tech. adv. bds. and coms. Contbr. over 60 articles to profl. jours. Pres. Verdandi Swedish Cultural Soc., Providence, 1992; bd. dirs. Verdandi Chorus Am. Union Swedish Singers, Providence, 1992. Recipient Civilian Navy Meritorious Svc. medal Dept. of Navy, 1991. Fellow Acoustical Soc. Am.; mem. IEEE (sr.), AIAA (svc. award 1977), Am. Geophys. Union. Achievements include research in ocean acoustic propagation, anti-submarine warfare acoustics, Arctic sea-ice ridges and lighter than air vehicles. Home: 20 Willow Ave Middletown RI 02842 Office: Marine Acoustics Inc 14 Pelham St Newport RI 02840

MELLOR, ARTHUR M(CLEOD), mechanical engineering educator; b. Elmira, N.Y., Jan. 1, 1942. BSE, Princeton U., 1963, MA, 1965, PhD in aerospace and mech. scis., 1968. From asst. prof. to assoc. prof. mech. engring. Purdue U., West Lafayette, Ind., 1967-75, prof., 1975—; prof. Coll. Engring. Drexel U., Phila., 1975-88; Centennial prof. mech. engring. Vanderbilt U., Nashville, 1988—. Mem. Air Polution Control Assn., Combustion Inst., Soc. Automotive Engrs., Sigma Xi. Research in chemical kinetics of combustion-generated air pollution, metal combustion, gas turbine combustor design. Office: Vanderbilt U Dept of Mech Engring Nashville TN 37203

MELLSTEDT, HÅKAN SÖREN THURE, oncologist, medical facility administrator; b. Lund, Sweden, Oct. 23, 1942; s. Ture and Stina (Carlstedt) M.; m. Eva Bengtsson, May 21, 1966. MD, Karolinska Inst., Stockholm, 1969, PhD, 1974. Asst. prof. medicine Karolinska Inst., Stockholm, 1975-86, assoc. prof., 1986—; sr. asst. physician Serafimer Hosp., Stockholm, 1978-80; head internal medicine, dept. oncology Karolinska Hosp. and Inst., Stockholm, 1980-85, head biotherapy and lymphoma unit, 1985—, dep. dir., 1986—; sci. adviser Centocor, Malvern, USA, 1988, E. Merck GMBH, Darmstadt, Germany, 1990, Kabipharmacia AB, Stockholm, 1992. Contbr. articles to profl. jours. Recipient Alfaferone prize Istituto Immunologico Italiano, Rome, 1989; grantee Swedish Cancer Soc., 1987—. Mem. Am. Assn. Cancer Rsch., N.Y. Acad. Sci., European Soc. Med. Oncology. Achievements include characterization of clonal tumor cells in myeloma; identification of a specific T cell tumor response in myeloma; development of treatment regimen with a-interferon of multiple myeloma, of a therapeutic concept of monoclonal antibodies and cytokines in colorectal carcinoma, and of a tumor vaccine in colorectal carcinoma based on anti-idiotypic antibodies and cloned antigens; analysis, in humans, of induction of a humoral and cellular idiotypic cascade after treatment with monoclonal antibodies. Office: Karolinska Hosp, Dept Oncology, S10401 Stockholm Sweden

MELMON, KENNETH LLOYD, physician, biologist, pharmacologist, consultant; b. San Francisco, July 20, 1934; s. Abe Irving and Jean (Kahn) M.; m. Elyce Edelman, June 9, 1957; children: Bradley S., Debra W. AB in Biology with honors, Stanford U., 1956; MD, U. Calif. at San Francisco, 1959. Intern, then resident in internal medicine U. Calif. Med. Ctr., San Francisco, 1959-61; clin. assoc., surgeon USPHS, Nat. Heart, Lung and Kidney Inst., NIH, 1961-64; chief resident in medicine U. Wash. Med. Ctr., Seattle, 1964-65; chief div. clin. pharmacology U. Calif. Med. Ctr., 1965-78; chief dept. medicine Stanford U. Med. Ctr., 1978-84, Arthur Bloomfield profl medicine, prof. pharmacology, 1978-86, prof. medicine and pharmacology, 1978—; dir. tech. transfer program Stanford U. Hosp., 1986—; mem. sr. staff Cardiovascular Rsch. Inst.; chmn. joint commn. prescription drug use Senate Subcom. on Health, Inst. Medicine and HEW-Pharm. Mfrs. Assn.; mem. Nat. Bd. Med. Examiners, 1987—; pres. Bio 2000, Woodside, Calif., 1983—85 co-founder, Immulogic, Boston, Palo Alto, Calif., 1988; sci. advisor Syntex, Hoffman LaRoche, Pharmetrix, Microprobe, LXR, others; cons. FDA, 1965-82, Office Tech. Assessment, 1974-75, Senate Subcom. on Health, 1975—; bd. dirs. Pharmatrix, San Mateo, Calif., Immulogic, Boston, Microprobe, Seattle; cons. to govt.; founder Inst. Biol. and Clin. Investigation, Ctr. for Molecular and Genetic; Medicine Stanford Community of Internists and Stanford Med. Group. Author articles, chpts. in books, sects. encys.; Editor: Clinical Pharmacology: Basic Principles in Therapeutics, 3d edit., 1992, Cardiovascular Therapeutics, 1974; assoc. editor: The Pharmacological Basis of Therapeutics (Goodman and Gilman), 1984; mem. editorial bd. numerous profl. jours. Surgeon USPHS, 1961-64. Burroughs Wellcome clin. pharmacology scholar, 1966-71; John Simon Guggenheim fellow Weizman Inst., Israel, 1971, NIH spl. fellow, Bethesda, 1971. Fellow AAAS (nat. coun. 1985-89); mem. Am. Fedn. Clin. Rsch. (pres. 1973-74), Am. Soc. Clin. Investigation (pres. 1978-79), Assn. Am. Physicians, Western Assn. Physicians (pres. 1983-84), Am. Soc. Pharmacology and Exptl. Therapeutics, Inst. Medicine of Nat. Acad. Sci., Am. Physiol. Soc., Calif. Acad. Medicine, Med. Friends of Wine, Phi Beta Kappa. Democrat. Jewish. Achievements include initiation of founding of Ctr. of Molecular and Genetic Medicine. Avocations: woodworking, photography, hiking, cycling, swimming. Home: 51 Cragmont Way Woodside CA 94062-2307 Office: Stanford U Med Ctr Dept Medicine S025 Stanford CA 94305

MELNIKOV, PAUL, analytical chemist, instrumentation engineer; b. Washington, June 29, 1951. BS, U. Md., 1972; BSEE, Va. Poly. Inst., 1973; postgrad., Mich. Tech. U., 1978-83. Rsch. asst. U. Md., College Park, 1973-78; ind. instrumentation systems engr., cons. Washington area, 1984-85; mass spectrometrist Velsicol Chem. Corp., Memphis, 1986-87, Argonne (Ill.) Nat. Lab. Chemistry Divsn., 1987—; mem. patent disclosure rev. com. Dept. Energy/U. Chgo./Argonne Nat. Lab., 1987—. Mem. AAAS, IEEE, Am. Soc. Mass Spectrometry, Am. Chem. Soc., Am. Vacuum Soc., Sigma Xi. Achievements include research and development of mass spectrometer ion source design, electron multiplier design. Office: Argonne Nat Lab 9700 Cass Ave Bldg 200 Lemont IL 60439-4831

MELTER, ROBERT ALAN, mathematics educator, researcher; b. N.Y.C., Mar. 20, 1935; s. George I. and Hattie (Eisenstein) M.; m. Therese Balavoine, Oct. 10, 1965; 1 child, Vanessa. AB, Cornell U., 1956; AM, U. Mo., 1960, PhD, 1962. Asst. prof. math. U. R.I., Kingston, 1962-64, U. Mass., Amherst, 1964-67, assoc. prof. math. U. S.C., Columbia, 1967-71; dir. sci. div. L.I. U., Southampton, N.Y., 1986-88, prof. math., 1971—. Assoc. editor: Math. Revs., 1973-74; translator Problems in Combinatorics, 1985, Combinatorial Configurations, 1988; contbr. articles to profl. jours. NAS Exch. fellow, 1981; Fulbright prof., 1985. Mem. Am. Math Soc. Office: LI U Dept Math Southampton NY 11968

MELTZER, YALE LEON, economist, educator; b. N.Y.C., Nov. 3, 1931; s. Benjamin and Ada (Luria) M.; BA, Columbia U., 1954, postgrad. Sch. Law, 1954-55; MBA, NYU, 1966; m. Annette Schoenberg, Aug. 7, 1960; children: Benjamin Robert, Philippe David. Asst. to chief patent atty. Beaunit Mills, Inc., Elizabethton, Tenn., 1955-56, prodn. mgr., 1956-58, rsch. chemist N.Y. Med. Coll., N.Y.C., 1958-59; rsch. chemist H. Kohnstamm & Co., Inc., mfg. chemists, N.Y.C., 1959-66, mgr. comml. devel., market rsch., patents and trademarks, 1966-68; sr. security analyst Harris, Upham & Co., Inc., 1968-70; instr. dept. econs. N.Y. U., 1972-79; adj. asst. prof. dept. acctg., fin. and mgmt. Pace U., N.Y.C., 1974-80, adj. assoc. prof., 1980-84; lectr. dept. polit. sci., econs. and philosophy Coll. S.I., CUNY, 1977-82, asst. prof. dept. polit. sci., econs. and philosophy, 1983—; lectr. bus., fin., econs., sci. and tech. Mem. AAAS, Am. Econ. Assn., Am. Chem. Soc. Author: Soviet Chemical Industry, 1967; Chemical Trade with the Soviet Union and Eastern European Countries, 1967; Chemical Guide to GATT, The Kennedy Round and International Trade, 1968; Phthalocyanine Technology, 1970; Hormonal and Attractant Pesticide Technology, 1971; Urethane Foams: Technology and Applications, 1971; Water-Soluble Polymers: Technology and Applications, 1972; Encyclopedia of Enzyme Technology, 1973; Economics, 1974; Foamed Plastics; Recent Developments, 1976; Water-Soluble Resins and Polymers: Technology and Applications, 1976; Putting Money to Work: An Investment Primer, 1976; (with W.C.F. Hartley) Cash Management: Planning, Forecasting, and Control, 1979; Water-Soluble Polymers: Recent Developments, 1979; Putting Money to Work: An Investment Primer for the '80s, 1981, updated edit., 1984; Water-Soluble Polymers: Developments since 1978, 1981; Expanded Plastics and Related Products: Developments Since 1978, 1983. Contbr. articles to profl. publs. Translator, Russian and German tech. lit. Home: 141-10 82d Dr Jamaica NY 11435 Office: Coll SI Dept Polit Sci Econs Philosophy 130 Stuyvesant Pl Staten Island NY 10301-1953

MELZACKI, KRZYSZTOF, physicist; b. Krakow, Poland, Jan. 29, 1932; came to U.S., 1984; s. Josef and Karolina (Wiecek) M.; m. Alicja W. Filipiak, Aug. 27, 1972. MSc, Jagiellonian U., Krakow, 1957, PhD in Physics, 1973. Rsch. asst., scientist Inst. Nuclear Physics, Krakow, 1956-74; sr. scientist, head of rsch. group Inst. Nuclear Rsch., Swierk, Poland, 1974-83; rsch. assoc. U. Wis., Madison, 1984-85; sr. scientist Interscience, Inc., Troy, N.Y., 1986-89; pres. KM Physics, Rensselaer, N.Y., 1990—; rsch. scientist Stevens Inst. Tech., Hoboken, N.J., 1992—; vis. scientist Uppsala

(Sweden) U., 1982-84. Translator: Introduction to Molecular Spectroscopy, 1968; contbr. over 30 articles to profl. jours. Mem. Am. Phys. Soc., Optical Soc. Am., The Planetary Soc. Achievements include patent in field of optics; design and construction of 13 devices for research in optical and atomic physics, optical spectroscopy, plasma diagnostics, and dense plasma pinches. Office: KM Physics PO Box 3021 Wallington NJ 07057

MEMOLI, MICHAEL ANTHONY, environmental engineer; b. Queens, N.Y., Apr. 21, 1950; s. Michael and Mildred (Cataldo) M.; m. Patti M. Porpora, May 6, 1973; children: Michael, Bryan. BSCE, SUNY, Buffalo, 1972; MSCE, Poly. Inst. N.Y., 1977. Registered profl. engr., N.Y., N.J., Pa., Conn.; diplomate Am. Acad. Environ. Engring. Project engr. Bowe Walsh & Assocs., Melville, N.Y., 1972-75; project mgr. Baldwin & Cornelius, Freeport, N.Y., 1975-77, Consoer Townsend & Assoc., Jericho, N.Y., 1977-82; assoc. Engring. Cons., Ft. Lee, N.J., 1982-83; sr. v.p. Camp Dresser & McKee, Woodbury, N.Y., 1983—. Bd. dirs. Kings Park (N.Y.) Soccer Club, 1987-92. Mem. Am. Water Works Assn., Water Environ. Fedn. Office: Camp Dresser & McKee 100 Crossways Park Woodbury NY 11797

MENDEL, MAURICE, audiologist, educator; b. Colorado Springs, Colo., Oct. 6, 1942; married; 3 children. BA, U. Colo., 1965; MS, Wash. U., 1967; PhD in Audiology, U. Wis., 1970. Asst. prof. audiology U. Iowa Hosp., 1970-74, assoc. rsch. scientist, 1975-76; assoc. prof. U. Calif., Santa Barbara, 1976-84, prof. audiology, 1984-88; chmn. dept. audiology and speech pathology Memphis State U., 1988—; program dir. speech and hearing sci. U. Calif., Santa Barbara, 1980-82. Fellow Am. Speech, Lang. & Hearing Assn., Soc. Ear Nose & Throat Advance in Children; mem. AAAS, Internat. Electrical Response Audiology Study Group, Internat. Soc. Audiology, Sigma Xi. Achievements include research in middle components of the auditory evoked potentials and their subsequent clinical applications to hearing testing. Office: Memphis State U CRISCI 807 Jefferson Ave Memphis TN 38152*

MENDELSOHN, AVRUM JOSEPH, psychologist; b. Chgo., June 22, 1940; s. Jack B. and Lena (Applebaum) Mendelsohn; divorced; children: Debra, Susan; m. Zehavah Whitney, June 27, 1982. BS, U. Ill., 1963; MS, IIT, Chgo., 1968; PhD, HEED, Hollywood, Fla., 1977. Lic. clin. psychologist, Ill. Psychologist Chgo. Police Dept., 1965-68; chief psychologist Ridgeway Hosp., Chgo., 1968-73; supervising psychologist Bur. of Testing Svcs., Hillside, Ill., 1973—; exec. dir. Police Cons., Inc., Westmont, Ill., 1973—. Grantee U.S. Justice Dept., 1978. Mem. Am. Psychol. Assn., Ill. Police Chiefs Assn. (police psychology com. 1989—), Ill. Fire and Police Commn. Assn., Ill. Psychol. Assn. Office: Police Cons Apt 210 825 N Cass # 210 Westmont IL 60559

MENDELSOHN, DENNIS, chemical pathology educator, consultant; b. Johannesburg, Transvaal, South Africa, Nov. 20, 1927; s. Max and Rachel (Sacks) M.; m. Leah Zar, Dec. 12, 1958; 1 child, Michaela. M.B.Ch.B, Witwatersrand U., Johannesburg, 1954, MD, 1963. Jr. lectr. dept. chem. pathology Witwatersrand U., Johannesburg, 1955-56, lectr. dept. chem. pathology, 1956-66, sr. lectr. dept. chem. pathology, 1967-72, assoc. prof. dept. chem. pathology, 1973-82, prof., head dept. chem. pathology, 1982—; chief cons. 5 teaching hosps., Johannesburg, 1982—; mem. bd. faculty medicine Witwatersrand U., Johannesburg, 1972—. Author: Fats in Food, 1988; contbr. numerous articles to profl. jours. USPHS postdoctoral fellow NIH, Bethesda, Md., 1962-64; recipient numerous rsch. grants Med. Rsch. Coun., Witwatersrand U., 1964—. Fellow Royal Coll. Pathologists (London); mem. Am. Assn. Clin. Chemists. Achievements include patent for cholesterol-free, polyunsaturated fat milk powder used to lower blood cholesterol levels. Home: 78 Linden Rd Bramley, Johannesburg 2090, South Africa Office: U Witwatersrand, Dept Chemical Pathology, York Rd Parktown, Johannesburg 2193, South Africa

MENDELSOHN, EVERETT IRWIN, history of science educator; b. Yonkers, N.Y., Oct. 28, 1931; s. Morris H. and May (Albert) M.; m. Mary B. Anderson, Sept. 14, 1974; children by previous marriage: Daniel Leeds, Sarah Ellicott, Joanna Moore; 1 stepson, Jesse Marshall Wallace. A.B., Antioch Coll., 1953; postgrad., Marine Biol. Lab., Woods Hole, Mass., 1957; Ph.D., Harvard U., 1960; D.H.L. (hon.), R.I. Coll., 1977. Research assoc. sci. and pub. policy Grad. Sch. Pub. Adminstrn. Harvard U., 1960-65, assoc. prof. history of sci., 1965-69, prof. history of sci., 1969—, chmn. dept., 1971-78; overseas fellow Churchill Coll., U. Cambridge, Eng.; fellow Van Leer Jerusalem Inst., Israel, 1978; vis. fellow Zentrum für Interdizsciplinare Forschung, Bielefeld, W. Ger., 1978; fellow Wissenschafts Kolleg, Berlin, 1983-84; dir. research group on bio-med. scis. Program on Tech. and Soc., Harvard U., 1966-68; mem. Soc. Fellows, 1957-60. Author/editor: Heat and Life: The History of the Theory of Animal Heat, 1964, Human Aspects of Biomedical Innovation, 1971, Topics in the Philosophy of Biology, 1976, The Social Production of Scientific Knowledge, 1977, The Social Assessment of Science, 1978, Sciences and Cultures, 1981, A Compassionate Peace: A Future for the Middle East, 1982, 89, Transformation and Transition in the Sciences, 1984; Nineteen Eighty Four: Science Between Utopia and Dystopia, 1984, Science, Technology and the Military, 1988, Technology and Pessimism, 1993; Editor: Jour. History Biology, 1967—; mem. editorial bd.: Sci, 1965-70, Social Studies of Sci, 1970-82, Ethics in Science and Medicine, 1973-80, Philosophy and Medicine, 1974-85, Sociology of Sciences, 1976—, Social Sci. and Medicine, 1981-92, Science in Context, 1986—, Social Epistemology, 1986—, Synthese, 1987—. Trustee Cambridge Friends Sch., The Sanctuary; bd. dirs. Inst. for Def. and Disarmament Studies; chmn. exec. com. Am. Friends Service Com. New Eng. regional office; mem. Commn. for Sci. and Cultural History of Mankind UNESCO; chmn. Harvard-Radcliffe Child Care Council.; mem. Cambridge Commn. for Nuclear Disarmament and Peace Edn.; pres. Inst. for Peace and Internat. Security. Recipient Bowdoin prize, 1957. Fellow AAAS (v.p., chmn. sect. L, com. on arms control and nat. security), Am. Acad. Arts and Scis. (chmn. program Middle East Security Studies; mem. Academie Internat. d'Historie des Scis., History Sci. Soc. (council), Internat. Acad. History Medicine, Internat. Council Sci. Policy Studies (pres.), Am. Scandinavian Found. (fellowship com.). Home: 26 Walker St Cambridge MA 02138-2404

MENDELSOHN, JOHN, oncologist, hematologist, educator; b. Cin., Aug. 31, 1936; s. Joe and Sarah (Feibel) M.; m. Anne Charles, June 23, 1962; children: John Andrew, Jeffrey Charles, Eric Robert. BA, Harvard U., 1958, MD, 1963. Diplomate Am. Bd. Internal Medicine, Am. Bd. Hematology, Am. Bd. Med. Oncology. Intern, resident Peter Bent Brigham Hosp., Boston, 1963-65, 67-68; fellow in hematology Washington U. Sch. Medicine, St. Louis, 1968-70; asst. prof. to prof. medicine U. Calif., La Jolla, 1970-85, Am. Cancer Soc. prof. clin. oncology, 1982-85, dir. Cancer Ctr., 1977-85; prof. medicine Cornell U. Med. Coll., N.Y.C., 1985—; chmn. dept. medicine Meml. Sloan Kettering Cancer Ctr., N.Y.C., 1985—; mem. bd. sci. counselors, div. cancer treatment, Nat. Cancer Inst., 1986-90; bd. dirs. Am. Assn. Cancer Rsch.; cons. Hybritech, Genentech, Immunex, Bristol-Myers; founder, 1st dir. U. Calif. San Diego Cancer Ctr. Editorial bd. Jour. Immunology, Blood, Cancer Rsch., Jour. Clin. Oncology, Growth Factors; contbr. over 100 articles in field of oncology to profl. jours. Mem. Gov.'s Cancer Adv. Coun., Calif., 1982-85; bd. dirs. Am. Cancer Soc., San Diego, 1981-85. Officer USPHS, 1965-67. Fulbright scholar U. Glasgow, Scotland, 1958-59; named Headliner of Yr. in Medicine, San Diego, 1985. Mem. Assn. Am. Physicians, Am. Soc. Clin. Investigation, Am. Soc. Clin. Oncology, Am. Assn. Cancer Rsch., Am. Soc. Hematology, Century Assn., Harvard Club N.Y., Phi Beta Kappa, Alpha Omega Alpha. Achievements include laboratory research establishing inhibition of tumor growth by antibodies against growth factor receptors. Avocations: tennis, music, history, hiking. Office: Meml Sloan-Kettering Cancer Ctr 1275 York Ave New York NY 10021-6094

MENDELSOHN, RAY LANGER, chemical engineering consultant; b. N.Y.C., Nov. 11, 1947; s. Randolph and Phyllis Janet (Langer) M.; m. Phyllis Hooper Gramlich, Dec. 8, 1978; children: Benjamin, David. B-SChemE, LeHigh U., 1969. Performance engr. Babcock & Wilcox, Barberton, Ohio, 1969-71; field svc. engr. Babcock & Wilcox, Atlanta, 1971-74; prodn. engr. Container Corp. Am., Brewton, Ala., 1974-78; tech. svc. engr. E.I. DuPont, Wilmington, Del., 1978-83; sr. engr. E.I. du Pont, Wilmington, Del., 1983-86, cons., 1986—. Contbr. papers to Afterburner Design. Youth advisor Congregation Beth Emeth, Wilmington, 1978-80, bd. dirs., 1982. Mem. AICE. Achievements include commercialization of 1st 3rd party co-generation plant in DuPont; evaluation of 1st low excess air fired organic media heaters, demonstrating no fluid degradation. Home: 8 Willing Way Wilmington DE 19807 Office: E I DuPont PO Box 6090 Newark DE 19717-6090

MENDELSON, SOL, physical science educator, consultant; b. Checonovska, Poland, Oct. 10, 1926; came to U.S., 1927; s. David C. and Frieda (Cohen) M. BME, CCNY, 1955; MS, Columbia U., 1957, PhD, 1961. Prof. engring. CCNY, 1955-58; sr. scientist Sprague Electric Co., North Adams, Mass., 1962-64, Airborne Instruments Lab., Melville, N.Y., 1964-65; phys. metallurgist Bendix Rsch. Lab., Southfield, Mich., 1966-67; cons., rschr., writer, N.Y.C. and Troy, Mich., 1968-72; adj. prof. phys. sci. CUNY, 1972-87. Contbr. numerous articles to sci. jours. Mem. Am. Phys. Soc., Fedn. Am. Scientists, Sigma Xi, Tau Beta Pi, Pi Tau Sigma. Achievements include research on theory and mechanisms of Martensitic transformations. Home: 515 E 7th St Brooklyn NY 11218

MENDEZ, C. BEATRIZ, obstetrician/gynecologist; b. Guatemala, Apr. 21, 1952; d. Jose and Olga (Sobalvarro) M.; m. Mark Parshall, Dec. 12, 1986. BS in Biology and Psychology, Pa. State U., 1974; MD, Milton Hershey Coll. Medicine, 1979. Diplomate Am. Bd. Ob-gyn. Resident in ob-gyn. George Washington U., Washington, 1979-83; pvt. practice Santa Fe, 1985—; chair perinatal com. St. Vincent's Hosp., Santa Fe, 1986-89, quality assurance mem., 1986—, chief ob-gyn., 1992; bd. dirs. Milton S. Hershey Coll. Medicine, Hershey, Pa., 1977-82. With USPHS, 1983-85. Mosby scholar Mosby-Hersey Med. Sch., Hershey, 1979. Fellow Am. Coll. Ob-Gyn. (Continuing Med. Edn. award 1989—); mem. AMA (Physician Recognition award 1986—), Am. Assn. Gynecol. Laparascopists, Internat. Soc. Gynecol. Endoscopy, Am. Fertility Soc., Am. Soc. Colposcopy and Cervical Pathology, N.Mex. Med. Soc., Santa Fe Med. Soc., Residents Assn. George Washington U. (co-founder 1981-83). Democrat. Office: Gallisteo Ob-gyn 539 Harkle Rd Santa Fe NM 87501

MENDEZ, CELESTINO GALO, mathematics educator; b. Havana, Cuba, Oct. 16, 1944; s. Celestino Andres and Georgina (Fernandez) M.; came to U.S., 1962, naturalized, 1970; BA, Benedictine Coll., 1965; MA, U. Colo., 1968, PhD, 1974, MBA, 1979; m. Mary Ann Koplau, Aug. 21, 1971; children: Mark Michael, Matthew Maximilian. Asst. prof. maths. scis. Met. State Coll., Denver, 1971-77, assoc. prof., 1977-82, prof., 1982—, chmn. dept. math. scis., 1980-82; adminstrv. intern office v.p. for acad. affairs Met. State Coll., 1989-90. Mem. advt. rev. bd. Met. Denver, 1973-79; parish outreach rep. S.E. deanery, Denver Cath. Community Svcs., 1976-78; mem. social ministries com. St. Thomas More Cath. Ch., Denver, 1976-78, vice-chmn., 1977-78, mem. parish council, 1977-78; del. Adams County Rep. Conv., 1972, 74, Colo. 4th Congl. Dist. Conv., 1974, Colo. Rep. Conv., 1982, 88, 90, 92, Douglas County Rep. Conv., 1980, 82, 84, 88, 90, 92; alt. del. Colo. Rep. Conv., 1974, 76, 84, 5th Congl. dist. conv., 1976, mem. rules com., 1978, 80, precinct committeeman Douglas County Rep. Com., 1976-78, 89—, mem. cen. com., 1976-78, 89—; dist. 29 Rep. party candidate Colorado State Senate, 1990; Douglas county chmn. Rep. Nat. Hispanic Assembly, 1989—; bd. dirs. Rocky Mountain Better Bus. Bur., 1975-79, Rowley Downs Homeowners Assn., 1976-78, Douglas County Leadership Program, 1990—; mem. Rep. Leadership Program, 1989-90; mem. exec. bd., v.p. Assoc. Faculties of State Inst. Higher Edn. in Colo., 1971-73; trustee Hispanic U. Am., 1975-78; councilman Town of Parker (Colo.), 1981-84, chmn. budget and fin. com. 1981-84; chmn. joint budget com. Town of Parker-Parker Water and Sanitation Dist. Bds., 1982-84. Recipient U. Colo. Grad. Sch. excellence in teaching award, 1965-67; Benedictine Coll. grantee, 1964-65. Mem. Math. Assn. Am. (referee rsch. notes sect. Am. Math. Monthly 1981-82, gov. Rocky Mountain section 1993—), Am. Math. Soc., Nat. Coun. Tchrs. of Math., Colo. Coun. Tchrs. of Maths., Colo. Internat. Edn. Assn., Assoc. Faculties of State Insts. Higher Edn. in Colo. (v.p. 1971-73). Republican. Roman Catholic. Assoc. editor Denver Metro. Jour. Math. and Computer Sci., 1993—; contbr. articles to profl. jours. including Am. Math. Monthly, Procs. Am. Math. Soc., Am. Math. Monthly, Jour. Personalized Instruction, Denver Met. Jour. Math. and Computer Sci., and newspapers. Home: 11482 S Regency Pl Parker CO 80134-7330 Office: 1006 11th St Denver CO 80204

MENDEZ, HERMANN ARMANDO, pediatrician, educator; b. Guatemala, Apr. 26, 1949; came to U.S., 1980; citizen of El Salvador; s. Hermann and Martha (Abularach) Mendez Fortun; m. Maria Elena Ortiz, Feb. 23, 1971; children: Natalia, Amalia. MD, U. El Salvador, 1977. Diplomate Am. Bd. Pediatrics. Asst. prof. pediatrics Health Sci. Ctr. SUNY, Bklyn., 1988-91, assoc. prof., 1991—; bd. dirs. Bklyn. Pediatric AIDS Network, 1989—, Kings County Hosp. Ctr.-Pediatric Maternal HIV Ctr., Bklyn., 1990—. Recipient Asst. Sec. for Health award HHS, 1990. Fellow Am. Acad. Pediatrics. Achievements include research in perinatal transmission of HIV, AIDS in children and their families. Office: SUNY HSCB Dept Pediatrics Box 49 450 Clarkson Ave Brooklyn NY 11201

MENDLOWSKI, BRONISLAW, retired pathologist; b. Tarnopol, Poland, June 28, 1914; came to U.S., 1948; s. Eugeniusz and Kazimiera (Zielinski) M.; m. Zita Pawlowski, Apr. 16, 1926 (dec. 1988); children: Jerry, Michael, Anna. DVM, U. Lwow, Poland, 1944; MRCVS, U. Edinburgh, Scotland, 1947; MS (hon.), U. Ill., Urbana, 1963. Pathologist W. Scotland Agr. Coll., 1945-46, Wis. State Diagnostic Lab., 1951-60, U. Ill., Urbana, 1960-63; sr. rsch. fellow in pathology Merck Inst. for Therapeutic Rsch., West Point, Pa., 1963-84; ret., 1984. Contbr. articles to profl. jours. Mem. N.Y. Acad. Sci. Roman Catholic. Home: 215 Cornwall Dr Chalfont PA 18914-2319

MENDOLIERA, SALVATORE, electrical engineer; b. Brolo, Messina, Italy, Aug. 28, 1952; s. Giuseppe and Maria (Bonfiglio) M.; m. Maria Opizzi, May 26, 1984; children: Giulia, Letizia. Degree in Electronics Engring., Pavia (Italy) U., 1977; degree in gen. mgmt., European Ctr. Permanent Edn., Fontainebleau, France, 1989. Cert. electronics engr. Jr. designer Italtel Spa, Milan, 1977-78, project leader, 1978-80; project leader Honeywell Info. Systems, Milan, 1980-82; firmware mgr. Cae Electronics, Milan, 1982-84; network mgr. Lombardia Info., Milan, 1984-85, tech. mgr., 1985-86, oper. mgr., 1986-88, deputy gen. mgr., 1988—; CEO TESEO Spa, Milan, Italy, 1992—. Mayor Town Coun., Villanterio, Italy, 1975-85. With Italian Army, 1977-78. Avocations: skiing, golf, reading. Home: Via delle novelle 9, Villanterio 27019, Italy also: TESEO Spa, Corso Sempione 32, 20100 Milan Italy

MENDOZA, GENARO TUMAMAK, mechanical engineering consultant; b. Villaba, Leyte, The Philippines, July 10, 1943; s. Benito Olitrez and Demetria E. (Tumamak) M.; m. Louella Apao; children: Genaro A., Khristine A. Degree in mech. engring., Cebu Inst. Tech., Cebu City, The Philippines, 1965, degree in elec. engring., 1966; postgrad., De La Salle U., Manila, The Philippines, 1972. Designing staff engr. Cebu Shipyard & Engring. Works, Lapulapu City, The Philippines, 1965-70; ops. engr. Esso Philippines Inc., Manila, 1970-71; aviation/marine/plant supr. Esso Philippines Inc., Visayas, Mindanao, 1974-76; plant engr. Esso Gasul, Pasig, The Philippines, 1971-72; relief supr. so. island dist. Esso Phils. Inc., Visayas, 1972-73; engring. cons. Philippine Nat. Oil Co., Makati, 1977-78; pres., gen. mgr. Marin Devel. Svcs., Cebu City, 1979—; engring. cons. Energy Devel. Corp., Makati, 1978. Avocations: tennis, basketball, table tennis. Office: Marin Devel Svcs, 13 JJ Paulino Sanches St, Cebu City The Philippines

MENG, WEN JIN, materials scientist; b. Beijing, June 24, 1962. BS, Calif. Inst. Tech., 1982, PhD, 1988. Postdoctoral rsch. fellow Argonne (Ill.) Nat. Lab., 1988-89; sr. rsch. scientist GM Rsch. Labs., Warren, Mich., 1989-92, staff rsch. scientist, 1992—. Contbr. articles to profl. jours. Mem. Material Rsch. Soc. (grad. rsch. award 1987), Sigma Xi., Tau Beta Pi. Achievements include first rsch. findings on ellucidating kinetic limitations on non-equilibrium crystal to glass transformations, structure and properties of crystals and superlattices formed by epitaxial growth. Office: GM Rsch & Devel Ctr Physics Dept Warren MI 48090

MENGEL, LYNN IRENE SHEETS, health science research coordinator; b. Toledo, Ohio, Jan. 21, 1955; d. William Burton and Darlene Ann (Ludwikowski) Paisie; m. Marvin C. Mengel, Apr. 15, 1989; children: Christopher David, Michael Patrick. Diploma in nursing, Riverside Meth. Sch. Nursing, Columbus, Ohio, 1976. RN, Fla. Staff nurse gen. patient care Doctor's Hosp. North, Columbus, Ohio, 1976, with orthopedics dept., 1976-77; with isolations unit Orlando (Fla.) Regional Med. Ctr., 1977-78, head nurse med. surg. unit, 1978-80, diabetes nurse educator, 1980-81; diabetes nurse educator, head nurse, rsch. coord. Diabetes and Endocrine Ctr. Orlando, 1981-88, coord. endocrine rsch., 1988—; speaker in field; researcher. Mem. NAFE, Am. Diabetes Assn., Am. Assn. Diabetes Educators (treas. local chpt. 1983-85). Republican. Roman Catholic. Office: 1118 S Orange Ave Ste 205 Orlando FL 32806-1235

MENIUS, ESPIE FLYNN, JR., electrical engineer; b. New Bern, N.C., Mar. 5, 1923; s. Espie Flynn and Sudie Grey (Lyerly) M.; BEE, N.C. State U., 1947; MBA, U. S.C., 1973; adopted children: James Benfield, Ruben Hughes, James Sechler, Steve Walden. With Carolina Power & Light Co., 1947-63, asst. to dist. mgr., Raleigh, Henderson, N.C., Sumter, S.C., 1947-50, elec. engr., Asheville, Southern Pines, Dunn, N.C., 1950-52, dist. engr. Hartsville, S.C., 1952-63; sr. elec. engr. Sonoco Products Co., Hartsville, 1963-74, engring. group leader, 1974-89, sr. profl. engr., 1989-91; profl. con., electrical engr., 1991—; instr. Florence-Darlington Tech. Ednl. Center. Mem. Hartsville Vol. Fire Dept., 1958—; Eagle Scout, Boy Scouts Am., 1938, scout troop leader New Bern, N.C., 1940-41, Raleigh, 1941-47, Henderson, 1948-49, Asheville, N.C., 1950, Southern Pines, N.C., 1951-52, Sumter, 1949-50, Hartsville, 1952-64; bd. mgrs. Nazareth Children's Home, Rockville, N.C., 1980—; chmn. bd. examiners City of Hartsville, 1980—. Served with AUS, 1943-46. Recipient Silver Beaver award Boy Scouts Am., 1959, Citizenship award S.C. State Firemen's Assn., 1993; named Hartsville's Citizen of Year, Rotary, 1960. Registered profl. engr., N.C., S.C., Tenn., Ga., Fla. Mem. IEEE, AAAS, Nat. Assn. Engrs., Am. Legion, Knight of St. Patrick, Scabbard and Blade, Eta Kappa Nu, Pine Burr, Phi Eta Sigma, Theta Tau, Beta Gamma Sigma. Presbyn. (elder, tchr. men's Bible class). Club: Civitan (past dir.). Author articles in field. Home and Office: 423 W Richardson Cir Hartsville SC 29550-5437

MENKE, JAMES MICHAEL, chiropractor; b. Greenville, Ohio, Oct. 1, 1951; s. Stewart Hume and Elizabeth Janette (Amburn) M.; m. Susan Christine Jameson, Sept. 6, 1986. Student, U. Cin., 1970-71, 1983-84; BS summa cum laude, Wright State U., 1975, MA, 1978; D of Chiropractic Medicine, Palmer U., 1987. Pvt. practice Menke-Jameson Clinic, Los Altos, Calif., 1987—; cons. Found. Chiropractic Edn. Rsch., Washington, 1987-91, Ford Motor Co., Batavia, Ohio, 1983-84; lectr. U. Cin., 1983-84; adj. faculty Palmer U., Sunnyvale, Calif., 1987—. Contbr. articles to profl. jours. Wright State U. scholar Dayton, Ohio, 1975. Mem. Am. Pub. Health Assn., Am. Chiropractic Scoliosis Found., Profl. Chiropractic Assn., Inst. Advancement Health, Calif. Chiropractic Assn., Mensa, Rotary. Avocations: reading, hiking, gardening. Office: Menke-Jameson Clinic 127 2nd St # 3 Los Altos CA 94022

MENNINGER, WILLIAM WALTER, psychiatrist; b. Topeka, Oct. 23, 1931; s. William Claire and Catharine Louisa (Wright) M.; m. Constance Arnold Libbey, June 15, 1953; children: Frederick Prince, John Alexander, Eliza Wright, Marian Stuart, William Libbey, David Henry. A.B., Stanford U., 1953; M.D., Cornell U., 1957; LittD (hon.), Middlebury Coll., 1982; DSc (hon.), Washburn U., 1982; LHD (hon.), Ottawa U., 1986. Diplomate: Am. Bd. Psychiatry and Neurology, Am. Bd. Forensic Psychiatry. Intern Harvard Med. Service, Boston City Hosp., 1957-58; resident in psychiatry Menninger Sch. Psychiatry, 1958-61; chief med. officer, psychiatrist Fed. Reformatory, El Reno, Okla., 1961-63; assoc. psychiatrist Peace Corps, 1963-64; staff psychiatrist Menninger Found., Topeka, 1965—, coordinator for devel., 1967-69, dir. law and psychiatry, 1981-85, dir. dept. edn., dean Karl Menninger Sch. Psychiatry and Mental Health Scis., 1984-90, exec. v.p., chief of staff, 1984-93; pres., chief exec. officer Menninger Found., 1993—; clin. supr. Topeka State Hosp., 1969-70, sect. dir., 1970-72, asst. supt., clin., dir. residency tng., 1972-81; pres., chief exec. officer Menninger Clinic, Topeka, 1991—; clin. prof. Kans. U. Med. Coll.; adj. prof. Washburn U., Wichita State U.; instr. Topeka Inst. for Psychoanalysis; mem. adv. bd. Nat. Inst. Corrections, 1975-88 , chmn., 1980-84; cons. U.S. Bur. Prisons; mem. Fed. Prison Facilities Planning Council, 1970-73; bd. dirs. Mercantile Bank Topeka (formerly Mchts. Nat. Bank, Topeka). Syndicated columnist: In-Sights, 1975-83; author: Happiness Without Sex and Other Things Too Good to Miss, 1976, Caution: Living May Be Hazardous, 1978, Behavioral Science and the Secret Service, 1981, Chronic Mental Patient II, 1987; editor: Psychiatry Digest, 1971-74; mem. editorial bd. Bull. Menninger Clinic, 1985—; contbr. chpts. to books, articles to profl. jours. Mem. nat. health and safety com. Boy Scouts Am., 1970-92, chmn., 1980-85, mem. nat. exec. bd., 1980-90, mem. nat. adv. coun., 1990—; mem. Kans. Gov.'s Adv. Commn. on Mental Health, Mental Retardation and Community Mental Health Services, 1983-90; bd. dirs. Nat. Com. for Prevention Child Abuse, 1975-83; mem. nat. adv. health council HEW, 1967-71; mem. Nat. Commn. Causes and Prevention Violence, 1968-69, Kans. Gov.'s Penal Planning Coun., 1970; chmn. Kans. Gov.'s Criminal Justice Adv. Commn., 1991—, rsch. adv. com. U.S. Secret Svc., 1990—; trustee Kenworthy-Swift Found., 1980—; ruling elder 1st Presbyn. Ch., Topeka, 1992—. Served with USPHS, 1959-64. Fellow ACP, Am. Psychiat. Assn. (chmn. com. on chronically mentally ill 1984-86, chmn. Guttmacher award bd. 1990—), Am. Coll. Psychiatrists; mem. AAAS, AMA, Group for Advancement of Psychiatry (chmn. com. mental health services 1974-77, 91—), Inst. Medicine NAS, Am. Psychoanalytic Assn. (chmn. com. on psychoanalysis, community, and society 1984—), Am. Acad. Psychiatry and Law, Stanford (Univ.) Assocs. Office: Menninger Found PO Box 829 Topeka KS 66601-0829

MENON, MAMBILLIKALATHIL GOVIND KUMAR, physicist; b. Mangalore, Aug. 28, 1928; s. Kizhekepat Sankara and Mambillikalathil Narayaniamma M.; m. Indumati Patel, 1955; 2 children. Ed. Jaswant Coll., Jodhpur, India, Royal Inst. Sci., Bombay, India, U. Bristol (Eng.); MSc, PhD, DSc (hon.), U. Jodhpur, U. Delhi, Sardar Patel U., U. Roorkee, Banaras Hindu U., Jadavpur U., Sri. Venkateswara U., Allahabad U., Andhra U., Utkal U. Aligarh Muslim U., North Bengal U., Indian Inst. Tech., Madras, Indian Inst. Tech., Kharagpur, Univ. Bristol, United Kingdom, hon. D of Eng. Stevens Inst. Tech. Dir. Tata Inst. Fundamental Rsch., Bombay, 1966-75; chmn. Electronics Commn. and sec. to Govt. India Dept. Electronics, 1971-78; sci. adviser to Min. of Def., dir.-gen. Def. Rsch. and Devel. Orgn., and sec. for def. rsch., 1974-78; dir.-gen. Coun. Sci. and Indsl. Rsch., 1978-81; sec. Dept. of Sci. and Tech., 1978-82; sec. Dept. Environ., 1980-81; chmn. Commn. for Additional Sources Energy, 1981-82; chmn. sci. adv. com. to Indian Cabinet, 1982-85; mem. Govt. Planning Commn., 1982-89; sci. adv. to Prime Min., 1986-89; min. state Sci. & Tech. Govt. India, 1989-90; pres. Internat. Coun. Sci. Unions, 1988-93; mem. Indian Parliament, 1990—. Recipient Sr. award Royal Commn. for Exhibition of 1851, 1953-55, Shanti Swarup Bhatnagar award for phys. Scis. Council Sci. and Indsl. Research, 1960, Repub. Day (nat.) awards Govt. India; Padma Shri, 1961; Padma Bhushan, 1968; Padma Vibhushan, 1985; numerous other awards. Fellow Nat. Acad. Scis. India (pres. 1987-88), Inst. Electronics and Telecommunications Engrs. India, Am. Acad. Arts and Scis. Russian Acad. of Scis., Asia Electronics Union (pres. 1973-75), IEEE (hon.); mem. Indian Nat. Sci. Acad. (pres. 1981-82), Royal Soc., Indian Acad. Sci. (pres. 1974-76), Indian Sci. Congress Assn. (pres. 1981-82), pontifical Acad Scis. Address: 47 Lodi Estate, New Delhi 110 003, India

MENON, PADMANABHAN, aerospace engineer; b. Bangalore, Karnataka, India, May 30, 1951; came to U.S., 1982; s. Achutha Aikara and Mani Kothenath M.; m. Prasanna Menon, Nov. 6, 1977; children: Jishnu P., Jayant P. BE, Osmania U., 1973; ME, Indian Inst. of Sci., Bangalore, 1975; PhD, Va. Polytechnic Inst., 1983. Rsch. asst. Ind. Inst. Sci., Bangalore, 1975; mission analyst Indian Space Rsch. Orgn., Trivandrum, India, 1976-81; rsch. asst. Va. Polytechnic Inst., Blacksburg, 1982-83; rsch. scientist Integrated Systems, Inc., Santa Clara, Calif., 1983-86; assoc. prof. Ga. Inst. Tech., Atlanta, 1986-88; pres., chief scientist Optimal Synthesis, Palo Alto, Calif., 1992—; vis. scientist NASA Ames Rsch. Ctr., Moffett Field, Calif., 1988-91; rsch. prof. Santa Clara U., 1989—; session organizer and chmn., Am. Control Conf., Pitts., Atlanta, Boston, San Diego, 1988-91; nat. chmn. workshop Internat. Fedn. Automatic Control Conf., Pitts., Atlanta, Boston, San Diego, 1988-91; nat. chmn. workshop Internat. Fedn. Automatic Control, Boston, 1989. Assoc. editor AIAA Jour. of Guidance, Control and Dynamics, 1991—; contbg. author: Integrated Technology Methods in Aerospace Design, 1992; contbr. articles to profl. jours., conf. publs. Team coach Odyssey of the Mind, Palo Alto, 1992; asst. scoutmaster Boy Scouts Am., Atlanta, 1987-88. Merit scholar Osmania U., Hyderabad, India, 1973. Fellow AIAA (assoc.); mem. Sigma Gamma Tau, Sigma Xi. Hindu. Achievements include development of new method for solving aircraft guidance problems, methods for machine vision-based helicopter guidance, missile guidance laws, advanced control systems for high performance aircraft, model for describing technology growth, hydraulic micromanipulator and intra-aortic heart assist device. Home: 908 E Meadow Dr Palo Alto CA 94303

MENTER, M(ARTIN) ALAN, dermatologist; b. Doncaster, Eng., Oct. 30, 1941; came to U.S., 1975; s. Harry Menter and Esme (Green) Behr; m. Pamela Mary Williams, Dec. 4, 1966; children: Keith, Colin, Kerith. MB Bch, U. Witwatersrand, 1966; MMed in Dermatology, U. Pretoria, 1971. Diplomate Am. Bd. Dermatology. Intern dept. medicine then dept. surgery Johannesburg (Republic South Africa) Gen. Hosp., 1967, sr. intern medicine and dermatology, 1968; resident in dermatology U. Pretoria and Pretoria Gen. Hosp., 1968-71; sr. resident in dermatology Guy's Hosp., London, 1972; sr. resident, tutor in dermatology St. John's Hosp. for Disease of Skin, London, 1972-73; cons. dermatologist Pretoria Gen. Hosp., 1973-75; dermatologist Baylor U. Med. Ctr., Dallas, 1975—; chmn. div. dermatology Baylor U. Med. Ctr., 1992—; med. dir. Nat. Psoriasis Found. Tissue Bank, Dallas, 1993—; fellow dept. dermatology U. Tex. Southwestern Med. Sch., Dallas, 1977-79, assoc. clin. prof. dermatology, 1977—; med. dir. Psoriasis Ctr., Baylor U. Med. Ctr., Dallas, 1979—; clin. assoc. prof. dept. periodontics Baylor Coll. Dentistry, Dallas, 1985—; presenter local, state and nat. dermatol. orgns. and teaching programs. Editorial bd. Jour. Am. acad. Dermatology; contbr. numerous articles to profl. jours., chpts. to books. Coach Rugby football team U. Pretoria, 1974; represented S. Africa Nat. Rugby football team, 1968; coach, commr. Boys Under 12 Classic League Soccer, Dallas, 1978-82; active various local civic organizations and coms. Recipient Clin. Rsch. award Imperial Chem. Industries, 1972-73. Mem. AMA, Am. Acad. Dermatology (com. on psoriasis 1988—, chmn. 1990—, com. on standards of care for psoriasis 1988-92, chmn. 1989-92, dir. Psoriasis Symposium 1991—), Am. Acad. Demratol. Surgery, Brit. Assn. Dermatology, Dallas County Med. Soc. (med. student rels. com. 1989—), Dallas Dermatol. Soc. (sec.-treas. 1979, pres. 1980, rep. to adv. coun. Am. Acad. Dermatology 1987-89), Dermatol. Therapy Assn. (pres. 1985), Tex. Dermatol. Soc. (program coord. 1987—), Tex. Med. Assn. (sub com. on joint sponsorship 1992), Dermatology Found. (rsch. award rev. com. 1992—). Home: 5230 Royal Ln Dallas TX 75229-5525 Office: Tex Dermatology Assocs 3409 Worth St Dallas TX 75246-2039

MENZEL, DAVID WASHINGTON, oceanographer; b. Bilasapur, India, Feb. 22, 1928; (parents Am. citizens); s. Emil W. and Ida T. Menzel; m. Dorothy Adamy, Sept. 7, 1951. BS, Elmhurst Coll., 1948; MS, U. Ill., 1951; PhD, U. Mich., 1958. Marine biologist Bermuda Biol. Sta., 1958-63, Woods Hole (Mass.) Oceanographic Instn., 1963-70; dir. Skidaway Inst. Oceanography, Savannah, Ga., 1970—. With U.S. Army, 1951-53. Recipient Disting. Assoc. award Dept. of Energy, 1990. Home: 20 Fallowfield Dr Savannah GA 31406-6420 Office: Skidaway Inst Oceanography McWorter Dr PO Box 13687 Savannah GA 31416-0687

MENZIE, DONALD E., petroleum engineer, educator; b. DuBois, Pa., Apr. 4, 1922; s. James Freeman and Helga Josephine (Johnson) M.; m. Jane Cameron Redsecker, Nov. 6, 1946; children: Donald, William Lee, John Peter, Thomas Freeman. B.S in Petroleum and Natural Gas Engring., Pa. State U., 1942, M.S, 1948, Ph.D., 1962. Marine engr. Phila. Navy Yard, 1943-46; rsch. asst. air-gas dr. recovery Pa. State U., 1946-48, instr. petroleum and natural gas engring., 1948-51; asst. prof. petroleum engring. U. Okla., Norman, 1951-55, assoc. prof., 1955-64, prof., 1964-91, Kerr-McGee Centennial prof. Petroleum and Geol. Engring., 1991—, Halliburton Disting lectr., 1982-84; disting. lectr. Okla. U., 1986-87; dir. Sch. Petroleum and Geol. Engring U. Okla., Norman, 1963-72, petroleum engr. rsch. info. systems program, 1979-88, assoc. exec. dir. Energy Resources Ctr., 1988; assoc. exec. dir. Microbial Enhanced Oil Recovery Rsch. Project, Norman, 1982-; microbial enhanced oil recovery rsch. project U. Okla., 1982—; pres., owner Petroleum Engring. Educators, Norman, 1971—; cons. in field. Author: Reservoir Mechanics, 1954, Waterflooding for Engineers, 1968, Applied Reservoir Engineering for Geologists, 1971, New Recovery Techniques, 1975, Microbial Enhanced Oil Recovery, 1987, Dispersivity As An Oil Reservoir Rock Characteristic, 1989; contbr. articles to profl. jours. Mem. enhanced oil com. Interstate Oil and Gas Compact Commn., 1982-; commr., scoutmaster Last Frontier Coun. Boy Scouts Am., 1951-81; mem. adminstrv. bd. McFarlin United Meth. Ch., Norman, also sunday sch. tchr., pres. fellowship class, treas.; pres. Jackson PTA, Norman, 1962-68; treas. Cleveland County Rep. Com.; mem. Norman Park Commn., 1974-80; co-chmn., dir. Norman Parks Found., 1983-; mem. Cen. Com. U. Okla., 1987-. Mem. AIME, Am. Assn. Petroleum Geologists, Okla. Soc. Profl. Engrs., Nat. Soc. Profl. Engrs., Am. Soc. Engring. Edn., Soc. Petroleum Engrs., Am. Petroleum Inst., AAAS, Okla. Engring. and Tech. Guidance Coun., Okla. Anthopol. Soc., Soc. Petroleum Engrs. (recipient Nat. Disting. Achievement award for petroleum engring. faculty 1989), Sigma Xi, Pi Epsilon Tau, Alpha Chi Sigma, Phi Lamda Upsilon, Phi Kappa Phi. Clubs: Sportsmen of Cleve. County, Sooner Swim (dir. 1966-78). Lodge: Masons. Home: 1503 Melrose Dr Norman OK 73069-5366 Office: U Okla F314 The Energy Ctr Norman OK 73019

MENZIES, CARL STEPHEN, agricultural research administrator, ruminant nutritionist; b. Menard, Tex., Mar. 6, 1932; s. Alex L. and Marguerite (Watson) M.; m. Shirley W. Martin, Sept. 2, 1952; children: John S., Linda D. Menzies Napier. BS, Tex. Tech Coll., 1954; MS, Kans. State U., 1956; PhD, U. Ky., 1965. Instr. animal sci. Kans. State U., Manhattan, 1955-58, asst. prof., 1958-65, assoc. prof., 1965-69; rsch. asst. animal sci. U. Ky., Lexington, 1961-62; head, prof. S.D. State U., Brookings, 1969-71; resident dir. of rsch., prof. animal sci. Tex. Agrl. Experiment Sta., San Angelo, 1971—; cons. Pakistan project King Ranch, Kingsville, Tex., 1975; bd. dirs. small ruminant CRSP project USAID, Davis, Calif., 1982—; mem. adv. com. sheep rsch. sta. USDA/ARS, Dubois, Idaho, 1985—; mem. adv. bd. Angelo State U. Agrl. Program, 1986—. Contbr. articles to profl. jours. Chmn. livestock com. Goals for San Angelo, 1988-90. Named to Manard Sch. Hall of Fame, 1973; recipient Appreciation for Svc. Kans. Purebred Sheep Breeders Assn., 1968, Silver Ram award Am. Sheep Producers Coun., 1988, award Dep. Chancelor for Agr. of TAMU for Disting. Performance in Adminstrn., 1983. Fellow Am. Soc. Animal Sci.; mem. Coun. Agrl. Sci. and Tech., San Angelo C. of C. (bd. dirs. 1990-93), Rotary, Sigma Xi, Gamma Sigma Delta, Phi Kappa Phi, Alpha Zeta. Avocations: ranching, producing registered sheep, outdoor activities. Office: Tex A&M U Agrl Rsch & Extension Ctr 7887 US Hwy 87 N San Angelo TX 76901-9714

MERAL, GERALD HARVEY, biologist; b. Detroit, Apr. 17, 1944; s. Abe and Sally (Rosenzweig) M.; m. Barbara Dale, Sept. 16, 1973. BS, U. Mich., 1965; PhD, U. Calif., Berkeley, 1973. Staff scientist Environ. Def. Fund, Berkeley, 1971-75; dep. dir. Calif. Dept. Water Resources, Sacramento, 1975-83; exec. dir. Planning and Conservation League, Sacramento, 1983—. Author, pub. chpt. California's Threatened Environment, 1992; author chpt. Achieving Consensus in California Water Policy, 1992. Founder Friends of the River, Berkeley, 1973; founder, bd. dirs. Tuolumne River Preservation Trust, Sacramento, 1981; organizer, campaigner wildlife and park bond act, 1988, wildlife protection measure, 1990, rail devel. bond act, 1990; developer tobacco tax measure, 1990. Recipient River Conservation award Perception, 1984, Pub. Svc. award Common Cause, 1984, Golden Spike award Nat. Assn. R.R. Passengers, 1990. Office: Planning and Conservation 926 J # 612 Sacramento CA 95814

MERCER, WILLIAM EDWARD, II, chemical research technician; b. Neubrücke, Fed. Republic Germany, Dec. 28, 1956; (parents Am. citizens); s. William Edward and Julia (Didio) M.; m. Debra Lynn Mokry, Nov. 3, 1979; children: Crystal Nichole, Lacey Lynn, Dustin Edward. Student, Tex. A&M U., 1974-75, 76-77, Del Mar Coll., Corpus Christi, Tex., 1975-76, 78-79. Asst. mgr., then dist. supr. Kwik Pantry Food Stores, Bryan and Beaumont, Tex., 1977-78; frozen foods mgr. H.E.B. Food Stores, Rockport, Tex., 1978-79; rsch. aide Tex. div. Dow Chem. Co., Freeport, 1979-81, rsch. aide II Tex. div., 1981-83, sr. chem. asst. researcher, 1983-87, chem. technician, 1987-90, sr. chem. technician, 1990—. Contbr. articles to profl. jours.; author scientific paper; patentee in field. Mem. South Brazoria County Bowling Assn. (bd. dirs. 1991-93, Dir. of Yr. 1992-93). Republican. Baptist. Avocations: bowling, stamp collecting, boating, fishing, horseshoe pitching. Home: 136 Cannon St Clute TX 77531-3612 Office: Dow Chem Co LJRC-212 Freeport TX 77541

MERCHANT, ROLAND SAMUEL, SR., hospital administrator, educator; b. N.Y.C., Apr. 18, 1929; s. Samuel and Eleta (McLymont) M.; m. Audrey Bartley, June 6, 1970; children—Orelia Eleta, Roland Samuel, Huey Bartley. B.A., N.Y.U., 1957, M.A., 1960; M.S., Columbia U., 1963, M.S.H.A., 1974. Asst. statistician N.Y.C. Dept. Health, 1957-60, statistician, 1960-63; statistician N.Y. TB and Health Assn., N.Y.C., 1963-65; biostatistician, adminstrv. coord. Inst. Surg. Studies, Montefiore Hosp., Bronx, N.Y., 1965-72; resident in adminstrn. Roosevelt Hosp., N.Y.C., 1973-74; dir. health and hosp. mgmt. Dept. Health, City of N.Y., 1974-76; from asst. adminstr. to adminstr. West Adams Community Hosp., L.A., 1976; spl. asst. to assoc. v.p. for med. affairs Stanford U. Hosp., Calif., 1977-82, dir. office mgmt. and strategic planning, 1982-85, dir. mgmt. planning, 1986-90; v.p. strategic planning Cedars-Sinai Med. Ctr., L.A., 1990—; clin. assoc. prof. dept. family, community and preventive medicine Stanford U., 1986-88, dept. health rsch. and policy Stanford U. Med. Sch., 1988-90. Served with U.S. Army, 1951-53. USPHS fellow. Fellow Am. Coll. Healthcare Execs., Am. Pub. Health Assn.; mem. Am. Hosp. Assn., Nat. Assn. Health Services Execs., N.Y. Acad. Scis. Home: 27335 Park Vista Rd Agoura Hills CA 91301-3639 Office: Cedars-Sinai Med Ctr 8700 Beverly Blvd Los Angeles CA 90048

MERCHENTHALER, ISTVAN JOZSEF, anatomist; b. Baja, Hungary, Apr. 29, 1949; came to U.S., 1988; s. Istvan and Maria (Gomori) M.; m. Agnes Katalin Major, Nov. 6, 1969; children: Istvan, Boglarka, Nora. MD, U. Med. Sch., Pecs, Hungary, 1974; PhD, Hungarian Acad. Sci., Budapest, 1986; DsC, Hungarian Acad. Sci., 1992. Asst. prof. Dept. Anatomy U. Med. Sch., Pecs, Hungary, 1974-86; assoc. prof. Dept. Anatomy U. Med. Sch., Pecs; vis. asst. prof. Dept. Anatomy U. N.C., Chapel Hill, 1981-83; vis. asst. prof. Hebert Rsch. Ctr. Tulane U., New Orleans, 1984; vis. scientist, head functional Morphology sect. NIEHS/NIH, Research Triangle Park, N.C., 1988—; adj. assoc. prof. Dept. Cell Biology and Anatomy U. N.C., 1991—. Contbr. articles to profl. jours. Recipient Lenhossek award Hungarian Assn. of Anatomists, 1984, Outstanding Young Scientist award Hungarian Acad. Scis., 1986. Office: LMIN NIEHS NIH MD C4-07 Research Triangle Park PO Box 12233 Durham NC 27709

MERCIER, DANIEL EDMOND, military officer, astronautical engineer; b. Biloxi, Miss., Dec. 7, 1950; s. Donald Lewis and Lillian Pearl (Long) M.; m. Diane Lynn Sawyer, June 12, 1976. BS in Astronautical Engring., Air Force Acad., Colorado Springs, Colo., 1972; MS in Astronautical Engring., Air Force Inst. Tech., Dayton, Ohio, 1978. Commd. 2d lt. Air Force, 1972, advanced through grades to lt. col., 1988; trajectory analyst Strategic Air Command, Omaha, 1972-77; acquisition officer Space Systems Div., L.A., 1978-80, Ballistic Missile Office, San Bernadino, Calif., 1980-82; asst. prof. Air Force Acad., Colorado Springs, 1982-86; test mgr. Air Force Operational Test and Evaluation Ctr., Albuquerque, 1986-89, Milstar test dir., 1989—. Contbr. papers to jours., symposium and confs. Named Outstanding Young Man of Am., U.S. Jaycees, 1982, 83. Mem. AIAA, Internat. Test and Evaluation Assn. Republican. Roman Catholic. Home: 940 Flaming Tree Way Monument CO 80132 Office: DET 4 AFOTEC/MIL 625 Suffolk St Peterson AFB CO 80914-1730

MERCIER, JOHN RENÉ, nuclear engineer, health physicist; b. Apr. 22, 1961; s. Joseph R.G. and Lucille (Trembley) M.; m. Christine Carol Chabai, Aug. 30, 1986. BS in Engring. Sci., U. Tex., 1984; ME in Nuclear Engring., Cornell U., 1991. Lic. nuclear plant sr. reactor operator. Commd. 2d lt. U.S. Army, 1985, advanced through grades to capt., 1987; nuclear med. sci. officer U.S. Army Environ. Hygiene Agy., Aberdeen Proving Ground, Md., 1985-86; chief radiation safety Darnall Army Community Hosp., Ft. Hood, Tex., 1986-89; project engr. Def. Nuclear Agy., Johnston Atoll, 1991-92; chief health physics Tripler Army Med. Ctr., Honolulu, 1992—. Vol. Habitat for Humanity, Hawaii. Decorated three Commendation medals U.S. Army; named U.S. Army grad. fellow, 1989-90. Mem. AAAS, IEEE Nuclear and Plasma Scis. Soc., N.Y. Acad. Scis., Am. Nuclear Soc., Health Physics Soc. (pres. Hawaii chpt.), Soc. Am. Mil. Engrs., Assn. U.S. Army, Sigma Xi, Tau Beta Pi, Tau Kappa Epsilon. Achievements include successful start-up and operation of world's first plutonium mining plant; research in radiologically contaminated soil; design of soil characterization of system. Home: 366-D Reno Rd Honolulu HI 96819

MERCURIO, ANTONINO MARCO, anthropologist, psychotherapist; b. Messina, Sicily, Italy, Nov. 8, 1930; s. Paolo and Maria (Urso) M.; m. Paola Sensini, Dec. 28, 1974. Doctor in Classical Literature, State U. Messina, 1958; Lic. in Philosophy, Jesuit Faculty of Messina, 1953; Doctor in Theology, Cath. U. Paris, 1964. Prof. theology Gregorian Pontifical U., Rome, 1969-70; founder, dir. Inst. Analytical Psychotherapy, Rome, 1970—; founder, pres. Associazione Psicoterapeuti Italiana, Rome, 1974-78; founder, rector Sophia U., Rome, 1978—; founder Sophia U., Geneva, 1980, Sophia U., Brussels, 1981, Sophia U., Paris, 1984, European Ctr. Research in Life as a Masterpiece of Art, 1986. sci. supr. several insts. psychotherapy, Italy, 1974—. Author: Amore e Persona, 1976, Teoria della Persona, 1978, Amore Libertà e Colpa, 1980, La vita come opera d'arte, 1988, Antropologia Esistenziale e Metapsicologia Personalistica, 1992; inventor sophianalysis, method of existential psychotherapy, 1970; inventor sophia-art, 1988; inventor Olimiadi Dello Fozza Amozosa, 1993; contbr. articles in field to profl. publs. Mem. European Assn. Humanistic Psychology (co-founder 1979), Acad. Psychoanalysis of Berlin. Home: 12 Via Pantanelle, 00043 Ciampino Rome Italy Office: 73 Via Potenza, 00043 Ciampino Rome Italy

MEREDITH, JULIA ALICE, nematologist, biologist, researcher; b. Asheboro, N.C., Aug. 21, 1943; d. Ralph Shanklin and Kathryn Virginia (Corder) M.; m. Renato N. Inserra, Aug. 31, 1985. BA, U. Ky., 1965, MS, 1966; DSc, U. Cen. Caracas, Venezuela, 1977. Instr. zoology Cen. U. Venezuela, Maracay, 1967-69, asst. prof., 1969-76, assoc. prof., 1976-79, prof. nematology, 1980-86, 89-90, head dept., 1980-90; rsch. assoc. U. Fla., Gainesville, 1986-87, vis. prof., 1988; rsch. assoc. U. Fla. and Agrl. Rsch. Svc. USDA, 1991—; cons. ministries agr. Colombia, 1967-71, Uruguay, 1971, Ecuador, 1977, 80, Venezuela, 1980-84; vis. prof. Va. Poly. Inst. and State U., Blacksburg, 1985. Author: (lab manuals) Nematologia Agricola, 1969, Field and Laboratory Methods, 1974; editor Nematropica, 1971-73; mem. editorial bd. Fitopatologia Venezolana, 1989-92. Rsch. grantee Nat. Coun. Sci. and Tech., Caracas, 1975, 84, Nat. Coun. for Sci. Devel., Caracas, 1988. Fellow European Soc. Nematologists; mem. Orgn. Nematologists of Tropical Am. (treas. 1974-75, pres. 1978-79, Disting. Svc. award 1988), Soc. Nematologists (edn. com. 1988-89), Venezuelan Phytopath. Soc., Internat. Soc. Chem. Ecology, Internat. Wildlife Fedn., Nat. Wildlife Fedn., Nat. Audubon Soc., Fla. Audubon Soc., Sigma Xi, Alpha Epsilon Delta, Phi Epsilon Phi. Methodist. Avocation: swimming. Office: USDA Agrl Rsch Svc IABBBRL 1700 SW 23rd Dr Gainesville FL 32604

MERENBLOOM, ROBERT BARRY, hospital and medical school administrator; b. Balt., July 13, 1947; Philip William and Florence Ruth (Surosky) M.;. B.A., U. Md., 1969; M.S., Morgan State U., 1973; M.B.A., U. Balt., 1980. Mem. staff Mayor Balt. Office Manpower Resources, 1972-73; assoc. staff mem. Office Dean, U. Md. Med. Sch., 1976-80; adminstrv. officer rsch. and devel. Balt. VA Med. Ctr., 1974-80; assoc. adminstr. dept. medicine Johns Hopkins U. Sch. Medicine, Balt., 1980-84, adminstr. dept. medicine Johns Hopkins Hosp., 1984-88, assoc. Sch. Hygiene and Pub. Health, 1984-88; lectr. dept. medicine Bowman Gray Sch. Medicine Wake Forest U., 1988—, asst. chmn. Dept. Medicine, 1988-91; assoc. chmn. Dept. Medicine, 1991—; instr. sociology U. Balt., 1973-76; adj. faculty Weekend Coll., Coll. Notre Dame, Balt., 1980—; assoc. mgmt. Babcock Grad. Sch. Bus. Wake Forest U. Exec. dir. J. Paul Sticht Ctr. on Aging. Recipient Hon. Corpsmen Leader award Office Mayor Balt., 1973; Outstanding Performance award Balt. VA Med. Ctr., 1975, Superior Performance award, 1980. Mem. Am. Gerontology Soc., So. Gerontology Soc., Soc. Rsch. Adminstrs., Nat. Coun. Univ. Rsch. Adminstrs., Adminstrs. Internal Medicine, Assn. Am. Med. Colls. (group on bus. affairs), Am. Hosp. Assn., Am. Pub. Health Assn., Am. Coll. Healthcare Adminstrs., Soc. Gen. Internal Medicine, John Hopkins Club, Piedmont Club.

MERGNER, HANS KONRAD, zoology educator; b. Lemgo, Fed. Republic Germany, May 8, 1917; s. Konrad Johannes and Luise Johanna (Tasche) M; m. Maria Theresia Zieger, July 16, 1933; children: Hans Joachim, Wolfgang Christian, Andreas. D in Natural Scis., U. Tubingen, Fed. Republic Germany, 1956. Researcher Max Planck Inst. Hirnforschung, Giessen, Fed. Republic Germany, 1956-58; from asst. to assoc. prof. Zool. Inst. U. Giessen, 1958-70; prof. Inst. Spezielle Zoologie, Bochum Ruhr U., Fed. Republic Germany, 1970-84, emeritus prof., 1984—; sci. dir. UNESCO Course for Ecology of Red Sea and Adjacent Waters, Gardagha, Egypt, 1977; dean faculty biology Ruhr U., Bochum, 1974-75, 78. Author: Schlechter: Orchideen, 1970—, Reverberi: Experimental Embryology of Marine and Fresh-Water Invertebrates; author, editor: Mergner: Orchideenkunde, 1992; contbr. numerous rsch. papers on embryology of hydroids, anatomy of brains, development physiology of sponges and ecology of coral reefs. Maj. German mil., 1936-45. Mem. Internat. Soc. Reef Studies, Deutsche Zoologische Gesellschaft, Gesellschaft Entwicklungsbiologie. Lutheran. Avocations: traveling, mountain climbing, diving, painting, classical music. Home: Hansstrasse 1, 44797 Bochum Germany Office: Inst Spez Zoology Ruhr U, Universitatsstrasse 150, 44797 Bochum Germany

MERHAUT, JOSEF, electroacoustics educator; b. Prague, Czech Republic, Nov. 5, 1917; s. Vaclav and Marcela (Jansenova) M.; m. Bezouskova Merhaut, June 26, 1942 (div. 1951); 1 child, Jan; m. Kristkova Merhaut, Dec. 29, 1951. Degree engring., T.U., Prague, 1946, Doctor, 1959. Techniquer Telegrafia Co., Pardubice, Czech Republic, 1940-45; audioengr. Tesla Co., Prague, 1945-58; prof. T.U., Prague, 1958-90, ret., 1990; chmn. Internat. Electrotechnic Commn., Geneva, Switzerland, 1976-84, v.p., 1978-84. Author: Theory of Electroacoustics, 1981; contbr. articles to Jour. of Audioengring., 1982-90. Fellow Acoustical Soc. of Am.; mem. Audio Engring. Soc. of Am. Home: Dvouletky 341, 10000 Prague Czech Republic

MERIKOSKI, JORMA KAARLO, mathematics educator; b. Helsinki, May 12, 1942; s. Antti Johannes Siimes and Ilona Marjatta Merikoski; m. Ulla Inkeri Kuhanen, Aug. 14, 1971; children: Marppa, Raisa, Tanja. MSc, U. Helsinki, 1964; PhLic, U. Jyväskylä, Finland, 1970; PhD, U. Tampere, Finland, 1976. Mathematician Finnish Cable Co., Helsinki, 1963-66; teaching asst. U. Helsinki, 1963-64; teaching asst. Tampere U. Tech., 1967, spl. tchr., 1974-75; lectr. math. U. Tampere, 1966—, docent, 1983—; vis. asst. prof. McGill U., Montreal, Que., Can., 1980; vis. rsch. fellow U. Coimbra, Portugal, 1985, Technion-Israel Inst. Tech., Haifa, 1988, Czechoslovak Acad. Scis., Prague, 1989, 93. Co-author 18 math. textbooks for primary and secondary schs.; contbr. articles to math. jours. 2d lt. inf. Finnish Army, 1965. Mem. Internat. Linear Algebra Soc., Finnish Math. Soc., Statis. Soc. Can., Am. Math. Soc., Soc. for Indsl. and Applied Math. Avocations: jogging, literature, chess. Home: Ylisenkatu 3 B 16, SF-33710 Tampere Finland Office: U Tampere Dept Math Scis, PO Box 607, SF-33101 Tampere Finland

MERILAN, CHARLES PRESTON, dairy husbandry scientist; b. Lesterville, Mo., Jan. 14, 1926; s. Peter Samuel and Cleo Sarah (Harper) M.; m. Phyllis Pauline Laughlin, June 12, 1949; children—Michael Preston, Jean Elizabeth. B.S. in Agr, U. Mo., 1948, A.M., 1949, Ph.D., 1952. Mem. faculty U. Mo., Columbia, 1950—; prof. dairy husbandry U. Mo., 1959—, chmn. dept., 1961-62; asso. dir. Mo. Agrl. Expt. Sta., 1962-63, asso. investigator space sci. research center, 1964-74, exec. sec., dir. grad. studies physiology area, 1969-72, chmn. univ. patent and copyright com., 1963-80. Served with USMC, 1944-45. Decorated Purple Heart. Fellow AAAS; mem. Am. Chem. Soc., Am. Dairy Sci. Assn., Am. Soc. Animal Sci., Soc. Cryobiology, Sigma Xi, Alpha Zeta, Gamma Sigma Delta, Phi Beta Pi. Achievements include research in biol. material preservation, computer assisted image analysis; patents in cryobiology and nuclear magnetic resonance fields. Home: 1509 Bouchelle Ave Columbia MO 65201-5920 Office: U Mo Columbia MO 65211

MERILAN, JEAN ELIZABETH, statistics educator; b. Columbia, Mo., Sept. 18, 1962; d. Charles Preston and Phyllis Pauline (Laughlin) M. AB summa cum laude, U. Mo., 1985, MA in Math., MA in Stats., 1987; post-grad., U. Ariz., 1987—. Grad. teaching asst. U. Mo., Columbia, 1985-87; grad. rsch. asst. U. Ariz., Tucson, 1988-89, grad. teaching asst., 1989—. Nat. Merit scholar, Univ. Curators scholar U. Mo., 1981-85, Grad. Acad. scholar U. Ariz., 1990—, Arts and Sci. Grad. scholar U. Mo., 1985-87; Gregory fellow U. Mo., 1985-87, Faculty of Sci. fellow U. Ariz., 1987-88. Mem. Am. Statis. Assn., Inst. Math. Stats., Soc. for Indsl. and Applied Math., Biometric Soc., Am. Math. Soc., Math. Assn. Am., Golden Key Nat. Honor Soc., Sigma Xi, Phi Beta Kappa, Phi Kappa Phi, Phi Eta Sigma, Pi Mu Epsilon. Office: U Ariz Dept of Stats Tucson AZ 85721

MERILAN, MICHAEL PRESTON, astrophysicist, educator; b. Columbia, Mo., Jan. 5, 1956; s. Charles Preston and Phyllis Pauline (Laughlin) M. B.S. summa cum laude in Physics, U. Mo., Columbia, 1978, M.S., 1980; Ph.D. in Astronomy, Ohio State U., 1985. Teaching asst. dept. physics and astronomy U. Mo., Columbia, 1976-78, grad. teaching asst., 1978-80; grad. teaching assoc., instr. dept. astronomy Ohio State U., Columbus, 1980-85; asst. prof. dept. physics and astronomy SUNY, Oneonta, 1985-91, assoc. prof., 1991—, chmn. dept. physics and astronomy, 1990-93, assoc. prof., 1991—; acting dean sci. and social sci., 1993—; astron. cons. Ohio Dept. Natural Resources, 1982-83; Oneonta smart node advisor Cornell Nat. Supercomputer Facility, 1987-92. Contbr. articles in field. O.M. Stewart fellow, 1979; U. Mo. Curators scholar, 1974-78; Mahan Writing award, 1975. Mem. AAAS, IAPPP, Am. Astron. Soc., Astron. Soc. Pacific, Sigma Xi, Phi Eta Sigma, Phi Kappa Phi, Phi Beta Kappa, Pi Mu Epsilon, Sigma Pi Sigma. Achievements include analytic and numeric investigation of protostellar hydrodynamics; determination of the properties of static and slowly rotating partially degenerate semirelativistic stellar structures. Office: SUNY Dept Physics and Astronomy Oneonta NY 13820

MERIN, ROBERT GILLESPIE, anesthesiology educator; b. Glens Falls, N.Y., June 16, 1933; s. Joseph Harold and Jessie Louisa (Gillespie) M.; m. Barbara R. Rothe, Mar. 1, 1958; children: Michael, Jan, Sarah. BA, Swarthmore Coll., 1954; MD, Cornell U., 1958. Diplomate Nat. Bd. Med. Examiners, Am. Bd. Anesthesiology. From asst. prof. to prof. anesthesiology U. Rochester (N.Y.) Med. Ctr., 1966-81; prof. anesthesiology U. Tex. Health Sci. Ctr., Houston, 1981-92; prof Anestheology Med. Coll. Ga., Augusta, 1992—; mem. anesthetic life support drug com. FDA, Washington, 1982-87, spl. cons., 1987—; Murray Mendolsohn Meml. lectr. U. Toronto Sch. Medicine, 1976, Harry M. Shields Meml. lectr., 1988; Litchfield lectr. Oxford U., 1977, William and Austin Friend Meml. vis. prof. Queens U., 1981, Joseph F. Artusio endowed lectr. Cornell U. Med. Coll., N.Y.C., 1991, and others. Editorial bd. Anesthesiology, 1977-86; contbr. articles to Anesthesiology, Jour. Pharmacology and Exptl. Therapeutics. Capt. U.S. Army, 1961-63. Recipient Rsch. Career Devel. award NIH, 1972-77. Mem. Assn. Univ. Anesthesiologists (pres. 1987-88), Am. Soc. Pharmacology and Exptl. Therapeutics. Achievements include pioneering work in demonstrating effects of anesthetics on myocardial perfusion and metabolism; cardioactive drug interactions with anesthetic drugs. Office: Med Coll Ga Dept Anesthesiology 6431 Fannin Augusta GA 30912-2700

MERIN, ROBERT LYNN, periodontist; b. L.A., Jan. 25, 1946; s. Marcus and Belle Merin; m. Barbara Rosen, June 27, 1971; children: Lori, Kimberly. DDS, UCLA, 1970; MS, Loma Linda U., 1972. Diplomate Am. Bd. Periodontology. Chief periodontal svc. Mather Air Force Hosp., Sacramento, 1972-74; pvt. practice, Woodland Hills, Calif., 1974—; chmn. dental staff Humana-West Hills (Calif.) Hosp., 1982-84; lectr. Sch. Dentistry, UCLA, 1970, 74—, dir. periodontal bd. cert. course, 1993—. Author: (with others) Glickman's Clinical Periodontics, 1978, 84, 90; contbr. articles to profl. jours. Active UCLA Dental Scholarship and Loan Com., 1984—; cons. L.A. Olympic Com., 1984. Mem. ADA, Am. Acad. Periodontics, Calif. Soc. Periodontists, San Fernando Valley Dental Soc. (mem. polit. action com. 1988), UCLA Dental Alumni Assn. (pres. 1979-80, bd. dirs. 1970—), UCLA Apollonians (pres. 1983-86). Avocations: windsurfing, sailing, magic. Office: 6342 Fallbrook Ave Ste 101 Woodland Hills CA 91367-1616

MERJAN, STANLEY, civil engineer, inventor; b. N.Y.C., Jan. 10, 1928; s. Morris and Rose (Katz) M.; m. Florence Louise Malone, June 5, 1954; children: Barbara, David, Alice. BCE, CCNY, 1948. Registered profl. engr., N.Y., Fla., Mass. Constrn. engr. N.Y.C. Bd. Water Supply, Downsville, N.Y., 1948-50, Gull Contracting Co., N.Y.C., 1950-51; structural designer Gen. Dynamics Corp., New London, Conn., 1951-53; exec. v.p. Underpinning and Found. Constructors, Inc., N.Y.C., 1955—. Pres. Port Washington Community Synagogue Brotherhood, 1978-80; mem. Port

Washington Community Chest, 1984—; chmn. United Jewish Appeal, Port Washington, 1986—. With U.S. Army, 1953-55. Recipient Sci. and Tech. award Am. ORT Fedn., N.Y.C., 1991. Mem. ASCE (life mem.), Am. Concrete Inst., N.Y. State Soc. Profl. Engrs., Deep Founds. Inst. (treas. 1992—), The Moles. Democrat. Achievements include U.S. and foreign patents for TPT piles (composite piles with pre-cast concrete bases) and pile driving appurtenances. Home: 96 Barkers Point Rd Sands Point NY 11050-1328 Office: Underpinning & Found Contrs 46-36 54th Rd Maspeth NY 11378

MERLINO, ANTHONY FRANK, orthopedic surgeon; b. Providence, Jan. 21, 1930; s. Anthony Frank and C. Mildred (Campagna) M.; m. Dolores Mary Aucello, Nov. 22, 1956; children: Christa Marianne, Paula Nicole. BS, Providence Coll., 1951; MS, U. Conn., 1952; MD, Jefferson Med. Coll., 1956. Diplomate Am. Bd. Orthopedic Surgery. Intern St. Joseph Hosp., Providence, 1956-57; resident orthopedic surgery VA Hosp., Phila., 1959-63; pvt. practice medicine specializing in orthopedic surgery, Phila., 1963-68, Providence, 1968—; attending orthopedic surgeon St. Joseph Hosp., Providence, pres. med. staff, 1974-75, trustee, 1973-76, med. staff/trustee joint conf. com. 1982; attending orthopedic surgeon Our Lady of Fatima Hosp., North Providence, R.I.; vis. orthopedic surgeon R.I. State Hosp., Howard, 1968-75; asst. orthopedic surgery Hahnemann Med. Coll., Phila., 1965-69; pediatric orthopedic surg. cons. Crippled Children's Program of R.I., 1968-86; cons. orthopedic surgeon Roger Williams Gen. Hosp., Providence, 1969-89; v.p. R.I. Orthopedic Group, Inc., Providence, 1969-83; pres., 1983—; team physician hockey and basketball teams Providence Coll., 1968-87; mem. R.I. Gov.'s Med. Malpractice Commn., 1975-77, R.I. Bd. Examiners in Chiropractic, 1977-80; mem. study commn. R.I. Med. Rev. Bd., 1977-85; mem. corp. Blue Cross/Shield R.I., 1976-87; physician-adv. R.I. Assn. Med. Assts., 1979-84; mem. R.I. Workers' Compensation Adv. Panel, 1978-88; mem. adv. bd. Cath. Social Svcs., 1981-84; police surgeon Am. Law Enforcement Officers' Assn., 1980; cons. orthopedic surgery Am. Assn. Medicolegal Cons., 1980-90; pres. Hindle Bldg. Assocs., 1983—. Contbr. articles to profl. jours. Mem. med. splty. adv. bd. Medical Malpractice Prevention, 1985-90. Capt., M.C., USAF, 1957-59. Recipient Dr. William McDonnell award Providence Coll. Alumni Assn., 1981. Fellow Am. Acad. Orthopedic Surgeons, ACS, (pres. R.I. chpt. 1982-84), Internat. Coll. Surgeons, Latin Am. Soc. Orthopedics and Traumatology; mem. AMA, Orthopaedic Rsch. and Edn. Found. (life), Am. Coll. Legal Medicine, Am. Fracture Assn., Pan-Pacific Surg. Assn., New Eng., R.I. (sec.-treas. 1978-80, v.p. 1980-82, pres. 1982-84), Ea. Orthopedic Socs., Jefferson Orthopaedic Soc., R.I. Med. Soc. (commr. profl. rels. 1976, ho. of dels. 1976-82, commr. internal affairs 1982), Providence Med. Assn., Am. Profl. Practice Assn., Am. Acad. Compensation Medicine, Am. Coll. Sports Medicine, Am. Orthopedic Soc. for Sports Medicine, Am. Med. Photography Assn., Internat. Soc. Orthopedics and Traumatology, Internat. Soc. Rsch. in Orthopedics and Trauma, Am. Soc. Law and Medicine, Thomistic Inst. Drs. Guild, R.I. Hist. Soc., Boston Orthopedic Club, Mal Brown Club, The 100 of R.I. Club. Roman Catholic. Home: 2 Countryside Dr Providence RI 02904-3419 Office: 655 Broad St Providence RI 02907-1444

MERMIN, N. DAVID, physicist, educator, essayist; b. New Haven, Mar. 30, 1935; s. John and Eva (Gordon) M.; m. Dorothy E. Milman, June 9, 1957; children—Jonathan George, Elizabeth Ruth. A.B. summa cum laude, Harvard U., 1956, A.M., 1957, Ph.D., 1961. NSF postdoctoral fellow U. Birmingham, Eng., 1961-63; postdoctoral fellow U. Calif., San Diego, 1963-64; asst. prof. physics Cornell U., Ithaca, N.Y., 1964-67; assoc. prof. physics Cornell U., 1967-72, prof. physics, 1972-90, Horace White prof. physics, 1990—, dir. Lab. Atomic and Solid State Physics, 1984-90; Loeb lectr. Harvard U., Cambridge, 1980, Emil Warburg prof. U. Bayreuth, Germany, 1981, Walker Ames prof. U. Washington, Seattle, 1984; Wunsch lectr. Technion, Haifa, Israel, 1992; Japan Soc. for Promotion of Sci. fellow Nagoya U., 1982. Author: Space and Time in Special Relativity, 1968, Solid State Physics, 1976, Boojums All the Way Through, 1990; contbr. articles to profl. jours. Sloan Found. fellow, 1966-68; Guggenheim Found. fellow, 1970-71. Fellow AAAS, Am. Acad. Arts and Scis., Am. Phys. Soc. (Julius Edgar Lilienfield prize 1989); mem. NAS. Avocation: piano. Home: 75 Hickory Rd Ithaca NY 14850 Office: Cornell U Lab Atomic and Solid State Phys Clark Hall Ithaca NY 14853

MEROLA, RAYMOND ANTHONY, engineering executive; b. Newark, Mar. 29, 1958; s. Raymond Anthony and Carolina (Allwell) M. Sr.; m. Sherri Jo Nelson, Dec. 30, 1983; children: Melinda, Andrea. BSCE, Valparaiso U., 1981. Registered profl. engr., Tex. Constrn. engr. Shell Oil Co., Indpls., 1981-83; project engr. Shell Oil Co., Houston, 1984-87, area engr., 1987-88; plant super. Shell Oil Co., Fall River, Mass., 1988-90; environ. and tech. engr. mgr. Shell Oil Co., Oakbrook, Ill., 1990—; coll. lectr. Nat. Louis U., Chgo., 1991—. High sch. coach various baseball programs, 1982—. Achievements include development of aggressive environmental cost management programs without sacrificing health, safety or regulatory compliance; pioneering concept of "commercial discipline" within environmental management field. Home: 6619 Muirwood Ct Lisle IL 60532 Office: Shell Oil Co 1415 W 22nd St Oak Brook IL 60522

MEROLLA, MICHELE EDWARD, chiropractor; b. Providence, Feb. 20, 1940; s. Joseph and Viola (Horne) M.; m. Ednamarie H.; children: Michele Edward II, Matthew Joseph, Samantha Joan, Alexandra Marie. BSc, Bryant Coll., 1961; DC, Chiropractic Inst. N.Y., 1965; LHD, Logan Chiropractic Coll., St. Louis, 1973. Owner chiropractic clinics, New Bedford, Taunton, Swansea, Seekonk, Attleboro and Westport, Mass., 1965—. Network radio talk show host Holistic Hotline. Mem. New Bedford City Coun., 1969-73, Airport Commn., 1972-75, Sch. Com., 1978-83, Recreation Commn., 1983-89; pres. New Bedford Aid Center, 1977; bd. dirs. Your Theatre Inc. Recipient Svc. award New England Chiropractic Coun., 1973. Mem. Southeastern Mass. Chiropractic Soc. (bd. dirs.), Mass. Chiropractic Soc., Am. Chiropractic Assn., N.Y. Acad. Sci., Fla. Chiropractic Soc., New Bedford Preservation Soc. (bd. dirs.). Editor: New England Jour. Chiropractic, 1965-75. Home: 62 Rear Manhattan Ave Fairhaven MA 02719 also: 3300 NE 23d Ave Lighthouse Point FL 33064 Office: 100 Bedford St New Bedford MA 02740-4896

MERRIAM, DANIEL F(RANCIS), geologist; b. Omaha, Feb. 9, 1927; s. Faye Mills and Amanda Frances (Wood) M. m. Annie Laura Young, Feb. 12, 1946; children: Beth Ann, John Francis, Anita Pauline, James Daniel, Judith Diane. BS in Geology, U. Kans., 1949, MS, 1953, PhD, 1961; MSc in Geology, Leicester U., England, 1969; DSc, Leicester U., 1975. Geologist Union Oil Co. Calif., 1949-51, 52; asst. instr. U. Kans., 1951-53, instr., 1954, rsch. assoc., 1963-71; geologist Kans. Geol. Survey, 1953-58, head divsn. basic geology, 1958-63, chief geol. rsch., 1963-71; Jessie Page Heroy prof. geology dept. geology Syracuse U., 1971-81, chmn. dept. geology, 1971-80; Endowment Assn. Disting. prof. natural scis. dept. geology Wichita State U., 1981—, chmn. dept. geology, 1981-87; vis. rsch. scientist Stanford U., 1963; dir. Internat. Field Inst. to Japan, Am. Geol. Inst., 1967; vis. prof. geology Wichita State U., 1968-70; vis. geol. scientist Am. Geol. Inst., 1969; cons. nat. gas survey Fed. Power Commn., 1972-75, 78, chmn. supply tech. adv. com., 1975-77; ad hoc panel earth resources survey NAS/NRC, 1972-73, chmn. U.S. Nat. Com. for Internat. Geol. Correlation program, 1976-79, ex-officio, 1979-80, 81-83, U.S. Nat. Com. on History of Geology, 1989—; Esso Disting. lectr. U. Sydney, Australia, 1979; mem. U.S. Nat. Commn. for UNESCO, U.S. Dept. State, 1979-85; vis. prof. Centre d'Informatique Geologique, Ecole des Mines de Paris, Fontainebleau, 1980; vis. sr. scientist Kans. Geol. Survey, 1990—. Author: The Geologic History of Kansas: Kansas Geological Survey , 1963, Computer Fundamentals for Geologists: COMPUTe, 1975, Bibliography of Computer Applications in the Earth Sciences, 1988; founder, editor-in-chief Jour. Math. Geology, 1968-76, Computers & Geosciences, 1975—; founder, editor Kansas Geological Survey, Computer Contributions, 1966-71, Syracuse University Geological Contributions, 1973-81; editor (series) Computer Applications in the Earth Sciences, 1969—, Computers and Geology, 1976— Computer Methods in the Geosciences, 1982—, (books and vols.) Mathematical Models of Sedimentary Processes, 1972, The Impact of Quantification on Geology, 1974, Random Processes in Geology, 1976, Geomathematics: Past, Present, and Prospects, 1978, Down-to-Earth Statistics: Solutions Looking for Geological Problems, 1981, Current Trends in Geomathematics, 1988, (colloquium) Geostatistics, 1970; translation editor Statistics for Geoscientists, 1987; co-editor Pacific Geology, 1971-83; editorial cons. Geosystems, 1971-83; mem. editorial rev. bd. Colo. Sch. Mines Quarterly, 1974-90; mem. editorial adv. bd. Geophysical Computer Programs, 1975-76, Applied Geochemistry, 1985-93; reviewer for nat. and internat. jours.; contbr. notes, articles to numerous jours. Bd. dirs. Kans Geol. Found., 1989-92. Fullbright-Hayes Sr. Rsch. fellow, U.K., 1964-65. Fellow AAAS (chairperson sect. E 1983-84, Sci. software adv. panel 1986—), Geol. Soc. Am. (com. on publs. 1973-76, chmn. com. geology dept. 1975-78), Geol. Soc. London (William Smith medal 1992), Sigma Xi; mem. Am. Assn. Petroleum Geologists (chmn. 1954, 57ednl. exhibits com., rsch. com., 1964-67, assoc. editor bulletin 1969-75, Geobyte 1985-92, computer applications in geology com. 1971-81, 86—, N.Y. Dist. rep. 1974-76, Kans. Dist. rep. 1956-57, 1985-91, chmn. 1989-91, Kans. rep. Midcontinent sect. 1988—, Disting. Svc. award 1987), Soc. Econ. Paleontologists and Mineralogists (chmn. organizer rsch. group in computer tech. 1970-75, 82-82, 89-90, publs. com. 1980-83, chmn. publs. 1981-82, chmn. Pa. Stratigraphy working group midcontinent sect. 1986—, ad. hoc. com. databases 1985-88, chmn databases 1986-88, organizer computer applications com. 1988—, chmn. computer applications 1988-92, spl. advisor headquarters and bus. com. 1988-91, chmn. 1991—), Nat. Assn. Geology Tchrs. (v.p. Kans.-Okla. sect. 1986, pres. 1987-89), Geoscience Info. Soc. (program com 1980-81), Internat. Assn. Math. Geology (mem. coun. 1968—, sec.-gen. 1972-76, preterim archivist 1989-92, archivist 1992—, William Christian Krumbein medal 1981), Leicester Geol. Soc. (hon. life), Sylvester-Bradley Geol. Soc. (hon. v.p. 1978-79), Classification Soc. (chmn. mem. com., bd. dirs. 1968-71), N.Y. State Geol. Assn. (exec. sec. 1972-77, pres. 1977-78, bd. dirs. 1987-83), Kans. Geol. Soc. (hon., bd. dirs., dir. many confs., presdl. citation 1989), Kans. Acad. Sci. (mem. coun. at large 1983-86, v.p. 1987, pres.-elect 1988, pres. 1989, chmn. com. for 2001, 1989-92, assoc. editor Transactions 1990-92, editor, 1992—), History of Earth Sci. Soc. (Earth Science History editorial bd. 1982—), Sigma Gamma Epsilon (pres. Alpha chpt. 1952-53, nat. coun. 1983—, nat. pres. 1990—, nat. editor The Compass, 1983-92), Phi Kappa Phi. Office: Kans Geol Survey U Kans Lawrence KS 66047

MERRIFIELD, ROBERT BRUCE, biochemist, educator; b. Ft. Worth, Tex., July 15, 1921; s. George E. and Lorene (Lucas) M.; m. Elizabeth Furlong, June 20, 1949; children: Nancy, James, Betsy, Cathy, Laurie, Sally. B.A., UCLA, 1943, Ph.D., 1949. Chemist Park Research Found., 1943-44; research asst. Med. Sch., UCLA, 1948-49; asst. Rockefeller Inst. for Med. Research, 1949-53, assoc., 1953-57; asst. prof. Rockefeller U., 1957-58, assoc. prof., 1958-66, prof., 1966—, John D. Rockefeller prof., 1984—; Developed solid phase peptide synthesis; completed (with B. Gutte) 1st total synthesis of an enzyme, 1969. Assoc. editor: Internat. Jour. Peptide and Protein Research; contbr. articles to sci. jours. Recipient Lasker award biomed. research, 1969, Gairdner award, 1970, Intra-Sci. award, 1970, Nichols medal, 1973, Alan E. Pierce award Am. Peptide Symposium, 1979, Nobel prize in chemistry, 1984, Chemical Pioneer award Am. Inst. Chemists, 1993. Mem. Am. Chem. Soc. (award creative work synthetic organic chemistry 1972, Hirschmann award in peptide chemistry 1990), NAS USA, Am. Soc. Biol. Chemists, Sigma Xi, Phi Lamda Upsilon, Alpha Chi Sigma. Office: Rockefeller U 1230 York Ave New York NY 10021

MERRILL, AUBREY JAMES, systems and electrical engineer; b. Dover-Foxcroft, Maine, Apr. 3, 1948; s. George A. and Gladys M. (Carey) M. Student, Colby Coll., 1966-68; BSEE, U. Maine, 1976, MSEE, 1978. Mem. tech. staff digital multiplex div. AT&T Bell Labs., Holmdel, N.J., 1978-86, mem. tech. staff electronic switch div., 1988-92, mem. tech. staff ops. planning div., 1992—; mem. tech. staff edn. div. AT&T Bell Labs., Middletown, N.J., 1986-88; chair adapted toys for disabled children com. Telephone Pioneers of Am., AT&T Labs., Holmdel, 1991—. With U.S. Army, 1968-71. Mem. IEEE, N.Y. Acad. Scis., Phi Kappa Phi, Tau Beta Pi, Sigma Xi (assoc.). Avocations: woodworking, metal working, golf, collecting antique tools. Home: PO Box 240 Holmdel NJ 07733-0240 Office: AT&T Bell Labs Rm 2M-609 101 Crawford Corner Rd Holmdel NJ 07733

MERRILL, EDWARD WILSON, chemical engineering educator; b. New Bedford, Mass., Aug. 31, 1923; s. Edward Clifton and Gertrude (Wilson) M.; m. Genevieve de Bidart, Aug. 19, 1948; children—Anne de Bidart, Francis de Bidart. A.B., Harvard, 1945; D.Sc., Mass. Inst. Tech., 1947. Research engr. Dewey & Almy div. W.R. Grace & Co., 1947-50; mem. faculty MIT, 1950—, prof. chem. engring., 1964—, Carbon P. Dubbs prof., 1973—; cons. in field, 1950—; cons. in biochem. engring. Harvard U. Health Services, 1982—. Author articles on polymers, rheology, med. engring. Pres. bd. trustees Buckingham Sch., Cambridge, 1969-74; trustee Browne and Nichols Sch., Cambridge, 1972-74, hon. trustee, 1974—. Fellow Am. Inst. for Med. and Biol. Engring., Am. Acad. Arts and Scis.; mem. Am. Chem. Soc., Am. Inst. Chem. Engrs., Soc. for Biomaterials. Patentee chem. and rheological instruments. Home: 90 Somerset St Belmont MA 02178-2010

MERRILL, RONALD THOMAS, geophysicist, educator; b. Detroit, Feb. 5, 1938; s. Robert Able and Freda (Havens) M.; m. Nancy Joann O'Byrne, Sept. 1, 1962; children: Craig Elliot, Scott Curtis. BS in Math., U. Mich., 1959, MS in Math., 1961; PhD in Geophysics, U. Calif., Berkeley, 1967. Asst. prof. oceanography U. Wash., Seattle, 1967-72, assoc. prof. geophysics and oceanography, 1972-77, prof. geophysics and geol. sci., 1977—, chmn. dept. geophysics, 1985-92. Author: (with M.W. McElhinny) The Earth's Magnetic Field, 1984; contbr. numerous articles to profl. jours. Recipient numerous rsch. grants from NSF, other founds. Fellow Am. Geophys. Union (pres. geomagnetism and paleomagnetism sect. 1988-90); mem. AAAS, Soc. Geomagnetism (Japan). Avocations: skiing, hiking, scuba diving, dancing. Office: U Wash Dept Geophysics AK-50 Seattle WA 98195

MERRILL, STEVEN WILLIAM, research and development executive; b. Oakland, Calif., Aug. 6, 1944; s. David Howard and Etha Nadine (Wright) M. BA in Chemistry, Calif. State U., 1987. Lic. pyrotechnic, Calif. Apprentice Borgman Sales Co., San Leandro, Calif., 1960-64; assembler Calif. Fireworks Display, Rialto, Calif., 1970; pyrotechnician Hand Chem. Industries, Milton, Ont., Can., 1972-74; dir. R&D Pyrospectaculars, Rialto, 1988—; experimenter in field, 1958—; chief chemist Baron Blakesly Solvents, Newark, Calif., 1987-88; court expert San Francisco Superior Ct., 1971, Victorville (Calif.) Superior Ct. Counselor Xanthos, Inc., Alameda, Calif., 1970. Mem. Am. Chem. Soc. Avocations: wood carving, sculpture, electronics. Home: PO Box 676 Crestline CA 92325-0676 Office: Pyrospectaculars 3196 N Locust Ave Rialto CA 92376-1414

MERRITT, DORIS HONIG, pediatrics educator; b. N.Y.C., July 16, 1923; d. Aaron and Lillian (Kunstlich) Honig; children: Kenneth Arthur, Christopher Ralph. B.A., CUNY, 1944; M.D., George Washington U., 1952. Diplomate Am. Bd. Pediatrics, Nat. Bd. Med. Examiners. Pediatric intern Duke Hosp., 1952-53; teaching and rsch. fellow pediatrics George Washington U., 1953-54; pediatric asst. resident Duke U. Hosp., 1954-55, cardiovascular fellow pediatrics, 1955-56, instr. pediatrics, dir. pediatric cardiorenal clinic, 1956-57; exec. sec. cardiovascular study sect., gen. medicine study sect. div. rsch. grants NIH, 1957-60; dir. med. rsch. grants and contracts Sch. Medicine Ind. U., 1961-62, asst. prof. pediatrics Sch. Medicine, 1961-68, asst. dean med. rsch. Sch. Medicine, 1962-65, asst. dir. med. rsch., aerospace rsch. application ctr. Sch. Medicine, 1963-65, assoc. dir. med. rsch. Sch. Medicine, 1965-68, asst. dean for rsch., office v.p. rsch. and dean advanced studies Sch. Medicine, 1965-67, dir. sponsored programs, asst. to provost Sch. Medicine, 1965-68, assoc. dean for rsch. and advanced studies, office v.p. and dean for rsch. and advanced studies Sch. Medicine, 1967-71, assoc. prof. pediatrics Sch. Medicine, 1968-73, prof. Sch. Medicine, 1973-80, prof. pediatrics, assoc. dean Sch. Medicine, 1988—; spl. asst. to dir. NIH, 1978-87, rsch. tng. and rsch. resource officer, 1980-87, acting dir. Nat. Ctr. Nursing Rsch., 1986-87; bd. dirs. Ind. Health Industry Reform; cons. USPHS, NIH div. rsch. grants Div. Health Rsch. Facilities and Resources, Nat. Heart Inst., 1963-78, Am. Heart Assn., 1963-67, Ind. Med. Assn. Commn. Vol. Health Orgns., 1964-67, Bur. Health Manpower, Health Profession's Constrn. Program, 1965-71, Nat. Library Medicine, Health Ctr. Libr. Constrn. Program, 1966-72; dir. office sponsored programs Ind. U.-Purdue U. Indpls. Office Chancellor, 1968-71, dean rsch. and sponsored programs, 1971-79; mem. Nat. Library Medicine biomed. communications rev. com., 1970-74; mem. com. to study rsch. capabilities acad. depts. ob-gyn Inst. Medicine, 1990-91. Contbr. articles to profl. jours. Chmn. Indpls. Consortium for Urban Edn., 1971-75; v.p. Greater Indpls. Progress Com., 1974-79; mem. Community Svc. Council, 1969-75; bd. dirs. Bd. for Fundamental Edn., 1973-77, Ind. Sci. Edn. Found., 1977-78, Community Addiction Svc. Agy., Inc., 1972-74; trustee Marian Coll., 1977-78; exec. com. Nat. Council U. Rsch. Adminstrs., 1977-78; bd. regents Nat. Library Medicine, 1976-80; chmn. adv. screening com. for life scis. Council Internat. Exchange of Scholars, 1978-81; bd. dirs. Community Svc. Coun. Cen. Ind., 1989—, Univ. Hosp. Consortium, Tech. Assessment Ctr., 1990-93; mem. Ind. Health Industry Forum, 1993—. Served to lt. (j.g.) USNR. Fellow Am. Acad. Pediatrics; mem. AAAS, George Washington U., Duke U. med. alumni assns., Phi Beta Kappa, Alpha Omega Alpha. Office: Ind U Sch Medicine FH312 Indianapolis IN 46202-5114

MERRITT, JAMES FRANCIS, biological sciences educator, administrator; b. Raleigh, N.C., July 21, 1944; s. Clifton and Emily (Rogers) M.; m. Sue Wall, Aug. 9, 1969; children—Ashley Grant, Bradley Gene, Carey Reid. B.S., E. Carolina U., 1966, M.S., 1968; Ph.D., N.C. State U., 1973. Asst. prof. biological scis. U. N.C., Wilmington, 1973-78, assoc. prof., chmn., 1978-89, dir. Ctr. Marine Scis. Rsch., Wilmington, 1989—. Contbr. articles to profl. jours. Chmn. PTA, Wilmington, 1983, Marine Expo Com., Wilmington, 1984—. E.G. Moss fellow N.C. State U., 1972. Mem. Am. Genetic Assn., N.C. Acad. Sci. (sect. chmn. 1975-76, vice pres. 1979-80), Chi Beta Phi (Outstanding Service award 1966). Avocations: fishing; woodworking. Home: 3523 Violet Ct Wilmington NC 28409-2541 Office: U NC at Wilmington Ctr. Marine Sci Rsch 7205 Wrightsville Ave Wilmington NC 28403

MERRITT, JOY ELLEN, chemist, editor; b. Flint, Mich., Mar. 15, 1943; d. Fred Clare and Leata Leone (Harder) Hutchins; m. Robert Edward Merritt, Mar. 17, 1973; children: Ronald Edward, Andrew Roy. BS in Chemistry, U. Mich., 1965, AMLS, 1967. Libr. trainee U.S. VA Hosp., Ann Arbor, Mich., 1966-67; asst. editor Chem. Abstracts Svc., Columbus, Ohio, 1967-73; assoc. editor Chem. Abstracts Svc., Columbus, 1973-77, sr. assoc. editor, 1977-84, sr. editor, 1984—; cons. AMA, WHO, U.S. Pharmacopeia, Nat. Formulary; mem. tech. adv. group Internat. Stds. Orgn., 1990—; com. mem. Nat. Acad. Scis. and NRC, 1967-73; mem. Nat. Formulary Adv. Panel on Chem. Data and Nonenclature, 1973-75; presenter various orgns. Contbr. articles to profl. jours. Mem. ASTM, Am. Chem. Soc., Spl. Librs. Assn., Drug Info. Assn., Internat. Soc. on Polycyclic Aromatic Compounds, Internat. Union of Pure and Applied Chemistry. Office: Chem Abstracts Svc PO Box 3012 2540 Olentangy River Rd Columbus OH 43210

MERSCH, CAROL LINDA, information systems specialist; b. Tulsa, Dec. 13, 1938; d. Forrest Delbert and Betty Clare (Kirk) Baker; 1 child, Teddy Melinda. BSBA, Okla. State U., 1960. Programmer, analyst Rockwell Internat., Tulsa, 1964-68; systems analyst Amoco Prodn. Co., Tulsa, 1968-73; dir., mgr. The Williams Cos., Tulsa, 1973-83; dir. info. svcs. Reading & Bates Corp., Tulsa, 1983-89; pres. Mersch-Bacher Assocs. (MBA) Inc., Tulsa, 1989—; chmn. info. tech. adv. bd. City of Tulsa, 1988—. Author: Systems Development Life Cycle, 1988; contbr. articles to profl. jours. Bd. dirs. United Way Venture Grant, Tulsa, 1987-88, chmn. adv. bd. Tulsa Info. Tech. 1992—. Avocations: jogging, art, writing. Office: Mersch-Bacher Assoc (MBA) Inc 810 S Cincinnati Ave Ste 105 Tulsa OK 74119-1601

MERSHIMER, ROBERT JOHN, chemical engineer; b. Youngstown, Ohio, Nov. 19, 1950; s. John A. and Evelyn M.; children: Matthew, Melinda. B-SChemE, Youngstown State U., 1973. Engr. in tng., W.Va. Technician GE, Niles, Ohio, 1970-73; engr. Firestone, Akron, Ohio, 1973-75; process engr. FMC, South Charleston, W.Va., 1975-79, USS Chemicals, Haverhill, Ohio, 1979-82; sr. process engr. South Point (Ohio) Ethanol, 1982—.

MERTA DE VELEHRAD, JAN, diving and safety engineer, scientist, psychologist, inventor, educator, civil servant; b. Stare Mesto, Czechoslovakia, Apr. 24, 1944; arrived in Can., 1968; s. Jan and Marie (Sebkova) M.; 1 child, Iveta. Diploma, Ucnovská Skola Technická, Slusovice, 1962, Coll. Social Law, Prague, 1968; BS, McGill U., Montreal, 1971; PhD in Psychology, U. Aberdeen, Scotland, 1978. Pres., pub. Jan's Pub. Co., Montreal, 1972-74; deep sea diver, diving supr. North Sea, Middle East, Africa, 1974-78; dir. R&D Wharton-Williams Ltd., Aberdeen, 1978-79, Oceaneering, Inc., Houston, 1979-81; chief inspector diving Govt. of Can., Ottawa, 1981—. Co-author: Exploring The Human Aura, 1976, Canadian Diving Regulations, 1987. Chmn. com. for survival suits Can. Gen. Standards Bd., 1983; br. chmn. Czech Assn. of Can., 1986; hon. appt. bd. Seneca Coll. Ont., 1983. Recipient Spl. Industry award Can. Assn. Diving Contrators, 1985, award for svc. to sub-sea industry, 1988, Internat. Cultural Diploma of Honour, 1989, Commemorative Medal of Honour, 1988; named Pursuivant, Spanish Coll. Arms, 1990, to Internat. Leadership Hall of Fame, 1988, Internat. Hall of Leaders, 1988;. Fellow Inst. Diagnostic Engrs., Inst. Petroleum; mem. Soc. Fire Protection Engrs., Brit. Psychol. Soc., Internat. Soc. Hyperbaric Medicine (v.p. 1990—), Undersea Med. Soc. Roman Catholic. Soc. rank: H.S.H. Prince of Armavir, Duke of Melk, Graf von Gratz, Count of Bavaria, Baron de Velehrad, Chevalier Ordre Royal de La Couronne de Boheme, Capt., Sea Eagle Legion. Achievements include 2 British patents; patents pending. Avocations: photography, exploration, literary investigation. Home: 308 MacKay St, Ottawa, ON Canada K1M 2B8 Office: Nat Energy Bd, 311 6th Ave S W, Calgary, AB Canada

MERTEN, UTZ PETER, physician; b. Cologne, Germany, July 22, 1942; s. Richard and Hildegard (Dewald) M.; m. Frania Sachs, 1973; children: Patrick Sinclair, Marc Leon. Student, Free U., Berlin, 1963-69, Oxford U., 1967-68; MD, Kiel U., Germany, 1970. Dir. Inst. Lab. Medizin, Cologne, Germany; cons. MCS-AG, DELAB, OKO-Control. Chmn. NAV-Virchowbund - LV Nordrhein, 1987—; bd. dirs., 1986—. Avocations: golf, tennis, field hockey, PPL-A. Office: Inst Lab Medizin, Stadtwaldgurtel 35, 5000 Cologne 41, Germany

MERTENS, JOSEF WILHELM, engineer; b. Aachen, Rheinland, Germany, Aug. 6, 1946; s. Joseph Peter and Maria (Schrouff) M.; m. Barbara Kuesters, Aug. 28, 1970; children: Yvonne, Birgit, Stephan, Monika. Diplomate aerospace, U. Tech., Aachen, 1972, PhD in Aerospace Engring., 1983. Asst. Lehrstuhl fuer Mechanik, U. Tech., Aachen, 1972-84; with theoretical aerodyn. Transport und Verkehrsflugzeuge Messerschmitt-Boelkow-Blohm GmbH, Bremen, Germany, 1984-86; coord. aerodyn. for high speed aircraft dept. EF1 Deutsche Airbus GmbH, Bremen, 1987—; mem. steering com. Minister for Rsch. and Tech. Study on High Speed Transport, Bonn, Germany, Munich, 1986-87, Brite-Euram: Supersonic Flow, Brussels, 1987—; tech. group mem. Supersonic Comml. Transport Internat. Study Working Group Sonic Boom, 1990—; mem. BMFT Working Group Hypersonics, Bonn, 1989—. Author, co-author publs., 1987—. Mem. AIAA, Deutsche Gesellschaft fuer Luft-und Raumfahrt. Roman Catholic. Office: Deutsche Aerospace Airbus GmbH, Huenefeldstr 1-5, D-28183 Bremen Germany

MERTENS, THOMAS ROBERT, biology educator; b. Fort Wayne, Ind., May 22, 1930; s. Herbert F. and Hulda (Burg) M.; m. Beatrice Janet Abair, Apr. 1, 1953; children—Julia Ann, David Gerhard. B.S., Ball State U., 1952; M.S., Purdue U., 1954, Ph.D., 1956. Research assoc. dept. genetics U. Wis.-Madison, 1956-57; asst. prof. biology Ball State U., Muncie, Ind., 1957-62, assoc. prof., 1962-66, prof., 1966—, dir. doctoral programs in biology, 1974—, disting. prof. biology edn., 1988—. Author: (with A.M. Winchester) Human Genetics, 1983; (with R.L. Hammersmith) Genetics Laboratory Investigations, 9th edit., 1991; contbr. numerous articles to profl. jours. Fellow NSF, 1963-64, Ind. Acad. Sci., 1969; co-recipient Gustav Ohaus award for Innovative Coll. Sci. Teaching, Nat. Sci. Tchrs. Assn., 1986; recipient Dist. Service to Sci. Edn. citation Nat. Sci. Tchrs. Assn., 1987. Fellow AAAS; mem. Nat. Assn. Biology Tchrs. (pres. 1985, hon. mem. 1988), Am. Genetic Assn., Genetics Soc. Am. Lutheran. Home: 2506 W Johnson Rd Muncie IN 47304-3066 Office: Ball State U Dept Biology Muncie IN 47306

MERTZ, EDWIN THEODORE, biochemist, emeritus educator; b. Missoula, Mont., Dec. 6, 1909; s. Gustav Henry and Louise (Sain) M.; m. Mary Ellen Ruskamp, Oct. 5, 1936; children: Martha Ellen, Edwin T.; m. Virginia T. Henry, Aug. 1, 1987. B.A., U. Mont., 1931, D.Sc. (hon.), 1979; M.S. in Biochemistry, U. Ill., 1933, Ph.D. in Biochemistry, 1935; D.Agr. (hon.), Purdue U., 1977. Research biochemist Armour & Co., Chgo., 1935-37; instr. biochemistry U. Ill., 1937-38; research assoc. pathology U. Iowa, 1938-40; instr. agrl. chemistry U. Mo., 1940-43; research chemist Hercules Powder Co., 1943-46; prof. biochemistry Purdue U., West Lafayette, Ind., 1946-76; emeritus Purdue U., 1976—; vis. prof. U. Notre Dame, South Bend, Ind., 1976-77; cons. in field. Author: Elementary Biochemistry, 1969; author, editor: Quality Protein Maize, 1992. Recipient McCoy award Purdue U., 1967; John Scott award City of Phila., 1967; Hoblitzelle Nat. award Tex. Research Found., 1968; Congressional medal Fed. Land Banks, 1968; Disting. Service award U. Mont., 1973; Browning award Am. Soc. Agronomy, 1974; Pioneer Chemist award Am. Inst. Chemists, 1976. Mem. AAAS, AAUP, Nat. Acad. Scis., Am. Soc. Biol. Chemists, Am. Inst. Nutrition (Osborne-Mendel award 1972), Am. Chem. Soc. (Spencer award 1970), Am. Assn. Cereal Chemists. Presbyterian. Co-discoverer high lysine corn, 1963. Office: Dept Agronomy Purdue U Lafayette IN 47907

MERTZ, PATRICIA MANN, dermatology educator; b. Normal, Ill., Dec. 11, 1939; d. John William and Paula Ernestine (Lester) Mann; m. John Sherman Mertz, Aug. 31, 1962; children: Kristin, Katria, Kamara. BA in Biology, U. Miami, 1961. Lab. technician U. Miami, Fla., 1961-62, lab. mgr. dept. dermatology, 1964-72, sr. rsch. assoc. dept. dermatology, 1972-80, rsch. assoc. prof. dept. dermatology, 1986—; high sch. tchr. Acad. of Assumption, Miami, 1962-63; rsch. asst. prof. dept. dermatology U. Pitts., 1980-86, rsch. assoc. prof. dept. dermatology, 1986; mem. com. instnl. rev. bd. U. Miami, 1988—; presenter workshops, lectr. in field; bd. dirs. Wound Care Info. Inst. Contbr. articles to profl. publs.; mem. editorial bd. Wounds Jour.; referee Archives of Dermatology, Wounds, Jour. Am. Acad. Dermatology, Jour. Investigative Dermatology. Mem. Am. Soc. Microbiology, Pseudomonas Soc., Mycological Soc. of Ams., Soc. for Investigative Dermatology, U.S. Fedn. for Culture Collections, Am. Acad. Dermatology, Am. Assn. for Lab. Animal Scis., Surg. Infection Soc., Bioelec. Repair and Growth Soc., Acad. Surg. Rsch., Applied Rsch. Ethics Nat. Assn. Achievements include the development of a wound healing and burn model. Office: U Miami Sch Medicine PO Box 016250 R-250 Miami FL 33101

MERTZ, WALTER, retired government research executive; b. Mainz, Germany, May 4, 1923; s. Oskar and Anne (Gabelmann) M.; m. Marianne C. Maret, Aug. 8, 1953. M.D., U. Mainz, 1951. Intern County Hosp., Hersfeld, Germany, 1952-53; resident Univ. Hosp., Frankfurt, Germany, 1953; vis. scientist NIH, Bethesda, Md., 1953-61; chief dept. biol. chemistry Walter Reed Army Inst. Research, Washington, 1961-69; mem. staff Nutrition Inst., Agrl. Research Service, Dept. Agrl., Beltsville, Md., 1969-72, chmn. inst., 1972-92; dir. Human Nutrition Research Ctr.; lectr. George Washington U. Med. Sch., 1963-73. Served with German Army, 1941-46. Recipient Osborne and Mendel award Am. Inst. Nutrition, 1971, Superior Performance award Dept. Agr., 1972, Lederle award in Human Nutrition, 1982, Internat. prize for Modern Nutrition, 1987, award for Disting. Svc., Dept. Agr., 1988. Mem. Am. Inst. Nutrition, Am. Soc. Biol. Chemists, Am. Soc. Clin. Nutrition.

MERVYN, LEONARD, biochemist; b. Liverpool, Eng., Aug. 26, 1930; s. William Henry and Esther (Tetlow) M.; m. Beryl Brown, Aug. 4, 1956; children: Gillian Lindsey, Timothy Michael, Adrian David Neil. BS, U. Liverpool, 1953, BS with honors, 1954, PhD, 1957. Pathology technician Nat. Health Svc., Liverpool, 1946-50; rsch. biochemist Glaxo Rsch., Greenford, Middy, Eng., 1957-75; tech. dir. Ferrosan Nutritional Products, Byfleet, Surrey, Eng., 1975—; lectr. Ewell Coll. Technol., 1970-76; external examiner Oxford U., 1972-78. Author: Vitamin Research, 1964-72; patentee in field; contbr. numerous articles to profl. jours. Sch. gov. various schs., Ruislip, 1968-76. Recipient A. Cressy-Morrison award N.Y. Acad. Scis., 1964, Ital-seber Gold medal U. Pavia, 1968. Fellow Royal Soc. Chemistry, Royal Soc. Health, Inst. Health Food Retailing; mem. N.Y. Acad. Scis., Brit. Soc. Nutritional Medicine, Whitefriars Club London. Anglican. Avocations: photography, bird-watching, walking, gardening, sailing. Home: 34 Roker Pk Ave, Ickenham Uxbridge UB108ED, England Office: Ferrosan Nutritional Products, York Close, Byfleet KT14 7HN, England

MERWIN, JUNE RAE, research scientist, cell biologist; b. Tonawanda, N.Y., Sept. 9, 1943; d. Raymond Francis and Vera Alice (Cornish) Brider; m. James Robert Merwin, June 3, 1967; children: Kristian, Kari. BS, So. Conn. State U., 1984, MS, 1985; PhD, MPh, Yale U., 1990. Postdoctoral scholar Yale U., New Haven, 1990-92, Bristol-Myers Squibb, Wallingford, Conn., 1990-92; rsch. scientist TargeTech, Inc., Meriden, Conn., 1992—; mem. grant study sect. NIH, 1992—; reviewer several scientific jours. Co-author 3 chpts. to books; contbr. articles to profl. jours. Pres. Luth. Ch., Madison, 1991—; vol. local sch. systems, 1980-85; active Boy Scouts Am., 1980—. With Fgn. Svc., 1964-66. Mem. Women in Cancer Rsch., Am. Soc. of Cell Biology, Yale Alumni Assn., Sigma Xi. Achievements include patent on T6F-B1/B2 A Novel Chimeric Transforming Growth Factor (U.S. and Fgn.), Patent on YEE(Ga1NAcAH)3 Ligand for Asor Receptor in Gene Therapy. Home: 19 Kenilworth Dr Clinton CT 06413 Office: TargeTech Inc 290 Pratt St Meriden CT 06450

MERZ, ANTONY WILLITS, aerospace engineer; b. Luanshya, Zambia, Nov. 25, 1932; s. Albert Russell and Virginia (Willits) H.; m. M. Margaret Milhoan, Sept. 10, 1964; children: Amy Rebecca, Alison. BS, MS in Aeronautics and Astronautics, MIT, 1956; MS in Math., Tex. Christian U., 1960; PhD in Aeronautics and Astronautics, Stanford U., 1971. Scientist Dynamics Rsch. Corp., Stoneham, Mass., 1960-63; sr. engr. Analytical Mechanics Assocs., Palo Alto, Calif., 1972-76; sr. scientist Aerophysics Rsch. Corp., Palo Alto, 1976-80; staff scientist Lockheed Missiles and Space, Palo Alto, Calif., 1980—; cons. Aerophysics Rsch. Corp., Bellevue, Wash., 1979-81, TAU Corp., Los Gatos, Calif., 1991-92; editorial bd. Jour. Dynamics and Control, 1990—. Contbr. articles to profl. jours., presented papers at profl. meetings. Recipient Fulbright scholarship U.S. Govt., Paris, 1956. Fellow AIAA (assoc.); mem. Sigma Xi. Home: 270 Willowbrook Dr Portola Valley CA 94028 Office: Orgn 92-20 Bldg 254E 3251 Hanover St Palo Alto CA 94304

MERZ, JAMES LOGAN, electrical engineering and materials educator, researcher; b. Jersey City, Apr. 14, 1936; s. Albert Joseph and Anne Elizabeth (Farrell) M.; m. Rose-Marie Weibel, June 30, 1962; children: Kathleen, James, Michael, Kimarie. BS in Physics, U. Notre Dame, 1959; postgrad., U. Göttingen, Fed. Republic Germany, 1959-60; MA, Harvard U., 1961, PhD in Applied Physics, 1967; PhD (hon.), Linghöping U., Sweden, 1993. Mem. tech. staff Bell Labs., Murray Hill, N.J., 1966-78; prof. elec. engring. U. Calif., Santa Barbara, 1978—, prof. materials, 1986—, chmn. dept. elec. and computer engring., 1982-84, assoc. dean for rsch. devel. Coll. Engring., 1984-86, acting assoc. vice chancellor, 1988, dir. semiconductor rsch. corp. core program on GaAs digital ICs, 1984-89, dir. Compound Semiconductor Rsch. Labs., 1986-92, dir. NSF Ctr. for Quantized Electronic Structures, 1989—; NATO Advanced Study Inst. lectr. Internat. Sch. Materials Sci. and Tech., Erice-Sicily, Italy, 1990; mem. exec. com. Calif. Microelectronics Innovation and Computer Rsch. Opportunities Program, 1986-92; mem. NRC com. on Japan, NAS/NAE, 1988-90; mem. internat. adv. com. Internat. Symposium on Physics of Semiconductors and Applications, Seoul, Republic of Korea, 1990, Conf. on Superlattices and Microstructures, Xi'an, China, 1992; participant, mem. coms. other profl. confs. and meetings. Contbr. numerous articles to profl. jours.; patentee in field. Fulbright fellow, Danforth Found. fellow, Woodrow Wilson Found. fellow. Fellow IEEE, Am. Phys. Soc.; mem. IEEE Lasers and Electo-Optics Soc. (program com. annual meeting 1980), Am. Vacuum Soc. (exec. com. electronic materials and processing div. 1988-89), Electrochem. Soc., Materials Rsch. Soc. (editorial bd. jour. 1984-87), Soc. for Values in Higher Edn., Inst. Electronics, Info. and Comm. Engrs. (overseas adv. com.), Sigma Xi, Eta Kappa Nu. Achievements include research in field of optoelectronic materials and devices: semiconductors and ionic materials; optical and electrical properties of implanted ions, rapid annealing; semiconductor lasers, detectors, solar cells, other optoelectronic devices; low-dimensional quantum structures. Office: U Calif 1413 Phelps Hall Santa Barbara CA 93106

MERZBACHER, EUGEN, physicist, educator; b. Berlin, Germany, Apr. 9, 1921; came to U.S., 1947, naturalized, 1953; s. Siegfried and Lilli (Wilmersdoerffer) M.; m. Ann Townsend Reid, July 11, 1952; children: Celia, Charles, Matthew, Mary. Licentiate, U. Istanbul, 1943; A.M., Harvard U., 1948, Ph.D., 1950; DSc (hon.), U. N.C., Chapel Hill, 1993. High sch. tchr. Ankara, Turkey, 1943-47; mem. Inst. Advanced Study, Princeton, N.J., 1950-51; vis. asst. prof. Duke U., 1951-52; mem. faculty U. N.C., Chapel Hill, 1952—; prof. U. N.C., 1961—, acting chmn. physics dept., 1965-67, 71-72, Kenan prof. physics, 1969-91, Kenan prof. physics emeritus, 1991—, chmn. dept., 1977-82; vis. prof. U. Washington, 1967-68, U. Edinburgh, Scotland, 1986; Arnold Bernhard vis. prof. physics, Williams Coll., 1993; vis. rsch. fellow Sci. and Engring. Rsch. Coun., U. Stirling, 1986; chmn. Internat. Conf. on the Physics of Electronic and Atomic Collisions, 1987-89. Author: Quantum Mechanics, 2d edit, 1970; also articles. NSF Sci. Faculty fellow U. Copenhagen, Denmark, 1959-60; recipient Thomas Jefferson award U. N.C., 1972; Humboldt sr. scientist award U. Frankfurt, Germany, 1976-77. Fellow Am. Phys. Soc. (pres. 1990), AAAS (sect. del. to coun., mem.-at-large sect. B); mem. Am. Assn. Physics Tchrs. (Oersted medal 1992), AAUP, Sigma Xi. Achievements include research on applications of quantum mechanics to study atoms and nuclei. Home: 1396 Halifax Rd Chapel Hill NC 27514-2724

MESCHAN, ISADORE, radiologist, educator; b. Cleve., May 30, 1914; s. Julius and Anna (Gordon) M.; m. Rachel Farrer, Sept. 3, 1943; children: David, Eleanor Jane Meschan Foy, Rosalind Weir, Joyce Meschan Lawrence. B.A., Western Res. U., 1935, M.A., 1937, M.D., 1939; Sc.D. (hon.), U. Ark. Med. Center, 1983. Instr. Western Res. U., 1946-47; prof., head dept. radiology U. Ark., Little Rock, 1947-55; prof., dir. dept. radiology Bowman Gray Sch. Medicine, Wake Forest U., Winston-Salem, N.C., 1955-77; now prof. emeritus Bowman Gray Sch. Medicine, Wake Forest U. Author: Atlas of Normal Radiographic Anatomy, 1951, Roentgen Signs in Clinical Diagnosis, 1956, (with R. Meschan) Synopsis of Roentgen Signs, 1962, Roentgen Signs in Clinical Practice, 1966, Radiographic Positioning Related Anatomy, 1969, 2d edit., 1978, Analysis of Roentgen Signs, 3 vols, 1972, Atlas of Anatomy Basic to Radiology, 1975, Synopsis of Analysis of Roentgen Signs, 1976, Synopsis of Radiographic Anatomy, 1978, 2d rev. edit., 1980, (with B.W. Wolfman) Basic Atlas of Sectional Anatomy, 2d edit.; co-author: Atlas of Cross-Sectional Anatomy, 1980, Roentgen Signs in Diagnostic Imaging, vol. 1, 1984, vol. 2, 1985, vol. 3, 1986, vol. 4, 1987; editor: The Radiologic Clinics of North America, 1965; contbr. articles to profl. jours. Recipient Disting. Alumnus award Case-Western Res. U. Sch. Medicine, 1984, Disting. Faculty Svc. Alumni award Wake Forest U. Bowman Gray Sch. Medicine, 1989. Fellow Am. Coll. Radiology (com. chmn., Gold medal 1978, Living Legends of Radiology 1986); mem. Am. Roentgen Ray Soc., AMA, Radiology Soc. N.Am., N.C. Radiol. Soc., So. Med. Assn., Soc. Nuclear Medicine, Assn. U. Radiologists, Phi Beta Kappa, Sigma Xi, Alpha Omega Alpha. Home: 305 Weatherfield Ln Kernersville NC 27284

MESCHAN, MRS. ISADORE See FARRER-MESCHAN, RACHEL

MESELSON, MATTHEW STANLEY, biochemist, educator; b. Denver, Col., May 24, 1930; s. Hymen Avram and Ann (Swedlow) M.; m. Jeanne Guillemin, 1986; children: Zoe, Amy Valor. Ph.B., U. Chgo., 1951, D.Sc. (hon.), 1975; Ph.D., Calif. Inst. Tech., 1957; Sc.D. (hon.), Oakland Coll., 1964, Columbia, 1971, Yale U., 1987, Princeton U., 1988. From research fellow to sr. research fellow Calif. Inst. Tech., 1957-60; assoc. prof. biology Harvard U., 1960—, prof. biology, 1964-76, Thomas Dudley Cabot prof. natural scis., 1976—. Recipient prize for molecular biology NAS, 1963, Eli Lilly award microbiology and immunology, 1964, Alumni medal U. Chgo., 1971; Lehman award 1975, Presidential award 1983, N.Y. Acad. Scis., 1975; Alumni Disting. Svc. award Calif. Inst. Tech., 1975; Leo Szilard award Am. Phys. Soc., 1978; MacArthur fellow, 1984-89. Fellow AAAS (Sci. Freedom and Responsibility award); mem. NAS, Inst. Medicine, Am. Acad. Arts and Scis., Fedn. Am. Scientists (chmn. 1986-88, Pub. Svc. award 1972), Coun. Fgn. Rels., Accademia Santa Chiara, Am. Philos. Soc., Royal Society (London), Académie des Sciences (Paris). Office: Harvard U Fairchild Biochem Bldg 7 Divinity Ave Cambridge MA 02138-2092

MESHII, MASAHIRO, materials science educator; b. Amagasaki, Japan, Oct. 6, 1931; came to U.S., 1956; s. Masataro and Kazuyo M.; m. Eiko Kumagai, May 21, 1959; children: Alisa, Erica. BS, Osaka (Japan) U., 1954, MS, 1956; PhD, Northwestern U., 1959. Lectr., rsch. assoc. dept. materials sci. and engring. Northwestern U., Evanston, Ill., 1959-60, asst. prof., assoc. prof., then prof., 1960-88, chmn. dept. materials sci. and engring., 1978-82, John Evans prof., 1988—; vis. scientist Nat. Rsch. Inst. Metals, Tokyo, 1970-71; NSF summer faculty rsch. participant Argonne (Ill.) Nat. Lab., 1975; guest prof. Osaka U., 1985; Acta/Scripta Metallurgica lectr., 1993—. Co-editor: Lattice Defects in Quenched Metals, 1965, Martensitic Transformation, 1978, Science of Advanced Materials, 1990; editor: Fatigue and Microstructures, 1979, Mechanical Properties of BCC Metals, 1982; contbr. 190 articles to tech. publs. and internat. jours. Recipient Founders award Midwest Soc. Electron Microscopists, 1987. Fellow ASM (Henry Marion Howe medal 1968), Japan Soc. Promotion of Sci.; mem. TMS-AIME, Japan Inst. Metals (Achievement award 1972). Home: 3051 Centennial Ln Highland Park IL 60035-1017 Office: Northwestern U Dept Materials Sci Engring Evanston IL 60208

MESIHA, MOUNIR SOBHY, industrial pharmacy educator, consultant; b. Alexandria, Egypt, Feb. 14, 1945; came to U.S., 1985; s. Madiha Sidhom, Sept. 4, 1972; 1 child, Mena. BS in Pharmacy with distinction, Pharmacy Coll., Alexandria, 1967; PhD in Pharmaceutics, Pharmacy Coll., Kharkov, Ukraine, 1977. Lic. pharmacist. Instr. Dept. of Pharms. U. Assiut, 1967-69, instr., 1969-72, sr. instr., 1972-77, asst. prof. indl. pharmacy, 1978-83, assoc. prof. indsl. pharmacy, 1983-85; prof. indsl. pharmacy U. P.R., San Juan, 1985-89, L.I. U., Bklyn., N.Y., 1990—; sci. com. mem. Industry U. Rsch. Ctr., P.R., 1988-89; pres. rsch. com. Coll. Pharmacy, San Juan, 1988-89. Contbr. articles to Jour. Pharmacy and Pharmacology, Die Pharmazie, Drug Devel. and Indsl. Pharmacy. Industry U. Rsch. Ctr. grantee, P.R., 1989. Mem. Am. Pharm. Soc., Am. Assn. Colls. Pharmacy, Am. Assn. Pharm. Scientists (chartered), Fedn. Internat. Pharmaceutique (assoc.). Christian Orthodox. Achievements include enhancement of gastrointestinal absorption of insulin in fatty acid media; research on interaction and incomplete release of drug substances incorporated with cellulose derivatives; acceleration of solubility of benzodiazepines by controlled crystallization, of mass transfer and properties of tablets during microwave drying of granules' energy and mass; optimization of extrusion and fluid bed granulation. Home: 239 Halsey Ave Jericho NY 11753-1625 Office: L I U Coll Pharmacy 75 Dekalb Ave Brooklyn NY 11201-5497

MESLOH, WARREN HENRY, civil, environmental engineer; b. Deshler, Nebr., Mar. 17, 1949; s. Herbert Frederick and Elna Florence (Petersen) M.; m. Barbara Jane Anderson, Sept. 7, 1969; children: Christopher Troy, Courtney James. BS, U. Kans., 1975; postgrad., Kans. State U., 1976-77. Registered profl. engr. Colo., Kans., Nebr. Project mgr. Wilson & Co. Engrs., Salina, Kans., 1975-80, process design dir., 1980-82; engring. dir. Taranto, Stanton & Tagge, Fort Collins, Colo., 1982-85; pres. The Engring. Co., Fort Collins, Colo., 1985—; mem. civil engring. adv. bd. Kans. U., Lawrence, 1982—. Contbg. author (book) Pumping Station Design, 1989, (water pollution control manual) Manual of Practice No. OM-2, 1991; contbr. articles to profl. jours. Cub master Boy Scouts Am., Salina, 1980-81; active Luth. Ch., 1982—; vol. Paralyzed Vets. Orgn., Fort Collins, 1985—; pres. Foothills Green Pool Assn., Fort Collins, 1987-88. Sgt. U.S. Army, 1971-73, Germany. Named Outstanding Engr.-In-Tng. NSPE, 1978. Mem. Am. Pub. Works Assn., Am. Water Works Assn., Water Pollution Control Fedn., Fort Collins Country Club. Republican. Avocations: golf, boating, snow skiing. Office: The Engring Co 2310 E Prospect Rd Fort Collins CO 80524

MESSENGER, GEORGE CLEMENT, engineering consultant; b. Bellows Falls, Vt., July 20, 1930; s. Clement George and Ethel Mildred (Farrar) M.; m. Priscilla Betty Norris, June 19, 1954; children: Michael Todd, Steven Barry, Bonnie Lynn. BS in Physics, Worcester Poly. U., 1951; MSEE, U. Pa., 1957; PhD in Engring., Calif. Coast U., 1986. Rsch. scientist Philco Corp., Phila., 1951-59; engring. mgr. Hughes Semicondr., Newport Beach, Calif., 1959-61; div. mgr. Transitron Corp., Wakefield, Mass., 1961-63; staff scientist Northrop Corp., Hawthorne, Calif., 1963-68; cons. engr., Las Vegas, Nev., 1968—; lectr. UCLA, 1969-75; v.p., dir. Am. Inst. Fin., Grafton, Mass., 1970-78; gen. ptnr. Dargon Fund, Anaheim, Calif., 1983—; v.p., tech. dir. Messenger and Assoc., 1987—, registered investment adviser, 1989—. Co-author: The Effects of Radiation on Electronic Systems, 1986; contbg. author: Fundamentals of Nuclear Hardening, 1972; contbr. numerous articles to tech. jours.; patentee microwave diode, hardened semicondrs. Recipient Naval Rsch. Lab. Alan Berman award, 1982; Best Paper award HEART Conf., 1983, Spl. Merit award HEART Conf., 1983; fellow IEEE, 1976, annual merit award 1986, Pete Haas award. HEART Conf., 1992. Mem. Rsch. Soc. Am., Am. Phys. Soc. Congregationalist. Home and Office: 3111 Bel Air Dr Apt 7F Las Vegas NV 89109-1510

MESSICS, MARK CRAIG, civil engineer; b. Allentown, Pa., June 8, 1960. BSCE, Lehigh U., 1982, MBA, 1987. Profl. engr., Pa. Civil engr. Pa. Dept. Transp., St. Davids, 1982-89; sr. project mgr. Waste Mgmt., Bensalem, Pa., 1989—. Mem. ASCE, NSPE. Libertarian. Home: 1117 Grove Dr Orefield PA 18069 Office: Pottstown Landfill 1425 Sell Rd Pottstown PA 19464

MESSIHA, FATHY S, pharmacologist, toxicologist, educator; b. Cairo, Egypt, Feb. 10, 1936. PhD in Physiological Biochemistry, U. Bern, Switzerland, 1965. Prof. pathology and psychiatry, dir. div. toxicology Tex. Tech. U. Health Scis. Ctr., 1980-87; prof. pharmacology and toxicology Sch. Medicine U. N.D., Grand Forks, 1987—. Recipient Internat. Görlich award for rsch. on alcoholism and substance abuse, 1990. Mem. Am. Acad. Clin. Toxicology, Acad. Pharm. Sci., Am. Soc. Pharmacology & Exp. Therapeutics. Achievements include rsch. in neuropharmacology, neurotoxicology, drug metabolism. Office: U ND Sch Med/Pharmacology and Toxicology PO Box 9037 Grand Forks ND 58202-9037

MESSING, CHARLES GARRETT, zoologist; b. Bronx, N.Y., July 5, 1948; s. Irving and Dollie (Friedman) M. BA in Biol. Sci., Rutgers U., 1970; MS in Biol. Oceanography, U. Miami, Coral Gables, Fla., 1975, PhD, 1979. Lectr. in marine sci. U. Miami, 1980-84; cons. Miami, Fla., 1985-88; rsch. assoc. Nova U. Oceanography Ctr., Dania, Fla., 1987-90, asst. prof., 1990-93, assoc. prof., 1993—; adj. asst. prof. dept. geology U. Miami, 1987-91. Contbr. articles to Bull. Marine Sci., Marine Biology, Jour. Crustacean Biology, Proceedings Biology Soc. Wash., Palaios. Postdoctoral fellow Smithsonian Instn., 1979, rsch. fellow Christensen Rsch. Found., 1991. Mem. Geol. Soc. Am., Biol. Soc. Wash., Sigma Xi. Office: Nova Oceanographic Ctr 8000 N Ocean Dr Dania FL 33004

MESSING, JOACHIM WILHELM, molecular biology educator; b. Duisburg, Germany, Sept. 10, 1946; came to U.S., 1978; s. Heinrich and Martha (Pfeifer) M; m. Rita C. Stremmer, Sept. 25, 1975; 1 child, Simon. MS, Free U., Berlin, 1971; Dr.Rer.Nat., LM U., Munich, 1975. Rsch. assoc. U. Calif., Davis, 1978-80; asst. prof. biochemistry U. Minn., St. Paul, 1980-82, assoc. prof., 1982-84, prof., 1984-85; univ. prof. molecular biology Waksman Inst. Rutgers U., Piscataway, N.J., 1985—, dir. rsch. Waksman Inst., 1985-88, dir. Waksman Inst., 1988—; mem. sci. adv. bd. Am. Cyanamid Co., Princeton, N.J., 1985-89, Metrigen, Inc., Piscataway, 1988-90; bd. dirs. Pharmacia PL Biochems., Inc., Milw. Contbr. articles to profl. jours. Trustee Rutgers Preparatory Sch., Somerset, N.J., 1992—. Ranked first among sci. leaders of the decade as most cited U.S. scientist under 45 yrs. old The Scientist, 1990. Mem. AAAS, Am. Soc. Biol. Chemistry, Am. Soc. Microbiology, Internat. Soc. Plant Molecular Biology. Achievements include development of DNA purification techniques using genetic rather than physical methods; analysis of gene structure and organization, transposition and somatic recombination on the DNA sequence level in plants; discovery of high methionine maize; research on post-transcriptional regulation of gene expression and genomic imprinting. Office: Rutgers U Waksman Inst PO Box 759 Piscataway NJ 08855-0759

MESTER, ULRICH, ophthalmologist; b. Troisdorf, Fed. Republic of Germany, Aug. 14, 1944. MD, U. Bonn, Fed. Republic of Germany, 1972; Habilitation, Bonn, Germany, 1978. Resident dept. ophthalmology U. Bonn, Bonn, Fed. Republic of Germany, 1972-75, sr. resident, 1975-78, Habilitation: head dept. retinal disorders, 1978; prof. Univ. Eye Clinic, Bonn, 1983; head dept. ophthalmology Bundesknappschafts Hosp., Sulzbach, Germany, 1985—, med. dir., 1989—. Contbr. over 90 articles to internat. ophthal. jours. Mem. numerous nat. and internat. ophthal. assns., including Am. Soc. Cataract and Refractive Surgery, Club Gonin, Rotary. Office: Bundesknappschaft's Hosp, Ander Keinik 1, 66280 Sulzbach Germany also: An der Klinik 10, 66280 Sulzbach Germany

MESTRE, S(OLANA) DANIEL, economics consultant; b. Barcelona, Catalonia, Spain, June 14, 1937; s. Lluis and Joaquima (Solana) M.; married, June 6, 1969; children: Oriol, Laia. Degree, Escuela de Altos Estudios Mercantiles, Barcelona, 1955; Programa Alta Direccion de Empresas, Instituto Estudios Superiores da la Empresa, Barcelona, 1972. V.p. Banco Condal, Barcelona, 1969-75, Banco de Madrid, Barcelona, 1975-77; pvt. practice mgmt. cons. Barcelona, 1977-81, Andorra la Vella, Andorra, 1981—; pres. Atlantis Assegurances, Andorra la Vella, 1982—. Author: Tax Planning Andorra, 1989. Sec. presidence Govt. of Andorra, 1985. Mem. Spanish Fin. Law Assn., Econ. Bus. Assn., Catalonia Auditors Assn., Horwath Internat. Office: Horwath Andorra SA, Bonaventura Armengol 15, Andorra la Vella Andorra

MESTRIL, RUBEN, biochemist, researcher; b. N.Y.C., Jan. 21, 1951; s. Fernando and Renee (Casanova) M.; m. Ilona Erika Brelewski, Dec. 16, 1984; 1 child, Sebastian. BA in Chemistry summa cum laude, St. Thomas U., 1981; PhD in Biochemistry, U. Miami, Coral Gables, Fla., 1986. Postdoctoral fellow German Cancer Rsch. Ctr., Heidelberg, 1986-88; asst. rsch. biochemist U. Calif., San Diego, 1988-92, asst. adj. prof., 1992—. Reviewer Circulation jour., San Diego, 1991—; contbr. revs., articles to profl. jours., chpts. to books. Grantee NSF, 1980, Am. Heart Assn., 1991. Mem. AAAS, Am. Inst. Chemists, Am. Soc. Biochemistry and Molecular Biology, Am. Heart Assn. (basic sci. coun. 1991—). Democrat. Achievements include research in heat shock and adaptive response to ischemia, regulation of heat shock genes in Drosophila, steroid hormone regulation. Office: U Calif San Diego Med Ctr 8412 200 W Arbor Dr San Diego CA 92103-8412

MÉSZÁROS, ERNÖ, meteorologist, researcher, science administrator; b. Budapest, Hungary, Apr. 12, 1935; s. Lajos and Julianna (Gréci) M.; m. Ágnes Nagy, Aug. 5, 1957 (dec. June 1986); 1 child, Lörinc. Degree in meteorology, L. Eötvös U., Budapest, 1957, Dr. rer. nat., 1961; D Earth Scis., Acad. Scis., Budapest, 1970; D H.C., U. Bretagne Occidente, Brest, France, 1983. Rsch. scientist Aerol. Observatory of Nat. Meteorol. Svc., Budapest, 1957-64, head div., 1964-71; dep. dir. Inst. Atmospheric Physics, Budapest, 1971-76, dir., 1976—; pres. geol. and mining scis. Magyar Tudományos Akadémia, Budapest; prin. Tng. Ctr. World Meterol. Orgn. on Air Pollution Monitoring, Budapest, 1978. Author: Atmospheric Chemistry, 1981, (manual) Air Pollution Monitoring, 1985; author, editor: Physical Meteorology, 1982; contbrg. author Environmental Warfare, 1984. Recipient Silver medal Order of Labor Presdl. Coun. Hungary, 1976, Gold medal Order of Labor, 1982, ProNatura award nat. Environ. Agy., 1987. Mem. Hungarian Acad. Scis. (chmn. meteorol. com. 1978—, v.p. dept. earth scis. 1985—, Acad. prize 1979), Meteorol. Soc. Hungary (v.p. sci. bd. 1980—, medallion 1975), European Assn. for Sci. of Air Pollution (v.p. 1985). Avocations: French lit., jazz. Home: 3 Peterhalmi, 1181 Budapest Hungary Office: Magyar Tudományos Akad, Roosevelt-Tér 9, 1051 Budapest Hungary*

METALA, MICHAEL JOSEPH, mechanical engineer; b. Homer City, Pa., Oct. 27, 1946; s. Michael D. and Ruth (Lydic) M.; m. Linda Ann Corbelli, May 17, 1969; children: Felecia, Ronna, Elizabeth. BSME, U. Pitts., 1979. Engr. steam turbine dept. Westinghouse Electric Co., Phila., 1979-82; sr. engr. combustion turbine dept. Westinghouse Electric Co., Concordville, Pa., 1982-84; fellow engr. R & D Ctr., Westinghouse Electric Co., Pitts., 1984-90; mgr. Power Generation Svc. Divsn. Westinghouse Electric Co., Pitts., Pa., 1990—. Contbr. articles to profl. publs. Fellow Am. Soc. for Nondestructive Testing (nat. cert. bd. 1989-92, chmn. Pitts. sect. 1990-91). Achievements include patents for apparatus and method for providing a combined ultrasonic and eddy current inspection of a tube, nondestructive testing for creep damage of a ferromagnetic work piece. Office: Westinghouse/PGSD 1061 Main St North Huntingdon PA 15642

METCALF, VIRGIL ALONZO, economics educator; b. Branch, Ark., Jan. 4, 1936; s. Wallace Lance and Luella J. (Yancey) M.; m. Janice Ann Maples, July 2, 1958; children: Deborah Ann, Robert Alan. BS in Gen. Agr., U. Ark., 1958, MS in Agrl. Econs., 1960; Diploma in Econs., U. Copenhagen, 1960; PhD in Agrl. Econs., U. Mo., 1964. Asst. prof. U. Mo., Columbia, 1964-65, asst. to chancellor, 1964-69, assoc. prof., 1965-69, prof., exec. asst. to the chancellor, 1969-71; prof. econs., v.p. administrn. Ariz. State U.,

Tempe, 1971-81, prof. Sch. Agribus. and Natural Resources, 1981-88, prof. internat. bus. Coll. of Bus., 1988—; asst. to the chancellor U. Mo., 1964-69, coord. internat. programs and studies, 1965-69, mem. budget com., 1965-71, chmn., co-chmn. several task forces; cons. Ford Found., Bogota, Colombia, 1966-67; mem. negotiating team U.S. Agy. for Internat. Devel., Mauritania, 1982, cons., Cameroon, 1983, agrl. rsch. specialist, India, 1984, agribus. cons., Guatemala, 1987, 88, asst. dir. Reform Coops. Credit Project, El Salvador, 1987-90; co-dir. USIA univ. linkage grant Cath. U., Bolivia, 1984-89; cons. World Vision Internat., Mozambique, 1989. Contbr. numerous articles to profl. jours. Mem. City of Tempe U. Hayden Butte Project Area Com., 1979; bd. commrs. Columbia Redevel. Authority; mem. workable project com. City of Columbia Housing Authority. Econs. officer USAR, 1963, econ. analyst, 1964-66. Fulbright grantee U. Copenhagen, 1959-60, U. Kiril Metodij, Yugoslavia, 1973. Mem. Am. Assn. Agrl. Economists, Soc. for Internat. Devel., Samaritans (chmn. 1976, bd. dirs. 1976, mem. task force of health svc. bd. trustees 1974, health svc. 1974-78, chmn. program subcom. 1975), Kiwanis, Blue Key, Gamma Sigma Delta, Alpha Zeta, Alpha Tau Alpha. Democrat. Home: 8415 S Kachina Dr Tempe AZ 85284-2517 Office: Ariz State U Tempe AZ 85287-3706

METDEPENNINGHEN, CARLOS MAURITS W., radiologist; b. Ghent, Belgium, May 13, 1935; s. Cesar Henri J. and Alice Maria (Bracke) M.; m. Claudine Agnes A. Vanderheyde, July 30, 1960; children: Patrick, Catheline, Annick, Chantal, Erwin. MD, State U., Ghent, 1961; postgrad., Inst. Tropical Medicine, Antwerp, Belgium, 1962; student in radiology, State U., 1963-66. Ged. officer Army Med. Svc., Belgium, 1954-75; army surgeon Army Med. Svc., Werl, Germany, 1962-63; radiologist Belgian Mil. Hosp., Cologne, Germany, 1966-68; idem radiologist Mil. Hosp., Antwerp, 1968-75, St. Camillus Clinic, Antwerp, 1969—; sec. Med. Bd. St. Camillus Clinic, 1978-88 (pres. 1988-89); pres. Med. Bd. Fusion-Hospital St. Augustinus-St. Bavo-St. Camillus, 1989—; mem. admission bd. for radiologists Ministry of Health, 1990—. Lt. Col. Army Med. Svc., 1954-75, Belgium. Named officer of the Crownorder, 1980, officer of the Leopoldorder, 1983; commdr. order Leopold II, 1993. Mem. Royal Belgian Radiological Soc., Belgian Profl. Union Radiology. Avocation: genealogy. Home: Magdalena Vermeeschlaan 19, 2540 Hove Belgium Office: St Camillus Clinic, Lockaertstraat 10, 2018 Antwerp Belgium

METEVIER, CHRISTOPHER JOHN, electrical engineer; b. New Haven, May 17, 1965; s. Ronald Arthur and Anne Marie (Banville) M. BSE, U. Ctrl. Fla., Orlando, 1988. Electronics engr. Naval Sea Combat Systems Engring. Sta., Norfolk, Va., 1989-90; mgmt. engr. Naval Sea Systems Command, Washington, 1990; project engr. Naval Tng. Systems Ctr., Orlando, 1990—. Mem. Fla. Mu Alumni Assn. (bd. dirs. 1991—), Sigma Xi (v.p. 1992—). Office: Naval Tng Systems Ctr Code 242 12350 Research Pky Orlando FL 32826

METIU, HORIA ION, chemistry educator; b. Cluj, Romania, Mar. 7, 1940; s. Ion and Erna (Weisser) M.; m. Jane Farrell, Oct. 8, 1971; children: Michael, Ion. BSChemE, Politechnic Inst., Bucharest, 1961; PhD in Theoretical Chemistry, MIT, 1974. Postdoctoral fellow MIT, Cambridge, Mass., 1974-75, U. Chgo., 1975-76; prof. U. Calif., Santa Barbara, 1976—; mem. exec. com. phys. chemistry div. Am. Chem. Soc., 1992. Mem. editorial bd. Jour. Phys. Chemistry, Jour. of Chem. Physics; contbr. over 200 articles to profl. jours. Named Sloan fellow, 1978, Dreyfus Tchr.-scholar, 1978, Solid State Chemistry Exxon fellow, 1979, Fellow of Japan Soc. for the Promotion of Sci., 1992. Fellow Am. Phys. Soc.; mem. Am. Chem. Soc. Achievements include devel. of Metastable Quenching Spectroscopy; devel. of theory of rate constants for chem. reactions; theory of photodissociation with short pulses; theory of crystal surface growth; theory of surface enhanced spectroscopy. Home: 1482 Crestline Dr Santa Barbara CA 93105 Office: U Calif Santa Barbara CA 93106

METLAY, MICHAEL PETER, nuclear physicist; b. Albany, N.Y., Mar. 6, 1962; s. Max and Katharine (Titof) M.; m. Suzanne Gene Traub, May 20, 1990. AB, Oberlin Coll., 1983; PhD, U. Pitts., 1992. Postdoctoral researcher U. Pitts., 1992-93, Fla. State U., 1993—. Mem. Am. Phys. Soc.

METSGER, ROBERT WILLIAM, geologist; b. N.Y.C., Apr. 27, 1920; s. William Martin and Mabel (Herbst) M.; m. Sylvia May Haff, Mar. 29, 1947 (dec. 1961); children: Mary Gwendolen Chong, Deborah Anne Warren; m. Barbara Lawrence Holbert, Aug. 2, 1963; 1 child, Robert Lawrence. AB in Geology, Columbia U., 1948. Cert. profl. geologist. Mine geologist N.J. Zinc Co., Franklin, 1949-54; resident geologist N.J. Zinc Co., Ogdensburg, 1954-64, northeast regional geologist, 1964-81; chief geologist N.J. Zinc Co., 1981-88; consulting geologist Horsehead Resource Devel. Co., Palmerton, Pa., 1988—; assoc. Ogdensburg Seismic Obs./Lamont-Doherty Geol. Obs., 1958-81; mem. adv. bd. N.J. Geol. Survey, Trenton, 1982-90. Contbr. articles and papers to profl. jours. Active Episcopal Diocese of Newark, past pres. western dist.; sr. warden Christ Ch., Newton, 1978-86. Lt. comdr. USNR, 1940-45. Fellow Geol. Soc. Am. (sr., chmn. N.E. sect. 1981-82, com. on geology and pub. policy 1981-82), Soc. Econ. Geologists (sr.); mem. AAAS, VFW, Am. Inst. Profl. Geologists, U.S. Naval Inst., Ret. Officers Assn. (life), Geol. Assn. N.J., Rotary. Republican. Episcopalian. Achievements include research in precambrian ore deposits, hydrogeology of karst areas.

METTINGER, KARL LENNART, neurologist; b. Helsingborg, Sweden, Nov. 1, 1943; came to the U.S., 1989; s. Nils Allan and Anna Katarina (Hallberg) M.; m. Chesne Maree Ryman, Jan. 27, 1979. MD, U. Lund, 1973; PhD, Karolinska Inst., 1982. Intern Stockholm Hosps., 1973-74; resident Karolinska Hosp., Stockholm, 1974-77, clin. neurologist, 1977-85; med. dir. Kabi Hematology, Stockholm, 1985-87; dep. gen. mgr. Kabi Cardiovascular, Stockholm, 1987-89; med. dir. Ivax/Baker Norton Pharms., Miami, Fla., 1989-93, sr. clin. rsch. dir., 1993—; assoc. prof. Karolinska Inst., Stockholm, 1983-91; cons. neurologist Odenplan Med. Ctr., Stockholm, 1984-89. Author: Cerebral Thromboembolism, 1982, Refaat-Myths and Billions in Biotech, 1987; editor: Coronary Thrombolysis: Current Answers to Critical Questions, 1988, Controversies in Coronary Thrombolysis, 1989. V.p. Fla. Festival Philharmonic, Fort Lauderdale, 1991—. Lt. Swedish Army, 1979. Recipient Silver award Spanish Health Ministry, 1989, Classical Langs. award King Gustav V Found., 1963. Mem. Swedish Stroke Soc. (bd. dirs. 1979-89, pres. 1984-86), Swedish Med. Soc., Swedish Christian Med. Soc. (bd. dirs. 1972-88, pres. 1983-88), Am. Heart Assn., N.Y. Acad. Scis., Nat. Found. for Advancement of Arts, Internat. Assn. Christian Physicians (exec. com. 1975-86). Achievements include 5 patents pending. Home: 1244 Sorolla Ave Coral Gables FL 30134 Office: IVAX 8800 NW 36 St Miami FL 30134

METTLER, RUBEN FREDERICK, former electronics and engineering company executive; b. Shafter, Calif., Feb. 23, 1924; s. Henry Frederick and Lydia M.; m. Donna Jean Smith, May 1, 1955; children: Matthew Frederick, Daniel Frederick. Student, Stanford U., 1941; BSEE, Calif. Inst. Tech., 1944, MS, 1947, PhD in Elec. and Aero. Engring, 1949; LHD (hon.), Baldwin-Wallace Coll., 1980; LLD, John Carroll U., 1986. Registered profl. engr., Calif. Assoc. div. dir. systems research and devel. Hughes Aircraft Co., 1949-54; spl. cons. to asst. sec. def. U.S. Dept. Def., 1954-55; asst. gen. mgr. guided missile research div., tech. supr. Atlas, Titan, Thor and Minuteman programs Ramo-Wooldridge Corp., 1955-58; exec. v.p., then pres. TRW Space Tech. Labs. (merger Thompson Products and Ramo-Wooldridge), 1958-65; pres. TRW Systems Group, 1965-68; asst. pres. TRW Inc., 1968-69, pres., chief operating officer, 1969-77, chmn. bd., chief exec. officer, 1977-88, also bd. dirs.; bd. dirs. Bank Am. Corp., Merck & Co., Japan Soc. Inc.; mem. Pres. Reagan's Commn. Exec. Exchange, Adv. Council on Japan-U.S. Econ. Rels., Pres.'s Blue Ribbon Def. Panel, 1969-70; vice chmn. Def. Industry Adv. Council, 1964-70, chmn. Pres.'s Task Force on Sci. Policy, 1969-70. Author: reports on airborne electronic systems; patentee interceptor fire control systems. Nat. campaign chmn. United Negro Coll. Fund, 1980-81; chmn. Nat. Alliance Bus., 1978-79; co-chmn. 1980 UN Day, Washington; chmn. bd. trustees Calif. Inst. Tech.; trustee Com. Econ. Devel., Cleve. Clinic Found.; bd. dirs. Nat. Action Council for Minorities in Engring. Served with USNR, 1942-46. Named one of Outstanding Young Men of Am., U.S. Jr. C. of C., 1955, So. Calif.'s Engr. of Year, 1964; recipient Meritorious Civilian Service award Dept. Def., 1969, Nat. Human Relations award NCCJ, 1979, Excellence in Mgmt. award Industry Week Mag., 1979, Disting. Service award Calif. Inst. Tech., 1966, Automotive Hall of Fame Leadership Award, Nat. Medal Honor (Electric Industries Assn.), Nat. Engring. award Am. Assn. Engring. Socs., 1993. Fellow IEEE, AIAA; mem. Sci. Research Soc. Am., Bus. Roundtable (chmn. 1982-84), Conf. Bd. (trustee 1982—), Bus. Council (vice chmn. 1981-82, chmn. 1986-87), Nat. Acad. Engring. (Arthur M. Bueche award 1992), The Japan Soc. (bd. dirs.), Sigma Xi, Eta Kappa Nu (Nation's Outstanding Young Elec. Engr. 1954), Tau Beta Pi, Theta Xi. Office: 11150 Santa Monica Blvd Los Angeles CA 90025-3314

METZ, ALAN, psychiatrist; b. Johannesburg, South Africa, Sept. 19, 1954; m. Alice McCall Smith, Mar. 29, 1985; 1 child, Deborah. BSc in Physiol. Chemistry, U. Witwatersrand, Johannesburg, South Africa, 1975, M.B.B.Ch., 1978. Diplomate Am. Bd. Psychiatry and Neurology; lic. physician N.C., Mass., Israel, N.Z., Eng., South Africa. Surg. intern Chaim Sheba Med. Ctr., Tel Hashomer Hosp., Tel Aviv, 1979-80; med. intern Royal Free Hosp., London, 1980; psychiat. registrar Oxford Rotational Tng. Scheme, Oxford, Eng., 1980-84; sr. registrar in gen. cons. and liaison psychiatry John Radcliffe and Warneford Hosps., Oxford, Eng., 1984-85; from psychopharmacology assoc. to dir. adult inpatient psycniatry svc. New Eng. Med. Ctr., Boston, 1985-88; assoc. dir. to dir. TMCA Psychopharmacology Assocs., Boston, 1987-88, 88-92; attending, cons./liaison psychiatry div. New Eng. Med. Ctr., Boston, 1990-92, dir. clin. psychopharmacology, 1990-92; asst. dir. CNS Clin. Rsch. Group, Glaxo, Inc., Research Triangle Park, N.C., 1992, assoc. dir., 1992—; vis. sr. lectr., cons. psychiatrist Christchurch Sch. Medicine, U. Otago, N.Z., 1990; cons. psychopharmacologist Quincy (Mass.) Mental Health Ctr., 1988-92; fellow in clin. and rsch. psychopharmacology Oxford U., Eng., 1984-85; psychopharmacology fellow Tufts U., Boston, 1985-86; sr. lectr. psychol. medicine Christchurch Sch. Medicine, 1990; asst. prof. psychiatry Tufts U., Boston, 1986-92; clin. assoc. prof. U. N.C. Med. Sch., Chapel Hill. Reviewer Jour. Clin. Psychopharmacology, Jour. of Clin. Psychiatry, Jour. Women's Health, Jour. Nervous and Mental Disease, Psychosomatics, Am. Psychiat. Press, Inc.; contbr. articles to profl. jours. Mem. AAAS, Am. Psychiat. Assn., Nat. Alliance for Mentally Ill, N.Y. Acad. Sci., N.C. Psychiat. Assn., Royal Coll. Psychiatrists, N.Z. Med. Soc. Home: 1100 Mt Carmel Church Rd Chapel Hill NC 27514 Office: Glaxo Inc Rsch Inst 5 Moore Dr Research Triangle Park NC 27709

METZ, DONALD JOSEPH, scientist; b. Bklyn., May 18, 1924; s. Emil Arthur and Madeline Margaret (Maas) M.; m. Dorothy Gorman, Aug. 30, 1947. BS, St. Francis Coll., Bklyn., 1947; DSc (hon.), St. Francis Coll., 1984; MS, N.Y. Poly. U., 1949, PhD, 1955. From lectr. to prof. St. Francis Coll., 1947-76; from assoc. to sr. scientist sci. edn. Brookhaven Nat. Lab., Upton, N.Y., 1954-93, ret., 1993; ednl. cons. Brrokhaven Nat. Lab., Upton, N.Y., 1993—. With U.S. Army, 1943-46, ETO. Roman Catholic. Home: 147 Southern Blvd East Patchogue NY 11772

METZGER, DARRYL EUGENE, mechanical and aerospace engineering educator; b. Salinas, Calif., July 11, 1937; s. August and Ruth H. (Anderson) M.; m. Dorothy Marie Castro, Dec. 16, 1956; children: Catherine Ann, Kim Marie, Lauri Marie, John David. BS in Mech. Engring., Stanford U., 1959, MS, 1960, PhD, 1963. Registered profl. engr., Ariz. Asst. prof. mech. engring Ariz. State U., Tempe, 1963-67, assoc. prof., 1967-70, prof., 1970-92, Regents' prof., 1992—, prof., chmn. dept., 1974-88, dir. thermosci. research, 1980-88; cons. Pratt & Whitney Aircraft, East Hartford, Conn., 1977—, Pratt & Whitney Aircraft Can., 1979—, United Techs. Corp., 1989—, Garrett Turbine Engine Corp., Phoenix, 1966-77, NASA Lewis Research Ctr., NASA Office of Aeronautics and Space Tech., USAF Aeropropulsion Lab., Worthington Turbine Internat., Solar Turbine Internat., Allied Chem. Corp., Office of Naval Research, Sundstrand Aviation, AT&T, Bell Labs., Calspan Advanced Tech. Ctr., Rocketdyne div. Rockwell Internat., Ishikawajima-Harima Heavy Industries Co., Ltd., Tokyo, United Tech. Corp.; keynote address NATO Adv. Group for Aerospace Research and Develop., Norway, 1985; U.S. del. U.S./China Binat. Workshop on Heat Transfer, Beijing, Xian, Shanghai, 1983, NSF U.S./China Program Dev. Meeting, Hawaii, 1983, NSF/Consiglio Nazionale delle Ricerche Italy Joint Workshop on Heat Transfer and Combustion, Pisa, Italy, 1982; gen. chmn. Symposium on Heat Transfer in Rotating Machinery, Internat. Centre for Heat and Mass Transfer, Yugoslavia, 1982, 92; mem. U.S. sci. com. Internat. Heat Transfer Conf., 1986; mem. NASA Space Shuttle Main Engine Rev. Team, 1986-87, NASA Space Engring. Program Externat. Task Team, 1987; chair prof. Office of Naval Tech. U.S. Naval Postgrad. Sch., Montery, Calif., 1989. Contbr. articles to profl. jours.; editor: Regenerative and Recuperative Heat Exchangers, 1981, Fundamental Heat Transfer Research, 1980, Heat and Mass Transfer in Rotating Machinery, 1983, Heat Transfer in Gas Turbine Engines, 1987, Compact Heat Exchangers, 1989, A Festschrift for A.L. London, 1990; mem. editorial bd. Internat. Jour. Exptl. Heat Transfer, 1987—. Ford Found. fellow, 1960, NSF fellow, 1981, ASEE/NASA fellow, 1964-65; recipient Alexander von Humboldt sr. rsch. scientist award Fed. Republic of Germany, 1985, 86, 87, Achievement award ASME, Japan Soc. Mech. Engrs., 1985, Faculty Achievement award Ariz. State U. Alumni Assn., 1987, Grad. Coll. Disting. Rsch. award Ark. State U., 1991; Sonderforschungsbereich grantee U. Karlsruhe, 1988, 89, 90. Fellow ASME (mem. gas turbine com., chmn. heat transfer div. 1982-84, mem. com. on faculty quality 1986), AIAA (assoc. fellow); mem. Soaring Soc. Am., Fed. Aero. Inst. (Internat. Diamond award), Phi Beta Kappa, Sigma Xi, Tau Beta Pi, Pi Tau Sigma, Phi Kappa Phi. Home: 8601 N 49th St Paradise Vly AZ 85253-2023 Office: Ariz State U Mech and Aerospace Engring Dept Tempe AZ 85287

METZGER, GERSHON, chemist, patent attorney; b. N.Y.C., June 25, 1935; s. Isidore and Jennie (Feigon) M.; m. Miriam Sylvia Resnikoff, Aug. 19, 1959; children: Tova, Shlomo, Rivka, Y'hoshua, Michael, Binyamin, Mordechai, Sarah, Daniel. DSc (hon.), Yeshiva U., N.Y.C., 1993, BA, 1955; MA, Columbia U., 1956, PhD, 1959; PhD (hon.), Russian Acad. Sci., 1992. Lic. patent atty., Israel, 1967. Alfred P. Sloan post-doctoral fellow Columbia U., N.Y.C., 1959-60; rsch. chemist Esso Rsch. and Engring. Co., Linden, N.J., 1960-64, Chems. and Phosphates Ltd., Haifa, Israel, 1964-65; from staff to dep. dir. Nat. Coun. for Rsch. and Devel., Jerusalem, Israel, 1965-86; dir. gen. Ministry Sci. and Tech., Jerusalem, Israel, 1991-93, chief scientist, 1993—; instr. in Chemistry Yeshiva U. 1960-62; adj. assoc. prof. chemistry Jerusalem Coll. Tech., 1973-77; bd. dirs. Weizmann Sci. Press, Israel Desalination Engring. Co., Ctr. for Indsl. Rsch. Recipient Commuity Svc. award Yeshiva U., 1992. Mem. AAAS, Am. Chem. Soc., Assn. Orthodox Jewish Scientists in Jerusalem (past pres.) Phi Lambda Upsilon, Sigma Xi. Achievements include patents for process for preparing aryl sulphonic acid esters, for preparing finely divided metals and alloys, for fuel cells. Office: Ministry of Sci and Tech, POB 18195, Jerusalem 96426, Israel

METZGER, HENRY, federal research institution administrator; b. Mainz, Germany, Mar. 23, 1932; came to U.S., 1938; naturalized, 1945; s. Paul Alfred and Anne (Daniel) M.; m. Deborah Stashower, June 16, 1957; children: Eran D., Renée V., Carl E. MD, Columbia U., 1957. Chief chem. immunology sect. Nat. Inst. Arthritis & Musculoskeletal & Skin Disease/NIH, Bethesda, Md., 1973—; br. chief USPHS, Bethesda, 1983—, sci. dir., 1987—, med. officer grade VI, 1975—; Carl Prausnitz Meml. lectr., 1982; Ecker Meml. lectr. Case Western Res. U., Cleve., 1984; Harvey Soc. lectr., 1984; Eli Nadel Meml. lectr. St. Louis U., 1987. Editor: Fc Receptors & the Action of Antibodies, 1990; assoc. editor Ann. Rev.Immunology, 1982—; contbr. numerous articles to profl. jours.; mem. editorial bd. numerous sci. jours. Recipient Meritorious Svc. award USPHS, 1978, Disting. Svc. award, 1985, Joseph Mather Smith prize Columbia U., 1984. Fellow AAAS, Am. Acad. Allergy and Immunology; mem. NAS, Am. Assn. Immunologists (pres. 1991-92), Am. Soc. Biol. Chemists, Am. Soc. Cell Biology, Am. Rheumatism Assn., Internat. Union Immunological Soc. (pres. 1992—), Found. for Advanced Edn. in the Scis. (pres. 1990-92), Alpha Omega Alpha. Home: 3410 Taylor St Bethesda MD 20815-4024 Office: NIH Bldg # 10 Rm # 9N228 Bethesda MD 20892

METZGER, WALTER JAMES, JR., physician, educator; b. Pitts., Oct. 30, 1945; s. Walter James Sr. and Marion Smith (Vine) M.; m. Carol Louise Hughes, Sept. 14, 1968; children: James Andrew, Joel Robert, Anne Elizabeth. BA, Stanford U., 1967; MD, Northwestern U., Chgo., 1971. Intern, resident Northwestern U., Chgo., 1971-74, rsch. fellow, 1974-76; asst. prof. U. Iowa, 1978-84; assoc. prof., sect. head East Carolina U., Greenville, N.C., 1984-91, prof., sect. head, 1993—, chmn. med. rsch. com., 1989-93; mem. study sect. merit rev. Nat. VA Rsch. Com., 1991-93. Mem. editorial bd. Allergy Procs., 1989-93. Forum leader Jarvis Meml. United Meth. Ch., Greenville, 1985-92. Maj. USAF, 1976-78. NIH grantee, Bethesda, Md., 1988-92. Fellow ACP, Am. Coll. Chest Physicians; mem. Am. Acad. Allergy Rsch. Coun. (vice chair 1988—), Am. Acad. Allergy Asthma (chair bronchoalveolar lavage com. 1993—), Rhinitis, Respiratory Diseases (chair interest sect. 1991-92), Chilean Lung Soc. (hon.). Achievements include 2 patents pending; research in allergic diseases: Dx and management, principles and practice in allergy, immunology and allergy clinics, immunopharmacology and investigation and classification of drugs. Office: East Carolina U Sch Medicine Moye Blvd Brody 3E-129 Greenville NC 27858-4354

METZLER, JERRY DON, nursing administrator; b. Mishawaka, Ind., Mar. 6, 1935; s. Gerald Donald and Cleota Christabell (Dowell) M.; m. Dorothy J. Masters, Aug. 18, 1962. BS, Ariz. State U., 1962, MEd, 1967; BSN, San Diego State U., 1973; MS, U. Ariz., Tucson, 1980. Sci. tchr. Washington Sch., Sanger, Calif., 1963-68; tchr. biology San Jacinto (Calif.) High Sch., 1968-70; staff nurse Maricopa County Hosp., Phoenix, 1973-76; staff nurse St. Luke's Hosp., Phoenix, 1976-77; nursing instr., dept. head Gila Pueblo Coll., Globe, Ariz., 1977-78; nurse educator, asst. dir. nursing USPHS Indian Hosp., Tuba City, Ariz., 1980-84; asst. nursing svc. mgr. Phoenix Indian Med. Ctr., 1984-85; pub. health educator Phoenix Indian Med. Ctr., 1985-88; dir. nursing USPHS Indian Hosp., Owyhee, Nev., 1988-90; sr. project officer USPHS, Dallas, 1990—. With USN, 1956-60, USPHS, 1980—. Mem. Res. Officers Assn., Am. Nurses Assn., Commd. Officers Assn. of USPHS, Masons, Sigma Theta Tau. Republican. Methodist. Home: 420 Shockley Ave De Soto TX 75115-3229 Office: PHS ROVI 1200 Main Tower Bldg Dallas TX 75202

MEUNIER, PIERRE JEAN, medical educator; b. Miribel, France, June 5, 1936; s. Marcel Jacques and Marie Ernestine (Ballufin) M.; m. Marie Aimée Loiseleur, July 16, 1961 (div. 1984); children: Gilles, Pascal, Francois; m. Annie Bernadette Mariés, Dec. 27, 1986. B, Lycée Ampère, Lyons, France, 1953; MD, Claude Bernard U., Lyons, France, 1967. Resident Lyons Hosps., 1963-67; asst. prof. Lyons Hosps. Claude Bernard U., 1967-71, prof., head dept. rheumatology and bone disease, 1971-79; head rsch. unit INSERM 234, 1979-92; pres. European Found. for Osteoporosis, 1989—, Groupe de Rsch. et d'Info. Osteoporoses, Paris, 1986-90; cons. in field. Editor-in-chief Bone, 1978-89, Osteoporosis Internat., 1989—; contbr. articles to profl. jours. Served with French Navy, 1962-63. Recipient Internat. League Against Rheumatism prize 1989, prize Paget's Disease Found., 1991. Mem. European Calcified Tissue Soc. (sec. 1985-91), Ordre des Palmes Académiques (officer). Avocations: music, tennis. Home: 31 rue du Bois De La Caille, 69004 Lyons France Office: Pavillon F Hosp, Ed Herriot Pl D'Arsonval, 69437 Lyons France

MEWIS, JOANNES J(OANNA), chemical engineering educator; b. Borgerhout, Belgium, Apr. 22, 1938; s. Frans and Irma (Bausmans) M.; m. Maria G. Suy, Apr. 23, 1966. Degree chem. engring., K.U. Leuven, 1961, PhDChemE, 1967. Asst. dir. IVP Lab., Leuven, 1964-69; docent part-time K.U. Leuven, 1967-69, docent, 1969-73, prof., 1974—, dept. chmn., 1989—; chmn. Internat. Commn. Rheology, 1992—; bd. dirs. BIG Emergency Info. Ctr., Geel, Belgium, 1982—; vis. prof. U. Del., 1981, Princeton U., 1982; cons. in field. Author: Gevaarlijke Stoffen, 1983; contbr. chpts. to books and articles to profl. jours. Postdoctoral fellowship NATO, U. Del., 1971-72. Mem. AICE, Belgian Group Rheology (pres. 1979-80, 1988-90), Plastics Soc. Royal Flemish Engrs. Assn. (chmn. 1976-84). Office: Chem Engring Dept, KU Leuven, de Croylaan 46, 3001 Heverlee Belgium

MEYER, DELBERT HENRY, organic chemist, researcher; b. Maynard, Iowa, Aug. 28, 1926. BA, Wartburg Coll., 1949; PhD in Chemistry, U. Iowa, 1953. Chemist Std. Oil Co., Ind., 1953-61; chemist Amoco Chem. Corp., 1961-67, rsch. supr. rsch. and devel. dept., 1967-77, dir. rsch., 1977—. Recipient Nat. Medal Tech., U.S. Dept. Commerce Tech. Adminstrn., 1992. Mem. Am. Chem. Soc. Achievements include research in esterification and oxidation reactions, polymerization, synthetic fibers, flame retardant polymers. Home: 1524 Clyde Dr Naperville IL 60565*

MEYER, DUANE RUSSELL, civil and cost engineer, consultant; b. Stambaugh, Mich., Mar. 31, 1948; s. Donald Gordon and Marilyn Lorraine (Zyskowski) Ml; m. Theresa Winifred Oliver, Aug. 21, 1971; children: Jennifer, Christopher, Jonathan, Melissa. BSCE, Mich. Technol. U., 1970. Registered profl. engr., Mich., Wis. Staff design engr. McNamee, Porter and Seeley, Ann Arbor, Mich., 1970-72, asst. resident engr., 1973-75, project mgr., 1976-81; mgr. cost estimating CH2M Hill, Milw., 1981-92; sr. cost estimator PSI Energy, 1993—; presenter papers on importance of ethics in engring. at profl. confs. Contbr. to profl. publs. Mem. Am. Assn. Cost Engrs. (pres. Wis. sect. 1985-87, Cost Engr. of Yr. 1991), Mich. Soc. Profl. Engrs. (Young Engr. of Yr. Ann Arbor chpt. 1981), Wis. Soc. Profl. Engrs. (chair ethics com. 1991—), Riverside Toastmasters (pres. 1986). Republican. Roman Catholic. Office: PSI Energy 1000 E Main St Plainfield IN 46168

MEYER, EDMOND GERALD, energy and natural resources educator, resources scientist, entrepreneur, former chemistry educator; b. Albuquerque, Nov. 2, 1919; s. Leopold and Beatrice (Ilfeld) M.; m. Betty F. Knobloch, July 4, 1941; children: Lee Gordon, Terry Gene, David Gary. B.S. in Chemistry, Carnegie Mellon U., 1940, M.S., 1942; Ph.D., U. N.Mex., 1950. Chemist Harbison Walker Refractories Co., 1940-41; instr. Carnegie Mellon U., 1941-42; asst. phys. chemist Bur. Mines, 1942-44; chemist research div. N.Mex. Inst. Mining and Tech., 1946-48; head dept. sci. U. Albuquerque, 1950-52; head dept. chemistry N.Mex. Highlands U., 1952-59; dir. Inst. Sci. Rsch., 1957-63; dean Grad. Sch., 1961-63; dean Coll. Arts and Sci., U. Wyo., 1963-75, v.p., 1974-80, prof. energy and natural resources, 1981-87, prof. and dean emeritus, 1987—; exec. cons. Diamond Shamrock Corp., 1980; bd. dirs. Carbon Fuels Corp., Am. Nat. Bank, Laramie; sci. adviser Gov. Wyo.; pres. Coal Tech. Corp., 1981—; cons. Los Alamos Nat. Lab., NSF, HHS, GAO, Diamond Shamrock Corp., Wyo. Bancorp.; contract investigator Research Corp., Dept. Interior, AEC, NIH, NSF, Dept. Energy, Dept. Edn.; Fulbright exch. prof. U. Concepcion, Chile, 1959. Co-author: Chemistry-Survey of Principles, 1963, Legal Rights of Chemists and Engineers, 1977, Industrial Research & Development Management, 1982; contbr. articles to profl. jours.; patentee in field. Lt. comdr. USNR, 1944-46, ret. Recipient Disting. Svc. award Jaycees; rsch. fellow U. N.Mex., 1948-50. Fellow AAAS, Am. Inst. Chemists (pres. 1992-93); mem. Assoc. Western Univs. (chmn. 1972-74), Am. Chem. Soc. (councilor 1962-90), Biophys. Soc., Coun. Coll. Arts and Scis. (pres. 1971, sec., treas. 1972-75, dir. Washington office 1973), Laramie C. of C. (pres. 1984), Laramie Regional Airport Bd. (chair 1989-93), Sigma Xi. Home: 1058 Colina Dr Laramie WY 82070-5015 Office: 2020 E Grand Ave Ste 440 Laramie WY 82070-4381 also: U Wyo Laramie WY 82071-3825

MEYER, FRANK HENRY, physicist; b. N.Y.C., July 11, 1915; s. Frank X. and Anna Helen (Wenzinger) M.; m. Winifred Josephine Duffy, Oct. 6, 1942; children: Frank Vincent, Vivian Pamela. BS in Natural Sci., CCNY, 1936; MS in Physics, N.Y. Poly. U., 1957; MA, U. Minn., 1968. Rsch. physicist Mt. Sinai Hosp., N.Y.C., 1948-51; rsch. physicist cen. rsch. div. Continental Oil Co., Ponca City, Okla., 1954-60; rsch. metallurgist Kaiser Aluminum and Chem. Corp., Spokane, 1960-63; sr. rsch. engr. Univac div. Sperry Rand, St. Paul, 1963-65; prof. physics and philosophy U. Wis., Superior, 1963-65; pres. sci. edn. corp. Internat. Soc. Unified Sci., Inc., Salt Lake City, 1978-89, editor Reciprocity, 1973—. Vol. Met. Sr. Fedn., Minn., 1970, First Unitarian Srs., Mpls., 1970. Mem. Am. Phys. Soc. (emeritus), Am. Assn. Physics Tchrs. (emeritus), Am. Cyrstallographic Assn. (founder, emeritus), Fedn. Am. Socs., Minn. Acad. Sci. Unitarian. Achievements include x-ray analysis cryogenic patent; discovery of two new solids FHCO3 and Fe8S8 (Kansite) and other determination of crystal structures; discovery of how motion is prior to rest; how time and space are inseparable; how motion, space and time are quantized. Home: 1103 15th Ave SE Minneapolis MN 55414 Office: Internat Soc Unified Sci 1680 E Atkins Ave Salt Lake City UT 84106

MEYER, GREGORY JOSEPH, power company executive; b. Euclid, Ohio, Mar. 11, 1955; s. Joseph Henry and Constance Mary (Atkinson) M. BA, U. Mich., Dearborn, 1980; MSA, Cen. Mich. U., 1989. Chartered indsl. gas cons.; cert. energy mgr. Resdl. energy cons. Consumers Power Co., Pontiac, Mich., 1981-83; mkt. svcs. cons. Consumers Power Co., Mt. Clemens, Mich., 1983-86; sr. mkt. svcs. cons. Consumers Power Co., Royal Oak, Mich., 1986-

90, indsl. account mgr., 1990—. Mem. ASM Internat., Assn. of Energy Engrs., Mich. Assn. Energy Engring., Nat. Assn. Energy Engring., Warren C. of C. Office: Consumers Power Co 4600 Coolidge Hwy # 369 Royal Oak MI 48068-0369

MEYER, HAROLD LOUIS, mechanical engineer; b. Chgo., June 25, 1916; s. Norman Robert and Martha (Stoewsand) M.; m. Charlotte Alene Tilberg, June 21, 1941 (dec. 1951); 1 child, John C. Nelson. Student, Armour Inst. Tech., Chgo., 1934-42, U. Akron, 1942-44; BA in Natural Sci., Southwestern Coll., Winfield, Kans., 1949; postgrad., Ill. Inst. Tech., 1988, 90; ME in Mech. Engring., 1988. Sales engr. Olsen & Tilgner, Chgo., 1938-39; project engr. Gen. Electric X-Ray, 1939-42, field engr., 1944-46; project engr. Goodyear Aircraft, 1942-44; chief x-ray technologist and therapist William Newton Meml. Hosp., Winfield, 1946-51; sr. design cons. Pollak and Skan, Chgo., 1952-58, cons. design specialist, 1963-68, 92—; project engr. Gaertner Scientific Co., Chgo., 1958-63; sr. design specialist Am. Steel Foundries, Chgo., 1969-74; cons. Morgen Design, Milw., 1974-76; propr. Meyersen Engring., Addison, Ill., 1981-92, also bd. dirs.; cons. dir. Miller Paint Equipment, Addison, 1976-87; design cons. R.R. Donnelley, Kraft Foods, Pollak and Skan, 1992—. Inventor: box sealing sta., 1939, chest x-ray equipment, 1941, G-2 airship, 1943, space program periscope, 1959-62, nuclear fuel inspection and measurement periscope, 1962, beer can filling machine, 1963, atomic waste handling vehicle, 1965, ry. freight car trucks, 1973, hwy. trailer 5th wheels, 1974, motorized precision paint colorant dispensing machines, 1986. Sponsered a family of Cambodian Chinese refugees; mem. Norwood Park (Ill.) Norwegian Old Peoples Home; mem. Family Shelter Svc., Glen Ellyn, Ill. With USNR, 1949-52. Recipient Appreciation award Lioness Club, Glendale Heights, 1985. Mem. AAAS, Chem. Engring. Product Rsch. Panel, Ill. Inst. Tech. Alumni Assn. (new student recruiter 1985—, Recognition award 1986, 87, honored for 50 yrs. of high standards of profl. activity and citizenship 1988, Emeritus Club award, 1990), Am. Registry of X-Ray Techs. Avocations: archery, golf, semi-precious gem stones, archaeology.

MEYER, HORST, physics educator; b. Berlin, Germany, Mar. 1, 1926. BS, U. Geneva, 1949; PhD in physics, U. Zurich, 1953. Fellow Swiss Assn. Rsch. Physics and Math. Studies, Oxford, Eng., 1953-55; Nuffield fellow Clarendon Lab. U. Oxford, 1955-57; lectr., rsch. assoc. dept. engring. and applied physics Harvard U., Cambridge, Mass., 1957-59; from asst. prof. to prof. Duke U., Durham, N.C., 1959-84, Fritz London prof. physics, 1984—; vis. prof. Technische Hochschule, Federal Republic of Germany, 1965, Tokyo U., 1980, 81, 83; traveling fellow Japanese Soc. for Promotion Sci., 1971, vis. scientist, 1979; guest scientist Inst. Laue-Langevin, France, 1974, 75; Yamada Found. fellow, Japan, 1986; guest scientist USSR Acad. Sci., 1988; chmn. Gordon Conf. on Solid H2, 1990. Editor Jour. Low Temperature Physics, 1992—, mem. editorial bd., 1988-92; contbr. articles to profl. jours. Alfred P. Sloan fellow, 1961-65. Fellow Am. Phys. Soc. (Jesse Beams prize, 1982, Fritz London prize 1993). Exptl. rsch. on the properties of liquid and solid helium, solid hydrogen and deuterium, magnetic insulators, critical phenomena. Office: Dept Physics PO Box 90305 Durham NC 27708-0305

MEYER, JAMES HENRY, meteorologist; b. St. Marys, Pa., July 20, 1928; s. Henry G. and Josephine C. (Hoehn) M.; m. Joan Elizabeth Mahoney, July 2, 1960; children: James Jr., John, Karin. BS, Pa. State U., 1953, MS, 1955. Cert. cons. meteorologist. Meteorologist USAF Rome (N.Y.) Air Devel. Ctr., 1954-55; atmospheric physicist MIT Lincoln Lab., Lexington, Mass., 1955-63, Tech. Ops., Burlington, Mass., 1963-64, Electromagnetic Rsch. Corp., College Park, Md., 1964-67; meteorologist Applied Physics Lab. Johns Hopkins U., Laurel, Md., 1967—; pres. Meteorol. Applications, Silver Spring, Md., 1976—. Com. chmn., leader Boy Scouts Am., Silver Spring, 1970-78. With USN, 1946-49. Mem. Am. Meteorol. Soc., Am. Geophys. Union, Marine Tech. Soc., Air and Waste Mgmt. Assn. Republican. Roman Catholic. Achievements include patents pending for a sea level evaluation device for study of air/water interface, a free floating air/water interface probe, yo-yo sounder for probing the lower atmosphere. Home: 12926 Allerton Ln Silver Spring MD 20904 Office: Johns Hopkins U Applied Physics Lab Johns Hopkins Rd Laurel MD 20723

MEYER, JAROLD ALAN, oil company research executive; b. Phoenix, July 28, 1938; s. Lester M. and Anita (Walker) M.; m. Diane Louise Wheeler; children: Ronald Alan, Sharon Lynne. BSChemE, Calif. Inst. Tech., 1960, MS, 1961. Mgr. process devel. Chevron Rsch., Richmond, Calif., 1978-82; tech. mgr. Chevron U.S.A., El Segundo, Calif., 1982-84; v.p. process rsch. Chevron Rsch., Richmond, 1984-86, pres., 1986—; sr. v.p. Chevron Rsch. and Tech., Richmond, 1990—; bd. dirs. Solvent Refined Coal Internat., Inc., San Francisco; adv. bd. Surface Sci. and Catalysis Program Ctr. for Advanced Materials Lawrence Berkeley Lab., 1988—; adv. coun. Lawrence Hall Sci., 1989—. Inventor petroleum catalysts; contbr. articles to profl. jours. Bd. visitors U. Calif., Davis, 1986—, trustee found., 1989—. Mem. Am. Chem. Soc., Nat. Petroleum Refining Assn., Indsl. Rsch. Inst., Conf. Bd. Internat. Rsch. Mgmt. Coun., Accreditation Bd. for Engring. and Tech. Indsl. Advisor, Tau Beta Pi, Sigma Xi. Avocations: electronics design and constrn., photography. Office: Chevron Rsch and Tech 100 Chevron Way Richmond CA 94802*

MEYER, JEAN-PIERRE, psychiatrist; b. Paris, Apr. 3, 1949; s. Henry Jules and Jacqueline Suzanne (Roux).; m. Marie Elisabeth Buisan, June 25, 1977; children: Arnaud Jean, Gauthier Henri. MD, Broussais U., Paris, 1975; Cert. of Maritime Medicine, 1976; Cert. of Med. Expertise, Cochin U., Paris, 1978; specialist in psychiatry, Necker U., Paris, 1978. Intern Fontainebleau (France) Hosp., 1974, Enfants Malades Hosp., 1975, Melun (France) Hosp., 1976, Mohamed V Hosp., Rabat, Morocco, 1976, Lagny (France) Hosp., 1977; intern psychiatrist infirmary of police Paris, 1977, sole practice medicine, specializing in psychiatry, 1979—; cons. Paris Hosp., 1986—; expert cons. Securite Sociale, Paris and Creil, 1984—; expert cons. Ct. of Appeals, Paris, 1988; archbishopric, Paris, 1979—. Author: Relaxation Therapeutique, 1986; co-author: Le Projet en Psychotherapie, 1983, Abrege de Neuro-Psychiatrie, Conduites Pratiques de Psychiatrie. Contbr. articles to profl. jours. V.p. Mutual Ins.'s, Paris, 1972—. Mem. Intergroupe de Formation en Relaxation, Med. Assn. Paris. Roman Catholic. Avocations: golf, skiing, surfing. Office: 9 rue du Général Delestraint, 75016 Paris France

MEYER, JOHN STIRLING, neurologist, educator; b. London, Feb. 24, 1924; came to U.S., 1940; s. William Charles and Alice Elizabeth (Stirling) M.; m. Wendy Haskell, July 20, 1947 (dec. 1986); children: Jane, Anne, Elizabeth, Helen, Margaret; m. Katharine Sumner, Aug. 2, 1987. BSc, Trinity Coll., Hartford, Conn., 1944; MD, CM, McGill U., Montreal, Que., 1948, MSc, 1949. Diplomate Am. Bd. Neurology and Psychiatry. Instr. rsch. assoc. Harvard Med. Sch., Boston, 1955-57; prof., chair dept. Wayne State U., Detroit, 1957-69; prof., chair dept. Baylor Coll. Medicine, Houston, 1969-75, prof. neurology, 1976—; demonstrator neuropathology and teaching fellow neurology Harvard U. Med. Sch., 1950-52; sr. rsch. fellow USPHS, 1952-54; instr. medicine Harvard U. Med. Sch., 1954-56; assoc. vis. physician neurology Boston City Hosp., 1956-57; cons. and lectr. neurology U.S. Naval Hosp., Chelsea, Mass., 1957; prof. neurology and chmn. dept. sch. medicine Wayne State U., 1957-69, chmn. dept., 1969-76; prof. neurology, dir. stroke lab. Baylor Coll. Medicine, Houston, 1976—; chair stroke panel Pres.' Commn. on Heart Disease Cancer & Stroke, Washington, 1964-65; mem. nat. adv. coun. Nat. Inst. Neurol. Diseases & Stroke, Bethesda, Md., 1965-69. Author 28 books; contbr. 768 articles to profl. jours. Mem. jury Albert Lasker Med. Rsch. Awards, N.Y.C., 1965-69. Lt. (s.g.) Med. Corps USN, 1953-55, Korea. Recipient Harold G. Woff award, Am. Assn. for Study of Head Ache, 1977, 79, Baylor Coll. Medicine award, Houston, 1980, 85, 90, Mihara award Mihara Found., Tokyo, 1987, Bertha Lecture award Salzburg Conf., Washington, 1992. Mem. Am. Heart Assn. (bd. dirs. 1968-70, chair coun. on stroke 1968-70). Republican. Episcopalian. Achievements include development of xenon contrast method for measuring cerebral blood flow using computerized toneography. Office: VA Med Ctr Rm 225 Bldg 110 2002 Holcombe Rd Houston TX 77030

MEYER, KARL V., civil engineer, consultant; b. Carlton, Kans., Oct. 16, 1926; s. Jesse E. and Thirza M. (Jones) M.; m. Patricia L. Talbot, June 5, 1949; children: Kyle P., Corey J. BSCE, U. Nebr., 1960. Structural engr. Nebr. Dept. Rds., Lincoln, 1956-67, B. H. Backlund & Assocs., Omaha, 1967-70; structural engr. The Schemmer Assocs. Inc., Omaha, 1970-82, v.p., 1982—; bd. dirs. Am. Concrete Inst., Nebr., 1990—. Mem. NSPE, ASCE, Am. Consulting Engrs. Coun. (pres. 1990-92, nat. dir. 1991-92), Soc. Am. Mil. Engrs., Omaha C. of C. (chmn. 1989, mem. transp. com. 1982-92). Democrat. Roman Catholic. Office: The Schemmer Assocs Inc 1044 N 115th St Omaha NE 68154

MEYER, KATHLEEN ANNE, school psychologist; b. Houston, Nov. 4, 1951; d. William Harry Delany and Clara Louise (Soland) Peltier; m. Michael Wayne Meyer; 1 child, Kristopher Rex. Student, Tex. Chiropractic Coll., 1968-70; AA in Elem. Edn., San Jacinto Jr. Coll., Pasadena, Tex., 1975; BS in Elem. Edn., Univ. Houston, 1980, MA in Psychology, 1985, post grad., 1988. Cert. tchr. Tex.; cert. state bd. of examiners of psychologists Tex., 1990; nat. cert. sch. psychologist, Tex., 1991. Instr. Pasadena I.S.D., Tex., 1980-84; tchr., and part time sch. psychologist St. Maur's Internat. Sch., Yokohama, Japan, 1985-87; assoc. sch. psychologist Santa Fe I.S.D., Tex., 1990-92; owner K & M Affiliates, Pasadena, Tex., 1993—. V.p. PTA, Pasadena, 1978, community rels., 1978-79; counselor Boy Scouts Am., Yokohama, Japan, 1986-87. Mem. APA, Coun. for Exceptional Children, S.E. Psychol. Assn., Nat. Assn. Sch. Psychologists, Am. Assn. of Suicidology, Tex. Psychol. Assn., Kappa Delta Pi. Avocation: ikebana, piano, japanese language, swimming. Home: 4211 Los Verdes Pasadena TX 77504-2415

MEYER, L. DONALD, agricultural engineer, researcher, educator; b. Concordia, Mo., Apr. 14, 1933; s. Lawrence Dick and Florence Malinda (Uphaus) M.; m. Loretta Lou Bush, Dec. 26, 1954; children: Dan W., James B., David J. Student, Cen. Coll., Fayette, Mo., 1950-51; BS in Agrl. Engring., U. Mo., 1954, MS in Agrl. Engring., 1955; PhD, Purdue U., 1964. Cert. profl. soil erosion and sediment control specialist; registered profl. engr., Ind. Agrl. engr. Agrl. Rsch. Svc., USDA, West Lafayette, Ind., 1955-73; agrl. engr. Nat. Sedimentation Lab., USDA, Oxford, Miss., 1973—; asst. prof., assoc. prof. Purdue U., West Lafayette, 1965-73; adj. prof. agr.-biol. engring. Miss. State U., Starkville, 1975—. Contbr. articles to profl. jours. Recipient Outstanding Performance award USDA Agrl. Rsch. Svc., 1959, 88, 89, 90, 91. Fellow Am. Soc. Agrl. Engrs. (dir. publs. 1968-69, chmn. soil and water div. 1972-73, Hancor award 1985), Soil and Water Conservation Soc.; mem. Soil Sci. Soc. Am. Office: USDA-ARS Nat Sedimentation Lab PO Box 1157 Oxford MS 38655-6004

MEYER, PAUL REIMS, JR., orthopaedic surgeon; b. Port Arthur, Tex.; s. Paul Reims and Evelyn (Miller) M.; children: Kristin Lynn, Holly Dee, Paul Reims III, Stewart Blair. BA, Va. Mil. Inst., 1954; MD, Tulane U., 1958; MA of Mgmt., J.L. Kellogg Grad. Sch. of Mgmt. (Northwestern U.), 1992. Dir. Spine Injury Ctr. Northwestern U., Chgo., 1972—, prof. orthopaedic surgery, 1981—; cons. Nat. Inst. Disability and Rehab. Rsch. VA, Washington, 1978—, chmn. spinal cord injury subcom., 1983—, also chmn. sci. merit com.; clin. prof. surgery, Dept. Surgery, USUHS; mem. adv. com. World Rehab. Fund, 1990—; orthopaedic cons. to Army Surgeon Gen. Author: Surgery of Spine Trauma, 1988; patentee cervical orthosis. Col. M.C., USAR. Fellow ACS; mem. Société Internationale de Chirurgie Orthopédique et de Traumatologie, Internat. Med. Soc. Paraplegia, Am. Acad. of Orthopaedic surgeons, Am. Trauma Soc. (bd. dirs. 1988—), Am. Ortho. Assn., Am. Spinal Injury Assn. (past pres.), Soc. Med. Cons. to Armed Forces, Mid-Am. Ortho. Assn. Roman Catholic. Avocations: photography, fishing, ham radio, aviation, boating. Office: Northwestern Meml Hosp 250 E Superior St Rm 619 Chicago IL 60611-2950

MEYER, PETER, physicist, educator; b. Berlin, Jan. 6, 1920; came to U.S., 1952, naturalized, 1962; s. Franz and Frida (Lehmann) M.; m. Luise Schützmeister, July 20, 1946 (dec. 1981); children: Stephan S., Andreas S.; m. Patricia G. Spear, June 14, 1983. Dipl.Ing., Tech. U., Berlin, 1942; Ph.D., U. Goettingen, Germany, 1948. Faculty U. Goettingen, 1946-49; fellow U. Cambridge, Eng., 1949-50; mem. sci. staff Max-Planck Inst. fuer Physik, Goettingen, 1950-52; faculty U. Chgo., 1953—, prof. physics, 1965-90, prof. emeritus, 1990—, chmn. Dept. Physics, 1986-89; dir. Enrico Fermi Inst., 1978-83; cons. NASA, NSF.; mem. cosmic ray commn. Internat. Union Pure and Applied Physics, 1966-72; mem. space sci. bd. Nat. Acad. Scis., 1975-78. Recipient Alexander von Humboldt Sr. U.S. Scientist award, 1984, Llewellyn John and Harriet Manchester Quantrell award, 1971. Fellow Am. Phys. Soc. (chmn. div. cosmic physics 1972-73), AAAS; mem. NAS, Am. Astron. Soc., Am. Geophys. Union, Max Planck Inst. fuer Physik und Astrophysik (fgn.), Sigma Xi. Office: 933 E 56th St Chicago IL 60637-1460

MEYER, RICHARD, psychiatrist; b. Sierentz, Alsace, France, June 15, 1942; s. Rodolphe and Suzanne (Sonntag) M.; m. Marianick Lemoine, Sept. 9, 1967 (div. 1988); children: Jérôme, Grégoire, Aurore; m. Veronique Fischer. MD, U. Strasbourg (France), 1973; PhD in Human Scis., U. Montpellier (France), 1964, U. Paris, 1978. Houseman in psychiatry Univ. Hosp., Strasbourg, 1971-75; asst. Univ. Hosp., Lausanne, Switzerland, 1973; pvt. practice Strasbourg, 1975—; lectr. psychiatry U. Strasbourg, 1975-80; pres., founder AJP/L'Innovation Psychiatrique, Strasbourg, 1981, Internat. Congress on Somatotherapy, Paris, 1988, Strasbourg, 1991. Author: Le Corps Aussi, 1982, Portrait de Groupe avec Psychiatres, 1984, Thérapies Corporelles, 1986, Reich ou Ferenczi, 1992, La Somatologie, 1992, Lieux du Corps en Psychotheapie, 1993, Somatologie tome II, 1993; editor: Somatotherapie, 1992; patentee in somatanalysis and socio and somat analytical psychotherapy. Mem. Internat. Assn. on Somatotherapy (founder, pres. 1992). Office: 20 Pl des Halles, 67000 Strasbourg France

MEYER, ROBERT VERNER, farmer; b. Rochester, Minn., May 13, 1954; s. Theodore John and Jennie Ann (Whiting) M. Farmer Elgin, Minn., 1954—; warehouse mgr. Meyer's Seeds, Inc., Elgin, 1981—. Designer 5 wooden-geared clocks, 1988-92; builder 23 wooden-geared clocks, 1988-92; author clock plans, 1988-92. Mem. Chesnut Found. Democrat. Lutheran. Achievements include designer and builder of 5 different wooden-geared, weight-driver, pendulum-regulated clocks; wrote plans for 1st design and sold over 400 to other woodworkers; built 23 clocks; bldg. giant clock with 4 foot gears. Home: 7347 Hwy 247 NE Elgin MN 55932 Office: Meyer's Seeds Inc 7813 Hwy 247 NE Elgin MN 55932

MEYER, STEVEN JOHN, electrical engineer; b. Glendale, Calif., Mar. 17, 1961; s. Albert John and Diane (Whitehead) M.; m. Wendy Dawn Fullmer, Apr. 23, 1988; children: Joshua David, Trevor John. BSEE, Brigham Young U., 1987; MSEE, Calif. State U., Northridge, 1991. Elec. engring. intern Radio Free Europe, Radio Liberty, Lisbon, Portugal, 1984; elec. engr. Weapons divsn. Naval Air Warfare Ctr., China Lake, Calif., 1987—. Contbr. articles to profl. jours. Asst. scoutmaster Boy Scouts Am., Ridgecrest, Calif., 1989—. Mem. Mercury Amateur Radio Assn. Republican. Mem. LDS Ch. Achievements include patent pending wire table procedure. Office: Naval Air Warfare Ctr Weapons Divsn Code C3923 (6424) China Lake CA 93555

MEYER, THOMAS ROBERT, television product executive; b. Buffalo, Apr. 20, 1936; s. Amel Robert and Mildred Lucille M.; m. Dawn E. Shaffer, 1985. Student Purdue U., 1953-55, Alexander Hamilton Inst. Bus., 1960-62, West Coast U., 1969-72; B in Math., Thomas Edison State Coll., 1988. Sect. chief wideband systems engring. Ground Elec. Engring. and Installation Agy., Dept. Air Force, 1960-66; product mgr., systems engr. RCA Corp., Burbank, Calif., 1966-71; systems cons. Hubert Wilke, Inc., L.A., 1971-72; product mgr. Telemation, Inc., Salt Lake City, 1972-77; v.p. engring. Dynair Electronics, San Diego, 1977-92; prin. Duir Assocs., San Diego, 1992—. Recipient Bronze Zero Defects award Dept. Air Force, 1966, Tau Beta Pi Eminent Eng. award, 1987. Fellow Soc. Motion Picture and TV Engrs. (mem. task force on TV/computer digital image architecture); mem. Soc. Broadcast Engrs. (sr.), Computer and Electronics Mktg. Assn. Rsch. and publs. on color TV tech. and optics, TV equipment and systems, application of computer to TV systems. Office: Duir Assocs 1220 Rosecrans St Ste 302 San Diego CA 92106

MEYERHOFF, ARTHUR AUGUSTUS, geologist, consultant; b. Northampton, Mass., Sept. 9, 1928; s. Howard Augustus and Anna Sophia (Theilen) M.; m. Kathryn Eleanor Laskaris, Jan. 2, 1951; children: James Charles, Richard Dietrich, Donna Kathryn. BA, Yale U., 1947; MS, Stanford U., 1950, PhD, 1952. Cert. profl. geologist; cert. petroleum geologist. Geologist U.S. Geol. Survey, Mont., 1948-52; sr. geologist Standard Oil Co. of Calif., Chevron Corp., 1952-65; publs. mgr. Am. Assn. Petroleum Geologists, Tulsa, 1965-74; pvt. practice geologist, cons. Tulsa, 1974—; guest lectr. U. Belgrade, 1973, USSR Acad. Scis., Inst. Physics of the Earth, 1974; hon. lectr. Can. Soc. Petroleum Geologists, 1975, U. Calgary, 1978, Petroleum Exploration Soc. Australia, 1982; vis. prof. Okla. State U., Stillwater, 1975-77, U. Calgary, 1978; overseas vis. scholar Gonville and Caius Coll., Cambridge U., 1979-80, others. Contbr. articles to profl. jours. Fellow AAAS, Geol. Soc. Am., Geol. Soc. London; mem. Tulsa Geol. Soc., Geol. Soc. of China, Can. Soc. of Petroleum Geologists, Am. Assn. Petroleum Geologists, Soc. Sedimentary Geology, Am. Geophys. Union, Assn. Earth Scis. Editors (pres. 1968-69), Assn. Geoscientists for Internat. Devel., Geol. Australia, Petroleum Exploration Soc. Australia, Geol. Soc. Malaysia, S.E. Asian Petroleum Explorationists, Soc. Econ. Geologists, Paleontol. Soc., Lafayette Geol. Soc., Houston Geol. Soc., Sigma Xi, more. Home: 3123 E 28th St Tulsa OK 74114

MEYERS, PAUL ALLAN, chemist; b. Rogers City, Mich., July 23, 1958; s. Raymond Leonard and Blanche Patricia M.; m. Michele Kay Kreft, May 16, 1981. AAS, Alpena Community Coll., 1979; B of Chemistry, Hayward State U., 1987. Chemist Dow Chem. Co., Walnut Creek, Calif., 1979-88; chemist III Landec Corp., Menlo Park, Calif., 1988—; guest lectr. Coll. of San Mateo, Calif., 1992—. Contbr. tech. papers to profl. publs. Recipient Jesse Besser award Besser Found., 1979, Inventor of Yr. award Dow Chem. Co., 1982. Mem. Am. Chem. Soc. Achievements include 5 patents for Mineral Filled Composites and 2 patents for Controlled Delivery Systems. Office: Landec Corp 3603 Haven Ave Menlo Park CA 94005

MEYERS, WAYNE MARVIN, microbiologist; b. Huntingdon County, Pa., Aug. 28, 1924; s. John William and Carrie Venca (Weaver) M.; m. Esther Louise Kleinschmidt, Aug. 26, 1953; children: Amy, George, Daniel, Sara. BS in Chemistry, Juniata Coll., 1947; diploma, Moody Bible Inst., 1950; M.S. in Med. Microbiology, U. Wis., 1953, Ph.D. in Med. Microbiology, 1955; M.D., Baylor Coll. Medicine, 1959; DSc (hon.), Juniata Coll., 1986. Instr. Baylor Coll. Medicine, 1955-59; intern Conemaugh Valley Meml. Hosp., Johnstown, Pa., 1959-60; staff physician Berrien Gen. Hosp., Berrien Ctr., Mich., 1960-61; missionary physician Am. Leprosy Missions, Burundi and Zaire, Africa, 1961-73; prof. pathology Sch. Medicine U. Hawaii, Honolulu, 1973-75; chief microbiology divsn. Armed Forces Inst. Pathology, Washington, 1975-89, chief mycobacteriology, 1989—, registrar leprosy registry, 1975—; mem. leprosy panel U.S.-Japan Coop. Med. Sci. Program, 1976-83; mem. sci. adv. bd. Leonard Wood Meml., 1981-85, sci. cons. dir., 1985-87, sci. dir., 1987-90; cons., 1990—, Leonard Wood Meml., 1990—; rsch. affiliate Tulane U., 1981—; corp. bd. dirs. Gorgas Meml. Inst. Tropical and Preventive Medicine, Inc. Bd. dirs. Internat. Jour. Leprosy, 1978—; contbr. numerous chpts. and articles on tropical medicine to textbooks and jours. Adv. bd. Damien-Dutton Soc. for Leprosy Aid, Inc., 1983—, Am. Leprosy Missions, Inc., 1979-88, chmn. bd., 1985-88, program cons. to bd., mem. bd. reference, 1988—; mem. Hansen's Disease Rsch. Adv. Com., Gillis W. Long Hansen's Disease Ctr., Carville, La., 1983-85; chmn., 1985—. With U.S. Army, 1944-46. Allergy Found. Am. fellow, 1957, 58; WHO rsch. grantee, 1978-87. Mem. Internat. Leprosy Assn. (councillor 1978-88, pres. 1988—), Internat. Soc. Tropical Dermatology, Am. Soc. Tropical Medicine and Hygiene, Am. Soc. Microbiology, Binford-Dammin Soc. Infectious Disease Pathologists (sec.-treas. 1988-91), Sigma Xi. Achievements include researching human and experimental leprosy. Office: Armed Forces Inst Pathology Washington DC 20306-6000

MEYERS-JOUAN, MICHAEL STUART, computer engineer; b. N.Y.C., May 12, 1948; s. Harold Hilliard and Ina Stuart (Szabad) Meyers; m. Sylvie G. Jouan, Dec. 24, 1980. BS in Humanities and Engring., MIT, 1969. Software engr. Digital Equip. Corp., Maynard, Mass., 1969-76, Hendrix Electronics, Inc., Manchester, N.H., 1976-78, Nashua (N.H.) Digital Scis., Inc., 1978-80; v.p. engring. Raster Graphics/Imteks, Mount Kisco, N.Y., 1980-86; mgr. strategic tech. Prodigy Svcs. Co., White Plains, N.Y., 1986—. Mem. MIT Alumni Assn. Office: Prodigy Svcs Co 445 Hamilton Ave White Plains NY 10601-1814

MEYERSON, SEYMOUR, retired chemist; b. Chgo., Dec. 4, 1916; s. Joseph and Rena (Margulies) M.; m. Lotte Strauss, May 22, 1943; children: Sheella, Elana. SB, U. Chgo., 1938, postgrad., 1938-39, 47-48; postgrad., George Williams Coll., 1939-40. Chemist Deavitt Labs., Chgo., 1941-42; inspector powder & explosives Kankakee Ordnance Works, Joliet, Ill., 1942; from chemist to rsch. cons. Standard Oil Co. Rsch. Dept., Whiting, Ind.-Naperville, Ill., 1946-84; mem. indsl. adv. coun. chemistry dept. U. Okla., Norman, 1967-69; charter mem. Editorial Organic Mass Spectrometry, 1968-87; mem. adv. bds. various mass spectrometry revs., 1980-87. Author or co-author 180 sci. publs. 2d lt. AUS, 1943-46, ETO. Mem. emeritus Am. Chem. Soc. (Frank H. Field and Joe L. Franklin award for outstanding achievement in mass spectrometry 1993), Am. Soc. for Mass Spectrometry. Achievements include many contributions to systematic chemistry of gas-phase organic ions; 2 patents in field. Home: 650 N Tippecanoe St Gary IN 46403

MEYR, SHARI LOUISE, information consultant; b. San Diego, Dec. 6, 1951; d. Herchell M. and Etta Louise (Bass) Knight; m. William Earl Groom, Oct. 22, 1977 (div. Sept. 1989), Herbert Carl Meyr Jr., Feb. 23, 1990. AS in Fire Scis., San Diego Mesa Coll., 1976. T.O.S.S. specialist Spectrum Scis. & Software, Mountain Home AFB, 1989—; equestrian instr. Summerwind Ctr., Mountain Home, 1979-91; Chow Chow breeder Meyr Kennels, Mountain Home, 1990—; multimedia P.C. cons. Access to Answers, Mountain Home, 1990—. Mem. NRA, U.S. Ski Assn. (competition lic., alpine ofcl., master's alpine racer 1991—), Summerwind Riding Club (founder, pres. 1981-89), Mountain Home Ski Club (founder, bd. dirs. 1991), Bogus Basin Ski Club, Sun Valley Ski Club, Amateur Trapshooting Assn. (life), Mensa. Avocations: ski racing, trap shooting, tae kwon do. Home: 570 E 16th N Mountain Home ID 83647-1717 Office: Spectrum Scis & Software PO Box 663 Mountain Home ID 83647

MEZIC, RICHARD JOSEPH, engineer, consultant; b. Bklyn., Dec. 29, 1968; s. Dino and Erna (Rom) M. BSME, Poly. U. Bklyn., 1991. CAD engr. Am. Power Tech., Inc., Lake Success, N.Y., 1991; design engr. Joseph R. Loring and Assocs., Inc., N.Y.C., 1991—. Asst. scoutmaster Queens (N.Y.) area Boy Scouts Am., 1986—. Mem. ASME (assoc.), Soc. Automotive Engrs. (assoc.). Home: 77-05 79th St Glendale NY 11385 Office: Joseph R Loring and Assocs 1 Penn Plz New York NY 10119

MEZZANOTTE, PAOLO ALESSANDRO, aeronautical engineer; b. Como, Italy, Sept. 25, 1943; s. Antonio and Alessandra (Favaro) M.; m. Ornella De Marchi, June 26, 1974; 1 child, Antonio. M in Aero. Engring. with honors, Poly. U., Milan, Italy, 1968. Rsch. asst. Poly. U., Milan, Italy, 1968-70; preliminary design engr. Aermacchi, Varese, Italy, 1970-86; dir. advanced studies Aermacchi, 1986-91, dir. new programs, 1990—. Home: Via Pozzi 14, 21020 Casciago Italy Office: Aermacchi, Via Sanvito 80, 21100 Varese Italy

MI, YONGLI, polymer engineer, researcher; b. Beijing, Feb. 10, 1961; came to U.S., 1987; s. Shengqing and Suyun (Yan) M.; m. Song Xiang, May 1, 1986; 1 child, Deborah J. BS, Chinese U. Sci. and Tech., Hefei, 1984; PhD, Syracuse U., 1991. Jr. rschr. Inst. Chemistry, Chinese Acad., Beijing, 1984-87; rsch. asst. Syracuse (N.Y.) U., 1987-91, rsch. assoc., 1992—. Contbr. articles to Jour. Polymer Sci., Macromolecules. Mem. Sigma Xi. Achievements include patent pending for synthesis of new polyimide materials for application of gas separation; developed new theoretical model of gas solution and transport in glassy polymers based on a "concentration-temperature superposition" principle. Home: 1169 Cumberland Ave Syracuse NY 13210 Office: Syracuse U 310 Hinds Hall Syracuse NY 13244

MIAN, FAROUK ASLAM, chemical engineer, educator; b. Lahore, Punjab, Pakistan, Aug. 10, 1944; came to U.S., 1969; s. Mohd Aslam and Qureshia Mian; m. Zahida Perveen, July 16, 1970; children: Shoaib F., Sophia F. BS in Chem. Engring., Inst. Chem. Tech., Punjab U., Lahore, 1964, MS in Chem. Engring., 1965; postgrad., Ill. Inst. Tech., Chgo., 1972-74. Registered profl. engr., Tex., Calif., Colo., La., Miss., Wis., Wyo.; registered environ. engr.; diplomate Am. Acad. Environ. Engrs. Chem. engr. Kohinoor/Didier-Werke, 1965-69, Nuclear Data, Inc., Palatine, Ill., 1969-71; prodn. supr. Searle Corp., Arlington Heights, Ill., 1971-74; lead process engr. Austin Co.,

Des Plaines, Ill., 1974-76; leac process engr. Crawford and Russell, Inc., Houston, 1976-77; supr. process Bechtel, Inc., Houston, 1977-80; process mgr. Litwin Corp., Houston, 1980; mgr. chems. Brown and Root, Inc., Houston, 1980—; chmn.'s adviser U.S. Congl. Adv. Bd., Am. Security Coun. Found., Washington, 1983, 84. Contbr. articles to profl. publs. Mem. AICE, NSPE, Instrument Soc. Am. (sr.), Inst. Food Technologists. Achievements include research in petrochemicals, inorganic and organic chemicals, specialty and fine chemicals, petroleum refining and coal gasification processes, chlor-alkali and electro-chemicals, food and pharmaceuticals. Office: Brown and Root Inc Box 3 Houston TX 77001

MIAN, GUO, electrical engineer; b. Shanghai, Feb. 6, 1957; came to U.S., 1987; s. Wenseng Mian and Guorong Sun; m. Ann Wang, Nov. 1, 1989. BS in Physics, Shanghai U. Sci. & Tech., 1982; MS in Physics, Western Ill. U., 1989; DSc in Elec. Engring., Washington U., 1992. Mgr. Rec. Media Lab. Magnetic Rec. Ctr., Shanghai (China) Ctrl. Chem. Ltd., 1982-85; vis. scientist materials sci. lab. Keio U., Yokohama, Japan, 1985-87; sr. rsch. elec. engring. Quantum Corp., Milpitas, Calif., 1992-93, Conner Peripherals, San Jose, Calif., 1993—. Contbr. articles to Jour. Materials Sci., IEEE Trans. Magnetics, Jour. Magnetism & Magnetic Materials, Jour. Applied Physics, Japanese Jour. Applied Physics. Recipient C & C Promotion award Found. for C & C Promotion, Tokyo, 1986. Mem. IEEE, IEEE Magnetics Soc., IEEE Computer Soc., Am. Phys. Soc. Achievements include discovered a transverse correlation length in magnetic thin film media, a linear relationship between correlation function of media noise and an off track displacement of a recording head; an algorithm to determine a signal to noise ratio for an arbitratry data sequence in time domain; inventor in field.

MIAO, SHILI, plant biology educator, plant ecology researcher; b. Chonging, Shichuan, People's Republic of China, Feb. 25, 1950; came to U.S., 1986; d. Ji Ming and Shenyun (Mao) M.; m. Zefu Chen, July 22, 1991; 1 child, Sonia. BS, S.W. China Normal U., 1982, MS, 1984; PhD, Boston U., 1990. Tchr. Sixtieth Primary Sch., Chonging, 1972-78; teaching asst., biology dept. S.W. China Normal Univ., Chonging, 1984-86, lectr., biology dept., 1987—; postdoctoral fellow Harvard Univ., Cambridge, Mass., 1990—; vis. scholar Harvard U., 1986-87. Contbr. chpt. to Ecological Studies on Evergreen Broadleaved Forests, 1988, Modern Ecology of Am., articles to Ecology and Oecologia. Recipient Presdl. fellowship Boston Univ., 1987-90. Mem. Ecol. Soc. Am., Ecol. Soc. China, Internat. Biologists Assn. Office: Harvard U Biol Lab 16 Divinity Ave Cambridge MA 02138-2097

MICHA, DAVID ALLAN, chemistry and physics educator; b. Argentina, Sept. 12, 1939; came to U.S., 1969, naturalized, 1974; s. Simon David and Catalina (Cohen) M.; m. Rebecca Stefan, 1991; children: Michael F., Anna K. MS, U. Cuyo, Bariloche, Argentina, 1962; DSc, U. Uppsala, Sweden, 1966. Rsch. assoc. Theoretical Chemistry Inst. U. Wis., Madison, 1965-66; asst. rsch. physicist Inst. Pure and Applied Sci. U. Calif., La Jolla, 1967-69; assoc. prof. chemistry and physics U. Fla., Gainesville, 1969-74, prof., 1974—, dir. Ctr. Chem. Physics, 1982-91; vis. prof. U. Gothenburg, Sweden, 1970, Harvard U., 1972, Max-Planck Inst., Gottingen, Fed. Republic Germany, 1976, Imperial Coll., London, 1977, U. Calif., Santa Barbara, 1982, U. Colo. and Weizmann Inst., Israel, 1983, U. Buenos Aires, 1988, Harvard U., 1990, supercomputer inst. Fla. State U., 1991. Mem. editorial bd. Internat. Jour. Quantum Chemistry, 1979-88, Few-Body Systems, 1985—; editor Finite Systems and Multiparticle Dynamics, 1990—, symposium procs.; contbr. several book chpts., over 130 articles to sci. jours. Recipient A. Von Humboldt Found. U.S. Sr. Scientist award, 1976; Alfred P. Sloan Found. fellow, 1971-74; Nat. Bur. Standards JILA fellow, 1983. Fellow Am. Phys. Soc.; mem. Am. Chem. Soc. Office: U Fla Williamson Hall 366 Gainesville FL 32611

MICHAEL, ALFRED FREDERICK, JR., physician, medical educator; b. Phila.; s. Alfred Frederick and Emma Maude (Peters) M.; children: Mary, Susan, Carol. M.D., Temple U., 1953. Diplomate: Am. Bd. Pediatrics (founding mem. sub-bd. pediatric nephrology, pres. 1977-80). Diagnostic lab. immunology and pediatric nephrology intern Phila. Gen. Hosp., 1953-54; resident Children's Hosp. and U. Cin. Coll. Medicine, 1957-60; postdoctoral fellow dept. pediatrics Med. Sch., U. Minn., Mpls., 1960-63; asso. prof. Med. Sch., U. Minn., 1965-68, prof. pediatrics, lab. medicine and pathology, 1968—, dir. pediatric nephrology, Regents' Prof., head Dept. Pediatrics, 1986—; established investigator Am. Heart Assn., 1963-68. Mem. editorial bd. Internat. Yearbook of Nephrology, Am. Jour. Nephrology, Clin. Nephrology, Am. Jour. Pathology; contbr. articles to profl. jours. Served with USAF, 1955-57. Recipient Alumni Achievement award in clin. scis. Temple U. Sch. Medicine, 1988; NIH fellow, 1960-63; Guggenheim fellow, 1966-67. Fellow Am. Acad. Pediatrics; mem. AMA, AAAS, Am. Assn. Immunologists, Am. Soc. Clin. Investigation, Assn. Am. Physicians, Am. Pediatric Soc., Soc. for Pediatric Research, Am. Assn. of Investigative Pathology, Am. Soc. Cell Biology, Central Soc. for Clin. Research, Midwestern Soc. for Pediatric Research, Am. Soc. Nephrology (coun., pres.-elect 1992—, pres. 1993), Internat. Soc. Nephrology, Soc. for Exptl. Biology and Medicine, Am. Fedn. Clin. Research, Soc. for Pediatric Nephrology, Northwestern Pediatric Soc., Mpls. Pediatric Soc., Minn. Med. Assn. Congregationalist. Office: U Minn Hosps & Clinic Dept Pediatrics Haward St at E River Rd Minneapolis MN 55455

MICHAELIDES, DOROS NIKITA, internist; b. Nicosia, Cyprus, Jan. 7, 1936; came to U.S., 1969; s. Nikita P. and Efipinike (Taliadorou) M.; m. Eutychia J. Loizides, Feb. 27, 1965; children: Nike-Elsie, Joanna-Doris. MD cum laude (Royal Greek Govt. scholar), U. Athens, 1962; D.T.M. and H. (Greek State Scholarship Found. scholar), U. Liverpool (Eng.), 1967; MSc in Clin. Biochemistry (Greek State Scholarship Found. scholar), U. Newcastle-upon-Tyne (Eng.), 1969. Clk., intern U. Uppsala (Sweden), 1962; resident Nicosia Gen. Hosp., 1963-66; fellow U. Liverpool Hosps., 1967; fellow internal and clin. medicine Royal Infirmary, U. Edinburgh, 1967-68; research fellow Royal Victoria Infirmary, U. Newcastle-upon-Tyne, 1968-69; resident internal medicine Bapt. Meml. Hosp., Memphis, 1969-72; fellow in chest diseases Western Okla. Chest Disease Hosp., 1970-71; chief clin. immunology and respiratory care center, Erie, Pa.; chief respiratory care center VA Med. Ctr., Erie, 1972-84, acting chief dept. medicine, 1980-81; asst. clin. prof. medicine Hahnemann Univ. Sch. Medicine, Phila., 1977—; asst. clin. prof. medicine Gannon U., Erie, 1977—; mem. staff internal medicine Hamot Med. Ctr., immunology & chest diseases Metro Health Ctr., Erie; preceptor medicine St. Vincent's Health Ctr.; affiliate staff Cleveland Clinic Found. Recipient citation for outstanding services to vets. DAV, 1975, citation Adminstr. U.S. Vets. Affairs, 1978. Diplomate Am. Bd. Family Practice, Am. Bd. Allergy and Immunology; cert. in infectious diseases and immunochemistry, Eng. Fellow ACP (life), Am. Assn. Cert. Allergists, Am. Coll. Allergy and Immunology (com. autoimmune diseases), Am. Assn. Clin. Immunology and Allergy (pulmonary com.). Am. Coll. Chest Physicians (life; critical care com.), Royal Soc. Medicine, Am. Coll. Angiology, N.Y. Acad. Scis., Am. Coll. Clin. Pharmacology, Am. Assn. Cert. Allergists. Democrat. Greek Orthodox. Author: The Occurrence of Proteolytic Inhibitors in Heart and Skeletal Muscle, 1969; Blood Gases, Acid-Base and Electrolytes Disturbances, 1980; Immediate Hypersensitivity: The Immunochemistry and Therapeutics of Reversible Airway Obstruction, 1980; The Equivalent Potency of Corticosteroid Preparations used in Reversible Airway Obstruction, 1981; contbr. articles to med. jours. Home: 4107 State St Erie PA 16508-3129 Office: Allergy Immunology & Chest Diseases 1611 Peach St Ste 220 Erie PA 16501-2172

MICHAELIS, ELIAS K., neurochemist; b. Wad-Medani, Sudan, Oct. 3, 1944; married, 1967; 1 child. BS, Fairleigh Dickinson U., 1966; MD, St. Louis U. Med. Sch., 1969; PhD in Physiology and Biophysics, U. Ky., 1973. Spl. fellow rsch. dept. physiology and biophysics U. Ky., 1972-73, from asst. prof. to prof. depts. human devel. and biochemistry, 1982-87; chair pharmacology and toxicology U. Kans., Lawrence, 1988—; dir. ctr. biomedical rsch. and Higuchi bioscience rsch. U. Kans., 1988—. Mem. AAAS, Am. Soc. Neurochemistry, Am. Soc. Biochemistry and Molecular Biology, Internat. Soc. Biomedical Rsch. on Alcoholism, Soc. Neuroscience, N.Y. Acad. Sci. Achievements include research in characterization of L-glutamate receptors in neuronal membranes, in membrane protein isolation and chemical analysis, in characterization of membrane transport systems for amino acids, sodium, potassium, and calcium, in neuronal membrane biophysics, in molecular neurobiology. •

MICHAELS, GORDON JOSEPH, metals company executive; b. Williamsport, Pa., May 9, 1930; s. Scott Joseph and Gloria Jean M.; m. Cleo Arlene Lela Tietbohl, June 12, 1954; children: Cathryn, Cheryl, Carole. BSEE, Bucknell U., 1959. Tool engr. Ternstedt div. Gen. Motors, Warren, Mich., 1950-59, sr. facilities engr., 1959-65; div. mgr. rectifiers M & T Chem. div. Am. Can Co., Rahway, N.J., 1965-71; v.p. mfg. and engring. Ullrich Copper Co., Kenilworth, N.J., 1971—; pres. Golld Truck Inc., Dorgo Products Inc.; treas. Tricor Am. Inc. Bd. dirs. Tech. Machinery Inst., Union, N.J.; active Jr. Achievement, Elizabeth, N.J.; mem. bd. advisors Kean Coll. N.J.; mem. Nat. Trust Hist. Preservation. Served with AUS, 1954-56. Mem. IEEE, Soc. Mining Engrs., Am. Electroplaters Soc., Copper Development Assn. (tech. com.), Soc. Mfg. Engrs., AAAS, Nat. Rifle Assn., Cryogenic Soc. Am., High Speed Rail Assn., Internat. Platform Assn., Am. Legion, Smithsonian Assn. (assoc.), Nat. Trust for Historic Preservation, Nat. Wildlife Fedn., Sierra Club. Republican. Lutheran. Home: Trout Run PA 17771 Office: HC64 PO Box 318 Trout Run PA 17771

MICHAELSON, HERBERT BERNARD, technical communications consultant; b. Washington, D.C., Dec. 29, 1916. B in Physics, N.Y. Univ., 1955. Head tech. info. Sylvania Elec., Bayside, N.Y., 1951-56; assoc. editor IBM Jour. Rsch. & Devel. IBM Corp. Headquarters, Armonk, N.Y., 1956-84; cons. tech. communications Jackson Heights, N.Y., 1984-92; cons. IBM Corp., 1985-91. Author: How to Write and Publish Engineering Papers and Reports, 1992; editor: IRE Transactions on Engineering Writing and Speech, 1961; assoc. editor: IBM Journal of Research and Development, 1956-84; contbr. articles to profl. jours. With U.S. Army Signal Corps., 1943-46. Recipient IEEE Goldsmith award, 1991, IBM Corp. Div. award, 1979. Fellow Soc. Tech. Communication; mem. Soc. Tech. Communication (bd. dirs), IRE Profls. Group (treas.), IEEE Profl. Communication Soc. (adminstr. bd.), IEEE (sr. mem.). Achievements include several published papers on periodicities of the electronic work function in the table of the elements. Home: 33-50 74 St Jackson Heights NY 11372

MICHAUD, GEORGES JOSEPH, astrophysics educator; b. Que., Can., Apr. 30, 1940; s. Marie-Louis and Isabelle (St. Laurent) M.; m. Denise Lemieux, June 25, 1966. BA, U. Laval, Que., 1961, BSc, 1965; PhD, Calif. Tech. Inst., Pasadena, 1970. Prof. Universite de Montreal, Can., 1969—; dir. Centre du Recherche en Calcul Appliqué, 1992—. Recipient Steacie prize NRC, 1980, Medaille Janssen, Academie des Sciences, Paris, 1982, Prix Vincent, ACFAS, 1979; Killam fellow Conseil des Arts, 1987-89. Office: Universite de Montreal, Dept de Physique, Montreal, PQ Canada H3C 3J7

MICHAUD, RICHARD OMER, financial economist, researcher; b. Salem, Mass., Nov. 18, 1941; s. Omer A. and Helene E. (Talbot) M.; m. Judith A. Slattery, July 20, 1968 (div. Sept. 1981); children: Robert, Christine. BA, Northeastern U., 1963; MA, U. Pa., 1966, Boston U., 1968; PhD, Boston U., 1971. Asst. prof. Boston U. Sch. Mgmt., 1976-79; dir. quantitative investment svcs. Prudential Securities, N.Y.C., 1979-83; dir. quantitative svcs. Lynch, Jones Ryan, N.Y.C., 1983-84; dir. valuation svcs. Zacks Investment Rsch., Chgo., 1984-86; head equity analytics Merrill Lynch, N.Y.C., 1986-90; dir. rsch. State St. Bank & Trust, Boston, 1990-91; sr. quantitative investment strategist Acadian Asset Mgmt., Boston, 1991—; adj. prof. Columbia U. Grad. Sch. Mgmt., 1982-87; bd. dirs. Inst. Quantitative Rsch. and Fin., N.Y.C., 1980—; editorial bd. Financial Analyst Journal, N.Y.C., 1980—. Contbr. articles to profl. jours. Recipient Graham and Dodd scroll Assn. Investment Mgmt. Rsch., N.Y.C., 1990. Office: Acadian Asset Mgmt 260 Franklin St Boston MA 02110

MICHEL, HARTMUT, biochemist; b. Ludwigsburg, Fed. Republic of Germany, July 18, 1948; m. Ilona S. Leger, 1979; 2 children. Doctorate, U. Wurzburg, 1977. With Max Planck Inst. Biochemistry, Martinsried, Federal Republic of Germany, 1979-87. Co-recipient Nobel prize for chemistry, 1988. Office: Max Planck Inst Biophysics, Heinrich-Hoffmann Str 7, 60528 Frankfurt/Main Germany

MICHEL, HENRI MARIE, medical educator; b. Saussan, Languedoc, France, Feb. 21, 1931; s. Gustave Gabriel and Adrienne Clemence (Monteils) M.; m. Elisabeth Germaine Lamouroux, July 8, 1961; children: Anne Pascale, Jacques Henri, Vincent Yves, Clara Anne. BS, Lycee of Montpellier, 1951; MD, UCLA, 1964. Externe Centre Hospitalier Regional, Montpellier, 1955-58, intern, 1958-62, medec, 1970—; resident Sch. of Medicine Lycee of Montpellier, 1970; prof. hepatology, gastroenterology, hydrology/climatology Centre Hospitalier Regional, Montpellier and Nimes, 1975—. Member conseil Adminstrn. of Hosp., Montpellier, 1974—; mem. med. com. Consultation Commn., Montpellier, 1974—; active Cours d'Appel Montpellier, 1985—. USPHS fellow, 1963-66. Mem. Nat. Gastroenterology French Soc. (pres. 1991—), Internat. Assn. for Study of Liver, French Parental Nutritional Soc. Republican. Roman Catholic. Home: Le Devois de Moulin, 34470 Murviel les, Montpellier France Office: Inst Mal Appareil Digestif, Rue Bertin Sans, 34059 Montpellier France

MICHELS, RICHARD STEVEN, microbiologist; b. Bridgeport, Conn., Dec. 27, 1951; s. Edward Joseph and Mary (Morawski) M.; m. Kathleen Bagoly (div. June 1979); 1 chld, Wade Emery; m. Linda Marie Wolak, Oct. 1984; children: Steven Douglas, Lauren Elisabeth. AA, Housatonic C.C., Bridgeport, 1973; BS in Biology, So. Conn. State U., New Haven, 1975. Microbiologist Chesebrough Ponds Inc., Clinton, Conn., 1975-80; rsch. microbiologist Chesebrough Ponds Inc., Oriskany Falls, N.Y., 1980-86; mgr. lab. control ctr. Sherwood Med., Oriskany Falls, 1986-89, microbial quality assurance mgr., 1989—. Mem. ASTM, AAAS, Am. Soc. Microbiology, N.Y. Acad. Scis., Trout Unltd (bd. dirs. 1985-91, pres. 1985, 87). Office: Sherwood Med 130 S Main St Oriskany Falls NY 13425

MICHELSEN, W(OLFGANG) JOST, neurosurgeon, educator; b. Amsterdam, Holland, Aug. 20, 1935; came to U.S., 1936; s. Jost Joseph and Ingeborg Mathilde (Dilthey) M.; m. Constance Richards, Sept. 21, 1963 (div. 1987); children: Kristina, Elizabeth, Ingrid; m. Claude Claire Grenier, Mar. 30, 1988. AB magna cum laude, Harvard U., 1959; MD, Columbia U., 1963. Diplomate Am. Bd. Neurol. Surgery. Intern in surgery Case Wester Res. U. Hosps., Cleve., 1963-64; asst. resident in neurology Mass. Gen. Hosp., Boston, 1964-65; asst. resident, then chief resident neurol. surgery Columbia-Presbyn. Med. Ctr., N.Y.C., 1965-69; from instr. to assoc. prof. neurosurgery Columbia U. Coll. Physicians and Surgeons, N.Y.C., 1969-89, prof. clin. surgery, 1990—; fellow in neurosurgery Presbyn. Hosp., N.Y.C., 1969-71, dir. neuro vascular surgery, 1989-90; dir. neurosurgery St. Luke's Roosevelt Hosp. Ctr., N.Y.C., 1990—; prof. and chmn. dept. neurological surgery Albert Einstein Coll. Medicine, Bronx, N.Y., 1992—; dir. neurosurgery Montefiore Med Ctr, Bronx, 1992—; asst. attending in neurosurgery, St. Luke's Hosp. Ctr., 1970—; cons. neurosurgeon Nyack (N.Y.) Hosp., 1972—, Englewood (N.J.) Hosp., 1972—; vis. prof. neurosurgery Tufts U., 1975, Emery U., 1977, Presbyn.-St. Luke's Hosp. Ctr., Chgo., 1978, Yale U., 1980; guest faculty Northwestern U., 1977, 78, U. Chgo., 1977, Colby Coll., 1980; mem. numerous panels on neurosurgery. Contbr. articles to profl. publs. 1st lt. U.S. Army, 1954-57. Grantee NIH, USPHS. Fellow ACS, Am. Heart Assn.; Mem. AMA, Am. Assn. Neurol. Surgeons (mem. sect. pediatric neurosurger), Neurosurg. Soc. Am. (v.p. 1984-85, pres. 1987-88), Congress Neurol. Surgeons, N.Y. Neurosurg. Soc., Neurosurg. Soc. State N.Y., N.Y. Acad. Scis., Assn. Rsch. in Nervous and Mental Diseases, Internat. Neurosurg. Soc., Internat. Pediatric Neurosurg. Soc., Explorers Club, N.Y. State Med. Soc., N.Y. County Med. Soc. Office: Montefiore Med Ctr 111 E 210th St Bronx NY 10467

MICHELSOHN, MARIE-LOUISE, mathematician, educator; b. N.Y.C., Oct. 8, 1941; d. Marcel and Lucy Friedmann; children: Didi, Michelle. BS, U. Chgo., 1962, MS, 1963, PhD, 1974. Asst. prof. U. Calif. San Diego, La Jolla, 1974-75; lectr. U. Calif., Berkeley, 1975-77; mem. Inst. des Hautes Études Scientifiques, Bures sur Yvette, France, 1977-78; asst. prof. SUNY, Stony Brook, 1978-82, assoc. prof., 1982-88, prof., 1988—; visitor Inst. Matematica Pura e Aplicada, Rio de Janeiro, 1980, Rsch. Inst. for Math. Scis., Kyoto, Japan, 1986, Tata Inst., Bombay, 1986-87; vis. mem. Inst. des Hautes Études Scientifiques, Bures-sur-Yvette, 1983-84, 93; dir. grad. program Dept. of Math SUNY, Stony Brook; rsch. prof. Math. Scis. Rsch. Inst., 1993-94. Author: Spin Geometry, 1989; contbr. articles to Am. Jour. Math., Acta Mathematica, Inventiones Mathematicae. Grantee NSF. Mem. Am. Math. Soc. Achievements include research in complex geometry, characterization of balanced spaces, Clifford and spinor cohomology, the geometry of spin manifolds and the Dirac operator, riemannian manifolds of positive curvature, the theory of algebraic cycles. Office: SUNY Dept Math Stony Brook NY 11794

MICHELSON, ALAN DAVID, pediatric hematologist; b. Melbourne, Victoria, Australia, Nov. 3, 1950; came to U.S., 1982; s. Fred Charles and Gisela (Schneider) M.; m. Lee Ann Simons, June 19, 1983; children: Daniel Simon, Sarah Kate. MB, BS, U. of Adelaide, Australia, 1973. Resident Royal Children's Hosp., Melbourne, Australia, 1975-76; registrar The Children's Hosp., Sydney, Australia, 1976-82; rsch. fellow Harvard Med. Sch., Boston, 1982-84; asst. prof. pediatrics U. Mass. Med. Sch., Worcester, 1984-88; lectr. Harvard Med. Sch., Boston, 1986-87; assoc. prof. pediatrics U. Mass. Med. Sch., Worcester, 1988—, assoc. prof. surgery, 1991—. Contbr. articles to profl. publs. and chpts. to books. Recipient Fulbright scholarship, 1982-85, First award Nat. Heart, Lung and Blood Inst., 1987-92, New Investigator Rsch. award, 1986-89, Basil O'Connor award March of Dimes, 1986-89, Grant-in-Aid, 1985-87. Mem. Soc. for Pediatric Rsch., Royal Coll. of Physicians, Am. Soc. Hematology, Am. Fedn. for Clin. Rsch. Achievements include appointment as prin. investigator of reference lab. and workshop leader for CD42 Vth Internat. Workshop on Human Leukocyte Differentiation Antigens. Office: U of Mass Med Sch 55 Lake Ave N Worcester MA 01655

MICHENER, CHARLES DUNCAN, entomologist, biologist, educator; b. Pasadena, Calif., Sept. 22, 1918; s. Harold and Josephine (Rigden) M.; m. Mary Hastings, Jan. 1, 1941; children: David, Daniel, Barbara, Walter. B.S., U. Calif., Berkeley, 1939, Ph.D., 1941. Tech. asst. U. Calif., Berkeley, 1939-42; asst. curator Am. Mus. Natural History, N.Y.C., 1942-46; assoc. curator Am. Mus. Natural History, 1946-48, research assoc., 1949—; assoc. prof. U. Kans., 1948-49, prof., 1949-89, prof. emeritus, 1989—, chmn. dept. entomology, 1949-61, 72-75, Watkins Disting. prof. entomology, 1959-89, acting chmn. dept. systematics, ecology, 1968-69, Watkins Disting. prof. systematics and ecology, 1969-89; dir. Snow Entomol. Museum, 1974-83, state entomologist, 1949-61; Guggenheim fellow, vis. research prof. U. Paraná, Curitiba, Brazil, 1955-56; Fulbright fellow U. Queensland, Brisbane, Australia, 1958-59; research scholar U. Costa Rica, 1963; Guggenheim fellow, Africa, 1966-67. Author: (with Mary H. Michener) American Social Insects, 1951, (with S.F. Sakagami) Nest Architecture of the Sweat Bees, 1962, The Social Behavior of the Bees, 1974, (with M.D. Breed and H.E. Evans) The Biology of Social Insects, 1982, (with D. Fletcher) Kin Recognition in Animals, 1987, also articles; editor: Evolution, 1962-64; Am. editor: Insectes Sociaux, Paris, 1954-55, 62-90; assoc. editor: Ann Rev. of Ecology and Systematics, 1970-90. Served from 1st lt. to capt. San. Corps AUS, 1943-46. Fellow Am. Entomol. Soc., Entomol. Soc. Am., Am. Acad. Arts and Scis., Royal Entomol. Soc. London, AAAS; mem. NAS, Linnean Soc. London (corr.), Soc. for Study Evolution (pres. 1967), Soc. Systematic Zoologists (pres. 1969), Am. Soc. Naturalists (pres. 1978), Internat. Union for Study Social Insects (pres. 1977-82), Bee Research Assn., Kans. Entomol. Soc. (pres. 1950), Brazilian Acad. Scis. (corr.). Home: 1706 W 2d St Lawrence KS 66044

MICHNOVICZ, JON JOSEPH, physician, research endocrinologist; b. Albuquerque, Nov. 9, 1953; s. John James and Mary Louise (Maloney) M.; m. Stephanie J. D'Ambra, May 19, 1990. PhD, Albert Einstein Coll. Medicine, 1982, MD, 1983. Assoc. physician Rockefeller U. Hosp., N.Y.C., 1985—, asst. prof., 1987-89; med. dir., v.p., founder Inst. for Hormone Rsch. Found. for Preventive Oncology, Inc., N.Y.C., 1989-92, pres., 1992—. Contbr. articles to profl. jours., chpts. to books. Grantee in field. Mem. Am. Soc. Preventive Oncology, Endocrine Soc., Am. Assn. Clin. Rsch. Achievements include co-discovery of plant antiestrogens as possible breast cancer prevention strategy, design of clinical programs in diet and cancer prevention; patentee on urinary hormone measurements in women. Office: Inst for Hormone Rsch 145 E 32d St New York NY 10016-6002

MICK, ELIZABETH ELLEN, medical technologist; b. Ashtabula, Ohio, Apr. 17, 1962; d. Okey Jr. and Alice Faye (Ashley) M. BS in Med. Tech., Bowling Green State U., 1984; postgrad., Northeastern U., 1986—. Registered med. technologist. Wire tuber WEK Industries, Inc., Jefferson, Ohio, 1978; student med. technologist Luth. Med. Ctr., Cleve., 1983, S.W. Gen. Hosp., Middleburg Heights, Ohio, 1983-84; bench med. technologist Brigham and Women's Hosp., Boston, 1984-85, weekend supr., 1985-90; bench med. technologist St. Elizabeth Med. Ctr. South, Edgewood, Ky., 1990—; safety technologist, chem. hygiene officer St. Elizabeth Med. Ctr. South, Covington, Edgewood, Ky., 1992—; mem. Lab. CQUI/QA Com., Covington, Edgewood, 1992—; mem. Lab. Safety Edn. Com., Covington, Edgewood, 1992—; workload recording officer St. Elizabeth Med. Ctr., Edgewood, 1991—. Named Most Outstanding 3d Shift Technologist Brigham and Women's Hosp., 1985. Mem. AAAS, Am. Soc. for Med. Technologists, Am. Soc. Clin. Pathologists, Am. Assn. Clin. Chemistry, Math. Assn. Am. Democrat. Methodist. Achievements include promotion of safe lab. practices as mandated by OSHA Standards and writing of lab. safety policy. Home: # 15 RR3 Rte 262 SR Box 266 AA Dillsboro IN 47018 Office: Saint Elizabeth Med Ctr S 1 Medical Village Dr Edgewood KY 41017

MICKELSON, ELLIOT SPENCER, quality assurance professional; b. Manti, Utah, Dec. 7, 1934; s. Rulon Spencer and Jessie (Chapman) M.; m. Zola G. Christensen, Sept. 6, 1957; children: Diana, Janet, Craig, Glenn, David. Degree in mech. engring. tech., U. Utah, 1977. Registered profl. engr., Calif. Quality engr. Hercules Inc., Salt Lake City, 1961-80; mgr. quality assurance Cambelt Internat., Salt Lake City, 1980-87, dir. tech., 1990-92; dir. quality IRECO Inc., Salt Lake City, 1987-90, BGA Internat., Salt Lake City, 1992—; cons. Quality Engring. and Mgmt. Systems, West Valley City, Utah, 1980—; chair program adv. com. Salt Lake Community Coll., West Valley City, 1979—. Author: Construction Quality Program Handbook, 1986, Quality Program Handbook, 1991, (with others) Quality Management for the Construction Project, 1987. Staff sgt. USAF, 1952-56. Recipient Dist. Award of Merit Boy Scouts Am., 1974. Fellow Am. Soc. for Quality Control (Internat. Inspector of Yr. 1974, Engr. of Yr. Salt Lake sect. 1982), Nat. Soc. Profl. Engrs., Am. Welding Soc., ASTM. Ch. Jesus Christ Latter Day Saints. Achievements include invention of mono-molded camflex and camwall belt tooling and process.

MIDAY, STEPHEN PAUL, quality assurance engineer; b. Canton, Ohio, Oct. 22, 1952; s. Albert Lawrence and Annebelle Louise (Moushey) M.; m. Linda Louise Smik, Oct. 24, 1980; children: Katrina Louise, Jacob Albert, Zachary Stephen. AS in Electronic Tech., U. Cin., 1975; BSME, Akron U., 1985. Registered profl. engr., Ohio; cert. quality engr., cert. quality auditor. Tech. supr. Telecomms., Inc., Cleve., 1976-77; technitian Bruel & Kjaer, Cleve., 1977-79; field technitian Honeywell Computers, Cleve., 1979; tech. asst. engr. Diebold, Inc., Canton, Ohio, 1979-81; quality assurance engineer Goodyear Aerospace/Loral Def., Akron, Ohio, 1981—. Mem. NSPE, Am. Soc. Quality Control, Akron Dist. Soc. Profl. Engrs. (sec. 1992-93, v.p. 93-94). Home: 13401 Crocus NW Hartville OH 44632

MIDDEN, WILLIAM ROBERT, chemist; b. Wood River, Ill., May 19, 1952. BS, St. Johns U., 1974; PhD in Biochemistry, The Ohio State U., 1978. Postdoctoral fellow The Ohio State U., 1978-79; postdoctoral fellow The Johns Hopkins U., Balt., 1980-81, rsch. assoc. dept. environ. health scis., 1981-83, asst. prof. dept. environ. health scis., 1983-87; adj. asst. prof. dept. pathology Med. Coll. of Ohio, Toledo, 1988—; asst. prof. dept. chemistry Ctr. for Photochemical Scis., Bowling Green (Ohio) State U., 1987—; vis. scientist Laboratoire Acide Nucleique, Departement de Recherche Fondamentale sur La Matiére Condensée, Service D'études des Systémes et Architectures Moléculaires, Centre D'Etudes Nucléaires, Grenoble, France, 1989; lectr. and seminar presenter in field; cons. Synergistic Lab., Libby, Mont., 1988, Fresh Products, Inc., Toledo, 1989, Calderon Automation, Inc., Bowling Green, 1989, Ricerca, Inc., Painesville, Ohio, 1988-91. Author: How to Use Chemistry, 1992; contbr. numerous articles to profl. jours. and monographs/. Nominated for Bowling Green State U. Master Tech. award, 1989, 90, 91, 92; grantee NIH, 1986, 87, Bowling Green State U., 1988, 1992-93, NSF, 1989; recipient Grad. Rsch. asst. award Bowling Green State U., 1992-93. Mem. AAAS, Am. Chem. Soc., Am. Soc. Photobiology, Environ. Mutagen Soc., Sigma Xi. Office: Bowling Green State U Dept Chemistry Bowling Green OH 43402

MIDDLEBROOK, ROBERT DAVID, electronics educator; b. England, May 16, 1929. BA, Cambridge U., England, 1952, MA, 1956; MS, Stanford U., 1953, PhD in Elec. Engring., 1955. Sr. tech. instr., mem. trade testing bd. Radio Sch. No. 3, Royal Air Force, Eng., 1947-49; asst. prof. electrical engring. Calif. Inst. Tech., Pasedena, 1955-58, assoc. prof., 1958-65, prof. electronics, 1965—; mem. hon. editorial adv. bd. Solid State Electronics, 1960-74; mem. WESCON tech. program com., 1964; lectr. 23 univs. and cos. in Eng., The Netherlands, Germany, 1965-66; mem. rsch. and tech. adv. coun. com. on space propulsion and power, NASA, 1976-77; gen. chmn. Calif. Inst. Tech. Indsl. Assocs. Conf. Power Electronics, 1982; cons. in field. Author: An Introduction to Transistor Theory, 1957, Differential Amplifiers, 1963, (with S. Cuk) Advances in Switched-Mode Power Conversion, Vols. I and II, 1981, 2d edit., 1983, Vol. III, 1983; mem. editorial bd. Internat. Jour. Electronics, 1976-82; presented 77 profl. papers; patentee in field. Recipient Nat. Profl. Group Indsl. Engrs. award, 1958, I•R 100 award Indsl. Rsch. Mag., 1980, award for the Best Use of Graphics Powercon 7, 1980, Powercon 8, 1981, William E. Newell Power Electronics award Inst. Elec. & Electronics Engrs., 1982, PCIM award for Leadership in Power Electronics Edn., 1990, Edward Longstreth Medal Franklin Inst., 1991. Fellow IEEE (exec com. San Gabriel Valley section, 1964-65, treas. 1977-78, gen. chmn. poer electronics specialists conf. 1973, AES-S electrical power/energy systems panel 1977-87, William E. Newell Power Electronics award 1982, program chmn. applied electronics conf. 1986, 87), Institution Electrical Engrs. (U.K.); mem. Inst. Radio Engrs. (Honorable Mention award 1958, subcom. 4.1 1956-62, chmn. L.A. chpt., 1960-61, vice chmn. Pasadena subsect. 1960-61), Sigma Xi. Achievements include research in new solid state devices, their development, representation and application; electronics education; power conversion and control. Office: Calif Inst Tech 116-81 Engring and Applied Sci Pasadena CA 91125

MIDDLEBUSHER, MARK ALAN, computer scientist; b. Springhill, La., Aug. 22, 1966; s. Jerry Almond and Karen Rae (York) M. BS in Computer Sci., U. South Ala., 1989. Systems analyst Internat. Paper, Vicksburg, Miss., 1989-90, sr. systems analyst, 1990-91, process systems analyst, 1991—. Bd. dirs. Vicksburg Chamber Choir, 1992-93. Mem. Digital Equipment Corp. Users Soc. Presbyterian. Home: 2501 Culkin Rd Apt C-7 Vicksburg MS 39180 Office: Internat Paper PO Drawer 950 Vicksburg MS 39181

MIDDLEDITCH, BRIAN STANLEY, biochemistry educator; b. Bury St. Edmunds, Suffolk, Eng., July 15, 1945; came to U.S., 1971; s. Stanley Stafford and Dorothy (Harker) M.; m. Patricia Rosalind Nair, July 18, 1970; 1 child, Courtney Lauren. BSc, U. London, 1966; MSc, U. Essex, 1967; PhD U. Glasgow, 1971. Rsch. asst. U. Glasgow, Scotland, 1967-71; vis. asst. prof. Baylor Coll. Medicine, Houston, 1971-75; asst. prof. U. Houston, 1975-80, assoc. prof., 1980-89, prof., 1989—; hon. prof. Eurotechnical Rsch. U. Author: Mass Spectrometry of Priority Pollutants, 1981, Analytical Artifacts, 1989, Kuwaiti Plants, 1991; editor: Practical Mass Spectrometry, 1979, Environmental Effects of Offshore Oil Production, 1981. Grantee Nat. Marine Fisheries Service 1976-80, Sea Grant Program, 1977-81, NASA, 1980-90, IBM, 1985-88, NIH, 1988—, Tex. Advanced Rsch. Program, 1988—, Nat. Dairy Coun., 1991—. Mem. Am. Chem. Soc., Am. Soc. Mass Spectrometry, World Mariculture Soc. Home: 4101 Emory Ave Houston TX 77005-1920 Office: U Houston Dept Biochemistry Houston TX 77204-5934

MIDDLETON, ANTHONY WAYNE, JR., urologist, educator; b. Salt Lake City, May 6, 1939; s. Anthony Wayne and Dolores Caravena (Lowry) M.; BS, U. Utah, 1963; MD, Cornell U., 1966; m. Carol Samuelson, Oct. 23, 1970; children: Anthony Wayne, Suzanne, Kathryn, Jane, Michelle. Intern, U. Utah Hosps., Salt Lake City, 1966-67; resident in urology Mass. Gen. Hosp., Boston, 1970-74; practice urology Middleton Urol. Assos., Salt Lake City, 1974—; mem. staff Primary Children's Hosp., staff pres., 1981-82; mem. staff Latter-Day Saints Hosp., Holy Cross Hosp.; assoc. clin. prof. surgery U. Utah Med. Coll., 1977—; vice chmn. bd. govs. Utah Med. Self-Ins. Assn., 1980-81, chmn. 1985-87. Bd. dirs. Utah chpt. Am. Cancer Soc., 1978-86; bishop, later stake presidency Ch. Jesus Christ Latter-day Saints; vice chmn. Utah Med. Polit. Action Com., 1978-81, chmn., 1981-83; chmn. Utah Physicians for Reagan, 1983-84; mem. U. Utah Coll. Medicine Dean's Search Com., 1983-84; bd. dirs. Utah Symphony, 1985—. Capt. USAF, 1968-70. Mem. ACS, Utah State Med. Assn. (pres. 87-88), Am. Urologic Assn. (socioecons. com. 1987—), AMA (alt. del. to House of Dels. 1989-92), Salt Lake County Med. Assn. (sec. 1965-67, pres. liaison com. 1980-81, pres.-elect 1981-83, pres. 1984), Utah Urol. Assn. (pres. 1976-77), Salt Lake Surg. Soc. (treas. 1977-78), Am. Assn. Clin. Urologists (bd. dirs. 1989-90, nat. pres. elect 1990-91, pres. 1991-92, nat. bd. chmn. urologic polit. action com. 1992—), Phi Beta Kappa, Alpha Omega Alpha, Beta Theta Pi (chpt. pres. Gamma Beta 1962). Republican. Contbr. articles to profl. jours. Home: 2798 Chancellor Pl Salt Lake City UT 84108-2835 Office: 1060 W 1st S Salt Lake City UT 84102-1501

MIDDLETON, DAVID, physicist, applied mathematician, educator; b. N.Y.C., Apr. 19, 1920; s. Charles Davies Scudder and Lucile (Davidson) M.; m. Nadea Butler, May 26, 1945 (div. 1971); children: Susan Terry, Leslie Butler, David Scudder Blakeslee, George Davidson Powell; m. Joan Bartlett Reed, 1971; children: Christopher Hope, Andrew Bartlett, Henry H. Reed. Grad., Deerfield Acad., 1938; AB summa cum laude, Harvard U., 1942, AM, 1945, PhD in Physics, 1947. Teaching fellow electronics Harvard U., Cambridge, Mass., 1942, spl. rsch. assoc., Radio Rsch. Lab., 1942-45, NSF predoctoral fellow physics, 1945-47, rsch. fellow electronics, 1947-49, asst. prof. applied physics, 1949-54; cons. physicist Cambridge, 1954—, Concord, Mass., 1957-71, N.Y.C., 1971—; adj. prof. elec. engring. Columbia U., 1960-61; adj. prof. applied physics and communication theory Rensselaer Poly. Inst., Hartford Grad. Ctr., 1961-70; adj. prof. communication theory U. R.I., 1966—, adj. prof. math. scis. Rice U., 1979-89, U.S. del. internat. conf. Internat. Radio Union, Lima, Peru, 1975; lectr. NATO Advanced Study Inst., Grenoble, France, 1964, Copenhagen, 1980, Luneburg, Fed. Republic Germany, 1984; Naval Rsch. Adv. Com., 1970-77, Inst. of Def. Analyses: Supercomputing Rsch. Ctr. Sci. Adv. Bd., 1987-91; cons. physicist since 1946 Johns Hopkins U., SRI Internat., Rand Corp., USAF, Cambridge Rsch. Ctr., Comm. Satellite Corp., Lincoln Lab., NASA, Raytheon, Sylvania, Sperry-Rand, Inst. of Def. Analyses, Office of Naval Rsch., Applied Rsch. Labs., U. Tex., GE, Honeywell Transp. Systems Ctr. of Dept. Transp., Dept. Commerce Office of Telecom., NOAA, Office Telecom. Policy of Exec. Office Pres., Nat. Telecom.and Info. Adminstrn., Sci. Applications Inc., Naval Undersea Systems Ctr., Lawrence Livermore Nat. Labs., others. Author: Introduction to Statistical Communication Theory, 1960, 87, Russian edit. Soviet Radio Moscow, 2 vols., 1961, 62, Topics in Communication Theory, 1965, 87, Russian edit., 1966; sci. editor: English edit. Statistical Methods in Sonar (by V.V. Ol'shevskii), 1978; mem. editorial bd. Info. and Control, Advanced Serials in Electronics and Cybernetics, 1971-82; contbr. articles to tech. jours. Recipient award (with W.H. Huggins) Nat. Electronics Conf., 1956; Wisdom award of honor, 1970; First prize 3d Internat. Symposium on Electromagnetic Compatibility Rotterdam, Holland, 1979; awards U.S. Dept. Commerce, 1978. Fellow AAAS., IEEE (life, awards 1977, 79), Am. Phys. Soc., Explorers Club, Acoustical Soc. Am., N.Y. Acad. Scis.; mem. Am. Math. Soc., Author's Guild Am., Electromagnetics Acad. MIT, Harvard Club (N.Y.C.), Cosmos Club (Washington), Dutch Treat (N.Y.C.), Phi Beta Kappa, Sigma Xi. Achievements include research in radar, telecommunications, underwater acoustics, oceanography, seismology, systems analysis, electromagnetic compatibility, communication theory; pioneering research in statistical communication theory. Home and Office: 127 E 91st St New York NY 10128-1601 also: MIND 48 Garden St Cambridge MA 02138 also: 13 Harbor Rd Harwich Port MA 02646

MIDDLETON, GERARD VINER, geology educator; b. Capetown, South Africa, May 13, 1931; s. Reginald Viner Cecil and Doris May (Hutchinson) M.; m. Muriel Anne Zinkewich, Apr. 4, 1959; children: Laurence, Teresa, Margaret. B.Sc., Imperial Coll., London U., 1952; Ph.D., Diploma, Imperial Coll., 1954. Geologist, Calif. Standard Oil Co., Calgary, Alta., Can., 1954-55; mem. faculty dept. geology McMaster U., Hamilton, Ont., Can., 1955—; assoc. prof. McMaster U., 1962-67, prof., 1967—, chmn. dept. geology, 1959-62, 78-84; vis. prof. Calif. Inst. Tech., 1964-65, Oxford (Eng.) U., 1971, Stanford U., 1978, Elf-Aquitaine Co., France, 1984-85, U. Washington, 1991-92; cons. oil cos. Author: (with H. Blatt and R. Murray) Origin of Sedimentary Rocks, 1972; editor: Primary Sedimentary Structures, 1965; contbr. numerous articles on sedimentology to profl. jours.; editor: Geosci. Can, 1974-78. Recipient Francis J. Pettijohn Sedimentary medal Soc. Sedimentary Geology, 1994. Fellow Royal Soc. Can.; hon. mem. Internat. Assn. Sedimentologists (v.p. 1972-82), Soc. Econ. Paleontologists and Mineralogists; mem. Geol. Assn. Can. (Logan medal 1980, v.p. 1986-87, pres. 1987-88), Am. Assn. Petroleum Geologists, Can. Soc. Petroleum Geologists. Roman Catholic. Club: Dundas Valley Curling.

MIDDLETON, MICHAEL JOHN, civil engineer; b. N.Y.C., May 14, 1953; s. Vincent Aloysius and Mary Hilda (Lehane) M. BS in Civil Engring., U. Calif., Davis, 1975. Registered profl. engr., Calif., Wash., Hawaii. Project mgr. G.A. Fitch & Assoc., Concord, Calif., 1975-78, v.p., 1978-80; project mgr. Santina & Thompson, Inc., Concord, 1980-83, dir. engring., 1983-88, sr. v.p., 1988—. scholar, Calif. Scholarship Fedn., 1971. Mem. ASCE, Nat. Soc. Profl. Engrs., Soc. Am. Mil. Engrs. Roman Catholic. Home: 1409A Bel Air Dr Concord CA 94521 Office: Santina & Thompson Inc 1355 Willow Way Ste 280 Concord CA 94520

MIDGLEY, A(LVIN) REES, JR., reproductive endocrinology educator, researcher; b. Burlington, Vt., Nov. 9, 1933; s. Alvin Rees and Maxine (Schmidt) M.; m. Carol Crossman, Sept. 4, 1955; children: Thomas, Debra, Christopher. B.S. cum laude, U. Vt., 1955, M.D. cum laude, 1958. Intern U. Pitts., 1958-59, resident dept. pathology, 1959-61; resident dept. pathology U. Mich., Ann Arbor, 1961-63, instr. pathology, 1963-64, asst. prof., 1964-67, assoc. prof., 1967-70, prof., 1970—, dir. Reproductive Scis. Program; chmn. BioQuant of Ann Arbor, Inc., 1985-89. Contbr. articles to med. jours. Recipient Parke-Davis award, 1970; Ayerst award Endocrine Soc., 1977; Smith Kline Bio-Sci. Labs. award, 1985; NIH grantee, 1960—; Mellon Found. grantee, 1979-91. Mem. Soc. Study Reprodn. (pres. 1983-84), Endocrine Soc., Am. Assn. Pathology, Am. Physiol. Soc. Home: 3600 Tubbs Rd Ann Arbor MI 48103-9437 Office: U Mich Reproductive Scis Program 300 N Ingalls St Fl 11 Ann Arbor MI 48109-2007

MIELE, ANGELO, engineering educator, researcher, consultant, author; b. Formia, Italy, Aug. 21, 1922; came to U.S., 1952, naturalized, 1985; s. Salvatore and Elena (Marino) M. D.Civil Engring., U. Rome, Italy, 1944, D.Aero. Engring., 1946; DSc (hon.), Inst. Tech., Technion, Israel, 1992. Asst. prof. Poly. Inst. Bklyn., 1952- 55; prof. Purdue U., 1955-59; dir. astrodynamics Boeing Sci. Research Labs., 1959-64; prof. aerospace scis., math. scis. Rice U., Houston, 1964-88, Foyt Family prof. engring., 1988—; cons. Douglas Aircraft Co., 1956-58, Allison div. Gen. Motors Corp., 1956-58, U.S. Aviation Underwriters, 1987, Boeing Comml. Airplane Co., 1989. Author: Flight Mechanics, 1962; editor: Theory of Optimum Aerodynamic Shapes, 1965, Math. Concepts and Methods in Sci. and Engring., 1974—; editor in chief Jour. Optimization Theory and Applications, 1966—; assoc. editor Jour. Astronautical Scis., 1964—, Applied Math. and Computation, 1975—, Optimal Control Applications and Methods, 1979—; mem. editorial bd. RAIRO-Ops. Rsch., 1990—; mem. adv. bd. AIAA Edn. Series, 1991—; contbr. numerous research papers in aerospace engring., windshear problems, hypervelocity flight, math. programming, optimal control theory and computing methods. Pres. Italy in Am. Assn., 1966-68. Decorated knight comdr. Order Merit Italy, 1972; recipient Levy Medal Franklin Inst. of Phila., 1974, Brouwer award AAS, 1980, Schuck award Am. Automatic Control Coun., 1988. Fellow AIAA (Pendray award 1982, Mechanics and Control of Flight award 1982, mem. adv. bd. Edn. Series 1991-94), Am. Astronautical Soc., Franklin Inst.; mem. Internat. Acad. Astronautics, Acad. Scis. Turin (corr.). Home: 3106 Kettering Dr Houston TX 77027-5504 Office: Rice U Aero-Astronautics Group PO Box 1892 Houston TX 77251-1892

MIELE, JOEL ARTHUR, SR., civil engineer; b. Jersey City, May 28, 1934; s. Jene Gerald Sr., and Eleanor Natale (Bergida) M.; m. Faith Roseann Trombetta, July 21, 1952 (div. 1954); m. 2d Josephine Ann Cottone, Feb. 14, 1959; children: Joel Arthur, Jr., Vita Marie, Janet Ann. B.C.E., Poly. Inst. Bklyn., 1955. Registered profl. engr., N.Y., N.J.; profl. planner, N.J. Civil engr. Yudell & Miele, Queens, N.Y., 1955-57; chief engr. Jene G. Miele Assocs., Queens, 1960-68; prin., chief exec. officer Miele Assocs., Queens, 1968—. Patentee masonry wall constrn. Commr. N.Y.C. Planning Commn., 1990—; pres. bd. visitors Creedmoor State Hosp., 1979—; pres., bd. dirs. Peninsula Hosp. Ctr., 1990—, Peninsula Nursing Home, 1990—; chmn. Community Bd. 10, Queens, 1978-90; trustee, treas. Queens Pub. Communications Corp., 1983—; trustee, past treas., sec. Queens Borough Pub. Libr., 1979—; pres., bd. dirs. Queen County Overall Econ. Devel. Corp., 1989—; exec. v.p. Queens coun. Boy Scouts Am., 1991—. Lt. (j.g.) USN, 1958-60; capt. USNR, 1960-88. Named Italian-Am. of Yr. Ferrini Welfare League, Queens, 1980; recipient Outstanding Community Leader award Boy Scouts Am., 1987, Pride of Queens award, 1990. Fellow ASCE; mem. ASTM, NSPE (trustee polit. action com. 1988—), N.Y. State Soc. Profl. Engrs. (v.p. 1984-86, pres. 1988-89, nat. dir. 1987-90, Engr. of Yr. 1983, pres. Queens chpt. 1980-82), Soc. Am. Mil. Engrs., N.Y. State Assn. of Professions (founding), Ozone Howard C. of C. (pres. 1980-84, 86-91), Am. Parkinson Disease Assn. (dir. 1985—, exec. com. 1987—). Democrat. Congregationalist. Office: Miele Assocs 81-01 Furmanville Ave Middle Village NY 11379

MIELKE, PAUL WILLIAM, JR., statistician; b. St. Paul, Feb. 18, 1931; s. Paul William and Elsa (Yungbauer) M.; m. Roberta Roehl Robison, June 25, 1960; children: William, Emily, Lynn. BA, U. Minn., 1953, PhD, 1963; MA, U. Ariz., 1958. Teaching asst. U. Ariz., Tucson, 1957-58; teaching asst. U. Minn., Mpls., 1958-60, statis. cons., 1960-62, lectr., 1962-63; from asst. to assoc. prof. dept. statistics Colo. State U., Fort Collins, 1963-72, prof. dept. statistics, 1972—. Contbr. articles to Am. Jour. Pub. Health, Jour. of Statis. Planning and Inference, Ednl. and Psychol. Measurement, Biometrika, Earth-Sci. Revs. Capt. USAF, 1953-57. Fellow Am. Statis. Assn.; mem. Am. Meteorol. Soc., Biometric Soc. Achievements include proposal that common statistical methods (t test and analysis of variance) were based on counter intuitive geometric foundations and provided alternative statistical methods which are based on appropriate foundations. Home: 736 Cherokee Dr Fort Collins CO 80525-1517 Office: Colo State U Dept Stats Fort Collins CO 80523

MIFFLIN, RICHARD THOMAS, applied mathematician; b. Havre de Grace, Md., Apr. 11, 1959; s. Thomas Roche and Carol Ann (Myers) M. BSChemE, Rice U., 1980; PhDChemE, Princeton (N.J.) U., 1986. Lectr. Dept. Chem. Engring. Princeton U., Princeton, N.J., 1984-85; mem. rsch. staff Exxon Prodn. Rsch. Co., Houston, 1985—. Contbr. articles to profl. jours. Hertz fellowship Fannie and John Hertz Found., 1981. Mem. AICE, Soc. of Petroleum Engrs., Soc. of Indsl. and Applied Math., Math. Assn. of Am. (mem. U.S. math team 1976, Putnam fellowship 1979), Soc. of Rheology. Achievements include rsch. on viscosity of spheres in viscoelastic fluids, resistance functions for spheres, formulation for thermal reservoir process simulation. Office: Exxon Prodn Rsch Co PO Box 2189 Houston TX 77252-2189

MIFSUD, LEWIS, electrical engineer, fire origin investigator, physicist; b. Zabbar, Malta, July 9, 1932; came to U.S., 1964; s. Felix and Josephine (Bonello) M.; m. Christine McFarland, July 14, 1989. BSc in Spl. Physics cum laude, U. London, 1963; MSEE, Rutgers U., 1968, PhD in Elect. Engring. and Physics, 1970. Lic. and registered, cert. elec. engr.; cert. quality engr., Am. Soc. Quality Control, cert. telecom. engr., London; cert. fire and explosion investigator, cert. fire investigation instr. Nat. Assn. Fire Investigators, cert. fire investigator Internat. Assn. of Arson Investigators. Apprentice electrician Elect. Installations Co., Ltd., London, 1947-49; equipment operator, asst. svc. engr. Rediffusion Broadcasting Co., London, 1949-51; svc. and maintenance engr. E.M.I. Industries, Staines, England, 1951-54; electronics product design, devel. engr., project group lead E.M.I. Electronics, Hayes, Middlesex, England, 1954-61; product designs and devel. engr., group leader Data Recording Instruments Co., Staines, Middlesex, England, 1961-64; electronics designs lead engr. and project leader R.C.A. Astro Electronics Div., N.J., 1964-68; sci. researcher and instr. Rutgers U., New Brunswick, N.J., 1968-70; tenured asst. prof. physics Pa. State U., Abington, 1970-88; pvt. practice as forensic engr., physicist, cons. Jenkintown, Pa., 1981—; presenter in field. Contbr. articles to Fire Jour. Recipient Cert. Achievement in fire investigation U.S. Fire Acad. the Fed. Emergency Mgmt. Agy., 1983. Mem. ASTM, AICE, Internat. Assn. Arson Investigators, Nat. Fire Protection Assn., Am. Acad. Foresic Scis., Am. Soc. Safety Engrs., Am. Soc. for Quality Control, Am. Soc. for Metals (1st and 3rd pl. award 1991), Soc. Automotive Engrs., Pa. Assn. Elect. Inspectors, Pa. Acad. Sci., N.J. Assn. Arson Investigators. Achievements include rsch. in criteria for ignition of flammable vapors by electrostatic discharge, elect. grounding & mech. protection of cables, a miniaturized gas/vapor chromatograph & hydrocarbon detector, gas & accelerant detection in arsons. Home and Office: 5 Township Line Rd Jenkintown PA 19046

MIGNANI, ROBERTO, physics educator and researcher; b. Messina, Italy, Jan. 28, 1946; s. Pietro and Maria Antonietta (Parise) M.; m. Rosa Amato, June 17, 1976; children: Ruggero, Diana. PhD in Physics, U. Palermo (Italy), 1970. Fellow, Inst. for Theoretical Physics, Catania U. (Italy), 1971-73, dept. physics, Università La Sapienza, Rome, 1974, assoc. prof. physics, 1977-80, prof. electrodynamics, 1981-92; assoc. prof. dept. physics L'Aquila U. (Italy), 1975-76; prof. theoretical physics Inst. for Basic Research, Palm Harbor, Fla., 1981-91; prof. theoretical physics dept. physics E. Amaldi III Univ. di Roma, 1993—; research assoc. Sicilian Ctr. for Nuclear Physics and Structure of Matter, Catania, 1971-74, Italian Nat. Inst. for Nuclear Physics (INFN), Rome, 1971—, Italian Nat. Orgn. for Nuclear Energy (ENEA), Rome, 1984. Contbr. articles on physics to profl. jours.; editor Hadronic Jour., 1981-91; (book) Selected Papers of Italian Physicists: Piero Caldirola vols. I-IV, 1986. Fellow Italian Phys. Soc. Office: I Universita di Roma, LaSapienza P le A Moro 2, 00185 Rome Italy

MIGNELLA, AMY TIGHE, environmental engineering researcher; b. Phoenix, Sept. 21, 1964; d. Michael Jr. and Eleanor (Tighe) M. Student, U. Ariz., 1991—, BA in Chemistry, 1987; MSCE, Stanford U., 1990. Sales engr. Photometrics, Ltd., Tucson, 1988, chemist, 1988-89; rsch. asst. dept. civil engrin. Stanford U., 1989-90; environ. engr. Edler & Kalinowski, San Mateo, Calif., 1991. Vol., law clk. Ariz. Ctr. for Law in Pub. Interest, Tucson, 1992. Mem. Am. Soc. Civil Engrs., Nat. Lawyers Guild.

MIHAILEANU, ANDREI CALIN, energy researcher; b. Arad, Romania, Mar. 5, 1923; s. Gheorghe M. and Cleopatra (Bestelei) Pascu; m. Ileana Dana Demetrescu, July 15, 1948 (div. 1954); 1 child, Serban Alexandru; m. Simona Niculescu, Jan. 7, 1957; 1 stepson, Dan Georgescu. Diploma Electromech. Engring., Polytech. Inst., Bucharest, Romania, 1945, D in Elec. Engring., 1972. Profl. elect. engr., energetics. Power plant engr. Concordia-Electrica, Câmpina, Romania, 1945-47; chief engr. Regional Electricity Utility, Bucharest, 1947-52; dep. dir. Tech. Div., Ministry of Elec. Energy/Electrotech. Industry, Bucharest, 1952-58, sr. rsch. officer, 1958-64; gen. mgr. Elec. Energy Dept. Ministry of Mines and Elec. Energy, Bucharest, 1964-65; researcher and dir. Elec. Energy and Co-Generation Rsch. Inst., Bucharest, 1967-74; rsearcher and gen. mgr. Cen. Energy Rsch. Inst., Bucharest, 1974-84, researcher and sci. sec., 1984-87; cons., sr. rsch. officer Energy Rsch. and Modernising Inst.-ICEMENERG, Bucharest, Romania, 1990—; asst. prof. Polytech. Inst., Bucharest, 1948-50, assoc. prof., 1950-70. Author five books, three inventions and over 80 scientific papers in field, 1946—; editor: Energetica, 1953-87. Recipient Traian Vuia prize Romanian Acad., 1979, Scientific Merit 1st Class Order, State Council, Bucharest, 1974. Mem. UNEcon. Commn. for Europe-Geneva (vice-chmn. Electric Power Com. 1956-58), Internat. Conf. on Large High Voltage Electric Systems, Romanian Nat. Com. CIGRE (chmn. 1976—, CIGRE SC 37 internat. study com. 1984-87, 91—), World Energy Coun. (conservation commn. 1974-87). Christian Orthodox. Avocations: breeding cocker spaniels and miniature pinschers. Home: Piata Alex Sahia 4 of post 22, RO 70203 Bucharest Romania Office: ICEMENERG, Bd Energeticienilor 8, R-74568 Bucharest Romania

MIHALOV, JOHN DONALD, aerospace research scientist; b. L.A., Dec. 28, 1937; s. John and Alice Alma Lydia (Wagner) M. BS, Calif. Inst. Tech., Pasadena, 1959, MS, 1961; degree in Space Sci. Engring., Stanford U., 1981. Mem. tech. staff Space Tech. Labs, Inc., El Segundo, Calif., 1959-60; asst. Cornell U., Ithaca, N.Y., 1960-61; mem. tech. staff The Aerospace Corp., El Segundo, 1961-66; rsch. scientist NASA, Moffett Field, Calif., 1966—. Contbr. articles to profl. publs. Mem. Am. Geophys. Union, Am. Phys. Soc., AAAS, Am. Astron. Soc. (Div. Planetary Sci.). Achievements include contbg. to energy-resolved proton measurements in earth's trapped radiation, to shock propagation in outer heliosphere, to measurements of solar wind proton parameter gradients in outer heliosphere, beyond major system planets. Office: Nasa-Ames Rsch Ctr 245-3 Moffett Field CA 94035-1000

MIHICH, ENRICO, medical researcher; b. Fiume, Italy, Jan. 4, 1928; came to U.S., 1957; s. Milan and Rosina (Lenaz) M.; m. Renata Marisa Mustacchi, Sept. 25, 1954; 1 child, Sylvia. B.S., U. Milan, Italy, 1944, M.D., 1951, docent, 1962; MD (honoris causa), U. Marseille, 1986. Research asst. Inst. Pharmacology U. Milan, Italy, 1951, asst. prof., 1954-56; vis. research fellow Sloan Kettering Inst. Cancer Research, N.Y.C., 1952-54; head pharmacology lab. Valeas Pharm. Industry, Milan, 1954-56; sr. cancer research scientist dept. exptl. therapeutics Roswell Park Cancer Inst., Buffalo, 1957-59, assoc. cancer research scientist, 1959-66, prin. scientist, 1966-71, dir. dept. exptl. therapeutics and Grace Cancer Drug Ctr., 1971—, assoc. dir. for sponsored programs, 1987—; prof. pharmacology SUNY-Buffalo, 1960—, research asst., 1960-66, research assoc., 1966-68, research prof. pharmacology, 1968-69, chmn. dept. pharmacology, 1969—; assoc. prof. biochem. pharmacology Sch. of Pharmacy, 1963-68, adj. prof. biochem. pharmacology, 1968—; cons., lectr. in field; participant numerous symposia; sci. advisor govt. agys., pvt. industry; mem. Nat. Cancer Adv. Bd., 1984-90. Author more than 230 books, articles, chpts. in books; editor in chief for N.Am. and Japan, Cancer Immunology and Immunotherap; mem. editorial bd. Advances in Cancer Chemotherapy, Internat. Jour. Immunopharmacology, Cancer and Metastasis Revs., J. Liposome Research, etc.; adv. editor Cancer Communications, Selective Cancer Therapeutics jours. Recipient numerous grants for med. research; Fulbright travel fellow, 1952-53; Sloan Found. fellow, 1953-54; Myron Karon Meml. Lectr., 1981. Mem. Am. Soc. Pharmacology and Exptl. Therapeutics, Am. Assn. Immunologists, Am. Assn. Cancer Research (pres. 1988), Am. Cancer Soc., Am. Soc. Clin. Oncology, Am. Coll. Clin. Pharmacology, Soc. Exptl. Biology and Medicine, Transplantation Soc., AAAS, Am. Chem. Soc. (div. medicinal chemistry), Internat. Soc. Biochem. Pharmacology, Internat. Soc. Immunopharmacology, European Assn. Cancer Research, N.Y. Acad. Scis., N.Y. State Soc. Med. Research, Sigma Xi. Office: Grace Cancer Drug Ctr Roswell Park Cancer Inst Elm and Carlton Sts Buffalo NY 14263*

MIHM, JOHN CLIFFORD, chemical engineer; b. Austin, Tex., July 28, 1942; s. Clifford Henry and Adeline (Cleary) M.; m. Janet Elanor Skales, May 29, 1964; 1 child, Mary Lynn. AA, Frank Phillips Coll., 1962; BSChemE, Tex. Tech. Engring., 1964. Registered profl. engr., Tex. With Phillips Petroleum Co., 1964—; v.p. R & D Phillips Petroleum Co., Bartlesville, Okla., 1992—; engr. mgr. E & P Phillips Petroleum Co., Stavanger, Norway, 1977-82; div. mgr. R & D Phillips Petroleum Co.; adv. bd. Tex. Tech. Engring., Lubbock, Tex., 1985—. Bd. dirs. Boy Scouts Am., Bartlesville, 1986—. Mem. ASME (ind. adv. bd. 1989—), NSPE, Am. Inst. Chem. Engring. (bd. dirs. 1989-93, chmn. 1992-93), Okla. Soc. Profl. Engrs. (Outstanding Engr. in Mgmt. award 1991), Soc. Profl. Engrs. Republican. Roman Catholic. Office: Phillips Petroleum Co 260 RF Bartlesville OK 74004

MIHNEA, TATIANA, mathematics educator; b. Bucharaest, Romania, May 24, 1951; came to U.S., 1984; m. Andrei Mihnea, May 29, 1976; 1 child, Radu. BS, U. Bucharest, 1975, MS, 1976. Community coll. instr. credential, Calif. Tchr. math. high sch., Bucharest, 1976-82, Lawrence Acad., Santa Clara, Calif., 1985-86; lectr. math. San Jose (Calif.) State U., 1986-87; instr. math. West Valley Coll., Saratoga, Calif., 1986—. Contbr. articles to Poetics. Mem. Math. Assn. Am., Am. Math. Soc. Home: 4967 Kenlar Dr San Jose CA 95124-5106 Office: West Valley Coll Dept Math 14000 Fruitvale Ave Saratoga CA 95070-5640

MIKALOW, ALFRED ALEXANDER, II, deep sea diver, marine surveyor, marine diving consultant; b. N.Y.C., Jan. 19, 1921; m. Janice Brenner, Aug. 1, 1960; children: Alfred Alexander, Jon Alfred. Student Rutgers U., 1940; MS, U. Calif., Berkeley, 1948; MA, Rochdale U. (Can.), 1950. Owner Coastal Diving Co., Oakland, Calif., 1950—, Divers Supply, Oakland, 1952—; dir. Coastal Sch. Deep Sea Diving, Oakland, 1950—; capt. and master rsch. vessel Coastal Researcher I; mem. Marine Inspection Bur., Oakland. marine diving contractor, cons. Mem. adv. bd. Medic Alert Found., Turlock, Calif., 1960—. Lt. comdr. USN, 1941-47, 49-50. Decorated Purple Heart, Silver Star. Mem. Divers Assn. Am. (pres. 1970-74), Treasury Recovery, Inc. (pres. 1972-75), Internat. Assn. Profl. Divers,

Assn. Diving Contractors, Calif. Assn. Pvt. Edn. (no. v.p. 1971-72), Authors Guild, Internat. Game Fish Assn., U.S. Navy League, U.S. Res. Officers Assn., Tailhook Assn., U.S. Submarine Vets. WWII, Explorer Club (San Francisco), Calif. Assn. Marine Surveyors (pres. 1988—), Masons, Lions. Author: Fell's Guide to Sunken Treasure Ships of the World, 1972; (with H. Rieseberg) The Knight from Maine, 1974. Office: 320 29th Ave Oakland CA 94601

MIKESELL, WALTER R., JR., mechanical engineer, engineering executive. BSChemE, MSChemE, Purdue U., 1957. Registered profl. engr., Calif., Tenn. Engr. Chgo. Bridge and Iron Co., Oakbrook, Ill., 1957-59, project engr. Operation Cryogenics, 1959-60, devel. engr., cons., 1960-66, mgr. stress analysis, 1968-72, from mgr. design engring. to asst. chief engr., 1976-83, from asst. chief engr. to chief mech. engr., 1983-86; mgr. engring. CBI Nuclear Co., Memphis, 1972-76; sr. cons. Robert L. Cloud & Assocs., Berkeley, Calif., 1986—; former mem. com. engring. Am. Bur. Shipping. Fellow ASME (boiler code coms., chmn. bd. accreditation, vice-chair coun. codes and standards, J. Hall Taylor medal 1989, Codes and Standards medal 1992), Materials Properties Coun. Office: Robert L. Cloud & Assoc Ste 1200 2150 Shattuck Ave Berkeley CA 94704*

MIKROPOULOS, ANASTASSIOS (TASSOS MIKROPOULOS), physicist, researcher; b. Kilkis, Greece, Feb. 13, 1961; s. Aristotelis and Androniki Mikropoulos; m. Ioanna Bellou, Jan 26, 1985; children: Aristotelis, Leonidas. BS in Physics, U. Ioannina, Greece, 1983; PhD in Physicis (hon.), U. Athens, Greece, 1990. Rsch. assoc. Nat. Hellenic Rsch. Found., Athens, 1984-89, 1991-92; tchr. physics Technol. and Ednl. Inst., Athens, 1990-91; lectr. Computers in Edn., dept. Edn. U. Ioannina, Greece; con. laser systems Fotoniki Ltd., Athens, 1988-90; lectr. dept. edn., U. Ioannina, Greece, 1992. Contbr. articles to profl. jours. Sgt. Hellenic Air Force, 1990-91. Royal Soc. scholar, 1986; Hellenic Orgn. for Small-Medium Cos. and Handicraft grantee, 1988. Mem. Am. Phys. Soc., Assn. Advancement of Computing in Edn., Soc. Tech. and Tchr. Edn., Hellenic Phys. Soc., Hellenic Lasers and Optronics Assn. Achievements include patent for flashlamp-pumped dye laser lithotripter. Home: PL Pargis 16, 453 32 Ioannina Greece Office: U Ioannina Dept Edn, Doboli 30, 45110 Ioannina Greece

MILAD, MOHEB FAWZY, consultant, urologist; b. Cairo, Arab Republic Egypt, Mar. 16, 1945; s. Fawzy Milad and Basima Helmi; m. Dalal Kamal; children: Nadine Moheb, Michael Moheb. MB. BCh., Ain Shams U., Cairo, 1969; diploma in urology, Inst. of Urology and U. London, 1985. Intern Ain Shams U. Hosp., Cairo, 1969-70; med. officer Amoco Oil Co., Cairo, 1970-71; sr. house officer Newcastle Area Health Authority, Newcastle, Eng., 1971-73; sr. house officer, jr. registrar Coventry (Eng.) Health Authority, 1974-76; registrar East Sussex Health Authority, Eastbourne, Brighton, Eng., 1976-78; asst. surgeon, urologist Internat. Hosp., Bahrain, 1978-80, chief of staff, 1978-80, cons. in charge, 1978-80; chmn. jour. club Dhahran (Saudi Arabia) Health Ctr. (Aramco), 1982-86, chmn. urology meeting, 1990—. Contbr. articles to profl. jours. Fellow Royal Coll. Surgeons Edinburgh, Internat. Coll. Surgeons; mem. Royal Coll. Surgeons Eng., Brit. Assn. Urol. Surgeons. Avocations: reading, squash, swimming, water skiing. Home: Saudi Aramco, Box 10005, Dhahran 31311, Saudi Arabia Office: Dhahran Health Ctr Saudi, Aramco Box 76, Dhahran 31311, Saudi Arabia

MILAM, JOHN DANIEL, pathologist, educator; b. Kilgore, Tex., May 22, 1933; s. Ott G. and Effie (White) M.; m. Carol Jones Milam, Aug. 1, 1959; children: Kay, Beth, John Jr., Julie. BS, La. State U., 1955, MS, 1957, MD, 1960. Attending pathologist St. Luke's Episcopal Hosp., Houston, 1967-89; cons. in pathology Tex. Children's Hosp., Houston, 1979—; prof. lab. medicine M.D. Anderson Cancer Ctr., U. Tex., Houston, 1990—; prof. pathology and lab. medicine Health Sci. Ctr. U. Tex., Houston, 1989—. Contbr. numerous articles to profl. jours., chpts., abstracts to books. Trustee Am. Bd. Pathology, 1985—; bd. dirs. Harris County chpt. ARC, 1978—. Mem. Am. Assn. Blood Banks (pres. 1984, Disting. Svc. award 1988), Tex. Soc. Pathologists (George T. Caldwell award 1981). Republican. Baptist. Home: 11927 Arbordale Houston TX 77024 Office: U Tex Houston Med Sch Dept Pathology 6431 Fannin Rm 2022 Houston TX 77030

MILANI-COMPARETTI, MARCO SEVERO, geneticist, bioethicist; b. Florence, Italy, May 15, 1926; s. Piero and Luisa (Fatichi) M.-C.; m. Donatella R. Riccitelli, Apr. 21, 1966; children: Alfredo P., Alessia V. PhD in Biology, Rome U., 1969. Dir. internat. rsch. Am. Inst. Mgmt., N.Y.C., 1953-56; geneticist The Gregor Mendel Inst., Rome, 1956-69; asst. prof. Rome U. Med. Sch., 1970-71; asst. prof. Ancona (Italy) U. Med. Sch., 1972-79, assoc. prof. human genetics and bioethics, 1980—; dir. med. sch. Inst. Biology & Genetics, Ancona, 1974—; asst. sec. gen. permanent com. Internat. Congresses of Human Genetics, 1961-91; sci. dir. Internat. Inst. for Ethical-Juridical Studies of New Biology, Milazzo, Italy, 1986—; sci. sec. Olympic Games Sci. Com., Rome, 1960. Author and co-author several books; contbr. articles to profl. jours.; co-editor Internat. Jour. Bioethics. Coun. mem. Provincial Coun., Ancona, 1983-85. Recipient Philip Noel Baker Rsch. prize Internat. Coun. Sports and Phys. Edn./UNESCO, 1977. Mem. AAAS, Am. Soc. Human Genetics, European Soc. Human Genetics, Italian Assn. Med. Genetics, N.Y. Acad. Scis., Lions. Christian Democrat. Roman Catholic. Home: La Miralunga, 55100 Arsina Lucca, Italy Office: Inst Biology & Genetics, Via Ranieri, 60131 Ancona Italy

MILANO, ANTONIO, engineering executive; b. Buenos Aires, Nov. 26, 1931; s. José and Maria Justina (De Gregorio) M.; divorced Mar. 1981; children: Marina Mercedes, Paula Beatriz, Flavia Silvina; m. Maria Rosario Ponsilglione, June 9, 1988. BSEE, U. Buenos Aires, 1956; postgrad., U. Catolica Argentina, 1969. Registered profl. engr., Argentina. Tech. advisor Ministry of Communications, Buenos Aires, 1953-57; R&D engr. Otis Elevator Co., Buenos Aires, 1957-60, Electrolux Corp., Conn., 1960-62; dir. human resources Eli Lilly & Co. of Argentina, Buenos Aires, 1962-73; mktg. dir. Fiplasto, S.A., Buenos Aires, 1973-76; mgr. Kepner Tregoe, S.A., Buenos Aires, 1976-81; pres. Bus. Processes Internat., S.A., Buenos Aires, 1981—; pub. rels. prof. U. Buenos Aires, 1965. Author: La Practica del Analisis de Problemas y Toma de Decisiones-Spain, 1987, Procesos Gerenciales Criticos, 1992. Conventional Christian Dem. Party, Argentina, 1958. Cpl. Argentinian Army, 1952-53. Mem. Tng. Cons. and Mgrs. Assn., Sporting Club (life), Maschwitz Country Club (pres. 1970-73). Avocations: tennis, golf, reading, music. Home: Sucre 1180-6th Fl Apt B, 1428 Buenos Aires Argentina Office: Bus Processes Internat, Av. Cordoba 659-5th. floor, 1054 Buenos Aires Argentina

MILANO, CHARLES THOMAS, obstetrician, legal medicine consultant; b. N.Y.C., July 18, 1951; s. Thomas F. Milano and Silvia (Tedescho) Perrotta; children: Rosanne, Teresa. MD, Mount Sinai Hosp. Med. Sch., N.Y.C., 1977; JD, Thomas Jefferson, L.A., 1986. Dir. gyn. dept. St. Francis Hosp., Poughkeepsie, N.Y., 1987—, coord. family practice residency gyn., 1987—; attending physician Vassar Bros. Hosp., Poughkeepsie, 1981—, pres. Cons. Ob.-Gyn., Poughkeepsie, 1986—; lectr. Mt. Sinai Hosp. Med. Sch., N.Y.C., 1990—; instr. N.Y. Med. Coll., Valhalla, 1991—; cons. Risk Mgmt. Svcs., San Rapheal, Calif., 1990—, St. Francis Hosp., Poughkeepsie, 1991—. Contbr. articles to jour. Ovarian Pregnancy, jour. Echinococcosis, jour. Phakomatosis, jour. Gynecol. Oncology, and jour. Prostaglandins. Fellow Internat. Coll. Surgery, ACS, Am. Coll. Legal Medicine, Am. Coll. Ob.-Gyn. Office: 1 Pine St Poughkeepsie NY 12601-3943

MILANOVICH, FRED PAUL, physicist; b. Rochester, Pa., Nov. 22, 1944; s. Fred Rade and Stephanie (Osowiecka) M.; m. Linda Elaine Bertram, Aug. 3, 1968; children: Robin, Scott. BS, USAF Acad., 1967; MS, U. Calif., Davis, 1968, PhD, 1974. Staff scientist, group leader Lawrence Livermore (Calif.) Nat. Lab., 1974—. Author: (with others) Instrumentation for Fiber Optic Chemical Sensors, 1991. Capt. USAF, 1967-71. Achievements include innovation in optical sensors, such as fast response, single fiber pH sensor, energy transfer based fiber sensor, highly sensitive solvent vapor fiber sensor. Home: 3383 Woodview Dr Lafayette CA 94549 Office: Lawrence Livermore Nat Lab 7000 East Ave Livermore CA 94550

MILAS, ROBERT WAYNE, neurosurgeon; b. Rock Island, Ill., Dec. 25, 1944; s. Peter and Josephine M.; m. Mary McCarron; children: Roberta, Maura. BS, U. Ill., 1966; MD, Loyola U., 1969. Resident gen. surgery V.A. Hosp., Hines, Ill., 1970-71, resident neurosurgery, 1971; lt. USN U.S.S. Intrepid, Quonset Naval Air Sta., 1971-72, lt. comdr., 1972-73; resident neurosurgery Mercy Med. Ctr., Chgo., 1973-74, U. Ill. Hosp., Chgo., 1974-77; neurosurgeon pvt. practice Franciscan Med. Ctr., Rock Island, Ill., 1977—. Fellow Am. Coll. Surgeons, Am. Heart Assn. Stroke Coun.; mem. AMA, Rock Island County Med. Soc. Roman Catholic. Office: Robert W Milas MD 2131 1st St A Moline IL 61265

MILBERG, MORTON EDWIN, chemist; b. N.Y.C., July 21, 1926; m. Helga Esther Weiss, June 29, 1962; children: David, Randall, Jonathan. BS, Rutgers U., 1946; PhD, Cornell U., 1949. Instr. chem. U. N.D., Grand Forks, 1950-52; prin. rsch. scientist Ford Motor Co., Dearborn, Mich., 1952-89; rsch. prof. U. Ariz., Tucson, 1989—; chmn. Gordon Rsch. Conf., 1971. Contbr. over 40 pubs. to profl. jours. Postdoctoral fellow U. Minn., Mpls., 1949-50. Mem. Am. Chem. Soc., Am. Ceramic Soc., Am. Crystallographic Assn., Phi Beta Kappa, Phi Lambda Upsilon, Sigma Xi. Office: Ariz Materials Labs 4715 E Ft Lowell Rd Tucson AZ 85712

MILBURN, DARRELL EDWARD, ocean engineer, coast guard officer; b. Ft. Belvoir, Va., Apr. 7, 1959; s. Henry Larry and Mary Ann (Ihnatko) M.; m. Mary Frances Pero, Apr. 9, 1983; children: Darcy Ann, Lauren Rose. BS, U.S. Coast Guard Acad., 1981; Ocean Engr., MIT, 1989. Commd. officer U.S. Coast Guard, 1981, advanced through grades to lt. comdr., 1992; comdg. officer USCG cutter Point Knoll, New London, Conn., 1983-85; sr. contr. USCG Atlantic Area Ops., N.Y.C., 1985-87; ocean engr. USCG R&D Ctr., Groton, Conn., 1989—. Author reports and articles. Cath. youth leader St. Mary's Youth Group, Groton, 1987—; asst. sailing coach USCG Acad., New London, 1989—. Mem. Soc. Naval Architects and Marine Engrs., Marine Tech. Soc., Ram Island Yacht Club, Sigma Xi. Achievements include development of mission data recorder system for USCG rescue boats; development of computer-aided mooring selection guide for navigation buoys; development of computer design tools for buoys. Home: 193 Brook St Noank CT 06340 Office: USCG R&D Ctr 1082 Shennecossett Rd Groton CT 06340

MILBURY, THOMAS GIBERSON, engineering executive; b. Rahway, N.J., Aug. 7, 1951; s. Wilmot Arthur and Helen Gertrude (Brendle) M.; m. Lorraine Alice Bruno, Dec. 15, 1984; children: Lydia Alison, Aaron Giberson. BS in Civil Engring. and Architecture, MIT, 1973; MS in Mech. Engring., Columbia U., 1983. Registered profl. engr., N.Y., N.J., Ohio. Structural designer, from engr. to sr. engr. Am. Electric Power, N.Y.C. and Columbus, Ohio, 1977-86; dir. electro-mech. engring. N.Y.C. Transit Authority, Bklyn., 1986—. Elder Germonds Presbyn. Ch., New City, N.Y., 1992. Mem. ASCE, NSPE, ASME, MIT Club of Westchester. Home: 26 Carolina Dr New City NY 10956 Office: NYC Transit Authority 370 Jay St Brooklyn NY 11201

MILCAREK, WILLIAM FRANCIS, marketing professional; b. Sterling, Ill., Nov. 22, 1947; s. Leonard Joseph and Caroline Marie (McGraw) M.; m. Kathleen Mary Flynn, July 12, 1969 (div. June 1980); children: Bonnie, William Jr.; m. Samantha Swanson, July 1980 (seperated); children: Kerri, Marri. BA, Tulane U., 1969. Home office underwriter Associated Aviation Underwriters, N.Y.C., 1974-76; asst. v.p. far east dept. Guy Carpenter & Co., Inc., N.Y.C., 1976-83; v.p. reins. worldwide mgr. Am. Internat. Underwriters, N.Y.C., 1983-85; v.p. aviation div. Frank B. Hall & Co., N.Y.C., 1985-87; v.p. mktg. AGF Reins. Corp. of the U.S., N.Y.C., 1987-92; v.p., aviation reins. broker Willcox, Inc., N.Y.C., 1992—; commanded nuclear alert facility, Dyess AFB, Tex., chief of safety. Vol. VA Hosp., Lyons, N.J., 1992. Capt. USAF, 1969-74, USAFR, 1975-79. Decorated D.F.C., Purple Heart, 5 Air medals. Mem. Chiselers Club of N.Y., John St. Club, Mil. Order of the Purple Heart (exec. com.), Am. Legion (comdr. chpt. 1870). Republican. Home: 99 Alexandria Way Basking Ridge NJ 07920

MILES, DONALD GEOFFREY, economist; b. Melbourne, Victoria, Australia, Aug. 26, 1952; s. Harry Raymond and Marian Edith (Lightfoot) M.; m. Judy E. Roberts, Dec. 14, 1991. B. Bus. with distinction, Curtin U. Tech., Muresk, Australia, 1981; MS in Econs., Iowa State U., 1983. Rsch. asst. Iowa State U., Ames, 1980-84; econs. lectr. Curtin U. Tech., Muresk, 1985-87; rsch. economist PRD Consulting Svcs., Pty., Ltd. & Max Christmas Pty. Ltd., Gold Coast, Australia, 1988-89; pres. Miles Internat., Australia and U.S., 1989—; econometric revenue forecaster State of Wash., 1990—. Inventor environ. wholistic and econ. models, Dept. of Licensing WA/U.S. growth index and transforms, trading day seasonality, copyrights for laws of human ecology, problem shifting analysis, systems econs., wholistic analysis, systems repair, systems improvement, quantifying inefficiency and waste, optimal rates of adjustment, adjustment boxes, events-prices and incomes analysis. Participant World Food Conf., Ames, 1976, Inst. World Affairs, Ames, 1981. Recipient Edwards Prize, Curtin U. Tech., 1987. Mem. World Future Soc. (life). Avocations: environmental economics, aboriginal history and trade, world affairs. Home and Office: 1015 7th Ave N Tumwater WA 98512-6315

MILES, EDWARD LANCELOT, marine studies educator, consultant, director; b. Port-of-Spain, Trinidad, W.I., Dec. 21, 1939; came to U.S., 1959; s. Cecil Bannister and Louise (DuFont) M.; divorced; children: Anthony Roger, Leila Yvonne. BA, Howard U., 1962; PhD, U. Denver, 1965. Instr. Sch. Internat. Studies U. Denver, 1965-66, asst. prof., 1966-70, assoc. prof., 1970-74; prof. Inst. Marine Studies & Grad. Sch. Pub. Affairs U. Wash., Seattle, 1974—, dir. Inst. for Marine Studies, 1982—; chief negotiator Micronesian Maritime Authority, 1980-90; chmn. ocean policy com. NAS/NRC, Washington, 1974-89; chmn. adv. com. on internat. programs NSF, Washington, 1990-92. Author: The Management of Marine Regions: The North Pacific, 1982. Mem. Coun. on Fgn. Rels., Inc., Seattle United Nations Assn., Harvard Club of N.Y.C., Sierra (adv. bd. N.Y.C. chpt. 1975-79). Avocations: literature, opera, theatre, hiking, skiing. Office: U Wash Sch Marine Affairs HF-05 Seattle WA 98195

MILES, PAULA EFFETTE, mechanical engineer; b. Victoria, Tex., Nov. 6, 1960; d. Milton D. and Robbie L. (Russell) Johnson; m. Edwin K. Miles, Jan. 26, 1985; 1 child, Patrick Isiah. BSME, U. Tex., 1984; MSPA, S.W. Tex. State U., 1991. Registered profl. engr., Tex. Engr. intern Union Carbide Corp., Texas City, Tex., 1979-82; admissions intern U. Tex., Austin, 1983-84; jr. engr. City Pub. Svc., San Antonio, 1984-87, project engr., 1987-91, supt., 1991—; charter mem. Women in Sci. and Engring. Collaborative, San Antonio, 1992—; founding mem. Pub. Adminstrn. Adv. Bd., San Marcos, Tex., 1989-92. Speaker, mentor various pub. sch. dists., San Antonio, 1985—; presenter Upward Bound., St. Mary's U., San Antonio, 1989; mem. Leadership San Antonio XVIII, 1992-93. Recipient Achievement award, 1979, Tex. Achievement award U. Tex., 1979. Mem. Profl. Engrs. Scholarship Fund (sec. 1992), Tex. Soc. Profl. Engrs., Nat. Assn. CorrosionEngrs. (sec. 1987-88), Pi Alpha Alpha Hon. Soc., Delta Sigma Theta Sorority. Office: City Pub Svc PO Box 1771 San Antonio TX 78296

MILES, THOMAS CASWELL, engineer; b. Atlanta, Mar. 21, 1952; s. Franklin Caswell and Eugenia Frances (Newsom) M.; m. Linda Susan Duggleby, Aug. 10, 1980. B of Mechanical Engring. Tech., So. Tech. Inst., 1977; postgrad., Troy State U., 1978-80. Assoc. engr., aircraft design Lockheed Aero. Systems Co., Marietta, Ga., 1980-82, engr., aircraft design, 1982-85, sr. engr., aircraft design, 1985-89, group engr., 1989-90, specialist engr., 1990—; mem. SAE-A-6 Mil. Aircraft & Helicopter Panel, 1987-91. Mem. AIAA (sr.), Nat. Mgmt. Assn., Soc. Automotive Engrs., Oxygen Standardization Coord. Group, Assn. Fraternity Advisors (affiliate), Tau Kappa Epsilon (dist. pres. 1987-88, dist. v.p. 1984—, chpt. advisor 1980-87, key leader 1985, 90, So. Order of Honor 1989). Avocations: sailing, scuba diving, screen printing. Home: 1205 Hickory Rd Canton GA 30114 Office: Lockheed Aeronautical Systems Co Dept 73-05 cc-34 Z-0199 Marietta GA 30063-0199

MILES, WILLIAM ROBERT, structural engineer; b. Cortland, N.Y., Feb. 23, 1951; s. Lawrence Robert and Anna Fay (Lefever) M.; m. Doreen Louise Wadsworth, July 12, 1975; children: Brooke Leigh, Valerie Anne. BS in Civil Engring., Syracuse U., 1973. Profl. engr. Md., N.Y., N.J. Structural engr. Kidde Cons., Inc., Towson, Md., 1973-78; project engr. Century Engring., Inc., Towson, 1978-89; prject mgr. Bergmann Assocs., Rochester, N.Y., 1989--. Partnership Sch. of the Arts, Rochester, 1991--. N.Y. State Regents scholar, 1969-73, L.C. Smith Coll. Engring. scholar, 1969-73. Mem. Am. Soc. Civil Engrs., Nat. Soc. Profl. Engrs., Institutional and Mcpl. Parking Congress, Am. Concrete Inst. Achievements include numerous design amoung which are USN Trident Submarine drydock in Kings Bay, Ga., rehabilitation of Troy Lock for Corps of Engrs., Lexington St. Parking Garage, Balt., Tampa Shipyards Drydocks Nos. 3 and 4., Sod Run Wastewater Treatment Plant, Harford County, Md., Red Dog Mine Seaport Facilities, Alaska, Hart-Miller Island diked disposal area, Balt., 500-bed medium/maximum security annex, Jessup, Md. Home: 1613 Beech Dr Walworth NY 14568 Office: Donald Bergmann Assocs 1 S Washington St Rochester NY 14614

MILEWSKI, STANISLAW ANTONI, ophthalmologist, educator; b. Bagrowo, Poland, June 16, 1930; s. Alfred and Sabina (Sicinska) M.; came to U.S., 1959, naturalized, 1967; BA, Trinity Coll., U. Dublin (Ireland), 1954, MA, 1959, B. Chir., M.B., B.A.O., 1956; m. Anita Dobiecka, July 11, 1959; children: Andrew, Teresa, Mark. House surgeon Hammersmith Hosp. Postgrad. Sch. London, 1958; intern St. Raffael Hosp., New Haven, 1960-61; resident in ophthalmology Gill Meml. Hosp., Roanoke, Va., 1961-64; practice medicine specializing in surgery and diseases of the retina and vitreous; mem. staff Manchester (Conn.) Meml. Hosp., 1964-71, chief of ophthalmology, sr. attending physician St. Francis Hosp., Hartford, Conn., 1971—; sr. attending St. Francis Hosp, Hartford, Conn.; asst. clin. prof. ophthalmology U. Conn., 1972—. Clin. fellow Montreal (Que., Can.) Gen Hosp., McGill U., 1971-72, Mass. Eye and Ear Infirmary, Harvard Med. Sch., Boston, 1974; diplomate Am. Bd. Ophthalmology. Fellow ACS; mem. Am., Conn. (sec.-treas.), Brit. med. assns., New Eng. Ophthal. Soc. Republican. Roman Catholic. Home: 127 Lakewood Cir S Manchester CT 06040-7018 Office: 191 Main St Manchester CT 06040-3556 also: 43 Woodland St Ste 100 Hartford CT 06105

MILEY, GEORGE HUNTER, nuclear engineering educator; b. Shreveport, La., Aug. 6, 1933; s. George Hunter and Norma Angeline (Dowling) M.; m. Elizabeth Burroughs, Nov. 22, 1958; children: Susan Miley Hibbs, Hunter Robert. B.S. in Chem. Engring., Carnegie-Mellon U., 1955; M.S., U. Mich., 1956, Ph.D. in Chem.-Nuclear Engring., 1959. Nuclear engr. Knolls Atomic Power Lab., Gen. Electric Co., Schenectady, 1959-61; mem. faculty U. Ill., Urbana, 1961—; prof. U. Ill., 1967—, chmn. nuclear engring. program, 1975-86, dir. Fusion Studies Lab., 1976—, fellow Ctr. for Advanced Study, 1985-86; dir. rsch. Rochford Tech. Assocs. Inc., 1990—; vis. prof. U. Colo., 1967, Cornell U., 1969-70, U. New South Wales, 1986, Imperial Coll. of London, 1987; mem. Ill. Radiation Protection Bd., 1988—; mem. Air Force Studies Bd., 1990—; chmn. tech. adv. com. Ill. Low Level Radioactive Waste Site, 1990—; chmn. com. on indsl. uses of radiation Ill. Dept. Nuclear Safety, 1989—. Author: Direct Conversion of Nuclear Radiation Energy, 1971, Fusion Energy Conversion, 1976; editor Jour. Fusion Tech., 1980—; U.S. assoc. editor Laser and Particle Beams, 1982-86, mng. editor, 1987-91, editor-in-chief, 1991—. With C.E. AUS, 1960. Served with C.E. AUS, 1960. Recipient Western Electric Teaching-Rsch. award, 1977, Halliburton Engring. Edn. Leadership award, 1990; NATO sr. sci. fellow, 1975-76; Guggenheim fellow, 1985-86. Fellow Am. Nuclear Soc. (dir. 1980-83, Disting. Svc. award 1980, outstanding achievement award Fusion energy Divsn. 1992), Am. Phys. Soc., IEEE; mem. Am. Soc. Engring. Edn. (chmn. energy conversion com. 1967-70, pres. U. Ill. chpt. 1973-74, chmn. nuclear div. 1975-76, Outstanding Tchr. award 1973), Sigma Xi, Tau Beta Pi. Presbyterian. Lodge: Kiwanis. Achievements include research on fusion, energy conversion, reactor kinetics. Office: U Ill 214 Nuclear Engring Lab 103 S Goodwin Ave Urbana IL 61801-2984

MILEY, GEORGE KILDARE, astronomy educator; b. Dublin, Ireland, Mar. 15, 1942; s. John Felix Miley and Josephine Mary Miley-Minch; m. Johanna Blomberg, Aug. 26, 1974; children: Helen Dorothy, Anne Christina. BS in Physics with first class honors, Nat. U. Ireland, Dublin, 1963; PhD in Radio Astronomy, U. Manchester, 1968. Rsch. asst. U. Manchester, Eng., 1963-68; rsch. assoc. Nat. Radio Astronomy Observatory, Charlottesville, 1968-70, asst. scientist, 1970; sr. astronomer European Space Agy. Space Telescope Sci. Inst., Balt., 1984-88, acting head acad. affairs dept., 1986-87; scientist Leiden (The Netherlands) Observatory U. Leiden, 1970-74, sr. scientist Leiden Observatory, 1974-88, prof. astronomy Leiden Observatory, 1988—; vis. prof. Lick Observatory, U. Calif., Santa Cruz, 1977-78; vis. scientist Kitt Peak Nat. Observatory, Tucson, 1981-82, Jet Propulsion Lab., Pasadena, Calif., 1982; adj. prof. dept. physics and astronomy Johns Hopkins U., Balt., 1986-88. Mem. Internat. Astron. Union, Am. Astron. Soc. Achievements include detection of ultra-compact components in several radio sources; establishment of the existence of angular diameter-redshift relation for quasars; explanation of head-tail radio sources as double sources distorted by motion through an intergalactic medium; first detection of radio emission from Cygnus X-1 and Cygnus X-3; optical identification of Cygnus X-1, a probable black hole; establishment of the several similarities between radio emission from galactic X-ray sources and that from active galaxies; measurement of alignment between inner and outer structure in 3C236, establishing that the ejection axis in a readio source remains fixed for at least 10(7) years; demonstration of widespread existence of rotational symmetry in radio source structure; discovery of radio jets in several radio sources establishing the importance of the jet phenomenon; discovery of a correlation between position angles of optical polarizatioon and radio elongation of quasars; detection of optical emission from several jets in extended radio sources; detection of optical emission lines from extended radio lobes showing for the first time a detailed spatial correlation between radio continuum and optical line emission; measurement of systematic asymmetry in line profiles showing that radial motion dominates the kinematics of narrow-line nuclear emission-line regions; meaurement of IRAS spectra of Seyferts suggesting that nuclear thermal infrared components are widespread in AGNs; establishment of a correlation between bending of radio structure and redshift for quasars providing evidence for cosmic evolution in the intergalactic and circumgalactic media; demonstration of alignment between the radio and optical axes fo high-redshift radio galaxies; discovery of the galaxy 4C40.46 with the largest known redshift; first 2.2 micron image of a galaxy of high redshift showing alignment between radio and infrared emission; discovery of the galaxy 4C41.17 with the largest known redshift. Office: Leiden U Observatory, Niels Bohring 2, 2300 Leiden The Netherlands

MILEY, HUGH HOWARD, retired physician; b. Wauseon, Ohio, Apr. 24, 1902; s. Howard Harland Miley and Edith (Martin) Esterline; m. Anna Horwitz, June 26, 1935; children: Ruth Eliza Jane, Howard Charles. BA, Ohio State U., 1924, MS, 1925, PhD, 1927; MD, Wayne State U., 1935. Intern Grace Hosp., Detroit, 1934-35; extern Delray Gen. Hosp., 1933-35; instr. physiology Detroit Coll. Medicine and Surgery, 1927-33; practice indsl. medicine Dodge Main Hosp., Chrysler Motor Co., Hamtramick, Mich., 1935-37; gen. practice indsl. medicine Marygrove Indsl. Clinics, Detroit, 1937-67; long-term care physician Wayne County Gen. Hosp., Eloise, Mich., 1967-72; gen. practice clinics Highland Park, Mich., 1972-82; gen. practice medicine Herman Kiefer Hosp. Bldg., Detroit, 1982-83; participant charity welfare programs North End Clinic, Grace Hosp., Detroit. Contbr. articles to profl. jours. Physician vaccination programs Detroit Pub. Schs., 1935-40, Neighborhood Health Svc. Ctr., Inc., 1972. Mem. AMA, Wayne County Med. Soc., Mich. State Med. Soc., Hon. Pathology Soc., Sigma Xi. Avocations: reading books, fishing, traveling, geriatrics. Home: 1038 Wimbleton Dr Birmingham MI 48009-7605

MILGROM, FELIX, immunologist, educator; b. Rohatyn, Poland, Oct. 12, 1919; came to U.S., 1958; naturalized, 1963; s. Henryk and Ernestina (Cyryl) M.; m. Halina Miszel, Oct. 15, 1941; children: Henry, Martin Louis. Student, U. Lwow, Poland, 1937-41, U. Lublin, Poland, 1945; MD, U. Wroclaw, Poland, 1947; MD (hon.), U. Vienna, Austria, 1976, U. Lund, Sweden, 1979, U. Heidelberg, Fed. Republic Germany, 1979, U. Bergen, Norway, 1980; DSc (hon.), U. Med. Dent., N.J., 1991. Rsch. assoc., prof. dept. microbiology Sch. Medicine U. Wroclaw, 1946-54, chmn. dept., 1954; prof., head dept. microbiology Sch. Medicine, Silesian U., Zabrze, Poland, 1954-57; rsch. assoc. Svc. de Chime Microbienne, Pasteur Inst., Paris, 1957; rsch. assoc., prof. dept. bacteriology, immunology Sch. Medicine, U. Buffalo, 1958-62; assoc. prof., then prof. and disting. prof. microbiology Sch. Medicine, SUNY, Buffalo, 1962—, chmn. dept., 1967-85. Author: Studies on the Structure of Antibodies, 1950; co-editor: International Convocations on Immunology, 1969, 75, 79, 85, Principles of Immunology, 1973, 2d edit., 1979, Principles of Immunological Diagnosis in Medicine, 1981, Medical Microbiology, 1982; editor in chief Internat. Archives of Allergy and Applied Immunology, 1965-91; contbg. editor Vox Sanguinis, 1965-76, Transfusion, 1966-73, Cellular Immunology, 1970-83, Transplation, 1975-78; contbr.

numerous articles to profl. jours. Recipient Alfred Jurzykowski Found. prize, 1986, Paul Ehrlich and Ludwig Darmstaedter prize, 1987. Mem. Am. Assn. Immunologists, Transplantation Soc. (v.p. 1976-78), Am. Acad. Microbiology, Coll. Internat. Allergologicum (v.p. 1970-78, pres. 1978-82), Sigma Xi. Achievements include research on the serology of syphillis, Tb, rheumatoid arthritis, organ and tissue specificity including blood groups, transplantation and autoimmunity. Home: 474 Getzville Rd Buffalo NY 14226-2555

MILHORAT, THOMAS HERRICK, neurosurgeon; b. N.Y.C., Apr. 5, 1936; s. Ade Thomas and Edith Caulkins (Herrick) M.; children: John Thomas, Robert Herrick. BA, Cornell U., 1957, MD, 1961. Intern, asst. resident in gen. surgery N.Y. Hosp.-Cornell Med. Ctr., 1961-63; clin. assoc., dept. surg. neurology Nat. Inst. Neurol. Diseases and Blindness, Bethesda, 1963-65; asst. resident, chief resident in neurosurgery N.Y. Hosp.-Cornell Med. Ctr., 1965-68, asst. neurosurgeon NIH, 1968-71; assoc. prof. neurol. surgery, assoc. prof. child health and devel. George Washington U. Sch. Medicine, Washington, 1971-74; prof. child health and devel. George Washington U., Washington, 1974-81, prof. neurol. surgery, 1974-81; chmn. dept. neurosurgery Children's Hosp. Nat. Med. Ctr., Washington, 1971-81; prof. neurol. surgery, dept. chmn. SUNY Health Sci. Ctr., Bklyn., 1982—; neurosurgeon-in-chief Kings County Hosp. Ctr.; regional chmn. neurol. surgery L.I. Coll. Hosp., 1986—, Coney Island Hosp., 1986—; program dir. Neurosurgery Rsch. Tng. Program, 1982—. Author: Hydrocephalus and Cerebrospinal Fluid, 1972, Pediatric Neurosurgery, 1978, Cerebrospinal Fluid and the Brain Edemas, 1987, (with M.K. Hammock) Cranial Computed Tomography in Infancy and Childhood, 1981; contbr. 175 articles to sci. publs. and chpts. to books. Chmn. bd. Internat. Neurosci. Found., pres., 1986—. Served to lt. comdr. USPHS, 1963-65. Awarded 1st prize in Pathology, Cornell U. Med. Sch. Dept. Ob-Gyn., 1960, Charles L. Horn prize Cornell Med. Sch., 1961, Best Paper award ann. combined meeting N.Y. Acad. Medicine/N.Y. Neurosurg. Soc., 1965, Best Doctors in N.Y. award New York mag., 1992. Mem. AAAS, Internat. Soc. Pediatric Neurosurgery, Am. Assn. Neurol. Surgery, Am. Assn. Neurol. Surgery (pediatric sect.), Am. Acad. Pediatrics (surg. sect.), Soc. Pediatric Research, N.Y. Acad. Medicine, N.Y. Soc. Neurosurgery (pres. 1989-91), Bklyn. Neurologic Soc. (pres. 1988-92), Soc. Neurosci., Internat. Soc. Neurosci., Soc. Neurol. Surgeons, Nat. Coun. Scientists NIH, Med. Club. Bklyn., Sigma Xi. Avocations: golf, billiards, baseball. Office: SUNY Health Sci Ctr Bklyn 450 Clarkson Ave PO Box 1189 Brooklyn NY 11203

MILHOUS, ROBERT THURLOW, hydraulic engineer; b. Indpls., Oct. 14, 1936; s. Russell Eugene and Mabel (Lee) M.; m. Marilyn Stout, Aug. 15, 1965; children:Sasha, Maya. BS in Civil Engring., Purdue U., 1961, MS in Civil Engring., 1964; PhD, Oreg. State U., 1973. Vol. Peace Corps, Tanganyika, 1961-63; water resources engr. Calif. Dept. of Water Resources, Sacramento, 1964-66; pub. works engr. U.S. AID, Kumasi, Ghana, 1966-68; hydraulic engr. Wash. Dept. of Ecology, Ol;mpia, 1972-78; rsch. hydraulic engr. U.S. Fish & Wildlife Svc., Ft. Collins, Colo., 1978—. Contbr. articles to profl. jours. Sgt. USMC, 1955-57. Office: NERC - USFWS 4512 McMurry Ave Fort Collins CO 80526

MILIC-EMILI, JOSEPH, physician, educator; b. Sezana, Yugoslavia, May 27, 1931; arrived in Canada, 1963; s. Joseph Milic-Emili and Giovanna Milic-Emili Perhavec; m. Ann Harding, Nov. 1, 1957; children—Claire, Anne-Marie, Alice, Andrew. M.D., U. Milan, 1955. Asst. prof. physiology and exptl. medicine McGill U., Montreal, Canada, 1963-65; assoc. prof. McGill U., Canada, 1965-69, prof. Dept. Physiology, Exptl. Medicine, 1970—, dir. Meakins-Christie Labs, 1979—; vis. prof. Laboratoire de Physiologie Faculte de Medecine Saint-Antoine, Paris and Service de Pneumologie Hopital Beujon, Paris, 1978-79, chmn. dept. physiology, 1973-78; vis. cons. medicine royal Postgrad. Med. Sch., London, 1969-70, aeronautics Imperial Coll. Tech., London, 1969-70; asst. prof. physiology U. Liege, Belgium, 1958-60; asst. prof. physiology U. Milan, Italy, 1956-58. Author of one of 100 most-cited articles in clin. research of 1960s; one of 1000 most-cited contemporary scientists, 1965-78; mem. editorial bd. Am. Jour. Physiology, 1970-76, Jour. Applied Physiology, 1970-76, Revue Franciase des Maladies Respiratoires, 1979—, Rivista di Biologia, 1979-86, Am. Rev. Respiratory Disease, 1982-89, Reanimation, soins intensifs, medicine d'urgence, 1984—; assoc. editor Clin. Investigative Medicine Can. Soc. Clin. Investigation, 1981-86. Mem. applied physiology and bioengring. study sec. NIH, 1975-78. Decorated Order of Can.; recipient Gold medal C. Forlanini u. Pavia, Italy, 1982; Am. Coll. Chest Physicians medalist, 1984; named Dr. Honoris Causa, U. Louvain, Belgium, 1987, Kunming Med. Coll., China, 1987; fellow Harvard U., Boston, 1960-63; Harry Wunderly medal Thoracic Soc. Australia, 1988. Fellow Royal Soc. Can., Slovenian Acad. Scis. (fgn. corr. mem. 1982); mem. Societe Belge de Physiologie, Assn. des Physiologistes de Langue Francaise, Am. Physiol. Soc., Can. Physiol. Soc., Can. Soc. for Clin. Investigation, Italian Physiol. Soc., Can. Thoracic Soc., Med. Rsch. Coun. (grants com. 1980). Home: 4394 Circle Rd, Montreal, PQ Canada H4W 1Y5 Office: McGill U Meakins-Christie Labs, 3626 St Urbain St, Montreal, PQ Canada H2X 2P2

MILJEVIC, VUJO I(LIJA), physicist, researcher; b. Topusko, Yugoslavia, July 2, 1931; s. Ilija and Vasilija (Vorkapic) M.; 2 children. BS, Physics Faculty, 1961, MS, 1964, PhD, 1970. Researcher Boris Kidric Inst. for Nuclear Scis., Vinca-Beograd, 1962-70; sr. researcher Boris Kidric Inst. for Nuclear Scis., Vinča-Beograd, 1970-84, head plasma dept., 1984—. Patentee of a new generation of the particle and radiation sources based on a hollow anode discharge; contbr. articles in fields of plasma physics, spectroscopy, lasers and ion-electron sources to profl. jours. Mem. Europe Physicists Soc. Office: Inst Nuclear Scis Vinca, Atomic Physics Lab PO Box 522, 11001 Belgrade Yugoslavia

MILKMAN, ROGER DAWSON, genetics educator, molecular evolution researcher; b. N.Y.C., Oct. 15, 1930; s. Louis Arthur and Margaret (Weinstein) M.; m. Marianne Friedenthal, Oct. 18, 1958; children: Ruth Margaret, Louise Friedenthal, Janet Dawson Lussenhop, Paul David. A.B., Harvard U., 1951, A.M., 1954, Ph.D., 1956. Student asst., instr., investigator Marine Biol. Lab., Woods Hole, Mass., 1952-72, 88—; instr., asst. prof. U. Mich., Ann Arbor, 1957-60; assoc. prof., prof. Syracuse U., N.Y., 1960-68; prof. biol. scis. U. Iowa, Iowa City, 1968—, chmn. univ. genetics PhD program, 1992-93; vis. prof. biology Grinnell Coll., 1990. Translator: Developmental Physiology, 1970; editor: Perspectives on Evolution, 1982, Experimental Population Genetics, 1983, Evolution jour., 1984-86; mem. editorial bd. Jour. Molecular Evolution, Molecular Phylogenetics and Evolution, Zool. Sci. (Japan); contbr. articles to profl. jours. Sec. Soc. Gen. Physiologists, 1963-65, Am. Soc. Naturalists, 1980-82; alumni rep. Phillips Acad., Andover, Mass., 1980—. NSF grantee, 1959—; USPHS grantee, 1984-87. Fellow AAAS; mem. Am. Soc. for Microbiology, Genetics Soc. Am., Corp. of Marine Biol. Lab., Soc. Study Evolution (NIH genetics study sect. 1986-87), Soc. of Molecular Biology and Evolution, Internat. Soc. for Molecular Evolution. Jewish. Avocations: travel; mountain hiking. Home: 12 Fairview Knoll NE Iowa City IA 52240 Office: U Iowa Dept Biol Scis 138 Biology Bldg Iowa City IA 52242-1324

MILLA GRAVALOS, EMILIO, industrial engineer; b. Zaragoza, Spain, Apr. 1, 1944; s. Pablo Milla Mallo and Emilia Grávalos Gil. Diploma in Indsl. Engring., Higher Tech. Sch. Indsl. Engrs., Madrid, 1972; Dr. in Indsl. Engring., Higher Tech. Sch. Indsl. Engrs. U. Politecnica, Madrid, 1977; diploma in Physics Sci., Complutense U., Madrid, 1984, diploma in History, 1989. Diplomate engring. divsn. Junta de Energia Nuclear, Madrid, 1973-78, inspector of nuclear power plants, 1978-82, project dir. air treatment Nuclear Systems divsn., 1982-86; safety engr. Nuclear Tech. Inst. Ctr. Energetic, Environ. Tech. Rsch., Madrid, 1986-93, project leader nuclear & radioactive facilities decommissioning Nuclear Tech. Inst., 1993—; mem. coord. com. on fast breeder nuclear reactors, Europe, 1991—; lectr. in atmosphereic pollution Trade and Indsl. Chamber, Madrid, 1992—. Contbr. articles and revs. to profl. jours. Mem. Internat. Assn. for Aerosol Rsch., Amnesty Internat., UNICEF. Avocations: lecturing, history, travel, mountains, music. Home: Maria Benitez 25, Pozuelo de Alarcon, 28224 Madrid Spain

MILLER, ALAN, software executive, management specialist; b. Bklyn., Apr. 20, 1954; s. Michael and Lillian Charlotte (Garment) M.; m. Zelda Sara Bochlin, Nov. 16, 1974; children: Michael Glenn, Dara Jennifer. BS in Computer Sci. magna cum laude, SUNY, 1975; MBA in Mgmt. with honors, Adelphi U., 1982. Tech. svcs. mgr. Guardian Life Ins. Co., N.Y.C., 1977-81; project mgr. Mfrs. Hanover Trust Co., N.Y.C., 1981-83; asst. v.p. Bankers Trust Co., N.Y.C., 1983-86; v.p., MIS dir. Bank Am. Trust Co. of N.Y., N.Y.C., 1986-87; assoc. John Diebold and Assocs., N.Y.C., 1987-89; mgr. banking practice AGS Info. Svcs., N.Y.C., 1989-90; v.p. bus. devel., product mgr. global trade fin. BIS Banking Systems, N.Y.C., 1990—. Chmn. Sch. Dist. Adv. Com., Plainview, N.Y., 1981-83; exec. producer Oklahoma prodn. Patio Players, Plainview, 1990-91; bd. dirs. men's club Plainview Jewish Ctr., 1986—. Mem. Delta Mu Delta. Jewish. Avocations: softball, theater, game shoes, volleyball. Home: 21 Beaumont Dr Plainview NY 11803-2507 Office: BIS Banking Systems 900 3d Ave New York NY 10022-4728

MILLER, ALEXANDRA CECILE, radiation biologist, researcher, educator; b. Carmel, Calif., Jan. 14, 1959; d. Joseph John and Elizabeth Cecilia (Economopoulos) M. BS, U. Md., 1981; PhD, U. N.Y., 1986. Rsch. affiliate Roswell Park Cancer Inst., Buffalo, 1981-85; rsch. assoc. Armed Forces Radiobiol. Rsch. Inst., Bethesda, Md., 1987-89; prin. investigator Armed Forces Radio Rsch. Inst., Bethesda, 1989—; cons. Navy Rsch. & Devel. Commn., Bethesda, 1991—; instr. med. effects nuclear weapons Dept. of Defense, 1991—; instr. continuing edn. George Washington U., Washington, 1990—. Contbr. articles to Molecular & Cell Biology, Internat. Jour. Cancer, Radiation Rsch., Jour. Nat. Cancer Inst., Jour. Investigative Dermatology, Free Radicals in Biology & Medicine. Mem. IEEE, AAAS, Am. Electrophoresis Soc., Radiation Rsch. Soc., Snow Hill Hist. Soc. Roman Catholic. Achievements include discovery of EJras oncogene associated with increased resistance to oxidative stress, DNA damage can occur after cellular exposure to ultraviolet radiation. Office: Armed Forces Radio Rsch Inst 8901 Wisconsin Ave Bethesda MD 20889

MILLER, ALLEN RICHARD, mathematician; b. Bklyn., Dec. 2, 1942; s. Hyman and Sylvia (Weitz) M. BS, Bklyn. Coll., 1965; MA, U. Md., 1971. Mathematician U.S. Naval Rsch. Lab., Washington, 1968-93; rsch. prof. George Washington U., Washington, 1992—. Contbr. rsch. articles to profl. jours. including Jour. Math. Analysis and Applications, Jour. Franklin Inst., Fibonacci Quar., Jour. Acoustical Soc. Am., Internat. Jour. Math. Edn. in Sci. and Tech., IEEE, Instn. Electrical Engrs. (U.K.), Navigation, Soc. Indsl. and Applied Math. Rev. Recipient Alan Berman Rsch. Publ. award Naval Rsch. Lab., 1986, 87. Served with U.S. Army, 1965-67. Mem. Math. Assn. Am., Soc. for Indsl. and Applied Math., London Math. Soc., Sigma Xi. Achievements include research in special functions, numerical analysis, applied mathematics, scattering theory. Office: George Washington U Dept Math Washington DC 20052

MILLER, BARRY ALAN, epidemiologist, cancer researcher; b. Manila, Oct. 1, 1955; s. Paul Albert and Joan (Leidner) M.; m. Allyson Bartlett, Apr. 24, 1987. BA, Johns Hopkins U., 1977; MS in Pub. Health, U. N.C., 1979. Scientist Nat. Inst. for Occupational Safety and Health Ctr. Disease Control, Cin., 1978; epidemiologist occupational studies sect. Nat. Cancer Inst., Bethesda, Md., 1979-88; sr. rsch. epidemiologist cancer statis. br. Nat. Cancer Inst., Bethesda, 1988—; advisor epidemiology and biostats. NCI, Bethesda, 1983-84, 88, Dept. of Labor, Washington, 1988-89; NCI rep. Nat. Acad. Scis. Workshop, Washington, 1990. Editor: Cancer Statistics Review: 1973-90, 1992; contbr. articles to profl. jours. Mem. Chesapeake Bay Found., Annapolis, Md., 1984—; With USPHS, 1979—. U.S. Pub. Health Svs ach. medal, 1988, commendation medal, 1991. Mem. Commd. Officers Assn., Soc. for Epidemiologic Rsch., Sierra Club (nat., D.C. chpt.). Achievements include first to report increased cancer mortality among professional artists exposed to paints, solvents and various dusts. Office: Nat Cancer Inst EPN/343J 9000 Rockville Pike Bethesda MD 20892

MILLER, BRUCE NEIL, physicist; b. N.Y.C., Dec. 12, 1941; s. Harold Charles and Shirley (Buchsbaum) M.; m. Etta Mendelson, Dec. 22, 1966; children: Ira Zevi, Bo Kim. BA, Columbia Univ., 1963; MsC, Univ. Chgo., 1965; PhD, Rice Univ., 1969. Asst. physicist Ill. Inst. Tech. Res. Inst., Chgo., 1965; postdoctoral fellow SUNY, 1970-71; asst. prof. physics Tex. Christian Univ., Fort Worth, Tex., 1971-77, assoc. prof. physics, 1977-85, prof. physics, 1985—; referee Physical Review Am. Physical Soc., N.Y.C., 1980—; prof. theoretical physics, Univ. New South Wales, Sydney, 1992. Contbr. articles to profl. jours. Bd. mem. B'nai B'rith. Grad. fellow Rice Univ., Houston, 1966-69; regents scholar N.Y.S., 1959-63. Mem. Am. Physical Soc., Am. Assn. Univ. Prof., Sigma Xi. Jewish. Achievements include designing method of measuring pore diameters by positronium decay, wedge billiard, path integral formulation of positron annihilation, diffusion model of one dimensional gravitational system. Home: 3932 Weyburn Dr Fort Worth TX 76109 Office: Tex Christian Univ Physics Dept Box 32915 Fort Worth TX 76129

MILLER, CATE, psychologist, educator; b. Baton Rouge, Apr. 1, 1964; d. Benjamin Robertson and Mertie Cate (Barnes) Miller, Jr. BA, Cath. U., 1985; MEd, Harvard U., 1986; MPhil, Columbia U., 1989, PhD, 1991. Lic. psychologist, N.Y.; cert. sec. edn. tchr. Rsch. asst. Harvard U., Cambridge, Mass., 1985-86; diagnostician Psychol. Testing Inc., Baton Rouge, summer 1986; vocat. counselor Ctr. for Psychol. Svcs., N.Y.C., 1986-87, psychotherapist, 1987-88; psychology intern NYU Med. Ctr., Rusk Inst., N.Y.C., 1988-89, rsch. scientist, 1990—; pvt. practice N.Y.C., 1991—; adj. asst. prof. Columbia U., N.Y.C., 1991—. Contbr. (textbook): Geometry, 1985. Vol. St. Francis Xavier Soup Kitchen, N.Y.C., 1991. Columbia U. Merit scholar, 1987-88. Mem. APA, Nat. Trust for Historic Preservation, Jr. League, Phi Beta Kappa, Sigma Xi, Phi Delta Kappa, Kappa Gamma Pi, Psi Chi,. Democrat. Roman Catholic. Avocations: tennis, aerobics, theatre. Office: NYU Med Ctr Rusk Inst 400 E 34th St RR506 New York NY 10016

MILLER, CHARLES GREGORY, biomedical researcher; b. Raleigh, N.C., May 14, 1960; s. Thurman Greene and Sally Lou (Everhart) M.; m. Kimberly Michelle Trivette, Aug. 14, 1982; 1 child, Charles Parker Trivette Miller. BS in Marine Biology, U. N.C. at Wilmington, 1983, MS in Biology, 1993. Rsch. asst. Inst. for Marine Biomedical Rsch., Wilmington, 1979-83; rsch. technician Ctr. for Marine Sci. Rsch., Wilmington, 1983-90; rsch. assoc. Duke U., Durham, N.C., 1990; rsch. scientist Burroughs Wellcome Co., Research Triangle Park, N.C., 1990—. Author several publs., 1984-93. Counselor Boy Scouts Am., Hampstead, N.C., 1988-90; mem. Action Network, Research Triangle Park, 1991—. Mem. AAAS, Izaak Walton League, Nature Conservancy, Sigma Xi. Democrat. Baptist. Achievements include development of deepsea retrieval equipment and high pressure aquaria; research in thermoregulation, biomineralization, and preclinical cancer drug development. Home: 1220 Trillium Cir Apt B Raleigh NC 27606-3444 Office: Burroughs Wellcome Co 3030 W Cornwallis Rd Durham NC 27709-2700

MILLER, CLIFF, engineer; b. Griffin, Ga., Aug. 8, 1958; s. Isaiah and Marline Miller; m. Charlotte Lenoir Free, Aug. 9, 1986. BS in Gen. Engring., U.S. Mil. Acad., 1982. Commd. 2d lt. U.S. Army, 1982, advanced through grades to capt., 1985; resigned, 1987; mgr. level 1, Procter & Gamble, Mehoopany, Pa., 1987-91, mgr. level 2, 1991—. Mem. Saddle Lake Homeowners Assn., Tunkhannock, Pa., 1987—; diversity trainer By Visions, Boston; leader Black Mgrs. Work Team, Mehoopany, 1988-89. Mem. Northeastern Networking, Tunkhannock Jaycees. Baptist. Avocations: reading, writing, jogging. Home: 1 Eastwoods Rd Tunkhannock PA 18657 Office: Procter & Gamble RR 87 Box 32 Mehoopany PA 18629

MILLER, DANIEL NEWTON, JR., geologist, consultant; b. St. Louis, Aug. 22, 1924; s. Daniel Newton and Glapha (Shuhardt) M.; m. Esther Faye Howell, Sept. 9, 1950; children: Jeffrey Scott, Gwendolyn Esther. B.S. in Geology, Mo. Sch. Mines, 1949, M.S., 1951; Ph.D., U. Tex., 1955. Intermediate geologist Stanolind Oil and Gas Co., 1951-52; sr. geologist Pan Am. Petroleum Corp., 1955-60, Monsanto Chem. Co., 1960-61; cons. geologist Barlow and Haun, Inc., Casper, Wyo., 1961-63; prof. geology, chmn. dept. So. Ill. U., 1963-69; state geologist, exec. dir. Wyo. Geol. Survey, 1969-81; adj. prof. geology U. Wyo., 1969-81; asst. sec. energy and minerals Dept. Interior, 1981-83; geol. cons., 1983-89; dir. Anaconda Geology Documents Collection, U. Wyo., 1989—; mem. Interstate Oil Compact Commn., 1969-81. Co-editor: Overthrust Belt of Southwestern Wyoming, 1960; gen. editor: Geology and Petroleum Production of the Illinois Basin, 1968; Contbr. articles to profl. jours. Served with USAAF, 1942-46. Decorated Air medal; recipient award merit So. Ill. U., 1967. Mem. Am. Assn. State Geologists (pres. 1979-80, chmn. govt. liaison com. 1970-71), Am. Assn. Petroleum Geologists (pres. Rocky Mountain sect. 1987, Pub. Svc. award), Am. Inst. Profl. Geologists (pres. 1992, Ben H. Parker Meml. medal 1993), Assn. Am. State Geologists (pres. 1979-80, hon. mem. 1982), Rocky Mountain Assn. Geologists (Disting. Pub. Svc. award). Achievements include patent for exhausto-port for automobiles. Home: 402 Colony Woods Dr Chapel Hill NC 27514-7908

MILLER, DAVID EDMOND, physician; b. Biscoe, N.C., June 6, 1930; s. James Herbert and Elsie Dale (McGlaughon) M.; m. Marjorie Willard Penton, June 4, 1960; children: Marjorie Dale, David Edmond. AB, Duke U., 1952, MD, 1956. Diplomate Am. Bd. Internal Medicine (subspecialty bd. cardiovasular disease). Internmed. ctr. Duke U., Durham, N.C., 1956-57, resident in internal medicine, 1957-58, 59, 60, research fellow cardiovascular disease, 1958-59, 61, assoc. internal medicine and cardiology, 1963-79, clin. asst., prof. medicine cardiology, 1979—; practice medicine specialising in internal medicine Durham, 1964—; attending physician internal medicine div. cardiology Watts Hosp., Durham, 1964-76, chief medicine, 1975-76; attending physician cardiology div. internal medicine Durham County Gen. Hosp., 1976—, chmn. dept. internal medicine, 1976-82, pres. med. staff, 1980-81; adv. com. Duke Med. Ctr. Contbr. articles to profl. jours. Council clin. cardiology N.C. chpt. Am. Heart Assn., 1963—. Served to lt. comdr. USNR, 1961-63. Fellow ACP, Am. Coll. Cardiology; mem. AMA, So. Med. Assn., N.C. Med. Soc. (del. ho. of dels. 1981, 82, 83), N.C. Durham-Orange County Med. Soc., Am. Soc. Internal Medicine, N.C. Soc. Internal Medicine (exec. council), Am. Fedn. Clin. Research. Methodist. Clubs: Capitol City, Hope Valley Country, University Club, Duke Faculty Club. Home: 1544 Hermitage Ct Durham NC 27707-1680 Office: 2609 N Duke St Ste 403 Durham NC 27704

MILLER, DAVID WILLIAM, geologist; b. Columbus, Ohio, July 5, 1959; s. Thomas Owen and Elaine (Brockmann) M. MS in Geology, Wright State U., 1987; postgrad., Ohio State U., 1989—. Instr. Urbana U., Dayton, 1987-90; teaching asst. Ohio State U., Columbus, 1990-92; instr. II Sinclair C.C., Dayton, 1987-92; instr. Jefferson C.C., Watertown, N.Y., 1992—; environ. specialist II Ohio EPA, Columbus, 1992; paleontologist Mur. of the Rockies, Bozeman, Mont., 1989-92. Contbr. articles to profl. jours. With USAF, 1976-78. Mem. Miami Valley Astron. Soc., Nat. Sci. Tchrs. Assn., Nat. Earth Sci. Tchrs. Assn., Geol. Soc. Am. Lutheran. Home: Hyde Lake Rd RR2 Box 62A Theresa NY 13691 Office: Jefferson Community Coll Outer Coffeen St Watertown NY 13601

MILLER, DAWN MARIE, meteorologist, product marketing specialist; b. Hartford, Conn., Sept. 17, 1963; d. Eugene E. Miller and Audrey E. (Flagg) Laurel; m. Dennis James Miller, Sept. 9, 1989; 1 child, Zackarey. BS in Meteorology, SUNY, Oneonta, 1985. Customer support specialist WSI Corp., Bedford, Mass., 1985-87; in media (TV) mktg. WSI Corp., Billerica, Mass., 1987-91, media (TV) and industry mktg. rep., 1991-92, mktg. communications specialist, 1992-93, product mktg. specialist-data svcs., 1993—. Mem. Am. Meteorol. Soc., Oneonta Alumni Assn., Nat. Arbor Day Found., Am. Film Inst. Republican. Episcopalian. Avocations: meteorology, astronomy, photography, gardening, cooking. Home: 10 Hartford Ln Nashua NH 03063-1904 Office: WSI Corp 4 Federal St Billerica MA 01821-3593

MILLER, DENNIS DIXON, economics educator; b. Chillicothe, Ohio, May 1, 1950; s. Kermit Baker and Martha (Ralston) M. BA, Heidelberg Coll., 1972; MA, U. Colo., 1979, PhD, 1985. Instr. in econs. Am. U., Cairo, Egypt, 1982-84; internat. economist USDA, Washington, 1985-86; assoc. prof. Baldwin-Wallace Coll., Berea, Ohio, 1987—; rsch assoc. Internat. Ctr. Energy and Econ. Devel., Boulder, Colo., 1979-82, Inst. Behavior Sci., U. Colo., Boulder, 1979-82, 84-85; .vis. scholar Hoover Instn., Stanford U., Palo Alto, fall 1986; acad. advisor Heartland Inst., Chgo., 1988—; book reviewer Choice mag., 1984—; pub. policy advisor Heritage Found.'s Listing, Washington, 1991—, econ. cons. gen., 1991—; vis. prof. Mithibai Coll., U. Bombay, India, summer and fall 1991; coord. agency Air Quality Public Adv. Task Force, 1993. Contbr. articles to profl. jours. on rsch. on height and longevity. Earhart Found. fellow, 1977-78. Mem. AAAS, Am. Econs. Assn., Cleve. Coun. on World Affairs, Internat. Joseph S. Schumpeter, Intertel, Nat. Assn. Bus. Economists, Assn. for the Study of Grants Econs., N.Am. Econ. and Fin. Assn., Sierra Club, Mensa. Avocations: running, tennis, reading. Home: 12 Adelbert St Apt 2 Berea OH 44017-1753 Office: Baldwin Wallace Coll Dept of Econs Berea OH 44017

MILLER, DON WILSON, nuclear engineering educator; b. Westerville, Ohio, Mar. 16, 1942; s. Don Paul and Rachel (Jones) M.; m. Mary Catherine Thompson, June 25, 1966; children: Amy Beth, Stacy Catherine, Paul Wilson Thompson. BS in Physics, Miami U., Oxford, Ohio, 1964, MS in Physics, 1966; MS in Nuclear Engring., Ohio State U., 1970, PhD in Nuclear Engring., 1971. Rsch. assoc. Ohio State U., Columbus, 1966-68, univ. fellow, 1968-69, tchg. assoc., 1969-71, asst. prof. nuclear engring., 1971-74, assoc. prof., 1974-80, chmn. nuclear engring. program, 1977—, prof., 1980—, dir. nuclear reactor lab., 1977—; sec., treas. Cellar Lumber Co., Westerville, Ohio, 1972-84, 85—; cons. Monsanto Rsch. Corp., Miamisburg, Ohio, 1979, NRC, Washington, 1982-84, Scantech Corp., Santa Fe, 1984—, Neoprobe Corp., Columbus, 1990, Electric Power Rsch. Inst., Palo Alto, Calif., 1992-93. Patentee in field; contbr. articles to profl. jours. Mem. Westerville Bd. Edn., 1976-91, pres., 1977-78, 86-88; mem. Ohio Sch. Bd.'s Assn., Columbus, 1976-91; mem. fed. rels. com. Nat. Sch. Bd.'s Assn., Washington, 1984-86. With USAR, 1960-68. Named Tech. Person of Yr. Columbus Tech. Coun., 1979; named to All Region Bd. Ohio Sch. Bd's. Assn., 1981, 86; recipient Coll. of Engring Rsch. award Ohio State U., 1984, Achievement award Mid Ohio Chpt Multiple Sclerosis Soc., 1988. Fellow Am. Nuclear Soc. (chmn. edn. div. 1986-87, bd. dirs. 1989-91, chair human factors div. 1993-94, cert. appreciation 1991); mem. IEEE, Am. Soc. Engring. Edn. (chmn. nuclear engring. div. 1978-79, Glenn Murphy award 1989), Nuclear Dept. Heads Orgn. (chmn. 1985-86), Westerville Edn. Assn. (Friend of Edn. award 1992), Rotary (Westerville chpt. (Courtright Community Svc. award 1989), Kiwanis, Hoover Yacht Club, Alpha Nu Sigma (chmn. 1991-93). Avocations: sailing, Am. history, traveling, amateur radio. Home: 172 Walnut Ridge Ln Westerville OH 43081-2464 Office: Ohio State U Dept Mech Engring Nuclear Engring Program 206 W 18th Ave Columbus OH 43210-1154

MILLER, DONALD KENNETH, engineering consultant; b. St. Louis, Oct. 18, 1925; s. Henry Edward and Ernestine Elizabeth (Schmeer) M.; m. Arline Louise Heckman, Feb. 27, 1953; children: Garry Edwin, Kristine Louise Miller Morris. BSChemE, Mo. U., 1950. Registered profl. engr., Pa. Application engr. York Corp., St. Louis, Houston, York, 1951-62; mgr. quality control York Div. Borg Warner Corp., 1962-65, chief engr., 1965-85; refrigeration specialist York Internat. Corp., 1985-88; pres. MDK Engring. Corp., York, 1988—. Author: (with others) Plant Engineering Handbook, 1959, ASHRAE, 1981-94, Applied Thermal Design, 1989; contbr. articles to profl. jours.; inventor desuperheater control in a refrigeration apparatus. With USNR, 1944-46. Mem. NSPE, ASHRAE (Cen. Pa. chpt., life, sec. 1972-73, treas. 1973-74, v.p. 1974-75, pres. 1975-76, Disting. Svc. award 1992), AICE, Pa. Soc. Profl. Engrs., Internat. Inst. Refrigeration. Republican. Presbyterian. Avocations: sketching, computers. Home: 1749 Prescott Rd York PA 17403-4607 Office: MDK Engring Corp 391 Greendale Rd York PA 17403-4635

MILLER, DONALD SPENCER, geologist, educator; b. Ventura, Calif., June 12, 1932; s. Spencer Jacob and Marquerite Rachael (Williams) M.; m. Carolyn Margaret Losee, June 12, 1954; children: Sandra Louise, Kenneth Donald, Christopher Spencer. BA, Occidental Coll., 1954; MA, Columbia U., 1956, PhD, 1960. Asst. prof. Rensselaer Poly. Inst., Troy, N.Y., 1960-64, assoc. prof., 1964-69, prof., 1969—, chmn. dept. geology, 1969-76, 80-90; research assoc. geology Columbia U., 1960-63; research fellow geochemistry Calif. Inst. Tech., Pasadena, summer 1963; NSF Sci. Faculty fellow U. Bern, Switzerland, 1966-67; sci. guest prof. Max-Planck Inst. Nuclear Physics, Heidelberg, Fed. Republic Germany, 1977-78, vis. prof., summer 1979, guest scientist, Aug. 1979, 80, 81, 82; vis. prof. Isotope Geology Lab., U. Berne, summer 1979; participant NATO exchange program Demokritos Inst., Athens, Greece, Sept. 1983, 85; vis. rsch. fellow U. Melbourne, Australia, summer 1988; mem. nat. screening com. Inst. Internat. Edn., 1988-91. Pres., treas. Troy Rehab. and Improvement, Inc., 1968-74; mem. Troy Zoning Bd. Appeals, 1970-85. Fellow Geol. Soc. Am.; mem. Am. Geophys. Union, Geochem. Soc., Nat. Assn. Geol. Tchrs., Sigma Xi, Sigma Pi Sigma. Home:

2198 Tibbits Ave Troy NY 12180-7015 Office: Rensselaer Poly Inst Dept Earth & Environ Scis Troy NY 12181

MILLER, DOUGLAS ALAN, civil engineer; b. Lyons, N.Y., Dec. 13, 1961; s. Alan Charles and Esther Gay Miller. BS, Syracuse U., 1984. Engr. N.Y. State Dept. Transp., Syracuse, 1984-85, Dayspring Constructors, Baldwinsville, N.Y., 1986-87, O'Brien's Gere Engrs., Syracuse, 1987—; mem. Town of Pompey (N.Y.) Planning Bd., 1990—, chmn., 1992—. Mem. Everson Art Soc., Syracuse, 1990. Mem. ASCE, Am. Water Works Assn., Water Pollution Control Fedn., Sigma Chi. Methodist. Achievements include research of City of Lockport composting facilities. Home: 7500 Broadfield Rd Manlius NY 13104 Office: OBrien & Gere Engrs 5000 Brittonfield Pky Syracuse NY 13221

MILLER, GEORGE ARMITAGE, psychologist, educator; b. Charleston, W.Va., Feb. 3, 1920; s. George E. and Florence (Armitage) M.; m. Katherine James, Nov. 29, 1939; children: Nancy, Donnally James. B.A., U. Ala., 1940, M.A., 1941; AM, Harvard U., 1944, Ph.D., 1946; Doctorat honoris causa, U. Louvain, 1976; D Social Sci. (hon.), Yale U., 1979; D.Sc. honoris causa, Columbia U., 1980; DSc (hon.), U. Sussex, 1984; LittD (hon., Charleston U., 1992. Instr. psychology U. Ala., 1941-43; research fellow Harvard Psycho-Acoustic Lab., 1944-48; asst. prof. psychology Harvard U., 1948-51, assoc. prof., 1955-58, prof., 1958-68, chmn. dept psychology, 1964-67, co-dir. Ctr. for Cognitive Studies, 1960-67; prof. Rockefeller U., N.Y.C., 1968-79; adj. prof. Rockefeller U., 1979-82; prof. psychology Princeton U., 1979-90, James S. McDonnell Disting. Univ. prof. psychology, 1982-90, James S. McDonnell Disting. Univ. prof. psychology emeritus, 1990—, program dir. McDonnell-Pew Program in Cognitive Neurosci., 1989—; assoc. prof. MIT, 1951-55; vis. Inst. for Advanced Study, Princeton, 1972-76, 82-83, mem., 1950, 70-72; vis. prof. Rockefeller U., 1967-68; vis. prof. MIT, 1976-79, group leader Lincoln Lab., 1953-55; fellow Ctr. Advanced Study in Behavioral Scis., Stanford U., 1958-59; Fulbright research prof. Oxford (Eng.) U., 1963-64; Sesquicentennial prof. U. Ala., 1981. Author: Language and Communication, 1951, (with Galanter and Pribram) Plans and the Structure of Behavior, 1960, Psychology, 1962, (with Johnson-Laird) Language and Perception, 1976, Spontaneous Apprentices, 1977, Language and Speech, 1981, The Science of Words, 1991; editor Psychol. Bulletin, 1981-82. Recipient Disting. Service award Am. Speech and Hearing Assn., 1976, award in behavioral scis. N.Y. Acad. Scis., 1982, Hermann von Helmholtz award Cognitive Neurosci. Inst., 1989, Nat. Medal Sci. NSF, 1991, Gold Medal Am. Psychological Found. 1990, Nat. Medal of Sci. 1991, Louis E. Levy medal Franklin Inst., 1991; Guggenheim fellow, 1986, William James fellow Am. Psychological Soc., 1989; Fondation Fyssen Priz Internat. for cognitive sci., 1992. Fellow Brit. Psychol. Assn. (hon.); mem. NAS, AAAS (chmn. sect. J 1981), Am. Psychol. Assn. (pres. 1968-69, Disting. Scientific Contbn. award 1963, William James Book award divsn. gen. psychology 1993), Eastern Psychol. Assn. (pres. 1961-62), Acoustical Soc. Am., Linguistic Soc. Am., Am. Statis. Assn., Am. Philos. Soc., Am. Physiol. Soc., Psychometric Soc., Soc. Exptl. Psychologists (Warren medal 1972), Am. Acad. Arts and Scis., Psychonomic Soc., Royal Netherlands Acad. Arts and Scis. (fgn.), Sigma Xi. Home: 753 Prospect Ave Princeton NJ 08540-4080 Office: Princeton Univ Dept Psychology Green Hall Princeton NJ 08544

MILLER, GEORGE MCCORD, electrical engineer; b. Abbeville, S.C., Dec. 14, 1919; s. Daisey (McCord) M.; m. Ethel Waldron, Aug. 8, 1945; 1 child George McCord, Jr. BEE, Clemson U., 1940; postgrad., Union Coll., 1946-47. Registred profl. engr., N.Y., Ala. Sponsor engr. apparatus sales div. GE Co., Schenectady, N.Y., 1948-55; application engr. user industry sales GE Co., Schenectady, 1955-58; application engr. elec. utility sales div. GE Co., Birmingham, Ala., 1958-71; mgr. application engring. transmission and distbn. sect. GE Co., Birmingham, 1971-75, sr. application engr. elec. utility sales div., 1975-84; pres. George M. Miller and Assocs., Birmingham, 1985—. 1st lt. U.S. Army, 1943-46, ETO. Mem. IEEE (sr. life), Nat. Soc. Profl. Engrs. Home and Office: 3369 Hermitage Rd Birmingham AL 35223-2003

MILLER, HAROLD WILLIAM, nuclear geochemist; b. Walton, N.Y., Apr. 21, 1920; s. Harold Frank and Vera Leona (Simons) M. BS in Chemistry, U. Mich., 1943; MS in Chemistry, U. Colo., 1948, postgrad. Control chemist Linde Air Products Co., Buffalo, 1943-46; analytical research chemist Gen. Electric Co., Richland, Wash., 1948-51; research chemist Phillips Petroleum Co., Idaho Falls, Idaho, 1953-56; with Anaconda (Mont.) Copper Co., 1956; tech. dir., v.p. U.S. Yttrium Co., Laramie, Wyo., 1956-57; tech. dir. Colo. div. The Wah Chang Co., Boulder, Colo., 1957-58; analytical chemist The Climax (Colo.) Molybdenum Co., 1959; with research and devel. The Colo. Sch. of Mines Research Found., Golden, 1960-62; cons. Boulder, 1960—; sr. research physicist Dow Chem. Co., Golden, 1963-73; bd. dirs. Sweeney Mining and Milling Corp., Boulder; cons. Hendricks Mining and Milling Co., Boulder; instr. nuclear physics and nuclear chemistry Rocky Flats Plant, U. Colo. Contbr. numerous articles to profl. jours. Recipient Lifetime Achievement award Boulder County Metal Mining Assn., 1990. Mem. Sigma Xi. Avocations: mineralogy, western U.S. mining history. Home and Office: PO Box 1092 Boulder CO 80306-1092

MILLER, HARTMAN CYRIL, JR., chemical hazardous material training specialist; b. Balt., Feb. 5, 1948; s. Hartman Cyril Sr. and Loretta Mae (Campbell) M.; m. Bonnie Mae Long, June 16, 1973. BA in Chemistry, Towson State Coll., 1970. Analytical chemist Joseph E. Seagram & Sons, Dundalk, Md., 1972-81; quality control mgr. Genstar Stone Products, Texas, Md., 1981-83; hazardous material tng. specialist, project team leader Hazmat Tng. Info. Svcs. Inc., Columbia, Md., 1991—. Mem. Second Amendment Found., Balt., 1990—. Nuclear biol. chem. non-commd. officer Md. Army NG, 1970-92; ret. Mem. ACS (analytical div.), Am. Inst. Chemists, Nat. Environ. Tng. Assn. Democrat. Roman Catholic. Achievements include research in reduction reactions in liquid sulfur, in neutralization chemicals for use in air purifying respirator filter cartridges, and in barrier fabrics and suit construction for chemical protective clothing. Home: 1533 Old Manchester Rd Westminster MD 21157-3835 Office: Hazmat Tng Info Svcs Inc 9017 Red Branch Rd Columbia MD 21045-2194

MILLER, HARVEY ALFRED, botanist, educator; b. Sturgis, Mich., Oct. 19, 1928; s. Harry Clifton and Carmen (Sager) M.; m. Donna K. Hall, May 9, 1992; children: Valerie Yvonne, Harry Alfred, Timothy Merk, Tanya Merk. B.S., U. Mich., 1950; M.S., U. Hawaii, 1952; Ph.D., Stanford U., 1957. Instr. botany U. Mass., 1955-56; instr. botany Miami U., 1956-57, asst. prof., 1957-61, assoc. prof., curator herbarium, 1961-67; prof., chmn. program in biology Wash. State U., 1967-69; vis. prof. botany U. Ill., 1969-70; prof., chmn. dept. biol. scis. U. Cen. Fla., 1970-75, prof., 1975—; v.p. Marine Research Assocs. Ltd., Nassau, 1962-65; assoc. Lotspeich & Assocs., natural systems analysts, Winter Park, Fla., 1979—; botanist U. Mich. Expdn. to Aleutian Islands, 1949-50; prin. investigator Systematic and Phytogeol. Studies Bryophytes of Pacific Islands, NSF, 1959, Miami U. Expdn. to Micronesia, 1960; dir. NSF-Miami U. Expdn. to Micronesia and Philippines, 1965; prin. investigator NSF bryophytes of So. Melanesia, 1983-86; research assoc. Orlando Sci. Ctr., Orlando; vis. prof. U. Guam, 1965; cons. tropical botany, foliage plant patents, also designs for sci. bldgs.; adj. prof. botany Miami U., 1985—; field researcher on Alpine meadows in Irian Jaya, 1991, 1992. Author: (with H.O. Whittier and B.A. Whittier) Prodromus Florae Muscorum Polynesiae, 1978, Prodromus Florae Hepaticarum Polynesiae, 1983; Field Guide to Florida Mosses and Liverworts, 1990; editor: Florida Scientist, 1973-78; contbr. articles to sci. jours. Mem. exec. bd. and chmn. scholarship and grant selection com. Mercury Seven Found., 1985—. Recipient Acacia Order of Pythagoras; recipient Acacia Nat. award of Merit; Guggenheim fellow, 1958. Fellow AAAS, Linnean Soc. London; mem. Pacific Sci. Assn. (chmn. sci. com. for botany 1975-83), Assn. Tropical Biology, Am. Inst. Biol. Scis., Am. Bryol. Soc. (v.p. 1962-63, pres. 1964-65), Brit. Bryol. Soc., Bot. Soc. Am., Internat. Assn. Plant Taxonomists, Internat. Assn. Bryologists, Mich. Acad. Sci. Arts and Letters, Hawaiian Acad. Sci., Am. Soc. Plant Taxonomists, Fla. Acad. Sci. (exec. sec. 1976-83, pres. 1980), Nordic Bryol. Soc., Acacia, Explorers Club, Sigma Xi, Phi Sigma, Beta Beta Beta. Home: PO Box 4413 Winter Park FL 32793-4413 Office: U Central Fla Dept Biology Orlando FL 32816

MILLER, HERBERT DELL, petroleum engineer; b. Oklahoma City, Sept. 29, 1919; s. Merrill Dell and Susan (Green) M.; BS in Petroleum Engring., Okla. U., 1941; m. Rosalind Rebecca Moore, Nov. 23, 1947; children: Rebecca Miller Friedman, Robert Rexford. Field engr. Amerada Petroleum Corp., Houston, 1948-49, Hobbs, N.Mex., 1947-48, dist. engr., Longview, Tex., 1949-57, sr. engr., Tulsa, 1957-62; petroleum engr. Moore & Miller Oil Co., Oklahoma City, 1962-78; owner Herbert D. Miller Co., Oklahoma City, 1978—. Maj., F.A., AUS, 1941-47; ETO. Decorated Bronze Star with oak leaf cluster, Purple Heart (U.S.); Croix de Guerre (France). Registered profl. engr., Okla., Tex. Mem. AIME, Petroleum Club. Republican. Episcopalian (pres. Men's Club 1973). Clubs: Oklahoma City Golf, Country. Home: 6708 NW Grand Blvd Oklahoma City OK 73116-6016 Office: 1236 First National Ctr W 120 N Robinson Oklahoma City OK 73102

MILLER, HERMAN LUNDEN, retired physicist; b. Detroit, Apr. 23, 1924; s. Josiah Leonidas and Sadie Irene (Lunden) M.; m. Dorothy Grace Sack, Sept. 15, 1951. BS in Engring. Physics, U. Mich., 1948, MS in Physics, 1951. Registered profl. engr., Mich. Physicist Ethyl Corp., Ferndale, Mich., 1948-49, Dow Chem. Co., Denver, 1950-55; mem. project rsch. staff Princeton (N.J.) U., 1955-65; physicist Bendix Aerospace, Ann Arbor, Mich., 1965-72; nuclear engr. Commonwealth Assocs., Jackson, Mich., 1973-80. Contbr. articles to profl. jours. With USAF, 1943-46, PTO, lt. col. Res. Mem. IEEE, Am. Phys. Soc., Am. Nuclear Soc.

MILLER, HUGH THOMAS, computer consultant; b. Indpls., Mar. 22, 1951; s. J. Irwin and Xenia S. Miller; m. Katherine Thorne McLeod, Sept. 17, 1988; 1 child, Jonathan William. BA, Yale U., 1976; SM in Mgmt., MIT, 1985. Owner Hugh Miller Bookstore, New Haven, 1976-83, Hugh Miller Cons., New Haven; ind. cons. microcomputers, 1983-85; supr. decision technologies div. Electronic Data Systems, Inc., Troy, Mich., 1985-86, supr. product and mfg. engring. div., 1986-90; product mgr. Indsl. Bus. Devel.; supr. Packard Electric Acct., Troy, 1990-92, acct. mgr. GM Chassis Systems Ctr., 1992—. Editor, ptnr. The Common Table, pub. firm. Bd. dirs. Irwin-Sweeney-Miller Found., Columbus, Ind., 1972—; bd. of govs. MIT Sloan Sch. Mgmt., 1989—. Mem. Am. Mktg. Assn., Assn. Computing Machinery, IEEE. Home: 1173 Lake Angelus Rd Lake Angelus MI 48326-1028 Office: EDS/Chassis Systems Ctr 1001 Charles H Orndorf Dr Brighton MI 48116

MILLER, JAMES EDWARD, computer scientist, educator; b. Lafayette, La., Mar. 21, 1940; s. Edward Gustave and Orpha Marie (DeVilbis) M.; m. Diane Moon, June 6, 1964; children—Deborah Elaine, Michael Edward. B.S., U. La.-Lafayette, 1961, Ph.D., 1972; M.S., Auburn U., 1964. Systems engr. IBM, Birmingham, Ala., 1965-68; asst. prof. U. West Fla., Pensacola, 1968-70, chmn. systems sci., 1972-86; grad. researcher U. La.-Lafayette, 1970-72; computer systems analyst EPA, Washington, 1979; prof., chmn. computer sci. and stats. U. So. Miss., Hattiesburg, 1986—; program evaluator Computer Sci. Accreditation Commn., 1986—, cons., lectr. in field; co-dir. NASA/Am. Soc. Engring. Edn. Summer Faculty Fellowship Program-Stennis Space Flight Ctr. Author numerous articles for tech. publs. Mem. Computer Soc. of IEEE, Assn. Computing Machinery (editor Computer Sci. Edn. spl. interest group bull. 1982—), Data Processing Mgmt. Assn. (dir. edn. spl. interest group 1985-86), Info. Systems Security Assn., Internat. Assn. Math. and Computer Modeling. Democrat. Methodist. Lodge: Rotary. Avocations: Research on computer crime, computer sci. edn. and optimal sensor deployment. Office: Univ of So Miss Comouter Science & Stat Box 5106 Southern Sta Hattiesburg MS 39406

MILLER, JAMES FREDERICK, geologist, educator; b. Davenport, Iowa, Feb. 18, 1943; s. Harry Earnest and Frances Elizabeth (Henry) M.; m. Louise L. Miersch, June 17, 1967; children: Jason F., Michelle L. BA cum laude, Augustana Coll., Rock Island, Ill., 1965; MA, U. Wis., 1968, PhD, 1971. Asst. prof. geology U. Utah, Salt Lake City, 1970-73, Lawrence U., Appleton, Wis., 1973, U. Pitts., Johnstown, Pa., 1974; asst. prof. geology S.W. Mo. State U., Springfield, 1974-77, assoc. prof. geology, 1977-82, prof. geology, 1982—, disting. scholar, 1988—; voting mem. Internat. Working Group on the Cambrian-Ordovician Boundary, 1974-93, sec., 1980-93, corr. mem., 1993—. Author pub. field trip guidebooks and abstracts; contbr. numerous articles to profl. jours. NSF grantee, 1977-80, 81-83, 85-88, 88-89, 88-92, faculty rsch. grantee S.W. Mo. State U., 1975, 76, 78-79, 80, 81-82, 83, 86, 88, 92, S.W. Mo. State U. Found. grantee, 1984. Mem. AAAS, Internat. Paleontol. Assn., Palaeontological Assn., Paleontol. Soc., Soc. Sedimentary Geology, Geol. Soc. Am., Nat. Assn. Geology Tchrs. (Ctrl. Sect. v.p. 1989-90, pres. 1990-91), Paleontol. Rsch. Instn., Pander Soc., Sigma Xi. Achievements include research in invertebrate paleontology, micropaleontology, biostratigraphy, stratigraphy, redefinition and correlation of the Cambrian-Ordovician Boundary, Upper Cambrian-Lower Ordovician conodonts. Office: Southwest Missouri State U Geography, Geology & Plng Springfield MO 65804-0089

MILLER, JAN DEAN, metallurgy educator; b. Dubois, Pa., Apr. 7, 1942; s. Harry Moyer and Mary Virginia (McQuown) M.; m. Patricia Ann Rossman, Sept. 14, 1963; children: Pamela Ann, Jeanette Marie, Virginia Christine. B.S., Pa. State U., 1964; M.S., Colo. Sch. of Mines, 1966, Ph.D., 1969. Research engr. Anaconda Co., Mont., 1966, Lawrence Livermore Lab., Calif., 1972; asst. prof. metallurgy U. Utah, Salt Lake City, 1968-72, assoc. prof., 1972-78, prof., 1978—; cons. on processing of mineral resources to various cos. and govt. agys. Editor: Hydrometalurgy, Research, Development, and Plant Practice, 1983. Contbr. over 200 articles to profl. jours. First commercial plant using air-sparged hydrocyclone tech. for deinking flotation in wastepaper recycling plant, 1992, patentee, 1983. Bethelehem Steel fellow, 1964-68; recipient Marcus A. Grossman award Am. Soc. Metals, 1974; Van Diest Gold medal Colo. Sch. Mines, 1977; Mellow Met award U. Utah, Salt Lake City, 1978, 82, Stefanko award coal divsn. Soc. Mining Engrs., 1988, Extractive Metallurgy Tech. award, Metall. Soc., 1988, Richards award Am. Inst. Mining, Metall. and Petroleum Engrs., 1990, Extractive and Processing Lectr. award The Minerals, Metals and Materials Soc., 1992. Mem. AIME (chmn. mineral processing div. 1980-81), Fine Particle Soc., AIME (Henry Krumb lectr. 1987), NAE, Am. Chem. Soc., Soc. Mining Engrs. (bd. dirs. 1980-83, program chmn. 1982-83, Taggart award, 1986), Metall. Soc., Soc. Mining, Metall. and Exploration (Disting. mem., Antoine M. Gaudin award 1992). Baptist. Clubs: Salt Lake Swim and Tennis; U. Utah Faculty. Office: U Utah Dept Metallurgy 412 William C Browning Bldg Salt Lake City UT 84112

MILLER, JANEL HOWELL, psychologist; b. Boone, N.C., May 18, 1947; d. John Estle and Grace Louise (Hemberger) Howell; B.A., DePauw U., 1969; postgrad. Rice U., 1969; M.A., U. Houston, 1972; Ph.D., Tex. A&M U., 1979; m. C. Rick Miller, Nov. 24, 1968; children—Kimberly, Brian, Audrey, Rachel. Asso. sch. psychologist Houston Ind. Sch. Dist., 1971-74; research psychologist VA Hosp., Houston, 1972; asso. sch. psychologist Clear Creek Ind. Sch. Dist., Tex., 1974-76; instr. psychology, counseling psychology intern Tex. A. and M. U., 1976-77; clin. psychology intern VA Hosp., Houston, 1977-78; coordinator psychol. services Clear Creek Ind. Sch. Dist., 1978-81, assoc. dir. psychol. services, 1981-82; pvt. practice, Houston, 1982—; faculty U. Houston-Clear Lake, 1984—; adolescent suicide cons., 1984—. DePauw U. Alumni scholar, 1965-69; NIMH fellow U. Houston, 1970-71; lic. clin. psychologist, sch. psychologist, Tex. Mem. Am. Psychol. Assn., Tex. Psychol. Assn., Houston Psychol. Assn. (media rep. 1984-85), Am. Assn. Marriage and Family Therapists, Tex. Assn. Marriage and Family Therapists, Houston Assn. Marriage and Family Therapists. Home: 806 Walbrook Dr Houston TX 77062-4030 Office: Southpoint Psychol Svcs 11550 Fuqua St Ste 450 Houston TX 77034

MILLER, JEFFREY STEVEN, hematologist, researcher; b. Cleve., Nov. 21, 1958; s. Alan Irving and Jean (Press) M.; m. Karen Behrens (div. 1993). BS in Engring., Northwestern U., 1981, MD, 1985. Diplomate Am. Bds. Internal Medicine. Med. resident U. Iowa, Iowa City, 1985-88; fellow in hematology and oncology U. Minn., Mpls., 1988-91, instr. medicine, 1991-92, asst. prof., 1992—. Contbr. articles to profl. jours. Recipient young investigator award Am. Soc. Clin. Oncology, 1991. Mem. ACP, Am. Soc. Hematology, Internat. Soc. Exptl. Hematology. Achievements include research on natural killer cell growth kinetics and immune ontogeny. Home: 59 Rosewood Dr Little Canada MN 55117 Office: Univ Minn Box 480 420 Delaware St SE Minneapolis MN 55455

MILLER, JOAN G., psychology educator; b. Cleve., Sept. 20, 1949. MA, U. Chgo., 1975, PhD, 1982. Rsch. asst. Michael Reese Med. Ctr., Chgo., 1977-79, rsch. assoc., 1980-82; predoctoral fellow The East-West Ctr., Honolulu, 1981; lectr. U. Chgo., 1982, postdoctoral fellow, 1982-84; asst. prof. Yale U., 1984-90, assoc. prof., 1990--. Contbr. articles to profl. jours. nat. Inst. Mental Health grantee 1980-81, 87-90, 91--, nat. Sci. Found. grantee, 1980-81, 91--; East-West Ctr. fellow, 1980-81. Mem. Soc. for Rsch. in Child Devel., Soc. Experimental Social Psychology, Soc. for Personality and Social Psychology, Soc. for Cross Cultural Rsch., Soc. for Psychological Anthropology, Am. Psychological Soc., Am. Psychological Assn., Am. Anthropologica Assn. Office: Yale U Box 11A Yale Sta New Haven CT 06520

MILLER, JOEL STEVEN, solid state scientist; b. Detroit, Oct. 14, 1944; s. John and Rose (Schpok) M.; m. Elaine J. Silverstein, Sept. 20, 1970; children: Stephen D., Marc A., Alan D. BS in Chemistry, Wayne State U., 1967; PhD, UCLA, 1971. Mgr. rsch. Occidental Rsch. Corp., Irvine, Calif., 1979-83; supr. rsch. Cen. R & D Lab. E. I. Du Pont Nemours & Co., Wilmington, Del., 1983-93; prof. chemistry U. Utah, 1993—; bd. dirs. Inorganic Synthesis Corp., Chgo.; vis. prof. U. Calif., Irvine, 1980, Weizmann Inst., Rehovot, Israel, 1985, U. Pa., Phila., 1988, U. Paris-Sud, 1991. Editor 8 books; mem. adv. bd. Jour. Chemistry Materials, 1990—, Jour. Materials Chemistry, 1991—; reporter rsch. news Advanced Materials, 1990—; contbr. over 200 articles to sci. jours. Mem. Am. Chem. Soc. (chmn. solid state subdiv. 1989). Achievements include discovery and development of molecular-based conductors and magnets. Office: U Utah Dept of Chemistry Salt Lake City UT 84108

MILLER, JOHN CAMERON, research chemist; b. Richmond, Va., Aug. 12, 1949; s. Edward McCarthy and Jean Elenor (Hudson) M.; m. Gray Marshall Staples, Sept. 13, 1975; 1 child, Marshall Gregory. BS in Chemistry, Ga. Inst. Tech., 1971; PhD in Chemistry, U. Colo., 1975. Instr. The Colo. Coll., Colorado Springs, 1976; postdoctoral fellow U. Va., Charlottesville, 1976-79; staff scientist Oak Ridge (Tenn.) Nat. Lab., 1979—, sect. head, 1989—. Contbr. 80 articles to jours., chpts. to books; editor 3 books. Recipient IR-100 award Indsl. Rsch. and Devel. Mag., 1983. Fellow Am. Phys. Soc., Optical Soc. Am.; mem. Am. Chem. Soc. Achievements include discovery of nonlinear quantum interference effects, first use of laser photoelectron spectroscopy, discovery of many novel van der Waals molecules. Office: Oak Ridge Nat Lab PO Box 2008 Bldg 4500-S MS 6125 Oak Ridge TN 37831

MILLER, JOHN RICHARD, scientist; b. Knoxville, Aug. 10, 1944; s. Charles Addison and Ruby Beulah (Bramlett) M.; m. Mary Ann Harris Miller, Mar. 25, 1967; children: Jennifer Nicole, John Richard II, Meghan Amanda. BS, East Tenn. State U., 1966, MS, 1968; PhD, U. Va., 1973. Sr. scientist Braddock, Dunn & McDonald, Inc., Vienna, Va., 1973-74; rsch. staff Oak Ridge (Tenn.) Nat. Lab., 1974-83; engr. Lawrence Livermore Nat. Lab. U. Calif., 1983-91; scholar, scientist Nat. High Magnetic Field Lab. Fla. State U., Tallahassee, 1991—; leader Magnet Design Unit, ITER Conceptual Design Activity, Garching, Fed. Republic of Germany, 1989-90; bd. dirs. Applied Superconductivity Conf., Inc., 1990—. Recipient Citation for notable contbr. to fusion energy rsch. Dir. Energy Rsch., U.S. Dept. Energy, 1991. Office: Nat High Magnetic Field Lab 1800 E Paul Dirac Dr Tallahassee FL 32306

MILLER, JOSEF M., otolaryngologist, educator; b. Phila., Nov. 29, 1937; married, 1960; 2 children. BA in Psychology, U. Calif., Berkeley, 1961; PhD in Physiology and Psychology, U. Wash., 1965; MD (hon.), U. Göteborg, Sweden, 1987. USPHS fellow U. Mich., 1965-67; rsch. assoc., asst. prof. dept. Psychology U. Mich., Ann Arbor, 1967-68, prof., dir. rsch. dept. Otolaryngology, dir. Kresge Hearing Rsch. Inst., 1984—; asst. prof. depts. Otolaryngology, Physiology and Biophysics U. Wash., Seattle, 1968-72, rsch. affiliate Regional Primate Rsch. Ctr, 1968-84, assoc. prof., 1972-76, acting chmn. dept. Otolaryngology, 1975-76, prof., 1976-84; study sect. Nat. Inst. Neurol. and Communiative Disorders and Stroke, NIH, 1978-84, ad hoc bd. dirs. sci. counselors, 1988; sci. review com. Deafness Rsch. Found., 1978-83, chair, 1983—; mem. faculty Nat. Conf. Rsch. Goals and Methods in Otolaryngology, 1982; adv. com. hearing, bio-acoustics and biomechanics Commn. Beharvioral and Social Scis. and Edn., Nat. Rsch. Coun., 1983—; hon. com. Orgn. Nobel Symposium 63, Cellular Mechanisms in Hearing, Karlskoga, Sweden, 1985; cons. Otitis Media Rsch. Ctr., 1985-89, Pfizer Corp., 1988; faculty opponent U. Göteborg, Sweden, 1987; rsch. adv. com. Galludet Coll., 1987; chair external scientific adv. com., House Ear Inst., 1988-91; author authorizing legis. NIDCD, NIH, 1988, co-chair adv. bd. rsch. priorities com., bd. dirs. Friends, adv. coun., 1989—, chair rsch. subcom., 1990—; grant reviewer Mich. State Rsch. Fund, NSF, VA; reviewer numerous jours. including Acta Otolanryngologica, Jour. Otology, Physiology and Behavior, Science. Mem. editorial bd. Am. Jour. Otolaryngology, 1981—, AMA, Am. Physiology Soc., Annals of Otology, Rhinology and Laryngology, 1980—, Archives of Oto-Rhino-Laryngology, 1985-93, Hearing Rsch., Jour. Am. Acad. Otolaryngology-Head and Neck Surgery, 1990—. Bd. dirs. Internat. Hearing Found., 1985—. Fellow U. Wash., 1962-65, Kresge Hearing Rsch. Inst., U. Mich., 1965-67; recipient award Am. Acad. Otolaryngology; grantee Deafness Rsch. Found., U. Wash., 1969-71; rsch. grantee NIH, 1969-73. Mem. AAAS, Am. Acad. Otolaryngology and Head and Neck Surgery (com. rsch. in otolaryngology 1971-82, continuing edn. com. 1975-79, NIH liaison com. 1988—, program steering com. jour. 1990—), Am. Auditory Soc., Am. Otological Soc., Am. Neurotological Soc., Am. Otologic Honor Soc., Acoustical Soc. Am. (com. rsch. psychol., physiol. acoustics 1969-78), Fedn. Am. Physiological Soc., Fedn. Am. Socs. Experimental Biology, Soc. Neurosci., Assn. Rsch. Otolaryngology (sec.-treas. 1979-80, pres. elect 1981, pres. 1982. program dir. mtg. 1983, award of merit com. 1985, chair 1988), Sigma Xi. Office: U Mich Kresge Hearing Rsch Inst 1301 E Ann St HR5032 Ann Arbor MI 48109-0506*

MILLER, JOSEPH ARTHUR, manufacturing engineer, educator, consultant; b. Brattleboro, Vt., Aug. 28, 1933; s. Joseph Maynard and Marjorie Antoinette (Hammerberg) M.; m. Ardene Hedwig Barker, Aug. 19, 1956; children: Stephanie L., Jocelyn A., Shana L., Gregory J. BS in Agrl., Andrews U., Berrien Springs, Mich., 1955; MS in Agrl. Mechs., Mich. State U., 1959; EdD in Vocat. Edn., UCLA, 1973. Constrn. engr. Thornton Bldg. & Supply, Inc., Williamston, Mich., 1959-63, C & B Silo Co., Charlotte, Mich., 1963-64; instr. and dir. retraining Lansing (Mich.) Community Coll., 1964-68; asst. prof./prog. coord./coop coord. San Jose State U., 1968-79; mfg. specialist Lockheed Missiles & Space Co., Sunnyvale, Calif., 1979-81, rsch. specialist, 1981-88, NASA project mgr., 1982-83, staff engr., 1988—, team leader Pursuit of Excellence award winning project, 1990—; agrl. engring. cons. USDA Poultry Expt. Sta., 1960-62; computer numerical control cons. Dynamechtronics, Inc., Sunnyvale, 1987—; machining cons. Lockheed, Space Sys. Div., 1986—; instr. computer numerical control DeAnza Coll., Cupertino, Calif., 1985-88, Labor Employment Tng. Corp., San Jose, Calif., 1988—. Author: Student Manual for CNC Lathe, 1990; contbr. articles to profl. jours. Career counselor Pacific Union Coll., Angwin, 1985—. UCLA fellow, 1969-73. Mem. Soc. Mfg. Engrs. (sr. mem.), Nat. Assn. Indsl. Tech. (pres. industry div. 1987-88, bd. cert. 1991-92), Calif. Assn. Indsl. Tech. (pres. 1974-75, 84-85), Am. Soc. Indsl. Tech. (pres. 1980-81). Seventh-day Adventist. Avocations: violin, camping, designing and building homes. Home: 338 Raccoon Rd Berry Creek CA 95916-9518 Office: Lockheed Missiles & Space 1111 Lockheed Way Sunnyvale CA 94089-3504

MILLER, KENNETH EDWARD, mechanical engineer, consultant; b. Weymouth, Mass., Dec. 24, 1951; s. Edward Francis and Lena Joan (Trotta) M.; m. Florence Gay Wilson, Sept. 18, 1976; children: Nicole Elizabeth, Brent Edward. BSME, Northeastern U., 1974; MS in Systems Mgmt., U. So. Calif., 1982. Registered profl. engr., N.Y., N.H., Ariz., Nev.; registered land surveyor, Ariz. Test engr. Stone & Webster Engring., Boston, 1974-76; plant engr. N.Y. State Power Authority, Buchanan, 1976-80; maintenance engr. Pub. Service Co. of N.H., Seabrook, 1980-82; cons. engr. Helios Engring. Inc., Litchfield Park, Ariz., 1982-87; sr. supervisory service engr. Quadrex Corp., Coraopolis, Penn., 1987-89; cons. engr. Helios Engring., Inc., Litchfield Park, Ariz., 1989—. Republican. Roman Catholic. Avocations: piloting, scuba diving. Office: 360 Ancora Dr S Litchfield Park AZ 85340-4639

MILLER, LEONARD DOY, army officer; b. Minden, La., Feb. 14, 1941; s. Doy Bracken and Jewel (Krouse) M.; m. Mona Carolyn Hall, Dec. 22, 1962; children: Doy Michael, Mark Leonard. BS, Northwestern State U. La., 1964. Commd. 2d lt. U.S. Army, 1964, advanced through grades to brig.

gen., 1989; student U.S. Army Command and Gen. Staff Coll., Ft. Leavenworth, Kans., 1973-74; ops. officer 9th Inf. Div. Arty., Ft. Lewis, Wash., 1977; bn. comdr. 1st Bn., 11th F.A., 9th Inf. Div., Ft. Lewis, 1977-79; student Nat. War Coll., Washington, 1979-80; action officer Orgn. Joint Chiefs of Staff, Washington, 1980-81; dep. div. chief, exec. officer Office Chief of Legis. Liaison, Washington, 1981-83; comdr. 5th Inf. Div. Arty., Ft. Polk, La., 1983-86; div. chief Office Chief of Legis. Liaison, Washington, 1986-87; chief of staff 5th Inf. Div., Ft. Polk, 1987-88; dep. chief Office Chief of Legis. Liaison, Washington, 1988-90; comdr. V Corps Arty., Frankfurt, Germany, 1990-92; comdr. field command Defense Nuclear Agy., Kirtland AFB, N.Mex., 1992—. Decorated Legion of Merit, Bronze Star, Meritorious Svc. medal, Def. Meritorious medal. Mem. Assn. U.S. Army. Baptist. Avocations: fishing, woodworking. Home: Box 16 Rt 5 Minden LA 71055 Office: Field Command Defense Nuclear Agy Kirtland AFB NM 87117

MILLER, LYNNE MARIE, environmental company executive; b. N.Y.C., Aug. 4, 1951; d. David Jr. and Evelyn (Gulbransen) M. AB, Wellesley Coll., 1973; MS, Rutgers U., 1976. Analyst Franklin Inst., Phila., 1976-78; dir. hazardous waste div. Clement Assocs., Washington, 1978-81; pres. Risk Sci. Internat., Washington, 1981-86, Environ. Strategies Corp., Vienna, Va., 1986—, Environ. Strategies Ltd., London, 1986—. Editor: Insurance Claims for Environmental Damages, 1989, editor-in-chief Environ. Claims Jour.; contbr. chpts. to books. Named Ins. Woman of Yr. Assn. Profl. Ins. Women, 1983. Mem. AAAS, Am. Cons. Engrs. Coun., N.Y. Acad. Sci., Washington Wellesley Club, Wellesley Bus. Coun. Office: Environ Strategies Corp 11911 Freedom Dr Ste 900 Reston VA 22090-5628

MILLER, MAYNARD MALCOLM, geologist, educator, research foundation director, explorer, state legislator; b. Seattle, Jan. 23, 1921; s. Joseph Anthony and Juanita Queena (Davison) M.; m. Joan Walsh, Sept. 15, 1951; children: Ross McCord, Lance Davison. BS magna cum laude, Harvard U., 1943; MA, Columbia U., 1948; PhD (Fulbright scholar), St. John's Coll., Cambridge U., Eng., 1957; student, Naval War Coll., Air War Coll., Oak Ridge Inst. Nuclear Sci.; D of Sci. (hon.), U. Alaska, 1990. Registered profl. geologist, Idaho. Asst. prof. naval sci. Princeton (N.J.) U., 1946; geologist Gulf Oil Co., Cuba, 1947; rsch. assoc., coordinator, dir. Office Naval Rsch. project Am. Geog. Soc., N.Y.C., 1948-52; staff scientist Swiss Fed. Inst. for Snow and Avalanche Rsch., Davos, 1952-53; instr. dept. geography Cambridge U., 1953-54, 56; assoc. producer, field unit dir. film Seven Wonders of the World for Cinerama Corp., Europe, Asia, Africa, Middle East, 1954-55; rsch. assoc. Lamont Geol. Obs., N.Y.C., 1955-57; sr. scientist dept. geology Columbia U., N.Y.C., 1957-59; asst. prof. geology Mich. State U., East Lansing, 1959-61, assoc. prof., 1961-63; prof. Mich. State U., Lansing, 1963-75; dean Coll. Mines and Earth Resources U. Idaho, Moscow, 1975-88, prof. geology, dir. Glaciological and Arctic Scis. Inst., 1975—; dir., state geologist Idaho Geol. Survey, 1975-88; elected rep. Legislature of State of Idaho, Boise, 1992—; prin. investigator, geol. cons. sci. contracts and projects for govt. agys., univs., pvt. corps., geographic socs., 1946—; geophys. cons. Nat. Park Svc., NASA, USAF, Nat. Acad. Sci.; organizer leader USAF-Harvard Mt. St. Elias Expdn., 1946; chief geologist Am. Mt. Everest Expdn., Nepal, 1963; dir. Nat. Geographic Soc. Alaskan Glacier Commemorative Project, 1964-74; organizer field leader Nat. Geographic Soc. Joint U.S.-Can. Mt. Kennedy Yukon Meml. Mapping Expdn., 1965, Muséo Argentino de Ciencias Naturales, Patagonian expdn. and glacier study for Inst.. Geologico del Peru & Am. Geog. Soc., 1949-50, participant adv. missions People's Republic of China, 1981, 86, 88, geol. expdns. Himalaya, Nepal, 1963, 84, 87, USAF mission to Ellesmere Land and Polar Sea, 1951; organizer, ops. officer USN-LTA blimp geophysics flight to North Pole area for Office Naval Rsch., 58; prin. investigator U.S. Naval Oceanographic Office Rsch. Ice Island T-3 Polar Sea, 1967-68, 70-73; dir. lunar field sta. simulation program USAF-Boeing Co., 1959-60; co-prin. investigator Nat. Geographic Soc. 30 Yr. Remap of Lemon & Taku Glaciers, Juneau Icefield, 1989-92; exec. dir. Found. for Glacier and Environ. Rsch., Pacific Sci. Ctr., Seattle, 1955—, pres., 1955-85, trustee, 1960—, organizer, dir. Juneau (Alaska) Icefield Rsch. Program (JIRP), 1946—; cons. Dept. Hwys. State of Alaska, 1965; chmn., exec. dir. World Ctr. for Exploration Found., N.Y.C., 1968-71; dir., mem. adv. bd. Idaho Geol. Survey, 1975-88; chmn. nat. coun. JSHS program U.S. Army Rsch. Office and Acad. Applied Sci., 1982-89; sci. dir. U.S. Army Rsch. Office-Nat. Sci. and Humanities Symposia program, 1991—; disting. guest prof. China U. Geoscis., Wuhan, 1981-88, Changchun U. Earth Scis., People's Republic of China, 1988—; affiliate prof. U. Alaska, 1986—. Author: Field Manual of Glaciological and Arctic Sciences; co-author books on Alaskan glaciers and Nepal geology; contbr. over 200 reports, sci. papers to profl. jours., ency. articles, chpts. to books, monographs; producer, lectr. 16 mm. films and videos. Past mem. nat. exploring com., nat. sea exploring com. Boy Scouts Am.; mem. nat. adv. bd. Embry Riddle Aero. U.; bd. dirs. Idaho Rsch. Found.; pres. state divsn. Mich. UN Assn., 1970-73; mem. Centennial and Health Environ. Commns., Moscow, Idaho, 1987-94. With USN, 1943-46, PTO. Decorated 11 battle stars; named Leader of Tomorrow Seattle C. of C. and Time mag., 1953, one of Ten Outstanding Young Men U.S. Jaycees, 1954; recipient commendation for lunar environ. study USAF, 1960, Hubbard medal (co-recipient t. Everest expdn. team) Nat. Geographic Soc., 1963, Elisha Kent Kane Gold medal Geog. So. Phila., 1964, Karo award Soc. Mil. Engrs., 1966, Franklin L. Burr award Nat. Geog. Soc., 1967, Commendation Boy Scouts Am, 1970, Disting. Svc. commendation plaque UN Assn. U.S.A., Disting Svc. commendation State of Mich. Legislature, 1975, Outstanding Civilian Svc. medal U.S. Army Rsch. Office, 1977, Outstanding Leadership in Minerals Edn. commendations Idaho Mining Assn., 1985, 87; recipient numerous grants NSF, Nat. Geographic Soc., others, 1948—. Fellow Geol. Soc. Am., Arctic Inst. N.Am., Explorers Club; mem. councilor AAAS (Pacific divsn. 1978-88), AIME, Am. Geophys. Union, Internat. Glaciological Soc. (past councilor), ASME (hon. nat. lectr.), Am. Assn. State Geologists (hon.), Am. Assn. Amateur Oarsmen (life), Am. Alpine Club (past councilor, life mem.), Alpine Club (London), Appalachian Club (hon. corr.), Brit. Mountaineering Assn. (hon., past v.p.), Himalyan Club (Calcutta), English Speaking Union (nat. lectr.), Naval Res. Assn. (life), Dutch Treat Club, Circumnavigators Club (life), Adventurers Club N.Y. (medalist), Harvard Club (N.Y.C. and Seattle), Sigma Xi, Phi Beta Kappa (pres. Epsilon chpt. Mich. State U. 1969-70), Phi Kappa Phi. Republican. Methodist. Avocations: skiing, mountaineering, photography. Home: 514 E 1st St Moscow ID 83843 Office: U Idaho Coll Mines & Earth Resources Mines Bldg Rm 204 Moscow ID 83843 also: House of Reps Idaho State House Boise ID 83720 also: Found for Glacier & Environ Rsch 4470 N Douglas Hwy Juneau AK 99801

MILLER, MERLE LEROY, retired manufacturing company executive; b. College Springs, Iowa, Dec. 20, 1922; s. Otis John and Hilda Edith (Reiff) M.; m. Doris Marie Young, June 4, 1944; children: Mark, Kevin, Jean. B.S. in Mech. Engring., U. Mo.-Columbia, 1944. Engr. Peerless Pump Div., Quincy, Ill., 1946; engr. John Deere Co., Waterloo, Iowa, 1946-55, sr. engr., 1955-59, project engr., 1959-65, sr. div. engr., 1965-84; dir. Council for Agr. Sci. and Tech., Ames, Iowa, 1981—. Patentee transmission valve, transmission housing. Served to lt. USN, 1942-53. Fellow Am. Soc. Agr. Engrs.; mem. Soc. Automotive Engrs., Pi Tau Sigma, Tau Beta Pi. Republican. Am. Baptist. Club: JD Suprs. (Waterloo). Lodge: Elks.

MILLER, MICHAEL BEACH, schizophrenia researcher; b. Clarksville, Tenn., May 19, 1958; s. Lester Earl and Nancy Anne (Beach) M.; m. Betty Violetta Muller, June 11, 1988; 1 child, Christopher Roman. BS, U. Mass., 1983; MS, U. Wis., 1991, 92, PhD, 1993. Mental retardation asst. Monson Devel. Ctr., Palmer, Mass., 1980-82; child and adolescent counselor Northampton (Mass.) Ctr. for Children & Families, 1982-86; direct care staff Community Support Program, Holyoke, Mass., 1985-86; rsch. asst. dept. psychology U. Wis., Madison, Wis., 1986—. Mem. AAAS, Am. Soc. of Human Genetics, Behavior Genetics Assn., Am. Psychol. Soc. Home: 105 A Eagle Heights Madison WI 53705 Office: Univ Wis Psychology Dept 1202 W Johnson St Madison WI 53706

MILLER, MICHAEL CARL, chemist, researcher; b. Chgo., July 8, 1955; s. Arthur Martin and Mary K. (Brunkala) M.; m. Debra Marie Jakimauskas, Aug. 30, 1981; children: Stephanie Anne, Brenda Michelle. BS in Forensic Chemistry, U. Ill., 1979; postgrad., Roosevelt U., 1983—. Quality assurance chemistry technician William Wrigley Jr. Co., Chgo., 1979-81; R & D analytical chemist Velsicol Chem. Corp., Chgo., 1981-85; R & D sr. chemist materials characterization Helene Curtis Industry, Inc., Chgo., 1985-87; chemistry mgr. Silliker Labs., Chicago Heights, Ill., 1987-88; cons. JEMS Assocs., Glenview, Ill., 1988; rsch. dir. CJR Processing Inc., Des Plaines, Ill., 1988-93, Regenex Rsch. Corp., Chgo., 1993—. Mem. AICE, ASTM (engine coolant subcom. 1990—, oil and insulation subcoms.), INDA, Am. Chem. Soc., Assn. Ofcl. Chemists, Am. Oil Chemists Soc., Soc. for Applied Spectroscopy, Chgo. Chromatography Discussion Group, Soc. Automotive Engrs., Plastics Inst. Am., Milw.-Chgo.-Madison Mass Spectrometry Discussion Group, U. Ill. Alumni Assn. (life), Trout Unltd. Roman Catholic. Avocations: fly fishing, racquetball, tennis, ping pong. Home: 1801 Concord Dr Downers Grove IL 60516-3149 Office: Regenex Rsch Corp 9400 W Foster Chicago IL 60656

MILLER, MILTON DAVID, agronomist, educator; b. Melmont, Wash., Nov. 27, 1911; s. Milton and Katie Virginia (Manney) M.; m. Mary Eleanor McGraw, July 24, 1932; children: Mary Lee Varone, Judith Marie McCullough. BS, U. Calif., Davis, 1935, MS, 1960. Extension agronomist emeritus U. Calif., Davis, 1936-74; tech. advisor Calif. Rice Research Bd., (disting. service award 1974), Yuba City, Calif., 1970-82; cons. World Bank, Romania, 1975, US Aid, Egypt, 1977. Served to lt. col. Q.M.C., 1941-46. Recipient Legion of Merit award U.S. Army, 1949, award of distinction Coll. Agriculture and Environ. Scis. U. Calif., Davis, 1990. Fellow AAAS, Am. Soc. Agronomy, Am. Crop Sci. Soc.; mem. Commonwealth Club of Calif., Masons, Sigma Xi, Alpha Zeta, Alpha Gamma Rho. Republican. Avocations: model rail roading. Home: 1111 Alvarado F-166 Davis CA 95616-5922

MILLER, MURRAY HENRY, soil science educator; b. Ont., Can., July 10, 1931; married, 1954; 3 children. BSA, Ont. Agrl. Coll., 1953; MS, Purdue U., 1955, PhD in Agrl., 1957. From asst. to assoc. prof. Ont. Agrl. Coll., U. Guelph, Guelph, Ont., Can., 1957-66, head dept. soil sci., 1966-71, prof. soil sci., 1966—. Recipient AIC Fellowship award Agrl. Inst. Can., 1991. Fellow Can. Soc. Soil Sci.; mem. Agrl. Inst. Can., Canadian Soc. Soil Sci., Internat. Soc. Soil, Soil Conservation Soc. Am., Ont. Inst. Agrologist. Achievements include research on soil fertility especially chemistry of nutrient elements in soils and their absorption by plants; plant nutrients, environmental quality and land productivity; soil physical factors and crop yield. Office: Land Resource Science, Prof Univ of Grelph, Guelph, ON Canada N1G 2W1*

MILLER, NICOLE GABRIELLE, clinical psychologist; b. N.Y.C., Apr. 19, 1962; d. Michael David and Merle Judith (Jablin) M. BA in Psychology, U. Pacific, 1984, MA in Psychology, 1988; PhD in Clin. Psychology, Calif. Sch. Profl. Psychology, 1990. Therapist various clinics, Stockton, Calif., 1984-86; psycholog. trainee Calif. Men's Colony State Prison, San Luis Obispo, 1987, Dept. Health, Fresno, Calif., 1987-88; intern VA Med. Ctr., Loma Linda, Calif., 1988-89; psycholog. technician and consultant VA Med. Ctr., Martinez, Calif., 1989-90, researcher, 1990, clin. psychologist, 1990—; clin. instr. psychiatry Sch. Medicine, U. Calif., Davis, 1991—, clin. dir. outpatient substance abuse program, 1990—. Mem. APA. Democrat. Office: VA Med Ctr Psychology Svcs 150 Muir Rd 116B Martinez CA 94553-4695

MILLER, PAUL ANDREW, editor; b. Tulsa, Okla., Nov. 11, 1959; s. Paul Melby and Emily Carol (Walker) M.; m. Emily Kay Riley, Apr. 18, 1992. BA, Williams Coll., 1982. Asst. editor British Indsl. Biol. Rsch. Assn., Carshalton, England, 1982-86; from news editor to editor Soc. Chem. Industry, London, 1986—. Contbr. numerous articles to Chemistry and Industry, Food and Chem. Toxicology. Sec. Dems. Abroad, UK, London, 1985. Mem. Assn. British Sci. Writers (com. mem. 1990-91). Office: Chemistry and Industry, 15 Belgrave Sq, London SW1X 8PS, England

MILLER, PAUL DEAN, breeding company executive, geneticist, educator; b. Cedar Falls, Iowa, Apr. 4, 1941; s. Donald Hugh and Mary (Hansen) M.; m. Nancy Pearl Huser, Aug. 23, 1965; children: Michael, Steven. BS, Iowa State U., 1963; MS, Cornell U., 1965, PhD, 1967. Asst. prof. animal breeding Cornell U., Ithaca, N.Y., 1967-72; v.p. Am. Breeders Service, DeForest, Wis., 1972—; adj. prof. U. Wis., Madison, 1980—. Contbr. articles to profl. jours. Mem. Beef Improvement Fedn. (disting. service award 1980), Am. Soc. Animal Sci., Am. Dairy Sci. Assn., Nat. Assn. Animal Breeders (dir. 1983, v.p. 1986). Republican. Home: 3665 Windsor Rd De Forest WI 53532-2727 Office: Am Breeders Service 6908 River Rd DeForest WI 53532

MILLER, PHILLIP EDWARD, environmental scientist; b. Waterloo, Iowa, May 29, 1935; s. Joe Monroe and Katherine Elva (Groom) M.; m. Cathy Ann Love, Sept. 15, 1962; children: Eric Anthony, Bryan Edward, Stefan Patrick, Gregory Joseph. BA in Sci. Edn., U. No. Iowa, 1961; MA in Sci. Edn., U. Iowa, 1964; postgrad., U. Wis., 1966-68. Physics and chemistry tchr. Millersburg (Iowa) Community High Sch., 1961-62; supervising tchr. NSF Insvc. Inst. U. Iowa, Iowa City, 1962-64; instr. biology, area coord. Office Equal Opportunity Western Ky. U., Bowling Green, 1964-66; sci. editor, journalism instr.-sci. and tech. Mich. State U., East Lansing, 1968-74; asst. prof. agr., forestry and home econs. U. Minn., St. Paul, 1974-77; sr. editor atomic energy div. E.I. du Pont de Nemours and Co., Aiken, S.C., 1977-89; sr. scientist environ. protection dept. Westinghouse Savannah River Co., Aiken, 1989—; panelist 26th Internat. Tech. Comm. Conf., L.A., 1979; participant Dept. Energy/Westinghouse Sch. for Environ. Excellence, Cin., 1991; invited contbr. to proceedings of the 1st Tatarstan Symposium on Energy, Environment and Econs., Kazan, Tatarstan, Russia, 1992. Mem. publs. com. Cen. Assn. Sci. and Math. Tchrs., Iowa City, 1969-72; editor Nat. Task Force on Agrl. Energy R&D, Washington, 1976; editor, contbr. Minn. Sci. Mag., 1974-77; contbr. several hundred med., sci. and engring. articles including to Procs. of Iowa Acad. Sci., Sch. Sci. and Math., Am. Biology Tchrs., Procs. of Internat. Communication Conf., and Procs. of Westinghouse Computer Symposium. Pres. Savannah River Rifle & Pistol Club, Aiken, 1981-82, Aiken Toastmasters, 1984; judge speech contests Optimist and 4-H Club Contests, Aiken, 1985-86. Sgt. U.S. Army, 1955-58. Decorated Disting. Marksman Badge gold medal; recipient 1st place sci. writing Argonne Labs. Assn., 1973, Profl. Achievement Permanent Profl. cert. Iowa State Bd. of Pub. Instrn., 1974, Blue Ribbon, Am. Assn. Agrl. Coll. Editors, Tex. A&M, 1976. Mem. AAAS, N.Y. Acad. Scis., Am. Chem. Soc., Phi Delta Kappa, Sigma Xi. Achievements include research in the causes and timing of pre-adolescent initial interest in science; discovery that low-zinc root environment causes delay of development and acceleration of senescence in tobacco plants. Office: Westinghouse Savannah River Co Environ Protection Dept Aiken SC 29801-0001

MILLER, RAYMOND JARVIS, agronomist, college dean, university official; b. Claresholm, Alta., Can., Mar. 19, 1934; came to U.S., 1957, naturalized, 1975; s. Charles Jarvis and Wilma Macy (Anderson) M.; m. Frances Anne Davidson, Apr. 28, 1956; children—Cheryl Rae, Jeffrey John, Jay Robert. B.S. (Fed. Provincial grantee 1954-56, Dan Baker scholar 1954-56), U. Alta., Edmonton, 1957; M.S., Wash. State U., 1960; Ph.D., Purdue U., 1962. Mem. faculty N.C. State U., 1962-65, U. Ill., 1965-69; asst. dir., then asso. dir. Ill. Agrl. Expt. Sta., 1969-73; dir. Idaho Agrl. Expt. Sta., 1973-79; dean U. Idaho Coll. Agr., 1979-85, v.p. for agr.; dean Coll. Agr. and Coll. Life Sci. U. Md., College Park, 1986-89, vice chancellor agr. and natural resources, 1989-91; pres. Md. Inst. for Agrl. and Natural Resources, 1991093, prof. agronomy, 1993—. Author numerous papers in field. Pres. Idaho Rsch. Found., 1980-85; bd. govs. Agrl. Rsch. Inst., 1979-80; chmn. legis subcom. Expt. Sta. Com. on Policy, 1981-82, chmn. bd. divsn. agr. Land Grant Assn., 1985-86; co-chmn. Nat. Com. Internat. Sci. Edn. Joint Com. USDA, 1991-93; bd. dirs. C.V. Riley Found., 1985-90; chmn. budget com. Bd. Agr., Nat. Assn. State Univs. and Land Grant Colls. Grantee Internat. Congress Soil Sci., 1960, Purdue U. Research Found., summers 1960, 61. Fellow Am. Soc. Agronomy, Soil Sci. Soc. Am.; mem. Internat. Soc. Soil Sci., Clay and Clay Minerals Soc., Am. Chem. Soc., Am. Soc. Plant Physiologists, Elks, Lions, Sigma Xi, Phi Kappa Phi, Gamma Sigma Delta, Alpha Zeta. Home: 3319 Gumwood Dr Hyattsville MD 20783-1934 Office: Elkins Bldg 3300 Metzerott Rd Adelphi MD 20783

MILLER, RICHARD KEITH, engineering educator; b. Fresno, Calif., June 12, 1949; s. Albert Keith and Gloria Mae (Pittman) M.; m. Elizabeth Ann Parrish, July 10, 1971; children: Katherine Elizabeth, Julia Ann. BS in Aerospace Engring., U. Calif., Davis, 1971; MS in Mech. Engring., MIT, 1972; PhD in Applied Mechanics, Calif. Inst. Tech., Pasadena, 1976. Asst. prof. mech. engring. U. Calif., Santa Barbara, 1975-79; assoc. prof. civil engring. U. So. Calif., L.A., 1979-85, prof., 1985-92, assoc. dean engring., 1989-92; prof., dean Coll. Engring., U. Iowa, 1992—; cons. Astro Aerospace Corp., The Aerospace Corp., Jet Propulsion Lab. Contbr. numerous articles to sci. and profl. jours. Mem. ASCE, Am. Soc. Engring. Edn. Office: U Iowa Coll Engring Office of Dean 3100 EB Iowa City IA 52242-1527

MILLER, RICHARD LEE, psychology educator; b. Houston, Oct. 14, 1945; s. Fred and Ida Lois (Kerby) M.; m. Beth Vivian Lyons, (div. 1972); 1 child, Nathaniel. BS, Weber State, 1968; MA, Northwestern U., 1970, PhD, 1974. Asst. prof. Georgetown U., Washington, 1973-75; dir. HumRRO, Heidelberg, Germany, 1975-82; adj. assoc. prof. U. Cologne, Germany, 1978-84; dir. Communit Learning Ctr., Mallorca, Spain, 1984-89; vis. prof. U. Ark., Monticello, 1989-90; prof., chmn. Psychology dept. U. Nebr., Kearney, 1990-. Co-editor: Social Comparison Process, 1977; contbr. articles to profl. jours. Ensign USNR, 1968-70. Mem. Sigma Xi (chpt. pres. 1991--). Office: U Nebr Dept Psychology Kearney NE 68849

MILLER, RICHARDS THORN, naval architect, engineer; b. Jenkintown, Pa., Jan. 31, 1918; s. Herman Geistweit and Helen Buckman (Thorn) M.; B.S. in Naval Architecture and Marine Engring., Webb Inst. Naval Architecture, 1940; Naval Engr., MIT, 1951; m. Jean Corbat Spear, Sept. 13, 1941; (dec.); children: Patricia (Mrs. Charles G. Fishburn), Linda (Mrs. John X. Carrier); m. 2d, Alice Johnson Houghton, May 19, 1984. Reg. profl. engr. Commd. ensign U.S. Navy, 1940, advanced through grades to capt., 1960; head preliminary design br. Bur. Ships, 1960-63; dir. Mine Def. Lab., Panama City, Fla., 1963-66; dir. ship design Naval Ship Engring. Ctr., 1966-68; specialized work design oceanographic research ships, mine sweepers, torpedo boats, destroyers; ret., 1968; mgr. ocean engring. Oceanic div. Westinghouse Electric Corp., 1969-75, adv. engr., 1975-79; cons. naval architect and engr., 1968—; arbitrator admiralty and ship building contract cases, 1978—; mem. com. naval architecture Am. Bur. Shipping, 1960-63, mem. tech. com., 1978-92; mem. ship structure com., 1966-68. Decorated Navy Legion of Merit; recipient William Selkirk Owen award Webb Alumni Assn., 1983. Fellow Soc. Naval Architects and Marine Engrs. (chmn. S.E. sect. 1965-66, chmn. marine systems com. 1970-77, chmn. tech. and rsch. steering com. 1977-78, chmn. small craft com. 1983-87, v.p. tech. and rsch. 1979-81, hon. v.p. (life), 1981—, mem. coun. 1976—, mem. exec. com. 1977-81; Capt. Joseph H. Linnard prize 1964, Disting. Svc. award 1988); mem. Am. Soc. Naval Engrs. (mem. council 1976-78), U.S. Naval Inst., N.Y. Yacht Club, Annapolis Yacht Club, Sailing Club of the Chesapeake, Sigma Xi. Author: (with R.G. Henry) Sailing Yacht Design, 1963, (with K.L. Kirkman) Sailing Yacht Design-A New Appreciation, 1990; also sects. in books, articles. Home and Office: 957 Melvin Rd Annapolis MD 21403-1315

MILLER, ROBERT ALAN, physicist; b. Montclair, N.J., Jan. 30, 1943; s. George U. and Florence Lahoma (Fairchild) M.; m. Mary K. Sheridan, Jan. 30, 1971; children: Brendan Alexander, Stacey Ann. BS, U. Ill., 1965, PhD, 1970. Rsch. assoc. Coll. of William and Mary, Williamsburg, Va., 1970-72, Rutgers U., New Brunswick, N.J., 1972-74; head theory program Fusion Energy Corp., Princeton, N.J., 1974-77; head applications software Princeton Gamma-Tech, 1977-81, sr. staff scientist, 1983-85; pres. Sci. Transfer Assocs., N.Y.C., 1981-83; mem. tech. staff AT&T Bell Labs., Holmdel, N.J., 1985—. Mem. IEEE. Achievements include founding and leading theory group for Fusion Energy Corporation. Home: 22 Evans Dr Cranbury NJ 08512 Office: AT&T Bell Labs HO 2B 438A 101 Crawfords Corner Rd Holmdel NJ 07733-3030

MILLER, ROBERT ALLEN, software engineering educator; b. Batavia, N.Y., Aug. 18, 1946; s. Wilford Earl and Mildred A. (Faith) M. BA, DePauw U., 1968. Cert. data processor, cert. systems profl. Software engr. AT&T, Alpharetta, Ga., 1972-88, instr. Bell Labs. Tech. Edn. Ctr., 1988—. Vol. ARC, Atlanta, 1992. Sgt. USAF, 1968-72. Mem. Assn. for Computing Machinery. Office: AT&T 300 Eastside Dr Alpharetta GA 30202

MILLER, ROBERT FRANCIS, physiologist, educator; b. Eugene, Oreg., Nov. 30, 1939; s. Irvin Lavere and Ettie (Graham) M.; m. Rosemary F. Fish, June 12, 1968; children: Derek, Drew. MD, U. Utah, 1967. Head neurophysiology Naval Aerospace Med. Rsch. Lab., Pensacola, Fla., 1969-71; asst. prof. physiology SUNY, Buffalo, 1971-76, assoc. prof. physiology, 1976-78; assoc. prof. ophthalmology Washington U. Sch. Medicine, St. Louis, 1978-83, prof. ophthalmology, 1983-88; 3M Cross prof., head physiology dept. U. Minn., Mpls., 1988—; Mem. VISA 2 NEI study sect. NIH, 1986—. Mem. editorial bd. Jour. Neurophysiology, 1986—. Lt. comdr. USN, 1969-71. Recipient Merit award NIH, 1988, award for med. rsch. Upjohn, 1967, Rsch. to Prevent Blindness Sr. Scientists award, 1988; James S. Adams scholar, 1982, Robert E. McCormick scholar, 1977-78. Mem. AAAS, Assn. Rsch. in Vision and Ophthalmology, Assn. Chmn. Depts. Physiology, N.Y. Acad. Scis., Neurosci. Soc. Achievements include discovery of major new excitatory amino acid receptor, discovery that the electroretinogram in generated by glia. Home: 4613 Golf Terr Edina MN 55424 Office: U Minn Dept Physiology 6-255 MIllard Hall 435 Delaware St SE Minneapolis MN 55455

MILLER, ROBERT FRANK, retired electronics engineer, educator; b. Milw., Mar. 30, 1925; s. Frank Joseph and Evangeline Elizabeth (Hamann) M.; m. La Verne Boyle, Jan. 10, 1948 (dec. 1978); children: Patricia Ann, Susan Barbara, Nancy Lynn; m. Ruth Winifred Drobnic, July 26, 1980. BSEE, U. Wis., 1947, MSEE, 1954, PhD in Elec. Engring., 1957. Profl. engr., Wis. Instr. physics Milw. Sch. Engring., 1949-53; sr. engr. semicondr. Delco Electronics/GMC, Kokomo, Ind., 1957-67, asst. chief engr., 1967-70, mgr. product assurance, 1970-73, dir. quality control, 1973-85; asst. prof. elec. engring. tech. Purdue U., Kokomo, 1986-90; ret., 1990; ind. cons., Kokomo, 1990—; mem. Ind. Microelectronics Commn., Indpls., 1987—. Author tech. papers; co-author lab. manuals. Bd. dirs. Howard Community Hosp. Found., Kokomo, 1974—; trustee YMCA, Kokomo, 1990—, bd. dirs., 1967-90. Named Disting. Alumnus U. Wis., Madison, 1980, 90. Mem. IEEE (life), Am. Soc. Quality Control (bd. dirs. sect. 0918, advisor Cen. Ind. sect. bd. 1988—), Sigma Xi, Tau Beta Pi, Phi Kappa Phi, Eta Kappa Nu. Presbyterian. Home: 3201 Susan Dr Kokomo IN 46902

MILLER, ROGER ALLEN, physicist; b. Chillicothe, Ohio, June 27, 1934; s. Joseph Perrin and Mary Josephine (Sowers) M.; m. Barbara Pauline Rice, Aug. 31, 1957; children: Erich Rice, Gretchen Rice, Carl Rice. BS, Ohio U., 1956; PhD, Case Inst., 1963. Rsch. assoc. Case Inst., Cleve., 1963-64; rsch. physicist Corning (N.Y.) Inc., 1964-71, sr. rsch. physicist, 1971-79, devel. assoc., 1979-87, sr. rsch. assoc., 1987—; spl. lectr. physics Elmira Coll., N.Y., 1966-69; mem. edit. bd. Fiber and Integrated Optics, Pasadena, Calif., 1976-86, mem. adv. bd. 1986-88. Contbr. articles to profl. jours. AART award Assn. for the Advancement Radiation Tech., 1990. Mem. Am. Phys. Soc., Optical Soc. Am., Am. Assn. Physics Tchrs., Sigma Xi, Phi Beta Kappa, Phi Kappa Phi. Achievements include patents in field. Office: Corning Inc Sullivan Pk SP FR 03 1 Corning NY 14831

MILLER, RONALD LEWIS, research hydrologist, chemist; b. Point Pleasant, N.J., Jan. 5, 1946; s. Normand Ralph and Janet Elizabeth (Kemmerer) M.; m. Joanne Divola, June 9, 1967; children: Timothy Normand, Rebecca Janet. BS, U. Fla., 1968, MS, 1970. Hydrologist U.S. Geol. Survey, Ocala, Fla., 1970-73; chemist U.S. Geol. Survey, Doraville, Ga., 1973-75; mem. grad. tng. program U.S. Geol. Survey, Gainesville, Fla., 1975-76; rsch. chemist U.S. Geol. Survey, Denver, 1976-79; hydrologist U.S. Geol. Survey, Tampa, Fla., 1979-86, rsch. hydrologist, 1986—; environ. svcs. tech. adv. com. Fla. Inst. Phosphate Rsch., Bartow, 1987—. Author reports; contbr. articles to profl. jours. Mem. AAAS, Am. Chem. Soc., Am. Water Resources Assn. Baptist. Achievements include development of equations to relate major ion concentration in natural waters to specific conductance, model for estimating estuarine flushing and residence times of water in river dominated estuaries. Office: US Geol Survey Ste B 5 4710 Eisenhower Blvd Tampa FL 33634

MILLER, SHELBY ALEXANDER, chemical engineer, educator; b. Louisville, July 9, 1914; s. George Walter and Stella Katherine (Cralle) M.; m. Jean Adele Danielson, Dec. 26, 1939 (div. May 1948); 1 son, Shelby Carlton; m. Doreen Adare Kennedy, May 29, 1952 (dec. Feb. 1971). B.S., U. Louisville, 1935; Ph.D., U. Minn., 1943. Registered profl. engr., Del., Kans., N.Y. Asst. chemist Corhart Refractories Co., Louisville, 1935-36; teaching, rsch. asst. chem. engring. U. Minn., Mpls., 1935-39; devel. engr., rsch. chem. engr. E.I. duPont de Nemours & Co., Inc., Wilmington, Del., 1940-46; assoc. prof.

chem. engring. U. Kan., Lawrence, 1946-50; prof. U. Kan., 1950-55; Fulbright prof. chem. engring. King's Coll. Durham U., Newcastle-upon-Tyne, Eng., 1952-53; prof., chem. engring. U. Rochester, 1955-69, chmn., 1955-68; assoc. lab. dir. Argonne (Ill.) Nat. Lab., 1969-74; dir. Ctr. Ednl. Affairs, 1969-79, sr. chem. engr., 1979-84; ret. sr. chem. engr., cons., 1984—; vis. prof. chem. engring. U. Calif., Berkeley, 1967-68; vis. prof. U. of Philippines, Quezon City, 1986. Editor: Chem. Engring. Edn. Quar, 1965-67; sect. editor: Perry's Chem. Engring. Handbook, 5th edit., 1973, 6th edit., 1984; contbr. articles to tech., profl. jours. Sec. Kans. Bd. Engring. Examiners, 1954-55; mem. adv. com. on tng. Internat. Atomic Energy Agy., 1975-79; treas. Lawrence (Kans.) League for Practice Democracy, 1950-52. Fellow AAAS, Am. Inst. Chemists, Am. Inst. Chem. Engrs. (past chmn. Kansas City sect.); mem. Am. Chem. Soc. (past chmn. Rochester sect.), Soc. Chem. Industry, Am. Soc. Engring. Edn. (past chmn. grad. studies div.), Am. Nuclear Soc., Filtration Soc., Triangle, Sigma Xi, Sigma Tau, Phi Lambda Upsilon, Tau Beta Pi, Alpha Chi Sigma. Presbyn. Home: 825 63d St Downers Grove IL 60516 Office: Chem Tech Div Argonne Nat Lab Argonne IL 60439

MILLER, STEPHEN DOUGLAS, immunologist, educator; b. Harrisburg, Pa., Jan. 22, 1948; s. Bruce Lloyd and E. Virginia (Perry) M.; m. Kimberley Ann Kohnlein, June 21, 1969; children: Jennifer R., Elizabeth K. BS, Pa. State U., 1969, PhD, 1975. Postdoctoral fellow U. Colo. Med. Ctr., Denver, 1975-78, instr., then asst. prof., 1978-81; asst. prof. Northwestern U. Med. Sch., Chgo., 1981-86, assoc. prof., 1986-92, prof. immunology and pathology, 1992—, dir. immunology and biology ctr., 1993—; mem. study sect. on neurology NIH, Bethesda, Md., 1991—. Editorial bd. Jour. Immunology, Bethesda, 1982-84, Regional Immunology, Miami, Fla., 1991—; contbr. chpts. to books, articles to profl. jours. With U.S. Army, 1970-72. Postdoctoral fellow NIH, 1975-78, recipient Young Investigator's award, 1978-81, grantee, 1981—; grantee Nat. Multiple Sclerosis Soc., 1988-91. Mem. AAAS, Am. Assn. Immunologists, Internat. Soc. Neuroimmunology, Phi Kappa Phi. Office: Northwestern Univ Med Sch Dept Microbiology 303 E Chicago Ave Chicago IL 60611-3008

MILLER, TERRY ALAN, chemistry educator; b. Girard, Kans., Dec. 18, 1943; s. Dwight D. Miller and Rachel E. (Detjen) Beltram; m. Barbara Hoffmann, July 16, 1966; children: Brian, Stuart. BA, U. Kans., 1965; PhD, Cambridge (Eng.) U., 1968. Disting. tech. staff Bell Telephone Labs, 1968-84; vis. asst. prof. Princeton U., 1968-71; vis. lectr. Stanford U., 1972; vis. fgn. scholar Inst. Molecular Sci., Okazaki, Japan, summer 1983; Ohio eminent scholar, prof. chemistry Ohio State U., Columbus, 1984—; chair Molecular Spectroscopy Symposium, Columbus, 1992—. Mem. editorial bd. Jour. Chem. Physics, 1978-81, Jour. Molecular Spectroscopy, 1982-87, Laser Chemistry, 1986—, Rev. of Sci. Instruments, 1986-89, Jour. Phys. Chemistry, 1989—, Jour. Optical Soc. Am., 1989—, Chemtracts, 1989-90, Annual Revs. Phys. Chemistry, 1989—; contbr. over 225 articles to profl. jours. Marshall fellow Brit. Govt., 1965-67, NSF fellow, 1967-68. Fellow Optical Soc. Am. (Meggars award 1993), Am. Phys. Soc.; mem. Am. Chem. Soc. (councilor). Office: Ohio State U 120 W 18th Ave Columbus OH 43210

MILLER, THOMAS ALBERT, entomology educator; b. Sharon, Pa., Jan. 5, 1940; s. Stephen Andrew and Amelia (Gorence) Miller (Chmeliar); m. Hollace Lee Gruhn, Dec. 18, 1965 (div. Nov. 1988); children: Remembrance L., Honor C.; m. Soo-ok Johnson, Dec. 13, 1991. BA in Physics, U. Calif., Riverside, 1962, PhD in Entomology, 1967. Rsch. assoc., NIH postdoctoral fellow U. Ill., Urbana, 1967-68; NATO postdoctoral fellow U. Glasgow, Scotland, 1968-69, vis. prof. zoology dept., 1973; NIH fellow U. Calif., 1964-67, asst. prof. entomology, 1969-72, assoc. prof., 1972-76, prof., entomologist, 1976—, acting head div. toxicology and physiology, 1979-80, head div., 1984-86; cons. in residence Wellcome Rsch. Labs., Berkhamsted, Eng., 1973-74, Australian Cotton Growers Rsch. Assn., 1983-84; vis. prof. U. Ariz., 1990; overseas cons. Wellcome Found., London, 1990-93; cons. AID, Ariz. Dept. Agr., Ciba-Geigy, Dow Chem. Co., DuPont Chem. Co., Food Machinery Corp., U. Calif., Berkeley, numerous others; organizer Symposium on Advances in Insect Neurobiology, Entomol. Congress, Hamburg, 1984, organizer, chmn. Symposium on Insect Autonomic Nervous System, Vancouver, 1988. Author: Insect Neurophysiological Techniques, 1979; editor 16 books; founder 2 book series; contbr. over 130 articles and revs. to sci. jours. and proc., including Jour. Analytical Chemistry, Annals Entomol. Soc. Am., Archives Insect Biochem. Physiology, Jour. Econ. Entomology, Jour. Neurochemistry, Pesticide Sci., also chpts. in books. Sgt. Calif. N.G., 1956-62. NAS exch fellow, Hungary, 1978-79, Czechoslovakia, 1986; grantee Nat. Inst. Neurol. Diseases and Stroke, 1969-72, Rockefeller Found., 1970-76, Nat. Inst. Environ. Health Scis., 1972-84; numerous others. Mem. Entomol. Soc. Am., Am. Chem. Soc. Achievements include discovered myogenicity of insect hearts, mode of action of pyrethroid insecticides, diapause protein in pink bollworm, resistance monitoring techniques for insects, automated recording of insect activity in crops; designed force transducers for insect muscles. Office: Entomology Dept U Calif Riverside CA 92521-0134

MILLER, THOMAS NATHAN, chemical engineer; b. Memphis, Oct. 20, 1954; s. Gene Warren and Florine (Holloway) M.; m. Carolyn Sue Spurlock, June 22, 1975; children: Derek Wesley, Reagan Thomas. BSChemE, U. Ark., 1977. Registered profl. engr., Ark. Process engr. PPG Industries, Lake Charles, La., 1977-78, Agrico Chem. Co., Blytheville, Ark., 1978-83; project engr. Riceland Foods, Inc., Stuttgart, Ark., 1983-91, dir. engring. and constrn., 1991—. Scoutmaster Boy Scouts Am., Holly Grove, Ark., 1992. Mem. Tri-State Oil Mill Supt.'s Assn., Rotary. Presbyterian. Office: Riceland Foods Inc 709 E 22d St Stuttgart AR 72160-0926

MILLER, TODD Q., social psychology educator; b. Salt Lake City, June 25, 1957; s. Ralph Howard and Janet Arlene (Mitchell) M. BS, U. Utah, 1983; PhD, Loyola U., Chgo., 1990. Acting assoc. dir. U. Ill., Chgo., 1988-91; asst. prof. U. Tex. Med. Br., Galveston, 1991—. Devel. grantee Ctr. for Disease Control, U. Ill., 1991, Sealy Smith Found, 1992, Am. Lung Assn., 1993, Am. Heart Assn. Tex. Affiliate, 1993. Mem. Am. Psychol. Soc. Achievements include identification of reasons for positive and negative findings in research on the relationship between Type A behavior and heart disease. Office: U Tex Med Br Dept Preventive Medicine 700 Strand Galveston TX 77555

MILLER, WALTER EDWARD, physical scientist, researcher; b. St. Johns, Ariz., Oct. 15, 1936; s. Walter Edward and Geraldine Marie (Sides) M.; m. Emma Lee Nelson, June 10, 1960; children: Carol Lynn, Brenda Kay Miller Flowers, Melissa Joy Johnson Williams. BS in Natural Sci. magna cum laude, Bethany Nazarene Coll., 1958; postgrad., Vanderbilt U., 1958-59, U. Ala., 1964-69. Cert. math. tchr., Tenn. Math. tchr. Trevecca High Sch., Nashville, 1958-59; physicist Electromagnetics Lab U.S. Army Missile Command, Redstone Arsenal, Ala., 1964-68; rsch. physicist RD & E Ctr. Army Missle Command, Redstone Arsenal, Ala., 1968-90, supr., phys. scientist, 1991—; quality control chemist S.W. Fertilizer Mfg., Bethany, Okla., 1957-58; tech. cons. NATO D-7 Panel, Brussels, 1972, 4 Power Working Group (internat.) Paris, 1983; test dir. Joint U.S./Fed. Republic Germany Laser Expt., Graffenwoehr, 1979; prin. investigator Hypervelocity Missile/LOSAT, Advanced Sensors, U.S. Army Missile Command, Redstone Arsenal, 1984-93; appointee Tri-Svc. Working Group U.S. Tech. Experts (optical guidance), 1986—; speaker at profl. confs. Contbr. numerous articles to scholarly and profl. jours.; patentee 32 U.S. and 3 fgn. patents; 12 patents pending. Speaker, mem. Gideons Internat., 1980—; mem. governing bd. Mastin Lake Nazarene Ch., 1975—. Capt. USMC, 1960-64, Cuba. Recipient Svc. award Mastin Lake Nazarene Ch., 1985, Army-wide R & D award, 1968, 74, 83. Mem. Phi Delta Lambda. Achievements include research in U.S. laser beam rider missile guidance (principal investigator), 1970-85; development of TOW2 missile IR guidance, 1976-79, and TOW2B remote target sensor, 1985-90. Office: US Army Missile Command AMSMI-RD-AS-OG Redstone Arsenal AL 35898

MILLER, WARNER ALLEN, physicist; b. Las Cruces, N.Mex., Dec. 10, 1959; s. Warner Haines and Kathryn Louise (Minars) M.; m. Catherine Sue Clark, Dec. 19, 1981; children: Cheryl Naomi, Anna Kathryn, Megan Diane. BS, U. Md., 1981; PhD, U. Tex., 1986. Commd. 2d lt. USAF, 1981, advanced through grades to capt., 1986; physicist Weapons Lab. USAF, Kirtland AFB, N.Mex., 1981-82, physicist Phillips Lab., 1986—; rsch. asst. Argonne (Ill.) Nat. Lab., 1979-81; teaching asst. U. Md., College Park, 1981; rsch. asst. U. Tex., Austin, 1982-86; J. Robert Oppenheimer fellow Los Alamos (N.Mex.) Nat. Lab., 1990-93, mem. staff theoretical divsn., 1993—; adj. prof. U. N.Mex., Albuquerque, 1988—; mem. external faculty Santa Fe Inst., 1990—. Author: Foundations of Null-Strut Calculus, 1986; editor: Between Quantum and Cosmos, 1988; contbr. articles to profl. jours. Decorated Meritorious Svc. medal; recipient Basic Rsch. award USAF, 1989. Mem. Am. Phys. Soc., Air Force Assn. (Ira C. Eaker fellow 1992), Los Alamos Astrophysics Soc. (bd. dirs. 1991—), Phi Kappa Phi. Achievements include development of mathematics of Null-Strut calculus to solve Einstein's equations, first relativistic smoothed particle hydrodynamics algorithm, introduction of the field of Spacetime Geodesy, design of first Ti-Fe polarizers for ultra-cold neutrons. Home: 45 Obsidian Loop Los Alamos NM 87544 Office: Los Alamos Nat Lab Tehoretical Div T6 Group Mail Stop B288 Los Alamos NM 87545

MILLER, WAYNE HOWARD, biochemist; b. Brevard, N.C., Oct. 5, 1952; s. Robert Allen and Barbara Ann (Medford) M.; m. Donna Faye Garren, Aug. 18, 1973. BS, N.C. State U., 1974. Sr. rsch. scientist Burroughs Wellcome Co., Research Triangle Park, N.C., 1974—. Mem. AAAS, Internat. Soc. Antiviral Rsch. Achievements include patents on antiviral nucleosides. Office: Burroughs Wellcome Co 3030 Cornwallis Rd Research Triangle Park NC 27709

MILLER, WENDELL SMITH, chemist, consultant; b. Columbus, Ohio, Sept. 26, 1925; s. Wendell Pierce and Emma Josephine (Smith) M.; m. Dorothy Marie Pagen, Aug. 18, 1949; children: William Ross, Wendell Roger. BA, Pomona Coll., 1944; MS, UCLA, 1952. Chemist U.S. Rubber Co., Torrance, Calif., 1944; sr. chemist Carbide & Carbon Chemicals Corp., Oak Ridge, 1944-48; ptnr. Kellogg & Miller, Los Angeles, 1949-56; patent coordinator Electro Optical Systems, Inc., Pasadena, Calif., 1956-59; v.p. Intertech. Corp. optical and optoelectronic system devel., North Hollywood, Calif., 1960-66, dir., 1966—; assoc. Ctr. for Study Evolution and Origin of Life, UCLA. Commr. Great Western Council Boy Scouts Am., 1960-65. Served with AUS, 1944-46. Decorated Army Commendation medal. Mem. Los Angeles Patent Law Assn., IEEE, AAAS, 20th Century Round Table, Sigma Xi, Phi Beta Kappa, Pi Mu Epsilon. Numerous patents in field. Home: 1341 Comstock Ave Los Angeles CA 90024-5314

MILLER, WILBUR HOBART, business diversification consultant; b. Boston, Feb. 15, 1915; s. Silas Reuben and Muriel Mae (Greene) M.; m. Harriett I. Harmon, June 20, 1941; children: Nancy Iber Miller Harray, Warren Harmon, Donna Sewall Miller Davidge. B.S., U. N.H., 1936, M.S., 1938; Ph.D., Columbia U., 1941. Rsch. chemist Am. Cyanamid Co., Stamford, Conn., 1941-49, Washington tech. rep., 1949-53, dir. food industry devel., 1953-57; tech. dir. products for agr. Cyanamid Internat. Am. Cyanamid Co., N.Y.C., 1957-60; sr. scientist Dunlap & Assos., Darien, Conn., 1960-63, sr. assoc., 1963-66; coord. new product devel. Celanese Corp., N.Y.C., 1966-67, mgr. comml. rsch., 1967-68, dir. corp. devel., 1969-84; bus. diversification cons., 1984—; lectr. on bus. and soc. Western Conn. State Coll., 1977-79. Contbr. sci. papers to profl. jours.; patentee in field. Chmn. Stamford Forum for World Affairs, 1954-87, hon. chmn. 1987—; mem. adv. bd. Ctr. for the Study of the Presidency, 1980—; bd. dirs. Stamford Symphony, 1974-80, v.p., 1978-80; bd. dirs. Stamford Hist. Soc., 1988, v.p., 1991—, pres. 1993—; pres. Coun. for Continuing Edn., Stamford, 1962, bd., 1960-70, Ch., nominating com., 1960-63; elder United Presbyn. Ch.; pres. Interfaith Coun. of Stamford, 1973; internat. fellow U. Bridgeport, 1985—; mem. pres.'s coun. U. N.H., 1982—. Recipient outstanding achievement award Coll. Tech., U. N.H., 1971, Am. Design award, 1948, Golden Rule Award J.C. Penney & Co., 1986; Univ. fellow Columbia U., 1940-41. Fellow AAAS, Am. Inst. Chemists (councillor N.Y. chpt. 1984-85); mem. Am. Chem. Soc. (news svc. adv. bd., 1948-53), N.Y. Acad. Scis., Société de Chimie Industrielle (v.p. fin. Am. sect. 1980-84, dir. 1984—), Inst. Food Tech., Soc. for Internat. Devel., Am. Acad. Polit. and Social Scis., Stamford Hist. Soc., Chemists Club (N.Y.C., treas. 1982-84), Sigma Xi, Alpha Chi Sigma, Phi Kappa Phi. Home: 19 Crestview Ave Stamford CT 06907-1906

MILLER, WILLIAM, science administrator; b. Kansas City, Mo., Feb. 21, 1940; m. Jeanne Wilmes Miller; children: Joyce, John, Daniel, Mitchell. BA, U. Mo., 1962, MD, 1966. Diplomate Am. Bd. Pathology. Intern U. Rochester, N.Y., 1966-67; clin. assoc. NIH, Bethesda, Md., 1967-69; resident and fellow U. Mo., 1969-71; asst. prof. dept. pathology U. Ky., 1971-73; clin. prof. Washington U., 1973-91, St. Louis U., 1973-91, U. Tex., 1992—; dir. histocompatibility lab. U. Mo. Med. Ctr., 1969-71; dir. Cen. Ky. Blood Ctr., Lexington, 1971-73, U. Ky. Hosp. Blood Bank, %; med. dir. U. Ky. Med. Ctr., 1971-73; dir. blood bank St. Louis U. Hosps., 1974-80; blood bank cons. Cardinal Glennon Hosp. for Children, 1977-80; dir. labs. St. Louis U. Hosps. for Children, 1977-81; mem. med. adv. bd. Primus Corp. Inc., Kansas City, Mo., 1991—, Biorelease Inc. 1991—, dir. 1989-91, Biocyte, Inc. 1992—. Lt. comdr. U.S. Pub. Health Svc., 1967-69. Recipient Dist. Citizen award St. Louis Jaycees, 1976, Dist. Alumnus award U. Mo., 1991. Mem. AMA, FDA (adv. com. on blood and blood products 1979-85, chmn. 1984-85, cons. 1986), Am. Nat. Red Cross (mem. pres. adv. com., chmn. nat. com. histocompatable blood transfusion, spl. com. bd. govs. plasma fractionation facility, chmn. spl. com. on blood resource sharing), Am. Assn. Blood Banks (chmn. com. tech. manual 1970-78, com. standards 1971-83), Am. Blood Commn. (bd. dirs. 1979-83, pres. 1983-85), Am. Soc. Clin. Pathologists, Am. Soc. Hematology, Am. Assn. Tissue Banks, Am. Coun. Transplantation (founding dir. 1984-85), Am. Soc. Histocompatibility and Immunogenetics (founding dir.), Internat. Soc. Blood Transfusion, South Central Assn. Blood Banks, AIDS Found. St. Louis (dir. 1988-91), U.S. Sailing Assn., Nat. Corvette Restorers Soc., Fort Worth Boat Club. Avocations: Sailing, Automobile Restoration, Folk and Bluegrass Music, Photography and Contemporary Art. Office: Wadley Inst Molecular Medicine 9000 Harry Hines Blvd Dallas TX 75235*

MILLER, WILLIAM HUGHES, theoretical chemist, educator; b. Kosciusko, Miss., Mar. 16, 1941; s. Weldon Howard and Jewel Irene (Hughes) M.; m. Margaret Ann Westbrook, June 4, 1966; children: Alison Leslie, Emily Sinclaire. B.S., Ga. Inst. Tech., 1963; A.M., Harvard U., 1964, Ph.D., 1967. Jr. fellow Harvard U., 1967-69; NATO postdoctoral fellow Freiburg (Germany) U., 1967-68; asst. prof. chemistry U. Calif., Berkeley, 1969-72, assoc. prof., 1972-74, prof., 1974—, dept. chmn., 1989-93; fellow Churchill Coll., Cambridge (Eng.) U., 1975-76. Alfred P. Sloan fellow, 1970-72; Camille and Henry Dreyfus fellow, 1973-78; Guggenheim fellow, 1975-76; recipient Alexander von Humboldt-Stiftung U.S. Sr. Scientist award, 1981-82, Ernest Orlando Lawrence Meml. award, 1985. Fellow AAAS, Am. Acad. Arts and Scis., Am. Phys. Soc. (Irving Langmuir award 1990); mem. NAS, Internat. Acad. Quantum Molecular Sci. (Ann. prize 1974). Office: U Calif Dept Chemistry Berkeley CA 94720

MILLETT, MERLIN LYLE, aerospace consultant, educator; b. East Moline, Ill., Dec. 29, 1923; s. Merlin Lyle Sr. and Erie Lucille (Hyland) m.; m. Glendola Mae Westlic, Feb. 23, 1945 (dec. 1968); 1 child, Debra Sue; m. Esther Lee Dayhuff, Aug. 21, 1970. BS, Iowa State Coll., 1945, MS, 1948, PhD, 1957. Registered profl. engr., Iowa. Draftsman Am. Machine and Metals, East Moline, Ill., 1941-42; instr. Iowa State Coll., Ames, 1946-48, asst. prof. aerospace engring., 1952-57, assoc. prof., 1957-61; prof. Iowa State U., Ames, 1961-75; flight test engr. Douglas Aircraft Co., Santa Monica, Calif., 1948-52; dean of faculty Parks Coll. St. Louis U., Cahokia, Ill., 1975-78; mgr. fighter aircraft Boeing Mil. Airplanes, Wichita, 1978-89; adj. prof. Oklahoma State U., Stillwater, 1979—; design cons. Architects Associated, Des Moines, 1970-72; aeronautical cons. Iowa Aeronautics Commn., Des Moines, 1972-74; co-prin. investigator U.S. Dept. Transp., Ames, 1973-75; power plant engr. Fed. Aviation Agy., Ames., 1971-75; cons. City of Ames., 1973. Patentee low cost drone. Pres. bd. dirs. Suntree East Home Owners Assn., 1993—. Lt. sr. grade SUNR, ret. Mem. AIAA (assoc. fellow, dep. dir. 1986—), Scottsdale Ranch Community Assn. (pres. 1993—, mem. arch. com. 1989—, bd. dirs. 1989—). Avocations: photography, music, woodworking, bicycling, tennis. Home: 10515 E Fanfol Ln Scottsdale AZ 85258-6032

MILLICAN, DAVID WAYNE, analytical chemist; b. Selma, Ala., Mar. 29, 1959; s. Raymond Charles and Marianne (Odom) M. BS, La. State U., 1981; PhD, Duke U., 1990. Chemist Dow Chem. Co., Plaquemine, La., 1981-85; rsch. and devel. chemist DuPont, Deepwater, N.J., 1990—. Mem. Am. Chem. Soc., Soc. Applied Spectroscopy. Achievements include demonstration of use of frequency-domain fluorescence in resolving component spectra from mixtures by multi-way array techniques, evolving factor analysis for resolution of component spectra from mixtures, also using frequency domain fluorescence. Office: E I DuPont Chambers Works QCL(M) Deepwater NJ 08023

MILLIGAN, VICTOR, civil engineer, consultant; b. Belfast, No. Ireland, Nov. 11, 1929; arrived in Can., 1956; s. Albert and Margaret (Walker) M.; m. Mary Ann Pelikan, July 20, 1955 (dec. 1988); children: Jeffrey, Michael; m. Audrey Morrow, Oct. 9, 1990. BS, Queen's U., No. Ireland, 1951, MS, 1952, DSc (hon.), 1993; D Engring. (hon.), Waterloo U., Ont., Can., 1990. Registered profl. engr. Ont., Alta., Nfld. Asst. engr. James Williamson & Ptnrs., Glasgow, Scotland, 1952-54; rsch. fellow Purdue U., Lafayette, Ind., 1954-55; tech. officer Imperial Chem. Industries Ltd., Cheshire, Eng., 1955-56; from dist. to asst. chief engr. Geocon, Ltd., Toronto, Ont., Can., 1956-60; prin. Golder Assocs., Toronto, 1960-74, pres., CEO, chmn., 1974-84, sr. prin., chmn., cons., 1984—; mem. faculty engring. sci. adv. com. U. Western Ont., 1973-76; adj. prof. dept. geol. engring. U. Toronto, 1980-83; pres. Consulting Engrs. Ont., 1982-83; chmn. assoc. com. on geotechnical rsch. NRC, 1984-89. Co-author: Stability in Open Pit Mining, 1971, Geotechnical Practice in Open Pit Mining, 1972; founding editor Can. Geotechnical Jour., 1963-68; contbr. 40 sci. papers. King George VI Meml. Rsch. fellow 1954-55; recipient Engring. Excellence medal Assn. Profl. Engrs. Ont., 1988. Fellow ASCE, Geol. Soc. Can., Engring. Inst. Can. (Julian C. Smith medal 1991); mem. Can. Geotechnical Soc. (R. F. Legget award 1973), Internat. Soc. Soil Mechanics and Found. Engring. Office: Golder Assocs Ltd, 2180 Meadowvale Blvd, Mississauga, ON Canada L5N 5S3

MILLIKAN, LARRY EDWARD, dermatologist; b. Sterling, Ill., May 12, 1936; s. Daniel Franklin and Harriet Adeline (Parmenter) M.; m. Jeanine Dorothy Johnson, Aug. 27, 1960; children: Marshall, Rebecca. B.A., Monmouth Coll., 1958; M.D., U. Mo., 1962. Intern Great Lakes Naval Hosp., Ill., 1962-63; housestaff in tng. U. Mich., Ann Arbor, 1967-69, chief resident, 1969-70; asst. prof. dermatology U. Mo., Columbia, 1970-74, assoc. prof., 1974-81; chmn. dept. dermatology Tulane U., New Orleans, 1981—; cons. physician Charity Hosp., New Orleans, Tulane U. Hosp., New Orleans, Huey P. Long Hosp., Pineville, St. Tammany Parish Hosp., Covington, La., Highland Park Hosp., Covington, AMI St. Jude Hosp., Kenner, La., Alexandria VA Hosp., La., New Orleans VA Hosp. Assoc. editor Internat. Jour. Dermatology, 1980—; mem. editorial bd. Current Concepts in Skin Disorders, Am. Jour. Med. Scis., Jour. Am. Acad. Dermatology, Postgraduate Medicine; contbr. articles to med. jours. Served with USN, 1960-67. Recipient Andres Bello awrd Govt. of Venezuela, 1989, citation of merit Sch. Medicine, U. Mo., 1993; named Disting. Alumnus, Monmouth Coll., 1990; Nat. Cancer Inst. grantee, 1976-84. Fellow ACP; mem. AAAS, AMA, Am. Acad. Dermatology (bd. dirs. 1986-90), Am. Dermatol. Assn., Am. Dermatol. Soc. for Allergy and Immunology (pres., bd. dirs.), Soc. for Investigative Dermatology (past pres. South sect.), So. Med. Assn. (vice chmn. dermatology sect. 1984, chmn.-elect 1993), Coll. Physicians Phila., Assn. Profs. Dermatology (bd. dirs. 1984-86), Orleans Parish Med. Soc., La. Med. Soc., Pan Am. medc. Assn., Internat. Soc. Dermatology (asst. sec. gen. 1989—), Mo. Allergy Assn. (past pres.), Am. Coll. Cryosurgery, Assn. Acad. Dermatologic Surgeons, Internat. Soc. Dermatologic Surgery, Dermatol. Found. Leaders Soc. (state chmn. 1993). Office: Tulane Univ Sch Medicine Dept of Dermatology 1430 Tulane Ave Ste 3551 New Orleans LA 70112-2699

MILLIMAN, JOHN D., oceanographer, geologist; b. Rochester, N.Y., May 5, 1938; married, 1963; 2 children. BS, U. Rochester, 1960; MS, U. Washington, 1963; PhD in Oceanography, U. Miami, 1966. Rsch. asst. radiation biology lab U. Washington, 1961; rsch. asst. Inst. Marine Sci, U. Miami, 1963-66, rsch. fellow, 1966, asst. scientist, 1966-71; assoc. scientist Woods Hole Oceanography Inst., 1971—. Alexander von Humboldt Found. Lab Sedimentology scholar U. Heidelberg, 1969-70; Recipient Francis P. Shepard medal Soc. Sedimentary Geology, 1992. Mem. AAAS, Geol. Soc. Am., Soc. Econ. Paleontologists Mineralogists. Achievements include research in deposition and diagenesis of marine sediments; continental shelf sedimentation; Holocene history and shallow structure; submarine precipitation and lithification of marine carbonates. Office: Woods Hole Oceanographic Inst Dept Geology & Geophysics Woods Hole MA 02543*

MILLIS, ROBERT LOWELL, astronomer; b. Martinsville, Ill., Sept. 12, 1941; married, 1965; 2 children. BA, Ea. Ill. U., 1963; PhD in Astronomy, U. Wis., 1968. Astronomer Lowell Obs., Flagstaff, Ariz., 1967-86, assoc. dir., 1986-90, dir., 1990—. Mem. Am. Astronomy Soc., Astronomy Soc. Pacific, Internat. Astronomy Union, Divsn. Planetary Sci. (sec.-treas. 1985-88). Achievements include research in planetary satellites and ring systems; occultation studies of solar system objects; research on comets. Office: Lowell Observatory PO Box 1269 1400 W Mars Hill Rd Flagstaff AZ 86001-4499*

MILLMAN, ROBERT BARNET, psychiatry and public health educator; b. N.Y.C., 1939. BA, Cornell U., 1961; MD, SUNY, 1965. Intern in internal medicine NYU div. Bellevue Hosp., N.Y.C., 1965-66; with rsch. and adminstrn NIH, Washington, 1966-68; resident in internal medicine N.Y. Hosp., 1968-70; resident in psychiatry Payne Whitney Psychiat. Clinic, 1974-77, assoc. attending psychiatrist, 1977-89, attending psychiatrist, 1989—; project dir. adolescent devel. program, dept. pub. health Dept. Pub. Health, 1970—, acting chmn., 1992—; clin. affiliate in medicine Payne Whitney Psychiat. Clinic, 1977—; assoc. prof. clin. psychiatry Cornell U., N.Y.C., 1980-86, Disting. prof. psychiatry and pub. health, 1986—; vis. assoc. physician Rockefeller U., 1972-79; dir. Substance Abuse Svcs., N.Y. Hosp., 1980—; chmn. tech adv. group on drugs, N.Y.C., 1989-92; chmn. com. on alcohol and drugs N.Y. Acad. Med., 1988—; med. advisor Major League Baseball, 1992—; adv. com. Robert Wood Johnson Found., 1990—. Mem. N.Y. Acad. Medicine, Am. Psychiat. Assn., Am. Pub. Health Assn., Acad. Med. Educators and Researchers in Substance Abuse, Am. Soc. on Addiction Medicine. Office: Cornell U Med Coll Dept Psychiatry Pub Health 411 E 69th St New York NY 10021-5697

MILLS, DAVID MICHAEL, physicist; b. Bremerton, Wash., Jan. 21, 1942; s. Thomas Olney and June T. (Weckwerth) M.; m. Carol Hardy, June 24, 1978; children: Kerry Kearns, Peter Kearns, Michael Jonathon Hardy. BSEE cum laude, U. Wash., 1964; MSEE, Stanford U., 1966, PhD in Applied Physics, 1971; MA in Clin. Psychology, Lone Mountain Coll., 1977; postdoctorate, U. Wash., 1989-91. Cert. marriage and family therapist, Wash. Rsch. asst. Stanford (Calif.) U., 1965-70; postdoctoral researcher Lick Obs., U. Calif., Santa Cruz, 1970-72; family therapist Mental Health North, Seattle, 1977-81; pvt. practice clin. psychology Seattle, 1982-90; dir. family therapy tng. program Montlake Inst., Seattle, 1984-87; postdoctoral rsch. scientist Virginia Merrill Bloedel Hearing Rsch. Ctr. U. Wash., Seattle, 1991—; adj. faculty mem. Seattle U., 1979, Antioch U., Seattle, 1979-82; curriculum developer, core faculty mem. Montlake Inst., Seattle, 1982-90; speaker in field. Contbr. articles to profl. publs. Mem. Acoustical Soc. Am. (assoc.), Soc. Neurosci. Office: U Wash Mail Stop RL-30 Seattle WA 98195

MILLS, JOHN JAMES, research director, mechanical engineering educator; b. Motherwell, Scotland, May 12, 1939; came to U.S., 1966; s. John Thompson King and Esther Houston (Leitch) M.; m. Dorothea Becker, Mar. 27, 1971; children: Jennifer, Julia, Janine, Ian. BS in Physics, U. Glasgow, Scotland, 1961; PhD in Applied Physics, U. Durham, Eng., 1965. Rsch. fellow dept. elec. engring. Imperial Coll., London, 1964-66; from scientist to sr. scientist IIT Rsch. Inst., Chgo., 1966-71; sr. fellow Inst. Silicatforschung der Fraunhofergesellschaft, Fed. Republic of Germany, 1972-73; sect. leader Inst. Werkstofftechnic, Fed. Republic of Germany, 1973-75; sr. scientist Martin Marietta Labs., Balt., 1975-79, mgr. aluminum fabrication R&D, 1979-84, mgr. mfg. tech. R&D, 1984-88, project dir., 1988-90; dir. Automation & Robotics Rsch. Inst., U. Tex. at Arlington, 1990—. Author: (chpt.) Properties of Pure Aluminum, 1984; contbr. articles to profl. jours. Recipient rsch. fellowship British Oxygen Co., 1964-66, A. von Humboldt fellowship Inst. Silicatforschung der Fraunhofergesellschaft, 1972-73, Outstanding Achievement award Martin Marietta Labs., 1981. Mem. ASME, Soc. Mfg. Engring., Am. Soc. Metals, Am. Inst. Mining and Metall. Engring., Am. Phys. Soc., Ft. Worth C. of C. (bd. dirs. East Area Coun. 1991—). Avocations: wine making, rowing, cabinet making. Office: Univ Tex Automation & Robotics Rsch 7300 Jack Newell Blvd S Fort Worth TX 76118

MILLS, LESTER STEPHEN, chemist; b. Sudbury, Eng., Oct. 18, 1958; came to U.S., 1990; s. Richard William and Doris Ellen (Carlo) M.; m. Silvia Salzmann, June 3, 1989. BA in Natural Sci., Caius Coll., Cambridge, Eng., 1980; MA, Cambridge U., 1985; PhD, U. East Anglia, Norwich, Eng., 1985. Rsch. chemist Glaxo, Ware, Eng., 1980-82, Lonza, Inc., Visp, Switzerland, 1986-90; mgr. new products Lonza, Inc., Fairlawn, N.J., 1990—. Contbr. to profl. publs. Mem. Royal Soc. Chemistry, Am. Chem. Soc., Am. Mktg. Assn. Republican. Roman Catholic. Achievements include patent for improved synthesis of minoxidil, synthesis of pyrimidines. Office: Lonza Inc 17-17 Rte 208 Fair Lawn NJ 07410

MILLS, ROBERT LAURENCE, physicist, educator; b. Englewood, N.J., Apr. 15, 1927; s. Frederick Cecil and Dorothy Katherine (Clarke) M.; m. Elise Ackley, July 21, 1948; children—Katherine, Edward, Jonathan, Susan, Dorothy. A.B., Columbia Coll., 1948; B.A., Cambridge (Eng.) U., 1950, M.A., 1954; Ph.D., Columbia, 1955. Research asso. Brookhaven Nat. Lab., Upton, N.Y., 1953-55; mem. Sch. Math. Inst. for Advanced Study, Princeton, 1955-56; asst. prof. physics Ohio State U., Columbus, 1956-59; asso. prof. Ohio State U., 1959-62, prof., 1962—. Author: Propagators for Many-Particle Systems, 1969, (with C.N. Yang) Rumford Premium, 1980, for devel. of a generalized gauge invariant field theory. Mem. Am. Phys. Soc., AAUP, Fedn. Am. Scientists. Home: 2825 Neil Ave Apt 816 Columbus OH 43202-2077

MILLS, RODNEY DANIEL, engineering company executive; b. Niagara Falls, N.Y., Nov. 26, 1942; s. Russell Thomas and Golda Marie (Rowe) M.; m. Alana Rae French, June 12, 1965 (div. 1971); children; Thomas Andrew, Robert Daniel; m. Sharon Louise Maggart, May 13, 1972; 1 child, Melinda Marie. Student, Tri-State U., 1961-64, Purdue U., 1966-72. Draftsman Kunkle Valve Co. Inc., Fort Wayne, Ind., 1965-67; draftsman Cen. Soya Co. Inc., Fort Wayne, 1967-74, sr. draftsman, 1974-76, staff engr., 1976-80, engr., 1980-84, project engr., 1984-90, engring. purchasing mgr., 1990—. Pres. Three Rivers Youth Soccer Assn., 1987-91; active United Soccer Boosters, 1988—. Recipient Rick Trowski Nat. Humanitarian award United Soccer Boosters, 1990. Home: 7131 Winnebago Dr Fort Wayne IN 46815 Office: Cen Soya Co Inc PO Box 2507 Fort Wayne IN 46801-2507

MILLS, TERRY, III, forensic chemist; b. Providence, Nov. 14, 1946; s. Terry and Gladys (Herbert) M. BS in Chemistry, Ga. Tech., 1970, MS in Analytical Chemistry, 1978. Forensic chemist Divsn. Forensic Scis., Atlanta, 1969-74, supr. drug chemistry, 1974-92; pres. Mills Forensics Svcs., Inc., Douglasville, Ga., 1985-92; lectr. Ga. Police Acad. Author: Instrumental Data for Drug Analysis, 1991; co-author chpt. The Analysis of Drugs of Abuse, 1991. Recipient Cert. of Appreciation Atlanta Bur. Pub. Svcs., 1984; named GBI Mgr. of Yr. State of Ga., 1988. Fellow Am. Acad. Forensic Scis.; mem. Mem. Am. Soc. for Testing Materials, So. Assn. Forensic Scis., Clandestine Lab. Investigating Chemists, Fourier Transform Infrared Forensic User's Group (bd. dirs.). Home: 6370 Shallowford Way Douglasville GA 30135 Office: 3121 Panthersville Rd Decatur GA 30034

MILO, FRANK ANTHONY, manufacturing executive; b. Bristol, Conn., Aug. 19, 1946; s. Frank Raymond and Helen Ellen Milo; BS in Indsl. Engring., Gen. Motors Inst., Flint, Mich. 1970. Abrasives supr. New Departure-Hyatt Bearings div. Gen. Motors Corp., Bristol, 1964-72; sales engr. air tools Hartford, Conn. br. Ingersoll Rand Co., Liberty Corners, N.J., 1972-74; process engr. Electric Boat div. Gen. Dynamics, Groton, Conn., 1974-75; regional sales mgr. Unbrako Chem. Products div. SPS Tech., FortWashington, Pa., 1975-78; nat. sales mgr. Permabond Internat., Englewood, N.J., 1978-80; v.p. sales and mktg. world-wide Pacer Tech. & Resources, Campbell, Calif., 1980-84; pres., founder Firecat Tech., Mountain View, Calif., 1984-89, founder, Ormond Beach, Fla., 1989—; mktg. mgr. Penn Internat. Chem., Mountain View, 1986-88; v.p. global ops. Miracle Workers Internat. Inc., San Diego, 1993—; exec. v.p. Global Dynamics Inc., Vista, Calif., 1993—. Mem. Sigma Nu. Republican. Roman Catholic.

MILONE, EUGENE FRANK, astronomer, educator; b. N.Y.C., June 26, 1939; arrived in Can., 1971; s. Frank Louis and Vera Christine (Joeckle) M.; m. Helen Catherine Louise (Ligor), Mar. 1, 1959; children: Bartholomew Vincenzo Llambro, Marie Christina Milone Jack. AB, Columbia U., 1961; MSc, Yale U., 1963, PhD, 1967. Astronomer space sci. div. rocket spectroscopy br. Naval Rsch. Lab., Washington, 1967-84; asst. prof. Gettysburg (Pa.) Coll., 1968-71; asst. prof. dept. physics and astronomy U. Calgary, Alta., Can., 1971-75, assoc. prof., 1976-81, prof., 1981—. Author: Infrared Extinction and Standardization, 1989, Challenges of Astronomy, 1991, Light Curve Modelling of Eclipsing Binary Stars, 1993; contbr. over 150 articles to profl. jours. Elected mem. com. for coll. and univ. svcs. Evang. Luth. Ch. in Can., Synod of Alberta and the Territories, Edmonton, Alta., 1989-93. Operating and Equipment grantee Natural Scis. and Engring. Rsch. Coun. Can., 1972—; Killam Resident fellow Killam Found. U. Calgary, 1982, 88. Mem. Internat. Astron. Union (mem. organizing com., commn. 25 1985-91), Am. Astron. Soc. (chmn. local organizing com. Calgary meeting 1981), Can. Astron. Soc., Sigma Xi (pres. U. Calgary chpt. 1979-80). Democrat. Lutheran. Achievements include development of Rothney Astrophysical Observatory, the Rapid Alternate Detection System, of light curve modeling techniques; research on the O'Connell Effect, on a new passband system for infrared photometry. Home: 1031 Edgemont Rd NW, Calgary, AB Canada T3A 2J5 Office: U Calgary Dept Physics and Astronomy, 2500 University Dr NW, Calgary, AB Canada T2N 1N4

MIL'SHTEIN, SAMSON, semiconductor physicist; b. Vinitza, USSR, Aug. 6, 1940; came to U.S., 1982; s. Khaim and Golda (Tzukerman) M.; married; children: Mark, Valery. MS, State U., Odessa, USSR, 1963; PhD, U. Jerusalem, 1976. Lectr. Jr. Coll., Kherson, USSR, 1963-65; mem. staff Inst. Marine Engrs., Odessa, 1965-67, Inst. Solid State Physics, Moscow, 1967-72; lectr. U. Jerusalem, 1974-76; sr. lectr. Ben-Gurion U., Beer-Shera, Israel, 1976-82; vis. scientist Bell Labs., Murray Hills, N.J., 1982-83; sr. scientist Semiconductors Group, Cabot Corp., Billeria, Mass., 1984-86; prof. elec. engring. U. Mass., Lowell, 1987—, dir. Advanced Electronic Tech. Ctr., 1990—. Contbr. over 100 articles to profl. jours., confs. Recipient lst prize for sci. achievements Inst. Solid STate Physics of Acad. Scis. USSR, 197l. Mem. IEEE, Materials Rsch. Soc., Elec. Micros. Soc. Jewish. Avocations: classical music, jazz, art, table tennis.

MILSTEIN, CÉSAR, molecular biologist; b. Oct. 8, 1927; s. Lázaro and Máxima Milstein; m. Celia Prilleltensky, 1953. Ed., Colegio Nacional De Bahia Blanca, U. Nacional de Buenos Aires, Fitzwilliam Coll., Cambridge. Brit. Council fellow, 1958-60; staff Instituto Nacional de Microbiologia, Buenos Aires, 1957-63; head Div. de Biologia Molecular, 1961-63; mem. staff M.R.C. Lab. of Molecular Biology, 1963—, dep. dir., 1988—. Contbr. articles to profl. jours. Recipient Royal medal Royal Soc., 1982, Nobel prize for medicine, 1984; Rozenberg prize, 1979, Mattia award, 1979, Gross Howit prize, 1980, Koch prize, 1980, Wolf prize in medicine, 1980, Wellcome Found. medal, 1980, Gimenez Diaz medal, 1981, Sloan prize Gen. Motors Cancer Research Found., 1981, Gardner award Gardner Found., 1981. Fellow Royal Coll. Physicians (hon.), royal Soc., Royal Coll. Pathologists (hon. 1987); mem. Nat. Acad. Scis. (fgn. assoc.). Avocation: cooking. Office: Med Rsch Coun Lab Mol Biol, Hills Rd, Cambridge CB2 2QH, England

MILU, CONSTANTIN GHEORGHE, physicist; b. Brasov, Romania, Feb. 16, 1943; s. Gheorghe and Paraschiva (Ovesea) M.; m. Marina Aurel Serbănescu, Sept. 11, 1971. MSc, U. Bucharest, Romania, 1966; PhD, Inst. Atomic Physics, Bucharest, 1987. Diplomate nuclear physics. Chief sci. investigator Inst. Hygiene and Pub. Health, Bucharest, 1967—; head secondary std. radiation dosimetry lab. WHO/IAEA, Bucharest, 1977—; mem. sci. adv. groups WHO, Montgomery, Ala., 1992, IAEA, Vienna, Austria, 1990-92. Co-author: (tech. report series) Calibration of Dose Meters Used in Radiotherapy, 1979; (handbook) Radiation Hygiene, 1985. Grantee Atominstitut, Vienna, 1978. Mem. Am. Assn. Physicists in Medicine, Romanian Soc. Radiol. Protection (pres. 1990—). Achievements include introduction of standard dosimetry in clinical and radioprotection dosimetry in Romania; evaluation of medical irradiation of the population in Romania; use of thermoluminescent dosimetry in tranzition zones, calculation of individual and collective doses after Chernobyl nuclear accident for the population in Romania. Office: Inst Hygiene and Pub Health, Str Dr Leonte No 1-3, R-76256 Bucharest Romania

MINAHEN, TIMOTHY MALCOLM, engineering educator; b. Vallejo, Calif., Nov. 21, 1955; s. Malcolm Edward and Betty Lee (Perkins) M.; m. Debra Susan Gillmer, Sept. 9, 1982. BS, U. Calif., Berkeley, 1977, MS, 1979; PhD, Calif. Inst. Tech., 1992. Mem. tech. staff Rocketdyne div. Rockwell Internat., Canoga Park, Calif., 1979-91; asst. prof. Okla. State U., Stillwater, 1992—. Contbr. articles Jour. Applied Mechanics, Internat. Jour. Solids and Structures. Recipient fellowship Calif. Inst. Tech., Pasadena, 1986. Mem. AIAA, ASME (assoc.). Office: Okla State Univ Sch Mech & Aerospace Engr 218 E N Stillwater OK 74078

MINASY, ARTHUR JOHN, aerospace and electronic detection systems executive; b. N.Y.C., July 19, 1925; s. John and Esther (Horvath) M.; B.S. in Adminstrv. Engring., N.Y. U., 1949, M.S. in Indsl. and Mgmt. Engring., 1952; postgrad. Case Inst. Tech., 1953-55; m. Jayne Marion Leary, June 29, 1946; children: Karen Lynn, Keith Leary, Kathy Jayne. Asst. gen. mgr. Def. div. Bulova Watch Co., Maspeth, N.Y., 1950-53; chief indsl. engr. Standard Products Co., Cleve., 1953-55; gen. mgr. ops. Gruen Industries, Cin., 1955-57; mgmt. cons. Booz-Allen and Hamilton, N.Y.C., 1957-60; mfg. mgr. Sperry Gyroscope Co., Great Neck, N.Y., 1960-62; v.p. ops. Belock Instrument Co., College Point, N.Y., 1962-64; pres. Detection Devices, Inc., Woodbury, N.Y., 1963—; chmn., chief exec. officer KNOGO Corp., Hauppauge, N.Y., 1966—; founder, pres. Internat. Electronic Articles Surveillance Mfrs. Assn., Brussels, 1989-93; bd. dirs. KNOGO Italia S.r.l., Milan, Italy, KNOGO SA, Belgium, KNOGO Caribe Inc., Cidra, P.R., KNOGO Australia, KNOGO The Netherlands B.V., KNOGO Switzerland S.A., KNOGO France S.A., KNOGO Denmark APS, KNOGO Deutschland GMBH, KNOGO Scandinavia AB, KNOGO UK Ltd., KNOGO Iberica SA; prin. Arthur J. Minasy Assocs., Mgmt. Cons., 1957-62; adv. bd. Abilities, Inc.; also lectr. in sci. law enforcement and detection systems. Dir., mem. adv. bd. Human Resources Found.; trustee Rehab. Inst. Served with AUS, 1943-46. Decorated Commdr. of Order of the Crown, King of Belgium, 1992; recipient Humanitarian of Yr. award Am. Cancer Soc.; named L.I.'s Entrepeneur of Yr., Inc. Mag., 1990; inductee to Smithsonian Nat. Mus. Am. History, 1991. Mem. Am. Inst. Indsl. Engrs., Internat. Electronic Article Surveillance Mfgrs. Assn. (founder, pres. 1989), Am. Ordnance Assn., Am. Mgmt. Assn., Tau Beta Pi, Alpha Pi Mu. Patentee in field. Home: 15 Hunting Hill Rd Woodbury NY 11797-1403 Office: KNOGO Corp 350 Wireless Blvd Hauppauge NY 11788-3927

MINEAR, ROGER ALLAN, chemist, educator; b. Seattle, June 19, 1939; s. Herbert Russell M. and Iris Ione (Merrill) Patterson; m. Carol Louise English, Aug. 12, 1966; children: Meredith Edin, Melinda Erin. BS, U. Wash., 1964, MS in Engring., 1966, PhD, 1971. Assoc. prof. dept. civil engring. U. Tenn., Knoxville, 1973-77, prof., 1977-82, Armour T. Granger prof., 1983-84; prof., dir. inst. for environ. studies U. Ill., Urbana, 1985—; sr. scientist Radian Corp., Austin, Tex., 1980-81; sr. devel. staff mem. environ. sci. div. Oak Ridge (Tenn.) Nat. Lab., 1983-84; dir. office solid waste rsch. U. Ill., 1987—; mem. scientific adv. com. hazardous waste rsch. ctr. La. State U., Baton Rouge, 1989-91; bd. scientific counselors Agy. for Toxic Substances and Disease Registry, Atlanta, 1988-93; mem. bd. environ. sci. and tech. Nat. Rsch. Coun., Washington, 1983-86. Editor: Water Chlorination, Vol. 6, 1989, Water Analysis, Vols. 1, 2, 3, 1982, 84; contbr. articles to profl. jours. Mem. Am. Water Works Assn., Assn. Environ. Engring. Profs. (pres. 1980-81, Disting. Svc. award 1984), Am. Chem. Soc. (councilor 1989—, Disting. Svc. award 1985), ASCE, Water Environ. Fedn. Home: 1003 Eliot Dr Urbana IL 61801 Office: U Ill Inst Environ Studues 1101 W Peabody Dr Urbana IL 61801

MINER, JOHN RONALD, agricultural engineer; b. Scottsburg, Ind., July 4, 1938; s. Gerald Lamont and Alice Mae (Murphy) M.; m. Betty Katheron Emery, Aug. 4, 1963; children—Saralena Marie, Katherine Alice, Frederick Gerald. B.S. in Chem. Engring, U. Kans., 1959; M.S.E. in San. Engring, U. Mich., 1960; Ph.D. in Chem. Engring. and Microbiology, Kans. State U., 1967. Lic. profl. engr., Kans., Oreg. San. engr. Kans. Dept. Health, Topeka, 1959-64; grad. research asst. Kans. State U., Manhattan, 1964-67; asst. prof. agrl. engring. Iowa State U., 1967-71, assoc. prof., 1971-72; assoc. prof. agrl. engring. Oreg. State U., 1972-76, prof., 1976—, head dept., 1976-86, acting assoc. dean Coll. Agrl. Sci., 1983-84, assoc. dir. Office Internat. Research and Devel., 1986-90, extension water quality specialist, 1991—; environ. engr. FAO of UN, Singapore, 1980-81; internat. cons.; cons. to livestock feeding ops., agrl. devel. firms. Co-author book on livestock waste mgmt.; author 3 books of children's sermons; contbr. numerous articles on livestock prodn., pollution control, control of odors associated with livestock prodn. to profl. publs. Mem. Am. Soc. Agrl. Engrs. (bd. dirs. 1985-87), Water Pollution Control Fedn., Sigma Xi, Gamma Sigma Delta, Alpha Epsilon, Tau Beta Pi. Presbyterian. Office: Dept Bioresource Engring Oreg State U Bioresource Eng Dept Corvallis OR 97331

MINES, RICHARD OLIVER, JR., civil and environmental engineer; b. Hot Springs, Va., July 23, 1953; s. Richard Oliver and Dreama Irene (Blankenship) M. BSCE, Va. Mil. Inst., 1975; ME in Civil Engring., U. Va., 1977; PhD, Va. Poly. Inst., 1983. Instr. Va. Mil. Inst., Lexington, 1977-79; project engr. William Matotan & Assocs., Albuquerque, 1979; grad. asst. Va. Poly. U., Blacksburg, 1980-83; asst. prof. U. South Fla., Tampa, 1983-85, Va. Mil. Inst., 1985-86; project engr. CH2M Hill, Gainesville, Fla., 1986-90; sr. process engr. Black & Veatch, Tampa, 1990-92; asst. prof. U. South Fla., Tampa, 1992—; adj. prof. Santa Fe Community Coll., Gainesville, 1989, U. South Fla., 1990—. Contbr. chpt. to book, articles to profl. jours. Capt. USAF, 1977. Engring. Found. grantee, 1985. Mem. ASCE, Water Pollution Control Fedn., Am. Water Works Assn., Alpha Kappa. Baptist. Achievements include research on oxygen transfer studies in the completely mixed activated sludge process. Office: U South Fla 4202 E Fowler Ave Tampa FL 33620

MINETTE, DENNIS JEROME, financial computing consultant; b. Columbus, Nebr., May 18, 1937; s. Lawrence Edward and Angela Ellen (Kelley) M.; B.S. in Elec. Engring., U. Nebr., 1970; M.B.A., Babson Coll., 1978; m. Virginia Rae Jordan, Oct. 27, 1961; children—Jordan Edward, Lawrence Edward II. Brokerage systems designer Honeywell Info. Systems, Mpls. and Wellesly, Mass., 1970-75; devel. mgr. Investment Info., Inc., Cambridge, Mass., 1975-77; product support mgr. Small Bus. Systems div. Data Gen. Corp., Westboro, Mass., 1977-81; pres. Minette Data Systems, Inc., Sarasota, Fla., 1981—. Capital improvement programs committeeman Town of Medway (Mass.), 1978-79, mem. town fin. com., 1979-80. With USN, 1956-60, 61-67, served to lt. commdr. res., 1967-87. Mem. IEEE, IEEE Computer Soc., Data Processing Mgmt. Assn. (cert.), Naval Res. Assn. (life), Res. Officers Assn., Am. Legion, U. Nebr. Alumni Assn. (life), Eta Kappa Nu, Sigma Tau. Republican. Roman Catholic. Office: Minette Data Systems Inc PO Box 15435 Sarasota FL 34277-1435

MING, SI-CHUN, pathologist, educator; b. Shanghai, China, Nov. 10, 1922; came to U.S., 1949, naturalized, 1964; s. Sian-Fan and Jan-Teh (Kuo) M.; m. Pen-Ming Lee, Aug. 17, 1957; children—Carol, Ruby, Stephanie, Michael, Jeffrey, Eileen. M.D., Nat. Central U. Coll. Medicine, China, 1947. Resident in pathology Mass. Gen. Hosp., Boston, 1952-56; assoc. pathologist Beth Israel Hosp., Boston, 1956-67; asst. prof. pathology Harvard U. Med. Sch., 1965-67; assoc. prof. U. Md., 1967-71; prof. Temple U., 1971—, acting chmn. dept. pathology, 1978-80, dep. chmn. dept. path., 1980-86; mem. Internat. Study Group on Gastric Cancer; U.S. rep. WHO Collaborating Ctr. for Primary Prevention, Diagnosis and Treatment of Gastric Cancer. Author: Tumors of the Esophagus and Stomach, 1973, supplement, 1985, Precursors of Gastric Cancer, 1984, Pathology of the Gastrointestinal Tract, 1992. Nat. Cancer Inst. sr. fellow Karolinska Inst., Stockholm, 1964-65. Mem. AAAS, U.S. Canadian Acad. Pathology, Am. Soc. Investigative Pathology, Am. Gastroenterol. Assn., N.Y. Acad. Scis. Achievements include development of classification method for stomach carcinoma based on the growth pattern of the cancer. Office: 3400 N Broad St Philadelphia PA 19140-5196

MININBERG, DAVID T., pediatric urology surgeon, educator; b. N.Y.C., May 28, 1936; s. Benjamin and Mildred (Zellermayer) M.; m. Anne Wikler, June 16, 1957; children: Gustav, Julien. BA, Yale U., 1957; MD, N.Y. Med. Coll., 1961. Intern Beth Israel Hosp., N.Y.C., 1961-62; resident in surgery East Orange (N.J.) Vets. Hosp., 1962-63; resident in urology N.Y. Med. Coll., N.Y.C., 1963-66; Ferdinand Valentine fellow N.Y. Acad. Medicine, 1966-67; instr. urology, pediatrics N.Y. Med. Coll., N.Y.C., 1967-69, asst. prof. urology, pediatrics, 1969-74, assoc. prof. urology, pediatrics, 1974-77; assoc. prof. surgery/urology Cornell U. Med. Coll., N.Y.C., 1977—, dir. pediatric urology, 1977—; bd. trustees Nat. Kidney Found.; bd. chmn. Nat. Enresis Soc.; mem. exec. com. Am. Acad. Pediatrics/Urology, 1985-91. Recipient Sprague Carleton award N.Y. Med. Coll., 1961, Ferdinand Valentine fellow N.Y. Adac. Medicine, 1966-67. Mem. Urology Am. Acad. Pediatrics (pres. sect. 1990). Home: 860 Fifth Ave New York City NY 10021 Office: Cornell Univ Med Coll 525 E 68th St New York NY 10021

MINIUTTI, ROBERT LEONARD, engineering company executive; b. Quincy, Mass., May 5, 1962; s. Leonard and Madeline M.; m. Elizabeth Fick, Aug. 19, 1984; children: Katherine, Anna, Emily. BS in Chem. Engring., Rensselar Poly. Inst., 1984. Process engr. Milliken and Co., La Grange, Ga., 1984-85; quality engr. Gentex Corp., Carbondale, Pa., 1985-86; project engr. Gentex Optics, Inc., Carbondale, 1986-90, tech. mktg. dir., 1990—; mem. ANSI Z136 Control Measures subcom., 1990--. Recipent 1992 Rsch. and Devel. 100 award Rsch. and Devel. mag., 1992. Mem. Am. Inst. Chem. Engrs. (sr.), Laser Inst. Am. (safety com. 1990--), Soc. Plastics Engrs., Optical Soc. Am., Soc. Photo Optical Instrumentation Engrs., Am. Vacuum Soc., Soc. Vacuum Coaters, Aerospace Lighting Inst., Am. Mgmt. Assn. Achievements include the development of laminated plastic lens (patent pending), tintable polysiloxane coatings for ophthalmic lenses, vacuum deposition compatible coatings for polymeric optics, Diamond Like Carbon (DLC) films for commercial sunwear, polymer dye absorption system for specialty filters. Home: 113 Upper Knapp Rd Clarks Summit PA 18411

MINKOWYCZ, W. J., engineering educator; b. Libokhora, Ukraine, Oct. 21, 1937; came to U.S., 1949; s. Alexander and Anna (Tokan) M.; m. Diana Eva Szandra, May 12, 1973; 1 child, Liliana Christine Anne. B.S. in Mech. Engring., U. Minn., 1958, M.S. in Mech. Engring., 1961, Ph.D. in Mech. Engring, 1965. Asst. prof. U. Ill., Chgo., 1966-68, assoc. prof., 1968-78, prof., 1978—; cons. Argonne Nat. Lab, Ill., 1970-82, U. Hawaii, Honolulu, 1974—. founding editor-in-chief Jour. Numerical Heat Transfer, 1978—; editor Internat. Jour. Heat and Mass Transfer, 1968—, Internat. Communications In Heat and Mass Transfer Jour., 1974—; editor book series: Computational and Physical Processes in Mechanics and Thermal Sciences, 1979—; editor: Rheologically Complex Fluids, 1972, Handbook of Numerical Heat Transfer, 1988; contbr. articles to profl. jours. Recipient Silver Circle for Excellence in Teaching, U. Ill.-Chgo., 1975, 76, 81, 86, 90, Harold A. Simon award Excellence in Teaching, 1986, Ralph Coats Roe Outstanding Tchr. award Am. Soc. Engring. Edn., 1988, U. Ill. Disting. Tchr. award, 1989. Fellow ASME; mem. Sigma Xi, Pi Tau Sigma. Republican. Ukrainian Catholic. Office: U Ill Dept Mech Engring Mail Code 251 PO Box 4348 Chicago IL 60680

MINN, YOUNG KEY, astronomer; b. Seoul, Korea, Jan. 1, 1938; s. Byung Chae and Hyun Sook (Lee) M.; m. Symyoung, May 12, 1967. BS, Seoul Nat. U., 1961; PhD, Rensselaer Poly. Inst., 1971. Rsch. scientist Max-Planck Inst. fur Radioastronomie, Bonn, Germany, 1971-73; asst. prof. U. Ala., Tuscaloosa, 1973-75; dir. Nat. Astron. Observatory, Seoul, Korea, 1975-85; assoc. prof. Seoul Nat. U., 1981-83; prof. Kyung Hee U., Yong-In, Korea, 1985—; dean coll. natural scis. Kyung Hee U., Yong-In, 1986-90. Mem. Am. Astron. Soc., Korean Astron. Soc., Internat. Astron. Union, Korean Sci. Writers Assn. (v.p. 1985—). Home: 409-163 Sillim-dong, Kwanuk-Ku, Seoul Republic of Korea Office: Kyung Hee U, Dept Astronomy & Space Sci, Yong-In, Kyunggi-Do 449-701, Republic of Korea

MINOR, MARK WILLIAM, allergist; b. Steubenville, Ohio, May 19, 1956; s. Garland Edgar Minor and Norma Jean McKenzie Shidock; m. Rachael Anne Hatfield, Aug. 15, 1987; children: Megan, Emily. BS in Biology, U. Miami, 1978; MD, W.Va. U., 1982. Intern W.Va. U., Charleston; residency in allergy/immunology U. So. Fla., Tampa; staff physician Holmes Regional Hosp., Melbourne, Fla.; clin. asst., prof. medicine U. South Fla.; physician Brevard Allergy Assocs., Melbourne, Fla., 1987—. Contbr. articles, referee Jour. Allergy and Clin. Immunology, So. Med. Jour. Fellow Am. Coll. Allergy and Immunology, Am. Coll. Physicians; mem. Am. Acad. Allergy and Immunology, Alpha Omega Alpha. Office: Brevard Allergy Assocs 1515 Airport Blvd Melbourne FL 32901

MINSHALL, GREG, computer programmer; b. Carmel, Calif., Apr. 21, 1952; s. Glenn Almon and Martha Jane (Hardesty) M.; m. Maria Concepci(ó)n Gonzalez, Dec. 30, 1976 (div. Jan. 1984); children: Matthew, Cecilia; m. Carol Ann Mendel, Oct. 4, 1987 (div. Feb. 1992); children: Oriana, Jacob. BA in Math., U. Calif., Berkeley, 1985. Computer programmer Stanford Linear Accelerator Ctr., Menlo Park, Calif., 1969-70, 72-73; computer programmer/engr. Inst. for Advanced Computation, Sunnyvale, Calif., 1978-80; computer programmer U. Calif., Berkeley, 1980-88; cons., 1984-88; computer programmer Novell, Inc., Walnut Creek, Calif., 1988—. Mem. IEEE, Assn. for Computing Machinery. Office: Novell Inc 1340 Treat Blvd Ste 300 Walnut Creek CA 94596

MINSKY, MARVIN LEE, mathematician, educator; b. N.Y.C., Aug. 9, 1927; s. Henry and Fannie (Reyser) M.; m. Gloria Anna Rudisch, July 30, 1952; children: Margaret, Henry, Juliana. B.A., Harvard U., 1950; Ph.D., Princeton U., 1954. Mem. Harvard Soc. Fellows, 1954-57; with Lincoln Lab., MIT, 1957-58, prof. math., 1958-61, prof. elec. engring., 1961—, Donner prof. sci., 1973, dir. artificial intelligence group MAC project, from 1958, dir. artificial intelligence lab. MAC project, from 1970. Author: Computation, 1967, Semantic Information Processing, 1968, Perceptrons, (with S. Papert), 1968, Robotics, 1985. Served with USNR, 1945-46. Recipient Turing award Assn. for Computing Machinery, 1970, Japan prize Sci. and Tech. Found. Japan, 1990. Fellow IEEE, NAE, NAS, Acad. Arts and Scis., N.Y. Acad. Scis. Office: MIT Dept Elec Engring - Computer Sci Cambridge MA 02139

MINTER, DAVID EDWARD, chemistry educator; b. Marshall, Tex., Aug. 26, 1946; s. Pete and Bettye Jeanne (Moore) M. BS, Stephen F. Austin Coll., 1968, MS, 1970; PhD, U. Tex., 1974. Chemist Dow Chemical Co., Freeport, Tex., 1970; teaching asst. U. Tex., Austin, 1970-72, instr., 1974-75; NIH postdoctoral fellow Health Sci. Ctr. U. Tex., San Antonio, 1975-77, rsch. scientist, 1977-80; asst. prof. Tex. Christian U., Ft. Worth, 1980-84, assoc. prof., 1985—. Contbr. articles to Tetrahedron Letters, Jour. Organic Chemistry, Jour. Chem. Edn., Magnetic Resonance in Chemistry. Mem. Am. Chem. Soc., Sigma Xi, Alpha Epsilon Delta. Office: Tex Christian U Dept Chemistry Box 32908 Fort Worth TX 76129

MINTZ, DANIEL HARVEY, endocrinologist, educator, academic administrator; b. N.Y.C., Sept. 16, 1930; s. Jacob A. and Fannie M.; m. Dawn E. Hynes, Jan. 15, 1961; children: David, Denise, Deborah. B.S. cum laude, St. Bonaventure Coll., 1951; M.D., N.Y. Med. Coll., 1956. Diplomate: Am. Bd. Internal Medicine. Intern Henry Ford Hosp., Detroit, 1956-57; resident Georgetown med. div. D.C. Gen. Hosp., Washington, 1957-59, Georgetown U. Hosp., Washington, 1958-59; fellow medicine Nat. Inst. Arthritis and Metabolic Diseases, 1959-60, Am. Diabetes Assn., 1960-61; practice medicine, specializing in internal medicine Miami, Fla.; asst. prof. medicine Georgetown U. Sch. Medicine, 1963-64; assoc. prof. medicine U. Pitts. Sch. Medicine, 1964-69; prof. medicine U. Miami Sch. Medicine, 1969—, Mary Lou Held prof. medicine, 1981—, chief div. endocrinology and metabolism, dept. medicine, 1969-80, Sci. dir. Diabetes Research Inst., 1980—; chief of service Georgetown U. Med. div. D.C. Gen. Hosp., Washington, 1963-64; chief of medicine Magee-Women's Hosp., Pitts., 1964-69; chief div. endocrinology and metabolism, dept. medicine Jackson Meml. Hosp., Miami; guest prof. U. Geneva, 1976-77. Contbr. articles to profl. jours. Fellow ACP; mem. Endocrine Soc., Am. Diabetes Assn. (program dir. 1972), Am. Fedn. Clin. Research, Am. Soc. Clin. Investigation, Central Soc. Clin. Investigation, So. Soc. Clin. Investigation., Am. Assn. Physicians. Office: U Miami Diabetes Rsch Inst PO Box 016960 (R-134) Miami FL 33101

MINTZER, PAUL, ophthalmologist, educator; b. Bklyn., Aug. 28, 1948; s. Alexander and Ellen Margaret (Brosnan) M.; m. Ellen A. Mintzer, Dec. 19, 1979; 1 child, Joshua. BA summa cum laude, Amherst Coll., 1970; MA, Cornell U., 1972; PhD, Northwestern U., 1977; MD, U. Calif., San Diego, 1981. Diplomate Am. Bd. Ophthalmology. Lectr. English lit. Northwestern U., Chgo., 1975-77, U. Ill., Chgo., 1976-77; ophthalmologist Med. West Community Health Plan, Chicopee, Mass., 1985-88, Hampshire Eye & Ear Assocs., Northampton, Mass., 1988—; v.p. student liaison bd. Calif. Med.

Assn., San Francisco, 1970-71. Author: Linguistics for the Foreign Language Student, 1970. Mem. Dem. Town Com., Southampton, Mass., 1989—. Fellow Cornell U., 1970-72, Northwestern U., 1972-77, Eastman fellow Amherst Coll., 1970-71. Mem. AMA, Royal Soc. Medicine, Physicians for Social Responsibility, Phi Beta Kappa. Jewish. Office: Hampshire Eye & Ear Assocs 61 Locust St Northampton MA 01060-2018

MIRABELLA, FRANCIS MICHAEL, JR., polymer scientist; b. New Haven, Dec. 27, 1943; s. Francis Michael and Mary Dorothy (DePrimo) M.; m. Doreen Lynn Lanier, Jan. 5, 1979; children: Andrea, Trent, Francis Michael III. AS, Norwalk State Tech. Inst., 1964; BA in Chemistry, U. Bridgeport, 1966; MS in Organic Chemistry, U. Conn., 1974, PhD in Polymer Sci., 1975. Chemist Olin, Corp., New Haven, 1966-69; rsch. scientist ARCO/Polymers, Monroeville, Pa., 1975-77; assoc. scientist Quantum Chem. Corp., Morris, Ill., 1977—. Author: Harrick Scientific Internal Reflection Spectroscopy, 1985; editor Marcel Dekker; contbr. 50 papers to profl. jours., including Polymer and Spectroscopic Sci. Jour. Dana scholar U. Bridgeport, 1966; fellow Inst. Material Sci. U. Conn., 1974-75. Mem. Am. Chem. Soc., Soc. for Applied Spectroscopy, Coblentz Soc. Office: Quantum Chemical Corp 8935 N Tabler Rd Morris IL 60450

MIRABITO, MICHAEL MARK, communications educator; b. N.Y.C., Apr. 15, 1956; s. Anthony J. and Jean (Cutrone) M.; m. Barbara Morgenstern, Oct. 17, 1980. BFA in Film/TV, NYU, 1977; MS in Comm., N.Y. Inst. Tech., 1979; PhD, Bowling Green State U., 1982. Asst. prof. comm. U. Tulsa, 1982-85, Ithaca (N.Y.) Coll., 1985—; cons. VITA, 1983. Author: Exploration of Space with Cameras, 1982, New Communications Technology, 1990, New Communications Technology II, 1993; contbr. articles to profl. jours., chpt. to book. Rsch. grantee Nat. Assn. Broadcasters, 1987, Digital Equipment Corp., 1988. Mem. Soc. Motion Picture and TV Engrs. (award com.), Nat. Space Soc., Nat. Space Club. Achievements include co-design of computer-television facilities. Office: Ithaca Coll Dept TV and Radio 953 Danby Rd Ithaca NY 14850

MIRACLE, MARIA ROSA, ecology educator; b. Barcelona, Spain, June 2, 1945. Licenciado, U. Barcelona, 1968, PhD, 1974. Teaching asst. U. Barcelona, 1968-71, researcher, 1974-76, adj. prof., 1976-79; postgrad. scholar for rsch. U. Calif., Davis, 1971-73; assoc. prof. ecology U. Valencia, Spain, 1979-81, prof. ecology, 1981—; vis. prof. U. Oreg., Corvallis, 1979. Author: Banyoles Lake Zooplankton, 1976, Ecologia, 1986; editor: Proc. of 6th Rotifer Symposium, 1992; contbr. articles to profl. jours. Recipient Premio a la Vocación, Fundación de la Vocación, 1975. Mem. Sci. Com. on Problems of Environ., Commn. Internat. pour l'Exploration de la Mer Mediterranée (pres. 1988—, com. coastal lagoons), Acad. Environ. Biology (pres. 1986-89), Spanish Limnology Assn. (pres. 1993—). Achievements include research in zooplankton community structure, limnology of meromictic lakes and coastal lagoons. Office: U Valencia, Facultad Biology Dept Ecol, 46100 Burjasot Valencia, Spain

MIRA GALIANA, JAIME JOSE JUAN, economist, consultant; b. Alicante, Spain, May 23, 1950; s. Jaime Mira Novella and Milagros Galiana Frances; m. Juana Camara Saez, May 5, 1974; children: Jaime, Cristian. Degree in econs., U. Valencia, Spain, 1972; computer diploma, IBM Edn. Ctr., Barcelona, Spain, 1973; diploma in indsl. mgmt., EADA, Barcelona, 1975. With orgn. dept., indsl. engring. systems and devel. dept. Unitransa Transport Co., Barcelona, 1972-76; mem. budget control staff, distbn. mgr., then logistic mgr. Frigo SA, Barcelona, 1977-84; logistic mgr. Henkel Iberica SA, Barcelona, 1984; cons. Tecnicas Logisticas SA, Barcelona, 1985; gen. dir. Credito Y Docks de Barcelona SA, 1986-87; logistics dir. Corbero SA (Electrolux Holding SA), Barcelona, 1987-90; cons. in orgn. and logistics Barcelona, 1991-92; gen. dir. COMBURSA (Comercial Burgos S.A.), Barcelona, 1991-92; cons. in orgn. and logistics Barcelona, 1992—; lectr. logistics ESMA Sch. Mktg., Barcelona, 1985—; lectr. total quality ops., comml. and phys. prodn. distbn. MBA and Operational and Superior Mktg. Sch.; lectr. Montpellier U., profl. confs. Contbr. articles to profl. publs.; editor: Manutencion y Almacenaje. Mem. Inst. Catalan de la Logistica Barcelona, Cen. Spain Logistica, CEL Madrid, Assn. Tecnicos Informatica (lectr.), Masnou Sport Club. Roman Catholic. Avocations: tennis, swimming, computers. Home and Office: Paseo Juan Carlos 1, Trebol 9 2-2, 08320 Masnou Barcelona Spain

MIRANDOLA, ALBERTO, mechanical engineering educator; b. Verona, Veneto, Italy, Feb. 11, 1942; s. Giovanni and Rina (Bacchiega) M.; m. Gabriella Minuzzi, June 30, 1967; children: Stefano, Marina. MS, U. Padua, Italy, 1967. Engr. Italian Railways, Florence, Italy, 1967; scholar Ministry of Edn., Padua, 1968-69; asst. prof. U. Padua, 1970-82, assoc. prof., 1983-86, prof., 1987—; dir. Dept. Mech. Engring., Padova, 1988—. Contbr. articles to profl. jours. Mem. ASME, Italian Thermotech. Com. Office: Dept Mech Engring, Via Venezia 1, 35131 Padua Italy

MIRDAMADI, HAMID REZA, structural engineering educator, researcher; b. Isfahan, Iran, Mar. 30, 1961; s. Seyed Ahmad and Mozaffar (Alaghemandan) Mirdamadi. BS in Structural Engring., Sharif U. Tech., Tehran, Iran, 1986, MS in Structural Engring., 1990. Researcher Sharif U. Tech., 1987-90, instr., 1990-91; design engr. Sanayee Havaee, Tehran, 1988-89; instr. structural engring. Azad U., Tehran, 1990—, Air Force U., Tehran, 1990—. Contbr. articles to Jour. Nonlinear Analysis Structures, Proc. 3d Internat. Congress Civil Engring. Fellow U. Victoria, B.C., Can., 1991; recipient letter commendation Minister Culture & Higher Edn. and V.p. Iran, 1990. Mem. AIAA. Achievements include research on application of BFGS algorithm to geometrically and materially nonlinear finite element analysis of continua, especially nonlinear vibrations and dynamics problems. Home: 224, Voroudi 8, Block 5-A3, Ekbatan, Tehran 13948, Iran Office: Office Sci and Internat Coops, Sharif U Tech, Tehran 11365, Iran

MIRICK, ROBERT ALLEN, military officer; b. Kingston, N.Y., June 26, 1957; s. Harry Lawrence and Jean Alice (Erickson) M.; m. Pamela Ann Warburton, July 24, 1982; children: Kristen E., Kathryn A., Meredith W., Abigail S. BS in Oceanography, U.S. Naval Acad., 1979; MS in Engring. Acoustics, Naval Postgrad. Sch., 1989. Commd. ensign USN, 1979, advanced through grades to lt. comdr.; navigator, propulsion asst. USS McCandless, 1979-82; diving and deck officer USS Pigeon, 1983-85; exec. officer, navigator USS Bolster, 1985-87; commdg. officer USS Hoist, 1990-92; vol. staff diver Monterey (Calif.) Bay Aquarium, 1987-89; field asst. Scripps Inst., San Diego, 1985. Contbr. article to Jour. of Acoustical Soc. Am. Pres. Parents Assn. of L.A., San Pedro, Calif., 1986. Decorated Meritorious Svc. medal USN, 1992. Mem. Acoustical Soc. Am., Am. Soc. Naval Engrs., U.S. Naval Inst., Soc. Colonial Wars. Republican. Achievements include research in sediment acoustics; development of apparatus to determine the complex mass of a viscous fluid contained in a rigid porous solid from acoustic pressure measurements; contributor to certification of USN MK2 Mod1 Deep Diving System to 850 feet. Office: Bur Naval Personnel Code 416 Washington DC 20370-0416

MIRMAN, MERRILL JAY, physician, surgeon; b. Darby, Pa., Jan. 24, 1941; s. Nathan and Rosalie (Applebaum-Yablotchnick) M. BS in Pharmacy, Phila. Coll. Pharmacy and Sci., 1962; D of Osteopathy, Phila. Coll. Osteo. Medicine, 1966. Diplomate Am. Acad. Pain Mgmt. Mem. faculty dept. osteo. principles and practice Pa. Coll. Osteo. Medicine, 1970-75; acting med. dir. S.E. Pa. Transp. Authority, Phila., 1970; mem. psychopharmacology rsch. unit Phila. Gen. Hosp., 1972-77; vice chmn. dept. gen. practice Tri-County Hosp., 1975, chmn. dept. gen. practice, 1976; mem., preceptor dept. family and community medicine Milton S. Hershey Med. Ctr. Pa. State U., 1978; founder, dir. Mirman Sch. Hypnosis, 1978-79; physician, surgeon Mirman Med. Ctr., Springfield, Pa., 1979—; osteo. med. cons. West Phila. Mental Health Consortium, 1971; mem. pvt. practice rsch. group U. Pa., 1972-77; chmn. osteo principles and practice com. Tri-County Hosp., 1976, chmn. pharmacy and therapeutics com., 1977, med. exec. com. of osteo. prin. and methods com., 1977, libr. com., 1977., med. exec. com., 1976-78; preceptor Phila. Coll. Osteo. Medicine, 1980-82, 87, U. Health Scis., Coll. Osteo. Medicine, Kansas City, 1985; physician reviewer region V Pa. Peer Review Orgn., 1985; reviewer Keystone Peer Review Orgn., 1986. Editor: Sclerotherapy, 1989; originator, prodr., dir. (cable TV series) To Your Health, 1982-86. Pres. planning comm., Glenolden, Pa., 1970-72; vol. radio and TV sta. operator Franklin Inst. Sci. Mus., Phila.; vol. examiner Am. Radio Relay League. Capt. U.S. Army Med. Corps., 1969, Vietnam. Recipient Four Chaplains Legion Honor award Chapel Four Chaplains, 1982, Air medal with oak leaf clusters. Fellow Am. Osteo. Acad. Sclerotherapy (sec.-treas. 1981-83, v.p. 1983-85, pres.-elect 1985-87, pres. 1987-89, bd. dirs. 1991-93), Acad. Psychosomatic Medicine, Am. Back Soc., Internat. Acad. Behavioral Medicine, Counseling, and Psychotherapy, Coll. Physicians Phila., Oculoplastic Fellowship Soc. N.Y.; mem. AAAS, Am. Bd. Sclerotherapy (cert., v.p. 1989-91, pres. 1991-93), Am. Coll. Gen. Practice (cert.), Am. Osteo. Assn., Am. Acad. Osteopathy (mem. hosp. com. 1973), Assn. to Advance Ethical Hypnosis (cert., pres. Pa. chpt. # 2 1972-73, 79-80), Acad. Psychosomatic Medicine, Internat. Assn. for Forensic Hypnosis (cert., exec. bd. 1982—, advisor 1979—), Can. Orthopedic Med. Assn., Nat. Aero. Assn., Am. Radio Relay League (vol. examiner), Pa. Osteo. Med. Assn. (chpt. # 2, alt. del. 1971-77), Del. County Osteo. Med. Assn., Del. County Pharm. Assn., Peer Review Orgn. Commonwealth Pa., Exptl. Aircraft Assn., Del. County Amateur Radio Assn., Airplane Owners and Pilots Assn., Chaverim Del. Valley, Holmesberg Amateur Radio Club, Philmont Mobile Radio Club., Rho Pi Phi. Office: Mirman Med Ctr 652 E Springfield Rd Springfield PA 19064-3644

MIRMIRAN, AMIR, civil engineer; b. Ghazvin, Iran, Feb. 20, 1961; s. Jafar and Asieh Mirmiran. BSCE, Tehran (Iran) U., 1984; MS in Structural Engring., U. Md., 1986, PhD in Structural Engring., 1991. Registered profl. engr., Md., Pa., Fla. Design engr. Gilan Civil Co., Tehran, 1980-83; project engr. J.M.T., Engrs., Balt., 1985-87, mgr. computer aided design and drafting, 1987-90; project mgr., CADD cons. Hurst-Rosche Engrs., Inc., Cockeysville, Md., 1990-93; asst. prof. Structural Engring U. Ctrl. Fla., Orlando, 1993—; part-time faculty mem. Towson (Md.) State U., 1990-93; postdoctoral fellow U. Md., College Park, 1992-93; pvt. practice computer cons., Cockeysville, 1990-93. Contbr. articles to profl. publs.; author software in field. Grantee Japan Dept. Higher Edn., 1984, U. Md., 1987, U.S. Dept. Edn., 1988-91. Mem. ASCE, Md. Soc. Profl. Engrs., Md. Cons. Engrs. Coun. (mem. CADD com. 1989-90). Islam. Achievements include development of method for the stability analysis of sequential fabrication of expandable systems. Home: # 305 4225 Thornbriar Ln Orlando FL 32822 Office: U Ctrl Fla Dept Civil & Environ Engring Orlando FL 32816-2450

MIRON, AMIHAI, electronic systems executive, electrical engineer; b. Ber-Shava, Israel, Feb. 21, 1953; came to U.S., 1981; s. Dov and Judith (Malovizka) M.; m. Ayala Loron, Mar. 26, 1980; children: Jonathan, Benjamin, Michelle. BSEE, Israel Inst. Tech., Haifa, 1981; MSEE, Poly. U., N.Y.C., 1984; Elec. Engr., Columbia U., 1986-88. Design engr. Philips Electonics N.V., Eindhoven, The Netherlands, 1981-82; sr. rsch. assoc. Philips Labs., Briarcliff Manor, N.Y., 1982-83, mem. tech. staff, 1983-85, sr. mem. tech. staff, 1985-88, dept. head Very Large Scale Integration and video, 1988-90, dir. electronic systems, 1990-93; with AMI Tech, Ossining, N.Y., 1993—; mem. advent Columbia U., N.Y.C., 1991—. Contbr. articles to profl. jours. Leader Youth Orgn., Israel, 1971. Capt. Israel Def. Army, 1972-76. Recipient Best in What is New grand prize Popular Sci., 1992, Video Tech. of Yr. grand prize Audio Video, 1992, R & D 100 award R & D Mag. Mem. IEEE, Soc. Motion Picture TV Engrs. Office: AMI Tech 56 Ganung Dr Ossining NY 10562

MIRON, MURRAY SAMUEL, psychologist, educator; b. Allentown, Pa., Aug. 7, 1932; s. Murray R. and Myrtle E. (Burton) M.; m. Helen Kutuchief, July, 1954 (div. June 1972); 1 child, Melinda; m. Cheryl Adamy, Aug. 5, 1973; 1 child, Murray Thomas. MA, U. Ill., 1956, PhD, 1960. From asst. prof. to assoc. prof. U. Ill., Urbana, 1960-70; prof. Syracuse (N.Y.) U., 1956; threat assessor FBI, Washington, 1972-92, U.S. Dept. Energy, Washington, 1980—, U.S. Dept. State, 1989-91. Author: Hostage, 1979; co-author: Cross-Cultural Universals of Affective Meaning, 1976; contbr. 50 articles to profl. jours. Negotiator Onondaga Sheriff's Office and Syracuse Police Dept., 1980-84; forensic cons. Onondaga Dist. Atty., Syracuse, 1992. Recipient Gold Badge Internat. Assn. of Chiefs of Police, 1977. Mem. Am. Psychol. Soc., Am. Soc. Indsl. Security. Achievements include pioneering rsch. in resolution of disputed communication origins, cross-cultural rsch. in meaning, psycholinguistic profiling of anonymous communications. Office: Psycholinguistic Rsch Ctr Ste C 2200 E Genesee St Syracuse NY 13210

MIRRA, CARLO, aerospace engineer, researcher; b. Naples, Italy, Dec. 18, 1963; s. Angelo G. and Maria I. (Silvestri) M. MSc in Aerospace Engring., Polytechs., Naples, 1987. Researcher U. Naples, 1985-87, Von Karman Inst., Bruxelles, 1987; project mgr. Mareco SpA, Aversa, Italy, 1987-88, Ciset SpA, Rome, 1988-89; head strategic devel. and advanced studies div., programme mgr. user support ctr. Mars Ctr., Naples, 1989—. Mng. editor Microgravity Qtr., 1990—; contbr. numerous articles to profl. jours. Mem. Internat. Astron. Fedn. (pub. com. 1989—, space sta. com. 1990—). Roman Catholic. Office: Mars Ctr, Via Diocleziano, 328, I-80125 Naples Italy

MIRZA, M. SAEED, civil engineering educator. Prof. civil engring. McGill U., Montreal, Que., Can. Recipient Le Prix A.B. Anderson award Can. Soc. Civil Engring., 1990. Office: McGill U Dept Civil Engineering, 817 Sherbrooke St W, Montreal, PQ Canada H3A 2K6*

MIRZA, SHAUKAT, engineering educator, researcher, consultant; b. Bhopal, India, Aug. 1, 1936; s. Mirza Afaq Beg and Birjees Jehan; m. Ferzana Beg, June 24, 1967; children—Sabah Jehan, Mazin. B.S. in Engring., Aligarh U., 1956; M.S. in Civil Engring., U. Wis.-Madison, 1960, Ph.D. in Engring. Mechanics, 1962. Sr. lectr. Delhi Coll. Engring., India, 1962-64; prof. Indian Inst. Tech., New Delhi, 1964-69; prof. mech. engring. U. Ottawa, Ont., Can., 1969—, vice dean rsch. and devel. faculty engring; vis. engr. Westinghouse Nuclear Europe, Brussels, 1976-77; vis. engr. Def. Rsch. Establishment, Ottawa, 1987-88; cons. Govt. of India, New Delhi, 1967-68, Atomic Energy Can., 1974-80, Bell No. Research, Ottawa, 1981-82. Invited keynote speaker various internat. profl. confs.; contbr. research articles, tech. reports to publs. Recipient Pres.'s gold medal, Roorkee U., India, 1958. Mem. ASME, Assn. Profl. Engrs. Ont. Office: U Ottawa, Faculty Engring, 770 King Edward Ave, Ottawa, ON Canada K1N 6N5

MIRZABEKOV, ANDREY DARYEVICH, molecular biologist; b. Baku, Russia, Oct. 19, 1937; m. Nataly Romanov, 1964; 1 child. Student, Inst. Fine Chem. Tech., Moscow. With W. Engelhardt Inst. Molecular Biology of USSR Acad. Scis., Moscow, 1961—, jr. rschr., chief dept., dep. dir., 1984, dir., 1985-90; vis. scientist MRC lab. molecular biology Cambridge, U.K., 1971, Calif. Inst. Tech., 1975, Harvard U., 1975; v.p. Internat. Human Genome Orgn., 1989—. Contbr. articles to profl. jours. Recipient State prize USSR, 1969, Anniversary prize Fedn. European Biochem. Socs., 1978, Gregor Mendel medal Deutsche Akad. der Naturforscher Leopoldina. Mem. Russian Acad. Scis. (former academician-secondary biophysics, biochemistry and chemistry of physiologically active compounds divsn.), Academia Europaea. Office: Engelgardt Inst Molecular Biology, Ulitsa Vavilova 32, 117984 Moscow Russia*

MISAWA, EDUARDO AKIRA, mechanical engineer, educator; b. Sao Paulo, Brazil, Oct. 4, 1956; came to U.S., 1983; s. Tomoaki and Hisae (Osaka) M.; m. Sonia Regina Midori Hayashida, July 26, 1981; 1 child, Erik. BSME, U. Sao Paulo, 1979, MSME, 1983; PhD in Mech. Engring., MIT, 1988. Rsch. engr. Heart Inst., lectr. U. Sao Paulo, 1980-83, asst. prof., 1988-90; grad. asst. MIT, Cambridge, 1983-88; researcher, then dept. head Inst. de Pesquisas Tecnologicas, Sao Paulo, 1988-90; asst. prof. mech. engring. Okla. State U., Stillwater, 1990—; cons. Holival R&D, Sao Paulo, 1980, Brazilian Found. for Teaching of Sci., Sao Paulo, 1983. Editor conf. procs. Advances in Nonlinear and Robust Control, 1992; contbr. articles to profl. jours. Office: Okla State U 218 Engineering North Stillwater OK 74078-0545

MISAWA, SUSUMU, physicist, educator; b. Menuma, Saitama, Japan, May 29, 1951; s. Masashi and Yoneko (Kakegawa) M.; m. Miyone Yamaoka, Nov. 18, 1984; children: Yoshihiro, Hideo. BS, Kyoto (Japan) U., 1975; MS, Tokyo U. of Edn., 1977; DSc, U. Tsukuba, Ibaraki, Japan, 1980. Asst. prof. Tokiwa U., Mito, Ibaraki, Japan, 1983-87, assoc. prof., 1987—. Contbr. articles to profl. jours. Mem. Am. Phys. Soc., Phys. Soc. of Japan. Office: Tokiwa U, Miwa, 1-430-1, Mito 310, Japan

MISCHEL, HARRIET NERLOVE, psychologist, educator; b. Chgo., Aug. 7, 1936; d. Samuel H. and Evelyn (Andelman) Nerlove; m. Walter Mischel, June 19, 1960; children: Judith, Rebecca, Linda. BA, Swarthmore Coll., 1958; PhD, Harvard U., 1963. Licensed psychologist, N.Y. With dept. psychology Stanford (Calif.) U., 1970-83; rsch. assoc. Columbia U., N.Y.C., 1983-85; fellow Inst. for Behavior Therapy, N.Y.C., 1983-85, Inst. for Rational Emotive Therapy, N.Y.C., 1983-85; pvt. practice N.Y., 1985—. Author: Essentials of Psychology, 1977; editor: Readings in the Theory of Personality, 1973; contbr. chpt.: Children, Feminism and the New Family, 1968; contbr. numerous articles to profl. jours. Mem. Am. Psychol. Assn., Assn. for Advancement Behavior Therapy, Phi Beta Kappa. Office: Behavioral Assocs 114 E 90th St 1A New York NY 10028

MISCHKE, CHARLES RUSSELL, mechanical engineering educator; b. Glendale, N.Y., Mar. 2, 1927; s. Reinhart Charles and Dena Amelia (Scholl) M.; m. Margaret R. Bubeck, Aug. 4, 1951; children: Thomas, James. BSME, Cornell U., 1947, MME, 1950; PhD, U. Wis., 1953. Registered mechanical engr. Iowa, Kans. Asst. prof. mech. engring. U. Kans., Lawrence, 1953-56; assoc. prof. mech. engring. U. Kans., 1956-57; prof., chmn. mech. engring. Pratt Inst., N.Y.C., 1957-64; prof. mech. engring. Iowa State U., Ames, 1964—, Alcoa Found. prof., 1974. Author: Introduction to Mechanical Analysis, 1963, Introduction to Computer-Aided Design, 1968, Mathematical Model Building, 1972; editor: Standard Handbook of Machine Design, 1986, Mechanical Engineering Design, 5th edit., 1989, eight Mechanical Designers Workbooks, 1990. Scoutmaster Boy Scouts Am., Ames. With USNR, 1944-75, mem. Res. ret. Recipient Ralph Teetor award Soc. Automotive Engrs., 1977, best book award Am. Assn. Pubs., 1986, Legis. Teaching Excellence award Iowa Assembly, 1990, Ralph Coates Roe award Am. Soc. for Engring. Edn., 1991. Fellow ASME (life, Machine Design award 1990); mem. ASEE (Centennial cert. 1993), Am. Gear Mfrs. Assn., Scabbard and Blade, Cardinal Key, Sigma Xi, Phi Kappa Phi, Pi Tau Sigma. Avocations: model bldg., railway history. Office: Iowa State U 3029 Black Ames IA 50011

MISCHKE, RICHARD EVANS, physicist; b. Bristol, Va., Aug. 19, 1940; s. Vernon Evans and Ruth (Shollenberger) M.; m. Alice Ruth Joyce, Dec. 27, 1962; children: Rachel Ellen, Rebecca Elise. Student, Lambuth Coll., 1957-59; BS, U. Tenn., 1961; MS, U. Ill., 1962, PhD, 1966. Instr., Princeton U., N.J., 1966-68, asst. prof., 1968-71; mem. staff Los Alamos Sci. Lab., N.Mex., 1971-73, asst. group leader, 1973-75, assoc. group leader, 1975-78, dep. group leader, 1978-86, mem. staff, 1986—; dir. Los Alamos Meson Physics Facility Users Group, Inc., Los Alamos, 1977; guest scientist Swiss Inst. Nuclear Research, 1978-79; program monitor U.S. Dept. Energy, 1987-88; mem. organizing com. Conf. on Intersections Between Particle and Nuclear Physics, 1984. Editor: Proceedings of Conference on Intersections Between Particle and Nuclear Physics, 1984—; contbr. physics articles to profl. jours. Patroller, Nat. Ski Patrol, 1974—; mem. central com. Democratic party, 1979-85, ward chmn., 1981-83; chmn. bicycle subcom. Los Alamos County, 1983-86, chmn. transp. bd., 1985-87. Nat. Methodist scholar, 1958-59; U. Ill. fellow, 1961-62, NSF fellow, 1962-64. Fellow Am. Phys. Soc.; mem. Phi Kappa Phi, Sigma Pi Sigma. Mem. Ch. of Christ. Home: 2172 Loma Linda Dr Los Alamos NM 87544-2769 Office: Los Alamos Nat Lab MS H846 Los Alamos NM 87545

MISFELDT, MICHAEL LEE, immunologist, educator; b. Davenport, Iowa, June 15, 1950; s. Melvin Lawrence and Wanda Irene (Dee) M.; m. Mary Alice Griffin, Aug. 4, 1973; children: Andrew Michael, Kristin Marie. BS, U. Ill., 1972; PhD, U. Iowa, 1977. Rsch. asst. U. Iowa, Iowa City, 1972-77; staff fellow NIH, Bethesda, Md., 1977-81; asst. prof. U. Mo. Sch. Medicine, Columbia, 1981-87, assoc. prof., 1987-92, prof. immunology, 1992—. Author: Veterinary Immunolgy Immunopathology, 1989, Infection Immunity, 1990, 92, Cellular Immunology, 1991, Encyclopaedia of Microbiology, 1992. Pres. Home Owners Assn., Columbia, 1988-90. Grantee, NIH, Bethesda, 1983, 86, 88, USDA, Washington, 1991. Mem. AAAS, Am. Soc. Microbiology (editorial bd. 1989—, lectr. 1991—), Am. Assn. Immunologists, Reticuloendothelial Soc. Episcopalian. Achievements include description of a unique microbial superantigen, isolated and characterized a swine T lymphocyte cell line. Home: 4102 S Wappel Columbia MO 65203 Office: U Mo Sch Medicine M642 Med Scis Columbia MO 65212

MISHKIN, FREDERIC STANLEY, economics educator; b. N.Y.C., Jan. 11, 1951; s. Sidney and Jeanne (Silverstein) M.; m. Sally A. Hammond; children: Matthew, Laura. Student, Oxford (Eng.) U., 1971-72; BS, MIT, 1973, PhD, 1976. Asst. prof. U. Chgo., 1976-81, assoc. prof., 1981-83; vis. assoc. prof. Northwestern U., Evanston, Ill., 1982-83; rsch. assoc. Nat. Bur. Econ. Rsch., Cambridge, Mass., 1980—; prof. econs. Grad. Sch. Bus., Columbia U., N.Y.C., 1983—; A. Barton Hepburn prof. econs. Columbia U. Grad. Sch. Bus., N.Y.C., 1991—; vis. scholar bd. govs. Fed. Res. System, Washington, 1977, Ministry of Fin., Japan, Tokyo, 1986; panel mem. Brookings Panel on Econ. Activity, Washington, 1977-78; mem. acad. adv. panel Fed. Res. Bank N.Y., N.Y.C., 1990—; vis. prof. Princeton U.. 1990-91. Author: A Rational Expectations Approach to Macroeconometrics, 1983, The Economics of Money, Banking and Financial Markets, 1986, 3d edit., 1992, Money, Interest Rates and Inflation, 1993, also articles. Mem. Am. Econ. Assn., Am. Fin. Assn., Phi Beta Kappa, Sigma Xi. Avocations: sailing, cross country skiing, long distance cycling, reading. Office: Grad Sch Bus Columbia U Uris Hall 619 New York NY 10027

MISHKIN, MORTIMER, neuropsychologist; b. Fitchburg, Mass., Dec. 13, 1926; married; 2 children. AB, Dartmouth Coll., 1946; MA, McGill U., Montreal, Can., 1949, PhD, 1951. Asst. in research and physiology and psychiatry Yale U. Med. Sch., New Haven, Conn., 1949-51; research assoc. Inst. of Living, Hartford-Conn. and NYU Bellevue Med. Ctr., N.Y.C., 1951-55; research psychologist, sect. on neuropsychology NIMH, Bethesda, Md., 1955-75, research physiologist, Lab. of Neuropsychology, 1976-78, chief, sect. on cerebral mechanisms, Lab. of Neuropsychology, 1979-80, chief Lab. of Neuropsychology, 1980—; part-time instr. psychology Howard U., 1956-58; vis. scientist Nencki Inst. Exptl. Biology, Warsaw, Poland, winter 1958, 68, Tokyo Met. Inst. Neurosci., summer 1978, Oxford U. Dept. Exptl. Psychology, summer 1979; mem. psychol. scis. panel NIH, 1959-61, exptl. psychology study sect., 1965-69; mem. NIMH Assembly of Scientists Council, 1962-64, 72-74; mem. NIMH Scientist Promotion Rev. Com., 1984-86; mem. adv. com. Cognitive Neurosci. Inst., 1982-86; mem. NIH Fogart Internat. Scholars-in-Residence Adv. Panel, 1985-89; adv. bd. McDonnell-Pew Program Cognitive Neurosci., 1989—; review com. Brain rsch., Human Frontier Sci. program, 1992—, chmn. 1993—. Cons. editor Jour. Comparative and Physiol. Psychology, 1963-73, Exptl. Brain Research, 1965—, Brain Research, 1974-78; cons. editor Neuropsychologia, 1963, mem. editorial bd., 1978-81; editorial bd. Human Neurobiology, 1981-87, Neuropsychologia, 1963-92, Exptl. Brain Rsch., 1965—, Brain Rsch., 1974-78, Human Neurobiology, Jour. Cognitive Neurosci., 1989—, Jour. NIH Rsch., 1989—, Cerebral Cortex, 1990—, Advances in Neurobiology, 1990—, Handbook Behavioral Neurology, 1990, Current Opinion in Neurobiology, 1991—, Behavioral and Neural Biology, 1992—; reviewing editors Sci., 1985-93; assoc. editor Neuroreport, 1990—; contbr. over 160 articles to profl. jours.; also abstracts and books revs. Served to lt. (j.g.) USNR. Fellow AAAS (chair-elect 1990-91, chair 1991-92, past chair 1992-93), Am. Psychol. Assn. (officer, mem. at large 1964-66, coun. rep. 1967-69, pres. 1968-69); mem. NAS (officer, chmn. 1989-92), Ea. Psychol. Assn., Internat. Brain Research Orgn. (officer, rep.-at-large governing coun. 1993—), Internat. Neuropsychol. Soc., Internat. Neuropsychol. Symposium, Internat. Primatological Soc., Internat. Soc. Neuroethology, Soc. Exptl. Psychologists, Soc. Neurosci. (officer, pres.-elect 1985-86, pres. 1986-87, past pres. 1987-88), Sigma Xi, Phi Beta Kappa. Achievements includes research in behavioral and cognitive neuroscience in primates. Office: NIMH Lab Neuropsychology Bldg 49 Rm 1B80 9000 Rockville Pike Bethesda MD 20892-0001

MISHRA, AJAY KUMAR, software engineer; b. Mirzapur, India, May 4, 1960; came to U.S., 1981; s. Sharda Prasad and Subhadra (Pandey) M.; m. Vrinda Mishra, June 30, 1989; children: Aditi and Ameeti (twins). BSc, Allahabad U., India, 1980; MS Edn. in Biology, Alcorn State U., 1983; MS in Computer Sci., U. So. Miss., 1985; MBA, U. Chgo., 1993. Teaching asst. U. So. Miss., Hattiesburg, 1983-85; weather software engr. NOW Weather Inc., Gulfport, Miss., 1986-87; project leader Universal Weather & Aviation, Houston, 1987-89; sr. staff analyst Computer People Unltd., Milw., 1989-90; sr. software engr. Motorola Cellular Inc., Libertyville, Ill., 1990—. Mem.

N.W. Suburb Toastmasters (pres. 1993). Hindu. Home: 110 George Town Dr Cary IL 60013 Office: Motorola Inc 600 N Hwy 45 Libertyville IL 60048

MISHRA, ARUN KUMAR, ophthalmologist; b. Motihari, Bihar, India, July 20, 1945; arrived in Eng., 1972; s. Vijayanand and Rajkali (Mishra) M.; m. Kalawati Mishra, Feb. 23, 1966; children: Suman, Vidya Bhusan, Shashi Bhushan. MB, BS, P.W. Med., Patna, Bihar, 1967; DO, Patna U., Bihar, 1970. Med. officer Coal Mines Welfare, Dhanbad, Bihar, 1970-72; ophthalmologist Southampton (Eng.) U., 1972-73; registrar Glasgow (Scotland) Royal Infirmary, 1973-74, U. Cardiff and Merthyr (Wales), 1974-76; registrar Nottingham (Eng.) U., 1976-78, sr. registrar and lectr., 1978-80; cons., head of dept. ophthalmology Dewsbury Dist. Hosp., Yorkshire, Eng., 1980—; chmn. surgery Dewsbury Health Authority, Yorkshire, 1981—; cons. Unit Mgmt. Bd., Dewsbury, 1987-91. Co-author: Brit. Jour. Ophthalmology, 1979. Justice of the peace Dewsbury Magistrates Ct., 1988—; sec. Yorkshire Indian Soc., Leeds, 1986—. Fellow Royal Coll. Surgeons, Coll. Ophthalmologists (U.K.); mem. U.K. Magistrates Assn., Brit. Med. Assn. (chmn. Dewsbury div. 1989-91), Overseas Doctors Assn. (v.p. 1985—), Bihar Ophthalmol. Soc., Rotary (pres. Batley chpt. 1989-92). Hindu. Avocations: photography, badminton, tropical fish keeping. Home: 214 Woodlands Rd, Batley WF17 0QS, England Office: Dewsbury Dist Hosp, Healds Rd, Dewsbury WF13 4HS, England

MISHRA, BRAJENDRA, metallurgical engineering educator, researcher; b. Muzaffarpur, Bihar, India, May 7, 1959; came to U.S., 1981; s. Jayamanta and Sushila (Roy) M.; m. Deepa Jha, June 1, 1983; 1 child, Ramya. BTech, Indian Inst. Tech., Kharagpur, 1981; MS, U. Minn., 1983, PhD, 1986. Rsch. asst. U. Minn., Mpls., 1981-86; product devel. engr. Tata Steel, Jamshedpur, India, 1986-87, R & D engr., 1987-90, rsch. asst. prof. metall. engring., 1990-93; rsch. asst. prof. metall. engring. Colo. Sch. Mines, Golden, 1990-93, rsch. assoc. prof. metall. engring., 1993—; mem. Tech. Information Forecasting and Assessment Coun., Govt. of India, Delhi, 1989-90. Contbr. over 65 articles to profl. jours. Doctoral fellow U. Minn., 1984; doctoral grantee Dow Chem. Co., 1984, rsch. grantee Schottglaswerke, 1992, Castolin Eutectic, 1992, Dept. of Energy, 1993. Mem. ECS, NACE, Metall. Soc., Am. Soc. for Metals (lectr. 1992), Reactive Metals Com. (program chmn. 1992—), Sigma Xi. Achievements include patent on iron powder production from sponge iron fires; development of process for actinide separation and waste minimization; steel development for corrosion resistance; thermomechanical control processed steel rolling in hot strip mill. Home: 116 Kimball Ave Golden CO 80401 Office: Colo Sch Mines Dept Metall Engring 1500 Illinois St Golden CO 80401

MISHRA, VISHNU S., research scientist; b. Ajmer, India, June 15, 1956; came to U.S., 1986.; s. Hari Shanker and Urmila Mishra; m. Rashmi Mishra, Apr. 23, 1985. PhD, Jawaharlal Nehru U., New Delhi, India, 1983. Rsch. assoc. Jawaharlal Nehru U., 1983-84, rsch. scientist, 1984-86; postdoctoral fellow U. Pa., Phila., 1986-88; postdoctoral fellow U. Fla., Gainesville, 1988-91, asst. scientist, 1992—. Published papers in field. Mem. AAAS. Achievements include first to clone and sequence the gene for a surface protein of Babesia bigemina, identify the multigene family of p58 protein of Babesia bigemina. Home: 507 NW 39th Rd # 303 Gainesville FL 32607 Office: U Fla PO Box 100277 Gainesville FL 32610-0277

MISKOVITZ, PAUL FREDERICK, gastroenterologist; b. Far Rockaway, N.Y., Sept. 18, 1949; s. Frederick William and Helen (Uram) M.; m. Leslie Joan Wagshol, Mar. 11, 1973; children: Sharyn, Steven. BS in Biology, SUNY, Stony Brook, 1971; MD, Cornell U. Med. Coll., 1975. Diplomate Am. Bd. Internal Medicine. Instr. Cornell U. Med. Coll., N.Y.C., 1980-81, asst. prof., 1981-87, assoc. prof., 1987—. Author: Diarrhea, 1993; contbr. articles to profl. jours. Mem. Am. Soc. Gastrointestinal Endoscopy, Am. Gastroent. Assn., N.Y. Acad. Gastroent. Office: Cornell U Med Coll Box 422 50 E 70th St New York NY 10021-4928

MISNER, ROBERT DAVID, electronic warfare and magnetic recording consultant, electro-mechanical company executive; b. Waynesville, Ill., May 1, 1920; s. Oscar and Elizabeth (Nyren) M.; student Ill. Wesleyan U., 1939-42; B.S. in Physics, George Washington U., 1946; postgr. U. Md., 1948; m. Virginia Fuehrer, June 4, 1949; children: Robin Beth, Christie Marie. Mem. staff U.S. Naval Rsch. Lab., Washington, 1942-44, 46—, br. head signal exploitation br., 1965-87; pres. MEMRE Co., 1987—; cons. Served in USNR, 1944-46. Recipient Disting. Civilian Service award USN, 1970; others. Mem. IEEE (sr.), Assn. Old Crows, Sigma Xi. Contbr. articles to profl. jours. Home: 7107 Sussex Pl Alexandria VA 22307-2006 Office: 4555 Overlook Ave Washington DC 20375

MISRA, JAYADEV, computer science educator; b. Cuttack, Orissa, India, Oct. 17, 1947; s. Sashibhusan and Shanty (Kar) M.; m. Mamata Das, Nov. 30, 1972; children: Amitav, Anuj. B Tech, Indian Inst. Tech., Kanpur, 1969; PhD, Johns Hopkins U., 1972. Staff scientist IBM, Gaithersburg, Md., 1973-74; from asst. prof. to prof. computer sci. U. Tex., Austin, 1974—, Regents chair in computer sci., 1992—; vis. prof. Stanford (Calif.) U., 1983-84; cons. on software and hardware design. Contbr. articles to profl. jours. Guggenheim fellow, 1988-89. Fellow IEEE; mem. Assn. Computing Machinery (Samuel N. Alexander Meml. award, 1970). Office: Univ Tex Dept Computer Sci Austin TX 78712

MISRA, PRASANTA KUMAR, physics educator; b. Kamakhyanagar, Orissa, India, Sept. 10, 1935; came to U.S., 1963; s. Jayakrishna and Parabati Misra; m. Swayamprava Sarangi, Dec. 3, 1961; children: Debasis, Moushumi, Sandeep. MS, Utkal U., Orissa, 1956; PhD, Tufts U., 1967. Reader in physics Utkal U., Bhabaneswar, 1968-70; rsch. assoc. U. Tex., Austin, 1970-72; prof. physics Berhampur (India) U., 1974-84, Mesa State Coll., Grand Junction, Colo., 1988—; vis. prof. Temple U., Phila., 1984-85, La. State U., Baton Rouge, 1985-86; lectr., vis. prof. U. R.I., Kingston, 1986-88; sec. Vignyan Prachar Samiti, Berhampur, 1979-84, referee Pramana physics jour., Bangalore, India, 1976-84, NSF, Washington, 1986-87. Contbr. numerous articles, rsch. papers to profl. publs. Grantee Indian Univ. Grants Commn., 1969-84, Coun. Sci. and Indsl. Rsch., India, 1982-84, Indian Dept. Atomic Energy, 1982-83; Fulbright Found. scholar, 1963-68. Mem. Am. Phys. Soc., Indian Physics Assn., Soc. Physics Students (advisor). Office: Mesa State Coll 1150 Texas Ave Grand Junction CO 81501

MISTRY, KISHORKUMAR PURUSHOTTAMDAS, biomedical scientist, educator; b. Vadodara, India, Dec. 25, 1953; came to U.S., 1987; s. Purushottamdas M. and Savitaben P. (Suthar) M.; m. Varsha B. Shah, Apr. 1, 1990. MS in Biochemistry, M.S. U., Baroda, India, 1977, PhD in Biochemistry, 1981. Res. officer Miles India Ltd., Vadodara, 1981-82, head quality assurance, 1982-83; chief biochemist M.P. Urol. Hosp., Nadiad, India, 1983-84; asst. prof. M.S. U., Vadodara, 1984-87; rsch. assoc. N.J. Med. Sch., Newark, 1987-90, instr. dept. physiology, 1990—; mem. admission com. M.S. U., 1984-87. Contbr. articles to profl. jours. Vol. Liberty Sci. Ctr., Jersey City, 1990—, Newark Literacy Campaign, 1989. Recipient scholarships and grant. Mem. Assn. Rsch. in Vision and Ophthalmology, Am. Assn. Clin. Chemistry. Hindu. Achievements include showing that there is a possibility of human fetus synthesizing vitamin C; research proving that hyperglycemic condition can inhibit lens myo-inositol uptake by 3 different mechanisms. Home: 38-C Donald St Bloomfield NJ 07003 Office: NJ Med Sch 185 S Orange Ave Newark NJ 07103

MISUREC, RUDOLF, physician, surgeon; b. Dobre Pole, Czechoslovakia, June 27, 1924; came to U.S., 1967; s. Gustav and Hilda (Safar) M.; m. Miluse Kisil, 1951 (div. 1978); children: Peter Clyde, Rudolph Carl; m. Stanislava Coufal, 1978. MD, Masaryk's U., Brno-Czechoslovakia, 1950. Diplomate Am. Bd. Urology, gen. surgery (Czechoslovakia), thoracic surgery (Czechoslovakia). Intern U. Ill., Chgo., 1967-68, resident in urology, 1968-71, clin. asst. prof. urology, 1975—. Mem. Rep. Presdl. Task Force, 1984, Rep. Presdl. Legion of Merit, 1992. Capt. Czechoslovakia Army, 1950-55. Recipient Cert. of Achievement U.S. Army, 1967, Letter of Appreciation, 1967. Fellow ACS; Internat. Coll. Surgeons, Am. Urol. Assn.; mem. AMA, Chgo. Med. Soc., N.Y. Acad. Scis., Czechoslovak Soc. Arts and Scis. (U.S.). Roman Catholic. Office: 3340 S Oak Park Ave Berwyn IL 60402

MITA, ITURA, polymer chemist; b. Tokyo, Aug. 20, 1929; s. Taizo and Takako (Fujimoto) M.; m. Hisako Matsuo, Nov. 17, 1962; children: Noriyuki, Reiko. BSc, U. Tokyo, 1954, MSc, 1956, DSc, 1971; DSc, U. Strasbourg, France, 1959. Asst. Inst. Sci., Tech., U. Tokyo, 1956-63; assoc. prof. Aero. Rsch. Inst. U. Tokyo, 1963-73, prof. Space and Aero. Rsch. Inst., 1973-88, prof. Rsch. Inst. Advanced Sci., Tech., 1988-90, prof. emeritus, 1990—; rsch. dir. Dow Corning Japan Ltd., Yamakita/Kanagawa, Japan, 1990—. Author: High Temperature Polymers, 1989, High Performance Composite Materials, 1990, Polymers for Microelectronics, 1991. Mem. Soc. Polymer Soc. Japan (pres. 1990-93, award 1989), Internat. Union Pure, Applied Chemistry (titular mem. 1980-89). Avocation: Go. Home: Komone 4-12-11, Itabashi, Tokyo 173, Japan Office: Dow Corning Japan Rsch Ctr, Kishi-603, Kanagawa, Yamakita 258-01, Japan

MITARAI, OSAMU, physics educator; b. Fukue, Nagasaki, Japan, Nov. 14, 1950; d.; s. Iwao and Kito Mitarai; m. Akiko Tani. BSME, Kyushu U., Fukuoka, Japan, 1974, MSME, 1976, D of Nuclear Engring., 1979. Postdoctoral fellow dept. physics U. Saskatchewan, Saskatoon, Can., 1981-83, rsch. assoc. dept. physics, 1983-84; lectr. Kumamoto (Japan) Inst. Tech., 1985-87, assoc. prof., 1987—. Author: Invention of "Alternating Current (AC) Tokamak Reactor: Nuclear Fusion, Fusion Technology, 1984, Invention of "Ignition Access Condition for a Tokamak Reactor" Fusion Technology, 1990. Mem. AAAS, IEEE, Am. PHys. Soc., Japan Phys. Soc., Japan Soc. Plasma, Sci. and Nuclear Fusion Rsch., Atomic Energy Soc. Japan. Achievements include pioneer work on D-3He tokamak reactor, first experimental demonstration of AC operation in the STOR-1M tokamak; design, building and experiments of first Canadian tokamak STOR-1M. Office: Kumamoto Inst Tech Dept EE, 4-22-1 Ikeda, Kumamoto 860, Japan

MITCHELHILL, JAMES MOFFAT, civil engineer; b. St. Joseph, Mo., Aug. 11, 1912; s. William and Jeannette (Ambrose) M.; BS, Northwestern U., 1934, CE, 1935; m. Maurine Hutchason, Jan. 9, 1937 (div. 1962); children: Janis Maurine Mitchelhill Johnson, Jeri Ann Mitchelhill Riney; m. 2d, Alicia Beuchat, 1982; Engring. dept. C., M., St. P. & P.R.R. Co., Chgo. and Miles City, Mont., 1935-45; asst. mgr. Ponce & Guayama R.R. Co., Aguirre, P.R., 1945-51, v.p., gen. mgr., 1969-70; mgr. Cen. Cortada, Santa Isabel, P.R., 1951-54; r.r. supt. Braden Copper Co., Rancagua, Chile, 1954-63; staff engr. Coverdale & Colpitts, N.Y.C., 1963-64; asst. to exec. v.p. Central Aguirre Sugar Co., 1964-67; v.p., gen. mgr. Coddea, Inc., Dominican Republic, 1967-68; asst. to gen. mgr. Land Adminstrn. of P.R., La Nueva Central Aguirre, 1970-71, for Centrals Aguirre Lafayette and Mercedita, 1971-72; asst. to gen. mgr. Corporacion Azucarera de P.R., 1973-76, asst. to exec. dir., 1977-79, asst. exec. dir. for environ., 1979-82; engring. cons., 1982-92; Kendall County engr., 1985—. Registered profl. engr., Mont., P.R., Tex. Fellow ASCE; mem. Am. Ry. Engring. Assn., Colegio de Ingenieros y Agrimensores de P.R., Explorers Club, Travellers Century Club, Sigma Xi, Tau Beta Pi. Home: PO Box 506 Boerne TX 78006-0506 Office: 12 Staudt St Boerne TX 78006-1745

MITCHELL, DEBORAH JANE, educator; b. Mound Bayou, Miss., Feb. 14, 1964; d. Robert Lee and Lillian E. (Sias) M. BS in Chemistry, Jarvis Christian, 1986; MS in Chemistry, Prairie View A&M, 1987. Cert. tchr. biology and chemistry, Tex. Chemist Argonne (Ill.) Nat. Lab., 1986-87; provisional tchr. Chgo. Bd. Edn., 1986-87; organic rsch. chemist Universal Energy Systems, Dayton, Ohio, 1988; corp. rsch. asst. Tex. Instruments, Inc., Dallas, 1989-90; student tchr. Houston Ind. Sch. Dist., 1990-91, tchr., 1991-92; sci. team coach Jefferson Davis Sr. High, Houston, 1991-92, sci. fair sponsor, 1991-92, ESL tchr., 1991-92. Editor, contbr. Waller County Tribune newspaper, 1989-92. Vol. Dem. Party, Chgo., 1987, Houston, 1992. Mem. NAFE, NAFE, Am. Chem. Soc., Nat. Orgn. Black Chemists/Chem. Engrs., Delta Sigma Theta (pres. 1989-91, Active Soror award 1990), Beta Kappa Chi, Kappa Delta Pi, Alpha Kappa Mu. Democrat. Mem. Christian Ch. (Disciples of Christ). Achievements include development of two methods for the removal of surface spot contamination on micro computer chips by laser desorption. Home: PO Box 155 Big Sandy TX 75755

MITCHELL, DENIS, civil engineer. Recipient Sir Casimir Gzowski medal Can. Soc. Civil Engring., 1991. Home: 115 Brock Cr, Pointe Claire, PQ Canada H9R 3B9*

MITCHELL, DONALD HEARIN, computer scientist; b. Bandung, Java, Indonesia, Mar. 13, 1959; came to U.S., 1959; s. Donald William and Kathryn Ann (Allen) M.; m. Lois Elaine Wright, Aug. 2, 1980; children: Laurel Ann, Donald Edward. BS in Psychology, Wheaton Coll., 1980; MS in Cognitive Psychology, Northwestern U., 1984, PhD in Cognitive Psychology, 1987. Pres. Mitchell Software Cons., Palatine, Ill., 1984-86; mem. tech. staff Allied Signal Aerospace, Columbia, Md., 1986-88; rsch. scientist Amoco Prodn. Co., Tulsa, 1988-93; pres. Proactive Solutions, Tulsa, 1993—; adj. prof. U. Tulsa, 1989-90, Okla. State U., Tulsa, 1990-91; mem. program com. Conf. on Artificial Intelligence in Petrol Exploration and Prodn., 1989—, Okla. Symposium in Artificial Intelligence, 1989—. Pres. Columbia Concert Band, 1987-88; coun. mem. Fellowship Luth. Ch., Tulsa, 1990—. Mem. Am. Assn. for Artificial Intelligence, Cognitive Sci. Soc., Judgement and Decision Making Soc. Democrat. Home: 10814 S Quebec Tulsa OK 74137

MITCHELL, EUGENE ALEXANDER, safety consultant; b. Chgo., Mar. 18, 1953; s. Frank and Delores (Langert) M.; m. Raulla Sue Sadoff, June 30, 1991. BS, U. Minn., 1977; JD, William Mitchell Coll. Law, 1984. Bar: Minn. 1984, Alaska 1985. Cons. Home Ins. Co., Mpls., 1977-84, Rollins Burdick Hunter, Anchorage, 1984-85, Marsh & McLennan, Mpls., 1986—; lectr. on workers' compensation, product safety, occupational safety. Mem. Alaska Bar Assn., Am. Soc. Safety Engrs. (profl., cert., chair regulatory affairs 1988—). Achievements include development of corporate safety programs. Office: Marsh & McLennan 90 S 7th St Ste 3600 Minneapolis MN 55402

MITCHELL, JAMES KENNETH, civil engineer, educator; b. Manchester, N.H., Apr. 19, 1930; s. Richard N. and Henrietta (Moench) M.; m. Virginia D. Williams, Nov. 24, 1951; children: Richard A., Laura K., James W., Donald M., David L. B.C.E., Rensselaer Poly. Inst., 1951; M.S., M.I.T., 1953; D.Sc., 1956. Mem. faculty U. Calif., Berkeley, 1958-93, prof. civil engring., 1968-89, chmn. dept., 1979-84, Edward G. and John R. Cahill prof. civil engring., 1989-92, Edward G. and John R. Cahill prof. civil engring. emeritus, 1993—; Berkeley citation, 1993; geotech. cons., 1960—. Author: Fundamentals of Soil Behavior, 1976, 2d edit., 1993; contbr. articles to profl. jours. Asst. scoutmaster Boy Scouts Am., 1975-82; mem. Moraga (Calif.) Environ. Rev. Com., 1978-80. Served to 1st lt. AUS, 1956-58. Recipient Exceptional Sci. Achievement medal NASA, 1973, Berkeley Citation, 1993. Fellow ASCE (hon., Huber prize 1965, Middlebrooks award 1962, 70, 73, Norman medal 1972, Terzaghi lectr. 1984, Terzaghi award 1985, pres. San Francisco sect. 1986-87); mem. Nat. Acad. Engring., Am. Soc. Engring. Edn. (We. Electric Fund award 1979), Geotech. bd. of NRC (chmn. 1990—), Transp. Rsch. Bd. (exec. com. 1983-87), Internat. Soc. Soil Mechanics and Found. Engring. (v.p. N.Am. 1989—), Earthquake Engring. Rsch. Inst., Brit. Geotech. Soc. (Rankine lectr. 1991), Sigma Xi, Tau Beta Pi, Chi Epsilon. Office: U Calif Dept Civil Engring Berkeley CA 94720

MITCHELL, JANET BREW, health services researcher; b. N.Y.C., Oct. 20, 1949; d. Robert Moscrip Mitchell and Dorothy Brennan; m. Jerry Lee Cromwell, June 15, 1980; children: Alexander, Genevieve. BA with highest honors, U. Calif., San Diego, 1971; MSW, UCLA, 1973; PhD, Brandeis U., 1976. Rsch. asst. Brandeis U./Worcester Tng. Program in Social Rsch. & Psych., Waltham, Mass., 1973-75; sr. analyst Abt Assocs., Cambridge, Mass., 1975-77; asst. prof. Boston U. Sch. Medicine, 1977-80; pres. Ctr. for Health Econs. Rsch., Waltham, Mass., 1980—; mem. com. on monitoring access to health care svcs. Inst. Medicine, 1989-92; mem. nat. adv. com. Robert Wood Johnson Health Care Fin. Fellows, 1988-93; cons. VA, 1982-85, NIH, 1983-85, Health Care Financing Adminstrn., 1979—; advisor Physician Reimbursement Study, Congl. Budget Office, 1984-85; mem. adv. panel on physicians & med. tech. Office of Tech. Assessment, 1984-85; mem. health care tech. study sect. Nat. Ctr. for Health Svcs. Rsch., 1984-88; psychiat. social worker UCLA Med. Ctr., 1971-72; med. social worker U. So. Calif., 1972-73, Univ. Hosp. San Diego, 1973. Author (with F.A. Sloan & J. Cromwell) Private Physicians and Public Programs, 1978; contbr. chpts. to 8 books; contbr. numerous articles to profl. jours. Thesis grantee VA, 1976-77. Office: Ctr for Hlth Econ Rsch 300 5th Ave Waltham MA 02154

MITCHELL, JERE HOLLOWAY, physiologist, researcher, medical educator; b. Longview, Tex., Oct. 17, 1928; s. William Holloway and Dorothea (Turner) M.; m. Pamela Battey, Oct. 1, 1960; children: Wendy O'Sullivan, Laurie Clemens M., Amy Dewing M. BS with honors, Va. Mil. Inst., 1950; MD, Southwestern Med. Sch., 1954. Intern Parkland Meml. Hosp., Dallas, 1954-55, resident in internal medicine, 1955-56; asst. prof. medicine and physiology U. Tex. Southwestern Med. Ctr., Dallas, 1962-66, dir. Weinberger Lab. for Cardiopulmonary Research, 1966—, assoc. prof., 1966-69, prof., 1969—, dir. Harry S. Moss Heart Ctr., 1976—, holder Frank M. Ryburn Jr. chair in heart research, 1982—; holder Carolyn P. and S. Roger Horchow Chair in Cardiac Rsch., 1989—; Pfizer vis. profl. Pa. State U., 1990; Percy Russo lectr. U. Sydney, Cumberland Coll., 1991. Established Investigator, Am. Heart Assn., 1962-67. Recipient Career Devel. award USPHS, 1968-73; Donald W. Seldin Research award U. Tex. Southwestern, 1978; recipient Carl J. Wiggers award Am. Physiology Soc. 1992. Attending physician Parkland Meml. Hosp., 1963—, St. Paul Med. Ctr., 1966—, VA Med. Ctr. Dallas, 1969—. Mem. Internat. Union Physiol. Soc. (commn. on cardiovascular physiology 1977—), Applied Physiol. Orthopedics Study Sect., NIH, 1979-81; Respirat. Appl. Physiol. Study Sect., NIH, 1981-82; Council of Cardiac Rehab. of Internat. Soc. & Fed. Cardiol., 1981—; Sci. Adv. Bd., USAF, 1986-90; mem. cardiology adv. com. NHLBI, 1988-92, rsch. review com. ANHLBI, 1992—; Med. Sci. Com. AAAS, 1988—. Mem. editorial bd. Am. Jour. Physiology, 1972-76, Circulation, 1978-81, 93—, Am. Jour. Cardiology, 1965-74, 82-84, Cardiovascular Research, 1979-87, Jour. Cardiopulmonary Rehab., 1981—, Clin. Physiology, 1981—, Experimental Physiology, 1993—, Jour. Applied Physiology, 1978-82, 84-89, assoc. editor, 1990-93. Fellow Am. Coll. Cardiology (Young Investigator award 1961), Am. Coll. Sports Medicine (Citation award 1983, Honor award 1988, Joseph B. Wolffe lectr. 1989); mem. Am. Heart Assn. (Award of Merit 1984, pres. Dallas div. 1977-78, pres. Tex. affiliate 1983-84, nat. v.p. 1990-91), Am. Fedn. Clin. Research (emeritus), Am. Soc. Clin. Investigation (emeritus), Assn. Am. Physicians, Am. Physiol. Soc. (cardiovascular sect.), Assn. Univ. Cardiologists, Alpha Omega Alpha. Office: U Tex Southwestern Med Ctr Harry S Moss Heart Ctr 5323 Harry Hines Blvd Dallas TX 75235-9034

MITCHELL, JERRY CALVIN, environmental company executive; b. Shubuta, Miss., Feb. 27, 1938; s. Joe Calvin and Elizabeth (Hudson) M.; s. Thelma Baker Becker, 1957; children: Jerry Jr., Charles Joseph, Stephen Thomas. BS in Chem. Engring., Miss. State U., 1960. Chem. engr. U.S. Gypsum Co., Greenville, Miss., 1960-61, quality engr., 1963-67; quality supt. U.S. Gypsum Co., Danville, Va., 1967-71; plant mgr. waste water treatment, City of Greenville, 1971-77; v.p. McCullough Environ., Murfreesboro, Tenn., 1977-83, pres., 1983-91; pres. Mitchell Tech. Svcs., Murfreesboro, 1991—; bd. dirs. AES, Inc. Charter organizer Sertoma Club, Greenville, 1972; pres. Rutherford County Assn. for Retarded Citizens, Murfreesboro, 1985-86; mem. Rutherford County Spl. Olmpics, Murfreesboro, 1986. Served to 2d lt. U.S. Army, 1961-62, 1st lt. Miss. Army Nat. Guard, 1962-67. Mem. Water Environment Fedn., Miss. Water and Wastewater Operators Assn., Miss. Water Environment Assn. (pres., v.p. 1974-76), Okla. Water and Pollution Control, Ky.-Tenn. Water Pollution Control Assn., Tenn. Water and Wastewater Operators Assn. Methodist. Avocations: fishing, reading, travel, woodworking. Office: Mitchell Tech Svcs Inc 325 W Mcknight Dr Murfreesboro TN 37129-2450

MITCHELL, JOAN LAVERNE, research scientist; b. Palo Alto, Calif., May 24, 1947; d. William Richardson and Doris LaVerne (Roddan) M. BS in Physics, Stanford U., 1969; MS in Physics, U. Ill., 1971, PhD in Physics, 1974. Rsch. staff mem. T.J. Watson Rsch. Ctr. IBM, Yorktown Heights, N.Y., 1974-88, mgr. T.J. Watson Rsch. Ctr., 1979-88; image tech. cons. mktg. IBM, White Plains, N.Y., 1989-91; rsch. staff mem. T.J. Watson Rsch. Ctr. IBM, Hawthorne, N.Y., 1991—, mgr. T.J. Watson Rsch. Ctr., 1992—; sole proprietor JLM's Bookcase, Ossining, N.Y.; del. CCITT Study Group XIV, 1978-79, ISO JPEG Com., 1987--. Co-author: JPEG Still Image Data Compression Standard, 1993; contbr. articles to profl. jours. Xerox Indsl. fellow, 1970-71. Mem. IEEE, Am. Info. and Image Mgmt., Am. Phys. Soc., Sigma Xi (sec. 1976, v.p. 1977, pres. 1978). Democrat. Achievements include co-inventor on numerous patents. Home: 7 Cherry Hill Cir Ossining NY 10562 Office: IBM 30 Saw Mill River Rd Hawthorne NY 10532

MITCHELL, JOHN NOYES, JR., electrical engineer; b. Pownal, Maine, Dec. 16, 1930; s. John Noyes and Frances (Small) M.; m. Marilyn Jean Michaelis, Sept. 1, 1956; children: Brian John, Cynthia Lynn Mitchell Tumbleson, Stephanie Lee Mitchell Judson. BSEE, Milw. Sch. Engring., 1957. Registered profl. engr., Ohio. Elec. rsch. engr. Nat. Cash Register Co., Dayton, Ohio, 1957-65; sr. engr. Xerox Corp., Rochester, N.Y., 1965-70, area mgr., 1970-73; area mgr. Xerox Corp., Dallas, 1973-76; area mgr. Xerox Corp., El Segundo, Calif., 1976-79, tech. program mgr., 1979-85, competitive benchmarking mgr., 1985—. With USN, 1949-53. Mem. IEEE, Mason. Republican. Episcopalian. Home: 11300 Providencia St Cypress CA 90630-5351 Office: Xerox Corp 701 S Aviation Blvd El Segundo CA 90245-4898

MITCHELL, JOSEPH PATRICK, architect; b. Bellingham, Wash., Sept. 29, 1939; s. Joseph Henry and Jessie Delila (Smith) M.; student Western Wash. State Coll., 1957-59; BA, U. Wash., 1963, BArch, 1965; m. Marilyn Ruth Jorgenson, June 23, 1962; children: Amy Evangeline, Kirk Patrick, Scott Henry. Asso. designer, draftsman, project architect Beckwith Spangler Davis, Bellevue, Wash., 1965-70; prin. J. Patrick Mitchell, AIA & Assoc./Architects/Planners/Cons., Kirkland, Wash., 1970—. Chmn. long range planning com. Lake Retreat Camp, 1965—; bldg. chmn. Northshore Baptist Ch., 1980—, elder, 1984-90; mem. bd. extension and central com. Columbia Baptist Conf., 1977-83; Northshore Bapt. Ch. del. World Bapt. Alliance Congress, Soul Korea, 1990. Recipient Internat. Architectural Design award St. John Vianney Parish, 1989. Cert. Nat. Council Archtl. Registration Bds. Mem. AIA, Constrn. Specification Inst., Interfaith Forum Religion, Art, and Architecture, Nat. Fedn. Ind. Bus., Christian Camping Internat., Wash. Farm Forestry Assn., Rep. Senatorial Inner Circle, Woodinville C. of C., Kirkland C. of C. Republican. Office: 12620 120th Ave NE Ste 208 Kirkland WA 98034-7511

MITCHELL, MAURICE MCCLELLAN, JR., chemist; b. Lansdowne, Pa., Nov. 27, 1929; s. Maurice McClellan and Agnes Stewart (Kerr) M.; m. Marilyn M. Badger, June 14, 1952. BS in Chemistry, Carnegie-Mellon U., 1951, MS in Chemistry, 1957, PhD in Phys. Chemistry, 1960. Group leader rsch. and devel. U.S. Steel Corp., Pitts., 1951-61; br. head phys. chemistry rsch. and devel. Melpar Inc., Falls Church, Va., 1961-64; group leader rsch. and devel. Atlantic Richfield Co., Phila., 1964-73; dir. rsch. and devel. Houdry div. Air Products and Chems., Inc., 1973-81; dir. rsch. and devel. Ashland (Ky.) Oil Inc., 1981-86, v.p. rsch. and devel., 1986-93; vis. lectr. dept. chem. Coll. Arts and Scis. Ohio U. Southern Campus, Ironton, 1993—. Contbr. articles to profl. jours.; patentee in field. Fellow Am. Inst. Chemists; mem. Am. Chem. Soc., Am. Inst. Chem. Engrs., Catalysis Soc. N.Am. (pres. 1985-89), AAAS, Sigma Xi. Republican. Presbyterian. Lodge: Kiwanis. Home: 2380 Hickory Ridge Dr Ashland KY 41101-3604 Office: Ohio U So Campus 1804 Liberty Ave Ironton OH 45638-2296

MITCHELL, NEIL CHARLES, geophysicist; b. Wokingham, Eng., Feb. 3, 1964. BA, Cambridge (Eng.) U., 1986, MA, 1990; DPhil, Oxford (Eng.) U., 1989. Adj. assoc. Columbia U., N.Y.C., 1990-91; sr. geophysicist Wimpol Ltd., Eng., 1992; rsch. fellow dept. geology Durham U., Eng., 1993—. Contbr. articles to profl. jours. Royal Commn. for Exhbn. of 1851 rsch. fellow, 1990-91. Fellow Geol. Soc. London; mem. Am. Geophys. Union, Acoustical Soc. Am. Achievements include research into application of underwater acoustics to sideseam sonar imagery to obtain quantitative information about the nature of the seabed.

MITCHELL, PAULA LEVIN, biology educator, editor; b. N.Y.C., Nov. 2, 1951; d. Louis X. and Jane (Schanfeld) Levin; m. Forrest Lee Mitchell, July 28, 1979; children: Robert, Evelyn. BA in Biology, U. Pa., 1973; PhD in Zoology, U. Tex., 1980. Rsch. assoc. dept. entomology La. State U., Baton Rouge, 1981-84; vis. prof. dept. biol. scis Tarleton State U., Stephenville, Tex., 1984-93; asst. prof. dept biology Winthrop U., Rock Hill, S.C., 1993—; adj. asst. prof. dept. biology Tex. Christian U., Fort Worth, 1985-88; editor Entomol. Soc. Am., Lanham, Md., 1986-93. Contbr. articles to profl. jours. Branch sec. AAUW, Stephenville, 1986-87. U. fellow U. Tex., Austin, 1973-76. Mem. Entomol. Soc. Am., Ga. Entomol. Soc., Southwestern Entomol.

Soc., Sigma Xi. Office: Winthrop U Box T-219 101 Sims Bldg Rock Hill SC 29733

MITCHELL, PAULA RAE, nursing educator; b. Independence, Mo., Jan. 10, 1951; d. Millard Henry and E. Lorene (Denton) Gates; m. Ralph William Mitchell, May 24, 1975. BS in Nursing, Graceland Coll., 1973; MS in Nursing, U. Tex., 1976; postgrad. N.Mex. State U. RN, Tex., Mo.; cert. childbirth educator. Commd. capt. U.S. Army, 1972; ob-gyn. nurse practitioner U.S. Army, Seoul, Korea, 1977-78; resigned, 1978; instr. nursing El Paso (Tex.) Community Coll., 1979-85, dir. nursing, 1985—, acting div. chmn. health occupations, 1985-86, div. chmn., 1986—, curriculum facilitator, 1984-86; ob-gyn. nurse practitioner Planned Parenthood, El Paso, 1981-86, mem. med. com., 1986-92. Author: (with Grippando) Nursing Perspectives and Issues, 1989, 93; contbr. articles to profl. jours. Founder, bd. dirs. Health-C.R.E.S.T., El Paso, 1981-85; mem. pub. edn. com. Am. Cancer Soc., El Paso, 1983-84, mem. profl. activities com., 1992—; mem. El Paso City-County Bd. Health, 1989-91; mem. Govt. Applications Rev. Com., Rio Grande Coun. Govts., 1989-91; mem. collaborative coun. El Paso Magnet High Sch. for Health Care Professions. Decorated Army Commendation medal, Meritorious Svc. medal. Mem. Nat. League Nursing (mem. resolutions com. Assocs. Degree coun. 1987-89, accreditation site visitor, AD coun. 1990—, mem. Tex. edn. com. 1991-92, Tex. 3rd v.p. 1992-93), Am. Soc. Psychoprophylaxis Obstetrics, Nurses Assn. Am. Coll. Obstetricians & Gynecologists (cert. in ambulatory women's health care; chpt. coord. 1979-83, nat. program rev. com. 1984-86, corr. 1987-89), Advanced Nurse Practitioner Group El Paso (coord. 1980-83 legis. committee 1984), Am. Phys. Therapist Assn. (commn. on accreditation, site visitor for phys. therapist assistant programs 1991—), Orgn. Assoc. Degree Nursing (Tex. membership chmn. 1985-89, chmn. long range goals com. 1989—, mem nat. bylaws com., 1990—), Am. Vocat. Assn., Am. Assn. Women Community & Jr. Colls., Tex. Orgn. Nurse Execs., Nat. Coun. Occupational Edn. (mem. articulation task force 1986-89, program standards task force 1991-93), Nat. Coun. Instructional Adminstrs., Sigma Theta Tau, Phi Kappa Phi. Mem. Christian Ch. (Disciples of Christ). Home: 4616 Cupid Dr El Paso TX 79924-1726 Office: El Paso C C PO Box 20500 El Paso TX 79998-0500

MITCHELL, PHILIP MICHAEL, aerospace engineer, air force officer; b. Mobile, Ala., Feb. 12, 1953; s. Philip Augustus and Betty J. (Hardy) M. BS in Aeros. magna cum laude, Embry-Riddle Aero. U., Daytona Beach, Fla., 1980, MS in Aeros., 1987. Radar systems engr. ITT, Van Nuys, Calif., 1980-82; commd. 2d lt. USAF, 1982, advanced through grades to maj., 1994; bomber br. chief 42d Orgnl. Maintenance Squadron, Loring AFB, Maine, 1983-86; officer-in-charge weapons br. 520th Aircraft Generation Squadron, RAF Upper Heyford, Eng., 1986; asst. maintenance supr. 20th Equipment Maintenance Squadron, RAF, RAF Upper Heyford, 1986-87, 88-90; weapons safety officer 20th Tactical Fighter Wing, RAF Upper Heyford, 1986-87; chief standardization and tng. div. 42d Bomb Wing, Loring AFB, 1990-91; chief of maintenance 42d Maintenance Squadron, Loring AFB, Maine, 1991-92; maintenance mgmt. officer 42d BMW, 1992—; adj. prof. European div. Embry-Riddle Aero. U., 1988-90; aerospace cons., 1987—. Recipient Meritorious Svc. medal, Commendation medal, Air Force Achievement medal. Fellow Brit. Interplanetary Soc.; mem. AIAA (sr.), Soc. Logistics Engrs., Air Force Assn., Royal Scottish County Dance Soc. Anglican. Avocations: flying, skiing, Scottish country dancing. Home: 304 Sweden St Caribou ME 04736 Office: 42d LSS/LSO PO Box 505 Loring AFB ME 04751

MITCHELL, ROBERT CURTIS, physicist, educator; b. Ft. Dodge, Iowa, Mar. 29, 1928; s. Curtis Bradshaw and Mabel Cecilia (Higgins) M.; m. Mary Jo Bennett, Aug. 30, 1949; children: Drake Curtis, John Douglas, Mari Cecilia. BS, N.Mex. State U., 1949; MS, U. Wash., 1952; PhD, N.Mex. State U., 1966. Researcher Anderson Labs., West Hartford, Conn., 1952-53; rsch. asst. U. Conn., Sunspot, N.Mex., 1953; solar observer Harvard Coll. Observatory, Sunspot, 1953-54; tchr. Colo. Rocky Mt. Sch., Carbondale, 1954-56, Gadsden High Sch., Anthony, N.Mex., 1956-62; assoc. prof. Cen. Wash. U., Ellensburg, 1966-71, prof. physics, 1971-93; dept. chair N000, 1990-93. Photographer: (slide set of stars) The Night Sky, 1981. Mem. AAAS, Am. Assn. Physics Tchrs. (chair com. on astronomy edn. 1990-92), Astron. Soc. of Pacific, Am. Assn. Variable Star Observers. Democrat. Presbyterian. Home: Rt 5 Box 880 Ellensburg WA 98926

MITCHELL, ROGER LOWRY, agronomy educator; b. Grinnell, Iowa, Sept. 13, 1932; s. Robert T. and and Cecile (Lowry) M.; m. Joyce Elaine Lindgren, June 26, 1955; children: Laura, Susan, Sarah, Martha. B.S. in Agronomy, Iowa State Coll., 1954; M.S., Cornell U., 1958; Ph.D. in Crop Physiology, Iowa State U., 1961. Mem. faculty Iowa State U., 1959-69, prof. agronomy, 1966-69, prof. charge farm operation curriculum, 1962-66; prof. agronomy, chmn. dept. U. Mo., Columbia, 1969-72, 81-83; dean agr., dir. expt. sta. U. Mo., 1983—, dean extension, 1972-75; v.p. agr. Kans. State U., Manhattan, 1975-80; exec. dir. Mid-Am. Internat. Agrl. Consortium, 1981; exec. bd. divsn. agr. Nat. Assn. State Univs. and Land Grant Colls., 1978-80, 85-90, chmn., 1988-89; mem. bd. agr. NRC/NAS, 1983-86. Author: Crop Growth and Culture, 1970; co-author: Physiology of Crop Plants, 1985. Served to 2d lt. USAAF, 1954-56. Danforth fellow, 1956-61; Acad. Adminstrn. fellow Am. Council Edn., 1966-67. Fellow AAAS (chmn. sect. O 1980-81), Am. Soc. Agronomy (pres. 1979-80), Crop Sci. Soc. (pres. 197); mem. Soil Sci. Soc. Am., Coun. Agrl. Sci. and Tech., Sigma Xi, Gamma Sigma Delta, Alpha Zeta, Phi Kappa Phi. Home: 502 W Lathrop Rd Columbia MO 65203-2804

MITCHELL, ROY DEVOY, industrial engineer; b. Hot Springs, Ark., Sept. 11, 1922; s. Watson W. and Marie (Stewart) M.; m. Jane Caroline Gibson, Feb. 14, 1958; children: Michael, Marilyn, Martha, Stewart, Nancy. BS, Okla. State U., 1948, MS, 1950; B of Indsl. Mgmt., Auburn U., 1960. Registered profl. engr., Ala., Miss. Instr. Odessa (Tex.) Coll., 1953-56; prof. engring. graphics Auburn (Ala.) U., 1956-63; field engr. HHFA, Community Facilities Adminstrn., Atlanta and Jackson, Miss., 1963-71; area engr. Met. Devel. Office, HUD, 1971-72, chief architecture and engring., 1972-75, chief program planning and support br., 1975, dir. archtl. br., Jackson, 1975-77, chief archtl. br. and engring. br., 1977-84, community planning and devel. rep., 1984-88; prin. Mitchell Mgmt. and Engring., 1988—; cons. Army Balistic Missile Agy., Huntsville, Ala., 1957-58, Auburn Research Found., NASA, 1963; mem. state tech. action panel Coop. Area Manpower Planning System; elected pub. ofcl., chmn. Bd. of Election Commrs., Rankin County, Miss. Mem. Cen. Miss. Fed. Personnel Adv. Council; mem. House and Home mag. adv. panel, 1977; trustee, bd. dirs. Meth. Ch., 1959-60. Served with USNR, 1943-46. Recipient Outstanding Achievement award HUD, Commendation by Sec. HUD. Mem. NSPE, Am. Soc. for Engring. Edn., Miss. Soc. Profl. Engrs., Nat. Assn. Govt. Engrs. (charter mem.), Jackson Fed. Execs. Assn., Cen. Miss. Safety Council, Am. Water Works Assn., Iota Lambda Sigma. Club: River Hills (Jackson). Home and Office: HUD 706 Forest Point Dr Brandon MS 39042-6220

MITCHELL, SANDRA LOUISE, biology educator, researcher; b. Clarksdale, Miss., Aug. 27, 1957; d. Melville E. and Gloria Ann (Woods) M. BA in Religion, Polit. Sci., U. of South, 1978; MS in Biology, Miss. State U., 1983; PhD in Biology, U. N.Mex., 1988. Biologist U. No. Colo., Greeley, 1988-89; biologist, instr. Western Wyo. Coll., Rock Springs, 1989-92, biologist, asst. prof., 1992—; researcher Fla. State U., Tallahassee, 1985-86, N.Mex. Game and Fish, Albuquerque, 1985-88, Nat. Fish and Wildlife Svc., Ft. Collins, Colo., 1988, Nat. Pk. Svc., Rock Springs, 1989-92. Author: (chpt.) Community Ecologies of Desert Amphibians, 1991; contbr. articles to profl. jours. 4-H leader, Rock Springs, 1991—. Mem. Herpetoloyist's League, Am. Soc. Naturalists, Animal Behavior Soc., Soc. for Study of Evolution, Behavioral Ecology Soc., Soc. for Study of Amphibians and Reptiles, Am. Soc. of Icthyologists and Herpetologists, Phi Beta Kappa. Office: Western Wyo Coll 2500 College Dr Rock Springs WY 82901

MITCHELL, WILLIAM AVERY, JR., orthodontist; b. Greenville, S.C., Apr. 26, 1933; s. William Avery and Eva (Rigdon) M.; m. Patricia Ann Scott, June 26, 1965; 1 child, William Avery III. BS, Furman U., 1955; DDS, Emory U., 1959; MS in Dentistry, 1967. Diplomate Am. Bd. Orthodontics. Pvt. practice Decatur, Ga., 1963-67, Greenville, S.C., 1967—; instr. Dental Sch., Emory U., 1965-66, Dept. Orthodontics, Atlanta, 1966-68; guest lectr. Greenville (S.C.) Tech. Coll., 1979—; councilor Coll. Diplomates of Am. Bd. Orthodontics, 1988-91, treas., 1991-92, sec., 1992-93, pres. elect, 1993-94. Bd. dirs. United Speech and Hearing, Greenville, 1972-75; chmn. dental divsn. United Way, Greenville, 1983, 84; fund raiser Roper Mt. Sci. Ctr., 1985; mem. adv. bd. Greenville Tech. Coll., 1985-88; pres. Booster Club, Christ Ch. Episcopal Sch., 1987-88, mem. bd. visitors, 1990—; deacon 1st Bapt. Ch. Capt. U.S. Army, 1959-61. Recipient Emil Eisenberg Scholarship, Ga. Dental Assn., Atlanta, 1958-59. Fellow Am. Coll. Dentists, Internat. Coll. Dentists; mem. ADA, Am. Assn. Orthodontists (del. 1985-91, 93—), S.C. Dental Assn., So. Assn. Orthodontists (pres.-elect 1992-93, trustee 1983, 84, 85, sr. dir. 1989-92), S.C. Orthodontic Assn. (pres. 1978), Piedmont Dist. Dental Soc. (bd. dirs. 1972, Greenville County Dental Soc. (pres. 1972), S.C. Acad. Dental Practice Adminstrn., Emory Orthodontoc Assn. (pres. 1972), Furman Paladin Club (bd. dirs. 1980-84), Commerce Club (bd. dirs. Greenville chpt. 1983—), Rotary, Greenville Country Club. Avocation: tennis. Office: 10 Cleveland Ct Greenville SC 29607-2414

MITCHELL, WILLIAM COBBEY, physicist; b. Rochester, Minn., Aug. 2, 1939; s. William Alexander and Ruth Chilla (Cobbey) M.; m. Carole Ann Hudson, June 16, 1962; children: William, James, Karen, Matthew. AB, Oberlin Coll., 1961; PhD, Washington U., St. Louis, 1967. Sr. rsch. physicist 3M, St. Paul, 1967-69; NBS-NRC fellow Nat. Bur. Standards, Gaithersberg, Md., 1969-71; cons. Thermoelectric Systems, 3M, St. Paul, 1971-72; rsch. physicist 3M, St. Paul, 1972-76, supr., mgr. Pioneering Lab., 1976-83, lab. mgr. Info. Storage Lab., 1983-88, staff scientist basic materials, 1988—. Nat. Merit scholar, 1957-61; NSF postdoctoral fellow, 1962-67. Mem. Am. Phys. Soc. Achievements include patents and/or publications in non equilibrium statistical mechanics, thermoelectric materials, electrochemistry, information storage. Home: 3568 Siems Ct Saint Paul MN 55112 Office: 3M Bldg 201-1C-18 3M Ctr Saint Paul MN 55144-1000

MITRA, SANJIT KUMAR, electrical and computer engineering educator; b. Calcutta, West Bengal, India, Nov. 26, 1935; came to U.S., 1958; MS in Tech., U. Calcutta, 1956; MS, U. Calif., Berkeley, 1960, PhD, 1962; D of Tech. (hon.), Tampere (Finland) U., 1987. Asst. engr. Indian Statis. Inst., Calcutta, 1956-58; from teaching asst. to assoc. Univ. Calif., Berkeley, 1958-62; asst. prof. Cornell U., Ithaca, N.Y., 1962-65; mem. tech. staff Bell Telephone Labs., Holmdel, N.J., 1965-67; prof. U. Calif., Davis, 1967-77; prof. elec. and computer engring. U. Calif., Santa Barbara, 1977—, chmn. dept. elec. and computer engring., 1979-82; cons. Lawrence Livermore (Calif.) Nat. Lab., 1974—; cons. editor Van Nostrand Reinhold Co., N.Y.C., 1977-88; mem. adv. bd. Coll. Engring. Rice U., Houston, 1986-89. Author: Analysis and Synthesis of Linear Active Networks, 1969, Digital and Analog Integrated Circuits, 1980; co-editor: Modern Filter Theory and Design, 1973, Two-Dimensional Digital Signal Processing, 1978, Miniaturized and Integrated Filters, 1989, Handbook for Digital Signal Processing, 1993. Named Disting. Fulbright Prof., Coun. for Internat. Exch. of Scholars, 1984, 86, 88, Disting. Sr. Scientist, Humboldt Found., 1989. Fellow AAAS, IEEE (Edn. award Circuits and Systems Soc. 1988), Internat. Soc. Optical Engring.; mem. Am. Soc. for Engring. Edn. (F. E. Terman award 1973, AT&T Found. award 1985), European Assn. for Signal Processing. Achievements include patents for two-port networks for realizing transfer functions; nonreciprocal wave translating device. Office: Univ Calif Dept Elec & Computer Engring Santa Barbara CA 93106

MITRY, MARK, chemist; b. Cairo, Aug. 1, 1945; came to U.S. 1971; s. Twefik and Widad (Soror) M.; m. Eva Futar, Sept. 28, 1973; children: Diana, Daniel. BS in Chemistry and Geology, Ain Shams U., Cairo, 1970; MSin Polymer Sci., U. Lowell, 1979; MBA, Fairleigh Dickinson U., 1986. Geologist Sonatrch, Algeria, 1970-71; chemist Dennison, Framingham, Mass., 1972-81; devel. assoc. Nat. Starch and Chem., Bridgewater, N.J., 1981--. Mem. Am. Shem. Soc., Am. Soc. for Testing and Materials (chmn. subcom. 1991--), Bio/Environ. Degradable Polymer Soc. Achievements include development of cyanoacrylate adhesives composition and ultraviolet curing adhesives, U.S. patent in laminating adhesives containing polymerizable surfactant, European patents pending. Home: 19 Fairview Dr Flemington NJ 08822

MITSUKE, KOICHIRO, chemistry educator; b. Tokuyama, Yamaguchi, Japan, Feb. 2, 1959; s. Tatsuo and Atsuko (Oyane) M. BS, U. Tokyo, Japan, 1981, MS, 1983, DSc, 1986. Rsch. assoc. U. Tokyo, Japan, 1986-91; assoc. prof. Inst. for Molecular Sci., Okazaki, Japan, 1991—. Mem. Japan Chem. Soc., Japan Phys. Soc., Am. Phys. Soc. Home: Tatsumi-Minami 2-5-1-1-42, Okazaki Aichi 444, Japan Office: Inst for Molecular Sci, Myodaiji, Okazaki Aichi 444, Japan

MITTAL, MANOJ, aerospace engineer; b. Lucknow, India, Nov. 21, 1964; came to U.S., 1986; s. Jagdish Kumar and Annapurna (Agrawal) M.; m. Swati Jain, July 15, 1991. BTech., Indian Inst. Tech., Kanpur, 1986; MS, Ga. Inst. Tech., 1989, PhD, 1991. Teaching asst. U. Minn., Mpls., 1986-87; rsch. asst. Ga. Inst. Tech., Atlanta, 1987-91, postdoctoral fellow, 1992—. Contbr. articles to profl. jours., chpts. to books. Vertical Flight Found. scholar, 1990; recipient Doctoral Dissertation award Sigma Xi, 1992. Mem. AIAA, AHS. Achievements include conducting first comprehensive modeling and control study of the twin-lift helicopter system.

MITTEL, JOHN J., economist, corporate executive; b. L.I., N.Y.; s. John and Mary (Leidolf) M.; 1 child, James C. B.B.A., CUNY. Researcher econs. dept. McGraw Hill & Co., N.Y.C.; mgr., asst. to pres. Indsl. Commodity Corp., J. Carvel Lange Inc. and J. Carvel Lange Internat., Inc., 1956-64, corp. sec., 1958-86, v.p., 1964-80, exec. v.p., 1980-86; pres. I.C. Investors Corp., 1972—, I.C. Pension Adv., Inc., 1977—; bd. dir. several corps.; plan adminstr., trustee Combined Indsl. Commodity Corp. and J. Carvel Lange Inc. Pension Plan, 1962-86, J. Carvel Lange Internat. Inc. Profit Sharing Trust, 1969-86, Combined Indsl. Commodity Corp. and J. Carvel Lange Inc. Employees Profit Sharing Plan, 1977-86. Mem. grad. adv. bd. Bernard M. Baruch Coll., CUNY, 1971-72. Mem. Conf. Bd., Am. Statis. Assn., Newcomen Soc. N.Am. Club: Union League (N.Y.C.). Co-author: How Good A Sales Profit Are You, 1961, The Role of the Economic Consulting Firm. Office: 10633 St Andrews Rd Boynton Beach FL 33436-4714

MITTER, SANJOY K., electrical engineering educator; b. Calcutta, India, Dec. 9, 1933. PhD, Imperial Coll., Eng., 1965. Prof. elec. engring. MIT, Cambridge, 1973—, dir. Lab. Info. and Decision Systems, 1981-86, co-dir Lab. Info. and Decision Systems, 1986—, dir. Ctr. Intelligent Control Systems, 1986—. Fellow IEEE; mem. Nat Acad. Eng. Office: MIT Lab Info & Decision Systems Bldg 35 Rm 308 Cambridge MA 02139-4307

MITTER, WERNER SEPP, physicist, researcher, educator; b. Leoben, Styria, Austria, Mar. 20, 1941; s. Josef and Anna (Wernig) M. Engr.'s diploma, Mining U., Leoben, 1965, inauguration decrete, 1986; PhD in Physics, U. Vienna, Austria, 1970. Staff mem. R & D dept. Studying Co. for Atomic Energy, Seibersdorf, Austria, 1966-69; staff mem. R & D dept. United Spl. Steelworks, Kapfenberg, Austria, 1975-77, group leader, 1977-86, head dept., 1986-88; staff mem. R & D dept. Bohler Spl. Steelworks, Kapfenberg, 1970-75, head dept., 1988-91, rsch. and devel. cons., 1992—; lectr. Mining U., 1986—. Author: Transformation Plasticity, 1987. Recipient Hans-Malzacher award Eisenhütte Österreich, 1980. Avocations: religions, astronomy. Office: Bohler Spl Steelworks, Mariazeller Strasse 25, A-8605 Kapfenberg Styria, Austria

MITTLEMAN, MARVIN HAROLD, physicist, educator; b. N.Y.C., Mar. 13, 1928; s. Sol and Jennie (Bookstein) M.; m. Sondra Ruth Koslow, Aug. 21, 1955; children: Richard, Joshua, Daniel. BS in Physics, Poly. Inst. of Bklyn., 1949; PhD in Physics, MIT, 1953. Instr. Columbia U., N.Y.C., 1952-55; rsch. physicist Lawrence Livermore Nat. Lab., Livermore, Calif., 1955-65, Space Sci. Lab., Berkeley, Calif., 1965-68; assoc. prof. physics CCNY, N.Y.C., 1968-70, prof., 1970—; cons. Lawrence Livermore Nat. Lab., 1985—, Los Alamos (N.Mex.) Sci. Lab., 1991—. Author: Introduction to the Theory of Laser-Atom Interactions, 1985, 2d edit. 1993; contbr. articles to Phys. Rev. AEC fellow, 1952. Fellow Am. Phys. Soc. Home: 74 George St Harrington Park NJ 07640 Office: CCNY 137th St at Convent Ave New York NY 10031

MITZ, VLADIMIR, plastic surgeon; b. Vielikoie Selo, USSR, Mar. 3, 1943; came to France, 1949, naturalized, 1957; s. Hersz Jumen and Maria (Mincberg) M.; 1 child, Illitch. M.D., 1973. Intern, Lariboisiere Hosp., Paris, 1967; resident Jackson Meml. Hosp., Miami, Fla., 1975; chief clinic Assistance Publique, Paris, 1973-78, mem. staff, 1980—; prof. anatomy Faculty Medicine, Paris, 1970-74; dir. micro vascular lab. Bovcicaut Hosp., Paris, 1980—. Author: Operation Beaute, 1984, Lambeaux Musculo Cutanes, 1984, Plaies et Bosses, 1988, Les Dessous de la Peau, 1990, le Choix d'être belle, 1992; producer movie Operation Verite, 1982. Pres. Mouvement de Recherche en Plastique Humaine, Paris, 1972. Mem. French Acad. Surgery, Brazilian Soc. Plastic Surgery (assoc.), French Soc. Plastic Reconstructive Aesthetic Surgery, Assn. des Chirurgiens Esthetiques et Plastiques au l'Hosp. Boucicaut. Avocation: painting, Sculpture. Home: 12 Rue du Renard, 75004 Paris France Office: 176 Blvd St Germain, 75006 Paris France

MIURA, TANETOSHI, acoustics educator; b. Aomori, Japan, Sept. 21, 1920; s. Taneyoshi and Toku Miura; m. Teru Kobayashi, Nov. 3, 1946; children: Shigemi, Midori Iwae. BS, Tohoku Imperial U., Sendai, Japan, 1943, D. of Engring., 1957. Researcher Japanese Telegraph and Telephone Pub. Corp., Tokyo, 1945-59, dir. transmission rsch., 1959-62; mgr. acoustics rsch. Cen. Rsch. Lab., Hitachi Ltd., Tokyo, 1962-77; prof. Tokyo Denki U., 1977—; cons. Audiolog. Soc. of Japan, 1988-91, Tokyo-To, 1985-93. Lt. Japanese Navy, 1944-45. Recipient award of Gov., 1951, 57, Tokyo-To, 1973, award of Invention, Inst. Invention, Tokyo, 1973. Fellow Acoustical Soc. of Am.; mem. Acoustical Soc. of Japan (life, pres. Tokyo chpt. 1985-87), Audio Engring. Soc. (life). Avocations: golf, gardening. Home: Minami-cho 1-11-20, Kokubunji Tokyo 185, Japan Office: Tokyo Denki U, 2-2 Kanda Nishiki-cho, Chiyoda-ku Tokyo 101, Japan

MIYACHI, IWAO, electrical engineering educator; b. Kochi, Japan, Sept. 27, 1916; s. Sakaki and Asami (Taniwaki) M.; m. Kazuko Nagano, Apr. 4, 1943; children: Yukiko Tanaka, Reiko Yamashita. B in Engring., U. Tokyo, 1940, D in Engring., 1953. From lectr. to prof. Nagoya U., Aichi, 1940-80, prof. emeritus, 1980—; prof. Aichi Inst. Tech., Toyota, 1980—. Author: Power Transmission and Distribution, 1958, Electrical Power Engineering, 1965, Electrical Power Generation, 1988. Decorated 2d Order of Nat. Merit, 1990. Mem. Inst. Elec. Engrs. of Japan (pres. 1977, Power Engring. award 1968, Best Treatise award 1973, Disting. Svcs. award 1980, Advanced Tech. award 1990), Elec. Coop. Rsch. Inst. Japan (adviser 1989-93), Soc. des Electriciens et des Electroniciens, Conf. Internat. des Grands Reseaux Electriques. Avocations: driving, worldwide sightseeing. Home: 3-6-5 Nishizakicho, Chikusa, Nagoya 464, Japan Office: Aichi Inst Tech, 1247 Yachigusa, Yakusa-cho, Toyota 470-03, Japan

MIYAJIMA, HIROSHI, aerospace engineer; b. Fukumitsu, Toyama, Japan, Feb. 5, 1940; s. Magohachi and Setsu (Furuike) M.; m. Michiko Iino, Aug. 30, 1964; children: Masafumi, Norifumi. M of Engring., Tokyo Inst. of Tech., 1964. Rsch. scientist Nat. Aerospace Lab., Kakuda, Japan, 1967-74; head propellant lab. Nat. Aerospace Lab., Kakuda Rsch. Ctr., Kakuda, Japan, 1974-80, head rocket altitude performance lab. Kakuda Rsch. Ctr., 1978-87, dep. dir., 1983-92, dir., 1992—; mem. tech. com. Nat. Space Devel. Agy. of Japan, Tokyo, 1990—. Contbr. articles to Jour. Spacecraft and Rockets, Jour. Propulsion and Power. Recipient Achievement award Sci. and Tech. Agy., Japan, 1983, Disting. Svc. award, 1992, Cert. of Merit AIAA, 1983. Mem. AIAA, Japan Soc. Aero. Space Sci. Achievements include devel. of a medium thrust cryogenic rocket engine; rsch. on performance potential of low thrust cryogenic rocket engine, bldg. a high altitude rocket engine test facility, bldg. a scramjet engine test facility. Home: 18-12 Tachimachi, Sendai 980, Japan Office: Nat Aerospace Lab, Kakuda Rsch Ctr, Kakuda, 981-15 Miyagi Japan

MIYAKADO, MASAKAZU, biochemist, chemist; b. Osaka, Japan, Feb. 22, 1947; s. Mitsugu and Tamaki Miyakado; m. Yoshiko Imahori, Feb. 25, 1973; children: Masanori, Masanao, Ran. BS, Kyoto U., 1970, MS, 1972; PhD, Nagoya U., 1980. Research chemist Sumitomo Chem. Co. Ltd., Takarazuka, Japan, 1972-79; asst. research assoc. Sumitomo Chem. Co. Ltd., 1979-85, research assoc., 1985—; faculty Cornell U., Dept. Chemistry, Ithaca, N.Y., 1984-86. Author: Natural Product Chemistry, 1983, Pesticides of Plant Origin, 1989, Development of New Agrochemicals, 1992. Recipient Award of Encouragement, Pesticide Sci. Soc. Japan, 1986. Mem. Internat. Soc. Chem. Ecology, AAAS, Agrl. Chem. Soc. Japan, Pesticide Sci. Soc. Japan, Kyuzan League. Avocations: building musical instruments, woodblock print. Home: 51 Arakusacho Kamigamo Kita, Kyoto 603, Japan Office: Sumitomo Chem Co Ltd, Agrl Sci Rsch Lab, Takarazuka Hyogo 665, Japan

MIYAKE, AKIO, biologist, educator; b. Kyoto, Japan, June 29, 1931; s. Yoshikazu and Yukie (Yamazaki) M.; m. Sadako Harada, Mar. 15, 1965 (dec. June 1986); children: Akiko, Toshio; m. Terue Harumoto, Dec. 30, 1988; 1 child, Yuka. BS, Kyoto U., 1953, D of Science, 1959. Asst. Osaka (Japan) City U., 1953-63; visiting scholar Ind. U., Bloomington, 1959-61; lectr. Kyoto (Japan) U., 1963-70; group leader Max-Planck Inst. for Molecular Genetics, West Berlin, 1970-74; visiting scholar U. Pisa, Italy, 1975-77, U. Münster, West Germany, 1978-83; prof. U. Camerino, Italy, 1983—. Contbr. articles on sexual reprodn. in microorganisms to profl. jours. Recipient Zool. Soc. of Japan Prize, 1981. Mem. Zool. Soc. Japan, Genetics Soc. of Japan, AAAS, Soc. Protozoologists. Avocations: origin and evolution of life, Italian opera music. Home: Corso Italia 150, Castelraimondo MC, Italy I-62022 Office: U Camerino Dept Cell Biol, Via F Camerini 2, Camerino MC, Italy I-62032

MIYAKE, AKIRA, physics educator; b. Yamaguchi-ken, Japan, July 31, 1925; s. Etsuo and Kiyo (Hayashi) M.; m. Kazuko Shinkai, Mar. 25, 1957 (dec. 1980); children: Tadashi, Hikaru, Iwao, Nozomu. BS, U. Tokyo, 1947; DSc, Hokkaido U., Sapporo, Japan, 1960. Cert. highsch. tchr. in sci. Asst. prof. faculty of liberal arts and sci. Shizuoka (Japan) U., 1951-61; assoc. prof. Coll. of Liberal Arts, Internat. Christian U., Tokyo, 1961-63, chmn. div. natural scis., 1967-69; dir. computer ctr. Internat. Christian U., Tokyo, 1967-69, acting pres., 1969-71, v.p. for acad. affairs, 1971-73, dir. Inst. for Ednl. Rsch. and Svc., 1976-82, dir. libr., 1978-82, chmn. div. natural scis. grad. sch., 1987-91, prof., 1963—; vis. fellow dept. chemistry Dartmouth Coll., Hanover, N.H., 1963-65; vis. scholar James Franck Inst., U. Chgo., 1982-83; vis. researcher Univ. Edn. Rsch. Ctr., Hiroshima (Japan) U., 1978-82. Co-editor, author (with others): Physics of Polymers, 1963, Structures and Properties of Polymers, 1963; contbr. articles to profl. jours.; mem. editorial bd. Reports on Progress in Polymer Physics in Japan, 1986—. Councillor Inter-Univ. Seminar House, Hachioji, Tokyo, 1969-71, 74—, exec. trustee, 1982—; councillor Tokyo High Sch. Alumni Assn., 1950—, trustee, 1979-81, 89-91; councillor Juridical Person, Internat. Christian U., Mitaka, Tokyo, 1968-73, 84-87, 90—. Rsch. grantee Ministry of Edn., Sci. and Culture, Japan, 1978, Pvt. Sch. Promotion Found., Tokyo, 1980-82. Mem. Phys. Soc. Japan (trustee, v.p. 1976-78, vice chmn., chmn. publ. bd. Japanese Jour. Applied Physics 1978-80, auditor 1980-82, exec. sec. 1985-87), Soc. Polymer Sci. Japan, Soc. Rheology Japan (award com. 1987-89), Inst. for Dem. Edn., Univ. Alumni Assn. Mem. United Ch. of Christ in Japan. Avocations: travel, reading. Home: 17-2 Chitosedai 2-chome, Setagaya-ku, Tokyo 157, Japan Office: Dept Physics Internat Christian U, 10-2 Osawa 3-chome, Mitaka-shi, Tokyo 181, Japan

MIYAMOTO, MAYLENE HU, computer science educator; b. Taipei, Republic of China, July 4, 1960; d. Whou-Cheng and Red-Found (Ching) Hu; m. Ken C. Miyamoto, Dec. 30, 1990. BA, Soochow U., Taipei, 1982; MA, Ball State U., 1986; postgrad., W.Va. U., 1988-90. System cons. Systemation Inc., Columbus, Ohio, 1986-87; West Liberty (W.Va.) State Coll., 1987-91. Avocations: music, travel, skiing, reading. Home: SBN 629 Princeton Theol Sem Box 5204 Princeton NJ 08543-5204

MIYAMOTO, RICHARD TAKASHI, otolaryngologist; b. Zeeland, Mich., Feb. 2, 1944; s. Dave Norio and Haruko (Okano) M.; m. Cynthia VanderBurgh, June 17, 1967; children: Richard Christopher, Geoffrey Takashi. BS cum laude, Wheaton Coll., 1966; MD, U. Mich., 1970; MS in Otology, U.So. Calif., 1978. Diplomate Am. Bd. Otolaryngology. Intern Butterworth Hosp., Grand Rapids, Mich., 1970-71, resident in surgery, 1971-72; resident in otolaryngology Ind. U. Sch. Medicine, 1972-75; fellow in otology and neurotology St. Vincent Hosp. and Otologic Med. Group, L.A., 1977-78; asst. prof. Ind. U. Sch. Medicine, Indpls., 1978-83, assoc. prof., 1983-88; prof. 1988—;chmn. 1987, chief Otology and Neurotology dept. Otolaryngology, Head and Neck Surgery, Ind. U., 1982—, chmn. dept. Otolaryngology, 1987—, Arilla DeVault prof., 1991; chief Otolaryngology, Head and Neck Surgery Wishard Meml. Hosp., 1979—. Mem. editorial bd.

Laryngoscope, Am. Jour. of Otology, Otolaryngology-Head and Neck Surgery; contbr. articles to profl. jours. Mem. adv. coun. Nat. Inst. Deafness and other communication disorders, 1989—; mem. med. adv. bd. Alexander Graham Bell Assn. for the Deaf, The Ear Found. Served to maj. USAF, 1975-77. Named Arilla DeVault Disting. investigator Ind. U., 1983. Fellow Am. Acad. Otolaryngology (gov. 1982—), ACS, Am. Otological, Rhinological, and Laryngological Soc. (Thesis Disting. for Excellence award), Am. Neurotology Soc. Am. Auditory Soc. (mem. exec. com. 1985—); mem. Otosclerosis Study Group (coun. 1993—), Am. Otol. Soc. (coun. 1992—), Marines Meml. Assn., Wheaton Coll. Scholastic Honor Soc., Cosmos Club of Washington, Columbia Club of Ind., Royal Soc. Medicine London, Alpha Omega Alpha. Avocation: tennis. Office: Riley Hosp 702 Barnhill Dr rm 0860 Indianapolis IN 46202-5230 also: Ind U Sch Medicine Indianapolis IN 46223

MIYAMOTO, SIGENORI, astronomy educator, researcher; b. Akashi, Hyogo, Japan, Oct. 20, 1931; s. Tsunezo and Tora (Yamasaki) M.; m. Shuko Hara, Apr. 23, 1961. MS, Osaka (Japan) U., 1956, DSc, 1961. Lectr. Osaka City U., 1963-66, assoc. prof., 1966-67; assoc. prof. U. Tokyo, 1967-77; prof. Osaka U., 1977—; mem. steering com. Inst. for Space and Astronautical Sci., Japan, 1980-91, guest prof., 1991-93. Author: (high sch. physics textbook) Butsuri, 1975-93. Recipient Nishina Meml. prize Nishina Meml. Soc., Tokyo, 1961, Asahi Culture prize Asahi Newspaper Co., Japan, 1963, Asahi prize, Asahi Newspaper Co., 1981. Mem. Phys. Soc. Japan, Astron. Soc. Japan, Royal Astron. Soc. (U.K.), Am. Astron. Soc., Internat. Astron. Union. Achievements include invention of Spark Chamber, of Hadamard Transform X-ray Telescope, of Hadamard X-ray Spectro-Telescope; determination of the position of Cyg X-1 (a black hole candidate); discovery of the phase lag of the time variations of the X-rays from Cyg X-1; discovery of Canonical time variations of X-rays from black hole candidates. Home: Kiyoshikojin 1-2-30-613, Takarazuka 665, Japan Office: Osaka U Faculty of Sci, Machikaneyama 1-1, Toyonaka 560, Japan

MIZE, CHARLES EDWARD, academic pediatrician; b. Tex., Mar. 3, 1934. PhD, John Hopkins U., 1961, MD, 1962. Diplomate Am. Bd. Pediatrics, Am. Bd. Nutrition. Intern, resident Johns Hopkins Hosp., Balt., 1962-64; from asst. prof. to assoc. prof. U. Tex., Southwestern Med. Ctr., Dallas, 1967—; vis. scientist I.N.S.E.R.M., Molecular Pathology, Paris, 1976-77; dir. metabolic clinic Children's Med. Ctr., Dallas, 1979-89, dir. nutrition support svc., 1980-92. Contbr. articles to profl. jours. Sr. surgeon USPHS, 1964-67. Spl. Rsch. fellowship USPHS, NIH, 1957-61; recipient Rsch. Career Devel. award USPHS, 1970-75. Mem. Am. Acad. Pediatrics, Am. Soc. of Human Genetics, Soc. for Magnetic Resonance in Medicine, Soc. for Pediatric Rsch., Soc. for Inherited Metabolic Disorders. Achievements include rsch. on hepatic glutathione reductase, phytanic acid metabolic pathway, modeling techniques and metabolic diagnosis, pediatric sequential NMR spectroscopy, mitochondrial DNA and pearson syndrome. Office: U Tex Southwestern Med Ctr 5323 Harry Hines Blvd Dallas TX 75235-9063

MIZE, JOE HENRY, industrial engineer, educator; b. Colorado City, Tex., June 14, 1934; s. Kelly Marcus and Birtie (Adams) M.; m. Betty Bentley, Mar. 16, 1966; 1 dau., Kelly Jean. B.S. in Indsl. Engring, Tex. Tech. Coll., 1958; M.S. (Research Found. grantee) in Indsl. Engring, Purdue U., 1963, Ph.D., 1964. Registered profl. engr., Ala., Okla. Indsl. engr. White Sands Missile Range, N.Mex., 1958-61; grad. research asst. Purdue U., Lafayette, Ind., 1961-64; asso. prof. engring. Auburn (Ala.) U., 1964-69; dir. Auburn (Ala.) U. (Computer Center), 1965-66; prof. engring. Ariz. State U., Tempe, 1969-72; prof., head Sch. Indsl. Engring. and Mgmt. Okla. State U., Stillwater, 1972-80; dir. Univ. Ctr. for Energy Research Okla. State U., 1980-83, Regents prof., 1982—; cons. to Air War Coll., 1968-69, U.S. Army, Ops. Analysis Standby Unit, U. N.C., 1965-69, various mfg. firms, 1964—; program adv. Office of Mgmt. and Budget, Exec. Office of the President, Washington, 1974-79; adv. to NSF, 1974-79, Nat. Center for Productivity and Quality of Work Life, 1973-78; chmn. tech. adv. council So. Growth Policies Bd., 1975-77; accrediting visitor Engrs. Council for Profl. Devel., 1973-80. Author: (with J.G. Cox) Essentials of Simulation (translated into Japanese 1970), 1968, Prosim V.: Instructor's Manual, 1971, Student's Manual, 1971, (with C.R. White and George H. Brooks) Operations Planning and Control, 1971, (with J.L. Kuester) Optimization Techniques with Fortran, 1973, (with W.C. Turner and K.E. Case) Introduction to Industrial and Systems Engineering, 3d edit., 1993 (named Book of Yr., Am. Inst. Indsl. Engrs. 1979), Guide to Systems Integration, 1991; contrbr. articles to profl. jours., more. Recipient Disting. Engring. Alumnus award Purdue U., 1978. Mem. Am. Inst. Indsl. Engrs. (exec. v.p. 1978-80, pres. 1981-82, H.G. Maynard Innovative Achievement award 1977, Gilbreth Indsl. Engring. award 1990), Am. Soc. for Engring. Edn. (sec. govt. rels. com. 1975-76), Nat. Soc. Profl. Engrs., Okla. Soc. Profl. Engrs. (Outstanding Engring. Achievement award 1977, Outstanding Engr. in Okla. 1981), Inst. Mgmt. Scis., Coun. Indsl. Engring. Acad. Dept. Heads (chmn. 1975-76), NAE, Nat. Rsch. Coun., Sigma Xi, Tau Beta Pi, Alpha Pi Mu. Office: Oklahoma State U Indsl Engring Dept Stillwater OK 74078

MIZOKAMI, IRIS CHIEKO, mechanical engineer; b. Honolulu, Oct. 19, 1953; d. Takeo and Muriel Yae (Maeda) M.; m. Joseph John Nainiger, Nov. 27, 1976 (div. June 1987). BSME, Case Inst. Tech., 1975. Project engr. Aluminum Co. of Am., Cleve., 1975-77; engr. machine, tool and die dept. Chevrolet-Parma, Ohio, 1977-81; engr. facilities Pearl Harbor (Hawaii) Naval Shipyard, 1988-90, acting drydock engr., 1990-91, design engr. facilites, 1991—. Vol. Manor Care Nursing Home, North Olmsted, Ohio, 1986-87; mem. Honolulu Community Band, 1987-90. Mem. ASME, Nat. Fire Protection Assn., Shetland Sheepdog Club of Hawaii (bd. dirs.), Chopin, Eta Kappa Nu. Mormon. Avocations: music, chess, quilts, reading, Shetland sheepdogs. Home: 94-323 Anaula St Mililani HI 96789 Office: Pearl Harbor Naval Shipyard PO Box 400 Honolulu HI 96860-0001

MIZRAH, LEN LEONID, physicist; b. Kiev, Ukraine, May 11, 1948; s. Benjamin and Maria (Gutin) M.; m. Zoya Gelvan, Jan. 6, 1989. MS, Inst. Electron Tech., Moscow, 1972; PhD, Inst. Physics, Kiev, 1985. Mem. rsch. staff The Inst. of Physics, Acad. of Scis., Kiev, 1972-89; lab. mgr. Fujitsu Microelectronics, Inc., San Jose, Calif., 1990—; lectr. Cogswell Poly. Coll., Cupertino, Calif., 1993—. Contbr. articles to profl. jours. Mem. Am. Phys. Soc., IEEE. Achievements include finding new mechanisms of Frenkel pair annihilation in irradiated diamond-like semiconductors; working out new methodology and making first direct measurements to estimate lifetime of primary radiation defects in covalent semiconductors; creation of an efficient method to define yield of CMOS intergrated circuits based on wafer parametric measurements. Home: 350 Sharon Park Dr # Q24 Menlo Park CA 94025 Office: Fujitsu Microelectronics 77 Rio Robles M/S 466 San Jose CA 95134-1807

MIZRAHI, ABRAHAM MORDECHAY, cosmetics and health care company executive, physician; b. Jerusalem, Apr. 16, 1929; came to U.S., 1952, naturalized, 1960; s. Solomon R. and Rachel (Haliwa) M.; m. Suzanne Eve Glasser, Mar. 15, 1956; children: Debra, Judith, Karen. B.S., Manchester Coll., 1955; M.D., Albert Einstein Coll. Medicine, 1960. Diplomate: Am. Bd. Pediatrics, Nat. Bd. Med. Examiners. Intern U. N.C., 1960-61; pediatric resident Columbia-Presbyn. Med. Center, N.Y.C., 1961-63; NIH fellow in neonatology Columbia-Presbyn. Med. Center, 1963-65; assoc. dir. Newborn Service Mt. Sinai Hosp., N.Y.C.; also dir. Newborn Service Elmhurst Med. Center, 1965-67; staff physician Geigy Pharm. Corp., N.Y.C., 1967-69; head cardio-pulmonary sect. Geigy Pharm. Corp., 1969-71; sr. v.p. corp. med. affairs USV Pharm. Corp., Tuckahoe, N.Y., 1971-76; v.p. health and safety Revlon, Inc., N.Y.C., 1976-89, sr. v.p. human resources, 1989—; assoc. in pediatrics Columbia U., 1963-67; cons. in neonatology Misericordia-Fordham Med. Ctr., 1967-89; clin. affiliate N.Y. Hosp.; clin. asst. prof. Cornell U. Med. Coll., 1982—. Contbr. articles to profl. jours. Trustee Westchester (N.Y.) Jewish Center. Mem. AMA, N.Y. State and County Med. Soc., Am., N.Y. acads. medicine, Am. Soc. Clin. Pharmacology and Therapeutics, Am. Pub. Health Assn., Am. Occupational Med. Assn. Home: 7 Jason Ln Mamaroneck NY 10543-2108 Office: 767 5th Ave New York NY 10022

MIZSAK, STEPHEN ANDREW, chemist; b. East Cleveland, Ohio, June 10, 1939; s. Stephen Joseph and Mary Ellen (Golden) M.; m. Marilyn Ann Clark, June 17, 1961; children: Stephen, Mark, Stephanie, Scott. BS in Chemistry, U. Dayton, 1961. Assoc. organic chemist Eli Lilly and Co., Indpls., 1961-65; assoc. rsch. engr. U.S. Steel, Monroeville, Pa., 1965-69; sr. rsch. chemist The Upjohn Co., Kalamazoo, 1969--. Contbr. articles to profl. jours. Mem. Am. Chem. Soc., N.Y. Acad. Sci. Republican. Roman Catholic. Achievements include patents. Home: 5717 Powderhorn Dr Kalamazoo MI 49009 Office: The Upjohn Co 7255-209-006 Kalamazoo MI 49001

MIZUTANI, HIROSHI, biogeochemist, researcher; b. Tokyo, Jan. 19, 1949; came to U.S., 1973; s. Senkichi and Haruno (Matsumoto) M.; m. Atsuko Anzai, Mar. 22, 1980; children: Wataru, Futoshi. BA, U. Tokyo, 1971, MS, 1973; PhD, U. Md., 1978. Rsch. assoc. U. Md., College Park, 1978-79; researcher Mitsubishi Kasei Inst. Life Scis., Tokyo, 1979-85, sr. researcher, 1985-93; prof. Senshu U. Ishinomaki, Japan, 1993—; lectr. Tokyo Gakugei U., 1983, U. Yamagata, Japan, 1984-85, Seijo U., Tokyo, 1985-86, 91, Juntendo U., Tokyo, 1988, Toho U., Chiba, Japan, 1989, Sophia U., Tokyo, 1989, U. Nagoya, Aichi, Japan, 1989-90, Sohka Univ., Tokyo, 1991-92, U. Tokyo, 1993; adv. bd. Soc. Japanese Aerospace Cos., 1984-87, 90, Ministry Internat. Trade and Industry, 1984-87, Nomura Rsch. Inst., 1985-86; scientist MV Rsch. Inst., 1984—. Author: Cycle of Life and Man on the Earth, 1987, Splendid Biotechnology, 1991; contbr. articles to profl. jours. Mem. Ecol. Soc. Am., Ecol. Soc. Japan, Internat. Soc. for Study Origin of Life, Japanese Soc. for Study Origin and Evolution of Life, Geochem. Soc. Japan (editor 1986-90), Assn. Study on Nature Matter and Value, Japanese Soc. for Biol. Scis. in Space (councilor 1987-89, 92—, editor 1987—), Future Earth Club (editor 1985—), CELSS Rsch. Group. Achievements include establishment of high atmospheric CO2 in Archean; formulation of sociogeochemical cyclings of elements; discovery of the latitudinal dependence of ammonia volatilization; invention of method to determine deserted animal colonies. Home: Meito 2-bankan 201, Kadonowaki Aza Nibanyachi 2-26 Ishinomaki Miyagi 986, Japan Office: Senshu U Ishinomaki Sch Sci and Engring Dept Basic Scis, Sinmito 1 Minamisakai, Ishinomaki Miyagi 986, Japan

MIZUTANI, JUNYA, chemist, educator; b. Fukagawa-shi, Hokkaido, Japan, Sept. 20, 1932; s. Ryuki and Tome (Eguchi) M.; m. Yukiko Nakagawa, May 3, 1964; children: Mahito, Yuka, Aki. BAgr, Hokkaido U., Sapporo, Japan, 1955, MAgr, 1957, DAgr, 1960. Instr. Hokkaido U., Sapporo, Japan, 1960; rsch. assoc. MIT, Cambridge, 1960-63; lectr., assoc. prof. Tokyo Noko U., Fuchu-shi, Japan, 1963-68; assoc. prof. Hokkaido U., Sapporo, 1969-71, prof., 1971—, dean faculty of agr., 1993—; project dir. Rsch. Devel. Corp. Japan, Eniwa-shi, 1988—. Co-author: Noyaku No Kagaku, 1979; editor: Agrl. Biol. Chemistry, 1990-91. Home: 7-6-14-22 Nishino Nishi-ku, Sapporo 063, Japan Office: Hokkaido U Dept Applied Biosci, Kita 9 Nishi 9 Kita-ku, Sapporo 060, Japan

MIZUTANI, SATOSHI, research administrator; b. Yokohama, Kanagawa, Japan, Nov. 19, 1937; s. Hideo and Hiroko (Kimura) M.; m. Kaoruko Kobayashi, June 17, 1966; 1 child, Takaharu. BS, Tokyo U. Agr. and Tech., 1962; PhD, U. Kans., 1969. Rsch. scientist Nippon Kayaku, K.K., Tokyo, 1962-65; postdoctoral fellow McArdle Lab. U. Wis., Madison, 1969-71, instr., 1971-72, asst. scientist, 1972-74, assoc. scientist, 1974-80; sr. molecular biologist Abbott Labs. North, Chgo., 1980-81; dir. Bethesda Rsch. Lab., Gaithersberg, Md., 1981-82; sr. rsch. fellow Merck Sharp & Dohme Rsch. Lab., West Point, Pa., 1982-88; assoc. dir. Wyeth-Ayerst Rsch., Radnor, Pa., 1988—. Fulbright scholar U.S.-Japan Fulbright Com., 1965-71; recipient Scholar award Leukemia Soc. Am., 1973-78. Mem. AAAS, Am. Soc. Microbiology, Internat. Soc. Antiviral Rsch., N.Y. Acad. Sci. Achievements include discovery of reverse transcriptase to prove provirus hypothesis, protein-nucleotidyl transferase activity of P-gene product of human hepatitis B virus human rhinovirus cDNA cloning and sequencing, infections RNA transcript synthesis in vitro. Office: Wyeth-Ayerst Rsch 145 King of Prussia Rd Radnor PA 19087

MJOLSNESS, ERIC DANIEL, computer science educator; b. Los Alamos, N.Mex., July 29, 1958; s. Raymond Charles and Patricia (McGeary) M.; m. Shelley Eulalie Scallan, Sept. 12, 1980; 1 child, Clare Eulalie. AB, Washington U., St. Louis, 1980; PhD, Calif. Inst. Tech., 1985. Asst. prof. computer sci. dept. Yale U., New Haven, 1985-90, assoc. prof. computer sic. dept., 1990—; co-dir. Ctr. for Computational Ecology, Yale U., 1992—, project dir. Ctr. for Theoretical and Applied Neurosci., 1992—. Contbr. articles to profl. pubs. Sr. faculty fellow Yale U., 1991-92. Mem. IEEE (assoc. editor transactions on neural networks 1991-92), AAAS, Am. Phys. Soc., Assn. for Computing Machinery. Democrat. Achievements include study of mathematical methods in artificial neural networks and in computational biology. Office: Yale U Computer Sci Dept 51 Prospect St New Haven CT 06520-2158

MLADENOVIC, NIKOLA SRETEN, mechanical engineer; b. Belgrade, Yugoslavia, Serbia, May 4, 1958; s. Sreten Nikola and Draginja Ljubica (Stankovic) M.; m. Jela Aleksandar Mirosavljevic, May 10, 1986. MS, U. Belgrade, 1982, PhD, 1986. Rschr. Aeronautical Inst., Belgrade, 1981-84; asst. prof. Faculty of Mech. Engring., Belgrade, 1984-90; cons. Aero. Inst., Belgrade, 1984-89; docent Faculty of Mech. Engring., Belgrade, 1990—. Contbr. articles to profl. jours. including Tehnika and Theoretical and Applied Mechanics. Recipient Oct. award City of Belgrade, 1983, Rastko Stojanovic award Yugoslav Assn. of Mechanics, 1986; Fulbright grant Fulbright Found., 1989. Mem. AIAA, Serbian Assn. of Mechanics, Yugoslav Assn. of Mechanics. Office: Faculty of Mech Engring, 27 Marta 80, 11000 Belgrade Yugoslavia

MO, JIAQI, mathematics educator; b. Zhenjiang, Jiangsu, People's Republic China, Mar. 30, 1937; s. Shanxiang and Yiying (Lu) M.; m. Xiufang Ou, Jan. 17, 1966; 1 child, Shengbin. Student, Fudan U., Shanghai, People's Republic China, 1955-60. Asst. Wannan U., Wuhu, People's Republic China, 1960-66; asst. Anhui Normal U., Wuhu, 1966-78, lectr., 1978-86, assoc. prof., 1986-87, prof., 1987—; prof. Shanghai U. Tech., 1988—. Author: Higher Mathematics, 2d edit., 1985; contbr. articles to profl. jours. Recipient Cert. of Merit, Anhui Govt., 1980, 86, 90, 93. Mem. Math. Soc. China, Wuhu Math. Soc. (pres. 1988-91), Anhui Math. Soc. (bd. dirs. 1985—, standing dirs. 1992—). Avocations: sports, bridge, tea, television, repairing and utilizing old or discarded things. Office: Anhui Normal U, Dept Math, Wuhu 241000, China

MO, ROGER SHIH-YAH, electronics engineering manager; b. Shanghai, Rep. of China, Mar. 10, 1939; s. Maurice Chun-Dat and Mary (Shen) M.; m. Amy Chun-Muh, June 21, 1964; 1 child, Karen Voong-Tsun. BSEE, MIT, 1962; MSEE, Northeastern U., Boston, 1964, PhD, 1967; MBA, Pepperdine U., 1980. Engr. Raytheon Corp., Sudbury, Mass., 1967-69; on tech. staff Xerox, El Segundo, Calif., 1969-74, mgr. memory, 1974-77, mgr. cirs. and subsystems, 1977-81, area mgr., 1981-87, program mgr., 1987-89, imaging systems mgr., 1989-92, systems design mgr., 1992—; sr. lectr. West Coast U., L.A., 1978-87, chmn. acad. standards com., 1981-82. Contbr. articles to profl. jours. Bd. dirs. The Wellness Community So. Bay Cities, 1989—. Mem. IEEE, Chinese Am. Assn. of So. Calif. (bd. dirs. 1983-87). Democrat. Roman Catholic. Lodge: Flip Flap (local chmn. 1974, nat. chmn. 1976). Avocations: golf, table tennis, photography. Home: 6852 Verde Ridge Rd Palos Verdes Peninsula CA 90274-4638 Office: Xerox Corp 701 S Aviation Blvd El Segundo CA 90245-4898

MOAWAD, ATEF, obstetrician-gynecologist, educator; b. Beni Suef, Egypt, Dec. 2, 1935; came to U.S., 1959; s. Hanna and Baheya (Hunein) M.; m. Ferial Fouad Abdel Malek, Aug. 22, 1966; children: John, Joseph, James. Student, Cairo U. Sch. Sci., 1951-52; MB, BCh, Cairo U. Sch. Medicine, 1957; MS in Pharmacology, Jefferson Med. Coll., 1963. Diplomate Am. Bd. Ob-Gyn; licentiate Med. Coun Can. Rotating intern Cairo U. Hosp., 1958-59, Elizabeth (N.J.) Gen. Hosp., 1959-60; resident in ob-gyn. Jefferson Med. Coll. Hosp., Phila., 1961-64; lect. dept. pharmacology U. Alta., Can., 1966; asst. prof. dept. ob-gyn. and pharmacology U. Alta., Can., 1967-70, assoc. prof., 1970-72; assoc. prof. dept. ob-gyn. and pharmacology U. Chgo., 1972-75, prof. dept. ob-gyn. and pediatrics, 1975—, co-dir. perinatal ctr., 1974-80; obstetrician-gynecologist, chief obstetrics, co-dir. perinatal ctr. The Chgo. Lying-in Hosp. U. Chgo., 1980—; vis. investigator dept. ob-gyn. U. Lund, Sweden, 1969. Co-author book chpts., jour. articles. Mem. perinatal adv. com. Chgo. March of Dimes, 1977—, health profl. adv. com., 1983—; mem. perinatal adv. bd. com. State of Ill., 1978—; mem. Chgo. Maternal Child Health Adv. Com., chmn., 1991—; mem. Mayor's Adv. Com. on Infant Mortality, 1991—. Fellow Jefferson Med. Coll., 1960-61, Case Western Reserve U., 1964-65; grantee Brush Found., 1966-67; recipient award Phila. Obstet. Soc., 1964. Fellow Am. Coll. Ob-Gyn. (Purdue-Frederick award 1978), Royal Coll. Surgeons (Can.); mem. Soc. for Gynecol Investigation, Pharmacol. Soc. Can., Am. Gynecol. and Obstet. Soc., Soc. Perinatal Obstetricians, N.Y. Acad. Scis., Chgo. Gynecol. Soc., Can. Med. Assn., Christian Med. Soc., Edmonton Obstetrics Soc. Office: U Chgo Dept Ob-Gyn 5841 S Maryland Ave Chicago IL 60637*

MOAZZAMI, SARA, civil engineering educator; b. Tehran, July 24, 1960; d. Morteza Moazzami and Ezzat Akbari. BS, George Washington U., 1981; MS, U. Calif., Berkeley, 1982, PhD, 1987. Rsch. asst. George Washington U., Washington, 1980-81; teaching asst. U. Calif., Berkeley, 1982-83, rsch. asst., 1983-87; prof. Univ. Conn., Stamford, 1987-91, Calif. Polytechnic State U., San Luis Obispo, 1991—; mem. 1989 Santa Cruz Earthquake Reconnaissance Team, Earthquake Engring. Rsch. Inst., Oakland, Calif., 1989; speaker internat. confs. in field. Author: (book) 3-D Inelastic Analysis of Reinforced Concrete Frame-Wall Structures, 1987. Recipient Genevieve McEnerney fellowship U. Calif., Berkeley, 1981-82, Martin Mahler prize in Materials Testing, George Washington U., 1981, Columbian Women Soc. scholarship, Washington, 1979-80. Mem. Am. Soc. Civil Engring. (scholarship 1980), Earthquake Engring. Rsch. Inst., Soc. Women Engrs. Avocations: biking, swimming, sewing, travel. Office: Calif Polytechnic State Univ Sch Engring San Luis Obispo CA 93407

MOBERG, CLIFFORD ALLEN, mold products company executive; b. Milw., Oct. 27, 1951; s. Clifford Clarence and Kathleen Marie (Dickinson) M.; m. Elaine M. Koepfle, July 23, 1951 (div. Aug. 1985); children: Scott Allen, Stephanie Ann; m. Lynette Rose Pavloski, Sept. 2, 1989; children: Shawna Rose McKee, Sara Eve McKee. Student, Milw. Tech. Coll., 1972-74, Waukesha (Wis.) Tech. Coll., 1980-82. Tool and die maker Interstate Tool and Engring., Muskego, Wis., 1970-78; shop foreman Dynamic Tool and Design, Menomonee Falls, Wis., 1978-80; toolroom supr. Sequist Plastics, div. of Pittway, Mukwonago, Wis., 1980-86; pres., owner Performance Alloys and Svcs., Menomonee Falls, Wis., 1987—; pres., ptnr. Performance Mold Products, Inc., Menomonee Falls, 1990—. Vol. Kettle Moraine Hosp., Oconomowoc, Wis., 1988, 89; trustee, chmn. fin. com. Village Bd., Nashotah, Wis. Mem. Soc. Plastics Engrs. (cons. 1990-91, dir. 1987—, sec. 1990-91, Cert. of Appreciation 1989), Engrs. and Scientists Milw. Achievements include patent for Injection Mold Sprue Bushing Design and Alloy Core Pins for Injection Molds. Home: 3284065wn Allendale Dr Nashotah WI 53058-9786 Office: Performance Alloys & Svcs W140 N9055 Lilly Rd Menomonee Falls WI 53051

MOBLEY, CLEON MARION, JR. (CHIP MOBLEY), physics educator, real estate executive; b. Reidsville, Ga., July 14, 1942; s. Cleon M. and Lucile (Anderson) M.; m. Martha Hewlett, 1962 (div. 1970); children: Lisa Anne, Arthur Marion. AS, So. Tech. Coll., 1961; BS, Oglethorpe U., 1963; MS, U. Mo., 1966; PhD The Union Inst. Ohio, 1987. Lic. airplane pilot. Faculty research assoc. Ga. Inst. Tech., Atlanta, 1963-65; instr. So. Coll. Tech., Marietta, Ga., 1965-67; faculty fellow NASA, 1967-68; from asst. to assoc. prof. physics Ga. So. U., Statesboro, 1968—, sec. Ga. Acad. Sci., 1991-93, dir. Ga. So. U. planetarium; pres. Assoc. Income Properties, Inc. Statesboro, 1982—; pres. Savannah Properties Mgmt., Inc., 1983-87; sci cons. AEC fellow, 1965; sec. Ga. Acad. Sci., 1990—; NASA-ASEE fellow, 1970. Contbr. articles to profl. jours. Mem. Statesboro Home Builders Assn., Am. Inst. Physics, Ga. Acad. Sci. Sigma Phi Epsilon. Methodist. Club: Optimist. Lodge: Elks. Office: PO Box 8031 Statesboro GA 30460

MOCHEL, MYRON GEORGE, mechanical engineer, educator; b. Fremont, Ohio, Oct. 9, 1905; s. Gustave A. and Rose M. (Minich) M.; m. Eunice Katherine Steinicke, Aug. 30, 1930 (dec. Dec. 1982); children: Kenneth R., David G., Virginia June. BSME, Case Western Res. U., 1929; MSME, Yale U., 1930. Registered profl. engr. N.Y., Mass., Pa. Devel. engr. nitrogen div. Allied Chem. Corp., Hopewell, Va., 1930-31; devel. engr. R&D dept. Mobil Corp., Paulsboro, N.J., 1931-37; design and devel. engr. gearing div. Westinghouse Electric Corp., Pitts., 1937-43; rsch. assoc. underwater sound lab. Harvard U., Cambridge, Mass., 1943-45; supr. of tng. steam turbine div. Worthington Corp., Wellsville, N.Y., 1945-49; prof. mech. engr. Clarkson U., Potsdam, N.Y., 1949-71; prof. emeritus Clarkson U., Potsdam, 1971—; lect. U. Pitts., 1938-43, N.Y. State U. Adult Edn., Wellsville, 1946-49, Oswego, 1965, N.Y. State High Sch. Enrichment Program, Potsdam, 1962-71; cons. Designers for Industry, Cleve., 1953, rsch. engr. Morris Machine Works, Baldwinsville, N.Y., 1954, design engr. Racquette River Paper Co., Potsdam, 1955. Author: Fundamentals of Engineering Graphics, 1960, Pre-Engineering and Applied Science Fundamentals, 1962, Fortran Programming, Programs and Schematic Storage Maps, 1971; co-author: (with Eunice S. Mochel) Funds For Fun, 1983, (with Donald H. Purcell) Beyond Expectations, 1985; contbr. articles to profl. jours. Officer, vol. St. Lawrence Valley Hospice, 1983; pres. Mayfield Tenants Assn., 1989-91. Mem. ASME, Am. Soc. Engring. Edn. (advt. mgr. Jour. Engring. Graphics 1963-66, sec. 1966-67, high schs. laision on engring. graphics 1962-65, awards com. chmn. 1965-66), Am. Assn. Ret. Persons (founder St. Lawrence County chpt., income tax counselor 1988-89, medicare/medicaid assistance program counselor 1988—, pres. 1989-90). Republican. Mem. Unitarian Universalist Ch. Home and Office: 9C Mayfield Dr Apt 1 Potsdam NY 13676-1309

MOCK, GARY NORMAN, textile engineering educator; b. Easton, Pa., Oct. 22, 1942; s. Norman James and Zelma (Donnaly) M.; m. Ruth Woodruff England, Sept. 9, 1964; children: Kevin, Darin, Brian. BS in Chem. Engring., Va. Poly. Inst. and State U., 1967; PhD in Chem. Engring., Clemson U., 1976. Process engr. Milliken & Co., Spartanburg, S.C., 1968-72; asst. prof. Coll. Textiles, N.C. State U., Raleigh, 1976-82, assoc. prof., 1982-88, prof., 1988—, asst. head textile chemistry, 1985-90, assoc. dean, 1991—; textile expert UN, India, 1991. Author, editor: Minicomputers and Microprocessors in the Textile Industry, 1983; contbr. articles to profl. jours. Mem. Am. Assn. Textile Chemists and Colorists (chmn. No. Piedmont sect. 1989), Am. Chem. Soc., Instrument Soc. Am. Presbyterian. Office: NC State U Coll Textiles 2401 Research Dr Raleigh NC 27695

MOCK, JOHN EDWIN, science administrator, nuclear engineer; b. Altoona, Pa., Sept. 29, 1925; s. Daniel Raymond and Sarah Adella (Lorenz) M.; m. Jeannette Daly, Oct. 25, 1947; children: Donna Jean, Susan Jean. BS, MS, Purdue U., 1950, PhD in Nuclear Engring., 1960; LLM in Patent Law, George Washington U., 1982. Bar: D.C. 1980, U.S. Ct. Appeals (D.C. cir.) 1981, U.S. Ct. Customs and Patent Appeals 1981, U.S. Ct. Appeals (fed. cir.) 1982, U.S. Supreme Ct. 1986, U.S. Tax Ct. 1987; registered profl. engr., Colo., Ga., N.Y. Commd. 2d lt. USAF, 1943, advanced through grades to lt. col., 1964; program mgr. Def. Atomic Support Agy., Dept. of Def., Washington, 1960-65, Advanced Rsch. Projects Agy., Dept. of Def., Washington, 1966-68; ret. USAF, 1968; dir. Ga. Sci. and Tech. Commn., Atlanta, 1968-71; chmn. Ga. Energy Commn., Atlanta, 1971-74; head R&D incentives program NSF, Washington, 1974; sr. sci. advisor U.S. Dept. Energy, Washington, 1975-80, dir. geothermal divsn., 1981—; commr. So. Interstate Nuclear Bd., Atlanta, 1968-74; chmn. Gov.'s Sci. Adv. Coun., Atlanta, 1970-74; advisor study com. on offshore oil exploration Pres.'s Coun. on Environ. Quality, 1973-74; advisor Nat. Acad. Engring., 1972-74, Office Sec. of Def. (ARPA), 1968-70, NASA, 1970-72, NSF, 1968-74, Nat. Acad. Pub. Adminstrn., 1968-71, Coun. State Govts., 1968-73, Coastal Plains Regional Commn., 1968-74, Dept. Interior's Outer Continental Shelf Rsch. Mgmt. Adv. Bd., 1973-74. Editor: Science for Society, 1971; co-author: Technology Assessment in State Government, 1972; author: Priorities for Technology Assessment, 1973; author: (with others) Encyclopedia of Architecture, 1989. Founder, dir. Coastal States Orgn., 1969; founder, chmn. Nat. Gov.'s Coun. for Sci. and Tech., 1970, 72; chmn. Ga. Seaward Lateral Boundary Commn., Atlanta, 1968-69; charter mem. U.S.Sr. Exec. Svc., 1979. Recipient Mark Mills award Am. Nuclear Soc., 1960, Lifetime Achievement award in tech. transfer Tech. Utilization Found., 1991, Spl. Achievement award Geothermal Resources Coun., 1991, Energy R&D Adminstrn., 1976. Fellow AAAS, Nat. Conf. Advancement Rsch. (chmn. 1976), Ga. Acad. Sci.; mem. Sci. and Tech. Coun., Am. Mgmt. Assn., Cosmos Club. Methodist. Home: 1326 Round Oak Ct McLean VA 22101 Office: US Dept Energy 1000 Independence Ave Washington DC 20585

MOCK, PETER ALLEN, hydrogeologist; b. Alamagordo, N.Mex., Oct. 26, 1959; s. Roy Edward and Greta Cynthia (Funk) M.; m. Mary Catherine White, Jan. 12, 1985. BS in Hydrology, U. Ariz., 1981. Registered profl. geologist, Ariz. Hydrologist Ariz. Dept. Water Resources, Phoenix, 1982-85; hydrogeologist CH2M Hill, Inc., Phoenix, 1985—. Author publs. in field including Siting of Spreading Basins for Underground Storage of Treated Municipal Wastewater, 1991. Mem. Am. Geophys. Union, Nat. Ground Water Assn., Ariz. Hydrol. Soc., ASTM, Assn. Groundwater Scientists and Engrs. Republican. Roman Catholic. Office: CH2m Hill Inc Ste 550 1620 W Fountainhead Pkwy Tempe AZ 85282

MOCK, ROBERT CLAUDE, architect; b. Baden, Fed. Republic of Germany, May 3, 1928; came to U.S., 1938, naturalized, 1943; s. Ernest and Charlotte (Geismar) M.; m. Belle Carol Bach, Dec. 23, 1952 (div.); children: John Bach, Nicole Louise; m. Marjorie Reubenfeld, Dec. 20, 1964. B.Arch., Pratt Inst., 1950; M. Arch., Harvard U., 1953. Registered architect, N.Y., Conn., N.J., Nat. Council Archtl. Registration Bds. Architect George C. Marshall Space Center, Huntsville, Ala., 1950-51; archtl. critic Columbia Sch. Architecture, N.Y.C., 1953-54; dir. facility design Am. Airlines, N.Y.C., 1955-60; founder Robert C. Mock & Assocs. (architects and engrs.), N.Y.C., 1960—; Mem. Mayor's Panel of Architects, N.Y.C. Prin. works include: Shine Motor Inn, Queens, N.Y., 1961 (recipient 1st prize motel category Queens C. of C. 1961), temporary terminal bldg. Eastern Air Lines , La Guardia Airport, N.Y.C., 1961, cargo bldgs United Airlines and Trans World Airlines, Kennedy Airport, N.Y.C., Bridgeport (Conn.) Airport, 1961, Eastern Air Lines Med. Ctr., Kennedy Airport, 1962, ticket office Trans World Airlines Fifth Ave., N.Y.C., 1962, terminal bldgs. Eastern Air Lines and Trans World Airlines , La Guardia Airport, N.Y.C., 1963, 7 bldgs. Mfrs. Hanover Trust Co. , 1964-66, kitchen and commissary bldg. Lufthansa German Airlines, 1964, Ambassador Club, La Guardia Airport, 1964, Happyland Sch., N.Y.C., 1965, cargo bldgs. Alitalia and Lufthansa German Airlines, Kennedy Airport, 1965, FAA-Nat. Prototype Air Traffic Control Tower, 1966; Lufthansa German Airlines; Irish Internat. Airlines, El Al Israel Airlines, Varig Brazilian Airlines; passenger terminals Kennedy Airport, 1970; Swiss Air Cargo Terminal, Lufthansa German Airlines, cargo terminals El Al Israel airline cargo terminal, Kennedy Airport, 1972, passenger terminal Aerolineas Argentina, 1974, N.Am. hdqrs. Aerolineas Argentinas, N.Y.C., 1974, corp. hdqrs. Am. Airlines, 1977, N.Am. hdqrs. Varig Brazilian Airlines, N.Y.C., 1977, Norel-Ronel Indsl. Pk., Hollywood, Fla., 1979, N.Am. hdqrs. Irish Internat. Airlines , N.Y.C., 1979, corp. hdqrs. Bankers Trust Co., N.Y.C., 1980, cargo terminal Air India, cargo terminal Flying Tiger, Kennedy Airport, 1982, 2 flight kitchen bldgs. Ogden Food Corp., Kennedy Airport, 1984, 88 and LaGuardia Airport, 1987, Greenwich Assn. Retarded Citizens Sch., 1983, passenger terminal extension Varig Brazilian Airlines , 1985, 3 restaurants La Guardia Airport, 1987, residences Palm Beach, Fla., 1989-92, Bethesda, Md., 1993. Recipient United Way Vol. of Yr. award, 1984. Mem. Am. Arbitration Assn. Clubs: City, Harvard, Admirals Cove. Office: 185 Byram Shore Rd Greenwich CT 06830-6909

MOCKAITIS, ALGIS PETER, mechanical engineer; b. Sakiai, Lithuania, Nov. 18, 1942; came to U.S., 1949; s. Petras and Marcele (Adomaitis) M.; m. Jolanda Maria Gliozeris, May 2, 1970; children Audra, Romas, Vitas. BS, Ill. Inst. Tech., 1964; MS, U. Pa., 1967, PhD, 1972. Aerospace engr. NASA, Cleve., 1964-66; instr. math. and physics Spring Garden Coll., Chestnut Hill, Pa., 1968-70; sr. engr. Continental Can Corp., Chgo., 1971-77; mgr. mech. engring. Continental Forest Industries, Lombard, Ill., 1977-84, Stone Container Corp., Willowbrook, Ill., 1984-86; dir. mech. engring. Stone Container Corp., Burr Ridge, Ill., 1986—; com. mem. U.S. Dept. Energy, Washington, 1987—. Contbr. articles to profl. jours. Dir. Lithuania Am. Community Sch., Lemont, Ill., 1980-82. Mem. ASME, TAPPI (com. mem.), Am. Paper Inst. Home: 813 Chestnut Dr Darien IL 60561 Office: Stone Container Corp 8170 S Madison Burr Ridge IL 60521

MOCKAITIS, JOSEPH PETER, logistics engineer; b. Hazleton, Pa., Sept. 9, 1942; s. Joseph Julius and Helen (Carl) M; m. Brigitte E. Maier, Apr. 29, 1971 (div. Mar. 1981); m. Karen E. Hess, Jan. 28, 1983; children: Peter, Monica, Tiffany, Curtis, Marlise, Michelle, Megan, Corrine. BSE, Bloomsburg U., 1964; MS in Logistics Mgmt., Air Force Inst. Tech., 1972; AA in Computer Info. Systems, El Camino Coll., 1989. Enlisted USAF, 1964, advanced through grades to maj., 1984; logistics engr. S-Systems, Inglewood, Calif., 1985-86, Northrop Aircraft Div., Hawthorne, Calif., 1986—. Pres. Young Men's Orgn. LDS Ch., Lawndale, Calif., 1991-92. Home: PO Box 1260 Hawthorne CA 90251-1260 Office: Northrop Aircraft Div 2522 Mailstop 89 1 Northrop Ave Hawthorne CA 90250

MOCZYGEMBA, GEORGE ANTHONY, research chemist; b. Panna Maria, Tex., Jan. 7, 1939; s. Edmund and Victoria (Keller) M.; m. Ann L. Franzetti, June 11, 1966; children: Michael, Richard, Kathryn. BS in Chemistry, U. Tex., 1962, PhD in Inorganic Chemistry, 1969. Sr. rsch. chemist Phillips Petroleum Co., Bartlesville, Okla., 1968-92, mem. devel./tech. staff, 1993—. Contbr. articles to profl. jours. Mem. Am. Chem. Soc., Sigma Xi. Achievements include 22 patents in polymers and related chemistry; advancements in polymerization, polymerization catalysis, technology transfers, quality improvement. Office: Phillips Petroleum 95-G PRC Bartlesville OK 74004

MODAK, CHINTAMANI KRISHNA, information and documentation officer; b. Panval, M.S., India, Jan. 25, 1947; s. Krishna Laxman and Mangala Krishna (Yamuna) M.; m. Vishali C. Sunita, May 23, 1976; children: Aditi C., Adwait C. BSc. in Libr. Sci., Poona U., Pune, Maharshtra, India, 1963. Libr. supr. Asian Paints Ltd., Bombay, India, 1971-72; libr. Cen. Sch., Uran, India, 1972-74; sr. info. and documentation officer Hidustan Organic Chem. Ltd., Rasayani, India, 1974—. Editor Spl. issue on Maintenance and Trouble Shooting, 1991, Spl. issue on Safety, 1991, Spl. issue on Mech. Engrs., 1992. Home: Opp Hotel Welcome Inn, 10 Sujata Society, Panvel 410 206, India Office: Hindustan Organic Chemicals, Post-Rasayani, MS Rasayani 410 207, India

MODEER, VICTOR ALBERT, JR., civil engineer; b. St. Joseph, Mo., Oct. 19, 1955; s. Victor Albert and Evelyn Mary (Borkowski) M.; m. Sheryll Sexton, Aug. 5, 1978; children: Michael Victor, Kathryn Lynn. BS, La. State U., 1978; MS, Purdue U., 1982. Registered profl. engr., Mo., Ill.; cert. hazardous materials mgr. Adj. prof. La. State U., Baton Rouge, 1980-81; project engr. Louis J. Capozzoli & Assocs., Baton Rouge, 1978-81, Woodward Clyde Cons., St. Louis, 1982-85; sr. project engr. Soil Cons. Inc., St. Louis, 1985-86; geotech. engr. Ill. Dept. Transp., Collinsville, 1986-91, materials engr., 1992, program implementation engr., 1992—; staff instr. civil engring. So. Ill. U.; mem. various coms. Nat. Acad. Sci.-Transp. Rsch. Bd., Washington, 1988—; mem. adv. bd. Ill. Dept. Transp., Ill. Assn. Asphalt Paving Contractors, Springfield, 1990—; cons. various firms and agys., St. Louis, 1986-91. Author: Groundwater, 1987; co-author: Landslides: Investigation and Mitigation, 1993. Bd. mem. various youth groups Boy Scouts, Soccer Clubs, St. Louis. With USNR, 1990-91, Desert Shield/Desert Storm. Mem. ASTM, ASCE, VFW, Inst. Hazardous Materials Mgrs., Hazardous Rsch. and Control Inst., Ill. Assn. Hwy. Engrs. Achievements include research in hazardous waste effects to transp. agys. and electric cone penetrometer deep foundations. Office: Ill Dept Transp 1100 Eastport Plz Dr Collinsville IL 62234-6198

MODESTINO, JAMES WILLIAM, electrical engineering educator; b. Boston, Apr. 27, 1940; s. William and Mary Elizabeth (Dooley) M.; m. Leone Marie MacDougall, Aug. 25, 1962; children: Michele Marie, Lee Ann. BS, Northeastern U., 1962; MS, U. Pa., 1966; MA, Princeton U., 1968, PhD, 1969. Mem. tech. staff Gen. Telephone Electronics Labs., Waltham, Mass., 1969-70; asst. prof. Northeastern U., Boston, 1970-72; prof. Rensselaer Polytechnic Inst., Troy, N.Y., 1972—; vis. prof. U. Calif., San Diego, 1981-82; vis. faculty fellow GE Corp. R&D Ctr., 1988-89; pres. Modcom Inc., Ballston Lake, N.Y., 1981—; v.p. ICUCOM Inc., Troy, N.Y., 1986—. Recipient Sperry Faculty award Sperry Corp., 1986. Fellow IEEE (S.O. Rice Prize Paper award 1984, gov. Info. Theory Soc. 1988—). Avocations: sailing, jogging, tennis, skiing. Office: Rensselaer Poly Inst 110 8th St Troy NY 12180-3522

MODLIN, JAMES MICHAEL, mechanical engineer, army officer; b. Long Beach, Calif., Nov. 6, 1955; s. James Kenneth and Pauline Sue (Hunt) M.; m. Cynthia Anne Maslak, June 11, 1977; children: James Michael, Matthew David, Daniel Christopher, Joshua Thomas. BS, West Point, 1977; MS, MIT, 1985; PhD, Ga. Inst. Tech., 1991. Registered profl. engr., Va. Commd. 2d lt. U.S. Army, 1977, advanced through grades to lt. col., 1993; exec. officer Co. A, 20th Engring. Bn., Ft. Campbell, Ky., 1978-79, engr. equip. mgr., 1979-80; civil engr. 34th Engr. Bn., Ft. Riley, Kans., 1981; comdr. 55th Engr. Co., Ft. Riley, Kans., 1981-83; asst. prof. mechanics U.S. Mil. Acad., West Point, N.Y., 1985-88; rsch./devel. coord. U.S. Army Strategic Def. Command, Huntsville, Ala., 1991-92; dep. missile engring. div. chief THAAD Project Office, Huntsville, Ala., 1992—; instr. mech. engring. U. Ala., Huntsville, 1992—. Contbr. articles to profl. jours. George W. Woodruff doctoral teaching fellow, 1988. Mem. AIAA, ASME, Soc. Am. Mil. Engrs., Sigma Xi. Roman Catholic. Office: Program Exec Office Missile Def Attn SFAE-MD-THA PO Box 1500 Huntsville AL 35807

MODLINSKI, NEAL DAVID, computer systems analyst; b. Milw., Feb. 8, 1958; s. Rudolph G. and Eugenia J. (Drewek) M.; m. Paula J. Modlinski, Nov. 7, 1981; 1 child, Alyssa B. AA in Elec. Engring., Milw. Sch. Engring., 1991; BS in Bus. Adminstrn., Cardinal Stritch Coll., Milw., 1993. Designer Harley-Davidson Motor Co., Milw., 1979-83; designer Erie Mfg. Co., Milw., 1983-87, CAD supr., 1987-92, systems analyst, 1992—; system adminstr. Word Art, Muskego, Wis., 1991—; cons. Oak Creek (Wis.) Tchrs. Assn., 1992. Mem. IEEE Computer Soc. Office: Erie Mfg Co 4000 S 13th St Milwaukee WI 53221-1791

MOE, MICHAEL K., physicist; b. Milw., Nov. 17, 1937; s. Kenneth Ingalls and Jane (Rettke) M.; m. Juanita Wolfram, Dec. 23, 1961; 1 child, Kimberly. BS, Stanford U., 1959; PhD, Case Inst. Tech., 1965. Postdoctoral fellow Calif. Inst. Tech., Pasadena, 1965-66; rsch. physicist U. Calif., Irvine, 1966—. Contbr. articles to profl. jours. Achievements include first lab. observation of double beta decay. U.S. patent holder. Office: U Calif Dept Physics Irvine CA 92717

MOE, OSBORNE KENNETH, physicist; b. L.A., Dec. 29, 1925; s. Osborne Rangvall and Mabel Agnes (Brenner) M.; m. Mildred Mary Minasian, Aug. 20, 1951; children: Robert Arthur, Karen Elizabeth. BA in Physics, UCLA, 1951, MA in Physics, 1953, PhD in Planetary and Space Sci., 1966. Analyst, engr. airborne radar div. RCA, L.A., 1955-56; mem. tech. staff Space Tech. Labs. TRW, Inc., Redondo Beach, Calif., 1956-59; pvt. practice cons. physicist L.A., 1960-67, Corona del Mar, Calif., 1975-87; sr. scientist McDonnell Douglas Astronautics, Huntington Beach, Calif., 1967-75; program mgr. Air Force Plant Rep. Office, L.A., 1987-89; sr. staff meteorologist Space and Missile Systems Ctr., USAF, L.A., 1989-93, electronics engr. devel. planning, 1993—; scientific panelist Climatic Impact Assessment Program, Dept. Transp., Washington, 1971-75; mem. Com. on Extensions to Standard Atmosphere, Washington, 1968—. Co-author: Flight Performance Handbook for Orbital Operations, 1963, The Natural Stratosphere of 1975, 1975, The U.S. Standard Atmosphere, 1976; contbr. articles to profl. jours. With USN, 1944-48. NASA fellow, 1962, 63, 64; recipient award NASA, 1991. Fellow AIAA (assoc., chmn. tech. com. on space sci. and astronomy 1971-75, Certs. of Appreciation), Explorers Club; mem. Am. Geophys. Union, Am. Meteorol. Soc. Achievements include measurement of satellite drag coefficients in orbit, analysis of gas-surface interactions in satellite instruments, thermospheric composition, effect of in-track winds on density and composition measurements. Home: 1520 Sandcastle Dr Corona del Mar CA 92625 Office: Space & Missile Systems Ctr XRF PO Box 92960 2420 Vela Way Ste 1467-A2 Los Angeles AFB CA 90245-4659

MOEHLE, JACK P., civil engineer, engineering executive. BSCE, U. Ill., 1977, MSCE, 1977, PhD, 1980. Registered civil engr., Calif. From asst. to assoc. prof. U. Calif., Berkeley, 1980-90, prof., 1990—, Roy W. Carlson Disting. prof. civil engring., vice-chair tech. svcs. civil engring., 1990-91, dir. earthquake engring. rsch. ctr., 1991—; tech. advisor Double Deck Peer Rev. Panel, Caltrans, 1990—; mem. sci. adv. com. Nat. Ctr. Earthquake Engring. Rsch.; proposal reviewer NSF; cons. in field; bd. dirs. Calif. Univs. Rsch. Earthquake Engring., Cooperating Orgns. No. Calif. Earthquake Rsch. and Tech. Contbr. articles to profl. jours.; reviewer tech. papers. Recipient Chi Epsilon Excellence Teaching award, 1986; Regents Jr. Faculty fellow, 1981. Fellow Am. Concrete Inst. (chmn. detail and proportion earthquake resisting structural elements and systems com. 1988—, mem. various coms.); mem. ASCE (publs. sec. com. seismic effects, Huber Rsch. prize 1990), Structural Engrs. Assn. Calif. (mem. seismology com., reinforced concrete com., bd. dirs.), Earthquake Engring. Rsch. Inst. Office: U Calif Berkeley Earthquake Engring Rsch Ctr 1301 S 46th St Richmond CA 94804-4698*

MOELLENBECK, ALBERT JOHN, JR., engineering executive; b. St. Louis, Aug. 4, 1934; s. Albert John and Josephine Marie (Fruth) M.; m. Charlotte Anne Zimmerman, Nov. 26, 1960; children: Albert, Mary, Cheryl, Joan. BSCE, U. Mo., Rolla, 1960; AMP in Bus., Harvard U., 1979. Registered prof. engr., Calif., N.Y., N.J., N.Mex. Engring. technician Calif. Divsn. Hwys., Sacramento, 1956-59; sr. engr. Calif. Dept. Water Resources, Sacramento, 1960-66, GE, San Jose, Calif., 1967-70; project mgr. R. Parsons Co., Pasadena, Calif., 1971-73; CEO, pres. Nuclear Power Svcs., N.Y.C., 1973-87; CEO, exec. v.p. BioTech Industries, Wyckoff, N.J., 1988-91; v.p. tech. Simon WTS BioTech, Santa Clara, Calif., 1991—; expert lectr. U. Calif., Berkeley, 1963-66; expert examiner Calif. P & Y Standards, Sacramento, 1964-68; expert instr. San Jose State U., 1967-69. Chief editor: Professional Engineering Examination Handbook, 1968; contbr. tech. articles to jours. With USN, 1952-55. Mem. ASME (nuclear codes com. 1972—, award 1990), ASCE, Am. Nuclear Soc. Achievements include patent for upflow biol. reactor treatment system. Home: 36 Wilderness Gate Santa Fe NM 87501

MOELLER, RONALD SCOTT, mechanical engineer; b. Teaneck, N.J., Nov. 12, 1963; s. Dennis Edward Moeller and Joyce (Anderson) Berger. AAS, County Coll. of Morris, Randolph, N.J., 1986; BSME, N.J. Inst. Tech., 1988. Devel. engr. I Lytel Corp., Somerville, N.J., 1986-88; staff engr. Ortel Corp., Alhambra, Calif., 1988—. Republican. Lutheran. Achievements include patent in field. Office: Ortel Corp 2015 W Chestnut St Alhambra CA 91803-1542

MOELLERING, ROBERT CHARLES, JR., internist, educator; b. Lafayette, Ind., June 9, 1936; s. Robert Charles and Irene Pauline (Nolde) M.; children: Anne Elizabeth, Robert Charles, Catherine Irene; m. Mary Jane Ferraro, July 11, 1987. BA, Valparaiso U., 1958, DSc, 1980; MD cum laude, Harvard U., 1962. Diplomate: Am. Bd. Internal Medicine. Intern Mass. Gen. Hosp., Boston, 1962-63, resident, 1963-64, postdoctoral fellow in infectious diseases, 1967-70, resident, 1966-67, mem. infectious disease unit and asst. physician, 1970-76, assoc. physician, 1976-83, hon. physician, 1983—, cons. bacteriology, 1972-87; instr. medicine Harvard U. Med. Sch., Boston, 1970-72, asst. prof., 1972-76, assoc. prof., 1976-80, prof., 1980—; chmn. dept. medicine, physician-in-chief New Eng. Deaconess Hosp., 1981—; Shields Warren-Mallinckrodt prof. clin. rsch. Harvard U. Med. Sch., Boston, 1981-89, Shields Warren-Mallinckrodt prof. med. rsch., 1989—; mem. subcom. on susceptibility testing Nat. Com. for CLin. Lab. Standards, 1976-88; mem. subcom. on antimicrobial agts. and chemotherapy, 1978-80; subcom. on antimicrobioi disc. diffusion suceptibility testing, 1980-88. Mem. editorial bd. Antimicrobial Agts. and Chemotherapy, 1977-81, editor, 1981-85, editor-in-chief, 1985—; editor European Jour. Clin. Microbial Infectous Diseases, 1990—; consulting editor Infectious Disease Clinics N.Am., 1986—; editor Les Infections, 1983; editorial bd. New Eng. Jour. Medicine, 1977-81, European Jour. Clin. Microbiology, 1981—, Jour. Infectious Diseases, 1981-85, 89—, Infectious Disease Alert, 1981-92, Pharmacotherapy, 1982—, Antimicrobial Agts. Ann., 1984-87, Zentralblatt Fur Bacteriologie, Microbiologie and Hygience, 1984—, Jour. of Infection, 1986—, Innovations, 1986—, Infectious Disease Clinics N. Am., 1986—, Residents Forum in Internal Medicine, 1988-90, Diagnostic Microbiology and Infectious Disease, 1989-90, Internat. Jour. Antimicrobial Agts., 1990—, Infectious Diseases in Clin. Practice, 1991-92. Served with USPHS, 1964-66. Grantee USPHS, NIH. Fellow ACP, Infectious Diseases Soc. Am. (v.p. 1988-89, pres. elect 1989-90, pres. 1990-91, past pres., 1991-92); mem. Am. Soc. Microbiology, Am. Clin. and Climatol. Assn., Internat. Soc. Chemotherapy, Am. Soc. Clin. Investigation, Assn. Am. Physicians, European Soc. Clin. Microbiology, Am. Fedn. Clin. Rsch., Roxbury Clin. Records Club, Mass. Med. Soc. (councilor), Brit. Soc. Antimicrobial Chemotherapy, Coun. Biology Editors, Alpha Omega Alpha, Phi Kappa Psi. Home: 49 Longfellow Rd Wellesley MA 02181-5220 Office: New Eng Deaconess Hosp Dept Medicine 110 Francis St Boston MA 02215-5501

MOESCHL, STANLEY FRANCIS, electrical engineer, management consultant; b. Cin., Mar. 14, 1931; s. Stanley F. and Matilda F. (Trenkamp) M.; m. Kathleen K. Koebel, Aug. 21, 1954; children: Stanley, Melissa, Deborah, Karen. BSEE, Purdue U., 1957. Engr. Honeywell Space Div., St. Petersberg, Fla., 1957-60; engring. mgr. Honeywell Space Div., St. Petersberg, 1960-69, program mgr., 1969-77; dir. engring. Honeywell Avionics Div., Mpls., 1977-80; v.p. gen. mgr. Honeywell Space Div., St. Petersberg, 1980-82, Honeywell Avionics Div., Mpls., 1982-88; pres. Sundstrand Data Control, Redmond, Wash., 1988-92; bd. mem. Com. of 100, St. Petersberg, 1980-82, Wash. Round Table, Seattle, 1989-92. Bd. dirs. Jr. Achievement, Mpls., 1983-86, Seattle, 1989-92. With USCG, 1951-54, Korea. Mem. IEEE, AIEE, Eta Kappa Nu, Tau Beta Pi. Home: 4575 S Landings Dr Fort Myers FL 33919

MOESE, MARK DOUGLAS, environmental consultant; b. Jersey City, Aug. 3, 1954; s. Harold Francis and Mary Frances (Wilk) M.; m. Elizabeth Renker Cozine, Apr. 20, 1991. BS, Fairleigh Dickinson U., 1976, MS, 1979; PhD, NYU, 1988. Rsch. asst. West Indies Lab., St. Croix, U.S.V.I., 1978-79, NYU Med. Ctr., Tuxedo, N.Y., 1980-86; staff scientist Hazen and Sawyer, P.C., N.Y.C., 1982-85; supr. risk assessment Ebasco Environ., Lyndhurst, N.Y., 1986—; cons. Taiwan Power Co., Taipei, Republic of China, 1987, 89, Hub River Power Co. , Fauji Corp., Karachi, Pakistan, 1991-92, Chinese Rsch. Acad. Environ. Scis., 1993; human and environ. risk assessments profl. Ebasco Environ., Lyndhurst, N.J., 1986—. Contbr. articles to profl. jours. Sigma Xi grantee-in-aid, 1978; grad. fellow NYU Med. Ctr., 1980-86. Mem. ASTM (voting mem., E-47 com., sediment toxicity subcom.), Soc. for Risk Analysis, Soc. Environ. Toxicology and Chemistry. Office: Ebasco Environ 160 Chubb Ave Lyndhurst NJ 07071-3586

MOESGEN, KARL JOHN, electronics engineer, science educator; b. Bonn, Fed. Republic Germany, Jan. 5, 1949; arrived in Peru, 1969; s. Martin Otto and Hermine (Dor) M.; m. Edith Badrian, Nov. 16, 1979; children: Deborah, Daniel. AS in Engring., Grantham Coll. Engrs., L.A., 1976; BS in Engring., Calif. Coast U., 1976, MS in Engring., 1977, PhD in Engring., 1983. Engring. technician Aero Peru, Lima, 1974-75; aux. prof. physics U. Lima, 1978-80; prof. flight theory Ministry of Aero., Lima, 1980; lectr. Peruvian Air Force Acad., Lima, 1978-79; asst. prof. physics U. Lima, 1981-82; lectr. relativity Cath. U. Peru, Lima, 1983; prof. physics electronics dept. San Marcos U., Lima, 1986, prof. physics faculty of physics, 1987; sci. educator and tutor Colegio Alexander von Humboldt, Lima, 1980—. Author: A Space-Time Model Concerning the Movement of Matter and Energy, 1986. Mem. Am. Phys. Soc., N.Y. Acad. Scis., German Phys. Soc., European Phys. Soc., Pilots Internat. Assn., Club de Regatas Lima, Club de la Union Lima. Office: Colegio Alexander von Humboldt, Casilla 18-1053, Lima 18, Peru

MOFFATT, HUGH MCCULLOCH, JR., hospital administrator, physical therapist; b. Steubenville, Ohio, Oct. 11, 1933; s. Hugh McCulloch and Agnes Elizabeth (Bickerstaff) M.; m. Ruth Anne Colvin, Aug. 16, 1958; children: David, Susan. AB, Asbury Coll., 1958; cert. in phys. therapy, Duke U., 1963. Lic. in phys. therapy and health care adminstrn., Alaska. Commd. officer USPHS, 1964, advanced through grades to capt.; therapist USPHS, N.Y.C., 1964-66, Sitka, Alaska, 1970-72; therapist cons. USPHS, Atlanta, 1968-70; clinic adminstr. USPHS, Kayenta, Ariz., 1972-73; hosp. dir. USPHS, Sitka, 1973-78; therapist cons. Idaho Dept. Health, Boise, 1966-68; contract health officer USPHS, Anchorage, 1978-89, ret., 1989; phys. therapy cons. Ocean Beach Hosp., Ilwaco, Wash., 1989—, Harbors Home Health Svcs., Aberdeen, Wash., 1990—; therapist cons. Our Lady of Compassion Care Ctr., Anchorage, 1979—, Alaska Native Med. Ctr., Anchorage, 1988—. With U.S. Army, 1955-57. Mem. Am. Phys. Therapy Assn., Commd. Officers Assn. USPHS, Res. Officers Assn., Ret. Officers Assn., Am. Assn. Individual Investors, Am. Assn. Ret. Persons, Eagles. Avocations: automobile repairs, woodworking, camping, fishing, church choir.

MOFFETT, MARK BEYER, physicist; b. Orrville, Ohio, Jan. 14, 1935; s. Rexford Wendell and Mary (Beyer) M.; m. Marilyn June Liechty, June 17, 1961; children: Alice Moffett Schultz, Bonnie Moffett, Mary Moffett, David Liechty Moffett. BS, MS, MIT, 1959; PhD, Brown U., 1970. Devel. engr. TRW Inc., Euclid, Ohio, 1959-65; rsch. asst. Brown U., Providence, 1966-69, rsch. assoc., 1970; asst. prof. U. R.I., Kingston, 1970-74; physicist Naval Underwater Systems Ctr., New London, Conn., 1974—; rsch. physicist Naval Underwater Systems Ctr., New London, 1971-74; adj. prof. U. R.I., Kingston, 1974-77, Hartford Grad. Ctr., 1979. Co-author, editor: NUSC Scientific and Engineering Studies-Nonlinear Acoustics 1954-83, 1984; contbr. numerous publs. to profl. jours. Mem. Waterford (Conn.) Community Band, 1980—; bd. trustees Waterford Pub. Libr., 1991—. Fellow Acoustical Soc. of Am.; mem. IEEE, Audio Engring. Soc. Achievements include patents for System for Detection of Transducer Defects; Lightweight Broadband Rayleigh Wave Transducer; Broadband, Acoustically Transparent, Nonresonant PVDF Hydrophone. Home: 12 Connshire Dr Waterford CT 06385 Office: Naval Undersea Warfare Ctr Code 3111 New London CT 06320

MOFFITT, CHRISTOPHER EDWARD, physicist; b. Boulder, Colo., Aug. 27, 1966; s. George Joseph and Susan Barbara (Hile) M. Bachelors in Physics and Math., Rockhurst Coll., 1988; Masters in Physics, U. Mo., Kansas City, 1991. Rsch. asst. U. Mo., Kansas City, 1989-92, lectr. physics, 1991--. Mem. Sigma Pi Sigma, Alpha Delta Gamma (chpt. v.p. 1987-88). Home: 1 NW 53rd Terr Gladstone MO 64118

MOFTAH, MOUNIR AMIN, engineering executive; b. Egypt, Nov. 22, 1922; arrived in Australia, 1968, naturalized 1972; m. Marcelle ElMahmoudy, Aug. 6, 1950; children: Magued, Medhat, Maha Prateley, Monica. B in Engring. with honors, Cairo U., 1948; MME, London U., 1970. Registered cons. engr., chartered engr. Workshop maintenance engr. Khanka Power, Water and Sewage Works, 1949-53; field supervising engr. for collective potable water project Ministry Mcpl. and Rural Affairs, Egypt, 1953-54; workshop engr. mech. and elect. power adminstrn. Ministry Housing and Utilities, Egypt, 1954-56, engr. in charge support svcs. br. mech./elec. power, 1956-60, from asst. dir. to dir. mech. fleet adminstrn., 1960-66, dir. mech. fleet sector planning dept., 1967-68; mgr. transport and vehicle depot United Distbn. Co., Cairo, 1966-67; engr. mech. and elect. design Water Authority West Australia, Leederville, 1969-86; deputy mgr. Kafr-el-Sheikh Water Supply Cons. Joint Venture KESCON, elManial, Cairo, Egypt, 1986-88; water and wastewater specialist Engring Consultants Group (ECG), Nasr City, Cairo, 1988—. Author: Pump Stations Design Manual, 1986, Water and Wastewater Treatment Plants, 1992; contbr. articles to profl. jours. Mem. ASME, Instn. Mech. Engrs., Instn. Engrs. Australia, Soc. Automotive Engrs., European Fedn. Nat. Engring. Assn. Home: 11 Syrinx Pl, Mullaloo 6025, Australia also: 8 Gameat elDowal elArabia, Apt 27 Mohandseen, 12411 Cairo Egypt

MOHAMMAD, SHAIKH NOOR, electronics engineer, educator; b. Paruldihi, India, Dec. 27, 1946; came to U.S., 1979; s. Shaikh Abdul and Shaikh Ayesha Jalil. PhD, Indian Inst. Tech., Kharagpur, India, 1977; DSc, Calcutta (India) U., 1982. Rsch. assoc. Indian Inst. Tech., Kharagpur, 1977-79, Case Western Res. U., Cleve., 1979-82; vis. prof. Indian Inst. Tech., Kharagpur, 1982-84; visitor Case Western Res. U., Cleve., 1985-87, U. Ill., Urbana, 1988-89; electronics engr. Semiconductor R&D Ctr., IBM, Hopewell Junction, N.Y., 1989—; prof. elec. engring. SUNY, New Paltz, 1990—. Contbr. over 100 articles to profl. jours. Mem. IEEE (sr., chmn. Electron Devices Soc. mid-Hudson sect. 1991—, Disting. Svcs. award 1992), Am. Phys. Soc. Achievements include patents and patents pending in field of bipolar transistors; research in bipolar transistors and related devices; in field effect transistors; in photovoltaics and optoelectronics; in solid-state electronics; in quantum mechanical study of materials structure. Office: IBM Semiconductor R&D Ctr 1580 Rt 52 Hopewell Junction NY 12533

MOHANAZADEH, FARAJOLLAH BAKHTIARI, chemist, educator; b. Masdjedsoliman, Khozestan, Iran, July 31, 1957; s. Fazel and Hamideh Mohanazadeh Bakhtiari; m. Atosa Mohanazadeh, Aug. 12, 1986; children: Merdad, Milad. PhD, U. Shiraz, Iran, 1989. Head faculty basic sci. U. Mazandaran, Babolsar, Iran, 1991—. Mem. Am. Chem. Soc., Iranian Chem. and Engring. Soc. Achievements include research in synthesis, characterization and application of polymer bond thiazolium. Office: U Mazanderan, PO Box 311, Babolsar Iran

MOHANTY, AJAYA K., physicist; b. Cuttack, Orissa, India, Feb. 21, 1952; came to U.S. 1976; s. Dibakar and Padmavati M.; m. Nivedita Mohanty, May 9, 1984. MS, SUNY, Albany, 1979; PhD, U. Md., 1984. Rsch. asst. SUNY, Albany, 1976-79; rsch. scholar NASA, Greenbelt, Md., 1980-84; rsch. asst. U. Md., 1979-84; rsch. assoc. NYU, 1984-85, Pa. State U., University Park, 1985-88; advisor rsch. and devel. IBM, Kingston, N.Y., 1988—; tech. group leader IBM Corp., 1989—. Contbr. articles to profl. jours., chpts. to books. Nat. Merit scholar, Govt. of India, 1969; NASA rsch. scholar, 1980; NSF postdoctoral fellow, 1984, Dept. of Energy postdoctoral fellow, 1985. Mem. Am. Phys. Soc., Indian Physics Assn. (chpt. sec. 1980-84). Achievements include original scientific research in theoretical atomic and molecular physics, quantum field theory and high energy physics; establishment of variational stability of relativistic basis set calculations in heavy atoms and extended the method to molecular structure. Office: IBM Corp Dept MLMA Power Parallel System 428 Neighborhood Rd Kingston NY 12401-1040

MOHANTY, UDAYAN, chemical physicist, theoretical chemist; b. Cuttack, India; s. Jitendra Nath and Sarbani Mohanty; m. Gail Fowler; children: Jayanta Fowler, Sudarsana Eldridge. BS, Cornell U., 1975; PhD, Brown U., 1981. Rsch. assoc. U. Calif., San Diego, 1980-82, U. Chgo., 1982-84; asst. prof. Boston Coll., Chestnut Hill, Mass., 1984-90; vis. scientist MIT, Cambridge, 1985-91, Harvard U., Cambridge, 1990-91, Calif. Inst. Tech., Pasadena, 1990-91; assoc. prof. Boston Coll., Chestnut Hill, 1990--; vis. scientist U. Calif., Santa Barbara, 1987. Contbr. articles to profl. jours. Grantee NSF, 1985-89, Petroleum Rsch. Funds of Am. Chem. Soc., 1985-90, Rsch. Corp., 1985-88, Cornell Theory Ctr., 1991-93, IBM Supercomputer Funds, 1989-90. Mem. Theoretical Chemistry Am. Chem. Soc., Am. Phys. Soc., Materials Rsch. Soc. Achievements include research on liquid-metal surface, liquid-vapor and liquid- solid transition, equilibrium and non-equilibrium statistical mechanics of supercooled and glassy states, complex fluids, protein folding and membranes, Brownian motion; re-normalization group approach to bonding of atoms and molecules; contributions to hydrodynamics of heat of transport, melted fluxiquid and quasicrystals. Office: Boston Coll Dept Chemistry Eugene F Merkert Chem Ctr Chestnut Hill MA 02167

MOHL, DAVID BRUCE, electronics engineer; b. Abington, Pa., Aug. 27, 1968; s. Elwood Ralph and Cleo Alma (Rudisill) M.; m. Bobbie Jo Painter, June 19, 1992. BSEE, Bob Jones U., 1990; MSEE, Drexel U., 1993. Lab. asst. Semcor, Inc., Warminster, Pa., 1989-90; electro-optics engr. Naval Air Warfare Ctr., Warminster, Pa., 1990—. Republican. Baptist. Achievements include development of sensor system to identify colors of ambient scene -- used to interface with active camouflage system on an unmanned aerial vehicle. Home: 432 Elm St Warminster PA 18974 Office: Naval Air Warefare Ctr Code 5013 PO Box 5152 Warminster PA 18974-0591

MOHL, JAMES BRIAN, aerospace engineer; b. Williston, N.D., Oct. 29, 1957; s. Keith LaVern and Olive Marlene (Zook) M.; m. Gwynne Rosanne Weiss, Sept. 1, 1978; children: Brianne Marie, Christine Elyse. BSEE with highest honors, Mont. State U., 1979; MS in Aero. Engring., U. Colo., 1991. Engr. Sperry Flight Systems, Phoenix, 1979-83; control system analyst Ball Aero. and Comm. Group, Boulder, Colo., 1983—. Vice pres. congl. coun. Christ the Servant Luth. Ch., Louisville, Colo., 1988-90. Mem. AIAA, IEEE, Am. Astron. Soc. Republican. Achievements include research in control of flexible structures. Home: 776 W Aspen Way Louisville CO 80027-9771 Office: Ball Aero and Comm Group CO 5 PO Box 1062 Boulder CO 80306-1062

MOHLENBROCK, ROBERT HERMAN, JR., botanist, educator; b. Murphysboro, Ill., Sept. 26, 1931; s. Robert Herman and Elsie (Treece) M.; m. Beverly Ann Kling, Oct. 19, 1957; children—Mark William, Wendy Ann, Trent Alan. B.A., So. Ill. U., 1953; M.S., 1954; Ph.D., Washington U., St. Louis, 1957. With dept. botany So. Ill. U., Carbondale, 1957—; chmn. dept. So. Ill. U., 1966-79, prof., 1966-85, disting. prof., 1985—. Author: A Flora of Southern Illinois, 1957, Plant Communities of Southern Illinois, 1963, Ferns of Illinois, 1967, Flowering Plants of Illinois: Flowering Rush to Rushes, 1970, Flowering Plants of Illinois: Lilies to Orchids, 1970, Grasses of Illinois, 1972, 73, Forest Trees of Illinois, 1973, Guide to the Vascular Flora of Illinois, 1975, Spring Wildflowers of Carlyle-Rend-Shelbyville Lakes, 1975, Summer and Fall Wildflowers of Carlyle-Rend-Shelbyville Lakes, 1975, Sedges of Illinois, 1976, Hollies to Loasas in Illinois, 1978, Distribution of Illinois Vascular Plants, 1978, Prairie Plants of Illinois, 1979, Hunter's Guide to Illinois Flowering Plants, 1980, Flowering Plants of Illinois: Willows to Mustards, 1980, Spring Woodland Wildflowers of Illinois, 1980, You Can Grow Tropical Fruit Trees, 1980, Flowering Plants of Illinois: Magnolias to Pitcher Plants, 1981, Wildflowers of Roadside Fields and Open Habitats in Illinois, 1981, Giant City State Park, An Illustrated Handbook, 1981, Flowering Plants of Illinois: Basswoods to Spurges, 1982, Where Have All the Wildflowers Gone?, 1983, The Field Guide to U.S. National Forests, 1984, New Guide to Illinois Vascular Plants, 1986, Flowering Plants of Illinois: Smartweeds to Hazelnuts, 1986, Macmillan's Wildflower Guide, 1987, MacMillan's Tree Guide, 1987, Field Guide to Illinois Wetlands, 1988, Midwestern Field Guide to Wetlands, 1988; monthly columnist Natural History mag.; also articles. Trustee Ill. Nature Conservancy, Mo. Native Plant Soc.; mem. Ill. Nature Preserves Commn. Mem. Am. Fern Soc., Assn. So. Biologists, So. Appalachian Bot. Club., other native plant socs. Home: 1 Birdsong Dr Carbondale IL 62901-9038

MOHLER, STANLEY ROSS, JR., aeronautical research engineer; b. Bethesda, Md., Feb. 28, 1961; s. Stanley Ross and Ursula Luise (Burkhardt) M.; m. Kathleen Joan Andre, Aug. 27, 1988; children: Samantha Marie, Kurt Andre. BA in Physics, Beloit Coll., 1983; MS in Aero. and Astro. Engring., Ohio State U., 1990. Teaching asst. Dept. Physics Wayne State Univ., Detroit, 1984-85; rsch. asst. Dept. Aero/Astro Engring Ohio State U., Columbus, 1986-89; aero. rsch. engr. Sverdrup Tech., Inc., Brook Park, Ohio. Author paper in field. Mem. AIAA, Exptl. Aircraft Assn., Aircraft Owners and Pilots Assn. Home: 11526 Harbour Light Dr North Royalton OH 44133

MOHLER, TERENCE JOHN, psychologist; s. Edward F. and Gertrude A. (Aylward) M.; m. Carol B. Kulczak; children: Renee, John, Timothy. BE, ME, EdS, Toledo U.; PhD, Walden U., Union Inst., 1979. Psychologist, Toledo Bd. Edn., 1969-89; sr. ptnr. Psychol. Assocs., Maumee, Ohio, 1970—; assoc. fellow Inst. for Advanced Study in Rational Psychotherapy, N.Y.C. Served with U.S. Army, 1951-53; Korea. Lic. psychologist, Ohio. Mem. Am., Ohio, Northwestern Ohio, Maumee Valley Psychol. Assns., Soc. Behaviorists, Toledo Acad. Profl. Psychology, Nat. Registry Mental Health Providers, Am. Pers. and Guidance Assn., Ohio Pers. and Guidance Assn., Coun. for Exceptional Children, Rotary (Paul Harris Fellow), Kappa Delta Phi. Home: 1904 Glen Ellyn Dr Toledo OH 43614-3256 Office: 5757 Monclova Rd Maumee OH 43537

MOHR, JOHN LUTHER, biologist, environmental consultant; b. Reading, Pa., Dec. 1, 1911; s. Luther Seth and Anna Elizabeth (Davis) M.; m. Frances Edith Christensen, Nov. 23, 1939; children: Jeremy John, Christopher Charles. A.B. in Biology, Bucknell U., 1933; student, Oberlin Coll., 1933-34; Ph.D. in Zoology, U. Calif. at Berkeley, 1939. Research asso. Pacific Islands Research, Stanford, 1942-44; research asso. Allan Hancock Found., U. So. Calif., 1944-46, asst. prof., 1946-47, asst. prof. dept. biology, 1947-54, asso. prof., 1954-57, prof., 1957-77; chmn. dept., 1960-62, prof. emeritus, 1977—; vis. prof. summers U. Wash. Friday Harbor Labs., 1956, '57; marine borer and pollution surveys harbors So. Calif., 1948-51, arctic marine biol. research, 1952-71; chief marine zool. group U.S. Antarctic research ship Eltanin in Drake Passage, 1962, in South Pacific sector, 1965; research deontology in sci. and academia; researcher on parasitic protozoans of anurans, crustaceans, elephants; analysis of agy. and industry documents, ethics and derelictions of steward agy., sci. and tech. orgns. as they relate to offshore oil activities, environ. effects of oil spill dispersants and offshore oil industry discharges. Active People for the Am. Way; mem. Biol. Stain Commn., 1948-80, trustee, 1971-80, emeritus trustee, 1981—, v.p., 1976-80. Recipient Guggenheim fellowship, 1957-58. Fellow AAAS (coun. 1964-73), So. Calif. Acad. Sci., Sigma Xi (exec. com. 1964-67, 68, 69, chpt. at large bd. 1968-69); mem. Am. Micros. Soc., Marine Biol. Assn. U.K. (life), Am. Soc. Parasitologists, Western Soc. Naturalists (pres. 1960-61), Soc. Protozoologists, Am. Soc. Tropical Medicine and Hygiene, Am. Soc. Zoologists, Ecol. Soc. Am., Planning and Conservation League, Calif. Native Plant Soc., Am. Inst. Biol. Scis., L.A. MacIntosh Group, Save San Francisco Bay Assn., Ecology Ctr. So. Calif., So. Calif. Soc. Toxicologists, So. Calif. Soc. Parasitologists, Common Cause, Huxleyan, Sierra Club, Phi Sigma, Theta Upsilon Omega. Democrat. Home: 3819 Chanson Dr Los Angeles CA 90043-1601

MOHR, SIEGFRIED HEINRICH, mechanical and optical engineer; b. Vöhrenbach, Baden, Fed. Republic Germany, Sept. 20, 1930; came to U.S., 1958.; s. Adolf and Luise (Faller) M.; m. Gloria P. Vauges, Apr. 25, 1959 (div. 1972); children: Michael S., Brigitte M.; m. Jeani Edith Hancock, Mar. 24, 1973; 1 child, Suzanne A. Diplom-Ingenieur, Universität Stuttgart, Fed. Republic Germany, 1957; MS in Optical Engring., SUNY, 1971. Thesis researcher Daimler Benz AG, Stuttgart, 1957; design engr. Russell, Birdsall & Ward B & Nut Co., Port Chester, N.Y., 1958-59; devel. engr. IBM Advanced Systems Devel. div., San Jose, Calif., 1960-64; rsch. engr., inventor Precision Instrument Co., Palo Alto, Calif., 1964-67; prin. engr. RCA Instructional Systems, Palo Alto, 1967-70; rsch. engr., scientist Singer Simulation Products, Sunnyvale, Calif., 1971-73; project leader Dymo Industries Tech. Ctr., Berkeley, Calif., 1973-77; leader, adv. rsch. & devel. NCR Corp. Micrographic Systems Div., Mountain View, Calif., 1977-89; advanced project mgr. electro-optics dept. (U.S.A.) Angénieux, Santa Clara, Calif., 1989—; translator for books in English, French and German; corr., writer for European jazz publs. Patentee, author in field. Bicycle activist League of Am. Wheelmen, Balt., 1975—; del. mem. U.S. Del. ISO Conf., Paris, 1988. Mem. Soc. Photo-Optical Instrumentation Engrs., Assn. Info. and Image Mgmt., Internat. Soc. Optical Engring. Avocations: bicycle touring, playing jazz piano, collecting jazz music, practicing fgn. langs., lit. and internat. rels. Home: 3311 Benton St Santa Clara CA 95051-4420

MOHRHERR, CARL JOSEPH, biologist; b. Buffalo, Sept. 29, 1944; s. Roderick Lowe Mohrherr and Beatrice (Shapiro) Lumpkin. BS in Zoology, Roosevelt U., Chgo., 1966; MS in Biology, DePaul U., 1969; PhD in Biology, TuLane U., 1973. Asst. prof. biology St. Mary's Dominican Coll., New Orleans, 1974-77, U. Eduardo Mond Lane, Maputo, Mozambique, 1977-84, Univ. S.W. La., Lafayette, 1984-85; rsch. assoc. Univ. West Fla., Pensacola, 1985—. Contbr. articles to profl. jours. Mem. Am. Zoologist Soc. Office: U West Fla Pensacola FL 32514

MOHRMANN, LEONARD EDWARD, JR., chemical engineer; b. Winston Salem, N.C., June 14, 1940; s. Leonard E. and Helen (Bean) M.; m. Sue Ross, June 18, 1966; children: Leonard III, Vaden, Nelwyn Ann. BS in Chemistry, U. Tex., 1963; PhD in Chemistry, Fla. State U., 1971; BSChemE, Tex. A&M U., 1980. Cert. profl. chemist. Tech. lab. coord. Tex. A&M U., College Sta., 1971-81, rsch. assoc., 1981-82; chemist, chemical engr. Tex. Dept. Health, Austin, 1982-92, Tex. Water Commn., Austin, 1992-93, Fugro McClelland, Houston, 1993—. Contbr. articles to sci. jours. Active Mathcounts program Tex. Soc. Profl. Engrs., Austin, 1991-92. Recipient Appreciation cert. Tex. Soc. Profl. Engrs., 1993. Mem. AICE (chmn., vice chmn. balcones fault subsect.). Achievements include development of Texas regulations for special waste disposal, medical waste handling and disposal, special waste program for municipal waste. Office: Fugro McClelland 6100 Hillcroft Ste 300 Houston TX 77081

MOILANEN, MICHAEL DAVID, civil engineer; b. Tampa, Fla., Oct. 30, 1968; s. David and Barbara (Jeannotte) M. BCE, Ga. Tech., 1991. Registered engr.-in-tng. Engr. II structures design office Fla. Dept. Transp., Tallahassee, 1992—. Mem. ASCE (assoc.). Home: 1515 Paul Russell Rd Tallahassee FL 32301 Office: Fla Dept Transp MS33 605 Suwannee St Tallahassee FL 32399

MOJCIK, CHRISTOPHER FRANCIS, internist; b. Bridgeport, Conn., Oct. 23, 1959; s. Michael George and Theresa Annette (Suich) M.; m. Robin Jean Guion, July 31, 1982; children: Daniel Stephen, Arla Elizabeth. BA, Washington U., St. Louis, 1981; MD, PhD, U. Conn., Farmington, 1988. Diplomate Am. Bd. Internal Medicine. Med. intern Roger Williams Gen. Hosp., Providence, 1988-89, med. resident, 1989-91; clin. assoc., fellow Nat. Inst. Arthritis, Muscoloskeletal and Skin Diseases, NIH, Bethesda, Md., 1991—; del. Nat. Inst. Diabetes, Digestive and Kidney Diseases, Nat. Inst. Arthritis, Muscoloskeletal and Skin Diseases, Bethesda, 1991—, Clin. Assocs. Com., Bethesda, 1992—. Mem. exec. bd. Sally Ride Elem. Sch. PTA, Germantown, Md., 1992—. Lt. comdr. USPHS, 1991—. Grad. student fellow U. Conn., Farmington, 1981-88. Assoc. ACP; mem. AMA, Am. Coll. Rheumatology. Roman Catholic. Achievements include reseach in phemophpic devel. and functional studies of RT6 Bearing Ret T Lymphocyte, immunology and immunopathology. Office: NIAMS NIH Bldg 10 Rm 9N216 9000 Rockville Pike Bethesda MD 20892

MOK, CARSON KWOK-CHI, structural engineer; b. Canton, China, Jan. 17, 1932; came to U.S., 1956, naturalized, 1962; s. King and Chi-Big (Lum) M.; B.S. in Civil Engring., Chu Hai U., Hong Kong, 1953; M.C.E., Cath. U. Am., 1968; m. Virginia Wai-Ching Cheng, Sept. 19, 1959. Structural designer Wong Cho Tong, Hong Kong, 1954-56; bridge designer Michael Baker Jr., Inc., College Park, Md., 1957-60; structural engr., chief design engr., asso. Milton A. Gurewitz Assos., Washington, 1961-65; partner Wright & Mok, Silver Spring, Md., 1966-75; owner Carson K.C. Mok, Cons. Engr., Silver Spring, 1976-81, pres., 1982—; facility engring. cons. Washington Met. Area Transit Authority, 1985-86; pres. Transp. Engring. and Mgmt. Assocs., P.C., Washington, 1986—; adj. asst. prof. Howard U., Washington, 1976-79, adj. assoc. prof., 1980-81; bd. dirs. U.S. Pan Asian Am. C. of C. Sec., N.Am. bd. trustees, China Grad. Sch. Theology, Wayne, Pa., 1972-74, pres., 1975-83, v.p., 1984-91; elder Chinese Bible Ch. Md., Rockville, 1978-80; chmn. Chinese Christian Ch. Greater Washington, 1958-61, 71, elder, 1972-76. Recipient Outstanding Standard of Teaching award Howard U., 1980; registered profl. engr., Md., D.C. Mem. ASCE, ASTM, Constrn. Specification Inst., Nat. Assn. Corrosion Engrs., Concrete Reinforcing Steel Inst., Am. Concrete Inst., Am. Welding Soc., Prestressed Concrete Inst., Post-Tensioning Inst., Soc. Exptl. Mechanics., Internat. Assn. Bridge and Structural Engring. Contbr. articles to profl. jours. Home: 4405 Bestor Ct Rockville MD 20853-2137 Office: 9001 Ottawa Pl Silver Spring MD 20910-2257

MOKHOV, OLEG IVANOVICH, mathematician; b. Orenburg, Russia, June 28, 1959; s. Ivan Vasiliyevich and Maya Nikolayevna (Khlebnikova) M. Student, Leningrad U., Russia, 1977; Honours Degree, Moscow U., 1981, PhD, 1984. Postgrad. Moscow U., Russia, 1981-84; scientist Nat. Sci. and Rsch. Inst. for Phys. Tech. Measurements, Mendeleevo, Russia, 1984-93; sr. scientist Vsesoyuzny Nauchno Issledovatelskiy Institut Fiziko Tekhnicheskikh i Radiotekhnicheskikh Izmereniy, Mendeleevo, 1990-93, Steklov Math. Inst., Moscow, 1993—. Contbr. articles to profl. jours. Mem. Moscow Math. Soc. Home: Ulitsa Kuybysheva 12 v ap 66, I41570 Mendeleevo Russia Office: Steklov Math Inst, ul Vavilova 42, Moscow 117966, Russia

MOKHTARZADEH, AHMAD AGHA, agronomist, consultant; b. Fassa, Fars, Iran, Oct. 23, 1933; came to U.S., 1989; s. Mohamad Hassan and Assieh (Kadivar) M.; m. Brigitte Becker, Nov. 12, 1980; 1 child, Mitra. BSc, BA, U. Tehran, Iran, 1956; MSc, U. Md., 1960; PhD, U. Paris, France, 1964. Assoc. prof. Shiraz U., Iran, 1964-80; cons. S. Pacifc regions Somalia, Vietnam, Namibia Food and Agrl. Orgn. U.N., Rome, 1981-89; cons. pvt. practice Leesburg, Va., 1990—; mem. Pres. Program Com. Shiraz U., Iran, 1977-79; advisor German Remote Sensing in Agrl., Somalia, 1984; officer-in-charge Food Agrl. Orgn., U.N., Western Samoa, 1985. Co-author: Bibliography of Natural History of Iran, 1965; contbr. articles to profl. jours. including Crop Sci., Der Züchter, Nematologia Mediterranea. Recipient 4-yr. scholarship Govt. of Iran, 1958-62, 2-yr. scholarship Govt. of France, 1962-64, Fulbright fellowship N.D. State U., 1968-69, Rsch. fellowship, Energy Rsch. and Devel., Oak Ridge, Tenn., 1976-77. Mem. Am. Soc. Agronomy, Sigma Xi. Home and Office: 141 Davis Ave SW Leesburg VA 22075-3405

MOLDENAERS, PAULA FERNANDE, chemical engineer, educator; b. Leuven, Belgium, May 2, 1957; d. Louis and Maria (Van Horenbeek) M.; m. Herman Van der Auweraer, Aug. 21, 1981. Degree in chem. engring., U. Leuven, 1980, PhD, 1987. Teaching asst. U. Leuven, 1980-87; rsch. assoc. U. Levuen, 1987-91, assoc. prof., 1991—. Contbr. over 30 articles to profl. jours. Recipient Ann. award British Soc. Rheology, 1991, Exxon Chem. Sci. and Engring. award Nat. Fonds voor Wetenschappelijk Onderzoek, 1992. Mem. Am. Chem. Soc., Belgian Group Rheology (sec. 1988—), Soc. Rheology, Polymer Proc. Soc., Plastics Soc., Royal Flemish Engrs. Assn. Office: U Leuven Dept Chem Engring, de Croylaan 46, 3001 Leuven Belgium

MOLIERE, JEFFREY MICHAEL, cardio-pulmonary administrator; b. San Pedro, Calif., Nov. 22, 1948; s. Dwight Hedrick and Geraldine Stabile. AA, L.A. Harbor Coll., 1968; postgrad., Calif. State U., Long Beach, 1968-69, Calif. Coll. for Health Sci., 1991—; Assoc. degree, Ind. U., Indpls., 1987, B. in Gen. Studies, 1990; postgrad., Calif. Coll. for Health Scis., 1992—. Registered respiratory therapist. Alt. supr. Good Samaritan Hosp., Vincennes, Ind., 1976-79; criticalcare technician Winona Meml. Hosp., Indpls., 1979-80; neonatal intensive care unit-critical care technician Mercy Hosp., Urbana, Ill., 1980-82; cardio-pulmonary supr. Winona Meml. Hosp., Indpls., 1982-92; dir. pulmonary svcs. MidWest Med. Ctr., Indpls., 1992—. Mem. advisory bd. allied health Ind. Vocat. Tech. Coll., 1987—. Mem. Am. Assn. for Respiratory Care, Ind. Soc. for Respiratory Care, Alpha Sigma Lambda (charter, Membership award 1990).

MOLINA, MARIO JOSE, physical chemist, educator; b. Mexico City, Mexico, Mar. 19, 1943; came to U.S., 1968; s. Roberto Molina-Pasquel and Leonor Henríquez; m. Luisa Y. Tan, July 12, 1973; 1 child, Felipe. Bachillerato, Acad. Hispano Mexicana, Mexico City, 1959; Ingeniero Químico, U. Nacional Autónoma de México, 1965; postgrad., U. Freiburg, Fed. Republic Germany, 1966-67; Ph.D., U. Calif., Berkeley, 1972. Asst. prof. U. Nacional Autónoma de México, 1967-68; research assoc. U. Calif.-Berkeley, 1972-73; research assoc. U. Calif.-Irvine, 1973-75, asst. prof. phys. chemistry, 1975-79, assoc. prof., 1979-82; sr. rsch. scientist Jet Propulsion Lab., 1983-89; prof. dept. earth, atom and planet sci., dept. chemistry MIT, Cambridge, 1989—, Martin prof. atmospheric chemistry. Recipient Tyler Ecology award, 1983, Esselen award for chemistry in pub. interest, 1987. Mem. Am. Chem. Soc., Am. Phys. Soc., Am. Geophys. Union, Nat. Acad. Scis., Sigma Xi. Achievements include discovering the theory that fluorocarbons deplete ozone layer of stratosphere. Home: 8 Clematis Rd Lexington MA 02173-7117 Office: MIT Dept of EAPS 54-1312 77 Massachusetts Ave Cambridge MA 02139

MOLINO-BONAGURA, LORY JEAN, neurobiologist; b. Quezon, The Philippines, Mar. 1, 1964; came to U.S., 1967; d. Lorenzo Daban and Carmelita (Jason) M.; m. Anthony Francis Bonagura, Aug. 10, 1991. BS, SUNY, Brockport, 1986; MA, NYU, 1989. Rsch. asst. dept. psychology SUNY, Brockport, 1985-86; rsch. asst. dept. psychology NYU, N.Y.C., 1986-87, rsch. asst. dept. biology, 1987-88; tech. rsch. specialist dept. psychiatry SUNY, Stony Brook, 1988-90; temp. regulatory affairs dept. Smithkline-Beecham, Phila., 1991-92; assoc. scientist cardiovascular biology dept. Rhone-Poulenc Rorer, Collegeville, Pa., 1992-93; sr. rsch. scientist Sterling-Winthrop PRD, Collegeville, 1993—. Presenter N.Y. Acad. Sci., 1989, Soc. for Neuroscience, 1989; contbr. articles to profl. jours. Mem. AAAS, Assn. for Women in Scis., N.Y. Acad. Scis., Soc. for Neurosci., Internat. Brain Orgn. Republican. Achievements include research in genetic predispositioning of learned helplessness in developmental neural circuitry; interactive capacity of monoaminerg system; vacillatory behavior in immature rats via a D-1 agonist. Home: 4562 Fleming St Philadelphia PA 19128-4719

MOLL, DAVID CARTER, civil engineer; b. Ames, Iowa, Aug. 5, 1948; s. Dale Curtis and Virginia (Carter) M.; m. Margaret E. Newman (div. 1989); 1 child, Megahn Elizabeth. BSCE, Iowa State U., 1971; cert. advanced study, Am. Grad. Sch. Internat. Mgmt., 1983; MBA with distinction, U. Mich., 1984. Engr. in tng. Iowa; field engr. Chgo. Bridge & Iron Co., 1971; subcontract supr., field engr. Morrison-Knudsen Internat. Co., Inc., Surinam and Panama, 1976; site supt. engring., asst. supt. constrn. Fluor Corp., Saudi Arabia, 1977-82; group mgr. Cummins Engine Co., Columbus, Ind., 1984-85; mgr. spl. projects Kerr-McGee Coal Corp., Oklahoma City, 1985-88; project mgr. Kerr-McGee Corp., Oklahoma City, 1989, U.K., 1989-90, Saudi Arabia, 1990—. Lt. USN, 1971-75. Mem. ASCE, Civil Engr. Corps. (Meritorious Svc. medal), Am. Soc. Quality Control (constrn. tech. com.), Am. Legion, Order of the Knoll, Chi Epsilon. Avocations: cross country skiing, jogging, golf, sailing, stamps.

MOLL, JOHN LEWIS, electronics engineer; b. Wauseon, Ohio, Dec. 21, 1921; s. Samuel Andrew and Esther (Studer) M.; m. Isabel Mary Sieber, Oct. 28, 1944; children: Nicolas Josef, Benjamin Alex, Diana Carolyn. B.Sc., Ohio State U., 1943, Ph.D., 1952; Dr. h.c., Faculty Engring., Katholieke U. Leuven, (Belgium), 1983. Elec. engr. RCA Labs., Lancaster, Pa., 1943-45; mem. tech. staff Bell Telephone Labs., Murray Hill, N.J., 1952-58; mem. faculty Stanford U., 1958-69, prof. elec. engring., 1959-69; tech. dir. optoelectronics Fairchild Camera and Instrument Corp., 1969-74; dir. integrated circuits labs. Hewlett-Packard Labs., Palo Alto, Calif., 1974-80; dir. IC structures research, sr. scientist Hewlett-Packard Labs., 1980-87, dir. Superconductivity Lab., 1987-90, mem. tech. staff, 1990—. Author: Physics of Semi Conductors, 1964; co-author Computer Aided Design and VLSI Device Development, 1985, rev. edit., 1988; inventor (with Ebers) first analytical transistor model, 1953, still valid and useful for circuit design. Guggenheim fellow, 1964; Recipient Howard N. Potts medal Franklin Inst., 1967, Disting. Alumnus award Coll. Engring., Ohio State U., 1970, Benjamin C. Lamme medal Coll. Engring. Ohio State U., 1988. Fellow IEEE (Ebers award 1971, Thomas A. Edison medal 1991), Am. Acad. Arts and Scis.; mem. Am. Phys. Soc., Nat. Acad. Engring., Nat. Acad. Scis., Sigma Xi, Sigma Pi Sigma, Tau Beta Pi. Home: 4111 Old Trace Rd Palo Alto CA 94306-3728 Office: 3500 Deer Creek Rd Palo Alto CA 94304-1392

MOLLARD, JOHN DOUGLAS, engineering and geology executive; b. Regina, Sask., Can., Jan. 3, 1924; s. Robert Ashton and Nellie Louisa (McIntosh) M.; m. Mary Jean Lynn, Sept. 18, 1952; children: Catherine Lynn, Jacqueline Lee, Robert Clyde Patrick. BCE, U. Sask., 1945; MSCE, Purdue U., 1947; PhD, Cornell U., 1952. Registered profl. engr., profl. geologist Sask., Alta. and B.C., Can. Resident constrn. engr. Sask. Dept. Hwys and Transp., 1945; grad. asst. Purdue U., West Lafayette, Ind., 1946-47, rsch. engr. Sch. Civil Engring., 1950-52; air surveys engr., soil and water conservation and devel. Prairie Farm Rehab. Adminstrn., Govt. of Can., 1947-50; chief, airphoto analysis and engring. geology div. Prairie Farm Rehab. Adminstrn., Govt. of Can., Regina, 1953-56; pres. J.D. Mollard and Assocs. Ltd., Regina, 1956—; aerial resource mapping surveys tech. adv. Colombo plan, Govts. Ceylon and Pakistan, 1954-56; advisor Shaw Royal Commn. on Nfld. Agr.; disting. lectr. series Ea. Can. Geotech. Soc., 1969; Cross Can. disting. lectr. Can. Geotech Soc., 1993; guest lectr., vis. lectr., instr. over 50 short courses on remote sensing interpretation aerial photos and satellite imagery numerous univs., cities and provinces in Can., also Cornell U., Ithaca, N.Y., Harvard U., Cambridge, Mass., U. Calif., Berkeley, U. Wis., Madison, U. Hawaii, 1952—. Author: Landforms and Surface Materials of Canada, 7 edits.; co-author: Airphoto Interpretation and the Canadian Landscape, 1986; contbr. over 100 articles to profl. pubs. Organizer, canvasser United Appeal campaigns; former bd. dirs. Regina Symphony Orch. Recipient Engring. Achievement award Assn. Profl. Engrs. Sask., 1983, Massey medal Royal Can. Geog. Soc., 1989. Fellow ASCE, Geol. Soc. Can., Geol. Soc. Am., Am. Soc. Photogrammetry and Remote Sensing (award for contbns. airphoto interpretation and remote sensing 1979), Internat. Explorers Club; mem. Engring. Inst. Can. (Keefer medal 1948), Assn. Cons. Engrs. Can., Can. Geotech. Soc. (1st R.M. Hardy Meml. Keynote lectr. 1987, Thomas Roy award with engring. geology div. 1989, R.F. Legget award 1992), Regina Geotech. Soc., Geol. Soc. Sask., Can. Soc. Petroleum Engrs., Regina YMCA (former dir.), Rotary (former dir. Regina club). Mem. United Ch. of Can. Avocations: jogging, reading, golf, tennis, nature study. Home: 2960 Retallack St, Regina, SK Canada S4S 1S9 Office: JD Mollard/Assoc 810 Avord Tower, 2002 Victoria Ave, Regina, SK Canada S4P 0R7

MOLLEGEN, ALBERT THEODORE, JR., engineering company executive; b. Meridian, Miss., Aug. 13, 1937; s. Albert Theodore and Harriette Ione (Rush) M.; m. Glenis Ruth Gralton, Feb. 16, 1962; children: Glenis Ione, Marion Anne. BEE, Yale U., 1961; postgrad. Poly. Inst. Bklyn., 1965-67. Cir. designer Melpar Inc., Falls Church, Va., 1958-59; system engr., group supr. Arma div. AMBAC Industries, N.Y.C., 1961-67; dept. mgr. Mystic Oceanographic Co. (Conn.), 1967-71; v.p., group mgr. Analysis & Tech. Inc., North Stonington, Conn., 1971-76, pres., CEO, 1976-91, chmn., CEO, bd. dirs., 1991—; bd. dirs. Continental Dynamics, Inc., Reston, Va., Automation Software, Inc., No. Kingston. R.I., Benthos, Inc., No. Falmouth, Mass., Philips Cir. Assemblies/Slatersville (R.I.) Inc., SEATECH, Groton, Conn. Applied Sci. Assocs., Inc., Butler, Pa., Gen. Systems Solutions, Inc., Groton, Ct., Naval Submarine League, Annandale, Va., Profl. Svcs. Coun., Vienna, Va. Inventor towed array rangefinder; writer numerous tech. reports. Chmn. Mystic Area Ecumenical Coun., 1970-72; treas. Seabury deanery Episcopal Diocese Conn., 1974-76, mem. diocesan fin. policy com., 1975, missionary strategy com., 1976-78, chmn. stewardship com., 1982-89, del. diocesan conv., 1974—, del. nat. gen. conv., 1991, alt. del. nat. gen. conv., 1982, 85, 88. Mem. Am. Soc. for Quality Control, IEEE, Am. Mgmt. Assn., Acoustical Soc. Am., Nat. Contract Mgmt. Assn., Ops. Rsch. Soc. Am., Inst. Mgmt. Scis., Navy League U.S., Am. Def. Preparedness Assn., Am. Soc. Naval Engrs., Surface Navy Assn., The Newcomer Soc. Home: 337 Pleasant St Willimantic CT 06226-3311 Office: Analysis & Tech Inc PO Box 220 North Stonington CT 06359-0220

MÖLLER, DETLEV, atmospheric chemist; b. Berlin, May 30, 1947; s. Heinz and Ingeborg (Barth) M.; m. Ursula Rahn, Mar. 6, 1972; children: Andreas, Stefan. Chemist degree, Humboldt U., Berlin, 1970, PhD, 1972; Habilitation, Acad. of Scis., Berlin, 1982. Chem. diplomate. Rsch. scientist Humboldt U., Berlin, 1972-74; rsch. scientist Inst. Geography & Geoecology, Berlin, 1974-82, head of a group, 1982-85; head of dept. Heinrich Hertz Inst., Berlin, 1986-91; br. head Fraunhofer Inst. for Atmospheric Environ. Rsch., Berlin, 1992—; lectr. Humboldt U., Berlin, 1988-93, Free U., Berlin, 1993—; mem. com. German EUROTRAC, Bonn, Fed. Republic Germany, 1990-92. Editor books; contbr. articles and revs. to profl. jours.; patentee in field; mem. editorial bd. Idöjaras, Jour. Hungarian Meteorol. Svc., 1991—. Mem. European Assn. on Scis. of Air Pollution (com. mem. 1990-91, v.p. 1992—), Commn. on Atmospheric Chemistry and Global Pollution. Avocations: painting, sailing. Home: Hackenbergstr, D-12489 Berlin Germany Office: Fraunhofer Inst, Br for Air Chemistry, D-12484 Berlin Germany

MÖLLER, GÖRAN, immunology educator; b. Vittangi, Sweden, Aug. 30, 1936; s. Sven and Edia (Falck) M.; m. Erna Lindell, June 10, 1960; children: Gunnar, Elisabeth. MD and PhD, Karolinska Inst., Stockholm, 1963. Lic. physician, Sweden. Scientist Swedish Cancer Soc., Stockholm, 1963-69; prof. Karolinska Inst., Stockholm, 1969-85, Stockholm U., 1985—; mem. Nobel Com., Stockholm, 1976-78, Nobel Assembly, Stockholm, 1977-85, vice chmn., 1985. Editor Jour. Immunological Revs., 1969—, Scandinavian Jour., 1990—. Recipient A. Jahres prize Oslo U., 1976. Mem. Am. Assn. Immunology (hon.), Scandinavian Soc. Immunology (hon.). Home: Morabergsvagen 14, 133 33 Saltsjobaden Sweden Office: Stockholm U, 106 91 Stockholm Sweden

MOLLOVA, NEVENA NIKOLOVA, chemist; b. Sofia, Bulgaria, Aug. 25, 1958; d. Nikola Mollov and Iordanka Anastasova (Hadjigrigorova) M. MSc, U. Sofia, 1981; PhD, Bulgarian Acad. Sci., 1988. Chemist Bulgarian Acad. Scis., Sofia, 1981-84, rsch. collaborator III, 1984-88, rsch. collaborator II, 1988-90; rsch. assoc. Coll. Pharmacy, U. Ariz., Tucson, 1990-93; asst. rsch. scientist Coll. Nursing, U. Ariz., Tucson, 1993—. Contbr. articles to mass spectrometry jours. Min. Rels. Exterieures fellow, Paris, 1985-86, Ctr. Internat. Echanges Scientifiques postdoctoral fellow, Paris, 1988-89. Mem. Am. Soc. Mass Spectrometry. Home: 6161 E Grant Rd Apt 20106 Tucson AZ 85712 Office: U Ariz Coll Pharm Dept Pharm & Tocxicology Tucson AZ 85721

MOLLOY, CHRISTOPHER JOHN, molecular cell biologist, pharmacist; b. N.Y.C., Feb. 18, 1954; s. James Francis and Dorothy Jean (Russell) M. BS in Pharmacy, Rutgers U., 1977, PhD in Pharmacology-Toxicology, 1987. Registered pharmacist, N.J. Pharmacist W.M. Weinstein Prescriptions, Livingston, N.J., 1977-81; teaching asst. Rutgers Coll. of Pharmacy, Piscataway, N.J., 1982-85; grad. asst. Rutgers U., Piscataway, 1985-86; postdoctoral fellow Nat. Cancer Inst., NIH, Bethesda, Md., 1986-90; sr. rsch. investigator Bristol-Myers Squibb Pharm. Rsch. Inst., Princeton, N.J., 1990—; adj. asst. prof., Robert Wood Johnson Med. Sch. Univ. Med. and Dentistry N.J., Robert Wood Johnson Med. Sch., Piscataway, N.J., 1991—; reviewer Cancer Rsch., 1987—, Carcinogenesis, 1988—. Contbr. articles to Nature, Molecular and Cellular Biology, Proc. Nat. Acad. Scis., Jour. of Biol. Chemistry, Defferentiation, Cancer Rsch. Recipient Grad. fellowship award Soc. of Cosmetic Chemists, 1983, Biotech. fellowship Nat. Cancer Inst., Bethesda, Md., 1986, Visiting fellowship, Spanish Health Ministry, Madrid, 1990, Selected Reference (Hot Paper) The Scientist Newspaper, Phila., 1991. Mem. AAAS, Am. Heart Assn., Am. Soc. Biochem. and Molecular Biol., N.Y. Acad. Scis., Am. Assn. Cancer Rsch., Sigma Xi. Achievements include description of a direct biochemical linkage between activated growth factor receptor typrosine kinases and the Ras signalling pathway, a possible critical step in cellular mitogenic signal transduction. Office: Bristol Myers Squibb PO Box 4000 Princeton NJ 08543-4000

MOLLOY, WILLIAM EARL, JR., systems engineer; b. Washington, May 19, 1949; s. William Earl Sr. and Margaret (Ballard) M.; m. Suzanne Lee Bailey, Aug. 9, 1969; children: Kathleen Erin, Mary Kee. BA in Psychology, U. South Fla., 1971; MS in Info. Systems, George Mason U., 1993. Commd. USN, 1971—; head mission dept. Spl. Projects Unit Two, Barbers Point, Hawaii, 1971-83; head air spl. projects Chief of Naval Ops., Washington, 1982-84; electronic warfare dept. head Fleet Air Reconnaissance Squadron 2, Rota, Spain, 1984-87; spl. tech. ops., comdr. Joint Chiefs of Staff, Washington, 1987-90; dir. system engring. Space and Naval Warfare System Comman, Washington, 1990—. Mem. Armed Forces Comms. and Electronics Assn.

MOLNAR, JOSEPH MICHAEL, plant physiologist, research director; b. Debrecen, Hungary, Dec. 12, 1931; arrived in Can., 1957; s. Jozsef and Erzsebet Maria (Farkas) M.; m. Susan E. Lancaster, July 8, 1982 (div. 1986); children: Richard, Michael. BSA, U. B.C., Vancouver, 1961; MSc, U. Alta., Edmonton, 1964; PhD, U. Man., Winnipeg, 1971. Credit advisor Farm Credit Corp., Edmonton, 1962-64; rsch. officer Agr. Canada Rsch. Br., Morden, Man., 1966-71; sr. scientist Agr. Canada Rsch. Br., Ottawa, Ont., 1971-77; dir. rsch. sta. Agr. Canada Rsch. Br., Sidney, B.C., 1977-85, Agassiz, B.C., 1985—. Contbr. articles to profl. jours. Scholar B.C. Fruit Growers Assn. Golden Jubilee, 1960, B.C. Govt., 1960, Alta. Govt., 1965. Mem. Can. Soc. Plant Physiologists, Can. Soc. Hort. Sci., Agrl. Inst. Can. Office: Agriculture Canada Research Stn, PO Box 1000, Agassiz, BC Canada V0M 1A0

MOLNAR-KIMBER, KATHERINE LU, immunologist, molecular biologist; b. East Cleveland, Ohio, July 19, 1953; d. Charles Joseph and Jean Elizabeth (Corser) Molnar; m. Robert Oliver Cope Kimber, May 19, 1979. BS, Slippery Rock State Coll., 1974; PhD in Immunology, U. Pa., 1980. Postdoctoral fellow Inst. for Cancer Rsch., Phila., 1980-83, rsch. assoc., 1983-84; rsch. scientist Damon Biotech, Needham, Mass., 1984-85, Wyeth-Ayerst Labs., Princeton, N.J., 1985—; temporary advisor World Health Orgn., Geneva, Switzerland, 1989; adj. prof. Delaware Valley Coll., Doylestown, Pa., 1991—. Contbr. articles to profl. jours. Del. People to People Amb. Program to Russia, 1992; vice-chmn. Cheltenham Tree Adv. Commn., Cheltenham Twp., Pa., 1992-93. Recipient Predoctoral Tng. grant NIH, U. Pa., 1978-79, Postdoctoral Tng. grant NIH, Inst. for Cancer Rsch., 1980-83. Mem. Am. Assn. Sci., Am. Soc. Microbiologists, Am. Soc. Virologists. Achievements include research in HBV replication, generated and characterized recombinant adenovirus for expression of heterologous proteins, vaccines, cytokine regulation, immunomodulators. Office: Wyeth-Ayerst Labs Inflammation Div 865 Ridge Rd Monmouth Junction NJ 08852

MOLZ, FRED JOHN, III, hydrologist, educator; b. Mays Landing, N.J., Aug. 13, 1943; s. Fred John Jr. and Viola Violet (MacDonald) M.; m. Mary Lee Clark, Dec. 17, 1966; children: Fred John IV, Stephen Joseph. BS in Physics, Drexel U., 1966, MCE, 1968; PhD in Hydrology, Stanford U., 1970. Hydraulic engr. U.S. Geol. Survey, Menlo Park, Calif., 1970; asst. prof. Auburn (Ala.) U., 1970-74, alumni asst. prof., 1974-76, alumni assoc. prof., 1976-80, asst. dean research, 1979-84, dir. Eng. exptl. sta., 1981-84, prof., 1980-84, Feagin prof., 1984-89, Huff eminent scholar, 1990—; cons. Battelle N.W., Richland, Wash., 1982-83, 84-85, Argonne (Ill.) Nat. Labs., 1983-85, Electric Power Rsch. Inst., Menlo Park, Calif., 1984—, U.S. Nuclear Regulatory Commn., 1991—. Author: (with others) Numerical Methods in Hydrology, 1971, Modeling Wastewater Renovation, 1981; contbr. articles to profl. jours. Recipient Disting. Faculty award Auburn U. Alumni Assn., 1987; grantee EPA, 1983, 86, 90, U.S. Dept. Edn., 1991, NSF, 1992. Mem. Am. Geophys. Union (Horton award 1992), Am. Soc. Agronomy, Nat. Ground Water Assn., Am. Inst. Hydrology. Avocations: reading, travel, investing. Home: 1224 Ferndale Ct Auburn AL 36830-6731 Office: Auburn Univ Dept of Civil Engring Harbert Engring Ctr Auburn AL 36849

MOMOKI, KOZO, analytical chemist, educator; b. Saitama Prefecture, Japan, June 12, 1921; s. Choji and Yoshi (Iino) M.; m. Moto Nakajima, Mar. 5, 1949; children: Akiko, Shiro. B in Engring., U. Tokyo, 1945, D in Engring., 1961. Asst. U. Tokyo, 1947-50; lectr. Yokohama Nat. U., Japan, 1950-55, asst. prof., 1955-65, prof. engring., 1965-87; prof. emeritus, 1987—. Author: (with H. Uchikawa) Analytical X-Ray Methods, 1965; editor: (series) Advances of X-Ray Analysis, 1964, 65, 66, 68; contbr. 45 rsch. papers on analytical chemistry. Mem. Am. Chem. Soc., Japan Soc. Analytical Chemistry, Chem. Soc. Japan. Avocations: gardening, bird feeding and watching, classical music. Home: 30-5 Shirahata Higashi Cho, Kanagawa Ku, Yokohama 221, Japan

MOMTAHENI, MOHSEN, oral and maxillofacial surgeon, academician, clinician; b. Tehran, Iran, Dec. 1, 1950; came to U.S., 1975; m. Esther Marry Michael, May 15, 1982; children: Adam, Ashley. DMD, Nat. U. Iran, Tehran, 1975; cert. in pathology, Harvard U., 1978; cert. in maxillofacial surgery, Albert Einstein Coll. Medicine, 1982. Diplomate Am. Bd. Oral Maxillofacial Surgery. Asst. prof. Albert Einstein Coll. of Medicine, N.Y.C., 1983-86; chief resident Montefiore Hosp. and Med. Ctr., N.Y.C., 1980-81; mem. surg. team Cranio Facial Ctr., asst. attending Albert Einstein Coll. of Medicine, N.Y.C., 1983—; dir. Temporomandibular Joint Clinic, TMJ Montefiore Med. Ctr., N.Y.C., 1984-90; asst. clin. prof. Sch. of Dentistry Columbia U., N.Y.C., 1986—; pvt. practice N.Y.C., 1986—; affiliate Albert Einstein Coll. Hosp., Montefiore Hosp. and Med. Ctr., Columbia Presbyn. Med. Ctr., Dr.'s Hosp., Cabrini Hosp.; surg. cons. Profl. Dental Reviewer, 1988-91. Co-author: (chpt.) Gastrointestinal Disorders of the Elderly, 1984; contbr. articles to Exposure of Impacted Teeth and Orthodontic Movements via Trans-osseous Forces, Newman Antibiotic in Dentistry. Mem. Rep., 1986. Nat. U. of Iran shcolar, 1975-78. Fellow Am. Assn. Oral and Maxillofacial Surgeons; mem. ADA, Internat. Soc. Plastic and Reconstructive Surgery, N.Y. Acad. Dentistry, Am. Acad. of Pain Mediicne, Am. Soc. Laser in Dentistry (founder, past pres.), Acad. of Osseointegration, 9th Dist. Dental Soc., Am. Coll. Oral and Maxillofacial Surgery, Am. Soc. Dental Anesthesiology, 1st Dist. Dental Soc., Montefiore Med. Ctr. Alumni Assn., Harvard U. Dental Alumni Assn., Tufts U. Dental Alumni Assn. Republican. Roman Catholic. Achievements include development of new treatment modalities for TMJ dysfunction, new bone graft techniques for atrophic jaw and dental implants; research on the effect of orthognathic surgery and condylar position, on alternative solutions for third molar removal, multiple new applications of C02 and KTP lasers in dentistry including implant exposure with C02 laser and TM joint arthroscopic surgery with KTP laser. Office: 630 5th Ave New York NY 10111-0002

MONAGHAN, RICHARD LEO, microbiologist; b. Boston, Dec. 15, 1948; s. Leo John and Marilyn (Freeley) M.; m. Joan Anne Farrell, Nov. 20, 1971; children: Anthony, Sean, Erin. BS, Boston Coll., 1970; PhD, Rutgers U., 1975. Sr. rsch. microbiologist Merck & Co. Inc., Rahway, N.J., 1974-76; research fellow Merck & Co. Inc., 1976-81, assoc. dir. basic microbiology, 1981-83, dir. basic microbiology, 1983-86, sr. dir. fermentation microbiology, 1986—; co-convener Internat. Conf. Biotech. of Microbial Products, 1988, 1990. Mem. editorial bd. Jour. Indsl. Microbiology, 1990-93; contbr. articles to profl. jours. and chpts. to books. Referee Mid. N.J. Soccer Assn., 1989-91. Recipient Thomas Alva Edison Patent award Rsch. Devel. Coun. N.J., 1989; named Inventor of Yr. Intellectual Property Owners Inc., 1988. Fellow Am. Acad. Microbiology; mem. Am. Soc. Microbiology, Soc. Indsl. Microbiology, Theobald Smith Soc. (edn. coms. 1983-91). Roman Catholic. Achievements include discovery of enzyme chitosanase; co-discovery of mevinolin, CCK antagonist asperlicin, antibacterials difficidin,L681,217, antifungals L671,329, L657,398 and L175,491, immunomodulating activity of ascomycin, tapeworm active L155,175 and cholesterol lowering compounds dihydocompactin and 1233A. Home: 48 Johnson Rd Somerset NJ 08873-2953 Office: Merck & Co PO Box 2000 Rahway NJ 07065-0900

MONAGHAN, W(ILLIAM) PATRICK, immunohematologist, retired naval officer, health educator, consultant; b. Ashtabula, Ohio, June 24, 1944; s. Paul E. and June E. (Sober) M. m. Mary Lou Gustafson, Mar. 15, 1976; children: Ian Patrick, Erin Kelly. BS, Old Dominion U., Va., 1968; MS in Biology, Bowling Green State U., Va., 1972, PhD, 1975. Enlisted U.S. Navy, 1961, commd. ensign Med. Service Corps, 1969, advanced through grades to comdr., 1983; staff med. technologist officer Nat. Naval Med. Ctr., Bethesda, Md., 1969; clin. lab. and blood bank officer USS Sanctuary (AH-17), S. Vietnam, 1969-70; clin. lab. officer Nat. Naval Med. Ctr., Charleston, S.C., 1970-72; blood bank fellow U.S. Army Med. Rsch. Lab., Ft. Knox, Ky., 1972-73; head blood bank Nat. Naval Med. Ctr., 1975-85, faculty and course dir. for immunohematology med. tech., 1976-84, dir. blood bank, 1976-84; asst. prof. pathology George Washington U. Sch. Medicine, Washington, S.C., 1976-83; assoc. prof. George Washington U. Sch. Medicine, Washington, 1983-88; mem. faculty Walter Reed Army Med. Ctr., Washington, 1976-88; dep. asst. dean grad. and continuing edn. Uniformed Svcs. U. of Health Sci., Washington, 1984-88, ret., 1988; v.p. Met. Washington Blood Banks, 1976-81, ex officio mem. bd. dirs. 1981-87; cons. D.C. chpt. Hemophiliac Found., 1977-78; spl. USN rep. Am. Soc. Med. Tech., 1976-88; dir. N.E. area blood system Navy Blood Program, 1978-88; mem. tri-service blood bank com. Dept. Def. Blood Program, 1978-88; faculty and program adv. com. ARC, Washington, 1978-84, Johns Hopkins Med. Sch., Balt., 1978-85; faculty U. Tenn. Center for Health Scis., Memphis, 1978, U. Ill. Sch. Medicine, Peoria, 1978-79; guest lectr. NIH Blood Bank, 1978-90; adj. assoc. prof. Bowling Green State U., Ohio, 1981-89; bd. dirs. Exam, Inc., Rockville, Md. Navy editor Procs. Armed Forces Med. Lab. Scientists, 1976, 79, 80, editor-in-chief, 1982-85; assoc. editor Am. Jour. Med. Tech., 1978—, Jour. Allied Health; Navy editor History of the Blood Program of the U.S. Military Services in Vietnam and S.E. Asia, 1976-84; contbr. articles to profl. jours. Active, Big Bros., 1976-85. Decorated several combat and svc. medals; USN grantee, 1977-79. Mem. Am. Soc. Med. Technologists (chmn. immunohematology task group 1976), Am. Blood Commn. (task force 1976, regionalization), Am. Assn. Blood Banks (sci. assembly 1976—, adminstrv. sect. 1976—, blood component therapy com. 1977-79, edn. com. 1976-83), AAAS, Am. Soc. Clin. Pathologists, Soc. Mil. Surgeons, Naval Inst., Sigma Xi, others. Home: 14116 Parkvale Rd Rockville MD 20853-2526

MONAHAN, EDWARD CHARLES, academic administrator, marine science educator; b. Bayonne, N.J., July 25, 1936; s. Edward C. and Helen G. (Lauenstein) M.; m. Elizabeth Ann Eberhard, Aug. 27, 1960; children: Nancy Elizabeth, Carol Frances, Eilis Marie. B of Engring. Physics, Cornell U., 1959; MA, U. Tex., 1961; PhD, MIT, 1966; DSc, Nat. U. Ireland, Dublin, 1984. Rsch. asst. Woods Hole (Mass.) Oceanographic Inst., 1964-65; asst. prof. physics No. Mich. U., Marquette, 1965-68; asst. prof. oceanography Hobart and William Smith Coll., Geneva, N.Y., 1968-69; asst. prof. dept. meteorology, oceanography U. Mich., Ann Arbor, 1969-71, assoc. prof. dept. atmosphere and ocean sci., 1971-75; dir. edn. and rsch. Sea Edn. Assn., Woods Hole, 1975-76; statutory lectr. phys. oceanography U. Coll., Galway, Ireland, 1976-86; prof. marine scis. U. Conn., Avery Point, 1986—; dir. Conn. Sea Grant Coll. Program, Avery Point, 1986—. Editor: Oceanic Whitecaps and Their Role in Air-Sea Exchange Processes, 1986, Climate and Health Implications of Bubble-Mediated Sea-Air Exchange, 1989; contbr. over 225 articles to profl. jours. Recipient more than 95 rsch. grants, 1966—. Fellow Royal Meteorol. Soc.; mem. AAUP, Am. Geophys. Union, Am. Meteorol. Soc. (profl.), Am. Soc. Limnology and Oceanography, Internat. Assn. Theoretical and Applied Limnology, Irish Meteorol. Soc., Acoustical Soc., The Oceanography Soc. (life). Avocation: recreational sculling.

MONAHAN, EDWARD JAMES, geotechnical engineer; b. Bayonne, N.J., Sept. 18, 1931; s. John Joseph Monahan and Anna Monahan; m. Mary-Jean Claire Thompson, Aug. 22, 1970. BSCE, Newark Coll. Engring., 1958, MSCE, 1961; PhD, Okla. State U., 1968. Registered profl. engr., N.J., N.Y. Instr. to prof. Newark Coll. Engring., 1958-84, prof. emeritus, 1984—; writer, cons. Bloomfield, N.J., 1984—; advisor Tau Beta Pi, Newark, 1958-78, Chi Epsilon, Newark, 1958-81. Author: Construction of and on Compacted Fills, 1986. Vol. performer, storyteller/singer at hosps., nursing homes, librs., schs. N.J. Storytellers Guild. Staff sgt. USAF, 1950-54. NSF summer grantee, Washington, 1960, NSF sci. faculty fellow, Washington, 1962. Mem. ASCE, Am. Soc. Engring. Edn. Baptist. Achievements include two patents on found. constrn. methodology. Home: 85 Newark Ave Bloomfield NJ 07003

MONEK, DONNA MARIE, pharmacist; b. New Brunswick, N.J., Aug. 9, 1947; d. James Frank and Angeline Eleanor (Marzella) M. BS, Phila. Coll. of Pharmacy, 1970; MBA, Fairleigh Dickinson U., East Rutherford, N.J., 1976. Reg. pharmacist, N.J. Staff pharmacist Freehold (N.J.) Area Hosp., 1971-72, dir. pharmacy, 1972-76; pharmacy administr. Rahway (N.J.) Hosp., 1976—; cons. home health care intravenous therapy, Rahway, N.J., 1985. Rep. committeewoman Middlesex County, 1972-86, 92—; mem. Bd. Health, Metuchen, N.J., 1987—. Mem. Am. Soc. Hosp. Pharmacists, N.J. Soc. Hosp. Pharmacists, N.J. Hosp. Assn. (group purchasing 1980-88, chairperson profl. standards 1989-90, vice chairperson state pharmacy com. 1990-91, chairperson 1992), Am. Pharm. Assn., N.J. Pharm. Assn., Metuchen Rep. Club, Cranford Dramatic Club, Kappa Epsilon. Roman Catholic. Avocation: photography. Office: Rahway Hosp 865 Stone St Rahway NJ 07065-2797

MONFORTON, GERARD ROLAND, civil engineer, educator; b. Windsor, Ont., Can., July 21, 1938; married, 1960; 4 children. BASc, Assumption U., 1961, MASc, 1962; PhD in Civil Engring., Case Western Reserve U., 1970. Lectr. civil engring. U. Windsor, Ont., Can., 1962-64; rsch. asst. solid mechanics Case Western Reserve U., 1964-68; from asst. prof. to assoc. prof. civil engring. U. Windsor, 1968-76, prof. civil engring., 1976—. Mem. Engring. Inst. Can., Can. Soc. Civil Engrs. Achievements include research in solid mechanics and structural design. Office: University of Windsor, Faculty of Engineering, Windsor, ON Canada N9B 3P4*

MONFRE, JOSEPH PAUL, mechanical engineer, consultant; b. Milw., Dec. 31, 1956; s. Peter Steven and Mary Monica (Miller) M.; m. Deborah Mae Mueller, Sept. 17, 1983; children: Caroline, Thomas, Daniel. BSME, Marquette U., Milw., 1979. Registered profl. engr., Wis. Sr. svc. engr., prin. application engr., prin. mktg. engr. The Trane Co., LaCrosse, Wis., 1980—; sr. prin. engr. The Trane co., La Crosse, Wis. Contbr. articles to profl. jours. Sec. Roncalli Newman Parish, LaCrosse, 1988. Mem. ASHRAE, ASME, Assn. Energy Engrs. Office: The Trane Co 3600 Pammel Creek Rd La Crosse WI 54650

MONISMITH, CARL LEROY, civil engineering educator; b. Harrisburg, Pa., Oct. 23, 1926; s. Carl Samuel and Camilla Frances (Geidt) M. BSCE, U. Calif., Berkeley, 1950, MSCE, 1954. Registered Civil Engr., Calif. From instr. to prof. of civil engring. dept. civil engring. U. Calif., Berkeley, 1951—, chmn. dept. civil engring., 1974-79, Robert Horonjeft prof. civil engring., 1986—; Henry M. Shaw lectr. in civil engring. N.C. State U., 1993; cons. Chevron Research Co., Richmond, Calif., 1957—, U.S. Army CE Waterways Expt. Sta., Vicksburg, Miss., 1968—, B.A. Vallerga, Inc., Oakland, Calif., 1980—, ARE, Austin, Tex. and Scotts Valley, Calif., 1978—; cons. Bechtel Corp., San Francisco, 1982-86. Contbr. numerous articles to profl. jours. Served to 2d lt. U.S. Army Corps Engrs., 1945-47. Scholar Fulbright Sr. U. New South Wales, 1971; Recipient Rupert Myers Medal U. New South Wales, 1976. Fellow ASCE (pres. San Francisco sect. 1979-80, ednl. activities com. 1989-91, State of the Art award 1977, James Laurie prize 1988); mem. Assn. Asphalt Paving Technologists (pres. 1968, W.J. Emms award 1961, 65, 85, hon. mem. 1989), Transp. Research Bd. (assoc., chmn. pavement design sect. 1973-79, K. B. Woods award 1972, 1st Disting. Lectureship 1992), Am. Soc. Engring. Edn., ASTM, NAE, Internat. Soc. for Asphalt Pavements (chmn. bd. dirs. 1988-90), The Asphalt Inst. (roll of honor 1990), Sigma Xi, Tau Beta Pi, Xi Epsilon. Avocations: swimming, stamp collecting. Office: U Calif Dept Civil Engring 115 McLaughlin Hall Berkeley CA 94720

MONIZ, ERNEST JEFFREY, physics educator; b. Fall River, Mass., Dec. 22, 1944; s. Ernest Perry and Georgina (Pavao) M.; m. Naomi Hoki, June 9, 1973; 1 child, Katya. B.S., Boston Coll., 1966; Ph.D., Stanford U., 1971. Prof. physics MIT, Cambridge, Mass., 1973—; dir. Bates Linear Accelerator Ctr. MIT, Middleton, Mass., 1983-91, head physics dept., 1991—; Cons. Los Alamos Nat. Lab., 1975—. Contbr. numerous articles to profl. jours. Office: MIT Dept Physics 6-113 77 Massachusetts Ave Cambridge MA 02139-4307

MONKE, EDWIN JOHN, agricultural engineering educator; b. Harvel, Ill., June 7, 1925; s. Edwin Herman and Emma Lillian (Prange) M.; m. Marian Mildred Hubb, Aug. 24, 1963; children: Karen Ruth, Sarah Elizabeth, Dianne Mildred. B.S., U. Ill., 1950, M.S., 1953, Ph.D., 1959. Registered profl. engr., Ind. Instr. agrl. engring. U. Ill., Urbana, 1951-58; asst. prof. agrl. engring. Purdue U., West Lafayette, Ind., 1958-62, assoc. prof., 1962-67, prof., 1967—; dir. Third Internat. Seminar for Hydrology Profs., 1971. Editor: Biological Effects in the Hydrological Cycle, 1971. Served with U.S. Army, 1944-46. Fellow Am. Soc. Agrl. Engrs. (chmn. soil and water div. 1978-79); mem. Soil and Water Conservation Soc. Am., Am. Geophys. Union, Nat. Soc. Profl. Engrs., Am. Soc. Engring. Edn., Ind. Soc. Profl. Engrs. (pres. A.A. Potter chpt. 1971-72, state dir. 1972-73), Sigma Xi, Tau Beta Pi, Alpha Zeta, Alpha Epsilon, Gamma Sigma Delta. Lutheran. Club: Optimists (West Lafayette) (pres. 1979-80). Home: 2501 Newman Rd West Lafayette IN 47906-4537 Office: Purdue U Agrl Engring West Lafayette IN 47907

MONKEWITZ, PETER ALEXIS, mechanical and aerospace engineer, educator; b. Berne, Switzerland, Nov. 9, 1943; came to U.S., 1977; s. Kurt and Marie M. (Krähenbühl) M.; m. Annelise Heidy Sturzenegger, Jan. 15, 1972; children: Serge M., Florence S., Cyril P. Diploma Physics, Fed. Inst. Tech., Zurich, 1967, D of Natural Scis., 1977. Software cons. Sperry Rand UNIVAC, Zurich, 1968; teaching asst. Fed. Inst. Tech., Zurich, 1969-77; rsch. assoc. U. So. Calif., L.A., 1977-80; asst. prof. UCLA, 1980-85, assoc. prof., 1985-91, prof., 1991—. Co-editor Jour. Applied Math. and Physics, 1991—; contbr. articles to Jour. Fluid Mechanics, Physics of Fluids, Ann. Revs. Fluid Mechanics, others. Recipient U.S. Sr. Scientist award Humboldt Found., Germany, 1988. Fellow Am. Phys. Soc.; mem. AIAA. Achievements include rsch. on self-excited oscillations and their control with regard to altering mixing rates or the drag of a body. Office: UCLA MANE Dept 405 Hilgard Ave Los Angeles CA 90024-1597

MONNINGER, ROBERT HAROLD GEORGE, ophthalmologist, educator; b. Chgo., Nov. 5, 1918; s. Louis Robert and Katherine (Lechner) M.; m. Anna Evelyn Turnen, Sept. 1, 1944; children—Carl John William, Peter Louis Philip. A.A., North Park Coll., 1939; B.S., Northwestern U., 1941, M.A., 1945; M.D., Loyola U., Chgo., 1953, Sc.D. (hon.), 1968. Diplomate Am. Bd. Cosmetic Plastic Surgery. Intern St. Francis Hosp., Evanston, Ill., 1953-54; resident Presbyterian-St. Luke's, U. Ill. Research and Eye, Va. hosps., 1954-57; mem. leadership council Ravenswood Hosp. Med. Ctr.; instr. chemistry Lake Forest Coll., Ill., 1946-47; instr. biochemistry, physiology Loyola U. Dental Sch., 1948-49; clin. assoc. prof. ophthalmology Stritch Sch. Medicine, Loyola U., Maywood, Ill., 1957-72; practice medicine specializing in ophthalmology Lake Forest, 1957—; prof., dir. acad. ophthalmology U. Health Scis.-The Chgo. Med. Sch.; guest lectr. numerous univs. med. ctrs. U.S., Can., Europe, Central and S.Am., Orient; resident lectr. Klinikum der Goethe-Universitat, Fed. Republic Germany, 1981; mem. panel Nat. Disease and Therapeutic Index; cons. Draize eve toxicity test revision HEW, cons. research pharm. cos. Nat. Assoc. Smithsonian Instn.; bd. dirs. Eye Rehab. and Research Found.; postgrad. faculty Internat. Glaucoma Congress; lectr. Hopital Dieu, Paris; lectr. postgrad. courses for developing nations physicians WHO; life mem. Postgrad. Sch. Medicine U. Vienna; cons. Nat. Acad. Sci.; adv. bd. Madera Del Rio Found. Cons. author Textbook of Endocrinology. Editorial bd. Clin. Medicine, 1958—, EENT Digest, 1958—, Internat. Surgery, 1972—, profl. jours. Served with USMCR, 1941-44. Recipient citation Gov. Bahamas, 1960, Ophthalmic Found. award, 1963, Sci. Exhibit award Ill. State Med. Soc., 1966, Franco-Am. Meritorious citation, 1967, Paris Post No. 1 Am. Legion award, 1967, citation Pres. Mexico, 1968, Sightsaving award Bausch & Lomb, 1968, exhibit award Western Hemisphere Congress Internat. Surgeons, 1968, Research citation Japanese Soc. Ophthalmology, 1969; Barraquer Gold Medallion; Physician's Recognition award AMA, Bicentennial citation Library of Congress Registration Book; Pres.'s medal of merit; meritorious citation Gov. Ill., citation and medal Lord Mayor of Rome, also Pres. of Italy, 1981, Civic Ctr. of Evanston, Ill., 1981, commendation and citation Ill. Gen. Assembly, 1982, cert. of accomplishment Loyola U. Alumni Assn., Chgo., 1983; Catherine White Scholarship fellow, 1945-46. Fellow Internat. Coll. Surgeons (postgrad. faculty continuing edn.), Am. Coll. Angiology, Oxford Ophthal. Congress and Soc. (lectr. 1960-61), Royal Soc. Health, Internat. Acad. Cosmetic Surgery (editorial bd.), Sociedad Mexicana Ortopedia (hon.), C. Puestow Surg. Soc.; mem. AAAs, Internat. Soc. Geog. Ophthalmology (program course coordinator, lectr. ocular electrophysiology VI Internat. Congress, Rio de Janiero), Pan Am. Assn. Ophthalmology, Assn. for Research Ophthalmology, Am. Assn. Ophthalmology, Am. Soc. Contemporary Ophthalmology, Internat. Glaucoma Soc., Ill. Soc. for Med. Research, Ill. Assn. Ophthalmology, Internat. Soc. clin. electrophysiology of Vision (hon., lectr. 1978), Brazilian Soc. Ophthalmology (hon. corr.), German Ophthal. Soc., Internat. Fedn. Clin. Chemists (lectr.), Primum Froum Ophthalmologicum (lectr.), European Ophthal. Soc. (lectr.), Internat. Congress Anatomists (lectr.), Assn. des Diabetologues Francise (lectr.), German Soc. for Internal Medicine (lectr.), Met. Opera Guild, Fedn. Am. Scientists, N.Y. Acad. Scis., Ill. Acad. Scis., AAUP, Nat. Soc. Lit. and Arts, Nat. Hist. Soc., Rush Med. Sch.-Presbyn. St. Luke's Alumni Assn., Sociedad Poblana Oftalmologia (hon, silver placue, commemorative prestige lectr. 1982) (Mex.), Internat. Platform Assn., Cousteau Soc., Sigma Xi, Sigma Alpha Epsilon, Phi Beta Pi, Theta Kappa Psi.

MONROCHE, ANDRÉ VICTOR JACQUES, physician; b. Saumur, France, May 31, 1941; s. Maurice and Thérèse (Chevreau-Rocheron) M.; m. Bodet-Pasquier, July 8, 1966; children: Benoît, Sabine, Hélène, Matthieu. MD, Faculté de Médecine, Angers, France, 1970; Degree in Rheumatology, Faculté de Médecine, Paris, 1972. Gen practice spa medicine Villa Forestier, Aix-les-Bains, France, 1970-72; gen. practice rheumatology and sports medicine Cabinet Med., Angers, 1973—. Author: Eléments de Rhumatologie, 1975, Eau et Sport pour votre Santé, 1988; editor Chiron-Paris; editor-in-chief Cinésiologie 1980, Sport Médecine Rev. Mem. Panathlon Club d'Angers (internat., founder, pres. 1990), Club of Paris (v.p. 1986, 89). Office: Cabinet Medical, 1 rue d'Alsace, 49100 Angers France

MONROE, DANIEL MILTON, JR., biologist, chemist; b. Memphis, Tenn., Sept. 4, 1943; s. Daniel M. and Vee T. M. BS, Christian Bros. Coll., 1965. Lab. asst. Christian Bros. Coll., Memphis, 1963-65; rsch. asst. U. Tenn., Memphis, 1967-75, rsch. assoc., 1975—; cons. in field. Contbr. chpts. to books, articles to profl. jours. Cited for continuing med. lab. edn. Am. Soc. Clin. Pathologists, 1989—. Mem. Am. Inst. Chemists (cert. profl. chemist 1982—). Lutheran. Achievements include development of diagnostic enzyme/liposome immunoassays, immuno-optrodes, disposable biosensing devices, antibody production, vaccines. Home: 6057 Knightsbridge Dr Memphis TN 38115-3317 Office: Univ Tenn VA Med Ctr 1030 Jefferson Ave Memphis TN 38163-0001

MONROE, FREDERICK LEROY, chemist; b. Redmond, Oreg., Oct. 13, 1942; s. Herman Sylvan and Mary Roberta (Grant) M.; B.S. in Chemistry, Oreg. State U., 1964; M.S. in Environ. Engring., Wash. State U., 1974. Control specialist Air Pollution Authority, Centralia, Wash., 1969-70; asst. chemist Wash. State U., 1970-74; environ. engr. Ore-Ida Foods, Inc., Idaho, 1974-77; cons., Idaho, 1977-78; applications engr. AFL Industries, Riviera Beach, Fla., 1979-80; mgr. chem. control PCA Internat., Matthews, N.C., 1980-85; quality assurance mgr. Stork Screens Am., Charlotte, N.C., 1985-93, environ. mgr., 1993—, grade IV N.C. wastewater treatment operator. Pres. Unity Ch., 1982-84. Served with USAF, 1964-68, maj. Res. ret.; served with N.G., 1973-78. Decorated Air Force Commendation medal; recipient Blue Thumb award Charlotte-Mecklenburg Utility Dist. Fellow Am. Inst. Chemists; mem. Am. Chem. Soc. Republican. Achievements include approved international shipment of hazardous wastes for recycling; avocation: languages including French, Spanish, Dutch and German, working to improve world health. Home: 4200 Nevin Rd Charlotte NC 28269-4366 Office: Stork Screens Am 3201 N I-85 Charlotte NC 28269

MONROE, JOHN ROBERT, electrical engineer; b. Madison, Wis., Nov. 23, 1960. B Engring., Vanderbilt U., 1982; JD, U. Wis., 1993. Registered profl. engr., Wis. Engr. Spectrum Planning, Dallas, 1983-87, Midwestern Relay Co., Milw., 1988, Wis. Power & Light Co., Madison, 1988—. Mem. IEEE, NSPE, Utilities Telecom. Coun., Wis. Soc. Profl. Engrs. Office: Wis Power & Light Co 222 W Washington Ave Madison WI 53703

MONSALVE, MARTHA EUGENIA, pharmacist; b. Bogotá, Colombia, Sept. 27, 1968; came to U.S., 1985; d. Oscar and Martha Lucia (Vidal) M.; m. Dale James Morphonios, Dec. 16, 1989. AA, Miami-Dade Community Coll., 1989; grad., Southeastern U., 1993. Lab. asst. William Harvey Co., Bogotá, 1981-82; pharmacy asst. Galloway Pharmacy, Miami, Fla., 1988; purchasing agt. FEPARVI, Ltd., Miami, 1985—; pharmacy intern Bapt. Hosp., Miami, 1992. Vol. Bapt. Hosp., Miami, 1988. Mem. Am. Soc. Hosp. Pharmacists, Fla. Soc. Hosp. Pharmacists. Roman Catholic. Avocations: reading, swimming, travel. Home: 8640 SW 84th Ave Miami FL 33143-6912

MONSON, RAYMOND EDWIN, welding engineer; b. Evanston, Ill., Apr. 1, 1957; s. Wilfred Raymond and Ruth Eleanor (Nelson) M.; m. Janelle Faye Johnson, Apr. 25, 1981. BS in Welding Engring., Le Tourneau U., 1979. Engr. in tng., Tex. Welding engr., mfg. engr., quality assurance unit mgr. Babcock & Wilcox Co., Barberton, Ohio and Paris, Tex., 1980-86; sr. welding engr. Precision Components Corp., York, Pa., 1986—; mem. welding adv. bd. York County Vocat./Tech. Sch., York, 1991—. Author Welder Tng. Manuals, 1988. Nadine scholar LeTourneau U., 1978. Mem. Welding Rsch. Coun., Am. Welding Soc. (chmn. York sect. 1991-92). Republican. Home: 1346 Prospect St York PA 17403 Office: Precision Components Corp 500 Lincoln St York PA 17405

MONTAG, MORDECHAI, mechanical engineer; b. Muran, Czechoslovakia, Oct. 30, 1925; came to U.S., 1955; s. Itzhak and Isabella (Frank) M.; m. Harriet Roth, May 3, 1953; children: Avram, Jonathan, Ethan, Benjamin. Ingenieur, Hebrew Inst. Tech., Haifa, Israel, 1951. Mech. engr. Brookhaven Nat. Lab., Upton, N.Y., 1965—. Jewish. Home: 3 Malton Rd Plainview NY 11803 Office: Brookhaven Nat Lab Bldg 510C Upton NY 19973-0510

MONTAGNA, WILLIAM, scientist; b. Roccacasale, Italy, July 6, 1913; s. Cherubino and Adele (Giannangelo) M.; m. Martha Helen Fife, Sept. 1, 1939 (div. 1975); children: Eleanor, Margaret, James and John (twins); m. Leona Rebecca Montagna, Apr. 19, 1980. A.B., Bethany Coll., 1936, D.Sc., 1960; Ph.D., Cornell U., 1944; D. B.S., Università di Sassari, 1964. Instr. Cornell U., 1944-45; asst. prof. L.I. Coll. Medicine, 1945-48; asst., assoc. prof. Brown U., 1948-52, prof., 1952-63, L. Herbert Ballou univ. prof. biology, 1960-63; prof., head exptl. biology U. Oreg. Health Scis. Ctr.; dir. Oreg. Regional Primate Rsch. Ctr., Beaverton, 1963-80, ret., affiliate scientist, 1985—. Author: The Structure and Function of Skin, 1956, 3d edit., 1974, Comparative Anatomy, 1959, Nonhuman Primates in Biomedical Research, 1976, Science Is Not Enough, 1980; co-author: Man, 1969, 2d edit., 1973, Atlas of Normal Human Skin, 1991; editor: The Biology of Hair Growth, 1958, Advances in Biology of Skin, 20 vols, The Epidermis, 1965, Advances in Primatology, 1970, Reproductive Behavior, 1974. Decorated Ordine di Cavaliere, 1963, Cavaliere Ufficiale, 1969, Commendatore della Repubblica Italiana, 1975; Italy; recipient spl. award Soc. Cosmetic Chemists, 1957; Gold award Am. Acad. Dermatology, 1958; gold medal for meritorious achievement Universitá di Sassari, 1964; Aubrey R. Watzek award Lewis and Clark Coll., 1977; Hans Schwarzkopf Research award German Dermatol. Soc., 1980. Mem. Acad. Dermatology and Syphilology (hon.), Soc. Investigative Dermatology (pres. 1969, recipient Stephen Rothman award 1972, ann. William Montagna lectr. 1975—), Sigma Xi (Pres. 1960-62). Research in biology mammalian skin with emphasis on primates and especially man. Office: Oregon Regional Primate Rsch Ctr 505 NW 185th Ave Beaverton OR 97006-3499

MONTAGUE, MICHAEL JAMES, plant physiologist; b. Flint, Mich., Dec. 25, 1947; s. Herbert Francis and Ella Margaret (O'Connor) M. AB, U. Mich., 1970, PhD, 1974. Dir. rsch. ops. Monsanto Co., St. Louis, 1975—; lectr. in biotech. field. Contbr. articles to profl. jours. Home: 2548 Town & Country Ln Saint Louis MO 63131 Office: Monsanto Co 800 N Lindbergh Saint Louis MO 63167

MONTANA, ANTHONY JAMES, analytical chemist; b. Bklyn., Apr. 1, 1950; s. Joseph James and Helen Theresa (Fazio) M. PhD, Columbia U., 1976; MBA, Fairleigh Dickinson U., 1984. Group leader Gaf Corp., Wayne, N.J., 1976-82; mgr. analytical R & D Diamond Shamrock, Morristown, N.J., 1982-88; dir. analytical R & D Atochem N.A., Branchburg, N.J., 1988-91; dir., materials chair Warner-Lambert Co., Morris Plains, N.J., 1991—. Editorial adv. bd. Office Tech. Mgmt. Mag., 1989-92; contbr. articles to profl. jours. Mem. Am. Oil Chemists Soc. (pres. 1988-90), Am. Soc. Quality Control, Am. Mgmt. Assn., N.J. Chromatography Topical Group (dir. 199-92, mem. com). Republican. Roman Catholic. Achievements include research in analytical chemistry, technical management, and quality process. Office: Warner Lambert Co 182 Tabor Rd Morris Plains NJ 07950

MONTANE, JEAN JOSEPH, mechanical engineer; b. N.Y.C., July 6, 1961; s. Claude John and Elizabeth Linda (Tedori) M. B. Engring., Stevens Inst., 1984. Registered profl. engr., Va. Design engr. Newport News (Va.) Shipbuilding, 1984—. Mem. ASME, NSPE, Soc. of Naval Architects and Marine Engrs. (Tidewater sect.). Home: 12970 Nettles Dr Apt A1 Newport News VA 23602

MONTAUDO, GIORGIO, chemistry educator, researcher; b. Catana, Italy, Aug. 9, 1934; s. Salvatore and Matilde (Prampolini) M.; m. Tudisco Paola, July 22, 1964; children: Maurizio, Matilde. D in Chemistry, U. Catania, Italy, 1959. Asst., lectr. dept. chemistry U. Catania, Italy, 1959-66; postdoctoral fellow Bklyn. Polytech., 1966-67; postdoctoral fellow U. Mich., Ann Arbor, 1967-68, rsch. assoc., 1971; Humboldt Found. fellow Mainz U., Germany, 1973; prof. macromolecular chemistry U. Catania, Italy, 1975; dir. Nat. Coun. Rsch. of Italy Inst. for Chemistry Tech. Polymers, Catania, Italy, 1980—; vis. prof. U. Mainz, Germany, 1980, U. Cin., 1988. Editorial bd. Jour. Polymer Sci., Polymer Degradation and Stabilization, Jour. Analytical and Applied Pyrolysis and Polymer Trends; contbr. numerous articles to profl. and internat. jours., chpts. to books. Recipient Italian Chem. Industry award, Milan, 1990. Office: U Catania, Viale A Doria 6, 95125 Catania Italy

MONTEALEGRE, JOSÉ RAMIRO, information systems consultant; b. Guatemala, Oct. 3, 1959; s. José Ramiro and Margarita (Villacorta) M.; m. Vilma Beatriz Cordón, Dec. 12, 1986; 1 child, José Ramiro. Degree in computer engring. magna cum laude, Francisco Marroquin U., Guatemala, 1982; M in Computer Sci., Carleton U., Ottawa, Ont., Can., 1984; postgrad., Harvard Bus. Sch. Teaching asst. Francisco Marroquin U., 1979-82; analyst/programmer Computo y Comunicaciones Avanzadas, Guatemala, 1979-80; field engr. Herrera y Llerandi Hosp., Guatemala, 1980-82; rsch. asst. Carleton U., Ottawa, 1982-84; univ. lectr. Francisco Marroquin U., San Carlos U., Guatemala, 1984-86; data processing cons. various cos., Guatemala, 1984-86; teleprocessing cons. Dept. Def., Republic of Guatemala, 1984-86; infosystems and data processing cons. Inst. Nutrition of Cen. Am. and Panama, 1986—; guest cons. Govt. of Taiwan, Taipei, 1988; examiner for Guatemalan univ. programs, 1984—; gen. coord. Nat. Computer and Info. Conv., Guatemala, 1985, 86; rsch. assoc. Harvard Bus. Sch., 1991; presenter at confs. in field. Mem. Nat. Dept. Pers. Devel., Republic of Guatemala Ministry of Fin., 1984-85. J.W. Kellogg Found. scholar, 1990; recipient Hon. Mention award Ministry of Interior, El Salvador, 1988. Mem. IEEE, Guatemalan Info. Assn. (v.p. 1985—), Panam. Integration of Engring. Assns. (sec. computer com.), Guatemalan Engring. Assn., Guatemalan Mgmt. Assn., Data for Devel. Assn. Roman Catholic. Avocations: softball, videotaping. Home: 18 Ave B 7-69 Zona 15, Vista Hermosa I, Guatemala Guatemala Mailing address: 128 Sycamore St Belmont MA 02178

MONTEIRO, RENATO DUARTE CARNEIRO, industrial engineer, educator, researcher; b. Rio De Janeiro, Brazil, Mar. 1, 1959; came to the U.S., 1984; s. Carlos Victorino M. Carneiro and Neyde (Duarte) M.; m. Monica Rasel, Aug. 30, 1986; children: Felipe Carneiro, Henrique Carneiro. MA in Math., Inst. de Matematica, Rio De Janeiro, 1984; PhD in Ops. Rsch., U. Calif., Berkeley, 1988. With AT&T Bell Labs., Holmdel, N.J., 1988-90, U. Ariz., Tucson, 1990—. Recipient $60,000 Rsch. grant NSF, Washington, 1991. Mem. Math. Programming Soc., Ops. Rsch. Soc. Am. Achievements include research on Morse's work about Schoenflies problems, interior path following primal-dual algorithms, an extension of Karmarkar type algorithm to a class of convex separable programming problems with global linear rate of convergence, a polynomial-time primal-dual affine scaling algorithm for linear and convex quadratic programming and its power series extension, limiting behavior of the affine scaling continuous trajectories for linear programming problems, convergence and boundary behavior of the projective scaling trajectories for linear programming, a geometric view of parametric linear programming, the continuous trajectories for a potential reduction algorithm for linear programming, a globally convergent interior point algorithm for convex programming, the global convergence of a class of primal potential reduction algorithms for convex programming. Office: U Ariz Systems and Indl Engring Dept Tucson AZ 85721

MONTEIRO, SERGIO LARA, physics educator; b. Rio de Janerio, Brazil, May 15, 1945; came to U.S., 1973; s. Manoel F.L. and Maria F.L. (Lima) M. BS, Fed. U., Rio de Janeiro, Brazil, 1968; MS, Univ. Wyo., 1975; PhD, Washington State Univ., 1981. Post-doctoral assoc. Rensselaer Polytech. Inst., 1982-83, asst. prof. Hudson Valley Community Coll., Troy, N.Y., 1983-84; dir. R&D Telecom Gen., Troy, N.Y., 1984-86; prof. Moorpark (Calif.) Coll., 1986—. Office: Moorpark Coll Dept Physics 7075 Campus Rd Moorpark CA 93021

MONTEIRO-RIVIERE, NANCY ANN, biologist, educator; b. New Bedford, Mass., Sept. 5, 1954; d. Theodore Leo and Jeannette Yvonne (Delage) Monteiro; m. Jim Edmond Riviere, May 31, 1976; children: Christopher, Brian, Jessica. BS in Biology cum laude, Stonehill Coll., 1976; MS in Anatomy, Purdue U., 1979, PhD, 1981; postgrad., Chem. Indsl. Inst. Toxicology, 1984. Asst. prof. coll. vet. medicine N.C. State U., Raleigh, 1984-90, assoc. prof., 1990—; rsch. assoc. prof. U. N.C. Med. Sch., Chapel Hill, 1991—; reviewer four profl. jours., 1981—; pathology and toxicology expert Internat. Life Scis. Inst., 1986; invited speaker several indsl. cos.; cons. several pharm. cos. Contbr. articles and abstracts to profl. jours. and chpts. to books. NIH grantee, 1987—, USAMRDC, 1984—, EPA, 1991-93. Fellow Soc. Investigative Dermatology, Soc. Toxicology (editorial bd. 1992—), Soc. Toxicol. Pathologists, Am. Assn. Anatomists, Am. Assn. Vet. Anatomists, World Assn. Anatomists, S.E. Electron Microscopy Soc., Am. Assn. Pharm. Scientists, Sigma Xi. Roman Catholic. Home: 5105 Lenoraway Dr Raleigh NC 27613 Office: NC State U 4700 Hillsborough St Raleigh NC 27606-1428

MONTEITH, DAVID KEITH BRISSON, toxicologist, researcher; b. Castro Valley, Calif., July 18, 1959; s. Lee E. and Gretchen (Surrey) M.; m. Marsha Candice Brisson; 1 child, Denton. BS, U. Wash., 1981; MS, U. Tex., 1983, PhD, 1985. Diplomate Am. Bd. Toxicology. Lab. supr. Indoor Air Rsch. Study U. Tex. Sch. Pub. Health, Houston, 1982-83; rsch. asst. M.D. Anderson Hosp., Houston, 1984-85; post doctoral fellow Duke U. Med. Ctr., Durham, N.C., 1985-87; asst. prof. Ind.-Purdue U., Ft. Wayne, Ind., 1987-90; rsch. assoc. Parke Davis Pharm. Rsch., Ann Arbor, Mich., 1990—. Author: (book chpts.) Indoor Air: Chemical Characterization and Personal Exposure, 1984, The Isolated Heptocyte: Use in Toxicology, 1987; contbr. articles to profl. jours. Grantee Purdue Rsch. Found., 1988, '89, Purdue Cancer Ctr., 1989-90. Mem. AAAS, Am. Assn. for Cancer Rsch., Soc. Toxicology, Sigma Xi. Episcopalian. Achievements include devel. human heptocyte culture techniques; metabolism and genotoxicity of aromatic amines and polycyclic aromatic hydrocarbons. Office: Parke Davis Pharm Rsch 2800 Plymouth Rd Ann Arbor MI 48104

MONTEMAYOR, JESUS SAMSON, physician; b. La Carlota City, The Philippines, May 17, 1939; came to U.S., 1972; s. Jesus Cavada and Elena Poticar (Samson) M.; m. Catalina Abellana, Sept. 26, 1971; children: Grace, Gail, Jennifer, Jessica. AA, U. San Carlos, Cebu City, The Philippines, 1958; MD, Southwestern U., Cebu City, Philippines, 1969. Adj. physician C.L. Montelibano Meml. Hosp., Bacolod City, The Philippines, 1969-70; resident physician Drs. Hosp., Bacolod City, 1970-72; resident in pathology Wyckoff Heights Hosp., Bklyn., 1975-80, emergency rm. physician, 1984-88; fellow in medicine NYU Med. Ctr.-Goldwater Meml. Hosp., Roosevelt Island, 1980-84, attending physician, 1984—; emergency rm. physician Astoria (N.Y.) Hosp., 1983-84. Fellow Am. Coll. Internat. Physicians; mem. AAAS, Am. Geriatric Soc., Assn. Philippine Physicians in Am., Am. Acad. Home Care Physicians, Am. Med. Dirs. Assn., N.Y. Acad. Scis. Republican. Roman Catholic. Achievements include research in electromagnetic energy in medicine. Home: 1073 Hart St Brooklyn NY 11237-3405 Office: NYU Med Ctr Goldwater Meml Hosp Roosevelt Island NY 10044

MONTESANO, ALDO MARIA, economics educator; b. Reggio, Calabria, Italy, July 14, 1939; s. Francesco M. and Maria I. (Gargiulo) M. Dr Engring., Polytechnic, Milan, Italy, 1962; DSc in Econ. Commerce, Bocconi U., Milan, Italy, 1966. Asst. prof. Bocconi U., Milan, 1966-76; assoc. prof. U. Venice, Italy, 1972-76, prof., 1976-83; prof. U. Milan, 1983-87; prof. econs. Bocconi U., Milan, 1987—. Editor Rivista Internat. Sci. Econ. Comml., 1983—; cons. editor Theory and Decision, 1987—; assoc. editor Economia Politica, 1986—; contbr. articles to profl. jours. Mem. Nat. Acad. Lincei (corr.), Soc. Ital. degli Economisti, Am. Econ. Assn., Royal Econ. Soc., Econometrica, European Econ. Assn., Internat. Soc. Intercommunication of New Ideas. Home: Via E Motta 6, I-20144 Milan Italy Office: Bocconi U, Via R Sarfatti 25, I-20136 Milan Italy

MONTGOMERY, ELIZABETH ANN, clinical research consultant; b. Springfield, Ohio, May 2, 1957; d. John Warwick and Joyce Ann (Bailer) M. Student, Bethany Luth. Coll., Mankato, Minn., 1975-76, Faculté de Médecine, Strasbourg, France, 1978-80; cert., Internat. Inst. Human Rights, Strasbourg, 1980; BS in Biol. Sci. with honors, U. Calif., Irvine, 1986. Lab. technician dept. psychobiology U. Calif., 1978, 80, lab. rsch. asst., 1983-84, lab. rsch. asst. dept. devel. and cell biology, 1984-86, clin. rsch. asst. dept. medicine, 1986; dispatcher Saddleback Chapel & Mortuary, Tustin, Calif., 1984; clin.. rsch. assoc. I, Baxter Edwards Labs., Irvine, 1987-88; engring. R & D biologist Baxter Bentley Labs., Irvine, 1988; clin. rsch. assoc. II, ICN Pharms., Costa Mesa, Calif., 1988-89; sr. clin. rsch. assoc. Berlex Labs. div. Schering AG, Alameda, Calif., 1989-92; clin. rsch. cons. Clinimetrics Rsch. Assocs., San Jose, Calif., 1992—. Vol. pediatrics and emergency room U. Calif. Med. Ctr., 1976-77. Mem. Assocs. Clin. Pharmacology and Drug Info. Assn., Internat. Regulatory Affairs Profls. Soc., Nat. Regulatory Affairs Profls. Soc., U. Calif.-Irvine Alumni Assn., Alpha Epsilon Delta. Lutheran. Home and Office: 313 S Overlook Dr San Ramon CA 94583-4542

MONTGOMERY, JOHN HAROLD, environmentalist; b. San Francisco, Mar. 16, 1955; s. Lloyd Cecil Montgomery and Liane Claire (Klein) Malinofsky; m. Patricia Elizabeth Keim, Sept. 13, 1986; 1 child, Kelly Elizabeth. AS in Chemistry, Brookdale C.C., Lincroft, N.J., 1978; BS in Geology, Stockton State Coll., Pomona, N.J., 1984. Lic. profl. geologist, Tenn.; lic. subsurface evaluator, N.J. Lab. technician Seals Eastern, Inc., Red Bank, N.J., 1977-78; rsch. technician Inter-Polymer Rsch. Corp., Farmingdale, N.J., 1978-79, CPS Chem. Co., Old Bridge, N.J., 1979-80; reporter Asbury Park Press, Neptune, N.J., 1974-86; rsch. chemist Union Carbide Corp., Bound Brook, N.J., 1985-86; prin. geologist N.J. Dept. Environ. Protection, Trenton, 1986-90; project mgr. Groundwater & Environ. Svcs. Inc., Wall, N.J., 1990—. Author: Groundwater Chemicals Desk Reference, 1990, Groundwater Chemicals Desk Reference - Vol. 2, 1991, Groundwater Chemicals Field Guide, 1991, Agrochemicals Desk Reference, Environmental Data, 1993. Mem. Nat. Ground Water Assn., Geol. Assn. N.J., Marina View Homeowner's Assn. (v.p. 1986-87). Republican. Methodist. Achievements include research in quantitative structure phys. relationships in determining aqueous solubilities of organic compounds, fate and transport of organics in groundwater. Home: 1115 Raymere Ave Wanamassa NJ 07712-4115 Office: Groundwater & Environ Svcs Inc 1340 Campus Pkwy Wall NJ 07719

MONTGOMERY, JOHN HENRY, electronics company executive; b. Greenwood, Miss., July 17, 1937; s. John Henry Montgomery and Sally Blanton (Williford) Montgomery Williamson; m. Elaine Marie Damicone, Dec. 17, 1971; children: John Henry IV, David Paul. BS in Elec. Engring., Miss. State U., Starkville, 1965. Field engr. Western Union Co., Atlanta, 1964-66, Codex Corp., Cambridge, Mass., 1966-72; v.p. MCI Telecommunications, Washington, 1972-80, No. Telecom, Inc., Richardson, Tex., 1980-81, Digital Switch Corp., Richardson, 1981-87; dir. Integrated Telecom Corp., 1986-90; pres., chief exec. officer Integrated Telecommunications, Inc., 1988-90; v.p., gen. mgr. CP Network System DSC Communication Corp., 1990—. Served with AUS 1958-60. Home: 2614 Woods Ln Garland TX 75044-2806 Office: DSC Communications Corp 1000 Colt Rd Plano TX 75075

MONTGOMERY, PHILIP O'BRYAN, JR., pathologist; b. Dallas, Aug. 16, 1921. BS, So. Meth. U., 1942; MD, Columbia U., 1945. Diplomate Am. Bd. Pathology, Am. Bd. Clin. Pathology and Forensic Pathology. Intern Mary Imogene Bassett Hosp., Cooperstown, N.Y., 1945-46; fellow in pathology Southwestern Med. Sch., Dallas, 1950-51, asst. prof. pathology, 1953-55, assoc. prof., 1955-61, prof., 1961—, assoc. dean, 1968-70, Ashbel Smith prof. pathology, 1991—; rsch. asst. pathology and cancer rsch. Cancer Rsch. Inst. New Eng. Deaconess Hosp., Boston, 1951-52; spl. asst. to chancellor U. Tex. System, 1971-75; exec. dir. Cancer Ctr. U. Tex. Health Sci. Ctr. Dallas, 1975-89; pathologist Parkland Meml. Hosp., Dallas, 1952—, Dallas City Zoo, 1955-68; med. examiner DallasCounty, 1955-58; cons. Navarro County Meml. Hosp., Corsicana, Tex., 1952-53, McKinney (Tex.) Vets. Hosp., 1952-65, Lisbons Vets. Hosp., Dallas, 1953—, St. Paul Hosp., Dallas, 1958—, Flow Meml. Hosp., Denton, Tex., 1958-65; pathologist Tex. Children's Hosp., Dallas, 1954-55. Contbr. numerous articles to profl. jours., sci. abstracts, jours. Bd. dirs. Planned Parenthood of Dallas, 1958-63, pres., 1958-60; trustee St. Mark's Sch. Tex., 1958—, v.p., chmn. exec. com. bd. trustee, 1966-68, v.p., 1968-69, pres. 1974-76; trustee Lamplighter Sch., 1967-70; chmn. Dallas Area Libr. Planning Coun., 1970-72, Goals for Dallas Health Task Force com., 1975-76, Fleet Adm. Nimitz Mus. commn., 1979-81; mem. adv. bd. Dallas Citizens coun., chmn. health com. 1988-89; bd. dirs. Met. YMCA, 1960-63, Dallas Coun. on World Affairs, 1962-65; pres., bd. dirs. Damon Runyon, Walter Winchell Cancer Fund, 1974-79; cord. Dallas Arts Dist., 1982—. Fellow Am. Soc. Clin. Pathologists; mem. Am. Assn. Pathologists and Bacteriologists, Am. Assn. Cancer Rsch., Internat. Acad. Pathology, Am. Acad. Forensic Scis., Soc. Exptl. Biology and Medicine, Internat. Soc. Cell Biology, Biophys. Soc., Am. Soc. Cell Biology, Am. soc. Exptl. Pathology, Tissue Culture Assn., Internat. Fedn. Med. Electronics, Profl. Group Med. Electronics of Inst. Radio Engrs., AAAS, Optical Soc. Tex. (founding), Pan-Am. Med. Assn., AMA, So. Med. Assn., Tex. Med. Assn., AAUP. Office: 5323 Harry Hines Blvd Dallas TX 75235-7200

MONTGOMERY, ROBERT RENWICK, medical association administrator, educator; b. New Castle, Pa., June 3, 1943. BS in Chemistry, Grove City Coll., 1965; MD, U. Pitts., 1969. Diplomate Am. Bd. Pediatrics. Intern Childrens Hosp. Phila. U. Pa., 1969-70; resident Harriet Lane Svc. Johns Hopkins Hosp., 1972-73, fellow, 1972-73; fellow U. Colo., 1973-76, Scripps Clinic and Rsch. Found., 1976-77; gen. med. officer USPHS, Chinle, Ariz., 1970-71; dep. chief pediatrics USPHS, Tuba City, Ariz., 1971-72; rsch. clin fellow in pediatric hematology U. Colo., 1973-76; rsch. fellow in molecular immunology Scripps Clinic and Rsch. Found., 1976-77; acting dir. Mountain States Regional Hemphilia Program U. Colo., 1977-78, asst. prof. dept. pediatrics, 1977-80, co-dir. coagulation rsch. labs., asst. dir. mountain sates regional hemophilia program, 1978-80; asst. prof. dept. pediatrics Med. Coll. Wis., 1980-81; dir. homeostasis program Milwaukee Children's Hosp., 1980-84; med. dir. Great Lakes Hemophilia Found., 1980-84; dir. regional homeostasis reference lab. The Blood Ctr. Southeastern Wis., 1981—; cons. hemostasis lab., dept. pathology The Children's Hosp. Wis., 1981—; assoc. prof. dept. pediatrics Med. Coll. Wis., 1981-84; sr. investigator The Blood Ctr. Southeastern Wis., 1982—, section head hemostasis rsch.; scientific dir. Great Lakes Hemophilia Found., 1984—; assoc. dir. rsch. The Blood Ctr. Southeastern Wis., 1984-86; assoc. clin. prof. dept. pediatrics Med. Coll. Wis., clin. prof. dept. pediatrics, 1986—; dir. rsch. The Blood Ctr. Southeastern Wis., 1986—; acting sect. head coagulation lab., dept. pathology Med. Coll. Wis., 1986-87; faculty med. tech. Marquette U., Milw., 1986-92; clin. prof. dept. pathology Med. Coll. Wis., 1987—; v.p., dir. rsch. The Blood Ctr.

Southeastern Wis., 1988—; mem. med. adv. com. Great Lakes Hemophilia Found., 1980—; mem. libr. com. The Blood Ctr. Southeastern Wis., 1981—, mem. human rsch. review com. The Blood Ctr. Southeastern Wis., 1981—, mem. rsch. mgmt. group, 1983—, rsch. strategic planning com., 1983—; mem. subcom. FVII and von Willebrand factor Internat. Congress Thrombosis and Haemostasis, 1984—; mem. radiation safety com. The Blood Ctr. Southeastern Wis., 1984—; ad hoc reviewer Heart Lung and Blood Inst., NIH, 1984—; mem. Inst. Biosafety com. The Blood Ctr. Southeastern Wis., 1985—; chmn. rsch. review com. Nat. Hemophilia Found., 1987—; ad hoc reviewer com. B Nat. Heart, Lung and Blood Inst., 1991—; chmn. von Willebrand subcom. Hemophilia Rsch. Soc., 1990—; pres. Hemophilia Rsch. Soc., 1990—; mem. med. scientific adv. com. Nat. Hemophilia Found., 1992—; mem. bd. dirs. Wis. Sickle Cell Disease Comprehensive Ctr., 1992—; chair med. adv. coun. Great Lakes Hemophilia Found., Milw., 1992—; mem. blood diseases and resources adv. com. Nat. Heart, Lung, and Blood Inst., 1992—. Sr. asst. surgeon USPHS, Indian Health Svc., 1971-72. Recipient Nat. Rsch. Svc. award Heart Lung and Blood Inst., NIH, 1975-77, Young Investigator award, 1978-81, Established Investigator award Am. Heart Assn., 1982-87, Jack Kennedy Alumni Achievement award Groce City Coll., 1985, Dr. Murray Thelin award Nat. Hemophilia Found., 1991. Mem. AAAS, Am. Soc. Clin. Investigators, Am. Soc. Pediatric Hematology/Oncology, Am. Soc. Hematology, Am. Fedn. Clin. Rsch., Am. Heart Assn., N.Y. Acad. Sci., Western Soc. Pediatric Rsch., Internat. Soc. Thrombosis and Hemostasis, Soc. Pediatric Rsch., Hemophilia Rsch. Soc. Office: Blood Center of SouthEastern Wis Blood Research Institute 1701 W Wisconsin Ave Milwaukee WI 53233*

MONTGOMERY, TERRY GRAY, textiles researcher; b. Ft. Knox, Ky., Aug. 25, 1953; s. Bobby Gray Montgomery and Shirley Mitchell Sutphin; m. Linda Stoltz, May 18, 1973; 1 child, Matthew Gray. BS, N.C. State U., 1974, PhD, 1980; MBA, Winthrop U., 1991. Asst. prof. of textile scis. Southeastern Mass. U., Dartmouth, Mass., 1979-81; product devel. mgr. Milliken & Co., Spartanburg, S.C., 1981-85; mgr. new bus. devel. Burlington Industries, Inc., Greensboro, N.C., 1985-87; product design mgr. Springs Industries, Inc., Ft. Mill, S.C., 1987-88, dir. R & D, 1988—; bd. dirs. Armor Techs., Houston. Author: (rsch. pub.) Textile Rsch. Jour., 1982-84; mem. adv. bd. (mag.) Safety & Protective Fabrics, 1992—. Mem. Am. Textile Mfg. Inst. (flammability com. 1992—, consumer affairs com. 1992—), Fiber Soc., Phi Kappa Phi, Beta Gamma Sigma, SAMPE. Republican. Moravian. Achievements include 3 U.S. patents. Home: Springs Industries Inc 220 Rose Arbor Ln Matthews NC 28105 Office: Springs Industries Inc 123 N White St Fort Mill SC 29715

MONTGOMERY, WILLARD WAYNE, physicist; b. Gamaliel, Ky., May 26, 1949; s. Willard Osborne Montgomery and Mildred Lucille (Parkhurst) Morse; m. Brenda Kaye Winberg, Aug. 3, 1973. BS, Mid. Tenn. State U., 1973; MS, Purdue U., 1975; PhD, The U. Tenn., 1980. Staff scientist Lockheed Missiles & Space Co., Huntsville, Ala., 1978—. Contbr. articles to profl. jours. Mem. Optical Soc. Am., Sigma Xi. Republican. Achievements include rsch. in coherent lidar systems; prin. investigator for Lockheed coherent pulsed laser radar independent devel. project. Office: Lockheed Missiles & Space PO Box 70017 Huntsville AL 35807-7017

MONTGOMERY-DAVIS, JOSEPH, osteopathic physician; b. Annapolis, Md., Aug. 27, 1940; s. John and Flonila Alice (Sutphin) Swontek. Student, U. Wis., Milw., 1967-70; DO, Chgo. Coll. Osteo. Medicine, 1974. Diplomate Nat. Bd. Examiners for Osteo. Physicians and Surgeons. Chief technologist nuclear medicine dept. Columbia Hosp., Milw., 1964-70; intern Richmond Heights (Ohio) Gen. Hosp., 1974-75; pvt. practice Raymondville, Tex., 1975—; med. care adv. com. Tex. Dept. Human Svcs., Austin, 1990—; health officer Willacy County Bd., Raymondville, 1984—. Contbr. articles to profl. jours. With USAF, 1959-63. Mem. Am. Osteo. Assn., Tex. Osteo. Med. Assn. (pres. 1989-90, Disting. Svc. award 1990), Am. Coll. Gen. Practitioners, Tex. Soc. Am. Coll. Gen. Practitioners (pres. 1985-86, Disting. Svc. award 1986), Tex. Med. Found., Tex. Coll. Osteo. Medicine Alumni Assn., Phi Eta Sigma, Sigma Sigma Phi. Office: Raymondville Med Clinic 5255 S 10th St Raymondville TX 78580

MONTI, LAURA ANNE, psychology researcher, educator; b. Evanston, Ill., Feb. 28, 1959; d. LeRoy John and Mary Alice (Foley) M. BA in Psychology, U. Ariz., 1981; MA in Cognitive Sci., Loyola U., Chgo., 1986, PhD, 1987; postgrad., Menninger Found., 1988. Mem. bd. dirs., co-owner Monti & Assocs. Inc., Arlington Heights, Ill., 1976—; v.p., co-owner MAM Imports and Creative Gifts, Kildeer, Ill., 1986-89; lectr. psychology Loyola U., Chgo., 1986-89; asst. prof. North Park Coll., Chgo., 1989-91; instr. Rush Presbyn. St. Luke's Med. Ctr., Chgo., 1992—; vis. rsch. specialist U. Ill. Inst. Devel. Disabilities, 1989-90; cons. Walter H. Sobel FAIA & Assocs., Chgo., 1987—, Yate and Auberle, Oakbrook, Ill., 1987-88; postdoctoral fellow Northwestern U., Evanston, Ill., 1990-92. Contbr. articles to profl. jours.; co-author tech. reports to various orgns. Tuition scholar Loyola U., Chgo., 1983-84; NIH fellow, 1992-94; Loyola U. grad. asst., 1986. Mem. APA (divsn. psychology of women 1989-92, gen. psychology, exptl. psychology 1989—), Psi Chi (faculty rep. for North Park Coll. 1989-91), Sigma Alpha Iota. Roman Catholic. Avocations: tennis, piano. Home: 632 Happfield Dr Arlington Heights IL 60004

MONTIJO, RALPH ELIAS, JR., engineering executive; b. Tucson, Oct. 26, 1928; m. Guillermina Paredes, Dec., 1947; children: Rafael (dec.), Suzanne, Felice. BSEE, U. Ariz., 1952; postgrad. in digital computer engring., U. Pa., 1953-57; postgrad. in mgmt., U. Calif., Los Angeles, 1958-60. Registered profl. engr., Tex. With RCA Corp., 1952-67; design and devel. engr. RCA Corp., Camden, N.J., 1952-55; mgr. West Coast EDP engring. RCA Corp., L.A., 1960-61, mgr. EDP systems engring., 1961-64; mgr. special systems and equipment planning, product planning divsn. RCA Corp., Cherry Hill, N.J., 1964-65; mgr. Calif. Dept. of Motor Vehicles program RCA Corp., Sacramento, 1965-66; mgr. spl. EDP programs RCA Corp., 1966-67; with Planning Rsch. Corp., 1967-72; dep. div. mgr. Eastern and European ops., computer systems div. Planning Rsch. Corp., Washington, 1968; dep. div. mgr. advanced systems planning, reservations systems, computer systems div. Planning Rsch. Corp., Moorestown, N.J., 1968-69; v.p., gen mgr. Internat. Reservations Corp. div. Planning Rsch. Corp., L.A., 1969-70, exec. v.p., 1970-71, pres., 1971-72, also bd. dirs.; v.p. Systems Sci. Devel. Corp. subs. Planning Rsch. Corp., L.A., 1972—; CEO Omniplan Corp., Culver City, Calif. and Houston, 1972—. Contbr. 37 articles to profl. jours.; patentee in field. Recipient Alumni Achievement award U. Ariz., 1985, Centennial medal U. Ariz., 1989. Mem. IEEE, NSPE, Am. Mgmt. Assn., Assn. for Computing Machinery, U. Ariz. Alumni Assn. (pres. So. Calif. chpt. 1980-81, Centennial Medallion award 1989, Scabbard and Blade, Alumni Achievement award 1985; pres. Houston chpt. 1992-93). Republican. Roman Catholic. Home: 2222 Gemini Ave Houston TX 77058-2049 also: 3651 Beverly Ridge Dr Sherman Oaks CA 91423 Office: Omniplan Corp 17041 El Camino Real Houston TX 77058-2617

MONTOLIU, JESUS, nephrologist, educator; b. Barcelona, Catalonia, Spain, Nov. 3, 1949; s. Jesus and Ana Maria (Duran) M.; m. Merja Kati Tirranen, Oct. 18, 1974 (div. Sept. 1990); children: Laura, Daniel, Sebastian, Jaume. MD, U. Barcelona, 1973. Diplomate in internal medicine and nephrology Am. Bd. Internal Medicine. Intern Mt. Sinai Hosp. Program, N.Y.C., 1973-74, med. resident, 1974-76; resident nephrology N.Y. Hosp., Cornell Med. Ctr., N.Y.C., 1976-78, Hosp. Univ. Barcelona, 1978-79; fellow nephrology U. Toronto (Can.), 1979-80; attending physician nephrology Hosp. Univ. Barcelona, 1980-88; chief nephrology svc. Hosp. Arnau de Vilanova, Lleida, Spain, 1988—; assoc. prof. medicine U. Barcelona, 1987-90, prof. medicine, 1990-92; prof. medicine, chmn. medicine U. Lleida, 1992—; dir. dept. medicineEstudi Gen. Lleida, 1990—; chief of studies Hosp. Arnau de Vilanova, Lleida, 1989-90. Author 2 books, chpts. in 20 books; contbr. articles to profl. jours. 2d lt. Spanish mil., 1973-74. Fellow ACP, Internat. Nephrology; mem. Am. Soc. Nephrology (corr.), EDTA-European Renal Assn., Spanish Soc. Nephrology (v.p. 1984-87). Convergencia Democràtica de Catalonia. Avocations: sailing, reading, sports, music, art. Home: Bisbe Irurita 12-7-3-1, 25006 Lleida Spain Office: Hosp Arnau de Vilanova, Rovira Roure 80, 25006 Lleida Spain

MONTONE, LIBER JOSEPH, engineering consultant; b. Apr. 21, 1919; s. Vito and Philomena (Carnicelli) M.; m. Clara Elisabeth Edwards, June 1, 1945; 1 child, Gregory Edwards. MS, Temple U., 1961. Registered profl. engr., Pa. Quality control supr. Haskell Electronic and Tool Corp., Homer, N.Y., 1950-53; researcher IBM Airborne Computer Lab., Vestal, N.Y., 1954-56; devel. engr. Western Electric Co. and Bell Labs., Laureldale, Pa., 1956-61; sr. devel. engr. Western Electric Co. Inc., Reading, Pa., 1961-65, sr. staff engr. R&D, 1965-82; cons. Naples, Fla., Fenwick Island, Del., 1983—; biomed. engr., cancer rsch. projects pathology and clin. lab. St. Joseph's Hosp., Reading, 1961-80; tech. cons. Reading Hosp., 1961-78. Contbr. articles to profl. jours. including The Engr., Am. Assn. Clin. Scientists Symposium. Capt. USAAF, 1942-45, ETO. Recipient Outstanding Paper award Engring. Rsch. Ctr., Princeton, N.J., 1963. Mem. NSPE, Fla. Engring. Soc., Res. Officer's Assn. (life). Achievements include patents in field; invention (with others) diagnostic cystic fibrosis capillary conduction test method. Office: 45 W Virginia Ave Fenwick Island DE 19944

MONTS, DAVID LEE, scientist, educator; b. Taylorville, Ill., Apr. 14, 1951; s. Robert Virgil and Margrett Nell (Lee) M.; m. Dana Rochelle Harrold, Nov. 10, 1984; children: Desiree Michelle, Derek Earl, Deidre Lee. BS, U. Ill., 1973; MA, Columbia U., 1974, MPhil, 1976, PhD, 1977. Postdoctoral fellow Rice U., Houston, 1977-79; lectr. Princeton (N.J.) U., 1979-81; asst. prof. U. Ark., Fayetteville, 1981-88, Miss. State U., 1988—. Co-author: Spectral Atlas of Nitrogen Dioxide, 1978; contbr. articles to jours. Chem. Physics, Optics Comms., Applied Spectroscopy, Rev. Sci. Instruments. Mem. Am. Phys. Soc., Am. Chem. Soc., Soc. Applied Spectroscopy, Sigma Xi. Achievements include research in molecular and atomic spectroscopy; development of new experimental techniques for identification, kinetic studies of transient species, science education. Office: Miss State U Dept Physics & Astronomy Mississippi State MS 39726-3574

MONTVILLE, THOMAS JOSEPH, food microbiologist, educator; b. Somerville, N.J., Jan. 10, 1953; s. Frank Vincent and Elisabeth (Para) M.; m. Nancy Helen Shiffner, June 6, 1976; children: Christopher, Rebecca, Matthew. BS cum laude, Rutgers U., 1975; PhD, MIT, 1979. Rsch. asst. food sci. dept. MIT, Cambridge, 1975-80; rsch. microbiologist USDA, Ea. Region Rsch. Ctr., Phila., 1980-84; assoc. prof. food sci. dept. Rutgers U., New Brunswick, N.J., 1984-91, prof. food sci. dept., 1991—, dir. grad. program in food sci., 1991—; cons., lectr. in field, 1985—; editorial bd. Jour. Food Protection, 1985—, Applied & Environ. Microbiology, 1980-83; bd. editors Food & Nutrition Press, 1988—, Process Biochemistry, 1993—; panel mgr. USDA CSRS Food Safety Grants Program, 1992. Editor: Food Microbiology, vol. 1, 2, 1987, Jour. Food Safety; contbr. articles to profl. jours. Recipient McGraff lectureship, Long Island U., 1991, Cert. Merit, USDA, 1983. Fellow Am. Acad. Microbiology; mem. AAAS, Internat. Assn. Milk Food and Environ. Sanitarians, Am. Soc. Microbiology, Inst. Food Technologists (chmn. biotech. div. 1991-92), Am. Soc. Indsl. Microbiology, Soc. Applied Bacteriology, Phi Tau Sigma. Office: Rutgers U Food Sci Dept PO Box 231 New Brunswick NJ 08903-0231

MONTY, RICHARD ARTHUR, experimental psychologist; b. Sanford, Maine, Apr. 2, 1935; s. Leo James and Evelyn J. (Delahunt) M.; m. Margaret E. Penniston, June 22, 1965. MA, Columbia U., 1957; PhD, U. Rochester, 1962. Engring. psychologist Missile & Space Vehicle dept. GE Co., Phila., 1961-62; rsch. psychologist Cornell Aero. Lab., Buffalo, 1962-65; team leader Human Engring. Lab., Aberdeen Proving Ground, Md., 1965-80, chief behavioral rsch., 1980-86, chief scientist, 1986-92; chief ergonomic br. Army Rsch. Lab., Aberdeen Proving Ground, Md., 1992—; cons. VA Hosp., Boston, Chgo., Syracuse, 1965—, Va. Poly. Inst. and State U., Blacksburg, 1972-81, Bowdoin Coll., 1969-72. Editor: Eye Movements and Psychological Processes, 1976, Eye Movements, Cognition and Visual Processes, 1981, Choice and Perceived Control, 1979, Eye Movements and Psychological Functions, 1983; contbr. over 90 articles to profl. jours. With U.S. Army, 1957-59. Fellow APA, Human Factors Soc., Am. Psychol. Soc.; mem. Cosmos Club. Republican. Home: 600 Harvest Ct Bel Air MD 21014 Office: Army Rsch Lab Aberdeen Proving Ground MD 21005

MONYAK, WENDELL PETER, pharmacist; b. Chgo., Sept. 14, 1931; s. Wendell and Mary Elizabeth M.; m. Lorraine Mostek, Aug. 29, 1964. BS in Chemistry, Roosevelt U., 1957; BS in Pharmacy, St. Louis Coll. Pharmacy, 1961. Asst. chief pharmacist Little Co. of Mary Hosp., Chgo., 1961-66; chief pharmacist MacNeal Meml. Hosp., Berwyn, Ill., 1966-72; dir. pharmacy Ill. Masonic Med. Ctr., Chgo., 1972, dir. pharm. services, 1972-87; dir. pharmacy services St. Anne's Hosp., Chgo., 1987-88; adminstr., 1989—; teaching assoc. U. Ill., 1972-87. Author: Hospital Formulary and Therapeutic Guide for Residents and Interns, 1974, 3d edit. 1986. Pres., chmn. bd. dirs. Bohemian Home for Aged, 1986—. With M.C., AUS, 1955-57. Mem. Am. Pharm. Assn., Am. Soc. Hosp. Pharmacists, Ill. Pharm. Assn. (Spl. Recignition award), No. Ill. Soc. Hosp. Pharmacists, Chgo. Hosp. Coun. Club: Oakbrook Exec. Home: 19 W 059 Chateau N Oak Brook IL 60521 Office: 5055 N Pulaski Rd Chicago IL 60630-2706

MOODY, BRIAN WAYNE, chemist; b. Lebanon, Pa., Dec. 23, 1955; s. Ralph Robert and Mary Louise (Brown) M.; children: Tiffany Ann, Derek Wayne. BS in Chem., Lebanon Valley Coll., 1977. Rsch. chemist Kerr, Lancaster, Pa., 1978-88; mgr. product tech. DSM Engring. Plastics, Evansville, Ind., 1988—. Contbr. articles to profl. jours. Mem. Soc. Plastics Engrs., Soc. Plastics Industry. Home: 7366 Tyring Rd Newburgh IN 47630 Office: DSM Engring Plastics 2267 West Mill Rd Evansville IN 47720

MOODY, GENE BYRON, engineering executive, small business owner; b. Calhoun, Ga., Aug. 29, 1933; s. Denzel Elwood and Mary Edna (Hughes) M.; m. Willie Earline Chauncey, Sept. 1, 1955; children: Byron Eugene, Iva Marie Levy. BSCE, U. Tenn., 1956. Registered profl. engr., Ala., Ark., Ga., La., Miss., Tex. V.p. S.I.P. Engring. Corp., Baton Rouge, 1968-70; project engr. S.I.P., Inc., Houston, 1970-73; dir. of engring. Jacus Assoc., Mpls., 1972-73; dir. of civil engring. Barnard & Burk, Baton Rouge, 1973-79; project mgr. Process Svcs., Baton Rouge, 1979-80, Salmon & Assoc., Baton Rouge, 1980-81; chief engr. Minton & Assoc., Lafayette, La., 1982; mgr. Assoc. Engr. Cons., Baton Rouge, 1982-86; owner Gene B. Moody, P.E., Baton Rouge, 1986—. Author: Good Homemakers, 1988, Deliverance Manual, 1989; contbr. articles to profl. jours. Deacon South Side Bapt. Ch., Baton Rouge, 1974; tchr. Hamilton Bible Camp, Hot Springs, Ark., 1981-91; trustee Manna Bapt. Ch., Baton Rouge, 1989-91. With U.S. Army, 1957. U. Chattanooga scholar, 1951, U. Tenn. scholar, 1953. Fellow ASCE; mem. Am. Soc. Safety Engrs., La. Soc. Profl. Surveyors, Soc. Automotive Engrs., Inst. Transp. Engrs., La. Engring. Soc., Transp. Res. Rsch. Bd. Mem. Christian Ch. (minister). Home and Office: 9852 Hillyard Ave Baton Rouge LA 70809-3109

MOODY, LAMON LAMAR, JR., civil engineer; b. Bogalusa, La., Nov. 8, 1924; s. Lamar Lamon and Vida (Seal) M.; B.S. in Civil Engring., U. Southwestern La., 1951; m. Eve Thibodeaux, Sept. 22, 1954 (div. 1991); children: Lamon Lamar III, Jennifer Eve, Jeffrey Matthew. Engr., Tex. Co., N.Y.C., 1951-52; project engr. African Petroleum Terminals, West Africa, 1952-56; chief engr. Kaiser Aluminum & Chem. Corp., Baton Rouge, 1956-63; pres., owner Dyer & Moody, Inc., Cons. Engrs., Baker, La., 1963—, also chmn. bd., dir. Chmn., Baker Planning Commn., 1961-63. Trustee La. Coun. on Econ. Edn., 1987-93. Served with USMCR, 1943-46. Decorated Purple Heart; registered profl. engr., La., Ark., Miss., Tex.; registered profl. land surveyor, La., Tex. Fellow ASCE, Am. Congress Surveying and Mapping (award for excellency 1972); mem. La. Engring. Soc. (dir., v.p. 1980-81, pres. 1982-83, Charles M. Kerr award for public relations 1971, A.B. Patterson medal 1981, Odom award for disting. svc. to engring. profession, 1986), Profl. Engrs. in Pvt. Practice (state chmn. 1969-70), La. Land Surveyors Assn. (pres. 1968-69, Land Surveyor of Yr. award 1975), Cons. Engrs. Coun., Pub. Affairs Rsch. Coun. of La. (exec. com., trustee 1983-93), Good Roads and Transp. Assn. (bd. dirs. 1984-92), Baker C. of C. (pres. 1977, Bus. Leader of Yr. award 1975), NSPE (nat. dir. 1982-83), Blue Key. Republican. Baptist. Clubs: Masons (32 deg., K.C.C.H. 1986), Kiwanis (dir. 1964-65). Home: 451 Ray Weiland Dr Baker LA 70714-3353 Office: 2845 Ray Weiland Dr Baker LA 70714

MOODY, MAXWELL, JR., retired physician; b. Tuscaloosa, Ala., Aug. 7, 1921; s. Maxwell and Jean Kilroy (Lahey) M.; m. Betty Alice Morrissey, May 10, 1946; children: Maxwell III, Susan, Elizabeth Sims. BA, U. Ala., 1941; MD, U. Pa., 1944. Diplomate Am. Bd. Internal Medicine. Intern Gorgas Hosp., Ancon, C.Z., 1944-45, Ancon, 1944-45; resident Grad. Sch. Medicine, U. Pa., Phila., 1957-58, Univ. Hosp., Birmingham, Ala., 1948-50; jr. and sr. resident in medicine U. Ala. Hosp., Birmingham, 1948-50; pvt. practice Tuscaloosa, 1950-87, ret., 1987; pres. Tuscaloosa County Med. Soc., Ala. Soc. Internal Medicine; pres., chn. bd. Ala. Heart Assn. Former state chmn., nat. trustee Ducks Unltd. Capt. U.S. Army, 1945-47. Fellow Am. Coll. Physicians. Republican. Episcopalian. Avocations: golf, hunting, fishing. Home: 7604 Mountbatten Rd NE Tuscaloosa AL 35406-1110

MOODY, ROGER WAYNE, civil engineer; b. Henderson, N.C., May 31, 1956; s. Calvin Wilton and Othela Pearl (Edwards) M.; m. Lynn Derisse Pergerson, Feb. 20, 1987; children: Bradley Dale, Hunter Wayne. BSCE, N.C. State U., 1979. Registered profl. engr., N.C. Design engr. Harland Bartholomew & Assocs., Raleigh, 1979-81, Tate Lanning & Assocs., Raleigh, 1981-84; project engr. Rummel, Klepper & Kahl, Raleigh, 1984-85; project mgr. Rummel, Klepper & Kahl, Virginia Beach, Va., 1988-91; office mgr. Am. Engrs., Raleigh, 1991-92, prin., 1993—; mem. N.C. Dept. of Transp./Cons. Engrs. Coun. Joint Transp. Com., Raleigh, 1992—, Am. Cons. Engrs. Coun/N.C. sect. Transp. Commn., Raleigh, 1992—. Mem. Community Devel. Adv. Bd., Henderson, N.C., 1992—. Mem. ASCE, Am. Soc. Hwy. Engrs., N.C. Cons. Engrs. Coun. (rep.), Lions. Presbyterian. Home: 223 Willowood Dr Henderson NC 27536 Office: American Engrs 3803-B Computer Dr Raleigh NC 27609

MOOERS, CHRISTOPHER NORTHRUP KENNARD, physical oceanographer, educator; b. Hagerstown, Md., Nov. 11, 1935; s. Frank Burt and Helen (Miner) M.; m. Elizabeth Eva Fauntleroy, June 11, 1960; children: Blaine Hanson MacFee, Randall Walden Lincoln. BS, U.S. Naval Acad., 1957; MS, U. Conn., 1964; PhD, Oreg. State U., 1969. Postdoctoral fellow U. Liverpool (Eng.), 1969-70; asst. prof. U. Miami (Fla.), 1970-72, assoc. prof., 1972-76; assoc. prof. U. Del., Newark, 1976-78, prof., 1978-79; prof., chmn. dept. oceanography Naval Postgrad. Sch., Monterey, Calif., 1979-86; dir. Inst. Naval Oceanography, Stennis Space Ctr., Miss., 1986-89; sci. advisor to dir. Inst. for Naval Oceanography, 1989; rsch. prof. U. N.H., Durham, 1989-91; prof., chmn. div. applied marine physics U. Miami, Fla., 1991-93, dir. ocean pollution rsch. ctr., 1992—. Editor Jour. Phys. Oceanography, 1991—; mng. editor Coastal and Estuarine Studies, 1978—. Served with USN, 1957-64. NSF fellow, 1964-67; NATO fellow, 1969-70; Sr. Queen Elizabeth fellow, 1980. Mem. AAAS, The Oceanography Soc. (interim councilor 1987-88), Am. Geophys. Union (pres. ocean sci. sect. 1982-84), Eastern Pacific Oceanic Conf. (chmn. 1979-86), Am. Meteorol. Soc., Marine Tech. Soc., Acoustical Soc. Am., Estuarine Rsch. Fedn., Sigma Xi. Achievements include pioneering direct observation of transient coastal ocean currents and fronts and mesoscale and coastal prediction research. Home: 2521 Inagua Ave Coconut Grove FL 33133-3811 Office: U Miami Div Applied Marine Physics RSMAS 4600 Rickenbacker Causeway Miami FL 33149-1098

MOON, BILLY G., electrical engineer; b. Louisville, Oct. 22, 1961; m. Catherine Alice Haunz, May 26, 1984; 1 child, William Victor. BSEE, U. Louisville, 1984. Elec. engr. coop. GTE of Ky., Lexington, 1981-82, Louisville (Ky.) Gas and Elec., 1982-83; elec. engr. II Motorola, Inc., Ft. Lauderdale, Fla., 1984-85, elec. engr. I, 1985-87; engring. sect. mgr. Americom Corp., Atlanta, 1987-89, advanced product planning mgr., 1989-90, dir. tech., 1990-91; v.p. engring. Uniden Am. Corp., Ft. Worth, 1992—; owner Moonsoft, Inc., Ft. Lauderdale, 1984-87. Deacon Duluth (Ga.) Seventh Day Adventist ch., 1989-92. Recipient 2nd place award elec. engring. design competition, U. Louisville, 1982. Mem. Nat. Honor Soc., Beta Club. Republican. Home: 600 Oak Hill Dr Southlake TX 76092 Office: Uniden Am Corp 4700 Amon Carter Blvd Fort Worth TX 76155

MOON, HARLEY WILLIAM, veterinarian; b. Tracy, Minn., Mar. 1, 1936; s. Harley Andrew Moon and Catherine Mary (Engesser) Lien; m. Irene Jeannette Casper, June 9, 1956; children: Michael J., Joseph E. Anne E. Teresa J. BS, U. Minn., 1958, DVM, 1960, PhD, 1965. Diplomate Am. Bd. Veterinary Pathologists. Instr. Coll. Vet. Medicine U. Minn., St. Paul, 1960-62, NIH postdoctoral fellow, 1963-65; vis. scientist Brookhaven Nat. Lab., Upton, N.Y., 1965-66; assoc. prof. Coll. Vet. Medicine U. Sask., Saskatoon, Can., 1966-68; rsch. vet. Nat. Animal Disease Ctr. Agrl. Rsch. Svc., USDA, Ames, Iowa, 1968-88, ctr. dir., 1988—; assoc. prof. Iowa State U., Ames, 1970-73, prof. 1973-74; cons. U. N.C., Chapel Hill, 1985-92, Pioneer Hy-Bred Internat., Johnson, Iowa, 1986-92. Contbr. articles reporting rsch. on animal diseases. Recipient Superior Svc. award USDA. Mem. NAS, Am. Coll. Vet. Pathologists, AVMA, Am. Soc. Microbiologists, AAAS, NAS, Sigma Xi, Phi Zeta. Avocation: farming. Home: 800 Shagbark Dr Nevada IA 50201-2702 Office: USDA ARS Nat Animal Disease Ctr PO Box 70 2300 Dayton Rd Ames IA 50010

MOON, MARLA LYNN, optometrist; b. Connellsville, Pa., July 31, 1956; d. George Donnelly and Pauline Harriet (Hough) M. BS, Pa. State U., 1978, Pa. Coll. Optometry, Phila., 1980; OD, Pa. Coll. Optometry, 1982. Cert. Nat. Bd. Examiners, Pa., N.J. Bds. of Optometric Examiners. Intern Gesell Inst. for Human Devel., New Haven, 1981, U.S. Mil. Acad., West Point, N.Y., 1981, Dr. William Moskowitz, Somerville, N.J., 1981-82, Elwyn Ins., Feinbloom Ctr., Phila., 1982; resident, pediatrics unit The Eye Inst., Phila., 1982-83; ptnr. Drs. Carlin and Moon, State College, Pa., 1983—; vis. lectr. Dominican Coll., Orangeburg, N.Y., 1985, Pa. State U., University Park, 1985-89, 91, 92; faculty Pa. Coun. Horseback Riding for Handicapped, State College, 1988-92; cons. JMS Mobility Assocs., Inc., Exton, Pa., 1983-89, Univ. Hosp. and Rehab. Ctr., Hershey, Pa., 1988—, John Heinz Rehab. and Med. Ctr., Wilkes-Barre, Pa., 1990—. Mem. adv. bd., v.p. Learning Disabilities Assn., State College, 1983—; com. chmn. Local Children's Team, State College, 1985-89; pres., bd. dirs. Cen.-Clear Child Svcs., Phila., 1984—; mem. Task Force Project Self Sufficiency, Bellefonte, Pa., 1988—; bd. dirs. Pa.-Del. Assn. for Educators and Rehab. of Blind and Visually Impaired, Harrisburg, 1988—. Recipient Phila. County Optometric Soc. award, 1982, Knight-Henry Meml. award Optometric Ext. Program, Phila., 1982, Disting. Svc. award Assn. Educators and Rehab. of Blind and Visually Impaired (Pa.-Del. chpt.), 1992. Mem. Am. Optometric Assn. (optometric recognition 1985-93), Pa. Optometric Assn. (chmn. 1989-91), Mid-Counties Optometric Soc. (pres. 1992—), Pa. State Alumni Assn. (life), Altrusa Club (sec., v.p., pres. 1987—), Omega Epsilon Phi. Avocations: spectator sports, golf, tennis. Office: 423 S Pugh St State College PA 16801-5308

MOONEY, DENNIS JOHN, forensic document examiner; b. Boston, June 25, 1949; s. Joseph H. and Theresa (Champia) M.; children: Dennis John Jr., Joseph C. BA, Shaw U., 1975. Diplomate Am. Bd. Forensic Document Examiners. Fingerprint tech. FBI, Washington, 1970-72; forensic document examiner N.C. State Bur. of Investigation, Raleigh, 1972-83; agt. in charge Colo. Bur. of Investigation, Denver, 1983—. Contbr. articles to profl. jours. With U.S. Army, 1967-70. Mem. Am. Acad. Forensic Scis., Southwestern Assn. Forensic Document Examiners (regional rep.). Office: Colo Bur Investigation 690 Kipling St Denver CO 80215

MOONEY, JOHN BRADFORD, JR., oceanographer, engineer, consultant; b. Portsmouth, N.H., Mar. 26, 1931; s. John Bradford and Margaret Theodora (Akers) M.; m. Martha Ann Huntley, Dec. 25, 1953 (dec. May 1990); children: Melinda Jean, Pamela Ann, Jennifer Joan; m. Jennie Marie Duca, Nov. 24, 1990. BS, U.S. Naval Acad., 1953; postgrad., George Washington U., 1970, 71, 76; grad. nat. and internat. security program, Harvard U., 1980. Commd. ens. USN, 1953, advanced through grades to rear adm., 1979; chief staff officer Submarine Devel. Group 1, 1971-73; commdr. Bathyscaphe Trieste II, 1964-66, Submarine Menhaden, 1966-68; comdg. officer Naval Sta., Charleston, S.C., 1973-75; dep. dir. Deep Submergence Systems Div., Office Chief Naval Ops., Washington, 1975-77; comdr. Naval Tng. Ctr., Orlando, Fla., 1977-78; dir. Total Force Planning Div., Office Chief Naval Ops., Washington, 1978-81; oceanographer USN, 1981-83, chief naval rsch., 1983-87, ret., 1987; pres. Harbor Br. Oceanographic Instn., Inc., Ft. Pierce, Fla., 1989-92, marine bd., 1991—; bd. dirs. Coltec Industries, 1992—; mem. marine programs adv. coun. U.R.I., Narragansett, 1989—. At controls of Trieste II when hull of Thresher was found on floor of Atlantic, 1964; coordinated deep search and recovery of hydrogen bomb lost off coast of Spain, 1966; condr. recovery operation from depth of 16,400 feet in Mid-Pacific, 1972. Decorated Legion of Merit with 1 gold star; recipient spl. citation Armed Forces Recreation Assn., 1975. Mem. Marine Tech. Soc. (pres. 1991-93), Nat. Acad. Engring., U.S. Naval Inst., Nat. Geog. Soc., Smithsonian Assocs., Masons, Shriners, Order of DeMolay. Avocations:

racquetball, sailing. Home and Office: 801 S Ocean Dr # 706 Fort Pierce FL 34949-3428

MOONEY, ROBERT MICHAEL, ophthalmologist; b. Mt. Vernon, N.Y., July 25, 1945; s. Robert Michael and Marie Evelyn (sabatini) M.; m. Dorothy May Kazmaier, Feb. 21, 1981. BS in Biology, Fordham U., 1966; MD, U. Bologna, Italy, 1972. Diplomate Am. Bd. Ophthalmology. Intern Grasslands Hosp., Valhalla, N.Y., 1972-73; resident in surgery Grasslands Hosp., 1973-74; resident in ophthalmology N.Y. Med. Coll., Valhalla, 1974-76; chief resident ophthalmology N.Y. Med. Coll., 1976-77; acting dir. dept. ophthalmology Westchester County Med. Ctr., Valhalla, 1980-86; pvt. practice Katonah-Mt. Kisco, N.Y., 1979—; asst. clin. prof. ophthalmology N.Y. Med. Coll., Valhalla, 1982—. Fellow Am. Acad. Ophthalmology, Am. Coll. Surgeons; mem. Med. Soc. State of N.Y., Westchester County Med. Soc., Westchester Acad. Medicine (chmn. sect. ophthalmology 1987-89), MENSA. Republican. Roman Catholic. Avocations: travel, photography. Office: 51 Bedford Rd Katonah NY 10536-2135

MOONEYHAN, ESTHER LOUISE, nurse, educator; b. Wabash, Ind., Oct. 2, 1920; d. Edward Lamont and Ina Louretta (Adams) Smithee; children: William Cecil, Mary Kathleen, Stephen Alan. Student (scholar), Ind. Wesleyan U., 1938-40; diploma (scholar), Meth. Hosp. Sch. Nursing, Indpls., 1943; Ind. Wesleyan U.; BS in Gen. Nursing, Ind. U., 1964, MS in Nursing Edn., 1965, Ed. D. (fellow), 1973. Staff nurse Putnam County (Ind.) Hosp., 1943-44, Meth. Hosp., Indpls., 1951; pvt. duty nurse Ind. U. Med. Ctr., Indpls., 1953-60; staff nurse, assoc. supr., ednl. dir. Bur. Pub. Health Nursing Health & Hosp. Corp., Indpls. and Marion County, Ind., 1960-67; nurse adviser AID, Haile Selassie I U. Coll. Pub. Health, Ethiopia, 1967-69; asst. prof. nursing Tex. Woman's U., Denton, 1972-73; assoc. prof. nursing Fla. Internat. U. Miami, 1973-79, chmn., 1974-76; prof., adminstr. Ind. U. Sch. Nursing, South Bend, Ind., 1979-88; prof. Ind. U. Sch. Nursing, Indpls., 1989-91; vis. prof. U. South Fla. Coll. Nursing, 1988-89, U. Ctrl. Fla., Orlando, 1991-92; community health nurse Manatee County Pub. Health unit Fla. Health and Rehab. Svcs., 1992—; charter mem. Nursing Rsch. Consortium North Ctrl. Ind.; speaker state workshops and nat. convs.; cons. in field. Contbr. articles to profl. jours. Mem. Am. Nurses Assn., Coun. Nurse Researchers, Fla. Nurses Assn. Dist. 4 , Nat. League Nursing, Am. Pub. Health Assn., Internat. Health Soc. (bd. dirs. 1977-79), Nat. Coun. Internat. Health, Am. Assn. for World Health, Pi Lambda Theta, Sigma Theta Tau. Baptist. Achievements include research interest in nursing curricula, international development, international health, international nursing, and family/community health.

MOORADIAN, GREGORY CHARLES, physicist; b. Tampa, Fla., Sept. 5, 1947; s. Richard Dick and Lucienne A. M.; m. Marsha Eileen Sterling, Dec. 29, 1973; children: Michelle, Michael. BS, Calif. State U., 1969; MS, U. Calif., San Diego, 1971, PhD, 1973. Ind. cons. Del Mar, Calif., 1969-72; mgr. laser lab. Gulf Gen. Atomic, La Jolla, Calif., 1971-72; chief scientist Naval Ocean Systems Ctr., San Diego, 1972-88; corp. v.p. Sci. Applications Internat. Corp., San Diego, 1988—; adv. bd. NSF, Washington, 1980-82; mem. Naval Studies Bd. Space Panel, Washington, 1982-83, Naval Rsch. Adv. Coun., Washington, 1983-84. Author: Blue-Green Pulsed Propagation Through Clouds, 1981; contbr. articles to Applied Optics, Jour. Applied Physics. Mentor, adviser Ctr. for Excellence in Edn., Washington, 1990-91. Recipient Outstanding Individual Performance award Def. Advanced Rsch. Project Agy., 1984, 86, medal for Meritorious Civilian Svc., Sec. of Def., 1987. Mem. Am. Assn. Physics Tchrs., Optical Soc. Am. (commendation), Soc. Photo-Optical Instn. Engrs. (commendation). Achievements include pioneering in development of submarine laser communications systems; patent for modulated sunlight communications to submerged submarines, pulsed laser measurement subsurface absorption, others. Home: 2140 Via Mar Valle Del Mar CA 92014 Office: Sci Applications Internat 11803 Sorrento Valley Rd San Diego CA 92121

MOORE, CARL GORDON, chemist, educator; b. Zanesville, Ohio, Feb. 7, 1922; s. Henry Carl and Hilda Marie (Oberfield) M.; m. Sheila Marie O'Toole, Nov. 2, 1951; children: Carl, Patrick, Martina, Michael, Maureen, Regina, Madeleine, Terence. BS in Chem. Engring., Ga. Inst. Tech., 1947; MS in Chem. Engring., Carnegie Mellon U., 1948, postgrad., 1948-51; postgrad., U. Newark, Del., 1973-74. Cert. tchr., Pa., Del. Chemist Manhattan Project, Oak Ridge, Tenn., 1944-46; chem. engr. Koppers Co., Pitts., 1946-47; rsch. chemist E.I. DuPont de Nemours & Co., Wilmington, Del., 1951-73; tchr. Chester (Pa.)-Upland Sch., 1974-78; tutor Del. Tutoring, Wilmington, 1981-88; instr. Del. Tech. and Community Coll., Wilmington, 1982-90, U. Del., Newark, 1984—. Author tech. reports on hydrogen over voltage of titanium and zirconium. Adult leader Wilmington area Boy Scouts Am., 1953-73; tchr. Sunday sch., Wilmington, 1962-67; group leader U.S. Census Bur., 1980. Sgt. U.S. Army, 1943-46. Mem. Am. Chem. Soc., Sigma Xi. Achievements include 13 patents, production of TiO_2 Rutile by chloride process, 100% oxygen oxidation of TiCl4, 100% anatase by chloride process. Home: 1913 Oak Lane Rd Wilmington DE 19803

MOORE, CARLETON BRYANT, geochemistry educator; b. N.Y.C., Sept. 1, 1932; s. Eldridge Carleton and Mabel Florence (Drake) M.; m. Jane Elizabeth Strouse, July 25, 1959; children—Barbara Jeanne, Robert Carleton. B.S., Alfred U., 1954, D.Sc. (hon.), 1977; Ph.D., Cal. Inst. Tech., 1960. Asst. prof. geology Wesleyan U., Middletown, Conn., 1959-61; mem. faculty Ariz. State U., Tempe, 1961—; prof., dir. Ctr. for Meteorite Studies Ariz. State U., Regents' prof., 1988—; vis. prof. Stanford U., 1974; Prin. investigator Apollo 11-17; preliminary exam. team Lunar Receiving Lab., Apollo, 12-17. Author: Cosmic Debris, 1969, Meteorites, 1971, Principles of Geochemistry, 1982, Grundzügeder Geochemie, 1985; editor: Researches on Meteorites, 1961, Jour. Meteoritical Soc.; contbr. articles to profl. jours. Fellow Ariz.-Nev. Acad. Sci. (pres. 1979-80), Meteoritical Soc. (life hon., pres. 1966-68), Geol. Soc. Am., Mineral. Soc. Am., AAAS (council 1967-70); mem. Geochem. Soc., Am. Chem. Soc., Am. Ceramic Soc., Sigma Xi. Home: 507 E Del Rio Dr Tempe AZ 85282-3764 Office: Ariz State U Ctr for Meteorite Studies Tempe AZ 85287

MOORE, CHARLES WILLARD, architect, educator; b. Benton Harbor, Mich., Oct. 31, 1925; s. Charles Ephraim and Nanette Kathryn (Almendinger) M. B.Arch., U. Mich., 1947; M.F.A., Princeton U., 1956, Ph.D., 1957; M.A. (hon.), Yale U., 1965. Architect Mario Corbett (Architect), 1947-48, Joseph Allen Stein (Architect), 1948-49; asst. prof. U. Utah, 1950-52; asst. prof. architecture Princeton U., 1957-59; asso. prof. U. Calif., Berkeley, 1959-62; chmn. dept. architecture U. Calif., 1962-65, Yale U., New Haven, 1965-69; dean Yale U., 1969-71, prof., 1971-75; prof. architecture UCLA, 1975-85, head dept., 1976-77, 77-80; architect Moore Lyndon Turnbull Whitaker (Architects), 1961-64, Moore Turnbull, San Francisco and New Haven, 1964-70, Charles Moore Assos., Essex, Conn., 1970-76, Moore Grover Harper, Essex, Conn., and Moore Ruble Yudell, Los Angeles, 1976—; O'Neil Ford Centennial prof. architecture U. Tex., Austin, 1985—. Author: The Place of Houses, 1974, Dimensions, 1975, Body Memory and Architecture, 1977, The Poetics of Gardens, 1988. Served to capt. U.S. Army, 1952-54. Recipient Topaz medallion for execellence in archtl. edn. AIA/Assn. Collegiate Schs. Architecture, 1989, 25 Yr. award Sea Ranch AIA, 1991; Nat. Endowment Arts grantee, 1975; Guggenheim grantee, 1976-77. Fellow AIA (Gold medal 1991). Democrat.

MOORE, CHRISTOPHER BARRY, industrial engineer; b. Deal, Kent, Eng., Feb. 25, 1938; s. Ernest Stanley and Millicent Lillian (Harris) M.; diploma mgmt. studies Barking Regional Coll. Tech., Eng., 1966; m. Jill Irene Porter, July 6, 1963; came to U.S., 1977; children—Andrew, Stephen, Jeremy, Jennifer. Prodn. unit mgr. Plessey Co. Ltd., Ilford, Eng., 1968-70, productivity mgr., Upminster, 1970-72, regional indsl. engr., Ilford, 1972-74; mgr. mfg. devel. Northern Telecom Ltd., Montreal, Que., Can., 1974-77, dir. mfg. engring., Nashville, 1977-88; dir. mfg. devel. No. Telecom Inc., Atlanta, 1988—. Served with RAF, 1956-59. Mem. Am. Inst. Indsl. Engrs. (sr.), Inst. Electronic and Radio Engrs., Brit. Inst. Mgmt., Soc. Mfg. Engrs. (sr.), Ravinia Club. Home: 5167 Killingsworth Trce Norcross GA 30092-1739

MOORE, DALTON, JR., petroleum engineer; b. Snyder, Tex., Mar. 25, 1918; s. Dalton and Anne (Yonge) M. Grad., Tarleton State U., 1938; BS, Tex. Agrl. and Mech. U., 1942; diploma, U.S. Army Command and Gen. Staff Sch., 1945. Field engr. Gulf Oil Corp., 1946; dist. engr. Chgo. Corp., 1947-48, chief reservoir engr., 1949; mgr. Burdell Oil Corp., N.Y.C. and Snyder, Tex., 1950-52; mgr. Wimberly Field Unit, 1953-55; profl. petroleum cons., Abilene, Tex., 1956—; pres. Dalton Moore Engring. Co., 1957-67, First Oil Co., 1960-67, Second Oil Co., 1960-72, Petroleum Engrs. Operating Co., 1967—, Evaluation Engr. for Investment Bankers Corp., 1968-89, Investment Bankers Oil Corp., Inc., 1968-89. Pres. Sweetwater (Tex.) Jr. C. of C., 1938; precinct chmn. Taylor County Dem. Com., 1956-76; bd. dirs. Taylor County chpt. ARC, 1956-62. Served to maj., AUS, 1940-46. Named Eagle Scout, Boy Scouts Am.; mem. Mus. Natural History. Mem. AIME (chmn. West Cen. Tex. sect. 1954), NRA (charter, Golden Eagle), Nat. Audubon Soc., Nat. Arbor Day Found., Nature Conservancy, Soc. Petroleum Engrs. (mem. legion of hon. 1989—), Abilene Geol. Soc., Smithsonian Assocs., Am. Mus. Natural History, Bass Anglers Sportsman Soc., N.Am. Hunting Club, VFW, KP. Address: 326 N Turner Ave Moore OK 73160

MOORE, DAVID AUSTIN, pharmaceutical company executive, consultant; b. Phoenix, May 8, 1935; s. Harry Theodore and Helen Ann (Newport) M.; m. Emily J. McConnell, Jan. 26, 1991; children by previous marriage: Austin Newport, Cornelia Christina, Christopher Robinson. Grad. high sch., Glendale, Ariz.; study opera and voice with Joseph Lazzarini, 1954, 55, 57-64; studies with Joseph Lazzarini, U.S., 1954-55, 57-64; studied opera and voice, Italy, 1955-56; study with Clarence Loomis, 1958-60. Pres., owner David A. Moore, Inc., Phoenix, 1969-71, Biol. Labs. Ltd., Phoenix, 1972-78; pres., co-owner Am. Trace Mineral Rsch. Corp., Phoenix, 1979-83; pres., owner Biol. Mineral Scis., Ltd., Phoenix, 1979-82; rsch. dir., pres., owner Nutritional Biols. Inc., Phoenix, 1979-83; nutritional dir.-owner Nutritional Biol. Rsch. Co., Phoenix, 1984-85; rsch. dir., product formulator, owner Nutrition and Med. Rsch., Scottsdale, Ariz., 1986—; biochem. cons. Nutripathic Formulas, Scottsdale, 1975-88; introduced di Calcium Phosphate free concept and 100 percent label disclosure, 1979-83. Pub. NMR Newsletter. Inventor first computerized comprehensive hair analysis interpretation, 1976. Recipient Plaque Am. Soc. Med. Techs., 1982, Mineralab Inc., 1976. Avocation: singing opera and Italian songs, teaching voice, coaching singers. Home and Office: PO Box 98 Barnesboro PA 15714

MOORE, DAVID SUMNER, forensic document examiner; b. Natick, Mass., Aug. 29, 1939; s. Sumner Lucian and Lillian Rose (Kirman) M.; m. Betty Edwina Morris, Feb. 3, 1962 (div. 1982); children: Ronald Anthony, Michael David; m. Beverly Ann Holt, Apr. 23, 1982; 1 child, Rachel Rene. BS in Law Enforcement and Corrections, U. Nebr., 1972; MEd in Adult Edn., Ga. So. Coll., 1976. Diplomate Am. Bd. Forensic Document Examiners. Enlisted pvt. U.S. Army, 1959, advanced through grades to CW4, chief questioned document sect., ret., 1979; sr. questioned document analyst U.S. Postal So. Region Crime Lab., Memphis, Tenn., 1980-82; sr. questioned document examiner Las Vegas (Nev.) Met. Crime Lab., 1982-84; sr. questioned document examiner II Calif. Dept. Justice, Sacramento, 1984—. Editorial staff: Southwestern Examiner, 1992; contbr. articles to Jour. Forensic Scis., Forensic Sci. Internat. Decorated Legion of Merit. Fellow Am. Acad. Forensic Scis.; m. Southwestern Assn. Forensic Document Examiners, So. Assn. Forensic Scientists, No. Calif. Laser Study Group, Phi Kappa Phi. Home: 9010 Barrhill Way Fair Oaks CA 95628 Office: Calif Dept Justice Bur Forensic Svcs Questioned Doc Sect 4949 Broadway Sacramento CA 95820

MOORE, DUNCAN THOMAS, optics educator; b. Biddeford, Maine, Dec. 7, 1946; s. Thomas Fogg Moore and Virginia Robinson Wing; m. Susan Marie Rocheleau, June 14, 1969 (div. Aug. 1986); 1 child, Matthew. BA in Physics, U. Maine, 1969; MS in Optics, U. Rochester, 1970, PhD in Optics, 1974. Asst. prof. U. Rochester, N.Y., 1974-79, assoc. prof., 1979-86, prof., 1986—, Kingslake prof. optics, 1993—; pres., founder Gradient Lens Corp., Rochester, 1980; vis. scientist Nippon Schlumberger, Tokyo, 1983; dir. Ctr. Advanced Optical Tech., Rochester, 1987—, Inst. Optics, Rochester, 1987-93; Congl. fellow Am. Phys. Soc., Washington, 1993—; bd. dirs. Am. Precision Optical Mfrs. Assn., 1988—, N.Y. State Photonics, Rome, N.Y. 1991-93. Contbr. 70 articles to profl. jours.; patentee in field. Chmn. Hubble Independent Review Panel, 1991-92; mem. adv. bd. high tech. Rochester C. of C., 1987—. Recipient Disting. Inventor of Yr. award Intellectual Property Law Assn., 1993, Grin Optics award Japanese Soc. Applied Physics, 1993, Sci. and Tech. award Greater Rochester C. of C., 1992. Mem. IEEE Lasers and Electro-Optics Soc., Am. Ceramic Soc., Am. Soc. Precision Engring., Optical Soc. Am. (editor Applied Optics 1989-92, bd. dirs. 1987-90, 92-93), Am. Assn. Engring. Soc., Nat. Rsch. Coun., Internat. Soc. Optical Engring. (bd. govs. SPIE 1986-88), Materials Rsch. Soc. Home: 4 Claret Dr Fairport NY 14450 Office: U Rochester Inst Optics 509 Wilmot Bldg Rochester NY 14627

MOORE, EDMUND HARVEY, materials science and engineering engineer; b. Newnan, Ga., Aug. 24, 1961; s. George Robert and Ruth Virginia (Harvey) M. BS, Fla. A&M U., 1983; SM in Materials Sci., MIT, 1987; MS in Materials Sci., U. Fla., 1989. Materials engr. Wright-Patterson AFB, Ohio, 1991—. Contbr. articles to profl. jours. Big Bro. Big Bros. Assn. of Boston, 1985-87; mem. Gainesville Pan Hellenic Coun., 1991—, Black Grad. Student Orgn., Gainesville, 1987—. Patricia Harris fellowship U. Fla., 1989, Grant Minority fellowship, 1987; scholarship Y-Band Civic Club, 1979, Playtex Corp., 1979, Omega Psi Phi, 1979. Mem. Am. Ceramic Soc./NICE, Soc. of Plastic Engrs., KREAMOS, Omega Psi Phi (vice basileus). Democrat. Baptist. Achievements include patent pending for microwave thermogravimetric analyzer. Home: 2363 Duncan Dr Apt 13 Fairborn OH 45324-2047

MOORE, FAY LINDA, systems analyst; b. Houston, Apr. 7, 1942; d. Charlie Louis and Esther Mable (Banks) Moore; m. Noel Patrick Walker, Jan. 5, 1963 (div. 1967); 1 child, Trina Nicole Moore. Student, Prairie View Agrl. and Mech. Coll., 1960-61, Tex. So. U., 1961, Our Lady Lake U., 1993—. Instr. Internat. Bus. Coll., Houston, 1965; keypunch operator IBM Corp., Houston, 1965-67, sr. keypunch operator, 1967-70, programmer technician, 1970-72, asst. programmer, 1972-73, assoc. programmer, 1973-84; sr. assoc. programmer, 1984-87, staff programmer, 1987-92; staff project specialist, 1992—; mem. space shuttle flight support team IBM, 1985-92; mem. space sta. team IBM, 1992—. Recipient Apollo Achievement award NASA, 1969, Quality and Productivity award, 1986, 1992. Mem. NAFE, Internat. Platform Assn., Booker T. Washington Alumni Assn., Ms. Found. for Women, Inc. Democrat. Roman Catholic. Avocations: personal computing, board games. Office: IBM Corp 3700 Bay Area Blvd # 5206 Houston TX 77058-1199

MOORE, FRANCIS DANIELS, JR., surgeon; b. Boston, Oct. 19, 1950; s. Francis Daniels and Laura (Bartlett) M.; m. Carla Dateo, Sept. 14, 1985; children: Kyle, Hadley, Lynsey, Colby, Alessandra. Cord. AB, Harvard Coll., 1972; MD, Harvard Med. Sch., 1976. Intern Peter Bent Brigham Hosp., Boston, 1976-77, resident, 1977-84; instr. in surgery Harvard Med. Sch., Boston, 1984-86, asst. prof., 1986-93, assoc. prof., 1993—; assoc. surgeon Brigham and Women's Hosp., Boston, 1983-90, staff physician, 1990—. Contbr. articles to profl. jours. Stuart fellow in infection Nat. Found. for Infectious Diseases, 1986; recipient Jr. Faculty Rsch. award Am. Cancer Soc., 1987. Fellow ACS; mem. Am. Assn. Immunologists, Am. Fedn. for Clin. Rsch., Surg. Infection Soc., Soc. Univ. Surgeons, Am. Assn. Endocrinology Surgeons. Office: Brigham Womens Hosp Dept Surgery 75 Francis St Boston MA 02115

MOORE, GARY THOMAS, aerospace architect; b. Calgary, Alta., Can., Mar. 13, 1945; came to U.S., 1962; s. John Thomas and Grace Irene (Bulmer) M.; children: Mindan Jennifer Gunther-Moore, Kelton Gary. BArch, U. Calif., Berkeley, 1968; MA, Clark U., Worcester, Mass., 1973, PhD, 1982. Asst. prof. dept. architecture U. Wis., Milw., 1976-83, coord. environ-behavior studies master's option, 1976-83, dir. Environ. Behavior Rsch. Inst., 1979-82, coord. PhD program in architecture, 1983-88, assoc. prof. architecture, 1983-89, prof., 1989—, rsch. coord. environ. professions and social scis. Office Indsl. Rsch and Tech. Transfer, Grad. Sch., 1985-86, dir. Ctr. for Architecture and Urban Planning Rsch., 1983-90, dir.Wis Space Grant Program and Consortium, 1991—; lectr. geography Clark U., Worcester, Mass., 1974-75; vis. lectr. man-environ. studies, vis. design critic archiecture Sydney (Australia) U., 1975; vis. lectr. architecture Queensland Inst. Tech., Brisbane, Australia, 1975; vis. lectr. social geography U. New South Wales, Sydney, 1975; vis. asst. prof. architecture U. Oreg., Eugene, 1980; vis. univ. fellow Victoria U. Wellington, New Zealand, 1983; vis. prof. architecture Jurusan Teknik Arsitektur, U. Gadjah Mada, Yogyakarta, Indonesia, 1986-87. Editor: (with others) Environmental Knowing: Theories, Research and Methods, 1976, Doctoral Education for Architectural Research: Questions of Theory, Methods and Implementation, 1984, Advances in Environment, Behavior and Design, Vols. 1, 2, 3, 1987, 89, 91; author: (with others) Designing Environments for Handicapped Children, 1979, Environmental Design Rsearch Directions: Process and Prospects, 1985; contbr. over 80 articles to profl. jours. Woodrow Wilson fellow, 1968-69, Can. Coun. doctoral fellow, 1969-72; recipient Design award Type Dirs. Club N.Y., 1971, Applied Rsch. Citation Progressive Architecture Ann. Awards Program, 1978, 79, Applied Design Rsch. award Progressive Architecture Ann. Awards Program, 1980, Rsch. award U. Wis.-Milw. Found., 1980, Wis. State Assembly Citation for Rsch. Achievements, 1980; grantee NSF, Can. Coun. Arts and Humanities, Nat. Endowment Arts, U.S. Army C.E., NASA, others. Fellow APA; mem. AIAA, AIA (com. on architecture for edn. 1991—, subcom. on child care and presch. design 1991—), Nat. Coun. Space Grant Dirs. (exec. com. 1991—), Environ. Design Rsch. Assn. (bd. dirs. 1978-81, chair bd. dirs. 1980-81), Internat. Assn. for People-Environ. Studies (Europe) (bd. dirs. 1988—), Nat. Space Soc., People and Phys. Environ. Rsch. Assn. (bd. dirs. 1983-86). Achievements include founder (with others) of field of environ.-behavior studies and for introducing rsch. into architecture. Office: U Wis-Milw Wis Space Grant Prog/Consor 333 AUP Bldg Milwaukee WI 53201-0413

MOORE, G(EORGE) PAUL, speech pathologist, educator; b. Everson, W. Va., Nov. 2, 1907; s. George B. and Emma (Ayers) M.; m. Gertrude H. Conley, June 10, 1929 (dec.); children—Anne Gertrude Moore Dooley, Paul David; m. 2d, Grace MacLellan Murphey, Mar. 1, 1981. A.B., W. Va. U., 1929, D.Sc. (hon.), 1974; M.A., Northwestern U., 1930, Ph.D., 1936. Faculty dept. communicative disorders, Sch. Speech, Northwestern U., 1930-62, dir. voice research lab., 1940-62, dir. voice clinic, 1950-62; lectr. in otolaryngology Northwestern Med. Sch., 1953-62, dir. research lab. Inst. Laryngology and Voice Disorders, Chgo., 1957-62; prof. speech U. Fla., Gainesville, 1962-77, chmn. dept. speech, 1962-73, dir. communicative scis. lab., 1962-68, disting. service prof., 1977, acting chmn. dept. speech, 1977-78, disting. service prof. emeritus, 1980, adj. prof. elec. engring., 1981—; vis. faculty U. Colo., summer 1948, 51, 67, U. Minn., 1963, U. Witwatersrand, Johannesburg, S.Africa, summer 1971; co-chmn. Internat. Voice Conf., 1957; mem. communicative scis. study sect. NIH, 1959-63; mem. speech pathology and audiology adv. panel, Vocat. Rehab. Adminstrn., HEW, 1962, 64; mem. rev. panel speech and hearing, Neurol. and Sensory Disease Service Program, Bur. State Services, HEW, 1963-66, adv. com., 1964-67; mem. communicative disorders research tng. com. Nat. Inst. Neurol. Diseases and Blindness, NIH, 1964-68; mem. communicative disorders program project rev. com. Nat. Inst. Neurol. diseases and Stroke, 1969-73, chmn. 1971-72, mem. nat. adv. neurol. and communicative disorders and stroke council, 1973-77; mem. Am. Bd. Examiners in Speech Pathology and Audiology, 1965-67. Recipient merit award Am. Acad. Ophthalmology and Otolaryngology, 1962; Gould award William and Harriet Gould Found., 1962; Barraquer Meml. award Smith, Miller and Patch, 1969; Disting. Faculty award Fla. Blue Key, 1975; Tchr. Scholar award U. Fla., 1976; honors Ill. Speech, Lang. and Hearing Assn., 1979; Fellow Am. Speech-Lang.-Hearing Assn. (pres. 1961; Honors of Assn. award 1966); mem. Fla. Speech and Hearing Assn. (honor award 1977), So. Speech Assn., Speech Communication Assn. (Golden Anniversary award for scholarship 1969), Internat. Coll. Exptl. Phonology, Internat. Assn. Logopedics and Phoniatrics, Am. Assn. Phonetic Scis., AMA (spl. affiliate), Sigma Xi. Republican. Presbyterian. Club: Kiwanis. Author: Organic Voice Disorders, 1971; patentee laryngoscope, 1975; contbr. chpts. to books, articles to profl. jours. Home: 2234 NW 6th Pl Gainesville FL 32603-1409 Office: U Fla 63 Dauer Hall Gainesville FL 32611-2005

MOORE, GREGORY JAMES, physicist; b. Harrisburg, Pa., Aug. 2, 1964; s. Wayne Herman and Priscilla Marie (Faulk) M.; m. Laura Lou Damrose, Aug. 23, 1986; children: Peter Gregory, Andrew James. BS in Physics and Biology, North Park Coll., 1986; MS in Nuclear Engring., MIT, 1988, PhD in Radiol. Scis., 1992. Rsch. asst., instr. MIT, Cambridge, Mass., 1986-92; vis. scientist VA Med. Ctr. U. N.Mex. Sch. Medicine, Albuquerque, 1992—; dir.'s postdoctoral fellow Los Alamos (N.Mex.) Nat. Lab., 1992—. Contbr. articles to profl. jours. NIH fellow, 1986-87; Internat. Enrico Fermi Sch. of Physics scholar Italian Phys. Soc., 1992. Mem. Am. Assn. Physicists in Medicine, Soc. Magnetic Resonance in Medicine, Am. Nuclear Soc. (chmn. pub. info. MIT chpt. 1987-89), Sigma Xi, Alpha Nu Sigma, Sigma Pi Sigma (pres. North Park Coll. chpt. 1985). Republican. Presbyterian. Achievements include patent for simultaneous multinuclear magnetic resonance imaging and spectroscopy. Office: Los Alamos Nat Lab Mail Stop M880 Biomedical NMR Facility Los Alamos NM 87545

MOORE, JAMES ALLAN, agricultural engineering educator; b. Fresno, Calif., Mar. 11, 1939; s. E. Dale and Anne Moore; m. Janet Rae Robles, Dec. 19, 1960; children: Catherine, Kimberly, Donald, Dennis, Korinna. BSAE, Calif. State Poly. U., 1962; MSAE, U. Ariz., 1964; PhD in Agrl. Engring., U. Minn., 1975. Registered profl. engr., Oreg. Research engr. agrl. engring. dept. U. Calif., Davis, 1964-68; from instr. to asst. prof. agrl. engring. U. Minn., Mpls./St. Paul, 1968-79; assoc. prof. agrl. engring. Oreg. State U., Corvallis, 1979-85, prof., 1985—, acting head dept. agrl. engring., 1983-84; forward planning com. Dept. Agrl. Engring., 1973, instruction com., 1973, 75. chmn. recruiting com, 1973-79, chmn. student activities, 1974-75; mem. steering com. Experiment Station Waste Mgmt., 1975-79, grad. com. 1979; mem. Midwest Plan Service Com. on Waste Mgmt., 1975-78; bd. dirs. Oreg. State U. Water Resources Research Inst., 1983-85, 89-91. Contbr. numerous articles to profl. jours. Recipient Superior State award USDA, 1985. Fellow Am. Soc. Agrl. Engrs. (tech. dir. 1979-81, various editorial positions, Blue Ribbon award 1983, 84, G.B. Gunlogson Countryside Engring. award, 1985), mem. Soil Conservation Soc. Am., AM. Soc. Agrl. Engrs., Council for Agrl. Sci. and Tech., Sigma Xi, Gamma Sigma Delta, Alpha Epsilon. Avocations: hunting, fishing.

MOORE, JAMES NEAL, design engineer; b. Delhi, La., May 18, 1959; s. Herbert A. and Fannie Lou (Kolb) M.; m. Kris Ferguson, Dec. 28, 1981; children: Andrea Nicole, James Neal Jr. BS, La. Tech. U., 1981. Registered profl. engr., La. Agrl. engr. USDA Soil Convervation Svc., Alexandria, La., 1982-84, planning engr., 1984-87, hydraulic engr., 1987-90, state design engr., 1990—. Developer various computer programs. Mem. Am. Soc. Agrl. Engring. (vice chair scholarship com. 1991—), La. Engring. Soc. (sec.-treas., v.p., pres.). Democrat. Baptist. Achievements include work on flood prevention and drainage projects, emergency watershed protection projects, irrigation project, wetland restoration programs. Office: USDA Soil Conservation Svc 3737 Government St Alexandria LA 71302

MOORE, JANICE KAY, biology educator; b. Waco, Tex., June 26, 1948; d. Doyle Liles and Matilda Mae (Spross) M. BA, Rice U., 1970; MA, U. Tex., 1974; PhD, U. N.Mex., 1981. Tech. editor Gov.'s Energy Adv. Coun., Austin, Tex., 1977; teaching asst. U. N.Mex., Albuquerque, 1977-81; lectr. U. Tex., Austin, 1982; rsch. assoc. Tall Timbers Rsch. Sta. Fla. State U., Tallahassee, 1983; prof. Colo. State U., Ft. Collins, 1983—; vis. scholar Glasgow (Scotland) U., 1991, Oxford (England) U., 1992; rep. Colo. State faculty Coun., Ft. Collins, 1990-93, Colo. Univs.' Faculty Leaders Assn. Conf., Mt. Vernon, Colo., 1992. Mem. editorial bd. Jour. Parasitology, 1990-92, assoc. editor, 1992-93; contbr. articles to sci. jours. Leader Sci. Clusters, Colo., 1989-91. Recipient Presdl. Young Investigator award NSF, 1985-90, Dir.'s Initiative Fund award WHO, 1990; Fogarty Sr. Internat. fellow NIH, 1991-92.; research grantee NSF. Mem. Am. Soc. Parasitologists (coun. 1987-90, 92—), Ecol. Soc. Am. (subco. 1987), S.W. Assn. Parasitologists (coun. rep. 1987-90), Soc. for Study of Evolution. Achievements include discovery that parasties alter the behavior of their hosts in ways that affect (often enhance) transmission in the field, which may have profound effects on disease/parasite transmission. Not only do parasitized animals frequently behav in ways that transmit parasites, but they also are distributed nonrandomly, thus causing anocunters with parasites/pathogens to be nonrandom, counter to many epidemiological models. Office: Colo State U Depr Biology Colorado State University CO 80523

MOORE, JOAN L., radiology educator, physician; b. Belmont, Mass., Oct. 26, 1935; d. Frank Joseph and Maria L. Mazzio; children: James Thomas, Edwin Stuart. BA in Chemistry and Theology, Emmanuel Coll., 1957; MA in Genetics and Physiology, Mass. Wellesley Coll., 1961; PhD in Genetics,

Bryn Mawr (Pa.) Coll., 1964; MD, Phila. Coll. of Medicine, 1977, MSc in Radiology, 1981. Instr. in biochemistry Gwynedd Mercy Coll., Springhouse, Pa., 1963-65; instr. in genetics Holy Family Coll., Phila., 1965-66; instr. in anatomy Phila. Coll. of Medicine, 1971-77, tchr., 1973-77, asst. prof., 1977-84; prof. W.Va. Sch. of Medicine, 1984—; rotating intern Phila. Coll. of Medicine Hosp., 1977-78, resident in radiology, 1978-81; lt. col. USAR, 1984—; prof. W.Va. Sch. of Medicine, Lewisburg, 1984—. Author: (with Dr. DiVirgilito) Essentials of Neuropathology, 1974. Lector St. Ann's Cath. Ch., Phoenixville, Pa., 1981-84; treas. Hist. Soc. of Frankford, Phila., 1968-75, Sch. Mother's Assn., Devon (Pa.) Prep., 1980-81. Lt. col. U.S. Army Med. Corps, 1992. Mem. AAUP, Am. Acad. Family Physicians, Am. Assn. Women Radiologists, Am. Med. Women's Assn., Am. Osteo. Coll. of Radiology, Am. Soc. Clin. Oncology, Am. Soc. Therapeutic Readiologists, Hist. Soc. of Lewisburg (life), Pa. Osteo. Med. Assn., Pa. Osteo. Gen. Practitioner's Soc., Radiol. Soc. N.Am., Radiation Rsch. Soc., Res. Officers Assn. (life), W.Va. Soc. Osteo. Medicine, Greenbrier River Hike and Bike Trail. Home: Rte # 1 Box 123 Anthony Creek Rd Frankford WV 24938 Office: WVa Sch of Medicine 400 N Lee St Lewisburg WV 24901-1196

MOORE, JOHN HAYS, chemistry educator; b. Pitts., Nov. 6, 1941; s. John Hays and Mary (Welfer) M.; m. Judy Ann Williams, Aug. 10, 1963; children: John H. IV, Victoria Inez. BS, Carnegie Tech, 1963; MS, Johns Hopkins U., 1965, PhD, 1967. Rsch. assoc. Johns Hopkins U., Balt., 1967-69; program officer NSF, Washington, 1980-81, 85-86; asst. prof. U. Md., College Park, 1969-73, assoc. prof., 1973-78, prof., 1978—. Author: Building Scientific Apparatus, 1982, 2d edit., 1989; contbr. 85 publs. to profl. jours. Named Joint Inst. for Lab. Astrophysics fellow, 1975, Am. Phys. Soc. fellow, 1990. Home: 3905 Commander Dr Hyattsville MD 20782 Office: U of Maryland Chemistry Dept College Park MD 20742

MOORE, JOHN JAMES CUNNINGHAM, neonatologist; b. West Point, N.Y., June 19, 1948; s. John James and Marguerite (Murray) M.; m. Nancy Rowland, Mar. 22, 1969; children: David J., Brian P., Michael K., Katherine. BS in Physics, Yale U., 1969; MD, U. Va., Charlottesville, 1976. Diplomate Am. Bd. Pediatrics, Sub-Bd. Neonatal-Perionatal Medicine. Intern Pitts. Children's Hosp., 1976-77, resident in pediatrics, 1977-79; fellow in neonatology Cin. Children's Hosp., 1979-82; attending neonatologist MetroHealth Med. Ctr., Cleve., 1982—, dir. neonatology, 1988—; asst. prof. pediatrics Case Western Res. U., Cleve., 1982-90, assoc. prof. pediatrics, 1990—; dir. Cleve. Regional Code Pink Program in Newborn Resuscitation, 1986—; cons. Cleve. Health Mus., 1992—. Contbr. articles to profl. jours. Lt. USN, 1969-73. Grantee NIH, 1986-89, March of Dimes, 1987-90. Fellow Am. Acad. Pediatrics; mem. Soc. Gynecol. Investigation, Soc. for Pediatric Rsch., Endocrine Soc. Achievements include establishment of code pink neonatal resuscitation training program. Office: MetroHealth Med Ctr Dept Pediatrics 2500 MetroHealth Dr Cleveland OH 44109

MOORE, KEVIN L., electrical engineering educator; b. Fayetteville, Ark., Feb. 2, 1960; s. Robin G. and Ruth S. (Smith) M.; m. Tamra Myers, June 9, 1979; children: Joshua, Julia. BSEE, La. State U., 1982; MSEE, U. So. Calif., 1983; PhD, Tex. A&M U., 1989. profl. engr., Idaho. Asst. prodn. engr. Tex. Instruments, Houston, 1981; mem. tech. staff Hughes Aircraft Co., El Segundo, Calif., 1982-85; lectr. elec. engring. Tex. A&M U., College Station, 1986-89; asst. prof. Coll. Engring., Idaho State U., Pocatello, 1989—. Author: Iterative Learning Control for Deterministic Systems, 1992; contbr. articles to Jour. of Robotic Systems, Automatica, IEEE Potentials, IEEE Transactions on Automatic Control. Guitarist 1st United Meth. Ch. Grantee Westinghouse Idaho Nuclear Corp., 1992, Air Force Office of Scientific Rsch., 1991. Mem. IEEE, Soc. for Indsl. and Applied Math., Am. Soc. of Engring. Educators (named 1993 Dow Outstanding Young Faculty/ Pacific N.W. sect.), Sigma Xi, Tau Beta Pi, Eta Kappa Nu. Achievements include rsch. on learning control theory for linear, time-invariant systems, with extensions to neural network based learning controllers for non-linear systems. Office: Coll of Engring Idaho State U 833 S 8th Box 8060 Pocatello ID 83209

MOORE, MARCUS LAMAR, mechanical engineer, consultant; b. Chattanooga, Mar. 6, 1965; s. Homer Douglas Jr. and Pansy Ann (Smith) M.; m. Deena Lee, May 20, 1989. BSME, U. Tenn., 1988. Registered prof. engr., Tenn. Mech. engr. David Talor R & D, Annapolis, Md., 1984-85, Acco-Cast, Inc., Chattanooga, 1985-86, Norton Chem., Soddy Daisy, Tenn., 1987-88, E.I. DuPont, Aiken, S.C., 1988-89; project engr. Nuclear Fuel Svcs., Erwin, Tenn., 1989-90; project and facility engr. Martin Marietta Energy Systems, Oak Ridge, Tenn., 1990—; mem. Nuclear Crticality Safety Div., Knoxville, Tenn., 1989—, fin. com. Knox County Solid Waste Bd., 1993; judge Chattanooga Regional Engring. Fair, 1990-91. Contbr. articles to profl. jours. Leader Bapt. Assn., Johnson City, Tenn., 1990. Mem. ASME, Am. Nuclear Soc., Chattanooga Engr. Club, Phi Eta Sigma. Achievements include patent pending for radioactive soil dryer; secondary coils applied to electromagnetic railgun for added efficiency; use of liquid metal for current carrier for railgun. Home: 121 Woodmont Cir Clinton TN 37716-3527 Office: Martin Marietta Energy Systems Bldg # 9204-1 MS # 8053 Oak Ridge TN 37830

MOORE, OMAR KHAYYAM, experimental sociologist; b. Helper, Utah, Feb. 11, 1920; s. John Gustav and Mary Jo (Crowley) M.; m. Ruth Garnand, Nov. 19, 1942; 1 child, Venn. BA, Doane Coll., 1942; MA, Washington U., St. Louis, 1946, PhD, 1949. Instr. Washington U., St. Louis, 1949-52; teaching assoc. Northwestern U., Evanston, Ill., 1950-51; rsch. asst., prof. sociology Tufts Coll., Medford, Mass., 1952-53; researcher Naval Rsch. Lab., Washington, 1953-54; asst. prof. sociology Yale U., New Haven, 1954-57, assoc. prof. sociology, 1957-63; prof. psychology Rutgers U., New Brunswick, N.J., 1963-65; prof. social psychology, sociology U. Pitts., 1965-71, prof. sociology, 1971-89, prof. emeritus, 1989—; scholar-in-residence Nat. Learning Ctr.'s Capital Children's Mus., Washington, 1989-90; pres. Responsive Environ. Found., Inc., Estes Park, Colo., 1962—; assessor of rsch. projects The Social Scis. and Humanities Rsch. Coun. Can., 1982—; adj. prof. U. Colo., Boulder, 1992—. Contbg. editor Educational Technology; contbr. numerous articles to profl. jours.; patentee in field; motion picture producer and director. Recipient Award The Nat. Soc. for Programmed Instruction, 1965, Award Doane Coll Builder Award, 1967, Ednl. Award Urban Youth Action, Inc., 1969, Award House of Culture, 1975, Cert. of Appreciation, 1986, Cert. of Appreciation D.C. Pub. Schs., 1987, da Vinci Award Inst. for the Achievement of Human Potential, 1988, Cert. of Appreciation Capital Children's Museum, 1988, award Jack & Jill of America Found., 1988, Cert. of Appreciation U.S. Dept. of Edn., 1988, Cert. of Appreciation D.C. Pub. Schs., 1990, Person of Yr. in Ednl. Tech. award Ednl. Tech. mag., 1990. Mem. AAAS, Am. Math. Soc., Am. Psychol. Assn., Internat. Sociol. Assn., Am. Sociol. Assn., Assn. for Symbolic Logic, Assn. for Anthrop. Study of Play, Philosophy Sci. Assn., Psychonomics Soc., Soc. for Applied Sociology, Soc. for Exact Philosophy, Math. Assn. Am. Republican. Avocation: mountaineering. Home and Office: 2341 Upper High Dr PO Box 1673 Estes Park CO 80517

MOORE, PENELOPE ANN, forensic scientist; b. Monroe, La., Oct. 1, 1944; d. James Byron and Mary Ellen (Conrad) M. MEd, Temple U., 1970; MS, John Jay Coll., 1988. Part-time lectr. Union Coll., Cranford, N.J., 1975-77; tchr. Cherry Hill (N.J.) Twp. Sch. Dist., 1966-68, Darby Twp. Sch. Dist., Glenolden, Pa., 1968-69, Black Horse Pike Sch. Dist., Blackwood, N.J., 1969-72; asst. to dir. N.J. Alcoholic Beverage Control, 1972-79; prin. forensic scientist N.J. State Police, 1979—. Contbr. articles to profl. jours. Mem. N.J. Assn. Forensic Scientists, Northeastern Assn. Forensic Scientists. Office: NJ State Police East Lab Sea Girt Ave Sea Girt NJ 08750

MOORE, RICHARD KERR, electrical engineering educator; b. St. Louis, Nov. 13, 1923; s. Louis D. and Nina (Megown) M.; m. Wilma Lois Schallau, Dec. 10, 1944; children: John Richard, Daniel Charles. B.S., Washington U. at St. Louis, 1943; Ph.D., Cornell U., 1951. Test equipment engr. RCA, Camden, N.J., 1943-44; instr. and research engr. Washington U., St. Louis, 1947-49; research asso. Cornell U., 1949-51; research engr., sect. supr. Sandia Corp., Albuquerque, 1951-55; prof., chmn. elec. engring. dept. U. N.Mex., 1955-62; Black and Veatch prof. U. Kans., Lawrence, 1962—; dir. remote sensing lab. U. Kans., 1964-74, 84-93; pres. Cadre Corp., Lawrence, 1968-87; cons. cos., govt. agys. Author: Traveling Wave Engineering, 1960; co-author: (with Ulaby and Fung) Microwave Remote Sensing, Vol. I, 1981, Vol. II, 1982, Vol. III, 1986; contbr. to profl. jours. and handbooks. Served to lt. (j.g.) USNR, 1944-46. Recipient Achievement award Washington U. Engring. Alumni Assn., 1978, Outstanding Tech. Achievement award Geosci. and Remote Sensing Soc., 1982, Louise E. Byrd Grad. Educator award U. Kans., 1984, Irving Youngberg Rsch. award U. Kans., 1989. Fellow IEEE (sect. chmn. 1960-61, Outstanding Tech. Achievement award coun. of oceanic engring. 1978); mem. NAE, AAUP, Am. Soc. Engring. Edn., Am. Geophys. Union, Internat. Sci. Radio Union (chmn. U.S. commn. F 1984-87, internat. vice chmn. commn. F 1990-93, chmn. 1993—), Sigma Xi, Tau Beta Pi. Presbyn. (past elder). Lodge: Kiwanis. Achievements include research in submarine communications, radar altimetry, radar as a remote sensor, radar oceanography; patent for polypanchromatic radar. Home: 1620 Indiana St Lawrence KS 66044-4046 Office: U Kans R S & Remote Sensing Lab 2291 Irving Rd Lawrence KS 66045-2969

MOORE, RICHARD LAWRENCE, structural engineer, consultant; b. Rocky Ford, Colo., Feb. 7, 1934; s. Lawrence and Margaret Kathryn (Bolling) M.; m. Donna St. Clair, Mar. 26, 1972 (div. 1983); 1 child, Andrew Trousdale; m. Margaret Ann Guthrie, May 4, 1984. BSCE, U. Colo., 1957; MS, Princeton U., 1963; PhD, Calif. Western U., Santa Ana, 1975. Registered profl. engr., Mass., Maine, Colo., Pa., Iowa, Nebr., N.Mex., Wyo., Ill., Ark., Mo., N.D., Mich., Okla. Structural engr. Cameron Engrs., Denver, 1964-66; v.p. Moore Internat., Jeddah, Saudi Arabia, 1967-78; asst. to pres. C.H. Guernsey Co., Oklahoma City, 1979-82; pres. R.L. Moore Co., Boston, 1983—; v.p., dir. Isolink Ing., Basel, Switzerland, 1990—; nat. chmn. Roof Cons. Inst., Raleigh, N.C., 1988—; prof. Episcopal Sch. Theology, Denver, 1967-71. Patentee in field. Member Mound City (Mo.) Libr. Bd., 1963-64; pres. Dist. Rep. Party, Boston, 1988—; sr. warden St. John Chrysostom Epis. Ch., Denver, 1966-71. Danforth Found. scholar, 1962. Mem. ASCE, NSPE, Am. Concrete Inst., Nat. Forensic Ctr. Avocations: golf, travel, antique pocket watch collecting. Home and Office: RL Moore Co 534 E Broadway Boston MA 02127-4407

MOORE, ROGER STEPHENSON, chemical and energy engineer; b. Wenatchee, Wash., Apr. 30, 1939; s. Roy Allen and Mildred (Stephenson) M.; m. Janet Anne Dragoo, Oct. 6, 1962; children: Sarah Beth (dec.), Amy Lynn. BS in Chemistry with honors, Wash. State U., 1961, BSChemE, 1962; MBA in Mgmt., Fairleigh Dickinson U., 1979. Registered profl. engr., N.Y. Process and indsl. engr. magnetic products divsn. 3M Co., Freehold, N.J., 1962-78, process engring. supr. magnetic products divsn., 1978-82, sr. vendor quality engr. magnetic products divsn., 1982-86, cost reduction coord. magnetic products divsn., 1983, 84; project devel. mgr. Cencogen, 1986, 87. Pres. Ind. Homeowners Assn., Freehold, 1972; mem. Freehold High Sch. PTO, 1981-82, pres., 1983-85. Recipient Cert. of Appreciation, Freehold High Sch., 1985. Mem. Assn. Energy Engrs. (cert. energy mgr., cert. cogeneration profl., sec. N.J. chpt. 1986-88, Engr. of Yr. 1986), Am. Chem. Soc., EcoVillage at Ithaca. Achievements include engineering assessments for technical and economic viability of new cogeneration power plants, fuel efficiency of operating electric power plants; development of computer programs utilized in such assessments. Home: 120 Roosevelt Rd Hyde Park NY 12538

MOORE, SALLY FALK, anthropology educator; b. N.Y.C., Jan. 18, 1924; d. Henry Charles and Mildred (Hymanson) Falk; m. Cresap Moore, July 14, 1951; children: Penelope, Nicola. B.A., Barnard Coll., 1943; LL.B., Columbia U., 1945, Ph.D., 1957. Asst. prof. U. So. Calif., Los Angeles, 1963-65, assoc. prof., 1965-70, prof., 1970-77; prof. UCLA, 1977-81; prof. anthropology Harvard U., Cambridge, Mass., 1981—, Victor Thomas prof. anthropology, 1991—, dean Grad. Sch. Arts and Scis., 1985-89. Author: Power and Property in Inca Peru, (Ansley Prize 1957), 1958, Law as Process, 1978, Social Facts and Fabrications, 1986, Moralizing States, 1993. Trustee Barnard Coll., Columbia U., 1991-92; master Dunster House, 1984-89. Research grantee Social Sci. Research Council, N.Y.C., 1968-69; research grantee NSF, Washington, 1972-75, 79-80, Wenner Gren Found., 1983. Fellow Am. Acad. Arts & Scis., Am. Anthrop. Assn., Royal Anthrop. Inst.; mem. Assn. Polit. and Legal Anthropology (pres. 1983), Am. Ethnological Soc. (pres. 1987-88). Democrat. Office: Harvard U 320 William James Hall Cambridge MA 02138

MOORE, SANDRA, architect, environmental designer, educator; b. Charleston, S.C., June 30, 1945. B.A. in Architecture, Tuskegee Inst., 1967; M.Environ. Design, Yale U., 1973; Doctorate, Harvard U., 1982. Architect Clauss and Nolan, Architects, Planners, Trenton, N.J., 1968-72; founder, exec. dir. Trenton Design Ctr., 1970-73; asst. prof. Schs. Architecture and Edn., U. Wis.-Milw., 1973-75; dir. ctrs. for environ. edn. Edn. Devel. Ctr., Cambridge, Mass., 1975-76; asst. prof. environ. design Mass. Coll. Art, Boston, 1975-76; assoc. Alexander Cooper & Assocs., N.Y.C., 1976; adminstr. Dept. Housing Preservation and Devel., N.Y.C., 1978-79; asst. prof., asst. dean Sch. Architecture, Fla. A&M U., Tallahassee, 1979-82; assoc. prof., assoc. dean Sch. Architecture, N.J. Inst. Tech., Newark, 1982-83, assoc. prof., 1983—; chmn. housing task force mayoral transistion team City of Newark, 1986-87; community advisor, Newark, 1987—. Co-editor Many Faces of Architecture, 1988; prodr. video documentary "Work-in-Progress", Black Women in Architecture: A Sense of Place, 1992. Mem. policy panels Nat. Endowment for Arts, Nat. League Cities; cons. design arts program N.J. State Council on the Arts, 1987—; mem. Nat. Def. Exec; mem. architecture adv. commn. Mercer County Community Coll., 1987—; mem. community reinvestment act adv. bd. Midlantic Nat. Bank, Edison, N.J., 1987—. Recipient Pub. and Inst. Svc. award N.J. Inst. Tech., 1989; named Alumnus of Yr. Tuskegee U. Dept. Architecture, 1987; Nat. Endowment Arts fellow, 1984, 85. Mem. N.J. Soc. Architects (bd. dirs., citation award 1983).

MOORE, SHARON PAULINE, biologist; b. Sutton, W.Va., Nov. 10, 1949; d. Arnold Creed and Zora Priscilla (Truman) M. MSc, Western Ky. U., 1974; PhD, U. Louisville, 1982. Rsch. technician, then rsch. assoc. Western Ky. U., Bowling Green, Ky., 1974-80; postdoctoral fellow Brookhaven Nat. Lab., Upton, N.Y., 1982-84; rsch. officer U. Bath, Eng., 1984-87; postdoctoral fellow Frederick (Md.) Cancer R&D Ctr., 1987-91, scientist assoc., 1991—. Home: 8232 Red Wing Ct Frederick MD 21701 Office: Frederick Cancer Rsch Advanced Bioscis Labs PO Box B Frederick MD 21702-1201

MOORE, THOMAS EDWIN, museum director, biology educator; b. Champaign, Ill., Mar. 10, 1930; s. Gerald Everitt and Velma (Lewis) M.; married, Feb. 4, 1951; children: Deborah Susan Moore-Yanchyshyn, Melinda Sifferd Kerr. BS, U. Ill., 1951, PhD, 1956. Tech. asst. Ill. Natural History Survey, Urbana, 1950-56; instr. U. Mich., Ann Arbor, 1958-59, asst. prof. zoology, 1959-63, prof. biology, 1966—, curator exhibit mus., 1956—, dir. exhibit mus., 1988—; vis. prof. Orgn. for Tropical Studies, Costa Rica and Brit. Honduras, 1970, 72; mem. steering com. U.S. Internat. Biol. Program, Ecosystems, 1969-75, Colloquium of Environ. Rsch. and Edn., Raleigh, N.C., 1991-92; mem. adv. com. Mich. High Sch. Accrediation, Ann Arbor, 1987-92; mem. planning com. Nat. Inst. for the Environment, 1991-93. Author, producer film: 17-Year Cicadas, 1975; editor: Cricket Behavior and Neurobiology, 1989, Lectures on Science Education, 1991-92, 93. Pres. Assn. for Tropical Biology, 1973-75; bd. dirs. Orgn. for Tropical Studies, 1968-79; mem. Huron River Watershed Coun., Washtenaw County, Mich., 1987-92. Fellow AAAS, Royal Encomol. Soc., Linnean Soc. London. Home: 4243 N Delhi Rd Ann Arbor MI 48103 Office: U Mich Mus Zoology 1109 Geddes Ave Ann Arbor MI 48109-1079

MOORE, THOMAS JOSEPH, research psychologist; b. N.Y.C., Aug. 5, 1939; s. Thomas Joseph and Marie Catherine (Flanagan) M.; m. Jane Theresa Ryan, Aug. 26, 1961; children: Thomas, Kerrie Ann, Beth Ann. BA, Manhattan Coll., Bronx, N.Y., 1961; MA, Ohio U., 1962; PhD, U. Mass., 1965. Rsch. psychologist Air Force Aerospace Med. Rsch. Lab., Wright-Patterson AFB, Ohio, 1968-90; sr. scientist Armstrong Lab., Wright-Patterson AFB, 1990-92, div. chief biodynamics and biocomm. div., 1992—; mem. adj. faculty U. Dayton, Ohio, 1965-74, Wright State U., Dayton 1980—; Air Force rep. interagy. coordinating com. NIH, Washington, 1991—. Contbr. over 50 articles to tech. and sci. jours., chpts. to books. Capt. USAF, 1965-68. Recipient sci. achievement award Air Force Systems Command, 1982, program mgmt. award Amrstrong Lab., 1991; mid-career fellow Princeton U. Woodrow Wilson Sch., 1982-83. Fellow Acoustical Soc. Am.; mem. Am. Psychol. Soc., Human Factors Soc. Achievements include pioneer in use of speech as an input signal in single cell neurophysiological studies; research on effects of infrasound on the auditory system, voice communications effectiveness and effects of environmental stressors on speech production and perception. Office: Armstrong Lab/CFB Wright Patterson AFB OH 45433

MOORE, W. JAMES, civil engineer, engineering executive; b. Sandusky, Mich., Nov. 21, 1916; s. W.A. and Mabel (Whitely) M.; m. Phyllis Kennedy, Sept. 21, 1946; children: Patricia, Susan, Nancy, James K. BSCE, U. Mich., 1939, MSCE, 1941. Registered profl. engr., Mich. Engr. Martin Fire Proofing Corp., Buffalo, 1946-49; engr., sales mgr. R.C. Mahon Co., Detroit, 1949-62; pres. Wayne Foundry and Stamping Co., Detroit, 1962—. With USNR, 1942-46. Mem. Engring. Soc. Detroit. Home: 785 Balfour St Grosse Pointe Park MI 48230 Office: Wayne Foundry and Stamping 3100 Hubbard St Detroit MI 48210-3293

MOORE, WALTER CALVIN, chemical engineer; b. Oklahoma City, Oct. 21, 1910; s. Walter Arthur and Mary Helen (Hingeley) M. Student, U. Okla., 1927-31; diploma in nuclear tech., Capitol Radio Engr. Inst., 1963, diploma in electronics tech., 1967; BS in Chemistry, U. State N.Y., 1986. Sanitary engr. U.S. Army Corps Engrs., Tallahassee, 1942-44; chem. engr. gaseous diffusion plant Union Carbide Corp., Oak Ridge, Tenn., 1944-50, asst. chief engr., 1950-52; project mgr. Oak Ridge Nat. Lab., 1953, asst. supt. tech. divs. plant Y-12, 1954-58; rsch. mgr. plastics in packaging Union Carbide Devel. Co., N.Y.C., 1958-59; EBOR nuclear reactor project mgr. Gen. Atomic, San Diego, 1959-62; v.p. engring. and rsch. York (Pa.) div. Borg-Warner Corp., 1962-76; cons. Borg-Warner Corp., York, 1976-81; coms. mgmt. and rsch., 1981—. Author: (with others) Our Western World's Most Beautiful Poems, 1985; contbr. articles to Supervisory Mgmt. Am. Mgmt. Assn. Asst. gen. chmn. corp. contbns. fund campaign York United Way, 1969-71; bd. dirs. Sheltered Workshop, 1971, York County Solid Waste and Refuse Authority, 1983. Mem. AICE, Am. Nuclear Soc. (charter), Am. Soc. Heating, Refrigerating, Air Conditioning Engrs., Internat. Inst. Refrigeration (mem. E-1 air conditioning commn. 1972-75), Internat. Solar Energy Soc., N.Y. Acad. Scis., Rotary Internat. (dist. chmn. group study exch. 1974), York Area C. of C. (chmn. nat. govtl. affairs com. 1967-69). Republican. Achievements include patents for control circuit and oil separator. Home: 360 Tri-Hill Dr York PA 17403-3839

MOORE, WALTER PARKER, JR., civil engineering company executive; b. Houston, May 6, 1937; s. Walter Parker Sr. and Zoe Alma (McBride) M.; m. Mary Ann Dillingham, Aug. 19, 1992; children: Walter P. III, Melissa Moore Magee, Matthew Dillingham. BA in Civil Engring., Rice U., 1959, BS in Civil Engring., 1960; MS in Civil Engring., U. Ill., 1962, PhD in Civil Engring., 1964. Registered profl. engr., Ark., Ariz., Colo., Fla., Ga., Idaho, Ill., Ind., Kans., Maine, Md., Mich., Minn., Mo., Nev., N.H., N.Mex., N.Y., N.C., Okla., Oreg., Pa., R.I., Tex., Utah, Wash., Wis., Wyo. Rsch. asst. U. Ill., Urbana, 1960-64; design engr. Walter P. Moore & Assocs., Inc., Houston, 1966-70, sec., treas., 1970-75, exec. v.p., 1975-83, pres., chmn., 1983—; engring. adv. coun. Rice U., 1970-74, adj. prof. architecture, 1975-82, archtl. adv. coun., 1988—; pres. Rice U. Alumni Assn., 1975; civil engring. vis. com. U. Tex. Austin, 1975-77; adv. com. effects of earthquake motions on reinforced concrete bldgs. U. Ill. Adv. Com., 1980; pres. Rice U. Engring. Alumni Assn., 1983, vis. lectr. Cornell U., 1986; mem. sesquicentennial com. State of Tex. Bus. and Fin. Com., 1986; vis. lectr. U. Ill., 1988, 89; engring. adv. coun. Tex. A&M U., 1990—. Group editor (monograph) Tall Building Systems and Concepts; mem. editorial adv. bd. Constrn. Bus. Rev., 1992—. Bd. dirs. Kiwanis Club Houston, 1974-76, Rice Design Alliance, 1976-79, Rice Ctr. for Community Design and Rsch., 1977-82, Harris County Heritage Soc., 1984-86, River Oaks Bank, 1983-91, Compass Bank, 1991-92; bd. dirs., v.p. The Forest Club, 1976-78; chmn. architects and engrs. United Way, 1985; mem. exec. com. River Oaks Bank, 1984-91. Capt. U.S. Army, 1964-66. Mem. NAE, NSPE, ASCE (activities chmn. 1968-69, sec. structures group Tex. sect. 1976-77, vice-chmn. 1977-78, structural standards divsn., exec. com. 1989—, keynote speaker La. conv. 1986), Am. Cons. Engrs. Coun. (bd. dirs. CEC-T 1981-83, v.p. 1990-92), Internat. Assn. Bridge and Structural Engrs., Am. Concrete Inst. (mem. com. # 318, bd. direction 1989-91, Alfred E. Lindau award 1992), Soc. Am. Mil. Engrs., Coun. on Tall Bldgs. and Urban Habitat (mem. steering group 1971—, editor Com. 3 1975—, chmn. final plenary session Hong Kong Fourth World Congress, 1990, Hong Kong Forth World Congress, 1990, internat. and regional conference 1990—), Consulting Engrs. Coun. Tex. (bd. dirs. 1974-76, v.p. 1976-77, pres. 1977-78), Tex. Soc. Profl. Engrs. (Young Engr. of Yr. 1969-70, region IV Engr. of Yr., 1985), Post Tensioning Inst. (juror nat. awards 1983—), Structural Engrs. Assn. Tex. Episcopalian. Office: 2d Fl 3131 Eastside Houston TX 77098

MOORE, WAYNE V., pediatrician, educator, endocrinologist; b. Wichita, Kans., May 3, 1942. BA, Friends U., 1964; PhD, U. Minn., 1969, MD, 1970. Diplomate Am. Bd. Pediatrics, Am. Bd. Pediatric Endocrine; lic. physician, Kans., Minn. Fellow in biochemistry U. Minn., Mpls., 1969-70, intern in pediatrics, 1970-71, resident in pediatrics, 1971-72; clin. assoc. NIH, Bethesda, Md., 1972-74; asst. prof. pediatrics U. Kans. Med. Ctr., Kansas City, 1974-78, assoc. prof., 1978-82, prof., 1982—; head. sect. pediatric endocrinology and metabolism U. Kans. Med. Ctr., 1974—, attending physician pediatric inpatient svc., 1976—, elected mem. grad. sch. faculty in biochemistry, 1977, departmental reviewer dept. diatetics and nutrition, 1974-75, acad. com. student promotions subcom., 1975—, physiology chmn. search com., 1975-76, ad-hoc com. for devel. tenure guidelines, 1976-77, faculty coord. for pediatric grand rounds, 1975-81, pediatric rsch. com., 1976-79, adminstrn. PKU and congenital hypothyroid follow-up clinic, 1976—, ad-hoc com. for rev. Nat. Intensive Care Unit, 1976-77, ENT chmn. search com., 1978, chmn. CRC adv. com., 1978-79, utilization rev. com., 1979-80, promotion and tenure com. dept. pediatrics, 1979-82, acting. chmn. pediatrics, 1981-83, instn. promotion and tenure com., 1984-86, dean search com., 1985, departmental awards com., 1986-87, dept. promotion and tenure com., 1987-88, chmn., 1988—, coll. promotion and tenure com., 1988-89, 89-91, telethon com., pediatric dept. rsch. com., 1988—; vis. prof. U. Okla-Tulsa, 1979, U. S.D., 1979, U. Minn., 1980, European Soc. for Pediatric Endocrinology, 1986; mem. Nat. Pituitary Agy., 1978-85, chmn. growth hormone subcom., 1984-89. Reviewer Jour. AMA, Endocrinology, Jour. Pediatrics, Diabetes, Jour. Clin. Endocrinology and Metabolism; contbr. over 150 articles and abstracts to profl. and sci. jours., chpt. to book. Am. Cancer Soc. scholar, 1970; grantee Nat. Pituitary Agy., 1975-85, State Dept. Child and Maternal Health, 1976—, NIH, 1980-85, 81-84, 87-90, Muscular Dystrophy Assn., 1980-82, Carey Endorsement, 1983—, Genetech, 1985—, Astrowe Found., 1986-87, 87-88, Juvenile Diabetes Found., 1986-88, Amoco Found., 1986—, Cosmopolitan Club, 1987—, Am. Heart Assn., 1987-89, 89-90, Ctr. for Disease Control., 1987-90, Mother Mary Ann Found., 1991—. Mem. Am. Acad. Pediatrics (v.p. Kans. chpt. 1988, pres. Kans. chpt. 1991), Am. Pediatric Soc., Am. Diabetes Assn. (grantee 1986-87), Endocrine Soc., Pediatric Endorcine Soc. (chmn. com. on future use of polypeptide hormones 1986-88), Human Growth Found., Soc. for Pediatric Rsch., Southwest Pediatric Soc., Kans. Med. Soc., Kans. Diabetes Assn., Greater Kansas City Diabetes Assn. (invited lectr. 1979—), Kansas City Endocrine Roundtable, Lawson Wilkins Pediatric Endocrine Soc., Sigma Xi, Alpha Omega Alpha. Achievements include research in mechanism of active growth hormones, early detection and treatment of chronic complications associated with diabetes, Islet transplantation, treatment of prediabetes with nicotinamide, role of intrathymic transplantation in development of tolerance, isolation and characterization of antigen presenting cells, early detection of vascular dysfunction in type 1 diabetes, treatment of short stature growth hormone, development of an implantable glucose sensor. Office: U Kansas Pediactric Endocrine Dept 3901 Rainbow Blvd Kansas City KS 66103

MOORE, WILLIAM GOWER INNES, biochemist; b. Phillipsburg, N.J., Dec. 18, 1951; s. Ralph Gower Davies and Joan Elizabeth Innes (Read) M.; m. Nancy Leo Moore, June 8, 1985. BA, SUNY, Binghamton, 1973; PhD, Purdue U., 1981; MS in Pub. Health, U. Ala., Birmingham, 1988. Asst. prof. Keuka Coll., 1981-83; visiting asst. prof. Birmingham-Southern Coll., 1983-84; research fellow U. Ala., Birmingham, 1984-88, research asst. prof., 1988—. Recipient Regents Scholarship SUNY, 1969. Mem. Am. Chem. Soc., AAAS, N.Y. Acad. of Scis., Sigma Chi (assoc.). Democrat. Episcopalian. Avocations: cooking, tailoring. Office: UAB Sch Dentistry 1919 7th Ave S Birmingham AL 35294-0001

MOORE, WILLIAM VINCENT, electronics and communications systems engineer; b. Derby, Conn., June 22, 1949; s. William Vincent and Mary Elizabeth (Palmer) M.; m. Kathleen Josephine Gavigan, July 7, 1979. BSEE, U. Conn., 1972; MSEE, U. So. Calif., 1980; cert. astronautical engring., UCLA, 1991. Project engr. U.S. Naval Air Test Ctr., Patuxent River, Md., 1973-75; sr. rsch. engr. Lockheed-Calif. Co., Burbank, Calif., 1975-81; sr. project engr. Hughes Space & Comm., El Segundo, Calif., 1981-85; tech. group supr. Jet Propulsion Lab., Pasadena, Calif., 1985—. Mem. IEEE (sr.), AIAA. Achievements include design and development of flight telecommunications system for Magellan, Topex/Podeidon, Mars observer, miniature seeker technology integration, Cassini, and MESUR Pathfinder. Home: 19661 Rosita St Tarzana CA 91356 Office: JPL MS 161-260 4800 Oak Grove Dr Pasadena CA 91109

MOORES, ANITA JEAN YOUNG, computer consultant; b. Poplar Bluff, Mo., Oct. 11, 1944; d. Joseph Samuel and Irene Anita (Sollars) Young; m. James Stephen Moores, June 5, 1965 (div. Jan. 1979); 1 child, Carolyn Terra. BS in Edn., So. Ill. U., 1972, MS in Edn., 1979. Cons. edn. and bus., sales Forsythe Computers, St. Louis, 1979-81; floor sales mgr., bus. cons., sales Computerland of Southwest Houston-Westheimer, 1981-82; bus. cons., sales Bus. Computer Systems and Software, Houston, 1982-83, MicroTask Computers, 1983-84; adminstrv. asst., tech. support Computerland-Techtron, 1984-85; distributer sales Cyber/Source, Houston, 1985; southwest regional sales mgr. Professions Info. Network, Houston, 1987; adminstrv. asst., computer specialist Human Affairs Internat. Inc., Houston; owner Moores' Consulting, Houston, 1986—. Author: (manuals) Choosing a Business Computer, 1983, Career Management, 1984, Training Manual-Computer, 1989; editor: Hounix Newsletter, 1988-89; artist oil paintings. Cons., trainer Meml. Luth. Ch., Houston, 1980-85; adminstr. Olympic Devel.-Soccer, Houston, 1988-89. So. Ill. U. Grad. fellow, 1975-76; named Outstanding Young Women Athlete, So. Ill. U., 1972.

MOORES, ELDRIDGE MORTON, geology educator; b. Phoenix, Oct. 13, 1938; s. Eldridge Morton Jr. and Geneva Pauline (Hofmann) M.; m. Judith Elizabeth Riker, June 5, 1965; children: Geneva, Brian, Kathryn. BS, Calif. Inst. Tech., 1959; PhD, Princeton U., 1963. From lectr. U. Calif., Davis, 1966-67, asst. prof., 1967-70, assoc. prof., 1970-76, prof., 1976—, chair, 1971-76, 88-89; chair tectonics panel Ocean Drilling Project, 1989—. Author: (with R.J. Twiss) Structural Geology, 1992; editor: Troodos 87 Proceedings - Ocean Lithosphere and Ophiolites, 1990, Shaping the Earth: Tectonics of Continents and Oceans, 1990. Fellow Geol. Soc. Am. (editor Geology 1981-88, sci. editor GSA Today 1990—, Disting. Svc. award 1988-91), AAAS; mem. Am. Geophys. Union. Home: 27033 Patwin Rd Davis CA 95616 Office: U Calif Geology Dept Davis CA 95616

MOORJANI, KISHIN, physicist, researcher; b. Karachi, Pakistan, Apr. 9, 1935; came to U.S., 1957; s. Santdas Gurmukhdas and Gomi (Jotsinghani) M.; m. Angela Bickel, Dec. 22, 1962. BSc with honours, Delhi (India) U., 1955; PhD, Cath. U. Am., 1964. Asst. prof. Cath. U. Am., Washington, 1965-66; vis. scientist Nat. Ctr. Sci. Rsch., Paris, 1966-67; mem. sr. staff Applied Physics Lab., Johns Hopkins U., Laurel, Md., 1967-74, prin. prof. staff, 1974—; group supr. Johns Hopkins U., Laurel, 1991—; chmn. applied physics program Johns Hopkins U., Balt., 1985—; cons. Melpar, Inc., Arlington, Va., 1961-67, NASA Goddard Space Flight Ctr., Greenbelt, Md., 1964-65; mem. adv. bd. Supercondr. Applications Assocs., Laguna Hills, Calif., 1989—. Co-author: (monograph) Magnetic Glasses, 1984; co-editor: Advances in Material Science and Applications of High Temperature Superconductors, 1991; editor-in-chief The Johns Hopkins APL Tech. Digest, 1993—; contbr. over 100 articles to profl. jours. Pres. Internat. Club, U. Md., 1958-59. Fellow NSF, 1963, Transfer Of Knowledge Through Emigrated Nats. fellow UN Devel. Program, India, 1990. Mem. Am. Phys. Soc., Commanderie de Bordeaux, Sigma Pi Sigma (chpt. pres. 1959-60). Achievements include patents on methods for detection and investigation of superconducting materials, for superconducting vector magnetometer. Home: 4000 N Charles St Baltimore MD 21218 Office: Johns Hopkins U Applied Physics Lab Johns Hopkins Rd Laurel MD 20723

MOORMAN, WILLIAM JACOB, agronomist, consultant; b. Nickerson, Kans., Jan. 15, 1923; s. Elmer O. and Abbie L. (Mood) M.; m. Mildred L. Morris, Nov. 1, 1947; children: David Morris, Margaret Jane. Student in Chem. Engring., Kans. U., 1941-43; student in Agronomy, Kans. State U., 1946-47. Owner, mgr. Moorman Feed & Seed Co., Inc., Nickerson, Kans., 1947-61; ter. sales mgr. Northrup King Co., Nickerson, Kans., 1961-65; regional sales mgr. Northrup King Co., Columbia, Mo., 1965-68; div. sales promotion, agronomic svcs. Northrup King Co., Richardson, Tex., 1968-73; br. mgr. Northrup King Co., St. Joseph, Mo., 1973-74; div. mktg. mgr. Northrup King Co., Columbus, Miss., 1974-82, sales agronomist, 1982-85; mktg. specialist Northrup King Co., Fredericksburg, Va., 1985-88; owner, mgr. Moorman Enterprises, Fredericksburg, Va., 1988—. Contbr. articles to profl. jours. Committeeman Rep. Party, Kans., 1948-61, Lions Club, 1948-62; adv. Svc. Corps Ret. Execs., Fredericksburg, 1991—; active United Meth. Ch., 1948—. With U.S. Army, 1942-46. Mem. Nat Agri-Mktg. Assn., Am. Soc. Agronomy, Am. Forage and Grasslands Coun., Am. Crop Sci. Soc., Am. Soil Sci. Soc. Home: 1127 James Madison Cir Fredericksburg VA 22405-1632 Office: Moorman Enterprises PO Box 885 Fredericksburg VA 22404-0885

MOOZ, ELIZABETH DODD, biochemist; b. Middletown, Conn., Nov. 22, 1939; d. J. Alfred C. and Lillian Hall (Potter) Dodd; m. R. Peter Mooz, Aug. 29, 1964; children: Ralph Peter Jr., Christopher Dodd. BA, Hollins Coll., 1961; PhD, Tufts U., 1967. Instr. U. Pa. Grad. Hosp., Phila., 1967-69; postdoctoral fellow U. Del., 1969-71; asst. prof. U. Del., Newark, 1971-73, rsch. assoc. Bowdoin Coll., Brunswick, Maine, 1973-76; asst. prof. Med. Coll. Va., Richmond, 1977-79; rsch. scientist Philip Morris U.S.A., Richmond, 1979-89; grants analyst Moody Found., Galveston, Tex., 1989-91; grants cons. Dallas, 1991-93; sci. rev. coord. Am. Heart Assn. Nat. Ctr., Dallas, 1993—; mem. state adv. com. New Bd. Higher Edn., Brunswick, 1975-77; mem. seminars com. Sci. Mus. Va., Richmond, 1985-89. Contbr. articles and abstracts to profl. jours. and chpts. to books. State del. Rep. Cen. Com. Tex., Galveston, 1990-91. NSF fellow, 1959-61, NIH fellow, 1961-67, 69-71; named one of Outstanding Young Women Am., 1970. Mem. Am. Chem. Soc. (Women's chemist com. 1980-87, James Lewis Howe award 1961), Rotary, Sigma Xi (pres. Med. Coll. Va. chpt. 1983). Episcopalian. Achievements include patent for process for increasing filling power of tobacco. Home: 3131 Maple Ave Apt 9-H Dallas TX 75201 Office: Am Heart Assn Nat Ctr 7272 Greenville Ave Dallas TX 75231

MORAFF, HOWARD, science foundation director; b. N.Y.C., Feb. 5, 1936; s. Frank and Sylvia Moraff; m. Connie J. McClure, Oct. 4, 1975; children: Kenneth, Judith, Steve. AB, Columbia U., 1956, BSEE, 1957, MSEE, 1958; PhD in Neurophysiology, Cornell U., 1967. Dir. vet. med. computing resources Cornell U., Ithaca, N.Y., 1967-82; dir. computing resources Merck Sharp & Dohme Rsch. Labs., Rahway, N.J., 1982-84; program dir. NSF, Washington, 1984—. Co-author: Electronics for Neurobiologists, 1973, Electronics for the Modern Scientist, 1982; contbr. articles to profl. jours. Capt. USAF, 1958-61. Mem. IEEE (sr.), Am. Assn. Artificial Intelligence, N.Y. Acad. Scis., Sigma Xi, Tau Beta Pi, Eta Kappa Nu, Phi Kappa Phi. Achievements include patent for transistor circuit; research in on-line laboratory computing and end-user computing. Office: NSF 1800 G St NW Rm 310 Washington DC 20550-0002

MORAHAN-MARTIN, JANET MAY, psychologist, educator; b. N.Y.C., Jan. 13, 1944; d. William Timothy and May Rosalind (Tarangelo) Morahan; m. Curtis Harmon Martin, June 2, 1979; 1 child, Gwendolyn May. AB, Rosemont (Pa.) Coll., 1965; MEd, Tufts U., 1968; PhD, Boston Coll., 1978. Asst. mkt. rsch. analyst Compton Advt. Co., N.Y.C., 1965-67; mkt. rsch. analyst Ogilvy & Mather Advt., N.Y.C., 1967; ednl. rsch. asst. Tufts U., Medford, Mass., 1968-69; counselor Psychol. Inst. Bentley Coll., Waltham, Mass., 1971-72; dir. counseling svcs. Bryant Coll., Smithfield, R.I., 1972-75, psychology instr., 1972-76, asst. prof. psychology, 1976-81, assoc. prof. psychology, 1981-91, prof. psychology, 1991—; bd. dirs. Multi-Svc. Ctr., Newton, Mass., 1980-82. Contbr. articles to profl. jours., chpts. to books; reviewer APA Conv., 1985—, Teaching of Psychology Jour., 1988—, Collegiate Micro-Computer Jour., 1991, 93, Nat. Soc. Sci. Jour., 1991—, Collegiate MicroComputer Jour., 1991—. Bd. dirs. Wellesley (Mass.) Community Children's Ctr., 1986-90, Coun. for Children, Newton, Mass., 1984-86. NIMH fellow, 1967-68; NSF grantee, 1974-76, U.S. Office Edn. grantee, 1980. Mem. Am. Psychol. Assn., Brewster (Mass.) Hist. Soc., Mass. Audubon Soc., Nat. Social Sci. Assn., Mass. Hort. Soc., N.E. Soc. for Behavioral Analysis and Therapy. Avocations: photography, antiques, gardening, literature. Home: 17 Fuller Brook Rd Wellesley MA 02181-7108 Office: Bryant Coll 1150 Douglas Pike Smithfield RI 02917-1220

MORALES, CYNTHIA TORRES, clinical psychologist, consultant; b. L.A., Aug. 13, 1952; d. Victor Jose and Lupe (Pacheco) Torres; m. Armando Torres Morales, June 30, 1989. BA, UCLA, 1975, M in Social Welfare, 1978, D in Counseling Psychology, 1986. Lic. psychologist, Calif. Clin. social worker VA, Brentwood, Calif., 1977-78; med. social worker Harbor-UCLA Med. Ctr., Carson, Calif., 1978-79; psychotherapist San Fernando Valley Child Guidance Clinic, Northridge, Calif., 1979-80; psychiat. social worker L.A. County Dept. Mental Health, 1980-81; child welfare worker L.A. County Dept. Children's Svcs., 1981-86; cons. psychologist, organizational devel. mgr. UCLA, 1986—; pvt. practice and consultation, 1992—; cons. Hispanic Family Inst., L.A., 1989—; mem. diversity com. UCLA, 1988—, mem. mental health emergency task force, 1986-89. Mem. Centro de Ninos Bd. Dirs., L.A., 1984-88; lobbyist self devel. people United Presbyn. Ch. Synod, L.A., 1982-88; chair Inner City Games Acad. Contest Hollenbeck Police Bus. Coun., L.A., 1992; co-chair Inner City Games Acad. Essay Contest, 1993. Recipient Cert. of Appreciation, Children's Bapt. Home, 1984, Cert. of Appreciation, Hollenbeck Police Bus. Coun. 1992, Spl. Recognition award Fed. Judge Takasugi, Pro Bono Bar Rev. and L.A. City Atty. 1993, Cert. of Appreciation, Hollenbeck Youth Ctr., 1992. Mem. APA. Office: Ste 1701 1100 Glendon Ave Westwood Village Los Angeles CA 90024

MORALES, RAUL HECTOR, physician; b. San Juan, P.R., Aug. 2, 1963; s. Raul and Sonia Margarita (Borges) M. BS, U. P.R., 1985; MD, San Juan Bautista Sch. Med., 1990. Diplomate P.R. Bd. Med. Examiners, Am. Bd. Internal Medicine. Aux. sales rep. Borges Warehouse of Textiles, Gurabo, P.R., 1981-85; tutor Computer Lab. U. P.R., Cayey, 1984, asst. researcher Ecol. Lab., 1985; intern Henry Ford Hosp., Detroit, 1990-91, resident, 1991-93; fellow, medical oncology Providence Hosp., Mich., 1993—; mem. prostate specific antigen clin. policy team Henry Ford Hosp., 1991-93, smoking cessation task force, 1993; summer tutor Edn. Dept. P.R., Gurabo, 1986-87; presenter in field. Achievements include reseach in peridontal analysis. Organizer Com. of Profl. Assn., San Juan, 1988-90, Com. of San Jose Marathon, 1985-91. Mem. ACP (assoc.); AAAS, N.Y. Acad. Scis., Henry Ford Med. Assn. Roman Catholic. Home: 27300 Franklin Rd Apt 223 Southfield MI 48034

MORALES, RICHARD, structural engineer; b. Albuquerque, Jan. 19, 1958; s. Alfonso and Nela (Pena); m. Diana Veronica Molina, Oct. 26, 1991; children: Yvonne Marianne, Yvette Adrianne. BSCE, U. N.Mex., 1985; MS in Structural Engring., U. Calif., Berkeley, 1986. Registered profl. engr., Calif. Structural rsch. engr. Nat. Bur. Stds., Washington, 1985-87; bridge design engr. Caltrans Dept. Transp., Sacramento, Calif., 1987-89; sr. project mgr. Chavez-Grieves Cons. Engrs., Albuquerque, 1989, Lockheed Engring. and Scis./NASA Johnson Space Ctr. WSTF, Las Cruces, N.Mex., 1989—. Sci. advisor vol. Las Cruces Pub. Schs., Sci. Advisor Program, 1992—; bd. dirs. So. N.Mex. Soc. Hispanic Profl. Engrs., 1991—; v.p. U. N.Mex. Young Dems., Albuquerque, 1984; lobbyist Associated Students of N.Mex., 1984. Nat. Hispanic fellowship, 1984. Mem. ASCE (pres. student chpt. 1984-85), Earthquake Engring. Rsch. Inst., Structural Engrs. Assn. of No. Calif., Kappa Mu Epsilon, Chi Epsilon. Roman Catholic. Home: 1095 Esplanada Circle El Paso TX 79932 Office: LESC PO Drawer MM Las Cruces NM 88004

MORALES-ACEVEDO, ARTURO, electrical engineering researcher, educator; b. Oaxaca, Mexico, Nov. 25, 1954; s. Arturo Morales-Cisneros and Ma del Carmen Acevedo-Ricardez. Elec. Engr., Inst. Politenico Nacional, Mexico City, 1977; MSc in Elec. Engring., Centro Investigacion Estudios Avanzados Instituto Politecnico Nacional, Mexico City, 1978, PhD in Elec. Engring., 1987; MSc in Physics, Purdue U., 1982. Prof. elec. engring. Centro Investigacion Estudios Avanzados Instituto Politecnico Nacional, Mexico City, 1983—; engring. cons. Electrosolar de Mexico, Mexico City, 1990-91, Opcion Solar, Mexico City, 1990-91; nat. investigator Edn. Ministry, Mexico, 1984—. David Ross fellow Purdue U., West Lafayette, Ind., 1981. Mem. IEEE, Internat. Solar Energy Soc., N.Y. Acad. Scis. Achievements include research in solar cells and computers in physics. Office: CINVESTAV-IPN, IPN 2508 Av, Mexico City 07360, Mexico

MORAN, JOSEPH JOHN, psychology educator; b. Hartford, Conn., Sept. 26, 1942; s. Joseph Peter and Margaret (McGrath) M.; m. Barbara Ann Burns, Sept. 4, 1965; children: Joseph Michael, Brian Matthew. BA in Psychology, Amherst (Mass.) Coll., 1964; MA in Edn., Trinity Coll., 1969; PhD in Psychology, Emory U., 1973. Lic. psychologist, N.C., N.Y. Tchr. Granby (Mass.) Pub. Schs., 1965-66, Hartford (Conn.) Pub. Schs., 1966-70; with SUNY, Buffalo, 1973—, asst. prof., assoc. prof.; cons. Mental Health Ctrs., N.C., N.Y., 1981-87, Pub. Sch. Dists., N.C., N.Y., 1982-92. Assoc. editor Child Study Jour., 1986—; contbr. articles to profl. jours. Mem. N.C. Botanical Garden. Recipient Spl. Merit award on Excellence in Cons. Project Innovation, 1988. Mem. APA, Am. Assn. of Adult and Continuing Edn. (chair, adult psychology unit 1992—). Democrat. Roman Catholic. Achievements include rsch. on measurement of personality and promoting mental health through educational policy. Office: SUNY 1300 Elmwood Ave Buffalo NY 14222

MORAN, RICARDO JULIO, economist; b. Havana, Cuba, Sept. 4, 1939; came to U.S., 1960; s. Jose Ricardo and Maria Luisa (Forcade) M.; m. Mayra Buvinic, Dec. 24, 1973 (div. 1981); m. Mary Louise Fox, Apr. 25, 1987. BA, Tulane U., 1963; MA in Econs., U. Calif., Berkeley, 1966, postgrad., 1968. Prof., sr. rsch. assoc. Universidad Catolica de Chile, Santiago, 1967-70; pvt. practice Washington, 1970-73, v.p. Moran Equities, Inc., Miami, Fla., 1973; economist The World Bank, Washington, 1973-90; prin. Moran Internat., Washington, 1990-92; sr. economist Interam. Devel. Bank, Washington, 1992—. Co-author: Declining Births in Chile, 1972, Brazil, 1981; contbr. articles to profl. jours. Fellow Latin Am. Teaching Program Tufts U., 1957, Ford Found., 1965, OAS, 1963-65. Mem. Am. Econ. Assn. Avocation: literature, music, travel, tennis. Home: 2910 Cortland Pl NW Washington DC 20008-3429 Office: Interam Devel Bank Washington DC 20577

MORAN, SHARON JOYCE, chemist; b. St. Louis, Apr. 2, 1946; d. Richard Francis and Olivette Leona (Kissel) M. BS in Chemistry, Webster U., 1968; MS in Chemistry, U. Mo., St. Louis, 1982. Rsch. chemist Monsanto Co., St. Louis, 1968-73, rsch. chemist II, 1973-74, sr. rsch. chemist, 1979-82, rsch. specialist, 1982-91, sr. rsch. specialist, 1991—. Mem. Am. Chem. Soc. Office: Monsanto Co 700 Chesterfield Pkwy N Saint Louis MO 63198

MORAND, PETER, research agency executive; b. Montreal, Que., Can., Feb. 11, 1935; s. Frank and Rose Alice (Fortier) M.; m. Dawn McKell, Oct. 10, 1957; children: Clifford, Tanya. BSc with honors, Bishop's U., Lennoxville, Que., 1956, DCL (hon.), 1991; PhD, McGill U., Montreal, 1959. NATO postdoctoral fellow Imperial Coll., London, 1959-61; sr. rsch. chemist Ayerst Labs., Montreal, 1961-63; asst. prof. chemistry U. Ottawa, Can., 1963-67, acad. asst. vice rector, 1968-71, dean sci.and engring., 1976-81, prof. chemistry, dir. rsch. svcs., 1981-87, vice rector univ. R&D, 1987-90; pres. Natural Scis. and Engring. Rsch. Coun., Ottawa, 1990—; bd. dirs. Ottawa-Carleton Econ. Devel. Corp., Can., Royal Ottawa Hosp. Contbr. articles to profl. jours.; patentee in field. Trustee B.C. Applied Systems Inst., Vancouver, Can., 1990—; bd. dirs. Ottawa Life Scis. Tech. Park. Natural Scis. and Engring. Rsch. Coun. grantee, 1964-90. Fellow Chem. Inst. Can.; mem. Soc. of Rsch. Adminstrs., Rideau Club, Cercle Univ. Office: Natural Scis & Engrng Rsch Coun. Can, 200 Kent St # 501, Ottawa, ON Canada K1A 1H5

MORARI, MANFRED, chemical engineer, educator; b. Graz, Austria, May 13, 1951; came to U.S., 1975; s. Manfred and Hilde (Florian) M.; m. Marina Korchynsky, May 12, 1984. Diploma Chem. Engring., Eidgenoessische Technische Hochschule, Zurich, Switzerland, 1974; PhD in Chem. Engring., U. Minn., 1977. Asst. prof. U. Wis., Madison, 1977-81, assoc. prof., 1981-83; prof. chem. engring. Calif. Inst. Tech., Pasadena, 1983—, McCollum-Corcoran prof., 1991—, exec. officer, 1990—; Gulf vis. prof. chem. engring. Carnegie Mellon U., 1987. Contbr. articles to profl. jours. Recipient D.P. Eckman award Am. Automatic Control Coun., 1980. Mem. IEEE (George S. Axelby Outstanding Paper award 1990), Am. Soc. for Engring. Edn. (Curtis W. McGraw rsch. award 1989), Am. Inst. Chem. Engrs. (A.P. Colburn award 1984), Nat. Acad. Engring., Am. Chem. Soc. Home: 2735 Ardmore Rd San Marino CA 91108-1768 Office: Calif Inst Tech Chem Engring 210-41 Pasadena CA 91125

MORAWETZ, CATHLEEN SYNGE, mathematician; b. Toronto, May 5, 1923; came to U.S., 1945, naturalized, 1950; d. John Lighton and Elizabeth Eleanor Mabel (Allen) Synge; m. Herbert Morawetz, Oct. 28, 1945; children: Pegeen Ann, John Synge, Lida Joan, Nancy Babette. B.A., U. Toronto, 1945; S.M., MIT, 1946; Ph.D., NYU, 1951; hon. degree, Eastern Mich. U., 1980, Smith Coll., 1982, Brown U., 1982, Princeton U., 1986, Duke U., 1988, N.J. Inst. Tech., 1988. Research assoc. Courant Inst., NYU, 1952-57, asst. prof. math., 1957-60, assoc. prof., 1960-65, prof., 1965—, assoc. dir., 1978-84, dir., 1984-88. Editor Jour. Math. Analytical Applications, Communications in PDE, Advanced Applications Math., Comm. in Pure and Applied Math.; author articles to applications of partial differential equations, especially transonic flow and scattering theory. Trustee Princeton U., 1973-78, Sloan Found., 1980—. Guggenheim fellow, 1967, 79; Office Naval Rsch. grantee, until 1990. Fellow AAAS; mem. NAS, Am. Math. Soc. (term trustee), Am. Acad. Arts and Scis., Soc. Indsl. and Applied Math. Office: 251 Mercer St New York NY 10012-1185

MORBEY, GRAHAM KENNETH, management educator; b. Birmingham, Eng., Apr. 5, 1935; came to U.S., 1956; m. Gillian M. Grist, Feb. 13, 1960; children: Alison J., Karen J. BScChemE with honors, U. Birmingham, 1956; MASc in Applied Chemistry, U. Toronto, 1957; MA, PhD, Princeton U., 1961. Rsch. scientist Dunlop Tire, Toronto, 1960-63; dir. product devel. Celanese, Charlotte, N.C., 1963-69; tech. dir. Hoechst Fibers Inc., Spartanburg, S.C., 1969-72; dir. R&D Johnson & Johnson Personal Products, Milltown, N.J., 1972-79; v.p. R&D Texon-Emhart, South Hadley, Mass., 1979-85; v.p. Lydall Inc., Manchester, Conn., 1989-92; assoc. prof. U. Mass., Amherst, 1985—; cons. Aminco, Amherst, 1985—. Contbr. articles to profl. jours. Hon. rsch. fellow Textile Rsch. Inst., Princeton. Mem. TAPPI, Product Devel. Mgmt. Assn., Princeton Club N.Y., Sigma Xi. Achievements include patents for absorbtive and thermal barrier products. Home: 81 Blossom Ln Amherst MA 01002

MORDECHAI, SHAUL, physicist; b. Bagdad, Iraq, Jan. 13, 1941; arrives in Israel, 1951; s. Salman and Nazima (De Gam) M.; m. Yaffa Cohen, June 24, 1969; children: Michal, Nurit, Merav, Amir. MS, Hebrew U., 1967, PhD, 1971. Sr. lectr. Ben-Gurion U., Beer-Sheva, Israel, 1971-75; rsch. investigator U. Pa., Phila., 1976-78; assoc. prof. Ben-Gurion U., 1982-89; rsch. faculty U. Tex., Austin, 1984-85; prof. physics Ben-Gurion U., 1990—; vis. scientist U. Tex., 1987-89; cons. Hebrew U., 1972-76, U. Pa., 1987-88; mem. adv. com. Israeli Acad. Scis. & Humanities, Jerusalem, 1989-91. Contbr. articles to profl. jours. Israel Nat. Acad. Sci. grantee, 1974-75, 81-83, U.S. -Israel Binat. Sci. Found., 1988-91; recipient Ben Sheva de Rothschild Found. award, 1973-74. Mem. Am. Phys. Soc., European Phys. Soc., Coun. Israeli Phys. Soc., LAMPF USers Group. Achievements include experimental observation of the phenomena of double giant resonances in the atomic nuclei. Office: Los Alamos Nat Lab MP-9 MS-H846 Los Alamos NM 87545

MORE, KANE JEAN, science educator; b. Norwalk, Ohio, Nov. 19, 1953; d. Charles Norman and Kazue (Miura) M.; m. Steven W. Gong, Nov. 24, 1979; children: Benjamin, Kaylyn. BS, Wright State U., 1976; MS in Zoology, U. Fla., 1982. Cert. tchr., Fla. Rsch. assoc. Dept. Zoology, U. Fla., Gainesville, 1979-86; sci. tchr. Spanish River High Sch., Boca Raton, Fla., 1987-91; sci. dept. chair Olympic Heights High Sch., Boca Raton, 1991—; sponsor, faculty Nat. Honor Soc., Boca Raton, 1991—. Contbr. articles to Jour. of Exptl. Biology. Recipient Outstanding Tchr. award, 1992-93; Tandy Tech. scholar. Mem. Mount Desert Island Biol. Lab., Fla. Assn. Sci. Tchrs., Sigma Xi. Democrat. Home: 9312 Neptunes Basin Ct Boca Raton FL 33434 Office: Olympic Heights High Sch 20101 Lyons Rd Boca Raton FL 33434

MORE, SYVER WAKEMAN, geologist; b. Washington, Jan. 27, 1950; s. John William and Virginia (Wakeman) M.; m. Judith Ann Bessler, May 25, 1974; children: Kristin Elisabeth, Andrew Alan. BS in Geoscis., U. Ariz., 1972, MS in Geoscis., 1980. Registered geologist, Ariz., Wyo., Ky. Asst. exploration geologist Continental Oil Co., Tucson, 1970, 71, 72-73; mine devel. geologist Conoco Minerals, Florence, Ariz., 1973-75; exploration geologist Exxon Co. U.S.A., 1976; exploration geologist Amax Exploration Inc., Tucson, 1979, Billiton Exploration U.S.A., Tucson, 1980-83; supervising geologist Billiton Polaris, Inc., Hibbing, Minn., 1984; geologist Billiton Exploration USA, Tucson and San Carlos, Ariz., 1985-86; sr. minerals geologist Billiton Minerals USA Inc., Tucson, 1986-87, Lancaster, Calif., 1987-88; sr. geologist Atlas Precious Metals, Sparks, Nev., 1988-89; cons. geologist, Tucson, 1989—. DuVal Corp. fellow, 1977. Fellow Soc. Econ. Geologists; mem. AIME Soc. Exploration, Mining and Metallurgy, Geol. Soc. Am., Am. Inst. Profl. Geologists (cert. profl. geologist), Nat. Rifle Assn. (life). Republican. Club: Tucson Rod and Gun. Research or work interests: Exploration and development of base-and precious-metal deposits; exploration program design and management (foreign and domestic). Subspecialties: Geology; Mineral exploration and development. Home: 11321 E Calle Vaqueros Tucson AZ 85749-9521

MOREAU, HUGUES ANDRE, physician; b. Limoges, France, July 17, 1948; s. Jacques and Colette (Hugues) M.; m. Martine Houles, Sept. 15, 1972 (div. 1980); children: Gilles, Eva; m. Josette Denise Roy, Sept. 24, 1984; 1 child, Martial. Grad., Lycee Gay Lussac, 1966, postgrad. in math., 1967; MD, U. Limoges, 1975. Intern Centre Hosp. Regional, Gueret, France, 1974; gen. practice medicine Limoges, 1975—. Served as chief med. officer French Air Force, 1974-75. Mem. Confedn. Syndicates Med. Francais (pres. dept. 1985), Assn. Conferal Formation Medicine (regional counsel 1984), Syndicat Autonome de Haute Vienne (pres. 1985—), Limoges Golf Club. Roman Catholic. Avocations: target practice, golf. Home: 197 Ave du General Leclerc, 87100 Limoges France Office: Cabinet Med La Bastide 2, 14 Allee Seurat, 87100 Limoges France

MOREHART, JAMES HENRY, mechanical engineer; b. Des Moines, July 31, 1959; s. Jonas Leroy and Constance Leontine (Paul) M.; m. Tammy Marie Short, May 31, 1986. BS, U. Md., 1986; MS, Calif. Inst. Tech., 1987, PhD, 1991. Rsch. mech. engr. Nat. Inst. Stds. and Tech., Gaithersburg, Md., 1990-91; assoc. rsch. engr. U. Dayton Rsch. Inst., Edwards AFB, Calif., 1991-92; sr. rsch. engr. SPARTA Inc., Edwards AFB, Calif., 1993—. Contbr. articles to profl. jours. Mem. AIAA, The Combustion Inst., Tau Beta Pi, Pi Tau Sigma, Phi Kappa Phi. Office: OLAC PL/RKFT 8 Draco Dr Edwards AFB CA 93524-7230

MOREHOUSE, DAVID FRANK, geologist; b. Charles City, Iowa, Dec. 8, 1943; s. Neal Francis and Florence E. (Schwendener) M. BS in Gen. Scis., State U. Iowa, 1967; MS in Geology, Iowa State U., 1970; postgrad., Pa. State U., 1970-74. Staff geologist Nat. Gas Survey and Planning and Spl. Projects Div., FPC, Washington, 1974-78; dir. Info. Processing and Interpretation and Analysis Divs. Oil and Gas Info. System, Energy Info. Adminstrn., Washington, 1978-80, sr. supervisory geologist, 1980—; advisor petroleum data system U. Okla., Norman, 1975-86; Energy Info. Adminstrn. rep. Am. Gas Assn. Com. on Natural Gas Res., Washington, 1991—, Potential Gas Com., Boulder, Colo., 1991—. V.p. Iowa Jr. Acad. Sci., 1961. Recipient awards for outstanding performance Fed. Govt., Washington, 1974-92. Fellow Nat. Speleological Soc.; mem. AAAS, AIME, Internat. Assn. Math. Geologists, N.Y. Acad. Scis. Congregationalist. Achievements include first evidence that sulfuric acid can be important to speleogenesis; exercising the prin. responsibility for design and establishment of fed. govts. domestic oil and gas reserves estimation and analysis program. Office: Energy Info Adminstrn EI443 1000 Independence Ave SW Washington DC 20585

MOREHOUSE, LAWRENCE GLEN, veterinarian, educational administrator; b. Manchester, Kans., July 21, 1925; s. Edwy O. and Ethel (Glenn)

M.; m. GeorgiaAnn Lewis, Oct. 6, 1956; children: Timothy, Glenn Ellen. BS in Biol. Sci., Kans. State U., 1952, DVM, 1952; MS in Animal Pathology, Purdue U., 1956, PhD, 1960. Veterinarian County Vet. Hosp., St. Louis, 1952-53; supr. Brucellosis labs. Purdue U., Lafayette, Ind., 1953-60; staff veterinarian lab. svcs. U.S. Dept. Agr., Washington, 1960-61; discipline leader in pathology and toxicology, animal health div. Nat. Animal Disease lab., Ames, Iowa, 1962-64; prof., chmn. dept. veterinary pathology Coll. Vet. Medicine U. Mo., Columbia, 1964-67, 84-86; dir. Vet. Med. Diagnostic Lab., 1968-88; prof. emeritus Coll. Vet. Medicine U. Mo., Columbia, 1986—; cons. to U.S. Dept Agr., Surgeon Gen. U.S. Army, Am. Inst. Biol. Scis. Nat. Acad. Sci., Miss. State U., St. Louis Zoo Residency Tng. Program, Miss. Vet. Med. Assn.; adv. com. med. rsch. and devel. U.S. Army. Co-editor: Mycotoxic Fungi, Mycotoxins, Mycotoxicoses: An Encyclopedic Handbook , 3 vols., 1977; contbr. numerous articles on diseases of animals to profl. jours. With USN, 1943-46, U.S. Army, 1952-56. Recipient Outstanding Svc. award U.S. Dept Agr., 1959, Merit Cert., 1963, 64, Disting. Svc. award Coll. Vet. Medicine U. Mo., 1987. Fellow Royal Soc. Health London; mem. Am. Vet. Med. Assn., Mo. Vet. Med. Assn., Nat. Assn. Vet. Lab. Diagnosticians (E.P. Pope award 1976, chmn. lab. accreditation bd. 1972-79, 87-90), N.Y. Acad. Sci., U.S. Animal Health Assn., Am. Assn. Lab. Animal Medicine, Mo. Soc. Microbiology, Am. Assn. Avian Pathologists. Presbyterian. Home: 916 Danforth Dr Columbia MO 65201-6164 Office: U Mo Vet Med Diagnostic Lab PO Box 6023 Columbia MO 65201

MOREL, FRANÇOIS M.M., civil and environmental engineering educator; b. Versailles, France, Oct. 11, 1944; came to U.S., 1967; s. André M.S. Morel and Elizabeth Penicaud; m. Nicole M.L. Laurens, July 8, 1967; children: Sébastien, Chantal. License-és-Scis. in Math. Appliquees, U. Grenoble, France, 1966, diplôme d'Ingénieur Hydraulicien, 1967; MS in Civil Engring., Calif. Inst. Tech., 1968, PhD in Engring. Scis., 1971. Rsch. fellow in environ. engring. scis. Calif. Inst. Tech., Pasadena, 1971-73; from asst. prof. to assoc. prof. to prof. MIT, Cambridge, 1973—, dir. R.M. Parsons Lab., 1991—; dir. R.M. Parsons Lab., MIT; vis. scientist Woods Hole Oceanographic Instn., 1982—; assoc. dir. for rsch. Ecole des Mines de Paris, 1986; vis. prof. Ecole Normale Supérieure, Paris, 1987—; panel on marine mineral tech. Nat. Rsch. Coun., 1976-77, com. on ocean waste transp., 1982-84, com. on irrigation-induced water quality problems, 1985-88; participant workshops Nat. Oceanic and Atmospheric Adminstrn., 1977, 78, 79, tech. reviewer, cons.; panel on ocean mining Nat. Adv. Com. for Oceans and Atmosphere, 1981-82; participant workshops, cons. EPA, Bur. Land Mgmt., Nat. Aerospace Adminstrn., Internat. Joint Commn.; environ. cons. Institut Nat. de l'Environnement Indsl. et des Risques, Environ. Def. Fund, Nashua Corp., Water Purification Assocs., Mass. Exec. Office Environ. Affairs, Procter and Gamble, Nat. Oceanic and Atmospheric Adminstrn., Ocean Mining Co., Pelican Pipe Co., Nat. Highway Adminstrn., Homestake Mining Co., Tetra Tech, MITRE Corp., Ciba-Geigy, Alcoa, Gradient Corp., AMOCO, other private and pub. interest groups. Author: Principles of Aquatic Chemistry, 1983, (with D.A. Dzombak) Surface Complexation Modeling: Hydrous Ferric Oxide, 1990, (with J.G. Hering) Principles and Applications of Aquatic Chemistry, 2d edit., 1993; co-editor vol. 36, No. 8 of Limnology & Oceanography, 1991; contbr. numerous chpts. to books including Ocean Margin Processes in Global Change, Aquatic Chemical Kinetics, Chemical Processes in Lakes, also articles to jours. including Deep-Sea Rsch., Biol. Oceanography, Water Rsch. Named Doherty prof. in ocean utilization, 1974-77; Govt. Ministry of Fgn. Affairs scholar, 1967, Fulbright Internat. travel scholar, 1967. Mem. AAAS, Am. Chem. Soc. (adv. bd. Environ. Sci. & Tech. 1978-83, assoc. editor Environ. Sci. & Tech. 1980), Am. Soc. of Limnology and Oceanography (organizer of symposium 1978, mem. editorial bd. Limnology & Oceanography 1980-83, editor spl. issue Limnology & Oceanography 1990), New Eng. Aquarium (sci. coun. 1988-90), Assn. of Environ. Engring. Profs., Am. Geophys. Union, Biogeochemistry (cons. editor 1991—). Achievements include research in theoretical and experimental studies on fate and effects of chemical pollutants; computer modeling of chemical characteristics of natural and polluted waters; coordination chemistry, surface chemistry, and photochemistry of trace metals in natural waters; interactions between the chemistry and microbiota in aquatic systems; trace metal nutrition and toxicity in phytoplankton. Home: MIT 75 Cambridge Pkwy Cambridge MA 02142 Office: MIT 77 Massachusetts Ave Cambridge MA 02139

MORELLO, JOSEPHINE A., microbiology educator, pathology educator; b. Boston, May 2, 1936; married, 1971. BS, Simmons Coll., 1957; AM, Boston U., 1960, PhD in Microbiology, 1962. Cert. med. microbiologist Am. Bd. Microbiology. Inst. microbiology Boston U., 1962-64; rsch. assoc. Rockefeller U., 1964-66; resident med. microbiologist Coll. Physicians & Surgeons, Columbia U., 1966-68, asst. prof. microbiology, 1968-69, asst. prof., 1970-73, assoc. prof. pathology and medicine, 1973-78; assoc. prof. pathology and medicine U. Chgo., 1978—, dir. clinical microbiology, 1970—; dir. microbiology Harlem Hosp. Ctr., 1968-69. Editor: Clinical Microbiology Review, Clinical Microbiology Newsletter. Recipient Sonnenwirth Meml. award Am. Soc. Microbiology,1991. Fellow Am. Acad. Microbiologist; mem. Acad. Clinical Lab. Physicians and Scientists, Am. Soc. Microbiology, Am. Soc. Clinical Pathologist, Sigma Xi. Achievements include research in inproved methods of clinical microbiology; epidemiology and characteristics of pathogenic neisseria. Home: 425 Luthin Rd Hinsdale IL 60521-2770*

MOREN, LESLIE ARTHUR, physician; b. Webster, Wis., Jan. 28, 1914; s. John Arthur and Jennie (Anderson) M.; m. Laurena Ann McBride, Sept. 9, 1939 (dec. Aug. 1987); children: Ann Nisbet, Allen, Kristin Madden, James A. BA, U. Minn., 1934, MD, 1938. Diplomate Am. Bd. Family Practice. Pvt. practice physician Eklo, Nev., 1938-40, St. Paul, 1940-42; pvt. practice physician Elko Regional Med. Ctr., 1946—. Maj. AUS, 1943-45, ETO. Mem. Nev. State Bd. Med. Examiners (pres. Reno chpt. 1950-77), Nev. State Med. Assn. (pres. Reno chpt. 1938-40, 46—), Rotary. Republican. Episcopalian. Avocation: reading. Home: 777 Court St Elko NV 89801-3330 Office: Elko Regional Med Ctr 762-14th St Elko NV 89801

MORENO-LOPEZ, JORGE, virologist, educator; b. Riobamba, Ecuador, Apr. 19, 1941; arrived in Sweden, 1965; s. Thelmo Elias and Dina Sabina (Lopez) M.; m. Lilian Margareta Pettersson, Apr. 2, 1969; children: Patric, Jessica, Daniel. Vet. degree, Faculty of Vet. Medicine, Quito, Ecuador, 1964, Royal Vet. Coll., Stockholm, 1971; PhD, Swedish U. Agr., Uppsala, 1977. Radio and tv journalist HCJB-TV 4, Quito, 1962-64; radio journalist Radio Sweden, Stockholm, 1967-74; scientist Nat. Vet. Inst., Stockholm, 1970-77; from scientist to assoc. prof. Swedish U. Agr. Scis., Uppsala, 1977-90; regional expert for Latin Am. Internat. Atomic Energy Agy., Vienna, Austria, 1990—; cons. Internat. Found. for Sci., Stockholm, 1985—, UN Food and Agr. Orgn., Stockholm, 1976—, Swedish Agy. for Rsch. Coop., Stockholm, 1985— (grant 1986—). Co-author: Animal Papilloma Viruses, 1985, Virus of Ruminants, 1990; editor: Diagnostic Virology I-II, 1989, 90; contbr. articles to profl. jours. Grantee Swedish Coun. for Forestry and Agrl. Rsch. 1980—. Mem. Swedish Vet. Assn., N.Y. Acad. Scis., Swedish Agy. for Rsch. Cooperation. Avocations: science, Latin American ancient cultures, antique collecting, archaeology. Home: Gernotgasse 74-76 Top 3, A-1220 Vienna Austria Office: Internat Atomic Energy Agy, Wagramerstrasse 5, PO Box 200, A-1400 Vienna Austria

MORESI, REMO P., mathematician; b. Lugano, Switzerland, Apr. 8, 1952. MPhil, U. Zurich, 1977, PhD, 1980. Prof. U. Catolica de Chile, Santiago, 1980-83, SMS TI, Switzerland, 1984—; researcher, admnstr. CERFIM, Locarno, Switzerland. Contbr. articles to profl. jours. Avocations: chess, sports, literature. Office: CERFIM, Via F Rusca 1, Locarno 6601, Switzerland

MOREY, PHILIP STOCKTON, JR., mathematics educator; b. Houston, July 11, 1937; s. Philip Stockton and Helen Holmes (Wolcott) M.; m. Jeri Lynn Snyder, Sept. 5, 1964; children: William Philip, Christopher Jerome. BA, U. Tex., 1959, Ma, 1961, PhD, 1967. Asst. prof. math. U. Nebr., Omaha, 1967-68; assoc. prof. Tex. A&I U., Kingsville, 1968-76, prof., 1976—; lectr. U. Tokyo, 1976, U. Hokkaido, 1977, 88. Contbr. articles to Tensor N.S., Internat. Jour. Engring. Sci, Tex. Jour. Sci. Recipient Researcher of Yr. awrd Tex. A&I Alumni Assn., 1985. Mem. Tex. Acad. Sci. (chmn. math. sect. 1982, '85), Am. Math. Soc., Tensor Soc., (Japan). Achievements include research in extensor analysis, tensor analysis, differential geometry, mathematical physics. Home: 1514 Lackey St Kingsville TX 78363-3199 Office: Tex A&I Univ Kingsville TX 78362

MORFOPOULOS, V., metallurgical engineer, materials engineer; b. Athens, Greece, Oct. 22, 1937. BS, Purdue U., 1958; MS, Columbia U., 1961, ScD in Engring. Sci., 1964. Rsch. assoc. metall. engring. Purdue U., 1957-60; rsch. engr. U.S. Steel Corp., 1961; instr. chem. CUNY, 1961-63; rsch. engr. Argonne Nat. Lab., 1963; rsch. engr. Am. Iron & Steel, Columbia U., 1964-65, sr. metall. sci., 1965-66; tech. dir. R&D testing Am. Standards Testing Bur., 1966—; cons. govt. and industry, 1966—; mem. Int. Commn. Chem. Thermodyn. & Kinetics; mem. Transp. Rsch. Bd., Nat. Rsch. Coun. Mem. AAAS, Am. Inst. Mining, Metall. Petroleum Engrs., Am. Soc. Engr. Edn., Assn. Cons. Chemists and Chem. Engrs., N.Y. Acad. Sci. Achievements include research and consulting in fields of corrosion and oxidation phenomena. low and high temperature thermodynamics, liquid metals and compounds, surface phenomena, electrometallurgy and electrode phenomena, electrical and magnetic properties of matter, failure and stress analysis, metal finishing, joining and working. Office: Am Standards Testing Bur Inc 40 Water St New York NY 10004-2605*

MORGAN, ALAN VIVIAN, geologist, educator; b. Barry, Glamorgan, Wales, Jan. 29, 1943; emigrated to Can., 1964, naturalized, 1977; s. George Vivian Williams and Sylvia Nesta (Atkinson) M.; m. Marion Anne Medhurst, June 14, 1966; children: Siân Kristina, Alexis John. B.Sc. with honors in Geology and Geography, U. Leicester, Eng., 1964; M.Sc. in Geography, U. Alta., Calgary, Can., 1966; Ph.D. in Geology, U. Birmingham, Eng., 1970. Postdoctoral fellow U. Western Ont. and U. Waterloo, Ont., Can., 1970-71; asst. prof. earth scis. and man-environ. studies U. Waterloo, 1971-78, assoc. prof. earth scis., 1978-85, prof., 1985—; assoc. dir. Quaternary Scis. Inst., U. Waterloo, 1992—; rep. Can. Geosci. Coun., 1977-83, exec. dir., 1988—; mem. Brit. Schs. Exploring Soc. Ctrl. Iceland Expn., 1960; coord. global change Geol. Survey Can., 1990-92; mem. com. on global change Royal Soc. Can., 1988—, com. on pub. awareness of sci., 1989—. Author 5 field guides; editor newsletter OYEZ, 1990—; contbr. articles to numerous profl. pubs.; dir., producer documentary film The Heimaey Eruption, 1974. Recipient award for M.Sc. thesis Can. Assn. Petroleum Geologists, 1967, Caroline Harrold Research award U. Birmingham, 1969, Disting. Tchr. award U. Waterloo, 1991; Charles Lapworth scholar, 1970; Nat. Scis. and Engring. Rsch. Coun. Can. grantee, 1971—. Fellow Geol. Assn. Can. (sec.-treas. 1975-83), Geol. Soc. Am.; mem. Am. Quaternary Assn. (pres. 1990-92), Can. Quaternary Assn. (pres. 1987-89), Brit. Quaternary Research Assn., Internat. Union Quaternary Research (sec. gen. XII congress 1983-87), Can. Entomol. Soc., Coleopterist's Soc., Sigma Xi. Office: U Waterloo, Dept Earth Scis, Waterloo, ON Canada N2L 3G1

MORGAN, AUDREY, architect; b. Neenah, Wis., Oct. 19, 1931; d. Andrew John Charles Hopfensperger and Melda Lily (Radtke) Anderson; m. Earl Adrian Morgan (div); children: Michael A., Susan Lynn Heiner, Nancy Lee, Diana Morgan Lucio. B.A., U. Wash., 1955. Registered architect, Wash., Oreg.; cert. NCARB. Project mgr. The Austin Co., Renton, Wash., 1972-75; med. facilities architect The NBBJ Group, Seattle, 1975-79; architect constrn. rev. unit Wash. State Health Divsn. Health, Olympia, 1979-81; project dir., med. planner John Graham & Co., Seattle, 1981-83; pvt. practice architecture, Ocean Shores, Wash., 1983—, also health care facility cons., code analyst. Contbg. author: Guidelines for Construction and Equipment of Hospitals and Medical Facilities; co-editor Design Considerations for Mental Health Facilities; co-editor: Design Considerations for Mental Health Facilities; contbr. articles to profl. jours. and govt. papers; prin. works include quality assurance coord. for design phase Madigan Army Med. Ctr., Ft. Lewis, Wash.; med. planner and code analyst Rockwood Clinic, Spokane, Wash., Comprehensive Health Care Clinic for Yakima Indian Nation, Toppenish, Wash.; code analyst S.W. Wash. Hosps., Vancouver; med. planner facilities for child, adult, juvenile and forensic psychiatric patients., States of Wash. and Oreg. Cons. on property mgmt. Totem council Girl Scouts U.S.A., Seattle, 1969-84, troop leader, cons., trainer, 1961-74; mem. Wash. State Bldg. Code Coun. Barier Free Com. Tech. adv. group for Ams. with Disabilities Act; assoc. mem. Wash State Fire Marshals Tech. Adv. Group. Mem. AIA (nat. acad. architecture for health 1980—, subcoms. codes and standards, health planning, chair mental health com., 1989-92, and numerous other coms., founding mem. Wash. council AIA architecture for health panel 1981—, recorder 1981-84, vice chmn., 1987, chmn. 1988, bd. dirs. S.W. Wash. chpt. 1983-84), Nat. Fire Protection Assn., Soc. Am. Value Engrs., Am. Hosp. Assn., Assn. Western Hosps., Wash. State Hosp. Assn., Wash. State Soc. Hosp. Engrs. (hon.), Seattle Womens Sailing Assn., Audubon Soc., Alpha Omicron Pi. Lutheran. Clubs: Coronado 25 Fleet 13 (Seattle) (past sec., bull. editor); GSA 25 Plus. Home and Office: PO Box 1990 Ocean Shores WA 98569-1990 also: 904 Falls of Clyde SE Ocean Shores WA 98569-1990

MORGAN, DANIEL CARL, civil engineer; b. Pitts., June 23, 1954; s. Walter and Gayle Lorraine (Vogel) Kohut; m. Patricia Sue Adair, Mar. 20, 1982; 1 child, Colin Adair. BSCE, U. Pitts., 1977. Engring. asst. John T. Boyd Co., Pitts., 1976; asst. project engr. Seymore S. Stein & Assoc. Inc., Pitts., 1977-78; quality control structural engr. Energy Cons., Inc. div. Schneider Enterprises, Inc., Pitts., 1978-81, sr. quality control structural engr., 1981-86; engr., designer Rollmec Inc., Pitts., 1986-88; staff engr. DLA div. Killam Assoc. Inc., Warrendale, Pa., 1988-91; sr. engr. UEC Environ. Systems, Inc., Pitts., 1991—. Home: 538 Oxford Blvd Pittsburgh PA 15243-1562

MORGAN, DONALD GEORGE, magnetic separation engineer; b. Milw., Nov. 7, 1931; s. Joseph George and Rose Terese (Podanovich) M.; m. Evelyn Marie Owsiany, June 13, 1953; children: Marilyn J., Daniel T., Thomas A., Andrew J. BSEE, Milw. Sch. Engring., 1958. Draftsman, then lab. technician Stearns Magnetic Products, Milw., 1947-58; prodn. engr., R&D engr. Ind. Gen. Corp., Milw., 1958-63, sales engr., then mgr. R&D, 1963-69, sales mgr., 1969-72; gen. mgr., div. v.p. Magnetics Internat., Stearns div., Milw., 1972-84; pres. Applied Magnetic Systems, Milw., 1984-91, O.S. Walker Co., Milw., 1991—. Contbr. articles on magnetic separation to sci. publs. Cpl. USMC, 1951-53. Mem. AIME, IEEE. Achievements include patent on high strength magnetic separators. Office: OS Walker Co PO Box 20911 Milwaukee WI 53220

MORGAN, DONNA JEAN, psychotherapist; b. Edgerton, Wis., Nov. 16, 1955; d. Donald Edward and Pearl Elizabeth (Robinson) Garey. BA, U. Wis., Whitewater, 1983, MS, 1985. Cert. psychotherapist, Wis., mental health alcohol and drug counselor; cert. alcohol and drug counselor; cert. counselor. Pvt. practice Janesville, Wis.; clin. supr. Stoughton (Wis.) Hosp.; prin. Morgan and Assocs., Janesville, Wis. Mem. underaged drinking violation alternative program Rock County, 1986—; co-chmn. task force on child sexual abuse, 1989—; mem. Rock County Multi-disciplinary Team on Child Abuse, 1990—; mem. speakers bur. Rock County C.A.R.E. House, 1990—. Mem. AACD, APA, Am. Profl. Soc. on the Abuse of Children, Rock County Mental Health Providers, Am. Assn. Mental Health Counselors, Wis. Assn. Mental Health Counselors, South Cen. Wis. Action Coalition, Am. Assn. Marriage and Family Therapy. Office: One Parker Pl Ste 625 Janesville WI 53545

MORGAN, EVAN, chemist; b. Spokane, Wash., Feb. 26, 1930; s. Evan and Emma Anne (Klobucher) M.; m. Johnnie Lu Dickson, Feb. 14, 1959; 1 child, James. BS, Gonzaga U., 1952; MS, U. Wash., 1954, PhD, 1956. Staff chemist IBM Corp., Poughkeepsie, N.Y., 1956-60; group supr. Olin Mathieson Co., New Haven, 1960-64; assoc. prof. chemistry High Point (N.C.) Coll., 1964-65; sr. research chemist Reynolds Metals Co., Richmond, Va., 1965-72; research assoc. Babcock & Wilcox, Lynchburg, Va., 1972—. Mem. Am. Chem. Soc., Cen. Va. Air Pollution Control Com. Home: 5128 Wedgewood Rd Lynchburg VA 24503-4208 Office: Babcock & Wilcox Corp PO Box 11165 Lynchburg VA 24506-1165

MORGAN, GARY PATRICK, energy engineer; b. Connellsville, Pa., Dec. 21, 1944; s. Frank John and Pauline Grace (Oglethorpe) M.; m. Linda Diane Carter, Dec. 17, 1967; 1 child, Tanya Lea Morgan. BSEE, Va. Poly. Inst. and State U., 1967; MS in Bus. Adminstrn., U. North Colo., 1978. Nuclear prodn. engr. USN Norfolk Naval Shipyard, Portsmouth, Va., 1972-74; dept. dir. of facilities Rocky Mountain Arsenal, Denver, 1974-76; tech. asst. PM EPA Region VIII, Denver, 1977-79; asst. to AM for space program Western Area Power Adminstrn., Loverland, Colo., 1980-84, converation program mgr., 1984-87, dir. energy svc., 1987-90; dir. sec. affairs/biomechs. Western Area Power Adminstrn., Loverland, 1991—. Co-author: Methane on the Move, 1978; producer (videotape) Palmers Pride, 1986, co-author/producer (videotape) Infared Thermography, 1990, Security Awareness. Co-chmn. Methane Gas Task Force, Denver, 1979; sponsor Weld County Spl. Olympics, Greeley, Colo., 1987. Mem. Nat. Guard Assn., Assn. of Engergy Engrs., Assn. of Indsl. Security. Home: 2332 S Dawson Way Aurora CO Office: Western Area Power Adminstn 1627 Cole Blvd Golden CO 80401

MORGAN, GEORGE EMIR, III, financial economics educator; b. Carmel, Calif., Jan. 2, 1953; s. George Emir Jr. and Dolores (Prydzial) M.; m. Donna Batts Vail, Dec. 31, 1977; 1 child, Abbie Vail. BS in Math., Georgetown U., 1973; MS in Stats., U. N.C., 1975, PhD in Fin., 1977. Sr. fin. economist Office of the Compt. of the Currency, Washington, 1978-79; asst. prof. U. Tex., Austin, 1979-84; assoc. prof. Va. Poly. Inst., Blacksburg, 1984-89, dir. PhD program, 1985-89, prof., 1989—; assoc. dir. bus. rsch. Ctr. Commercial Space Communications, 1990—; cons. Investment Co. Inst., Washington, 1982-83. Editor (newsletter) 90 Day Notes, 1986—. Mem. Am. Econ. Assn., Am. Fin. Assn., Am. Statis. Assn., So. Fin. Assn., Fin. Mgmt. Assn., Macintosh Blacksburg Users Group (faculty advisor 1989—), Beta Gamma Sigma. Avocations: soccer, squash, Macintosh personal computers. Office: Va Poly Inst Dept Fin 1016 Pamplin Hall Blacksburg VA 24061-0221

MORGAN, J. RONALD, aerospace company executive; b. Xenia, Ohio, July 9, 1952; s. James Pierce and Phylis Lou (Compton) M.; m. Cheryl D. Morgan, Jan. 24, 1981; children: James L., Christopher P. AA in Electronics, Sinclair C.C., 1986; BA in Computer Sci., Capital U., Columbus, Ohio, 1988. Engring. technician Universal Energy System, Wright Patterson AFB, Ohio, 1973-76; test technician Huffman Mfg. Co., Celina, Ohio, 1977-80; sales engr. Allied-Signal Airsupply, Dayton, Ohio, 1981-87, mem. corp. sales staff, 1987-90; regional mgr. Allied-Signal Airsupply, Torrance, Calif., 1990-92; dir. eastern region Allied-Signal Airsupply, Tempe, Ariz., 1992—. Office: Allied Signal Airsupply 2441 W Erie Dr Tempe AZ 85252

MORGAN, JACOB RICHARD, cardiologist; b. East St. Louis, Ill., Oct. 10, 1925; s. Clyde Adolphus and Jennie Ella Henrietta (Van Ramshorst) M.; m. Alta Eloise Ruthruff, Aug. 1, 1953; children: Elaine, Stephen Richard. BA in Physics, BBA, U. Tex., 1953; MD, U. Tex., Galveston, 1957. Diplomate Am. Bd. Internal Medicine, Am. Bd. Cardiology. Ensign USN, 1944, advanced through grades to capt., 1969; intern U.S. Naval Hosp., Oakland, Calif., 1957-58; chief medicine U.S. Naval Hosp., Taipei, Republic of China, 1962-64; internal medicine staff San Diego, 1964-67, chief cardiology, 1969-73; ret., 1973; dir. medicine R.E. Thomas Gen. Hosp., El Paso, Tex., 1973-75; asst. clin. prof. medicine U. Calif., San Diego, 1970-73; prof. medicine, assoc. chmn. dep. Tex. Tech U. Sch. Medicine, Lubbock and El Paso, 1973-75; pvt. practice National City, Calif., 1976—; dir. cardiology Paradise Valley Hosp., National City, 1976-88; presenter in field. Contbr. articles on cardiology to sci. jours. Recipient Casmir Funk award, 1972. Fellow ACP, Am. Coll. Cardiology, Am. Coll. Chest Physicians, Am. Heart Assn. (coun. on clin. cardiology). Avocation: golf. Home: 9881 Edgar Pl La Mesa CA 91941-6833 Office: 2409 E Plaza Blvd National City CA 91950-5101

MORGAN, JAMES C., electronics executive; b. 1938. BSME, Cornell U., MBA. With Textron Inc., 1963-72, West Ven Mgmt., San Francisco, 1972-76; chmn. bd., pres., CEO Applied Materials, Inc., Santa Clara, Calif., 1976-87, chmn. bd., CEO, 1987—. Office: Applied Materials Inc 3050 Bowers Ave Santa Clara CA 95054-3201

MORGAN, JEFF, research engineer; b. Salt Lake City, Sept. 3, 1954; s. David Nyle and Dene Huber (Olsen) M.; m. Linda Mae Marquez, May 28, 1982 (div.). BS, U. Calif., San Diego, 1976; MS, U. Hawaii, 1978, PhD, 1982. Rsch. assoc. U. Hawaii, Honolulu, 1982-85; sr. rsch. assoc. Stanford U., Palo Alto, Calif., 1985-91; rsch. engr. U. Wash., Seattle, 1991—. Mem. Am. Astron. Soc. Office: U Wash Dept Astronomy FM-20 Seattle WA 98195

MORGAN, JOHN DAVIS, physicist; b. Washington, Apr. 10, 1955; s. John Davis Jr. and Leta (Bretzinger) M. BS, George Washington U., 1974; MSc, Oxford (Eng.) U., 1978; PhD, U. Calif., Berkeley, 1978. Lectr., postdoctoral fellow dept. physics Princeton (N.J.) U., 1978-81; asst. prof. U. Del., Newark, 1981-86, assoc. prof., 1986—; vis. scholar dept. chemistry Harvard U., Cambridge, Mass., 1988-90, vis. assoc. prof. dept. physics, 1990, assoc. dept of classics, 1993; vis. scientist Harvard-Smithsonian Ctr. for Astrophysics, Cambridge, 1989-90, 92, Inst. for Nuclear Theory U. Wash., 1993; vis. mem. sch. hist. studies Inst. for Advanced Study, 1994. Contbr. articles to Phys. Rev. A., Phys. Rev. Letters, Jour. Chem. Physics, Theoretica Chimica Acta. Marshall scholar Brit. Govt., 1974-76; grantee Rsch. Corp., 1983-84, NSF, 1984—, Precision Measurement award grantee Nat. Inst. Standards and Tech., 1987-91; fellow NEH, 1993. Mem. Am. Phys. Soc. (counsellor topical group on few-body systems 1989-92), Am. Philological Assn. Achievements include research on high-precision calculations of helium atom energy levels, elucidation of convergence rates of Rayleigh-Ritz variational calculations, performance of large-order Rayleigh-Schroedinger perturbation theory, application of mathematical physics to problems in atomic and molecular structure, ancient astronomy and chronology. Office: U Del Dept Physics and Astronomy Newark DE 19716

MORGAN, KENNETH, civil engineering educator, researcher; b. Llanelli, Wales, U.K., June 9, 1945; s. Idris and Megan Elizabeth (Richards) M.; m. Elizabeth Margaret Harrison, Apr. 7, 1969; children: Gareth, David Kenneth. BSc. in Math., U. Bristol, Eng., 1966, PhD in Math., 1970, DSc, 1987. Chartered engr., chartered mathematician, sci. officer U.K. A.E.A., Aldermaston, Eng., 1969-72; lectr. U. Exeter, Eng., 1972-75; from lectr. to prof. U. Wales, Swansea, U.K., 1975-89, 1991—, head dept. civil engring., 1991—; Zaharoff prof. of aviation Imperial Coll., London, 1989-91; vis. rsch. prof. U. Va., 1989—. Co-author: Finite Elements and Approximation, 1983; contbr. articles to profl. jours. Recipient Group Achievement award, Langley Rsch. Ctr., 1989. Fellow Inst. Civil Engrs., Inst. of Math. and Its Applications; mem. AIAA. Home: 137 Pennard Dr, Pennard, Swansea Wales SA3 2DW Office: Univ Wales, Dept Civil Engring, Swansea Wales SA2 8PP

MORGAN, LINDA CLAIRE, industrial engineer; b. Plainfield, N.J., Mar. 7, 1958; d. Luther William and Dolores Odessa (Minor) Hammond; m. Bruce A. Morgan, July 3, 1982; children: William, Eric. BS in Indsl. Engring., N.J. Inst. Tech., 1982. Assoc. indsl. engr. Ortho Diagnostics, Inc., Raritan, N.J., 1982-84; indsl. engr. Johnson & Johnson Internat., New Brunswick, N.J., 1984-85, assoc. project coord., 1985-89; sect. mgr. corp. indsl. engrin. Ethicon, Inc., New Brunswick, N.J., 1989-91; mgr. corp. indsl. engring. Johnson & Johnson Internat., New Brunswick, N.J., 1991-93; staff mfg. mgr. Ethicon, Inc., Somerville, N.J., 1992—. Editor Raritan Valley Newsletter, 1989-91. Recipient N.J. Black Achievers award, 1988, Tribute to Women and Industry award, 1993. Mem. NAFE, Am. Inst. Indsl. Engrs.

MORGAN, MICHAEL JOSEPH, advanced information technology executive; b. Canton, Ohio, Nov. 28, 1953; s. Edward A. and Mary Arlene (Maier) M.; m. Deborah Mae Beach, Mar. 2, 1979 (div. Sept. 1986). BSBA, Old Dominion U., 1982, MBA, 1989. Lic. journeyman, aircraft mechanic. Aircraft mechanic Naval Air Rework Facility, Norfolk, Va., 1976-82; resource analyst Newport News (Va.) Shipbuilding, 1982-84; sr. planner Bendix Electronic Controls Div., Newport News, 1984-85; program mgr. Superior Engring., Norfolk, 1985-88; dept. mgr. CACI, Virginia Beach, Va., 1988—. With USN, 1971-75. Mem. Am. Prodn. and Inventory Control Soc. Republican. Roman Catholic. Avocations: diving, swimming, jogging, carpentry. Home: 512 Concord Dr Hampton VA 23666-2205 Office: CACI 1300 Diamond Springs Rd Ste 300 Virginia Beach VA 23465-0001

MORGAN, ROBERT STEVE, mechanical engineer; b. Oklahoma City, Oct. 10, 1945; s. Chester Steve and Madelein Ruth (Stowers) M.; m. Margaret Ann Groves, June 7, 1971; children: Jerri Dianna, Jamie Deann. Diploma, S.W. Tech. Inst., 1967. Chief draftsman R.L. Gilstrap Inc., Oklahoma City, 1966-67; mgr. print dept. Phelps-Spitz-Ammerman-Thomas Inc., Oklahoma City, 1967-68; sr. drafter GE, Oklahoma City, 1968-70; mech. designer

Honeywell Inc., Oklahoma City, 1970-75; pvt. practice Oklahoma City, 1970—; sr. designer Control Data Corp., Oklahoma City, 1970-75; sr. designer, drafting coord., cad adminstr. BTI Systems Inc., Oklahoma City, 1984—. Patentee overlap document detector, double document detector, ribbon cartridge. Voter organizer Rep. Party, Oklahoma City, 1967; asst. leader Girl Scouts of Am., Yukon, Okla., 1984. Mem. Confederate Air Force (col.). Republican. Avocations: gun collecting, off road auto touring, coin collecting, bowling. Home: 704 Victoria Dr Yukon OK 73099-5341 Office: BTI Systems Inc PO Box 83439 Oklahoma City OK 73148-1439

MORGAN, ROGER JOHN, research scientist; b. Manchester, England, Nov. 2, 1942; came to U.S., 1968; s. Leslie Budworth and Hilda May (Bevins) M.; m. Anne Christine Cheetham, Sept. 23, 1967; children: Jacqueline, Nicholas, Melissa. BS in Chemistry with honors, U. of London, 1965; PhD in Polymer Phyics, U. of Manchester, 1968. Asst. rsch. prof. Washington U., St. Louis, 1968-72; scientist McDonnell Douglas Rsch. Labs., St. Louis, 1972-78; group leader Lawrence Livermore (Calif.) Nat. Lab., 1978-85; mem. tech. staff Rockwell Internat., Thousand Oaks, Calif., 1985-86; head of composites Mich. Molecular Inst., Midland, 1986—. Co-editor Advanced Composites Bull., 1989—; mem. editorial adv. bd. Jour. Composite Materials, 1985—, SAMPE Quar., 1991—; contbr. over 110 articles on composites and polymer sci. Mem. Am. Chem. Soc., Soc. for Advancement of Materials and Process Engrs. Achievements include research in composites/polymer science. Office: Mich Molecular Inst 2203 Eastman Ave Midland MI 48640

MORGAN, ROSE MARIE, biology educator, researcher; b. Minot, N.D., Nov. 17, 1935; d. Clinton Edward and Clara Adlyn (Fedje) M.; m. J.L. Parsons, Sept. 23, 1967 (div. 1970). BS, Minot (N.D.) State U., 1963; MS, N.D. State U., 1968; PhD, Tex. Woman's U., 1981. Rsch. microbiologist N.D. State U., Fargo, 1965-75; teaching asst. Tex. Woman's U., Denton, 1977-81; assoc. prof. Minot State U., 1983—; bench med. technologist Trinity Med. Ctr., 1960-65, adj. prof. clin. lab. sci., 1989-90; extension lectr. in anatomy, physiology, microbiology and biochemistry Minot State U., 1981-83; adj. prof. clin. lab. sci. St. Joseph's Hosp., Minot, 1989-90. Contbr. articles to Jour. Environ. Sci. Health, Tex. Jour. Sci., Am. Jour. Med. Tech., Jour. Am. Sci., Pollution; textbook reviewer Times Mirror Mosby, 1986, West Pub. Co., 1990, Prentice-Hall, 1993, Wm. C. Brown, 1993, Macmillan, 1993. Recipient numerous grants. Mem. AAUW (internat. fellowships panel 1993—), NEA, Am. Soc. Clin. Pathologists (assoc.), Am. Soc. Med. Technologists, N.W. Dakota Sci. Tchrs. Assn., N.D. Acad. Sci. (chairperson Dennison com. 1988-89), N.D. Higher Edn. Assn., N.D. Edn. Assn., N.Y. Acad. Sci., Minot State U. Edn. Assn., Sigma Xi, Phi Delta Kappa, Phi Delta Gamma (Alpha Theta chpt.), Delta Kappa Gamma. Lutheran. Achievements include research in co-insult effects on physiological mechanisms, physiological changes in vital organ systems following interactions of cadmium and gamma radiation; pioneer in effects on interactions of environmental pollutants; comparison and report of differential effects of three different radiations on seed germination and cadmium lethalities. Home: 823 6th St SW Minot ND 58701-4581 Office: Minot State U 8500 University Ave Minot ND

MORGAN, STEVEN GAINES, marine ecology researcher, educator; b. Portsmouth, Va., Sept. 2, 1952; s. Walter Nathaniel and Lois (Gaines) M. BS in Biology, Randolph-Macon Coll., 1974; MS in Biol. Oceanography, Old Dominion U., 1977; PhD in Zoology, U. Md., 1986. Postdoctoral fellow Smithsonian Tropical Rsch. Inst., Balboa, Panama, 1987, Marine Environ. Scis. Consortium, Dauphin Island, Ala., 1988-90; rsch. assoc. U. South Ala., Mobile, 1989-90; adj. prof. dept. ecology and evolution SUNY, Stony Brook, 1990—, asst. prof. Marine Scis. Rsch. Ctr., 1990—. Contbr. articles to profl. publs. Grantee NSF, 1989, NOAA, 1989, N.Y. Sea Grant, 1991, Hudson River Found., 1991. Mem. AAAS, Ecol. Soc. Am., Crustacean Soc., Omicron Delta Kappa. Achievements include research in predation by planktivorous fishes as a dominant selective force shaping life histories of brachyuran crabs, including the timing of larval release, dispersal patterns, larval morphologies and pigmentations. Home: PO Box 819 Miller Place NY 11764 Office: Marine Scis Rsch Ctr State Univ NY Stony Brook NY 11774-5000

MORGAN, WALTER, retired poultry science educator; b. Ledyard, Conn., Dec. 22, 1921; s. Walter Clifford and Margaret (Allyn) M.; m. Marcella Hodge, Dec. 28, 1948 (div. 1960); children: Nancy, Peggy, Beth; m. Helen Naden, May 14, 1966. BSc, U. Conn., 1946; MSc, George Washington U., 1949, PhD, 1953. Animal husbandman Nat. Cancer Inst., NIH, Bethesda, Md., 1946-49; rsch. assoc. Nevis Rsch. Sta., Columbia U., Irvington-on-Hudson, N.Y., 1950-53; asst. prof. U. Tenn., Knoxville, 1953-54; assoc. prof., prof. S.D. State U., Brookings, 1954-85; ret., 1985; fgn. expert on English and genetics People's Republic China, 1991; researcher biology div. Nuclear Energy Ctr., Mol, Belgium, 1968-69, genetics div. Commonwealth Sci. and Indsl. Rsch. Orgn., Sydney, Australia, 1975-76; cons., Kuala Lumpur, Malaysia, 1988. Author: (poetry) Now and Then, 1982, Down Under, 1983, Hitchin' Around, 1985, Here and There, 1990, What's Good About China, 1992; contbr. over 100 articles to profl. jours. Former scoutmaster Boy Scouts Am., Brookings; pres. Men's Brotherhood, Brookings, 1974, U.S. Friends Fgn. Students, Brookings, 1986-88. Sgt. USAAF, 1942-45, ETO. Fellow AAAS; mem. Am. Genetics Assn., World Poultry Assn., N.Y. Acad. Scis., S.D. Acad. Sci. (pres.). Home: 1610 1st St Brookings SD 57006-2617

MORGENSTERN, BRIAN D., civil engineer. With Buckland & Taylor Ltd., North Vancover, B.C., Can. Recipient Le Prix P.L. Pratley award Can. Soc. Civil Engring., 1991. Office: Buckland & Taylor Ltd, 1591 Bowser Ave, North Vancouver, BC Canada V7P 2Y4*

MORGENSTERN, NORBERT RUBIN, civil engineering educator; b. Toronto, Ont., Can., May 25, 1935; s. Joel and Bella (Skornik) M.; m. Patricia Elizabeth Gooderham, Dec. 28, 1960; children: Sarah Alexandra, Katherine Victoria, David Michael Gooderham. BASc, U. Toronto, 1956, DEng h.c., 1983; DIC, Imperial Coll. Sci., 1964; PhD, U. London, 1964; DSc h.c., Queen's U., 1989. Research asst., lectr. civil engring. Imperial Coll. Sci. and Tech., London, 1958-68; prof. civil engring. U. Alta., Edmonton, Can., 1968-83, Univ. prof., 1983—; cons. engr., 1961—. Contbr. articles to profl. jours. Bd. dirs. Young Naturalists Found., 1977-82, Edmonton Symphony Soc., 1978-85. Athlone fellow, 1956; recipient prize Brit. Geotech. Soc., 1961, 66, Huber prize ASCE, 1971, Legget award Can. Geotech. Soc., 1979, Alta. order of Excellence, 1991. Fellow Royal soc. Can., Can. Acad. Engring.; mem. U.S. Nat. Acad. Engring. (fgn. assoc.), Cancian Geosci. Coun. (pres. 1983), Can. Geotechnical Soc. (pres. 1989-91), Internat. Soc. for Soil Mechanics and Found. Engring. (pres. 1989—), Royal Glenora Club, Athenaeum (London), various other profl. assns. Home: 106 Laurier Dr, Edmonton, AB Canada T5R 5P6 Office: U Alta, Edmonton, AB Canada T6G 2G7

MORGENTALER, ABRAHAM, urologist, researcher; b. Montreal, Quebec, Can., May 14, 1956; came to U.S., 1974; s. Henry Morgentaler and Chawa Rosenfarb; m. Susan Deborah Edbril, June 12, 1982; children: Maya Edbril, Hannah Edbril. AB, Harvard U., 1978, MD, 1982. Diplomate Am. Bd. Urology. Intern Harvard Surg. Svc.-N.E. Deaconess Hosp., Boston, 1982-83; resident Harvard Program in Urology, Boston, 1984-88; instr. surgery Harvard Med. Sch., Boston, 1988-92, asst. prof. surgery (urology), 1993—; staff urologist, dir. male infertility program and impotency Beth Israel Hosp., Boston, 1988—; dir. andrology lab. Beth Israel Hosp., 1990—. Mem. AMA, Am. Urologic Assn., Am. Fertility Soc., Am. Soc. Andrology, Boston Fertility Soc., Am. Assn. Clin. Urologists. Achievements include detection of protein abnormalities in infertile sperm, detection of temperature dependent protein expression in rat testis, use of investigational stents for treatment of benign prostatic hypertrophy. Office: Beth Israel Hosp 330 Brookline Ave Boston MA 02215-5491

MORGENTHALER, ANN WELKE, electrical engineer; b. Boston, Sept. 5, 1962; d. Frederic Richard and Barbara (Pullen) M.; m. Carey Milford Rappaport, Nov. 12, 1989. AB in Physics, Harvard Coll., 1984; MSEE, MIT, 1988. Teaching asst. MIT, Cambridge, 1986-87, rsch. asst., 1988—. Author: (with Staelin, Kong) Electromagnetic Waves, 1993. NSF fellow, 1984-87, Unisys Corp. grad. fellow, 1988-91. Mem. IEEE, Sigma Xi. Home: 53 E Quinobequin Rd Newton MA 02168 Office: MIT 38-280 77 Massachusetts Ave Cambridge MA 02139

MORGENTHALER, JOHN HERBERT, chemical engineer; b. Cleve., Jan. 5, 1929; s. Frederick Herman and Anna Margarethe (Welke) M.; m. Kathleen Ann Merriman, June 23, 1956 (dec. Oct. 1986); children: John David, Jennifer Ann, Jeffrey Paul; m. Susan Kay Braaten, Dec. 27, 1988. SB, MIT, 1951, SM, 1952; PhD, U. Md., 1965. Group leader Procter & Gamble Co., Cin., 1954-58; project mgr. Atlantic Rsch. Corp., Alexandria, Va., 1958-62; sr. staff engr. Applied Physics Lab. Johns Hopkins U., Silver Spring, Md., 1962-65; project scientist Marquardt's Gen. Applied Sci. Labs., Westbury, N.Y., 1965-67; rsch. dir. Textron's Bell Aerospace Co., Buffalo, 1967-74; sect. mgr. Stauffer Chem. Co., Richmond, Calif., 1974-77; mgr. comml. ventures Bechtel Corp., San Francisco, 1977-78; pres. JHM Assocs., Tacoma, Wash., 1978—; cons. Moore Rsch. Labs., Inc., Bethesda, Md., 1959-65; mem. adv. bd. U. Tenn. Space Inst., Tullahoma, 1967-68, Assn. Bay Area Govts., Oakland, Calif., 1976-77; chmn. membership com. Nat. Capitol sect. Am. Inst. Chem. Engrs., 1959-61; treas. Buffalo sect. AIAA, 1974. Contbr. articles to Internat. Jour. Heat and Mass Transfer, Jour. Fluids Engring., Jour. Spacecraft and Rockets. Chmn. Joe Berg Sci. Soc., Niagara Falls, N.Y., 1967-71, com. chair Lewiston (N.Y.) Cub Scouts., 1970-71, Walnut Creek (Calif.) Boy Scouts, 1980-82; v.p. Homeowners Assn., Walnut Creek, 1983-85. 1st lt. chem. corps U.S. Army, 1952-54. Scholar Westinghouse Corp., 1947, MIT, 1947-51. Mem. AAAS, Elks, Sigma Xi, Kappa Kappa Sigma (hon.). Republican. Unitarian. Achievements include patents in Process for the Regeneration of Spent Sulfuric Acid using controlled combustion, Process for Making a Bleach Composition using fluidized bed technology. Home and Office: 46 Bonney St Steilacoom WA 98388-1502

MORHARDT, JOSEF EMIL, IV, environmentalist, engineering company executive; b. Bishop, Calif., Aug. 19, 1942; s. J. Emil III and G.H. Morhardt; m. Sylvia Staehle, Jan. 23, 1965; 1 child, Melissa Camille. BA, Pomona Coll., 1964; PhD, Rice U., 1968. Asst. prof. Wash. U., St. Louis, 1967-75; chief scientist Henningson, Durham & Richardson, Santa Barbara, Calif., 1975-78; sr. v.p., dir. western div. EA Engring. Sci. & Tech., Inc., Lafayette, Calif., 1978—. Home: 4520 Canyon Rd Lafayette CA 94549-2709 Office: EA Engring Sci & Tech Inc 11019 McCormick Rd Cockeysville Hunt Valley MD 21031

MORHARDT, SIA S., environmental scientist. BA in Botany, Pomona Coll., 1965; MA in Biology, Rice U., 1968; PhD in Biology, Washington U., 1971. Dir. environ. assessment and mgmt., v.p. EA Engring. Sci. & Tech. Inc., Cockeysville Hunt Valley, Md.; panel moderator Western Social Sci. Assn., 1991, Am. River Mgmt. Soc., 1992; panelist Nat. Hydropower Assn., 1991. Planning commr. City of Lafayette, Calif., 1983-89, chair. Achievements include studies in developement of water resources in the central and western states. Office: EA Engring Sci & Tech Inc 11019 McCormick Rd Cockeysville Hunt Valley MD 21031

MORI, SHIGEJUMI, mathematician educator. Prof. Rsch. Inst. Math. Scis., Kyoto U., Kyoto, Japan. Recipient Frank Nelson Cole Algebra prize Am. Math. Soc., 1992. Office: Kyoto Univ, Res Inst For Math Sciences, Kyoto 606, Japan*

MORI, SHIGEYA, economist, educator; b. Kamigori-cho Ako-gun, Japan, Feb. 15, 1926; s. Ryusaku and Hana Mori; m. Takako Tsuge, Mar. 20, 1954; children: Hiroshige, Keiko, Junko, Kyoko, Tsuneo, Yuko. BA, Kobe U. Econs., 1952, D in Econs., 1984; PhD, Cath. U. Am., 1961. With Bank of Kobe, Ltd., Japan, 1945-48; rsch. asst. dept. econs. Kobe U., 1952-54; rsch. asst. Nanzan U., Nagoya, Japan, 1954-55, instr., 1955-60, asst. prof., 1960-68, prof., 1968—, dean, 1970-71, dean faculty econs., 1975-84, v.p., 1985-91. Author: The Theory of Supply and Demand in the English Classical School, 1961, Introduction to Economics, 1970, History of price Theory, 1982, The Theory of Economic Growth in the English Classical School, 1992. Roman Catholic. Avocations: travel, movies. Home: 92 Yamazato-cho Showa-ku, Nagoya 466, Japan Office: Nanzan U, 18 Yamazato-cho Showa-ku, Nagoya 466, Japan

MORIER, DEAN MICHAEL, psychology educator; b. Detroit, Jan. 8, 1960; s. Bruce John and Dorothy M. (Brausseur) M.; m. Susan Marie Bergmann, Aug. 17, 1986; children: Evan Philippe, Corinne Elise. BA in Psychology, Hope Coll., 1982; PhD in Psychology, U. Minn., 1987. Vis. asst. prof. Union Coll., Schenectady, N.Y., 1987-89; asst. prof. Mills Coll., Oakland, Calif., 1989—. Contbr. articles to profl. jours. Mem. APA, Am. Psychology and Law Soc., Soc. for Personality and Social Psychology, We. Psychol. Assn. Achievements include research on conjunction fallacy; jury research in capital murder trials; friendship and personality studies. Home: 13 Faculty Rd Oakland CA 94613 Office: Mills Coll 5000 MacArthur Blvd Oakland CA 94613

MORIKIS, DIMITRIOS, physicist; b. Athens, Attiki, Greece, June 6, 1960; s. Vasilios and Ekaterini (Vastaki) M.; m. Maria Gloria González-Rivera, July 27, 1986; 1 child, Vasilios Aris. BS, Aristotle U. of Thessaloniki, Thessaloniki, Greece, 1983; MS, Northeastern U., 1985, PhD, 1990. Teaching asst. Northeastern U., Boston, 1983-85, rsch. asst., 1985-90; postdoctoral fellow The Scripps Rsch. Inst., La Jolla, Calif., 1990—; referee for publs. in field. Contbr. articles to profl. jours. Named IAESTE Student trainee, 1981, Fulbright Exch. Student, 1983-84. Mem. Am. Phys. Soc. Greek Orthodox. Achievements include original basic rsch. in fields of biophysics and structural biology studying HEME proteins with a variety of spectroscopic techniques. Office: Scripps Rsch Inst MB2 10666 N Torrey Pines Rd La Jolla CA 92037

MORIMOTO, RODERICK BLAINE, research engineer; b. Fresno, Calif., Dec. 27, 1961; s. Hiromi and Martha Miyo (Toyama) M. BS in Gen. Engring., Harvey Mudd Coll., 1984. Cert. engr. Rsch. engr. SRI Internat., Menlo Park, Calif., 1984—. Mem. IEEE. Office: SRI Internat 333 Ravenswood Ave Menlo Park CA 94025

MORIN-POSTEL, CHRISTINE, international operations executive; b. Paris, Oct. 6, 1946. Inst. d'Etudes Polit., Paris, 1965-68, Inst. de Controle, 1972-74. Mgr. econs. C. of C. and Industry of Normandie, Rouen, 1968-73; mgr. small bus. dept. Agence Nat. de Valorisation de la Recherche, Paris, 1973-76; assoc. SOFINNOVA-SOFININDEX, Paris, 1976-79; sr. v.p. for corp. devel. and internat. operations Lyonnaise des Eaux Dumez, Paris, 1979-91, with, 1979-82, v.p. corp. devel., 1982-84, sr. v.p. corp. devel. and internat. opers., 1984-90, exec. v.p. corp. devel. and internat. opers., 1990—; bd. dirs. Ufiner, Degremont S.A., Sita, Sofinco, Shimizu France, Aqua Chem Inc., U.S., Trigen Energy Corp., U.S., Gen. Waterworks Corp., U.S., Dic Degremont, Japan, Aguas De Barcelona, Spain, Plantsbrook, U.K., and others. Mem. internat. adv. bd. Cranfield U. Decorated Chevalier de l'Ordre Nat. du MErite. Mem. Polo de Paris, Reform Club. Office: Lyonnaise des Eaux Dumez, 72 avenue de la liberte, Nanterre 92000, France

MORISHIGE, FUKUMI, surgeon; b. Fukuoka, Japan, Oct. 24, 1925; s. Fukumatsu and Teruko M.; m. Fumie Osada, Apr. 18, 1954; children: Kyoko, Hisakazu, Noritsugu. MD, Kurume U., 1952, DMS, 1962; PhD, Fukuoka U., 1983. Intern Kurume U., 1951-52; asst. Kurume (Japan) U., Dept. Pathology, 1952-55, Kyoto (Japan) U. Inst. Chest Disease, 1955-58; v.p. Tachiarai Hosp., Fukuoka, Japan, 1959-67, Torikai Hosp., Fukuoka, Japan, 1968-80; chmn. Tachiarai Hosp., Fukuoka, Japan, 1980-84; dir. Nakamura Hosp., Fukuoka, Japan, 1984-86, supreme advisor, 1987—; dir. Morishige Cancer Clinic, Chiba, Japan, 1992—; resident fellow Linus Pauling Inst. of Sci. and Medicine, Palo Alto, Calif., 1976—; chemistry advisor Nissan Chem. Industries Ltd., Tokyo, 1983—. Author: Nutrition of Nucleic Acid, 1983, Brain Blood Circulation, 1986; contbr. articles to profl. jours. Fellow Linus Pauling Inst. Sci. and Medicine; mem. Japan Soc. Magnetic Resonance (founder, bd. dirs. 1978—, exec. sec. 1979—), Internat. Assn. for Vitamin & Nutritional Oncology (exec. com. 1983—), Japanese Cancer Assn., Japanese Assn. for Thoracic Surgery, Japan Surg. Soc. Democrat. Buddhist. Home and Office: 190-1 Komagome, Ooami-Shirasato-Machi, Sambu-Gun Chiba 299-32, Japan Office: Sta Pla Hotel 1401, 2-1-1 Hakata-Eki-Mae, Hakata-ku Fukuoka, Japan

MORISHIMA, AKIRA, physician, director, educator, consultant; b. Tokyo, Apr. 18, 1930; came to U.S., 1955; s. Azusa and Toshiko (Tezuka) M.; m. Hisayo Oda, June 3, 1961; children: Amy, Alyssa. MD, Keio U., Tokyo, 1954, PhD, 1963. Postdoctoral fellow Columbia U., N.Y.C., 1958-61, instr. in pediatrics, 1961-63, assoc. in pediatrics, 1963-65, asst. prof. pediatrics, 1965-66; asst. prof. pediatrics U. Calif., San Francisco, 1966-68; assoc. prof. pediatrics Columbia U., N.Y.C., 1968—, dir. div. pediatric endocrinology, 1969—; assoc. attending physician Presbyn. Md. Ctr., N.Y.C., 1968—, Englewood (N.J.) Hosp., 1991—; cons. Overlook Hosp., Summit, N.J., 1978—; hon. cons. St. Luke's Roosevelt Hosp. Ctr., N.Y.C., 1988—. Contbr. over 55 articles to profl. jours. Mem. Health Planning Bd., Bronx, 1974-76; v.p., treas. Dist. 10 Community Sch. Bd., N.Y.C., 1974-80; chmn. com. Community Bd. # 8, N.Y.C., 1977-85. Fulbright scholar U.S. Edn. Commn., 1955-57; grantee NIH; recipient Citation, City Coun. of N.Y., 1985. Mem. AAAS, AAUP, Am. Pediatrics Soc., Acad. Pediatrics (Japan), Environ. Mutagen Soc., Soc. for Study of Reprodn., Soc. Human Genetics, Endocrine Soc., Soc. for Pediatric Rsch., N.Y. Japanese/Am. Lions. Republican. Presbyterian. Office: Columbia U Coll Physicians & Surgeons 622 W 168th St New York NY 10032-3702

MORISHIMA, ISAO, biochemistry educator; b. Kyoto, Japan, July 11, 1942; s. Masao and Miyako Morishima; m. Chieko Kaneda, July 9, 1972; children: Masaki, Tomoki. BS, Kyoto U., 1966, MS, 1968, PhD, 1971. Rsch. assoc. faculty of agr. Tottori (Japan) U., 1971-75; assoc. prof., 1976-89; prof. Tottori U., 1989—; vis. rsch. assoc. Med. Sch. U. Tex., Houston, 1975-76. Mem. Internat. Soc. Devel. and Comparative Immunology, Japanese Soc. Biosci., Biotech., and Agrochemistry. Avocations: computer programming, wood crafts. Home: 176 Mihagino 2-chome, Tottori 689-02, Japan Office: Tottori U, Faculty of Agriculture, Tottori 680, Japan

MORISHITA, ETSUO, aeronautical, mechanical engineering educator; b. Ueno, Japan, Aug. 26, 1949; s. Katsuaki and Tazuko (Aoki) M.; m. Midori Shigematsu, Apr. 16, 1977; children: Masao, Yuko, Saiko. BS, U. Tokyo, 1972, ME, 1974, D of Engring., 1985; MSc, U. Cambridge, U.K., 1983. Rsch. engr. Mitsubishi Electric Corp., Amagasaki, Japan, 1974-87; assoc. prof. U. Tokyo, 1987—. Co-author: Applied Aerodynamics (in Japanese), 1991; contbr. articles to profl. jours. Mem. AIAA, Japan Soc. Aero. and Space Scis., Japan Soc. Mech. Engrs. Buddhist. Achievements include developing scroll compressor, co-rotating scroll vacuum pump. Home: 3-5-9 Minamigaoka, Ryugasaki Ibaraki 301, Japan Office: Univ Toyko Aero & Astro Dept, 7-3-1 Hongo Bunkyo-ku, Tokyo 113, Japan

MORISHITA, TERESA YUKIKO, veterinarian, consultant, researcher; b. Honolulu; d. Yasuyuki and Doris M. MS, U. Hawaii; DVM, MPVM, U. Calif., Davis. Diplomate Am. Coll. Poultry Veterinarians. Avian veterinarian U. Calif., Davis, 1989-91; pvt. practice avian veterinarian Davis, 1991—; asst. dir. Raptor Ctr., Davis, 1991—. Author poems; contbr. articles to periodicals and profl. jours. Mem. Spl. Olympics Com., Honolulu, 1978-85. Recipient award for rsch. excellence in clin. avian medicine Sch. Vet. Medicine U. Calif., Davis; grantee Pacific Egg and Poultry Assn., 1990-92, Calif. Turkey Industry Assn., 1990-91, Nat. Wildlife Rehab. Assn., 1992-93; Pacific Egg and Poultry Assn. scholar, 1985-93. Mem. AVMA (environ. affairs com.), Am. Assn. Avian Pathologists (enteric com.), Assn. Avian Veterinarians, Am. Assn. Zoo Veterinarians, Calif. Vet. Med. Assn. Roman Catholic.

MORITA, KAZUTOSHI, psychology educator, consultant; b. Tokyo, Nov. 25, 1937; s. Kyoichi and Tomiko Morita; m. Yoshie Nakahashi, Oct. 10, 1968; children: Toshio, Hideyo. BA in Psychology, U. Tokyo, 1961. Lectr. Sanno Jr. Coll., Tokyo, 1968-71, asst. prof., 1972-79, mgr. Inst. Orgnl. Behavior, 1973-77, asst. v.p., 1978-79; asst. prof. indsl. and orgnl. psychology and behavioral sci. Sangyo Noritsu U., Isehara, Japan, 1980-83; prof. Sanno U. Sch. Mgmt. and Informatics, Kanagawa, Japan, 1984—, mgr. adminstrv. dept., 1988—; vis. scholar U. Mich., Ann Arbor, 1971, Harvard U., Cambridge, Mass., 1972; com. mem. human resource devel. project Ministry Internat. Trade and Industry, 1973-76; com. mem. long range plan for edn. Ministry Edn., 1975-77; vice chmn. Conf. Commerce Activities Adjustment, Fujisawa, Japan, 1988—. Author, editor: Cases of Organizational Development in Japan, 1978, Formats for Human Resource Management, 1990; author: Behavioral Sciences in Business, 1984. Recipient Ueno Godo prize 18th All Japan Mgmt. Conf., 1966. Mem. Japanese Assn. Indsl. and Orgnl. Psychology (editorial com. 1986—). Home: 14-3 Matsukazedai, Chigasaki Kanagawa 253, Japan Office: Sanno U, 1573 Kamikasuya, Isehara Kanagawa 259-11, Japan

MORIZUMI, SHIGENORI JAMES, applied mathematician; b. San Francisco, Nov. 13, 1923; s. Mohei and Hatsue (Kawaharada) M.; m. Hiroko Kimura, Nov. 20, 1956; 1 child, Miachel N. BS, U. Calif., Berkeley, 1955; MS, Calif. Inst. Tech., 1957; PhD, UCLA, 1970. Sr. aerodynamicist Douglas Aircraft, Santa Monica, Calif., 1955-60; sr. staff engr. TRW Space Tech. Lab., Redondo Beach, Calif., 1960-81; dir. HR-Textron, Irvine, Calif., 1981-82; sr. scientist Hughes Aircraft, El Segundo, Calif., 1982-89. Author: An Investigation of Infrared Radiation by Vibration-Rotation Bands of Molecular Gases, 1970; speaker, presenter in field; contbr. articles to AIAA Jour., Jour. of Quantitative Spectroscopy and Radiative Transfer, Jour. of Spacecraft and Rockets. Mem. Sigma Xi. Achievements include rsch. in rocket exhaust gas plume heat modeling, pitch-yaw couple trajectory radio guidance equations, application of thermo electric converter to A/C engine for power generation, simulation of molecular gas radiation by wide band absorp. model. Home: 29339 Stadia Hill Ln Rancho Palos Verdes CA 90274

MORKOC, HADIS, electrical engineer, educator; b. Senkaya, Erzurum, Turkey, Oct. 2, 1947; came to U.S., 1971; s. Mustafa and Saadet (Metin) M.; m. Amy C. Ahlberg, Sept. 5, 1975; 1 child, Erol Taner. MS, Tech. U., Istanbul, 1969; PhD, Cornell U., 1975. Postdoctoral fellow Cornell U., Ithaca, N.Y., 1975-76, mem. tech. staff Varian Assocs., Palo Alto, Calif., 1976-78; prof. elec. engring. U. Ill., Urbana, 1978—; disting. vis. scientist Calif. Inst. Tech., Pasadena, 1987-88; bd. advisers Kopin Corp., Tounton, Mass., 1985—; cons., Motorola, IBM, AT&T, others. Author: Principles and Technology of Modfets, 1991; contbr. chpts. to 2 books, 700 articles to profl. jours. Fellow IEEE, AAAS, Am. Phys. Soc.; mem. Math. Rsch. Soc., Optical Soc. Am., Sigma Xi, Eta Kappa Nu. Achievements include investion of fastest transistor in world. Home: 1803 Golfview Dr Urbana IL 61801 Office: Univ IOll 104 S Goodwin St Urbana IL 61801

MORONI, ANTONIO, chemist, consultant, international coordinator; b. Barga, Toscany, Italy, Oct. 25, 1953; came to U.S., 1980; PhD, Univ. Pisa, Italy, 1979; MBA, Seton Hall Univ., 1993. Postdoctoral fellow Univ. Mass., Amherst, 1980-81, Univ. of Pisa, Italy, 1982-83, Polytechnic Univ., Bklyn., 1983-85; sr. rsch. chemist Pennwalt Corp., King of Prussia, Pa., 1985-88; rsch. assoc. Warner Lambert, Morris Plains, N.J., 1988—. Contbr. articles to profl. jours.; patentee in field. Mem. Am. Chemical Soc., Sigma Chi. Home: 43 Burnham Rd Morris Plains NJ 07950

MOROWITZ, HAROLD JOSEPH, biophysicist, educator; b. Poughkeepsie, N.Y., Dec. 4, 1927; s. Philip Frank and Anna (Levine) M.; m. Lucille Rita Stein, Jan. 30, 1949; children: Joanna Lynn, Eli David, Joshua Alan, Zachary Adam, Noah Daniel. B.S., Yale U., 1947, M.S., 1950, Ph.D., 1951. Physicist Nat. Bur. Standards, 1951-53, Nat. Heart Inst., Bethesda, Md., 1953-55; mem. faculty Yale, 1955—, assoc. prof. biophysics, 1960-68, prof. molecular biophysics and biochemistry, 1968-88, master Pierson Coll., 1981-86; faculty George Mason U., Fairfax, Va., 1988—, Robinson prof. biology and natural philosophy, 1988—; chmn. com. on models for biomed. rsch. NRC, 1983-85, mem. bd. on basic biology, 1986-92. Author: Life and the Physical Sciences, 1964, (with Waterman) Theoretical and Mathematical Biology, 1965, Energy Flow in Biology, 1968, Entropy for Biologists, 1970, (with Lucille Morowitz) Life On The Planet Earth, 1974, Ego Niches, 1977, Foundations of Bioenergetics, 1978, The Wine of Life, 1979, Mayonnaise and the Origin of Life, 1985, Cosmic Joy and Local Pain, 1987, The Thermodynamics of Pizza, 1991, Beginnings of Cellular Life, 1992, (with James Trefil) The Facts of Life, 1992, Entropy and the Magic Flute, 1993; contbr. articles to profl. jours. Mem. sci. adv. bd. Santa Fe Inst., 1991—. Mem. Biophys. Soc. (mem. exec. com. 1965), Nat. Ctr. for Rsch. Resources (mem. coun. 1987-92). Office: George Mason Univ 207 E Bldg Fairfax VA 22030

MORREL-SAMUELS, PALMER, experimental social psychologist; b. Hartford, Conn., Oct. 9, 1951; s. Harold and Rita (Palmer) Samuels; children: Ana, Elias, Eva. MA, U. Chgo., 1980; MPhil, Columbia U., 1986, PhD, 1989. Rsch. asst. psychology dept. Yale U., New Haven, 1981-82; rsch. asst. IBM Watson Rsch. Ctr., Yorktown, N.Y., 1986-87; researcher Kewalo Basin Lab., Honolulu, 1987-91; dir. group dynamics EDS Ctr. for Advanced Rsch., Ann Arbor, Mich., 1991—; asst. rsch. scientist Cognitive Sci./Machine Intelligence Lab, U. Mich., Ann Arbor, 1993—. Columbia U. faculty fellow, N.Y.C., 1982-87. Mem. AAAS, APA, Psychonomic Soc., Am. Psychol. Soc. Achievements include research on the role of visual info. in communication, with particular attention to the relation between gesture, word and tone of voice, gesture lang. signs and gesture-driven computer interfaces. Office: EDS Ctr for Advanced Rsch 2001 Commonwealth Rd Ann Arbor MI 48105

MORREY, WALTER THOMAS, electronics engineer; b. Berkeley, Calif., Dec. 27, 1946; s. Charles B. and Frances E. (Moss) M.; m. Lori Swartworth, Dec. 19, 1976 (div. 1993); children: Jeremy M., Zachary M. BSEE, MIT, 1968. Staff engr. MIT Instrumentation Lab, Cambridge, Mass., 1968-69; chief engr. Logitron Inc., Cambridge, 1969-71; chief engr., ptnr. D B Systems, Jaffrey Center, N.H., 1975-77; staff engr. Burr-Brown Corp., Tucson, 1977-78; cons. Electronic Design Cons., New London, N.H., 1978—; sr. staff scientist Household Data Svcs. Inc., Reston, Va., 1985-90. Patentee TV Scrambling Sys., 1989; contbr. articles to profl. jours. Mem. IEEE, Sigma Xi. Office: EDC PO Box 489 New London NH 03257

MORRIN, THOMAS HARVEY, engineering research company executive; b. Woodland, Calif., Nov. 24, 1914; s. Thomas E. and Florence J. (Hill) M.; m. Frances M. Von Ahn, Feb. 1, 1941; children: Thomas H., Diane, Linda, Denise. B.S., U. Calif., 1937; grad., U.S. Navy Grad. Sch., Annapolis, Md., 1941. Student engr. Westinghouse Electric Mfg. Co., Emeryville, Calif., 1937; elec. engr. Pacific Gas & Electric Co., 1938-41; head microwave engring. div. Raytheon Mfg. Co., Waltham, Mass., 1947-48; chmn. elec. engring. dept. Stanford Research Inst., 1948-52, dir. engring., research, 1952-60, gen. mgr. engring., 1960-64, vice pres. engring., sci., 1964-68; pres. University City Sci. Inst., Phila., 1968-69; pres., chmn. bd. Morrin Assos., Inc., Wenatchee, Wash., 1968-72. Trustee Am. Acad. Transp. Served as officer USNR, 1938-58, comdr. USN, 1945-48. Decorated Bronze Star; recipient Bank Am. award for automation of banking during 1950's, 1992. Fellow IEEE, AAAS; mem. Sci. Research Soc. Am., U.S. Naval Inst., Navy League, Marine Meml. Club (San Francisco). Address: 654 23d Ave San Francisco CA 94121

MORRIS, ALVIN LEE, retired consulting corporation executive, meteorologist; b. Kim, Colo., June 7, 1920; s. Roy E. and Eva Edna (James) M.; BS in Meteorology (U.S. Weather Bur. fellow), U. Chgo., 1942; MS, U.S. Navy Postgrad. Sch., 1954; m. Nadean Davidson, Jan. 16, 1979; children: Andrew N., Nancy L., Mildred M., Ann E., Jane C. Meteorologist Pacific Gas and Electric Co., San Francisco, 1947-50; commd. U.S. Navy, 1942, advanced through grades to capt., USNR, 1962, assignments including staff, comdr. 7th Fleet; dir. rsch. Navy Weather Research Facility, Norfolk, Va., 1958-62; facilities coord., mgr. sci. balloon facility, Nat. Ctr. for Atmospheric Rsch. Boulder, Colo., 1963-75; pres. Ambient Analysis Inc., Internat. Cons., Boulder, 1975-86. Treas. Home Hospitality for Fgn. Students Program, U. Colo., 1969-70; del. People to People Del. on Environment; Peoples Republic of China, 1984. Served with USN, 1942-46, 50-58. Mem. Am. Meteorol. Soc. (cert. cons. meteorologist), Am. Geophys. Union, ASTM, Ret. Officers Assn., N.Y. Acad. Scis., Boulder County Knife and Fork Club. Editor Handbook of Scientific Ballooning, 1975; assoc. editor Jour. Oceanic and Atmospheric Tech., 1984-88; contbr. articles to profl. jours.; convenor, editor proceedings ASTM conf. Home: 880 Sunshine Canyon Dr Boulder CO 80302-9727

MORRIS, DAVID BRIAN, chemical engineer; b. Marlboro, Mass., Apr. 25, 1946; s. John Paul and Jean (Zarozinski) M.; m. E. Diane Morris, Oct. 8, 1973; children: Christopher P., Christina F., Collin D. BSChemE, Worcester Poly. Inst., 1969, MS in Mgmt.&Engring., 1979. Process engr. Stone & Webster Engring., Boston, 1969-72; devel. engr. Abcor, Inc., Cambridge, Mass., 1972-76; account mgr. Betz Labs., Trevose, Pa., 1976-82; area gen. mgr. Continental Water Systems, San Antonio, Tex., 1982-88; pres. N.Eng. Water Systems, Auburn, Mass., 1988-90; cons. Fla. Progress Corp., St. Petersburg, 1990—; mem. program adv. bd. Worcester Poly. Inst., 1991—; cons. in field. Troop leader Boys Scouts Am., Wilbraham, Mass., 1988-89. NASA fellow, 1970. Mem. Am. Inst. Chem. Engrs. Roman Catholic. Achievements include development of trademarked product name for laboratory high purity water systems, developed and administered AWWA award winning water conservation program for Fla. Power Corp. Home: 267 Millstone Dr Palm Harbor FL 34863

MORRIS, DON MELVIN, surgical oncologist; b. Longview, Tex., Jan. 4, 1946; s. John Raymond and Martha Ann (Walker) M.; m. Judy Ray Miller, Sept. 1970 (div. 1978); 1 child, Curtis John Walker; m. Katherine Crowe, June 26, 1982. BA in Biology, U. Tex., 1968; MD, U. Tex. Med. Br., Galveston, 1972. Intern Bexor County Hosp. Bexor County Hosp., U. Tex. Med. Sch., San Antonio, 1972-73; resident gen. surgery U. Md. Hosp., Balt., 1973-75, surg. resident, 1976-77, chief resident dept. surgery, 1977-78; clin. assoc. Nat. Cancer Inst. Balt. Cancer Rsch. Ct., 1975-76; instr. surg. oncology U. Md. Sch. Medicine, Balt., 1978-80, asst. prof., 1980-81; asst. prof. surgery La. State U. Med. Sch., Shreveport, 1981-84, assoc. prof. surgery and oral and maxillofacial surgery, 1984-88, prof., 1988-90; prof. surgery U. N.Mex. Sch. Medicine, Albuquerque, 1992—; dir. surg. oncology U. N.Mex. Cancer Ctr., Albuquerque, 1990—; chief surg. svc. VA Med. Ctr., Shreveport, 1984-88, Albuquerque, 1990-92. Contbr. chpt. to book, numerous articles to profl. jours. Stiles grantee, 1986; VA Rsch. Group grantee, 1985-86, 91. Fellow ACS (mem. commn. on cancer 1991—); mem. Soc. Univ. Surgeons, So. Surg. Assn., Soc. for Surgery of the Alimentary Tract, numerous others. Republican. Mem. Disciples of Christ Ch. Home: 5720 Bartonwood Pl NE Albuquerque NM 87111 Office: U NMex Cancer Ctr 900 Camino de Salud NE Albuquerque NM 87131

MORRIS, JAMES BRUCE, internist; b. Rochester, N.Y., May 13, 1943; s. Max G. and Beatrice Ruth (Becker) M.; B.A., U. Rochester, 1964; M.D., Yale U., 1968; m. Susan Carol Shencup, July 31, 1966; children—Carrie, Douglas, Deborah, Rebecca. Intern, SUNY, Buffalo, 1968-69, resident, 1969-70, 72-73, chief resident, 1973; fellow U. Miami, 1974; practice medicine specializing in internal medicine and infectious diseases, Plantation, Fla., 1974—; chmn. infection control com. Lauderdale Lakes Gen. Hosp., 1974-76; chmn. infection control com. Plantation Gen. Hosp., 1976-80, 83-85, chmn. pharmacy com., 1980-81, chmn. tissue com., 1982; sec., program chmn. dept. medicine Bennett Community Hosp., 1978-80, chmn. dept. medicine, 1980-81, vice chief staff, 1981-83; chmn. infection control com. Fla. Med. Center, 1980-82; chief staff Humana Hosp. Bennett, 1983-85, trustee, 1983-88, chmn. infection control com., 1985-87 ; clin. asst. prof. U. Miami Med. Sch., 1975—. Served with USAR, 1970-72. Diplomate Am. Bd. Internal Medicine, Am. Bd. Infectious Diseases. Mem. ACP, AMA, Am. Soc. Microbiology, Infectious Diseases Soc. Am., Am. Soc. Internal Medicine, Fla. Med. Assn., Broward County Med. Assn. Office: Sachs Morris & Sklaver 7353 NW 4th St Fort Lauderdale FL 33317-2241

MORRIS, JAY KEVIN, chemical engineer; b. Tulsa, May 21, 1959; s. James Milton and Gloria Jean (Sanders) M.; m. Lynn Marie Scherrer, June 29, 1985; children: John William, Melanie Jean. BS in Chem. Engring. with distinction, U. Okla., 1981, MS in Chem. Engring., 1982. Prodn. engr. Gulf Oil Exploration & Prodn., Houston, 1982-85; project engr. Gulf Oil Internat., Houston, 1985; facilities engr. Chevron Overseas Petroleum, Inc., San Ramon, Calif., 1985-88; lead facilities engr. Zaire Gulf Oil Co., Kinshasa, Zaire, 1988-91; mech. engr. Chevron (U.K.) Ltd., London, 1991—. Advisor Jr. Achievement, Houston, 1983-85; commr. Kinshasa Am. Softball League, 1990-91. Mem. NSPE (com. chmn. 1982-85), Am. Inst. Chem. Engrs., Tau Beta Pi, Pi Kappa Phi. Avocations: car restoration, golf, softball, international travel. Home: Orchehill Ave, Gerreards Cross UK SL9 8PT Office: Chevron London PO Box 5046 San Ramon CA 94583

MORRIS, JOHN MICHAEL, energy technology educator; b. Chester, Pa., Aug. 31, 1949; s. John Arthur and Joan Ann (O'Hara) M.; m. Lois Ann Rouleau, Apr. 18, 1970; children: Tracy Noel, Andrew, Katie, Matt. BS, U. Wis., Kenosha, 1976; MS, Calif. State U., 1977; PhD, Wayne State U., 1987. Cert. energy mgr. Asst. dir. phys. plant Waukesha County Tech. Inst., Sewaukee, Wis., 1976-77; dir. phys. plant Monroe County C.C., Monroe, Mich., 1977-81; prof. energy tech. Henry Ford C.C., Dearborn, Mich., 1981—; cons. engr. Techni-Serve, Inc., Monroe, 1987—. Scoutmaster Boy Scouts Am., Monroe, 1977—, dir. cub scout resident camp, Gregory, Mich.; bd. dirs. Monroe City Libr., 1980—. With USN, 1968-76, USNR, 1976-90. Mem. ASHRAE, Am. Inst. Plant Engrs. (bd. dirs. info. resources 1991-92), Mich. Assn. Energy Engrs. (chair publs. 1990-91, Energy Profl. of Yr. 1990). Roman Catholic. Home: 50 E Grove St Monroe MI 48161 Office: Henry Ford CC Searle Tech Bldg 13020 Osborn Dearborn MI 48124

MORRIS, JUSTIN ROY, food scientist, enologist, consultant; b. Nashville, Ark., Feb. 20, 1937; s. Roy Morris; m. Ruby Lee Blackwood, Sept. 5, 1956; children: Linda Lee, Michael Justin. BS, U. Ark., 1957, MS, 1961; PhD, Rutgers U., 1964. Rsch. asst. Rutgers U., New Brunswick, N.J., 1957-61, instr., 1961-64; extension horticulturist U. Ark., Fayetteville, 1964-67, from asst. to assoc. prof., 1967-75, prof., 1975-85; univ. prof., 1985—; cons. viticulture sci. program Fla. A&M U., Tallahassee, 1979-81; cons. viticulture and enology program Grayson City Coll., Denison, Tex., 1987—; cons. J. M. Smucker Co., Orrville, Ohio, 1982-91. Co-author: Small Fruit Crop Management, 1990, Quality and Preservation of Fruits, 1991; assoc. editor: Am. Jour. Enology and Viticulture, 1985; contbr. more than 240 articles to sci. jours. Recipient rsch. award Nat. Food Processors Assn., 1982. Fellow Am. Soc. for Hort. Sci. (assoc. editor 1985, Gourley award 1979, Outstanding Rsch. award 1983); mem. Ozark Food Processors Assn. (exec. v.p. 1988—), Coun. for Agrl. Sci. and Tech. (bd. dirs. 1987—, chmn. nat. concerns 1987-91, pres.-elect 1993), Inst. Food Technologists (steering com. divsn. fruit and vegetable 1987—), Gamma Sigma Delta (Outstanding Rsch. award 1982, Faculty Achievement award for rsch. & svc. 1993). Achievements include development of mechanical cane fruit harvester, of mechanical strawberry harvester, of modified grape harvester for wine grapes, of mechanical fruit positioner for grapes; development of systems for the production, harvesting, handling, utilization, and marketing of grape juice and wine. Office: U Ark Dept Food Sci 272 Young Ave Fayetteville AR 72703-5584

MORRIS, MICHAEL DAVID, chemistry educator; b. N.Y.C., Mar. 27, 1939; s. Melvin M. and Rose (Pollock) M.; m. Leslie Tuttle, June 5, 1961; children: Susannah, David, Rebecca, Ari. BA in Chemistry, Reed Coll., 1960; PhD in Chemistry, Harvard U., 1964. Asst. prof. Penn State U., University Park, Pa., 1969; assoc. prof. U. Mich., Ann Arbor, 1969-82, prof., 1982—; mem. editorial bd. Spectrochim Acta Rev., 1987-92, editor, 1993. Editor: (book) Spectroscopic and Microscopic Imaging of the Chemical State, 1993. Mem. Am. Chem. Soc., Soc. for Applied Spectroscopy, Microbeam Analysis Soc. Office: Univ of Michigan Dept of Chemistry Ann Arbor MI 48109

MORRIS, ROBERT DUBOIS, epidemiologist; b. New Haven, Conn., Dec. 8, 1956; s. John Mclean and Majorie Stout (Austin) M. BA, Yale U., 1978; PhD, U. Wis., Milw., 1986; MS, Med. Coll. Wis., 1988, MD, 1991. Rsch. asst. Marine Biol. Lab., Woods Hole, Mass., 1975; prodn. mgr. Windworks, Inc., Mukwonago, Wis., 1980-81; rsch. asst. U. Wis., Milw., 1982-85, lectr., 1986; rsch. assoc. Med. Coll. Wis., Milw., 1986-90; vis. scientist Harvard U. Sch. Pub. Health, Boston, 1990; asst. prof. Med. Coll. Wis., Milw., 1991--; expert advisor Nat. Rsch. Coun. Com. on Environ. Epidemiology; mem. health effects working group Great Lakes Protection Fund; mem. interant. collaborative group on clin. trial registries. Contbr. articles to profl. jours. U. Wis. fellow, 1982-85. Mem. AMA, Am. Chem. Soc., Soc. for Epidemiological Rsch., Internat. Soc. for Environ. Epidemiology, Phi Kappa Phi. Achievements include meta analysis to define assn. between water chlorination by products and cancer. Office: Med Coll Wis 8701 Watertown Plank Rd Milwaukee WI 53217

MORRIS, SAMUEL CARY, environmental scientist, consultant, educator; b. Summit, N.J., Dec. 16, 1942; s. Samuel Cary Jr. and Roberta Ann (Griffiths) M.; m. Stephanie Margaret Rose, Aug. 13, 1966; children: Jennifer, Daniel, Laura. BSCE, Va. Mil. Inst., 1965; MS in Sanitary Engring., Rutgers U., 1966; ScD in Environ. Health, U. Pitts., 1973. Asst. prof. environ. sci. Ill. State U., Normal, 1971-72; rsch. assoc. U. Pitts., 1972-73; asst. scientist, then assoc. scientist Brookhaven Nat. Lab., Upton, N.Y., 1973-77, scientist, 1977—, dep. head div. analytic scis., 1990—; adj. prof. Carnegie Mellon U., Pitts., 1976—; editorial bd. Environment Internat., 1983—; lectr. SUNY-Stony Brook, 1992—. Author: Cancer Risk Assessment, 1990; contbr. chpts. to books, articles to profl. jours. Capt. U.S. Army, 1966-68. Mem. ASCE, Inst. Mgmt. Scis., Air and Waste Mgmt. Assn., Soc. Risk Analysis (coun. mem. 1984-86), Delta Omega. Office: Brookhaven Nat Lab Upton NY 11973

MORRIS, STANLEY M., research and engineering executive; b. Burlington, Vt., Jan. 15, 1942; s. I. Philip Morris and Elizabeth (Kershner) Marcus; m. Sandra L. Harris, 1963 (div. 1981); children: Jeffrey, Seth; m. 2d, Linda Hecht, 1983; stepchildren: Andrew Zinnes, Alexandra Zinnes. BSChemE, Carnegie-Mellon U., 1964; MSChemE, Lehigh U., 1966, PhD, 1967. Sr. engr. Exxon Rsch. and Engring. Inc., Florham Pk., N.J., 1967-73; v.p. tech. and bus. devel. chems. Air Products and Chemicals Inc., Allentown, Pa., 1973—; bd. dirs., UTI Inc., Milpitas, Calif., 1984-87. Patentee in field. Bd. dirs. Big Bros./Big Sisters of Lehigh County, Allentown, Pa., 1983, Planned Parenthood N.E. Pa., 1990—. Mem. Am. Inst. Chem. Engrs., Sigma Xi. Democrat. Unitarian. Avocations: sailing, hiking, raquetball. Home: 2820 W Gordon St Allentown PA 18104-4851 Office: Air Products & Chems Inc 7201 Hamilton Blvd Allentown PA 18195*

MORRIS, STEPHEN BLAINE, clinical psychologist; b. Logan, Utah, Aug. 22, 1951; s. Blaine and Helen (Bradshaw) Morris, Jr.; m. Marilyn Smith, Sept. 12, 1974; children: David Stephen, Angela, Michael Andrew. BA magna cum laude, Utah State U., 1976; MA, Brigham Young U., 1979; PhD, U. Utah, 1986. Lic. psychologist, Utah. Psychol. cons. U. Utah Speech and Hearing Clinic, 1982-83; psychology intern Primary Children's Med. Ctr., Salt Lake City, 1983-86; postdoctoral resident in clin. psychology Comprehensive Psychol. Svcs. of Utah, Salt Lake City, 1986, clin. psychologist, 1988-89; dir. children's program Charter Summit Hosp., Midvale, Utah, 1989-90; cons. children's program Charter Summit Hosp., Midvale, 1991-93; clin. psychologist and dir. outpatient child and family svcs Charter Counseling Ctr., Midvale, 1990; clin. psychologist in pvt. practice Salt Lake City, 1990—; instr. dept. psychology U. Utah, summer 1986; lectr. in field. Contbr. articles to profl. jours. Mem. APA, Utah Psychol. Assn. (sec.-treas. div. hosp. practice 1989-92, pres. practice divsn. 1992-93), Assn. of Mormon Counselors and Psychotherapists, Phi Kappa Phi, Phi Eta Sigma. LDS. Office: 195 W 7200 S Midvale UT 84047

MORRIS, WILLIAM ALLAN, engineer; b. Cardiff, Wales, June 24, 1933; s. William Charles and Agnes (Gertrude) M.; m. Elizabeth Mary Miller, Aug. 23, 1959; children: Lisa, Adrian William. M. in Engring., U. Birmingham (Eng.), 1963; full tech. cert., Cardiff Tech. Coll., 1964; chartered engr., U. Wales, Cardiff, 1966; postgrad. diploma in mgmt. studies, Mgmt. DMS, Cardiff, 1974. Registered European engr.; chartered engr./surveyor, U.K.; licentiate in engr. Inst. Maintenance, Ireland; cert. quality assurance lead assessor, U.S. Indentured craft apprentice Curran Engring., Cardiff, 1950-55, design engr., 1957-60; instrumentation engr. Firth Cleveland Instrumentation, Treforest, Wales, 1960-63; sr. lectr. Gwent Polytechnic, Newport Gwent, 1963-76; chief engr. G.K.N., Treforest, 1970-72; cons. engr. Eur. Ing. William A. Morris M.Sc., Cardiff, 1972—. Author: Workshop Technology, 1965. Jr. technician Royal Air Force, 1955-57. Fellow Inst. Mech. Engr., Inst. Marine Engrs., Welding Inst.; mem. Inst. of Quality, Brit. Inst. Non-Destructive Testing, Inst. Elec./Electronic Engrs., Am. Welding Soc. Avocations: gardening, painting, decorating, swimming, studying. Home: The Tee Squares 9 padarn C1, Cardiff CF2 6ER, Wales Office: Inspection Bur Tee Squares, 9 Padarn Cl, Cardiff CF2 6ER, Wales

MORRISON, ADRIAN RUSSELL, veterinarian educator; b. Phila., Nov. 5, 1935; married, 1958; 5 children. DVM, Cornell U., 1960, MS, 1962; PhD in Anatomy, U. Pa., 1964. Spl. fellow neurophysiology U. Pisa, 1964-65, asst. prof., 1966-70; assoc. prof. anatomy Sch. Vet. Medicine, U. Pa., Phila., 1970—. Recipient AAAS Sci. Freedom and Responsibility award AAAS, 1991. Mem. Am. Vet. Med. Assn., Am. Assn. Vet. Anatomy, Am. Assn. Anatomy, Assn. Psychophysiol Sleep Study. Achievements include research in neuroanatomical and neurophysiological bases of mammalian behavior. Office: Univ of Penn Sch of Veterinary Philadelphia PA 19104*

MORRISON, ANGUS CURRAN, aviation executive; b. Toronto, Ont., Can., Apr. 22, 1919; s. Gordon Fraser and Mabel Ethel (Chalcraft) M.; m. Carlotta Townsend Munoz, Mar. 1, 1947; children—Sandra, James, Christian, Mark. Student, Upper Can. Coll., Bishop's Coll. Sch. Pres. Atlas Aviation, Ltd., Ottawa, Ont., 1946-51; sec. Air Industries and Transport Assn. Can., Ottawa, 1951-62; pres., chief exec. officer Air Transport Assn. Can., 1962-85; ret., 1985—; v.p., dir. Munoz Corp., Montclair, N.J., 1969-73. Councillor Town of Almonte, Ont., 1960-65. Served with Royal Can. Armoured Corps, 1939-46. Recipient Diplome Paul Tissandier Fedn. Aeronautique Internationale, 1977, Casi C.D. Howe award, 1987, Companion of Order of Flight, 1989; named to Can.'s Aviation Hall of Fame, 1989. Assoc. fellow Can. Aeronautics and Space Inst.; mem. Internat. N.W. Aviation Coun., Chartered Inst. Transport. Anglican. Clubs: Rideau, Wings. Home: Burnside, PO Box 609, Almonte, ON Canada K0A 1A0

MORRISON, GEORGE HAROLD, chemist, educator; b. N.Y.C., Aug. 24, 1921; s. Joseph and Beatrice (Morel) M.; m. Annie Foldes, Oct. 19, 1952; children—Stephen, Katherine, Althea. B.A., Bklyn. Coll., 1942; Ph.D., Princeton, 1948. Instr. chemistry Rutgers U., 1948-50; research chemist AEC, 1949-51; head inorganic and analytical chemistry Gen. Tel. & Electronic Labs., 1951-61; prof. chemistry Cornell U., 1961—; chmn. com. analytical chemistry NAS-NRC, 1965-77; Internat. Francqui chair U. Antwerp, Belgium, 1989. Editor Analytical Chemistry, 1980-91; contbr. articles to profl. jours. Served with AUS, 1943-46. Recipient Benedett-Pichler award Am. Microchem. Soc., 1977, Ea. Analytical Symposium Jubilee award, 1986, Pitts. Analytical Chemistry award Soc. for Analytical Chemists of Pitts., 1990; NSF sr. fellow, U. Calif, San Diego, 1967-68, Guggenheim fellow, U. Paris, Orsay, 1974-75, NIH sr. fellow, Harvard Med. Sch., 1982-83. Mem. Am. Chem. Soc. (award analytical chemistry 1971), Soc. Applied Spectroscopy (award 1975), Sigma Xi. Office: Cornell Univ Baker Lab Chemistry Ithaca NY 14853

MORRISON, GERALD LEE, mechanical engineering educator; b. Bartlesville, Okla., Oct. 11, 1951; s. Edward Ranz and Alice Eileen (Gresham) M.; m. Nelda Fayrene Nelson, May 26, 1973; children: Jennifer Eileen, Jeffery Edward. BS, Okla. State U., 1973, MS, 1974, PhD, 1977. Registered profl. engr., Tex. Asst. prof. mech. engring. Tex. A&M U., College Station, 1977-81, assoc. prof., 1981-88, prof., 1988—; assoc. dir. Ctr. for Space Power, College Station, 1987-90; adv. bd. mem. Turbomachinery Lab. Pump Symposium, College Station, 1989—; mem. Gas Metering Rsch. Coun. Am. Gas Assn., Arlington, Va., 1991—. Author: Fluid Mechanics Laboratory, 1992; editor: (with others) Third International Symposium on Laser Anemometry, 1987; contbr. articles to profl. jours. Asst. scoutmaster Boy Scouts Am., College Station, 1991—, den leader Cub Scouts, 1989-91; mem. adminstrv. bd. Aldersgate United Meth. Ch., College Station, 1981-83. Gulf Oil fellow, 1974; sr. Tex. Engring. Exptl. Sta. fellow Tex. A&M U., 1986; named Nelson Jackson prof., 1993; recipient Cert. of Recognition NASA, 1984. Assoc. fellow AIAA (aeroacoustics tech. com. 1983-91); mem. ASME, Sigma Xi, Phi Kappa Phi. Achievements include patents pending for slotted orifice flowmeter and method and apparatus for calibration of high volume gas flowmeters; recognized expert in laser anemometry and flow meters; performance of unique measurements of the flow field inside labyrinth and annular seals. Home: 2807 Mescalero Ct N College Station TX 77845 Office: Tex A&M U Mech Engring Dept College Station TX 77843-3123

MORRISON, HARRY, chemistry educator, university dean; b. Bklyn., Apr. 25, 1937; s. Edward and Pauline (Sommers) M.; m. Harriet Thurman, Aug. 23, 1958; children—Howard, David, Daniel. B.A., Brandeis U., 1957; Ph.D., Harvard U., 1961. NATO-NSF postdoctoral fellow Swiss Fed. Inst., Zurich, 1961-62; research asso. U. Wis., Madison, 1962-63; asst. prof. chemistry Purdue U., West Lafayette, Ind., 1963-69, assoc. prof., 1969-76, prof., 1976—, dept. head, 1987-92, dean Sch. Sci., 1992—. Assoc. editor Photochemistry and Photobiology, 1986—; contbr. numerous articles to profl. jours. Bd. fellows Brandeis U. Mem. Am. Chem. Soc., Am. Soc. Photobiology, Interam. Photochem. Soc., Phi Beta Kappa, Sigma Xi. Office: Purdue U Sci Adminstrn Math Bldg West Lafayette IN 47907-1390

MORRISON, ROBERT THOMAS, engineering consultant; b. Manson, Iowa, June 4, 1918; s. Charles Henry and Ida Magdeline (Fuessley) M.; m. Callie Louise Warren, July, 25, 1942; children: Linda Ann, Allan Charles, Janis Lou. BS in Mech. Engring., Iowa State U., 1942; MS in Engring., U. Calif., Los Angeles, 1961. Engr. Gen. Elecric Co., Schenectedy, N.Y., 1942-45; sales engr., inventory supr. Gen. Electric Supply Corp., Omaha, 1945-50; pres. Morrison Mfg. Co., Omaha, 1950-52; elec. system designer Douglas Aircraft, Long Beach and Santa Monica, Calif., 1952-58; system engr., proposal mgr. Rockwell Internat., Downey, Seal Beach, Anaheim, Calif., 1958-81; freelance cons. Garden Grove, Calif., 1981—; originator, coord. system engring. program West Coast U., L.A., 1963-71, assoc. dir. devel., 1972; moderator Rockwell System Engring. Seminar, 1964. Author: Proposal Manager's Guide, 1972, Proposal Style Guide, 1988, Proposal Publications Guide, 1988. Lay minister Crystal Cathedral, Garden Grove, 1980—, dir. New Hope Telephone Counseling Ctr., 1990—; Garden Grove Energy Commn., 1982-85, Garden Grove Planning Commn., 1960-61; com. chmn. March of Dimes, Orange County, 1973-75; pres. Meth. Men, Garden Grove, 1964, 65. Recipient Apollo Achievement award NASA, 1970, Apollo-Soyuz Test Project award NASA, 1975, Space Shuttle Approach and Landing Test award NASA, 1978, Profl. Achievement citation in Engring. Iowa State U., 1984. Mem. Assn. Profl. Cons., World Future Soc., Inst. Mgmt. Scis., Ops. Rsch. Soc. Am., Assn. Proposal Mgmt. Profls., Tech. Mktg. Soc. Am., Masons, Toastmasters, Palm Springs Tennis Club. Republican. Avocations: travel, photography, gardening, shooting, municipal bands.

MORRISON, ROGER BARRON, geologist; b. Madison, Wis., Mar. 26, 1914; s. Frank Barron and Elsie Rhea (Bullard) M.; BA, Cornell U., 1933, MS, 1934; postgrad. U. Calif., Berkeley, 1934-35, Stanford U., 1935-38; PhD, U. Nev., 1964; m. Harriet Louise Williams, Apr. 7, 1941 (deceased Feb. 1991); children: John Christopher, Peter Hallock and Craig Brewster (twins). Registered profl. geologist, Wyo. Geologist U.S. Geol. Survey, 1939-76; vis. adj. prof. dept. geoscis. U. Ariz., 1976-81, Mackay Sch. Mines, U. Nev., Reno, 1984-86; cons. geologist; pres. Morrison and Assocs., 1978—; prin. investigator 2 Landsat-1 and 2 Skylab earth resources investigation projects NASA, 1972-75. Fellow Geol. Soc. Am.; mem. AAAS, Internat. Assn. Quaternary Research (past mem. Holocene and pedology commns.), Am. Soc. Photogrammetry, Am. Soc. Agronomy, Soil Sci. Soc. Am., Internat. Soil Sci. Soc., Am. Quaternary Assn., Am. Water Resources Assn., Colo. Sci. Soc., Sigma Xi, Colorado Mountain Club. Author 2 books, co-author one book, co-editor 2 books; editor: Quaternary Nonglacial Geology, Conterminous U.S., Geol. Soc. Am. Centennial Series, vol. K-2, 1991; mem. editorial bd. Catena, 1973-88; contbr. over 150 articles to profl. jours. Research includes Quaternary geology and geomorphology, hydrogeology, environ. geology, neotectonics, remote sensing of Earth resources. Office: 13150 W 9th Ave Golden CO 80401-4201

MORRISON, ROLLIN JOHN, physicist, educator; b. Akron, Ohio, Oct. 8, 1937; s. Charles Harmon and Dorothy (Sechrist) M.; m. Anne Ellen Kranek, June 10, 1964; children: Susan Anne, Rollin Wesley. BA, Ohio Wesleyan U., 1959; MS, U. Ill., 1961, PhD, 1964. Volkswagen fellow Deutsches Elecktronen Synchnotron, Hamburg, Germany, 1964-67; asst. prof. U. Calif., Santa Barbara, 1967-71, assoc. prof., 1971-78, prof., 1978—. Contbr. articles to profl. jours. Fellow Am. Phys. Soc. Achievements include research in measurement of charmed meson lifetimes and branching ratios. Home: 760 Kristen Ct Santa Barbara CA 93111 Office: U Calif Santa Barbara Dept Physics Santa Barbara CA 93106

MORRISSETTE, JEAN FERNAND, electronics company executive; b. Montreal, Que., Can., Sept. 3, 1942; s. Gaétan C. and Liliane (Bilodeau) M.; m. Margot Michaud, 1965; children: Sylvie, Natalie. Chmn. of bd. Parcap Elec. & Electronics Inc., Montreal, 1983—, Parcap Foods Inc., Montreal, 1983—, Parcap Fin. Inc., Montreal, 1983—, Parcap Real Estate Inc., Montreal, 1983—, Parcap Technologies Inc., Montreal, 1983—; chmn. of bd., chief exec. officer Parcap Holdings Inc., Boston, 1983—; chmn. of bd. Parcap Elec. Inc., Boston, 1983—. Named Man of the Month, Electric League,

Can., 1976. Mem. Club St-Denis. Office: Parcap Mgmt Inc 1800 McGill, College Ave Ste 2800, Montreal, PQ Canada H3A 3J6

MORROW, BRUCE W., business executive, consultant; b. Rochester, Minn., May 20, 1946; s. J. Robert and Frances P. Morrow; m. Jenny Lea Morrow; B.A., U. Notre Dame, 1968, M.B.A. with honors in Mgmt., 1974, M.A. in Comparative Lit., 1975; grad. U.S. Army Command and Gen. Staff Coll., 1979. Adminstrn. mgr. Eastern States Devel. Corp., Richmond, Va., 1977; v.p. JDB Assos., Inc., Alexandria, Va., 1976-78; owner Aardvark Prodns., Alexandria, 1978-80, Servital Foods, Alexandria, Va., 1980-82; sr. cons. Data Base Mgmt., Inc., Springfield, Va., 1982—; dir. continuing edn. Southside Va. Community Coll., Alberta, Va., 1989-91. Active Boy Scouts Am., 1960-69; chmn. elem. German, U. Notre Dame, 1973-75; mem. Roanoke Wildwood Vol. Fire Dept., 1991-93. Lt. col. USAR. Decorated Bronze Star, Army Commendation medals, Army Achievement medal, Meritorious Svc. medal, Parachutist's badge. Mem. VFW (life), Nat. Eagle Scout Assn., Lake Gaston C. of C. (bd. dirs.), Am. Legion, Lions (v.p. local club), Beta Gamma Sigma, Delta Phi Alpha. Clubs: Friends Internat. (Am. v.p. 1969-71, Boeblingen, Germany), Order of DeMolay. Contbg. columnist Notre Dame Mag., 1974-86; composer songs. Office: Commonwealth Dominion Corp PO Box 400 Gasburg VA 23857-0400

MORROW, GRANT, III, genetecist; b. Pitts., Mar. 18, 1933; married, 1960; 2 children. BA, Haverford Coll., 1955; MD, U. Pa., 1959. Intern U. Colo., 1959-60; resident in pediatrics U. Pa., 1960-62, fellow neonatology, 1962-63, assoc. prof., 1964-65, assoc. prof. metabolic disease, 1965-71; from assoc. prof. to prof. neonatology and metabolism U. Ariz., 1972-78, assoc. chmn. dept., 1976-78; med. dir. Columbus (Ohio) Children's Hosp., 1978—; prof. neonatology and metabolism, chmn. dept. Ohio State U., 1978—. Mem. Am. Pediatric Soc., Am. Soc. Clin. Nutrition, Soc. Pediatric Rsch. Achievements include research on children suffering inborn errors of metabolism, mainly amino and organic acids, patients on total parental nutrition. Office: Children's Hosp Rsch Found 700 Childrens Dr Columbus OH 43205-2696*

MORROW, JOSEPH EUGENE, psychology educator; b. Paragould, Ark., Sept. 18, 1935; s. James Andrew Morrow and Edith Hortense (Higgins) Stouder. PhD, Washington State U., 1965. Asst. prof. psychology Calif. State Coll., Fullerton, 1965-68; asst. prof., chmn. dept. psychology Ind. U., South Bend, 1968-70; prof. psychology Calif. State U., Sacramento, 1970--; owner and clin. administr. Applied Behavior Cons., Sacramento, 1988--. Editor: (jour.) Behaviorists for Social Action, 1982. Mem. Am. Psychologist Assn., Assn. Behavior Analysis, Soc. Am. magicians. Achievements include being the first to discover a learning to learn phenomonon in an invertebrate organism. Home: 878 Woodside East #1 Sacramento CA 95825 Office: Calif State U 6000 J St Sacramento CA 95819

MORROW, PAUL LOWELL, forensic pathologist; b. N.Y.C., Mar. 30, 1949; s. Rufus Clegg and Dorothy Bell (Jackson) M.; m. Emily G. Rubenstein, Jan. 13, 1978; 1 child, Lillian Elaine. BA, Haverford Coll., 1971; MD, U. Vt., 1976. Diplomate Am. Bd. Pathology. Resident in pathology SUNY, Syracuse, 1976-78, Evanston (Ill.) Hosp., 1978-80; asst. chief med. examiner N.C. Office of Chief Med. Examiner, Chapel Hill, 1980-81; dep. chief med. examiner Vt. Office Chief Med. Examiner, Burlington, 1981-90, chief med. examiner, 1990—; asst. clin. prof. pathology U. of Vt., Burlington, 1981-92, assoc. clin. prof. pathology, 1992—. Contbr. articles to profl. jours. Fellow Am. Acad. Forensic Scis.; mem. Nat. Assn. Med. Examiners. Soc. of Friends. Office: Office of Chief Med Examin 18 East Ave Burlington VT 05401

MORROW, ROY WAYNE, chemist; b. Hopkinsville, Ky., Sept. 28, 1942; s. Roy Campbell and Gladys (Smith) M.; m. Margaret Kelly, Dec. 28, 1966; children: Shawn Alison, Brian Wayne. BS, Murray (Ky.) State U., 1964; MS, U. Tenn., 1968, PhD, 1970. Devel. chemist Y-12 Union Carbide, Oak Ridge, Tenn., 1971-75, sect. mgr., 1976-79, lab. mgr. K-25, 1979-80; lab. mgr. K-25 Martin Marietta, Oak Ridge, 1981-92, tech. dir., 1992—; mem. analytical support team U.S. Dept. Energy, Washington, 1990—. Mem. ASTM (subcom. chair 1988—), Am. Chem. Soc. (membership com. local sect. 1980), Soc. Applied Spectroscopy, Sigma Xi (nominations com. local sect. 1984). Home: 102 Morningside Dr Oak Ridge TN 37830 Office: Martin Marietta PO Box 2003 Oak Ridge TN 37831-7279

MORROW, WALTER EDWIN, JR., electrical engineer, university laboratory administrator; b. Springfield, Mass., July 24, 1928; s. Walter Edwin and Mary Elizabeth (Ganley) M.; m. Janice Lila Lombard, Feb. 25, 1951; children—Clifford E., Gregory A., Carolyn F. S.B., M.I.T., 1949, S.M., 1951. Mem. staff Lincoln Lab., MIT, Lexington, Mass., 1951-55, group leader, 1956-65; head div. communications MIT Lincoln Lab., 1966-68, asst. dir., 1968-71, asso. dir., 1972-77, dir., 1977—. Contbr. articles to profl. publs. Recipient award for outstanding achievement Pres. M.I.T., 1963, Edwin Howard Armstrong Achievement award IEEE Communications Soc., 1976. Fellow IEEE, Nat. Acad. Engring. Achievements include patent for synchronous satellite, electric power plant using electrolytic cell-fuel cell combination. Office: MIT Lincoln Lab 244 Wood St PO Box 73 Lexington MA 02173

MORROW, WILLIAM JOHN WOODROOFE, immunologist; b. Sept. 12, 1949; came to U.S., 1982; s. Albert William and Elfreda Elizabeth (Woodroofe) M.; m. Anna Maria Abai, Oct. 25, 1980; 1 child, Olivia. BSc, Univ. Coll., Cardiff, Wales, United Kingdom, 1974; PhD, Plymouth Polytech., United Kingdom, 1978. Postdoctoral rsch. fellow Middlesex Hosp. Med. Sch., London, 1977-82; vis. scientist U. Calif. Sch. Medicine, San Francisco, 1982-85, asst. rsch. immunologist, 1985-87; immunologist, sr. scientist IDEC Pharm. Corp., La Jolla, Calif., 1987—. Co-author: (book) Autoimmune Rheumatic Disease, 1987; editor BBA: Reviews on Cancer Spl. issue on Human Immunodeficiency Virus, 1989; contbr. 60 articles to profl. jours. Grantee: Royal Coll. of Physicians, London, 1982, The Royal Soc., London, 1982, The Am. Found. for AIDS Rsch., 1986, Calif. U.-Wide Task Force on Aids, 1986, NIH, 1988, '90. Mem. Am. Assn. Immunologists, British Soc. Immunology, Internat. AIDS Soc., Marine Biol. Assn. of U.K. Achievements include research in autoimmune diseases and HIV infection. Office: IDEC Pharm Corp 11099 N Torrey Pines Rd La Jolla CA 92037-1081

MORSE, AARON HOLT, chemist, consultant; b. Batavia, N.Y., 1946; s. Chauncey D. and Roberta (Holt) M.; m. Jeannine E. Morse, May 1967; children: Aaron P., Matthew P., Sarah A. BS, Roberts Wesleyan Coll., North Chili, N.Y., 1969, postgrad., 1969-75. Chemist Eastman Kodak, 1971-72; sales rep. Dearborn Chem. Co., Lake Zurich, Ill., 1975-78, Noxell Corp., 1973-75; account exec. Betz Indsl., Trevose, Pa., 1978-91; gen. mgr. WWM, Sioux Falls, S.D., 1991—. Pres. 4-H Coun., Kenai, Alaska, 1988, 90; mem. ch. coun. Luth. Ch. With U.S. Army, 1969-71. Mem. Am. Chem. Soc., Nat. Assn. Corrosion Engrs., Kenai C. of C. Home: PO Box 2813 Kenai AK 99611-2813

MORSE, F. D., JR., dentist; b. Glen Lyn, Va., Apr. 5, 1928; s. Frank D. and Ida Estell (Davis) M.; B.S., Concord Coll., 1951; D.D.S., Med. Coll. Va., 1955; m. Patsy Lee Apple, Feb. 4, 1967; children—Fortis Davis, Pamela Marie. Free lance photographer, 1950-56; practice dentistry, Pearisburg, Va., 1958—; mem. staff Giles Hosp., Pearisburg, 1958-86. Served from asst. dental surgeon to sr. asst. dental surgeon USPHS, 1955-57; assigned to USCG, 1957-58. Mem. Am., S.W. Va. dental assns., Assn. Mil. Surgeons, AAAS Nat. Assn. Advancement Sci., Fedn. Dentaire Internat., Internat. Platform Assn., W.Va. Collegiate Acad. Sci., Beta Phi. Kiwanian. Achievements include research in dental ceramics and roof coatings. Home: Bicuspid Acres Pearisburg VA 24134 Office: Giles Profl Bldg Pearisburg VA 24134

MORSE, MARTIN A., surgeon; b. Louisville, June 25, 1957; s. Marvin Henry and Betty Anne (Hess) M. BS in Zoology with distinction, Duke U., 1979, MD, 1983. Diplomate Nat. Bd. Med. Examiners. Intern, jr. resident dept. surgery Barnes Hosp./Washington U., St. Louis, 1983-85; rsch. fellow dept. pediatric surgery Children's Hosp./Harvard Med. Sch., Boston, 1985-87; sr. resident dept. of surgery U. Rochester, N.Y., 1987-89, chief resident, 1989-90; rsch./clin. fellow in transplantation dept. pediatric surg. Children's Hosp. Med. Ctr., Cin., 1990-92; clin. fellow hand and upper extremity surgery dept. orthopedic surgery U. Pitts. Med. Ctr., 1992-93; fellow in plastic and reconstructive surgery, dept. surgery U. Fla. Coll. Medicine, Gainesville, 1993—; lab. investigator Lab. Exptl. Pathology div. cancer cause and prevention, Nat. Cancer INst./NIH, Rockville, Md., summers, 1974-80; invited prof. dept. grad. nursing Simmons Coll., Boston, 1986-87. Contbr. articles to profl. jours. Vol. Cystic Fibrosis, Am. Cancer Soc., Am. Heart Assn., March of Dimes, Am. Lung Soc.; founding mem. Statue of Liberty/Ellis Isle Found., N.Y.C., 1985, JFK Libr. Found., Boston, 1987, Challenger Ctr. for Space Sci. Edn., Washington, 1987, U.S. Naval Meml. Found., Washington, 1990; mem. Friends of Nat. Libr. of Medicine; mem. Col. Williamsburg Found., Met. Mus. of Art., Boston Mus. Fine Arts, Carnegie Mellon Mus. Lt. comdr. USNR Med. Corps, 1989—. Farley Found. fellow Children's Hosp., Harvard Med. Sch., 1986; recipient Outstanding Svc. award Nat. Cancer Inst., NIH, 1977. Mem. AMA, AAAS, ACS (assoc. fellow), Am. Soc. Artificial Internal Organs, Am. Trauma Soc., Pa. State Med. Soc., Assn. for Acad. Surgery, Aerospace Med. Assn., Assn. Mil. Surgeons of U.S., Am. Soc. Cell Biology and Tissue Culture Assn.,So. Med. Assn., Physicians for Social Responsibilty, N.Y. State Med. Soc., Ohio State Med. Soc., Rochester Surg. Soc., N.Y. Acad. Scis., Phi Beta Kappa, Alpha Omega Alpha, Phi Lambda Epsilon. Achievements include patent for Controlled Cellular Implantation Using Artificial Matrices; first to describe long-term growth of established human extrahepatic biliary epithelial cells in culture; first to describe a specific chemoattractant neutral proteinase in whole human skin, fibroblasts, lymphocytes, and granulocytes. Office: U Fla Coll Medicine Div Plastic Surgery Dept Surgery/Shands Hosp Gainesville FL 32610

MORSE, PHILIP DEXTER, II, chemist, educator; b. Bakersfield, Calif., Oct. 17, 1944; s. Philip Dexter and Constance (Brown) M.; m. Kiyo Ann Akaba, Sept. 3, 1966; children: Emiko, Keiko, Tami. BA, U. Calif., Davis, 1967, PhD, 1972. Postdoctoral fellow U. Bern, Switzerland, 1971-73, Pa. State U., University Park, Pa., 1973-75; asst. prof. Wayne State U., Detroit, 1975-82; rsch. assoc. U. Ill. Coll. Medicine, Urbana, 1982-88; assoc. prof. Ill. State U., Normal, 1988—; owner, operator Scientific Software Svcs., Bloomington, Ill., 1987—; mem. bd. dirs. Steppingstone, Inc., Plymouth, Mich. Contbr. chpt. to Structure and Properties of Cell Membranes, 1984, Methods in Enzymology, 1985; contbr. articles to profl. jours. Ciba-Geigy fellow, 1972. Mem. Biophys. Soc., AAAS, Am. Chem. Soc. Home: 42583 Five Mile Rd Plymouth MI 48170 Office: Ill State U Dept Chemistry # 4160 Normal IL 61761

MORSE, STEPHEN SCOTT, virologist, immunologist; b. N.Y.C., Nov. 22, 1951; s. Murray H. and Phyllis Morse; m. Marilyn Gewirtz, Feb. 1991. BS, CCNY, 1971; MS, U. Wis., 1974, PhD, 1977. NSF trainee dept. bacteriology U. Wis., Madison, 1971-72, rsch. asst., 1972-77; rsch. fellow Nat. Cancer Inst.-Med. Coll. Va./Va. Commonwealth U., Richmond, 1977-80, instr., 1980-81; asst. prof. microbiology Rutgers U., New Brunswick, N.J., 1981-85; rsch. assoc. Rockefeller U., N.Y.C., 1985-88, asst. prof., 1988—; cons. U.S. Congress Office Tech. Assessment, Washington, 1989—; chair conf. on emerging viruses NIH, 1989; cons. Inst. Medicine-Nat. Acad. Scis., com. mem. microbial threats to health, chair subcom. on viruses, 1990—. Editor: Emerging Viruses, 1993, Evolutionary Biology of Viruses, 1993. Mem. Am. Soc. Microbiology, Am. Assn. Pathologists, Am. Assn. Immunologists, Marine Biol. Lab., Sigma Xi. Office: Rockefeller U 1230 York Ave New York NY 10021-6399

MORTAZAWI, AMIR, electrical engineering educator; b. Yazd, Iran, June 7, 1962; came to U.S., 1983; s. Ali and Khanom (Salahi) M. BSEE, SUNY, Stony Brook, 1987; MSEE, U. Tex., 1988, PhDEE, 1990. Teaching asst. U. Tex., Austin, 1987-88, rsch. asst., 1987-90; asst. prof. elec. engring. U. Cen. Fla., Orlando, 1990—; reviewer IEEE Trans. on Microwave Theory and Techniques, IEEE Microwave and Guided Wave Letters; presenter profl. confs. Contbr. articles to profl. jours. Recipient rsch. initiation award NSF, 1991. Mem. IEEE, Tau Beta Pi, Eta Kappa Nu. Achievements include research in microwave circuits and devices, quasi optical techniques. Office: Dept Elec and Computer Eng U Cen Fla Orlando FL 32816-2450

MORTENSEN, KENNETH PETER, materials engineer, water treatment engineer; b. Kansas City, Mo., Nov. 18, 1955; s. Tage Alfred and Else (Eie) M.; m. Sandra Ferenz Lawber, Sept. 11, 1982 (div. 1986); m. Mary Lue Mason, Apr. 27, 1991. Student, Northwestern U., 1973-75; BSChemE, MIT, 1977; postgrad., Washburn U., 1977-78. Registered profl. engr., Kans. Project engr. Marley Cooling Tower Co., Mission, Kans., 1978-87, sr. engr. materials, 1987-92, supr. materials and processes, 1992—; presenter at profl. confs. Dir. security Lake of the Forest, Inc., Bonner Springs, Kans., 1988, treas., 1989-90. Mem. Internat. Ozone Assn. (rsch. mem.), Nat. Fire Protection Assn. (mem. water cooling towers com. 1987—), Soc. Plastics Engring. (treas. Kansas City chpt. 1987-89, pres. 1990-91), Kans. Assn. Profl. Engrs. Achievements include development of low-pressure pipe gaskets for cooling tower spray systems, virgin/regrind extrusion compound for cooling tower film fill, application of ozone water treatment to cooling, low combustibility fill configurations; development and mechanism research of film fill fouling in counter flow cooling towers. Home: 527 Lake Forest Bonner Springs KS 66012 Office: Marley Cooling Tower Co 5800 Foxridge Dr Mission KS 66202

MORTENSEN, RICHARD EDGAR, engineering educator; b. Denver, Sept. 29, 1935; s. Edgar Steele and Frieda Amalie (Boecker) M.; m. Sarah Jean Raulston, Oct. 12, 1974 (div. 1978). BSEE, MIT, 1958, MSEE, 1958; PhD, U. Calif., Berkeley, 1966. Co-op. engr. GE Co., Schenectady, N.Y., 1955-57; mem. tech. staff Space Tech. Labs., L.A., 1958-61; rsch. asst. U. Calif., Berkeley, 1961-65; prof. engring. UCLA, 1965-91, prof. emeritus, 1991—; cons. TRW, Inc., Redondo Beach, Calif., 1966-70, Aerojet-Gen. Corp., Azusa, Calif., 1970-72, Applied Sci. Analytics, Inc., Canoga Park, Calif., 1980-82. Author: Random Signals and Systems, 1987; contbr. to profl. publs. Team mem. Beyond War, Topanga, Calif., 1986-89; alcoholism counselor. Grantee NSF, 1987-90. Mem. IEEE, Soc. Indsl. and Applied Math., Sigma Xi, Tau Beta Pi, Eta Kappa Nu. Avocations: hiking, yoga. Office: Dept Elec Engring 405 Hilgard Ave Los Angeles CA 90024-1594

MORTIMER, J. THOMAS, biomedical engineering educator; b. Las Vegas, Oct. 12, 1939; s. John Thomas and Frances Harriet (Sample) M.; m. Sarah Van Gorder Watterson, Oct. 28, 1967; children—Elizabeth Van Gorder, Katherine Sample, Anne Watterson. B.S. in Elec. Engring., Tex. Tech Coll., Lubbock, 1964; M.S. in Engring., Case Inst. Tech., Cleve., 1965; Ph.D. in Engring., Case-Western Res. U., Cleve., 1968. Mem. faculty Case-Western Res. U., Cleve., 1969—; prof. biomed. engring. Case-Western Res. U., 1981—, dean engring., 1985-86, chmn. computer engring. and sci., 1986-88; vis. prof. U. Karlsruhe, Fed. Republic Germany, 1977-78; cons. Medtronic, Inc., Mpls., 1974—, NIH, Bethesda, Md., 1975-77, VA, Washington, 1979—, LaJolla Tech, Inc., Calif., 1983-84. Recipient numerous grants, NIH, VA. Mem. Biomed. Engring. Soc., Soc. Neurosci., N.Y. Acad. Scis., Am. Assn. Engring. Edn. Office: Case Western Res U 10900 Euclid Ave Cleveland OH 44106-4901

MORTIMER, JAMES WINSLOW, analytical chemist; b. Mt. Kisco, N.Y., Mar. 11, 1955; s. James Winslow and Eileen Ruth (Cutting) M.; m. Dawn Romay Kania, Apr. 30, 1977. BA, Washington and Jefferson U., 1976. Tech. sales rep. Waters Assocs., Milford, Mass., 1978-82; dir. nat. accounts Zymark Corp., Hopkinton, Mass., 1982-89; v.p. Microflex Tech., Triadelphia, W.Va., 1989-90; mgr. mktg. Berthold Systems, Inc., Aliquippa, Pa., 1990—; speaker at profl. confs. Author: Laboratory Robotics, 1987; cons. editor Lab. Robotics Jour., Hershey, Pa., 1990—; assoc. editor Lab. Robotics and Automation, 1988, 90; contbr. articles to tech. publs. Mem. TAPPI, Soc. Analytical Chemists (speaker 1978, 87), Masons. Achievements include development of cleavastat surgical instrument, beaker that will not cause vortexing action. Home: 113 Little John Dr McMurray PA 15317

MORTLOCK, ROBERT PAUL, microbiologist, educator; b. Bronxville, N.Y., May 12, 1931; s. Donald Robert and Florance Mary (Bellaby) M.; m. Florita Mary Welling, Sept., 1954; children—Florita M., Jeffrey R., Douglas P. B.S., Rensselaer Poly. Inst., N.Y., 1953; Ph.D., U. Ill., Urbana, 1958. Asst. prof. microbiology U. Mass., Amherst, 1963-68, assoc. prof. microbiology, 1968-73, prof. microbiology, 1973-78; prof. microbiology Cornell U., Ithaca, N.Y., 1978—. Editor: Microorganisms as Model Systems for Studying Evolution, 1984, The Evolution of Metabolic Function, 1992. Served to 1st lt. U.S. Army, 1959-61. Fellow Am. Acad. Microbiology; mem. AAAS, Am. Soc. Microbiology, Northeastern Microbiologists, Physiology, Ecology and Taxonomy (pres. 1984-91). Office: Cornell U Sect Microbiology Wing Hall Ithaca NY 14852

MORTON, DONALD CHARLES, astronomer; b. Kapuskasing, Ont., Can., June 12, 1933; s. Charles Orr and Irene Mary (Wightman) M.; m. Winifred May Austin, Dec. 12, 1970; children: Keith James, Christine Elizabeth. BA, U. Toronto, 1956; PhD, Princeton U., 1959. Astronomer U.S. Naval Rsch. Lab., Washington, 1959-61; from rsch. assoc. to sr. rsch. astronomer with rank of prof. Princeton (N.J.) U., 1961-76; dir. Anglo-Australian Obs., Epping and Coonabarabran, Australia, 1976-86; dir. gen. Herzberg Inst. Astrophysics, NRC of Can., Ottawa, Ont., 1986—. Contbr. numerous articles to profl. jours. Fellow Australian Acad. Sci.; mem. Internat. Astron. Union, Royal Astron. Soc. (assoc. 1980), Astron. Soc. Australia (pres. 1981-83, hon. mem. 1986), Royal Astron. Soc. Can., Am. Astron. Soc. (councilor 1970-73), Can. Astron. Soc. Australian Inst. Physics (Pawsey Meml. lectr. 1985), Can. Assn. Physicists, U.K. Alpine Club, Am. Alpine Club, Alpine Can. Clubs: U.K. Alpine, Am. Alpine, Alpine Can. Avocations: mountaineering, rock climbing, ice climbing, marathon running. Office: Herzberg Inst Astrophysics, 100 Sussex Dr, Ottawa, ON Canada K1A 0R6

MORTON, HAROLD S(YLVANUS), JR., retired mechanical and aerospace engineering educator; b. Mpls., Oct. 30, 1924; s. Harold S(ylvanus) and Rhoda Ethelwyn (Cowin) M.; m. Margaret Elizabeth James, July 21, 1951; children: Elizabeth Anne, Harold S. III, Rhoda Sue, Nancy James, Joseph Henry. BS in Mech. Engring., U. Minn., 1947; PhD in Physics, U. Va., 1953. R & D engr. N.Mex. Sch. Mines, Albuquerque, 1947-48; engr. Nat. Bur. Stds., Washington, 1949, Applied Physics Lab., Johns Hopkins U., Silver Spring, Md., 1950, 51; rsch. physicist Linde Co. (Union Carbide), Tonawanda, N.Y., 1953-55; physicist U.S. Atomic Energy Commn., Oak Ridge, Tenn. and Wahsington, 1956-62; assoc. prof. Mech. and Aero. Engring. U. Va., Charlottesville, 1962—. Bd. mem. Piedmont Va. C.C., Charlottesville, 1970-80 (chmn. 1978-80). Mem. AIAA (sr.), Am. Physical Soc., Am. Soc. for Engring. Edn., Va. Acad. Sci. Presbyterian. Achievements include research in engineering R & D in areas of plasma dynamics, low density gas dynamics, space flight mechanics, rigid-body rotational dynamics. Home: 370 Mallard Ln Earlysville VA 22936 Office: U Va Dept Mech Aero Nuclear Engring Thornton Hall McCormick Rd Charlottesville VA 22903

MORTON, STEPHEN DANA, chemist; b. Madison, Wis., Sept. 7, 1932; s. Walter Albert and Rosalie (Amlie) M.; B.S., U. Wis., 1954, Ph.D., 1962. Asst. prof. chemistry Otterbein Coll., Westerville, Ohio, 1962-66; postdoctoral fellow water chemistry, pollution control U. Wis., Madison, 1966-67; water pollution research chemist WARF Inst., Madison, 1967-73; head environ. quality dept., 1973-76; mgr. quality assurance Raltech Sci. Services, 1977-82; pres. SDM Cons., 1982—. Served to 1st lt. Chem. Corps, AUS, 1954-56. Mem. Am. Chem. Soc., Am. Water Works Assn., Am. Soc. Limnology and Oceanography, Water Pollution Control Fedn., AAAS. Author: Water Pollution—Causes and Cures, 1976. Home: 1126 Sherman Ave Madison WI 53703-1620

MORTON, TERRY WAYNE, architect; b. Wichita, Kans., July 1, 1957; s. Charlie Edmon and Geneva (Cooper) M.; m. Christine Ann Dumais, July 31, 1976 (div. 1990); children: Terrance Wayne,. Elizabeth Laurel; m. Cheri Renee Jorgenson, Apr. 11, 1992. BArch, U. Ark., 1981. Registered architect, Mo., Tenn. Intern Saunders-Thalden, Inc., St. Louis, 1981-83; project architect Henderson Gantz Architects, St. Louis, 1983-87; assoc. Grieve and Ruth Architects, Knoxville, Tenn., 1987-92; architect project mgr. Barber & McMurry Architects, Knoxville, Tenn., 1992—; vol. architect East Tenn. Community Design Ctr., Knoxville, 1989-90; guest thesis critic U. Tenn., Knoxville, 1991. Vol. profl. Vols. of Am. Homeless Shelter, Knoxville, 1990-91. Mem. AIA, Habitat for Humanity Internat., Nat. Trust Historic Preservation, Sports Car Club Am. Methodist. Avocations: hiking, making furniture, sports car restoration and racing. Home: 629 Deaderick Rd Knoxville TN 37920 Office: Barber & McMurry Architects 623 Lindsay Pl Knoxville TN 37919

MOSCOWITZ, ALBERT JOSEPH, chemist, educator; b. Manchester, N.H., Aug. 20, 1929; s. Mark and Sarah (Kavesh) M. B.S., City Coll. N.Y., 1950; M.A., Harvard, 1954, Ph.D., 1957. NRC-Am. Chem. Soc.-Petroleum Research Fund postdoctoral fellow Harvard, 1957-58, Washington, 1958-59; mem. faculty U. Minn., 1959—, prof. chemistry, 1965—; vis. prof. U. Copenhagen, Denmark, 1961-62, 67-68, 76; Seydel-Woolley vis. prof. Ga. Inst. Tech., 1966; cons. to industry, 1960—, Chmn. nat. screening com. Fulbright-Hays awards for, Scandinavia and Iceland, 1966; vice chmn. Gordon Conf. in Theoretical Chemistry, 1968, chmn., 1970. Adv. editorial bd.: Chem. Physics Letters, 1967-80; assoc. editor: Jour. Chem. Physics, 1970-73; Contbr. profl. jours. Fulbright lectr., 1961; Sloan fellow, 1962-66. Fellow AAAS, Am. Phys. Soc., N.Y. Acad. Scis.; mem. Am. Chem. Soc., Faraday Soc., Royal Soc. Chemistry (Eng.), Royal Danish Acad. Scis. and Letters (fgn. mem.), Harvard Club Minn., Skylight Club, Sigma Xi, Phi Beta Kappa. Home: The Towers 19 S First St Minneapolis MN 55401 Office: U Minn Dept Chem Minneapolis MN 55455

MOSELEY, JOHN MARSHALL, nurseryman; b. New Canton, Va., Aug. 8, 1911; s. John Marshall and May Baxter (Staehlin) M.; m. Edith Batchelor Hancock, Dec. 31, 1942 (div. Jan. 1966); children: John Marshall, James Hancock, William Rogers. BS, U. Richmond, Va., 1930. Analytical chemist Va. State Dept. Hwys, Richmond, 1930-31; chemist Am. Tobacco Co., Richmond, 1931-56, asst. to dir. rsch., 1956-59, from asst. to dir. rsch. to asst. to v.p., 1959-64; dir. agrl. rsch. Am. Tobacco Co., Hopewell, Va., 1964-69, mgr. basic materials rsch., 1969-71, leaf svcs. mgr., 1971-72; farmer Dillwyn, Va., 1973-78; English boxwood nurseryman Dillwyn, 1978—; mem. industry adv. com. tobacco divsn. Agrl. Stabilization and Conservation Svc., mem. industry adv. com. tobacco divsn. Agrl. Rsch. Svc., USDA, 1955-72; mem. state adv. com. Sch. Agriculture, Va. Poly. Inst. and State U., Blacksburg, 1955-72. Contbr. articles to profl. jours. Mem. Agrl. Rsch. Inst., Nat. Acad. of Scis. Washington, 1955-72, Am. Chem. Soc., 1935-72, Va. Acad. Sci., 1939-72; mem. Fishing Bay Yacht Club, 1939-72, commodore, 1947-48. Capt. USAAF, 1942-46, PTO. Recipient Citation, Am. Men of Sci., 1961, 66, Tobacco Sci., 1967, Am. Men & Women of Sci., 1972, The Daily Progress, 1991. Mem. Va. Farm Bur., Richmond Nursery Assn., Am. Tree Farm System, Buckingham Ruritan (sec. 1989-91, v.p. 1993—), Theta Chi. Baptist. Avocations: sailing, gardening. Home: RR 2 Box 93 Dillwyn VA 23936-9525

MOSER, KENNETH MILES, physician; b. Balt., Apr. 12, 1929; s. Simon and Helene Joyce M.; m. Sara Falk, June 17, 1951; children—Gregory, Kathleen, Margot, Diana. B.A., Haverford Coll., 1950; M.D., Johns Hopkins U., 1954. Diplomate: Am. Bd. Internal Medicine. Intern, resident in medicine D.C. Gen. Hosp., Georgetown Hosp., 1954-59; chief pulmonary and infectious disease service Nat. Naval Med. Center, Bethesda, Md., 1959-61; dir. pulmonary div. Georgetown U. Med. Center, Washington, 1961-68; prof. medicine, dir. pulmonary and critical care med. div. U. Calif., San Diego Sch. Medicine, 1968—; dir. Specialized Ctr. Rsch. U. Calif-San Diego NHLBI, 1978—. Author 10 books in field of pulmonary medicine and thrombosis.; Contbr. articles to med. jours. Bd. dirs. Am. Lung Assn. of San Diego and Imperial Counties, 1969-76, Am. Lung Assn. of Calif., 1976-80; mem. manpower com. Nat. Heart, Lung and Blood Inst., bd. dirs. 1978—. Served with U.S. Navy, 1959-61. Fellow A.C.P., Am. Coll. Chest Physicians; mem. Am. Thoracic Soc. (exec. bd., pres. 1985-86), Am. Heart Assn. Coun. on Thrombosis, Am. Physiol. Soc. Office: U Calif San Diego Med Ctr 200 W Arbor Dr San Diego CA 92103-8372*

MOSER, ROYCE, JR., physician, medical educator; b. Versailles, Mo., Aug. 21, 1935; s. Royce and Russie Frances (Stringer) M.; m. Lois Anne Hunter, June 14, 1958; children: Beth Anne Moser McLean, Donald Royce. BA, Harvard U., 1957, MD, 1961; MPH, Harvard Sch. Pub. Health, Boston, 1965. Diplomate Am. Bd. Preventive Medicine (trustee), Am. Bd. Family Practice. Commd. officer USAF, 1962, advanced through grades to col., 1974; resident in aerospace medicine USAF Sch. Aerospace Medicine, Brooks AFB, Tex., 1965-67; chief aerospace medicine Aerospace

Def. Command, Colorado Springs, Colo., 1967-70; comdr. 35th USAF Dispensary Phan Rang, Vietnam, 1970-71; chief aerospace medicine br. USAF Sch. Aerospace Medicine, Brooks AFB, 1971-77; comdr. USAF Hosp., Tyndall AFB, Fla., 1977-79; chief clin. scis. div. USAF Sch. Aerospace Medicine, Brooks AFB, 1979-81, chief edn. div., 1981-83, sch. comdr., 1983-85; ret., 1985; prof., vice chmn. Dept. Family and Preventive Medicine U. Utah Sch. Medicine, Salt Lake City, 1985—, dir. Rocky Mountain Ctr. for Occupational and Environ. Health, 1987—; cons. in occupational, environ. and aerospace medicine, Salt Lake City, 1985—; presenter nat. and internat. med. meetings. Author: Effective Management of Occupational and Environmental Health and Safety Programs, 1992; contbr. book chpts. and articles to profl. jours. Mem., past pres. First Bapt. Ch. Found., Salt Lake City, 1987-89; mem., chmn. numerous univ. coms., Salt Lake City, 1985—; bd. dirs. Hanford Environ. Health Found., 1990-92; mem. preventive medicine residency review commn. Accredation Coun. Grad. Med. Edn., 1991—; mem. ednl. adv. bd. USAF Human Systems Ctr., 1991—. Decorated Legion of Merit (2). Fellow Aerospace Med. Assn. (pres. 1989-90, Harry G. Mosely award 1981, Theodore C. Lyster award 1988), Am. Coll. Preventive Medicine (regent 1981-82), Am. Coll. Occupational and Environ. Medicine, Am. Acad. Family Physicians; mem. Internat. Acad. Aviation and Space Medicine (selector 1989—), Soc. of USAF Flight Surgeons (pres. 1978-79, George E. Schafer award 1982), Phi Beta Kappa. Avocations: photography, fishing. Home: 664 Aloha Rd Salt Lake City UT 84103-3329 Office: Dept Family & Preventive Medicine 50 N Medical Dr Salt Lake City UT 84132-0001

MOSES, ELBERT RAYMOND, JR., speech and dramatic arts educator; b. New Concord, Ohio, Mar. 31, 1908; s. Elbert Raymond Sr. and Helen Martha (Miller) M.; m. Mary Miller Sterrett, Sept. 21, 1933 (dec. Sept. 1984); 1 child, James Elbert (dec.); m. Caroline Mae Entenman, June 19, 1985. AB, U. Pitts., 1932; MS, U. Mich., 1934, PhD, 1936. Instr. U. N.C., Greensboro, 1936-38; asst. prof. Ohio State U., Columbus, 1938-46; assoc. prof. Ea. Ill. State U., Charleston, 1946-56; asst. prof. Mich. State U., E. Lansing, Mich., 1956-59; prof. Clarion (Pa.) State Coll., 1959-71, chmn. dept. speech and dramatic arts, 1959—, emeritus prof., 1971—; Fulbright lectr. State Dept. U.S. Cebu Normal Sch., Cebu City, Philippine Islands, 1955-56; vis. prof. phonetics U. Mo., summer 1968; hon. sec.'s advocate dept. of aging State of Pa., Harrisburg, 1980-81. Author: Guide to Effective Speaking, 1957, Phonetics: A History and Interpretation, 1964, Three Attributes of God, 1983, Adventure in Reasoning, 1988, Beating the Odds, 1992; poems included in Best Poems of the 90s, 1992; contbr. articles to profl. jours. Del. 3d World Congress Phoneticians, Tokyo, 1976; mem. nat. adv. com. fng. students and tchrs. HEW; del. to Internat. Congress Soc. Logopedics and Phoniatre, Vienna, 1965; liaison rep. to Peace Corps; pres. County Libr. Bd.; past exec. dir. Clarion County United Way; commr. Boy Scouts Am., 1976-77; pres. Venango County Adv. Coun. for Aging, 1978-79. Maj. AUS, 1942-46, lt. col. AUS, ret. Recipient Ret. Sr. Vol. Program Vol. of Yr. award No. Ariz. Coun. Govts., 1989, Spl. award Speech Communication Assn., 1989, Endowment Benefactor award, 1991; 6 Diamond Pin of Melvin Jones Found., Internat. Lions. Fellow United Writers Assn.; mem. Hospitalier Order of St. John of Jerusalem, Knights Hospitalier, Knightly and Mil. Order of St. Eugene of Trebizond (chevalier), Soverign and Mil. Order of St. Stephen the Matyr (comdr.), Knightly Assn. of St. George the Matyr, Ordre Chevaliers du Sinai, Hist. File, VFW (comdr.), Am. Legion (comdr.), Rotary (pres. 1966-67, dist. gov. 1973-74), Order of White Shrine of Jerusalem, Niadh Nask (Marshall of Kilbonane), Internat. Chivalric Inst., Confedn. of Chivalry (life, mem. grand coun.), Ordre Souverain et Militaire de la Milice du Saint Sepulcre (chevalier grand cross), Sovereign World Order of White Cross (lord of knights, dist. commdr. Ariz.), Prescott High Twelve Club (pres. 1990), Phi Delta Kappa (Svc. Key 1978). Republican. Methodist. Avocation: ham radio. Home: 2001 Rocky Dells Dr Prescott AZ 86303-5685

MOSES, GREGORY ALLEN, engineering educator; b. Kalamazoo, Mich., Apr. 7, 1950; s. John Stuart and Blanche (Marousek) M.; m. Sharon Louise Hallquist, Apr. 16, 1983; children: Laurel Erin, Lindsey Elizabeth. BS in Engring., U. Mich., 1972, PhD, 1976. Asst. prof. U. Wis., Madison, 1976-80, assoc. prof., 1980-84, prof. of engring., 1984-89, assoc. dean rsch., 1989—. Co-author: Inertial Confinement Fusion, 1982; author: Engineering Applications Software Development, 1989; contbr. over 50 articles to profl. jours. Mem. Am. Nuclear Soc., Am. Phys. Soc. Lutheran. Home: 5 Mt Rainier Ln Madison WI 53705-2453 Office: U Wis Engring Expt Sta 1500 Johnson Dr Madison WI 53706-1687

MOSES, HAMILTON, III, neurology educator, hospital executive; b. Chgo., Apr. 29, 1950; s. Hamilton Jr. and Betty Anne (Theurer) M.; m. Elizabeth Lawrence Hormel, 1977 (dec. 1988); m. Alexandra MacCollough Gibson, 1992. BA in Psychology, U. Pa., 1972; MD, Rush Med. Coll., Chgo., 1975. Intern in medicine Johns Hopkins Hosp., Balt., 1976-77, resident in neurology, 1977-79, chief resident, 1979-80, assoc. prof. neurology, 1986—, vice chmn. neurology and neurosurgery, 1980-86, v.p., 1988—, dir. Parkinson's Ctr., 1984—. Editor, major author: Principles of Medicine, 1985-88; editor newsletter Johns Hopkins Health, 1988—; contbr. numerous articles to med. jours. Mem. com. on med. ministries Episcopal Diocese Md., Balt., 1987; bd. dirs. Valleys Planning Ct. Mem. Am. Acad. Neurology (sec. 1989—), Am. Neurol. Assn., Md. Neurol. Soc. (pres. 1984-86), Movement Disorders Soc., Md. Club, Green Spring Valley Hunt Club (Garrison, Md.). Republican. Avocations: landscape photography, sailing. Office: Johns Hopkins Hosp 600 N Wolfe St Baltimore MD 21205-2104

MOSES, JOHNNIE, JR., microbiologist; b. Kinston, N.C., May 24, 1939; s. Johnnie Moses and Lillie Ann (Williams) Dillahunt; m. Mirian Louise Mosely, Aug. 16, 1958; children: Nicholas G., Adrianne D. BA, Fordham U., 1978; MA, NYU, 1982. Lic. clin. lab. tech. Lab. technologist Harlem Hosp. Ctr., N.Y.C., 1962-68, sr. lab. technologist, 1968-80, lab. microbiologist, 1980-90, lab. microbiology cons., 1980—, assoc. microbiology, 1990—; hematology isntr. Mandl Med. Asst. Sch., N.Y.C., 1983-85; Am. history prof. Malcolm-King Coll., N.Y.C., 1983-88. Mem. editorial rev. bd. Black Chronicle. Adv. mem. N.Y. State Assembly, N.Y.C., 1988; treas., exec. Addicts Rehab. Ctr., 1975—; treas. Manhattan Christian Reformed Ch., 1975—. Recipient Cert. of Appreciation, Harlem Hosp. Pathology Dept., 1990. Mem. Internat. Soc. Clin. Lab. Tech. (certs. of merit 1982, 91), Nat. Sickle Cell Anemia Found., Guitar Found. Am., Fordham U. Alumni Assn., NYU Alumni Assn. Democrat. Avocations: microphotography, classical guitar, art, African-American research. Home: 990 Tinton Ave Bronx NY 10456-7106 Office: Harlem Hosp Ctr 506 Lenox Ave New York NY 10037-1802

MOSHER, ALAN DALE, chemical engineer; b. Concordia, Kans., Aug. 23, 1963; s. Robert Dean and Helen Imogene (Starr) M.; m. Jacqueline Marie Rohr, Apr. 23, 1988; children: Michelle Leigh, Jonathan Robert. AA, Cloud County Community Coll., Concordia, Kans., 1983; BS in Chem. Engring., U. Kans., 1986. Registered profl. engr., Kans. Process engr. J.C. Butler Assocs., Salina, Kans., 1986-89, chief chemist of metals lab., 1986-89; asst. process engr. The C.W. Nofsinger Co., Kansas City, Mo., 1989-90, staff process engr., 1990-92; sr. process engr. Strato Inc., Leawood, Kans., 1992—. Bd. dirs. Country Oaks Homes Assn., Stanley, Kans., 1992. Mem. Am. Inst. Chem. Engrs., Tau Beta Pi, Phi Lambda Upsilon. Methodist. Home: 15299 Newton Stanley KS 66223 Office: Stratco Inc 4601 College Blvd #300 Leawood KS 66211

MOSHER, DONALD RAYMOND, chemical engineer, consultant; b. Mpls., Jan. 7, 1930; s. Cleveland Bert and Rose (Alkofer) M.; m. Jane Lucille Ryan, June 20, 1954 (div. Dec. 1989); children: Leslie Renee Mosher Goode, Lee David, Laura Ann Mosher Flanders, Jennifer Lynn Mosher Konzen, Jill Teresa, Jody Lavonne; m. Lurlie Elizabeth Amsler, Dec. 19, 1992. BSChemE, U. Minn., 1953, MS, 1954. Prodn. area engr. U.S. Chem. Corps., Edgewood, Md., 1955-57; mng. engr. Union Carbide Corp., South Charleston, W.Va., 1957-68; cons. in new ventures Union Carbide Linde Div., Tanawhonda, N.Y., 1968-69; chief engr. design Hess Oil Corp., Woodbridge, N.J., 1969-70; plant mgr. Stearns Roger Ops., Rapid City, S.D., 1970-72; mgr. engring. sect. Stearns Roger Corp., Denver, 1972-82; mgr. gas process and environ. Allis Chalmers Coal Gas Corp., Milw., 1982-88; cons. Ralston Internat. Trading Assn. Inc., Kingsport, Tenn., 1988—, EG & G Analytical Svcs. Inc., Morgantown, W.Va., 1989—. Author conf. reports; contbr. articles to profl. jours. Chmn. South Hills Community Assn., Charleston, 1967; mem. sponsor's com. So. Ill. U., 1986. With U.S. Army, 1955-57. Univ. and AEC scholar, 1952-54. Mem. Am. Inst. Chem. Engrs. (treas. local sect. 1963), Am. Chem. Soc., Assn. Cons. Chemists and Chem. Engrs. (cons. 1989—), Police Star Found. Tex., Rotary, Phi Kappa (v.p. 1952-54), Phi Beta Epsilon. Republican. Roman Catholic. Achievements include patent for oxidation of butane; invention of system for conserving energy while cleaning dirty gas, of system for removing oil and tar from water with no discharge. Office: D R Mosher Corp PO Box 820295 Houston TX 77282-0295

MOSHER, FREDERICK KENNETH, engineering consultant; b. Middletown, N.Y., Aug. 25, 1943; s. Fred J. and Ruth M. (Werlau) M.; student N.Y.U., 1970-72, Lafayette U., 1973-74; m. Gail J. Berry, Jan. 24, 1968; children—Scott, Kerri, Dean. With Mayo, Lynch & Assos., Architect & Engrs., Hoboken, N.J., 1962-64, designer, 1964-69; mech. designer Louis Goldberg & Assos., Metuchen, N.J., 1969-74, assoc., 1975; partner Brownworth, Mosher & Doran, Piscataway, N.J., 1976-90, Mosher & Doran, Edison, N.J., 1990—. Pres., St. Luke's Luth. Ch., Washington, N.J., 1975-81; mem. Warren County Uniform Constrn. Code Bd. Appeals. Served with Security Agy., U.S. Army, 1965-71. Recipient Mem. Recognition award Cons. Engrs. Coun. N.J., 1990. Fellow Am. Soc. Cert. Engring. Technicians, Am. Cons. Engring. Coun. (Nat. Award for Engring. Excellence 1979); mem. N.J. Cons. Engrs. Coun. (chmn. engring. excellence com.), Am. Soc. Mil. Engrs., IEEE, ASHRAE (3d pl. award for alternative or renewable energy utilization 1982), Nat. Soc. Profl. Engrs., Constrn. Specification Inst. Lutheran. Home: 21 Oak Ridge Rd Washington NJ 07882-1503 Office: Ste 300 3090 Woodbridge Ave Edison NJ 08818

MOSHER, HARRY STONE, chemistry educator; b. Salem, Oreg., Aug. 31, 1915; s. Daniel Harris and Maude Aurelia (Stone) M.; m. Carol Beth Walker, June 23, 1944; children: Janet Lee, Stephen Eric, Leslie Jean. BA, Willamette Univ., 1937; MS, Oreg. State Coll., 1939; PhD, Pa. State Coll., 1942; DSc, Willamette U., 1981. Asst. prof. Willamette U., Salem, Oreg., 1939-40, Pa. State Coll., State College, 1943-46; asst. prof. to prof. Stanford (Calif.) U., 1947-52; vis. prof. F.U. of Amsterdam, 1975. Co-author: Asymmetric Organic Reactions; contbr. articles to profl. jours. Bd. trustees Palo Alto Meth. Ch., 1987-90. Named Disting. Alumni Citation, Willamette U., 1966. Fellow AAAS, Calif. Acad. Sci., Chem. Soc. London; mem. Am. Chem. Soc. (councilor 1966-86, com. on profl. tng. 1970-80). Achievements include research on stereochemistry, natural products, animal toxins, organic peroxides and synthetic methods. Home: 713 Mayfield Ave Stanford CA 94305 Office: Dept Chemistry Stanford U Stanford CA 94305

MOSIER, STEPHEN RUSSELL, college program director, physicist; b. San Rafael, Calif., Nov. 14, 1942; s. Russell Glenn and Marjorie Jean (Carhart) M.; m. Catherine Priscilla Spindle, June 14, 1964; children: Catherine Priscilla, Roger Carhart. BS, Coll. William & Mary, 1964; PhD, U. Iowa, 1970. Rsch. scientist NASA/Goddard Space Flight Ctr., Greenbelt, Md., 1971-78; dir. U.S.-Japan programs NSF, Washington, 1978-81, dir. U.S.-France program, 1981-83; assoc. v.p. internat. affairs U. Houston System, 1983-86; dir. rsch. svcs. U. N.C., Greensboro, 1986—; vice chmn., bd. dirs. N.C. Assn. for Biomed. Rsch., Raleigh, 1989—; bd. dirs. Ctr. for Applied Tech., Houston, 1984-86; rsch. cons. various univs. Contbr. articles to Jour. Geophys. Rsch., Solar Physics, Nature, Transactions (IEEE). Mem. exec. bd. Gen. Greene Coun. Boy Scouts Am., Greensboro, 1987—; Scout leader Washington, Houston, Greensboro, 1970—. Mem. AAAS, Am. Geophys. Union, Soc. Rsch. Adminstrs., Sigma Xi. Methodist. Achievements include research in magnetospheric physics, solar physics, international science and technology policy. Office: U NC at Greensboro Rm 100 McIver Bldg Greensboro NC 27412

MOSJIDIS, CECILIA O'HARA, botanist, researcher; b. Lima, Peru, Aug. 15, 1960; d. Felix M. and Aurea M.Y. (Gaberscik) O'Hara; m. Jorge A. Mosjidis, Mar. 23, 1986; 1 child, Alexis. BS in Biology, U. Cayetano Heredia, Lima, 1982, MS in Biochemistry, 1985; MS in Botany, Auburn U., 1992. Student aide Auburn (Ala.) U., 1987-88, rsch. technician II, 1988-90, rsch. technician VII, 1990—. Contbr. articles to profl. jours.

MOSKAL, JOSEPH RUSSELL, biochemist; b. Saginaw, Mich., June 10, 1950; s. Robert James M. and Lavonne Bernice (Brandt) McIntyre; m. Danielle Marie Sedilleau, July 21, 1980. BS, U. Notre Dame, 1972, PhD, 1977. Dir. rsch. Chgo. Inst. for Neurosurgery and Neurorsch., 1990—; adj. assoc. prof. dept. cell, molecular and structural biology med. sch. Northwestern U., Chgo., 1991—, dept. biomed. engring., 1991—; cons. Chgo. Children's Mus., 1991—. Editor: New Directions in Neuro-Oncology, 1992. NSF fellow, 1971. Mem. Am. Soc. Biochemistry and Molecular Biology, Internat. Soc. Neurochemistry, Soc. Neurosci., AAAS. Achievements include development of monoclonal antibodies that enhance learning and memory by interactin with NMDA receptor. Office: Chgo Inst Neurosurgery 428 W Deming Pl Chicago IL 60614

MOSKOWITZ, RICHARD, physician; b. Passaic, N.J., Dec. 16, 1938; s. Harry and Sylvia (Margoles) M.; m. Carol Ann Dukes, Feb. 15, 1963 (div. 1965); 1 child, Richard; m. Linda Ruth Reissman, Oct. 20, 1984; 1 child, Jennifer. BA, Harvard U., 1959; MD, NYU, 1963; fellow in philosophy, U. Colo., 1963-65. Diplomate Am. Bd. Homeotherapeutics. Intern St. Anthony's Hosp., Denver, 1966-67; staff physician Beth Israel Hosp., Denver, 1967-68, West Side Neighborhood Health Ctr., Denver, 1968-69, Red Hook Neighborhood Health Ctr., Bklyn., 1969-70; pvt. practice, 1971; mem. faculty Nat. Ctr. Homeopathy, Washington, 1978—, bd. dirs., 1980-87, pres., 1985-86; mem. faculty Internat. Found. Homeopathy, Seattle, 1992—. Author: Dissent in Medicine (with others), 1989, Homeopathic Medicines in Pregnancy and Childbirth, 1992; contbr. articles to Jour. Am. Inst. Homeopathy, Brit. Homeopathic Jour., others. Mem. ACLU, Am. Inst. Homeopathy, Am. Holistic Med. Assn., Amnesty Internat., Phi Beta Kappa. Democrat. Jewish. Achievements include investigation of homeopathy as a species of medical reasoning, malpractice as a subspecies of iatrogenic illness and injury, vaccinations as a hidden source of chronic diease. Office: 173 Mount Auburn St Watertown MA 02172

MOSKOWITZ, RONALD, electronics executive; b. N.Y.C., Feb. 15, 1939; m. Phyllis Lenes, June 30, 1957; 4 children. BEE, CUNY, 1961; MSEE, Rutgers U., 1963, PhD in Engring., 1966. Registered profl. engr., Mass. Prin. investigator RCA Labs. and Astroelectronics div., 1961-67; instr. elec. engring. Rutgers U., 1965-67; prof. elec. engring. U. Miss., 1967; sr. cons. engr., programs mgr. Avco. Corp., 1967-68; founder, chmn. bd., chief exec. officer Ferrofluidics Corp., Nashua, N.H., 1968—; adj. faculty Lowell Tech. Inst., 1967. Contbr. articles to profl. jours.; patentee in field. RCA fellow MIT, 1963, AEC fellow Princeton U., 1964, NSF fellow Rutgers U., 1964-65. Mem. IEEE, AAAS, Am. Vacuum Soc., Tech. Assn. Graphic Arts, ASME, Am. Soc. Lubrication Engrs., Sigma Xi, Eta Kappa Nu. Office: Ferrofluidics Corp 40 Simon St Nashua NH 03061-3027

MOSS, BERNARD, virologist, researcher; b. N.Y.C., July 26, 1937; s. Jack and Goldie (Altman) M.; m. Toby Frima Lieberman, Dec. 25, 1960; children: Robert, Jennifer, David. BA, NYU, 1957, MD, 1961; PhD, MIT, 1966. Diplomate Am. Bd. Medical Examiners. Intern Children's Hosp., Boston, 1961-62; investigator, sect. head NIH, Bethesda, Md., 1966—, lab. chief, 1984—; mem. adv. bd. Virus Res., 1984—, Current Opinion Biotech., 1989—. Assoc. editor Virology Jour., 1976-92, editor., 1992—; mem. editorial bd. Jour. of Virology, 1972—, Antimicrobial Agts. and Chemotherapy, 1973-79, Jour. Biol. Chemistry, 1982-87; AIDS rsch. Human Retroviruses, 1989—; contbr. more than 400 articles to profl. jours. Mem. adv. com. Am. Cancer Soc., N.Y.C., 1983-86; bd. dirs. Found. Advanced Edn. in Scis., Bethesda, 1985-91; mem. NIH AIDS vaccine selection com., 1989—. Served as med. dir. USPHS, 1966—. Named one of 100 Most Innovative Scientists of 1986, Sci. Digest; recipient Solomon A. Berson Alumni Achievement award Sch. Medicine, NYU, Meritorious Svc. medal USPHS, Disting. Svc. medal USPHS, Dickson prize in medicine, Invitrogen award for eukaryotic gene expression. Mem. AAAS, Am. Soc. Biol. Chemists, Am. Soc. Microbiology, Am. Soc. Virology, Nat. Acad. Sci., Phi Beta Kappa, Sigma Xi, Alpha Omega Alpha. Office: NIH 9000 Rockville Pike Bldg Rm 229 Bethesda MD 20814-1436

MOSS, CHARLES NORMAN, physician; b. L.A., June 13, 1914; s. Charles Francis and Lena (Rye) M.; A.B., Stanford U., 1940; M.D., Harvard U., 1944; cert. U. Vienna, 1947; M.P.H., U. Calif.-Berkeley, 1955; Dr.P.H., UCLA, 1970; m. Margaret Louise Stakias; children—Charles Eric, Gail Linda, and Lori Anne. Surg. intern Peter Bent Brigham Hosp., Boston, 1944-45, asst. in surgery, 1947; commd. 1st lt. USAF, M.C., USAAF, 1945, advanced through grades to lt. col., USAF, 1956; Long course for flight surgeon USAF Sch. Aviation Medicine, Randolph AFB, Tex., 1948-49, preventive medicine div. Office USAF Surgeon Gen., Washington, 1955-59; air observer, med., 1954, became sr. flight surgeon 1956; later med. dir., Los Angeles div. North Am. Rockwell Corp., Los Angeles; chief med. adv. unit Los Angeles County, now ret. Decorated Army Commendation medal (U.S.); Chinese Breast Order of Yun Hui. Recipient Physicians Recognition award AMA, 1969, 72, 76, 79, 82. Diplomate in aerospace medicine and occupational medicine Am. Bd. Preventive Medicine. Fellow Am. Pub. Health Assn., AAAS, Am. Coll. Preventive Medicine, Royal Soc. Health, Am. Acad. Occupational Medicine, Western Occupational Med. Assn., Am. Assn. Occupational Medicine; mem. AMA, Mil. Surgeons U.S., Soc. Air Force Flight Surgeons, Am. Conf. Govt. Hygienests, Calif. Acad. Preventive Medicine, (dir.), Aerospace Med. Assn., Calif., Los Angeles County med. assns., Assn. Oldetime Barbell and Strongmen. Research and publs. in field. Home: 7714 Cowan Ave Los Angeles CA 90045-1135

MOSS, HERBERT IRWIN, chemist; b. Bklyn., Mar. 8, 1932; s. Jacob and Marcelle (Garblik) M.; m. Geraldine Georgia Germek, Sept. 10, 1960; children: Jennifer, David, Jessica. BS, U. Louisville, 1953; PhD, Ind. U., 1960. Mem. tech. staff RCA Labs., Princeton, N.J., 1959-87; freelance cons. Point Pleasant, Pa., 1987-; mem. steering com. Thin Film div. Am. Vacuum Soc., N.Y., 1965-68. Co-author: Refractory Coatings, 1981; contbr. articles to profl. jours. Bd. dirs. Bucks Tourist Commn., Doylestown, Pa., 1988-91. Mem. Am. Ceramic Soc. (program com. 1966-67, session chmn. 1967), Am. Chem. Soc., Sigma Xi. Achievements include patents on materials for magnetic recording heads, pressure sintering and video disc styli. Home and Office: PO Box 569 Point Pleasant PA 18950-0569

MOSS, JOEL, medical researcher; b. Bklyn., Nov. 27, 1946. BA, Brandeis U., 1967; MD and PhD, NYU, 1972. Intern, then resident Johns Hopkins Hosp., Balt., 1972-74; rsch. assoc., pulmonary fellow Nat. Heart, Lung and Blood Inst., Bethesda, Md., 1974-77, staff investigator, 1977-79, head molecular mechanisms sect., 1979—, dep. chief, Lab. Cellular Metabolism, 1986—. Office: NIH Rm 5N-307 Bldg # 10 Bethesda MD 20892

MOSS, KENNETH WAYNE, neurologist; b. Teaneck, N.J., June 19, 1950; s. Joseph Daniel and Doris Marie (Smith) M.; m. Audrey Arnold, June 16, 1976 (div. June 1991); 1 child; Kathryn Elizabeth-Dolan; m. Dena Ann Avery, Oct. 20, 1991. BS, Georgetown U., 1972, MS, 1974, MD, 1979, PhD, 1980. Diplomate Am. Bd. Psychiatry and Neurology. Neurologist Andrews AFB, Md., 1983-87, Neurology Svcs., Fairfax, Va., 1987-88, 98th Gen. Hosp., Nurenberg, Germany, 1988-89, Womack Army Med. Ctr., Fort Bragg, N.C., 1989—. Lt. col. U.S. Army, 1988—. Mem. Am. Acad. Neurology. Republican. Lutheran. Office: Womack Army Med Ctr Dept Psych & Neuro Fort Bragg NC 28307-5000

MOSS, SIMON CHARLES, physics educator; b. Woodmere, N.Y., July 31, 1934; married, 1958; 4 children. SB, Mass. Inst. Tech., 1956, SM, 1959, ScD in Metallurgy, 1962. Mem. rsch. staff metallurgy Raytheon Mfg. Co., 1956-57; from asst. to assoc. prof. Mass. Inst. Tech., 1962-64; dir. sci. dept. Energy Conversion Devices, Inc., 1970-72; prof. physics U. Huston, 1972—. Ford Engring. fellow Mass. Inst. Tech., 1962-64, Guggenheim fellow, 1968-69; Alexander von Humboldt Sr. Scientist grantee U. Munich, 1979; Recipient David Adler Lectureship award Am. Physical Soc., 1993. Fellow Am. Physical Soc. Achievements include research in X-ray and neutron diffraction; structure of disordered and defective solids; crystallography and thermodynamics of phase transformations; amorphous semiconductors; hydrogen in metals; biological structures. Office: Univ of Houston Dept of Physics 1 Main St Houston TX 77002*

MOSSAVAR-RAHMANI, BIJAN, oil and gas company executive; b. Tehran, Iran, June 14, 1952; came to U.S., 1978; s. Morteza and Fatemeh (Mohtashem-Nouri) Mossavar-R.; m. Sharmin Batmanghelidj, Oct., 1980. BA, Princeton (N.J) U., 1974; MS, U. Pa., 1975; MPA, Harvard U., 1982. Oil and energy columnist Kayahan Group of Newspapers, Iran, 1975-78; energy policy analyst Govt. of Iran, 1976-78; vis. rsch. fellow The Rockefeller Found., N.Y., 1978-80; rsch. coord. internat. natural gas study Harvard U., Mass., 1982-85, asst. dir. internat. energy studies, 1985-87; pres. Apache Internat., Inc., Houston, 1988—; bd. dirs. Apache Internat., Inc., Tex., Compagnie des Énergies Nouvelles de Côte d'Ivoire; sr. exec. cons., dir. oil and gas studies Temple, Barker & Sloane, Inc., Mass., 1983-87; chmn. Assocs. Harvard Internat. Energy Program, Cambridge, Mass., 1988-91; mem. Internat. Consultative Group on Mid. East. Author: Energy Policy in Iran, 1981; co-author: OPEC and the World Oil Outlook, 1983, World Natural Gas Outlook, 1984, The OPEC Natural Gas Dilemma, 1986, Energy Security Revisited, 1987, Natural Gas in Western Europe, 1987, Lower Oil Prices: Mapping the Impact, 1988; mem. editorial adv. bd. Offshore mag., 1992—. Bd. dirs. U.S.-Angola C. of C., 1990-92; mem. coun. Internat. Exec. Svc. Corps, 1991—. Mem. Internat. Assn. of Energy Economists, Denver U. Club, Nassau Club, Ivy Club, Harvard Club of N.Y. Avocation: art collecting. Office: Apache Internat Inc 2000 Post Oak Blvd Houston TX 77056-4400

MÖSSBAUER, RUDOLF LUDWIG, physicist, educator; b. Munich, Jan. 31, 1929; s. Ludwig and Erna M.; 3 children. Ed., Technische Hochschule, Munich; D.Sc. (hon.), Oxford U., 1973, U. Leicester, Eng., 1975; Dr. honoris causa, U. Grenoble, France, 1974. Research asst. Max-Planck Inst., Heidelberg, Fed. Republic Germany, 1955-57; research fellow Technische Hochschule, Munich, 1958-60; research fellow Calif. Inst. Tech., 1960, sr. research fellow, 1961, prof. physics, 1961; prof. exptl. physics Tech. U. Munich, 1964-72, 77—; dir. Inst. Max von Laue, Grenoble, France and German-French-Brit. High Flux Reactor, 1972-77. Author publs. on recoilless nuclear resonance absorption and neutrino physics. Recipient Research Corp. award, 1960; Röntgen prize U. Giessen, 1961; Elliott Cresson medal Franklin Inst., Phila., 1961; Nobel prize for physics, 1961; Guthrie medal Inst. Physics (London), 1974; Lomonossovmedal Acad. Sci. USSR, 1984; Einstein medal Albert Einstein Soc., Bern, 1986. Mem. Deutsche Physikalische Gesellschaft, Deutsche Gesellschaft der Naturforscher, Leopoldina, Am. Phys. Soc., European Phys. Soc., Indian Acad. Scis., Am. Acad. Sci. (fgn.), Am. Acad. Arts Scis. (fgn), Nat. Acad. Scis. (fgn. assoc.), Bavarian Acad. Scis., Academia Nazionale dei XL Roma, Pontifical Acad. Scis., Acad. Sci. USSR (fgn.), Acad. European Scis. Arts des Letters France, Hungarian Acad. of Scis., Internat. Acad. Scis., ICSD Munich, Acad. Europe U.K. Office: Tech U Munich, Dept Physics, 8046 Garching Germany

MOSSELMANS, JEAN-MARC, physician; b. Sorengo, Switzerland, Mar. 7, 1963; arrived in Belgium, 1977; s. George and Nicole (Seret-Fontaine) M.; m. N. Cox, Apr. 4, 1987. MD, Cath. U. Louvain en Woluwe, Brussels, 1987; postgrad., Inst. Cath. Hautes Etudes Commerciales, Brussels, 1991—. Freelance journalist ICS-Med. Time Communication, Brussels, 1988; med. dir. Jour. Medicin/Artsenkrant IMS Communications Systems, Brussels, 1989-90; med. dir. ICS, Brussels, 1991—. Contbr. over 100 articles to profl. jours., also booklets. Mem. Assn. Medecins Generalistes Bruxelles Sud-Est (admnistr. 1989-90). Home: Ave Rogier 387, B-1030 Brussels Belgium Office: ICS, 935-937, chaussée de Waterloo, B-1180 Brussels Belgium

MOSSER, HANS MATTHIAS, radiologist; b. Feldkirchen, Kärnten, Austria, Feb. 24, 1955; s. Heinz and Gertrud (Kopper) M. MD, U. Vienna, Austria, 1982. Med. resident Hainburg (Austria) Hosp., 1983, Poliklinik, Vienna, 1983-84; resident in radiology Rudolfstiftung Hosp., Vienna, 1984-90, radiologist, 1990-91; radiologist SMZO Hosp., Vienna, 1991—; presenter in field, 1988—. Author: Ultrasound, 1991, PACS Research, 1991; contbr. articles to profl. jours. Mem. Austrian Roentgen Soc., Radiol. Soc. N.Am., Am. Inst. Ultrasound, Soc. Computer Assisted Radiology, Internat. Soc. for Optical Engring. Avocations: mountaineering, music. Office: SZMO Danube Hosp, Langobardenstrasse 122, A-1220 Vienna Austria

MÖSSNER, JOACHIM, internist, gastroenterologist; b. Würzburg, Bavaria, Fed. Republic Germany, Nov. 17, 1950; s. Franz Emil and Ursula Amalie (Gunder) M.; m. Karin Sigrid Neidhardt, July 22, 1978; children: Felix Oskar, Lone Dorothea, Flora Eleonore. Student, U. Würzburg, Fed.

Republic Germany, 1970-76, MD, 1978, Habilitation, 1987. Intern Dept. of Surgery, Tauberbischofsheim, Fed. Republic Germany, 1977, Medizinische Poliklinik, U. Würzburg, 1978; resident Med. Poliklinik, U. Würzburg, 1978-82, 85-86, chief gastroenterology unit, 1986-93; rsch. assoc. dept. physiology U. Calif., San Francisco, 1983-85; chief dept. internal medicine, gastroenterology U. Leipzig, 1993—; full prof. medicine U. Würzburg, 1989—. Assoc. editor Internat. Jour. Pancreatology. Grantee Deutsche Forschungsgemeinschaft, 1982—. Mem. German Soc. Internal Medicine, Am. Soc. Gastroenterology, Internat. Soc. Pancreatology, Am. Soc. Pancreatology, Internat. Gastro Surg. Club. Home: Guttenbergerstr 20, 97082 Würzburg Germany Office: U Leibzig Medizinische Klinik II, Liebigstrasse, 04103 Leipzig Germany

MOSTELLER, FREDERICK, mathematical statistician, educator; b. Clarksburg, W.Va., Dec. 24, 1916; s. William Roy and Helen (Kelley) M.; m. Virginia Gilroy, May 17, 1941; children: William, Gale. ScB, Carnegie Inst. Tech. (now Carnegie-Mellon U.), 1938, MSc, 1939, DSc (hon.), 1974; AM, Princeton U., 1942, PhD, 1946; DSc (hon.), U. Chgo., 1973, Wesleyan U., 1983; D. of Social Scis. (hon.), Yale U., 1981; LLD (hon.), Harvard U., 1991. Instr. math. Princeton U., 1942-44; research assoc. Office Pub. Opinion Research, 1942-44; spl. cons. research br. War Dept., 1942-43; research mathematician Statis. Research Group, Princeton, applied math. panel Nat. Devel. and Research Council, 1944-46; mem. faculty Harvard U., 1946—, prof. math. stats., 1951-87, Roger I. Lee prof., 1978-87, prof. emeritus, 1987—, chmn. dept. stats., 1957-69, 75-77, chmn. dept. biostats., 1977-81, chmn. dept. health policy and mgmt., 1981-87; dir. Tech. Assessment Group, 1987—; vice chmn. Pres.'s Commn. on Fed. Stats., 1970-71; mem. Nat. Adv. Council Equality of Ednl. Opportunity, 1973-78, Nat. Sci. Bd. Commn. on Pre-coll. Edn. in Math., Sci. and Tech., 1982-83; Fund for Advancement of Edn. fellow, 1954-55; nat. tchr. NBC's Continental Class-room TV course in probability and stats., 1960-61; fellow Center Advanced Study Behavioral Sciences, 1962-63, bd. dirs., 1980-86; Guggenheim fellow, 1969-70; Miller research prof. U. Calif. at Berkeley, 1974-75; Hitchcock Found. lectr. U. Calif., 1985. Co-author: Gauging Public Opinion (editor Hadley Cantril), 1944, Sampling Inspection, 1948, The Pre-election Polls, 1948, 1949, Stochastic Models for Learning, 1955, Probability with Statistical Applications, 1961, Inference and Disputed Authorship, The Federalist, 1964, The National Halothane Study, 1969, Statistics: A Guide to the Unknown, 3d edit., 1988, On Equality of Educational Opportunity, 1972, Sturdy Statistics, 1973, Statistics By Example, 1973, Cost, Risks and Benefits of Surgery, 1977, Data Analysis and Regression, 1977, Statistics and Public Policy, 1977, Data for Decisions, 1982, Understanding Robust and Exploratory Data Analysis, 1983, Biostatistics in Clinical Medicine, 1983, 2d. edit. 1986, Beginning Statistics with Data Analysis, 1983, Exploring Data Tables, Trends and Shapes, 1985, Medical Uses of Statistics, 1986, 2d edit., 1992, Quality of Life and Technology Assessment, 1989, Fundamentals of Exploratory Analysis of Variance, 1992; author articles in field. Trustee Russell Sage Found.; mem. bd. Nat. Opinion Research Center, 1962-66. Recipient Outstanding Statistician award Chgo. chpt. Am. Statis. Assn., 1971, Boston chpt., 1989, Myrdal prize Evaluation Research Soc., 1978, Paul F. Lazarsfeld prize Council Applied Social Research, 1979, R.A. Fisher award Com. of Pres.'s of Statis. Socs., 1987, Medallion of Ctrs. for Disease Control, 1988. Fellow AAAS (chmn. sect. U 1973, dir. 1974-78, pres. 1980, chmn. bd. 1981), Inst. Math. Statistics (pres. 1974-75), Am. Statis. Assn. (v.p. 1962-64, pres. 1967, Samuel S. Wilks medal 1986), Social Sci. Research Council (chmn. bd. dirs. 1966-68), Math. Social Sci. Bd. (acad. governing bd. 1962-67), Am. Acad. Arts and Scis. (council 1986-88), Royal Statis. Soc. (hon.); mem. Am. Philos. Soc. (council 1986-88), Internat. Statis. Inst. (v.p. 1986-88, pres.-elect 1989, pres. 1991-93), Math. Assn. Am., Psychometric Soc. (pres. 1957-58), Inst. Medicine of Nat. Acad. Scis. (council 1978), Nat. Acad. Scis., Biometric Soc. Office: 1 Oxford St Cambridge MA 02138-2901

MOSTILLO, RALPH, medical association executive; b. Newark, Apr. 11, 1944; s. Joseph and Antoinette (Cipriano) M. BA in Chemistry magna cum laude, Rutgers U., Newark, 1972; MA in Biochemistry, Princeton U., 1974, PhD in Biochemistry, 1978. NIH rsch. fellow Princeton (N.J.) U., 1972-78; sr. scientist drug regulatory affairs Hoffmann-La Roche, Inc., Nutley, N.J., 1979-85; founder, chmn., chief exec. officer Am. Cancer Assn., Nutley, 1986—. Assoc. editor U.S. Pharmacopoeia XX-Nat. Formulary XV, 1980-85. With USN, 1962-66, Vietnam. Mem. Am. Chem. Soc., Am. Mgmt. Assn., Am. Mktg. Assn., N.Y. Acad. Scis., Am. Legion, Phi Beta Kappa. Achievements include research on molecular transport systems in E. coli as general models for drug delivery into cells. Home: P O Box 505 Nutley NJ 07110-0505 Office: Am Cancer Assn PO Box 87 Nutley NJ 07110-0087

MOSTOW, GEORGE DANIEL, mathematics educator; b. Boston, July 4, 1923; s. Isaac J. and Ida (Rotman) M.; m. Evelyn Davidoff, Sept. 1, 1947; children: Mark Alan, David Jechiel, Carol Held, Jonathan Carl. B.A., Harvard U., 1943, M.A., 1946, Ph.D., 1948; DSc (hon.), U. Ill., Chgo., 1989. Instr. math. Princeton U., 1947-48; mem. Inst. Advanced Study, 1947-49, 56-57, 75, trustee, 1982-92; asst. prof. Syracuse U., 1949-52; asst. prof. math. Johns Hopkins U., 1952-53, assoc. prof., 1954-56, prof., 1957-61; prof. math. Yale U., 1961-66, James E. English prof. math., 1966-81, Henry Ford II prof. math., 1981—, chmn., 1971-74; vis. prof. Conselho Nat. des Pesquisas, Inst. de Matematica, Rio de Janiero, Brazil, 1953-54, 91, U. Paris, 1966-67, Hebrew U., Jerusalem, 1967, Tata Inst. Fundamental Rsch., Bombay, 1970, Inst. des Hautes Etudes Scientifiques, Bures-Sur-Yvette, 1966, 71, 75, Japan Soc. for Promotion of Sci., 1985, Eidgenossische Technische Hochschule, Switzerland, 1986; chmn. U.S. Nat. Com. for Math , 1971-73, 83-85, Office Math. Scis., NRC, 1975-78; mem. sci. adv. coun. Math. Scis. Rsch. Inst., Berkeley, Calif., 1988-91, Weizmann Inst., Israel, Tel Aviv U.; mem. vis. com. dept. math. Harvard U., 1975-81, MIT, 1981—. Assoc. editor Annals of Math, 1957-64, Trans. Am. Math. Soc, 1958-65, Am. Scientist, 1970-82; editor Am. Jour. Math, 1965-69; asso. editor, 1969-79; author research articles. Fulbright rsch. scholar, Utrecht U., The Netherlands; Guggenheim fellow, 1957-58. Mem. AAAS, NAS (chmn. sect. math. 1982-84), Am. Math. Soc. (pres. 1987-88), Internat. Math. Union (chmn. U.S. del. to gen. assembly Warsaw 1982, mem. exec. com. 1983-86, Ritt lectr. Columbia U. 1982, Bergman lectr. Stanford U. 1983, Sachar lectr. Tel Aviv 1985, Karcher lectr. U. Okla. 1986, Markert lectr. Pa. State U. 1993), Harvard Grad. Coun., Phi Beta Kappa, Sigma Xi. Home: Beechwood Rd Woodbridge CT 06525-1331 Office: Yale U New Haven CT 06520

MOTE, CLAYTON DANIEL, JR., mechanical engineer, educator; b. San Francisco, Feb. 5, 1937; s. Clayton Daniel and Eugenia (Isnardi) M.; m. Patricia Jane Lewis, Aug. 18, 1962; children: Melissa Michelle, Adam Jonathan. BSc, U. Calif., Berkeley, 1959, MS, 1960, PhD, 1963. Registered profl. engr., Calif. Asst. specialist U. Calif. Forest Products Labs., 1961-62; asst. mech. engr., 1962-63; lectr. mech. engring. U. Calif., Berkeley, 1962-63, asst. prof., 1967-69, asst. research engr., 1968-69, assoc. prof., assoc. research engr., 1969-73, prof., 1973—, vice chmn. mech. engring. dept., 1976-80, 83-86, chmn. mech. engring. dept., 1987-91, vice chancellor univ. rels., FANUC chair mech. systems, 1991—; research fellow U. Birmingham, Eng., 1963-64; asst. prof. Carnegie Inst. Tech., 1964-67; vis. prof. Norwegian Inst. Wood Tech., 1972-73, vis. sr. scientist, 1976 '78, '80 '84 '85; cons. in engring design and analysis; sr. scientist Alexander Von Humboldt Found., Fed. Republic Germany, 1988, Japan Soc. for Promotion of Sci., 1991. Mem. editorial bd. Soma Jour. Sound and Vibration, Machine Vibration; contbr. articles to profl. jours.; patentee in field. NSF fellow, 1963-64; recipient Disting. Teaching award, U. Calif., 1971, Pi Tau Sigma Excellence in Teaching award, U. Calif., 1975, Humboldt Prize, Fed. Republic Germany, 1988, Frederick W. Taylor Rsch. medal. Soc. Mfg. Engrs., 1991, Hetenyi award Soc. Exptl. Mechanics, 1992. Fellow NAE, AAAS, ASME (Blackall award 1975, v.p. environ. and transp. 1986-90, nat. chmn. noise control and acoustics 1980-84, chmn. San Francisco sect. 1978-79, Disting. Svc. award 1991), Internat. Acad. Wood Sci., Acoustical Soc. Am.; mem. ASTM (com. on snow skiing F-27 1984-87, chmn. new projects subcom.), Am. Acad. Mechanics, Am. Soc. Biomechanics, Orthopaedic Rsch. Soc., Internat. Soc. Skiing Safety (v.p., sec. 1977-85, bd. dirs. 1977—, chmn. sci. com. 1985—), Sigma Xi, Pi Tau Sigma, Tau Beta Pi. Office: U Calif Dept Mech Engring Berkeley CA 94720

MOTHEO, ARTUR DE JESUS, chemist, educator; b. Santos, Sao Paulo, Brazil, Nov. 16, 1952; s. Arthur dos Santos and Rosa de Jesus Motheo; m. Angela Maria Ferguson Cavichiolli, July 8, 1978; children: Daniel, Tathiana, Stephanie. B in Chemistry, U. Sao Paulo, 1976, M in Phys. Chemistry, 1981, DSc, 1986. Cert. chemist. From instr. to asst. prof. U. Sao Paulo, Sao Carlos, Brazil, 1977-86; assoc. prof. U. Sao Paulo, Sao Carlos, 1986—; assoc. researcher U. Calif., Davis, 1988-90. Author: Laboratory Manual for General Chemistry, 1991; contbr. articles to profl. jours. Grantee U. Guelph, Can. 1982, U. Sao Paulo 1991. Mem. AAAS, Am. Chem. Soc., Internat. Soc. Electrochem., Electrochem. Soc., Inc., Brazilian Chem. Soc., Brazilian Soc. for Advancement of Sci., Brazilian Assn. Surface Treatment, Brazilian Assn. Tech. Rules. Roman Catholic. Avocations: tennis, soccer, chess, home videos. Office: U Sao Paulo Dept Fisico-Quimica, Av Dr Carlos Botelho 1465, 13560 Sao Carlos Sao Paulo, Brazil

MOTHKUR, SRIDHAR RAO, radiologist; b. Mothkur, India, Oct. 5, 1950; came to U.S., 1975; s. Venkat Rao and Laxmi Bai (Gundepally) M.; m. Sheila Rama Rao Paga, Nov. 30, 1973; children: Swathi, Preethi, Venkat Krishna. PVC, Osmania U. Arts & Sci. Coll., Siddipet, India, 1966; MBBS, Osmania U & Inst. Med. Scis., Hyderabad, India, 1972; DPH, Osmania U., Hyderabad, India, 1974. Diplomate Am. Bd. Radiology. Rotating intern Osmania Gen. Hosp., Hyderabad, 1972-73, internal medicine intern, 1973, resident in surgery, 1974-75; resident Resurrection Hosp., Chgo., 1975-76; resident in radiology Luth. Gen. Hosp., Park Ridge, Ill., 1976-79, chief resident radiology, 1978-79; with rotations in nuclear medicine, angiography and neuroradiology Rush-Presbyn. St. Luke's Med. Ctr., Chgo., 1978; chmn. and med. dir. dept. radiology Louise Burg Hosp., Chgo., 1979-85, Shriner's Hosp., Chgo., 1986-88; fellow in ultrasound and computered tomography U. Ill., Chgo., 1988-89, fellow in magnetic resonance imaging, 1988-89; staff radiologist St. Anthony Hosp., Meml. Hosp. and Kingwood Hosp., Michigan City, Ind., 1989—; spl. staff radiologist Christ Hosp. Med. Ctr., Oaklawn, Ill., 1988-89; med. dir. magnetic resonance imaging and interventional radiology St. Anthony Hosp. and Meml. Hosp., Michigan City, Ind., 1989—; clin. asst. prof. radiology U. Ill., Chgo., 1990—. Fellow Am. Coll. Internat. Physicians, Am. Coll. Angiology, Internat. Coll. Angiology; mem. AMA, Am. Coll. Radiology, Soc. Magnetic Resonance Imaging, Ill. State Med. Soc., Chgo. Med. Soc., Radiol. Soc. of N.Am., Am. Roentgen Ray Soc., India Med. Assn. N.W. Ind., Am. Assn. Physicians from India, Am. Coll. Internat. Physicians, Am. Diabetes Assn., Am. Coll. Emergency Physicians, Soc. Cardiovascular and Interventional Radiology, Susruta Radiol. Soc., Soc. Magnetic Resonance in Medicine, Am. Soc. Head and Neck Radiology, Telugu Assn. Greater Chgo., Am. Telugu Assn., Tristate Telugu Assn., Telugu Assn. N.Am., Internat. Soc. of Krishna Consciousness. Hindu-Brahmin. Home: 1018 Prestwick Dr Frankfort IL 60423-9539 Office: Michigan City Radiologists Inc 916 Washington St Michigan City IN 46360-3593

MOTIN, REVELL JUDITH, data processing executive; b. Bayonne, N.J., July 24, 1941; d. Charles and Belle (Laks) Motin; children from a previous marriage: Laura Mantell, Deborah Mantell. BS in Psychology cum laude, CUNY, 1969. Systems analyst Univac div. Sperry Corp., N.Y.C., 1961-66; programmer, analyst J.C. Penney Co., N.Y.C., 1966-67; systems and programming cons. Automated Concepts, Inc., N.Y.C., 1968-72; ind. systems and programming cons. N.Y.C., 1972-76; mgr. systems and programming Citibank, NA, N.Y.C., 1976-83; v.p. data processing Columbia Savs. Bank S.L.A., Fair Lawn, N.J., 1983—. Mem. Fin. Mgrs. Soc., Mensa. Jewish. Home: 3 Jockey Ln New City NY 10956-6608 Office: Columbia Savs Bank SLA 25-00 Broadway Fair Lawn NJ 07410

MOTLOCH, CHESTER GEORGE, nuclear engineer; b. Hamtramck, Mich., Jan. 4, 1948; s. Chester and Valentine Victoria (Majewski) M.; m. Andrea Christine Logan, June 1970 (div. June 1978); 1 child, Elizabeth Christine; m. Vicki Marie Bowen, Aug. 20, 1983; 1 child, Jacob Chester. BS in Physics, UCLA, 1970; MS in Physics, Oakland U., 1972. Sound, vibration engr. Chrysler Corp., Highland Park, Mich., 1972-74; nuclear safety analyst Bettis Atomic Power Lab., West Mifflin, Pa., 1974-76; nuclear quality assurance Shippingport (Pa.) Atomic Power Sta., 1976-77; quality assurance supr. Naval Reactor Facility, Idaho Falls, Idaho, 1977-79; gen. mgr. Energy Inc., Idaho Falls, Idaho, 1979-89; prin. project engr. Idaho Nat. Engring. Lab., Idaho Falls, 1989—; Contbr. articles to profl. jours. NSF scholar, 1966. Mem. AIAA, Am. Nuclear Soc., Idaho Acad. Sci. Achievements include research in nuclear propulsion and advanced reactor concepts. Office: Idaho Nat Engring Lab PO Box 1625 MS 3511 Idaho Falls ID 83415

MOTOBA, TOSHIO, physics educator; b. Iwata, Shizuoka, Japan, May 29, 1944; s. Jouji and Mura (Hakamata) M.; m. Setsuko Shima, Mar. 21, 1970; 1 child, Atsushi. BS in Physics, Kyoto U., Japan, 1967, MS in Physics, 1969, DSc in Nuclear Physics, 1975. Rsch. fellow Kyoto U., Faculty Sci., Japan, 1972-76; postdoctoral fellow Japan Soc. for Promotion of Sci., Kyoto, Japan, 1973-74, Yukawa Meml. Found., Kyoto, Japan, 1972, 75; lectr. Osaka Electro-Communication U., Japan, 1976-77, assoc. prof., 1978-83, prof., 1984—; organizing com. Internat. Conf. Hypernuclear and Strange Particle Physics, 1989-91; rsch. coun. mem. Kyoto U., Yukawa Inst. for Theoretical Physics, 1990-91. Author: Fundamentals of Physics II, 1985, (jour.) Production, Structure and Decay of Hypernuclei, 1991. Mem. Phys. Soc. Japan (sec. for theoretical nuclear physics 1985-86). Home: 50 Umenoki-cho Ichijoji, Kyoto 606, Japan Office: Osaka Electro-Comm U, 18-8 Hatsu-cho, Neyagawa 572, Japan

MOTT, CHARLES DAVIS, civil engineer; b. Phila., Aug. 30, 1914; s. Charles Hillard and Emma (Davis) M.; m. Ellen Mary Hooge, Aug. 13, 1938 (dec.); children: Ellen H., Charles H., Joseph W. H. BSc in Civil Engring., U. Pa., Phila., 1932; M Engring. Adminstrn., George Washington U., 1967. Engr. Cruse Kemper Co., Ambler, Pa., 1936-37; flight leader Am. Vol. Group, Burma, China, 1941-45; tech. staff/mgr. Analytic Svcs., Arlington, Va., 1963—; mem. staff Rsch. and Devel. Bd., Office Sec. of Def., Washington, 1952-55. Pres. Lakevale Ct. Citizens Assn., Vienna, Va., 1972, 87. Capt. USN, 1937-41, 46-63. Recipient Cloud and Banner medal Chinese Air Force, 1958, POW medal USN, 1990. Mem. AIAA (mem. coun. Nat. Capitol sect. 1984-86), Am. Def. Preparedness Assn. Baptist. Achievements include membership in concept formulation team, F-15 and participant Project Forecast; research in air to surface guided weapons. Home: 2522 Rocky Branch Rd Vienna VA 22181-4068 Office: Analytic Svcs CG-3 Ste 800 1215 Jeff Davis Hwy Arlington VA 22202

MOTT, SIR NEVILL (FRANCIS MOTT), physicist, educator, author; b. Leeds, Eng., Sept. 30, 1905; s. C.F. and Lilian Mary (Reynolds) M.; m. Ruth Horder, Mar. 21, 1930; children: Elizabeth, Alice. Student, Cambridge (Eng.) U., 1924-29; MA, St. John's Coll., Cambridge, 1929; DSc (hon.), Univs. of Sheffield, London, Louvain, Grenoble, Paris, Poitiers, Bristol, Univs. of Ottawa, Liverpool, Reading, Warwick, Lancaster, Heriot Watt, Bordeaux, Univs. of St. Andrews, Essex, Stuttgart, Sussex, William and Mary, Marburg, Univs. of Bar Ilan, Lille, Rome, Lisbon; D Tech, Linköping. Lectr. math. Cambridge U., 1930-33, Cavendish prof. exptl. physics, 1954-71; master Gonville and Caius Coll., 1959-66; prof. physics U. Bristol, 1933-54, bd. dir. H.H. Wills Phys. Lab., 1948-54; Page-Barbour lectr. U. Va., 1956. Author: An Outline of Wave Mechanics, 1930, (with H.S.W. Massey) The Theory of Atomic Collisions, 1933, (with H. Jones) The Theory of the Properties of Metals and Alloys, 1936, (with R.W. Gurney) Electronic Processes in Ionic Crystals, 1940, Wave (with I.N. Snedden) Wave Mechanics and its Applications, 1948, Elements of Wave Mechanics, 1952, Atomic Structure and the Strength of Metals, 1956, (with E.A. Davis) Electronic Processes in Noncrystalline Materials, 1971, 2d edit., 1979, Elements of Quantum Mechanics, 1972, Metal-Insulator Transitions, 1974, 2d edit., 1991, Conduction in Noncrystalline Materials, 1987, 2d edit., 1993, (autobiography) A Life in Science, 1986, Can Scientists Believe?, 1991. Mem. cen. adv. coun. Ministry of Edn., 1956-59; chmn. com. physics edn. Nuffield Found., 1965-75. Sci. adviser to Anti Aircraft Command, also supt. theoretical rsch. in armaments Armament Rsch. Dept., World War II. Decorated knight bachelor; recipient Nobel prize for physics, 1977. Fellow Royal Soc. (Hughes medalist 1941, Royal medal 1953, Copley medal 1972), Phys. Soc. of Great Brit. (pres. 1956-58); mem. NAS, AAAS (corr.), Inst. Physics (hon. fellow), Internat. Union Pure and Applied Physics (pres. 1951-57), Modern Langs. Assn. (pres. 1955), Société Française de Physique (hon.).

MOTT, PETER ANDREW, biologist; b. Shawinigan, Quebec, Canada, Oct. 21, 1959; came to the U.S., 1963; s. Edgar Andrew and Anna Elizabeth (Muggah) M.; m. Mary Elizabeth Reed, Apr. 14, 1984; children: Justin Andrew, Lauren Elizabeth. BS, East Stroudsburg State U., 1982. Scientist Schering Plough Corp., Lafayette, N.J., 1982-86; sr. group leader Marypaul Labs., Inc., Sparta, N.J., 1986—. Mem. Am. Water Works Assn., Am. Pub. Health Assn., N.J. Soc. Indsl. Microbiology. Office: Marypaul Labs Inc PO Box 952 Sparta NJ 07871

MOTTELSON, BEN R., physicist; b. Chgo., July 9, 1926; naturalized Danish citizen, 1971; s. Goodman and Georgia (Blum) M.; m. Nancy Jane Reno, 1948 (dec. 1975); 3 children. B.Sc., Purdue U., 1947; Ph.D., Harvard U., 1950; hon. degrees, Purdue U., U. Heidelberg, Fed. Republic Germany. Fellow Inst. Theoretical Physics, Copenhagen, 1950-51, U.S. AEC fellow, 1951-53; with theoretical study group European Orgn. for Nuclear Research, Copenhagen; prof. Nordic Inst. for Theoretical Atomic Physics, Copenhagen, 1957—; physicist Bohr Inst.; bd. dirs. Nordita; vis. prof. U. Calif., Berkeley, 1959. Author: Nuclear Structure, vol. 1, 1969, vol. 2 (with A. Bohr), 1975; numerous other publs. in field. Recipient Nobel prize for physics, 1975. Mem. Nat. Acad. Scis. (fgn. assoc.). Address: Nordita, Blegdamsvej 17, DK-2100 Copenhagen Denmark

MOUCHATY, GEORGES, physicist; b. Aleppo, Syria, June 9, 1950; came to U.S., 1980; s. Elias and Jeanette (Dib) M. Advanced studies diploma, U. Paris VI, 1975, doctorate, 1977. Rsch. assoc. Tex. A&M U., College Station, 1980-84, accelerator physicist, 1985—; devel. engr. Schlumberger, Houston, 1984-85. Contbr. articles to profl. jours. Lt. Mil. Acad., 1977-80, Aleppo. French Govt. scholar Acad. of Paris, 1974-77. Mem. Am. Phys. Soc., Am. Chem. Soc., Tex. Acad. Sci. Achievements include design of successful new ECR ion source; development of better operations of ECR-K500 accelerator system and better understanding of heavy ion nuclear reactions and nuclear structure. Office: Tex A&M Univ The Cyclotron Inst College Station TX 77843

MOULIJN, JACOB A., chemical technology educator; b. Harlingen, Friesand, The Netherlands, Mar. 13, 1942. D, U. Amsterdam, The Netherlands, 1967, Lab. Chem. Tech., The Netherlands, 1974. Asst. prof. U. Amsterdam, 1972-86, full prof. chem. tech., 1986-90; full prof. chem. tech. Tech. U. Delft, The Netherlands, 1990—. Editor jour. Fuel Process Tech., 1987—; mem. editorial bd. jours. Fuel, 1982, Applied Catalysis, 1992; editor 4 books; contbr. over 200 articles to profl. jours. Recipient Nat. BP Energy prize, 1982. Mem. AICE, Royal Chem. Soc. (The Netherlands). Achievements include patents in combustion, catalysis, zeolite-based membrane; research in coal gasification, coal combustion, petroleum conversion, exhaust gas catalysis, mass transfer in porous materials, catalytic reactors, catlytic testing. Office: U Delft, Julianalaan 136, 2628 BL Delft The Netherlands

MOULTHROP, JAMES SYLVESTER, research engineer, consultant; b. DuBois, Pa., Feb. 24, 1939; s. John Oliver Jr. and Pauline Margaret (Allen) M.; m. Martha Sue Mulloy, Sept. 14, 1963; children: James S. Jr., Margaret Winn, Mary Mulloy. BA, St. Joseph's Coll., Rensselaer, Ind., 1960; MS, Kans. state U., 1963. Registered profl. engr., Pa. Soils engr. Pa. Dept. Transp., Franklin and Harrisburg, 1963-72; materials engr. Pa. Dept. Transp., Harrisbvurg, 1972-81, dir. maintenance, 1981-83; regional engr. Chemkrete Techs., Inc., Wyckliff, Ohio, 1983-85; sr. applications engr. Exxon Chem. Co., Houston, 1985-87; rsch. engr. U. Tex., Austin, 1987-93; co-founder, sr. ptnr. Asphalt Rsch. and Delvel. Internat., Austin, 1993—; mem. Trasnp. Rsch. Bd., Washington, 1970—. Co-author manual: Statistical Quality Control of Highway Construction in Pennsylvania, 1975, Practical Applications of Statistical Quality Control in Highway Construction, 1979. With U.S. Army, 1963. Recipient Creativity award Nat. Univ. Extension Assn. Mem. ASTM (subcom. chair 1987—), Assn. Asphalt Paving Tech., Am. Soc. Hwy. Engrs. (sr.), Am. Concrete Inst. Achievements include chairing committee responsible for implementation of roadway management system in Pennsylvania, directing technical program for $50 million strategic highway research program in asphalt. Home: 7400 Anaqua Dr Austin TX 78750 Office: Asphalt Rsch Devel Internat Ste 210 8240 N Mopac Austin TX 78759

MOUNTAIN, CLIFTON FLETCHER, surgeon, educator; b. Toledo, Apr. 15, 1924; s. Ira Fletcher and Mary (Stone) M.; m. Marilyn Isabelle Tapper, Feb. 28, 1945; children: Karen Lockerby, Clifton Fletcher, Jeffrey Richardson. AB, Harvard U., 1947; MD, Boston U., 1954. Diplomate Am. Bd. Surgery. Dir. dept. statis. rsch. Boston U., 1947-50; cons. rsch. analyst Mass. Dept. Pub. Health, 1951-53; intern U. Chgo. Clinics, 1954, resident, 1955-58, instr. surgery, 1958-59; sr. fellow thoracic surgery Houston, 1959; mem. staff M.D. Anderson Hosp. and Tumor Rsch. Inst.; asst. prof. thoracic surgery U. Tex., 1960-63, assoc. prof. surgery, 1963-76, prof., 1976—, chief sect. thoracic surgery, 1970-79, chmn. thoracic oncology, 1979-84, chmn. dept. thoracic surgery, 1980-85, chmn. program in biomath and computer sci., 1962-64, Mike Hogg vis. lectr. in S.Am., 1967; mem. sci. mission on cancer USSR, 1970-78, and Japan, 1976-84; mem. com. health, rsch. and edn. facilities Houston Community Coun., 1964-78; cons. Am. Joint Com. on Cancer Staging and End Result Reporting, 1964-74, mem. Am. Joint Com. on Cancer, 1974-86, chmn. lung and esophagus task force; mem. working party on lung cancer and chmn. com. on surgery, Nat. Clin. Trials Lung Cancer Study Group, NIH, 1971-76; mem. plans and scope com. cancer therapy Nat. Cancer Inst., 1972-75, mem. lung cancer study group, 1977-89, chmn. steering com., 1973-75, mem. bd. sci. counselors div. cancer treatment, 1972-75; hon. cons. Shanghai Chest Hosp. and Lung Cancer Ctr., Nat. Cancer Inst. of Brazil. Editor The New Physician, 1955-59; mem. editorial bd. Yearbook of Cancer, 1960-88, Internat. Trends in Gen. Thoracic Surgery, 1984-91; contbr. articles to profl. jours., chpts. to textbooks. Chmn. profl. adv. com. Harris County Mental Health Assn.; bd. dirs. Harris County chpt. Am. Cancer Soc. Lt. (j.g.) USNR, 1942-46. Recipient award Soviet Acad. Sci., 1977, Garcia Meml. medal Philippine Coll. Surgeons, 1982, Disting. Alumni award Boston U., 1988, Disting. Achievement U. Tex. M.D. Anderson Cancer Ctr., 1990, Disting. Svc. award Internat. Assn. for the Study of Lung Cancer, 1991, Disting. Alumnus award Boston U. Sch. of Medicine, 1992. Fellow ACS, Am. Coll. Chest Physicians (chmn. com. cancer 1967-75), Am. Assn. Thoracic Surgery, Inst. Environ. Scis., N.Y. Acad. Sci., Assn. Thoracic and Cardiovascular Surgeons of Asia (hon.), Hellenic Cancer Soc. (hon.), Chilean Soc. Respiratory Diseases (hon., hon. pres. 1982); mem. AAAS, Am. Assn. Cancer Rsch., AMA, So. Med. Assn., Am. Thoracic Soc., Soc. Thoracic Surgeons, Soc. Biomed. Computing, Biomed. Info. Processing Orgn., Am. Fedn. Clin. Rsch., Internat. Assn. Study Lung Cancer (pres. 1976-78), Am. Radium Soc., Pan-Am. Med. Assn., Am. Congress Rehab. Medicine, Houston Surg. Soc., Soc. Surg. Oncology, James Ewing Soc., Sigma Xi. Achievements include conception and development of program for application of mathematics and computers to the life sciences, of resource for experimental designs, applied statistics and computational support; first clinical use of physiologic adhesives in thoracic surgery; demonstration of clinical behavior of undifferentiated small cell lung cancer; first laser resection of lung tissue at thoracotomy; development of international system for staging of lung cancer. Home: 1612 South Blvd Houston TX 77006-6338 Office: 6723 Bertner St Houston TX 77030-4095

MOUSA, ALYAA MOHAMMED ALI, neurobiologist, researcher; b. Makkah AL-Mukkaramah, Saudi Arabia, Nov. 19, 1964; d. Mohammed Ali Hussein Mousa and Mariam (Abbas) Afandi. B in Sci., Edn., Girls' Edn. Coll., Jeddah, Saudi Arabia, 1986; MS in Neurobiology, King Saud U., Riyadh, Saudi Arabia, 1991. Researcher King Faisal Specialist Hosp. and Rsch. Ctr., Riyadh, Saudi Arabia, 1989—; team mem. detoxification and decontamination King Faisal Specialist Hosp. and Rsch. Ctr., Riyadh, 1990-91. Avocation: reading. Address: Karolinska Inst-Geriatrics, Huddinge Univ Hosp B56, S-141 86 Huddinge Sweden Office: King Faisal Specialist Hosp, Box 3354, Riyadh Saudi Arabia

MOUSSA, KHALIL MAHMOUD, polymer photochemist, researcher; b. Jib-Jinnine, Lebanon, Jan. 19, 1959; came to U.S., 1989; s. Mahmoud Khalil and Najibe Ali M.; m. Narges Khalil Abdul-Baki, Aug. 23, 1983; children: Racha, Lama, Mohammad. MS, Univ. Saint Jerome, 1984; PhD, Univ. Haute-Alsace, 1988. Rsch. scientist Ecole Nationale Superieure de Chimie, Mulhouse, France, 1984-88; post doctorate Laboratoire de Photochimie Generale, Mulhouse, France, 1988-89, Univ. Southern Miss., Hattiesburg, 1989-90; sr. rsch. assoc. Polychrome Corp., Columbus, Ga., 1990—. Contbr. articles to profl. jours. Recipient Advances in Radiation Curing award Radcure Europe, 1987, Founders award for Outstanding Paper Radtech Internat., 1992. Mem. Am. Chem. Soc., Internat. Union Pure and Applied Chem. Achievements include patents, real time study of laser induced polymerization, established a new method for monitoring the ultra-fast photopolymerization by real time infrared spectroscopy. Home: 4312 Old

Macon Rd Columbus GA 31907 Office: Polychrome Corp One Polychrome Pk Corp Ridge Industrial Pk Columbus GA 31907

MOUTAERY, KHALAF REDEN, neurosurgeon; b. Dhomer, Syria, July 1, 1949; s. Reden Bayan Moutaery and Fatima (Asaad) Khalaf; m. Majda Hassan, Dec. 12, 1985; children: Anoud, Abdullah, Najla, Haifa. PhD, Med. High Sch., Hanover, Fed. Republic Germany, 1984. Cons. neurosurgeon Riyadh (Saudi Arabia) Armed Forces Hosp., 1988-89, head neuroscis. dept., 1989—, chmn. med. staff, 1989—. Author: Solitary Brain Metastasis, 1985. Col. Saudi Med. Svc., 1989—. Fellow Royal Coll. Surgeons; mem. German Neurosurg. Assn., European World Fed. Neurosurgeons, Saudi Neurosurgery Club. Muslim. Avocations: swimming, skiing. Home and Office: Riyadh Armed Forces Hosp, PO Box 7897, Riyadh 11159, Saudi Arabia

MOUTON, PETER RANDOLPH, neuroscientist, biologist; b. Houston, June 4, 1958; s. Francis Edward and Lynne Marie (Moore) M. BS, U. So. Fla., 1983, PhD, 1990. Guest scientist Karolinska Inst., Stockholm, 1987-88; adj. prof. U. So. Fla., Tampa, 1988-90; postdoctoral fellow Neurol. Rsch. Lab, Copenhagen, 1990-92; rsch. assoc. Johns Hopkins Sch. Medicine, Balt., 1992—; cons. Neurosearch A/S, Copenhagen, 1991, Microbrightfield, Inc., Balt., 1993—. Contbr. articles to profl. jours. Vol. Bill Clinton Presdl. Campaign, Balt., 1992. Postdoctoral fellow Am. Scandinavian Found., N.Y., 1987. Mem. AAAS, N.Y. Acad. Scis., Soc. Neuroscis. Democrat. Achievements include research in symbiosis and evolution of the coral reef, the safety of diving and pregnancy. Office: Johns Hopkins Sch Medicine 720 Rutland Ave Baltimore MD 21205

MOW, VAN C., engineering educator, researcher; b. Chengdu, China, Jan. 10, 1939. B. Aero. Engring., Rensselaer Poly. Inst., 1962, Ph.D., 1966. Mem. tech. staff Bell Telephone Labs., Whippany, N.J., 1968-69; assoc. prof. mechanics Rensselaer Poly. Inst., Troy, N.Y., 1969-76, prof. mechanics and biomed. engring., 1976-82, John A. Clark and Edward T. Crossan prof. engring., 1982-86; prof. mechanical engring. and orthopedic bioengring. Columbia U., N.Y.C., 1986—; dir. Orthopedic Research Lab., Columbia-Presbyn. Med. Ctr., N.Y.C., 1986—; vis. mem. Courant Inst. Math. Sci., NYU, 1967-68; vis. prof. Harvard U., Boston, 1976-77; chmn. orthopaedics and musculoskeletal study sect. NIH, Bethesda, Md., 1982-84; hon. prof. Chengdu U. Sci. Tech., 1981, Shanghai Jiao Tong U., 1987; mem. grants rev. bd. Orthopaedic Rsch. Edn. Found., 1992-96; bd. dirs. Hoar Rsch. Found., 1993—; cons. in field. Assoc. editor Jour. Biomechanics, 1981—, Jour. Biomech. Engring., 1979-86; chmn. editorial adv. bd. Jour. Orthopedic Rsch., 1983-90; contbr. numerous articles to profl. jours. Founder Gordon Research Conf. on Bioengring. and Orthopedic Sci., 1980. NATO sr. fellow, 1978; recipient William H. Wiley Disting. Faculty award Rensselaer Poly. Inst., 1981; Japan Soc. for Promotion Sci. Fellow, 1986, Fogarty Sr. Internat. fellow, 1987; Alza disting. lectr. Biomed. Engring. Soc., 1987; H.R. Lissner award ASME, 1987, Kappa Delta award AAOS, 1980, Giovani Borelli award, 1991. Fellow ASME (chmn. biomechanics div. 1984-85, Melville medal 1982), Am. Inst. Med. Biol. Engring.; mem. NAE, Orthopaedic Rsch. Soc. (pres. 1982-83), Am. Soc. Biomechanics (founding), Internat. Soc. Biorheology, U.S. Nat. Com. on Biomechanics (sec.-treas. 1985-90, chmn. 1991—). Office: Columbia-Presbyn Med Ctr 630 W 168th St BB-1412 New York NY 10032

MOWBRAY, ROBERT NORMAN, forest ecologist, government agricultural and natural resource development officer; b. Warren, Pa., Feb. 26, 1935; s. Leonard Kelly and Jean Elizabeth (Lowes) M.; m. Sonia de los Angeles Baquerizo, June 7, 1969; children: Norma Mercedes, Elizabeth Laning. BA, Dartmouth Coll., 1957; M of Forestry, Yale U., 1963; postgrad., Duke U., 1966-68. Rsch. asst. forest ecology Duke U., Panama, 1967, Ecuador, 1968-70; research asst. forest ecology U. Tenn., Knoxville, 1970-71; research asst. ecology Oak Ridge (Tenn.) Nat. Labs., 1971-72; reclamation crew chief Tenn. Mountain Mgmt., Knoxville, 1972; assoc. dir. Peace Corps, Asunción, Paraguay, 1972-78; agrl. devel. officer A.I.D., San Jose, Costa Rica, 1978-80, Kingston, Jamaica, 1980-83, Washington, 1983-88, 90-91, Quito, Ecuador, 1988-90; sr. forest ecologist and natural resource mgmt. specialist A.I.D., Washington, 1990—; forestry vol. Peace Corps, Ecuador, 1963-66, editor tech. newsletter, 1964-66. Author: (with others) Natural Resource Management and Conservation of Biodiversity and Tropical Forests in Ecuador-A Strategy for USAID, 1989; contbr. articles to profl. jours. 1st lt. USMC, 1958-61. Mem. World Wildlife Fund, Nature Conservancy, Assn. for Tropical Biology, Internat. Soc. Tropical Foresters, Friends of the Nat. Zoo, Nat. Coun. Returned Peace Corps Vols. Avocations: gardening, photography. Home: 2218 Wheelwright Ct Reston VA 22091-2313 Office: Agy Internat Devel R&D/ENR Rm 503I SA-18 Washington DC 20523-1812

MOWREY, TIMOTHY JAMES, management and financial consultant; b. Lewiston, N.Y., Oct. 18, 1958; s. William Ronald and Joan (Cupp) M.; m. Karrie Rae Kaminske, Sept. 9, 1978; children: Christin R., Andrea M., Ryan T. BA of Profl. Studies in Mgmt., Data Proc., SUNY, Buffalo, 1982. Registered investment advisor. Rsch. technician Carborundum Co., Niagara Falls, N.Y., 1978-79; programmer KVS Info. Systems, Kenmore, N.Y., 1979-80; systems analyst Moore Bus. Forms Inc., Niagara Falls, 1980-83; telecommunications project coord. Marine Midland Bank, N.A., Buffalo, 1983-85; pres., owner Micro-Tec, Niagara Falls, 1982-86; telecommunications specialist Electronic Data Systems, Lockport, N.Y., 1985-86; mng. cons. Computer Task Group Inc., Buffalo, 1986-88, mng. communications cons., 1988-91, practice mgr. enterprise cons., 1992-93, practice dir. mgmt. cons., 1993—; owner Mowrey Investment Mgmt., 1992—; Registered investment advisor, SEC. Scholar N.Y. State Bd. of Regents, 1976. Mem. Soc. Mfg. Engrs. Democrat. Roman Catholic. Avocations: hiking, reading, golf, Tae Kwon Do. Office: Mowrey Investment Mgmt 6211 Baer Rd Sanborn NY 14132-2094

MOWREY-MCKEE, MARY FLOWERS, biochemist; b. Warren, Pa., Feb. 21, 1941; d. Harold Thomas and Florence (Irwin) F.; m. Thomas Wesley Mowrey, June 9, 1962 (div. Mar. 1976); children: Susan Ariel, Elisabeth Jennifer, Juliet Deirdre; m. Leonard William McKee, Mar. 10, 1978; stepchildren: Kevin Sean, William Scott, Kathleen Margaret. AB in Chemistry, U. Rochester, 1963, MS in Biophysics, 1975, PhD in Biophysics, 1977. Assoc. chemist Xerox Corp., Webster, N.Y., 1963-64; program analyst Svc. Bur. Corp. (IBM), N.Y.C., 1964-67; sr. biologist Bausch & Lomb, Rochester, N.Y., 1977-80, sr. rsch. specialist, 1980-83, project mgrr., 1983-85, mgr. tech. sect., 1985-87, sr. scientist, 1987-90; mgr. Wesley-Jessen, Chgo., 1990-92; dir. of solutions development Ciba Vision, Duluth, Ga., 1992—; expert Internat. Standards Orgn., Geneva, 1986—; scientist, chmn. microbiology com. Contact Lens Inst., 1992—; unit chmn. Am. Nat. Standards Inst., 1991—; presenter at profl. meetings, 1976-90; lectr. Am. Type Culture Collection, Rockville, Md., 1987. Contbr. articles to sci. jours. Violinist Penfield (N.Y.) Symphony Orch., 1979-90, Greece (N.Y.) Symphony, 1982, Finger Lakes Orch., Canandaigua, N.Y., 1990, DuPage (Ill.) Symphony Orch., 1991-92, Cobb Symphony, 1993. Scholar Aeroflow Dynamics, Inc., 1959-63; fellow ERDA, 1973-75; rsch. grantee NIH, 1975-77. Mem. Assn. for Rsch. in Vision and Ophthalmology, Tissue Culture Assn., Am. Soc. for Microbiology. Achievements include 2 patents pending relating to antimicrobial activity of disinfecting solutions for contact lenses; determination that protein uptake by contact lenses is increased by increasing concentrations of methacrylic acid leading to grouping of lenses by the FDA, that patient-handling is a primary source of microbial contamination of contact lenses and that a significant reduction in concentration of microbes on the lens occurs during lens wear; confirmation that adaptation of bacteria to growth at high temperature before infection greatly increased production of T4D bacteriophage at that temperature. Home: 210 Mirrowood Dr Alpharetta GA 30202 Office: Ciba Vision 11460 Johns Creek Parkway Duluth GA 30136-1518

MOWRY, ROBERT WILBUR, pathologist, educator; b. Griffin, Ga., Jan. 10, 1923; s. Roy Burnell and Mary Frances (Swilling) M.; m. Margaret Neilson Black, June 11, 1949; children: Janet Lee, Robert Gordon, Barbara Ann. B.S., Birmingham So. Coll., 1944; M.D., Johns Hopkins U., 1946. Rotating intern U. Ala. Med. Coll., 1946-47, resident pathology, 1947-48; sr. asst. surgeon USPHS-NIH, Bethesda, Md., 1948-52; fellow pathology Boston City Hosp., 1949-50; asst. prof. pathology Washington U., St. Louis, 1952-53; asst. prof. pathology U. Ala. Med. Ctr., Birmingham, 1953-54, assoc. prof. pathology, 1954-57; prof. U. Ala. Med. Center, Birmingham, 1958-89, prof. emeritus, 1989—, prof. health svcs. adminstrn., 1976-84, dir. Anat. Pathology Lab., 1960-64, dir. grad. programs in pathology, 1964-72; sr. scientist U. Ala. Inst. Dental Research, 1967-72, dir. autopsy services, 1975-79; vis. scholar dept. pathology U. Cambridge, Eng., 1972-73; cons. FDA, 1975-81. Author: (with J.F.A. McManus) Staining Methods: Histologic and Histochemical, 1960; mem. editorial bd. Jour. Histochemistry and Cytochemistry, 1960-75, Stain Tech., 1965-90, AMA Archives of Pathology, 1967-76, Biotechnics and Histochemistry, 1991—. Served with USPHS, 1948-52. Mem. Am. Assn. Pathologists, Internat. Acad. Pathology, Biol. Stain Commn. (v.p. 1974-76, pres. 1976-81, trustee 1966—), Soc. for Complex Carbohydrates, Am. Assn. Univ. Profs. Pathology, Phi Beta Kappa, Sigma Xi, Delta Sigma Phi, Alpha Kappa Kappa. Presbyterian. Achievements include perfection of staining methods for complex carbohydrates (Alcian blue and colloidal iron) and insulin (Alcian blue-aldehyde fuchsin); showed the utility of these in diagnostic histopathology. Home: 4165 Sharpsburg Dr Birmingham AL 35213-3234

MOY, RONALD LEONARD, dermatologist, surgeon; b. Stuttgart, Germany, June 10, 1957; s. Howard Leonard Stephen and Jenny (Yee) M.; m. Lisa Wing Lan Lin, Aug. 10, 1986; children: Lavren, Erin. Grad., Rensselaer Poly. Inst., 1977, Albany Med. Coll., 1981. Dir. Mohs micrographic surgery div. dermatology UCLA, 1988-93, dir. dermatologic surgery div. dermatology, 1988-93, co-chief div. dermatology, 1992-93. Author: Atlas of Cutaneous Flaps and Grafts, 1990; editor: Principle and Practice of Dermatologic Surgery, 1993; contbr. articles to profl. jours. Bd. dirs. L.A. Costal unit Am. Cancer Soc., 1988. Recipient J. Lewis Piplan award in dermatology Nat. Student Rsch. Forum, 1981, Henry Christian award Am. Fedn. Clin. Rsch., T-cell and Cytolcine Patterns in Skin Cancer award NIH, 1992. Fellow Am. Acad. Dermatology (Gold award 1986); mem. Am. Soc. Dermatologic Surgery (bd. dirs. 1993—), Am. Coll. Mohs Micrographic Surgery and Cutaneous Oncology (bd. dirs. 1992—), Assn. Acad. Dermatologic Surgeons (bd. dirs. 1992—). Roman Catholic. Office: UCLA Div Dermatology 100 UCLA Med Plz #590 Los Angeles CA 90024

MOYA DE GUERRA, ELVIRA, physics educator; b. Albacete, La Mancha, Spain, Feb. 19, 1947; d. Augusto and Maria Luisa (Valgañon) M.; m. Jesus Guerra, Jan. 29, 1971; children, Elvira, Andres. Bachiller, Inst. Miguel Servet, Zaragoza, Spain, 1961, Bachiller Superior, 1964; M in Physics, U. Zaragoza, 1969, PhD in Physics, 1974. Asst. prof. U. Zaragoza, 1970-74; vis. scientist MIT, Cambridge, Mass., 1974-76; rsch. assoc. MIT, Cambridge, 1976-79; assoc. prof. U. Autonoma Madrid, 1979-82; physics prof. U. Extremadura, Badjoz, Spain, 1982-86; rsch. prof. Consejo Superior Investigaciones Cientificas, Madrid, 1986—; vice dir. I.E.M. Consejo Superior Investigaciones Cientificas, Madrid, 1990—; pres. Nuclear Physics Div. RSEF, Madrid, 1990—; mem. editorial bd. Nuclear Physics News in Europe jour., 1990. Contbr. articles to profl. jours. Grantee NSF (U.S.), 1979, '89, Juan March Found., Spain, 1972. Mem. Am. Physical Soc., Spanish Physics Soc. (pres. physics div. 1990—), Nuclear Physics European Collaboration Com. Roman Catholic. Avocations: antiques, restoration, decoration, sports. Office: IEM CSIC, Serrano 119-123, 28006 Madrid Spain

MOYER, DAVID LEE, veterinarian; b. Reading, Pa., May 7, 1940; s. Robert Bretz and Helen Vera (Lutz) M.; m. Ann Marie DeGarmo, Aug. 8, 1964 (div. Sept. 1979). BS, Pa. State U., 1962; postgrad., Temple U., 1967; VMD, U. Pa., 1968; psychology, Kutztown U., 1980. Lic. doctor vet. medicine, Pa. Veterinarian Antietam Valley Animal Hosp., Reading, 1969-76, Kutztown Animal Hosp., Dryville, Pa., 1976-83; meat and poultry inspection supr., vet. med. officer U.S. Dept. Agr., Food Safety and Inspection Svc., Fredericksburg, Pa., 1983-88; veterinarian Exeter Animal Hosp./Cat Clinic/Pet Care Ctr., Birdsboro, Pa., 1983—; veterinarian Reading Police Dept. K-9 Corps, City of Reading, 1969-80, Humane Soc. Berks County, Reading, 1970-72, Animal Rescue League SPCA, Reading, 1970—. Advisor Boy Scouts Am., Reading, 1969-74. Mem. Am. Vet. Med. Assn., Am. Animal Hosp. Assn., Pa. Vet. Med. Assn., Keystone Vet. Med. Assn., Schulykill Vet. Med. Assn., Lehigh Vet. Med. Assn., Western Vet. Conf., Vet. Orthopedic Soc., Am. Soc. Vet. Ophthalmology, Am. Assn. Feline Practitioners (charter), Cornell Feline Health Ctr., Acad. Feline Medicine, Internat. Vet. Ophthalmology, Am. Soc. Vet. Nutrition, Am. Assn. Food Hygiene, Am. Assn. Avian Pathologists, Vet. Cardiology Soc. (pres. 1983), Am. Vet. Dental Soc., Toastmasters (pres. 1983). Democrat. Lutheran. Avocations: skiing, gardening, fine dining, travel. Home and Office: 6800 Daniel Boone Rd Birdsboro PA 19508

MOYER, JAMES WALLACE, physicist, consultant; b. Syracuse, Aug. 16, 1919; s. Wallace Earl and Viola (Hook) M.; m. Nedra Blake, Sept. 10, 1940; children—Jeffry Mark, Elaine, Virginia, Julia; m. Ruth Pierce Hughes, Jan. 23, 1993. A.B., Cornell U., 1938; postgrad. Rutgers U., 1938-41; Ph.D., U. Rochester, 1948. Insp. ordnance U.S. Army, 1941-42; physicist Radiation Lab., U. Calif., Berkeley, 1942-43; sr. physicist Tenn. Eastman, Oak Ridge, 1943-46; research assoc. Gen. Electric Research, Knolls Atomic Power Lab., Schenectady, N.Y., 1948-55; cons. engr. Gen. Electric Microwave Lab., Palo Alto, Calif., 1955-57; mgr. phys. sci. Gen. Electric Tempo, Santa Barbara, Calif., 1957-60; research dir. Sperry Rand Research Ctr., 1960-61; research dir. Servo Mechanisms, Inc., 1961-63; dir. applied research Autonetics div. N.Am. Aviation, Anaheim, Calif., 1963-65; dir. phys. scis. Northrop Space Lab., 1965-67; dir. engring. Northrop Corp., Beverly Hills, Calif., 1967-70; mgr. phys. systems So. Calif. Edison, Rosemead, 1976-84; cons., 1984—; cons. Nat. Bur. Standards, 1956-62; mem. panel Nat. Acad. Sci., 1967-70, 71. Mem. IEEE (sr.), Am. Phys. Soc., Sigma Xi. Home: 1520 Stone Brooke Rd Ames IA 50010

MOYER, JOHN HENRY, III, physician, educator; b. Hershey, Pa., Apr. 1, 1917; s. John Henry and Anna Mae (Gruber) M.; m. Mary Elizabeth Hughes; children: John Henry IV, Michael, Carl, Anna Mary, Nancy Elizabeth, Mary Louise, Matthew Timothy. BS, Lebanon Valley Coll., 1939, DSc (hon.), 1968; MD, U. Pa., 1943. Diplomate Am. Bd. Internal Medicine, Nat. Bd. Med. Examiners; lic. physician Mass., Pa., Tex. Intern Pa. Hosp., Phila., 1943; resident in Tb and contagious diseases Belmont Hosp., Worcester, Mass., 1944-45; asst. instr. Tb and contagious diseases U. Vt., 1944-45; chief resident in medicine Brooke Gen. Hosp., San Antonio, 1947; fellow in pharmacology and medicine Sch. Medicine, U. Pa., Phila., 1948-50; attending physician, then. sr. attending physician Jefferson Davis Hosp., Houston, 1950-57, Meth. Hosp., Houston, 1950-57; from asst. prof. to prof. internal medicine and pharmacology Coll. Medicine, Baylor U., Houston, 1950-56, prof., 1956-57; prof., chmn. dept. medicine Hahnemann Med. Coll. and Hosp., Phila., 1957-74, exec. v.p. acad. affairs, 1971-73; sr. v.p., dir. profl. and ednl. affairs Conemaugh Valley Meml. Hosp., Johnstown, Pa., 1974-88; emeritus dir. profl. and ednl. affairs Conemaugh Valley Meml. Hosp., Johnstown, 1988—; prof. Temple U., 1977—; dir. regional affairs Sch. Medicine Temple U., 1977-88; clin. prof. Coll. Medicine Pa. State U., Hershey, 1986—; adj. prof. natural scis. U. Pitts. at Johnstown, 1982—; adj. prof. physician asst. sci. St. Francis Coll., 1983—, sr. cons. physician asst. program adv. com., 1985—; vis. prof., lectr. various ednl. instns.; mem. Pa. State Bd. Med. Edn. and Licensure, 1977-86, sec. to bd., 1982-86; mem. task force on profl. edn., mem. hypertension info. and edn. adv. com. U.S. HEW, 1972-75; chmn. high blood pressure control adv. bd. to sec. health, State of Pa., 1980-86; cons. numerous profl. orgns. Editorial cons. Am. Jour. Cardiology, 1960-72; editor-in-chief Cyclopedia of Medicine, Surgery and Specialties, 1963-65; mem. editorial adv. bd. Internal Medicine News, 1969-92; editor 16 multi-authored textbooks; contbr. more than 600 articles to profl. jours. Mem. bd. trustees Pa. Heart Assn., 1959-65, v.p. bd. trustees, 1965; mem. bd. govs. Heart Assn. Southeastern Pa., 1958-64, 67-72; bd. dirs. Houston Heart Assn., 1952-57; mem., then emeritus fellow med. adv. bd. coun. for high blood pressure Am. Heart Assn., 1954—, chmn., 1964-65, mem., then emeritus fellow cen. adv. bd. coun. on circulation. Maj. U.S. Army, 1945-48. Recipient Susan and Theodora R. Cummings Humanitarian award, 1962, 65, 66, Presdl. citation Cultural Exchg. Program, U.S. State Dept., 1964, Honors Achievement award Angiology Rsch. Found., 1965; named Alumni of Yr., Lebanon Valley Coll., Annville, Pa., 1967. Fellow ACP (Laureate award for western Pa. 1986), Am. Coll. Cardiology (trustee 1961-68), N.Y. Acad. Scis. (emeritus), Am. Coll. Chest Physicians (emeritus); mem. AMA (emeritus, mem. ho. dels. 1966-72, cons. coun. on drugs 1968-72, mem. sect. coun. on clin. pharmacology and therapeutics), AAAS (emeritus), Am. Soc. Clin. Pharmacology and Therapeutics (emeritus dir., hon. dir.), Am. Fedn. Clin. Rsch. (emeritus), Am. Soc. Pharmacology and Exptl. Therapeutics (emeritus), Assn. Am. Med. Colls., Am. Acad. Med. Dirs., Am. Soc. Internal Medicine, Pa. Soc. Internal Medicine (pres. 1992—), Sems. and Symposia (pres.), Assn. Hosp. Med. Edn., Assn. Former Chmn. Medicine, Sigma Xi, many others. Republican. Achievements include extensive research in cardiovascular diseases. Address: 1090 Miller Rd Palmyra PA 17078

MOYER, RALPH OWEN, JR., chemist, educator; b. New Bedford, Mass., May 19, 1936; s. Ralph Owen and Annie (Brown) M. BS, U. Mass., Dartmouth, 1957; MS, U. Toledo, 1963; PhD, U. Conn., 1969. Devel. engr. Union Carbide Corp., Fostoria, Ohio, 1957-64; asst. prof. chemistry Trinity Coll., Hartford, Conn., 1969-76, assoc. prof., 1976-86, prof., 1986-91, Scovill prof. chemistry, 1991—; vis. lectr. U. West Indies, Kingston, Jamaica, 1985; rsch. collaborator Brookhaven Nat. Lab., Upton, N.Y., 1977-78. Contbr. articles and chpt. to profl. pubs. With U.S. Army, 1959. Mem. Am. Chem. Soc. (chmn. Connecticut Valley sect. 1984), N.Y. Acad. Scis., Sigma Xi (pres. Hartford chpt. 1990-91). Achievements include two patents for carbon fibers. Home: 9 Grandview Ter Wethersfield CT 06109-3240 Office: Trinity Coll 300 Summit St Hartford CT 06106-3100

MOYER, THOMAS PHILLIP, biochemist; b. Dearborn, Mich., Nov. 4, 1946; s. John Burroughs and Charlotte Mary (Smith) M.; m. Diane Marie Prescott, Dec. 21, 1971; children: Michael Andrew, Peter Benjamin, Thomas Prescott. BA, U. Minn., 1970; PhD, N.D. State U., 1976. Cons. lab. medicine Mayo Clinic, Rochester, Minn., 1979-90, chair divsn. clin. biochemistry, 1990—. Editor: Applied Therapeutic Drug Monitoring, Vol. 1, 1984, Vol. 2, 1985; contbr. more than 130 papers and abstracts to profl. jours. Founding chair Rochester Area Math/Sci. Partnership, 1991. Sgt. U.S. Army, 1970-72. NIH grantee, 1981, 83, 91. Mem. Acad. Clin. Lab. Physicians and Scientists, Am. Assn. Clin. Chemistry (bd. dirs. 1990-92, presdl. citation 1989), Rotary (bd. dirs. Rochester chpt. 1987-90, Paul Harris fellow 1990), Sigma Xi. Republican. Roman Catholic. Office: Mayo Clinic 200 SW 1st St Rochester MN 55905

MOZAFFARI, MOJTABA, computer science educator; b. Tehran, Mar. 30, 1952; s. Abolhasan Mozaffari and Aqdasolmolok Gouya; m. Tahmineh Khaledi, May 21, 1982; children: Mohammad Hassan, Mohammad Ali, Maryam, Mostafa. BS in Math. and Computer Sci., Sharif U. Tech., Tehran, 1976; MS in Computer Sci., U. Mich., 1979; PhD in Computer Sci., Hokkaido U., Sapporo, Japan, 1989. Instr. Ahwaz (Iran) Shahid Chamran U., 1976-77, systems analyst, 1976-77; lectr. Kerman (Iran) Shahid Bahonar U., 1979-89, assoc. prof., 1989-91; assoc. prof. Sharif U. Tech., 1991-93, Amir Kabir U., 1993—. Mem. IEEE, ACM, Iranian Info. Soc. Home: Shahrak Ekbatan Block A4, Entrance 8 # 287, Tehran Iran Office: Amir Kabir U, Math and Computer Dept Hafez Ave, Tehran Iran

MOZDZIAK, PAUL EDWARD, growth biologist; b. New Haven, May 14, 1967; s. Edward and June C. (Libby) M.; m. Lori Anne Pogmore, Aug. 5, 1989. BS with honors and distinction, Cornell U., 1989; MS, U. Wis., 1991. Rsch. asst. U. Wis., Madison, 1989—. Daniel W. Kops scholar, 1988. Mem. AAAS, Am. Meat Sci. Assn., Inst. Food Technologists, Poultry Sci. Assn., Cornell U. Alumni Assn. Roman Catholic. Office: Univ of Wis 1805 Linden Dr Madison WI 53706

MOZENA, JOHN DANIEL, podiatrist; b. Salem, Oreg., June 9, 1956; s. Joseph Iner and Mary Teresa (Delaney) M.; m. Elizabeth Ann Hintz, June 2, 1979; children: Christine Hintz, Michelle Delaney. Student, U. Oreg., 1974-79; B in Basic Med. Scis., Calif. Coll. Podiatric Medicine, D in Podiatric Medicine, 1983. Diplomate Am. Bd. Podiatric Surgery. Resident in surg. podiatry Hillside Hosp., San Diego, 1983-84; pvt. practice podiatry Portland, Oreg., 1984—; dir. residency Med. Ctr. Hosp., Portland, 1985-91; lectr. Nat. Podiatric Asst. Seminar, 1990, Am. Coll. Gen. Practitioners, 1991. Contbr. articles to profl. jours.; patentee sports shoe cleat design, 1985. Fellow Am Coll. Ambulatory Foot Surgeons, Am. Coll. Foot Surgeons. Republican. Roman Catholic. Avocations: softball, basketball, piano, jogging, piano. Office: Town Ctr Foot Clinic 8305 SE Monterey Ave Ste 101 Portland OR 97266-7728

MRUK, CHARLES KARZIMER, agronomist; b. Providence, Sept. 23, 1926; s. Charles and Anna (Pisarek) M. BS in Agr., U. R.I., 1951, MS in Agronomy, 1957. Soil scientist soil conservation svc. Dept. Agr., Sunbury, Pa., 1951; insp. Charles A. McGuire Co., Providence, 1952; claims insp. R.R. Perishable Inspection Agy., Boston, 1953-55; asst. in agronomy U. R.I., 1955-57; agronomist Hercules Inc., 1957-79, tech. salesman, 1957-79; sr. tech. sales rep. BFC Chems., Inc., 1981-82; are devel. supr. Ea. States, 1982-84, ret., 1984; cons. turf maintenance Olympic Stadium and grounds, Mexico City, 1968, Fenway Park, Boston, 1963-70. author and editor articles on turf culture and fertilizers, 1960-81. Mem. Rep. Ward Com., Providence, 1963-76. With USN, 1944-46. U.S. Golf Assn. Green Sect. grantee, 1955-57. Mem. Am. Soc. Agronomy, New Eng. Sports Turf Mgrs. Assn., R.I. Golf Course Supts. Assn., Mass. Turf and Lawn Grass Coun. (dir., mem. planning com., chmn. fin. com., 1987, pres., 1987-89), VFW, Am. Registry Cert. Profls. in Agronomy (cert. agronomist), Sigma Xi, Alpha Zeta. Mem. Polish National Ch. Home: 75 Burdick Dr Cranston RI 02920-1517

MU, EDUARDO, electrical engineer; b. Lima, Peru, Dec. 15, 1957; s. Enrique and Maria (Hoyos) Mu. MSEE, U. Pitts., 1987, PhD in Electrical Engring., 1991. Chief maintenance Panamerican TV Network, Lima, 1981-85; teaching fellow dept. electrical engring. U. Pitts., 1985-91; electrical engr. Amatrol, Clarksville, Ind., 1992--; cons. Cardiac Telecom Corp., Pitts., 1991, Dept. Cardiology U. Pitts., 1991. Contbr. articles to profl. jours. Recipient Outstanding Master of Sci. Thesis award Dept. Electrical Engring. U. Pitts., 1987; U. Pitts. scholar, 1986-91. Mem. IEEE, Sigma XI. Roman Catholic. Home: 812 Applegate Ln 817 Applegate Ln Clarksville IN 47129

MUATHEN, HUSSNI AHMAD, chemistry educator; b. Makkah, Saudi Arabia, Nov. 8, 1956; s. Ahmad Ibraheem and Zainab (Sabbagh) M.; m. Maatuka Mohammad Mokhtar, Aug. 15, 1979; children: Summer, Sahhar, Sommayah, Youssif. Grad. diploma in chem. scis., East Anglia U., Norwich, Eng., 1980; BSc in Chemistry, Umm Al-Qura, Makkah, Saudi Arabia, 1978; PhD in Chemistry, Bristol (Eng.) U., 1985. Demonstrator Umm Al-Qura U., 1978-79, asst. prof. organic chemistry, 1985—; head Applied Scis. and Engring. Rsch. Ctr., Umm Al-Qura, 1989-92. Contbr. articles to profl. jours. in Jour. of Organic Chemistry and Indian Jou. of Chemistry. Mem. Am. Chem. Soc. Muslim. Achievements include rsch. in synthesis and application of new halogen complexes. Office: U Umm Al-Qura, PO Box 7283, Makkah Saudi Arabia

MUBAYI, VINOD, physicist; b. Lahore, Punjab, Pakistan, Sept. 27, 1941; s. Baikunth Nath and Kunwar Rani (Razdan) M.; m. Joan Ilsa Feldman, Aug. 25, 1969 (div. July 1988); 1 child, Suneel; m. Felicitas Figueroa Santiago, Sept. 11, 1989. BS, Delhi (India) U., 1961; PhD, Brandeis U., 1968. Rsch. assoc. Cornell U., Ithaca, N.Y., 1967-69; rsch. fellow Tata Inst. Fund Rsch., Bombay, 1969-74; rsch. assoc. Brandeis U., Waltham, Mass., 1974-75; physicist Brookhaven Nat. Lab., Upton, N.Y., 1976—; adj. prof. SUNY, Stony Brook, 1978-81; cons. UN, N.Y.C., 1981-85, Idea, Inc., Washington, 1984-87. Mem. AAAS, Am. Nuclear Soc., Am. Solar Energy Soc., N.Y. Acad. Sci. Achievements include research in calculated risk of radiation exposure during low power operation of reactors, oper. risk of a high level nuclear waste repository, benefit-cost analysis of radiation protection; devel. of approach to modeling oil/gas exploration under uncertainty; developed solar thermal technology research program for developing countries; obtained exact numerical solution of Heisenberg spin systems. Office: Brookhaven Nat Lab Bldg 130 Upton NY 11973

MUCHA, JOHN FRANK, data processing professional; b. Ludlow, Mass., Sept. 12, 1950; s. Joseph Walter and Sophie (Chrusciel) M.; m. Anne Virginia Casey, Sept. 1, 1973 (div. Feb. 1989). BA in Polit. Sci., U. Mass., 1972; MBA in Tech. and Profl. Communications, Frostburg State U., 1985. Cert. in data processing. Computer programmer IRS, Washington, 1974-79, computer systems programmer, 1979-81; computer systems programmer IRS, Martinsburg, W.Va., 1981-86; staff systems programmer fed. systems div. IBM, Gaithersburg, Md., 1986-87; chief tech. support IRS Martinsburg Computing Ctr., 1987-91; staff asst. to projects dir. info. systems devel. IRS, Washington, 1991-92, computer specialist transition mgmt. office, 1992—. Contbr. articles to profl. jours. Team mem., v.p. Beginning Experience of

Balt., 1989—; pres. Cath. Single Again Coun. of Balt., Inc., 1991—. Mem. Data Processing Mgmt. Assn. (bd. dirs.), Assn. of the Inst. for Certification of Computer Profls., Moose. Democrat. Roman Catholic. Avocations: reading, instrumental music, hiking, travel, single again ministry. Home: 2482 Warm Spring Way Odenton MD 21113-1542

MUCHMORE, JOHN STEPHEN, endocrinologist; b. L.A., Jan. 1, 1945; s. Allan Winner and Lyntha Carol (Weed) M.; m. Susan Jill Crawford, June 13, 1968; children: Adam Ian, Rachel Kathlene. AB, Knox Coll., 1967; MD, U. Okla., 1975; PhD, U. Rochester, 1976. Diplomate Am. Bd. Internal Medicine. Intern U. Ohio Health Sci. Ctr., 1975-76, resident in medicine, 1976-78, fellow endocrinology, metabolism and hypertension, 1978-80; Pvt. practice Oklahoma City, 1980—; cardiac transplant physician Okla. Transplantation Inst., Bapt. Med. Ctr. of Okla., Oklahoma City, 1989—. Recipient Nat. Rsch. Svc. award Nat. Heart, Lung and Blood Inst., Bethesda, Md., 1978. Mem. AMA, AAAS, Am. Coll. Physicians, Alpha Omega Alpha. Office: Plaza Med Assocs 3433 NW 56th St Oklahoma City OK 73112-4444

MUCHNICK, RICHARD STUART, ophthalmologist; b. Bklyn., June 21, 1942; s. Max and Rae (Kozinsky) M.; BA with honors. Cornell U., 1963, MD, 1967; m. Felice Dee Greenberg, Oct. 29, 1978; 1 child, Amanda Michelle. Intern in medicine N.Y. Hosp., N.Y.C., 1967-68, now assoc. attending ophthalmologist, chief Pediatric Ophthalmology Clinic; resident in ophthalmology, 1970-73; practice medicine, specializing in ophthalmology, notably strabismus and ophthalmic plastic surgery N.Y.C., 1974—; attending surgeon, chief Ocular Motility Clinic, Manhattan Eye, Ear and Throat Hosp., N.Y.C.; clin. assoc. prof. ophthalmology Cornell U., N.Y.C., 1984—. Served with USPHS, 1968-70. Recipient Coryell Prize Surgery Cornell U. Med. Coll., 1967. Diplomate Am. Bd. Ophthalmology, Nat. Bd. Med. Examiners. Fellow A.C.S., Am. Acad. Ophthalmology; mem. Am. Soc. Ophthalmic Plastic and Reconstructive Surgery, Am. Assn. Pediatric Ophthalmology and Strabismus, Internat. Strabismological Assn., N.Y. Soc. Clin. Ophthalmology, AMA, N.Y. Acad. Medicine, Manhattan Ophthal. Soc., N.Y. Soc. Pediatric Ophthalmology and Strabismus, Alpha Omega Alpha, Alpha Epsilon Delta. Clubs: Lotos, 7th Regt. Tennis. Clin. researcher strabismus, ophthalmic plastic surgery, 1973—. Office: 69 E 71st St New York NY 10021-4213

MUCI KÜCHLER, KARIM HEINZ, mechanical engineering educator; b. Valencia, Carabobo, Venezuela, May 22, 1964; came to Mexico, 1981; s. Moussa and Luise Gertrud (Küchler) Muci Abraham; m. Alejandra Castañeda González, June 25, 1983; children: Karim Ibrahim, Moses Alejandro, Claudia María. BS, Inst. Tech. y Estudios Superiores Monterrey, Mexico, 1985; MS, Inst. Tech. y Estudios Superiores Monterrey, 1988; PhD, Iowa State U., 1992. Mech. maintenance Altos Hornos de Mexico, Monclova, 1986; teaching asst. Iowa State U., Ames, 1989, rsch. asst., 1989-92; prof. mech. engring. Inst. Tech. y Estudios Superiores Monterrey, 1993—. Mem. ASME (assoc.), Internat. Soc. Boundary Elements, Sigma Xi (assoc.), Tau Beta Pi, Phi Kappa Phi. Achievements include development of higher order boundary elements for three dimensional problems. Office: ITESM Campus Monterrey, Dept Ing Mecanica, Sucursal de Correos J Monterrey 64849, Mexico

MUDAR, M(ARIAN) J(EAN), biologist, environmental scientist; b. Albany, N.Y.; d. Michael and Jean (Hnyda) M. BS in Biology, SUNY, Albany, 1973, MS in Biology Edn., 1976; MS Urban/Environ Studies summa cum laude, Rensselaer Poly. Inst., 1981, PhD in Urban/Environ. Studies cum laude, 1991. Cert. tchr., N.Y. With N.Y. State Environ. Facilities Corp., Albany, 1981—, indsl. waste program analyst, 1981-86, program analyst, 1986-91, environ. scientist, 1991—; mem. adv. panel Congl. Office Tech. Assessment, 1986; mem. editorial adv. bd. Govt. Insts., Inc., Rockville, Md., 1986-91; mem. adv. bd. Nat. Roundtable of State Pollution Prevention Programs, Mpls., 1990-2; mem. coord. com. Forum on State and Tribal Toxics Action, Washington, 1991-93; project lead Pollution Prevention Project, Washington, 1991-93; mem. team of scientists and engrs. to conduct joint Indo-U.S. workshop on indsl. wastewater recycling and reuse Govt. of India., 1986. Author, editor materials in field; editor: (pamphlet) Guide to Household Hazardous Wastes, 1988; author: Reducing Plastic Contamination of the Marine Environment Under MARPOL Annex V, A Model for Recreational Harbors and Ports. Mem. Cousteau Soc. (founding), USA Track and Field, Internat. Marina Inst. (rsch. assoc.), Boat/U.S., World Wildlife Orgn., Air and Waste Mgmt. Assn. Avocations: sculling, skiing, racewalking, cycling, photography.

MUDD, JOHN BRIAN, biochemist; b. Darlington, U.K., Aug. 31, 1929; came to U.S., 1955; s. John Curry and Clara Kennedy (Bell) M.; m. Monika Ittig, Aug. 10, 1974; 1 child, Simon Marius. BA, Cambridge (Eng.) U., 1953; MSc, U. Alta., Edmonton, Can., 1955; PhD, U. Wis., Madison, 1958. Postdoctoral fellow U. Calif., Davis, 1958-60; asst. prof., assoc. prof., then prof. U. Calif., Riverside, 1961-81; group leader Plant Cell Rsch. Inst., Dublin, Calif., 1981-86, v.p. rsch., 1987-90; dir. Statewide Air Pollution Rsch. Ctr., Riverside, 1990-93; bd. dirs. Vector Labs., Burlingame, Calif.; vis. prof. plant rsch. lab. Mich. State U., East Lansing, 1979-80. Editor: Responses of Plants to Air Pollution, 1975, Metabolism, Structure and Function of Plant Lipids, 1987; editor Archives Biochem. Biophys. Jour. Jane Coffin Childs Fund fellow, 1959-60. Mem. Am. Chem. Soc., Am. Soc. Plant Physiologists, Biochem. Soc., Sigma Xi. Office: U Calif Dept Botany Riverside CA 92521

MUDEK, ARTHUR PETER, automation engineer, consultant; b. Milw.. BSEE, Marquette U., 1970; MBA, Loyola U., Chgo., 1973; PhD, Calif. Coast U., 1993. Automation engr. Automatic Electric Co., Nrothlake, Ill., 1970-73; engring. mgr. McGraw-Edison Co., Geneva, Ill., 1973-75; mgr. mfg. A.O. Smith Co., Elkhorn, Wis., 1975-79; mgr. mfg. engring. Johnson Controls Co., Watertown, Wis., 1980-87; applications mgr. Machinery Systems Co., Milw., 1987-90; mgr. mfg. engring. Alloy Products Co., Waukesha, Wis., 1990-93; pres., automation cons. A.C. Automation, Delafield, Wis., 1972—, bd. dirs. Waukesha Aviation Club, 1982—, mem. Fedn. Environ. Tech., Milw., 1991—; inspection authority FAA, Milw., 1992. Autohr: Electronic Specifications, 1971, Environmental Management Plan, 1992; contbr. to profl. publs. Mem. Soc. Mech. Engrs. (sr.), Inst. Hazardous Materials Mgmt. (master cert.). Achievements include work in automation, robotic automation, electronic automation, computer-controlled machines and electro-chemical automation. Home: PO Box 112 Delafield WI 53018 Office: A C Automation Inc PO Box 112 Delafield WI 53018

MUELLER, DENNIS WARREN, physicist; b. Moline, Ill., Aug. 2, 1946; s. Warren Rudolph and Margie Irene (Fitzpatrick) M.; m. Linda Jean Klee, June 19, 1971; children: Luke Klee, Colby Dennis. BA in Physics and Math., MacMurray Coll., 1968; MS in Physics, Mich. State U., 1974, PhD, 1976. Postdoctoral researcher cyclotron lab. Mich. State U., East Lansing, 1976; instr., researcher cyclotron lab. Princeton (N.J.) U., 1976-78, mem. rsch. staff plasma physics lab., 1978-82, rsch. physicist plasma physics lab., 1982-91; br. head ops. Tokamak Fusion Test Reactor, 1990—; prin. rsch. physicist plasma physics lab. Princeton (N.J.) U., 1991—; cons. for radioactive remediation Middlesex (N.J.) Borough Coun., 1978-79. Contbr. over 130 articles to profl. publs. Mem. AAAS, Am. Phys. Soc. Office: Princeton U Plasma Physics Lab PO Box 451 Princeton NJ 08541

MUELLER, DONALD SCOTT, chemist; b. Cleve., May 8, 1947; s. Robert Leo and Edith Marie (Somershield) M.; m. Linda J. Meredith; children: Scott, Cyndi, Tracey, Laura. BA in Chemistry, Hiram Coll., 1969; PhD, U. Ill., 1973. Chemist Rohm and Haas Co., Phila., 1973-77, group leader, 1978-81, mktg. mgr., 1981-82, rsch. mgr., 1983-85; rsch. dir. Johnson and Johnson, New Brunswick, N.J., 1985-91; v.p. rsch. and devel. Ashland Chem., Columbus, 1991--. Rep. Twp. Civic Assn., Plainsboro, N.J., 1990-91, Ch. Stewardship Comm., Dublin, Ohio, 1992. Rsch. fellow U. Ill., 1972-73. Mem. Indsl. Rsch. Inst., Am. Chem. Soc. Achievements include 4 patents. Office: Ashland Chem Inc 5200 Blazer Parkway Dublin OH 43017

MUELLER, GEORGE E., corporation executive; b. St. Louis, July 16, 1918; m. Maude Rosenbaum (div.); children: Karen, Jean; m. Darla Hix, 1978. B.S. in Elec. Engring., Mo. Sch. Mines, 1939; M.S. in Elec. Engring., Purdue U., 1940; Ph.D. in Physics, Ohio State U., 1951; hon. degrees, Wayne State U., N.Mex. State U., U. Mo., 1964, Purdue U., Ohio State U., 1965. Mem. tech. staff Bell Telephone Labs., 1940-46; prof. elec. engring. Ohio State U., 1946-57; cons. electronics Ramo-Wooldridge, Inc., 1955-57; from dir. electronic lab. to v.p. research and devel. Space Tech. Labs., 1958-62; assoc. adminstr. for manned space flight NASA, 1963-69; corporate officer, sr. v.p. Gen. Dynamics Corp., N.Y.C., 1969-71; chmn., pres. System Devel. Corp., Santa Monica, Calif., 1971-83, chmn., chief exec. officer, 1983; sr. v.p. Burroughs Corp., 1981-83; pres. Jojoba Propagation Labs, from 1981, George E. Mueller Corp., from 1984, Internat. Acad. Astronautics, from 1983. Author: (with E.R. Spangler) Communications Satellites. Recipient 3 Disting. Service medals NASA, 1966, 68, 69; Eugen Sanger award, 1970; Nat. Medal Sci., 1970; Nat. Transp. award, 1979. Fellow AAAS, IEEE, AIAA (Goddard medal 1983, Sperry award 1986), Am. Phys. Soc., Am. Astronautical Soc. (Space Flight award), Am. Geophys. Union, Brit. Interplanetary Soc.; mem. Internat. Acad. Astronautics (pres. 1982—), Nat. Acad. Engring., N.Y. Acad. Scis. Patentee in field. Home: Santa Barbara CA Office: PO Box 5856 Santa Barbara CA 93150-5856

MUELLER, KENNETH HOWARD, pathologist; b. Harvey, Ill., Mar. 27, 1940; s. Walde Herman and Joann Rose (Rohrschneider) M.; m. Carol Ann Lulonic, June 16, 1962; children: Margaret, Joan, Peter, Sharon. BS, Carroll Coll., 1961; MD, Harvard U., 1965. Intern & resident pediatrics, resident pathology U. Cin. Children's Hosp., 1965-70; with U. Cin./Children's Hosp., 1965-70; commd. 2d lt. USAF, 1970, advanced through grades to lt. col., ret., 1980; pathologist, assoc. med. examiner State of Mont., Billings, 1980—. Contbr. articles to profl. jours. Bass trombonist Billings Brass Soc., 1980—. capt. USNR, 1980-92. Decorated Commendation medal, Def. Commendation. Fellow Nat. Assn. Med. Examiners, Am. Acad. Forensic Scis., Coll. Am. Pathology, Am. Soc. Clin. Pathology. Office: Saint Vincent Hosp 1233 N 30th St Billings MT 59101

MUELLER, PAUL, chemist, educator; b. Ennetbaden, Switzerland, Feb. 23, 1939; s. Franz and Paula (Mueller) M.; m. Ruth Helen Bochsler, July 17, 1965; children: Michael, Susanne Ruth, Corinne Maya, Yolanda. Diploma, Swiss Fed. Inst. Tech., Zurich, 1963; PhD, ETH, Zurich, 1966. Rsch. assoc. U. Ill., Chgo., 1966-68, U. Chgo., 1968-69; chef de travaux U. Geneva, Switzerland, 1969-73, asst. prof., 1973-78, assoc. prof., 1978-82, prof., 1982—; chmn. IUPAC Com. III, 2, 1985-93. Mem. Swiss Chem. Soc. (award 1975), Am. Chem. Soc. Office: U Geneva, 30 Quai E Ansermet, 1211 Geneva Switzerland

MUELLER, PETER STERLING, psychiatrist, educator; b. N.Y.C., Dec. 28, 1930; s. Reginald Sterling and Edith Louise (Welleck) M.; m. Ruth Antonia Shipman, Aug. 9, 1958; children: Anne Louise, Peter Sterling, Paul Shipman, Elizabeth Ruth. A.B., Princeton U., 1952; M.D., U. Rochester, 1956. Am. Cancer Soc. student fellow Francis Delafield Hosp., N.Y.C., summer 1955; intern Bellevue Hosp., Columbia U., N.Y.C., 1956-57; asst. resident in psychiatry Henry Phipps Psychiat. Clinic, Johns Hopkins Hosp., Balt., 1963-66; asst. prof. psychiatry Sch. Medicine, Yale U., New Haven, 1966-72; asso. prof. psychiatry Coll. Medicine and Dentistry of N.J., Rutgers Med. Sch., Piscataway, 1972-76; clin. prof. psychiatry Coll. Medicine and Dentistry of N.J., Rutgers Med. Sch., 1976-82; cons. for Rehab. Unit and Center for Indsl. Human Resources, Community Mental Health Center, 1973—; mem. courtesy staff dept. psychiatry Princeton Med. Center, 1976—; cons. in psychotherapy Conn. Valley Hosp., Middletown, 1966-72; cons. in psychiatry Carrier Clinic, Belle Mead, N.J., 1973—, VA Hosp., Lyons, N.J., 1975-78. Contbr. writings in field to profl. publs. U.S. and Brit., papers to profl. confs. on the use patents in U.S. and fgn. countries for direct dopamine agonists in the treatment of tobacco addiction. Served with USPHS, 1957-63. Mem. Am. Psychosomatic Soc., Am. Psychiat. Assn., AAAS, Amyotrophic Lateral Sclerosis Found. (adv. bd.), Sigma Xi. Episcopalian. Home: 182 Snowden Ln Princeton NJ 08540-3915 Office: 601 Ewing St Princeton NJ 08540-2754

MUELLER, RAYMOND JAY, software development executive; b. Denver, Nov. 16, 1959; s. Frank Joseph and JoAnn A. (Seib) M.; m. Hiro K. Abeyta; 1 child, Michael Raymond. A in Acctg. and Computer Sci., Metro State Coll., 1981; cert. in computer sci., Denver U., 1983. cert. data processor, 1985—. Data processing mgr. Bailey Co., Denver, 1980-86; pres. MIS, Inc., Lakewood, Colo., 1986—; cons. to local high schs., 1985-88; speaker in field. Contbr. articles to computer mags.; inventor Touch 2000; patentee in field. Mem. Assn. for Inst. Cert. of Computer Profls., Data Processing Mgmt. Assn., Greater Denver C. of C., Colo. Inst. Artificial Intelligence. Republican. Roman Catholic. Avocations: tennis, golf, scuba diving, motorcycling, public speaking. Office: MIS Inc 355 Union Blvd # 300 Lakewood CO 80228-1500

MUELLER, ROBERT WILLIAM, process engineer; b. Louisville, Aug. 18, 1964; s. Harris Clinton and Elizabeth Anne (Scholz) M.; m. Marilyn Jean Williams, May 22, 1987; children: Christopher, Jaclyn. BSChemE, U. Louisville, 1986, MSChemE, 1987. Grad. asst. Protein Techs. Internat., Louisville, 1986-87, process engr., 1987-89, sr. process engr., 1989—. Mem. AICE (local sect. chair), Instrument Soc. Am. Republican. Home: 9427 Fairground Rd Louisville KY 40291 Office: Protein Techs Internat 2441 S Floyd St Louisville KY 40217

MUELLER, RUDHARD KLAUS, toxicologist; b. Glauchau, Saxony, Fed. Republic Germany, Aug. 20, 1936; s. Rudhard Otto and Hildegard Dora (Krasselt) M.; m. Ursula Hanni Rossberg, May 5, 1961; children: Cornelia, Beatrix, Mildred. Diploma in chemistry, U. Leipzig, German Dem. Republic, 1960, D in Nat. Scis., 1965, D in habil., 1977. Mem. staff forensic medicine U. Leipzig, German Dem. Republic, 1960-75, venia legendi toxicological chemistry, 1975-82, assoc. prof. forensic toxicology, 1982—, head postgrad. study program toxicology faculty of medicine, 1987, prof., 1989—. Author, editor: (book) Toxicological Analysis, (German) 1976, (English) 2d edit. 1991. Recipient Rudolf Virchow award Min. Pub. Health, Berlin, 1977, Gottfried Wilhelm Leibniz award U. Leipzig, 1979. Mem. Internat. Assn. Forensic Toxicologists (co-found., mem. com. Systematic Toxicological Analysis, 1978—, chmn. 1991—), Soc. Legal Medicine, Soc. Toxicol. and Forensic Chem. Avocations: chamber music, organ playing. Home: 92 Kurt Eisner St, D-04275 Leipzig Germany Office: U Forensic Medicine, 28 Johannisallee, D-04103 Leipzig Germany

MUELLER, STEPHAN, geophysicist, educator; b. Marktredwitz, Ger., July 30, 1930; s. Hermann Friedrich and Johanna Antonie Fanny (Leuze) M.; m. Doris Luise Pfleiderer, July 31, 1959; children: Johannes Christoph, Tobias Ulrich. Dipl.-Phys., Inst. Tech. Stuttgart, 1957; M.Sc. in Elec. Engring., Columbia U., 1959; Dr.rer.nat., U. Stuttgart, 1962. Lectr. geophysics U. Stuttgart, 1962-64; vis. prof. S.W. Center Advanced Studies, Richardson, Tex., 1964-65; prof. geophysics U. Karlsruhe, 1964-71, dean Faculty Natural Scis., 1968-69; vis. prof. U. Tex., Dallas, 1969-70; prof. Swiss Fed. Inst. Tech., 1971—, U. Zurich, 1977—; dean Sch. Natural Scis., Swiss Fed. Inst. Tech., 1978-80; dir. Swiss Earthquake Service, 1971—; pres. Swiss Geophys. Commn., 1972-93, European Seismol. Commn., 1972-76, Internat. Commn. Controlled Source Seismology, 1975-83; chmn. governing council Internat. Seismol. Centre, 1975-85; chmn. European-Mediterranean Seismol. Centre, 1976-82. German Acad. Interchange scholar, 1954-55. Fellow Royal Astron. Soc. (hon. fgn. assoc.), Am. Geophys. Union, Geol. Soc. London (hon.); mem. Internat. Assn. Seismology and Physics of Earth's Interior (pres. 1987-91), European Geophys. Soc. (pres. 1978-80, hon. mem. 1984—), European Union Geoscis. (Alfred Wegener medal 1993), German Geophys. Soc., Soc. Exploration Geophysicists, European Assn. Exploration Geophysicists, Seismol. Soc. Am., Seismol. Soc. Japan, Acoustical Soc. Am., Swiss Geophys. Soc. (pres. 1977-80), Swiss Acad. Scis. (life), Academia Europaea (founding mem.), German Acad. Leopoldina Researchers in Natural Scis., Sigma Xi. Co-editor Pure and Applied Geophysics, 1974-83; editor-in-chief Annales Geophysicae, 1982-87; editorial bd. Jour. Geophysics, 1969-87, Tectonophysics, 1971-77, 84—, Bolletino di Geofisica Teorica ed Applicata, 1978—, Jour. Geodynamics, 1983—, Tectonics, 1988—. Office: ETH-Geophysics, CH-8093 Zurich Switzerland

MUELLNER, WILLIAM CHARLES, computer scientist, physicist; b. Chgo., Jan. 7, 1944; s. Frank Joseph and Catherine (Mika) M.; m. Marilyn Joy Bocan, Dec. 20, 1969; children: Kimberly Ann, Kevin Alan. BS in Physics, DePaul U., 1966; MS in Physics, Purdue U., 1968; PhD, U. Ill., 1973. Prof. Morton Coll., Cicero, Ill., 1969-76; engring. group leader Motorola Corp., Schaumburg, Ill., 1977-78; engring. mgr., sr. scientist Perkin-Elmer Corp., Oak Brook, Ill., 1978-80; prof., dept. chmn. computer sci. Elmhurst (Ill.) Coll., 1980—; cons. Perkin-Elmer Corp., Oak Brook, 1980-86. Author: Advanced C Language Programming, 1992; contbr. articles to Physics Rev. B., Solid State Communications. Mem. Am. Soc. Engring. Edn., Am. Phys. Soc., Assn. for Computing Machinery, Phi Kappa Phi. Achievements include 4 patents in field. Office: Elmhurst College 190 Prospects Elmhurst IL 60126

MUFSON, MAURICE ALBERT, physician, educator; b. N.Y.C., July 7, 1932; s. Max and Faye M.; m. Diane Cecile Weiss, Apr. 1, 1962; children: Michael Jeffrey, Karen Andrea, Pamela Beth. A.B., Bucknell U., 1953; M.D., NYU, 1957. Intern Bellevue Hosp., N.Y.C., 1957-58; resident Bellevue Hosp., 1958-59; chief resident Cook County Hosp., Chgo., 1965-66; sr. surgeon USPHS Lab. Infectious Diseases, NIH, 1961-65; asst. prof. medicine U. Ill., 1965-69, assoc. prof., 1969-73, prof., 1973-76; prof., chmn. dept. medicine Marshall U., 1976—; vis. scientist Karolinska Inst., 1984-85. Contbr. articles to profl. jours. Served with U.S. Navy, 1959-61. WHO grantee, 1967; recipient Meet-the-Scholar award Marshall U., 1986, Researcher of Yr. award Sigma Xi, Marshall U., 1989. Fellow ACP (traveling scholar 1987); fellow Infectious Diseases Soc. Am.; mem. Soc. Exptl. Biology and Medicine, Central Soc. Clin. Research, So. Soc. Clin. Investigation, AMA, W.Va. State Med. Assn., Assn. Profs. Medicine (counselor 1992—), Alpha Omega Alpha. Office: Dept Medicine Marshall U Sch Medicine Huntington WV 25701

MUFTI, AFTAB A., civil engineering educator; b. Sukkur, Sind, Pakistan, Apr. 24, 1940; arrived in Canada, 1963; s. Abdul Wahid D. and Shah Jahan M.; m. Janet Cockayne, Dec. 19, 1964 (div. April 1981); 1 child, Javed Christopher; m. Zahara Omar, Jan. 2, 1992. BCE, NED Engring. U., Karachi, Sind, 1962; MCE, McGill U., Montreal, 1965, PhD, 1969. Asst. prof. McGill U., 1969-72; assoc. prof., head dept. comp. sci. Acadia U., Wolfville, Nova Scotia, 1972-76, prof., dir. Sch. Comp. Sci., 1976-80; prof. civil engring., dir. Tech. U. Nova Scotia, Halifax, 1980—; judge Canadian Cons. Engring. Awards, 1987. Author: (book) Elementary Computer Graphics, 1982; editor: Advanced Composite Materials in Bridges and Structures, 1992, Finite Element Method in Civil Engring., 1993. Vol. fireman Wolfville Fire Dept., 1976-77. Recipient Phelp Johnson prize Engring Inst. Can., Montreal, 1969. Fellow Engring. Inst. Can., Canadian Soc. Civil Engring. (recipient Whitman Wright award 1990); mem. ASCE. Achievements include a Bridge Deck patent. Office: Tech U of NS, PO Box 1000, Halifax, NS Canada B3L 2X4

MUFTI, NAVAID AHMED, network design engineer; b. Karachi, Pakistan, July 6, 1967; arrived in Can., 1970; s. Iqbal Ahmed and Hamida (Memon) M.; m. Rashida Lodhi, Apr. 23, 1992. BEng, McGill U., Montreal, Que., Can., 1990. Computer operator Control Data Can., Montreal, 1985-87; systems engr. Electronic Data Systems, Toronto, Ont., Can., 1990; network designer Bell Mobility Cellular, Toronto, 1988, 90—; part time instr. Sheridan Coll. Mem. IEEE. Islam. Avocations: baseball, hockey, stamp collecting. Home: 1394 Grist Mill Ct, Mississauga, ON Canada L5V 1S5 Office: Bell Cellular, 20 Carlson Ct, Etobicoke, ON Canada M9W 6V4

MÜHE, ERICH, surgical educator; b. Bad Windsheim, Bavaria, Germany, May 23, 1938; s. Karl and Friedel (Schmidt) M.; m. Hannelore Strebel, Oct. 6, 1952; children: Christian, Michael. MD, U. Erlangen, Fed. Republic Germany, 1966. Diplomate in surgery, Fed. Republic Germany. First resident Surg. Clinic, U. Erlangen, 1973-82, lectr., 1973-79, prof., 1979—; chief surg. clinic Kreiskrankenhaus, Böblingen, Fed. Republic Germany, 1982—; prof. surgery U. Tübingen (Fed. Republic Germany), 1982—. Contbr. over 250 articles to med. jour., chpts. to books; inventor bed bicycle for prevention thrombosis and pulmonary embolism, 1971, 1st laparoscopic cholecystectomy in the world (with-and without pneumoperitoneum), 1985. Recipient L. Rehn prize Mittelrheinische Chirurgen, 1974. Fellow ACS; mem. Deutsche Gesellschaft für Chirurgie (Jubilee prize 1992), Bayerische Gesellschaft für Chirurgie (J.H. Nussbaum prize 1965), Deutsche Krebsgesellschaft. Avocation: bicycling in Black Forest, Alps and Pyrenees. Office: Kreiskrankenhaus, Bunsenstrasse 120, D-7030 Böblingen Germany

MUHLENBRUCH, CARL W., civil engineer; b. Decatur, Ill., Nov. 21, 1915; s. Carl William and Clara (Theobald) M.; m. Agnes M. Kringel, Nov. 22, 1939; children: Phyllis Elaine (Mrs. Richard B. Wallace), Joan Carol (Mrs. Frederick W. Wenk). BCE, U. Ill., 1937, CE, 1945; MCE, Carnegie Inst. Tech., 1943. Research engineer Aluminum Research Labs., Pitts., 1937-39; cons. engring., 1939-50; mem. faculty Carnegie Inst. Tech., 1939-48; assoc. prof. civil engring. Northwestern U., 1948-54; pres. TEC-SEARCH, Inc. (formerly Edni. and Tech. Consultants Inc.), 1954-67, chmn. bd., 1967—; Pres. Profl. Centers Bldg. Corp., 1961-77. Author: Experimental Mechanics and Properties of Materials; Contbr. articles engring. publs. Treas., bd. dirs. Concordia Coll. Found.; dir. Mo. Lutheran Synod, 1965-77, vice chmn. 1977-79. Recipient Stanford E. Thompson award, 1945. Mem. Am. Econ. Devel. Council (certified indsl. developer), Am. Soc. Engring. Edn. (editor Educational Aids in Engring.), Nat. Soc. Profl. Engrs., ASCE, Sigma Xi, Tau Beta Phi, Omicron Delta Kappa. Club: University (Evanston). Lodge: Rotary (dist. gov. 1980-81, dir. service projects Ghana and the Bahamas). Home: 4071 Fairway Dr Wilmette IL 60091-1005

MUIR, HERMAN STANLEY, III, lawyer; b. San Antonio, Tex., Apr. 11, 1949; s. Herman Stanley Jr. and Frances Jane (Insley) M.; m. Lucinda Adams, Aug. 15, 1970; children: Carrie Sloan, Herman Stanley IV, Devon Whitfield. BS in Indsl. Engring., Va. Poly., 1971; JD, Coll. of William and Mary, 1977. Bar: Va. 1977, Ohio 1991, U.S. Patent Office 1976. Engr. Nuclear Power Generation div. Babcock & Wilcox, Lynchburg, Va., 1971-74; corporate lawyer Babcock & Wilcox, Lynchburg, 1977-83; ind. contractor NASA Langley Field, Hampton, Va., 1975-77; corporate lawyer, internat. counsel McDermott Inc., New Orleans, 1983-85; corporate lawyer NCR Corp., Dayton, Ohio, 1985-91; ptnr. Graham & Muir, Dayton, 1991—; adj. prof. Lynchburg Coll., 1981-83; lectr. Govtl. Controls in Internat. Bus., 1991—; mem. bd. advisors Internat. Law Quarterly, 1991—. Campaign chmn. United Way, Lynchburg, 1981. With USAR, 1971-77. Mem. ABA. Office: Graham & Muir 15 W Dorothy Ln Ste 202 Dayton OH 45429

MUIR, PATRICIA SUSAN, biology educator, researcher; b. Madison, Wis., Apr. 4, 1953; d. William Howard and Elizabeth Ann (Townsend) M.; m. Bruce Pettit McCune, Aug. 18, 1979; children: Myrica Muir McCune, Sara Muir McCune. BA, U. Mont., 1975; PhD, U. Wis., 1984. Rsch. scientist Holcomb Rsch. Inst., Indpls., 1984-87; asst. prof. dept. gen. sci. Oreg. State U., Corvallis, 1987-91, asst. prof. dept. botany and plant pathology, 1991—. Contbr. articles to profl. jours. Mem., vol. PTO, Corvallis, 1991. Recipient NSF fellowship, 1979, rsch. grant USAF Office of Sci. Rsch., 1985, rsch. grant Nat. Pk. Svc., 1986, rsch. grant U.S. EPA Office of Exploratory Biology, 1992. Mem. AAAS, Am. Inst. Biol. Sci., Soc. for Conservation Biology, Sigma Xi. Achievements include discoveries related to effects of disturbance history on Pinus contorta, high concentrations of pollutants in fog water, effects of air pollutants on plants. Home: 1840 NE Seavy Ave Corvallis OR 97330 Office: Oreg State Univ Dept Botany Plant Pathology Cordley Hall 2082 Corvallis OR 97331-2902

MUKHERJEE, TRISHIT, reproductive infertility specialist; b. Konnagar, India, Jan. 1, 1934; came to U.S., 1970; s. Anath Nath and Sarala (Banerjee) M.; m. Annapurna Chatterjee, June 8, 1962; children: Tanmoy, Kushal, Kingshuk. MD, Calcutta U., 1956; PhD, London U., 1970. Asst. prof. Columbia U., N.Y.C., 1972-77; clin. assoc. prof. Downstate Med. Ctr., N.Y.C., 1978, N.Y. Med. Coll., 1981—; chief gyn.-endocrinology Lenox Hill Hosp., N.Y.C., 1981—, Our Lady of Mercy Med. Ctr., Bronx, 1989—. Editorial bd. Infertility, 1990—. Named One of Best Drs. in N.Y.C., N.Y. Mag., 1991. Fellow Royal Coll. Surgeons, Royal Coll. of Ob-Gyn., Am. Fertility Soc. Office: 895 Park Ave New York NY 10021

MULASE, MOTOHICO, mathematics educator; b. Kanazawa, Japan, Oct. 11, 1954; came to U.S., 1982; s. Ken-Ichi and Mieko (Yamamoto) M.; m. Sayuri Kamiya, Sept. 10, 1982; children: Kimihico Chris, Paul Norihico, Yurika. BS, U. Tokyo, 1978; MS, Kyoto U., 1980, DSc, 1985. Rsch. assoc. Nagoya (Japan) U., 1980-85; JMS fellow Harvard U., Cambridge, Mass., 1982-83; vis. asst. prof. SUNY, Stony Brook, 1984-85; Hedrick asst. prof. UCLA, 1985-88; asst. prof. Temple U., Phila., 1988-89; assoc. prof. U. Calif., Davis, 1989-91, prof., 1991—; mem. Math. Scis. Rsch. Inst., Berkeley, Calif.,

1982-84, Inst. for Advanced Study, Princeton, N.J., 1988-89; vis. prof. Max-Planck Inst. for Math., Bonn, Germany, 1991-92. Contbr. articles to profl. jours. Treas. Port of Sacramento Japanese Sch., 1990-91. Mem. Math. Soc. Japan, Am. Math. Soc. Avocation: music. Office: U Calif Dept Math Davis CA 95616

MULCKHUYSE, JACOB JOHN, energy conservation and environmental consultant; b. Utrecht, the Netherlands, July 21, 1922; came to U.S., 1982; s. Lambertus D. and Aagje (Van Geyn) M.; m. Cornelia Jacoba Wentink, Jan. 17, 1953; children: Jacobien, Hans, Dieuwke, Linda, Marlies. MSc, U. Amsterdam (the Netherlands), 1952, PhD, 1960. Dir. Chemisch-Farmaceutische Fabriek Hamu, the Netherlands, 1951-57; tech. asst. mgr. Polak & Schwarz (now IFF), the Netherlands, 1957-60; asst. tech. mgr. Albatros Superphosphate Fabrieken, the Netherlands, 1960-61; tech. mgr. for overseas subsidiaries Verenigde Kunstmestfabrieken, the Netherlands, 1961-64, gen. mgr. process engring. dept., 1964-70; dept. head process engring. dept. Unie van Kunstmestfabrieken, the Netherlands, 1970-82; sr. chem. engr. World Bank, Washington, 1982-83, sr. cons. chem. engr., 1983-87; ind. cons. World Bank and several cons. firms, 1987—. Author: (with Heath and Venkataraman) The Potential for Energy Efficiency in the Fertilizer Industry, 1985, (with Gamba and Caplin) Industrial Energy Rationalization in Developing Countries and Constraints in Energy Conservation, 1990, Process Safety Analysis: Incentive for the Identification of Inherent Process Hazards; editor: Environmental Balance of the Netherlands, 1972. Mem. AICE, Royal Dutch Chem. Soc., Fertilizer Soc. (pres. 1969-70), Internat. Inst. for Energy Conservation (bd. dirs. 1990-93), Rotary. Avocations: philosophy, tennis, advising developing countries. Home: 53 Ponderosa Ln Palmyra VA 22963-2405

MULHEARN, CYNTHIA ANN, industrial/organizational psychology researcher; b. New Britain, Conn., Oct. 23, 1963; d. Robert John and Virginia Ann (Ryiz) Domurat; m. Thomas James Mulhearn, Dec. 16, 1988; 1 child, Alaina Ann. BA, U. Conn., 1986; MA, U. West Fla., Pensacola, 1990. Analyst Life Ins. Mktg. and Rsch. Assn., Farmington, Conn., 1990—. Contbr. articles to Market Facts and Mgr.'s Mag. Mem. Soc. for Indsl./Orgnl. Psychology, Am. Psychol. Soc. Office: Life Ins Mktg & Rsch Assn 8 Farm Springs Farmington CT 06032

MULICH, STEVE FRANCIS, safety engineer; b. Kansas City, Mo., Apr. 23, 1934; s. Stephen Francis and Mary Margret (Mish) M.; m. Apr. 5, 1974 (div.); children: Michael Francis, Mischelle Marie, Merko Mathew, Cheri Regina, Michael Klaus, Gary John, Josette Marie. BS in Gen. Sci., U. Notre Dame, 1956. Phys. chemist high altitude combustion Army Rocket and Guided Missile Agy., Huntsville, Ala., 1957-59; ballistics facility mgr. Aerojet Gen. Corp., Sacramento, 1960-65; chief engr. minute man penetration aids MB Assoc., Bollinger Canyon, Calif., 1968-72; lab mgr. hazardous materials and ballistics Martin Marietta, Waterton, Colo., 1966-75; plant mgr. smog sampler collectors mfg. Gen. Tex. Corp., Santa Clara, Calif., 1976-77; chief engr. auto airbag plant mgr. Talley Industries, Mesa, Ariz., 1978-84; prin. engr. hypervelocity guns FMC Corp., Mpls., 1984—. Author: Solid Rocket Technology, 1967; inventor stun gun, combustion augemented plasma gun, semiconductor initiator. With U.S. Army, 1957-59. Recipient Acad. Achievement award Bausch & Lomb, Kenosha, Wis., 1952. Mem. IEEE, AIAA (assoc.), Am. Def. Preparedness Assn., Navy League, Old Crows. Avocations: skiing, sailing, climbing, camping, hiking. Home: 1325 104th Pl NE Minneapolis MN 55434-3620 Office: FMC Corporation 4800 E River Rd Minneapolis MN 55421-1498

MULJADI, EDUARD BENEDICTUS, electrical engineer; b. Surabaya, Indonesia, Nov. 28, 1957; came to U.S., 1982; s. Johanes Joseph and Hartatik (Ong) Harijanto; m. Lili Gunawidjaja, Dec. 26, 1981; children: Anthony Mahardhika, Patrick Mahardhika. BS, Surabaya Inst. Tech., 1981; MS, U. Wis., 1984, PhD, 1987. Lab. instr. Surabaya Inst. Tech., 1979-81, U. Wis., Madison, 1983-87; asst. prof. Calif. State U., Fresno, 1988-91, assoc. prof., 1991-92; elec. engr. Nat. Renewable Energy Lab., Golden, Colo., 1992—. Contbr. articles on power electronics, control of electric machines, reactive power, other topics to profl. publs. Mem. IEEE (indsl. drives com.), Soc. Mfg. Engrs., Sigma Xi, Eta Kappa Nu. Achievements include research on power electronics applications for photovoltaics, wind power, power systems and electric vehicle. Office: Nat Renewable Energy Lab 1617 Cole Blvd Golden CO 80401

MULLEN, ALEXANDER, information scientist, chemist; b. Glasgow, Scotland, Nov. 23, 1945; s. John and Barbara (Sloan) M.; m. Irmela Bodemer, July 21, 1972; children: Peter, Anika. BSc, Glasgow U., 1967, PhD, 1971. Info. scientist Ruhrchemie AG, Germany, 1975-80, Goedecke AG, Germany, 1980-82, Bayer AG, Germany, 1982—; mem. scientific adv. panel Current Drugs, London, 1990—. Author: Information Sources in Pharmaceuticals: Chemical and Physicochemical Data, 1991, Information Sources in Chemistry: Pharmaceutical Industry, 1993, Carbonylation Reactions: Ring Closure Reactions, Reppe Reactions, 1981. Mem. Am. Chem. Soc. (info. divsn.), Royal Soc. of Chemistry (chartered chemist 1975), Pharm. Documentation Ring (v.p. 1988-91, pres. 1992—), German Chem. Soc. Home: Flehenberg 72, 42489 Wülfrath Germany

MULLEN, KEN IAN, chemist; b. Detroit, Jan. 7, 1955; s. Lewis O. and Carol Sue (Spiller) M.; m. Mary Anna Laspe, Aug. 19, 1989; children: Stefan, Kari. BS in Chemistry, Ft. Lewis Coll., 1989; PhD in Analytical Chemistry, U. Wyo., 1992. Hydrologist Bur. Reclamation, Durango, Colo., 1977-89; rsch. asst. dept. chemistry U. Wyo., Laramie, 1989-92; chemist Los Alamos (N.Mex.) Nat. Lab., 1993—. Contbr. articles to profl. publs. Mem. Am. Chem. Soc., Soc. Applied Spectroscopy. Achievements include patent for molecular specific optical fibers for surface enhanced raman spectroscopy. Home: 352 Bryce Los Alamos NM 87544

MULLENAX, CHARLES HOWARD, veterinarian, researcher; b. Sterling, Colo., Feb. 5, 1932; s. Guy William and Evelyn Irene (Simpson) M.; m. Phyllis Jean Brown, June 11, 1954 (div. 1972); children: Mark David, Craig Collins, Jean Gail, Nancy Alba; m. Lidia Hincapie, Nov. 7, 1974. BS, Colo. State U., 1953, DVM, 1956; MS, Cornell U., 1961; diploma de honor, Ctrl. U., Quito, Ecuador, 1966. Owner Mt. Pks. Vet. Hosp., Evergreen, Colo., 1956-59; teaching and rsch. asst. Cornell U., Ithaca, N.Y., 1959-61; rsch. veterinarian Nat. Animal Disease Ctr., Agrl. Rsch. Svc., USDA, Ames, Iowa, 1961-64; Fulbright prof., dir. of clinics Ctrl. Univ., Quito, 1964-66; pathologist, tng. leader Rockefeller Found., Internat. Ctr. for Tropical Agriculture, Bogotá and Cali, Colombia, 1966-71; project dir., cons. World Bank, Washington, 1971-74; prof., rschr. Tech. U. Llanos, Villavicencio, Bogotá, 1974-84; mem. acad. staff Univ. Calif. and Univ. Mo., 1985-87; rsch. assoc. Rural Devel. Inst., Univ. Wis., River Falls, 1988—; cons. Bahamas Livestock Co., Eleuthera, 1956-57, U.S. Agy. Internat. Devel., Washington, 1985-87; ofcl. rep. Colombian Ministry Agr., Bogotá, 1972-73, Livestock Prodrs. Assn. of Meta, Villavicencio, 1982-83. Contbr. more than 50 articles to sci. jours. Lay min. Presbyn. Bd. Nat. Missions, Lapwai, Idaho, 1955; pres. sch. bd. Am. Sch., Quito, 1965. Recipient Medal of Merit, Colombian Vet. Pharm. Inst. and Livestock Prodrs. Assn. of Meta, Bogotá, 1985. Mem. Am. Soc. Tropical Vet. Medicine, Vet. Hemoparasite Rsch. Workers, N.Y. Acad. Scis. Achievements include first discovery and reporting correlation between changes of the earth's geomagnetic field strength, soil pH, availability of mineral soil nutrients for plant uptake, plant productivity and quality and impact of this on agricultural and livestock production and ecosystem function. Home: N7003 710th St Beldenville WI 54003

MULLENNEX, RONALD HALE, geologist, consultant; b. Harrisonburg, Va., Nov. 12, 1949; s. Roy Hale and Josie (Harman) M.; m. Linda Jean Miller, June 29, 1974; children: Rory, Megan, Sarah Beth. BS, W.Va. U., 1971, MS, 1975. Cert. profl. geologist; cert. ground water profl. Coal geologist W.Va. Geol. and Econ. Survey, Morgantown, 1974-77; project geologist Geol. Cons. Svcs. Inc., Bluefield, Va., 1977-79; v.p. Geol. Cons. Svcs. Inc., Bluefield, 1979-90; sr. v.p. Marshall Miller and Assocs., Bluefield, 1990—; tech. adv. bd. mem. W.Va. Dept. Environ. Health, Charleston, 1991—. Contbr. articles to profl. jours. Comprehensive Plan Com. mem. Tazewell (Va.) County Planning Commn., 1991—. Mem. AIME, Assn. Ground Water Scientists and Engrs., Geol. Soc. Am., Am. Inst. Profl. Geologists. Presbyterian. Office: Marshall Miller & Assocs PO Box 848 Bluefield VA 24605

MULLER, ACHIM, chemistry educator; b. Detmold, Germany, Feb. 14, 1938. PhD in Natural Scis., U. Göttingen, Germany, 1965. Lectr. U. Göttingen, 1967-71; assoc. prof. U. Dortmund (Germany), 1971-77; full prof. U. Bielefeld (Germany), 1977—; v.p. European Congress on Molecular Spectroscopy, 1980—. Editor: (with others) Spectroscopy in Chemistry and Physics, 1980, Transition Metal Chemistry, 1981, Electron and Proton Transfer in Chemistry and Biology, 1992, Matrix Isolation Spectroscopy, 1981, Nitrogen Fixation: The Chemical-Biochemical-Genetic Interface, 1982, Sulfur, Its Significance for Chemistry, for the Geo- and Bio- and Cosmosphere and Technology, 1984. Mem. AAAS, N.Y. Acad. Scis., Am. Chem. Soc., Royal Soc. Chemistry, Gesellschaft Deutscher Chemiker. Achievements include research on transition metal sulfur clusters, on new class of metal oxides. Home: Altenberndstr 1, 4930 Detmold Germany Office: U Bielefeld, UniversitÄtsstrasse 25, 4800 Bielefeld 100131, Germany

MULLER, DANIEL, biomedical scientist; b. Bronx, N.Y., Nov. 28, 1953; s. Oscar and Lilian (Otelsberg) M.; m. Roberta Riportella, June 5, 1976; children: Scott Jacob, Benjamin Sean. PhD, U. Wis., 1981; MD, SUNY, Stony Brook, 1985. Diplomate Am. Bd. Internal Medicine, Am. Bd. Rheumatology. Resident in medicine N.C. Meml. Hosp., Chapel Hill, 1985-88; postdoctoral fellow in microbiology and immunology U. N.C., Chapel Hill, 1988-91, rheumatology fellow, 1989-91; asst. prof. medicine U. Wis., Madison, 1991—. Contbr. articles to jour. Sci. Recipient Nat. Rsch. Svc. award NIH, 1989. Mem. ACP, AAAS, Am. Coll. Rheumatology, Am Soc. Microbiology. Achievements include research in molecular biology of the cellular immune response, microbial host-parasite relationships. Office: U Wis Dept Medicine 2605 MSC 1300 University Ave Madison WI 53706

MULLER, DIETRICH ALFRED HELMUT, physicist, educator; b. Leipzig, Germany, Sept. 14, 1936; came to U.S., 1968; s. Herbert and Johanna (Potzger) M.; m. Renate Runkel, Mar. 23, 1968; children: Georg, Michael, Agnes. Student, U. Leipzig, German Dem. Republic; diploma in physics, U. Bonn, Fed. Republic Germany, 1961, PhD in Physics, 1964. Rsch. assoc. U. Bonn., 1965-68; rsch. assoc. U. Chgo., 1968-70, asst. prof., 1970-77, assoc. prof., 1977-84, prof. physics, 1985—, dir. Enrico Fermi Inst., 1986-92; prin. investigator on rsch. in high energy astrophysics using baloons and spacecraft, NASA; rsch. in exptl. physics and particle physics. Contbr. numerous articles to profl. jours. Fellow Am. Phys. Soc.; mem. AAAS, Am. Astron. Soc. Office: U Chgo Enrico Fermi Inst 933 E 56th St Chicago IL 60637-1460

MULLER, JEAN-CLAUDE, nuclear engineer; b. Nilvange, France, Apr. 29, 1948; s. Jean-Pierre and Catherine (Gradel) M.; m. Iris Gaudenzi, Apr. 3, 1971; 1 child, Marie-Eve. PhD in Nuclear Engring., U. Louis Pasteur, Strasbourg, France, 1976, PhD in Physics, 1982. Rsch. engr. Ctr. Nat. de la Recherche, Strasbourg, 1978—. Contbr. articles to profl. publs.; author over 100 rsch. papers in field. Mem. Coopération Française pour l'Etude et le Développement de l'Energie Solaire. Achievements include patent for new doping technique using molecular ion-implantation for semiconductor material; development of new technologies for manufacturing solar cells as dry processing, including ion implantation. Office: CNRS-Laboratoire PHASE, 23 rue du Loess - BP 20, 67037 Strasbourg France

MULLER, JULIUS FREDERICK, chemist, business administrator; b. Bklyn., Sept. 5, 1900; s. Edward Jefferson and Julia (Lang) M.; m. Ethel Mae Johnson, June 18, 1927; children: Richard J., J. Edward, James H. BS, Rutgers U., 1922, MSc, 1928, PhD, 1930. Mgr. Princewick Farm, Franklin Park, N.J., 1922-26; rsch. fellow Walker-Gordon Labs., Plainsboro, N.J., 1930-32; bacteriologist Nat. Oil Products Co., Harrison, N.J., 1932-35; owner, mgr. Muller Labs., Balt., 1935-41; div. mgr. Borden, Inc., N.Y.C., 1941-65; cons. Borden, Inc., N.Y.C., 1965-68. Contbr. articles to profl. jours. Fellow Am. Inst. Chemists, mem. Am. Chem. Soc. (emeritus), Alpha Zeta, Sigma Xi. Republican. Presbyterian. Achievements include 3 patents on uses of SO_2 in oils; invention of mull-soy first commercial canned liquid soybean milk. Home: # 313 5700 Williamsburg Landing Dr Williamsburg VA 23185-3779

MÜLLER, KARL ALEXANDER, physicist, researcher; b. Apr. 20, 1927. PhD in Physics, Swiss Fed. Inst. Tech., 1958; DSc (hon.), U. Geneva, 1987, Tech. U. Munich, 1987, U. Studi di Pavia, Italy, 1987. Project mgr. Battelle Inst., Geneva, 1958-63; lectr., U. Zurich, Switzerland, from 1962, titular prof., from 1970, prof., 1987—; researcher solid-state physics IBM Zurich Research Lab., Rüschlikon, Switzerland, 1963-73, mgr. dept. physics, 1973-82, fellow, 1982-85; researcher Switzerland, 1985—. Contbr. over 200 articles to tech. publs. Recipient Marcel-Benoist Found. prize, 1986, Nobel prize in physics, 1987, (with J. Georg Bednorz) Fritz London Meml. award, 1987, Dannie Heineman prize Acad. Scis. Göttingen, Fed. Republic of Germany, 1987, Robert Wichard Pohl prize German Phys. Soc., 1987, Europhysics prize Hewlett-Packard Co., 1988. Fellow Am. Phys. Soc. (Internat. prize for new materials research 1988); mem. European Phys. Soc. (mem. ferroelectricity group), Swiss Phys. Soc., Zurich Phys. Soc. (pres. 1968-69), Groupement Ampère, Nat. Acad. Scis. (fgn. assoc.). Office: IBM Zurich Rsch Lab, Saumerstrasse 4, CH8803 Rueschlikon Switzerland

MULLER, PAUL-EMILE, electrical engineer; b. Reutlingen, Germany, Jan. 10, 1926; came to U.S., 1956; s. Paul-Emile and Maria-Elisa (Borel) M.; children: Paul Emile, Bernard, Roland, Yves, Wanda. BSEE, Geneva Inst. Tech., 1946; MSEE, U. Pa., 1962; diploma in energy, Swiss Inst. Tech., Lausanne, 1980, diploma in European Engring., 1991. With Remington Rand Univac RCA Corp. Am., Phila., Venezuela, 1956-62; prin. engr. N.Am. Aviation, Downey, Calif., 1962-65; chief dept. Noordwijk (Hollande) Agence Spatiale Européene, 1965-68; mgr. info. dept. McDonnell-Douglas Astronautics Co., Huntington Beach, Calif., 1968-70; cons. engr. Prospective Engring. Gestion, Geneva, 1970-72; dir. Ecole d'ingénieurs de St.-Imier, 1972-76; dir. dept. edn. Geneva Inst. Tech., 1976-91; dir. World ORT Union High Tech Inst., Geneva, 1992—. Mem. Swiss Astronomy Soc. (cen. sec. 1991—). Home: 10 Chemiu du Marais-Long, 1217 Meyrin Geneva Switzerland

MULLER, RICHARD STEPHEN, electrical engineer, educator; b. Weehawken, N.J., May 5, 1933; s. Irving Ernest and Marie Victoria Muller; m. Joyce E. Regal, June 29, 1957; children: Paul Stephen, Thomas Richard. ME, Stevens Inst. Tech., Hoboken, N.J., 1955; MSEE, Calif. Inst. Tech., 1957, PhD in Elect. Engring. and Physics, 1962. Engr.-in-tng., 1955. Test engr. Wright Aero/Curtiss Wright, Woodridge, N.J., 1953-54; mem. tech. staff Hughes Aircraft Co., Culver City, Calif., 1955-61; instr. U. So. Calif., L.A., 1960-61; asst. prof., then assoc. prof. U. Calif., Berkeley, 1962-72, prof., 1973—; guest prof. Swiss Fed. Inst. Tech., 1993; dir. Berkeley Sensor and Actuator Ctr., 1985—. Co-author: Device Electronics for Integrated Circuits, 1977, 2d rev. edit., 1986, Microsensors, 1990; contbr. more than 200 articles to profl. jours. Vice chmn. Kensington (Calif.) Mcpl. Adv. Coun. Fellow Hughes Aircraft Co. 1955-57, NSF 1959-62, NATO postdoctoral 1968-69, Fulbright 1982-83, Alexander von Humboldt 1993. Fellow IEEE; mem. NAE, Electron Devices Soc. (adv. com. 1984—), Internat. Sensor and Actuator Meeting (chmn. steering com.). Achievements include 12 patents; construction of first operating micromotor. Office: U Calif Dept EECS 401 Cory Hall Berkeley CA 94720

MULLER, ROBERT JOSEPH, gynecologist; b. New Orleans, Dec. 5, 1946; s. Robert Harry and Camille (Eckert) M.; m. Susan Philipsen, Aug. 22, 1974; children: Ryan, Matt. BS, St. Louis U., 1968; BS, MSc, Emory U., 1976; MD, La. State U., New Orleans, 1981. Intern Charity Hosp., New Orleans, 1981-82; resident La. State U. Affiliate Hosp., 1982-85; resident staff physician La. State U. Med. Ctr., New Orleans, 1981-85; pvt. practice Camellia Women's Ctr., Slidell, La., 1985—; staff physician Tulane Med. Ctr., New Orleans, 1986—; med. dir. Northshore Regional Med. Ctr., Slidell, 1987—, New Orleans Police Dept., 1981—, S.W. La. Search and Rescue, Covington, La., 1986—, St. Tammany Parish Sheriff Dept., Covington, 1989—, commdr., 1990— Camellia City Classic, Slidell, 1989—, Crawfishman Triathlon, Mandeville, La., 1988—, Res-Q-Med Laser Team, 1984—. Contbr. articles to profl. jours. Recipient Commendation Medal New Orleans Police Dept., 1986, 87, 89, Medal Valor St. Tammany Parish Sheriff Office, Covington, 1990, Cert. Valor S.E. La. Search and Rescue, Mandeville, 1990; named one of Outstanding Young Men of Am., 1984. Mem. Am. Coll. Ob-Gyn., La. State Med. Soc., Profl. Assn. Diving Instrs. (divemaster 1991), So. Offshore Racing Assn. (med. dir. 1982—), Offshore Profl. Racing Tour (med. dir. staff 1990—), Am. Power Boat Assn. (med. staff 1984-89). Roman Catholic. Avocations: scuba diving, boating, shooting. Home: 128 Golden Pheasant Dr Slidell LA 70461-3104 Office: Camellia Womens Ctr 105 Smart Pl Slidell LA 70458-2039

MÜLLER-ISBERNER, JOACHIM RÜDIGER, psychiatrist, educator; b. Leipzig, Saxon, Germany, Mar. 25, 1952; s. Joachim Roland and Ursula (Köhler) M.; m. Petra Isberner, Sept. 28, 1984; 1 child, Nora Isberner. MD, Liebig U., Giessen, Germany, 1976. Intern Univ. Hospital, Giessen, Germany, 1976-77; resident Psychiat. Hospital Giessen, 1979-83; med. dir. Forensic Hospital, Haina, Germany, 1987—; instr. Justus Liebig U., 1988—. Contbr. articles to med. jours. Capt. German Army, 1977-78. Mem. Internat. Acad. Law and Mental Health, German Soc. for Psychiat. and Nerve Medicine. Home: An der Liebigshöhe 2, D-35394 Giessen Hessen, Germany Office: Forensic Psychiat Hospital, Landgraf-Philip-Platz, D-35114 Haina Hessen, Germany

MULLETTE, JULIENNE PATRICIA, television personality and producer, astrologer, author, health center administrator; b. Sydney, Australia, Nov. 19, 1940; came to U.S., 1953; d. Ronald Stanley Lewis and Sheila Rosalind Blunden (Phillips) M.; m. Fred Gillette Sturm, Nov. 24, 1964 (div. Dec. 1969); children: Noah Khristoff Mullette-Gillman, O'Dhaniel Alexander Mullette-Gillman. B.A., Western Coll. for Women, Oxford, Ohio, 1961; postgrad., Harvard U., 1964, U. Sao Paulo, Brazil, 1965, Inst. do Filosofia, Sao Paulo, 1965, Miami U., Oxford, 1967-69. Tchr. English, High Mowing Sch., Wilton, N.H., 1962-64, Stoneleigh-Prospect Hill Sch., Greenfield, Mass., 1964; seminar dir. Western Coll., Oxford, Ohio, 1967-69; pres. Family Tree, The Home Univ., Montclair, N.J., 1978-80; dir. Pleroma Holistic Health Ctr., Montclair, 1980—; dir. Astrological Rsch. Ctr., Sydney, Australia, 1983; hostess (radio talk show) You and the Cosmos Sta. WFMU, East Orange, N.J., 1985, Sta. WJFF, Jeffersonville, N.Y., 1992—, The Juliette Mullette Show, Connections TV, Newark, 1985—, The Juliette Mullette Show Sta. WFDU, Fairleigh Dickinson U., N.J., 1986—, (TV program) You and the Cosmos, Woodstock, N.Y., 1992—; founder Spiritual Devel. Rsch. Group 1986—; pvt. astrology counselor, 1962—; lectr., speaker worldwide, 1968—; guest on radio and TV shows, U.S. and Can., 1962—; host syndicated radio talk show The Juliette Mullette Show, N.Y., N.J., 1987—; owner, pres. Moonlight Pond, Woodbourne, N.Y., 1988—; founder The Spiritual Devel. Ctr., 1986—, Pleroma Found. for Astrological Rsch. and Studies, 1990; breeder, trainer llamas, alpacas and other exotic animals; The Juliette Mullette Show, You and the Cosmos, WJFF pub. radio, Jeffersonville, N.Y., 1992. Author: The Moon-Understanding the Subconscious, 1973; also articles, 1968—; founding editor KÓSMOS mag., 1968-78, The Jour. of Astrological Studies, 1970; contbg. columnist I Love Cats, 1988—. Founder local chpt. La Leche League, Montclair, 1974. Mem. AAUW (chair cultural affairs Montclair chpt.), Spiritual Devel. Group (founder 1987), Internat. Soc. Astrological Research (founding pres. 1968-78), Am. Fedn. Astrologers (cert.), Société Belge d'Astrologie, Am. Assn. Humanistic Psychology, AAUW (dir. cultural affairs 1987—), NAFE, Internat. Llamas Assn. Avocations: competitive tennis, local theatre, singing. Home: PO Box 65 Bearsville NY 12409

MULLIGAN, RICHARD C., molecular biology educator. Prof. molecular biology MIT, Cambridge, Mass. Recipient AMGEN award Am. Soc. Biochemistry and Molecular Biology, 1992. Office: MIT Dept Biochemistry 77 Massachusetts Ave Cambridge MA 02139*

MULLIN, GERARD EMMANUEL, physician, educator, researcher; b. Pequannock, N.J., Nov. 5, 1959; s. Gerard Vincent Jr. and Frances Rita (Magnanti) M. BS in Biology and Chemistry, William Patterson Coll., 1981; MD, U. Medicine and Dentistry N.J., 1985. Diplomate Am. Bd. Internal Medicine, Am. Bd. Gastroenterology, Am. Bd. Med. Examiners. Intern Mt. Sinai Hosp., N.Y.C., 1985-86, resident, 1986-88; fellow in gastroenterology Johns Hopkins Hosp., Balt., 1988-91; fellow NIH, Bethesda, Md., 1989-91; instr., scientist Cornell U. Med. Coll., N.Y., 1991-93, asst. prof., 1993—. Cotbr. articles to med. and sci. jours. Grantee Nat. Found. Ileitis and Colitis, 1990, North Shore U. Hosp., 1991. Fellow ACP; mem. AMA, Crohn's and Colitis Found. Am. (Young Investigator of Yr. 1991, grantee 1991, 93), N.J. Med. Sch. Alumni Assn. (bd. dirs.), Fieri No. N.J. (bd. dirs.), Alpha Omega Alpha, Phi Beta Sigma. Roman Catholic. Achievements include discovery that suppressor cell function is elevated in AIDS, lymphomas are increased in inflamatory bowel disease, helicobacter pylori gastritis causes increased gastric acid secretion and hypergastrinemia, intestinal T cells are activated and make IL2 in Crohn's disease but not ulcerative colitis. Office: Cornell U Med Coll 300 Comminuty Manhasset NY 11030

MULLIN, SHERMAN N., engineering executive. Pres. Lockheed Advanced Devel., Burbank, Calif. Recipient Aircraft Design award Am. Inst. Aeronautics and Astronautics, 1992; Wright Brothers Aeronautics lectureship Am. Inst. Aeronautics and Astronautics, 1992. Office: Lockheed Advanced Development PO Box 551 Burbank CA 91520-0001*

MULLINS, MICHAEL DREW, environmental educator; b. Tampa, Fla., Nov. 24, 1947; s. Hugh Clifford Jr. and Ida (Dunbar) M.; m. Barbara McManeus, Aug. 15, 1970; 1 child, David Edward. BA, U. South Fla., 1970, MEd, 1977, cert. supr., 1985. Cert. tchr., supr., Fla. Biol. aid U.S. Dept. Interior, 1967-69; sci. tchr. Sch. Bd. Hillsborough County, 1970-74; instr. energy edn. Sch. Bd. Pasco County, 1974-75; high sch. sci. tchr. Sch. Bd. Hillsborough County, 1975-78, summer instr. marine biology, 1978, dir. U.S. youth conservation corps, 1980-81, supr. environ. edn., mid. sch. instrumental tech., 1978-92, supr. mid. sch. instrnl. tech., 1992—, acting gen. dir. secondary edn., 1992; assoc. dir. West Ctrl. Fla. Regional Environ. Edn. Project, 1990—; pres. Mangrove Pencil Publs., Tampa, 1985-89; prof. community edn. Hillsborough Community Coll., 1987—; acting gen. dir. secondary edn. Sch. Bd. Hillsborough County, 1992. Co-author: A Field Guide to the Copeland Park Nature Trail, 1990, Marine Biology: A Laboratory Text, 1986, The Florida Environmental Guide, 1985, Pine Woods Communities, 1993; editor: The Estuary: A Balance of Forces, 1986, Florida's Freshwater Wetlands, 1985, Florida's Wildlife: A Habitat Approach, 1984, The Original Floridians, 1983, Nature's Classroom: Classroom Teacher's Guide, 1982, Florida's Energy Use: A Historical Perspective, 1981, Coastal Zone Management, 1981, An Introduction to Aquaculture, 1981, Pine Woods Communities, 1993, others. Mem. adv. com. Teco Corp. Stewartship, 1989—; active Tampa Bay Area Environ. Action Team, 1988—; adv. com. Fla. Conservation Corps, 1981-82, Hillsborough C.C. Environ. Studies Ctr., 1973-83, chmn., 1979-82; steering com. Earth Day Tampa, 1990—; vice chair recycling com. County of Hillsborough, 1989-91; active Mayor's Water Adv. Com., Tampa, 1990-91; active edn. com. S.W. Fla. Water Mgmt. Dist., Tampa, 1980—. Mem. Oceanic Soc., Nat. Marine Edn. Assn., League Environ. Educators Fla. (pres. 1986-87, chmn. membership 1981-84, workshop planning com. 1981-88), North Am. Assn. Environ. Edn., Audubon Soc., Fla. Audubon Soc., Tampa Audubon Soc., Cousteau Soc., Fla. Marine Sci. Edn. Assn. (pres. 1974-75, chmn. membership, treas. 1976, 1982-86, bd. dirs. 1975-84, chmn. workshop planning com. 1974-75, 83-84), Phi Kappa Phi. Methodist. Avocations: skin diving, nature photography, reading. Office: Sch Bd Hillsborough County PO Box 3408 Tampa FL 33601-3408

MULLINS, OBERA, microbiologist; b. Egypt, Miss., Feb. 15, 1927; d. Willie Ree and Maggie Sue (Orr) Gunn; B.S., Chgo. State U., 1974; M.S. in Health Sci. Edn., Governors State U., 1981; m. Charles Leroy Mullins, Nov. 2, 1952; children—Mary Artavia, Arthur Curtis, Charles Leroy, Charlester Teresa, William Hellman. Med. technician, microbiologist Chgo. Health Dept., Chgo., 1976—. Mem. AAUW, Am. Soc. Clin. Pathologists (cert. med. lab. technician), Ill. Soc. Lab. Technicians. Roman Catholic. Home: 9325 S Marquette Ave Chicago IL 60617-4131 Office: Daley Cte Lower Level 82 50 W Washington St Chicago IL 60623

MULLINS, RICHARD AUSTIN, chemical engineer; b. Seelyville, Ind., Apr. 22, 1918; s. Fred A. and Ethel (Zenor) M.; B.S. in Chem. Engring., Rose Poly. Inst., 1940; postgrad. Yale, 1942-43; m. Margaret Ann Dellacca, Nov. 27, 1946 (dec. Nov. 1982); children—Scott Alan, Mark Earl. Chemist, Ayrshire Collieries Corp., Brazil, Ind., 1940-49; chief chemist Fairview Collieries Corp., Danville, Ill., 1949-54; preparations mgr. Enos Coal Mining Co., Oakland City, Ind., 1954-72, Enoco Collieries, Inc., Bruceville, Ind., 1954-62; mining engr. Kings Station Coal Corp.; mgr. analytical procedures Old Ben Coal Corp., 1973-84; ret., 1984. Am. Mining Congress cons. to Am.

Standards Assn. and Internat. Orgn. for Standards, 1960-74; mem. indsl. cons. com. Ind. Geol. Survey, 1958-72; mem. organizing com. 5th Internat. Coal Preparation Congress, Pittsburgh, 1966. Mem. exec. bd. Buffalo Trace council Boy Scouts Am., also mem. speakers bur. Bd. dirs. Princeton Boys Club. Served with AUS, 1942-46; ETO. Decorated Medaille de la France Liberee (France); recipient Eagle Scout award, Boy Scouts Am., 1935, Silver Beaver award, 1962, Wood Badge Beads award, 1960; Outstanding Community Svc. award Princeton Civitan Club, 1964; Engr. of Year award S.W. chpt. Ind. Soc. Profl. Engrs., 1965; Prince of Princeton award Princeton C. of C., 1981, Sagamore of the Wabash award Ind. gov. R.D. Orr, 1984. Registered profl. engr., Ind., Ill. Mem. AIME (life mem.), ASTM (sr. mem., R.A. Glenn award 1985), Am. Chem. Soc., Nat. Soc. Profl. Engrs. (life mem.), Ind., Ill. mining insts., Ind. Coal Soc. (pres. 1958-59), Am. Mining Congress (chmn. com. coal preparation 1964-68), Am. Legion (past commn. chmn.), VFW (40 & 8 coms.), Ind. Soc. Profl. Land Surveyors, Rose Tech. Alumni Assn. (pres. 1976-77, Honor Alumnus 1980), Order of Ring, Sigma Nu. Methodist (lay speaker). Mason, Elk. Contbr. articles to profl. jours. Home: RR 4 Box 310 Princeton IN 47670-9412

MULLIS, KARY BANKS, biochemist; b. Lenoir, N.C., Dec. 28, 1944; s. Cecil Banks Mullis and Bernice Alberta (Barker) Fredericks; children: Christopher, Jeremy, Louise. BS in Chemistry, Ga. Inst. Tech, 1966; PhD in Biochemistry, U. Calif., Berkeley, 1973. Lectr. biochemistry U. Calif., Berkeley, 1972; postdoctoral fellow U. Calif., San Francisco, 1977-79, U. Kans. Med. Sch., Kansas City, 1973-76; scientist Cetus Corp., Emeryville, Calif., 1979-86; dir. molecular biology Xytronyx, Inc., San Diego, 1986-88; cons. Specialty Labs, Inc., Amersham, Inc., Chiron Inc. and various others, Calif., 1988—; chmn. StarGene, Inc., San Rafael, Calif. Contbr. articles to profl. jours.; patentee in field. Recipient Preis Biochemische Analytik award German Soc. Clin. Chem., 1990, Allan award Am. Soc. of Human Genetics, 1990, award Gairdner Found. Internat., 1991, Nat. Biotech. award, 1991, Robert Koch award, 1992, Chiron Corp. Biotechnology Rsch. award Am. Soc. Microbiology, 1992, Japan prize Sci. and Tech. Found. Japan, 1993, Nobel Prize in Chemistry, Nobel Foundation, 1993; named Calif. Scientist of Yr., 1992, Scientist of Yr., R&D Mag., 1991. Mem. Am. Chem. Soc., Inst. for Further Study (dir. 1983—). Achievements include invention of Polymerase Chain Reaction (PCR); avocations: computers, gardening, surfing, skiing, rollerblading. Office: 6767 Neptune Pl Apt # 5 La Jolla CA 92037

MUMFORD, DAVID BRYANT, mathematics educator; b. Worth, Sussex, Eng., June 11, 1937; came to U.S., 1940; s. William Bryant and Grace (Schiott) M.; m. Erika Jentsch, June 27, 1959 (dec. July 30, 1988); children: Stephen, Peter, Jeremy, Suchitra; m. Jenifer Moore, Dec. 29, 1989. B.A., Harvard U., 1957, Ph.D., 1961; D.Sc. (hon.), U. Warwick, 1983. Jr. fellow Harvard U., 1958-61, asso. prof., 1962-66, prof. math., 1966-77, Higgins prof., 1977—, chmn. dept. math, 1981-84; v.p. Internat. Math. Union, 1990—. Author: Geometric Invariant Theory, 1965, Abelian Varieties, 1970, Introduction to Algebraic Geometry, 1976. Recipient Fields medal Internat. Congress Mathematicians, 1974; MacArthur Found. fellow, 1987-92. Fellow Tata Inst. (hon.); mem. Accad. Nazionale dei Lincei, Nat. Acad. Scis., Am. Acad. Arts and Scis. Home: 26 Gray St Cambridge MA 02138-1510 Office: Harvard U 1 Oxford St Cambridge MA 02138-2901

MUMMA, MICHAEL JON, physicist; b. Lancaster, Pa., Dec. 3, 1941; s. John Henry and Violet Lyndell (Baxter) M.; m. Sage Bailey Tower, Aug. 20, 1966; children: Peter Robb, Amy Elizabeth. A.B. in Physics with honors, Franklin and Marshall Coll., 1963; Ph.D. in Physics, U. Pitts., 1970. Grad. research asst. U. Pitts., 1963-70; astrophysicist NASA Goddard Space Flight Center, Greenbelt, Md., 1970-76; head br. Infrared and Radio Astronomy NASA Goddard Space Flight Center, 1976-84, assoc. chief Lab. Extraterrestrial Physics, 1984-85, head Planetary Systems br., 1985-90, chief scientist Lab. Extraterrestrial Physics, 1990—; adj. research assoc. in physics Pa. State U., 1978-81, prof. physics, 1981—; mem. numerous working groups and adv. coms. NASA, Nat. Bur. Standards, NSF, Nat. Acad. Scis., 1973—; lectr. in field. Contbr. numerous articles to profl. publs., 1970—; editor: The Study of Comets, Vols. 1, 2, 1976, Vibrational-Rotational Spectroscopy for Planetary Atmospheres, vols. 1, 2, 1982, Astrophysics from the Moon, 1990. Recipient NASA medal for Exceptional Sci. Achievement, 1986; Kershner award for physics, 1962; Coll. Trustee's scholar Franklin and Marshall Coll., 1963. Fellow Am. Phys. Soc., Washington Acad. Sci.; mem. AAAS, Am. Astron. Soc., Am. Geophys. Union, Internat. Astron. Union, Sigma Pi Sigma. Achievements include discovery of natural lasers in atmospheres of Mars, Venus, and Jupiter; first detection of water vapor, formaldehyde and methanol in comets; first definitive measurements of deuterium and hydrogen on Mars and Venus; first absolute wind measurements on Venus and Mars; invention of tunable diode laser heterodyne spectrometer and other advanced instruments; development of Doppler-limited infrared spectroscopy for laboratory and astrophysical applications, of absolute calibration procedures in vacuum ultraviolet, of molecular branching ratio technique for intensity calibration in vacuum ultraviolet; measurement of many absolute cross sections in vacuum ultraviolet; research on atomic and molecular physics and chemistry, on comets, on planetary atmospheres, on infrared astronomy, on high-resolution spectroscopy, and in the field of dissociative excitation of molecules. Office: Code 690 Goddard Space Flight Ctr Greenbelt MD 20771

MUMTAZ, MOHAMMAD MOIZUDDIN, toxicologist, researcher; b. Hyderabad, India, Aug. 25, 1949; s. Mohammad Abdur Raheem and Khairunnisa Begum; m. Farzana Begum, Nov. 2, 1978; 1 child, Mohammad Nabeeluddin. BSc with honors, Osmania U., Hyderabad, 1970; MS, Oreg. State U., 1977; PhD, U. Md., 1984. Rsch. scientist Ill. State Psychiat. Inst., Chgo., 1978-79; instr. Med. Coll. Va., Richmond, 1979-81; rsch. asst. U. Md., College Park, 1981-84; rsch. assoc. U. Tex. Med. Br., Galveston, 1984-87; toxicologist office R&D U.S. EPA, Cin., 1987-92; sr. toxicologist, sci. adviser ATSDR, USPHS, Atlanta, 1992—; mem. guidelines work group U.S. EPA, 1989-92, pharmocokinetics work group OHEA, 1990-92, interagy. work group Kuwati Crude Oil Fires, 1991-92. Contbr. to profl. publs. Mem. Am. Coll. Toxicology, Fedn. Am. Socs. Exptl. Biologists, Soc. Toxicology. Office: USPHS Div Toxicology Ste E 29 1600 Clifton Rd Atlanta GA 30333

MUMZHIU, ALEXANDER, machine vision systems engineer; b. St. Petersburg, Russia, June 6, 1937; came to U.S., 1979; s. Mikhail and Henrietta (Rosenblum) M.; m. Natalya Takjas, Sept. 22, 1967; children: Jennifer, Daniel. Engirng. diploma, Leningrad Tech. Inst., 1959. Engr. Petzochemical Inst., St. Petersburg, Russia, 1959-65; sr. engr. optical dept. Mendeleev Inst. Metrology, St. Petersburg, Russia, 1965-79; prin. engr. Mid-West Instr., Troy, Mich., 1980-83; electro-optical engr. Perceptron Co., Farm Hills, Mich., 1983-84; cons. Arthur D. Little Co., Washington, 1984-85; sr. rsch. scientist Hunterlab., Reston, Va., 1985—. Contbr. articles to profl. jours. Recipient awards Moscow Indsl. Show. Mem. IEEE, OSA. Achievements include 2 U.S. patents for fiber optic sensor of rotation; application of color TV cameras with image processor and personal computers for measurement of color and appearance; design of first scanning spectrophotometer, automatic lensmeter, thermoelectric flowmeter. Office: Hunter Lab Inc 11491 Sunset Hills Rd Reston VA 22090

MUNASINGHE, MOHAN, development economist; b. Colombo, Sri Lanka, July 25, 1945; s. Peter Munasinghe and Flower Wickramasinghe; m. Sria Gooneratne, May 8, 1970; children: Anusha, Ranjiva. BA with honors, Cambridge (Eng.) U., 1967, MA, 1968; SM, MIT, 1968, EE, 1970; PhD in EE, McGill U., Montreal, Can., 1973; MA in Econs., Concordia U., 1975. Rsch. officer Ceylon Inst. Sci. and Indsl. Rsch., Colombo, 1968-70; asst. dir. Internat. Inst. Quantitative Econs., Montreal, 1973-75; div. chief World Bank, Washington, 1975—; vis. prof. Am. U., Washington, 1977-81, Inst. Tech. Policy in Devel., SUNY, 1982-88, Energy Ctr. U. Pa., Phila., 1988—; fellow Beiger Inst., Royal Swedish Acad. Scis., 1993—; sr. advisor to pres. Office of Pres. Sri Lanka, Colombo, 1982-87, chmn. computer and info. tech. coun., 1983-86; sr. rsch. fellow Ctr. Internat. Devel. and Conflict Mgmt., U. Md., College Park, 1987-90; pres.-emeritus Sri Lanka Energy Mgmt. Assn., Colombo, 1985—, pres., 1983-85. Author: 40 books including Economics of Power System Reliability and Planning, 1979, Energy Economics, Demand Management and Pricing, 1983, Rural Electrification for Development, 1987, Integrated National Energy Planning and Management, 1988, Computers and Informatics in Developing Countries, 1989, Energy Analysis and Policy, 1990, Electric Power Economics, 1990, Water Supply and Environmental Management, 1992, Energy Modelling and Policy, 1992, Towards Sustainable Development, 1993; author over 200 tech. papers. Recipient Prize for Outstanding Achievement Latin Am. and Caribbean Energy Conf., 1988, Exceptional Contributions award Internat. Assn. of Energy Econs., 1987, Outstanding Scientists Gold medal Lions Internat., 1985; Grass fellowship MIT, 1968. Fellow Nat. Acad. Scis. (Sri Lanka), Royal Soc. Arts, (U.K.) Inst. Elec. Engrs. (U.K., Beauchamp prize 1967), Inst. Engrs. (Sri Lanka); mem. IEEE, Am. Econ. Assn., Am. Phys. Soc., Sri Lanka Assn. Adv. Sci., Sri Lanka Econ. Assn. Home: 4201 E West Hwy Bethesda MD 20815-5910 Office: World Bank 1818 H St NW Washington DC 20433-0002

MUNCHMEYER, FREDERICK CLARKE, naval architect, marine engineer; b. Washington, Mar. 26, 1922; s. Frederick Stephenson and Corinne (Clarke) M.; m. Janet Six, Feb. 4, 1945 (div. May 1964); children: Leslie, Kurt Wayne, Kristen; m. Denise Hirshfield, Sept. 4, 1964; 1 child, Katherine Anne. BS in Engring., USCG Acad., 1942; MS in Naval Constrn., MIT, 1948; PhD in Naval Architecture and Marine Engring., U. Mich., 1977. From asst. prof. to prof. U. Hawaii, Honolulu, 1963-82, chmn. mech. engring. dept., 1980-82; prof., chmn. Sch. Naval Architecture and Marine Engring., U. New Orleans, 1982-88; cons., New Orleans, 1988—; sec. Ship Structure Com., Washington, 1955-59; cons. prof. Northwestern Poly. Univ., X'ian, People's Republic of China, 1983. Comdr. USCG, 1942-63. Ford Found. fellow, 1968-69, Deutsche Acad. Austauch Dienst fellow, 1977-78; teaching and rsch. grantee NSF, 1965-82. Mem. Am. Soc. Naval Engrs., Soc. Naval Architects and Marine Engrs., Sigma Xi. Achievements include research on math. of surfaces, geometric modeling, ocen sci. and engring., marine tech. and computer aided ship design. Home: 226 Rue Saint Peter Metairie LA 70005 also: Rt 2 Box 167 Waterford OH 45786

MUNDELL, JOHN ANTHONY, environmental engineer, consultant; b. Frankfort, Ind., Apr. 11, 1957; s. Loren Sherman-Sheridan Mundell and Juanita Fern (Thompson) Rogers; m. Julia Ann Mooney, Aug. 4, 1979; children: Sarah Marie, Andrew Jacob, Daniel Isaac, James Thomas. BSCE with highest distinction, Purdue U., 1979, MSCE with highest distinction, 1980; postgrad., U. Notre Dame, 1984-88. Registered profl. engr., Ind. Grad. teaching asst. Purdue U., West Lafayette, Ind., 1979-80; staff engr., then project engr. Am. Testing and Engring. Corp. (ATEC) Assocs., Inc., Indpls., 1981-84, corp. dir. environ. svcs., 1988-89, v.p., corp. dir. tech. svcs., 1989—; pvt. cons. Notre Dame, Ind., 1984-88; rsch. assoc. U. Notre Dame, 1984-88. Contbr. articles to Jour. Geotech. Engring., Jour. Environ. Engring., Jour. Geophys. Rsch., Soils and Founds., other profl. publs. Music group leader, guitarist, liturgist St. Elizabeth Seton Parish, Carmel, Ind., 1981-84, 88—, Christ the King Parish, South Bend, Ind., 1985-88. Mem. ASCE, ASTM, Nat. Ground Water Assn., Am. Geophys. Union. Achievements include first to engineer the use of on-site waste fixation/stabilization approved by U.S. EPA at a Superfund site; development of multicomponent geochemical model for lead mobility analysis, of first predictive models to determine effect of waste on hydraulic properties of clay barriers at disposal sites; research in laboratory testing and field control guidelines to achieve clay liner compaction conditions to meet EPA standards of waste isolation, three-dimensional visualization techniques for assessing organic contamination at industrial sites. Home: 10411 White Oak Dr Carmel IN 46033-3975 Office: ATEC Assocs Inc PO Box 501970 Indianapolis IN 46250-6970

MUNDLAK, YAIR, agriculture and economics educator; b. Pinsk, Poland, June 6, 1927; arrived in Israel, 1927; s. Lipa and Batia (Bodankin) M.; m. Yaffa Mundlak; children: Tal, Yaelle, Guy. BS in Agrl. Econs. with highest honors, U. Calif., Davis, 1953; MS in Stats., U. Calif., Berkeley, 1956, PhD in Agrl. Econs., 1957. Prof. agrl. econs. Hebrew U., Jerusalem, 1956-85, head dept. agrl. econs., 1965-73, dir. research Ctr. for Agrl. Econs. Research, 1968-85, dean faculty agriculture, 1972-74; vis. prof. econs. U. Chgo., 1966-67, 78-85, F.H. Prince prof. econs., 1978—; vis. prof. agrl. econs. U. Calif., Berkeley, 1961-63; vis. prof. econs. Harvard U., Cambridge, Mass., 1974-76; research fellow Internat. Food Policy Research Inst., Washington, 1976—; vis. fellow CORE, Louvain, Belgium, 1973; vis. rsch. fellow World Bank, 1990-91; cons. Ministry Agriculture, Devel. Research Ctr. The World Bank, Washington, 1972; mem. research coms. Nat. Council for Research and Devel., 1964-74. Co-author: (with Ben-Shahar, Berglas and Sadan) The West Bank and Gaza Strip—Economic Structure and Development Prospects, 1971; editor: Research in Agricultural Economics, 1976, (with F. Singer) Arid Zone Management, 1977; contbr. articles to profl. jours.; assoc. editor Jour. Econometrics, 1973-77. Pres. bd. trustees Israel Found., 1977-88. Recipient Bareli prize, 1965, Rothschild prize, 1972, Quality of Research Discovery award Am. Agrl. Econ. Assn., 1980, 82; Ford Found. research fellowship, 1966-67. Fellow Econometric Soc.; mem. Phi Beta Kappa. Office: Univ of Chgo Dept Econs 1127 E 59th St Chicago IL 60637-1539 also: Faculty Agriculture, PO Box 12, Rehovot 76100, Israel

MUNDY, PHILLIP CARL, engineer. Pres. Can. Posture and Seating Ctr., Fitchner, Ont., Can. Recipient Young Engr. Achievement award Can. Coun. Profl. Engrs., 1991. Office: Canadian Posture & Seating Ctr, PO Box 1473 Station C, Kitchener, ON Canada N2G 4P2*

MUNGAN, NECMETTIN, petroleum consultant; b. Mardin, Turkey, Mar. 1, 1934; came to U.S., 1953, came to Can., 1966; s. Kerim and Saide M.; m. Gunilla Ersman, May 19, 1962; children—Carl Edward, Christina Deniz, Nils Kerim, Tanya Katerina. BS Petroleum Engring., U. Tex., 1956, BA Math., 1957, MS Petroleum Engring., 1958, PhD Petroleum Engring., 1961. Head rsch. Sinclair Research, Tulsa, 1961-66; chief research officer Petroleum Recovery Inst., Calgary, Alta., Can., 1966-78; pres. Mungan Petroleum Cons., Ltd., Calgary, 1978-86; chief tech. advisor AEC Oil and Gas Co., Calgary, 1986—. Contbr. numerous articles to profl. publs. Named hon. citizen State of Tex., 1956, hon. prof. chem. and petroleum engring. U. Calgary, 1993; recipient Excellence in Presentation of Tech. Paper award Am. Inst. Chem. Engrs., 1966. Mem. Soc. Petroleum Engrs. (Uren Award, 1990), AIME (C.K. Ferguson award 1966, Disting. lectr 1969-70, Disting. Svc. award 1992), Petroleum Soc., Assn. Profl. Engrs., Geologists and Geophysicists Alta., Sigma Xi, Tau Beta Pi, Sigma Gamma Epsilon, Pi Epsilon Tau, Kappa Mu Epsilon. Office: AEC Oil and Gas Co, 3900 421-7 Ave SW, Calgary, AB Canada T2P 4K9

MUNGER, ELMER LEWIS, civil engineer, educator; b. Manhattan, Kans., Jan. 4, 1915; s. Harold Hawley and Jane (Green) M.; m. Vivian Marie Bloomfield, Dec. 28, 1939; children: John Thomas, Harold Hawley II, Jane Marie. B.S., Kans. State U., 1936, M.S., 1938; Ph.D., Iowa State U., 1957. Registered profl. engr., Nebr., Kans., Iowa, Vt. Rodman St. Louis-Southwestern Ry., Ark., Mo., 1937-38; engr. U.S. Engr. Dept., Ohio, Nebr., 1938-46; missionary engr. Philippine Episcopal Ch., 1946-48; engr. Wilson & Co., Salina, Kans., 1948; tchr. Iowa State U., 1948-51, 54-58; engr. C.E., U.S. Army, Alaska, 1951-54; from tchr. to dean Norwich U., Northfield, Vt., 1958-69; prof. gen. engring. U. P.R., Mayagüez, 1969-75; prof. civil engring. Mich. Tech. U., 1975-80; cons. engr., 1980—; Mem. spl. com. on engring. Inter-Am. Devel. Bank, U. W.I., 1971. Author: (with Clarence J. Douglas) Construction Management, 1970. Fellow ASCE; mem. Soc. Am. Mil. Engrs., Nat., Vt. socs. profl. engrs., Am. Soc. Engring. Edn., Phi Kappa Phi, Sigma Tau, Tau Beta Pi, Chi Epsilon. Episcopalian. Clubs: Masons, Shriners. Home: 21028 Tucker Ave Port Charlotte FL 33954

MUNGER, HAROLD CHARLES, architect; b. Toledo, July 25, 1929; s. Harold Henry and Lela Marie (Hoffman) M.; m. Patricia Ann Billeter, Oct. 2, 1954; children: Hal Peter, Peter Charles, David James. B.Arch., U. Notre Dame, 1951; cert., Davis Bus. Coll., 1947, U. Toledo, 1949, 50, Toledo Mus. Art, 1954, Leica Sch., 1982. Registered architect, Ohio, Mich., Ind., cert. Nat. Council Arctl. Registration Bds. Draftsman atomic energy br. Giffels & Vallet, Architects and Engrs., Detroit, 1951-52; chief designer, assoc. Britsch and Munger, Architects, Toledo, 1952-55; chief architect, ptnr. Munger, Munger and Assocs., Architects, Toledo, 1955-70, owner, proprieter, 1970-83; pres. Munger, Munger and Assocs., Inc., Toledo, 1983—; mem. nat. exam. evaluation com. Nat. Council Archtl. Registration Bds., Washington, 1983; presenter Ohio Assn. Sch. Officials, 1985; charter mem. Historic Dist. Design Rev. Bd., City of Perrysburg, Ohio, 1982—; mem. mayors com. planning, City of Perrysburg, mem. downtown task force, 1985; mem. archtl. jury Nat. Sch. Bds. Assn., 1961—; mem. pres.'s council Toledo Mus. Art, 1985—; mem. archtl. ann. awards jury Ind. Masonry Inst., 1988. Author: Housing Physically Disabled Elderly, 1964; co-author: Lucas County Bldg. Code, 1955-56; assoc. editor: Ohio Architect, 1955-58, Architectural Graphics Standard, 8th edit., 1986—. Dist. officer, merit badge counselor, committeeman Toledo Area Boy Scouts Am., 1953—; mem. Vocat. Tech. High Sch. Bldg. Trades Adv. Com., Toledo, 1956-79, Perrysburg 1st City Charter Commn., 1960-62; co-chmn. Chase Park Urban Renewal Adv. awards Com., Toledo, 1962-65; pres. St. Rose Bd. of Edn., Perrysburg, 1966-69; trustee Way Pub. Library, 1960-80, Historic Perrysburg, Inc., 1977-81. Recipient Pub. and Comml. award Toledo Area Concrete Assn., 1965, Indsl. award Toledo Area Concrete Assn., 1973, Boss of Yr. award Per Ro Ma chpt. Am. Bus. Women's Assn., 1977, Bus. Assoc. of Yr. award, 1985, St. George award Cath. Com. on Scouting, Diocese of Toledo, 1978, Masonry Honor award Masonry Inst. Northwestern Ohio, 1981, 85 (2 awards), 86, 87, 89 (2 awards); Excellence in Masonry Design award Ohio Masonry Coun., 1986, 88, Man of Yr. award U. Notre Dame Alumni Club of Toledo, 1987, Best Project award Internat. Union Bricklayers, 1989, Toledo Design Forum award of excellence in architecture, 1991. Fellow AIA (nat. committeeman design, inquiry, housing, architecture for edn. 1965—, pres., past. bd. dirs., Devoted Service award 1963, Architect of Yr. award Toledo chpt. 1991); mem. Constrn. Specifications Inst. (charter mem. cert. of recognition), AIA Architects Soc. Ohio (pres., past dir. Silver Gavel award, 1969, honor award 1981, 86, 88, Gold medal award 1987 Ohio chpt., 25-Yr. Bldg. award of excellence 1991), Toledo Club, Rotary (charter, cert. of recognition). Home: 446 W Front St Perrysburg OH 43551-1433 Office: Munger Munger & Assocs Architects Inc 225 N Michigan St Toledo OH 43624-1648

MUNGER, PAUL R., civil engineering educator; b. Hannibal, Mo., Jan. 14, 1932; s. Paul Oettle and Anne Lucille (Williams) Munger; m. Frieda Ann Mette, Nov. 26, 1954; children: Amelia Ann, Paul David, Mark James, Martha Jane Munger Cox. BSCE, Mo. Sch. Mines and Metallurgy, 1958, MSCE, 1961; PhD in Engring. Sci., U. Ark., 1972. Registered profl. engr., Mo., Ill., Ark. Instr. civil engring Mo. Sch. Mines and Metallurgy, Rolla, 1958-61, asst. prof., 1961-65; assoc. prof. U. Mo., Rolla, 1965-73, prof., 1973—; dir. Inst. River Studies, U. Mo., Rolla, 1976—; exec. dir. Internat. Inst. River and Lake Systems, U. Mo., Rolla, 1984—. Mem. Nat. Soc. Profl. Engrs., Mo. Soc. Profl. Engrs., Am. Soc. Engring. Edn., ASCE, Nat. Coun. Engring. Examiners (pres. 1983-84), Mo. Bd. Architects, Profl. Engrs. and Land Surveyors (chmn. 1978-84). Office: U Mo 111 Civil Engring Rolla MO 65401

MUNIZ, BENIGNO, JR., aerospace engineer, space advocate; b. Mayaguez, P.R., Nov. 22, 1958; came to U.S., 1958; s. Beningo and Anna Margareta (Rüdinger) M. BSME, Clarkson Coll., 1982. Asst. engr. Grumman Aerospace Corp., Bethpage, N.Y., 1982-84, assoc. engr., 1984-87, engr., 1987-90; sr. engr. Rocketdyne Div., Rockwell Internat., Canoga Park, Calif., 1990-92, engring. specialist, 1992—; bd. dirs. Calif. Space Devel. Coun. Recpient Clarkson Trustee award Clarkson Coll., Potsdam, N.Y., 1976, Group Achievement award NASA, 1992; named John Bardeen scholar, 1977. Mem. AIAA (sr.), Am. Astronautical Soc. (sr.), Space Studies Inst., Nat. Space Soc. (founder, pres. Rocketdyne employees chpt. 1991—), Orgn. for Advancement of Space Industrialization and Settlement, Theta Chi. Achievements include participation in first experimental validation of phenomenon of body-freedom flutter for forward-swept-wing aircraft and first flight of full scale forward-swept wing aircraft; instructor for company-sponsored course in computer-aided structural analysis using NASA Structural Analysis. Home: 1319 Gonzalez Rd Simi Valley CA 93063 Office: Rocketdye Div Rockwell Internat 6633 Canoga Ave Mail Code LB25 Canoga Park CA 91309

MUNK, WALTER HEINRICH, geophysics educator; b. Vienna, Austria, Oct. 19, 1917; came to U.S., 1933; m. Edith Kendall Horton, June 20, 1953; children: Edith, Kendall. BS, Calif. Inst. Tech., 1939, MS, 1940; PhD in Oceanography, U. Calif., 1947; PhD (hon.), U. Bergen, Norway, 1975, Cambridge (Eng.) U., 1986. Asst. prof. geophysics Scripps Inst. Oceanography, U. Calif., San Diego, 1947-54, prof., 1954—; dir. Inst. Geophysics and Planetary Physics, U. Calif., La Jolla, 1960-82; prof. geophysics, dir. heard island expt. Scripps Inst., U. Calif. Author (with Mac Donald) The Rotation of the Earth: A Geophysical Discussion, 1960; Contbr. over 200 articles to profl. jours. Recipient Gold medal Royal Astron. Soc., 1968, Nat. medal Sci., 1985; Marine Tec. Soc. award, 1969, Capt. Robert Dexter Conrad award Dept. Navy, 1978, G. Unger Vetlesen prize Columbia U., 1993; named Calif. Scientist of Yr. Calif. Mus. Sci. and Industry, 1969; fellow Guggenheim Found., 1948, 55, 62, Overseas Found., 1962, 81-82, Fulbright Found., 1981-82; Sr. Queen's fellow, 1978. Fellow Am. Geophys. Union (Maurice Ewing medal 1976, William Bowie medal 1989), AAAS, Am. Meteorol. Soc. (Sverdrup Gold medal 1966), Accoustical Soc. Am., Marine Tech. Soc. (Compass award 1991); mem. Nat. Acad. Scis. (Agassiz medal 1976, chmn. ocean studies bd. 1985-88), Am. Philos. Soc., Royal Soc. London (fgn. mem.), Deutsche Akademie der Naturforscher Leopoldina, Am. Acad. Arts and Scis. (Arthur L. Day medal 1965), Am. Geol. Soc. Office: U Calif San Diego Scripps Inst Oceanography 0225 La Jolla CA 92093

MUNSINGER, ROGER ALAN, marketing executive; b. Wichita, Kans., June 24, 1948; s. Dorus M. and Betty Gene (Frailey) M. BA in Chemistry, Kans. U., 1970; MS in Biochemistry, Okla. State U., 1973. Lab. coord. Okla. State U., Stillwater, 1973-75; lab. supr. U. Tenn., Knoxville, 1975-78; product mgr. to mktg. mgr. Brinkmann Inst., Westbury, N.Y., 1978—. Contbr. articles to profl. jours. Mem. Assn. of Analytical Chemists Internat., Am. Chem. Soc., Am. Seminar Leaders Assn. Home: 565 Avenue A #203 Uniondale NY 11553 Office: Brinkmann Instruments One Cantiague Rd Westbury NY 11590

MUNSON, JANIS ELIZABETH TREMBLAY, engineering company executive; b. Beverly, Mass., Dec. 17, 1948; d. Louis Story Tremblay and Doroth Ellen (Burnham) Tonkin; divorced. BS in Geology summa cum laude, Boston U., 1976, M in Urban Planning, 1982. Tech. libr. United Engrs. & Constructors, Boston, 1971-73, land use planner, 1973-76, lead land use planner, 1976-80, supervising lic. engr., 1980—, environ./scientific cons., 1980-85, head mktg. analysis svcs. group power div., 1987-89, mgr. land use planning group, 1989-92, sr. ptnr., 1992—. Bd. dirs. Ctr. City Residents Assn., Phila., 1986; mem. Multiple Sclerosis Soc.; vol. for disabled. Mem. Internat. Platform Assn., Internat. Biog. Inst., Am. Planning Assn., Am. Inst. Cert. Planners (assoc.), World Affairs Coun., Smithsonian Assn. Republican. Congregationalist. Achievements include research on transmission line site selection process, on crime control through environmental design, on emotion exercise and nutrition fro those labeled chronic/progressive. Home: 2401 Pennsylvania Ave Apt 30-50 Philadelphia PA 19130-3061 Office: United Engrs and Constructors 30 S 17th St Philadelphia PA 19103-4118

MUNSON, NORMA FRANCES, biologist, ecologist, nutritionist, educator; b. Stockport, Iowa, Sept. 22, 1923; d. Glenn Edwards and Frances Emma (Wilson) M.; BA, Concordia Coll., 1946; MA, U. Mo., 1955; PhD (NSF fellow 1957-58, Chgo. Heart Assn. fellow 1959), Pa. State U., 1962; postgrad. Ind. U., 1957, Western Mich. U., 1967, Lake Forest Coll., 1971, 72, 78; student various fgn. univs., 1964-71. Tchr., Aitkin (Minn.) High Sch., 1946-48, Detroit Lakes (Minn.) High Sch., 1948-54, Libertyville (Ill.) High Sch., 1955-79; researcher Nutrition, Alzheimer's, Hypoglycemia and Multiple Sclerosis, Libertyville, 1965—; lectr. counseling and nutrition. Author biology lab. manual; contbr. articles to profl. jours. Ruling elder 1st Presbyn. Ch., Libertyville, 1971-77; pres. Lake County Audubon Soc., 1975-79, 82-86, 88-89, Libertyville Edn. Assn., 1964-67; active Rep. Party Ill., Citizens to Save Butler Lake, Citizens Choice, Defenders; mem. U.S. Congl. Adv. Bd., 1985—; bd. dirs. Holy Land Christian Mission Internat.; mem. Heritage Found., Citizens Lake County for Environ. Action Reform, Wilderness Soc. Recipient Hilda Mahling award, 1967, C. of C. award, 1971, Ill. Best Tchr. award, 1974; Biology Tchr. of Yr. award, 1971; NSF fellow, 1957, 58, 60-62, 70-71. Fellow Am. Biog. Inst. Rsch., Internat. Biog. Assn.; mem. Nat. Biology Tchrs. Assn. (rsch. in degenerate diseases, award 1971), AAAS, Am. Inst. Biol. Sci., Ill. Environ. Coun., Nat. Audubon Soc., Lake County Audubon Soc. (pres. 1982-89, 91—), Ill. Audubon Coun., Nat. Health Fedn., Internat. Platform Assn., Internat. Profl. and Bus. Women, Nat. Wildlife Fedn., N.Y. Acad. Scis., Chgo. Acad. Sci., Parks and Conservation Assn., Concerned Women Am. Nature Conservation, Evanston North Shore Bird Club, Delta Kappa Gamma. Contbr. research articles to publs. Home and Office: 206 W Maple Ave Libertyville IL 60048-2172

MUNTWYLER, URS WALTER, electronics engineer; b. Solothurn, Switzerland, Mar. 22, 1958; s. Theo and Erika (Flury) M.; m. Sigrid Kleindienst, 1986; children: Barbara, Lea, Anna. Degree in engring., Ing. schule Biel, Switzerland, 1982; postgrad., U. Freiburg, 1989-90. Indsl. electronic monteur FEAM Autophon, Solothurn, 1974-78, FEAM Telephonie, Geneva, 1979; devel. engr. Hasler AG, Bern, 1982-84; application engr. Jenni Energietechnik AG, Oberburg, Switzerland, 1985; dir. Tour de Sol Orgn., Berne, 1985-92, Engring. Bur. Muntwuyler, Berne, 1985—, Muntwyler Energietechnik AG, Berne, 1988—. Author: Solarmobile im Alltag, 1985, Praxis mit Solarzellen, 1992, Leicht-Elektromobile im Alltag, 1993. program mgr. Swiss Fed. Office Energy, Bern, 1992—. Mem. Solar Fedn. (v.p. 1987—), Swiss. Assn. for Solar Energy (adv. bd. 1983—). Home: Vereinsweg 7, CH 3012 Berne Switzerland Office: Engineer Bureau Muntwyler, PO Box 512, 3052 Zollikofen Switzerland

MUNTZ, ERIC PHILLIP, aerospace engineering and radiology educator, consultant; b. Hamilton, Ont., Can., May 18, 1934; came to U.S., 1961, naturalized, 1985; s. Eric Percival and Marjorie Louise (Weller) M.; m. Janice Margaret Furey, Oct. 21, 1964; children: Sabrina Weller, Eric Phillip. B.A.Sc., U. Toronto, 1956, M.A.Sc., 1957, Ph.D., 1961. Halfback Toronto Argonauts, 1957-60; group leader Gen. Electric, Valley Forge, Pa., 1961-69; assoc. prof. aerospace engring. and radiology U. So. Calif., Los Angeles, 1969-71, prof., 1971-87, chmn. aerospace engring., 1987—; cons. to aerospace and med. device cos., 1967—; mem. rev. of physics (plasma and fluids) panel NRC, Washington, 1983-85. Contbr. numerous articles in gas dynamics and med. diagnostics to profl. publs., 1961—; patentee med. imaging, isotope separation, nondestructive testing net shape mfg., transient energy release micromachine. Mem. Citizens Environ. Adv. Council, Pasadena, Calif., 1972-76. U.S. Air Force grantee, 1961-74, 82—; NSF grantee, 1970-76, 87—; FDA grantee, 1980-86. Fellow AIAA (aerospace Contbn. to Soc. award 1987); mem. Am. Assn. Physicists in Medicine, Nat. Acad. Engring. Episcopalian. Home: 1560 E California Blvd Pasadena CA 91106-4104 Office: U So Calif Univ Pk Los Angeles CA 90089-1191

MUNTZ, RICHARD ROBERT, computer scientist, educator; b. Jersey City, Mar. 6, 1941; s. Benjamin Augustus and Ruth (Wise) M.; m. Carma Jewl Celestino, Feb. 15, 1964 (dec. 1973); m. Alice Hwei-Yuan Meng, Mar. 28, 1981. PhD, Princeton U., 1969. Mem. tech. staff AT&T Bell Labs., Holmdel, N.J., 1964-66; asst. prof. dept. computer sci. UCLA, 1969-73, assoc. prof., 1974-78, prof. computer sci., 1979—; mem. corp. tech. adv. bd. Teradata Corp., L.A., 1989-91; presenter at profl. confs. Contbr. articles to tech. jours. Fellow IEEE; mem. Assn. Computing Machinery, Sigma Xi, Tau Beta Pi. Office: UCLA Dept Computer Sci 3732 Boelter Hall Los Angeles CA 90024

MUNYER, EDWARD A., zoologist, museum adminstrator; b. Chgo., May 8, 1936; s. G. and M. (Carlson) M.; m. Marianna J. Munyer, Dec. 12, 1981; children: Robert, William, Richard, Laura, Cheryl. BS, Ill. State U., 1958, MS, 1962. Biology tchr. MDR High Sch., Minonk, Ill., 1961-63; instr. Ill. State U., Normal, 1963-64; curator zoology Ill. State Mus., Springfield, 1964-67, asst. dir., 1981—; assoc. prof. Vincennes (Ill.) U., 1967-70; dir. Vincennes U. Mus., 1968-70; assoc. curator Fla. Mus. Natural History, Gainesville, 1970-81; mem. Mus. Accreditation Vis. Com. Roster, 1976—; cons. Am. Assn. Mus., 1987—. Contbr. articles to profl. jours. Mem. Am. Assn. Mus. (cons. 1976—, bd. dirs.1992-95), Midwest Mus. Conf. (pres. 1990-92), Ill. Mus. Congress (bd. dirs. 1981-86), Wilson Ornithol. Soc. (life). Office: Ill State Mus Spring & Edward Sts Springfield IL 62706

MUNZER, MARTHA EISEMAN, writer; b. N.Y.C., Sept. 22, 1899; d. Samuel and Stella (Stettheimer) Eiseman; m. Edward M. Munzer, June 15, 1922 (dec. 1960); children: Edward Munzer Jr. (dec.), Martha Amato, Stella Loeb; m. Isaac Corkland, Apr. 30, 1980 (dec. 1986). BS in Electrochem. Engring., MIT, 1922. Tchr. chemistry, chairperson community services Fieldston Sch., N.Y.C., 1930-54; writer Conservation Found., N.Y.C., 1954-68, Wave Hill Ctr. for Environ. Studies, Riverdale, N.Y., 1968-73; freelance writer Ft. Lauderdale, Fla., 1978—. Author: Teaching Science Through Conservation, 1960, Unusual Careers, 1962, Planning Our Town, 1964, Pockets of Hope, 1966, Valley of Vision, 1969, Block by Block, 1973, New Towns, 1974, Full Circle, 1978, The Three R's of Ecology: A Personal Collection, 1986, Lauderdale by the Sea: A Living History, 1989, Friends of the Everglades: A Living History, 1993; contbr. articles to profl. jours. Mem. conservation adv. commn. Town of Larchmont, N.Y., 1964-78, planning and zoning bd. Lauderdale by the Sea, Fla., 1985—. Recipient Oscar R. Foster award Chemistry Tchrs. Club, 1947. Fellow Soc. Women Engrs.; mem. AAUW (chair lit. group 1985—), Fla. Engring. Soc. (award for engring. journalism 1985), LWV (mem. natural resources com.), Fla. Hist. Soc. Democrat. Home and Office: 4411 E Tradewinds Ave Fort Lauderdale FL 33308-4410

MUNZNER, ROBERT FREDERICK, biomedical engineer; b. Balt., July 3, 1936; s. Robert F. Munzner and Catherine E. (Appel) Gay; m. Jo Ann Goettee, Sept. 2, 1960 (div. 1980); children: Elizabeth Mae, Robert Victor, Ann Catherine. BS in Physics, Loyola Coll., Balt., 1963; PhD in Biomed. Engring., U. Va., 1976. Aerospace engr. Westinghouse Def. and Space, Balt., 1963-69; rsch. assoc. Johns Hopkins U., Balt., 1975-77; chief, neurol. devices br. U.S. FDA, Rockville, Md., 1977—; exec. sec. neurol. device adv. panel. Contbg. author: Cerebellar Stimulation for Spasticity, 1984; contbr. articles to profl. jours. Recipient postdoctoral fellowship Johns Hopkins U., Balt., 1975, Univ. fellowship U. Va., Charlottesville, 1972-73, Thornton fellowship, 1971. Mem. IEEE, Biomed. Engring. Soc., AAAS, Sigma Xi. Achievements include demonstration of atrial mechanical stimulation producing vasomotor reflex. Office: FDA Device Evaluation 1390 Piccard Dr HFZ-450 Rockville MD 20850

MURADIN-SZWEYKOWSKA, MARIA, physicist; b. Poznan, Poland, Mar. 28, 1952; arrived in The Netherlands, 1979; d. Jerzy and Alicja Maria (Stabecka) Szweykowska; m. Joshi Arnold Muradin, Sept. 18, 1978; children: Marvick, Stefan. MSc, Warsaw U., Poland, 1977; PhD, Leiden U., Holland, 1984. Asst. Warsaw U., 1976-78, Leiden U., 1979-83; program mgr. NOVEM, Utrecht, Holland, 1986-91, advisor, 1992—. contbr. articles to profl. jours. Roman Catholic. Achievements include development of long term strategies of energy research in The Netherlands to Netherlands agency for energy and the environment; management of the Dutch National Photovoltaic Programme; research on bacteriorhedopsins with cehmically modified chromophones. Office: Novem, St Jacosstraat 61, 3503RE Utrecht The Netherlands

MURAI, SHINJI, chemistry educator; b. Osaka, Japan, Aug. 24, 1938; s. Takeshi and Masako (Nishikawa) M.; m. Masako Noguchi, Feb. 20, 1966; children: Kanoko, Gyo. BS, Osaka U., 1961, D, 1966. From asst. to assoc. prof. Osaka U., 1966-87, prof. chemistry, 1987—; cons. Dow Corning Co., Midland, Mich., 1988-91. Contbr. numerous articles to profl. jours. Mitsubishi Found. grantee, 1988. Mem. Chem. Soc. Japan (bd. dirs. 1991—, co-chmn. publicity of chemistry 1989, award 1984). Avocations: oil painting, pottery making. Home: 5-1-97 Midoridai, Kawanishi Hyogo 666-01, Japan Office: Osaka U Chemistry Dept, Faculty Engring, Suita Osaka 565, Japan

MURAKAMI, EDAHIKO, chemistry educator; b. Kyoto, Japan, Jan. 11, 1922; s. Sanji and Nobuko M.; m. Hiroko, Aug. 28, 1949; children: Akihiko, Jiro. BS, Nagoya Imperial U., 1944; PhD, Nagoya U., 1967. Asst. Nagoya Imperial U., Japan, 1944-49; asst. prof. Aichi Kyoiku U., Japan, 1950-68, prof., 1968-85, honored prof., 1985—; prof. Nagoya Women's Jr. Coll. of Commerce, Japan, 1985—; mgr. Japanese Study Group for Tryptophan Rsch., 1974—. Author: Kagaku Koyomi, Numon Seikagaku, 1977, Seikagaku Numon, 1989. Recipient Award for Disting. Svc., Assn. of the Sci. Edn., 1985. Avocations: painting. Home: 3-13 Kamioka Cho Meito-ku, Nagoya 465, Japan

MURAMATSU, ICHIRO, chemistry educator; b. Ikeda, Osaka, Japan, Aug. 21, 1928; s. Bunzo and Etsuko Muramatsu; m. Nobuko Tomiyama, Mar. 29, 1958; children: Megumi Takimoto, Mariko. BA, Osaka U., 1951, PhD, 1961. Asst. Osaka U., 1954-58; assoc. prof. chemistry Rikkyo U., Tokyo, 1958-70, prof., 1970—, dean Coll. Sci., 1979-83, Registration Office, 1986-90; postdoctoral researcher Case Western Res. U., Cleve., 1967-69. Contbr. articles to profl. jours. Mem. Chem. Soc. Japan, Pharm. Soc. Japan, Biochem. Soc. Japan. Avocations: photography, classical music. Home: 2-8-3 Midorigaoka, Oi-machi, Saitama 356, Japan Office: Rikkyo U Dept Chemistry, Nishi Ikebukuro Toshima-ku, Tokyo 171, Japan

MURANAKA, KEN-ICHIRO, biophysicist; b. Nagoya, Aichi, Japan, Sept. 7, 1958; s. Megumi and Takeko (Ohtsuka) M. BS in Math., U. Ill., Chgo., 1982; MS in Biophysics, U. Ill., Urbana, 1985. Teaching asst. U. Ill., Chgo., 1982-83, rsch. asst., summer 1983; teaching asst. U. Ill., Urbana, fall 1984, rsch. asst., 1984-85; grad. teaching assoc. Ohio State U., Columbus, 1985-87; coop. editor in patent svcs. Chem. Abstracts Svc., Columbus, 1986-87; quantitative analyst Yasuda Trust and Banking, Tokyo, 1987-91; asst. prof. Mizuho Jr. Coll., Nagoya, Japan, 1991-93; rsch. scientist Nippon Shinyaku Pharm. Co., Kyoto, Japan, 1993—; v.p. Pre-Health Profl., U. Ill., Chgo., 1982-83. Author: (in Japanese) Theory of Options, 1988, (poems) Foreigners, 1987 (Golden Poet award). Mem. AAAS, N.Y. Acad. Scis., Biohpys. Soc. Japan, Soc. for Econ. Studies of Securities, Chem. Soc. Japan, Crystallographic Soc. Japan, Phi Eta Sigma, Alpha Lambda Delta. Roman Catholic. Avocations: classical guitar, music composition and arrangement.

MURAO, KENJI, chemist; b. Niihama, Ehime, Japan, Sept. 28, 1946; s. Masahito and Sadako (Sogame) M.; m. Motoe Kan; children: Tetsushi, Tadashi. BS, Osaka (Japan) U., 1969, MS, 1971, PhD, 1987. Rsch. scientist rsch. lab. Hitachi Ltd., Ibaraki, Japan, 1972-85; sr. rsch. scientist advanced rsch. lab. Hitachi Ltd., Saitama, Japan, 1985-91, sr. rsch. scientist Hitachi rsch. lab., 1991—; vis. scientist chemistry dept. MIT, Cambridge, Mass., 1980-81. Contbr. articles to profl. jours. Mem. Am. Chem. Soc., Chem. Soc. Japan, Phys. Soc. Japan, Japan Soc. Applied Physics. Avocations: tennis, fishing, skiing. Home: Daihara-cho 3-3-4, Hitachi-shi Ibaraki 316, Japan Office: Hitachi Rsch Lab Hitachi Lt, Ohmika-cho, Hitachi-shi, Ibaraki 319-12, Japan

MURATA, YASUO, economics educator; b. Osaka, Japan, Jan. 26, 1931; s. Masao and Sadae (Morii) M.; m. Hiroko Sakurai, Feb. 7, 1960; 1 d., Akiko. B.A., Kobe U. (Japan), 1953, M.A., 1955, D. Econs., 1970; Ph.D., Stanford U., 1965. Lectr., Kobe U. Commerce, 1958-60, assoc. prof., 1960-68, prof., 1968-71; prof. Dalhousie U., Halifax, N.S., Can., 1971-74; prof. econs. Nagoya City U. (Japan) 1974-86. Author: (with Michio Morishima) Working of Econometric Models, 1972; Mathematics for Stability and Optimization of Economic Systems, 1977; Optimal Control Methods for Discrete-Time Economic System, 1982; Modern Macroeconomics (in Japanese), 1984; Finance, Exchange, Prices and Investment (in Japanese), 1992; editorial bd. Optimal Control Applications and Methods, 1984-92. Grantee Fulbright Commn., 1962, Japan Econ. Research Found., 1979, Mishima Meml. Found., 1982. Mem. Econometric Soc., Japan Assn. Econs. and Econometrics (trustee 1980-92), Japan Assn. Automatic Control. Home: 1-16-17 Karatodai Kitaku, Kobe 651-13, Japan Office: Kansai U, Dept Econs, Suita Osaka 564, Japan

MURAYAMA, MAKIO, biochemist; b. San Francisco, Aug. 10, 1912; s. Hakuyo and Namiye (Miyasaka) M.; children: Gibbs Soga, Alice Myra. B.A., U. Calif., Berkeley, 1938, M.A., 1940; Ph.D. (NIH fellow), U. Mich., 1953. Research biochemist Children's Hosp. of Mich., Detroit, 1943-48, Harper Hosp., Detroit, 1950-54; research fellow in chemistry Calif. Inst. Tech., Pasadena, 1954-56; research asso. in biochemistry Grad. Sch. Medicine, U. Pa., Phila., 1956-58; spl. research fellow Nat. Cancer Inst. at Cavendish Lab., Cambridge, Eng., 1958; research biochemist NIH, Bethesda, Md., 1958—. Author: (with Robert M. Nalbandian) Sickle Cell Hemoglobin, 1973; discovered DIPA (decompression-inducible platelet aggregation, 1975; discovered DIPA causes vascular occlusion in both acute mountain sickness and diver's syndrome. Fellow Am. Inst. Chemists; mem. AAAS, Am. Chem. Soc., Am. Soc. Biol. Chemists, Assn. Clin. Scientists, Undersea and Hyperbaric Med. Soc., Aerospace Med. Assn., Internat. Platform Assn., West African Soc. Pharmacology (hon.), N.Y. Acad. Sci., Sigma Xi. Achievements include patent for automatic amperometric titration apparatus; development of molecular mechanism of human red cell sickling and prevention of sickle cell crisis by oral prophylactic carbamide; discovery that simulation of diving in frogs and mice indicated that diver's disease could be alleviated by piracetam and thymol, antiplatelet agents. Home: 5010 Benton Ave Bethesda MD 20814-2804 Office: NIH Bldg 6 Room 129 Bethesda MD 20892

MURAZAWA, TADASHI, mathematics educator; b. Seki-Kabuto, Suzukagun, Mie, Japan, July 31, 1940; s. Shougoro and Tosie Murazawa; m. Tamiko Murazawa, Sept. 12, 1970; 2 children. BA, Kyoto (Japan) U. of Edn., 1964; MA, Kobe (Japan) U., 1968, Dr.Sci., 1992. Instr. Osaka (Japan) U., 1968-73; vis. researcher U. Ill., 1974; lectr. Kyoto Prefectural U., 1974-77, asst. prof., 1978—; reviewer Math. Rev. of Am. Math. Soc., 1980—. Author: Introduction of Mathematics, 1985; translator: Introduction of Par. diff. Equation, 1975. Advisor A Coop. of Kyoto, 1989—. Mem. Japanese Math. Soc., Japan Soc. of Home Econs. Avocations: driving, traveling, tennis. Home: 3-choume 12-12, Kitaouji Otsu 520, Japan Office: Kyoto Prefectural U, Shimogamo-Sakyo, Kyoto 606, Japan

MURCRAY, FRANK JAMES, physicist; b. Stillwater, Okla., Mar. 12, 1950; s. David Guy and Evelyn May (Hannigan) M.; m. Marci Jo Everson, Aug. 21, 1986; children: Matthew, Meghan, Melissa. BS, U. Denver, 1972; MA, Harvard U., 1973, PhD, 1978. Tech./rsch. asst. U. Denver, 1966-72, teaching asst., 1971-72, rsch. physicist, 1978-81, rsch. physicist/asst. rsch. prof., 1981-86, rsch. physicist/assoc. rsch. prof., 1986-89, sr. rsch. physicist/assoc. prof., 1989—; rsch. asst. Harvard U., Cambridge, 1973-77; assoc. dir. Chamberlin Obs., Denver, 1986—. Contbr. articles to profl. jours. NSF grad. fellow, 1972-75, Antarctic Svc. medal, 1981; Harvard U. teaching fellow, 1976, 77. Mem. Am. Geophys. Union, Optical Soc. Am., Sigma Xi, Phi Beta Kappa. Office: Univ of Denver Dept Physics Denver CO 80209

MURDOCH-KITT, NORMA HOOD, clinical psychologist; b. Clinton, S.C., May 16, 1947; d. Bernard Constantine and Martha Grace (Hood) Murdoch; m. Jonathan Michael Murdoch-Kitt, Mar. 23, 1974; children: Kelly Michelle, Mark Jason, Sabrina Brittany, Laura Kristina. BA, Wake Forest U., 1969; MS, U. Pitts., 1971, PhD, 1975 . Psychology intern Eastern Pa. Psychiat. Inst., 1972-73; asst. prof., therapist campus counseling center Coll. William and Mary, Williamsburg, Va., 1973-74; staff psychologist child psychiatry dept. Med. Coll. Va., 1974-75; pvt. practice individual psychotherapy and family and marital therapy, Richmond, Va., 1975—. Mem. Richmond Dem. Com., 1976-79, 82-85, 88-89, 91—; v.p. govtl. relations com. Ginter Park Residents Assn., 1987, pres., 1988, 89; mem. Richmond Human Relations Adv. Commn., 1976-80, Richmond Mayor's Com. on Concerns of Women, 1987—, chair, 1989-93; mem. Richmond Citizens Crime Commn., 1985-88; founder, 1st state chmn. polit. action com. ERA, 1977-78; chief lobbyist ERA Ratification Council, 1977-79. USPHS fellow, 1969-72. Mem. Am. Psychol. Assn. (steering com. State Leadership Conf. 1986-91, chair 1991), Va. Psychol. Assn. (state legis. lobbyist 1978-79, chmn. legis. com. 1981-83, bd. profl. affairs 1981-85, pres. 1986), Va. Acad. Clin. Psychologists (chmn. profl. affairs com. 1982-84), Va. Breast Cancer Found. (rsch. chair 1992—), Richmond Area Psychol. Assn., LWV, ACLU. Presbyterian. Club: Richmond First (chmn. edn. com. 1979-80, dir. 1980-81). Office: Murdoch-Kitt Profl Bldg 3217 Chamberlayne Ave Richmond VA 23227-4806

MURDOCK, BRUCE, physicist; b. Hackensack, N.J., June 20, 1956; s. Simon and Estelle (Goldstein) M.; m. Bridget Hyatt, June 10, 1978; children: Dylan, Meagan. BS in Physics, SUNY, Stony Brook, 1980; MS in Physics, MIT, 1983. Physicist Tektronix, Inc., Wilsonville, Oreg., 1983-88, sr. sci. program mgr., 1988-91, dir. electronics rsch., 1991—; cons. Jet Propulsion Lab., NASA, Pasadena, 1989; prin. investigator DARPA, Washington, 1988, 92—; bd. dirs. Coun. on Superconductivity for Am. Competitiveness, Washington, 1989—. Contbr. articles to profl. jours. Mem. AAAS, Am. Phys. Soc. Achievements include patents in information input technology, electronics and measurement techniques; significant novel work in high speed superconductive measurement techniques and cryogenic systems design for superconductive digital microelectronic. Office: Tektronix Inc 14150 Karl Braun Dr Beaverton OR 97005

MURESANU, VIOLETA ANA, civil, structural engineer; b. Bucharest, Romania, May 1, 1942; came to U.S., 1981; d. Romulus and Elena (Murgescu) M.; m. Lucian Popescu, Oct. 28, 1960; 1 child, Diana Rodica. M in Civil Engring., Inst. for Construction, Bucarest, 1965. Civil engr. Inst. for Design Standard Structures, Bucharest, 1965-79; draft person Muesser, Ruthledge de Desimone, N.Y.C., 1981-83; civil engr. The L.I. RR, Jamaica, N.Y., 1983--. Achievements include the design of a bridge and numerous other structures. Office: The LI RR Hillside Support Facility 93-59 183rd St Jamaica NY 11435

MURFET, IAN CAMPBELL, botany educator; b. Launceston, Tasmania, Australia, Apr. 2, 1934; s. Campbell Vernon and Pretoria Clara (Rutherford) M.; m. Barbara Janet Forward, Feb. 1, 1958; children: Gregory, Andrew, Robyn. BSc, U. Tasmania, Hobart, 1957, BSc with first class honors, 1958, PhD, 1971. Temporary lectr. botany U. Tasmania, Hobart, 1961, lectr. botany, 1962-70, sr. lectr. botany, 1971-76, reader botany, 1977—, sub dean faculty sci., 1977-91, assoc. prof. plant sci., 1992—, head dept. plant sci., 1989; vis. fellow Cornell U., Ithaca, N.Y., 1974-75, 78. Editor Pisum Genetics, 1989—; contbr. articles to profl. jours., chpts. to books. Mem. Australian Soc. Plant Physiologists, Scandinavian Soc. Plant Physiology, Crop Sci. Soc. Am., Pisum Genetics Assn. (mem. coordinating com. 1975—, chmn. 1989—). Avocations: tennis, swimming, bush walking. Office: U Tasmania Dept Plant Sci, Hobart TAS 7001, Australia

MURNIK, MARY RENGO, biology educator; b. Manistee, Mich., Aug. 30, 1942; d. John Everett and Lorraine P. (ReVolt) R.; m. James M. Murnik, July 30, 1970; 1 child, John. Student Marquette U., 1960-62; B.S., Mich. State U., 1964, Ph.D., 1969. Asst. prof. Fitchburg State Coll., Mass., 1968-70; from asst. prof. to prof. Western Ill. U., Macomb, 1970-80; prof., head biol. sci. dept. Ferris State U., Big Rapids, Mich., 1980-92, prof. 1992—. articles to profl. jours. Author two lab. manuals. NIH fellow HEW, Mich. State U., 1965-68; NIH grantee Western Ill. U., 1976; grantee Environ. Mutagen Soc., Edinburgh, Scotland, 1977, Western Ill. U., 1972-79. Mem. AAAS, Genetics Soc. Am., Behavior Genetics Soc., Mich. Acad. Sci., Arts and Letters, Sigma Xi. Roman Catholic. Home: 331 W Slosson Ave Reed City MI 49677-1167 Office: Ferris State U Dept Biol Scis Big Rapids MI 49307-2225

MUROFF, LAWRENCE ROSS, nuclear medicine physician; b. Phila., Dec. 26, 1942; s. John M. and Carolyn (Kramer) M.; m. Carol R. Savoy, July 12, 1969; children: Michael Bruce, Julie Anne. AB cum laude, Dartmouth Coll., 1964, B of Med. Sci., 1965; MD cum laude, Harvard U., 1967. Diplomate Am. Bd. Radiology, Am. Bd. Nuclear Medicine. Intern Boston City Hosp., Harvard, 1968; resident in radiology Columbia Presbyn. Med. Ctr., N.Y.C., 1970-73, chief resident, 1973; instr. dept. radiology, asst. radiologist Columbia U. Med. Ctr., N.Y.C., 1973-74; dir. dept. nuclear medicine, computed tomography and magnetic resonance imaging Univ. Community Hosp., Tampa, Fla., 1974—; clin. asst. prof. radiology U. South Fla., 1974-78, clin. assoc. prof., 1978-82, clin. prof., 1982—; clin. prof. U. Fla., 1988—. Contbr. articles to profl. jours. Pres. Ednl. Symposia, Inc., 1975—. Lt. comdr. USPHS, 1968-70. Fellow Am. Coll. Nuclear Medicine (disting. fellow, Fla. del.), Am. Coll. Nuclear Physicians (regents 1976-78, pres.-elect 1978, pres. 1979, fellow 1980), Am. Coll. Radiology (councilor 1979-80, 91—, chancellor 1981-87, chmn. commn. on nuclear medicine 1981-87, fellow 1981); mem. Am. Assn. Acad. Chief Residents Radiology (chmn. 1973), AMA, Boylston Soc., Fla. Assn. Nuclear Physicians (pres. 1976), Fla. Med. Assn., Hillsborough County Med. Assn., Radiol. Soc. N.Am., Soc. Nuclear Medicine (coun. 1975-90, trustee 1980-84, 86-89, pres. Southeastern chpt. 1983, vice chmn. correlative imaging coun. 1983), Fla. Radiol. Soc. (exec. com. 1976-91, treas. 1984, sec. 1985, v.p. 1986, pres. elect 1987, pres. 1988-89), West Coast Radiol. Soc., Soc. Mag. Resonance Imaging (bd. dirs. 1988-91, chmn. ednl. program 1989, chmn. membership com. 1989-93), Phi Beta Kappa, Alpha Omega Alpha. Office: 1527 S Dale Mabry Hwy Tampa FL 33629-5808

MURPHY, DOUGLAS BLAKENEY, cell biology educator; b. Hartford, Conn., Jan. 25, 1945; s. Robert Blakeney and Dorothy Lou (Brown) M.; m. Christine VanWegen, Aug. 26, 1967; children: Ann V., Blake M. AB, U. Rochester, 1967; MS, Syracuse U., 1969; PhD, U. Pa., 1973. Asst. prof. biology Kans. State U., Manhattan, 1976-78; asst. prof. cell biology Johns Hopkins U., Balt., 1978-84, assoc. prof. cell biology, 1984-88, prof. cell biology, 1988—; mem. NIH cell biology study sect., Bethesda, Md., 1988-92; exch. scientist Russia, Nat. Acad. Scis., Balt., 1984, 87. Mem. editorial bd. Jour. Cell Biology, 1986-89. Recipient Rsch. Career Devel. award NIH, 1980-85; Fulbright scholar 1985. Mem. Am. Soc. for Cell Biology. Home: Johns Hopkins Med Sch Dept Cell Biology/Anatomy 725 N Wolfe St Baltimore MD 21205

MURPHY, EDWARD THOMAS, engineering executive; b. Boston, Nov. 20, 1947; s. Edward William and Eleanor Catherine (Brown) M.; m. Marianne Scheid, May 1, 1976; children: Edward Robert, Cynthia Kathrine. BS, Calif. Inst. Tech., 1969; MS in Nuclear Sci. and Engring., Carnegie Mellon U., 1971. Registered profl. engr., Pa., Md. Containment system engr. Westinghouse Electric Corp., Monroeville, Pa., 1969-74, fuel projects engr., 1974-80; mgr. licensing ops. Westinghouse Electric Corp., Bethesda, Md., 1980-84; acting regulatory com. supr. Westinghouse Electric Corp., Avila Beach, Calif., 1984-85; mgr. control system analysis and support Westinghouse Electric Corp., Monroeville, 1985-88; spl. project mgr. reactor restart div. Westinghouse Savannah River Co., Aiken, S.C., 1989-92, site configuration mgmt. regulatory affairs/starting projects mgr. Engring. & Projects Divsn., 1992—. Mem. Am. Nuclear Soc. Roman Catholic. Office: Westinghouse Savannah River Co Savannah River Site Aiken SC 29808

MURPHY, EUGENE F., aerospace, communications and electronics executive; b. Flushing, N.Y., Feb. 24, 1936; s. Eugene P. and Delia M.; m. Mary Margaret Cullen, Feb. 20, 1960. BA, Queens Coll., 1956; JD, Fordham U., 1959; LLM, Georgetown U., 1964. Bar: N.Y. With RCA Global Communications Inc., N.Y.C., 1964-81, v.p. and gen. counsel, 1969-71, exec. v.p. ops., 1972-75, pres., chief operating officer, 1975-76, pres., chief exec. officer, 1976-81; chmn., chief exec. officer RCA Communications Inc., N.Y.C., 1981-86, sr. v.p. communications and info. svcs., 1986-91; pres., chief exec. officer GE Aerospace, King of Prussia, Pa., 1992-93; pres., CEO GE Aircraft Engines, Cin., 1993—; bd. dirs. Martin Marietta Corp.; mem. Pres. Reagan's Nat. Sec. Telecommunications Adv. Com.; bd. govs. Aerospace Industries Assn. Bd. Served with USMCR, 1959-60. Mem. Armed Forces Communications and Electronics Assn. (past nat. chmn.), ABA. Clubs: Marco Polo, Plandome Country, Plandome Field and Marine. Office: GE Aircraft Engines Maildrop 101 1 Neumann Way Cincinnati OH 45215

MURPHY, JOHN CARTER, economics educator; b. Ft. Worth, July 17, 1921; s. Joe Preston and Rachel Elsie (Carter) M.; m. Dorothy Elise Haldi, May 1, 1949; children: Douglas C., Barbara E. Student, Tex. Christian U., 1939-41; BA, North Tex. State U., 1943, BS, 1946; AM, U. Chgo., 1949, PhD, 1955; postgrad., U. Copenhagen, 1952-53. Instr. Ill. Inst. Tech., 1947-50; instr. to assoc. prof. Washington U., St. Louis, 1950-62; vis. prof. So. Meth. U., Dallas, 1961, prof., 1962-90, prof. emeritus, 1990—, dir. grad. studies in econs., 1963-68, chmn. dept., 1968-71, faculty summer program in Oxford, 1982-91, dir., 1991, pres. faculty senate, 1988-89, co-dir. Insts. on Internat. Fin., 1982-87; vis. prof. Bologna (Italy) Center, Sch. Advanced Internat. Studies, Johns Hopkins U., 1961-62; UN tech. assistance expert, Egypt, 1964; vis. prof., spl. field staff Rockefeller Found., Thammasat U., Bangkok, 1966-67; sr. staff economist Council Econ. Advisers, 1971-72; mem. U.S. dels. econ. policy com. and working party III OECD, 1971-72, U.S. del. 8th meeting Joint U.S.-Japan Econ. Com., 1971; cons. Washington U. Internat. Econs. Research Project, 1950-53, U.S. Treasury, 1972; referee NSF; witness and referee congl. coms.; lectr. USIA Program, Fed. Republic Germany, 1961-62, 84, Philippines, South Viet Nam, Thailand, 1972, France, Belgium, 1984; lectr. Southwestern and Midwestern Grad. Sch. Banking; adj. scholar Am. Enterprise Inst. for Pub. Policy Research, 1976—. Author: The International Monetary System: Beyond the First Stage of Reform, 1979; (with R.R. Rubottom) Spain and the U.S.: Since World War II, 1984; editor: Money in the International Order, 1964; contbr. articles to profl. books and jours. Chmn. rsch. com. on internat. conflict and peace Washington U., 1959-61; lectr. mgmt. tng. programs Southwestern Bell Telephone Co., 1961-66, St. Louis Coun. on Econ. Edn., 1958-61; mem. regional selection com. H.S. Truman Fellowships, 1976-89; pres. Dallas Economists, 1981, Town and Gown of Dallas, 1980-81; mem. Dallas Com. on Fgn. Rels. Lt. USNR, 1943-46. Decorated Silver Star; Fulbright scholar to Denmark, 1952-53; Ford Found. Faculty Research fellow, 1957-58; U.S.-Spanish Joint Com. for Cultural Affairs fellow, 1981; Sr. Fulbright lectr. Italy, 1961-62. Mem. Am.

Econ. Assn., So. Econ. Assn. (bd. editors Jour. 1969-71), Midwest Econ. Assn., Am. Fin. Assn., Soc. Internat. Devel., Peace Rsch. Soc., Southwestern Social Sci. Assn. (pres. econs. sect. 1971-72), AAUP (chpt. pres. 1964-65). Home: 10530 Somerton Dr Dallas TX 75229-5323 Office: So Meth U Dept Econs Dallas TX 75275

MURPHY, KATHLEEN JANE, psychologist; b. Worcester, Mass., Nov. 9, 1962; d. Frederick George and Dorothy Jane (McGuiness) M.; m. Gary Lee Tatum, July 3, 1991. BA cum laude, Holy Cross Coll., 1984; MA, Assumption Coll., 1987; PhD, Tex. A&M U., 1991. Lic. profl. counselor, marriage and family therapist. Counselor Tex. Rehab. Commn., College Station, 1988-89, psychometrician, 1989; intern clin. psychology Worcester (Mass.) State Hosp., 1989-90; psychotherapist Sandstone Ctr., College Station, 1991-92, Luth. Social Svc., Bryan, Tex., 1992—. Editor: Report on Inquiry newsletter, 1988. Mem. Am. Psychol. Assn., Am. Counseling Assn., Nat. Register Health Svc. Providers in Psychology, Phi Beta Kappa, Psi Chi, Phi Kappa Phi. Democrat. Roman Catholic. Avocations: swimming, walking, gardening, travel. Home: 614 Abbey Ln College Station TX 77845-5853

MURPHY, MARTIN JOSEPH, JR., cancer research center executive; b. Colorado Springs, Dec. 29, 1942; s. Martin J. Sr. and Gertrude F. (Heffting) M.; m. Ann. A. Flesher, May 29, 1965; children: Siobhan, Deirdre, Martin III, Sean, Brendan. BS, Regis Coll., 1964; MS, N.Y.U., 1967, PhD, 1969. Vis. fellow Inst. de Pathologie Cellulaire, Paris, 1969-71, Christie Hosp., Manchester, Eng., 1971-72; visiting fellow John Curtin Sch. of Med. Rsch., Canberra, Australia, 1972-74; asst. prof. Sch. Medicine Cornell U., N.Y.C.; dir. Bob Hipple Lab for Cancer Rsch., N.Y.C., 1977-85; dir. hematology tng. program Sloan-Kettering Inst. for Cancer Rsch., N.Y.C., 1978-79; prof. Sch. of Medicine Wright State U., Dayton, 1984--; pres., chief exec. officer Hipple Cancer Rsch. Ctr., Dayton, 1985--; bd. chmn. AlphaMed Press, Inc., Dayton, 1983--; bd. dirs. Dayton Clin. Oncology Program. Editor: In Vitro Aspects of Erythropoleis, 1978, Blood Cell Growth Factors: Their Biology and Clinical Applications, 1990, Blood Cell Growth Factors: Their Utility in Hematology and Oncology, 1991; editor-in-chief Internat. Jour. Cell Cloning, 1982-93, Stem Cells, 1993—; contbr. articles to profl. jours. NIH Postdoctoral fellow Nat. Inst. of Health, 1969-70, Damon Runyon fellow, 1970-71, Spl. fellow Leukemia Soc. Am., 1971-71, Pro America award, Dayton Exec. Club. mem. Assn of Am. Cancer Inst., Am. Soc. of Hematology, Am. Soc. of Oncology, Am. Assn. for Cancer Research, Internat. Soc. for Exptl. Hematology, Dayton Area Cancer Assn. (trustee). Avocation: photography. Office: Hipple Cancer Rsch Ctr 4100 Kettering Blvd Dayton OH 45439-2092

MURPHY, RICHARD ALAN, physiology educator; b. Twin Falls, Idaho, July 4, 1938; s. Albert M. and S. Elizabeth (McClain) M.; m. Genevieve M. Johnson, Dec. 16, 1961; children: Hayley McClain Murphy Parrish, Wendy Louisa Murphy Bossong. AB in Biology cum laude, Harvard U., 1960; PhD in Physiology, Columbia U., 1964. NIH postdoctoral fellow Max Planck Inst. Physiology, Heidelberg, Germany, 1964-66; rsch. assoc. U. Mich., 1966-68; asst. prof. U. Va., Charlottesville, Va., 1968-72; assoc. prof. U. Va., Charlottesville, 1972-77, prof. physiology, 1977—; sabbatical Open U. Rsch. Unit, Oxford, Eng., 1974-75. Assoc. editor Am. Jour. Physiology, 1985, Cell Physiology, 1990—; spl. topics editor Annual Revs. Physiology, 1989; contbr. chpts. to books and articles to profl. jours. and revs. Bd. dirs. Scientists Ctr. for Animal Welfare, Bethesda, Md., 1986-92; chair rsch. com. Va. affiliate Am. Heart Assn., Richmond. Recipient Prin. Investigator award NIH, 1977-94. Mem. Am. Soc. Biochemistry and Molecular Biology, Am. Physiol. Soc., Biophys Soc., Internat. Soc. Heart Rsch. Office: U Va Dept Molecular Physiology Box 449-HSC Charlottesville VA 22908

MURPHY, RICHARD DAVID, physics educator; b. Omaha, May 30, 1938; s. Francis Allan and Anne (Pettinger) M.; m. Donna Marie Caldwell, July 16, 1965; children: Michael, Susan. BA, U. Colo., 1961; MA, U. Minn., 1964, PhD, 1968. Physicist Nat. Bur. Stds., Boulder, 1958-61; rsch. asst. U. Minn., Mpls., 1961-68; physicist Honeywell Rsch., Mpls., 1962-68; postdoctoral physicist IBM Rsch. Labs., San Jose, Calif., 1968-70; asst. prof. Meml. U., St. John's, Nfld., Can., 1970-74; assoc. prof. to prof. U. Mo., Kansas City, 1974—; summer rschr. Ballistic Rsch. Labs., Aberdeen, Md., 1983, Seiler Lab., Air Force Acad., Colo., 1984, Naval Ocean Systems Ctr., San Diego, 1989-91, Phillips Labs., Albuquerque, 1992; cons. and reviewer fed. agencies. Contbr. articles to profl. jours. Bd. dirs., v.p. Friends of Kansas City Symphony Chorus, 1988-92. Recipient Sci. Faculty Profl. Devel. award NSF, 1978-82. Achievements include research on large scale scientific computation, condensed matter and plasma physics. Office: U Mo Physics Dept Kansas City MO 64110

MURPHY, ROBERT EARL, scientist, government agency administrator; b. Yakima, Wash., Sept. 24, 1941; s. William Barry and Caroline Norbeth (Boyd) M.; m. Nancy Jane Hybner, June 26, 1965; children: Kimberly Elizabeth, Mark Hybner. B.S. in Math, Worcester (Mass.) Poly. Inst., 1963; M.A. in Astronomy, Georgetown U., 1966; Ph.D., Case Western Res. U., 1969. Astronomer U.S. Army Map Service, 1965-66; asst. prof. astronomy Inst. Astronomy, U. Hawaii, 1969-73; exec. dir. Md. Acad. Scis., Balt., 1973-76; pres. Scientia Inc., Balt., 1976—; chief planetary atmospheres programs and program scientist Galileo project NASA, Washington, 1977-81; head earth resources br., project scientist heat capacity mapping mission Goddard Space Flight Center, Greenbelt, Md., 1981-84; chief land processes br. NASA Hdqrs., Washington, 1984-89, chief biochemistry, geophysics bur., 1989—. Author curriculum materials, articles in field. Mem. AAAS, Am. Astron. Soc., Am. Geophys. Union. Republican. Baptist. Office: NASA Hdqrs Land Processes Br Code SE Washington DC 20546

MURPHY, ROBERT FRANCIS, biology educator; b. Bklyn., Aug. 25, 1953; s. Robert Francis and Marguerite Ann (McClean) M.; m. Vivian Mathilde Grosswald, Aug. 15, 1981 (div. May 1990); children: Robert Emile, Charles Francis; m. Cynthia Ann Miller, Nov. 23, 1991. BA, Columbia U., 1974, PhD, Calif. Inst. Tech., 1979. Rsch. assoc. Columbia U., N.Y.C., 1979-83; asst. prof. dept. biol. sci. Carnegie Mellon U., Pitts., 1983-89, assoc. prof., 1989—; cons. Becton Dickinson Immunocytometry Systems, San Jose, Calif., 1982-92; assoc. Pitts. Cancer Inst., 1986—; mem. cell biology study panel NSF, Washington, 1989-92, biol. scis. study sect. NIH, 1993—. Co-editor: Applications of Fluorescence in the Biomedical Sciences, 1986, Endosomes and Lysosomes: A Dynamic Relationship, 1993; contbr. over 50 articles to profl. publs. Damon Runyon-Walter Winchell Cancer Found. postdoctoral fellow, 1979; grantee NIH, 1983, 87, 91, NSF, 1989; recipient Presdl. Young Investigator award NSF, 1984. Mem. AAAS, Internat. Soc. Analytical Cytology, Am. Soc. Cell Biology, Sigma Xi. Achievements include development of use of flow cytometry for the study of endocytosis, development of computational biology curriculum; research in analysis of ligand/receptor processing and lysosome biogenesis, application of computers in biology. Home: 161 Kingsdale Rd Pittsburgh PA 15221-3909 Office: Carnegie Mellon U 4400 5th Ave Pittsburgh PA 15213-2683

MURPHY, STEPHAN DAVID, electrical engineer; b. Cin., July 12, 1948; s. James Martin and Oswalda (Magalli) M.; m. Nancy Elizabeth Benton, Apr. 20, 1979; children: Colleen B., Brian B. BSEE, Case Western Res. U., 1971. Design engr. Gould Ocean Systems, Cleve., 1971-74; project engr. Victoreen Inst. div. Sheller-Globe, Cleve., 1974-78, TRW, Inc., Euclid, Ohio, 1978-85; engring. mgr. Textron, Inc., Danville, Pa., 1985—. Author, editor: In-Process Measurement of Control, 1990; contbr. tech. articles to profl. publs. Community chmn. Cleve. unit Am. Heart Assn., 1984; coach Am. Youth Soccer Orgn., Danville, Pa., 1992, Danville Little League, 1992. Mem. AAAS, IEEE, Soc. Mfg. Engrs. (sr.). Republican. Presbyterian. Achievements include patent in area of non-contact gaging and ultrasonic defect detection, pioneering in development of non-contact gaging. Office: Textron Inc Rte 11 at Woodbine Ln Danville PA 17821-0400

MURPHY, THOMAS JOSEPH, chemistry educator; b. Pitts., Oct. 4, 1941; s. Hugh Joseph and Margaret Marie (Rohalley) M.; m. Cecilia Schmuttenmaer, Dec. 6, 1969. BS, U. Notre Dame, 1963; PhD, Iowa State U., 1967. Prof. DePaul U., Chgo., 1985—, faculty mem., 198—; rsch. adv. com. Ill. and Ind. Sea Grant, Champaign, 1990—, Hazardous Substance Rsch. Ctr., Baton Rouge, 1991—; mem. rev. panel U.S. EPA, Washington, 1991—. Editor Jour. Great Lakes Rsch., 1991—. Dir. Lake Mich. Fedn., Chgo., 1972—; pres. Becker House, Chgo., 1991—. Recipient Chandler-Meisner award Internat. Assn. Great Lakes Rsch., Ann Arbor, Mich., 1976, Environ. Quality award U.S. EPA, Chgo., 1976; grantee U.S. EPA, Chgo., 1985, 89, Hazardous Waste Rsch. and Info. Ctr., Champaign, 1987. Mem. AAAS, Am. Chem. Soc., Am. Geophys. Union. Achievements include first documented fact that the atmosphere was a significant source of nutrients and semi volatile toxic organic compounds to the Great Lakes. Office: DePaul U Chemistry Dept 1036 W Belden Chicago IL 60614

MURRAY, ARTHUR JOSEPH, engineering consultant, researcher; b. Portsmouth, Va., Jan. 12, 1954; s. Arthur Patrick and Regina Agneta (Lescavage) M.; m. Deborah Marie Moyer, Sept. 6, 1975; 1 child, Arthur III. BSEE, Lehigh U., 1975; MEA, George Washington U., 1982, DSc, 1989. Electronics engr. USN Ordnance Sta., Indian Head, Md., 1975-81; rsch. engr. Inst. for Artificial Intelligence, Washington, 1985-87; sr. tech. staff The Titan Corp., Vienna, Va., 1982-89; assoc. professorial lectr. Sch. Engring. and Applied Sci., The George Washington U., Washington, 1985—; mgr. advanced technology McDonnell Douglas Electronic Systems Co., McLean, Va., 1989-91; sr. tech. cons. Gemini Industries, Inc., Vienna, Va., 1991—; conf. com. AI Systems in Govt. Conf., Washington, 1986, 90, Am. Soc. Info. Sci., Atlanta, 1988; referee Interfaces, 1993—. Named First Titan fellow Titan Systems, Inc., 1985. Mem. Am. Assn. for Artificial Intelligence, Am. Soc. Naval Engrs., Engring. Mgmt. Soc., Nat. Bus. Incubation Assn., Agile Mfg. Enterprise Forum, Am. Soc. for Performance Improvement (bd. dirs. 1978-79), Lambda Chi Alpha. Republican. Roman Catholic. Achievements include development of virtual enterprise corporate knowledge server and architecture. Home: 203 S Fillmore St Arlington VA 22204-2079 Office: Gemini Industries Inc 8391 Old Courthouse Rd # 170 Vienna VA 22182-3819

MURRAY, DELBERT MILTON, engineer; b. Fordland, Mo., Aug. 22, 1941; s. Chester Augustus and Iris Morene (Hamilton) M.; m. Orilla Maxine Stoaks, Sept. 15, 1962; children: Cynthia Ann, Norman Lee, Orilla Mae, Delbert Lynn. BS, S.W. Mo. State U., 1963. Prodn. planner McDonnell Douglas Corp., St. Louis, 1963-65; tool planning engr. The Boeing Corp., Wichita, Kans., 1965-70; indsl. engr. NCR Corp., Wichita, 1972-77; sr. mfg. engr. Emerson Electric Co., Ava, Mo. Chmn. Mt. Zion Ch. of God., Mo., 1977—. Mem. NRA, N.Am. Hunting Club, Gideons. Republican. Avocations: hunting, fishing, photography, woodworking. Home: RR 1 Box 305 Ava MO 65608-9720 Office: Emerson Electric Co 1400 NW 3D St Ava MO 65608

MURRAY, JEANNE MORRIS, scientist, educator, consultant; b. Fresno., Calif., July 6, 1925; d. Edward W. and Augusta R. (French) Morris; m. Thomas Harold Murray, June 19, 1964; children: Jeanne, Margaret, Barbara, Thomas, William. B.S. in Math., Morris Harvey Coll., 1957; M.S. in Info. and Computer Sci., Ga. Inst. Tech., 1966; Ph.D., in Pub. Adminstrn., Tech. Mgmt., Am. U., 1981. Research scientist Ga. Inst. Tech., Atlanta, 1959-68; adj. prof. Am. U., Washington, 1968-73; computer scientist U.S. Dept. Def., Washington, 1968-69; staff scientist Delex Systems, Inc., Arlington, Va., 1969-70; mgmt. analyst GSA, Washington, 1971-74; assoc. prof. No. Va. Community Coll., 1975-76, U. Va., 1976—; guest lecturer, Computers and Soc., U. Md., 1986—, Govtl. Rels., Marymount U. Arlington Va., 1990; cons., TechDyn Systems, ABA Corp., OrKand Corp., 1978-80; pres. Sequoia Assocs., Arlington, Va., 1981—; panelist Inst. Agr., Akadamgorodok, Siberia, 1991, Inst. Nuclear Physics, 1991, M.Ulughbek Inst., Samarkand, Uzbekistan, 1991. Author: Development and Testing of a System of Encoding Visual Information Based on Optimization of Neural Processing in Man--with Application to Pattern Recognition in the Computer, 1966, Cybernetics and the Management of the Research and Development Function in Society, 1971, Cybernetics as a Tool in the Control of Drug Abuse, 1972, Development of a General Computerized Forecasting Model, 1971, Political Humankind and the Future of Governance, 1974, The Doctrine of Management Planning, 1973, Policy Design, 1980, Computer Futures, 1982, A Search for Positive Response Level Indicators (PRLI's) Under Stress, 1987, Strategic Planning: Pathfinder to the Future, Beijing, 1988, Strategic Planning: A Systems Perspective, Shanghai PRC 1988, Electronic Control Systems for Railroads, Wuhan PRC 1988, Technology Forecasting Methodologies for Use on Personal Computers, 1989, Japan's Burgeoning Rates of Economic Expansion in the U.S. and other Western Countries, 1990, Technology Transfer and National Security, 1992, Privatization Mechanism for the Former Soviet Union and Central European Countries, 1992, Curriculum Development for Privatization Training of Entrepreneurs in Siberia, 1993. Mem. Carter transition team, 1976-77, Arlington (Va.) Civil Def. Com., 1983—, Washington Met. Area Emergency Assistance Com., Arlington County Com. on Sci. and Tech. Mem. IEEE (sr. mem., vice chmn. Washington sect., chmn. panel on internat. mktg. high tech. in the persence of def. tech. controls 1983, nat. com. on a tech. transfer policy for the U.S. 1986), AAAS, N.Y. Acad. Scis., Assn. Computing Machinery, Washington Evolutionary Systems Soc., Inst. Noetic Scis., Soc. Gen. Research, Am. Soc. Pub. Adminstrn., World Future Soc., Better World Soc., Acad. Polit. Sci., Personality Assessment System Found., Soc. for the Advance of Socio-Econs.; Episcopalian. Home and Office: 2915 27th St N Arlington VA 22207-4922

MURRAY, JOSEPH, chemistry educator; b. Pocatello, Idaho, Dec. 14, 1916; s. Oscar and Nettie Muriel (Cotham) M.; m. Irma Leontina Chaix, Jan. 4, 1941; children: Carl, JoAnn. BA in Chemistry, U. Puget Sound, 1948; MS in Chemistry, Oreg. State U., 1950; PhD in Chemistry, SUNY, Buffalo, 1956. Accredited profl. chemist. Various university and profl. chemist positions, 1948-51; organic researcher Trojan Powder Co., Allentown, Pa., Wolf Lake, Ill., 1951-52; instr. chemistry SUNY, Buffalo, 1952-56; sr. evaluations engr. Olin Corp., 1956-59, dir. nuclear computer ctr., 1959-61, corp. mgr. tech. data ctr., 1961-62; asst. prof. chemistry Mont. Tech., Butte, 1962-64, assoc. prof. chemistry, 1964-70, prof. chemistry, 1970-80, prof. emeritus chemistry, 1980—; acting dept. head Mont. Tech., Butte, 1971-75, chmn., 1975; chem. cons. local police and fire depts., Def. Rsch. Inst., Am. Assn. Railroads, local lawyers. Contbr. articles to profl. jours. Fellow Am. Inst. Chemists (life), Am. Chem. Soc. (past program chmn. Mont. sect., past sect. chmn.), Mu Sigma Delta, Soc. of Sigma Xi, Phi Lambda Upsilon. Achievements include developement of computer tutorial program for hich school and college students, programs to assist chemists in determining the relationships between temperature, pressure, and concentrations, computer program to assist environmental chemists in determining the relative concentrations of various chemical species in aqueous solutions. Office: 803 W Granite St Butte MT 59701-9053

MURRAY, JOSEPH EDWARD, plastic surgeon; b. Milford, Mass., Apr. 1, 1919; s. William Andrew and Mary (DePasquale) M.; m. Virginia Link, June 2, 1945; children—Virginia, Margaret, Joseph Link, Katharine, Thomas, Richard. A.B., Holy Cross Coll., 1940, D.Sc., 1965; M.D., Harvard, 1943; D.Sc., Rockford (Ill.) Coll., 1966, Roger Williams Coll., 1986; hon. degree, Anna Marie Coll., 1993, SUNY, Albany, 1993, U. Suffolk, 1993. Diplomate: Am. Bd. Surgery, Am. Bd. Plastic Surgery (chmn. 1969). Chief plastic surgeon Peter Bent Brigham Hosp., Boston, 1951-86, chief plastic surgeon emeritus, 1986—; chief plastic surgeon Children's Hosp. Med. Center, Boston, 1972-85, emeritus, 1985; prof. surgery Harvard Med. Sch., 1970—. Served to maj. M.C. AUS, 1944-47. Recipient Gold medal Internat. Soc. Surgeons, 1963, hon. award Am. Acad. Arts and Sci., 1962, Nobel prize for medicine or physiology, 1990. Fellow AAAS (hon.), AMA, Royal Australasian Coll. Surgeons, Royal Coll. Surgeons of Eng., Royal Coll. Surgeons Ireland, Royal Coll. Surgeons Edinburgh; mem. ACS (regent 1970-79, v.p. 1983), NAS, Am. Surg. Assn. (v.p. 1979), New Eng. Surg. Assn. (pres. 1986-87), Boston Surg. Soc. (pres. 1975), Soc. U. Surgeons, Am. Assn. Plastic Surgeons (hon. award 1969, pres. 1964-65), Am. Acad. Arts and Sci., Harvard Med. Sch. Alumni Coun. (pres. 1984), Alpha Omega Alpha. Clubs: Badminton and Tennis, Wellesley Country. Home: 108 Abbott Rd Wellesley MA 02181-6104

MURRAY, JOSEPH JAMES, JR., biology educator, zoologist; b. Lexington, Va., Mar. 13, 1930; s. Joseph James and Jane Dickson (Vardell) M.; m. Elizabeth Hickson, Aug. 24, 1957; children—Joseph James III, Alison Joan, William Lister. B.S., Davidson Coll., 1951; B.A., Oxford U., Eng., 1954, M.A., 1957, D.Phil., 1962. Instr. biology Washington & Lee U., Lexington, Va., 1956-58; asst. prof. biology U. Va., Charlottesville, 1962-67, assoc. prof., 1967-73, prof., 1973-77, Samuel Miller prof. biology, 1977—, chmn. dept. biology, 1984-87; co-dir. Mountain Lake Biol. Sta., Pembroke, Va., 1963-91. Author: Genetic Diversity and Natural Selection, 1972; contbr. articles to profl. jours. Served with U.S. Army, 1955-56. Rhodes scholar, 1951-54. Fellow AAAS, Va. Acad. Sci.; mem. Am. Soc. Naturalists, Genetics Soc. Am., Soc. Study Evolution, Am. Soc. Ichthyologists and Herpetologists, Va. Acad. Sci. (pres. 1986-87), Va. Soc. Ornithology (pres. 1976-79). Avocations: walking; mountaineering; shooting. Office: U Va Dept Biology Gilmer Hall Charlottesville VA 22901

MURRAY, MARY KATHERINE, reproductive biologist, educator; b. Evergreen Park, Ill., Oct. 21, 1956; d. William Anthony and Phyllis Katherine (Grevas) M. BS, U. Ill., 1978; PhD, U. Ill. Med. Ctr., Chgo., 1984. Lab. tutor, anatomical minority student program U. Ill. Med. Ctr., Chgo., 1979-80; instr. anatomical scis. Dr. W. Scholl. Coll. Pediatric Medicine, Chgo., 1980-83; teacing asst. anatomical scis. U. Ill. Med. Ctr., Chgo., 1980-84; instr. anatomical scis. Northwestern U., Chgo., 1982; asst. prof. U. N.H., Durham, 1986-87, Tufts U., Boston, 1987—; reviewer in field. Contbr. to 1 book and articles to profl. jours. Rsch. grantee NIH, 1991—. Mem. AAAS, AAVA, Mass. Soc. Med. Rsch. (speakers bur. 1992—), Soc. Study Reproduction. Achievements include the detection and purification and assessment of biologicalfunction of oviduct and uterine steroid regulated proteins. Office: Tufts U Dept Anat & Cell Biology 136 Harrison Ave Boston MA 02111

MURRAY, PAMELA ALISON, quality data processing executive; b. Phila., Sept. 22, 1955; d. Everett Hickman Jr. and Laura Frances (Lautenbach) M. BA in Math. cum laude, Gettysburg (Pa.) Coll., 1977; MBA in Mktg., Temple U., 1986; postgrad., U. Del., 1978-80. Asst./assoc. systems programmer Burroughs Corp., Paoli, Pa., 1977-79, cost acctg. analyst, gen. acctg. analyst, 1979-80, sr. tech. analyst, 1980-82, mgr. data ctr. tech. support, 1982-83; mgr. software product assurance Unisys Corp./Burroughs Corp., Paoli, Pa., 1983-86; mgr. product assurance Unisys Corp., Devon, Pa., 1986-88; sr. staff engr. Unisys Corp., Blue Bell, Pa., 1988-90, system quality mgr., 1990-91; mgr. quality assurance Unisys Corp., Paoli and Blue Bell, 1991-92; mgr. requirements assurance Unisys Corp., Paoli, 1992—; instr. after-hours tng. Burroughs Corp./Unisys, 1983-86. Fundraiser Am. Heart Assn., 1987—, Am. Cancer Soc., 1989, March of Dimes, 1987, 89; active Phila. Soc. Preservation Landmarks, Phila. Mus. Art. Recipient Young Alumni Achievement award Gettysburg Coll., 1989, Phila. Alumni Club Silver Plate, 1988, Svc. award, 1989. Mem. Am. Soc. Quality Control, Gettysburg Coll. Alumni Assn. (chmn. alumni reunion com. 1992—, treas. 1991-93, chmn. alumni ctr. task force 1990—, career rep. 1988—), Jr. League Phila. (chmn. cookbook sales com. 1990-92, v.p. 1984-86), Temple U. Alumni Assn. Republican. Mem. Christian Ch. Avocations: travel, classical music, antiques, decorating. Home: 404 Danor Ct Wayne PA 19087-1232

MURRAY, PATRICK ROBERT, microbiologist, educator; b. L.A., Jan. 15, 1948; married, 1970; 3 children. BS, St. Mary's Coll., Calif., 1969; MS, U. Calif., L.A., 1972, PhD in Microbiology, 1974. Rsch. fellow clinical microbiology Mayo Clinic & Mayo Found., 1974-76; asst. prof. medicine Wash. U. Sch. Med., 1976-82, assoc. dir., 1976; dir. clinical microbiology Barnes Hosp., St. Louis, 1977—, dir. postdoctoral training program, 1982—; assoc. prof. medicine Wash. U. Sch. Medicine, 1983—; cons. St. Luke's Hosp., 1985—; mem. Nat. Com. Clinical Lab. Standards, 1980—; chmn. exam com. Am. Bd. Medical Microbiologist, 1982—, mem. joint standards and exam com., 1985—; bd. dirs. Southwestern Assn. Clinical Microbiologist; chmn. Clinical Microbiologist Divsn. Am. Soc. Microbiologist. Becton Dickinson Co. Clinical Microbiology award Am. Soc. Microbiology, 1993. Fellow Am. Acad. Med. Microbiologist, Infectious Disease Soc; mem. Am. Assn. Pathologist, Am.Soc. Microbiologist, Med. Mycol Soc. Am., Am. Federation Clinical Rsch., Sigma Xi. Achievements include research in new diagnostic and therapeutic tests for clinical microbiology. Office: Washington Univ Div of Lab Medicine PO Box 8118 660 S Euclid Saint Louis MO 63110*

MURRAY, PETER, metallurgist, manufacturing company executive; b. Rotherham, Yorks, Eng., Mar. 13, 1920; came to U.S., 1967, naturalized, 1974; s. Michael and Ann (Hamstead) M.; m. Frances Josephine Glaisher, Sept. 8, 1947; children: Jane, Paul, Alexander. BSc in Chemistry with honors, Sheffield (Eng.) U., 1941, postgrad., 1946-49; PhD in Metallurgy, Brit. Iron and Steel Research Bursar, Sheffield, 1948. Research chemist Steetley Co., Ltd., Worksop, Notts, Eng., 1941-45; with Atomic Energy Research Establishment, Harwell, Eng., 1949-67; head div. metallurgy Atomic Energy Research Establishment, 1960-64, asst. dir., 1964-67; tech. dir., mgr. fuels and materials, advanced reactors div. Westinghouse Electric Corp., Madison, Pa., 1967-74; dir. research Westinghouse Electric Europe (S.A.), Brussels, 1974-75; chief scientist advanced power systems divs. Westinghouse Electric Corp., Madison, Pa., 1975-81; dir. nuclear programs Westinghouse Electric Corp., Washington, 1981-92; sr. cons. Nuclear Programs, 1992—; mem. divisional rev. coms. Argonne Nat. Lab., 1968-73; Mellor Meml. lectr. Inst. Ceramics, 1963. Contbr. numerous articles to profl. jours.; editorial adv. bd.: Jour. Less Common Metals, 1968—. Recipient Holland Meml. Research prize Sheffield U., 1949. Fellow Royal Inst. Chemistry (Newton Chambers Research prize 1954), Inst. Ceramics, Am. Nuclear Soc.; mem. Brit. Ceramics Soc. (pres. 1965), Am. Ceramic Soc., Nat. Acad. Engring. Roman Catholic. Home: 20308 Canby Ct Gaithersburg MD 20879-4014 Office: Westinghouse Electric Corp One Montrose Metro 11921 Rockville Pike Ste 450 Rockville MD 20852

MURRAY, ROBERT WALLACE, chemistry educator; b. Brockton, Mass., June 20, 1928; s. Wallace James and Rose Elizabeth (Harper) M.; m. Claire K. Murphy, June 10, 1951; children: Kathleen A., Lynn E., Robert Wallace, Elizabeth A., Daniel J., William M., Padraic O'D. AB, Brown U., 1951; MA, Wesleyan U., Middletown, Conn., 1956; PhD, Yale U., 1960. Mem. tech. staff Bell Labs., Murray Hill, N.J., 1959-68; prof. chemistry U. Mo., St. Louis, 1968-81; chmn. dept. U. Mo., 1975-80, curators' prof., 1981—; vis. prof. Engler-Bunte Inst. U. Karlsruhe, Fed. Republic Germany, 1982, dept. chemistry Univ. Coll., Cork, Ireland, 1989; cons. to govt. and industry. Co-editor: Singlet Oxygen, 1979; contbr. articles to profl. jours. Mem. Warren (N.J.) Twp. Com., 1962-63, mayor, 1963; mem. Planning Com. and Bd. Health, 1962-64, Bd. Edn., 1966-68. Served with USNR, 1951-54. Grantee EPA, NSF, NIH, Office of Naval Research. Fellow AAAS, Am. Inst. Chemists, N.Y. Acad. Scis.; mem. Am. Soc. Photobiology, Am. Chem. Soc., The Oxygen Soc., Sigma Xi. Home: 1810 Walnutway Dr Saint Louis MO 63146-3659 Office: Univ Mo Dept Chemistry Saint Louis MO 63121

MURRAY, ROYCE WILTON, chemistry educator; b. Birmingham, Ala., Jan. 9, 1937; s. Royce Leeroy and Justina Louisa (Herd) M.; m. Judith Studinka, 1957 (div.); children: Katherine, Stewart, Debra, Melissa, Marion; m. Mirtha X. Umana, Dec. 11, 1982. BS in Chemistry, Birmingham So. Coll., 1957; PhD in Analytical Chemistry, Northwestern U., 1960. Instr. U. N.C., Chapel Hill, 1960-61, asst. prof., 1961-66, assoc. prof., 1966-69, prof., 1969—, vice chmn., 1970-75, acting chair dept. chemistry, 1970-71, dir. undergrad. studies, 1978-80, dept. chmn., 1980-85, adj. prof curriculum of applied scis., 1987—, div. chmn., 1987—, Kenan prof., 1980—. Contbr. articles to jours. in field. Recipient award Japanese Soc. for Promotion Sci., 1978, Electrochem. Group medal Royal Soc. Chemistry, 1989; Alfred P. Sloan fellow, 1969-72, Guggenheim fellow, 1980-82. Fellow Am. Inst. Chemists, Am. Acad. Arts and Scis.; mem. NAS, AAAS, Soc. for Electroanalytical Chemistry (bd. dirs., co-founder 1982-84, Charles N. Reilley award 1988, pres. 1991-93), Am. Chem. Soc. (Electrochemistry award 1990, Analytical Chemistry award 1991, editor in chief Jour. Analytical Chemistry 1991—), Electrochem. Soc. (hon. life, Carl Wagner Meml. award 1987). Presbyterian. Office: The Univ of NC Dept of Chemistry Chapel Hill NC 27599-3290

MURRAY, RUSSELL, II, aeronautical engineer, defense analyst, consultant; b. Woodmere, N.Y., Dec. 5, 1925; s. Herman Stump and Susanne Elizabeth (Warren) M.; m. Sally Tingue Gardiner, May 22, 1954; children: Ann Tingue, Prudence Warren, Alexandria Gardiner. BS in Aero. Engring, MIT, 1949, MS, 1950. Guided missile flight test engr. Grumman Aircraft Engring. Corp., Bethpage, N.Y., 1950-53, asst. chief operations analysis, 1953-62; prin. dep. asst. sec. of def. for systems analysis The Pentagon, Washington, 1962-69; dir. long range planning Pfizer Internat., N.Y.C., 1969-73; dir. review Center for Naval Analyses, Arlington, Va., 1973-77; asst. sec. of def. for program analysis and evaluation Dept. of Def., The Pentagon, Washington, 1977-81; prin. Systems Research & Applications

Corp., Arlington, Va., 1981-85; spl. counsellor Com. on Armed Services U.S. Ho. of Reps., 1985-89, nat. security cons., 1989—. Served with USAAF, 1944-45. Recipient Sec. of Def. Medal for meritorious civilian service, 1968; Disting. Public Service medal Dept. Def., 1981. Home: 210 Wilkes St Alexandria VA 22314-3839

MURRAY, STEVEN NELSEN, marine biologist, educator; b. L.A., Sept. 7, 1944; s. Arthur Willard and Andrea Jensina (Nelsen) M.; m. Nancy Mary DiGerolami, June 25, 1966; children: Timothy Daniel, Andrea Patrice. BA, U. Calif., Santa Barbara, 1966, MA, 1968; PhD, U. Calif., Irvine, 1971. Asst. prof. Calif. State U., Fullerton, 1971-74, assoc. prof., 1974-78, prof. marine biology, 1978—; cons. rsch. and edn., Calif. Contbr. articles to profl. jours. Mem. Internat. Phycological Soc., Ecol. Soc. Am., Phycological Soc. Am., So. Calif. Acad. Scis., Western Soc. Naturalists (treas. 1982-89, pres. 1992). Office: Calif State Univ Dept Biol Sci Fullerton CA 92634

MURRAY, THOMAS REED, aerospace engineer; b. Logan, Utah, Aug. 20, 1924; s. Thomas Bailey and Modena (Affleck) M.; m. Elaine Marshall, Sept. 10, 1947; children: William Reed, Eric Alan, Paul Raymond, Kent Douglas, Brian Lane, Carla Morene. BSME, U. N.Mex., 1945; MS in Aero. Engring., Stanford U., 1947, ME in Engring. Mechanics, 1949. Rsch. assist, assoc. Stanford (Calif.) U., 1947-49; structures analyst Douglas Aircraft Co., Santa Monica, Calif., 1949-51, stress analyst, supr., 1953-63; engring. mgr. stress analysis McDonnell Douglas Astro Co., Huntington Beach, Calif., 1963-78, mgr., chief engr. structures, 1978-87, dep. dir. engring., 1987-90; ret., 1990; cons. Lear Corp., Santa Monica, 1960; bd. dirs. Sci. Fair, Irvine, Calif., 1980-85; presenter at profl. confs. Author tech. reports. Chair architect com. College Park Home Owners' Assn., Irvine, 1985-87; pres. Mo. Independence mission Ch. of Jesus Christ of Latter-Day Saints, 1991—. With USN, 1943-46, PTO, 1951-53, Atlantic. Mem. AIAA, Sigma Tau, Kappa Mu Epsilon. Achievements include Titan II fairing failure analysis, application of suppressive device to mitigate launch vehicle longitudinal oscillations, analysis of longitudinal oscillation of Thor launch vehicles. Home: 5787 Bower St Kansas City MO 64133 Office: Ch Jesus Christ Latter Day Saints 517 W Walnut St Independence MO 64051

MURRAY, WILLIAM, food products executive; b. 1936. With UN Relief Works Agy, Lebanon, 1960-65; head budget divsn. Internat. Labor Orgn., 1965-70; with Philip Morris Inc., 1970—, v.p., 1976-78, exec. v.p. internat. divsn., 1978-84, pres. and CEO internat. divsn., 1984-87, vice chmn. bd., 1987-91, pres., COO., 1991—; pres. Benson & Hedges Canada Ltd., 1974. Office: Phillip Morris Co Inc 120 Park Ave New York NY 10017-5523*

MURRAY, WILLIAM JAMES, anesthesiology educator, clinical pharmacologist; b. Janesville, Wis., July 20, 1933; s. James Arthur and Mary Helen (De Porter) M.; m. Therese Rose Dooley, June 25, 1955; children: Michael, James, Anne. BS, U. Wis., 1955, PhD, 1959; MD, U. N.C., 1962. Diplomate Am. Bd. Anesthesiology. Rsch. asst. U. Wis., Madison, 1955-59; instr. pharmacology U. N.C., Chapel Hill, 1959-62, resident and fellow in surgery (anesthesiology), 1962-64, instr., 1964-65, asst. prof., 1965-68; asst. to dir. for drug availability FDA, Washington, 1968-69; assoc. prof. pharmacology, clin. pharmacology and anesthesiology U. Mich., Ann Arbor, 1969-72; assoc. prof. anesthesiology Duke U., Durham, N.C., 1972-81, prof., 1981—; assoc. dir. Upjohn Ctr. for Clin. Pharmacology, Ann Arbor, 1969-72. Mem. AMA, Am. Soc. Anesthesiologists, Internat. Anesthesia Rsch. Soc., Soc. for Ambulatory Anesthesia, Am. Pharm. Assn., N.Y. Acad. of Sci., N.C. Soc. Anesthesiologists, Am. Soc. Hosp. Pharmacists. Republican. Roman Catholic. Home: 135 Pinecrest Rd Durham NC 27705-5822 Office: Duke U Med Ctr Dept Anesthesiology Box 3094 Durham NC 27710

MURRELL, KENNETH DARWIN, research administrator, microbiologist; b. Burley, Idaho, Jan. 19, 1940; s. Kenneth Leland and Margaret (Madelin) .; m. Joyce Elizabeth, July 10, 1965; children: Duncan, Amy. BA, Chico (Calif.) State Coll., 1957; MSPH, U. N.C., 1963, PhD, 1969. Rsch. assoc. U. Chgo., 1967-71; resch. sci. Naval Med. Rsch. Inst., Bethesda, Md., 1971-78; resch. sci. Agrl. Rsch. Svc., Beltsville, Md., 1978-87, assoc. area dir., 1987-89; area dir. Beltsville Agrl. Rsch. Ctr., 1989—. Lt. USN, 1966-69. Recipient ARS Outstanding Sci., USDA, Beltsville, Md., 1984; named Disting. Par asitologist, Am. Assn. Vet. Parasitologist., Orlando, Fla., 1987. Mem. Am. Soc. Parasitologist, Am. Assn. Veterinary Parasitology, AAAS. *

MURTAGH, JOHN EDWARD, chemist, consultant; b. Wallington, Surrey, Eng., Sept. 12, 1936; came to U.S., 1982; s. Thomas Henry and Elsie (Kershaw Paterson) M.; m. Eithne Anne Fawsitt, July 18, 1959; children: Catherine, Rhoda, Sean, Aidan, Doreen. BSc, U. Wales, 1959, MSc, 1970, PhD, 1972. Rsch. coord. House of Seagram, Long Pond, Jamaica, 1959-63; whisky distillery mgr. House of Seagram, Beaupre, Que., Can., 1963-64; rum distillery mgr. House of Seagram, Richibucto, N.B., Can., 1965-68; rsch. mgr. House of Seagram, Montreal, Que., 1968-70; distillery cons. Murtagh & Assocs., Buttevant, Ireland, 1972-77, 79-82, Winchester, Va., 1982—; vodka distillery mgr. Iran Beverages, Tehran, 1977-79; ethanol tech. cons., adv. bd. Info. Resources, Inc., Washington, 1988—; lectr. Alltech Ann. Alcohol Sch., Lexington, Ky., 1982—. Author: Glossary of Fuel-Ethanol Terms, 1990; contbr. articles to profl. jours. Adv. bd. Byrd Sch. Bus., Shenandoah U., Winchester, Va., 1989—. Recipient Millers Mutual prize, U. Wales, 1959. Fellow Am. Inst. Chemists, Inst. Chemistry of Ireland, Inst. Food Sci. and Tech. of Ireland; mem. Royal Soc. Chemistry (chartered), Am. Arbitration Assn. (arbitrator nat. comml. panel 1990—). Achievements include development of proprietary process for production of ethanol from cheese whey and the design of whey-ethanol production plants. Home and Office: 160 Bay Ct Winchester VA 22602

MURTAZA, GHULAM, chemist, consultant; b. Lahore, Punjab, Pakistan, Oct. 27, 1955; s. Mustafa Ghulam and Surraya Khanum; married; children: Zoya, Zara. BSc, Punjab U., 1977, MS in Chemistry, 1980. Mgr. sales Hoechst (Pakistan) Ltd., Karachi, 1981-83; br. mgr. Chemtech Svcs. Ltd., Lahore, 1983-88; tech. dir. Seena Chems., Lahore, 1989—. U.S. AID scholar, Mo. U., 1991. Mem. Am. Soc. Metals, Am. Chem. Soc. Achievements include development of low temperature, sludge free phosphatizing processes as paint base for anodic and cathodic paints, water based anticorrosion coatings for ferrous and nonferrous metals, corrosion inhibitors for steel pickling in strong acid solutions, various additives for corrosion and scale prevention of metals used in water based cooling systems. Office: Seena Chems Khokhar Rd, 26-Nishat Park, Badamibagh Lahore 54000, Pakistan

MURTHY, KAMALAKARA AKULA, chemical engineer, researcher; b. Kadiri, India, Dec. 15, 1950; came to U.S., 1987; s. Venkatappa and Sheshamma Akula; m. Sudha Akula, Sept. 11, 1981. BSc, S.V. Arts Coll., Tirupati, India, 1970; MSc, S.V.U. Coll., Tirupati, India, 1972; PhD, Jadavpur U., Calcutta, India, 1977. Postdoctoral fellow Cen. Leather Rsch. Inst., Madras, India, 1977-79; asst. prof. Sri Krishnadevaraya U., Anantapur, India, 1979-86; dept. asst. U. Mass., Amherst, 1983-86; vis. scientist U. Wash., Seattle, 1987-90; rsch. assoc. U. Del., Newark, 1990-92; rsch. dir. KAI Sci. & Tech., Inc., 1992—; cons. Du Pont, Wilmington, Del., 1989-90, A.E. Staley, Decatur, Ill., 1990—. Contbr. articles to profl. jours. Mem. Am. Chem. Soc., Am. Phys. Soc., N.Y. Acad. Scis. Achievements include patents in Spontaneous Vesicles from Aqueous Mixtures of Cationic and Anionic Single-Tailed Surfactants; in Hot Melt Inks for Use in Ink-Jet Printers. Home: 423 Old Blue Rock Rd Lancaster PA 17603-9415

MURTHY, KRISHNA KESAVA, infectious desease and immunology scientist; b. Bangalore, Karnataka, India, June 5, 1950; came to U.S., 1975; s. Shamana Kesava and Tumkur (Kamalama) M. DVM, U. Agriculture Sci., Bagalore, 1971, MS, 1975; PhD, Cornell U., 1979. Sr. rsch. assoc. U. Ga., Athens, 1979-81, asst. prof., 1982-84; asst. prof. U. Basel, Switzerland, 1981-82, La. State U. Med. Ctr., New Orleans, 1984-90; staff scientist S.W. Fedn. for Biomed. Rsch., San Antonio, 1990-91, assoc. scientist, 1992—. Contbr. articles to profl. jours. Mem. AIDS Rsch. and Edn. Com., San Antonio, 1992—. Grantee NIH, NIAID, NHLBI. Mem. Assn. Assoc. Immunologists. Democrat. Home: 4423 Huntington Woods San Antonio TX 78249 Office: SW Fedn Biomedical Rsch 7620 NW Loop 410 San Antonio TX 78227-5301

MURTHY, SRINIVASA K., engineering corporation executive; b. Bangalore, Karnataka, India, June 12, 1949; came to U.S., 1979; s. Ramaswamy and Gowramma Kadur. BS in Physics, Bangalore U., India, 1967; MS in Physics, Bangalore U., 1969; MSEE, Mysore U., India, 1971. Mgr. project engring. Indian Space Rsch. Orgn., Bangalore, 1971-79; asst. prof. Calif. State U., Pomona, 1979-80, Fullerton, 1981-82; mgr. project Systems and Applied Scis. Corp., Anaheim, Calif., 1980-83; dir. div. IMR Systems Corp., Arlington, Va., 1983-84; mgr. systems engring. GE, Portsmouth, Va., 1984-85; bus. mgr. AT&T Bell Labs, Holmdel, N.J., 1985—; bd. advisors IMR Systems Corp., Roslyn, Va., 1988—. Contbr. articles to profl. jours. Recipient Disting. Achievement award Dept. Space, Indian Govt., 1975. Mem. IEEE (sr. mem., bd. dirs. 1986—, standards bd. 1986—, bd. dirs. Electronics and Aerospace Systems conf. 1983-84, editorial bd. dirs. Network Jour 1986—, area activities bd. and tech. activities bd. 1988, Computer Soc. 1987—, lectr. India, Singapore, Austalia 1989, South Am. 1990, numerous other coms.), Engring. Mgmt. Soc. of IEEE (bd. govs. 1986—), Nat. Rsch. Coun., Advt. Rsch. Found. (coun. mem.). Home: 5 Polo Club Dr Eatontown NJ 07724-3823 Office: AT&T Bell Labs Rm 1K-219 Crawford Corner Rd Holmdel NJ 07733

MURTHY, SUDHA AKULA, physicist, consultant; b. Vijayawada, India, June 5, 1957; came to U.S., 1984; d. Pakeeriah and Kalyani (Nagisetty) Ponnaganti; m. Kamalakara A. Murthy, Sept. 11, 1981. MSc in Edn., U. Mysore, India, 1980; MS, U. Mass., 1986, PhD, 1990. Asst. prof. Lincoln U., Oxford, Pa., 1990-91, Millersville (Pa.) U., 1992—; v.p., CFO KAI Sci. and Tech., Redwood City, Calif., 1991—. Contbr. articles to profl. jours. Mem. Am. Phys. Soc., N.Y. Acad. Scis. Office: 441 Poplar Ave # 226 Redwood City CA 94061

MURTHY, VADIRAJA VENKATESA, biochemist, researcher, educator; b. Bombay, Mar. 27, 1940; came to U.S., 1969; s. Ramanathpur Venkatesa and Saroja (Bai) M.; m. Jayashree Deshpande, Sept. 21, 1969; children: Deepti, Seema. BSc with honors, U. Bombay, 1959, MSc, 1961; PhD in Biochemistry, U. Md., 1968. Clin. chemist Nat. Registry of Clin. Chemists; lic. dir. clin. chemistry, N.Y.C. and N.Y. state. Sr. rsch. biochemist, asst. group leader USV Pharms., Yonkers, N.Y., 1970-71; rsch. assoc. Toxicology Ctr., U. Iowa, Iowa City, 1971-72; vis. scientist NIH-Environ., Research Triangle Park, N.C., 1972-74; sr. rsch. assoc. Emory U. Sch. Medicine, Atlanta, 1974-75; adj. asst. prof. Atlanta (Ga.) Univ., 1975-76; prof., co-dir. Talladega (Ala.) Coll., 1976-83; asst. prof., co-dir. Albert Einstein Coll. Medicine, Bronx, N.Y., 1983—; chmn. sci. rev. San Diego (Calif.) Conf. on Nucleic Acids, 1992; chmn. sci. symposium on molecular diagnostics Am. Assn. Clin. Chemistry Nat. Meetings, N.Y., 1993. Contbr. chpt. to book and articles to profl. jours. Hon. sec. Vishwa Kalyana Trust, Washington, 1989—. Named Fogarty Internat. Vis. Scientist Nat. Inst. Environ. Health Scis., NIH, 1972-74; rsch. grantee NIH, 1976-83, Resource Ctr. grantee NSF, 1981-83. Mem. Am. Assn. for Clin. Chemistry (chmn. elect molecular pathology div. 1992-93, chair 1993—), Am. Assn. Cancer Rsch., Am. Chem. Soc., Assn. Clin. Lab. Physicians and Scientists. Achievements include U.S. patent for spectrophotometric attachment for analyzing two-phase systems. Home: 100 Lindbergh Blvd Teaneck NJ 07666 Office: Albert Einstein College Medicine 1300 Morris Park Ave Bronx NY 10461

MURTHY, VANUKURI RADHA KRISHNA, civil engineer; b. Hyderabad, India, June 20, 1928; came to U.S., 1963; s. Rama Vannkuri and Sita (Dittakavi) Rao; m. Lakshmi Gruha Gadiraju, Mar. 12, 1952; children: Siva, Prabha, Jyothy, Lata. BE in Civil, Osmania U., 1949; MS in Engring., U. Fla., 1964; PhD, U. Pa., 1967. Registered profl. engr., Tex. Engr. Govt. of Andhra Pradesh, Hyderabad, 1949-62; grad. rsch. asst. U. Fla., Gainesville, 1963-64, U. Del., Newark, 1964-67; grad. rsch. asst. U. Pa., Phila., 1964-67, instr., 1967-68; hydrologist Tex. Water Rights Commn., Austin, 1968-71, head hydraulics design sect., 1972-77; head basin modeling Tex. Water Commn., Austin, 1978-90; chief engr. Tex. Water Commn., 1990-91; cons. water resources, 1991—; engr. advisor to commr. Pecos River Compact Commn., Austin, 1987-91, chmn. engring. adv. com., 1987-91; prin. expert witness in interstate law suit for Tex. against N.Mex., 1977-90. Author: Allocation of Pecos River Basin Water, 1991; also articles. Travel grantee NSF, 1975. Mem. ASCE (editor Jour. Pipeline Div. 1969-74), Internat. Water Resources Assn., Sigma Xi. Achievements include development of water availability models for all major river basins in Texas. Home: 3011 Val Dr Austin TX 78723-2317 Office: 600 W 28th St Austin TX 78705-3708

MUSCHENHEIM, FREDERICK, pathologist; b. N.Y.C., July 9, 1932; s. Carl and Haroldine (Humphreys) M.; m. Linda Alexander, Mar. 29, 1958; children: Alexandra Lydia, Carl William, David Henry. AB, Harvard U., 1953; MDCM, McGill U., Montreal, Can., 1963. Intern Santa Clara County Hosp., San Jose, Calif., 1963-64; resident pathology U. Colo. Med. Ctr., Denver, 1964-68, chief resident clin. pathology, 1968-69; pathologist Freeman, Hanske, Munkittrick & Foley PA, Mpls., 1969-77; clin. pathologist Union-Truesdale Hosp., Fall River, Mass., 1977-78; chief pathologist St. Clare's Hosp., Denville, N.J., 1978-83, Oneida (N.Y.) City Hosps., 1984—; clin. asst. prof. SUNY Health Sci. Ctr., Syracuse, 1984-90, clin. assoc. prof., 1990—; chief med. staff Oneida City Hosps., 1991. Mem. choir 1st Presbyn. Ch. of Cazenovia (N.Y.), 1984—, bd. trustees, 1985-89. Mem. Assn. Clin Scientists (pres. 1990, Diploma of Honor 1991), Coll. Am. Pathologists (key contact 1990—), N.Y. State Soc. Pathologists (com. chmn.), Med. Soc. State of N.Y. (mem. legis. com. 1991—), Med. Soc. Madison County (v.p. 1990-91, pres. 1991-93—), N.Y. State Assn. Pub. Health Labs. (v.p. 1992-93, pres. 1993—). Home: 5257 Owera Point Cazenovia NY 13035-9804 Office: Oneida City Hosps 321 Genesee St Oneida NY 13421-2611

MUSCI, TERESA STELLA, developmental biologist, researcher; b. Chgo., Apr. 21, 1960; d. Stanislaw and Genowefa (Magnus) Cichocki; m. Anthony Gerard Musci, Apr. 29, 1989. BS in Biology, Loyola U., Chgo., 1982; PhD in Anatomy, U. Utah, 1991. Rsch. specialist Howard Hughes Med. Inst., Salt Lake City, 1991—. Contbr. articles to profl. jours. Grad. rsch. fellow U. Utah, 1987-88. Mem. AAAS, Am. Assn. Anatomists, Soc. Neuroscience, Soc. Devel. Biology. Achievements include research in cell mixing in the spinal cord of mouse chimeras, the developmental defects of the ear, cranial nerves, hindbrain from a gene targeted disruption of HOX 1.6; discovered swaying is a mutant allele of WNT-1. Home: 625 E 9th Ave Salt Lake City UT 84103 Office: U Utah, H Hughes Med Inst Eccles Inst Human Genetics Bldg 533, Rm 5440 Salt Lake City UT 84112

MUSHA, TOSHIMITSU, physicist, educator; b. Tokyo, June 29, 1931; s. Genjirou and Masa Musha; m. Tazuko Iwano, Mar. 16, 1961; children: Yuniko, Mitsuru, Neriko. BS, U. Tokyo, 1954, DSc, 1964. Rsch. staff Elec. Communication Lab. NTT, Musashino, Japan, 1954-62, group leader, 1962-64; mem. staff rsch. lab. elec. MIT, Boston, 1964-65; rsch. staff Royal Inst. Tech., Stockholm, 1965-66; rsch. staff Tokyo lab. RCA, 1966; assoc. prof. Tokyo Inst. Tech., 1966-71, full prof. electronics, 1971-92; prof. sci. U. Tokyo, 1992—; pres. Brain Functions Lab., Inc., Kawasaki, Japan, 1992—. Author: Fluctuations in Nature, 1981; editor and chmn. proceedings; inventor in dipole tracing system for human brain; discovered biol. rhythm subject to 1/f fluctuations. Bd. dirs. Assn. Supportors for Machida City Orch., Tokyo, 1975—; chmn. Local Community Coun., Machida, 1986-88. Grantee Casio Found., 1984, NHK, 1985, Mazda Found., 1986, Nissan Found., 1989-91, New Energy Devel. Orgn., 1990—. Mem. Inst. Electronics Engrs. Japan (chmn. rsch. com. on electromagnetism 1976-79, chmn. com. phys. electronics 1980-82), Acad. Tech. Scis. of Russia. Avocations: travel, playing hand-made harpsichord. Home: 2-13-17 Minami-Tsukushino, Machida Tokyo 194, Japan

MUSIELAK, ZDZISLAW EDWARD, physicist, educator; b. Piwonice, Kalisz, Poland, Feb. 13, 1950; came to U.S., 1983; s. Edward and Irena (Bak) M.; m. Grazyna Roszkiewicz, Aug. 23, 1975; 1 child, Agnieszka Maria. MS in Physics, A. Mickiewicz U., 1976; PhD in Physics, U. Gdansk, 1980, hon. scholar, 1990. Cert. tchr., Poland, 1976. Teaching asst. A. Mickiewicz Univ., Poznan, Poland, 1976-77; rsch. asst. U. Gdansk, Poland, 1977-80, instr., 1980-82; vis. scientist U. Heidelberg, Fed. Republic Germany, 1982-83; rsch. staff MIT, Cambridge, 1983-86; NAS/NRC fellow NASA Marshall Space Flight Ctr., Huntsville, Ala., 1986-89; assoc. prof. U. Ala., Huntsville, 1989—; vis. scientist Harvard-Smithsonian Ctr. for Astrophy., Cambridge, 1985; cons. Univ. Chgo., Ill., 1988—; vis. prof. Univ. Heidelberg, 1990. Contbr. articles to Acta Astron., Astron. & Astrophys. Jour., Astrophys. Jour., Jour. Plasma Phys. Mem. Solidarity, Gdansk, 1980-83, deputy, 1980; mem. Nat. Geographic Soc., Washington, 1984. Recipient Sonnenforschungsbereich award U. Heidelberg, 1985, NAS/NRC award U.S. Nat. Acad. Scis., Huntsville, 1986, Rsch. award Smithsonian obs., Cambridge, 1987, NASA theory grant NASA Marshall Space Flight Ctr., Huntsville, 1989; named Rosat Guest Observer, NASA, German Space Agy., Washington, 1990; grantee NSF, 1991, 92. Mem. Am. Astron. Soc., Internat. Astron. Union (grants 1982, 88), Polish Astron. Soc. (travel grant 1979, 80). Achievements include research on theoretical models of stellar chromospheres, general and unique form of wave equation in inhomogeneous media, acoustic and MHD wave energy fluxes for main sequence stars, non-decaying MHD bending waves, stellar X-ray emissions, stellar wind acceleration. Office: U Ala EB-117H Huntsville AL 35899

MUSIHIN, KONSTANTIN K., electrical engineer; b. Harbin, China, June 17, 1927; s. Konstantin N. and Alexandra A. (Lapitsky) M.; m. Natalia Krilova, Oct. 18, 1964; 1 child, Nicholas; came to U.S., 1967, naturalized, 1973; student YMCA Inst., 1942, North Manchurian U., 1945, Harbin Poly. Inst., 1948. Registered profl. engr., Calif., Colo., N.Y., N.J., Pa., Ill., Wash. Asst. prof. Harbin Poly. Inst., 1950-53; elec. engr. Moinho Santista, Sao Paulo, Brazil, 1955-60; constrn. project mgr. Caterpillar-Brazil, Santo Amaro, 1960-61; mech. engr. Matarazzo Industries, Sao Paulo, 1961-62; chief of works Vidrobras, St. Gobain, Brazil, 1962-64; project engr. Brown Boveri, Sao Paulo, 1965-67; sr. engr. Kaiser Engrs., Oakland, Calif., 1967-73; sr. engr. Bechtel Power Corp., San Francisco, 1973-75; supr. power and control San Francisco Bay Area Rapid Transit, Oakland, 1976-78; chief elec. engr. L.K. Comstock Engring. Co., San Francisco, 1978-79; prin. engr. Morrison Knudsen Co., San Francisco, 1979-84; prin. engr. Brown and Caldwell, Cons. Engrs., Pleasant Hill, Calif., 1984-85; cons. engr. Pacific Gas and Electric Co., San Francisco, 1986-89; sr. engr. Bechtel Corp., San Francisco, 1989—. Mem. IEEE (sr.), Instrument Soc. Am. (sr.), Am. Mgmt. Assn., Nat., Calif. socs. profl. engrs., Nat. Assn. Corrosion Engrs., Instituto de Engenharia de Sao Paulo. Mem. Christian Orthodox Ch. Clubs: Am.-Brazilian, Brit.-Am. Home: 320 Park View Ter Apt 207 Oakland CA 94610-4653

MUSIKAS, CLAUDE, chemical researcher; b. Paris, May 1, 1937; s. Dimitros and Myrsina (Valana) M.; m. Anne-Marie Osouf, Sept. 24, 1964; children: Helene, Nicolas. Engr. in Chemistry, Conservatoire des Arts/Metiers, Paris, 1962; DSc, U. Orsay, France, 1977. Technician French Atomic Energy, Fontenay, 1955-57, engr., 1962-74, basic researcher, 1975-80, group leader basic rsch., 1980-91; rsch. dir. French Atomic Energy, Fotnenay, 1992; technician French Atomic Energy, Saclay, 1957-62; vis. scientist Oak Ridge (Tenn.) Nat. Lab., 1974-75; dir. doctorate thesis progs. various French univs., 1972-91; cons. in field. Author, editor: Principles and Practices of Solvent Extraction, 1992; contbr. articles to profl. jours.; patentee in field. Avocations: tennis, skiing, cross country walking, reading, theatre. Home: 5 Avenue Moissan, 91440 Bures/Yvette France Office: Cogema, 15 rue Paul Dautier, 78140 Velizy-Villacoublay France

MUSILLO, JOSEPH H., electrical engineer; b. Paterson, N.J., Nov. 23, 1949; s. Rocco and Mary (Zito) M.; m. Jennifer Mary Janet Pucci, July 10, 1976; children: Joy Margret Ann, Jenna Nancy Rose. BE in Elec. Engring., Stevens Inst. Tech., Hoboken, N.J., 1971. Cert. hazardous materials mgr.; registered environ. profl.; lic. first class engr., N.J. Asst. engr. oper. dept. Linden Generating Sta. Pub. Svc. Gas and Electric Co., Newark, 1971-72, maintenance supr., 1973-74, gas turbine supr. Linden, Essex and Bayonne Generating Stas., 1975-78, sta. performance engr. Kearny, Essex and Bayonne Generating, 1979-84, sta. oper. engr. Sewaren and Edison Generating Stas., 1985-86, supervising engr. prodn. dept. projects group, 1986-88, sr. staff engr. prodn. dept. environ. engrs. group, 1988-90, sr. staff engr. prodn. dept. Linden Generating Sta., 1990-92, sr. environ. engr., 1992—. Vice chmn. Environ. Com., Cranford, 1990-92; mem. Mayor's Refuse Com., Cranford, 1989-91, Clean Communities Com., Cranford. Mem. Performance Engrs. Assn. (chmn. 1982-84), Inst. Hazardous Materials Mgrs., Water Pollution Control Assn. Home: 8 Venetia Ave Cranford NJ 07016-1934 Office: Pub Svc Elec and Gas Co Linden Generating Sta 4001 Wood Ave S Linden NJ 07036-6591

MUSMANNI, SERGIO, chemist, researcher; b. San Jose, Costa Rica, May 13, 1960; came to U.S. 1988; s. Jorge and Maria Eugenia (Sobrado) M. B.Chemistry, Universidad de Costa Rica, 1987; PhD in Chemistry, U. Tex.-Dallas, Richardson, 1993. Rsch. asst. Electrochemistry Rsch. Ctr., San Jose, 1983-85, Natural Products Rsch. Ctr., San Jose, 1985-87; chemist Formulaciones Quimicas S.A., San Jose, 1987-88; teaching asst. U. Tex. at Dallas, Richardson, 1988-91, rsch. asst. chemistry, 1992—; chemist Shell Devel. Co., Houston, 1991-92. Contbr. articles to profl. jours. State of Tex. Pub. Edn. grantee, 1989. Mem. Am. Chem. Soc., Royal Soc. Chemistry (Eng.), Am. Inst. Chem. Engrs. Achievements include research in organic synthesis; heterogeneous catalysis; organic conducting polymers; degradable polymers; magnetic resonance imaging contrast agents; alternative sources of energy.

MUSSENDEN, GEORG ANTONIO, electronics engineer; b. San Juan, P.R., Aug. 25, 1959; s. Gustavo Adolfo and Christa-Maria (Gotsch) M. Student U. P.R.-Rio Piedras, 1977-79; B.S. in E.E. with honors, U. Fla.-Gainesville, 1982, postgrad. in elec. engring. Electronics technician Radiotelephone Communicators of P.R. (Motorola), 1976; computer systems programmer and operator U. P.R., Rio Piedras, 1978-79, research asst. dept. physics, 1978-79; computer programmer Regional Electrocardiogram Analysis Ctr., J. Hillis Miller Health Ctr., U. Fla., Gainesville, 1981; pre-profl. engr. IBM Corp. Devel. Lab., Endicott, N.Y., 1981, sr. assoc. engr./scientist entry systems tech. SPD, CPD and ESD design and devel. labs., Boca Raton, Fla., 1982—. Contbr. 7 articles to profl. jours. Scholar San Jose Alumni, 1973-77, Fonalledas Found., 1977-79, Procter and Gamble, 1980; U.F. Sr. Honors scholar, 1980; scholar Nat. Fund Minority Engring., 1980, Du Pont, 1981; Nat. Consortium for Grad. Degrees for Minorities in Engring. fellow, 1981. Mem. N.Y. Acad. Scis., IEEE, SMPTE, AES, Golden Key, Eta Kappa Nu, Tau Beta Pi, Phi Kappa Phi. Roman Catholic. Clubs: Audio-Visual, Amateur Radio. Achievements include 8 technological invention disclosures. Home: 5150 E Club Circle Apt 204 Boca Raton FL 33487-3757 Office: IBM Entry Systems Divsn 1000 NW 51 St Internal Zip 1300 Boca Raton FL 33431

MUSSENDEN, GERALD, psychologist; b. N.Y.C., June 1, 1941; s. Geraldo and Adele (Gimenez) M.; m. Iris Manuela Prado, Aug. 11, 1967; children: Gerald, Ricardo-Antonio, Gina. BA, Tarkio Coll., 1968; MS, Brigham Young U., 1971, PhD, 1974. Diplomate Am. Bd. Profl. Disability Cons. Dir. child program Albert Einstein Coll. Medicine, N.Y.C., 1974-76; psychologist Mental Health Ctr., Bartow, Fla., 1976-77, Norside Community Mentala Health Ctr., Tampa, Fla., 1977-80; pvt. practice Brandon (Fla.) Counseling Ctr., 1980—; criminal ct. psychologist Fla. Cts., Hillsborough, Fla., 1978—; with children's svcs. State Rehab., Hillsborough, 1977—; rehab. psychologist Vocat. Rehab., Hillsborough; psychologist Div. Blind Svcs., Hillsborough. Fellow Ford Found., 1972-73. Mem. APA, Fla. Psychol. Assn., Bay Area Psychol. Assn. Home: 317 Cactus Road Seffner FL 33584 Office: Brandon Counseling Ctr 134 N Moon Ave Brandon FL 33510-4420

MUSSENDEN, MARIA ELISABETH, psychologist, substance abuse counselor; b. San Juan, P.R., Oct. 4, 1949; came to U.S., 1979; d. Gustavo Adolfo and Christa-Maria (Gotsch) M. BA in Music magna cum laude, U. P.R., Rio Piedras, 1971, MA in Gen. Exptl. Psychology, 1978; MS in Counseling Psychology, U. Fla., 1989, postgrad. in counseling psychology, 1979-89. Cert. tchr., P.R. Counselor alcohol dependency treatment program VA Hosp., Gainesville, Fla., fall 1980, counselor cardiac rehab. unit, fall 1981; counseling psychology intern Psychol. & Vocat. Counseling U. Fla., Gainesville, 1982-83; alcohol specialist Alcothon House, Gainesville, 1983-84, counselor I, 1984—; rehab. therapist North Fla. Evaluation and Treatment Ctr., Gainesville, 1985, human svcs. counselor III, 1985-89; substance abuse counselor Vista Pavilion, Gainesville, 1987—; psychol. specialist Marion Correctional Instn., Lowell, Fla., 1989—; pvt. practice tchr., 1966-79; part-time lectr. Inter-Am. U., Hato Rey, P.R., 1971-78; part-time music tchr. Music Acad., 1972-78; part-time music therapist Centro Didactico del Lenguaje, Hato Rey, 1978-79; tchr. music P.R. Dept. of Edn., San Juan, 1974-79; presenter in field. Ford Found. fellow, 1972-73, 74-75, Angel Ramos Found. fellow, 1973-74, Fonalledas Found. fellow, 1982-83; recipient Presdl. award U. Fla., 1985. Mem. APA, Fla. Assn. for Behavior Analysis, Mental Health Assn., Nat. Alliance for the Mentally Ill, Puerto Rican Music

Therapy Assn. (co-founder), Mu Alpha Phi (hon. Gold medal 1967). Roman Catholic. Avocations: music, reading, sewing, cooking. Home: 125 SW 40th Ter Gainesville FL 32607-2754

MUSSHOFF, KARL ALBERT, radiation oncologist; b. Wuppertal, Germany, June 11, 1910; s. Gustav and Amalie (Kauls) M.; m. Margarethe Herbst, Aug. 25, 1951; children: Stephan, Renate. MD, U. Munich, 1936. Univ. lectr. medicine and radiology U. Freiburg, Germany, 1960; prof., 1966—, emeritus dir. sect. radiotherapy Radiol. Centre Albert-Ludwigs-U., Freiburg, 1977—. Recipient W.C. Röntgen Plakette of Remscheid-Lennep, Birth-town of Röntgen, 1983; Leopold Freund Medaille of Österreichische Gesellschaft für Radiologie, Radiobiologie, Medizinishe Radiophysik, 1985. Mem. Group European Radiotherapists (hon., pres. 1973), European Soc. Therapeutic Radiology and Oncology, German Cancer Soc. (hon.), German Radiation Soc., S.W. German Radiation Soc. (pres. 1972), German Soc. Internal Medicine, German Soc. Hematology and Oncology, Soc. German Physicists and Doctors, Am. Coll. Radiology (hon. fellow), Beirat der Deutschen Krebshilfe. Co-editor Jour. Cancer Rsch. Clin. Oncology, Strahlentherapie, Jour. Radiation Oncology Biol. Physics, Pneumonologie-Pneumonology, Kösener Senioren Convent. Home: 33 Eichenweg, 76 547 Sinzheim-Vormberg, Baden Germany

MUSSON, DONALD GEORGE, chemist; b. Pomona, Calif., June 7, 1950; s. Richard George and Ada Pearl (Hensely) M.; m. Jaqueline Joyce Roque, June 19, 1971; children: Wesley George, Stephanie Ann. BS, U. Calif., Riverside, 1972, MS, 1973; PhD, U. Calif., San Francisco, 1979. Postdoctoral fellow U. Kans., Lawrence, 1978-79; sr. rsch. chemist Merck Sharp & Dohme, West Point, Pa., 1979-84, rsch. fellow, 1984-88; group leader IDLAB Corp. div. of Johnson & Johnson, Claremont, Calif., 1988—. Contbr. articles to Biomed & Environ., Mass Spectroscopy, Jour. Med. Chemistry, Jour. of Chromatography, Pharm. Rsch. and others. Mem. Am. Chem. Soc., Am. Assn. Pharm. Scientists, Controlled Release Soc. Achievements include patents (with others) in aqueous ophthalmic microemulsion of tepoxalin, (with other)in stabilizing preparation for thymoxamine. Home: 2073 Redgrove Way Upland CA 91786-7972 Office: IOLAB Corp 500 W Iolab Dr Claremont CA 91711-4881

MUSTAKALLIO, KIMMO KALERVO, dermatologist; b. Helsinki, Finland, July 7, 1931; s. Martti Joeli and Heddi Manghild (Sjöberg) M.; m. Marita Nordberg, Dec. 21, 1956; 1 child, Sami. MD, U. Helsinki, 1956, hon. degree, 1966. Intern Helsinki U. Cen. Hosp., 1952-56, resident in dermatology, 1957-65, MD, 1966, assoc. prof. dermatology, 1966-68, prof., head dermatology dept., 1968—. Editor-in-chief: Annals of Clin. Rsch., 1969-75; contbr. over 200 articles to med. jours.; patentee in field. Pres. Finnish CP Assn., Helsinki, 1965-69, Finnish Med. Soc. Duodecim, Helsinki, 1978-81. Named Comdr. of the Order of Finnish Lion. Mem. European Soc. Dermatol. Rsch. (hon.), Finnish Acad. Scis., Dermatol. Socs: Finland (hon.), Germany (hon.), Israel (hon.), Poland (hon.), Denmark (hon.), France (corr.), Norway (corr.), and Sweden (corr.). Lutheran. Avocations: classical music, gastronomy. Office: Helsinki U Cen Hosp, Meilahdentie 2, 00250 Helsinki Finland

MUSTARD, JAMES FRASER, research institute executive; b. Toronto, Oct. 16, 1927; s. Allan Alexander and Jean Anne (Oldham) M.; m. Christine Elizabeth Sifton, June 4, 1952; children: Cameron, Ann, Jim, Duncan, John, Christine. MD, U. Toronto, 1953; PhD, Cambridge U., 1956. Asst., then assoc. prof pathology U. Toronto, 1963-66, asst. prof. medicine, 1965-66, hon. prof. pathology, 1990—; prof. pathology McMaster U., Hamilton, Ont., Can., 1966-88, prof. emeritus, 1988—, chmn. pathology, 1966-72, dean faculty health. scis., 1972-80, v.p. faculty health scis., 1980-82; bd. dirs., pres., sr. fellow Can. Inst. Advanced Rsch., Toronto, 1982—; mem. Ont. Premier's Coun. on Econ. Renewal, 1991—; mem. Ont. Coun. Health, 1966-72; founder, chmn. Health Rsch. Com., 1966-73 ; chmn. Spl. Task Force Future Arrangements Health Edn., 1970-71, Task Force Health Planning for Ont., MInistry Health and Govt. Ont., 1973-74; mem. Ont. Coun. Univ. Affairs, 1975-81; chmn. Adv. Coun. Occupational Health and Safety, 1977-83; mem. Royal Commn. Matters Health and Safety Arising Use of Asbestos in Ont., 1980-83; mem. Bovey Commn. Study Future Devel. Univ. in Ont., 1984-85; mem. Premier's Coun. Ont., 1986-91, Premier's Coun. Health Strategy, 1988-91; chmn. ctrs. excellence com. Govt. Ont., 1987-91; mem. Prime Minister's Nat. Adv. Bd. on Sci. and Tech., Ottawa, vice chmn., 1988-91; mem. adv. com. Networks of Ctrs. of Excellence Govt. Can., 1988—; rsch. and analysis adv. com. Stats. Can., Ottawa, 1986—; nat. adv. coun. Can. Adv. Tech. Assn., Ottawa, 1988—; adv. coun. Ctr. Health Econs. and Policy Analysis, McMaster U., 1990—; chmn. bd. dirs. Ont. Workers' Compensation Inst., 1990—; mem. advisory bd. Man. Ctr. for Health Policy and Evaluation, 1991—; pres. Assn. Can. Med. Colls., 1975-76; bd. dirs. Steel Co. Can., Hamilton, 1985, Atomic Energy Can. Ltd., 1990. Bd. dirs. Heart and Stroke Found. Can., 1971-82, Heart and Stroke Found. Ont., 1982-87; bd. govs. McMaster U., 1978-82; bd. dirs. McMaster U. Med. Centre, 1972-82; trustee Advanced Systems Inst. Found., Vancouver, 1986—, Aga Khan U., Karachi, Pakistan, 1985—; mem. chancellor's commn. Aga Khan U., 1992—. Decorated Officer Order of Canada, 1986; recipient Disting. Svc. award Can. Soc. Clin. Investigation, J. Allyn Taylor Internat. Prize, 1988, Internat. award Gairdner Found., 1967, James F. Mitchell award, 1972, Izaak Walton Killam prize in health sci. Can. Council, 1987, Disting. Career award for contbns. Internat. Soc. Thrombosis and Haemostasis, 1989, Pvt. Sector Leadership award Can. Advanced Tech. Assn., 1989, R & D Mgmt. award Can. Rsch. Mgmt. and Assn., 1989, Man of Yr. award Can. High Tech. Coun., 1989, Xerox Can.-Forum award Corp. Higher Edn. Forum, 1990. Fellow Royal Coll. Physicians Can., Royal Soc. Can., Internat. Soc. Thrombosis and Haemostasis (pres. 1965-66); mem. Am. Soc. Clin. Investigation, Assn. Am. Physicians, Am. Assn. Pathologists, Am. Soc. Hematology (pres 1970), Can. Atherosclerosis Soc. (councillor exec. com. 1986—), Can. Soc. Clin. Investigation (pres. 1965-66), Internat. Congress on Thrombosis and Haemostasis (Robert P. Grant award for contbns. to progress 1987). Home: 422 Sumach St, Toronto, ON Canada M4X 1B5 Office: Can Inst Advanced Rsch, 179 John St Ste 701, Toronto, ON Canada M5T 1X4

MUSTARD, JOHN FRASER, geologist, educator; b. Toronto, Ont., Can., Mar. 13, 1961; s. James Fraser and Christine Elizabeth (Sifton) M. BSc, U. B.C., Vancouver, Can., 1983; MS, Brown U., 1986, PhD, 1990. Postdoctoral rsch. asst. Brown U., Providence, 1990-91, asst. prof. geology, 1991—; panelist Planetary Astronomy Rev. Panel, Washington,1 992; presenter at profl. confs. Assoc. editor Procs. Lunar and Planetary Sci. Conf., Houston, 1990-92; contbr. articles to refereed jours. Fellow Sigma Xi; mem. Am. Geophys. Union. Achievements include participation in experiment for proposed Lunar Scout mission by NASA. Home: 71 Edgehill Rd Providence RI 02906 Office: Brown Univ Dept Geol Sci PO Box 1846 Providence RI 02912

MUTCH, JAMES DONALD, pharmaceutical executive; b. Portland, Oreg., Mar. 6, 1943; s. Keith William and Dorothy (Wones) M.; m. Judith Ann Thompson, June 12, 1965; children: William James, Alicia Kathleen. BS in Pharmacy, Oreg. State U., 1966. Registered pharmacist, Calif., Oreg.; cert. regulatory affairs profl. Mgr. regulatory affairs Syntex Labs., Palo Alto, Calif., 1970-72, assoc. dir. regulatory affairs, 1972-76, dir. regulatory affairs, 1976-80; dir. regulatory affairs and clin. devel. Cooper Vision, Inc., Mt. View, Calif., 1980-86; dir. regulatory affairs and pre-clin. devel. Salutar, Inc., Sunnyvale, Calif., 1987-89, v.p. product devel., 1990-91; pres. Altos Biopharm. Inc., Los Altos, 1991-92; v.p. regulatory affairs and product devel. Pharmacyclics, Inc., Mountain View, Calif., 1992—. Pres. bd. Woodland Vista Swim & Racquet Club, Los Altos, Calif., 1982-83. With USPHS, 1966-68. Mem. AAAS, Am. Pharm. Assn., Regulatory Affairs Profl. Soc. Democrat. Achievements include co-development of Naprosyn, Lidex, Polycon, CSI, Clerz, Clerz-2. Office: Pharmacyclics Inc 265 N Whisman Rd Mountain View CA 94043

MUTCHLER, CALVIN KENDAL, hydraulic research engineer; b. Oceola, Ohio, Jan. 25, 1926; s. Jesse Cleveland and Bessie Dell (Cox) M.; m. Margaret Norens Hopper, June 9, 1951; children: Sue, Beth, Jan, Leigh, Don, Kathryn. B in Agrl. Engring., Ohio State U., 1951, MS, 1952; PhD, U. Minn., St. Paul, 1970. Registered profl. engr. Instr. dept. agrl. engring. U. Conn., Storrs, 1952-53; tech. mgr. B. F. Goodrich Co., Akron, Ohio, 1953-54; agrl. engr. Agrl. Rsch. Svc. USDA, State College, Miss., 1954-58, St. Paul, 1958-60, Morris, Minn., 1960-72; hydraulic engr. Agrl. Rsch. Svc., Nat. Sedimentation Lab. USDA, Oxford, Miss., 1972—. Contbr. articles to profl. jours. Col. USAR, 1944-81, ETO. Mem. Am. Soc. Agrl. Engrs., Soil and Water Conservation Soc. Achievements include research in soil erosion and sediment yield. Home: 216 Carol Ln Oxford MS 38655-3404 Office: USDA Box 1157 Nat Sedimentation Lab Oxford MS 38655

MUTZIGER, JOHN CHARLES, physician; b. Natchez, Miss., June 23, 1949; s. Dudley Henson and Marie Louise (Eyrich) M.; m. Sarah Edwards Meyer, Nov. 17, 1973 (div. Apr. 1979); m. Janis Merle Averyt, Sept. 13, 1980; children: John Charles Jr., William Westley. BS in Anthropology with honors, Tulane U., 1971; DO, U. Health and Sci., 1982. Owner Family Med. Clinic of Philadelphia, Miss., 1983-90, Family Med. Clinic of Decatur, Miss., 1990-91, Poplar Springs Family Med. Clinic, 1992—; emergency room dir. Laird Hosp., Union, 1990-92. Mem. Rotary, Phila., 1983-89; bd. dirs. Meridian Symphony Orch., 1992. Mem. AMA (Phys. recognition award 1991), Am. Coll. Coll. Practitioners, Am. Acad. Family Practitioners, Am. Osteo. Assn., Am. Osteo. Acad. Addictionology (charter mem. 1986—), Miss. State Med. Assn., Miss. Osteo. Med. Assn. (pres. 1986-87, program chmn. ann. coast conv. 1986-93), So. Med. Assn. Republican. Presbyterian. Avocations: golf, tennis, sailing, gourmet cooking, bridge. Home: 5503 13th Pl Meridian MS 39305-1446 Office: Poplar Springs Family Med Ctr 4707 Poplar Springs Dr Meridian MS 39305

MYERS, ANDREW GORDON, chemistry educator; b. Fort Bragg, N.C., Aug. 14, 1959. SB in Chemistry, MIT, 1981; PhD in Chemistry, Harvard U., 1985. Asst. prof. chemistry Calif. Inst. Tech., Pasadena, 1986-91, assoc. prof. chemistry, 1991—. Postdoctoral Rsch. fellow Harvard U., 1985-86; Arthur C. Cope Scholar award, Am. Chem. Soc., 1993. Office: Cal Inst Tech Dept of Chemistry 1201 E California Blvd Pasadena CA 91125*

MYERS, DALE DEHAVEN, government, industry, aeronautics and space agency administrator; b. Kansas City, Mo., Jan. 8, 1922; s. Wilson and Ruth (Hall) M.; m. Marjorie Williams, Sept. 18, 1943; children—Janet Louise Myers Westling, Barbara Toby Myers Curtis. Student, Kansas City Jr. Coll., 1939-40; B.S. in Aero. Engring, U. Wash., Seattle, 1943; Ph.D. (hon.), Whitworth Coll., 1970. Chief engr. missile devel. div. N. Am. Aviation, 1946-57, v.p., weapons systems mgr., 1957-63; asst. div. dir. advanced systems Rockwell Internat. Corp., El Segundo, Calif., 1963-64; v.p., program mgr. Rockwell Internat. Corp. (Apollo CSM programs), 1964-69, Rockwell Internat. Corp. (space shuttle program), 1969-70; asso. adminstr. manned space flight NASA, 1970-74; pres. N.Am. Aircraft ops. Rockwell Internat. Corp., corporate v.p.; Rockwell Internat., 1974-77; under-sec. Dept. Energy, Washington, 1977-79; pres., chief oper. officer Jacobs Engring. Group Inc., Pasadena, Calif., 1979-84; pres. Dale D. Myers & Assoc. Cons. Aerospace & Energy, 1984-86; dep. adminstr. NASA, Washington, 1986—; mem. adv. com. NASA, Washington, 1984-86; responsible for Apollo Command and Svc. Module constrn. and launch, 1960's, for overall Apollo program Apollo 13 through 17, for complete Skylab program, for concepts and initiation of the Shuttle program. Contbr. articles to profl. jours. Dist. mgr. United Way, Greater L.A., 1983; U.S. rep. 22d conf. Internat. Atomic Energy Agy., Geneva, 1979; bd. dirs. Internat. Aero. Hall of Fame, San Diego, 1989. Recipient Meritorious Service award Compton (Calif.) Schs., 1977, Achievement award Los Angeles City Schs., 1976, Public Service award, 1969, Disting. Service medal Dept. Energy, 1979, Von Karman award Mus. of Sci. and Industry, L.A., 1987. Fellow AIAA (nat. dir.), Am. Astronautics Soc.; mem. Nat. Acad. Engring., Newcomen Soc. in N.Am., Calif. C. of C. (dir.), Calif. Roundtable, Sigma Alpha Epsilon. Presbyterian (elder). Clubs: Century Flying, Calif., (Los Angeles). Achievements include rsch. in and development of aerospace vehicles and mgmt. of large high tech. projects. Office: NASA 400 Maryland Ave SW Washington DC 20546-0001

MYERS, DANIEL LEE, manufacturing engineer; b. South Bend, Ind., Nov. 15, 1961; s. Harold Lee and Gerda Martha (Schulz) M.; m. Vanessa Rea Burkhart, Aug. 27, 1986; children: Wayne Anthony, Amberli Rea. BS in Indsl. Engring., Purdue U., 1984. Indsl. engr. Anderson-Bolling Mfg. Co., Goshen, Ind., 1985; mfg. engr. Johnson Controls Inc., Goshen, 1985-88, Siemens Energy and Automation Inc., Bellefontaine, Ohio, 1988—. Mem. NSPE, Inst. Indsl. Engrs., Ind. Soc. Profl. Engrs., Toastmasters. Republican. Home: 3384 Bell Rd Huntsville OH 43324 Office: Siemens Energy & Automation 811 N Main St Bellefontaine OH 43311-2362

MYERS, DARYL RONALD, metrology engineer; b. Denver, July 12, 1948; s. James Elmer Myers and Betty Mae (Gannon) Welborn; m. Donna Lee Olsen, Oct. 3, 1990. BS in Applied Math., U. Colo., 1970, postgrad., 1974-75. Staff physicist Smithsonian Radiation Biology Lab., Rockville, Md., 1974-78; metrology engr. Solar Energy Rsch. Inst. (now Nat. Renewable Energy Lab.), Golden, Colo., 1978—. Contbr. articles to Solar Energy, Solar Cells, profl. publs. Russian and German interpreter, Cultural Diversity Com., Arvada, Colo., 1992. With U.S. Army, 1970-74. Mem. ASTM, Precision Measurement Assn. (John Quincy Adams award 1979-80), Am. Solar Energy Soc., Coun. Optical Radiation Measurement, Math. Assn. Am. Democrat. Achievements include contribution of significant algorithms for interpolation and corrections to measured precipitable water, aerosol optical depth, and radiometric data used in developing national solar radiation data base. Office: Nat Renewable Energy Lab 1617 Cole Blvd Golden CO 80401

MYERS, DAVID FRANCIS, chemical engineer; b. Flemington, N.J., June 11, 1959; s. Francis Stanley and Elizabeth Anne (Vahle) M.; m. Christina Joyce Hodgin, Aug. 7, 1982; children: Andrew, Sophie, Brett. SB, SM, MIT, 1982; PhD, Princeton U., 1988. Mgr. bldg. products rsch. W.R. Grace & Co., Columbia, Md., 1987—. Mem. Am. Ceramic Soc. (Brunauer award 1992). Achievements include 6 patents in field of specialty chemicals (specifically in area of cement additives).

MYERS, DONALD RICHARD, biomedical engineer; b. Lancaster, Pa., June 12, 1951; s. Richard Cooper and Faye Ellen (Longnecker) M. BA in Biology, Swarthmore Coll., 1973; BS in Bioengring. summa cum laude, Temple U., 1976; MS, Drexel U., 1978, PhD, 1980. Instr. dept. biology Temple U., Phila., 1974-75; rsch. assoc. rehab. engring. ctr. Einstein Med. Ctr., Phila., 1976-80, cons., 1980-82; NRD postdoctoral rsch. fellow Nat. Bur. Standards, 1981-82; project leader robot systems div. Nat. Inst. Standards & Tech., Gaithersburg, Md., 1983-86; program mgr. corp. rsch. ctr. Lord Corp., Cary, N.C., 1986-89; sr. rsch. assoc. Triangle R&D Corp., Research Triangle Park, N.C., 1989—; adj. assoc. prof. dept. elec. engring. Duke U., Durham, N.C., 1988-91. Contbr. chpt. to International Encyclopedia of Robotics: Applications and Automation, 1988; contbr. articles to Jour. Dynamic Systems, Measurement & Control, Internat. Jour. Robotics and Computer-Integrated Manufacturing, Jour. Robotic Systems, Internat. Jour. Robotics Rsch., Indsl. Robot, IEEE Transactions on Systems, Man & Cybernetics. NIH grantee, 1976. Republican. Lutheran. Home: 414 Glen Bonnie Ln Cary NC 27511

MYERS, EDWARD E., retired biology and anthropology educator; b. Santa Cruz, Calif., Jan. 22, 1925; s. Earl S. Myers and Alice Reidel Schmidt; m. Pauline Sine Cleven, Aug. 26, 1945; children: Earl Tobias, Susan E. BA, Adelphi Coll., 1948; MA, Calif. State U., L.A., 1957. Cert. sec./elem. tchr. Tchr. Pomona Elem. Sch., 1949-54; tchr. sci. and math. Pomona (Calif.) City Sch. Dist., 1954-59; prof. Chaffey C.C., Alta Loma, Calif., 1949-86. Nature trail guide L.A. County Parks. With USN, 1943-45. NSF grantee, 1957, 58, 59, others. Mem. Nat. Acad. Sci., Nat. Assn. Phys. Anthropologists, Phi Delta Kappa. Democrat. Home: 10210 Baseline Rd D229 Alta Loma CA 91701

MYERS, ERIC ARTHUR, physicist; b. Salem, Oreg., Jan. 5, 1958; s. Ray Arthur and Delores Ellen (Miller) M. BA in Math. and Physics, Pomona Coll., 1980; MPhil in Physics, Yale U., 1983, PhD in Physics, 1984. Rsch. assoc. physics dept. Brookhaven Nat. Lab., L.I., N.Y., 1984-86; postdoctoral fellow math. dept. Dalhousie U., Halifax, N.S., 1986-89; vis. rsch. assoc. physics dept. Boston U., 1987-89; postdoctoral fellow Ctr. for Relativity, U. Tex., Austin, 1989-92; asst. prof. physics dept. sci. and math. Pks. Coll. St. Louis U., Cahokia, Ill., 1992-93; vis. prof. of physics Vassar Coll., Poughkeepsie, N.Y. Mem. Am. Phys. Soc. Achievements include demonstration that intersecting cosmic strings will intercommute; computation of gravitational Casimir energy in Kaluza-Klein theory.

MYERS, GREGORY EDWIN, aerospace engineer; b. Harrisburg, Pa., Jan. 1, 1960; s. Bernard Eugene and Joyce (Calhoun) M.; m. Susan Ann Hayslett, Dec. 30, 1983; children: Kimberly, Benjamin. BS in Aerospace Engring., U. Mich., 1981; MS in Aerospace Engring., Air Force Inst. Tech., 1982. Aerospace engr. Sperry Comml. Flight Systems group Honeywell, Inc., Phoenix, 1987-90; sr. project engr. satellite systems ops. Honeywell, Inc., Glendale, Ariz., 1990-92; sr. project engr. air transport systems Honeywell, Inc., Phoenix, 1992-93, prin. engr., 1993—; presenter in field. Contbr. articles to profl. jours. Mem. Aviation Week Rsch. Adv. Panel, 1990-91. Recipient Certs. of Recognition and Appreciation Lompoc Valley Festival Assn., Inc., 1983, Arnold Air Soc. (comdr. 1979), Cert. of Appreciation Instrument Soc. Am., 1991. Mem. AIAA (sr.). Lutheran. Avocations: softball, tennis, reading, computer programming. Office: Honeywell Air Transport Systems Divsn 21111 N 19th Ave Phoenix AZ 85036

MYERS, JERRY ALAN, computer engineer; b. Dayton, Ohio, May 11, 1958; s. Gleason Harold and Loretta P. (Phillips) M. BS in Computer Engring., Wright State U., 1983, MS in Computer Engring., 1993. Electronics engr. USAF, Wright Patterson AFB, Ohio, 1983-84; firmware engr. Digital Tech., Inc., Dayton, Ohio, 1984-85; computer engr. Comml. Flight Systems Div. Honeywell, Phoenix, 1985-88; computer engr. USAF, Wright Patterson AFB, 1989—. Office: USAF WL/AAA-2 Bldg 146 Rm 122 Wright Patterson AFB OH 45433

MYERS, KENNETH ALAN, air force officer, aerospace engineer; b. Waynesboro, Pa., Sept. 11, 1942; s. Richard Dudley and Esta (Decker) M.; m. Leanor Ann Littlefield, Sept. 25, 1965; children: Kenneth Jr., Lisa Louise, Leanna Lynne, Keith Austin, Kevin Andrew. BS in Aerospace Engring., Pa. State U., 1964; MS in Astronautics, Air Force Inst. Tech., 1967; Phd in Aerospace Engring., U. Tex., Austin, 1974; MA in Bus. Mgmt., U. Nebr., 1980; MPS in Polit. Sci., Auburn U., 1986. Commd. 2d lt. USAF, 1964, advanced through grades to col., 1985; astronautical engr. Def. Systems Applications Program, L.A. AFB, 1967-71; program mgr. Global Positioning System Tech., Wright-Patterson AFB, Ohio, 1974-77; chief systems engring. and analysis Space Ops. Group, Offutt AFB, Nebr., 1977-80; dep. systems and tech. Sec. of the Air Force/Space Systems, Pentagon, Va., 1980-85; dir. flight systems engring. Air Force Systems Command, Wright-Patterson AFB, Ohio, 1986-87; dep. dir. space ops. U.S. Space Command, Peterson AFB, Colo., 1987-90; faculty advisor Indsl. Coll. of Armed Forces, Nat. Def. U., Washington, 1990-91; dep. dir. nat. systems Def. Intelligence Agy., Arlington, Va., 1991-92; cons. Booz Allen & Hamilton, Inc. for Advanced Rsch. Projects Agy., War Breaker Program, 1992-93; cons. Nat. Systems Simulation and Analysis Program, 1993—; panel chmn. U.S. Space Command/Industry Space Ops. Workshop, Peterson AFB, 1988. Contbr. articles to Airpower jour. (Ira C. Eaker award 1988), Decision Scis., AIAA jour., Jour. of Spacecraft and Rockets, IEEE Transactions on Aerospace and Electronic Systems. Asst. cubmaster Boy Scouts Am., Colorado Springs, Colo., 1988-90, asst. scoutmaster, Fairfax, Va., 1993—; usher coord. and chmn. lay ministry bd. Lord of Life Luth. Ch., Fairfax, 1990-93. Mem. AIAA, Am. Astron. Soc., The Planetary Soc., Nat. Space Soc., Air Force Assn., Sigma Gamma Tau. Home: 6067 Burnside Landing Dr Burke VA 22015 Office: Booz Allen & Hamilton Inc 8283 Greensboro Dr Mc Lean VA 22102

MYERS, LAWRENCE STANLEY, JR., radiation biologist; b. Memphis, Apr. 29, 1919; s. Lawrence Stanley and Jane (May) M.; m. Janet Vanderwalker, June 13, 1942; children: David Lee, Frederick Lawrence, Lee Scott. BS, U. Chgo., 1941, PhD, 1949. Jr. chemist Metall. Lab. of Manhattan Engring. Dist., U. Chgo., 1942-44; asst. chemist Clinton Labs. of Manhattan Engring. Dist., Oak Ridge, Tenn., 1944-46; chemist Inst. for Nuclear Studies, U. Chgo., 1947-48; assoc. chemist Argonne (Ill.) Nat. Lab., 1948-52; asst. prof. radiology UCLA, 1953-70, assoc. rsch. phys. chemist Atomic Energy project, 1952-59, lectr. in radiol. scis., 1970-76, adj. prof. radiol. scis., 1976-82; rsch. radiobiologist, chief radiobiology div. UCLA Lab. Nuclear Medicine and Radiation Biology, 1959-76; prof. radiology and nuclear medicine Uniformed Svcs. Univ. of Health Scis., 1982-88; sci. advisor Armed Forces Radiobiology Rsch. Inst., 1982-87; cons. Oak Ridge Assoc. Univs., 1987—; vis. scientist AFRRI, 1987—; co-organizer UCLA Internat. Conf. on Radiation Biology, 1957, 59; participant in three major Fed. Govt. planning exercises related to energy rsch. and devel. in U.S., 1973-74; mem. adv. com. Ctr. for Fast Kinetic Rsch. U. Tex., Austin, 1975-81, chmn., 1977-81; mem. adv. bd. Radiation Chemistry Data Ctr., U. Notre Dame, 1976-84, sec. 1979-81, chmn. 1981-83; chmn. Long Range Planning Com., Radiation Rsch. Soc., 1976-78; dir. Issues and Requirements Workshop for Analysis of the 1976 "Inventory of Fed. Energy Related Environ. and Safety Rsch.", 1977. Contbr. more than 100 sci. articles and abstracts to profl. jours. Com. mem. Boy Scouts of Am., Pacific Palisades and Malibu, Calif., 1956-67. Fellow AAAS; mem. Radiation Rsch. Soc., Biophys. Soc., N.Y. Acad. Sci., Am. Inst. Biol. Scis., Am. Soc. for Photobiology, Soc. for Free Radical Rsch., European Soc. for Photobiology, Sigma Xi. Home: 11810 Coldstream Dr Rockville MD 20854-3612 Office: Armed Forces Radiobiology Rsch Inst Bethesda MD 20889-5603

MYERS, MARCUS NORVILLE, research educator; b. Boise, Idaho, May 30, 1928; s. Marcus Lemuel and Lucille (Sabin) M.; m. Mary Ann Storrs, Aug. 4, 1950; children: Mary Ann, Mark Art, Dora Jean. BS, Brigham Young U., 1950, MS, 1952; PhD, U. Utah, 1965. Engr. Hanford Works, GE, Richland, Wash., 1951-57; sr. chemist ANP, GE, Idaho Falls, 1957-61; sr. scientist Valecios Lab., GE, Pleasanton, Calif., 1961-62; rsch. assoc. Chemistry Dept. U. Utah, Salt Lake City, 1965-67, assoc. rsch. prof., 1967—; cons., adv. bd. FFFractionation, Inc., Salt Lake City, 1986—. Contbr. numerous papers for profl. publs. Mem. Am. Chem. Soc., Sigma Xi. Mem. LDS. Achievements include patents for field-flow factionation. Office: Chemistry Dept U Utah Salt Lake City UT 84112

MYERS, RONALD EUGENE, research chemist; b. Hanover, Pa., Aug. 12, 1947; s. Ivan Elmer and Betty Jane (Gibbons) M.; m. Ewha Chun, June 18, 1972; children: Michele, Jennifer. BA in Chemistry, Gettysburg (Pa.) Coll., 1969, PhD in Inorganic Chemistry, Purdue U., 1977. Advanced R & D chemist B.F. Goodrich Co., Brecksville, Ohio, 1977-80, sr. R & D chemist, 1980-83, R & D assoc., 1983-88, sr. R & D assoc., 1988—; vis. scholar Ohio Acad. Sci., Brecksville, 1990-92. Instr. sci. Strongsville (Ohio) Assn. for Gifted and Talented Students, 1987—. Sgt. U.S. Army, 1970-72, Korea. Fellow Am. Inst. Chemists; mem. AAAS, Am. Chem. Soc., Am. Ceramic Soc. (presdl. com. on pre-coll. edn. 1990—), Phi Lambda Upsilon, Sigma Xi. Achievements include 10 U.S. and foreign patents; inventor electrically conducting polymers, flame retardants, preceramic polymers; current research and development in area of high temperature polymers and composites. Office: BF Goodrich Co 9921 Brecksville Rd Cleveland OH 44141-3289

MYERSON, ALBERT LEON, physical chemist; b. N.Y.C., Nov. 14, 1919; s. Myer and Dora (Weiner) M.; m. Arline Harriet Rosenfield, May 10, 1953; children: Aimee Lenore, Lorraine Patrice, Paul Andrew. BS, Pa. State U., 1941; postgrad., Columbia U., 1942-45; PhD, U. Wis., 1948. Rsch. asst. Manhattan Project Columbia U., N.Y.C., 1941-45; sr. rsch. chemist Franklin Inst. Labs., Phila., 1948-56; mgr. phys. chemistry Gen. Electric Co., Phila., 1956-60; prin. phys. chemist Aero. Lab. Cornell U., Buffalo, 1960-68; rsch. assoc. Exxon Rsch. and Engring. Co., Linden, N.J., 1969-79; head phys. chemistry sect. Mote Marine Lab., Sarasota, Fla., 1979-85; sr. scientist Princeton (N.J.) Sci. Enterprises, Inc., 1985—; cons. in field. Violinist in area symphonies; co-editor: Physical Chemistry in Aerodynamics and Space Flight, 1961; contbr. articles in field to profl. jours.; patentee in field. Mem. Am. Chem. Soc., Am. Phys. Soc., Combustion Inst., Pa. State U. Alumni Assn., Sigma Xi, Phi Lambda Upsilon. Home and Office: 4147 Rosas Ave Sarasota FL 34233-1614

MYERSON, ALLAN STUART, chemical engineering educator, university dean; b. Bklyn., Nov. 17, 1952; s. Jules Myerson and Tilda (Rogalsky) Herman; m. Nancy Winget, June 15, 1979; 1 child, Megan. BS, Columbia U., 1973; MS, U. Va., 1975, PhD, 1977. Asst. prof. chem. engring. U. Dayton, Ohio, 1977-79; asst. prof. chem. engring. Ga. Inst. Tech., Atlanta, 1979-83, assoc. prof., 1983-85; assoc. prof. Polytech. U., Bklyn., 1985-88, prof., 1988-90, Joseph J. and Violet J. Jacobs prof., 1990—, head dept., 1985-92, dean Sch. Chem. and Materials Sci., 1992—; cons. E.I Du Pont de Nemours & Co., Wilmington, Del., 1988—, Ajinomoto, Inc., Tokyo, 1990—, Molecular Simulations, Cambridge, Eng., 1992—. Editor: Crystallization as a Separation Process, 1990, Handbook of Industrial Crystallization, 1992; also over 60 articles. Mem. AICE, Am. Chem. Soc., Sigma Xi (faculty disting. rsch. award Poly. U. chpt. 1992), Tau Beta Pi. Achievements include patents on Purification of Terephthalic Acid by Supercritical Fluid Extraction, (with W. Ernst) Removal of Inorganic Contaminants from Catalysts and Regeneration of HDS Catalysts; research on area of crystallization from solution, metastable solution structure and impurity crystal interactions. Office: Poly U Six Metrotech Ctr 333 Jay St Brooklyn NY 11201

MYKYTIUK, ALEX P., chemist. Fellow Chem. Inst. Can. (Norman and Marion Bright Meml. award 1991). Home: 1592 Meadowfield Pl, Gloucester, ON Canada K1C 5V6*

MYLAR, J(AMES) LEWIS, psychologist, consultant; b. Kansas City, Mo., Feb. 15, 1940; s. James Owern and Martha Jeanne (James) M.; m. Carol Sue Rinard, Feb. 22, 1983; children: Mark Jeffrey, Michelle Anne, Doug, Mara. MDiv, Fuller Theol. Sem., 1967; PhD, Fuller Grad. Sch. Psychology, 1970. Lic. pasychologist, Colo., Tex.; marriage, family counselor, Calif. Chief psychologist Pike's Peak Mental Health Ctr., Colo. Springs, Colo., 1970-72; prof. U. Colo., Colo. Springs, 1972-74; psychologist private practice, Colo. Springs, 1972-92, private practice Colo. Ctr. for Psychology, Colo. Springs, 1980-92, private practice, San Antonio, Tex., 1991—; chmn. Colo. Bd. Psychologist Examiners, Denver, 1974-81. Mem. Am. Psychol. Assn., El Paso Psychol. Soc. (Annual award 1975). Home: 29327 Summit Ridge Dr Fair Oaks Ranch TX 78006 Office: 7979 Broadway San Antonio TX 78209

MYLES, KEVIN MICHAEL, metallurgical engineer; b. Chgo., July 18, 1934; s. Michael J. and Ursula (May) M.; m. Joan Christine Ganczewski, Dec. 16, 1967; children: Kathleen, Gary, Jennifer. BS in Metallurgical Engring., U. Ill., 1956, PhD in Metallurgical Engring., 1963. Asst. mgr. nuclear fuel repossing program Argonne (Ill.) Nat. Lab., 1977-79, dep. dir. fossil energy program, 1982-87, mgr. fuel cell program, 1987-88, mgr. electrochemical tech. program, 1988—, assoc. dir. chem. tech. div., 1992—; adj. prof. materials sci. U. Ill., Chgo., 1968-70; prof. materials sci. Midwest Coll. Engring., Lombard, Ill., 1970-82. Contbr. articles to Jour. Phys. Chemistry, Chem. Engring. Sci., Jour. Electrochemical Soc., Jour. Fusion Energy, Jour. Power Sources. Mem. Sch. Bd. Dist. #58, Downers Grove, Ill., 1964-70. Capt. USAR, 1956-68, Korea. Mem. Am. Soc. for Metals, Am. Inst. Metallurgical Engrs., Alpha Sigma Mu. Achievements include 8 patents in field. Office: Argonne Nat Lab 9700 S Cass Ave Argonne IL 60439

MYLONAKIS, STAMATIOS GREGORY, chemist; b. Athens, Greece, Aug. 18, 1937; came to U.S., 1963; s. Gregory and Vassiliki (Charalampopoulos) M.; m. Pamela H. Morton, May 15, 1965 (dec. Mar. 1978); 1 son., Gregory John (dec. June 1992). BS in Chemistry, U. Athens, 1961; MS in Phys. Organic Chemistry, Ill. Inst. Tech., 1964; PhD in Phys. Organic Chemistry, Mich. State U., 1971. Research scientist Brookhaven Nat. Lab., Upton, N.Y., 1965-68; instr. U. Calif., Berkeley, 1971-73; group leader Rohm and Haas Co., Springhouse, Pa., 1973-76; supr. DeSoto Inc., Des Plaines, Ill., 1976-79; staff scientist Borg-Warner Chems., Inc. Des Plaines, 1979-81, research and devel. mgr., 1981-87, dept. head EniChem Ams. Inc., Princeton, N.J., 1988— . Author numerous research papers; assoc. editor Jour. Applied Polymer Sci.; patentee in polymer synthesis and applications fields. Mem. tech. adv. bd. Case Western Res. U.; mem. PhD thesis adv. com. Lehigh U., mem. adv. bd. NSF Ctr. Polymer Interfaces; mem. rev. panel NSF. Served as lt., Greek Army, 1961-63. Ill. Inst. Tech. fellow, 1963-64; Mich. State U. fellow, 1968-71. Mem. Am. Chem. Soc., N.Y. Acad. Scis., Sigma Xi. Office: EniChem Ams Inc 2000 Princeton Pk Monmouth Junction NJ 08852

MYODA, TOSHIO TIMOTHY, microbiologist, consultant, educator; b. Mukden, China, Mar. 17, 1929; arrived in Japan, 1946; came to U.S., 1956; s. Kihach and Hisako (Kanda) M.; m. Lois Johnson, June 9, 1963; children: Samuel Peter, Paul Timothy. BS, Hokkaido Coll., Sapporo, Japan, 1949; MS, Hokkaido U., Sapporo, Japan, 1952, PhD, 1954; PhD, Iowa State U., 1959. Instr. Hokkaido U., Sapporo, 1954-59; rsch. asst. Iowa State U., Ames, 1956-59; rsch. fellow Nat. Rsch. Coun. Can., Saskatoon, 1959-60, Western Res. U., Cleve., 1960-64; sr. rsch. fellow Inst. Microbial Chemistry, Tokyo, 1964-66; instr., rsch. assoc. La Rabida-U. Chgo. Inst., Chgo., 1966-67; various positions Alfred I. DuPont Inst. of The Nemours Found., Wilmington, Del., 1967-86; pres. Uni-Tech Assocs., USA, Inc., Rockland, Del., 1985—; prof. Tokai U. Sch. Medicine, Isehara, Japan, 1989-91; dir. sci. affairs Margaronics, Inc., East Brunswick, N.J., 1990—; vis. prof. Valparaiso (Ind.) U., 1967; adj. prof. U. Del. Sch. Life and Health Scis., Newark, 1986—; cons., speaker to various symposia, univs. and govt. Co-author: Perspectives in Culture Collections Number 1: Procurement of Patent Cultures, 1992; contbr. numerous articles to profl. jours. Waksman Found. grantee, 1965; recipient Japan Soc. scholarship, 1957, 58, Disting. Svc. medal Internat. Union Microbiol. Socs., 1982. Mem. AAAS, Am. Acad. Microbiology, Am. Chem. Soc. (biol. chemistry div., fermentation div., Del. div.), Am. Soc. Microbiology, The Biochem. Soc. (U.K.), Internat. Fedn. Advancement of Genetic Engring. and Biotech., The Japan Assn. Microbiology, The Soc. Biosci., Biotech. and Agrochemistry, N.Y. Acad. Sci., The Soc. Actinomycetes Japan, Soc. Indsl. Microbiology, Inst. Food Tech., U.S. Fedn. Culture Collections, World Fedn. Culture Collections, Rotary, Rodney Square Club. Home: 1224 Evergreen Rd Wilmington DE 19803-3514 Office: Uni-Tech Assocs USA Inc PO Box 354 Rockland DE 19732-0354

MYSLINSKI, NORBERT RAYMOND, medical educator; b. Buffalo, Apr. 14, 1947; s. Bernard and Amelia Joan (Lesniak) M.; m. Patricia Ann Byrne, June 19, 1970 (dec. 1980); m. René Carter, Nov. 21, 1993. BS in Biology, Canisius Coll., Buffalo, 1965-69; PhD in Pharmacology, U. Ill., Chgo., 1973. Research assoc. Tufts U., Boston, 1973-75; asst. prof. U. Md., Balt., 1975-80; assoc. prof. physiology U. Md., 1980—, co-dir. Facial Pain Clinic, 1980-84, instr. nursing, 1982-84; research fellow U. Bristol, Eng., 1984-85; instr. Community Coll. Balt., 1980-82; dir. grad. prog. dept. physiology U. Md., 1981—; cons. in field; reviewer profl. jours. Editor newsletter Med. Soc. Med. Rsch., 1977-82; contbr. articles to profl. jours. and books on pharmacology and neurosci.; inventor in field. Rep. Task Force on Aging, U. Md., 1979-84; instr. Am. Heart Assn., Balt., 1978—, ARC, Balt., 1977-83. Capt. U.S. Army, 1969-77. Grantee, NIH, various drug cos., founds. Mem. European Brain and Behavior Soc. (hon.), Internat. Brain Rsch. Orgn., Md. Soc. Med. Rsch. (exec. com., bd. dirs. 1978-86), Internat. Assn. Dental Rsch. (advisor 1980-81), Am. Physiol. Soc., Soc. for Neurosci. (pres. Balt. chpt. 1990-92), Sigma Xi. Democrat. Roman Catholic. Home: 108 Rockrimmon Rd Reisterstown MD 21136-3214 Office: U Md Physiology Dept 666 W Baltimore St Baltimore MD 21201-1586

NABHOLZ, JOSEPH VINCENT, biologist, ecologist; b. Memphis, Nov. 3, 1945; s. Martin Peter and Helen Kathleen (Garbacz) N.; m. Sue Ann Winterburn, Aug. 12, 1972; children: Karen Stacey, Pamela Michelle. BS, Christian Bros. U., Memphis, 1968; MS, U. Ga., 1973, PHD, 1978. Sr. biologist U.S. EPA, Washington, 1979—; reviewer NSF and profl. jours., 1973—, Standards Methods Com., Am. Water Works Assn., Denver 18th and 19th edits.; evaluator Office Exptl. Learning U. Md., College Park, Md., 1984-86. Co-author (with others) Methods of Ecological Toxicology, 1981, Testing for Effects of Chemicals on Ecosystems, 1981; author: Estimating Toxicity of Industrial Chemicals to Aquatic Organisms Using Structure Activity Relationships, 1988; contbr, 17 articles to profl. jours. Bd. dirs. Community Assn. Rollingwood Village (4th sect.), Woodbridge, Va., 1981-90, v.p. 1981-82, pres. 1983-90, maintaience chmn. 1990—. Decorated Army Commendation medal with oak leaf cluster, U.S. Army, Vietnam, 1969, '70. Mem. AAAS, Am. Inst. Biol. Scis., Assn. Southeastern Biologists, Internat. Assn. Ecology, Ecol. Soc. Am. (life), Phi Kappa Phi (life). Roman Catholic. Achievements include pragmatic application of theory of chemical structure activity relationships for routine risk assessment of industrial chemicals for environmental toxicity. Home: 13627 Bentley Cir Woodbridge VA 22192-4340 Office: Office Pollution Prevention and Toxics 401 M St SW Washington DC 20460-0001

NABIRAHNI, DAVID M.A., chemist, educator; b. Tehran, Iran, July 10, 1956; came to U.S., 1979; Dakhil and Shokat (Hajizadeh) N.; m. Fariba Eshaghzadeh, Mar. 14, 1979; children: Ozzie, Bobby. BS, Nat. U. Iran, 1979; MS with honors, Ea. N.Mex. U., 1980; PhD, U. New Orleans, 1985. Chmn. and English tchr. Armaghan Tarbiat Sch., Tehran, 1975-79; teaching and rsch. asst. Ea. N.Mex. U., Portales, 1979-80; from teaching and rsch. asst. to sr. rsch. chemist U. New Orleans, La., 1981-86; asst. prof. analytical chemistry Pace U., Pleasantville, N.Y., 1986-89, assoc. prof., 1990-92, prof., 1992—, dir. ctr. applied analytical chemistry, 1988—, adj. prof. environ. law, 1992—; vis. prof. II U. Rome, 1987; adj. rsch. chemist Am. Health Found., Valhalla, N.Y., 1987; adj. prof. chemistry Manhattanville Coll., Purchase, N.Y., 1990—; vis. prof. Inst. Analytical Chemistry, U. Florence, Italy, 1990; vis. scientist IBM T. J. Watson Rsch. Ctr., Yorktown Heights, N.Y., 1991-92, Ciba Geigy Corp., 1993—; Fulbright sr. rsch. scholar Tech. U. Denmark, 1993; vis. faculty fellow Oxford U., U.K., 1994. Contbr. 60 articles/presentations to profl. jours. Environ. advisor to Congresswoman Nita M. Lowey, 1990-92. Mem. Am. Chem. Soc., Am. Assn. Clin. Chemistry, Am. Assn. U. Profs., Perisan Am. Chemists Assn. (founder 1990-92, chmn., sec. bd. dirs. 1988—, ran for chmn. N.Y. sect. 1991, 93, chmn. Westchester div. 1991), Soc. Electroanalytical Chemistry, N.Y. Acad. Sci., Sigma Xi. Muslim. Avocations: soccer, swimming, hiking, pro-environ. quality, computers. Office: Pace U Dept Chemistry Pleasantville NY 10570

NACHT, SERGIO, biochemist; b. Buenos Aires, Apr. 13, 1934; came to U.S., 1965; s. Oscar and Carmen (Scheiner) N.; m. Beatriz Kahan, Dec. 21, 1958; children: Marcelo H., Gabriel A., Mariana S., Sandra M. BA in Chemistry, U. Buenos Aires, 1958, MS in Biochemistry, 1960, PhD in Biochemistry, 1964. Asst. prof. biochemistry U. Buenos Aires, 1960-64; asst. prof. medicine U. Utah, Salt Lake City, 1965-70; rsch. scientist Alza Corp., Palo Alto, Calif., 1970-73; sr. investigator Richardson-Vicks Inc., Mt. Vernon, N.Y., 1973-76; asst. dir., dir. rsch. Richardson-Vicks Inc., Mt. Vernon, 1976-83; dir. biomed. rsch. Richardson-Vicks Inc., Shelton, Conn., 1983-87; sr. v.p. rsch. and devel. Advanced Polymer Systems, Redwood City, Calif., 1987—; lectr. dermatology dept. SUNY Downstate Med. Ctr., Blkyn., 1977-87. Contbr. articles to profl. jours.; patentee in field. Mem. Soc. Investigative Dermatology, Soc. Cosmetic Chemists (award 1981), Dermatology Found., Am. Physiological Soc., Am. Acad. Dermatology. Democrat. Jewish. Home: 409 Wembley Ct Redwood City CA 94061-4308

NADEL, NORMAN ALLEN, civil engineer; b. N.Y.C., Apr. 10, 1927; s. Louis and Bertha (Julius) N.; m. Cynthia Esther Jereski, July 6, 1952; children: Nancy Sarah Frank, Lawrence Bruce. B.C.E., CCNY, 1949; postgrad., Columbia U., 1949-50. Registered profl. engr., N.Y., Conn. Engr. Arthur A. Johnson Corp., N.Y.C., 1950-53; engr. Slattery Contracting Corp., N.Y.C., 1953-56; mgr., estimator Hartsdale Constrn. Corp., Hartsdale, N.Y., 1956-59; engr. MacLean Grove & Co., Inc., Greenwich, Conn., 1959-63, project mgr., 1963-66, v.p., 1966-70, pres., 1970—; chmn. Nadel Assocs., Inc., Greenwich, 1988—; cons. tunnel and underground constrn.; bd. dirs. United Am. Energy Corp., Karl Koch Erecting Co., Inc., PB-KBB Inc.; mem. com. on tunneling Transp. Rsch. Bd., Washington, 1974-75; mem. U.S. Nat. Com. on Tunneling Tech., Washington, 1976-82, chmn., 1980-81; chmn. adv. com. Superconducting Super Collider Underground Tech., 1992—. Trustee Tunnel Workers Welfare Fund, N.Y.C., 1976-88; mem. exec. coun. Pace U. , N.Y.C., 1984—; mem. CCNY Engring. Sch. bd. adv., 1992—. Served with USNR, 1945-46. Named Heavy Constrn. Man of Yr., United Jewish Appeal, 1984; Benjamin Wright award Conn. Soc. Civil Engrs., 1984, Townsend Harris medal City Coll. of N.Y. Alumni Assn., 1987. Fellow ASCE (Constrn. Mgmt. award 1986); mem. Nat. Acad. Engring., Conn. Acad. Sci. and Engring., The Moles (pres. 1982-83, Outstanding Achievement in Constrn. award 1985), Am. Arbitration Assn., Tau Beta Pi, Chi Epsilon. Office: MacLean Grove & Co Inc 14 Reynwood Mnr Greenwich CT 06831-3145

NADIM, ALI, aerospace and mechanical engineering educator; b. Teheran, Iran, May 18, 1962; came to U.S., 1979; s. Jafar and Simin (Sadjadpour) N.; m. Mitra Kutchemeshgi, Dec. 26, 1992. BSChemE summa cum laude, U. Calif., Davis, 1982; MSChemE, MIT, 1985, DSChemE, 1986. Asst. prof. applied math. MIT, Cambridge, 1986-91; assoc. prof. aerospace and mech. engring. Boston U., 1991—; dir. ad interim Div. of Engring. and Applied Sci., Boston U., 1992—; chair grad. com. Coll. Engring., 1992—; student advisor Boston U., MIT, 1986—. Contbr. articles to profl. jours. including Phys. Fluids, Jour. Fluid Mech., Jour. Chem. Phys., Physica A, others; referee Jour. Fluid Mech., Phys. Fluids, AICE Jour., Internat. Jour. Multiphase Flow, Chem. Engring. Sci., Jour. Colloid Interface Sci. Mem. ASME, AICE, Am. Phys. Soc., Sigma Xi. Achievements include research and teaching in transport phenomena, fluid dynamics and applied mathematics. Office: Dept Aero & Mech Engring Boston U 110 Cummington St Boston MA 02215

NAESER, MARGARET ANN, linguist, medical researcher; b. Washington, June 22, 1944; d. Charles Rudolph and Elma Mathilda (Meyer) N. BA in German, Smith Coll., 1966; PhD in Linguistics, U. Wis., 1970. Chief speech pathology sect. Martinez (Calif.) VA Med. Ctr., 1972-74, Palo Alta (Calif.) VA Med. Ctr., 1974-77; rsch. linguist Boston VA Med. Ctr., 1977—; dir. CT scan/MRI scan aphasia rsch. lab. Boston U. Sch. Medicine, asst. rsch. prof. neurology, 1978-84, assoc. rsch. prof., 1984—; mem. adv. bd. CT scan/aphasia VA Nat. Task Force, Washington, 1990-91. Contbr. articles to Neurology, Archives of Neurology, Brain; author: Outline Guide to Chinese Herbal Patent Medicines in Pill Form, 1990. NDEA fellow, 1967, AAUW fellow, 1970. Mem. Acoustical Soc. Am., Am. Speech, Lang., Hearing Assn., Acad. Aphasia, AAAS, Am. Assn. Acupuncture and Oriental Medicine. Office: Boston VA Med Ctr 150 S Huntington Ave Boston MA 02130

NAESSENS, JAMES MICHAEL, biostatistician; b. St. Louis, Mich., July 25, 1952; s. Henry Joseph and Josephine Ann (Gredys) N.; m. Susan Carla Waltz, Sept. 6, 1975; children: Lauren, Kathryn. BS, U. Mich., 1974, MPH, 1979. Mgr. rsch. and info. systems Cen. Mich. Med. Care Rev., Lansing, 1979-82; biostatistician Mayo Clinic, Rochester, Minn., 1982—. Contbr. articles to profl. publs. Mem. Biometrics. Home: 5804 NW 26th Ave Rochester MN 55901 Office: Mayo Clinic Harwick 7-Statistics Rochester MN 55905

NAG, ASISH CHANDRA, cell biology educator; b. Jamalpur, Bengal, India, Jan. 1, 1932; came to U.S., 1964; s. Kshitish Chandra and Nalinibala Nag; m. Nilima Banerjee, Dec. 3, 1961; children: Debasish, Koushik. MS, U. Hawaii, 1966; PhD, U. Alta., Edmonton, Can., 1970. Postdoctoral fellow U. Pa., Phila., 1971-72; rsch. assoc. U. Chgo., 1973-75; asst. prof. biology Oakland U., Rochester, Mich., 1975-81, assoc. prof., 1982-87, prof., 1987—; mem. peer rev. com. for rsch. Mich. affiliate Am. Heart Assn., Southfield, 1989-92. Contbr. articles to Jour. Cell Biology, Jour. Molecular Cellular Cardiology, Biochem. Jour., Circulation Rsch., Jour. Cell. Biochemistry. Rsch. grantee Am. Heart Assn., 1977-88, 91-92, NIH, 1981-84, NSF, 1987-91. Mem. AAAS, Am. Soc. for Cell Biology, Electron Microscopy Soc. Am., N.Y. Acad. Scis. Achievements include research on ultrastructural quantitations and biochemical determination of ATPase activity of red and white muscle cells, establishment of long-term culture of adult cardiac muscle cells, expression of myosin heavy chain isoforms in cultured cardiac muscle cells, effect of Amiodarone (antiarrythmic drug) on cultured cardiac muscle cells; TPA (a tumor promotor) has no influence on the myosin heavy chains. Office: Oakland U University Dr Rochester MI 48309-4401

NAGAI, TSUNEJI, pharmaceutics educator; b. Gumma, Japan, June 10, 1933; s. Ushinosuke and Take Nagai; m. Kiyoko Usui, May 5, 1964. BS, U. Tokyo, 1956, MS, 1958, PhD, 1961. Lic. pharmacist, Japan. Postdoctoral fellow Columbia U., N.Y.C., 1965-66, U. Mich., Ann Arbor, 1966-67; rsch. and teaching assoc. U. Tokyo, 1961-71; prof. pharmaceutics Hoshi U., Tokyo, 1971—; pres. Acad. Pharm. Sci. and Tech., Tokyo, 1965-67. Contbr. articles to profl. jours.; patentee in field. Adv. mem. Ministry of Health and Welfare, Tokyo, 1968-83; trustee Hoshi U., Tokyo, 1979-91, Iwaki Found., Tokyo, 1985—. Recipient Japan Nat. Invention Prize, The Invention Assn., Tokyo, 1984; named hon. prof. Beijin (China) Med. U., 1987, China Pharm. U., 1990. Fellow Am. Assn. Pharm.; mem. Internat. Pharm. Fedn. (v.p. 1986—, Hoest-Madsen Gold medal for rsch. 1986), Pharm. Soc. Japan (pharm. tech. award 1982, acad. prestige prize 1988), Japan Soc. Drug Delivery System (pres. 1988-89). Home: 1-23-10-103 Hon-Komagome, Bunkyo-ku Tokyo113, Japan Office: Hoshi U Dept Pharms, 2-4-41 Ebara, Tokyo 142 Shinagawa-ku, Japan

NAGAMIYA, SHOJI, physicist, educator; b. Mikage-shi, Hyogo-Ken, Japan, May 24, 1944; came to U.S., 1973; s. Takeo and Masako Nagamiya; m. Tae Nagamiya, Nov. 9, 1969; children: Masahiko, Kenji. BSc, U. Tokyo, 1967; DSc, Osaka (Japan) U., 1972. Rsch. assoc. U. Tokyo, 1972-75; staff scientist Lawrence Berkeley (Calif.) Lab., 1975-82; assoc. prof. U. Tokyo, 1982-88; prof. Columbia U., N.Y.C., 1986—, chmn. dept. physics, 1991—; chmn. Internat. Conf. on Quark Matter, Lennox, Mass., 1987-88. Author: Advances in Nuclear Physics, 1984, Annual Review of Nuclear & Particle Science, 1984; mem. editorial bd. Il Nuovo Cimento, 1989—, Jour. Phys. Soc. Japan, 1984-86; mem. adv. com. Jour. Phys. G., 1992—, Internat. Jour. Modern Phys. E., 1992—. Mem. evaluation com. Swedish Natural Rsch. Coun., 1987; mem. vis. com. Lawrence Berkeley Lab., Oak Ridge (Tenn.) Nat. Lab., Brookhaven Nat. Lab. and others; chmn. Nuclear Physics Com. of Japan, 1985-87. U.S. Dept. of Energy grantee, 1986—, Yamada Sci. Found. grantee, 1989-91; recipient Inoue prize, 1992. Fellow Am. Phys. Soc. Home: 24 Victor Dr Irvington NY 10533-1923 Office: Columbia U Dept of Physics W 120th St New York NY 10027

NAGAO, MAKOTO, electrical engineering educator; b. Ise City, Japan, Oct. 4, 1936; s. Kaoru and Yukie Nagao; m. Mikiko Nagao, Dec. 5, 1964; children: Noriko, Fumiko, Takashi. BS, Kyoto U., Japan, 1959, MS, 1961, D in Engring., 1965. Asst. prof. elec. engring. Kyoto U., 1961-68, assoc. prof. elec. engring., 1968-73, prof. elec. engring., 1973—; prof. Nat. Mus. Ethnology, Osaka, Japan, 1976—. Author: Knowledge and Inference, 1988, English transl., 1990, Machine Translation, 1986, English transl., 1990; editor Ency. Computer Sci., 1990. Recipient IEEE Emanuel R. Piore award, 1993. Mem. IEEE, IEICE (v.p. 1993—), Internat. Assn. Machine Translation (pres. 1991—), Cognitive Sci. Soc. Japan (pres. 1988-90), Internat. Assn. Pattern Recognition (v.p. 1986-88). Avocations: book reading, golf, swimming, classical music, walking. Home: 9-11 Asahigaoka, Hirakata Osaka 573, Japan Office: Kyoto U Dept Elec Engring, Yoshida-Hon Machi Sakyo-Ku, Kyoto 606, Japan

NAGARAJAN, SUNDARAM, metallurgical engineer; b. Valayapatty, India, Aug. 17, 1962; came to U.S., 1986; s. M.N. and Saraswathy Sundaram; m. Anandhi Subbiah, May 25, 1989. BS, S.V. Regional Engring. Coll., Surat, Gujarat, India, 1986; MS, Auburn U., 1988, PhD, 1991. Rsch. assoc. Auburn (Ala.) U., 1986-91; metall. engr. Tri-Mark, Inc., Piqua, Ohio, 1991—; mem. peer rev. com. Jour. Engring. for Industry Am. Soc. Mech. Engrs. Contbr. articles to profl. jours., including Welding Jour., Materials Evaluation. Mem. Am. Welding Soc. (vice chmn. 1990-91), Am. Soc. for Metals (exec. com. 1992-93), Sigma Xi. Achievements include development of IR sensing techniques to automatically track joint contours in gas tungsten arc welding processes. Office: Tri-Mark Inc 8585 Industry Park Dr Piqua OH 45356

NAGARKATTI, JAI PRAKASH, chemical company executive; b. Hyderabad, India, Feb. 18, 1947; came to U.S., 1970; s. Surendranath and Shakuntala (Bai) N.; m. Linda Susan Slaughter, Mar. 14, 1975; 1 child, Shanti. BS, Osmania U., 1966, MS, 1968; MS, East Tex. State U., 1972, EdD, 1976. Group leader Aldrich Chem. Co. Inc., Milw., 1977-78, supr. prodn., 1978-79, mgr. prodn., 1979-84, dir. prodn., 1985, v.p., 1985-87, pres., 1987—, also bd. dirs.; lectr. chemistry V.V. Coll., Hyderabad, 1969-70. Contbr. articles to profl. jours. Robert A. Welch fellow East Tex. State U., Commerce, 1974-76. Fellow Indian Chem. Soc.; mem. Am. Chem. Soc. (chmn. membership Milw. chpt. 1981). Avocations: philately, tennis. Office: Aldrich Chem Co Inc 940 W St Paul Ave Milwaukee WI 53233-2625

NAGASAKA, KYOSUKE, chemical engineer; b. Shimizu, Shizuoka, Japan, Nov. 22, 1955; s. Masami and Yasuko (Hosaka) N. PhD, Kyoto U., 1990. Sr. staff rschr. Mitsubishi Rsch. Inst., Inc., Tokyo, 1990—. Office: Mitsubishi Rsch Inst Inc, 2-3-6 Otemachi Chiyoda-ku, Tokyo 100, Japan

NAGASAWA, YUKO, aerospace psychiatrist. Researcher JASDF Aeromedical Lab., Tokyo, Japan. Recipient Raymond F. Longacre award Aerospace Med. Assn., Va., 1991. Office: JASDF Aeromedical Lab, 1-2-10 Sakae-cho Tachidawa, Tokyo Japan*

NAGATA, HIROSHI, psycholinguistics educator; b. Kamoto-gun, Kumamoto, Japan, Apr. 10, 1949; m. Yuriko Kohhara; children: Naoko, Tomoko. BA in Psychology, Okayama (Japan) U., 1972, MA, 1975. Instr. Okayama U., 1975-86, asst. prof., 1986-87, assoc. prof. Sch. Health Sci., 1987—. Author: Language Acquisition, 1987; contbr. articles to profl. jours. Ministry of Edn. grantee, 1981, 90, 93. Office: Okayama U Sch Health Scis, Okayama 700, Japan

NAGATA, ISAO, chemist; b. Tokyo, Jan. 1, 1942; came to U.S., 1966; s. Shigeto and Takako (Watari) N.; m. Hiroko Hayashi, June 8, 1969; children: Nobuko, Yuko. BS in Chemistry, Doshisha U., Kyoto, Japan, 1966; MS in Organic Chemistry, NYU, 1971; MS in Polymeric Material, Polytech. Inst., Bklyn., 1976, PhD in Polymer Chemistry, 1980. Postdoctoral fellow Polytechnic Inst., Bklyn., 1980-82; editor Chem. Abstract, Columbus, Ohio, 1982-87; chemist E.I. DuPont, Troy, Mich., 1987—. Co-author jour. articles on macromolecules, 1981-83. Mem. Am. Chem. Soc., Detroit Soc. Coating Tech. Home: 3745 Horseshoe Dr Troy MI 48083-5652 Office: EI DuPont 945 Stephenson Hwy Troy MI 48083-1185

NAGATA, MINORU, electronics engineer; b. Shinjuku-ku, Tokyo, Japan, June 4, 1933; s. Kikushiro and Sigeko Nagata; m. Yasuko Fukukita, Mar. 31, 1963; children: Noboru, Mitsuru. BS, U. Tokyo, 1952, PhD, 1966. Rsch. assoc. Stanford (Calif.) Electronics Lab., Stanford U., 1964; head VLSI dept. Cen. Rsch. Lab., Hitachi Ltd., Kokubunji, Tokyo, 1972-75, dir., sr. chief scientist, 1985—. Recipient Gov.'s award Metropolis of Tokyo, 1979, Nat. Medal Honor with purple ribbon, 1993; named Person of Sci. and Tech. Merits, Sci. and Tech. Agy., 1986. Fellow IEEE; mem. Engring. Acad. Japan. Avocations: swimming, golf, amateur radio. Home: 4-5-3 Jyosuihoncho, Kodaira-shi, Kokubunji, Tokyo 187, Japan Office: Hitachi Ltd Cen Rsch Lab, 1-280 Higachi-koigakubo, Tokyo 185, Japan

NAGEL, JOACHIM HANS, biomedical engineer, educator; b. Haustadt, Saarland, Feb. 22, 1948; came to U.S., 1986; s. Emil and Margarethe Nagel; m. Monika Behrens. MS, U. Saarbruecken, Fed. Republic Germany, 1973; DSc, U. Erlangen, Fed. Republic Germany, 1979. Rsch. assoc., lectr., instr. U. Saarbruecken, 1973-74; rsch. assoc., lectr., instr. Dept. Biomed. Engring., U. Erlangen-Nuernberg, 1974-75, asst. prof., 1975-79, dir. med. electronics and computer div., 1976-85, acad. councillor, 1980-86; assoc. prof. radiology Med. Sch. U. Miami, Coral Gables, Fla., 1990-91, assoc. prof. psychology Sch. Arts and Scis., 1988-91, assoc. prof. biomed. engring. Coll. Engring., 1986-91, prof. biomed. engring, radiology and psychology, 1991—. Editor Annals of Biomedical Engineering, Section Instrumentation, 1989—; contbr. numerous articles to profl. jours. NIH grantee since 1986. Mem. IEEE (sr.), IEEE/Engring. in Medicine and Biology Soc. (chmn. Internat. Conf. 1991, chmn. Internat. Progr. Com. Conf. 1989, 90, 92), IEEE/Acoustics, Speech, and Signal Processing Soc., Biomed. Engring. Soc. (sr.), N.Y. Acad. Scis., Internat. Soc. Optical Engring., Romanian Soc. for Clin. Engring. and Med. Computing (hon.), Sigma Xi. Roman Catholic. Achievements include numerous U.S., German and European patents; invention and development of procedure for Sub-Nyquist Sampling of signals for statistic signal processing, NMR imaging of electric currents, passive telemetry for analogue signals, ECG detection; invention of Macro programming; portable drug infusion systems; new techniques for impedance cardiography and perinatal monitors. Office: U Miami PO Box 248294 Miami FL 33124-0621

NAGEL, MAX RICHARD, retired, applied optics physicist; b. Bautzen, Saxony, Germany, Dec. 15, 1909; s. Karl August and Auguste Anna (Rich) N.; m. Angelika von Braun, June 1, 1940 (dec. Mar. 1979); children: Michael von Braun, Stefan von Braun; m. Sieglinde Albertine Jobst, Dec. 16, 1983. Diploma in Engring., Tech. U., Dresden, Germany, 1935, D. in Engring., 1939; D. in Engring. Habilitation, Tech. U., Berlin, 1942. Engr. Dept Optics Engring. Bosch AG, Stuttgart, Germany, 1935-36; sci. co-worker Deutsche Versuchanstalt f. Luftfahrt, Berlin-Adlershof, Germany, 1936-42; chief engring. office Hamburg, Germany, 1945-49; sci. co-worker Photo Reconnaissance Lab., Aerial Reconnaissance Lab. Airforce Rsch. and Devel. Command, Dayton, Ohio, 1949-58; chief Thermal Radiation Lab. Cambridge Air Force Rsch. Ctr., Bedford, Mass., 1958-60, NATO SHAPE Air Defence Tech. Ctr., The Hague, Netherlands, 1960-62; acting chief, sr. scientist Space Optics Lab. NASA Cambridge (Mass.) Electronics Rsch. Ctr., 1963-70; sci. co-worker Deutsche Forschungs- und Versuchsanst f. Luft- u. Raumfahrt, Oberpfaffenhofen, Fed. Republic Germany, 1971-78. Co-author: Tables...Graphs on Planck Radiation, 1952, Tables of Blackbody Radiation Function, 1961, Daylight Illumination...Tables, 1978; contbr. articles to profl. jours. With German Armed Forces, 1942-45. Recipient Sustained Superior Performance award, 1957, Spl. Act Svc. award Dept. Air Force, 1961. Mem Optical Soc. Am., Deutsche Gesellschaft für angewandte Optik, Oberlausitzische Gesellschaft der Wissenschaften.

NAGEY, DAVID AUGUSTUS, physician, researcher; b. Cleve., Oct. 14, 1950; s. Tibor Franz and Patricia Ann (Griffin) N.; m. Elaine Traicoff, Aug. 7, 1971; children: Stefan Anastas, Nicholas Tibor. Student Cornell U., 1966-67; BS with distinction, Purdue U., 1969; PhD in Bioengring., Duke U., 1974, MD, 1975. Diplomate Am. Bd. Obstetrics and Gynecology, Am. Bd. Maternal-Fetal Medicine; registered profl. engr., N.C., Md. Resident in ob/gyn., Duke U. Sch. Medicine, Durham, N.C., 1975-79, fellow in maternal-fetal medicine, 1979-81; asst. prof. U. Md. Sch. Medicine, Balt., 1981-84, assoc. prof., 1984—, asst. dir., div. maternal-fetal medicine, 1981-85, dir., div. maternal-fetal medicine, 1985—, rsch. assoc. prof. Dept. of Epidemiology and Preventive Medicine, 1992—; adj. assoc. prof. dept. elec. engring. U. Md., 1986—; adj. assoc. prof. dept. maternal and child health Sch. of Hygiene and Pub. Health, Johns Hopkins U., 1986—; rsch. assoc. Nat. Inst. Child Health & Human Devel., 1991-92; assoc. examiner Am. Bd. Obstetrics and Gynecology, 1991—. Assoc. editor: Computers in Medicine and Biology, 1984—; mem. editorial bd. Jour. Maternal-Fetal Investigation, 1991—; contbr. articles to med. jours. ACOG/Syntex grantee, 1987. Fellow Am. Coll. Obstetricians and Gynecologists (com. sci. program 1988—); mem. AAAS, IEEE (healthcare engring. policy com. 1987-91), N.Y. Acad. Scis., So. Perinatal Assn. (pres. 1987), Nat. Perinatal Assn. (bd. dirs. 1986-90), Bayard Carter Assn. Ob-Gyn., Md. Ob-Gyn. Soc. (pres. 1990-91, exec. com. 1987—). Avocation: sailing. Office: U Md Med System/Hosp Dept Obstetrics and Gynecology 22 S Greene St Baltimore MD 21201-1544

NAGYS, ELIZABETH ANN, environmental issues educator; b. St. Louis; d. Dallas and Miriam (Miller) Nichols; m. Sigi Nagys, Feb. 7, 1970; children: Eric M., Jennifer R., Alex M. BS., So. Ill. U. Extenstion, Edwardsville, 1970. Cert. tchr., Mo. Announcer Sta. KMTY, Clovis, N.Mex., 1970-71; substitue tchr. Ritneour Sch. Dist., Overland, Mo., 1977-78; instr. biology Southwestern Mich. Coll., Dowagiac, 1988—; reviewer textbooks Harcourt, Brace & Co., 1993. Bd. dirs. United Meth. Ch., Marvin Park, 1979-84; coord. United Meth. Women, 1980-87; mem. Hazardous Waste Com. for Elkhart County, Ind., 1991—. Mem. AAUW, Sierra Club, Welcome Wagon Club. Avocations: reading, gardening.

NAHAR, SULTANA NURUN, research physicist; b. Dacca, Bangladesh, Jan. 10, 1955; came to U.S., 1979; d. Abdur Razzak and Shamsun Nahar Khanam; m. Lutfur Rahman, July 15, 1979. BSc in Physics with honors, Dacca U., 1977, MSc, 1979; MA, Wayne State U., 1987, PhD in Atomic Physics, 1987. Grad. teaching asst. Wayne State U., Detroit, 1979-84, 86-87, Univ. fellow, 1984-86; rsch. assoc. Ga. State U., Atlanta, 1987-90; rsch. fellow in physics Ohio State U., Columbus, 1990—. Contbr. articles to profl. jours. Recipient Gustafson teaching award Wayne State U., 1984, Rumble fellow, 1984, 85. Mem. Am. Phys. Soc., Islamic Soc. N.Am., Assn. Bangladesh. Achievements include new computational method for total radiative and dielectronic electron-ion recombination of atoms and ions, accurate results on photoionization cross sections and oscillator strengths for atoms and ions such as for carbon sequence, silicon sequence, Fe II, relativistic calculation for (electron-positron, atom) scattering in model potential approach. Office: Ohio State U Dept Astronomy 174 W 18th Ave Columbus OH 43210

NAIL, PAUL REID, psychology educator; b. Kansas City, Mo., July 7, 1952; s. Lowell Thomas and Lee Ann (Reid) N.; m. Jenifer Jane Ellinger, May 21, 1978; children: Anthony Reid, Lance Thomas Walker. MS, Southwestern Okla. State U., 1978; PhD, Tex. Christian U., 1981. Instr. Mustang (Okla.) Elem. Sch., 1974-76; prof. psychology Southwestern Okla. State U., Weatherford, 1980—. Contbr. articles to profl. publs. Mem. Am. Psychol. Soc., Soc. for Personality and Social Psychology, Southwestern Psychol. Assn., Soc. Southwestern Social Psychologists (pres. 1989-90). Republican. Baptist. Home: 1709 Chisholm Trail Weatherford OK 73096 Office: Southwestern Okla State U 100 Campus Dr Weatherford OK 73096-3098

NAIMARK, GEORGE MODELL, marketing and management consultant; b. N.Y.C., Feb. 5, 1925; s. Myron S. and Mary (Modell) N. B.S., Bucknell U., 1947, M.S., 1948; Ph.D., U. Del., 1951; m. Helen Anne Wythes, June 24, 1946; children: Ann, Richard, Jane. Rsch. biochemist Brush Devel. Co., Cleve., 1951; dir. quality control Strong, Cobb & Co., Inc., Cleve., 1951-54; dir. sci. svcs. White Labs., Inc., Kenilworth, N.J., 1954-60; v.p. Burdick Assocs., Inc., N.Y.C., 1960-66; pres. Rajah Press, Summit, N.J., 1963—; pres. Naimark and Barba, Inc., N.Y.C., 1966—. With USNR, 1944-46. Fellow AAAS, Am. Inst. Chemists; mem. Am. Chem. Soc., N.Y. Acad. Scis., Am. Mktg. Assn., Pharm. Advt. Coun., Med. Advt. Agy. Assn. (bd. dirs.). Author: A Patent Manual for Scientists and Engineers, 1961, Communications on Communication, 1971, 3d edit., 1987; patentee in field; contbr. articles in profl. jours. Home: 87 Canoe Brook Pky Summit NJ 07901-1404 Office: Naimark & Barba Inc 248 Columbia Tpke Florham Park NJ 07932-1210

NAIMI, SHAPUR, cardiologist; b. Tehran, Iran, Mar. 28, 1928; s. Mohsen and Mahbuba (Naim) N.; came to U.S., 1959, naturalized, 1968; M.B., Ch.B., Birmingham (Eng.) U., 1953; m. Amy Cabot Simonds, May 11, 1963; children—Timothy Simonds, Susan Lyman, Cameron Lowell. House physician Royal Postgrad Med. Sch. London, 1955; sr. house officer Inst. Diseases of the Chest, London, 1956; fellow in grad. tng. New Eng. Med. Center and Mass. Inst. Tech., 1961-64; cardiologist Tufts New Eng. Med. Center, Boston, 1966—, dir. intensive cardiac care unit, 1973—, assoc. prof. 1970-93, prof. 1993—. Recipient Distinguished Instr. award, 1972, Teaching citation, 1976, Excellence in Teaching award, 1982 (all Tufts Med. Sch.), diplomate Royal Coll. Physicians London, Royal Coll. Physicians Edinburgh, Am. Bd. Internal Medicine (subsplty. bd. cardiovascular disease). Fellow Royal Coll. Physicians (Edinburgh), A.C.P., Am. Coll. Cardiology; mem. Am. Soc. Exptl. Biology and Medicine, Am. Heart Assn., Mass. Med. Soc. Clubs: Country Brookline; Cohasset Yacht. Contbr. to profl. jours. Home: 265 Woodland Rd Chestnut Hill MA 02167-2204 Office: 750 Washington St Boston MA 02111-1854

NAIR, CHANDRA KUNJU PILLAI, internist, educator; b. Trichur, India, May 20, 1944; married; three children. BS with honors, Bombay U., India, 1964, MD, 1972; MBBS, Armed Forces Med. Coll., Poona, India, 1968. Diplomate Am. Bd. Internal Med., Am. Bd. Cardiology. Resident in internal medicine Bombay (India) U. Affiliated Hosp., 1969-70, registrar in cardiology, 1970-73; resident in internal medicine Med. Coll. Affiliated Hosp., Toledo, 1973-76, chief resident in internal medicine, 1975; clin. instr. in internal medicine Creighton U. Affiliated Hosp., Omaha, Nebr., 1977-78, asst. prof., 1978-85, assoc. prof., 1985-90, prof. medicine, 1990—; bd. dirs. Cardiology Outpatient Clinic Creighton U. Affiliated Hosp., 1978-81, Exercise Electrocardiographic Lab., 1981-83, Ambulatory Electrocardiographic Lab., 1982-90; active staff St. Joseph Hosp., Omaha, 1978—; attending physcian VA Adminstrn. Hosp., Omaha, 1977—; cons. cardiology Crawford County Meml. Hosp. Denison, Iowa, 1979-87, 1989—, Saunders County Community Hosp., Wahoo, Nebr., 1979-86, Boone County Community Hosp., Albion, Nebr., 1986-87, Genoa (Nebr.) Community Hosp., 1987—, Antelope Meml. Hosp., Neligh, Nebr., 1987—, Fulerton (Nebr.) Meml. Hosp., 1988-90. Contbr. articles to profl. jours.; mem. editorial bd. Chest, 1988—, Am. Jour. Cardiology, 1989—, Cardiovascular Review and Reports, 1991. Fellow ACP, Am. Coll. Cardiology (mem. coun. geriatric cardiology), Am. Coll. Chest Physicians (mem. coun. on criteria care), Am. Heart Assn. (Nebr. affiliate, mem. coun. clin. cardiology), Am. Coll. Angiology; mem. Am. Fedn. Clin. Rsch., Am. Soc. Internal Medicine, Midwest Clin. Soc., Cardiovascular Soc. Omaha, Metro Omaha Med. Soc., Coun. Clin. Cardiology, N.Y. Acad. Scis., Coun. Geriatric Cardiology. Achievements include research in coronary heart disease, cardiac arrhythmias, valvular heart disease, echocardiology and Dopplers, exercise testing, Holter monitor. Home: 9929 Devonshire Omaha NE 68114 Office: Creighton U. Affiliated Hosps 3006 Webster St Omaha NE 68131-2044

NAIR, K. MANIKANTAN, materials scientist; b. Vaikam, India, May 25, 1933; came to the U.S., 1962; s. Kumaran and Karthiyani (Amma) N.; m. Xina Nair, Sept. 21, 1970; 1 child, M. Nathan. BS, U. Kerala, 1959, MS, 1961; MS, Pa. State U., 1964; PhD, U. Wash., 1969. Assoc. Inst. Chemists, Royal Inst. Chemists. Rsch. scientist CGCRI, CSIR, Calcutta, India, 1961-62; researcher Pa. State U., University Park, 1962-65, U. Wash., Seattle, 1965-74; faculty mem. U. Cin., 1974-78; scientist DuPont Co., Wilmington, Del., 1978—; pres. Ideas, Inc., East Amherst, N.Y., 1989—; adj. prof. SUNY, 1992—. Editor: Processing for Improved Productivity, 1984, Super Conductivity, Vols. I and II, 1990, 91, Microelectric Systems, 1990, Electro Optics and Nonlinear Optics Materials, 1990, Glasses for Electronics, 1991, Ferroelectric Films, 1992, Dielectric Ceramics: Preparation, Properties and Applications, 1993; author: Two Dramas in Malayalam, 1955. Active various polit. groups, Kerala, 1944-62. Recipient many internat. awards in materials sci. Fellow Am. Ceramic Soc. (chair, program chair 1986-92, trustee 1993). Republican. Hindu. Achievements include 21 U.S. patents and numerous patents in Europe, Canada, and Japan. Home: 3350 Westminster Ln Doylestown PA 18901 Office: DuPont Co PO Box 80334 Wilmington DE 19880-0334

NAIR, MADHAVAN PUTHIYA VEETHIL, immunologist, consultant; b. Kattachira, Kerala, India, Nov. 18, 1943; came to U.S., 1977; s. Narayanan Pillai and Kutyamma K. N.; m. Rema Devi, Nov. 30, 1961; children: Narayanan, Harikrishnan. MS, U. Bombay, India, 1973, PhD, 1977. Research Cancer Inst., Bombay, 1973-77; postdoctoral scholar Sloan-Kettering Cancer Inst., N.Y.C., 1977-79; postdoctoral scholar U. Mich., Ann Arbor, 1979-82, rsch. immunologist, 1982-84, asst. rsch. scientist, 1984-89, dir. AIDS Rsch. Lab., 1988—, assoc. rsch. scientist, 1989—; cons. Hoffman LaRoche Inc., Nutley, N.J., 1990—; rev. mem. Nat. Inst. Drug Abuse (NIDA), Bethesda, Md., 1988—, NIDA AIDS Tng. program, 1990—. Contbr. articles to profl. jours. Mem. PTA, Ann Arbor, 1991—. Recipient Disting. Rsch. Soc. award, U. Mich., 1990; grantee faculty rsch. U. Mich., 1986, Children's Leukemia Found., 1986, Hoffman LaRoche Inc., 1988—, NIMH, 1990—. Mem. AAAS, Am. Assn. for Cancer Rsch., Am. Assn. Immunologists, N.Y. Acad. Scis., Rsch. Soc. Alcoholism. Achievements include discovery of a serum factor which suppresses our defense against cancer and may aid in the diagnosis of cancer susceptibility, that smoking and alcohol consumption are frequently associated with infections (including HIV), that proteins from HIV suppress the immune response of lymphocytes, and that alcohol supresses our defense mechanism against HIV infection. Home: 105 Mill Valley East Amherst NY 14051 Office: SUNY-Buffalo Buffalo Gen Hosp 100 High St Buffalo NY 14203

NAIR, SUDHAKAR EDAYILLAM, mechanical and aerospace engineering educator; b. Kasaragod, Kerala, India, Apr. 12, 1944; came to U.S., 1969; s. Krishnan and Meenakshi Nair; m. Celeste Churchill, Dec. 18, 1983. BS in Mechs., U. Kerala, 1967; M of Aero. Engring., Indian Inst. Sci., Bangalore, India, 1969; PhD in Applied Mechs., U. Calif.-La Jolla, 1974. Rsch. engr. U. Calif., San Diego, 1974-77; from asst. prof. to assoc. prof. Ill. Inst. Tech., Chgo., 1977-91, prof., 1991—; chmn. faculty coun. Ill. Inst. Tech., Chgo., 1992-93; cons. in field. Contbr. over 35 articles to sci. jours. Mem. ASME, AIAA, Am. Acad. Mechs. Achievements include rsch. in theory of shells, residual stresses in structures, nonlinear vibrations. Home: 6157 N Sheridan Chicago IL 60660 Office: Ill Inst Tech Dept Mech Aerospace Engr Chicago IL 60616

NAIR, VELAYUDHAN, pharmacologist, medical educator; b. India, Dec. 29, 1928; came to U.S., 1956, naturalized, 1963; s. Parameswaran and Ammini N.; m. Jo Ann Burke, Nov. 30, 1957; children: David, Larry, Sharon. Ph.D. in Medicine, U. London, 1956, D.Sc., 1976. Research assoc. U. Ill. Coll. Medicine, 1956-58; asst. prof. U. Chgo. Sch. Medicine, 1958-63; dir. lab. neuropharmacology and biochemistry Michael Reese Hosp. and Med. Center, Chgo., 1963-68; dir. therapeutic research Michael Reese Hosp. and Med. Center, 1968-71; vis. asso. prof. pharmacology Chgo. Med. Sch., 1963-68, vis. prof., 1968-71, prof. pharmacology, 1971—, vice chmn. dept. pharmacology and therapeutics, 1971-76; dean Chgo. Med. Sch. (Sch. Grad. and Postdoctoral Studies), 1976—. Contbr. articles to profl. publs. Recipient Morris Parker award U. Health Scis./Chgo. Med. Sch., 1972. Fellow AAAS, N.Y. Acad. Scis., Am. Coll. Clin. Pharmacology; mem. AAUP, Internat. Brain Rsch. Orgn., Internat. Soc. Biochem. Pharmacology, Am. Soc. Pharmacology & Exptl. Therapeutics, Am. Soc. Clin. Pharmacology & Therapeutics, Radiation Rsch. Soc., Soc. Toxicology, Am. Chem. Soc., Brit. Chem. Soc., Royal Inst. Chemistry (London), Pan Am. Med. Assn. (council on toxicology), Soc. Exptl. Biology & Medicine, Soc. Neurosci., Internat. Soc. Chronobiology, Am. Coll. Toxicology, Internat. Soc. Developmental Neurosci., Sigma Xi, Alpha Omega Alpha. Club: Cosmos (Washington). Office: UHS/Chgo Med Ctr 3333 Green Bay Rd North Chicago IL 60064-3095

NAIR, VELUPILLAI KRISHNAN, cardiologist; b. Kerala, India, Dec. 30, 1941; came to U.S., 1973; s. Veupillai and Bharathy Nair; m. Sathy C. Nair, Apr. 22, 1971; children: Parvathy, Pradeep. BSc, Kerala U., Trivandum, India, 1961, MB BS, 1965, MD, 1971. Diplomate Am. Bd. Internal Medicine, Am. Bd. Cardiology. Asst. prof. N.Y. Med. Coll. Lincoln Hosp., Bronx, 1978-80; cardiologist, dir. cardiology svc. Somerset (Pa.) Hosp., 1980—, chief of med. dental staff, 1990—. Former pres. Somerset County divsn. Am. Heart Assn. Fellow Am. Coll. Cardiology; mem. AMA, Pa. Med. Soc., Somerset County Med. Soc. (former pres.), Soc. Hypertension, Soc. Echocardiography, Cardiac Club (advisor). Avocations: reading, tennis, travel. Office: 223 S Pleasant Ave Somerset PA 15501

NAISMITH, JAMES POMEROY, civil engineer; b. Dallas, Aug. 4, 1936; s. James S. and Frances (Pomeroy) N.; m. Beverly Mozeney, Feb. 2, 1957; children: Anne Elizabeth, James Mozeney, Robert Alan, Margaret Lynn. B of Civil Engring., Cornell U., 1958, MS, 1959. Registered profl. engr. Tex., N.Y., Mo.; registered pub. land. surveyor Tex. Instr. Cornell U., Ithaca, N.Y., 1959-60; asst. engr. Calif. Water Pollution Control Bd., San Luis Obispo, 1960-61; engr. to chief exec. officer Naismith Engrs. Inc., Corpus Christi, Tex., 1961-89; mgr., dist. engr. San Patricio Mcpl. Water Dist., Ingleside, Tex., 1989—. Trustee Calallen Ind. Sch. Dist., Corpus Christi, 1985-91; chmn. Zoning Bd. of Adjustment, Corpus Christi, 1970. Mem. NSPE, ASCE (dir. 1982-85), ASCE Tex. Sect. (pres. 1973), Rotary. Office: San Patrico Mcpl Water Dist PO Drawer S Ingelside TX 78362

NAJARIAN, JOHN SARKIS, surgeon, educator; b. Oakland, Calif., Dec. 22, 1927; s. Garabed L. and Siranoush T. (Demirjian) N.; m. Arlys Viola Mignette Anderson, Apr. 27, 1952; children: Jon, David, Paul, Peter. AB with honors, U. Calif., Berkeley, 1948; MD, U. Calif., San Francisco, 1952; LHD (hon.), Univ. Athens, 1980; DSc (hon.), Gustavus Adolphus Coll., 1981; LHD (hon.), Calif. Luth. Coll., 1983. Diplomate Am. Bd. Surgery. Surg. intern U. Calif., San Francisco, 1952-53, surg. resident, 1955-60, asst. prof. surgery, dir. surg. research labs., chief transplant service dept. surgery, 1963-66, prof., vice chmn., 1966-67; spl. research fellow in immunopathology U. Pitts. Med. Sch., 1960-61; NIH sr. fellow and asso. in tissue transplantation immunology Scripps Clinic and Research Found., La Jolla, Calif., 1961-63; Markle scholar Acad. Medicine, 1964-69; prof., chmn. dept. surgery U. Minn. Hosp., Mpls., 1967—, chief hosp. staff, 1970-71, Regents' prof., 1985—, Jay Phillips Disting. Chair in Surgery, 1986—; spl. cons. USPHS, NIH Clin. Rsch. Tng. Com., Inst. Gen. Med. Scis., 1965-69; cons. U.S. Bur. Budget, 1966-68; mem. sci. adv. bd. Nat. Kidney Found., 1968; mem. surg. study sect. A div. rsch. grants NIH, 1970; chmn. renal transplant adv. group VA Hosps., 1971; mem. bd. sci. cons. Sloan-Kettering Inst. Cancer Rsch., 1971-78; mem. screening com. Dernham Postdoctoral Fellowships in Oncology, Calif. div. Am. Cancer Soc. Editor: (with Richard L. Simmons) Transplantation, 1972; co-editor: Manual of Vascular Access, Organ Donation, and Transplantation, 1984; mem. editorial bd. Jour. Surg. Rsch., 1968—, Minn. Medicine, 1968—, Jour. Surg. Oncology, 1968—, Am. Jour. Surgery, 1967—, assoc. editor, 1982—; mem. editorial bd. Year Book of Surgery, 1970-85, Transplantation, 1970—, Transplantation Procs, 1970—, Bd. Clin. Editors, 1981-84, Annals of Surgery, 1972—, World Jour. Surgery, 1976—, Hippocrates, 1986—, Jour. Transplant Coordination, 1990—; assoc. editor: Surgery, 1971; editor-in-chief: Clin. Transplantation, 1986—. Bd. dirs., v.p. Variety Club Heart Hosp., U. Minn.; trustee, v.p. Minn. Med. Found. Served with USAF, 1953-55. Hon. fellow Royal Coll. of Surgeons of Eng., 1987; hon. prof. Univ. Madrid, 1990; named Alumnus of Yr. U. Calif. Med. Sch. at San Francisco, 1977; recipient award Calif. Trudeau Soc., 1962, Ann. Brotherhood award NCCJ, 1978, Disting. Achievement award Modern Medicine, 1978, Internat. Gt. Am. award B'nai B'rith Found., 1982, Uncommon Citizen award 1985, Sir James Carreras award Variety Clubs Internat., 1987, Silver medal IXth Centenary, Univ. of Bologna, 1988, Humanitarian of Yr., U. Minn., 1992. Fellow ACS; mem. Soc. Univ. Surgeons, Soc. Exptl. Biology and Medicine, AAAS, Am. Soc. Exptl. Pathology, Am. Surg. Assn. (pres. 1988-89), Am. Assn. Immunologists, AMA, Transplantation Soc. (v.p. western hemisphere 1984-86, pres. 1994-96), Am. Soc. Nephrology, Internat. Soc. Nephrology, Am. Assn. Lab. Animal Sci., Assn. Acad. Surgery (pres. 1969), Internat Soc. Surgery, Soc. Surg. Chairmen, Soc. Clin. Surgery, Central Surg. Assn., Minn., Hennepin County med. socs., Mpls., St. Paul, Minn., Howard C. Naffziger, Portland, Halsted surg. socs., Am. Heart Assn., Am. Soc. Transplant Surgeons (pres. 1977-78), Council on Kidney in Cardiovascular Disease, Hagfish Soc., Italian Research Soc., Minn. Acad. Medicine, Minn. Med. Assn., Minn. Med. Found., Surg. Biology Club, Sigma Xi, Alpha Omega Alpha, others. Office: U Minn Surgery Dept Mayo Meml Bldg Box 195 Minneapolis MN 55455

NAJERA, RAFAEL, virologist; b. Córdoba, Spain, Feb. 19, 1938; s. Luis and María (Morrondo) N.; m. Margarita Vazquez de Parga, July 1, 1965; children: Isabel, Gonzalo. M.B., Madrid U., 1962, M.D., 1967; M.Sci., Birmingham (Eng.) Med. Sch., 1967, postgrad., 1967-68; D.P.H., Sch. Public Health Madrid, 1965. Assoc. chief, service respiratory and exanthematic viruses Nat. Center for Virology, Majadahonda, Madrid, 1963-72, chief service, 1972-80; med. officer virus diseases unit WHO, Geneva, 1980-81; assoc. prof. virology Madrid Faculty Medicine, 1970-73; dir. Nat. Center Microbiology, 1982-86; dir. gen. NIH, Ministry of Health, Madrid, 1986-92; dir. Dept. Retrovirus Rsch. Instituto de Salud Carlos III, Madrid, Spain, 1992—. Contbr. articles to profl. jours. Dir. WHO Collaborating Ctr. for AIDS; pres. Spanish Soc. AIDS, 1989—. Decorated knight comdr. Civil Order Sanidad. Mem. Spanish Soc. Microbiology (sec., pres. virology group 1970—), Spanish Soc. Virology (pres. 1988—), Soc. Gen. Microbiology, Am. Soc. Microbiology, Internat. Assn. Biol. Standardization, European Teratology Soc., Internat. Epidemiol. Assn., Am. Public Health Assn., others. Office: Instituto de Salud Carlos III, Majadahonda, 28220 Madrid Spain

NAJM, ISSAM NASRI, environmental engineer; b. Lebanon, Sept. 1, 1963; came to U.S., 1985; s. Nasri Majeed and Georgette (Francis) N. MS in Environ. Engring., U. Ill., 1987, PhD in Environ. Engring., 1990. Registered engr.-in-tng., Calif. Supervising engr. applied rsch. dept. Montgomery Watson, Pasadena, Calif., 1990—. Contbr. articles to profl. publs. Mem. Am. Water Works Assn. (com. on organic contaminants rsch. 1992—, Acad. Achievement award 1990), Internat. Assn. Water Quality. Office: Montgomery Watson 301 N Lake Ave Ste 600 Pasadena CA 91101

NAKAGAKI, MASAYUKI, chemist; b. Tokyo, Apr. 19, 1923; s. Sengoro and Shizue N.; m. Hisako Yoshitake, May 3, 1951. BSc, Imperial U., Tokyo, 1945; DSc, U. Tokyo, 1950. Instr. Imperial U., Tokyo, 1945-51; lectr. U. Tokyo, 1951-54; prof. Osaka (Japan) City U., 1954-60; prof. Kyoto (Japan) U., 1960-87, emeritus prof., 1987—; vis. prof. Wayne State U., Detroit, 1955-57, 68-69; dean faculty pharm. sci. Kyoto U., 1978-80; prof. Hoshi U., Tokyo, 1987-92; dir. Tokyo Inst. Colloid Sci., 1992—. Regional editor: Colloid and Polymer Sci., Darmstadt, Germany, 1982—. Recipient rsch. award Takeda Found. 1971. Mem. Membrane Soc. Japan (inspector 1988—, pres. 1978-88), Pharm. Soc. Japan (rsch. award 1970). Home: 354 Kotokujicho Teramachi, Kamigoryo, Kyoto 602, Japan Office: Tokyo Inst Colloid Sci, 502 Higashi-Nakano 4-4-3, Tokyo 164, Japan

NAKAGAWA, JOHN EDWARD, aeronautical engineer; b. Honolulu, Nov. 24, 1962; s. David and Laura Nakagawa. BS in Aero. Engring., U. So. Calif., 1985. Trajectory simulations mission analyst Space Vector Corp., Northridge, Calif., 1986; advanced systems aero. engr. Space Vector Corp., Fountian Valley, Calif., 1986—. Mem. steering com. Seal Beach (Calif.) Citizens United, Sensible Traffic Planning, Seal Beach. Mem. AIAA, Union Concerned Scientists, Greenpeace. Home: 320 12th St # 5 Seal Beach CA 90740-6438 Office: Space Vector Corp 17330 Brookhurst #150 Fountain Valley CA 92708

NAKAGAWA, KIYOSHI, communications engineer; b. Himeji, Japan, Apr. 24, 1945; s. Minoru and Hisano (Shiraoka) N.; m. Tomoko Tanaka, Oct. 11, 1970; children: Mizuho, Akihisa. BS, Osaka U., Japan, 1968, MS, 1970, PhD, 1980. Rsch. engr. Elec. Communication Lab., NTT, Musashino, Japan, 1970-84; sect. head Yokosuka ECL, NTT, Yokosuka, Japan, 1985-86; rsch. group leader NTT Transmission System Labs., NTT, Yokosuka, Japan, 1987-88, exec. rsch. engr., 1989—; lectr. Yokohama Nat. U., 1984-85, Osaka U., 1992-93; assoc. editor The IEEE JLT, N.Y.C., 1990-93; gen. co-chair OSA/IEEE Meeting on Optical Amplifiers, 1991; guest co-editor IEEE JLT/JSAC Coherent Communications, N.Y., 1987; speaker in field. Co-author of several books on optical communications; patentee in field; contbr. numerous articles to profl. jours. Mem. IEEE, Optical Soc. Am., Laser Soc. Japan, Inst. Electronics, Info. and Communication Engrs. (Young Engr. award 1976, Achievement award 1984, 91). Buddist. Avocations: amateur radio, hiking, fishing, reading. Office: NTT Transmission Systems Labs, 1-2356 Take, Yokosuka 238-03, Japan

NAKAGAWA, YUZO, laboratory administrator; b. Wakayama, Japan, Mar. 27, 1932; s. Masaharu and Shizuka (Yoshimatsu) N.; m. Atsuko Torii, May 26, 1967; children: Kumiko, Misako. BS, Osaka U., 1955, M Pharm. Sci., 1957, PhD, 1960. Lic. pharmacist, Japan. Rsch. assoc. Kyoto (Japan) U., 1960-61; lectr. Mukogawa Women's U., Nishinomiya, Japan, 1960-61; rsch. assoc. Stanford (Calif.) U., 1961-64; rsch. assoc. Shionogi Rsch. Labs., Osaka, Japan, 1961-69, rsch. dir., 1970-92; rsch. mgr. Matsushita Technoresearch, Osaka, 1993—, prof. Osaka Bioengineering Coll., 1992—, lectr. Kyoto U., 1982-83. Author books on organic chemistry and mass spectrometry; editor-in-chief Spectroscopy, An Internat. Jour., 1986—; contbr. numerous articles on organic chemistry and mass spectrometry to profl. jours.; patentee in field. Pres. Neighborhood Community Assn., Kobe, 1988-90; advisor Kobe Nishi Police Sta., 1991—. Mem. Mass. Spectroscopy Soc. Japan (bd. dirs. 1970—, chmn., organizer ann. conf. 1972, 90, pres. 1993—), Liquid Chromatography/Mass Spectrometry Soc. Japan (pres. 1991—), Am. Soc. Mass Spectrometry, Am. Chem. Soc., Royal Soc. Chemistry, Pharm. Soc. Japan, Japanese Soc. Med. Mass Spectrometry (bd. dirs. 1977—), Stanford Alumni Assn. (pres. 1975—). Avocations: tennis, camera work, gardening, listening to music. Home: Takenodai 5-18-20 Nishi-ku, Kobe 651-22, Japan Office: Technoresearch, 3-1-1 Yagumo Moriguchi, Osaka 570, Japan

NAKAHARA, MASAYOSHI, chemist; b. Tokyo, Jan. 19, 1927; s. Kyusaku and Hatsu Nakahara; m. Kikue Nagumo, May 20, 1956; children: Kaori, Haruka. BA, Tokyo Inst. Tech., 1952; DSc, Nagoya (Japan) U., 1962. Asst. prof. Rikkyo U., Tokyo, 1962-71, prof., 1971-92, emeritus prof., 1992—. Author: Electron. Structure and Period, 1976, Color Science, 1985; editor: Chem. of Noble Metals, 1984; chief editor Jour. Japan Spectroscopy Soc., 1981-83. Mem. Japan Spectroscopy Soc. (pres. Tokyo chpt. 1989-91). Office: Rikkyo U, 3 Nishiikebukuro, Toshimaku, Tokyo 171, Japan

NAKAI, HIROSHI, civil engineering educator; b. Mitsukaido, Ibaragi, Japan, Dec. 14, 1935; s. Ichiro and Kikue (Yoshida) N.; m. Yoshiko Sawa, Oct. 26, 1968; children: Fuyuko, Makiko. M in Engring., Osaka City (Japan) U., 1961; DEng, Osaka U., 1972. Rsch. assoc. Osaka City U., 1961-66, asst. prof., 1966-70, assoc. prof., 1970-73, prof. bridge engring., 1973—; vis. prof. U. Md., 1976, San Paulo (Brazil) U., 1983; academic advisor Hanshin Expy. Pub. Coop., Osaka, 1969, Japan Civil Engring. Cons. Assn., Kinki br., Osaka, 1977, Osaka Mcpl. Office. Author: (with C. H. Yoo) Analysis and Design of Curved Steel Bridges, 1988; contbr. articles to profl. jours. Mem. ASCE, Internat. Assn. Bridge and Structural Engring., Japan Soc. Civil Engrs., Japanese Soc. Steel Constrn., Japan Rd. Assn., Japanese Soc. Material Sci. Avocations: fishing, traveling, farming. Office: Osaka City U Civil Engring, Sugimoto 3 3 138, Sumiuosyi-Ku, Osaka 558, Japan

NAKAJIMA, AMANE, computer engineer, researcher; b. Nagoya, Aichi, Japan, Mar. 25, 1961; s. Tsutomu and Masako (Furuhashi) N.; m. Misako Suzuki, Mar. 27, 1988; 1 child, Takeshi. B in Engring., U. Tokyo, 1983, M in Engring., 1985. Researcher Tokyo Rsch. Lab. IBM Japan Ltd., 1985—. Contbr. articles to profl. jours. Mem. Inst. Electronics, Info. and Comm. Engrs. Japan (Best Paper award 1987), IEEE (feature editor comm. mags.

1991—, Asia Pacific tech. co. 1990—), Assn. for Computing Machinery, Info. Processing Soc. of Japan. Avocation: reading. Office: IBM Tokyo Rsch Lab, 1623-14 Shimotsuruma Yamato, Kanagawa 242, Japan

NAKAMOTO, TETSUO, nutritional physiology educator; b. Kure, Japan, Dec. 20, 1939; came to U.S. 1964; s. Takamori and Masae Nakamoto; m. Lynda G. Ward, May 14, 1980; children: Andrew T., Christopher W.T. DDS, Nihon U. Sch. Dentistry, Tokyo, 1964; MS, U. Mich., 1966, 71, U. N.D., 1969; PhD, MIT, 1978. Lic. dentist. Teaching asst. U. N.D., Grand Forks, 1967-68, rsch. asst., 1968-69; teaching fellow U. Mich., Ann Arbor, 1971-72; asst. prof. La. State U. Med. Ctr., New Orleans, 1978-84, assoc., 1984-91, prof., 1991--; vis. prof. Nihon U., Tokyo, 1987, 92. Contbr. articles to profl. jours. Mem. Am. Inst. Nutrition, The Am. Physiological Soc., Am. Assn. for Dental Rsch. (pres. New Orleans sect. 1987-88), Soc. for Experimental Biology and Medicine, Omicron Kappa Upsilon. Roman Catholic. Office: La State U Med Ctr 1100 Florida Ave New Orleans LA 70119-2799

NAKAMURA, HIROSHI, urology educator; b. Tokyo, Mar. 22, 1933; s. Yataroh and Hideko (Tanaka) N.; m. Miyoko Kodachi, Aug. 13, 1966. MD, Keio U., Tokyo, 1960; PhD, Grad. Sch. Medicine, Keio U., 1966. Med. diplomate. Asst. resident Mt. Sinai Hosp., N.Y.C., 1962-63; rsch. fellow Cornell U. Med. Coll., N.Y.C., 1966-68; asst. Sch. Medicine Keio U., Tokyo, 1968-70; chmn. urology dept. Tokyo Elec. Power Hosp., 1970-73; vis. asst. prof. surgery Cornell U. Med. Coll., N.Y.C., 1973; chmn. urology Kitasato Inst. Hosp., Tokyo, 1973-77; chmn. dept., prof. urology Nat. Def. Med. Coll., Tokorozawa, Saitama, Japan, 1977—. Author: New Clin. Urology, 1982, Practice of Renal Transplantation, 1985, Bedside Urology, 1991; editor: Up-to-date Urology, 1983. Recipient Tamura award Keio U. Sch. Medicine, 1967, All-around Med. award, Igaku-Shoin, Ltd., Tokyo, 1967. Buddhist. Avocations: jazz, audiophile, travel, fishing, baseball. Home: 4-403 Boei Idai 3-2 Namiki, Tokorozawa Saitama 359, Japan Office: Nat Def Med Coll Dept Urol, 3-2 Namiki, Tokorozawa Saitama 359, Japan

NAKANISHI, KOJI, chemistry educator, research institute administrator; b. Hong Kong, May 11, 1925; came to U.S., 1969; s. Yuzo and Yoshiko (Sakata) N.; m. Yasuko Abe, Oct. 25, 1947; children: Keiko, Jun. B.Sc., Nagoya U., Japan, 1947; Ph.D., Nagoya U., 1954; DSc (hon.), Williams Coll., 1987, Georgetown U., 1992. Asst. prof. Nagoya U., 1955-58; prof. Tokyo Kyoiku U., 1958-63, Tohoku U., Sendai, Japan, 1963-69; prof. chemistry Columbia U., N.Y.C., 1969-80; Centennial prof. chemistry Columbia U., 1980—; dir. research Internat. Ctr. Insect Physiology and Ecology, Nairobi, Kenya, 1969-77; dir. Suntory Inst. for Bioorganic Research, Osaka, Japan, 1979-91. Author: Infrared Spectroscopy-Practical, 1962, rev. edit., 1977, Circular Dichroic Spectroscopy-Exciton Coupling in Organic Stereochemistry, 1983, A Wandering Natural Products Chemist, 1991. Recipient Asahi Cultural award Asahi Press, Tokyo, 1968, E.E. Smissmann medal U. Kans., 1979, H.C. Urey award Phi Lambda Upsilon chpt. Columbia U., 1980, Alcon award in ophthalmology, 1986, Pual Karrer gold medal U. Zurich, 1986, Egbert Havinga medal Havinga Found., Leider, 1989, Imperial Prize of Japan Acad., 1990, Japan Acad. prize, 1990, R.T. Major medal U. Conn., 1991, L.E. Harris award U. Nebr., 1991. Mem. Chem. Soc. Japan (soc. award 1954, 79), Am. Chem. Soc. (E. Guenther award 1978, Remsen award 1981, A.C. Cope award 1990, Nichols medal 1992, Mosher award 1993), Brit. Chem. Soc. (Centenary medal 1979), Swedish Acad. Pharm. Scis. (Scheele award 1992), Am. Acad. Arts and Sci., Am. Soc. Pharmacognosy (rsch. achievement award 1985), Pharm. Soc. Japan (hon.), Acad. Nazionale d'Scienze, Italy (fgn. fellow). Home: 560 Riverside Dr New York NY 10027-3202 Office: Columbia U Dept Chemistry 116th St & Broadway New York NY 10027

NAKANISHI, TSUTOMU, pharmaceutical science educator; b. Osaka, Japan, Mar. 4, 1939; s. Noboru and Shizuko (Kurata) N.; m. Naoko Nishihara, May 3, 1969; children: Wataru N., Nozomi N., Hiromu N. Diploma in pharm. scis., Osaka U., 1961, PhD, 1968; diploma in organic chemistry, Imperial Coll., London. Lectr. pharm. scis. Osaka U., 1970-78, assoc. prof., 1978-83; prof. faculty pharm. sci. Setsunan U., Osaka, 1983—; rsch. fellow in organic chemistry, Imperial Coll., London, 1973-74. Author; contbr. articles to English and Japanese lang. sci. publs. Mem. Pharm. Soc. Japan, Chem. Soc. Japan, Royal Soc. Chemistry. Office: Pharm Scis Setsunan U, 45-1 Nagaotoge cho, Hirakata 573-01, Japan

NAKANO, TATSUHIKO, chemist, researcher, educator; b. Osaka, Japan, Feb. 4, 1925; s. Denichi and Shie (Kubo) N.; m. Toshiko Kitagawa, Apr. 23, 1965. Licenciado, Kyoto U., 1950, PhD, 1955. Rsch. fellow Kyoto (Japan) U., 1950-55, assoc. prof., 1960-65; postdoctoral fellow Wayne State U., Detroit, 1956-59, Stanford (Calif.) U., 1959-60; investigador titurar Inst. Venezolano de Investigaciones Centificas, Caracas, Venezuela, 1965—; prof. U. Cen. de Venezuela, Caracas, 1981—; disting. prof. Inst. de Tecnologia y Estudios Superiores, Monterrey, Mex., 1979; vis. prof. U. N.C., Chapel Hill, 1990; regional editor Revista Latinoamericana de Química, Mex., D.F., 1970—. Contbr. articles to Jour. Chem. Soc., Jour. Chem. Rsch., Tetrahedron, Tetrahedra Letters; author: Studies in Natural Products Chemistry, 1989, 90. Recipient Spl. Rsch. fellowship IKUEIKAI, 1950, Rsch. award Rockefeller Found., 1961, NIH, 1962, Order Andres Bello II and III Ministry of Edn. Venezuela, 1979, 89. Mem. Am. Chem. Soc., Royal Soc. Chemistry, N.Y. Acad. Scis. Office: IVIC Centro de Quimica, Apartado 21827, Caracas 1020-A, Venezuela

NAKATANI, ALAN ISAMU, physicist; b. L.A., Aug. 26, 1957. BA, U. Calif., San Diego, 1978; MS, U. Wis., 1981, U. Conn., 1984; PhD, U. Conn., 1987. Sr. rsch. technician McGaw Labs., Irvine, Calif., 1978-80; physical scientist Nat. Inst. of Standards and Tech., Gaithersburg, Md., 1987—. Contbr. articles to Polymer Communications, Macromolecules, Jour. Chem. Physics, Phys. Rev. Letters, Phys. Rev. B. Nat. Rsch. Coun. fellow, 1987. Mem. Am. Chem. Soc., Am. Physical Soc., Soc. for Rheology, Soc. of Plastics Engrs. Office: Nat Inst Standards Tech Bldg 224 Rm B210 Gaithersburg MD 20899

NAKATSUJI, NORIO, biologist; b. Hashimoto, Wakayama, Japan, Mar. 26, 1950; s. Rikichi and Akiko Nakatsuji; m. Takako Nakatsuji. BS, Kyoto (Japan) U., 1972, MS, 1974, DSci, 1977. Rsch. asst. Umea (Sweden) U., 1978; postdoctoral assoc. MIT, Cambridge, 1978-80; rsch. assoc. George Washington U., Washington, 1980-81, rsch. scientist, 1981-83; vis. scientist Med. Rsch. Coun. U.K., London, 1983-84; sr. scientist Meiji Inst. Health Sci., Odawara, Japan, 1984-88, div. head, 1988-91; prof. Nat. Inst. Genetics, Mishima, Japan, 1991—. Contbr. articles to profl. jours. Mem. Am. Soc. for Cell Biology, British Soc. Developmental Biologists, Japanese Soc. Developmental Biologists, Japanese Soc. Molecular Biologists. Office: Nat Inst Genetics, Mishima 411, Japan

NAKAYAMA, WATARU, engineering educator; b. Kamakura, Kanagawa, Japan, Jan. 7, 1936; s. Shiroh and Haru N.; m. Michiko Aoyagi, Jan. 8, 1967. BS, Defense Acad., Yokosuka, Japan, 1958; MS, Tokyo Inst. Tech., 1963, DEng, 1966. Lectr. U. Sherbrooke, Que., 1969-70; rschr. Hitachi, Ltd., Tokyo, 1970-71; chief rschr. Hitachi, Ltd., Tsuchiura, Japan, 1971-78, sr. rschr., 1978-88, sr. chief rschr., 1988-91, hon. engr., 1991-92; Hitachi chair prof. Tokyo Inst. Tech., 1989-92, prof., 1992—; lectr. in field. Author: (with others) Heat Transfer in Electronic and Microelectronic Equipment, 1990, High Performance Computing in Japan, 1992, Computers and Computing in Heat Transfer Science and Engineering, 1993; contbr. articles to profl. jours. Recipient New Tech. Innovation award Ichimura Found., 1978, Best Paper award Gas Turbine Soc. of Japan, 1984. Fellow ASME (K-16 com. 1981 , chmn. Japanese chpt. 1990-92, Best Paper award 1981, Heat Transfer Meml. award 1992); mem. IEEE (sr.), Japanese Soc. Mech. Engrs. (vice chmn. thermal engring. divsn. 1989-90, chmn. 1990-91, Best Paper award 1965, 80, Tech. award 1978). Achievements include patents for industrial application of heat transfer enhancement techniques to heat exchangers, rotating machinery, cooling systems of computers. Home: 920-7 Higashi Koiso OH-ISO Machi, Naka Gun Kanagawa 255, Japan Office: Tokyo Inst Tech, 2-12-1 OH-Okayama, Meguro-ku 152, Japan

NAKAZATO, HIROSHI, molecular biologist; b. Osaka, Japan, July 25, 1941; s. Goro and Yaye (Kijima) N.; m. Michiyo Matsuda; children: Mari, Yuri. BS, Tokyo U., 1965, MS, 1967, PhD, 1970. Postdoctoral fellow U. Pitts., 1970-74, rsch. assoc., 1974-76; cancer expert NCI, Bethesda, Md., 1976-80; molecular biologist Suntory Inst. for Biomed. Rsch., Osaka, Japan, 1981—. Contbr. articles to profl. jours. Mem. AAAS, Metastasis Rsch. Soc., Japanese Cancer Assn., Japanese Soc. Immunology, Molecular Biology Soc. Japan, Japanese Biochem. Soc., Japanese Soc. for Bone and Mineral Rsch. Avocations: fishing, bonsai, gardening. Home: G306 3-9 Higashinara, Ibaraki-shi, Osaka 567, Japan Office: Suntory Inst Biomed Rsch, 1-1-1 Wakayamadai, Shimamoto-cho Osaka 618, Japan

NAKAZAWA, MITSURU, ophthalmologist, educator; b. Hakodate, Hokkaido, Japan, Jan. 20, 1956; s. Mitsutake and Chie (Kanazawa) N.; m. Junko Oizumi, Apr. 8, 1980; 1 child, Yuki. MD, Tohoku U., Sendai, Japan, 1980, MSD, 1989. Diplomate Japanese Bd. Ophthalmology. Resident in ophthalmology Tohoku Univ. Hosp., Sendai, 1980-82, fellow in ophthalmology, 1982-85, mem. ophthalmology staff, 1985, 88-89, asst. prof. ophthalmology, 1989—; postdoctoral asst. U. Cin., 1985-88; bd. dirs. Tohoku U. Eye Bank, Miyagi Ophthalmologists Assn. Contbr. articles to profl. jours. Grantee Ministry of Edn., 1989, 91, 93, Japan Eye Bank Assn., 1989, Japan Soc. for Prevention of Blindness, 1993. Mem. AAAS, Japanese Soc. Ophthalmology, Japan Ophthalmologists Assn., Assn. of Rsch. for Vision and Ophthalmology, Japanese Soc. of Ophthalmic Surgeons. Avocations: travel, swimming, social dance, sumo watching. Office: Tohoku U Sch Medicine, 1-1 Seiryo-machi Aoba-ku, Sendai Miyagi 980, Japan

NAKHLA, ATIF MOUNIR, biochemist; b. Cairo, Oct. 23, 1946; came to the U.S., 1981; s. Mounir and Afifa (Nagib) N.; 1 child, Ashraf. BS (hon.) in Biochemistry, Cairo U., 1967, MS, 1971, PhD, 1975. Instr., lectr. Cairo U., 1967-80, assoc. prof., 1980-85; rsch. scientist Coll. Physicians and Surgeons Columbia U., N.Y.C., 1985—; postdoctoral fellow Aarhus (Denmark) U., 1976-79; fellow in residence Rockefeller U., N.Y.C., 1981-85. Contbr. over 40 articles to profl. jours. Fellow Danish Internat. Devel. Agy., 1976, World Health Orgn., 1981. Mem. AAAS, Am. Soc. Biochemistry and Molecular Biology, Endocrine Soc. U.S.A., Am. Recorder Soc., Egyptian Biochem. Soc., Sigma Xi. Avocations: music, drawing, horseback riding, swimming, tennis, chess. Home: PO Box 7917 Jersey City NJ 07307-0917

NAKICENOVIC, NEBOJSA, economist, interdisciplinary researcher; b. Belgrade, Serbia, Yugoslavia, July 1, 1949; arrived in Austria, 1973; s. Slobodan and Dobrila (Dajlevic) N. BA in Econ., Princeton U., 1971; MA in Econ., U. Vienna (Austria), 1980, PhD in Econ. and Computer Sci., 1984. Rsch. asst. dept. econ. Princeton (N.J.) U., 1969-70; scientist Applied Systems Analysis and Reactor Physics Inst., Karlsruhe, Fed. Republic of Germany, 1971-73; rsch. scholar energy program Internat. Inst. for Applied Systems Analysis, Laxenburg, Austria, 1974-83, prin. investigator Dynamics of Tech. Project, 1984-88, project leader Environ. Compatible Energy Strategies Project, 1989—; dir. studies project on energy prospectives World Energy Coun., 1993—; instr. Tech. U. Graz, 1992—; cons. World Bank, Washington, 1991—, Shell Internat. Petroleum Co., Ltd., London, 1989-90, Ministry of Sci. and Tech., Brazil, 1986-87. Co-author: Energy in a Finite World, 1981, Technological Progress, Structural Change and Efficient Energy Use, 1989, Diffusion of Technologies and Social Behavior, 1991; contbr. chpts. to books; mem. editorial bd. Technol. Forecasting and Social Change Jour., 1989—, Jour. Evolutionary Econ., 1989—, Energy, The Internat. Jour., 1992—; contbr. articles to profl. jours. Grantee Volkswagenwerk Found., 1976, Ministry Sci. and Tech., Fed. Republic of Germany, 1981, Austrian Electricity Bd., 1987, Global Indsl. and Social Progress Rsch. Inst., Japan, 1990, Tokyo Electric Power Co., Inc., 1991. Avocations: cosmology, commericial aviation. Home: Muehlengasse 42, A-2362 Biedermannsdorf Austria Office: Internat Inst Applied Systems, Schlossplatz 1, A-2361 Laxenburg Austria

NAM, JUNG WAN, mathematics educator; b. Chinju, Korea, Apr. 21, 1927; s. Eok Man Nam and Uh Ik Choi; m. W. Jae Ok Park; children: Soo-kyung, Yi-Kyung, Hyun-chul, Mee-kyung. BS, Seoul Nat. U., 1956; MS, Kyungpook Nat. U., 1971; PhD, Pusan Nat. U., 1979. chief dept. math. Gyeongsang Nat. U., 1970-80, dean coll., 1980-84, dean grad. sch. edn., 1984-87; com. of edn. Gyeongnam Edn. Bd., Changwon, Korea, 1984—. Recipient Edn. Merit award Korean Edn. Fedl., 1983, Nation award Govt. of Korea, 1986. Home: 297-7 Sangdae-dong, 660-320 Chinju Republic of Korea Office: Gyeongsang Nat U, 900 Kazoa-dong, 660-701 Chinju Republic of Korea

NAM, TIN (TONNY NAM), chemical engineer; b. Kowloon, Hong Kong, Apr. 16, 1961; came to U.S., 1979; s. Ching and Po Kin (Yu) N.; m. Wing-sze Nam, Aug. 24, 1986; children: Tiffany, Stacy. BS, Iowa State U., 1982; PhD, U. Wis., 1990. Sr. devel. engr. Olin Corp., Lake Charles, La., 1990-92, assoc. devel. engr., 1992—. Office: Olin Corp PO Box 2896 Lake Charles LA 70602

NAMBA, TATSUJI, physician, researcher; b. Changchun, China, Jan. 29, 1927; came to U.S., 1959, naturalized, 1968; s. Yosuke and Michino (Hinata) N. M.D., Okayama U., Japan, 1950, Ph.D., 1955. Asst., lectr. medicine Okayama U. Med. Sch. and Hosp., 1955-62; research assoc. Maimonides Med. Ctr., Bklyn., 1959-66; dir. neuromuscular labs. Maimonides Med. Ctr., 1966-70, dir. neuromuscular disease div., head electromyography clinic, 1966—; instr., asst. prof., assoc. prof. medicine State U. N.Y., Bklyn., 1959-76; prof. State U. N.Y., 1976—; mem. med. adv. bd. Myasthenia Gravis Found., 1968—. Recipient commendation for rsch. and clin. activities on insecticide poisoning Minister Health and Welfare, Japanese Govt., 1958; Fulbright scholar, 1959-62. Fellow ACP, Royal Soc. Medicine; mem. AMA, Am. Acad. Neurology, Am. Soc. Pharmacology and Exptl. Therapeutics, Am. Soc. Clin. Pharmacology and Therapeutics, Am. Assn. Electrodiagnostic Medicine. Office: 4802 10th Ave Brooklyn NY 11219-2999

NAMBOODIRI, KRISHNAN, sociology educator; b. Valavoor, Ind., Nov. 13, 1929; s. Narayanan and Parvathy (Kutty) N.; m. Kadambari Kumari, Sept. 7, 1954; children: Unni (dec.), Sally. B.Sc., U. Kerala, 1950, M.Sc., 1953; M.A., U. Mich., 1962, Ph.D., 1963. Lectr. U. Kerala, India, 1953-55, 58-59; tech. asst. Indian Statis. Inst., Calcutta, 1955-58; reader demography U. Kerala, 1963-66; asst. prof. sociology U. N.C., Chapel Hill, 1966-67; asso. prof. U. N.C., 1967-73, prof., 1973-84, chmn. dept., 1975-80; Robert Lazarus prof. population studies Ohio State U., Columbus, 1984—, chmn. dept. sociology, 1989—. Author: (with L.F. Carter and H.M. Blalock) Applied Multivariate Analysis and Experimental Designs, 1975; editor: Demography, 1975-78, Survey Sampling and Measurement, 1978, Auth. Matrix Algebra: An Introduction, 1984, (with C.M. Suchindran) Life Table Techniques and Their Applications, 1987, (with R.G. Corwin) Research in Sociology of Education and Socialization: Selected Methodological Issues, 1989, Demographic Analysis: A Stochastic Approach, 1991; contbr. articles to profl. jours. Fellow Am. Statis. Assn.; mem. Population Assn. Am. (dir. 1975-76), Internat. Union Sci. Study Population, Am. Sociol. Assn., Indian Sociol. Assn., Am. Statis. Assn., Sociol. Research Assn. Home: 3107 N Star Rd Columbus OH 43221-2366

NAMBOODIRI, KRISHNAN, chemist; b. Calicut, Kerala, India, Jan. 25, 1953; came to U.S., 1983; s. Krishnan Kalpakasseri Pattathil Namboodiri and Savithri (Paduthol) Namboodiripad; m. Indiradevi Vasudevan Elayath, Feb. 2, 1985; children: Arya K., Soorya K. MS, Sardar Patel U., 1978; PhD, Madras U., 1983. Rsch. asst. Indian Inst. of Sci., Bangalore, 1982, Case Western Res. U., Cleve., 1983-84; postdoctoral rsch. asst. Mt. Sinai Med. Ctr., N.Y.C., 1984-86, N.J. Inst. Tech., Newark, 1986-87; postdoctoral rsch. assoc. Georgetown Univ. Med. Ctr. and Naval Rsch. Lab., Washington, 1987-90, asst. prof., 1990-92; sr. sci. analyst Martin Marietta Tech. Svcs. & Nat. Environ. Supercomputing Ctr., Bay City, Mich., 1992—; cons. Nat. Biomed. Rsch. Found., Washington, 1989-92, Nat. Ctr. for Biotech. Info. NIH, Bethesda, Md., 1990-92. Contbr. articles to profl. jours. Mem. AAAS, Am. Crystallographic Assn., Am. Chem. Soc., Sigma Xi. Achievements include linkage of 2 important biomolecular databases; explanation of toxicity of important polymer precursers. Home: 5111 Loganberry Dr Saginaw MI 48603

NAMBOODRI, CHETTOOR GOVINDAN, mechanical engineer; b. Greensboro, N.C., Nov. 1, 1968; s. Chettoor Govindan and Sandra Marie (Neal) N.; m. Shannon Mary Leahy, Dec. 30, 1989. BSME, Clemson U., 1990; MSME, Va. Tech., 1992. Mfg. mgmt. engr. GE, Salem, Va., 1993—. Big Bro. Big Bros.-Big Sisters, Inc., New River Valley, Va., 1991-93; coach Youth Soccer League, Recreation Dept., North Augusta, S.C., 1988. Nat. Def. Sci. and Engring. fellow, 1990. Mem. ASME, Am. Soc. Engring. Edn., Tau Beta Pi (treas. 1988-89), Pi Tau Sigma (v.p. 1988-89). Achievements include research articles on recent advances in adaptive materials and nuclear waste mgmt.

NAMDARI, BAHRAM, surgeon; b. Oct. 26, 1939; s. Rostam and Sarvar Namdari; M.D., 1966; m. Kathleen Diane Wilmore, Jan. 5, 1976. Resident in gen. surgery St. John's Mercy Med. Ctr., St. Louis, 1969-73; fellow in cardiovascular surgery with Michael DeBakey, Baylor Coll. Medicine, Houston, 1974-75; practice medicine specializing in gen. and vascular surgery and surg. treatment of obesity, Milw., 1976—; mem. staff St. Mary's, St. Luke's, St. Michael, St. Francis hosps. (all Milw.); founder, pres. Famous Mealwaukee Foods Enterprises. Diplomate Am. Bd. Surgery. Fellow ACS, Internat. Coll. Surgeons; mem. Med. Soc. Milw. County, Milw. Acad. Surgery, Wis. Med. Soc., Wis. Surg. Soc., Royal Soc. Medicine Eng. (affiliate), Am. Soc. for Bariatric Surgery, AMA, World Med. Assn., Internat. Acad. Bariatric Medicine (founding mem.), Michael DeBakey Internat. Cardiovascular Soc. Contbr. articles to med. jours.; patentee med. instruments and other devices. Office: Great Lakes Med and Surg Ctr 6000 S 27th St Milwaukee WI 53221-4805

NAMINI, AHMAD HOSSEIN, structural engineer, educator; b. Montreal, Que., Can., July 1, 1961; s. Hassan and Roshanak (Dara) N.; m. Geralyn Smariga, Nov. 7, 1987; children: Sarah, Troy. BS, U. Md., 1982, MS, 1984, PhD, 1989. Cert. engr. Civil engr. Sys Corp., Santa Monica, Calif., 1984-85, Best Ctr., College Park, Md., 1985-87, EMTEC Corp., Bethesda, Md., 1987-88; rsch. fellow Fed. Highway Adminstn., Washington, 1988-89; prof. U. Miami, Coral Gables, Fla., 1989--; strutural engr. Baker Engring., Pitts., 1990-91. Recipient Rsch. Initiation award Nat. Sci. Found., 1991; named Knight fellow J.L. Knight Found., 1991. Mem. ASCE, Internat. Assn. for Bridge and Structural Engring. Achievements include computer aided design of cable stayed bridges; algorithm for aeroelastic analysis of long span bridges. Office: U Miami 303 McArthur Coral Gables FL 33124

NANCE, RICHARD DAMIAN, geologist; b. St. Ives, Cornwall, U.K., Oct. 25, 1951; came to U.S., 1980; s. Richard William Morton and Edith Eleanor (Leach) N.; m. Rita Felice Carpenter, Aug. 28, 1982; children: André Bernard Carpenter, Sarah Marie Eleanor, Christopher Louis Morton. BS in Geology, U. Leicester, U.K., 1972; PhD, U. Cambridge, U.K., 1978. Asst. prof. St. Francis Xavier U., Antigonish, Nova Scotia, 1976-80; from asst. prof. to full prof. Ohio U., Athens, 1980—; W.F. James Prof. of Pure and Applied Sci. St. Francis Xavier U., 1993—; rsch. cons. La. State U., Baton Rouge, 1977-81, Argonne Nat. Lab., Chgo., 1984-88; sr. rsch. geologist Exxon Prodn. Rsch. Co., Houston, 1982-83. Contbr. articles to profl. jours. Grantee Geol. Survey of Can., 1987-88, NSF, Oxford, 1990, New Brunswick, 1990, Earthwatch, Mt. Olympus, Greece, 1991-93, Nat. Geographic, Mt. Olympus, 1993—, Ohio U., New Brunswick, 1981-92. Fellow Geol. Soc. Am., Royal Geol. Soc. Cornwall; mem. Am. Geol. Assn. Canada, Am. Geophys. Union, Am. Assn. Petroleum Geologists, Ussher Soc., Trevithick Soc., Soc. for Indsl. Archaeology, Internat. Stationary Steam Engine Soc., Sigma Xi. Avocations: cornish mining history, stationary steam engines, jogging. Home: 3 Julian Dr Athens OH 45701-3661 Office: Ohio University Dept Geological Sciences Athens OH 45701

NANDAGOPAL, MALLUR R., engineer; b. Kolar, Karnataka, India, May 14, 1938; came to U.S., 1976; s. M. Ramanuja Iyengar and Garudammal; m. Sreedharani K. Ramamurthy; children: Radha, Meena, Sudha. BS, Cen. Coll., Bangalore, India, 1958; B of Tech., Indian Inst. Tech., Bombay, 1962; ME, Indian Inst. Sci., Bangalore, 1963, PhD, 1974. Registered profl. engr., Wash. Mem. faculty Indian Inst. Sci., 1963-77; engr. City of Spokane, Wash., 1977—; coord. summer sch. Indian Inst. Sci., 1974-75. Contbr. articles to profl. jours. Mem. IEEE (sr.), Inst. Sci. (sec. Staff Club 1972-74), Fed. Emergency Mgmt. Agy. (mitigation com.). Hindu. Avocations: tennis, astrology, reading, movies. Home: 410 E Shiloh Hills Dr Spokane WA 99208-5819

NANDIVADA, NAGENDRA NATH, biochemist, researcher; b. Bhimavaram, India, Aug. 3, 1950; came to U.S., 1982; s. Krishna Murty and Putali Bai Nandivada; m. Vijaya Lakshmi, Aug. 14, 1980; children: Pratima, Sandeep. BS, Osmania Univ., Hyderabad, India, 1975; MS in Biochemistry, Bombay U., 1978, PhD, 1981. Postdoctoral fellow Population Coun., N.Y.C., 1982-84; rsch. assoc. Med. Ctr. La. State U., New Orleans, 1984-86; supr. RIA lab. L.I. Jewish Med. Ctr., N.Y.C., 1986-87; scientist Med. Ctr. SUNY, Bklyn., 1988—. Postdoctoral fellow Melan Found. Australia, 1982-84. Mem. N.Y. Acad. Scis., Endocrine Soc. (India). Republican. Hindu. Home: 21 Locust Ct Staten Island NY 10309 Office: SUNY Health Sci Ctr 450 Clarkson Ave Brooklyn NY 11203

NANDY, SUBAS, chemical engineer, consultant; b. Calcutta, India, Feb. 14, 1957; came to U.S., 1979; s. Sachindra Nath and Padma (Roy) N.; m. Sushmita De, July 19, 1981. B Tech with honors, Indian Inst. Tech., Kharagpur, 1979; MS, Colo. State U., 1981; PhD, Pa. State U., 1986; MBA, U. Mass., 1992. Registered profl. engr., Colo. Rsch. assoc. dept. chem. engring. MIT, Cambridge, 1986-87; sr. engr. Polaroid Corp., Waltham, Mass., 1988—; cons. Nat. Coun. Examiners Engring. and Surveying, Clemson, S.C., 1990—; adj. faculty Northeastern U., Boston, 1992—. Contbr. articles to Jour. Biomech. Engring., Biorheology, other profl. jours. Sec. Bangla-O-Biswa, Bengalee Club Boston, 1987-88. Mem. AIChE, Am. Chem. Soc., Sigma Xi, Tau Beta Pi. Achievements include demonstration that growth of carboxylic acid chains on polymer latex surface provides steric stabilization preventing coagulation during high shear coating operation. Home: Apt 2516 25 Francis Ave Mansfield MA 02048-1511 Office: Polaroid Corp Bldg 4 1265 Main St Waltham MA 02254

NANNERY, MICHAEL ALAN, civil engineer; b. Norfolk, Va., Nov. 10, 1967; s. James Michael and Barbara Ann (Selig) N. BS, Old Dominion U., 1992. Asst. engr. St. Mary's County Met. Commn., Lexington Park, Md., 1992—. Mem. ASCE (assoc.), NSPE. Republican. Roman Catholic. Home: PO Box 1393 Lexington Park MD 20653 Office: St Mary's County Met Commn 191-B Shangri-La Dr Lexington Park MD 20653

NANNICHI, YASUO, engineering educator; b. Sendai, Miyagi, Japan, Aug. 7, 1933; s. Minoru and Kiyo (Tsuneda) N.; m. Ikuko Utsumi, May 5, 1967; children: Ken, Shoji, Tomo. B. in Engring., U. Tokyo, 1956; postgrad., Stanford U., 1960-61; PhD, U. Tokyo, 1966. Physicist Nippon Electric, Tokyo, 1956-79; rsch. assoc. Stanford U., Palo Alto, Calif., 1967-68; sr. expert Internat. Telecommunication Union, Geneva, 1977-78; project mgr. UN Devel. Program/Internat. Telecomm. Union, Brasilia, Brazil, 1978; prof. U. Tsukuba, Japan, 1979—, chmn. Inst. Materials Sci., 1981-85, dean Coll. Engring. Systems, 1991-92, v.p., 1992—. Editor: Frontier of Materials Science, 1981. Recipient Medal of Honor laser life improvement, 1977. Mem. Am. Phys. Soc., Japan Soc. Applied Physics, IEEE, Internat. Conf. on Solide State Devices (chmn. 1990-92). Avocation: classic camera collection. Office: U Tsukuba, 1-1-1 Ten-nodai, Tsukuba 305, Japan

NANZ, CLAUS ERNEST, economist, management consultant; b. Stuttgart, June 14, 1934. B. Commerce (Dipl. rer. pol.) U. Frankfurt; Doctor's degree (Dr. rer. oec.), U. Mannheim. Gen. mgr., bd. dirs. various German holding cos.; pres., owner Eurofound Internat. Mgmt. Cons., 1976—. Author: Fluktuation: Das Problem und betriebliche Massnahmen seiner Minderung. Recipient Best Thesis award Univ. Mannheim and Rhein-Main C. of C. Avocations: climbing, skiing, boating, scientific travelling, languages. Address: Landhaus Valbrava, D87569 Mittelberg Germany

NAQVI, SARWAR, aerospace engineer; b. New Delhi, India, Oct. 20, 1943; came to U.S., 1967; s. Syed Manzoor Ali and Qamar Jehan (Agha) N.; m. Patricia Engel, July 25, 1970 (div. Jan. 1981); children: Arif Ali Naqvi, Zafar Ali Naqvi; m. Ghizala Islam, Mar. 22, 1983; children: Mustafa Ali Naqvi, Asad Ali Naqvi. BS in Engring. with honors, U. London, 1965; MS, Brown U., 1971, Rice U., 1971; PhD, Rice U., 1972. Sr. rsch. engr. Esso Rsch. Co., Houston, 1972-73; sr. rsch. engr. Honeywell, Inc., Mpls., 1973-74; tech. specialist McDonnell Douglas Corp., Houston, 1974-81; scientist IBM Fed. Systems Div., Houston, 1981-85; dir. BCCI Found. for Advancement of Sci.

and Tech., Karachi, Pakistan, 1985-86; project mgr. Rockwell Space Operation Co., Houston, 1987—. Author: (book chpt.) Advances in Control Systems, 1971; (publs.) Journal of Mathematical Analysis, 1972, Journal of Astronautical Sciences, 1973; contbr. articles to profl. jours. Pres. Pakistan Assn. Greater Houston (elected 9 times). Recipient rsch. fellowship U. London, Brown U., Rice U.; awarded full colors for batting performance U. London cricket team, 1966. Fellow Inst. Engrs. Pakistan (vice chmn. 1984-85); assoc. fellow AIAA (vice chmn. Houston 1980-81, 1991-92, councilor 1981-82); mem. IEEE, Instrument Soc. Engrs. (sr.), Pakistan Engring. Coun. (profl. engr.), Sigma Xi. Home: 16014 Lost Rock Ct Webster TX 77598 Office: Rockwell Space Ops 600 Gemini Houston TX 77058

NARAHASHI, TOSHIO, pharmacology educator; b. Fukuoka, Japan, Jan. 30, 1927; came to U.S., 1961; s. Asahachi and Itoko (Yamasaki) Ishii; m. Kyoko Narahashi, Apr. 21, 1956; children: Keiko, Taro. BS, U. Tokyo, 1948, PhD, 1960. Instr. U. Tokyo, 1951-65; research assoc. U. Chgo., 1961, asst. prof., 1962; asst. prof. Duke U., Durham, N.C., 1962-63, 65-67, assoc. prof., 1967-69, prof., 1969-77, head pharmacology div., 1970-73, vice chmn. dept. physiology and pharmacology, 1973-75; prof., chmn. dept. pharmacology Northwestern U. Med. Sch., Chgo., 1977—; Alfred Newton Richards prof. Northwestern U., Evanston, Ill., 1983—, John Evans prof., 1986—; mem. pharmacology study sect. NIH, 1976-80; mem. research rev. com. Chgo. Heart Assn., 1977-82, vice chmn. research council, 1986-87, chmn., 1988-90; mem. Nat. Environ. Health Scis. Council, 1982-86; rev. com. Nat. Inst. Environ. Health Scis., 1991—. Editor: Cellular Pharmacology of Insecticides and Pheromones, 1979, Cellular and Molecular Neurotoxicology, 1984, Insecticide Action: From Molecule to Organism, 1989; editor Ion Channels, 1988—; contbr. articles to profl. jours. Recipient Javits Neurosci. Investigator award NIH, 1986. Fellow AAAS; mem. Am. Soc. for Pharmacology and Exptl. Therapeutics, Am. Physiol. Soc., Soc. for Neurosci., Biophys. Soc. (Cole award 1981), Soc. Toxicology (DuBois award 1988, Merit award 1991), Agrochem. Div. Am. Chem. Soc. (Burdick L. Jackson Internat. award 1989). Home: 175 E Delaware Pl Apt 7911 Chicago IL 60611-1732 Office: Northwestern U Med Sch Dept Pharmacology 303 E Chicago Ave Chicago IL 60611-3008

NARASAKI, HISATAKE, analytical chemist; b. Fukuoka, Kyushu, Japan, Nov. 13, 1933; s. Kumeo and Tsutae Narasaki; m. Satoko Masuzawa, Nov. 1965; children: Jun-ichi, Masako. BS, Kyushu U., Fukuoka, Japan, 1956; MS, U. Tokyo, Japan, 1962, PhD, 1965. Lectr. Saitama U., Urawa, Japan, 1965-67, asst. prof., 1967-87, prof. analytical chemistry, 1988—; vis. rsch assoc. N.E. London Polytech., 1975-76. Author: Talanta: The Use of Approximation Formulae in Calculations of Acid-Base Equilibria, 1980, Analytical Chemistry: Automated Hydride Generation Atomic Absorption Spectrometry, 1986. Mem. Chem Soc. Japan, Japan Soc. Analytical Chemistry. Home: 67-9 Kami-Okubo, Urawa 338, Japan Office: Saitama U Faculty Sci Dept Chem, 255 Shimo-Okubo, Urawa 338, Japan

NARASIMHA, RODDAM, laboratory director, educator; b. Bangalore, Karnataka, India, July 20, 1933; s. R. L. Narasimhaiya and R. N. Leela Devi; m. Neelima R. Narasimha; children: Maithreyi, Aditi. BE in Mech. Engring., U. Mysore, Bangalore, 1953; DIISc, Indian Inst. Sci., Bangalore, 1955, AIISc, 1957; PhD in Aero. and Physics, Calif. Inst. Tech., 1961. Asst. prof. Indian Inst. Sci., Bangalore, 1962-70, prof., 1970—, dean engring., 1980-82, chmn. aerospace engring., 1983-84; chief project coord. Hindustan Aeros. Ltd., Bangalore, 1977-79; Sherman Fairchild scholar Calif. Inst. Tech., Pasadena, 1982-83, Clark B. Millikan prof., 1985—; J. Nehru prof. Cambridge U., Eng., 1989-90; dir. Nat. Aero. Lab., Bangalore, 1984-93; mem. sci. adv. coun. to Prime Minister, India, 1984-89; mem. governing body Aero. Devel. Agy., New Delhi, 1984—; mem. gen. assembly Internat. Union Theoretical & Applied Mechs., Paris, 1984—; bd. dirs. Hindustan Aeros. Ltd., Bangalore. Coun. mem. Karnataka State Coun. Sci. & Tech., Bangalore, 1983—. Minta Martin Nat. Student Inst. Aerospace Scis., U.S., 1960; recipient Bhatngar prize Coun. Sci. & Indsl. Rsch., New Delhi, 1976; named Disting. Alumnus Calif. Inst. Tech., 1986, Padma Bhushan Govt. India, 1987. Fellow Indian Acad. Scis. (pres. 1992—), Indian Nat. Sci. Acad. (coun. 1982-86, Bhatnagar medal 1985), Aero. Soc. India (Burmah-Shell medal 1970), Third World Acad. Scis., Royal Soc. (London); mem. NAE (fgn. assoc.). Avocations: history, walking. Home: 72 Jaladarshini Layout, Rajmahal Vilas (II Stage), Bangalore 560 094, India Office: Nat Aero Lab, PB No 1779 Kodihalli, Bangalore 560 017, India

NARATH, ALBERT, national laboratory director; b. Berlin, Mar. 5, 1933; came to U.S., 1947; s. Albert Narath and Johanna Agnes Anne (Bruggemann) Bruckmann; m. Worth Haines Scattergood, (div. 1976); children: Tanya, Lise, Yvette; m. Barbara Dean Camp, Aug. 8, 1976; 1 child, Albert. BS in Chemistry, U. Cin., 1955; PhD in Phys. Chemistry, U. Calif., Berkeley, 1959. Mem. tech. staff, mgr. phys. sci. Sandia Nat. Labs., Albuquerque, 1959-68, dir. solid state sci., 1968-71, mng. dir. phys. sci., 1971-73, v.p. rsch., 1973-82, exec. v.p. rsch. and adv. weapons systems, 1982-84, pres., 1989—; v.p. govt. systems AT&T-Bell Labs, Whippany, N.J., 1984-89; vice chmn. basic energy scis. adv. com. Dept. Energy, 1987—; cons. in field. Contbr. sci. articles to profl. jours. Fellow Am. Phys. Soc. (George E. Pake prize 1991); mem. NAE, AAAS. Office: Sandia Nat Labs PO Box 5800 Albuquerque NM 87185-5800

NARAYAN, JAGDISH, materials science educator; b. Kanpur, India, Oct. 15; came to U.S., 1969; s. Shri Sheo Nath and Radha Prasad; m. Ratna (Katyar) Narayan, Nov. 27, 1973; 1 child, Roger. BS in Materials Sci. with highest honors, Indian Inst. Tech., Kanpur, 1969; MS, U. Calif., Berkeley, 1970, PhD, 1971. Group leader solid state dvsn. Oak Ridge (Tenn.) Nat. Lab., 1971-84; dir. div. materials rsch. NSF, Washington, 1990-92, maj. facilities reviewer; disting. univ. prof. materials sci. and engring. N.C. State U., Raleigh, 1984—; dir. Microelectronic Ctr. N.C., Raleigh, 1984-86; tech. adv. bd. Kopin Corp., Taunton, Mass., 1988—; mem. exec. coun. electronic, magnetic and photonic materials dvsn. TMS, awards chair. Editor: Defects in Semiconductors, 1981, Laser-Solid Interactions and Transient Thermal Processing of Metals, 1984, High-Temperature Superconductors, 1990, others; contbr. over 400 sci. papers to profl. jours. Vice-pres. bd. trustees Hindi Vikas Mandal, Raleigh, 1990. Recipient award for outstanding sustained rsch. Dept. of Energy. Fellow AAAS, Am. Phys. Soc. (life), Nat. Acad. Scis. India (life), Am. Soc. Metals Internat.; mem. Materials Rsch. Soc. (fall meeting co-chair 1984, councillor 1984-87, long-range planning com. 1987-89), Bôhmische Physikalische Gesellschaft. Achievements include 14 patents in fields of laser processing of materials, high performance ceramics, novel methods for materials synthesis and processing, laser processing and patterning of diamond films. Office: NC State Univ Dept Materials Sci/Engring Raleigh NC 27695-7916

NARAYAN, K(AVASSESY) SURESWARAN, physicist; b. Madras, India, Jan. 26, 1964; came to U.S., 1986; s. K.K. and V. Sureswaran. MS, The Ohio State U., 1988, PhD, 1991. Grad. teaching asst. dept. physics Ohio State U., Columbus, 1986-88, grad. rsch. asst., 1988-91; sr. postdoctoral fellow Wright Patterson AFB, Ohio, 1992; scientist Systran Corp., Dayton, Ohio, 1992—. Contbr. articles to profl. jours. Pres. SPIC-Macay, The Ohio State U., 1991-92; rep. Grad. Student Orgn., 1988. Grantee Ecole Normal Superieure, 1991. Mem. Am. Phys. Soc., Materials Rsch. Soc., Sigma Xi. Achievements include discovery of the role of defects in a magnetic chain; anamolous behavior in certain molecular magnets; heating effect of ladder-type polymer BBL.

NARAYAN, RAMESH, astronomy educator; b. Bombay, India, Sept. 25, 1950; came to U.S., 1983; s. G.N. and Rajalakshmi (Sankaran) Ramachandran; m. G.V. Vani, June 6, 1977. BS in Physics, Madras U., 1971; MS in Physics, Bangalore U., 1973, PhD in Physics, 1979. Rsch. scientist Raman Rsch. Inst., Bangalore, India, 1978-83; postdoctoral fellow Calif. Inst. Tech., 1983-84, sr. rsch. fellow, 1984-85; assoc. prof. U. Ariz., Tucson, 1985-90, prof. astronomy, 1990-91; prof. astronomy Harvard U., Cambridge, Mass., 1991—; sr. astronomer Smithsonian Astrophys. Obs., Cambridge, 1991—. Contbr. articles to profl. jours. Named NSF Presdl. Young Investigator, 1989. Mem. AAAS, Am. Astronomical Soc., Internat. Astronomical Union, Astromonical Soc. India. Achievements include research in the general area of theoretical astrophysics, specializing in accretion disks, collapsed stars, gravitational lenses, hydrodynamics, image processing and scintillation. Office: Harvard-Smithsonian Ctr Astrophysics 60 Garden St MS 51 Cambridge MA 02138

NARDONE, ROBERT CARMEN, cell biologist; b. Wilkes-Barre, Pa., May 27, 1953; s. Lucas Louis and Anna Ursula (Rusavage) N.; m. Marie Rose Koval, Oct. 11, 1980. BS in Biology, Kings Coll., 1975; postgrad., U. N.C., Chapel Hill, N.C., 1977-82. Rsch. technician U.S. Govt., Research Triangle Park, N.C., 1975-82; rsch. technician dept. immunology Duke U., Durham, N.C., 1982-83; rsch. assoc. dept. genetic toxicology Pharmakon Rsch., Waverly, Pa., 1984-87; rsch. technician biotech. group DuPont Co., Wilmington, Del., 1987-89; rsch. assoc. dept. cell culture Terumo Med. Corp., Elkton, Md., 1989-90; assoc. scientist dept. immunobiology Centocor, Inc., Malvern, Pa., 1990—. Author: (poster presentation) Mutaganicity of a Petroleum Oil Extract to Salmonella Typhhmurium and Mouse Lymphoma Cells, for Pharmakon Rsch., Internat., Waverly, Pa., 1986. Mem. AAAS, Tissue Culture Assn., N.Y. Acad. Sci. Home: 158 S Kingscroft Dr Bear DE 19701-1443 Office: Centocor Inc Immunobiology 220 Great Valley Pky Malvern PA 19355-1307

NARENDRA, KUMPATI SUBRAHMANYA, electrical engineering educator, association administrator; b. Madras, India, Apr. 14, 1933; came to U.S., 1954, naturalized, 1974; s. Subrahmanya and Sarada (Alladi) Kumpati; m. Barbara Lamb, Nov. 3, 1961. BEE with honors, U. Madras, 1954; MS, Harvard U., 1955, PhD, 1959; MA (hon.), Yale U., 1968. Lectr., postdoctoral asst. Harvard U., Cambridge, Mass., 1959-61, asst. prof., 1961-65; assoc. prof. Yale U., New Haven, Conn., 1965-68, prof. elec. engring., 1968—, chmn. dept. elec. engring., 1984-87; cons. to comml. firms, 1961—; dir. Ctr. for Systems Sci., 1980—. honorary vis. prof. Anna Univ., Madras, India, 1993; mem. adv. bd. Inst. Advanced Engring, Korea. Author: Frequency Domain Criteria For Absolute Stability, 1973, Stable Adaptive Systems, 1989, Learning Automata: An Introduction, 1989; editor: Applications of Adaptive Control, 1980, Adaptive and Learning Systems: Theory and Applications, 1987, Advances in Adaptive Control, 1991; editor issue on learning automata Jour. Cybernetics and Info. Sci., vol. I, 1977. Recipient Edn. award Am. Automatic Control Coun., 1990. Fellow AAAS, Inst. Elec. Engrs., IEEE United Kingdom (Franklin V. Taylor award 1973, George S. Axelby award, 1988, Outstanding Paper of neural network coun. 1991); mem. Sigma Xi. Home: 35 Old Mill Rd Woodbridge CT 06525-1523 Office: Yale U Ctr Systems Sci PO Box 2157 New Haven CT 06520-2157

NARIN, FRANCIS, research company executive; b. Phila., May 10, 1934; s. Bernard E. and Anna L. (Lipsius) N.; m. Carole Shapiro, July 6, 1958; children: Shari, Cindy. BS in Chemistry, Franklin and Marshall U., Lancaster, Pa., 1955; MS in Nuclear Engring., N.C. State U., 1957; PhD in Bibliometrics, Walden U., 1980. Sr. staff mem. Ill. Inst. Tech. Rsch. Inst., Chgo., 1957-59, '64-68; staff mem. Los Alamos Nat. Lab., N. Mex., 1960-63; pres. Chi Rsch. Inc., Haddon Heights, N.J., 1968—. Contbr. articles to profl. jours. Numerous grants: NSF, NIH, Washington. Fellow AAAS; mem. World Future Soc., Assn. for Computing Machinery, Am. Soc. for Info. Sci., N.Y. Acad. Scis. Office: Chi Rsch Corp 10 White Horse Pike Haddon Heights NJ 08035

NARRAMORE, JIMMY CHARLES, aerospace engineer; b. Mountain Home, Ark., Mar. 13, 1949; s. Charles Augustus Narramore and LaVern (Tolliver) Kerr; m. Deborah Sue Campbell, June 5, 1971; 1 child, Keri Elizabeth. BS, U. Tex., 1972, MS, 1973. Engr., scientist McDonnell Douglas Aircraft Co., Long Beach, Calif., 1973-75; advance design engr. Cessna Aircraft Co., Wichita, 1975-79; sr. engr. Bell Helicopter Textron, Fort Worth, 1979-89, prin. engr., 1989—. Contbr. articles to profl. jours. Musician Trinity Arts Coun., Bedford, Tex., 1991-92; choir pres. First United Meth. Ch., Hurst, Tex., 1984-85, tchr. Sunday sch., 1991—. Mem. AIAA, Am. Helicopter Soc. (chmn. aerodynamics com. 1991, aerodynamics com. 1985-91). Achievement includes a patent for flaperon system for tilt rotor wings; designed rotor and wing airfoils for V-22 Osprey tilt rotor aircraft. Home: 2505 Rollingshire Ct Bedford TX 76021 Office: Bell Helicopter Textron PO Box 482 Fort Worth TX 76101

NARUTIS, VYTAS, chemist, researcher; b. Germany, May 25, 1950; came to U.S., 1952.; s. Pilypas and Elvra Narutis; m. Jolita, 1982; 1 child, Aras. PhD, Ill. Inst. Tech., 1982. Chemist Humko-Sheffield/Kraft Co., Glenview, Ill., 1972-75; scientist EPA, Chgo., 1980-82; g. leader Nalco Chem., Naperville, Ill., 1982—. Contbr. articles to profl. jours. Active Chgo. Coun. Fgn. Affairs, 1988—; bd. dirs. Lithuanian Fund., Chgo., 1990—. Mem. Am. Chem. Soc. (polymer divsn.), Internat. Soc. Magnetic Resonance, Sigma Xi. Achievements include two patents in phosphinate compounds; development of microstructural and sequence analysis of polymers by NMR; research in conformational analysis of polymers, hydrogen exchange and peptide conformational analysis, NMR conformational analysis of dipeptides, stereoelectronic effects on hydroboration. Office: Nalco Chem 1 Nalco Ctr Naperville IL 60566

NASER, NAJIH A., chemistry educator, researcher; b. Zaita, West Bank, Oct. 10, 1962; s. Abdelrahman and Amneh A. Naser. BSc, Yarmouk U., Irbid, Jordan, 1985; MSc, U. Bridgeport, 1987; postgrad., N.Mex. State U., 1992—. Analytical chemist Danbury (Conn.) Pharmacal, 1987; rsch. asst. N.Mex. State U., Las Cruces, 1988-92, teaching asst., 1989-92; pres.'s assoc., N.Mex. State U., 1990-91. Contbr. articles to profl. jours. Achievements include research in design and characterization of novel portable electrochemical biosensors for bioanalysis and biotechnology, and for on-line monitoring and screening of environmentally significant compounds. Office: NMex State U Dept Chem Box 3C Las Cruces NM 88003

NASH, HOWARD ALLEN, biochemist, researcher; b. N.Y.C., Nov. 5, 1937; s. Harvey and Harriet (Ratner) N.; m. Dominie Maria Shortino, Aug. 31, 1963; children: Janet Elisabeth, Emily Julia. BS, Tufts U., 1957; MD, U. Chgo., 1961, PhD, 1963. Intern U. Chgo. Clinics, 1963-64; rsch. assoc. NIMH, Bethesda, Md., 1964-68, med. officer (res), 1968-84, chief, sec. molecular genetics, 1984—; chmn. Gordon conf. on Nucleic Acids, 1988; vice-chair FASEB Conf. on Genetic Recombination, 1993. Assoc. editor: Cell Jour., 1985-91; editorial bd.: Current Biology Jour., 1993—. Lt. comdr. USPHS, 1964-66. Recipient Superior Svc. award USPHS, 1983, Disting. Svc. award HHS, 1990. Fellow Am. Acad. of Arts & Sci.; mem. NAS, Am. Soc. for Biochemistry and Molecular Biology, Am. Soc. for Microbiology. Avocations: choral singing, birdwatching. Office: Lab Molecular Biology 9000 Rockville Pike Bethesda MD 20892-0036

NASH, LILLIAN DOROTHY, gynecologist, reproductive endocrinologist; b. Lyndhurst, N.J., Jan. 27, 1931; d. Wilfrid Joseph and Lillian Bernadette (Rogers) N. BS, Chestnut Hill Coll., Phila., 1953; MD, Med. Coll. Pa., Phila., 1957. Diplomate Am. Bd. Ob-Gyn. Rotating intern Orange (N.J.) Meml. Hosp., Orange, 1957-58; asst. resident St. Clare's Hosp., N.Y.C., 1958-60, chief resident, 1960-61; asst. resident Sloane Hosp., N.Y.C.; asst. physician The Roosevelt Hosp., N.Y.C., 1965-71, Community Mem. Hosp., N.J., 1971-73; gynecologist Fertility Rsch., N.Y.C., 1977-89; assoc. attending physician St. Barnabas Med. Ctr., 1983—, St. Lukes Roosevelt Hosp., 1984—; asst. prof. SUNY, Stony Brook, 1973-76; clin. instr. ob-gyn. Columbia U. Coll. Phys. and Surg., N.Y.C., 1984—. Barnes Foster fellow Sloane Hosp. Women, N.Y.C., 1962-65, NIH Trainee grantee, 1962-65, Ford Found. fellow, 1962-65. Fellow Am. Coll. Ob-gyn.; mem. Am. Fertility Soc., Am. Assn. Gyn Laparoscopists, Women's Med. Assn. N.Y., N.Y. Gynecol. Soc. Roman Catholic. Avocations: reading, traveling, exercise. Home: 425 E 72d St New York NY 10021 Office: 15 James St Florham Park NJ 07932-1346 also: 230 Central Park South New York NY 10019

NASH, WILLIAM WRAY, JR., retired city planning educator; b. N.Y.C., Nov. 25, 1928; s. William Wray and Janet Caroline (Fobes) N.; m. Dorothy Elaine Westerberg, Dec. 23, 1950; children: Meryl Elaine, Wendy Wray, Janet Amanda, Joseph Adamson. BA, Harvard U., 1950; M City Planning, U. Pa., 1956, PhD, 1961. Rsch. analyst Am. Coun. To Improve Our Neighborhood, Phila., 1956-58; mem. faculty Harvard U., Cambridge, Mass., 1958-71, chmn. dept. city planning, 1964-69; advisor on urban affairs Office of Gov., State of Ga., Atlanta, 1970-74; mem. faculty Ga. State U., Atlanta, 1971-90, dean Coll. Urban Life, 1976-81, Regents prof., 1983; ret., 1990; prin. Nash-Vigier, Inc., Cambridge, 1962-71; advisor UN Tech. Assistance Bd., Bandung, Indonesia, 1961-62; traveling lectr. USIA, Japan, 1980. Author: Residential Rehabilitation, 1959, (with B. Frieden) Shaping an Urban America, 1969; contbg. author: Taming the Metropolis, 1967, Four Days: Forty Hours, 1970. Mem. Atlanta Mayor's Com. To Reorganize City Govt., 1973, Atlanta Zoning Rev. Bd., 1974-75, Atlanta 2000, 1982; pres. Ga. Planning Assn., Atlanta, 1973-74; chmn. Atlanta Regional Forum, 1976; vestryman Ch. of Incarnation, 1973-76, 83-85, sr. warden, 1986-88; mem. Atlanta Multiple Sclerosis Support Group, 1986. With U.S. Army. 1950-53. Decorated Bronze Star. Mem. Am. Inst. Planners, Am. Inst. Cert. Planners, Sigma Xi, Tau Sigma Delta, Phi Kappa Phi. Democrat. Episcopalian. Home: 3086 Cascade Rd SW Atlanta GA 30311

NASHMAN, ALVIN ELI, computer company executive; b. N.Y.C., Dec. 16, 1926; s. Joseph and Fay (Portnoy) N.; m. Honey Weinstein, May 29, 1960; children—Jessica Rachel, Pamela Wynne, Stephanie Paige. B.E.E., CUNY, 1948; M.E.E., NYU, 1951; Sc.D. (hon.), Pacific U., 1968, George Washington U., 1986. With Ketay Mfg. Corp., N.Y.C., 1951-52; sr. project engr., exec. engr., assoc. lab. dir., lab. dir. ITT Fed. Labs., Nutley, N.J., 1952-62; dir. ops. ITT Intelcom, Inc., Falls Church, Va., 1962-65; pres. System Scis. Corp., Falls Church, 1965-67; with Communications & Systems, Inc., Falls Church, 1967-69; pres. Systems div., corp. v.p. Computer Scis. Corp., Falls Church, 1969-77; pres. Systems Group, corp. v.p., 1977-91, also dir. Patentee in field; contbr. articles to profl. jours. Trustee Fairfax Hosp. System Found. With USN, 1944-46. Fellow IEEE; mem. Armed Forces Communications and Electronics Assn. (dir., internat. v.p. 1976-79, chpt. pres. 1979-80, exec. com. 1980-84, chmn. bd. 1984-86), AIAA, Nat. Space Club, Nat. Security Indsl. Assn., Tau Beta Pi, Eta Kappa Nu. Republican. Jewish. Home: 3609 Ridgeway Ter Falls Church VA 22044-1308 Office: Computer Systems Group 3170 Fairview Park Dr Falls Church VA 22042-4501 also: Computer Sciences Corp 2100 E Grand Ave El Segundo CA 90245

NASON, DOLORES IRENE, computer company executive, counselor, eucharistic minister; b. Seattle, Jan. 24, 1934; d. William Joseph Lockinger and Ruby Irene (Church) Gilstrap; m. George Malcolm Nason Jr., Oct. 7, 1951; children: George Malcolm III, Scott James, Lance William, Natalie Joan. Student, Long Beach (Calif.) City Coll., 1956-59; cert. in Religious Edn. for elem tchrs., Immaculate Heart Coll., 1961, cert. teaching, 1962, cert. secondary teaching, 1967; attended, Salesian Sem., 1983-85. Buyer J. C. Penney Co., Barstow, Calif., 1957; prin. St. Cyprian Confraternity of Christian Doctrine Elem. Sch., Long Beach, 1964-67; prin. summer sch. St. Cyprian Confraternity of Christian Doctrine Elem. Sch., Long Beach, 1965-67; pres. St. Cyprian Confraternity Orgn., Long Beach, 1967-69; dist. co-chmn. L.A. Diocese, 1968-70; v.p. Nason & Assocs., Inc., Long Beach, 1978—; pres. L.A. County Commn. on Obscenity & Pornography, 1984—; eucharistic minister St. Cyprian Ch., Long Beach, 1985—; bd. dirs. L.A. County Children's Svcs., 1988—; part-time social svcs. counselor Disabled Resources Ctr., Inc., Long Beach, 1992—; vol. Meml. Children's Hosp., Long Beach, 1977—; mem. scholarship com. Long Beach City Coll., 1984-90, Calif. State U., Long Beach, 1984-90. Mem. Sunland-Tujuna (Calif.) Citizens-Police Coun., 1987—; mem. adv. bd. Pro-Wilson 90 Gov., Calif., 1990; mem. devel. bd. St. Joseph High Sch., 1987—; pres. St. Cyprian's Parish Coun., 1962—; mem. Long Beach Civic Light Opera, 1973—, Assistance League of Long Beach, 1976—. Mem. L.A. Fitness Club, U. of the Pacific Club, K.C. (Family of the Month 1988). Republican. Roman Catholic. Avocations: physical fitness, theater, choir, travel.

NASSER, ESSAM, electrial engineer, physicist; b. Cairo, Feb. 3, 1931; s. Abdelaziz Hassan and Aida (Darwish) N.; m. Fawkeya Shaker; children: Nadya, Mona. B.E.E., Cairo U., 1952; Dipl-Ing., Tech. U. West Berlin, 1955, Dr.-Ing., 1959. Research engr., group leader Siemens Co., Fed. Republic Germany, 1958-61, 62-63; asst. prof. physics U. Calif.-Berkeley, 1961-62; prof. elec. engring. Engring Research Inst., Iowa State U., Ames, 1963-71, 73-77; vis. prof. elec. engring. Am. U. Beirut, 1971-73; vis. prof. power engring. Tech. U. Denmark, Lyngby, 1972; vis. prof. physics Am. U., Cairo, 1974-76; head Dar Al-Handash Cons., Cairo, 1977-80; pres. Middle East Cons., Cairo, 1980—; cons. U.S. Dept. Interior, Washington, Ministry Electric Power, Cairo; nat. chmn. Working Group on Insulator Contamination, N.Y.C., 1967-74; rep. Klockner-Moeller GmbH, Egypt, 1979—, Teledyne Systems, Egypt, 1988—. Author: Fundamentals of Gaseous Ionization and Plasma Electronics, 1971; Transmission Line Corona Effects, 1972; The Structure of the electric Energy System, 1974; name Paul Harris Fellow, 1984. Contbr. articles to profl. jours. Patentee in field. Recipient award Senate of West Berlin, 1958; grantee Office Naval Research, Washington, 1961, NSF, Washington, 1968-70. Fellow IEEE, Power Engring. Soc.; mem. Am. Phys. Soc., Conf. on Large Electric Networks, Paris, AAUP, Sigma Xi. Lodge: Rotary. Home: 10 Geziret Elarab Str, Dokki Egypt Office: PO Box 181, Dokki Egypt

NASSIRHARAND, AMIR, systems engineer; b. Tehran, Iran, Jan. 12, 1961; came to U.S. 1977; BS in Mech. Engring., Okla. State U., 1980, MS, 1981, PhD, 1986. Asst. prof. U. Ky., 1986-90, Mich. Tech. U., 1990-91; pres., dir. Sci. Inst. of Scholars, Reno, Nev., 1992—; vis. asst. prof. Okla. State U., 1986. Contbr. articles to profl. jours., contbr. to Voice Market, 1989—. Recipient Best Paper award Am. Control Conf., 1989, Intergovtl. Pers. Act award Dept. Def., 1990; Battle summer faculty fellow, 1988. Mem. ASME, Inst. Aeronautics and Astronautics, Inst. Elec. Electronics Engrs., Am. Soc. Engring. Edn. (summer faculty fellow 1989), Sigma Xi. Office: Sci Inst of Scholars PO Box 3095 Reno NV 89505-3095

NASTASE, ADRIANA, aerospace scientist, educator, researcher; b. Galati, Moldova, Romania; arrived in Germany, 1973; d. Constantin and Maria N. Diploma in Mech. Engring., Polytech. U., Bucharest, Romania, 1956; Diploma in Math., U. Bucharest, 1957; DEng in Aerodynamics, Romanian Acad., 1968; Doctorate in Applied Math., Université Paris VI, 1970. Rschr. dept. high speed aerodynamics Inst. Fluid Mechanics, Bucharest, 1956-73, master rsch., 1968—; asst. lectr. Polytech. Inst. Bucarest, 1957-73, lectr., 1968—; rsch. high speed aerodynamics Deutsche Forschungsgemeinschaft Tech. Inst. Braunschweig, Germany, 1973-74; prof. aerodynamics, head dept. Aerodynamics of Flight Rheinisch Westfäische Technische Hochschule Aachen, Germany, 1975—. Author: (with E. Carafoli and D. Mateescu) Wing Theory in Supersonic Flow, 1969, Forme Aerodinamice Optime, Prin Metoda Variationala, 1969, Contribution a L'etude des Formes Aerodynamiques Optimales, 1970, Utilizarea Calculatorelor in Optimizarea Formelor Aerodinamice, 1973, Analytische und Numerische Untersuchungen Für Optimum Optimorum Modelle von Deltaflügeln in Überschallströmungen, 1974; editor: Proceedings of High Speed Aerodynamics I, 1987, Proceedings of High Speed Aerodynamics II, 1990; contbr. articles to profl. jours., papers to profl. proceedings. Recipient Boursière Scientifique de la Centre Nat. Recherche Scientifique, 1956-57, Humboldt Forschungs-Stipendiatin, 1973-74. Mem. AIAA, Deutsche Gesellschaft für Luft und Raumfahrt, Gesellschaft für Angewandte Mathematik und Mechanik. Greek Orthodox. Achievements include patent in optimal shape of the wing for supersonic transport aircraft of second generation, research in optimal shape of the wing-fuselage configuration for supersonic aircraft and for horizontal space vehicles in two stages, the shape of the wing-fuselage-flaps configuration of variable geometry optimized at two cruising speeds, and spectral solutions for three-dimensional compressible boundary layer. Office: Rheinisch Westfäisch Techische Hochscule, Templergraben 55, 5100 Aachen Germany

NASU, SHOICHI, electrical engineering educator; b. Sendai-Shi, Miyagi-Ken, Japan, July 10, 1933; s. Nobuyuki and Yuri N.; m. Masako Hoshinami, Oct. 8, 1967; children: Masayuki, Akiko. BS, Kyoto U., Japan, 1959, DSc, 1967. Chief fuel property lab. Japan Atomic Energy Rsch. Inst., Tokai, Ibaraki, Japan, 1960-81; rsch. assoc. U. Pitts., 1969-70; exec. dir., gen. mgr. rsch. and devel. Ushio Inc., Tokyo, 1981-91, cons., 1991; prof. dept. elec. engring. Kanazawa Inst. Tech., Japan, 1992—; mem. internat. adv. com. Internat. Symposium in Electronic Structure of Actinides, Argonne, Ill., 1973-74; Japanese del. IAEA Advising Group, Vienna, Austria, 1979. Japanese compiler of Actinides News Letters, 1976-80; contbr. revs., handbooks and articles to profl. jours.; patentee in field. Mem. AAAS, Am. Phys. Soc. (life), N.Y. Acad. Scis., Am. Assn. Physics Tchrs., Planetary Soc., Atomic Energy Soc. Japan (life), Laser Soc. Japan, Japan Soc. Applied Physics. Avocations: jogging, swimming, mountain climbing, golf. Home: Higashi Itchoda 7-11, Mishima-shi, Shizooka-ken 411, Japan Office: Ohgigaoka 7-1, Nonoichi, Kanazawa Minami-kyoku, Ishikawa 921, Japan

NATAN, BENVENISTE, aeronautical engineer, educator, researcher; b. Thessaloniki, Greece, Jan. 1, 1954; arrived in Israel, 1971; s. Moses and Louiza (Balestra) N.; m. Dalia Bar-Ner, July 21, 1983; children: Irene L.,

Daniel M. BSc in Aero. Scis., Technion-Israel Inst. Tech., Haifa, 1977, MSc in Aero. Scis., 1982, DSc in Aero. Scis., 1988. Rsch. assoc. Naval Postgrad. Sch., Monterey, Calif., 1989-91; instr. aero. engring. Technion-Israel Inst. Tech., 1977-89, rsch. and teaching assoc., 1991—; cons. Israel Aircraft Industries, Lod, 1986. Contbr. articles to profl. jours. Recipient Rsch. award Ben-Gurion Found., 1988; rsch. grantee NRC, 1989, 90. Mem. AIAA (sr.), Combustion Inst., Inst. Liquid Atomization and Spray Systems. Jewish. Achievements include research in rocket and ramjet propulsion, combustion and thermodynamics. Office: Faculty Aerospace Engring, Technion-Israel Inst Tech, Haifa 32000, Israel

NATARAJ, CHANDRASEKHAR, mechanical engineering educator, researcher; b. Chikmagalur, Karnataka, India, Nov. 26, 1959; came to U.S., 1982.; s. A. S. Chandrasekhar and K. S. Mahalakshmi; m. Latha Ramamurthy, July 8, 1991; 1 child, Chiraag. MS, Ariz. State U., 1984, PhD, 1987. Instr. Ariz. State U., Tempe, 1986-87; v.p. Trumpler Assocs., Inc., West Chester, Pa., 1987-89; asst. prof. Villanova (Pa.) U., 1988—; bd. dirs. Turbo Rsch. Found., West Chester, 1987—. Contbr. articles to Jour. of Trbiology. Mem. ASME (assoc. 1989—, contbr. articles to Jour. Vibration and Acoustics 1984, 87, 90, Jour. Applied Mechanics 92), Am. Acad. Mechanics, Am. Soc. Engring. Educators, Sigma Xi. Office: Villanova U Villanova PA 19085

NATARAJAN, PARAMASIVAM, chemistry educator; b. Madras, Tamilnadu, India, Sept. 17, 1940; s. Somanampatti Kailasam and Chellammal Paramasivam; m. Sivabagyam Natarajan, Jan. 17, 1971; children: Shiva Suganti, Sakthi. BSc, Madras (India) U., 1959; MSc, Banaras U., Varanasi, India, 1963; PhD, U. So. Calif., 1971. Reader U. Madras, 1974-77, prof. chemistry, 1977-91, head chemistry dept., 1982-91; dir. Cen. Salt and Marine Chems. Rsch. Inst., Bhavnagar, India, 1991—; mem. Dept. Sci. and Tech., New Delhi, 1980-88, 91—, Coun. Sci. and Indsl. Rsch., New Delhi, 1987-91, U. Grants Commn. Chemistry Panel, New Delhi, 1988-90. Recipient Shanti Swarup Bhatnagar prize in chem. sci. CSIR, 1984, Best Tchr. award Tamil Nadu Govt., 1984. Fellow Indian Acad. Scis., Indian Nat. Sci. Acad., Tamil Nadu Acad. Scis., N.Y. Acad. Scis., Gujarat Acad. Scis. Achievements include development of electrode materials of polymeric dyes which show a new type of photoelectrochemical behavior, investigation of flash photolysis behavior of molecules. Office: CSMCRI, Gijubhai Badheka Marg, Bhavnagar Gujarat 364002, India

NATARAJAN, THYAGARAJAN, civil engineer; b. Chidambaram, India, Sept. 30, 1942; came to U.S. 1971; s. Ramaswamy and Janaki (Venketraman) T.; m. Indira Natarajan, May 17, 1971; 1 child, Sripriya. BE in Civil Engring. with hons., Coll. Engring., Guindy, India, 1963; MSc in Structural Engring., Coll. Engring., Guindy, 1965. Design engr. Atomic Energy Commn. of India, Bombay, 1965-71; sr. designer Balt. Gas & Elec. Co., 1971-76, engr., 1976-80, sr. engr., 1980—. Mem. ASCE. Achievements include research on power plant waterfront structures, microwave towers and analysis of concrete pipes. Home: 8808 Valleyfield Rd Timonium MD 21093 Office: Balt Gas & Elec Co 1000 Brandon Shores Rd Baltimore MD 21226

NATH, RAVINDER, physicist; b. Jullundar, Punjab, India, Apr. 9, 1942; came to U.S., 1967; d. Kedar Nath and Rajrani (Malmotra) Katyal; m. Rashmi Duggal Nath, Oct. 27, 1971; children: Anjali, Sameer. BS, Delhi U., 1963, MS, 1965; PhD, Yale U., 1971. Diplomate Am. Bd. Radiology. Staff physicist Yale U., New Haven, 1971-75, asst. prof., 1976-79, assoc. prof., 1979-85, chief physicist, 1992—; mem. radiation study sect. NIH. Contbr. over 100 articles to profl. jours. Fellow Am. Coll. Radiology, Am. Assn. Med. Physicists (chm. radiation therapy com. 1988-90); mem. Am. Endocurietherapy Soc. (chmn. physics com.), Am. Assn. Physicists in Medicine (pres.-elect), Inter-Soc. Coun. Radiation Oncology. Office: Yale Univ 333 Cedar St New Haven CT 06510

NATHAN, DAVID GORDON, physician, educator; b. Boston, May 25, 1929; s. E. Geoffrey and Ruth (Gordon) N.; m. Jean Louise Friedman, Sept. 1, 1951; children: Deborah, Linda, Geoffrey. BA, Harvard U., 1951, MD, 1955. Diplomate Am. Bd. Internal Medicine, Am. Bd. Pediatrics. Intern dept. medicine Peter Bent Brigham Hosp., Boston, 1955-56, sr. resident, 1958-59; jr. assoc. in medicine Brigham and Women's Hosp., Boston, 1961-67, sr. assoc. in medicine, 1967—; assoc. in medicine, hematology Childrens Hosp., Boston, 1963-68, chief, div. hematology, 1968-73, chief div. hematology and oncology, 1974-84; pediatrician-in-chief Dana Farber Cancer Inst., Boston, 1974-85; Robert A. Stranahan prof. pediatrics Harvard Med. Sch., Boston, 1977—; physician-in-chief Childrens Hosp., Boston, 1985—. Editor: Hematology in Infancy and Childhood, 4th edit., 1993. With USMC, 1948-49. Recipient Nat. medal Sci. NSF, 1990. Mem. Inst. of Medicine of NAS, Am. Acad. Arts and Scis., Am. Pediatric Soc., Soc. Pediatric Rsch., Assn. Am. Physicians, Am. Soc. Clin. Investigators, Am. Soc. Hematology (pres. 1986), Phi Beta Kappa (hon.). Avocations: tennis, hiking. Office: Childrens Hosp 300 Longwood Ave Boston MA 02115-5737

NATHANIELSZ, PETER WILLIAM, physiologist; b. Colombo, Sri Lanka, Jan. 31, 1941; came to the U.S., 1976; s. Arthur Holman and Constance Ethel (Mouncy) N.; m. Diana Joyce Crawford-Smith, Mar. 19, 1966; children: Helen Julie, David William. PhD, Cambridge U., 1969, MD, 1977, ScD, 1993. Physician Brit. Med. Assn., London, 1964-76; intern Univ. Coll. Hosp., London; resident Cambridge (Eng.) U.; dir. lab. for pregnancy and newborn rsch. Coll. Vet. Medicine, Cornell U., Ithaca, N.Y.; chairperson maternal-child health rsch. com. Nat. Inst. of Child Health and Human Devel., Washington, 1992—. Author: Fetal Endocrinology, 1975, Life Before Birth and a Time To Be Born, 1992. Achievements include discovery of signal giving ability of fetal brain to start the birth process. Office: Cornell U Dept Physiology Lab Pregnancy/Newborn Rsch Ithaca NY 14853-6401

NATHANS, DANIEL, biologist; b. Wilmington, Del., Oct. 30, 1928; s. Samuel and Sarah (Levitan) N.; m. Joanne E. Gomberg, Mar. 4, 1956; children: Eli, Jeremy, Benjamin. B.S., U. Del., 1950; M.D., Washington U., 1954. Intern Presbyn. Hosp., N.Y.C., 1954-55; resident in medicine Presbyn. Hosp., 1957-59; clin. assoc. Nat. Cancer Inst., 1955-57; guest investigator Rockefeller U., N.Y.C., 1959-62; prof. microbiology Sch. Medicine, Johns Hopkins, 1962-72, prof., dir. dept. microbiology, 1972-82, Univ. prof. molecular biology and genetics, 1982—; sr. investigator Howard Hughes Med. Inst., 1982—. Recipient Nobel prize in physiology or medicine, 1978, Nat. Medal of Sci., Nat. Sci. Found., 1993. Fellow Am. Acad. Arts and Scis.; mem. NAS, Pres.'s Coun. Advisers on Sci. and Tech. Office: Johns Hopkins U-Sch of Medicine Dept of Mo-Bio & Genetics 725 N Wolfe St Baltimore MD 21205-2105

NATHANSON, JAMES A, neurologist; b. Hartford, Conn., Jan. 21, 1947; m. Barbara Hastings. BS, Trinty Coll., Hartford, 1968; MD, Yale Med. Sch., New Haven, 1973; PhD, Yale, 1973. Lic. physician Mass.; cert. neurology, psychiatry. Dir. neuro-pharmacology res. lab. Mass. Gen. Hosp., Boston, 1979; assoc. prof. Harvard Med. Sch., Boston, 1979. Office: Dept Neurology Mass Gen Hosp Res CNY-6 Boston MA 02114

NATHANSON, LINDA SUE, technical writer, software training specialist; b. Washington, Aug. 11, 1946; d. Nat and Edith (Weinstein) N.; m. James F. Barrett. BS, U. Md., 1969; MA, UCLA, 1970, PhD, 1975. Tng. dir. Rockland Research Inst., Orangeburg, N.Y., 1975-77; asst. prof. psychology SUNY, 1978-79; pres. Cabri Prodns., Inc., Ft. Lee, N.J., 1979-81; research supr. Darcy, McManus & Masius, St. Louis, 1981-83; mgr. software tng., documentation On-Line Software Internat., Ft. Lee, 1983-85; pvt. practice cons. Ft. Lee, 1985-87; founder, exec. dir. The Edin. Group, Inc., Gillette, N.J., 1987—. Author: (with others) Psychological Testing: An Introduction to Tests and Measurement, 1988; contbr. articles to mags., newspapers and profl. jours. Recipient Research Service award 1978; Albert Einstein Coll. Medicine Research fellow, 1978-79. Jewish. Home and Office: 102 Sunrise Dr Gillette NJ 07933-1944

NATHWANI, BHARAT NAROTTAM, pathologist, consultant; b. Bombay, Jan. 20, 1945; came to U.S., 1972; s. Narottam Pragji and Bharati N. (Lakhani) N. MBBS, Grant Med. Coll., Bombay, 1969, MD in Pathology, 1972. Intern Grant Med. Coll., Bombay U., 1968-69; asst. prof. pathology Grant Med. Coll., 1972; fellow in hematology Cook County Hosp., Chgo., 1972-73; resident in pathology Rush U., Chgo., 1973-74; fellow in hematopathology City of Hope Med. Ctr., Duarte, Calif., 1975-76, pathologist, 1977-84; prof. pathology, chief hematopathology U. So. Calif., L.A., 1984—; cons. Norris Cancer Hosp., L.A., 1986—. Contbr. numerous articles to profl. jours. Recipient Grant awards Nat. Libr. Medicine, Bethesda, Md., Nat. Cancer Inst., 1991. Mem. AAAS, Internat. Acad. Pathology, Am. Soc. Clin. Pathology, Am. Soc. Hematology, Am. Soc. Oncology. Office: U So Calif Sch Medicine HMR 204 2025 Zonal Ave Los Angeles CA 90033-4526*

NATION, JOHN ARTHUR, electrical engineering educator, researcher; b. Bridgwater, Eng., Aug. 8, 1935; came to U.S., 1965; naturalized, 1991.; s. Arthur John and Doris Edith (Rides) N.; m. Sally Gillian Leeds, May 31, 1961; children—Philip David, Robert James. B.Sc., Imperial Coll., London, 1957; Ph.D., Imperial Coll., 1960. Cons. Comitato Nazionale per L'Energia Nucleare, Frascati, Italy, 1960-61; staff physicist Central Electricity Generating Bd., Leatherhead, Eng., 1962-65; elec. engring. faculty Cornell U., Ithaca, N.Y., 1965—; prof. Cornell U., 1978—, dir. Sch. Elec. Engring., 1984-89; cons. miscellaneous cos., 1965—. Contbr. articles to profl. jours. Fellow Am. Phys. Soc., IEEE (plasma scis. exec. com. 1985-89, chair com. 1987-88, Centennial medal 1984). Avocation: golf, tennis. Home: 1041 Hanshaw Rd Ithaca NY 14850-2741 Office: Cornell U 325 Engring & Theory Ctr Ithaca NY 14853

NATION, LAURA CROCKETT, electrical engineer; b. Ft. Worth, Dec. 1, 1957; d. Donald Ray and Cora Lee (Holt) Crockett; m. David Hunter Nation, Aug. 23, 1986. BA, U. Tex., Arlington, 1980, BSEE, 1984. Registered profl. engr., Tex. With engring. coop. TU Electric, Dallas, 1981-84; assoc. engr. TU Electric, Euless, Tex., 1984-85, Irving, Tex., 1985-87; engr. TU Electric, Dallas, 1987-91; staff engr. TU Electric, Ft. Worth, 1991—. Vol. United Way, Ft. Worth, 1992, Muscular Dystrophy Assn., Dallas, 1987-91. Scholar Mogul Corp., Canton, Ohio, 1976. Mem. IEEE (vol. Discover "E"), NSPE. Republican. Missionary Alliance. Home: 1303 Brittany Ln Arlington TX 76013 Office: TU Electric PO Box 970 Fort Worth TX 76101-0970

NATOWITZ, JOSEPH B., chemistry educator, administrator; b. Saranac Lake, N.Y., Dec. 24, 1936. BS in Chemistry, U. Fla., 1958; Cert. in Meteorology, UCLA, 1959; PhD in Nuclear Chemistry, U. Pitts., 1965. Staff meteorologist, 1st lt. USAF, 1958-61; grad. teaching asst. U. Pitts., 1961-62, grad. rsch. asst., 1962-65; postdoctoral rsch. assoc. SUNY, Stony Brook, 1965-67; rsch. collaborator Brookhaven Nat. Lab., 1965-67; asst. prof. Tex. A&M U., College Station, 1967-72, assoc. prof., 1972-76, prof., 1976—, head dept. chemistry, 1981-85, dir. Cyclotron Inst., 1991—; part-time instr. SUNY-Stony Brook, 1966-67; rsch. collaborator Lawrence Radiation Lab., Berkeley, Calif., 1966, Los Alamos (N.Mex.) Nat. Lab., 1973-74; Alexander Von Humboldt sr. scientist Max Planck Inst. für Kernphysik, Heidelberg, Germany, 1978; vis. prof. Inst. for Nuclear Studies, U. Tokyo, 1979, U. Claude Bernard, Inst. de Physique Nucleaire, 1983, U. de Caen, 1985, Ctr. des Etudes Nucleaires de Saclay, 1986, U. Cath. de Louvain, 1987. Contbr. over 80 articles to profl. jours., also to approx. 30 books & procs. Chmn. Cub Scout Pack 802, 1973-75; v.p. College Hills PTO, 1974-75; mem. A&M Consol. Sch. Bd., 1975-78, pres., 1977-78; pres. A&M Consol. Band Boosters, 1980-81. NSF summer fellow, 1962; NASA predoctoral fellow, 1964-65; recipient Disting. Achievment award-rsch. Tex. A&M U., 1988. Fellow Am. Phys. Soc.; mem. Am. Chem. Soc. (vice chmn. Divsn. Nuclear Chemistry & Tech., divsn. chmn.-elect), Chem. Inst. Can., Sigma Xi, Phi Lambda Upsilon. Office: Texas A & M University Cyclotron Institute College Station TX 77843

NATSOULAS, THOMAS, psychology educator; b. N.Y.C., Mar. 1, 1932; s. Anthony and Helen (Theodorou) N.; m. Prokopia Levenderis, Jan. 18, 1953; children: Anthony, John. BA, Bklyn. Coll., 1953; PhD, U. Mich., 1960. Asst. prof. Wesleyan U., Middletown, Conn., 1959-62, U. Wis., Madison, 1962-64; prof. psychology U. Calif., Davis, 1964—. Assoc. editor Jour. Mind. and Behavior, 1984—; mem. editorial bd. New Ideas in Psychology, 1982—, Consciousness and Cognition: Internat. Jour., 1991—, Advances in Consciousness Rsch. (book series), 1991—; mem. editorial adv. bd. Philos. Psychology, 1990—; contbr. articles to Am. Jour. of Psychology, Imagination, Cognition, and Personality, Jour. Mind. and Behavior, Jour. for Theory Social Behaviour, Psychol. Rsch., Psychoanalysis and Contemporary Thought. Fellow APA, Am. Psychol. Soc.; mem. Internat. Soc. for Ecol. Psychology, Phi Beta Kappa. Home: 1030 Fordham Dr Davis CA 95616-0925 Office: U Calif Psychology Dept Davis CA 95616-8686

NATSUKARI, NAOKI, psychiatrist, neurochemist; b. Tokyo, July 16, 1952; came to U.S., 1991; s. Masao and Hideko (Kobayashi) N.; m. Ikuko Ueno, May 28, 1989; 1 child, Shunya. BS, Waseda U., Tokyo, 1976, MS, 1978; MD, Hamamatsu (Japan) U., 1986, PhD, 1991. Researcher Nikkiso Corp., Shizuoka, Japan, 1978-79; resident Hamamatsu U. Hosp., 1986, Okada Hosp., Okazaki, Japan, 1986-87; rsch. assoc. Nat. Inst. Physiol. Scis., Okazaki, 1991-92; postdoctoral fellow Med. Coll. Pa., Phila., 1991-92; vis. fellow NIMH, Washington, 1992—. Recipient award Ichiro Kanehara Found., Tokyo, 1991. Mem. Soc. Neurosci., Japanese Neurochem. Soc., Japanese Biol. Soc., Japanese Neurology and Psychiatry Soc. Achievements include rsch. on involvement of calcium/calmodulin in the mechanisms of brain dopaminergic and B-adrenergic transmission, interaction between calmodulin and its target proteins, calmodulin-binding proteins, pathogenesis of schizophrenia. Office: NIMH at St Elizabeth Hosp Neuropsychiatry Br 2700 Martin Luther King Jr Washington DC 20032

NAUERT, ROGER CHARLES, healthcare executive; b. St. Louis, Jan. 6, 1943; s. Charles Henry and Vilma Amelia (Schneider) N.; B.S., Mich. State U., 1965; J.D., Northwestern U., 1969; M.B.A., U. Chgo., 1979; m. Elaine Louise Harrison, Feb. 18, 1967; children: Paul, Christina. Bar: Ill. 1969. Asst. atty. gen. State of Ill., 1969-71; chief counsel Ill. Legis. Investigating Commn., 1971-73; asst. state comptroller State of Ill., 1973-75; dir. adminstrn. and fin. Health and Hosps. Governing Commn. Cook County, Chgo., 1975-79; nat. dir. health care services Grant Thornton, Chgo., 1979-88; exec. v.p. Detroit Med. Ctr., 1988-91; exec. v.p. Columbia-Presbyn. Med. Ctr., N.Y.C., 1991-93; sr. v.p. Mt. Sinai Med. Ctr., N.Y.C., 1993—; vis. lectr. healthcare mgmt. & fin. Columbia U., Vanderbilt U., U. Chgo., 1978—; preceptor Wharton Sch., U. Pa. Ford Found. grantee, 1968-69. Mem. Am. Hosp. Assn., Am. Public Health Assn., Am. Coll. Healthcare Execs., Nat. Health Lawyers Assn., Health Care Fin. Mgmt. Assn. (faculty mem.), Alpha Phi Sigma, Phi Delta Phi, Delta Upsilon. Clubs: N.Y. Athletic, Mt. Kisco Country. Author: A Sociology of Health, 1977; The Demography of Illness, 1978; Proposal for a National Health Policy, 1979; Health Care Feasibility Studies, 1980; Health Care Planning Guide, 1981; Health Care Strategic Planning, 1982; Overcoming the Obstacles to Planning, 1983; Principles of Hospital Cash Management, 1984; Healthcare Networking Arrangements, 1985; Strategic Planning for Physicians, 1986; HMO's: A Once and Future Strategy, 1987, Mergers, Acquisitions and Divestitures, 1988, Tax Exempt Status Under Seige, 1989, Governance in Multi-Hospital Systems, 1990, Planning Alternative Delivery Systems, 1991, Direct Contracting: The Future is Now, 1992, The Rise and Fall of the U.S. Healthcare System, 1993. Home: 461 Haines Rd Mount Kisco NY 10549 Office: Mt Sinai Med Ctr 1 Gustave Levy Pl New York NY 10029

NAUGHTON, JOHN PATRICK, cardiologist, medical school administrator; b. West Nanticoke, Pa., May 20, 1933; s. John Patrick and Anne Frances (McCormick) N.; m. Margaret Louise Fox; children: Bruce, Marcia, Lisa, George, Michael, Thomas. A.A., Cameron State Coll., Lawton, Okla., 1952; B.S., St. Louis U., 1954; M.D., Okla. U., 1958. Intern George Washington U. Hosp., Washington, 1958-59; resident U. Okla. Med. Center, 1959-64; asst. prof. medicine U. Okla., 1966-68; assoc. prof. medicine U. Ill., 1968-70; prof. medicine George Washington U., 1970-75, dean acad. affairs, 1973-75, dir. div. rehab. medicine and Regional Rehab. Research and Tng. Center, 1970-75; dean Sch. Medicine, SUNY, Buffalo, 1975—; prof. medicine and physiology Sch. Medicine, SUNY, 1975—, lectr. in rehab. medicine, 1975; acting v.p. for health scis. SUNY, 1983-84, v.p. clin. affairs, 1984—; dir. Nat. Exercise and Heart Disease Project, 1972—; chmn. policy adv. bd. Beta-blocker heart attack trial Nat. Heart, Lung and Blood Inst., 1977-82; pres. Western N.Y. chpt. Am. Heart Assn., 1983-85, v.p. N.Y. State affiliate, 1985, pres. N.Y. state affiliate, 1988-90; chmn. clin. applications and preventions adv. com. Nat. Heart, Lung and Blood Inst., 1984; mem. N.Y. Gov.'s Commn. on Grad. Med. Edn., 1985; mem. N.Y. State Coun. on Grad. Med. Edn., 1988-90; pres. Assoc. Med. Schs. N.Y., 1982-84, mem. adminstrv. com. Council of deans, 1983-89; mem. N.Y. State Dept. of Health Adv. Com. on Physician Recredentialing; mem. exec. coun. Nat. Inst. on Disability and Rehab. Rsch., 1991-92. Author: Exercise Testing and Exercise Training in Coronary Heart Disease, 1973, Exercise Testing: Physiological, Biomechanical, and Clinical Principles, 1988. Career Devel. awardee Nat. Heart Inst., 1966-71; recipient Brotherhood-Sisterhood award in medicine NCCJ, N.E. Minority Educators award, 1990, Acad. Alumnus of Yr. award Okla. U., 1990, award for svc. to minorities in med. edn., 1991. Fellow ACP, Am. Coll. Cardiology, Am. Coll. Sports Medicine (pres. 1969-73), Am. Coll. Chest Physicians; mem. N.Y. State Heart Assn. (pres.). Office: SUNY Buffalo Sch Medicine Biomed Scis 3435 Main St Buffalo NY 14214

NAUMANN, ROBERT BRUNO ALEXANDER, chemist, physicist, educator; b. Dresden, Germany, June 7, 1929; came to U.S., 1932, naturalized, 1951; s. Eberhard Bruno and Elsa Henriette (Haege) N.; m. Marina Grot Turkevich, Sept. 16, 1961; children: Kristin Ragnhild, Andrew John Bruno. B.S., U. Calif., Berkeley, 1949; M.A., Princeton U., 1951, Ph.D., 1953. Mem. faculty Princeton U., 1953—, prof. chemistry and physics, 1973-92, prof. emeritus chemistry and physics, 1992—; mem. vis. staff Los Alamos Nat. Lab., 1970-86; rsch. collaborator Brookhaven Nat. Lab., 1984-87; sci. assoc. CERN, Geneva, 1985—; vis. prof. physics dept. Tech. U. Munich, 1988; vis. scholar physics Dartmouth Coll., 1992—. Author articles electromagnetic isotope separation, nuclear structure via radioactive and charged particle nuclear spectroscopy, implantation radioactive isotopes into solids, formation and properties of muonic atoms. Recipient Alexander von Humboldt Stiftung Sr. U.S. Scientist award, 1978, 83; Allied Chem. and Dye Corp. fellow, 1951-52, Procter and Gamble faculty fellow, 1959-60; Deutsche Forschungsgemeinschaft grantee, 1988. Fellow Am. Phys. Soc., AAAS; mem. Am. Chem. Soc. (chmn. Princeton U. sect. 1975, Chmn. Div. Nuclear Chemistry and Technology 1984), Sierra Club, Phi Beta Kappa, Sigma Xi (chmn. Princeton, N.J. sect. 1986-87). Episcopalian. Home: RR 1 Box 483A 54 Hawk Pine Hills Norwich VT 05055-9516 Office: Dartmouth Coll Physics Dept Hanover NH 03755

NAURATH, DAVID ALLISON, engineering psychologist, researcher; b. Houston, Mar. 11, 1927; s. Walter Arthur and Joy Frances (Bradbury) N.; m. Barbara Ellen Coverdell; children: Kathleen Ann, David Allen, Cynthia Ellyn, Randall Austin. BA, Simpson Coll., Indianola, Iowa, 1948; MA, Southern Meth. U., 1949; postgrad., U. Denver, 1955-57. Job analyst U.S. Air Force, San Antonio and Denver, 1951-55; rsch. psychologist U.S. Air Force, Lowry AFB, Colo., 1955-60, Navy, Life Scis. & Systems div., Point Mugu, Calif., 1960-76; engring. psychologist Navy Systems Engring., Point Mugu, 1976-83; ret.; presenter at profl. socs. and orgns. in field. Contbr. articles to Jour. Engring. Psychology, jour. Soc. for Info. Display, jour. Soc. Photo-optical Instrument Engrs. With USAAF, 1944-46. Mem. AAAS (life), IEEE (sr.), Am. Psychol. Assn., Human Factors Soc. (panel mem. Certification of Human Factors Engrs. 1976), Soc. Engring. Psychologists, Soc. for Info. Display (life). Methodist. Home: 5633 Pembroke St Ventura CA 93003-2200

NAVRATIL, GERALD ANTON, plasma physicist; b. Troy, N.Y., Sept. 5, 1951; s. Lloyd George and Frances Mary (Scalise) N.; m. Joan Frances Etzweiler, Sept. 4, 1976; children—Frances, Alexis, Paula. B.S., Calif. Inst. Tech., 1973; M.S., U. Wis., 1974, Ph.D., 1976. Project assoc. dept. physics U. Wis.-Madison, 1976-77; asst. prof. engring. sci. Columbia U., N.Y.C., 1977-78, asst. prof. applied physics, 1978-83, assoc. prof., 1983-88, prof. and chmn., 1988—; vis. fellow Princeton U., 1985-86; cons. MIT, 1984-86, Fusion Systems, Inc., 1988, Inst. Def. Analysis, 1992-93, com.; mem. Nat. Adv. Coun. TPX Tokomak Project and chair TPX Program Adv. Com., 1993—. Contbr. articles to profl. jours. Patentee in field. Cottrell Research grantee, 1978; U.S. Dept. Energy High Beta Tokomak Research contract, 1982—; NSF grantee, 1978—; Alfred P. Sloan research fellow, 1984. Fellow Am. Phys. Soc.; mem. Univ. Fusion Assn. (sec., treas., pres. 1991), Sigma Xi. Office: Columbia U Dept Applied Physics 500 W 120th St Rm 206 New York NY 10027

NAVROTSKY, ALEXANDRA, geophysicist. BS, U. Chgo., 1963, MS, 1964, PhD, 1967. Rsch. assoc. Technische Hochschule, Clausthal, Germany, 1967-68, Pa. State U., Pitts., 1968-69; mem. geology and chemistry faculty Ariz. State U., 1969-85, dir. Ctr. Solid State Sci., 1984-85; prof. geol. and geophys. scis. Princeton (N.J.) U., 1985—, chair geol. and geophys. scis., 1988-91; Albert G. Blanke, Jr. chmn., 1992—. Editor: Physics and Chemistry of Minerals, 1986-91; contbr. articles to profl. jours. Alfred P. Sloan fellow, 1973. Fellow Am. Geophys. Union; mem. NAS, Mineral. Soc. Am. (pres. 1992-93, v.p. 1991-92, award 1981), Princeton Materials Inst. Office: Princeton U Dept Geol & Geophys Scis 114 Guyot Ave Princeton NJ 08544-1003

NAWROCKI, H(ENRY) FRANZ, propulsion technology scientist; b. Pueblo, Colo., Dec. 10, 1931; s. Henry Vincent and Verna Ella (Weyand) N.; m. Marlene Charlotte Kryak, Sept. 1, 1973. BS Aero. Engring., U. Colo., 1953; MS Aerospace Engring., U. So. Calif., 1968. Group supr. for RJ43 qualification Marquardt, Van Nuys, Calif., 1953-64; flight test analyst engr. for L-1011 cert. Lockheed, Palmdale, Calif., 1964-72; flight test program coord. for B-1 qualification Rockwell Internat., El Segundo, Calif., 1972-77; propulsion group supr. for CL600 cert. Canadair Ltd., Montreal, Que., Can., 1977-80; design mgr. for LF2000 devel. Lear Fan Ltd., Reno, 1980-82; staff scientist Gulfstream Aerospace Corp., Savannah, Ga., 1982—; engring. rep. FAA L.A., 1980-82, Atlanta, 1985—; engring. cons. Savannah Art Assn., 1982-88, Gallery 209, 1984—. Contbr. articles to profl. jours. Recipient La Verne Noyes scholarship U. Colo., 1952. Mem. AIAA, Soc. Automotive Engrs., Soc. Flight Test Pilots, Aerospace Industries Assn., Soc. North Am. Goldsmiths, Fla. Soc. Goldsmiths, Golden Isle Gem and Mineral Club, Sherwood Homeowners Assn. Avocations: mineralogy, model railroading, lapidary, metal working. Home: 18 Landon Ln Savannah GA 31410-3830

NAWROCKY, ROMAN JAROSLAW, research electrical engineer; b. Peremysl, Poland, Apr. 30, 1932; came to U.S., 1950; s. Stefan and Maria (Lominsky) N.; m. Marta Maria Malynowska, Aug. 20, 1966; 1 child, Anne Myroslawa. BSEE, U. Man., Winnipeg, Can., 1956; MSEE, Poly. Inst. N.Y., 1963, PhDEE, 1975. Design engr. Can. GE, Toronto, Ont., 1956-58, National Co., Medford, Mass., 1958-60; design engr. Instrumentation Lab., MIT, Cambridge, 1960-61; sr. rsch. engr. Brookhaven Nat. Lab., Upton, N.Y., 1964—. Contbr. over 35 articles to profl. jours. Mem. IEEE (sr.), Ukrainian Engrs. Soc. Am., Sigma Xi. Office: Bldg 725B Brookhaven Nat Lab Upton NY 11973

NAWY, EDWARD GEORGE, civil engineer, educator; b. Baghdad, Iraq, Dec. 21, 1926; came to U.S., 1957, naturalized, 1966; s. George M. and Ava (Marshall) N.; m. Rachel E. Shebbath, Mar. 23, 1949; children: Ava Margaret, Robert M. DIC, Imperial Coll. Sci. and Tech., London, 1951; CE, MIT, 1959; D of Engring., U. Pisa, Italy, 1967. Registered profl. engr. (P.E.), N.J., N.Y., Pa., Calif., Fla. Head structures Israel Water Planning Authority, Tel-Aviv, 1952-57; mem. faculty Rutgers U., New Brunswick, N.J., 1959—; mem. grad. faculty Rutgers U., 1961—, prof. civil engring., 1966-72, Distinguished prof. (prof. II), 1972—, chmn. dept. civil and environ. engring., dir. grad. programs, 1980-86; chmn. Coll. Engring. Del. Assembly, 1969-72; mem. Univ. Senate, 1973-80, also mem. exec. com.; also faculty rep. mem. bd. govs. and trustee; Guest prof. Nat. U. Tucuman, Argentina, summer 1963, Imperial Coll. Sci. and Tech., summer 1964; vis. prof. Stevens Inst. Tech., Hoboken, 1968-72; hon. prof. Nanjin Inst. Tech., China, 1987; mem. N.J. Chancellor Higher Edn. Com. for Higher Edn. Master Plan; mem. bridge com., Rutgers U. rep. Transp. Research Bd.; cons. to industry; U.S. mem. commn. on cracking Comité EuroInternat. du Beton; mem. Civil Engring. Tech. Adv. Council N.J., 1966-72; concrete systems cons. FAA, Washington; cons. energy div. U.S. Gen. Accounting Office, Washington; gen. chmn. Internat. Symposium on Slabs and Plates, 1971; mem. hon. presidium RILEM Internat. Conf., Budapest, 1977. Author: Reinforced Concrete, 1985, 2d edit., 1990, Simplified Reinforced Concrete, 1986, Prestressed Concrete, 1989; contbr. over 115 articles to profl. publs. Vice pres. Berkeley Twp. Taxpayers Assn., Ocean City, N.J., 1966-70. Recipient merit citation and award N.J. Concrete Assn., 1966; C. Gulbenkian Found. fellow, 1972. Fellow ASCE (mem. joint com. on slabs), Instn. Civil Engrs. (London), Am. Concrete Inst. (pres. N.J. chpt. 1966, 77-78, chmn. nat. com. on cracking 1966-73, bd. com. chpts. 1969-72, ACI rep. internat. commn. fractures, H.L. Kenneday award 1972, award of recognition N.J. chpt. 1972, chpt. activities award 1978, chmn. nat. com. on deflection 1989—); mem. NSPE, AAUP (chmn. budget and priorities com Rutgers U. chpt. 1972), Am. Soc. Engring. Edn., Prestressed Concrete Inst. (Bridge Competition award 1971, mem. tech. activities com.), N.Y. Acad. Scis., N.J. Bldg., Tall Bldgs. Coun., N.J. Contractors Assn. (cons. ednl. com., tall bldgs. coun.), Rotary, Sigma Xi, Tau Beta Pi, Chi Epsilon (hon.). Office: Rutgers State U of NJ Civil Engring Dept New Brunswick NJ 08855

NAYFEH, JAMAL FARIS, mechanincal and aerospace engineering educator; b. Kuwait, Mar. 2, 1960; came to U.S., 1985; s. Faris Abdulrahman and Bahiya Hasan N.; m. Stephanie Leigh Rouse, July 23, 1990; 1 child, Haythem Jamal. BSCEin Engring. Mechanics, Kuwait U., 1983; MSME, Yarmouk U., Jordan, 1985; PhD, Va. Tech. Inst., 1990. Rsch., teaching asst. Yarmouk U., Jordan, 1983-85; rsch. asst. Va. Tech., Blacksburg, 1985-90; prof. U. Ctrl. Fla., Orlando, 1990—. Contbr. articles to Internat. Jour. Nonlinear Mechanics, Acta Mechanica, Physics of Fluids, Jour. Applied Mechanics, Nonlinear Dynamics. Mem. ASEM (presenter tech. papers at regional , nat. and internat. confs.), ASCE, AIAA, Am. Acad. Mechanics. Office: U Ctrl Fla Mech & Aerospace Engring PO Box 162450 Orlando FL 32816-2450

NAYFEH, MUNIR HASAN, physicist; b. Tulkarem, Jordan, Dec. 13, 1945; came to U.S. 1970; s. Hasan Ahmad and Khadra Said (Sirhan) N.; m. Hutaf Mahmud Abu Hantash, Sept. 31, 1973; children: Hasan, Maha, Ammar, Osama, Mona. BSc, Am. U., Beirut, 1968, MSc, 1970; PhD, Stanford U., 1974. Rsch. physicist/postdoctoral fellow Oak Ridge (Tenn.) Nat. Lab., 1974-76; lectr. Yale U., New Haven, 1976-78; prof. physics U. Ill., Urbana, 1978—; cons. Argonne (Ill.) Nat. Lab., 1985-88, Teepak, Inc., Danville, Ill., 1989—. Co-author: Electricity and Magnetism, 1985; co-editor: Atomic Excitation and Recombination in External Fields, 1985, Atomic Spectra and Collisions in External Fields, 1988, Atoms in Strong Fields, 1990. Recipient IR-100 award, 1977. Mem. Am. Phys. Soc., Am. Electrochem. Soc. Achievements include co-development of resonance ionization spectroscopy. Office: Univ of Ill 1110 W Green St Urbana IL 61801

NAZMY, ALY SADEK, structural engineering educator, consultant; b. Port Said, Egypt, Nov. 14, 1955; came to U.S., 1983; s. Sadek Adly Nazmy and Bahigah Sayed Ibrahim; m. Hala Esmat Elnagar, July 15, 1983; children: Rania, Mohamed. BSc with honors, Ain Shams U., Cairo, 1977, MSc, 1982; MA, Princeton U., 1985, PhD, 1987. Instr. Ain Shams U., 1977-83, asst. prof., 1989-91; rsch. asst. Princeton U., 1983-87; asst. prof. structural engring. Poly. U., Bklyn., 1987-89, 91-93; assoc. prof. structural engring. U. Maine, Orono, 1993—. Contbr. articles to profl. jours. Recipient George Van Ness Lothrop Honorific award Princeton U., 1986. Mem. ASCE, Earthquake Engring. Rsch. Inst., N.Y. Acad. Scis., Sigma Xi. Achievements include research to formulate the procedure for performing nonlinear seismic analysis of cable-supported bridges under the effect of non-synchronous ground motion. Office: U Maine 5711 Boardman Hall Orono ME 04469-5711

NAZOS, DEMETRI ELEFTHERIOS, obstetrician, gynecologist; b. Mykonos, Greece, July 20, 1949; came to U.S., 1967, naturalized, 1983; s. Eleftherios D. and Anousso (Grypari) N.; m. Dorothea A. Lazarides, Dec. 3, 1977; children: Anna D., Elliot D. BS, Loyola U., Chgo., 1971; student in psychol. scis., Pierce Am. Coll., Athens, 1972; MD, U. Athens, 1976; postgrad. biol. med. scis., N.D. State U., 1976; postgrad. cert. Am. Acad. Family Physicians, Washington U., St. Louis, 1982, Am. Yag Laser Assn., 1986; postgrad. med. cert. in microsurgery, Tulane Med. Sch., 1989. Diplomate Am. Bd. Ob-Gyn.; cert. Ednl. Commn. Fgn. Med. Grads.; Nat. Bds. cert.; cert. Mich. Med. Practice Bd.; cert. State of Ill. Extern in internal medicine U. Minn., Mpls., 1974-75; intern U. Athens Hosps., 1975-76; resident Harper Grace Hosp., Wayne State U., Detroit, 1976-80; practice medicine specializing in ob-gyn., Livonia, Mich., 1980-81, Joliet, Ill., 1981—; mem. staff St. Joseph Med. Ctr., Silver Cross Hosp.; pres., chief exec. officer Amsurg Surg. Ctr., Demetri E. Nazos, Ltd. Med. Profl. Corp., 1982—; pres., chief exec. officer, exec. dir. Joliet Diagnostic Imaging, Ltd. 1982. Sustaining mem. Rep. Nat. Com. Presdl. Task Force; mem. Inner Circle, Washington. Named Man of Yr. 1990, ABI, REcognition and Appreciation Cert., Joliet Jr. Woman's Club, 1986. Fellow Am. Coll. Ob-Gyn., Am. Fertility Soc.; mem. AMA Physician's Med. award), Internat. Med. Specialists Assn., Royal Soc. Medicine-Eng., Greek Med. Assn., Am. Biog. Inst., Inc. (Man of Yr. in Medicine 1991), Am. Assn. Laparoscopists, Am. Inst. Ultrasound in Medicine (cert.), Ill. State Med. Assn., Southeastern Surg. Soc. Mich., Will-Grundy County Med. Soc., Am. Mgmt. Inst., Harper/Grace Hosps. Med. Postgrad. Physicians' Assn., Mich. State Med. Assn., Mykonos Soc., NRA, Ill. State Rifle Assn., Hellenic Med. Soc. Chgo., Inst. Profl. Bookkeepers, Assn. Reproductive Health Profls. Clubs: Senatorial; Joliet Country; Lincoln Skeet and Trap. Greek Orthodox. Lodge: Rotary. Avocations: photography; hunting; gun collecting. Home: 24436 Woodridge Way # 1 Joliet IL 60436-8608 Office: 330 Madison St Joliet IL 60435-6565

NEAL, STEPHEN WAYNE, electrical engineer; b. Mildenhall, Eng., Aug. 26, 1964; came to U.S., 1964; s. Charles Houston and Ruby (Puckett) N.; m. Sandra Hanshew, June 14, 1986. BSEE, W.Va. Inst. Tech., 1986; MSEE, Johns Hopkins U., 1990. Electronic engr. Dept. Def., Ft. Meade, Md., 1986—. Mem. IEEE, Tau Beta Pi (pres. student chp.t 1986), Eta Kappa Nu (sec. student chpt. 1986), Alpha Chi.

NEALE, MICHAEL CHURTON, behavior geneticist; b. Amersham, Eng., Mar. 10, 1958; came to U.S., 1986; s. Albert Henry and Sylvia Strangways (Collins) N.; m. Lynne Neale, July 22, 1980; children: Matthew, Benjamin, Catherine, Zoë. BSc, London U., 1980, PhD, 1983. Mem. nursing aux. staff Coney Hill Mental Hosp., Gloucester, Eng., 1976-77; lectr. London U., 1983-86; asst. prof. genetics Med. Coll. Va., Richmond, 1988-92, assoc. prof. genetics, 1992—; mem. Health Scis. Computing Adv. Com., Richmond, 1989-92; lectr., coord. Twin Methodology Workshops, Leuven, Belgium, 1989-91; cons. Mx Software Devel., Richmond, 1990-92. Author: Methodology for Genetic Studies of Twins and Families, 1992, (software and manual) Mx: A Package for Statistical Analysis, 1991; editor spl. issue Behavioral Genetics, 1989; contbr. articles to profl. publs. Pres. Queens Point Homeowners Assn., Richmond, 1988-90; mem. exec. bd. Citizens for Neighborhood Integrity, Richmond, 1988-92. Mem. APA, Behavior Genetics Assn. (chair profl. tng. com. 1989-91, Thompson Meml. award 1983). Achievements include development of mathematical models for age at onset of disease, rater bias, multivariate genetic and cultural transmission, ascertainment and volunteer bias, applications to anxiety, depression, schizophrenia, alcoholism, etc. Office: Med Coll Va Box 710 MCV Richmond VA 23298-0710

NEASE, ALLAN BRUCE, research chemist; b. Wichita, Kans., Dec. 2, 1954; s. Lewis Edward and Margarette (Douglass) N. BS in Chemistry, Wichita State U., 1975; PhD, U. Mo., 1979. Chemist U.S. EPA, Kansas City, Kans., 1976-77; postdoctoral fellow U. S.C., Columbia, 1979-81; sr. rsch. chemist Monsanto Rsch. Corp., Miamisburg, Ohio, 1981-88; rsch. specialist EG&G-Mound Applied Techs., Miamisburg, Ohio, 1988—. Mem. Am. Chem. Soc. (chmn. profl. practices 1983), Soc. Applied Spectroscopy (sect. chmn. 1984-85). Home: 175 Golfwood Dr West Carrollton OH 45449 Office: EG&G PO Box 3000 Miamisburg OH 45343

NEBELKOPF, ETHAN, psychologist; b. N.Y.C., June 13, 1946; s. Jacob and Fannie (Carver) N.; m. Karen Horrocks, July 27, 1976; children: Demian David, Sarah Dawn. BA, CCNY, 1966; MA, U. Mich., 1969; PhD, Summit U., 1989. Social worker Project Headstart, N.Y.C., 1965; coord. Project Outreach, Ann Arbor, 1968-69; program dir. White Bird Clinic, Eugene, Oreg., 1971-75; counseling supr. Teledyne Econ. Devel. Corp., San Diego, 1976-79; dir. planning and edn. Walden House, San Francisco, 1979-89, dir. tng., 1990—; adj. prof. Dept. Social Work, San Francisco State U., 1982-87; cons. Berkeley Holistic Health Ctr., Berkeley, 1979-84, Medicine Wheel Healing Co-op, San Diego, 1976-79; alternate del. Nat. Free Clinic Coun., Eugene, Oreg., 1972-74. Author: White Bird Flies to Phoenix, 1973, The New Herbalism, 1980, The Herbal Connection, 1981, Hope Not Dope, 1990. Mem. Mayor's Task Force on Drugs, San Francisco, 1988; mem. treatment com. Gov.'s Policy Coun. on Drugs, Sacramento, 1989; task force Human Svcs. Tng., Salem, Oreg., 1972; organizer West Eugene Bozo Assn., 1973; founder Green Psychology, 1993. Named Outstanding Young Man of Am., U.S. Jaycees, 1980; recipient Silver Key, House Plan Assn., 1966. Fellow Am. Orthopsychiat. Assn.; mem. Calif. Assn. Family Therapists, World Fedn. of Therapeutic Communities, Nat. Writer's Club, N.Y. Acad. Scis., Internat. Assn. for Human Rels. Lab. Tng., Calif. Assn. of Drug Programs and Profls. (pres. 1988-90), Phi Beta Kappa. Avocations: herbs, rocks, cactus, yoga, baseball cards. Office: 6641 Simson St Oakland CA 94605-2220

NEBERGALL, ROBERT WILLIAM, orthopedic surgeon, educator; b. Des Moines, Dec. 31, 1954; s. Donald Charles and Shirley (Williams) N.; m. Teresa Rae Fawell, May 27, 1978; children: Nathaniel Robert Baird, Bartholomew William Campbell. BS in Biology, Luther Coll., 1977; DO, U. Osteo. Health Scis., 1981. Intern Des Moines Gen. Hosp., 1981-82; resident orthopedic surgery Tulsa Regional Med. Ctr., 1982-86; trauma fellow Assn. Osteosynthesis/Assn. Study of Internal Fixation Fellowship Program, Stuttgart and Mainz, West Germany, 1986; sports medicine fellow U. Oreg. Orthopedic and Fracture Clinic, Eugene, 1986; orthopedic surgeon Tulsa Orthopedic Surgeons, 1987—; team physician Tulsa Ballet Theatre, 1987—, Internat. Pro Rodeo Assn., Pauls Valley, Okla., 1987—, Nathan Hale High Sch. Football, 1992, Tulsa Roughnecks Soccer, 1993; clin. asst. prof. surgery Okla. State U. Coll. Osteo. Medicine. Reviewer Okla. Found. for Peer Rev., Oklahoma City, 1988—; pres. Culver (Ind.) Summer Sch. Alumni Assn., 1991; trustee Culver Ednl. Found. of Culver Mil Acad., 1991. Recipient Vol. award Tulsa Ballet Theatre, 1990, Physicians Recognition award AMA, 1990; named Outstanding Young Man in Am., U.S. Jaycees, 1983. Mem. Am. Osteo. Acad. Orthopedics, Am. Coll. Sports Medicine, Nat. Strength and Conditioning Assn., Am. Osteo. Orthopedic Soc. Sports Medicine, (pres.), Sigma Sigma Phi. Methodist. Avocations: strength tng., bicycling, hunting. Home: 2116 S Detroit Ave Tulsa OK 74114-1208 Office: Tulsa Orthopedic Surgeons 802 S Jackson Ave Ste 130 Tulsa OK 74127-9010

NECHES, RICHARD BROOKS, cardiologist, educator; b. Long Branch, N.J., Nov. 22, 1955; s. Jacob and Evelyn (Brooks) N. BS summa cum laude, Wofford Coll., 1976; MD, Med. U. S.C., Charleston, 1979. Diplomate Am. Bd. Internal Medicine, cardiovascular diseases; lic. physician, N.Y., N.J. Resident internal medicine Brookdale Hosp. Med. Ctr., Bklyn., 1979-82; fellow cardiovascular diseases Beth Israel Med. Ctr., N.Y.C., 1982-84, chief fellow cardiology, 1984; pvt. practice cardiology, 1984—; co-chief div. cardiology dept. medicine St. John's Episcopal Hosp., Far Rockaway, N.Y., 1990—; clin. instr. medicine SUNY, Stony Brook, 1987-88; clin. asst. prof. medicine U. Medicine and Dentistry N.J., Robert Wood Johnson Med. Sch., New Brunswick, 1989-91, SUNY Health Sci. Ctr., Bklyn., 1991—; participant internat. cardiology studies. Contbr. articles to profl. jours. Mem. Am. Heart Assn. Fellow Am. Coll. Cardiology, Am. Coll. Angiology, N.Y. Cardiological Soc.; mem. Phi Beta Kappa. Office: St John's Episcopal Hosp Dept Medicine 327 Beach 19 St Far Rockaway NY 11691

NECULA, NICHOLAS, electrical engineering educator, researcher; b. Campulung, Arges, Romania, July 1, 1940; came to U.S., 1985; s. Nicolae and Svetlana (Stepanov) N.; m. Maria-Ana Parau, July 1, 1964; 1 child, Maria-Cristina. MSEE, Poly. Inst., Bucharest, Romania, 1961, PhD in Elec. Engring., 1968. Asst. prof. Sch. Electronics, Poly. Inst., 1961-72, assoc. prof., 1972-85; dean asst. Sch. Electronics Poly. Inst., 1972-76, head dept. telecommunications, 1981-85; prin. systems engr. Contel I.P.C., Stamford, Conn., 1986-91; cons., 1992—; reviewer Zentralblatt fur Mathematik, Heidelberg, Fed. Republic Germany, 1970-90. Author: Logic Circuits: Automatic Synthesis, 1972, also 3 textbooks; mem. editorial bd. Internat. Jour. on Digital Processes, 1971-77; contbr. numerous articles to profl. jours.; inventor in field. Recipient award Romanian Acad. Scis., 1964, 72. Mem. IEEE (sr.), Internat. Neural Network Soc., N.Y. Acad. Scis. (assoc.).

NEDOLUHA, ALFRED KARL FRANZ, physicist; b. Vienna, Austria, Sept. 13, 1928; came to U.S., 1957; s. Franz and Petronella (Forstner) N.; m. Auguste Stahl, Mar. 13, 1957; 1 child, Gerald. PhD, Univ., Vienna, Austria, 1951. Head high voltage testing lab. Felten & Guileaume, Vienna, Austria, 1951-57; missile analyst White Sands Missile Range, N.Mex., 1957-59; rsch. physicist Naval Weapons Ctr., Corona, Calif., 1959-70, Electronics Lab. Ctr., San Diego, 1970-75; chief electronics br. European Rsch. Office, U.S. Army, London, 1975-79; rsch. physicist Naval Ocean Systems Ctr., San Diego, 1979-82, head electronic materials sci. div., 1982-88, emeritus, 1988—. Contbr. numerous articles to profl. jours. Mem. Am. Phys. Soc.

NEEDHAM, CHARLES WILLIAM, neurosurgeon; b. Bklyn., Oct. 14, 1936; s. William and Jeanne (Studioso) N.; m. Constance Taft, June 15, 1958; children—Susan, Andrew, Jennifer, Sarah, Benjamin. B.S. cum laude Wagner Coll., 1957; M.D., Albany Med. Coll., 1961; M.Sc., McGill U., 1969. Cert. Am. Bd. Neurol. Surgery; lic. physician Conn. Asst. prof. neurol. surgery UCLA Sch. Medicine, 1969-71; clin. assoc. prof. neurol. surgery U. Ariz., Tucson, 1971-84; staff neurosurgeon Norwalk (Conn.) Hosp., Greenwich (Conn.) Hosp., 1984—; clin. instr. neurosurgery Yale U. Sch. Medicine, 1989—; postdoctoral fellow Nat. Inst. Neurol. Diseases & Blindness, 1967-69. Author: Neurosurgical Syndromes of the Brain, 1973, Cerebral Logic, 1978, Principles of Cerebral Dominance, 1982, Neurosurgical Signs, 1986; contbr. articles to profl. jours. Served to capt. USAF MC, 1963-65. Recipient numerous awards for excellence in medicine including AMA Continuing Edn. awards, 1978—, Yale U. Sch. Medicine award, 1986. Fellow ACS; mem. AAAS, Am. Assn. Neurol. Surgeons, Congress Neurol. Surgeons, Brain and Behavioral Scis. Assn., N.Y. Acad. Scis., New Eng. Neurosurg. Soc., Conn. State Neurosurg. Soc. (pres. 1992—), Fairfield (Conn.) County Med. Soc. Avocations: philosophy, physics, anthropology, writing. Home: 1 Sipperleys Hill Rd Westport CT 06880-1245 Office: # 5 Elmcrest Terr Norwalk CT 06850

NEEDHAM, JOSEPH, biochemist, historian of science, orientalist; b. 1900; s. Joseph and Alicia N.; M.A., Ph.D., Sc. D., Cambridge U.; F.R.S., F.B.A., hon. F.R.C.P.; D.Sc. (hon.), U. Brussels, U. Norwich, Chinese U. Hong Kong; LL.D. (hon.), U. Toronto, U. Salford; Litt.D. (hon.), U. Hong Kong, U. Newcastle upon Tyne, U. Hull, U. Chgo., U. N.C., Wilmington, U. Cambridge, U. Peradeniya, Sri. Lanka; D.Univ. (hon.), U. Surrey; Ph.D. (hon.), U. Uppsala; Companion of Honor awarded Queen's Birthday Honors, 1992; m. Dorothy Mary Moyle, 1924 (dec. 1979); m. Lu Gwei-Djen, 1989 (dec. 1991). Fellow, Gonville and Caius Coll., Cambridge, 1924-66, librarian, 1959-60, pres., 1959-66 master, 1966-76, sr. fellow, from 1976, emeritus sr. fellow; univ. demonstrator in biochemistry Cambridge U., 1928-33, Sir William Dunn reader in biochemistry, 1933-66, now emeritus; vis. prof. biochemistry Stanford U., 1929; Hitchcock prof. U. Calif., 1950; vis. prof. U. Lyon, 1951, U. Kyoto, 1971, Collège de France, Paris, 1973, U. B.C., Vancouver, 1975; hon. prof. Inst. History of Sci., Peking, 1982, Grad. Sch., Nat. Acad. Social Scis., Peking, 1984; head Brit. Sci. Mission, China, and sci. counsellor Brit. Embassy, Chungking, adv. to Chinese Nat. Resources Commn., Chinese Army Med. Adminstrn., and Chinese Air Force Research Bur., 1942-46; dir. dept. natural sci. UNESCO, 1946-48; chmn. Ceylon Govt. Univ. Policy Commn., 1958. Recipient Sir William Jones medal Asiatic Soc. Bengal, 1963; George Sarton medal Soc. History of Sci., 1968; Leonardo da Vinci medal Soc. History of Tech., 1968; Dexter award for History of Chemistry, 1979; Nat. award 1st class Chinese Sci. and Tech. Commn., 1988, Fukuoka gold medal, 1990. Mem. Internat. Union History of Sci. (pres. 1972-74), Nat. Acad. Sci. U.S. (fgn.), Am. Acad. Arts and Scis. (fgn.), Nat. Acad. Scis. China (fgn.), Royal Danish Acad. (fgn.), Internat. Acads. History of Sci. and Medicine, Sigma Xi (hon.). Author: Man a Machine, 1927; The Skeptical Biologist, 1929; Chemical Embryology (3 vols.), 1931; The Great Amphibium, 1932; A History of Embryology, 1934; Order and Life, 1935; Adventures before Birth, 1936; Biochemistry and Morphogenesis, 1942; Time, the Refreshing River, 1943; History is on Our Side, 1945; Chinese Science, 1946; Science Outpost, 1948; Science and Civilisation in China (7 vols. in 25 parts), 1954—; The Development of Iron and Steel Technology in China, 1958; Heavenly Clockwork, 1960; Within the Four Seas, 1970; The Grand Titration, 1970; Clerks and Craftsmen in China and the West, 1970; Moulds of Understanding, 1976; Celestial Lancets, a history and rationale of Acupuncture and Moxa, 1980; Trans-Pacific Echoes and Resonances—Listening Once Again, 1985; The Hall of Heavenly Records; Korean Clocks and Astronomical Instruments, 1380 to 1780, 1986; editor: Christianity and the Social Revolution, 1935; Background to Modern Science, 1938; Hopkins and Biochemistry, 1949; The Chemistry of Life, 1970, (with Mansel Davies) Selections from the Writings of Joseph Needem, 1990; contbr. articles to profl. jours. Home: 2 A Sylvester Rd, Cambridge England CB3 9AF Office: East Asian History of Sci Libr, 8 Sylvester Rd, Cambridge England

NEEL, LOUIS EUGENE FELIX, physicist; b. Lyons, France, Nov. 22, 1904; s. Louis Antoine and Marie Antoinette (Hartmayer) N.; m. Hélène Hourticq, Sept. 14, 1931; children: Marie-Francoise, Marguerite Guély, Pierre. Agrégé de l'Université, Ecole Normale Supérieure, 1928; Docteur es-Sciences, Strasbourg, France, 1932. With Faculté des Sciences, Strasbourg, 1928-45, prof., 1937-45; prof. Faculté des Sciences Grenoble, France, 1945-76; dir. Lab. Electrostatics and Physics of Metal, 1940-71; pres. Institut Nat. Polytechnique, Grenoble, 1970-76; dir. Centre d'Etudes Nucléaires, Grenoble, 1957-71; French rep. sci. council NATO, 1960-82; pres. Conseil Sup. Sûreté nucléaire, 1973-86. Decorated grand croix Legion of Honor, Gold medal Nat. Center Sci. Research; Nobel prize in physics, 1970. Mem. French Acad. Sci., acads. sci. Moscow, Halle, Royal Soc. London, Romanian Acad., Royal Netherlands Acad. Scis., Am. Acad. Arts and Scis., French Soc. Physics (hon. pres.), Internat. Union Pure and Applied Physics (hon. pres.). Research and numerous publs. on magnetic properties of solids; introduced sci. ideas of ferrimagnetism and antiferromagnetism; discoveries of certain magnetic properties of fine grains and crystals, directional order of magnetism, magnetic after effect. Home: 15 rue Marcel-Allégot, 92190 Meudon-Bellevue France

NEELAKANTAN, GANAPATHY SUBRAMANIAN, geotechnical engineer; b. Dindigul, T.N., India, Mar. 15, 1965; came to U.S., 1986; s. Eswara Ganapathy and Jayam (Srinivasan) Subramanian; m. Chitra Meenakshi, Dec. 13, 1990. B Tech., Indian Inst. Tech., Madras, 1986; MS, SUNY, Buffalo, 1988; PhD, U. Ariz., 1991. Teaching/rsch. asst. dept. civil engring. SUNY, Buffalo, 1986-88, U. Ariz., Tucson, 1988-91; geotech. engr. Geotech. Cons., Inc., San Francisco, 1991—. Mem. ASCE (assoc.). Hindu. Achievements include performance of seismic tests on tied-back retaining walls; proposal of a new method of designing tied-back retaining walls to efficiently withstand earthquakes. Home: 9 Ponmeni Jayanagar, Madurai, Tamil Nadu 625010, India Office: Geotech Cons Inc 111 New Montgomery # 600 San Francisco CA 94105

NEER, CHARLES SUMNER, II, orthopaedic surgeon, educator; b. Vinita, Okla., Nov. 10, 1917; s. Charles Sumner and Pearl Victoria (Brooke) N.; m. Eileen Meyer MacFarlane, June 12, 1990; children from previous marriage: Charlotte Marguerite, Sydney Victoria. BA, Dartmouth Coll., 1939; MD, U. Pa., 1942. Intern U. Pa. Hosp., Phila., 1942-43; asso. in surgery N.Y. Orthopedic-Columbia-Presbyn. Med. Center, N.Y.C., 1943-44; instr. in surgery Coll. Physicians and Surgeons, Columbia U., N.Y.C., 1946-47; instr. orthopaedic surgery Coll. Physicians and Surgeons, Columbia U., 1947-57, asst. prof. clin. orthopaedic surgery, 1957-64, asso. prof., 1964-68, prof. clin. orthopaedic surgery, 1968-90, prof. clin. orthopaedic surgery emeritus, spl. lectr. orthopaedic surgery, 1990—; attending orthopaedic surgeon Columbia-Presbyn. Med. Ctr., N.Y.C.; chief adult reconstructive svc. N.Y. Orthopaedic Hosp.; chief shoulder and elbow clinic Presbyn. Hosp.; cons. Orthopaedic Surgeon emeritus, N.Y. Orthopaedic Columbia Presbyterian Med. Ctr., 1991—, chmn. 4th internat. Congress Shoulder Surgeons. Founder, chmn. bd. trustees Jour. Shoulder and Elbow Surgery, 1990—; contbr. articles to books, tech. films, sound slides. Served with U.S. Army, 1944-46. Fellow ACS (sr. mem. nat. com. on trauma), Am. Acad. Orthopaedic Surgeons (com. on upper extremity, shoulder com.); mem. AMA, Am. Bd. Orthopaedic Surgeons (Disting. Svc. award 1975, bd. dirs. 1970-75), Am. Shoulder and Elbow Surgeons (inaugural pres.), Am. Assn. Surgery of Trauma, Am. Orthopaedic Assn., Am. Coll. Surgeons (mem. com. trauma), Mid-Am. Orthopaedic Assn. (hon.), N.Y. Acad. Medicine, Allen O. Whipple Surg. Soc., N.Y. State Med. Soc., N.Y. County Med. Soc., Pan Am. Med. Assn., Am. Trauma Soc., Sociedad Latino Americana de Ortopedia y Traumatologia, Internat. Soc. Orthopaedic Surgery and Traumatology, Va. Orthopaedic Soc. (hon.), Carolina Orthopaedic Alumni Assn. (hon.), Conn. Orthopaedic Club (hon.), Houston Orthopaedic Assn. (hon.), Soc. Française de Chirurgie Orthopédique et Traumatologique (honneur), Soc. Italiana Ortopedia Etravmatologia (Onorario); patron, Shoulder and Elbow Soc. of Australia, Internat. Bd. of Shoulder Surgery (chmn. 1992—), Alpha Omega Alpha, Phi Chi. Home and Office: 231 S Miller St Vinita OK 74301-3625

NEFF, GREGORY PALL, manufacturing engineering educator, consultant; b. Detroit, Nov. 23, 1942; s. Jacob John and Bonnie Alice (Pall) N.; m. Bonita Jean Dostal, Apr. 27, 1974; 1 child, Kristiana Dostal Neff. BS in Physics, U. Mich., 1964, MA in Math., 1966, MS in Physics, 1967; MSME, Mich. State U., 1982. Registered profl. engr., Ind.; cert. mfg. engr.; cert. mfg. technologist; cert. sr. indsl. technologist. Rsch. asst. cyclotron lab U. Mich., Ann Arbor, 1968-72, teaching fellow physics dept., 1973; instr. sci. dept. Lansing (Mich.) C.C., 1976-82; guest lectr. Purdue U. Calumet, Hammond, Ind., 1982-83, asst. prof., 1984-91, assoc. prof. mfg. engring. technologies and supervision, 1991—; cons. Inland Steel Co., Indsl. Engring., East Chicago, Ind., 1984-86, Polyurethane div. Pinder Industries, East Chicago, 1990-92, Elevated div. Pitts. Tank & Tower, Henderson, Ky., 1990-91. Contbr. articles to profl. jours. County commr. Ingham County Bd. of Commr., Mason, Mich., 1977-80, Tri-County Regional Planning Commn., Lansing, 1978-80, chair, non-motorized adv. coun. Mich. Dept. Transp., Lansing, 1982-83. Mem. ASME, AAUP, Soc. Mfg. Engrs. (chpt. 112, bd. dirs. 1986—, Appreciation award 1990, 92), Ind. Soc. of Profl. Engrs., Am. Soc. for Engring. Edn., Nat. Assn. of Indsl. Tech., Order of the Engr. Democrat. Roman Catholic. Office: Purdue U Calumet 2200 169th St Hammond IN 46323

NEFF, HAROLD PARKER, environmental engineer, consultant; b. Williamsport, Pa., July 21, 1933; s. James Bell and Alberta (Parker) N.; m. Joanne Miller, June 5, 1954; children: David M., Denise J. Shipman, Deborah L. Chamberlin. BS, Lycoming Coll., 1955; MPH, U. N.C., 1960. Diplomate Am. Acad. Sanitarians. Various positions Pa. Dept. Health, 1955-64; dir. environ. health rsch. Avco Corp., Williamsport, 1964-68; v.p. ops. Lyco Systems Inc., Williamsport, 1968-70; v.p., engring. mgr. Lyco Inc., Williamsport, 1971-84; applications engr. Bendlin-Du Hamel Assocs. Inc., Montclair, N.J., 1984-85; pres. Fluid Systems Internat. Inc., Williamsport, 1985—; process cons. Graver Water, Union, N.J., 1992—. Mem. profl. adv. com. Lycoming County Career Consortium, 1992—. Mem. Pa. Environ. Health Assn. (pres. 1971-72), Water Environment Fedn., Am. Water Works Assn. Republican. Presbyterian. Achievements include patent for Apparatus and Method for Denitrification of Wastewater. Home: RD # 3 1933 Mountview Ave Montoursville PA 17754 Office: Fluid Systems Internat Inc 2010 Northway Rd Williamsport PA 17701

NEFF, RAY QUINN, electric power consultant; b. Houston, Apr. 29, 1928; s. Noah Grant and Alma Ray (Smith) N.; m. Elizabeth McDougald, Sept. 4, 1982. Degree in Steam Engring., Houston Vocat. Tech., 1957; BSME, Kennedy Western U., 1986. Various positions Houston Lighting & Power Co., 1945-60, plant supr., 1960-70, plant supt. asst., 1970-80, tech. supr., 1980-85, tng. supr., 1985-87; owner, operator Neff Enterprises, Bedias, Tex., 1987—; tng. supr. Tex. A&M U., 1991—; cons. Houston Industries, 1987-89. Author: Power Plant Operation, 1975, Power Operator Training, 1985, Power Foreman Training, 1986. Judge Internat. Sci. and Engring. Fair, Houston, 1982, Sci. Engring. Fair Houston, 1987. Mem. ASME, Assn. Chief Operating Engrs. Republican. Methodist. Lodge: Masons. Avocations: farming, ranching, classic cars. Home: Hwy 90 Rte 2t Box 193-A Bedias TX 77831 Office: Power Plant Cons PO BX CL College Station TX 77841

NEFZGER, CHARLES LEROY, astronautics company executive; b. Twin Falls, Idaho, May 26, 1944; s. Alvin Elroy Netzger and Eileen (Lorain) Roberts; m. Betty Ann Anderson, Aug. 20, 1966; children: Mark LeRoy, Treena Suzanne. BSEE, Purdue U., 1966; MS, AF Inst. Tech., 1968. Commd. 2d lt. USAF, 1966, advanced through grades to lt. col., 1982; system analyst Cen. Inertial Guidance Test Facility USAF, Holloman AFB, N.Mex., 1973-76; project mgr. Space and Missile Test Ctr. USAF, Vandenberg AFB, Calif., 1973-76; program mgr. Office Sci. Rsch. USAF, Bolling AFB, Washington, 1976-80; space systems resource mgr. HQ, Pentagon USAF, Washington, 1980-84; dir. chief SDI Laser Weapons Lab. USAF, Kirkland AFB, N.Mex., 1984-88; ret. USAF, 1988; v.p. ops. Lentec Corp., Albuquerque, 1988-91; tech. mgr. Martin Marietta Astronautics Group, Denver, 1991—; cons. optical engring. W. J. Schafer Assocs.,

Calabassas, Calif., 1991. Author various govt. publs. Home: 8081 Sweetwater Rd Littleton CO 80124

NEGELE, JOHN WILLIAM, physics educator, consultant; b. Cleve., Apr. 18, 1944; s. Charles Frederick and Virgil Lea (Wettich) N.; m. Rose Anne Meeks, June 18, 1967; Janette Andrea, Julia Elizabeth. B.S., Purdue U., 1965; Ph.D., Cornell U., 1969. Research fellow Niels Bohr Inst., Copenhagen, 1969-70; vis. asst. prof. MIT, Cambridge, 1970-71, faculty mem., 1971—, prof. physics, 1979—, William A. Coolidge prof., 1991—, assoc. dir. Ctr. for Theoretical Physics, 1988-89, dir. Ctr. for Theoretical Physics, 1989—; cons. Los Alamos Sci. Lab., Brookhaven Nat. Lab., Lawrence Livermore Nat. Lab., Oak Ridge Nat. Lab.; mem. physics div. rev. com. Argonne Nat. Lab., (ILL.), 1977-83; mem. nuclear sci. div. rev. com. Lawrence Berkeley Lab., (Calif.), 1982—; mem. adv. bd., steering com. Inst. for Theoretical Physics, U. Calif.-Santa Barbara, 1982-86; mem. adv. bd. inst. for Nuclear Theory U. Washington, 1990—, chair, 1992—; program adv. com. Tandem Van de Graaff Accelerator, Brookhaven Nat. Lab., 1977-78, Bates Linear Accelerator, 1977-80, Los Alamos Meson Prodn. Facility, 1986-89, Brookhaven Alternating Gradient Synchraton, 1987-90. Author: Quantum Many Particle Systems, 1987; contbr. articles to profl. jours.; editor: Advances in Nuclear Physics, 1977—. Grantee NSF, 1965-69; grantee Danforth Found., 1965-69, Woodrow Wilson Found., 1965, Alfred P. Sloan Found., 1979, Japan Soc. for Promotion Sci., 1981, John Simon Guggenheim Found., 1982. Fellow Am. Phys. Soc. (exec. com. 1982-84, program com. 1980-82, editorial bd. Phys. Rev. 1980-82, exec. com. topical group on computational physics 1990—, chair div. computational physics 1992-93, exec. com. 1992—, Bonner prize Com. 1984-85), AAAS (nominating com. 1987-91, mem. physics sect. com. 1991—), Fedn. Am. Scientists. Home: 70 Buckman Dr Lexington MA 02173-6000 Office: MIT Dept Physics 6-308 77 Massachusetts Ave Cambridge MA 02139-4307

NEHER, ERWIN, biophysicist; b. Landsberg, Bavaria, Germany, Mar. 20, 1944; s. Franz Xaver and Elisabeth (Pfeiffer) N.; m. Eva-Maria Ruhr, Dec. 26, 1978; children: Richard, Benjamin, Carola, Sigmund, Margret. MS, U. Wis., 1967; Dr. rer. nat., Tech. U., Munich, 1970; Doctorate honoris causa, U. Limburg, Belgium, 1988; doctorate honoris causa, U. Alicante, Spain, 1993—; postgrad., U. Wis., 1993. Rsch. assoc. Max Planck Inst. for Psychiatry, Munich, 1970-72; rsch. assoc. Max Planck Inst. for Biophys. Chemistry, Göttingen, 1972-75, 1976-83, rsch. dir., 1983—; rsch. assoc. Yale Univ., 1975-76; Fairchild Scholar, Calif. Inst. of Tech., 1988-89. Author: Elektronische Messtechnik, 1974; editor: Single Channel Recording, 1983; contbr. numerous articles to profl. jours. Co-recipient Nobel Prize physiology or medicine, 1991; recipient Louisa Gross-Horwitz award Columbia U., N.Y.C., 1986, Leibniz award Deutsche Forschungsgemeinschaft, Bonn, 1986, Gairdner Found. award, 1989, H. HellmutVits prize U. Münster, Germany, 1990. Mem. NAS (fgn. assoc.), Bavarian Acad. Scis. (corr. mem.), Academia Europea, Akademie d. Wissensch. zu Goettinger. Roman Catholic. Office: Max Planck Inst Biophys Chemistry, Am Fassberg Pf 2841, 3400 Göttingen Germany

NEHER, LESLIE IRWIN, engineer, former air force officer; b. Marion, Ind., Sept. 15, 1906; s. Irvin Warner and Lelia Myrtle (Irwin) N.; m. Lucy Marion Price; 1 child, David Price; m. Cecelia Marguerite Hayworth, June 14, 1956; BS in Elec. Engring., Purdue U., 1930. Registered profl. engr., Ind., N.Mex. Engr. high voltage rsch., 1930-32; engr. U.S. Army, Phila., 1933-37; heating engr. gas utility, 1937-40; commd. 2d lt. U.S. Army, 1929, advanced through grades to Col., 1947; dir. tng., Tng. Command, Heavy Bombardment, Amarillo (Tex.) AFB, 1942-44; dir. mgmt. tng., 15th AF, Colorado Springs, Colo., 1945-46; mgr. Korea Electric Power Co., Seoul, 1946-47, ret., 1960; engr. Neher Engring. Co., Gas City, Ind., 1960—; researcher volcanic materials, 1948-49. Chmn. Midwest Indsl. Gas Coun., 1969; historian Grant County, Ind., 1985—. Named Outstanding Liaison Officer, Air Force Acad., 1959; Ambassador for Peace, Republic of Korea, 1977; recipient Republic of Korea Svc. medal, 1977. Mem. Ind. Soc. Profl. Engrs. (Outstanding Engr. 1982, Engr. of Yr. Ind. 1986), Nat. Soc. Profl. Engrs., Midwest Indsl. Gas Assn. (chmn. 1969), Am. Assn. of Retired Persons (pres. Grant County chpt. 1986, 87, 89, 90, dir. dist. 5 1992-93), NAUS (pres. Grissom chpt. 1992-93). Republican. Methodist. Lodge: Kiwanis (Disting. sect. 1979-85, lt. gov. 1964; Disting. Svc. award 1962).

NEIDELL, NORMAN SAMSON, oil and gas exploration consultant; b. N.Y.C., Mar. 11, 1939; s. Harry and Eva (Dermansky) N.; m. Elizabeth Joy Reay, May 17, 1963; children: shani E.R., H. Penelope C., R. Alexander R., Nicholas H. T., Victoria A. C. BA, NYU, 1959; postgrad., Brown U., 1960; diploma, Imperial Coll., 1961; PhD, Cambridge (U.K.) U., 1964. Rsch. geophysicist Gulf R&D Co. (now Chevron), Harmarville, Pa., 1964-68; tech. asst. to pres. Seiscom-Delta (now Grant-Norpac), Houston, 1968-71; co-founder GeoQuest Internat. Inc. (now Petroleum Info.), Houston, 1973-82; co-founder, chmn., pres., bd. dirs. Zenith Exploration Co., Houston, 1977-87; mng. ptnr. Delphian Signals, Ltd., Houston, 1978—; exec. dir. Gandalf Explorers Internat., Ltd., Houston, 1989—; cons. N. S. Neidell and Assoc., Houston, 1971—; adj. prof. geophysics U. Houston, 1971-87; adv. bd. Rice U. Geophysics Program, Houston, 1984—, Ind. Geol. Survey Computer, 1975—; chmn. SEG Tech. Program for OTC, Houston, 1974-77. Contbr. articles to profl. jours. including Geophysics, spl. issue Geophysics on Interpretation, others. Fellowship Woods Hole, Brown U. 1960, NSF, 1961-64. Mem. Geophysics Soc. Houston (pres. 1984), Soc. Exptl. Geophysics (assoc. editor 1976-79, Disting. lectr. 1986, Best Presentation award 1973), Am. Assn. Petroleum Geologists, Soc. Petroleum Engrs. Achievements include patents for echolocation and naviagtion systems; development of first practical seismic modeling system and amplitude analysis for thin bed thickness estimation, theory of dolphin echolocation; discovery of three sand/shale reflectivity zones and their seismic characteristics; negotiation of international oil and gas producing sharing agreements as an independent operator. Home: 315 Vanderpool Ln Houston TX 77024 Office: NS Neidell and Assocs 2929 Briarpark Dr #125 Houston TX 77042

NEIDHARDT, JEAN SLIVA, systems integration professional, sales executive; b. Canastota, N.Y., Oct. 29, 1956; d. William Rudolf and Marian Jean (Cady) Sliva; m. Gil E. Neidhardt, Aug. 20, 1982; children: Christopher S., Ryan P. BA in Sociology, LeMoyne Coll., Syracuse, N.Y., 1978. Programmer, analyst Digital Equipment Corp., Maynard, Mass., 1978-79, bus. systems analyst, 1979-81, sr. telecommuns. analyst, 1981-85, mgr. systems support and telecommuns., 1985-86; cons. Digital Equipment Corp., Liverpool, N.Y., 1986-90, sales support mgr., 1990—. Mem. NAFE. Democrat. Roman Catholic. Home: 99 West Main St Morrisville NY 13408 Office: Digital Equipment Corp 290 Elwood Davis Rd Liverpool NY 13088

NEIER, REINHARD WERNER, chemistry educator; b. Basel, Switzerland, Dec. 11, 1950; s. Reinhard and Anna (Angehrn) N.; m. Christiane Bobillier; children: Roland, Carine. Diploma in chemistry, U. Basel, 1972; PhD in Chemistry, Eidgenössisch Technische Hochschule-Zurich, Switzerland, 1978. Postdoctoral fellow U. Cambridge, Eng., 1978-79; maitre asst. U. Geneva, 1979-80; chef des travaux U. Fribourg, Switzerland, 1980-90, assoc. prof., 1991; prof. chemistry U. Neuchâtel, Switzerland, 1991—; co-organizer seminar 3-ème Cycle, Les Diablerets, 1984; co-organizer XIII European Colloguium on Heterocyclic Chemistry, Fribourg, 1988. Recipient Fuiji-Otsuka prize Otsuko Co., Tokushima, Japan, 1991. Mem. German Chem. Soc., Swiss Chem. Soc., Royal Chem. Soc., Am. Chem. Soc. Achievements include research in organic chemistry. Office: Inst Chemistry, Bellevaux 51, CH-7000 Neuchâtel Switzerland

NEIFELD, JAMES PAUL, surgical oncologist; b. Paterson, N.J., June 5, 1948; s. Herbert S. and Elinor (Charney) N.; m. Ramona S. Simmons, Apr. 27, 1985; children: Emily Claire, Jillian Rose. Student, Lafayette Coll., 1965-68; MD, Med. Coll. Va., 1972. Asst. prof. surgery Med. Coll. Va., Richmond, 1978-82, assoc. prof., 1982-86, prof., 1986—. Lt. comdr. USPHS, 1974-76. Office: PO Box 11 M C VSta Richmond VA 23298

NEIL, GARY LAWRENCE, pharmaceutical company research executive, biochemical pharmacologist; b. Regina, Sask., Can., June 13, 1940; came to U.S., 1962; s. Bert Lawrence and Barbara Jessie (Robinson) N.; m. Beverly May Hendry, Apr. 16, 1939; children: Deborah Nadine, Michael Lawrence. BS with honors, Queen's U., Kingston, Ont., Can., 1962; PhD, Calif. Inst. Tech., 1966. Rsch. scientist The Upjohn Co., Kalamazoo, Mich., 1966-73, rsch. head, 1973-79, rsch. mgr., 1979-82, group mgr., 1982-83, exec. dir., 1983-85, v.p., 1985-89; sr. v.p. Wyeth-Ayerst, Radnor, Pa., 1989-90, exec. v.p., 1990-93; pres., CEO Therapeutic Discovery Corp., Palo Alto, Calif., 1993—. Editor Investigational New Drugs, 1983-88; contbr. over 50 articles to profl. jours. Mem. Am. Chem. Soc., Am. Assn. Cancer Rsch., Am. Soc. Clin. Pharmacology and Exptl. Therapeutics. Presbyterian. Avocation: sailing. Home: Apt 417 1830 Oak Creek Dr Palo Alto CA 94304 Office: Therapeutic Discovery Corp PO Box 10051 1290 Page Mill Rd Palo Alto CA 94303*

NEIL, GEORGE RANDALL, physicist; b. Springfield, Mo., Apr. 11, 1948; s. George and Doris (Neely) N.; m. Doreen Osowski, May 6, 1978; children: Rebecca T., Daniel L. BS, U. Va., 1970; PhD, U. Wis., 1977. Program mgr. TRW, Redondo Beach, Calif., 1977-90; dept. mgr. CEBAF, Newport News, Va., 1990—; bd. dirs. Peninsula High Tech. Coun., Newport News, 1991-92. Contbr. articles to profl. publs. With U.S. Army, 1970-72, Vietnam. Mem. Am. Phys. Soc., Southeast Photon Consortium (bd. dirs., pres. 1990—). Achievements include patents in the field of high energy lasers, particle accelerators, and the development of controlled thermonuclear fusion. Office: CEBAF 12000 Jefferson Ave Newport News VA 23606

NEILL, ROBERT D., engineering executive; b. Fredericton, N.B., Can., Aug. 8, 1932; s. Wallace Raymond and Marjorie Haines (Fletcher) N.; m. Joey J. Coates, June 25, 1954; children: Katherine Josephine, Kimberly Robert. BS, Univ. N.B., 1954, DSc (hon.), 1985. Design engr. N.B. Power, 1954-56, sr. mech. engr., 1957-61, chief design engr., 1962-64; exec. v.p. Neill and Gunter Ltd., Fredericton, N.B., 1964-78, pres., CEO, 1979-84, chmn., CEO, 1984—; pres. Neill and Gunter Inc., 1976-84; pres. NGM Internat., 1978—; v.p. Mfg. Tech. Ctr. of N.B., CADMI Microelectronics Ctr., Univ. N.B.; chmn. Wood Sci. Tech. Inst. Mem. ASME, Assn. Profl. Engrs. of N.B., Can. Coun. Profl. Engrs. (chmn., CEO, Gold Medal award 1991), Forest Product Rsch. Soc., Fredericton Garrison Club Inc. Home: 505 Golf Club Rd, Fredericton, NB Canada E3B 5Z5 Office: Neill & Gunter Ltd, 191 Prospect St W, Fredericton, NB Canada E3B 5B4*

NEILSON, ALASDAIR HEWITT, microbiologist; b. Glasgow, Scotland, June 26, 1932; s. Alexander and Margaret (Johnston) N. BSc, U. Glasgow, 1954; PhD, U. Cambridge, Eng., 1958. Tech. officer ICI Pharms., 1962-65; univ. lectr. U. Sussex, Brighton, 1965-68; rsch. bacteriologist U. Calif., Berkeley, 1968-72; rsch. biochemist U. Stockholm, 1972-75; sect. leader Swedish Environ. Rsch. Inst., Stockholm, 1975-87, rsch. mgr., 1987-92, prin. scientist, 1992—. Contbg. author: Handbook of Environmental Chemistry; contbr. articles to profl. jours. including Applied and Environ. Microbiology, Jour. Applied Bacteriology, Ecotoxicology and Environ. Safety, Jour. Chromatog. Mem. Am. Chem. Soc., Am. Soc. for Microbiology, Soc. of Gen. Microbiology, Soc. of Applied Bacteriology. Home: Baldersvägen 3, S-141 44 Huddinge Sweden Office: IVL, Box 21060, S-10031 Stockholm Sweden

NEILSON, RONALD PRICE, ecology educator; b. Portland, Oreg., Feb. 21, 1949; s. Ronald Price and Marian (Schwichtenberg) N.; m. Avery Sedgwick, June 8, 1974; children: Elizabeth Ann, Charles Robert. BA in Biology, U. Oreg., 1971; MS in Biology, Portland State U., 1975; PhD in Biology, U. Utah, 1981. Asst. prof. N.Mex. State U., Las Cruces, 1982-84; asst. prof. U. Ariz., Tucson, 1984-85; sr. rsch. scientist U. Utah Rsch. Inst., Salt Lake City, 1985-87; assoc. prof. Oreg. State U., Corvallis, 1987—; bioclimatologist USDA Forest Svc., 1992—. Author: (book chpts.) Landscape Boundaries, 1992, Ecotones, 1991; contbr. articles to Jour. of Biogeography, Sci., Landscape Ecology, Ecol. Applications. Lectr. U.S. Forest Svc., 1992, World Affairs Coun. of Oreg., 1990, Solar Energy Assn. Oreg., 1990. Recipient W.S. Cooper award Ecol. Soc. Am., 1987, Tech. Contbn. awards U.S. EPA, 1988, 90; rsch. grantee U.S. EPA, 1987-93, Nat. Park Svc., 1992—. Mem. AAAS, U.S. Internat. Assn. Landscape Ecologists (councillor at large 1991-93), Am. Inst. Biol. Sci., Am. Geophys. Union. Achievements include rsch. in biotic effects of global climate change; development of computer models of coterminous U.S. and global vegetation predicting future distribution and biosphere feedbacks. Office: USDA Forest Svc Forest Sci Lab Corvallis OR 97331

NEIMARK, VASSA, interior architect; b. Miami, Fla., Dec. 9, 1954; d. William Rolla and Bettijean (Davison) Meyer; m. Philip John Neimark, Oct. 29, 1982; 1 child, Dashiel Charles. Student, Art Inst. Ft. Lauderdale, 1974, Art Inst. Chgo., 1980. Owner, prin. Vassa Inc., Chgo., 1979—. Contbr. articles to local mag. Bd. dirs. M.R.I.C. Michael Med. Found., Chgo., Expressways Mus., Chgo., Orchard Village Home for Retarded Adults, Park Ridge Youth Campus, Des Plaines, Ill. Recipient Star on Horizon award Chgo. Mdse. Mart-Chgo. Design Sources, 1985, Spl. Recognition in Design award, 1987. Mem. Internat. Soc. Interior Designers (bd. dirs. 1985-86), Women in Design Industry, IFA Found. N. Am. (v.p.), Internat. Inst. for Bau-Biologie and Ecology, Inc., Carlton Club, Club Internat. Avocation: designing and manufacturing one-of-a-kind art furniture pieces. Office: Vassa Inc 1923 N Halsted St Chicago IL 60614-5008

NEISHLOS, ARYE LEON, chemist; b. St. Petersburg, Russia, Jan. 8, 1955; s. Yefim and Sophia (Zhukov) N.; m. Yanina Zilberfarb, July 25, 1984; 1 child, Maurice Ronald. MS in Chemistry, St. Petersburg Tech. Inst., Russia, 1977; BSChE, Technion-Israel Inst. Tech., Haifa, 1981. Chem. engr. Israel Def. Forces, 1980-82; chem. project engr. Haifa Oil Refinery, 1982-84; chem. engr. Impala Platinum Ltd. Refineries, Springs, Republic of South Africa, 1985-87; plant engr. Anhydrides and Chems., Newark, 1987-88; sr. devel. chemist Castrol North Am., Piscataway, N.J., 1988—. Mem. Am. Chem. Soc., Israel Assn. Engrs. and Architects, South African Inst. Mining and Metallurgy, STLE. Jewish. Office: Castrol North Am 240 Centennial Ave Piscataway NJ 08854

NEITZ, DAVID ALLAN, information scientist; b. Paterson, N.J., Feb. 16, 1961; s. Robert Eugene Neitz and Betty Louise (Jones) Washington; m. Meta Rose Bonfadini, Aug. 13, 1983 (div. 1988); m. Sharon Jean Shifflet, Sept. 21, 1991. B Computer Sci., Kennedy Western U., 1993. Computer oper. Artistic Identifications Systems, Pompton Lakes, N.J., 1978-79; customer engr. Computer Scies. Corp., Suitland, Md., 1983-84; field engr. Data Gen. Corp., Vienna, Va., 1984-86; sr. programmer/analyst Dexel Systems, Vienna, Va., 1989-90; pres., computer cons. Software Express, Inc., Redding, Calif., 1987-88; ind. cons. Washington, 1990-92; mgr. MIS H&K, Inc., Sterling, Va., 1992—. With U.S. Army, 1979-83. Republican. Achievements include design of methodologies for fourth generation code generation for IBM midrange computers. Home: 4339 Cub Run Rd Chantilly VA 22021

NELKIN, DOROTHY, sociology and science policy educator, researcher; b. Boston, July 30, 1933; d. Henry and Helen (Fine) Wolfers; m. Mark Nelkin, Aug. 31, 1952; children: Lisa, Laurie. B.A., Cornell U., 1954. Research assoc. Cornell U., Ithaca, N.Y., 1963-69, sr. research assoc., 1970-72, assoc. prof., 1972-76, prof. sci. tech. sociology program, 1976-90, prof. sociology, 1977-90; Univ. prof., prof. sociology, affiliate prof. law NYU, 1990—, Clare Boothe Luce vis. prof., 1988-90; cons. OECD, Paris, 1975-76, Inst. Environ., Berlin, 1978-79; maitre de conference U. Paris, 1975-76; maitre de recherche Ecole Polytechnique, Paris, 1980-81. Author: The Creation Controversy, 1982, Workers at Risk, 1984, The Atom Besieged, 1981, Science as Intellectual Property, 1983, Selling Science: How the Press Covers Science and Technology, 1987, Dangerous Diagnostics: The Social Power of Biological Information, 1989, A Disease of Society: Cultural Impact of AIDS, 1991, The Animal Rights Crusade, 1991, Controversy: Politics of Tecnical Decision, 3d edit., 1992. Adviser Office Tech. Assessment, 1977-79, 82-83; expert witness ACLU, Ark., 1982; mem. Nat. Adv. Coun. to NIH Human Genome Project, 1991—. Vis. scholar Resources for the Futures, 1980-81; vis. scholar Russell Sage Found., N.Y.C., 1983; Guggenheim fellow, 1983-84. Fellow AAAS (bd. dirs.) Hastings Inst. Soc. Ethics and Life Scis.; mem. NAS Inst. of Medicine, Medicine in the Pub. Interest (bd dirs.), Soc. for Social Studies Sci. (pres. 1978-79). Home: 3 Washington Square Vlg New York NY 10012-1836 Office: NYU Dept Sociology 269 Mercer St New York NY 10003-6633

NELL, JANINE MARIE, metallurgical and materials engineer; b. Milw., Jan. 15, 1959; d. Joseph Frank (Gabrhel) and Joyce Cecelia (Jans) Clendening; m. Michael Paul Nell, Aug. 19, 1978. SB in Materials Sci. and Engring., MIT, 1981, PhD in Metallurgy, 1989. Rsch. asst. MIT, Cambridge, 1981-89; sr. engr., asst. to pres. Failure Analysis Assocs., Inc., Menlo Park, Calif., 1989-91, sr. engr. exec. office, 1991-92, materials and mechanical lab. supr., 1992--. Author: Progress in Powder Metallurgy, 1986, Superalloys 92, 1992; contbr. articles to profl. jours. Recipient Karl Taylor Compton award MIT, 1986; Cabot Corp. fellow, 1981-85. Mem. Am. Soc. Mechanical Engrs., ASM Internat., The Metallurgical Soc. Am. Inst. Mining, Metallurgical and Petroleum Engrs., Soc. Plastics Engrs., Am. Welding Soc., Sigma Xi, Tau Beta Pi. Achievements include designed, manufactured and tested new high temperature alloys for gas turbine applications based on multiphase strengthening and engineered grain structures. Office: Failure Analysis Assocs Inc 149 Commonwealth Dr Menlo Park CA 94025

NELSON, ALFRED JOHN, retired pharmaceutical company executive; b. Dalmuir, Scotland, Jan. 24, 1922; came to U.S., 1972; s. John and Mary Catherine (Duncan) N.; m. Frances C. Hillier, Dec. 5, 1952; children: J. Stuart, Andrew D. MBChB, U. Glasgow, Scotland, 1945, MD with commendation, 1957; DPH, Royal Inst. Pub. Health and Hygiene, London, 1948. Resident Ayr County Hosp., 1945, Belvidere Fever Hosp., Glasgow, Scotland, 1948; cons. N.Y. State Dept. Health, Albany, 1950-51; dir. venereal disease control B.C. Dept. Health and Welfare, Vancouver, Can., 1952-55, cons. epidemiology, 1954-55; asst. dean medicine, assoc. prof. pub. health U. B.C., Can., 1955-57; dir. health services B.C. Hydro and Power Authority, Vancouver, Can., 1957-70; v.p. Hoechst-Roussel Pharm., Inc., Somerville, N.J., 1972-81, sr. v.p., med. dir., 1981-87, ret., 1987; hon. mem. staff Vancouver Gen. Hosp. Served with RCAF, 1953-56. Recipient John J. Sippy Meml. award Am. Pub. Health Assn., 1959. Fellow ACP, Royal Coll. Physicians and Surgeons Can., Am. Coll. Preventive Medicine, N.Y. Acad. Medicine; mem. Can. Med. Assn. (sec. com. 1954-57). Presbyterian. Home: 29436 Port Royal Way Laguna Niguel CA 92677

NELSON, CARL MICHAEL, construction executive; b. Hazleton, Pa., July 12, 1956; s. Carl Oke and Loretta Adeline (Ravina) N.; m. Wendy Lynn LeBrun, Oct. 24, 1987; 1 child, Carley Lynn. BS in Civil Engring., Pa. State U., 1978; MBA, U. Md., 1987. Engr. in tng. Constrn. engr. Burns and Roe, Inc., Oradell, N.J., 1981; constr. supr. Potomac Electric Power Co., Washington, 1978-80, sr. constrn. supr., 1981-85, constrn. supt., 1985-88; mgr., constrn. Potomac Capital Investment Corp., Washington, 1988—; tech. com. Kramer Junction Co., Washington, 1991—. Author: (manual) Segs Project Mgmt. Manual, 1989. Mem. Glenwick Homeowners Assn., Alexandra, Va., 1985; key person United Way Campaign, Washington, 1986-88. Mem. ASCE, Constrn. Industry Inst., Pa. State U. Alumni Assn. Republican. Methodist. Home: 5120 Morningside Ln Ellicott City MD 21043 Office: Potomac Capital Investment 900 19th St NW Washington DC 20006

NELSON, CHAD MATTHEW, chemical engineer, researcher; b. Redwood City, Calif., Feb. 8, 1965; s. Jerome Charles and Sandra (Christl) N.; m. Nike Victoria Agman, Feb. 6, 1990. BSChemE, Calif. Inst. Tech., 1987; PhDChemE, U. Calif., Berkeley, 1991. Lab. asst. Stanford U. Linear Accelerator Ctr., Menlo Park, Calif., 1984; rsch. asst. dept. chemistry Calif. Inst. Tech., Pasadena, 1986; rsch. asst. dept. chem. engring. U. Calif., 1987-91, instr., 1988-89; sr. chem. engr. Advanced Fuel Rsch., Inc., East Hartford, Conn., 1991—. Grad. fellow NSF, 1988. Mem. AICE, AAAS, Am. Chem. Soc., Am. Soc. for Microbiology. Republican. Roman Catholic. Office: Advanced Fuel Rsch Inc 87 Church St East Hartford CT 06108

NELSON, DARRELL WAYNE, university administrator; b. Aledo, Ill., Nov. 28, 1939; s. Wayne Edward and Olive Elvina (Peterson) N.; m. Nancyann Hyer, Aug. 27, 1961; children: Christina Lynne, Craig Douglas. BS in Agriculture, U. Ill., 1961, MS in Agronomy, 1963; PhD in Agronomy, Iowa State U., 1967. Cert. profl. soil scientist. Div. chief U.S. Army Chem. Corps., Denver, 1967-68; asst. prof. Purdue U., West Lafayette, Ind., 1968-73, assoc. prof., 1973-77, prof. agronomy, 1977-84; dept. head U. Nebr., Lincoln, 1984-88, dean for agr. rsch. and dir. Nebr. Agrl. Experiment Sta., 1988—; cons. U.S. EPA, Washington, 1977-79, Ind. Bd. of Health, Indpls., 1977-83, Eli Lilly Co., Indpls., 1976. Editor: Chemical Mobility and Reactivity in Soils, 1983. Served to capt. U.S. Army, 1967-68. Fellow AAAS, Am. Soc. Agronomy (bd. dirs., CIBA-Geigy award 1975, Agronomic Achievement award 1983, Environ. Quality Rsch. award 1985), Soil Sci. Soc. Am. (bd. dirs., pres. elect 1992, pres. 1993); mem. Internat. Soil Sci. Soc., Lions Lodge (treas. 1980-83 Lafayette, Ind. chpt.). Presbyterian. Avocations: fishing, skiing, jogging. Office: Univ of Nebr Agrl Rsch Div Lincoln NE 68583-0704

NELSON, DAVID BRIAN, physicist; b. Lincoln, Nebr., Oct. 23, 1940; s. Carl Leroy and Charlotte Mary (Butler) N.; m. Kathryn Muriel Puester, May 12, 1964; children: Mark, Laura. AB, Harvard U., 1962; MA, NYU, 1965, PhD, 1967. Mem. research staff in engring. physics Oak Ridge (Tenn.) Nat. Lab., 1966-71, mem. research staff in plasma physics, 1971-79; chief fusion theory and computer services Office Fusion Energy, Dept. Energy, Washington, 1979-84, dir. applied plasma physics, 1984-87; dir. sci. computing Office Energy Research, 1987, exec. dir., 1987—; mem. adv. com. on civil def. Nat. Research Council, Washington, 1968-71; vis. mem. Courant Inst., NYU, 1975-76. Contbr. articles to profl. jours. Council pres. bd. trustees Christ Luth. Ch., Bethesda, Md., 1986-87. Recipient Appreciation award Def. Electric Power Adminstrn., Washington, 1973. Mem. Am. Phys. Soc. Avocations: tennis, skiing, music, drama. Office: US Dept Energy Office Energy Research Washington DC 20585

NELSON, DAVID LOREN, geneticist, educator; b. Washington, June 25, 1956; s. Erling Walter and Marlys Joan (Jorgenson) N.; m. Claudia Jane Hackbarth, July 31, 1982; children: Jorgen William, Erik Alexander. BA, U. Va., 1978; PhD, MIT, 1984. Staff fellow NIH, Bethesda, Md., 1985-86; sr. assoc. Baylor Coll. Medicine, Houston, 1986-89, instr., 1989-90, asst. prof., 1990—. Editor Genome Data Base, 1992—; exec. editor Genetic Analysis, 1990—; contbr. articles to profl. jours. Achievements include development of Alu PCR; discovery of fragile X syndrome gene (FMR-1), new form of genetic mutation (simple repeat expansion). Office: Baylor Coll Medicine Inst Molecular Genetics 1 Baylor Pla Houston TX 77030

NELSON, DENNIS GEORGE ANTHONY, dental researcher, life scientist; b. New Plymouth, New Zealand, Dec. 25, 1954; came to U.S., 1983; s. Hugo and Johanna Katherina (Dekker) N.; m. Joanne Elizabeth Dick; 1 child, Kathryn Sarah. BS with honors, Victoria U., Wellington, New Zealand, 1977, PhD, 1981. Postdoctoral fellow Med. Rsch. Coun. of New Zealand, Wellington, 1981-82; rsch. assoc. Materia Technica Rijksuniversiteit, Groningen, Netherlands, 1982-83; Fogarty Internat. fellow Eastman Dental Ctr., Rochester, N.Y., 1983-85; sr. fellow Med. Rsch. Coun. of New Zealand, Wellington, 1985-88; staff scientist Procter & Gamble Co., Cin., 1988—; rev. cons. NIH, Washington, 1991—; sci. reviewer for various jours., 1983—. Contbr. articles to profl. jours. Recipient Cogate-Palmolive Travel award Internat. Assn. for Dental Rsch., 1980, Cogate-Palmolive prize Internat. Assn. for Dental Rsch., 1980, Edward H. Hatton award, 1981, Hamilton Meml. prize Royal Soc. of New Zealand, 1983. Mem. AAAS, Internat. Assn. for Dental Rsch., European Orgn. for Caries Rsch. Achievements include patents in field and patents pending; rsch. in high resolution TEM of hydroxyapatites; rsch. in interaction of laser radiation with dental enamel; rsch. in elucidation of fluoridation mechanisms of dental enamel and apatites. Office: Procter & Gamble 1180 E Miami River Rd Ross OH 45061

NELSON, DON JEROME, electrical engineering and computer science educator; b. Nebr., Aug. 17, 1930; s. Irvin Andrew and Agnes Emelia (Nissen) N. BSc, U. Nebr., 1953, MSc, 1958; PhD, Stanford U., 1962. Registered profl. engr., Nebr. Mem. tech. staff AT&T Bell Labs., Manhattan, N.Y., 1953, 55; instr. U. Nebr., Lincoln, 1955-58, from asst. to assoc. prof., 1960-63, dir. computer ctr., 1963-72, prof. electrical engring., 1967—, prof. computer sci., 1969—, co-dir. Ctr. Comm. & Info. Sci., 1988-91; cons. Union Life Ins., Lincoln, 1973, Nebr. Pub. Power Dist., Columbus, 1972-83, Taiwan Power Co., Taipei, 1974. 1st lt. USAF, 1953-55. Mem. IEEE (sr., Outstanding Faculty award 1989), Assn. Computing Machinery. Republican. Office: U Nebr 209N WSEC Lincoln NE 68588-0511

NELSON, FREDERICK CARL, mechanical engineering educator, academic administrator; b. Braintree, Mass., Aug. 8, 1932; s. Carl Edwin and Marjorie May (Miller) N.; m. Delia Ann Dwaresky; children: Jeffrey, Karen, Richard, Christine. BSME, Tufts U., 1954; MS, Harvard U., 1955, PhD, 1961. Registered profl. engr., Mass. Instr. Tufts U., Medford, Mass., 1955-57, asst. prof. mech. engring., 1957-64; assoc. prof. mech. engring. Tufts U., Medford,

1964-71, prof. mech. engring., 1971—, dean engring., 1980—; bd. dirs. Mass. Tech. Park Corp., Westboro. Translator: (book) Mechanical Vibrations for Engineers, 1983. Fellow ASME (Centennial Medal award 1980), Acoustical Soc. Am.; mem. AAAS, Nat. Inst. Applied Scis. of Lyons (medal 1988), Korea Advanced Inst. Sci. and Tech. (medal 1988), Tufts U. Alumni Assn. (medal 1991). Office: Tufts Univ Coll of Engring Office of the Dean Medford MA 02155-5555

NELSON, GARY ROHDE, computer systems executive; b. Charleston, S.C., Oct. 19, 1942; s. Louis August Rohde Jr. and Ruth Ann (Hynes) N.; m. Carol Lynn Starr, Dec. 31, 1967 (div. Mar. 1990); children: Catherine, Sara; m. Holly Roos, Nov. 23, 1990. AB, Duke U., 1964; PhD, Rice U., 1972. Economist Inst. Def. Analyses, Arlington, Va., 1969-72; staff economist Rand Corp., Santa Monica, Calif., 1972-75; prin. analyst Congressional Budget Office, Washington, 1975-77; deputy asst. sec. U.S. Dept. Defense, Arlington, Va., 1977-79; assoc. dir. U.S. Office Personnel Mgmt., Washington, 1979-81; prin. Systems Rsch. and Applications Corp., Arlington, 1982-85, sr. v.p., 1985-88, exec. v.p., 1988—; cons. in field; mem. com. on Social Security Modernization, Washington, 1988-92; charter mem. Sr. Exec. Svc. of U.S., Washington, 1979. Contbr. to books and articles to profl. jours. Pres. Montgomery County chpt. Md. Ornithological Soc., Bethesda, 1985-87, bd. dirs., 1988—. Ford Found. fellow, 1967-68, Samuel Fain Carter fellow Rice U., 1965-67. Mem. Am. Birding Assn., Audubon Naturalist Soc., Duke U. Alumni Assn. Avocations: birding, golf, tennis. Office: SRA Corp 2000 15th St N Arlington VA 22201

NELSON, GEORGE DRIVER, astronomy and education educator, former astronaut; b. Charles City, Iowa, July 13, 1950; s. George Vernon and Evelyn Elenor (Driver) N.; m. Susan Lynn Howard, June 19, 1971; children: Aimee Tess, Marti Ann. BS, Harvey Mudd Coll., 1972; MS, U. Wash., 1974, PhD, 1978. Astronaut NASA, Houston, 1978-89; mission specialist Space Shuttle flight, 1984, 86, 88; asst. provost, assoc. prof. astronomy U. Wash., Seattle, 1989—, adj. assoc. prof. edn., 1989—. Unitarian. Avocations: reading; athletics; guitar. Office: U Wash Office of the Provost AH 20 Seattle WA 98195

NELSON, GORDON LEIGH, chemist, educator; b. Palo Alto, Calif., May 27, 1943; s. Nels Folke and Alice Virginia (Fredrickson) N. BS in Chemistry, U. Nev., 1965; MS, Yale U., 1967, PhD, 1970; DSc (hon.), William Carey Coll., 1988. Staff research chemist corp. research and devel. Gen. Electric Co., Schenectady, N.Y., 1970-74; mgr. combustibility tech. plastics div. Gen. Electric Co., Pittsfield, Mass., 1974-79, mgr. environ. protection plastics div., 1979-82; v.p. materials sci. and tech. Springborn Labs. Inc., Enfield, Conn., 1982-83; prof., chmn. dept. polymer sci. U. So. Miss., Hattiesburg, 1983-89; dean Coll. Sci. and Liberal Arts, prof. chemistry Fla. Inst. Tech., Melbourne, 1989—, mem. coun. sci., soc. pres., sec., 1989-90, chair-elect, 1991, chair, 1992; cons. in field. Author: Carbon-13 Nuclear Magnetic Resonance for Organic Chemists, 1972, 2d edit., 1980; co-author: Polymer Materials--Chemistry for the Future, 1989, Carbon Monoxide and Human Lethality, 1993; editor: Fire and Polymers--Hazard Identification and Prevention, 1990; editor books on coatings sci. tech.; contbr. articles to profl. jours. Mem. Am. Inst. Chemists (Mems. and Fellow lectr. award 1989), Soc. Plastics Engrs., Am. Chem. Soc. (pres. 1988, bd. dirs. 1977-85, 87-89, 92—, Henry Hill award 1986), Computer and Bus. Equipment Mfrs. Assn. (chmn. Plastics Task Group), ASTM (E5 cert. of appreciation 1985), So. Soc. for Coatings Technology, IEC (U.S. tech. adv. group on info. processing equipment), Structural Plastics div. Soc. of the Plasics Industry (Man of Yr. 1979), Coun. Colls. Arts and Scis., Yale Chemists Assn. (pres. 1981—), Nev. Hist. Soc., Sigma Xi. Republican. Presbyterian. Avocations: travel, western U.S. history. Office: Fla Inst Tech Coll Sci & Liberal Arts 150 W University Blvd Melbourne FL 32901-6988

NELSON, GORDON LEON, agricultural engineering educator; b. Chippewa County, Minn., Dec. 28, 1919; s. John Anton and Hilda (Weberg) N.; m. Florence Jeanne Wise, June 7, 1942; children: Gordon Leon, Carol Nelson Earl, Linda Nelson Ochsner, Janet (dec.), David, Barbara Nelson Pumphrey. B.Agrl. Engring., U. Minn., 1942; certificate naval engring. design, U.S. Naval Acad. Postgrad. Sch., 1945; M.Sc., Okla. State U., 1951; Ph.D., Iowa State U., 1957. Sr. agrl. engr. Portland Cement Assn., Chgo., 1946-47; asso. prof. to prof. agrl. engring. Okla. State U., 1947-69; prof. agrl. engring. Ohio State U., also Ohio Agrl. Research and Devel. Center, 1969-81, chmn. dept., 1969-81; dir. Ohio State U.-Ford Found. project Coll. Agrl. Engring., Punjab (India) Agr. U., 1969-72; cons. in field.; Mem. 7 engring. edn. and accreditation ad hoc visitation teams to evaluate agrl. engring. curricula Engrs. Council Profl. Devel. Author and co-author 2 agrl. engring. textbooks, 1988; contbr. articles to profl. jours. Chmn. bd. dirs. Stillwater (Okla.) Municipal Hosp., 1956-60; mem. grad. council Ohio State U., 1970-74; bd. dirs. Council for Agrl. Sci. and Tech., 1975-81. Served to comdr. USNR, 1942-68. NSF Sr. Postdoctoral fellow U. Calif., Berkeley and Davis, 1964, 65-66. Fellow Am. Soc. Agrl. Engrs. (dir. awards dept., bd. dirs., dir. edn. and research dept. 1979, Metal Bldg. Mfg. award 1960, 8 outstanding Paper awards, Massey-Ferguson Gold Medal award 1986, Cyrus Hall McCormick-Jerome Increase Case Medal 1990); mem. Am. Soc. Engring. Edn., Am. Assn. Engring. Socs. (chmn. continuing edn. com.), Sigma Xi, Tau Beta Pi, Sigma Tau, Alpha Epsilon, Phi Kappa Phi, Phi Tau Sigma, Gamma Sigma Delta. Republican. Baptist (chmn. deacons 1971). Home: 6000 Sedgwick Rd Columbus OH 43235-3319

NELSON, GORDON LEON, JR., aeronautical engineer; b. N.Y.C., June 29, 1943; s. Gordon L. and Florence Jeanne (Wise) N.; m. Carolyn Stover, Aug. 2, 1968 (div. Feb. 1990). BS in Physics, Okla. State U., 1966, MS, 1969; M of Aeronautics and Astronautics, U. Washington, 1977. Sr. engr. Boeing Comml. Aircraft Co., Renton, Wash., 1977-80; sr. engr. MSE, Inc., Butte, Mont., 1980-83, engr., 1988-90, supr. test and evaluation br., 1990—; instr. aero. engring. Ariz. State U., Tempe, 1983-88; adj. prof. Mont. Coll. Mineral Sci. & Tech., Butte, 1989—. Contbr. articles to profl. jours. Lt. (j.g.) USNR, 1969-72. Mem. AIAA, Sigma Pi Sigma. Achievements include patent pending for pressurization of the liquid metal tundish in a nozzle aspirated thermal spray device, MHD generator performance, fluid dynamics of Magneto-Hydro-Dynamics generators, plasma physics. Home: 1200 W Diamond Butte MT 59701 Office: MSE Inc PO Box 3767 Butte MT 59701

NELSON, JAMES ALONZO, radiologist, educator; b. Cherokee, Iowa, Oct. 20, 1938; s. Joe George and Ruth Geraldine (Jones) N.; m. Katherine Metcalf, July 16, 1966; children: John Metcalf, Julie Heaps. AB, Harvard U., 1961, MD, 1965. Asst. prof. radiology U. Calif. San Francisco, 1972-74; assoc. prof. U. Utah, Salt Lake City, 1974-79, prof., 1979-86; prof. U. Wash., Seattle, 1986—; dir. radiol. rsch. U. Calif.-San Francisco/Ft. Miley VAH, 1973-74, U. Utah, 1984-85, U. Wash., 1986—. Contbr. articles to Aj. Jour. Roentgeneology, Radiology, Investigative Radiology, others; contbr. chpts. to books. Capt. USAF, 1967-69. John Harvard scholar, 1957-61, James Picker Found. scholar, 1973-77; recipient Mallinkrodt prize Soc. Body Computerized Tomagraphy, 1990, Roscoe Miller award Soc. Gastrointestinal Radiology, 1991. Fellow Am. Coll. Radiology (diplomate); mem. Radiol. Soc. N.Am., Assn. Univ. Radiology. Achievements include patents for Non-surgical Peritoneal Lavage, Recursive Band-Pass Filter for Digital Angiography; patent (with others) for Unsharp Masking for Chest Films, Improved Chest Tube, Oral Hepatobiliary MRI Contrast Agent. Office: U Wash Dept Radiology SB-05 Diagnostic Imaging Sci Ctr Seattle WA 98195

NELSON, JAMES HAROLD, health sciences administrator; b. Gosnell, Ark., Apr. 26, 1936; s. J.D. and Louise (Gann) N.; m. Betty Sue Leonard, Sept. 21, 1974; children: Amelia Rebecca, Rachel Louise. BS, Ark. State U., 1961, MS, 1969; PhD, Okla. State U., 1972. Br. chief U.S. Army Environ. Hygiene Agy., Edgewood, Md., 1972-76; from rsch. area mgr. to div. chief U.S. Army Biomed. R & D Lab., Frederick, Md., 1976-92; project mgr. applied med. systems U.S. Army Med. Materiel Devel. Activity, Fort Derick, Md., 1992—; mem. Fed. Work Group Pest Mgmt., Washington, 1977-81; chmn. equipment com. Armed Forces Pest Mgmt. Bd., Washington, 1979-83; cons. dir. engrs. Ft. Detrick, Frederick, 1976—; guest lectr. Acad. Health Scis., U.S. Army, Ft. Sam Houston, Tex., 1986-88. Contbr. articles to profl. jours.; assoc. editor: Jour. Am. Mosquito Control Assn., 1982-88; chmn. editorial bd.: Equipment & Insecticides-Mosquito Control, 1989. With USN, 1954-58. Recipient numerous commendations U.S. Army, Ft. Detrick, 1981-93, R&D Achievement award Asst. Sec. of the Army, 1988, Order of Mil. Med. Merit, 1992. Mem. AAAS, Am. Pub. Health Assn., Entomol. Soc. Am., Assn. Mil. Surgeons U.S., N.Y. Acad. Scis., Internat. Platform Assn., AMVETS, Am. Legion, Sigma Xi (pres. 1987-88). Republican. Episcopalian. Achievements include patent pending for lemark far-forward surg. table. Home: 2419 Tabor Dr Middletown MD 21769-9006 Office: US Army Med Material Devel Activity Fort Detrick MD 21702-5009

NELSON, JEANNE FRANCESS, mathematics educator; b. Ottumwa, Iowa, May 19, 1933; d. Arthur Bartine and Opal Irene (Elliott) Tyler; m. Robert Koren Nelson, June 7, 1953; children: Shelaine, Kirsten, Teresa, Robert Jr., Michael Kalani. AS, U. No. Iowa, 1983; BE, U. Iowa, 1986; MA in secondary edn., U. Hawaii, 1989. Tchr. math. Kamehameha Schs., Honolulu, 1970—. Recipient Presdl. award State of Hawaii, 1985, 86, 91, Presdl. award for Excellence in Math. Teaching, Washington, 1991; grantee NSF, 1991. Mem. Nat. Coun. Tchrs. Math., Hawaii Coun. Tchrs. Math. (pres. 1986-88, chairperson Western Regional Conf. 1990), Pacific Region Edn. Lab. (math. and sci. consortium), Mu Alpha Theta (nat. gov. 1989-92, chairperson nat. conv. 1993). Episcopalian. Avocations: reading, golf, tennis, swimming, dancing. Home: 686 Ainapo St Honolulu HI 96825-1042 Office: Kamehameha Schs Kapalama Heights Honolulu HI 96817

NELSON, JOHN HOWARD, food company research executive; b. Chgo., May 29, 1930; s. Harold Eugene and Zoe (Peters) N.; m. Jacqueline Raff, Apr. 30, 1952; children: Keith E., Kevin E., Kristen E. BS in Horticulture and Food Tech., Purdue U., 1952, MS in Food Tech. and Microbiology, 1953; PhD in Biochemistry and Microbiology, U. Minn., 1961. From rsch. biochemist to head R&D dept. Gen. Mills., Mpls., 1955-67; dir. R&D to v.p. R&D Peavey Co., Mpls., 1968-76; ptnr. Johnson Powell & Co., Mpls., 1976-78; v.p. R&D, then v.p. mktg. and product devel. Am. Maize Products Co., Hammond, Ind., 1978-82; v.p. corp. devel., then chief oper. officer Roman Meal Co., Tacoma, Wash., 1982-86; corp. dir. R&D, then v.p. R&D McCormick & Co. Inc., Hunt Valley, Md., 1986-88, v.p. sci. and tech., 1988—; sci. program advisor Charles F. Kettering Rsch. Lab., Dayton, Ohio, 1976-77. Trustee St. Joseph Hosp., Towson, Md., 1989—; mem. chancellor's adv. com. U. Md., 1989—; chmn. Ind./Acad. bd. Towson State U., 1990-92. Visking fellow Visking Corp., 1958. Fellow League for Internat. Food Edn. (pres. 1981-82, chmn. Project SUSTAIN 1988-89); mem. Am. Assn. Cereal Chemists (pres. 1974-75, William F. Geddes award 1979), Am. Chem. Soc., Inst. Food Technologists, Elks, Sigma Xi, Gamma Sigma Delta, Alpha Zeta. Republican. Avocations: travel, golf, gardening. Office: McCormick & Co Inc 18 Loveton Cir Sparks MD 21152-6000

NELSON, KAY YARBOROUGH, author, columnist; b. Atlanta, Feb. 16, 1945; d. Newell Dudley and Katie Wilmouth (Durrwachter) Yarborough; m. Raymond A. Nelson, June 3, 1978; children: Holly, Eric, Todd, Karin, Carol. BA, U. Chgo., 1965. Editorial mgr. Sci. Rsch. Assocs., 1966-81; freelance editor various high tech. publs., 1981-85. Author: WordPerfect Macro Handbook, 1989, 2nd edit., 1990, Advanced Techniques in WordPerfect, 1987, 2nd edit., 1988, The Little Windows Book, 1991, 2nd edit., 1992, The Little System 7 Book, 1991, 2nd edit., 1992, Voodoo Windows, 1992, Voodoo DOS, 1992, Voodoo Mac, 1993, Friendly Windows, 1993, The Little OS/2 Book, 1993, numerous others; contbg. editor various high tech. mags.; columnist Home Office Computing mag., DOS Resource Guide, others; also articles. Home and Office: 5751 Pescadero Rd Pescadero CA 94060

NELSON, LARRY KEITH, questioned document examiner; b. Frederick, Okla., Feb. 26, 1948; s. Bernard Leroy and Una Lee (Greeson) N.; m. Barbara Sue Stout, Feb. 26, 1972; children: Shawn Keith, Jeffery Ryan. BS in Forestry, Okla. State U., 1972; MS in Bus. Adminstrn., Boston U., 1986. Apptd. Warrant Officer 1 U.S. Army, 1975, advanced through grades to Chief Warrant Officer 4, 1992; criminal investigator U.S. Army Criminal Investigation Div., Ft. Riley, Kans., 1975-78; questioned document student U.S. Army Criminal Investigation Lab., Ft. Gordon, Ga., 1978-80, questioned document examiner, 1980-82; questioned document examiner U.S. Army Criminal Investigation Lab., Frankfurt, Fed. Republic of Germany, 1982-83, chief questioned document div., 1983-86; questioned document tng. officer U.S. Army Criminal Investigation Lab., Ft. Gillem, Ga., 1986-91, chief questioned document div., 1991-93; v.p. Carney & Nelson Forensic Document Lab. Inc., Norcross, Ga., 1993—; Diplomate Am. Bd. Forensic Document Examiners; cert. profl. instr. Ga. peace officer standers and tng. coun. Contbr. articles to profl. jours. Fellow Am. Acad. Forensic Scis.; mem. Southeastern Assn. Forensic Document Examiners (charter, treas. 1988-91, sec. 1991-93, pres. 1993—). Office: Carney & Nelson Forensic Document Lab. Inc. Ste 240 5855 Jimmy Carter Blvd Norcross GA 30071

NELSON, LAWRENCE MERLE, reproductive endocrinologist; b. Topeka, Kans., Oct. 20, 1947; s. Paul Wharton and Charlotte Irene (Ament) N.; m. Nancy Karen Palmer, June 2, 1973; children: Lance, Christina, Lauren. BS, Westminster Coll., 1969; MD, U. Pitts., 1973. Diplomate Am. Bd. Obstetrics and Gynecology. Resident U. So. Calif., L.A., 1973-77; pvt. practice Lynchburg, Va., 1977-85; rsch. fellow U. London, 1985; fellow George Washington U., Washington, 1986-88; adj. scientist NIH, Bethesda, Md., 1988-90, sr. clin. specialist, 1990—; mem. clin. rsch. subpanel Nat. Inst. Child Health and Human Devel., Bethesda, 1990—. Contbr. articles to profl. jours. Lt. comdr. USPHS, 1990—. Fellow Am. Coll. Obstetricians and Gynecologists; mem. Am. Fertility Soc., Endocrine Soc., Soc. for Study of Reprodn. Presbyterian. Achievements include demonstration of association of premature ovarian failure with increased immune activation; demonstration that tubal salpingotomy closed by secondary intention functions as well as one closed primarily. Office: NIH Bldg 10 Rm 10N262 Bethesda MD 20892

NELSON, MICHAEL GORDON, mining engineer, educator, consultant; b. Afton, Wyo., Aug. 18, 1951; s. Thomas Gordon and Shirley LaRee (Allred) N.; m. Judith Irvine, July 3, 1973; children: Ana, Michael P., Jacob C., Mary J., Thomas I., Marta M. BS, U. Utah, 1975, MS, 1983; PhD, W.Va. U., 1989. Asst. metallurgist U.S. Steel Corp., Geneva, Utah, 1974; metall. engr. Kennecott Copper Corp., Salt Lake City, 1975-78; sr. metall. engr. Western Zirconium, Inc., Ogden, Utah, 1978-81; metall. process engr. Torkelson-Rust, Inc., Salt Lake City, 1981-82; sr. rsch. engr. Consolidation Coal Co., Inc., Pitts., 1983-89; asst. prof. U. Ala., Fairbanks, 1989-91, assoc. prof., 1992—; v.p. Fairbanks Exploration, Inc., 1991—; pres. Ala. Mining Svcs., Fairbanks, 1990—; bd. dirs. Ala. Miners Assn., Anchorage. Editor Procs. 2d Internat. Symposium on Mining in Arctic, 1992; contbr. articles to profl. jours. Pres. Hunter Sch. PTA, Fairbanks, 1990. Recipient Meritorious Svc. award Sch. Mineral Engring., 1990. Mem. Soc. Mining, Metallurgy and Exploration, Instrument Soc. Am. (sr., Outstanding Mining Engring. Faculty award 1991), Greater Fairbanks C. of C. Achievements include 7 patents in mine automation and coal processing. Office: Univ Ala Dept Mining and Geol Engring Fairbanks AK 99775

NELSON, NANCY ELEANOR, pediatrician, educator; b. El Paso, Apr. 4, 1933; d. Harry Hamilton and Helen Maude (Murphy) N. B.A. magna cum laude, U. Colo., 1955, M.D., 1959. Intern, Case Western Res. U. Hosp., 1959-60, resident, 1960-63; practice medicine specializing in pediatrics, Denver, 1963—; assoc. clin. prof. U. Colo. Sch. Medicine, Denver, 1977-88, clin. prof., 1988—, asst. dean Sch. Medicine, 1982-88, assoc. dean, 1988—. Mem. Am. Acad. Pediatrics, AMA, Denver Med. Soc. (pres. 1983-84), Colo. Med. Soc. (bd. dirs. 1985-88, judicial coun. 1992—). Home: 1265 Elizabeth St Denver CO 80206-3241 Office: 4200 E 9th Ave Denver CO 80262

NELSON, PETER EDWARD, energy engineer; b. San Francisco, Nov. 23, 1953; s. William A. and Andre Lucille (Wilson) N. BA in Physics, U. Calif., Santa Cruz, 1977; MS in Solar Energy, Trinity U., 1982. Rschr. Trinity U., San Antonio, 1979-82; cons. The Energy Ctr., Ft. Collins, Colo., 1982-84; solar specialist Idaho Dept. Water Resources, Boise, 1984-87; engr., rschr. Lambert Engring., Bend, Oreg., 1987-90; mgr., cons. Aloha Systems, Seattle, 1990-91; engr. Puget Sound Power and Light Co., Bellevue, 1991—; adv. bd. Internat. Solar Coordinating Coun., 1985-87. guest (TV show) Canyon Forum, 1987; contbr. articles to profl. jours. Past master counselor Order of DeMolay, Burlingame, Calif., 1971; past chmn. bd. Snake River Alliance, Boise, 1986. Grantee Tex. Energy and Natural Resources Adv. Coun., 1980. Mem. ASHRAE, Assn. of Energy Engrs., Assn. of Demand Side Mgmt. Profls., Oreg. Solar Energy Assn. Democrat. Achievements include development of directory of Idaho's energy industry; design of low cost, 3-tank, solar breadbox water heater; research in feasibility of solar-regenerated dessicant dehumidiers for residences.

NELSON, PHILIP EDWIN, food scientist, educator; b. Shelbyville, Ind., Nov. 12, 1934; s. Brainard R. and Alta E. (Pitts) N.; m. Sue Bayless, Dec. 27, 1955; children: Jennifer, Andrew, Bradley. BS, Purdue U., 1956, PhD, 1976. Plant mgr. Blue River Packing Co., Morristown, Ind., 1956-60; instr. Purdue U., West Lafayette, Ind., 1961-76, head dept. food sci., 1984—; cons. PEN Cons., West Lafayette, 1974; chair Food Processors Inst., Washington, 1990—. Editor: Fruit Vegetable Juice Technology, 1980, Principles of Aseptic Processing and Packaging, 1987. Fellow Inst. Food Techs. (Indsl. Achievement award 1976); mem. AAAS, Sigma Xi, Phi Tau Sigma (pres. 1976-77). Achievements include 10 U.S. and foreign patents. Office: Purdue U Dept Food Sci 1160 Smith Hall West Lafayette IN 47907-1160

NELSON, RICHARD DAVID, electro-optics professional; b. Detroit, Feb. 28, 1945; s. Richard Harold and Lorraine (Simo) N.; m. Leigh S. Nelson, Aug. 28, 1969; children: Brett, Sarah. BS, W.Va. State U., 1967; PhD in Physics, Mich. State U., 1972. Scientist, mgr. Rockwell Internat., Anaheim, Calif., 1972-80; mgr. Loral Corp., Newport Beach, Calif., 1980-92; pres. NT Corp., Santa Ana, Calif., 1992—; bd. dirs. Summa Internat. Rsch. Inst., Huntsville, Ala., WIT Corp., Huntsville; adj. prof. U. Calif. Contbr. articles on electro-optics to tech. jours. Achievements include 4 patents in electro-optics and semiconductors. Office: NT Corp 12122 Red Hill Ave Santa Ana CA 92705

NELSON, ROY LESLIE, cardiac surgeon, researcher, educator; b. N.Y.C., May 3, 1941; s. Sam and Anna (Kaminetsky) N.; m. Anne Judith Sachs, Jan. 6, 1973; children: Samuel Philip, Amy Joy, Jill Heather. BS, Lafayette Coll., Easton, Pa., 1963; MD magna cum laude, Free U. Brussels, Belgium, 1971. Cert. MD Am. Bd. Surgery, Am. Bd. Thoracic Surgery; cert. in laser surgery. Intern surgery Bronx Mcpl. Hosp., A.E.C.O.M., 1971-72; resident surgery NYU Med. Ctr. Bellevue Hosp., N.Y.C., 1972-74; thoracic rschr. UCLA Med. Ctr., L.A., Calif., 1974-76; resident surgery NYU Med. Ctr. Bellevue Hosp., 1976-78, fellow cardiothoracic surgery NYU Med. Ctr., 1978-80; asst. attending cardiothoracic surgeon Dept. Surgery, Divsn. Cardiovascular Sugery North Shore U. Hosp., Manhasset, N.Y., 1980-83, assoc. attending cardiothoracic surgeon, 1984—, asst. dir. Dept. Surgery, 1990—; rschr. Bureau Biological Rsch., New Brunswick, N.J., 1963-64, Surg. Rsch. Lab., St. Pierre Hosp., Free U. Brussels, 1969-71, Divsn. Thoracic Sugery, UCLA, 1974-76; physician-in-charge Cardiovascular Rsch. Lab., North Shore U. Hosp., 1980—; teaching asst. Dept. Surgery, Albert Einstein Coll. Medicine, 1971-72, NYU Med. Ctr., 1972-74, 76-77; clin. instr. surgery NYU Med. Ctr., 1977-80; asst. prof. surgery Cornell U. Med. Coll., 1980—. Author: (with others) Plasmapheresis, 1982, Pathophysiology and Techniques of Cardiopulmonary Bypass II, 1983; contbr. articles to profl. jours. Recipient Barnett Meml. prize NYU, 1974, Physician's Recognition award AMA, 1986. Fellow ACS, Am. Coll. Angiology, Am. Coll. Cardiology, Am. Coll. Chest Physicians (coun. critical care 1990—), Am. Soc. for Laser Medicine and Surgery, N.Y. Cardiological Soc.; mem. AAAS, Am. Heart Assn. (rsch com. 1982—, coun. cardiovascular surgery 1984—), Am. Soc. Artificial Internal Organs, Am. Soc. Extra-Corporeal Tech., Internat. Soc. for Artificial Organs (reviewer artificial organs 1984—), Internat. Soc. for Heart Transplantation, Med. Soc. State N.Y., Nassau County Med. Soc., N.Y. Acad Scis., N.Y. Soc. for Thoracic Surgery, N.Am. Society for Pacing and Electrophysiology, Soc. Critical Care Medicine, Soc. Thoracic Surgeons, Spencer Surg. Soc., Undersea Med. Soc. Achievements include research in radial transplantation of the lungs studying different experimental procedures, the importance of alkalosis in maintenance of "ideal" blood pH during hypothermia, the effects of profound topical cardiac hypothermia on myocardial blood flow, metabolism, compliance and function, myocardial preservation during cardiopulmonary bypass, citrate reperfusion of ischemic hearts on cardiopulmonary bypass, improved myocardial performance after aortic cross clamping by combining pharmacologic arrest with topical hypothermia, the effects of hypothermia on regional myocardial blood flow and metabolism during cardiopulmonary bypass, optimizing myocardial supply/demand balance with adrenergic drugs during cardiopulmonary resuscitation, hemoconcentration by ultrafiltration, following cardiopulmonary bypass, intra-aortic balloon rupture, cocaine induced acute aortic dissection, the role of cardioplegia oxygen concentration in limiting myocardial reperfusion injury, the role of morbid obesity and diabetes in the outcome of coronary bypass surgery, isolated intra-thoracic trauma following deployment of an air bag. Office: North Shore U Hosp 300 Community Dr Manhasset NY 11030

NELSON, SANDRA LYNN, biochemist; b. Moline, Ill., May 15, 1953; d. Theodore Edward and Alice Ruth (Scott) N.; m. Rodney Robert Walters, Mar. 1, 1975 (div. Apr. 1986); children: Alexander Michael Walters, Theodore James Walters; m. Donald Paul Zimmerman, Sept. 28, 1990; children: Ethan R. Zimmerman, Jacob Nelson Zimmerman. BS, Iowa State U., 1975; PhD, U. N.C., 1980. Rsch. biochemist Ames (Iowa) Lab. U.S. Dept. Energy, 1980-84; rsch. bov. resp. dis. Nat. Animal Disease Ctr., USDA, Ames, 1984-88; anal. biochem. rsch. and devel. Procter & Gamble Pharms., Cin., 1988—; presenter in field. Contbr. sci. papers to profl. publs. Den leader Boy Scouts Am., Fairfield, Ohio, 1990-91. Mem. N.Y. Acad. Scis. Democrat. Lutheran. Office: Procter & Gamble Co Miami Valley Labs Cincinnati OH 45239

NELSON, THOMAS JOHN, research biochemist; b. Pitts., July 26, 1954; s. Robert Eugene N. PhD, U. R.I., 1984. Sr. staff fellow NIH, Bethesda, Md., 1989-91, collaborative rschr., 1991—. Office: NIH NINDS Park 5 Bldg Rm 435 Bethesda MD 20892

NELSON, WALTER WILLIAM, computer programmer, consultant; b. Seattle, May 7, 1954; s. Arne A. and Helen R. (Truitt) N.; m. Paula E. Truax, Dec. 21, 1985. BA in Zoology, U. Wash., 1976, BS in Psychology, 1977; PhC in Psychology, U. Minn., 1982. Systems analyst Dept. of Social and Health Svcs., State of Wash., Seattle, 1986-89; computer info. cons. Dept. of Health, State of Wash., Seattle, 1989-90; pres. Data Dimensions, Inc., Seattle, 1990—; cons. The Heritage Inst., Seattle, 1989—; pres. Tech Alliance, Renton, Wash., 1990-91. Contbr. articles to profl. jours. Mem. Tech Alliance, Berkeley Macintosh Users Group, Seattle Downtown Macintosh Bus. Users Group, 4th Dimension Spl. Interest Group (founder, pres. 1990—). Avocations: tennis, golf, thoroughbred horse racing. Office: Data Dimensions Inc 1100 NW Elford Dr Seattle WA 98177-4129

NÊME, JACQUES, economist; b. St. Laurent de Maroni, France, June 13, 1930; s. Marcel Pierre and Aimée Jeanne (Janin) N.; m. Marine Collignon, Jan. 29, 1957 (div. 1958); children: Jean Pol, Richard, Sylvie; m. Colette Jeanne Cordebas, Aug. 30, 1963; children: Christiane, Charles Henri, Isabelle, Vincent Neme-Peyron. Diploma, Inst. Etudes Polit., Paris, 1951; lic. en droit, Faculte de Droit, Paris, 1952. Chief svc. economist soc. Gen. de Presse, Paris, 1951-58; dir. Europe Svc., Paris, 1958-80, Afrique Svc., Paris, 1960-80; sec. gen. for econs. Soc. Generale de Presse, Paris, 1980-92; lectr. Inst. du Commerce Internat., 1959-79, U. Paris II, 1973-92, U. Paris IX Dauphine, 1992—, Ecole de Guerre, 1979; cons. Syndicat de la Margarine, Paris, 1970, Assn. Francaise des Banques, 1979. Author: European Economies, 1970, International Economic Organizations, 1972, Compared Economic Policies, 1977, 2d edit., 1989, (with C. Nême) The European Economic Community, 1992; contbr. articles to profl. jours. With French Air Force, 1954-55. Recipient Officer award Order of Merit, 1988, Laureat de l'Institut de France, 1972. Roman Catholic. Avocation: gardening. Home: 83 Rue de Rome, 75017 Paris France

NEMECZ, GEORGE, biochemist; b. Hungary, Nov. 15, 1952; s. György and Maria (Szaghmeister) N.; m. Zsuzsanna Kovacs, Apr. 23, 1977; children:Attila, Akos. MS, U. Szeged, 1978, PhD, 1981. Post doctoral fellow Biological Rsch. Ctr., Szeged, Hungary, 1981-82, rsch. scientist, 1982-86; post doctoral rsch. assoc. St. Louis U., 1986-87, U. Cin., 1987-91; asst. prof. Campbell U., Buies Creek, N.C., 1991--. Contbr. articles to profl. jours. Fed. Support grantee, Hungary, 1985, Ind. Coll. Fund grantee, 1992, Burroughs Wellcome Fund grantee, Braswell Milling Co. Support grantee. Mem. ASBMB, AAAS, AACP, N.Y. Acad. Scis. Achievements include demonstration that showed b-adrenergic agonist sensitivity can be modified by phospholipase-C, that showed linoleic acid-rich diet has a protective effect on the acute phase of coronary occlusion; research on LH hormone regulation of phospholipase-C which controls Ca2 mobilization in ovarian granulosa cells; characterization of liver fatty acid binding protein cholesterol binding capacity; development of new fluorescent technique to measure cholesterol exchange. Office: Campbell U PO Box 1090 Buies Creek NC 27506

NEMETH, EDWARD JOSEPH, process research specialist; b. Glassport, Pa., Oct. 1, 1938; s. John M. and Ann (Breza) N.; m. Eleanor Zakrajsek, Aug. 3, 1968; 1 child, Michael John. BSChemE, U. Pitts., 1960, MSChemE, 1962; PhD, U. Pitts., 1967. Engr., textile fibers I.E. DuPont, Richmond, Va., 1963-64; sr. rsch. engr. U.S. Steel, Monroeville, Pa., 1967-75; rsch. supr. U.S. Steel, Monroeville, 1976-81; mgr. facility planning U.S. Steel, Pitts., 1981-82; dir. rsch. USS Chems./Aristech, Monroeville, pa., 1982—; Contbr. articles to profl. jours.; patentee in field. Mem. Am. Inst. Chem. Engrs. Avocations: youth baseball, theoretical physics, ethnic studies. Office: Aristech Chem Corp 1000 Tech Center Dr Monroeville PA 15146*

NEMETZ, PETER NEWMAN, policy analysis educator, economics researcher; b. Vancouver, B.C., Can., Feb. 19, 1944; s. Nathan Theodore and Bel Nemetz. BA in Econs. and Polit. Sci., U. B.C., 1966; AM in Econs., Harvard U., 1969, PhD in Econs., 1973. Teaching fellow, tutor Harvard U., Cambridge, Mass., 1971-73; lectr. Sch. Planning, U. B.C., Vancouver, 1973-75, asst. prof., assoc. prof. policy analysis, 1975—, chmn., 1984-90; postdoctoral fellow Westwater Research Centre, Vancouver, 1973-75; vis. scientist, dept. med. stats. and epidemiol. Mayo Clinic, 1986-88, sr. visiting scientist Dept. of Health Scis. Research Mayo Clinic, 1988—; cons. consumer and corp. affairs, Can., 1977-80; program chmn. The Vancouver Inst., 1990—; mem. rsch. mgmt. com. Ctr. Health Svcs and Policy Rsch., U. B.C., 1990—, mgmt. com. S.E. Asia Ctr., 1992—. Mem. bd. mgmt. BC-Yukon divsn. Can. Nat. Inst. for Blind. Editor Jour. Bus. Adminstrn., 1978—. Contbr. articles to sci. jours. Grantee Natural Scis. and Engring. Research Council Can., 1976-92, Consumer and Corp. Affairs Can., 1978-80, Econ. Council of Can., 1979-80, Max Bell Found., 1982-84. Mem. Am. Econ. Assn., AAAS, Assn. Environ. and Resource Economists, Internat. Epidemiol. Assn. Liberal. Jewish. Clubs: Harvard of B.C. (pres. 1986—), University (Vancouver). Avocations: swimming; photography. Office: Univ British Columbia, Faculty of Commerce, Vancouver, BC Canada V6T 1Z2

NEMZEK, THOMAS ALEXANDER, nuclear engineer; b. Fargo, N.D., Mar. 22, 1926; s. Alexander Jerome and Anne Jane (Hagen) N.; m. Margaret Clare Peters; children: Paula, Alexandra, Thomas, Michael. BS, U.S. Naval Acad., 1949; MS in Nuclear Engring., N.C. State U., Raleigh, 1952. Mgr. tech. ops. Chgo. Ops. Office, U.S. AEC, Argonne, Ill., 1959-64; dept. mgr. San Francisco Ops. Office, U.S. AEC, Berkeley, 1964-69; mgr. Richland (Wash.) Ops. Office, U.S. AEC, 1969-73; div. dir. reactor devel. AEC/ERDA, Washington, 1973-76; pres. J.A. Jones Applied Rsch. Co., Charlotte, N.C., 1976—. Capt. USAF, 1949-57. Recipient Disting. Svc. award, U.S. AEC, 1971, Spl. Achievement award, U.S. ERDA, 1976. Mem. Am. Nuclear Soc., Tech. Transfer Soc., Rotary (pres. 1989). Office: JA Jones Applied Rsch Co 1300 W Harris Blvd Charlotte NC 28262-8557

NEOU, IN-MEEI CHING-YUAN, mechanical engineering educator, consultant; b. Wuhing, Chekiang, China, Jan. 19, 1917; came to U.S., 1945; s. Yah-son and Wai-sen (Wu) N.; m. Katherine Neou, Aug. 30, 1952; 1 child, Vivian. BS, Chekiang U., 1941; MS, MIT, 1947; PhD in Mech. Engring., Stanford U., 1950. Teaching asst. Chekiang U., Hangzhow, China, 1941-45, MIT, Cambridge, Mass., 1946; rsch. asst. Stanford (Calif.) U., 1947-50; asst. prof. Syracuse (N.Y.) U., 1951-55; from assoc. prof. to prof. U. Bridgeport, Conn., 1955-66; prof. mech. engring. W.Va. U., Morgantown, 1966-82; engring. cons. in pvt. practice Palo Alto, Calif., 1982—. Contbr. articles to Jour. of Applied Mechanics. Mem. ASME, AAAS, Am. Soc. Engring. Edn. Achievements include five patents for calculating and graphic mechanisms, environmental pollution control devices, Evaporative Emission Control of Liquid Storage Tank Using Bellow Sealing Systems. Home: 3444 Murdoch Ct Palo Alto CA 94306-3633

NEPPE, VERNON MICHAEL, neuropsychiatrist, author, educator; b. Johannesburg, Transvaal, Rep. South Africa, Apr. 16, 1951; came to U.S., 1986; s. Solly Louis and Molly (Hesselsohn) N.; m. Elisabeth Selima Schachter, May 29, 1977; children: Jonathan, Shari. BA, U. South Africa, 1976; MB, BCh, U. Witwatersrand, Johannesburg, 1973, D in Psychol. Medicine, 1976, M in Medicine, 1979, PhD in Medicine, 1981; MD, U.S., 1982. Diplomate Am. Bd. Psychiatry and Neurology, Am. Bd. Geriatric Psychiatry; registered psychiatry specialist U.S., Rep. South Africa, Can. Specialist in tng. dept. psychiatry U. Witwatersrand, Johannesburg, 1974-80; sr. cons. U. Witwatersrand Med. Sch., Johannesburg, 1980-82, 83-85; neuropsychiatry fellow Cornell U., N.Y.C., 1982-83; div. dir. U. Wash. Med. Sch., Seattle, 1986-92; dir. Pacific Neuropsychiat. Inst., Seattle, 1992—; mem. clin. faculty dept. psychiatry and behavioral scis. U. Wash. Med. Sch.; attending physician Northwest Hosp.; neuropsychiatry cons. South African Brain Rsch. Inst., Johannesburg, 1985—; chief rsch. cons. Epilepsy Inst., N.Y.C., 1989; mem. faculty lectr. Epilepsy: Refining Medical Treatment, 1993. Author: The Psychology of Deja Vu, 1983, Innovative Psychopharmacotherapy, 1990, (text) BROCAS SCAN, 1992; author (with others) 31 book chpts.; editor 14 jours. issues; contbr. articles to profl. jours. Recipient Rupert Sheldrake Prize for Rsch. Design (2d prize) award New Scientist, 1983, Marius Valkhoff medal South African Soc. for Psychical Rsch., 1982, George Elkin Bequest for Med. Rsch. U. Witwatersrand, 1980, Overseas Travelling fellow, 1982-83. Fellow Psychiatry Coll. South Africa (faculty), Royal Coll. Physicians of Can., North Pacific Soc. for Neurology, Neurosurgery and Psychiatry, Coll. Internat. Neuropharmacologicum; mem. Parapsychologic Assn., Am. Psychiatric Assn. (U.S. transcultural collaborator diagnostic and statis. manual 1985-86, cons. organic brain disorders 1988—), Am. Epilepsy Soc., Soc. Biol. Psychiatry, Can. Psychiat. Assn., Soc. Sci. Exploration, Am. Soc. Clin. Psychopharmacology. Jewish. Avocations: chess, table tennis, tennis, computers. Office: Pacific Neuropsychiat Inst 10330 Meridian Ave N Ste 380 Seattle WA 98133

NERDAL, WILLY, research chemist; b. Bergen, Norway, Sept. 28, 1954; s. Ludvik Bernhard and Inga Johanne (Svardal) N. Cand. Sci., U. Bergen, 1985, PhD, 1990. Fellow Royal Norwegian Coun. Scientific and Indsl. Rsch., Seattle, Wash., 1985-88; fellow U. Bergen, 1988-92, rsch. assoc. dept. chemistry, 1992-93, asst. prof., 1993—. Contbr. articles to profl. jours. Mem. Am. Chem. Soc., Norwegian Chem. Soc., Norwegian Mountaineering Club. Avocations: mountaineering, sport fishing, sailing. Office: Univ Bergen, Dept Chemistry, Allegaten 41, Bergen Norway

NEREM, ROBERT MICHAEL, engineering educator, consultant; b. Chgo., July 20, 1937; s. Robert and Borghild Guneva (Bakken) N.; m. Jill Ann Thomson, Dec. 21, 1958 (div. 1977); children: Robert Steven, Nancy Ann Nerem Chambers; m. Marilyn Reed, Oct. 7, 1978; stepchildren: Christina Lynn Maser, Carol Marie Teasley. BS, U. Okla., 1959; MS, Ohio State U., 1961, PhD, 1964; D (honoris causa), U. Paris, 1990. Asst. prof. Ohio State U., Columbus, 1964-68, assoc. prof., 1968-72, prof., 1972-79, assoc. dean Grad. Sch., 1975-79; prof. mech. engring., chmn. dept. U. Houston, 1979-86; Parker H. Petit prof. Ga. Inst. Tech., Atlanta, 1987—, Inst. prof., 1991—; mem. Ga. Gov.'s Adv. Coun. on Sci. and Tech. Devel., Atlanta, 1992—; ALZA disting. lectr. Biomed. Engring. Soc., 1991. Contbr. over 80 articles to profl. jours. Recipient Lissner award ASME, 1989. Fellow Am. Inst. Med. and Biol. Engring. (founding pres. 1992—), ASME, AAAS; mem. Biomed. Engring. Soc., NAE, Inst. Medicine, Internat. Union for Phys. and Engring. Scis. in Medicine (pres. 1991—), Internat. Fedn. for Med. and Biol. Engring. (immediate past pres.), U.S. Nat. Com. on Biomechanics (immediate past chmn.). Home: 2950 Waverly Ct Atlanta GA 30339-4200 Office: Ga Inst Tech Sch Mech Engring Ferst and Cherry Sts Atlanta GA 30332-0405

NERODE, ANIL, mathematician, educator; b. Los Angeles, June 4, 1932; s. Nirad Ranjan and Agnes (Spencer) N.; m. Sondra Raines, Feb. 12, 1955 (div. 1968); children: Christopher Curtis, Gregory Daniel; m. Sally Riedel Sievers, May 16, 1970; 1 child, Nathanael Caldwell. B.A., U. Chgo., 1949, B.S., 1952, M.S., 1953, Ph.D., 1956. Group leader automata and weapons systems Lab. Applied Sci., U. Chgo., 1954-57; mem. Inst. for Advanced Study, Princeton, 1957-58, 62-63; vis. asst. prof. math. U. Calif. at Berkeley, 1958-59; mem. faculty Cornell U., 1959—, prof. math., 1965—, Goldwin Smith prof. math., 1990—, chmn. dept. math., 1982-87, dir. Math. Sci. Inst., 1986—; acting dir. Center for Applied Math., 1965-66; vis. prof. Monash U., Melbourne, Australia, 1970, 74, 78, 79, U. Chgo., 1976, M.I.T., 1980, U. Calif., San Diego, 1981; disting. vis. scientist EPA, 1985-87; prin. investigator numerous grants; mem. sci. adv. bd. EPA, 1988—, chair tech. adv. panel Global Change, 1990-92; mem. policy adv. bd. High Performance Computing Ctr., U. Minn., 1990—; mem. policy adv. bd. U. Pa. AI Ctr.; mem. sci. adv. bd. Ctr. for Intelligent Control, Harvard-MIT-Brown U., 1988—; cons. to govt. and industry. Author: (with John Crossley) Combinatorial Functors, 1974, (with Richard Shore) Logic for Applications, 1993; editor: Advances in Mathematics, 1967-70, Jour. Symbolic Logic, 1967-82, Annals of Pure and Applied Logic, 1983—, Future Generation Computing Systems, 1983—, Annals of Math. and Artificial Intelligence, 1989—, Logical Methods in Computer Sci., 1991—, Models and Simulation, 1991—. Mem. IEEE Computer Soc., Am. Math. Soc. (assoc. editor procs. 1962-65, v.p. 1992—), Soc. Indsl. and Applied Math., Math. Assn. Am., Assn. for Computing Machinery, Assn. Symbolic Logic, European Assn. for Theoretical Computer Sci. Home: 406 Cayuga Heights Rd Ithaca NY 14850-1402 Office: Cornell U Math Sci Inst 409 College Ave Ithaca NY 14853

NESBITT, LLOYD IVAN, podiatrist; b. Toronto, Ont., Can., Sept. 24, 1951; s. Allan Jay and Rose (Shuster) N.; m. Marlene Cindy Wegler, May 13, 1984; children: Hilary Liza, Andrea Eve, Jeffrey Ryan. D in Podiatric Medicine, Calif. Coll. Podiatric Medicine, San Francisco, 1975. Diplomate Internat. Soc. Podiatric Laser Surgery. Residency program Vancouver (B.C.) Gen. Hosp., Can., 1975-76; pvt. practice podiatric medicine Toronto; cons. podiatry Alan Eagleson Sports Medicine Clinic, Toronto, 1979—; lectr. numerous colls., fitness ctrs. and sports medicine confs., Ont., 1979—. Contbr. numerous articles to sports medicine books and jours; editor Canadian Podiatrist Jour., 1979-88. Fellow Can. Podiatric Sports Medicine Acad. (pres. 1979-89, editor newsletter 1977-89); mem. Internat. Soc. Podiatric Laser Surgery (diplomate), Am. Podiatric Med. Assn., Sierra Club. Avocations: skiing, skating, sailing, cycling, gardening. Home: 122 Argonne Crescent, Willowdale, ON Canada M2K 2K1 Office: Madison Ctr Office Tower, 4950 Yonge St Ste 2414, Toronto, ON Canada

NESPOR, JERALD DANIEL, electrical engineer, educator; b. Chgo., July 20, 1958; s. John Frank and Helen Ann (Ciz) N.; m. Dawn Marie Nixon, Dec. 22, 1984; 1 child, Velika Ann. BSEE, U. Ill., Chgo., 1980, MSEE, 1983. Specialist engr. Boeing Aerospace, Kent, Wash., 1984-88; staff scientist U. Ill., Urbana, 1988-90; prin. mem. engring. staff govt. electronic div. Martin Marietta, Moorestown, N.J., 1990—; instr. North Seattle C.C., 1986-88, Burlington County Coll., Pemberton, N.J., 1992—; presenter profl. confs., 1982—. Contbr. articles to profl. publs. Mem. IEEE, Am. Meteorol. Soc., Sigma Xi. Achievements include patent on Doppler Tolerant Range Side Lobe Suppression with Time Domain Determination of Spectral Moments, Correction of Radar Transmitter Distortion. Home: 73 Boothby Dr Mount Laurel NJ 08054 Office: Martin Marietta Govt Electronic Systems 199 Borton Landing Rd Moorestown NJ 08057

NESS, NORMAN FREDERICK, astrophysicist, educator, administrator; b. Springfield, Mass., Apr. 15, 1933; s. Herman Hugo and Eva (Carlson) N.; children: Elizabeth Ann, Stephen Andrew. B.S., Mass. Inst. Tech., 1955, Ph.D., 1959. Space physicist, asst. prof. geophysics UCLA, 1959-61; Nat. Acad. Sci.-NRC post doctoral research assoc. NASA, 1960-61; research physicist in space scis. Goddard Space Flight Center, Greenbelt, Md., 1961—; head extraterrestrial physics br. Goddard Space Flight Center, 1968-69; chief Lab. for Extraterrestrial Physics, 1969-86; pres., prof. Bartol Research Inst., U. Del., 1987—; lectr. math. U. Md., 1962-64, assoc. research prof., 1965-67. Contbr. articles profl. jours. Recipient Exceptional Sci. Achievement award NASA, 1966, 81, 86, Arthur S. Flemming award, 1968, Space Sci. award AIAA, 1971, Disting. Svc. medal NASA, 1986, Nat. Space Club Sci. award, 1993, Emil Wiechert medal German Geophys. Soc., 1993. Fellow Am. Geophys. Union (John Adam Fleming award 1968), Royal Astron. Soc.; mem. NAS, Academia Nazionale dei Lincei, Royal Ocean Racing Club. Achievements include research, experimental studies of interplanetary and planetary magnetic fields by satellites and space probes. Home: 9 Wilkinson Dr Landenberg PA 19350-9359 Office: U Del Bartol Research Inst Newark DE 19716-4793

NESSMITH, H(ERBERT) ALVA, dentist; b. Miami, Fla., Nov. 27, 1935; s. William Boyd and Florence Editha (Lowe) N.; m. Paula Ann Fox, Oct. 1, 1960 (div. 1984); children: Amy Susan, Lynn Margaret, Mark Alva. Student, U. Miami, Fla., 1953-56; DDS, Northwestern U., 1960. Gen. practice dentistry Tequesta, Fla., 1963—; dental cons. Palm Beach-Martin County Med. Ctr., Jupiter, Fla., 1970—. Mem. adminstrv. bd. United Meth. Ch. Tequesta, Jupiter, 1970—, chmn., 1988-90; pres. Meth. Men, 1982; chmn. Coun. on Ministries, 1992—; pres. Jupiter Elem. PTO, 1972; clarinetist Symphonic Band of Palm Beaches; pianist and clarinetist United Meth. Ch.; mem. Village of Tequesta Hist. Commn., 1992—. Mem. ADA, North Palm Beach County Dental Soc., Fla. Dental Assn., Jupiter-Tequesta-Juno Beach C. of C. Democrat. Lodge: Kiwanis (pres. Jupiter/Tequesta chpt. 1980-81). Avocations: sailing, gardening, music. Home: 196 River Dr Jupiter FL 33469-1934 Office: Inlet Profl Bldg 175 Tequesta Dr Jupiter FL 33469-2721

NESTER, EUGENE WILLIAM, microbiology educator, immunology educator; b. Johnson City, N.Y., Sept. 15, 1930; married, 1959; 2 children. BS, Cornell U., 1952; PhD, Western Reserve U., 1959. Am. Cancer Soc. rsch. fellow genetics Stanford U., 1959-62, instr. microbiology, 1962-63, from asst. to assoc. prof. microbiology and genetics, 1963-72; prof. microbiology and immunology U. Wash., Seattle, 1972—. Chiron Corp. Biotechnology Rsch. award Am. Soc. Microbiology, 1991. Mem. Am. Soc. Microbiology. Achievements include research in genetics and biochemistry of enzyme regulation; bacterial-plant relationships. Office: Univ of Wahington Microbiology Dept AC-42 Seattle WA 98195*

NESTVOLD, ELWOOD OLAF, oil service company executive; b. Minot, N.D., Mar. 19, 1932; came to Netherlands 1979; s. Ole Enevold and Ragnhilda (Quanbeck) N.; m. Simone Chriqui, Dec. 6, 1955 (dec. Jan. 1990); children: Rebecca Lynn, Paul Stephen; m. Jeannette Garvin, Mar. 23, 1991; stepchildren: Michele Marie, Jennifer Ann, Michael Dennis. BA, Augsburg Coll., Mpls., 1952; postgrad., U. Wash., 1952-53; MS, U. Minn., 1959, PhD, 1962. Physics instr. U. Minn., Mpls., 1956-61; physicist and section leader Shell EP Rsch. Lab., Houston, 1962-68, mgr. geophysics rsch., 1968-71; mgr. geophysics Shell Western Div., Denver, Houston, 1971-74, Pecten Internat., Houston, 1974-77; chief geophysicst Woodside Petroleum, Perth, Australia, 1977-78; mgr. EP processing ctr. Shell EP Rsch. Lab., Rijswijk, Netherlands, 1979-81; chief geophysicst Shell Internat. Petroleum, The Hague, 1981-86, dir. geophysics and topography, 1986-92; chief geophysicist GECO-Prakla div. Schlumberger Ltd., Paris, 1992—; v.p. mktg. GECO-Prakla div. Schlumberger Ltd., Paris, 1993—; cons. Lighting and Transients Rsch. Inst., Mpls., 1957-61. Contbr. articles to profl. jours. 1st lt. USAF, 1952-56. Mem. IEEE, Am. Assn. Physics Tchrs., European Assn. Petroleum Geoscientists, European Assn. Exploration Geophysicists, Soc. Exploration Geophysicists, N.Y. Acad. Scis., Soc. Petroleum Engrs., Sigma Xi. Avocations: hiking, museums. Home: 6423 Genstar Dallas TX 75252-5403 Office: 42, rue Saint-Dominque, 75340 Paris CEDEX 07, France

NETTING, ROBERT M., anthropology educator; b. Racine, Wis., Oct. 14, 1934; s. Robert Jackson and Martha Marie (McCorckle) N.; m. Rhonda Marie Gillett, Mar. 13, 1993; children: Robert Frazier, Jessa Forte, Laurel Marthe; 1 child from previous marriage, Jacqueline Ann Frazier. BA English summa cum laude, Yale U., 1957; MA Anthropology, U. Chgo., 1959, PhD Anthropology, 1963. From asst. prof. to assoc. prof. U. Pa., 1963-72; prof. anthropology U. Ariz., Tucson, 1972-91, Regents' prof. anthropology, 1991—; field researcher Ft. Berthold Reservation, N.D., 1958, Jos Plateau, Northern Nigeria, 1960-62, 66-67, 84, Törbel, Valais, Switzerland, 1970-71, 77, Senegal, Ivory Coast, 1977, Portugal, 1982; cons. AID project Stanford U., USDA, USAID Agrl. Devel. Program.; mem. adv. coun. Wenner-Gren Found. Anthropological Rsch., 1982-86, search com. new dir. rsch., 1985-86; mem. com. human dimensions global change commn. behavioral and social scis. and edn. Nat. Rsch. Coun., 1989-91; pres. Internat. Assn. Study Common Property, 1991-92. Author: Documentary History of the Fox Project, 1948-59, 1960, Hill Farmers of Nigeria, 1968; Cultural Ecology of the Kofyar of the Jos Plateau, 1968, Cultural Ecology, 1977, 2d edit., 1986, Balancing on an Alp: Ecological Change and Continuity in a Swiss Mountain Community, 1981, Smallholders, Householders: Farm Families and the Ecology of Intensive Sustainable Agriculture, 1993; editorial com. Annual Rev. Anthropology, 1976-81; bd. editors Ethnohistory, 1983-88; editor Ariz. Studies in Human Ecology, 1984—; contbr. numerous articles to profl. jours. Recipient Robert F. Heizer prize best jour. article ethnohistory Am. Soc. Ethnohistory, 1987; Ctr. Advanced Study Behavioral Scis. fellow, 1986-87, Guggenheim fellow, 1970-71, NSF grantee, 1984-87, 77-78, 71, 58-60, Nat. Inst. Child Health and Human Devel. Ctr. Population Rsch. grantee, 1974-76, Social Sci. Rsch. Coun. grantee, 1966-67, Ford Found. Fgn. Area Studies fellow, 1960-62, Woodrow Wilson fellow, 1957-58. Fellow Am. Anthropological Assn. (exec. bd. 1981-84); mem. NAS, Am. Ethnological Soc. (councillor 1976-79), Am. Anthrop. Assn. (exec. bd. 1981-84), Soc. Ethnohistory, Phi Beta Kappa. Avocations: hiking, fishing, photography. Office: Dept Anthropology Univ of Arizona Tucson AZ 85721

NETTLES, JOSEPH LEE, dentist; b. Fairhope, Ala., Jan. 11, 1954; s. Arthur and Gladys (Fore) N.; m. Dana Renea Samuels, July 11, 1981; children: Joseph Jr., Nicholas, Kimberly. Cert., Meharry Med. Coll., 1973-75, Brookhaven Nat. Lab., 1975; BS, Jarvis Christian Coll., 1976; DMD, U. Ala., 1981. Asst. dental officer, lt. USNR, San Diego, 1981-83, Subic Bay, Philippines, 1983-85, Orlando, Fla., 1985-87; dental officer, lt. comdr. USNR, Mobile, Ala., 1987-89, Individual Ready Reserves, New Orleans, 1989—; gen. dentist pvt. practice Mobile, 1987—. Recipient Award for Patriotism, USN, 1986-89, Plaque of Appreciation, Philippines Dental Assn., 1984, Cert. of Appreciation, 1982, Cert. of Appreciation, Baker Elem. Sch., 1989, 90, 91, Navy scholarship, 1976. Mem. 1st Dist. Dental Soc., Ala. Dental Assn., Am. Dental Assn., Naval Res. Assn., PTA, Am. Fund for Dental Health, Republic Philippines Dental Assn., Tri-Svc. Dental Soc. of Japan, Alpha Kappa Mu, Beta Kappa Chi, Oak Leaf Dental Corps. Avocations: fishing, travel, basketball, reading, swimming. Home and Office: 103 Holly St Mobile AL 36608-4512

NEUBAUER, WERNER GEORGE, physicist; b. White Plains, N.Y., Apr. 18, 1930; s. Paul and Lilli (Reinhardt) N.; m. Jean Harding Quigg, Feb. 27, 1952; children: Steven Werner, Dale Jean. BS, Roanoke Coll., 1952; PhD with distinction, Cath. U. Am., 1969. Physicist Sound div. Naval Rsch. Lab., Washington, 1953-58; head micracoustics sect. Acoustics div. Naval Rsch. Lab., Washington, 1958-68, acting head phys. acoustics br., 1968-70, head microacoustics sect., 1970-79, spl. asst., 1979-82; cons. Werner G. Neubauer, Cons., Annandale, Va., 1982—. Assoc. editor Linear Acoustics Jour. of Acous. Soc. Am., 1986—; author: Acoustic Reflection from Surfaces and Shapes, 1986; contbr. numerous articles to profl. jours., chpts. to books. Walter Cotress-Quincy scholar, Johns Hopkins U., 1953, Edison Meml. scholar, 1966-68; named Sesquicentenial Disting. Alumnus Roanoke Coll., 1992. Fellow Acoustical Soc. Am. (D.C. chpt. pres. 1979-80), Washington Acad. Sci., Sigma Xi (Applied Sci. award 1981). Presbyterian. Achievements include first direct observation and measurement of creeping waves in underwater acoustics; measurement of most accurate sound speed in water than extant; theoritical development and exptl. verification of acoustical reflection from spheres and cylinders. Home and Office: 4603 Quarter Charge Dr Annandale VA 22003

NEUFELD, MURRAY JEROME, aerospace scientist/engineer, consultant; b. N.Y.C., Oct. 20, 1930; s. Solomon and Dora (Faleck) N.; m. Arlene C. Selzer, Jan. 23, 1966 (div. Mar. 1988); m. Jan Hersh, Sept. 16, 1988; stepchildren: Mathew Ellis, Todd Ellis. BS in Physics, CCNY, 1953; MS in Astronautics, MIT, 1955. Electronics scientist U.S. Naval Air Devel. Ctr., Johnsville, Pa., 1953-54; weapon systems engr. U.S. Navy Bur. Aeronautics, Washington, 1955-58; mem. tech. staff, sr. staff engr., chief scientist, mgr. Hughes Aircraft Co., L.A., 1958-92; cons. orbit control Goddard Space Flight Ctr. NASA, Greenbelt, Md., 1962-72; cons. navigation satellites USAF, Dept. Transp., Washington, 1973-75; adv. affiliate physics industry forum MIT, Cambridge, Mass., 1989-92. Recipient Cert. of Appreciation, U.S. Dept. Transp., 1977. Mem. AIAA, N.Y. Acad. Scis., CCNY Alumni Assn. (v.p. programs 1992, pres. So. Calif. chpt. 1993). Achievements include patents for stabilization and control of spin-stabilized communication satellites, communication network synchronization; first team to demonstrate the controllable geostationary spacecraft enabling low cost global communication.

NEUFELD, RONALD DAVID, environmental engineering consultant, researcher, educator; b. N.Y.C., Feb. 10, 1947; s. Milton and Norma Neufeld; m. Toby Heringer, Aug. 31, 1968; children: Steven, Todd, Jennifer. B Engring. in Chem. Engring., Cooper Union, 1967; MS in Chem. Engring., Northwestern U., 1969, PhD in Civil Engring. and Environ. Health Engring., 1973. Registered profl. engr., Pa.; diplomate Am. Acad. Environ. Engrs. Asst. prof. U. Pitts., 1973-77, assoc. prof., 1977-82, prof. civil engring., 1982—, prof. environ. health engring., Grad. Sch. Pub. Health, 1985—, also mem. energy resources faculty Sch. Engring.; cons. in field to govt. and industry; rsch. cons. indsl. waste treatment, waste mgmt. environ. systems for energy devel.; chmn. Mid-Atlantic Indsl. Waste Conf., 1991; mem. hazardous waste in highway rights of way com. NAS/Transp. Rsch. Bd.; chmn. environ. sci. com. Coun. Internat. Exchange Scholars; mem. affluent guidelines task force EPA; mem. indsl. waste com. WEF, 1992—, program com.; mem. edn. com. Am. Acad. Environ. Engring, 1992. Contbr. articles to profl. jours. Fulbright sr. scholar, 1983-84. Mem. ASCE (bd. dirs. Pitts. sect. 1984-87, chmn. environ. effects of energy div., chmn. energy div. specialty conf. 1991), Am. Inst. Chem. Engrs., Internat. Assn. Water Pollution Research, Assn. Environ. Engring. Profs., Assn. Engring. Educators, Water Environ. Fedn. Air and Waste Mgmt. Assn., Sigma Xi, Chi Epsilon. Research using biotechnology for remediation of wastes and soils containing fuel hydrocarbons, heavy metals, and PCBs, environmental process engineering, environmental engineering for energy development. Home: 6558 Bartlett St Pittsburgh PA 15217-1834 Office: Univ Pittsburgh Dept Civil Engring 939 BEH Pittsburgh PA 15261

NEUGEBAUER, GERRY, astrophysicist, educator; b. Gottingen, Germany, Sept. 3, 1932; came to U.S., 1939; s. Otto E. and Grete (Brück) N.; m. Marcia MacDonald, Aug. 26, 1956; children: Carol, Lee. B.S., Cornell U., 1954; Ph.D., Calif. Inst. Tech., 1960. Mem. faculty Calif. Inst. Tech., Pasadena, 1962—, prof. physics, 1970—, Howard Hughes Prof. Physics, 1985—, chmn. divsn. physics, math and astronomy, 1988-93; mem. staff Hale Obs., 1970-80; acting dir. Palomar Obs., 1980-81, dir., 1981—. Served with AUS, 1961-63. Fellow Am. Acad. Arts and Scis.; mem. NAS, Am. Philos. Soc. Office: 320-47 Pasadena CA 91125

NEUGER, SANFORD, orthodontics educator; b. Cleve., Aug. 17, 1925; s. Samuel and Ethel (Manheim) N.; m. Marjorie Odess, Sept. 8, 1963; 1 child, Howard Michael. BS, Western Res. U., 1947, DDS, 1953; MS in Orthodontics, Ind. U., 1957. Diplomate Am. Bd. Orthodontics. Orthodontics demonstrator Western Res. U., Cleve., 1957-58; asst. prof., assoc. prof. orthodontics Western Res. U./Case Western Res. U., Cleve., 1958-75; clin. prof. orthodontics Case Western Res. U., Cleve., 1975—, acting chmn. Orthodontics Dept., 1969-71; asst. dental surgeon U. Hosp., Cleve., 1967—. Author: (syllabus) Contemporary Edgewise Mechanics-Sliding Mechanics, 1973, Limited Tooth Movement, 1970; author-presenter: (videotape) Orthodontics Soldering, 1970. Vol. United Way, 1988, Case Western Res. U. Alumni Assn., Jewish Nat. Fund. Comdr. USNR (ret. 1972). Named Man of Yr. Case Western Res. U. Orthodontics alumni, 1982. Fellow Am. Coll. Dentists; mem. Am. Dental Soc., Cleve. Dental Soc. (bd. dirs. 11965-90), Cleve. Soc. Orthodontists (pres. 1969)., Great Lakes Assn. Orthodontists Assn., Am. Assn. Orthodontists, Pierre Fauchard Soc., Alpha Omega (pres. Cleve. chpt. 1984-85), Omicron Kappa Upsilon. Jewish. Avocations: replicar building. Home: 24850 Hilltop Dr Cleveland OH 44122-1350 Office: 1500 S Green Rd Cleveland OH 44121-4086

NEUMAIER, GERHARD JOHN, environment consulting company executive; b. Covington, Ky., July 27, 1937; s. John Edward and Elli Anna (Raudies) N.; m. Ellen Elaine Klepper, Oct. 24, 1959; children: Kevin Scott, Kirsten Lynn. BME, Gen. Motors Inst., 1960; MA in Biophysics, U. Buffalo, 1963. Research ecologist, project mgr. Cornell Aero. Lab., Buffalo, 1963-70; pres., chief exec., chmn. bd. Ecology and Environment Inc., Buffalo, 1970—. Mem. Am. Pub. Health Assn., Air Pollution Control Assn., Internat. Assn. Gt. Lakes Research, Inst. Environ. Scis., Ecol. Soc. Am., Am. Inst. Biol. Scis., Urban Land Inst., Arctic Inst. N.Am., Nat. Parks and Conservation Assn., Defenders of Wildlife, Nat. Wildlife Fedn., Wilderness

Soc., Am. Hort. Soc., Smithsonian Assocs., Nat. Audubon Soc. Home: 284 Mill Rd East Aurora NY 14052-2805 Office: Ecology & Environment Inc 368 Pleasant View Dr Lancaster NY 14086-1397

NEUMAN, CHARLES P., electrical and computer engineering educator, consultant; b. Pitts., July 26, 1940; s. Daniel and Frances G. Neuman; m. Susan G. Neuman, Sept. 4, 1967. B.S. in Elec. Engring. with honors, Carnegie Inst. Tech., 1962; S.M., Harvard U., 1963, Ph.D. in Applied Math., 1968. Teaching fellow Harvard U., Cambridge, Mass., 1962-64, research asst., 1964-67; mem. tech. staff Bell Telephone Labs., Whippany, N.J., 1967-69; asst. prof. elec. engring. Carnegie-Mellon U., Pitts., 1969-71, assoc. prof., 1971-78, prof. elec. engring., 1978-83, prof. elec. and computer engring., 1983—. Mem. editorial bd. Internat. Jour. Modelling and Simulation, Control and Computers; contbr. numerous articles to profl. jours. Mem. IEEE (sr., assoc. editor Trans. on Systems, Man and Cybernetics), Inst. Mgmt. Scis., AAAS, Instrument Soc. Am. (sr.), Soc. Harvard Engrs. and Scientists, Soc. Indsl. and Applied Math., Sigma Xi, Phi Kappa Phi, Tau Beta Pi, Eta Kappa Nu. Office: Carnegie-Mellon U Dept Elec and Computer Engring Pittsburgh PA 15213

NEUMAN, MICHAEL ROBERT, biomedical engineer; b. Milw., Nov. 25, 1938; s. Robert B. and Jane G. N.; m. Judith H. Borton, Aug. 2, 1973; 1 child, Elizabeth Ann. BSEE, Case Inst. Tech., 1961, PhD, 1966; MD, Case Western Res. U., 1974. Asst. prof., then assoc. prof. Case Western Res. U., Cleve., 1966-74, assoc. prof. reproductive biology, 1974—; guest prof. U. Zurich (Switzerland) Women's Hosp., 1980; Lilly vis. prof. Duke-N.C. ERC, Durham, 1990; adv. bd. Wash. Regional Primate Ctr., Seattle, 1989—; mem. device adv. panel FDA, Rockville, Md., 1992—; presenter at profl. confs. Contbg. author: Biomedical Instrumentation, 1978, 2d edit., 1992; contbr. articles to sci. publs. U.S. Steel Found. fellow, 1966; recipient Career Devel. award NIH, 1970-74; grantee NIH, NSF. Fellow Am. Inst. Med. and Biol. Engrs.; mem. IEEE (sr., v.p. engring. in medicine and biology soc. 1987-88), Internat. Soc. Biotelemetry (pres. 1984-88), Biomed. Engring. Soc. Achievements include development of biotelemetry for fetal monitoring, infant monitoring devices, biomedical sensors for infant monitoring and for feedback control of paralyzed hand, implantable electrochemical pH sensors; 2 patents in field. Office: Case Western Res U MetroHealth Med Ctr 2500 MetroHealth Dr Cleveland OH 44109-1998

NEUMANN, ANDREW CONRAD, geological oceanography educator; b. Oak Bluffs, Mass., Dec. 21, 1933; s. Andrew Conrad Neumann and Faye Watson (Gilmore) Gilmour; m. Jane Spaeth, July 7, 1962; children: Jennifer, Christopher, Jonathan. BS in Geology, Bklyn. Coll., 1955; MS in Oceanography, Tex. A&M U., 1958; PhD in Geology, Lehigh U., 1963. Asst. prof. marine geology Lehigh U., Bethlehem, Pa., 1963-65; asst. prof. marine sci. U. Miami, Fla., 1965-69, assoc. prof. marine sci., 1969-72; prof. marine sci. U. N.C., Chapel Hill, 1972-85, Bowman and Gray prof. geol. oceanography, 1985-88; program dir. NSF, Washington, 1969-70; Kenan prof. U. Edinburgh, Scotland, 1978; summer vis. investigator U.S. Geol. Survey, Woods Hole, Mass., 1981—, Woods Hole Oceanographic Inst., 1981—; vis. prof. U. Naples, Italy, 1984, Eötvös U., Budapest, Hungary, 1991. Contbr. articles to profl. jours. Trustee Bermuda Biol. Sta. for Research Inc., 1972-76. Recipient Disting. Alumni award Bklyn. Coll., 1987. Fellow Geol. Soc. Am.; mem. Soc. Econ. Paleontologists and Mineralogists, N.C. Acad. Sci. Avocations: fishing, gardening, sailing. Office: U NC Dept Marine Scis Curriculum 1205 Venable Hall Chapel Hill NC 27599-3300

NEUMANN, BERNHARD HERMANN, mathematician; b. Berlin-Charlottenburg, Germany, Oct. 15, 1909; s. Richard and Else (Aronstein) N.; m. Hanna von Caemmerer, Dec. 22, 1938 (dec. Nov. 1971); children: Irene Brown, Peter, Barbara Cullingworth, Walter, Daniel; m. Dorothea Zeim, Dec. 24, 1973. Student U. Freiburg, Germany, 1928-29; Dr.phil., U. Berlin, 1932; PhD, Cambridge U., Eng., 1935; D.Sc., U. Manchester, Eng., 1954; D.Sc. (hon.), U. Newcastle, Australia, 1974, Monash U., Australia, 1982; D. Math. (hon.), U. Waterloo, 1986., Dr.rer.nat. (hon.) Humboldt U. (Berlin), 1992. Lectr., Univ. Coll., Hull, Eng., 1946-48; faculty U. Manchester, Eng., 1948-61; prof., head dept. math. Inst. Advanced Studies, Australian Nat. U., Canberra, 1962-74, hon. univ. fellow, 1975—; hon. research fellow div. Math. and Stats. Commonwealth Sci. and Indsl. Rsch. Orgn., Canberra, 1978—. Editor Houston Jour. Math., 1974—; editor, pub. IMU Canberra Circular, 1972—; other editorships; contbr. numerous articles to math. jours. Served with Brit. Armed Forces, 1940-45. Recipient prize Wiskundig Genootschap, Amsterdam, Netherlands, 1949; Adams prize U. Cambridge, 1952-53. Fellow Royal Soc., Australian Acad. Sci. (v.p. 1969-71, Matthew Flinders lectr. 1984), Inst. Combinatorics and Its Applications (hon.); mem. London Math. Soc. (v.p. 1959-61, editor proc. 1959-61), Australian Math. Soc. (v.p. several terms, pres. 1964-66, hon. mem. 1981—, editor bull. 1969-79, hon. editor 1979—), many other profl. orgns., also chess and musical clubs and socs. Avocations: classical music (cello), chess, cycling. Home: 20 Talbot St, Forrest ACT 2603, Australia Office: Australian Nat U, Canberra ACT 0200, Australia also: CSIRO-DMS, GPO Box 1965, Canberra ACT 2601, Australia

NEUMANN, DONALD LEE, civil engineer, environmental specialist; b. St. Louis, Apr. 10, 1945; s. Homer George and Rose (Bieler) N.; m. Barbara Ann Decker, Oct. 27, 1973; children: Kimberly, Jennifer, Margaret. BSCE, St. Louis Univ., 1968. Cert. profl. engr. Hwy. engr. Fed. Hwy. Adminstrn., 1968-71; project mgr. Fed. Hwy. Adminstrn., Arlington, Va., 1971-75; area engr. Fed. Hwy. Adminstrn., Lincoln, Nebr., 1975-77; safety engr. Fed. Hwy. Adminstrn., Washington, 1977-80, planning engr., 1980-82; program review engr. Fed. Hwy. Adminstrn., Jefferson City, Mo., 1982—; adv. com. Mo. Hwy. Transp. Dept., Jefferson City, 1988-93; Mo. Wetlands adv. coun. Dept. Nat. Resources, 1991-93. Contbr. articles to profl. jours. Mem. Cole County Traffic and Safety Bd., Jefferson City, 1990-93, charity art auctioneer Jefferson City West Rotary Club, 1992, interdenominational com. First United Meth. Ch., Jefferson City, 1993. Recipient Outstanding Performance award Fed. Hwy. Adminstrn., 1989, Exceptional Ach. award, 1990, 91, Superior Accomplishment award, 1992, Exceptional Achievement award, 1992. Mem. Mo. Prof. Engrs. In Govt. (chmn., pres. 1989-90, Dist. Svc. award 1990), Jefferson City Engrs. Club (pres. 1990-91 leadership, 1991), Mo. Soc. Prof. Engrs. (com. chmn. 1993-94, Outstanding Achievement award 1991, pres.-elect adv. com. 1992-93), Rotary Club (chmn. 1991-93). Home: 5314 Foxfire Ln Lohman MO 65053 Office: Jefferson City MO 65053

NEUMANN, GERHARD, mechanical engineer. Recipient R. Tom Sawyer award ASME, 1991. Home: 53 Ocean View Rd Swampscott MA 01907

NEUMANN, PETER GABRIEL, computer scientist; b. N.Y.C., Sept. 21, 1932; s. J.B. and Elsa (Schmid) N.; 1 child, Helen K. AB, Harvard U., 1954, SM, 1955; Dr rerum naturarum, Technisch Hochschule, Darmstadt, Fed. Republic Germany, 1960; PhD, Harvard U., 1961. Mem. tech. staff Bell Labs, Murray Hill, N.J., 1960-70; Mackay lectr. Stanford U., 1964, U. Calif., Berkeley, 1970-71; computer scientist SRI Internat., Menlo Park, Calif., 1971—. Contbr. articles to profl. jours. and chpts. to books. Fulbright fellow, 1958-60. Fellow IEEE; mem. AAAS (mem.-at-large sect. com. on info., computing and communications 1991—), Assn. for Computing Machinery (editor jour. 1976—, chmn. com. on computers and pub. policy 1985—). Avocations: music, tai chi, holistic health. Office: SRI Internat EL-243 333 Ravenswood Ave Menlo Park CA 94025-3493

NEUMANN, ROBERT WILLIAM, engineer; b. Plainfield, N.J., Feb. 21, 1952; s. Robert Thomas and Lucille Francis (Perry) N. BS in Engring., Lafayette Coll., 1974; MS in Mgmt., Stevens Inst., 1984. Cert. plant engr. Marine tech. Larry Smith Electronics, Absecon, N.J., 1974; plant mgr. Neumann Sheet Metal, Plainfield, N.J., 1974-75; plant engr. Allison Corp., Garwood, N.J., 1976-77; engr. Midland Ross Corp., Somerset, N.J., 1977-81; plant engr. Container Corp. of Am., Matawan, N.J., 1981-85, Eastern Steel Barrel, Piscataway, N.J., 1985-91, Elizabeth (N.J.) Gen. Med. Ctr., 1991—. Pres., treas. South Plainfield (N.J.) Jaycees, 1975— (outstanding citizen 1982-83); v.p. N.J. Jaycees, 1977— (keyman 1982-83, 85); sec., treas. N.J. J.C.I. Senate, 1984—, pres., 1989-90; pres. N.J. Leadership Seminars, 1986-87. Mem. IEEE, Inst. Indsl. Engrs., Am. Inst. Plant Engrs., Assn. Advancement Med. Instrumentation, Exec. Hosp. Engrs. Avocations: collecting comic books, coin operated vending machines, music. Home: 1507 Central Ave South Plainfield NJ 07080-3704

NEUMANN, THOMAS WILLIAM, archaeologist; b. Cin., Aug. 30, 1951; s. William Henry and Virginia Marie (Walz) N.; m. Mary Louise Spink, Sept. 3, 1988. BA in Anthropology, U. Ky., 1973; PhD in Anthropology, U. Minn., 1979. Instr. U. Minn., Mpls., 1977-79; asst. prof. Syracuse U., 1979-86, dir. archaeology field program, 1979-86; sr. ptnr. Neumann & Sanford Cultural Resource Assessments, Syracuse, 1985-87; sr. scientist R. Christopher Goodwin & Assocs., Inc., Frederick, Md., 1987-92; rsch. assoc. Terrestrial Environ. Specialists, Phoenix, N.Y., 1980-83, SUNY Rsch. Found., Potsdam, 1985-87; external reviewer NSF, Washington, 1982-85; dir. Ctr. for Archaeol. Rsch. and Edn., Houston, Minn., 1982-84; vis. assoc. prof. Emory U., 1991—; independent cons. 1991—; bd. dirs. Georgia Coun. of Profl. Archaeologists, 1992—. Author/co-author 70 monographs including 2 winners of The Anne Arundell County Historic Preservation award; asst. editor Amanuensis, 1972-73; contbr. articles to profl. jours. Nat. Trust Historic Preservation honor award. Grantee, Am. Philos. Soc., 1981, Appleby-Mosher Found., 1983, Landmarks Assn. Cen. N.Y., 1984; recipient Oswald award, U. Ky., 1973. Mem. AAAS, N.Y. Acad. Sci., Soc. for Am. Archaeology, Ea. States Archaeol. Fedn., Phi Beta Kappa. Roman Catholic. Achievements include development of use of vegetation successional stages for cultural resource assessments; identification of cause of passenger pigeon extinctions, microlithic compound tool industry in the eastern prehistoric U.S. Home: 3859 Wentworth Ln SW Lilburn GA 30247-2260 Office: Ind Archeol Cons 3859 Wentworth Lane Lilburn GA 30247

NEUMANN, WILHELM PAUL, chemistry educator; b. Würzburg, Bavaria, Fed. Republic of Germany, Oct. 29, 1926; s. Wilhelm A.E. and Margarete (Bertram) m. Mechtild Maier, Feb. 7, 1959, (wid. 1978); children: Brigitte, Albrecht, Doris; m. Gerda Deutskens, Mar. 21, 1983. Diploma in Chemistry, U. Würzburg, 1949, D degree, 1952. Cert. in Habilitation, Chemistry, U. Giessen, Fed. Republic of Germany. Rsch. and tng. specialist U. Würzburg, 1949-55; asst. rsch. assoc. Max-Planck-Inst. of Coal Research, Mülheim, Ruhr, Fed. Republic Germany, 1955-59; lectr. U. Giessen, Hessen, 1959-65, prof., 1965-68; prof. organic chemistry U. Dortmund, Fed. Republic Germany, 1968—; pres. of convent, U. Dortmund, head Chem. Dept., 1975-76. Author: The Organic Chemistry of Tin, 1970; contbr. several handbook articles, reviews and over 265 articles to profl. jours. Fellow Japan Soc. for the Promotion of Sci., Tokyo, 1988, Hon. Mem. Soc. Argentina de Investigaciones en Quimica Organica, Argentina, 1987. Mem. Soc. German Chemists (chmn. local section), Rotary (pres. Dortmund chpt. 1986-87). Avocations: gliding, travel, Baroque music. Office: U Dortmund Organic Chem, Otto-Hahn-Str 6, D-4600 Dortmund 50, Germany

NEUMEYER, JOHN LEOPOLD, research company administrator, chemistry educator; b. Munich, Germany, July 19, 1930; came to U.S., 1945, naturalized, 1950; s. Albert and Martha (Stern) N.; m. Evelyn Friedman, June 24, 1956; children: Ann Martha, David Alexander, Elizabeth Jean. BS, Columbia U., 1952; PhD, U. Wis., 1961. Rsch. chemist Ethicon Inc., New Brunswick, N.J., 1952-57, FMC Corp., Princeton, N.J., 1961-63; sr. staff chemist Arthur D. Little Inc., Cambridge, Mass., 1963-69; prof. medicinal chemistry, chemistry Northeastern U., Boston, 1969-91, dir. grad. sch., 1978-85, disting. emeritus prof., 1992—; chmn. bd., co-founder Rsch. Biochem. Internat., Natick, Mass., 1981—; cons. in field. Patentee in field. Contbr. articles to profl. jours., also chpts. to books in field. Mem. Bd. Health, Wayland, Mass., 1968-75, Pesticide Bd., Mass., 1972-75. Served to cpl. U.S. Army, 1953-55. Recipient Lunsford Richardson award, 1961, Marie Curie award in Nuclear Medicine, 1992; Sr. Hayes Fulbright fellow, 1975-76. Fellow AAAS (mem. at large 1983-87, chmn. pharm. sci. sect. 1992-93), Am. Assn. Pharm. Scis., Acad. Pharm Scis. (rsch. achievement award in medicinal chemistry 1982, Northeastern U. faculty lectr. award 1978, U. disting. prof. 1982-92); mem. Am. Soc. Neurosci., Am. Soc. Exptl. Pharm. & Exptl. Therapeutics, Am. Chem. Soc. (councilor 1985—, trustee 1989-93, bd. editors Jour. Medicinal Chemistry 1974-88, chmn. div. med. chem. 1982). Office: Rsch Biochems Internat 1 Strathmore Rd Natick MA 01760

NEUMILLER, PHILLIP JOSEPH, III, research scientist; b. Kenosha, Wis., Mar. 27, 1939; s. Phillip II and Stella (Kozlik) N.; m. Loraine A. Neumiller, July 16, 1960; children: Sheryl, Phillip IV, Judith. Sr. rsch. scientist S.C. Johnson Wax, Racine, Wis., 1958—; with Sta. KC91S. Lutheran. Achievements include over 50 patents in field. Home: 10017 Spring St Racine WI 53406 Office: SC Johnson Wax 1525 Howe St Racine WI 53403

NEUSPIEL, DANIEL ROBERT, pediatrician, epidemiologist; b. Haifa, Israel, May 15, 1952; came to U.S., 1953; s. William and Miriam (Schwerstein) N.; m. Cathy Canepa, Apr. 12, 1987; children: Juliana, Samuel. BA, Rutgers U., 1975; MD, N.J. Med. Sch., 1979; MPH, U. Pitts., 1984. Diplomate Nat. Bd. Med. Examiners, Am. Bd. Pediatrics, Am. Bd. Preventive Medicine. Resident in pediatrics Children's Hosp., Pitts., 1979-82; fellow in epidemiology U. Pitts., 1982-84; asst. prof. Albert Einstein Coll. of Medicine, Bronx, 1984-90, assoc. prof., 1990—; founding dir. early family outreach program North Cen. Bronx Hosp. Contbr. articles to Jour. AMA, Am. Jour. Pub. Health, Neurotoxicol. Teratol, Devel. Behavior Pediatrics. Office: North Central Bronx Hosp 3424 Kossuth Ave Bronx NY 10467-2489

NEVA, FRANKLIN ALLEN, physician, educator; b. Cloquet, Minn., June 8, 1922; s. Lauri Albin and Anna (Lahti) N.; m. Alice Hanson, July 5, 1947; children: Karen, Kristin, Erik. SB, U. Minn., 1944, MD, 1946; AM (hon.), Harvard U., 1964. Diplomate Am. Bd. Internal Medicine. Intern Harvard Med. Services, Boston City Hosp., 1946-47, resident, 1949-50; research fellow Harvard Med. Sch., 1950-53; asst. prof. U. Pitts. Med. Sch., 1953-55; mem. faculty Harvard Sch. Pub. Health, 1955-69, John LaPorte Given prof. tropical pub. health, 1964-69; chief Lab. Parasitic Diseases, Inst. Allergy and Infectious Diseases, NIH, 1969—; mem. commn. parasitic diseases, assoc. mem. commn. virus infections Armed Forces Epidemiol. Bd., 1963-68; mem. Latin Am. sci. bd. Nat. Acad. Scis.-NRC, 1963-68; bd. sci. counselors Inst. Allergy and Infectious Diseases, NIH, 1966-69. Served to lt. (j.G.) USNR, 1947-49. Mem. Soc. Exptl. Biology and Medicine, Infectious Diseases Soc. Am. (Joseph Smadel lectr. 1985), Am. Soc. Tropical Medicine and Hygiene (Bailey K. Ashford award 1965), Assn. Am. Physicians (Presdl. Meritorious Exec. Rank award 1985). Achievements include special research infectious diseases especially tropical, parasitic and virus infections. Home: 10851 Glen Rd Potomac MD 20854-1401 Office: NIH Inst Allergy and Infectious Diseases Bethesda MD 20892

NEVITT, MICHAEL VOGT, materials scientist, educator; b. Lexington, Ky., Sept. 7, 1923; s. Charles Alonzo and Cecelia (Vogt) N.; m. Jean Meierdirks, Sept. 14, 1946; children: Carole Davis, Michael C., Timothy J., Robert M. BS, U. Ill., 1944, PhD, 1954; MS, Va. Poly. Inst., 1950. Rsch. metallurgist Olin-Matheson Corp., East Alton, Ill., 1946-48; asst. prof., assoc. prof., head dept. materials Va. Poly. Inst., Blacksburg, 1948-55; scientist, div. dir., dep. lab. dir. Argonne (Ill.) Nat. Lab., 1955-90; adj. prof. physics Clemson (S.C.) U., 1990—. Editor 2 handbooks, also procs. vols.; contbr. more than 100 articles to sci. and tech. jours. Lt. (j.g.) USN, 1943-46; PTO. Union Carbide fellow U. Ill., 1952-53; Sci. Rsch. Coun. fellow, London, 1965-66; Am. Soc. Metals fellow, 1970. Mem. AAAS, Materials Soc. of AIME, Am. Soc. Metals Internat., Alpha Sigma Mu. Home: 1 Boatswain Way Salem SC 29676 Office: Clemson U Dept Physics and Astronomy 107 Kinard Lab Clemson SC 29634

NEWBERGER, BARRY STEPHEN, physicist, research scientist; b. Huntington, W.Va., June 19, 1945; s. Reuben I. and Sara Mae (Gaffin) N.; m. Betsy Lenore Pobanz, Nov. 28, 1975. BSEE, Carnegie Inst. Tech., 1967; PhD, Princeton U., 1976. Staff mem. Los Alamos (N.Mex.) Nat. Lab., 1971-84; rsch. scientist Mission Rsch. Corp., Albuquerque, 1984-87, U. Tex., Austin, 1987—; cons. Mission Rsch. Corp., Albuquerque, 1987-90; guest scientist Los Alamos Nat. Lab., 1984—; vis. scientist Nat. Lab. for High Energy Physics, Tsukuba, Japan, 1990-91. Contbr. articles to profl. jours. Mem. permanent program com. N.Mex. Coal Surface Mining Commn., 1983-87; sec. bd. dirs. Barton Creek West Homeowners Assn., Austin, 1992. NASA fellow, 1967, 68. Mem. Am. Phys. Soc. Achievements include research on sum rules for series of Bessel functions. Home: 10037 Circleview Dr Austin TX 78733 Office: Inst for Fusion Studies U Tex RLM 11.218 26th and Speedway Austin TX 78712-1060

NEWBERN, LAURA LYNN, forestry association executive, editor; b. Valdosta, Ga., Feb. 25, 1954; d. Jefferson Lamar Jr. and Laura Helen (Downs) N. BA in Art History, Emory U., 1977. Adminstrv. asst. Ga. Forestry Assn., Norcross, 1977—, editor TOPS, Ga. Forestry Assn. News, 1987—; state coord. Project Learning Tree, Norcross, 1988—; officer, bd. dirs. Turn In Poachers, Inc., Atlanta, 1987—. Recipient Natl. Outstanding Program award Project Learning Tree, Washington, Outstanding Svc. award Ga. Tree Farm Com., 1987, Outstanding Tree Farm Svc. award Am. Forest Coun., Atlanta, 1988, Nat. Outstanding New Program award Project Learning Tree, Washington, 1988, Outstanding Svc. award Coalition for Green Ga., Macon, 1989, Friend of Ga. Wildlife award Turn In Poachers, Inc., 1990; named Outstanding Facilitator of Yr., Ga. Project Learning Tree, 1988, Outstanding Educator of Yr., Ga. Urban Forest Coun., 1992. Avocations: scuba diving, photography, travel. Office: Ga Forestry Assn 500 Pinnacle Ct Ste 505 Norcross GA 30071-3662

NEWBERRY, CONRAD FLOYDE, aerospace engineering educator; b. Neodesha, Kans., Nov. 10, 1931; s. Ragan McGregor and Audra Anitia (Newmaster) N.; m. Sarah Louise Thonn, Jan. 26, 1958; children: Conrad Floyde Jr., Thomas Edwin, Susan Louise. AA, Independence Jr. Coll., 1951; BEME in Aero. Sequence, U. So. Calif., 1957; MSME, Calif. State U., Los Angeles, 1971, MA in Edn., 1974; D.Environ. Sci. and Engring., UCLA, 1985. Registered profl. engr., Calif., Kans., N.C., Tex. Mathematician L.A. div. N.Am. Aviation Inc., 1951-53, jr. engr., 1953-54, engr., 1954-57, sr. engr., 1957-64; asst. prof. aerospace engring. Calif. State Poly. U., Pomona, 1964-70, assoc. prof. aerospace engring., 1970-75, prof. aerospace engring., 1975-90, prof. emeritus, 1990—; staff engr. EPA, 1980-82; engring. specialist space transp. systems div. Rockwell Internat. Corp., 1984-90; prof. aeronautics, astronautics Naval Postgrad. Sch., Monterey, Calif., 1990—, acad. assoc. space systems engring., 1992—. Recipient John Leland Atwood award as outstanding aerospace engring. educator AIAA/Am. Soc. Engring. Edn., 1986. Fellow AIAA (dep. dir. edn. region VI 1976-79, dep. dir. career enhancement 1982-91, chmn. L.A. sect. 1989-90, chmn. Point Lobos sect. 1990-91, dir. tech.-aircraft systems 1990-93), Inst. Advancement Engring, Brit. Interplanetary Soc.; mem.IEEE, AAAS, ASME, NSPE, Royal Aeronautical Soc., Calif. Soc. Profl. Engrs., Am. Acad. Environ. Engrs. (cert. air pollution control engr.), Am. Soc. Engring. Edn. (chmn. aerospace div. 1979-80, div. exec. com. 1976-80, 89-93, exec. com. ocean and marine engring. div. 1982-85, 90—, program chmn. 1991-93, chmn. 1993—), Am. Soc. Pub. Adminstrn., Am. Meteorol. Soc., U.S. Naval Inst., Am. Helicopter Soc., Soc. Naval Architects and Marine Engrs., Air and Waste Mgmt. Assn., Inst. Environ. Scis., Exptl. Aircraft Assn., Water Pollution Control Fedn., Soc. Automotive Engrs., Soc. Allied Weight Engrs., Assn. Unmanned Vehicle Systems, Calif. Water Pollution Control Assn., Nat. Assn. Environ. Profls., Am. Soc. Naval Engrs., Planetary Soc., Tau Beta Pi, Sigma Gamma Tau, Kappa Delta Pi. Democrat. Mem. Christian Ch. (Disciples of Christ). Achievements include research on aircraft, space, missile and engine design and related impacts on exergy, quality, concurrent engineering, cost and environmental controls. Home: 9463 Willow Oak Rd Salinas CA 93907-1037 Office: Naval Postgrad Sch Dept Aeronautics and Astronautics AA/Ne Monterey CA 93943-5000

NEWBOLD, HERBERT LEON, JR., psychiatrist, writer; b. High Point, N.C., Nov. 3, 1921; s. Herbert Leon and Mary Temperance (Sherrod) N.; m. Susan Deena Hecht; children: Lucile, Susan. Student, U. Chgo., 1941, Coll. William and Mary, 1941; B.S., Duke U., 1945, M.D., 1945; postgrad., Northwestern U., 1951, New Sch. Social Research, 1960-61. Intern U. Chgo. Clinics, 1945-46, U. Minn., 1949-50; resident Woodlawn Hosp., Chgo., 1946; resident in internal medicine Vanderbilt U. and associated VA Hosp., Nashville, 1946-47; resident in psychiatry U. Ill. and associated VA Hosp., Hines, Ill., 1955-58; practice medicine specializing in internal medicine Newton, N.C., 1947-48; practice medicine specializing in psychiatry Chgo., 1950-55, 1958-60; Asheville, N.C., 1961-70; N.Y.C., 1970—, pvt. practice specializing in psychiatry and neurology, 1976—; instr. neurology and psychiatry Sch. Medicine, Northwestern U., Chgo., 1958-61. Freelance writer, 1950—; novels include 1/3 of an Inch of French Bread, 1961, Long John, 1979, Dr. Cox's Couch, 1979; others under pseudonym, 1950-60; sci. books include text Psychiatric Programming of People, 1972, Mega-Nutrients for Your Nerves, 1975, Doctor Newbold's Revolutionary New Discoveries about Weight Loss, How to master hidden allergies that make you fat, 1977, Physicians Handbook on Orthomolecular Medicine, 1977, Vitamin C Against Cancer, 1979, Mega-Nutrients, 1987, Dr. Newbold's Type A/Type B Weight Loss Book, 1991, Dr. Newbold's Nutrition for Your Nerves, 1993; author: (with others) The New Chemotherapy in Mental Illness, 1958; contbr. articles to profl. jours.; numerous appearances radio and TV. Served with U.S. Army, 1943-45, 46-47. Mem. AMA. Address: 151 E 31st St New York NY 10016

NEWCOMB, ROBERT WHITNEY, biotechnologist, neuroscience researcher; b. Redwood City, Calif., Apr. 19, 1956; s. Robert Wayne and Sarah Elenor N.; m. Marcelle Nguyen, Oct. 12, 1985. BA in Chemistry, Molecular Biology, U. Colo., 1977; PhD in Biochemistry, Biophysics, U. Hawaii, 1983. Teaching asst. dept. chemistry U. Hawaii, Honolulu, 1977-78, rsch. asst. dept. biochemistry, 1978-83, postdoctoral researcher Bekesy Lab. Neurobiology, 1983-85, asst. prof. Pacific Biomed. Rsch. Ctr., 1987-89; postdoctoral researcher in biol. scis. Stanford (Calif.) U., 1985-87; INSERM fellow Ctr. Neurochemistry, U. Louis Pasteur, Strasbourg, France, 1987; scientist, project coord. analytical biochemistry Neurex Corp., Menlo Park, Calif., 1989—. Contbr. articles to Jour. Neurosci., Jour. Neurochemistry, other profl. pubs. Grantee NIH, 1985, 86. Mem. Phi Beta Kappa. Office: Neurex Corp 3760 Haven Ave Menlo Park CA 94301

NEWELL, NANETTE, biotechnologist; b. Pensacola, Fla., Aug. 20, 1951; d. Floyd Alex and Irene Olga (Martin) N. BS in Chemistry, Lewis and Clark Coll., 1973; PhD in Biochemistry, Johns Hopkins Med. Sch., 1978; MBA, U. N.C., 1992. Postdoctoral fellow U. Wis., Madison, 1978-80; asst. prof. Reed Coll., Portland, Oreg., 1980-81; project dir. Office Tech. Assessment U.S. Congress, Washington, 1981-84; dir. rsch. adminstrn. Calgene, Davis, Calif., 1984-85; pres. Newell Cons. Group, San Francisco, 1985-87, Apex, N.C., 1987-92, Portland, Oreg., 1993—; prin. Synertech Group, Research Triangle Park, N.C., 1987-91; exec. dir. Oregon Biotech. Assn. and Found., 1992—; bd. dirs. EcoTech., Langhorne, Pa. Contbr. chpt. to books; co-author: International Biotechnology, 1984. Named Outstanding Young Alumnae Lewis and Clark Coll., 1988; fellow NIH, 1973-81. Mem. AAAS, Am. Chem. Soc., Assn. Women in Sci. (pres. local chpt. 1989-91), Beta Gamma Sigma. Home and Office: 2523 SE Harrison Portland OR 97214

NEWELL, PHILIP BRUCE, physicist, lighting engineer; b. Lincoln, Nebr., Sept. 25, 1937; s. Laurence Cutler and Lois (Milbourne) N.; m. Joan Betty Hutchinson, Mar. 21, 1962; children: Laurence Jeffrey, Gwendolyn Jo. BS, MIT, 1959; MS, Boston U., 1965, PhD, 1969. Devel. engr. EG&G Inc., Salem, Mass., 1959-76, GTE Sylvania, Danvers, Mass., 1976-86; ind. cons. Carlisle, Mass., 1986-87; devel. engr. GTE Sylvania, Danvers, 1987—. Contbr. articles to profl. jours. Mem. Am. Physical Soc., Illumination Engring. Soc. Achievements include 4 patents in circuits and lamps for photography; development of criteria for design, fabrication and application of noble gas flashlamps and cw lamps for photography reprography and lasers and of high intensity metal halide lamps for specialty and general illumination. Office: OSRAM Sylvania Inc 100 Endicott St Danvers MA 01923

NEWHOUSE, ALAN RUSSELL, federal government executive; b. N.Y.C., Feb. 27, 1938; s. Russell Conwell and Clara Lucille (Scovell) N.; m. Margo Stiles Hicks, Feb. 3, 1960; children: Daryl, Jeffrey, William. BEE, Cornell U., 1960. Engr. Bur. of Ships, Washington, 1964-66; nuclear power engr., chief West Milton field office AEC, Schenectady, N.Y., 1966-69; sr. exec. AEC, ERDA, U.S. Dept. Energy, Washington, 1969-1993; dep. asst. sec. Space and Def. Power Systems Office Nuclear Energy, Washington, 1993—. Composer numerous musical works. Bd. trustees River Road Unitarian Ch., Bethesda, Md., 1973-75; mem. McLean Symphony Orchestra, Washington Men's Camerata, Musica Antiqua, Interamerican Chamber Singers, Continuum Chamber Singers, U. Md. Chorus. Lt. USN, 1960-64. Mem. IEEE, AIAA, Am. Nuclear Soc., Am. Soc. Naval Engrs., Soc. Naval Architects and Marine Engrs., Am. Astronomical Soc. Republican. Unitarian. Home: 11108 Deborah Dr Rockville MD 20854-2721 Office: Deputy Asst Sec for Space and Defense Power Systems Office of Nuclear Energy NE-50 Washington DC 20585

NEWHOUSE, JOSEPH PAUL, economics educator; b. Waterloo, Iowa, Feb. 24, 1942; s. Joseph Alexander and Ruth Linnea (Johnson) N.; m. Margaret Louise Locke, June 22, 1968; children: Eric Joseph, David Locke. BA, Harvard U., 1963, PhD, 1969; postgrad (Fulbright scholar), Goethe U., Frankfort, Germany, 1963-64. Staff economist Rand Corp., Santa Monica, Calif., 1968-72, dep. program mgr., health and biosci. rsch., 1971-88, sr. staff economist, 1972-81, head econs. dept., 1981-85, sr. corp. fellow, 1985—; John D. MacArthur prof. health policy and mgmt., dir. div. Health Policy Rsch. and Edn., Harvard U., 1988—; lectr. UCLA, 1970-83, adj. prof., 1983-88; mem. faculty Rand Grad. Sch., 1972-88; dir. Rand-UCLA Ctr. for Study Health Care Fin. Policy, 1984-88, co-dir., 1988—; prin. investigator health ins. study grant HHS, 1971-86; chmn. health svcs. rsch. study sect. HHS-Agy. for Health Care Policy and Rsch., 1989-93; mem. Nat. Commn. Cost Med. Care, 1976-77; mem. health svcs. devel. grants study sect. HEW, 1978-82, Inst. Medicine of NAS, 1978—, mem. coun., 1992—; mem. Physician Payment Rev. Commn., 1993. Author: The Economics of Medical Care, 1978, The Cost of Poor Health Habits, 1991, A Measure of Malpractice, 1993, Free for All?, 1993; editor Jour. Health Econs., 1981—; assoc. editor Jour. Econ. Perspectives, 1992—; contbr. articles to profl. jours. Recipient David Kershaw award and prize Assn. Pub. Policy and Mgmt., 1983, Baxter Am. Found. prize, 1988, Adminstr.'s citation Health Care Fin. Adminstrn., 1988. Mem. Assn. for Health Svcs. Rsch. (Article of Yr. award 1989, bd. dirs. 1991—, pres. 1993), Am. Econ. Assn., Royal Econ. Soc., Econometric Soc., Phi Beta Kappa. Office: Harvard U Health Policy Rsch and Edn 25 Shattuck St # B Boston MA 02115-6092

NEWHOUSE, JOSEPH ROBERT, plant pathologist, mycologist; b. Greensburg, Pa., June 13, 1955; s. Joseph John and Lucy Leona (Welty) N.; m. Noreen Halgas, June 25, 1977; 1 child, Samantha Anne. BS in Biology, St. Vincent Coll., 1977; MS in Biology, California U. of Pa., 1980; PhD in Plant Pathology, W.Va. U., 1988. Grad. rsch. asst., dept. biol. scis. California U. of Pa., 1978-80; grad. rsch. asst., dept. plant pathology/agrl. microbiology W.Va. U., Morgantown, 1980-84, grad. teaching asst., dept. agronomy, 1984-85; postdoctoral rsch. plant pathologist USDA Agrl. Rsch. Svc., Frederick, Md., 1987-90; pvt. practice Belle Vernon, Pa., 1990-91; faculty mem. dept. biol. scis. California U. of Pa., California U. Pa., 1991—. Contbr. articles to profl. jours. and mags. Mem. Mycol. Soc. Am., Am. Phytopath. Soc. (program com. 1988), Sigma Xi, Beta Beta Beta (James H. Long hon. award 1989), Gamma Sigma Delta. Democrat. Roman Catholic. Achievements include discovery of virus-like particles in hyphae of double-stranded RNA containing isolates of the chestnut blight fungus, Cryphonectria parasitica; discovery of double-stranded RNA in European and Peruvian isolates of the potato late blight fungus, Phytophthora infestans, and application of freeze-substitution technique to the preservation of gypsy moth larvae infected by various microfungi for observation using scanning electron microscopy. Home: 935 Brown St Belle Vernon PA 15012-1625 Office: California U of Pa Dept Biol Scis California PA 15419

NEWHOUSE, NORMAN LYNN, mechanical engineer; b. Lincoln, Nebr., Apr. 4, 1951; s. Keith Norman and Betty Jane (Douglass) N. BS, U. Nebr., 1973, MS, 1975, PhD, 1984. Registered profl. engr., Nebr. Design engr. Brunswick Corp., Lincoln, Nebr., 1975-84; product design mgr. Brunswick Corp., Lincoln, 1984-92, mgr. engring., 1992—; vis. asst. prof. U. Nebr., Lincoln, 1978-86; engring. alumni bd. mem. U. Nebr., 1991—. Mem. AIAA, ASME, Sigma Xi. Office: Brunswick Corp 4300 Industrial Ave Lincoln NE 68504

NEWHOUSE, QUENTIN, JR., social psychologist, educator, researcher; b. Washington, Oct. 20, 1949; s. Quentin Sr. and Berlene Delois (Byrd) N.; m. Brenda Joice Washington, Feb. 17, 1973 (div. Mar. 1984); m. Debra Ann Carter, July 7, 1984; 1 child, Alyse Elizabeth Belinda. BA in Psychology, Marietta (Ohio) Coll., 1971; MS in Psychology, Howard U., 1974, PhD in Psychology, 1980. Asst. prof. Antioch U., Balt., 1976-79; pres. Quentin Newhouse Jr. and Assocs., Inc., Washington, 1981-84; computer systems analyst U.S. Army, Washington, 1984, 85, Alexandria, Va., 1987-88; asst. prof. Howard U., Washington, 1982-88; adj. prof. U. D.C., 1984, 91—, Bowie State U., 1982-86, 91—; mentor Prince George's Community Coll., Largo, Md., 1986-89; statistician Bur. of the Census, Ctr. for Survey Methods Rsch., Suitland, Md., 1991—; computer specialist Bur. of the Census, Ctr. for Survey Methods Rsch., 1988-91; v.p. Bureautots, Inc., Largo, 1989-91. Commr. Prince George's County Children and Youth, Upper Marlboro, Md., 1991, Prince George's County Commn. for Children, Youth and Families, 1992—; bd. dirs. Prepare Our Youth, Inc., Tacoma Park, Md., 1990—, Shiloh Bapt. Ch. Nursery, Washington, 1988-90; mem. State of Md. Adv. Com. for Children, Youth and Families, 1992-93. Recipient Community Svc. award U. D.C., 1982, 84; named Outstanding Young Man of Am., 1982, 86. Mem. APA, Social Sci. Computing Assn., Census SAS Users Group (co-chair 1991-93), Tau Epsilon Phi (life). Democrat. Avocations: walking, reading, working with computers. Office: Bur of the Census Washington Pla Rm 433 Washington DC 20233

NEWLING, DONALD WILLIAM, urological surgeon; b. Nottingham, Eng., Feb. 12, 1941; s. Gilbert Wood Wilkerson and Elizabeth Mary (Allan) N.; m. Rosemary Judith Cracknell, Sept. 7, 1968; children: Benedict, Nicholas, Barnaby. BA in Natural Sci., U. Cambridge, Eng., 1963, MBB Chir, 1967. House physician St. Thomas's Hosp., London, 1966-67; house surgeon Essex County Hosp., Colchester, Eng., 1967; sr. house officer intensive care unit St. Thomas's Hosp., London, 1967-68, sr. house officer, demonstrator in anatomy, 1968-69; surg. registrar Warwick (Eng.) Hosp., 1969-71; registrar urology dept. urology Hull (Eng.) Hosp., 1971-73; sr. registrar urology Guy's Hosp., London, 1973-74; cons. urologist Hull and East Yorkshire Health Authority, 1974-90; prof., chmn. dept. urology Acad. Hosp. of Free U. of Amsterdam, The Netherlands, 1990—; chmn. Yorkshire Urol. Cancer Rsch. Group, 1982-86; chmn. Genito-Urinary Group of European Orgn. Rsch. and Treatment Cancer, Brussels, 1988—; chmn. Chmns. Conf. Coop Groups European Orgn. Rsch. and Treatment Cancer, Brussels, 1989—, coord. clin. cooop. groups, 1989—; bd. examiners Diploma Urology, Inst. Urology, London, 1986—. Contbr. articles to profl. jours. Chmn. Acute Svcs. Planning Team, Hull and East Yorkshire Health Authority, 1984—, chmn. Humberside Area Med. Com., 1980-84; pres. East Yorkshire div. British Med. Assn., Hull, 1981-82; mem. Exec. of Yorkshire Regional Cancer Orgn., Cookridge, Leeds, Eng., 1986—. Fellow Royal Coll. Surgeons; mem. Superficial Bladder Cancer Subgroup MRC, Prostatic Cancer Subgroup MRC, British Prostate Group, European Orgn. Rsch. and Treatment Cancer. (mem. bd. coun. 1988—). Mem. Ch. of Eng. Avocation: music. Home: Nassaulaan 47, 1213 BB Hulversun The Netherlands Office: A Z V U Atd Urologie, De Boelelaan 1117 Postbus 7057, 1007 MB Amsterdam The Netherlands

NEWMAN, BRETT, aerospace engineer; b. Oklahoma City, Oct. 23, 1961; s. Gene E. and Virginia A. (Harding) N.; m. Amanda L. Dinger, May 16, 1987. MS, Okla. State U., 1985; PhD, Purdue U., 1992. Rsch. assoc. Aerospace Rsch. Ctr. Ariz. State U., Tempe, 1989-91; flight dynamics and control engr. Orbital Scis. Corp., Chandler, Ariz., 1991—. Contbr. articles to profl. jours. Mem. AIAA, NSPE. Republican. Home: 728 S Kenwood Ln Chandler AZ 85226 Office: Orbital Scis Corp 3380 S Price Rd Chandler AZ 85248

NEWMAN, ELSIE LOUISE, mathematics educator; b. Bowling Green, Ohio, Mar. 25, 1943; d. Carroll E. and Grace G. (Underwood) Frank; m. Lawrence J. Newman, Sept. 15, 1962; children: Timothy, Jennifer. BS cum laude, Bowling Green (Ohio) State U., 1968; MEd, U. Toledo, 1992. Study supr. After Sch. Study Tutorial Program, Bowling Green, 1983-85; instr. Owens Tech. Coll., Toledo, 1987—; office mgr. K.C. Ins. Co., Bowling Green, 1984; tutor in maths. Bowling Green City Schs., 1984-88, Bur. of Vocat. Rehab., Oregon, Ohio, 1988-91. Advisor 4-H Club, Bowling Green, 1985—; asst. Christmas clearing bur. Voluntary Action Ctr., United Way, Bowling Green, 1982-86; residential crusade chmn. Am. Cancer Soc., Bowling Green, 1981-82. Bowling Green U. scholar, 1966-68. Mem. Nat. Coun. Tchrs. of Maths., Phi Kappa Phi, Kappa Delta Pi, Pi Lambda Theta. Home: 328 S Summit St Bowling Green OH 43402-3017

NEWMAN, JAN BRISTOW, surgeon; b. Butte, Mont., Mar. 15, 1951; d. M. Jack and Joan (Bristow) N. AS, Westchester Community Coll., Valhalla, N.Y., 1972; BS, SUNY, Bklyn., 1974, MD, 1980. Diplomate Am. Bd. Surgery. Resident in surgery U. Tex. Med. Br., Galveston, 1980-84; surg. asst. Christian Hosp., St. Louis, 1984-85; resident in surgery U. Vt., Burlington, 1985-88; pvt. practice Butte, 1988-93. Contbr. chpts. to book: Gastrointestinal Endocrinology, 1987. Fellow ACS, Internat. Coll. Surgeons; mem. AMA, N.Y. Acad. Scis., Mont. Med. Assn. Avocations: skiing, dressage, jumping, hiking, photography. Home: 3436 Mountain Dr Clinton MT 59825 Office: 900 N Orange Missoula MT 59802

NEWMAN, JOHN DENNIS, neuroethologist, biomedical researcher; b. Newark, N.J., Sept. 28, 1940; s. Richard A. and Elsie (Brown) N.; m. Judith Osler Scaffidi, Aug. 18, 1962; children: Peter John, Matthew Paul. BS, Cornell U., 1962; PhD, U. Rochester, 1970. Staff fellow Behavioral Biology Branch, Nat. Inst. Child Health & Human Devel., NIH, Bethesda, Md., 1970-74; vis. scientist Max Planck Inst. Psychiatry, Munich, Germany, 1974-75; physiologist Lab. Devel. Neurobiology, Nat. Inst. Child Health & Human Devel., NIH, Bethesda, 1975-80, rsch. physiologist, 1980-84; rsch. physiologist Lab. Devel. Neurobiology, Nat. Inst. Child Health & Human Devel., NIH, Poolesville, Md., 1984-88; head unit on Neuroethology Comparative Behavioral Genetics Sect., Lab. Comparative Ethology, Nat. Inst. Child Health & Human Devel., NIH, Poolesville, 1988-90; chief sect. on Neuroethology Nat. Inst. Child Health & Human Devel., NIH, Poolesville, 1990—; consulting editor Am. Jour. Primatology, 1988—; mem. editorial bd. Bioacoustics, 1989-92. Contbr. chpts. to books: Psychopharmocology of Anxiolytics and Antidepressents, 1991, Language Origin: a Multi-disciplinary Approach, 1992; co-contbr. (with Paul Mac Lean) Brain Research Vol. 450 pp 111-123; editor: The Physiological Control of Mammalian Vocalization, 1988. Recipient Humboldt award, Humboldt Found., Bad Godesburg, Germany, 1974. Office: NIH Bldg 112 Elmer School Rd Poolesville MD 20837-0529

NEWMAN, JOHN HUGHES, medical educator; b. Balt., Oct. 1, 1945; s. Elliot Voss and Ailsa (MacKay) N.; m. Rebecca Lyford, May 13, 1978; children: Katherine MacKay, Alexander Lyford. AB, Harvard U., 1967; MD, Columbia U., 1971. Intern Columbia-Presbyn. Hosp., N.Y.C., 1971-72; resident Johns Hopkins Hosp., Balt., 1973-74; asst. prof. Vanderbilt Med. Sch., Nashville, 1979-84, assoc. prof., 1984-91, prof. medicine, 1991—, Elsa S. Kanigan chair in pulmonary medicine, 1984—; chief pulmonary St. Thomas Hosp., Nashville, 1984—. Maj. U.S. Army, 1974-76. Mem. So. Soc. Clin. Investigation, Am. Thoracic Soc. (chmn. pulmonary circulation sect. 1992—). Office: Vanderbilt Univ Sch Medicine B1308 MCN Nashville TN 37232

NEWMAN, LISA ANN, speech pathologist, educator; b. Chgo., Apr. 11, 1952; d. Nathan Sidney and Shirley (Gilbert) N.; m. David A. Dickson, Mar. 4, 1989; 1 child, Jacob. BS, Northwestern U., 1973, MA, 1974; ScD with distinction, Boston U., 1992. Cert. clin. competence speech pathology. Speech pathologist Mercy Hosp., Chgo., 1978-84, Mass. Gen. Hosp., Boston, 1985-88; rsch. speech pathologist Boston, 1988-91; lectr. Boston U., 1990-92, clin. instr., 1987-89; instr. Northeastern U., Boston, 1990; vis. asst. prof. Bridgewater (Mass.) State Coll., 1991; asst. prof. U. Wis., Whitewater, 1992—; mem. alumni admissions coun. Northwestern U., 1990-92. Contbr. articles to profl. jours.; author profl. videotapes. Mem. West Roxbury Chaurah (bd. dirs. 1991-92). Achievements include pioneering research of maturational effects on infant swallowing. Home: 357 Spring Ln Delavan WI 53115 Office: U Wis Roseman Bldg Rm 1020 Whitewater WI 53190

NEWMAN, MARC ALAN, electrical engineer; b. Jasper, Ind., Nov. 21, 1955; s. Leonard Jay and P. Louise (Shainberg) N.; m. Shelley Jane Martin, Aug. 13, 1977; 1 child, Kelsey Renée. BSEE, Purdue U., 1977, MSEE, 1979. Sr. elec. engr. Sperry Corp. Flight Systems, Phoenix, 1979-85; staff engr. Motorola Inc., Tempe, Ariz., 1985-88, Quincy St. Corp., Phoenix, 1988-89; prin. staff scientist Motorola Inc., Chandler, Ariz., 1989-91, Scottsdale, Ariz., 1991—; Prolog and artificial intelligence expert Motorola Inc., Tempe, Chandler and Scottsdale, 1985—. Mem. IEEE, The Assn. for Logic Programming (London), Am. Assn. Artificial Intelligence, Ariz. Artificial Intelligence Assn. (founder), Internat. Platform Assn., Phi Sigma Kappa, Eta Kappa Nu. Avocations: fine music, photography, astronomy, bicycling, traveling. Home: 1539 N Hobson St Mesa AZ 85203-3653 Office: Motorola Inc 8201 E Mcdowell Rd Scottsdale AZ 85257-3893

NEWMAN, PAUL RICHARD, physicist; b. N.Y.C., June 8, 1947; s. Bernard and Edna Ethel (Albert) N.; m. Sharan Elizabeth Hill, June 21, 1971; 1 child, Allison Catherine. BS, Antioch Coll., Yellow Springs, Ohio, 1970; PhD, Mich. State U., 1975. Post-doctoral rschr. U. Pa., Lab. for Rsch. on the Structure of Matter, Phila., 1975-77; mem. tech. staff Rockwell Internat. Sci. Ctr., Thousand Oaks, Calif., 1977-83; mgr. Rockwell Internat. Sci. Ctr., Thousand Oaks, 1983-88, prin. scientist, 1988—; vis. prof. Technion-Israel Inst. Tech., Haifa, 1982. Contbr. articles to profl. jours. Recipient Bauch & Lomb award Bauch & Lomb Co., 1965. Mem. Nat. Assn. for the Advancement Sci., Am. Phys. Soc., Am. Optical Soc., Materials Rsch. Soc. Achievements include patents for Process for Producing Electrically Conducting Composites and Composites Produced Therein, Method of Stabilizing Conductive Polymers, Laser Generated Electricity Conducting Pattern, others. Office: Rockwell Internat Sci Ctr 1049 Camino Dos Rios Thousand Oaks CA 91360

NEWMAN, PAUL WAYNE, dentist, cattle rancher; b. Houston, Feb. 26, 1955; s. Emmett Wayne and Grace Uvaughn (Love) N. BS in Pre-Med, Houston Bapt. U., 1979; DDS, U. Tex. Dental Br., Houston, 1989. Lic. dentist, Tex. Gen. dentist in pvt. practice Houston; cattle rancher. Mem. ADA, Tex. Dental Assn., Greater Houston Dental Soc., Acad. Gen. Dentistry, Alumni U. Tex. Houston, Alumni L.D. Pankey Inst., Galleria C. of C. (mem., cons. health com. 1991—), Wilderness Soc., Sierra Club, Audubon Soc. Avocations: hunting, fishing, horseback riding, travel, performing arts. Office: 1011 Augusta Dr Ste 106 Houston TX 77057-2035

NEWMAN, RAYMOND MELVIN, biologist, educator; b. New Castle, Pa., June 10, 1956; s. Raymond Melvin and Sarah L. (Lawton) N.; m. Patricia Ann Scott, Nov. 22, 1989. BS in Biology, Slippery Rock (Pa.) U., 1978; MS, U. Minn., 1982, PhD in Fisheries, 1985. Grad. asst. U. Minn., St. Paul, 1979-84, rsch. specialist forest resources, 1985-86, asst. prof. fisheries, 1988—; postdoctoral fellow natural resources U. Conn., Storrs, 1986-88; investigator U. Mich. Biol. Sta., Pellston, 1987-88; acad. coord. Electric Power Rsch. Inst./Instream Flow Needs Project, St. Paul, 1990—; mem. exotics task force Nat. Sea Grant, Silver Spring, Md., 1991—; mem. interagy. exotics species com. Minn. Dept. Natural Resources, St. Paul, 1992. Mem. editorial bd. Ecology Freshwater Fish, 1992—; contbr. articles to profl. jours. Dir. Twin Cities Trout Unltd., Mpls., 1982-87. Mem. Am. Fisheries Soc. (chair river com. Minn. chpt. 1989-91), Am. Inst. Fishery Rsch. Biologists, Ecol. Soc. Am., North Am. Benthological Soc. Achievements include documentation of chemical defense from herbivory by aquatic plants. Office: U Minn Fisheries Wildlife 1980 Folwell Ave Saint Paul MN 55108-1037

NEWMAN, ROBERT WYCKOFF, research engineer; b. St, John's, Mich., Nov. 14, 1951; s. Richard Allen and Joan Christine (Ballantine) N.; married June 18, 1983. BS, USAF Acad., Colo. Springs, 1974; MBA, U. Balt., 1983; MS Clin. Engring., Johns Hopkins Sch. of Medicine, 1983. Pilot USAF, various locations, 1974-80; applications devel. engr. G.E. Med. Magnetic Resonance, Milw., 1983-90; mgr. applications devel. G.E. CGR Magnetic Resonance, Paris, 1990-91, G.E. Med., Minimally Invasive Therapy, Milw., 1991—. Author: Clinical Magnetic Resonance Imaging, 1990. (jours.) Revue d'Imagerie Medicale, 1991, Jour. of Pediatrics, 1988; writer in field. Capt. USAF, 1969-80. Mem. Am. Assn. Physicists in Medicine, Soc. Minimally Invasive Therapy, Soc. Magnetic Resonance in Medicine. Achievements include patents for MR Phantom for Quality Assurance Testing, 1987, Phantom for Evaluating Signal to Noise, 1988. Office: GE Med Systems PO Box 414 W 801 Milwaukee WI 53201

NEWMAN, SIMON LOUIS, immunologist, educator; b. Jacksonville, Fla., Oct. 31, 1947; s. Melvin and Jeanette J. (Marks) N.; m. Diana S. Bayar, Mar. 19, 1972; 1 child, Andrew David. BA, Emory U., 1969; MS, U. Ala., Tuscaloosa, 1971; PhD, U. Ala. Med. Ctr., Birmingham, 1978. Rsch. asst. prof. U. N.C. Coll. Medicine, Chapel Hill, 1982-87; assoc. prof. U. Cin. Coll. Medicine, 1987—. Contbr. articles to sci. publs. Grantee Nat. Arthritis Found., 1982-85, NIH, 1983-86, 87-91, 90-94, 91-94, Am. Found. AIDS Rsch., 1991-92. Mem. AAAS, Am. Soc. Microbiology, Am. Assn. Immunologists. Office: Univ Cin Coll Medicine 231 Bethesda Ave ML 560 Cincinnati OH 45267

NEWMAN, STANLEY RAY, oil refining company executive; b. Milo, Idaho, Mar. 5, 1923; s. Franklin Hughes and Ethel Amelda (Crowley) N.; student Tex. A&M U, 1944-45; B.S., U. Utah, 1947, Ph.D., 1952. m. Rosa Klein, May 27, 1961 (div. Mar. 1980); children: Trudy Lynn, Susan Louise, Karen Elizabeth, Paul Daniel, Phillip John; m. Madelyn Wycherly, Jan. 10, 1991; children: Heidi, Heather, Amy. With Texaco Res. Ctr., Beacon, N.Y., 1951-82, technologist, 1973-77, sr. technologist research mfg.-fuels, 1977-82, profl. cons. on fuels and chems., 1983—. Chmn., Planning Bd., Village of Fishkill, N.Y., 1973- 77; village trustee, 1990-92; mem. Dutchess County Solid Waste Mgmt. Bd., 1974-76. With inf. Signal Corps U.S. Army, 1944-46. Mem. AAAS, N.Y. Acad. Sci., Dutchess County Geneal. Soc. (pres. 1981-87, exec. v.p. 1987-88), N.Y. Fruit Testing Assn., Sigma Xi (pres. Texaco Res. Ctr. br. 1980-81). Republican. Mormon. Patentee in field. Home: 285 Plantation Cir Idaho Falls ID 83404

NEWMARK, EMANUEL, ophthalmologist; b. Newark, May 25, 1936; s. Charles Meyer and Bella (Yoskowitz) N.; m. Tina Steinberg, Aug. 25, 1957; children: Karen Beth, Heidi Ellen, Stuart Jeffry. BS in Pharmacy, Rutger's U., 1959; postgrad., Univ. Amsterdam, The Netherlands, 1960-63, Armed Forces Instn. Pathology, Washington, 1971; MD, Duke U., 1966; Lancaster course in ophthalmology, Harvard U., 1967. Diplomate Am. Bd. Ophthalmology. Intern George Washington U. Hosp., Washington, 1966; trainee NIH rsch. Univ. Fla., Gainesville, 1967-70; resident ophthalmology U. Fla. Hosp., 1967-70; instr. dept. ophthalmology Univ. Fla., 1970; cons. ophthalmology Gainesville VA Hosp., 1970; clin. instr. ophthalmology U. Tex. Med. Sch., San Antonio, 1971-72; cons. ophthalmology Kerrville (Tex.) VA Hosp., 1971-72; asst. chief ophthalmology svc. Brooke Army Gen. Hosp., Fort Sam, Tex., 1971-72; assoc. clin. prof. Bascom Palmer Eye Inst., Miami, Fla., 1973—; clin. asst. prof. ophthalmology Bexar County Hosp. and Clinics, San Antonio, 1971-72; teaching faculty Joint Com. Allied Health Pers. Ophthalmology, St. Paul, 1981—; chief ophthalmology svc. JFK Med. Ctr., Atlantis, Fla., 1982; cons. Bascom Palmer Eye Inst., 1983—; dir., sec., treas. Palm Beach Eye Assocs., Atlantis, 1973—; mem. pharm. adv. com. Fla. Dept. Profl. Regulation, 1991; mem. med. adv. bd. Fla. east coast chpt. Nat. Sjorgren's Syndrome Assn., 1990—; bd. dirs. Fla. Eye Injury and Disease Registry. Contbr. articles and presentations to profl. jours. and chpts. to books. Alumni assoc. Rutger's Coll. Pharmacy, 1960—; chmn. reunion 1986 Duke U. Med. Alumni Assn., N.C., 1967—; centurian Davison Club--Duke U. Med. Sch., N.C., 1982—; campaign chmn., nat. vice chmn. Israel Bond's, Palm Beach County, Fla., 1988—; participant charitable orgns.; treas. Palm Beach Liturgical Culture Found., 1987—, trustee, 1987—. Decorated Lion of Judea State of Israel, 1984, recipient Gates of Jerusalem medal, 1991. Fellow ACS, Am. Acad. Ophthalmology, Am. Castroveiejo Cornea Soc.; mem. AMA, Internat. Platform Assn., Assn. for Rsch. in Vision and Ophthalmology, Contact Lens Assn. Ophthalmologists, Am. Orgn. for Rehab. Through Tng. Fedn. (nat. exec. com.-campaign cabinet 1987, pres. 1990—, Palm Beach Men's Achievement award 1988, Presdl. award 1989), So. Med. Assn., Fla. Med. Assn. (Ho. Dels. 1993), Palm Beach County Ophthal. Soc. (pres. 1984-85), Fla. Soc. Ophthalmmology (ethics chmn. 1985-90, pres. 1990-91). Jewish. Avocations: travel, radio broadcasting, teaching. Home: 335 Glenbrook Dr Atlantis FL 33462-1009 Office: Palm Beach Eye Assocs 140 JFK Cir Atlantis FL 33462-6690

NEWNHAM, ROBERT EVEREST, materials scientist, department chairman; b. Amsterdam, N.Y., Mar. 28, 1929. BS, Hartwick Coll., 1950; MS, Colo. State U., 1952; PhD in Physics, Pa. State U., 1956; PhD in Crystallography, Cambridge U., 1960. Assoc. prof. elec. engring. MIT, 1959-66; assoc. prof. solid state sci. Pa State U., University Park, 1966-71, prof., 1971—, sect. head, 1977—. Recipient International Cermaics prize, Acad. Ceramics, 1992. Mem. Am. Phys. Soc., Am. Crystallography Assn., Am. Ceramic Soc. (John Jeppson medal 1991), Mineral Soc. Am. Achievements include research in crystal and solid state physics, x-ray crystallography. Office: Penn State U Solid State Sci Program University Park PA 16802*

NEWPORT, BRIAN JOHN, physicist; b. Papakura, New Zealand, Aug. 10, 1958; came to U.S., 1980; s. Dennis Bradley and Marjorie Ann (Hill) N. BSc, Victoria U., Wellington, New Zealand, 1979; PhD, Calif. Inst. Tech., 1986. McCormick fellow U. Chgo., 1986-88, rsch. assoc., 1988-89, sr. rsch. assoc., 1989—; aux. faculty U. Utah, Salt Lake City, 1988—. Mem. Am. Phys. Soc. Office: U Utah Dept Physics Salt Lake City UT 84112

NEWTON, CRYSTAL HOFFMAN, mechanical engineer; b. York, Pa., Dec. 5, 1956; d. Martin James and Mary Jane (Blake) Hoffman. BSME, Carnegie-Mellon U., 1978; MS in Applied Mechanics, Lehigh U., 1980, PhD in Mechanical Engring., 1985. Rsch. engr. Lehigh U., Materials Rsch. Ctr., Bethlehem, Pa., 1984-86; project engr. Materials Scis. Corp., Ft. Washington, Pa., 1986—; Editor: Manual on the Building of Material Property Databases, 1993, (handbook) MIL-Handbook-17 Polymer Matrix Composites, 1988—. Mem. Am. Soc. Testing and Materials (subcom. chmn. 1990—, 2d vice chmn., 1992—), Pi Tau Sigma, Sigma Xi. Achievements include research on material property databases, standardization of composite materials testing, expert systems. Office: Materials Scis Corp 500 Office Center Dr Fort Washington PA 19034

NEWTON, JEFFREY F., project engineer. AAS in Tech., Acad. Aero.; BS in Engring., Widener U. Design engr. Boeing Helicopters, Ridley Park, Pa., 1973-80, 84-87, engring. team leader, 1987-89, mgmt. trainee, 1989-90, project engr., 1990—; staff process engr. West Co., Phoenixville, Pa., 1980-84. mem. AIAA, Am. Helicopter Soc.

NEWTON, RHONWEN LEONARD, microcomputer consultant; b. Lexington, N.C., Nov. 13, 1940; d. Jacob Calvin and Mary Louise (Moffitt) Leonard; children: Blair Armistead, Allison Page, William Brockenbrough III. AB, Duke U., 1962; MS in Edn., Old Dominion U., 1968. French tchr. Hampton (Va.) Pub. Schs., 1962-65, Va. Beach (Va.) Pub. Schs., 1965-66; instr. foreign lang. various colls. and univs., 1967-75; foreign lang. cons. Portsmouth (Va.) Pub. Schs., 1973-75; dir. The Computer Inst., Inc., Columbia, S.C., 1983; pres., founder The Computer Experience, Inc., Columbia, 1983-88, RN Enterprises, Columbia, 1991—. Author: WordPerfect, 1988, All About Computers, 1989, Microsoft Excel for the Mac, 1989, Introduction to the Mac, 1989, Introduction to DOS, 1989, Introduction to Lotus 1-2-3, 1989, Advanced Lotus 1-2-3, 1989, Introduction to WordPerfect, 1989, Advanced WordPerfect, 1989, Introduction to DisplayWrite 4, 1989, WordPerfect for the Mac, 1989, Introduction to Microsoft Works for the Mac, 1990, Accountant, Inc for the Mac, 1992, Introduction to Filemaker Pro, 1992, Quicken for the MAC, 1993, Quicken for Windows, 1993, WordPerfect for Windows, 1993, Advanced WordPerfect for Windows, 1993, Lotus 1-2-3 for Windows, 1993. Mem. Columbia Planning Commn., 1980-87; bd. dirs. United Way Midlands, Columbia, 1983-86; bd. dirs. Assn. Jr. Leagues, N.Y.C., 1980-82; trustee Heathwood Hall Episcopal Sch., Columbia, 1979-85. Republican. Episcopalian. Avocations: golf, walking. Home and Office: 1635 Kathwood Dr Columbia SC 29206-4509

NEWTON, RICHARD WAYNE, food products executive; b. Baytown, Tex., Aug. 26, 1948; s. Cecil J.V. and Selma Ruth (Waldrep) N.; m. Natalee Gay Wood, Feb. 16, 1968; children: Nathan Wayne, Natalee Christine. BSEE, Tex. A&M U, 1970, MSEE, 1971, PhD, 1977. Registered profl. engr., Tex.; lic. comml. pilot. Engr., scientist Lockheed Electronics Co., Inc., Clear Lake City, Tex., 1971-73; program mgr. remote sensing ctr. Tex. Engring. Expt. Sta., Tex. A&M U., College Station, 1973-77, assoc. dir. remote sensing ctr., asst. prof. elec. engring., 1977-80, dir. remote sensing ctr., assoc. prof., 1980-84, dir. microwave microelectronics lab., assoc. prof., 1985-86; prin. Aerial Surveys Inc., 1977-84, Innovative Devel. Engring. Assocs., Inc., Cannon Bear, Inc., 1980-86; group mgr. rsch. and devel. Frito-Lay, Inc., Plano, Tex., 1986-88, group mgr. corp. engring., 1988-89, sr. group mgr. corp. engring., 1989—; prin. investigator over 19 contracts and grants; mem. Internat. Union Radio Sci., Commn. F., 1985-89; fellow Tex. Engring. Expt. Sta., 1983-84. Contbr. articles to profl. jours. Bd. trustees Univs. Space Rsch. Assn., 1983-86, 86-89, Alliance for Higher Edn., 1991—; organizer, pres. Highland Meadows Homeowners Assn., 1988-90; mem. Mayor's Commn. on Internat. Trade, Sci. and Tech. Com., City of Dallas, 1987-88; mem. Dallas Internat. Initiative, 1989-92; mem. Colleyville City

Coun., Place 2, 1989-93; mayor City of Colleyville, 1992—. Named Young Engr. of Yr. Tex. Soc. Profl. Engrs., Brazos chpt., 1983. Achievements include 2 patents in field. Home: 6107 Lansford Ln Colleyville TX 76034

NEWTON, SEAN CURRY, cell biologist; b. Rockford, Ill., Sept. 9, 1959; s. Delbert Russell and Cathrine Ann (Curry) N.; m. Ann Marie Summer, May 21, 1988. MS, So. Ill. U., 1984, PhD, 1989. Postdoctoral fellow U. S.C., Columbia, 1990-91, rsch. asst. prof., 1992-93; rsch. assoc. prof. Creighton U., Omaha, 1993—. Contbr. articles to profl. jours. NIH postdoctoral fellow, 1992; So. Ill. U. grad. fellow, 1988; Sigma Xi grantee, 1988. Mem. Am. Assn. of Anatomists, Am. Soc. Cell Biology, Soc. for Study of Reprodn., Am. Soc. for Andrology. Republican. Roman Catholic. Office: Creighton U Dept Phys Therapy Omaha NE 68178

NEWTON, V. MILLER, medical psychotherapist, writer; b. Tampa, Fla., Sept. 6, 1938; s. Virgil M. Jr. and Louisa (Verri) N.; m. Ruth Ann Klink, Nov. 9, 1957; children: Johanna, Miller, Mark. BA, U. Fla., 1960; MDiv, Princeton Theol. Sem., 1963; postgrad., U. Geneva, Switzerland, 1962; PhD, The Union Inst., 1981. Min. dir. Flectcher Pl. Urban Social Ministry, Indpls., 1963-65; coord. staff tng. and community rels. Breckinridge Job Corps Ctr., Ky., 1965-66; asst. prof., program dir. social scis. Webster Coll., St. Louis, 1966-69; assoc. prof., program dir. edn. U. South Fla., Tampa, 1969-73; clk. of the cir. ct. Pasco County, Fla., 1973-76; exec. dir. Fla. Alcohol Coalition, Inc., 1979-80; program and nat. clin. dir. Straight, inc., St. Petersburg, Fla., 1980-83; dir. KIDS of North Jersey, Inc., 1983—; mem. Sec. Task Force Confidentiality and Client Info. System, Fla. Dept. of Health and Rehab. Svcs., 1979-80; chmn. pres.'s adv. Coun. Webster Coll., 1968-69; guest lectr. at the Grad. Inst. of Community Devel. So. Ill. U., 1968-69; cons. Tampa Model Cities Program, 1969-70; chmn. planning com. Tchr. Corps. Nat. Conf.; faculty mem. Internat. U. for Pres., Munich; co-chmn. Mayor's com. on Pre-sch. Edn., Indpls., 1964-65; speaker in field. Author: Gone Way Down: Teenage Drug-Use is A Disease, 1981, Kids, Drugs, and Sex, 1986; co-author: Not My Kid: A Parent's Guide to Kids and Drugs, 1984; appeared on TV programs NBC Mag., 1982, 1986 NBC, 1986, Drugs: A Plague upon America with Peter Jennings ABC, 1988; contbr. articles to profl. jours. Member drug abuse adv. coun. State of N.J., 1985-91; chmn., bd. dirs. Adjustment Madeira Beach, Fla., 1981—, Alcohol Community Treatment Svcs., Inc., Tampa, 1979; pres. Pasco County Coun. on Aging, 1977-79; chmn. bd. San Antonio Boys Village, 1975-76; chmn. Pasco County Data Ctr. Bd., 1973-75, Cen. Pasco Urban Planning Commn., 1972-73; adult del. White House Conf. on Youth, 1971; chmn. Nat. Tchr. Corps Field Coun., 1970-71; pres. Christian Inner City Assn., Indpls., 1964-65; mem. Gov. Ashew's Adv. Com., Pasco County, 1974-76. Aldersgate fellow, 1962; recipient Honor award Nat. LWV, 1963, Cert. Appreciation Pinellas County Bd. of County Commrs., 1982; named Outstanding Young Man of Yr., Indpls. Jaycees, 1965, Outstanding Govt. Leader, Dade City Jaycees, Fla., 1973-74. Mem. Am. Anthrop. Assn., Soc. Med. Anthropology, Psychol. Anthropology, AACD, APHA, Soc. Behavioral Medicine, Phi Delta Theta, Rotary Internat., Order of DeMolay (state master counselor 1957). Democrat. Methodist. Home: Apt 4-14K 7003 Boulevard E Guttenberg NJ 07093 Office: KIDS of North Jersey PO Box 2455 Secaucus NJ 07096-2455

NEY, RONALD ELLROY, JR., chemist; b. Harrisonburg, Va., July 4, 1936; s. Ronald Ellroy and Evelyn (Ash) N.; m. Sue McClure, June 12, 1959; children: Ronald Ellroy III, Dorothy Ellen. BS, James Madison U., 1959; PhD, Pacific Western U., 1989. Tchr. chemistry, sci. Fairfax (Va.) County Schs., 1959-62; chemist U.S. FDA, Washington, 1962-65; supervisory chemist USDA U.S. FDA, 1965-70; chemist, sci. adviser U.S. EPA, 1970-86; real estate broker, pres. Environ. Realty Cons. Svc., Fairfax, 1986—; instr. No. Va. C.C., Fairfax, 1991—. Author: Where Did That Chemical Go?, 1990, Your Guide to Safety, 1992. Sec. Fairfax Archtl. Rev. Bd., 1991—. Recipient letters of commendation U.S. Dept. Justice, 1977, 78. Achievements include initiation of U.S. program on environmental chemistry dealing with fate and transport of chemicals, pesticides in environment. Home: 3818 Charles Stewart Dr Fairfax VA 22033 Office: Environ Realty Cons Svc Inc 3818 Charles Stewart Dr Fairfax VA 22033

NEYLAN, JOHN FRANCIS, III, nephrologist, educator; b. Chgo., Feb. 20, 1953; s. John Francis and Mary Alice (Coogan) N.; m. Cynthia Barnes, May 17, 1981; 1 child, John Francis IV. BS, Duke U., 1975; MD, Rush Med. Coll., Chgo., 1979. Intern in medicine Vanderbilt U., Nashville, 1979-80, resident, 1980-82; fellow in nephrology Brigham and Women's Hosp., Boston, 1983-84; fellow in immunogenetics Harvard U. Med. Sch., Boston, 1984-86, clin. preceptor, 1986; asst. prof. medicine U. Calif., Davis, 1986-88; asst. prof. medicine Emory U., Atlanta, 1988-93, assoc. prof., 1993—, med. dir. renal transplantation, 1988—; vis. cons. Wanless Hosp., Miraj, India, 1982-83; assoc. med. dir. Lifelink of Ga. Organ Procurement Orgn., Atlanta, 1989—; bd. govs. Lifelink Found., Tampa, Fla., 1988—. Contbr. articles and abstracts to med. jours., chpts. to books. Vol. Nat. Kidney Found., N.Y.C., 1990—, ARC, Atlanta, 1991, Spl. Olympics, Atlanta, 1991—, Habitat for Humanity, 1993—. Recipient Physician's Recognition award AMA, 1989. Mem. ACP, Am. Fedn. Clin. Rsch. (councillor 1988), Am. Soc. Transplant Physicians (co-chmn. patient care com. 1988-90, chmn. 1991-93, councillor-at-large exec. coun. 1993—), Am. Soc. Nephrology, Internat. Soc. Nephrology, Transplantation Soc., United Network for Organ Sharing, Circumnavigators Club, Alpha Omega Alpha. Avocations: windsurfing, tennis, cycling. Office: Emory U 1365 Clifton Rd NE Atlanta GA 30307-1013

NEZHAD, HAMEED GHOLAM, energy management educator; b. Gorgan, Iran, Aug. 21, 1941; came to U.S., 1969; s. Ghorban and Sultaneh (Meighani) Gholamnezhad; m. Madeline Bunsu, Mar. 15, 1977 (div. 1984); m. Sandee L. Wendeling, Feb. 14, 1987; 1 child, Sheila June. BS in Physics, Tehran U., 1963; MS in Physics, Eastern Ill. U., 1972; PhD, U. Pa., 1979. Prodn. engr. Oil Svc. Co., Ahwaz, Iran, 1972-74; rsch. asst. U. Pa., Phila., 1976-79; dir. energy mgmt. program Eastern Ill. U., Charleston, 1979-84; dir. energy inst. Moorhead (Minn.) State U., 1984—; pres., founder Energy Educators Assn., 1988—; cons. in field. Author: How to Make Decisions in a Complex World, 1989, (software) DECIDE: A Decision Support System, 1990, STRUCTURE: A Problem Solving Software, 1992; contbr. articles to profl. jours. Recipient Energy Conservation award State of Minn., 1989, Merit award Moorhead State U., 1988. Mem. Moorhead Rotary Club. Home: 910 4th Ave N Moorhead MN 56560-2010 Office: Moorhead State U Energy Inst Moorhead MN 56563

NEZU, CHRISTINE MAGUTH, clinical psychologist, educator; b. Passaic, N.J., June 18, 1952; d. Frank Joseph and Alice Anna (Hingstmann) Maguth; m. Arthur Maguth Nezu, June 12, 1983; children: Frank, Alice, Linda. BA, Fairleigh Dickingson U., Rutherford, N.J., 1977; MA, Fairleigh Dickingson U., Teaneck, N.J., 1981, PhD, 1987. Lic. psychologist, N.Y., N.J., Pa. Psychology intern Beth Israel Med. Ctr., N.Y.C., 1985-86; coord. rsch. and clin. supervision Project NSTM, Fairleigh Dickinson U., Teaneck, 1986-87; clin. asst. prof. and supervising psychologist Beth Israel Med. Ctr., Mt. Sinai Sch. Medicine, N.Y.C., 1987-89; asst. prof., dir. intern tng. Hahnemann U., Phila., 1989—; dir. Phila. mental retardation-sex offender project Hahnemann U., 1991—. Co-author: Problem Solving Therapy, 1989; co-editor Clinical Decision-Making, 1989, Psychotherapy of Persons with Mental Retardation: Clinical Guidelines for Assessment and Treatment, 1992; contbr. articles to profl. jours. Mem. Phila. Mus. of Art, Mus. of the U. of Pa.; vol. Winter Shelter Program for the Homeless, Phila., 1991—. Recipient Bd. Trustees fellowship award Fairleigh Dickinson U., 1985, rsch. fellowships, 1983-86. Fellow Pa. Psychol. Assn.; mem. Am. Psychol. Assn., Am. Assn. on Mental Retardation, Assn. for the Advancement of Behavior Therapy (chair com. on acad. tng. 1990-93, coord. acad. and profl. issues 1993—). Lutheran. Avocations: music, sailing, jogging, theatre, travel. Home: 2426 Fitler Walk Philadelphia PA 19103 Office: Hahnemann U Broad and Vine Philadelphia PA 19102

NG, KHENG SIANG, cardiologist, educator, researcher; b. Singapore, Apr. 5, 1960; s. Siak Jinng and Wang Hong (Tan) N.; m. Lau Yung Sang, Oct. 24, 1991; 1 child, Alyssa. MBBS, Nat. U. Singapore, 1984. House officer Ministry Health, Singapore, 1984-85; med. officer Ministry Defense, Artillery, Singapore, 1985-87; med. trainee Singapore Gen. Hosp. and Tan Tock Seng Hosp., Singapore, 1987-89; registrar Singapore Gen. Hosp., 1989-92; registrar Nat. U. Hosp., 1992-93, sr. registrar, 1993—; clin. tutor Nat. U. of Singapore, 1989—; vis. cardiologist Med. Classification Ctr., Cen. Manpower Base, 1990—. Contbr. articles to profl. jours. Vol. Community Health Svc., Singapore, 1985-90, Singapore Coronary Club, 1992—. Recipient Pub. Svc. Commn. Local Merit award, 1979. Mem. Royal Coll. Physicians and Surgeons (Glasgow), Singapore Cardiac Soc., Singapore Nat. Heart Assn. Office: Nat U Hosp, 5 Lower Kent Ridge Rd, Singapore 0511, Singapore

NG, KWOK-WAI, physics educator; b. Hong Kong, Aug. 15, 1958; came to U.S., 1981; s. Wan-Fu and Kam-Har (Sin) N.; m. Grace Mun Yan, Dec. 28, 1987. BSc, U. Hong Kong, 1981; PhD, Iowa State U., 1986. Postdoctoral fellow U. Tex., Austin, 1986-88; asst. prof. U. Ky., Lexington, 1988—. Contbr. articles to Phys. Rev. Letter, Phys. Rev. B, Japanese Jour. Applied Physics. Mem. IEEE, Am. Phys. Soc., Phi Kappa Phi. Achievements include gap anisotropy of high Tc superconductors; superconducting tunneling spectroscopy. Office: Univ Ky Dept Physics & Astronomy Lexington KY 40506-0055

NG, LEWIS YOK-HOI, civil engineer; b. Chaozhou, Guangdong, People's Republic of China, Aug. 28, 1955; came to U.S., 1976; s. Shing Yee and Sukyau Ng; m. Sandra Jane Moy, June 19, 1982; children: Andrea, Gabriel. BS, U. Wis., Platteville, 1980; MS, U. Minn., 1984. Cert. profl. engr., Minn., Ohio, Wis. Sr. engr. Twin City Testing Corp., St. Paul, 1980-86; project mgr. Walker Engrs., Mpls., 1986-88; chief engr., mgr. Twin City Testing Corp., 1988-91; mgr., prin. STS Cons., Mpls., 1992—. Chmn. deacons bd. Twin City Chinese Christian Ch., 1992-93. Mem. ASCE, Am. Concrete Inst. (tech. com. 1985—, planning com. 1989—), Constrn. Specifier Inst. (tech. com. 1990—). Home: 3107 Evelyn St Saint Paul MN 55113-1214

NGAI, KA-LEUNG, biochemist, researcher; b. Kowloon, Hong Kong, Aug. 31, 1950; came to U.S., 1969; s. Chung-Yin and Wen-Yue (Sun) N. BSChemE, U. Wis., 1972; PhD, U. Pa., Phila., 1981. Rsch. fellow Yale U., New Haven, Conn., 1981-86; dir. biotech. facility Northwestern U., Evanston, Ill., 1986-92, Northwestern U. Med. Sch., Chgo., 1990-92; dir. genetic engring. facility U. Ill., Urbana, 1992—. Author: (with others) Methods in Enzymology, 1990; ad hoc reviewer Jour. Bacteriology, 1985—; mem. adv. bd. Who's Who in Technology, 6th edit., 1988; contbr. articles to sci. jours. Mem. AAAS, Am. Chem. Soc., Am. Soc. for Microbiology, Tau Beta Pi. Roman Catholic. Achievements include research in biodegradation of aromatics via Beta-Ketoadipate pathway. Office: U Ill 200 Noyes Lab Box 62-1 Urbana IL 61801

NGHIEM, LONG XUAN, computer company executive; b. Saigon, Vietnam, July 26, 1952; arrived in Can., 1970.; s. Huynh Xuan Nghiem and Chinh Thi Tran; m. Hanh Dgoc Duong, Apr. 12, 1980; children: Sarah B., Kristen D. BASc in Chem. Engring., U. Montreal, Can., 1975; MASc in Chem. Engring, U. Waterloo, Ont., Can., 1976. Dir. rsch. and devel. Computer Modelling Group, Calgary, Alta., 1986—; cons. U.N., Inst. Reservoir Studies, Oil and Natural Gas Commn., Ahmedabad, India, 1992-93; lectr. numerous confs., forums, seminars;. Mem. Assn. Profl. Engrs., Geologists and Geophysicists of Alta., Soc. Petroleum Engrs. (organizer comparative solution project, symposium project, tech. editor Resevoir Engring.), Petroleum Soc. CIM. Achievements include research includes enhancement of user's friendliness of reservoir simulators, improvement of simulator speed, investigation of reservoir/wellbore interactions, characterization of reservoir heterogeneities for simulation, modelling bypassing due to heterogeneities/viscous fingering, modelling asphaltene precipitation in compositional simulation. Office: Computer Modeling Group, 200-3512-33rd St NW, Calgary, AB Canada T2L 2A6

NGHIEM, SON VAN, electrical engineer; b. Saigon, Vietnam; came to U.S. 1982; s. Bui and Minh Van N.; m. Anh Lieu. BS summa cum laude, Tex. A&M U., 1985; SM, MIT, 1988, PhD, 1991. Rsch. asst. MIT, Cambridge, 1986-91; mem. tech. staff Calif. Inst. Tech. Jet Propulsion Lab., Pasadena, Calif., 1991--; com. mem. Progress in Electromagnetic Rsch. Symposium, Boston, 1989, Cambridge, 1991, Pasadena, 1993. Co-author: Polarimetric Active Remote Sensing, 1990. Recipient Dean's Honor award Tex. A&M U., 1983, 84, Earl Graham scholar, Teagle Found. scholar Exxon, 1984-85; travel grantee Internat. Union Radio Sci., 1986. Mem. IEEE, Phi Kappa Phi, Sigma Xi. Achievements include electromagnetic wave models for scattering and emission from geophysical media; first measurements of polarimetric emission including third stokes parameters from periodic soil surface; derivation of relationships among polarimetric scattering coefficients from symmetry groups. Office: Jet Propulsion Lab 4800 Oak Grove Dr MS300-235 4800 Oak Grove Dr Pasadena CA 91109

NGUYEN, ANN CAC KHUE, pharmaceutical and medicinal chemist; b. Sontay, Vietnam; came to U.S., 1975; naturalized citizen; d. Nguyen Van Soan and Lu Thi Hieu. BS, U. Saigon, 1973; MS, San Francisco State U., 1978; PhD, U. Calif., San Francisco, 1983. Teaching and research asst. U. Calif., San Francisco, 1978-83, postdoctoral fellow, 1983-86; research scientist U. Calif., 1987—. Contbr. articles to profl. jours. Recipient Nat. Research Service award, NIH, 1981-83; Regents fellow U. Calif., San Francisco, 1978-81. Mem. AAAS, Am. Chem. Soc., N.Y.Acad. Scis., Bay Area Enzyme Mechanism Group, Am. Assn. Pharm. Scientists. Roman Catholic. Home: 1488 Portola Dr San Francisco CA 94127-1409 Office: U Calif Lab Connective Tissue Biochemistry Box 0424 San Francisco CA 94143

NGUYEN, HAO MARC, aerospace engineer; b. Washington, July 7, 1967; s. Hung Michel and Loc Nguyen. BS, U. Va., 1989. Analytical engr. UTC Pratt & Whitney, GSP, West Palm Beach, 1989-91; space systems engr. BDM Internat., Washington, 1991—; chairperson Discover E Program, Pratt & Whitney, West Palm Beach, 1990-91, coord., 1991; advisor U. Va., SEA, Charlottesville, Va., 1988-89. Mem. AIAA, Sigma Gamma Tau. Office: BDM Internat Ste 340 409 3rd St NW Washington DC 20024

NGUYEN, PHILONG, electrical engineer; b. Cantho, Vietnam, Aug. 31, 1956; came to U.S., 1985; s. Ngoc The and Thi Dihn (Dinh) N.; m. Bachlan Thi, Sept. 2, 1978; children: Hoangvi Nguyen T., Phivu Nguyen. AS in Elec. Sci., Westchester Community Coll., Valhalla, N.Y., 1987; BS in Elec. Sci., Manhattan Coll., 1990. Registered profl. engr., N.Y. Owner, mgr. Elec. Power Workshop, Vungtau, Vietnam, 1978-83; operator elec. generator and mech. pump Indonesia Refugee Camp, 1983-84; work study-researcher, jr. engr. IBM T.J. Watson Rsch. Ctr., Yorktown, N.Y., 1987-89, assoc. engr., 1989—. Mem. IEEE. Roman Catholic. Home: 313 E Blackwell St Dover NJ 07801 Office: IBM Corp TJ Watson Rsch Ctr PO Box 218 Yorktown Heights NY 10598-0218

NGUYEN, QUAN A., medical physicist; b. Saigon, Vietnam, May 24, 1960; came to U.S., 1975; s. Suy D. and Tra T. (Nguyen) N.; m. Anh-Nguyen Ho, June 15, 1985; children: Tammy, Vincent. MS, U. So. Calif., 1985. Radiation physicist U. Calif. Irvine Med. Ctr., Orange, 1990—. Mem. Am. Assn. Med. Physicists. Home: 4 Sorrento Irvine CA 92714 Office: U Calif Irvine Med Ctr Radiation Oncology 101 City Blvd W Orange CA 92668-2901

NGUYEN, TIEN MANH, communications systems engineer; b. Saigon, Vietnam, Apr. 5, 1957; came to the U.S., 1975; s. Hung The and Bi Thi (Luu) N.; m. Thu Hang Thi, Dec. 28 1986. BS in Engring., Calif. State U., Fullerton, 1979, MS in Engring., 1980; postgrad., Calif. State U., Long Beach, 1991—, MA in Engring. Math., 1993; MEE, U. Calif., San Diego, 1982; PhD in Engring., Columbia Pacific U., 1986. Cert. EMC engr., Mfg. tech. Teaching asst. U. Calif., San Diego, 1982-83; chief automated mfg. dept. ITT Ednl. Svcs., West Covina, Calif., 1983-85; mem. tech. staff Jet Propulsion Lab., Pasadena, Calif., 1985—; prin. tech. advisor Internat. Consultative Com. for Space Data Systems, Pasadena, 1985-90. Editor: Proceedings of CCSDS RF & Modulation, 1989; contbr. more than 18 articles to profl. jours. Grad. rep. EECS dept. U. Calif., San Diego, 1982-83; NASA del. to internat. CCSDS, 1986—. San Diego fellow, 1980-82, Long Beach Found. scholar Calif. State U.; recipient Bendix Mgmt. Club award, 1987, NASA Honor award, 1988, over 12 NASA monetary awards, 1989—. Mem. IEEE (vice chmn. 1987—, session chmn. internat. symposium on EMC 1986, award 1986), AIAA (sr.), Soc. Mfg. Engrs., Am. Math. Soc., Armed Forces Communications and Electronics Assn., N.Y. Acad. Scis., U.S. Naval Inst., Internat. Platform Assn. Republican. Buddhist. Achievements include patent for technique to resolve phase ambiguity for QPSK systems; development of new algorithms to design communications systems for space applications. Home: 1501 W Maxzim Ave Fullerton CA 92633 Office: Jet Propulsion Lab 4800 Oak Grove Dr Pasadena CA 91109-8099

NGUYEN, TINH, materials scientist; b. Tuyhoa, Vietnam, Dec. 20, 1948; came to U.S., 1973; s. Tang and Don T. N.; m. Susan Tran, Jan. 4, 1976; children: Richard, Derrick. Diploma in engring., U. Saigon, Vietnam, 1970; MS, U. Philippines, Los Banôs, 1974; PhD, U. Calif., Berkeley, 1979. Rsch. asst. U. Calif., Richmond Exptl. Sta., 1974-79; rsch. chemist Atlantic Richfield Chem Co., Newtown Square, Pa., 1979-82; phys. scientist Nat. Inst. Standards and Tech., Gaithersburg, Md., 1982—. Contbr. articles to profl. jours. Mem. Adhesion Soc., Ceramic Soc., Am. Nondestructive Soc., Am. Chem. Soc. (surface and colloid and polymeric materials div.). Republican. Achievements include patent for UV stabilizers for polyurethane sealants, new binders for woodboard products, internal release agents; development of techniques for surface and interface studies of polymer-coated metals, understanding of degradation mechanism of organic protective coatings. Home: 13219 Dodie Dr Darnestown MD 20878 Office: Nat Inst Standards/Tech 226/B348 Gaithersburg MD 20899

NGUYEN, TRUC CHINH, analytical chemist; b. Saigon, Vietnam, Apr. 21, 1960; came to the U.S., 1981; s. Duc Huu Nguyen and Cam Thi Doan. BS, U. Tex., 1987. Rsch. asst. Univ. Tex., San Antonio, 1984-87; assoc. scientist Radian Corp., Austin, Tex., 1987-88; scientist Radian Corp., Austin, 1988-90, staff scientist, project dir., 1991—; assistance dir. Southeast Asia Internat. Trade Assocs., Inc., 1992—; pres. L'Expression Internat., 1993. Contbr. articles to profl. jours. Pres. Young Vietnamese-Am. Assn., Tex. 1990—; bd. dirs. Vietnam TV, Austin, 1988-90. Recipient Minority Biomed. Rsch. Support grant NIH, 1984-87. Mem. Am. Chem. Soc. (analytical chemistry div.), Internat. Platform Assn. Home: 11914 Snow Finch Rd Austin TX 78758-3008 Office: Radian Corp 8501 N Mo Pac Expy Austin TX 78720-1088

NGUYEN-DINH, THANH, internist, geriatrician; b. Saigon, Vietnam, Nov. 7, 1950; s. Bam and Chanh Thi (Duong) Nguyen-Dinh; m. Kim-Chi Nguyen-Dinh, Oct. 28, 1950; children: Trung, Kim-Trang, Kim-Trinh, Trong. MD, Free U. Brussels, 1974; Tropical MD, Antwerp Tropical Med. Inst., 1975. Diplomate Am. Bd. Internal Medicine, Am. Bd. Geriatric Medicine. Asst. prof. medicine Howard Med. Svc., Washington, 1981—; physician dir. St. Elizabeth Unit, D.C. Gen. Hosp., Washington, 1983—; co-dir. Howard U. Md. Clinics, D.C. Gen. Hosp., 1990—. Contbr. articles to profl. jours. Fellow ACP. Avocations: chess, swimming. Office: 611 S Carlin Springs Rd Ste 211 Arlington VA 22204-1071

NGUYEN-TRONG, HOANG, physician, consultant; b. Hue, Republic of Vietnam, Sept. 4, 1936; s. Nguyen-Trong Hiep and Nguyen-Phuoc Ton-nu-Thi Sung. B in Math., Lycée d'Etat Michel Montaigne, Bordeaux, 1956; state diploma of medicine, Sch. Medicine, Paris, 1966, also cert. aeronautical medicine and health and sanitation. Resident surgeon Compiegne State Hosp., 1963-64, Meaux State Hosp., 1964-66, Lagny State Hosp., 1966; specialist in health and sanitation Paris Sch. Medicine, 1965—; specialist in family planning French Action of Family Planning, Paris, 1968—; practice medicine, Nanterre, France, 1969—; cons. physician various pharm. labs., Paris, 1987; investigator physician WHO regional office for Europe, 1991. Contbr. articles to profl. jours. Mem. French Soc. Aviation and Space Physiology and Medicine (titulary, specialist in aviation medicine), Assn. Nanterre Physicians, Assn. Vietnamese Practitioners in France, Chambre Syndicale des Medecins des Hauts de Seine, Ordre des Medecins des Hauts de Seine, Les Ex du XIV Shooting Club. Avocations: painting, poetry, classical and modern jazz music, riflery, martial arts. Home: 3 Rue Gazan, 75014 Paris France Office: Cabinet Med Privé, 38 Rue des Fontenelles, 92000 Nanterre France

NIBLACK, TERRY L., nematologist, plant pathology educator; b. Lubbock, Tex.. MS, U. Tenn., 1982; PhD, U. Ga., 1985. Postdoctoral assoc. Iowa State U., Ames, 1986-87; asst. prof. plant pathology dept. U. Mo., Columbia, 1988—. Editor Annals of Applied Nematology, 1992—. Recipient Disting. Svc. award So. Soybean Disease Workers, 1991. Mem. Soc. Nematologists, Am. Phytopath. Soc., Orgn. Nematologists of Tropical Am., Sigma Xi. Achievements include recognition of symptomless yield losses due to soybean cyst nematode in the North Central region. Office: U Mo Dept Plant Pathology 108 Waters Hall Columbia MO 65211

NICASTRO, DAVID HARLAN, forensic engineer, consultant, author; b. L.A., Mar. 12, 1961; s. Leo and Ruth Elizabeth (Moody) N. BA, Pomona Coll., 1983; MS, U. Tex., 1985. Registered profl. engr., Tex. Staff engr. Law Engring. Inc., Houston, 1985-86, project engr., 1986-90, sr. engr., 1990-91; prin. engr., 1992—. Contbr. articles to profl. jours. Recipient Tileston prize Pomona Coll., 1983. Mem. ASCE (chmn. materials divsn. Houston chpt. 1989, tech. coun. for forensic engring. control group com. on dissemination of failure info. 1991—), ASTM (vice chmn. com. C24). Democrat. Mem. Christian Ch. Achievements include development of taxonomy for failure classification, matrix method of objective failure mechanism evaluation. Office: Law Engring Inc Ste # 450 13831 NW Freeway Houston TX 77040

NICASTRO, FRANCESCO VITO MARIO, agricultural science educator; b. Ginosa, Italy, Jan. 19, 1953; s. Giuseppe and Anna (Rizzi) N.; m. Filomena Nunziante, Apr. 30, 1981; children: Annarita, Tiziana. Diploma, Inst. Tech. Agrario, Matera, Italy, 1972; PHD in Agr., U. Bari, Italy, 1978. Researcher dept. animal prodn. U. Bari, 1980—; prof. U. Molise, Campobasso, Italy, 1991—; leader Unity of Rsch., RAISA/Coun. Nat. Rsch., Bari, 1991—, chmn. rsch. groups, Bari, 1991—; lectr. in field; cons. Minister Fgn. Affairs, Rome, 1989, Food and Agrl. Orgn. of UN, Rome, 1990, IRI-Italstat, Rome, 1990; rep. Italy Internat. Congress of Meat Sci. and Tech., 1988—. Co-author: International Dictionary of Meat Science, 1988; contbr. articles to profl. publs. Coun. Nat. Rsch. grantee, 1978-80, 83, 84-90; recipient Best Program for Rsch. Animal Sci. award Formez-Italia, 1986. Fellow Italian Soc. Animal Prodn. (chmn. histological com. 1989), Italian Soc. Sheep and Goat Phatology, Italian Soc. Vet. Sci., Am. Soc. Animal Sci., Am. Meat Sci. Assn., N.Y. Acad. Scis. Roman Catholic. Avocations: reading, music, tennis. Home: Via Oberdan # 24, 70126 Bari Italy Office: U Bari Dept Animal Prodn, Via Amendola 165/A, 70126 Bari Italy

NICCOLINI, DREW GEORGE, gastroenterologist; b. Rockville Center, N.Y., July 27, 1945; s. George D. and Elaine A. (Augsbury) N.; m. Martha Dodge, Jan. 3, 1971; children: Alyssa, Rachael, Lesley, Matthew, Adam. BA, Johns Hopkins U., 1967; MD, Tufts U., 1971. Diplomate Am. Bd. Internal Medicine, Am. Bd. Gastroenterology. Intern St. Elizabeth Hosp., Boston, 1971-72, resident, 1972-74, gastrointestinal fellow, 1974-75; gastrointestinal fellow Faulkner Hosp., Boston, 1975-76; clinician Pentucket Med. Assocs., Haverhill, Mass., 1976—; staff physician Hale Hosp., Haverhill, 1976—, Lawrence (Mass.) Gen. Hosp., 1976—; cons. Holy Family Hosp., Methuen, Mass., 1976—; chief medicine Hale Hosp., haverhill, Mass., 1987-88. Capt. U.S. Army, 1971-77. Fellow Am. Coll. Gastroenterology; mem. ACP, AMA, New Eng. Endoscopy Soc., Am. Soc. Gastroent. Endoscopy, Alpha Omega Alpha. Avocations: skiing, tennis. Office: Pentucket Med Assocs 1 Parkway Haverhill MA 01830

NICEWANDER, WALTER ALAN, psychology educator; b. Eagle River, Wis., Sept. 23, 1939; s. B. Walter and Dorothy (Shirley) N.; children: David, Brent, Karen. BS, Purdue U., 1961, PhD, 1971. Prof. psychology U. Okla., Norman, 1970—. Recipient Regent's Award Superior Teaching, 1988. Mem. Soc. Multivariate Exptl. Psychology. Democrat. Achievements include research on latent trait based reliability estimate and upper bound and symmetric, invariant measures of multivariate association. Home: 2525 Hollywood Norman OK 73072 Office: U Okla Dept Psychology Norman OK 73019

NICHOL, DOUGLAS, geologist; b. Glasgow, Scotland, U.K., June 21, 1947; s. Donald and Mary (Semple) N.; m. Christine Judith Cannon, June 20, 1970; children: Sancha Kay, Craig Douglas, Philip John. BSc with honors, U. Glasgow, 1970; MSc, U. N.S.W., Sydney, 1981, PhD, 1988. Chartered geologist; chartered engr.; European engr. Geologist S. Australian Geol. Survey, Adelaide, 1970-74; sr. geologist Newbold Gen. Refractories

Ltd., Sydney, 1974-76; chief geologist Minerals Pty. Ltd., Sydney, 1976-79, Comml. Minerals Ltd. formerly Steetley Industries Ltd., Sydney, 1979-87; geologist, geotechnical engr. Wrexham, North Wales, 1987—. Contbr. over 40 articles to profl. jours. Fellow Geol. Soc., Instn. of Mining and Metallurgy, Australasian Inst. Mining and Metallurgy; mem. Brit. Geotechnical Soc., Geol. Soc. Glasgow, Am. Inst. Mining Engrs., Clay Minerals Soc. Presbyterian. Home: 39 Buckingham Rd, Wrexham Clwyd, Wales LL112RH Office: Shire Hall Ste 4359, Mold, Clwyd CH7 6NF, Wales

NICHOLAS, JAMES THOMAS, electronics engineer; b. Buckingham, Va., Feb. 25, 1956; s. Emmett Andersona nd Lelia Ethel (Coles) N.; m. Magdline Elizabeth Rollins, June 25, 1986; stepchildren: Stephanie, James. BS in Elec. Engring. Tech., Va. Poly. Inst. and State U., 1974-78; student in computer info. systems, New River C.C., Dublin, Va., 1975-79. Quality control engr. Poly-Sci. Litton, Blacksburg, Va., 1978-79; engr. Radford Army Ammunition Plant/Hercules, Inc., Radford, Va., 1979-84, maint. supr., 1984-93; pres. Nicholas Systems, Roanoke, Va., 1990—. Mem. Instrument Soc. Am. (sr.). Home and Office: 2320 Staunton Ave NW Roanoke VA 24017-3942

NICHOLAS, NICKIE LEE, industrial hygienist; b. Lake Charles, La., Jan. 19, 1938; d. Clyde Lee and Jessie Mae (Lyons) N.; B.S., U. Houston, 1960, M.S., 1966. Tchr. sci. Pasadena (Tex.) Ind. Sch. Dist., 1960-61; chemist FDA, Dallas, 1961-62, VA Hosp., Houston, 1962-66; chief biochemist Baylor U. Coll. Medicine, 1966-68; chemist NASA, Johnson Spacecraft Center, 1968-73; analytical chemist TVA, Muscle Shoals, Ala., 1973-75; indsl. hygienist, compliance officer OSHA, Dept. Labor, Houston, 1975-79, area dir., Tulsa, 1979-82, mgr., Austin, 1982—; mem. faculty VA Sch. Med. Tech., Houston, 1963-66. Recipient award for outstanding achievement German embassy, 1958, Suggestion award VA, 1963, Group Achievement award Skylab Med. Team, NASA, 1974, Personal Achievement award Dept. Labor Fed. Women's Program, 1984, Career Achievement award Federally Employed Women, Inc., 1988, Meritorious Performance award DOL-OSHA, 1990, Disting. Career Svc. award Dept. Labor, 1991, Sec.'s Exceptional Acievement award Dept. Labor, 1991, Cert. Appreciation, OSHA, 1991. Mem. Am. Chem. Soc. (dir. analytical group Southeastern Tex. and Brazosport sects. 1971, chmn. elect 1973), Am. Assn. Clin. Chemists, Am. Conf. Govtl. Indsl. Hygenists, Am. Ind. Hygiene Assn., Am. Soc. Safety Engrs., Am. Harp Soc., Fed. Exec. Assn. (pres. 1984-85), Kappa Epsilon. Home: 1305 Shannon Oaks Trl Austin TX 78746-7342 Office: 611 E 6th St Ste 303 Austin TX 78701-3786

NICHOLAS, RALPH WALLACE, anthropologist, educator; b. Dallas, Nov. 28, 1934; s. Ralph Wendell and Ruth Elizabeth (Oury) N.; m. Marta Ruth Weinstock, June 13, 1963. B.A., Wayne U., 1957; M.A., U. Chgo., 1958, Ph.D., 1962. Asst. prof. to prof. Mich. State U., East Lansing, 1964-71; prof. anthropology U. Chgo., 1971—, chmn. dept., 1981-82, dep. provost, 1982-87, dean of coll., 1987-92, dir. Ctr. Internat. Studies, 1984—, William Rainey Harper prof. of Anthropology and Social Scis., 1992—; cons. Ford Found., Dhaka, Bangladesh, 1973. Author: (with others) Kinship Bengali Culture, 1977; editor: Jour. Asian Studies, 1975-78. Trustee Am. Inst. Indian Studies, v.p., 1974-76; trustee Bangladesh Found. Ford Found. fgn. area tng. fellow, India, 1960-61; Sch. Oriental and African Studies research fellow, London, 1962-63; sr. Fulbright fellow, West Bengal, India, 1968-69. Fellow AAAS, Am. Anthrop. Assn., Royal Anthrop. Inst. (Eng.); mem. Assn. Asian Studies, India League of Am. Found. (trustee). Home: 5473 S Ellis Ave Chicago IL 60615-5058 Office: U Chgo 5828 S. University Ave. Chicago IL 60637-1513

NICHOLLS, RALPH WILLIAM, physicist, educator; b. Richmond, Surrey, Eng., May 3, 1926; s. William James and Evelyn Mabel (Jones) N.; m. Doris Margaret McEwen, June 28, 1952. B.Sc., Imperial Coll., U. London, 1945, Ph.D., 1951, D.Sc. in Spectroscopy, 1961. Sr. demonstrator in astrophysics Imperial Coll., U. London, 1945-48; instr. U. Western Ont. (Can.), London, 1948-50; lectr. U. Western Ont. (Can.), 1950-52, asst. prof. physics, 1952-56, assoc. prof., 1956-58, prof., 1958-63, sr. prof., 1963-65; prof. York U., Toronto, Ont., 1965—, Disting. Research prof. physics, 1983—; chmn. dept. physics York U., 1965-69; dir. Centre for Rsch. in Earth and Space Scis. (formerly Centre for Rsch. in Exptl. Space Scis.), 1965-92; dir. atmospheric physics lab. Inst. for Space and Terrestrial Sci., 1987—; vis. scientist Nat. Bur. Standards, 1959; vis. prof. Stanford U., 1964, 68, 73, 90. Author: (with B.H. Armstrong) Emission, Absorption and Transfer of Radiation in Heated Atmospheres, 1972; editor: Can. Jour. Physics, 1986-92. Walter Gordon rsch. fellow, York U., 1982-83. Fellow Royal Soc. Can., Optical Soc. Am., Am. Phys. Soc., Can. Aero. and Space Inst. Home: 9 Pinevale Rd, Thornhill, ON Canada L3T 1J5 Office: York U, 4700 Keele St, North York, ON Canada M3J 1P3

NICHOLLS, RICHARD AURELIUS, obstetrician, gynecologist; b. Norfolk, Va., Aug. 12, 1941; s. Richard Beddoe and Aurelia (Gill) N.; m. Geri Bowden, Feb. 24, 1986. BS in Biology, Stetson U., 1963; MD, Med. Coll. Va., 1967. Diplomate Am. Bd. Ob-Gyn. Intern, Charity Hosp., Tulane div., New Orleans, 1967-68, resident in ob-gyn, 1968-71; asst. prof. ob-gyn Tulane Med. Sch., New Orleans, 1973-74, clin. asst. prof., 1974-83; practice medicine specializing in ob-gyn, Pascagoula, Miss., 1974-89; pvt. practice medicine, Ocean Spring, Miss., 1989—; mem. staff Singing River Hosp., chmn. surg. and ob-gyn depts., 1979-80, chmn. Ob-Gyn Dept., 1984, mem. staff Ocean Springs Hosp., laser com., pharmacy com., and theraputics com., chmn. OB-Gyn dept., mem. exec. bd., 1990-91; sec., treas. staff Ocean Springs Hosp., 1991-92, exec. bd., 1991-92, chief of staff elect, 1992-93. Bd. dirs. Miss. Racing Assn. Maj. US. Army, 1971-73. Fellow Am. Coll. Ob-Gyn, ACS; mem. Miss. State Med. Soc., Singing River Med. Soc., Am. Fertility Soc., Am. Assn. Gynecol. Laparoscopists, Am. Med. Soc., So. Med. Soc., New Orleans Grad. Med. Assembly, New Orleans Ob-Gyn Soc., Gulf Coast Ob-Gyn Soc., Conrad Collins Ob-Gyn Soc., Am. Venereal Disease Soc., Am. Cancer Soc. (bd. dirs Jackson County Br.). Contbr. articles to med. jours.

NICHOLS, BUFORD LEE, JR., physiologist; b. Ft. Worth, Dec. 12, 1931; married; 3 children. BA, Baylor U., 1955, MS, 1958; MD, Yale U., 1960. Diplomate Am. Bd. Pediatrics, Am. Bd. Nutrition. Instr. pediatrics Baylor U. Coll. Medicine, Houston, 1956-57, instr. physiology and pediatrics, 1964-66, from asst. prof. to assoc. prof. pediatrics, 1966-67, instr. physiology, 1967-74, chief sect. nutrition and gastroenterology, dept. Pediatrics, 1970-78, assoc. prof. community medicine, 1975—, prof. physiology and pediatrics, 1977—, head sect. nutrition and physiology, 1979—; intern in pediatrics Yale-New Haven Med. Ctr., 1960-61, chief resident in pediatrics, 1963-64; resident in pediatrics Johns Hopkins Hosp., 1961-63; instr. pediatrics Yale U. Sch. Medicine, 1963-64; dir. USDA Children's Nutritional Rsch. Ctr., Houston, 1979—. Recipient award Bristol-Myers, 1984. Mem. Am. Acad. Pediatrics, Am. Soc. Clin. Nutrition, Am. Coll. Nutrition (v.p. 1975-76, pres. 1977-79). Achievements include research in environmental effects upon growth and development in the infant especially alterations in body composition and muscle physiology in malnutrition, chronic diarrhea and malnutrition. Office: Baylor Coll Medicine Childrens Nutrition Rsch Ctr 1100 Bates Ave Houston TX 77030-2600*

NICHOLS, C(LAUDE) ALAN, mechanical engineer; b. Tucson, Mar. 11, 1947; s. David Arthur and Dorothy May (Pritchett) N.; m. Gina Burton, June 1, 1969 (div. 1975); 1 child, Wendy Ann; m. Susan Frances Toth, Oct. 12, 1978; children: Alisha Janelle, Christiann Michelle, David Michael. BSME, U. Ariz., 1969; postgrad., Va. Poly. Inst. and State U., 1972. Registered profl. engr., Ariz., Calif., Nev.; cert. energy mgr. Project engr. Western Elec., Greensboro, N.C., 1969-73; plant engr. Ponderosa Paper, Flagstaff, Ariz., 1974-77; process engr. W.L. Gore, Flagstaff, Ariz., 1977-79; process engr./ptnr. Maxlight Phoenix, 1979-80; project engr. Tierney Mfg., Phoenix, 1980-82, Goetting Engring., Tucson, 1982-84, A.D.P., Tucson, 1984-85; sr. engr. Cella Barr, Tucson, 1985-89; prin. Burnside Nichols Engring., Tucson, 1989—; mem. adv. com. Tucson Metro. Energy Commn., 1992, Joint Tucson/Pima County Ad Hoc Com. for building energy code, 1992. Author: The Easy Does It Solar Water Still, 1985; author pamphlets: The Tracking Solar Cooking Oven, 1989, Reflections on a Solar Cooker, 1992. Recipient Clifford C. Sawyer Achievement award Am. Cons. Engrs. Assn., 1991, 93, Engring. Excellence award, 1992, Recognition for Exceptional Design Ariz. Gov.'s award of Spl. Recognition for Residence, 1986; projects recognized with 16 additional energy and engring. excellence awards including one nat. award for energy conservation, 1986-91. Mem. ASHRAE (chpt. past pres. 1986, bd. dirs.), Assn. Energy Engrs. (sr. mem., founder So. Ariz. chpt.), Am. Soc. Plumbing Engrs., Elec. League of Ariz., Ariz. Solar Energy Assn., Citizens for Solar. Achievements include research in setting of energy and construction standards for Tucson area school districts; deisgner the tracking solar cooking oven.

NICHOLS, EUGENE DOUGLAS, mathematics educator; b. Rovno, Poland, Feb. 6, 1923; came to U.S., 1946, naturalized, 1951; s. Alex and Anna (Radchuk) Nichiporuk; m. Alice Bissell, Mar. 31, 1951. BS, U. Chgo., 1949, postgrad., 1949-51; MEd, U. Ill., 1953, MA, 1954, PhD, 1956. Instr. math. Roberts Wesleyan Coll., North Chili, N.Y., 1950-51, U. Ill., 1951-56; assoc. prof. math. edn. Fla. State U., 1956-61, prof., head dept., 1961-73; dir. Project for Mathematical Devel. of Children, 1973-77; dir. math program NSF, 1958-61; dir. Math. Inst. Elem. Tchrs., 1961-70; pres. Nichols Schwartz Pub., 1992—; Chmn. U. Ill. Com. on Sch. Math., 1954-55; cons. editor math McGraw-Hill Book Co., summer 1956. Co-author: Modern Elementary Algebra, 1961, Introduction to Sets, 1962, Arithmetic of Directed Numbers, 1962, Introduction to Equations and Inequalities, 1963, Introduction to Coordinate Geometry, 1963, Introduction to Exponents, 1964, Understanding Arithmetic, 1965, Elementary Mathematics Patterns and Structure, 1966, Algebra, 1966, Modern Geometry, 1968, Modern Trigonometry, 1968, Modern Intermediate Algebra, 1969, Analytic Geometry, 1973, Holt Algebra 1, 1974, 78, 82, 86, 92, Holt Algebra 2, 1974, 78, 82, 86, 92, Holt Geometry, 1974, 78, 82, 86, Holt School Mathematics, 1974, 78, 81, Holt Mathematics, 1981, 85, Pre-Algebra Mathematics, 1980, 86, Holt Pre-Algebra, 1992, Elementary School Mathematics and How to Teach It, 1982, Geometry, 1991; author: Pre-Algebra Mathematics, 1970, College Mathematics for General Education, rev. edit, 1975, Introductory Algebra for College Students, 1971, Mathematics for the Elementary School Teacher, 1971, College Mathematics, 1975. Named Fla. State U. Disting. Prof., 1968-69; recipient Disting. Alumni award U. Ill. Coll. Edn., 1970. Mem. Am. Math. Soc., Math. Assn. Am., Sch. Sci. and Math. Assn., Nat. Coun. Tchrs. Math., Coun. Basic Edn., Pi Mu Epsilon, Phi Delta Kappa. Home: 3386 W Lakeshore Dr Tallahassee FL 32312-1305

NICHOLS, FREDERIC HONE, oceanographer; b. Boston, Nov. 28, 1937; s. Lorrel Brayton and Carolyn Merriam (Hone) N.; m. Kirstin Margaret Clark, June 27, 1971; children: Matthew Clark, John Hone, Andrew Larsen. AB, Hamilton Coll., 1960; MS, U. Wash., 1968, PhD, 1972. Teaching asst., rsch. asst. U. Wash., Seattle, 1965-72; oceanographer U.S. Geol. Survey, Menlo Park, Calif., 1971—, assoc. chief br. Pacific marine geology, 1979-82, chief br. regional rsch., water resources div., 1990—; vis. scientist U. Kiel, Fed. Republic Germany, 1975-76; governing bd. Tiburon Ctr. Environ. Studies, 1977-86, chmn., 1982-83; panelist Nat. Rsch. Coun.-Nat. Acad. Scis., 1987-89; mem. San Francisco Bay Conservation and Devel. Commn. Sci. and Tech. Adv. Com., 1986—; tech. adv. com. San Francisco Estuary Project, EPA, 1987—. Editor: Temporal Dynamics of an Estuary: San Francisco Bay, 1985; editorial bd. Estuaries jour., 1987—. Mem. Palo Alto (Calif.) Planning Commn., 1977-82, chmn., 1980-81. Lt. (j.g.) USNR, 1960-63. Fellow Calif. Acad. Scis.; mem. AAAS (exec. com. Pacific div. 1988-92, pres. 1989-90), Estuarine Rsch. Fedn. (chmn. organizing com. internat. conf. 1991, pres. 1993-95), Am. Soc. Limnology and Oceanography (sec.-treas. Pacific sect. 1978-82), Biol. Soc. Wash., Ecol. Soc. Am., Oceanography Soc., Pacific Estuarine Rsch. Soc., Western Soc. Naturalists, Sigma Xi. Office: US Geol Survey MS 472 345 Middlefield Rd Menlo Park CA 94025

NICHOLS, JAMES ROBBS, university dean; b. Jackson, Tenn., May 30, 1926; s. William Ed and Buelha (Robbs) N.; m. Johnnie Jones; 1 dau., Tina Jean Nichols Benson. BS, U. Tenn., 1949; MS, U. Minn., 1955, PhD, 1957. Former mem. faculty Pa. State U., U. Tenn.; mem. faculty Va. Poly. Inst. and State U., Blacksburg, 1964-71, 73—; head dairy sci. dept. Va. Poly. Inst. and State U., 1964-69; assoc. dean Va. Poly. Inst. and State U. (Coll. Agr.), 1969-71, 73-75; dean Coll. Agr. & Life Scis., dir. Va. Agr. Exptl. Sta., 1975—; exec. v.p., gen. mgr. Select Sires, Inc., Columbus, Ohio, 1971-73. Served with USAAF, World War II. Named Man of Yr. in Agr. in Va. Progressive Farmer mag., 1975; hon. state farmer Tenn. Mem. AAAS, Am. Dairy Sci. Assn., Sigma Xi, Phi Kappa Phi, Alpha Zeta. Methodist. Clubs: Rotary. Office: Va Poly Inst and State U Coll Agr and Life Scis 104G Hutchenson Hall Blacksburg VA 24061-2903

NICHOLS, RONALD LEE, surgeon, educator; b. Chgo., June 25, 1941; s. Peter Raymond and Jane Eleanor (Johnson) N.; m. Elsa Elaine Johnson, Dec. 4, 1964; children: Kimberly Jane, Matthew Bennett. M.D., U. Ill., 1966, M.S., 1970. Diplomate: Am. Bd. Surgery (assoc. cert. examiner, New Orleans, 1991), Nat. Bd. Med. Examiners. Intern U. Ill. Hosp., Chgo., 1966-67, resident in surgery, 1967-72, instr. surgery, 1970-72, asst. prof. surgery, 1972-74; assoc. prof. surgery U. Health Scis. Chgo. Med. Sch., 1975-77, dir. surg. edn., 1975-77; William Henderson prof. surgery Tulane U. Sch. Medicine, New Orleans, 1977—, prof. microbiology and immunology, 1979—, vice chmn. dept. surgery, 1982-91; staff surgeon Tulane U. Med. Ctr. Hosp., New Orleans; sr. vis. surgeon Charity Hosp. La., New Orleans, 1990—; cons. surgeon VA Hosp., Alexandria, La., 1978-93, Huey P. Long Hosp., Pineville, La., 1978—, Lallie Kemp Charity Hosp., Independence, La., 1977-85, Touro Infirmary, New Orleans, Monmouth Med. Ctr., Long Branch, N.J.; mem. VA Coop. Study Rev. Bd., 1978-81, VA Merit Rev. Bd. in Surgery, 1979-82; mem. sci. program com. 3d Internat. Conf. Nosocomial Infections, Ctr. Disease Control; bd. dirs. Nat. Found. Infectious Diseases, 1989—; hon. fellow faculty Kasr El Aini Cairo U. Sch. Medicine, 1989; mem. adv. com. on infection control Ctrs. for Disease Control, 1991—; disting. quest, vis. prof. Royal Coll. Surgeons Thailand 17th Ann. Clin. Congress, 1992; mem. infectious deseases adv. bd. Roche Labs., 1988—, Abbott Labs., 1990-92, Kimberly Clark Corp., 1990—, Smith Klein Beecham Labs., 1990—, Fugispura Parms., chmn. 1990—. Author: (with Gorbach, Bartlett and Nichols) Manual of Surgical Infection, 1984; (with Nichols, Hyslop Jr. and Bartlett) Decision Making in Surgical Sepsis, 1991; mem. editorial bd. Current Surgery, 1977—, Hosp. Physician, 1980—, Infection Control, 1980-86, Guidelines to Antibiotic Therapy, 1976-81, Am. Jour. Infection Control, 1981—, Internat. Medicine, 1983—, Confronting Infection, 1983-86, Current Concepts in Clin. Surgery, 1984—, Fact Line, 1984—, Host/Pathogen News, 1984—, Infectious Diseases in Clin. Practice, 1991—, surg. sect. editor, 1992—, Surgical Infections: Index and Reviews; mem. adv. bd. Physicians News Network, 1991—. Elected faculty sponsor graduating class Tulane Med. Sch., 1979-80, 83, 85, 87, 88, 91-92. Served to major USAR, 1972-75. Recipient House Staff teaching award U. Ill. Coll. Medicine, 1973, Rsch. award Bd. Trustees U. Health Scis.-Chgo. Med. Sch., 1977, Owl Club Teaching award, 1980-86, 90; named Clin. Prof. of Yr. U. Health Scis., Chgo. Med. Sch., 1977, Clin. Prof. of Yr., Tulane U. Sch. Medicine, 1979; Douglas Stubbs Lectr. award Surg. Sect. Nat. Med. Assn., 1987, Prix d'Elegance award Men of Fashion, New Orleans, 1993, Disting. Guest/vis. prof. Royal Coll. Surgeon Thailand, 14th Ann. Clin. Congress, 1989. Fellow Infectious Disease Soc. Am. (mem. FDA subcom. to develop guidelines in surg. prophylaxis 1989-93, co-recipient Joseph Susman Meml. award 1990), Am. Acad. Microbiology, Internat. Soc. Univ. Colon and Rectal Surgeons, ACS (chmn. operating room environment com. 1981-83, sr. mem. 1984-87, internat. relations com. 1987—); mem. AMA, Nat. Found. for Infectious Diseases (bd. dirs. 1988—), Joint Commn. on Accreditation of Health Care Orgn. (Infection Control adv. group, 1988—, sci. program com. 3d internat. conf. nosocomial infections CDC/Nat. Found. Infectious Diseases 1990, FDA Subcom. to Develop Guidelines in Surg. Prophylaxis), 5th Nat. Forum on AIDS (sci. program com.), Assn. Practitioners in Infection Control (physician adv. coun. 1991—), Internat. Soc. Anaerobic Bacteria, So. Med. Assn. (vice chmn. sect. surgery 1980-81, chmn. 1982-83), Assn. Acad. Surgery, N.Y. Acad. Sci., Warren H. Cole Soc. (pres.-elect 1988, pres. 1989-90), Assn. VA Surgeons, Soc. Surgery Alimentary Tract, Inst. Medicine Chgo., Midwest Surg. Assn., Cen. Surg. Assn., Ill. Surg. Soc., European Soc. Surg. Rsch., Collegiuae Digestivae, Chgo. Surg. Soc. (hon.), New Orleans Surg. Soc. (bd. dirs. 1983-87), Soc. Univ. Surgeons, Surg. Soc. La., Southeastern Surg. Soc., Phoenix Surg. Soc. (hon.), Hellenic Surg. Soc. (hon.), Cen. N.Y. Surg. Soc. (hon.), Tulane Surg. Soc., Alton Ochsner Surg. Soc., Am. Soc. Microbiology, Soc. Internat. de Chirugie, Surg. Infection Soc. (sci. study com. 1982-83, fellowship com. 1985-87, ad hoc sci. liaison com. 1986-89, program com. 1986-87, chmn. ad hoc com. rels. with industry 1990-93), Soc. for Intestinal Microbial Ecology and Disease, Soc. Critical Care Medicine, Am. Surg. Assn., Kansas City Surg. Soc., Bay Surg. Soc. (hon.), Cuban Surg. Soc. (hon.), Panhellenic Surg. Soc. (hon.), Sigma Xi, Alpha Omega Alpha. Episcopalian. Home: 1521 7th St New Orleans LA 70115-3322 Office: 1430 Tulane Ave New Orleans LA 70112-2699

NICHOLS, STEVEN PARKS, mechanical engineer, academic administrator; b. Cody, Wyo., July 1, 1950; s. Rufus Parks Nichols and Gwen Sena (Frank) Keyes; m. Mary Ruth Barrow, Aug. 5, 1990; 1 child, Nicholas Barrow Nichols. PhD, U. Tex., Austin, 1975, JD, 1983. Assoc. dir. Tex. Space Grant Consortium, Austin, 1989-91; dep. dir. ctr. for energy studies U. Tex., Austin, 1988-91, dir. ctr. for energy studies, 1991—. Patentee (with others) railgun igniter, inert burner, other patents pending. Mem. NSPE, ASME, ABA, Am. Soc. Engring. Edn., Nat. Inst. Engring. Ethics (bd. govs. 1987—). Home: 105 Laurel Ln Austin TX 78705 Office: U Tex Ctr for Energy Studies 10100 Burnet Rd Austin TX 78758

NICHOLS, THOMAS ROBERT, biostatistician, consultant; b. Green Bay, Wis., 1948. BS, U. Wis., Green Bay, 1975; MS, Med. Coll. Wis., 1986. Sr. researcher Arthritis Rsch. Lab., Med. Coll. Wis., Milw., 1978-83; biostatis. cons. Cleve. Clinic Found., 1986-88; sr. statis. cons. Trilogy Cons. Corp., Waukegan, Ill., 1989-91; sr. clin. rsch. statistician Hollister, Inc., Libertyville, Ill., 1991—. Contbr. articles to Neurology, Investigative Radiology, Jour. of AMA, Annals of Allergy, Am. Jour. Preventive Medicine, others. With U.S. Army, 1968-71. Rsch. scholar Med. Coll. Wis., 1984. Mem. AAAS, Am. Statis. Assn., Assocs. in Clin. Pharmacology, N.Y. Acad. Scis., Soc. for Clin. Trials. Office: Hollister Inc 2000 Hollister Dr Libertyville IL 60048-3746

NICHOLSON, JAMES ALLEN, orthodontist, inventor; b. Richton, Miss., June 30, 1948; s. Edward A. and Glady S. (Guthrie) N.; m. Patricia Tatum, 1969; children: Angela M., Michael B. BS, U. So. Miss., 1973; DDS, La. State U., 1981; MS, U. Tenn., 1983. Registered radiologic and nuclear medicine technologist. Orthodontist in pvt. practice Hattiesburg, Miss., 1983—. Patentee in field of orthodontics. So. Soc. Orthodontists grantee, 1982. Mem. ADA, Am. Assn. Orthodontists, So. Soc. Orthodontists, Miss. Dental Assn., South Miss. Dental Soc., 3d Dist. Dental Soc., Miss. Assn. Orthodontists. Baptist. Office: 120 S 28th Ave Hattiesburg MS 39401-7151

NICHOLSON, WILLIAM JOSEPH, forest products company executive; b. Tacoma, Aug. 24, 1938; s. Ferris Frank and Athyleen Myrtle (Fesenmaier) N.; m. Carland Elaine Crook, Oct. 10, 1964; children: Courtney, Brian, Kay, Benjamin. SB in ChemE, MIT, 1960, SM in ChemE Practice, 1961; PhD in ChemE, Cornell U., 1965; MBA, Pacific Luth. U., 1969. Registered profl. chem. engr., Wash. Sr. devel. engr. Hooker Chem. Co., Tacoma, 1964-69, Battelle N.W., Richland, Wash., 1969-70; planning assoc. Potlatch Corp., San Francisco, 1970-75, mgr. corp. energy service, 1976—; chmn. electricity com. Am. Paper Inst., 1977—, mem. solid waste task force, 1988-91, mem. air quality com., 1989—; mem. adv. bd. Univ. Calif. Forest Products Lab., 1992—, chmn., 1993—. Mem. Am. Chem. Soc., Am. Inst. Chem. Engrs. (assoc.), Tech. Assn. Pulp and Paper Industry, AAAS, Sigma Xi. Democrat. Clubs: Commonwealth (San Francisco), Cornell (N.Y.). Avocation: industrial history. Home: PO Box 1114 Ross CA 94957-1114 Office: Potlatch Corp 244 California St Ste 610 San Francisco CA 94111

NICHOLSON, WILLIAM MAC, naval architect, marine engineer, consultant; b. Napa, Calif., June 15, 1918; s. William John and Hazel (McIlmoil) N.; m. Lynda Bishop, Feb. 10, 1947 (div. Aug. 1959); children: Lynda Joanne, Samuel Bishop; m. Leslie Marie Earle-Thomas, Apr. 1, 1964; children: Richard Thomas, Glenn Thomas, Ronald Thomas. BS, U.S. Naval Acad., 1941; MS, MIT, 1948. Commd. ensign USN, 1941, advanced through grades to capt., 1960; ret., 1971; design supt. Boston Shipyard, 1951-52; project officer minesweeping, hydrofoils Bur. of Ships, Washington, 1955-59; comptr. Puget Sound Naval Shipyard, Bremerton, Wash., 1959-62; prof. naval constrn. MIT, Cambridge, 1962-65; dir. ship design div. Bur. of Ships, Washington, 1965-66; program mgr. deep submergence Chief of Naval Material, Washington, 1966-71; assoc. dir. Nat. Ocean Survey, NOAA, Washington, 1971-81; retired, 1981. chmn. U.S.-Japan panel on marine facilities under U.S.-Japan Natural Resource Agreement., 1972-81; vice-chmn. marine bd. Acad. Engring., NAS, Washington, 1984-85; pres., 1983, chmn., bd. dirs., 1984, Sea-Space Symposium. Decorated Legion of Merit. Mem. Soc. Naval Architects and Marine Engrs. (sect. chmn. N.E. 1964-65), Soc. Naval Engrs. (coun. 1966-68), Marine Tech. Soc. (sect. chmn. 1974-75), Tau Beta Pi, Sigma Xi. Republican. Home: 4000 Towerside Ter # 2307 Miami FL 33138

NICKEL, HORST WILHELM, psychology educator; b. Spangenberg, Hessen, Germany, Sept. 30, 1929; s. Konrad and Marie Nickel; m. Irmtraud Ulm, Sept. 26, 1959; children: Wolfram, Cordula. Tchr. exam., Tchrs. Coll., Weilburg/Lahn, Fed. Republic Germany, 1953; diploma in psychology, U. Marburg/Lahn, Fed. Republic Germany, 1961; PhD, U. Erlangen-Nuernberg, Fed. Republic Germany, 1965. Tchr. primary sch., Hessen, 1953-57; lectr. Tchrs. Coll., Weilburg, Lahn, 1957-62, U. Bayreuth, Fed. Republic Germany, 1962-65; sr. lectr. U. Hamburg, Fed. Republic Germany, 1965-67; prof. Tchrs. Coll., Flensburg, Fed. Republic Germany, 1967-69, Bonn, Fed. Republic Germany, 1969-72; prof. psychology U. Düsseldorf, Fed. Republic Germany, 1972—, chmn. dept.; cons. prof. dept. psychology U. Ga., Athens, 1987-91. Author 25 books including Entwicklunspsychologie des Kindes- und Jugendalters, vol. 2, 3d edit., 1981, vol. 1, 4th edit., 1982, Psychologie des Lehrerverhaltens, 2d edit., 1978, Begriffsbildung im Kindesalter, 1984, Sozialisation im Vorschulalter, 1985, Psychologie der Entwicklung und Erziehung, 1993; co-author: Psychologie in der Erziehungswissenschaft 4 Vol., 1976-78, Erzieher-und Elternverhalten im Vorschulbereich, 1980, Sozialverhalten von Vorschulkindern, 1980, Modelle und Fallstudien der Schul- und Erziehungsberatung, 1982, Oekopsychologie der Entwicklung im Freuhen Kindesalter, 1987, Vom Kleinkind zum Schulkind, 4th edit., 1991, Begabung und Hochbegabung, 1992; editor Psychologie in Erziehung und Unterricht, also over 150 articles. Mem. German Soc. for Psychology, Internat. Soc. Pre- and Perinatal Psychology and Medicine, Internat. Soc. for Study Behavioural Devel. Home: Berliner Strasse 25, D-55340 Meckenheim Germany Office: Heinrich-Heine U, Universtiätsstrasse 1, D-40225 Düsseldorf Germany

NICKERSON, RAYMOND STEPHEN, psychologist; b. Bangor, Maine, Nov. 5, 1931; s. Raymond W. and Velma Katherine (Rand) N.; m. Doris Van Sant, Aug. 1, 1953; children: Daniel (dec.), Nathan, Betsy, Sheri. MA in Exptl. Psychology, U. Maine, 1959; PhD in Exptl. Psychology, Tufts U., 1965. Sr. scientist Bolt Beranek & Newman, Cambridge, Mass., 1966-69; div. v.p. Bolt Beranek & Newman, Cambridge, 1970-75, v.p., 1975-79, sr. v.p., 1979-91; ret., 1991; chmn. com. on human factors Nat. Rsch. Coun., 1991—. Author: Using Computers: Human Factors in Information Systems, 1986, Reflections on Reasoning, 1986, Looking Ahead: Human Factors Challenges in a Changing World, 1992; co-author: The Teaching of Thinking, 1985. Bd. dirs. Mass. Assn. for Retarded Citizens, Waltham, Mass., 1974-85, pres., 1976-79; mem. Mass. Devel. Disabilities Coun., Boston, 1982-90. With U.S. Army, 1955-57. Fellow APA (Franklin V. Taylor award Engring. Psychology div. 1991), AAAS, Am. Psychol. Soc., Human Factors Soc., Soc. Exptl. Psychologists. Baptist. Home: 5 Gleason Rd Bedford MA 01730

NICKLAS, JOHN G., systems analyst; b. Atlantic City, Mar. 9, 1944; s. John Henry and Caroline (Gibson) N.; m. Kay Susanne McDowell, June 15, 1968; children: Andrew, Alexander, Melissa. BA, Syracuse U., 1966; MA, U. No. Colo., 1978. Comd. 2d lt. USAF, 1967, advanced through grades to lt. col., 1987; instr. 90 SMU, Few, Wyo., 1971-76; chief test divsn. 1 STRAD, Vandouleus AFB, Calif., 1976-81; chief MX instrumentation divsn. 6595 MTG, Vandouleus AFB, Calif., 1981-83; divsn. chief Joint Cruise Missile Office, Crystal City, Va., 1983-87; systems analyst RCI, Vienna, Va., 1987—. Founder acadmeic booster club Thomas Jefferson High Sch., 1986, St. Peter's Episcopal Ch., Fairfax Sation, VA., 1988; trustee Burke Centre Housing Corp. Episcopalian. Home: 5904 Vernons Oak Ct Burke VA 22015-2514 Office: RCI 1960 Gallowd Rd Vienna VA 22182-3824

NICKLE, DENNIS EDWIN, electronics engineer, church deacon; b. Sioux City, Iowa, Jan. 30, 1936; s. Harold Bateman and Helen Cecilia (Killackey) N. BS in Math., Fla. State U., 1961. Reliability mathematician Pratt & Whitney Aircraft Co., W. Palm Beach, Fla., 1961-63; br. supr. Melpar Inc., Falls Church, Va., 1963-66; prin. mem. tech. staff Xerox Data Systems, Rockville, Md., 1966-70; sr. tech. officer WHO, Washington, 1970-76; software quality assurance mgr. Melpar div. E-Systems, Inc., Falls Church, 1976—; ordained deacon Roman Catholic Ch., 1979. Chief judge for computers Fairfax County Regional Sci. Fair, 1964-88; mem. Am. Security Council; scoutmaster, commr. Boy Scouts Am., 1957-92; youth custodian Fairfax County Juvenile Ct., 1973-87; chaplain No. Va. Regional Juvenile Detention Home, 1978-88; moderator Nocturnal Adoration Soc.; parochial St. Michael's Ch., Annandale, Va., 1979-89, Christ the Redeemer, Sterling, Va., 1990—. Served with U.S. Army, 1958-60. Recipient Eagle award, Silver award, Silver Beaver award, other awards Boy Scouts Am.; Ad Altare Dei, St. George Emblem, Diocese of Richmond. Mem. Assn. Computing Machinery, Computer Soc., Am. Soc. For Quality Control, CODSIA (mem., chmn working groups), ORLANDO II (Govt./industry working group), Old Crows Assn., Rolm Mil-Spec Computer Users Group (internat. pres.), San Antonio I (select industry coordination group), Nat. Security Indsl. Assn. (convention com. 1985—, software quality assurance subcom., regional membership chmn. 1981-89, nat. exec. vice-chmn. 1989—), Am. Security Coun., IEEE (sr., mem. standards working group in computers 1983—), Defence Software Devel. Standards (chmn. adv. bd. 1991—), Soc. Software Quality, Hewlett Packard Users Group, Smithsonian Assn., Internat. Platform Assn., NRA (endowment), Nat. Eagle Scout Assn. (life), Alpha Phi Omega (life), Sigma Phi Epsilon. Club: KC (4 deg.). Author: Stress in Adolescents, 1986; co-author: Handbook for Handling Non-Productive Stress in Adolescence, Standard For Software Life Cycle Processes, IMPEESA Junior Leader Training Guide, Standard for Software Quality Assurance, 1984-91; contbr. to profl. jours. Office: 7700 Arlington Blvd Falls Church VA 22042-2902

NICOL, MARJORIE CARMICHAEL, research psychologist; b. Orange, N.J., Jan. 6, 1929; d. Norman Carmichael and Ethel Sarah (Siviter) N. BA, Upsala Coll., MS, 1978; MPh, PhD, CUNY, 1988. Mgr. advt. prodn. RCA, Harrison, N.J., 1950-58; advt. mgr., writer NPS Advt., East Orange, N.J., 1960-67; pres. measurement and eval., chief exec. officer, psychol. evaluator Nicol Evaluation System, Millburn, N.J., 1967—; chief exec. officer., dir. Rafiki, Essex County, N.J., 1965—. Author: Nicol Index, Nicol Evaluation System, 1991. Officer Montclair Rehab. Orgn., 1981—; founder, patron Met. Opera at Lincoln Ctr. Republican. Presbyterian. Home: 85 Linden St Millburn NJ 07041-2160 Office: PO Box 111 Millburn NJ 07041-0111

NICOLADIS, MICHAEL FRANK, engineering company executive; b. New Orleans, Aug. 15, 1960; s. Frank and Pagona (Gemelos) N. B Engring. magna cum laude, Vanderbilt U., 1982; MBA (Fuqua scholar, Conoco scholar), Duke U., 1984. Assoc. N-Y Assocs., Inc., Metairie, La., 1984-85, v.p., 1985—. Mem. St. Katherine's Greek Orthodox Ch., West Palm Beach. Mem. ASCE, Soc. Am. Mil. Engrs., Order of Engrs., Tau Beta Pi, Chi Epsilon. Avocations: boating, tennis, reading, travel. Office: N Y Assocs Inc 2750 Lake Villa Dr Metairie LA 70002

NICOLAOU, KYRIACOS COSTA, chemistry educator; b. Karavas, Kyrenia, Cyprus, June 5, 1946; came to U.S., 1972; s. Costa and Helen (Yettimi) N.; m. Georgette Karayianni, July 15, 1973; children; Colette, Alexis, Christopher, Paul. BSc, Bedford Coll., London, 1969; PhD, U. Coll., London, 1972. Rsch. assoc. Columbia U., N.Y.C., 1972-73, Harvard U., Cambridge, Mass., 1973-76; from asst. prof. to Rhodes-Thompson prof. chemistry U. Pa., Phila., 1976-89; Darlene Shiley prof. chemistry, chmn. dept. Rsch. Inst. Scripps Clinic, La Jolla, Calif., 1989—; prof. chemistry U. Calif., La Jolla, 1989—; vis. prof. U. Paris, 1986; mem. exec. com. Biann. Cyprus Conf. on Drug Design; mem. med. study sect. B, NIH, 1988-90. Author: (with N. A. Petasis) Selenium in Natural Products Synthesis, 1984; co-editor: Synthesis, Germany, 1984-90; editorial bd. Prostaglandins, Leukotrienes and Medicine, 1978-88, Synthesis, 1990—, Accounts of Chem. Rsch., 1992—, Carbohydrate Letters, 1993—; mem. bd. consulting editors Tetrahedron Publs., 1992—; mem. adv. bd. Contemporary Organic Synthesis, 1993—; mem. regional adv. bd. J. C. S. Chem. Comm., 1989—, J. C. S. Perkin I, 1991—. ; contbr. numerous articles to profl. jours.; patentee in field. Recipient Japan Soc. for Promotion Sci. award, 1987-88, U.S. Sr. Scientist award Alexander von Humboldt Found., 1987-88, Alan R. Day award Phila. Organic Chemisyts Club, 1993; fellow A. P. Sloan Found., 1979-83, J. S. Guggenhiem Found., 1984; Camille and Henry Dreyfus scholar, 1980-84, Arthur C. Cope scholar, 1987. Fellow N.Y. Acad. Scis.; mem. AAAS, Am. Chem. Soc. (Creative Work in Synthetic Organic Chemistry award 1993), Chem. Soc. London, German Chem. Soc., Japanese Chem. Soc. Office: Rsch Inst Scripps Clinic 10666 N Torrey Pines Rd La Jolla CA 92037-1027

NICOLAY, JEAN HONORÉ, cardiologist; b. Nice, France, July 24, 1920; s. Gabriel Claude and Rose (Baossa) N.; m. Evelyne Lecointe, Oct. 21, 1961; children: Laetitia, Jean-Gabriel. P.C.B., Marseille Faculte des Scis., France, 1940; MD, Faculte de Medecine, Marseille, 1945. Intern Hopitaux de Marseille, France, 1945-48, cardiologist, ex-chef clinique, cardiologique et experimentale, 1960-62; attache de cardiologie Hopitaux Pasteur, Nice, France, 1962-90; pvt. practice Nice, 1950-90; cardiologist neurol. dept. Pasteur Hosp. Contbr. articles to profl. jours. Lt. French Army, 1945-46. Mem. Medical and Surgical Soc. (pres. 1985-86), Rotary. Roman Catholic. Avocations: swimming, golf. Home: 54 Boulevard Victor Hugo, Nice 06, France 06000

NICOLIS, GREGOIRE, science educator; b. Athens, Greece, Sept. 11, 1939; s. Stamatios and Catherine (Siganou) N.; m. Catherine Rouvas, Aug. 21, 1966; children: Helen, Stamatios. BEE, Nat. Tech. U., Athens, 1962; PhD in Physics, U. Brussels, 1965. Research asst. U. Brussels, 1964-66, asst. prof., 1968-72, assoc. prof., 1973-86, prof., 1986—; postdoctoral fellow, research assoc. U. Chgo., 1966-67; vis. prof. U. Tex., Austin, 1972, U. Oreg., 1976, U. Paris, 1975-77, 86, 91, U. Geneva, 1979. Author: Self-organization in Nonequilibrium Systems, 1977, Exploring Complexity, 1989; co-editor: Membranes, Dissipative Structures and Evolution, 1975, Order and Fluctuations in Equilibrium and Nonequilibrium Statistical Mechanics, 1981, Aspects of Chemical Evolution, 1984, Chemical Instabilities, 1984, Irreversible Phenomena and Dynamical Systems Analysis in Geosciences, 1987, From Chemical to Biological Organization, 1988, Spatial Inhomogeneities and Transient Behaviour in Chemical Kinetics, 1990; assoc. editor: Bull. Math. Biology, 1975-86, Biophys. Chemistry, 1975-91, Jour. Nonequilibrium Thermodynamics, 1978, Advances in Chem. Physics, 1978, Jour. Statis. Physics, 1982-91, Dynamics and Stability of Systems, 1986, Bifurcation and Chaos, 1990, Chaos, Solitons and Fractals, 1991. Mem. Am. Phys. Soc., European Phys. Soc., Belgian Acad. Sci. (fgn. assoc.), European Acad. Sci., Academia Europaea, Acad. Athens (corr.). Home: 26 ave de l'Uruguay, 1050 Brussels Belgium Office: ULB CP, 231 Blvd du Triomphe, 1050 Brussels Belgium

NICOSON, STEVEN WAYNE, geotechnical engineer; b. Aurora, Ill., Oct. 2, 1959; s. Jack O. and Dorothy J. Nicoson; m. Anita L. McCarthy, July 30, 1983; children: Ashley M. and Christie J. (twins). BS in Civil Engring., U. Ill., 1981, MS in Geotech. Engring., 1982. Engr. Fed. Hwy. Adminstrn., Concord N.H. and Albany, N.Y., 1987-89, GZA GeoEnviron. Inc., Newton Upper Falls, Mass., 1989—. Mem. ASCE (assoc.), Boston Soc. Civil Engrs., Tau Beta Pi, Chi Epsilon. Home: 31 Whitney St Milford MA 01757

NIEBUHR, CHRISTOPHER, chemical engineer; b. Des Plaines, Ill., Nov. 8, 1965; s. Kenneth and Vivian (Wilson) N. BS in Chem. Engring., Iowa State U., 1987. Process engr. Monsanto Chem. Co., Muscatine, Iowa, 1988-89; maintenance engr. Monsanto Chem. Co., Muscatine, 1989-90; systems engr. Nalco Chem. Co., Naperville, Ill., 1990-92; with chem. mktg. dept. Pulp & Paper Industry at Nalco, Naperville, 1993—. Achievements include flow meter for pulsing flows (patent pending). Office: One Nalco Ctr Naperville IL 60563

NIEBUR, ERNST DIETRICH, computational neuroscientist; b. Lipperode, West Germany, Apr. 7, 1955; came to U.S., 1989; s. Ernst and Helene (Brand) N.; m. Dagmar Kryn, July 4, 1986. MS, U. Dortmund, Germany, 1982; PhD, U. Lausanne, Switzerland, 1988. Asst. diplomé U. Lausanne, 1983-88, premier asst., 1988-89; rsch. fellow Calif. Inst. Tech., Pasadena, 1989-92, sr. rsch. fellow, 1992—. With German Navy, 1974-76. Recipient Seymour Cray award Cray Rsch., Switzerland, 1989. Mem. AAAS, Soc. Neurosci., Internat. Neural Network Soc. Office: Calif Inst Tech Mail Stop 216-76 Pasadena CA 91125

NIEDERDRENK, KLAUS, mathematician, educational administrator, educator; b. Velbert, Germany, Sept. 8, 1950; s. Horst and Doris (Broich) N.; m. Jutta Kaczor, May 31, 1974; children: Maren, Lisa, Laura. Diploma math., RWTH, Aachen, Germany, 1977, D of natural sciences, 1980. High sch. asst. RWTH, Aachen, 1977-80, scientist, 1980-86; scientist Berufsbildungsstätte Westmünsterland, Ahaus, Germany, 1986-88; acad. dir. Tech. Acad., Ahaus, 1989-93; prof. Fachhochschule Münster, Steinfurt, Germany, 1993—; tchr. Fachhochschule Aachen, 1981-86, Fachhochschule Münster, 1992-93. Author: Fourier-and Walsh-Transform, 1982, 84; co-author: Functions of One Variable, 1987, Ordinary Differential Equations, 1987; contbr. articles to profl. jours. Mem. Greenpeace. Avocations: music, sports. Home: Kaland 28, D-48683 Ahaus Germany Office: Fachhochschule Münster, Stegerwaldstr 39, D-48565 Steinfurt Germany

NIEDERREITER, HARALD GUENTHER, mathematician, researcher; b. Vienna, June 7, 1944; s. Simon and Erna (Emig) N.; m. Gerlinde Hollweger, Aug. 30, 1969. PhD, U. Vienna, 1969. Asst. prof. So. Ill. U., Carbondale, 1969-72, assoc. prof., 1972-73; mem. Inst. Advanced Study, Princeton, N.J., 1973-75; vis. prof. UCLA, 1975-76; prof. U. Ill., Urbana, 1976-78, U. W.I., Kingston, Jamaica, 1978-81; researcher Austrian Acad. Scis., Vienna, 1981-89, dir. Inst. Info. Processing, 1989—. Author: Uniform Distribution of Sequences, 1974, Russian transl., 1985, Finite Fields, 1983, Russian transl. 1988, Introduction to Finite Fields and Their Applications, 1986, Random Number Generation and Quasi-Monte Carlo Methods, 1992; contbr. over 180 rsch. articles to math. jours.; assoc. editor Mathematics of Computation, 1988—, ACM Trans. Modeling and Computer Simulation, 1990—; mem. editorial bd. Caribbean J. Math., 1982—, Applicable Algebra, 1990—, Stochastic Optimization and Design, 1991—, J. Ramanujan Math. Soc., 1991—, Acta Arithmetica, 1992—, Monatshefte Math., 1993—, Finite Fields and Their Applications, 1993—. Named hon. prof. U. Vienna, 1986. Mem. Am. Math. Soc., Austrian Math. Soc., Austrian Acad. Scis. (elected corr. 1993), Internat. Assn. Cryptologic Rsch., Austrian Computer Soc., Gesellschaft für Informatik. Home: Sieveringer Str 41, A1190 Vienna Austria Office: Austrian Acad Scis, Sonnenfelsgasse 19, A1010 Vienna Austria

NIELSEN, FORREST HAROLD, research nutritionist; b. Junction City, Wis., Oct. 26, 1941; s. George Adolph and Sylvia Viola (Blood) N.; m. Emily Joanne Currie, June 13, 1964; children: Forrest Erik, Kistin Emily. BS, U. Wis., 1963, MS, 1966, PhD, 1967. NIH grad. fellow, dept. biochemistry U. Wis., Madison, 1963-67; rsch. chemist, Human Nutrition Rsch. Inst. USDA, Beltsville, Md., 1969-70; rsch. chemist Human Nutrition Rsch. Ctr., USDA, Grand Forks, N.D., 1970-86, ctr. dir. and rsch. nutritionist, 1986—; rsch. assoc. dept. biochemistry, U. N.D., Grand Forks, 1971—, speaker in field. Assoc. editor Magnesium and Trace Elements Jour., 1990-93; mem. editorial bd. Biol. Trace Element Rsch. Jour., 1979—, Jour. Nutrition, 1984-88; contbr. articles to profl. jours. Capt. U.S. Army, 1967-69. Recipient Klaus Schwarz Commemorative medal and award Internat. Assn. of Bioinorganic Scientists. Mem. Internat. Soc. Trace Element Rsch. in Humans (gov. bd. 1989—, pres. 1992—), Internat. Assn. Bioinorganic Scis., Soc. for Exptl. Biology and Medicine, Am. Inst. Nutrition, Am. Soc. Magnesium Rsch., N.D. Acad. Sci. (pres. 1988-89), Sigma Xi (pres. U. N.D. chpt. 1976-77). Lutheran. Achievements include patent for use of Boron Supplements to Increase in ViVo Production of Hydroxylated Steroids; discovery of the nutritional essentiality of the trace elements boron and nickel. Office: USDA ARS GFHNRC PO Box 9034 Grand Forks ND 58202-9034

NIELSEN, JAKOB, user interface engineer; b. Copenhagen, Oct. 5, 1957; came to U.S., 1990; s. Gerhard and Helle (Hopfner) N.; m. Hannah Kain, Feb. 18, 1984. MS in Computer Sci., Aarhus (Denmark) U., 1983; PhD in Computer Sci., T.U. of Denmark, 1988. Rsch. fellow Aarhus U., 1983-84; vis. scientist IBM User Interface Inst., Yorktown Heights, N.Y., 1985; adj. asst. prof. U. Denmark, Lyngby, 1986-90; mem. rsch. staff Bell Comm. Rsch., Morristown, N.J., 1990—. Author: Hypertext and Hypermedia, 1990, Usability Engineering, 1993; editor: Coordinating User Interfaces for Consistency, 1989, Designing User Interfaces for International Use, 1990, Usability Inspection Methods, 1993; editorial bd. Behavior and Info. Tech., 1989—, Hypermedia Jour., 1989—, Interacting with Computers, 1989—, Internat. Jour. Human-Computer Interaction, 1989—, Internat. Jour. Man-Machine Studies, 1991—; contbr. 45 articles to profl. jours. Mem. Assn. for Computing Machinery (spl. interest group on computer human interaction, papers co-chair internat. conf., 1993). Achievements include founding of discount usability engineering approach; invention (with R. Molich) of heuristic evaluation method for cost-effective improvement of user interfaces; demonstration (with T.K. Landauer) that user testing and heuristic evaluation both follow same mathematical model; definition of the parallel design method for rapidly exploring user interface alternatives. Home: 72 Derby Ct Madison NJ 07940 Office: Bellcore 445 South St Morristown NJ 07960

NIELSEN, KENNETH ANDREW, chemical engineer; b. Berwyn, Ill., Oct. 10, 1949; s. Howard Andrew and La Verne Alma (Wentzer) N.; m. Linda Kay Miller, Aug. 20, 1970; children: Annette Marie, Kirsten Viola. BS in Chem. Engring., Iowa State U., 1971, MS in Chem. Engring., 1974, PhD in Chem. Engring., 1977. Sr. engr. Union Carbide Corp., Charleston, W.Va., 1976-80, project scientist, 1980-87, rsch. scientist, 1987—. Contbr. articles to profl. jours. Co-founder Forest Hills Asns., Charleston, 1981; advisor Boy Scout Explorer Post, Charleston, 1992. Recipient Fellowships NDEA Title IV, Procter and Gamble Co., Am. Oil Co., Elias Singer award Troy Chem. Co., 1990, Kirkpatrick Chem. Engring. Achievement award Chem. Engring. mag., 1991, Profl. Progress in Engring. award Coll. Engring. Iowa State U., 1992. Mem. Am. Inst. Chem. Engrs., Soc. Rheology. Achievements include invention of UNICARB system for spray coating, a recognized major new pollution-prevention tech.; co-inventor of SERT process for applying mold release agents in polyurethane foam manufacture; discovery of fundamentally new type of spray atomization, known as a decompressive spray. Holder of 20 U.S. patents and 10 U.S. patents pending, also foreign patents. Home: 108 Stratford Pl Charleston WV 25303 Office: Union Carbide Corp PO Box 8361 South Charleston WV 25303

NIELSEN, PETER EIGIL, biochemist, educator; b. Copenhagen, Aug. 7, 1951; s. Eigil Louis and Linda (Rumessen) N.; m. Christina Maria Bonçalves. MSc, U. Copenhagen, 1976, PhD, 1980. Mem. faculty U. Copenhagen, 1983—, assoc. prof. biochemistry, 1991—. Editor: Photochemical Probes in Biochemistry, 1989; adv. bd. Jour. Molecular Recognition; contbr. articles to profl. publs. Niels Bohr fellow, 1983, Hallas Møller fellow Novo Found., 1985, Alfred Benzon Found. fellow, 1990. Mem. Am. Chem. Soc. Achievements include development of uranyl ions as probes in molecular biology, co-invention of peptide nucleic acid chimerae (PNA). Office: Panum Inst Dept Biochemistry, Blegdamsvej 3, DK2200N Copenhagen Denmark

NIELSEN-BOHLMAN, LYNN TRACY, neuroscientist; b. L.A., July 9, 1958; d. Robert L. and Mary A. (Nielsen) Bohlman. BA, Calif. State U., Long Beach, 1983; PhD, U. Calif., Davis, 1993. Rsch. assoc. U. Calif., Davis, 1987—, mem. human rsch. rev. bd., 1988-89. Contbr. articles to profl. pubs. NIMH fellow, 1991-93. Mem. AAUW, Soc. Neurosci., Sigma Xi, Phi Sigma. Democrat. Episcopalian. Office: U Calif Davis VA Med Ctr 150 Muir Rd Martinez CA 94553

NIEMOLLER, ARTHUR B., electrical engineer; b. Wakefield, Kans., Oct. 4, 1912; s. Benjamin Henry and Minnie Christine (Carlson) N.; m. Ann Sochor, May 29, 1937 (dec. June 1982); children: Joanna Matteson, Arthur D. BSEE, Kans. State U., 1933. Registered profl. engr., N.Y., N.J., Pa., Ill., Ohio. Engr. Westinghouse, Newark, N.J., 1937-48, Hillside, N.J., 1948-59, Chgo., 1959-61, Pitts., 1961-65, Cin., 1965-77; pvt. practice engr. Montgomery, Ohio, 1977—. Patentee in field. Elder Presbyterian Church. Served with USN, 1933-37. Mem. AAAS, IEEE, NSPE, N.Y. Acad. Scis. Republican. Home and Office: 7888 Mitchell Farm Ln Cincinnati OH 45242-6410

NIENHUIS, ARTHUR WESLEY, physician, researcher; b. Hudsonville, Mich., Aug. 9, 1941; s. Willard M. and Grace (Prince) N.; m. Sheryl Ann Kalmink Nienhuis, Sept. 20, 1968; children: Carol Elizabeth, Craig Wesley, Kevin Robert, Heather Grace. Student, Cornell Coll., 1959-61; MD, U. Calif., L.A., 1963-68. Am. Bd. Internal Medicine, Am. Bd. Hematology. Intern Mass. Gen. Hosp., Boston, 1968-69, asst. resident, 1969-70; clin. assoc. NHLBI, NIH, Bethesda, Md., 1970-72; clin. fellow hematology Children's Hosp., Boston, 1972-73; chief. clin. svc. Molecular Hematology NIH, Bethesda, Md., 1973-77; dept. clin. dir. NHLBI, NIH, Bethesda, Md., 1976—, chief clin. Hematology Branch, 1976-93; dir. St. Jude Children's Rsch. Hosp., Memphis, 1993—; editor BLOOD-J Am. Soc. Hematology, Bethesda, Md., 1988-92; chmn. Hematology Bd. Am. Bd. Internat Med., Phila., 1988-92; mem. bd. dirs. Am. Bd. Internat Med., Phila., 1988-92. Editor: Molecular Basis of Blood Diseases, 1986, 93. Mem. Am. Soc. Hematology, Am. Soc. Clin. Investigation, Assn. Am. Physicians. Office: St Jude Children's Rsch Hosp 332 N Lauderdale Memphis TN 38101

NIENOW, JAMES ANTHONY, biologist, educator; b. San Diego, Dec. 31, 1953; s. Elmer A. and Frances C. (Kuettel) N.; m. Karen A. Meagher, Jan.13, 1980; children: Tatyana E., Adam J., Anastasia V.F. BA in Math., U. Calif., San Diego, 1975, MA in Math., 1977; PhD in Biology, Fla. State U., 1987. Asst. prof. biology Waycross (Ga.) Coll., 1988—. Contbr. articles to profl. publs., chpt. to book. Mem. Am. Soc. Microbiology, Phycological Soc. Am., Limnology and Oceanography, Ecol. Soc. Am., Sigma Xi. Roman Catholic. Office: Waycross Coll Dept Biology Waycross GA 31501

NIER, ALFRED OTTO CARL, physicist; b. St. Paul, May 28, 1911; s. August Carl and Anna J. (Stoll) N.; m. Ruth E. Andersen, June 19, 1937; children—Janet, Keith; m. Ardis L. Hovland, June 21, 1969. B.S., U. Minn., 1931, M.S., 1933, Ph.D., 1936. Nat. Research fellow Harvard, 1936-38; asst. prof. physics U. Minn., 1938-40, asso. prof., 1940-43, prof., 1946-80, prof. emeritus, 1980—; physicist Kellex Corp., N.Y.C., 1943-45. Mem. NAS, AAAS, Am. Phys. Soc., Minn. Acad. Sci., Geochem. Soc., Am. Geophys. Union (William Bowie medal 1992), Am. Philos. Soc., Geol. Soc. Am., Am. Soc. Mass Spectrometry, Soc. Applied Spectroscopy, Am. Acad. Arts and Scis., Royal Swedish Acad. Scis., Max Planck Soc. (Fed. Republic Germany), Sigma Xi. Achievements include research activities include devel. of mass spectrometer and its application to problems in physics, chemistry, geology, medicine, space sci.; first to separate rare isotope of uranium, U-235, 1940; with J.R. Dunning, E.T. Booth and A.V. Grosse of Columbia, demonstrated it was source of atomic energy when uranium is bombarded with slow neutrons. Home: 2001 Aldine St Saint Paul MN 55113-5601 Office: Univ Minn Sch Physics Dept Physics & Astronomy 116 Church St SE Minneapolis MN 55455

NIETO-ROIG, JUAN JOSE, mathematician, educator, researcher; b. Madrid, Spain, Sept. 27, 1958; s. Ruperto and Maria (Concepción) N.; m. Angela Torres-Iglesias, July 13, 1985; 1 child, Juan. M.Math., U. Santiago, Spain, 1980; D.Math., 1983. Teaching asst. U. Santiago, Spain, 1980, prof. math., 1983—, vice dean faculty math. U. Santiago, 1989, acting dean, 1990; vis. prof. U. Tex.-Arlington, 1981-83; dir. several rsch. projects in math. Reviewer math. revs.; contbr. articles, referee to math. jours. Referee NSF USA, sci. div. NATO. Recipient Best Student award Matematicas Univ., Santiago, 1980. Research fellow U. Santiago, 1981-83; recipient Rsch. prize Ingeniero Comerma, Spain, 1983, 90. Mem. Am. Math. Soc., Am. Biog. Inst. (rsch. bd advisors), Internat. Biog. Centre (adv. coun.), Soc. Indsl. and Applied Math., Assn. Computing Machinery, Soc. Math. Biology, Applied Math. Soc. Spain. Office: Univ Santiago, Facultad Matematicas, Santiago La Coruna, Spain

NIEWIAROWSKI, STEFAN, physiology educator, biomedical research scientist; b. Warsaw, Poland, Dec. 4, 1926; came to U.S., 1972, naturalized, 1978; s. Marian and Janina (Sledzinska) N.; m. Marta Ciswicka (div. 1974); children: Agata, Tomasz. MD, Warsaw U., 1952, PhD, 1960, Dozent, 1961. Lic. physician, Pa.; cert. Ednl. Coun. Fgn. Med. Grads. Intern, med. resident Inst. Hematology, Warsaw, 1951-54; rsch. fellow, rsch. assoc. dept. physiol. chemistry Warsaw U. Med. Sch., 1948-54; rsch. assoc., sr. rsch. assoc. Lab. Clin. Biochemistry, Inst. Hematology, Warsaw, 1951-61; physician in charge Outpatient Dept. for Hemophiliacs, Warsaw, 1957-61; head dept., prof. physiol. chemistry Med. Sch., Bialystok, Poland, 1961-68; assoc. prof. pahtology dept. pathology McMaster U., Hamilton, Ont., Can., 1970-72; rsch. prof. medicine, head coagulation sect. Specialized Ctr. Thrombosis Rsch., Temple U. Sch. Medicine, Phila., 1972-78; prof. physiology Temple U. Sch. Medicine, Phila., 1975—, prof. physiology Thrombosis Rsch. Ctr., 1978—; cons. dept. infectious diseases Warsaw U. Med. Sch., 1954-60; vis. scientist Centre Nat. de Transfusion Sanguine, Paris, 1959; cons. dept. pediatrics Warsaw U. Med. Sch., 1961-65; vis. scientist Vascular Lab., Lemeul Shattuck Hosp., Boston, 1965, 68-70; vis. prof. medicine Tufts U. Sch. Medicine, Boston, 1968-70; dir. Blood Components Devel. Lab., Hamilton Red Cross and McMaster U., 1971-72; mem. sr. coun. Internat. Com. on Haemostatis and Thrombosis, 1973—; mem. NIH rsch. rev. coms., 1975—. Editor Thrombosis Rsch., 1972-80; mem. editorial com. Procs. of Soc. of Exptl. Biology and Medicine, 1980-82, mem. editorial bd. 1990—; reviewer Jour. Clin. Investigation, Jour. Lab. and Clin. Medicine, Blood, Biochimica et Biophysica Acta, Archives of Biochemistry and Biphysics, Jour. Biol. Chemistry, Am. Jour. Physiology; author, co-author 250 articles in the field of blood coagulation, platlet physiology and all adhesion; contbr. articles to profl. jours. Ont. Heart Found. fellow, 1970-71; recipient Jurzykowski Found. award, 1990, rsch. awards NIH, 1972—. Mem. Internat. Soc. Hematology, Internat. Soc. Thrombosis and Hemostasis, Am. Physiology Soc., Am. Soc. Hematology, Coun. of Thrombosis of Am. Heart Assn., Soc. Exptl. Biology and Medicine, Polish Inst. Arts and Scis. in Am., Am. Soc. Exptl. Pathology, Polish Am. Med. Soc. (hon.). Achievements include patent for trigramin a platelet aggregation inhibiting polypeptide. Home: 445 Woodbine Narberth PA 19072 Office: Temple U Sch Medicine 3400 N Broad St Philadelphia PA 19140

NIGAM, BISHAN PERKASH, physics educator; b. Delhi, India, July 14, 1928; came to U.S., 1952; s. Rajeshwar Nath and Durga (Vati) N.; m. Indira Bahadur, Nov. 14, 1956; children—Sanjay, Shobhna, Ajay. B.S., U. Delhi, 1946, M.S., 1948; Ph.D., U. Rochester, N.Y., 1955. Research fellow U. Delhi, 1948-50; lectr. in physics, 1950-52, 55-56; postdoctoral fellow Case Inst. Tech., Cleve., 1954-55; postdoctoral research fellow NRC, Ottawa, Can., 1956-59; research assoc. U. Rochester, 1959-60, asst. prof. physics, part-time 1960-61; prin. scientist Gen. Dynamics/Electronics, Rochester, N.Y., 1960-61; assoc. prof. physics SUNY, Buffalo, 1961-64; prof. physics Ariz. State U., Tempe, 1964—, U. Wis., Milw., 1966-67. Author: (with R.R. Roy) Nuclear Physics, 1967; also articles. Govt. of India scholar U. Rochester, 1952-54. Fellow Am. Phys. Soc. Office: Dept Physics Ariz State U Tempe AZ 85287

NIGG, BENNO M., biomechanics educator; b. Walenstadt, St. Gallen, Switzerland, Apr. 10, 1938; s. Josef B. Nigg and Edwina Nigg-Widrig; m. Margaretha J. Bolleter, Aug. 28, 1965; children: Andreas Reto, Claudio, Sandro. Diploma in physics, ETH, Zurich, Switzerland, 1965, Dr. sci. nat., 1975. Instr. Lyceum Alpinum Zuoz, Switzerland, 1965-71; researcher Biomechs. Lab. ETH, Zurich, 1971-76, dir., 1976-81; prof. U. Calgary, 1981—, dir. Human Performance Lab., 1981—; cons. Adidas, Germany, 1976—, Nike, 1981-85; mem. steering com. World Congress on Biomechs., 1988—. Recipient Michael Jaeger award GOTS, Munich, 1986, Wartenweiler Meml. award ISB, UCLA, 1989, NOVEL award, Vienna, 1991. Mem. Internat. Soc. Biomechanics (pres. 1983-85). Office: U Calgary, Human Performance Lab, Calgary, AB Canada T2N 1N4

NIGRO, ALDO, physiology and psychology educator; b. Scigliano, Italy, Mar. 27, 1927; s. Francesco and Mariannina (Pagliuso) N.; m. Rosa Siss Merro, Sept. 8, 1969; 1 child, Francesco. MD, U. Messina, Italy, 1950. Pvt. dozent U. Messina, 1959, prof. psychology, 1969; dir., mgr. Polidadattico, Messina, 1990—. Author: Dialogo proteico, 1974, Modello Psicolinguistico, 1988, Human Light: A Cybernetic Theory of Consciosness, 1992. Pres. Universita Tempo Libero, Messina, 1990. Mem. Accademia Peloritana Pericolanti. Roman Catholic. Home: N Panoramica 1330, Messina Italy 98168 Office: Cattedra Psychology, T Cannizzaro 278, Messina Italy 98100

NIITU, YASUTAKA, pediatrician, educator; b. Nagano Prefecture, Japan, Aug. 3, 1920; s. Shusuke and Tsuruji N.; m. Tokiko Komuro, Dec. 8, 1952; children: Hidetaka, Iwayasu, Munetaka. Igakushi, Tohoku Imperial U., 1943, M.D. (Igakuhakase), 1949. Research fellow Research Inst. for TB and Leprosy, Tohoku U., Sendai, Japan, 1943-54, asst. prof. dept. pediatrics, 1954-63, Univ. prof. pediatrics Research Inst. for TB and Cancer, 1964-84, prof. emeritus, 1984—; dir. Miyagi br. Japan Anti TB Assn., 1984—; pres. Sendai Kosei Hosp., 1973-80, 86-89; adviser Japan Sarcoidosis Com., Soc. Paediatrica Japonica, Japan Soc. Pediat. Pulmonology (pres. 1980-88), Japan Soc. Pediat. Infectious Diseases; tech. adviser Shenyang First Tuberculosis Hosp., Shenyang Anti-tuberculosis Ctr., Shenyang Chest Hosp., People's Republic of China. Capt. M.C., Japanese Army, 1943-45. Mem. Internat.

Orgn. Mycoplasmology, World Assn. Sarcoidosis and Other Granulomatous Diseases, Japan Soc. Chest Diseases, Soc. Japanese Virologists, Japanese Soc. TB, Am. Soc. Microbiology, Internat. Union Against TB and Lung Disease, Japan Soc. Internat. Medicine, Japan Assn. Infectious Diseases, Japanese Assn. Child Health. Co-author books in Japanese, including: Virus and Diseases, 1969; Routine Pediatric Diagnosis and Treatment, 1971; New Virology, 1972; Pediatric X-Ray Diagnosis, 1972; Handbook of Clinical Pneumology, 1977; Present Pediatrics, 1978; Clinical Virology, 1978; Sarcoidosis, 1979; Practice of Pediatric Infections, 1980; Mycoplasma, 1981; Illustrations of Mycoplasma, 1981; Illustrations of Pediatric Diagnosis and Treatment, 1981; Illustrated Pediatric Chart 1982; Topics of Infections, 1983. Mem. editorial bd. Pediatric Pulmonology. Contbr. 300 articles to profl. jours. Home: Higashikatsuyama 1-5-6, Aobaku, Sendai 981, Japan Office: Japan Anti TB Assn, Miyagi Br Miyamachi 1-1-5, Aobaku Sendai 980, Japan

NIKI, KATSUMI, chemist, educator; b. Atsugi, Japan, Mar. 8, 1931; s. Yuji and Yae Kanzaki; m. Hisae Niki, Dec. 8, 1961; children: Yuko Morishima, Jun. M in Engring., Tokyo Inst. Tech., 1957, D in Engring., 1978; PhD in Chemistry, U. Tex., 1966. Project engr. Sumitomo Chem. Co., Osaka, Japan, 1957-68; lectr. Yokohama (Japan) Nat. U., 1968-79, asst. prof., 1979-82, prof., 1982—; sec., chmn. electrochem. commn. Internat. Union Pure and Applied Chemistry, Oxford, England; adv. bd. Langmuir, Washington, 1988-90; organizer U.S.-Japan seminar NSF, 1982, 89, Internat. Soc. Electrochemistry, 1989, 90, v.p., 1993—. Editorial bd.: Jour. Electroanalytical Chemistry, 1992—; author: Encyclopedia of the Electrochemistry of the Elements, 1986; editor: Redox Mechanisms and Interdacial Behaviors of Molecules of Biological Importance, 1988; contbr. articles to Jour. Electroanalytical Chemistry, Jour. Electrochem. Soc., Langmuir. Grantee Office Naval Rsch., 1987—. Home: 1-11-2 Nagahama Kanazawa-ku, Yokohama 236, Japan Office: Yokohama Nat U, 156 Tokiwadai Hodogaya-ku, Yokohama 240, Japan

NIKIFORUK, PETER NICK, university dean; b. St. Paul, Alta., Can., Feb. 11, 1930; s. DeMetro N. and Mary (Dowhaniuk) N.; m. Eugenie F. Dyson, Dec. 21, 1957; children: Elizabeth, Adrienne. B.Sc., Queen's U., Ont., Can., 1952; Ph.D., Manchester U., Eng., 1955, D.Sc., 1970. Engr. A.V. Roe Ltd., Toronto, Ont., 1951-52; def. sci. service officer Def. Research Bd., Quebec, Que., Can., 1956-57; systems engr. Canadair Ltd., Montreal, Que., 1957-59; asst. prof. U. Sask., Saskatoon, 1960-61; assoc. prof. U. Sask., 1961-65, prof., 1965—, chmn. div. control engring., 1964-69, head mech. engring., 1966-73, dean engring., 1973—; cons. in field.; mem. council NRC, 1973-78. Contbr. articles to profl. jours. Bd. dirs. Sask. Research Council, Can. Inst. Indsl. Tech. Fellow Royal Soc. Arts, Inst. Physics, Inst. Elec. Engrs. (Kelvin Premium), Can. Acad. Engring., Engring. Inst. Can., Can. Soc. Mech. Engr. (past v.p.); mem. IEEE (Centennial medal), Assn. Profl. Engrs. Sask (chmn. bd. dirs., Disting. Svc. award). Home: 31 Bell Crescent, Saskatoon, SK Canada S7J 2W2

NIKISHENKO, SEMION BORIS, chemical research engineer; b. St. Peterburg, Russia, Oct. 20, 1946; s. Boris Samuel Bregman and Lidiya Sergey Nikishenko; m. Irene Alexander Frolova, Oct. 20, 1970; children: Natalia, Maxim. BS in Phys. Engring., Moscow Phys. Engring. Inst., 1970; PhD in Phys. Chemistry, Inst. Organic Chemistry of Acad. Sci., Moscow, 1980. Rsch. engr. All-Union Oil Refining R & D Inst., Moscow, 1970-76, All-Union Oil & Gas R & D Inst., Moscow, 1976, Inst. Organic Chemistry of Acad. Sci., Moscow, 1976-83; sr. rsch. engr. Sector of Ingomog Media Mechanics, Moscow, 1983-84; rsch. engr. Inst. Gas & Oil Industry, Moscow, 1984-85; sr. rsch. engr. Chem. Tech. R&D Inst., Moscow, 1985-88, All-Union Cable R & D Inst., Moscow, 1988—. Editor, reviewer 4500 abstracts Russian Jour. Abstracts, 1980-92; reviewer All-Union R & D Inst. Sci. & Engring. Infos., 1980-91, editor, 1991-92; contbr. 26 articles to Kinetics & Catalysis. Mem. Am. Chem. Soc., N.Y. Acad. Scis. Achievements include research in theoretical and experimental studies of characteristics of surface and structure of Co-Mo/AL203 HDS catalysts; nature of interactions between HTCS ceramics and noble metals. Home: Pionerskaya 2-44, 142105 Podolsk Russia

NIKKARI, TAPIO URHO, medical biochemistry educator; b. Lumivaara, Finland, Mar. 13, 1934; s. Urho and Aino (Hirvonen) N.; m Marja Särkilä, 1959; children: Seppo, Simo. MD, U. Turku, Finland, 1962, D in Med. Sci., 1965. Instr. med. chemistry U. Turku, 1959-63, assoc. prof., 1969-70; rsch. asst. Finnish Med. Rsch. Coun., Turku, 1963-65, investigator, 1967-69; asst. prof. Rockefeller U., N.Y.C., 1971-72; prof. med. biochemistry U. Tampere, Finland, 1973—; dean of med. faculty U. Tampere, 1979-80; chmn. dept. biomed. scis., U. Tampere, 1984-85. Contbr. articles to profl. jours. Lt. Med. Corps, Finnish Army, 1953-54, 61. Decorated Knight First Class of the Order of the White Rose of Finland, 1976. Mem. Finnish Med. Assn., Finnish Med. Assn. Duodecim, N.Y. Acad. Scis. Home: Kuninkaankatu 19 A 17, Tampere SF-33210, Finland Office: U Tampere Dept Biomed Scis, PO Box 607, Tampere SF-33101, Finland

NIKOLAI, TIMOTHY JOHN, aerospace engineer; b. Marshfield, Wis., Jan. 24, 1964; s. Thomas Frederik and Hildegard J. (Stiglmaier) N. B Aerospace Engring and Mechanics, U. Minn., 1986. Engring. intern Army Materiel Command, Red River Army Depot, Tex., 1986-87; project engr. Human Systems Program Office, Brooks AFB, Tex., 1987—. Mem. AIAA. Home: 7914 Misty Forest San Antonio TX 78239-3542 Office: HSC/YAED Bldg 809 Brooks AFB TX 78235-5000

NIKOLIC, GEORGE, cardiologist, consultant; b. Belgrade, Yugoslavia, Feb. 7, 1945; came to Australia, 1964; s. Ilija and Sofija (Rajic) N.; m. Annette Courtney Smith, Feb. 7, 1978; children: Alexandra Courtney, John George. MB, BS with honors, Sydney U., 1971. Diplomate Am. Bd. Internal Medicine, Am. Bd. Cardiovascular Med., Critical Care Medicine resident St. Vincent's Hosp., Sydney, Australia, 1971-72, med. registrar, 1973-74, cardiology registrar, 1975; med. and cardiology registrar Woden Valley, Canberra, Australia, 1976-77, dir. intensive care, 1978, 1982-91, sr. specialist, 1991—; cardiology fellow St. Vincent Hosps., S.I., Worcester, 1979-82; asst. prof. medicine U. Mass., Worcester, 1981-82.Contbr. articles in field to med. jours. Fellow Royal Australasian Coll. Physicians (sec. Australian capital Territory br., Canberra 1983), Am. Coll. Cardiology, Am. Coll. Chest Physicians; mem. Australia-N.Z. Intensive Care Soc., Critical Care Soc., Cardiac Soc. Australia and New Zealand. Office: Woden Valley Hosp, PO Box 11, Woden ACT, Australia 2605

NIKSA, STEPHEN JOSEPH, research chemical engineer, consultant; b. Bridgeport, Pa., Nov. 28, 1953; s. Steven Niksa and Theresa Eleanor Granese Marconi; m. Patricia Lynn Andersen, Nov. 23, 1986 (div. May 1989); m. Ellen Zhi Ping Chen, Feb. 21, 1993. BS in Chem. Engring. summa cum laude, Case Western Res. U., 1975; PhD, Princeton U., 1982. Rsch. staff assoc. dept. chem. engring. Princeton U., 1981-82; mem. tech. staff combustion rsch. facility Sandia Nat. Labs., Livermore, Calif., 1982-85; asst. prof. dept. mech. engring. Stanford U., 1985-92; sr. chem. engr. Molecular Physics Lab. SRI Internat., Menlo Park, Calif., 1992—; cons. in field, 1987-92. Contbr. articles to Energy Fuels, Am. Inst. Chem. Engrs. Jour., Combustion Flame, others. Mem. Combustion Inst. (exec. bd. of we. states sect. 1989—), Am. Inst. Chem. Engrs., Am. Chem. Soc., Sigma Xi. Achievements include numerous contributions to our understanding of pulverized coal combustion; formation mechanisms for noxious gases and organic mutagens; mechanisms that generate sub-micron particulates, rates of char oxidation and thermal behavior of catalytic converters for automobiles. Office: SRI Internat Molecular Physics Lab 333 Ravenswood Ave Menlo Park CA 94025

NILSSON, BO INGVAR, hematologist; b. Helsingborg, Sweden, Sept. 15, 1947; s. Jonny Ingvar and Sonja Eugenia (Akesson) N.; m. Elsa Sigbritt Warkander, Aug. 20, 1971; children: Victoria, Filip, Sofia. MD, U. Gothenburg, 1973; PhD, U. Lund, 1985. Lic. physician, 1973. Resident Dept. Radiology, Univ. Hosp., Gothenburg, 1973-74, Dept. Internal Medicine, County Hosp., Angelholm, 1974-76; resident Dept. Internal Medicine, Univ. Hosp. Lund, 1976-78, amanuensis, 1978-83; sr. house officer Dept. Internal Medicine, County Hosp., Helsingborg, 1983-86; med. dir. oncology Pharmacia Leo Therapeutics, Helsingborg, 1986-88; rsch. dir. oncology Pharmacia Leo Therapeutics, 1989-90; rsch. dir. oncology Kabi Pharmacia Therapeutics, 1991—, dir. med. affairs, 1992—. Contbr. articles to profl. jours. U. Lund, Segerfalk Found. grantee, 1978-86. Mem. Am. Soc. Hematology, Internat. Soc. Regional Chemotherapy, Swedish Soc. Medicine, Swedish Soc. Hematology-Oncology (edn. com. 1985-89), N.Y. Acad. Scis., European Soc. Med. Oncology, Oncology of So. Sweden (bd. dirs. 1980-83), Wig. Home: Kristinehamnsg 12 4 Weaver Dr Martinsville NJ 08836 Office: 800 Centennial Ave Piscataway NJ 08866

NILSSON, DAN CHRISTER, plant ecology educator, researcher; b. Umea, Sweden, May 7, 1951; s. Dan Lennart and Ingrid Alice (Engelmark) N.; m. Gunnel Agneta Grelsson, Oct. 30, 1976; children: Frida Susanna, Tobias Vidar. PhD, Umea U., 1981. Assoc. prof. dept. ecol. botany Umea U., 1986—; reviewer Jour. Ecology, Conservation Biology, Jour. Biogeography, Oikos, Lund, Sweden. Contbr. articles to Biol. Conservation, Can. Jour. Fish and Aquatic Sci., Jour. Ecology, Ecology Jour., Am. Naturalist mag., Can. Jour. of Botany, Conservation Biology. Grantee: Swedish Natural Sci. Rsch. Coun., 1991—, Swedish Coun. for Forestry and Agrl. Rsch., 1992—, Swedish Soc. for Conservation of Nature, 1985-90, Swedish EPA, 1979—, WWF, Sweden, 1991—, European Community, 1992—. Mem. Internat. Assn. Vegetation Sci., Soc. Conservation Biology, Brit. Ecol. Soc., Swedish Oikos Soc., Ecol. Soc. Am. Office: Umeå U, 90187 Umeå Sweden

NILSSON, KURT GÖSTA INGEMAR, biotechnology and enzyme technology scientist; b. Göteryd, Kronberg, Sweden, June 25, 1953; s. Sven Gösta and Ebba Birgit (Karlsson) N.; m. Pia Ingrid Elisabeth, July 30, 1983; children: Jenny Ingrid Maria, Johan Martin Ingemar. MSci. in Chemistry, U. Lund, Sweden, 1976, DrSci. in Applied Chemistry, 1984, postgrad., 1976-84. Sr. scientist Swedish Sugar Co., Ltd., Malmö, Arlöv, Sweden, 1985-87; mgr. biotech. Carbohydrate Internat., Arlöv, 1987-88; assoc. prof. Chem. Ctr. U. Lund, 1989—, docent Chem. Ctr., 1991; pres. Glycorex AB, Lund, 1992—; cons. in field, 1988—. Contbr. articles to profl. jours. Mem. Swedish Chem. Soc., Am. Chem. Soc., Biotransformation Club. Achievements include patents for antibody immobilisation, antibody conjugates for enzyme immunoassay synthesis of complex carbohydrates. Office: Glycorex AB, Ideon Sci Park, Solveg 41, S-22370 Lund Sweden

NIMMAGADDA, RAO RAJAGOPALA, materials scientist, researcher; b. Donepudi, Andhra Pradesh, India, July 1, 1944; came to U.S., 1967; s. Suryaprakasa Rao and Bullemma (Venigalla) N.; m. Usha Rani Chava, Nov. 7, 1965 (div. Nov. 1980); children: Sandhya Rani, Pramada Shree; m. Jhansi Rani Talluri, Dec. 18, 1980; children: Sai Chandra and Sri Spandana. B Tech. with honors, Indian Inst. Tech., Bombay, 1965; MS, Mich. Tech. U., 1970; PhD, UCLA, 1975. Jr. sci. officer Def. Metall. Rsch. Labs., Hyderabad, India, 1965-67; postdoctoral scholar UCLA, 1975-78, rsch. engr., 1978-81; rsch. scientist Smith Tool, Irvine, Calif., 1981-83, Burroughs, Westlake Village, Calif., 1983-84; rsch. engr. Memorex Corp., Santa Clara, Calif., 1984-86; staff scientist Lockheed Missiles & Space Co., Palo Alto, Calif., 1986-93; staff engr. Akashic Memories Corp., San Jose, Calif., 1993—. Contbr. articles to profl. jours. Pres. Telugu Assn. So. Calif., 1977, dir. Hindu Temple Soc. So. Calif., 1977-80, pres., 1980. Recipient of Outstanding Tech. Achievement award Strategic Def. Initiative Orgn., Washington, 1989. Mem Am. Vacuum Soc., Materials Rsch. Soc. Republican. Hindu. Achievements include patents for titanium nitride coatings to improve wear resistance of oil drilling bit components, dual squeeze seal gland for oil drilling bits, wear resistant coatings for o-ring seals of oil drilling bits. Home: 120 Gilbert Ave Santa Clara CA 95051-6705 Office: Akashic Memories Corp 305 W Tasman San Jose CA 95134

NIMOITYN, PHILIP, cardiologist; b. Phila., Mar. 6, 1951; s. Benjamin Solomon and Edith (Ornstein) N.; m. Hillary Rachel Saul, June 11, 1989. BS in Biology with distinction, Phila. Coll. Pharmacy and Sci., 1972; MD, Thomas Jefferson U., 1976. Cert. Nat. Bd. Med. Examiners, Am. Bd. Internal Medicine, Am. Bd. Cardiovascular Disease. Intern Hahnemann U. Hosp., Phila., 1976-77; resident in internal medicine Thomas Jefferson U. Hosp., Phila., 1977-79, cardiovascular disease fellow, 1979-81, instr. medicine, 1981-90, clin. asst. prof., 1990—, attending physician, 1981—; cons. physician Wills Eye Hosp., Phila., 1981—. Author: (with others) Artificial Cardiac Pacing, 1984, Quick Reference to Cardiovascular Disease, 1987, Cardiac Emergency Care, 1991; contbr. articles to profl. jours. Recipient Cert. of Merit for Sci. Exhibits AMA, 1974, 2d prize for sci. exhibits Ind. State Med. Assn., 1974. Fellow Am. Coll. Cardiology; mem. AMA, Pa. Med. Soc., Phila. County Med. Soc. Office: 1128 Walnut St Ste 401 Philadelphia PA 19107

NING, CUN-ZHENG, physicist; b. Xianyang, Shaanxi, China, Oct. 21, 1958; arrived in Germany, 1988; s. Xi-Wu and Shu-Xian (Cheng) N.; m. Ya-E Zhang, Jan. 1, 1983; 1 child, Feng-Tao. BS, Northwestern U., 1982, MS, 1985; D in Natural Sci., U. Stuttgart, Fed. Republic of Germany, 1991. Vis. scientist U. Stuttgart, 1986-87, rsch. scientist, 1988—; lectr. Northwestern U., Xian, 1987-88. Co-editor: Lectures in Synergetics, 1987; contbr. numerous papers to profl. jours. Mem. Am. Phys. Soc., Chinese Phys. Soc., German Phys. Soc., N.Y. Acad. Scis. Home: Saarstr 15, D-70374 Stuttgart Germany Office: U Stuttgart Inst Theoretical Physics & Synergetics, Pfaffenwaldring 57/4, D-70550 Stuttgart Germany

NING, TAK HUNG, physicist, microelectric technologist; b. Canton, China, Nov. 14, 1943; came to U.S., 1964; s. Hong and Kwai-Chan (Lee) N.; m. Yin Ngao Fan; children: Adrienne, Brenda. BA in Physics, Reed Coll., 1967; MS in Physics, U. Ill., 1968, PhD in Physics, 1971. IBM Rsch. Div., Yorktown Heights, N.Y., 1973-78, Mgr. bipolar devices and cirs., 1978-82, mgr. Advanced Silicon Technology Lab., 1982-83, mgr. silicon devices and technology, 1983-90; mgr. VLSI design and tech. IBM Rsch. Div., 1990-91; IBM fellow, 1991—. Patentee in field. Fellow IEEE (assoc. editor Trans. on Electron Devices 1988-90, J.J. Ebers award 1989, Jack a. Morton award 1991); mem. Nat. Acad. Engring. Home: 3085 Weston Ln Yorktown Heights NY 10598-1962 Office: IBM T J Watson Research Ct Yorktown Heights NY 10598

NING, XUE-HAN (HSUEH-HAN NING), physiologist, researcher; b. Peng-Lai, Shandong, People's Republic of China, Apr. 15, 1936; came to U.S., 1984; s. Yi-Xing and Liu Ning; m. Jian-Xin Fan, May 28, 1967; 1 child, Di Fan. MD, Shanghai 1st Med. Coll., People's Republic of China, 1960. Investigator and head cardiovascular unit Shanghai Inst. Physiology, 1960-84, prof. and chair hypoxia dept., 1988-90, vice chairperson academic com., 1988-90; NIH internat. rsch. fellow U. Mich., Ann Arbor, 1984-87, vis. prof. dept. physiology, 1991—; prof. and dir. Hypoxia Physiology Lab. Academia Sinica, Shanghai, 1989—; acting leader High Alt. Physiology Group, Chinese mountaineering and sci. expdn. team to Mt. Everest, 1975, Dept. Metall. Industry of China and Ry. Engring. Corps, 1979; vis. prof. dept. physiology Mich. State U., East Lansing, 1989-90. Author: High Altitude Physiology and Medicine, 1981, Environment and Ecology of Qinghai-Xizong (Tibet) Plateau, 1982; contbr. articles to Med. and Biol. Engring. and Computing, Am. Jour. Physiology; mem. editorial bd. chinese Jour. Applied Physiology, 1984—, Acta Physiologica, 1988—. Recipient Merit award Shanghai Sci. Congress, 1977, All-China Sci. Congress, Beijing, 1978, Upper Class award Acadademia Sinica, Beijing, 1986, 1st Class award Nat. Natural Scis., Beijing, 1987. Mem. Am. Physiol. Soc., Internat. Soc. Heart Rsch., Royal Soc. Medicine, Shanghai Assn. Physiol. (bd. dirs. 1988—), Chinese Assn. Physiol. (com. applied physiology 1984—, com. cardiovascular physiology 1988—), Chinese Soc. Medicine, Chinese Soc. Biomed. Engring. Achievements include research in predictive evaluation of mountaineering performance, paradox phenomenon of cardiac pump function injury after climbing or giving oxygen, blood flow-metabolism-function relationship of heart during hypoxia and ischemia, effect of Chinese medicinal herbs on cardiac performance during hypoxia, cardiovascular adaptation to hypoxia and ischemia; first electrocardiograph recording at summit of Mt. Everest. Office: Univ Mich Thoracic Rsch Lab B 564 MSRBII Ann Arbor MI 48109 also: Apt 1-7, 850 Cao-Xi Bei Rd, Shanghai China

NINNIE, EUGENE DANTE, civil engineer; b. Endicott, N.Y., Mar. 18, 1957; s. Eugene Robert and Elizabeth Ann (Lang) N. BSCE, U. Wyo., 1982. Project mgr. Ninnie Constrn. Corp., Beacon, N.J., 1982-88; sr. design engr. Hayward & Pakan Assocs., Poughkeepsie, N.Y., 1988-90; mng. prin. Civil Technologies and Engring., Wappingers Falls, N.Y., 1990—. Mem. NSPE, ASCE, Am. Inst. Steel Cos., N.Y. State Soc. Profl. Engrs., Concrete Reinforcing Steel Inst., Builders Assn. Hudson Valley, Steel New Force. Republican. Roman Catholic. Home: 9 Cottam Hill Rd Wappingers Falls NY 12590 Office: Civil Technologies and Engring Rt 9D Wappingers Falls NY 12590

NIORDSON, FRITHIOF IGOR, mechanical engineering educator; b. Johannesburg, Republic South Africa, Aug. 1, 1922; s. Johan Fredrik Niord and Helena de Makeff; m. Ann-Marie Odqvist (div. 1973); children: Ulrika Marie, Johan Niord; m. Hanne Oerregaard, Jan. 15, 1975; children: Christian Niord, Mathilde Antonina. MSc, Royal Inst. Tech., Stockholm, 1947; PhD, Brown U., 1952. Head Nilson & Niordson, Cons. Engrs., Sweden, 1949-59; prof. mech. engring. Tech. U. Denmark, 1958-92, chmn. faculty mech. engring., 1975-91; pres. Inernat. U. Tehoretical Applied Mechanics, 1976-80; g. Author monograph: Shell Theory, 1985. Named Knight of Dannebrog, 1st class, Danish Queen; recipient Euler medal USSR Acad. Scis. Mem. Royal Swedish Acad. Scis., Polish Soc. Theoretical and Applied Mechanics, ASME (hon. mem.), German Soc. for Applied Math. and Mechanics (mem. coun.), Sigma Xi. Lutheran. Avocations: sailing, skiing, horseback riding. Home: Geelsvej 19, Holte Denmark 2840

NIRENBERG, LOUIS, mathematician, educator; b. Hamilton, Ont., Can., Feb. 28, 1925; came to U.S., 1945, naturalized, 1954; s. Zuzie and Bina (Katz) N.; m. Susan Blank, Jan. 25, 1948; children: Marc, Lisa. BSc, McGill U., Montreal, 1945, DSc (hon.), 1986; MS, NYU, 1947, PhD, 1949; DSc (hon.), U. Pisa, Italy, 1990, U. Paris Dauphine, 1990. Mem. faculty N.Y. U., 1949—, prof. math., 1957—, dir. Courant Inst., 1970-72; visitor Inst. Advanced Study, 1958; hon. prof. Nankai U., Zhejiang U. Author research articles. Recipient Crafoord prize Royal Swedish Acad., 1982; NRC fellow, 1951-52; Sloan Found. fellow, 1958-60; Guggenheim Found. fellow, 1966-67, 75-76; Fulbright fellow, 1965. Mem. NAS, Am. Acad. Arts and Scis., Am. Math. Soc. (v.p. 1976-78, M. Bôcher prize 1959), Am. Philos. Soc., French Acad. Scis. (fgn. assoc.), Accademia dei Lincei (fgn. mem.), Istituto Lombardo, Accad. di Scienze e Lettere (fgn. mem.). Home: 221 W 82nd St New York NY 10024-5406 Office: Courant Inst 251 Mercer St New York NY 10012-1185

NIRENBERG, MARSHALL WARREN, biochemist; b. N.Y.C., N.Y., Apr. 10, 1927; s. Harry Edward and Minerva (Bykowsky) N.; m. Perola Zaltzman, July 14, 1961. B.S. in Zoology, U. Fla., 1948, M.S., 1952; Ph.D. in Biochemistry, U. Mich., 1957. Postdoctoral fellow Am. Cancer Soc. at NIH, 1957-59; postdoctoral fellow USPHS at NIH, 1959-60; mem. staff NIH, 1960—; research biochemist, chief lab. biochem. genetics Nat. Heart, Lung and Blood Inst., 1962—; researcher mechanism protein synthesis, genetic code, nucleic acids, regulatory mechanisms in synthesis macromolecules, and neurobiology. Recipient Molecular Biology award Nat. Acad. Scis., 1962, award in biol. scis. Washington Acad. Scis., 1962, medal HEW, 1964, Modern Medicine award, 1963, Harrison Howe award Am. Chem. Soc., 1964, Nat. Medal Sci. Pres. Johnson, 1965, Hildebrand award Am. Chem. Soc., 1966, Research Corp. award, 1966, A.C.P. award, 1967, Gairdner Found. award merit Can., 1967, Prix Charles Leopold Meyer French Acad. Scis., 1967, Franklin medal Franklin Inst., 1968, Albert Lasker Med. Research award, 1968, Priestly award, 1968; co-recipient Louisa Gross Horowitz prize Columbia, 1968, Nobel prize in medicine and physiology, 1968. Fellow AAAS, N.Y. Acad. Scis.; mem. Am. Soc. Biol. Chemists, Am. Chem. Soc. (Paul Lewis award enzyme chemistry 1964), Am. Acad. Arts and Scis., Biophys. Soc., Nat. Acad. Scis., Washington Acad. Scis., Soc. for Study Devel. and Growth, Harvey Soc. (hon.), Leopoldina Deutsche Akademie der Naturforscher, Pontificial Acad. Scis. Office: Nat Inst Health Lab Biochemical Genetics 9000 Rockville Pike Bethesda MD 20892-0001

NISBETT, RICHARD EUGENE, psychology educator; b. Littlefield, Tex., June 1, 1941; s. R. Wayne and Helen (King) N.; m. Susan Ellen Isaacs, June 29, 1969; children: Matthew, Sarah. A.B. summa cum laude, Tufts U., 1962; Ph.D., Columbia U., 1966. Asst. prof. psychology Yale U., New Haven, 1966-71; assoc. prof. psychology U. Mich., Ann Arbor, 1971-77, prof., 1977—, Theodore M. Newcomb prof. psychology, 1989-92, Theodore M. Newcomb disting. prof. of the social scis., 1992—. Author: (with others) Attribution: Perceiving the Causes of Behavior, 1972, Induction: Processes of Inference, Learning, and Discovery, 1986, Rules for Reasoning, 1992; editor (with L. Ross) Human Inference: Strategies and Shortcomings of Social Judgment, 1980, The Person and the Situation, 1991; editor (with others) Social Psychology: Explorations in Understanding, 1974—. Recipient Donald T. Campbell award for disting. research in social psychology, 1982, Disting. Sci. Contbn. award APA, 1991; President's fellow, 1963-65; NSF fellow, 1965-66; fellow Ctr. for Advanced Studies in Behavioral Scis. Office: U Mich 5261 ISR Rsch Ctr Group Dynamics Ann Arbor MI 48109

NISHI, KENJI, electrical engineer; b. Shingu, Wakayama, Japan, Apr. 28, 1949; s. Hiroshi and Yomi Nishi; m. Atsuko Taya, May 6, 1980; children: Yoko, Junko. BS, U. Tokyo, 1973, PhD, 1988. Researcher OKI Electric Industry Co. Ltd., Hachioji, Tokyo, 1973-82, sr. researcher, 1984-90, mgr., 1990—, chief researcher, 1992—; vis. scientist MIT, Cambridge, 1982-84. Co-author: Modern Semiconductor Simulation Technology, 1990; contbr. articles to profl. jours. Mem. IEEE, Info. and Communication Engrs. of IEEE, Japan Soc. Applied Physics, Materials Rsch. Soc. Avocations: piano playing, mountaineering, classical music, Go playing. Home: 1-4-23 Nishihashimoto, Sagamihara Kanagawa 229, Japan Office: OKI Electric Industry Co Lt, 550-1 Higashiasakawa, Hachioji Tokyo 193, Japan

NISHI, MICHIHIRO, mechanical engineering educator; b. Fukuoka, Japan, July 4, 1943; s. Hiroshi and Junko (Inamochi) N.; m. Yukiko Abe, Nov. 28, 1971; 1 child, Mariko. D. Engring., Kyushu U., Fukuoka, 1976. Cert. tchr. Rsch. assoc. Rsch. Inst. Indsl. Sci. Kyushu U., Fukuoka, 1965-71; lectr. Kyushu Inst. Tech., Kitakyushu, Japan, 1971-75, assoc. prof., 1975-84, prof., 1984—; vis. prof. Stanford (Calif.) U., 1987-88. Author: Book Series on Hydraulic Machinery, 1991; reviewer Transactions Japan Soc. Mech. Engrs.; contbr. articles to Transactions Japan Soc. Mech. Engrs., Transactions ASME, Proceedings IAHR Symposium. Advisor Div. Commerce and Industry, Fukuoka Prefecture Office, 1985—. Grantee Harada Found., 1983, 86, 92. Mem. Japan Soc. Mech. Engrs. (Hatakeyama prize 1965, prize 1986), Internat. Assn. Hydraulic Rsch. (Delft, the Netherlands), AIAA. Achievements include research on fluid mechanics of internal flow; on boundary layer control; on diffuser performance; on draft tube surging; on flow measurement. Office: Kyushu Inst Tech, Sensui-cho 1 Tobata, Kitakyushu 804, Japan

NISHIMOTO, NOBUSHIGE, pharmacognosy educator; b. Osaka, Japan, Oct. 19, 1929; s. Gihei and Tsurue (Yoshida) N.; m. Itsuyo Iwakiri, Dec. 28, 1959; children: Soh, Koh. Dr.Pharm.Sci., Tohoku U., Sendai, 1968. Asst. researcher Osaka U., 1951-58; hygienic engring. ofcl. Osaka br. Nat. Inst. Hygienic Sci., Osaka, 1958-63; chief researcher Research Lab. of Rhoto Pharm. Co. Ltd., Osaka, 1963-83; temp. instr. Tohoku U., Sedai, 1975-79; researcher Research Inst. Oriental Medicines, Toyama U., 1977-78; prof. pharmacognosy Tokushima Bunri U., 1983—. Contbr. articles to profl. jours.; inventor in field. Mem. Pharm. Soc. Japan, Japanese Soc. Pharmacognosy, Med. and Pharm. Soc. for Wakan-yaku, Japanese Assn. for Plant Tissue Culture. Avocations: travel, plant collecting. Home: 5-7-209 Shinkanaoka-cho, Sakai, Osaka Japan 591 Office: Tokushima Bunri Univ, Yamashiro-cho, Tokushima Japan 770

NISHIMURA, CHIAKI, molecular virologist educator; b. Obama, Fukui, Japan, Nov. 17, 1928; s. Kikugoro and Natsuko N.; m. Kyoko Tanaka, Feb. 3, 1955; children: Akira, Masahiko. B of Pharmacy, Kyoto U., Japan, 1953, PharmD, 1960. Tech. ofcl. Nat. Inst. Hygienic Scis., Tokyo, Japan, 1954-57; sect. chief NIH (Japan), Tokyo, Japan, 1957-71; vis. fellow The Rockefeller Inst., N.Y.C., 1960-61; vis. assoc. NIH, Washington, 1961-62; prof. Kitasato U., Tokyo, 1971—. Patentee in field; contbr. articles to profl. jours. Mem. Japanese Pharm. Soc., Japanese Virological Soc., Japanese Immunological Soc., Japanese Cancer Soc. Democrat. Buddhist. Avocations: music, reading, travel. Home: 2-47-4-103 Mitsuwadai, Wakaba-ku Chiba 264, Japan

NISHIMURA, MANABU, nephrologist; b. Kitakyusyu, Fukuoka, Japan, Aug. 20, 1950; s. Noboru N. and Fujiko K. Nishimura; m. Hitomi Nishimura, Sept. 15, 1980; children: Sayaka, Misaki, Megumi, Takashi. MD, Kurume U. Sch. Medicine, Japan, 1977. Med. diplomate. Resident Kokura Nat. Hosp., Kitakyusyu, 1977-79; physician dept. internal medicine Asou Iizuka Hosp., Iizuka City, Japan, 1979-82, dir. dept. Kidney

Ctr., 1982-85, dir. dept. nephrology, 1985—. Patentee in field. Mem. Soc. Japanese Internal Medicine, Japanese Soc. Dialysis Therapy (coun. mem. 1989—), Soc. Nephrology, Soc. Transplantation. Mem. Am. Soc. Nephrology, Internat. Soc. Peritoneal Dialysis. Avocations: jogging, music, photography. Office: Asou Iizuka Hosp, 3-83 Yoshio-cho, Fukuoka, 820 Iizuka City Japan

NISHIMURA, SUSUMU, biologist; b. Taito-ku, Tokyo, Japan, Apr. 7, 1931; s. Kiyoshi and Chie (Watabe) N.; m. Michiko Uzawa, Nov. 14, 1961; children: Tomoo, Kazuo. BSc in Chemistry, U. Tokyo, 1955, PhD in Biochemistry and Biophysics, 1960. Rsch. assoc. Cancer Inst., Tokyo, 1960-62; postdoctoral fellow Biol. Div. Oak Ridge (Tenn.) Nat. Lab., 1961-63; rsch. assoc. Inst. for Enzyme Rsch. U. Wis., Madison, 1963-65; sect. chief Virology Div. Nat. Cancer Ctr. Rsch. Inst., Tokyo, 1965-68, div. chief Biology Div., 1968-92; exec. dir. Banyu Tsukuba Rsch. Inst., Banyu Pharm. Co., Ltd., Japan, 1992—. Recipient Naito Sci. award Naito Found., 1980, Acad. and Imperial prize, Japan Acad., 1988, Princess Takamatsu Cancer Rsch. prize, 1988, Fujiwara Found. prize, 1990. Mem. Am. Soc. for Biochemistry and Molecular Biology (hon.), N.Y. Acad. Sci. Avocations: radio-controlled model airplane, roses. Home: 3-10-7 Hanbatake, Tsukuba-City Ibraki 330-32, Japan Office: Banyu Tsukuba Rsch Inst, Tsukuba Techno-Park Oho, Okubo 3 Tsukuba 300-33, Japan

NISHIYAMA, TOSHIYUKI, physics educator; b. Osaka, Japan, Jan. 31, 1922; s. Ushinosuke and Ko-o (Yamanaka) N.; m. Sawako Handa, Mar. 15, 1953; 2 daus., Ayami Nishiyama Kobayashi, Yuri Nishiyama Kamino-o. B.Sc., Osaka U., 1945, D. Sc., 1954. Research assoc. Osaka U., 1945-55, assoc. prof., 1955-58, prof., 1958-85, prof. emeritus, 1985—; prof. Osaka Inst. Tech., 1985-92; asst. prof. U. Md., College Park, 1961-62; prof. grad. sch. Osaka U., 1959-85, Inst. Plasma Physics Nagoya U., Japan, 1963; mem. governing body Baika Gakuen, 1988—. Contbr. articles to profl. jours. Yukawa Meml. fellow Osaka U., 1951; Weizmann Meml. fellow Weizmann Inst., Israel, 1960. Mem. Phys. Soc. Japan (dir. Osaka br. 1977-78), Am. Phys. Soc., N.Y. Acad. Scis., Sigma Xi. Lodge: Rotary Internat. Home: 2-14-55 Minami, Okamachi, Toyonaka Osaka 560, Japan Office: INS Co Ltd 2-14-55 Minami, Okamachi 560, Osaka 560, Japan

NISHIZAWA, TEIJI, computer engineer; b. Osaka, Japan, Oct. 4, 1949; s. Kiyota and Harue Nishizawa; m. Junko Nishizawa, June 18, 1978; children: Yoshiko, Tomoko, Kouji. BS in Engring., Osaka U., 1972, MS in Engring., 1974. Engr. Matsushita Electric Ind. Co. Ltd., Kadoma, Osaka, 1974-83, sr. engr., mgr., 1986—; vis. researcher Carnegie-Mellon U., Pitts., 1983-85. Author: Parallel Processing for Computer Graphics, 1991. Mem. IEEE, Computer soc. of IEEE, Info. Processing Soc. Japan, Inst. of Electronics, Info. and Communication Engrs. Office: Matsushita Electric Ind Co, 1006 Kadoma, Kadoma-shi Osaka 571, Japan

NISLY, KENNETH EUGENE, chemical engineer, environmental engineer; b. Canton, Ohio, Aug. 8, 1950; s. Alvin Leo and Catherine (Kurtz) N.; m. Suellen Jo Wyler, July 29, 1972; children: Daniel Alan, Rebecca Ann. BS in Chem. Engring., U. Akron, 1973. Registered profl. engr., Ohio, Pa., Ill. Environ. engr. Republic Steel Co., Canton, Ohio, 1973-82, lubrication engr., 1982-86; plant environ. engr. LTV Steel Co., Canton, 1986; prin. environ. engr. Babcock & Wilcox Co., Barberton, Ohio, 1987-89; mgr. engring. Envirite Corp., Canton, 1989—. Trustee Lake Center Christian Sch., Hartville, Ohio, 1987-89, Hartville Meadows Mentally Retarded/Devel. Disabled Home, 1990—. Mem. Am. Inst. Chem. Engrs., Air and Waste Mgmt. Assn. Office: Envirite Corp 2050 Central Ave SE Canton OH 44707

NISOLLE, ETIENNE, industrial engineer; b. Mons, Hainaut, Belgium, Aug. 28, 1964; s. Rene and Lise-Marie (Rousseau) N.; m. Marina Frittaion, Nov. 17, 1990; 1 child, Cindy. BS in Electronics, Ecole Centrale des Arts et Metiers, Brussels, 1986. Project engr. Siemens, Brussels, 1986—. Mem. Brit. Amateur TV Club, Union Belge of Amateurs. Avocation: amateur radio. Office: Siemens, Chaussee de Charleroi 116, B1060 Brussels Brabant, Belgium

NISPEROS, ARTURO GALVEZ, engineering geologist, petrographer; b. San Fernando, Luzon, The Philippines, Nov. 19, 1941; came to U.S., 1965; s. Bartolome Apilado Sr. and Florencia (Galvez) N.; m. Lourdes Medrano, July 4, 1970; children: Arthur, Arnez, Arnold. BS in Geology, U. of The Philippines, Quezon City, 1962; postgrad., Utah State U. and U. Wyo., 1965-66; MBA in Tech. Mgmt., Ill. Inst. Tech., Chgo., 1989. Registered profl. geologist, Ind. Geologist Philippine Bur. of Pub. Works, Manila, 1963-68; geologist, supervising geochemist Kenting Earth Scis., Toronto, Ont., Can., 1968-72; lab. technician Flood Testing Lab., Chgo., 1972-75; sr. rsch. engr., head quality control lab. Material Svc. Corp., Chgo., 1975-88; dir. quality assurance and tech. svc. Prairie Group, Bridgeview, Ill., 1988-89; sr. rsch. petrographer Constrn. Tech. Labs., Inc., Skokie, Ill., 1989-90, supr. petrographic svcs., 1990—. Co-editor: Internat. Conf. on Cement Microscopy Proceedings, 1981—; contbr. articles to profl. publs. Bd. dirs. Cagayan Cir. of Chgo. and Suburbs, 1988—. Recipient Chgo's. Most Outstanding New Citizen award Citizenship Coun. Met. Chgo., 1979, Outstanding Achiever of Industry award YMCA of Met. Chgo., 1982, Most Outstanding Filipino in Midwest in Field of Tech. Cavite Assn. of Am., Chgo., 1982, one of 20 Most Outstanding Filpino-Ams., Fil-Am. Image, 1991. Mem. ASTM, Am. Concrete Inst., Assn. of Engring. Geologists, Geol. Soc. Am., Philippine Engrs. and Scientists Orgn. (past pres. 1984-85, bd. advisers 1985—, outstanding leadership 1985, outstanding scientist, 1986), Am. Inst. Profl. Geologists (cert.); U. of The Philippines Club of Am., Chgo. (pres. 1992-93), LA Union Club of Chgo. (pres. 1990—). Roman Catholic. Home: 1260 68th St Downers Grove IL 60516-3329 Office: Construction Tech Labs Inc 5420 Old Orchard Rd Skokie IL 60077-1060

NISSIM, ELIAHU, aerospace engineer, researcher; b. Jerusalem, June 9, 1933; s. Isaac and Rachel (Hayon) N.; m. Miriam Michelson, July 25, 1961; children: Rachel, Isaac, Dina. BSc, U. Bristol, Eng., 1957, PhD, 1963; MSc, Technion-Israel Inst. Tech., Haifa, Israel, 1961. Registered profl. engr., Israel. Lectr. in aerospace engring. Technion-Israel Inst. Tech., 1963-66, sr. lectr. in aerospace engring., 1966-71, assoc. prof. aerospace engring., 1971-78, dean faculty, 1971-73, 78-80, prof. aerospace engring., 1978—, v.p. for acad. affairs, 1983-86; mem. bd. govs. Israel Aircraft Industries, Lod, 1984-87; mem. Coun. for higher Edn., Jerusalem, 1990—; sr. rsch. assoc. U.S. NRC, Washington, 1969, 75, 87, 88. Contbr. over 40 articles to profl. jours. Fellow Royal Aero. Soc. (George Taylor award 1966); mem. AIAA. Achievements include patent in field. Home: 62 Bikurim St, Haifa 34577, Israel Office: Faculty Aerospace Engring, Technion City, Haifa 32000, Israel

NISTRI, ANDREA, pharmacology educator; b. Florence, Italy, Jan. 15, 1947; s. Carlo and Anna (Baroncelli) N.; m. Sandra Breckon, Aug. 4, 1972; children: Giampaolo, Christian. MD, U. Florence, 1971. Registered physician, Eng., Italy. Intern, hon. registrar in clin. toxicology St. Maria Nuova Hosp., Florence, 1972-73; resident, hon. registrar in clin. pharmacology St. Bartholomew's Hosp., London, 1973-75; asst. prof. U. Florence, 1975-76, assoc. prof., 1983-91; rsch. fellow Med. Rsch. Coun. McGill U., Montreal, Que., Can., 1977-78; lectr. St. Bartholomew's Hosp. Med. Coll., London, 1979-84, sr. lectr., 1985-87, reader, 1987-90; reader Queen Mary and Westfield Coll., London, 1990-92; prof. cellular and molecular pharmacology Internat. Sch. Advanced Studies, Trieste, Italy, 1992—. Contbr. numerous articles to profl. jours. Lt. Italian mil., 1972-74. Fellow Royal Soc. Medicine, Ciba Found., London, 1974-75; rsch. fellow NATO, Montreal, 1976. Mem. Internat. Brain Rsch. Orgn., Internat. Soc. Neurochemistry, European Soc. Neurochemistry, Physiol. Soc., Soc. Neurosci., Brit. Pharm. Assn., Brit. Med. Assn. Avocations: classic cars, mountain hiking, art history. Office: Internat Sch Advanced, Via Beirut 2, 34013 Trieste Italy

NISWENDER, GORDON DEAN, physiologist, educator; b. Gillette, Wyo., Apr. 21, 1940; s. Rex Lel and Inez Irene (Dillinger) N.; m. Joy Dean Thayer, June 14, 1964; children: Kevin Dean, Kory Dean. B.S., U. Wyo., 1962; M.S., U. Nebr., 1964; Ph.D., U. Ill., 1967. NIH postdoctoral fellow U. Mich., 1967-68, asst. prof. physiology, 1968-72; mem. faculty Colo. State U., Ft. Collins, 1972—; prof. physiology Colo. State U., 1975—; assoc. dean research Coll. Veterinary Medicine and Biomed. Scis., 1982—, dir. animal reproduction and biotech. lab., 1986—, disting. prof., 1987—; mem. rev. panels NIH; cons. FDA. Recipient Merit award NIH, grantee, 1968—. Mem.Am. Assn. Animal Scientists (Outstanding Young Scientist award western sect. 1974, Animal Physiology and Endocrinology award 1983), Soc. Study Reprodn. (treas. 1972-75, pres. 1981-82, rsch. award 1988). Office: Colo State U Animal Reprod & BiotechnologyLab College of Veterinary Medicine Fort Collins CO 80523

NITHIANANDAM, JEYASINGH, physicist; b. Virudhunagar, Tamil Nadu, India, Apr. 24, 1950; came to U.S., 1979; s. Daniel and Etta Sophie Vedamanickam; m. Vinitha Angeline Chandrasekaran, Sept. 6, 1978; children: Christopher Daniel, Divya Christine. MA in Physics, U. Scranton, 1980; PhD in Exptl. Physics, U. Va., 1989. Assoc. prof. Am. Coll., Madurai, India, 1971-79; grad. fellow U. Scranton, Pa., 1979-80; grad. tchr. asst. U. Va., Charlottesville, 1980-82, grad. rsch. asst., 1983-89; asst. prof. Brevard (N.C.) Coll., 1982-83; rsch. assoc. NRC, Washington, 1989-91; prin. scientist Hughes STX Corp., Lanham, Md., 1991—; faculty mem. Summer Sch. on Vaccum and Thiin Films Physics, 1978; mem. microwave humidity sounder design proposal rev. panel NASA Goddard Space Flight Ctr., Greenbelt, Md., 1991. Contbr. articles to Phys. Rev. Letters, Applied Physics Letters and Phys. Rev. B. Mem. Am. Phys. Soc., Am. Geophys. Soc., Materials Rsch. Soc., Indian Assn. for Physics Tchrs. Methodist. Achievements include discovery of dipole forbidden core excitonic transition in diamond x-ray absorption and emission spectra, surface core excitons at internal interfaces of polycrystalline diamond films, anomalous life-time effects on valence band of doped amorphous hydrogenated silicon. Home: 4626 Smokey Wreath Way Ellicott City MD 21042 Office: Hughes STX Corp 4400 Forbes Blvd Lanham MD 20706

NITKA, HERMANN GUILLERMO, hospital administrator; b. Buenos Aires, Nov. 28, 1926; s. Natalio and Emma (Wille) N.; m. Ingeburg Sigrid Imkamp, Apr. 29, 1953; children: Cristina Brígida Nitka de Laube, Federico Germán. Cert. in pub. acctg., U. Buenos Aires, 1949, Cert. Translator/Interpreter of German, 1950, D Econ. Scis., 1956. CPA, Argentina. Chief internal auditing Siam di Tella Ltda., Buenos Aires, 1950-56; acctg. and fin. mgr. SIAT S.A., Lanus, Argentina, 1956-67; mgr. adminstrn. and fin. Establecimentos Klöckner S.A., Buenos Aires, 1967-87; dir. Hosp. Alemán, Buenos Aires, 1987—; pres. Gelia S.A., Buenos Aires, 1987—. Treas. Argentine chpt. of Ecumenical Ch. Loan Fund, a div. of the World Coun. of Chs., Geneva, 1987—. Mem. Coun. of German Evangelical Congregation of Buenos Aires. Home: Dr Lisandro de la Torre, 1.555, 1638 Vicente López BA, Argentina Office: Hospital Alemán, Avda Pueyrredón 1640, 1118 Buenos Aires Argentina

NITSCHE, JOHANNES CARL CHRISTIAN, mathematics educator; b. Olbernhau, Fed. Republic Germany, Jan. 22, 1925; came to U.S., 1956; s. Ludwig Johannes and Irma (Raecke) N.; m. Carmen Dolores Mercado Delgado, July 1, 1959; children: Carmen Irma, Johannes Marcos and Ludwig Carlos (twins). Diplom für Mathematik, U. Göttingen, 1950; PhD, U. Leipzig, 1951; Privatdozent, Tech. U. Berlin, 1955. Asst. U. Göttingen, 1948-50; rsch. mathematician Max Planck Institut für Strömungsforschung Göttingen, 1950-52; asst. Privatdozent Tech. U., Berlin, 1952-56; vis. assoc. prof. U. Cin., 1956-57; assoc. prof. U. Minn., Mpls., 1957-60; prof. math. U. Minn., 1960—, head Sch. Math., 1971-78; vis. prof. U. P.R., 1960-61, U. Hamburg, 1965, Tech. Hochschule Vienna, 1968, U. Bonn, 1971, 75, 77, 80, 81, U. Heidelberg, 1979, 82, 83, U. Munich, 1983, U. Florence, 1983; keynote speaker Festive Colloquium, U. Ulm, 1986; co-organizer workshop statis. thermodynamics and differential geometry U. Minn., 1991; keynote speaker Meml. Colloquium Tech. U. Berlin, 1991, speaker Internat. Workshop on Geometry and Interfaces, Aussois, France, 1990. Author: Vorlesungen uber Minimalflachen, Springer-Verlag, 1975, Lectures on Minimal Surfaces, 1989; mem. editorial bd. Archive of Rational Mechanics and Analysis, 1967-91; editor: Analysis, 1980—; assoc. editor: Contemporary Math., 1980-88, Zeitschrift für Analysis und ihre Anwendungen, 1993—; contbr. articles to profl. jours. Mem. Am. del. joint Soviet-Am. Symposium on Partial Differential Equations, Novosibirsk, 1963, U.S.-Japan Seminar on Differential Geometry, Tokyo, 1977; speaker 750th Berlin Anniversary Colloquium, Free U. Berlin, 1987. Recipient Lester R. Ford award for outstanding expository writing, 1975, George Taylor Disting. Svc. award U. Minn. Found., 1980, Humboldt prize for sr. U.S. scientists Alexander von Humboldt Found., 1981; Fulbright rsch. fellow Stanford, 1955-56. Fellow AAAS; mem. Am. Math. Soc., Circolo Matematico di Palermo, Deutsche Mathematiker-Vereinigung, Edinburgh Math. Soc., Gesellschaft für Angewandte Mathematik und Mechanik, Math. Assn. Am., N.Y. Acad. Scis., Österreichische Mathematische Gesellschaft, Soc. Natural Philosophy. Home: 2765 Dean Pky Minneapolis MN 55416-4382

NITTA, KYOKO, aerospace engineering educator; b. Nanao, Ishikawa, Japan, Sept. 22, 1962; d. Hiroshi and Eiko (Nakamura) N. BEng, Tokyo U., 1985. Rsch. engr. Computervision Japan Co., Atsugi/Kanagawa, 1985-86; rsch. assoc. Nogoya U., 1986—. Contbr. articles to profl. jours. Mem. AIAA, Japan Soc. Aero. and Space Scis. (sec. 1991—). Home: 40 Inoue Chikusa, Hoshigaoka Hogi Bldg 2C, Nagoya 464, Japan Office: Nagoya Univ, 1 Furo Chikusa, Nagoya Japan 464-01

NITTOLI, THOMAS, chemist, consultant; b. Newark, Sept. 21, 1963; s. Jerry and Angelina (Gioino) N. BA in Chemistry, Rutgers U., Newark, 1987; postgrad., SUNY, Stony Brook, 1992—. Per diem tchr. Newark Bd. Edn., 1986-87; info. scientist Merck and Co., Rahway, N.J., 1987-88; scientist CIBA-GEIGY Pharm., Summit, N.J., 1988-92; grad. asst. SUNY, Stony Brook, 1992—; chem. cons. Amalthea Wine Cellers, Atco, N.J., 1987—. Mem. Am. Chem. Soc. Home: 29 Martin Rd West Caldwell NJ 07006-7431

NITZ, FREDERIC WILLIAM, electronics company executive; b. St. Louis, June 22, 1943; s. Arthur Carl Paul and Dorothy Louise (Kahm) N.; m. Kathleen Sue Rapp, June 8, 1968; children: Frederic Theodore, Anna Louise. AS, Coll. Marin, 1970; BS in Electronics, Calif. Poly. State U., San Luis Obispo, 1972. Electronic engr. Sierra Electronics, Menlo Park, Calif., 1973-77, RCA, Somerville, N.J., 1977-79; engring. mgr. EGG-Geometrics, Sunnyvale, Calif., 1979-83; v.p. engring. Basic Measuring Insts., Foster City, Calif., 1983-91; exec. v.p. Reliable Power Meters, Los Gatos, Calif., 1991—; cons. in field, Boulder Creek, Calif., 1978—. Patentee in field. Bd. dirs. San Lorenzo Valley Water Dist., Boulder Creek, 1983—, Water Policy Task Force, Santa Cruz County, Calif., 1983-84. With U.S. Army, 1965-67. Democrat. Lutheran. Home: 12711 East St Boulder Creek CA 95006-9148 Office: Reliable Power Meters 400 Blossom Hill Rd Los Gatos CA 95032-4511

NITZSCHE, FRED, aeronautical engineer; b. Sao Paulo, Brazil, June 21, 1953; s. Frederico Eriberto and Alice (Trindade) N.; m. Regina Fuser Pillis, Dec. 17, 1977. MS, Stanford U., 1980, PhD, 1983. Researcher Brazilian Space Inst., S.J. Campos, Brazil, 1977-85; sr. engr. Avibras S.A., S.J. Campos, 1985-86; mgr. asst. Embraer S.A., S.J. Campos, 1986-91; assoc. prof. Inst. Tech. de Aeronautica, S.J. Campos, 1984-90; guest scientist DLR-German Aerospace Rsch. Establishment, Gottingen, 1991-92, researcher, 1992—; cons. Avibras S.A., 1983, D.F. Vasconcellos S.A., Sao Paulo, 1984-85, Promon Engring. S.A., Sao Paulo, 1985. Contbr. articles to Jour. of Aircraft. Lt. Brazilian Air Force, 1972-73. Scholar Brazilian Govt., 1974-75, 78-82, fellow, 1991-92. Mem. AIAA. Achievements include patents pending. Office: DLR Inst fur Aeroelastik, Bunsenstr 10, W-37073 Göttingen Germany

NIU, KEISHIRO, science educator; b. Katsuragi, Ito, Wakayama, Japan, Aug. 5, 1929; s. Takayoshi and Chie Niu; m. Hiromi Niu, Oct. 6, 1956; children: Naoki, Hiroki Ishikawa. BS, Kyoto (Japan) U., 1951, MS, 1953, DSc, 1961. Assoc. prof. Osaka (Japan) U., 1963-69; prof. Tokyo Inst. Tech., 1969-90, Teikyo U. Tech., Ichihara, Chiba, Japan, 1991—; vis. scientist Columbia U., 1965-66; vis. prof. Paris U., Padova (Italy) U., Darmstadt (Germany) Tech. U., Sevelle (Spain) U., Prague (Czechoslovakia) Tech. U. Author: Nuclear Fusion, 1989; mem. editorial bd. Laser and Particle Beams, 1983—; author books in Japanese. Avocations: tennis, painting, travel. Office: Teiko U Tech, Uruido, Ichihara Chiba 290-01, Japan

NIX, MARTIN EUGENE, engineer; b. Winston-Salem, N.C., Sept. 26, 1951; s. Richard Nix and Margaret Searborn-Collins. B, U. N.Mex., 1973, postgrad., 1976. Engr. Boeing, Seattle, 1980—; chief exec. officer Solarshack, Seattle, 1989—.

NIX, WILLIAM DALE, materials scientist, educator; b. King City, Calif., Oct. 28, 1936; s. Earl Thomas and Ethel Lula (Jackson) W.; m. Jean Aldene Telford, June 15, 1958; children—Cynthia Lea, Jeffrey Alan, Rebecca Jean. B.S., San Jose State Coll., 1959; M.S., Stanford U., 1960, Ph.D., 1963. Asst. prof. materials sci. Stanford U., 1962-66, asso. prof., 1967-72, prof., 1972—, dir. Ctr. Materials Research, 1968-70, asso. chmn. dept., 1975-86, acting chmn. dept., 1986-87, Lee Otterson prof. Sch. of Engring., 1989—, chmn. dept., 1991—; asst. to dir. technology Stellite div. Union Carbide Corp., Kokomo, Ind., 1966; cons. to govt. and indsl. orgns. Author: (with C.R. Barrett and A.S. Tetelman) The Principles of Engineering Materials, 1973; contbr. articles to profl. jours. Recipient Teaching Excellence award Western Electric Fund, 1964, Disting. Alumni award San Jose State U., 1980, Gold medal Acta Metallurgica, 1993. Fellow Am. Soc. Metals (Bradley Stoughton award 1970, Edward DeMille Campbell lectr. 1989), Metall. Soc.; mem. AIME (Mathewson Gold medal 1979, R.F. Mehl medal 1988, Inst. Metals lectr 1988), NAE, Am. Soc. Engring. Edn., Sigma Xi. Republican. Presbyterian. Home: 1001 Persimmon Ave Sunnyvale CA 94087 Office: Stanford U Dept Materials Sci Stanford CA 94305

NIXON, CHARLES WILLIAM, bioacoustician; b. Wellsburg, W.Va., Aug. 15, 1929; s. William E. and Lenora S. (Treiber) N.; m. Barbara Irene Hunter, May 19, 1956; children: Timothy C., Tracy Scott. BS, Ohio State U., 1952, MS, 1953, PhD, 1960. Tchr. spl. edn. Ohio and W.Va. Pub. Schs., Wheeling, 1954-56; rsch. audiologist Aeromed Lab., Wright Patterson AFB, Ohio, 1956-67; supervisory rsch. audiologist Armstrong Lab., Wright Patterson AFB, 1967—; chair W4 Am. Nat. Stds. Inst., N.Y.C., 1968—; U.S. rep. hearing protection Internat. Stds. Orgn., Geneva, 1968—; USAF rep. NRC-NAS Hearing Com., Washington, 1976—; chair robotics panel Joint Dirs. Labs., Washington, 1987-88. Author reports and book chpts. Cpl. U.S. Army, 1953-55. Recipient Meritorious Svc. medal U.S. Dept. Def., Dayton, Ohio, 1986. Fellow Acoustical Soc. Am.; mem. Rsch. Soc. Am. Achievements include research on noise exposure, voice communications, hearing protection, sonic boom, others. Home: 4316 Sillman Pl Dayton OH 45440

NIZAMUDDIN, KHAWAJA, structural engineer; b. Hyderabad, India, May 22, 1960; came to U.S., 1983; s. Khawaja and Fatima (Qureshy) Qumaruddin; m. Noorunisa Shahnaz Razzack, Jan. 26, 1989; 1 child, Rehan. BSCE, Ned U., Karachi, Pakistan, 1983; MS in Structural Engring., U. Calif., Berkeley, 1985; MBA in Fin., DePaul U., 1991. Registered profl. engr., Wis.; lic. structural engr., Ill. Structural engr. Kolbjorn Saether & Assocs., Chgo., 1985-89, chief engr., 1990—. Recipient Best Structures award Structural Engrs. Assn. Ill., 1988; Pres.'s fellow Ministry of Edn., Govt. of Pakistan, 1978-82; Aga Khan Found. fellow, Geneva, 1983. Mem. ASCE (assoc.), Fin. Mgmt. Assn., Delta Mu Delta (Eta chpt.). Office: Kolbjorn Saether & Assocs 1062 W Chicago Ave Chicago IL 60622-5416

NIZIN, JOEL SCOTT, surgeon; b. N.Y.C., Mar. 14, 1953; s. Harry and Florence (Silverstein) N.; m. Vickie Rene Johnson, Sept. 2, 1979; children: Zia Tahirih, Anisa Jamal. BA cum laude, Amherst Coll., 1974; MD, Howard U., 1978. Intern, then resident, then chief resident in surgery St. Luke's Roosevelt Hosp. Ctr., N.Y.C., 1978-83; fellow in colon rectal surgery U. Minn., Mpls., 1983-84; colon rectal surgeon N.J. Colon Rectal Surg. Assocs., Fairlawn, N.J., 1984—; mem. cancer com. Valley Hosp., 1987—; mem. various coms. Chilton Hosp., 1985—; presenter in field. Contbr. articles to Jour. Biol. Chemistry, Am. Jour. Surgery, Contemporary Surgery, Disease Colon Rectal Surgery. Judge City Wide Sci. Fair, N.Y.C., 1990. Fellow ACS, Am. Coll. Colon Rectal Surgeons (pres. N.J. chpt. 1990-91, mem. various nat. coms.); mem. AMA, AAAS, Am. Cancer Soc. (pres. Passaic chpt. 1985-86), Am. Soc. Gastroenterology, Nat. Colitis Ileitis Coun., N.J. Med. Soc., N.Y. Surg. Soc., N.Y. Colon Rectal Soc., N.J. Colon Rectal Soc., N.J. Surg. Soc., Bergen County Med. Soc., Bergen County Stoma Soc., Sigma Xi. Office: N Jersey Colon Rectal Surg 4-14 Saddle River Rd Fair Lawn NJ 07410

NOAR, MARK DAVID, internist, gastroenterologist, therapeutic endoscopist, consultant, inventor; b. Boston, Sept. 10, 1953; s. Myron Theodore and Phyllis (Krinsky) N.; m. Martine Denise Motard, May 15, 1983; children: Emmanuelle, Ariane. BS in Biology, Ursinus Coll., Collegeville, Pa., 1975; MPH in Internat. Health, Tulane U., 1977; MD, U. Cen. del Este, Dominican Republic, 1980. Intern 5th Pathway program Coll. Medicine and Dentistry N.J.-Newark Beth Israel Hosp., 1980-81; resident in internal medicine U. Nebr. Med. Ctr., Omaha, 1981-84; fellow in gastroenterology SUNY Downstate Med. Ctr., Bklyn., 1984-86; fellow in therapeutic and surg. endoscopy, vis. staff Univ. Hosp. Hamburg, Germany, 1986-87; pvt. practice, Balt., 1988—; CEO Chesapeake Bay Brewing Co., Balt., 1992—; clin. cons. in therapeutic endoscopy Bklyn. VA Med. Ctr., 1987; dir. project devel., v.p. med. devel. Ixion, Inc., Seattle, 1987—; staff physician dept. gastroenterology St. Joseph Hosp., Balt., Franklin Square Hosp., Balt.; bd. dirs., dir. ops. Disaster Support Network, Balt., 1990—; session co-chmn. World Congress Gastroenterology Sydney, Australia, 1990, IX European Workshop on Therapeutic Digestive Endoscopy, Brussels, 1991; CEO, med. dir. EMA-Ambulatory Surgery Ctr., Inc., Balt., 1990—; CEO, bd. dirs. Capitol City Brewing Co. Md., 1992—. Author: (with N. Soehendra and H. Grimm) A Compendium of Therapeutic Endoscopy for the General Practitioner, 1991; editor-in-chief Internat. Video Jour. Therapeutic and Diagnostic Endoscopy; assoc. editor Endoscopy Rev.; contbr. articles and abstracts to med. jours., chpts. to books; inventor robotic interactive endoscopy simulation. Pub. lectr. Am. Cancer Soc., Balt., 1988—; physician educator Doctor and Lawyer Coalition Against Drugs, Balt., 1991-92. Fellow Royal Soc. Tropical Medicine and Hygiene; mem. ACP, AMA, Am. Coll. Gastroenterology, Am. Soc. Gastrointestinal Endoscopy (instr. regional advanced endoscopy 1992—, award for achievement and edn. in diagnostic/therapeutic biliary and pancreatic endoscopy 1992), Baltimore County Med. Soc., Sigma Xi. Avocations: guitar, banjo, sailing, orchid culture, gourmet cooking. Office: Endoscopic Microsurgery Assocs 7402 York Rd Ste 100 Baltimore MD 21204-7519

NOBEL, JOEL J., physician; b. Phila., Dec. 8, 1934; s. Bernard D. and Golda R. (Nobel) Judovich; m. Bonnie Sue Goldberg, June 19, 1960 (div.); children—Erika, Joshua; m. Loretta Schwartz, Oct. 28, 1979; 1 child, Adam. A.B., Haverford Coll., 1956; M.A., U. Pa., 1958; M.D., Thomas Jefferson Med. Coll., Phila., 1963. Intern Presbyn. Hosp., Phila., 1963-64; resident in surgery Pa. Hosp., Phila., 1964-65; resident in neurosurgery U. Pa. Hosp., 1965-66; practice medicine specializing in biomed. engring. research Phila., 1968—; dir. research Emergency Care Research Inst., Plymouth Meeting, Pa., 1968-71; dir., pres. Emergency Care Research Inst., 1971—; pres. Plymouth Inst., 1979—; cons. in field; bd. dirs. Consumers Union, 1976-79, 80—, chmn. tech. policy com., exec. bd. Publisher Health Devices, 1971—, Health Devices Alerts, 1977—. Contbr. articles to profl. jours. Served with USNR, 1966-68. Smith, Kline & French fgn. fellow, 1962; grantee HEW, 1968-72; grantee Am. Heart Assn., 1965-66. Mem. Assn. Advancement Med. Instrumentation, Critical Care Med. Soc., Am. Public Health Assn., AMA, Pa. Med. Assn., Am. Def. Preparedness Assn. Clubs: Union League, Sunday Breakfast. Home: 1434 Monk Rd Gladwyne PA 19035-1315 Office: ECRI 5200 Butler Pike Plymouth Meeting PA 19462-1241

NOBLE, JAMES KENDRICK, JR., media industry consultant; b. N.Y.C., Oct. 6, 1928; s. James Kendrick and Orrel Tennant (Baldwin) N.; m. Norma Jean Rowell, June 16, 1951; children: Anne Rowell, James Kendrick III. Student, Princeton U., 1945-46; BS, U.S. Naval Acad., 1950; postgrad., USN Gen. Line Sch., 1955-56; MBA, NYU, 1961; postgrad., Sch. Edn., 1962-68. Chartered fin. analyst. Commd. ensign USN, 1950; transferred to USNR, 1957; advanced through grades to capt. USNR, 1973; asst. gunnery officer in U.S.S. Thomas E. Fraser, 1950-51; student naval aviator USNR, 1951-52, pilot asst. ops. officer, 1952-55; instr. U.S. Naval Acad., 1956-57, Officer Candidate Sch., Newport, R.I., 1958; asst. to pres. Noble & Noble, Pub., Inc., N.Y.C., 1957-60; dir. spl. projects Noble & Noble, Pub., Inc., 1960-62, v.p., 1962-65, exec. v.p., 1965-66, dir., 1957-65; dir., v.p. Transl. Pub. Co., N.Y.C., 1958-65; cons. Transl. Pub. Co., 1965-66; v.p., dir. Elbon Realty Corp., Bronxville, N.Y., 1960-65; cons. Elbon Realty Corp., 1965-66; comdg. officer NAIRU R2, 1968-70; staff NARS W2, 1970-71, NRID 3-1, 1971-74; comdg. officer NRCSG 302, 1974-76; sr. analyst F. Eberstadt & Co., 1966-69; sr. analyst Auerbach, Pollak & Richardson, 1969-75, v.p., 1972-75, mgr. spl. rsch. projects, 1973-75, dir., 1975; v.p. rsch. Paine, Webber, Jackson & Curtis, Inc., 1975-77, assoc. dir. rsch., 1976-77; v.p.

Paine Webber, N.Y.C., 1977-79; 1st v.p. Paine Webber, 1979-91; pres. Noble Cons. Inc., 1991—; bd. dirs. Curriculum Info. Center, Inc., Denver, 1972-78; instl. investor All Am. Rsch. Team, 1972-90. Author: Ploob, 1949, rev., 1956; editor pub.: The Years Between, 1966; also articles in various publs. V.p. Bolton Gardens Assn., 1959-61; mem. Bronxville Bd. Edn., 1968-74, pres., 1970-72; Republican co-leader 21st Dist., Eastchester, N.Y., 1961-65; dir. Merit; cons., dir. Space and Sci. Train, 1962-63; trustee St. John's Hosp., Yonkers, N.Y., 1972-92, com. chmn., 1980-92. Fellow AAAS; mem. Info. Industry Assn. (disting. profl. mem.), Nat. Inst. Social Scis., N.Y. Soc. Security Analysts (mem. com. 1971-91, dir. 1975-84, v.p. 1977-81, exec. v.p. 1981-82, pres. 1982-83), Am. Textbook Pub. Inst. (com. chmn. 1964-66), AIAA (mem. com. 1957-61), Media and Entertainment Analysts Assn. (pres. 1969-71), Fin. Analysts Fedn. (dir. 1984-87), Naval Res. Assn. (v.p. N.Y. Navy chpt. 1968-76), Wings Club, Siwanoy Country Club. Mem. Reformed Ch. Home: 45 Edgewood Ln Bronxville NY 10708-1946 Office: Noble Cons Inc 45 Edgewood Ln Bronxville NY 10708-1946

NOBLIN, CHARLES DONALD, clinical psychologist, educator; b. Jackson, Miss., Dec. 16, 1933; s. Charles Thomas and Margaret (Byrne) N.; m. Patsy Ann Beard, Aug. 12, 1989. BA, Miss. Coll., 1955; MS, Va. Commonwealth U., 1957; PhD, La. State U., 1962. Lic. psychologist, N.J., N.C. Instr. to asst. prof. La. State U., Baton Rouge, 1961-63; asst. to assoc. prof. U. N.C., Greensboro, 1963-66; assoc. prof. Rutgers Med. Sch., New Brunswick, N.J., 1966-69; dir. clin. training Va. Commonwealth U., Richmond, 1969-72; chmn. dept. psychology Va. Tech., Blacksburg, 1972-82; dir. clin. training U. So. Miss., Hattiesburg, 1982-85, chmn. dept. psychology, 1985-91, prof., 1991-93, dir. clin. tng., 1993—. Contbr. over 60 articles and presentations. Recipient Clin. Tng. grant NIMH, 1983-86, Victim Behavior & Personal Space rsch. grant U.S. Dept. Justice, 1970-71, Trubeck Found. Rsch. award, 1968-69. Baptist. Avocation: antique art glass. Home: 2501 Sierra Cir Hattiesburg MS 39402-2540 Office: U So Miss Dept Psychology Hattiesburg MS 39406

NOBLITT, NANCY ANNE, aerospace engineer; b. Roanoke, Va., Aug. 14, 1959; d. Jerry Spencer and Mary Louise (Jerrell) N. BA, Mills Coll., Oakland, Calif., 1982; M.S. in Indsl. Engring., Northeastern U., 1990. Data red specialist, Universal Energy Systems, Beaver Creek, Ohio, 1981; aerospace engr. turbine engine div. components br. turbine group aero-propulsion lab. Wright-Patterson AFB, Ohio, 1982-84, engine assessment br. spl. engines group, 1984-87; lead analyst cycle methods computer aided engr. Gen. Electric Co., Lynn Mass., 1987-90, Lynn PACES project coord., 1990-91; software systems analyst Sci. Applications Internat. Corp., artificial intelligence Mc Lean, Va., 1991-92, software engring. mgr., Intelligence Applications Integration, Hampton, Va., 1992-93, mgr. test engring. and systems support, 1993—. Math and sci. tutor Centerville Sch. Bd., Ohio, 1982-86, math. and physics tutor Marblehead Sch. Bd., Mass., 1988-90; rep. alumnae admissions Mills Coll., Boston area, 1987-91. Recipient Notable Achievement award U.S. Air Force, 1984; recipient Special award Fed. Lab. Consortium, 1987. Avocation: book collecting, weight training. Home: 109 Signature Way # 626 Hampton VA 23666 Office: Sci Applications Internat Corp Hampton VA 23666

NOCE, ROBERT HENRY, neuropsychiatrist, educator; b. Phila., Feb. 19, 1914; s. Rev. Sisto Julius and Madeleine (Saulino) N.; m. Carole Lee Landis, 1987. A.B., Kenyon Coll., 1935; M.D., U. Louisville, 1939; postgrad., U. Pa. Sch. Medicine, 1947, Langley-Porter Neuropsychiat. Inst., 1949, 52. Rotating intern Hamot Hosp., Erie, Pa., 1939-40; resident psychiatrist Warren (Pa.) State Hosp., 1940-41, staff psysician, 1946-48; staff physician Met. State Hosp., Norwalk, Calif., 1948-50; dir. clin. services Pacific State Hosp., Spadra, Calif., 1950-52; dir. clin. services Modesto (Calif.) State Hosp., 1952-58, asst. supt. psychiat. services, 1958-64; pvt. practice medicine specializing in neuropsychiatry, 1965-73; Mem. faculty postgrad. symposiums in psychiatry for physicians U. Calif., 1958, 66. Author: film Reserpine Treatment of Psychotic Patients; Contbr. articles to profl. jours. Served from lt. (j.g.) to lt. comdr. M.C. USNR, 1941-46. Recipient Albert and Mary Lasker award for integration reserpine treatment mentally ill and mentally retarded, 1957; Wisdom award of honor, 1970. Life fellow Am. Psychiat. Assn. (sec. 1954, 55); fellow Royal Soc. Health; mem. Phi Beta Kappa, Delta Psi. Episcopalian. Home: 407 E Colgate Dr Tempe AZ 85283-1809

NOCI, GIOVANNI ENRICO, electronics engineer; b. Lucca, Italy, July 30, 1955; s. Remo N.; m. Stefania Baggiani; children: Claudia, Enrico. BS in Elec. Engring., Pisa (Italy) U., 1980. System engr. OTD Melara, La Spezia, Italy, 1982-86; programs and systems head Proel Tech., Florence, Italy, 1986—; researcher, cons. Pisa U., 1980-81. Contbr. articles to sci. publs. Mem. AIAA.

NOCKS, JAMES JAY, psychiatrist; b. Bklyn., Apr. 17, 1943; s. Henry and Pearl (Klein) N.; m. Ellen Jane Leblang, June 21, 1964; children: Randy, Jason. BA in English Lit., U. Pa., 1964, MD, 1968. Diplomate Nat. Bd. Med. Examiners, Am. Bd. Psychiatry and Neurology; Cert. in adminstrv. psychiatry Am. Psychiatric Assn. Rotating intern Chgo. Wesley Meml. Hosp., Northwestern U., 1968-69; resident psychiatry U. Pa., Phila., 1969-73; chief alcoholism program VA Med. Ctr., West Haven, Conn., 1975-87; asst. chief psychiatry svc. VA Med. Ctr., 1978-87; chief staff VA Med. Ctr., Coatesville, Pa., 1987—; sr. resident psychiatry U. Pa., 1971-72, chief resident psychiatry, 1972-73, asst. instr. psychiatry, 1971-73; clin. asst. prof. psychiatry Health Sci. Ctr. U. Tex., San Antonio, 1973-75; asst. clin. prof. psychiatry Yale U. Sch. Medicine, New Haven, 1975-80, assoc. clin. prof. psychiatry, 1980-87; clin. prof. psychiatry and human behavior Jefferson Med. Coll., Phila., 1987-91; prof. psychiatry Temple U. Sch. Medicine, Phila., 1991—. Contbr. chpts. in Psychiatry: Pre-Test, Self-Assessment and Review, 2d edit., 1982, 3d edit., 1984, Alcoholic Liver Disease, 1985; contbr. articles to profl. jours. Mem. exec. com. Med. Alumni Soc., U. Pa., 1989—. Major Med. Corps, USAF, 1973-75. Fellow Am. Psychiat. Assn.; mem. Pa. Psychiat. Soc., Phila. Psychiat. Soc., Am. Soc. of Addiction Medicine (cert.), Assn. for Med. Edn. & Rsch. in Substance Abuse, Am. Assn. for Social Psychiatry, Am. Acad. Psychiatrists in Alcoholism & Addictions. Jewish. Avocations: bicycling, tennis, music. Office: VA Med Ctr Blackhorse Hill Rd Coatesville PA 19320-3313

NOCKS, RANDALL IAN, systems engineer; b. Chgo., Apr. 5, 1969; s. James J. and Ellen J. (Leblang) N.; m. Lisa R. Sprague, Oct. 19, 1991. BS in Aerospace Engring., Boston U., 1991; postgrad., Drexel U., 1992—. Systems engr. mgmt. and data systems Martin Marietta, King of Prussia, Pa., 1988—. Founding sponsor Challenger Ctr., 1988—.

NODA, YUTAKA, physician, otolaryngologist; b. Toyonaka, Osaka, Japan, Sept. 22, 1937; s. Masayuki and Yukiko (Yuasa) N.; m. Hiroko Tamura, Nov. 19, 1963; children: Maki, Miki. BA, Med. Faculty Keio U., Tokyo, 1962; MD, Nihon U., Tokyo, 1972. Intern Keio U. Hosp., Tokyo, 1962-63; resident Nihon U. Hosp., Tokyo, 1963-65, asst., 1965-68, asst. lectr., 1972-73; asst. Hamburg (Fed. Republic of Germany) U. Hosp., 1968-72; assoc. prof. Ryukyu U. Hosp., Naha-Shi, Okinawa, Japan, 1973-81, prof., 1981-83; prof. medicine Ryukyu U., Nishihara, Okinawa, 1983—; vis. prof. Guangxi (Republic of China) Med. Coll., 1989—. Contbr. articles to profl. jours. Mem. Japanese Soc. Stomatopharyngology (bd. dirs. 1988—), Otorhinolaryngology Soc. Japan (councilor 1978—), German Soc. Throat, Nose and Ear Medicine, Soc. Head and Neck Surgery (corr.), Soc. for Promotion Rsch. in Otorhinolaryngology in Ryukyus (chief dir. 1983—). Avocations: Japanese archery, travel. Office: U Ryukyus, Aza-Uehara 207, Nishihara-Cho, Okinawa 903-01, Japan

NOEBE, RONALD DEAN, materials research engineer; b. Canton, Ohio, July 17, 1961; s. Donald Richard and Ellen Marie (Makley) N.; m. Anita Diane Tenteris, June 8, 1985. BS, Case Western Reserve U., 1983, MS, 1986; MS in Materials Engring., U. Mich., 1986. Technician Rep. Steel, Ctrl. Alloy Div., Canton, Ohio, 1980; asst. turn metallurgist Rep. Steel, Ctrl. Alley Divsn., Canton, Ohio, 1981; lab. mgr. Chase Brass and Copper, Solon, Ohio, 1982; materials rsch. engr. NASA Lewis Rsch. Ctr., Cleve., 1987—; symposium organizer MRS Conf., Boston, 1994. Contbr. over 70 tech. papers to profl. publs. Mem. Am. Powder Metallurgy Inst., The Metall. Soc., Materials Rsch. Soc., Alpha Sigma Mu, Sigma Xi. Independent. Home: 2510 1/2 Medina Rd Medina OH 44256 Office: NASA Lewis Rsch Ctr 21000 Brookpark Rd MS 49-3 Cleveland OH 44135

NOEL, DALE LEON, chemist; b. Wichita, Kans., May 21, 1936; s. Marvin D. and L. Frances (Knowles) N.; m. E. Jean Stovall, Apr. 21, 1962; children: Brian, Susan. BA in Chemistry, Friends U., 1958; MS in Analytical Chemistry, Wichita State U., 1960; PhD in Analytical Chemistry, Kans. State U., 1970. Asst. prof. chemistry Friends U., Wichita, 1961-65, Eastern Nazarene Coll., Quincy, Mass., 1967-69; assoc. prof. chemistry Kearney (Nebr.) State Coll., 1970-74; rsch. assoc. Internat. Paper, Tuxedo, N.Y., 1974-80, sr. rsch. assoc., 1980-90, prin. scientist, 1990-. Mem. Am. Chem. Soc. (chmn. 1964-65), Tech. Assn. Pulp and Paper Industry. Office: Internat Paper Long Meadow Rd Tuxedo Park NY 10987

NOGA, MARIAN, mechanical engineer; b. Lwow, Poland, Feb. 2, 1939; s. Antoni and Maria (Dobrzyniecka) N.; m. Maria Drozdowska June 3, 1967; 1 child, Anna. MS in Engring., Acad. of Mining and Metallurgy, AGH/ Kraków, Poland, 1961, Doctor, 1969, Doctor Habilitation, 1975. Asst. Acad. Mining and Metallurgy, Kraków, 1961-69, asst. prof., 1976-82, prof., 1982—; asst. prof. The Sielesian Univ., Katowice, Poland, 1976-82, prof., 1982-89; dir. The Computer Ctr. AGH, Kraków, 1988-89, The Acad. Computer Ctr./Cyfronet, Kraków, 1989—; cons. Project Office for Electric Machines, Katowice, 1973-75, The Katowice Steelmill, 1975-88. Co-author two books in field; contbr. articles to profl. jours; patentee in field. Recipient II Degree prizes Minister of Higher Edn., Warsaw, 1980, 83, III Degree prize, 1970. Mem. Assn. of Polish Electricians (expert, bd. dirs. sect. of computer sci. 1989), Coun. of the Users of Polish Acad. Computer Networks (chmn. 1991). Avocations: skiing, tourism. Office: Cyfronet, Nawojki 11 PO Box 386, 30-950 Kraków 61, Poland

NOGUCHI, CONSTANCE TOM, molecular biologist, researcher; b. Canton, China, Dec. 8, 1948; came to U.S., 1949; m.; Philip David Noguchi, Sept. 7, 1969; children: Jamie K., Matthew S. AB, Univ. Calif., 1970; PhD, George Washington Univ., 1975. Rsch. physicist Nat. Inst. Health, Bethesda, Md., 1975—; study section mem. Hematology, Nat. Heart, Lung and Blood Inst., Bethesda, 1990—. Mem. Am. Soc. Hematology, Am. Physicial Soc., Biophysicial Soc., Sigma Xi. Achievements include biophysicial studies in sickle cell anemia; understanding expression of erythroid genes. Office: Nat Inst Health Bldg 10 Rm 9N307 Bethesda MD 20892

NOGUCHI, HIROSHI, structural engineering educator; b. Tokyo, Aug. 9, 1946; s. Kou and Kimie (Ohtake) N.; m. Yoriko Ito, Jan. 8, 1982; children: Mariko, Eriko. B in Engring., U. Tokyo, 1970, M in Engring., 1972, D in Engring., 1976. Registered architect 1st class. Rsch. assoc. U. Tokyo, 1976-77; asst. prof. Chiba (Japan) U., 1977-79, assoc. prof., 1979-90; prof. Chiba U., 1991—; vis. researcher U. Toronto, 1984-85. Author: (with others) Shear Analysis of Reinforced Concrete Structures, 1983, Finite Element Analysis of Reinforced Concrete Structures, 1986, Shear Resistance Mechanisms of Beam-Column Joints Under Reversed Cyclic Loading, 1987, Guidelines for Application of FEM to RC Design, 1989, Development of Mixed Structures in Japan, 1990, Experimental Studies on Shear Performances of RC Interior Column-Beam Joints with High-Strength Materials, 1992; author: State-of-the-Art of Theoretical Studies in Membrane Shear Behavior in Japan, 1991, Recent Developments of Researches and Applications of RCFEM in Japan, 1991. Mem. ASCE, Archtl. Inst. Japan, Japan Concrete Inst. (Meritorious Paper award, 1985), Tokyo Soc. Architects, Am. Concrete Inst., Internat. Assn. Bridges and Structural Engring. Avocations: swimming, skiing, classical music. Home: 1-5-3-103 Eifuku, Suginami-ku, Tokyo 168, Japan Office: Chiba U Dept Architecture, 1-33 Yayoi-cho, Inage-ku Chiba 263, Japan

NOGUCHI, SHUN, chemistry educator; b. Tokyo, Jan. 28, 1930; s. Teiji and Yaeno Noguchi; m. Etsuko Noguchi, Nov. 16, 1956; children: Chifumi Watanabe, Tokimi, Wakane. Degree, U. Tokyo, 1952, D in Chemistry, 1961. Rsch. worker Mitsuwa Soap Co., Tokyo, 1952-65; asst. U. Tokyo, 1965-70; asst. prof. Kyoritsu Women's U., Tokyo, 1970-74, prof., 1974—; lectr. Tokyo U. Agriculture, 1968-70, 88—, Shizuoka (Japan) Women's U., 1969-71. Author: Water Behavior on Cooking, 1978, Science of the Relation between Water and Foods, 1992; contbr. articles to profl. jours. Min. Edn. grantee, 1972—; recipient prize for excellent treatises Japan Assn. Fats and Oils Industry, 1964. Mem. Chem. Soc. Japan, Japan Soc. for Bioscience, Biotechnology and Agrochemistry, Japan Soc. Home Econs. (bd. dirs. 1978—, prize for excellent treatises 1989). Avocations: tennis, music. Home: 3-13-7 Higashiyama Meguro, Tokyo 153, Japan Office: Kyoritsu Womens U, 1-710 Motochachioji-machi, Hachioji-shi Tokyo 193, Japan

NOGUEIRA E SILVA, JOSE AFONSO, engineering executive; b. Lisbon, June 30, 1946; s. Afonso Lourenco Dias Da and Maria Jose Barata (Nogueira) Silva; m. Isabel Rivera, Sept. 17, 1990. Diploma, Inst. Superior Technico, Lisbon, 1972. Student travel guide Student Travel Svc., Europe, 1966-69; nat. travel guide Portuguese Travel Agys., Europe, 1970-72; engr. Prof. Costa Lobo/Prof. Johnson Marshall, Regional Porto Plan, 1973-74; minister cons. Internal Affairs Minister, Lisbon, 1976; engr. cons. Portuguese Town Halls, Portugal, 1977-78; travel engr. cons. Nat. Travel Assn., Lisbon, 1978-79; co. exec. Nogueira e Silva, Portugal, 1978-91, Luxembourg, 1991—; founder, pres. Zona do Pinhal Rural Bank, Serta, 1982-84; pres. Agr. Coop., 1984-91. Avocation: stamp collection. Home: Rua Artilharia Um 46-2 Dt, 1500 Lisbon Portugal Office: Rua Prof R. Santos 50-B, 1500 Lisbon Portugal also: 53 rue de Beggen, L-1221 Luxembourg Luxembourg

NOLAN, PETER JOHN, physics educator; b. N.Y.C., Mar. 25, 1934; s. Peter John and Nora (Gleeson) N.; divorced 1978; children: Thomas, James, John, Kevin. BS in Physics, Manhattan Coll., 1956; cert. in meteorology, UCLA, 1958; MS in Physics, Adelphi U., 1966, PhD in Physics, 1974. Engr. various corps., N.J., N.Y., 1956-63; systems analysis engr. Gruman Aircraft Engring. Corp., Bethpage, N.Y., 1963-66; asst. prof. Physics SUNY, Farmingdale, 1966-68, assoc. prof. Physics, 1968-71, prof. Physics, 1971—; chmn. Physics dept. SUNY, Farmingdale, 1970-77. Author: Experiments in Physics, 1982, Fundamentals of College Physics, 1993; contbr. articles to profl. jours. Mem. Am. Assn. Physics Tchrs. Home: 47 Fairdale Dr Brentwood NY 11717-1337 Office: SUNY Farmingdale NY 11735

NOLAN, ROBERT PATRICK, chemistry educator; b. Paterson, N.J., Jan. 28, 1955; s. Charles Vincent and Catherine Elizabeth (Hutchinson) N. BA, Rutgers U., 1978; MPhil, CUNY, 1985, PhD, 1986. Assoc. dir. environ. sci. lab. Bklyn. Coll., 1988—, asst. prof. chemistry, 1988—; mem. doctoral faculty Grad. Sch. and Univ. Ctr. CUNY, N.Y.C., 1992—; cons. U.S. Consumer Product Safety Commn., Washington, 1988; advisor WHO, Geneva, 1992. Contbr. over 30 sci. reports, articles and monographs to profl. publs., chpts. to books. Am. Microchem. Soc. scholar; Stony Wold-Herbert Fund fellow, 1986-88. Mem. N.Y. Acad. Scis., Mineralogical Soc. Am., Harvey Soc., Chemists' Club N.Y. Achievements include research on definition of physico-chemical properties of minerals, e.g. silica, asbestos, asbestos substitutes which have been associated with a health hazard. Office: Bklyn Coll Environ Sci Lab Ave H and Bedford Ave Brooklyn NY 11210

NOLAN, STANTON PEELLE, surgeon, educator; b. Washington, May 29, 1933; s. James Parker and Ellen Dubose (Peelle) N.; m. Marion Faro, June 16, 1955; children—Stanton Peelle Jr., Tiphanie Ravenel. B.A., Princeton U., 1955; M.D., U. Va., 1959, M.S., 1962. Cert. Am. Bd. Surgery, Am. Bd. Thoracic Surgery. Intern U. Va. Med. Ctr., Charlottesville, 1959-60, asst. resident gen. surgery, 1960-61, research fellow surgery, 1961-62, sr. asst. resident gen. surgery, 1962-64, chief resident gen surgery, 1964-65, chief resident thoracic cardiovascular surgery, 1965-66; sr. rsch. assoc. Clinic of Surgery Nat. Heart Inst., NIH, Bethesda, Md., 1966-68; asst. prof. surgery U. Va. Med. Ctr., Charlottesville, 1968-70, assoc. prof. surgery, 1970-74, surgeon in charge div. thoracic cardiovascular surgery, 1970—, prof. surgery, 1974—, Claude A. Jessup prof. surgery, 1981—, med. dir. Thoracic Cardiovascular post-operative unit, 1989—; established investigator Am. Heart Assn., 1969-74; cons. thoracic cardiovascular surgery VA Hosp., Salem, Va., 1968—, Am. Bd. Surgery to qualifying examination com., 1988-91; surg. cons. Bur. Crippled Children, Charlottesville, 1968—; mem. surgery A study sect. NIH, 1972-76, surgery and bioengring. study sect., 1984-87, chmn., 1985-87. Mem. editorial bd. Jour. Surg. Rsch., 1973-79, Annals of Thoracic Surgery, 1979-88; mem. sci. adv. bd. Jour. for Heart Valve Disease; contbr. numerous articles to profl. jours., chpts. to textbooks. Recipient John Horsley Meml. prize U. Va. Med. Sch., 1962; Merit award Research Forum of Am. Coll. Chest Physicians, 1968; research fellow Va. Heart Assn., 1961-62, Am. Cancer Soc., 1963-64; grantee NIH, 1968-84, Am. Heart Assn., 1970-73, Medtronic Corp., 1975-81. Fellow ACS, Am. Coll. Cardiology, Am. Surg. Assn.; mem. Am. Assn. Thoracic Surgery, Am. Heart Assn. (coun. on cardiovascular surgery 1969—, anesthesiology, radiology and surgery study com. 1991—), Andrew G. Morrow Soc., Assn. Acad. Surgery, Assn. Advancement of Med. Instrumentation (co-chmn. cardiac valve prostheses standards com. 1974—, mem. internat. standards com. 1989—, bd. dirs. 1990—, standards bd. 1991—, edn. com. 1992—), Internat. Stds. Orgn. (chmn. subcom. on cardiovascular surg. implants 1982—), Assn. Clin. Cardiac Surgeons, Halsted Soc. (exec. com. 1985-89), Coordinating Com. on Perfusion Affairs (chmn. 1990—), Internat. Cardiovascular Soc., Muller Surg. Soc. (pres. 1979), Societé Internat. de Chirurgie, Soc. Vascular Surgery, Soc. Thoracic Surgeons (ad hoc com. on industry rels. 1992—, stds. and ethics com. 1993—), Soc. Univ. Surgeons, Southeastern Surg. Congress, So. Surg. Assn. (2d v.p. 1982), Thoracic Surgery Dirs. Assn. (coun. 1989—, rep. to coun. of Acad. Socs. of Assn. of Am. Med. Colls. 1991-93), Va. Surg. Soc. (v.p. 1980-83, pres. 1984), Va. Vascular Soc. (exec. coun. 1985-86), Soc. Critical Care Medicine, Raven Soc., Assn. Am. Med. Colls. (rep. coun. acad. socs. 1992—), Alpha Omega Alpha, Omicron Delta Kappa. Clubs: Chevy Chase (Md.); Farmington Country (Va.); Princeton (N.Y.C.). Office: Univ Va Dept Surgery Charlottesville VA 22908

NOLAND, ROBERT EDGAR, dentist; b. Tampa, Fla., Oct. 15, 1930; s. John Edgar and Myrtle Irene (Murray) N.; m. Lena Vermeille Lide, June 16, 1956; children: Douglas Eugene Noland, Thomas Edgar Noland, Edward Alan Noland, Robyn Lee Condra. BS, U. Fla., 1951; DDS, Emory U., 1956. Lt. dental div. officer USN, Pensacola, Fla., 1956-58; pvt. practice Pensacola, Fla., 1958—. Recipient Fellowship award Acad. Gen. Dentistry, 1991, 30 yr. in Good Standing award, Greater Pensacola (Fla.) Dental Assn., 1988, Merit Club Officer award Kiwanis Internat., 1989. Mem. ADA, Fla. Dental Assn., N.W. Dist. Dental Assn., Greater Pensacola (Fla.) Dental Assn., Acad. Gen. Dentistry, Kiwanis Club (West Pensacola pres. 1974-75, 88-89). Baptist. Avocations: swimming, boating, travel, music. Home: 60 Highpoint Dr Gulf Breeze FL 32561-4014 Office: 2220 N Palafox St Pensacola FL 32501-1749

NOLLER, HARRY FRANCIS, JR., biochemist, educator; b. Oakland, Calif., June 10, 1939; s. Harry Francis and Charlotte Frances (Silva) N.; m. Betty Lucille Parnow, Nov. 25, 1964 (div. 1969); children: Maria Irene; m. Sharon Ann Sussman; children: Eric Francis. AB, U. Calif., Berkeley, 1960; PhD, U. Oreg., 1965. Postdoctoral fellow MRC Lab. of Molecular Biology, Cambridge, Eng., 1965-66, Inst. Molecular Biology, Geneva, Switzerland, 1966-68; asst. prof. biology U. Calif., Santa Cruz, 1968-73, assoc. prof., 1973-79, prof. biology, 1979—, Robert Louis Sinsheimer prof. molecular biology, 1987—; Sherman Fairchild Disting. scholar, Calif. Inst. Tech., 1990. Mem. NAS. Office: Sinsheimer Labs U Calif Santa Cruz High St Santa Cruz CA 95064-1099

NOLTE, MARTY DEE, nuclear power plant training manager; b. Lake City, Iowa, Feb. 6, 1963; s. Maurice Dee and Patricia Joyce (Miller) N.; m. Faith Ann Gallo, June 21, 1988; 1 child, Makenna Leigh. BSME, U. Nebr., 1986. Assoc. engr. S1W Plant Westinghouse, Idaho Falls, Idaho, 1987-88; nuclear plant engr. SIW Plant Westinghouse, Idaho Falls, Idaho, 1988-89, A1W Plant Westinghouse, Idaho Falls, Idaho, 1989-90; shift supr. AIW Plant Westinghouse, Idaho Falls, Idaho, 1990-92, staff tng. mgr., 1992—. Lutheran.

NOMISHAN, DANIEL APESUUR, science and mathematics educator; b. Gboko, Benue, Nigeria, June 1, 1940; came to U.S., 1978; s. Akpam and Mary Senave (Gakem) N.; m. Bridget Mernan Iyongovihi, Jan. 28, 1972; children: Yadoo, Ioryina, Terngu, Sesugh. Student, U. Lagos (Nigeria); BS, Morgan State U., Balt.; MS, Morgan State U., 1980; EdD, Morgan State U., Balt.; DEd, Indiana U. Pa. Cert. tchr. Tchr. sci. Bristow Secondary Sch., Gboko, 1965-66, Oduduwa Coll., Ile-Ife, Nigeria, 1967-69, Ahmadiya Grammar Sch., Lagos, Nigeria, 1969-71, Christ Apostolic Grammar Sch., Iperu, Nigeria, 1971-72, Benue State Schs. Bd., Makurdi, Nigeria, 1973-77, Balt. City Pub. Sch., 1979-86; teaching assoc. Indiana U. Pa., 1986-88; asst. prof. Fla. Atlantic U., Boca Raton, 1988-91, Fitchburg (Mass.) State Coll., 1991—; cons. Pine Jog Environ. Edn. Ctr., West Palm Beach, Fla., 1988-90, Broward Sch. Dist., Ft. Lauderdale, Fla., 1990, Partnerships Advancing Learning of Math. and Sci., 1993—; dir. Nopento Sci. and Edn. Svc., Ft. Lauderdale, 1990; bd. dirs. Profl. Devel. Ctr., Fitchburg. Author: Science Activities for Elementary and Middle Schools, 1990. Chmn. Conf. Nigerian People and Orgns., Boca Raton, 1989-90. Mem. NSTA, Assn. Tchr. Educators (instnl. rep. 1987—). Democrat. Methodist. Avocations: reading, soccer, tennis, community svc. Home: 250 Whitney St # 610 Leominster MA 01453-3263 Office: Fitchnurg State Coll Fitchburg MA 01420

NOMURA, SHIGEAKI, aerospace engineer; b. Tokyo, Jan. 6, 1936; s. Shigeji and Chou Nomura; m. Keiko Tezuka, Oct. 19, 1963; ;children: Kyougo, Takahiko. BS, Sci. U. Tokyo, 1959; M in Engring. Sci., NYU, 1971; D in Engring., Tokyo U., 1976. Researcher Nat. Aerospace Lab., Tokyo, 1960-70, sr. researcher, 1970-71, sect. chief, 1971-89, dir. aerodynamics divsn., 1989—; bd. mem. of specialists Space Activity Commn. Japanese Govt., Tokyo, 1991—. Author: Handbook of Aerospace Engineering, 1974, 2d edit., 1991; contbr. articles to Jour. AIAA. Recipient Disting. Rsch. award Min. Sci. and Tech. Agy. Japan, 1989. Fellow (assoc.) AIAA; mem. Japan Soc. for Aero. and Space Sci. (div. head com. 1992—). Achievements include patent for heat protection material. Home: 5-13-13 Miyamae, Suginami-ku, Tokyo 168, Japan

NONEMAN, EDWARD E., engineering executive. V.p. TRW Inc., Cleve. Recipient Space Systems award Am. Inst. Aeronautics and Astronautics, 1992. Office: TRW Inc 1900 Richmond Rd Cleveland OH 44124-3719*

NONNEMAN, ARTHUR J., psychology educator; b. Chgo., Jan. 17, 1943; s. Oscar R. and Helen Marthena (Watson) N.; m. Kathleen Ann Marie Moran, Sept. 11, 1965; children: Cheryl Lynn Nonneman Murphy, Wendy Sue Nonneman Hollis. BA, Northwestern U., Evanston, Ill., 1965; MS, U. Mich., 1968; PhD, U. Fla., 1970. Asst. prof. psychology U. Ky., Lexington, 1973-75, assoc. prof., 1975-81, prof., 1981-91, dir. grad. study psychology, 1983-89, chmn. psychology dept., 1988-91; prof. psychology Asbury Coll., Wilmore, Ky., 1991—, dir. Inst. Rsch. and Planning, 1991—; cons. VA Med. Ctr., Lexington, 1981-83. Contbr. articles and revs. to profl. jours., chpts. to books. Dist. coun. Explorers Boy Scouts Am., Lexington, Ky., 1975-78; asst. Brownie troop leader Wilderness Rd. Girl Scouts Coun., Lexington, 1974-77; pres. PTA, Garden Springs Elem. Sch., 1977-78; pres. Lafayette High Sch. Band Parents Assn., Lexington, 1984-85. Rsch. grantee NIMH, U. Ky., 1974-75, 75-78, NIH, 1978, 79, 83, 84, 85, 87, 88. Fellow APA, Am. Psychol. Soc.; mem. Ky. Psychol. Assn. (bd. mem. 1991—), Soc. Neurosci., Psychonomic Soc., Phi Beta Kappa, Sigma Xi (pres. 1987-88), Phi Kappa Phi. Achievements include training 14 doctoral students through completion of the PhD, training masters students through completion of the MA or MS. Home: 701 Corbitt Dr Wilmore KY 40390 Office: Asbury Coll 1 Macklem Dr Wilmore KY 40390

NORA, JAMES JACKSON, physician, author, educator; b. Chgo., June 26, 1928; s. Joseph James and May Henrietta (Jackson) N.; m. Barbara June Fluhrer, Sept. 7, 1949 (div. 1963); children: Wendy Alison, Penelope Welbon, Marianne Leslie; m. Audrey Faye Hart, Apr. 9, 1966; children: James Jackson Jr., Elizabeth Hart Nora. AB, Harvard U., 1950; MD, Yale U., 1954; MPH, U. Calif., Berkeley, 1978. Intern Detroit Receiving Hosp., 1954-55; resident in pediatrics U. Wis. Hosps., Madison, 1959-61, fellow in cardiology, 1962-64; fellow in genetics McGill U. Children's Hosp., Montreal, Can., 1964-65; assoc. prof. pediatrics Baylor Coll. Medicine, Houston, 1965-71; prof. genetics, preventive medicine and pediatrics U. Colo. Med. Sch., Denver, 1971—; dir. genetics Rose Med. Ctr., Denver, 1980—; dir. pediatric cardiology and cardiovascular tng. U. Colo. Sch. Medicine, 1971-78; mem. task force Nat. Heart and Lung Program, Bethesda, Md., 1973; cons. WHO, Geneva, 1983—; mem. U.S.-U.S.S.R. Exchange Program on Heart Disease, Moscow and Leningrad, 1975. Author: The Whole Heart Book, 1980, 2d rev. edit., 1989), (with F.C. Fraser) Medical Genetics, 4th rev. edit., 1993, Genetics of Man, 2d rev. edit., 1986, Cardiovascular Diseases: Genetics Epidemiology and Prevention, 1991; (novels) The Upstart Spring, 1989, The Psi Delegation, 1989. Com. mem. March of Dimes, Am. Heart Assn., Boy Scouts Am. Served to lt. USAAC, 1945-47. Grantee Nat. Heart, Lung and Blood Inst., Nat. Inst. Child Health and Human Devel., Am. Heart Assn., NIH; recipient Virginia Apgar Meml. award. Fellow Am.

Coll. Cardiology, Am. Acad. Pediatrics, Am. Coll. Med. Genetics; mem. Am. Pediatric Soc., Soc. Pediatric Rsch., Am. Heart Assn., Teratology Soc., Transplantation Soc., Am. Soc. Human Genetics, Authors Guild, Authors League, Acad. Am. Poets, Mystery Writers Am., Rocky Mountain Harvard Club. Democrat. Presbyterian. Avocations: writing fiction, poetry. Home: 3110 Fairweather Ct Olney MD 20832 Office: Parklawn Bldg 5600 Fishers Ln Rm 18-05 Rockville MD 20857

NORBEDO, ANTHONY JULIUS, engineering executive; b. Trieste, Italy, Feb. 18, 1951; s. Joseph and Paula Norbedo. BSEE, Northeastern U., 1975. Electronic engr. S.H. Couch div., Quincy, Mass., 1972; engring. aide Draper Labs, Cambridge, Mass., 1973-75; electronic engr. New England Instrument Co., Natick, Mass., 1976-78; lead engr. Mobil Solar Energy Corp., Billerica, Mass., 1978-89; engring. mgr. Integrated Power Corp., Rockville, Md., 1990—. Contbr. articles to profl. jours. Achievements include patent for PV module disk; designed first middle eastern standalone photvoltaic desalination plant. Office: Integrated Power Corp 7524 Standish Pl Rockville MD 20855

NORBERG, RICHARD EDWIN, physicist, educator; b. Newark, Dec. 28, 1922; s. Arthur Edwin and Melita (Roefer) N.; m. Patricia Ann Leach, Dec. 27, 1947 (dec. July 1977); children—Karen Elizabeth, Craig Alan, Peter Douglas; m. Jeanne C. O'Brien, Apr. 1, 1978. B.A., DePauw U., 1943; M.A., U. Ill., 1947, Ph.D., 1951. Research assoc., control sytems lab. U. Ill., 1951-53, asst. prof., 1953; vis. lectr. physics Washington U., St. Louis, 1954—; mem. faculty Washington U., 1955—, prof. physics, 1958—, chmn. dept., 1962-91; dir. Technilab Inc. Mem. editorial bd. Magnetic Research Rev. Served with USAAF, 1943-46. Fellow Am. Phys. Soc., Internat. Soc. Magnetic Research. Home: 7134 Princeton Ave Saint Louis MO 63130-2308 Office: Washington U Dept Physics Box 1105 Saint Louis MO 63130

NORCROSS, DAVID WARREN, physicist, researcher; b. Cin., July 18, 1941; s. Gerald Warren and Alice Elizabeth (Downey) N.; m. Mary Josephine Boudrias, Aug. 26, 1967, children—Joshua David, Sarah Elizabeth. A.B., Harvard Coll., 1963; M.Sc., U. Ill., 1965; Ph.D., Univ. Coll., London, 1970. Research assoc. U. Colo., Boulder, 1970-74; physicist Nat. Bur. Standards, Boulder, 1974—; cons. Lawrence Livermore Lab., Calif., Los Alamos Lab., N.Mex.; fellow Joint Inst. Lab. Astrophysics, Boulder, 1976—. Contbr. articles to profl. jours. Recipient Bronze medal Nat. Bur. Standards, 1982. Fellow Am. Phys. Soc. Office: U Colo Joint Inst Lab Astrophysics Boulder CO 80309-0440

NORCROSS, MARVIN AUGUSTUS, veterinarian, government agency official; b. Tansboro, N.J., Feb. 8, 1931; s. Marvin A. and Katherine V. (McGuigan) N.; m. Diane L. Tuttle, Nov. 22, 1956 (div. 1991); children: James, Janet. Student, Rutgers U., 1954-55; V.M.D., U. Pa., 1959, Ph.D., 1966. Pathologist Merck Sharp & Dohme Research Labs., Rahway, N.J., 1966-69; dir. clin. research Merck Sharp & Dohme Research Labs., 1969-72, sr. dir. domestic vet. research, 1972-75; dir. div. vet. med. research Ctr. Vet. Medicine, FDA, Rockville, Md., 1975-78; assoc. dir. for research Ctr. Vet. Medicine, FDA, 1978-82, assoc. dir. for human food safety, 1982-84, assoc. dir. for new animal drug evaluation, 1984-87; asst. dep. adminstr., then dep. adminstr. Sci. and Tech., Food Safety and Inspection Svc. USDA, Washington, 1987-93, exec. asst. to adminstr., 1993—; adj. prof. faculty Va.-Md. Regional Coll. Vet. Medicine, Blacksburg, Va., 1980-85. Contbr. articles to profl. jours. Trustee Scotch Plains (N.J.) Community Fund, 1969-72. Served to lt. AUS, 1952-54; col. Res., 1954-83 (ret.). Recipient FDA Merit award, 1978, Meritorious Presdl. Rank award, 1989. Mem. Am. Vet. Med. Assn., AAAS, Am. Assn. Avian Pathologists, Soc. Toxicologic Pathologists, Assn. Mil. Surgeons U.S., Civil Affairs Assn., Inst. Food Technologists, Nat. Assn. Fed. Veterinarians, N.J. Acad. Sci., N.Y. Acad. Scis., Res. Officers Assn., Sigma Xi. Home: 14304 Brickhowe Ct Germantown MD 20874-3431 Office: Food Safety and Inspection Service USDA 14th and Independence Ave SW Washington DC 20250-0001

NORDBY, EUGENE JORGEN, orthopedic surgeon; b. Abbotsford, Wis., Apr. 30, 1918; s. Herman Preus and Lucille Violet (Korsrud) N.; m. Olive Marie Jensen, June 21, 1941; 1 child, Jon Jorgen. B.A., Luther Coll., Decorah, Iowa, 1939; M.D., U. Wis., 1943. Intern Madison Gen. Hosp., Wis., 1943-44, asst. in orthopedic surgery, 1944-48; practice medicine specializing in orthopedic surgery Madison, Wis., 1948—; pres. Bone and Joint Surgery Assocs., S.C. 1969-91; chief staff Madison Gen. Hosp., 1957-63; assoc. clin. prof. U. Wis. Med. Sch., 1961—; chmn. Wis. Physicians Svcs., 1979—; dir. Wis. Regional Med. Program, Chgo. Madison and No. RR; bd. govs. Wis. Health Care Liability Ins. Plan; chmn. trustees S.M.S. Realty Corp.; mem. bd. attys. Profl. Responsibility of Wis. Supreme Ct., 1992—. Assoc. editor Clin. Orthopaedics and Related Research, 1964—. Pres. Vesterheim Norwegian Am. Mus., Decorah, Iowa, 1968—. Served to capt. M.C., AUS, 1944-46. Decorated knight 1st class Royal Norvegian Order St. Olav; recipient Disting. Svc. award Luther Coll., 1964, Sr. Svc. award Internat. Rotary, 1987. Mem. Am. Acad. Orthopaedic Surgeons (dir. 1972-73), Clin. Orthopaedic Soc., Assn. Bone and Joint Surgeons (pres. 1973), Internat. Soc. Study Lumbar Spine, State Med. Soc. Wis. (chmn. 1968-76, treas. 1976—, Coun. award 1976), N.Am. Spine Soc., Internat. Intradiscal Therapy Soc. (sec. 1987—), Wis. Orthopaedic Soc., Dane County Med. Soc. (pres. 1957), Nat. Exchange Club, Madison Torske Klubben (founder, pres. 1978—), Phi Chi. Lutheran. Home: 6234 S Highlands Ave Madison WI 53705-1115 Office: 2704 Marshall Ct Madison WI 53705-2297

NORDÉN, BENGT JOHAN FREDRIK, chemist; b. Lund, Sweden, May 15, 1945; s. Jan Gunnar and Marit Birgitta (Strömberg) N.; married, 1970; children: Anna, Birgitta, Catharina. BSc with honors, U. Lund, 1967, PhD, 1971, DSc, 1972. Docent Dept. Inorganic Chemistry, U. Lund, 1972-79; acting prof. U. Lund, 1979; chair phys. chemistry Dept. Phys. Chemistry, Chalmers U. Tech., Gothenburg, 1979—; chmn. Dept. of Phys. Chemistry, Chalmers U. Tech., Gothenberg, 1982—, dean Sch. of Chemistry, 1983-85; advisor of sci. and tech. The Swedish Govt., 1991—; fellow The Swedish Nat. Sci. Rsch. Coun., 1986-92. Editor: Interaction Mechanisms of Low-Level Electromagnetic Fields in Living Systems, 1992; editor The Nat. Ency. of Sweden: Physical Chemistry; editorial bd. Jour. Crystallographic and Spectroscopic Rsch. (U.K.) 1981—, Internat. Jour. of Genome Rsch., 1990—, Spectroscopy Letters, 1976—; contbr. over 250 articles to profl. jours. Fellow Royal Acad. Scis. (life, Göran Gustafsson Prize of Chemistry 1992), Royal Soc. of Arts and Scis. of Gothenburg (life), Royal Physiographic Soc. of Lund (life, Fabian Gyllenbergs Prize, 1972), Royal Soc. of Chemistry of U.K.; mem. Am. Chem. Soc., European Biophys. Chemistry. Achievements include devel. of the linear dichriosm spectroscopy as a tool for determining molecular structure; rsch. on methodologies for determining spectrscopic transition mements, novel non-intercalated DNA structures in solution, including complex with recombination enzyme. Home: Dörjeskärgatan 15, S-42160 Västra Frölunda Sweden Office: Dept Phys Chemistry, Chalmers U of Tech, S-41296 Gothenburg Sweden

NORDGREN, RONALD PAUL, civil engineering educator, research engineer; b. Munising, Mich., Apr. 3, 1936; s. Paul A. and Martha M. (Busse) N.; m. Joan E. McAfee, Sept 12, 1959; children: Sonia, Paul. BS in Engring., U. Mich., 1957, MS in Engring., 1958; PhD, U. Calif., Berkeley, 1962. Rsch. asst. U. Calif. at Berkeley, 1959-62; mathematician Shell Devel. Co., Houston, 1963-68, staff rsch. engr., 1968-74, sr. staff rsch. engr., 1974-80, rsch. assoc., 1980-90; Brown prof. civil engring. Rice U., Houston, 1989—; lectr. mech. engring. U. Houston, 1980; mem. U.S. nat. com. on theoretical and applied mechanics NRC, 1984-86, U.S. nat. com. for rock mechanics, 1991—. Contbr. tech. papers to profl. jours.; assoc. editor Jour. Applied Mechanics, 1972-76, 81-85; patentee in field. Fellow ASME; mem. NAE, ASCE, Soc. Industrial and Applied Math., Soc. Engring. Sci., Sigma Xi. Office: Rice U PO Box 1892 Houston TX 77251-1892

NORDIN, PAUL, physicist, system engineer; b. Kansas City, Mo., Feb. 12, 1929; s. Paul Sr. and Marguerite Edith (Desmond) N.; m. Barbara Boardman, Mar. 15, 1953 (div. 1972); children: Sandra Nordin Dixon, Paul III; m. Muriel Brown Greenwood, Nov. 18, 1989. BA in Physics with honors, U. Calif., Berkeley, 1956, MA in Physics, 1957, PhD in Physics, 1961. Engring. draftsman Northrop Aircraft, Hawthorne, Calif., 1952-53; teaching asst., then rsch. asst. U. Calif., Berkeley, 1956-61; sr. scientist aero. div. Ford Motor Co., Newport Beach, Calif., 1961-63; system engring. mgr. Aerospace Corp., San Bernardino, Calif., 1963-68, TRW Inc., Redondo Beach, Calif., 1968-90, Grumman Aerospace Corp., Bethpage, N.Y., 1990—. Contbr. articles to Phys. Rev., Applied Physics Letters, others. With USCG, 1946-48. Mem. AAAS, IEEE, Am. Phys. Soc., Sigma Xi, Phi Beta Kappa. Home: 77 Maple Ave Bethpage NY 11714-2231 Office: Grumman Aerospace Corp B29-025 S Oyster Bay Rd Bethpage NY 11714

NORDLANDER, JAN PETER ARNE, physicist, educator; b. Stockholm, Sweden, Nov. 21, 1955; came to U.S., 1985; s. Arne Nils Ludwig and Blenda Mimmi (Sjosell) N.; m. Nancy Jean Halas, Aug. 1, 1986. MS in Engring. Physics, Chalmers U., Gothenburg, Sweden, 1980, PhD, 1985. Postdoctoral IBM, Thomas J. Watson Rsch. Ctr., Yorktown Heights, N.Y., 1985-86; rsch. asst. prof. Vanderbilt U., Nashville, 1987-88; sr. postdoctoral Rutgers U., Piscataway, N.J., 1988-89; asst. prof. Rice U., Houston, 1989-93, assoc. prof., 1993—; cons. AT&T Bell Labs., Murray Hill, N.J., 1987-89; adj. asst. prof. Vanderbilt U., 1988-91; vis. prof. Univ. of Paris, Orsay, France, 1992. Lt. Swedish Army, 1976-78, Sweden. Mem. Am. Phys. Soc., Am. Chem. Soc. Achievements include rsch. in electronic structure of surfaces, charge transfer processes in atom-surface scattering and non-equilibrium processes. Office: Rice Univ Dept Physics 6100 S Main Houston TX 77251

NORDLIE, ROBERT CONRAD, biochemistry educator; b. Willmar, Minn., June 11, 1930; s. Peder Conrad and Myrtle (Spindler) N.; m. Sally Ann Christianson, Aug. 23, 1959; children: Margaret, Melissa, John. B.S. St. Cloud State Coll., Minn., 1952; M.S., U. N.D., 1957, Ph.D., 1960. Teaching and research asst. biochemistry U. N.D. Med. Sch., Grand Forks, 1955-60, Hill research prof. biochemistry, 1962-74, Chester Fritz disting. prof. biochemistry, 1974—, Cornatzer prof., chmn. dept. biochemistry and molecular biology, 1983—; hon. prof. San Marcos U., Lima, Peru, 1982—; NIH fellow Inst. Enzyme Rsch., U. Wis., 1960-61; mem. biochemistry study sect. NIH; cons. enzymology Oak Ridge, 1961—; vis. prof. Tokyo Biomed. Inst., 1984. Mem. editorial bd.: Jour. Biol. Chemistry, Biochimca et Biophysica Acta. Research publs. on enzymology relating to metabolism of various carbohydrates in mammalian livers, regulation blood sugar levels. Served with AUS, 1953-55. Recipient Disting. Alumnus award St. Cloud State U., 1983; recipient Sigma Xi Rsch. award, 1969, Golden Apple award U. N.D., 1968, Edgar Dale award U. N.D., 1983, Burlington No. Faculty Scholar award, 1987, Thomas J. Clifford Faculty Achievement award for excellence in rsch. U. N.D. Found., 1993. Mem. Am. Soc. Biol. Chemistry and Molecular Biology, Am. Chem. Soc., AAAS, Internt. Union Biochemists, Am. Soc. Microbiology, Soc. Exptl. Biology and Medicine, Am. Inst. Nutrition, Brit. Biochem. Soc., Sigma Xi, Alpha Omega Alpha. Home: 162 Columbia Ct Grand Forks ND 58203-2947

NORDMARK, GLENN EVERETT, civil engineer; b. Sioux City, Iowa, Jan. 12, 1929; s. Klas Everett and Laura Evangeline (Medalen) N.; m. Mary Lou Scott, Aug. 23, 1952; children: Bruce Scott, Craig Eric, Scott Douglas. BSCE, S.D. State U., 1951; MSCE, U. Ill., 1955. Registered profl. engr., Pa. From rsch. engr. to tech. specialist Alcoa Tech. Ctr., Alcoa Center, Pa., 1955-93. Contbr. articles to profl. jours. Scoutmaster Boy Scouts Am., Lower Burrell, Pa., 1957-90. Capt. USAF, 1951-53. Mem. ASCE (fatigue com. 1958-62, 73-78), SAE (fatigue design evaluation com.), Toastmasters (pres. 1980-90, Outstanding Club Pres. 1980), Am. Welding Soc., Sigma Xi. Presbyterian. Home: 331 Claremont Dr Lower Burrell PA 15068 Office: Alcoa Tech Ctr Alcoa Center PA 15069

NORDSTROM, JAMES WILLIAM, nutritionist, educator; b. Eagle Bend, Minn., Feb. 21, 1929; s. Albin Carl and Anna Effie (Sandburg) N.; m. Paula Wankum, Oct. 6, 1986. BS in Agrl. Edn., U. Minn., 1957, MS in Animal Sci. (Nutrition), 1961, PhD in Animal Sci. (Nutrition), 1964. Postdoctoral trainee U. Calif., Berkeley, 1964-66; asst. prof. U. Ill., Urbana, 1966-68, U. Minn., St. Paul, 1968-72; assoc. prof. to prof. Lincoln U., Jefferson City, Mo., 1972—; adj. prof., U. Mo., Columbia, 1972—. Author Nutrition Education And The Elderly, 1991; co-author: Generation In The Middle, 1988, Human Nutrition A Comp Treatise, 1989; contbr. articles to profl. jours. bd. mem., Am. Heart Assn., Jefferson City, 1992—. Sgt. U.S. Army, 1952-54. Recipient rsch. grant, AID, Belize, C.Am., 1986-88, USDA, 1988—, AID, Nigeria, 1991-93. Mem. Am. Soc. for Clin. Nutrition, Am. Inst. Nutrition, Sierra Club, Nature Conservancy, Audubon Soc. Methodist. Avocations: gardening, nature study, conservation, carpentry. Home: 1020 Primrose Jefferson City MO 65109 Office: Lincoln Univ Nutrition Lab Jefferson City MO 65101

NORELL, MARK ALLEN, paleontology educator; b. St. Paul, July 26, 1957; s. Albert Donald Norell and Helen Louise Soltau; m. Vivian Pan, Nov. 1, 1991. BS, Long Beach State U., 1980; MS, San Diego State U., 1983; PhD, Yale U., 1988. Asst. curator Am. Mus., N.Y.C., 1989—; adj. asst. prof. biology dept. Yale U., New Haven, 1991—. Author: All You Need to Know About Dinosaurs, 1991; contbr. articles to profl. jours. Fellow Willi Hennig Soc., Explorers Club; mem. Soc. Vertebrate Paleontology (Romer prize 1987). Office: Am Mus Natural History 79th at CPW New York NY 10024

NORGARD, MICHAEL VINCENT, microbiology educator, researcher; b. Glenridge, N.J., Oct. 5, 1951; s. Bernard Raymond and Marion C. (Testa) N.; m. Gabriella Rosella Lombardo, July 18, 1976; 1 child, Gina Gabriella. AB, Rutgers U., 1973; PhD, U. Medicine and Dentistry, N.J., 1977. Postdoctoral fellow Roche Inst. Molecular Biology, Nutley, N.J., 1977-79; asst. prof. U. Tex. Southwestern Med. Sch., Dallas, 1979-86, assoc. prof., 1986-91, prof., 1991—; cons. U.S. Justice Dept., Washington, 1986-92; mem. NIH Study Sect. Bacteriology and Mycology, Bethesda, Md., 1990—; mem. editorial bd. Jour. Sexually Transmitted Diseases, 1988—, Jour. Infection and Immunity, 1993—; panelist Congenital Syphillis Policies Ctrs. for Disease Control, Atlanta, 1987, Trepenol Vaccines, WHO, Birmingham, Eng., 1989, Internat. Sexually Transmitted Diseases Diagnostics Network, NIH, Bethesda, 1990—. Contbr. articles to Infection and Immunity, Jour. Infectious Diseases, Current Opinion Infectious Diseases, 1989—. Grantee NIH, 1980—, Robert A. Welch Found., 1982—, Austin, Tex., 1982—, Serex Internat., Van Nuys, Calif., 1983, 86, 89, Dallas Biomed. Corp., 1988-89, Tex. Higher Edn. Coordinating Bd., Austin, 1990. Mem. AAAS, Am. Venereal Disease Assn., Am. Soc. Microbiology, Tex. Infectious Diseases Soc. Roman Catholic. Achievements include U.S. patents for monoclonal antibodies against Treponema, methods for Diagnosing Syphilis, cloning of the 47-KDa antigen of Treponema pallidum; first to develop monoclonal antibodies against the syphilis bacterium; discoverer of membrane lipoproteins in T pallidum. Office: U Tex Southwestern Med Ctr 5323 Harry Hines Blvd Dallas TX 75235

NORGREN, RALPH, neuroscientist; b. Washington, Mar. 22, 1943. BA, U. Pa., 1965; PhD, U. Mich., 1969. Postdoctoral Rockefeller U., N.Y.C., 1969-71, from asst. to assoc. prof., 1971-83; prof. Pa. State U. Coll. Medicine, Hershey, Pa., 1983—. Office: Pa State U Coll Medicine Dept Behavioral Sci 500 University Dr Hershey PA 17033

NORINS, ARTHUR LEONARD, physician, educator; b. Chgo., Dec. 2, 1928; s. Russell Joseph and Elsie (Lindemann) N.; m. Mona Lisa Wetzer, Sept. 12, 1954; children: Catherine, Nan, Jane, Arthur. B.S. in Chem. Engring, Northwestern U., 1951, M.S. in Physiology, 1953, M.D., 1955. Diplomate: Am. Bd. Dermatology; subcert. in dermatopathology. Intern U. Mich., Ann Arbor, 1955-56; resident in dermatology Northwestern U., Chgo., 1956-59; asst. prof. Stanford U., 1961-64; prof., chmn. dept. dermatology, prof. pathology Ind. U. Sch. Medicine, Indpls., 1964—; chief pediatric dermatology Riley Children's Hosp.; chief dermatology Wishard Hosp.; mem. staff Univ. Hosp.; cons. VA Hosp. Contbr. articles to profl. jours. Capt. M.C. U.S. Army, 1959-61. Recipient Pres.' award Ind. U., 1979. Fellow ACP; mem. Am. Acad. Dermatology (bd. dirs.), Am. Dermatol. Assn., Soc. Pediatric Dermatology (founder, past pres.), Am. Soc. Dermatopathology, Am. Soc. Photobiology (founder), Soc. Investigative Dermatology. Home: 1234 Kirkham Ln Indianapolis IN 46260-1637 Office: Ste 3240 550 W University Blvd Indianapolis IN 46202-5267

NORMAN, ARNOLD MCCALLUM, JR., project engineer; b. Little Rock, May 1, 1940; s. Arnold McCallum and Ann Carolyn (Gibson) N.; m. Sylvia Burton, July 1, 1962 (div. 1967); m. Marisha Irene Malin, June 7, 1969; children: Frank Lee, Paul James. BS in Physics, Ga. Inst. Tech., 1962. Test engr. Rocketdyne div. Rockwell Internat., Canoga Park, Calif., 1962-64, engr. in charge of various programs, 1964-75, engr. in charge, project engr. large chem. lasers, 1975-85, project engr. space sta. propulsion system, 1985-87, project engr. nat. launch system health mgmt. systems, 1987-92, project engr. kinetic energy weapons, 1993—; mem. ops. com. health mgmt. ctr. U. Cin., 1988—; mem. program com. Am. Internat. Conf. on Engring. Applications of Artificial Intelligence, 1988-90; presenter in field. Mem. editorial bd. Jour. Applied Intelligence, 1990—; author numerous papers in field. Fellow AIAA (assoc., chair San Fernando Valley sect. 1989-91, chair sr. adv. com. 1991—), Inst. Advancement Engring.; mem. Tau Beta Pi. Home: 20238 Mobile St Canoga Park CA 91306 Office: Rocketdyne div Rockwell Int 6633 Canoga Ave Canoga Park CA 91303

NORMAN, E. GLADYS, business computer educator, consultant; b. Oklahoma City, June 13, 1933; d. Joseph Eldon and Mildred Lou (Truitt) Biggs; m. Joseph R.R. Radeck, Mar. 1, 1953 (div. Aug. 1962); children: Jody Matti, Ray Norman, Warren Norman, Dana Norman; m. Leslie P. Norman, Aug. 26, 1963; 1 child, Elayne Pearce. Student, Fresno (Calif.) State Coll., 1951-52, UCLA, 1956-59, Linfield Coll., 1986—. Math. aid U.S. Naval Weapons Ctr., China Lake, Calif., 1952-56, computing systems specialist, 1957-68; systems programmer Oreg. Motor Vehicles Dept., Salem, 1968-69; instr. in data processing, dir. Computer Programming Ctr., Salem, 1969-72; instr. in data processing Merritt-Davis Bus. Coll., Salem, 1972-73; sr. programmer, analyst Teledyne Wah Chang, Albany, Oreg., 1973-79; sr. systems analyst Oreg. Dept. Vets. Affairs, Albany, 1979-80; instr. in bus. computers Linn-Benton Community Coll., Albany, 1980—; presenter computer software seminars, State of Oreg., 1991-92, Oreg. Credit Assoc. Conf., 1991, Oreg. Regional Users Group Conf., 1992; computer cons. in field. Mem. Data Processing Mgmt. Assn. (bd. dirs. 1977-84, 89—, region 2 pres. 1985-87, assoc. v.p. 1988, Diamond Individual Performance award 1985). Democrat. Avocations: family, sewing, crocheting. Office: Linn-Benton CC Bus Mgmt Dept 6500 Pacific Blvd SW Albany OR 97321-3755

NORMAN, JAMES HAROLD, mathematician, researcher; b. Foster, Okla., Mar. 15, 1942; s. Vica Buren and May Elizabeth (Whitehead) N.; m. Nancy Anne Carmona, Aug. 21, 1982; children: Erik Andrew, Alesha Elizabeth. BA in History, East Cen. State U., Ada, Okla., 1964; BS in Math., Cen. State U., Edmond, Okla., 1971. Mathematician Missile Electronic Lab., White Sands Missile Range, N.Mex., 1973-84, Nat. Range Ops., White Sands Missile Range, N.Mex., 1984—; mem. comdr.'s bd. White Sands Missile Range, 1986-87. With U.S. Army, 1967-69, Vietnam. Recipient spl. act award Missile Warfare Lab., 1983, superior performance award Data Scis. Div., 1987; performance award Dept. Army, 1990, 92, customer svc. award, 1992. Mem. Digital Computer Users Assn., Am. Mensa. Republican. Mem. Ch. of Christ. Achievements include developer computer software real time data acquisition, statistical analysis. Home: 1208 Akers St Las Cruces NM 88005 Office: Nat Range Ops 1600 Headquarters Ave White Sands Missile Range NM 88002

NORRIS, DOLORES JUNE, computer specialist; b. Belmore, N.Y., Feb. 10, 1938; d. Abe and Doris Cyril (Stahl) Wanser; m. William Dean Norris, June 11, 1960; children: William Dean II, Ronald Wayne, Darla Cyrille. BS in Elem. Edn., So. Nazarene U., 1959; MS in Computer Edn., Nova U., 1988, EdS in Computer Applications, 1990. Cert. elem. edn. and computer sci. tchr., Fla. Tchr. 4th and 5th grades Ruskin (Fla.) Elem. Sch., 1959-61; tchr. 5th grade Emerson Elem. Sch., Kansas City, Kans., 1961-63; tchr. 1st grade Hickman Mills, Mo., 1964-65; tchr. 3d and 4th grades Lake Mary Elem. Sch., Sanford, Fla., 1968-72; tchr. 1st grade St. Charles Cath. Sch., Port Charlotte, Fla., 1976-77; primary tchr. Meadow Park Elem. Sch., Port Charlotte, 1977-89; computer specialist Vineland Elem. Sch., Rotanda West, Fla., 1989-90, Myakka River Elem. Sch., Port Charlotte, 1990—; reading coun. Charlotte County Schs., Port Charlotte, 1987—, rep., 1989-90, in-svc. com. 1990-93; program planner Meadow Park Elem. Sch., 1988-89; program planner Myakka River Elem. Sch. 1991-93. Mem. Rotary, Punta Gorda, Fla., 1982-86; co-dir. teens Touring Puppet Group, Punta Gorda, 1982-86; puppet co-dir. NOW Teens, Punta Gorda, 1976-80. Mem. Fla. Assn. Computers in Edn. Avocations: piano, swimming, travel. Home: 1171 Richter St Port Charlotte FL 33952

NORRIS, FRANKLIN GRAY, surgeon; b. Washington, June 30, 1923; s. Franklin Gray and Ellie Narcissus (Story) N.; m. Sara Kathryn Green, Aug. 21, 1945; children: Gloria Norris Sales, F. Gray III. BS, Duke U., 1947; MD, Harvard U., 1951. Resident, Peter Bent Brigham Hosp., Boston, 1951-54, Bowman Gray Sch. Medicine, 1954-57; practice medicine specializing in thoracic and cardiovascular surgery, 1957—; pres. Norris Assocs., Orlando, 1985—; mem. staff Brevard Meml. Hosp., Melbourne, Fla., Waterman Meml. Hosp., Eustis, Fla., West Orange Meml. Hosp., Winter Garden, Fla., Orlando Regional Med. Ctr., Fla. Hosp., Lucerne Hosp., Arnold Palmer Children Hosp., Princeton, Fla. Hosp. N.E. and South (all Orlando). Bd. dirs. Orange County Cancer Soc., 1958-64, Central Fla. Respiratory Disease Assn., 1958-65. Served to capt. USAAF, 1943-45. Decorated Air medal with 3 oak leaf clusters. Diplomate Am. Bd. Surgery, Am. Bd. Thoracic and Cardiovascular Surgery, Am. Bd. Gen. Vascular Surgery. Mem. Fla. Heart Assn. (dir. 1958—), Orange County Med. Soc. (exec. com. 1964-75, pres. 1971-75), Cen. Fla. Hosp. Assn. (bd. dirs., 1980-85), ACS, Soc. Thoracic Surgeons, So. Thoracic Surg. Assn., Am. Coll. Chest Physicians, Fla. Soc. Thoracic Surgeons (pres. 1981-82), Am. Coll. Cardiology, So. Assn. Vascular Surgeons, Fla. Vascular Soc., Phi Kappa Psi. Presbyterian (elder). Clubs: Citrus, Orlando Country. Home: 1801 Bimini Dr Orlando FL 32806-1515 Office: Norris Assocs 55 W Columbia St Orlando FL 32806-1180

NORRIS, JAMES RUFUS, JR., chemist, educator, consultant; b. Anderson, S.C., Dec. 29, 1941; s. James Rufus and Julia Lee (Walker) N.; m. Carol Anne Poetzsch, Dec. 28, 1963; children: Sharon Adele, David James. BS, U. N.C., 1963; PhD, Washington U., St. Louis, 1968. Postdoctoral appointee Argonne (Ill.) Nat. Lab., 1968-71, asst. chemist, 1971-74, chemist, 1974-79, photosynthesis group leader, 1979—, sr. chemist, 1981—; prof. chemistry U. Chgo., 1984—; chmn. internat. organizing com. 7th Internat. Conf. on Photochemical Conversion and Storage of Solar Energy, Northwestern U., Evanston, Ill., 1988. Co-editor: Photochemical Energy Conversion, 1989; mem. editorial bd. Applied Magnetic Resonance Jour., 1989—. Recipient Disting. Peformance award U. Chgo., 1977, 2 R&D 100 awards R&D mag., 1988, E.O. Lawrence Meml. award Dept. of Energy, 1990, Rumford Premium AAAS, 1992, Humboldt Rsch. award for Sr. Scientists, 1992. Mem. Am. Chem. Soc., Biophysical Soc. Achievements include discovery that the primary donor of photosynthesis is a dimeric special pair of chlorophyll molecules. Office: Univ of Chicage Dept of Chemistry 5735 Ellis Ave Chicago IL 60637

NORRIS, JOHN ROBERT, design engineer; b. St. Louis, July 5, 1967; s. John Charles and Judy Louise (Feddersen) N. BS in Engring. Mechanics, U. Mo., Rolla, 1989. Teaching asst. U. Mo., Rolla, 1987-89; structural steel detailer E-M-E, Inc., St. Louis, 1987-89; engr. Mitek Industries, Inc., Chesterfield, Mo., 1989—. Mem. Intermural Mgrs. Assn., Rolla, 1987; commr. North County Fast-Pitch League,Florissant, Mo., 1991-92. Mem. NSPE, Mo. Soc. Profl. Engrs., Beta Sigma Psi Alumni Assn. (sec. 1989-92, treas. 1992—), Beta Sigma Psi (pres., 1st v.p., rush chmn., steward, athletic mgr., sgt.-at-arms). Lutheran. Avocations: fast-pitch softball, fishing, soccer. Home: 9208 Hallock Dr Saint Louis MO 63123

NORRIS, KARL HOWARD, optics scientist, agricultural engineer; b. Glen Richey, Pa., May 23, 1921; married, 1948; 2 children. BS, Pa. State U., 1942. Radio engr. Airplane & Marine Instruments Co., 1945-46; electronic engr. U. Chgo., 1946-49; lab. dir. Instrumentation Rsch. Lab., Agrl. Rsch. Svc., USDA, 1950-77; chief instrument rsch. lab. Sci. & Edn. Adminstr., 1977-88; retired. Recipient McCormick Gold medal Am. Soc. Agrl. Engrs., 1974; Recipient Alexander von Humboldt award, 1978, Maurice F. Hasler award Soc. Applied Spectroscopy, 1991. Mem. Nat. Acad. Engrs., Am. Soc. Agrl. Engrs., Soc. Applied Spectros. Achievements include research in instrumentation for the measurement of quality factors of agricultural products. Office: 11204 Montgomery Rd Beltsville MD 20705*

NORRIS, PAMELA, school psychologist; b. Springfield, Mass., May 11, 1946; d. William Henry Jr. and Loretta Agnes (Houck) N. BA in English, Keuka Coll., 1968; MA in Philosophy, U. Mass., 1969; MEd in Guidance and Counseling, Westfield (Mass.) State Coll., 1973; MA in Clin. Psychology with distinction, Am. Internat. Coll., 1991. Nat. cert. sch. psychologist; cert. sch. psychologist, Mass., Conn.; lic. ednl. psychologist, Mass.; lic. real estate broker, Mass.; cert. guidance counselor, Mass.; cert. English tchr., Mass., Conn. English tchr. Agawa Pub. Schs., Feeding Hills, Mass., 1970-87, sch. psychologist, 1987—. Telephone operator Springfield (Mass.) Hotline, 1970-72; telephone operator, counselor Falmouth (Mass.) Emergency and Referral Svc., 1971, SPAN Ctr., Feeding Hills, Mass, 1972-73, CHEC-Line, West Springfield, Mass., 1974-75; occupational therapist Monson (Mass.) State Hosp., 1976-77. Mem. Am. Psychol. Assn. (assoc.), Nat. Assn. of Sch. Psychologists, Mass. Sch. Psychology Assn., Western Mass. Sch. Psychology Assn., Sigma Tau Delta, Pi Delta Epsilon. Avocations: swimming, hiking, travel, reading, gardening. Office: Agawa Pub Schs 1305 Springfield St Feeding Hills MA 01030

NORRIS, PETER EDWARD, semiconductor physicist; b. Bklyn., Nov. 27, 1942; s. Sidney and Priscilla (Goldman) N.; m. Fern Dee Olin (div. July 1976); 1 child, Rebecca; m. Amy H. Rugel, May 30, 1982. BSEE, MIT, 1965, MSEE, 1967, PhD, 1979. Registered profl. engr., Mass. Mem. tech. staff RCA Labs., Princeton, N.J., 1968-73; sr. mem. tech. staff GTE Labs., Waltham, Mass., 1980-81, prin. mem. tech. staff, 1981-87; v.p. process tech. Emcore Corp., Somerset, N.J., 1987-92; pres., CEO NZ Applied Techs., Cambridge, Mass., 1992—; mem. tech. exch. com. N.J. Commn. Sci. and Tech., 1991-92; head tech. adv. bd. United Laser Techs., Providence, 1992—; invited speaker, session chmn. confs. Japan, U.S., Eng., Russia, Germany, 1984-93. Contbr. over 60 articles to tech. jours. Nat. Honor Soc. scholar, 1960; David Sarnoff fellow, 1972-74. Mem. Metallurgical Soc., Material Rsch. Soc., Sigma Xi. Achievements include patents in process control, device fabrication and CVD equipment design. Home: 12 Whitney Ave Cambridge MA 02139 Office: NZ Applied Techs c/o Bus Innovation Ctr 100 Jersey Ave New Brunswick NJ 08901

NORSTRAND, IRIS FLETCHER, psychiatrist, neurologist, educator; b. Bklyn., Nov. 21, 1915; d. Matthew Emerson and Violet Marie (Anderson) Fletcher; m. Severin Anton Norstrand, May 20, 1941; children: Virginia Helene Norstrand Villano, Thomas Fletcher, Lucille Joyce. BA, Bklyn. Coll., 1937, MA, 1965, PhD, 1972; MD, L.I. Coll. Medicine, 1941. Diplomate Am. Bd. Psychiatry and Neurology, Am. Bd. EEG. Med. intern Montefiore Hosp., Bronx, N.Y., 1941-42; asst. resident in neurology N.Y. Neurol. Inst.-Columbia-Presbyn. Med. Ctr., N.Y.C., 1944-45; pvt. practice Bklyn., 1947-52; resident in psychiatry Bklyn. VA Med. Ctr., 1952-54, resident in neurology, 1954-55, staff neurologist, 1955—, asst. chief neurol. svc., 1981-91; neurol. cons. Indsl. Home for Blind, Bklyn., 1948-51; clin. prof. neurology SUNY Health Sci. Ctr., Bklyn., 1981—; attending neurologist Kings County Hosp., Bklyn., State U. Hosp., Bklyn. Contbr. articles to med. jours. Recipient spl. plaque Mil. Order Purple Heart, 1986, also others. Fellow Am. Psychiat. Assn., Am. Acad. Neurology, Internat. Soc. Neurochemistry, Am. Assn. U. Profs. Neurology, Am. Med. EEG Soc. (pres. 1987-88), Nat. Assn. VA Physicians (pres. 1989-91, James O'Connor award 1987), N.Y. Acad. Scis., Sigma Xi. Republican. Presbyterian. Avocations: writing, piano, travel, reading. Home: 7624 10th Ave Brooklyn NY 11228 Office: Bklyn VA Med Ctr 800 Poly Pl Brooklyn NY 11209-7198

NORTH, DAVID LEE, medical physicist; b. Phila., May 4, 1947; s. Leon Levi and Margaret Grater (Brunner) N.; m. Karen Ann Blanchard, Nov. 7, 1985; 1 child, Victoria. BA, Grinnell (Iowa) Coll., 1969; ScM, Brown U., 1972. Diplomate Am. Bd. Radiology. Med. physicist The Miriam Hosp., Providence, 1974—, R.I. Hosp., Providence, 1992—; cons. physicist R.I. Magnetic Resonance Imaging Network, Providence, 1988—. Contbr. articles to profl. jours. including Physics in Medicine and Biology, Health Physics, Radiology and Am. Jour. of Cardiology. Mem. Am. Assn. of Physicists in Medicine, Am. Coll. Med. Physics, Health Physics Soc., Am. Coll. Radiology. Achievements include devel. of use of ventricular stroke volume ratio as a measure of cardiac ventricular volume overload. Office: RI Hosp 593 Eddy St Providence RI 02903

NORTH, RICHARD ALAN, neuropharmacologist; b. Halifax, England, May 20, 1944; came to U.S., 1975; BSc, MB in Chem. Biology, U. Aberdeen, Scotland, 1969, PhD, 1973. Registered med. practitioner, U.K. Assoc. prof. Loyola U., Chgo., 1975-81; prof. MIT, 1981-86; sr. scientist Vollum Inst., Portland, Oreg., 1987—. Contbr. articles to profl. jours. Recipient Schweppe award Schweppe Found., Chgo., 1978, Boehringer prize, Ingelheim, Germany, 1986. Mem. Physiol. Soc., Soc. for Neurosci., Soc. Gen. Physiologists, British Pharmacol. Soc. (Gaddum prize 1988). Achievements include rsch. on physiology and pharmacology of nervous tissue at cellular and molecular level. Office: Vollum Institute Oregon Health Sciences Univ Portland OR 97201

NORTH, ROBERT JOHN, biologist; b. Bathurst, Australia, Aug. 22, 1935; s. Herbert John North and Loraine (Grace) Lamrock. BS, Sydney U., Australia, 1958; PhD, Nat. U., Canberra, Australia, 1967; DSc (hon.), SUNY, 1992. Vis. investigator Trudeau Inst., Saranac Lake, N.Y., 1967-70; assoc. mem. Trudeau Inst., 1970-74, mem., 1974-76, dir., 1976—. Mem. editorial bd. Jour. Exptl. Medicine, Infection and Immunity, Cancer Immunology and Immunotherapy; contbr. articles on immunity to infections and cancer to profl. jours. Recipient Friedrich Sasse sci. prize, 1984, rsch. award Soc. of Leukocyte Biology, 1990; grantee NIH, Am. Cancer Soc. Mem. Am. Assn. Immunologists, Reticuloendothelial Soc. (pres. 1983), Transplantation Soc., AAAS, Am. Soc. Microbiologists. Avocation: classical music. Home and Office: Trudeau Inst Inc PO Box 59 Saranac Lake NY 12983*

NORTHINGTON, DAVID K., research center director, botanist, educator. BA in Biology, U. Tex., 1962, PhD in Systematic Botany, 1971. Prof. Texas Tech U., 1971-84; exec. dir. Nat. Wildflower Rsch. Ctr., Austin, Tex., 1984—; vis. assoc. prof. Southwest Tex. State U., 1985—; adj prof. dept. botany U. Tex., 1984—; curator E.L. Reed Herbarium Tex. Tech. U.; dir. Tex. Tech. Ctr., Junction. Co-author 3 sci. books; contbr. numerous articles to profl. jours., mags., newspapers, newsletters. Mem. AAAS, Am. Soc. Plant Taxonomists, Nature Conservancy (bd. dirs. Tex. chpt.). Office: Nat Wildflower Research Ct 2600FM 973 N Austin TX 78725

NORTHUP, BRIAN KEITH, ecologist; b. Plainville, Kans., May 22, 1958; s. Keith Elvin and Katherine Joan (Kollman) N. BS in Botany, Ft. Hays State U., 1984, MS in Biology, 1990; PhD in Agronomy, U. Nebr., 1993. Soil conservationist USDA Soil Conservation Svc., Osborne, Kans., 1984, Salina, Kans., 1985-86; teaching asst. dept. Biology Ft. Hays State U., Hays, Kans., 1986-87; rsch. asst. dept. Agronomy U. Nebr., Lincoln, 1989-93. Contbr. articles to profl. jours. With USN, 1976-80. Mem. AAAS, Soc. for Range Mgmt., Southwestern Assn. Naturalists, Sigma Xi. Office: U Nebr WCREC Rt 4 Box 46A North Platte NE 69101

NORTHUP, T. EUGENE, nuclear engineer. With GA Technologies, Inc., San Diego, Calif. Recipient Bernard F. Langer Nuclear Codes and Standards award ASME, 1991. Office: GA Technologies Inc Torry Pines Po Box 85608 San Diego CA 92186*

NORTON, DESMOND ANTHONY, economics educator; b. Dublin, Ireland, Aug. 22, 1942; s. Ernest Gerard and Sarah Elizabeth (Collins) N.; m. Shirley Jean Renfro, Jan. 4, 1969; children: Chantelle, Shevawn, Emma. BA in Economics, Univ. Coll., Dublin, 1964; MA, Univ. N.H., Durham, 1966; PhD, Univ. Calif., Berkeley, 1973. Cert. Economist. Asst. Univ. N.H., 1964-66; asst. lectr. Univ. Coll., Dublin, 1967-69; asst. Univ. Calif., Berkeley, 1970-73; asst. lectr. Univ. Coll., Cork, Ireland, 1973-74; lectr. Univ. Coll. Dublin, 1974-78; visiting prof Univ. N.H., 1978; visiting prof. Cornell Univ., 1978-79; lectr. Univ. Coll., Dublin, 1979-86, sr. lectr., 1986—; economics correspondent The Cork Examiner 1974-76; cons. Statistical Office of European Commission, 1985-86, World Bank, Washington, 1986-87, Hungarian Acad. Scis., Budapest, 1988-91. Author: 7 books, approximately 35 acad. articles, 100 popular articles. Recipient: Univ. N.H., Univ. Calif., Economic & Social. Rsch Inst. and Inst. of Pub. Adminstrn. fellowships. Mem. European Economic Assn., Royal Economic Soc., Development Studies Assn., Irish Economic Assn. Avocation: postal history. Home: 4 Eaton Pl Monkstown, Dublin Ireland Office: U Coll, Belfield, Dublin 4, Ireland

NORTON, HARRY NEUGEBAUER, aerospace engineer; b. Vienna, Austria, June 28, 1922; came to U.S., 1940; s. Richard Neugebauer and Edith (Prager) N.; m. Hanrietta A. Eckstrom, 1953 (div. 1973); children: Cynthia E., Andrew D., Paul E. BSEE, NYU, 1950. Prodn. engr. Air King Radio, Bklyn., 1940-47; elec. engr. various cos., N.Y.C., 1947-53; avionics engr. Lockheed Aircraft Svc. Internat., JFK Airport, N.Y., 1953-56; instrumentation engr. Convair Aerospace, San Diego, 1956-68; mem. tech. staff Jet Propulsion Lab., Pasadena, Calif., 1968-88, cons. spacecraft systems, 1988—; tech. author, editor pvt. practice, Altadena, Calif., 1973. Author: 7 books, 1969-92; editor: 2 books, 1984; contbr. articles to profl. jours. Fellow AIAA (assoc.), Instrument Soc. Am. (Recognition award 1974, standards dir. 1965-74). Democrat. Achievements include developed error band concept (for transducer specifications); developed comet nucleus sampling mechanisms; directed development of I.S.A./A.N.S.I transducer standards. Home and Office: 604 Devonwood Rd Altadena CA 91001

NORTON, JERRY DON, biology educator; b. Baytown, Tex., Jan. 18, 1958; s. David Lawrence and Naomi Elaine (Davis) N. BA, Tex. Tech U., 1979, MS, 1985; postgrad., U. Tex. McLaughlin pre-doctoral fellow U. Tex. Med. Br., Galveston, 1985-88; asst. instr. U. Tex., Austin, 1990-91; part-time faculty Austin Community Coll., 1992—; biology instr. Tyler (Tex.) Jr. Coll., 1992—. Asst. scoutmaster Troop 108 Boy Scouts Am., Midland, Tex., 1977-87. Named Eagle Scout Boy Scouts Am., Lion of Yr. Hub Lions Club, Lubbock, 1981; recipient Outstanding Svc. award Tatanka Lodge, Boy Scouts Am., Buffalo Trail Coun., 1979. Mem. Tex. Jr. Coll. Tchrs. Assn., Nat. Sci. Tchrs. Assn., Am. Soc. Microbiology, Nat. Assn. for Rsch. in Sci. Teaching, Sigma Xi, Phi Delta Kappa, Phi Kappa Phi. Home: 2908 Rebel Dr Midland TX 79707 Office: U Tex at Austin Sci Edn Ctr EDB340 Austin TX 78712 also: Tyler Jr Coll Math and Sci Div Tyler TX 75701

NORTON, ROBERT LEO, SR., mechanical engineering educator, researcher; b. Boston, May 5, 1939; s. Harry Joseph and Kathryn (Warren) N.; m. Nancy Auclair, Feb. 27, 1960; children: Robert L., Jr., MaryKay, Thomas J. BS, Northeastern U., 1967; MS, Tufts U., 1970. Registered profl. engr., Mass, N.H. Engr. Polaroid Corp., Cambridge, Mass., 1959-67; project engr. Jet Spray Cooler, Inc., Waltham, Mass, 1967-69; rsch. assoc. N.E. Med. Ctr., Boston, 1969-74; prof. Tufts U., Medford, Mass., 1974-79; sr. engr. Polaroid Corp., Waltham, 1979-81; prof. mech. engring., Worcester Poly., Mass., 1981—; pres. Norton Assocs., Norfolk, Worcester, 1970—. Patentee (13) in field; contbr. articles to profl. jours; author engring. textbooks. Mem. ASME, Am. Soc. Engring. Edn. (J.F. Curtis award 1984, Merle Miller award 1987, 92), Pi Tau Sigma, Sigma Xi. Democrat. Avocations: sailing, computers. Office: WPI 100 Institute Rd Worcester MA 01609

NORTON, STEVEN DAVID, immunologist, researcher; b. Endicott, N.Y., Aug. 14, 1965; s. David Gibbons and Ro Jean (Reeves) N.; m. Lynette Dayton, Sept. 30, 1987; children: Jonathan Steven, Gregory Dayton. BS, Brigham Young U., 1988; PhD, U. Minn., 1992. Predoctoral fellow U. Minn., Mpls., 1988-92; postdoctoral fellow U. Chgo., 1992—. Contbr. articles to sci. jours.; contbr. chpt. to Encyclopedia of Immunology. Recipient Rsch. award 20th Midwest Immunology Conf., 1991. Mem. Phi Kappa Phi. Office: U Chgo 920 E 58th St Chicago IL 60637

NORUP, KIM STEFAN, civil engineer; b. Copenhagen, Mar. 13, 1945; s. Erik and Ulla (Helgesen) N.; m. Loimata Faleafa, Dec. 7, 1974; children: Stewart Magnus, Signe Louise. MSc, Tech. U. Denmark, Copenhagen, 1968. Asst. engr. Port Authority Copenhagen, 1968-72; assoc. expert UN Devel. Programme, Apia, Western Samoa, 1972-73; project mgr. Spl. Project Devel. Corp., Apia, 1973-75; asst. project engr. Sir Alex Gibb & Ptnrs. Australia, Adelaide, 1975-77; resident engr. Sir Alex Gibb & Ptnrs. Australia, Lautoka, Fiji, 1977-85, Cowiconsult, Copenhagen, 1985-86; project mgr. Cowiconsult, Thimphu, Bhutan, 1986-90; supervision cons. Cowiconsult, Bandung, Indonesia, 1990—. Mem. Instn. Civil Engrs. U.K., Dansk Ingenior Forening Denmark. Office: Cowiconsult, Jalan Cimanuk 34, Bandung 40115, Indonesia

NORVELLE, NORMAN REESE, environmental and industrial chemist; b. Wewoka, Okla., Oct. 5, 1946; s. Norman Geneese and Aldulia Maud (Furney) N.; m. Carolyn E. Sewell, Jan. 24, 1970; children: Alex Reese, Heather Lynn. BS, Ea. N.Mex. U., 1970, M Natural Scis., 1974. Registered hazardous substance profl., environ. mgr., environ. profl. cert. hazardous material mgr.; cert. environ. trainer, N.Mex. Lab. scientist N.Mex. Health Dept., Farmington, 1970-74; tchr. Farmington Pub. Sch. System, 1975-76; sales mgr. Sales Assocs., Inc., Albuquerque, 1976-77; sr. chemist Pub. Svc. Co. N.Mex., Farmington, 1977-88; regional chemist El Paso Natural Gas Co., Farmington, 1989—; adv. com. water utility tech. Dona Ana br. N.Mex. State U., Las Cruces, 1980-84. Asst. scoutmaster Farmington area Boy Scouts Am., 1983—, asst. scoutmaster Nat. Jamboree, Washington, 1989. Mem. Nat. Assn. Corrosion Engrs. (treas. 1990, sec. 1991, vice-chmn. 1992, chmn. 1993, south cen. region conf. sec. 1991), Am. Chem. Soc., Am. Water Works Assn., Am. Indsl. Hygiene Assn., Nat. Environ. Health Assn., N.Mex. Water and Wastewater Assn. (sec.-treas. 1978-82, Double IV Club 1989). Achievements include invention of chlorine leak detection bottle. Home: 3510 Kayenta Dr Farmington NM 87402 Office: El Paso Natural Gas Co 770 W Navajo St Farmington NM 87401

NORWICK, BRAHAM, textile specialist, consultant, columnist; b. N.Y.C., July 6, 1916; s. Mark and Rose (Ungar) N.; m. Thérèse Thoisy, May 7, 1939; 1 child, Noel Alex. BS in Chemistry, Rensselaer Polytech. Inst., 1938; cert. d' Etudes, U. de Besançon, France, 1945. Tech. dir. Beaunit Mills, N.Y.C., 1938-78; v.p. Joseph Bancroft & Sons Co., Wilmington, Del., 1973-78; expert witness TAG, N.Y.C., 1980—; columnist Wirkerei-und Strickerei-Technik, Bamberg, Fed. Republic of Germany, 1983—; vis. prof. Cornell U., Ithaca, N.Y., 1982. Author: (mag.) Indsl. & Engring. Chemistry, 1942; contbg. author: Developments in Applied Spectroscopy, 1968, Analytical Methods for a Textile Laboratory, 1968, 84; columnist Daily News Record Newspaper, 1979-86. Cons. Jacques Marchais Mus., S.I., 1975—. With U.S. Army, 1943-46, ETO. Recipient Bronze medal Am. Assn. Textile Techs., 1977. Fellow ASTM (spokesman Internat. Standardization Orgn. 1960—, Gold Harold de Witt Smith medal, 1984), Am. Soc. for Quality Control; mem. Am. Chem. Soc. (life), Am. Assn. Textile Chemists and Colorists, N.Y. Acad. Scis., Chemists Club. Achievements include first to make heat set nylon parachutes; first characterization and standardization of dyestuffs for industrial computerized dyeing; development of wash and wear finishes; utilization of small sample statistical methods and rare events statistics in textile processing, chromatographic methods; employment of infrared methods in textile chemicals analysis. Home: 200 E 57th St New York NY 10022-2899

NORWOOD, CAROL RUTH, research laboratory administrator; b. N.Y.C., Oct. 10, 1949; d. John Theodore and Ruth Arnold (Shields) Gundlach; m. Christian K.-H. Schneider; 1 child, from previous marriage, Jonathan Blair. BA, U. Colo., 1971, PhD in Biophysics, 1975. Rsch. assoc. div. biochemistry MIT, Cambridge, 1974-76; rsch. assoc. dept. cardiac surgery Children's Hosp., Boston, 1976-83; instr. dept. surgery Harvard U. Med. Sch., Boston, 1980-83; asst. prof. U. Pa. Sch. Medicine, Phila., 1984-87; dir. Cardiothoracic Surgery Rsch. Labs. Children's Hosp. Phila., 1984—. Contbr. articles to sci. jours., chpts. to books. Mem. Am. Chem. Soc. (biochemistry divsn. pub. outreach), Am. Heart Assn. Avocations: science education, vernacular and landscape architecture. Office: Children's Hosp Phila 34th St and Civic Center Blvd Philadelphia PA 19104

NORWOOD, KEITH EDWARD, civil engineer; b. Shreveport, La., Mar. 13, 1963; s. Charles M. and Marlene T. (Hagedorn) N. BSCE, La. State U., 1986. Registered profl. engr., La. Engr.-in-tng. Aillet, Fenner, Jolly & McClelland, Inc., Shreveport, 1986-91, civil engr., 1991; civil engr. City of Shreveport/Dept. Pub. Works, Engring. Div., 1991—. Mem. adv. coun. Community Cultural Plan, Shreveport, 1992-93; chmn. Downtown Revitalization Com., 1993—; media chmn. Artbreak, Shreveport Regional Arts Coun., 1990. Mem. ASCE, La. Engring. Soc. Republican. Achievements include sitework for Schumpert Med. Ctr., South Wing, Phase I Expansion. Office: City of Shreveport Dept Pub Works/Engring PO Box 31109 Shreveport LA 71130

NOSÉ, YUKIHIKO, surgeon, educator; b. Inamisawa, Hokkaido, Japan, May 7, 1932; came to U.S., 1962; s. Minoru and Haru (Murakami) N.; m. Bonnie Jean MacDonald, Mar. 15, 1965 (div. 1987); children: Kimi Wilhelmina, Ken Willem, Kevin Scott; m. Ako Funakoshi, May 5, 1990. MD, U. Hokkaido, Sapporo, Japan, 1957, PhD, 1962. Surgeon in charge sect. artificial organs U. Hokkaido Sch. Medicine, 1961-62; rsch. assoc. Maimonides Hosp., Bklyn., 1962-64; postgrad. fellow dept. artificial organs Cleve. Clinic Found., 1964-66; mem. staff dept. artificial organs Cleve. Clinic, 1966-67, chmn. dept. artificial organs, 1967-89, chmn. emeritus, 1989-90; prof. surgery Baylor Coll. Medicine, Houston, 1991—; v.p. Internat. Ctr. Artificial Organs and Transplantation, Cleve., 1979—; adj. prof. surgery Tokyo Med. Coll., Tsukuba (Japan) U., Keio (Japan) U.; cons., mem. surgery and bioengring. study sect. NIH, 1981-87; assoc. dean Asian region Internat. Faculty Artificial Organs, 1992—; prof. Bologna (Italy) U. Sch. Medicine, 1992—; congress pres. 1994 Houston Congress of World Apheresis Assn. Author: Manual on Artificial Organs: Volume I-The Artificial Kidney, 1969, Volume II-The Oxygenator, 1973, Cardia Engineering, 1970, Die Kunstliche Niere, 1974, Plasmapheresis, Historical Perspective, Therapeutic Applications and New Frontiers (with Kambic), 1983, Future Perspective for the Development of Artificial Organs (with Kolff), 1988; contbr. to numerous profl. publs. Fellow Am. Inst. Med. and Biol. Engring., N.Y. Acad. Sci.; mem. AMA, AAAS, Internat. Soc. Artificial Organs (trustee, past pres.), Am. Soc. Artificial Internal Organs (past pres.), World Apheresis Assn. (gen. sec., congress pres.), Am. Soc. Testing Materials (chair subcom. on cardiovascular prosthesis in med. and surg. materials and devices, Moses award 1979), Am. Heart Assn., Am. Soc. Apheresis, Am. Soc. Artificial Internal Organs, Am. Soc. Biomaterials, Assn. Advancement Med. Instrumentation. Achievements include development of various types of artificial organs including cardiac prosthesis, artificial kidney, hepatic assist, respiratory assist, plasmapheresis, biomaterials. Home: 1400 Hermann St Houston TX 77004 Office: Baylor Coll Medicine Dept Surgery 1 Baylor Plz Houston TX 77030

NOSEK, THOMAS MICHAEL, physiologist, educator; b. Bklyn., Apr. 25, 1947; s. Stanley M. and Regina A. (Bernatowicz) N.; m. Claudia M. Wilchek, Aug. 16, 1968; children: Thomas A., Christopher M. BS in Physics, U. Notre Dame, 1969; PhD in Biophysics, Ohio State U., 1973. Instr. Bowman Gray Sch. Medicine, Winston-Salem, 1973-76; asst. prof. Med. Coll. Ga., Augusta, 1976-82, assoc. prof., 1982-93, prof. physiology, 1993—; bd. dirs. MCG Rsch. Inst., Augusta, 1992—. Contbr. articles to profl. jours. Scoutmaster Boy Scouts Am., Augusta, 1982-89; pres. PTA, Warren Rd. Elem. Sch., Augusta, 1981-82; coach Augusta Recreation Dept., 1976-85. Rsch. grantee NIH, 1978-82, 88—, Ga. Heart Assn., 1979-88, Dept. of Navy, 1991—; recipient Outstanding Rsch. award Grad. Sch./Med. Coll. Ga., 1990, Excellence in Teaching award, 1992. Mem. AAAS, AAUP, N.Y. Acad. Sci., Am. Heart Assn., Biophys. Soc. Unitarian. Office: Med Coll of Ga Dept Physiology/Endocrinol Augusta GA 30912-3000

NOTARIO, VICENTE, molecular biology educator, researcher; b. Ciudad Real, La Mancha, Spain, Nov. 28, 1952; came to U.S., 1980; s. Vicente Notario-Garcia and Maria (Ruiz) de Notario; m. Sofia P. Becerra, July 31, 1982; children: Patricia, Rafael. BS, U. Salamanca, Spain, 1974, PhD in Biol. Scis., 1977. Rsch. instr. Faculty Scis. and Faculty Pharmacy U. Salamanca, 1974-77; postdoctoral fellow dept. biochemistry Cambridge (Eng.) U., 1978-79, Nuffield Found. fellow, 1979; fellow sect. enzymes and cellular biochemistry Nat. Inst. ADDK, NIH, Bethesda, Md., 1980, vis. assoc., 1980-81; vis. assoc. Lab. Cellular and Molecular Biology, Nat. Cancer Inst., Bethesda, 1981-86; rsch. asst. prof. biochemistry Sch. Medicine Georgetown U., Washington, 1986-89, assoc. prof., head exptl. carcinogenesis div. Sch. Medicine, 1989—, mem. Lombardi Cancer Rsch. Ctr. Sch. Medicine, 1988—; guest worker somatic cell genetics sect. Lab. Biology, DCE, NIH, 1986-89, adj. scientist, 1988-90; referee Analytical Biochemistry, Cancer Rsch.; sci. cons. molecular biology BioSynthesis Labs., Denton, Tex.; presenter at profl. meetings, workshops and symposia, 1975—. Contbr. numerous articles to sci. jours. Fellow Spanish Ministry Edn. and Sci., 1975-77. Mem. AAAS, Am. Assn. for Cancer Rsch., Spanish Soc. for Biochemistry, Spanish Soc. for Microbiology, Radiation Rsch. Soc., Inst. Estudios Manchegos (hon.), Sigma Xi. Achievements include establishment of role of yeast cell wall glucan in resistance to antifungal drugs; isolation of human oncogenes (RAS, SIS) and their products (FGR), isolation and purification of yeast cell-wall hydrolases; reproducible activation of oncogenes with chemical carcinogens in animal models; cloning of a novel oncogene from hamster tumor cells initiated chemically; research in molecular basis for cell growth, differentiation and morphogenesis, mechanisms of oncogenesis, molecular evolution, molecular genetics with emphasis on yeast genetics. Office: Georgetown Univ Radiation Med Dept 3800 Reservoir Rd NW Washington DC 20007-2196

NOTHNAGEL, EUGENE ALFRED, plant cell biology educator; b. Litchfield, Minn., June 5, 1952; s. Andrew H. and Rose (Schindele) N.; m. Janet Marie Luber, Aug. 9, 1980; children: Sarah Lynn, Alex Leon. BA in Physics, Math., U. Minn., Morris, 1973; MA in Physics, So. Ill. U., 1975; PhD in Applied Physics, Cornell U., 1981. Rsch. assoc. dept. chemistry U. Colo., Boulder, 1981-83; asst. prof. dept. botany and plant scis. U. Calif., Riverside, 1983-89, assoc. prof., 1989—; mem. adv. panel biol. energy rsch. Dept. Energy, Washington, 1986; mem. adv. panel cell biology NSF, Washington, 1988-92. Editor: Plant Senescence: Its Biochemistry and Physiology, 1987; contbr. articles to profl. jours. Fellow for young biophysicists Japanese Govt., Kyoto, Japan, 1978, postgrad. fellow Rockefeller Found., N.Y.C., 1981-83. Mem. Am. Soc. Plant Physiologists, Am. Soc. Cell Biology, Am. Soc. Hort. Sci., Biophys. Soc. Achievements include patents for NBD-acidic phallotoxins and their use in the fluorescence staining of F-actin and acidic phallotoxin derivatives and methods of preparation. Office: U Calif Dept Botany & Plant Scis 072 Riverside CA 92521-0124

NOTKINS, ABNER LOUIS, physician, researcher; b. New Haven, May 8, 1932; s. Louis Adolph and Sair Rosalyn (Mathog) N.; m. Susan Woodward, Aug. 9, 1969. BA Yale U., 1953; MD, NYU, 1958. Intern Johns Hopkins U., Balt., 1959, asst. resident, 1960; research assoc. Nat. Cancer Inst., NIH, Bethesda, Md., 1960-61; prin. investigator Nat. Inst. Dental Research, NIH, Bethesda, Md., 1961-67, sect. chief, microbiology lab., 1967-73, chief, oral medicine lab., 1973—, dir., intramural research program, 1985—; cons. Nat. Commn. on Diabetes, 1975, Virology Task Force, 1976, Nat. Diabetes Research Interchange, 1981—, Wistar Inst. Persistent Virus Program, 1983—, NASA, 1983; lectr. various univs. and orgns. Mem. editorial bd. Infection and Immunity, 1970-73, Jour. of Virology, 1984—, Microbial Pathogenesis, 1985—; contbr. numerous articles to profl. jours.; editor 3 books. Served to med. dir. PHS, 1960-85. Recipient Meritorious Service medal Dept. HHS, 1973, David Rumbough Sci. award Juvenile Diabetes Found., 1980, Distinguished Service award Dept. HHS, 1981, Paul E. Lacey research award Nat. Diabetes Research Interchange, 1982, Paul Ehrlich prize, Paul Ehrlich Found., Fed. Republic of Germany, 1986. Mem. Assn. U. Pathologists (Pluto Soc.), Assn. Am. Physicians, Alpha Omega Alpha. Home: 1179 Crest Ln Mc Lean VA 22101-1805 Office: HHS Nat Inst of Dental Research 9000 Rockville Pike Bldg 30 Bethesda MD 20892-0001

NOTOWIDIGDO, MUSINGGIH HARTOKO, information systems executive; b. Indonesia, Dec. 9, 1938; s. Moekarto and Martaniah (Brodjonegoro) N.; m. Sihar P. Tambunan, Oct. 1, 1966 (dec. Nov. 1976); m. Joanne S. Gutter, June 3, 1979; children: Matthew Joseph, Jonathan Paul. BME, George Washington U., 1961; MS, NYU, 1966, postgrad., 1970. Cons. Dollar Blitz & Assocs., Washington, 1962-64; ops. research analyst Am. Can Co., N.Y.C., 1966-69; prin. analyst Borden Inc., Columbus, Ohio, 1969-70, mgr. ops. research, 1970-71, mgr. ops. analysis and research, 1972-74, asst. gen. contr., officer, 1974-77, corp. dir. info. systems/econ. analysis, officer, 1977-83; v.p. info. systems Wendy's Internat., 1983-90; founder, pres. World Techs. Group, 1990-91; chmn., co-founder AMG Inc., 1991—; adj. lectr. Grad. Sch. Adminstrn. Capital U.; mem. gov.'s tech. adv. bd. State of Ohio. Contbr. articles to profl. jours. Mem. Gov.'s Tech. Adv. Bd.; mem. bd. dirs. Optimum Tech., Inc. Mem. Fin Execs. Inst. (trustee), Inst. Mgmt. Sci., Am. Mgmt. Assn., Nat. Assn. Bus. Economists, Long Range Planning Soc., Am. Statis. Assn., World Future Soc., Data Processing Mgmt. Assn., Soc. Info. Mgmt., N.Y. Acad. Scis. Republican. Clubs: Capital, Racquet. Home: 1965 Brandywine Dr Columbus OH 43220-4421 Office: 6400 Riverside Dr Bldg E Dublin OH 43017

NOURI-MOGHADAM, MOHAMAD REZA, mathematics and physics educator; b. Tehran, Iran, July 1, 1949; s. Gholam Reza and Zahra (Razeghian) Nouri-Moghadam; m. Vida Nouri-Moghadam, May 18, 1968; children: Yasaman, Ardavan. BS, U. London, 1972, PhD, 1975. Postdoctoral research fellow U. London, King's Coll., 1975-76; Fulbright sr. research fellow and vis. asst. prof. Princeton U., 1979-80; prof. Sharif U.

Tech., Tehran, 1976-87; rsch. fellow Kyoto U., Japan, 1986-87; vis. prof. U. Toledo, Ohio, 1987-88; prof. math., sci. tech. and soc. Pa. State U., 1991—; assoc. rschr. Internat. Ctr. for Theoretical and Math. Physics, Trieste, Italy, 1977-80, 88-93; chair faculty senate Pa. State U., Wilkes Barre Campus, 1992-93. Editor-in-chief Iranian Math. Soc. Bull.; author: Vector and Tensor Analysis. Mem. SIAM, Am. Math. Soc., Internat. Neural Network Soc., London Math. Soc. Office: Pa State U E-Mail # MNM1 PO Box PSU Lehman PA 18627 also: Pa State U 126 Willard University Park PA 16802

NOVAK, CHARLES R., computer scientist; b. Saint Augustine, Fla., Aug. 8, 1957; s. Ralph Charles and Wanda (Vermulen) N.; m. Wanda Marie Thompson, Nov. 30, 1976; children: Gabriel, Erin, Beth, Joey, Marianne, Jordan, Melanie. BSBC, Nat. U., 1984. Instrumention technician Fed. Civil Svc., San Diego, 1979-81; prodn. controller Fed. Civil Svc., Oceanside, Calif., 1981-84; sr. programmer Gen. Dynamics Corp., Fort Worth, Tex., 1984-87; cons. Digital Equip. Corp., Houston, 1987-88; sr. cons. Info. Industries Inc., Richmond, Va., 1988-90; staff analyst Westinghouse Savannah River Co., Aiken, S.C., 1990--; mem. Digital Mass Storage Product Directions Forum, Malboro, Mass., 1991--. Sgt. USMC, 1975-79. Achievements include development of highly efficient algorithms for advanced avionic systems. Home: 611 Kimball Pond Rd Aiken SC 29803 Office: Westinghouse Savannah Co Blg 773-51A Aiken SC 29808

NOVAK, RAYMOND FRANCIS, research institute director, pharmacology educator; b. St. Louis, July 26, 1946; s. Joseph Raymond and Margaret A. (Cerutti) N.; m. Frances C. Holy, Apr. 12, 1969; children: Jennifer, Jessica, Janelle, Joanna. BS in Chemistry, U. Mo., St. Louis, 1968; PhD in Phys. Chemistry, Case Western Res. U., 1973. Assoc. in pharmacology Northwestern U. Med. Sch., Chgo., 1976-77, asst. prof. pharmacology, 1977-81, assoc. prof., 1981-86, prof., 1986-88; prof. pharmacology Wayne State U. Sch. Medicine, Detroit, 1988—; dir. Inst. Chem. Toxicology Wayne State U., Detroit, 1988—; mem. toxicology study sect. NIH, Bethesda, Md., 1984-88; adj. sci. Inhalation Toxicology Rsch. Inst., Lovelace Biomed. and Environ. Rsch. Inst., 1991-92. Assoc. editor Toxicol. Applied Pharmacology, 1992—; editor Drug Metabolism and Disposition, 1994—; mem. editorial bd. Jour. Toxicology and Environ. Health, 1987-92, In Vivo, 1986—; contbr. articles to profl. jours. Recipient Disting. Alumni award U. Mo., St. Louis, 1988; grantee Nat. Inst. Environ. Health Sci., 1979—, Gen. Medicine sect. NIH, 1979—. Mem. Am. Soc. for Biochem. and Molecular Biology, Soc. Toxicology, Am. Assn. for Cancer Rsch., Am. Soc. for Pharmacology and Exptl. Therapeutics, Am. Soc. Hematology, Am. Chem. Soc., Biophys. Soc., Internat. Soc. for Study of Xenobiothics. Office: Wayne State U Inst Chem Toxicology 2727 2nd Ave Rm 4000 Detroit MI 48201-2654

NOVAK, ROBERT LOUIS, civil engineer, pavement management consultant; b. Chgo., Feb. 29, 1928; s. Louis and Frances (Kucera) N.; m. Virginia Staas, Jan. 22, 1955 (div. 1962); children: Susan Grace, Nina Louise; m. Joyce Eloise Keen, May 7, 1966; stepchildren: Robert John Moore, William Keen Moore, Marilyn Joyce Moore, James Clifford Moore. BCE, Ga. Inst. Tech., 1948. Soils engr. Soil Testing Svc., Chgo., 1952-54; dir. field invest. USAF Acad., Colorado Springs, 1954-58; asst. dir. engring. Naess & Murphy, Chgo., 1958-60; pres. Novak, Dempsey & Assocs., Palatine, Ill., 1960-85; ptnr. Infrastructure Mgmt. Svcs., Arlington Heights, Ill., 1985-89, cons., 1989—. Contbr. articles to profl. jours. With U.S. Army, 1950-52. Mem. ASTM, Am. Pub. Works Assn. (meritorious svc. award 1990), Transp. Rsch. Bd. Achievements include development of one of the first pavement management programs. Home: 813 Auburn Woods Dr Palatine IL 60067

NOVAKOVIĆ, BRANKO MANE, engineering educator; b. Sanski Most, Bosnia and Hercegovina, July 8, 1940; s. Mane Lazo and Boja Petar (Tonković) N.; m. Borka Branko Milinović, Dec. 16, 1967; children: Dario, Alen. BSME, Tech. U., Novi Sad, Yugoslavia, 1966; MSME, U. Zagreb, Croatia, 1974, D of Engring. Sci., 1978. Project designer Rade Koncăr, Zagreb, 1966-67; dept. chief Termomehahika, Zagreb, 1967-72; offering engr. Hladjenje, Zagreb, 1972-74; asst. prof. FSB-Univ. of Zagreb, 1974-79, docent prof., 1979-87, prof., 1987—; dept. dir. FSB-Univ. Zagreb, 1975-78, pres. gen. coun., 1986-88, pres. edn. coun., 1989—. Author: Control Methods of Technical Systems, 1990 (J.J. Strossmayer award 1991), numerous sci. papers. Fellow Jurema (mgmt. mem. 1985—); mem. IEEE, N.Y. Acad. Scis., Etan, Infomatika, Interklima. Achievements include research in control systems, robotics; design of model for artificial neural networks. Home: Cesariceva 33, 41000 Zagreb Croatia Office: FSB-U Zagreb, Salajeva 5, 41000 Zagreb Croatia

NOVARA, MAURO, space system engineer; b. Turin, Italy, Sept. 13, 1957; arrived in Netherlands, 1984; s. Erminio and Elsa (Lovisone) N. Degree in Aero. Engring., Politecnico di Torino, Turin, Italy, 1981. Thermal engr. Aeritalia, Turin, 1982-84; thermal engr. ESA/ESTEC, Noordwijk, Netherlands, 1984-88, life support engr., 1988-93, space system engr., 1993—. Douglas Marsh fellow European Space Agy., 1987. Mem. AIAA. Roman Catholic. Office: ESA/ESTEC, Keplerlaan 1, 2201 AZ Noordwijk The Netherlands

NOVELLO, ANTONIA COELLO, U.S. surgeon general; b. Fajardo, P.R., Aug. 23, 1944; d. Antonio and Ana D. (Flores) Coello; m. Joseph R. Novello, May 30, 1970. BS, U. P.R., Rio Piedras, 1965; MD, U. P.R., San Juan, 1970; MPH, Johns Hopkins Sch. Hygiene, 1982. Diplomate Am. Bd. Pediatrics. Intern in pediatrics U. Mich. Med. Ctr., Ann Arbor, 1970-71, resident in pediatrics, 1971-73, pediatric nephrology fellow, 1973-74; pediatric nephrology fellow Georgetown U. Hosp., Washington, 1974-75; project officer Nat. Inst. Arthritis, Metabolism and Digestive Diseases NIH, Bethesda, Md., 1978-79, staff physician, 1979-80; exec. sec. gen. medicine B study sect., div. of rsch. grants NIH, Bethesda, 1981-86; dep. dir. Nat. Inst. Child Health & Human Devel., NIH, Bethesda, 1986-90; surgeon gen. HHS, Washington, 1990—; clin. prof. pediatrics Georgetown U. Hosp., Washington, 1986, 89, Uniformed Svcs. U. of Health Scis., 1989; mem. Georgetown Med. Ctr. Interdepartmental Rsch. Group, 1984—; legis. fellow U.S. Senate Com. on Labor and Human Resources, Washington, 1982-83; mem. Com. on Rsch. in Pediatric Nephrology, Washington, 1981—; participant grants assoc. program seminars Nat. Inst. Arthritis, Diabetes and Digestive and Kidney Diseases, NIH, Bethesda, 1980-81; pediatric cons. Adolescent Medicine Svc., Psychiat. Inst., Washington, 1979-83; nephrology cons. Met. Washington Renal Dialysis Ctr. affiliate Georgetown U. Hosp., Washington, 1975-78; phys. diagnosis class instr. U. Mich. Med. Ctr., Ann Arbor, 1973-74; chair Sec.'s Work Group on Pediatric HIV Infection and Disease, DHHS, 1988; cons. World Health Orgn., Geneva, 1989; mem. Johns Hopkins Soc. Scholars, 1991. Contbr. numerous articles to profl. jours. and chpts. to books in field; mem. editorial bd. Internat. Jour Artificial Organs, Jour. Mexican Nephrology. Served to capt. USPHS, 1978—. Recipient Intern of Yr. award U. Mich. Dept. Pediatrics, 1971, Woman of Yr. award Disting. Grads. Pub. Sch. Systems, San Juan, 1980, PHS Commendation medal HHS, 1983, PHS Citation award HHS, 1984, Cert. of Recognition, Div. Research Grants NIH, 1985, PHS Outstanding medal HHS, 1988, PHS Unit Commendation, 1988, PHS Surgeon Gen.'s Exemplary Svc. medal, 1989, PHS Outstanding Unit citation, 1989, DHHS Asst. Sec. for Health cert. of commendation, 1989, Surgeon Gen. Medallion award, 1990, Alumni award U. Mich. Med. Ctr., 1991, Elizabeth Blackwell award, 1991, Woodrow Wilson award for disting. govt. svc., 1991. Fellow Am. Acad. Pediatrics; mem. AMA, Internat. Soc. Nephrology, Am. Soc. Nephrology, Latin Am. Soc. Nephrology, Soc. for Pediatric Rsch., Am. Pediatric Soc., Assn. Mil. Surgeons U.S., Am. Soc. Pediatric Nephrology, Pan Am. Med. and Dental Soc. (pres.-elect, sec. 1984), D.C. Med. Soc. (assoc.), Alpha Omega Alpha. Avocation: collecting antique furniture. Home: 1315 31st St NW Washington DC 20007-3334 Office: Office of Surgeon Gen 200 Independence Ave SW Washington DC 20201

NOVICH, BRUCE ERIC, materials engineer; b. Phila., Mar. 15, 1957; s. Samuel David and Vivian Rose Novich; m. Susan S. Novich, Sept. 5, 1982; children: Scott, Spencer, Cory. BA, Colgate U., 1979; BSChemE, MIT, 1980, MS in Geology, 1982, MS in Civil Engring, 1982, PhD, 1984. V.p. rsch., devel., engring. Ceramics Process System, Milford, Mass., 1984—. Contbr. over 30 articles to profl. jours. Achievements include 15 patents in particulate processing. Home: 55 Pleasant St Lexington MA 02173 Office: Cermics Process Systems 8 Ave E Hopkinton MA 01748

NOVICK, RONALD PADROV, electrical engineer; b. N.Y.C., Jan. 5, 1957; s. Sylvan R. and Pearl (Codner) N.; m. M. Wendy Carp, June 2, 1985; children: Jessica, Ari, Leora. BSEE, U. Rochester, 1979; postgrad., U. New Haven, 1991—. Design engr. Tex. Instruments, Dallas, 1979-81; F-16 field svc. engr. Gen. Dynamics, Ft. Worth, 1981-83; electronics countermeasures test engr. ITT, Nutley, N.J., 1983-84; cons. avionics design Sikorsky Aircraft, Stratford, Conn., 1985-86; cons. systems integration Norden Systems, Norwalk, Conn., 1986-88; sr. systems engr. Barnes Engring. div. EDO Corp., Shelton, Conn., 1988—. Contbr. sci. papers to conf. and symposium proceedings. Bd. dirs. Congregation or Shalom, Orange, Conn., 1989-90, com. chmn., 1991-92. Mem. IEEE, AIAA. Democrat. Achievements include first to use field programmable gate arrays for space applications; design of space qualified MIL-STD 1750 computer and high speed data acquistion subsystem. Office: Barnes Engring div EDO Corp 88 Longhill Crossroads Shelton CT 06484

NOVITSKI, CHARLES EDWARD, biology educator; b. Rochester, N.Y., Oct. 3, 1946; s. Edward and Esther Ellen (Rudkin) N.; m. Margaret Thornton Sime, June 15, 1968; children: Nancy Ellen, Linda Nicole, Elise Michelle. BA in Biology, Columbia Coll., 1969; PhD in Biophysics, Calif. Inst. Tech., 1979. Rsch. fellow and assoc. City of Hope Nat. Med. Ctr., Duarte, Calif., 1977-80; sr. tutor in biochemistry Monash U., Victoria, Australia, 1980-82, lectr. in biochemistry, 1982-84; program leader and rsch. scientist in nematode control Agrigenetics Advanced Sci. Co., Madison, Wis., 1985-88; assoc. prof. molecular biology Cen. Mich. U., Mt. Pleasant, 1989—. Author: (with others) Manipulation and Expression of Genes in Eukaryotes, 1983; contbr. articles to Annals of Human Genetics, Genetics, Cytobios, Biochem. Internat., Trends Biochem., Nucleic Acids Rsch., Current Genetics, European Jour. Biochem., Ann. Applied Biology. Mem. Soc. of Nematologists, Internat. Soc. of Plant Molecular Biology. Achievements include research in the molecular genetics of mitochondria and of nematodes. Home: 1208 E Preston Rd Mount Pleasant MI 48858 Office: Cen Mich U Dept Biology Mount Pleasant MI 48859

NOVOTNY, JIRI, biophysicist; b. Kladno, Czechoslovakia, Dec. 15, 1943; came to U.S., 1979; s. Jaroslav and Eva (Foustkova) N.; m. Jarmila Novotny, Feb. 26, 1967; 1 child, Paula. RNDr, Charles U., Prague, Czechoslovakia, 1970, PhD, 1970. Scientist Inst. Organic Chemistry & Biochemistry, Prague, 1970-77; sr. scientist Inst. Molecular Genetics, Prague, 1971-79; from asst. to assoc. prof. Harvard Med. Sch., Boston, 1980-88; dir. dept. Bristol-Myers Squibb Rsch. Inst., Princeton, N.J., 1988—; vis. scientist Oxford U., Eng., 1979, dept. molecular biology Princeton U., 1993—; dir. rsch. Pasteur Inst., Paris, 1987; adj. prof. U. Pa., Phila., 1989—; cons. Creative Biomolecules Inc., Hopkinton, Mass., 1984-88, Immulogic, Inc., Cambridge, Mass., 1987-88; mem. sci. adv. bd. Procept, Inc., Cambridge, 1989—. Contbr. articles to jours. Procs. NAS, Nature, Biochemistry. Recipient prize of the Czechoslovak Acad. of Scis., 1979. Mem. Am. Phys. Soc., Am. Soc. Biol. Chemists, Am. Biophys. Soc., Princeton Chess Club, Belle Meade Friends of Music. Achievements include patent for Single-Chain T Cell Receptor Fragments; development of empirical free energy potential, of molecular theory of protein antigenicity; identification of C1q binding site on antibody molecules; research on secondary structure prediction algorithm. Home: 101 Red Hill Rd Princeton NJ 08540-1307 Office: Bristol-Myers Squibb Rsch Inst Princeton NJ 08543-4000

NOVOTNY, MILOS V., chemistry educator. BS, U. Brno, Czechoslovakia, 1962, PhD, 1965. With Ind. U., Bloomington, 1971—, now Rudy prof. of chemistry. Office: Ind U Dept of Chemistry Bloomington IN 47405

NOVY, DIANE MARIE, psychologist; m. Stanley B. Novy, July 23, 1967; children: Mara, Meredith. BS, U. Tex., 1967; MA, Houston Bapt. U., 1984; PhD, U. Houston, 1990. Intern VA Hosp., Houston, 1989-90; asst. prof. psychiatry, asst. prof. anesthesiology U. Tex. Health Sci. Ctr., Houston, 1990—; adj. prof. U. Houston, 1990—; bd. dirs. Alumni Assn. U. Houston Coll. Edn., 1992. Bd. mem. Tex. Rehab. Comm., State of Tex., 1992—. Mem. APA, Tex. Psychol. Assn., Am. Ednl. Rsch. Assn., Am. Pain Soc. Office: Univ Tex Med Sch Dept Anesthesiology Houston TX 77030

NOWAK, CHESTER JOHN, JR., infosystems specialist; b. Chgo., Nov. 20, 1941; s. Chester John and Ann Barbara (Pluta) N. BS in Archtl. Engring., Chgo. Tech. Coll., 1963. Inventory control, internal expiditer Nat. Starch & Chem., Chgo., 1963-69; dir. records storage ops. Exchange Nat. Bank, Chgo., 1970-83; records mgmt. analyst Sargent & Lundy Engrs., Chgo., 1984-88; records info. mgr. II Labat-Anderson Inc., Chgo., 1988—; cons. Gangi Marine Bldrs., Naperville, Ill. Vol. Dem. Com. to elect Richard Daley Mayor of Chgo., 1989. Mem. Assn. Records Mgrs. and Adminstrs., Model Railroading Club (LaGrange, Ill.), KC. Roman Catholic. Avocation: model railroading. Home: 5618 S Neva Ave Chicago IL 60638 Office: Labat-Anderson Inc US EPA WMD Records Ctr H-75 77 W Jackson 7th Fl Chicago IL 60604

NOWAK, FELICIA VERONIKA, endocrinologist, molecular biologist, educator; b. Camden, N.J.; d. Walter Ignatius and Felicia Valeria (Krukowska) N. AB cum laude, Trinity Coll., Washington, D.C., 1970; PhD, U. Wis., 1975; MD, Washington U., 1978. Resident in internal medicine Washington U. Med. Ctr., St. Louis, 1978-80; fellow in endocrinology Harbor/UCLA Med. Ctr., L.A., 1980-83; asst. prof. medicine Columbia U., N.Y.C., 1985-87; asst. prof. molecular cell biology U. Conn., Storrs, 1988-92; asst. prof. medicine div. endocrinology St. Louis U., 1992—. Contbr. articles to Endocrinology, Steroids, Molecular Endocrinology, Molecular and Cellular Neuroscis. Tutor Ctr. for Acad. Programs U. Conn., 1991-92. Recipient Basil O'Connor Starter Scholar Rsch. award March of Dimes Rsch. Found., 1985, New Investigator award NIH, 1986. Mem. AAAS, Sigma Xi. Achievements include discovery of the porf-1 and porf-2 genes. Office: Saint Louis U Sch Medicine 1402 S Grand Blvd Saint Louis MO 63104

NOWAK, ROMUALD, physicist; b. Wroclaw, Poland, Jan. 17, 1957; came to U.S. 1985; s. Eugeniusz Czeslaw and Teresa (Janowska) N. MS in Physics, Tech. U. Wroclaw, 1980, PhD in Phys. Chemistry, 1984. Grad. rsch. asst. Tech. U. Wroclaw, 1980-84; rsch. assoc. Colo. State U., 1985-88; process engr. Applied Materials Inc., Santa Clara, Calif., 1988-89, mem. tech. staff, 1989-91, sr. mem. tech. staff, 1991-92, mgr. process tech., 1991-93, dir. tech., 1993—. Contbr. articles to profl. jours. Mem. Electrochem. Soc., Materials Rsch. Soc. Office: Applied Materials Inc 3100 Bowers Ave MS 0204 Santa Clara CA 95054

NOWLAN, GODFREY S., geologist. BA in Geology with honors, Trinity Coll., Dublin, Ireland, 1971; MSc in Geology, Meml. U. Newfoundland, 1973; PhD in Biology, U. Waterloo, 1976. Postdoctoral fellow NSERC, 1976-77; rsch. micropaleontologist Geol. Survey Can., Ottawa, 1977-85, head eastern paleontology sect., 1985-88, chief paleontologist, 1988-92, sr. rsch. scientist, 1992—; head paleontology subdivsn. Inst. Sedimentary and Petroleum Geology, Calgary, Can., 1988-92; adj. prof. U. Ottawa, 1984-88, U. Calgary, 1988-91; mem. steering com. Can. Continental Drilling Program, 1987—; mem. Alta. Paleontol. Adv. Com., 1988-91. Editor GEOLOG, 1982-85; asst. editor Geosci. Can., 1983—; contbr. over 77 publs. to sci. jours. Recipient Bancroft award Royal Soc. Can., 1992. Mem. Geol. Assn. Can. (chmn. paleontology divsn. 1980-81, mem. exec. com., chair publs. com. 1985-88), Can. Soc. Petroleum Geologists (assoc. editor), Paleontol. Soc. (Golden Tribute award 1993), -Calgary Sci. Network (co-founder, pres. 1988—), Sigma Xi. Achievements include research in Lower Paleozoic biostratigraphy, paleontology and regional geology. Office: Inst Sediment & Petrol Geology, 3303-33rd St NW, Calgary, AB Canada T2L 2A7*

NOWOTNY, JANUSZ, materials scientist; b. Moscice, Poland, Jan. 2, 1936; s. Stanislaw and Irena (Olszanik) N.; m. Janina Smoczek, May 3, 1969; 1 child, Maria Karolina. MS, Tech. U., Gliwice, Poland, 1958; PhD, Polish Acad. Scis., 1967; DSc, Acad. Mining and Metallurgy, Krakow, Poland, 1974. Rsch. assoc. Tech. U., Gliwice, 1957-62; rsch. assoc. Polish Acad. Scis., 1963-69, prin. investigator, 1970-73, prof. materials sci., 1974-86; post doctoral rsch. fellow Northwestern U., Evanston, Ill., 1969-70; vis. scientist Max Planck Inst. for Solid State Rsch., Stuttgart, Germany, 1986-89; vis. prof. U. Nancy, France, 1989-90; sr. prin. rsch. scientist Australian Nuclear Sci. and Tech. Orgn., Lucas Heights, 1990—; rsch. cons. Inst. Heavy Organic Syntheses, Blachownia, Poland, 1960-63, Inst. Indsl. Electronics, Krakow, 1982-86; vis. prof. U. Bordeaux, France, 1972-73, U. Grenoble, France, 1980-81, Tokyo Inst. Tech., 1982-83; vis. scientist U. Dijon, France, 1985-86; dir. internat. confs. on Non-stoichiometric Compounds, Poland, 1980, Interfaces of Ceramic Materials, Australia, 1993; co-chmn. of NATO sponsored meetings: Surfaces Grain Boundaries and Structural Defects, Germany, 1988, Surfaces and Interfaces of Ceramics, France, 1988; rsch. coord. BRITE/EURAM Project sponsored by the European Community on Interface Segregation in Ceramic Materials and its Effect on Processing and Properties, 1990-94. Editor: Solid State Chemistry, 1981, Transport in Nonstoichiometric Compounds, 1982, Reactivity of Solids, 1982, Surface and Near-Surface Chemistry of Oxide Materials, 1988, External and Internal Surfaces in Metal Oxides, 1988, Nonstoichiometric Compounds, 1989, Interface Segregation and Related Processes in Materials, 1991, Science of Ceramic Interfaces, 1991, Electronic Ceramics, 1991, Diffusion in Solids and High Temperature Oxidation of Metals, 1991, Chemical Gas Sensors, 1993, Science of Ceramic Interfaces II, 1993; contbr. 150 articles to profl. jours. on defects and diffusion in solids, ceramic interfaces, ceramic gas censors and gas/solid heterogeneous processes. Recipient Rsch. award Japan Soc. for Promotion of Sci., 1982. Home: 30 Fretus Ave, Woonona, NSW 2517, Australia Office: Australian Nuclear Sci and Tech Orgn, Lucas Heights 2234, Australia

NOYES, RICHARD MACY, physical chemist, educator; b. Champaign, Ill., Apr. 6, 1919; s. William Albert and Katharine Haworth (Macy) N.; m. Winninette Arnold, July 12, 1946 (dec. Mar. 1972); m. Patricia Jean Harris, Jan. 26, 1973. A.B. summa cum laude, Harvard U., 1939; Ph.D., Calif. Inst. Tech., 1942. Research assoc. rocket propellants Calif. Inst. Tech., 1942-46; mem. faculty Columbia U., 1946-58, assoc. prof., 1954-58; Guggenheim fellow, vis. prof. U. Leeds, Eng., 1955-56; prof. chemistry U. Oreg., 1958—, head dept., 1963-68, 75-78, ret., 1984—. Editorial adv. com.: Chem. Revs, 1967-69; editorial adv. com.: Jour. Phys. Chemistry, 1973-80; assoc. editor: Internat. Jour. Chem. Kinetics, 1972-82, Jour. Phys. Chemistry, 1980-82; Contbr. to profl. jours. Fulbright fellow; Victoria U. Wellington, New Zealand, 1964; NSF sr. postdoctoral fellow Max Planck Inst. für Physikalische Chemie, Göttingen, Fed. Republic Germany, 1965; sr. Am. scientist awardee Alexander von Humboldt Found., 1978-79. Fellow Am. Phys. Soc.; mem. Nat. Acad. Scis., Am. Acad. Arts and Scis., Hungarian Acad. Scis. (hon.), Am. Chem. Soc. (chmn. div. phys. chemistry 1961-62, exec. com. div. 1960-75, mem. coun. 1960-75, chmn. Oreg. sect. 1967-68, com. on nominations and elections 1962-68, com. on publs. 1969-72), Chem Soc. (London), Wilderness Soc., ACLU, Sierra Club (past chmn. Atlantic and Pacific N.W. chpts., N.W. regional v.p. 1973-74), Phi Beta Kappa, Sigma Xi. Achievements include research mechanisms chemical reactions, developing general theories, intrepretation physical properties chemicals. Home: 2014 Elk Dr Eugene OR 97403-1734 Office: U of Oregon Dept Chemistry Eugene OR 97403

NOYES, RONALD T., agricultural engineering educator; b. Leedey, Okla., Jan. 4, 1937; s. Johnnie Lyle and Anna Madeline (Allen) N.; m. Zona Gail McMillen, Apr. 16, 1960; children: Cynthia Gail, Ronald Scott, David Eric. BS in Agrl. Engring., Okla. State U., 1961, MS in Agrl. Engring., 1964; postgrad., Purdue U., 1966-68, U. Okla., 1988-93. Profl. engr., Ind., Okla. Asst. prof. Purdue U., West Lafayette, Ind., 1964-68; chief engr. Beard Industries, Inc., Frankfort, Ind., 1968-81, v.p. engring., 1981-85; assoc. prof. Okla. State U., Stillwater, 1985-88, prof., 1988—; cons. Ronald T. Noyes, Profl. Agrl. Engr., Stillwater, 1988—. Co-author: Designing Pesticide and Fertilizer Containment Facilities, 1991; contbr. chpts. to books. 1st lt. U.S. Army, 1961-63. Recipient Disting. Svc. award U.S. Dept. Agr., 1992, Outstanding Ext. Faculty award Okla. State U., 1991. Fellow Am. Soc. Agrl. Engrs.; mem. Aircraft Owners & Pilots Assn., Am. Soc. for Engring. Edn., Nat. Agrl. Aviation Assn. (assoc.). Achievements include 6 patents in field; developed new aeration management procedure for controlling insects in stored grain that reduces chemical use. Home: 1116 Westwood Dr Stillwater OK 74074 Office: Oklahoma St Univ Agr Engineering Dept 224 Ag Hall Stillwater OK 74078

NOYES, WALTER OMAR, tree surgeon; b. Brookton, Maine, Aug. 4, 1929; s. Vinal Lloyd and Gladys May (Craig) N.; m. Anne Elizabeth Prout; children: Andrew W., Cynthia A.; m. Lorraine Pearle Gay, June 18, 1983. Grad. high sch., Lee, Maine. Tree climber Bartlett Tree Expert Co., Washington, 1949-50, R.E. Tillgren Tree Co., Brockton, Mass., 1953-55, Hartney Tree Co., Dedham, Mass., 1963-64, Town of Bridgewater, Mass., 1964-69; tree climber, foreman Davey Tree Expert Co., Las Vegas, 1969-70, Maltby Tree Co., Stoughton, Mass., 1970-79; tree lift operator Asplundh Tree Expert Co., Brockton, 1979—; devel. tng. and safety manual for entry-level jobs in field. Contbr. articles to profl. jours. Cpl. U.S. Army, 1951-53. Avocations: writing poetry, philosophy. Home: 71 Maplewood Ave Holbrook MA 02343

NOZ, MARILYN EILEEN, radiology educator; b. N.Y.C., June 17, 1939. BA summa cum laude, Marymount Coll., 1961; MS in Physics, Fordham U., 1963, PhD in Physics, 1969. Diplomate Am. Bd. Med. Physics, Am. Bd. Radiology, Am. Bd. of Sci. in Nuclear Medicine. Asst. prof., chmn. Dept. Physics, Marymount Coll., Tarrytown, 1964-69; assoc. prof. physics Dept. Physics, Ind. U. of Pa., 1969-74; physicist Div. Nuclear Medicine, Dept. Radiology, Bellevue Hosp. Ctr., 1974—, Div. Nuclear Medicine, Dept. Radiology, U. Hosp., 1974—; asst. prof. Dept. Radiology NYU, 1974-77, assoc. prof., 1977-92, prof., 1992—; adj. asst. prof. Dept. Physics, Iona Coll, New Rochelle, N.Y., 1967-69; adj. assoc. prof. Dept. Radiol. and Health Scis., Manhattan Coll., Bronx, N.Y., 1975—; cons. Bartech Med. Systems, Ltd., U.K., Radiology Dept., Middlesex U. Hosp., New Brunswick, N.J. Referee Am. Jour. of Physics, Founds. of Physics, Internat. Jour. of Microcomputer Applications; contbr. numerous articles to profl. jours. N.Y. State Regents scholarship, 19657-61, 61-63, Nat. Def. Edn. Act fellowship, 1961-64, sr. internat. rsch. fellowship Fogarty Internat. Ctr. of NIH, 1983-84; recipient Excellent in Rsch. award Soc. of Computed Body Tomography, 1989; recipient numerous grants. Mem. Am. Phys. Soc. (N.Y. state sect., exec. com. 1977-81), Am. Assn. of Physicists in Medicine (nuclear medicine com. 1978—), Radiol. and Med. Physics Soc., Am. Coll. Radiology, Am. Teihard Soc. for the Future of Man (adv. bd. 1981—), Soc. of Nuclear Medicine (rep. mem. to Am. Bd. Sci. in Nuclear Medicine 1989—), Soc. for Computer Applications in Radiology, Internat. Soc. for Optical Engring., Sigma Xi. Achievements include rsch. on relativistic quantum mechanics and the symmetric quark model, quantum optics, med. image processing, picture archiving and communication systems for med. images. Office: Div Nuclear Medicine Dept Radiology NYU 550 1st Ave New York NY 10016

NOZAKI, SHINJI, engineering educator; b. Otaru, Hokkaido, Japan, Sept. 16, 1953; came to U.S. 1976.; s. Hideo and Yuriko (Tomoda) N.; m. Kaori Maki, Nov. 28, 1992. BSEE, Tokyo Inst. Tech., 1976; MSEE, Wichita State U., 1980; PhD in Elec. Engring., Carnegie Mellon, 1984. Sr. device physicist Intel Corp., Santa Clara, Calif., 1984-92; vis. assoc. prof. Tokyo Inst. Tech., 1991-93; assoc. prof. U. Electro-Communications, Japan, 1993—; vis. scholar Nagoya Inst. Tech., 1988-89, Tokyo Inst. Tech., 1989-91; mem. adv. bd. Japanese IEE, Tokyo, 1991—, mem. editorial bd., 1992—. Mem. Am. Phys. Soc., IEEE, Materials Rsch. Soc. Achievements include patent for technique to reduce si autodoping during growth of GaAs on Si. Office: U Electro-Comm Dept Comm/Systems, 1-5-1 Chofugaoka, Chofu 182, Japan

NOZATO, RYOICHI, metallurgy educator, researcher; b. Amagasaki, Hyogo, Japan, May 27, 1926; s. Mitsujiro and Kokame (Ashida) N.;m. Kazuko Ando, Nov. 15, 1955 (dec. 1981); children: Kimiko, Kohei; m. Harumi Miura, May 23, 1985. BEng, Osaka (Japan) U., 1950, DEng, 1962. Instr. metallurgy U. Osaka Prefecture, Sakai, Japan, 1963-64, assoc. prof., 1964-83; prof. metallurgy Himeji (Japan) Inst. Tech., 1983-92, chmn. dept. material engring., 1984-85, 89-90, prof. emeritus, 1992—. Contbr. articles on phase diagram of alloys and precipitation in alloys to profl. jours. Recipient Hofmann Meml. prize Lead Devel. Assn., 1977. Mem. Japan Inst. Metals, Japan Soc. Applied Physics, Atomic Energy Soc. Japan, Hyogo Indsl. Assn. (hon.). Buddhist. Avocations: painting, golf. Home: 15-13 Yamate-cho, Tondabayashi 584, Japan

NOZOE, TETSUO, organic chemist, research consultant; b. Sendai, Japan, May 16, 1902; s. Juichi Kinoshita and Toyoko Nozoe; m. Kyoko Horiuchi, Nov. 2, 1927; children: Takako Masamune, Shigeo Nozoe, Yoko Ishikura, Yuriko Higashihara. BS in Chemistry, Tohoku Imperial U., 1926; DSc in

Chemistry, Osaka Imperial U., 1936. Asst. prof. Taihoku Imperial U., 1929-37, prof., 1937-45; prof. Nat. Tawian U., Taipei, 1945-48; prof. Tohoku U., Sendai, Japan, 1948-66, prof. emeritus, 1966—; rsch. cons. Kao Corp., Tokyo, 1966—, Takasago Perfumery Co., Tokyo, 1966—, Sankyo (pharm.) Co., Tokyo, 1966—; mem. Found. IUPAC Symposium in Chemistry Nonbenzenoid Aromatic Compounds, 1970—. Author: Nonbenzenoid Aromatic Compounds, 1960, Tamkang Chair Lecture, 1991, American Chemical Society Autobiography (Seventy Years in Organic Chemistry), 1991; editor: Topics in Nonbenzenoid Aromatic Chem. (2 vols.) 1972, Organic Chemistry (2 vols.) 1970; contbr. article to profl. jours. Named hon. citizen Sendai City, 1959, Taipei City, 1982; recipient Majima award Chem. Soc. Japan, 1944, Asahi Cultural award Asahi Newspaper, 1952, Order of Cultural Merit Japanese Govt., 1958, Von Hofman Meml. medal German Chem. Soc., 1981, Grand Prize Soc. Synthetic Org. Chem. Japan, 1984. Mem. AAAS, Royal Swedish Acad. Scis. (fgn. mem.), Japan Acad. (award 1953), N.Y. Acad. Scis. (hon.), Chinese Chem. Soc. Taiwan, Swiss Chem. Soc.; hon. mem. Chem. Soc. Japan, Japan Soc. Bioscience, Biotechnology, and Agrochemistry, Pharm. Soc. Japan, Soc. Synthetic Organic Chemistry. Achievements include discovery of hinokitiol, having 7-membered aromatic system, structural elucidation of hederagenin and oleanol, novel constituents of wool wax, establishment of a wide area of nonbenzenoid aromatic chemistry, synthesis of representative troponoid compounds, one-pot synthesis of various azulene derivatives, of tropocoronands, others. Home: 2-5-1-811 Kamiyoga, Setagaya-ku, Tokyo 158, Japan Office: Tokyo Rsch Labs Kao Corp, 103 Bunka 2-chome Sumida-ku, Tokyo 131, Japan

NUCCITELLI, SAUL ARNOLD, civil engineer, consultant; b. Yonkers, N.Y., Apr. 25, 1928; s. Agostino and Antoinette (D'Amicis) N.; m. Concetta Orlandi, Dec. 23, 1969; 1 child, Saul A. BS, NYU, 1949, MCE, 1954; DCE, MIT, 1960. Registered profl. engr., N.Y., Mo., Colo., Conn., Mass.; lic. land surveyor, Mo., Colo., Conn., Mass. Asst. civil engr. Westchester County Engrs., N.Y.C., 1949-51, 53-54; project engr. H.B. Bolas Enterprises, Denver, 1954-55; asst. prof., rsch. engr., U. Denver, 1955-58; mem. staff MIT, 1958-60; asst. prof. engring. Cooper Union Coll., N.Y.C., 1960-62; pvt. practice cons. engring., Springfield, Mo., 1962—; organizer, bd. dirs. Met. Nat. Bank, Springfield; former adviser, bd. dirs. Farm & Home Savs. and Loan Assn. Contbr. articles to profl. jours. Past chmn. Adv. Council on Mo. Pub. Drinking Water; chmn. Watershed Com. of the Ozarks; bd. dirs. YMCA; Greene County Mus. of Ozarks; past chmn. Bd. City Utilities, Springfield; past pres. Downtown Springfield Assn. Served with U.S. Army, 1951-53. Recipient Cert. of Appreciation, Mo. Mcpl. League, 1981; named Mo. Cons. Engr. of Yr., 1973. Fellow ASCE; mem. Nat. Soc. Profl. Engrs., Mo. Soc. Profl. Engrs. (past pres. Ozark chpt.), Boston Soc. Civil Engrs., Am. Concrete Inst., Am. Inst. Steel Constrn., Am. Welding Soc., ASTM, Am. Soc. Mil. Engrs., Springfield C. of C. (past v.p.), Rotary (past pres.). Home: 2919 S Brentmoor Ave Springfield MO 65804-3925 Office: 122 Park Central Sq Springfield MO 65806-1394

NUCKOLLS, JOHN HOPKINS, physicist, researcher; b. Chgo., Nov. 17, 1930; s. Asa Hopkins and Helen (Gates) N.; m. Ruth Munsterman, Apr. 21, 1952 (div. 1983); children—Helen Marie, Robert David; m. Amelia Aphrodite Liaskas, July 29, 1983. B.S., Wheaton Coll., 1953; M.A., Columbia U., 1955; D.Sc. (hon), Fla. Inst. Tech., 1977. Physicist U. Calif., Lawrence Livermore Nat. Lab., 1955—, assoc. leader thermonuclear design div., 1965-80, assoc. leader laser fusion program, 1975-83, div. leader, 1980-83, assoc. dir. physics, 1983-88, dir., 1988—. Recipient E.O. Lawrence award AEC, 1969, Fusion Leadership award Fusion Power Assocs., 1983, Edward Teller medal Internat. Workshop Laser Interaction and Related Plasma Phenomena, 1991. Fellow AAAS, Am. Phys. Soc. (J.C. Maxwell prize 1981); mem. NAE. Office: Lawrence Livermore Nat Lab PO Box 808 Livermore CA 94551-0808

NUCKOLS, FRANK JOSEPH, psychiatrist; b. Akron, Ohio, Apr. 7, 1926; s. William Alexander Jr. and Jean (Harrison) N.; m. Jane Fleetwood McIntosh, June 16, 1948; children: Claud Alexander, John Andrew. BA, U. Louisville, 1946; MD, U. Ala., 1951. Diplomate Am. Bd. Psychiatry and Neurology. Intern Holy Name Jesus Hosp., Gadsden, Ala., 1951; ward physician Ala. State Hosp., Tuscaloosa, 1951-52; resident USPHS Hosp., Lexington, Ky., 1953-56; mem. faculty dept. psychiatry U. Ala. Med. Ctr., Birmingham, 1958-68, dir. tng. psychiat. residents, 1964-68, head div. community psychiatry, 1964-68, head continguing psychiat. edn. for physicians, 1964-68; chief psychiat. staff in-patient svc. U. Hosp., Birmingham, 1966-68; dir. tng. Hill Crest Hosp., Birmingham, 1975-79; pvt. practice Birmingham, 1968—; staff Med. Ctr. East Hosp., Birmingham, Bapt. Med. Ctr. Montclair, Birmingham; cons. staff St. Vincent's Hosp., Birmingham, Lloyd Noland Hosp., Birmingham, south Highland Hosp., Birmingham; vis. faculty Harvard Univ., Boston, 1963-66, Baylor Univ. Med. Sch., Houston, Tex., 1967-71. Vol. faculty family practice residency Med. Ctr. East Hosp. Ensign USNR, 1941-43; sr. surgeon USPHS, 1956—. Fellow APA (life), So. Psychiat. Assn.; mem. Med. Assn. State Ala., So. Med. Assn., Jefferson County Mental Health Assn. (v.p. 1960), Jefferson County Med. Soc., Mental Health Assn. State Ala. (chair profl adv. com. 1961), Phi Beta Pi, Tau Kappa Epsilon. Home: 3741 River Oaks Cir Birmingham AL 35223-2117 Office: Psychiatry Assocs PC 6900 6th Ave S Birmingham AL 35212-1902

NUGENT, MATTHEW ALFRED, biochemist; b. Boston, Sept. 30, 1961; s. Alfred Emmanuel and Louise Mary (Roche) N.; m. Nancy Ann Kruszona, June 23, 1984. BA, Brandeis U., 1983, PhD, 1990. Postdoctoral fellow MIT, Cambridge, 1989-93; rsch. scientist Harvard U.-MIT, Cambridge, 1993—; asst. prof. Boston U., 1993--. Contbr. articles to profl. jours. Recipient Nat. Rsch. Svc. award NIH, 1990-92; Hoffman-LaRoche Rsch. fellow, 1987-89, Pub. Health fellow, 1984-87. Mem. AAAS, N.Y. Acad. Scis., Materials Rsch. Soc., Tissue Culture Assn., Inc. Home: 6 Jeffrey Cir Bedford MA 01730 Office: Boston U Sch Medicine 80 E Concord St Boston MA 02118

NUMAI, TAKAHIRO, scientist, electrical engineer; b. Kurashiki, Japan, Jan. 1, 1961; s. Yasumasa and Aiko Inoue;m. Reiko Numai, Mar. 8, 1986. BS, Keio U., 1983, MS, 1985, PhD, 1992. Mem. tech. staff NEC Corp., Kawasaki, 1985-90; asst. mgr. NEC Corp., Tsukuba, 1990—. Contbr. articles to profl. jours. Mem.The Japan Soc. Applied Physics, Inst. Electronics Info. and Communication Engrs., IEEE, Optical Soc. Am. Home: 7-5-205 Ikuta-cho, Tsuchiura Ibaraki 300, Japan Office: NEC Corp, 34 Miyukigaoka, Tsukuba Ibaraki 305, Japan

NUMBERE, DAOPU THOMPSON, petroleum engineer, educator; b. Buguma, Nigeria, Mar. 30, 1951; came to the U.S., 1975; s. Thompson and Norah Alfred (West) N.; m. Tonye Eugenia Higgwe, Dec. 29, 1987. BS in Mech. Engring., U. Coll. Swansea, 1975; MS in Petroleum Engring., Stanford U., 1977; PhD, U. Okla., 1982. Asst. prof. U. Mo., Rolla, 1982-88, assoc. prof., 1988—; cons. Sigma Cons., Mattoon, Ill., 1987—. Author: Petroleum Reservoir Mechanics, 1991, Introduction to Water Flooding, 1989. Recipient Shell-BP award, 1971-75, Caswell Massey prize U. Coll. Seansea, 1975, Okla. Rsch. award Okla Rsch. Coun., 1981. Mem. Soc. Petroleum Engrs., Sigma Xi. Achievements include development of a new method for streamline generation for oil recovery prediction, simultaneous prediction of oil recovery and water influx for oil and gas reservoirs. Office: U Mo Rolla 119 McNutt Hall Rolla MO 65401

NUNEZ, STEPHEN CHRISTOPHER, aerospace technologist; b. Presque Isle, Maine, Dec. 8, 1961; s. Harold Claude and Sadie Elizabeth (Thomson) Frost; m. Cynthia Marlene Cuevas, Nov. 17, 1984. AD, Miss. Gulf Coast Jr. Coll., 1981; BS in Civil Engring., Miss. State U., 1984. Registered profl. engr., Miss. Jr. engr. Pan Am World Svcs., Inc., Stennis Space Center, Miss., 1984-85, engr., 1986-88; civil design engr. Simpkins and Costelli, Inc., Gulfport, Miss., 1985-86; aerospace technologist liquid propulsion systems NASA, Stennis Space Center, 1989—. Vol., Spl. Olympics, Stennis Space Center, 1990, 91. Mem. AIAA, Nat. Soc. Profl. Engrs., Chi Epsilon. Avocations: golf, swimming, tennis, softball. Office: NASA Bldg 1100 Mail Code FA30 Stennis Space Center MS 39529-6000

NUNEZ-CENTELLA, RAMON ANTONIO, chemist, science educator, journalist; b. La Coruna, Spain, Aug. 16, 1946; s. Ramon and Josefa (Centella) N.; m. Mercedes Visos, June 28, 1970. BS in Chemistry, Santiago U., 1969; MA in Sci. Edn., NYU, 1977. Tchr. sci. High Sch., La Coruna, Spain, 1970-76; faculty NYU, N.Y.C., 1976-77; head dept. sci. High Sch., La Coruna, 1977-83; dir. edn. City Coun., La Coruna, 1983-85; dir. Casa de las Ciencias, la Coruna, 1985—; cons. U. de la Coruna, 1990-91; dir. planetarium shows, including This is Astronomy, Moon and Seven Stars, El Zodiaco, Milky Way, others. Contbr. over 500 articles to profl. jours. Recipient Periodismo Cientifico Consejo Superior de Investigaciones Cientificas, 1989, 90, 92, Silver medal CSIC, 1992. Mem. Padres y Maestros, Inst. Jose Cornide, A.E. Periodismo Cientifico, Internat. Sci. Writers Assn. Office: Casa de las Ciencias, La Coruna Spain 15005

NUNN, JAMES ROSS, engineering executive; b. Clayton, N. Mex., Aug. 21, 1940; s. James Arthur and Emma Thelma (Morrison) N.; m. Nina Yvonne Kenney, Nov. 22, 1961; children: Stephanie, Shawn. BS in Math & Physics, W. Tex. State U., 1962; MS in Engring. Physics, Air Force Inst. Tech., 1971. Commd. 2lt throu OTS USAF, 1963, advanced through grades to col., 1985; program mgr. munitions SPO USAF, Eglin AFB, Fla., 1979-83; program dir. AGM-130/GBU-15 USAF, Eglin AFB, 1983-84; dir. engine test Arnold Engring. Devel. Ctr., Arnold AFB, Tenn., 1984-86; comdr. USAF Rocket Propulsion/Astronautics Lab., Edwards AFB, 1986-89; retired USAF, 1989; v.p., gen. mgr. Sverdrup Tech., Stennis Space Ctr., Miss., 1989—. Decorated Legion of Merit USAF, Edwards AFB, Calif., 1989; recipient Laurels 1988 Avation Week Mag. Mem. AIAA (sr., charter chmn.), DAV, Air Force Assn. Republican. Baptist. Office: Sverdrup Tech Space Ctr Stennis Space Center MS 39529

NUNN, JENNY WREN, pharmacist; b. Atlanta, May 5, 1944; d. Joshua Hugh and Jenny Wren (Scott) N. Student, Gulf Park Coll.; BS, U. Tenn., 1967; AS, Dyersburg State Community Coll, 1975; BS in Law Enforcement, Samford U., 1980, BS in Pharmacy, 1981. Registered Pharmacist, Tenn. Pvt. practice Ripley, Tenn., 1981—; planter, land owner Wren's Flight Plantation, Chestnut Bluff, Tenn., 1980—. Founder, bd. dirs. Mid South chpt. Greyhound Pets of Am., Memphis, 1985—; pres. Lauderdale County Humane Soc., bd. dirs. 1981—; bd. dirs. Nostalgia U.S.A. Fellow Am. Soc. of Cons. Pharmacists, Internatl Inst. of History of Pharmacy Assn.; mem. NRA (life), Am. Pharm. Assn., Am. Soc. Hosp. Pharmacists, Christian Med. Soc., Bus. and Profl. Women's Assn., Daughters of Am. Revolution, Dames of Ct. of Honor, Zeta Tau. Alpha. Methodist. Avocations: horseback riding, raising dogs, writing, sporting clubs, archeology. Home and Office: RR 1 Box 1 Scottlawn Plantation Ripley TN 38063-9709

NUNNALLY, STEPHENS WATSON, civil engineer; b. Gadsden, Ala., Nov. 30, 1927; s. John Marshall and Mae Louise (Watson) N.; m. Joan Marie Arel, May 29, 1957; children: Stephens Jr., Janine, John. BS, U.S. Mil. Acad., 1949; MS, Northwestern U., 1958, PhD, 1966. Cert. profl. engr. Fla., Ala. Commd. lt. U.S. Army, 1949, advanced through grades to lt. col., retired, 1970; asst. prof. U. Fla., Gainesville, 1971-75; prof. N.C. State U., Raleigh, 1975-84; freelance cons. Satellite Beach, Fla., 1984--. Author: Managing Construction Equipment, 1977, Construction Methods and Management, 1980, 87, 93; co-author: Residential and Ligt Building Construction, 1990. Recipient Outstanding Extension Svc. award N.C. State U. 1982. Fellow Am. Soc. Civil Engrs. (exec. com. COlo. div. 1990--), Am. Soc. for Engring. Edn. (com. chmn. 1980-81), Soc. Am. Mil. Engrs. Republican. Episcopaian. Home: 474 St Lucia Ct Satellite Beach FL 32937

NUNZ, GREGORY JOSEPH, aerospace engineer, mathematics educator, technical manager; b. Batavia, N.Y., May 28, 1934; s. Sylvester Joseph and Elizabeth Marie (Loesell) N.; m. Georgia Monyea Costas, Mar. 30, 1958; children: Karen, John, Rebecca, Deirdre, Jaimie, Marta. BSChemE, Cooper Union, 1955; postgrad., U. So. Calif., Calif. State U., 1991; MS in Applied Math., Columbia Pacific U., 1991, PhD in Mgmt. Sci., 1993. Adv. design staff, propulsion mgr. U.K. project Rocketdyne div. Rockwell, Canoga Park, Calif., 1955-65; mem. tech. staff Aerospace Corp., El Segundo, Calif., 1965-70; mem. tech. staff propulsion div. Jet Propulsion Lab., Pasadena, Calif., 1970-72; chief. monoprop. engring. Bell Aerospace Corp., Buffalo, N.Y., 1972-74; group supr. comb. devices Jet Propulsion Lab., Pasadena, 1974-76; asst. div. leader, program mgr. internat. HDR geothermal energy program, project mgr. space-related projects Los Alamos (N.Mex.) Nat. Lab., 1977—; assoc. prof. electronics Los Angeles Pierce Coll., Woodland Hills, Calif., 1961-72; instr. No. N.Mex. Community Coll., Los Alamos, 1978-80; div. head scis. U. N.Mex., Los Alamos, 1980-92, adj. prof. math., 1980—. Author: Electronics Lab Manual I, 1964, Electronics in Our World, 1972; co-author: Electronics Mathematics, vol. I, II, 1967; contbg. author Prentice-Hall Textbook of Cosmetology, 1975, Alternative Energy Sources VII, 1987; contbr. articles to profl. jours.; inventor smallest catalytic liquid N2H4 rocket thrustor, co-inventor first monoprop/biprop bimodal rocket engine, tech. advisor internat. multi-prize winning documentary film One With the Earth. Mem. Aerial Phenomena Research Orgn., L.A., 1975. Fellow AIAA (assoc.); mem. Tech. Mktg. Soc. Am., Math. Assn. Am., ARISTA, Shrine Club, Masons, Ballut Abyad Temple. Avocations: travel, archaeology, fgn. langs., golf. Office: Los Alamos Nat Lab PO Box 1663 MS D 460 Los Alamos NM 87545

NURHUSSEIN, MOHAMMED ALAMIN, internist, geriatrician, educator; b. Adwa, Ethiopia, Apr. 4, 1942; came to U.S., 1972; s. Hagos and Teberih (Yusuf) N.; m. Zahra Said, June 10, 1972; children: Nadia, Siham, Safy. BS, Haile Selasie Mil. Acad., Harar, Ethiopia, 1961; MD, Zagreb (Yugoslavia) U., 1968. Intern, resident, then fellow Bklyn.-Cumberland Med. ctr., 1972-77; emergency rm. physician Cumberland Hosp., Bklyn., 1977-79; attending physician in medicine Kings County Hosp. Ctr., Bklyn., 1979—; faculty practice medicine, geriatrics SUNY Univ. Hosp., Bklyn., 1983—; instr., then asst. prof. SUNY Health Sci. Ctr., bklyn., 1979-92; med. cons., dir. drug abuse treatment Coney Island Hosp., Bklyn., 1982-84; adv. bd. Bklyn. Alzheimer's Disease Assistance Ctr., 1992—. Fellow ACP; mem. Am. Geriatric Soc., Am. Lung Assn., N.Y. Acad. Scis., Amnesty Internat., Physicians for Human Rights. Democrat. Moslem. Office: SUNY Health Sci Ctr 450 Clarkson Ave Brooklyn NY 11203

NÜRNBERGER, GÜNTHER, mathematician; b. Marktredwitz, Bavaria, Germany, Jan. 13, 1948; s. Rudolf and Lotte (Schodner) N.; m. Gudrun Tiller, Apr. 7, 1982; 1 child, Sandra. Mathematician, Erlangen U., Nuremberg, Germany, 1974, D of natural scis., 1975, D of natural scis. habilitation, 1979. Asst. prof., assoc. prof. Erlangen U., Nuremberg, 1974-83, 85-89; assoc. prof Mannheim U., Germany, 1983-85; prof. Mannheim U., 1989—, dean, 1993—; cons. Siemens Inc., Nuremberg, 1985-88; organizer internat. sci. confs. Author: Approximation by Spline Functions, 1989; editor: Delay Equations, Approximation and Application, 1985, Numerical Methods of Approximation Theory, 1987, Jour. of Approximation Theory, 1988; contbr. 50 articles to profl. jours. Fiebiger prof. Munich 1985. Avocations: music, sports. Office: U Mannheim, Faculty Math and Computers, 6800 Mannheim Germany

NURNBERGER, JOHN I., JR., psychiatrist, educator; b. N.Y.C., July 18, 1946; married; 3 children. BS in Psychology magna cum laude, Fordham U., 1968; MD, Ind. U., 1975, PhD, 1983. Diplomate Am. Bd. Psychiatry and Neurology. Resident in psychiatry Columbia Presbyn. Med. Ctr., N.Y.C., 1975-78, med. officer sect. psychogenetics, 1977-78; sr. staff fellow, outpatient clinic adminstr. sect. psychogenetics NIH, Bethesda, Md., 1978-83, staff psychiatrist, chief NIMH Outpatients Clinic, 1983-86, acting chief sect. clin. genetics, 1986; prof. psychiatry, dir. Inst. Psychiatric Rsch., rsch. coord. dept. psychiatry Ind. U. Med. Ctr., Indpls., 1986—; prof. med. neurobiology Ind. U. Grad. Sch., Indpls., 1987—; clin. cons. Cold Spring VA Hosp., 1986—; cons., lectr. in field. Editor-in-chief: Psychiatric Genetics; field editor: Neuropsychiatric Genetics; contbr. articles to profl. jours. NSF fellow, 1968; recipient NAMI Exemplary Psychiatrist award Nat. Alliance Mentally Ill, 1992. Fellow Am. Psychiatric Assn., Am. Psychpathological Assn.; mem. AAAS, Am. Soc. Human Genetics, Internat. Soc. Psychiatric Genetics (bd. dirs.), Soc. Light Treatment and Biol. Rhythms, Soc. Neursci., Assn. Rsch. in Nervous and Mental Disease, Soc. Biol. Psychiatry, Sigma Xi. Office: Ind U Sch Medicine Psychiatric Rsch Inst 791 Union Dr Indianapolis IN 46202-4887

NURRE, JOSEPH HENRY, engineering educator; b. Cin., Jan. 15, 1960; s. John William and Dorothy Mary (Balastra) N.; m. Michelle Doench, Oct. 8, 1988; 1 child, Joseph Gerard. BSEE, U. Cin., 1983, MS, 1985, PhD in Mech. Engring., 1989. Registered profl. engr., Ohio. Asst. prof. Ohio U., Athens, 1990—; cons. Procter & Gamble Co., Cin., 1987, GE, 1989. Contbr. articles to jours. IEEE. Mem. IEEE, Soc. Mfg. Engrs., Sigma Xi. Achievements include patent pending for General Electric Aircraft Engines. Office: Ohio U Dept Elec Engring Athens OH 45701

NUSHOLTZ, GUY SAMUEL, research engineer; b. Detroit, Nov. 4, 1948; s. Phillip and Shirley Fern (Altschuler) Nusholtz; m. Pat S. Kaiker, June 5, 1988. BS, Antioch Coll., 1972; MS, U. Mich., 1974. Fire chief Antioch Coll. Fire Dept., Yellow Springs, Ohio, 1970-72; rsch. assoc. U. Mich., Ann Arbor, 1974-81, rsch. scientist, 1981-88; rsch. engr. Chrysler, Auburn Hill, Mich., 1988—; bio-engring. Washington State U., Detroit, 1993—; cons. mech. engring. U. Va., Charlottesville, 1989—; mem. adv. bd. STAPP, Warrendale, Mich., 1991—, Motor Vehicle Rsch. Adv. Com., Washington, 1991—. Mem. AAAS, Soc. Exptl. Mechanics, Soc. Automobile Engrs., Sigma Xi. Achievements include devel. of motion-tracking system using linear accelerometers, x-ray cinematographic, crash impact signal processing procedures, non linear shift varient filters, fire fighting procedures using high pressure fog. Office: Chrysler CIMS 483-05-10 800 Chrysler Dr E Auburn Hills MI 48326-2757

NUSIM, STANLEY HERBERT, chemical engineer, executive director; b. N.Y.C., Oct. 2, 1935; s. Seymour and Ranna T. (Weiner) N.; m. Marcia Anne Borsig, Feb. 21, 1960; children: David Mark, Jill Wendi. BSChemE, CCNY, 1957; MSChemE, N.Y. U., 1960, PhD, 1967. Rsch. engr. Battelle Meml. Inst., Columbus, Ohio, 1956; researcher, chem. engring. Merck & Co. Rsch. Labs. Div., Rahway, N.J., 1957-68; sect. mgr., chem. engring R & D Merck Rsch. Labs. Div., Rahway, 1968-70; tech. svcs. mgr. Merck Chem. Mfg. Div., Rahway, 1970-73, mfg. mgr., 1973-80; dir. subsidiary projects Merck Internat. Div., Rahway, 1981-82, exec. dir. Latin Am., Far East, Near East ops., 1982-88; exec. dir. licensee, Latin Am., Far East, Asia ops. Merck Pharm. Mfg. Div., Rahway, 1989—; exec. dir. licensee ops. mfg. divsn. Merck & Co. Inc., Whitehouse Station, N.J., 1992—; adv. bd. CCNY Sch. Engring., 1982—. Author: Kinetic Studies on C4 Hydrocarbon Systems, 1967. V.p. men's club Temple Beth Shalom, Livingston, N.J., 1975-78; rep. to bd. edn. Livingston Home and Sch. Assn., 1982-83; bd. govs. Turnberry Isle Yacht & Racquet Club, Aventura, Fla., 1992—. Mem. Am. Inst. Chem. Engrs. (bd. dir. N. Jersey sect. 1968-71, scholarship award 1955), Am. Chem. Soc., Tau Beta Pi, Garden State Yacht Club (bd. govs. 1987-88). Jewish. Achievements include U.S. and foreign. patents on the continous manufacture of halogenated acetone, on the first commercial process, on the development of sophisticated recovery systems for complex organic compounds from catalytic processes. Home: 454165 Prospect Ave West Orange NJ 07052-3214 Office: Merck & Co Inc One Merck Dr PO Box 100 WS 3D-30 Whitehouse Station NJ 08889-0100

NUSSENBAUM, SIEGFRIED FRED, chemistry educator; b. Vienna, Austria, Nov. 21, 1919; came to U.S., 1939; s. Marcus and Susan Sara (Rothenberg) N.; m. Celia Womark, Feb. 20, 1951; children; Deborah M., Evelyn R. BS in Chemistry, U. Calif., Berkeley, 1941, MS in Food Tech., 1948, PhD in Comparative Biochemistry, 1951. Analytical chemist Panam. Engring. Co., Berkeley, 1942-43; asst. chief chemist Manganese Ore Co., Las Vegas, 1943-45; rsch. assoc. U. Calif., Berkeley, 1951-52; dir. master clin. lab. sci. program U. Calif., San Francisco, 1969-87; from instr. to prof. Calif. State U., Sacramento, 1952-90, chair dept. chemistry, 1958-65; cons. biochemist Sacramento County Hosp., 1958-70; lectr. U. Calif. Davis Med. Ctr., 1970—. Author: Organic Chem-Principles and Application, 1963; contbr. articles to profl. jours. Sgt. U.S. Army, 1945-47. Fellow AAAS; mem. Am. Chem. Soc., Am. Assn. Clin. Chemistry (Outstanding Contbn. in Edn. award no. sect. 1991), Nat. Acad. Clin. Biochemistry. Achievements include research in pectic enzymes, mechanism of amylopectin formation and differentiation from amylose, phenotyping of lipemias. Home: 2900 Latham Dr Sacramento CA 95864-5644

NÜSSLEIN-VOLHARD, CHRISTIANE, medical researcher; b. Magdeburg, Germany, Oct. 20, 1942; d. Rolf Volhard and Brigitte (Haas) Volhard. Diploma in Biochemistry, U. Tübingen, 1968, PhD, 1973; ScD (hon.), Yale U. Rsch. assoc. lab. of Dr. Schaller Max-Planck Inst. for Devel. Biology, Tübingen, 1972-74; postdoctoral fellow lab. of Dr. W. Gehring, Biozentrum, Basel, Switzerland, 1975-76; postdoctoral fellow lab of Dr. K. Sander U. Freiburg, 1977; head rsch. group European Molecular Biology Lab., Heidelberg, 1978-80; rsch. group leader Friedrich-Miescher Laboratorium Max-Planck-Gesellschaft, Tübingen, 1981-85; sci. mem. Max-Planck-Gesellschaft, dir. Max-Planck Inst. for Devel. Biology, Tübingen, 1985-90, dir. genetics dept., 1990—; hon. prof. U. Tübingen. Contbr. numerous articles to profl. jours. Recipient Albert Lasker Basic Med. Rsch. award Albert and Mary Lasker Found., 1991, Louisa Gross Horowitz prize Columbia U., 1992, Forderpreis award Deutschen Forschungsgemeinschaft, 1986, Leibnizpreis der Deutschen Forschungsgemeinschaft, Franz Vogt prize U. Giessen, 1986, Carus prize City of Schweinfurth, 1989, Rosenstiel medal Brandeis U.; Brooks lecturer Harvard Med. Sch., 1988, Stilliman lecturer Yale U., 1989. Mem. European Molecular Biology Orgn., Deutsche Gesellschaft fur Entwicklungsbiologie, Academia Europeaea. Achievements include research in using embryos, in which she created a series of genetic screens that led to the identification of most of the genes responsible for the organism's body segment development; she also established that genes encode signaling molecules that tell cells where they are in the organism's overall structure and what their function is to be. Office: Max Planck Inst, Spemannstr 35 Postfach 2109, D7400 Tübingen Germany*

NUTHMANN, CONRAD CHRISTOPHER, civil engineer, consultant; b. Palo Alto, Calif., Oct. 22, 1958; s. Conrad Francis Nuthmann and Anne (McDaniel) Matthes; m. Lucinda Seabury. BS in Civil Engring., U. Mass., 1983. Registered profl. engr. Mass. Staff engr. Somerville (Mass.) Engring. Inc., 1984-86; v.p. I.E.S. Inc., Somerville, 1986; staff engr. GHR Engring., Lexington, Mass., 1986-88; sr. engr. Fay, Spofford & Thorndike Inc., Lexington, 1988—. Author newspaper U. Mass Daily Collegian, 1983. Mem. ASCE, Mass. Assn. Conservation Comms., Inst. Transp. Engrs., Delta Epsilon (pres. 1982-83). Home: 94 Sunset Ln Lunenburg MA 01462 Office: Fay Spofford & Thorndike 191 Spring St PO Box 9117 Lexington MA 02173

NUTTER, DALE E., mechanical engineer; b. Borger, Tex., July 20, 1935; s. I. Earl and Martha (Crowell) N.; m. Donna K. Duggar, Oct. 22, 1983; children: Mark E., Michael E. BSME, Okla. State U., 1958. Registered profl. engr., Okla., Tex. Project engr. Douglas Aircraft & Space Systems, Santa Monica, Calif., 1958-61; sales engr. Nutter Engring., Tulsa, 1961-68, sales mgr., R&D mgr., 1968-78; R&D cons. Nutter Engring. div. Harsco, Tulsa, 1978—; licensor Mitsui Engring. & Shipbldg., Tokyo, 1971—; mem. chem. engring. adv. bd. Tulsa U., 1987. Contbr. articles to profl. jours. Mem. Am. Inst. Chem. Engrs. (adv. bd. 1992), Fractionation Rsch. Inc. (bd. dirs., tech. com., v.p.), Acad. Model Aeronautics, Tulsa Soaring Club (pres.). Achievements include patents in U.S., U.K., France, Italy, Netherlands, Japan and Korea. Office: Nutter Engring PO Box 700489 Tulsa OK 74170

NUWAYSIR, LYDIA MARIE, analytical chemist; b. Pottsville, Pa., Feb. 8, 1963; d. Fuad Sami Nuwaysir and Willa Anne (Sechler) Traub. BS in Chemistry with distinction, Stockton State Coll., 1985; PhD in Analytical Chemistry, U. Calif., Riverside, 1990. Water chemist Stockton Environ. Lab., Pomona, N.J., 1984-85; teaching asst. Stockton State Coll., Pomona, 1985; teaching asst. U. Calif., Riverside, 1985-86, rsch. asst., 1986-90, postdoctoral fellow, 1990-91; postdoctoral fellow Genentech, Inc., South San Francisco, Calif., 1991—. Contbr. chpt. to book Lasers and Mass Spectrometry, 1990; contbr. articles to Analytical Chemistry, Mass Spectrometry Revs. Vol. Dem. party, Riverside, 1988. Rsch. fellow Std. Oil Ohio, 1985. Mem. Am. Chem. Soc. (contbr. to jour.), Am. Soc. for Mass Spectrometry (contbr. to jour.), Bay Area Mass Spectrometry, U. Calif. Riverside Grad. Student Assn. (pres. 1987-89). Home: 741 40th Ave San Francisco CA 94121 Office: Genentech Inc 460 Pt San Bruno Blvd South San Francisco CA 94080

N'VIETSON, TUNG THANH, civil engineer; b. Bien Hoa, Vietnam, Nov. 12, 1949; came to U.S., 1979; s. Cu Van and Dao Thi Nguyen; m. Phan Thi, Oct. 12, 1973; children: Trucle C., Leyna C., Tung Thanh Jr. BS in Civil Engring., Nat. Inst. Tech., Saigon, Vietnam, 1973. Registered profl. engr., La., Va., D.C., Md., Fla., Tex. Asst. province engr. Darlac Province Pub. Works, Banmethuot, Vietnam, 1973-75; design engr. Daklak Province, Dept.. of Transp., Banmethuot, 1975-79, Burk & Assocs. Inc., New Orleans, 1980-85; project engr. David Volkert & Assocs. Inc., Metairie, La., 1985-87,

Cervantes & Assocs. P.C., Fairfax, Va., 1987-88, Progressive Engring. Cons. Inc., Fairfax, 1988-89; sr. project mgr. Sheladia Assocs. Inc., Rockville, Md., 1989-93; v.p. Guillot-Vogt Assocs., Inc., Metairie, La., 1993—. Mem. ASCE. Roman Catholic. Home: 2601 Metairie Lawn Dr # 14-118 Metairie LA 70002 Office: Guillot-Vogt Assocs Inc 2720 Metairie Lawn Dr Metairie LA 70002

NYBERG, STANLEY ERIC, cognitive scientist; b. Boston, Jan. 30, 1948; s. Leroy Milton and Anna Maria (Olson) N. PhD, SUNY, Stony Brook, 1975; M of Pub. and Pvt. Mgmt., Yale U., 1984. Postdoctoral fellowship U. Calif., Berkeley, 1975-76; asst. prof. North Pk. Coll., Chgo., 1976-79, Barnard Coll., Columbia U., N.Y.C., 1979-82; systems mgmt. Interactive Data Corp., Lexington, Mass., 1984-88, Dept. of Revenue, Commonwealth of Mass., Boston, 1988—; bd. dirs. Children's Home of Cromwell, Conn. Co-author: Human Memory: An Introduction to Research and Theory, 1982. V.p. "L" St. Running Club, South Boston, 1993; mem. ch. coun. Luth. Ch. of Redeemer, Woburn, Mass., 1989—. Fellow Am. Psychol. Soc.; mem. APA. Home: PO Box 1849 GMF Boston MA 02205

NYCE, DAVID SCOTT, electronics company executive; b. Norristown, Pa., Jan. 25, 1952; s. Jonathan J. and Emma R. (Dusza) N.; m. Gwen Ann Gordon, Apr. 26, 1975; children: Timothy S., Christopher D., Megan S. BSEE, Temple U., 1973. Project engr. Robinson-Halpern Co., Plymouth Meeting, Pa., 1973-77; sr. devel. engr. Honeywell, Ft. Washington, Pa., 1977-78; chief engr. Chatlos Systems, Inc., Whippany, N.J., 1978-79; mgr. engring. Environ. Tectonics Corp., Southampton, Pa., 1979-80; v.p., dir. engring. Neutronics, Inc., Exton, Pa., 1980-90; dir. engring. measurement and automation group MTS Systems Corp., Research Triangle Park, N.C., 1990—; cons. in field; proprietor Nyce Sporting Equipment, Apex, N.C., 1985-91, Nyce Sounds, Trappe, Pa., 1987-90. Tchr. aerodynamics Apex Elem. Sch. System, 1990, 92. Recipient Vaaler award Chem. Engring. Mag., 1988. Mem. AAAS, Instrument Soc. Am. (sr.), NRA (life, cert. instr., sharpshooter), Lower Providence Rod and Gun Club (capt. pistol team 1987-89, high on team 1988-89), U.S. Hang Gliding Assn., U.S. Parachute Assn., U.S. Judo Assn. (brown belt), Nat. Assn. Rocketry, Nat. Trappers Assn., Acad. Model Aero., The Planetary Soc., Tripoli Rocketry Assn. Achievements include patents for low power magnetostriction, threshold compensating detector, bandwidth limiting, densimeter. Office: MTS Systems Corp Sensors Divsn 3001 Sheldon Dr Cary NC 27513-2007

NYCZEPIR, ANDREW PETER, nematologist; b. Englewood, N.J., Feb. 25, 1952; s. Andrew and Lillian Elizabeth (Gudz) N.; married. BS, U. Ga., 1974; MS, Clemson U., 1976, PhD, 1980. Rsch. assoc. USDA, ARS, Prosser, Wash., 1980-82; rsch. nematologist USDA, ARS, Byron, Ga., 1982—. Contbr. articles to profl. publs., chpt. to book. Baptist. Achievements include demonstration that the ring nematode (Criconemella xenoplax) was the biological agent responsible for predisposing peach trees to peach-tree-short-life disease complex and that it can be managed with non-chemical means such as ground cover-planting winter wheat. Office: USDA ARS SE Fruit & Tree Nut Rsch Lb PO Box 87 Byron GA 31008

NYHAN, WILLIAM LEO, pediatrician, educator; b. Boston, Mar. 13, 1926; s. W. Leo and Mary (Cleary) N.; m. Christine Murphy, Nov. 20, 1948; children: Christopher, Abigail. Student, Harvard U., 1943-45; M.D., Columbia U., 1949; M.S., U. Ill., 1956, Ph.D., 1958; hon. doctorate, Tokushima U., Japan, 1981. Intern Yale U.-Grace-New Haven Hosp., 1949-50, resident, 1950-51, 53-55; asst. prof. pediatrics Johns Hopkins U., 1958-61, assoc. prof., 1961-63; prof. pediatrics, biochemistry U. Miami, 1963-69, chmn. dept. pediatrics, 1963-69; prof. U. Calif., San Diego, 1969—; chmn. dept. pediatrics U. Calif., 1969-86; mem. FDA adv. com. on Teratogenic Effects of Certain Drugs, 1964-70; mem. pediatric panel AMA Council on Drugs, 1964-70; mem. Nat. Adv. Child Health and Human Devel. Council, 1967-71; mem. research adv. com. Calif. Dept. Mental Hygiene, 1969-72; mem. med. and sci. adv. com. Leukemia Soc. Am., Inc., 1968-72; mem. basic adv. com. Nat. Found. March of Dimes, 1973-81; mem. Basil O'Connor Starter grants com., 1973—; mem. clin. cancer program project rev. com. Nat. Cancer Inst., 1977-81; vis. prof. extraordinario U. del Salvador (Argentina), 1982. Author: (with E. Edelson) The Heredity Factor, Genes, Chromosomes and You, 1976,Genetic & Malformation Syndromes in Clinical Medicine, 1976, Abnormalities in Amino Acid Metabolism in Clinical Medicine, 1984, Diagnostic Recognition of Genetic Disease, 1987; editor: Amino Acid Metabolism and Genetic Variation, 1967, Heritable Disorders of Amino Acid Metabolism, 1974; mem. editorial bd. Jour. Pediatrics, 1964-78, King Faisal Hosp. Med. Jour., 1981-85, Western Jour. Medicine, 1974-86, Annals of Saudi Medicine, 1985-87, mem. editorial com. Ann. Rev. Nutrition, 1982-86; mem. editorial staff Med. and Pediatric Oncology, 1975-83. Served with U.S. Navy, 1944-46; U.S. Army, 1951-53. Nat. Found. Infantile Paralysis fellow, 1955-58; recipient Commemorative medallion Columbia U. Coll. Physicians and Surgeons, 1967. Mem. AAAS, Am. Fedn. Clin. Research, Am. Chem. Soc., Soc. Pediatric Research (pres. 1970-71), Am. Assn. Cancer Research, Am. Soc. Pharmacology and Exptl. Therapeutics, Western Soc. Pediatric Research (pres. 1976-77), N.Y. Acad. Sci., Am. Acad. Pediatrics (Borden award 1980), Am. Pediatric Soc., Am. Inst. Biol. Scis., Soc. Exptl. Biology and Medicine, Am. Soc. Clin. Investigation, Am. Soc. Human Genetics (dir. 1978-81), Inst. Investigaciones Citologicas (Spain; corr.), Biochem. Soc., Société Française de Pediatrie (corr.), Sigma Xi, Alpha Omega Alpha. Office: U Calif San Diego Dept Pediatrics 0609A La Jolla CA 92093-0609

NYMAN, BRUCE MITCHELL, electrical engineer; b. Paterson, N.J., July 4, 1960; s. Burton Harold and Barbara Ann (Levine) N.; m. Eva A. Gutterson, Aug. 11, 1985; children: Robert, Michael. BS, Columbia U., 1982, PhD, 1988. Mem. tech. staff Western Electric Engring. Rsch. Ctr., Princeton, N.J., 1982-88; & AT&T Bell Labs., Holmdel, N.J., 1988—. Contbr. articles to profl. jours. Achievements include patent for method of inspecting ceramic substrates, participation in demonstration of first long-distance communications using optical amplifiers using both NRZ and Solitons; development of optical measurement methods for optical amplifier systems.

NYSTROM, GUSTAV ADOLPH, mechanical engineer; b. Chgo., May 10, 1948; s. Halvard A. and Gloria (Alvaro) N.; children: G. David, Christina E. BS in Aerospace Engring., U. Ill., 1970; PhD in Applied Mechanics, Stanford (Calif.) U., 1975. Registered profl. engr., Calif. Rsch. engr. Uniroyal Rsch. Ctr., Middlebury, Conn., 1976-77; rsch. specialist Exxon Prodn. Rsch. Co., Houston, 1977-86; staff scientist Peda Corp., Palo Alto, Calif., 1986-89; sr. engr. Failure Analysis Assocs., Menlo Park, Calif., 1989-92, Modeling and Computing Svcs., Newark, Calif., 1992—; community work group Lawrence Livermore Nat. Lab. Ground Water Superfund Project, 1987—. Contbr. 19 articles to profl. jours. Draft counselor Ctrl. Com. of Cons. Objectors, San Francisco, 1992; with fgn. delegation Pax Christi, Nicaragua, 1987; jail min. Cath. Ch., Alameda County Jail, 1986—. Recipient Advocate for Social Justice award St. Augustine's Ch., 1992, Cost Savs. award Lockheed Missiles and Space, 1975. Mem. ASME, Soc. Automotive Engrs., Sigma Tau (past pres.), Sigma Xi. Home: 1387 Greenwood Rd Pleasanton CA 94566 Office: Modeling & Computing Svcs 39675 Cedar Blvd #290 Newark CA 94560

OAK, RONALD STUART, health and safety administrator; b. Fargo, N.D., Dec. 20, 1956; s. Duane Lowel and Beverly Alice (Anderson) O. BS in Environ. Health, Colo. State U., 1979. Cert. indsl. hygienist Am. Bd. Indsl. Hygiene, cert. hazardous materials mgr. Inst. Hazardous Materials Mgmt. Compliance officer Wyo. Occupational Health and Safety Dept., Cheyenne, 1980-82, OSHA cons., 1982-84; indsl. hygienist Hager Labs., Inc., Denver, 1984-86; from assoc. to sr. indsl. hygienist Ecology and Environment, Inc., Denver, 1987-91; sr. indsl. hygienist Harding Lawson Assocs., Inc., Santa Ana, Calif., 1991-92; health and safety mgr. IT Corp., San Jose, Calif., 1993—. Mem. adv. com. Wyo. Gov.'s Com. on Hazardous Materials Response, Cheyenne, 1983-84; nat. mem. Smithsonian Assocs., Washington, 1990—. Mem. AAAS, Am. Indsl. Hygiene Assn. Achievements include research on hazardous waste sites and their remediation, on occupational health and safety management, on industrial hygiene evaluations and indoor air quality assessments. Office: IT Corp 2055 Junction Ave San Jose CA 95131

OAKES, CARLTON ELSWORTH, physicist; b. Mineola, N.Y., June 18, 1966; s. Donald K. O. BS in Physics, U. Mass., Amherst, 1990. Project mgr. Donald K. Oakes Gen. Contractor, Hobe Sound, Fla., 1990; staff scientist Radiation Monitoring Devices, Inc., Watertown, Mass., 1991—. Mem. Am. Phys. Soc., Materials Rsch. Soc., Internat. Soc. Optical Engring. Office: Radiation Monitoring Devices Inc 44 Hunt St Watertown MA 02172

OAKES, ELLEN RUTH, psychotherapist, health institute administrator; b. Bartlesville, Okla., Aug. 19, 1919; d. John Isaac and Eva Ruth (Engle) Harboldt; m. Paul Otis Oakes Sr., June 12, 1937 (div. April 1974); children: Paul Otis Jr., Deborah Ellen, Nancy Elaine Masters; m. Siegmar Johann Knopp, Nov. 24, 1975. BA in Sociology, Psychology summa cum laude, Oklahoma City U., 1961; MS in Clin. Psychology, U. Okla., 1963, PhD, 1967. Lic. clin. psychologist, Okla. Chief psychometrist Okla. U. Guidance Ctr., Norman, 1962; psychology trainee VA Hosp., Oklahoma City, 1962-64, Cerebral Palsy Ctr., Norman, Okla., 1964-65; psychology intern Guidance Service, Norman, 1965-66, staff psychologist, 1966-67; asst. prof. psychology Okla. U. Med. Sch., Oklahoma City, 1967-70; supr. psychology interns Okla. Univ. Health Scis. Ctr., 1967-80; founder, dir. Timberridge Inst., Oklahoma City, 1970-90, pres., 1980-90; pvt. practice clin. psychologist Oklahoma City, 1970-92; instr. Okla. U. extension course, Tinker AFB, Oklahoma City, 1963, U. Okla., 1965-66; discussion leader Inst. for Tchrs. of Disadvantaged Child Oklahoma City Sch. System, 1966; leader group therapy sessions Asbury Meth. and Westminster Presbyn. Chs., Oklahoma City, 1966; mem. psychology team confs. for hearing disorders, Okla. U. Med. Sch., 1967-70; cons. Oklahoma City Pub. Schs., 1970-72; cons., group leader halfway house, 1972; lectr. chs., PTAs, hosps.; reviewer Am. Psychol. Assn. Civilian Health and Med. Program of the Uniformed Svcs., 1978-89. Workshop conductor on Shame & Sexuality, Zurick Jungian Inst. winter seminar, 1992; attended Européen Congrès de Gestalt Thérapie in Paris, 1992; contbr. articles to profl. jours. Speaker Okla. County Mental Health Assn. Annual Worry Clinic, St. Luke's Ch., Oklahoma City, 1968-92, psychology dept. Sorosis Club, St. Luke's Ch.. Mem. Am. Psychol. Assn. (peer rev. project with CHAMPUS, 1978-89), Okla. Psychol. Assn. (pres. 1975-76). Avocations: art, travel, poetry, photography, walking.

OAKES, MELVIN ERVIN LOUIS, physics educator; b. Vicksburg, Miss., May 11, 1936; married, 1963; three children. Student, Fla. State U., 1958-64, PhD in Plasma Physics, 1964. Physicist USAR Guided Missile Agy., Redstone Arsenal, 1960; asst. in physics Fla. State U., Tallahassee, 1958-60, 60-64; asst. prof. physic U. Ga., Athens, 1964, rsch. assoc., 1964-65, from asst. prof. to assoc. prof., 1965-70; prof. of physics U. Tex., Austin, 1975—, now William David Blunk meml. prof. of physics. Mem. Am. Phys. Soc., Am. Assn. Physics Tchrs. Achievements include rsch. on electromagnet interaction with plasmas, plasma waves and radio frequency heating. Office: U Tex Dept Physics Austin TX 78712-1157

OAKES, THOMAS WYATT, environmental engineer; b. Danville, Va., June 14, 1950; s. Wyatt Johnson and Relia (Sceacre) O.; m. Terry Lynn Jenkins, June 15, 1974; 1 child, Travis Wyatt. BS in Nuclear Engring., Va. Polytechnic U., 1973, MS in Nuclear Engring., 1975; MS in Environ. Engring., U. Tenn., 1981. Health physics asst. Va. Polytechnic U., Blacksburg, 1972-74; radiation engr. Babcock and Wilcox Co., Lynchburg, Va., 1974-75; dept. mgr. Oak Ridge (Tenn.) Nat. Lab., 1975-78, environ. mgr., 1978-85; corp. environ. coord. Martin Marietta, Oak Ridge, 1985-87; asst. v.p. Sci. Applications Internat. Corp., Oak Ridge, 1987-90; environ. mmgr. Westinghouse Environ. and Geotech. Svcs., Knoxville, Tenn., 1990-91; mgr. S.E. region environ. svcs. ATEC & Assocs., Inc., Marietta, Ga., 1991—. Contbr. over 107 articles to scholarly and profl. jours. Recipient Spl. Recognition award Union Carbide Corp., 1980, Best Paper award Nat. Safety Coun., 1982, Tech. Publs. award Soc. Tech. Communications, 1987. Mem. AAAS, Am. Indsl. Hygiene Assn., N.Y. Acad. Scis., Health Physics Soc. (sec.-treas. environ. sect. 1984-85), Am. Naval Soc., Am. Soc. for Quality Control. Office: ATEC Assocs Inc 1300 Williams Dr Ste A Marietta GA 30066-6299

OAKESHOTT, GORDON B(LAISDELL), geologist; b. Oakland, Calif., Dec. 24, 1904; s. Philip S. and Edith May (Blaisdell) O.; m. Beatrice Clare Darrow, Sept. 1, 1929 (dec. 1982); children: Paul Darrow, Phyllis Joy Oakeshott Martin, Glenn Raymond; m. Lucile Spangler Burks, 1986. BS, U. Calif., 1928, MS, 1929; PhD, U. So. Calif., 1936. Asst. field geologist Shell Oil Co., 1929-30; instr. earth sci. Compton Coll., 1930-48; supervising mining geologist Calif. Div. Mines, 1948-56, dep. chief, 1956-57, chief, 1958; dep. chief Calif. Div. Mines and Geology, 1959-72; cons. geologist, 1973-85, ret.; lectr. geology Calif. State U., Sacramento, 1972-73, Calif. State U., San Francisco, 1975. Author: California's Changing Landscapes—A Guide to the Geology of the State, 1971, 2d edit., 1978, Volcanoes and Earthquakes-Geologic Violence, 1975, Japanese edit., 1981, My California: Autobiography of a Geologist with a Tribute to Don Tocher, 1989; contbr. articles to profl. jours. Fellow AAAS, Geol. Soc. Am., Calif. Acad. Sci.; mem. Seismol. Soc. Am., Nat. Assn. Geology Tchrs. (pres. 1970-71, Webb award 1981), Am. Assn. Petroleum Geologists (hon., Michel T. Halbouty Human Needs award 1993), AIME, Mining and Metall. Soc. Am., Peninsula Geol. Soc. (past pres.), Engrs. Club San Francisco, Geol. Soc. Sacramento (past pres.), Peninsula Gem and Geol. Soc. (hon.), Assn. Engring. Geologists (hon.), Earthquake Engring. Research Inst. (past dir.), Am. Inst. Profl. Geologists (emeritus). Home and Office: Byron Park # 443 1700 Tice Valley Blvd Walnut Creek CA 94595

OAKESON, DAVID OSCAR, aerospace engineer; b. Charleston, W.Va., Apr. 1, 1965; s. Gary Oscar and Barbara Ann (Hyde) O. BS, Purdue U., 1988; MS, Tex. A&M U., 1992. Coop. student GE Aircraft Engines, Evendale, Ohio, 1984-87; aerospace engr. Allison Gas Turbines div. GM, Indpls., 1988—. Adult advisor Student Venture, Brownsburg, Ind., 1992-93. Recipient Rsch. in Aero. Propulsion Tech. fellowship USAF, 1989-92. Mem. AIAA. Home: 6417 Cotton Bay Dr N Indianapolis IN 46254

OAKESON, RALPH WILLARD, pharmaceutical and materials scientist; b. Salt Lake City, Nov. 5, 1963; s. Willard LeRoy and Nelda (Haderlie) O.; m. FaNae Rock, Aug. 26, 1986; children: Nicole, Audrey. BS in Chemistry, U. Utah, 1988, postgrad., 1991—. Researcher Dept. Materials Sci. U. Utah, Salt Lake City, 1987-91, researcher Dept. Pharmaceutics, 1991--; mem. adv. coun. Dept. materials Sci. U. Utah, 1988--. Participant Community Coun., Salt Lake City, 1987--. Mem. Controlled Release Soc., Inc., Am. Assn. Pharm. Scientists, ASM Intrnat., Ctr. for Controled Chem. Delivery. Office: U Utah 421 Wakara Way Salt Lake City UT 84108-1210

OAKFORD, LAWRENCE XAVIER, electron microscopist, laboratory administrator; b. Cleve., Jan. 6, 1953; s. Gerald Frederick and Mary Elizabeth (Pestak) O.; m. Dorothy Jean Savage, Aug. 18, 1985; children: Tony, Jamil. MS, Calif. State Poly. U., 1977; PhD, Wash. State U., 1986. Teaching asst., technician Calif. State Poly. U., Pomona, 1975-77; rsch. asst. Wash. State U., Pullman, 1977-79, teaching asst., 1977-82; lab. dir. Tex. Coll. Osteo. Medicine, Ft. Worth, 1982—. Contbr. 15 articles to profl. jours. Com. chmn. Cowtown Marathon and 10k Run, Ft. Worth, 1984-93; mentor, judge Ft. Worth Ind. Sch. Dist. Adopt-a-Sch. Program, Ft. Worth, 1984-93; mem. pastoral coun. St. Andrews Cath. Ch., 1993—. Mem. AAAS, Electron Microscopy Soc. Am., Societé de Microscopie du Can., Tex. Soc. for Electron Microscopy. Office: U Tenn Health Sci Ctr Dept Anatomy & Cell Biology 3500 Camp Bowie Blvd Fort Worth TX 76107-2644

OAKLEY, MARTA TLAPOVA, psychologist; b. Prague, Czechoslovakia, Feb. 7, 1948; came to U.S. 1964; d. Jan Tlapa and Jirina Valerie (Vesela) Rick; m. Daniel Taylor, June 20, 1969; children: Daniel Jan, Jennifer Lee. MA, U. N.C., 1984, PhD, 1987. Teaching ast., lab. asst. U. N.C., Greensboro, 1980-82, 84-87; lab. mgr. A&T State U., Greensboro, 1983-84; researcher Los Alamos (N.Mex.) Nat. Lab., 1987-90, tng. specialist, 1990-; sec. N.Mex. Psychol. Assn., 1990—. Contbr. articles to profl. jours. Bd. dirs. Santa Fe Rape Crisis Ctr., N.Mex., 1990-92; mem. Gov.'s Task Force for the Prevention Sexual Crimes in N.Mex., 1990-91. Republican. Achievements include presentation of evidence for subcortical gating mechanism during selective attention in the human visual system correlated with the intention to make a motor response. Office: Los Alamos Lab MS M589 Los Alamos NM 87545

OANA, HARRY JEROME, optical engineer; b. Sibiu, Romania, Feb. 6, 1957; arrived in U.S., 1985; s. Ironim and Eugenia (Ciora) O.; m. Lillian Maria Campeanu, July 26, 1980; 1 child, Alexandra. BS in Physics, Bucharest (Romania) U., 1981; MS in Optics, 1982. Registered profl. engr., N.J. Physicist Inst. Electrotechnics, Bucharest, Romania, 1982-83; rsch. physicist Inst. Mechanics, Bucharest, 1983-85, Huth Labs. Inc., Culver, Ind., 1986-89; optical systems specialist Instruments S.A., N.J., 1989-90; optical engr. Datacolor Internat., Lawrenceville, N.J., 1990—. Achievements include development of miniature spectrometer for spectro colorimetry. Home: 48 Sagamore Ave Edison NJ 08820 Office: Datacolor Internat 5 Princess Rd Lawrenceville NJ 08648

OAS, JOHN GILBERT, neurologist, researcher; b. Baton Rouge, May 29, 1958; s. Reynold Gilbert and Donna Jean (Billington) O.; m. Kimberly Jane Hall, Nov. 29, 1986; 1 child, Molly Lucille. BS in Engring. magna cum laude, U. Mich., 1982; MD, U. Tex., Galveston, 1986. Project engr. NASA-Johnson Space Ctr., Houston, 1979-82; intern St. Elizabeth's Hosp., Boston, 1986-87; neurology resident U. Tex. Med. Br. Hosps., Galveston, 1987-90; NIH fellow UCLA Med. Ctr., 1990-92, clin. instr., 1990-92; clin. instr. Harvard Med. Sch., Boston, 1992—; med. dir. Jenks Vestibular Lab. Mass. Eye & Ear Infirmary, Boston, 1992—. Contbr. articles to profl. jours. Recipient Cooley Writing award U. Mich., 1982. Mem. AMA, AIAA, Am. Acad. Neurology (assoc. mem.), Tau Beta Pi. Achievements include research in disorders of balance and dizziness, otolith function and disorders of neurological nature as they refer to the vestibular system. Office: Mass Eye & Ear Infirmary 243 Charles St Boston MA 02114-3096

OBADIA, ANDRE ISAAC, surgeon; b. Oran, Algeria, Mar. 21, 1927; s. Sam and Marthe Marie (Bouchara) O.; m. Annie Judith Ziza, June 5, 1961; children—Dominique, Laurence, Olivia. S.P.C.N., Faculty of Scis., Algiers, 1945; B. Ethnology, Faculty of Scis., Paris, 1959, B. Anthropology, 1960; med. qualification Ednl. Council Fgn. Med. Grads., Evanston, Ill., 1961. Externe des hopitaux Assistance Publique, Paris, 1949-54, intern des hopitaux, 1957-61; attaché CNRS, Hopital Broussais, Paris, 1964; chef de clinique Faculty of Med., Paris, 1962-66; surgeon Clinique Sully, Maisons-Laffitte, 1964—, Poissy Hosp., France, 1965—; cons. Centre Hospitalier des Courses, 1978—; med. expert Cour d'Appel, Versailles, 1984—; prof. Nurses Tng. Sch., 1969—. Contbr. articles to profl. publs. Donateur, Appel Unifie Juif de France, Paris; hon. mem. Orphelinat Mutualiste de la Police Nationale Pris. Served to lt. French Armed Forces, 1955-57. Ministere des Affaires Etrangeres grantee, 1958; Fulbright fellow, 1962. Mem. Assn. Francaise de Chirurgie, Coll. de Pathologie Vasculaire, Compagnie des Experts de Versailles, Union Nationale des Medecins de Reserve, Coll. Nat. des Chirurgiens Francais, Assn. des Membres de l'Ordre Nat. du Merite. Jewish. Avocations: ethnology, minerology, chess. Home: 20 rue Euler, 75008 Paris France Office: Clinique Sully, 2 Place Sully, 78600 Maisons-Lafitte France

OBENAUER, JOHN CHARLES, physicist; b. Columbus, Ohio, Oct. 1, 1967; s. Charles Edward and Jean (Nuhfer) O. BA, St. John's Coll., Annapolis, Md., 1990. Physicist Sachs Freeman Assocs. at Naval Rsch. Lab., Washington, 1990—; teaching asst. SUNY, Buffalo, 1991-92. Mem. Am. Phys. Soc., Am. Assn. for the Advancement Sci. Home: 1954 Columbia Rd NW #710 Washington DC 20009-5040 Office: Code 7650 Naval Rsch Lab 4555 Overlook Ave SW Washington DC 20375

OBENOUR, JERRY LEE, scientific company executive; b. Waterloo, Iowa; s. Fred Jerry Obenour and Marian E. (Miller) Swank; m. Mary Ann Dieckmann, June 27, 1965. BSEE, U. Iowa, 1972. Engr. Tex. Instruments, Inc., Dallas, 1972-79, mgr., 1979-92; v.p. Sci. Applications, Internat., San Diego, 1992—; mem. Statis. Process Control Working Group, Washington, 1992—; mem. Electronic Mfg. Productivity Facility, Tech. Adv. Bd., Indpls., 1990—; speaker on quality mgmt. Mem. AIAA, IEEE, Am. Soc. Quality Control, Tau Beta Pi. Republican. Achievements include patent for auditory response detection method apparatus. Home: 3481 Overpark Rd San Diego CA 92130 Office: SAIC 10260 Campus Point Dr MS/D7 San Diego CA 92121

OBENSON, PHILIP, computer science educator; b. Mamfe, Cameroon, Apr. 22, 1948; s. Philip and Veronica (Oben) O.; m. Francoise Chambon, Nov. 21, 1981; children: Michael, Olivier, Anais. BSc, Fourah Bay Coll., Freetown, 1972; MSc, U. Grenoble (France), 1976; PhD, U. Montpellier (France), 1981. Rsch. scientist IBM, Paris, 1981-82; assoc. prof. African Inst. Informatics, Libreville, Gabon, 1982-87, prof., 1987-89; prof. Univ. Centre, Douala, Cameroon, 1989—; cons. UNESCO, Nairobi, Kenya, 1985, Govt. of Gabon, Libreville, 1984-89, PanAfrican Inst. for Devel., Douala, 1989—, USAID Agrl. Project, Dschang, Cameroon, 1991. Author: Logic Programming, 1991, Databases, 1986; editor-in-chief Organisational Computing, 1991; contbr. articles to profl. jours. Mem. Assn. for Computing Machinery, IEEE Computer Soc. (sr.), Am. Assn. for Artificial Intelligence, Brit. Computer Soc. Roman Catholic. Home: PO Box 12426, Douala Cameroon Office: Univ Centre, PO Box 1872, Douala Cameroon

OBERDORFER, FRANZ, chemist; b. Aalen, Germany, Nov. 8, 1951; m. Ute Reichmann, Dec. 28, 1977; children: Agnes, Georg, Konrad. Diplom Chemiker, U. Heidelberg, 1979, Dr.rer.nat., 1982. Rsch. asst. German Cancer Rsch. Ctr., Heidelberg, 1979-82; hon. rsch. asst. Med. Rsch. Coun., London, 1982; rsch. fellow Commissariat a L'Energie Atomique, Saclay, France, 1983-84; sr. rsch. chemist German Cancer Rsch. Ctr., Heidelberg, 1984—; lectr. U. Faculty of Pharmacy, Heidelberg. Fellow German Chem. Soc., Am. Chem. Soc. (div. fluorine chemistry), European Assn. Nuclear Medicine, Soc. Nuclear Medicine. Office: German Cancer Rsch Ctr, Im Neuenheimer Feld 280, Heidelberg Germany 6900

ÖBERG, KJELL ERIK, physician, educator, researcher; b. Kalix, Norbotten, Sweden, Apr. 15, 1946; s. Emil K. and Carin I. (Nordlund) Ö.; m. Anita I. Eriksson, May 24, 1969; children: Anna, Andreas. Med kand, Med. Faculty U., Gothenburg, 1967; Med lic, Med. Faculty U., Umeå, 1972; PhD, specialist in endocrinology, Med. Faculty U., Uppsala, 1982; specialist in internal medicine, Boden, 1976. Cons. internist Dept. Internal Medicine, Boden, 1975-77; sr. registrar Endocrine Unit, Uppsala, 1977-79, rsch. fellow, 1979-82, cons. endocrinologist, 1982—; head Nat. Referal Ctr. for Neuroendocrine Tumours Univ. Hosp., Uppsala, 1987—; clin. assoc. dir. Ludwig Inst. for Cancer Rsch., Uppsala, 1986-92; head of the endocrine unit Univ. Hosp., Uppsala, 1990—. Contbr. over 250 articles to profl. jours. Recipient European award for Interferon Rsch., Germany, 1991. Mem. Swedish Med. Assn., Swedish Med. Rsch. Assn., Lion Cancer Rsch. Found. (sci. bd. 1988—), European Soc. for Med. Oncology, European Sch. Oncology, European Orgn. for Rsch. on Treatment of Cancer. Avocations: cross country skiing, hiking, jogging, fishing. Home: Ekolnsväg 15A, S 756 53 Uppsala Sweden Office: Dept Internal Medicine, Endocrine Unit, 75185 Uppsala Sweden

OBRAMS, GUNTA IRIS, health facility administrator; b. Düsseldorf, Germany, Sept. 2, 1953; came to U.S., 1961; d. Robert and Olga (Baltins) O.; m. Malcolm DeWitt Patterson, Dec. 22, 1975; 1 child, Andrew McDougal Patterson. BS in Biology cum laude, Rensselaer Poly. Inst., 1977; MD, Union U., Albany, N.Y., 1977; MPH, Johns Hopkins U., 1982, PhD, 1988. Resident in obstetrics and gynecology Ea. Va. Grad. Sch. Medicine, Norfolk, 1977-78; community physician Southampton Meml. Hosp., Franklin, Va., 1978-81; resident in gen. preventive medicine sch. hygiene and pub. health Johns Hopkins U., Balt., 1981-84, project dir., 1983-85, med. dir., 1985-86; med. officer divsn. cancer etiology Nat. Cancer Inst., Bethesda, Md., 1986-89, dep. chief, 1989-90, chief, 1990—. Editor: (with M. Potter): The Epidemiology and Biology of Multiple Myeloma, 1991; contbr. articles to profl. jours. With USPHS, 1987—. Decorated Achievement medal; recipient Nat. Rsch. Svc. award, 1981, Rsch. Career award NIOSH, 1985; scholar Am. Med. Women's Assn., 1977. Mem. Phi Beta Kappa, Delta Omega, Alpha Omega Alpha. Office: National Cancer Institute 6130 Executive Blvd Ste 535 Bethesda MD 20892

O'BRIEN, ELLEN K., hydrologist; b. Hartford, Conn., Feb. 11, 1954; d. Roger W. and Virginia (Dolliver) Knight; m. Richard John O'Brien, July 1, 1954; children: Flann Christopher, Madelyn MacKenzie, Torin Skye. BS summa cum laude, U. N.H., 1976; MSCE, Northeastern U., 1982. Cert. geologist, Maine. Hydrologist Frederic R. Harris, Inc., Boston, 1976-84; lectr. Bates Coll., Lewiston, Maine, 1987-92; sr. hydrologist Acheron Engring. Svcs., Winthrop, Maine, 1987—. Chair Mechanic Falls (Maine) Planning Bd., 1984-88; mem. exec. bd. Androscoggin Valley Coun. Govts., Auburn, Maine, 1984-88; classroom vol. Winthrop (Maine) Grade Sch., 1988—. Mem. Am. Inst. Hydrology (cert.), Assn. Groundwater Hydrologists. Unitarian Universalist. Office: Acheron Engring Svcs PO Box 173 Winthrop ME 04364

O'BRIEN, JAMES JOSEPH, meteorology and oceanography educator; b. N.Y.C., Aug. 10, 1935; s. Maurice J. and Beatrice (Cuddihy) O'B.; m. Sheila O'Keeffe, Nov. 29, 1958; children: Karen, Kevin, Sean, Dealyn, Denis. BS in Chemistry, Rutgers U., 1957; MS in Meteorology, Tex. A&M U., 1964, PhD in Meteorology, 1966. Meteorologist USAF, 1957-58; chemist E.I. duPont, 1958-62; mgr. Office Naval Res., Washington, 1974-76; prof. Fla. State U., Tallahassee, 1972-82, sec. navy prof., 1982—, disting. rsch. prof., 1991—; pres. IAPSO, IUGG, 1987-91. Editor: The Sea, Vol. 6, 1976, Advanced Physical Modelling, 1986; contbr. over 100 papers to profl. jours. Mem. bd. advisors Naval Postgrad. Sch., Monterey, Calif., 1990—. Capt. USAF, 1957-60. Recipient Liege Medal of Honor, Liege U., Disting. Ocean Educator awaard Office Naval Rsch., 1989, Disting. Rsch. Prof. award Fla. state U., 1991. Fellow AAAS, Am. Meteorol. Soc. (Sverdrups Gold medal), Am. geophys. Union, Royal Meteorol. Soc. Democrat. Roman Catholic. Achievements include expertise on Upper Ocean Modellings, forecasting; model to forecast El Nino; expertise in Ocean Models, climate models, physical oceanography, statistics. Office: Fla State U Dept Meteorology Tallahassee FL 32306

O'BRIEN, JOHN CONWAY, economist, educator, writer; b. Hamilton, Lanarkshire, Scotland; s. Patrick and Mary (Hunt) O'B.; m. Jane Estelle Judd, Sept. 16, 1966; children: Kellie Marie, Kerry Patrick, Tracy Anne, Kristen Noël. B.Com., U. London, 1952, cert. in German lang., 1954; tchr.'s cert., Scottish Edn. Dept., 1954; AM, U. Notre Dame, 1959, PhD, 1961. Tchr. Scottish High Schs., Lanarkshire, 1952-56; instr. U. B.C., Can., 1961-62; asst. prof. U. Sask., Can., 1962-63, U. Dayton, Ohio, 1963-64; assoc. prof. Wilfrid Laurier U., Ont., Can., 1964-65; from asst. to full prof. Econs. and Ethics Calif. State U., Fresno, 1965—; vis. prof. U. Pitts., 1969-70, U. Hawaii, Manoa, 1984; keynote speaker Wageningen Agrl. U., The Netherlands, 1987; presenter papers 5th, 6th, 10th World Congress of Economists, Tokyo, 1977, Mexico City, 1980, Moscow, 1992; presenter Schmoller Symposium, Heilbronn am Neckar, Fed. Republic Germany, 1988, paper The China Confucius Found. and "2540" Conf., Beijing, 1989, 6th Internat. Conf. on Cultural Econs., Univ. Umeä, Sweden, 1990, Internat. Soc. Intercommunication New Ideas, Sorbonne, Paris, 1990, European Assn. for Evolutionary Polit. Economy, Vienna, Austria, 1991, World Cong. Economists, Tokyo, 1977, Mexico City, 1980, Moscow, 1992; active rsch. U. Göttingen, Fed. Republic Germany, 1987; acad. cons. Cath. Inst. Social Ethics, Oxford; presenter in field. Author: Karl Marx: The Social Theorist, 1981, The Economist in Search of Values, 1982, Beyond Marxism, 1985, The Social Economist Hankers After Values, 1992; editor: Internat. Rev. Econs. and Ethics, Internat. Jour. Social Econs., Ethical Values and Social Economics, 1981, Selected Topics in Social Economics, 1982, Festschrift in honor of George Rohrlich, 3 vols., 1984, Social Economics: A Pot-Pourri, 1985, The Social Economist on Nuclear Arms: Crime and Prisons: Health Care, 1986, Festschrift in honor of Anghel N. Rugina, parts I and II, 1987, Gustav von Schmoller: Social Economist, 1989, The Eternal Path to Communism, 1990, (with Z. Wenxian) Essays from the People's Republic China, 1991, Festschrift in Honor of John E. Elliott, parts I and II, 1992, Communism Now and Then, 1993; translator econ. articles from French and German into English; contbr. numerous article to profl. jours. With British Royal Army Service Corps, 1939-46, ETO, NATOUSA, prisoner of war, Germany. Recipient GE Corp. award Stanford U., 1966. Fellow Internat. Inst. Social Econs. (mem. coun., program dir. 3d World Cong. Social Econs. 1983, keynote speaker 4th conf. 1986), Internat. Soc. for Intercommunication New Ideas (disting.); mem. Assn. Social Econs. (dir. west region 1977—, pres.-elect 1978-89, program dir. conf. 1989, pres. 1990, presdl. address, Washington 1990), Western Econ. Assn. (organizer, presenter 1977-93), History Econs. Soc., Soc. Reduction Human Labor (exec. com.), European Assn. Evolutionary Polit. Econ. Roman Catholic. Avocations: jogging, collecting miniature paintings, soccer, tennis, photography. Home: 2733 W Fir Ave Fresno CA 93711-0315 Office: Calif State U Econs and Ethics Dept Fresno CA 93740

O'BRIEN, JOHN F(RANCIS), civil engineer; b. Kalamazoo, Oct. 30, 1962; s. Kyran and Loretta (Lemmer) O'B.; m. Elizabeth Ann Carpenter, Oct. 8, 1984; 1 child, Matthew. BSCE, BS in Chemistry, Mich. Tech. U., 1986. Registered profl. engr., Mich. Staff engr. City of Battle Creek, Mich., 1986-89, pub. utilities engr., 1989-90, mgr. pub. utilities, 1990—; presenter at profl. confs. Mem. Ann. Groundwater Contamination Conf., Nat. Well Water Assn., Am. Water Works Assn. (regulatory adv. com. Mich. sect. 1990—, chmn. ground water com. Mich. sect. 1992—). Republican. Roman Catholic. Home: 60 Chestnut St Battle Creek MI 49017 Office: City of Battle Creek PO Box 1717 Battle Creek MI 49016

O'BRIEN, JOHN JOSEPH, physicist; b. Fermoy, Ireland, Sept. 10, 1948; came to the U.S., 1970; s. John J. and Catherine (Power) O'B.; m. Mary Beth Maus, June 26, 1976; children: Margaret, Sean. BSc, Nat. U. Ireland, Cork, 1970; PhD, U. Pitts., 1976. Mem. rsch. staff Yale U., New Haven, Conn., 1977-79; rsch. geophysicist Gulf Oil R & D, Harmarville, Pa., 1979-83; sr. geophysicist Sohio/B P Exploration, Anchorage and Houston, 1983-92; prof. S.D. State U., Brookings, 1992—. Editor: Dynamical Geology of Salt and Related Structures, 1988; contbr. articles to sci. jours. Mem. Am. Phys. Soc., Soc. Exploration Geophysicists. Achievements include applications of geophysics in oil exploration, oilfield development, and groundwater studies; research in exploration geophysics, geology, nuclear physics and atomic physics, geodynamic modeling and stratigraphic prediction from seismic data. Office: SD State U PO Box 2219 Brookings SD 57007-0395

O'BRIEN, JOHN STEININGER, clinical psychologist; b. Lewisburg, Pa., June 3, 1936; s. Peck Zanders and Esther (Steininger) O'B.; m. Joan Irene Romanos, Nov. 1, 1976; children: Peck David, Timothy. AB, Pa. State U., 1967; MA, So. Ill. U., 1969; PhD, Boston U., 1980. Diplomate Internat. Acad. Profl. Psychotherapists, Internat. Acad. Behavioral Medicine/Psychotherapy. Asst. tchr. educable retarded children Selin's Grove (Pa.) State Sch., 1964-66; clin. rsch. asst. Pa. State U., State Coll., 1966-67; rsch. technician Anna (Ill.) State Hosp., 1968; intern Boston City Hosp., 1968-69, from coord. alcohol study unit to psychologist, 1969-73; clin. instr. psychiatryi Sch. Medicine Tufts U., St. Elizabeth's Hosp., Brighton, Mass., 1973-87; dir. psychol. svcs. Baldpate Hosp., Georgetown, Mass., 1981—, dir. outpatient substance abuse rehab. program, 1991—; bio-behavioral cons. Bhavioral Medicine Inst., Quincy, Mass., 1985-88; clin. dir. Social Learning Ctr., Quincy, 1971—; behavioral therapist, clin. coord. TAP Boston Childrens Svc., 1973-76. Author: Moments with Peck, 1982; contbr. articles to profl. jours. Mem. Am. Psychol. Assn., Nat. Register Health Svcs. in Psychology, Soc. Study of Addiction, Assn. Advancement Behavioral Therapy, Am. Assn. Clin. Counselors, Biofeedback Soc. Am., Internat. Acad. Profl. Counselors and Psychotherapists. Avocations: ocean cruising, deep sea fishing, photography, gardening. Home: 72 Kenneth Rd Scituate MA 02066-2950 Office: Baldpate Psychiat Assocs Baldpate Rd Georgetown MA 01833-2301

O'BRIEN, JOSEPH EDWARD, laboratory director; b. N.Y.C., Feb. 2, 1933; s. Joseph and Sarah Elizabeth (Kerrigan) O'B.; m. Yvonne Joan, Feb. 22, 1958; children: Deborah, Douglas. Med. degree, Columbia Coll., 1957, N.Y. State Coll. medicine, 1961. Asst. attending in pathology Englewood (N.J.) Hosp., 1967-69; dir. tissue Met. Pathology, N.Y.C., 1968-69; lab. dir. MetPath, Inc., Teterboro, N.J., 1969—, sr. v.p. lab. medicine, 1976—; asst. prof. pathology Columbia U. Coll. Physicians and Surgeons, N.Y.C., 1971—. Contbr. articles to profl. jours. Fellow Am. Soc. Clin. Pathologists, Coll. Am. Pathologists. Office: MetPath 1 Malcolm Ave Teterboro NJ 07608

OBRINSKI, VIRGINIA WALLIN, retired school psychologist; b. Stanton, Iowa, Sept. 4, 1915; d. John Edward Wallace and Frances Geraldine (Tinsley) Wallin; m. Peter James Obrinski, May 2, 1981 (dec. Mar. 26, 1989). BA, U. Del., 1936; MEd, Duke U., 1941. Lic. psychologist, Del. Sch. psychologist Del. Dept. Pub. Instrn., Dover, 1936-64, various suburban sch. dists., Wilmington, Del., 1964-67, Mt. Pleasant (Del.) Sch. Dist., 1967-68, Stanton (Del.) Sch. Dist., 1968-69, Conrad (Del.) Area Sch. Dist., 1969-78; ret., 1978. Recipient plaque Del. Coun. for Exceptional Children Fedn., 1985. Mem. AAUW, APA, DAR, Del. Psychol. Assn., Coun. for Exceptional Children. Baptist. Avocations: gardening, travel, reading. Home: 311 Highland Ave Lyndalia Wilmington DE 19804

O'BRYAN, SAUNDRA M., clinical chemist, biosafety officer. BS, Duquesne U., 1969; MS, Northeastern U., 1979. Cert. chemist Am. Soc. Clin. Pathologists. Supr. clin. lab. SmithKline Beecham, Waltham, Mass., 1973-86; mgr. quality control Biocraft Labs., Elmwood Park, N.J., 1986-87; clin. chemist ICI Americas Inc., Wilmington, Del., 1987-91; biosafety officer Zeneca Pharm. (formely ICI Americas Inc.), Wilmington, Del., 1991—. Mem. Am. Assn. for Clin. Chemistry, Am. Soc. Clin. Pathologists, Sigma Xi. Office: Zeneca Pharm 1800 Concord Pike Wilmington DE 19897

OCH, MOHAMAD RACHID, psychiatrist, consultant; b. Damascus, Syria, Apr. 1, 1956; came to U.S., 1981; s. Seifeddine and Souad (Oubari) O.; m. Marianne Noonan, July 24, 1960; children: Seifeddine, Adam. MD, Aleppo (Syria) U., 1980. Psychiat. cons. Human Resource Inst., Brookline, Mass., 1985; med. dir. Spectrum House, Westboro, Mass., 1986-87; assoc. med. dir. Boston Rd. Clinic, Shrewsbury, Mass., 1985—, v.p., 1989—; med. dir. mental health unit Holden (Mass.) Hosp., 1988-90; med. dir. Basic Health Mgmt., Worcester, Mass., 1988-90; asst. med. dir. Boston Rd. Clinic, Shrewsbury, 1986—, Holden Hosp., 1988—, Basic Health Mgmt., Worcester, 1988—; attending psychiatrist, asst. prof. U. Mass. Med. Ctr., Worcester; dir. mental health unit Milford Whitinsville Hosp., 1990—, chmn. dept. psychiatry, 1991-92; med. dir. Seven Hills Intensive Residential Treatment Program, 1990—. Mem. Am. Psychiat. Assn., AMA. Moslem. Office: Boston Road Clinic 108 Belmont St Worcester MA 01605-2937

OCHOA, AUGUSTO CARLOS, immunologist, researcher; b. Medellin, Colombia, Sept. 29, 1955; came to the U.S., 1982; s. Bernardo and Alina (Gautier) O. MD, U. de Antioquia, Colombia, 1981. Rsch. asst. Tansplantation Lab., Medellin, 1976-81; intern Hosp. San Vicente de Paul, Medelin, 1981-82; postdoctoral assoc. Immunobiology Rsch. Ctr. U. Minn., Mpls., 1982-85, asst. prof., 1985-89; sr. scientist Nat. Cancer Inst. Frederick (Md.) Cancer Rsch. Ctr., 1985-89; co-founder Onco Therapeutics Inc., Mpls., 1988. Contbr. articles to profl. jours. Mem. Assn. Medica de Antioquia, Am. Assn. Immunology. Achievements include development of description of anti-CO3 monoclonal antibodies to grow cells for immunotherapy of cancer, IL2 in liposomes to decrease toxicity of IL2. Office: Frederick Cancer Rsch Ctr PO Box B NCI-FCRDC Frederick MD 21702

OCHOA, SEVERO, biochemist; b. Luarca, Spain, Sept. 24, 1905; came to U.S., 1940, naturalized, 1956; s. Severo and Carmen (Albornoz) O.; m. Carmen G. Cobian, July 8, 1931. A.B., Malaga (Spain) Coll., 1921; M.D., U. Madrid, Spain, 1929; D.Sc., Washington U., U. Brazil, 1957, U. Guadalajara, Mexico, 1959, Wesleyan U., U. Oxford, Eng., U. Salamanca, Spain, 1961, Gustavus Adolphus Coll., 1963, U. Pa., 1964, Brandeis U., 1965, U. Granada, Spain, U. Oviedo, Spain, 1967, U. Perugia, Italy, 1968, U. Mich., Weizman Inst., Israel, 1982; Dr. Med. Sci. (hon.), U. Santo Tomas, Manila, Philippines, 1963, U. Buenos Aires, 1968, U. Tucuman, Argentina, 1968; L.H.D., Yeshiva Univ., 1966; LL.D., U. Glasgow, Scotland, 1959. Lectr. physiology U. Madrid Med. Sch., 1931-35; head physiol. div. Inst. for Med. Research, 1935-36; guest research asst. in physiology Kaiser-Wilhelm Inst. for Med. Research, Heidelberg, Germany, 1936-37; Ray Lankester investigator Marine Biol. Lab., Plymouth, Eng., 1937; demonstrator Nuffield research asst. biochemistry Oxford (Eng.) U. Med. Sch., 1938-41; instr., research asso. pharmacology Washington U. Sch. of Medicine, St. Louis, 1941-42; research asso. medicine NYU Sch. Medicine, 1942-45, asst. prof. biochemistry, 1945-46, prof. pharmacology, chmn. dept., 1946-54, prof., chmn. dept. biochemistry, 1954-76, prof., 1976—; distinguished mem. Roche Inst. Molecular Biology. Author publs. on biochem. of muscles, glycolysis in heart and brain, transphosphorylations in yeast fermentation, pyruvic acid oxidation in brain and role of vitamin B1; RNA and Protein biosynthesis; genetic code. Decorated Order Rising Sun Japan; recipient (with Arthur Kornberg) 1959 Nobel prize in medicine, Albert Gallatin medal N.Y. U., 1970, Nat. Medal of Sci., 1980. Fellow N.Y. Acad. Scis., N.Y. Acad. Medicine, Am. Acad. of Arts and Sci., AAAS; mem. NAS, Am. Philos. Soc., Soc. for Exptl. Biology and Medicine, Soc. of Biol. Chemists (pres. 1958, editor jour. 1950-60), Internat. Union Biochemistry (pres. 1961-67), Biochem. Soc. (Eng.), Harvey Soc. (pres. 1953-54), Alpha Omega Alpha (hon.); fgn. mem. German Acad. Nat. Scis., Royal Spanish, USSR, Polish, Pullian, Italian, Argentinian, Barcelona (Spain), Brazilian acads. sci., Royal Soc. (Eng.), Pontifical Acad. Sci., G.D.R. Acad. Scis., Argentinian Nat. Acad. Medicine.

OCHS, WALTER J., civil engineer, drainage adviser; b. Springfield, Minn., May 20, 1934; s. Walter Minrod and Cleo (Schultz) O.; m. Connie Mae Strate, Sept. 15, 1956; children: Julie, Brian. BS in Agrl. Engring., South Dakota U., 1957. Registered profl. civil engr., Mich. Engr. in training USDA, Soil Conservation Svc., Watertown, S.D., 1957-58; project engr. USDA, Soil Conservation Svc., Britton, S.D., 1958-61; area engr. USDA, Soil Conservation Svc., Sioux Falls, S.D., 1961-63; asst. state conservation engr. USDA, Soil Conservation Svc., East Lansing, S.D., 1963-66, state conservation engr., 1966-69; asst. state conservationist USDA, Soil Conservation Svc., Saint Paul, Minn., 1969-71; nat. drainage engr. USDA, Soil Conservation Svc., Washington, 1971-86; drainage adviser World Bank, Washington, 1986—; bd. dirs. Internat. Inst. for Land Reclamation and Improvement Postgrad Land Drainage Course, The Netherlands; cons. for over 25 countries; mem. Internat. Commn. Irrigation and Drainage. Contbr. to profl. jours. Named Federal Engr. Of The Year, Nat. Soc. Profl. Engrs., 1982; recipient Outstanding Alumnus award South Dakota State Univ., 1977, Outstanding Contributions award Corrugated Plastic Tubing Assn., 1981. Fellow Am. Soc. Agrl. Engrs., 1991; Mem. Am. Soc. Civil Engrs. (chmn. drainage com. 1975-76). Home: 6731 Fern Ln Annandale VA 22003 Office: The World Bank 1818 H St NW Washington DC 20433

OCHSNER, SEYMOUR FISKE, radiologist, editor; b. Chgo., Nov. 29, 1915; s. Albert Henry Ochsner and Fleda Fiske; m. Helen Keith, Sept. 8, 1945 (dec. Jan. 1976; children: Anne, Diana, Lida; m. Bobbie Sue Mercer, Dec. 31, 1981. AB, Dartmouth Coll., 1937; MD, U. Pa., Phila., 1947. Diplomate Am. Bd. Radiology. Staff radiologist Ochsner Clinic, New Orleans, 1953-89, also chmn. dept., 1969-77; clin. prof. radiology Tulane Med. Sch., New Orleans, 1955-75; editor Orleans Parish Med. Bulletin, New Orleans, 1985—. Contbr. articles to profl. jours. Pres. PTA, Metairie, La., 1964. Recipient Disting. Svc. medal So. Med. Assn., 1972, Disting. Svc. award AMA, 1992. Mem. Radiol. Soc. La. (pres. 1965), So. Radiol. Conf. (pres. 1968), Am. Coll. Radiology (pres. 1972, Gold medal 1982), Am. Rowgen Roy Soc. (pres. 1975, Gold medal 1986), Rex Orgn., So. Yacht Club, Candlewood Club. Republican. Episcopalian. Avocations: reading, gardening, travel, sailing. Home: 107 Holly Dr Metairie LA 70005-3915 Office: Orleans Parish Med Soc 1800 Canal St New Orleans LA 70112-3078

O'CONNELL, DANIEL CRAIG, psychology educator; b. Sand Springs, Okla., May 20, 1928; s. John Albert and Letitia Rutherford (McGinnis) O'C. B.A., St. Louis U., 1951, Ph.L., 1952, M.A., 1953, S.T.L., 1960; Ph.D., U. Ill., 1963. Joined Soc. of Jesus, 1945; asst. prof. psychology St. Louis U., 1964-66, asso. prof., 1966-72, prof., 1972-80, trustee, 1973-78, pres., 1974-78; prof. psychology Loyola U., Chgo., 1980-89; prof. psychology Georgetown U., Washington, 1990—, chmn., 1991—; vis. prof. U. Melbourne, Australia, 1972, U. Kans., 1978-79, Georgetown U., 1986; Humboldt fellow Psychol. Inst. Free U. Berlin, 1968; sr. Fulbright lectr. Kassel U., W. Ger., 1979-80. Author: Critical Essays on Language Use and Psychology, 1988; contbr. articles to profl. jours. Recipient Nancy McNeir Ring award for outstanding teaching St. Louis U., 1969; NSF fellow, 1961, 63, 65, 68; Humboldt Found. grantee, 1973; Humboldt fellow Tech. U. of Berlin, 1987. Fellow Am., Mo. psychol. assns., Am. Psychol. Soc.; mem. Midwestern, Southwestern, Eastern psychol. assns., Psychologists Interested in Religious Issues, Psychonomic Soc., Soc. for Scientific Study of Religion, N.Y., Mo. acads. sci., AAUP, AAAS, Phi Beta Kappa. Home and Office: Georgetown U Dept Psychology 37th & 0 Sts NW Washington DC 20057

O'CONNELL, WALTER EDWARD, psychologist; b. Reading, Mass., Aug. 2, 1925; s. Walter Edward and Margaret Cecilia (Turner) O'C.; m. Gloria June Kane, Aug. 5, 1960. BA, U. Mass., 1950; MA, U. Tex., 1952, PhD, 1958. Diplomate Am. Bd. Profl. Psychology; licensed psychologist, Tex. Rsch. and clin. psychologist VA Med. Ctr., 1955-86; clin. assoc. prof. Baylor Coll. Medicine, Houston, 1967—, U. Houston, 1967-86; dir. Natural High Ctr., Bastrop, Tex., 1986—; adj. prof. U. St. Thomas, Houston, 1966-86, Baylor U., Waco, Tex., 1959-66, Union Inst., Cin., 1984—; pvt. practice in psychotherapy, Houston, 1973-86; cons. to hosps., univs., chs., schs., and bus., 1965-86. Author: Action Therapy and Adlerian Theory, 1975, 80; contbr. chpts. to books and 350 articles to Voices, The Art and Sci. of Psychotherapy, Individual Psychology, Adlerian Jour. Rsch., Theory and Practice; columnist: Bastrop Advertiser, 1991—. Legislator dist. II Tex. Silver-Haired Legislature, Austin, 1992—; officer, chairperson Bastrop County Retirement Orgns., 1987—. With U.S. Army, 1943-45. Recipient award Roth Foun., 1977, Annual Nat. VA Profl. Svc. award for therapeutic svc. to vets., 1983. Fellow APA, Internat. Acad. Eclectic Psychotherapists; mem. N.Am. Soc. Adlerian Psychology (life, pres. 1972, bd. dirs. 1966-72, 74-77, Outstanding Performance awards 1977, 92), Am. Acad. Psychotherapists (life, editorial bd. 1965-70, 76-83). Democrat. Achievements include research in natural high theory and practice, psychospiritual approach to wellness and interactive health, theory and practice of self-esteem, sense of humor, encouragement, community connectedness, didactic experiential methods. Home and Office: 106 Kelley Rd Bastrop TX 78602

O'CONNOR, DENIS, mechanical engineer; b. Boston, Oct. 14, 1954; s. Denis and Margaret (O'Neil) O'C.; m. Jane Margaret Diggins, Sept. 13, 1981. AS in Mech. Power Engring., Wentworth Inst. Tech., Boston, 1979, BSMET, 1981. Plant engr. Semline, Braintree, Mass., 1981-83; sales engr. Flow Controls, Needham, Mass., 1983-86; field engr. Romicon Inc., Woburn, Mass., 1986-90; project engr. EUA Cogenex, Lowell, Mass., 1990—. With USN, 1972-76. Mem. ASME, Assn. Energy Engrs. Office: EUA Cogenex Boott Mills S 100 Foot of John St Lowell MA 01852

O'CONNOR, DIANE GERALYN OTT, engineer; b. Chgo., June 23, 1967; d. Roy James and Lucille Delores (Steggers) Ott; m. John Thomas O'Connor, Nov. 23, 1991. BS in Engring., U. Ill., 1989; postgrad., Harvard U., 1992. Mktg. specialist electronics GE Plastics, Pittsfield, Mass., 1989-90, application devel. engr., 1990, programs specialist recycle, 1990; regional materials engr. GE Plastics, Atlanta, 1991-92. Advisor Jr. Achievement, Pittsfield, 1990-91; coord. Minority Introduction to Engring., Atlanta, 1991; group leader GE Plastics Outreach Program, Pittsfield, 1990. Mem. Soc. Plastic Engrs., Am. Bus. Women's Assn. Roman Catholic.

O'CONNOR, JAMES JOHN, utility company executive; b. Chgo., Mar. 15, 1937; s. Fred James and Helen Elizabeth O'Connor; m. Ellen Louise Lawlor, Nov. 24, 1960; children: Fred, John (dec.), James, Helen Elizabeth. BS, Holy Cross Coll., 1958; MBA, Harvard U., 1960; JD, Georgetown U., 1963. Bar: Ill. 1963. With Commonwealth Edison Co., Chgo., 1963—, asst. to chmn. exec. com., 1964-65, comml. mgr., 1966, asst. v.p., 1967-70, v.p., 1970-73, exec. v.p., 1973-77, pres., 1977-87, chmn., 1980—, chief exec. officer, also bd. dirs.; bd. dirs. Corning, Inc., Chgo. Stock Exch., Tribune Co., United Air Lines, Scotsman Industries, Am. Nat. Can., First Chgo. Corp., First Nat. Bank Chgo., Am. Nuclear Energy Coun.; past chmn. Nuclear Power Oversight Com.; chmn. Advanced Reactor Corp., U.S. Coun. for Energy Awareness; bd. dirs., past chmn. Edison Electric Inst. Mem. The Bus. Coun.; bd. dirs. Assocs. Harvard U. Grad. Sch. Bus. Adminstrn., Leadership Coun. for Met. Open Communities, Lyric Opera, Mus. Sci. and Industry, St. Xavier Coll., Reading is Fundamental, Helen Brach Found.; past chmn. Met. Savs. Bond Campaign; trustee Adler Planetarium, Northwestern U.; bd. dirs., past chmn. Chgo. Urban League; bd. advisors Mercy Hosp. and Med. Ctr.; past chmn. bd. trustees Field Mus. Natural History; trustee Mus. Sci. and Industry; exec. bd. Chgo. area coun. Boy Scouts Am.; chmn. Cardinal Bernardin's Big Shoulders Fund; exec. v.p. The Hundred Club Cook County. With USAF, 1960-63. Mem. ABA, Ill. Bar Assn., Chgo. Bar Assn., Chgo. Assn. Commerce and Industry (bd. dir., chmn.), Ill. Bus. Roundtable. Home: 9549 Monticello Ave Evanston IL 60203-1119 Office: Commonwealth Edison Co PO Box 767 1 First Nat Pla Chicago IL 60690-0767

O'CONNOR, KEVIN NEAL, psychologist; b. Mineola, N.Y., July 30, 1953; s. Neal Evan and Marianne (Mueller) O'C. BA, Muhlenberg Coll., 1975; PhD, Columbia U., 1988. Rsch. assoc. U. Rochester, N.Y., 1986-90, U. Mass., Amherst, 1990—. Mem. N.Y. Acad. Scis., Soc. for Neurosci., Am. Psychol. Soc. Achievements include research in auditory asymmetries in rats and echoic memory in rats. Office: U Mass Dept Psychology Tobin Hall Amherst MA 01003

O'CONNOR, KEVIN PATRICK, civil engineer; b. Buffalo, Mar. 24, 1966; s. Frank O'Connor and Maureen Vogt. BS, U. Ctrl. Fla., 1992. Host Walt Disney World, Kissimmee, Fla., 1987-90; drafter Fred Humphrey & Assocs., Orlando, 1990-92; engr. Glace & Radcliffe Inc., Orlando, 1992—. Mem. ASCE, Fla. Hwy. Engring. Soc., Pi Kappa Alpha Frat. Roman Catholic. Achievements include development of computer software for drainage design and geometric design. Home: PO Box 678399 Orlando FL 32867 Office: Glace & Radcliffe 800 S Orlando Ave Maitland FL 32751

O'CONNOR, KIM CLAIRE, chemical engineering and biotechnology educator; b. N.Y.C., Nov. 18, 1960; d. Gerard Timothy and Doris Julia (Bisagni) O'C. BS magna cum laude, Rice U., Houston, 1982; PhD, Calif. Inst. Tech., Pasadena, 1987. Postdoctoral rsch. fellow chemistry dept. Calif. Inst. Tech., Pasadena, 1987-88; postdoctoral rsch. fellow chem. engring., biochemistry, molecular biology, and cell biology depts. Northwestern U., Evanston, Ill., 1988-90; asst. prof. chem. engring. Tulane U., New Orleans, 1990—, faculty molecular and cellular biology grad. program, Newcomb fellow, 1991—; mem. steering com. molecular and cellular biology grad. program Tulane U., 1993—; cons. in field. Reviewer of profl. jours. Mem. Am. Chem. Soc., Am. Inst. Chem. Engrs., Am. Soc. Engring. Edn., European Soc. Animal Cell Tech., Sigma Xi, Tau Beta Pi, Phi Lambda Upsilon. Achievements include interdisciplinary research in engineering, animal- and microbial-cell culture, genetics, and protein chemistry; specifically research in enzyme structure and function with immobilization and protein engineering, oxygenation and cell metabolism of animal cells in bioreactors, bioseparation of DNA fragments with HPLC, and large-scale production of recombinant protein. Office: Tulane U Dept Chem Engring Dept Chem Engring Lindy Boggs Ctr Rm 300 New Orleans LA 70118

O'CONNOR, RAYMOND JOSEPH, wildlife educator; b. Dublin, Ireland, Jan. 20, 1944; came to U.S., 1987; s. Kevin and Nora (Crowley) O'C.; m. Deirdre Máire Mageean, Apr. 26, 1979; 1 child, Eóin Hugh. BSc with honors, U. Coll. Dublin, 1965; DPhil, Oxford (Eng.) U., 1973. Lectr. in zoology Queen's U., Belfast, Northern Ireland, 1971-75, Univ. Coll. North Wales, Bandor, Gwynned, 1975-78; dir. Brit. Trust for Ornithology, Tring, Eng., 1978-87; prof. wildlife U. Maine, Orono, 1987—; mem. Internat. Ornithol. Com., 1978-86; mem. adminstrv. bd. Lab. of Ornithology, Cornell U., 1988-92; sr. rsch. associateship Nat. Rsch. Coun., 1993—. Author: The Growth and Development of Birds, 1984; co-author: Farming and Birds, 1986; contbr. numerous articles to profl. jours. Fellow Inst. Biology; mem. Am. Ornithologists Union, Ecol. Soc. Am., Wildlife Soc., Brit. Ornithologists Union (mem. coun. 1982-87). Roman Catholic. Home: 17 College Heights Orono ME 04473 Office: U Maine 238 Nutting Hall Orono ME 04469

OCRANT, IAN, pediatric endocrinologist; b. Buffalo, N.Y., Nov. 13, 1954; s. Lawrence and Nancy Jean (Harris) O.; m. Peggy Ann Kondo, Apr. 21, 1978. BA, Univ. Colo., 1975, MD, 1979. Diplomate Am. Bd. Pediatrics, Am. Bd. Pediatric Endocrinology. Pediatric residency Stanford (Calif.) Univ., 1979-82; pediatrician U.S. Air Force, 1982-86; pediatric endocrin fellowship Stanford Univ., 1986-89; pediatric endocrinologist Brown Univ., Providence, R.I., 1989—; dir. pediatric endocrine clinic Rhode Island Hosp., Providence, 1989—; dir.-at-large Human Growth Found., Falls Church, Va., 1990—. Contbr. articles to profl. jours. With USAF, 1982-86. Recipient Individual Nat. Rsch. Svc. award Nat. Inst. Health, 1987-89. Mem. Endocrine Soc., Lawson Wilkins Pediatric Endocrine Soc., Am. Federation Clinical Rsch. Office: Rhode Island Hosp 593 Eddy St Providence RI 02903

OCVIRK, ANDREJ, Slovene federal official; b. Teharje, Celje, Slovenia, Nov. 30, 1942. D in Chem. Sci. Boris Kidric Chem. instr., 1970; dir. Belinka, 1970-84; mgr. Yulon Co., 1984-86; minister Dept. Power and Industry, 1986; asst. dir. Petrol Co., 1990; v.p. Exec. Coun. Republic of Solvenia. Mem. Democratic Party. Office: Fed Com Energy & Industry, Omladinskih brigada 1, 11070 Belgrade Yugoslavia also: Fed Comm Energy & Industry, bul Avnoj-a 104, 11070 Belgrade Yugoslavia*

ODA, TAKUZO, biochemist, educator; b. Shinichicho, Japan, Oct. 20, 1923; s. Ryoichi and Misu Oda; M.D., Okayama U., 1947, Ph.D., 1953; m. Kazue Matsui, Dec. 8, 1946; children: Mariko, Yumiko. Intern, Okayama U. Hosp., 1947-48, mem. faculty, 1949—, prof. biochemistry, 1965—, dir. Cancer Inst., from 1969; dean Okayama U. Med. Sch., 1985-87; prof. Okayama U. Sci., 1989—; pres. and prof. Niimi Women's Coll., 1993—; postdoctoral trainee Inst. Enzyme Research, U. Wis., Madison, 1960-62. Fellow Rockefeller Found., 1959-60, grantee, 1964; USPHS grantee, 1963-68. Mem. Japanese Soc. Electron Microscopy (Seto prize 1966), Japanese Soc. Biochemistry, Am. Soc. Cell Biology, Japanese Soc. Cancer, Japanese Soc. Cell Biology, Japanese Soc. Histochemistry, Japanese Soc. Virology, Am. Soc. Microbiology. Author: Biochemistry of Biological Membranes, 1969, Mitochondria, Handbook of Cytology; editor, writer: Cell Biology series, 1979—; chief editor Acta Medica Okayama, 1975-85 . Home: 216-38 Maruyama, Okayama 703, Japan Office: Niimi Women's Coll, 1263-2 Nishikata, Niimi City Okayama 718, Japan

ODABASI, HALIS, physics educator; b. Gurun, Sivas, Turkey, Dec. 10, 1931; came to U.S., 1960; s. Mevlut Turan and Zemhanur (Engin) O.; m. Marian Jeanne Maticka, July 10, 1965; 1 child, Turan Paul. BSME, Turkish Merchant Marine Acad., Istanbul, 1954; MSME, U. Colo., 1962, MS in Physics, 1965, PhD in Physics, 1968. Assoc. prof. dept. physics Bogazici U., Istanbul, 1973-79, prof., 1979-82, chmn. dept. physics, 1980-82; vis. faculty dept. physics and chemistry, Cen. Mo. State U., Warrensburg, 1982-84, dept. physics, E. Carolina U., Greenville, N.C., 1985-87, 90—, dept. physics and astronomy, The U. Miss., Oxford, 1987-89. Co-author: Atomic Structure, 1980, Korean edit. 1990; editor: Topics in Modern Physics, 1971, Environmental Problems and Their International Implications, 1973, Topics in Mathematical Physics, 1979, Selected Scientific Papers of E.U. Condon, 1990; contbr. articles to profl. jours. Lt. Turkish Navy, 1955-57. Mem. Am. Phys. Soc., Am. Assn. Physics Tchrs., N.Y. Acad. Scis., Turkish Phys. Soc. Office: Dept Physics East Carolina U Greenville NC 27858

ODAVIC, RANKO, physician, pharmaceutical company executive; b. Vienna, June 1, 1931; s. Simo J. and Angelina (Ljubibratic) O.; m. Katharina Elisabeth Delz, Oct. 17, 1986. MD, U. Sarajevo, 1956, DSc, 1964. Asst. prof. biochemistry U. Sarajevo, 1958-67, sr. lectr., 1967-72, assoc. prof., 1972-74; rsch. fellow dept. biochemistry Royal Coll. Surgeons, London, 1962-63; rsch. fellow Cen. Hematol Lab. Inselpital, Bern, Switzerland, 1974-80, Bur. Radiol. Health, FDA, HEW, Rockville, Md., 1972-73, Dept Hematol./Oncology Children's Hosp., Bern, 1981-82; sect. head cardiology Boehringer Ingelheim Schweiz, Basle, Switzerland, 1983—. Contbr. over 80 articles to profl. publs. Mem. Internat. Soc. Exptl. Hematology, Internat. Soc. Thrombosis Haemostasis, Internat. Union Angiology, N.Y. Acad. Scis. Avocations: gardening, classical music, sports. Home: Steinenbuehlstrasse 173, 4232 Fehren Switzerland Office: Boehringer Ingelheim Scheiz, Peter Merian St 19/21, 4002 Basel Switzerland

ODAWARA, KEN'ICHI, economist, educator; b. Tokyo, Mar. 8, 1933; s. Tsuneo and Kimie (Nagazumi) O.; m. Tsuneko Kurosawa, Sept. 25, 1965; children: Jun'ichi, Nobuo. B.A., Jochi (Sophia) U., 1955, M.A., 1957; M.A., Boston Coll., 1958; postgrad. Columbia U., 1957-58. Rsch. asst. Jochi U., Tokyo, 1956-59, instr., 1959-65, assoc. prof., 1965-70, prof. econs., 1970—; lectr. U. Tokyo, Komaba campus, 1973-84, 90-93, Waseda U., Tokyo, 1988—, Hitotsubashi U., Tokyo, 1989-91; cons. UN. Bankok, Thailand, 1965-68, 69-70, 74; vis. prof. U. Pitts., 1977, U. Hawaii, 1978; vis. scholar Yale U., 1972, MIT, 1975, Hitotsubashi U., Tokyo, 1985-86, Columbia U., 1988. Author: Found. Paper on Regional Payments Arrangements for ECAFE Countries, 1970; UN ESCAP documents, Preliminary Draft Agreement for Asian Payments Arrangements; UN ESCAP Documents, 1970. Author: The Great American Disease, 1980, The Economic Friction between the U.S., Europe and Japan, 1981; author, co-editor: The Textbook of World Economy, 1981, International Political Economy, 1988; author, editor: Report of Symposium on U.S.-Japan Telecommunications Trade, 1986; contbr. profl. jours. including Japan Quar., Sophia Econ. Review, MITI Jour., Japan Econ. Jour. Chief instr. Inst. Money & Banking for Indonesian Govt. Trainees, Japan Internat. Coop. Agy., 1961-62; expert advisor Sci. Council, Ministry Edn., Tokyo, 1980-82; interviewer Econs. com. Fulbright Program, 1985, 91; mem. internat. econs. research coord. com. Sci. Council Japan, 1985-91; chmn. Islamic Econ. Research Council, Japan Indsl. Policy Research Inst., Ministry of Internat. Trade & Industry, 1979-81. Japan Found. grantee, 1978, 92; Union Nat. Econ. Assns. Japan grantee, 1983. Mem. Japan Soc. Internat. Econs. (councillor 1972-78, editor proceedings 1974-76. dir., 1978-85, exec. dir. 1985—, sec.-treas. 1980-91), Am. Econ. Assn. Roman Catholic. Home: 1-3-4 Fujigaya, Kugenuma, Fujisawa-shi, Kanagawa-ken 251, Japan Office: Jochi (Sophia) U, 7-1 Kioi-Cho, Chiyoda-ku, Tokyo 102, Japan

O'DAY, KATHLEEN LOUISE, food products executive; b. Chgo., July 10, 1951; d. Alfred Anton and Maria (Weidinger) Schuld; m. Gary Michael O'Day, Oct. 25, 1975; children: Colleen Marie, Daniel Michael. BS in Biology, Mundelein Coll., Chgo., 1973; Assoc. Mid-Mgmt. & Mktg., McHenry County Coll., Crystal Lake, Ill., 1986. Food techologist Park Corp., Barrlington, Ill., 1973-84; sr. product devel. scientist J.W. Allen & Co., Wheeling, Ill., 1984—. Mem. v.p. Transfiguration Sch. Bd., Wauconda, Ill., 1983-85; sponsor Transfiguration RCIA Orgn., Wauconda, 1992. Mem. Am. Assn. Cereal Chemists, Inst. Food Technologists, Food Exec. Women. Republican. Roman Catholic. Home: 3287 Kings Point Ct Island Lake IL 60042

ODDERSHEDE, JENS NORGAARD, chemistry educator; b. Hillerslev, Thy, Denmark, Aug. 19, 1945; s. Harry Thomas and Marie Kirstine (Norgaard) O.; m. Bente Broeng, Dec. 29, 1967; children: Lene, Poul. Grad., Aarhus (Denmark) U., 1970, DSc, 1978. Postdoctoral fellow Aarhus U., 1970-71, sr. postdoctoral fellow, 1973-77; postdoctoral fellow U. Utah, Salt Lake City, 1971-73; lectr. Odense (Denmark) U., 1977-88, prof. chemistry, 1988—; dean sci. Odense U., Denmark, 1992—. Author: Problems in Quantum Chemistry, 1983. Recipient Rsch. prize Fyens Stiftstidende, 1981. Mem. Am. Phys. Soc., Danish Phys. Soc. (div. atomic physics chmn. 1987-89), Danish Chem. Soc. (bd. dirs. 1989-92, theoretical chemistry div. chmn. 1981-85), Danish Natural Sci. Acad. Home: Fyrrehøjen 7, DK-5330 Munkebo Denmark Office: Odense Univ Dept Chemistry, Campusvej 55, DK-5230 Odense Denmark

ODELL, WILLIAM DOUGLAS, physician, scientist, educator; b. Oakland, Calif., June 11, 1929; s. Ernest A. and Emma L. (Mayer) O.; m. Margaret F. Reilly, Aug. 19, 1950; children: Michael, Timothy, John D., Debbie, Charles. AB, U. Calif., Berkeley, 1952; MD, MS in Physiology, U. Chgo., 1956; PhD in Biochemistry and Physiology, George Washington U., 1965. Intern, resident, chief resident in medicine U. Wash., 1956-60, postdoctoral fellow in endocrinology and metabolism, 1957-58; sr. investigator Nat. Cancer Inst., Bethesda, Md., 1960-65; chief endocrine service NICHD, 1965-66; chief endocrinology Harbor-UCLA Med. Center, Torrance, Calif., 1966-72; chmn. dept. medicine Harbor-UCLA Med. Center, 1972-79; vis. prof. medicine Auckland Sch. Medicine, New Zealand, 1979-80; prof. medicine and physiology, chmn. dept. medicine U. Utah Sch. Medicine, Salt Lake City, 1980—. Mem. editorial bds. med. jours.; author 6 books in field; contbr. over 300 artilces to med. jours. Served with USPHS, 1960-66. Recipient Disting. Svc. award U. Chgo., 1973, Pharmacia award for outstanding contbns. to clin. chemistry, 1977, Gov.'s award State of Utah Sci. and Tech., 1988, also rsch. awards, Mastership award ACP, 1987. Mem. Am. Soc. Clin. Investigation, Am. Physiol. Soc., Assn. Am. Physicians, Am. Soc. Andrology (pres.), Endocrine Soc. (v.p., Robert Williams award 1991), Soc. Study of Reprodn. (bd. dirs.), Pacific Coast Fertility Soc. (pres.), Western Assn. Physicians (pres.), Western Soc. Clin. Rsch., Soc. Pediatric Rsch., Alpha Omega Alpha. Office: U of Utah Med Center 50 Medical Dr Salt Lake City UT 84132-0002

ODEN, JOHN TINSLEY, engineering mechanics educator, consultant; b. Alexandria, La., Dec. 25, 1936; s. John James and Sara Elizabeth (Lyles) O.; m. Barbara Clare Smith, Mar. 19, 1965; children: John Walker, Elizabeth Lee. BS, La. State U., 1959; MS, Okla. State U., 1960, PhD, 1962; doctorate in sci. (hon.), Tech. U. Lisbon, Portugal, 1986. Registered profl. engr., Tex., La. Teaching asst. La. State U., Baton Rouge, 1959; asst. prof. Okla. State U., Stillwater, 1961-63; sr. structures engr. Gen. Dynamics, Fort Worth, 1963-64; prof., head dept. engring. mechanics U. Ala., Huntsville, 1964-73; prof. U. Tex., Austin, 1973—; Carol and Henry Groppe prof. engring. U. Tex., Ernest and Virginia Cockrell chair in engring., 1987; prof. Coope U. Fed., Brazil, 1974; dir. Tex. Inst. Computational Mechanics, Austin, 1974-93, Tex. Inst. Computational and Applied Math., Austin, 1993—; Sci. Rsch. Coun. vis. scholar Brunel U., Eng., 1981; mem. com. on computational mechanics NRC; mem. U.S. Nat. Com. on Theoretical and Applied Mechanics. Author, editor 38 books; contbr. numerous articles to profl. jours. Decorated medal Order de Palms Admique (France); recipient Rsch. award Southeastern Conf. on Theoretical and Applied Mechanics, 1978, Lohmann medal Okla. State U., 1991. Fellow ASCE (Outstanding Svc. award 1968, Walter Huber Rsch. award 1973, Theodore Von Karman medal 1992), ASME (Worcester Reed Warner medal 1990), Am. Acad. Mechanics (pres.), U.S. Nat. Acad. Engring.; mem. Soc. Engring. Sci. (pres. 1978, Eringen medal 1991), Soc. Indsl. and Applied Math., Internat. Assn. Computational Mechanics (pres.), U.S. Assn. Computational Mechanics (pres. 1990-92, John Von Neumann medal 1993), Soc. Natural Philosophy, Nat. Acad. Engring. Mex. Home: 7403 W Rim Dr Austin TX 78731-2044 Office: ASE-EM Dept U Tex WRW Bldg 305A Austin TX 78712

ODER, ROBIN ROY, physicist; b. Jefferson City, Tenn., Sept. 26, 1934; s. Charles R.L. and Nell (Bramham) O.; m. Peggy Powers, Aug. 3, 1956 (div. 1971); m. Marcia Randlett, Oct. 2, 1971; children: Karin Oder Colin, Richard P. BS, MIT, 1959, PhD, 1965. Rsch. physicist MIT Nat. Magnet Lab., Cambridge, 1965-70; group leader rsch. J.M. Huber Corp., Macon, Ga., 1970-74; engring. specialist Bechtel Corp., San Francisco, 1974-77; dir. coal tech. Gulf Oil Corp., Pitts., 1977-82; pres. Exportech Co., Inc., New Kensington, Pa., 1982—; tech. adv. com. EPRI Coal Cleaning Test Faculty, 1981-82; program com. Ill. Ctr. for Rsch. on Sulfur in Coal, Champaign, 1986-88. Contbr. over 100 articles to profl. jours. Kennecott Copper fellow in physics, 1961; Case Western Res. U. Centennial scholar, 1980. Mem. Am. Phys. Soc., Am. Chem. Soc., Am. Inst. Chem. Engrs. (local officer 1983-91). Achievements include 12 patents. Office: Exportech Co Inc PO Box 588 New Kensington PA 15068

ODEYALE, CHARLES OLAJIDE, biomedical engineer; b. Ile-Ife, Nigeria, Dec. 16, 1950; came to U.S., 1972; s. Samuel Olagunju and Clara Olatundum (Aina) O.; m. Claudette Renee Wright, Dec. 20, 1986; 1 child, Elani Joy. BSc, Bowie State U., 1976; MS, U. N.Mex., 1980; PhD, Walden U., 1993. Rsch. asst. Washington U., St. Louis, 1981; Rsch. asst. Naval Med. Rsch. Inst., Bethesda, Md., 1985-87, biomed. engr., 1987—; pres., founder Knowledge Engring. & Rsch., Inc., 1993—. Contbr. articles to profl. jours. With USN, 1982-85. Mem. IEEE, Computer Soc., Engring. Mgmt. Soc. Achievements include system and method for quantitation of macrophages phagocytosis (patent), knowledge representation and inference strategies for effective military biomedical R&D management; artificial neural networks and expert systems synergy in knowlege based systems; design & development of implantable coated wire clucose sensor for an artificial beta cell. Office: Naval Med Rsch Inst 8901 Wisconsin Ave Bethesda MD 20889

ODINK, DEBRA ALIDA, chemist, researcher; b. Milw., Nov. 2, 1963; d. Harry Jean and Betty Ann (Agnew) O. BS in Chemistry, Calif. State U. Stanislaus, Turlock, 1986; PhD, U. Calif., Davis, 1991. Postdoctoral researcher Inst. des Matériaux, Nantes, Frances, 1991—. Contbr. articles to Jour. Am. Chem. Soc., Chemistry of Materials, Jour. Power Sources. Mem. Am. Chem. Soc., Materials Rsch. Soc.

ODMAN, MEHMET TALAT, mechanical engineer, researcher; b. Uskudar, Turkey, July 20, 1960; came to U.S., 1983; s. Mustafa Nejat and Zeliha Mesadet (Saltug) O.; m. Deniz Ozuslu, May 26, 1992. BS, Bogazici U., Istanbul, Turkey, 1983; MS, Fla. Inst. Tech., 1986; PhD, Carnegie Mellon U., 1992. Teaching asst. Fla. Inst. Tech., Melbourne, 1983-86, Carnegie Mellon U., Pitts., 1988-92; rsch. scientist N.C. Supercomputing Ctr., Research Triangle Park, N.C., 1992—. Contbr. articles to Atmospheric Environ., JAWMA, Jour. Geophys. Rsch. Mem. Am. Geophys. Union, Sigma Xi. Islam. Office: NC Supercomputing Ctr PO Box 12889 Research Triangle Park NC 27709

O'DONNELL, BRENDAN JAMES, naval officer; b. Detroit, Sept. 7, 1949; s. John James and Margaret Mary (Shupe) O'D.; m. Barbara Jean Frinder, June 12, 1971; children: Brendan Neil, Sean Michael, Daniel James. AB cum laude in French and Russian, Holy Cross Coll., 1971; MS with distinction in Aero. Engring., Naval Postgrad. Sch., 1978; student, Nat. War Coll., 1989-90. Commd. USN, 1971, advanced through grades to capt.; standardization officer Patrol Squadron 46, Mountain View, Calif., 1973-76; anti-sub. warfare officer USS John F. Kennedy, Norfolk, Va., 1978-80; evaluation dept. head Air Test and Evaluation Squadron One, Patuxent River, Md., 1980-82; maintenance/safety officer Patrol Squadron Eight, Brunswick, Maine, 1983-85; staff officer commdr. Patrol Wings Atlantic, Brunswick, 1985-87; exec. officer Tng. Squadron Two, Milton, Fla., 1987-88, commanding officer, 1988-89; div. chief Joint Staff C4 Directorate, Washington, 1990—. Asst. scoutmaster Boy Scouts Am. Troop 1544, Burke, VA., 1989—, treas., 1991—; den leader, treas. Cub Scouts Am. Pack 648, Brunswick, 1985-87. Mem. AIAA, Armed Forces Communications and Electronics Assn., Assn. Naval Aviation, U.S. Naval Inst. Roman Catholic. Office: Joint Staff J6E The Pentagon Washington DC 20318-6000

O'DONNELL, MATTHEW, electrical engineering, computer science educator; b. Bronxville, N.Y., Dec. 18, 1950; s. Hubert Bernard and Loretto Anne (Schmidt) O'D.; m. Catharine Gleason, Jan. 3, 1976; children: Brendan Gleason, Sean Gleason. BS in Physics with honors, U. Notre Dame, 1972, PhD in Solid State Physics, 1976. Grad. rsch. and teaching fellow in physics U. Notre Dame, Ind., 1972-76; postdoctoral fellow in physics Washington U., St. Louis, 1976-78, sr. rsch. assoc. in physics, rsch. instr. medicine, 1978-80; physicist Rsch. and Devel. Ctr., GE Co., Schenectady, N.Y., 1980-90; prof. elec. engring. and computer sci. U. Mich., Ann Arbor, 1990—; rsch. fellow elec. engring. Yale U., New Haven, Conn., 1984-85; presenter in field. Editorial bd.: Ultrasonic Imaging; contbr. articles to Phys. Rev. Letters, Solid State Communications, Jour. Acoustical Soc. Am., Ultrasound in Medicine, Am. Jour. Physiology, Circulation Rsch., Jour. Applied Physics, Am. Jour. Cardiology, Ultrasonic Imaging, others. Fellow IEEE; mem. Am. Phys. Soc., Sigma Xi. Achievements include 29 patents incuding Collimation of Ultrasonic Linear Array Transducer, Method for Imaging Blood Flow Using Multiple-Echo, Phase-Contrast NMR, Method and Apparatus for Fully Digital Beam Formation in a Phased Array Ultrasonic. Office: EECS Dept Univ Mich Ann Arbor MI 48109-2122

O'DONNELL, SEAN, zoologist; b. Abbington Twp., Pa., Nov. 20, 1964; s. John Elwood and Barbara Ann (Kelley) O'D. BS in Biology cum laude, St. Joseph's U., Phila., 1986; MS in Entomology, U. Wis., 1989, PhD in Zoology, Entomology, 1993. Analytical technician gas chromatography/mass spectroscopy Uniroyal Inc., Naugatuck, Conn., 1984-86; teaching asst. Gen. Biology Lab., Coral Gables, Fla., 1986-87; rsch. asst. NSF Grant, Madison, Wis., 1987-92, Hatch Grant, Madison, 1992; teaching asst. gen. entomology U. Wis., Madison, 1989, teaching asst. BIOCORE biology honors program, 1992; mem. resource faculty Orgn. Tropical Studies, Republic of Costa Rica, summer 1992; field trip leader The Nature Conservancy, Madison, 1991. Recipient Pilot Rsch. award Orgn. Tropical Studies, Republic of Costa Rica, 1988, Orgn. Tropical fellow, 1991, Villas Fellowship U. Wis. Grad. Sch., Republic of Venezuela, 1990, Short-term fellowship Smithsonian Tropical Rsch. Inst., Republic of Panama, 1990. Mem. Animal Behavior Soc., Entomol. Soc. Am., Internat. Union Study of Social Insects, Am. Entomol. Soc., Internat. Behavioral Ecology Soc. Achievements include research (published in refereed jours.) in physiol. entomology, animal behavior, insectes sociaux, insect behavior and behavioral ecology and sociobiology. Office: U Calif Davis Dept Entomology Davis CA 95616

O'DONNELL, TERESA HOHOL, software development engineer, antennas engineer; b. Springfield, Mass., Nov. 25, 1963; d. Marion Henry and Lena Ann (Zajchowski) Hohol. BS in Computer Engring., MIT, 1985, BSEE, 1985, MSEE, MS in Computer Sci., 1986. Rsch. asst. MIT Rsch. Lab for Electronics, Cambridge, 1985-86; lead VHSIC insertion engr. USAF Electronic Systems Div., Hanscom AFB, Mass., 1986-88; intelligent antennas engr. USAF Rome Lab., Hanscom AFB, Mass., 1988-91; software devel. engr. Arcon Corp., Waltham, Mass., 1991—. Composer: (choral mass setting) Mass of Rejoicing, 1989. Performer Zbeide's Harem, Tewksbury, Mass., 1986—; organist/composer St. Theresa's Choir, Billerica, Mass., 1987—. Capt. USAF, 1986-91, selective svc. officer, 1992—. Decorated 2 Commendation medals, Nat. Def. medal; recipient 2 Air Force Orgnl. Excellence awards USAF Rome Lab., 1989, 91. Mem. IEEE, Nat. Assn. Pastoral Musicians, Am. Guild Organists, Assn. Computing Machinery, Air Force Assn., Eta Kappa Nu (v.p., 1985-86), Sigma Xi Rsch. Soc. Roman Catholic. Avocations: music, dancing, equistrianship, composing, roller skating. Office: Arcon Corp 260 Bear Hill Rd Waltham MA 02154-1018

O'DONNELL, WILLIAM JAMES, engineering executive; b. Pitts., June 19, 1935; s. William James and Elizabeth (Rau) O'D.; m. Joanne Mary Kusen, Jan. 31, 1959; children—Suzanne, Janice, William, Thomas, Kerry, Amy. B.S.M.E., Carnegie Inst. Tech., 1957; M.S.M.E., U. Pitts., 1959, Ph.D., 1962. Jr. engr. Westinghouse Research Lab., 1957-58, asso. engr., 1958; with Westinghouse Bettis Atomic Power Lab., West Mifflin, Pa., 1961-70, adv. engr., 1966-70; pres., chmn. bd. O'Donnell & Assos., Inc., Pitts., 1970—. Contbr. numerous articles on engring. and mechanics to profl. jours.; holder patents on processes and devices. Served with C.E. AUS, 1963-64. Recipient Machinery's Achievement award as outstanding mech. designer, 1957, Pi Tau Sigma Gold medal for achievements in engring., 1967. Fellow ASME (Nat. award for outstanding contbn. to engring. profession 1973, Internat. award for best publ. in pressure vessels and piping 1988, Engr. of Yr. 1988); mem. AAAS, ASTM, Soc. Exptl. Stress Analysis, Am. Soc. Metals, Am. Nuclear Soc., Nat. Soc. Profl. Engrs., Sigma Xi, Tau Beta Pi, Pi Tau Sigma. Home: 3611 Maplevue Dr Bethel Park PA 15102-1423 Office: O'Donnell & Assoc 241 Curry Hollow Rd Pittsburgh PA 15236-4696

O'DOR, RON, physiologist, marine biologist; b. Kansas City, Mo., Sept. 20, 1944; s. Claude Marvin O'Dor and Opal LaMoyne (Sears) Mathes; m. Janet Ruth Spiller, Dec. 30, 1967; children: Matthew Arnold, Stephen Roderick. Student, El Camino Coll., 1963-65; AB, U. Calif., Berkeley, 1967; PhD, U. B.C., 1971. From asst. prof. to prof. Dalhousie U., Halifax, N.S., 1973—; dir. Aquatron Lab., Halifax, 1986—; summer scientist Laboratoire Arago, Banyuls-sur-Mer, France, 1979-85; vis. scientist Pacific Biol Sta., Nanaimo, B.C., 1980, U. Papua New Guinea, Motupore Island, 1989-91; vis. prof. U. B.C., Vancouver, 1986-87; seasonal prof. Bamfield (B.C.) Marine Sta., 1987; cons. UN, Rome, 1987—; adv. space stas. Can. Space Agy., Ottawa, 1992—; mgmt. com. Ocean Prodn. Enhancement Network, Halifax, 1993—. Author, editor: Cephalopod Fishery Biology, 1993; contbg. author Cephalopod Life Cycles, Vol. I, 1983, Vol. II, 1987; contbr. articles, reviews to profl. jours. Scholar Med. Rsch. Coun. Can., 1968; fellow MRC, Cambridge U., Eng., 1971-73, Stazione Zoologica, Naples, Italy, 1971-73; recipient Hon. Mention Rolex Awards Enterprise, 1987. Mem. Can. Soc. Zoologists (councillor 1989-92), Can. Fedn. Biological Socs.,Am. Soc. Zoologists, Am. Soc. Gravitational and Space Biology, Cephalopod Internat. Adv. coun. (councillor, pres. 1988-91), Phi Beta Kappa. Achievements include co-development of acoustic pressure transducer/transmitters and radio-linked tracking systems for monitoring cephalopod bioenergetics in nature, co-organization of expeditions to the Azores and Papua New Guinea to monitor squid and nautilus, first international cephalopod research conference in Japan, investigation for Aquatic Research Facility for Space Shuttle. Home: 1181 South Park St, Halifax, NS Canada B3H 2W9 Office: Dalhousine University, Biology Dept, Halifax, NS Canada B3H 4J1

ODREY, NICHOLAS GERALD, industrial engineer, educator; b. Kittanning, Pa., Mar. 22, 1942; s. Nicholas and Katherine (Ogrodowczyk) O.; m. Saundra Lee Sidora, Sept. 18, 1971. BS, Pa. State U., 1964, MS in Aerospace Engring., 1968, PhD in Indsl. Engring., 1978. Devel. engr. Goodyear Aerospace Corp., Akron, 1967-70; rsch. asst. Pa. State U., University Park, 1971-76; asst. prof. U. R.I., Kingston, 1976-81; assoc. prof. W.Va. U., Morgantown, 1981-83; assoc. prof. Lehigh U., Bethlehem, Pa., 1983-91, dir. Inst. Robotics, 1987-91, prof. indsl. engring., 1991—. Co-author: Industrial Robotics: Technology, Programming and Applications, 1986; contbr. chpts. to engring. books; contbr. articles, reports, monographs to profl. publs.; editorial bd. Internat. Jour. Flexible Automation and Integrated Mfg., 1992—. Grad. trainee NSF, 1964; faculty fellow Air Force Office Sci. Rsch., 1979; grantee NSF, Nat. Inst. for Standards and Technology, others. Mem. ASME, Soc. Mfg. Engrs. (sr.), Robotics Internat. of Soc. Mfg. Engrs. (bd. dirs. 1987-89, chair R&D div. 1985-89, cert. mfg. engr.), Inst. Indsl. Engrs. (pres. Lehigh Valley chpt. 77 1986-87, award of excellence 1987), Sigma Xi, Alpha Pi Mu. Home: 1665 Pleasant Dr Bethlehem PA 18015 Office: Lehigh U 200 W Packer Ave Bethlehem PA 18015

O'DRISCOLL, JEREMIAH JOSEPH, JR., safety engineer, consultant; b. Galveston, Tex., Dec. 8, 1925; s. Jeremiah Joseph and Nate Alice (Jones) O'D.; m. Marie McCormack, Feb. 7, 1953; children: Maureen, Sharon, Maria, Jeremiah Joseph III, Kathleen, Paul, Jonathan. BBA, U. Tex., 1950; student, Southwestern La. Inst., 1943-44. Registered profl. engr., Calif. Mgr. safety and security Catalyst Rsch. Inc., Balt., 1955-58; mgr. safety Atlas Powder Co., Wilmington, Del., 1959-65; corp. mgr. safety Atlas Chem. Industries, Wilmington, 1965-69; supt. safety So. Ry. System, Atlanta, 1969-72, mgr. safety planning, 1972-75, dir. safety and hazardous materials, 1975-82, dir. hazardous materials and loss prevention, 1982-87; cons. Jody Inc. and Assocs., Atlanta, 1987—; steering com. Bur. Explosives; panelist Bd. Radioactive Waste Mgmt. for Waste Internment Pilot Plant, Nat. Rsch. Coun. Author: Emergency Action Plan, 1969, (guides) Emergency Action, Hazardous Materials, 1975, 83, 88, 90; contbr. to profl. publs. Lt. USNR, 1943-47, 50-54. Recipient George L. Wilson award Hazardous Materials Adv. Coun., 1988. Mem. Assn. Am. R.R.'s (emeritus). Republican. Roman Catholic. Home and Office: 505 Valley Hall Dr Atlanta GA 30350-4632

ODUM, JEFFERY NEAL, mechanical engineer; b. Bristol, Tenn., Sept. 11, 1956; s. Herschel S. and Minnie Lee (Carrier) O.; m. Stacy Elaine Ferrell, mar. 18, 1989; 1 child, Charles Wesley Ferrell. BSME, Tenn. Technol. U., 1978; MS in Engring., U. Tenn., 1983. Sr. project engr. TVA, Knoxville, 1978-81; sr. constrn. engr. Stone & Webster Engring. Corp., Boston, 1981-84; div. engr. E.I. DuPont de Nemours & Co., Aiken, S.C., 1984-89, engring. mgr. Flour Daniel, Greenville, S.C., 1989-92; mgr. of projects, Pharmaceutical Bus. Group CRS Sirrine Engrs., Inc., Raleigh, N.C., 1992-93; sr. project mgr. Gilbaje Bldg. Co., Raleigh, 1993—. Contbr. articles to profl. jours. Vol. Spl. Olympics. Recipient DuPont Engring. Achievement award 1986, 88, 89, Nat. Svc. Alumni award Univ. Tenn. Mem. Soc. Mfg. Engrs., Internat. Soc. Pharm. Engrs. (bd. dirs. Carolina Chpt.), U. Tenn. Nat. Alumni Assn. (pres. Augusta chpt. 1987-89), Order of Engr., Kappa Sigma. Republican. Presbyterian. Avocations: sports, biking, cooking, physical fitness, writing. Office: Gilbaje Bldg Co 4815 Emperor Blvd Ste 230 Morrisville NC 27560

OEHLERT, WILLIAM HERBERT, JR., cardiologist, educator; b. Murphysboro, Ill., Sept. 11, 1942; s. William Herbert Sr. and Geneva Mae (Roberts) O.; m. L. Keith Brown, Mar. 14, 1976; children: Emily Jane, Amanda Elizabeth. BA, So. Ill. U., 1967; MD, Washington U., St. Louis, 1967. Diplomate Nat. Bd. Med. Examiners, Am. Bd. Internal Medicine, Am. Bd. Cardiovascular Disease, North Am. Soc. Pacing and Electrophysiology. Intern Union Meml. Hosp., Balt., 1967-68, resident, 1968-69; resident U. Iowa, Iowa City, 1969-70, cardiology fellow, 1970-72; asst. prof. medicine, dir. coronary care units U. Okla. Health Sci. Ctr., Oklahoma City, 1972-74, asst. clin. prof. medicine, 1974-82, assoc. clin. prof. medicine, 1982-88, clin. prof. medicine, 1988—; chmn. dept. cardiology Baptist Med. Ctr., 1992—; pres. Cardiovascular Clinic, Oklahoma City, 1987-91, chmn. exec. com. 1987-91; pres., med. dir. Cardiovascular Imaging Svcs. Corp., Oklahoma City, 1987-92; v.p. Plaza Med. Group, 1992—. Author: Arrhythmias, 1973, Cardiovascular Drugs, 1976; contbr. articles to profl. jours. Fellow Am. Heart Assn. (bd. dirs. 1974-88, nat. program com. 1979-82, pres. Okla. affiliate 1985-86, ACLS nat. affiliate faculty 1987-90), Am. Coll. Cardiology; mem. AMA, Nat. Assn. Residents and Interns, Am. Soc. Internal Medicine, Am. Coll. Physicians, Okla. County Med. Assn. (chmn. quality of care com. 1990-91), Okla. State Med. Assn., Okla. City Clin. Soc., Okla. Cardiac Soc. (pres. 1978-79), Osler Soc., Soc. Nuclear Medicine, Wilderness Med. Soc., Stewart Wolf Soc., Phi Eta Sigma, Phi Kappa Phi. Home: 3017 Rockridge Pl Oklahoma City OK 73120-5713 Office: Cardiovascular Clinic 3433 NW 56th St Ste 400 Oklahoma City OK 73112-4450

OEHME, FREDERICK WOLFGANG, medical researcher and educator; b. Leipzig, Germany, Oct. 14, 1933; came to U.S., 1934; s. Friedrich Oswald and Frieda Betha (Wohlgamuth) O.; m. Nancy Beth MacAdam, Aug. 6,

1960 (div. June 1981); children: Stephen Frederick, Susan Lynn, Deborah Ann, Heidi Beth; m. Pamela Sheryl Ford, Oct. 2, 1981; 1 child, April Virginia. BS in Biol. Sci., Cornell U., 1957, DVM, 1958; MS in Toxicology and Medicine, Kans. State U., 1962; DMV in Pathology, Justus Liebig U., Giessen, Germany, 1964; PhD in Toxicology, U. Mo., 1969. Diplomate Am. Bd. Toxicology, Am. Bd. Vet. Toxicology, Acad. Toxicological Scis. Resident intern, Large Animal and Ambulatory Clinic Cornell U., 1957-58; gen. practice vet. medicine, 1958-59; from asst. to assoc. prof. medicine Coll. Vet. Medicine Kans. State U., 1959-66, 69-73, dir. comparative toxicology labs., 1969—, prof. toxicology, medicine and physiology Coll. Vet. Medicine, 1974—; postdoctoral research fellow in toxicology, NIH U. Mo., 1966-69; cons. FDA, Washington, Ctr. for Vet. Medicine, Rockville, Md., Animal Care com., U. Kans., Lawrence, 1969-76, Syntex Corp., Palo Alto, Calif., 1976-77; mem. sci. adv. panel on PBB Gov.'s Office, State of Mich., 1976-77, Council for Agrl. Sci. and Tech. Task Force on Toxicity, Toxicology and Environ. Hazard, 1976-83; cons., mem. adv. group on pesticides EPA, Cin., 1977—; expert state and fed. witness; advisor World Health Orgn., Geneva, Switzerland; presenter over 500 papers to profl. meetings; numerous other activities. Author over 600 books and articles on toxicology and vet. medicine; editor, pub. Vet. Toxicology, 1970-76, Vet. and Human Toxicology, 1977—; assoc. editor Toxicology Letters, mem. editorial bd. Am. Jour. Vet. Research, 1975-83, Toxicology, 1979-85, Clin. Toxicology, Jour. de Toxicologie Medicale, Toxicology and Indls. Health, Poisindex, Jour. Analytical Toxicology, Companion Animal Practice; reviewer Toxicology and Applied Pharmocology, Jour. Agrl. and Food Chemistry, Spectroscopy, numerous others. Mem. council Luth. Ch. Am., sr. choir, numerous coms., adv. council Cub Scouts Am., Eagle Scouts, mgr., coach Little League Baseball; council rep., treas. area council, various coms. PTA; mem. Manhattan Civic Theatre; bd. trustees Manhattan Marlin Swim Team; dir. meet Little Apple Invitational Swim Meet, 1984. Morris Animal Found. Project fellow, 1967-69. Fellow Am. Acad. Clin. Toxicology (charter, past pres., numerous coms.), Am. Acad. Vet. and Comparative Toxicology (past sec.-treas., numerous coms.); mem. Soc. Toxicology (past pres., numerous coms.), World Fedn. Clin. Toxicology Ctrs. and Poison Control Ctrs. (past pres.), Soc. Toxicologic Pathologists, N.Y. Acad. Scis., Am. Vet. Med. Assn. (com. on environmentology 1971-73, adv. com. council on biol. and therapeutic agts. 1971-74, various others), Nat. Ctr. Toxicological Rsch. (vet. toxicology rep. sci. adv. bd.), Nat. Rsch. Coun. (subcom. on organic contaminants in drinking water, safe drinking water com., adv. ctr. on toxicology assembly life scis. 1976-77, panel on toxicology marine bd., assembly of engring. 1976-79, com. on vet. med. scis. assembly life scis. 1976-78), Nat. Ctr. for Toxicological Rsch. (grad. edn. subcom., sci. adv. bd. 1974-77), Cornell U. Athletic Assn, Omega Tau Sigma, Phi Zeta, Sigma Xi, numerous others. Republican. Clubs: Cornell U. Crew; Manhattan Square Dance. Avocations: hist. readings, sci. writings, nature tours and walks, travel. Home: 148 S Dartmouth Dr Manhattan KS 66502-3079 Office: Kans State U Comparative Toxicology Labs Manhattan KS 66506-5606

OELBERG, DAVID GEORGE, neonatologist, educator, researcher; b. Waukon, Iowa, May 26, 1952; s. George Robert and Elizabeth Abigail (Kepler) O.; m. Debra Penuel, Aug. 4, 1979; 1 child, Benjamin George. BS with highest honors, Coll. William and Mary, 1974; MD, U. Md., 1978. Diplomate Am. Bd. Pediatrics, Am. Bd. Neonatal-Perinatal Medicine. Intern U. Tex. Med. Br., Galveston, 1978-79, resident, 1979-81, pediatric house staff, 1978-81; postdoctoral fellow in neonatal medicine U. Tex. Med. Sch., Houston, 1981-84, asst. prof. dept. pediatrics, 1984-90, assoc. prof., 1990—; mem. hosp. staff Hermann Hosp., Houston, 1983—; physician Crippled Children's Services Program, Houston, 1985—; mem. hosp. staff Lyndon B. Johnson County Hosp., 1990—; visiting prof. Wyeth-Ayerst Labs., 1992. Mem. editorial adv. bd. jour. Neonatal Intensive Care; contbr. articles to profl. jours. Physician cons. Parents of Victims of Sudden Infant Death Syndrome, Houston, 1984. Recipient award in analytical chemistry Am. Chem. Soc., 1974, NIH Clin. Investigator award Nat. Heart, Lung and Blood Inst., 1989—; rsch. grantee Am. Lung Assn., 1989-90, NIH, 1989—. Fellow Am. Acad. Pediatrics; mem. AMA, NAS, Soc. Exptl. Biology and Medicine, So. Soc. Pediatric Research, Harris County Med. Soc., Tex. Pediatric Soc. (fetus and newborn com.), Tex. Med. Assn., Houston Pediatric Soc., Soc. for Pediatric Rsch. Patent pending method for optical measurement of bilirubin in tissue. Avocations: sailing, gardening. Home: 1624 W Little Neck Rd Virginia Beach VA 23452 Office: Ea Va Med Sch Ctr Pediatric Rsch 855 W Brambleton Ave Norfolk VA 23510

OELCK, MICHAEL M., plant geneticist, researcher; b. Münster, Fed. Republic of Germany, July 26, 1954; s. Max and Ursula (Swiderski) O.; m. Elke Steinmann, Oct. 15, 1983; children: Fabian, Alexandra, Florian, Victoria. Diploma in Plant Genetics and Agronomy, U. Agr., Bonn., Fed. Republic of Germany, 1981; PhD in Molecular Genetics, U. Cologne, Fed. Republic of Germany, 1984. Leader rsch. lab. Hoechst AG, Frankfurt, Fed. Republic of Germany, 1984-88; vis. scientist Agriculture Can., Ottawa, Ont., 1988-90; group leader biotech. Hoechst Can., Saskatoon, Sask., Can., 1990—. Contbr. articles to Jour. Plant Breeding, Jour. Plant Physiology, Biotech., Internat. Rapeseed Congress. Emergency ambulance officer Malteser Hilfsdienst, Kelkheim, Germany, 1987-88. Grad. grantee Max Planck Inst., Cologne, 1980-83. Mem. Deutsche Landwirtschaftsgesellschaft, Verein für Angestellte Akademiker, Deutsche Gesellschaft für Angewandte Botanik. Achievements include mesophyll protoplast regeneration of numerous plant species; discovery of origin of somatic embryogenesis triggered by glufosinate-ammonium; genetic engineering of Canadian canola for herbicide tolerance. Office: Hoechst Can Inc, 106 Research Dr, Saskatoon, SK Canada S7N 0W9

OELLERICH, MICHAEL, chemistry educator, chemical pathologist; b. Heidelberg, Germany, July 15, 1944; s. Friedrich W. and Roswitha (Kamner) O.; m. Pushpa Singh, Mar. 12, 1976; children: Mark, Thomas, Diana. MD, U. Heidelberg, 1970; Habilitation, Med. U. Hannover, 1978. Intern U. Heidelberg, 1971-72; resident dept. internal medicine U. Düsseldorf, Fed. Republic of Germany, 1973-75; resident Inst. for Clin. Chemistry, Med. U. Hannover, Fed. Republic of Germany, 1972-73, 75-79, head physician, 1979-82, dep. chmn., 1982-91; prof. clin. chemistry U. Hannover/Göttingen, Fed. Republic of Germany, 1982—; dir. dept. clin. chemistry, ctrl. lab. Ctr. Internal Medicine U. Göttingen, Fed. Republic Germany, 1991—; Mem. DFG Commn. for Clin. Toxicol. Analysis, 1984-89. Contbr. over 170 articles on hypoglycemia inducing substances, liver function tests in clin. transplantation, therapeutic drug monitoring, and non-isotopic immunoassay techniques to sci. jours., chpts. to books; mem. adv. bd. Jour. Clin. Chemistry and Clin. Biochemistry, 1985-88; mem. editorial bd. Therapeutic Drug Monitoring, 1987—; inventor liver function test, hypoglycemic hydrazonopropionic acid. Recipient Ludolf-Krehl prize S.W. German Soc. for Internal Medicine, 1971. Mem. German Soc. for Clin. Chemistry (bd. dirs. 1984-86). Lutheran. Achievements include invention of hypoglycemic 2-(phenylethylhydrazono)-propionic acid and of the MEGX liver function test. Office: Georg-August U Dept Clin Chemistry, Robert-Koch-Str 40, D-37070 Göttingen Germany

OELOFSE, JAN HARM, game rancher, wildlife management consultant; b. Burgersdorp, Cape Province, South Africa, July 12, 1934; s. Andries and Johanna (Vorster) O. Diploma in agriculture, Agrl. Coll., Cradock Cape, South Africa, 1952. Animal and trapping staff Tanganykia Game Ltd., Arusha Tanganykia, 1954-63; sr. game warden game translocation Natal Parks Bd., Hlahluwe Gamereserve, Natal, South Africa, 1964-72; game rancher Farm Okonjati, Kalkfeld, Namibia, South Africa, 1974—; mem. Wilderness Leadership Sch., Durban, Natal, 1967—; mem. tourist bd. Namibia Adminstrn., Windhoeck, 1984—. Discoverer technique to capture wild animals with plastic material; patentee in field. Named Most Outstanding Hunter of Yr., 1982; recipient Internat. Order of Merit, 1990. Mem. Safari Club Internat., Internat. Profl. Hunters Assn. Lodge: Etosha Otjiwarongo. Avocations: sculpture; wildlife photography. Home and Office: Mount Etjo Safari Lodge, PO Box 81, Kalkfeld 9000, Namibia

OEMLER, AUGUSTUS, JR., astronomy educator; b. Savannah, Ga., Aug. 15, 1945; s. Augustus and Isabelle Redding (Clarke) O.; children: W. Clarke, Bryan S. AB, Princeton U., 1969; MS, Calif. Inst. Tech., 1970, PhD, 1974. Postdoctoral assoc. Kitt Peak Nat. Obs., Tucson, 1974-75; instr. astronomy Yale U., New Haven, 1975-77, asst. prof., 1977-79, assoc. prof., 1979-83, prof., 1983—, chmn. dept., 1988—. Contbr. articles to profl. jours. Alfred P. Sloan fellow, 1978-80. Mem. Am. Astronom. Soc., Internat. Astronom. Union. Republican. Roman Catholic. Home: 93 Water St Guilford CT 06437-2863 Office: Yale U Obs 260 Whitney Ave New Haven CT 06511-3748

OERTEL, GOETZ K. H., physicist, professional association administrator; b. Stuhm, Germany, Aug. 24, 1934; came to U.S., 1957; s. Egon F.K. and Margarete W. (Wittek) O.; m. Brigitte Beckmann, June 17, 1960; children: Ines M.H. Oertel Downing, Carsten K.R. Abitur, Robert Mayer, Heilbronn, Fed. Republic Germany, 1953; vordiplom, U. Kiel, Fed. Republic Germany, 1956; PhD, U. Md., 1963. Aerospace engr. Langley Ctr. NASA, Hampton, Va., 1963-68; chief solar physics NASA, Washington, 1968-75; analyst Office of Mgmt. and Budget, Washington, 1974-75; head astronomy div. NSF, Washington, 1975; dir. def. and civilian nuclear waste programs U.S. Dept. Energy, Washington, 1975-83; acting mgr. sav. river ops. office, Aiken, S.C., 1983-84; dep. mgr. ops. office, Albuquerque, 1984-85; dep. asst. sec. EH, Washington, 1985-86; pres. Assn. Univs. for Rsch. in Astronomy, Inc. (AURA, Inc.), Washington, 1986—, also bd. dirs.; cons. Los Alamos Lab., N.Mex., 1987-92, Westinghouse Electric, 1988—; bd. dirs. AURA, Inc., Inst. for Sci. and Soc., Ellensburg, Wash., IUE Corp. Patentee in field. Fulbright grantee, 1957. Fellow AAAS; mem. Am. Assn. for Advancement of Tech. (bd. dirs.), Am. Phys. Soc., Am. Astron. Soc., Internat. Astron. Union, N.Y. Acad. Scis., Internat. Univ. Exch., Inc. (bd. dirs.), Cosmos Club, Sigma Xi. Lutheran. Avocations: fitness, computing. Home: 9609 Windcroft Way Potomac MD 20854-2864 Office: Assn Univs for Rsch in Astronomy 1625 Massachusetts Ave NW Ste 701 Washington DC 20036-2212

OESTREICHER, MICHAEL CHRISTOPHER, architect; b. Columbus, Ohio, Dec. 24, 1947; s. Robert T. and Jane Ellen (Michaels) O.; m. Jean C. Gowdey, June 13, 1975. M Bus., Ohio U., 1971; MSc, U. Mass., 1986. Owner, dir. Challenges Unltd., Springfield, Mass., 1980—; cons., expert witness lawyers, 1991—. Author, editor newsletter Recreation Access 2000, 1989-92; contbr. articles to profl. jours. Com. mem. Habitat for Humanity, Springfield, 1988—. With U.S. Army, 1967-71. Mem. ASTM (com. mem. 1991—). Achievements include designing over 500 handicapped integrated playgrounds nationwide, Can. and Mideast, designing 5 special items of play for these playgrounds. Office: Challenges Unltd 136 William St Springfield MA 01105

O'FARRELL, TIMOTHY JAMES, clinical psychologist; b. Lancaster, Ohio, Apr. 22, 1946; s. Robert James and Helen Loretta (Tooill) O'F.; BA, U. Notre Dame, 1968; PhD in Psychology, Boston U., 1975; m. Jayne Sara Talmage, May 19, 1973; 1 child, Colin. Instr. Harvard U. Med. Sch., Boston, 1977-82, asst. prof., 1982-86, assoc. prof., 1986—; staff psychologist VA Med. Ctr., Brockton, Mass., 1975-78, dir. Alcoholism Clinic, 1978-83, dir. Counseling for Alcoholics' Marriages Project, 1978—, chief Alcohol and Family Studies Lab., 1981—, assoc. chief psychology svc., 1988—. VA predoctoral grantee, 1969-72; rsch. grantee VA, 1978—, Nat. Inst. on Alcohol Abuse and Alcoholism, 1991—, Smithers Found., 1991—, Guggenheim Found., 1993-94; fellow Behavior Therapy and Research Soc., Mem. NIAAA (psychosocial rsch. rev. group 1989-93, editorial bd. Jour. Family Psychology, editorial bd. Family Dynamics Addition Quarterly 1990-92), Am. Psychol. Assn., Assn. for Advancement Behavior Therapy, Eastern Psychol. Assn. Contbr. articles on marriage and alcoholism to profl. publs. Home: 260 High St Duxbury MA 02332-3406 Office: VA Med Center 116B1 Brockton MA 02401

OFFNER, FRANKLIN FALLER, biophysics educator; b. Chgo., Apr. 8, 1911; s. I.H. and Jennie (Faller) O.; m. Janine Y. Zurcher, Sept. 22, 1956; children—Laurens, Alexandra, Sylvia, Robin. B.Chemistry, Cornell U., 1933; M.S., Cal. Inst. Tech., 1934; Ph.D., U. Chgo., 1938. Pres. Offner Electronics Inc., Chgo., 1939-63; prof. biophysics Northwestern U., 1963—; Laureate in tech. Lincoln Acad. Ill. Author: Electronics for Biologists, 1967, research papers biophysics, electronics. Recipient achievement citation U. Chgo. Fellow IEEE (Centennial medal), Am. Inst. for Med. and Biol. Engring.; mem. NAE, Am. Electroencephalographic Soc., Am. Phys. Soc., Biophys. Soc., Sigma Xi. Achievements include patents in field of electronics, control, and hydraulics; invention of modern electrocardiograph, of electroencephalograph, of operating room monitors, of infrared guided missiles, of electronic fuel controls for jet engines, of phase synchronizer for propeller aircraft, of electronic aids for rehabilitation; research on basic theory of axonal excitation and conduction. Home: 1890 Telegraph Rd Bannockburn Deerfield IL 60015 Office: Northwestern U Technol Inst Northwestern U Evanston IL 60208

O'FLAHERTY, GERALD NOLAN, secondary school educator, consultant; b. Bradley, Ill., Sept. 24, 1933; s. John Nolan and Dorothy Ann (Toohey) O'F.; m. Marilyn Ann McFarland, July 24, 1955; children: Jeffery Nolan, Mark Edward, Heather Maureen. BS in Edn., Ea. Ill. U., 1961; MS in Biol. Scis., St. Mary's Coll., Winona, Minn., 1965; cert. advanced study in biol. scis., No. Ill. U., 1970; cert. advanced study in ednl. adminstrn., Ill. State U., 1986. Cert. secondary tchr., ednl. adminstr., Ill. Chemist Bradley Roper Co., 1954-56; tchr. St. Anne (Ill.) Elem. Sch., 1958-61; tchr. sci. Kankakee (Ill.) Sch. Dist. 111, 1961-63; tchr. sci., chmn. dept., dir. curriculum mgmt. Bradley-Bourbonnais Community High Sch., 1963—; rsch. cons. NSF, Washington, 1962-64, spl. seminar participant, New Orleans, 1982; cons. Ill. Office Edn., Springfield, 1965-71, North Cen. Accrediting Assn., Chgo., 1975—, numerous others; presenter in field. With U.S. Army, 1956-58. Recipient Outstanding Tchr. award Bradley-Bourbonnais Community High Sch., 1986; grantee NSF, 1962-75. Mem. NEA, NSTA (secondary curriculum com. 1982-85, spl. seminar participant 1978), Internat. Tissue Culture Assn., Nat. Assn. Biology Tchrs., Ill. Edn. Assn., Sigma Xi. Avocations: target shooting, photography, reading, writing. Office: Bradley-Bourbonnais Community High Sch 700 W North St Bradley IL 60915-1099

OFOSU, MILDRED DEAN, immunogeneticist; b. Gloster, Miss., Oct. 9, 1946; d. Willie and Leola (Robinson) Huff; m. Gustav Atta Ofosu, Aug. 17, 1968; children: Asua, Asi, Jasper. BS, Alcorn State U., 1967; MS, Tuskegee U., 1969; PhD, Howard U., Washington, 1984. Cert. phlebotomy technologist. Lab. technician Tulane U., New Orleans, 1967; teaching asst. Tuskegee (Ala.) Inst., 1967-69; tchr. Detroit Pub. Schs., 1969-72, Atlanta Pub. Schs., 1972-77; instr. Del. State Coll., Dover, 1977-85, asst. prof., 1985-91, assoc. prof., 1991—, dir. sponsored programs, 1992—, assoc. dean grad. studies & rsch., 1993—; pres. Minority Biomed. Rsch. Program Dirs. Orgn., Bethesda, Md., 1992—; dir. Saturday Acad. Del. State Coll., Dover, 1988—. Author: (with others) Saturday Academic Process for Accessing Math, 4 vols., 1990. Named Outstanding Black Delawarean Umfibuzi-Del. State Coll., Dover, 1986, Outstanding Vol. Del. Tech. and C.C., Dover, 1991; recipient Point of Light award White House, Washington, 1992. Mem. Nat. Phelbotomy Assn., Am. Soc. Hematology, Am. Soc. Human Genetics, Assn. Black Women in Higher Edn., Ministry Biomed. Rsch. Support Program Dir. (pres. 1992—). Achievements include research in clin. immmunology and immunopathology, hematology and dermatology. Home: 347 Mayberry Ln Dover DE 19901

ÖFVERHOLM, STEFAN, electrical engineer; b. Ludvika, Sweden, Nov. 21, 1936; s. Håkan and Ragnhild Gudrun (Andersson) Ö.; m. Eva Guy Tiselius, Sept. 9, 1966 (div. 1971); m. Ulrika Mathilda Marie Skaar, Oct. 10, 1975; children: Harald, Ingegerd. MSEE, Chalmers U. Tech., Gothenburg, Sweden, 1963. Engr. microwave systems devel. Ericsson AB, Mölndal, Sweden, 1963-66; mgr. microwave lab. Trelleborgplast AB, Ljungby, Sweden, 1966-67; mgr. process and prodn. control projects Ericsson AB, Stockholm, Sweden, 1967-71; head computer devel. dept. Ericsson AB, Mölndal, Sweden, 1974-77; head testing methods and tech. Ericsson AB, Stockholm, Sweden, 1977-81; mgr. design and devel. Asea Lme Automation AB, Västerås, Sweden, 1971-74; v.p., gen. mgr. hybrid divsn. Rifa AB, Stockholm, Sweden, 1981-85; gen. mgr. controls divsn. Tour & Andersson AB, Västerhaninge, Sweden, 1985-89; dir. ops. mobile telephone systems divsn. Ericsson Radio Systems AB, Stockholm, 1989-90, v.p., gen. mgr. ops. divsn., 1990-93, v.p. bus. unit mobile telephone systems European stds., 1993—. Avocations: classical music, modern art. Home: Erik Dahlbergsallen 11, S-11520 Stockholm Sweden

OGAARD, LOUIS ADOLPH, environmental administrator, computer consultant; b. Sacramento, Jan. 22, 1947; s. Adolph T. and Borghild (Braafladt) O.; children: Erik Allen, Kirk Anders. BA, St. Olaf Coll., 1968; MA, U. No. Iowa, 1974; PhD, N.D. State U., 1979. Rsch. assoc. dept. agrl. econs. N.D. State U., Fargo, 1976-81; environ. scientist Pub. Svc. Commn., Bismarck, N.D., 1981-84, supr. reclamation div., 1984-85, dir. AML div., 1985—; gen. ptnr. OK Systems, Bismarck, 1987—; adj. prof. Bismarck State Coll., 1985-92; state rep. AML R&D panel Bur. of Mines, Washington, 1987-92. Varsity soccer coach Bismarck High Sch., 1987-92. With USN, 1968-72, Spain. Mem. Sigma Xi (assoc.), Beta Beta Beta. Home: PO Box 7433 Bismarck ND 58507 Office: Pub Svc Commn State Capitol Bismarck ND 58505

OGAWA, SEIICHIRO, chemistry educator; b. Tokyo, Sept. 18, 1937; s. Yukio and Hatsu (Yahagi) O.; m. Tomiko Fujita, May 21, 1964; children: Hiroya, Takayuki. BS, Keio U., Tokyo, 1961, MS, 1964; PhD in Engring., Keio U., 1967. With Tejin Ltd., Tokyo, 1961-67; lectr. Keio U., Tokyo, 1967-71, asst. prof., 1971-75, assoc. prof., 1975-84, prof., 1984—. Editor-in-chief Jour. of Synthetic Organic Chemistry, 1992-93. Mem. Soc. of Synthetic Organic Chemistry (Japan, bd. dirs. 1990-93), Chem. Soc. of Japan. Home: 16-8 Kichijojiminami, 3-chome, Musashino-shi Tokyo 180, Japan Office: Keio U, Faculty Sci and Tech, 14-1 Hiyoshi 3-chome, Kohoku-ku, Yokohama 223, Japan

OGAWA, TOMOYA, chemist, veterinary medical science educator; b. Komazawa, Setagaya, Japan, Mar. 19, 1939; s. Hiroshi and Kaoru (Kato) O.; m. Sachiko Ogita, Nov. 10, 1969; children: Kyoko, Tatsu. BS, U. Tokyo, 1962, DSc, 1967. Asst. U. Tokyo, 1967-68; rsch. assoc. U. Montreal, Que., Can., 1972-74; scientist Riken Inst., Saitama, Japan, 1968-79, head lab., 1979—; prof. vet. med. sci. U. Tokyo, 1990—; counsilor Ciba-Geigi Found. for Promotion of Sci., Osaka, 1987—. Author: Synthetic Studies on Glycoconjugates, 1975—; mem. editorial bd. Jour. Carbohydrate Chemistry, 1982—, Glycoconjugate Jour., 1984—, Carbohydrate Rsch., 1990—. Recipient award Sci. and Tech. Agy., Tokyo, 1982, Okochi Found. prize, Tokyo, 1984, Internat. Carbohydrate award Internat. Carbohydrate Orgn., Utrecht, 1984, Sci. Rsch. award Upjohn Co., Tsukuba, 1988, Alexander von Humboldt Found. sr. scientist award for sr. scientist, Bonn, 1991. Fellow Royal Soc. Chemistry (Harworth Meml medal 1993); mem. Am. Chem. Soc., Japan Soc. Biosci., Biotech. and Agrochemistry (dir. 1989—), Japanese Soc. Carbohydrate Rsch. (dir. 1988—), Japanese Biochem. Soc., Chem. Soc. Japan, Sigma Xi. Avocations: reading, tennis, travel, music, golf. Home: 3-6-3-101, Kichijojikitamachi, Musashinoshi Tokyo 180, Japan Office: Riken Inst, 2-1 Hirosawa, Saitama Wako 351-01, Japan

OGBONNAYA, CHUKS ALFRED, entomologist, agronomist, educator; b. Akoli-Imenyi, Imo, Nigeria, June 30, 1953; came to U.S., 1975; s. Alfred Agbaeze and Christy (Agubuche) O.; m. Joyce Elizabeth Belgrave, Mar. 30, 1985; children: Latoya, Oluchi, Kelechi. BS, U. Nebr., 1979, PhD, 1985; MS, N.W. Mo. State U., 1981. Cert. profl. crop scientist, profl. agronomist. Lab. asst. U. Nebr., Lincoln, 1976-78, rsch. asst., 1978-80, 82-85, postdoctoral fellow, 1985; asst. prof., postdoctoral fellow Mountain Empire Coll., Big Stone Gap, Va., 1985-90, assoc. prof., 1990—, coord. environ. sci. dept., 1986—; Disting. scholar-in-residence Pa. State U., summer 1990, vis. prof., 1990. Soccer coach Parks and Recreation, Big Stone Gap, 1989; mem. Va. Water Resources Statewide Adv. Bd. Recipient Times Teaching award Community Coll. Times, 1990, Chancellor's award Va. Community Coll. System, 1990, Outstanding Contbn. to the Community award; Fulbright scholar, 1993-94. Mem. Am. Soc. Agronomy, Crop Sci. Soc. Am., Entomol. Soc., Va. Acad. Sci., Va. Mining Assn., Internat. Platform Assn., Phi Beta Kappa. Methodist. Avocations: tennis, soccer. Home: 1000 University Blvd Apt 23B Kingsport TN 37660-1041 Office: Mountain Empire Coll Drawer 700 Big Stone Gap VA 14219

OGDEN, MICHAEL RICHARD, civil engineer, consultant; b. Elmira, N.Y., Mar. 30, 1965; s. Richard Edwin and Ruth Ann (Codney) O.; m. Susan Marie Klingler, Feb. 14, 1987; 1 child, Richard Earl. BSCE, Lehigh U., 1987, MSCE, 1991. Registered profl. engr., Pa. Rsch. asst. Lehigh U., Bethlehem, Pa., 1989-90, teaching asst., 1988-89; asst. project engr. Langan Engring. and Environ. Svcs., Elmwood Park, N.J., 1987—. Recipient Clayton Fund scholarship Anderson Clayton, 1983. Mem. ASCE (assoc.), Am. Shore and Beach Preservation Assn., Am. Water Resources Assn. Republican. Office: Langan Engring River Drive Ctr 2 Elmwood Park NJ 07407

OGG, JAMES ELVIS, microbiologist, educator; b. Centralia, Ill., Dec. 24, 1924; s. James and Amelia (Glammeyer) O.; m. Betty Jane Ackerson, Dec. 27, 1948; children—James George, Susan Kay. B.S., U. Ill., 1949; Ph.D., Cornell U., 1956. Bacteriologist Biol. Labs., Ft. Detrick, Md., 1950-53; cons. Biol. Labs., 1953-56, med. bacteriologist, 1956-58; prof. microbiology Colo. State U., Ft. Collins, 1958-85, prof. emeritus, 1985—; asst. dean Grad. Sch., 1965-66, head dept. microbiology, 1967-77; dir. Advanced Sci. Edn. Program div. grad. edn. in sci. NSF, Washington, 1966-67; Fulbright-Hays sr. lectr. in microbiology, Nepal, 1976-77, 81; acad. adminstrn. advisor Inst. Agr. and Animal Sci., Tribhuvan U., Nepal, 1988-91; cons. NASA, 1968-69, NSF, 1968-73, Martin Marietta Corp., 1970-76; cons.-evaluator North Central Assn. Colls. and Secondary Schs., 1974-89; cons. Consortium for Internat. Devel., 1990—, Winrock Internat. Inst. for Agrl. Devel., 1992—. Contbr. articles to profl. jours. Served with AUS, 1943-46, 50-51. Fellow AAAS, Am. Acad. Microbiology; mem. Am. Soc. Microbiology (chmn. pub. service and adult edn. com. 1975-80), Fulbright Alumni Assn., Sigma Xi, Phi Kappa Phi, Gamma Sigma Delta. Home: 1442 Ivy St Fort Collins CO 80525-2348

OGILVIE, KELVIN KENNETH, chemistry educator; b. Windsor, N.S., Nov. 6, 1942; s. Carl Melbourn and Mabel Adelia (Wile) O.; m. Emma Roleen, May 7, 1964; children: Kristine, Kevin. B.Sc. with honors, Acadia U., 1964, D.Sc. honoris causa, 1983; Ph.D., Northwestern U., 1968; D.Sc. honoris causa, U. N.B., Can., 1991. Assoc. prof. U. Man., Winnipeg, 1968-74; prof. chemistry McGill U., Montreal, 1974-88, Can. Pacific prof. biotech., 1984-87; bd. dirs. Sci. Adv. Bd., Biologicals, Toronto, Ont., 1979-84; dir. Office of Biotech. McGill U., 1984-87; v.p. acad. affairs, prof. chemistry Acadia U., Wolfville, N.S., 1987-93, pres., 1993—; invited lectr. on biotechnology Tianjin, People's Republic of China, 1985; Snider lectr. U. Toronto, 1991; Gwen Leslie Meml. lectr., 1991; mem. Coun. of Applied Sci. and Tech. for N.S., 1988—; mem. Nat. Biotech. Adv. Com., 1988—; mem. Fisher (Can.) Biotech. Adv. Ctr., 1989-92; bd. dirs. N.S. Rsch. Found. Corp., 1990-91; mem. sci. adv. bd. Allelix Biopharms., 1991-93; chair adv. bd. NRC Inst. for Marine Biosci., 1990-93; mem. steering com. on biotech. labor Can., 1990-92. Mem. editorial bd. Nucleosides and Nucleotides, 1981-82; contbr. over 150 articles to profl. jours.; holder 14 patents. Bd. dirs. Plant Biotech. Inst., 1987-90. Decorated Knight of Malta, 1985, Order of Can., 1991; recipient Commemorative medal for 125th Anniversary of Confedn. Can., 1992, Buck-Whitney medal, 1983, Manning Prin. award, 1992; named to McLean's Honor Roll of Canadians Who Made a Difference, 1988; E.W.R. Steacie Meml. fellow, 1982-84. Fellow Chem. Inst. of Can.; mem. Am. Chem. Soc., Ordre des Chemists of Que., Assn. Canadienne Française pour l'Advancement des Sciences. Achievements include inventing of BIOLF-62 (ganciclovir), antiviral drug used worldwide; developed general synthesis of RNA; developed original 'gene machine'; developer complete chemical synthesis of large RNA molecules. Home: PO Box 307, Canning, NS Canada B0P 1HO Office: Acadia U, Academic VP, Wolfville, NS Canada B0P 1XO

OGNIBENE, FREDERICK PETER, internist; b. Jamestown, N.Y., Aug. 30, 1953; s. Vincent Larry and Alma Linda (Martinelli) O. BA, U. Rochester, 1975; MD, Cornell U., 1979. Diplomate Am. Bd. Internal Medicine, Am. Bd. Internal Medicine-Critical Care. From intern to resident N.Y. Hosp./Cornell Med. Ctr., 1979-82; from med. to sr. staff fellow Critical Care Medicine Dept. NIH, Bethesda, Md., 1982-87; sr. investigator NIH, Bethesda, 1987—; asst. clin. prof. George Washington U., Washington, 1990—. Manuscript reviewer; contbr. articles and chpts. Mem. Washington Project of the Arts. comdr. Pub. Health Svcs., 1985—. Fellow ACP, Am. Coll. Critical Care Medicine (chair credentials com. 1992—); mem. Cornell U. Med. Coll. Alumni Assn. (bd. dirs.), Am. Fedn. Clin. Rsch. (sec.-treas. ea. sect. 1987-91, chair-elect 1991-92, chair 1992—), Am. Fedn. Clin. Rsch. Found. (trustee), Alpha Omega Alpha. Democrat. Roman Catholic. Avocations: travel, studying Italian language, collecting contemporary American art. Home: 2227 20th St NW Washington DC 20009-5075 Office: NIH 9000 Rockville Pike Bldg 10 Bethesda MD 20892-0001

OGUNTOYE, FERDINAND ABAYOMI, economist, statistician, computer consultant; b. Epe, Lagos, Nigeria, Mar. 4, 1949; s. Mathias Okeowo and Veronica (Adenowo) O.; m. Francisca Omosola Adesanya, May 19, 1950; children: Frederick Arayemi, Francis Adeyele, Faustina Omolola, Felicia Omobusola, Fidelis Abayomi Jr. BSc in Social Scis., U. Ife, Oyo State, Nigeria, 1975; MA in Devel. Econs., Williams Coll., 1982; PhD, U. Ife Oyo State, Nigeria, 1984. Tchr. Lagos State Teaching Commn., Epe, Lagos, Nigeria, 1968-72; econ. planner Ogun State Civil Svc., Abeokuta, Nigeria, 1976-81, chief planning officer, 1980-81, 82-84; commodity economist, statistician Assn. Tin Producing Countries, Kuala Lumpur, Malaysia, 1984—; computer applications trainer, software designer, developer and engr.; cons. Indsl. Devel. Corp., Abeokuta, 1982-84. Area commr. Boy Scouts Movement, Abeokuta, Ogun State, Nigeria, 1976—. Mem. Am. Econs. Assn., Nigerian Econ. Assn., Am. Inst. Mgmt., Malaysia Nat. Council Comp. Educ, Malaysian Econ. Assn., Singapore Microcomputer Soc., Anunsa Club (pres. 1973-74). Office: Assn Tin Producing Country, 4th Fl Menara Dayabumi, Jln Sultan Hishamuddin, Kuala Lumpur 50050, Malaysia

OGUT, ALI, mechanical engineering educator; b. Pazarcik, Turkey, Dec. 7, 1949; came to U.S., 1972; s. Hasan Demirci and Done (Karakiz) O.; m. Ozden Ceyhan, Nov. 18, 1991; 1 child, Sefkan. BSChemE, Hacettepe U., Ankara, Turkey, 1972; MSChemE, U. Md., 1976, PhD, 1979. Rsch. engr. Westvaco Laurel (Md.) Rsch. Lab., 1979-81, Mixing Equipment Co., Rochester, N.Y., 1981-83; asst. prof. St. John Fisher Coll., Rochester, 1983-85; vis. asst. prof. Rochester Inst. Tech., 1985-86, asst. prof., 1986-90, assoc. prof. dept. mech. engring., 1990—; cons. The Pfendler Co., Rochester, 1990—, Gould Pumps, Inc., Seneca Falls, N.Y., 1991-92, Dow Corning Co., Midland, Mich., 1992-93; creator/developer industrial profl. devel. seminar. Contbr. articles to profl. jours. Fellow NASA/ASEE; grantee NSF, 1987-89, 91-93, 92-94, Mobil Chem., 1990-91, NASA Lewis Rsch. Ctr., 1990-94. Mem. ASME, Am. Soc. Engring. Edn., Am. Inst. Chem. Engrs. Achievements include development of a mechanical device to remote control small fluid pumps and authoring of a fluid mixing reference book. Home: 6 Winchester Dr Pittsford NY 14534 Office: Rochester Inst Tech Mech Engring Dept 1 Lomb Memorial Dr Rochester NY 14623-5640

OH, KOOK SANG, diagnostic radiologist, pediatric radiologist; b. Korea, Jan. 3, 1936; came to U.S., 1964; m. Oct. 6, 1961; children: Jean, Eugene, Young. MD, Seoul (Korea) Nat. U., 1961. Diplomate Am. Bd. Radiology. From asst. prof. to assoc. prof. radiology Johns Hopkins U. Sch. Medicine, Balt., 1971-76; assoc. prof. radiology U. Pitts. Sch. Medicine, 1976-78, prof. radiology, 1978-90; prof. radiologic scis. Med. Coll. Pa., 1992—. Fellow Am. Coll. Radiology; mem. AMA, N.Y. Acad. Scis. (life), Sci. Rsch. Soc. N.Am., Soc. Pediat. Radiology, Radiol. Soc. N.Am., Soc. Thoracic Radiology. Home: 107 Wynnwood Dr Pittsburgh PA 15215 Office: Med Coll Pa Allegheny Gen Hosp Pittsburgh PA 15212

OH, NAM HWAN, food scientist; b. Seoul, Korea, July 1, 1950; came to U.S., 1979; s. Wang-keun and Ki-soon (Nam) O.; m. Myung-hee Kim, July 15, 1979; children: Elizabeth, Samuel. MS, Kans. State U., 1978, PhD, 1984. Rsch. coord. Office of Rural Devel., Suweon, Korea, 1975-79; grad. rsch. asst. Kans. State U., Manhattan, 1979-84; sr. food scientist Best Foods Rsch. and Engring. Ctr., Union, N.J., 1984-87, prin. food scientist, 1987—. Author: (with Paul A. Seib) Introduction to Oriental Noodles, 1982; contbr. articles to profl. jours. 1st lt. Republic of Korea Army, 1972-74. Recipient First Best Paper award Internat. Gluten Mfrs. Assn., 1984. Mem. Am. Assn. of Cereal Chemists, Inst. of Food Technologists. Mem. Seventh Day Adventist. Achievements include research on methodology to quantify attributes critical in oriental noodles, impacts of raw material attributes on oriental noodle quality, optimization of processing conditions, ingredient functionality and processing impacts on pasta product quality, peanut butter processing and processing technology development. Office: Best Food Rsch and Engring Ctr 150 Pierce St Somerset NJ 08873-6710

OHADI, MICHAEL M., mechanical engineering educator; b. Kerman, Iran, Oct. 23, 1954; came to U.S., 1975; s. Mehdi M. and Fathy (Bagehri) O.; m. Mahnia Vakil, June 15, 1985; 1 child, Nicholas. D in Mech. Engring., Northeastern U., 1982; PhD, U. Minn., 1986. Project engr. Atmosphere Cons. Engrs., Tehran, Iran, 1975-77; rsch. asst. So. Ill. U., Carbondale, 1978-80; rsch. asst., instr. Northeastern U., Boston, 1980-84; rsch. assoc. U. Minn., Mpls., 1984-86; asst. prof. mech. engring. Mich. Tech. U., Houghton, 1986-89, assoc. prof., 1989-90; assoc. prof. U. Md., College Park, 1990—; faculty advisor Nat. Capital chpt. ASHRAE, Washington, 1990—, Blue Key Nat. Honor Soc., Houghton, 1987-89. Contbr. articles to profl. jours. Recipient Ralph Teetor Disting. Faculty award Soc. Automotive Engrs., 1989, Mich. Assn. Governing Bds. award State of Mich., 1989. Mem. ASME, ASHRAE, Am. Waterjet Assn., Am. Soc. Engring. Educators. Achievements include patents pending for enhanced gas-to-gas recuperator, compound-enhanced electrohydrodynamic boiler. Home: 9533 Clocktower Ln Columbia MD 21046 Office: Dept Mech Engring U Md College Park MD 20742

OHAMA, YOSHIHIKO, architectural engineer, educator; b. Shimonoseki, Japan, Mar. 29, 1937; s. Yoshihiro and Ritsuko O.; m. Ikuko Ohama, Mar. 24, 1966; children: Akiko, Noriko, Hiroshi. BE, Yamaguchi U., Japan, 1959; PhD, Tokyo Inst. Tech., 1974. Rsch. engr. cen. rsch. lab. Onoda Cement Co., Ltd., Koto-ku, Tokyo, 1959-66; rsch. engr. and dept. head Bldg. Rsch. Inst. Japanese Govt., Shinjuku-ku, Tokyo, 1966-76; from assoc. prof. to prof. Nihon U., Koriyama, Fukushima-ken, Japan, 1976—; rsch. engr. Inst. Indsl. Sci., U Tokyo, 1972-81; adv. prof. Tongji U., Shanghai China, 1990; guest prof. Shandong Inst. Bldg. Materials, Jinan, China, 1991; v.p. Internat. Congress on Polymers in Concrete, Detroit, 1987—; chmn. Polymers in Concrete Com., Japan Tech. Transfer Assn., Tokyo, 1987—. Author: Polymeric Waterproofing Work, 1972; co-author: Plastics Concrete, 1965, Building Materials, 1981, Concrete Admixtures Handbook, 1984, Construction Materials, 1990. Mem. ASTM, Am. Concrete Inst., Internat. Union Testing and Rsch. Labs. for Materials and Structures (chmn. tech. com. 113-CPT, 151-APC), Soc. for Advancement of Material and Process Engring. (Japan chpt., chmn. advanced constn. tech. com. 1988—), Soc. Materials Sci. Japan, Architectural Inst. Japan, Japan Concrete Inst. Avocations: travel, mountaineering, ham. Home: 14-10-402 Hiyoshi 2-chome, Kohoku-ku, Yokohama Kanagawa-ken 223, Japan Office: Nihon U Coll Engring, Koriyama Fukushima-ken 963, Japan

O'HANDLEY, DOUGLAS ALEXANDER, astronomer; b. Detroit, May 7, 1937; s. Malcolm Joseph and Georgie Roberta (MacPherson) O'H.; m. Christine Jeannette Stube, July 20, 1991; 1 child, Douglas Alexander, Jr. AB, U. Mich., 1960; MS, Yale U., 1964, PhD, 1967. Astronomer U.S. Naval Obs., Washington, 1960-67; scientist Jet Propulsion Lab., Pasadena, Calif., 1967-85; dir. space station Ames Rsch. Ctr., Moffett Field, Calif., 1985-86; mgr. TRW Space Tech. Group, Redondo Beach, Calif., 1986-88; dep. asst. adminstr. office exploration NASA, Washington, 1988-91; special asst. Center for Mars Exploration Ames Rsch. Ctr., Moffett Field, Calif.,1991—; chmn. com. for protection of human subjects in med. rsch., 1982-85; lectr. grad. sch. Georgetown U., Washington, 1964-67; speaker at med. soc. meetings. Contbr. articles to profl. jours. Bd. dirs. Cath. Big Bros.; extraordinary minister St. Bede's Roman Cath. Ch. Recipient NASA Group Achievement award Planetary Ephemeris Devel. Team, 1982. Fellow Royal Soc. Medicine, Internat. Astronomical Union, Internat. Acad. Astronautics, Aerospace Med. Assn. (fellow), AIAA (assoc.). Republican. Home: 1580 Grackle Way Sunnyvale CA 94087 Office: Ctr Mars Exploration Ames Rsch Ctr Moffett Field CA 94035-1000

OHARA, AKITO, biology educator; b. Aki, Kohchi, Japan, Feb. 11, 1948; s. Hideo and Fumie Ohara; m. Junko Yamamoto, June 14, 1981; children: Takayasu, Hiroto. BSc, Kyoto (Japan) U., 1972; MSc, Nagoya (Japan) U., 1974, PhD, 1982. Rsch. assoc. Emory U. Sch. Medicine, Atlanta, 1989-91; rsch. instr. Shiga U. Med. Sci., Ohtsu, Japan, 1977—. Contbr. articles Wilhelm Roux's Archives, Develop., Growth and Differ., Med. Sci. Rsch., Devel. Biology, The FASEB Jour. Mem. Zool. Soc. Japan, Japanese Soc. Devel. Biologists, Physiol. Soc. Japan, Biophys. Soc. (U.S.A.). Achievements include research which clarified relationships between neural competence of presumptive ectoderms and inducing time during neural induction process; discovered that Na/K pump activity appears in new membranes formed at the first cleavage in newt eggs, that a pertussis toxin-sensitive G protein inactivates highly-Na selective, amiloride-blockable epithelial sodium channels in apical membranes of A6 cells. Home: 3-39-146 Higashi-Yakura, Kusatsu 525, Japan Office: Shiga Univ Med Sci, Otsu 520-21, Japan

O'HARA, ROBERT JAMES, evolutionary biologist; b. Arlington, Mass., Nov. 21, 1959. BA, U. Mass., 1981; PhD, Harvard U., 1989. Tutor Dudley House, Harvard Coll., Cambridge, 1983-89; fellow Smithsonian Instn., Washington, 1990-91, Ctr. for Critical Inquiry, U. N.C., Greensboro, 1992—; NSF fellow and adj. curator U. Wis., Madison, 1991-92; disting. vis. prof. Transylvania U., Lexington, Ky., 1992. Mem. Soc. of Systematic Biologists, Soc. for History of Natural History, Sigma Xi, Phi Beta Kappa. Office: Univ of NC Ctr for Critical Inquiry Greensboro NC 27412

OHAYON, ROGER JEAN, aerospace engineering educator, scientific deputy; b. Meknes, Morocco, Jan. 3, 1942; s. David and Lily (Sassoon) O.; m. Ingrid Schmischke, Oct. 5, 1968; children: Carine, Philippe. D in Engring., U. Orsay, Paris, 1971; Habilitation Diriger Recherches, U. Paris 6, Marie Curie, 1990. Registered profl. engr., Chatillon. With Office Nat. d'etudes et de Recherches Aerospatiales, Chatillon, France, 1970—; sci. dep. Office Nat. d'études et de Recherches Aérospatiales, Chatillon, France, 1991—; prof., chair mechanic Conservatoire National des Arts et Métiers, Paris, 1992—; prof. Ecole Cen. Arts and Mfrs., Chatenay, France, 1978-86, Ecole Nat. Tech. Avancées, Paris, 1986—; external prof. U. Paris 6, 1986—; mem. sci. coun. Lab. Cen. des Ponts et Chaussées, Paris, 1987—. Co-author: Fluid-Structure Interaction, 1992, English edit., 1993); co-editor 4 books; mem. editorial bd. 6 hours.; contbr. numerous articles to profl. jours. Mem. and organizer of several Internat. Congresses. Recipient Acad. Scis. Price du Gen. Muteau award, 1989. Mem. ASME, French Computational Structural Mechanics Assn. (pres. 1991), Spanish Soc. Computational Mechanics, Internat. Assn. Computatonal Mechanics, Groupe pour Avancement Méthodes Njmériques Ingénieur. Home: 22 Kellermann Blvd, 75013 Paris France

OHE, SHUZO, chemical engineer, educator; b. Tokyo, Mar. 31, 1938; s. Kunio and Chizu (Tabata) O.; m. Nobuko Motegi, Oct. 31, 1975; 1 child, Kenzo. BS, Sci. U. Tokyo, 1962; D of Engring., Tokyo Met. U., 1971. Chem. engr. Ishikawajima-Harima Heavy Indsl. Co. Ltd., Tokyo, 1962-65, reseracher, then rsch. mgr., 1966-80; assoc. prof. Tokai U., Tokyo, 1980-82; prof. chem. engring. Tokai U., 1982-91; prof. indsl. engring. Sci. U. of Tokyo, 1991—; vis. researcher Fractionation Rsch., Inc., South Pasadena, Calif., 1973; cons. Chiyoda Engring. Co. Ltd., 1986-89, NKK, Tokyo, 1988—. Inventor angle tray distillation tower; author, editor: Computer-Aided Data Book of Vapor Pressure, 1976, Vapor-Liquid Equilibrium Data, 1988, Vapor-Liquid Equilibrium Data at High Pressure, 1990, Vapor-Liquid Equilibrium Data-Salt Effect, 1991; author: Chemical Engineering Design By P.C., 1985. Mem. Kanagawa Micon Club (pres. 1988—), Japan Info. Ctr. Sci. and Tech., Am. Inst. Chem. Engrs., Japan Soc. Chem. Engring. Avocation: golf. Office: Sci Univ Tokyo, 1-3 Kagurazaka Shinjuku-ku, Tokyo 162, Japan

OHLROGGE, JOHN B., botany and plant pathology educator; b. June 7, 1949. BA in Chemistry, Earlham Coll., 1971; PhD in Biochemistry, U. Mich., 1975. Post doctoral fellow in biochemistry U. Calif., Davis, 1976-79; rsch. biochemist U.S. Dept. Agriculture, Peoria, Ill., 1980-86; assoc. prof. botany and plant pathology Mich. State U., East Lansing, 1987-91, prof., 1991—; mem. peer review panel USDA Competitive Grants Program. Mem. editorial bd. Archives Biochemistry and Biophysics, Plant Physiology; contbr. over 50 articles to sci. jours. Grantee Agrl. Rsch. Svc. Rsch. Assoc. Program, 1982-84, USDA, 1984-86, 86-89, Am. Soybean Assn., 1987-89, DOE, 1987-90, 91—, USDA/DOE.NSF, 1988-93, 92—, NSF, 1989, 90—, Mich. State U. Rsch. Excellence Funds, 1993, Midwest Plant Pathology Consortium, 1993—. Achievements include research in plant fatty acid synthesis, organization of lipid metabolism in plant cells, molecular biology of genes controlling plant lipid synthesis. Office: Mich State U Dept Botany & Plant Pathology East Lansing MI 48824-1312

OHLSSON, STELLAN, scientist, researcher; b. Karlskrona, Blekinge, Sweden, Jan. 7, 1948; came to U.S., 1983; s. Sven Axel Harald and Margit Stella Maria (Haggstrom) O. M, U. Stockholm, 1972, PhD, 1980. Rsch. assoc. The Robotics Inst. Carnegie-Mellon U., Pitts., 1983-84; rsch. scientist U. Pitts. Learning Rsch. & Devel. Ctr., 1985-92, sr. scientist, 1992—. Assoc. editor Jour. Artificial Intelligence in Edn., 1992—; contbr. articles to profl. jours. Rsch. grantee Office of Naval Rsch., 1985-92, Office Ednl. Rsch. & Improvement, 1987—, NIH, 1990—. Mem. Am. Psychol. Soc., Artificial Ednl. Rsch. Soc. (assoc.), Cognitive Soc. Home: 835 N St Clair St Pittsburgh PA 15266 Office: Learning Rsch & Devel Ctr 3939 O'Hare St Pittsburgh PA 15260

OHMAN, MARIANNE, medical technologist; b. Lansing, Mich., Apr. 15, 1952; d. Erwin and Bette Jean (Spinde) Kuhn; m. Dennis Edward Ohman, Dec. 11, 1976. BS in Microbiology and Pub. Health, Mich. State U., 1974; BS in Med. Tech., Northwestern U., 1980. Lic. med. technologist, Calif., Tenn., histocompatibility specialist. High sch. sci. tchr. Joseph Merrick Bapt. Coll., Cameroon, Africa, 1974-76; rsch. asst., lab. mgr. dept. dermatology Oregon Health Sci. U., Portland, 1976-79; med. technologist I tissue typing lab. Northwestern U. Med. Sch., Chgo., 1980-81; clin. lab. technologist immunogenetics & transplantation lab. dept. surgery U. Calif., San Francisco, 1981-89; rsch. lab. specialist dept. pathology St. Jude Children's Rsch. Hosp., Memphis, 1989-90, rsch. lab. specialist dept. hematology/oncology, 1990; sr. rsch. assoc. dept. surgery-transplant div. U. Tenn., Memphis, 1991—; seminar speaker on current tech. of kidney transplantation immune monitoring, Porto, Portugal, 1988. Participant in fund-raising event St. Jude Children's Rsch. Hosp., Memphis, 1992. Mem. Am. Soc. Clin. Pathologists, Am. Soc. Histocompatibility & Immunogenetics (co-organizer regional edn. workshop 1987, presenter at ann. meetings). Home: 2318 Hickory Forest Memphis TN 38119 Office: Univ Tenn Dept Surgery 956 Ct Ave Rm H221 Memphis TN 38163

OHNAMI, MASATERU, mechanical engineering educator; b. Kyoto, Japan, Apr. 6, 1931; s. Eijiro and Hisae O.; m. Hiroko Ohnami, Oct. 10, 1959; 1 child, Masahiro. B in Engring., Ritsumeikan U., Kyoto, Japan, 1954; D in Engring., Kyoto U., 1960. Asst. prof. Kyoto U., 1955-61; assoc. prof. Ritsumeikan U., Kyoto, 1961-67, prof., 1967—, dean acad. affairs, 1978-80, dean faculty sci. & engring., 1988-90, pres., 1991—; vis. rsch. prof. Columbia U., N.Y.C., 1963-64; mng. dir. Jpan Assn.Prvt. Colls. and Univs., Tokyo, 1991—; dir. Japanese Univ. Accreditation Assn., Tokyo, 1991—; mem. Sci. Coun. Ministry Edn., 1984-86. 1989-91. Author: Plasticity and High Temperature Strength of Materials, 1988, Fracture and Society, 1992. Mem. Deutscher Verband für Materialforschung und prüfung e.V. (hon. 1992—), Soc. of Materials Sci. Japan (bd. dirs. 1971-74, 81-84, 85-88, Prize 1971), Japanese Soc. for Strength and Fracture of Materials (bd. dirs. 1984—), Sci. Coun. Japan (material rsch. liaison com. 1988—). Avocations: oil painting, reading. Home: 8-10 Hyugacho, Takatsuki Osaka Prefecture 569, Japan Office: Ritsumeikan Univ 56-1, Tojiin Kitamachi, Kita-ku Kyoto 603, Japan

OHNING, BRYAN LAWRENCE, neonatologist, educator; b. Evansville, Ind., Sept. 8, 1952; s. Victor G. and Florence E. (Pollock) O.; m. Diane L. Goforth; children: Gavin Nicholas, Collin Richard, Erin Elaine. BS in Biochemistry, Purdue U., 1974; PhD in Biochemistry, Case Western Reserve U., 1981, MD, 1981. Resident in pediatrics Duke U. Med. Ctr., Durham, N.C., 1981-84; fellow neonatal/perinatal Children's Hosp. Center, 1984-87; asst. prof. Med. U. S.C., Charleston, 1988—; bd. dir. Medacare Pediatric Transport. Contbr. articles to profl. jours. Chmn. March of Dimes, Charleston, 1980—, grants review com., 1989—. Rsch. scholar Inst. Developmental Rsch., Cin., 1986-88; recipient Clinician/Scientist award Am. Heart Assn., 1986. Mem. Am. Acad. Pediatrics, Am. Thoracic Soc., Nat. Perinatal Assn. Achievements include research in the biophysical activity of the hydrophobic low molecular weight pulmonary surfactant apoproteins SP-B and SP-C and their interactions with phospholipids. Office: Med U SC Dept Pediatrics 171 Ashley Ave Charleston SC 29425

OHNISHI, STANLEY TSUYOSHI, biomedical director, biophysicist; b. Ohtsu City, Japan, Dec. 17, 1931; came to U.S., 1966; s. Teruhiko and Miyoko (Tomoda) O.; m. Tomoko Kirita, Mar. 25, 1958; children: Hiroshi, Noriko. MS in Chemistry, Kyoto (Japan) U., 1956; PhD in Biophysics, Nagoya (Japan) U., 1959. Rsch. prof. dept. hemtology Hahnemann U., Phila., 1984, rsch. prof. dept. biochemistry, 1984; dir. Membrane Rsch. Inst., Phila., 1989—; dir. developing chemotherapeutic drugs Phila. Biomed Rsch. Inst., King of Prussia, Pa., 1989—. Editor: Experimental Techniques in Biomembranes, 1967, Mechanisms Gated Calcium Transport, 1981, Membrane-Linked Diseases, 1993; contbr. 150 articles to Biochem. Biophys. Pharmacology; editor 3 books. Buddhist. Achievements include 5 patents in field. Office: Phila Biomed Rsch Inst 502 King Of Prussia Rd Radnor PA 19087

OHRN, NILS YNGVE, chemistry and physics educator; b. Avesta, Sweden, June 11, 1934; came to U.S., 1966; s. Nils E. and Gerda M. (Akerlund) O.; m. Ann M.M. Thorsell, Aug. 24, 1957; children: Elisabeth, Maria. M.S., Uppsala U., 1958, Ph.D., 1963, F.D., 1966. Research assoc. Uppsala (Sweden) U., 1963-66; assoc. prof. U. Fla., Gainesville, 1966-70, prof. chemistry and physics, 1971—, assoc. dir. Quantum Theory Project, 1976-77, dir. Quantum Theory Project, 1983—, chmn. dept. chemistry, 1977-83. Editor: Internat. Jour. Quantum Chemistry, 1970—. Fulbright grantee Com. for Internat. Exchange of Scholars, Washington, 1961-63; recipient Bicentennial Gold medal King of Sweden, 1980; Fla. Acad. Scis. medal, 1984. Fellow Am. Phys. Soc.; mem. Am. Chem. Soc., Royal Acad. Scis. Sweden (fgn.), Finnish Acad. Scis. (fgn.), Royal Danish Acad. Scis. (fgn.), Sigma Xi, Phi Beta Kappa. Home: 1823 NW 11th Rd Gainesville FL 32605-5323 Office: U Fla Quantum Theory Project 362 Williamson Hall Gainesville FL 32611

OHRT, JEAN MARIE, chemist; b. Buffalo, N.Y., Dec. 7, 1923; d. Raymond William and Helen Mary (Hayes) O. Diploma, Mount Mercy Acad., 1943; student, D'Youville Coll., 1943-45; BA, U. Buffalo, 1949. Ptnr. Worth & Ohrt, 1945-49; tchr. Holy Family Sch., Buffalo, 1949-50; asst. med. records Children's Hosp., Buffalo, 1951-53; with Dr. E.H. De Kleine, 1953-55; cancer rsch. scientist x-ray crystallography Roswell Park Meml. Inst., Buffalo, 1955-79; tchr. Gaelic Trocaire Coll., Buffalo Irish Ctr. Contbr. over 50 articles to profl. jours. Mem. Am. Crystallographic Assn., AAAS, An Teanga Mharthanach, Irish-Am. Cultural Inst., Sigma Xi. Roman Catholic. Home: 33 Tamarack St Buffalo NY 14220-1730

OHTA, HIDEAKI, economics and finance researcher; b. Hiroshima, Japan, June 10, 1955; s. Isao and Nobuko (Matsuda) O.; m. Michiko Fujibayashi, Nov. 4, 1985; 1 child, Masako. B Econs., U. Tokyo, 1980; diploma, U. Stockholm, 1981; MPhil in Devel. Econs., Cambridge (Eng.) U., 1982. Researcher, economist Japan Devel. Inst. (formerly Engring. Cons. Firms Assn.), Tokyo, 1982-84; indsl. devel. officer UN Indsl. Devel. Orgn., Vienna, Austria, 1984-86, area program officer, 1986-90; assoc. sr. researcher Asia and Pacific rsch. dept. Nomura Rsch. Inst., Tokyo, 1990—; cons., investment advisor UN Indsl. Devel. Orgn. Investment Promotion Svc., Tokyo, 1983-84. Contbr. articles to profl. jours. Mem. Assn. Grads. Major Nat. Univs., Assn. Grad. Faculty Econs. U. Tokyo. Home: 6 Brookvale Walk #01-09, Brookvale Park, Singapore 2159, Singapore Office: 6 Battery Rd #40-02, Singapore 0104, Singapore

OHTA, HIROBUMI, aeronautical engineer, educator; b. Nagoya, Japan, Nov. 23, 1945; s. Hiroyuki and Toshiko O.; m. Mariko Kaneko, Dec. 5, 1970; 1 child, Mayumi. M Engring., Nagoya U., 1970, D Engring., 1975. Rsch. assoc. Nagoya U., 1973-78, assoc. prof. dept. aeronautical engring., 1979—; researcher Chubu Internat. Airport Rsch. Found., Nagoya, 1987—; vis. researcher Nat. Aerospace Lab., Tokyo, 1989-91. Contbr. articles on aircraft control design to profl. publs. Home: 3-3-5 Miyoshigaoka Miyoshi-cho, Nishikamo gun Aichi 470-02, Japan Office: Nagoya Univ Aeronautical Engring, Furo Cho Chikusa Ku, 464-01 Nagoya Japan

OHTA, HIROSHI, economics educator; b. Hamamatsu, Shizuoka, Japan, Oct. 2, 1940; s. Seiichi and Fumiko (Ohishi) O.; m. Yuko Narita, June 3, 1966; children: Isamu, Yasuo. BA in Econs., Aoyama Gakuin U., 1964, MA in Econs., 1966; PhD in Econs., Tex. A&M U., 1971; D of Econs., Aoyama Gakuin U., 1978. Lectr. Aoyama Gakuin U., Tokyo, 1971-73, assoc. prof., 1973-82; adj. prof. U. Houston, 1980—; prof. Aoyama Gakuin U., Tokyo, 1982—; Tokyo correspondent ISSJ, Paris, 1979—; assoc. editor The Annals of Regional Science, Umea U., 1989—, Jour. of Regional Sci., U. Pa., 1989—. Author: Spatial Price Theory of Imperfect Competition, 1988; editor: (with Jacques Thisse) Does Economic Space Matter?, 1993. Mem. Am. Econ. Assn., Royal Econ. Soc., Regional Sci. Assn. Japan (dir.), Regional Sci. Assn. Internat., Japan Assn. Econs. and Econometrics, Japan Fedn. Econs. Assns. (councilor 1991—). Home: 4-19-10 Kugayama, Suginami, Tokyo 168, Japan Office: Aoyama Gakuin U, 4-4-25 Shibuya, Tokyo 150 Shibuya-ku, Japan

OI, RYU, chemist; b. Fukusaki, Hyogo, Japan, Feb. 7, 1958; s. Susumu and Asako (Yasumoto) O.; m. Keiko Okada, June 15, 1984; children: Makoto, Takahiro. MS, Okayama U., 1984, PhD, 1989. Rsch. chemist Mitsui Toatsu Chems., Inc., Omuta, Japan, 1984-89; chief chemist Mitsui Toatsu Chems., Inc., Yokohama, Japan, 1992—; vis. scientist MIT, Boston, 1989-91, Scripps Rsch. Inst., La Jolla, Calif., 1991-92. Contbr. articles to profl. jours. Mem. Japan Chem. Soc., Assn. Francophile de Omuta (vice chmn. 1984-85), Kamakura English Club (vice chmn. 1992). Buddhist. Achievements include research in industrial application of electroorganic reduction, electrosynthesis of m-hydroxybenzyl alcohol for new insecticide ethofenprox, asymmetric dihydroxylation, stable equivalents of enantiopure glycer- and glycialdehyde. Home: 2882 Iijimi #17, Sakae-ku Yokohama 244, Japan Office: Mitsui Toatsu Chems Inc, 1190 Kasama-cho Sakae-ku, Yokohama 248, Japan

OIKAWA, ATSUSHI, pharmacology educator; b. Tokyo, Jan. 27, 1929; s. Yasuo and Tsuya (Yamagishi) O.; m. Masako Hirota, Apr. 4, 1931; children: Hikaru, Kahoru, Akira. BS, U. Tokyo, 1953; PhD, Osaka U., 1961. Research asst. Osaka U., 1957-61, lectr., 1961-62; sect. head Nat. Cancer Ctr. Research Inst., Tokyo, 1962-74; prof. pharmacology Tohoku U., Sendai, 1974-92, prof. emeritus, 1992—; dir. cancer cell repository Tohoku U., 1988-90; cons. Nihon Bioservice, 1992 . Editor: Pigment Cells (in Japanese), 1982; author: Cancer: Fundamentals (in Japanese), 1986, What is the Biological Science (in Japanese), 1992. Recipient Seiji Meml. award for pigment research Lydia O'Leary Meml. Fund, 1984. Mem. Japanese Cancer Assn. (councilor), Japanese Biochem. Soc. (councilor), Molecular Biology Soc. Japan., N.Y. Acad. Scis., Planetary Soc. Office: Rm 405 Bldg 2 Asahi Pla, Kuzuokashita Goroku Aobaku, Sendai 989 31, Japan

OIKAWA, HIROSHI, materials science educator; b. Sakhalin, Japan, Oct. 15, 1933; s. Torao and Tomi (Kumagai) O.; m. Ayako Otomo, May 4, 1963; children: Makoto, Junko. BE, Tohoku U., Sendai, Japan, 1956, ME, 1958, D in Engring., 1961. Instr. Tohoku U., Sendai, 1961-63, lectr., 1963-64, assoc. prof., 1964-82; rsch. fellow U. Fla., Gainesville, 1966-68; prof. Tohoku U., Sendai, 1982—, councilor, 1993—. Co-editor: Metals Handbooks, 1990, Metals Databook, 1993. Mem. Japan Inst. Metals (bd. dirs. 1992—, bd. dirs. Tohoku chpt. 1991-93), Iron and Steel Inst. of Japan (bd. dirs. 1990-92), Japan Inst. Light Metals (bd. dirs. 1989—), Minerals, Metals and Materials Soc., ASM Internat., Inst. Materials. Office: Tohoku U Dept Materials Sci, Faculty Engring, Sendai 980, Japan

OJCIUS, DAVID MARCELO, biochemist, researcher; b. Porto Alegre, Brazil, Sept. 30, 1957; s. Alex and Eva (Koval) O.; m. Louise Jeanne Burgaud, Nov. 25, 1988. BA, U. Calif., Berkeley, 1979, PhD in Biophysics, 1986. Rsch. fellow Harvard U., Cambridge, Mass., 1985-87; rsch. fellow Rockefeller U., N.Y.C., 1988-91, asst. prof., 1991—; guest investigator Pasteur Inst., Paris, France, 1991—; referee Jour. of Exptl. Medicine, N.Y.C., 1988-91. Contbr. articles to Trends Biochem. Sci., Proceedings Nat. Acad. Sci., Jour. Immunology, Exptl. Cell Rsch., Cancer Cells. Democrat. Achievements include rsch. on pore-forming proteins, mechanisms of cytotoxicity of cytolytic lymphocytes, programmed cell death and antigen presentation by the major histocompatibility complex. Home: 20A rue des Laitières, 94300 Vincennes France Office: Institut Pasteur, 25 rue du Dr Roux, 75724 Paris France

OJWANG, JOSHUA ODOYO, molecular biologist; b. Mfangano, Nyanza, Kenya, Mar. 5, 1960; came to U.S., 1985; s. Owuor and Louise (Akuku) O.; m. Rose Anyango, June 30, 1984; children: Desmond Omondi, Beryl Adhiambo, Audrey Akinyi. BSc, U. Nairobi, Kenya, 1985; PhD in Chemistry,

Boston U., 1990. Teaching fellow Boston U., 1985-88, rsch. asst., 1988-90; molecular biologist U. Calif., San Diego, 1990-92; rsch. sci. Triplex Pharm. Corp., The Woodlands, Tex., 1992—. Scholarship Govt. of Kenya, 1981-85; fellowship World Lab., 1991. Achievements include research in cancer and AIDS. Home: 2218 Pincher Creek Dr Spring TX 77386 Office: Triplex Pharm Corp 9391 Grogaws Mill Rd The Woodlands TX 77380

OKA, KUNIO, chemistry educator; b. Shimoaso, Gifu, Japan, May 28, 1945; s. Akitaro and Kanayo (Kanai) O.; m. Sachiko Kawata, May 2, 1971; children: Tetsunori, Michiko. BS, Toyama U., 1968; M in Engring., U. Osaka Prefecture, 1973, D in Engring., 1990. Researcher Osaka Prefectural Radiation Rsch. Inst., Sakai, Osaka, 1975-85, sr. researcher, 1986-90; assoc. prof. U. Osaka Prefecture, Sakai, 1990—; mem. drafting com. Tech. Innovation Assn., Osaka, 1984-86. Author: Chitin, Chitosan and Related Enzymes, 1984; contbr. articles to profl. jours. Ministry of Edn., Sci. and Culture grantee, 1981. Mem. Am. Chem. Soc., Chem. Soc. Japan, Kinki Chem. Soc., Organosilicon and Related Material Assn. Japan, Silicon Photochem. Soc. Japan, Nakamozu Lawn Tennis Club. Avocations: tennis, golf, gardening. Home: 4-34-3 Niwashirodai, Sakai, Osaka 59001, Japan Office: U Osaka Prefecture RIAST, 1-2 Gakuencho, Sakai Osaka 593, Japan

OKA, TETSUO, medical educator; b. Okayama, Japan, Jan. 9, 1938; s. Hiroshi and Naoko (Takai) O.; m. Yasuko Otsuka, Mar. 24, 1964; children: Keiko, Taro. MD, Keio U., Tokyo, 1963, Phd, 1970. Instr. dept. pharmacology Sch. of Medicine, Keio U., 1964-71, asst. prof. dept. pharmacology, 1972-74, assoc. prof. dept. pharmacology, 1974; prof., chmn. dept. pharmacology Sch. of Medicine Toaki U., Isehara, Japan, 1974—; lectr. Keio U., 1974—, Yamanashi Med. Coll., Kofu, Japan, 1988—, U. Nagoya, Japan, 1990—. Author: Essentials of Pharmacology, 1988. Mem. N.Y. Acad. Scis., Japanese Pharm. Soc. (trustee 1989-92). Liberal Democrat. Buddhist. Avocation: swimming. Home: 3-16-1007 Sengen-cho, Hiratsuka 254, Japan Office: Tokai U Sch of Medicine, Dept Pharmacology, Isehara 259-11, Japan

OKABE, MITSUAKI, economist; b. Shirotori, Kagawa, Japan, May 8, 1943; s. Teruo and Kazue (Tsuchiya) O.; m. Michiko Shinohara, Mar. 24, 1970; children: Koichiro, Akiko. BA in Econs., U. Tokyo, 1968; MBA, U. Pa., 1973. Economist Bank of Japan, Tokyo, 1968-85, chief rsch. div., 1986-90, sr. advisor, 1990—; prof., dir. Ctr. for Japanese Econ. Studies Macquarie U., Sydney, Australia, 1992—; vis. lectr. Wharton Sch. Bus., U. Pa., 1990-91; vis. lectr Princeton U., 1991-92. Author: Thoughts on Monetary Policy, 1988; editor: Toward a World of Economic Stability, 1988, Evolution of International Monetary System, 1990. Mem. Am. Econ. Assn., Japanese Assn. Theoretical Econs. Avocations: studying butterflies, studying Mozart, tennis. Home: Miyazono 2-20-9, Nagareyama-Shi 270-01, Japan Office: Sch Econ and Fin Studies, Macquarie U, Sydney New South Wales 2109, Australia

OKADA, AKANE, chemist; b. Kyoto, Japan, July 2, 1946; s. Takeo and Yaeko (Yoshimatsu) O.; m. Hisayo Sakakibara, Jan. 14, 1981; 1 child, Keitaroh. BS, Kyoto U., 1969; PhD, Tohoku U., Sendai, Japan, 1974. Researcher Mitsubishi Chem. Ind., Yokohama, Japan, 1974-85; researcher Toyota Cen. R & D Labs., Nagakuta, Japan, 1985-89, lab. mgr. organic materials, 1989—. Contbr. to tech. publs. Mem. Am. Chem. Soc., Material Rsch. Soc., Chem. Soc. Japan, Soc. Polymer Sci. Japan. Achievements include industrialization of first polymeric molecular composite, related patents. Home: 2 333 Momoyama, Ohbu 474, Japan Office: Toyota Cen R&D Labs, Nagakute, Aichi 480-11, Japan

OKADA, FUMIHIKO, psychiatrist; b. Obihiro, Hokkaido, Japan, Nov. 6, 1940; s. Michimaro and Chieko (Saida) O.; m. Junko Takeda, Apr. 28, 1978; children: Takabumi, Mona. MD, Hokkaido U., Sapporo, 1964, postgrad., 1965-73. Intern Hokkaido U. Hosp., 1964-65, asst. prof. Health Adminstrn. Ctr., 1976-81, assoc. prof. Health Adminstrn. Ctr., 1981—. Contbr. articles to profl. jours. Research fellow Vanderbilt U., 1981-82. Fellow Japanese Assn. Autonomic Nerve; mem. Japanese Assn. Psychiat. Neurology, N.Y. Acad. Scis. Home: Chuo-ku S 10 W 18 1-3-304, Sapporo Hokkaido 064, Japan Office: Hokkaido U Health Adminstrn Ctr, North 8 West 5, Sapporo Hokkaido 060, Japan

OKADA, ROBERT DEAN, cardiologist; b. Seattle, Sept. 18, 1947; m. Carolyn Okada. BA summa cum laude, U. Wash., 1969; MD, U. Pa., 1973. Intern U. Ariz. Health Scis. Ctr., Tucson, 1973-74, resident in internal medicine, 1974-76, clin. fellow in cardiology, 1976-78; clin. and rsch. fellow in medicine Mass. Gen. Hosp., Boston, 1978-79; fellow in medicine Harvard Med. Sch., Boston, 1978-79, instr., 1979-81, asst. prof., 1981-85; prof. U. Okla. Med. Sch., Tulsa, 1985—; staff cardiologist St. Francis Hosp., Tulsa, 1985—; asst. in medicine Mass. Gen. Hosp., 1982-86, cons. in nuclear medicine, 1981—, sr. staff cardiac catheterization lab., 1979-86, clin. asst. in medicine, 1979-82; established investigator Am. Heart Assn., 1982-87; clin. prof. Tulsa Med. Coll., 1985. Contbr. articles to profl. jours. Recipient Am. Legion award U. Wash., 1966, Neisei Vets. award, 1966. Fellow Am. Coll. Cardiology, ACP, Am. Coll. Chest Physicians, N.Y. Acad. Scis.; mem. Am. Fedn. for Clin. Rsch., Paul Dudley White Soc., Soc. Nuclear Medicine, Am. Heart Assn. Coun. Clin. Cardiology, Soc. Magnetic Resonance in Medicine, AAAS, Mass. Med. Soc., Soc. for Magnetic Resonance Imaging, Suffolk County Med. Soc., Tulsa County Med. Soc., Okla. State Med. Soc., AMA, Am. Soc. Internal Medicine, AAUP, Phi Beta Kappa, Alpha Xi Sigma. Office: Cardiology of Tulsa Inc 6585 S Yale Ave Ste 800 Tulsa OK 74136-8374

OKADA, SHIGERU, pathology educator; b. Okayama, Japan, Feb. 15, 1940; s. Keizo and Moyoko (Nishigaki) O.; m. Naoko Kobashi, Nov. 7, 1965; children: Satoru, Rie, Mari. MD, Okayama U. Med. Sch., Japan, 1964, PhD, 1969. Asst. Okayama U. Med. Sch., Japan, 1969-71, lectr., 1971-77; chief pathologist Kyoto City Hosp., Japan, 1977-80; lectr. Kyoto U. Sch. Medicine, Japan, 1980-90; prof. Okayama U. Med. Sch., Japan, 1990; head Radiation Protection Com., Okayama U., 1991. Contbr. articles to profl. jours. Mem. Japan Pathol. Soc. Tokyo, Japan Haematological Soc. Kyoto, Internat. Soc. Hematology, Japanese Cancer Assn. Tokyo, N.Y. Acad. Sci. Office: Okayama U Med Sch, 2-5-1 Shikata, Okayama 700, Japan

OKAFOR, MICHAEL CHUKWUEMEKA, chemical engineer; b. Port-Harcourt, Nigeria, Apr. 25, 1955; came to the U.S., 1977; s. Samuel Iwenofu and Josephine Onarife (Ilonze) O.; m. Linda Mae Bornkessel, Mar. 28, 1980 (div. 1984); children: Toney, Kodilli. BSChE, U. Wyo., 1980, MBA in Fin., 1986. Registered profl. engr., Tex. Process engr. Darenco Inc., Casper, Wyo., 1981-83; pres. Smid Air Control Co., Denver, 1987-88; sr. process engr. M.W. Kellogg Co., Houston, 1988—; mem. safety bd. M.W. Kellogg Co., 1990-91. Contbr. articles to profl. jours. Fundraiser United Way, Houston, 1991. Mem. AIChE. Home: PO Box 741345 Houston TX 77274-1345 Office: MW Kellogg PO Box 4557 Houston TX 77210-4557

OKAMOTO, YOSHIO, chemistry educator, researcher; b. Osaka, Japan, Jan. 10, 1941; s. Takashi and Sawae O.; m. Kozue Misumi, Apr. 26, 1970; 1 child, Kazumasa. BS, Osaka U., 1964, MS, 1966, DSc, 1969. Asst. prof. Osaka U., Toyonaka, Japan, 1969-83, assoc. prof., 1983-90; rsch. assoc. U. Mich., Ann Arbor, 1970-72; prof. Nagoya U., Japan, 1990—. Inventor Synthesis of helical polymer, 1979, Resolution by liquid chromatography, 1981. Mem. Soc. Polymer Sci. Japan (award 1982), Chem. Soc. Japan (award 1991), Am. Chem. Soc., Japan Soc. for Analytical Chemistry. Office: Dept Applied Chem Nagoya Univ, Furo-cho, Chikusa-ku, Nagoya 464-01, Japan

OKAMURA, KIYOHISA, mechanical engineer, educator; b. Gwangyon, Korea, Feb. 8, 1935; s. Shundo and Tomoye (Mikami) O. BSME, Kyushu Inst. Tech., Japan, 1957; MSME, U. Tokyo, 1959; PhD, Purdue U., 1963. Mem. rsch. staff Japan Atomic Energy Rsch. Inst., 1959-60; sr. project engr. Allison div. GMC, Indpls., 1963-66; asst. prof. Rensselaer Poly. Inst., Troy, N.Y., 1966-68; assoc. prof. N.D. State U., Fargo, 1968-85; prof. dept. mech. engring. Bradley U., Peoria, Ill., 1985—; cons. Yomiuri-Shimbun, Boeing Co., Catepillar Co. Contbr. articles to profl. publs. Mem. ASME. Office: Bradley Univ Dept Mech Engring Peoria IL 61625

OKAWA, YOSHIKUNI, computer science educator; b. Tokyo, July 7, 1934; s. Zensaku and Kimi Okawa; m. Akiko Okawa; children: Yoshitaka, Kunihiko. BS, Tokyo U., 1959, MS, 1961, PhD, 1964. Assoc. prof. Yamagata U., Yonezawa, Japan, 1967-70; prof. Gifu (Japan) U., 1970-85, Osaka (Japan) U., 1985—. Author numerous books, papers; contbr. numerous articles to profl. jours. Recipient award Japan Soc. for Finishings Tech., 1993. Mem. Computer Soc. of IEEE, Am. Assn. Artificial Intelligence, Assn. for Computing Machinery. Avocations: swimming, walking, mountain climbing, tennis, reading. Home: 3-1-39-201 Kita MidorigaOka, Toyonaka, Osaka 560, Japan Japan Office: Faculty of Engring, Osaka U, 2-1 Yamada Oka, Suita, Osaka 565, Japan

OKAY, JOHN LOUIS, information scientist; b. Emmett, Mich., Mar. 27, 1942; s. Stanley John and Mildred Isabell (Little) O.; m. Judith Ann Gerlach, Aug. 22, 1964; children: Stephen, Christopher, Douglas. BS in Agr., Mich. State U., 1964, PhD in Resource Econs., 1974. Agrl. economist U.S. Soil Conservation Svc., East Lansing, Mich., 1967-73; program analyst U.S. Soil Conservation Svc., Washington, Mich., 1974-83; dir. info. systems U.S. Soil Conservation Svc., Washington, 1983-85; assoc. dir. info. systems USDA, Washington, 1985-91, dir. info. systems, 1991—. Recipient Meritorious Exec. award Pres. of U.S., 1989. Mem. Sr. Execs. Assn. (bd. dirs. 1989—). Office: Info Resources Mgmt/USDA 14th & Independence Ave SW Washington DC 20250

O'KEEFE, JOSEPH KIRK, systems engineer; b. Paxton, Ill., Sept. 30, 1933; s. Richard Joseph and Ruth Louise (Shinn) O'K.; m. Mary Helen Waters, Oct. 4, 1958; children: Richard Kirk, Catherine Elizabeth. B Aero. Engring., Rensselaer Poly. Inst., 1955, MS in Mgmt., 1963. Aerodynamist N.Am. Aviation, Inglewood, Calif., 1955, McDonnell Aircraft Corp., St. Louis, 1957-59; systems analyst Lockheed Missiles & Space Co., Inc., Sunnyvale, Calif., 1959-61, program mgr., 1963-90; test engr. AVCO Rsch. & Advanced Devel., Wilmington, Mass., 1961-62; dir. ARGOSystems-Tex., Corinth, 1990—; lectr. applied math. U. Santa Clara, Calif., 1963-65. Contbr. articles to profl. jours. 1st lt. USAF, 1955-57. Mem. Found. for N.Am. Wild Sheep, Safari Club Internat. Republican. Roman Catholic. Achievements include invention of innovative techniques to nearfield verification of microwave autotracking and very low bit error rate verification, mini-remote powered vehicle; design of small satellite that forms the basis for an on-going program. Home: 3207 Fairview Dr Corinth TX 76205-2627 Office: 7801 S Stemmons Corinth TX 75065

O'KEEFFE, HUGH WILLIAMS, oil industry executive; b. Ft. Smith, Ark., May 23, 1905; s. Patrick Francis and Elizabeth Ann (Williams) O'K.; m. Marrisa Lucylle Davis Durkee, Mar. 27, 1949 (dec. Dec. 1965); m. Josephine Helen Loughmiller, June 10, 1969 (dec. May 1980); m. Grace H. Freeny, Sept. 3, 1983 (dec. Jan. 1990). BS with honors, U. Ark., 1928. Jr. geologist Phillips Petroleum of Okla., Tex., Kans. Colo. and, Okla., 1928-33; party chief on surface Phillips Petroleum of Okla., Shawnee, 1933, dist. geologist, 1934-37; div. geologist Phillips Petroleum of Okla., Bartlesville, 1937-40, asst. chief geologist, 1940-45; exploration mgr. Davon Oil Co., Oklahoma City, 1946-52; co-owner Davon Oil & Gas, Oklahoma City, 1952-55; pvt. practice cons. Oklahoma City, 1955-58; co-owner Wyant & O'Keeffe, Oklahoma City, 1930—. Mem. Am. Assn. Petroleum Geologists, Oklahome City Geol. Soc., Soc. Ind. Petroleum and Earth Scientists, Shawnee Geol. Soc. (v.p. 1936-37), Tulsa Geol. Soc., Petroleum Club of Oklahoma City (charter), Beacon Club of Oklahoma City, Oklahoma City Golf and Country Club. Republican. Roman Catholic. Avocations: tennis, hunting, fishing. Home: 6511 Avondale Dr Oklahoma City OK 73116-6405 Office: Wyant & O'Keeffe 222 NE 50th St Oklahoma City OK 73105-1893

O'KEEFFE, MARY KATHLEEN, psychology educator; b. Sewickley, Pa., May 14, 1959. BA in Psychology with honors, Indiana U. Pa., 1981; PhD in Med. Psychology, Uniformed Svcs. U. Health Sci., 1991. Alcoholism counselor Gateway Rehab. Ctr., 1980, 81, Balt. City Hosps., 1981-82; rsch. asst. behavioral pharmacology rsch. unit Johns Hopkins U. Sch. Medicine, 1982-85; rsch. and teaching asst. dept. med. psychology Uniformed Svcs. U. Health Scis., 1985-91, rsch. assoc., 1991; asst. prof. dept. psychology Providence Coll., 1991—; cons. Cumberland (R.I.) Substance Abuse Prevention Task Force, Brown U. Ctr. for Alcohol Studies project; adminstrv. aide Acad. Behavioral Medicine rsch. conf., 1988, 89, div. 38 rsch. conf., 1988, Soc. Behavioral Medicine conf., 1986. Ad hoc reviewer Health Psychology, Jour. Applied Social Psychology, Jour. Exptl. Analysis of Behavior; contbr. articles to profl. jours. Grantee Com. to Aid Faculty Rsch., Providence Coll., 1992, Instrumentation and Lab. Improvement Grant, NSF, 1992. Mem. APA (student travel award 1989). Achievements include research in deterring preadolescent drug and AIDS risky behavior, Cathecholamine and blood pressure response to stress in coffee drinkers, chronic stress in victimized communities. Office: Providence Coll Dept Psychology Providence RI 02918

OKHI, SHINPEI, biophysicist; b. Kawagoe, Saitama, Japan, Jan. 1, 1933; s. Akira and Shizue (Hotta) O.; m. Catherine Balasz, May 30, 1972; children: Elise, Thomas. BS, Kyoto U., 1956, MS, 1957, PhD, 1965. Instr. Tokyo Met. U., 1962-67; asst. prof. biophysics SUNY, Buffalo, 1969-74, assoc. prof., 1974-85, prof., 1986—. Editor: Molecular Mechanisms of Membrane Fusion, 1988; Cell and Model Membrane Interactions, 1991. Grantee NIH, 1978—. Mem. United Ch. of Christ. Office: SUNY Dept Biophysics 224 Cary Hall Buffalo NY 14214

OKITANI, AKIHIRO, food chemistry educator; b. Yotsuya-ku, Tokyo, Japan, Nov. 15, 1940; s. Jinzo and Hisako Okitani; m. Chieko Aoki, Mar. 6, 1967; children: Yuko, Makiko. BS, U. Tokyo, 1963, PhD, 1968. Rsch. assoc. U. Tokyo, 1968-84; assoc. prof. Nippon Vet. and Animal Sci. U., Musashino-shi, Tokyo, 1984-85, prof., 1985—; vis. researcher Iowa State U., Ames, 1975-76. Mem. Japan Soc. Nutrition and Food Sci. (mem. coun. 1986—, editor 1988-92), Japan Soc. Zootech. Sci. (mem. coun. 1991—), Japan Soc. Biosci. and Biotech. Agrochemistry (assoc. editor 1986—), Japan Soc. Meat and Meat Products (mem. coun. 1987—, v.p. 1993—). Avocation: oil painting. Home: 19-23 Inogata 2-chome, Komae-shi Tokyo 201, Japan Office: Nippon Vet and Animal Sci U, 7-1 Kyonan-cho 1 chome, Musashino-shi Tokyo 180, Japan

OKRENT, DAVID, engineering educator; b. Passaic, N.J., Apr. 19, 1922; s. Abram and Gussie (Pearlman) O.; m. Rita Gilda Holtzman, Feb. 1, 1948; children—Neil, Nina, Jocelyne. M.E., Stevens Inst. Tech., 1943; M.A., Harvard, 1948, Ph.D. in Physics, 1951. Mech. engr. NACA, Cleve., 1943-46; sr. physicist Argonne (Ill.) Nat. Lab., 1951-71; regents lectr. UCLA, 1968, prof. engring., 1971-91, prof. emeritus, rsch. prof., 1991—; vis. prof. U. Wash., Seattle, 1963, U. Ariz., Tucson, 1970-71; Isaac Taylor chair Technion, 1977-78. Author: Fast Reactor Cross Sections, 1960, Computing Methods in Reactor Physics, 1968, Reactivity Coefficients in Large Fast Power Reactors, 1970, Nuclear Reactor Safety, 1981; contbr. articles to profl. jours. Mem. adv. com. on reactor safeguards AEC, 1963-87, also chmn., 1966; sci. sec. to sec. gen. of Geneva Conf., 1958; mem. U.S. del. to all Geneva Atoms for Peace Confs. Guggenheim fellow, 1961-62, 77-78; recipient Disting. Appointment award Argonne Univs. Assn., 1970, Disting. Service award U.S. Nuclear Regulatory Commn., 1985. Fellow Am. Phys. Soc., Am. Nuclear Soc. (Tommy Thompson award 1980, Glenn Seaborg medal 1987), Nat. Acad. Engring. Home: 439 Veteran Ave Los Angeles CA 90024-1956

OKU, AKIRA, chemical engineer, educator; b. Ueda, Japan, June 19, 1938; s. Masami and Etsuko Oku; children: Wakako, Yusuke. D in Engring., Kyoto U., 1966. Postdoctoral fellow Mich. State U., East Lansing, 1966-68; asst. prof. Kyoto (Japan) Inst. Tech., 1968-70, assoc. prof., 1970-79, prof., 1979—; mem. com. Ministry Internat. Trade and Industry, 1989—; chmn. Internat. Symposium on Carbene Chemistry, 1989. Co-author: Organic Medicinal Chemistry, 1989; contbr. articles to profl. jours. Nissha scholar Inst. for Prodn. Tech., 1988; grantee Kashima Found., 1990. Mem. Chem. Soc. Japan, Am. Chem. Soc., Synthetic Organic Chemistry, Soc. Polymer Sci., Soc. Environ. Sci. Achievements include patent for chemical destruction of CFCs; research in synthesis of polyparabanic acid precursors, reductive destruction of PCBs, synthetic reactions utilizing carbene-type reactive molecules, organic electron transfer reactions. Office: Kyoto Inst Tech, Matsugasaki Sakyo-ku, Kyoto 606, Japan

OKU, TATSUO, mechanical engineering educator; b. Kagoshima, Japan, Aug. 28, 1934; s. Mitsuo and Shizu (Matsumoto) O.; m. Hiroko Nakauchi, Mar. 30, 1964; children: Takeo, Yukari. B in Engring., Waseda U., Tokyo, 1958, M in Engring., 1960; DEng, Tokyo Inst. Tech., 1975. Researcher Nissan Chem. Co., Ltd., Tokyo, 1960-61; researcher Japan Atomic Energy Rsch. Inst., Tokai-mura, Japan, 1961-72, sr. engr., 1972-77, prin. engr., head materials strength lab., 1977-90; prof. dept. mech. engring. Ibaraki U., Hitachi, Japan, 1990—; mem. ad hoc investigation com. on Soviet Nuclear Power Plant Accident, 1986-87; coop. researcher Nat. Inst. Fusion Sci., Nagoya, Japan, 1990—; rsch. cons. Japan Atomic Energy Rsch. Inst., Tokai-mura, 1991—. Co-author: (book) Nuclear Materials, 1972, Fusion Reactor Materials, 1986, Ceramic Engineering Handbook, 1989, also co-editor: Reactor Materials Handbook, 1977; editor-in-chief: Carbon Soc. of Japan, 1985-87. Recipient original paper award Atomic Energy Soc. Japan, 1981, 90. Mem. Japan Soc. Promotion of Sci. Home: 2920-217 Mawatari Mukaino, Katsuta Ibaraki 312, Japan Office: Ibaraki U Faculty Engring, 12-1 Nakanarusawa 4 chome, Hitachi 316, Japan

OKUDA, KUNIO, emeritus medical educator; b. Tokyo, May 21, 1921; s. Kinmatsu and Hatsue (Hashida) O.; m. Hinae Katsumata, May 24, 1947; children: Hiroaki, Keiko. MD, Manchuria Med. Coll., 1944; DSc, Chiba Med. Coll., Japan, 1951; PhD, Fukuoka U., 1983. With med. staff Chiba (Japan) Nat. Hosp., 1945-51; asst. prof. Yamaguchi Med. Coll., Ube, Japan, 1951-53, assoc. prof., 1953-63; asst. prof. Johns Hopkins Sch. Medicine, Balt., 1958-60; prof. medicine Kurume U. Sch. Medicine, Japan, 1963-71; prof. medicine Chiba U. Sch. Medicine, 1971-87, emeritus prof., 1987—; past pres. Japanese Soc. Hepatology, Tokyo, 1977-78; v.p. Orgn. de Gastro-Enterologie, 1982-86, hon. pres., 1986—. Contbr. articles to profl. jours. Recipient Spl. award Japanese Med. Assn., 1967, Abbott award Japanese Nuclear Medicine Soc., Publ. and Translation Cultural award Japanese Publ. Soc., 1977, Disting. Svc. award Am. Assn. for Study of Liver Diseases, 1990. Mem. Japan Vitamin Soc. (pres. 1987-91, Spl. award 1963), Orgn. Mondiale de Gastro-Enterologie (hon. pres. 1986—), Internat. Assn. Study of Liver (past pres. 1978-80), Asian Pacific Assn. Study of Liver (past pres. 1980-82, hon. pres. 1990—), Am. Gastroenterological Assn., Soc. Scholars Johns Hopkins U. Lodge: Rotary. Avocations: violin, music, stamp collecting.

OKUI, KAZUMITSU, biology educator; b. Ohta, Japan, July 8, 1933; s. Sadajiroh and Ume (Tanaka) O.; m. Mizue Aoki Okui, May 6, 1961; children: Teiichiroh, Ari. B Agr., Tokyo U. Agr., 1962, M Agr., 1964, D Agr., 1967. Lectr. biology Denki-Tsushin U., Tokyo, 1968-70, Kitasato U., Sagamihara, Japan, 1970-73; asst. prof. biology Kitasato U., 1973-82, prof. ethology, 1982—; v.p. Internat. Centre of Wild Silkworm, 1990—; vis. rsch. prof. Waikato U., New Zealand, 1991; mem. book rev. com. Yomiuri Shimbun, Tokyo, 1981-85, Sankei Shimbun, Tokyo, 1990—. Author: Entomology, 1976, Ethology, 1976, General Zoology, 1984, General Zoology, 1985, General Entomology, 1992, Human Ethology, 1992. Mem. Internat. Soc. Wild Silkworm (sec. 1988—, v.p. 1990—), Japan Cosmo-Biol. Soc. (councilor 1987-89). Home: 972-7 Yumoto-machi, 250-03 Kanagawa-ken Japan Office: Sch Liberal Arts Kitasato U, 1-15-1 Kitasato, Sagamihara 228 Kanagawa, Japan

OKUN, DANIEL ALEXANDER, environmental engineering educator, consulting engineer; b. N.Y.C., June 19, 1917; s. William Howard and Leah (Seligman) O.; m. Elizabeth Griffin, Jan. 14, 1946; children: Michael Griffin, Tema Jon. BS, Cooper Union, 1937; MS, Calif. Inst. Tech., 1938; ScD, Harvard U., 1948. Registered profl. engr., N.C., N.Y. With USPHS, 1940-42; teaching fellow Harvard, 1946-48; with Malcolm Pirnie (cons. environ. engrs.), N.Y.C., 1948-52; asso. prof. dept. environ. scis. and engring. U. N.C. at Chapel Hill, 1952-55, prof., 1955-73, Kenan prof., 1973-87, Kenan prof. emeritus, 1987—, head dept. environ. scis. and engring., 1955-73; bd. dirs. Water Resources Research Inst., 1965-84, chmn. faculty, 1970-73; vis. prof. Tech. U. Delft, 1960-61, Univ. Coll., London, 1966-67, 73-75, Tianjin U., 1981; editor environ. scis. series Acad. Press., 1968-75; cons. to industry, cons. engrs., govtl. agys. World Bank, WHO, UNDP, with spl. svc. in Switzerland, Israel, Jordan, Peru, Egypt, Colombia, Brazil, Venezuela, Thailand, Indonesia, Kenya, Zambia, Tunisia, Togo, Morocco, People's Republic China; mem. environ. adv. coun. Rohm & haas Co., Inc., 1985-92; chmn. expert panel on N.Y.C. water supply EPA, 1992-93. Author: (with Gordon M. Fair and John C. Geyer) Water and Wastewater Engineering, 2 vols., 1966-68, Elements of Water Supply and Wastewater Disposal, 1971; (with George Ponghis) Community Wastewater Collection and Disposal, 1975; Regionalization of Water Management—A Revolution in England and Wales, 1977; editor: (with M.B. Pescod) Water Supply and Wastewater Disposal in Developing Countries, 1971; (with C.R. Schulz) Surface Water Treatment for Communities in Developing Countries, 1984; contbr. to publs. in field. Chmn. Chapel Hill Fellowship for Sch. Integration, 1961-63; mem. adv. bd. Ackland Meml. Art Mus., 1973-78; bd. dirs. Warren Regional Planning Corp., 1971-77, Inter-Ch. Council Housing Corp., 1975-83, N.C. Water Quality Council, 1975-77; adv. com. for med. research Pan Am. Health Orgn., 1976-79; chmn. Washington Met. Area Water Supply Study Com., 1976-80, NAS-NRC; bd. sci. and tech. for internat. devel. NRC, 1978-81, vice chmn. environ. studies bd., 1980-83, chmn. water sci. and tech. bd., 1991—; mem. com. on human rights NAS, 1988—; pres. Chapel Hill chpt. N.C. Civil Liberties Union, 1991-93. Served from lt. to maj. AUS, 1942-46. Recipient Harrison Prescott Eddy medal for research Water Pollution Control Fedn., 1950, Gordon Maskew Fair award Am. Acad. Environ. Engrs., 1973, Thomas Jefferson award U. N.C. at Chapel Hill, 1973, Gordon Y. Billard award N.Y. Acad. Scis., 1975; 1st Thomas R. Camp Meml. lectr. Boston Soc. Environ. Engrs.; Gordon Maskew Fair medal Water Pollution Control Fedn., 1978; Friendship medal Inst. Water Engrs. and Scientists (Gt. Britain), 1984; NSF fellow, 1960-61; Fed. Water Pollution Control Adminstrn. fellow, 1966-67; Fulbright-Hayes lectr., 1973-74. Mem. NAE, AAUP (U. N.C. chpt. pres. 1963-64), ASCE (chmn. environ. engring. div. 1967-68, 1st Simon W. Freese award 1977), Am. Water Works Assn. (hon., N.C. Fuller award 1983, best paper award ednl. div. 1985, Abel Wolman award of excellence 1991), Inst. Medicine, Water Pollution Control Fedn. (hon., chmn. rsch. com. 1961-66, dir.-at-large 1969-72), Am. Acad. Engring. (pres. 1969-70, hon. diplomate), Order of Golden Fleece, Sigma Xi (U. N.C. pres. chpt. 1968-69). Home: 900 Linden Rd Chapel Hill NC 27514-9162

OKUNIEFF, PAUL, radiation oncologist, physician; b. Chgo., Mar. 8, 1957; s. Michael and Beverly Okunieff; m. Debra Trione, Sept. 7, 1989. SB in Elec. Engring. & Biology, MIT, 1978; MD, Harvard U., 1982. Diplomate Nat. Bd. Med. Examiners; cert. in therapeutic radiology Am. Bd. Radiology. Intern in medicine Beth Israel Hosp., Boston, 1982-83; resident in radiation oncology Mass. Gen. Hosp., Boston, 1983-86, fellow in radiation oncology, 1986-87, asst. prof. radiation oncology, 1987-93; chief dept. radiation oncology NIH, Bethesda, Md., 1993—; instr. in radiation medicine Harvard Med. Sch., Boston, 1987, asst. prof. radiation oncology, 1988—; asst. radiation therapist Waltham (Mass.) Hosp. Med. Ctr., 1989—, Mt. Auburn Hosp., Cambridge, Mass., 1989—; assoc. radiation oncologist Univ. Hosp., Boston, 1990—, co-dir. dept. radiation therapy, 1990-91; chief radiation neuro-oncology Mass. Gen. Hosp., Boston, 1991-93; chief radiation oncology br. Nat. Cancer Inst., Bethesda, 1993—; invited lectr. various med. schs., congresses & confs. Contbr. over 50 articles to profl. jours., also chpts. to books. Recipient Essay award IEEE, 1978, Young Oncologist Essay award Am. Radium Soc., 1987, travel award Am. Coll. Radiology, 1987, Young Investigator travel award VIth Internat. Meeting on Chem. Modifiers of Cancer Treatment, 1988, Basic Sci. travel grant ASTRO Am. Meeting, 1989, USNC/UICC travel grant UICC Meeting, Hamburg, Germany, 1990, Melvin H. Knisely award Internat. Soc. Oxygen, 1991; grantee/fellow Am. Cancer Soc., 1985-86, 86-87, 89-92, Mass. Gen. Hosp., 1986, NIH, 1988-93. Office: NCI-Radiation Oncology Branch Bldg 10 9000 rockville Pike Bethesda MD 20892

OKUYAMA, SHINICHI, physician; b. Yamagata, Japan, Dec. 4, 1935; s. Kinzo Okuyama and Asayo Hasegawa; m. Masako Fujii, Dec. 4, 1966 (dec.); children: Yuriko, Izumi, Takashi, Jun; m. Junko Hsun Chen, Mar. 21, 1983; children: Midori, Shaw. MD, Tohoku (Japan) U., 1961, PhD, 1966. Intern Saiseikan Hosp., Yamagata, 1961-62; resident in radiology Tohoku U. Research Inst., Sendai, Japan, 1962-66; research assoc. Tohoku U. Inst. Tb, Leprosy and Cancer, 1966-73, assoc. prof. radiology, 1974-80; dir. radiology Tohoku Rosai Hosp., Sendai, 1980—. Author: Diagnostic Bone Scintigraphy, 1974, Compton Radiography, 1979, Induction of Cancer Redifferentiation, 1983, Evolution of Cancer, 1990, Origin of Reed-Sternberg cell in Hodgkin's Disease, 1991, Evolution of Human Diseases, 1991. Recipient

Compton Tomography-Radiotherapy Planning award Japanese Ministry of Edn., 1980. Mem. AAAS, Japanese Soc. Radiology, Japanese Soc. Nuclear Medicine, Japanese Soc. Reticuloendothelial Systems, Japanese Soc. Internal Medicine, N.Y. Acad. Scis. Achievements include reinforcing aerosol cisplatin for radiotheraphy of laryngeal cancer, pasting chemotherapy for raditherapy of uterine, rectal and esophageal cancers. Home: Kamo 4-4-5, Izumi-ku, Sendai 981-31, Japan Office: Tohoku Rosai Hosp, Dainohara 4-3-21, Aoba-ku, Sendai 981, Japan

OLAGUNJU, DAVID OLAREWAJU, mathematician, educator; b. Nsuta, Ghana, Sept. 8, 1948; came to the U.S., 1976; s. Elijah Adeyeye and Deborah (Arinpe) O.; m. Elizabeth O. Olawole, Dec. 13, 1975; children: Deji, Yemisi, Segun, Adeola. BS, Ahmadu Bello U., Zaria, Nigeria, 1974, MS, 1977; PhD, Northwestern U., 1981. Asst. lectr. Ahmadu Bello U., 1976-79, lectr. II, 1979-82, lectr. I, 1982-85; sr. lectr. Nigerian Def. Acad., Kaduna, 1985-88; asst. prof. U. Del., Newark, 1989—; vis. scholar Northwestern U., Evanston, Ill., 1987, rsch. fellow, 1986; acting head math. Nigerian Def. Acad., 1985-88. Contbr. articles to Siam Jour. Applied MAth., Jour. Non-Newtonian Fluid Mechanics, Quar. Jour. Applied Math. Nat. Merit scholar Govt. of Nigeria, 1972-74; Afgrad fellow African-Am. Inst., 1977-80, Royal Cabell fellow Northwestern U., 1980-81; grantee U. Del. Rsch. Found., 1991; recipient math. prize Internat. Computers Ltd., 1974. Mem. Soc. for Indsl. and Applied Math., Soc. Rheology, Tau Beta Pi. Office: U Del 501 Ewing Hall Newark DE 19716

OLAH, GEORGE ANDREW, chemist, educator; b. Budapest, Hungary, May 22, 1927; came to U.S., 1964, naturalized, 1970; s. Julius and Magda (Rasznai) O.; m. Judith Agnes Lengyel, July 9, 1949; children: George John, Ronald Peter. PhD, Tech. U. Budapest, 1949, hon. degree, 1989; DSc honoris causa, U. Durham, 1988, U. Munich, 1990. Mem. faculty Tech. U. Budapest, 1949-54; asso. dir. Cen. Chem. Rsch. Inst., Hungarian Acad. Scis., 1954-56; rsch. scientist Dow Chem. Can. Ltd., 1957-64, Dow Chem. Co., Framingham, Mass., 1964-65; prof. chemistry Case-Western Res. U., Cleve., 1965-69, C.F. Mabery prof. rsch., 1969-77; Donald P. and Katherine B. Loker disting. prof. chemistry, dir. Hydrocarbon Rsch. Inst., U. So. Calif., L.A., 1977—; vis. prof. chemistry Ohio State U., 1963, U. Heidelberg, Germany, 1965, U. Colo., 1969, Swiss Fed. Inst. Tech., 1972, U. Munich, 1973, U. London, 1973-79, L. Pasteur U., Strasbourg, 1974, U. Paris, 1981; hon. vis. lectr. U. London, 1981; cons. to industry. Author: Friedel-Crafts Reactions, Vols. I-IV, 1963-64, (with P. Schleyer) Carbonium Ions, Vols. I-V, 1969-76, Friedel-Crafts Chemistry, 1973, Carbocations and Electrophilic Reactions, 1973, Halonium Ions, 1975, (with G.K.S. Prakash and J. Somer) Superacids, 1984; (with Prakash, R.E. Williams, L.D. Field and K. Wade) Hypercarbon Chemistry, 1987, (with R. Malthotra and S.C. Narang) Nitration, 1989, Cage Hydrocarbons, 1990, (with Wade and Williams) Electron Deficient Boron and Carbon Clusters, 1991, (with Chambers and Prarash) Synthetic Fluorine Chemistry, 1992, also chpts. in books, numerous papers in field; patentee in field. Recipient Leo Hendrik Baekeland award N.J. sect. Am. Chem. Soc., 1966, Morley medal Cleve. sect., 1970; Alexander von Humboldt Sr. U.S. Scientist award, 1979, Pioneer of Chemistry award Am. Inst. Chemists, 1993. Fellow AAAS, Chem. Inst. Can.; mem. NAS, Italian NAS, European Acad. Arts, Scis. and Humanities, Italy Chem. Soc. (hon.), Hungarian Acad. Sci. (hon.), Am. Chem. Soc. (award petroleum chemistry 1964, award Synthetic organic chemistry 1979, Roger Adams award in organic chemistry 1989), German Chem. Soc., Brit. Chem. Soc. (Centenary lectr. 1978), Swiss Chem. Soc., Sigma Xi. Home: 2252 Gloaming Way Beverly Hills CA 90210-1717 Office: U So Calif Dept Chemistry Los Angeles CA 90007

OLANOW, C(HARLES) WARREN, neurologist, educator; b. Toronto, Ont., Can., Aug. 15, 1941; s. Max Ronald and Betty (Bright) O.; m. Mariana Faroyh, Apr. 2, 1970; children: Edward, James, Alessandra, Andrew. Student, U. Toronto, 1961, MD, 1965. Asst. in neurology Columbia U., N.Y.C., 1970-71; lectr. McGill U., Montreal, Que., Can., 1971-77; asst. prof., then assoc. prof. Duke U., Durham, N.C., 1977-86; prof. neurology U. South Fla., Tampa, 1986—, prof. psychiatry, 1992—, prof. pharmacology and exptl. therapeutics, 1991—; med. adv. bd. United Parkinson Found., Chgo., 1988—, Am. Parkinson Disease Assn., N.Y., 1991—; editorial bd. Jour. Neural Transmission. Editor Annals of Neurology Supplement, 1992; contbr. over 30 chpts. to books, over 100 articles to profl. jours. Fellow Am. Acad. Neurology; mem. Am. Neurol. Assn., Internat. Soc. Movement Disorders (v.p., pres. 1990-92), Movement Disorder Soc. (treas. 1992). Achievements include development of pergolide and deprenyl, original study of central pharmacokinetics of L-dopa in Parkinson's Disease, studies of iron and oxidant stress in pathogenesis of Parkinson's Disease. Office: Movement Disorder Ctr Ste 410 4 Columbia Dr Tampa FL 33606

OLBRANTZ, DON LEE, chemist; b. Camp Pendleton, Calif., Oct. 27, 1951; s. Julius John Jr. and Aurelia Lee (Kosmickie) O.; m. Roberta Gwendolyn Longacre, June 21, 1975; 1 child, Justin Matthew. AA, Santa Ana Coll., 1971; BA, Calif. State U., Long Beach, 1977. Chemist Michelson Labs., Inc., Commerce, Calif., 1976-80, mgr. dept. chemistry, 1980-82, dir. lab. svcs., 1982-89, mgr. quality assurance, 1989-90, analytical chemist in instrumentation, 1990-92; tech. dir., 1992—. Chair Whittier area chpt. Crusade for Life, 1987—. Mem. ASTM, Am. Chem. Soc. Operation Rescue. Republican. Christian Evangelical. Home: 15033 Neartree Rd La Mirada CA 90638 Office: Michelson Labs Inc 6280 Chalet Dr Commerce CA 90040

OLCOTT, RICHARD JAY, data processing professional; b. Chgo., Sept. 20, 1940; s. Irwin and Sylvia Jane (Gutstadt) O.; m. Valerie Gale Hirst, June 30, 1962; children: Lorian Elizabeth, Timothy Jason. BS in Chemistry, Beloit Coll., 1961; PhD in Inorganic Chemistry, U. Wis., 1972. Asst. prof. chemistry Southwestern Coll., Memphis, 1968-72; rsch. fellow St. Jude Children's Rsch. Hosp., Memphis, 1972-74; programmer-analyst Cook Industries, Memphis, 1974-77; systems programmer Blue Cross/Blue Shield Memphis, 1977-80; sr. performance analyst Schering-Plough Corp., Memphis, 1980—. Editor: Computer Mgmt. and Evaluation Newsletter, 1983-88, CMG Transactions, 1993—; contbr. articles to profl. jours. Bd. dirs. Friends of the Orpheum, 1986-92; mem. exec. com. Vollentine-Evergreen Community Assn., Memphis, 1987—. NSF rsch. fellow, 1965-68. Mem. SHARE, Inc., Computer Mgmt. Group Inc., Assn. for Computing Machinery, Fedn. Am. Aquarium Socs. (founding pres. 1973). Achievements include research in computer systems management. Home: 1853 Snowden Memphis TN 38107 Office: Schering-Plough Corp 3030 Jackson Memphis TN 38151

OLCZAK, PAUL VINCENT, psychologist; b. Buffalo, N.Y., May 25, 1943; s. Vincent Henry and Helen (Babula) O.; m. Marie Rose Oliveri, Oct. 20, 1973; children: Paul V. II, Patrick J., Drew M. MA, No. Ill. U., 1969, PhD, 1972. Clin. psychologist Family Ct. Psychiat. Clinic, Buffalo, 1975-77, cons. supervisory psychologist, 1977—; supr. psychol. svcs. Hopevale, Inc., Hamburg, N.Y., 1977-89; clin. psychologist Amherst (N.Y.) Police Dept., 1989—; asst. prof. psychology SUNY, Geneseo, 1977-83, assoc. prof. psychology, 1983-90, prof. psychology, 1990—. Co-editor: Community Mediation, 1991; contbg. author: The POI in Clinical Situations: A Review, 1991, Self-actualization-Polemics Surrounding Its Use, 1991; contbr. articles to profl. jours./publs. Mem. Am. Psychol. Assn., Ea. Psychol. Assn., Midwestern Psychol. Assn., Psychonomic Soc., Soc. Exptl. Social Psychology, Psi Chi, Sigma Xi. Home: 185 N Long St Buffalo NY 14221-5313 Office: SUNY Dept Psychology Geneseo NY 14454

OLDAKER, BRUCE GORDON, physicist, military officer; b. Albuquerque, June 3, 1950; s. Marion Joseph and Minerva Rae (Rogers) O.; m. Patricia Rose Cooney, Feb. 17, 1973; children: Ian Joseph, Kathleen Marie. B Math., U. Minn., 1972; MS, U. Colo., 1981; PhD, MIT, 1990. Commd. 2d lt. U.S. Army, 1972, advanced through grades to lt. col., 1990; asst. prof., now assoc. prof. physics U.S. Mil. Acad. U.S. Army, West Point, N.Y., 1981—, dir. Photonics Rsch. Ctr., U.S. Mil. Acad., 1990—. Contbr. articles to sci. jours. Hertz Found. fellow, 1985-89. Mem. Am. Phys. Soc., Assn. U.S. Army, Sigma Xi, Phi Kappa Phi. Achievements include proving that momentum distribution of atoms after interaction with a photon field can be used to study properties of the photon field. Home: 256 A Beauregard Pl West Point NY 10996 Office: Photonics Rsch Ctr US Mil Acad Bldg 753 Rm B 21 West Point NY 10996

OLDEN, KENNETH, public health service administrator, researcher; b. Newport, Tenn., July 22, 1938; s. Mack L. and Augusta (Christmas) O.; m. Sandra L. White; children: Rosalind, Kenneth, Stephen, Heather. BS, Knoxville Coll., 1960; MS, U. Mich., 1964; PhD, Temple U., 1970. Rsch. fellow, physiology instr. Harvard U., Cambridge, Mass., 1970-74; sr. staff fellow NIH, Nat. Cancer Inst., Bethesda, Md., 1974-77, biochemistry expert, 1977-78, rsch. biologist, 1978-79; assoc. dir. rsch. Howard U. Med. Sch. Cancer Ctr., Washington, 1979-82, dep. dir., 1982-85, dir., 1985-91; dir. Nat. Inst. Environ. Health Scis. and Nat. Toxicology Program NIH, Rsch. Triangle Park, N.C., 1991—. Assoc. editor: Cancer Rsch., 1990—, Jour. Nat. Cancer Inst., 1990—, Molecular Biology of the Cell, 1991-93, Environmental Health Perspectives, 1992—. Mem. awards bd. Gen. Motors Cancer Rsch. Found., Detroit, 1992—. Porter Devel. Postdoctoral fellow Am. Physiol. Soc., 1970, Postdoctoral fellow NIH, 1970-73, Macy Faculty fellow Macy Found., 1973-74. Mem. Am. Soc. Cell Biology, Am. Soc. Biol. Chemistry, Am. Assn. Cancer Rsch. (bd.dirs.), So. Biol. Response Modifiers, Metastasis Rsch. Soc., Internat. Soc. Study Comparative Oncology. Baptist. Avocations: tennis, hiking, cycling, cooking. Home: 19 Quail Ridge Rd Durham NC 27705-1870 Office: Nat Inst Environ Health Scis PO Box 12233 Research Triangle Park NC 27709-2233

OLDFIELD, JAMES EDMUND, nutrition educator; b. Victoria, B.C., Can., Aug. 30, 1921; came to U.S., 1949; s. Henry Clarence and Doris O. Oldfield; m. Mildred E. Atkinson, Sept. 4, 1942; children: Nancy E. Oldfield McLaren, Kathleen E. Oldfield Sansone, David J., Jane E. Oldfield Imper, Richard A. BSA, U. B.C., 1941, MSA, 1949; PhD, Oreg. State U., 1951. Faculty Oreg. State U., Corvallis, 1951—, head dept. animal sci., 1967-83, dir. Nutrition Research Inst., 1986-90; mem. nat. tech. adv. com. on water supply U.S. Dept. Interior, Washington, 1967-68; bd. dirs. Coun. for Agrl. Sci. and Tech., Ames, Iowa, 1978-84; mem. nutrition study sect. NIH, Bethesda, Md., 1975-80, 85-87. Editor: Selenium in Biomedicine, 1967, Sulphur in Nutrition, 1970, Selenium in Biology and Medicine, 1987; author: Selenium in Nutrition, 1971. Served to maj. Can. Army, 1942-46, ETO. Fulbright research scholar U.S. Dept. State, 1974, Massey U., New Zealand. Fellow Am. Soc. Animal Sci. (pres. 1966-67, Morrison award 1972), Am. Inst. Nutrition; mem. Am. Chem. Soc., Am. Registry Profl. Animal Scientists (pres. 1990), Fed. Am. Socs. Exptl. Biol. Republican. Episcopalian. Lodge: Kiwanis (pres. 1964, lt. gov. 1986). Home: 1325 NW 15th St Corvallis OR 97330-2604 Office: Oreg State U Dept Animal Sci Corvallis OR 97331

OLDHAM, TIMOTHY RICHARD, physicist; b. St. Louis, Jan. 14, 1947; s. Richard Thomas and Gladys Marion (Althen) O.; m. Kathleen V. McKnight, Feb. 11, 1984; children: Kristin, Katherine, Karoline, Kimberly. BS in Physics, Mich. State U., 1969; Ms in Physics, Am. U., 1975; PhD in Physics, Cath. U. Am., 1982. Physicist U.S. Army-Harry Diamond Lab., Adelphi, Md., 1969-85; supervisory physicist U.S. Army-Harry Diamond Lab., Adelphi, 1985—; cons. Lab. for Phys. Scis., College Park, Md., 1981—, Def. Advanced Rsch. Projects Agy., Arlington, Va., 1991—, U.S. Army Strategic Def. Command, Huntsville, Ala., 1991—; tech. program chmn. IEEE Nuclear and Space Radiation Effects Conf., Reno, Nev., 1990, gen. conf. chmn., Tucson, 1994; program com. mem. IEEE Semiconductor Interface Specialists Conf., 1990—. Contbr.: (book) Ionizing Radiation Effects in MOS Devices & Circuits, 1989; contbr. articles to profl. jours. Mem. AAAS, IEEE, Am. Phys. Soc., Toastmasters Internat. (Adelphi). Achievements include experimental and theoretical contributions to understanding field dependent recombination in microelectronic insulators, crucial to understanding differences in test results obtained in different radiation sources.

OLDSHUE, JAMES Y., chemical engineering consultant; b. Chgo., Apr. 18, 1925; s. James and Louise (Young) O.; m. Betty Ann Wiersema, June 14, 1947; children: Paul, Richard, Robert. B.S. in Chem. Engring., Ill. Inst. Tech., 1947, M.S., 1949, Ph.D. in Chem. Engring., 1951. Registered engr., N.Y. With Mixing Equipment Co., Rochester, N.Y., 1950-92, dir. research, 1960-63, tech. dir., 1963-70, v.p. mixing tech., 1970-92; pres. Oldshue Techs. Internat., Inc., Fairport, N.Y., 1992—; adj. prof. chem. engring. Beijing Inst. Chem. Tech., 1992—. Author: Fluid Mixing Technology, 1983; contbr. chpts. and articles to books and jours. Chmn. budget com. Internat. div. YMCA; bd. dirs. Rochester YMCA. Served with AUS, 1945-47. Recipient 1st Disting. Svc. award NE YMCA Internat. Com., 1979; named Rochester Engr. of Yr., 1980. Fellow Am. Inst. Chem. Engrs. (pres. 1979, treas. 1983-89, Founders award 1981, Eminent Chem. Engr. award 1983, Svc. to Soc. award 1989, chmn. internat. activities com. 1989-92); mem. NAE, Am. Assn. Engring. Socs. (chmn. 1985, K.A. Roe award 1987), Am. Chem. Soc., Internat. Platform Assn., World Congress Chem. Engrs. (v.p. 1986), N.Am. Mixing Forum (chmn. 1990-93, Mixing Achievement rsch. award 1992), Interam. Confederation Chem. Engrs. (sec. gen. 1991-93, v.p. 1993—, Victor Marquez award 1983), Rochester Engring. Soc. (pres. 1992-93). Mem. Reformed Ch. in Am. (gen. program coun.). Achievements include design and scale-up procedures in field of fluid mixing. Home: 141 Tyringham Rd Rochester NY 14617-2522 Office: 811 Ayvault Rd Fairport NY 14450

OLEA, RICARDO ANTONIO, geological researcher, engineer; b. Santiago, Chile, July 1, 1942; came to U.S., 1985; s. Ricardo Humberto and Celinda Antonia (Meneses) O.; m. Lucila Amelia Valencia, Sept. 2, 1966; children: Claudia, Ricardo, Pablo. Mining Engr., Universidad de Chile, Santiago, 1967; MS, U. Kans., 1972, DEng, 1982. Prof. Universidad Técnica del Estado, Punta Arenas, Chile, 1968-70; seismologist Empresa Nacional del Petróleo, Punta Arenas, Chile, 1967-72; math. geologist Empresa Nacional del Petróleo, Santiago, Chile, 1972-82, asst. to v.p., 1982-85; vis. indsl. scientist Kans. Geol. Survey, Lawrence, 1970-72, 74-75, rsch. assoc., 1979—; cons. Texaco, Houston, 1983-86, EPA, Las Vegas, 1988, Marathon Oil Co., Denver, 1988—, GE, Albany, N.Y., 1990-91, Davis Cons., Baldwin City, Kans., 1990—. Author: Optimum Mapping Techniques, 1975, Measuring Spatial Dependence with Semivariograms, 1975, LOG II, 1981, Systematic Sampling of Spatial Functions, 1984, CORRELATOR, 1988; editor: Geostatistical Glossary and Multilingual Dictionary, 1991; author and co-author 40 papers and 120 reports. Adv. panel Celebration of Cultures, Lawrence, 1990-91. Recipient Juan Brüggen prize Instituto Ingenieros de Minas, Santiago, 1966, Task Com. Excellence award ASCE, 1990; grantee Texaco Inc., Union Oil Co., Marathon Oil Co., NSF. Mem. Am. Assn. Petroleum Geologists, N.Am. Coun. Geostats. (exec. officer 1988-91), Internat. Geostatis. Assn., Internat. Assn. Math. Geologists (coun. mem. 1989-92, sec.-gen. 1992—), Soc. Petroleum Engrs. Roman Catholic. Achievements include design of the CORRELATOR method for geological correlation of wireline logs. Home: 507 Abilene St Lawrence KS 66049-2239 Office: Kans Geological Survey 1930 Constant Ave Lawrence KS 66047-3726

OLEARCHYK, ANDREW S., cardiothoracic surgeon, educator; b. Peremyshl, Ukraine, Dec. 3, 1935; s. Simon and Anna (Kravéts) O.; m. Renata M. Sharan, June 26, 1971; children: Christina N., Roman A., Adrian S. Grad., Med. Acad., Warsaw, Poland, 1961. Diplomate Am. Bd. Surgery, Am. Bd. Thoracic Surgery. Chief div. anesthesiology, asst. dept. surgery Provincial Hosp., Kielce, Poland, 1963-66; resident in gen. surgery Geisinger Med. Ctr., Danville, Pa., 1968-73; resident in thoracic, cardiac surgery Allegheny Gen. Hosp., Pitts., 1980-82; pvt. practice medicine specializing in cardiac, thoracic and vascular surgery Phila.; assoc. div. cardiothoracic surgery Episc. Hosp., Phila., 1984—. Contbr. articles to med. jours., also 6 monographs. Fellow ACS; mem. Soc. Thoracic Surgeons, Ukrainian Med. Assn. N. Am. Achievements include design of Olearchyk R Triple Ringed Cannula Spring Clip to secure vein grafts over blunted cannulas in coronary artery bypass surgery; demonstration of safety of simultaneous use of fluothane and curare as general anesthesia; introduction of endarterectomy and external prosthetic grafting of ascending and transverse aorta under hypothermic circulatory arrest; pioneering promotion of grafting of the left anterior descending coronary artery system during resection of cardiac aneurysms, and of diffusely diseased coronary arteries with the internal thoracic artery; first to combine insertion of the inferior vena cava filter with iliofemoral venous thrombectomy. Home: 129 Walt Whitman Blvd Cherry Hill NJ 08003-3746

O'LEARY, DENNIS SOPHIAN, medical organization executive; b. Kansas City, Mo., Jan. 28, 1938; s. Theodore Morgan and Emily (Sophian) O'L.; m. Margaret Rose Wiedman, Mar. 29, 1980; children: Margaret Rose, Theodore Morgan. BA, Harvard U., 1960; MD, Cornell U., 1964. Diplomate Am. Bd. Internal Medicine, Am. Bd. Hematology. Intern U. Minn. Hosp., Mpls., 1964-65, resident, 1965-66; resident Strong Meml. Hosp., Rochester, N.Y., 1966-68; asst. prof. medicine and pathology George Washington U. Med. Ctr., Washington, 1971-73, assoc. prof., 1973-80, prof. medicine, 1980-86, assoc. dean grad. med. edn., 1973-77, dean clin. affairs, 1977-86; pres. Joint Commn. on Accreditation Healthcare Orgns., Chgo., 1986—; acting med. dir. George Washington U. Hosp., 1974-85, v.p. Univ. Health Plan, 1977-85; pres. D.C. Med. Soc., 1983. Chmn. editorial bd. Med. Staff News, 1985-86; contbr. articles to profl. jours. Founding mem. Nat. Capital Area Health Care Coalition, Washington, 1982; co-chmn. Washington Physicians for Reagan-Bush '84; trustee James S. Brady Found., Washington, 1982-87; bd. dirs. D.C. Polit. Action Com., 1982-84. Maj. U.S. Army, 1968-71. Recipient Community Service award D.C. Med. Soc., 1981, Key to the City, Mayor of Kansas City, Mo., 1982. Fellow Am. Coll. Physician Execs.; mem. ACP, AMA (resolution commendation 1981), Am. Soc. Internal Medicine, Soc. Med. Adminstrs, Am. Hosp. Assn. (del. 1984-86, resolution commendation 1981), Internat. Club, DuPage Club (Chgo.). Avocation: tennis. Office: Joint Commn Accreditation Healthcare Orgns 1 Renaissance Blvd Villa Park IL 60181-4813

OLER, WESLEY MARION, III, physician, educator; b. N.Y.C., Mar. 8, 1918; s. Wesley Marion and Imogene (Rubel) O.; m. Virginia Carolyn Craemer, Dec. 8, 1951; children: Helen Louise (dec.), Wesley Marion IV, Stephen Scott. Grad., Phillips Andover Acad., 1936; AB, Yale U., 1940; MD, Columbia U., 1943. Intern Bellevue Hosp., N.Y.C., 1944; resident Bellevue Hosp., 1948-50; fellow Hosp. U. Pa., 1951; practice medicine specializing in internal medicine Washington, 1952—; sr. adv. staff vice chmn. dept. medicine Washington Hosp. Center, 1962-64, v.p. med. bd., 1971-72, trustee, 1973-81; clin. prof. medicine Med. Sch., Georgetown U. Contbr. articles on old musical instruments to jours. Founder, past pres. Washington Recorder Soc.; bd. dirs. Am. Recorder Soc. Maj. M.C. U.S. Army (paratroops), 1944-47. Fellow ACP (gov. 1980-84); mem. Mensa, Osler Soc. Washington (past pres.), Met. Club, Cosmos Club, Chevy Chase Club. Republican. Episcopalian. Home: 8101 Connecticut Ave Apt 612N Bethesda MD 20815-2805 Office: 3301 New Mexico Ave NW Washington DC 20016-3622

OLESEN, DOUGLAS EUGENE, research institute executive; b. Tonasket, Wash., Jan. 12, 1939; s. Magnus and Esther Rae (Myers) O.; m. Michaele Ann Engdahl, Nov. 18, 1964; children: Douglas Eugene, Stephen Christian. B.S., U. Wash., 1962, M.S., 1963, postgrad., 1965-67, Ph.D., 1972. Research engr. space research div. Boeing Aircraft Co., Seattle, 1963-64; with Battelle Meml. Inst., Pacific NW Labs., Richland, Wash., 1967-84, mgr. water resources systems sect., water and land resources dept., 1970-71, mgr. dept., 1971-75, dep. dir. research labs., 1975, dir. research, 1975-79, v.p. inst., dir. NW div., 1979-84; exec. v.p., chief operating officer Battelle Meml. Inst., Columbus, Ohio, 1984-87, pres., chief exec. officer, 1987—. Patentee process and system for treating wast water. Trustees Capital Univ., Columbus Mus. of Art, Riverside Hosp., INROADS/Columbus Inc., Franklin County United Way; bd. dirs. Ohio State U. Found. Mem. Ohio C. of C. (trustee). Office: Battelle Meml Inst 505 King Ave Columbus OH 43201-2693*

OLESEN, MOGENS NORGAARD, mathematics educator, traffic researcher; b. Frederiksvaerk, Denmark, May 10, 1948; s. Niels Norgaard and Katrine (Rasmussen) O.; m. Lilly Groen Pedersen, July 14, 1974; children—Thomas Norgaard, Eva Groen, Morten Norgaard. Candidate Sci., U. Copenhagen, Denmark, 1976. Vis. researcher U. Calif., San Diego, 1977-78; assoc. prof. math. Coll. of Elsinore, Denmark, 1978—, Sch. Forests, Fredensborg, Denmark, 1979— , U. Copenhagen, 1982—. Author: The Exponential Functions and their Applications, 1982, Mathematical Problems and Their Solutions, 1983, The Great Belt Crossing During 100 Years, 1983, Our Solar System, 1984, Planar Curves, 1984, Practical Mathematics, 1985, Applied Mathematics, 1985, Halley's Comet and the History of Cosmology, 1985, The Ferries on the Kattegat, 1986, Danish Railway Steam Ferries, 1987, The History of the Rodby-Puttgarden Crossing, 1988, The Railwayferry Crossing Copenhagen-Malmo, 1989, The History of Astronomy, 1989, The History of the Kalundborg-Aarhus Crossing, 1989, The Ferry Sjaelland, 1990, Everything is Numbers, 1990, The History of the Danish Railways, 1990, Linear Algebra with Applications, 1991, The History of the Elsinore-Helsingborg Crossing, 1992. Copenhagen travelling grantee for mathematicians, 1976. Mem. Danish Math. Soc., Am. Math. Soc., European Math. Soc., Danish Ferry Hist. Soc., World Ship Soc. Lutheran. Avocations: history of mathematics, science, shiping ships and railways, classical music. Home: H P Christensensvej 18, DK-3300 Frederiksvaerk Denmark Office: U Copenhagen, Studiestraede 6, DK-1455 Copenhagen Denmark

OLESKOWICZ, JEANETTE, physician; b. N.Y.C., Oct. 10, 1956; d. John Francis and Helen (Zielinski) O. BA, NYU, 1977; D of Chiropractic, N.Y. Chiropractic Coll., 1982; MS, U. Bridgeport, 1984; MD, U. Medicine & Dentistry N.J., 1990. U.S. immigration officer U.S. Dept. Justice, N.Y.C., 1977; commd. med. officer USAR, 1983; advanced through grades to capt. HPSP, 1990—. Am. sponsor for a cripples child's health care in Mid. East. Mem. AMA, Am. Psychiat. Assn., Ga. Phychiat. Assn. Roman Catholic. Avocations: volunteer ch. nursery, adult education instructor. Home: 2714B Woodcrest Dr Augusta GA 30909-4200 Office: Eisenhower Army Med Ctr Fort Gordon Augusta GA 30905

OLEVSKII, VICTOR MARCOVICH, chemical engineer, educator; b. Charkov, Ukrain, Sept. 7, 1925; s. Marc and Ludmila (Zelenko) O.; m. Podeiko Galina, June 7, 1953; 1 child, Olevskii Vitalii. Degree in chem. engring., U. Moscow, 1947, 1st sci. degree, 1952, D in Tech. Sci., 1969. Rsch. engr. State Inst. for Nitrogen Industry, Moscow, 1947-56, head lab. for chem. engring., 1956-63, vice dir., 1963—; prof. Superior Testifying Commn., Moscow, 1970. Author ten books for chem. engring. caprolactam and fertilizer prodn., 1967-90; contbr. articles to profl. jours.; patentee in field. Mem. Acad. Sci. Russia (state coun. for chem. engring. 1970—, spl. counsel), State Com. for Technics and Sci. (commn. for mass-transfer tower equipment 1973—), Agrochim Assn. (sci. coun. 1982—), Soc. Knowledge (head 1980—). Avocations: lawn tennis, travel, skiing, boating. Home: Leningradski Prospect 75A, Apt 91, 125057 Moscow Russia Office: State Inst for Nitrogen Industry, Zemlianoy Val 50, 109815 Moscow Russia

OLILA, OSCAR GESTA, soil and water scientist; b. Cayang, Bogo, Cebu, The Philippines, Sept. 9, 1955; came to U.S., 1986; s. Escolastico Mangubat Olila and Gregoria (Jagdon) Gesta; m. Prima Carreon, July 11, 1987; 1 child, Glen Charles. BS in Agr., Cen. Mindanao U., Musuan, Bukidnon, The Philippines, 1977; MS in Soil Sci., U. Philippines, Los Baños, Laguna, 1983; postgrad., U. Mass., 1986-87; PhD, U. Fla., 1987-92. Instr. soil sci. Cen. Mindanao U., 1977-83, asst. prof., chmn. dept., 1984-86, head Soil and Plant Testing Lab., extension worker, 1985-86; grad. rsch. asst. U. Fla., Gainesville, 1987-92, post doctoral rsch. assoc., 1993—. Contbr. articles to profl. and sci. jours. Scholar Philippine Coun. for Agr. and Resources Rsch., 1979-81, Philippine Devel. scholar, 1983, Cen. Mindanao U. faculty scholar, 1986-89. Mem. AAAS, Soil Sci. Soc. Am., Am. Soc. Agronomy, Am. Soc. Limnology and Oceanography, Internat. Soc. Soil Sci., Cen. Mindano U. Alumni Assn. (v.p. 1985-86), Alpha Zeta, Gamma Sigma Delta. Roman Catholic. Achievements include research on liming and soil exchangeable aluminum, raw phosphate rocks as P fertilizers, fractionation of inorganic P in sediments, P reactivity and kinetics of P sorption/desortion in sediments of eutrophic lakes, biogeochemistry of wetlands, mobility of P and associated cations as influenced by redox potential and pH. Office: U Fla Soil Sci Dept 106 Newell Hall Gainesville FL 32611

OLINGER, CHAD TRACY, physicist; b. Ponca City, Okla., Oct. 9, 1962; s. Eugene C. and P. Darlene (Jones) O.; m. Tammy Lu Black, Aug. 14, 1985; children: Brandon Drake, Miranda Louise. BA in Physics magna cum laude, Gustavus Adolphus Coll., 1985; MA, Washington U., 1987, PHD, 1990. Postdoctoral fellow Los Alamos (N.Mex.) Nat. Lab., 1990-92, staff mem., 1992—. Contbr. articles to Nature and other geochem. and space sci. jours. Cons. Navajo Sci. Edn. Cons. Program, Shiprock, N.Mex., 1991-93. Fellow NASA, 1987-90, McDonnell Douglas, 1985-87, Hughes fellow Washington U., 1985-86; recipient Nininger Meteorite prize Ctr. for Meteorite Studies Ariz. State U., 1990. Mem. AAAS, Am. Phys. Soc., Am. Geophys. Union. Republican. Achievements include support in U.S.-Russian disarmament agreements, international safeguards; proof of extraterrestrial origin of suspected micrometeorites found in Greenland and Antarctica. Home: 2122

Candelero St Santa Fe NM 87505 Office: Los Alamos Nat Lab N-4 MS E541 Los Alamos NM 87545

OLIPARES, HUBERT BARUT, biological safety officer; b. Honolulu, Mar. 7, 1957; s. Micolas Pascua and Victorina (Barut) O.; m. Nancy Marie Linnan, Oct. 5, 1991. BS in Med. Technology, U. Hawaii, 1981, MSPH, 1984. Cert. med. technologist, clin. lab. scientist. Environ. technologist U. Hawaii, Honolulu, 1982-84, grad. asst., 1984-85, biol. safety officer, 1988—; com. researcher Hawaii State Ho. of Reps., Honolulu, 1984-87; solid-hazardous wastes cons. Trust Territory of the Pacific, Saipan, 1984-85; med. technologist Castle Med. Ctr., Kailua, Hawaii, 1982-88; student jr. researcher U. Hawaii, 1978-88. Mem. Young Dems., Honolulu, 1985, Hawaii Filipino Jaycees, Honolulu, 1980. Traineeship USPHS, Honolulu, 1985, 86. Mem. Am. Indsl. Hygiene Assn., Hawaii Soc. Med. Tech., Hawaii Infection Control Assn., Hawaii Soc. Microbiology. Democrat. Roman Catholic. Office: Environmental Health/Safety Univ Hawaii 2040 East-West Rd Honolulu HI 96822

OLIPHANT, CHARLES ROMIG, physician; b. Waukegan, Ill., Sept. 10, 1917; s. Charles L. and Mary (Goss) R.; student St. Louis U., 1936-40; m. Claire E. Canavan, Nov. 7, 1942; children: James R., Cathy Rose, Mary G., William D. Student, St. Louis U., 1936-40, MD, 1943; postgrad. Naval Med. Sch., 1946. Intern, Nat. Naval Med. Ctr., Bethesda, Md., 1943; pvt. practice medicine and surgery, San Diego, 1947—; pres., chief exec. officer Midway Med. Enterprises; former chief staff Balboa Hosp., Doctors Hosp., Cabrillo Med. Ctr.; chief staff emeritus Sharp Cabrillo Hosp.; mem. staff Mercy Hosp., Children's Hosp., Paradise Valley Hosp., Sharp Meml. Hosp.; sec. Sharp Sr. Health Care, S.D.; mem. exec. bd., past. comdr. San Diego Power Squadron. Charter mem. Am. Bd. Family Practice. Served with M.C., USN, 1943-47. Recipient Golden Staff award Sharp Cabrillo Hosp. Med. Staff, 1990. Fellow Am. Geriatrics Soc. (emeritus), Am. Acad. Family Practice, Am. Assn. Abdominal Surgeons; mem. AMA, Calif. Med. Assn., Am. Acad. Family Physicians (past pres. San Diego chpt., del. Calif. chpt.), San Diego Med. Soc., Public Health League, Navy League, San Diego Power Squadron (past comdr.), SAR. Clubs: San Diego Yacht, Cameron Highlanders. Home: 4310 Trias St San Diego CA 92103-1127

OLIU, RAMON, chemical engineer; b. Cantonigros, Catalunya, Spain, May 30, 1923; s. Francesc and Concepcio (Carol) L.; m. Isabel Herberg, Oct. 27, 1951; children: Elisabeth, Jorge, Ramon, Edward, Michael, John, Paul. Degree in chem. scis., U. Barcelona, 1947; MS of Petroleum Engring., U. Tulsa, 1957; MS in Chem. Engring., CCNY, 1963. Prof., dean Sch. Petroleum Engring. U. Bucaramanga, Colombia, 1954-57; process engr. Foster Wheeler Corp., N.Y.C., 1957-59; devel. engr. FMC Corp., Balt. and N.Y.C., 1959-62; assoc. engr. FMC Corp., Princeton, 1972-77, mgr. engring svcs., 1984-91; sr. assoc. engr. Mobil Oil Corp., N.Y.C., 1962-72; dir. tech. Foret S.A. div. FMC, Barcelona, Spain, 1977-84; cons. Esso Barrancabermeja, Colombia, 1954-57. Contbr. articles to profl. jours. Founder, dir. Barcelona Marathon, 1977-84; fundraiser St. Ambrose Parish, Old Bridge, N.J., 1961-62. Mem. Am. Assn. Chem. Engrs., N.Y. Acad. Scis. Roman Catholic. Achievements include patents for purification of terephthalic acid, for simple manufacture of SO2-O2 free (U.S., Spanish); patent related to improving manufacturing of phosphoric acid, and for phosphoric acid. Home: 97 Robin Dr Trenton NJ 08619-1158

OLIVARES DEL VALLE, FRANCISCO JAVIER, physical chemistry educator; b. Sevilla, Andalucia, Spain, Jan. 6, 1950; s. Joaquin and Ines Del Valle Lopez; m. Maria Del Carmen Marin Murga, Apr. 13, 1975; children: Mara, Maria Del Carmen, Francisco Javier. Degree, Extremadura U., Badajoz, Spain, 1974, Dr. degree, 1980. Prof. U. de Extremadura, 1987—. Contbr. articles to Jour. Chem. Physics, Chem. Physics, and J. Comp. Chem. Office: Extremadura U, Avd Elvas s/n, 06071 Badajoz Spain

OLIVER, BERNARD MORE, electrical engineer, technical consultant; b. Soquel, Calif., May 27, 1916; s. William H. and Margaret E. (More) O.; m. Priscilla June Newton, June 22, 1946; children: Karen, Gretchen, Eric. AB in Elec. Engring., Stanford U., 1935; MS, Calif. Inst. Tech., 1936, PhD, 1940. Mem. tech. staff Bell Telephone Labs., N.Y.C., 1940-52; dir. R&D, Hewlett-Packard Co., Palo Alto, Calif., 1952-57, v.p. R&D, 1957-81, tech. adviser to pres., 1981—; lectr. Stanford U., 1957-60; cons. Army Sci. Adv. Com., from 1966; mem. sci. and tech. adv. com. Calif. State Assembly, 1970-76. Contbr. articles on electronic tech. and instrumentation to profl. jours.; patentee electronic circuits and devices. Mem. Pres.'s Commn. on the Patent System, 1966, State Senate Panel for the Bay Area Rapid Transit System, 1973; trustee Palo Alto Unified Sch. Dist., 1961-71. Recipient Disting. Alumni award Calif. Inst. Tech., 1972, Nat. Medal of Sci., 1986, Exceptional Engring. achievement medal, NASA, 1990, Pioneer award Internat. Found. Telemetering, 1990; ann. Symposium series est. in his name Hewlett-Packard Labs., 1991. Fellow IEEE (Lamme medal 1977); mem. AAAS, Astron. Soc., Nat. Acad. Sci., Nat. Acad. Engring. Republican. Club: Palo Alto. Office: Hewlett-Packard Co MS 109 PO Box 10490 Palo Alto CA 94303-0969

OLIVER, BRUCE LAWRENCE, information systems specialist, educator; b. Westfield, Mass., Nov. 20, 1951; s. Ernest Lawrence and Elizabeth (Welchek) O. AS, Greater Hartford Commmunity, 1972; BS, U. Mass., 1974; MBA, U. Hartford, 1989. Cert. tchr. sec. and vocat. edn., Mass., Conn. Comml. sales Gordon Realty, Enfield, Conn., 1972-75; forestry tech. research Dept. Environ. Protection, State of Conn., Hartford, 1973, 1974; res. sales Forsman Realty, Enfield, 1975-77; substitute sec. tchr. Enfield Sch. Systems, 1975-78; collections mgr. New Eng. Bank & Trust, Enfield, 1978-79; ops. CCEC/McCullahg Leasing, Inc., S. Windsor, Conn., 1979-81; pres. Ollie & Ike's, Inc., Enfield, 1985-86; MBA Adj. U. Hartford, W. Hartford, Conn., 1988-89; workstation engr. Travelers, Hartford, 1982-89; v.p. 1st Class Expert Systems, Inc., Wayland, Mass., 1989-90, Microsoft Corp., Boston, 1991—; cons., pres. Profl. Office Solutions, Enfield; del. leader Comparative Studies Assn.; Internat. Cultural Exch. with China, Washington; pub. sepaker Speakers Bus., U. Hartford; vis. mem. faculty mgmt. info. sci. U. Hartford, Conn., 1989-91. Author: A Novice's Guide to Personal Computer Buying. Gubernatorial appointee Conn. bd. trustees Reg. Community Colls., 1985-89; vice chmn. Student Affairs and Acad. Policies Com. Hartford, 1987; chmn., trustee Conn. Data Processing Curriculum Com., Hartford, 1989; elected com. mem. Enfield Dem. Com., 1975; chmn. regional adv. coun. Asnuntuck C.C.; notary pub. Conn., 1972-92; gubernatorial appointee Conn. bd. trustees Community Tech. Colls., 1990-93. Recipient CTM Degree, Toastmasters Internat., Hartford, 1987, State Farmer Degree Conn. Future Farmers Am., DeKalb Agrl. Accomplishment award, cert. of recognition, Bicentennial (USA) Commn., Enfield, 1976, Vigil Hon. BSA: Order of the Arrow, Hartford, 1972, Eagle Sc. Boy Sc. of Am., Hartford, 1967. Mem. World Affairs Coun. of Hartford, Computer Soc. of IEEE, Am. Assn. for Artificial Intelligence, Assn. Community Coll. Trustees, Am. Assn. Community & Tech. Colls., Internat. Platform Assn., Oldefield Farms Homeowners' Assn. (residence com. sec. 1990-91), Hartford County Soil and Water Conservation Dist., Nat. Press Club Found., Robert Schueller's Eagles Club, Masons. Democrat. Roman Catholic. Avocations: travel, refinishing antiques, tennis, hiking, real estate investment. Home: 71 Oldefield Farms Enfield CT 06082-4565

OLIVER, DANIEL ANTHONY, electrical engineer, company executive; b. Camden, N.J., Sept. 15, 1939; s. James D. and Mary Catherine (Duda) O.; m. Judith Anne Hatch, Oct. 15, 1966; children: Jeanne Marie, James Daniel II, Joseph, J. Andrew, Jerome, Judith, Anne, Justin, Joy, Jackson. BSEE, U. Colo., 1960; MS, BA, Drexel U., 1970. Design engr. def. space and spl. systems group Burroughs Corp., Paoli, Pa., 1963-67; prin. programmer Univac div. Sperry Rand Corp., Blue Bell, Pa., 1967-73; mgr. Can. ops. Fairchild Test Systems Group, San Jose, Calif., 1973-81; pres., founder Dan Oliver & Assocs., Inc., Haddon Heights, N.J., 1981—; prin. Source Sci., Inc., Haddon Heights, 1988—. Contbr. articles to profl. jours. With U.S. Army, 1960-63. Recipient best tech. paper award Hanover (Germany) Fair, 1982. Mem. Assn. Test Engrs. (sr.), Am. Aviation Hist. Soc., U.S. Naval Inst. (life), Assn. Old Crows (adv. bd. 1987—), Am. Legion, Polish Legion Am. Vets. Republican. Roman Catholic. Achievements include patent for design or fixuring for a 50 OHM environment. Office: PO Box 397 Haddon Heights NJ 08035-0397

OLIVER, DAVID EDWIN, physicist; b. Horsham, Sussex, U.K., June 13, 1956; came to U.S. 1986; s. Edwin Charles and Laurette Henrietta (Taylor) O.; m. Evelyn Mary Foster, June 30, 1979; children: Amy Evelyn, Thomas David. Degree in Physics, U. Leicester, Eng., 1977. Chartered physicist. Rsch. officer Sira Ltd., Chislehurst, Kent, U.K., 1978-82, sr. rsch. officer, 1982; applications mgr. Ometron Ltd., Chislehurst, Kent, U.K., 1982-83; tech. sales mgr. Ometron Inc., Herndon, Va., 1984-85; applications mgr. Ometron Ltd., Chislehurst, 1985; gen. mgr. Ometron Inc., Herndon/Sterling, Va., 1986-91; product mgr. Polytec Optronics, Auburn, Mass., 1991—. Contbr. articles to profl. jours. Coach Nike Children's Running Devel. Program, Reston, Va., 1991. Mem. ASME, ASM Internat., Brit. Inst. Physics (optical group com. 1979-81), Soc. for Exptl. Mechanics (thermal methods com.), SAWG, Brit. Am. Bus. Assn., Optical Soc. Am. Achievements include development of method for applying laser doppler vibrometry to non-destructive testing; thermal technique for full field not-contact stress analysis; pushed frontiers in high frequency, high temperature measurement; others. Home: 30 Thomas Newton Dr Westborough MA 01581

OLIVER, DAVID JARRELL, electronic and systems engineer; b. Denton, Tex., June 23, 1954; s. Dellis Baker and Vera Belle (Garrison) O.; m. Marsha Ann Michaelson, June 19, 1976; children: Kristina Marie, Kevin Michael. B in Tech., Electronics Engring. Tech., Oreg. Inst. Tech., 1976. Integrated circuit design engr. Intersil, Inc., Cupertino, Calif., 1976-78, laser trim/test engr., 1978-82; test engr. LTX, Corp., San Jose, Calif., 1982-85; systems engr. Attain, Inc., Milipitas, Calif., 1985-88, Hewlett-Packard, Santa Clara, Calif., 1988—; mem. EIA/JEDEC JC-25 com. on transistors, 1988—. Contbr. article to Internat. Power Conversion Conf. Proceedings, 1989. Bd. dirs. Camp Fire, Inc., Santa Clara, 1989-92. Mem. Optical Soc. No. Calif. Republican. Roman Catholic. Achievements include development of optical system, electronics and software for automatic alignment of wafers on a laser trim system. Office: Hewlett-Packard 5301 Stevens Creek Blvd Santa Clara CA 95052-8059

OLIVERO, JOHN JOSEPH, JR., physics educator; b. Yonkers, N.Y., Jan. 18, 1941; s. John Joseph and Adele Marie (Iwanski) O.; m. Linda Kathleen Platt, Aug. 19, 1961; children: Lisa Marie, Mary Frances, Michael John, David Andrew. BS in Physics, Fla. State U., 1962; MS in Physics, Coll. William and Mary, 1966; PhD in Aeronomy, U. Mich., 1970. Aerospace technologist NASA-Langley Rsch. Ctr., Hampton, Va., 1962-70; instr. physics and astronomy U. Fla., Gainesville, 1970-72; prof. dept. meterology Pa. State U., University Park, 1972—. Office: Pa State U 509 Walker Bldg University Park PA 16802

OLKOWSKI, ZBIGNIEW LECH, physician, educator; b. Wilno, Poland, Nov. 24, 1938; came to U.S., 1970; s. Joseph and Jane (Poplawska) O.; m. Krystyna Nardelli, Apr. 15, 1963. MD, Silesian U., Zabrze, Poland, 1963, ScD, 1969. Cert. lab. dir., Ga. Assoc. prof. dept. radiation oncology Emory U. Med. Sch., Atlanta, 1972-75, assoc. prof., 1975—; dir. lab. tumor biology Emory U. Clinic, Atlanta, 1973-85; resident radiation oncology Emory U. Med. Sch., Atlanta, 1976-78, fellow radiation oncology, 1979-80; mem. spl. study sect. on cancer Nat. Cancer Inst., Bethesda, Md., 1980; mem. senate com. on environ. Emory U., Atlanta, 1992-94. Contbr. 50 articles on oncology and clin. immunology to nat. and internat. sci. jours., 1970-92, chpts. to books. Recipient contracts and grants NIH, Bethesda, Md., 1976-85. Fellow Internat. Acad. Cytology; mem. Royal Microscopical Soc., Brit. Inst. Radiology, Am. Soc. Therapeutic Oncology, Am. Rifle Assn. (life), Nat. Skeet Shooting Assn (life), Ga. Skeet Shooting Assn. (life). Achievements include discovery of dopamine transporter system on human lymphocytes, method of labeling and edn. of human lymphocytes with technesium 99m to make them selectively cytotoxic to human malignant tumor cells. Office: Emory U Clinic 1365 Clifton Rd NE Atlanta GA 30322

OLMSTEAD, WILLIAM EDWARD, mathematics educator; b. San Antonio, June 2, 1936; s. William Harold and Gwendolyn (Littlefield) O.; m. Adele Cross, Aug. 14, 1957 (div. 1967); children: William Harold, Randell Edward. BS, Rice U., 1959; MS, Northwestern U., 1962, PhD, 1963. Mem. research staff S.W. Research Inst., San Antonio, 1959-60; Sloan Found. postdoctoral fellow Johns Hopkins, 1963-64; prof. applied math. Northwestern U., Evanston, Ill., 1964—, chmn. dept. engring. scis. and applied math., 1991-93; vis. mem. Courant Inst. Math. Scis., NYU, 1967-68; faculty visitor Univ. Coll. London, Eng., 1973, Calif. Inst. Tech., 1987, 90. Contbr. articles to profl. jours. Recipient Teaching Excellence Alumni award Northwestern U., 1993; named Technol. Inst. Tchr. of Yr., 1980. Mem. Am. Acad. Mechanics, Am. Math. Soc., Am. Phys. Soc., Soc. Indsl. and Applied Math., Am. Contract Bridge League (life master), John Evans Club (Northwestern U.), Sigma Xi, Tau Beta Pi, Sigma Tau. Episcopalian. Home: 141 Lockerbie Ln Wilmette IL 60091-2947 Office: Northwestern U Dept Engring Scis and Applied Math Evanston IL 60208

OLMSTED, ROBERT AMSON, civil engineer; b. N.Y.C., Nov. 7, 1924; s. Harold McLain and Sophia (Amson) O.; m. Pauline Weiner, June 25, 1949; children: Elizabeth, Alan, Lawrence. BCE, Cornell U., 1946; MCE, Poly. Inst., Bklyn., 1953. Registered profl. engr., N.Y. Jr. civil engr. Triborough Bridge and Tunnel Authority, N.Y.C., 1946-49; civil engr. Port Authority of N.Y. and N.J., N.Y.C., 1949-51; engr. TAMS Engrs., N.Y.C., 1951-62; assoc. transp. engr. Office of Transp. State of N.Y., N.Y.C., 1962-67; asst. dir. planning Met. Transp. Authority, N.Y.C., 1967-89; transp. cons. N.Y.C., 1989—; adj. prof. Poly. Inst. Bklyn., 1966-70, Manhattan Coll., 1975-78. Contbr. articles to profl. jours. Sgt. U.S. Army, 1943-46, PTO. Mem. ASCE (bd. dirs. N.Y. sect. 1974-76, chmn. History and Heritage com. N.Y. sect., 1992—, Thomas Kavanaugh award N.Y. Met. sect. 1985), Inst. Transp. Engrs., Am. Planning Assn., City Club; affiliate Transp. Rsch. Bd. Address: 33-04 91st St Jackson Heights NY 11372

OLSAK, IVAN KAREL, civil engineer; b. Nitra, Czechoslovakia, Apr. 30, 1933; came to the U.S., 1970; s. Innocenc and Jolana (Rutkovska) O.; m. Renata Gabriela Franclova, Sept. 26, 1959; children: Ruth E., Patricia L. Degree in civil and sanitary engring., Slovak Tech. U., 1958. Registered profl. engr., Pa., Fla. Chief engr. Keramoproject, Bratislava, Czechoslovakia, 1961-63; chief of commune hygien Slovak Dept. Health, Bratislava, 1963-68; draftsman Crippen Acres Ltd., Winnipeg, Canada, 1969-71; design engr. Bouguard & Assocs., Harrisburg, Pa., 1971-72; chief dept. engring. Adair & Brady, Inc., West Palm Beach, Fla., 1972-74; profl., engr., ptnr. Weimer & Co., Inc., West Palm Beach, 1974-75; pvt. practice West Palm Beach, 1975—. Active Rep. Cen. Com., Tallahassee, Fla., 1984—, Rep. Presdl. Task Force, Washington, 1986—, Rep. Nat. Com., Washington, 1989—, Citizens Against Govt. Waste, Washington, 1988—; hon. mem. Fla. Sherrif's Assn., 1992. Recipient Cert. of Appreciation Palm Beach County Bar Assn., 1990, Fla. Assn. State Troopers, 1987. Mem. NSPE, Fla. Engring. Soc., Profl. Engrs. in Pvt. Practice. Roman Catholic. Achievements include research in oil refinery construction, influence of oil exfiltration on ground water system. Home: 308 Greymon Dr West Palm Beach FL 33405 Office: PO Box 6727 West Palm Beach FL 33405

OLSAVSKY, JOHN GEORGE, aerospace engineer; b. Quantico, Va., Nov. 2, 1966; s. John Andrew and Patricia Ann (Lewis) O. BS Aero. Engring., Purdue U., 1989. Aerospace engr. Naval Air Systems Command, Washington, 1989—. Vol. Orange Park (Fla.) Community Theatre, 1990, Nat. Aquarium, Balt., 1992—. Mem. AIAA (treas. Purdue chpt. 1988-89). Office: Naval Air Systems Command Code PMA205-211 Washington DC 20361-1205

OLSEM, JEAN-PIERRE, economics educator; b. Bar-Le-Duc, Meuse, France, Sept. 6, 1942; s. André and Renée (Lamy) O. Degree in bus. adminstrn., U. Nancy, France, 1962; D. in Econ., U. Reims, France, 1972. Asst. U. Nancy, U. Paris, U. Reims, U. Besançon, France, 1967-89; prof. U. Besançon, 1989—; vis. scholar U. Calif., Berkeley, 1979-89, U. Sydney, Australia, 1984; vis. prof. Karl Franzen U., Graz, Austria, 1991, 93. Author: Pour Ordre Concurrentiel, 1983, L'energie dans le monde, 1984-85, Economie Industrielle, 1991, La Strategie du Gagneur, 1993. Home: BP 313, F 25017 Besançon Cedex, France Office: Faculte de Droit, 25030 Besançon France

OLSEN, ALLEN NEIL, civil engineer; b. Council Grove, Kans., Sept. 28, 1934; s. Archie B. and La Vona (Carr) O.; m. Inger Pedersen, July 14, 1977; children: Mette, Joy, Nina, Angela, Jan Fredrik. BSCE, Kans. State U., 1959; MS in Fin. Mgmt., Naval Postgrad. Sch., 1973. Profl. engr. Tex., 1965. Design engr. R.J. Cummins Co, Houston, 1959-62; capt. civil engr. corps USN, worlwide, 1962-86; gen. mgr. Worlwide Tech. Construction, Inc., Vienna, Va., 1986-87; project mgr. Perland Environ., Inc., Burlington, Mass., 1987--; mem. devel. group Civil Engring. Rsch. Found., Washington, 1991. Mem. ch. coun. Pearl Harbor, Hawaii, 1967, Framingham, Mass, 1989-92. Decorated Bronze Star with "v", 1969, Def. Superior Svc. medal, 1986. Fellow Am. Soc. Civil Engrs. (pres. 1957-58); mem. Soc. Am. Mil. Engrs. (v.p. 1969-70, commendation 1970), U.S Naval Inst., Houston Engring. and Scientific Soc. Lutheran. Home: 16 Grove St Framingham MA 01701 Office: Perland Environ 8 New England Executive Pk Burlington MA 01803

OLSEN, CLIFFORD WAYNE, physical chemist; b. Placerville, Calif., Jan. 15, 1936; s. Christian William and Elsie May (Bishop) O.; m. Margaret Clara Gobel, June 16, 1962 (div. 1986); children: Anne Katherine Olsen Cordes, Charlotte Marie; m. Nancy Mayhew Kruger, July 21, 1990. AA, Grant Tech. Coll., Sacramento, 1955; BA, U. Calif.-Davis, 1957, PhD, 1962. Physicist, project leader, program leader, task leader Lawrence Livermore Nat. Lab., Calif., 1962—; mem. Containment Evaluation Panel, U.S. Dept. Energy, 1984—, mem. Cadre for Joint Nuclear Verification Tests, 1988; organizer, editor procs. for 2nd through 7th Symposiums on Containment of Underground Nuclear Detonations, 1983-93. Contbr. articles to profl. jours. Recipient Chevalier Degree, Order of DeMolay, 1953. Mem. AAAS, Am. Radio Relay League, Seismol. Soc. Am., Sigma Xi, Alpha Gamma Sigma, Gamma Alpha. Democrat. Lutheran. Avocations: gardening, amateur radio, music, cooking. Office: Lawrence Livermore Nat Lab PO Box 808 M/S L-221 Livermore CA 94551

OLSEN, DAREN WAYNE, electrical engineer; b. Johnstown, Pa., Aug. 9, 1965; s. Clarence Daren Olsen and Donna Lee (Horwitz) Murray; m. Beth Ann Holtschneider, Aug. 20, 1988; 1 child, Nicole Renee. BEE, U. Pitts., 1987. Engr. in tng., Md. Systems engr. Vitro Corp., Silver Spring, Md., 1987-89; elec. engr. Airflow Co., Frederick, Md., 1989-93, Air Tech Systems, Frederick, Md., 1993—. Mem. NSPE, IEEE, Armed Forces Comm. and Electronics Assn., Tau Beta Pi, Eta Kappa Nu. Democrat. Methodist. Home: 6270 N Steamboat Way New Market MD 21774 Office: Air Tech Systems 4510 Metropolitan Ct Frederick MD 21701

OLSEN, DAVID MAGNOR, science educator; b. Deadwood, S.D., July 23, 1941; s. Russell Alvin and Dorothy M. Olson; m. Muriel Jean Bigler, Aug. 24, 1963; children: Merritt, Chad. BS, Luther Coll., 1963; MS in Nat. Sci., U. S.D., 1967. Instr. sci., math. Augustana Acad., Canton, S.D., 1963-66; instr. chemistry Iowa Lakes Community Coll., Estherville, Iowa, 1967-69; instr. chemistry Merced (Calif.) Coll., 1969—, instr. astronomy, 1975—, div. chmn., 1978-88, coord. environ. hazardous materials tech., 1989—. Trustee Merced Union High Sch. Dist., 1983—, pres., 1986-87. Mem. NEA, Am. Chem. Soc., Astron. Soc. of the Pacific, Calif. Tchrs. Assn., Planetary Soc., Calif. State Mining & Mineral Mus. Assn. (bd. dirs. 1988—, sec. 1990-93), Nat. Space Soc., Merced Coll. Faculty Assn. (pres. 1975, 93, treas. 1980-90, bd. dirs. 1986—, sec. 1990-91), Merced Track Club (mem. exec. bd. 1981), M Star Lodge, Sons of Norway (v.p. 1983), Rotary Internat. Democrat. Lutheran. Home: 973 Idaho Dr Merced CA 95340-2513 Office: Merced Coll 3600 M St Merced CA 95348-2806

OLSEN, DONALD BERT, biomedical engineer, experimental surgeon, research facility director; b. Bingham, Utah, Apr. 2, 1930; s. Bertram Hansen and Doris (Bodel) O.; m. Joyce Cronquist; children: Craig, Kathy, Debbie, Jeff, Gary. BS, Utah State U., 1952; DVM, Colo. State U., 1956. Gen. practice vet. medicine Smithfield, Utah, 1956-63; extension veterinarian U. Nev., Reno, 1963-65, researcher Deseret Rsch. Inst., 1965-68; postdoctoral fellow U. Colo. Med. Sch., Denver, 1968-72; researcher U. Utah, Salt Lake City, 1972—, rsch. prof. surgery, 1973—, dir. artificial heart lab., 1976—, rsch. prof. pharmaceutics, 1981—, rsch. prof. biomed. engring., 1986—, dir. Inst. Biomed. Engring., 1986—, prof. surgery, 1986—; mem. sci. adv. bd. Link Resources, Inc., 1987. Contbr. articles to profl. jours.; patentee in field. Recipient Clemson award Soc. Biomaterials, N.Y.C., 1987, Centennial award for outstanding alumni Utah State U., Logan, 1988, Gov.'s medal for Sci. and Tech., 1988; named Alumnus of Yr. Colo. State U., 1986. Mem. Am. Soc. Artificial Internal Organs (trustee 1985—, mem. fellowship rev. com. 1983, chmn. program com. 1988-89, sec., treas. 1989-90, pres. elect. 1990-91, pres. 1991-92), Internat. Soc. Artificial Organs (v.p. 1987—), Am. Coll. Vet. Surgeons (hon.), Utah Vet. Med. Assn. (trustee 1985-88), Alpha Zeta. Mem. LDS Ch. Avocations: fishing, hunting, camping, hiking. Office: U Utah Inst Biomed Engring 803 N 300 W Salt Lake City UT 84103-1414

OLSEN, DONALD PAUL, communications system engineer; b. Glasgow, Mont., July 19, 1938; s. Arthur Leif and Frances Lucille (Beachler) O.; m. Carol Buck, Sept. 3, 1964; children: David Tanner, Paul Tanner, Julia Elizabeth. BSEL, Calif. State Poly. U., 1963; MSEE, Brigham Young U., 1964; PhD, Purdue U., 1969. With Advanced Comm. Inc., Chatsworth, Calif., 1962-71, Tech. Svc. Corp., Santa Monica, Calif., 1971-74; comm. systems engr. Aerospace Corp., El Segundo, Calif., 1974—. Scoutmaster Granada Hills (Calif.) area Boy Scouts Am., 1985-87. Mem. AIAA (sr., comm. com. 1989-92, chair comm. standards com. 1990-92), Highlander Band Parents Assn. (pres.). Republican. Mormon. Achievements include development of hybrid interleaver for survivable packet compatible communication waveforms. Home: 10918 Chimineas Ave Northridge CA 91326 Office: Aerospace Corp 2350 E El Segundo Blvd El Segundo CA 90245

OLSEN, GARY ALVIN, design engineer; b. Ypsilanti, Mich., Feb. 13, 1949; s. Reinhart Alvin and Marian (Losee) O. AS in Archtl. Drafting Tech., Washtenaw Community Coll., Ann Arbor, Mich., 1972. Draftsman, rodman Washtenaw Engring., Inc., Ann Arbor, 1969; archtl. draftsman Campbell Engring., Inc., Detroit, 1972-73, Garity Constrn. Co., Southfield, Mich., 1973; engring. draftsman Bechtel Power Corp., Ann Arbor, 1974; archtl. draftsman engring. dept. U. Mich., Ann Arbor, 1974-76; engring. detailer Ford Motor Co., Dearborn, Mich., 1977-90, engring. designer, 1990—. Republican. Lutheran. Avocations: bowling, golf, horseback riding, pewter figurines. Home: 41445 Elsa Ct Canton MI 48187-3815

OLSEN, HAROLD WILLIAM, research civil engineer; b. Casper, Wyo., Aug. 12, 1931; s. Jens and Anna (Hytmo) O.; m. Virginia Evans, July 15, 1961 (div. Sept. 1986); children: Nina Elizabeth, Karl Nephi, Matthew Daniel; m. Charlotte Blomquist Jensen, Dec. 27, 1986. SBCE, MIT, 1954, SMCE, 1956, ScDCE, 1961. Rsch. hydrologist water resources divsn. U.S. Geol. Survey, Washington, 1961-66; rsch. civil engr. br. engring. geology U.S. Geol. Survey, Menlo Park, Calif., Lakewood, Colo., 1966-83; rsch. civil engr. br. geologic risk assessment U.S. Geol. Survey, Golden, Colo., 1983—; geotech. cons. tech. assistance programs U.S. Geol. Survey, Peru, 1974, Indonesia, 1984, Bangladesh, 1988-90, geotech. cons. UN UNDP Project in People's Republic of China, 1988-90; adj. prof. dept. engring. Colo. Sch. Mines, 1986—; prof. adjoint dept. civil engring. U. Colo., 1983—. Author about 80 rsch. papers. Named Disting. Lectr., La. State U. Dept. Civil Engring., 1988; rsch. and study grantee Fulbright, U. Naples, 1959-60. Mem. ASCE (various coms., chair soil property com. 1975-80), Assn. Engring. Geologists, Assn. Groundwater Scientists and Engrs., Soil Sci. Soc. Am. Achievements include research on the fundamentals of soil behavior, on laboratory methodologies for soil property measurements. Home: 467 Columbine St Denver CO 80206 Office: US Geol Survey Box 25046 Mail Stop 966 Denver CO 80225

OLSEN, KATHIE LYNN, neuroscientist, administrator; b. Portland, Oreg., Aug. 3, 1952; d. Roland Berg and Gladys Elizabeth (Eldreth) O. BS, Chatham Coll., 1974; PhD, U. Calif., Irvine, 1979. Postdoct. fellow Harvard Med. Sch., Boston, 1979-80; rsch. scientist Long Island Rsch. Inst., Stony Brook, N.Y., 1980-83; rsch. asst. prof. SUNY, Stony Brook, 1982-85, asst. prof., 1985-89; assoc. program dir. NSF, Washington, 1984-86, program dir., 1988—, leader neurosci., 1991—; adj. assoc. prof. George Washington U., Washington, 1989—; cons. editor Hormones and Behavior, 1988—. Contbr. articles to profl. jours, chapters to books. Mem. Soc. Neurosci., Endocrine Soc., Women in Neurosci., Sod. Study of Reproduction, Internat. Acad. Sex Rsch. Office: Nat Sci Found 1800 G St NW Washington DC 20550

OLSEN, KENNETH HAROLD, geophysicist, astrophysicist; b. Ogden, Utah, Feb. 20, 1930; s. Harold Reuben and Rose (Hill) O.; m. Barbara Ann Parson, June 15, 1955; children: Susan L., Steven K., Christopher P., Richard Scott. BS, Idaho State Coll., 1952; MS, Calif. Inst. Tech., 1954,

PhD, 1957. Grad. rsch. asst. Calif. Inst. Tech., Mt. Wilson and Palomar Obs., Pasadena, 1952-57; staff member, group leader Los Alamos (N.Mex.) Nat. Lab., 1957-89; lab. assoc. Los Alamos Nat. Lab., 1989—; geophys. cons. GCS Internat., Lynnwood, Wash., 1989—; vis. rsch. fellow Applied Seismol. Group, Swedish Nat. Def. Inst., Stockholm, Sweden, 1983; sr. vis. scientist fellow Norwegian Seismic Array, Oslo, Norway, 1983; vis. scholar Geophysics Program, Univ. Wash., Seattle, 1989-91. Author and editor: Continental Rifts: Evolution, Structure, Tectonics, 1993; conr. articles to profl. jours. Recipient Nat. Medal of Technology, U.S. Dept. Commerce Technology Admin., 1993. Mem. Am. Geophys. Union, Geol. Soc. Am., Seismol. Soc. Am., Royal Astron. Soc. Home: 1029 187th Pl SW Lynnwood WA 98037 Office: GCS Internat PO Box 1273 Lynnwood WA 98046

OLSEN, RICHARD GALEN, biomedical engineer, researcher; b. Colorado Springs, Colo., Aug. 10, 1945; s. Floyd Edwin and Ruth Elizabeth (Robinson) O.; m. Karen Fidler Brubaker, June 17, 1973; children: Kathryn Elizabeth, Nickolas Robert. BSEE, U. Mo., Rolla, 1968; MS, U. Utah, 1970, PhD, 1975. Registered profl. engr., Fla. Engr. Bendix Corp., Kansas City, Mo., 1968-69; elec. engr. Naval Aerospace Med. Rsch. Lab., Pensacola, Fla., 1975-79, chief engring. systems div., 1979-82, head bioengineering div., 1982—; tech. cons. Armstrong Lab., USAF, 1991—, Naval Surface Warfare Ctr., Dahlgren, Va., 1989-91, Naval Sea Systems Command, Arlington, Va., 1989-91, Naval Spl. Warfare Ctr., Coronado, Calif., 1990—. Contbr. artilces to profl. jours and books. With U.S. Army, 1970-72. Recipient NDEA fellowship U. Utah, 1969, Fred A. Hitchcock award Aerospace Physiologist Soc. of Aerospace Med., 1987; named Engr. of the Yr., N.W. Fla. Engrs. Coun., 1991. Mem. IEEE (sr. chmn. Pensacola sect. 1982-83, mem. radio frequency and microwave measuring methods 1982—, mem. nonionizing radiation hazards 1983—, mem. SCC-28 com., cert. appreciation 1983), Bioelectromagnetics Soc. (charter, editorial bd. 1990—), Aerospace Med Assn. (editorial cons. 1986—), Fla. Engring. Soc., Rotary (bd. dirs. Suburban West chpt. 1980-81), Bream Fishermen Assn. (bd. dirs. 1982—), Sigma Xi, Eta Kappa Nu, Tau Beta Pi (sec. Mo. Beta chpt. 1967), Phi Kappa Phi. Republican. Adventist. Achievements include patents in RF coil for hypothermia resusitation, RF dosimetry system and RF warming of submerged extremeties. Home: 1503 N Baylen St Pensacola FL 32501-2101 Office: Aerospace Med Rsch Lab Naval Air Sta Pensacola FL 32508-1046

OLSEN, RICHARD GEORGE, microbiology educator; b. Independence, Mo., June 25, 1937; s. Benjamin Barth and Ruth Naomi (Myrtle) O.; m. Melinda J. Tarr; children—Cynthia Olsen-Noll, David G., Susan B., John D. B.A., U. Mo., Kansas City, 1959; M.S., Atlanta U., 1963; Ph.D., SUNY, Buffalo, 1969. Tchr. Rustin High Sch., Kansas City, Mo., 1960-61; instr. Met. Jr. Coll., Kansas City, 1963-67; from asst. prof. to prof. microbiology Ohio State U., Columbus, 1969-89; prof. emeritus Ohio State U., 1989—; dir. Parhelion Corp. Ohio State U., Columbus; v.p. R & D Parhelion Corp., Columbus. Author: Immunology and Immunopathology, 1979, Feline Leukemia, 1981; also numerous papers. Inventor method of recovering cell antigen and preparation of feline leukemia vaccine. NIH fellow SUNY-Buffalo, 1967-68; Nat. Cancer Inst. grantee, 1973—. Mem. Am. Soc. Microbiology, Internat. Assn. Research Leukemia, Am. Assn. Cancer Research, Internat. Soc. Immunopharmacology, Am. Soc. Virology. Avocation: farming. Home: 2255 State Route 56 London OH 43140-9628 Office: Ohio State U 1925 Coffey Rd Columbus OH 43210-1093

OLSEN, RICHARD SCOTT, geotechnical engineer; b. Portland, Oreg., Mar. 5, 1954; s. Irving E. and Gerry (Wiley) O.; m. Christine B. Olsen, Apr. 14, 1984; children: Erin, Jonathan. BS, Oreg. State U., 1976; MS, U. Calif., Berkeley, 1977, PhD, 1993. Profl. engr. Calif. Engr. Harding Lawson Assocs., San Francisco, 1978-80; staff engr. FURGO/Earth Tech. Corp., Long Beach, Calif., 1980-82; rsch. civil engr. Waterways Experiment Sta., Vicksburg, Miss., 1983—; reviewer NSF, 1989—. Mem. ASCE, Earthquake Engr. Rsch. Inst. Achievements include development and computerization of the in-situ fiber optic fluorescence cone penetrometer test system; developed method for prediction of static and dynamic geotechnical properties using both Cone Penetrometer Test measurements; developed the failure envelope curature and cavity expansion normalization technique for the CPT and SPT; co-developer of the seismic CPT; developer of massive CPT database for statistical evaluation; stratigraphic techniques for CPTs and borings; developer of several geophysical and CPT field data acquisition computer systems. Office: WES-GG-X 3909 Halls Ferry Rd Vicksburg MS 39180

OLSEN, THOMAS WILLIAM, geologist; b. Omaha, Aug. 25, 1952; s. L.C. and Dorlene N. (Bornholdt) O.; m. Janet Kaye Fulks, Sept. 2, 1978; children: Lindy Kaye, Timothy William. BS in Geol. Engring., Colo. Sch. of Mines, Golden, 1974. Geologist Cities Svc. Oil., Oklahoma City, 1974-76; geol. engr. Tenneco Oil, Oklahoma City, 1976-80; geologist Ind. Cons., Oklahoma City, 1980-85, Quintin Little Co., Ardmore, Okla., 1985—. Mem. Am. Assn. Petroleum Geologists, Oklahoma City Geol. Soc., Ardmore Geol. Soc. (sec.-treas. 1989, v.p. 1990, pres. 1991). Office: Quintin Little Co 2007 N Commerce Ardmore OK 73402-1509

OLSON, CARL MARCUS, chemist retired; b. Chgo., Sept. 18, 1911; s. Oscar Nils and Ida Wilhelmina (Peterson) O.; m. Loraine Bernita Swanson, Oct. 2, 1937; children: Erik, Marcia, Nicholas. AB, Augustana Coll., 1932; PhD, U. Chgo., 1936. Rsch. chemist E.I. du Pont de Nemours, Inc., Wilmington, Del., 1936-42, rsch. supr., 1945-50; lab. dir. Pigments Dept. Exptl. Sta., Wilmington, Del., 1950-68; v.p., treas. Micron, Inc., Wilmington, Del., 1966-80; ret., 1980—; supr. U. Chgo. Metallurgy Lab., 1943—, Clinton Lab., Oak Ridge, Tenn., 1944, Hanford Engring. Works., Richland, Wash., 1945. Achievements include patents for Titanium Pigments; Extractive Metallurgy; Super Pure Silicon for Electronics. Home: 43 Winslow Rd Newark DE 19711

OLSON, DARIN S., research scientist; b. Seattle, Nov. 12, 1964; s. Donald A. and Ruth I. (Pierce) O.; m. Emily R. Dinan, Aug. 6, 1988. BS, MIT, 1987; PhD, Stanford U., 1992. Rsch. engr. Boeing Aerospace Co., Seattle, 1983-87; rsch. scientist Spectra Diode Labs., Sunnyvale, Calif., 1988; rsch. asst. Stanford (Calif.) U., 1987-92; rsch. scientist Lockheed, Palo Alto, 1992—; vis. scientist Nat. Inst. for Rsch. on Inorganic Materials, Tsukuba, Japan, 1991. Contbr. articles to profl. jours. Mem. Sigma Xi (assoc.), Tau Beta Pi. Achievements include development of process to convert graphite directly into diamond thin films. Office: Stanford Univ Stanford CA 94305-2205

OLSON, DAVID P., physicist; b. Meriden, Conn., Nov. 22, 1956; s. Hilding Paul and Ruth Ann (Reppy) O.; m. Lauren L. Hiestand, June 5, 1982. MS, U. Conn., 1980; PhD, U. Ill., 1988. Cons. AT&T Info. Svcs., Oakbrook, Ill., 1986-87; dir. med. computing rsch. lab dept. pediatrics St. Francis Hosp., Hartford, Conn., 1988-91; instr. dept. pediatrics U. Conn. Sch. Medicine, Farmington, 1989—, cons., 1992—; assoc. rsch. scientist dept. physics U. Conn., Storrs, 1989—; cons. Yale U. Sch. Medicine, New Haven, 1990—, Dept. Health Svc., State of Conn., Hartford, 1989—; mem. adv. bd. Conn. Hosp. Rsch. Edn. Found., Wallingford, 1991—; mem. trauma adv. group St. Francis Hosp. and Med. Ctr., Hartford, 1989-91. Contbr. articles to profl. publs. Mem. AMA, Am. Phys. Soc., Am. Assn. Physicists in Medicine, Am. Med. Informatics Assn., Sigma Xi. Office: U Conn Health Ctr Farmington Ave Farmington CT 06248

OLSON, FERRON ALLRED, metallurgist, educator; b. Tooele, Utah, July 2, 1921; s. John Ernest and Harriet Cynthia (Allred) O.; m. Donna Lee Jefferies, Feb. 1, 1944; children: Kandace, Randall, Paul, Jeffery, Richard. BS, U. Utah, 1953, PhD, 1956. Consecrated Bishop Ch. LDS, 1962. Research chemist Shell Devel. Co., Emeryville, Calif., 1956-61; assoc. research prof. U. Utah, Salt Lake City, 1961-63, assoc. prof., 1963-68, chmn. dept mining, metall. and fuels engring., 1966-74, prof. dept. metallurgy and metall. engring., 1968—; cons. U.S. Bur. Mines, Salt Lake City, 1973-77, Ctr. for Investigation Mining and Metallurgy, Santiago, Chile, 1978; dir. U. Utah Minerals Inst., 1980-91. Author: Collection of Short Stories, 1985; contbr. articles to profl. jours. Del. State Rep. Conv., Salt Lake City, 1964; bishop, 1962-68, 76-82, missionary, 1988. With U.S. Army, 1943-46, PTO. Named Fulbright-Hayes lectr., Yugoslavia, 1974-75, Disting. prof. Fulbright-Hayes, Yugoslavia, 1980, Outstanding Metallurgy Instr., U. Utah, 1979-80, 88-89, Disting. Speaker U. Belgrade-Bor, Yugoslavia, 1974. Mem. Am. Inst. Mining, Metall. and Petroleum Engrs. (chmn. Utah chpt. 1978-79), Am. Soc. Engring. Edn. (chmn. Minerals div. 1972-73), Fulbright Alumni Assn., Am. Bd. Engring. and Tech. (bd. dirs. 1975-82). Republican. Achievements include research on explosives ignition and decomposition; surface properties of thoria, silica gels, silicon monoxide in ultra high vacuum; kinetics of leaching of Chrysocolla, Malachite and Bornite; electrowinning of gold; nodulation of copper during electrodeposition. Home: 1862 Herbert Ave Salt Lake City UT 84108-1832 Office: U Utah Dept Metallurgy 412 Browning Bldg Salt Lake City UT 84112

OLSON, HECTOR MONROY, research support engineer; b. Chgo., Oct. 26, 1959; s. Hector Olson and Sonia (Monroy) Cerda; m. Christina Anne Klock, June 7, 1985 (div. 1992); children: Ileah Rhiannon, Adam Monroy, Karissa Elyse. BS in Psychology, Duke U., 1981; BS in Engring., U. Ill., 1986. Rsch. support engr., svc. mgr. SLMA Instruments, Urbana, Ill., 1986—; rsch. support engr. Monroy Cons., Madison. Mem. Horological Soc. (sec. 1984-86), Soc. Nuclear Engrs. (sec. 1981-82), Parapsychology Soc., Clock Tower Restoration Group, L-5 Soc. Home: 2483 Fiedler Ln Madison WI 53713 Office: Monroy Cons 76 Bolder Rod Ln Ste 3 Madison WI 53719

OLSON, JAMES HILDING, software engineer; b. Phila., Nov. 15, 1965; s. Christopher Hilding Olson and Sandle Lillian (Williams) Brennan. BSEE, Drexel U., 1988. Elec. engr. Naval Ship Systems Engring. Sta., Phila., 1985-90; software engr. Telecomm. Techniques Corp., Germantown, Md., 1990—. Author: (computer program) Icon Editor, 1988. Mem. Nat. Computer Graphics Assn., Apple Programmers and Developers Assn. Republican. Office: Telecomm Techniques Corp 20410 Observation Dr Germantown MD 20876

OLSON, MARIAN EDNA, nurse, social psychologist; b. Newman Grove, Nebr., July 20, 1923; d. Edwrd and Ethel Thelma (Hougland) Olson; diploma U. Nebr., 1944, BS in Nursing, 1953; MA, State U. Iowa, 1961, MA in Psychlogy, 1962; PhD in Psychology, UCLA, 1966. Staff nurse, supr. U. Tex. Med. Br., Galveston, 1944-49; with U. Iowa, Iowa City, 1949-59, supr. 1953-55, asst. dir. 1955-59; asst. prof. nursing UCLA, 1965-67; prof. nursing U. Hawaii, 1967-70, 78-82; dir. nursing Wilcox Hosp. and Health Center, Lihue, 1970-77; chmn. Hawaii Bd. Nursing, 1974-80; prof. nursing No. Mich. U., 1984-88. Mem. Am. Nurses Assn. (mem. nat. accreditation bd. continuing edn. 1975-78), Nat. League Nursing, Am. Hosp. Assn., Am. Public Health Assn., LWV. Democrat. Roman Catholic. Home and Office: 6223 County 513 T Rd Rapid River MI 49878-9595

OLSON, NEIL CHESTER, physiologist, educator; b. Starbuck, Minn., Aug. 23, 1950; s. Chester Edward Olson and Esther (Pederson) Helgeson; m. Peggy Sue Wittenberg, Sept. 14, 1974; children: Jennifer JoAnn, Stephanie Sarah. BS, U. Minn., 1973, DVM, 1975; PhD, Mich. State U., 1982. Intern Vet. Coll., Cornell U., Ithaca, N.Y., 1975-76; resident Vet. Coll., Mich. State U., 1976-79; asst. prof. Vet. Coll., N.C. State U., Raleigh, 1982-85, assoc. prof., 1985-90, prof., 1990—; dir. interdepartmental grad. physiology program N.C. State U., Raleigh, 1991—; cons. Am. Inst. Biol. Scis., Washington, 1992—; rsch. grant study sect. mem. Am. Heart Assn., Raleigh, 1992—. Contbr. chpt. to book and articles to profl. jours. Recipient Nat. Rsch. Svc. award NIH, Bethesda, Md., 1979-82, RO1 Rsch. Grant award NIH, Bethesda, 1985—. Mem. Am. Physiol. Soc., Am. Thoracic Soc., N.Y. Acad. Scis., Comparative Respiratory Soc. (bd. dirs.). Home: 9509 Donegal Ct Raleigh NC 27615 Office: NC State Univ Coll Vet Medicine 4700 Hillsborough St Raleigh NC 27606

OLSON, NORMAN FREDRICK, food science educator; b. Edmund, Wis., Feb. 8, 1931; s. Irving M. and Elva B. (Rhinerson) O.; m. Darlene Mary Thorson, Dec. 28, 1957; children: Kristin A., Eric R. BS, U. Wis., 1953, MS, 1957, PhD, 1959. Asst. prof. U. Wis.-Madison, 1959-63, assoc. prof., 1963-69, prof., 1969—, dir. Walter V. Price Cheese Research Inst., 1976-93; dir. Ctr. Dairy Research, 1986-93; disting. prof. U. Wis., 1993—. Author: Semi-soft Cheeses; inventor enzyme microencapsulation. Served to lt. U.S. Army, 1953-55. Fellow Inst. Food Technologists (Macy award 1986); mem. Am. Dairy Sci. Assn. (v.p. 1984-85, pres. 1985-86, Pfizer award 1971, Dairy Research Fedn. award 1978, Borden Found. award 1988), Inst. Food Technologists. Democrat. Lutheran. Avocation: cross-country skiing. Home: 114 Green Lake Pass Madison WI 53705-4755 Office: U Wis Dept Food Sci Babcock Hall Madison WI 53706

OLSON, REX MELTON, oil and gas company executive; b. Alva, Okla., Mar. 6, 1940; s. Harrison and Zylpha Mable (Redgate) O.; m. Sandra Florence Barker, June 5, 1959; children: Brenda Carol Sikes, Brian Rex. Student, Abilene Christian U., 1958-60; BS, Okla. State U., 1963, DVM, 1965. Veterinarian Waynoka (Okla.) Animal Clinic, 1965-83; pres. Zoroco Petroleum, Inc., Waynoka, 1979-88, Santo Resources, Inc., Waynoka, 1982—; owner/operator ranch Waynoka, 1958—; co-founder Waynoka Mfg. Co., 1987; bd. dirs. First State Bank, Waynoka. Mem. Waynoka Airport Commn.; past mem. Waynoka City Coun., Waynoka Indstrl. Authority; v.p. Woods County Indsl. Com.; bd. dirs. Waynoka Sch. Found., York Coll., Nebr., trustee, 1988—; pres. Waynoka Hist. Soc., 1992—. Mem. AVMA, Okla. Vet. Med. Assn. (past bd. dirs.), Okla. Mineral Owners Assn., Aircraft Owners and Pilots Assn., Waynoka C. of C. (pres. 1977, 87). Republican. Mem. Ch. of Christ. Avocations: flying, skiing, fishing, golfing. Home: 604 E Maple St Waynoka OK 73860 Office: Santo Resources Inc 101 Missouri Waynoka OK 73860

OLSON, RICHARD DEAN, researcher, pharmacology educator; b. Rupert, Idaho, June 22, 1949; s. Emerson J. and Thelma Maxine (Short) O.; m. Carol Ann Dyba, Jan. 5, 1974; children: Stephan Jay, David Richard, Jonathan Philip. BS, Coll. Idaho, 1971; postgrad., Idaho State U., 1972-74; PhD, Vanderbilt U., 1978. Instr. Vanderbilt U., Nashville, 1980-81, asst. prof., 1982, head pediatric clin. pharmacology unit, 1982; asst. prof. U. S. Ala., Mobile, 1982-83; acting asst. prof. U. Wash., Seattle, 1984-85, research assoc. prof., 1985—; v.p. Olson, Wong and Walsh Labs., Inc., Lindenhurst, Ill., 1987-90, pres., 1990—; chief cardiovascular pharmacology rsch. VA Med. Ctr., Boise, Idaho, 1984—; hon. dir. cardiovascular rsch. lab. Capital Inst. Medicine Beijing, People's Republic of China, 1986; investigator Am. Heart Assn., Nashville, 1981, NIH, Mobile, 1982; bd. dirs. Idaho Affailiate Am. heart Assn.; pres. JB Internat. Inc., Boise, 1990; sr. cons. Longwood Cons. Group, Boston, 1990—. Contbr. articles to profl. jours. Pres. Fellowship Crusade for Christ, Inc., Nampa, Idaho, 1986—. Grantee Am. Heart Assn., 1981, 83, 84, 86, 88, Am. Fedn. Aging Research, Inc., Boise, 1985; VA Merit Review grantee, 1985, 88, 91; NIH trainee, 1975-78; NIH New Investigator 1982; fellow, Am. Fedn. Aging Research, Inc. 1985—, NIH U. Colo., Denver, 1978-80. Mem. AAAS, N.Y. Acad. Scis., Am. Soc. Pharmacology and Exptl. Therapeutics, Am. Heart Assn., Am. Fedn. for Clin. Research, Sigma Xi. Avocations: camping, radio-controlled model airplanes. Home: 605 Crocus Ct Nampa ID 83651 Office: VA Med Ctr #151 500 W Fort St Boise ID 83702-4501

OLSON, ROY EDWIN, civil engineering educator; b. Richmond, Ind., Sept. 13, 1931; s. Roy Edwin and Hester Elizabeth (Nelson) O.; children: Sandra Lee Christian, Cheryl Ann Jann, Chresten Edwin. BS in Engring., U. Minn., 1953, MS in Civil Engring., 1955; PhD, U. Ill., 1960. Registered profl. engr., Tex. Prof. U. Ill., Urbana, 1960-70; prof. civil engring. U. Tex., Austin, 1970-86, 87—; program dir. geomechanics NSF, Washington, 1986-87; bd. trustees Deep Founds. Inst., Springfield, N.J., 1983-86. Mem. ASCE (exec. com. geotech. engring. div. 1969-79, Huber rsch. prize 1971, Norman medal 1975, Croes medal 1984), ASTM (Hogentogler award 1973, 87), Transp. Rsch. Bd. Home: 4901 Ridge Oak Dr Austin TX 78713 Office: U Tex Dept Civil Engring Austin TX 78712

OLSON, WILLIAM HENRY, neurology educator, administrator; b. Haxtun, Colo., Sept. 2, 1936; s. William Henry and Burdene (Anderson) O.; m. Shirley Gorden, July 24, 1967; children: Erik, Marnie. B.A., Wesleyan U., 1959; M.D., Harvard U., 1963. Diplomate: Am. Bd. Psychiatry and Neurology. Intern Beth Israel Hosp., Boston, 1963-65; resident Children's Hosp. Med. Ctr., Boston, 1965-67; staff assoc. NIH, Bethesda, Md., 1969-70; asst. prof. neurology and anatomy Vanderbilt U., Nashville, 1970-73, assoc. prof. neurolog and anatomy, 1973-75; prof., chmn. dept. adult neurology U. N.D., Fargo, 1975-80; chmn., prof. dept. neurology U. Louisville, 1980—. Co-author: Practical Neurology and the Primary Care Physician, 1981. Fulbright scholar Tubingen, Germany, 1958-59. Fellow Am. Acad. Neurology; mem. Phi Beta Kappa. Home: 331 Zorn Ave # 1 Louisville KY 40206-1542 Office: Dept Neurology Univ Louisville Louisville KY 40292

OLSSON, NILS URBAN, research manager; b. Västeras, Sweden, Sept. 1, 1959; s. Nils Leopold Ossian and Elly Gabriella Karolina (Gustavsson) O.; m. Kathryn Anne Beale, Feb. 14, 1987; children: Nils Erik Alexander, Laila Elizabeth, Samuel Leopold. BS in Chemistry, Stockholm (Sweden) U., 1987, PhD in Analytical Chemistry, 1991. Project asst. Thoracic Clinic, Karolinska Hosp., Stockholm, 1986-87; rsch. chemist Karlshamns LipidTeknik AB, Stockholm, 1987-91, rsch. mgr., 1991—; bd. dirs. Karlshamns (Sweden) Rsch. Coun.; chmn. 2d Meeting on Contemporary Lipid Analysis, Stockholm, 1992. Contbr. articles to profl. jours. Mem. Am. Chem. Soc. Achievements include patent for herbal extract composition; rsch. for analysis and charactarisation of polar lipids with chromatographic and spectroscopic methods. Home: Nordkapsgatan 5 4tr, 16436 Kista Sweden Office: Karlshamns LipidTeknik AB, PO Box 6686, 11384 Stockholm Sweden

OLSSON, RONALD ARTHUR, computer science educator; b. Huntington, N.Y., Nov. 16, 1955; s. Ronald Alfred and Dorothy Gertrude (Hofmann) O. BA and MA, SUNY, 1977; MS, Cornell U., 1979; PhD, U. Ariz., 1986. Teaching asst. Cornell U., Ithaca, N.Y., 1977-79, rsch. asst., 1979; lectr. SUNY, Brockport, 1979-81; rsch. assoc. U. Ariz., Tucson, 1981-86; prof. Computer Sci. U. Calif., Davis, 1986—. Author (book) The SR Programming Language: Concurrency in Practice, 1993; contbr. articles to profl. jours. MICRO grantee U. Calif., 1987, 92, NSF grantee, 1988, Dept. Energy grantee, 1988-92. Mem. Assn. for Computing Machinery. Avocations: bicycling, hiking, cross-country skiing, movies. Home: 1333 Arlington Blvd Apt 31 Davis CA 95616-2664 Office: U Calif Dept Computer Sci Davis CA 95616-8562

OLSTOWSKI, FRANCISZEK, chemical engineer, consultant; b. N.Y.C., Apr. 23, 1927; s. Franciszek and Marguerite (Stewart) O.; A.A., Monmouth Coll., 1950; B.S. in Chem. Engring., Tex. A. and I. U., 1954; m. Rosemary Sole, May 19, 1952; children—Marguerita Antonina, Anna Rosa, Franciszek, Anton, Henryk Alexander. Research and devel. engr. Dow Chem. Co., Freeport, Tex., 1954-56, project leader, 1956-65, sr. research engr., 1965-72, research specialist, 1972-79, research leader, 1979-87; dir. Tech. Cons. Services, Freeport, 1987—. Lectr. phys. scis. elementary and intermediate schs., Freeport, 1961-85. Vice chmn. Freeport Traffic Commn., 1974-76, chmn., 1976-79, vice-chmn. 1987-89, chmn., 1989-92. With USNR, 1944-46. Fellow Am. Inst. Chemist; mem. AAAS, Am. Chem. Soc., Electrochem. Soc. (sec. treas. South Tex. sect. 1963-64, vice chmn. 1964-65, chmn. 1965-67, councillor 1967-70), N.Y. Acad. Sci, Velasco Cemetery Assn. (sec.-treas. 1992—). Patentee in synthesis of fluorocarbons, natural graphite products, electrolytic prodn. magnesium metal and polyurethane tech.

OLSZEWSKI, JAMES FREDERICK, metallurgical engineer; b. Buffalo, Aug. 4, 1948; s. Frederick and Helen (Stelmach) O.; m. Susan Linda Wesolowski, June 30, 1970; children: Jeffrey, Jacob. AAS, Erie Community Coll., Buffalo, N.Y., 1971; BA, SUNY, Buffalo, 1972. Cert. quality engr. Am. Soc. Quality Control. Chemist PM Refining, Buffalo, 1973-76; quality control supr., metallurgist Anaconda Co., Buffalo, 1976-80; rsch. engr. ARCO Metals, Arlington Heights, Ill., 1980-84; sr. metallurgist Gibraltar Steel, Buffalo, 1985-87; sr. staff metallurgist Alcoa, Massena, N.Y., 1987--. Mem. Am. Soc. Metallurgists, Am. Chem. Soc., Soc. Mfg. Engrs., Minerals Metals and Material Soc. Democrat. Roman Catholic. Home: PO Box 355 Waddington NY 13694

OLSZEWSKI, JERZY ADAM, electrical engineer; b. Nieswiez, Poland, Feb. 2, 1929; came to the U.S., 1955; s. Wincenty and Jadwiga (Malyszewicz) O.; m. Lidia Cegielnik, Sept. 28, 1958 (div. 1984); 1 child, Renata L.; m. Stefania Lonski, Nov. 10, 1988. MSc, Loughborough U. Tech., England, 1954. Rsch. engr. Gen. Cable Co., Bayonne, N.J., 1955-72; asst. dir. rsch. Gen. Cable Co., Union, N.J., 1972-80; tech. dir. fiber div. Gen. Cable Co., Edison, N.J., 1986-87, sr. scientist, 1987—; tech. dir. Okonite Co., Providence, 1980-81, Gen. Cable Internat. Inc., Edison, 1981-86. Contbr. articles to profl. jours. Mem. IEEE (sr.), Optical Soc. Am., Wire Assn. Internat. Republican. Roman Catholic. Achievements include 17 U.S. and 6 foreign patents; development of theory of structural return loss in coaxial cables. Home and Office: 27 Mechanic St Bayonne NJ 07002-4513

OLSZEWSKI, LEE MICHAEL, instrument company executive; b. East St. Louis., Ill., July 21, 1935; s. Leon Ignatius and Mary Elizabeth (Ryznek) O.; m. Elizabeth Ann Schnelly, De. 22, 1990; children: Theresa, Anna. BA in Chemistry, So. Ill. U., 1963; PhD in Biology, Washington U., 1967. Rsch. chemist Monsanto Co., St. Louis, 1953-67; sales mgr. Barber Coleman Co., Rockford, Ill., 1967-70, Bendix Corp., Ronceverte, W.Va., 1970-78; pres. Delta Instrument Co., Inc., Mountain View, Mo., 1978—; cons., advisor Am. Nat. Can Co., Des Moines, Iowa, 1980-92; cons. Union Electric Co., St. Louis, 1984-92; ptnrs. in edn. Mountain View Sch. Dist., Mo., 1992. Named Outstanding Rep. Shimadzu Scientific, 1987-92. Fellow Instrument Soc. Am.; mem. Am. Chem. Soc., Mountain View C. of C., NRA, Aircraft Owners and Pilots Assn. Democrat. Achievements include assisting in the development of the first programmed temperature thermal conductivity gas chromatogrph. Home: Rt 3 Box 3677 Mountain View MO 65548 Office: Delta Instrument Co PO Box 570 Mountain View MO 65548

O'MALLEY, BERT WILLIAM, cell biologist, educator, physician; b. Pitts., Dec. 19, 1936; s. Bert Alloysius O'M.; m. Sally Ann Johnson; children: Sally Ann, Bert A., Rebecca, Erin K. BS, U. Pitts., 1959, MD summa cum laude, 1963; DSc (hon.), N.Y. Med. Coll., 1979, Nat. U. Ireland, 1985; MD (hon.), Karolinska Inst., Stockholm, 1984. Intern, resident Duke U., Durham, N.C., 1963-65; clin. assoc. Nat. Cancer Inst., NIH, Bethesda, Md., 1965-67, head molecular biology sect., endocrine br., 1967-69; Lucius Birch prof., dir. Reproductive Biology Ctr. Vanderbilt U. Sch. Medicine, Nashville, 1969-73; Tom Thompson prof., chmn. dept. cell biology Baylor Coll. Medicine, Houston, 1973—, Disting. Svc. prof., 1985, dir. Baylor Ctr. for Reproductive Biology, 1973—; mem. endocrine study sect., NIH, 1970-73, chmn., 1973-74; chmn. CETUS-UCLA Symposium on Gene Expression, 1982; con., mem. coun. rsch. and clin. investigation awards Am. Cancer Soc., 1985-87. Author: (with A.R. Means) Receptors for Reproductive Hormones, 1973, (with L. Birnbaumer) Hormone Action, vols. I and II, 1977, vol. III, 1978, (with A.M. Gotto) The Role of Receptors in Biology and Medicine, 1986; contbg. author to over 400 publs. Lt. comdr. USPHS, 1965-69. Recipient Ernst Oppenheimer award Am. Endocrine Soc., 1975, Gregory Pincus medal, 1975, Lila Gruber Cancer award, 1977, Disting. Achievement in Modern Medicine award, 1978, Borden award Assn. Am. Med. Colls., 1978, Dickson prize for Basic Med. Rsch., 1979, Philip S. Hench award U. Pitts., 1981, Axel Munthe Reproductive Biology award, Capri, Italy, 1982, Bicentennial Medallion of Distincton U. Pitts., 1987. Mem. Am. Soc. Biol. Chemists, Endocrine Soc. (pres. 1985—, Fred Conrad Koch medal 1988), Am. Soc. Clin. Investigation, Am. Inst. Chemists, Fedn. Clin. Rsch., Harvey Soc., AAAS, Alpha Epsilon Delta, Phi Beta Kappa, Alpha Omega Alpha. Democrat. Roman Catholic. Office: Baylor Coll Medicine Dept Cell Biology One Baylor Pla Houston TX 77030

O'MALLEY, WILLIAM DAVID, architect; b. Kansas City, Mo., Oct. 12, 1947; s. Richard J. and Clare Alphild (Eide) O'M.; m. Constance Jeanne Rabideau, June 21, 1969; children: Derek, Kevin. BArch, U. Minn., 1971. Registered architect, Minn. Plan review engr. City of Mpls., 1972-79; architect Hammel Green & Abrahamson, Mpls., 1979-82, sr. project architect, 1982-85, sr. project mgr., 1985-90, dir. info. systems, 1990—; adv. com. archtl. tech. Mpls. Tech. Coll., 1991—. Archtl. works include Mall of America, Carnegie Inst. Omnimax Theater, Boston's Mus. of Sci. Omnimax Theater, Chgo. Mus. of Sci. and Industry Crown Space Ctr., Plaza 7 Hotel. Recipient Honor award AIA Minn., 1984. Mem. AIA, AIA Minn., Constrn. Specifications Inst. Office: HGA Architects 1201 Harmon Pl Minneapolis MN 55403

OMAN, HENRY, retired electrical engineer, engineering executive; b. Portland, Oreg., Aug. 29, 1918; s. Paul L. and Mary (Levonen) O.; m. Winifred Eleanor Potter, June 17, 1944 (dec. Nov. 1950); m. Earlene Mary Boot, Sept. 11, 1954; children: Mary Janet, Eleanor Eva, Eric Paul. BSEE, Oreg. State U., 1940, MSEE, 1951. Registered profl. engr., Wash. Application engr. Allis Chalmers Mfg. Co., Milw., 1940-48; rsch. engr. Boeing Co., Seattle,

1948-63, engring. mgr., 1963-92; Author: Energy Systems Engineering Handbook, 1986; contbr. 35 tech. articles to profl. jours. Mem. team that restarted amateur radio communication to the outside world from the People's Republic of China, 1981. Recipient prize paper award Am. Inst. Elec. Engrs., 1964. Fellow IEEE (v.p. Aerospace and Electronics Systems Soc. 1984-88, Harry Mimno award 1989), AIAA (assoc.); mem. AAAS (bd. dir. Pacific divsn. 1992—). Republican. Methodist. Achievements include development of concepts for solar power satellite which generates power in geo-synchronous orbit 24 hours per day and beams it to the Earth surface with a microwave beam. Home: 19221 Normandy Park Dr SW Seattle WA 98166

OMAN, PAUL RICHARD, entrepreneur; b. Manchester, N.H., Sept. 5, 1956; s. Ted and Mildred (Lindquist) O. BS in Geolgoy, Humboldt State U., 1977; BS in Oceanography, Calif. State U., Arcata, 1977; MBA, U. Houston, 1984, MS in Phys. and Geol. Sci., 1986. Petroleum logging engr. Schlumberger Offshore, Pearland, Tex., 1978; geologist J. Rose & Assoc. & Total Petroleum, Pearland, Tex., 1979-83; pres., founder, PROTEC, Pearland, Tex., 1983—; chmn. adv. bd. Entrepreneurial Devel. Ctr., Software Bus. Ctr.; instr. small bus. devel. ctr. Alvin C.C., Leisure Learning Unltd. Columnist Allied Feature Syndiacte, Mgr.'s Monthly. Achievements include research in communications and marketing. Home and Office: 4607 Linden Pl Pearland TX 77584

OMAR, HUSAM ANWAR, civil engineering educator; b. Kuwait, Aug. 8, 1959; s. Anwar A.J. and Abla Mohamad (Dahleh) O.; m. Hayfa Jameel, Sept. 13, 1990; 1 child, Ahmad. BS in Civil Engring., U. Tex., Arlington, 1981, MS in Civil Engring., 1982; PhD in Civil Engring., U. Man., Winnipeg, 1988. Rsch. asst. U. Man., Winnipeg, 1984-88; asst. prof. civil engring. U. South Ala., Mobile, 1988—. Recipient Cert. of Appreciation NASA, Huntsville, Ala., 1990, 91, Grants NASA, Huntsville, 1990, 91, 92; named Outstanding Faculty Haliburton Found., Mobile, 1991. Mem. ASCE, Am. Concrete Inst., Am. Soc. Engring. Educators, Tau Beta Pi. Moslem. Achievements include performance of structural design of components of NASA's large lunar telescope, design of a lunar lander for NASA's lunar telescope; study of concrete prodn. on the moon which is supported by NASA. Office: U South Ala 307 University Ave Mobile AL 36688

O'MEARA, ONORATO TIMOTHY, university administrator, mathematician; b. Cape Town, Republic of South Africa, Jan. 29, 1928; came to U.S., 1957; s. Daniel and Fiorina (Allorto) O'M.; m. Jean T. Fadden, Sept. 12, 1953; children—Maria, Timothy, Jean, Kathleen, Eileen. B.Sc., U. Cape Town, 1947, M.Sc., 1948; Ph.D., Princeton U., 1953; LLD (hon.), U. Notre Dame, 1987. Asst. lectr. U. Natal, Republic South Africa, 1949; lectr. U. Otago, New Zealand, 1954-56; mem. Inst. for Advanced Study, Princeton, N.J., 1957-58, 62; asst. prof. Princeton, 1958-62; prof. math. U. Notre Dame, Ind., 1962-76; chmn. dept. U. Notre Dame, 1965-66, 68-72, Kenna prof. math., 1976—, provost, 1978—; vis. prof. Calif. Inst. Tech., 1968; Gauss prof. Göttingen Acad. Sci., 1978; mem. adv. panel math. scis. NSF, 1974-77, cons., 1960—. Author: Introduction to Quadratic Forms, 1963, 71, 73, Lectures on Linear Groups, 1974, 2d edit., 1977, 3d edit., 1988, Russian translation, 1976, Symplectic Groups, 1978, 82, Russian translation, 1979, The Classical Groups and K-Theory (with A.J. Hahn), 1989; contbr. articles on arithmetic theory of quadratic forms and isomorphism theory of linear groups to Am. and European profl. jours. Mem. Cath. Commn. Intellectual and Cultural Affairs, 1962—; bd. govs., trustee U. Notre Dame Australia, 1990—. Recipient Marianist award U. Dayton, 1988; Alfred P. Sloan fellow, 1960-63. Mem. Am. Math. Soc., Am. Acad. Arts and Scis., Collegium (bd. dirs. 1992—). Roman Catholic. Home: 1227 E Irvington Ave South Bend IN 46614-1417 Office: U Notre Dame Office of Provost Notre Dame IN 46556

OMER, GEORGE ELBERT, JR., orthopaedic surgeon, educator; b. Kansas City, Kans., Dec. 23, 1922; s. George Elbert and Edith May (Hines) O.; m. Wendie Vilven, Nov. 6, 1949; children: George Eric, Michael Lee. B.A., Ft. Hays State U., 1944; M.D., Kans. U., 1950; M.Sc. in Orthopaedic Surgery, Baylor U., 1955. Diplomate Am. Bd. Orthopaedic Surgery, 1959, re-cet. orthopaedics and hand surgery, 1983 (bd. dirs. 1983-92, pres. 1987-88), cert. surgery of the hand, 1989. Rotating intern Bethany Hosp., Kansas City, 1950-51; resident in orthopaedic surgery Brooke Gen. Hosp., San Antonio, 1952-55, William Beaumont Gen. Hosp., El Paso, Tex., 1955-56; chief surgery Irwin Army Hosp., Ft. Riley, Kans., 1957-59; cons. in orthopaedic surgery 8th Army Korea, 1959-60; asst. chief orthopaedic surgery, chief hand surgeon Fitzsimons Army Med. Center, Denver, 1960-63; dir. orthopaedic residency tng. Armed Forces Inst. Pathology, Washington, 1963-65; chief orthopaedic surgery and chief Army Hand Surg. Center, Brooke Army Med. Center, 1965-70; cons. in orthopaedic surgery to Surgeon Gen. Army, 1967-70, ret., 1970; prof. orthopaedics, surgery and anatomy, chmn. dept. orthopaedic surgery, chief div. hand surgery U. N.Mex., 1970-90, med. dir. phys. therapy, 1972-90, acting asst. dean grad. edn. Sch. Medicine, 1980-81; mem. active staff U. N.Mex. Hosp., chief of med. staff, 1984-86, Albuquerque; cons. staff other Albuquerque hosps.; cons. orthopaedic surgery USPHS, 1966-85, U.S. Army, 1970—, U.S. Air Force, 1970-78, VA, 1970—, Carrie Tingley Hosp. for Crippled Children, 1970—, interim med. dir., 1970-72, 86-87. Bd. editors: Clin. Orthopaedics, 1973-90, Jour. AMA, 1973-74, Jour. Hand Surgery, 1976-81; contbr. over 200 articles to profl. jours., also 2 books, over 50 chpts. to books. Commd. 1st lt. M.C. U.S. Army, 1949; advanced through grades to col. 1967. Decorated Legion of Merit, Army Commendation medal with 2 oak leaf clusters; recipient Alumni Achievement award Ft. Hays State U., 1973, Recognition plaque Am. Soc. Surgery Hand, 1989, Recognition plaque N.Mex. Orthopaedic Assn., 1991. Fellow ACS, Am. Orthopaedic Assn. (pres. 1988-89, exec. dir. 1989-93), Am. Acad. Orthopaedic Surgeons, Assn. Orthopaedic Chmn., N.Mex. Orthopaedic Assn. (pres. 1979-81), La. Orthopaedic Assn. (hon.), Korean Orthopaedic Assn. (hon.), Peru Orthopaedic Soc. (hon.), Caribbean Hand Soc., Am. Soc. Surgery Hand (pres. 1978-79), Am. Assn. Surgery of Trauma, Assn. Bone and Joint Surgeons, Assn. Mil. Surgeons U.S., Riordan Hand Soc. (pres. 1967-68), Sunderland Soc. (pres. 1981-83), Soc. Mil. Orthopaedic Surgeons, Brazilian Hand Soc., S.Am. Hand Soc. (hon.), Groupe D'Etude de la Main, Brit. Hand Soc., Venezuela Hand Soc. (hon.), South African Hand Soc. (hon.), Western Orthopaedic Assn. (pres. 1981-82), AAAS, Russell A. Hibbs Soc. (pres. 1977-78), 38th Parallel Med. Soc. (Korea) (sec. 1959-60); mem. AMA, Phi Kappa Phi, Phi Sigma, Alpha Omega Alpha, Phi Beta Pi. Home: 316 Big Horn Ridge Rd NE Sandia Heights Albuquerque NM 87122

OMIROS, GEORGE JAMES, medical foundation executive; b. Uniontown, Pa., Oct. 26, 1956; s. Chris George and Alice (Zervoudi) O.; m. Sophia Florent, June 28, 1980; children: Christopher George, Alicia Helene. BS in Polit. and Philosophy, U. Pitts., 1978; M, Cen. Mich. U., 1982. Campaign coordinator, program assoc. SW Pa. chpt. Am. Heart Assn., Greensburg, 1979, fundraising dir., 1979-80, dir. devel., 1980-84; v.p. devel., ops. Western Pa. chpt. Am. Heart Assn., Pitts., 1984-85, dep. exec. v.p., 1985-87, exec. v.p., 1987-88, exec. dir., 1988—, nat. mktg. rep., 1988—, asst. v.p. nat. office, 1991-93; sr. exec. dir., nat. dir. donor devel. Western Pa. chpt. Am. Heart Assn., Pitts., 1993—. Cons. devel. Greek Orthodox Archdiocese, Pitts., 1982—, v.p. 1987—; mem. council, rev. com. Health Systems Agy. Southwest Pa., Pitts., 1983-87; mem. Parish Council, St. Spyridon Greek Orthodox Ch., Monessen, Pa., 1982—. Mem. Nat. Soc. Fundraising Execs. (cert., founder 1980, pres. 1985-87, Outstanding Fundraising Exec. 1990), Pitts. Planned Giving Coun. (founding com. 1983—), Friends of George C. Marshall (steering com.), Uniontown Country Club, Uniontown Rotary (local treas. 1985, sec. 1986, v.p. 1987, pres. 1988), Pitts. Rotary, Masons. Republican. Greek Orthodox. Avocations: stained glass work, art collections, gardening, stamp collecting. Office: Leukemia Soc Am 13 North 2 Gateway Ctr Pittsburgh PA 15222

OMLAND, TOV, physician, medical microbiologist; b. Kviteseid, Telemark, Norway, May 15, 1923; s. Hans Omland and Torbjorg Lid; m. Ellen-Margrethe Soderstrom, Aug. 13, 1949 (dec. Mar. 1981); children: Anne Katerine, Hans Harald. MD, Oslo U., 1949, specialist med. microbiology, 1959. Med. officer Internat. Tb Campaign, Greece, 1950-51; gen. practice medicine, Norway, 1951-52; med. officer WHO, Egypt and Turkey, 1952-53; trainee internal medicine and surgery Oslo City Hosp., 1954-55; trainee, asst. prof. Oslo U., 1955-63; dir. State Microbiology Lab., Lillehammer, Norway, 1963-68; dir. Norwegian Def. Microbiology Lab., Oslo, 1968-88; bd. dirs. Ellen-Margrethe and Tov Omland's Lab. for Med. Microbiology, Ski, Norway, 1973—; adv. on disarmament biol. weapons Ministry of Fgn. Affairs, Norway. Contbr. chpts. books and articles to profl. jours. Served to col. Norwegian Med. Corps, 1978. Recipient prize Schering Corp. U.S.A., 1986. Mem. Norwegian Soc. Microbiology (pres. 1973-75), Norwegian Soc. Pathology, Norwegian Soc. Infectious Diseases, Soc. Gen. Microbiology (U.K.), Am. Soc. Microbiology. Conservative. Lutheran. Club: Oslo Militaere Samfund, Norwegian Reserve Officers' Club (pres. regional divsn. 1992—). Home: Gamlevegen 55, 1400 Ski Norway Office: EM & T Omland's Lab Med Microbiology, 1400 Ski Norway

OMURA, TSUNEO, medical educator; b. Shizuoka, Japan, July 29, 1930; s. Bunzo and Yasu Omura; m. Yone Tominaga, Nov. 16, 1957; children: Shigeru, Minoru, Kaoru. BS, U. Tokyo, 1953, DSc, 1962. Instr. Shizuoka U., 1953-60; asst. prof. Osaka (Japan) U., 1960-70; prof. Kyushu U., Fukuoka, Japan, 1970—. Editor, author: Cytochrome P-450, 1978, 2d edit., 1993. Mem. Japanese Biochem. Soc., Am. Soc. Biochemistry and Molecular Biology (hon.). Home: 7-17-7 Hinosato, Munakata Fukuoka 811-34, Japan Office: Kyushu U Med Sch, Maidashi 3-1-1, Fukuoka 812, Japan

ONARAL, BANU KUM, electrical/biomedical engineering educator; b. Istanbul, Turkey, June 15, 1949; came to U.S., 1974; d. Mehmet Serhan and Sukufe (Demirag) Kum; m. Ibrahim Etem, Sept. 21, 1973; 1 child, Mutlu Can. MSEE, Bogazici U., Istanbul, 1974; PhD, U. Pa., 1978. Postdoctoral fellow U. Pa., Phila., 1979-80; vis. asst. prof. Drexel U., Phila., 1980, asst. prof., 1981-85, assoc. prof., 1985-91, prof., 1991—; vis. asst. prof. Bogazici U., 1980-81; vis. prof. Bogarici U., 1987-88; cons. TOKTEN/UN, Istanbul, 1989. Contbr. articles to profl. jours. Fulbright scholar Inst. Internat. Edn., 1974-78. Fellow AAAS, IEEE (gen. chair 12th ann. internat. conf. IEEE/EMBS Phila. 1990, tech. meetings coun./TAB Piscataway, N.J. 1991—), Am. Inst. Med. and Biol. Engring.; mem. Soc. Women Engrs. (sr.), Am. Soc. Engring. Edn., Sigma Xi. Achievements include research on scaling dynamics in signals and systems engineering, bioelectrodes, and scaling in non-destructive testing; patent in Biocompatible Electrode and Use in Orthodontic Electo Osteogenesis. Home: 2301 Cherry St # 6D Philadelphia PA 19103-1029 Office: Drexel U 32d and Chestnut St Philadelphia PA 19104

ONAYA, TOSHIMASA, internal medicine educator; b. Tokyo, May 7, 1935; s. Toshinobu and Kou (Ide) O.; m. Michiko Yamamoto, Mar. 7, 1965; children: Jun, Tina. MD, Gunma U., Maebashi, Japan, 1962, PhD, 1971. Intern U.S. Army Hosp. Zama, Sagamihara, Japan, 1962-63; postgrad. Gunma U., 1963-65, 69-71; rsch. fellow U. Calif., Berkeley, 1965-66; rsch. assoc. U. Iowa, Iowa City, 1966-68; postgrad. rsch. endocrinologist UCLA, 1968-69; asst. prof. Shinshu U., Matsumoto, Japan, 1971-73; assoc. prof. Shinshu U., Matsumoto, 1973-83; prof. medicine U. Yamanashi (Japan) Med. Sch., 1983—; prof., chmn. Third Dept. Internal Medicine, U. Yamanashi Med. Sch., 1983—. Contbr. articles to profl. jours. Recipient Shichijo prize Japan Thyroid Assn., 1971, Daiichi prize Asia and Oceania Thyroid Assn., 1991. Mem. Japanese Soc. Internal Medicine, Japan Endocrine Soc., Japan Diabetes Soc., Japan Atherosclerosis Soc., Japan Thyroid Assn., Japan Geriatrics Soc. Home: 12-3 Arigasakidai, Matsumoto 390, Japan Office: U Yamanashi Med Sch, 1110 Tamaho, Yamanashi-ken 409-38, Japan

ONCLEY, JOHN LAWRENCE, biophysics educator, consultant; b. Wheaton, Ill., Feb. 14, 1910; s. Lawrence and Emma Arena (Hunsche) O.; m. Genevieve Reese, June 14, 1933 (dec. May 1971); children: Genevieve Louise Oncley August, Nancy Anne Oncley Thyng; m. 2d, Lephia French Giles, Mar. 3, 1972. A.B., Southwestern Coll., Winfield, Kans., 1929, Sc.D. (hon.), 1954; Ph.D., U. Wis., 1933; A.M. (hon.), Harvard U., 1946. Instr. U. Wis., Madison, 1934-35; with research lab. Gen. Electric Co., Schenectady, N.Y., summer 1935; instr. MIT, Cambridge, 1935-43; research assoc. to prof. Harvard U. Med. Sch., Boston, 1939-62; prof. biophysics, dir. U. Mich., Ann Arbor, 1962-80, emeritus prof. chemistry and biol. chemistry, 1980—; cons. Gen. Radio Co., Cambridge, Mass., 1941-44; biophysics cons. Army Chem. Corps, Ft. Detrick, Frederick Md., 1954-67; mem. study sect./tng. grant coms. NIH, Gen. Med. Sci., Bethesda, Md., 1954-65, Gen. Med. Sci. Council, 1968-72. Editor-in-chief: Biophysicial Science—A Study Program, 1958-59; editor Biophysics Jour., 1964-66. Fellow Coffin Found., 1931-32, Fulbright Found., 1953, Guggenheim Found., 1953; Nat. Research Council grantee, 1932-34. Fellow Nat. Acad. Scis., AAAS, Am. Heart Assn. (High Blood Pressure Soc.; Stauffer prize 1972); mem. Am. Chem. Soc. (councilor 1956-62, Pure Chemistry award 1942), Am. Acad. Arts and Scis. (fellow; sec. 1959-62), Biophys. Soc. (pres. 1962-63), Internat. Union Pure and Applied Chemistry (com.; pres. com. on protein standards 1951-55, 57-59), Am. Soc. Biol. Chemists, Sigma Xi, Phi Kappa Phi, Gamma Alpha, Phi Lambda Upsilon, Alpha Chi Sigma. Lodge: Rotary (sec. 1982-85, 88-89). Avocations: travel; stamp collecting. Home: 9 Heatheridge St Ann Arbor MI 48104-2757 Office: U Mich Biophysics Rsch Divsn Rm 4046 Chemistry Bldg Ann Arbor MI 48109-1055

ONDETTI, MIGUEL ANGEL, chemist, consultant; b. Buenos Aires, Argentina, May 14, 1930; came to U.S., 1960, naturalized, 1971; s. Emilio Pablo and Sara Cecilia (Cerutti) O.; m. Josephine Elizabeth Garcia, June 6, 1958; children: Giselle Christine, Gabriel Alexander. Licensiate in Chemistry, U. Buenos Aires, 1955, D.Sc., 1957. Prof. chemistry Inst. Tchrs., Buenos Aires, 1957-60; instr. organic chemistry U. Buenos Aires, 1957-60; rsch. scientist Squibb Inst. Med. Rsch., Buenos Aires, 1957-60; rsch. investigator Squibb Inst. Med. Rsch., Princeton, N.J., 1960-66; rsch. supr. Squibb Inst. Med. Rsch., 1966-73, sect. head, 1973-76, dir. biol. chemistry, 1976-79; assoc. dir. Squibb Inst., 1980-82, v.p. rsch. cardiopulmonary disease, 1982-86, sr. v.p. cardiovascular rsch., 1987-91; pharm. cons., 1991—; ad-hoc cons., sculptor NIH; mem. adv. com. dept. chemistry Princeton U., 1982-86. Patentee in field (115); contbr. articles to sci. jours. Served with Argentine Army, 1950-51. Recipient Thomas Alva Edison Patent award R&D Coun. N.J., 1983, Ciba award for hypertension rsch. Am. Heart Assn., 1983, Perkins medal Soc. Chemistry Industry, 1991, Warren Alpert Found. award, 1991; scholar Brit. Coun., 1960, Squibb, 1956. Mem. AAAS, Am. Chem. Soc. (Alfred Burger award 1981, Creative Invention award 1992, Perkin medal 1992), Am. Soc. Biol. Chemists. Home: 79 Hemlock Circle Princeton NJ 08540-5405

ONDREJACK, JOHN JOSEPH, mechanical engineer; b. Newark, Jan. 27, 1954; s. John Joseph and Marie Rose (Bistak) O.; m. Karen Madaline Sasileo, June 8, 1980; children: Nicole Janelle, Jonathan Michael. BSME, Lehigh U., 1976. Registered profl. engr., Fla. Internat. mktg. mgr. Worthington Pump Corp., East Orange, N.J., 1976-79; Caribbean sales mgr. Dresser Pump Divsn., Ft. Lauderdale, Fla., 1980-87; regional mgr. Dresser Pump Divsn., Ft. Lauderdale, 1987—; cons. in field, 1990—. Contbr. articles on applying pumps to hydraulic systems to profl. jours. Dir. St. Jude Hunger Program, Boca Raton, Fla., 1990-92, Fishin Kids Stay Clean, Lake Worth, Fla., 1992. Mem. ASME, Nat. Soc. Mech. Engrs., Fla. Engring. Soc. Roman Catholic. Achievements include research in applying rotating equipment on different hydraulic systems. Office: Dresser Pump Divsn # 220 2701 W Oakland Park Blvd Fort Lauderdale FL 33311

O'NEAL, DENNIS LEE, mechanical engineering educator; b. Ft. Worth, Sept. 1, 1951; s. John Earl and Melba Pearl (Reed) O'N.; m. Sondra Kaye Walter, Jan. 11, 1975; children: Justin Earl, Stephen Walter, Sean Michael. BS, Tex. A&M U., 1973; MS, Okla. State U., 1977; PhD, Purdue U., 1982. Registered profl. engr., Tex. Project engr. Fluid Power Rsch. Ctr., Stillwater, Okla.; rsch. staff mem. Oak Ridge (Tenn.) Nat. Lab., 1977-80; rsch. asst. dept. mech. engring. Purdue U., West Lafayette, Ind., 1980-82; assoc. prof. dept. mech. engring. Tex. A&M U., College Station, 1983—; cons. Lawrence Berkeley (Calif.) Lab., 1989-91, N.W. Power Planning Coun., Portland, 1991-92; mem. peer rev. group EPA, Research Triangle Park, N.C., 1992. Editor conf. procs. in field; contbr. articles to profl. publs. Mem. Energy Mgmt. Com., College Station, 1984-88. Exxon Ednl. Found. faculty asst. grantee, 1986; Tex. Engring. Exptl. Sta. rsch. fellow, 1991. Mem. ASME, Am. Soc. Engring. Edn., Am. Soc. Heating, Ventilating and Air-Conditioning Engrs., Sigma Xi, Phi Kappa PHi. Achievements include development of models for characterizing frost growth on heat exchanger surfaces, flows models for refrigerants in orifices. Office: Tex A&M U Dept Mech Engring College Station TX 77843

O'NEAL, KATHLEEN LEN, financial administrator; b. Ft. Riley, Kans., May 24, 1953; d. Leonard Arthur and Mary (Modlin) O'N. BS with honors, U. Mo., 1975; MBA, Calif. Coast U., Santa Ana, 1991. Cert. secondary tchr. Instr. math. Killian Sr. High Sch., Miami, Fla., 1975-78; mfg. supr. Western Electric Co., Lee's Summit, Mo., 1978-79; prodn. control supr. Western Electric Co., 1979-81; dept. mgr. Lee Wards Co., Independence, Mo., 1981-83; materials mgmt. specialist Northrup-Wilcox Electric, Kansas City, Mo., 1983-84; bus. resource planning mgr. AT&T, Lee's Summit, 1984-85; product mgr. AT&T, Berkeley Heights, N.J., 1985-87; fin. mgr. AT&T, Bedminster, N.J., 1987-89; info. sys. devel. mgr. AT&T, Piscataway, N.J., 1989-90; sr. fin. mgr. AT&T, Jacksonville, Fla., 1990-91, asst. treas., 1992—. Recipient Spl. Recognition award United Way, 1980. Mem. Am. Prodn. and Inventory Control Soc. (v.p. membership 1987-88, regional del. 1988-89, instr. inventory mgmt. 1987-88), NOW. Avocations: aerobics, reading, sailing, scuba diving. Office: AT&T Co 8775 Baypine Rd Jacksonville FL 32256-7569

O'NEAL, STEPHEN MICHAEL, information systems consultant; b. Boulder, Colo., Oct. 4, 1955; s. Eldon Ellsworth and Ruth (Law) O'N.; m. Rebecca Susan Miller, Oct. 26, 1985; children: Margaret McKamie, Michael Patrick. BBA, U. Colo., 1977. Software salesman Burroughs Corp., Denver, 1977-79; systems analyst ESCOM, Denver, 1979-80; sr. systems analyst Gathers Software, Inc., Denver, 1980-84, computer data ctr. mgr., Dallas, 1984-86, oil and gas trust product mgr., 1986; cons. software and mgmt., Dallas, 1986—. Asst. master Mile High Coun. Boy Scouts Am., Denver, 1981-84; pres. South Cen. Prime Users Group, 1987-88. Mem. Inst. Cert. Computer Profls. (cert. systems profl.), Data Processing Mgmt. Assn., Nat. Soc. Fundraising Execs., Ind. Computer Cons. Assn. Republican. Methodist. Home and Office: 1400 Wisteria Way Richardson TX 75080-4134

O'NEIL, DANIEL JOSEPH, academic administrator, research executive; b. Boston, June 5, 1942; s. Daniel Joseph and Grace Veronica (Francis) O'N.; m. Elizabeth Noone, Nov. 14, 1964; children: Elizabeth Grace, Daniel Joseph, Dara Veronica. BA, Northeastern U., 1964; MS, So. Con.. State U., 1967; PhD, U. Dublin, 1972. Sr. rsch. chemist Raybestos-Manhattan Advanced Rsch. Lab., Stratford, Conn., 1964-67; unit leader Hitco Materials Sci. Ctr., Gardena, Calif., 1967-68; tech. dir. Euroglas Ltd., Middlesex, Eng., 1970-72, Kildare, Ireland, 1970-72; dir. external liaison and coop. edn., lectr. polymer sci. U. Limerick, Ireland, 1972-75; chief exec. European Rsch. Inst. Ireland, Limerick, 1981-83; sr. rsch. scientist Ga. Tech. Rsch. Inst., Atlanta, 1975-78, prin. rsch. scientist, 1978—, dir. energy and materials sci. lab., 1988-90, group dir. office of dir., 1990-91; v.p. and dean grad. coll., dir. Sarkeys Energy Ctr. U. Okla., Norman, 1991—, prof. chemistry, 1991-92; pres. rsch. corp. Univ. Okla. Rsch. Corp., Norman, 1992—; founder, mng. dir. Okla. Energy Rsch. Ctr., Atlanta, 1992—; pres. Univ. Systems Corp., Atlanta, 1993—; bd. dirs. U. Okla. Rsch. Corp., Okla. Ctr. for Advancement of Sci. and Tech., Okla. Exptl. Program Stim. Comp. Res.; mem. adv. bd. Gov.'s Energy Coun.; mem. Okla. Higher Edn. State Regents Coun. on Rsch. and Grad. Edn., 1991—; panelist bd. on sci. and tech. for internat. devel. NAS/NRCĄ Washington, 1978-79, 86-87; cons. EEC, Brussels, 1982, 87, 89, U.S. rep., 1989; mem. nat. policy rev. panel U.S. Dept. Energy, Washington, 1986; witness, cons. energy R & D com. U.S. Senate, Washington, 1986-88; reviewer small bus. innovation rsch. program U.S. Dept. Energy, Washington, 1988-90. Author and co-author of 100 reports and publs. including High Flux Materials Treatment, 1990, USDOE Solar Theraml Tech., 1989, Energy From Biomass and Wastes XIII, 1989, Internat. Conf. Pyrolysis/Gasification, 1989. Pres. U. Okla Res. Corps., 1992—, bd. dirs., 1992—; mgn. dir., Okla. Energy Res. Ctr., 1992—; bd. dirs. ; expert evaluator NBS Office Energy-Related Inventions, Gaithersburg, Md., 1978-86; adv. bd. dirs. tech. utilization USDOC Office Minority Bus. Enterprises, Washington, 1977-86; U.S. del. U.S.-Brazil energy workshop U.S. Dept. State, Washington, Brazil, 1980; mem. Okla. Higher Edn. State Regents Coun. on Rsch. and Grad. Edn., 1991—. Mem. AAAS, FAIC, CoGR, ORAU, URA, MAGS, NASULGC, NCURA, Cortech, Am. Inst. Chem., Am. Chem. Soc., Assn. Big Eight Univs., Biomass Energy Res. Assn. (bd. dirs. 1990—), Coun. on Grad. Schs., Ga. Acad. Sci. (councillor 1990-91), Ga. Inst. Chem. (pres 1988-90), Internat. Club of Atlanta (founder), Japan-Okla. Soc., Petroleum Club of Okla., Assn. Western U. (bd. dirs. 1991—), Century Club, Northeastern U., Univ. Okla. Assn., Trinity Coll. Dublin Alumni Assn. Office: U Okla Grad Coll 2660 Goodfellow Rd Tucker GA 30084

O'NEIL, HAROLD FRANCIS, JR., psychologist, educator; b. Columbia, S.C., Jan. 26, 1943; s. Harold Francis Sr. and Margaret Mary (Ryan) O'N.; m. Eva L. Baker, Sept. 15, 1984; children: Tristan, Christopher. PhD, Fla. State U., 1969; MS, Hollins Coll., 1970. Asst., assoc. prof. U. Tex., Austin, 1971-75; program mgr. Def. Advanced Rsch. Projects Agy., Arlington, Va., 1975-78; from team chief to dir. Tng. Rsch. Lab. Army Rsch. Inst., Alexandria, Va., 1979-85; prof. U. So. Calif., L.A., 1985—; cons. Army Rsch. Inst., Alexandria, 1985—, Inst. for Def. Analyses, Alexandria, 1985—, Nat. Acad. Scis., Washington, 1989—. Editor: Academic Press Education and Technology Series, 1977-92; contbr. book chpts. and articles to profl. jours. Fellow APA, Am. Psychol. Soc.; mem. Cosmos Club. Achievements include research in role of cognition and affect in computer-based instruction; development of measures for metacognition, effort, and anxiety; founding editor of Japanese journal in education. Office: Univ So Calif 600 WPH University Park Los Angeles CA 90089-0031

O'NEIL, THOMAS MICHAEL, physicist, educator; b. Hibbing, Minn., Sept. 2, 1940; married; 1 child. BS, Calif. State U., Long Beach, 1962; MS, U. Calif., San Diego, 1964, PhD in Physics, 1965. Rsch. physicist Gen. Atomic, 1965-67; prof. physics U. Calif., La Jolla, 1967—; mem. adv. bd. Inst. Fusion Studies, 1980-83, Inst. Theoretical Physics, 1983—. Assoc. editor Physics Review Letters, 1979-83; correspondent Comments Plasma Physics & Controlled Fusion, 1980-84. Recipient Plasma Physics Rsch award Am. Physical Soc., 1991. Fellow Am. Physical Soc. Achievements include research in theoretical plasma physics with emphasis on nonlinear effects in plasmas and on non-neutral plasmas. Office: Univ of California Dept of Physics 9500 Gilman Dr La Jolla CA 92093*

O'NEILL, JAMES FRANCIS, chemist; b. Syracuse, N.Y., Mar. 9, 1943; s. John F. and Kathleen M. (Edinger) O'N.; m. Marianne Custozzo, Oct. 3, 1970; 1 child, Tyler James. AAS, Onondaga C.C., 1967; BS, U. Rochester, 1970. Technician GE, Syracuse, 1966-67; technician Xerox Corp., Webster, N.Y., 1967-70, chemist, 1970—. Contbr. articles to sci. jours. Mem. IEEE, Am. Vacuum Soc. Achievements include 13 patents in field; rsch. in magnetic properties of iron-borosilicate glass, FEBO3 solid solutions: synthesis, chemistry, magnetic and optical properties, method of improving step coverage, preparation and crystal chemistry of some new lanthanum iron borates, thermal ink jet printer in a recess. Home: 60 Pine Brook Cir Penfield NY 14526 Office: Xerox Corp MS 0201-08L 800 Phillips Rd Webster NY 14580

O'NEILL, JAMES WILLIAM, structural engineer; b. St. Louis, Nov. 28, 1958; s. James W. and Mary K. (Stuart) O'N.; m. Joan Marie Thomas, Apr. 25, 1992. BSCE, U. Mo., 1981, MSCE, 1982. From engr. to unit mgr. McDonnell Douglas Corp., St. Louis, 1981-92, unit chief, 1992—. Mem. AIAA, Mo. Evans Scholar Alumni Assn. (pres. 1992). Roman Catholic. Achievements include research in stress ratio effects on crack initiation. Office: McDonnell Douglas Corp Mailcode 2702355 PO Box 516 Saint Louis MO 63166-0516

O'NEILL, MARK JOSEPH, solar energy engineer; b. Maysville, Ky., Mar. 10, 1946; s. Stanley Lawrence and Margaret Walton (Brannon) O'N.; m. Natalie Frances Degolian, June 7, 1968. BS in Aerospace Engring., Notre Dame U., 1968. Project engr. Lockheed Missiles & Space Co., Huntsville, Ala., 1969-74; v.p. engring. Northrup, Inc., Dallas, 1974-75; dir. energy programs E-Systems, Inc., Dallas, 1975-83; exec. v.p. engring. and ops. Entech, Inc., Dallas-Ft. Worth Airport, 1983—, also dir. Recipient New-Tech. awards NASA, 1969-92. Mem. IEEE, Am. Solar Energy Soc., Am. Quarter Horse Assn., Am. Paint Horse Assn., Keller Horse Owners Assn., Richland Hills Riding Club, Notre Dame Alumni Assn. Democrat. Achievements include U.S. and fgn. patents on 7 different solar energy devices, one of which is licensed to 3M, and another of which has been used to establish 13 world records for performance of various photovoltaic (solar cell) devices.

Home: 812 Belinda Dr Keller TX 76248 Office: Entech Inc PO Box 612246 1015 Royal Ln Dallas TX 75261

O'NEILL, MICHAEL JAMES, medical physicist; b. Seattle, Feb. 18, 1958; s. Layton James O'Neill and Barbara (Butler) Kellam; m. Natalie Schaffer, Sept. 4, 1983; children: Benjamin James, Jonathan Shea. BA, Western Md. Coll., 1980; MS, Johns Hopkins U., 1985. Diplomate Am. Bd. Radiology, Am. Bd. Med. Physics. Med. dosimetrist Johns Hopkins U., Balt., 1980-83, med. physicist, 1983-88; med. physicist St. Joseph Hosp., Houston, 1988—; cons. Cheung Labs., Inc., Washington, 1983-87, Gamma Mgmt. Inc., Lancaster, Calif., 1991-92. Contbr. articles to profl. publs. Vol. Galveston Bay Found., Houston, 1991-92; mem. Chesapeake Bay Found., Balt., 1986-89. Mem. IEEE, Am. Assn. Physicists in Medicine, N.Am. Hyperthermia Soc., Am. Coll. Radiology, Beta Beta Beta, Kappa Mu Epsilon. Achievements include patent pending for microwave treatment device for cancer; project consultant for first solid state microwave hyperthermia device. Office: St Joseph Hosp 1919 La Branch Houston TX 77002

O'NEILL, MICHAEL WAYNE, civil engineer, educator; b. San Antonio, Feb. 17, 1940; s. Wayne Jackson and Delores Hazel (Shaw) O'N.; m. Jerilyne Arleen Busse, Jan. 22, 1972; 1 child, Ronald Christopher. PhD, U. Tex., 1970. Registered profl. engr., Tex. Rsch. assoc. U. Tex., Austin, 1970-71; div. mgr. Southwestern Labs., Inc., Houston, 1971-74; prof. U. Houston, 1974—. Author: Design of Structures and Foundations for Machines, 1979, (with others) Construction and Design of Drilled Shafts, 1988; contbr. articles to profl. jours. Capt. U.S. Army, 1965-67. Fellow ASCE (chmn. deep founds. com. 1982-86, John B. Hawley award 1975, 81, 90, Walter L. Huber Rsch. prize, 1986); mem. Transp. Rsch. Bd., NSPE, Internat. Soc. for Soil, Mechanics and Found. Engrs. Lutheran. Achievements include research in reliability of load transfer on drilled shafts, interaction among piles in a group. Office: U Houston 4800 Calhoun Rd Houston TX 77204-4791

O'NEILL, PATRICK J., pharmacologist; b. Columbus, Ohio, July 12, 1949; s. Timothy J. and Martha F. (Reed) O'N.; m. Brenda Jane Brown, June 30, 1973; children: Kristin Ann, Michael Patrick. BS in Pharmacy with distinction, Ohio State U., 1972, PhD, 1976. Various rsch. positions McNeil Pharmaceutical, Spring House, Pa., 1976-82, dir. dept. drug metabolism, 1982-85, exec. dir. project mgmt., 1985-86, exec. dir. new products, 1986-88, exec. dir. devel. rsch., 1988-89; v.p. R & D Ethicon Inc., Somerville, N.J., 1989—. Office: Ethicon Inc PO Box 151 Somerville NJ 08876-0151

O'NEILL, PETER THADEUS, software engineer; b. Brentwood, N.Y., Feb. 10, 1957; s. Jerome F. and Noreen (Connolly) O'N. Grad. high sch., Smithtown, N.Y. Sr. software engr. Anorad Co., Hauppauge, N.Y., 1978-87, Linotype Co., Hauppauge, 1987-88, Metco Peskin Elmer, Westbury, N.Y., 1988—. Office: Metco Peskin Elmer 1101 Prospect Ave Westbury NY 11590-2724

O'NEILL, RUSSELL RICHARD, engineering educator; b. Chgo., June 6, 1916; s. Dennis Alysious and Florence Agnes (Mathurin) O'N.; m. Margaret Bock, Dec. 15, 1939; children: Richard A., John R.; m. Sallie Boyd, June 30, 1967. BSME, U. Calif., Berkeley, MSME, 1940; PhD, UCLA, 1956. Registered profl. engr., Calif. Design engr. Dowell, Inc., Midland, Mich., 1940-41; design engr. Dow Chem. Co., Midland, 1941-44, Airesearch Mfg. Co., Los Angeles, 1944-46; lectr. engring. UCLA, 1946-56, prof. engring., 1956, asst. dean engring., 1956-61, assoc. dean, 1961-73, acting dean, 1965-66, dean, 1974-83, dean emeritus, 1983—; staff engr. NAS-NRC, 1954; dir. Data Design Labs., 1977-86, dir. emeritus, 1986—; mem. engring. task force Space Era Edn. Study Fla. Bd. Control, 1963; mem. regional Export Expansion Coun. Dept. Commerce, 1960-66, Los Angeles Mayor's Space Adv. Com., 1964-69; mem. Maritime Transp. Rsch. Bd., 1974-81; bd. advisers Naval Postgrad. Sch., 1976-84; mem. Nat. Nuclear Accreditation Bd., 1983-88; mem. accrediting bd. Dept. Energy, 1992—. Trustee West Coast U., 1981-90; bd. dirs. Western region United Way, 1982-90. Mem. NAE, Am. Soc. Engring. Edn., Sigma Xi, Tau Beta Pi. Home: 15430 Longbow Dr Sherman Oaks CA 91403-4910 Office: 405 Hilgard Ave Los Angeles CA 90024-1301

O'NEILL, SHAWN THOMAS, waste water treatment executive; b. Providence, Sept. 7, 1958; s. Martin James O'Neill Jr. and Paula Anne (Zuromski) O'Neill Larivee; m. Kimberly Anne Correia, May 11, 1985; children; Joshua David, Lealah Catherine, Chelsea Elizabeth, Ian James. A in Liberal Arts, C.C. R.I., 1979. Cert. grade 4 waste water operator, R.I. Supt., chief ops. waste water treatment facility Town of Jamestown, R.I., 1987-91; asst. supt. waste water treatment facility Town of East Greenwich, R.I., 1991—. Coach Jamestown Little League, 1990—, Youth Basketball, Jamestown, 1991—, Youth Soccer, Jamestown, 1992—. Mem. Narragansett Water Pollution Control Assn. (exe. bd. 1991-93). Democrat. Home: 5 Colubmia Ave Jamestown RI 02835 Office: Town of East Greenwich 111 Pierce St East Greenwich RI 02818

ONET, VIRGINIA, veterinary parasitologist, researcher, educator; b. Sarmasag, Salaj, Romania, Mar. 17, 1939; came to U.S., 1986; d. Virgil and Eugenia (Marinescu) Constantinescu; m. Gheorghe Emil Onet, Sept. 3, 1981. DVM, Coll. Vet. Med., Cluj-Napoca, Romania, 1966; PhD, Coll. Vet. Med., Bucharest, Romania, 1974. Assoc. prof. Coll. Vet. Medicine, Cluj-Napoca, 1966-81, lectr., 1981-85; pvt. researcher Fed. Republic Germany, 1985-86; ind. cons. Detroit, 1986-87; project leader Grand Labs., Inc., Larchwood, Iowa, 1987-88, rsch. group leader, 1988-92, mgr. R&D dept. parasitology, 1992—; mem. profl. bd. Coll. Vet. Medicine, Cluj-Napoca, 1970-72, mem. faculty com., 1980-81; mem. Exam. Bd. for Screening Vet. Medicine Candidates, Cluj-Napoca, 1974-85. Author: Diagnosis Guide for Parasitic Disease, 1983; co-author: Laboratory Diagnosis in Veterinary Medicine, 1978; author 7 textbooks; contbr. 45 articles to profl. jours. Merit scholar Coll. Vet. Medicine, Bucharest, 1964. Mem. AVMA, AAAS, Am. Assn. Vet. Parasitologists, World Vet. Poultry Assn., World Assn. for Advancement Vet. Parasitology, Romanian Vet. Medicine Soc., Romanian Soc. Biologists, World Assn. Buiatrics. Avocations: music, poetry, travel, crocheting. Home: 4509 Mountain Ash Dr Sioux Falls SD 57103-4959 Office: Grand Labs Inc Box 193 Larchwood IA 51241

ONG, BOON KHENG, electronics executive; b. Penang, Malaysia, Aug. 18, 1935; m. Lean Eng Cheong; 1 child, Eng Jin. Degree in elec. engring., U. Canterbury, New Zealand, 1963. Dep. gen. mgr. Guthrie Engring. Malaysia, 1970-77; mng. dir. Electcoms Sdn Bhn, Selangor, Malaysia, 1977—. Mem. Instn. Elec. Engrs. London, Instn. Elec. Engrs. Malaysia, Profl. Engrs. Bd. Singapore. Avocations: swimming, reading. Home: Endah, Damansara Heights, 2A Persiaran Damansara, 50490 Kuala Lumpur Malaysia Office: Electcoms Sdn Bhn, 12 A Jalan 13/4, 46200 Petaling Jaya, Malaysia

ONG, SAY KEE, environmental engineering educator; b. Kuala Lumpur, Malaysia, Nov. 8, 1956; came to U.S., 1984; s. Heng Moon Ong and Puah Lan Tan; m. Siew Huang Tan, Aug. 28, 1982; children: L. Lin Ong, L. Wern Ong. MS, Vanderbilt U., 1987; PhD, Cornell U., 1990. Lic. profl. engr., Ohio. Project engr. Paterson Candy, Kuala Lumpur, 1980-84; teaching asst. Vanderbilt U., Nashville, 1984-85; rsch. asst. Cornell U., Ithaca, N.Y., 1985-90; prin. rsch. scientist Battelle Meml. Inst., Columbus, Ohio, 1990-92; asst. prof. Environ. Engring. Poly. U., Bklyn., 1992—. Contbr. articles to profl. jours. Recipient Hargill Centennial scholarship U. Malaysia, 1978-80, Cornell U. summer fellowship, 1989. Mem. ASCE, Am. Chem. Soc., Water Environment Fedn. (hazardous waste com. 1992—). Home: 75-35B 217th St Bayside NY 11364 Office: Poly U 333 Jay St Brooklyn NY 11201

ONISHI, AKIRA, economics and global modeling educator, academic administrator; b. Tokyo, Jan. 5, 1929; s. Tatsunosuke and Tomi (Fusegawa) O.; m. Noriko Shimizu, Sept. 27, 1963; children: Kimihiro, Masahiro. BA in Econs., Keio U., Tokyo, 1954, MA in Econs., 1958, PhD in Econs., 1963. Rsch. officer Inst. Developing Economies, Tokyo, 1963-65; assoc. prof. Chuo U., Tokyo, 1965-67; econ. affairs officer UN Econ. Commn. for Asia and Far East, Bangkok, Thailand, 1967-68; econ. affairs officer ILO, Geneva, 1968-70; chief economist Japan Econ. Research Ctr., Tokyo, 1970-71; prof. econs. Soka U., Tokyo, 1971-90, dir. Inst. Applied Econ. Research, 1976-90, dean Grad. Sch. Econs., 1978-90, v.p., 1991—; bd. dirs. Inst. Systems Sci., dean Faculty of Engring., 1991—. Author: Japanese Economy in Global Age, 1974; numerous articles on global econ. modeling, 1977—. Grantee Japan Found., 1958-60, Japan Econ. Rsch. Found., 1974-75, Australia-Japan Found., 1981-82; recipient SGI Culture award, 1985, Supreme Article award Japan Assn. Planning Adminstrn., 1991. Fellow Japan Soc. Internat. Econs., Japan Soc. Econ. Policy, Japan Assn. Simulation and Gaming (v.p. 1989-92, pres. 1993—). Buddhist. Futures of Global Interdependence global model used by UN, 1982—. Home: 4-9-4 Seijyo, Setagaya-Ku, Tokyo 157, Japan Office: Soka U, 1-236 Tangi-cho, Hachiji-shi, Tokyo 192, Japan

ONISHI, RYOICHI, aeronautical engineer, designer; b. Fukoka, Japan, Feb. 15, 1955; s. Mitsuji and Yoshiko O.; m. Midori Onishi, Apr. 11, 1987. BS in Mech. Engring., Kyushu U., Fukuoka, Japan, 1979. Engr. automobile CAD system Suzuki Motor Co., Ltd., Hamamatsu, Japan, 1979-82; engr. turbine mfg. Fuji Elec. Co., LTd., Kawasaki, Japan, 1982-84; sr. rschr., asst. mgr. aircraft design, computer aided design Mitsubishi Rsch. Inst., Inc., Tokyo, 1984—; bd. dirs. software re-use com. Japan Info. Svc. Industries Assn., Tokyo, 1985, bd. dirs. SIGMA project com., 1988. Mem. AIAA (CAD/CAM tech. com. 1993), Royal Aeronautical Soc., Japan Aero Engr. Assn. Office: Mitsubishi Rsch Inst Inc, Arco Tower 8-1 Shimomeguro 1 Meguro, Tokyo 153, Japan

ONISHI, YASUO, environmental researcher; b. Osaka, Japan, Jan. 25, 1943; came to U.S., 1969; s. Osamu and Tokiko (Domukai) O.; m. Esther Anna Stronczek, Jan 22, 1972; children: Anna Tokiko and Lisa Michiyo. BS, U. Osaka Prefecture, 1967, MS, 1969; PhD, U. Iowa, 1972. Rsch. engr. U. Iowa, Iowa City, 1972-74; sr. rsch. engr. Battelle Meml. Inst., Richland, Wash., 1974-77, staff engr., 1977-92, mgr. rsch. program office, 1984-92, sr. program mgr., 1992—. Co-author: Principles of Health Risk Assessment, 1985, several other environ. books; contbr. articles to profl. jours.; featured in TV program NOVA. Recipient Best Platform Presentation award ASTM, 1979. Mem. ASCE (chmn. task com. 1986—), IAEA (advisor on environ. issues, U.S. coord. water and soil assessment bilateral joint work on Chernobyl nuclear accident), Nat. Coun. Radiation Protection and Measurements (adj., mem. task com. 1983—), Sigma Xi. Lutheran. Achievements include coordingation of bilateral joint soil and environmental assessment of Chernobyl accident. Avocations: camping, skiing. Home: 144 Spengler Rd Richland WA 99352-1971 Office: Battelle Pacific NW Labs Batelle Blvd Richland WA 99352

ONOKPISE, OGHENEKOME UKRAKPO, agronomist, educator, forest geneticist, agroforester; b. Lagos, Nigeria, May 10, 1951; came to U.S., 1981; s. Jerome Esagwu and Margaret E. (Agbanobi) O.; m. Lucy Omotaka Edemo, Jan. 31, 1977; children: Oghenemaro, Omurhu, Oghogho. BS, U. Ife, Ile-Ife, Nigeria, 1974; MS, U. Guelph, Ont., Can., 1980; PhD, Iowa State U., 1984. Tutor Sch. Agrl., Yandev, Nigeria, 1974-75; rsch. officer Rubber Rsch. Inst. Nigeria, Benin City, 1975-81; rsch. asst. Iowa State U., Ames, 1981-85; postdoctoral researcher Ohio State U., Wooster, 1985-86; asst. prof. Fla. A&M U., Tallahassee, 1986-91, assoc. prof., 1991—; mem. Germplasm Collection Team Internat. Rubber R&D Bd. London, Eng. in Brazil, 1981; team leader Cocoyam Breeding Team USAID, Cameroon, Republic of West Africa, 1988-90. Contbr. articles to Annals Applied Biology, Plant Breeding, Acta Agronomca, Seed Sci. and Tech., African Jour. Genetics, Am. Jour. of Enology and Viticulture, Indian Jour. of Plant Breeding and Genetics, Silvae Genetica, Agronomie, Jour. of Plantation Crops, African Tech. Forum. Editor Pack 104 Cub Scouts Newsletter, Boy Scouts Am., Tallahassee, 1986-88. mem. Parish Coun. St. Louis Parish, Tallahassee, 1987-88; tutor Bapt. High Sch., Buea, Cameroon, 1988-89. Recipient Sci. Paper award, Assn. Rsch. Dirs., Washington, 1987; named Best Agrl. Instr., Agrl. Sci. Club FAMU Student, Tallahassee, 1988; grantee USAID FAMU, Cameroon, 1988-90, NASA FAMU, 1988-91, USDA. Fellow Indian Soc. Genetics and Plant Breeding; mem. Am. Soc. Agronomy, Commonwealth Forestry Assn., Soc. of Am. Foresters (campus faculty rep.). Achievements include development of concepts of moving forest for the tropical rain forest; growth of carrots in hydroponic system within growth chambers. Home: 2810 Kennesaw Pl Tallahassee FL 32303-1202 Office: Fla A&M U Martin Luther King Blvd Tallahassee FL 32307

ONSAGER, JEROME ANDREW, research entomologist; b. Northwood, N.D., Apr. 8, 1936; s. Alfred and Anne Marie (Kielbauch) O.; m. Bette Lynn Stanton, Aug. 16, 1958. B.S., N.D. State U., 1958, M.S., 1960, Ph.D., 1963. Research entomologist USDA Agrl. Research Service, Bozeman, Mont., 1963—. Contbr. articles on biology, ecology, population dynamics of rangeland grasshoppers to profl. jours. Fellow NSF, 1960. Mem. Entomol. Soc. Am., Entomol. Soc. Can., Orthopterist Soc., Sigma Xi, Alpha Zeta. Lutheran. Avocations: fishing; big game hunting; equestrian activities. Home: 4141 Blackwood Rd Bozeman MT 59715-9130 Office: USDA ARS Rangeland Insect Lab Bozeman MT 59717

ONSTOTT, EDWARD IRVIN, research chemist; b. Moreland, Ky., Nov. 12, 1922; s. Carl Ervin and Jennie Lee (Foley) O.; m. Mary Margaret Smith, Feb. 6, 1945; children: Jenifer, Peggy Sue, Nicholas, Joseph. BSChemE, U. Ill., 1944, MS in Chemistry, 1948, PhD in Inorganic Chemistry, 1950. Chem. engr. Firestone Tire & Rubber Co., Paterson, N.J., 1944, 46; research chemist Los Alamos Nat. Lab., 1950—, now guest scientist. Patentee in field. Served with C.E., AUS, 1944-46. Fellow AAAS, Am. Inst. Chemists; mem. Am. Chem. Soc., Electrochem. Soc., N.Y. Acad. Scis., Internat. Assn. Hydrogen Energy, Rare Earth Research Confs., Izaak Walton League, N.Mex. Acad. Scis., Los Alamos Hist. Soc. Republican. Methodist. Home: 225 Rio Bravo Dr Los Alamos NM 87544-3848

ONTTO, DONALD EDWARD, environmental analytical chemist, wastewater treatment consultant; b. Carlsbad, N.Mex., Sept. 18, 1952; s. Edward Donald and Dorothy Mae (Johnson) O.; m. Patricia Ann Baldassari, Apr. 12, 1975; 1 child, John. BS in Chemistry with honors, Mich. Tech. U., 1974. Lic. wastewater treatment plant operator, Ind. Rsch. asst. Midland (Mich.) Macromolecular Inst., 1974, Mich. Tech. U., Houghton, 1974-79; chem. foreman No. Ind. Pub. Svc. Co., Wheatfield, 1980-87; chemist Nat. Steel Corp., Portage, Ind., 1988, Tenco Environ. Labs., Schererville, Ind., 1988—; lab. hygiene officer Tenco Environ. Labs., Schererville, 1991—. Contbr. articles to profl. jours. Vol. Red Cross, Valparaiso, Ind., 1985—; soundman Vineyard Christian Fellowship, Valparaiso, 1991—. Achievements include development, tested and/or adapted numerous methods of physical and/or chemical analysis of environmental samples on materials ranging from drinking water to very hazardous waste materials. Home: 2114 E Glendale Blvd Valparaiso IN 46383 Office: Tenco Labs 1152 Junction Ave Schererville IN 46392

ONUFROCK, RICHARD SHADE, pharmacist, researcher; b. Colorado Springs, Colo., July 5, 1934; s. Frank and Mildred Joy (Overstreet) O.; m. Karen Faye Larson, June 15, 1958 (div. 1980); children: Richard Alan (dec.), Amy Mildred. BS in Pharmacy, U. Colo., 1961; diploma, Famous Artists Schs., 1963. Registered pharmacist, Colo., Ariz., South Africa. Pharmacist Aley Drug Co., Colorado Springs, 1961-75, St. Joseph Hosp., Denver, 1976-77, Navajo Nation Health Found., Ganado, Ariz., 1977-81, Kearny (Ariz.) Kenecot-Samarital Hosp., 1984-85, NIH, Warren G. Magnuson Clin. Ctr., Bethesda, Md., 1988—; dir. pharmacy, chief pharmacist Tintswalo Hosp., South Africa, 1981-84; pharmacist, chief pharmacist Miami (Ariz.)-Inspiration Hosp., 1985-88; instr. Coll. of Ganado, 1979-80; asst. in textbook revision and illustration U. Colo., 1961; cons. Heritage Health Care Ctr., Globe, Ariz., 1988. Illustrator Pharmacy for Nurses, 1961, Colo. Jour. of Pharmacy, 1962-64; illustrations exhibited Colo. Springs Fine Art Ctr., 1964-66, Gilpin County Art Assn., Central City, Colo., 1968-74, 1st Nat. Space Art Show, Denver, 1969. dem. precinct committeeman, 1974-76; den leader Boy Scouts Am., com. mem., 1975-76; fireman, lt. Ganado Vol. Fire Dept., 1977-81; compassionate med. missionary Nazarene Ch., Tintswalo Hosp. Gazankulu, South Africa, 1981-84;bd. dirs. Friends of Libr., Kearny, 1985-87; active Grace Episcopal Ch. Mem. Am. Pharm. Assn., Am. Soc. Hosp. Pharmacists, D.C. Soc. Hosp. Pharmacists, Phi Delta Chi, Delta Sigma Phi. Avocations: traveling, bicycling, hiking, skiing, computers. Home: 4831 36th St NW 202 Washington DC 20008 Office: NIH Clin Ctr Pharmacy Dept Bldg 10 Rm 1N257 9000 Rockville Pike Bethesda MD 20892

ONUKI, HIDEO, physicist; b. Tokyo, Aug. 11, 1943; s. Hide and Tsune O.; m. Rutsu Hirano, Apr. 2, 1976. BS, St. Paul U., Tokyo, 1966; MS, Tokyo U. Edn., 1969, ScD, 1973. Researcher Electrotech. Lab., Tokyo, 1974-78, prin. researcher, 1978-87; chief planning section Lab., Sortec Corp., Tsukuba, 1987-88; chief optical radation sect. Electrotech. Lab., Tsukuba, 1988—. Editor-in-chief The Spectroscopical Soc. Japan, 1991—; author: Synchrotron Radiation, 1986, Techniques in Synchrotron Radiation, 1990; patentee on new synchrotron radiation source, 1984. Mem. Am. Phys. Soc., Phys. Soc. Japan, Spectroscopical Soc. Japan, Japan Soc. Applied Physics, Inst. Elec. Engrs. Japan, Illuminating Engring. Inst. Japan. Avocations: music, baseball, Japanese chess. Home: 2-820-5 Azuma, Tsukuba Ibaraki, Japan 305 Office: Electrotech Lab, 1-1-4 Umezono, Tsukuba Ibaraki, Japan 305

ONYONKA, ZACHARY, federal agency administrator; b. Feb. 28, 1939; s. Godrico Deri and Kerobina (Kebati) O.; married; 6 children. Student, Mosocho Sch., Nyaburu Sch., St. Mary's Sch.; D. in Pub. Service (hon.), Syracuse U., 1981. Lectr. dept econs. U. Coll., Nairobi, Kenya, 1968-69; M.P. Govt. of Kenya, 1969—, minister econ. planning and devel., 1969-70, 79-83, 88—, minister of info. and broadcasting, 1970-73, minister of health, 1973-74, minister of edn., 1974-76, minister of housing and social services, 1976-78, minister of fgn. affairs, 1987-88, now minister of Research, Technical Training, and Technology; mem. nat. exec. com. Kenya African Nat. Union; pres. African-Caribbean and Pacific Group of Countries, 1981. Research fellow U. Coll., Nairobi, 1967; recipient Elder Golden Heart medal. Fellow Internat. Bankers Assn. Office: Min of Rsch Sci & Tech, Utali House-Uhuru Hwy-POB 30623, Nairobi Kenya

OORT, ABRAHAM HANS, meteorologist, researcher, educator; b. Leiden, The Netherlands, Sept. 2, 1934; came to U.S., 1961; s. Jan Hendrik and Johanna Maria (Graadt Van Roggen) O.; m. Bineke Pel, May 20, 1961; children: Pieter Jan, Michiel, Sonya. MS, MIT, 1963; PhD in Meteorology, U. Utrecht, The Netherlands, 1964. Rsch. meteorologist Koninkiyk Nederlands Meteorologisch Instituut, De Bilt, The Netherlands, 1964-66; rsch. meteorologist Geophysical Fluid Dynamics Lab/NOAA, Washington, 1966-68; rsch. meteorologist Geophysical Fluid Dynamics Lab/NOAA, Princeton, N.J., 1968-77, sr. rsch. meteorologist, 1977—; prof. dept. geological and geophysical scis., Princeton U., 1971—. Author: Physics of Climate, 1992; contbr. monographs in field. 2d lt. Netherlands Air Force, 1959-61. NATO Sci. fellow, MIT, Cambridge, Mass., 1961-63; 10th Victor P. Starr Meml. Lectr. MIT, 1988; recipient Gold medal U.S. Dept. Commerce, Washington, 1979. Fellow N.Y. Acad. Scis., Am. Meteorol. Soc. (Jule G. Charney award 1993, asst. editor 1966—), Royal Meteorol. Soc.; mem. Am. Geophys. Union. Democrat. Avocations: sculpture, shiatsu, meditation. Office: Ntl Ocean & Atmospheric Admin Princeton Univ Box 308 Princeton NJ 08542

OPAVA-STITZER, SUSAN CATHERINE, physiologist, researcher; b. N.Y.C., Mar. 14, 1947; d. Joseph Francis and Beatrice Agnes (Fitzgerald) O.; m. Louis Kent Stitzer, Sept. 2, 1973 (div. Sept. 1989); children: Joshua Kent Stitzer, Matthew Louis Stitzer. BS in Biology, Coll. Mt. St. Vincent, N.Y.C., 1968; PhD Physiology, U. Mich., 1972. Post doctoral fellow Dartmouth Med. Sch., Hanover, N.H., 1972-74; asst. prof. U. P.R. Med. Sch., San Juan, 1974-81, assoc. prof., 1981-86, prof., 1986—; chairperson Dept. Physiology, U. P.R. Med. Sch., San Juan, 1989—; mem. NIH Rsch. Resources Adv. Com., Washington, 1988-90. Author: J. Physiol. 280:487-498, 1978, J. Physiol. 305:97-106, 1980, Annals N.Y. Acad. Scis., 194:278-285, 1982, J. Histochem. Cytochem. 31:956-959, 1983, Am. Jour. Med. Sci. 303:1-7, 1992. Mem. Sci. and Tech. Com., P.R., 1990-92; participant Surgeon General's Workshop on Hispanic/Latino Health, Washington, 1992. Recipient Rsch. Ctrs. in Minority Inst. award NIH, U. P.R., 1986, Rsch. Career Devel. award, NIH, 1982-87; fellow Am. Assn. for Advancement of Sci., 1992. Mem. Am. Physiological Soc., Am. Soc. of Nephrology, Am. Assn. for Advancement of Sci., Inter-Am. Soc. Hypertension. Office: U PR Dept Physiology PO Box 365067 San Juan PR 00936-5067

OPIE, JANE MARIA, audiologist; b. N.Y.C., Jan. 14, 1961; d. Everett George and Maria Beata (Kals) O. MA, Northwestern U., Evanston, Ill., 1984, PhD, 1992. Cert. audiologist. Audiologist Albert Einstein Coll. of Medicine, Bronx, N.Y., 1984-86; rsch. asst. Northwestern U., 1986-90; cons., pvt. practice Chgo., 1988-90; rsch. asst. Loyola U., Chgo., 1988-89; rsch. scientist U. Iowa, Iowa City, 1991-92; faculty rsch. assoc. Ariz. State U., Tempe, 1992—; organizing com. mem. Conf. on Assistive Devices for the Hearing Impaired, Iowa City, 1991-92. Contbr. articles to profl. jours. Tutor Programmed Activities for Correctional Edn., Chgo., 1989. Dissertation Yr. grantee Northwestern U., 1990-91, Northwestern fellow, 1983, 86-90. Mem. Acoustical Soc. Am., Am. Speech-Lang.-Hearing Assn., Am. Auditory Soc. Office: Ariz State U Speech/Hearing Dept Tempe AZ 85287-1908

OPPENHEIM, IRWIN, chemical physicist, educator; b. Boston, June 30, 1929; s. James L. and Rose (Rosenberg) O.; m. Bernice Buresh, May 18, 1974; 1 child, Joshua Buresh. A.B. summa cum laude, Harvard U., 1949; postgrad., Calif. Inst. Tech., 1949-51; Ph.D., Yale, 1956. Physicist Nat. Bur. Standards, Washington, 1953-60; chief theoretical physics Gen. Dynamics/Convair, San Diego, 1960-61; asso. prof. chemistry MIT, Cambridge, 1961-65; prof. MIT, 1965—; lectr. physics U. Md., 1953-60; vis. assoc. prof. physics U. Leiden, 1955-56, Lorentz prof., 1983; vis. prof. Weizmann Inst. Sci., 1958-59, U. Calif., San Diego, 1966-67; Van der Waals prof. U. Amsterdam, 1966-67. Author: (with J.G. Kirkwood) Chemical Thermodynamics, 1961; editor: Phys. Rev. E, 1992—. Fellow Am. Phys. Soc., Am. Acad. Arts and Scis., Washington Acad. Sci.; mem. Phi Beta Kappa, Sigma Xi. Research in quantum statis. mechanics, statis. mechanics of transport processes, thermodynamics. Home: 140 Upland Rd Cambridge MA 02140-3623 Office: MIT 77 Massachusetts Ave Cambridge MA 02139

OPPENHEIM, VICTOR EDUARD, consulting geologist; b. Riga, Latvia, Oct. 31, 1906; came to U.S., 1947; s. Hert Moritz and Mathilde (Trius) O.; m. Dorothy Fay Allport, May 18, 1949. BS, Ind. Coll., Riga, 1922; student, U. Toulouse, France, 1923; MS, U. Caen, France, 1927. Engr., geologist Engr. Cons. Ludovic Barreaux, Paris, 1928-29, engring. geologist Argentine Govt. Oilfields, Buenos Aires, 1931-32; geologist Govt. of Brazil, Rio de Janeiro, 1933-36, Royal Dutch Shell Oil Co., The Hague, The Netherlands, 1936-39, Govt. of Colombia, Bogota, 1940-41; cons. geologist Socony Vacuum Oil Co., N.Y.C., 1941-43; chief geologist Govt. of Peru, Lima, 1943-45; cons. geologist, Dallas, 1947-52; ptnr., geologist Oppenheim & Briscoe, Dallas, 1953-59; pres. Internation Minerals Co., Dallas, 1960-75, V. Oppenheim & Assocs., cons., Dallas, 1975-93. Author: Exploration East of the Andes, 1958; also 130 articles. Recipient Legion of Honor, AIME, 1984, Human Needs award Am. Assn. Petroleum Geologists, 1988, Explorer award Explorers Club, 1990, Outstanding Geologist award Tex. sect. Am. Inst. Profl. Geologists, 1992. Mem. Dallas Geol. Soc. (hon. life). Mem. Am. Inst. Profl. Geologists, Nat. Soc. Profl. Engrs. Achievements include mapping the geology of entire continent of South America from 1931-44; discovery of one of largest coal deposits in world, "El Cerrjon" mine, Colombia; co-discovery of Cusiana oilfield; made geological maps of southern Brazil, Bolivia, Colombia, Uruguay. Home and Office: 6106 Averill Way Dallas TX 75225

OPPENHEIMER, JOEL K., civil engineer; b. Balt., June 20, 1957; s. Morton S. and Paula (Straus) O.; m. Karen Eve Doroshow, July 7, 1984; children: Bradley Edward, Lauren Michele. B in Engring. and Sci. summa cum laude, U. Pa., 1979; M in Engring. Adminstrn., George Washington U., 1988. Registered profl. engr., Md., Del., Va., Pa., N.C., Ky., W.Va., Fla. Designer Whitman, Requardt and Assocs., Balt., 1979-84, engr., 1984; project engr. Century Engring., Inc., Towson, Md., 1984-87, asst. dir. hwys., 1987-89, v.p. transp., 1989—. Advisor Mid-Atlantic Fedn. Temple Youth, Balt., 1981-84. Mem. ASCE (Annette Estrada award Women's Aux. 1978, Md. sect. Young Engr. of Yr. 1993), NSPE, Am. Soc. Hwy. Engrs., Md. Soc. Profl. Engrs. (v.p. 1991-92, pres. Chesapeake chpt. 1992—, Young Engr. of Yr. 1993), Inst. Transp. Engrs., Tau Beta Pi. Office: Century Engring Inc 32 West Rd Towson MD 21204

OPPENHEIMER, LARRY ERIC, physical chemist; b. N.Y.C., Aug. 18, 1942. BS in Chemistry, Clarkson Coll., 1963, PhD in Chemistry, 1967. Sr. rsch. chemist Eastman Kodak Co., Rochester, N.Y., 1966-80, rsch. assoc., 1980-92. Contbr. articles to profl. jours., chpts. to books. Lever Bros. fellow, 1963-66. Mem. AAAS, Am. Chem. Soc., Sigma Xi. Achievements include rsch. in analytical chemistry, photgraphic, polymer and colloid sci.

OPPENHEIMER, PRESTON CARL, psychotherapist, counseling agency administrator, psychodiagnostician; b. Jacksonville, Fla., Sept. 17, 1958; s. Dawson N. and Audrey Muriel (Pressman) O.; m. Pilar Apodaca, Oct. 28, 1989; 1 child, Leo Morris. BA in Psychology, U. Calif., San Diego, 1980;

MS in Edn., U. So. Calif., 1982, Phd in Counseling, 1988. Registered psychol. asst. Contract counselor Glendale (Calif.) Family Svc. Assn, 1987-88, intake coord., 1988-89, adminstrv. dir., 1989—; psychol. asst. Curtis Psychol. Assocs., L.A., 1988—; assoc. clin. prof. UCLA, 1988—. Narrator, co-editor (documentary) Time Out for Judy--It Beats a Belt, 1984. Active community rels. com. Ea. Region Jewish Fedn. Coun., Covina, Calif., 1987-88, Temple Sinai Glendale, 1989—, bd. dirs., 1991—. Named One of Outstanding Young Men Am., 1986, 87. Mem. Am. Psychol. Assn., Pasadena Jaycees (recipient William D. Sloss award for creative inspiration 1988, project chmn. yr. 1987)/. Republican. Jewish. Avocations: fundraising, community service, adult education, writing, tennis. Office: Glendale Family Svc Assn 3436 N Verdugo Rd Glendale CA 91208-1595

OPPERMAN, DANNY GENE, packaging professional, consultant; b. Fostoria, Ohio, June 29, 1938; s. Roy and Iva Ann (Dotson) O.; m. Dorothy Rae Bugner, Dec. 30, 1957; children: Carrie Rae Opperman Hammond, Melissa Ann Opperman Lee, Jon Aaron, Christopher Douglas. Assoc., ICS, 1960. Tool engr. Ford Motor Co., Fostoria, 1957-68; packaging engr. Allied-Signal Corp., Fostoria, 1968-86; machine designer Interconnect, Inc., Toledo, 1987; prodn. engr. TRW, Elyria, Ohio, 1987-88; pres. packaging consulting firm Opperman/Assocs., Inc., Fostoria, 1988—; cons. PacTech Engring., Inc., Cin., 1990—. Pres. Fostoria Jaycees, 1970-71; advisor Fostoria Teen Ctr., Inc., 1960-66. Mem. ASTM (D-10 packaging com), Inst. Packaging Profls. (cert., chpt. bd. dirs. 1984-92), Packaging Cons. Coun., Elks (exalted ruler 1984-85), Masons.

OPPLT, JAN JIRI, pathologist, educator; b. Czechoslovakia, Nov. 25, 1921; came to U.S., 1969; s. Jan Opplt and Anna (Vochocova) O.; m. Marie A. Koudelkova, Dec. 22, 1960. MD, Charles U., Prague, Czechoslovakia, 1949, PhD in Biochemistry, 1952, PhD in Med. Scis., 1966, Docent Habilit, 1969. Asst. prof. med. and clin. chemistry Charles U., 1948-51, dir., chmn. dept. clin. chemistry, 1951-69; vis. prof., rsch. fellow lab. medicine Clev. Clinic Found., 1969-70; dir. clin. chemistry Cleve. Met. Gen. Hosp., 1971-88; asst. prof. Case Western Res. U., Cleve., 1971-72, assoc. prof. pathology, 1973—; clin. prof. chemistry Cleve. State U., 1970-92. Co-author: Diabetic Nephropathy, 1969, Diabetes Mellitus, 1970, Handbook of Electrophoresis, 1980; patentee in field. Recipient Gold award Am. Soc. Clin. Pathologists, 1971, Svc. award Am. Heart Assn., 1978, Internat. Recognition award IFCC, Holland, 1987; Fellow WHO, 1959, Cleve. Clinic Found., 1970. Fellow Nat. Acad. Clin. Biochemistry, Coun. on Atherosclerosis; mem. Am. Assn. Clin. Chemsitry, N.Y. Acad. Sci., Cleve. Acad. Medicine, Czechoslovak Soc. Arts and Scis. Achievements include development of courses in differential diagnostic procedures and clinico-pathologic correlations as well as in clinical chemical pathology, used in therapeutic and preventative medicine, photometric analytical scanning of proteinograms, lipoproteinograms, permanent standard for electrophoretic analyses, isolation and description of medium-density lipoprotein, anti-atherogenic anti-hypertensive drug Quanadrel, correlations of metabolic and conformational changes in lipoproteins, ultra centrifugal determination of natural subclasses of serum lipoproteins, molecular phoresis, new screening project for atherosclerosis, research of physiological and pathological metabolism of plasma proteins and lipoproteins. Home: 17364 Falling Water Rd Strongsville OH 44136

OPRANDY, JOHN JAY, research scientist, molecular biologist; b. White Plains, N.Y., Sept. 20, 1956; s. Harold and Dale (Pozza) O. BS, Boston Coll., 1978; PhD, U. R.I., 1982. Postdoct. fellow Mt. Sinai Sch. Medicine, N.Y.C., 1982-83; postdoct. fellow Yale U. Sch. Medicine, New Haven, Conn., 1983-85, instr., 1985-86; naval officer/scientist USN Med. Rsch. Inst., Washington, 1986-89; program dir. USN Med. Rsch. Inst., Bethesda, Md., 1989-91; dir. rsch. and devel. Integrated Diagnostics, Inc., Balt., 1991-92; group leader IGEN, Inc., Rockville, Md., 1992—; cons. Nat. Acad. Scis., Washington, 1987-88, Millipore Corp., Bedford, Mass., 1988-91; speaker U.S. Agy. for Internat. Devel., Washington, 1988-91. Contbr. articles to profl. jours. Decorated Navy Achievement medal, Mertiorious Unit citation, Superior Svc. award, Commendation medal; recipient Nat. Rsch. Svc. award U.S. Public Health Svc., 1983. Mem. Am. Soc. Tropical Medicine and Hygiene, Am. Soc. Microbiology, Sigma Xi. Achievements include patents in the area of novel techniques for membrane based immunoassay development and clinical spciaman processing for DNA/RNA purification and isolation.

ORAN, GARY CARL, computer scientist; b. Kearney, Nebr., July 5, 1948; s. Carl Franklin and Bernice Merideth (Grosh) O.; m. Janet Marie Petersen, Sept. 9, 1969; children: Jenny Marie, Ryan Alan. BSBA, U. Nebr., 1978; postgrad., U. So. Calif., 1985—. Unit mgr. Sears Roebuck & Co., Springfield, Mo., 1980-83; dep. G 4 35th Inf. Div. Mechanized, Fort Leavenworth, Kans., 1983-85; chief customer svc. Br. Info. Mgmt. Agy., Washington, 1985-86; dep. chief Info. Mgmt. Agy. Nat. Guard Bur., Washington, 1986-88; liaison officer army standard systems Info. Mgmt. Agy., Washington, 1988; project dir. Army Pers. Div. Nat. Guard Bur., Washington, 1988-89; mgmt. analyst Fed. Ins. Adminstrn., Washington, 1989—; cons. Nat. Security Agy., Balt., 1992—, Roland Holland and Co., Alexandria, Va., 1992—. Dist. membership chmn. Boy Scouts Am., Montgomery County, Md., 1990—, Pack 1760 com. chmn., Gaithersburg, Md., 1989—; troop leader Girl Scouts U.S., Gaithersburg, 1988—; den leader Cub Scouts, Gaithersburg, 1988—. Mem. Nat. Guard Assn. U.S. (life), Am. Legion, Nat. Guard Assn. Nebr. (pres. 1982-83), Nat. Guard Assn. Kans. (life). Achievements include developing Equipment Readiness Analysis Computer System recognized in the Army Green Book as one of the outstanding achievements of 1984, 85. Home: 819 Diamond Dr Gaithersburg MD 20878-1807

O'RANGERS, JOHN JOSEPH, biochemist; b. Phila., June 11, 1936; s. John Joseph and Mabel V. (Cope) O'R.; m. Eleanor Mary Jablonksi, June 12, 1965; children: Eleanor Ann, John Robert. BS, St. Josephs U., 1958; PhD, Hahnemann U., 1972. Rsch. instr. dept. medicine Hahnenn U., Phila., 1973-74; asst. prof. biochemistry Phila. Coll. Osteo Medicine, 1974-75; sr. scientist Princeton (N.J.) Labs., 1975-76; nat. expert biochemistry U.S. FDA, Phila., 1976-78; nat. expert biochemistry U.S. FDA, Washington, 1978-85, sr. regulatory scientist, 1985—; chmn. analyt. method U.S.-Can. Trade Talks, Washington, 1990—; cons. Internat. Union Pure and Applied Chemistry Food Chem. Commn., Washington, 1988—. Contbr. articles to profl. jours. Lt. (j.g.) USNR, 1959-62. Recipient Award of Merit, USDA, Washington, 1989. Fellow Assn. Official Analytical Chemists; mem. AAAS, Sigma Xi (FDA chpt. pres. 1985, chmn. centenial com. 1986). Achievements include research in validation of immunoassap, biochemical method diagnostic testing, analytical methods for drug residues, drug screening tests. Office: US FDA 7500 Standish Pl Rockville MD 20855

ORAZEM, MARK EDWARD, chemical engineering educator; b. Ames, Iowa, June 7, 1954; s. Frank and Slava (Furlan) O. BSChemE, Kans. State U., 1976, MSChemE, 1978; PhD in Chem. Engring., U. Calif., Berkeley, 1983. Asst. prof. dept. chem. engring. U. Va., Charlottesville, 1983-88; assoc. prof. dept. chem. engring. U. Fla., Gainesville, 1988-92, prof. dept. chem. engring., 1992—. Contbr. articles to profl. jours. Mem. AIChE (area chmn. 1990-92), Nat. Assn. Corrosion Engrs. (rsch. com. 1992—), Electrochem. Soc. Achievements include development of measurement model techniques for electrochemical impedance spectroscopy and optically stimulated deep-level impedance spectroscopy for characterization of semiconductors. Office: Univ Fla Dept Chem Engring. Gainesville FL 32611

ORBELL, JOHN DONALD, chemist, educator; b. London, May 30, 1949; s. William George and Ann Mabel (Watson) O.; m. Gaik Beng Kok, July 20, 1985; children: Winston John, Penelope Ann. BSc, U. Auckland, 1971, MSc, 1973, PhD, 1980. Lectr. in chemistry U. Auckland, New Zealand, 1974-78; postdoctoral fellow Johns Hopkins U., Balt., 1980-81, U. Trieste, Italy, 1981-82; lectr. in chemistry Auckland Tech. Inst., 1983; lectr., rsch. fellow Flinders U. South Australia, 1984-85; rsch. assoc. U. Md., College Park, 1985-86; sr. rsch. officer St. Vincent's Inst. Med. Rsch., Victoria, Australia, 1986-88; assoc. prof. Victoria U. Tech., 1988—. Mem. Am. Chem. Soc., Royal Australian Chem. Inst., Soc. Crystallographers in Australia. Achievements include research on interaction of metal ions with nucleic acid, mode of action of DNA-binding platinum antitumor agents. Office: Victoria U Tech, McKechnie St, Saint Albans 3021, Australia

ORBISON, DAVID VAILLANT, clinical psychologist, consultant; b. Hartford, Conn., Apr. 2, 1952; s. Theodore Tucker and Edith Vaillant (Julier) O.; m. Beth Lynne Fruendt, June 19, 1981; children: Henry Douglas, Charles Vaillant, Samuel Tucker. BA, Yale U., 1974; MA in Psychology, Duquesne U., 1976, PhD in Clin. Psychology, 1986. Psychologist Harmarville Rehab. Ctr., Pitts., 1975-84; pvt. practice Pitts., 1981—; mem. adj. profl. staff Shadyside Hosp., Pitts., 1981—. Fellow Pa. Psychol. Assn. (chmn. program and edn. bd. 1988—); mem. Greater Pitts. Psychol. Assn. (treas. 1984-86, chmn. bd. dirs. 1986-88). Office: 401 Shady Ave Ste 102C Pittsburgh PA 15206-4409

ORBISON, JAMES GRAHAM, civil engineer, educator; b. Cleve., Oct. 27, 1953; s. James Lowell and Olga Andrea (Dianich) O.; m. Nancy Anne Miller, June 11, 1977; children: Ryan Brantly, Eric James. BSCE, Bucknell U., 1975; MEC, Cornell U., 1976, PhD, 1982. Project engr. English Engring. Corp., Williamsport, Pa., 1976-77; lectr. Bucknell U., Lewisburg, Pa., 1977-78, asst. prof. civil engring., 1982-87, assoc. prof. engring., 1987-93, prof. engring. 1993—; reviewer ASME, ASTM, Am. Inst. Steel Construction, Prestressed Concrete Inst, Pa. Dept. Commerce, Harper & Row Pubs; contbr. articles to profl. jours. Mem. ASCE (Lindback award for Disting. Teaching 1988), Pa. Soc. Profl. Engrs. (Engr. of Yr. 1985), Am. Acad. Mechanics, Am. Inst. Steel Constrn., Am. Soc. for Engring. Edn. (Excellence in Instrn. Engring. Students award AT&T Found. 1990), Am. Concrete Inst., Sigma Xi. Office: Bucknell U Civil Engring Dept Lewisburg PA 17837

ORDONEZ, NELSON GONZALO, pathologist; b. Bucaramanga, Santander, Colombia, July 20, 1944; came to U.S., 1972; s. Gonzalo and Itsmenia Ordonez; m. Miranda Lee Ferrell, Dec. 18, 1976 (div. June 1983); 1 child, Nelson Adrian; m. Catherine Marie Newton, Nov. 6, 1987; 1 child, Sara Catherine Itsmenia. BA and Sci., Instituto Daza Dangond, Bogota, Colombia, 1962; MD, Nat. U. Colombia, Bogota, 1970. Resident pathology U. N.C., Chapel Hill, 1972-73; resident pathology U. Chgo., 1974-76, asst. prof. pathology, 1977-78; asst. prof. pathology U. Tex. M.D. Anderson Cancer Ctr., Houston, 1978-82, assoc. prof., 1983-85, prof., 1985—, dir. immunocytochemistry sect., 1981—. Author: (with others) Renal Biopsy Pathology and Diagnostic and Therapeutic Implications, 1980, Tumors of the Lung, 1991; contbr. chpts. to books, numerous articles to med. jours. Nat. Kidney Found. fellow, 1977-78. Mem. AMA, Am. Assn. Pathologists, Internat. Acad. Pathology, Am. Soc. Clin. Pathologists, Am. Soc. Cytology, Internat. Acad. Cytology, Arthur Purdy Stout Soc. Surg. Pathologists, Latin-Am. Soc. Pathology. Office: U Tex MD Anderson Cancer Ctr 1515 Holcombe Blvd Houston TX 77030

ORDORICA, STEVEN ANTHONY, obstetrician/gynecologist, educator; b. N.Y.C., Jan. 4, 1957; s. Vincent and Rose (Goiricelaya) O. BA magna cum laude, NYU, 1979; MD, Stony Brook U., 1983. Diplomate Am. Coll. Obstetrics and Gynecology, speciality cert. maternal-fetal medicine; lic. Nat. Bd. Med. Examiners. Resident obstetrics and gynecology NYU-Bellevue Hosp. Ctr., 1983-87, fellow maternal-fetal medicine, 1987-89, instr. obstetrics-gynecology, 1989-91; clin. instr. obstetrics-gynecology NYU, 1986-89, asst. prof. ob/gyn., 1991—; dir. perinatal clinics and prenatal diagnostic unit Gouverneur Hosp., N.Y.C., 1989—; perinatal cons. Bellevue Hosp. Ctr., N.Y.C., 1989—; faculty mem. perinatal div. NYU Med. Ctr., 1989—; presenter in field. Contbr. articles to Surgery, Am. Jour. Obstetrics and Gynecology, Am. Jour. Perinatal, Surgery, Obstetrics and Gynecology, Jour. Reproductive Medicine, Acta Geneticae Medicae et Gemellologiae, Jour. Rheumatology. Mem. Am. Coll. Obstetrics and Gynecology, Soc. Perinatal Obstetricians, N.Y. Acad. Scis., N.Y. State Perinatal Soc., AMA, Phi Beta Kappa, Beta Lambda Sigma. Achievements include research in investigating aspects of maternal-fetal physiology. Office: NYU Med Ctr 530 1st Ave Ste 5F New York NY 10016-6402

O'REAR, EDGAR ALLEN, III, chemical engineering educator; b. Jasper, Ala., Feb. 24, 1953; s. Edgar Allen Jr. and Edith (Idzorek) O'R. BSChemE., Rice U., 1975; SM in Organic Chemistry, MIT, 1977; PhD, Rice U., 1981. Rsch. engr. Exxon Rsch. and Engring., Baytown, Tex., 1975; asst. prof. U. Okla., Norman, 1981-86, assoc. prof., 1986-91, prof., 1991—, Conoco Disting. lectr., 1987-92; vis. sr. researcher Hitachi Cen. Rsch. Lab., Kokubunji, Japan, 1988; vis. scientist RIKEN-Inst. for Phys. and Chem. Rsch., Wakoshi, Japan, 1992; cons. Kerr-McGee Corp., Oklahoma City, Boehringer-Mannheim Corp., Indpls., Baxter-Travenol, Inc., Deerfield, Ill., Associated Metallurgists, Norman. Contbr. articles to profl. jours.; co-author books; inventor, patentee producing polymeric films from a surfactant template, method and composition for treatment of thrombosis in a mammal; reviewer for sci. pubs. and funding agencies. Mentor Big Bros./Big Sisters, Norman, 1984-86; usher, mem. parish coun. St. Thomas More Cath. Ch., Norman; mem. Japan-Okla. Soc., Oklahoma City, U. Okla. Faculty Senate, Norman, 1984-86. Recipient Sigma Xi Faculty Rsch. award U. Okla. chpt. Sigma Xi, 1986. Mem. AAAS, NSF (separations and purification program dir. 1993—), Internat. Soc. Biorheology (sec. gen.), North Am. Soc. Biorheology (charter, membership com.), Am. Chem. Soc., Biomed. Engring. Soc., Soc. Rheology, Okla. Acad. Scis., MIT Chemists' Club, Tau Beta Pi, Sigma Xi, Phi Lambda Upsilon. Roman Catholic. Office: U Okla Dept Chem Engring 100 E Boyd St Rm 335T Norman OK 73019-0001

O'REILLY, JOHN JOSEPH, engineer; b. Santa Cruz, Calif., May 12, 1959; s. Richard Carroll and Naydene Joyce (Hensley) O'R.; m. Michelle Marie Johnson, Sept. 27, 1986. BS in Mech. Engring., U. Nebr., 1982; MS in Engring. Mgmt., So. Meth. U., 1988. Registered profl. engr., Tex. Facilities engr. Texas Instruments, Dallas, 1982-85; internat. facilities engring. mgr. United Techs. MOSTEK, Carrollton, Tex., 1985; mech./contamination control engr. Henningson, Durham & Richardson, Dallas, 1985-86; with Digital Equipment Corp., Dallas and Austin, Tex., 1986-91; program mgr. reliability SEMATECH, Austin, 1988-91; equipment reliability engr. Motorola Advanced Products R&D Lab., Austin, Tex., 1991—; cons. Sematech's U. Nat. Lab. Program, 1990-91, Sandia Nat. Labs., Albuquerque. Contbr. articles to profl. jours. Charter mem. Diocesan Forum Cath. Profls., Austin, 1988—; co-chair Diocesan ForuHomeless Com., 1989-92; vol. Our Lady's Youth Ctr., Austin, 1989-92; mem. Sematech Vol. Coun., 1990-91; bd. dirs., chair individual-giving subcom. Caritas Devel. Adv. Bd., 1991—; bd. dirs. Sts. of Hope. Mem. Am. Soc. Quality Control, Inst. Environ. Scis. (newsletter editor 1985-86, annual tech. meeting facilities co-chair 1986, Achievement award 1986). Avocations: reading, running, gardening, softball, tennis.

OREL, ANN ELIZABETH, applied science educator, researcher; b. Lake Charles, La., Oct. 26, 1955; d. Bernard Anthony and Bernice Josephine (Toplikar) O.; m. W. Hugh Woodin, Aug. 15, 1985; children: Christine, Alexandria. BS, Calif. Inst. Tech., 1977; PhD, U. Calif., Berkeley, 1981. Staff scientist Lawrence Livermore (Calif.) Nat. Lab., 1980-84; mem. tech. staff Aerospace Corp., El Segundo, Calif., 1985-88; assoc. prof. dept. applied sci. U. Calif. Davis, Livermore, 1988-90, assoc. prof. dept. applied sci., 1990—. Contbr. articles to profl. jours. Recipient Anna Louise Hoffman award Iota Sigma Pi, 1980. Mem. Am. Phys. Soc. Office: Univ Calif Davis Dept Applied Sci PO Box 808 L794 Livermore CA 94550

ORESKES, NAOMI, earth sciences educator, historian; b. N.Y.C., Nov. 25, 1958; d. Irwin Oreskes and Susan Eileen Nagin Oreskes; m. Kenneth Belitz, Sept. 28, 1986; 1 child, Hannah Oreskes Belitz. BSc with honors, Imperial Coll., London, 1981; PhD, Stanford U., 1990. Geologist Western Mining Corp., Adelaide, Australia, 1981-84; rsch. and tng. asst. Stanford (Calif.) U., 1984-89; vis. asst. prof. Dartmouth Coll., Hanover, N.H., 1990-91, asst. prof., 1991—; consulting geologist Western Mining Corp., 1984-90; consulting historian Am. Inst. Physics, N.Y.C., 1990—. Contbr. articles to profl. jours. Recipient Lindgren prize Soc. Econ. Geologists, 1993; fellow NEH, 1993. Mem. Geol. Soc. Am., History Sci. Soc. Jewish. Home: HC 61 Box 265 Etna NH 03750 Office: Dartmouth Coll Dept Earth Scis 6105 Fairchild Hall Hanover NH 03755

ORHON, NECDET KADRI, physician; b. Manisa, Turkey, Aug. 13, 1928. MD, U. Istanbul, Turkey, 1951. Lic. to practice medicine in Ill., Iowa, Minn., Ohio, Va., Fla. Intern. U. Istanbul, Turkey, 1952-53; asst. residency in internal medicine Erlanger Hosp., Chattanooga, Tenn., 1953-54; resident in internal medicine Lloyd Noland Hosp., Birmingham-Fairfield, Ala., 1954-55; chief med. resident Good Samaritan Hosp., Lexington, Ky., 1955-56; internist Army Hosp., Turkey, 1956-58; chief resident in internal medicine, chief house doctor Glenwood Hills Hosps., Mpls., 1958-65, internist, chief house officer, 1968-69; fellow oncology Tuft U., Boston, 1966-67; internist Hillcrest Hosp., Birmingham, Ala., 1967-68; dir. emergency room. Mercy Med. Ctr., Springfield, Ohio, 1969-82; practice medicine specializing in internal medicine Springfield, Ohio, 1982—. Mem. Clark County Med. Soc., Ohio Med. Assn. Home: 2066 Northridge Dr Springfield OH 45504-1055 Office: 411 West Handing Springfield OH 45504

ORI, KAN, political science educator; b. Osaka, Japan, Jan. 28, 1933; s. Tazo and Tamaji (Moriki) O.; m. Teruko Horie, Apr. 27, 1961; children: Akemi, Harumi, Noriko. BA cum laude, Taylor U., Upland, Ind., 1956; MA, Ind. U., Bloomington, 1958; PhD, Ind. U., 1961. Instr. Jochi (Sophia) U., Tokyo, 1965-67; assoc. prof. Jochi (Sophia) U., 1967-70, prof. polit. sci., 1970—, mem., faculty Inst. Internat. Relations, 1969—, dean grad. sch. Faculty Fgn. Studies, 1985-87; vis. prof. U. Malaya, Kuala Lumpur, Malaysia, 1974-75, U. Mo., Columbia, 1975-76 (winter semester), U. Minn., Mpls., summer 1978, 88, Princeton (N.J.) U., 1982-83 (fall semester). Author (with Roger Benjamin): Tradition and Change in Postindustrial Japan, 1981; editor Jour. Internat. Studies, 1978-82; editorial bd. Internat. Studies Quar., 1980-84. Ind. U. fellow, 1960; Matsunaga Sci. Found. grantee, 1970; Japan Found. grantee, 1974-75. Mem. Japanese Polit. Sci. Assn., Japanese Assn. Internat. Relations (councilor 1975—), Japanese Assn. Am. Studies (councilor 1986—). Office: Sophia U Inst Internat Rel, 7-1 Kioi-cho Chiyoda-ku, Tokyo Japan 102

ORIAKU, EBERE AGWU, economics educator; b. Atlanta, Sept. 16, 1951; (parents Am. citizens); s. Benjamin Agwu and Nwachi (Benjamin) O.; m. Ngozi Okonkwo, Sept. 3, 1983; children: Chiduzie, Uzoechi. BBA in Econs. and Fin., U. Ark., Little Rock, 1976; MS in Econs., Atlanta U., 1977; JD in Internat. Law, Antioch U., 1983; PhD in Econs., Howard U., 1984. Asst. pres. NSEEO, Washington, 1980-83; dir. Ctr. for Econ. Edn., Elizabeth City, N.C., 1985—; prof. Elizabeth City State U., 1990—; vice-chmn. Dept. of Bus. Econs., Elizabeth City State U., N.C., 1985—; coord. Black Exec. Exch. Prog., Elizabeth City State U., 1985—, Bank Cent. Elizabeth City State U., 1989—; cons. internat. trade law, Washington, 1982-85, internat. trade regulations, Washington, 1985—. Author: Role of Multi National Corporations in Developing Nations, Outstanding Young Men of America, 1978, Role Multinational Corporations in Economic Development of Nigeria, Analysis of Large Scale Irrigation Schemses. Analyst Elizabeth City (N.C.) Comml. Devel. Budget, 1985—. Named Prof. of Yr., Elizabeth City State U., 1985-88; nominee for Marx Garner Disting. Prof. award, 1991, 1993, Gov.'s Excellence award N.C., 1991; grantee Am. Assn. Retired Persons, 1990; honoree for excellent rsch. achievement, NAFEO. Fellow Omicron Delta Epsilon; mem. Am. Mgmt. Assn., Southern Econs. Assn. Home: 1604 Camellia Dr Elizabeth City NC 27909-6314 Office: Ctr for Econ Edn Box 916 Elizabeth City NC 27909

ORIN, DAVID EDWARD, electrical engineer; b. Portsmouth, Ohio, Mar. 26, 1949; s. Edward Everett and Sophia Eileen (Bond) O.; m. Katrina Loy Bentley, June 16, 1984; children: Jeffrey Lake, Kari Lake, Kimberly Lake. BSEE, The Ohio State U., 1972, MS, 1973, PhD, 1976. Asst. prof. Case Western Res. U., Cleve., 1976-80; asst. prof. The Ohio State U., Columbus, 1980-84, assoc. prof., 1984-89, prof., 1989—; sec. IEEE Robotics and Automation Soc., Piscataway, N.J., 1991—; cons. Unique Mobility, Golden, Colo., 1991—; mem. editorial bd. Robotics Rev., MIT Press, Cambridge, 1990—. Contbr. articles to profl. jours. Pastor Reorganized Ch. of Jesus Christ of LDS, Columbus, 1991-92. Grantee NSF, 1987, 84. Fellow IEEE; mem. Sigma Xi, Tau Beta Pi, Eta Kappa Nu. Reorganized Ch. of Jesus Christ of Latter-day Saints. Achievements include devel. of dexterous grasping robotic system called DIGITS; being first to develop efficient robot dynamics algorithms. Office: The Ohio State U 2015 Neil Ave Columbus OH 43210

ORKWIS, PAUL DAVID, aerospace engineering educator; b. Jamaica, N.Y., Oct. 27, 1962; s. John Steven and Jane (Kasprzak) O.; m. Pauline Nina Giovanniello, July 11, 1987; 1 child, Alexander Sterling. PhD in Aerospace Engring., N.C. State U., 1990. Rsch. asst. N.C. State U., Raleigh, 1986-90; asst. prof. U. Cin., 1991—. Contbr. articles to profl. jours. Bd. dirs. Richland Runs Homeowners Assn., Raleigh, 1989-90. Fellowship U.S. Army Rsch. Office, 1986-90, grant, 1992—, U.S. Air Force Office of Scientific Rsch., 1991-92. Mem. AIAA, Sigma Xi. Achievements include first to demonstrate quadratic convergence of a computational algorithm for computing solutions of the Navier-Stokes equations for high speed viscous flow fields. Office: U Cin Dept Aero Engring & Engring Mechanics ML 70 Cincinnati OH 45221

ORLAND, FRANK JAY, oral microbiologist, educator; b. Little Falls, N.Y., Jan. 23, 1917; s. Michael and Rose (Dorner) O.; m. Phyllis Therese Mrazek, May 8, 1943; children: Frank R., Carl P., June Rose, Ralph M. AA, U. Chgo., 1937, SM, 1945, PhD, 1949; BS, U. Ill., 1939, DDS, 1941. Diplomate Am. Bd. Med. Microbiology. With U. Chgo., 1941—; intern Zoller Meml. Dental Clinic, U. Chgo., 1941-42; Zoller fellow, asst. in dental surgery U. Chgo., 1942-49, instr., asst. prof., assoc. prof., prof. dental sci., 1949-88, prof. emeritus, 1988—, from instr. to assoc. prof. microbiology, 1950-58, rsch. assoc. prof., 1958-64; dir. Zoller Meml. Dental Clinic, 1954-66; prof. Fishbein Ctr. for Study History Sci. and Medicine, 1980-88, prof. emeritus, 1988—; attending dentist Country Home for Convalescent Children; past cons. Nat. Inst. Dental Rsch., NIH, Bethesda, Md.; mem. panel on dental drugs The Nat. Formulary; past chmn. dental adv. bd. Med. Heritage Soc. Author: The First Fifty-Year History of the International Association for Dental Research, 1973, Microbiology in Clinical Dentistry, 1982, William John Gies-His Contributions to the Advancement of Dentistry, 1992; editor: Jour. Dental Rsch., 1958-69, (Centennial brochure) Loyola U. Sch. Dentistry, 1983; editor, contbr. Microbiology in Clinical Dentistry, 1982; writer, prodr. 50th anniversary booklet Zoller Meml. Dental Clinic U. Chgo., 18997; contbr. articles to profl. jours. Past chmn. adv. coun. Forest Park (Ill.) Bd. Edn.; mem. Forest Park Citizens Com. for Better Schs.; past pres. Garfield Sch. PTA, 1953-55; chairperson heritage com. Bicentennial Commn. on Forest Park, 1983-85; editor Chronicles of Forest Park, 1976—. Recipient Rsch. Essay award Chgo. Dental Soc., 1955, Cook County Sheriff Medal of Honor award, 1993; named Citizen of Yr., Forest Park, 1989. Fellow AAAS, Inst. Medicine Chgo. (chmn. com. publ. communications), Am. Acad. Microbiology, Am. Coll. Dentists; mem. ADA (past chmn. coun. dental therapeutics), Internat. Assn. Dental Research (pres. 1971-72, past councilor Chgo. sect., past chmn. program com., past chmn. com. on history), Am. Assn. Dental Schs. (past chmn. conf. oral microbiology past chmn. com. on advanced edn.), Am. Assn. Dental Editors (William Gies Editorial award 1968), Ill. State Dental Soc. (chmn. com. on history), Fedn. Dentaire Internat. (Commn. on History Rsch.), Am. Acad. History Dentistry (pres.1976-77, Hayden-Harris award 1980), Hist. Soc. Forest Park (pres.), Soc. Med. History Chgo. (past pres.), Chgo. Lit. Club, Sigma Xi, Gamma Alpha. (pres. Chgo. chpt.). Home: 519 Jackson Blvd Forest Park IL 60130-1896 Office: 521 Jackson Blvd Forest Park IL 60130-1896

ORLANDO, CARL, medical research and development executive; b. Palermo, Italy, Sept. 26, 1915; came to U.S., 1928; s. Peter and Maria (Bongiorno) O.; m. Ann Bovè, June 29, 1943; children: Ann Marie, Francine, Patricia, Charleen, Joan. BS, Columbia U., 1941; postgrad., Rochester U., 1943. Chief photo optics U.S. Army Elec. Commd., Ft. Monmouth, N.J., 1945-75; cons. pvt. practice, New Shrewsbury, N.J., 1975-79; v.p. rsch. & devel. Analytical R&D Inc., Eatontown, N.J., 1979-88; cons. rsch. & devel. Engring. Devel. Co., Tinton Falls, N.J., 1986-88; pres., rsch. & devel. dir. Sens-O-Tech Indsutries Inc., Eatontown, N.J., 1988—; chmn. bd. dirs. Sens-O-Tech Industries, 1988-92. Contbr. articles to profl. jours. Bd. dirs. Monmouth Regional High Sch., Tinton Falls, 1974; com. mem. Tinton Falls Environ. Unit, 1984; chmn. Entertainment Activities St. Dorothaas Ch., Eatontown, 1983. With USN, 1945. Recipient Monetary Suggestion award Signal Corp. Engring. Lab., 1948. Mem. Soc. Photographic Scientist & Engrs. (sr. mem.), Soc. Imaging Sci. & Tech., Elks, Battle Ground Country Club. Republican. Roman Catholic. Achievements include over 20 patents in various fields: non-invasive heart and breathing alarm monitors, moving target indicator, photographic reproduction in 0.2 second, one step photographic technic, image stabilization system. Home and Office: 47 Willow Rd Tinton Falls NJ 07724

ORLEN, JOEL, association executive; b. Holyoke, Mass., Aug. 1, 1924; s. Barnet and Fannie (Fuchs) O.; m. Yana Sorra Edmundson, Nov. 24, 1963; 1 stepson, Charles. BA, U. Chgo., 1950, MA, 1952. Fgn. svc. officer U.S. Dept. State, Washington, 1952-56; fgn. affairs officer AEC, Washington, 1956-58; officer NAS, Washington, 1958-63; asst. dir. Desert Rsch. Inst., Reno, 1963-65; exec. officer, provost MIT, Cambridge, 1965-80; v.p. Sci. Mus. Minn., St. Paul, 1980-86; exec. officer Am. Acad. Arts and Letters, Cambridge, 1986—; dir. Mounds Park Acad., St. Paul, 1983-86; cons. MIT, 1980-82; exec. sec. Mass. Tech. Devel. Corp., Boston, 1978-80; advisor U.S. del. to UN, N.Y.C. and Geneva, 1952-63. Contbr. chpts. to several books. Staff sgt. U.S. Army, 1943-46, ETO. Home: 130 Mt Auburn St Cambridge MA 02138-5757 Office: Am Acad Arts and Scis 136 Irving St Cambridge MA 02138-1996*

ORLIDGE, LESLIE ARTHUR, electrical engineer; b. Johnstown, Pa., Nov. 11, 1953; s. Arthur Eugene and Mary Kail (Huffman) O.; m. Kimber Leigh Volkmar, Mar. 26, 1988; children: Jessica Leigh, Julia Leigh. BSEE, BA, Pa. State U., 1977; M.Adminstrv. Sci., Johns Hopkins U., Balt., 1984. Systems engr. AAI Corp., Hunt Valley, Md., 1977-78, lead systems engr., 1979-82, engring. mgr., 1983-85, IR&D program mgr., 1985-87, IR&D tech. dir., 1988-90, prin. devel. engr., 1990-91, mgr. new bus. devel. electronics div., 1991—. Contbr. articles to profl. jours. Mem. IEEE (vice chmn. SCC-20 AI subcom. 1990—), Assn. Computing Machinery, AIAA, Computer Soc. of IEEE, IEEE Instrument and Measurement Soc., Masons (past master 1988), Shriner. Methodist. Avocations: golf, tennis, hiking, electronics, family. Home: 1613 Walker Rd Freeland MD 21053-9541 Office: AAI Corp PO Box 126 Cockeysville Hunt Valley MD 21030-0126

ORLITZKY, ROBERT, engineer; b. Nadrag, Romania, Dec. 5, 1960; came to U.S., 1965; s. Josef and Anna (Steingasser) O.; m. Barbar Ann Piazza, July 12, 1989; 1 child, Michael J. Student prof. aero., Embry Riddle Aero U., 1985-89. Master technician, USAF. Engr. product support Bendix Aerospace, Teterboro, N.J., 1983-85; engr. level V Test Tech., Inc., Hauppauge, N.Y., 1985-89; v.p. Bus. Svcs. Network, Balt., 1989-90; mgr. computer group Kohler & Co., Greenbelt, Md., 1990-91; mgr. systems group Sacks, McGibney & Trotta, P.A., Balt., 1992—; conf. speaker Kohler Healthcare Cons., Balt., 1992—. Co-author, editor: Computerization in the Physicians Office, 1992; author: (software and documents) Nursing Employee Tracking System, 1990, MDCASH Cash Flow Projection, 1992. Community svc. Balt. (Md.) Bd. Edn., 1988-90; fundraiser Am. Heart Assn., Balt., 1991—, Am. Cancer Soc., Balt., 1991—. Sgt. USAF, 1978-82. Recipient Commendation, Gen. Dynamics Corp., 1981, Tech. Excellence award Strategic Air Command, 1981, Merit/Maintenance Support award Strategic Air Command, 1981. Mem. AIAA, IEEE , Air Force Assn. (life), Microcomputer Mgrs. Assn. Republican. Home: 1933 Kelly Ave Baltimore MD 21209

ORLOWSKI, STANISLAW TADEUSZ, architect; b. Skarzysko, Poland, Sept. 24, 1920; s. Tadeusz and Irena (Malawczyk) O.; m. Krystyna Joanna Przyborowska, July 23, 1949; children: Alexandra Maria Izabela, Irena Krystyna, Helena Victoria. Reader in econs., Leicester (Eng.) Sch. of Architecture, 1945-46, Diploma Architecture, 1951; M.Sc. in Architecture with hons., U. London, 1954. Architect Ont. Dept. Pub. Works, 1952-55, Page & Steel, Architects, Toronto, 1955-57; Allward & Gouinlock, Toronto, 1965-67; area architect Pub. Works of Can., Toronto, 1957-65; chief rsch. architect Ont. Ministry of Edn., 1967-73, assoc. chief architect, 1980-85, chief architect, 1985; chief architect Ont. Ministry of Colls. and Univs., 1973-80; cons. on ednl. facility design, 1968—; vis. prof., lectr. various internat. Univs. and profl. orgns., 1967—; lectr. Ont. Inst. Studies & Edn., 1967—. Researcher, editor planning and design studies. Contbr. articles to Can., Brit. and Italian jours. World v.p. Polish Girl Guides and Boy Scouts Assn., 1970-88; 1st v.p. Polish-Can. Congress, 1980-86, pres. 1986-90; chmn. World Polonia Coord. Coun., 1986—; mem. Royal Canadian Legion; prisoner concentration camp Piechorlag, 1940-41. Capt. Polish Army, NATOUSA, 1941-45. Decorated Polish and British mil. medals, 1945; Field Marshal Alexander scholar, U.K., 1945; recipient citation Coun. Ednl. Facility Planners Internat., 1991. Fellow Royal Archtl. Inst. of Can.; mem. Royal Inst. Brit. Architects, Polish Engrs. Assn. Can., Internat. Union Architects (ednl. facilities com. UNESCO 1975-80), Polish Engrs. (nat. pres. 1967-70), AIA (architecture for edn. com. 1983-86), Internat. Energy Commn. (coun. ednl. facilities planners 1979-86), Ontario Assn. architects (hon.), Polish Inst. Arts and Scis., Royal Can. Mil. Inst., Empire Club. Roman Catholic. Home: 42 Braeside Rd, Toronto, ON Canada M4N 1X7

ORMASA, JOHN, retired utility executive, lawyer; b. Richmond, Calif., May 30, 1925; s. Juan Hormaza and Maria Inocencia Olondo; m. Dorothy Helen Trumble, Feb. 17, 1952; children: Newton Lee, John Trumble, Nancy Jean Davies. BA, U. Calif.-Berkeley, 1948; JD, Harvard U., 1951. Bar: Calif. 1952, U.S. Supreme Ct. 1959. Assoc. Clifford C. Anglim, 1951-52; assoc. Richmond, Carlson, Collins, Gordon & Bold, 1952-56, ptnr., 1956-59; with So. Calif. Gas Co., L.A., 1959-66, gen. atty., 1963-65, v.p., gen. counsel, 1965-66; v.p., system gen. counsel Pacific Lighting Service Co., Los Angeles, 1966-72; v.p., gen. counsel Pacific Lighting Corp., Los Angeles, 1973-75, v.p., sec., gen. counsel, 1975. Acting city atty., El Cerrito, Calif., 1952. Served with U.S. Navy, 1943-46. Mem. ABA, Calif. State Bar Assn., Kiwanis (v.p. 1959). Republican. Roman Catholic.

ORMISTON, TIMOTHY SHAWN, mechanical engineer; b. Fort Wayne, Ind., Jan. 11, 1958; s. Rodney Earl and Shirley Hope (Kessler) O.; m. Kathleen C. Stewart, June 7, 1980. BS in Mech. Engring., Ind. Inst. Tech., 1980. Registered profl. engr., Pa. Project engr. to engring. PPG Industries, Inc., Carlisle, Pa., 1980-89, 91—; plant engr. Guangdong Float Glass Co., Ltd., China, 1989-91. Mem. NSPE (chpt. dir. 1989-90). Avocations: bicycling, raquetball, camping, woodworking.

ORNELLAS, DONALD LOUIS, chemist researcher; b. San Leandro, Calif., July 7, 1932; s. Louis Donald and Anna (Gerro) O.; children: Timothy Donald, Kathryn Ann, Melinda Dawn. BS in chemistry, Santa Clara U., 1954. Chemist Kaiser Gypsum Co, Redwood City, Calif., 1954-55, Kaiser Aluminum & Chem. Co., Permanente, Calif., 1957-58, Lawrence Livermore (Calif.) Nat. Lab., 1958—; presenter at tech. meetings. Contbr. 16 articles to profl. jours. Capt. U.S. Army, 1955-57. Recipient Annual medal award, Am. Inst. Chemist, 1954. Mem. Parents Without Ptnrs. (chpt. 53 pres. 1971-73, chpt. 458 bd. dirs. 1982-91). Democrat. Roman Catholic. Achievements include research in high explosives, development specializing in detonation and combustion calorimetry; holder 2 patents. Home: 559 S N St Livermore CA 94550-4365 Office: Lawrence Livermore Nat Lab PO Box 808 L-282 Livermore CA 94550

O'ROURKE, RONALD EUGENE, computer and electronics industry executive; b. Loma Linda, Calif., Jan. 31, 1957; s. Eugene Lawrence and Marilyn Jean (Rickert) O'R.; m. Victoria Lippincott, Apr.16, 1988; 1 child: Ryan Lawrence. BS in Computer Systems Engring., Stanford U., 1980; postgrad. in computer sci., San Diego State U., 1993. Registered engr.-in-tng., Calif.; lic. pvt. pilot. Asst. computer programmer Fluor Corp., Irvine, Calif., 1977; systems engr. intern IBM Corp., San Francisco, 1978; computer programmer Ford Aerospace & Communications Corp., Newport Beach, Calif., 1979; mktg. rep. IBM Corp., San Francisco, 1980-82; program mgr. Gen. Atomics, Sorrento Electronics, San Diego, 1989—. Chmn. reconstrn. com. La Jolla (Calif.) Village Homeowners' Assn., Inc., Southpointe, 1991, dir., 1991—, U.S. Navy Nuclear Submarine Program, 1983-89. Lt. USN, 1987-89, USNR, 1989—. Mem. IEEE, Am. Nuclear Soc., Naval Res. Assn., Naval Submarine League, U.S. Naval Inst., Stanford Alumni Assn. (life), Mensa, Theta Delta Chi (v.p. 1977-78). Republican. Presbyterian. Home: 3388 Caminito Vasto La Jolla CA 92037-2919 Office: Sorrento Electronics 10240 Flanders Ct San Diego CA 92121-3990

O'ROURKE, THOMAS DENIS, civil engineer, educator; b. Pitts., July 31, 1948; s. Lawrence Robert and Adel Mildred (Moloski) O'R.; m. Patricia Ann Lane, Aug. 12, 1978; B.S.C.E., Cornell U., Ithaca, N.Y., 1970; M.S.C.E., U. Ill., 1973, Ph.D., 1975. Soils engr. Dames & Moore, N.Y.C., 1970; research asst. U. Ill., Urbana, 1970-75, asst. prof., 1975-78; asst. prof. Cornell U., 1978-80, assoc. prof., 1981-87, prof, 1987—. Mem. ASCE (pres. Ithaca sect. 1981-82, Collingwood prize 1983, Huber prize 1988), ASME, NAE, ASTM (C.A. Hogentogler award 1976), Earthquake Engring. Research Inst., Internat. Soc. Engring. Geology, Internat. Soc. Rock Mechanics, U.S. Nat. Com. on Tunnelling Tech. (chmn. 1987-88). Home: 10 Twin Glens Rd Ithaca NY 14850-1041 Office: Cornell U Sch Civil Environ Engring 265 Hollister Hall Ithaca NY 14853

O'ROURKE, THOMAS JOSEPH, aerospace engineer, consultant; b. N.Y.C., Apr. 20, 1931; s. Thomas Joseph and Margaret Catherine (Broderick) O'R.; married; children: Thomas, Sean. BS in Civil Engring., Manhattan Coll., 1953. Sr. engr. N.Am. Aviation, Columbus, Ohio, 1953-56; staff engr. Pan Am, N.Y.C., 1959-66; staff engr. Grumman, 1966-74, sect. chief, 1974-87, prin. engr. ground ops.-space station, 1987—. With USN, 1956-59; lt. comdr. USNR, 1959-76. Mem. Soc. Am. Mil. Engrs., Ret. Officers Assn., KC. Roman Catholic. Home: 3 West Ct Sterling VA 20165

ORPHANIDES, GUS GEORGE, chemical company official; b. N.Y.C., Jan. 27, 1947; s. Gus G. and Savesta (Agapetus) O.; m. Jeanne Wood, Feb. 3, 1968; children: Alyson, Paul, Lindsay. BS with honors, Hobart Coll., 1967; PhD, Ohio State U., 1972. Chemist E.I. Du Pont de Nemours & Co., Wilmington, Del., 1974-79, Beaumont, Tex., 1979-81; chemist Air Products, Allentown, Pa., 1981-84, applications mgr., 1984-85, comml. mgr., 1985-88, rsch. mgr., 1988-91, sr. comml. devel. mgr., 1991—. Contbr. articles to profl. publs.; patentee in field. 1st lt. U.S. Army, 1972-74. Decorated Army Commendation medal; recipient Army Cert. of Achievement. Mem. TAPPI, Am. Chem. Soc. Republican. Presbyterian. Achievements include technical and commercial development of new polymer and monomer products for the water based polymer industry; basic and applied research in polymer chemistry, synthesis, process, applications technology and manufacturing; expertise in vinyl polymers and polyurethanes for adhesives, coatings and elastomers, and emulsion polymerization. Home: 3460 W Highland St Allentown PA 18104-2673 Office: Air Products 7201 Hamilton Blvd Allentown PA 18195-9642

ORR, MARCIA, child development researcher, child care consultant; b. Anamosa, Iowa, Mar. 2, 1949; d. Harold Edward Eiben and Clara Elizabeth (Hubbard) E.; m. Robert J. Orr, Sept. 6, 1969; 1 child, Jennifer. Student, U. Iowa, 1977; BS, St. Xavier U., Chgo., 1981; postgrad., Nat. Lewis U., 1993—. Edn. Bookkeeper Monticello State Bank, 1967-69; exec. sec. Davenport Bank and Trust, 1969-73; asst. educator Elisabeth Ludeman Devel. Ctr., Park Forest, Ill., 1979; tchr. Flossmoor Hills (Ill.) Elem. Sch., 1980-1984; exec. dir. Co-Care, Inc., Park Forest, 1984-89; child development researcher Flossmoor, Ill., 1989—; educator Nazarene Nursery Sch. and Kindergarten, Chicago Heights, Ill., 1991; child care ctr. cons. Matteson Sch. Dist. 162, Park Forest, 1991—; founder, pres. Before and After Sch. Enrichment, Inc. Matteson Sch. Dist. 162, Park Forest, Ill., 1991—, adv. mem. project early start, 1991—; home-sch. coord. Matteson Sch. Dist. 162, Park Forest, 1992—; officer Boleo Childcare Ctr., Iowa City, 1975-77. Tchr. religion Infant Jesus of Prague Ch., Flossmoor, Ill., 1982—; mem. Flossmoor Chorus, 1987-89; music chmn. Dist. 161 PTO, 1988-90. Mem. Nat. Secs. Assn., Women Employed Orgn., Internat. Platform Assn., Women's Club, Southland C. of C. Democratic. Catholic. Avocations: piano, classical music, travel. Home: 2250 Marston Ln Flossmoor IL 60422-1336 Office: Matteson Sch Dist 162 210 Illinois St Park Forest IL 60466-1100

ORSAK, JOSEPH CYRIL, civil engineer; b. Port Arthur, Tex., Aug. 28, 1928; s. Adolph and Marie Veronica (Repka) O.; m. Patricia Ann Viau, May 30, 1951; children: Joseph Cyril Jr., David, Darlene, Karen, James, Patricia Elaine. BSCE, Syracuse U., 1957. Project engr. Firestone Tire & Rubber Co., Akron, Ohio, 1957-69; project mgr. Firestone Brema, Bari, Italy, 1969-76, Firestone Australia Pty., Sydney, 1976-78, Firestone Can. Ltd., Hamilton, Ont., 1978-80, Firestone Fire & Rubber Co., Memphis, 1980-81; prin. Orsak Project & Constrn. Mgmt., Bartlett, Tenn., 1981-83; constrn. adminstr. Askew Nixon Ferguson Wolfe Architects, Memphis, 1983-90; project mgr. Memphis-Shelby County Airport Authority, 1990-91. Sgt. U.S. Army, 1950-51, Japan. Mem. ASCE (life), Assn. Profl. Engrs. Ont. Roman Catholic. Home: 5932 Spruce Hollow Cove Bartlett TN 38134

ORSILLO, JAMES EDWARD, computer systems engineer, company executive; b. Elmira, N.Y., Oct. 30, 1939; s. Giacomo and Irene (Heppy) O.; 1 child, June Lynne. BEE, RCA Insts., 1962; BS in Elec. Engring. and Math., Ind. Inst. Tech., 1964; MS, Rensselear Poly., 1968; BS in Nuclear Engring., Capital Radio Electronic Inst., 1974. Communications engr. Bell Telephone Labs., Holmdel, N.J., 1962-63; video engr. Westinghouse, Elmira, N.Y., 1965-66; computer engr. GE, Pittsfield, Mass., 1966-67; systems specialist Control Data Corp., Mpls., 1968-70; software specialist Computer Sci. Corp., Morristown, N.Y., 1970-72; prin. cons. Computer Cons. Assocs., Elmira, 1972-78; pres. ORTHSTAR, Inc., Elmira, 1974—; owner, pres. O-K Properties Co., Elmira, 1984—, Thundering Hooves Stables, Elmira, 1985—. Mem. IEEE, Am. Nuclear Soc., Soc. Indsl. and Applied Math., Am. Helicopter Soc., Army Aviation Assn. Am., Internat. Flying Engrs., USAF Assn., U.S. Naval League, U.S. Polo Assn. Republican. Achievements include invention of Integrated Data Acquisition System (IDAS), of Thread Algebra used in simulation development, of Extended Sentient Non-linear Ensemble (ESNE). Office: ORTHSTAR Inc 1890 W Water St PO Box 3430 Elmira NY 14905-0430

ORSZAG, STEVEN ALAN, applied mathematician, educator; b. N.Y.C., Feb. 27, 1943; s. Joseph and Rose (Siegel) O.; m. Reba Karp, June 21, 1964; children—J. Michael, Peter Richard, Jonathan Marc. B.S., M.I.T., 1962; postgrad. (Henry fellow), St. John's Coll., Cambridge (Eng.) U., 1962-63; Ph.D., Princeton U., 1966. Mem. Inst. Advanced Study, Princeton, N.J., 1966-67; prof. applied math. M.I.T., 1967-84; prof. applied and computational math. Princeton U., 1984—, dir., 1990-92, Hamrick prof. engring., 1989—; chmn. CHI, Inc. 1975—; vice chmn. Nektonics Inc., 1985—; cons. in field. Author: Studies in Applied Mathematics, 1976, Numerical Analysis of Spectral Methods, 1977, Advanced Mathematical Methods for Scientists and Engineers, 1978; editor: Springer Series in Computational Physics, 1977—, AIP Lecture Series Computational Physics, 1993—, Jour. Sci. Computing, 1986—; numerous research publs. in field. A.P. Sloan Found. fellow, 1970-74, Guggenheim fellow, 1989-90. Fellow Am. Inst. Physics (Otto Laporte award 1991), AIAA (Fluid and Plasmadynamics award, 1986), Soc. Indsl. and Applied Math. Office: 218 Fine Hall Princeton NJ 08544

ORTEGA-RUBIO, ALFREDO, ecologist, researcher; b. Mexico City, Apr. 16, 1956; s. Rodolfo G. and Sara C. (Rubio) Ortega; m. Laura Arriaga, Dec. 20, 1986; 1 child, Paloma O. BS in Biology, Nat. Sch. Biol. Scis., Mex., 1979, D in Ecology, 1986, PhD in Ecology, 1986; MS in Biology, Nat. Poly. Inst., Mex., 1981, grad. with distinction, 1989. Expert in biology Infraestructure Resources and Svcs., Mexico City, 1977-78; titular prof. Nat. Autonomous U. Mex., Mexico City, 1978-93, researcher, 1978-79; titular prof. Nat. Poly. Inst., Mexico City, 1979-86; researcher Inst. Ecology, Mexico City, 1979-86, mgr. biosphere res., 1984-85; div. dir. Biol. Rsch. Ctr., La Paz, Baja Calif. Sur, Mex., 1986-93, Biol. Rsch. Ctr. Northwest, La Paz, Baja Calif. Sur, 1993—; nat. expert & project evaluator Nat. Coun. Sci. and Tech., Mexico City, 1978—; advisor Ctr. of Rsch. in Ecology and Arid Zones, Coro Venezuela, 1989—, Rschs. Group for the Study of Calif. Gulf, Mexico City, 1990—; gen. dir. ecol. mgmt. natural resources Nat. Ministry Social Devel., Mexico City, 1992—; researcher Ricardo J. Zevada Found., Mexico City, 1989, Fed. Govt. Mex., 1986-91; nat. rschr. Nat. Rschs. System Fed. Mex. Govt., 1985—. Co-editor: LaSierra de la Laguna at Baja Calif., 1988, Environmental Impact Assessment, 1989; editor 7 books, 15 book chpts., 55 tech. reports; contbr. over 54 articles to profl. jours., 35 abstracts to symposia and congress. Recipient Biology award Third World Acad. Scis., 1987, Merit on Sci. Rsch. award Govt. Baja Calif., 1990. Mem. Ecol. Soc. Am., Sci. Rsch. Acad. Mex., Spanish Herpetological Assn. (founder), Ariz.-Nev. Acad. Sci. Achievements include co-establishment of world's largest Biosphere Reserve. Home: Calle de la Langosta # 110, La Paz 23000, Mexico Office: Biol Rsch Ctr, Apt Postal # 128, La Paz 23000, Mexico

ORTIZ, JOSEPH VINCENT, chemistry educator; b. Bethpage, N.Y., Apr. 26, 1956; s. José Vicente and Mary Davies (Bryant) O.; m. Karen Fagin, Dec. 16, 1979. BS, U. Fla., 1976, PhD, 1981. Rsch. fellow Harvard U., Cambridge, Mass., 1981-82; postdoctoral fellow Cornell U., Ithaca, N.Y., 1982-83; asst. prof. U. of N.Mex., Albuquerque, 1983-89, assoc. prof., 1989—; cons. Los Alamos (N.Mex.) Nat. Lab., 1986—. Contbr. 50 articles to profl. jours. Recipient Sr. Rsch. fellowship Am. Soc. Engring. Edn., 1986, 87, Rsch. fellowship Associated Western Univs., 1991, Rsch. grant NSF, 1985-91, 91-94, Grant-in-Aid, Petroleum Rsch. Fund, 1991-93, 84-86, Sandia Univ. Rsch. Program grant, 1983-85. Mem. Am. Chem. Soc., Am. Phys. Soc. Achievements include devel. and applications of propagator theory in quantum chemistry; elucidation of electronic structure of anions, polymers, organometallics and clusters. Office: U NMex Chemistry Dept Clark Hall Albuquerque NM 87131

ORTIZ, MARY THERESA, biomedical engineer, educator; b. N.Y.C., Mar. 25, 1957; d. Henry and Viola (Rega) O. BS, Wagner Coll., 1979; MS, Rutgers U., 1981, PhD, 1987. Emergency med. technician, N.Y. Adj. lectr. N.Y.C. Tech. Coll., Bklyn., 1981-89; teaching/rsch. asst. Rutgers U., New Brunswick, N.J., 1982-86; rsch. scientist N.Y. State Inst. for Basic Rsch., S.I., 1988-93; asst. prof. Kingsboro C.C., Bklyn., 1993—; adj. asst. prof. Coll. S.I., 1989—. Contbr. articles to sci. jours. Mem. youth adv. coun. N.Y.C. Youth Bd. Beame Adminstrn., 1970's; participant N.Y.C. Tech. Coll. Access for Women, Bklyn., 1980's, Rutgers U. Coll. Engrs. Open House, Piscataway, 1985-86. Grad. Prof. Opportunities Program fellow Rutgers U., 1979-82, Grad. Student Dissertation and Research Support grantee, 1986; Women's Rsch. and Devel. Fund grantee CUNY, 1988. Mem. IEEE, N.Y. Acad. Scis. (judge city and boro sci. fairs), Nat. Engrs. Honor Soc., Kappa Mu Epsilon, Beta Beta Beta. Democrat. Roman Catholic. Home: 31 Ruth Pl Staten Island NY 10305-2430 Office: Kingsboro C C Inst for Basic Rsch 2001 Oriental Blvd Brooklyn NY 11235

ORTIZ-ARDUAN, ALBERTO, nephrologist; b. Madrid, Spain, July 20, 1963; came to U.S., 1992; s. Arturo and Josefa (Arduan) O.; m. Neus Martin, Mar. 26, 1988; children: Natalia, Alejandro. MD, U. Autonoma, Madrid, 1987, PhD, 1992. Nephrology fellow Fundacion Jimenez Diaz, Madrid, 1988-91; rsch. fellow U. Pa., Phila., 1992--; vocal Spanish Med. Specialty Fellowships Coun., Madrid, 1988. Author: Nefrologia Jlropicis, 1992; contbr. articles to profl. jours. Recipient Extraordinary Licenciature award U. Autonoma, 1988; Consejo Superior Investigaciones Cientifcas grantee, 1986-87, Cajamadrid grantee, 1986, Fondo de Investigaciones Sanitarias dela Seguridad Soc. grantee, 1992, Fundacion Conchita Rabago grantee, 1992. Mem. Spanish Soc. Nephrology, Interant. Soc. Nephrology, N.Y. Acad. Scis. Roman Catholic. Achievements include the discovery of local secretion of lipid mediators and cytokines by intrinsic glomerular cells and infiltrating leucocytes that have a key role in the pathogenesis of several experimental glomerulopathies, ranging from minimal change nephrotic syndrome to proliferative glomerulonephritis; research in cellular and molecular mechanisms of interstitial inflammation and fibrosis in primary glomerular diseases and the relationship to progression of renal disease. Office: U Pa 422 Curie Blvd Philadelphia PA 19104

ORTIZ-LEDUC, WILLIAM, plant biologist, educator, researcher; b. Humacao, P.R., Oct. 21, 1951; s. William and Carmen Maria (Leduc) Ortiz. BS in Biology, Univ. P.R., 1973; MS, Rutgers U., 1975, PhD, 1979. Swiss Fonds Nat. Sci. Rsch. postdoctoral fellow U. Neuchâtel, Switzerland, 1979-81; postdoctoral U. Calif., Berkeley, 1981-85; asst. prof. U. Okla., Norman, 1985-91, assoc. prof., 1991—. Postdoctoral fellow NSF, 1980. Mem. Am. Soc. Plant Physiologists, Internat. Soc. Plant Molecular Biology, Phycol. Soc. Am., Sigma Xi. Office: U Okla Dept Botany Microbiology Norman OK 73019

ORTMEYER, THOMAS HOWARD, electrical engineering educator; b. Hampton, Iowa, Nov. 10, 1949; s. Howard William and Jane Elizabeth (Albright) O.; m. Ann Smits, July 27, 1974; children: Corrine Anna, Karl Thomas. PhD, Iowa State U., 1980. Gen. engr. Commonwealth Edison Co., Chgo., 1972-76; asst. prof. elec. engring Clarkson U., Potsdam, N.Y., 1979-85, assoc. prof., 1985-92, prof. elec. engring., 1992—; cons. Alcoa, Kaman Scis. Corp. Contbr. articles to profl. jours.; author chpt. to book: Power Systems Harmonics. Chmn. safety com. PTA, Potsdam, 1991-93. Recipient Prize Paper award Ind. Applications Soc., IEEE, 1985, Commonwealth Edison Co., 1975; NASA/Lewis Rsch. Ctr. summer faculty fellow, 1981, USAF summer faculty fellow, 1982. Mem. IEEE (sr., St. Lawrence subsect. chmn. 1988-92), Sigma Xi. Methodist. Home: 15 Lawrence Ave Potsdam NY 13676 Office: Clarkson Univ Elec & Computer Engring Dep Potsdam NY 13699-5720

ORTOLANO, RALPH J., engineering consultant. BS in Marine Engring., U.S. Mcht. Marine Acad., 1954; MBA, Santa Clara U., 1969. Registered profl. engr. Engring. watch officer USN, 1954-56; with marine divsn. Westinghouse, 1956-69; mgr. project engring. corp. cost recovery dept. Litton Ship Systems, Inc., 1969-72; consulting engr., sci. So. Calif. Edison Co., Rosemead, Calif., 1978-92, chief cons., 1993—; formed Turbine RESCUE, 1984. Contbr. articles to profl. jours.; patentee in field. Mem. ASME (past dir. ASME-SCAC power chpt., past chmn. steam turbine com., chmn. EEI prime movers com. task force on steam turbine crack prevention, past chmn. power divsn., mem. exec. com., George Westinghouse Gold medal 1991. *

ORTON, GEORGE FREDERICK, aerospace engineer; b. Flushing, N.Y., Aug. 8, 1941; s. Harry and Evelyn (Brostrom) O.; m. Susan K., Dec. 21, 1962; children: Karen, Kevin, Kristen. BS in Aeron. Engring., U. Md., 1964; MS in Engring. Mechanics, St. Loius U., 1971. Engr. propulsion McDonnell Douglas, St. Louis, 1964-73, sr. engr. propulsion, 1973-77, unit chief propulsion, 1977-81, sect. chief propulsion, 1981-86, chief br. nat. aerospace plane, 1986-90, staff dir. nat. aerospace plane, 1990-92, dir. space programs, 1992-93, mgr. programs nat. aerospace plane, 1993—. Contbr. articles to profl. jours. Advisor Exlporer Post 9005, St. Louis, 1980-87. Fellow AIAA (assoc., liquid propulsion tech. com., 1980-84, 91—, Best Paper award 1987) St. Louis Head Injury Assn. Methodist. Achievements include patent for propellant acquisition device for zero-g engine starts, patent for propellant resupply system, NASA technology cash award for work on shuttle OMS tankage. Office: McDonnell Douglas Corp Mailcode 1064150 P O Box 516 Saint Louis MO 63166

ORVIN, GEORGE HENRY, psychiatrist; b. Columbia, S.C., Aug. 6, 1922; s. Jesse Wright and Ruth Veril (Walton) O.; m. Rosalie Greer Salvo, Sept. 16, 1944; children: Candace, Jay Scott, Debra Anne, Nancy Lee. BS, The Citadel, 1943; MD, Med. U. S.C., Charleston, 1946. Diplomate Am. Bd. Psychiatry. Pvt. practice Charleston, S.C., 1948-57; resident psychiatry Med. U. S.C., 1957-60; clin. asst. U. London, 1960-61; instr. Med. U.S.C., 1961; chief adolescent psychiatry Med. U. S.C., 1967-89, pres. faculty senate, 1977-79, prof. psychiatry, 1977-89; founder, chmn. New Hope, Inc., Charleston, 1984—. Sr. editor Annals Adolescent Psychiatry, 1985; contbr. chpts. to books, articles to profl. jours. Vice-chmn. S.C. Com. Alcohol/Drug Abuse, 1973-89; mem. Gov.'s Cabinet for Children, 1984, Gov.'s Task Force Adolescent Pregnancies, 1985. Fellow Am. Psychiatric Assn. (life), Am. Soc. Adolescent Psychiatry (life), Royal Soc. Medicine; mem. St. Andrews Soc., Citadel Brigadier Club (founder, pres. 1948-53). Episcopalian. Achievements include development of new treatment modalities for adolescents. Home: 126 Rutledge Ave Charleston SC 29401-1333 Office: New Hope Inc 225 Midland Pky Summerville SC 29485-8104

OSBERG, TIMOTHY M., psychologist, educator, researcher, clinician; b. Buffalo, Aug. 11, 1955; s. John Carlton and Adeline Rose (Weichsel) O.; m. Debra A. Morreale, July 14, 1990; children: John Peter, Erika Evelyn. BA, SUNY, Buffalo, 1977, MA, 1980, PhD, 1982. Lic. psychologist, N.Y. Intern VA Med. Ctr., Buffalo, 1981-82; from asst. prof. to prof. Niagara (N.Y.) U., 1982—; pvt. practice Niagara Falls, N.Y., 1985—; psychologist Optifast Weight Loss Program, Niagara Falls, 1989-92; editorial bd. Jour. Personality and Social Psychology, 1988-92, Teaching of Psychology, 1991—, Jour. Correctional Edn., 1993—; instr. Attica Correctional Facility, 1980—; presenter in field. Contbr. articles to profl. jours. Vol. group leader prerelease program Attica (N.Y.) Correctional Facility, 1984-90, exec. com. Psychol. Assn. Western N.Y., Buffalo, 1982-87. Recipient Feldman-Cohen Meml. award SUNY, Buffalo, 1977. Mem. Am. Psychol. Assn., Eastern Psychol. Assn., Soc. for Personality Assessment, Assn. Advancement Behavior Therapy, Correctional Edn. Assn., Phi Beta Kappa. Democrat. Roman Catholic. Avocations: spectator sports, golf, hockey, football, tennis. Home: 2652 David Dr Niagara Falls NY 14304-4619 Office: Niagara U Dept Psychology Niagara University NY 14109

OSBORN, ANN GEORGE, retired chemist; b. Nowata, Okla., Aug. 1, 1933; d. David Thomas and Alice Audrey (Giles) George; m. Charles Wesley

Osborn, Nov. 8, 1958 (dec. Dec. 1977); 1 child, Charles David. BA in Chemistry, Okla. Coll. Women, 1955. Rsch. chemist thermodynamics rsch. lab. Bartlesville (Okla.) Energy Rsch. Ctr., U.S. Dept. Energy, 1957—; ret., 1983. Contbr. articles to profl. jours. Mem. AAAS (emeritus), Am. Chem. Soc. Republican. Mem. Christian Ch. (Disciples of Christ). Home: 647 S Pecan Nowata OK 74048

OSBORN, JAMES HENSHAW, operations research analyst; b. Carbondale, Pa., May 22, 1941; s. Daniel Cargill Jr. and Marguerite Isabel (Henshaw) O.; m. Mary Inez Tompkins, Nov. 26, 1966; children: Kevin Daniel, James Clifton. BA, Northeastern U., 1964; MA, U. Rochester, 1966; PhD, U. Wis., 1972. Computer programmer Sylvania Applied Rsch. Lab., Waltham, Mass., 1962-64, sr. engr., 1966-67; vis. asst. prof. Wright State U., Dayton, Ohio, 1972-73; systems analyst Computer Scis. Corp., Moorestown, N.J., 1973-75; ops. rsch. analyst Hdqs. Tactical Air Command, Langley AFB, Va., 1975-90, dir. ops. analysis, 1990-91; chief scientific analyst Air Combat Command, Joint Studies Group, Langley AFB, Va., 1991-92, analyst ops. rsch., 1992—; lectr. George Washington U., Hampton, Va., 1977-79, Golden Gate U., Langley AFB, Va., 1982-87; mem. Air Force Chief Scientists Group, Washington, 1990-92. Treas. York River Community Orch., Yorktown, Va., 1989-90. Mem. Math. Assn. of Am., Ops. Rsch. Soc. of Am. Episcopalian. Home: 404 Old Dominion Rd Yorktown VA 23692-4733 Office: Joint Studies Group HQ ACC/XP-JSG Langley AFB VA 23665-5520

OSBORNE, DANIEL LLOYD, electronics engineer; b. San Mateo, Calif., July 19, 1946; s. B. Lloyd and Sybil C. Osborne; m. Rita M. Stewart, July 26, 1986; 1 child, Lars S. Student, Coll. San Mateo, 1965-67, U. Alaska, 1967-71. Engr. Geophys. Inst., U. Alaska, Fairbanks, 1970-85, project engr., 1985—; bd. dirs., treas. Golden Valley Electric Assn., Fairbanks. Producer specialty videos. Recipient Antartic Svc. medal NSF, 1978. Mem. Am. Alpine Club, Alaska Alpine Club (pres.), Sierra Club (life, local pres. 1970-71). Achievements include first to ascend new routes on Alaska mountains; first ultra-low light level imaging; first true color and true speed moving images of the Northern Lights; research in peripatetic engineering from the South Pole to Northern Pole, Poker Flat research range, major sounding rockets. Office: U Alaska Geophys Inst Fairbanks AK 99775-0800

OSBORNE, MASON SCOTT, mathematician, educator; b. Rupert, Idaho, Sept. 2, 1946; s. Mason Scott and Olive May (Williamson) O. BS, U. Wash., 1968; PhD, Yale U., 1972. Vis. asst. prof. U. Mich., Ann Arbor, 1972-73; Dickson instr. U. Chgo., 1973-75; prof. math. U. Wash., Seattle, 1975—. Co-author: Theory of Eisenstein Systems, 1981; contbr. articles to sci. jours. Mem. Am. Math. Soc., Math. Assn. Am., Am. Phys. Soc. Office: Univ Wash Dept Math GN 50 Seattle WA 98195

OSBORNE, MICHAEL PIERS, surgeon, health facility administrator; b. Sutton, Surrey, Eng., Jan. 6, 1946; came to U.S., 1980; s. Arthur Frederick and Leonora Kate Hope (Miller) O.; m. Carolyn Patricia Malkinson, June 22, 1974; children: James, Simon, Andrew, Emma. MB, BS, London U., 1970, MS, 1980. Diplomate Royal Coll. Surgeons of Eng., Am. Coll. Surgeons. Intern Charing Cross Group of Hosp., England, 1970-71; resident Brompton Hosp., London, St. James Hosp., London, West Herts Hosp., Eng.; hon. lectr. surgery Royal Marsden Hosp., London, 1977-81; fellow in surg. oncology Meml. Sloan-Kettering Cancer Ctr., N.Y.C., 1980-81, fellow, attending surgeon, 1981-91, head breast cancer rsch. lab., 1984-91; chief breast surgery N.Y. Hosp.-Cornell Med. Ctr., N.Y.C., 1991—; prof. surgery Cornell U. Med. Coll., N.Y.C., 1991—; dir., CEO Strang Cancer Prevention Ctr., N.Y.C., 1991—; dir. Strang- Cornell Breast Ctr., N.Y.C., 1991—; cons. Meml. Sloan-Kettering Cancer Ctr., N.Y.C., 1991—; vis. physician Rockefeller U. Hosp., N.Y.C., 1991—; sci. adv. com. Am.-Italian Found., N.Y.C., 1991—; bd. trustees Nat. Consortium of Breast Ctrs., 1992—. Contbr. 12 chpts. to textbooks; contbr. over 100 articles to profl. jours. Recipient Gov.'s Clin. Gold medal Charing Cross Hosp. Med. Sch., 1970, Prize in Surgery, Charing Cross Hosp. Med. Sch., 1970, Raven prize British Assn. Surg. Oncologist, 1978; Wellcome Trust fellow, 1975. Mem. N.Y. Acad. Scis., N.Y. Surg. Soc., Brit. Assn. Surg. Oncology, Soc. Surg. Oncology, Royal Soc. Medicine, Soc. For The Study Of Breast Disease (bd. trustees 1993—). Office: Strang Cancer Prevention 428 E 72d St New York NY 10021

OSBORNE, ROBIN WILLIAM, civil engineer, educator; b. Roseau, Dominica, W.I., Apr. 9, 1944; s. John William and Josephine Elisabeth (Roberts) O.; m. Annette Phillips, July 29, 1972; children: Roanne, Robin Jr., Peter. BSCE with honors, U. W.I., St. Augustine, Trinidad, 1972, PhD in Civil Engring., 1990. Civil engr., head materials tech. and testing div. Caribbean Indsl. Rsch., St. Augustine, 1972-79; lectr. dept. civil engring. U. W.I., St. Augustine, 1979—; attached worker, UNIDO fellow Bldg. Rsch. Establishment, Garston, Watford, Hertfordshire, U.K., 1974-75. Contbr. articles to tech. jours. Chmn. bd. dirs. Emmaus Bible Schs. Trinidad and Tobago, Port of Spain, 1991—; vice-chmn. Intersch./Intervarsity Christian Fellowship of Trinidad and Tobago, 1992—. Mem. ASTM, Am. Concrete Inst. Mem. Christian Brethren Ch. Office: Dept Civil Engring, U West Indies, Saint Augustine Trinidad and Tobago

OSBORNE, WILLIAM GALLOWAY, JR., chemical engineer; b. Longview, Tex., Dec. 30, 1940; s. William Galloway and Edna Wesson (Webb) O.; m. Susan Anne Cline, Jan. 20, 1965; children: Scott W., Anne E., William G. III. BSChemE, Tex. A&M, 1962; MSChemE, Okla. State U., 1964, PhD in Chem. Engring., 1967. Rsch. engr. E. I. DuPont, Parlin, N.J., 1966-69; chem. engr. Kaiser Aluminum & Chem., Oakland, Calif., 1969-71, environ. engr., 1972-75, purchasing mgr., 1976-81, bus. mgr., 1981-87; gen. mgr. LaRoche Chems., Baton Rouge, La., 1988-91, v.p., 1991—. Mem. AICE, Am. Chem. Soc., Sigma Xi. Presbyterian. Home: 19622 Creekround Ave Baton Rouge LA 70817 Office: LaRoche Chems Inc 1200 Airline Hwy Baton Rouge LA 70805

OSBOURN, GORDON CECIL, materials scientist; b. Kansas City, Mo., Aug. 13, 1954; 2 children. BS, U. Mo., 1974, MS, 1975; PhD in Physics, Calif. Inst. Tech., 1979. Tech. staff Sandia Nat. Labs., 1979-83, divsn. supr., 1983—; rsch. scientist, dept. elec. engring. Columbia U., NYC. Recipient E.O. Lawrence award Dept. Energy, 1985; Internat. prize For New Materials Am. Physical Soc., 1993. Fellow Am. Physical Soc. Office: Columbia University Dept of Electrical Engineering 116th St & Broadway New York NY 10027*

OSBURN, CARLTON MORRIS, electrical and computer engineering educator; b. Lansing, Mich., Nov. 16, 1944; s. Carlton M. and Ramona Alice (Taylor) O.; m. Mary Anna Sampsell, Aug. 25, 1965. BS, Purdue U., 1966, MS, 1967, PhD, 1970. Mem. rsch. staff IBM, Yorktown Heights, N.Y., 1970-73, mgr. silicon process studies, 1973-76, mgr. fabrication tech., 1976-83; dir. advanced semicondr. tech. Microelectronics Ctr. N.C., Research Triangle Park, N.C., 1983-92, dir. device fabrication tech., 1992-93; prof. elec. and computer engring. N.C. State U., Raleigh, 1983—; mem. univ. adv. bd. Semicondr. Rsch. Corp., Research Triangle Park, 1985-90; chmn. 1st Internat. Workshop on Measurement of Doping Profiles in Semicondrs., Research Triangle Park, 1991. Contbg. author: Rapid Thermal Processing, 1992; editor 8 symposium procs.; contbr. over 90 articles to tech. jours. Mem. Croton-on-Hudson (N.Y.) Planning Bd., 1978-80. Recipient Maurice Simpson award Inst. Environ. Scis., 1988. Mem. IEEE (sr.), Electrochem. Soc. (chmn. electronics div. 1991-93, Callinan award 1975, electronics div. award 1991), N.C. Roadrunners Club (exec. bd. 1987, 91), Sigma Xi. Achievements include numerous patents in field of materials and device processing. Home: 103 Homestead Dr Cary NC 27511 Office: NC State U Box 7911 Raleigh NC 27695-7911

OSCHERWITZ, STEVEN LEE, internist, infectious disease physician; b. Ft. Worth, June 13, 1961. BA in Biochemistry, U. Tex., Austin, 1982; MD, U. Tex., Dallas, 1986. Diplomate Am. Bd. Infectious Dieases, Am. Bd. Intrnal Medicine. Resident in internal medicine U. Tex. Health Sci. Ctr., San Antonio, 1986-89; chief resident U. Tex. Health Sci. Cgtr., San Antonio, 1989-90, fellow in infectious diseases, 1990-92; pvt. practice Tempe, Ariz., 1992—; mem. teaching faculty internal medicine, infectious diseases U. Tex. Health Sci. Ctr., San Antonio, 1989-92; parasitology del. People to People Program, People's Republic of China, 1991; physician for Maya Lowlands Archaeol. Project, Guatemala, 1990. Contbr. articles to profl. publs. Mem. ACP, AMA, Infectious Disease Soc. Am., Am. Soc. Tropical Medicine and Hygiene, Internat. Soc. Travel Medicine, Phi Beta Kappa. Achievements include research in antifungal agents in humans, fungal meningitis. Office: Ste 22 2501 E Southern Ave Tempe AZ 85282

OSGOOD, RICHARD M., JR., applied physics and electrical engineering educator, research administrator; b. Kansas City, Mo., Dec. 28, 1943; s. Richard Magee and Mary Neff (Russell) O.; m. Alice Rose Dyson, June 25, 1966; children—Richard Magee, III, Nathaniel David, Jennifer Anne. B.S. in Engring., U.S. Mil. Acad., 1965; M.S. in Physics, Ohio State U., 1968; Ph.D., MIT, 1973. Rsch. assoc. dept. physics MIT, Cambridge, 1969-72, mem. rsch. staff Lincoln Lab., 1973-80, project leader Lincoln Lab., 1980-81; assoc. prof. applied physics and elec. engring. Columbia U., N.Y.C., 1981-82, prof., 1982-91, Higgins prof., 1989—, dir., 1984-90; bd. dirs. Microelectronics and Scis. Labs., Army Sci. and Tech., Basic Energy Scis. Adv. Com. Materials Rsch. Coun.-DARPA; cons. Los Alamos Nat. Lab. Editor: Laser Diagnostics and Photochemical Processing of Semiconductor Devices, 1983; assoc. editor: Applied Physics; contbr. articles to profl. jours.; patentee in field. Served to capt. USAF, 1965-69. Recipient Samuel Burka award USAF Avionics Lab., 1968, Leos Travelling Lectr. award, 1986-87, Disting. Travelling Lectr. APS; John Simon Guggenheim fellowship, 1989. Fellow IEEE, Optical Soc. Am. (R.W. Wood award, 1991); mem. Am. Chem. Soc., Materials Rsch. Soc. (councillor 1983-86), Optical Device Assn. (Japanese hon. lectr. 1990), Am. Phys. Soc. (travelling lectureship 1992). Home: 345 Quaker Rd Chappaqua NY 10514-2615 Office: Columbia U Microelectronics Scis Labs New York NY 10027

O'SHEA, JOHN ANTHONY, consulting engineer; b. Caherciveen, Ireland, July 23, 1939; s. Michael K. and Bridget (Courtney) O'S.; children—Bjorn Michael Karl, Charlotte Courtney. B.Engring., Univ. Coll. Cork, 1966. Telegrapher, Western Union Internat., Valentia, Ireland, 1958-62; elec. engr. Brown, Boveri, A.C., Baden, Switzerland, 1966-67, English Electric Co. Ltd., Stafford, 1967-68; chief elec. engr. Imperial Continental Gas Assn., Brussels, Belgium, 1968-72; mng. dir. Integrated Energy Systems, Ltd., Preston, U.K., 1972—; sr. ptnr. Paterson, Beal & O'Shea, Dublin, Ireland, 1975-90; chmn. Euro Container Shipping, Dublin; bd. dirs. D&F Fellows Ltd., Welte & Lasser Ltd., Dublin, IES, Harare, Zirbabwi, IES Internat. London. Author: The Efficient Use of Energy, 1977, Energy Centers, 1978, Energy Use in the Eighties, 1978. Contbr. articles to profl. jours. Mem. IEEE, Instn. Elec. Engrs. London. Mem. Fianna Fail Party. Club: Officers Mess. Home: 52 Glenageary Woods, Dublin Ireland Office: Integrated Energy Systems Ltd, 11 Lune St, Preston PR1 2NL, England

OSHIRO, YASUO, medicinal chemist; b. Oita-prefecture, Japan, Nov. 23, 1944; s. Tetsuo and Yoshiko (Tamaki) O.; m. Itsuko Morimitsu, Dec. 16, 1974 (dec. Mar. 1983); children: Takafumi Oshiro, Naoto Oshiro, Erina Oshiro. BS, Toyama (Japan) U., 1967, MS, 1970; PhD, Osaka (Japan) U., 1974. Fellow dept. med. chemistry Jonie Miller Health Ctr./U. Fla., Gainesville, 1979-81; chief scientist Otsuka Pharm. Co. Ltd., Tokushima, Japan, 1982—. Contbr. articles to profl. jours. Achievements include patents for new anti-psycotic agent (Japan and U.S.), novel cerebroprotective agents and agents for treatments of disturbance of consciousness (Japan), anti-histamine agents (Japan, U.S.), beta blocker agents (Japan). Home: Yoshino-Honchou 6 chome, 61-banchi, Tokushima 770, Japan Office: Otsuka Pharm Co Ltd, 463-10 Kawauchi Kagasuno, Tokushima 771-01, Japan

OSHIYOYE, ADEKUNLE EMMANUEL, physician, realtor; b. Lagos, Nigeria, Jan. 5, 1951; s. Alfred and Grace (Apena) O.; m. m. Olwwatoyin, Dec. 28, 1991. BS, N.Y.S.U., 1974; MD, American U., Monsterrat, Wis., 1979. Real Estate, Physician lic. Staff physician South Chgo. Hosp., Chgo., 1980-81, Cook County Hosp., Chgo., 1981-85, Mercy Hosp., Chgo., 1985-89; health physician City of Chgo. Dept. of Health, 1989—; cons. physician Dept. of Health, Chgo., 1989—. Med. Editor: African Connections, 1990—, Newsbreed Mag., 1990—. Organizer Harold Washington Coalition, Chgo., 1983—, operation push, Chgo. Mem. Am. Med. Assn., Cook County Physician Assn., Knights of Rose Croix #28, Eureka Lodge #64, Ancient Arabic Order of Mystic Shrine, Alpha Phi Alpha, Nigerian American Forum. Democrat. Apostolic. Avocations: basketball, skiing, ping pong, fishing, golf. Home: 4800 S Chicago Beach Dr Chicago IL 60615-2009

OSHTRAKH, MICHAEL IOSIFOVICH, physicist, biophysicist; b. Sverdlovsk, USSR, Sept. 18, 1956; s. Iosif Z. and Maya Z. (Yantovsky) O. MS engring. physics, Ural Poly. Inst., Sverdlovsk, 1979, PhD molecular physics and biophysics, 1990. Engr. Div. Applied Biophysics, Physico Tech. Dept. Ural Poly. Inst., Sverdlovsk, 1979-87; jr. rsch. worker Physico-Tech. Dept., Ural Poly. Inst., Sverdlovsk, 1987-91; sr. rsch. worker Ural Poly. Inst., 1991—; sci. cons. Sci.-Prodn. Bus. Uralcomplex, Sverdlovsk, 1990. Contbr. numerous articles to profl. jours. Organizer Sverdlovsk Soc. Jewish Culture, 1988-89, bd. dirs., 1989-90, v.p., 1990—; mem. organizing com. and del. I and II III Congresses of Jewish Orgns. and Communities, Moscow, 1989, 91, Odessa, 1992, I Congress of Fedn. Jewish Orgns. and Communities of Russia, N. Novgorod, 1992; organizer Sverdlovsk Assn. for Jewish Studies, 1991, pres., 1991—. Mem. Sverdlovsk Phys. Soc., Fedn. Jewish Orgns. and Communities of Russia (mem. presidium of coun., del. I Congress). Achievements include research in biophysical and biomedical applications of Mössbauer spectroscopy, the Mössbauer study of normal hemoglobins with different molecular structure and hemoglobins from patients with leukaemias and erythremia, the Mössbauer and position annihilation studies of hemoglobin radiolysis and iron-containing pharmaceutical compounds. Home: Zavodskaya str 32/3 Apt 45, Sverdlovsk Russia 620131 Office: Ural State Tech U-UPI, Div Applied Biophysics, Sverdlovsk Russia 620002

OSIPOV, YURII, mathematician, mechanical scientist, educator; b. July 7, 1937; 1 child, Nataliya. PhD in Physics and Maths., Ural State U., Sverdlovsk, 1971. Pres. Russia Acad. of Scis., Moscow; dir. Inst. of Maths. and Mechs. Ural Br. Acad. of Scis. of U.S.S.R., Sverdlovsk; head chair of optimal control Moscow State U.; assoc. prof. Ural State U., 1965-70; researcher Inst. Math. and Mechanics, Acad. Scis. USSR, 1970-72, head sect. differential equations, 1972—, dir., 86—; prof. Moscow U., 1990—. Mem. editorial bd. Jour. Doklady Acad. Scis., Izvegtiya Acad. Sci., Jour. Computational Maths. and Math. Physics, Automation and Telemechanics; contbr. article to profl. jours. Recipient Lenin prize Govt. of U.S.S.R., 1976. Mem. Am. Math. Soc., Acad. of Scis. of U.S.S.R. (corr. 1984—, academician 1987—). Home: Leninskie Gory, MGU korp L kv 3, 117234 Moscow USSR also: ul Marshala Zhukova 11, 62, 620077 Sverdlovsk Russia Office: Moscow State U, Faculty Computational Math/Mechs, 119899 Moscow Russia also: Inst Math and Mechanics, ul Kovalevskoi 16, 620066 Sverdlovsk USSR*

OSIPYAN, YURI ANDREYEVICH, physicist; b. Feb. 15, 1931. Student, Moscow Steel Inst. Sr. rschr. Ctrl. Inst. Metallophysics, 1955-62; dep. dir. inst. crystallography USSR (now Russian) Acad. Scis., 1962-63, with staff inst. solid state physics, 1963—, prof., 1970—, dep. dir., then dir., 1973—; dean, prof., head of faculty Moscow Inst. Physics and Tech., mem. pres. coun., 1990-91; pres. Internat. Union Theoretical and Applied Physics, 1990—. Named Hero of Socialist Labour, 1986, USSR People's Dep., 1989-91; recipient P. N. Lebedev Gold medal 1984, A. P. Karpinsky prize City of Hamburg, 1991. Office: Russian Acad Scis-Inst Sld St Physics, Chernogolovka, 142432 Moscow Russia*

OSIYOYE, ADEKUNLE, obstetrician/gynecologist, educator; b. Lagos, Nigeria, Jan. 5, 1951; came to U.S., 1972; s. Alfred and Grace (Apena) Oshiyoye; m. Toyin Osinowo Oshiyoye, Dec. 28, 1991; 1 child, Adekunle Jr. Student, Howard U., 1972-73; BS, U. State of N.Y., 1974; postgrad., Columbia U., 1974-78; MD, Am. U., Montserrat, West Indies, 1979. Intern South Chgo. Community Hosp., 1980-81; intern dept. obstetrics-gynecology Cook County Hosp., Chgo., 1981-82, resident physician, 1982-84, chief resident physician dept. obstetrics-gynecology, 1984-85; assoc. prof. dept. obstetrics-gynecology Chgo. Osteo. Coll. Medicine, 1986—; health physician, cons. physician City of Chgo. Dept. Health, 1989—; attending physician St. Bernard Hosp., Chgo., 1985—, Hyde Park Hosp., Chgo., 1986—, Mercy Hosp., Chgo., 1987—, Roseland Hosp., Chgo., 1985—, Columbus Hosp., Chgo., 1985—, Jackson Park Hosp., Chgo., 1985—; coord. emergency rm. Cook County Hosp., 1983-85. Med. editor African Connections, 1990—; med. columnist Newsbreed Mag., 1990—; founding mem. Ob-Gyn Video Jour. Am. Organizer Harold Washington Coalition, Chgo., 1983-87; operation mem. Operation P.U.S.H., Chgo., 1987—; active Chgo. Urban League, 1989—, Cook County Dem. Party, 1988—; mem. Mayor's Commn. on Human Rels., Chgo., 1990—, State of Ill. Inaugaural Com., 1991. Shell scholar, 1965-69; recipient Fed. Govt. scholarship award, 1972, Howard Univ. scholarship award, 1973, Fed. Govt. Nigeria grad. med. scholarship award, 1975-79, Cerebral Palsy rsch. award, 1977, Ob-gyn. Video Jour. award, 1989, Role Model award Chgo. Police Dept., 1991, 92, Chgo. Bd. Edn., 1991, Chgo. 100 Black Men, 1991, Gov.'s Recognition award, 1992; named one of Best Dressed Men in Chgo., Chgo. Defender, 1990, 91. Fellow Am. Coll. Internat. Physicians, Am. Coll. Obstetricians & Gynecologists; mem. AMA (physician recognition award 1986), Am. Coll. Glegal Medicine (edn. com.), Am. Soc. Law Medicine, Am. Pub. Heart Assn., Nat. Med. Assn., Ill. Med. Soc., Chgo. Med. Assn., Chgo. Gynecol. Soc., Cook County Physician Assn., Nigerian Am. Forum (chmn. health com., chmn. election com.), Cook County Hosp. Surg. Alumni Assn., Howard U. Alumni Assn. (regent, chmn. scholarship com. Chgo. chpt.), Eureka Lodge (investigating com.), Masons, Shriners, Order of Eastern Star, Alpha Phi Alpha (life mem., mem. Labor Day com., dir. ednl. programs Xi Lambda chpt. 1990—, co-chmn. courtesy Black & Gold com. 1989, 90, Recognition award 1991), Pan Hellenic Action Coun. (chmn. pub. rels. com.), Ill. Maternal and Child Health Coalition, Beta Kappa Chi. Apostolic. Avocations: ping pong, fishing, golf, basketball, swimming. Home: PO Box 49477 Chicago IL 60649-0477 Office: Dept Health 37 W 47th St Chicago IL 60609-4657

OSOWIEC, DARLENE ANN, psychologist, consultant, educator; b. Chgo., Feb. 16, 1951; d. Stephen Raymond and Estelle Marie Osowiec; m. Barry A. Leska. BS, Loyola U., Chgo., 1973; MA with honors, Roosevelt U., 1980; postgrad. in psychology, Saybrook Inst., San Francisco, 1985-88; PhD in Clin. Psychology, Calif. Inst. Integral Studies, 1992. Mental health therapist Ridgeway Hosp., Chgo., 1978; mem. faculty psychology dept. Coll. Lake County, Grayslake, Ill., 1981; counselor, supr. MA-level interns, chmn. pub. rels. com. Integral Counseling Ctr., San Francisco, 1983-84; clin. psychology intern Chgo.-Read Mental Health Ctr. Ill. Dept. Mental Health, 1985-86; mem. faculty dept. psychology Moraine Valley C.C., Palos Hills, Ill., 1988-89; lectr. psychology Daley Coll., Chgo., 1988-90; cons. Gordon & Assocs., Oak Lawn, Ill., 1989-93; child, adolescent and family therapist Orland Twp. Youth Svcs., Orland Park, Ill., 1993—; lectr. psychology Daley Coll., Chgo., 1988-90; mem. faculty dept. psychology Moraine Valley C.C., Palos Hills, Ill., 1988-89. Ill. State scholar, 1969-73; Calif. Inst. Integral Studies scholar, 1983. Mem. AACD, APA, Am. Women in Psychology, Am. Statis. Assn., Ill. Psychol. Assn., Calif. Psychol. Assn., Am. Assn. Marriage and Family Counselors, Am. Soc. Clin. Hypnosis, Internat. Platform Assn., Chgo. Soc. Clin. Hypnosis, NOW (chairperson legal advocate corps, Chgo., 1974-76). Avocations: playing piano, gardening, reading, backpacking, writing. Home: 6608 S Whipple St Chicago IL 60629-2916

OSSIPYAN YURIY, ANDREW, physicist, metallurgist, educator; b. Moscow, Feb. 15, 1931; s. Andrew Jacob Ossipyan and Bella Joseph Ossipyan Pugatch; m. Alla Nicholas Matantseva, Mar. 20, 1957 (div. 1964); 1 child, Olga; m. Ludmilla Nicholas Golenko, Nov. 19, 1965; children: Alexander, Sergey. BS in Physics, Steel and Alloy U., Moscow, 1955; MS in Math., Moscow State U., 1957, PhD in Physics, 1960. Jr. rschr. Inst. Metallphysics, Moscow, 1955-58, sr. rschr., 1958-62; prof. Inst. Phys. Problems, Moscow, 1962—; dep. dir. Inst. Cristallography, Moscow, 1962-63; dep. dir. Inst. Solid State Physics, Moscow, Chernogolovka, 1963-71, dir., 1971—; chmn. Solid State Physics Sci. Coun., Moscow, 1982—; mem. Internat. COSPAR Com.; lectr. in field. Author: Dislocations and Electronic Properties of Semiconductor, 1982; editor-in-chief Physics, Chemistry, and Mechanics Surfaces, 1990, Quantum, 1990, Physica C; contbr. articles to profl. jours. M.P. Parlament USSR, Moscow, 1988-91; mem. Presdl. Coun. USSR, Moscow, 1989-91. Recipient 7 Highest State medals, USSR, 1970-89, Latin Am. Order Liberty and Unity, 1990, Internat. Karpinsky prize Internat. Karp Found., 1991. Mem. Nat. Acad. Engring. USA, Russian Acad. Scis. (mem. divsn. gen. physics and astronomy bd., chmn. sci. coun. condensed matter 1982—, v.p. 1988-92, mem. presidium 1991—, P.N. Lebedev golden medal 1984), Hungarian Acad. Scis., Czechoslovakian Acad. Scis., Bulgarian Acad. Scis., Indian Metall. Soc., Internat. Acad. Astronautics, Internat. Union Pure and Applied Physics (pres. 1990—), Internat. Coun. Sci. Unions (mem. gen. com.). Achievements include 45 patents in physics and technology; discovery of photoplastic effect; research in the structure and physical property of cristalline materials with dislocations, in high temperature superconductivity, in cuprates, and in fullerenes at high pressure. Office: Institute of Solid State Physics, 64-A Leninskii prospect, 117290 Moscow Russia

OSTAPENKO, ALEXIS, civil engineer, educator; b. Ukraine, Oct. 1, 1923; came to U.S. 1951; s. Peter and Natalia O.; married; 3 children. Dipl.Ing., Tech. U. Munich, Germany, 1951; ScD, MIT, 1957. Structural engr. Fay, Spofford & Thorndike, Boston, 1952, Thomas Worcester, Boston, 1952-54, various firms, 1955-57; from asst. prof. to prof. civil engring. Lehigh U., Bethlehem, Pa., 1957—. Rsch. grantee USN, PennDOT, USCG, U.S. Dept. of Transp., U.S. Dept Interior, U.K. Dept. Energy, Exxon, Chevron, many others, 1958—. Mem. ASCE, Sigma Xi. Office: Lehigh Univ Bldg 13 Fritz Engring Lab Bethlehem PA 18015

OSTBY, FREDERICK PAUL, JR., meteorologist, government official; b. New Haven, Jan. 20, 1930; s. Frederick Paul and Edna Maria (Kruckenberg) O.; m. Joanne Bernice Sorvig, Jan. 1, 1955; children: Paul, Neil, Karen, Lynn. B.S. in Meteorology, NYU, 1951, M.S., 1960. Meteorologist TWA, N.Y.C., 1954-55, Kansas City, Mo., 1955-56; Meteorologist N.E. Weather Service, Lexington, Mass., 1955, Travelers Weather Service, Hartford, Conn., 1956-60; research scientist Travelers Research Center, Hartford, 1960-70; meteorologist Nat. Weather Service, Silver Spring, Md., 1970-72; dep. dir. Nat. Severe Storms Forecast Center, Dept. Commerce, Kansas City, Mo., 1972-80; dir. Nat. Severe Storms Forecast Center, 1980—. Contbr. papers to profl. lit. Served with USAF, 1951-53. Fellow Am. Meteorol. Soc. (council 1977-80, 84-87). Home: 9720 Reeder St Overland Park KS 66214-2577 Office: Nat Severe Storms Forecast Center 601 E 12th St Kansas City MO 64106-2808*

OSTENSO, NED ALLEN, oceanographer, government official; b. Fargo, N.D., June 22, 1930; s. Nels Andres and Estella (Temple) O.; m. Grace Elaine Laudon, June 29, 1963. BS, U. Wis., 1952, MS, 1953, PhD, 1962; postgrad., Johns Hopkins U., 1975. Scientist Arctic Inst. N.Am., Washington, 1956-66; asst. prof. geology and geophysics U. Wis., Madison, 1962-66; dir. marine geol. and geophys. programs Office Naval Rsch., Washington, 1966-69, sr. oceanographer, 1970-77; asst. Presdl. sci. advisor White House, Washington, 1969-70; dir. nat. sea grant coll. program NOAA, Washington, 1977-83, dir. Office Oceanic Rsch., 1983-89, chief scientist, 1989-90, asst. adminstr. for rsch., 1990—; founder Joy/Mac Petroleum, Madison, 1959-65; fellow Fed. Execs. Inst., Charlottesville, Va., 1974, Am. Polit. Sci. Assn., Washington, 1975-76. Cotnbr. over 60 articles on polar regions, oceanography and geophysics to sci. jours., chpts. to books. 1st lt. Signal Corps, U.S. Army, 1953-56. Recipient Antarctic Svc. medal Dept. of Def., 1958, Meritorius Svc. citation NAS, 1959, Superior Accomplishment award USN, 1968; Mt. Ostenso named in his honor, 1963, Ostenso Seamount (Arctic Ocean) named in his honor, 1978. Fellow Geol. Soc. Am., Arctic Inst. N.Am., Marine Tech. Soc., Explorers Club; mem. Acad. Polit. Sci., Am. Geophys. Union, UN Assn. U.S.A., Cosmos Club. Home: 2871 Audubon Ter NW Washington DC 20008-2309 Office: Dept of Commerce Oceananic & Atmospheric Research 1335 East-West Hwy Silver Spring MD 20910

OSTENSON, JEROME EDWARD, physicist; b. Elbow Lake, Minn., Mar. 10, 1938; s. Tilmer Orville and Edith Evelyn (Paulson) O.; m. Janet Sue Borts, June 28, 1986. BA cum laude, Concordia Coll., Moorhead, Minn., 1962; MS, Iowa State U., 1969. Jr. physicist Iowa State U., Ames Lab., 1962-69, asst. physicist I, 1969-71, asst. physicist II, 1971-78, assoc. physicist, 1978-92, physicist, 1992—. Contbr. over 55 articles to profl. jours. With U.S. Army, 1958-60. Mem. Am. Phys. Soc., Sigma Xi. Achievements include patentee in field. Office: Iowa State U/Ames Lab US Dept Energy/ Physics Ames IA 50011

OSTER, LUDWIG FRIEDRICH, physicist; b. Konstanz, Ger., Mar. 8, 1931; came to U.S., 1958, naturalized, 1963; s. Ludwig Friedrich and Emma Josefine (Schwarz) O.; m. Cheryl M. Oroian, Oct. 10, 1987; children from previous marriage: Ulrika, Mattias. B.S., U. Freiburg, 1951, M.S., 1954;

Ph.D., U. Kiel, 1956. Fellow German Sci. Council, U. Kiel, 1956-58; research asso. Yale U., 1958-60, asst. prof., then asso. prof. physics, 1960-67; mem. faculty U. Colo., Boulder, 1967-83; prof. physics and astrophysics U. Colo., 70-83; program mgr. Nat. Radio Astronomy Obs., NSF, 1983—; fellow Joint Inst. Astrophysics, Boulder, 1967-83; NRC Sr. assoc., 1981-82; vis. prof. U. Bonn, W.Ger., 1966, Johns Hopkins, 1981; cons. to govt. and industry. Author: Modern Astronomy, 1973 (also Spanish and Polish edits); editor: Scripta Technica, 1960-70; contbr. articles profl. jours. Recipient Humboldt Stiftung, 1974. Mem. Am. Phys. Soc., Am. Astron. Soc., Internat. Astron. Union, German Astron. Soc., Sigma Xi.

OSTER, PAMELA ANN, radiologic technologist; b. Springfield, Mo., Dec. 11, 1951; d. Charles James and Barbara Ann (Springer) Plowman; m. Ellis Oster, June 3, 1977; children: Andrew Ellis, Charles James. Diploma, St. John's Hosp., 1971. Staff technician St. Joseph's Med. Ctr., Ponca City, Okla., 1971-76, back-up technician, 1977—; radiation therapy technician U. Hosp., Oklahoma City, 1976-77. Pres. Kay County Med. Aux., Ponca City, 1980-81, Okla. State Med. Aux., Oklahoma City, 1984-85; bd. dirs. Drug and Alcohol Abuse Prevention, Ponca City, 1984; mem. Boy Scouts Am., Ponca City, 1985—; fundraiser Ponca City Sch. Found, 1988—. Recipient Vol. of Month award Sta. WBBZ Radio, Ponca City, 1989, Dist. award merit Boy Scouts Am., Ponca City, 1990, Silver Beaver award Boy Scouts Am., 1992. Mem. Am. Registry Radiol. Technicians, Am. Soc. Radiol. Technicians, Okla. Soc. Radiol. Technicians. Republican. Methodist. Avocations: calligraphy, music, reading, volunteering in school. Home: 917 E Overbrook Ave Ponca City OK 74601-3419

OSTERBERGER, JAMES SHELDON, JR., internist, educator; b. Baton Rouge, June 16, 1952; s. James Sheldon Osterberger and Mary Virginia (Wilcox) Lathrop; m. Patricia JoAnn Holcomb, May 26, 1973; children: Shelley JoAnn, Virginia Grace, Patricia Ford, Sarah Glen. BS, La. State U., 1974, MD, 1977. Diplomate Am. Bd. Internal Medicine. Intern, resident U. Ala. Hosps., Birmingham, 1977-80; assoc. Med. Assocs., Baton Rouge, 1980—; team physician La. State U., Baton Rouge, 1982—; clin. instr. La. State U. Sch. Medicine, New Orleans, 1984—; med. dir. Community Health Network, Inc., Baton Rouge, 1987-89, La. Commn. on HIV/AIDS, 1993—; co-investigator La. Community AIDS Rsch. Program, Baton Rouge, 1989—; cons. physician U.S. Olympic Com., Baton Rouge, 1985; pres. La. State Nutrition Coun., Baton Rouge, 1984-86; mem. La. State Nutrition Adv. Com., Baton Rouge, 1984. Contbr. articles to profl. jours. Recipient Spl. Recognition award La. Dietetic Assn., 1984. Fellow La. Heart Assn.; mem. ACP, AMA, AAAS, Am. Soc. Internal Medicine, Am. Coll. Sports Medicine, N.Y. Acad. Sci., So. Med. Assn., Soc. Critical Care Medicine, Rotary. Democrat. Roman Catholic. Office: Med Assocs 7777 Hennessy Blvd Ste 1000 Baton Rouge LA 70808

OSTERBROCK, DONALD E(DWARD), astronomy educator; s. William Carl and Elsie (Wettlin) O.; m. Irene L. Hansen, Sept. 19. 1952; children: Carol Ann, William Carl, Laura Jane. Ph.B., U. Chgo., 1948, B.S., 1948, M.S., 1949, Ph.D., 1952; DSc (hon.), Ohio State U., 1986, U. Chgo., 1992. Mem. faculty Princeton, 1952-53, Calif. Inst. Tech., 1953-58; faculty U. Wis.-Madison, 1958-73, prof. astronomy, 1961-73, chmn. dept. astronomy, 1966-67, 69-72; prof. astronomy U. Calif., Santa Cruz, 1972-92, prof. emeritus, 1993—; dir. Lick Obs., 1972-81; mem. staff Mt. Wilson Obs., Palomar Obs., 1953-58; vis. prof. U. Chgo., 1963-64, Ohio State U., 1980, 86; Hill Family vis. prof. U. Minn., 1977-78. Author: Astrophysics of Gaseous Nebulae, 1974, James E. Keeler, Pioneer American Astrophysicist and the Early Development of American Astrophysics, 1984, (with John R. Gustafson and W.J. Shiloh Unruh) Eye on the Sky: Lick Observatory's First Century, 1988, Astrophysics of Gaseous Nebulae and Active Galactic Nuclei, 1989, Pauper and Prince: Ritchey, Hale, and Big American Telescopes, 1993; editor: (with C.R. O'Dell) Planetary Nebulae, 1968, (with Peter H. Raven) Origins and Extinctions, 1988, (with J.S. Miller) Active Galactic Nuclei, 1989; Stars and Galaxies: Citizens of the Universe, 1990; letters editor Astrophys. Jour., 1971-73. Served with USAAF, 1943-46. Recipient Profl. Achievement award U. Chgo. Alumni Assn., 1982; Guggenheim fellow Inst. Advanced Studies, Princeton, N.J., 1960-61, 82-83, Ambrose Monnell Found. fellow, 1989-90; NSF Sr. Postdoctoral Rsch. fellow U. Coll., London, 1968-69. Mem. NAS (chmn. astronomy sect. 1971-74, sec. class math. and phys. sci. 1980-83, chmn. class math and phys. sci. 1983-85, councilor 1985-88), Am. Acad. Arts and Scis., Internat. Astron. Union (pres. commn. 34 1967-70), Royal Astron. Soc. (assoc.), Am. Astron. Soc. (councilor 1970-73, v.p. 1975-77, pres. 1988-90, vice chmn. hist. astronomy div. 1985-87, chmn. 1987-89, Henry Norris Russell lectr. 1991), Astron. Soc. Pacific (chmn. history com. 1982-86, Catherine Wolfe Bruce medal 1991), Wis. Acad. Scis. Arts and Letters, Am. Philos. Soc. Congregationalist. Home: 120 Woodside Ave Santa Cruz CA 95060-3422

OSTERHOUDT, WALTER JABEZ, geophysical and geological exploration consultant; b. Bramans (Stalkers), Pa., Jan. 21, 1906; s. Walter Osterhoudt and Grace May Bailey; m. Gretchen Marie Zierath, May 14, 1935; children: Hans Walter, Peter Gerard. BA in Physics, U. Wis., 1930. Registered profl. engr., Tex., Colo. Exploration geologist and geophysicist Gulf Rsch. and Devel. Corp., Pitts., 1930-49, Gulf Oil Corp., Houston, 1930-49, Estate of Wm. G. Helis, New Orleans, 1950-72; cons. in exploration for oil and gas in Greece, Venezuela, Tunisia, others. Created emblem of Houston Geol. Soc. Fellow AAAS; mem. Pi Kappa Alpha. Republican. Achievements include patents for vessels for submarine navigation, cable clamps and seal, shock absorber. Home and Office: 3403 County Rd 250 Durango CO 81301

OSTERLE, JOHN FLETCHER, mechanical engineering educator; b. Pitts., July 31, 1925. BS, Carnegie Inst. Tech., 1946, MS, 1949, DSc in mech. engring., 1952. From instr. to assoc. prof. Carnegie-Mellon U., Pitts., 1946-58, chmn. nuclear sci. and engring., 1975-85, acting head mech. engring. dept., 1985-87, Theodore Ahrens prof. mech. engring., 1958—; vis. prof. Delft U. Tech., The Netherlands, 1957-58, Oxford U., 1971. Recipient Walter D. Hodson award Am. Soc. Lubrication Engrs., 1956. Mem. ASME, Am. Nuclear Soc. Research in fluid mechanics, thermodynamics. Office: Carnegie Mellon U Dept Mech Engring Pittsburgh PA 15213-3890

OSTERLOH, EVERETT WILLIAM, county official; b. Luxemburg, Mo., June 7, 1919; s. Fred and Esther (Miller) O.; m. Eunice Gramann, Oct. 20, 1940 (dec. Apr. 1983); m. Herta Anna Emery, Oct. 25, 1987. BSME, Washington U., St. Louis, 1958. Registered profl. engr., Mo.; cert. code ofcl. Plant engr. Jasper-Blackburn Corp., St. Louis, 1958-60; equipment engr. Monsanto Chem. Co., St. Louis, 1960-68; pres. Caribean Beach Club, Antigua, West Indies, 1968-73; dep. dir. pub. works St. Louis County Govt., Clayton, Mo., 1973—; engring. instr. St. Louis Community Coll., 1986-91; exec. sec. Profl. Code Com. St. Louis, 1975—, Met. Area Code Com., St. Louis, 1982—. Staff sgt. USAAF, 1942-53. Mem. Mo. Soc. Profl. Engrs. (emergency response task force 1990-92, pres. St. Louis chpt. 1991-92, Outstanding Engr. in Govt. 1987). Lutheran. Home: 283 Spring Oaks Dr Ballwin MO 63011 Office: St Louis County Govt 41 S Central Ave Clayton MO 63105

OSTERMAN, LISA ELLEN, geologist, researcher; b. Cin., July 11, 1953; d. Clifford Paul Osterman and Norma Lee (Bledsoe) Gould; m. Mark Bykowsky, June 18, 1977; children: Marcus Burton, Spencer Paul. BS, U. Dayton, 1975; MS, U. Maine, Orono, 1977; PhD, U. Colo., 1982. Rsch. assoc. Inst. Arctic and Alpine Rsch., Boulder, Colo., 1982-85; postdoctoral fellow Smithsonian Inst., Washington, 1985-86, rsch. collaborator dept. paleobiology, 1986—; asst. prof. George Washington U., Washington, 1986—; presenter in field. Contbr. articles to profl. jours. Rsch. grantee NSF, 1981, 83, 85, Petroleum Rsch. Fund Am. Chem. Soc., 1989-90. Fellow Cushman Found.; mem. AAAS, Geol. Soc. Am. (J. Hoover Mackin award 1978), Am. Quaternary Assn., Paleontol. Soc. Washington (pres. 1991-92), Am. Women in Sci., Sigma Xi. Achievements include rsch. in polar paleo-oceanography and foraminiferal paleoecology. Home: 4513 Chase Ave Bethesda MD 20814

OSTMO, DAVID CHARLES, engineering executive; b. Mason City, Iowa, Jan. 16, 1959; s. Gene Charles and Charlene Lucille (Evans) O. Diploma, Brown Inst., 1979. Maintenance engr. Sta. KJRH-TV, Tulsa, 1981-84; chief news editor Sta. KOTV, Tulsa, 1984-86; chief engr. Sta. KXON-TV/Rogers State Coll., Claremore, Okla., 1986-90; dir. engring. Fairview AFX, Tulsa, 1990-92, KABB-TV, San Antonio, 1992—; columnist TV Tech., Falls Church, Va., 1988-90. Producer, dir. (TV show) Theodore Kirby Show, 1988; dir. (TV show) Zebra Sports Review, 1987, RSC Spotlight, 1987; writer mag. News From the Back Porch, 1988-90, Tulsa TV Tidbits, 1988-91. Named Eagle Scout Boy Scouts Am., 1974. Mem. Soc. Broadcast Engrs. (sr. broadcast engr., chmn. Tulsa chpt. 1989-90, chmn. San Antonio chpt. 1993—). Lutheran. Avocations: photography, genealogy, creative writing, computer science, golf. Home: 3500 Oakgate Dr # 2706 San Antonio TX 78230 Office: KABB-TV 520 N Medina San Antonio TX 78207

OSTRACH, SIMON, engineering educator; b. Providence, Dec. 26, 1923; s. Samuel and Bella (Sackman) O.; m. Gloria Selma Ostrov., Dec. 31, 1944 (div. Jan. 1973); children: Stefan Alan, Louis Hayman, Naomi Ruth, David Jonathan, Judith Cele; m. Margaret E. Stern, Oct. 29, 1975. B.S. in Mech. Engring, U. R.I., 1944, M.E., 1949; Sc.M., Brown U., 1949, Ph.D. in Applied Math, 1950; D.Sci. (h.c.) Technion, Israel Inst. Tech., 1986. Rsch. scientist NACA, 1944-47; rsch. assoc. Brown U., 1947-50; chief fluid physics br. Lewis Rsch. Ctr. NASA, 1950-60; prof. engring., head div. fluid, thermal and aerospace scis. Case Western Res. U., Cleve., 1960-70; Wilbert J. Austin Distinguished prof. engring. Case Western Res. U., 1970—; home sec. Nat. Acad. Engring., 1992—; disting. vis. prof. City Coll. CUNY, 1966-67, Fla. A&M U., Fla. State U. Coll. Engring., 1990; Lady Davis fellow, vis. prof. Technion-Israel Inst. Tech., 1983-84; cons. to industry, 1960—; mem. rsch. adv. com. fluid mechanics NASA, 1963-68, mem. space applications adv. com., 1985—; hon. prof. Beijing U. Aeronautics and Astronautics, 1991; mem. space studies bd. NRC, 1992. Contbr. papers to profl. lit. Fellow Japan Soc. for the Promotion of Sci., 1987; recipient Conf. award for best paper Nat. Heat Transfer Conf., 1963, Richards Meml. award Pi Tau Sigma, 1964, Disting. Svc. award Cleve. Tech. Socs. Coun., 1987, Disting. pub. svc. medal NASA, 1993. Fellow Am. Acad. Mechanics, AIAA, ASME (hon., Heat Transfer Meml. award 1975, Freeman scholar 1982, Thurston lectr. 1987, Max Jacob meml. award 1983, Heat Transfer div. 50th Anniversary award 1988); mem. NAE (chmn. com. on membership 1986, chmn. nominating com. 1989, chmn. awards com. 1990, sec., mem. space studies bd. 1992), Univs. Space Rsch. Assn. (trustee 1990), Soc. Natural Philosophy, Sigma Xi (nat. lectr. 1978-79), Tau Beta Pi. Home: 28176 Belcourt Rd Cleveland OH 44124-5618 Office: Case Western Res U Cleveland OH 44106

ØSTREM, GUNNAR MULDRUP, glaciologist; b. Oslo, Mar. 25, 1922; s. Sigurd and Alfhild (Paulsen) Ø; m. Britta Louise Nyström, June 30, 1952; children: Hans Peter, Anne Christine, Eva Louise, Karin Johanne. BA, U. Oslo, 1948, MA, 1954; Fil. lic., U. Stockholm, 1961, Fil. Dr., 1965. Lectr. various high schs. and colls., Koping and Tranas, Sweden, 1950-58; asst. prof. U. Stockholm, 1958-62, assoc. prof., 1966-81, prof., chmn. dept. phys. geography, 1981-83; chief glaciology div. Fed. Dept. Mines and Tech., Surveys, Ottawa, Ont., 1965-66; sr. glaciologist Norwegian Water Resources and Energy Adminstrn., Oslo, 1983-92, sr. cons., 1993—. Author: Glacier Atlas of Northern Scandinavia, 1973, Bathymetric Maps of Norwegian Lakes, 1984, Atlas of Glaciers in South Norway, 1988 (rev. edit.); editor Geografiska Annaler Jour., 1976-92; contbr. articles to profl. jours. Recipient Hans Egede medal, 1982, Gold medal King of Norway, 1992, J. A. Wahlberg's Silver medal, 1993. Fellow Arctic Inst. N.Am., Norwegian Geophys. Soc.; mem. Internat. Glaciological Soc., Swedish Geog. Soc., European Assn. Remote Sensing Labs. (treas. 1976-85), Can. Remote Sensing Soc. Home: Ovenbakken 4, N-1345 Osteras Norway Office: Norwegian Water Resources and Energy Adminstrn, PO Box 5091-Mj, Oslo 3, Norway

OSTRIKER, JEREMIAH PAUL, astrophysicist, educator; b. N.Y.C., Apr. 13, 1937; s. Martin and Jeanne (Sumpf) O.; m. Alicia Suskin, Dec. 1, 1958; children—Rebecca, Eve, Gabriel. A.B., Harvard, 1959; Ph.D. (NSF fellow), U. Chgo., 1964; postgrad., U. Cambridge, Eng., 1964-65; hon. degree, U. Chgo., 1992. Rsch. assoc., lectr. astrophysics Princeton (N.J.) U., 1965-66, asst. prof., 1966-68, assoc. prof., 1968-71, prof., 1971—, chmn. dept. astronomy, dir. obs., 1979—, Charles A. Young prof. astronomy, 1982—. Editorial bd., trustee Princeton U. Press; Contbr. articles to profl. jours. Alfred P. Sloan Found. fellow, 1970-72. Fellow AAAS; mem. Am. Astron. Soc. (councilor 1978-80, Warner prize 1972, Russell prize 1980), Internat. Astron. Union (bd. govs. 1993—), Nat. Acad. Scis. (counselor 1992—), Am. Acad. Arts and Scis. Home: 33 Philip Dr Princeton NJ 08540-5409 Office: Dept of Astrophysical Princeton Univ Obs Peyton Hall Princeton NJ 08544

OSTROFF, NORMAN, chemical engineer; b. Bklyn., Feb. 20, 1937; s. Samuel and Lee (Kacer) O.; m. Barbara Jane Pollack, June 24, 1961; children: Lawrence, William, Julie, Jonathan. BSChemE, Poly. Inst. of Bklyn., 1958, MSChemE, 1962, PhD in Chem. Engring., 1970. Registered profl. engr., Conn., Mass., Va. Math. tchr. Bridgeport Engring. Inst., Fairfield, Conn., 1963-80; chem. engr. Am. Cyanamid, Stamford, Conn., 1960-73, Engelhand Corp., Newark, 1973-74; rsch. mgr. UOP Corp. (now Wheelabrator), Darien, Conn., 1974-78; engring. supr. Peabody Process Systems (now A.B.B.), Norwalk, Conn., 1978-86; chem. engring. tchr. Ctr. Profl. Advancement, New Brunswick, N.J., 1988—; cons. engr. Stamford, 1986—. Contbr. articles to profl. jours. Mem. Environ. Protection Bd., Stamford, 1983-90. Rsch. fellow Poly. Inst. of Bklyn., 1958-60. Mem. ASME, Am. Assn. Environ. Engrs. (diplomate), Conn. Bus. and Industry Assn., Conn. Cogeneration Soc. (bd. dirs.), Air and Water Mgmt. Assn., Sigma Xi, Phi Lambda Upsilon. Avocation: photography. Home and Office: 87 Fishing Trl Stamford CT 06903-2422

OSTROGORSKY, ALEKSANDAR GEORGE, mechanical engineer, educator; b. Belgrade, Yugoslavia, Sept. 6, 1952; came to U.S., 1980; s. George and Fanula (Papazoglu) O.; m. Natacha DePaola, Mar. 11, 1989; 1 child, George. MS, Rensselaer Poly. Inst., 1981; ScD, MIT, 1986. Rsch. engr. Inst. Gosa Industries, Belgrade, 1977-82; rsch. asst. MIT, Cambridge, 1983-85, postdoctoral assoc., 1986-87, asst. prof. Columbia U., N.Y.C., 1987-92; assoc. prof. mech. engring. Rensselaer Poly. Inst., Troy, N.Y., 1993—. Contbr. chpt. to books, articles to profl. jours. Fulbright scholar, 1980; recipient Initiation award NSF, 1988; Alexander von Humboldt fellow, Bonn, Germany, 1990. Mem. Am. Soc. Crystal Growth, N.Y. Acad. Scis. Achievements include patent of method and apparatus for single crystal growth, patent for method for directional solidification of single crystals. Office: Rensselaer Poly Inst Dept Mech Engring Troy NY 12180-3590

OSTROM, CHARLES CURTIS, financial consultant, former military officer; b. Rockford, Ill., July 24, 1962; s. Curtis Earl and Carol Ann (Schleger) O.; m. Teri Lei Govig, Nov. 17, 1984; 1 child, Nicholas Andrew. BS, So. Ill. U., 1984. Lic. airline transport pilot/multi-engine land. Fin. cons. IDS Fin. Svcs., Inc., Rockford, Ill., 1992—. Airshow participant Midwest Airfest, Rockford, 1991. Capt. USAF, 1984-91, Desert Storm. Mem. Sigma Pi. Republican. Office: IDS Fin Svcs Inc Ste 200 6833 Stalter Dr Rockford IL 61108

OSTROVSKY, ALEXEY, geophysicist, researcher; b. Moscow, Oct. 2, 1953; s. Alexey Emeljan and Valentina (Gultjaeva) O.; m. Natalija Chmirjeva, 1979 (div. 1982); m. Ekaterina Mironova, July 14, 1984; 1 child, Nikita. MS in Geophysics, Moscow State U., 1977; PhD in Physics and Math, Inst. Oceanology, Moscow, 1980. Rsch. scientist USSR Acad. Scis., Moscow, 1980-85, sr. rsch. sci., 1985—. Contbr. articles to profl. jours. Alexander Von Humboldt Found. fellow, Germany, 1991-93. Achievements include design of ocean bottom seismic noise spectral model; research in situ OBS coupling characteristic estimation, fractality of the ocean bottom, fatigue triggering of earthquakes, single airgun Moho sounding under the Baltic Sea. Home: Kachalova str 16-18, Moscow 121069, Russia Office: Inst Oceanology, Krasikova str 23, Moscow 117218, Russia

OSTROW, JAY DONALD, gastroenterology educator, researcher; b. N.Y.C., Jan. 1, 1930; s. Herman and Anne Sylvia (Epstein) O.; m. Judith Fargo, Sept. 9, 1956; children: Geroge Herman, Bruce Donald, Margaret Anne. B.S. in Chemistry, Yale U., 1950; M.D., Harvard U., 1954; M.Sc. in Biochemistry, Univ. Coll., London, 1970. Diplomate Am. Bd. Internal Medicine, Am. Bd. Gastroenterology. Intern Johns Hopkins Hosp., Balt., 1954-55; resident Peter Bent Brigham Hosp., Boston, 1957-58; NIH trainee in gastroenterology, 1958-59; NIH trainee in liver disease Thorndike Mem. Lab. Boston City Hosp., 1959-62; instr. in medicine Harvard U., Boston, 1959-62; asst. prof. medicine Case-Western Res. U., Cleve., 1962-70; assoc. prof. U. Pa., Phila., 1970-76, prof., 1977-78; Sprague prof. medicine Northwestern U., Chgo., 1978-89, prof. medicine, 1989—, chief gastroenterology sect., 1978-87; med. investigator VA Hosp., Phila., 1973-78, VA Med. Ctr. Lakeside, Chgo., 1990—. Editor, contbg. author: Bile Pigments and Jaundice, 1986. Asst. scoutmaster Valley Forge council Boy Scouts Am., Merion, Pa., 1972-78; asst. scoutmaster Northeast Ill. council Boy Scouts Am., 1978-81; vestryman St. Matthew's Episcopal Ch., Evanston, Ill., 1979-82; treas. Classical Children's Chorale, Evanston, 1982. Served to lt. comdr. M.C. USN, 1955-57. Recipient Gastroenterology Rsch. award Beaumont Soc., El Paso, 1979, Sr. Disting. Scientist award Alexander von Humboldt Found., Germany, 1989-90; NIH fellow, 1958-62, grantee, 1962—; VA grantee, 1970—. Mem. Am. Assn. Study Liver Diseases (councillor 1983-85, v.p. 1985-86, pres. 1987), Am. Gastroent. Assn. (chmn. exhibit com. 1969-72), Am. Soc. Clin. Investigation, Am. Physiol. Soc. (asst. editor 1979-84), Internat. Assn. Study Liver, Am. Soc. Photobiology. Club: Peripatetic (Bethesda, Md.). Office: Rsch Svc (151) DVA Lakeside Med Ctr 400 E Ontario St Chicago IL 60611-4493

OSTROWSKI, PETER PHILLIP, aerospace engineer, researcher; b. Dover, N.J., May 17, 1940; s. Alexander Joseph and Delia Ann (Rages) O.; m. Carol Anne Carr, Aug. 31, 1968; children: Stephen T., Elizabeth A. BS, U. Md., 1963, MS, 1969; PhD, U. Conn., 1977. Group mgr. Geo-Ctrs. Inc., Washington, 1976-84; v.p. Applied Combustion Tech. Inc., Alexandria, Va., 1984-87; prin. engr. Arinc Rsch. Corp., Annapolis, Md., 1987-88; pres. Energetic Materials Tech., Alexandria, 1988—. Mem. AIAA, Internat. Pyrotechnics. Achievements include rsch. in lasers, pyrotechnics, explosives, combustion and laser ignition of pyrotechnics and energetic materials. Home: 1205 Potomac Ln Alexandria VA 22308 Office: Energetic Materials Tech PO Box 6931 Alexandria VA 22306-0931

OSTRY, SYLVIA, Canadian public servant, economist; b. Winnipeg, Man., Can.; d. Morris J. and B. (Stoller) Knelman; m. Bernard Ostry; children: Adam, Jonathan. BA in Econs., McGill U., 1948, MA, 1950; PhD, Cambridge U. and McGill U., 1954; also 17 hon. degrees. Lectr., asst. prof. econs. McGill U.; research officer Inst. Stats., U. Oxford, Eng.; assoc. prof. U. Montreal, Can.; with dept. stats. Econ. Coun. Can., Ottawa, 1964-72, chmn., 1978-79; chief statistician Stats. Can., Ottawa, 1972-75; dep. minister consumer and corp. affairs Govt. Can., Ottawa, 1975-78, dep. minister internat. trade, coordinator internat. econ. relations, 1984-85, ambassador for multilateral trade negotiations, personal rep. of Prime Minister for Econ. Summit, 1985-88; chmn. Centre for Internat. Studies U. Toronto, Can., 1990—; chancellor U. Waterloo, 1991—; head dept. econs. and stats. OECD, Paris, 1979-83; lectr. Per Jacobssen Found., 1987; chmn. nat. coun. Can. Inst. Internat. Affairs, 1990—; western co-chmn. The Blue Ribbon Commn. for hungary's Recovery, 1990—; chmn. internat. adv. bd. Bank of Montreal; bd. dirs. Power Fin. Corp., mem. internat. adv. coun.; bd. dirs. Kellog Can. Inc., UN Univ. World/World Inst. Devel. Econs. Rsch. Helsinki; expert adv. Commn. Transnat. Corps., UN; mem. internat. com. InterAm. Devel. Bank/Econ. Commn. Latin Am. & Caribbean project. Author: Governments and Corporations in a Shrinking World: The Search for Stability, 1990; co-author: The Political Economy of Policy Making in the Triad, 1990, Regional Trading Blocs: Pragmatic or Policy?; editor: Authority and Academic Scribblers: The Role of Research in East Asian Policy Reform, 1991; contbr. articles on empirical and policy-analytic subjects to over 80 profl. publs. Decorated companion Order of Can.; recipient Outstanding Achievement award Govt. of Can., 1987, Hon. Assoc. award Conf. Bd. of Can., 1992; Disting. vis. fellow Volvo, 1989-90, U. Toronto fellow, 1989-90. Fellow Royal Soc. Can., Am. Statis Assn.; mem. Am. Econ. Assn., Can. Econ. Assn., Royal Econ. Soc. (founding), Ctr. for European Policy Studies (adv. bd.), Group of Thirty, Inst. for Internat. Econs. (adv. bd.). Avocations: films, theatre, contemporary reading. Office: U Toronto Ctr Internat, Studies Office of Chmn, 170 Bloor St W Ste 500, Toronto, ON Canada M5S 1T9

O'SULLIVAN, MICHAEL ANTHONY, civil engineer; b. N.Y.C., Sept. 7, 1960; s. Edward J. and Mary T. O'Sullivan; m. Erin Diamond; children: Patrick, Nora. BSCE, U. Notre Dame, 1982; MBA, U. Chgo., 1987. Registered profl. engr., Ill. Unit supr. engr., constrn. mgr., staff engr. Commonwealth Edison Co., Chgo., 1982-89; devel. mgr. Homart Devel. Co., Chgo., 1989-91; v.p. Indeck Energy, Buffalo Grove, Ill., 1991—. Mem. ASCE, Assn. Energy Engrs. Office: Indeck Energy 1130 Lake Cook Rd Buffalo Grove IL 60089

OSUMI, MASATO, utility company executive; b. Osaka, Japan, Aug. 20, 1942; s. Masahiro and Sachiko Osumi; m. Masako Nakajima, Apr. 21, 1968; children: Masanori, Koji, Yuko. BS, Yokohama Nat. U., 1966; MS, NYU, 1971; D in Engring., Kyoto U., 1978. Engr. Sanyo Elec. Co. Ltd., Hirakata, Osaka, 1966-71, chief researcher rsch. ctr., 1978-85, mgr., 1985-86, mgr. control systems rsch. ctr., 1987-89, div. mgr., 1989-92; gen. mgr. Sanyo Elec. Co. Ltd., 1992—; engr. Tokyo Sanyo Elec. Co. Ltd., Ohizumi-Cho, Gunma, Japan, 1971-73. Contbr. articles to profl. jours. Mem. Japanese Soc. Mech. Engrs. (bd. dirs. 1986-87), Japanese Soc. Precision Engrs. Avocation: golfing. Home: 3-33 Takiimotomachi, Moriguchi Osaka 570, Japan Office: Sanyo Elec Co Ltd 1-18-13, Hashiridani, Hirakata, Osaka 573, Japan

OSWAKS, ROY MICHAEL, surgeon, educator; b. Bklyn., Dec. 9, 1945; m. Jill S. Detty, June 4, 1982; children: Aaron, David. BA cum laude, CUNY, Bklyn., 1967; MD, SUNY, Buffalo, 1971. Intern in gen. surgery SUNY-Millard Fillmore Hosp., Buffalo, 1971-72, resident in surgery, 1974-76; resident in surgery Maricopa County Gen. Hosp., Phoenix, 1972-73, Brookdale Hosp. Med. Ctr., Bklyn., 1973-74; surgeon U.S. Army Hosp., Ft. Monmouth, N.J., 1976-78, Sentara Bayside Hosp., Virginia Beach, Va., 1982—; chief surgery Sentra Bayside Hosp., Virginia Beach, Va., 1989-91, USPHS, Norfolk, Va., 1979-81; chmn. operating rm. com. Sentara Bayside Hosp., 1989-91, sec.-treas. med. staff, 1991-92. Active United Jewish Fedn. of Tidewater, Norfolk, Va. Mem. AMA, IPA of Southeastern Va. (bd. dirs., mem. exec. com., reimbursement and grievance com.), Med. Soc. Va., Va. Surg. Soc., So. Med. Assn., N.Y. Acad. Scis., Maimonides Med. Soc., Virginia Beach Med. Soc., S.E. Surg. Congress. Office: Roy M Oswaks MD Ltd PC 816 Independence Blvd # 1-A Virginia Beach VA 23455

OSZUREK, PAUL JOHN, industrial chemist; b. Hartford, Conn., May 31, 1959; s. Benjamin and Honorata (Wröbel) O. BS in Chemistry, U. Hartford, 1981; MS in Chemistry, St. Joseph Coll., West Hartford, Conn., 1986; MS in Ops. Mgmt., Rensselaer Poly. Inst., 1991. Sr. devel. chemist Amplex Corp., Bloomfield, Conn., 1982—. Mem. Am. Chem. Soc. Republican. Roman Catholic. Achievements include development and refining processes for manufacture of electroplated brazed abrasive coated grinding tools; co-development of alternative braze paste with resultant 70% decrease in cost. Home: 16 Harding St Wethersfield CT 06109 Office: Amplex Corp 16 Britton Dr Bloomfield CT 06002

OTANI, MIKE, optical company executive; b. Atsumi, Aichi, Japan, July 25, 1945; s. Yuichi and Miyako (Suzuki) O.; m. Jane Ashley Campbell, Aug. 25, 1976; 1 child, Michael Taro. Degree in Internat. Fin. and Econs., Shiga U., Japan, 1967. Office mgr. Kumagai-Gumi Ltd, Osaka, Japan, 1968-73; dude rancher Tumbling River Ranch, Grant, Colo., 1974-77; merchandiser Nobel, Inc., Denver, 1978-82; v.p. Charmant Eyewear, Inc., Morris Plains, N.J., 1983-88, pres., chief exec. officer, 1988—; bd. dirs. Charmant Optical Co., Ltd., Fukui, Japan, Charmant Optical GMRH Europe, Munich, Optical Manufacturers Assn., Falls Church, Va. Donator Project Literacy U.S., Pitts., 1990, Pa. Coll. Optometry, Phila., 1990; organizer N.J. Fukui Sister State Activity, 1990; sponsor Big Bros. and Little Sisters Morris County, 1993—. Recipient Vendor of Yr. award Walmart, Inc., 1991. Mem. Optical Mfgrs. Assn., Fukui/New York Club (v.p. 1992), Big Bros. and Big Sisters of Morris County (spon.), 1993—. Avocations: golf, horseback riding, reading, traveling, fishing. Home: 9 October Hill Rd Oak Ridge NJ 07438-9194 Office: Charmant Eyewear Inc 400 American Rd Morris Plains NJ 07950-2451

OTHMER, DONALD FREDERICK, chemical engineer, educator; b. Omaha, May 11, 1904; s. Frederick George and Fredericka Darling (Snyder) O.; m. Mildred Jane Topp, Nov. 18, 1950. Student, Ill. Inst. Tech., Chgo., 1921-23; B.S., U. Nebr., 1924, D.Eng. (hon.), 1962; M.S., U. Mich., 1925, Ph.D., 1927; D.Eng. (hon.), Poly. U., Bklyn., 1977, N.J. Inst. Tech., 1978. Registered profl. engr., N.Y., N.J., Ohio, Pa. Devel. engr. Eastman Kodak Co. and Tenn. Eastman Corp., 1927-31; prof. Poly. U., Bklyn., 1933; disting. prof. Poly. U., 1961—, sec. grad. faculty, 1948-58; head dept. chem. engring., 1937-61; hon. prof. U. Conception, Chile, 1951; cons. chem. engr., licensor of process patents to numerous cos., govtl. depts., and countries, 1931—; developer program for chem. industry of Burma, 1951-54; cons. UN, UNIDO, WHO, Dept. Energy, Office Saline Water of U.S. Dept. Interior, Chem. Corps. and Ordnance Dept. U.S. Army, USN, WPB, Dept. State, HEW, Nat. Materials Advisory Bd., NRC Sci. Adv. Bd., U.S. Army Munitions Command; mem. Panel Energy Advisers to Congress, also other U.S. and fgn. govt. depts.; sr. gas officer Bklyn. Citizens Def. Corps.; lectr. Am. Swiss Found. Sci. Relations, 1950, Chem. Inst. Can., 1944-52, Am. Chem. Soc., U.S. Army War Coll., 1964, Shri RAM Inst., India, 1980, Royal Mil. Coll. Can., 1981; plenary lectr. Peoples Republic of China; hon. del. Engring. Congresses, Japan, 1983; plenary lectr., hon. del. Fed. Republic of Germany, Greece, Mex., Czechoslovakia, Yugoslavia, Poland, P.R., France, Can., Argentina, India, Turkey, Spain, Rumania, Kuwait, Iran, Iraq, Algeria, China, United Arab Emirates; designer chem. plants and processes for numerous corps., U.S., fgn. countries. Holder over 150 U.S. and fgn. patents on methods, processes and engring. equipment in mfg. of pharms., sugar, salt, acetic acid, acetylene, fuel-methanol, synthetic rubber, petro-chems., pigments, zinc, aluminum, titanium, also wood pulping, refrigeration, solar and other energy conversion, water desalination, sewage treatment, peat utilization, coal desulfurization, pipeline heating, etc.; contbr. over 350 articles on chem. engring., chem. mfg., synthetic fuels and thermodynamics to tech. jours.; co-founder/co-editor: Kirk-Othmer Ency. Chem. Tech., 17 vols., 1947-60, 24 vols., 2d edit., 1963-71, 26 vols., 3d edit., 1976-84, 4th edit., 25 vols., 1992, Spanish edit., 16 vols., 1960-66; editor: Fluidization, 1956; co-author: Fluidization and Fluid Particle Systems, 1960; mem. adv. bd.: Perry's Chem. Engr.'s Handbook; tech. editor: UN Report, Technology of Water Desalination, 1964. Bd. regents L.I. Coll. Hosp., bd. dirs. numerous ednl. and philanthropic instns., engring. and indsl. corps. Recipient Golden Jubilee award Ill. Inst. Tech., 1975, Profl. Achievement award Ill. Inst. Tech., 1978, Award of Honor for Sci. and Tech. Mayor of N.Y.C., 1987, Outstanding Alumnus award U. Nebr., 1989, Citation for Improvement of Quality of Life, Pres. Borough Bklyn., 1989, award for significant contbns. to life Polytechnic U., 1989; named to Hall of Fame, Ill. Inst. Tech., 1981. Fellow AAAS, Am. Inst. Cons. Engrs., Am. Inst. Chemists (Honor Scroll 1970, Chem. Pioneer award 1977), ASME (hon. life, chmn. chem. processes div. 1948-49), N.Y. Acad. Scis. (hon. life, chmn. engring. sect. 1972-73), Instn. Chem. Engrs. (London) (hon. life), Am. Inst. Chem. Engrs. (Tyler award 1958, chmn. N.Y. sect. 1944, dir. 1956-59, Founders award 1991); mem. Am. Chem. Soc. (council 1945-47, E.V. Murphree-Exxon award 1978, hon. life mem.), Soc. Chem. Industry (Perkin medal 1978), Am. Soc. Engring. Edn. (Barber Coleman award 1958), Engrs. Joint Council (dir. 1957-59), Societe de Chimie Industrielle (pres. 1973-74), Chemurgic Council (dir.), Japan Soc. Chem. Engrs., Assn. Cons. Chemists and Chem. Engrs. (award of Merit 1975), Newcomen Soc., Am. Arbitration Assn. (panel mem. or sole arbitrator numerous cases), Deutsche Gesellscaft für Cheme. Apparatewesen (hon. life), Norwegian Club Bklyn., Chemists Club N.Y.C. (pres. 1974-75), Rembrandt Club Bklyn., Sigma Xi (citation disting. research 1983), Tau Beta Pi, Phi Lambda Upsilon, Iota Alpha, Alpha Chi Sigma, Lambda Chi Alpha. Home: 140 Columbia Heights Brooklyn NY 11201-1631 Office: Poly U 333 Jay St Brooklyn NY 11201-2990

OTIS, JOHN JAMES, civil engineer; b. Syracuse, N.Y., Aug. 5, 1922; s. John Joseph and Anna (Dey) O.; m. Dorothy Fuller Otis, June 21, 1958; children: Mary Eileen Dawn, John Leon Jr. B of Chem. Engring., Syracuse U., 1943, MBA, 1950, postgrad., 1951-55. Registered profl. engr., Ala., Tex. Jr. process engr. GM, Syracuse, 1951-53, prodn. engr., 1954-58, process control engr., 1958-59, process engr., 1960-61; engr., writer GE, Syracuse, 1961-63; configuration control engr. GE, Phila., 1969; assoc. research engr. Boeing Co., Huntsville, Ala., 1963-65; assoc. Planning Rsch. Corp., Huntsville, 1965-67; prin. engr. Brown Engring. Co. subs. Teledyne Co., Huntsville, 1967-69; mech. designer Drever Co., Beth Ayres, Pa., 1970-71; civil engr. U.S. Army Corps Engrs., Mobile, Ala., 1971-74, Galveston, Tex., 1974—. Lector, lay minister Roman Cath. Ch. Served with USNR, 1944-46. Mem. Am. Inst. Indsl. Engrs. (past v.p. Syracuse and Huntsville chpts.), Tex. Soc. Profl. Engrs. (dir. Galveston County chpt. 1976-79, sec.-treas. 1979-80, v.p. 1980-81, pres. 1982-83), Tau Beta Pi, Phi Kappa Tau, Alpha Chi Sigma, Chi Eta Sigma. Home: 2114 Yorktown Ct N League City TX 77573-5056 Office: US Army Corps Engrs Jadwin Bldg 2000 Fort Point Rd Galveston TX 77550-3038

OTOKPA, AUGUSTINE EMMANUEL OGABA, JR., research scientist, consultant; b. Agila town, Izote, Nigeria, Sept. 8, 1945; s. Otokpa and Ochanya (Obande) Okpekwu; m. Grace Onefeli, Jan. 3, 1969; children—Evelyn, Christabel, Loretta, Paul, Isaac, Jocelyn (dec.). A.B., Tenn. Christian Coll., 1974; M.B.A., Western Colo. U., 1978; S.J.D., Heed U., Fla., 1984; Ph.D. in Bus. Adminstrn. and Mgmt., Columbia Pacific U., 1982; cert. Am. Inst. Mgmt., 1974; cert. Assn. Cost and Exec. Accts., London, 1980; cert. and diploma Inst. Mktg., London, 1982; cert. fundamental and applied econs. Henry George Sch. of Social Sci., N.Y.C., 1968; CAM in Profl. Design., Acad. Adminstrv. Mgmt., Washington, 1989; CMC in Profl. Desig., Inst. Mgmt. Cons., N.Y., 1990; cultural doctorate in parapsychology World U., Tucson, 1980; diploma in Parapsychology, Am. Parapsycho. Rsch. Found., Calif. Pres., BRMC, Kano, Nigeria, 1976-81; dir. BIMS, Mangalore, India, 1981-84; personnel dir. DMB Internat., Belgium, from 1973; cons. prof. Faculty Accts. Pakistan, 1981-86; chair doctoral com. Asian Research Ctr., UNIDO, 1983-86; chair council econ. affairs World U., Benson, Ariz, 1982; faculty adviser N.Am. Regional Coll., Ariz., 1984, Columbia Pacific U., 1985. Corr. Interavia Daily Air Letter, U.K., 1981-86; appointed to West African Editorial Bur. for African Air Transport and World AirNews. Author: Scientific Evidence of the Proof of Immortality of the Human Soul, 1982; The Truth about Parapsychology in Ultimate Reality and Meaning, 1982; also articles; assoc. rev. editor Leadership & Orgn. Devel. Jour., 1980-86; freelance writer Bus. and Comml. Aviation, N.Y., Air and Space/Smithsonian Inst., Washington. Dep. mem. gen. assembly Internat. Parliament Safety and Peace. Fellow Inter-Univ. Seminar on Armed Forces and Soc., Coll. Preceptors (cert.) (London); mem. Internat. Inst. Adminstrv. Scis. (Belgium), Am. Soc. Inernat. Law, Am. Arbitration Assn. (internat. panel), Irish Mgmt. Inst. (cert.; council 1982); AFRICA: Council of Reprographics Execs. and Orgn. Planning Mgmt. Assn. (v.p. 1979-86), Acad. Mgmt., Acad. Internat. Bus., Am. Cons. League (cert.), Young Pres. Orgn., Internat. Council Psychologists, Assn. Behavior Analysis (area resource person 1982-86), Thomas Jefferson Research Ctr. (assoc.), Am. Mgmt. Assn., Nat. Mil. Intelligence Assn., Aviation-Space Writers Assn., Internat. Assn. Religion and Parapsychology (Japan). Emergency World Council in The Hague, Inst. Jud. Adminstrn., Fed. Jud. Ctr., Comml. Law League Am., Inst. of Sci. and Tech. Communicators (Eng.), Internat. Indsl. Relations Assn. (Geneva), Geneva Consultants Registry Ltd. (U.S.), Ctr. for Devel. of Industries, Internat. Peace Research Assn., Internat. Registry of Orgn. Devel. Profls., Nat. Council Tchrs. of English, Speech Communication Assn., Dignity of Man Found. (African rep. 1973-86), Internat. Assn. Chiefs of Police (assoc.). Roman Catholic. Club: London (Topeka). Avocations: writing; travel; photography; correspondence; music. Home: 54B Airport Rd, PO Box 117-2384, Kano Nigeria Office: BRMC, PO Box 117, Kano Nigeria

OTOONI, MONDE A., materials scientist, physicist; b. Broudjerd, Lorestan, Iran, Aug. 17, 1933; came to U.S., 1951; s. Abdel H. and Robob J. Otooni; m. Beatrice Azar, Sept. 8, 1962; children: Christopher K., Paul C. BS in Physics, Tehran (Iran) U., 1950; BS Geol. Engring., BS Geophys. Engring., Colo. Sch. Mines, 1955; MA in X-ray Physics, Columbia U., 1959; MSc in Solid State Physics, Syracuse U., 1970. Instr. phys. scis. NYU, N.Y.C., 1960-61, Bklyn. Coll., N.Y.C., 1961-65; asst. prof. physics S.I. Coll., N.Y.C., 1965-68; asst. prof. phys. sci. Glassboro (N.J.) State Coll., 1971-75; metallurgist, materials scientist USN/Philadel Shipyard, Phila., 1978-80, U.S. Army/Picatinny Arsenal, Dover, N.J., 1980—; vis. scholar, tchr. Stanford Rsch. Inst., Menlo Park, Calif., 1975—; postdoctoral fellow Ariz. State U., Tempe, 1976-78; vis. rsch. scientist U.S. Army Rsch. Office, Durham, N.C., 1985-86; sr. scientist, Undersec. of Army fellow MIT, 1989-91; rsch. fellow Armament Tech. Directorate, 1987; exec. fellow Office Tech. Dir., 1987-88; rsch. assoc. Army Rsch. Office, 1985-86. Contbr. articles to profl. jours. Chmn. UN Assn. U.S., Glassboro State Coll., 1973-75; chmn. Energy Symposium, Glassboro State Coll., 1973-75. Mem. Am. Phys. Soc., Materials Rsch. Soc., Sigma Xi. Roman Catholic. Achievements include patents for appartus for rapid solidification, Collison Centrifugal Atomization Unit; processing path for Collision Centrifugal Atomization Processing. Home: 4 Fairview Dr Glassboro NJ 08028

OTOSHI, TOM YASUO, electrical engineer; b. Seattle, Sept. 4, 1931; s. Jitsuo and Shina Otoshi; m. Haruko Shirley Yumiba, Oct. 13, 1963; children: John, Kathryn. BSEE, U. Wash., 1954, MSEE, 1957. With Hughes Aircraft Co., Culver City, Calif., 1956-61; mem. tech. staff Jet Propulsion Lab., Calif. Inst. Tech., Pasadena, 1961—; cons. Recipient NASA New Tech. awards. Mem. Wagner Ensemble of Roger Wagner Choral Inst., L.A. Bach Festival Chorale. Mem. IEEE, Sigma Xi. Contbr. articles to profl. jours; patentee in field. Home: 3551 Henrietta Ave La Crescenta CA 91214-1136 Office: Jet Propulsion Lab 4800 Oak Grove Dr Pasadena CA 91109-8099

OTSUKA, HIDEAKI, chemistry educator; b. Oomuta, Fukuoka, Japan, Aug. 2, 1947; s. Eiji and Sayoko (Ogata) O.; m. Junko Hashimoto, June 3, 1986. BS, Tokyo U., 1971, MS, 1973, PhD, 1976. Asst. prof. Tokyo U., 1976-81; assoc. prof. pharm. phytochemistry Hiroshima (Japan) U., 1981—; vis. assoc. prof. Ohio State U., Columbus, 1986. Contbr. rsch. papers to profl. publs. mem. Am. Chem. Soc. Buddhist. Home: 2-6-26-402 Shinonomehonmachi, Minami-ku Hiroshima 734, Japan Office: Hiroshima U Sch Medicine, 1-2-3 Kasumi, Minami-ku Hiroshima 734, Japan

OTSUKA, KANJI, information science educator; b. Kyoto, Japan, June 2, 1935; s. Shirou and Miyo Otsuka; m. Keiko, June 7, 1965; children: Kayoko, Hiroko, Atsuko, Noriko. BS, Kyoto Inst. Tech., 1958; D Material Sci., Tokyo Inst. Tech., 1987. Asst. tech. Kyoto (Japan) Inst. Tech., 1958-59; engr. Hitachi, Ltd., Semiconductor Group, Tokyo, 1959-69, sr. engr., 1970-84; project leader Hitachi Ltd., Device Devel. Ctr., Tokyo, 1985—; 1985-93; prof. information sci. Meisei U., Tokyo, 1993—. Author: Ceramic Multilayer Substrate, 1987. Home: Higashiyamoto-shi, Higashiyamoto-shi, Tokyo 207, Japan Office: Meisei U Faculty Info Sci, Omeshi, 2-590 Nagafuchi,, Tokyo 198, Japan

OTT, DAVID MICHAEL, engineering company executive; b. Glendale, Calif., Feb. 24, 1952; s. Frank Michael and Roberta (Michie) O.; m. Cynthia Dianne Bunce. BSEE, U. Calif., Berkeley, 1974. Electronic engr. Teknekron Inc., Berkeley, 1974-79; chief engr. TCI, Berkeley, 1979-83; div. mgr. Integrated Automation Inc., Alameda, Calif., 1983-87, Litton Indsl. Automation, Alameda, 1987-92; founder, chmn. Picture Elements Inc., Berkeley, 1992—. Inventor method for verifying denomination of currency, method for processing digited images, automatic document image revision. Mem. IEEE, AAAS, Assn. Computing Machinery, Union of Concerned Scientists. Office: Picture Elements Inc 777 Panoramic Way Berkeley CA 94704

OTT, KARL OTTO, nuclear engineering educator, consultant; b. Hanau, Germany, Dec. 24, 1925; came to U.S., 1967, naturalized, 1987; s. Johann Josef and Eva (Bergmann) O.; m. Gunhild G. Göring, Sept. 18, 1958 (div. 1986); children: Martina, Monika. BS, J. W. von Goethe U., Frankfurt, Germany, 1948; MS, G. August U., Göttingen, Fed. Republic Germany, 1953, PhD, 1958. Physicist Nuclear Rsch. Ctr., Karlsruhe, Fed. Republic Germany, 1958-67, sect. head, 1962-67; prof. Sch. Nuclear Engring. Purdue U., West Lafayette, Ind., 1967—; cons. Argonne (Ill.) Nat. Lab., 1967—. Author: Nuclear Reactor Statics, 1983, 2nd edit., 1989, Nuclear Reactor Dynamics, 1985. Fellow Am. Nuclear Soc. Office: Purdue U Sch Nuclear Engring Lafayette IN 47907-1290

OTT, WAYNE ROBERT, environmental engineer; b. San Mateo, Calif., Feb. 2, 1940; s. Florian Funstan and Evelyn Virginia (Smith) O.; m. Patricia Faustina Bertuzzi, June 28, 1967 (div. 1983). BA in Econs., Claremont McKenna Coll., 1962; BSEE, Stanford U., 1963, MS in Engring, 1965, MA in Communications, 1966, PhD in Environ. Engring., 1971. Commd. lt. USPHS, 1966, advanced to comdr., 1982; chief lab. ops. br. U.S. EPA, Washington, 1971-73, sr. systems analyst, 1973-79, sr. rsch. engr., 1981-84, chief air toxics and radiation monitoring rsch. staff, 1984-90; vis. scientist dept. stats. Stanford (Calif.) U., 1979-81; vis. scholar Ctr. for Risk Analysis and dept. stats., civil engring., 1990-93; sr. environ. engr., EPA Atmospheric Rsch. and Exposure Assessment Lab, 1993—; dir. field studies Calif. Environ. Tobacco Smoke Study, 1993—. Author: Environmental Indices: Theory and Practice, 1976, Environmental Statistics: Probability Theory Applied to Environmental Problems; contbr. articles on indoor air pollution, total human exposure to chems., stochastic models of indoor exposure, motor vehicle exposures, personal monitoring instruments, and environ. tobacco smoke to profl. jours. Decorated Commendation medal. Mem. Internat. Soc. of Exposure Analysis (v.p. 1989-90), Am. Statis. Assn., Am. Soc. for Quality Control, Air and Waste Mgmt. Assn., Phi Beta Kappa, Sigma Xi, Tau Beta Pi, Kappa Mu Epsilon. Democrat. Clubs: Theater, Jazz, Sierra. Avocations: hiking, photography, model trains, jazz recording. Developer nationally uniform air pollution index, first total human exposure activity pattern models. Home: 1008 Cardiff Ln Redwood City CA 94061-3678 Office: Stanford U Dept Stats Sequoia Hall Stanford CA 94305

OTTENSMEYER, DAVID JOSEPH, neurosurgeon, health care executive; b. Nashville, Tenn., Jan. 29, 1930; s. Raymond Stanley and Glenda Jessie (Helpingstein) O.; m. Mary Jean Langley, June 30, 1954; children: Kathryn Joan, Martha Langley. BA, Wis. State U., Superior, 1951; MD, U. Wis., Madison, 1959; MS in Health Svcs. Adminstrn., Coll. St. Francis, 1985. Diplomate Am. Bd. Neurological Surgery. Intern then resident in gen. surgery Univ. Hosps., Madison, Wis., 1959-61; resident in neurol. surgery Univ. Hosps., 1962-65; staff neurosurgeon Marshfield Clinic, Wis., 1965-76; from instr. of neurol. surgery to clin. asst. prof. U. Wis. Med. Sch., Madison, 1964-77; chief exec. officer Lovelace Med. Ctr., Albuquerque, 1976-86; chmn. Lovelace Med. Ctr., 1986-91; clin. prof. community medicine U. N.Mex., Albuquerque, 1977-79; clin. prof. neurol. surgery U. N.Mex., 1979—; exec. v.p., chief med. officer Equicor, 1986-90; v.p. Marshfield Clinic, 1970-71, pres., chief exec. officer, 1972-75; chmn. bd. dirs. Lovelace Med. Ctr., Albuquerque, 1986-91; pres., chief exec. officer Lovelace Med. Ctr. and Clinic, Albuquerque, 1976-86, Lovelace Med. Found., 1991—; sr. v.p., chief med. officer Travelers Ins. Co., 1990-91; served on numerous adv. and com. posts. Contbr. articles to profl. jours. Col. USAR, 1960-90. Fellow ACS, Am. Coll. Physician Execs. (pres. 1985-86); mem. Am. Group Practice Assn. (pres. 1983-84), Am. Bd. Med. Mgmt. (bd. dirs. 1989-93). Republican. Episcopalian. Avocations: flying; golf; travel. Home: 2815 Ridgecrest Dr SE Albuquerque NM 87108-5132

OTTO, FRED BISHOP, physics educator; b. Bangor, Maine, Aug. 17, 1934; s. Carl Everett and Edna Rosena (Bishop) O.; m. Alma Merrill, June 23, 1957; children: Janet, Nancy, Robert, Kathryn. BS, U. Maine, 1956; MA, U. Conn., 1960, PhD, 1965. Registered profl. engr. Maine. Asst. prof. physics Colby Coll., Waterville, Maine, 1964-68; asst. prof. elec. engirng. U. Maine, Orono, 1968-74; design engr. Eaton W. Tarbell & Assocs., Bangor, Maine, 1974-79; quality control engr. Pyr-ApLavm, Brewer, Maine, 1979-80; consulting engr. pvt. practice, Orono, 1980-84; instr. physics Maine Maritime Acad., Castine, 1982-89, asst. prof. physics, 1989-93; assoc. prof. Maine Maritime Acad., 1993—. Editor: Instructor's Resource Manual to Accompany Wilson's Physics, 1992, (contbg. editor) The Castle Electrical Curriculum, 1992; contbr. articles to profl. jours. Mem. Am. Phys. Soc., Am. Assn. Physics. Tchrs., Illaminating Engring. Soc. Methodist. Home: 430 College Ave Orono ME 04473 Office: Maine Maritime Acad Castone ME 04420

OTU, JOSEPH OBI, mathematical physicist; b. Ntamante, Boki, Nigeria, Aug. 10, 1957; came to U.S., 1987; s. Michael Nku and Bridgitte Barong (Bisong) O.; m. Immaculata Obi Otu, Dec. 19, 1980; children: Michael Aguribe, Bridgitte Bangkpang. MS, Simon Fraser U., Vancouver, Can., 1986; PhD, U. Ala., Brimingham, 1990. Instr. U. Cross River State, Nigeria, 1978-82; teaching and rsch. asst. Simon Fraser U., 1983-86; teaching and rsch. asst. U. Ala., 1987-90, postdoctoral fellow, 1990, staff mem., 1991-92; prof. physics U. Wis., Waukesha, 1992—. Contbr. publs. on elem. particle physics relativity and magnetic resonance imaging. Mem. Am. Phys. Soc., Ala. Acad. Sci., Internat. Ctr. for Theoretical Physics (Italy), Phi Kappa Phi. Roman Catholic. Achievements include research on the exact metric for a static gravitational ring mass with a quadrupole moment and studied its gravitational effect on electromagnetic waves, high frequency magnetic resonance imaging (MRI) for clinical applications; demonstration that one form of effective potential for chiral symmetry breaking porposed by Casalbuoni has stable solutions that are bounded from below. Office: U Wis Dept Physics and Astronomy Waukesha WI 53188

OTVOS, LASZLO ISTVAN, JR., organic chemist; b. Szeged, Csongrad, Hungary, May 17, 1955; came to U.S., 1985; s. Laszlo and Ilona (Elekes) O.; m. Elisabeth Papp, Aug. 6, 1977; children: Balint, Judy. MS, Eotvos L. U., Budapest, 1979; PhD, Hungarian Acad. Sci., Budapest, 1985. Assoc. scientist Chem. Works of Gedeon Richter, Budapest, 1979-82; assoc. scientist Wistar Inst., Phila., 1985-89, rsch. assoc., 1990-91, asst. prof., 1992—; asst. prof. pathology and lab. medicine U. Pa., Phila., 1992— . Contbr. 61 articles to sci. jours. Recipient Weil award Am. Assn. Neuropathologists, 1988; grantee, NIH, NSF, AHAF, 1990—. Mem. Am. Chem. Soc., Am. Peptide Soc. (charter). Achievements include patent for Stabilizing Peptide Structure. Home: 801 Mockingbird Ln Audubon PA 19403 Office: Wistar Inst 3601 Spruce St Philadelphia PA 19104

OVAERT, TIMOTHY CHRISTOPHER, mechanical engineering educator; b. Chgo., Apr. 30, 1959; s. Walter Allen and Joyce Ann (Collins) O.; m. Valerie Mora, July 16, 1988; 1 chld, Teresa Noel. BSME, U. Ill., 1981; MEM, Northwestern U., 1985, PhD, 1989. Plant engr. Wells Mfg. Co.-Dura Bar Div., Woodstock, Ill., 1981-85; mech. engr. Nat. Inst. of Standards and Tech., Gaithersburg, Md., 1986; asst. prof. Penn State U., 1989—. Contbr. articles to profl. jours. Traffic safety com. Borough of State College, Pa., 1992. Named Nat. Young Investigator, NSF, 1992, Engring. Rsch. grant NSF, 1990. Mem. ASME, Soc. Tribologists and Lubrication Engrs. Office: Penn State U 313-B Mech Engring Bldg University Park PA 16802

OVER, JANA THAIS, program analyst; b. Sangley Point, USN Sta., Philippines, Aug. 4, 1956; d. John James Jr. and Betty June (Pugh) O.; m. David Paul Harrington, Mar. 23, 1985. BA cum laude, Randolph-Macon Woman's Coll., 1978; MBA, Marymount Coll. Va., 1986. Rsch. asst., typist U.S. Dept. Treasury, Washington, 1978-82, computer programmer, analyst, Office Sec. Domestic Fin., 1982-85, computer specialist, Office Fiscal Asst. Sec., 1985-91; computer specialist (telecommunication project) U.S. Dept. Treasury, 1991-92; program analyst (datacomm.) Office Asst Sec. Mgmt., 1992—. Docent, U.S. Dept. Treasury, Office of the Curator, Washington, 1989—; mem. Choctaw Nation of Okla. Recipient Asian Studies award, Randolph-Macon Woman's Coll., Va., 1978, Award of Distinction in Cash Mgmt. U.S. Dept. Treasury, 1988. Mem. NAFE, Randolph-Macon Woman's Coll. Alumnae (treas. Washington area chpt. 1982-83, pres. 1991-93), Treasury Hist. Assn. (bd. dirs., treas. 1989-90). Office: US Dept Treasury 1500 Pennsylvania Ave NW Washington DC 20220

OVERBY, VERITI PAGE, chemist, environmental protection specialist; b. Ann Arbor, Mich., Oct. 18, 1964; d. Quentin Barry Smith and Janet Marie (Gatzka) Egri; m. Lawrence Overby Jr., May 18, 1990. BS in Chemistry, Mary Washington Coll., 1986; MS in Chemistry, Old Dominion U., 1990; student, Tidewater C.C., 1991—. Phys. sci. technician Norfolk Naval Shipyard, Portsmouth, Va., 1989, chemist, 1990, sr. chemist, 1991; sr. chemist Naval Supply Ctr., Norfolk, Va., 1991—; presenter at profl. confs. Contbr. articles to profl. publs. Mem. Am. Chem. Soc., Va. Acad. Scis. Home: Apt 74 7621 Bondale Ave Norfolk VA 23505 Office: Naval Supply Ctr Fuel Lab Bldg W 388 Code 702 Norfolk VA 23511-3392

OVERCASH, MICHAEL RAY, chemical engineering educator; b. Kannapolis, N.C., July 17, 1944; s. Ray Leonard and Ruth (Crabbe) O.; m. Mary C.N. Yoong; 1 child, Rachael. BS, N.C. State U., 1966; MS, U. New South Wales, Sydney, Australia, 1967; PhD, U. Minn., 1972. Prof. biol. and agrl. engring. N.C. State U., Raleigh, 1972—, prof. chem. engring. dept., 1981—, ctr. dir. process improvement and pollution prevention ctr., 1989—. Author: Design Industrial Land Treatment Systems, 1979, Industrial Pollution Prevention, 1986, Prevention, Management and Compliance for Hazardous Wastes, 1987. Com. mem. N.C. Govs. Waste Mgmt. Bd., 1984-88, Air Force Sci. Adv. Bd., 1984-86. Fulbright scholar, 1966-67; disting. vis. scientist EPA, 1986-89. Mem. Am. Inst. Chem. Engrs., Am. Soc. Agrl. Engrs. (young researcher award 1984), Inst. Chem. Engrs., N.C. Acad. Scis., History of Sci. Soc., Nat. Acad. of Sci. Office: Pollution Prevention Rsch Ctr NC State U Dept Chem Engring PO Box 7905 Raleigh NC 27695-7905

OVERHAUSER, ALBERT WARNER, physicist; b. San Diego, Aug. 17, 1925; s. Clarence Albert and Gertrude Irene (Pehrson) O.; m. Margaret Mary Casey, Aug. 25, 1951; children—Teresa, Catherine, Joan, Paul, John, David, Susan, Steven. A.B., U. Calif. at Berkeley, 1948, Ph.D., 1951; D.Sc. (hon.), U. Chgo., 1979. Research asso. U. Ill., 1951-53; asst. prof. physics Cornell U., 1953-56, asso. prof., 1956-58; supr. solid state physics Ford Motor Co., Dearborn, Mich., 1958-62; mgr. theoret. scis. Ford Motor Co., 1962-69, asst. dir. phys. scis., 1969-72, dir. phys. scis., 1972-73; prof. physics Purdue U., West Lafayette, Ind., 1973-74; Stuart disting. prof. physics Purdue U., 1974—. With USNR, 1944-46. Recipient Oliver E. Buckley Solid State Physics prize Am. Phys. Soc., 1975, Alexander von Humboldt sr. U.S. scientist award, 1979. Fellow Am. Phys. Soc., Am. Acad. Arts and Scis.; mem. Nat. Acad. Scis. Home: 236 Pawnee Dr West Lafayette IN 47906-2115 Office: Purdue U Dept of Physics West Lafayette IN 47907

OVERMYER, ROBERT FRANKLYN, astronaut, marine corps officer; b. Lorain, Ohio, July 14, 1936; s. Rolandus and Margaret (Fabian) O.; m. Katherine Ellen Jones, Oct. 17, 1959; children: Carolyn Marie, Patricia Ann, Robert Rolandus. B.S. in Physics, Baldwin Wallace Coll., 1958; M.S. in Aeros., U.S. Naval Postgrad. Sch., 1964. Commd. 2d lt. U.S. Marine Corps., 1958, advanced through grades to lt. col., 1972; completed aerospace research pilot sch. Edward AFB, 1966; astronaut with Manned Orbiting Lab. NASA, 1966-69; astronaut with Manned Spacecraft Ctr. NASA, Houston, from 1969; mem. support crew Appolo-Soyuz Test Project, 1973-75; co-pilot on 5th mission Columbia NASA, Houston, 1982; comdr. Spacelab 3 mission NASA, 1985. Recipient Alumni Merit award Baldwin Wallace Coll., 1967. Mem. Soc. Exptl. Test Pilots (assoc.), Sigma Xi. Office: 18510 Point Lookout Dr Houston TX 77058-4027

OVERSKEI, DAVID, physicist; b. Brookings, S.D., Dec. 26, 1948. BA, U. Calif., Berkeley, 1971; PhD in Physics, MIT, 1976. Operation mgr. plasma physics ctr. ptrogram, Francis Bitter Nat. Magnet Lab., plasma fusion ctr., MIT, 1976-81; mgr. Tokamak Physics Program, Fusion Divsn. Gen. Atomic Co., 1981—. Mem. Am. Phys. Soc. Achievements include research in controlled thermonuclear fusion research. Office: General Atomics PO Box 85608 San Diego CA 92186*

OVERSTREET, JAMES WILKINS, obstetrics and gynecology educator, administrator. BA in Biology magna cum laude, U. South, 1967; BA in Natural Scis., U. Cambridge, Eng., 1970, PhD in Reproductive Physiology, 1973, MA in Natural Scis., 1974; MD, Columbia U., 1974. Diplomate Nat. Bd. Med. Examiners; lic. physician, Calif. NIH Med. Scientist Tng. fellow dept. anatomy coll. physicians ans surgeons Columbia U., 1970-72, NIH Med. Scientist Tng. fellow Internat. Inst. for Study Human Reproduction, 1972-74; asst. resident in ob-gyn. Presbyn. Hosp., N.Y.C., 1974; Ford Found. Postdoctoral Rsch. fellow dept. ob-gyn. Cornell U. Med. Coll., 1975-76; asst. prof. human anatomy and ob-gyn. sch. medicine U. Calif., Davis, 1976-80, assoc. prof. human anatomy, 1980-84, assoc. prof. ob-gyn. sch. medicine, 1980-85, prof. ob-gyn. sch. medicine, 1985—, chief divsn. reproductive biology and medicine dept. ob-gyn. sch. medicine, 1983-86, dir. lab. for energy=related health rsch., 1985-88, dir. Inst. Toxicology and Environ. Health, unit leader devel. and reproductive biology Primate Rsch. Ctr., 1988—; mem. sci. rev. panel for health rsch., reviewer test rules dept. and reproductive abd devel. toxicology brs. U.S. EPA; mem. ad hoc study sect., cons. Nat. Inst. for Occupational Safety and Health; chair AIDS and related rsch. rev. group NIH; chmn. spl. study sect., mem. site visit team, mem. ad hoc reproductuve endocrinology study sect. Nat. Inst. Child Health and Human Devel.; reviewer, mem. site visit team Nat. Inst. Environ. Health Scis./NIH, Med. Rsch. Coun. Can.; mem. tech. adv. com. and site visit team contraceptive rsch. and devel. project Agy. for Internat. Devel./Ea. Va. Med. Sch.; temp. advisor spl. program rsch. devel. and rsch. tng. in human reproduction WHO; mem. reproductive and devel. toxicology program rev. panel Chem. Industry Inst. Toxicology; mem. exec. com. systemwide toxic substances rsch. and tng. program U. Calif.; reviewer NSF, Office Health and Environ. Rsch., U.S. Dept. Energy, March of Dimes Reproductive Hazards in the Workplace Rsch. Grants Program, Mt. Sinai Hosp., Alta. Heritage Cancer Grants, U.S.-Israel Binational Agrl. Rsch. and Devel. Fund; clin. cons. lab. surveys program Coll. Am. Pathologists; cons. Ctr. for Drugs and Biologics, U.S. FDA, Inst. for Internat. Studies in Natural Family Planning, Georgetown U., Internat. Devel. Rsch. Ctr. Assoc. editor

Molecular Reproduction and Devel.; mem. editorial bd. Biology of Reproduction, 1983-86, Jour. In Vitro Fertilization and Embryo Transfer, 1984-89, Reproductive Toxicology, Fertilty and Sterility, 1984-92, Jour. Andrology, 1990-92; referee Jour. Reproduction and Fertilty, Jour. Exptl. Zoology, Am. Jour. Physiology, Am. Jour. Ob-Gyn., Jour. Urology, Science, Jour. Clin. Endocrinology and Metabolism, Archives Internal Medicine, Internat. Jour. Andrology, Reproduction, Nutrition, Devel., Western Jour. Medicine, Contraception, Human Reproduction, Proceedings Royal Soc. Series B.; invited lectr. in field. Georgia Fulbright scholar, 1967-68; recipient Rsch. Career Devel. award NIH, 1978-83, Disting. Career in Clin. Investigation award Columbia Presbyn. Med Ctr., 1992; grantee Syntex Rsch. Divsn., 1978-81, NIH, 1978-81, 78-83, 79-90, 81-90, 85-88, 86—, 87—, 88—, 89—, 90— (two grants), 91—, 92—, 93—, U. Calif., 1981-82, U.S. EPA, 1981-84, 82-84, Nat. Inst. for Occupational Safety and Health, 1982-84, U.S. Dept. Energy, 1985-88, Georgetown U., 1987-89, 92—, Merck Rsch. Labs., 1988-90, 92, 92—, March of Dimes, 1988-90, Semiconductor Industry Assn., 1989-92, Tobacco Related Disease Rsch. Program, 1990-93, Andrew W. Mellon Found., 1993—,. Mem. Am. Fertility Soc., Am. Soc. Andrology (exec. coun. 1986-89), Soc. for Study Fertility, Soc. for Study Reproduction, Phi Beta Kappa. Achievements include research in physiology of mammalian spermatozoa, sperm transport in the female reproductive tract, in vivo and in vitro mammalian fertilization, diagnosis and therapy of human male infertility, reproducti, environmental and occupational hazards to male and female fertility, contraceptive development, reproductive endocrinology. Office: U Calif Davis Sch Medicine Divsn Reproductive Biology & Medicine Dept Ob-Gyn Suber House Davis CA 95616

OVERTON, SANTFORD VANCE, applications chemist; b. Rocky Mount, N.C., Sept. 15, 1949; s. Levy Lemuel and Irma Mae (Jenkins) O.; m. Joan Ann Lasota, Nov. 28, 1987. MS in Biology, East Carolina U., 1979; PhD in Plant Pathology, Va. Poly. Inst. and State U., 1986. Staff scientist Organogenesis, Inc., Cambridge, Mass., 1986-88; cons. Agri-Diagnostics Assocs., Cinnaminson, N.J., 1988-89; product mgr. Sci. Instrument Svcs., Ringoes, N.J., 1989—. Contbr. articles to profl. publs. Mem. Am. Chem. Soc., Am. Soc. Mass Spectrometry, Am. Phytopathol. Soc., Sigma Xi, Gamma Sigma Delta, Phi Kappa Phi. Achievements include patent for short-path thermal desorption apparatus for use in gas chromatography techniques, patent for injection assembly for transferring a component to be tested. Home: 8 Country Club Dr Ringoes NJ 08551 Office: Sci Instrument Svcs 1027 Old York Rd Ringoes NJ 08551

OVERWEG, NORBERT IDO ALBERT, physician; b. Enschede, The Netherlands; s. Ido and Bella Theresa (Lievenboom) O.; MD, U. Amsterdam, 1957; m. Angelique de Gorter; children: Eleonore, Elizabeth, Harold. Intern, Univ. Amsterdam Hosp., 1958-60; resident Rochester (N.Y.) Gen. Hosp., 1961-62; postdoctoral fellow dept. pharmacology Columbia U. Coll. Physicians and Surgeons, 1962-65; instr. dept. public health Columbia U., 1965-66; rsch. assoc. dept. surgery Columbia U., 1967-71; rsch. collaborator, asst. attending physician Brookhaven Nat. Lab., 1966-67; asst. prof. dept. physiology and pharmacology N.Y. U., 1971-78; cons. Lung Rsch. Ctr., Yale U. Sch. Medicine, 1972-73; pvt. practice medicine specializing in internal medicine, N.Y.C., 1967—; attending staff St. Clare's Hosp. and Health Center, Cabrini Med. Ctr., Med. Arts Center Hosp.; clin. investigator antihypertension, anti-depressant, anti-anxiety and gastro-intestinal drugs. NIH fellow, 1964-65. Mem. Am. Soc. Pharmcology and Exptl. Therapeutics, Am. Physiol. Soc., Am. Soc. Hypertension, Am. Coll. Clin. Pharmacology, N.Y. Acad. Scis., AAAS, AAUP, Royal Dutch Soc. Advancement of Medicine, Harvey Soc., Netherlands Am. Med. Soc., Eastern Hypertension Soc., N.Y. County Med. Soc., Med. Soc. of N.Y., Sigma Xi. Club: Netherlands of N.Y., Inc. Contbr. articles to profl. jour. Office: 133 E 73d St New York NY 10021

OWEN, BRUCE DOUGLAS, animal physiologist; b. Edmonton, Alta., Can., Oct. 1, 1927. BSc, U. Alta., 1950, MSc, 1952; PhD in Animal Nutrition, U. Sask., 1961. Animal nutritionist Lederle labs. divsn. Am. Cyanamid Co., N.Y., 1952-54; animal husbandman Can. Dept. Agr., 1954-57; from lectr. to prof. animal sci. U. Sask., 1957-77; prof., chmn. dept. animal sci. U. B.C., Vancouver, 1977—. Mem. Am. Soc. Animal Sci., Can. Soc. Animal Sci., Nutrition Soc. Can., Agr. Inst. Can. (AIC Fellowship award 1991). Achievements include research in passive immunity, fat soluble vitamin transport, vitamin requirements of swine, protein quality evaluation, swine nutrition and physiology, factors influencing composition of feed grains and forages. Office: Univ of British Columbia, 2357 Main Mall Ste 248, Vancouver, BC Canada V6T 1Z4*

OWEN, DUNCAN SHAW, JR., physician, medical educator; b. Fayetteville, N.C., Oct. 24, 1935; s. Duncan S. and Mary Gwyn (Hickerson) O.; m. Irene Lacy Rose, Oct. 22, 1966; children: Duncan Shaw III, Robert Burwell, Frances Gwyn. BS, U. N.C., 1957, MD, 1960. Diplomate Am. Bd. Internal Medicine. Intern Med. Coll. Va., Richmond, 1960-61; jr. asst. resident in medicine N.C. Meml. Hosp., Chapel Hill, 1961-62; asst. resident in medicine Med. Coll. Va., Richmond, 1964-65, fellow in rheumatic diseases, 1965-66; practice medicine specializing in internal medicine and rheumatology Richmond, Va., 1966—; instr. in medicine Med. Coll. Va., Richmond, 1966-67, asst. prof., 1967-71, assoc. prof., 1971-78, prof. dept. internal medicine, 1978—; Taliaferro/Scott Disting. prof. internal medicine Med. Coll. Va., Va. Commonwealth U., 1989—; mem. staff Med. Coll. Va. Hosp., McGuire VA; bd. dir. clin. tng. div. rheumatology, allergy, immunology, chmn. clin. activities comm., dept. internal medicine; chmn. med. adv. com. Richmond Br. Arthritis Found., 1966-75, bd. dirs., 1966—, mem. nat. patient edn. com., 1979-80; med. adv. Social Security Adminstrn., HHS, 1967—; bd. dirs. Blue Shield Va., 1975-77; co-chmn. Arthritis Project Va. Regional Med. Program, 1975-76; bd. dirs. U. Internal Medicine Found., 1979—; producer Your Health TV Series, Va. Ednl. TV, 1978-79; producer Update in Medicine, Good Morning Virginia TV Show, 1980; mem. various coms. in field; proctor Am. Bd. Internal Medicine, 1977—. Contbr. numerous papers, chpts. in books, articles to profl. jours.; assoc. editor: Va. Med., 1978—; editorial reviewer Jour. AMA, 1979—, Arthritis Rheumatism, 1981—, Jour. Rheumatology, 1984—. Mem. usher's guild First Presbyn. Ch., Richmond, Va., 1966-70, deacon, 1974-77, chmn. of diaconate, 1976-77, elder, 1978—, chmn. witness com., 1978-80; co-chmn. physicians statewide capital funds campaign Va. Commn. U., 1986-87; bd. dirs. Mooreland Farms Assn., 1971-73, 77-81, Va. chpt. Arthitis Found., 1970-85; mem. Va. Mus., Richmond Symphony; bd. dirs. Richmond Area Health Care Coalition, 1980-84. Served to capt. MC, 1962-64. Recipient Army Commendation medal, 1964. Nat. Inst. Arthritis and Metabolic Diseases fellow, 1965-66; recipient Gerard B. Lambert award, 1974-75, Disting. Service award Arthritis Found., 1971. Fellow ACP, Am. Coll. Rheumatology; mem. AMA (expert on diagnostic and therapeutic tech. assessment program), Am. Rheumatism Assn. (exec. com. 1979-80), Richmond Acad. Medicine (pres. 1982, chmn. bd. 1983, parliamentarian 1988—), Med. Soc. Va. (mem. com. on aging 1980-89, v.p. 1973, 75, del. 1972—, scholarship com. 1980-89), Richmond Soc. Internal Medicine (bd. dirs. 1971-73), Jr. Clin. Club (emeritus), Met. Richmond C. of C. (bd. dirs. 1981-84), Country Va. Club, Alpha Omega Alpha. Avocations: hunting, fishing, photography, amateur radio. Home: 8910 Brieryle Rd Richmond VA 23229-7704 Office: Med Coll Va Nelson Clinic Rm 742 PO Box 647 Richmond VA 23298-0647

OWEN, KENNETH DALE, orthodontist; b. Charlotte, N.C., May 9, 1938; s. Olin Watson and Ruth (Watlington) O.; m. Lura Aven Carnes, Feb. 14, 1958; children: Kenneth Dale, Aven Anna. BS, Davidson Coll., 1959; DDS, U. N.C., 1963, MSc in Orthodontics, 1967. Diplomate Am. Bd. Orthodontics. Pvt. practice dentistry specializing in orthodontia, Charlotte, 1966—; asst. clin. prof. U. N.C. Sch. Dentistry, 1969-72. Bd. dirs. N.C. Dental Found., 1973-81, 89-90, exec. com., 1974-80, v.p. 1976-77, pres., 1978-79; bd. dirs. Holiday Dental Conf. Found., 1989—, v.p., 1990—, conf. chmn. 1990—. Adminstrv. bd. Meyers Park United Meth. Ch., 1976-79. Served with Dental Corps AUS, 1963-65. Fellow Internat. Coll. Dentists (dep. regent N.C. 1986, 87), Am. Coll. Dentists; mem. ADA (ho. of dels. 1981-92, 16th trustee dist. caucus vice chmn. 1986-89, chmn. 1989-92), Am. Assn. Orthodontists (ho. of dels. 1980-88, 90-93) So. Assn. Orthodontists (trustee 1983-85, dir. 1987-93, pres. 1991-92), N.C. Assn. Orthodontists (bd. dirs. 1976-88, sec.-treas. 1976-88, pres. 1979-80), N.C. Dental Soc. (ho. dels. 1969-77, 81-93, trustee 1980-91, sec. treas. 1987-88, pres. elect 1988-89, pres. 1989-90), 2d Dist. Dental Soc. (editor 1967-69, sec.-treas. 1971-74, pres. 1975-76, exec. coun. 1971-77, 80-87), Charlotte Dental Soc. (chmn. various coms., dir. 1978-79, v.p. 1980-81), Stanly County Dental Soc., Coll. Diplomates Am. Bd. Orthodontics, U. N.C. Orthodontic Alumni Assn. (sec.-treas. 1971, v.p. 1972-73, pres. 1974-75, exec. com. 1971-76), U. N.C. Gen. Alumni Assn. (life), U. N.C. Dental Alumni Assn., Orthovista Orthodontic Study Group, Delta Sigma Delta (life; pres. N.C. grad. chpt. 1970-71), Omicron Kappa Upsilon, Kappa Sigma, Alpha Epsilon Delta. Home: 3724 Pomfret Ln Charlotte NC 28211-3726 Office: 497 N Wendover Rd Charlotte NC 28211-5001 also: 119 Yadkin St Albemarle NC 28001

OWEN, MICHAEL, agronomist, educator. Prof. dept. agronomy Iowa State U., Ames. Recipient CIBA-GEIGY/Weed Sci. Soc. Am. award CIBA-GEIGY Corp., 1992. Office: Iowa State U Dept Agronomy 2104 Agronomy Hall Ames IA 50011

OWEN, RAY DAVID, biology educator; b. Genesee, Wis., Oct. 30, 1915; s. Dave and Ida (Hoeft) O.; m. June J. Weissenberg, June 24, 1939; 1 son, David G. B.S., Carroll Coll., Wis., 1937, Sc.D., 1962; Ph.D., U. Wis., 1941, Sc.D., 1979; Sc.D., U. of Pacific, 1965. Asst. prof. genetics, zoology U. Wis., 1944-47; Gosney fellow Calif. Inst. Tech., Pasadena, 1946-47; assoc. prof. div. biology Calif. Inst. Tech., 1947-53, prof. biology, 1953-83, also chmn., v.p. for student affairs, dean of students, prof. emeritus, 1983—; research participant Oak Ridge Nat. Lab., 1957-58; Cons. Oak Ridge Inst. Nuclear Studies; mem. Pres.'s Cancer Panel. Author: (with A.M. Srb) General Genetics, 1952, 2d edit. (with A.M. Srb, R. Edgar), 1965; Contbr. articles to sci. jours. Mem. Genetics Soc. Am. (pres., Thomas Hunt Morgan medal 1993), Am. Assn. Immunologists, Am. Soc. Human Genetics, Western Soc. Naturalists, Am. Soc. Zoologists, Am. Genetics Assn., Nat. Acad. Scis., Am. Acad. Arts and Scis., Am. Philos. Soc., Sigma Xi. Home: 1583 Rose Villa St Pasadena CA 91106-3524 Office: Calif Inst Tech 156-29 Pasadena CA 91125

OWEN, THOMAS EDWARD, environmental engineer; b. Storm Lake, Iowa, July 12, 1954; s. Elmer Edward and Jayn Elizabeth (Mittelstadt) O.; m. Connie Gail Vinson, Jan. 1, 1993; 1 child, Jayn Elizabeth. BS, U. Wyo., 1977, MSCE, 1985. Environ. engr. U.S. Dept. Energy, Laramie, Wyo., 1977-83, Western Rsch. Inst., Laramie, 1983-86; environ. scientist Ariz. Pub. Svc. Co., Phoenix, 1986-89; sr. environ. engr. Intel Corp., Rio Rancho, N.Mex., 1989—, vice chmn. environ. team, 1991, chmn. environ. team, 1992; cons. Synthetic Fuels Corp., 1981-83. Contbr. articles to profl. pubs. Mem. Sigma Xi (assoc.), Omicron Delta Kappa. Achievements include documentation of exhaust emission rates from semiconductor process equipment; development of semiconductor processing equipment environmental standards and performance criteria. Home: 5413 Purcell Dr NE Albuquerque NM 87111

OWEN, THOMAS EDWIN, electrical engineer; b. Alexandria, La., Jan. 6, 1931; s. Edwin Lewis and Carrie Lou (Tullos) W.; m. Dorothy Ann Nelken, Nov. 15, 1953; children: Michael Scott, Carol Lynn Owen. BSEE, Southwestern La. U., 1952; MSEE, U. Tex., 1957, PhD in Elec. Engring., 1964. Registered profl. engr., Tex. Rsch. engr. Def. Rsch. Lab.-Univ. Tex., Austin, 1954-60; asst. prof. elec. engring. U. Tex., Austin, 1957-60; sr. rsch. engr. S.W. Rsch. Inst., San Antonio, 1960-65, mgr. earth sci. applications, 1965-74, dir. geoscis., 1974-89, inst. engr., 1989—; adv. com. for rsch. S.W. Rsch. Inst., San Antonio, 1989—, chmn. com. for rsch., 1991-92. Author: (chpt.) Development in Geophysical Experimental Methods, 1982, (vol. 3) Geotechnical and Environmental Geophysics, 1990; editor Jour. of Applied Geophy sics, 1991—. V.p. Helotes (Tex.) Vol. Fire Dept., 1992. With USN, 1952-54. Mem. IEEE, Acoustical Soc. Am., Soc. Profl. Well Log Analysts, Soc. Exploration Geophysicists, Sigma Xi. Republican. Methodist. Achievements include 18 patents in fields of biomedical, electrrostatics, micro-particle prodn. and sorting, borehole geophys. instrumentation, environ. geophyscis, geophys. exploration, underwater sound sources. Home: 10914 Bar X Trail Helotes TX 78023 Office: Southwest Rsch Inst 6220 Culebra Rd San Antonio TX 78228

OWEN, THOMAS J., environmental engineer; b. Hunt Park, Calif., Nov. 17, 1949; s. Edwin C. and Adelle F. (Kaplan) O. BA, U. Calif., Davis, 1974; MSCE, Calif. State U., Long Beach, 1993. Cert. sanitation engr., site assessment and remediation, hazardous materials mgmt., environ. inspector; registered environ. mgr., hazardous substance profl.; lic. gen. engring. contractor, hazardous substances remediation/ removal. Project engr. TPE Environ., Santa Ana, Calif., 1984-87; sr. project mgr. ENSR Corp., Irvine, Calif., 1987-90; sr. engr. Ebasco Environ., Santa Ana, 1990-91; prin. engr. Geraghty & Miller, Inc., Industry, Calif., 1991, Burlington Environ., Inc., Huntington Beach, Calif., 1991—. Contbr. articles to profl. jours. Recipient Citation for Achievements in Environ. Industry Nat. Environ. Registry, 1992. Mem. NSPE (fellow), Nat. Environ. Health Assn. Democrat. Roman Catholic. Achievements include devel. of on-site remediation treatment of hazardous waste with a soil washing process, bioreactor, in-situ air-scarging. Office: Burlington Environ Inc 18377 Beach Blvd Ste 211 Huntington Beach CA 92648

OWEN, THOMAS JOSEPH, acoustical engineer; b. Louisville, Ky., Apr. 27, 1946; s. Wilbur Thomas and Mary Anita (Meagher) O.; m. Linda Mary Kitchin, Oct. 13, 1972 (div. 1984); children: Michael, Jennifer, Ben; m. Sandy Kay Metheny, June 10, 1989. BA in History, Bellarmine U., 1969. Sound engr. CTI Records, Inc., N.Y.C., 1972-73; chief engr. Springboard Apex Records, Rahway, N.J., 1973-78, Rodger & Hammerstein Archive of Recorded Sound, N.Y.C., 1979-89; pres. Owl Investigations, Inc., New York, NY, 1989—. Contbr. articles to Audio Engring. Soc. Jour., Internat. Assn. for Identification Jour. Bd. dirs. Warren County Arts Commn., Bowling Green, 1991—; county chmn. Corns for Lt. Gov., Bowling Green, 1990; com. mem. Downtown Bus. Assn., Bowling Green, 1990—. Nominated for two Grammy awards; winner Ampex Golden Reel award. Mem. Audio Engring. Soc. (bd. dirs. 1989-91, chmn. WG-12 forensic audio 1991—), Acoustical Soc. Am., Internat. Assn. for Identification (bd. cert. 1987—). Office: Owl Investigations Inc 500 Fifth Ave Ste 2300 New York NY 10110

OWENS, ALBERT HENRY, JR., oncologist, educator; b. Staten Island, N.Y., Aug. 27, 1926; married, 1949; 4 children. BA, MD, Johns Hopkins U., 1949. Diplomate Am. Bd. Internal Medicine. Staff internal medicine Johns Hopkins U. Hosp., 1949-50, 52-53, 55-56; from instr. to assoc. prof. medicine Johns Hopkins U. Sch. Medicine, 1956-68, instr. pathobiology sch. hygiene and pub. health, 1957-76, prof. oncology and medicine, 1968—, dir. oncology ctr., 1976—; fellow pharmacology and exptl. therapeutics, sch. medicine Johns Hopkins U., 1953-55; vis. physician Balt. City Hosp., 1957—. Fellow ACP; mem. Am. Soc. Pharmacology and Exptl. Therapeutics, Am. Soc. Clin. Oncology, Am. Assn. Cancer Insts., N.Y. Acad. Sci. Achievements include research in neoplastic diseases. Office: Johns Hopkins U Oncology Ctr 600 N Wolfe St Baltimore MD 21205-2104*

OWENS, EDWARD HENRY, geologist, consultant; b. Stamford, Eng., Mar. 25, 1945; came to U.S., 1975; s. Charles Edward and Gladys Elizabeth (Sauntson) O.; m. Beti Wyn Williams, July 25, 1966 (div. 1973); 1 child, Sasha; m. Linda Marie Zimlicki, Oct. 2, 1974; children: Jennifer, Gareth, Tristan. BS in Geology, U. Wales, 1967; MS, McMaster U., Hamilton, Ont., Can., 1969; PhD, U. S.C., 1975. expert cons. to Internat. Maritime Orgn. in Africa, S.Am. and Caribbean; tech. adviser for oil spills including Arrow, Nestucca, Exxon Valdez, Arabian Gulf Coast. Rsch. scientist Geol. Survey of Can., Bedford, 1971-75; asst. prof. Coastal Studies Inst., La. State U., Baton Rouge, 1975-79; v.p. Woodward-Clyde Cons., Victoria, B.C., Can., 1979-83; Europe/Africa mgr. Woodward-Clyde Oceaneering, Aberdeen, Eng., 1983-85; gen. mgr. Oceaneering Geoscience, Aberdeen, Eng., 1985-86; mng. dir. Geoscience Svcs. Ltd., Aberdeen, Eng., 1986-87; sr. cons. Woodward-Clyde Cons., Seattle, 1987—. Chmn. Environ. Adv. Bd., Saanich, B.C., Can., 1981-83. Mem. Geol. Soc. Am., Nat. Acad. Scis. (marine bd. oil R&D com. 1991—), Wash. Environ. Industry Assn. (bd. dirs. 1990—), Nat. Assn. Environ. Profls. (bd. dirs. 1988-89), Am. Assn. Petroleum Geologists. Office: Woodward Clyde Cons 3440 900 4th Ave Seattle WA 98164-1001

OWENS, EDWIN GEYNET, mathematics educator; b. Clearfield, Pa., Jan. 26, 1953; s. Geynet Harold and Priscilla Jane (Stevens) O.; m. Patti Ann Lingle, Oct. 18, 1975; children: Steven, Lisa, Kristy. BS cum laude, Lock Haven State Coll., 1974; MEd, Pa. State U., 1978, cert. in Ednl. Administrn., 1985. Math. tchr. Penn Manor High Sch., Millersville, Pa., 1974; math. tchr., computer coord. Hollidaysburg (Pa.) Area Sch. Dist., 1974-87; math. instr. Susquehanna U., Selinsgrove, Pa., 1987-92; math. dept. chair, math., computer tchr. Solanco Area Sch. Dist., Quarryville, Pa., 1989-90; asst. prof., chair math dept. Pa. Coll. Tech., Williamsport, Pa., 1990—; adj. math. instr. Pa. State U., York, 1989-90; math. tchr. Upward Bound program St. Francis Coll., Loretto, Pa., 1980-83; part-time math. instr. Williamsport Area C.C., 1988-89; part-time math. and computer instr. Pa. State U., Altoona, 1981-87; instr. sci. edn. and math. Lycoming Coll., Williamsport. Mem. ASCD, NEA, Nat. Coun. Tchrs. Math., Pa. State Edn. Assn. Presbyterian. Avocations: computer technology, swimming, hunting, travel. Home: 258 Delaware Dr Watsontown PA 17777-9714 Office: Pa Coll Tech 1 College Ave Williamsport PA 17701-5799

OWENS, GARY MITCHELL, family physician; b. Salisbury, Md., July 31, 1949; s. Avery Donovan and Elizabeth (Mitchell) O.; B.A., U. Pa., 1971; M.D., Thomas Jefferson U., 1975; children: Aaron David, Scott Christopher, Stefanie Erin. Resident in family medicine Wilmington (Del.) Med. Center, 1975-78, chief resident, 1978, teaching assoc. dept. family medicine, 1978-91; practice medicine specializing in family practice, Wilmington, 1978-91; teaching assoc. dept. family medicine St. Francis Hosp., Wilmington, 1978-91; med. dir. Phoenix Steel Co., 1980-87; med. dir. HMO of Delaware Valley, Delaware Plan, 1985-91, Delaware Valley HMO, 1991, assoc. med. dir. Quality Assurance, 1987-91, also chmn. credentials com.; med. dir. Keystone Health Plan East, Phila., 1991—; staff, coun. mem., chmn. reappointment com. Med. Ctr. Del.; cons. NorAm. Chem. Co., 1984-91. Diplomate Am. Bd. Family Practice. Fellow Am. Acad. Family Physicians; mem. AMA, Am. Coll. Physician Execs., Med. Soc. Del. (congress of del. 1986-90), New Castle County Med. Soc., Phila. Acad. Family Physicians , Am. Occupational Med. Assn., Alpha Omega Alpha. Roman Catholic. Home: 19 Circle Dr West Grove PA 19390 Office: PO Box 7516 1901 Market St Philadelphia PA 19101-7516

OWENS, GREGORY RANDOLPH, physician, medical educator; b. Glendale, W. Va., Oct. 3, 1948; s. Elmer Herman and Anne Elizabeth (Kroggel) O.; m. Jane Marie Fleming, June 1, 1974; children: Gregory R. Jr., Allison Fleming. AB cum laude, Princeton U., 1970; MD, U. Pa., 1974. Diplomate Am. Bd. Internal Medicine, Am. Bd. Pulmonary Medicine, Am. Bd. Critical Care Medicine. Intern in internal medicine Hosp. U. Pa., Phila., 1974-75, residency in internal medicine, 1975-77, fellowship pulmonary disease, 1977-78, chief med. resident, 1978-79, fellowship pulmonary disease, 1979-80; asst. prof. medicine U. Pitts., 1980-86, assoc. prof. medicine, 1986-93, prof. medicine, 1993—, assoc. chief div. pulmonary medicine, 1984—; chief Montefiore U. Hosp., Pitts, 1991—; dir. Pulmonary Exercise Physiology Lab., Presbyn.-Univ. Hosp., Pitts, 1980—, co-dir. Pulmonary Function Lab., 1980—; co-dir. Occupational Lung Clinic, Falk Clinic, Pitts., 1983—. Contbr. articles to New Eng. Jour. Medicine, Am. Jour. Medicine, Jour. Lab. Clin. Medicine and others, presenter at profl. confs. Coach Oakmont (Pa.) Athletic Assn., 1988—, bd. dirs. Am. Lung Assn. Named awardee Future Leaders of Pulmonary Medicine, Chgo., 1984, Preventive Pulmonary Acad. award NIH, 1988—; grantee, Health Rsch. and Svc. Found., 1980-82, 83-85, NIH Lung Health Study P.I., 1984-89, 1987—. Mem. ACP, Am. Coll. Chest Physicians, Am. Thoracic Soc., Am. Fedn. for Clin. Rsch., Pa. Thoracic Soc. (pres. 1991). Avocations: golfing, gardening. Office: U Pitts Sch Medicine 1117 Kaufman Bldg 3471 Fifth Ave Pittsburgh PA 15213

OWENS, JOSEPH FRANCIS, III, physics educator; b. Syracuse, N.Y., Sept. 23, 1946; s. Joseph Francis Jr. and Georgianna (Borst) O.; m. Linda Nancy Baker, May 30, 1970; children: Jeffrey Forrest, Susan Melinda. BS, Worcester Polytechnic Inst., 1968; PhD, Tufts U., 1973. Rsch. assoc. Case Western Res. U., Cleve., 1973-76; rsch. assoc. Fla. State U., Tallahassee, 1976-79, rsch. asst. prof., 1979-80, asst. prof., 1980-82, assoc. prof., 1982-85, prof., 1985—, assoc. chmn. physics dept., 1988-91, chmn. physics dept., 1991—. Contbr. articles to profl. jours. Named Outstanding Jr. Investigator, U.S. Dept. Energy, 1979; recipient Developing Scholar award Fla. State U., 1982. Mem. Am. Phys. Soc., Acad. Model Aeronautics, Aircraft Owners and Pilots Assn. Achievements include rsch. in theoretical high energy physics using perturbation theory applied to quantum chromodynamics, scaling violations, structure functions, high-pt. processes, higher order calculations. Office: Physics Dept B-159 Florida State Univ Tallahassee FL 32306*

OWENS, JUSTINE ELIZABETH, cognitive psychology educator; b. New Brunswick, N.J., Oct. 1, 1953; d. William Thomas and Monica (LuBera) O.; m. Kirk Roberts, June 16, 1989; 1 child, Samuel Evan. BA in Psychology with highest honors, Rutgers U., 1975; PhD in Psychology, Stanford U., 1987. Rsch. asst. psychology dept. Stanford (Calif.) U., 1975-76, 84-85, rsch. asst. biochemistry dept., 1979-81, rsch. collaborator Sleep Lab., 1979-83, rsch. asst. Ctr. for Advanced Study in Behavioral Scis., 1985-86; postdoctoral scholar Stanford Med. Ctr., Palo Alto VA Hosp., Menlo Park VA Hosp., Calif., 1987-89; rsch. asst. prof. dept. behavioral medicine and psychiatry U. Va. Health Scis. Ctr., Charlottesville, 1989—, cons. dept. neurosurgery, 1992; statis. cons. IBM, Palo Alto, Hewlett Packard, Palo Alto, Ctr. for Econ. and Monetary Affairs, Redwood City, Calif., Ctr. for Chicano Rsch. at Stanford, 1986-92; presenter in field. Contbr. articles to Jour. Ednl. Psychology, Sleep Rsch., Memory and Cognition, Brain and Lang., Lancet, Jour. Abnormal Psychology. NIMH scholar Stanford U.; fellow U. Va. Commonwealth Ctr. for Lit. and Cultural Change, U. Va., 1992—. Mem. AAAS, Am. Psychol. Soc., Soc. for Sci. Exploration, Nat. Honor Soc. Achievements include research on effects of repression-sensitization on memory for pleasant and unpleasant information in panic disorder patients, near death experiences. Home: 1401-B Short 18th St Charlottesville VA 22902 Office: Box 152 Health Scis Ctr U Va Charlottesville VA 22908

OWENS, MARK JEFFREY, geotechnical engineer; b. Springfield, Ill., May 13, 1955; s. Harry Grant and Anniett Marion (Collins) O. B in Civil Engring. Tech., Western Ky. U., 1978; MSE, U. Tex., Austin, 1981. Registered profl. engr., Nev., Calif., Ariz., Utah. Staff engr. Dames & Moore, Boca Raton, Fla., 1981-82; project mgr. J.H. Kleinfelder, Las Vegas, Nev., 1982-84; prin. Western Techs., Inc., Las Vegas, 1984-93; dir. geotech. engring. Terracon Cons. Western Inc., Las Vegas, 1993—; mem. govt. affairs com. Cons. Engrs. Coun. Nev., Las Vegas, 1991—. Mem. ASCE (br. pres. 1989-90, sect. bd. dirs. 1990—), NSPE, Nev. Soc. Profl. Engrs. (So. Nev. Young Engr. of Yr. 1987, Nev. Young Engr. of Yr. 1987). Office: Western Techs Inc 3611 W Tompkins St Las Vegas NV 89103 Office: Terracon Cons Western Inc 3711 Regulus Dr Las Vegas NV 89102

OWENS, TYLER BENJAMIN, chemist; b. Norfolk, Va., Aug. 28, 1944; s. Arthur Samuel and Julia Tyler (Downs) O.; m. Brenda Anne Coates, Sept. 5, 1980; children: Brooks Downs, Elizabeth Tyler. BA in Chemistry, Campbell U., Buies Creek, N.C., 1967; postgrad., N.C. State U., 1967-69. Sanitarian State of Va. Health Dept., Manassas, Va., 1971-72; chief chemist Goodmark Foods, Raleigh, N.C., 1972-75; real estate broker Nadine Hodge Realty, Raleigh, 1976-77; sales engr. Hewlett Packard Co., Palo Alto, Calif., 1977-80; sales rep. Sperry Univac Corp., Blue Bell, Pa., 1980-81; sales engr. Spectra Physics Corp., San Jose, 1981-83; pres. Batchelor & Owens, Inc., Raleigh, 1983-88; territory mgr. Extrel Corp., Pitts., 1988-90; sales mgr. Delsi Nermag Instruments, Paris, 1990-91, Viking Instruments Corp., Reston, Va., 1991—. Active YMCA, Raleigh, 1989—; bd. dirs. Stonebridge Homeowners Assn., Raleigh, 1990-93, pres., 1993; bd. govs. Friends of the Children, Wake Meml. Hosp., Raleigh, 1990-93; precinct del. Wake County Rep. Party, Raleigh, 1985; vestry Episcopal Ch. of the Nativity, Raleigh, 1988-90, sr. warden, 1988-89. Mem. Am. Soc. Mass Spectrometry, N.C. Real Esate Commn., Triangle Mass Spectrometer Discussion Group, Wake County Rep. Men's Club. Episcopalian. Avocations: running, flying, ham radio, bridge. Home and Office: 1009 Carrington Dr Raleigh NC 27615-1212

OWUSU-ANSAH, TWUM, mathematics educator; b. Kumasi, Ghana, July 23, 1935; s. Kofi and Emma Twumwaa (Assoku) O.-A.; m. Faustina Forson, Jan. 6, 1966; children: Kwame Osei, Amma Twumwaa, Kwadwo Asare Ntim. BS (Lond.) Math., U. Ghana, Accra, 1960; MS, U. Alta., Edmonton, Can., 1968; PhD in Math., U. Toronto, Ont., Can., 1972. Math. tchr. St. Augustine's Coll., Cape Coast, Ghana, 1960-63; lectr. math. U. Sci. and Tech., Kumasi, 1963-74, sr. lectr., 1974-78, assoc. prof. math., 1978-84; prof. math. , 1984—, head dept. math., dean bd. postgrad. studies, 1990—; prof. math U. Ghana, Legon, part-time 1984-86. Editor: Proc. 2d Internat. Symposium on Functional Analysis and Its Applications, 1979; mem. editorial bd. Afrika Mathematika, Brazzaville, Congo, 1978-86. Contbr. articles to profl. jours. and books. Govt. of Ghana scholar, 1957-60; Commonwealth scholar Govt. of Can., 1966-68; fellow U. Toronto, 1969-71. Mem. Am. Math. Soc., Can. Math. Soc., African Math. Union (exec. com. 1976-86), Governing Coun. of CSIR Ghana (rep. coun. mgmt. bd. to Bldg. and Road Rsch. Inst.), African Network of Sci. and Tech. Instns. (vice chmn. 1985—), Ghana Sci. Assn. (nat. pres. 1989-91), Sr. Staff Club (U. Sci. and Tech.). Anglican. Avocations: music, reading, volleyball, cycling, swimming. Office: U Sci and Tech, Dept Math, Kumasi Ghana

OXER, JOHN PAUL DANIELL, civil engineer; b. Atlanta, Sept. 7, 1950; s. Robert B. Sr. and Leila Marie (Hammond) O.; m. Catherine Ann Stevens, Jan. 8, 1977. BCE, Ga. Inst. Tech., 1973; postgrad., U. Tex., Arlington, 1982-83. Registered profl. engr., Ala., Ark., Calif. Colo., Fla., Ga., Ky., La., Miss., Nev., N.J., N.Mex., N.C., Okla., S.C., Tenn., Tex, Utah, Wyo. Project engr. J.S. Ross & Assocs. Inc., Smyrna, Ga., 1973-75, Welker & Assocs. Inc., Marietta, Ga., 1976-78; sr. project. engr. Claude Terry & Assocs., Inc., Atlanta, 1978-79; chief environ. engr. Bernard Johnson Inc. (SE), Atlanta, 1979-80; chief civil engr. region VI Ecology & Environ., Inc., Dallas, 1981-84; S.E. regional mgr. Ecology & Environ., Inc., Tallahassee, Fla., 1984-88; dir. program devel. Ecology & Environ., Inc., Dallas, 1988-91, exec. asst. to the pres., 1991-92; exec. asst. to the pres. Ecology & Environ., Inc., Houston, 1992—; guest lectr. U. Fla., Gainesville; mem. Industry Functional Adv. Com. on Standards for Trade Policy Matters, 1993—. Executive producer video documentary; writer, producer video documentary; co-author, dir. video prodn. Mem. indsl. bd. advisors Speed Sci. Sch. U. Louisville, 1991—. Named Young Engr. of Yr., Ga. Soc. Profl. Engrs., 1980. Mem. NSPE, ASCE, Am. Soc. Agrl. Engrs., Nat. Def. Exec. Reserve, Fla. Bar (assoc.), Masons, FEMA. Republican. Avocations: piano, biathlon, photography, gardening. Office: Ecology & Environ Inc 4801 Woodway # 280-W Houston TX 77056 also: Ecology & Environ Inc 1999 Bryan St Ste 2000 Dallas TX 75201

OXNARD, CHARLES ERNEST, anatomist, anthropologist, human biologist, educator; b. Durham, Eng., Sept. 9, 1933; arrived in Australia, 1987; s. Charles and Frances Ann (Golightly) O.; m. Eleanor Mary Arthur, Feb. 2, 1959; children: Hugh Charles Neville, David Charles Guy. B.Sci. with 1st class honors, U. Birmingham, Eng., 1955, M.B., Ch.B. in Medicine, 1958, Ph.D., 1962, D.Sc., 1975. Med. intern Queen Elizabeth Hosp., Birmingham, 1958-59; rsch. fellow U. Birmingham, 1959-62, lectr., 1962-65, sr. lectr., 1965-66, court govs., 1958-66; assoc. prof. anatomy, anthropology and evolutionary biology U. Chgo., 1966-70, prof., 1970-78, gov. biology collegiate div., 1970-78, dean coll., 1973-77; dean grad. sch. U. So. Calif., Los Angeles, 1978-83; univ. rsch. prof. biology and anatomy U. So. Calif., 1978-83, univ. prof., prof. anatomy and cell biology, prof. biol. scis., 1983-87; prof. anatomy and human biology, head dept. of anatomy and human biology U. Western Australia, 1987-90, 93-95, dir. ctr. for human biology, 1989—, head div. agr. and sci., 1990-92; rsch. assoc. Field Mus. Natural History, Chgo., 1967—; overseas assoc. U. Birmingham, 1968—; Lo Yuk Tong lectr. U. Hong Kong, 1973, hon. prof., 1978— , Chan Shu Tzu lectr., 1980, Octagon lectr. U. Western Australia, 1987, Latta lectr. U. Nebr., Omaha, 1987; Stanley Wilkinson orator, 1991; rsch. assoc. L.A. County Natural History Mus., 1984—, George C. Page Mus., L.A., 1986—; bd. dirs. U. Western Australia Press, 1993—. Author: Form and Pattern in Human Evolution, 1973, Uniqueness and Diversity in Human Evolution, 1975, Human Fossils: The New Revolution, 1977, The Order of Man, 1983, Humans, Apes and Chinese Fossils, 1985, Fossils, Teeth and Sex, 1987, Animal Anatomies and Lifestyles, 1990; mem. editorial bd. Annals of Human Biology; cons. editor: Am. Jour. Primatology, Jour. Human Biology, Jour. Human Evolution; Australia com. mem. Ency. Britannica, 1991—; contbr. articles to anat. and anthrop. jours. Mem. Pasteur Found., 1988; bd. dirs. West Australian Inst. for Child Health, 1991—, electoral bd. Freemantle Hosp., 1991—. Recipient Book award Hong Kong Coun., 1984, S.T. Chan Silver medal, Univ. of Hong Kong, 1980; grantee USPHS, 1960-71, NIH, 1974-87, NSF, 1971-87, Raine Found., 1988-91. Fellow N.Y. Acad. Sci., AAAS, So. Calif. Acad. Sci. (bd. dirs. 1985); mem. Chgo. Acad. Soc. (hon. life), Australasian Soc. for Human Biology (pres. 1987-90), Australia and New Zealand Anat. Soc. (pres. 1989-90), Anat. Soc. Gt. Britain and Ireland (councillor 1992—), Nat. Health and Med. Rsch. Coun. (grantee), Australian Rsch. Coun. (grantee), Soc. for Study Human Biology (treas. 1962-66), Sigma Xi (pres., nat. lectr. 1987—), Phi Beta Kappa (pres. chpt.), Phi Kappa Phi (pres., Book award 1984). Office: U Western Australia, Nedlands WA 6009, Australia

OYEKAN, SONI OLUFEMI, chemical engineer; b. Aba, Nigeria, June 1, 1946; came to U.S., 1966; s. Theophilous Adebayo Oyekan; m. Emiliu Uduak (Inyang) O.; m. Priscilla Ann Parker, June 6, 1970; children: Ranti, Ima, Femi, Arit. BSChemE, Yale U., 1970; MSChemE, Carnegie Mellon U., 1972, PhD in Chem. Engring., 1977. Coord., instr. U. Pitts., 1972-77; rsch. engr. Exxon, Baton Rouge, 1977-78, sr. engr., 1979-80; rsch. sect. head Engelhard, Edison, N.J., 1980-84, rsch. mgr., 1985-86, sr. engr., cons., 1986-90; rsch. assoc. DuPont Chems., Orange, Tex., 1991—; cons. Ghaip, Tema, Ghana, 1986; instr. N.J. Inst. Tech., Newark, 1984. Mem. AICHE (meeting program chair, spring nat. meeting, 1994, chmn. tech. programming com. 1991-93, lectr. continuing edn. program 1992—), NAACP (life), Carnegie Mellon Admission Coun., Yale Alumni Admission Coun., Nat. Orgn. for Profl. Advancement of Black Chemists, Phi Kappa Phi, Sigma Xi. Democrat. Episcopalian. Achievements include 9 patents in petroleum refining. Home: 1804 Longfellow Rd Orange TX 77630

OYIBO, GABRIEL A., aeronautics and astronautics educator; b. Idah, Benue, Nigeria, Aug. 27, 1950; came to U.S., 1977; s. Sulaiman and Hadisatu (Alhaji) O.; m. Leslie M. Ford; children: Usman Dan, Ejima Sulaiman, Hassana Kia. B of Engr./Aeronautics, ABU/Imperial, London, 1970; MS in Engring./Aeronautics, Renssaeler Polytechnic U., Troy, N.Y., 1978, PhD in Aero/Astronautics, 1981. Prin. rsch. scientist Fairchild Republic Aerospace Co., Farmingdale, N.Y., 1981-87; assoc. prof. of aeronautics/astronautics Polytechnic Univ., Farmingdale, 1986—; cons. Quest Tech McClean, Va., 1988, Goodyear Aerospace, Akron, Ohio, 1984-85, NASA, Langley, Va., 1983—; Sidney Goldstein lectr. Technion Univ. Isreal, 1985. Author: (book) New Group Theory for Mathematical Physics, Gas Dynamics and Turbulence, 1993; book reviewer in field; contbr. articles to profl. jours. Named Educator of Yr., Network, Columbia U., 1990, Keynote Lectr. in Physics, Lawrence Livermore Labs/NSBP, 1990; grantee Dept. of Def./AFSOR, 1983—. Fellow AIAA (assoc., jour. award 1983). Office: Polytechnic U Dept Aerospace Engring Farmingdale Ctr Farmingdale NY 11735

OZAKI, SATOSHI, physicist; b. Osaka, Japan, July 4, 1929; married, 1960; 2 children. BS, Osaka U., 1953, MS, 1955; PhD in Physics, MIT, 1959. Rsch. asst. physics MIT, 1956-59; rsch. assoc. Brookhaven Nat. Lab., 1959-61, asst. physicist, 1961-63, assoc. physicist, 1963-66, physicist, 1966-72, group leader physics, 1970—, sr. physicist, 1972—; dir. Relativistic Heavy Ion Collider (RHIC), 1993—; vis. prof. Osaka U., 1975-76; vis. prof., dept. physics Nat. Lab. High Energy Physics, Japan, 1981—. Fellow Am. Phys. Soc., Phys. Soc. Japan. Achievements include the study of high energy particle interactions, particle spectoscopy, high energy physics instrumentation. Office: Brookhaven National Laboratory Relativistic Heavy Ion Collider Upton NY 11973*

OZAWA, KEIYA, hematologist, researcher; b. Tatsuno-machi, Nagano-Ken, Japan, Feb. 23, 1953; s. Iwao and Taka (Ono) O.; m. Masami Inokuchi, Oct. 16, 1983; children: Sayaka, Tomo-o. MD, U. Tokyo, 1977, PhD, 1984. Resident in medicine Tokyo U. Hosp., 1977-79; with U. Tokyo, 1977—; rsch. assoc. Jichi Med. Sch., Tochigi, 1980-82; staff fellow faculty medicine U. Tokyo, 1984-87; asst. prof. dept. hematology-oncology Inst. Med. Sci., 1987-90, assoc. prof., 1990—; Fogarty fellow, clin. hematology br. Nat. Heart Lung and Blood Inst., NIH, Bethesda, Md., 1985-87. Mem. editorial bd. Internat. Jour. Hematology, Kyoto, 1990-92. Mem. AAAS, Am. Soc. Hematology, Am. Fedn. for Clin. Rsch., N.Y. Acad. Scis., Japanese Cancer Assn., Japanese Soc. Hematology (councilor 1990—), Japanese Soc. Internal Medicine, Japan Soc. Clin. Hematology (councilor 1991—). Achievements include research in hematopoiesis, hematopoietic growth factors, leukemogenesis, and gene therapy. Home: 1-2-12-302 Yushima Bunkyo, 113 Tokyo Japan Office: U Tokyo Inst Med Sci, 4-6-1 Shirokanedai Minato, 108 Tokyo Japan

OZELLI, TUNCH, economics educator, consultant; b. Ankara, Turkey, May 18, 1938; came to U.S., 1962; s. Sufyan and Saziye (Ozmorali) O.; m. Lale A. Baymur, Dec. 30, 1960 (div. Mar. 1972); children: Selva, Kerem; m. Nancy Ann Goldschlager, Feb. 3, 1974 (div. Dec. 1984); m. Meral Ozdemir, May 9, 1992. MBA, Fla. State U., 1963; PhD, Columbia U., 1968. Rsch. fellow Harvard U., Cambridge, Mass., 1969-70; econ. advisor Office Prime Minister, Ankara, 1970-72; assoc. prof. mgmt. N.Y. Inst. Tech., N.Y.C., 1972—; spl. advisor State Planning Orgn., Ankara, 1989—. Contbr. articles to profl. jour. Ford Found. scholar, 1963-64, Found. for Econ. Edn. fellow, 1968. Mem. Am. Econ. Assn., Middle East Studies Assn., Turkish Mgmt. Assn., Delta Mu Delta. Avocation: equestrian activities. Office: NY Inst Tech Old Westbury NY 11568

OZERNOY, LEONID MOISSEY, astrophysicist; b. Moscow, May 19, 1939; came to U.S., 1986, naturalized citizen; m. Marianne Rosen; 2 children. BS in Physics, Moscow State U., 1961, MS in Astronomy, 1963; PhD in Astrophysics, Shternberg Astron. Inst., Moscow, 1966; DSc in Astrophysics, Lebedev Phys. Inst., USSR Acad. Scis., Moscow, 1971. Rsch. scientist, dept. theoretical physics Lebedev Phys. Inst., 1966-71, sr. rsch. scientist, 1971-86; asst. then assoc. prof. Moscow Physics and Tech. Inst., 1968-79, fired after applying for emigration from USSR; prof. astrophysics, disting. vis. scientist Ctr. for Astrophysics, Harvard Coll. Obs. and Smithsonian Astrophys. Obs., Cambridge, Mass., 1986-89; prof. astrophysics, disting. vis. scholar Inst. Geophysics and Planetary Physics, Los Alamos (N.Mex.) Nat. Lab., 1989-91; sr. assoc. Nat. Rsch. Coun.-NAS Goddard Space Flight Ctr., NASA, Greenbelt, Md., 1991-92, sr. rsch. scientist Univ. Space Rsch. Assn., 1993—; prof. computational sci. and space sci. Inst. Computational Sci. and Informatics, George Mason U., Fairfax, Va., 1993—; vis. prof. Boston U., 1986-87, Harvard U., 1987; mem. sci. coun. on plasma astrophysics, Presidium of USSR Acad. Scis., 1971-75. Co-editor Astrophysics and Space Scis. mag., 1976-86; author several books and over 200 sci. papers in theoretical astrophysics and related fields. Recipient Silver medal USSR Exhbn. of Achievements in Nat. Economy, 1968; NRC/NAS fellow, 1991. Fellow Am. Phys. Soc. (com. on internat. freedom of scientists 1988-90); mem. AAAS, Internat. Astron. Union, N.Y. Acad. Scis., Am. Astron. Soc. (exec. com. high energy astrophysics div. 1989-90), Astron. Soc. Pacific, COSPAR (Com. Space Rsch., Internat. Coun. Sci. Unions), The Planetary Soc., Internat. Platform Assn. Achievements include establishment of mass-radius and mass-angular momentum relationships for galaxies, of methods for measuring/constraining the black hole mass at the center of the galaxy; research on theory for evolution of cosmological turbulence; and on theory for structure and evolution of supermassive, rotating, magnetized bodies (magnetoids). Office: NASA Goddard Space Flight Ctr Lab Astronomy and Solar Physics Code 685 Greenbelt MD 20771 also: George Mason U Inst Computat Sci & Informatics Sci and Tech Bldg I Fairfax VA 22030-4444

OZKAN, UMIT SIVRIOGLU, chemical engineering educator; b. Manisa, Turkey, Apr. 11, 1954; came to U.S., 1980; d. Alim and Emine (Ilgaz) Sivrioglu; m. H. Erdal Ozkan, Aug. 13, 1983. BS, Mid. East Tech. U., Ankara, Turkey, 1978, MS, 1980; PhD, Iowa State U., 1984. Registered profl. engr., Ohio. Grad. rsch. assoc. Ames Lab. U.S. Dept. Energy, 1980-84; asst. prof. Ohio State U., Columbus, 1985-90, assoc. prof. chem. engring., 1990—. Contbr. articles to profl. jours. Recipient Women of Achievement award YWCA, Columbus, 1991, Outstanding Engring. Educator Ohio award Soc. Profl. Engrs., 1991, Union Carbide Innovation Recognition award, 1991, 92, NSF Woman Faculty award in Sci. and Engring., 1991. Fellow Am. Inst. Chemists; mem. NSPE (edn. trustee Franklin County chpt. 1989—), Am. Inst. Chem. Engring. (chair career guidance com. cen. Ohio sect. 1986—), Am. Soc. Engring. Edn., Am. Chem. Soc., Combustion Inst., Sigma Xi. Achievements include research in selective oxidation, catalytic incineration, NO reduction, hydrodesulfurization, hydrodenitrogenation, in-situ spectroscopy. Office: Ohio State U Chem Engring 140 W 19th Ave Columbus OH 43210-1180

ÖZMEN, ATILLA, federal agency administrator, physics educator; b. Göksun, Turkey, Jan. 21, 1941; s. M. Necati and Fetihiye (Türkkan) Ö; m. Irene Joyce Saunders, 1961 (div. 1978); children: Suzan, Ilhan; m. Senay Ozan, July 23, 1984; children: Güliz, Gökhan, Gözde. BS, Mid. East Tech. U., Ankara, Turkey, 1967, MS, 1969, PhD, 1977. Asst. lectr. physics dept. Mid. East Tech. U., 1967-69, lectr., 1969-81, v.p., 1977-79; assoc. prof., dir. Inst. Sci. and Tech. Gazi U., Ankara, 1982-88, prof., 1988—; now pres., dir. gen. Turkish Atomic Energy Authority Turkish Fed. Govt., Ankara; Pres. Turkish Atomic Energy Authority, Ankara, 1987—; mem. Nat. Bd. Edn., Ministry Edn., Ankara, 1985-87. Editor Jour. Sci. Tech. Gazi U., 1983-88, Turkish Jour. Nuclear Sci., 1987—. Served to lt. Turkish mil., 1973-75. Avocations: chess, bridge, backgammon, table tennis. Home: Cevre Sokak 52/30, Cankaya Turkey Office: Turkish Atomic Energy Authority, Alaçum Sok 9, Cankaya Ankara, Turkey*

PÄÄBO, SVANTE, molecular biologist, biochemist. Achievements include pioneering in ancient DNA extraction studies. Office: U Munich Dept of Zoology, Postfach 202136, D-80021 Munich Germany

PAALMAN, MARIA ELISABETH MONICA, public health executive; b. Zwolle, Overyssel, The Netherlands, Dec. 7, 1951; d. Herman J. and Wieb (Lievestro) P. MS in Clin. Psychology, U. Groningen, 1977; postgrad. in psychotherapy, Pesso Psychomotor, Boston, 1982. Staff mem. Hoog Hullen-Ther. Comm., Eelde, Netherlands, 1977-79; asst. dir. Krauweelhuis-Ther.Comm., Amsterdam, 1979-82; dir. STD Found., Utrecht, 1983-93; cons. WHO, Geneva, 1990; mem. Nat. Com. AIDS Control, The Netherlands, 1983-93; nat. coord. campaigns on STDs, AIDS, safer sex, 1985-93; tech. asst. STD/AIDS EC, Tanzania, 1993—. Editor: Dutch Std. Bull., 1983-93; editor: Promoting Safer Sex, 1990; mem. bd. editors Internat. Jour. on STD and AIDS, 1989—, AIDS Health Promotion Exch., 1991—, AIDS Edn. and Prevention, 1992—; contbr. articles to profl. jours. Mem. Internat. Soc. for STD Rsch. (bd. dirs. 1989—). Home: PR Beatrixlaan 24, 4001 AH Tiel The Netherlands Office: Ministry of Health, Dar es Salaam Tanzania

PAARMANN, LARRY DEAN, electrical engineering educator; b. Maquoketa, Iowa, Feb. 3, 1941; s. Arthur Herman Paarmann and Blanche Caroline (Lozenzen) Earles. BS, No. Ill. U., 1970; MS, U. Ill., 1977; PhD, Ill. Inst. Technology, 1983. Instr. Kishwaukee Coll., Malta, Ill., 1970-75; rsch. asst. U. Ill., Urbana, 1975-76; engr. IIT Rsch. Inst., Chgo., 1977-78; instr. IIT, Chgo., 1978-83; asst. prof. Drexel U., Phila., 1983-90; assoc. prof. Wichita (Kans.) State U., 1990—; co-organizer IEEE ednl. seminar, Phila., 1988, session of 15th N.E. Bioengring. Conf., Boston, 1989, session of 34th Midwest Symposium on Cirs. and Systems, Monterey, Calif., 1991; reviewer IEEE Transactions on Signal Processing. Contbr. articles to profl. jours. Rsch. grantee NSF, 1984, 85, AT&T Found., 1984, Am. Heart Assn., 1985, Kans. Electric Utilities Rsch., 1992. Mem. IEEE (sr.; chpt. chmn. 1989-90), European Assn. for Signal Processing. Achievements include contbns. to modeling and signal processing techniques. Home: 3224 Longfellow Ct Wichita KS 67226 Office: Wichita State U Dept Elec Engring Wichita KS 67260-0044

PAASWELL, ROBERT EMIL, civil engineer, educator; b. Red Wing, Minn., Jan. 15, 1937; s. George and Evelyn (Cohen) P.; m. Rosalind Snyder, May 31, 1958; children: Judith Marjorie, George Harold. B.A. (Ford Found. fellow), Columbia U., 1956, B.S., 1957, M.S., 1961; Ph.D., Rutgers U., 1965. Field engring. asst. Spencer White & Prentis, Washington, 1954-56; engr. Spencer White & Prentis, N.Y.C., 1957-59; rsch. scientist Davidson Lab., N.J., 1964; rsch. fellow Greater London Council, 1971-72; rsch. and teaching asst. Columbia U., 1959-62; asst. prof. civil engring. SUNY, Buffalo, 1964-68; chmn. bd. govs. Urban Studies Coll., 1973-76, assoc. prof., 1968-76, prof. civil engring., 1976-82; dir. Center for Transp. Studies and Research, 1979-82, chmn. dept. environ. design and planning, 1980-82; prof. transp. engring. U. Ill., Chgo., 1982-86, 89-90, dir. Urban Transp. Ctr., 1982-86; exec. dir. Chgo. Transit Authority, 1986-89; dir. transp. rsch. consortium, prof. civil engring. CCNY, 1990—, disting. prof., 1991—; faculty-on-leave Dept. Transp., 1976-77, cons., 1981—; v.p. Faculty Tech. Cons., Inc., Midwest Systems Scis., Inc., 1982-86; dir. Urban Mass Transp. Adminstrn. Summer Faculty Workshop, 1980, 81; cons. transp. planning, energy and soil mechanics; spl. cons. to Congressman T. Dulski, 1973; vis. expert lectr. Jilin U. Tech., Changchun, Peoples Republic of China, 1985, hon. prof. transp., 1986—; bd. dirs. D'Escuto Architects & Engrs., Chgo., Hickling Co., Ottawa, Can., Transit Devel. Corp. Author: Problems of the Carless, 1977; editor: Site Traffic Impact Assessment, 1992; contbg. author: Decisions for the Great Lakes, 1982, World Book Encyclopedia, 1993; bd. editors: Jour. Environ. Systems, 1971—, Jour. Urban Systems, 1974—, Transp., 1978—; contbr. articles to profl. jours. Mem. Buffalo Environ. Mgmt. Commn., 1972-74; mem. Area Com. for Transit, Mayor's Energy Adv. Bd., 1974, Block Grant Rev. Com., City of Buffalo; chmn. com. on transp., mem. rev. adv. bd. Rsch. and Planning Coun. Western N.Y.; mem. transp. com. Chgo. 1992 Worlds Fair; mem. citizens' adv. bd. Chgo. Transit Authority, 1985—; mem. strategic planning com. Regional Transp. Authority, 1985; mem. steering com. Nat. Transit Coop. Rsch. Program, 1991—, Borough pres. (Manhattan) Trans. Adv. Bd., Bronx Ctr. Devel. Project; bd. dirs. Transit Devel. Corp., 1992—. Recipient Dept. Transp. award, 1977; SUNY faculty fellow, 1965-66. Fellow ASCE (past pres. Buffalo sect., chmn. steering com. 1992 specialty conf. traffic impact analysis); mem. AAAS, Transp. Rsch. Bd. (chmn. com. on transp. disadvantaged, mem. exec. com., peer rev. com. nat. transp. ctrs. 1988—), Inst. Transp. Engrs. (transit coun., exec. com., chmn. legis. policy com.), Sigma Xi. Office: CCNY Inst Transp Systems Rm 220-Y 135th St and Convent Ave New York NY 10031

PABST, MICHAEL JOHN, immunologist, dental researcher; b. Washington, Oct. 19, 1945. BS in Chemistry, Boston Coll., 1967; PhD in Biochemistry, Purdue U., 1972. Asst. prof. U. Colo. Sch. Dentistry, Denver, 1974-77; assoc. prof. Nat. Jewish Hosp., Denver, 1977-86; prof. U. Tenn., Memphis, 1986—; study section mem. Nat. Inst. Health, Bethesda, Md., 1990—. Author: Handbook of Inflammation, 1989. Grantee Nat. Inst. Health, 1975-78, 80—. Mem. Am. Assn. Immunologists, Soc. for Leukocyte Biology, Internat. Soc. Immunopharmacology, Am. Soc. Biochemistry and Molecular Biology. Achievements include research in examination of enzyme regulation and kinetics by mutational analysis; phagocyte biochemistry. Office: U Tenn 894 Union Ave Memphis TN 38163

PACHTER, JONATHAN ALAN, biochemist, researcher; b. Phila., Nov. 16, 1957; s. Irwin Jacob and Elaine Anna (White) P.; m. Wendy Jane Cole, July 12, 1987; children: Barbara Danielle, Jeremy David. BA in Biology, U. Rochester, 1979; MS in Pharmacology, Baylor Coll. Medicine, 1981, PhD in Neurosci., 1985. Postdoctoral fellow Yale U. Sch. Medicine, New Haven, Conn., 1985-89; sr. scientist Schering-Plough Corp., Bloomfield, N.J., 1989-92; prin. scientist Schering-Plough Rsch. Inst., Kenilworth, N.J., 1992—. Contbr. articles to profl. jours. Recipient Nat. Rsch. Svc. award NIH, 1987-88. Mem. AAAS, N.Y. Acad. Scis. Achievements include research in interregulation of signal transduction pathways in mammalian cells. Office: Schering Plough Rsch Inst 2015 Galloping Hill Rd Kenilworth NJ 07033-0539

PACKARD, DAVID, manufacturing company executive, electrical engineer; b. Pueblo, Colo., Sept. 7, 1912; s. Sperry Sidney and Ella Lorna (Graber) P.; m. Lucile Salter, Apr. 8, 1938 (dec., 1987); children: David Woodley, Nancy Ann Packard Burnett, Susan Packard Orr, Julie Elizabeth Stephens. B.A., Stanford U., 1934, EE, 1939; LLD (hon.), U. Calif., Santa Cruz, 1966, Catholic U., 1970, Pepperdine U., 1972; DSc (hon.), Colo. Coll., 1964; LittD (hon.), So. Colo. State Coll., 1973; D.Eng. (hon.), U. Notre Dame, 1974. With vacuum tube engring. dept. Gen. Electric Co., Schenectady, 1936-38; co-founder, ptnr. Hewlett-Packard Co., Palo Alto, Calif., 1939-47, pres., 1947-64, chief exec. officer, 1964-68, chmn. bd., 1964-68, 72—; U.S. dep. sec. defense Washington, 1969-71; dir. Genetech, Inc., 1981-92; bd. dirs. Beckman Laser Inst. and Med. Clinic; chmn. Presdl. Commn. on Def. Mgmt., 1985-86; mem. White House Sci. Coun., 1982-88. Mem. President's Commn. Pers. Interchange, 1972-74, President's Coun. Advisors on Sci. and Tech., 1990-92, Trilateral Commn., 1973-81, Dirs. Coun. Exploratorium, 1987-90; pres. bd. regents Uniformed Svcs. U. of Health Scis., 1975-82; mem. U.S.-USSR Trade and Econ. Coun., 1975-82; mem. bd. overseers Hoover Instn., 1972—; bd. dirs. Nat. Merit Scholarship Corp., 1963-69, Found. for Study of Presdl. and Congl. Terms, 1978-86, Alliance to Save Energy, 1977-87, Atlantic Coun., 1972-83, vice chmn., 1972-80, Am. Enterprise Inst. for Public Policy Rsch., 1978—, Nat. Fish and Wildlife Found., 1985-87, Hitachi Found. Adv. Coun., 1986—; vice chmn. The Calif. Nature Conservancy, 1983-90; trustee Stanford U., 1954-69, pres., 1958-60, David and Lucile Packard Found., pres., chmn. 1964—, Herbert Hoover Found., 1974—, Monterey Bay Aquarium Found., chmn., 1978—, The Ronald Reagan Presdl. Found., 1986-91, Monterey Bay Aquarium Rsch. Inst., chmn., pres. 1987—. Decorated Grand Cross of Merit Fed. Republic of Germany, 1972, Medal Honor Electronic Industries, 1974; numerous other awards including Silver Helmet Def. award AMVETS, 1973, Washington award Western Soc. Engrs., 1975, Hoover medal ASME, 1975, Gold Medal award Nat. Football Found. and Hall of Fame, 1975, Good Scout award Boy Scouts Am., 1975, Vermilye medal Franklin Inst., 1976, Internat. Achievement award World Trade Club of San Francisco, 1976, Merit award Am. Cons. Engrs. Council Fellows, 1977, Achievement in Life award Ency. Britannica, 1977, Engring. Award of Distinction San Jose State U., 1980, Thomas D. White Nat. Def. award USAF Acad., 1981, Disting. Info. Scis. award Data Processing Mgmt. Assn., 1981, Sylvanus Thayer award U.S. Mil. Acad., 1982, Environ. Leadership award Natural Resources Def. Council, 1983, Dollar award Nat. Fgn. Trade Council, 1985, Gandhi Humanitarium Award, 1988, Roback Award Nat. Contract Mgmt. Assn., 1988, Pub. Welfare Medal NAS, 1989, Chevron Conservation Award, 1989, Doolittle Award Hudson Inst., 1989, Disting. Citizens Award Commonwealth Club San Francisco, 1989, William Wildback award, Nat. Conf. Standards Labs., Washington, 1990; named to Silicon Valley Engring. Hall of Fame, Silicon Valley Engring. Coun., 1991, Pueblo (Colo.) Hall of Fame, 1991. Fellow IEEE (Founders medal 1973); mem. Nat. Acad. Engring. (Founders award 1979), Instrument Soc. Am. (hon. lifetime mem.), Wilson Council, The Bus. Roundtable, Bus. Council, Am. Ordnance Assn. (Crozier Gold medal 1970,) Henry M. Jackson award 1988, Nat. Medal Tech. 1988, Presdl. Medal of Freedom 1988, Sigma Xi, Phi Beta Kappa, Tau Beta Pi, Alpha Delta Phi (Disting. Alumnus of Yr. 1970). Office: Hewlett-Packard Co PO Box 10301 Palo Alto CA 94304-1112

PACKER, KAREN GILLILAND, cancer patient educator, researcher; b. Washington, Apr. 27, 1940; d. Theodore Redmond and Evelyn Alice (Johnson) Gilliland; m. Allan Richard Packer, Sept. 27, 1962; 1 child, Charles Allan. Student, Duke U., 1957-59, U. Ky., 1959-60, 61-62, U. P.R. Sch. Medicine, 1960-61. Genetics researcher U. Ky., Lexington, 1959-60, 61-62; biologist Melpar Inc., Nat. Cancer Inst., Springfield, Va., 1964-66; rsch. assoc., epidemiology rsch. ctr. U. Iowa Coll. Medicine, Iowa City, 1981-85; founder, dir. Marshalltown (Iowa) Cancer Support Group, 1987—; mem. County Health Planning Commn., Marshalltown, 1989—; adv. bd. Community Nursing Svc., Marshalltown, 1990—; v.p. Community Svcs. Coun., Marshalltown, 1992-93. Editor The Group Gazette, 1988—. bd. mem. 1st United Ch. Christ, Hampton, Va., 1973-75; corresponding sec. DAR, Marshalltown, 1988—; chmn. cancer and rsch. aux. VFW, Marshalltown, 1990—. Genetics Rsch. grantee NSF, 1959-60, NIH, 1961-62; recipient Leadership award Marshalltown Area C. of C., 1988; spl. recognition Nat. Coalition for Cancer Survivorship, 1990; 1st place in state award Community Cancer Edn. VFW Aux., 1990-91, 91-92. Mem. AAAS, Nat. Guard Bur. Officers Wives Club (publ. editor 1965-68), Nat. Alliance Breast Cancer Orgns., Nat. Tumor Registrars Assn., Iowa Tumor Registrars Assn., N.Y. Acad. Scis. Mem. Congregational Ch. Achievements include establishment of regional orgn. for cancer info. and edn. Home and Office: 1401 Fairway Dr Marshalltown IA 50158-3825

PACKER, KENNETH JOHN, spectroscopist; b. Kettering, Northants, Eng., May 18, 1938; s. Harry James and Alice Ethel (Purse) P.; m. Christine Frances Hart, Aug. 18, 1962; children: James Vernon, Alison Frances. BSc 1st class with honors, London U., 1959; PhD, Cambridge (Eng.) U., 1962. Chartered chemist. Postdoctoral researcher E.I. duPont de Nemours, Wilmington, Del., 1962-63; SERC rsch. fellow U. East Anglia, Norwich, Eng., 1963-64, lectr. chemistry, 1964-71, sr. lectr., 1971-78, reader in chemistry, 1978-82, prof. chemistry, 1982-84; chief rsch. assoc. BP Rsch., Sunbury-on-Thames, Eng., 1984—; mem. sci. bd. UK Sci. and Engring. Rsch. Coun., London, Swindon, Eng., 1988-91. Editor Molecular Physics jour., 1982-85; contbr. articles to profl. jours. Rsch. grantee UK Sci. and Engring. Rsch. Coun., 1964-84. Fellow Royal Soc. Chemistry London (coun. Faraday div. 1988-91), Royal Soc. London for the Improvement of Natural Knowledge. Avocations: fly-fishing, skiing, music, gardening, sailing. Home: Periwinkle Cottage, Perry Hill Worplesdon, Guildford GU3 3RG, England Office: BP Rsch, Chertsey Rd, Sunbury on Thames TW16 7LN, England

PACKEY, DANIEL J., economist, researcher; b. Detroit, Mar. 5, 1952; s. Kenneth F. and Mary (Dyplowski) P.; m. Jeanne Marie Smiley, Nov. 1, 1986; children: Rechelle Anne, Andrea Marie. BS in Econs., BA in Bus., Cen. Mich. U., 1976; MS, U. Oreg., 1985, PhD, 1986. Economist Bonneville Power Adminstrn., Portland, Oreg., 1982; asst. prof. Calif. State U., Fresno, 1982-86; sr. analyst Potomac Electric Power Co., Washington, 1986-88; sr. energy economist Pub. Svc. Commn. D.C., Washington, 1988-90; sr. economist Solar Energy Rsch. Inst., Golden, Colo., 1990-92, Nat. Renewable Energy Lab., Golden, Colo., 1992—; guest lectr. renewable energy numerous univs., 1990—. Author: Market Penetration of New Energy Technologies, 1992; contbr. articles to profl. jours. Mem. Trout Unltd., Duck Unltd., Denver, 1992. Mem. Am. Econs. Assn., Internat. Assn. Energy Econ., Assn. Demand Side Mgmt. Profls. Democrat. Office: Nat Renewable Energy Lab 1617 Cole Blvd Golden CO 80401

PACUMBABA, R.P., plant pathologist, educator; b. Los Banos, Laguna, Philippines, July 15, 1935; came to U.S. 1964; s. Prudencio O. P.; m. Teresita P. Osi, Feb. 24, 1960; children: Marjorie Pacumbaba-Devlin, Gina Pacumbaba-Watson, Rudy O. Jr. BS in Agr., U. Philippines, 1958; MS, Kans. State U., 1966, PhD, 1970. Vocat. tchr. agr. Roxas Meml. Agrl. Sch., Guinobatan, Albay, Philippines, 1956-58; plant pathologist Dept. Botany U. Philippines, Quezon City, 1958-60; plant pathologist Dept. Agr. and Natural Resources Bur. of Plant Industry, Guinobatan Expt. Sta., 1960-64; grad. rsch. asst. plant pathology Kans. State U., Manhattan, 1966-70, postdoctoral traineeship, 1970, rsch. assoc. molecular virology, 1971-76; plant pathologist Internat. Inst. Tropical Agr., Kinshasa, Zaire, 1976-79; assoc. prof. plant virology Riyadh U., Saudi Arabia, 1980-81; assoc. prof. plant pathology Ala. A&M U., Normal, 1981-92, prof. plant pathology, 1992—; cons. in field; hon. advisor Rsch. Bd. Advisors, Am. Biog. Inst., Inc., Raleigh, 1992. Contbr. numerous articles to profl. jours. Recipient Lifetime Achievement award Am. Biog. Inst., 1992; named Internat. Man of the Yr. Internat. Biog. Centre, Cambridge, Eng., 1991-92; Rockefeller Found. fellow, 1964-66. Mem. Am. Phytopathol. Soc. (ad hoc com. for minorities 1991—, virology com. 1989-92, tropical plant pathology com. 1989-92), Philippine Phytopathol. Soc. (charter), So. Assn. Agrl. Scientists, So. Soybean Disease Workers, Soybean Breeders-Pathologists Assn., Internat. Soc. Plant Pathologists. Roman Catholic. Office: Ala A&M U Dept Plant/Soil Scis Normal AL 35762

PACZKOWSKI, JERZY, chemistry educator; b. Mroczno, Torun, Poland, Nov. 26, 1946; s. Feliks and Jadwiga (Cegielska) P.; m. Bozena Kwintera, Feb.17, 1973. Master, N. Copernicus U., 1971, PhD, 1979. Asst. Tech. and Agrl. U., Bydgoszcz, 1971-79, asst. prof., 1979-83, 1986-90, prof. chemistry, 1991—, rector, 1993—; postdoctoral fellow Bowling Green (Ohio) State U., 1984-85, sr. rsch. assoc., 1990-91. Contbr. articles to profl. jours. Home: Ogrody 25/219, 85-870 Bydgoszoz Poland Office: Tech and Agrl U, Seminaryjna 3, 85-326 Bydgoszcz Poland

PACZYNSKI, BOHDAN, astrophysicist, educator; b. Wilno, Poland, Feb. 8, 1940; came to U.S., 1981; s. Jan and Helena (Milkowska) P.; m. Hanna Adamska, Aug. 25, 1965; children—Agnieszka, Marcin. M.A., Warsaw U., 1962, Ph.D., 1964. Prof. astrophysics Polish Acad. Sci., Warsaw, 1976-82; Lyman Spitzer Jr. prof. theoretical physics Princeton (N.J.) U., 1982—. Recipient State prize in Sci., Govt. of Poland, 1980, Alfred Jurzykowski Found. award, 1982, Dannie Heineman prize for astrophysics Am. Astron. Soc., 1992. Mem. Polish Acad. Sci. (corr.), U.S. Nat. Acad. Sci. (fgn. assoc.), Deutsche Academie Leopoldina, Royal Astron. Soc. (U.K., Eddington medal 1987), Internat. Astron. Union (invited lectr. 1979). Office: Princeton U 124 Peyton Hall Princeton NJ 08544

PADGETT, BOBBY LEE, II, chemist; b. Gastonia, N.C., Apr. 8, 1966; s. Bobby Lee and Elizabeth Ann (Pate) P. BS in Chemistry, U. N.C., 1988. Cert. EMT-defibrillation, N.C. Chemist Applied Analytical, Inc., Wilmington, N.C., 1988-89, sr. analyst, 1989-90, lab supr., 1990-91; plant chemist Saber Internat. Ltd., Warrenton, N.C., 1991—; vice chmn. Warren County Local Emergency Planning Com., Warrenton, 1991-93. Del. Warren County Dem. Conv., 1992; candidate Wilmington City Coun., 1989. Mem. U. N.C. Gen. Alumni Assn., U.N.C. Ednl. Found., Gen. Matt Ransom Camp 861 SCV (compatriot), Mason. Presbyterian. Home: 43 Beech Ct Littleton NC 27850 Office: Saber Internat Ltd PO Box 860 Hwy 58 Warrenton NC 27589

PADMANABHAN, MAHESH, food and chemical engineer, researcher; b. Nagercoil, Tamil Nadu, India, Nov. 24, 1962; came to U.S., 1985; s. Saraswathi Padmanabhan. BTech in Agrl. Engring., Indian Inst. Tech., Kharagpur, 1986; MS in Agrl. Engring., U. Minn., St. Paul, 1989, PhD in Agrl. Engring., 1992, MSChemE, 1993; Phd, 1992. Cert. engr.-in-tng., Minn. Postdoctoral rsch. assoc. U. Minn., 1985—; presenter at profl. meetings. Contbr. articles to Jour. Food Sci., Jour. Food Engring., Cereal Foods World, Jour. Rheology, Trends in Food Sci. and Tech., Ency. Food Sci. and Tech., chpt. to book. Grantee U. Minn., 1990. Mem. Am. Inst. Chem. Engrs. (presenter), Soc. Rheology (presenter), Brit. Soc. Rheology, Inst. Food Technologists (presenter), Sigma Xi, Alpha Epsilon, Gamma Sigma Delta. Home: Apt 8 1057 Everett Ct Saint Paul MN 55108-1510

PAEK, JAMES JOON-HONG, civil engineer, construction management educator; b. Seoul, July 29, 1958; s. Sang H. and Kang (Lee) P.; m. Jennifer Miae Jeong, Dec. 21, 1985; children: Leah, Catherine. MS, U. Tex., 1984, PhD, 1987. Rsch. asst. civil engring. U. Tex., Austin, 1982-87; rsch. assoc. Constrn. Industry Inst., Austin, 1987-88; asst. prof. constrn. mgmt. U. Nebr., Lincoln, 1988—; civilian rsch. contractor Army-Corps of Engrs., Champaign, Ill., 1990; cost estimator Kiewit Engring. Co., Omaha, 1991; maker cons. Kiewit Western Co., Omaha, 1992. Contbr. articles to profl. jours. Recipient Faculty Scientific Svc. award U.S. Army-Corps Engrs., Champaign, 1990, Cooperative Rsch. Program award P.K. Constrn. Co., Taegu, Korea, 1992. Mem. Am. Soc. Civil Engrs., Am. Arbitration Assn., Assoc. Gen. Contractors, Korean Scientists and Engrs. Assn. Republican. Presbyterian. Home: 2832 Sedalia Dr Lincoln NE 68516

PAGALA, MURALI KRISHNA, physiologist; b. Sri Kalahasti, Andhra, India, Oct. 2, 1942; came to U.S., 1970; s. Lakshmaiah and Radhamma (Bhimavaram) P.; m. Vijaya Bhimavaram, Dec. 12, 1969; children: Sobhan, Suresh. PhD in Zoology, S. V. Univ., Tirupati, A.P., India, 1969; MS in Computer Sci., Pratt Inst., N.Y., 1985. Postdoctoral fellow Inst. for Muscle Disease, N.Y.C., 1970-73, asst. mem., 1974; assoc. rsch. scientist NYU, N.Y.C., 1974-75; asst. to dir. Neuromus Disease Div. Maimonides Med. Ctr., Bklyn., 1975-89, dir. neuromuscular rsch., 1990—; vis. scientist II Physiol. Inst., U. Saarlandes, Hamburg, Germany, 1981, 82; sci. cons. UNDP/TOKTEN Program, Calcutta, India, 1990, NIGMS/FASEB MARC Program Grambling State U., La., 1992. Contbr. articles to profl. jours. Life mem. Telugu Assn. of North Am., 1989; sci. fair judge N.Y. Acad. Scis., N.Y.C., 1986—. Named Best speaker Zool. Soc. of S. V. Univ., 1964, Best Basic Rsch. paper Maimonides Med. Ctr., 1983, 88; recipient Fatigue Rsch. grant Maimonides Rsch. Devel. Found., 1986—, Drug Rsch. grant Maimonides Rsch. Devel. Found., 1989—. Mem. N.Y. Acad. Scis., Am. Physiol. Soc., Assn. Scientists of Indian Origin in Am. (pres. 1993-94). Democrat. Hindu. Achievements include devel. in vitro electromyographic and electrocardiographic chambers to evaluate neuromuscular and cardiac functions of exptl. animals; awaiting patent for unique design of fan called Global Fan. Home: 82 Pacific Ave Staten Island NY 10312 Office: Maimonides Med Ctr 4802 Tenth Ave Brooklyn NY 11219

PAGANO, ALPHONSE FREDERICK, obstetrician/gynecologist; b. Bklyn., Nov. 23, 1909; s. Domenico Antonio Pagano and Amalia Rose (Foresta) Cundari; m. Adele Marie Savarese, May 20, 1939; children: Althea, Amelia, Alphonse Frederick. Ph Ch, Columbia U., 1930, BS, 1934; MD, SUNY, Bklyn., 1935. Diplomate Am. Bd. Ob.-Gyn. Asst. attending obgyn. Luth. Med. Ctr., Bklyn., 1938-48, assoc. attending ob-gyn., 1948-52, attending ob-gyn., 1952-88; mem. Malpractice Panel Supreme Ct. N.Y., 1965-85. Cons. Selective Svc. U.S.A., 1948-57; pres. Bay Ridge Med. Soc., 1952; chmn. membership com., 1952—; maternity mortality com. Kings County Med. Soc., 1955-85, chmn. adv. com. ob-gyn., 1965-88. 1st lt. USAR Med. Corps., 1935-42. Recipient Selective Svc. medal, Pres. Truman, Washington, 1948, Cert. Appreciation, Supreme Ct. Appellate Div., Kings County Med. Soc.; named Physician of Yr., 1966, Bay Ridge Med. Soc., Bklyn. Fellow ACS, Am. Coll. Ob.-Gyn. (founding). Republican. Roman Catholic. Achievements include supervision of delivery of over 10,000 babies. Home: 2 Westgate Dr Sayville NY 11782-3054

PAGANO, JOSEPH STEPHEN, physician, researcher, educator; b. Rochester, N.Y., Dec. 29, 1931; s. Angelo Pagano and Marian (Vinci) Signorino; m. Anna Louise Reynolds, June 8, 1957; children: Stephen Reynolds, Christopher Joseph. A.B. with honors, U. Rochester, 1953; M.D., Yale U., 1957. Intern Mass. Meml. Hosp., 1957-58; resident Peter Bent Brigham Hosp. Harvard U., Boston, 1960-61; fellow Karolinska Inst., Stockholm, 1961-62; mem. Wistar Inst., Phila., 1962-65; asst. prof., then assoc. prof. U. N.C., Chapel Hill, 1965-73, prof. medicine, 1974—, dir. div. infectious diseases, 1972-75; dir. U. N.C. Lineberger Comprehensive Cancer Ctr., Chapel Hill, 1974—; attending physician N.C. Meml. Hosp., Chapel Hill; vis. prof. Swiss Inst. Cancer Rsch., Lausanne, 1970-71, Lineberger prof. cancer rsch., 1986—; mem. virology study sect. NIH, Bethesda, Md., 1973-79; spl. lectr. European Molecular Biology Assn., Oxford, Eng., 1979; cons. Burroughs Wellcome Co., Research Triangle Park, N.C., 1978—, Am. Inst. Biologic Sci., 1985-90; mem. Recombinant DNA adv. com. USPHS, 1986-90; mem. chancellor's adv. com. U. N.C., 1985-91, chair 1990-91. Mem. editorial bd. Jour. Virology, 1974-90; bd. assoc. editors Cancer Rsch., 1976-80; assoc. editor Jour. Gen. Virology, 1979-84, Antimicrobial Agts. and Chemotherapy, 1984—; contbr. numerous articles to profl. publs., chpts. to books. Bd. dirs. Am. Cancer Soc., N.C., 1980—; trustee Chapel Hill Preservation Soc., 1982-86; mem. commn. Episcopal Diocese N.C., Raleigh, 1984-87. Nat. Found. fellow, Stockholm, 1961-62; recipient Sinsheimer Found. award, 1966-68, USPHS Research Career award NIH, 1968-73. Mem. AAAS (Newcomb Anderson prize selection com. 1984-88), Infectious Disease Soc., Am. Soc. Microbiology, Am. Soc. Clin. Investigation (emeritus), Am. Soc. Virology, Internat. Assn. for Study and Prevention of Virus Assn., Am. Assn. Physicians, Internat. Assn. for Rsch. in Epstein-Barr Virus and Associated Diseases (pres. 1991, First Gertrude & Werner Henle Lectureship in Viral Oncology 1990). Episcopalian. Clubs: Chapel Hill Country, Chapel Hill Tennis (pres. 1980-82). Avocations: tennis; squash. Home: 114 Laurel Hill Rd Chapel Hill NC 27514-4323 Office: U NC Lineberger Comp Cancer Ctr Box 7295 Chapel Hill NC 27599-7295

PAGDON, WILLIAM HARRY, optical systems technician; b. Elizabeth, N.J., Sept. 7, 1962; s. Daniel Joseph and Virginia Mary (Mason) P.; m. Faith Irene Cohen, Aug. 20, 1990. Grad., U.S.N. Opticalman Class "A" Sch., Great Lakes, Ill., 1980. Optical test technician Instruments S.A., Edison, N.J., 1983-90; optical systems technician Spex Industries, Edison, N.J., 1990—. With USN, 1979-82. Roman Catholic. Home: 39 Idlewild Rd Edison NJ 08817

PAGE, ARMAND ERNEST, data processing director, consultant; b. Attleboro, Mass., June 19, 1955; s. Ernest Orville and Doris Clarice (Achin) P.; m. Rita Gail Hyder, Nov. 21, 1979; children: Rachel, Eric. BA in Natural Sci., U. So. Fla., 1977. Dir. data processing Bass & Swaggerty, Holly Hill, Fla., 1980—; owner/dir. Caywest Software Svcs., Port Orange, Fla., 1985—. Code enforcement bd. Port Orange, 1986-88; co-founder Coalition of Port Orange Homeowners Assn., 1987; city coun. mem. Port Orange Municipality, 1988; planning commn. mem. Port Orange, 1989—. Recipient Cert. of Appreciation, Port Orange, 1988. Mem. Data Processing Mgmt. Assn. Democrat. Avocations: sailing, boating, fishing, motorcycling, camping. Home: 1520 Rusty Circle Port Orange FL 32119 Office: Bass & Swaggerty 330 Carswell Ave Daytona Beach FL 32117-4437

PAGE, ARTHUR ANTHONY, astronomer; b. Yokohama, Japan, Aug. 3, 1922; arrived in Australia, 1941; s. Anthony Elephthere and Elena Artemiovna (Orbinsky) Pappadopoulos; m. Muriel Nancy Woods, Dec. 7, 1946 (div. Apr. 1963); children: Meredith Ann, Robert Arthur; m. Berenice Rose, Feb. 15, 1964 (dec. 1970); m. Aileen Sturgess Goddard, Oct. 8, 1971. Diploma in physiotherapy, U. Queensland, Brisbane, Australia, 1950. Registered physiotherapist, Queensland. Staff physiotherapist Brisbane (Australia) Gen. Hosp., 1950, from physiotherapist to chief physiotherapist Commonwealth Rehab. Svc., 1950-83; hon. rsch. cons. dept. of physics U. Queensland, Brisbane, 1987—; optical astron. observer radiophysics div. CSIRO, 1964—; dir. Mt. Tamborine Obs., 1973—; convenor planning and design com. Australian Physiotherapy Assn., 1979-84; observational researcher internat. photog. and photoelec. coop Flare Star programs, 1965—. Author: Spectrum and Magnitude Data Bank of B(e), B(p) and B(pe) Stars, 1982, Atlas of Flare Stars in the Solar Neighbourhood, 1988, Nova Search, 1989; co-author: Planning and Designing Physiotherapy Departments, 1981; contbr. articles to profl. jours. Col. 2d Australian Imperial Force, 1942-45, Australian Army Reserve, PTO. Recipient Copernicus Quincentennial medal, 1973; grantee in field. Fellow Royal Astron. Soc., Australian Inst. Physics (hon.); mem. Am. Astron. Soc. (full), Am. Inst. Physics, Astron. Soc. Australia, Internat. Astron. Union. Achievements include discovery of flare characteristics of 66 Oph and other associated phenomena; co-discovery of faint Open Cluster region flare stars; photographic and photoelectric astronomer of chromospherically active stars and B(e) stars; observational photographic and photoelectric research of Flare Stars. Home: 4/70 Herries St, Toowoomba QLD 4350, Australia Office: U Queensland Dept Physics, Mt Tamborine Observatory, Saint Lucia QLD 4072, Australia

PAGE, DENNIS, coal scientist, consultant; b. Darjeeling, India, Aug. 2, 1932; arrived in U.K., 1934; s. Arthur and Doris Harriet (Dixon) P.; m. Mary Bernadette Leonard, June 10, 1957; children: Carol, Julie, Geraldine, Denise, Christopher. Grad. Royal Inst. of Chemistry, Rutherford Coll. Tech., Newcastle Upon Tyne, 1957; M in Philosophy, Sunderland Poly, 1972. Chartered chemist. Asst. analyst Nat. Coal Bd., Durham County, Eng., 1949-60; rsch. asst. Imperial Chem. Industries, Billingham, Eng., 1961-64; rsch. asst. Nat. Coal Bd., Newcastle upon Tyne, 1964-77, head investigations, 1977-82; hdqrs. scientist Nat. Coal Bd., Harrow, Eng., 1982-87; sr. scientist Brit. Coal, Burton on Trent, Eng., 1987-91; mem. tech. com. Brit. Standards Inst., London, 1986-90, chmn., 1990—; adviser radiation protection Brit. Coal, London, 1986-91. Patentee in field; contbr. articles to profl. jours. Fellow Royal Soc. Chemistry; mem. Soc. Radiol. Protection. Roman Catholic. Achievements include research in on-line analysis of coal, radon in mining industry. Avocations: photography, study of natural history. Home and Office: 10 The Green, Barton under Needwood, Burton-on-Trent DE13 8JB, England

PAGE, DON NELSON, theoretical gravitational physics educator; b. Bethel, Alaska, Dec. 31, 1948; s. Nelson Monroe and Zena Elizabeth (Payne) P.; m. Catherine Anne Hotke, June 28, 1986; children: Andrew Nicolaas Nelson, John Paul Weslie. AB in Physics and Math., William Jewell Coll., 1971; MS in Physics, Calif. Inst. Tech., 1972, PhD in Physics and Astronomy, 1976; MA, U. Cambridge, Eng., 1978. Rsch. asst., assoc. Calif. Inst. Tech., Pasadena, 1972-76, 87; rsch. asst., NATO fellow U. Cambridge, 1976-79; rsch. fellow Darwin Coll., Cambridge, 1977-79; asst. prof. physics Pa. State U., University Park, 1979-83, assoc. prof., 1983-86, prof., 1986-90; prof. U. Alta., Edmonton, Can., 1990—; vis. rsch. faculty, assoc. U. Tex., Austin, 1982, 83, 86; vis. rsch. assoc. Inst. Theoretical Physics, U. Calif., Santa Barbara, 1988; vis. prof. U. Alta, Edmonton, Can., 1989-90; mem. Inst. Advanced Study, Princeton, N.J., 1985; assoc. Can. Inst. Advanced Rsch., Toronto, 1987-91, fellow, 1991—; cons. Time-Life, Alexandria, Va., 1990—. Editorial bd. jour. Classical and Quantum Gravity, 1988-91; assoc. editor Can. Jour Physics, 1992—; contbr. articles to The Phys. Rev., Physics Letters, Phys. Rev. Letters, Nuclear Physics. U.S. Presdl. scholar, 1967; Danforth Found. grad. fellow, 1971-76; Alfred P. Sloan Found. rsch. fellow, 1982-86; John Simon Guggenheim Found. fellow, 1986-87. Fellow Am. Sci. Affiliation, Can. Inst. Advanced Rsch.; mem. Am. Phys. Soc. Baptist. Achievements include calculation of rates of black-hole emission; discovery of first inhomogeneous compact Einstein metric and other gravitational instantons, of approximate weight of radiation near a black hole; disproof of semiclassical gravity; research on size and entropy of quantum universe. Home: 5103-126 St, Edmonton, AB Canada T6H 3W1 Office: U Alta, 412 Physics Lab, Edmonton, AB Canada T6G 2J1

PAGE, EDWARD CROZER, JR., chemical engineer, consultant; b. Bryn Mawr, Pa., Nov. 14, 1919; s. Edward Crozer and Elizabeth (Griffith) P.; m. Barbara Benson Jefferys; children: Barbara, Carol, Edward. BSE, Princeton U., 1942, MSE, 1946. Rsch. chem. engr. E.I. duPont de Nemours & Co., Wilmington, Del., 1946-52; v.p., tech. dir. Bower Chem. Co., Phila., 1951-60; div. mgr. J.T. Baker Chem. Co., Phillipsburg, N.J., 1960-65; dir. corp. chem. control Polaroid Corp., Cambridge, Mass., 1965-82; pres. UCI, Inc.-Environ. Engring., Newton, Mass., 1983-89; v.p., sr. engring. assoc. David Gordon Assocs., Newton, Mass., 1989—; cons. Synthetic Organic Chem. Mfg. Assn., Washington, 1978-85. Contbr. articles to profl. jours. Pres. Boston Guild for Hard of Hearing, 1985-90; pres. PTA, Friends' Sch., Media, Pa., 1950-53. Lt. comdr. USNR, 1943-55. Grantee Calco, Inc., 1942, Phillips Petroleum, 1946. Mem. NSPE, Am. Inst. Chem. Engrs., Am. Chem. Soc., Am. Electroplaters and Surface Finishers Soc. (Pres.'s award 1988), Phi Beta Kappa, Tau Beta Pi. Episcopalian. Achievements include patents in fungicidal chemicals; electrochemical methods for recovering copper and silver from etching solutions; novel materials and methods for thin film coatings. Home: 305 Haley Rd PO Box 18 Hollis Center ME 04042

PAGE, LAWRENCE MERLE, ichthyologist, educator; b. Fairbury, Ill., Apr. 17, 1944; s. Elmo Merle and Morseta Vivian (Ellis) P. MS, U. Ill., 1968, PhD, 1972. Asst. profl. scientist Ill. Natural History Survey, Champaign, 1972-77, assoc. profl. scientist, 1977-80, profl. scientist, 1980—; prof. U. Ill., Champaign, 1983—; adv. Ill. Nature Preserves Commn., Springfield, 1987—, Ill. Endangered Species Bd., Chgo., 1985—; dir. ctr. biodiversity Ill. Natural Hsitory Survey, Champaign, 1989—. Author: Handbook of Darters, 1983, Peterson Field Guide to Freshwater Fishes, 1991. Named Assoc. in Ichthyology U. Kans. Mus. Natural History, Lawrence, 1979—. Mem. Am. Soc. Ichthyologists and Herpetologists (bd. givs. 1987-91, treas. 1991—). Achievements include research on systematics of darters (family Percidae) and development of classification of breeding behaviors in darters. Office: Ill Natural History Survey 607 E Peabody Dr Champaign IL 61820

PAGE, LORNE ALBERT, physicist, educator; b. Buffalo, July 28, 1921; s. John Otway and Laura (Stewart) P.; m. Muriel Emily Jamieson, Sept. 7, 1946; children: J. Douglas, Kenneth L., James F., Donald S., David K. B.Sc., Queen's U., Can., 1944; Ph.D., Cornell U., 1950. Faculty U. Pitts., 1950—, prof. physics, 1958-86, prof. emeritus, 1987—. Contbr. articles to profl. jours. Served to lt. Royal Canadian Navy, 1944-45. Guggenheim fellow Upsala U., Sweden, 1957-58; Alfred P. Sloan research fellow, 1961-63. Fellow Am. Phys. Soc. Episcopalian. Achievements include measurement of the position's mass; identification of positronium in condensed matter; development of method for analyzing circular polarization of high energy x-rays; measurement of first inherent polarization of positive beta particles. Home: 157 Lloyd Ave Pittsburgh PA 15218-1645

PAGE, PHILIP RONALD, chemist; b. Rochford, Essex, Eng., Feb. 21, 1951; s. Arthur Leonard and Joyce Jean (MacFarlane) P.; m. Penny Alison Gambrill, Apr. 8, 1978 (div. 1992); children: Sarah Dawn, Karen Louise. B.S., Victoria U. of Manchester (Eng.), 1972, M.S., 1973, Ph.D., 1975. Mgmt. trainee T.B.A. Indsl. Products Ltd., Rochdale, Eng., 1975-76, tech. rep., 1976-77, asst. market devel. mgr., 1977-78; asst. rsch. dir. Hovione-Sociedade Quimica Lda., Loures, Portugal, 1978-80, rsch. dir., 1980-84, exec. and tech. dir., 1984—. Patentee in field. Fellow Royal Soc. Chemistry. Mem. Ch. of Eng.

PAGE, RICHARD ALLEN, materials scientist; b. Washington, May 30, 1952; s. Franklin Farnsworth Page and Eunice Marguerite (Larson) Foster; m. Cynthia Louise du Menil. B of Metallurgical Engring., U. Minn., 1974, MS, 1976; PhD, Northwestern U., 1980. Rsch. engr. Zimpro Inc., Rothschild, Wis., 1976-77; sr. rsch. engr. Southwest Rsch. Inst., San Antonio, 1980-87, staff scientist, 1987-89, mgr., 1989—. Contbr. articles to profl. jours. Am. Can Co. fellow, 1978. Mem. Am. Ceramic Soc., Am. Crystallographic Assn., Nat. Assn. Corrosion Engrs., Metallurgical Soc., Tau Beta Pi, Alpha Sigma Mu, Delta Soc. Independent. Lutheran. Achievements include pioneering the use of small-angle scattering for creep cavitation studies in ceramics; developed series of wear resistant, self-lubricating ceramic composites; developed a stress corrosion cracking model. Office: Southwest Rsch Inst PO Drawer 28510 6220 Culebra Rd San Antonio TX 78228-0510

PAGE, ROY CHRISTOPHER, periodontist, educator; b. Campobello, S.C., Feb. 7, 1932; s. Milton and Anny Mae (Eubanks) P. B.A., Berea Coll., 1953; D.D.S., U. Md., 1957; Ph.D., U. Wash., 1967; Sc.D. (hon.), Loyola U., Chgo., 1983. Cert. in periodontics. Pvt. practice periodontics Seattle, 1963—; asst. prof. U. Wash. Schs. Medicine and Dentistry, Seattle, 1967-70, prof., 1974—, dir. Ctr. Research in Oral Biology, 1976—; dir. grad. edn. U. Wash. Sch. Dentistry, 1976-80, dir. research, 1976—; dir. clin. dental rsch. ctr., 1990—; vis. scientist MRC Labs., London, 1971-72; cons., lectr. in field. Author: Periodontal Disease, 1977, 2d edit., 1990, Periodontitis in Man and Other Animals, 1982. Recipient Gold Medal award U. Md., 1957; recipient Career Devel. award NIH, 1967-72. Fellow Internat. Coll. Dentists, Am. Coll. Dentists, Am. Acad. Periodontology (Gies award 1982); mem. ADA, Am. Asssn Dental Rsch. (pres. 1982-83), Am. Soc. Exptl. Pathology, Internat. Assn. Dental Rsch. (pres. 1987, basic periodontal rsch. award 1977). Home: 8631 Inverness Dr NE Seattle WA 98115-3935

PAGE, TONYA FAIR, cogeneration facility coordinator; b. Greenville, S.C., Oct. 29, 1956; d. Warren John and Georgia (Combs) Fair; m. Stanley Wilson, Aug. 18, 1979; children: Anthony Austin, Kyle Martin. BS in Humanities, U. S.C., 1978, BS in Geology, 1983, MS in Geology, 1985. Rsch. asst. U. S.C., Columbia, 1983-85; cons. Earth Sci. Rsch. Inst., Columbia, 1984-85; v.p. tech. svcs. United Devel. Group, Columbia, 1988—; mem. adv. bd. opportunity scholars program U. S.C., Columbia, 1990—. Recipient Alumni Merit award U. S.C., Columbia, 1990. Mem. NOW, S.C. Wildlife Fedn., U. S.C. Alumni Assn. Republican. Presbyterian. Achievements include research in SO2 absorption capability of dolomitic limestones and beneficial use applications of coal combustion by-products. Home: 2836 Burney Dr Columbia SC 29205 Office: United Devel Corp 1401 Main St Ste 1115 Columbia SC 29201

PAGE, VALDA DENISE, epidemologist, researcher, nutritionist; b. Houston, Jan. 23, 1958; d. Ulysses and Dorothy Lee (Sells) P. BS in Food Sci. and Tech., Tex. A&M U., 1981; BS in Nutrition and Dietetics, U. Tex., Houston, 1985, MPH, 1991. Registered and lic. dietitian. Clk. various temp. svcs., Houston, 1980-86; nutritionist, nutrition coord., asst. dir. W.I.C., pub. health analyst Harris County Health Dept., Houston, 1986-93; planner Harris County Juvenile Probation Dept., Houston, 1993—; Dietary cons. Deer Park (Tex.) Hosp., 1987-89; clin. instr. U. Tex., Houston, 1986—. Chmn. Health and Welfare ministry Windsor Village United Meth. Ch., 1989—; mem. Windsor Village AIDS Ministry, Houston, 1989—; team capt. March of Dimes Walk Am., Houston, 1988—; mem. Pres.'s Former Student Adv. Com. on Black Issues, College Station, Tex., 1989—. Mem. APHA, Am. Dietetic Assn., Am. Diabetes Assn. (affiliate, minority initiative task force), Am. Heart Assn. (Speaker's Bur. 1986—), Inst. Food Technologists, Soc. Epidemiologic Rsch., Nat. Perinatal Assn., Network Blacks in Dietetics and Nutrition. Democrat. Avocations: swimming, traveling, reading volleyball, sports. Home: 9421 Rosehaven Dr Houston TX 77051-3128 Office: Harris County Juvenile Probation Dept Rsch Planning And Evaluation 3540 W Dallas Houston TX 77019-1796

PAIK, JOHN KEE, structural engineer; b. Seoul, Feb. 17, 1934; came to U.S., 1955; s. Nam Suk and Kyong Ock (Yun) P.; m. Aine Fenoula Ievers, Feb. 20, 1970; 1 child, Brian Ievers Paik. BSCE, So. Meth. U., 1961; PhD, NYU, 1975. Lic. profl. engr. N.Y., N.J., Conn., Pa., Md., Mass., Vt., Ga., Fla. Chief engr. T.Y. Lin and Assocs., N.Y.C., 1960-67; chief structural engr. Soros Assocs., N.Y.C., 1967-68; sr. project engr. Stauffer Chem. Co., Dobbs Ferry, N.Y., 1975-77; prin., founder Paik and Assocs., Westchester County, N.Y., 1977—; chmn. The Future Home Tech. Inc., Port Jervis, N.Y., 1986—; chmn., pres. J.K.P. Constrn. Co. Inc., Mohegan Lake, N.Y., 1989—; adj. assoc. prof. Grad. Sch. Engring. Manhattan Coll., Bronx, 1985; lectr. Grad. Sch. Engring. Polytech. U., Bklyn., 1973-85, Cooper Union, N.Y.C., 1972. Mem. ASCE, Am. Inst. Steel Constrn., Prestressed Concrete Inst., N.Y Acad. Scis., Am. concrete Inst., Post Tensioning Inst., So. Meth. U. Alumni Club (pres. 1964), Chi Epsilon. Republican. Methodist. Achievements include the design of over 100 million sq. feet of comml., residential, indsl. and instnl. structures including several highrise bldgs. over 40 stories in N.Y.C. Home: Dyckman Dr Mohegan Lake NY 10547 Office: Paik and Assocs 2861 Lexington Ave Mohegan Lake NY 10547-1835

PAIK, YOUNG-KI, biochemist, nuclear biologist; b. Daejeon, ChungNam, Korea, Jan. 8, 1953; s. Nam-Chul and Bok-Im (Yoo) P.; m. Jung-Im Cho, June 10, 1979; children: Kyung-Min, Yoon-Jung, Sang-Jin. BS, Yonsei U., Seoul, 1975; PhD, U. Mo., 1983. Rsch. fellow Agency for Defence Devel., Daejeon, Korea, 1977-79; postdoctoral fellow Gladstone Found. Labs., U. Calif.-San Francisco, 1983-86; staff rsch. investigator Gladstone Found. Labs., San Francisco, 1986-89; assoc. prof. dept. biochemistry Hanyang U., Ansan, Korea, 1989-93; vis. scientist E.I. duPont Co., Wilmington, Del., 1981-83. Daewoo Found. grantee, Seoul, 1989, Ministry of Edn. Rsch. grantee, Seoul, 1990-94, Korean Sci. and Engring. Found. Rsch. grantee, Daejeon, 1991, Ministry of Sci. & Tech. grantee, 1992-94. Mem. Am. Soc. Biochemistry and Molecular Biology, Am. Soc. Microbiology, Biochem. Soc. Republic of Korea (sec. gen. affairs 1991, editorial bd. mem., assoc. editor 1993-94), Korean Soc. Molecular Biology, Korean Soc. Lipidology (planning sec. 1991-92, editorial bd. mem., v.p. 1993-94). Office: Yonsei U Coll Sci, Dept Biochem Suddemoon-ku, 425-791 Seoul 120-749, Republic of Korea

PAIS, ABRAHAM, physicist, educator; b. Amsterdam, Holland, May 19, 1918; s. Jesaja and Kaatje (van Kleeff) P.; m. Lila Atwill, Dec. 15, 1956 (div. 1962); 1 child, Joshua; m. Agnes Ida Benedicte Nicolaisen, Mar. 15, 1990. B.Sc., U. Amsterdam, 1938; M.Sc., U. Utrecht, 1940, Ph.D., 1941. Research fellow Inst. Theoretical Physics, Copenhagen, Denmark, 1946; prof. Inst. Advanced Study, Princeton, N.J., 1950-63; prof. physics Rockefeller U., N.Y.C., 1963-81, Detlev Bronk prof., 1981—, prof. emeritus, 1988—; Balfour prof. Weizmann Inst., Israel, 1977. Author: Subtle is the Lord (Am. Book award 1983, Am. Inst. Physics award 1983), Inward Bound, 1986, Niels Bohr's Times, 1991. Decorated officer Order of Oranje Nassau (The Netherlands); recipient J.R. Oppenheimer Meml. prize, 1979, Physica prize The Netherlands, 1992, Gemant award Am. Inst. Physics, 1993; Guggenheim fellow, 1960. Fellow Am. Phys. Soc.; mem. Royal Acad. Scis. Holland (corr., medal of sci. 1993), Royal Acad. Scis. and Letters, Denmark, Am. Acad. Arts and Scis., Am. Philos. Soc., Nat. Acad. Scis., Council on Fgn. Relations. Home: 1161 York Ave New York NY 10021 Office: Rockefeller Univ New York NY 10021

PAKES, STEVEN P., medical school administrator; b. St. Louis, Jan. 19, 1934; married; 4 children. BSc, Ohio State U., 1956, DVM, 1960, MSc, 1964, PhD in Vet. Pathology, 1972. Vet. pathologist U.S. Army, Ft. Detrick, Md., 1960-62; chief animal colonies Pine Bluff Arsenal, Ark., 1964-66; chief comparative pathology Naval Aerospace Med. Inst., 1966-69; dir. lab. animal medicine Coll. Vet. Medicine Ohio State U., 1969-72; assoc. prof. Southwestern Med. Sch. U. Tex. Southwestern Med. Ctr., Dallas, 1972-80, prof. comparative medicine, chmn. dept. Southwestern Med. Sch., 1980—; mem. exec. com. Inst. Lab. Animal Resources NAS-Nat. Rsch. Coun.; chmn. coun. accreditation Am. Assn. Accreditation Animal Care, 1974-76, treas., bd. trustees, 1983—; adv. bd. Vet. Specialities, 1981-82, chmn., 1985; chmn. lab. guide review com. Inst. Lab. Animal Resources NAS, 1983-85; mem. Inst. Lab. Animal Resources Coun., 1985—, chmn., 1987—. Mem. AAAS, Am. Assn. Lab. Animal Sci. (bd. trustees 1985-88), Am. Coll. Lab. Animal Medicine (pres. 1973, exam com. 1968-69), Am. Vet. Medicine Assn., Am. Soc. Microbiology, Sigma Xi. Achievements include research in infectious diseases of laboratory animals, effect of spontaneous diseases of laboratory animals on biomedical research. Office: U Tex Dept Comparative Medicine 5323 Harry Hines Blvd Dallas TX 75235*

PAL, SIBTOSH, mechanical engineer, researcher; b. Lagos, Nigeria, Feb. 25, 1963; Indian citizen; came to U.S., 1978; s. Sambhu Nath and Malina Pal. PhD, Penn State U., 1990. Rsch. assoc. Penn State U., University Park, 1990—. Contbr. articles to Physics of Fluids, Atomization and Sprays. Mem. AIAA, ASME, ASTM. Office: Pa State U Propulsion Engring Rsch Ctr 129 Bigler Rd University Park PA 16802

PALACIOS, JOAQUIN ALQUISIRA, polymer scientist; b. Huamantla, Mex., Nov. 30, 1943; s. Joaquin Palacios Escudero and Maria Alquisira Mendez. MS in Chemistry, UNAM, Mexico City, 1973, PhD in Chemistry, 1976. R&D engr. Plastiglass, Toluca, 1968-70; asst. prof. UNAM, Mexico City, 1972-73, phys.-chem. prof., 1976—; vis. scientist Akron U., 1973-76; postdoctoral MMI, Midland, Mich., 1981; ind. cons. IRSA, Toluca, Mex., 1978-81, Fisisa, Mexico City, 1981-82, Henkel, Mexico City, 1988, Reichhold, Mexico City, 1989-90. Author: Copolymers Characterization, 1982, Polyester Resins, 1981; contbr. articles to profl. jours. Fellow Sistema Nal. Investigadores, 1984. Mem. Am. Chem. Soc., Assn. Quimica Analitica, Sociedad Quimica de Mex. (trustee), Sociedad Polimerica de Mexico (trustee 1984-87). Achievements include patent for low temperature aromatic polyesters reaction, microwave initiated copolymerization of acrylic monomers; rsch. on new natural source of trans poly (isoprene) microstructure of guayule rubber, microstructure determination of opuntia mucilage. Office: UNAM Ciudad Universitaria, Fac Quimica, 04510 Mexico City Mexico

PALADE, GEORGE EMIL, cell biologist, educator; b. Jassy, Romania, Nov. 19, 1912; came to U.S., 1946, naturalized, 1952; s. Emil and Constanta (Cantemir) P.; m. Irina Malaxa, June 12, 1941 (dec. 1969); children—Georgia Teodora, Philip Theodore; m. Marilyn G. Farquhar, 1970. Bachelor, Hasdeu Lyceum, Buzau, Romania; M.D., U. Bucharest, Romania. Instr., asst. prof., then assoc. prof. anatomy Sch. Medicine, U. Bucharest, 1935-45; vis. investigator, asst. assoc., prof. cell biology Rockefeller U., 1946-73; prof. cell biology Yale U., New Haven, 1973-83; sr. research scientist Yale U., 1983-89; prof.-in-residence, dean sci. affairs Med. Sch., U. Calif., San Diego, 1990—. Author sci. papers. Recipient Albert Lasker Basic Research award, 1966, Gairdner Spl. award, 1967, Horwitz prize, 1970, Nobel prize in Physiology or Medicine, 1974, Nat. Medal Sci., 1986. Fellow Am. Acad. Arts and Scis.; mem. Nat. Acad. Sci., Pontifical Acad. Sci., Royal Soc. (London), Leopoldina Acad. (Halle), Romanian Acad., Royal Belgian Acad. Medicine. Research interests correlated biochem. and morphological analysis cell structures.

PALADINO, JOSEPH ANTHONY, clinical pharmacist; b. Utica, N.Y., May 5, 1953; s. Paul Francis and Jacqueline Ann (Monaco) P.; m. Carol Ann Jenny, June 5, 1976; children: Nicholas Joseph, Matthew Jerome, Kathryn Elizabeth. BS in Biology, Siena Coll., Loudonville, N.Y., 1975; BS in Pharmacy, Mass. Coll. Pharmacy, 1977; D Pharmacy, Med. U. S.C., 1982. Acting dir. pharmacy Utica (N.Y.) Psychiat. Ctr., 1978-80; decentralized pharmacist Med. U. Hosp., Charleston, S.C., 1980-82; asst. dir. pharmacy Rochester (N.Y.) Gen. Hosp., 1982-87, adj. med. staff, 1986-87; clin. instr. pediatrics U. Rochester Sch. Medicine, 1985-87; clin. asst. prof. pharmacy SUNY, Buffalo, 1989—; dir. pharmacokinetics Millard Fillmore Suburban Hosp., Williamsville, N.Y., 1987—; editorial adv. bd. Jour. of Infections Disease Pharmacotherapy, 1993; mem. adv. bd. several major pharm. cos.; cons. numerous major pharm. cos.; 1st vis. prof. pharmacy Tayside Health Bd., Dundee, Scotland, 1992. Big Bro., Big Bros. and Big Sisters, Albany, N.Y., 1972-80; bd. mem., den leader Cub Scouts, Clarence, N.Y., 1989-91; founding bd., coach Clarence Little League Football, 1992—. Mem. Am. Coll. Clin. Pharmacy, Am. Soc. Microbiology, Soc. Infectious Diseases Pharmacists, N.Y. Acad. Scis. Achievements include pioneering a method of treating certain infectious diseases with an early switch from intravenous to oral antibiotics, thus changing a standard of practice; co-development of a standardized format for dosing charts for critical intravenous medications; discovery of drug interaction between secobarbital and theophylline; integration of switch therapy with pharmacology, outcomes therapy, and pharmacoeconomics. Office: Millard Fillmore Suburban Hosp 1540 Maple Rd Williamsville NY 14221

PALANCE, DAVID M(ICHAEL), electrical engineer; b. Hicksville, Ohio, Oct. 5, 1960; s. matthew Palance and Jeanette (White) Deardorff; m. Liisa Victoria Ekker, Sept. 12, 1981; children: Katrina Rose, Benjamin Michael. AS in Indsl. Electronics, Keene State Coll., 1981; postgrad., Daniel Webster Coll. Tech. MFE Inc., Salem, N.H., 1978-79, U. N.H., Keene, 1980-81; engr. assoc. Sanders Lockhead, Nashua, N.H., 1981-86; electronics engr. Pneutronics Corp., Hollis, N.H., 1986-88, advanced devel. engr., 1988—. Active Amherst Town Band, 1984—, Willard Family Assn., 1986—; sec. bd. trustees Messiah Luth. Ch., 1989—. Mem. Am. Vacuum Soc., Instrument Soc. of Am., United We Stand Am. Republican. Achievements include patents for electronic variable orifice, electronically controlled variable device; patent pending for electronically controlled loading device. Office: Pneutronics Corp 26 Clinton Ste 103 Hollis NH 03049

PALAY, SANFORD LOUIS, retired scientist, educator; b. Cleve., Sept. 23, 1918; s. Harry and Lena (Sugarman) P.; m. Victoria Chan Curtis, 1970 (div. Nov. 1990); children: Victoria Li-Mei, Rebecca Li-Ming. A.B., Oberlin Coll., 1940; M.D. (Hoover prize scholar 1943), Western Res. U., 1943. Teaching fellow medicine, rsch. assoc. anatomy Western Res. U., Cleve., 1945-46; NRC fellow med. scis. Rockefeller Inst., 1948, vis. investigator, 1953; from instr. anatomy to assoc. prof. anatomy Yale U., 1949-56; chief sect. neurocytology, lab. neuroanatomical scis. Nat. Inst. Neurol. Diseases and Blindness, NIH, Washington, 1956-61; chief lab. neuroanatomical scis. Nat. Inst. Neurol. Diseases and Blindness, NIH, 1960-61; Bullard prof. neuroanatomy Harvard, Boston, 1961-89, prof. emeritus, 1989—; Linnean Soc. lectr., London, 1959; vis. investigator Middlesex Hosp. (Bland-Sutton Inst.), London, Eng., 1961; Phillips lectr. Haverford Coll., 1959; Ramsay Henderson Trust lectr. U. Edinburgh, Scotland, 1962; George H. Bishop lectr.. Washington U., St. Louis, 1990; Disting. Scientist lectr. Tulane U. Sch. Medicine, 1969, 75; vis. prof. U. Wash., 1969; Rogowski Meml. lectr. Yale, 1973; Disting. lectr. biol. structure U. Miami, 1974; Disting. Scientist lectr. U. Ark., 1977; other Disting. lectureships; vis. prof. U. Osaka, Japan, 1978, Nat. U. Singapore, 1983; spl. vis. prof. U. Osaka, 1988; chmn. study sect. on behavioral and neural scis. NIH, 1984-86; mem. fellowship bd. NIH, 1958-61, cell biology study sect., 1959-65, adv. com. high voltage electron microscope resources, 1973-80, mem. rev. com. behavioral and neurol. scis. fellowships, 1979-86; chmn. Gordon Research Conf. Cell Structure and Metabolism, 1960; asso. Neuroscis. Research Program, 1962-67, cons. assoc., 1975—; mem. anat. scis. tng. com. Nat. Inst. Gen. Med. Scis., 1968-72; mem. sci. adv. com. Oreg. Regional Primate Research Center, 1971-76. Author: The Fine Structure of the Nervous System, 1970, 3d edit., 1991, Cerebellar Cortex, Cytology and Organization, 1974; editor: Frontiers of Cytology, 1958, The Cerebellum, New Vistas, 1982; mem. sci. coun. Progress in Neuropharmacology and Jour. Neuropharmacology, 1961-66; mem. editorial bd. Exptl. Neurology, 1959-76, Jour. Cell Biology, 1962-67, Brain Research, 1965-71, Jour. Comparative Neurology, 1966—, Jour. Ultrastructure Research, 1966-86, Jour. of Neurocytology, 1972-87, Exptl. Brain Research, 1965-76, Neurosci, 1975—, Anatomy and Embryology, 1968; co-mng. editor, 1978-88; editor in chief Jour. Comparative Neurology, 1981-93; mem. adv. bd. editors Jour. Neuropathology and Exptl. Neurology, 1963-82, Internat. Jour. Neurosci, 1969-74, Tissue and Cell, 1969-86; contbr. articles to profl. jours. Served to capt. M.C. AUS, 1946-47. Recipient 50 Best Books of 1974 award Internat. Book Fair, Frankfurt, Fed. Republic Germany, Best Book in Profl. Readership award Am. Med. Writers Assn., 1975, Biomed. Rsch. award Assn. Am. Med. Colls., 1989, Lashley award Am. Philos. Soc., 1991, Camillo Golgi award Fidia Rsch. Found., 1992; Guggenheim fellow, 1971-72; Fogarty scholar-in-residence NIH, Bethesda, 1980-81. Fellow Am. Acad. Arts and Scis.; mem. NAS, Am. Assn. Anatomists (chmn. nominating com. 1964, mem. exec. com. 1970-74, anat. nomenclature com. 1975-78, pres. 1980-81, Henry Gray award 1990), Histochem. Soc., Electron Microscope Soc. Am., AAAS, Am. Soc. Cell Biology (program com. 1975), Internat. Soc. Cell Biology, Soc. for Neurosci. (Gerard award 1990), Washington Soc. Electron Microscopy (organizing com., sec.-treas. 1956-58), Soc. Francaise de Microscopie Electronique (hon.), Royal Microscopical Soc. (hon.), Golgi Soc. (hon.), Anat. Soc. Gr. Britain and Ireland (hon.), Cajal Club (pres. 1973-74), Phi Beta Kappa, Sigma Xi, Alpha Omega Alpha. Home: 78 Temple Rd Concord MA 01742-1520

PALEOS, CONSTANTINOS MARCOS, chemist; b. Chios, Greece, June 6, 1941; s. Marcos J. and Matrona M. (Vagianos) P.; m. Maria Maistrali, Dec. 26, 1971; children: Matrona, Evagellia, Marcos. BS in Chemistry, U. Athens, Greece, 1964; PhD in Chemistry, Drexel U., 1970. Project chemist Amoco Chems., Naperville, Ill., 1970-71, Motor Oil (Hellas), Athens, 1971-73; rsch. dir. NRC Demokritos, Athens, 1973—; vis. prof. L. Pasteur U., Strasbourg, France, 1991-92; cons. in field; mem. adv. bd. for molecular crystals and liquid crystals Gordon and Breach, Phila; mem. adv. bd. Chimica Chronica. Editor: Polymerization in Organized Media, 1992; contbr. articles to profl. pubis. Greek Ministry for Tech. and Rsch. grantee, 1973—. Mem. Union of Hellenic Chemists, Am. Chem. Soc., Liquid Crystalline Soc. Office: Nat Rsch Ctr Demokritos, Aghia Paraskevi Attikis, 15310 Athens Greece

PALERMO, CHRISTOPHER JOHN, aerospace engineer; b. Ft. Leavenworth, Kans., Sept. 15, 1962; s. B.G. Frank J. and Irmgard M. (Sadlemaier) P. BS in Aerospace Engring., U. Va., 1984; MS in Engring. Mgmt., U. Dayton, 1989. Commd. 2d lt. USAF, 1984, advanced through grades to capt., 1990; acquisition logistics staff officer Aero. Systems Div., Wright-Patterson AFB, Ohio, 1984-87, F-15 support equip. mgr., 1987-89; satellite ops. officer 1st Space Ops. Squadron, Falcon AFB, Colo., 1990-92; satellite ops. test mgr. 50 Ops. Support Squadron USAF, Falcon AFB, 1992—. Mem. AIAA, Air Force Assn. Roman Catholic. Home: 6740 Holt Dr Colorado Springs CO 80922 Office: 50 OSS/AFOTEC Falcon AFB CO 80912

PALEY, STEVEN JANN, engineering executive; b. Rutherford, N.J., Jan. 7, 1955; s. Edward and Florence (Janowitz) P.; m. Laura Fern Mironov, Sept. 16, 1979; children: Rachel, Shoshana. BS, BA, Tufts U., 1977; MSE in Product Design, Stanford U., 1983. Engr. IBM, Rochester, Minn., 1977-80; design engr. AT&T Bell Labs., Holmdel, N.J., 1985-87; dir. R&D Texwipe Co., Upper Saddle River, N.J., 1987-90, exec. v.p., 1990—. Editorial bd. Chinese Contamination Control Jour., 1992—; contbr. articles to profl. pubis. Bd. dirs. Congregation Bet Tefillah, Paramus, N.J., 1990—. Recognized for invention in semiconductor industry Semiconductor Internat. mag., 1990. Mem. IEEE, Inst. Environ. Scis. (Maurice Simpson Tech. Editors award 1992), Am. Vacuum Soc. Achievements include patents in field, development of measurement techniques and test methods. Office: Texwipe Co 650 E Crescent Ave Upper Saddle River NJ 07458

PALFFY-MUHORAY, PETER, physicist, educator; b. Nyireqyhaza, Hungary, May 20, 1944; arrived in Can., 1957; s. Zoltan and Eva (Meqyery) Palffy-Muhoray; m. Eunice Louise Boyd, Dec. 18, 1985; 1 child, Nicole Marie. MASc in Elec. Engring., U. B.C., Vancouver, Can., 1969, PhD in Physics, 1977. Instr. Capilano Coll., Vancouver, 1977-87; sr. rsch. fellow Liquid Crystal Inst., Kent, Ohio, 1987—, assoc. dir., 1990—; rsch. fellow U. Sheffield, Eng., 1981-82; hon. asst. prof. U. B.C., Vancouver, 1983-90; vis. prof. Hungarian Acad. Sci., Budapest, 1987; cons. AT&T Bell Labs., Murray Hill, N.J., 1989-90. Guest editor: Procs. 13th Internat. Liquid Crystal Conf., 1990, Procs. 4th Internat. Meeting on Optics of Liquid Crystals, 1991; editorial adv. bd. Molecular Crystals and Liquid Crystals, 1992, regional editor; assoc. editor World Sci. Series Liquid Crystals; regional editor Liquid Crystals Today, Sheffield, 1992. Mem. Am. Phys. Soc., Optical Soc. Am., Can. Assn. Physicists, Internat. Soc. Optical Engrs. Achievements include research contributions in areas of liquid crystal physics, pattern formation and nonlinear optics. Office: Kent State U Liquid Crystal Inst Kent OH 44242

PALIK, ROBERT RICHARD, mechanical engineer; b. Iowa City, Iowa, Mar. 10, 1923; s. Frank and Maria (Pavco) P.; m. Wanita Slaughter, Dec. 19, 1945; children: Andrea Denise, Stephen Brett, Robert Neil. BSME, State U. Iowa, 1949, MSME, 1950. Registered profl. engr. Va. Lab. tech. State U. Iowa Physics Dept., Iowa City, 1941-42; chief engr. Keokuk (Iowa) Steel Casting Co., 1950-54; mgr. rsch. engring. Reynolds Aluminum Co., Richmond, Va., 1954-70; v.p. Crown Aluminium Inds., Roxboro, N.C., 1970-75; div. chief City of Richmond, 1975-79; gen. engr. U.S. Dept. Housing and Urban Devel., Richmond, 1979—. Pres. West Wistar Civic Assn., Richmond, 1957; legis. chmn., treas. PTA, Richmond, 1957. Lt. Col. USAF, 1942-45. Fellow ASME; mem. Pi Tau Sigma. Achievements include method of treating metal patent. Home: 9318 Westmoor Dr Richmond VA 23229

PALISANO, JOHN RAYMOND, biologist, educator; b. Buffalo, Mar. 6, 1947; s. John P. and Marie L. Palisano; m. Peg A. Sutherland, Dec. 26, 1970; 1 child, Benjamin N. BS cum laude, U. Tenn., 1969, PhD, 1975. Instr. dept. medicine Case Western Res. U., Cleve., 1976-85, prof. dept. cell biology and anatomy, 1980-85; asst. prof. biology Emory (Va.) & Henry Coll., 1985-90; McConnell profl. biology, 1987; assoc. prof. Emory (Va.) & Henry Coll., 1990-93; assoc. prof. biology U. of the South, Sewanee, Tenn., 1993—; adj. asst. prof. Cleve. State U., 1978-80. Contbr. articles to Jour. Cell Biology, European Jour. Cell Biology, Thorax Nova Hedwigia, Jour. Phycology. Grantee Jeffress Meml. Trust, 1990-91, 88-89, 93—, Nat. Assn. Advisors for Health Professions, 1991; fellow Mabel Pew Myrin Trust, 1991. Mem. Am. Soc. for Cell Biology, Am. Soc. for Microbiology, Electron Microscopy Soc. Am., AAAS. Office: U of the South Dept Biology Sewanee TN 37383-1000

PALL, DAVID B., manufacturing company executive, chemist; b. Ft. William, Ont., Can., Apr. 2, 1914; came to U.S., 1939, naturalized, 1942; s. Adolph and Mary (Donner) P.; m. Josephine H. Blatt, Feb. 4, 1940 (dec. 1959); children: Stephanie (Mrs. Wendel Kincaid, Jr.), William, Ellen, Abigail; m. Helen R. Stream, July 10, 1960; 1 stepchild, Jane Block. B.Sc. in Chemistry, McGill U., 1936, Ph.D. in Phys. Chemistry, 1939, D.S. (hon.), 1987; postgrad., Brown U., 1936-37; DSc (hon.), L.I. U., 1985. Group leader research labs. Interchem. Corp., 1939-44; founder Pall Corp. (successor to Micrometallic Corp.), Glen Cove, N.Y., 1944; then pres., now chmn. bd. Pall Corp. (successor to Micrometallic Corp.). Inventor biomedical filters. Trustee North Shore Hosp., Manhasset, N.Y., cold Spring Harbor Lab., 1987. Recipient Nat. medal Tech. U.S. Dept. Commerce Tech., 1990. Mem. Am. Chem. Soc., Am. Soc. Metals, AAAS. Home: 5 Hickory Hill Roslyn NY 11576-1713 Office: Pall Corp 2200 Northern Blvd Greenvale NY 11548-1289

PALLAS, CHRISTOPHER WILLIAM, cardiologist; b. Chattanooga, Mar. 27, 1956; s. William Charles and Katherine (Rigas) P. Student, Vanderbilt U., 1974-75; BA in Biology, U. Tenn., 1978; MD, Wake Forest U., 1982. Diplomate Am. Bd. Internal Medicine, Am. Bd. Cardiology. Intern Med. Coll. Ga., Augusta, 1982-83, resident, 1983-85, chief med. resident, 1985-86, clin. fellow cardiology, 1986-88, instr. in cardiology, 1988-89, attending physician, 1988—, asst. prof. cardiology, 1989—, researcher clin. and basic cardiology, 1989—; cons. cardiovascular diseases VA Med. Ctr., Dublin, Ga., 1988-89, dir. coronary care unit, Augusta, 1988—. Contbr. articles to profl. jours. Fellow Am. Coll. Chest Physicians, Am. Coll. Cardiology; mem. AMA, Med. Assn. Ga., Richmond County Med. Soc. Greek Orthodox. Avocations: golf, collecting antiques. Home: 950 Stevens Creek Rd L-3 Augusta GA 30907 Office: Med Coll Ga Cardiology BA-A535 1120 15th St Augusta GA 30912-3150

PALM, WILLIAM JOHN, mechanical engineering educator; b. Balt., Mar. 1, 1944; s. William John Jr. and Lillian Mary (Hartmann) P.; m. Mary Louise Palm, Aug. 17, 1968; children: Aileen, William IV, Andrew. BS in Engring. Physics, Loyola Coll., Balt., 1966; PhD in Mech. Engring. and Astron. Sci., Northwestern U., 1971. Engr. Ballistic Rsch. Labs., Aberdeen, Md., 1966-67; prof. mech. engring. U. R.I., Kingston, 1971—, dir. Robotics Rsch. Ctr., 1985—; cons. Analysis and Tech. Inc., North Stonington, Conn., 1977—, Vitro Inc., Silver Spring, Md., 1989—. Author: Modeling, Analysis and Control of Dynamic Systems, 1983, Control Systems Engineering, 1986; patentee in field. Asst. cubmaster Boy Scouts Am., Kingston, 1986-89, asst. scoutmaster, 1989—. Mem. ASME, IEEE, AIAA, Tau Beta Pi, Pi Tau Sigma. Avocations: cycling, sailing. Office: U RI Wales Hall Kingston RI 02881

PALMER, ALAN MICHAEL, neuroscientist; b. Neath, Wales, U.K., Jan. 4, 1958; came to U.S., 1989; s. John Hugh and Joan Mary (Richards) P.; m. Susan Elizabeth Hawton, children: Dean Richard, Steven Russell, Gavin Hugh. BS, Warwick U., Coventry, U.K., 1980; MS, London U., 1981, PhD, 1987. Asst. prof. psychiatry U. Pitts., 1989, asst. prof. pharmacology, 1990; neuroscientist Western Psychiatric Inst. and Clin., Pitts., 1989—. Office: Western Psychiatric Inst W-1642 Biomedical Sci 3811 O'Hara St Pittsburgh PA 15213-2593

PALMER, ASHLEY JOANNE, aerospace engineer; b. Chilliwack, B.C., Can., June 1, 1951; d. Roland Jack and Alice Jean (Gavin) P. BSME, Mich. Tech., 1978; MSME, U. Mich., 1981. Assoc. engr. Remington Arms Co., Ilion, N.Y., 1979-80; sr. engr. Boeing Co., Seattle, 1981-89, Rohr Industries, Inc., Chula Vista, Calif., 1989—; cons. Howland Enterprises, Santee, Calif., 1990; co. rep. Nat. Mil. Standards Body, Washington, 1990. Inventor in field. Recipient Pres.'s 100, Dir. Civilian Marksmanship, Camp Perry, Ohio, 1985, Disting. Rifleman, Dir. Civilian Marksmanship, Washington, 1986. Avocations: competitive shooting, bridge, gourmet cooking.

PALMER, BRENT DAVID, microanatomy educator, reproductive biologist; b. Burbank, Calif., May 13, 1959; s. Warren Thayer and Yvonne Lita (McKelvey) P.; m. Sylvia Irena Karalius, June 26, 1982. BA, Calif. State U., Northridge, 1985; MS, U. Fla., 1987, PhD, 1990. Grad. asst. U. Fla., Gainesville, 1985-90; asst. prof. Wichita (Kans.) State U., 1990-91, Ohio U., Athens, 1991—; cons. Fla. Game and Fresh Water Fish Commn., Tallahassee, 1987—, U.S. EPA, 1992—; reviewer Biology of Reproduction, Champaign, Ill., 1988—, Harper Collins Pubrs., N.Y.C., 1991—. Contbr. chpts. to books, articles on vertebrate reproductive biology and toxicology to profl. jours. Coord. Sci. Olympiad, Wichita, 1991, So. Ohio Dist. Sci. Day, 1992—, Sci. and Engring. Fair, 1992—. Recipient Best Student Paper award Herpetologists League, 1988, Stoye award Am. Soc. of Ichthyologists & Herpetologists, 1988, Student Rsch. award Soc. Study of Amphibian & Reptiles, 1989, Grants-in-aid of Rsch., Sigma Xi, 1989. Mem. AAAS, Am. Soc. Zoologists, Electron Microscopy Soc. Am., Soc. Study Reproduction, Phi Beta Kappa. Achievements include description of functional morphology, physiology, and biochemistry of reptilian and amphibian oviducts; establishment of the evolution of an archosaurian reproductive system that may have implications for dinosaur reproduction. Office: Ohio U Dept Biol Scis Athens OH 45701-2979

PALMER, C(HARLES) HARVEY, electrical engineer, educator; b. Milw., Dec. 8, 1919; s. Charles Harvey and Grace Hambleton (Ober) P.; m. Elizabeth Hall Machen, Sept. 11, 1948; children: Charles Harvey III, Helen Palmer Stevens. SB in Physics magna cum laude, Harvard U., 1941, MA in Physics, 1946; PhD in Physics, Johns Hopkins U., 1951. Staff mem. MIT Radiation Lab., Cambridge, 1942-45; asst. prof., then assoc. prof. physics Bucknell U., Lewisburg, Pa., 1950-54; rsch. assoc. Lab. Astrophysics and Phys. Meteorology, Balt., 1954-60; asst. prof. elec. engring. Johns Hopkins U., Balt., 1960-64, assoc. prof., 1964-79, prof. elec. engring., 1979-92, prof. emeritus, 1992—; cons. Caterpillar Tractor Co., Aberdeen Proving Ground, Bendix Radio. Author: Optics: Experiments and Demonstrations, 1962; contbr. papers on ultrasonics, anomalous behavior of diffraction gratings, infrared spectrum of water vapor and measurement of ultrasmall angles to profl. pubis. Fellow Optical Soc. Am.; mem. IEEE, Am. Phys. Soc., Sigma Xi. Democrat. Office: Johns Hopkins U Dept Elec and Computer Engr 3400 N Charles St Baltimore MD 21218

PALMER, DAVID MICHAEL OLIVER, astronomer; b. Oakland, Calif., Oct. 17, 1962; s. Leigh Hunt and Evelyn Jean (Teyssier) P. BE in Computer Sci., U. Wash., 1984, BSc in Physics, 1984; PhD in Physics, Calif. Inst. Tech., 1992. Systems programmer Microsoft, Redmond, Wash., 1983-84; rsch. asst. Calif. Inst. Tech., Pasadena, 1984-92; rsch. assoc. Goddard Space Flight Ctr., NASA, Greenbelt, Md., 1992—. Mem. AAAS, Am. Phys. Soc., Am. Astron. Soc. Office: Goddard Space Flight Ctr NASA Code 661 Greenbelt MD 20771

PALMER, EDWARD EMERY, economics educator; b. Denver, Sept. 18, 1945; arrived in Sweden, 1970; s. Edward Emery and Helen Marie (Michael) P.; m. Anna Christina Castberg, June 19, 1969; children: Joanna Rebecca, Andrea Katarina. BA, U. Colo., 1967; MSc, U. Stockholm, 1971, PhD, 1982. Researcher Swedish Nat. Inst. of Econ. Rsch., Stockholm, 1971-81, sr. researcher, 1982-84; head dept. rsch. and analysis Swedish Nat. Social Ins. Bd., Stockholm, 1985—; lectr. U. Stockholm, 1979-82; adj. prof. of social ins. econs. U. Gothenburg, Sweden, 1988—; advisor, expert Swedish Ministry of Fin., Stockholm, 1982—, Swedish Ministry of Health and Social Affairs, Stockholm, 1985—; expert Internat. Social Security Assn., Geneva, 1986—; expert spl. studies Rehab. Internat., N.Y.C., 1989—. Author: Determination of Personal Consumption, 1981; co-author: Swedish Commodity Exports and Imports, 1985, (in Swedish) Social Security Pension, Medium and Long-term Financial Prospects, 1987, Pharmaceutical Costs and the Role of Social Insurance, 1987; contbr. articles to profl. jours. William Gambill scholarship William Gambill Found., 1967; fellowship Food Rsch. Inst., Stanford U., 1969; grantee Am. Coun. of Life Ins., 1977, Swedish Coun. for Social Rsch., 1988—. Mem. Swedish Econ. Soc., Am. Econ. Soc., Econometric Soc., Swedish Pharm. Cos. Coun. for the Rational Use of Medicine, Gothenburg Planning Group for Health Care Rsch. Home:

Klotmurklevägen 8, 18157 Lidingö Sweden Office: Swedish Nat Social Ins Bd, 10351 Stockholm Sweden Office: Gothenburg Sch Econs, Viktoriagatan 30, 41125 Gothenburg Sweden

PALMER, JAMES DANIEL, information technology educator; b. Washington, Mar. 8, 1930; s. Martin Lyle and Sarah Elizabeth (Hall) P.; m. Margret Kupka, June 21, 1952; children: Stephen Robert, Daniel Lee, John Keith. AA, Fullerton Jr. Coll., 1953; BS (Alumni scholar), U. Calif., Berkeley, 1955, MS, 1957; PhD, U. Okla., 1963; DPS (hon.), Regis Coll., Denver, 1977. Chief engr. Motor vehicle and Illumination Lab. U. Calif., Berkeley, 1955-57; assoc. prof. U. Okla., Norman, 1957-63; prof. U. Okla., 1963-66, asst. to dir. Rsch. Inst., 1960-63, cons. Rsch. Inst., 1966-69, dir. Sch. Elec. Engring., 1963-66, dir. Systems Rsch. Center, 1964-66; dean sci. and engring., prof. elec. engring. Union Coll., Schenectady, 1966-71; pres. Met. State Coll., Denver, 1971-78; rsch. and spl. programs adminstr. Dept. Transp., Washington, 1978-79; v.p., gen. mgr. rsch. and devel. div. Mech. Tech., Inc., Latham, N.Y., 1979-82; exec. v.p. J.J. Henry Co., Inc., Moorestown, N.J., 1982-85; BDM internat. prof. info. tech. George Mason U., Fairfax, Va., 1985—; bd. dir. J.J. Henry Co., Inc.; cons. Sym Mgmt. Co., Boston, Higher Edn. Exec. Assocs., Denver, PERI, Princeton; adj. prof. U. Colo. Co-author: (with A.P. Sage) Software Systems Engineering, (with Aseltine, Beam and Sage) Introduction to Computer Systems, Analysis, Design and Application. Bd. dirs., exec. v.p. adv. com. U.S.A. Vols. for Internat. Tech. Assistance, 1967-83, exec. v.p., 1970-71, chmn. exec. com.; trustee, vice chmn. Nat. Commn. on Coop. Edn.; mem. exec. policy bd. Alaska Natural Gas Pipeline, 1978-79; trustee Auraria Higher Edn. Program, Denver; mem. Fulbright fellow Selection Com., Colo.; bd. mgrs., mem. exec. com. Hudson-Mohawk Assn. Colls. and Univs., trustee, chmn. bd., 1970-71; adv. com. USCG Acad., 1972-82, chmn. adv. com., 1979-82; mem. Colo. Gov.'s Sci. and Tech. Adv. Council; pres. Denver Cath. Community Services Bd.; mem. Archdiocesan Catholic Charities and Community Services; mem. bd. U. Okla. Rsch. Inst.; mem. adv. com. Mile-Hi Red Cross. With USMC, 1950-51. Case-Western Res. Centennial scholar, 1981; recipient U.S. Coast Guard award and medal for meritorious pub. service, 1983. Fellow IEEE (exec. and adminstrv. coms., v.p. long-range planning and finance, chmn. com. on large scale systems); mem. Systems, Man and Cybernetics Soc. (pres., Outstanding Contbns. award 1981), alumni assns. U. Calif. and U. Okla., Inst. Internat. Edn. (bd. dir. Rocky Mt. sect.), Soc. Naval Architects and Marine Engrs., Am. Soc. Engring. Edn., Am. Mil. Engrs., N.Y. Acad. Sci., Navy League, Sigma Xi, Eta Kappa Nu, Pi Mu Epsilon, Alpha Gamma Sigma. Home: 12903 New Parkland Dr Herndon VA 22071-2646 Office: George Mason U Sch of Info Tech & Engring Fairfax VA 22030

PALMER, MARTHA JANE, computer specialist; b. Sellersburg, Ind., Jan. 5, 1947; d. James William and Dorotha (Townsend) Peyton; m. John Edward Palmer, Mar. 10, 1973; children: Rebecca Lin, Elizabeth Jane. BA in Math., Coll. of St. Francis, Joliet, Ill., 1986. Cert. secondary math. and comprehensive computer sci. tchr., Ill. With tng. and devel. dept. Silver Cross Hosp., Joliet, 1967-73; instr. Lockport (Ill.) Twp. Park Dist., 1975-77; tchr. math. Deer Creek Jr. High Sch., University Park, Ill., 1987-88; tchr., head coach math. team Crete (Ill.)-Monee High Sch., 1988-89; supr. & instr. Computer Lab., adj. instr. math. Joliet Jr. Coll., 1989—, adminstr. sr. and alumni programs, 1989-91; computer specialist, regional dir. support staff Info. Systems Ctr. VA Hosps., 1992—; oralist judge River Valley Conf., 1989—; sch. del. Ill. Gifted Conf., 1987. Sec. Lockport Planning Commn., 1989-91, Lockport Heritage and Architecture Commn., 1989—; mem. bd. advisors Lockport Main Street Program, 1991—. Recipient Disting. Svc. award Taft Dist. 90 PTA, Lockport, 1981. Mem. Nat. Coun. Tchrs. Math., Ill. Coun. Tchrs. Math. (del. 1988), Nat. Soc. DAR (registrar Louis Joliet chpt. 1983—, pres. 1987-89), Nat. Soc. Daus. of Union, Nat. Soc. Daus. War of 1812, Nat. Soc. Colonial Dames XVII Century (pres., bd. dirs. Sarah Hodsdon Morrill chpt. 1989-91, mem. Thomas Hooker chpt.), Peyton Soc. Va. (life), Nat. Soc. New Eng. Women (life, corr. sec. 1990—), Nat. Soc. Children Am. Revolution (sr. pres. Ill. Prairie Soc. 1991—, sr. state treas. 1993—), Nat. Soc. Dames of Ct. of Honor, Nat. Soc. Daus. Colonial Wars, Lockport Women's Club (scholarship com. 1992—), Towsend Soc. Am. (life), Nat. Soc. Daus. Am. Colonists, Kappa Mu Epsilon. Avocations: genealogy, fashion design and construction, historic preservation, archaeology. Home: 1520 Johnson St Lockport IL 60441-4483 Office: Region Direct Support Group care Hines ISC PO Box 7008, 162-4 Bldg 37 Hines IL 60141

PALMER, MELVILLE LOUIS, retired agricultural engineering educator; b. Dobbinton, Ont., Can., Aug. 30, 1924; came to U.S., 1953, naturalized, 1960; s. Louis Grange and Laura Lavina (Peacock) P.; m. Shirley Adams, Aug. 2, 1952; children: Laura, Melanie, Bradley. B.S., U. Toronto, 1950; M.S., Ohio State U., 1955. Asst. dean men Ont. Agrl. Coll., Guelph, 1950-52; asst. mgr. farm machinery United Coop., Toronto, 1952-53; research asst. Ohio Agrl. Expt. Sta., Wooster, 1953-55; extension agrl. engr. Ohio State U., Columbus, 1955-87, prof. agrl. engring., 1970-87, prof. emeritus, 1987—; cons. Ford Found., 1967. Contbr. numerous articles to tech. jours. Served with RCAF, 1943-46. Inducted into Internat. Drainage Hall of Fame, 1987. Fellow Am. Soc. Agrl. Engrs. (chmn. Ohio sect. 1979, recipient Hancor Soil and Water Engring. award, Gunlogson Countryside Engring. award 1986), Soil and Water Conservation Soc. (pres. All-Ohio chpt. 1973); mem. Ohio Extension Profs. Assn. (pres. 1964), Water Mgmt. Assn. Ohio (Disting. Svc. award 1985), Ohio Land Improvement Contractors (hon. mem., ednl. advisor 1956-87), Gamma Sigma Delta, Epsilon Sigma Phi (Disting. Svc. award 1989). Home: 4731 Goose Lane Rd Alexandria OH 43001-9730

PALMER, MILES R, engineering scientist, consultant; b. Roby, Tex., Dec. 10, 1953; s. Jim H. and Sylva M. (Russell) P.; m. Mary K. Wallace June 21, 1976; 1 child, Oliver James. BSEE, MIT, 1976, PhD, U. Calif., San Diego, 1980. Sr. staff scientist Sci. Applications Internat., Albuquerque, 1986—. Contbr. articles to profl. jours. Pilot Civil Air Patrol, Albuquerque, 1992. Capt. USAF, 1981-86. Recipient NSF fellowship, 1977; named Astronaut nominee NASA, 1980, Astronaut selectee USAF, 1984. Mem. AIAA (tech. com. 1992—), IEEE (applications panel 1992—). Achievements include patents for laser devices; discovery of non-biological mechanisms for early evolution of earth's primordial atmosphere; national leader for gun launch to space movement. Office: SAIC 2109 Air Park Rd Albuquerque NM 87106

PALMER, RAYETTA J., computer educator; b. Tribune, Kans., Dec. 9, 1949; d. Raymond H. and Helen Jean (Whittle) Helm; children: Carol Lynn, Eric Lee. BA in Bus. Edn., U. No. Colo., 1970; MA in Computer Edn., Lesley Coll., 1990. Bus./computer tchr. Dept. Def. Schs., Mannheim, Germany, 1983-87; computer tchr./coord. Cheyenne County Sch. Dist., Cheyenne Wells, Colo., 1987—; part-time instr. Lamar Community Coll., 1987—; part-time edn. industry specialist IBM, 1990—. Treas. Cheyenne County Rep. Cen. Com., Colo., 1989—. Mem. Internat. Soc. for Tech. in Edn., Pi Omega Pi. Republican. Avocations: reading, needlework, bridge, travel. Home: PO Box 771 Cheyenne Wells CO 80810-0771 Office: Cheyenne County Sch Dist RE-5 PO Box 577 Cheyenne Wells CO 80810

PALMER, RONALD ALAN, materials scientist, consultant; b. Ft. Knox, Ky., Oct. 21, 1950; s. John Curtis and Rosa Marie (Jury) P.; m. Ellen Bess Goldberg, Jan. 19, 1975; children: Michael Aaron, Alexander Steven. BS, Alfred (N.Y.) U., 1972; MS, U. Fla., 1979, PhD, 1981. Quality control engr. Metro Containers Corp., Jersey City, 1972-76; sr. chemist, unit mgr. Rockwell Hanford Ops., Richland, Wash., 1979-84; sr. ceramist 3M Co., St. Paul, 1984-88; lab. mgr. Keramont Corp., Tucson, 1988-89; mgr. vitrification process devel. West Valley (N.Y.) Nuclear Svcs. Co., 1989—; pvt. practice cons., nationwide, 1977—. Contbr. articles to profl. jours. Mem. Human Rights Commn., Golden Valley, Minn., 1985-88; del. Minn. League of Human Rights Commns., Mpls., 1987; legis. chair PTA, Williamsville, N.Y., 1990-92. Mem. AAAS, ASTM, Am. Ceramic Soc. (chair Eastern Wash. chpt. 1984), Materials Rsch. Soc., Soc. of Glass Tech. Home: 60 Pin Oak Dr Buffalo NY 14221-1640 Office: West Valley Nuclear Svcs Co PO Box 191 West Valley NY 14171-0191

PALMER, WILLIAM ALAN, entomologist; b. Melbourne, Victoria, Australia, Mar. 6, 1947; s. Alan Nicol and Joan Mary (Penny) P.; m. Jennifer Alice Bartlett, Dec. 16, 1972; children: Katherine Alice, Thomas William, Eleanor Charlotte. B in Rural Sci., U. New Eng., 1970; PhD, Tex. A&M, 1979. Rsch. officer N.S.W. Dept. Agr., Lismore, 1970-76; sr. entomologist N.S.W. Dept. Agr., Narrabri, 1976-81; entomologist in charge Queensland Dept. of Lands, Temple, Tex., 1982—. Contbr. over 35 articles to profl. jours. Recipient Merideth prize Robb Coll., Armidale, N.S.W., 1970; scholar Australia Meat Rsch. Coun., Sydney, 1976. Mem. Entomol. Soc. Am., Australian Entomol. Soc., Entomol. Soc. Washington, Colepterists Soc. Achievements include development of methods of examination for cattle tick, numerous successful agents; research in foreign exploration for biological control agents for various Australian weeds. Home and Office: 2801 Arrowhead Circle Temple TX 76502

PALMISANO, PAUL ANTHONY, pediatrician, educator; b. Cin., Dec. 30, 1929; s. William Robert and Lillian Rita (Schwarz) P. BS, Xavier U., 1952; MD, U. Cin., 1956; MPH, U. Calif.-Berkeley, 1979. Diplomate Am. Bd. Pediatrics, 1962. Intern, resident Children's Hosp., Cin., 1956-59; dep. dir. Bur. Medicine, FDA, 1963-66; mem. faculty U. Ala. Sch. Medicine, Birmingham, 1966—, prof. pediatrics, 1973—, asst. dean, 1975-79, assoc. dean, 1979-90; dir. Jefferson County Poison Control Ctr., 1968-83; Ala. del. U.S. Pharmacopeial Conv. Capt. AUS, 1959-61. Mem. AMA, Am. Acad. Pediatrics (chmn. com. accidental poisoning), Am. Soc. Pharmacology and Exptl. Therapeutics, Med. Assn. Ala. Author numerous articles on toxicology, pharmacology and pub. health. Office: 1600 7th Ave S Birmingham AL 35233-1711

PALOVCIK, REINHARD ANTON, research neurophysiologist; b. Dornheim, Hessen, W. Ger., June 30, 1950; came to U.S. 1956; s. Anton and Elfriede (Lankus) P. BS, U. Mich., 1973; MA, Wayne State U., Detroit, 1979, PhD, 1982. Rsch. asst. E.B. Ford Inst. Med. Rsch., Detroit, 1973-78; teaching asst. Dept. Psychology, Wayne State U., Detroit, 1978-79, grad. trainee, 1979-81, grad. asst., 1981-82; postdoctoral assoc. dept. physiology U. Fla., Gainesville, 1982-86, postdoctoral assoc. dept. neurosci., 1986-89, postdoctoral assoc. dept. neurosurgery, 1989-90, postdoctoral assoc. neurology dept., 1991—; postdoctoral assoc. neurology svc. VA Med. Ctr., Gainesville, 1990-91. U. Mich. Regents Alumni scholar, 1969; NIMH predoctoral tng. grantee, 1979; NIH Nat. Rsch. Svc. awardee, 1983; Epilepsy Rsch. Found. Fla. postdoctoral grantee, 1990. Mem. AAAS, APA, Soc. for Neurosci., Am. Statis. Assn., Internat. Neural Network Soc. Avocations: classical and modern music, Korean karate, creative photography, paleontology. Home: 2209 NE 15th Ter Gainesville FL 32609-3978

PALTAUF, RUDOLF CHARLES, aerospace engineer; b. Vienna, Austria, Aug. 26, 1930; s. Rudolf M. and Mary A. (Motz) P.; m. Shirley Jean Crabtree, June 8, 1955; children: Rosemarie, Roxanne (dec.), Rudolf, Ronald, Renee, Robert, Raymond. BS in Chemistry, Georgetown U., 1952; MBA in R & D Mgmt., U. Conn., 1965. Commd. to 2d lt. USAF, 1952, advanced through grades to lt. col., 1968, ret., 1980; chief surveillance divsn. Rome Air Devel. Ctr., Griffiss AFB, N.Y., 1973-76; chief test divsn. Airforce Phys. Security Systems, Hanscom AFB, Mass., 1977-80; systems engr. Norden Systems, Norwalk, Conn., 1981-89; sr. systems engr. Realtime Systems, Waterbury, Conn., 1990; mil. applications specialist Cyberchron, Cold Springs, N.Y., 1990; sr. engring. writer Norden Systems, Norwalk, 1991; pvt. practice as cons. Redding, Conn., 1992—; cons. Philips Med. Systems, Shelton, Conn., 1993—; exec. officer Airforce Avionics Lab., Wright-Patterson AFB, Ohio, 1960-64; scientist Hdqrs. Def. Atomic Support Agy., Pentagon, Washington, 1965-70; chief C3 divsn. Hdqrs. Airforce Systems Command, Andrews AFB, Mass., 1970-73; presenter papers Georgetown U., Aerial Reconnaissance Lab., Wright-Patterson AFB, 15th nat. aerospace electronics conf. Sessions on Electronic Device Tech. and Electric Propulsion, Ad Hoc Laser Comms. Working Group, others. Contbr. articles to profl. jours. Past mem. rules and by-lawd com. Sch. Bd. Nominating Conv. of Charles County Md.; past del. Charles County Md. Coun. Parents and Tchrs., chmn. nominating com.; past pres. and del. Indian Head (Md.) Elem. Sch. PTA; past v.p. Gen. Smallwood Mid. Sch. PTA, Indian Head. Mem. IEEE (sr.), AAAS, Air Force Assn. (membership chmn. Charles A. Lindbergh chpt.), Ret. Officers Assn. (pres., we. Conn. chpt.), Nat. Officers Assn., Nat. Assn. for Uniformed Svcs., N.Y. Acad. Scis., Am. Legion, Sigma Xi. Republican. Roman Catholic. Achievements include identification of certification problem with launch rocket system; resolution of system design issues for marine integrated fire and air support system; consolidation of phys. security equipment tests of Army and Navy with Air Force; instigation of first U.S.A. engineering feasability study of spaceborne radar surveillance system; integration of technology development efforts of DARPA by transistion mechanism to Aerospace Defense Command. Home and Office: 95 Great Oak Ln West Redding CT 06896 Office: Philips Med Systems 710 Bridgepart Ave Shelton CT 06484

PAMPEL, ROLAND D., computer company executive; b. Mystic, Conn., Nov. 16, 1935; s. Alban and Doris (Denison) P.; m. Carol Patricia Clay, Apr. 5, 1958; children: Lynne Pampel Albertini, Jean Pampel Losier, Sandra. BSEE, U. Conn., 1956. Product mgr., then gen. mgr. IBM Corp., 1961-82; gen. mgr., lab dir. IBM Corp., Kingston, N.Y., until 1982; v.p. research and devel. Prime Computer, Natick, Mass., 1982-85; sr. v.p. mktg. and devel. AT&T Computer Div., Morristown, N.J., 1985-86; pres. Apollo Computer Inc., Chelmsford, Mass., 1986-87, CEO, 1987-90; CEO Bull HN Onformation Systems, Inc., Billerica, Mass., 1990-91; pres., CEO Nicolet Instrument Corp., Madison, Wis., 1991—. Lt. USN, 1956-59. Mem. Nat. Computer Graphics Assn. (vice chmn. 1988—), Braeburn Country Club (Newton, Mass.). Office: Nicolet Instrument Corp 5225 Verona rd Madison WI 53711

PAN, CODA H. T., mechanical engineering educator, consultant, researcher; b. Shanghai, China, Feb. 10, 1929; came to U.S., 1948; s. Ming H. Pai and Chih S. Ling; m. Vivian Y.C. Chang, June 2, 1951; children—Lydia Codetta, Philip Daniel. Student, Tsing Hwa U., Beijing, China, 1946-48; B.S. in Mech. Engring., Ill. Inst. Tech., 1950; M.S. in Aero. Engring., Rensselaer Poly. Inst., 1958, Ph.D., 1961. Engr. Gen. Electric Co., Schenectady, 1950-61; dir. research Mech. Tech. Inc., Latham, N.Y., 1961-73; tech. dir. Shaker Research Corp., Ballston Lake, N.Y., 1973-81; prof. mech. engring. Columbia U., N.Y.C., 1981-87; sr. cons. engr. Digital Equipment Corp., Shrewsbury, Mass., 1987-92; adj. prof. Rensselaer Poly. Inst., 1961-81; vis. prof. Tech. U. Denmark, Copenhagen, 1971, U. Poitiers, France, 1987; mem. adv. panel rand Corp., 1974; engring. cons., 1992—, v.p. Indsl. Tribology Inst., Troy, N.Y., 1982—; co-prin. investigator Spacelab I, 1984. Contbg. author: Tribology, 1980, Structural Mechanical Software Series 3, 1980; contbr. articles to profl. jours. Recipient IR-100 award, 1967; NIH fellow, 1972. Fellow Am. Soc. Lubrication Engrs., ASME; mem. AAAS, Am. Phys. Soc., Am. Acad. Mechanics. Achievements include 3 patents in field; patent pending; avocation: classical music. Home and Office: 5 Pinehurst Cir Millbury MA 01527-3361

PAN, DAVID HAN-KUANG, polymer scientist; b. Chia-Yi, Taiwan, Jan. 2, 1954; s. Chu-Tsai and Yuen-Ling (Lau) P.; m. Jenny Kuo-Jen Lee, Jan. 16, 1982; children: Joyce Ming Pan, Brian Way Pan. BS, Nat. Taiwan U., 1976; PhD in Chem. Engring., U. Wis., 1983. Rsch. assoc. Argonne (Ill.) Nat. Lab., 1981-82; rsch. asst. U. Wis., Madison, 1982-83, lectr. and postdoctoral fellow, 1983; mem. rsch. staff Xerox Webster Rsch. Ctr., Webster, N.Y., 1983—; speaker Internat. Symposium Polymers, Japan, 1992, Fuji Photo Film, Inc., Soc. Plastics Engrs., Boston, 1986, Dallas, 1990. Contbr. articles to profl. jours. Mem. Am. Phys. Soc. (speaker 1985-93), Materials Rsch. Soc. (speaker 1990), Am. Chem. Soc. (speaker 1992). Achievements include patents for novel, high-toughness, low surface energy elastomers for high temperature applications, modification of polymer particle surface for imaging applications; developments include segregation and enrichment at surface of polymer blends, compressive yielding and fracture in aged copolymer glasses, supermagnetic properties and microstructures in ion-containing polymers, crystal bending for focussing X-ray beam. Home: 10 Westfield Commons Rochester NY 14625 Office: Xerox Webster Rsch Ctr 800 Phillips Rd 0114-39A Webster NY 14580

PAN, HENRY YUE-MING, clinical pharmacologist; b. Shanghai, China, Dec. 27, 1946; came to U.S. 1969; s. Chia-Liu and Siu-Ging (Sung) P.; m. Mary Agnes Tse; children: Lincoln Jonathan, Gregory Kingsley. BSc (hon.), McGill U., Montreal, 1969; MS in Toxicology, U. Hawaii, 1973, PhD in Pharmacology, 1974; MD, U. Hong Kong, 1979. Rsch. asst. U. Hawaii, Honolulu, 1969-74, teaching asst., 1970-74; med. officer Queen Mary Hosp., Hong Kong, 1979-81; asst. prof. medicine U. Hong Kong, 1981-85; vis. asst. prof. Stanford (Calif.) U., 1983-85; asst. clin. pharmacology dir. Squibb Inst. Med. Rsch., Princeton, N.J., 1985-87, assoc. clin. pharmacology dir., 1987-88, clin. pharmacology dir., 1988-89, exec. dir. clin. rsch., 1989-91; v.p. clin. rsch. Bristol-Myers Squibb Pharm. Rsch. Inst., Princeton, 1991-92; v.p. clin. rsch. and devel. Du Pont Merck Pharm. Co., Wilmington, Del., 1992—. Contbr. articles to profl. jours. Stanford Asian Med. Fund grantee, 1983-85. Fellow Am. Coll. Clin. Pharmacology, Am. Heart Assn. Coun., Am. Coll. Cardiology; mem. AAAS, Am. Soc. Clin. Pharmacology and Therapeutics, Am. Soc. for Pharmacology and Exptl. Therapeutics, Am. Fedn. Clin. Rsch. Roman Catholic. Avocations: tennis, golf, distance running, cycling, baseball, classical music, plays. Office: Du Pont Merck Pharm Co PO Box 80026 Wilmington DE 19880-0026

PAN, HUO-HSI, mechanical engineer, educator; b. Fuzhou, Peoples Republic of China, Nov. 11, 1918; came to the U.S., 1948; s. Bai-ming and Won-ching (Chen) P.; m. Chao Pan, June 4, 1960; children: Lillian, Nina. BS in Mech. Engring., Nat. S.W. Associated U., Kunming, Peoples Republic of China, 1943; MS in Mech. Engring., Tex. A&M U., 1949; MS in Applied Mechanics, Kans. State Coll., 1950; PhD, U. Calif., Berkeley, 1954. Asst. engr. Yunnan Smelting Plant, Peoples Republic of China, 1942-43; from mem. tech. staff to head inspection dept. 21st Arsenal, Peoples Republic of China, 1943-47; from teaching asst. to assoc. mech. engring. U. Calif., Berkeley, 1950-53; rsch. engr. Portland Cement Assn., 1954; asst. prof. U. Toledo, 1954-55, U. Ill., Champaign, 1955-57; asst. prof. engring. mechanics NYU, 1957-59, from asst. prof. to prof. applied mechanics, 1957-73; prof. applied mechanics, mech. engring. Poly. U., 1973-90, prof. emeritus, 1990—; cons. Frankford Arsenal, Picatinny Arsenal, Petro-Chem Devel. Co.; referee Jour. Applied Mechanics, AIAA Jour., Internat. Jour. Mech. Sci., Internat. Jour. Solid and Structures, NSF; reviewer Applied Mechanics Revs.; sect. chmn. Internat. Modal Analysis Confs.; lectr. Kunming Inst. Tech., Tsinghua U., Jilin U. Tech., Jilin U., 1984. Contbr. numerous articles to Jour. Applied Mechanics, AIAA Jour., Jour. Mecanique, Jour. Engring. Mechanics, Jour. Applied Math. and Physics, Quar. Jour. Mechanics and Applied Math., Quar. Applied Math., Internat. Jour. Mechanics and Sci., Bull. Acad. Polonaise des Scis., Jour. Sound and Vibration, many others. Grantee NSF, 1964-67, NASA, 1966-68. Mem. ASME, AIAA, Am. Acad. Mechanics, Soc. Engring. Sci., Soc. for Indsl. and Applied Math., U.S. Assn. for Computational Mechanics, Internat. Assn. for Computational Mechanics, Phi Kappa Phi, Sigma Xi, Tau Beta Pi, Pi Tau Sigma, Pi Mu Epsilon. Achievements include development of method for reduction of vibrational systems, general method of modal analysis, solution for ordinary differential equation containing symbolic functions, eigenfunction expansion method in vibration problems of viscoelestic bodies. Home: 76 Edgars Ln Hastings-on-Hudson NY 10706-1122 Office: Poly U Dept Mech Engring 6 Metrotech Ctr 333 Jay St Brooklyn NY 11201-2990

PAN, TZU-MING, biotechnologist; b. Ching-Shui, Taiwan, Republic of China, Jan. 9, 1947; s. Wan-Chih and Teng (Yang) P.; m. Mei-Lan Chang, Dec. 25, 1973; children: Chia-Ying, Chia-Yu, Chia-Yueh. BS, Nat. Taiwan U., 1969, MS, 1972, PhD, 1978. Assoc. prof. Chinese Culture U., Taipei, 1978-82, prof., 1982—, chmn., dir., 1988-92; dir. dept. bacteriology Nat. Inst. Protective Medicine, Taipei, 1993—; com. of patent screening Ministry of Econ. Affairs, Taipei, Taiwan, 1987—, com. of waste reduction task force, 1991—; com. of standard method EPA, Republic of China, 1990—;. Author: Organic Chemistry, 1984, Experiment in Chemistry, 1984, 86, Experiment in Analytical Chemistry, 1984, Experiment in Organic Chemistry, 1990. Recipient Award of Rsch. and Writings Ministry of Edn., 1977, 79, Outstanding Tchrs. award, 1989, Youth Medal award Ministry of Youth, 1978, Outstanding Teaching Material award Ministry of Edn., 1992. Mem. Chinese Biochem. Soc., Chinese Tchrs. Assn. (exec. dir.), Biotech. Indsl. Assn. (exec. dir.), Chinese Agrl. Chem. Assn. (exec. dir.), Am. Chemistry Soc. Achievements include rsch. on new recovery method of glutamic acid using ion-exch. resin, new microorganism to produce dedecanedic acid from n-dodecane, new microorganism to produce vitamin B12 from methanol, method of immobilization of papain and bromelain for beer chill-proofing. Office: Nat Inst Protective Medicine, 161 Kuin Yang St, Nan Kang Taipei 11513, Taiwan

PAN, ZHENGDA, physicist, researcher; b. Shanghai, China, Jan. 26, 1944; came to U.S., 1986; s. Xing-Huang and Xigu (Zhu) P.; m. Haiyun Liu, May 1, 1978; 1 child, Robert. BS, Qing Hua U., Beijing, 1967; MS, Okla. State U., 1989, PhD, 1992. Lectr. Huazhong U. of Sci. and Tech., Wuhan, China, 1977-86; rsch. and teaching asst. Ctr. for Laser Rsch. Okla. State U., Stillwater, 1986-92; rsch. assoc. Fisk Ctr. for Photonic Materials and Devices, Nashville, 1992—. Contbr. articles to profl. jours. Mem. Am. Phys. Soc., Chinese Inst. Electronics. Achievements include developing a rapid method for measuring C-V characteristics of MOS structures and the photoluminescence and multiphonon transition in poly (p-phenylene sulfide) films; studied the optical losses of heavy metal fluoride glasses for fiber materials; current research includes Raman scattering, nonlinear optical properties in glass and II-VI semiconductor materials. Home: 511 Chesterfield Ave Apt 7D Nashville TN 37212 Office: Fisk U Physics Dept Nashville TN 37208-3051

PANAGIOTOPOULOS, PANAGIOTIS DIONYSIOS, mechanical engineering educator; b. Thessalonike, Greece, Jan. 1, 1950; s. Dionysios and Kalliopi Panagiotopoulos. Diplom. Ingenieur, Aristotle U., 1972, Dr. Ingenieur, 1974; dozent habilitation Tech. U. Aachen, 1977. Head research group Tech. U. Aachen, Fed. Republic Germany, 1977-78, privatdozent (research), 1977-81, hon. prof., 1981—; full prof., dir. Inst. Steel Structures, Aristotle U., Thessalonike, 1978—; vis. prof. U. Hamburg, Fed. Republic Germany, 1981; vis. prof. PUC Rio de Janeiro, 1986; Fulbright scholar MIT, Cambridge, Mass., 1984. Author: Inequality Problems in Mechanics and Applications, 1985, Topics in Non-smooth Mechanics, 1988, Non-smooth Mechanics and Applications, 1988, A Boundary Integral Approach to the Static and Dynamic Contract Problem, 1992, Hemivariational Inequalities, 1993. Contbr. numerous articles on convex and nonconvex energy functions and spl. structural analysis software to profl. jours. Recipient award for grad. students Tech. Chamber Greece, 1971, 72; Alexander von Humbolt fellow, 1974-77, sr. fellow, 1978. Mem. ASCE, Am. Math. Soc., N.Y. Acad. Sci., Soc. Indsl. Applied Math., Math. Programming SoCo, Internat. Soc. for Computational Mechanics, Gesellschaft fur Angewandte Math. Mechanik, Acad. of Athens (corr. 1989), Acad. Europaea London. Office: Aristotle Univ, Thessaloniki Greece other: Tech U Aachen, Faculty of Math, Aachen 51 Federal Republic of Germany

PANARETOS, JOHN, mathematics and statistics educator; b. Kythera, Lianianika, Greece, Feb. 23, 1948; s. Victor and Fotini (Kominu) P.; m. Evdokia Xekalaki; 1 child, Victor. First degree, U. Athens, 1972; MSc, U. Sheffield, Eng., 1974; PhD, U. Bradford, Eng., 1977. Lectr. U. Dublin, Ireland, 1979-80; asst. prof. U. Mo., Columbia, U.S, 1980-82; assoc. prof. U. Iowa, Iowa City, U.S., 1982-83, U. Crete, Iraklio, Greece, 1983-84; assoc. prof. div. applied math., Sch. Engring. U. Patras, Greece, 1984-87, assoc. dean sch. engring., chmn. div. applied math., 1986-87, vice-rector, 1988-91; prof. Athens U. Econs., 1991—, chair dept. stats., 1993—; sec.-gen. Ministry Edn. and Religious Affairs, Greece, 1988-89. Contbr. articles to profl. jours. Mem. Sci. Coun. of Greek Parliament, 1987—; chmn. rsch. com., pers. com. U. Patras, 1988-91. Mem. Am. Statis. Assn., Inst. Math. Statistics, Bernoulli Soc. for Probability and Math. Statistics, Greek Math. Soc., Greek Statis. Inst., Internat. Statis. Inst. Office: Athens U Econs, PO Box 31466, 10035 Athens Greece

PANAYIOTOU, CONSTANTINOS, chemical engineering educator; b. Valcano, Trikala, Greece, June 26, 1951; s. George and Vassiliki (Papadogeorgos) P.; m. Kalliopi Pazaitou, Feb. 27, 1982; children: George, Vassiliki, Stella, Gregory. Diploma in chem. engring., Nat. Tech. U. Athens, Greece, 1974; Maitrise es Sciences Appliquees, U. Sherbrooke, Can., 1979; PhD in Chem. Engring., McGill U., Montreal, Can., 1982. Lic. engring., Greece. Lectr. U. Thessaloniki, Greece, 1983-86, asst. prof., 1986-92, assoc. prof., 1992—; vis. scholar U. Tex., Austin, 1990-91; rsch. scientist Chem. Processing Engring. Rsch. Inst., Thessalonoiki; founding mem. Hellacel Ltd./Cons. Co., Kavala, Greece, 1991—. Editor (proceedings) 4th International IUPAC Conf., 1989; author: (book notes) Physical Chemistry/Polymers, 1989; contbr. articles to profl. jours. Senator U. Thessaloniki, 1987-89. Res. officer Greek Air Force, 1974-77. Hon. scholar Nat. Scholarship Found., Greece, 1970-74; fellow Francohellenic Coun., 1974, Nat. Rsch. Coun. of Can., 1978-81; rsch. grantee Gen. Sec. Rsch. Tech., Greece, Econ. European Community, Brit. Coun., French Embassy, Local Industry, 1985—. Mem.

Am. Chem. Soc., Internat. Union Pure and Applied Chem., Tech. Chamber of Greece, Greek Chem. Assn., Greek Polymer Soc., Am. Phys. Soc. Avocations: soccer, swimming, chess, music. Home: 42 Vasil Georgiou Ave, 54640 Thessaloniki Greece Office: U Thessaloniki, Dept Chem Engring, 54006 Thessaloniki Greece

PANCHANATHAN, VISWANATHAN, product development specialist; b. Madras, Tamil Nadu, India, Aug. 12, 1939; came to U.S., 1980; naturalized, 1985; s. Puvanoor Arunachalam and Saraswathi V.; m. Ananda, July 4, 1969; 1 child, Anita. M of Engring., Indian Inst. Sci., 1961, PhD in Engring., 1964; MS in Mfg. Mgmt., GMI Engring. & Mgmt. Inst., 1987. Rsch. engr. Mukand Iron & Steel Works Ltd., Bombay, India, 1965-69; asst. prof., prof. Indian Inst. Tech., Madras, India, 1969-80; mgr. process engring. Marko Materials Inc., North Billerica, Mass., 1980-83; product engr. Magnequench-Delco Remy Div. of GM, Anderson, Ind., 1983—; cons. Sivananda Steel Ltd., MAdras, 1975-80, Die Casting Foundry, Madras, 1976-78. Mem. Metals and Materials Soc. Hindu. Achievements include 12 patents on ferrous and nonferrous alloy devel. by rapid solidification process and magnetic properties and processing of rare earth-transition metal-boron compositions. Office: Magnequench-Delcoremy Div GM 6435 S Scatterfield Rd Anderson IN 46013

PANCHERI, EUGENE JOSEPH, chemical engineer; b. South Bend, Ind., Jan. 23, 1947; s. Raymond Albert and Dora Lugenia (Martin) P.; m. Janice Edwina Sutton, Mar. 9, 1986; children: Brent, Ayrie, Joseph. BSChE, Purdue U., 1969. Staff mem. Procter & Gamble, Cin., 1969-74, group leader, 1974-92, prin. engr., 1993—. Mem. Am. Inst. Chem. Engrs., Am. Oil Chemists Soc., Phi Eta Sigma, Alpha Tau Omega. Achievements include 11 U.S. patents and 13 foreign patents for dishwashing and laundry cleaning products. Office: Procter & Gamble ITC 5299 Spring Grove Ave Cincinnati OH 45217

PANDA, MARKANDESWAR, chemistry researcher; b. Goudagaon, Orissa, India, May 5, 1952; came to U.S., 1981; s. Satyabadi and Sukanti (Satpathy) P.; m. Bidyut Prava Nanda, Aug. 20, 1981; children: Shantayan, Shree Lekha. PhD, Berhampur U., Orissa, India, 1983, U. Calif., Santa Cruz, 1987. Predoctoral fellow U. Grants Com., New Delhi, India, 1976-80; teaching asst. U. Calif., 1981-87; postdctoral fellow Miami U., Oxford, Ohio, 1987-89; postdoctoral assoc. U. Tex. Health Sci. Ctr., San Antonio, 1989-92, sr. rsch. assoc., 1992—. Contbr. articles to Internat. Jour. Chem. Kinetics, Jour. Am. Chem. Soc., Jour. Organic Chemistry. Mem. Am. Chem. Soc., Biophys. Soc. Hindu. Achievements include research in structure reactivity relationship in nucleophilic addition reactions, detection of long-lived intermediate in the carcinogenic path of amine-n-sulfonates mechanism of HCN binding of cytochrome oxidase. Office: U Tex Health Sci Ctr Dept Biochemistry 7760 Floyd Curl Dr San Antonio TX 78284-7760

PANDELIDIS, IOANNIS O., engineering manager; b. Thessaloniki, Greece, Mar. 15, 1957; s. Onoufrios and Kiveli (Tzarta) P.; married; 1 child, Alexander Thomas. BS in Elec. & Computer Engring., U. Wis., 1976, MS in Elec. & Computer Engring., 1978, PhD in Mech. Engring., 1983. Asst. prof. mech. engring. U. Md., College Park, 1983-88, asst. rsch. scientist, 1988-90; group mgr. Gillette Co., Boston, 1990—; cons. Airpax, Westvaco, Nat. Tech. U., Md., 1984-90. Contbr. articles to profl. jours., chpt. to book. Mem. Sharon Arts Coun., Sharon Chamber Music Assn., 1992—. Office: Gillette Co 1 Gillette Park 6D Boston MA 02127-1096

PANDEY, LAKSHMI NARAYAN, physicist, researcher; b. Varanasi, India, July 19, 1956; came to U.S., 1984; s. Ganapati Pandey and Phoolmati Devi; m. Nirmala Devi, June 11, 1973; children: Mata Prasad, Renu, Suman. BS, Banaras Hindu U., Varanasi, 1976, MS, 1978, PhD, 1984. Sr. fellow Banaras Hindu U., Varanasi, 1982-84; post-doctoral fellow SUNY, Buffalo, 1984-88, rsch. instr., 1988-91; scientist Wash. State U., Pullman, 1991—. Contbr. articles to Phys. Rev. Letters, Physics Letters A, Applied Physics Letters. Mem. Am. Phys. Soc., Material Rsch. Soc. Hindu. Achievements include theoretical prediction of feasibility of intrinsic bistability in quantum confined semiconductor heterstructures; theoretical studies of tailoring of infra-red transitions in quantum well systems. Home: 125 NW Larry St # 6 Pullman WA 99163 Office: Wash State U Dept Physics Pullman WA 99164-2814

PANDEY, RAGHVENDRA KUMAR, physicist, educator; b. Bath, India, Jan. 7, 1937; s. U.N. and Praja (Choudhary) P.; m. Christa U. Pandey, June 1, 1967. MS in Physics, Patna U., 1959; DSc in Physics, U. Cologne, 1967. Assoc. prof. elec. engring. Tex. A&M U., College Station, 1977-82, prof., 1982-89, Halliburton prof. elec. engring., 1989-90, dir. ctr. for electronic materials, 1990—; cons. Tex. Instruments, Inc., Dallas, 1986—, Microelectronics Consortium, Austin, Tex., 1989-90, Sandia Nat. Labs., Albuquerque, 1978-80. Contbr. 2 chpts. to books and 60 articles to profl. jours. Democrat. Hindu. Achievements include patents for new ferroelectric high-T superconductors and semiconductor materials and original work in crystal growth. Home: 1907 Amber Ridge Dr College Station TX 77845-5536 Office: Tex A&M U Elec Engring Dept College Station TX 77843-3128

PANDEY, RAMESH CHANDRA, chemist; b. Naugaon, India, Nov. 5, 1938; came to U.S., 1967; s. Gauri Dutt and Jivanti Pandey. B.Sc., U. Allahabad (India), 1958; M.Sc., U. Gorakhpur (India), 1960; Ph.D., U. Poona (India), 1965. Jr. research fellow C.S.I.R. Nat. Chem. Lab., Poona, India, 1960-64, research officer, 1965-67, scientist organic div., 1970-72; research assoc. dept. chemistry U. Ill., Urbana, 1967-70, vis. scientist, 1972-77; sr. scientist fermentation program Nat. Cancer Inst. Frederick (Md.) Cancer Research Facility, 1977-82, head chem. sect., 1982-83; sr. scientist Abbott Labs., North Chicago, Ill., 1983-84; pres. Xechem, Inc., Melrose Park, Ill., 1984-90, pres., chief exec. officer, dir. tech. devel., New Brunswick, N.J., 1990—; cons. Washington U. Sch. Medicine, St. Louis, 1976-85, LyphoMed, Inc., Melrose Park, 1984-85; vis. prof. Waksman Inst. Rutgers U., Piscataway, N.J., 1984-86. Mem. editorial bd. Internat. Jour. Antibiotics, 1986—; patentee graft thin layer chromatography. Fellow Am. Inst. Chemists; mem. Am. Chem. Soc., Am. Soc. Microbiology, Am. Soc. Mass Spectrometry, Am. Assn. Cancer Rsch., Am. Soc. Hosp. Pharmacists, Am. Soc. Pharmacognosy, Soc. Indsl. Microbiology, N.Y. Acad. Scis., Indian Sci. Congress Assn. Office: Xechem Inc 100 Jersey Ave Ste 310 New Brunswick NJ 08901-3279

PANDEYA, NIRMALENDU KUMAR, plastic and flight surgeon, U.S. Air Force officer; b. Bihar, India, Feb. 9, 1940; came to U.S., 1958, naturalized, 1965; s. Balbhadra and Ramasawari (Tewari) P.; children: by previous wife Alok, Kiran; m. Haripriya Pradhan, June 15, 1988; 1 stepchild, Bibek. BSc, MS Coll., Bihar U.-Motihari, 1958; MS, U. Nebr., 1962; postgrad. U. Minn., 1959, Ft. Hays State Coll., 1961, D.O., Coll. Osteo. Medicine and Surgery, Des Moines, 1969, Hamilton Co. Pub. Hosp.; grad. Sch. Aerospace Medicine, U.S. Air Force, 1979. Diplomate Nat. Bd. Osteo. Med. Examiners. USPHS fellow dept. ob-gyn Coll. Medicine, U. Nebr., Omaha, 1963-65; intern Doctors Hosp., Columbus, Ohio, 1969-70; resident in gen. surgery Des Moines Gen. Hosp., 1970-72, Richmond Heights Gen. Hosp. (Ohio), 1972-73; fellow in plastic surgery Umea U. Hosp. (Sweden), 1973, Karolinska Hosp., Stockholm, 1974-75; mil. cons. in plastic surgery, USAF surgeon gen.; clin. prof. scis. Coll. Osteo. Medicine and Surgery, Des Moines, 1975-76, also adj. clin. prof. plastic and reconstructive surgery; state air surgeon Iowa Air Nat. Guard; practice in reconstructive and plastic surgery, Des Moines, 1975—; mem. staff Des Moines Gen. Hosp., Mercy Hosp. Med. Ctr., Charter Community Hosp., Davenport Osteo. Hosp., Franklin Gen. Hosp., Ringgold County Hosp., Madison County Meml. Hosp., Winterset, Iowa, Mt. Ayr Surgery Ctr. of Des Moines, Hamilton County Hosp., Webster City Decatur County Hosp., Leon, Story County Hosp., Nev. Served to col. M.C., USAF; sr. flight surgeon Iowa Air N.G. Regents fellow U. Nebr., Lincoln, 1961-62. Fellow Internat. Coll. Surgeons, Interam. Coll. Surgeons, 1990; mem. Assn. Plastic Surgeons of India (life), Assn. Surgeons of India (life), Assn. Mil. Surgeons of U.S. (life), Assn. Mil. Plastic Surgeons, AMA, Am. Osteo. Assn., Polk County Med. Soc., Iowa Soc. Osteo. Physicians and Surgeons, Polk County Soc. Osteo. Physicians and Surgeons (pres. 1978), Soc. U.S. Air Force Clin. Surgeons, Aerospace Med. Assn., Air N.G. Alliance of Flight Surgeons, AAUP, Am. Coll. Osteo. Surgeons, Am. Acad. Osteo. Surgeons (cert.), Soc. U.S. Air Force Flight Surgeons. Hindu. Club: Army Navy, DMGCC. Contbr. numerous articles to profl. jours. Home: 4405 Mary Ann Cir West Des Moines IA 50265 Office: Cosmetic Surgery Ctr 1000 73rd St # 21 Des Moines IA 50311-1321

PANDIARAJAN, VIJAYAKUMAR, industrial engineer; b. Coimbatore, India, May 9, 1961; came to U.S., 1989; s. Pandiarajan Ramasamy and Nagarathinam Pandiarajan; m. Punithavathy Vijayakumar, June 25, 1990. BE with honors, Madras (India) U., 1983; M of Tech. Prodn., Indian Inst. Tech., Madras, 1985; PhD in Mfg., W.Va. U., 1992. Mgmt. trainee Hindustan Aeronautics Ltd., Bangalore, India, 1983-84, engr., 1984-88, shop mgr., 1988-89; teaching asst. W.Va. U., Morgantown, 1989-90; rsch. assoc. Concurrent Engring. Rsch. Ctr., Morgantown, 1990-91; instr. W.Va. U., Morgantown, 1991-92; tech. specialist Concurrent Techs. Corp., Johnstown, Pa., 1992—. Contbr. articles to profl. jours. Mem. Nat. Cadet Corps, India, 1981-83. Mem. Soc. Mfg. Engrs. (sr.), Inst. Indsl. Engrs., Sigma Xi, Alpha Pi Mu. Achievements include development of system to alter and evaluate complex geometries to improve their manufacturability. Home: 16 Sarada Mill Rd, Podanur Tamil Nadu South India 641023 Office: Concurrent Techs Corp 1450 Scalp Ave Johnstown PA 15904

PANDINA, ROBERT JOHN, neuropsychologist; b. Rochester, N.Y., July 19, 1945; s. Jack John and Jane (Presevento) P.; 1 child, GAhan. BA in Psychology, Hartwick Coll., 1967; MA, U. Vt., 1969, PhD in Psychology, 1973. Prof. N.J. Sch. Alcohol and Drug Abuse Studies Rutgers U., Piscataway, 1976—, clin. psychology and neuroscis., 1976—, grad. faculty clin. psychology and neuroscis., 1976—, sci. dir. Ctr. Alcohol Studies, 1983—, prof. clin. psychology grad. sch. applied & profl. psych., 1990—, prof. psychology Ctr. Alcohol Studies, 1990—, dir. Ctr. Alcohol Studies, 1992—; lectr. Rutgers/Prodential Alcohol Edn. Workshops, Rutgers U., 1984—; cons. in field; mem. adv. bd. N.J. Collegiate Consortium for Health in Edn., 1991; mem. sci. adv. bd. Ctr. for Edn. and Drug Abuse Rsch., Western Psychiat. Inst. and Clinic, Pitts., 1991. Reviewer, mem. editorial bd. Am. Psychologist, Jour. Studies on Alcohol; reviewer Psychol. Bull., Jour. Abnormal Psychology; contbr. articles to profl. publs., chpts. to books. Trustee Alcohol Rsch. Documentation, Inc., Piscataway, 1982—, v.p., 1982-91, pres., 1991—. Grantee Nat. Inst. Drug Abuse, 1982-83, 83-86, 86-88, 89-91, 91-94, Nat. Inst. Alcohol Abuse and Alcoholism, 1983-86, 86-88, 88-91, Dept. Health, Human Svcs. Pub. Health Svc., 1978-81, Nat. Inst. Justice, 1981-83. Mem. APA, Rsch. Soc. on Alcoholism, Soc. Psychologists in Addictive Behavior. Achievements include research on animal and human psychopharmacology, physiological and behavioral mechanisms in alsohol/drug related problems, experimental and clinical neuropsychology, neuropsychological models of mental disorders. Office: Rutgers U Ctr Alcohol Studies Busch Campus-Smithers Hall Piscataway NJ 08855-0969

PANDINI, DAVIDE, electrical engineer; b. Ferrara, Emilia Romagna, Italy, Mar. 21, 1961; s. Alberto Pandini and Mercedes Malossi. First cert. in English, U. Cambridge, Eng., 1990; diploma in elec. engring., U. Bologna, Italy, 1991; cert. of proficiency in English, U. Cambridge, Eng., 1991; PhD in Electronics, U. Padova, Italy, 1991. Tchr. Centro di Formazione Profl. Regionale, Ferrara, Italy, 1986-87; elec. engr. Sirti SpA, 1991-92. Served with Italian Army, 1988-89. Mem. English Speaking Club (sec. 1989-90). Roman Catholic. Avocations: playing piano. Home: Via Polonia 12, 44100 Ferrara Italy

PANDIT, SUDHAKAR MADHAVRAO, engineering educator; b. Gherdi, India, Dec. 3, 1939; came to U.S., 1968; s. Madhavrao Dhondopant and Ramabai P.; m. Maneesha Sudhakar Mangala Nulkar, May 12, 1966; children: Milind, Devarat. MS in Indsl. Engring., Pa. State U., 1970; MS in Statistics, U. Wis., 1972, PhD in Mech. Engring., 1973. Trainee engr. Kirloskar Oil Engines Ltd., Pune, India, 1961-62; engr. East Asiatic Co., Bombay, India, 1962; asst. engr. Heavy Engring. Corp., Ranchi, India, 1962-68; teaching asst. Pa. State U., State College, 1968-70; rsch. asst. U. Wis., Madison, 1970-73, lectr., 1973-76; prof. Mich. Tech. U., Houghton, 1976—; faculty rep. Nat. Tech. U., Ft. Collins, Colo., 1991—; ASA/NSF/NIST sr. rsch. fellow, 1993-94. Author: Time Series and System Analysis with Applications, 1983, Modal and Spectrum Analysis: Data Dependent Systems in State Space, 1991; contbr. articles to profl. jours. Mem. ASME, Soc. Mfg. Engrs., Inst. Indsl. Engrs., Sigma Xi. Achievements include developed a new philosophy and methodology of system analysis, prediction and control called data dependent systems. Home: Rt 1 Box 41 Royalewood Houghton MI 49931 Office: Mich Tech U ME-EM Dept 1400 Townsend Dr Houghton MI 49931-1295

PANDURANGI, ANANDA KRISHNA, psychiatrist; b. Dharwar, Karnataka, India, July 8, 1951; came to U.S., 1979; s. Krishna Tamanna and Susheela (Kusuma) P.; m. Rama Ananda, Mar. 2, 1979; children: Abhinav, Ashvin. MBBS, Jipmer, Madras U., Pondicherry, India, 1975; MD, Nimhans, Bangalore U., 1978. Resident in psychiatry Upstate Med. Ctr., SUNY, Syracuse, 1979-82; rsch. Columbia U., N.Y.C., 1982-84; lectr. in psychiatry Nimhans Bangalore U., Bangalore, India, 1978-79; asst. instr. SUNY Upstate Med. Ctr., Syracuse, 1979-82; rsch. fellow psychiatry Columbia U., N.Y.C., 1982-84; asst. prof. psychiatry Med. Coll. of Va., Richmond, 1984-90, assoc. prof. psychiatry, 1990—; chmn. Div. of Inpatient Psychiatry, Med. Coll. of Va., Richmond, 1990—; dir. Schizophrenia Program, Med. Coll. of Va., Richmond, 1984—; cons. Ctrl. State Hosp., Petersburg, Va., 1985-90. Contbr. articles to profl. jours. Bd. dirs. Gateway Found., Richmond, 1987-91, Bharat Swa-Mukti Found., Mansfield, Ohio, 1989-92. Named Young Investigator, NIMH, 1987, Exemplary Psychiatrist Nat. Alliance for Mentally Ill, 1992, Clin. Svc. award Med. Coll. Va.-Psychiatry, 1991. Fellow Am. Psychiat. Assn.; mem. Soc. for Biol. Psychiatry, N.Y. Acad. Scis., Indian Psychiat. Soc. Achievements include NIMH grant for study of bio-behavioral heterogeneity in schizophrenia. Office: Med Coll Va PO Box 710 Richmond VA 23298

PANETH, THOMAS, retired physicist; b. Vienna, Austria, Apr. 14, 1926; arrived in Argentina, 1938; s. Erwin and Anne (Deutsch) P. Degree in engring., U. Buenos Aires. Joined Soc. of Jesus, Cath. Ch., 1948. Researcher in solar physics Commn. Nat. Study Geo-Heliophysics, San Miguel, Buenos Aires, 1970-78, Commn. Nat. Investigations Espaciales, San Miguel, 1979-91; ret., 1991. Contbr. articles to profl. jours. Mem. IEEE (sr.), Argentine Astronomy Assn., Argentine Assn. Electrotech. Avocations: solving problems of maintaining large houses. Home: Av Mitre 3226, 1663 San Miguel Buenos Aires Argentina

PANG, HERBERT GEORGE, ophthalmologist; b. Honolulu, Dec. 23, 1922; s. See Hung and Hong Jim (Chuu) P.; student St. Louis Coll., 1941; BS, Northwestern U., 1944, MD, 1947; m. Dorothea Lopez, Dec. 27, 1953. Intern Queen's Hosp., Honolulu, 1947-48; postgraduate course ophthalmology N.Y.U., Med. Sch., 1948-49; resident ophthalmology Jersey City Med. Ctr., 1949-50, Manhattan Eye, Ear, & Throat Hosp., N.Y.C., 1950-52; practice medicine specializing in ophthalmology, Honolulu, 1952-54, 56—; mem. staffs Kuakini Hosp., Children's Hosp., Castle Meml. Hosp., Queen's Hosp., St. Francis Hosp.; asst. clin. prof. ophthalmology U. Hawaii Sch. Medicine, 1966-73, now asso. clin. prof. Cons. Bur. Crippled Children, 1952-73, Kapiolani Maternity Hosp., 1952-73, Leahi Tb. Hosp., 1952-62. Capt. M.C., AUS, 1954-56, Diplomate Am. Bd. Ophthalmology. Mem. AMA, Am. Acad. Ophthalmology and Otolaryngology, Assn. for Rsch. Ophthalmology, ACS, Hawaii Med. Soc. (gov. med. practice com. 1958-62, chmn. med. speakers com. 1957-58), Hawaii Eye, Ear, Nose and Throat Soc. (pres. 1960), Pacific Coast Oto-Ophthalmological Soc., Pan Am. Assn. Ophthalmology, Mason, Shriner, Eye Study Club (pres. 1972—). Home: 346 Lewers St Honolulu HI 96815-2345

PANGRAZI, RONALD JOSEPH, chemist; b. Natrona, Pa., Oct. 11, 1961; s. Louis Gene and Evelyn Lee (Gallo) P.; m. Dana Lorraine Shields, Sept. 22, 1984; children: Michael Joseph, Marc Vincent. BA, St. Vincent Coll., Latrobe, Pa., 1984. Cert. profl. chemist. Chemist Nat. Starch and Chem. Co., Bridgewater, N.J., 1984-86, devel. chemist, 1986-89, project supr., 1989-93; tech. mgr. R & D Garden State Tanning, Inc., Fleetwood, Pa., 1993—. Contbr. articles to profl. publs. Commentator St. Gertrude Roman Cath. Ch., Vandergrift, Pa. Mem. ALCA, Am. Inst. Chemists, Am. Chem. Soc. Republican. Roman Catholic. Achievements include five patents in field. Home: 24 E Barkley St Topton PA 19562-9999 Office: Garden State Tanning Inc 16 S Franklin St Fleetwood PA 19522

PANICKER, MATHEW MATHAI, nuclear engineer, educator; b. Kundara, Kerala, India, Aug. 19, 1945; s. Mathai M. and Thankama (Daniel) P.; m. Annamma Varghese, Feb. 2, 1976; 1 child, Dinesh M. BSc, U. Kerala, 1965, MSc, 1967; MS in Nuclear Engring., N.C. State U., 1983; PhD in Nuclear Engring., U. Fla., 1989. Lectr. in physics Carmel Poly., Alleppey, Kerala, 1967-68; lectr. in physics Catholicate Coll. U. Kerala, Pathanamthitta, 1968-69; asst. prof. Cuttington U. Coll., Monrovia, Liberia, 1969-75, assoc. prof., chmn. sci. div., 1977-80; grad. asst. N.C. State U., Raleigh, 1980-83; grad. asst. U. Fla., Gainesville, 1983-89, rsch. assoc., 1989-91; asst. prof. nuclear engring. Vogtle Nuclear Plant Campus, Am. Tech. Inst., Waynesboro, Ga., 1992—; cons. Innovative Nuclear Space Power and Propulsion Inst./U. Fla., 1989-91. Contbr. articles to profl. publs. Nat. Merit scholar Govt. of India, 1966-67. Mem. AIAA, Am. Nuclear Soc., Am. Phys. Soc., Sigma Xi, Alpha Nu Sigma, Tau Beta Pi, Phi Beta Delta. Achievements include development of coupled core neutron kinetics model to study time dependent neutronic behavior of multiple cavity gas core nuclear reactors. Home: 9 Marseilles Ct Savannah GA 31419 Office: Am Tech Inst PO Box 8 8760 Baylor Rd Brunswick TN 38014

PANIDES, ELIAS, mechanical engineer; b. Kastoria, Greece, Aug. 27, 1959; came to the U.S., 1969; s. Simeon and Thomai (Pinou) P. BS, Columbia U., 1981, MS, 1983, M in Philosophy, 1985, PhD, 1987. Tutor, translator Columbia U., N.Y.C., 1981-82, teaching asst. dept. mech. engring., 1981-83, grad. rsch. asst., 1983-87; HVAC engr. Flack & Kurtz Cons. Engrs., N.Y.C., 1982-84; rsch. scientist Grumman Aerospace Corp., Bethpage, N.Y., 1987-88, Xerox Corp., North Tarrytown, N.Y., 1988—; adj. prof. dept. mechan. engring. Columbia U., 1988—. Contbr. articles to Proceedings Am. Phys. Soc., Jour. Fluid Mechanics. Mem. ASME, KRIKOS, Sigma Xi. Achievements include patent for UV transfuse. Home: 18-06 Parsons Blvd Whitestone NY 11357

PANIN, FABIO MASSIMO, mechanical engineer; b. Milan, Italy, Mar. 30, 1957; m. Vana Secardin, July 8, 1987. Degree in Mech. Engring., Polytechnic of Milan, Italy, 1982. Mech. engr. IBM, Italy, 1982-85; mech. engr. European Space Agy., Holland, 1985-89, mech. systems engr., 1989—. Contbr. articles to profl. jours. Achievements include patents for Self-Reversing Ratchet; Fine Positioning Mechanism; Jettisoning Mechanism; Latching Mechanism. Office: European Space Agy, Keplerlaan 1, 2202 AG Noordwijk The Netherlands

PANIZZA, MICHAEL, civil engineer; b. Herrsching, Bavaria, Fed. Republic Germany, Aug. 13, 1946; s. Wolfgang and Rosemarie (Schafer) P.; m. Ursula Jaxt, Oct. 30, 1970; 1 child, Andreas. Diploma in Engring., U. Munich., 1973. Br. mgr. Doka, Heinsberg, Fed. Republic Germany, 1971-74; engr. Goetz, Greiner, Munich, Fed. Republic Germany, 1974-76; sales mgr. Spectra Physics, Darmstadt, Fed. Republic Germany, 1976-80; br. mgr. Laser Alignment, Landsberg, Fed. Republic Germany, 1980—. Diplom, Melioratia, Moscow, 1979, 1983; recipient gold medal Sympomech, Bratislava, Czechoslovakia, 1985. Avocations: windsurfing; caravaning; gardening. Home: Johann Michael Fischer Str 29, 86908 Diessen Germany Office: Laser Alignment Inc, Breslauer Str 42-46, 86884 Landsberg Germany

PANJEHPOUR, MASOUD, research scientist, educator; b. Kermanshah, Iran, May 22, 1958; came to U.S., 1977; s. Javad and Esmat Sadat (Emami) P.; m. Pamela Spray Davis, Aug. 24, 1986; children: Sara Marie, Emily Suzon. BSEE, U. N.C., 1981; MSEE, U. Toledo, 1983, PhD, 1988. Rsch. scientist Thompson Cancer Survival Ctr., Knoxville, Tenn., 1988—; asst. rsch. prof. U. Tenn. Coll. Vet. Medicine, Knoxville, 1989—; cons. Davol, Inc., Cranston, R.I., 1983-84, Oak Ridge (Tenn.) Nat. Lab., 1989—; mem. Rev. Bd. Com., Lasers in Surgery and Medicine Jour., Boston, 1991—. Contbr. articles to profl. jours. Do-Dad Girl Scouts of Am., Knoxville, Tenn., 1992. Recipient grant for laser hyperthermia NIH, Bethesda, Md., 1990. Fellow Internat. Soc. for Optical Engring., Am. Soc. for Laser Medicine & Surgery; mem. Am. Soc. for Photobiology. Achievements include development of balloon for gastrointestinal photodynamic therapy of cancer; development of laser-induced fluorescence for early detection of cancer in gastrointestinal tract. Home: 1222 Reston Ct Knoxville TN 37923 Office: Thompson Cancer Survival Ctr 1915 White Ave Knoxville TN 37916

PANOFSKY, WOLFGANG KURT HERMANN, physicist, educator; b. Berlin, Germany, Apr. 24, 1919; came to U.S., 1934, naturalized, 1942; s. Erwin and Dorothea (Mosse) P.; m. Adele Du Mond, July 21, 1942; children: Richard, Margaret, Edward, Carol, Steven. A.B., Princeton U., 1938, DSc (hon.), 1983; Ph.D., Calif. Inst. Tech., 1942; D.Sc. (hon.), Case Inst. Tech., 1963, U. Sask., 1964, Columbia U., 1977, U. Hamburg, Fed. Republic Germany, 1984, Yale U., 1985; hon. degree, U. Beijing, Peoples Republic China, 1987; DSc (hon.), U. Rome, 1988; hon. degree, Uppsala U., Sweden, 1991. Mem. staff mem. radiation lab. U. Calif., 1945-51, asst. prof., 1946-48, asso. prof., 1948-51; prof. physics Stanford U., 1951-89, prof. emeritus, 1989—; dir. Stanford (High Energy Physics Lab., Stanford Linear Accelerator Center), 1962-84, dir. emeritus, 1984—; Am. del. Conf. Cessation Nuclear Tests, Geneva, 1959; mem. President's Sci. Adv. Com., 1960-65; cons. Office Sci. and Tech., Exec. Office Pres., 1965—, U.S. ACDA, 1968-81; mem. gen. adv. com. to White House, 1977-81; mem. panel Office of Sci. and Tech. Policy, 1977—; nat. def. research Calif. Inst. Tech. and Los Alamos, 1942-45; chmn. bd. overseers Superconducting Super Collider Univs. Rsch. Assn., 1984—. Decorated officer Legion of Honor; recipient Lawrence prize AEC, 1961; Nat. Medal Sci., 1969; Franklin medal, 1970; Ann. Pub. Service award Fedn. Am. Scientists, 1973; Enrico Fermi award Dept. Energy, 1979; Shoong Found. award for sci., 1983, Hilliard Roderick prize Sci. AAAS, 1991; named Calif. Scientist Yr., 1966. Mem. Nat. Acad. Scis. (chmn. Com. on Internat. Security and Arms Control 1985-93), Am. Phys. Soc. (v.p., pres. 1974), Am. Acad. Arts and Scis., Phi Beta Kappa, Sigma Xi. Home: 25671 Chapin Rd Los Altos CA 94022-3413 Office: Stanford Linear Accelerator Ctr PO Box 4349 Palo Alto CA 94309-4349

PANOUTSOPOULOS, BASILE, electrical engineer; b. Amalias, Ilias, Greece, Dec. 1, 1956; came to U.S. 1979; s. Theodoros and Poulia (Kotsifas) P. ME in Elec. Engring., CCNY, 1985; MS in Applied Math., N.J. Inst. Tech., 1987; PhD, CUNY, 1991. Instr. elec. engring. Manhattan Coll., N.Y.C., 1986-89; adj. asst. prof. dept. elec. engring. CCNY, 1985—; electronics engr. Naval Undersea Warfare Ctr., New London, Conn., 1992—; adj. asst. prof. math. York Coll./CUNY, 1991-92; adj. lectr. physics Queensborough Community Coll./CUNY, 1991; computer cons. Contbr. articles to profl. jours. CUNY Grad. Ctr. fellow, 1990. Mem. IEEE, Am. Soc. Engring. Edn., Order of Engr., Eta Kappa Nu. Achievements include research on computer aided design - new educational approaches; computer simulation in physics, mathematics and applications of electromagnetics and electro-optics. Office: CCNY/CUNY Dept Elec Engring New York NY 10031

PANSKY, BEN, anatomy educator, science researcher; b. Milw., Feb. 18, 1928; s. Abraham and Leah (Namerosfky) P.; m. Julie Beverly Gossin, May 3, 1953; 1 child, Jonathan Hugh. BA, U. Wis., 1948, MS, 1950, PhD, 1954; MD, N.Y. Med. Coll., 1968. Diplomate Nat. Bd. Med. Examiners. Instr. U. Wis., Madison, 1950-53; asst. prof. N.Y. Med. Coll., N.Y.C., 1953-60, assoc. prof., 1960-65; intern Presbyn. Hosp., N.Y.C., 1968-69, resident in pathology, 1969-70; prof. Med. Coll. Ohio, Toledo, 1970—. Author/illustrator: (books) Functional Approach to Neuroscience 3rd edit., 1979, Review of Embryology 1st edit., 1982, Review of Gross Anatomy 5th edit., 1984, Review of Neuroscience 2nd edit., 1988. 1st lt. ROTC, 1944-48. Mem. AAAS, AAUP, AMA, Am. Assn. Anatomists, Sigma Xi, Phi Chi. (med. fraternity). Avocations: writing and illustrating children's books, writing science fiction. Home: 2809 Manchester Blvd Toledo OH 43606-2827 Office: Med Coll Ohio CS # 10008 Toledo OH 43699

PANTELIS, JOHN ANDREW, JR., civil engineer; b. Pitts., May 16, 1956; s. John Andrew and Audrey Jean (Kenny) P.; m. M. Catherine McLaughlin, June 18, 1983; 1 child, Ursula Leigh. BS in Civil Engring., U. Pitts., 1978, postgrad., 1990—. Registered profl. engr., Pa., W.Va., Ohio. Engr. in tng. unit leader W.Va. Dept. Natural Resources, Charleston, 1978-79; project engr.-estimator John A. Pantelis Painting Contractor, Carnegie, Pa., 1979-85; project engr. Cerrone & Assocs., Inc., Wheeling, W. Va., 1985-89, Killam Assocs. DLA Div., Warrendale, Pa., 1989-92; staff engr. Mcpl. Sewer and Water Authority Cranberry Twp., 1992—. Sec., Little League mgr., Overbrook Boys Club, Pitts., 1973-75; basketball referee and coach, St. Norbert

Sch. League, 1975, Pitts., 1975; basketball referee, YMCA League, Charleston, W.Va., 1979. Mem. ASCE, Water Works Operators' Assn. Pa., Western Pa. Water Pollution Control Assn., Order of the Engr., U. Pitts. Civil Engring. Alumni Club (sec. 1993—). Democrat. Roman Catholic. Home: 116 Chadborne Ct Zelienople PA 16063-1702 Office: Mcpl Sewer and Water Authority Cranberry Twp 2700 A Rochester Rd Mars PA 16046

PANTOS, WILLIAM PANTAZES, mechanical engineer, consultant; b. Ann Arbor, Mich., May 15, 1957; s. William Van and Lillian William (Skinner) P. BS in Mech. Engring., Northwestern U., Evanston, Ill., 1979; MS in Mech. Engring., San Diego State U., 1991. Registered profl. engr., Calif. Owner Signs & Symbols, Niles, Ill., 1975-80; engr. Hughes Aircraft, El Segundo, Calif., 1980-83, Gen. Dynamics, San Diego, 1983-85; staff engr. TRW, San Diego, 1985-90; pres. Tekton Industries, Carlsbad, Calif., 1990—. NROTC scholar USN, 1975. Mem. Am. Soc. Mech. Engrs., Nat. Soc. Profl. Engrs., Alpha Delta Phi. (pres. 1978). Greek Orthodox. Home: 6857 Seaspray Ln Carlsbad CA 92009-3738

PANUSCHKA, GERHARD, civil and sanitary engineer, inventor; b. Linz, Austria, Apr. 2, 1959; came to U.S., 1987; s. Johann and Anneliese (Lettmayr) P.; m. Tracey Lu Segur, Jan. 17, 1986; children: Richard, Michael. Diplom ingenieur, Univ. Vienna, 1987. Registered engr., Calif. Field engr. Ingenieurbuero Rosinak, Vienna, 1977-87; engr. I & II Black & Veatch Engr. and Arch., Santa Ana, Calif., 1988-89; project engr. Harding Lawson Assocs., Tustin, Sacramento, Calif., 1989—. Competitor in the 1984 Olympic Games in L.A. Mem. ASCE, Nat. Soc. Profl. Engrs. Roman Catholic. Home: 6115 Laguna Vale Way Elk Grove CA 95758 Office: Harding Lawson Assocs 3247 Ramos Circle Sacramento CA 95827

PANZA, GIULIANO FRANCESCO, seismologist, educator; b. Faenza, Ravenna, Italy, Apr. 27, 1945; s. Giuseppe and Giuseppina (Liverani) P.; m. Rita Zoccoli, Dec. 12, 1970. Maturita Classica, Liceo M. Minghetti, Bologna, Italy, 1963; laurea in Fisica, U. Bologna, 1967. Postdoctoral fellow U. Bologna, 1968-70; vis. postdoctoral fellow U. Uppsala, Sweden, 1969; asst. prof. geodesy U. Bari, Italy, 1970-80; postdoctoral fellow UCLA, 1971-74; assoc. prof. geodesy U. Bari, 1973-80; prof. seismology U. Della Calabria, Cosenza, Italy, 1975-77; vis. prof. Politechnic of Zurich, 1977; prof. geophysics U. Trieste, Italy, 1980-88, prof. seismology, 1988—; pres. Sch. Geology, U. Trieste, 1983-86; dir. Istituto di Geodesia e Geofisica Universita, Trieste, 1985-91; mem. commn. Strong Motion Seismology Internat. Assn. Seismology and Physics of Earth's Interior; cons. Internat. Ctr. Theoretical Physics, Trieste, 1991—, Consiglio Nazionale Richerche, Rome, 1990—, others; pres. Cent.European Initiative Com. Earth Scis., 1991—; chmn. Task Group Inter-Union Commn. for Lithosphere, 1991—. Editor spl. vol. Pure and Applied Geophysics: Generation and Propagation of Seismic Waves, 1979, Digital Seismology and Fine Modeling of the Lithosphere, Plenum, 1989, The Structure of the Alpine-Mediterranean Area; Contribution of Geophysical Methods, Terra Nova, 1990, Synthetic Seismograms: Generation and Use Geophys. Jour., 1985. Fulbright fellow, 1970; recipient Premio Ettore Cardani, U. Torino, 1968, Premio Linceo, Accademia Nazionale dei Lincei, 1990. Fellow Academia Europaea of London, Am. Geophys. Union, Royal Astron. Soc. London; mem. European Union Geoscis. (coun. mem., v.p.), European Geophys. Soc. (coun. mem.), Accademia Nazionale Lincei Roma (corr. fellow). Home: Via di Scorcola 4, Trieste Italy 34134 Office: Istituto di Geodesia, Via dell Universita 7, Trieste Italy 34100

PANZARELLA, JOHN EDWARD, water quality chemist; b. Freeport, N.Y., Dec. 5, 1960; s. Angelo John Panzarella and Irma (Schunrock) Cohen; m. Kathleen Huysman, Mar. 29, 1988; children: Leora-Anne, John. BS in Chemistry, SUNY, Old Westbury, 1986. Minority Biomed Rsch. Support rsch. asst. SUNY, 1983-86, biology technician, 1986-87; water quality chemist Suffolk County Water Authority, Oakdale, N.Y., 1988-91, lab. supr., 1991—; cons. chemistry and physics dept. SUNY, 1989—. Recipient certificate of merit Chem. Rubber Co., 1984. Home: 11 Bowdoin Rd Centereach NY 11720-2331 Office: Suffolk County Water Authority Pond Rd & Sunrise Hwy Oakdale NY 11769

PANZER, HANS PETER, chemist; b. Ratingen, Rhineland, Germany, July 26, 1922; came to U.S., 1957; s. Peter Franz and Margarete Ernestine (Schwellenbach) P.; m. Ursula Hedwig Stratmann, Oct. 22, 1956; children: Peter Hermann, Maria Margaret. MS in Chemistry, U. Muenster, Germany, 1954, PhD, 1957. Postdoctoral fellow Purdue U., Lafayette, Ind., 1957-58; sr. rsch. chemist Am. Machine and Foundry Co., Springdale, Conn., 1959-63; rsch. specialist Gen. Foods Corp., Tarrytown, N.Y., 1963-65; from various supervisory positions to assoc. rsch. fellow Am. Cyanamid Co., Stamford, Conn., 1965-91; cons. in field, Stamford, 1991—. Contbr. entries to: Ency. Chem. Tech., 3rd edit., 1980, Ency. Polymer Sci. and Engring., 2nd edit., 1987; contbr. articles to profl. jours.; patentee in field. Mem. Am. Chem. Soc. (divsn. polymer chemistry, divsn. polymeric materials, sci. and engring.), German Chem. Soc. Roman Catholic. Avocations: tennis, hiking. Office: Am Cyanamid Co 1937 W Main St Stamford CT 06902-4580

PANZONE, RAFAEL, mathematics educator; b. Buenos Aires, Apr. 4, 1932; s. Felix Antonio and Maria Luisa (Musante) P.; m. Agnes Ilona Benedek, Jan. 2, 1959; children: Susanne, Pablo, Pedro, Ines. Dr. Math., U. Buenos Aires, 1958. Assoc. prof. Nat. U. Buenos Aires, 1962-67; prof. Nat. U. del Sur, Bahía Blanca, Argentina, 1967—, head dept. math., 1968-70, head inst. math., 1968-71, 75-79, 83-88; researcher Conicet, Buenos Aires, 1962—; cons. Conicet, Buenos Aires, 1967-73, 76-85, 89—. Recipient Odol Prize, Conicet, Buenos Aires, 1966, Konex diploma, Buenos Aires, 1983; J.S. Guggenheim fellow, N.Y.C., 1978. Mem. Unión Matemática Argentina (editor-in-chief jour. 1968-69, editor 1970—), Am. Math. Soc. Office: Nat U del Sur, Alem 1253, 8000 Bahia Blanca Argentina

PAPADAKIS, CONSTANTINE N., engineering educator, dean; b. Athens, Greece, Feb. 2, 1946; came to U.S., 1969; s. Nicholas and Rita (Masciotti) P.; m. Eliana Apostolides, Aug. 28, 1971; 1 child, Maria. Diploma in Civil Engring., Nat. Tech. U. Athens, 1969; MS in Civil Engring., U. Cin., 1970; PhD in Civil Engring., U. Mich., 1973. Cert. profl. engr., Mich., Va., Ohio, Greece. Engring. specialist, geotechnical group Bechtel, Inc., Gaithersburg, Md., 1974-76; supr. and asst. chief engr. geotechnical group Bechtel, Inc., Ann Arbor, Mich., 1976-81; v.p., bd. dirs. water resources div. STS Cons. Ltd., Ann Arbor, 1981-84; v.p. water and environ. resources dept. Tetra Tech-Honeywell, Pasadena, Calif., 1984; head dept. civil engring. Colo. State U., Ft. Collins, 1984-86; dean Coll. Engring. U. Cin., 1986—, dir. Groundwater Rsch. Ctr., 1986—; dir. Ctr. Hill Solid and Hazardous Waste Rsch. Ctr. EPA, Cin., 1986—; adj. prof. civil engring. U. Mich., 1976-83; cons. Gaines & Stern Co., Cleve., 1983-84, Honeywell Europe, Maintal, Fed. Republic of Germany, 1984-85, Arthur D. Little, Boston, 1984-85, Camargo Assocs., Ltd., Cin., 1986, King Fahd Univ. Rsch. Inst., Dhahram, Saudi Arabia, 1987, King Abdulaziz City for Sci. and Tech., Riyadh, Saudi Arabia, 1991, Henderson & Bodwell Cons. Engrs. Inc., 1991, Cin. Met. Sewer Dist., 1992; acting pres. Ohio Aerospace Inst., 1988-90; interim pres. Inst. Advanced Manufacturing Scis. Ohio Edison Tech. Ctr., 1989-90; mem. bd. govs. Edison Materials Tech. Ctr., 1988—; mem. adv. bd., founding mem. Hamilton County Bus. Incubator, 1988—; bd. dirs. NES Inc. subs. Penn Cen. Co., 1991—. Author: Problems on Strength of Materials, 1968, Sewer Systems Design, 1969; editor: Fluid Transients and Acoustics, 1978, Pump-Turbine Schemes, 1979, Small Hydro Power Fluid Machinery, 1982; Megatrends in Hydraulics, 1987; contbr. over 55 articles to profl. jours. Mem. Great Cin. C. of C. Blue Chip Campaign for Econ. Devel. Task Force, 1988, Ohio Coun. on Rsch. and Econ. Devel., 1988, Ohio Sci. and Tech. Commn. Adv. Group, 1989—; coun. mem. St. Nicholas Ch. Parish, Ann Arbor, 1981-84; mem. City of Ft. Collins Drainage Bd., 1984-86. Recipient Horace W. King scholarship civil engring. dept. U. Mich., 1971-73, 6 Bechtel Merit awards, 1974-79, Young Engr. of Yr. award Mich. Soc. Profl. Engrs., Ann Arbor, Mich., 1982, Disting. Engr. award Engrs. and Scientists Cin. Tech. Socs. Coun., 1989. Fellow ASCE (pres. Ann Arbor br. 1980-81, pres.-elect Mich. sect. 1983-84, hydraulics div. publ. com. 1980-83), ASME (chmn. fluid transients com. 1978-80, mem. fluids engring. div. awards com. 1981-84); mem. NSPE, Am. Soc. Engring. Edn., Order of the Engr., Internat. Assn. for Hydraulic Rsch., Ohio Engring. Dean's Coun. (chmn.-elect 1989-91), Sigma Xi, Chi Epsilon, Tau Beta Pi. Greek Orthodox. Avocations: photography, classical music, travel, swimming, racquetball. Home: Indian Hill 7354 Sanderson Pl Cincinnati OH 45243 Office: U Cinn Coll Engring Mail Location 18 Cincinnati OH 45221-0018

PAPADAKOS, PETER JOHN, critical care physician; b. Bklyn., Feb. 4, 1957; s. John and Irene (Vahaviolos) P. BA, NYU, 1979; MD, CUNY, 1983. Intern, then resident in surgery The Roosevelt Hosp., N.Y.C., 1983-85; resident in anesthesiology Mt. Sinai Hosp., N.Y.C., 1985-87, fellow in critical care medicine, 1987-88; clin. dir. surg. ICU U. Rochester, N.Y., 1988—. Contbr. articles to profl. jours. Trustee Incurable Illness Found., N.Y.C., 1986-88. Recipient USN Rsch. award, 1975. Mem. Shock Soc., Soc. Critical Care. Achievements include research on affect of inverse ratio ventilation on sepitic shock and oxygen delivery. Office: U Rochester 601 Elmwood Ave Rochester NY 14642-9999

PAPADIMITRIOU, CHRISTOS, computer science educator. Endowed chair computer and info sci. U. Calif.-San Diego, La Jolla. Office: U Calif San Diego Dept Computer Sci La Jolla CA 92093

PAPAIOANNOU, EVANGELIA-LILLY, psychologist, researcher; b. Thessaloniki, Greece, Mar. 22, 1963; came to U.S. 1984; d. Nicholas and Ekaterini (Goulias) P. Bus. studies certificate with high honors, Anatolia Coll., Thessaloniki, Greece, 1983; BA in Psychology magna cum laude, Smith Coll., 1986; postgrad., Am. Univ., Washington, 1989. Guest researcher NIH, Bethesda, Md., 1986—. Author articles in press and profl. jours. Active in Hellenic Soc. for the Health Scis., Bethesda, 1987—. Recipient: scholarships Smith Coll. and Anatolia Coll. Mem. Jean Piaget Soc., Am. Psychol. Assn., Internat. Platform Assn., Alliance Francaise, Smith Coll. Alumnae Assn., Anatolia Coll. Alumni Assn., Nat. Mus. of Women in the Arts, Brazilian-Am. Cultural Inst., Phi Beta Kappa, Psi Chi, Smith Coll. First Group Scholars. Greek Orthodox. Avocations: modern dance, classical ballet, horseback riding, travel. Home: Promenade Towers 5225 Pooks Hill Rd Apt 1711N Bethesda MD 20814-2052 Office: NIH 9000 Rockville Pike Bethesda MD 20892-0001

PAPAKYRIAKOU, MICHAEL JOHN, biomechanical engineer; b. Hamilton, Ont., Can., June 3, 1958; arrived in U.S., 1985; s. John and Gertrude (Stocker) P.; m. Peg Bailey, Feb. 23, 1985; children: Ben, Matt. BE, McMaster U., Hamilton, 1982; MSE, U. New Brunswick, 1985. Biomech. engr. Nat. Coll. Chiropractic, Lombard, Ill., 1985—. Contbr. articles to profl. jours. Engr. Ont. scholar, 1977. Mem. IEEE. Achievements include research in muscle fiber conduction velocity, EMG changes during exercise. Office: Nat Coll Chiropractic 200 E Roosevelt Rd Lombard IL 60148

PAPALEXOPOULOS, ALEX DEMOCRATES, electrical engineer; b. Greece, June 27, 1957; came to U.S., 1980; s. Democrates and Maria (Nikolopoulos) P.; m. Deirdre Jan Canepa, Jan. 7, 1989; children: Alexis, Arianna. BSEE, Nat. Tech. U. Athens, Greece, 1980; MSEE, Ga. Inst. Tech., 1982, PhD in Elec. Engring., 1985. Rsch. asst. Ga. Inst. Tech., Atlanta, 1980-85; with Pacific Gas and Electric Co., San Francisco, 1986—, supr. systems engring., 1992—; industry adviser Electric Power Rsch. Inst., Palo Alto, Calif., 1990—. Contbr. articles to profl. publs. Vicepres. Hellenic Soc., Ga. Inst. Tech., Atlanta, 1983-85. Mem. IEEE (pres. student br. Ga. Inst. Tech. 1984-85, sr. mem. power engring. soc. 1990—). Office: Pacific Gas and Electric Co Mail Code T25A PO Box 770000 San Francisco CA 94117

PAPANASTASIOU, TASOS CHARILAOU, chemical engineering educator; b. Holetria, Paphos, Cyprus, Apr. 22, 1953; s. Charilaos and Galatia (Christofi) P.; m. Androula Michael Papanastasiou, Oct. 20, 1979; children: Charilaos, Yiangos. PhD, U. Minn., 1984. Asst. prof. U. Mich., Ann Arbor, 1985-91, assoc. prof., 1991-92; assoc. prof. Aristotle U. of Thessalonia, Greece, 1992—. Author: Applied Fluid Mechanics, 1993; contbr. over 35 articles to profl. jours. Pres. PTO, Saint Nicolas Greek Sch., Ann Arbor, 1990-92; soccer coach Logan Elem. Sch., Ann Arbor, 1988-90; mem. higher edn. com. Dem. Party, 1988. Fellowship IKY, 1973-78, U. Minn., 1982-83, faculty fellowship Rohms Haas Co, 1985. Mem. AICE, Am. Chem. Soc., Am. Phys. Soc., Internat. Polymer Processing Soc. Achievements include rsch. on constitutive equation for viscoelastic liquids, the continuous viscoplastic equation for viscoplastic liquids, the streamlined finite elements method, the inverse finite element method, the free boundary condition, frequent citation in CSI. Office: Aristotle U of Thessaloniki, Faculty of Chem Engring, 54006 Thessaloniki Greece

PAPANDREOU, CONSTANTINE, tele-informatics scientist, educator; b. Messolongi, Greece, Aug. 28, 1939; s. Andreas and Vassiliki (Georgouli) P.; m. Antonia Tsafoulia, May 22, 1971; 1 child: Andreas. M Engring., Tech. U. Munich, 1965, M Econ. Engring., 1972; PhD, U. Munich, 1984. Chief ops. rsch. dept. Rsch. Ctr. Nat. Def., Athens, Greece, 1968-70; systems analyst, designer, EDP mgr., then dept. coord. Greek Telecom Orgn., Athens, 1970-81; mem. rsch. staff U. Munich, 1981-84; dist. tech. mgr. Greek Telecom Orgn., 1985-86, cons., 1986—; asst. prof. informatics Athens U. of Econs. and Bus., 1990—; bd. dirs. EDP Ctr., Social Svcs. of Greece, Athens, 1989—, Govt. Com. Informatics, Athens, 1989—, Orgn. and Info. S.A., 1991—; cons. in EDP, 1970—; mem. rsch. staff Ctr. Econ. Rsch., 1986—; vice dir. Informatics Greek Telecom Orgn., Athens, 1993—. Author: Introduction to Electronic Computers and Information Systems, 1973, The Development of New Forms of Telecommunications (in German), 1984; sci. editor: Teleinformatics (in Greek), 1989; contbr. over 50 articles on telematics, telecommunications and informatics. 2d lt. Greek army, 1968-70. Grantee Govt. of Bavaria, Munich, 1962; prize winner 23d conf. Fedn. Telecommunications Engrs. of European Community, Rome, 1984; decorated Cross of Knight Ordo Sancti Constantini Magni-Hellenic Exarhate; recipient Cert. Recognition of Holy City Messolongi. Mem. Ops. Rsch. Soc. Greece (coun. mem. 1973-76), Soc. Econ. Engrs. of Greece (coun. mem. 1984—), Greek Computer Soc. (coun. mem. 1989-90, v.p. 1990—), Union Profs. of Bus. Univs., Union German Engrs., German Soc. of Informatics, Engring. Chamber of Greece, Ordo Santi Constantini Magni (pres. elect 1993). Greek Orthodox. Home: Kalliga Str 17, GR 11473 Athens Greece

PAPARAZZO, ERNESTO, chemist; b. Rome, July 15, 1950; s. Tancredi and Enrica (Bruschini) P. PhD in Chemistry, U. Rome, 1976. Postdoctoral fellow Nat. Rsch. Coun. Italy, Rome, 1979-82, staff researcher, 1982-83, 86—; assoc. researcher Lawrence Berkeley (Calif.) Lab., 1983-85; salesman Italian Br. of Rank Xerox, Rome, 1977-78; vis. scientist dept. chmistry U. Namur, Belgium, 1980; vix. prof. dept. physics U. Merida, Venezuela, 1993; cons. Ilva-Nat. Steel Co. Italy, Rome, 1990-92; responsible Electron Spectroscopy for Chem. Analysis Ctr. of Italian Rsch. Coun., Montelibretti, Rome, 1983; presenter internat. confs. Europe, U.S.A., Italy, France, Germany, England, Holland, 1980—; author invited seminars depts. Chemistry and Metallurgy U.Calif., Berkeley, Nat. Inst. Standards and Tech., Gaithersburg, Md., dept. physics Johns Hopkins U., Balt., AT&T Bell Labs., Murray Hill, N.J., Europe, various Italian Univs., Nat. Rsch. Coun. France, Paris, 1980—; organizing com. 3d Internat. Conf. Formation Semiconductor Interfaces, 1991; reviewer refereeing papers submitted for publ. in internat. jours., 1991—. Contbr. over 60 articles to internat. jours. and over 30 papers to internat. confs. Italian Com. for Atomic Energy fellow, Rome, 1978; recipient Stampacchia Prize, U. Rome, 1979. Mem. Italian Vacuum Soc., Am. Vacuum Soc., Italian Chem. Assn. Roman Catholic. Avocations: classical and jazz music, checkers, comedy movies. Home: Via San Silvestro 35, Montecompatri, I-00040 Rome Italy Office: Istituto Struttura della Materia CNR, Via E Fermi 38, I-00044 Frascati Italy

PAPAS, ANDREAS MICHAEL, nutritional biochemist; b. Kato Moni, Cyprus, Oct. 29, 1942; s. Michael and Maria (Hadjikyriakou) P.; m. Kalliopi Tsirozidou, June 29, 1969; 1 child, Konstantinos. MSc, U. Ill., 1971, PhD, 1973. Rsch. assoc. U. Man., Winnipeg, 1975, asst. prof., 1976; chemist biochemistry lab. Eastman Chem. Co., Rochester, N.Y., 1977, sr. chemist, 1978-82, lab. mgr., 1983-85; mgr. animal nutrition supplements Eastman Chem. Co., Kingsport, Tenn., 1985-88, tech. dir., 1989—; adj. prof. East Tenn. State U., Johnson City, 1989—. Contbr. articles to profl. jours. Pres. Tricities Greek Orthodox Ch., Kingsport, 1991; bd. dirs. Hellenic Cultural Soc., Rochester, 1984; mem. biotech. com. Cornell U., Ithaca, N.Y., 1983. Fulbright scholar, 1969; Wright scholar U. Ill., 1971; U. Ill. fellow, 1972. Mem. Natural Source Vitamin E Assn. (pres. tech. com. 1993—). Office: Eastman Chem Co Kingsport TN 37662

PAPATHANASIOU, ATHANASIOS GEORGE, mechanical engineer; b. Athens, Attica, Greece, Mar. 3, 1957; s. George Athanasios and Margaret (Matsoukas) P.; m. Georgia Aharidis, Sept. 2, 1990. Diploma, U. Salomica, Greece, 1980; PhD, N.C. State U., 1986. Chartered engr., Greece. Apprentice designer State Aircraft Factory, Greece, 1976-77; rsch. asst. N.C. State U., Raleigh, 1981-86, teaching asst., 1982-84; aerospace engr. Hellenic Aerospace Industries, Greece, 1988-89; rsch. engr. Nat. Tech. U. Athens, Greece, 1989—; mech. engr. Martedec S.A., Athens, 1992-93; project mgr. R&D Marine Applied Rsch. and New Techs. Ltd., Athens, 1993—; cons., Athens, 1989—. Contbr. articles to profl. jours. Sgt. Hellenic Air Force, 1986-87. Mem. AIAA, ASME, Union of Greek Mech. Engrs., Tech. Chambers of Greece (univ. student award 1980). Home: 23 Ipponaktos St, 11744 Athens Greece

PAPATHEOFANIS, FRANK JOHN, biochemist; b. Melrose Park, Ill., Apr. 20, 1959; s. John and Rina (Alexopoulos) P.; m. Julie Ann Teskey. BA in Biochemistry, Northwestern U., 1981; PhD, U. Ill., 1986, MD, 1991. NIH predoctoral fellow U. Ill., Chgo., 1983-86, chief bone metabolism lab., 1986-90, dir. biomaterials and bioengring. lab. dept. orthopaedics, 1990—, asst. prof., 1990—. Author: Bioelectromagnetics: Biophysical Principals in Medicine and Biology, 1986, Orthopaedics Made Ridiculously Simple, 1992, Magnetic Properties of Biological Systems and Tissues, 1993. Coord. Spl. Olympics, Evanston, Ill., 1977; mem. sch. bd. Plato Elem. Sch., Chgo., 1979. Warren and Clara Cole Found. rsch. fellow, 1991; recipient Edwin L. Shumann prize for excellence in English Northwestern U., 1979. Mem. AAAS, Am. Inst. Physics, Am. Assn. Physicists in Medicine, Internat. Soc. for Bioelectricity, Orthopaedic Rsch. Soc. Office: U Ill Dept Orthopaedics MC 844 Chicago IL 60680

PAPATZACOS, PAUL GEORGE, mathematical physicist, educator; b. Cairo, Aug. 16, 1941; arrived in Norway, 1966; s. Alexander and Helen (Chrysanthou) P.; m. Ruth Jorunn Hagen, Dec. 29, 1965; 1 child, David Alexander. Degree in civil and aero. engring., Ecole Nat. Superieure Aero., Paris, 1965; PhD in Physics (hon.), U. Trondheim, Norway, 1975. Researcher in physics U. Trondheim, 1966-75, Nordita, Copehhagen, 1975-77; researcher in seismics Geophys. Co. of Norway, Stavanger, 1977-79; petroleum engr. Statoil, Stavanger, 1979-82; researcher in petroleum Rogalandsforskning, Stavanger, 1982-83; assoc. prof. Høgskolesenteret i Rogaland, Stavanger, 1983-93, prof., 1993—; cons. Rogalandsforskning, 1983—. Contbr. articles to profl. publs. Mem. Norwegian Phys Soc., Assn. for Computing Machinery. Office: Høgskolesenteret i Rogaland, PO Box 2557, 4001 Stavanger Norway

PAPAVASSILIOU, ATHANASIOS GEORGE, molecular biologist, researcher; b. Thessaloniki, Greece, Mar. 13, 1961; s. George A. and Ioanna T. (Gelitsalis) P.; m. Effie K. Basdra, Feb. 18, 1984; children: George A., Konstantinos A. MD summa cum laude, Aristotelian U. Thessaloniki, 1984; PhD in Cellular, Molecular, and Biophys. Studies, Columbia U., 1989. Grad. rsch. assoc. dept. microbiology Columbia U. Coll. Physicians and Surgeons, N.Y.C., 1984-89, postdoctoral rsch. scientist, 1989-90; postdoctoral fellow European Molecular Biology Lab., Heidelberg, Germany, 1990-91, staff scientist, 1991—, instr., 1990—. Author: (monograph) Parenteral Nutrition-Therapeutic Administration of Fluids, 1989, (book) Pediatric Neurology, 1994; contbr. articles to profl. publs. Fellow Ministry Nat. Edn., 1978-84. Fellow European Molecular Biology Orgn.; mem. AAAS, Am. Chem. Soc., Am. Soc. for Microbiology (Young Investigator award 1990), Am. Soc. for Biochemistry and Molecular Biology (assoc.), Biochem. Soc., Soc. for Gen. Microbiology, N.Y. Acad. Scis. Christian Orthodox. Avocations: poetry, quantum physics and chemistry, soccer. Office: European Molecular Biol Lab, Meyerhofstr 1 PO 10.2209, D-69012 Heidelberg Germany

PAPERT, SEYMOUR AUBREY, mathematician, educator, writer; b. Pretoria, South Africa, Mar. 1, 1928; came to U.S., 1964; s. Jack and Betty P.; m. Androula Christofides, Apr. 10, 1963 (div.); 1 dau., Artemis; m. Sherry Turkle, Dec. 18, 1977. B.A., U. Witwatersrand, S. Africa, 1949, Ph.D., 1952; Ph.D., Cambridge U., Eng., 1959. Co-dir. artificial intelligence lab. MIT, Cambridge, 1967-73, dir. Logo Group, 1970-81, prof. math. and edn., 1964—, Cecil and Ida Green prof. edn., 1974-80; vis. prof. math. Rockefeller U., N.Y.C., 1980-81; sci. dir. Centre Mondial Informatique et Ressource Humaine, Paris, 1982-83. Author: (with Marvin Minsky) Perceptrons, 1969, Artificial Intelligence, 1974, (with McNaughton) Non-Counting Automata, 1971, Mindstorms: Children, Computers..., 1980. Marconi Internat. fellow, 1981; J.S. Guggenheim fellow, 1980. Office: MIT Dept Media Tech Cambridge MA 02139

PAPET, LOUIS M., federal official, civil engineer; b. White Castle, La., June 14, 1933; s. Leonce A. and Corrine C. (Comedux) P.; m. Lee Anna Blanchard, May 30, 1959; children: Louis M. Papet Jr., Benjamin J. Papet. BSCE, La. State U., 1954-57. Registered profl. civil engr. La.; land surveyer La. Project engr. La. Dept Hwys., Baton Rouge, La., 1957-61; hwy. engr. U.S. Bur. Pub. Roads, Ala., N.C., Ga., Md., 1963-81; divsn. adminstr. U.S. Fed. Hwy. Adminstrn., Harrisburg, Pa., 1981-85, Atlanta, Ga., 1985-88; chief, pavement divsn. U.S. Fed. Hwy. Adminstrn., Washington, 1989—. With Army Civil Engrs., 1952-54. Home: 2785 Rudder Dr Annapolis MD 21401

PAPIKE, JAMES JOSEPH, geology educator, science institute director; b. Virginia, Minn., Feb. 11, 1937; s. Joseph John and Sistine Marie (Tassi) P.; m. Pauline Grace Maras, Sept. 6, 1958; children: Coleen, Coreen, Jimmy, Heather. BS in Geol. Engring. with high honors, S.D. Sch. Mines and Tech., 1959; PhD, U. Minn., 1964. Rsch. geologist U.S. Geol. Survey, Washington, 1964-69; assoc. prof. dept. earth and space scis. SUNY, Stony Brook, 1969-71, prof., 1971-82, chmn. dept., 1971-74; prof. dept. geology and geol. engring. S.D. Sch. Mines and Tech., Rapid City, 1982-87, Disting. prof., 1987-90, dir. Inst. for Study Mineral Deposits, 1982-90, dir. Engring. and Mining Expt. Sta., 1984-90; Regents' prof. dept. geology, dir. Inst. Meteoritics, U. N.Mex., Albuquerque, 1990—; mem. adv. com. for earth scis. NSF, 1985-89, Continental Sci. Drilling Rev. Group, Dept. Energy, 1986-87, Lunar and Planetary Sample Team, 1990—, Lunar Outpost Site Selection Com., 1990-91, organizing com. for FORUM for Continental Sci. Drilling, 1990—, adv. com. Inst. Geophysics and Planetary Physics, Los Alamos Nat. Lab., 1991—. Assoc. editor procs. 4th Lunar Sci. Conf., 1973, Jour. Geophys. Rsch., 1975-77, 82-84; editor procs. Internat. Conf.: The Nature of Oceanic Crust, 1976, Luna 24 Conf., 1977, Conf. on Lunar Highlands Crust, 1980; guest editor spl. issue Geophys. Rsch. Letters, 1991; mem. editorial bd. Procs. Lunar and Planetary Sci. Confs., 1987; contbr. numerous articles to profl. jours. Recipient NASA medal, 1973, Centennial Alumni award S.D. Sch. Mines and Tech., 1985; grantee NSF, NASA, Dept. Energy, 1969—. Fellow Geol. Soc. Am., Mineral. Soc. Am. (life, MSA medal 1974, past mem. coun.), Soc. Econ. Geologists (mem. coun.); mem. Am. Geophys. Union (past sec.), Geochem. Soc. (v.p. 1988-89, pres. 1989-91), Meteoritical Soc., Mineral. Assn. Can. Roman Catholic. Home: 103 La Mesa Placitas NM 87043 Office: U NMex Inst Meteoritics Northrop Hall Rm 313 Albuquerque NM 87131-1126

PAPINI, MAURICIO ROBERTO, psychologist; b. Buenos Aires, Argentina, Dec. 28, 1952; came to U.S. 1988; s. Victorio Roberto and Rosa Elina (Bianchi) P.; m. Mirta Toledo, Jan. 7, 1977; children: Santiago, Angel. Licenciate Psychology, U. Buenos Aires, 1976; PhD in Psychology, U. San Luis, 1985. Fellow Nat. Rsch. Coun., Argentina, 1977-85; asst. prof. U. Buenos Aires, 1986; asst. rschs. Nat. Rsch. Coun., Argentina, 1987; postdoctoral fellow U. Hawaii, Honolulu, 1988-90; asst. prof. psychology Tex. Christian U., Ft. Worth, 1990—. Contbr. numerous articles to profl. jours. Nat. Rsch. Coun. Argentina fgn. fellow, 1980-82. Mem. Psychonomic Soc., Internat. Soc. for Comparative Psychology (sec. 1984-88). Achievements include comparative analysis of learning processes in a variety of species including octopuses, fish, amphibians, reptiles, birds, mammals; research in brain mechanisms of learning in amphibians; others. Office: Tex Christian U Dept Psychology Fort Worth TX 76129

PAPP, LASZLO GEORGE, architect; b. Debrecen, Hungary, Apr. 28, 1929; came to U.S., 1956; m. Judith Liptak, Apr. 12, 1952; children: Andrea, Laszlo-Mark (dec. 1978). Archtl. Engr., Poly. U. Budapest, 1955; MArch, Pratt Inst., 1960. Designer Harrison & Abramovitz, Architects, N.Y.C., 1958-63; ptnr. Whiteside & Papp, Architects, White Plains, N.Y., 1963-67;

pres. Papp Architects, P.C., White Plains, N.Y., 1967—. Mem. Pres.'s Adv. Com. on Pvt. Sector Initiatives; mem. adv. com. Westchester Comunity Coll., 1971—, Iona Coll., New Rochelle, N.Y., 1982—, Norwalk State Tech. Coll., 1983—; v.p. Clearview Sch., 1985-89, pres., 1990-91; councilman, New Canaan, Conn. Fellow AIA (reg. dir. 1983-85); mem. Internat. Union Architects (rep. habitat com. 1986-90), N.Y. State Assn. Architects (v.p. 1977-80, pres. 1981), Am.-Hungarian Engrs. Assn. (bd. dirs. 1978-90), Hungarian Univ. Assn. (pres. 1958-60), Weschester County C. of C. (bd. dirs. 1968-71, vice chmn. bd. for area devel. 1983-89, chmn. bd. dirs. 1989-90), Am.-Hungarian C. of C. (charter 1989—). Home: 1197 Valley Rd New Canaan CT 06840-2428 Office: Papp Architects PC 7-11 S Broadway White Plains NY 10601-3531

PAPPAS, COSTAS ERNEST, aeronautical engineer, consultant; b. Providence, Oct. 14, 1910; s. Ernest and Sofie (Rose) P.; m. Thetis Hero, June 9, 1940; children: Alceste, Conrad. B.S., NYU, 1933, M.S., 1934. Registered profl. engr., N.Y., Calif. Stress analyst, aerodynamicist Republic Aviation Corp. (formerly Seversky Aircraft Corp.), 1935-39, chief aerodynamics, 1939-54, chief aerodynamics and thermodynamics, 1954-57, asst. dir. sci. research, 1957-59, asst. to v.p. research and devel., 1959-64; cons. to aerospace industry, 1964—; cons. sci. adv. bd. USAF-Aero Space Vehicles Panel. Author: Design Concepts and Technical Feasibility Studies of an Aerospace Plane, 1961, (with Thetis H. Pappas, memoirs) To the Rainbow and Beyond, 1992; contbr. articles to profl. jours.; patentee in field. Mem. NYU Alumni Vis. Com.; mem. San Mateo County Devel. Assn.; mem. grievance com. San Mateo-Burlingame Bd. Realtors, 1982-83, mem. legis. com., 1985-86; planning commr. City of Redwood City (Calif.), 1984-87, vice chmn. planning commn., 1986; mem. adv. council Growth Policy Counc., 1985-90. Recipient Wright Bros. award Soc. Automotive Engr., 1943; award Rep. Aviation Corp., 1944; Certificate of Distinction NYU Coll. Engring., 1955. Mem. NASA (subcom. high speed aerodynamics 1947-53, spl. subcom. research problems transonic aircraft design 1948), Inst. Aero Scis. (asso. fellow, mem. adv. bd. and membership com.), Inst. Aerospace Scis. (chmn. vehicle design panel 1960), Am. Inst. Aeros. and Astronautics (mem. ram jet panel, chmn. workshop profl. unemployed), Calif. Soc. Profl. Engrs., Air Force Assn., Tau Beta Pi (founder, 1st pres. San Francisco Peninsula alumnus chpt. 1971, asst. dir. Dist. 15 1979—), Redwood City-San Mateo County C. of C. (econ. devel. and govtl. rels. com. 1987—), Iota Alpha. Club: Commonwealth of Calif. Address: PO Box 5633 San Mateo CA 94402

PAPPAS, JOHN, clinical psychologist; b. N.Y.C., Aug. 24, 1957; s. Dimitrios and Agnes (Tunis) P.; m. Lois M. McNally, Aug. 17, 1986. BS summa cum laude, U. Bridgeport, 1980; MA, New Sch. for Social Rsch., PhD, 1991. Lic. psychologist, N.Y. State. Psychology intern Hall-Brooke Found., Westport, Conn., 1986-87; clin. psychologist N.Y.C. Health Dept., 1987-88; clin. psychotherapist Washington Sq. Inst., N.Y.C., 1988-91; cons. Assessment Systems, Inc., N.Y.C., 1988-92; staff psychologist N.Y.C. Police Dept., 1987-92; dir. Downtown Profl. Cons., N.Y.C., 1990—; grad. teaching asst. New Sch., N.Y.C., 1985-86; rsch. asst. Yale U., New Haven, 1983-84; attending psychologist, Regent Hosp. Author: Social Network Orientation and Psychotherapy, 1991; adv. bd. Healthway Mag. Recipient Herbert award in psychology U. Bridgeport, 1980. Mem. APA, AAAS, N.Y. Acad. Sci., N.Y. State Psychol. Assn., Mus. Modern Art. Office: Downtown Profl Cons 150 Broadway 23d Floor New York NY 10038

PAPROSKI, RONALD JAMES, structural engineer; b. Danbury, Conn., Aug. 30, 1966; s. James Donald and Esther Ernestine (Kuhne) P. BSCE, U. Conn., 1988, MSCE, 1990; MBA, Marist Coll., 1992. Surveyor Dept. Transp. State of Conn., New Milford, 1985; engr.'s asst., surveyor Arthur Howland, PELS, New Milford, 1985-88; rsch. asst. U. Conn., Storrs, 1988-90; design, field engr. Greenman-Pedersen Inc., East Hartford, Conn., 1990-92; engr. State of N.Y., Poughkeepsie, 1992—. Recipient Barton Wellar scholarship Vitramon Inc., 1984, Conn. State Police scholarship, 1984. Mem. ASCE, N.Y. Soc. Transp. Engrs., Am. Alpine Club, Appalachian Mountain Club, Chi Epsilon.

PARADA, JAIME ALFONSO, mechanical engineer, consultant; b. Linares, Chile, Oct. 11, 1957; s. Leandro Parada and Mariana Ibauez; m. Maria Martha Dalpiaz, June 1, 1984; 1 child, Bruno Parada Dalpiaz. MS, U.F.R.G.S., Porto Alegre, Brazil, 1987; PhD, U. Politecnica de Madrid, 1992. Project engr. Climatic Ltda., Santiago, Chile, 1981-82, Aerolite S.A.C.I., Santiago, 1982-83; rschr. Universidade Federal Do Rio Grande Do Sul, 1984-87, Universidad Politecnica de Madrid, 1987-92; head of non conventional energy and environment Nat. Energy Commn., Santiago, 1992—; tchr. Unisinos U., Porto Alegre, 1986-87; cons. Tau Environment, Madrid, 1989-92. Contbr. articles to profl. jours. Recipient study grants CNPQ, 1984, 87, Ministry of Edn./Spain, 1991. Fellow Brazilean Solar Energy Assn., Chilean Solar Energy Assn. Achievements include development of a photovoltaic module with static concentration. Office: Comision Nacional Energia, Teatinos 120 7 Piso, Santiago Chile

PARADIES, HASKO HENRICH, educator; b. Bremen, Germany, Feb. 18, 1940; s. Henry J. and Erna (Poppinga) P.; m. Gundrun K. Patzelt, June 28, 1973; children: Gesa-Kundry, Jan-Henry, Felix-Benjamin. Diploma chem., MD, U. Munster, 1966, PhD in Medicine, 1967, postgrad., 1970; PhD in Chemistry, U. Uppsala, 1969; postgrad. King's Coll., 1969-71, MIT, 1970-71, D in Biochem. Sci. (hon.), Royal Crown Spain, 1986; PhD in Biochem. (hon.), Albert Einstein-Found., 1990. Rsch. assoc. U. Munster, 1966; postdoctoral fellow King's Coll., 1969-71, Boston, 1971-72; rsch. assoc. Max Planck Inst., Berlin, 1971; prof. biochemistry Free U. Berlin, 1974-83, chmn. dept. chemistry and biochemistry, 1974-77, chmn. dept. plant physiology and cell biology, 1980-82; guest and vis. prof. chemistry Cornell U., Ithaca, N.Y., 1977-79; dir. rsch. and devel. Medice-Corp., Ltd., 1984-86, dir. 1986— ; adj. prof. chemistry U. Mo.-Rolla, 1985; sr. lectr. tech. chemistry dept. engring. U. Hagen, 1985-86; scientific cons., counsel Sherex, Inc., Dublin, Ohio, Medice, Inc., Federal Republic of Germany, Octapharma, Federal Republic of Germany; guest prof. Biotech. and Physical Chem. Märkische Hochschule Iserlohn, 1987-88, Chaired Prof., 1988—; vis. prof. La. State U., 1991-92, Ohio State U. Chemistry Dept., 1992—. Author 3 books. Contbr. articles to profl. jours. Recipient Albert Einstein Internat. Acad. Found. Bronze Medal for Peace, III class. Deutsche Forschungsgem fellow, 1967-71, grantee, 1974-79; Umweltbundesant grantee, 1979, 82, colloid and surface sci., 1982, Internat. Copper Rsch. Assn., 1987, European Community of Microbiology, 1991. Mem. Am. Chem. Soc., N.Y. Acad. Scis., Gesellschaft Deutscher Chemiker, Gesellschaft Deutsche Naturforscher und Arzte. Club: Stadler (Ithaca). Achievements include over 60 patents in field. Home: 38 Goerresstrasse, D-58636 Iserlohn Germany Office: Märkische Hochschule, Frauenstuhlweg 37, D 5860 Iserlohn Germany

PARADIS, RICHARD ROBERT, energy conservation engineer; b. Fall River, Mass., Oct. 17, 1956; s. Arthur A. and Joan M. (Kuttner) P.; m. Katherine M. Mackie, Feb. 26, 1983; children: Jason Ryan, Kyle Richard, Eric James. BA in Geography/Geology, Clark U., 1978. Cert. energy mgr. Energy conservation coord. Norwood (Mass.) Mcpl. Light Dept., 1978-79; analyst Shooshanian Engrins. & Assoc., Boston, 1979-84; facility mgr. Cooperative Corp., Waltham, mass., 1984-89; energy conservation specialist R.W. Beck & Assocs., Waltham, 1989; group mgr. Energy Investment, Inc., Boston, 1989-93; regional energy engr. systems and svcs. divsn. NE Region Johnson Ctrls., Inc., Lynnfield, Mass., 1993—; panelist Illuminating Engring. Soc. Award Com., N.Y.C., 1992. Co-author: Conservation & Load Management Program Design for Fall River Gas Company, 1991, Evaluation of Natural Gas C&LM Program for Residential New Construction, 1992. Coach Franklin (Mass.) Youth Soccer Assn., 1990—; musical minister Paulist Ctr., Boston, 1982-92. Recipient Region 1 award NASA Skylab Project, 1970. Mem. AAAS, Assn. Energy Engrs., Illuminating Engring. Soc. (bd. mgrs. New England sect., Internat. Illuminating Design award com. chmn.). Democrat. Roman Catholic. Home: 51 Dale St Franklin MA 02038 Office: Johnson Ctrls Inc 39 Salem St Lynnfield MA 01940

PARADOWSKI, ROBERT JOHN, history of science educator; b. South River, N.J., June 13, 1940; s. Adolph Edward and Helen Mary (Czechowicz) P. MA in Chemistry, Brandeis U., 1967; PhD in History of Sci., U. Wis., 1972. Asst. prof. Bklyn. Coll., 1972-77, Eisenhower Coll., Seneca Falls, N.Y., 1977-79; vis. prof. Linus Pauling Inst. Sci. and Medicine, Menlo Park, Calif., 1979-80, 81-82; assoc. prof. Eisenhower Coll. of RIT, Seneca Falls, 1980-81, 82-83; assoc. prof. history of sci. Rochester (N.Y.) Inst. of Tech., 1983—; rsch. phys. chemist Panametrics, Inc., Waltham, Mass., 1969. Introducer, contbr.: The Nobel Prize Winners: Chemistry, 3 vols., 1990; primary contbr.: Linus Pauling: A Man of Intellect and Action, 1991; major contbr.: The Great Scientists, 12 vol. set, 1989. Woodrow Wilson Nat. Found. fellow, 1965-66, NSF grad. fellow Brandeis U., 1965-67, U. Wis., 1970-72; NSF rsch. grantee Linus Pauling Inst., 1980-82. Mem. Am. Chem. Soc., History of Sci. Soc., Brit. Soc. for History of Sci., Sigma Xi. Democrat. Roman Catholic. Achievements include research in great events from history, sci. and tech. Home: 680 North Rd Apt 2 Scottsville NY 14546 Office: Rochester Inst Tech 1 Lomb Memorial Dr Rochester NY 14623

PARASKEVOPOULOS, NICHOLAS GEORGE, electrical engineering educator; b. Plainfield, N.J., July 15, 1960; s. George Nicholas and Helen (Antonopoulos) P. BSE, U. Pa., 1982; PhD, Rutgers U., 1991. Prof. Rutgers U., New Brunswick, N.J., 1992--; sci. editor Rsch. and Edn. Assn., Piscataway, N.J., 1990--. Contbr. articles to profl. jours. Mem. IEEE, Sigma Xi. Office: Rutgers U Brett and Bowser Rds Piscataway NJ 08854

PARAVATI, MICHAEL PETER, medico-legal investigator, forensic sciences program administrator, correction specialist; b. Rochester, N.Y., Feb. 4, 1949; s. Michael Gerard and Antoinette (Cappelli) P.; 1 child, James Michael. AA, Herkimer Community Coll., 1969; BA, Syracuse U., 1972; postgrad. in Pub. Mgmt., Rockefeller Coll., 1984. Dep. sheriff Oneida County Sheriff's Dept., Oriskany, N.Y., 1971-73; sr. tng. adminstr. N.Y. State Commn. of Correction, Albany, 1973-77, sr. forensic investigator med. review bd., 1977—. Contbr. articles to profl. jours. Committeeman Albany County Dem. Com., 1982-90. Recipient Gov.'s Productivity award Gov. State of N.Y., 1988. Office: NY State Commn Correction 60 S Pearl St Albany NY 12207

PARBERRY, EDWARD ALLEN, mathematician, consultant; b. Phila., Feb. 24, 1941; s. Edward and Blanche Mary (Greenway) P.; children: Christina Park, Stella. BS in Physics, Pa. State U., 1965, MA in Math., 1967; PhD in Math., 1969. Asst. prof. math. Wells Coll., Aurora, N.Y., 1969-75; sr. analyst Ketron Inc., Arlington, Va., 1975-78; sr. corp. scientist Analysis & Tech. Engring. Tech. Ctr., Mystic, Conn., 1978—. Contbr. articles to profl. jours. Mem. Acoustical Soc. Am. Home: 3 Gallup Ct Pawcatuck CT 06379 Office: Engring Tech Ctr Ste 105 240 Oral School Rd Mystic CT 06355-1208

PARDEE, ARTHUR BECK, biochemist, educator; b. Chgo., July 13, 1921; s. Charles A. and Elizabeth B. (Beck) P.; m. Ruth Sager; children by previous marriage: Michael, Richard, Thomas. B.S., U. Calif. at Berkeley, 1942; M.S., Calif. Inst. Tech., 1943, Ph.D., 1947. Merck postdoctoral fellow U. Wis., 1947-49; mem. faculty U. Calif. at Berkeley, 1949-61, assoc. prof., 1957-61; NSF fellow Pasteur Inst., 1957-58; prof. biology, chmn. dept. biochem. scis. Princeton, 1961-67; prof. biochemistry Princeton U., 1961-75; Donner prof. sci. Princeton, 1966; prof. Dana Farber Cancer Inst. and biochem. pharmacology dept. Harvard Med. Sch., Boston, 1975—; Mem. research adv. council Am. Cancer Soc., 1967-71. Co-author: Experiments in Biochemical Research Techniques, 1957; Editor: Biochemica et Biophysica Acta, 1962-68, 74—. Trustee Cold Spring Harbor Lab. Quantitative Biology, 1963-69. Recipient Young Biochemists travel award NSF, 1952, Krebs Medal Fedn. European Biochem. Socs., 1973, Rosenstiel award Brandeis U., 1975, 3M award Fedn. Am. Socs., Exptl. Biology, 1980, CIIT Prize, 1993; Princess Takamatu lectr., 1990. Fellow AAAS; mem. Nat. Acad. Sci. (editorial bd. proc. 1971-73, com. on scis. and pub. policy 1973-76), Am. Chem. Soc. (Paul Lewis award 1960), Am. Soc. Biol. Chemists (treas. 1964-70, pres. 1980-81), Am. Assn. Cancer Research (pres. 1985-86), Am. Soc. Microbiologists, Japanese Biochem. Soc., Ludwig Inst. Cancer Res. (sci. com. 1988—), Chem. Industry Inst. Toxicology (Founder's award), Phi Beta Kappa, Sigma Xi. Home: 30 Codman Rd Brookline MA 02146-7555 Office: Dana-Farber Cancer Inst 44 Binney St Boston MA 02115-6084

PARDES, HERBERT, psychiatrist, educator; b. Bronx, N.Y., July 7, 1934; s. Louis and Frances (Bergman) P.; m. Judith Ellen Silber, June 9, 1957; children: Stephen, Lawrence, James. BS, Rutgers U., 1956, MD, SUNY, Bklyn., 1960; DSc (hon.) SUNY, 1990. Straight med. intern Kings County Hosp., 1960-61, resident in psychiatry, 1961-62, 64-66; asst. prof. psychiatry Downstate Med. Ctr., Bklyn., 1968-72, prof., chmn. dept., 1972-75; dir. psychiat. services Kings County Hosp., Bklyn., 1972-75; prof., chmn. dept. psychiatry U. Colo. Med. Sch., 1975-78; dir. psychiat. services Colo. Psychiat. Hosp., Denver, 1975-78; dir. NIMH, Rockville, Md., 1978-84; asst. surgeon gen. USPHS, 1978-84; prof. psychiatry Columbia U., N.Y.C., 1984—, chmn. dept., 1984—, dir. Psychiat. Inst., 1984-89, v.p. for health scis. and dean faculty of medicine, 1989—. Committeeman, Kings County Dem. Com., 1972-75; pres. sci. bd. Nat. Alliance for Research on Schizophrenia and Depression. Capt. M.C., AUS, 1962-64. Decorated Army Commendation medal; ann. hon. lectr. Downstate Med. Ctr. Alumni Assn., 1972, recipient Alumni Achievement medal, 1980, William Menninger award ACP, 1992, Dorothy Dix award Mental Illness Fedn., 1992, Vester Mark award, 1993. Mem. Am. Psychiat. Assn. (v.p. 1986-88, pres. 1989-90, Disting. Svc. award 1993), Inst. Medicine, Am. Psychoanalytic Assn., Coun. of Deans (adminstrv. bd.), Phi Beta Kappa, Alpha Omega Alpha. Contbr. articles to med. jours. Home: 15 Claremont Ave Apt 93 New York NY 10027-6814 Office: Columbia U Coll Phys & Surgeons Dept Psychiatry 630 W 168th St New York NY 10032 also: NY State Psychiat Inst 722 W 168th St New York NY 10032

PAREDES, EDUARDO, medical physicist; b. Guayaquil, Ecuador, Aug. 2, 1957; came to U.S., 1984; s. Victor Manuel and Lupe Yolanda (Gonzalez) P.; m. Yolanda Aviles, Dec. 20, 1980; children: Eduardo Javier, Fernando Andres. BSEE, Politechnica Nacional, Quito, Ecuador, 1982; MS, U. Mass., 1990. Pres., owner Sistemas Avanzados de Electromedicina, Quito, 1980-84; regional specialist Data Scan Inc., Ann Arbor, Mich., 1984-86; nat. support specialist Kontron Instruments, Everett, Mass., 1986-89, prin. engr., 1989-91; pres. Med. Imaging Specialists, Shrewsbury, Mass., 1991—; ultrasound specialist Kontron S.A., Paris, 1986-91; nat. svc. mgr. Westminster Med., Orangeburg, N.Y., 1991-92; pres., owner Medical Equipment Specialists, Inc., Shrewsbury, Mass., 1992—. Contbr. articles to profl. publs. Office: Med Imaging Specialists 6 Kinglet Dr Shrewsbury MA 01545

PAREJA-HEREDIA, DIEGO, mathematics educator, bookseller consultant; b. San Pablo, Colombia, Dec. 23, 1939; s. José Euclides and Rosa Victoria (Heredia) Pareja; m. Neira Cerón de Pareja, July 30, 1966; children: Sandra Natalia, Leslie Sofía, Mauricio. Licenciado, U. Libre, Bogotá, Colombia, 1966; MS, U. Colo., 1972. Prof. math. U. Quindío, Armenia, Colombia, 1967—; vis. prof. NYU, 1981-82; cons. Librería Primavera, Armenia, Colombia, 1990—. Author: El Análisis Matemático, 1973, Historia de las Matemáticas, 1979; editor: Jour. Memorias del Seminario Interno, 1986. Recipient ICETEX scholarship, 1970, NSF scholarship, 1973. Mem. Math. Assn. Am., Am. Math. Soc., Nat. Geog. Soc. Home: Cra 15 No 12N32, Armenia Quindio Columbia Quindío, South America Office: Librería Primavera, PO Box 1004, Armenia Colombia

PARENT, EDWARD GEORGE, biochemist; b. Queens, N.Y., Mar. 10, 1957; s. George F. and Anna S. (Shidlaskas) P.; m. Lucia C. Burbano, Aug. 11, 1990; 1 child, Hans. BS in Biology, SUNY, Old Westbury, 1983. Sr. rsch. asst. Biomatrix Inc., Ridgefield, N.J., 1983—. Contbr. articles to profl. jours. including Jour. Orthopedic Rsch., Jour. Biomed. Materials Rsch., Jour. Cosmetic Pharmac. Appl. of Polymer, Jour. Biomed. Mat. Rsch., Soc. for Biomaterials. Home: 105 Vreeland Rd West Milford NJ 07480

PARENT, RICHARD ALFRED, toxicologist, consultant; b. Lynn, Mass., Jan. 16, 1935; s. Alfred Joseph and Lillian Anita (Gagnon) P.; m. Eileen Frances Crutti, June 26, 1985. BS, U. Mass., 1957; MS, Northeastern U., 1959; PhD, Rutgers U., 1963. Diplomate Am. Bd. Toxicology; cert. regulatory affairs, 1992. Sr. rsch. chemist Am. Cyanamid Co., Bound Brook, N.J., 1959-69; staff toxicologist Xerox Corp., Rochester, N.Y., 1969-79; v.p., dir. Food & Drug Rsch. Labs., Waverly, N.Y., 1979-82; dir. life scis. Gulf South Rsch. Inst., New Iberia, La., 1982-84; pres., chief exec. officer Consultox Ltd., Baton Rouge, 1984—; cons. Delta Labs., Rochester, 1969-70, Internat. Union Airline Flight Attendants, Rochester, 1977-79; founder, chmn. Roundtable of Toxicology Cons., Washington, 1984—. Editor: (series) Comparative Biology of the Normal Lung, 1992; editor, founder Acute Toxicity Data jour., 1989—; N.Am. editor Jour. of Applied Toxicology, 1980—; mem. editorial bd. Toxicology Methods jour. 1990—; sect. editor Jour. of Am. Coll. of Toxicology jour., 1988—; contbr. over 60 articles to profl. jours. Recipient NSF fellowships, 1958-59, 60-61. Mem. ASTM (chmn. com. occupational health and safety 1979-84, Outstanding Svc. award 1983)), Soc. Toxicology, Am. Coll. Toxicology, European Soc. Toxicology, French Soc. Toxicology, Internat. Soc. Study of Xenobiotics, Soc. Biomaterials, Regulatory Affairs Profl. Soc. Achievements include 17 patents. Home: 135 Knox Hill Dr Baton Rouge LA 70810 Office: Consultox Ltd 7920 Wrenwood Blvd St D Baton Rouge LA 70809

PARIENTE, RENÉ GUILLAUME, physician, educator; b. La Marsa, Tunisia, Sept. 1, 1929; s. Jules and Vera (Guttieres) P.; m. Dominique Savary, Dec. 26, 1971; children: Pierre, David, Benjamin. MD, U. Paris, 1962. Prof. medicine U. Paris, 1966—, head dept. intensive care and chest disease; bd. dirs. Inst. Nat. du Recherche Med. Mem. Soc. Française de Cardiologie, Soc. de Pathologie Respiratoire, Soc. de la Tuberculose, Soc. Française de Microscopie Electronique, Am. Soc. Chest Physicians, N.Y. Acad. Sci., Am. Thoracic Soc. Home: 12 rue de la Neva, 75008 Paris France Office: 100 G LeClerc Hosp, Beaujon Clichy, France

PARIKH, DILIP, quality assurance engineer; b. Bombay, India, Oct. 26, 1948; came to U.S., 1968; s. Jamnadas H. and Savita J. P. BSChE, Lowell Tech. Inst., 1973; MS in Pulp & Paper Engring., U. Lowell, 1980. Quality engr. Courier Corp., Lowell, Mass., 1973-80; rsch. staff BASF Corp., Holland, Mich., 1982; quality assurance mgr. Queen's Group, Indpls., 1982-85; dir. quality & tech. svcs. & safety Ft. Dearborn Lithograph Co., Niles, Ill., 1985—; bd. dirs. Statis. Process Control Users Group, Chgo.; presenter in field. Bd. dirs. Sci. Profls. Soc., Chgo., 1991-92; vol. ARC, Indpls., 1982-85, Jaycees, Holland, 1981-82. Mem. AICE, ASTM, Am. Soc. Quality Control, Tech. Assn. Pulp & Paper. Office: Ft Dearborn Lithograph 6035 Gross Point Rd Niles IL 60714

PARIKH, HEMANT BHUPENDRA, chemical engineer; b. Baroda, India, May 11, 1951; s. Bhupendra M. and Nirmala B. (I. Chokshi) P.; m. Sushma H., May 1, 1977; children: Deepa, Riki. BE in Chem. Engring., MSU. of Baroda, India, 1973; MS in Chem. Engring., Bklyn. Poly., 1980; MS in Energy Sci. (hon.), NYU, 1980; MS in Computer Sci. (hon.), Fairleigh Dickinson U., 1987. Profl. engr. N.J. Process engr. GSFC, Baroda, 1973-83; process, project engr. Stepan Co., Maywood, N.J., 1983-90; sr. process engr. Henkel Corp., Harrison, N.J., 1990-91, Internat. Splty. Products (subs. GAF Chems.), Wayne, N.J., 1991--. Mem. Am. Inst. Chem. Engring., Inst. Soc. Am. Achievements include development of new sunscreen and new lubricants. Home: 35 Oxford Ln Harriman NY 10926

PARIKH, MANOR MADANMOHAN, mechanical engineer; b. Hydrabad, India, Aug. 12, 1964; came to U.S., 1988; s. Madanmohan V. Parikh and Malati P. Dalal; m. Neelima H. Shah, Feb. 13, 1990. BSME, L.D. Coll. Engring., Ahmedabad, India, 1988; MSME, Stevens Inst. Tech., 1991. Rsch. asst. Stevens Inst. Tech., Hoboken, N.J., 1988-91; sr. devel. engr. Wallace and Tiernan div. Northwest Water Plc., Belleville, N.J., 1991—. Mem. ASME, Soc. Mfg. Engrs., Alumni Gujarat Law Soc. (v.p. 1987). Hindu. Achievements include design for assembly of analysis software. Home: 37D Colfax Manor Roselle Park NJ 07204 Office: Wallace and Tiernan 25 Main St Belleville NJ 07109

PARIS, DAVID ANDREW, dentist; b. Milw., Jan. 16, 1962; s. John Baptistia and Geraldine Louella (Grosso) P. BA, UCLA, 1985, DDS, 1989. Oral surgery extern VA, Phoenix, 1989; primary practitioner Aids Project L.A. Dental Clinic, 1990—; assoc. M. Marchese D.D.S., Sun Valley, Calif., 1990-92, D. Pickrell DMD, West Hollywood, Calif., 1992—. Mem. ADA, Calif. Dental Assn., San Fernando Valley Dental Assn., Acad. Gen. Dentistry, Delta Sigma Delta. Avocations: cello, Italian language.

PARIS, TANIA DE FARIA GELLERT, information technology service executive; b. Santo André, Brazil, Nov. 26, 1951; d. Rubens and Dirce (Zuccon) Faria; m. Robert Gellert Paris Jr., July 8, 1972; children: Tabata, Natasha. BS in math., Fundação Santo André, Brazil, 1972. Jr. programmer GM Brazil, São Caetano, Brazil, 1970-72, sr. programmer, 1972, systems analyst, 1972-75, sr. systems analyst, 1975, systems analyst mgr., 1975-82, mgr. info. ctr., 1982-83, mgr. info. ctr. and office, 1983-85; dir. GM account EDS Brazil, São Caetano, 1985-88; mktg. dir. EDS Brazil, São Paulo, Brazil, 1988-93, v.p. sales and mktg. south and cen. Am., 1993—. Home: Rua Alm Tamandaré 177 Apt 51, 09040 Santo André SP Brazil Office: EDS Brazil Ltd, Ave Juscelino Kubitschek 1830 Torre I, 04543 São Paulo Brazil

PARISER, ROBERT JAY, dermatologist; b. Norfolk, Va., July 22, 1948; s. Harry and Alice (Plon) P.; m. Mary Margaret Shenkenberger, May 13, 1978; children: Kenneth Charles, Marlene Sara, Andrew Phillip. BA, Princeton U., 1970; MD, Med. Coll. Va., 1974. Intern Med. Coll. Va., Richmond, 1974-75; pvt. practice Norfolk, 1978—; resident in dermatology U. Miami, Fla., 1975-78; asst. prof. dept. pathology Ea. Va. Med. Sch., Norfolk, assoc. prof. dept. medicine. Editor: Primary Care, 1989. Mem. Am. Acad. Dermatology, Am. Soc. Dermatopathology (assoc.), Va. Dermatological Soc. (v.p. 1992-93), Phi Beta Kappa, Alpha Omega Alpha. Office: 601 Medical Tower Norfolk VA 23507

PARISH, RICHARD LEE, engineer, consultant; b. Kansas City, Mo., May 31, 1945; s. Charles Lee and Ruth (Duncan) P.; m. Patricia Ann Erickson, June 2, 1968; children: Christie Lynn, Kerry Anne. BS in Agrl. Engring., U. Mo., 1967, MS in Agrl. Engring., 1968, PhD, 1970. Registered profl. engr., Ohio. Asst., then assoc. prof. engring. Univ. Ark., Fayetteville, 1969-74; mgr. mech. research and devel. O.M. Scott & Sons Co., Marysville, Ohio, 1974-83; assoc. prof., then prof. La. State U., Baton Rouge, 1983—; pvt. practice engring. cons. Baton Rouge, 1984—; cons. O.M. Scott, Marysville, Ohio, 1988-90, Sierra Chem. Com., Milapas, Calif., 1989-93, Lesco, Rocky River, Ohio, Am. Plant Food Corp., Houston, Marvin Williams Assocs., Columbus, 1986-87, Dickey Machine Works, Pine Bluff, Ark., 1974. Contbr. over 100 articles to profl. jours. and trade pubs.; patentee in field (3). Bd. dirs. Agrl. Missions Found. Recipient Quality award ITT, 1979, Rsch. Dirs. award O.M. Scott Co., 1978; NSF fellow, 1967-69; rsch. grantee Cotton Inc., Raleigh, N.C., 1970-74, 91-93, La. Dept. Natural Resources, 1985-87, Italian Trade Commn., 1988-90. Mem. Am. Soc. Agrl. Engrs. (chmn. agrl. chem. application com. 1982-83, power and machinery div. program com. 1986-87), La. Vegetable Growers Assn., Am. Soc. Hort. Sci. Republican. Baptist. Avocations: gardening, woodwork, bicycling. Home: 834 Forge Ave Baton Rouge LA 70808 Office: La State U Dept Biol and Agrl Engring Baton Rouge LA 70803-4505

PARISI, GIORGIO, physicist, educator. Prof. dept physics U. Rome, Italy. Recipient Boltzmann medal Internat. Union Pure and Applied Physics, Sweden, 1992. Office: Univ of Rome-Dept of Physics, Piazzale Aldo Moro 5, I-00185 Rome Italy*

PARIZA, MICHAEL WILLARD, research institute executive, microbiology and toxicology educator; b. Waukesha, Wis., Mar. 10, 1943; married; 3 children. BSc in Bacteriology, U. Wis., 1967; MSc in Microbiology, Kans. State U., 1969, PhD in Microbiology, 1973. Postdoctoral trainee McArdle Lab. for Cancer Rsch. U. Wis., Madison, 1973-76, asst. prof. Food Rsch. Inst. dept. Food Microbiology and Toxicology, 1976-81, assoc. prof., 1981-84, prof., 1984—, assoc. dept. chmn., 1981-82, dept. chmn. food microbiology and toxicology, 1982—, dir. Food Rsch. Inst., 1986—, disting. prof., 1993—; with Wis. Clin. Cancer Ctr., Environ. Toxicology Ctr., Dept. Nutritional Scis., Dept. Food Sci.; mem. Inst. Medicine's Food Forum; mem. com. on comparative toxicity naturally occurring carcinogens NAS; trustee Internat. Life Scis. Inst.-N.Am., 1986—. With U.S. Army, 1969-71. Office: U Wis Food Rsch Inst 1925 Willow Dr Madison WI 53706-1187

PARK, BYEONG-JEON, engineering educator; b. Chill-Jeon, Chindo, Korea, May 5, 1934; s. Ho-June and Joo-Hyon (Kim) P.; m. Sung-Tcho, Oct. 27, 1956; children: Kwang-Yeol, Kwang-Hee, Kyu-Yeol, Yoo-Hee. BE, Seoul Nat. U., 1956; ME, Chonbuk Nat. U., 1975; D Engring., Chosun U., 1978. Archiect diplomate. Lectr. Mokpo (Chonnam) Tech. High Sch., 1957-58, Coll. Engring., Chosun U., Kwangju, Chonnam, 1958-61, Chonnam Nat. U., 1962-63, Coll. Engring., Chonbuk Nat. U., Chonju, 1963-64; asst.

prof. Coll. Engring., Chonbak Nat. U., Chonju, 1965-68; assoc. prof. Coll. Engring., Chonbuk Nat. U., Chonju, 1968-71, prof., 1971—; dean grad. sch. environ. studies Chonbuk Nat. U., Chonju, 1991—; guest prof. Tokyo U., 1981-82; dir. Archtl. Inst. Korea, Seoul, 1966-76, Korea Inst. Fire Sci. and Engring., Seoul, 1989—; mem. Archtl. Inst. Japan, Tokyo, 1979—; head Inst. Urban and Environ. Studies, Chonbuk Nat. U., Chonju, 1990—; spl. work prof. Ministry of Constrn., Republic of Korea, 1989—; mem. adv. com. Structures of Sound Insulation Ministry of Constrn., Republic of Korea, 1990—. Author: Building Equipments, 1981, Enivronmental Science for Architect, 1985, Science of Dewelling, 1987, Building Acoustics, 1989. Mem. com. Chonju City Planning Com., 1983—, Com.for Energy Conservation, Chonbuk Provincial Govt., 1987—, Traffic Policy Com. of Chonju City, 1990—, Chonbuk Provincial Govt. Adv. Com., 1980. Mem. Acoustical Soc. Korea (v.p., dir. 1985—), Soh-Woo Club. Home: 167-186 2Ga Tokjindong, Tokjinku Chonju, Chonbuk 560-190, Republic of Korea Office: Chonbuk Nat U, 664-14 1 Ga Tokjindong, Tokjinku Chonju 560-756, Republic of Korea

PARK, BYIUNG JUN, textile engineer; b. Seoul, Republic of Korea, Feb. 28, 1934; came to U.S., 1954; s. Kyung Hak and Tansil (Kiu) P.; m. Chunghi Hong, June 28, 1958. BS, R.I. Sch. Design, 1958; MS, MIT, 1961, Mech. Engr., 1963; PHD, Leeds U., Eng., 1966. Rsch. asst. MIT, Cambridge, 1961-63; v.p. Consumer Testing Labs., Inc., Canton, Mass., 1966-84, sr. v.p., 1984-86; pres. Merchandise Testing Labs., Brockton, Mass., 1986—. Contbr. articles to profl. jours. Recipient Bronze medal No. Textile Assn., Boston, 1958; rsch. fellow Internat. Wool Secretariat, Leeds, 1964-66. Mem. ASTM, Textile Inst., Am. Assn. Textile Chemists and Colorists, Am. Assn. Textile Technology, Am. Soc. for Quality Control, Indls. Fabrics Assn. Internat. Achievements include research in understanding mechanical behavior of textile assemblages. Office: Merchandise Testing Labs 244 Liberty St Brockton MA 02401-5522

PARK, BYUNG-SOO, chemistry educator, dean; b. Kumi, Kyungbuk, Korea, Oct. 16, 1930; s. So-Joon and Duk-Im (Yu) P.; m. Young-Hee Choi, May 12, 1952 (dec. July 1978); children: Kyung-Ae, Ki-Jung, Kun-Jung, Dong-Jung, Jung-Ae; m. Hwa-Ja Youn. BS, Kyungpook Nat. U., Taegu, Republic of Korea, 1955, PhD, 1989; MS, Yeungnam U., Kyungsan, Republic of Korea, 1982. Tchr. Andong (Republic of Korea) Girl's High Sch., 1961-66; prof. Andong Tchr.'s Coll., 1966-78, Andong Jr. Coll., 1978-79; prof. Andong Nat. U., 1979—, dean of students, 1979-80, head dept. of chemistry, 1983-86, dean of student affairs, 1983-84, dean Grad. Sch., 1991—; exch. prof. Kyungpook Nat. U., 1987-88. Advisor Andong City Policy Adv. Com., 1981-83. Mem. Chem. Soc. Korea (councillor 1987-90, bd. dirs. Kyungpook chpt. 1987-88, chmn. Taegu and Kyungbuk br. 1992—). Avocations: tennis, floriculture. Home: 1-602, Hyundai Apt, Taehwa-dong, Andong 760-290, Republic of Korea Office: Andong Nat U Dept Chemistry, 388, Songchun-dong, Andong 760-749, Republic of Korea

PARK, CHANG HWAN, physicist, educator; b. Taegu, Korea, Jan. 11, 1946; came to U.S., 1975; s. Moosik and Pilsoo (Kim) P.; m. Young Ai, May 31, 1975; children: Soonki, Eunhe, Inki. BS in Physics, Younsei U., Seoul, South Korea, 1974; MS in Physics, U. Oreg., 1978; PhD in Physics, U. Nebr., 1984. Postdoctoral fellow dept. radiology Med. Sch. Yale U., New Haven, 1985-87; med. physicist dept. radiation oncology Ohio Valley Med. Ctr., Wheeling, W.Va., 1987-89; asst. prof. physics Med. Sch. U. Louisville, 1989—. Contbr. articles to profl. publs. Mem. Am. Assn. Physics in Medicine. Presbyterian. Home: 7413 Timjoe Dr Louisville KY 40242 Office: U Louisville Med Sch 529 S Jackson St Louisville KY 40292

PARK, JON KEITH, dentist, educator; b. Wichita, Kans., May 26, 1938; s. William Ray and Eleanor Jeanette (Cunningham) P.; DDS, U. Mo., 1964; BA, Wichita State U., 1969; MS in Dental Hygiene Edn., U. Mo., 1971; MS in Oral Pathology, U. Md., 1982; cert. in dental radiology U. Pa. Sch. Dental Medicine, 1982. Diplomate Am. Bd. Oral and Maxillo-facial Radiology. Pvt. practice dentistry, Wichita, 1964-67; chmn. dept. dental hygiene Wichita State U., 1967-72; assoc. prof. oral diagnosis, dir. oral radiology Balt. Coll. Dental Surgery, U. Md., 1972—; program dir. U. Md. dental externship, 1974-77; lectr. Essex C.C., Harford County C.C.; cons. in radiology VA Hosp.; mem. Md. State Radiation Control Adv. Bd., 1981—; mem. Ute Pass Hist. Soc.; chmn. devel. com. Introduction to Basic Concepts in Dental Radiography, Dental Assisting Nat. Bd., Inc., Am. Dental Assts. Assn., 1991; editor Am. Acad. Oral and Maxillofacial Radiology Newsletter. Recipient U. Md. Media Achievement award, 1977, 78. Fellow Am. Acad. Dental Radiology; mem. ADA, Md. State Dental Assn., Balt. City Dental Soc. (ad hoc com. radiation safety), Am. Acad. Oral Pathology, Am. Acad. Oral and Maxillofacial Radiology (editor newsletter), Orgn. Tchrs. Oral Diagnosis, Am. Theater Organ Soc., Kans. Dental Hygienists Assn. (hon.), Balt. Music Club, Am. Acad. Dental Radiology (ednl. standards com.), Am. Assn. Dental Schs., Internat. Assn. Dental and Maxillofacial Radiology, Balt. Opera Guild, Engring. Soc. Balt., Met. Opera Guild, Balt. Symphony Orch. Assn., Ute Pass Community Assn., Univ. Club, Omicron Kappa Upsilon, Psi Omega. Episcopalian. Patentee pivotal design dental chair.

PARK, KWAN IL, energy engineer; b. Seoul, South Korea, Apr. 20, 1950; s. Do Jin and Young Kil (Bank) P.; m. Mi Ja Lim, May 7, 1983; 1 child, Chan Mi. Degree in phys. engring., Korea U., Seoul, 1973; postgrad., South Bailo U., L.A., 1991. Researcher dept. clean and new energy Korean Inst. Energy and Resources, Seoul, 1976-81; communal pres. Korea Wind Energy System, Inc., Seoul, 1981-86; rep. L.A. br. Korea Wind Energy System, Inc., 1986—; v.p., tech. advisor Unltd. Sci. Am., Inc., Garden Grove, Calif., 1988—; tech. advisor Inst. Am., San Diego, 1989—. Author: Future Energy, 1986; contbr. articles to profl. publs. Co-chmn. Korea Solar Energy Assn., Seoul, 1979-89. Sgt. South Korean Air Def. Army, 1973-76. Mem. Assn. Energy Engrs. (sr.). Republican. Roman Catholic. Achievements include study on the analysis of energy use and technological effect of energy, flow measurement using piezoelectric vortex sensors, mathematical modeling and simulation of biological conversion of solar energy, electric storage system's economic effect on load control of power generation, others; patents for Paper Pot, Sectional Brick, others. Office: 6734-5 Los Verdes Dr Rancho Palos Verdes CA 90274

PARK, MIN-YONG, human factors and safety engineer, educator; b. Kimcheon, South Korea, Feb. 21, 1955; came to the U.S., 1981; s. Heechang and JeongBoon (Park) P.; m. Mihwan Kim, June 28, 1981; children: Stephen, Veronica, John, Caroline. MS, Hayang U., Seoul, 1981, Okla. State U., 1986, Va. Poly. Inst., 1989; PhD, Va. Poly. Inst., 1991. Grad. rsch. and teaching assoc. Okla. State U., Stillwater, 1983-86, Va. Poly. Inst., Blacksburg, 1987-91; asst. prof. N.J. Inst. Tech., Newark, 1991—; assoc. dir. occupational safety and health program N.J. Inst. Tech., 1991—. Contbr. articles to profl. jours. Recipient Rsch. awards Va. Poly. Inst., 1990, 91. Mem. Acoustical Soc. Am., Am. Soc. for Quality Control, Am. Soc. for Safety Engrs., Nat. Hearing Conservation Assn., Human Factors Soc., Inst. Indsl. Engrs. (Rsch. award 1990), Alpha Pi Mu, Phi Kappa Phi. Roman Catholic. Avocations: music, photography, tennis. Home: 7 Boynton Dr Livingston NJ 07039 Office: NJ Inst Tech Dept Indsl Engring Newark NJ 07102

PARK, O OK, chemical engineering educator; b. Pohang, Republic of Korea, Mar. 10, 1954; s. Jae-dong and Byung-nam (Kim) P.; m. Dong Hwa Lee, June 2, 1979; children: Young Gul Park, Young Sun Park, Young Yoon Park. MSChemE, Kaist, Seoul, 1978; PhDChemE, Stanford (Calif.) U., 1985. Asst. dir. Ministry of Commerce and Industry, Seoul, 1978-81; asst. prof. Dept. of Chem. Engring., Kaist, Seoul, 1985-90, assoc. prof., 1990—; vis. prof. Dept. of Applied Phsycis, Nayoya (Japan) U., 1990; vis. scientist Lucky Polymer Rsch. Ctr., Taeduk, Korea, 1991, IBM Almaden Rsch. Ctr., San Jose, 1990-91; planning dir. The Polymer Soc. of Korea, Seoul, 1992—; sec. Polymen Processing Soc. Regional Meeting '90, Seoul, 1990, Internat. Union of Pure and Applied Chemistry Polymer Symposium '89, Seoul, 1988-89. Contbr. articles to profl. jours. Fellow Korean Inst. of Chem. Engring.; mem. The Polymer Soc. of Korea (bd. dirs. 1986—, editorial com. 1987-90), Am. Chem. Soc., Soc. of Rheology. Achievements include devel. of a new finisher for PET polymerization; rsch. on MWD in condesation polymerization, irreversible annealing of liquid crystalline polymers, electrohydrodynamic response of polymer solutions. Home: Hyundai Apt 101-905, 431-6 Doyongdong Yusongku, Dorjon 305-340, Korea Office: Dept Chem Engring Kaist, 373-1 Kusung-dong Yusung-gu, Taejon 305-701, Republic of Korea

PARK, RODERIC BRUCE, university official; b. Cannes, France, Jan. 7, 1932; came to U.S., 1932; s. Malcolm Sewell and Dorothea (Turner) P.; m. Marijke DeJong, Aug. 29, 1953; children: Barbara, Marina, Malcolm. AB, Harvard U., 1953; PhD, Calif. Inst. Tech., 1958. Postdoctoral fellow Calif. Inst. Tech., 1958, Lawrence Radiation Lab., Berkeley, Calif., 1958-60; prof. botany U. Calif., Berkeley, 1960-89, prof. plant biology, 1989-93, prof. emeritus, 1993—; chmn. dept. instrn. in biology U. Calif., 1965-68; provost, dean U. Calif. (Coll. Letters and Sci.), 1972-80, vice chancellor, 1980-90; pres. Brickyard Cove Harbors, Inc., 1975-77; dir. William Kaufmann, Inc., 1976-86; mem. corp. Woods Hole Oceanographic Instn., 1974-80; mem. Harvard Vis. Com. on Biochemistry and Molecular Biology, 1990—. Co-author: Cell Ultrastructure, 51967, Papers on Biological Membrane Structure, 1968; Biology editor, W.H. Freeman & Co., 1966-74; Contbr. articles to profl. jours. Trustee Athenian Sch., 1980—, U. Calif.-Berkeley Found., 1986-90, Jepson Endowment, 1992—; bd. dirs. Assoc. Harvard Alumni, 1976-79; bd. overseers Harvard U., 1981-87; mem. exec. com. Coun. Acad. Affairs, 1986-90, chmn., 1988-89; mem. exec. com. Nat. Assn. State Univs. and Land Grant Colls., 1988-90; mem. vis. com. Arnold Arboretum, 1981-88, chmn., 1986-88; acting dir. Univ. and Jepson Herbaria, 1991-93. Recipient New York Bot. Gardens award, 1962. Fellow AAAS; mem. Am. Soc. Plant Physiologists, Am. Bot. Soc., Am. Soc. Photobiology, Danforth Assn. (pres. San Francisco chpt. 1972), Richomond Yacht Club (commodore 1972, dir. Found. 1992—), Transpacific Yacht Club, Explorers Club. Home: 531 Cliffside Ct Richmond CA 94801-3766 Office: U Calif Koshland Hall Berkeley CA 94720

PARK, SANG-CHUL, molecular biologist, educator; b. Kwang-Ju, Korea, Mar. 25, 1949; s. Son-Hong and Young-Rye (Kang) P.; m. Soo-Hyang Lee, June 2, 1974; children: Kyung-Hee, Chang-Hee. MD, Seoul Nat. U., 1973, MS, 1975, PhD, 1980. Asst. Seoul Nat. U., 1973-77, lectr., 1980-84, asst. prof. Coll. of Medicine, 1984-88, assoc. prof., 1988—; chief Naval Hosp., Seoul, 1977-80; vis. fellow NIH, Bethesda, Md., 1982-84; div. chief Rsch. Inst. Molecular Biology, Seoul, 1985-89, Rsch. Inst. Health Sci., 1988—; bd. dirs. Korean Rsch. Found. Health Sci., Seoul, 1990—, Internat. Trade Rsch., 1991—. Contbr. articles to profl. jours.; patentee in field. Lt. comdr. Korean Navy, 1977-80. Named Best Scientist of the Yr. Nat. Press Club, Seoul, 1990, Best Achievement Ministry of Sci., Seoul, 1990; Receipent KumHo award, 1991. Mem. Korean Soc. Gerontology (treas.), Korean Soc. Biochemistry (gen. sec. 1990), Korean Soc. Molecular Biology (gen. sec. 1991), Korean Environ., Mutagen Soc. (gen. sec. 1987-89), Korean Genetic Soc. (bd. dirs. 1985—), Korean Soc. Microbiology (bd. dirs. 1985—), Korean Soc. Toxicology (bd. dirs. 1985—), Korean Cancer Soc. (bd. dirs. 1987—). Avocation: climbing. Home: 30-1205 Kaenari Apt, Ryuk-Sam Dong Kang Nam Ku, Seoul Republic of Korea Office: Seoul Nat U Coll Medicine, 28 Yun Kun Dong Chongnoku, Seoul Republic of Korea 110-744

PARK, THOMAS JOSEPH, biology researcher, educator; b. Balt., June 8, 1958; s. Lee Crandall and Barbara Ann (Merrick) P.; m. Stephanie Suzanne Reynolds, June 22, 1985. BA, Johns Hopkins U., 1982; PhD, U. Md., 1988. Vis. scientist Coll. of France, Paris, 1988-89; rsch. fellow U. Tex., Austin, 1989—. Contbr. chpt. to book, articles to Jour. of Neurosci., Jour. of Comparative Psychology, Hearing Rsch. Grantee NIMH, 1986, Nat. Ctr. Sci. Rsch., Paris, 1988, NIH. Mem. AAAS, Soc. for Neurosci., Assn. for Rsch. in Otolaryngology. Office: U Tex Dept Zoology Austin TX 78712

PARK, WON-HOON, chemical engineer; b. Seoul, Korea, Feb. 10, 1940; s. Hyo-Hum and Soon-Ok (Yoo) P.; m. Oksoo Han, Sept. 25, 1971; children: Suzanne, Thomas. BE, Seoul Nat. U., 1964; PhD, U. Minn., 1971. Postdoctoral fellow U. Houston, 1971-72; vis. fellow SUNY, Buffalo, 1974-75; head lab. Korea Inst. Sci. & Tech., Seoul, 1972-81; prof. Sung Kyun Kwan U., Seoul, 1981-83; v.p. Korea Inst. Energy & Resource, Taejon, 1983-86; dir. Korea Inst. Sci. & Tech., Seoul, 1988-93; pres. Sci. and Tech. Policy Inst., Seoul, 1993—; sec. gen. World Acad. Conf. of Seoul Olympiad '88, Seoul, 1987-88; chmn. Korea Sci. and Engring. Found., 1984-88; advisor Presdl. Commn. on Sci. and Tech., Seoul, 1989-90. Author: Fluidization - Japan and Korea, 1987; contbr. articles to profl. jours. With Korean Army, 1961-63. Recipient Nat. Medal (Sokryu) Korean Govt., 1980, Acad. Grand Prize, Kyunghyang Daily News, 1986, Citation, Pres. of Korea, 1989. Mem. Korean Inst. Chem. Engrs., Korean Soc. Energy Engring., Korean Chem. Soc., Korean Solar Energy Soc. (pres. 1990-91). Home: 17-29 Kuki-Dong Chongro-Ku, Seoul Republic of Korea 110-011 Office: Korea Inst Sci & Tech, Sungbuk-Ku 39-1 Hawolgog-Dong, Seoul Republic of Korea 136-791

PARK, YONG-TAE, chemistry educator; b. Taegu, Korea, Nov. 30, 1941; s. Kwang-Jun and Young-Jo (Kim) P.; m. Kyung Ja Lee, Dec. 19, 1968; children: Young A., Won-Ho, Han-Ho. BS, Kyungpook Nat. U., 1965; PhD, U. Nev., 1977. Asst. prof. Kyungpook Nat. U., Taegu, Korea, 1969-86, prof., 1986-92, chmn. dept. chemistry, 1986-88, vice dean Coll. of Natural Sci., 1992—. Author: Basic Science Laboratory, 1980, General Chemistry, 1990. Rsch. grant Ministry of Sci., Seoul, 1988, 91. Mem. Korean Chem. Soc. (bd. trustees 1989-92). Achievements include photoisomerization of bilirubin model compounds, photooxygenation of bilirubin model compounds, photocyclization of 2-halopyridinium salts and benzanilides, hydrogen and oxygen generation with artificial solar energy storage. Office: Kyungpook Nat U, Sankyuk dong, Taegu 702-701, Republic of Korea

PARKE, DAVID W., II, ophthalmologist, educator, healthcare executive; b. Columbus, Ohio, May 19, 1951; s. David William Parke and Eunice Joyce Erikson; m. Julie Diane Thorne, Sept. 15, 1975; children: David W. III, Laura Thorne, Lindsey Diane. AB, Stanford U., 1973; MD, Baylor U., 1977. Diplomate Am. Bd. Ophthalmology. Asst. prof. Baylor Coll. Medicine, Houston, 1983-90, assoc. prof., 1990-92; prof., chair dept. ophthalmology U. Okla., Oklahoma City, 1992—; pres., CEO McGee Eye Inst., Oklahoma City, 1992—. Active Okla. Econ. Devel. Found., 1992, Okla. Health Ctr. Found., 1992. Fellow Am. Acad. Ophthalmology (assoc. sec. 1983-92, Honor award 1980); mem. Assn. Univ. Profs. Ophthalmology, Retina Soc., Alpha Omega Alpha. Office: Dean A McGee Eye Institute 608 Stanton L Young Blvd Oklahoma City OK 73104

PARKE, ROSS DUKE, psychology educator; b. Huntsville, Ont., Can., Dec. 17, 1938. BA, U. Toronto, 1962, MA, 1963; PhD, U. Waterloo, 1965. From asst. to prof. psychology dept. U. Wis., Madison, 1965-71; chief social devel. sect. Fels Rsch. Inst., Yellow Springs, Ohio, 1971-75; prof. psychology U. Ill., Champaign-Urbana, 1975-90; prof. psychology U. Calif., Riverside, 1990—, dir. Ctr. for Family Studies, 1992—. Author: Fathers, 1981; co-author: Child Psychology, 1993; co-editor: Family - Peer Relationships, 1992; contbr. over 100 articles and chpts. to jours. and books. Grantee Office of Child Devel., 1978-80, NICHD, 1980-90, NSF, 1990—, Spencer Found., 1991. Fellow APA (pres. div. devel. psychology 1991, assoc. editor Child Devel. 1973-77, editor Devel. Psychology 1987-92), Am. Psychol. Soc.; mem. Soc. for Rsch. in Child Devel. (exec. com. 1978-83). Office: U Calif Dept Psychology Riverside CA 92521

PARKER, CHARLES DEAN, JR., engineering company executive; b. Gaffney, S.C., June 25, 1954; s. Charles Dean and Bobbie Jo (Humphries) P.; m. Cheryl Maude Thomas, Feb. 29, 1992; 1 child, Dustin Beau. BS in Engring., U. S. Merchant Marine Acad., 1976; MBA, The Citadel, 1984. Registered profl. engr., S.C. Test engr. Ingalls Shipbldg., Pascagoula, Miss., 1976-77; with Life Cycle Engring., Inc., Charleston, S.C., 1978—, program mgr., dir. contracts, corp. treas. Mem. NSPE, S.C. Soc. Profl. Engrs., Treasury Mgmt. Assn. Office: Life Cycle Engring Inc Ste 300 1 Poston Rd Charleston SC 29407

PARKER, DAVID, chemistry educator; b. Consett, Durham, Eng., July 30, 1956; s. Joseph William and Mary (Hill) P.; m. Fiona Mary MacEwan, July 27, 1979; children: Eleanor, Julia Rose, Philip. BA, Oxford (Eng.) U., 1978, D. Philosophy, 1980. NATO fellow U. Louis Pasteur de Strasbourg (France), 1980-81; univ. lectr. U. Durham (Eng.), 1982-89, sr. lectr., 1989-92, prof., 1992—; cons. Celltech Ltd., Slough, U.K., 1985—. Contbr. numerous articles to profl. jours.; patentee in field. Recipient Rsch. prize in organic chemistry Imperial Chem. Industries, 1991. Fellow Royal Soc. Chemistry (Perkin coun. 1990-93, Corday-Morgan medal and prize 1987, Hickinbottom fellow 1988, 89); mem. Am. Chem. Soc. Avocations: cricket, soccer, golf. Office: U Durham Dept Chemistry, South Rd, Durham DH1 3LE, England

PARKER, DAVID HIRAM, physicist, electrical engineer; b. Dothan, Ala., July 10, 1951; s. Marvin and Lois (Beckham) P.; m. Ginger Lois Beck, Dec. 14, 1975; children: Natalie Nicole, Elizabeth Ann. BEE, Auburn U., 1974, MS in Physics, 1982. Registered profl. engr., Ala. With So. Railway System, Atlanta, 1972-74; project engr. Owens-Corning Fiberglas, Anderson, S.C., 1976-78; constrn. engr. Owens-Corning Fiberglas, Amarillo, Tex., 1978-79; physicist U.S. Army Metrology Standards, Redstone Arsenal, Ala., 1983; sr. R & D engr. B.F. Goodrich R & D Ctr., Brecksville, Ohio, 1983-87; sr. engr. Quantex, Rockville, Md., 1988-90, Nat. Radio Astronomy Observatory, Green Bank, W.Va., 1990—. Contbr. articles to profl. jours. Mem. IEEE, Optical Soc. Am., Soc. Photo-Optical Instrumentation Engrs., Sigma Xi (Rsch. award 1983). Achievements include patents for method of simulating tire tread noise and apparatus, improving tread noise by relative rotation of a rib and simulating the effect. Home and office: Radio Astronomy Observatory Rte 28 Box 2 Green Bank WV 24944

PARKER, EUGENE NEWMAN, physicist, educator; b. Houghton, Mich., June 10, 1927; s. Glenn H. and Helen (MacNair) P.; m. Niesje Meuter, 1954; children—Joyce, Eric. BS, Mich. State U., 1948; PhD, Calif. Inst. Tech., 1951; DSc, Mich. State U., 1975; Doctor Honoris Causa in Physics and Math., Univ. Utrecht, The Netherlands, 1986; Doctor of Philosophy Honoris Causa in Theoretical Physics, U. Oslo, 1991. Instr. math. and astronomy U. Utah, 1951-53, asst. prof. physics, 1953-55; mem. faculty physics U. Chgo., 1955—, prof. dept. physics, 1962—, prof. dept. astronomy and astrophysics, 1967—. Author: Interplanetary Dynamical Processes, 1963, Cosmical Magnetic Fields, 1979. Recipient Space Sci. award AIAA, 1964; Chapman medal Royal Astron. Soc., 1979, Gold medal, 1992; Disting. Alumni award Calif. Inst. Tech., 1980; James Arthur Prize Lecture Harvard-Smithsonian Ctr. Astrophysics, 1986; Karl Schwarzschild award Astronomische Gesselschaft, 1990, Gold medal, British Royal Astronomical Scoeity, 1992. Mem. NAS (H. K. Arctowski award 1969, U.S. Nat. Medal of Sci. award 1989), Am. Astron. Soc. (Henry Norris Russell lectr. 1969, George Ellery Hale award 1978), Am. Geophys. Union (John Adam Fleming award 1968, William Bowie medal 1990), Am. Acad. Arts and Scis., Norwegian Acad. Sci. and Letters. Achievements include development of theory of the origin of the dipole magnetic field of Earth; of prediction and theory of the solar wind and heliosphere; of theoretical basis for the X-ray emission from the Sun and stars. Home: 1323 Evergreen Rd Homewood IL 60430-3410 Office: Lab Astrophysics 933 E 56th St Chicago IL 60637-1460

PARKER, JACK ROYAL, engineering executive; b. N.Y.C., Apr. 25, 1919; s. Harry and Clara (Saxe) P.; m. Selma Blossom, Dec. 8, 1946 (dec. Dec. 1991); children: Leslie Janet, Andrew Charles. Student, Bklyn. Poly. Inst., 1943; D.Sc. (hon.), Pacific Internat. U., 1956. Instr. Indsl. Tng. Inst., 1938-39; engr. Brewster Aero Corp., 1939-40; pres. Am. Drafting Co., 1940; design engr., plan div. Navy Dept., 1941-44; also supr. instr. N.Y. Drafting Inst.; instr. Gasoline Handling Sch., also Inert Gas Sch., U.S.N.T.S., Newport, R.I., 1944-46; cons. Todd Shipyards Corp., 1947-54; tech. adviser to pres. Rollins Coll., 1949-50; v.p. Wattpar Corp., 1947-54; pres. Parco Co. Can. Ltd., 1951-55; pres., dir., chief project mgr. Royalpar Industries, Inc., N.Y.C., 1947-75; chmn. Med. Engrs. Ltd., Nassau, Bahamas, 1957-60; pres. Parco Chem. Systems, Inc., 1965-69; pres., dir. Parco Internat., Inc., 1965-69; pres. Guyana Oil Refining Ltd., S.A., 1966-69, Refineria Peruana del Sur S.A., 1965-68; v.p., dir., founder Refinadora Costarricense de Petreoleo, S.A., 1963-73; pres. Oleoducto Trans Costa Rica, 1970-73, Trans Costa Rica Pipeline Operating Co., 1970-73, Due Diligence, Inc., 1985—; gen. mgr. Kellex power services Pullman Kellogg Co., 1975-77; past pres., sec., dir. Vernitron Corp., Amsterdam Fund, European securities Pub. Co.; Cons., Dominican Republic, 1964, Republic Costa Rica, 1964-65; Cons. Malta Indsl. Devel. Study Co. Ltd., 1965-67, Hambros Bank Ltd., London, 1967-68, Stone & Webster Engring. Corp., Boston, 1957-75, Hambro Am. Bank & Trust Co., N.Y.C., 1968-70; pres. J. Royal Parker Assos., Inc., 1975—, pres. Due Diligence, Inc., 1985—; Internat. Mfg. Centers, Inc., 1977-81, Delaware Valley Fgn. Trade Zone, Inc., 1977-80, Brown & Root (Delaware Valley) Inc., 1980-84; chmn. Summa Engring. Ltd., 1985-86; lectr. One World Club, Cornell U., 1963; chmn. project fin. panel Global Energy Forum '84; chmn., chief exec. officer Export Refinery Western Hemisphere Ltd., Castries, St. Lucia, W.I., 1991., 1991—; chmn., CEO Euro-Siberia Export Refining Corp., Siberia, Russia, 1992, chmn. Russian Trading Corp., Wilmington, Del., 1993—; rep. U.S. Merchant Marine Acad., Kings Point, N.Y., 1993—. Author: Gasoline Systems, 1945; also articles; developer Due Diligence process for project fin. analysis; patentee in field. Mem. adv. bd. Drafting Ednl. Adv. Commn. N.Y.C. Bd. Edn., 1968; founder museum U.S. Mcht. Marine Acad., 1977 (Disting. Service award 1987); pres. Am. Mcht. Marine Mus. Found., U.S. Mcht. Marine Acad., Kings Point, N.Y., 1981-87, pres. emeritus 1987—; Trustee Coll. Adv. Sci., Canaan, N.H. Decorated knight Order St. John of Jersusalem.; recipient Humanitarian award Fairleigh Dickinson U., 1974, Disting. Service award U.S. Merchant Marine Acad. Alumni Found., 1987. Fellow A.A.A.S.; mem. Inst. Engring. Designers (London), Am. Petroleum Inst., Soc. Am. Mil. Engrs., Presidents Assn., Am. Mgmt. Assn., Am. Inst. Chem. Engrs., Am. Inst. Dsgn. and Drafting. Republican. Clubs: Masons; Marco Polo (N.Y.C.); Royal Automobile (London). Achievements include invention of Lazy Golfer; development of Operation Centraport (Limon Costa Rica Internat. Freezone Trade). Home: 1117 Society Hill Blvd Cherry Hill NJ 08003-2421 Office: PO Box 945 Woodcrest Sta Cherry Hill NJ 08003

PARKER, JAMES ROGER, chemist; b. L.A., July 19, 1936. BS, Pomona Coll., 1958; PhD, Iowa State U., 1964. Lab. asst. Ames (Iowa) Lab Atomic Energy Commn., 1958-61; analytical supr. PPG Industries, Natrium, W.Va., 1964-73, Corpus Christi, Tex., 1973-82; agrl. chemist PPG Industries, Barberton, Ohio, 1982-89; infrared spectroscopist PPG Industries, Monroeville, Pa., 1989—. Contbr. articles to profl. jours. Mem. Am. Chem. Soc., Soc. for Applied Spectroscopy, Spectroscopy Soc. Pitts., Soc. for Analytical Chemists Pitts., Phi Lambda Upsilon. Achievements include research in analytical chemistry of metal halides, iodine compounds, alkali metal oxides, coordination chemistry of phosphine oxides, qualititative identifications with proton magnetic resonance spectroscopy. Office: PPG Industries 440 College Park Dr Monroeville PA 15146

PARKER, JOHN R., physician, radiologist; b. Rochester, Minn., Apr. 29, 1967; s. Joseph Corbin and Patricia (Singleton) P. BA, MD, U. Mo., Kansas City. Rsch. asst. U. Tenn. Meml. Hosp., Knoxville, 1985; autopsy technician Truman Med. Ctr., Kansas City, 1990-93. Contbr. articles to Annals of Clin. and Lab. Sci., Archives of Pathology. Organizer 4-H Summer Scholars Med. Terminology, Lakewood Hosp., 1989-91; co-chmn. Impaired Med. Student Coun., 1990-91. Recipient Richardson K. Noback Clin. Excellence award, 1993; Pathology Student fellow Truman Med. Ctr., 1989-90. Mem. AMA, Okla. State Med. Assn., Med. Student Adv. Com., N.Y. Acad. Sci., Mortar Board, Golden Key, Alpha Omega Alpha, Omicron Delta Kappa. Democrat. Office: Okla Meml Hosp Rm INP606 Dept Radiology Oklahoma City OK 73190

PARKER, JON IRVING, ecologist; b. Danville, Pa., Jan. 1, 1944; s. Byron S. and Hilda B. (Cooper) P.; m. Danielle I. Koury, July 2, 1966; children: Jonathan B., Koury A. BS, Bloomsburg State U., 1965; MS, U. Idaho, 1972; PhD, U. N.H., 1974. Tchr. Union Springs (N.Y.) Cen. High Sch., 1965-70; ecologist Argonne (Ill.) Nat. Lab., 1974-81; asst. prof. Lehigh U., Bethlehem, Pa., 1981-86; contr. Parker Contracting, Allentown, Pa., 1986-91; asst. project scientist Woodward-Clyde Cons., Plymouth Meeting, Pa., 1991—. Contbr. articles to profl. jours. Recipient numerous rsch. grants U.S. EPA, Dept. of Energy, NRC, 1974-81. Mem. Am. Chem. Soc., Soc. of Environ. Toxicology and Chemistry. Achievements include documentation of biogeochemical silicon cycle in Lake Michigan; biogeochemical cycles of Zn and Cd in Lake Michigan. Home: 319 Congo Niantic Rd Barto PA 19504 Office: Woodward-Clyde Cons 5120 Butler Pike Plymouth Meeting PA 19462

PARKER, JOSEPHUS DERWARD, corporation executive; b. Elm City, N.C., Nov. 16, 1906; s. Josephus and Elizabeth (Edwards) P.; m. Mary Wright, Jan. 15, 1934 (dec. Dec. 1937); children: Mary Wright (Mrs. Henry Avon Perry), Josephus Derward; m. Helen Hodges Hackney, Jan. 24, 1940; children: Thomas Hackney, Alton Person, Derward Hodges, Sarah Helen

(Mrs. Robert Seavey). AB, U. of the South, 1928; postgrad., Tulane U., 1928-29, U. N.C., 1929-30, Wake Forest Med. Coll., 1930-31. Founder, chmn. bd. J.D. Parker & Sons, Inc., Elm City, 1955—, Parker Tree Farms, Inc., 1956—; founder, pres. Invader, Inc., 1961-63; pres. dir. Brady Lumber Co., Inc., 1957-62; v.p., dir. Atlantic Limestone, Inc., Elm City, 1970—; owner, operator Parker Airport, Eagle Springs, N.C., 1940-62; founder, pres. Parkhurst Plantation, Inc., Elm City; pres. Toisnot Farms Inc., Elm City. Author: The Parker Family, 1987, The Red Wedding Gown, 1992. Served to capt. USAF, 1944-47. Mem. Moose, Lions, Wilson (N.C.) Country Club. Episcopalian. Home: PO Box 905 Elm City NC 27822-0905

PARKER, KEVIN JAMES, electrical engineer educator. BS in Engring. Sci. summa cum laude, SUNY, Buffalo, 1976; MSEE, MIT, 1978, PhD, 1981. Rsch. assoc. lab. for med. ultrasound MIT, Cambridge, 1977-81; asst. prof. dept. electrical engring. U. Rochester, N.Y., 1981-85, assoc. prof., 1985-91, prof., chair, 1992—, assoc. prof. dept. radiology, 1989-91, prof., 1992—; dir. Rochester Ctr. Biomedical Ultrasound, 1990—; com. mem. Internat. Symposium on Ultrasound Imaging, 1989—. Editorial bd. Ultras. Med. Biology, 1989—; contbr. numerous articles to profl. jours., chpts. to books. Fellow NIH, 1979, Lilly Teaching fellow, 1982; named IBM Supercomputing Contest Finalist, 1989; recipient Ultrasound in Medicine and Biology prize World Fed., 1991, Outstanding Innovation award Eastman Kodak Co., 1991. Mem. IEEE (sr. mem., Ultrasound Symposium Tech. Com. 1985—), Acoustical Soc. Am., Am. Inst. Ultrasound in Medicine (ethics com. 1987—, standards com. 1990—). Achievements include three patents in field. Office: Univ of Rochester Ctr for Biomedical Ultrasound 309 Hopeman Engineering Bldg Rochester NY 14627

PARKER, LARRY LEE, electronics company executive, consultant; b. St. Paul, Oct. 21, 1938; s. Clifford Leroy and Evelyn Elaine (McArtor) P.; m. Esperanza Victoria Delgado, Aug. 7, 1965; children: Sean Lawrence, Nicole Kathleen. AA in Engring., Antelope Valley Coll., Lancaster, Calif., 1964; BS in Indsl. Engring., U. Calif., Berkeley, 1966, MS in Ops. Rsch., 1968. Prin., cons. Ted Barry & Assocs., L.A., 1968-73; v.p. mfg. Pacific div. Mark Controls, Long Beach, Calif., 1973-79; v.p. world ops. ARL div. Bausch & Lombe, Sunland, Calif., 1979-84; pres. control products div. Leach Corp., Buena Park, Calif., 1984-88, exec. v.p., chief operating officer parent co., 1988-90, pres., COO, 1990—; also bd. dirs.; advisor engring. coun. U. Calif., Long Beach, 1990—; bd. dirs. So. Calif. Tech. Exec. Network. With USN, 1956-59. Recipient Outstanding Achievement award Los Angeles County Bd. Suprs., 1964. Mem. Am. Prodn. Inventory Control Soc. (mem. exec. com. mfg.), Am. Electronics Assn. (pres.'s roundtable). Avocations: fishing, reading, promoting world glass manufacturing in the United States. Home: 2711 Canary Dr Costa Mesa CA 92626-4747 Office: Leach Corp PO Box 5032 Buena Park CA 90622-5032

PARKER, MICHAEL ANDREW, materials scientist; b. Chgo., Oct. 1, 1949; s. Andrew Stanley and Elizabeth Rose (Jendrzejczyk) P.; m. Mary Ellen Donahue, June 19, 1982; children: Andrew Walter, Thaddeus Joseph. BS in Physics, Ill. Inst. Tech., Chgo., 1974, MS in Metall. Engring., 1978; PhD in Materials Sci., Stanford U., 1988. Rsch. asst. Metall. Dept. IIT, Chgo., 1975-77; staff engr. Systems Products div. IBM, Rochester, Minn., 1977-82; IBM resident study fellow Materials Sci. dept. Stanford (Calif.) U., 1982-88; adv. scientist Advanced Storage and Retrieval, San Jose, Calif., 1986—. Contbr. articles to Nature, 1986, Proceedings Elec. Micros. Soc. Am., 1991, IEEE Transactions on Magnetics, 1991. Resident study fellow IBM, White Plains, N.Y., 1982. Mem. IEEE and Magnetics Soc., Am. Inst. Metall. Engrs., Am. Phys. Soc., Materials Rsch. Soc. (award 1985), Electron Microscopy Soc. Am. Achievements include development of technique for observing atomic motions attending phase transformations within the transmission electron microscope, of a new technique for depth profiling the crystallographic structure of thin films by elongated probe micro-diffraction in the TEM, and of models for thin film growth and epitaxy relevant to magnetic recording media and magnetic recording head technology; research on the kinetics of solid phase epitaxial growth of silicon and development of a universal model for it relevant to 3 dimensional very large scale integration schemes for integrated circuits and implications of other film growth processes for magnetic recording technology. Office: IBM Fellow Program Dept 808/028 5600 Cottle Rd San Jose CA 95193-0001

PARKER, NORMAN NEIL, JR., software engineer, mathematics educator; b. Chgo., June 23, 1949; s. Norman Neil and Sarah Anne (Dodds) P.; m. Rowena Ubaldo Robles, June 27, 1987. BS with honors, Iowa State U., 1971, MS with honors, 1974. Cert. secondary math. tchr., Ill. Grad. teaching asst. math. dept. Iowa State U., Ames, 1971-72; tchr. math. dept. Thornwood High Sch., South Holland, Ill., 1972-81; software engr. IBM, Houston, 1981—; cons. Atomic Energy Commn., Iowa State U., Ames, 1970-72; Iowa State U. rep. NSF Regional Conf., Northfield, Minn., 1972. Contbr. articles to profl. jours. Life mem. Order of Demolay, 1963—; gymnastics judge Ill. High Sch. Assn., Nat. Gymnastics Judges Assn., Internat. Gymnastics Fedn., 1971-81; gymnastics coach Thornwood High Sch., South Holland, Ill., 1972-80; gymnastics program dir. South Holland Park Dist., 1976-80; devel. coord. Spaceweek Corp., Houston, 1983-87; officer Filipino-Internat. Families Tex., Houston, 1989—. Mem. AIAA (sr.), Clear Lake Area Spl. Interest Group Ada, Johnson Space Ctr. Employees Activities Assn. (assoc.), Gong Yuen Chuan Fa Fedn. (sr.). Republican. Roman Catholic. Home: 3300 Pebblebrook # 9 Seabrook TX 77586 Office: IBM 3700 Bay Area Blvd Rm 6406 Houston TX 77058

PARKER, PETER D.M., physicist, educator, researcher; b. N.Y.C., Dec. 14, 1936; s. Allan Ellwood and Alice Francis (Heywood) P.; m. Judith Maxfield CUrren, Dec. 27, 1958; children: Stephanie, Gregory, Gretchen. BA, Amherst Coll., 1958; PhD, Calif. Tech., 1963. Physicist Brookhaven Nat. Labs., Upton, N.Y., 1963-66; prof. Yale U., New Haven, Conn., 1966—. Office: Yale U Physics Dept Wright Nuclear Structure 272 Whitney Ave New Haven CT 06511-8124

PARKER, ROBERT ALLAN RIDLEY, astronaut; b. N.Y.C., Dec. 14, 1936; s. Allan Elwood and Alice (Heywood) P.; m. Joan Audrey Capers, June 14, 1958 (div. 1980); children: Kimberly Ellen, Brian David Capers; m. Judith S. Woodruff, Apr. 2, 1981. A.B., Amherst Coll., 1958; Ph.D., Calif. Inst. Tech., 1962. NSF postdoctoral fellow U. Wis., 1962-63, asst. prof., then assoc. prof. astronomy, 1963-74; astronaut NASA, Johnson Space Center, 1967-91; dir. policy plan Office Space Flight, NASA Hdqs., Washington, 1991-92, dir. ops. program, 1992—; mem. support crew Apollo XV and XVII, mission scientist Apollo XVII, program scientist Skylab program, mission specialist for Spacelab 1, ASTRO-1, 1967-91. Mem. Am. Astron. Soc., Phi Beta Kappa, Sigma Xi. Office: NASA-HQ Code M-6 Washington DC 20546

PARKER, ROBERT ANDREW, psychologist, computer consultant; b. Jacksonville, Fla., Mar. 21, 1946; s. Farrand Drake and Laurel (Cook) P. BA with distinction in Psychology, Ohio State U., 1969; MA, U. Regina, Sask., Can., 1976, PhD, 1983. Lic. psychologist, N.Y. Psychologist Weyburn (Can.) Psychiat. Ctr., 1981-85; psychologist No. Westchester Guidance Clinic, Mt. Kisco, N.Y., 1985-87, cons., 1988—; sr. psychologist Abbott House, Irvington, N.Y., 1988-89, cons., 1989—; pvt. practice, White Plains, N.Y., 1988—; assoc. psychologist Rockland Children's Psychiat. Ctr., Orangeburg, N.Y., 1989—; cons. Westchester Mental Health Assn., White Plains, 1990. Author: Waiting for Cargo, 1993; also articles; author computer programs Klopfer I, 1983, Exner I, 1986. Mem. client svcs. com. So. N.Y. chpt. Nat. Multpile Sclerosis Soc., White Plains, 1990—. Mem. APA, Soc. for Personality Assessment, Internat. Rsch. Soc. Avocations: music, camping, flying. Home: 16 Newcomb Pl White Plains NY 10606-2004

PARKER, ROBERT HALLETT, ecologist; b. Springfield, Mass., Feb. 14, 1922; s. Ralph Coy and Mildred (Hallett) P.; m. Harriet Elizabeth Logan, Dec. 23, 1945; foster children: Joycelene Bryan, Alfred Freeman Bryan, Sandee Jones. Student, Duke U., 1941-43, 49-50; BS, U. N.Mex., 1948, MS, 1949; PhD, U. Copenhagen, 1963. Cert. sr. ecologist, registered profl. geologist. Marine biologist Tex. Game and Fish Commn., Rockport, 1950-57; rsch. ecologist Scripps Inst. Oceanography, U. Calif., LaJolla, 1951-63; resident ecologist Marine Biol. Lab., Woods Hole, Mass., 1963-66; assoc. prof. biology and geology dept. Tex. Christian U., Ft. Worth, 1966-70; pres., chmn. bd. Coastal Ecosystems Mgmt., Inc., Ft. Worth, 1970—; cons. Humble Oil Co., Houston, 1956-58, Standard Oil Co. N.J., N.Y.C., 1958. Author: Zoo Geography and Ecology of Macro-Invertebrates, 1964, The Study of Benthic Communities, 1975, Benthic Invertebrates in Tidal Estuaries and Coastal Lagoons, 1969; co-author: Marine and Estuarine Environments, Organisms and Geology of the Cape Cod Region, 1967, Sea Shells of the Texas Coast, 1972; contbr. articles to profl. jours. With U.S. Army, 1943-45, ETO. Nat. Acad. Sci. fellow, 1959; recipient Best Abstract award Moscow Oceanographic Congress Nat. Acad. Sci., Moscow, 1966. Fellow Geol. Soc. Am., Explorer's Club N.Y., Tex. Acad. Sci.; mem. AAAS, Am. Assn. Petroleum Geologists (presidential award 1956), Sigma Xi. Republican. Avocations: woodwind music, swimming, photography, ornithology. Home: 3601 Wren Ave Fort Worth TX 76133-2130 Office: Coastal Ecosystems Mgmt 4850 McCart Ave Ste A Fort Worth TX 76115

PARKER, THEODORE CLIFFORD, electronics engineer; b. Dallas, Oreg., Sept. 25, 1929; s. Theodore Clifford and Virginia Bernice (Rumsey) P.; B.S.E.E. magna cum laude, U. So. Calif., 1960; m. Jannet Ruby Barnes, Nov. 28, 1970; children: Sally Odette, Peggy Claudette. V.p. engring. Telemetrics, Inc., Gardena, Calif., 1963-65; chief info. systems Northrop-Nortronics, Anaheim, Calif., 1966-70; pres. AVTEL Corp., Covina, Calif., 1970-74, Aragon, Inc., Sunnyvale, Calif., 1975-78; v.p. Teledyne McCormick Selph, Hollister, Calif., 1978-82; sr. staff engr. FMC Corp., San Jose, Calif., 1982-85; pres. Power One Switching Products, Camarillo, Calif., 1985-86; pres. Condor D.C. Power Supplies, Inc., 1987-88, pres. Intelligence Power Tech. Inc., Camarillo, 1988—. Mem. IEEE (chmn. autotestcon '87), NRA (life), Am. Prodn. and Inventory Control Soc., Am. Def. Preparedness Assn., Armed Forces Communications and Electronics Assn., Tau Beta Pi, Eta Kappa Nu. Home: 1290 Saturn St Camarillo CA 93010-3520 Office: Intelligence Power Tech Inc 829 Flynn Rd Camarillo CA 93012-8702

PARKER, THOMAS SHERMAN, civil engineer; b. Auburn, N.Y., Jan. 12, 1945; s. Sherman and Helen (Hodgson) P.; m. Katherine Ann Murphy, June 14, 1969. BS in Civil Engring., Clarkson U., 1968. Registered profl. engr., N.Y., Vt. Jr. engr. Hwy. div. State of Calif., San Francisco, 1968-69; jr. engr. Erdman, Anthony, Assocs., Rochester, N.Y., 1969-70, engr., 1970-74, sr. engr., 1974-78, project engr., 1978-84, assoc., project engr., 1984-90, assoc., project mgr., 1990—. Town supr. Town of Sweden, Brockport, N.Y., 1977-85; county legislator County of Monroe Dist 2, N.Y., 1985-89; v.p. Monroe County Legis., 1988-89. Mem. ASCE, NSPE. Democrat. Baptist. Home: 78 South Ave Brockport NY 14420 Office: Erdman Anthony Assocs 259 Monroe Ave Rochester NY 14607

PARKEY, ROBERT WAYNE, radiology and nuclear medicine educator, research radiologist; b. Dallas, July 17, 1938; s. Jack and Gloria Alfreda (Perry) P.; m. Nancy June Knox, Aug. 9, 1958; children: Wendell Wade, Robert Todd, Amy Elizabeth. BS in Physics, U. Tex., 1960; MD, S.W. Med. Sch., U. Tex., Dallas, 1965. Diplomate Am. Bd. Radiology, Am. Bd. Nuclear Medicine. Intern St. Paul Hosp., Dallas, 1965-66; resident in radiology U. Tex. Health Sci. Ctr., Dallas, 1966-69, asst. prof. radiology, 1970-74, assoc. prof., 1974-77, prof., chmn. dept. radiology, 1977—; chief nuclear medicine Parkland Meml. Hosp., Dallas, 1974-79, chief dept. radiology, 1977—. Contbr. numerous chpts., articles and abstracts to profl. publs. Served as capt. M.C., Army N.G., 1965-72. NIH fellow Nat. Inst. Gen. Med. Sci., U. Mo., Columbia, 1969-70; Nat. Acad. Scis.-NRC scholar in radiol. research James Picker Found., 1971-74. Fellow Am. Coll. Cardiology, Am. Coll. Radiology; mem. Am. Coll. Nuclear Physicians (charter, ho. of dels. 1974—), Council on Cardiovascular Radiology of Am. Heart Assn., AMA, Assn. Univ. Radiologists, Dallas County Med. Assn., Dallas Ft. Worth Radiol. Soc., Radiol. Soc. N.Am., Soc. Chairmen of Acad. Radiology Depts., Soc. Nuclear Medicine (acad. council, rep. to Inter-Soc. Com. on Heart Diseases 1975—, trustee S.W. chpt. 1975-78), Tex. Assn. Physicians in Nuclear Medicine (charter, sec.-treas. 1973-74), Tex. Med. Assn., Tex. Radiol. Soc., Sigma Xi, Alpha Omega Alpha. Avocations: gardening, golf, tennis. Academic research interests: nuclear cardiology, development of new imaging technologies, medical education. Office: U Tex Southwestern Med Ctr Dallas Dept Radiology 5323 Harry Hines Blvd Dallas TX 75235-8896

PARKHURST, CHARLES LLOYD, electronics company executive; b. Nashville, Aug. 13, 1943; s. Charles Albert Parkhurst and Dorothy Elizabeth (Ballou) Parkhurst Crutchfield; m. Dolores Ann Oakley, June 6, 1970; children: Charles Thomas, Deborah Lynn, Jere Loy. Student, Hume-Fogg Tech. Coll., 1959-61; AA, Mesa Community Coll., 1973; student, Ariz. State U., 1973-76. Mem. design staff Tex. Instruments, Dallas, 1967-68; mgr. design Motorola, Inc., Phoenix, 1968-76; pres. LSI Cons., Inc., Tempe, Ariz., 1976-85, LSI Photomasks, Inc., Tempe, 1985—. Mem. Rep. Congl. Leadership Coun., Washington, 1988; life mem. Rep. Presdl. Task Force, 1990. Served as cpl. USMC, 1961-64. Mem. Bay Area Chrome Users Soc., Nat. Trust Hist. Preservation, Ariz. State U. Alumni Assn. (life). Baptist. Avocations: genealogy, coin collecting, scuba diving, book collecting. Office: LSI Photomasks Inc 406 S Price Rd Ste 5 Tempe AZ 85281-3195

PARKIN, GERARD FRANCIS RALPH, chemistry educator, researcher; b. Middlesbrough, Cleveland, Eng., Feb. 15, 1959; s. Ralph and Clementine (Gill) P.; m. Rita K. Upmacis. BA with honors, Oxford (Eng.) U., 1981, MA, 1984, PhD, 1985. NATO/SERC (U.K.) postdoctoral rsch. fellow Calif. Inst. Tech., 1985-88; asst. prof. Columbia U., N.Y.C., 1988-91, assoc. prof., 1991—. Contbr. more than 60 articles to profl. jours. A.P. Sloan Rsch. fellow; recipient Presdl. Faculty fellowship NSF, 1992—, Camille and Henry Dreyfus Tchr.-Scholar award, 1991—. Roman Catholic. Achievements include discovery that bond stretch isomerism in an artifact. Office: Columbia U 116th St and Broadway New York NY 10027

PARKINSON, CLAIRE LUCILLE, climatologist; b. Bay Shore, N.Y., Mar. 21, 1948; d. C. V. and Virginia (Hafner) P. BA, Wellesley Coll., 1970; MA, Ohio State U., 1974, PhD, 1977. Rsch. asst. Inst. Polar Studies, Columbus, 1972-74; teaching asst. Ohio State U., Columbus, 1973-76; rsch. asst. Nat. Ctr. Atmospheric Rsch., Boulder, Colo., 1976-78; rsch. scientist Goddard Space Flight Ctr., NASA, Greenbelt, Md., 1978—; sci. colloquium com. mem. Goddard Space Flight Ctr., 1986—; project scientist Earth Observing System PM Mission, NASA, 1993—. Author: Breakthroughs, 1985; co-author: Antarctic Sea Ice, 1983, Three-Dimensional Climate Modeling, 1986; lead author: Arctic Sea Ice, 1987 (Peer award 1988); assoc. editor: Internat. Glaciological Soc., Cambridge, Eng., 1989-92; contbr. articles to profl. jours. Vol. Spl. Olympics, Annapolis, Md., 1989; tutor Greenbelt Cares, 1989—; sci. speaker, sci. fair judge local schs., 1989—. Recipient Goddard Lab. Atmospheric Scis., 1984, group award for co-editorship of atlas of satellite observations related to global change, 1993. Mem. Am. Polar Soc., Am. Meteorol. Soc. (history com. chmn. 1990), Nat. Oceanographic and Atmospheric Adminstrn. (adv. panel on climate and global change 1990—), Internat. Glaciological Soc., Assn. for Philosophy of Math., Oceanography Soc., Phi Beta Kappa. Achievements include research in global change, satellite remote sensing, sea ice/climate connections, climate modeling, history of science. Home: 8345 Canning Ter Greenbelt MD 20770-2701 Office: NASA Goddard Space Flight Ctr Code 971 Greenbelt MD 20771

PARKINSON, WILLIAM QUILLIAN, paleontologist; b. Henderson, Nev., Apr. 18, 1957; s. Preston Allen and Janet (Quillian) P. BS in Biology, SUNY, Albany, 1990; MA in History, So. Calif. Coll., 1992; MS in Paleontology, Loma Linda U., 1993. Rsch. asst. Loma Linda (Calif.) U., 1990-91. Mem. AAAS, Paleontol. Assn. Eng., Soc. Sedimentary Geologists, Geol. Soc. Am., Paleontol. Soc. Am., N.Y. Acad. Scis.

PARKS, HAROLD RAYMOND, mathematician, educator; b. Wilmington, Del., May 22, 1949; s. Lytle Raymond Jr. and Marjorie Ruth (Chambers) P.; m. Paula Sue Beaulieu, Aug. 21, 1971 (div. 1984); children: Paul Raymond, David Austin; m. Susan Irene Taylor, June 6, 1985; 1 stepchild, Kathryn McLaughlin. AB, Dartmouth Coll., 1971; PhD, Princeton U., 1974. Tamarkin instr. Brown U., Providence, 1974-77; asst. prof. Oreg. State U., Corvallis, 1977-82, assoc. prof., 1982-89, prof. math., 1989—; vis. assoc. prof. Ind. U., Bloomington, 1982-83. Author: Explicit Determination of Area Minimizing Hypersurfaces, vol. II, 1986, (with Steven G. Krantz) A Primer of Real Analytic Functions, 1992; contbr. articles to profl. publs. Cubmaster Oregon Trail Coun. Boy Scouts Am., 1990-92. NSF fellow, 1971-74. Mem. Am. Math. Soc., Math. Assn. Am., Soc. Indsl. and Applied Math., Phi Beta Kappa. Republican. Mem. Soc. of Friends. Home: 33914 Dorset Ln Philomath OR 97370-9555 Office: Dept Math Oreg State Univ Corvallis OR 97331-4605

PARKS, JAMES EDGAR, physicist, consultant; b. Morganton, N.C., Jan. 12, 1939; s. Thomas Edgar and Ada Conley (Wakefield) P.; m. Barbara Frances Catron, Sept. 15, 1962; children: Sharon Kaye, James Edgar II, Kermit Hunter, Christine Carol. BS in Physics, Berea (Ky.) Coll., 1961; MS in Physics, U. Tenn., 1965; PhD in Physics, U. Ky., 1970. Instr. Berea Coll., 1964-66; grad. asst. U. Ky., Lexington, 1966-70; prof. physics Western Ky. U., Bowling Green, 1970-81; tech. dir. Atom Scis., Oak Ridge, Tenn., 1981-88; dir. inst. resonance ionization spectroscopy U. Tenn., Knoxville, 1988—. Editor Procs. Internat. Symposia on RIS and its Applications; contbr. articles to profl. publs. Achievements include patents for Electrical Sensor of Plane Coordinates, Method and Apparatus for Noble Gas Detection with Isotopic Selectivity, Sputter Initiated Resonance Ionization Spectroscopy. Office: Inst RIS 10521 Research Dr Ste 300 Knoxville TN 37932

PARKS, ROBERT EDSON, optical engineer; b. Lakewood, Ohio, Jan. 22, 1942; s. James Gordon and Helen Marie (Edson) P.; m. Bertha Helen Tomlinson, June 14, 1965 (div. 1985); 1 child, Margery Helen. BS in Physics, Ohio Wesleyan U., 1964; MA in Physics, Williams Coll., 1966. Optical engr. Eastman Kodak Co., Rochester, N.Y., 1966-68, Itek Corp., Lexington, Mass., 1968-72; asst. mgr. Frank Cooke, Inc., North Brookfield, Mass., 1972-76; optical engr. U. Ariz., Tucson, 1976-83, adj. lectr., 1983-91, asst. rsch. prof., 1991—; leader U.S. Delegation to ISO/TC172, Harrison, N.Y., 1980—; chmn. ANSI/NAPM IT.11 Com., Harrison, 1991—. Recipient NASA Pub. Svc. medal NASA, 1991. Fellow SPIE (bd. dirs. 1990-92); mem. Optical Soc. Am., Am. Soc. Precision Engrs. Home: 4149 E Holmes St Tucson AZ 85711 Office: U Ariz Optical Scis Ctr Tucson AZ 85721

PARKS, ROBERT EMMETT, JR., medical science educator; b. Glendale, N.Y., July 29, 1921; s. Robert Emmett and Carolyn M. (Heinemann) P.; m. Margaret Ellen Ward, June 15, 1945; children: Robert Emmett III, Walter Ward, Christopher Carr. AB, Brown U., 1944; MD, Harvard U., 1945; PhD, U. Wis., 1954. Intern Boston's Children's Hosp., 1945-46; rsch. assoc. Amherst (Mass.) Coll., 1948-51; postdoctoral fellow Enzyme Inst., Madison, Wis., 1951-54; mem. faculty U. Wis. Med. Sch., 1954-63, prof. pharmocology, 1961-63; prof. med. sci. Brown U., Providence, 1963-91, prof. emeritus, 1991—, dir. grad. program in pharmacology and exptl. pathology, 1978-81, chmn. sect. biochem. pharmacology, 1963-78, 83-91; cons. in field. Contbr. articles to profl. jours. With AUS, 1943-45, 46-48. Acad. medicine scholar John and Mary Markle Found., 1956-61. Mem. Am. Soc. Pharmacology and Exptl. Therapeutics, Am. Soc. Biol. Chemists, Am. Assn. Cancer Rsch. (dir. 1982—), Sigma Xi. Home: 62 Alumni Ave Providence RI 02906-2310 Office: Brown U 429 Biomed Ctr Providence RI 02912

PARKS, RODNEY KEITH, mechanical engineer, consultant; b. Birmingham, Ala., Aug. 22, 1962; s. Rodney Howard and Ondrea Shaelene (Stewart) P.; m. Patricia Michelle Hawsey, Aug. 31, 1985; 1 child, Rebekah Michelle. BSME, Auburn Univ., 1984; postgrad., Univ. Ala., 1990—. Process engr. 3M Corp, Guin, Ala., 1985-86; resident project engr. 3M Corp, —, 1986-88; area leader Rust Internat, Birmingham, 1988-89, staff engr., 1989-90, staff chemical engr., 1990-91, design leader, 1991—. Active United Cerebal Palsey of Ala., 1992. Mem. ASME (assoc.), Tech. Assn. Pulp Paper Industries (assoc.). Home: 500 Sunrise Blvd Hueytown AL 35023 Office: Rust Internat 100 Corporate Pkwy Birmingham AL 35201

PARKS, STEVEN JAMES, aerospace engineer; b. Detroit, July 13, 1965; s. Donald W. and Arlene B. (Judd) P. B of Aero. Engring., U. Mich., 1987. Sr. assoc. engr. Lockheed Engring. and Scis. Co., Houston, 1988-92; project engr. Optimal Computer Aided Engring. Co., Novi, Mich., 1992—. Mem. AIAA. Achievements include design of methodology providing intersecting grid components within chimera overlapped grid interpolation scheme, full Navier-Stokes flow field calculation on space shuttle launch vehicle in ascent configuration. Office: Optimal CAE Ste 200 39555 Orchard Hill Pl Novi MI 48375

PARKS, VINCENT JOSEPH, civil engineering educator; b. Chgo., May 5, 1928; s. Joseph and Nora (Carr) P.; m. Julia Catherine Pyles, Feb. 12, 1955; children: Sean, Michael, David, Nora, Joseph, Gregory, Laurence. AS, Lewis Coll. Sci., 1948; BSME, Ill. Inst. Tech., 1953; MCE, Cath. U. Am., 1963, PhD, 1968. Engr. Andrew Corp., Orland Park, Ill., 1953-55; rsch. engr. Armour Rsch. Found., Chgo., 1955-61; rsch. assoc. Cath. Univ. Am., Washington, 1961-65, asst. prof., 1965-68, assoc. prof., 1968-73, prof., 1973—; rsch. cons. U.S. Naval Rsch. Lab., Washington, 1973-90, Sandia Nat. Lab., Albuquerque, 1980—. Co-author: Moire Analysis of Strain, 1970; editor: Progress in Experimental Mechanics, 1975. Recipient Resident Rsch. Associateship, Nat. Rsch. Coun., Washington, 1971-72, Faculty Rsch. grant Am. Soc. for Engring. Edn., Washington, 1983-89. Fellow ASME, Soc. Experimental Mechanics (Hetenyi 1974, Frocht 1981); mem. Am. Acad. Mechanics, Sigma Xi. Democrat. Roman Catholic. Office: The Cath U Am Dept Civil Engring Washington DC 20064

PARLOS, ALEXANDER GEORGE, systems and control engineer; b. Istanbul, Turkey, July 12, 1961; came to U.S., 1980; s. George Alexander and Helen (Stavridis) P.; m. Dalila Marcia Vieira, Aug. 25, 1985. MS, MIT, 1985, DSc, 1986. Rsch. asst. Tex. A&M U., College Sta., 1982-83, MIT, Cambridge, Mass., 1984-86; sr. rsch. assoc. The U. N.Mex., Albuquerque, 1986-87; asst. prof. Tex. A&M U., College Sta., 1987-92, assoc. prof., 1993—; cons. engr. BDM Internat., Inc., McLean, Va., 1988—; sr. engring. assoc. API, Albuquerque, 1990—. Contbr. articles to Internat. Jour. Control, IEEE Transactions on Nuclear Sci., AIAA Jour. Propulsion and Power, Space Nuclear Power Systems, Nuclear Tech., Nuclear Sci. Engring., IEEE Transactions of Automatic Control, others. Treas. S.W. Crossing Assn., College Sta., 1990-91; chair program planning Am. Nuclear Soc. Remote Systems Div., La Grange Pk., Ill., 1989-90; advisor Hellenic Student Assn., College Sta., 1988-91. Grantee NASA, 1988—, Dept. of Energy, 1989—, Lockheed MIss. Co., 1989, Electric Power Rsch. Inst., 1988. Mem. IEEE, AIAA, ASME, Am. Nuclear Soc., Internat. Neural Networks Soc. Achievements include 1 patent pending in the area of neural information processing. Office: Tex A&M U 129 Zachry Bldg College Station TX 77843-3133

PARMELEE, WALKER MICHAEL, psychologist; b. Grand Haven, Mich., Apr. 26, 1952; s. Walker Michael and Evelyn Mae (Essengerg) P.; m. Gayle Ann Klempel, Jan. 11, 1975; children: Morgan Christine, Kathryn Ann, Elizabeth Mae. BS, Cen. Mich. U., 1974, MA, cert. specialist in psychology, 1977; D Counseling Psychology, Western Mich. U., 1986. Lic. psychologist, Mich. Sch. psychologist Oakridge Pub. Schs., Muskegon, Mich., 1977-82, Ravenna (Mich.) Schs., Muskegon Heights (Mich.) Schs., 1982-84; sr. staff therapist Steelcase Counseling Svcs., Grand Rapids, Mich., 1984-90; prin., psychologist Parmelee Psychology Ctr., Grand Haven, 1989—; consulting psychologist Chem. Dependency Clinic, Grand Haven, 1989—. Contbr. articles to profl. jours. Bd. dirs. Planned Parenthood, Muskegon, 1979-82, Harbinger Inc., Grand Rapids, 1986-90; elder 2d Ref. Ch., Grand Haven, 1989-92; mem. women and families adv. group Allegan, Muskegon, Ottawa Substance Abuse Agy., 1992—. Mem. Am. Psychol. Assn., Am. Group Psychotherapy Assn., Nat. Assn. Child Alcoholics, Mich. Psychol. Assn., Mich. Sch. Psychologists. Avocations: woodworking, skiing, running, tennis, camping. Home: 215 Howard St Grand Haven MI 49417-1806 Office: Parmelee Psychology Ctr 220 Franklin St Grand Haven MI 49417-1336

PARMENTER, CHARLES STEDMAN, chemistry educator; b. Phila., Oct. 12, 1933; married, 1956; 3 children. BA, U. Pa., 1955; PhD in Phys. Chemistry, U. Rochester, 1963. Tech. rep. photo prodn. E.I. du Pont de Nemours & Co., 1958; NSF fellow chemistry Harvard U., Boston, 1962-63, NIH rsch. fellow, 1963-64, from asst. prof. to prof., 1964-88; Disting. prof. chemistry Ind. U., Bloomington, 1988—; Simon H. Guggenheim fellow U. Cambridge, 1971-72; vis. fellow Joint Inst. Lab. Astrophysics, Nat. Bur. Standards and U. Colo., 1977-78, 92. Lt. USAF, 1956-58. Recipient Humboldt Sr. Scientist award Tech. U. Munchen, 1986; Fulbright Sr. Scholar Griffith U., Australia, 1980. Fellow AAAS, Am. Phys. Soc.; mem. Am. Chem. Soc. Research in photochemistry, spectroscopy, energy transfer. Office: Ind U Dept of Chemistry Bloomington IN 47405

PARMER, DAN GERALD, veterinarian; b. Wetumpka, Ala., July 3, 1926; s. James Lonnie and Virginia Gertrude (Guy) P.; student L.A. City Coll., 1945-46; DVM, Auburn U., 1950; m. Donna Louise Kesler, June 7, 1980; 1 son, Dan Gerald; 1 dau. by previous marriage, Linda Leigh. Gen. practice vet. medicine, Galveston, Tex., 1950-54, Chgo., 59—; vet. in charge Chgo. Commn. Animal Care and Control, 1974-88; med. dir. food protection div. Chgo. Dept. Health, 1988-93; ret. 1993; chmn. Ill. Impaired Vets. Com.; tchr. Highlands U., 1959. Served with USNR, 1943-45, PTO; served as staff vet. and 2d and 5th Air Force vet. chief USAF, 1954-59. Decorated 9 Battle Stars; recipient Vet. Appreciation award U. Ill., 1971, Commendation, Chgo. Commn. Animal Care and Control, 1987. Mem. Ill. Vet. Medicine Assn. (chmn. civil def. and package disaster hosps. 1968-71, Pres.' award 1986), Chgo. Vet. Medicine Assn. (bd. govs. 1969-72, 74-13, pres. 1982), South Chgo. Vet. Medicine Assn. (pres. 1965-66), Am. Animal Hosp. Assn. (dir.), AVMA (nat. com. for impaired vets., coun. pub. health and regulatory medicine 1990—), Ill. Acad. Vet. Practice (pres. 1993), Nat. Assn. of Professions, Am. Assn. Zoo Vets., Am. Assn. Zool. Parks and Aquariums, VFW, Midlothian Country Club, Valley Internat. Country Club, Masons, Shriners, Kiwanis. Democrat. Discoverer Bartonellosis in cattle in N.Am. and Western Hemisphere, 1951; co-developer bite-size high altitude in-flight feeding program USAF, 1954-56. Address: 6720 Post Oak Ln Montgomery AL 36117

PARMER, EDGAR ALAN, radiologist, musician; b. N.Y.C., Sept. 14, 1928; s. Nathan and Selma (Benett) P.; m. Nina Ash (div. 1964); children: Vicki, Robert; m. Judith Rae Parmer, Nov. 22, 1969. AA in Music, UCLA, 1950, BA, 1951, MA; MD, N.Y. Med. Coll., 1958. Lic. physician, N.Y., Calif., Conn. Intern Grasslands Hosp., Valhalla, N.Y., 1958-59; resident Vets. Hosp., Bronx, N.Y., 1959-62; assoc. radiologist Francis Delafield Hosp. Columbia Presbyn. Med. Ctr., 1964-82; assoc. radiologist, dir. nuclear medicine Mt. Vernon (N.Y.) Hosp., 1964-82; pvt. practice New Rochelle, N.Y., 1982-84; instr. radiology Columbia U.; vis. fellow dept. radiology Columbia-Presbyn. Med. Ctr.; cons. tech. affairs HEW; mem. staff Radiologic Technicians Program Westchester Community Coll., chmn. adv. com.; cons., staff mem. St. Barnabas Radiology, 1987—; med. dir. Ultrasound Diagnostic Sch., 1985-87; assoc. dir. radiology, dir. ultrasound Strang Clinic; lectr. in field. Contbr. articles to profl. jours.; soloist, recordings include N.Y.C. Symphony Orch., 1940, Burbank (Calif.) Orch., Glendale (Calif.) Orch., MGM Symphony, Westchester Symphony Orch., Westchester Philharm. Orch. Mem., past pres. Doctor's Symphony Orch. N.Y.; sec. bd. dirs., pres. Westchester Philharm. Orch.; past dir., mem. Am. Cancer Soc.; pres. bd. dirs. Premium Point Pk. With U.S. Army. Am. Cancer Soc. grantee. Fellow Am. Coll. Angiology, Am. Inst. Ultrasound Medicine (sr.), Royal Soc. Health, ; mem. AMA, AAAS, Am. Assn. Advancement Boxing (bd. dirs.), Am. Coll. Radiology (past pres. Westchester div.), Am. Coll. Nuclear Medicine, Am. Coll. Med. Imaging, Soc. Nuclear Medicine, N.Y. State Med. Soc., Westchester County Med. Soc., Westchester Radiol. Soc. (past pres.), N.Y. Acad. Sci. Avocations: tennis, golf, scuba diving. Home: 7 Shore Dr New Rochelle NY 10801-5331

PARMESANI, ROLANDO ROMANO, engineer; b. Canazei, Trento, Italy, June 11, 1960; s. Enrico and Luciana (Proto) P. Grad. engr., U. Trieste (Italy), 1987. Stage mem. Fiat Aviazione, Torino, Italy, 1986-87; tech. dir. Soc. I.T. Canazei, Canazei, Italy, 1988-89; pres. Aerostudi, Trieste, 1989—; cons. Penal Ct., Trento, 1989—. 2d lt. Italian Air Force, 1988-89. Office: Aerostudi SpA, Colombara di Vignano 7, 34015 Muggia Trieste, Italy

PARMLEY, LOREN FRANCIS, JR., medical educator; b. El Paso, Tex., Sept. 19, 1921; s. Loren Francis and Hope (Bartholomew) P.; m. Dorothy Louise Turner, Apr. 4, 1942; children—Richard Turner, Robert James, Kathryn Louise. B.A., U. Va., 1941, M.D., 1943. Diplomate Am. Bd. Internal Medicine, Am. Bd. Internal Medicine-Cardiovascular Disease. Commd. 1st lt. U.S. Army, 1944; advanced through grades to col., 1968; intern Med. Coll. Va., 1944; resident in internal medicine Brooke Gen. Hosp., San Antonio, 1948-49, U. Wis. Gen. Hosp., Madison, 1949-51; asst. prof. mil. med. sci. Med. Coll. U. Wis., Madison, 1949-51; asst. attache (med.) U.S. Embassy, New Delhi, 1953-55; fellow in cardiovascular disease Walter Reed Gen. Hosp., Washington, 1956-57; chief medicine and cardiology Letterman Gen. Hosp., San Francisco, 1958-63; med. and cardiology cons. U.S. Army Europe, Heidelberg, Germany, 1963-64; chief medicine Walter Reed Gen. Hosp., Washington, 1965-68; prof. medicine, asst. dean Med. U. S.C., Spartanburg, 1968-75; dir. med. edn. Spartanburg Gen. Hosp., Spartanburg, 1968-75; prof. medicine U. South Ala., Mobile, 1975-87, chief div. cardiology, 1980-87, prof. emeritus medicine, 1988—; lectr. medicine U. Calif.-San Francisco, 1959-63; clin. assoc. prof. medicine Georgetown U., Washington, 1967-68; clin. prof. medicine Med. Coll. Ga., Augusta, 1969-75; cons. internal medicine Surgeon Gen. U.S. Army, Washington, 1966-68. Contbg. author: The Heart, 1966, 70, 74, 78 Cardiac Diagnosis and Treatment, 1976, 80, The Heart in Industry, 1960, 70. Recipient Gold award sci. exhibit Am. Soc. Clin. Pathologists and Coll. Am. Pathologists, 1959; Certificate of Achievement in cardio-vascular disease Surgeon Gen. U.S.A., Washington, 1962; Bronze Medallion Meritorious Service, Am. Heart Assn., S.C., Columbia, 1969, 73; decorated Legion of Merit. Fellow ACP, Am. Coll. Cardiology (bd. govs. U.S. Army 1967, S.C. 1969-73), Am. Coll. Chest Physicians; mem. Am. Heart Assn. (fellow coun. on clin. cardiology), Soc. Med. Cons. to Armed Forces, Kiwanis. Republican. Episcopalian. Avocations: golf, swimming. Home: 549 Fairway Dr Kerrville TX 78028-6440 Office: U South Ala Coll Medicine Dept Medicine 2451 Fillingim St Mastin Bldg Mobile AL 36617

PARR, ROBERT GHORMLEY, chemistry educator; b. Chgo., Sept. 22, 1921; s. Leland Wilbur and Grace (Ghormley) P.; m. Jane Bolstad, May 28, 1944; children: Steven Robert, Jeanne Karen, Carol Jane. AB magna cum laude with high honors in Chemistry, Brown U., 1942; PhD in Phys. Chemistry, U. Minn., 1947; D (hon.), U. Leuven, 1986. Asst. prof. chemistry U. Minn., 1947-48; mem. faculty Carnegie Inst. Tech., 1948-62, prof. chemistry, 1957-62; prof. chemistry Johns Hopkins U., 1962-74, chmn. dept., 1969-72; William R. Kenan, Jr. prof. theoretical chemistry U. N.C., Chapel Hill, 1974-90, Wassily Hoeffding prof. chem. physics, 1990—; vis. prof. chemistry, mem. Center Advanced Study U. Ill., 1962; distinguished vis. prof. State U. N.Y. at Buffalo, also Pa. State U., 1967; vis. prof. Japan Soc. Promotion Sci., 1968, 79, U. Haifa, 1977, Free U. Berlin, 1977; Firth prof. U. Sheffield, 1976; Chmn. com. postdoctoral fellowships in chemistry Nat. Acad. Sci.-NRC, 1961-63; chmn. panel theoretical chemistry Westheimer com. survey chemistry Nat. Acad. Sci., 1964; mem. council Gordon Research Conf., 1974-76; mem. Commn. on Human Resources, NRC, 1979-82. Author: Quantum Theory of Molecular Electronic Structure, 1963, Density-Functional Theory of Atoms and Molecules, 1989, also numerous articles.; Asso. editor: Jour. Chem. Physics, 1956-58, Chem. Revs, 1961-63, Jour. Phys. Chemistry, 1963-67, 77-79, Am. Chem. Soc. Monographs, 1966-71, Theoretica Chimica Acta, 1966-69, 92—; bd. editors: Jour. Am. Chem. Soc, 1969-77; adv. editorial bd.: Internat. Jour. Quantum Chemistry, 1967—, Chem. Physics Letters, 1967-79. Recipient Outstanding Achievement award U. Minn., 1968, N.C. Disting. Chemist award, 1982; fellow U. Chgo., 1949; research asso., 1957; Fulbright scholar U. Cambridge, Eng., 1953-54; Guggenheim fellow, 1953-54; NSF sr. postdoctoral fellow U. Oxford (Eng.) and Commonwealth Sci. and Indsl. Research Orgn., Melbourne, Australia, 1967-68; Sloan fellow, 1956-60. Fellow AAAS, Am. Phys. Soc. (chmn. div. chem. physics 1963-64); mem. AAUP, Am. Chem. Soc. (chmn. div. phys. chemistry 1978), Am. Acad. Arts and Sci., Nat. Acad. Sci., Indian Nat. Sci. Acad., Internat. Acad. Quantum Molecular Sci. (v.p. 1973-79, pres. 1991—), Phi Beta Kappa, Sigma Xi, Phi Lambda Upsilon, Pi Mu Epsilon. Home: 701 Kenmore Rd Chapel Hill NC 27514-2019 Office: U NC Dept Chemistry Chapel Hill NC 27599-3290

PARRA-DIAZ, DENNISSE, biophysical chemist; b. San Juan, P.R., Feb. 19, 1961; d. Jacinto and Ada I. (Colón) Parra; m. Carlos E. Diaz, Jan. 6, 1987; 1 child, Sophia. BS cum laude, U. P.R., 1982; PhD, U. Miami, Fla., 1990. Teaching asst. U. P.R., Rio Piedras, 1982, U. Miami, Fla., 1984, 89; analytical chemist Upjohn Chem. Co., Barceloneta, P.R., 1983; rsch. assoc. Temple U., Phila., 1991; adj. prof. St. Joseph's U., Phila., 1992; rsch. assoc. USDA, Phila., 1992—; speaker The Franklin Inst., Phila., 1991. Contbr. articles to sci. publs. Sci. fair judge South Miami High Sch., 1989; speaker at nat. meeting Girl Scouts U.S., Phila., 1991. Mem. Am. Chem. Assn., Am. Assn. Molecular Biology and Biochemistry, Grad. Student Assn. (chemistry rep. 1988), Grad. Assn. Fee Allocation Com. (rep. 1988), Sigm Xi, Phi Lambda Upsilon (v.p. 1987). Achievements include immobilization of lipoxygenase-1 on a solid support that could survive organic solvents. Home: 11547 Z SW 109th Rd Miami FL 33176

PARRIS, C(HARLES) DEIGHTON, television engineer, consultant; b. Bridgetown, Barbados, W.I., Oct. 11, 1922; s. Charles Reginald and Bernal (Black) P.; m. Erma Elizabeth Harriott, Nov. 23, 1963 (div. 1972); children: Deighton Keith, Selena Dawn. Student, Liverpool (Eng.) Coll. Tech., 1958-59, Borough Poly., London, 1959-60; grad. in elec. engring., No. Poly., London, 1960. Chartered engr., U.K. Tech. author, TV engr. Associated Rediffusion, London, 1958-60; sr. engr. Radio Jamaica, Jamaica, 1960-62; asst. chief engr. Trinidad & Tobago TV Co. Ltd., 1962-63, chief engr., 1965-80, gen. mgr., chief engr., 1980-82; TV broadcast cons. Trinidad, 1982—; bd. dirs. Valpark Shopping Plaza Ltd., Crown Point Hotel, Tobago; cons. Caribbean Broadcast Corp., Barbados, 1984-87, Caribbean Comm. Network, Trinidad, 1990. Contbr. articles to profl. jours. Mem. Inst. Elec. Engrs. U.K., Coun. Engring. Instns. U.K. (chartered), Royal TV Soc. U.K., Assn. Profl. Engrs. Trinidad & Tobago, Internat. Solar Soc., St. James Tennis Club. Home: 8 Bimitti Rd, Valsayn Park South, Trinidad Trinidad and Tobago

PARRIS, LUTHER ALLEN, gemologist, goldsmith; b. Presque Isle, Maine, Apr. 20, 1955; s. Luther King and Constance Marie (Brewer) P.; m. Teresa Ellen Cleeremans, May 5, 1984; children: Luke Aaron, Charles Maxfield. Student, Mesa (Ariz.) Community Coll., 1981, Boulder Sch. of Urantia, 1983. Cert. gemologist diamonds, diamond grading, colored stones. Silk screen printer, mgr. Safety Engring. and Supply, Tempe, Ariz., 1977-78; silversmith Argentinair Silversmith Shop, Mesa, 1978-79; prodn. mgr. Palmos Del Sol Semingson, Mesa, 1979-83, John Hay and Assocs., Boulder, 1983-86; mgr., gemologist Golden's Designer Jewelry, Rogers, Ark., 1986-92; owner Parris Designs Jewelry, Rogers, 1992—. Contbr. articles to profl. jours., newspapers. Speaker Springdale (Ark.) Jr. High Sch., 1989, Rogers Gem and Mineral Soc., 1988' artist Rogers Ark. Mus., 1988. Mem. Gemological Inst. Am. (student, cert.). Libertirian. Urantian. Home: 743 N 5th St Rogers AR 72756 Office: Parris Designs 728 N 2nd St Rogers AR 72756

PARRISH, CLYDE ROBIN, III, electronics engineer; b. Peoria, Ill., Feb. 18, 1952; s. Clyde R. and Alma Lucielle (Ellis) P.; m. Marie Theresa Betz, May 10, 1972 (div. Nov. 1974); m. Marla Esther Goodman, Sept. 28, 1985; children: Clyde Robin IV, Robbie. Aerospace engr., Ind. Inst. Tech., Ft. Wayne, 1972, U. Ill., 1974. Packaging engr. Signode, Chgo., 1975-78; design engr. UIP Engring., Chgo., 1978-80; svc. engr. Topaz Electronics, San Diego, 1980-83, Emerson Electric, Santa Ana, Calif., 1983-85; customer svc. mgr. Elgar Corp., San Diego, 1985-91; engring. mgr. Ultimate Power Solutions, San Diego, 1991—. Mem. IEEE. Office: Ultimate Power Solutions 13875 Davenport Ave San Diego CA 92129-3107 Mfg Office: Ultimate Power Solutions 12544 Kirkham Ct # 19 Poway CA 92064

PARRISH, EDWARD ALTON, JR., electrical engineering educator, researcher; b. Newport News, Va., Jan. 7, 1937; s. Edward Alton and Molly Wren (Vaughan) P.; m. Shirley Maxine Johnson, Oct. 26, 1963; children—Troy Alton, Gregory Sinton. B.E.E., U. Va., 1964, M.E.E., 1966, D.Sc. in Elec. Engring., 1968. Registered profl. engr., Tenn., Va. Group leader Amerad Corp., Charlottesville, Va., 1961-64; asst. prof. elec. engring. U. Va., Charlottesville, 1968-71, assoc. prof. elec. engring., 1971-77, prof. elec. engring., 1977-86, chmn. dept. elec. engring., 1978-86; dean, centennial prof. electrical engring. Vanderbilt U., Nashville, 1987—; cons. U.S. Army, Charlottesville, Va., 1971-77, ORS, Inc., Princeton, N.J., 1973-74, Sperry Marine Systems, Charlottesville, 1975-76, Hajime Industries Ltd., Tokyo, 1978-84. Contbr. numerous articles to profl. jours. Served with USAF, 1954-58. Recipient numerous grants, contracts from industry, govt. agys. Fellow IEEE (bd. dirs. 1990-91, v.p. ednl. activities 1992-93); mem. Accreditation Bd. Engring. Tech. (engring. accreditation commn. 1989—, exec. com. 1991—, officer, 1993—), IEEE Computer Soc. (sec. 1977, v.p. 1978-81, pres. 1988), Pattern Recognition Soc., Sigma Xi, Eta Kappa Nu, Tau Beta Pi. Baptist. Avocations: music; woodworking. Office: Vanderbilt U Sch Engring Nashville TN 37235

PARRISH, JOHN ALBERT, dermatologist, research administrator; b. Louisville, Ky., Oct. 19, 1939; Children: Lynn, Susan, Mark. BA, Duke U., 1961; MD, Yale U., 1965. Diplomate Am. Bd. Dermatology. Medicine intern U. Mich., Ann Arbor, 1965-67; dermatology resident Harvard Med. Sch., Boston, 1969-72; dermatologist Mass. Gen. Hosp., Boston, 1972-87, dir. Wellman labs., 1975—, dir. cutaneous biology rsch. lab. Harvard, 1987—; chief dermatology Harvard Med. Sch., Boston, 1987—; dermatology cons. Beth Israel Hosp., Boston, 1973—. Author: A Doctor's Year in Vietnam, 1972, Dermatology and Skin Care, 1975, Effects of Ultraviolet Radiation on the Immune System, 1983; co-author: Science of Photomedicine, 1982, Photoimmunology, 1983. Lt. Commdr. USN, 1968-89. Decorated Vietnamese Cross Gallantry with gold; recipient Outstanding Gen. Med. Officer award USN, 1969; Dohi lectr. Japanese Soc. Dermatology, 1990. Mem. Am. Soc. Dermatology (photobiology task force 1972—, Marion B. Sulzberger award 1988), Am. Soc. Lasers in Surgery and Medicine (pres. 1987-88), Am. Soc. Photobiology (coun. 1978-82), Soc. Investigative Dermatology (Wm. Montagna award 1982). Achievements include developing novel and safe effective treatment of psoriasis. Office: Mass Gen Hosp Derm Wel 2 32 Fruit St Boston MA 02114

PARRISH, JOHN WESLEY, JR., physiologist, biology educator; b. Dennison, Ohio, Mar. 5, 1941; s. John Wesley Parrish Sr. and Dorothy Irene (Dickinson) Price; m. Paula Schmanke, July 9, 1966; children: Corinne Danelle, Wesley Allen. BS, Denison U., 1963; MA, Bowling Green (Ohio) State U., 1970, PhD, 1974. Sci. tchr. Northwood Jr. High Sch., Norfolk, Va., spring 1967; vis. instr. dept. biology Kenyon Coll., Gambier, Ohio, 1973-74; postdoctoral fellow dept. zoology U. Tex., Austin, 1974-76; asst. prof. dept. biol. sci. Emporia (Kans.) State U., 1976-82, assoc. prof. dept. biol. sci., 1982-88, assoc. chair dept. biol. sci., 1987-88; prof., chair biology dept. Ga. So. U., Statesboro, 1988—; vis. rsch. assoc. prof. dept. physiology Cornell U., Ithaca, N.Y., fall 1986. Author: (with others) Field and Laboratory Biology, 1985; editor: Activities in the Environmental and Life Sciences, 1986; mem. editorial bd. Oriole, 1991—, contbr. numerous articles to profl. jours. With USNR, 1964-67. Rsch. grantee, 1977-88. Mem. AAAS, Am. Inst. Biol. Sci., Am. Ornithologists' Union (life, student award 1972), Kans. Ornithol. Soc. (life), Ga. Ornithol. Soc. (life), Soc. for Study of Reproduction, So. Cyclists (v.p.), Sigma Xi (life). Avocations: computers, photography, tennis, bird watching. Office: Ga So U Dept Biology LB 8042 Statesboro GA 30460

PARRISH, THOMAS DENNISON, computer systems engineer; b. Gardner, Mass., Dec. 13, 1935; s. Frank T. and Harriet G. (Wilder) P.; m. Pamela M. Despres, Feb. 20, 1969; children: Michelle, Simone, Denise. B of Engring. Physics, Cornell U., 1958. Commd. ensign USN, 1958, advanced through grades to lt., 1963, resigned, 1967; computer systems scientist, engr. Planning Rsch. Corp., McLean, Va., 1967—. Democrat. Achievements include first known application of Quicksort algorithm in production compiler environment; development of operational context-based message retrieval system for U.S. Navy, of generalized Fibonacci series and applied to computer sorting system; designer automated patent system for U.S. Patent Office. Home: 7104 45th St Chevy Chase MD 20815 Office: PRC Inc 1500 PRC Dr Mc Lean VA 22102

PARRY, ROBERT WALTER, chemistry educator; b. Ogden, Utah, Oct. 1, 1917; s. Walter and Jeanette (Petterson) P.; m. Marjorie J. Nelson, July 6, 1945; children: Robert Bryce, Mark Nelson. BS, Utah State Agr. Coll., 1940; MS, Cornell U., 1942; PhD, U. Ill., 1946; DSc (hon.), Utah State U., 1985. Research asst. NDRC Munitions Devel. Lab., U. Ill. at Urbana, 1943-45, teaching fellow 1945-46; mem. faculty U. Mich., 1946-69, prof. chemistry, 1958-69; Distinguished prof. chemistry U. Utah, 1969—; indsl. cons., 1952—. Chmn. com. teaching chemistry Internat. Union Pure and Applied Chemistry 1968-74. Recipient Mfg. Chemists award for coll. teaching, 1972, Sr. U.S. Scientist award Alexander Von Humboldt-Stiftung (W. Ger.), 1980, First Govs. Medal of Sci. State Utah, 1987. Mem. Am. Chem. Soc. (Utah award Utah Sect. 1978, past chmn. inorganic div. and div. chem. edn. award for distinguished service to inorganic chemistry 1965, for chem. edn., 1977, dir. 1973-83, bd. editors jour. 1969-80, pres.-elect 1981-82, pres. 1982-83, Priestly medal 1993), Internat. Union Pure and Applied Chemistry (chmn. U.S. nat. com.), AAAS, Sigma Xi. Founding editor Inorganic Chemistry, 1960-63. Research, publs. on some structural problems of inorganic chemistry, and incorporation results into theoretical models; chemistry of phosphorus, boron and fluorine. Home: 5002 Fairbrook Ln Salt Lake City UT 84117 Office: U Utah Dept Chemistry Henry Eyring Bldg Salt Lake City UT 84112

PARSA, BAHMAN, nuclear chemist; b. Tehran, Iran, May 16, 1940; came to U.S. 1984; s. Seifollah and Mahrokhsar (Razmara) P.; m. Sima Kermanshahi, Sept. 15, 1972; children: Pantea, Parham. BS, U. Calif., Berkeley, 1963; PhD, MIT, 1967. Asst. prof. nuclear chemistry Tehran U. Nuclear Ctr., 1967-71, asst. dir. rsch., 1968-69; chmn. dept. nuclear sci. U. Tehran, 1969-71; dir. Tehran U. Nuclear Ctr., 1973-74; dep. minister sci. rsch. Ministry of Sci. and Higher Edn., Iran, 1974-78; assoc. prof. nuclear chemistry U. Tehran, 1971-83; rsch. scientist Dept. Environ. Protection, Trenton, N.J., 1984—. Contbr. articles to profl. jours. Fulbright scholar, 1972-73, James Flack Norris fellow, MIT, 1965-66. Mem. Sigma Xi, Alpha Gamma Sigma. Achievements include discovery of K-46 (a new nuclide); determination of lead in atmosphere; trace elements analysis via neutron activation analysis; radon in drinking water. Home: 319 Flint Rd Langhorne PA 19047-8209 Office: NJ Dept Environ Protection and Energy CN411 Environ Labs Trenton NJ 08625

PARSEGIAN, V. ADRIAN, biophysicist; b. Boston, May 28, 1939; s. Voscan Lawrence and Varsenig (Boyajian) P.; m. Valerie Phillips, Mar. 2, 1963; children: Andrew, Homer, Adam. AB, Dartmouth Coll., Hanover, N.H., 1960; PhD, Harvard U., 1965. Rsch. physicist Phys. Scis. Lab., DCRT, NIH, Bethesda, Md., 1967—. Editor Biophys. Jour., 1977-80, Biophys. Discussions, 1978—; contbr. over 100 articles to profl. jours.; editorial bd. 10 jours. NIH Dirs. award, 1974. Mem. Biophys. Soc.

PARSHALL, GEORGE WILLIAM, research chemist; b. Hackensack, Minn., Sept. 19, 1929; s. George Clarence and Frances (Virnig) P.; m. Naomi B. Simpson, Oct. 9, 1954; children: William, Jonathan, David. B.S., U. Minn., 1951; Ph.D., U. Ill., 1954. Research chemist E.I. duPont de Nemours & Co., Wilmington, Del., 1954-65; research supr. E.I. duPont de Nemours & Co., 1965-79, dir. chem. sci., 1979-92, cons., 1992—; bd. chem. sci. NRC, Washington, 1983-86; Reilly lectr. Notre Dame U., 1980-92. Author: Homogeneous Catalysis, 1980, 2d rev. edit. 1992; editor: Inorganic Syntheses, 1974, Jour. Molecular Catalysis, 1977-80. Recipient Bailar medal in inorganic chemistry U. Ill., 1976. Mem. Am. Chem. Soc. (award in inorganic chemistry 1983, award for leadership in chem. research mgmt. 1989), Nat. Acad. Sci., Am. Acad. Arts Scis., Catalysis Soc. (Phila Catalyst Club award 1978). Episcopalian. Home: 2504 Delaware Ave Wilmington DE 19806-1220 Office: E I du Pont de Nemours & Co Exptl Sta Wilmington DE 19880-0328

PARSHALL, KAREN VIRGINIA HUNGER, mathematics and science historian; b. Virginia Beach, Va., July 7, 1955; d. Maurice Jacques and Jean Kay (Wroton) Hunger; m. Brian J. Parshall, Aug. 6, 1978. BA, U. Va., 1977, MS, 1978; PhD, U. Chgo., 1982. Asst. prof. math. Sweet Briar (Va.) Coll., 1982-87, U. Ill., Urbana, 1987-88; asst. prof. math. and history U. Va., Charlottesville, 1988-93, assoc. prof. math. and history, 1993—. Year's ago editor Mathematical Intelligencer, N.Y.C., 1990-93; mng. editor Historia Mathematica, San Diego, 1994—, book rev. editor, 1990-93; author: (with David Rowe) Emergence of American Mathematics Research Community, 1994; (with others) The History of Modern Mathematics, 1989; contbr. articles to profl. jours. including Archive for History of Exact Scis., History of Sci. Scholars award NSF, 1986-87, 90-93. Mem. Am. Math. Soc., History of Sci. Soc., Math. Assn. of Am., Am. Hist. Assn., Phi Beta Kappa. Office: U Va Depts Math and History Math/Astro Bldg Charlottesville VA 22903-3199

PARSONS, BRUCE ANDREW, biomedical engineer, scientist; b. Kingston, Jamaica, Mar. 13, 1960; came to U.S., 1961; s. Roy Dixon and Lupé Claudette (Bruce) P.; m. Jacquéline González, Dec. 19, 1987. BS in Elec. Engring., Fla. Atlantic U., 1983; MS in Biomed. Engring., U. Miami, 1985; PhD in Bioengring., Clemson U., 1990. Elec. engr. Trackmaster, Inc., Pompano, Fla., 1983-85, 90-91; sr. R&D engr. Becton Dickinson Vascular Access, Sandy, Utah, 1991-92; postdoctoral rschr. Dept. Cellular Pharmacology, Miami, Fla., 1992—; indsl. fellow Coulter Corp., Hialeah, Fla. Contbr. articles to Biomed. Sci. and Tech., Circulation. Grantee State of Fla., 1984; NIH postdoctoral fellow, 1992; rsch. fellow Am. Heart Assn., 1993—. Achievements include direct proof that catecholamine induced calcium overload is a contributary cause of myocardial necrosis. Home: 1951 NE 59th Pl Fort Lauderdale FL 33308

PARSONS, DANIEL LANKESTER, pharmaceutics educator; b. Biscoe, N.C., Sept. 10, 1953; s. Solomon Lankester and Doris Eva (Bost) P. BS in Pharmacy, U. Ga., 1975, PhD, 1979. Asst. prof. pharmaceutics U. Ariz., Tucson, 1979-82; asst. prof. Auburn (Ala.) U., 1982-86, assoc. prof., 1986-91, prof., 1991—, chmn. divsn., 1990—; cons. Wyeth-Ayerst, Phila., 1989-93, Technomics, Ardsley, N.Y., 1990—; presenter in field. Contbr. articles to Drug Devel. and Indsl. Pharmacy, Drug Metabolism Revs., Archives Internat. Pharmacodynamics, Materials Rsch. Soc., Jour. Pharmacy and Pharmacology, Biochem. Pharmacology. Named Disting. Alumni, Sandhills Coll., 1990. Mem. Am. Pharm. Assn., Am. Assn. Pharm. Scientists, Am. Coll. Clin. Pharmacology (mem. coun. 1990-93), Phi Kappa Phi, Kappa Psi (advisor 1990—, Svc. award 1990, Advisor award 1992). Achievements include research on plasma protein binding of drugs and effects of perfluorochemical blood substitutes on such binding. Office: Auburn U Sch Pharmacy Auburn AL 36849

PARSONS, DONALD OSCAR, economics educator; b. Pitts., Oct. 22, 1944; s. Leonard J. and Marian (Williams) P.; m. Deborah Arneson, Dec. 27, 1967; children: Donald Williams, Christopher Milne. AB, Duke U., 1966; PhD, U. Chgo., 1970. Asst. prof. econs. Ohio State U., Columbus, 1970-73, assoc. prof., 1973-77, prof., 1977—; Fulbright disting. prof. econs., Siena, Italy, 1991. Author: Poverty and the Minimum Wage, 1980; bd. editors Jour. Econs. and Bus., 1979-91, contbr. articles to Jour. Polit. Economy, Am. Econ. Rev. Harry Scherman rsch. fellow Nat. Bur. Econ. Rsch., 1975-76; grantee NIH. Mem. Am. Econ. Assn., Econometric Soc., Western Econ. Assn., So. Econ. Assn. Achievements include findings in modelling and estimation of relationship between job turnover and training in the employment contract, assessment of job search models of quit behavior, measurement of impact of social insurance programs, especially the Social Security Disability Program, on labor force participation. Office: Ohio State U Econs Dept 1945 High St Columbus OH 43210

PARSONS, GLENN RAY, biology educator; b. Fairfield, Ala., Oct. 30, 1953; s. Thomas Edward and Marion (Morris) P.; m. Cheryl Kaye Bouler, Oct. 25, 1980; 1 child, Erin Nicole. BS, U. Ala., 1977; MS, U. South Ala., 1980; PhD, U. South Fla., 1987. Rsch. technician So. Rsch. Inst., Birmingham, Ala., 1975-77; rsch. technician U. Ala., Birmingham, 1977-78; rsch. fellow Dauphin Island (Ala.) Sea Lab, 1980-81; rsch. asst. Marine Rsch. Lab., St. Petersburg, Fla., 1984-87; asst. prof. U. Miss., University, 1987—; cons. Nat. Geographic film crew Long Key, Fla., 1991. Contbr. articles to profl. jours. Recipient grants U.S. Army Corps Engrs., 1991, 92, Miss. Dept. Wildlife, Fisheries, 1992; rsch. fellowships Dauphin Island Sea Lab, 1979, Gulf Oceanographic Rsch. Found., 1985. Mem. Am. Soc. Ichthyologists, Am. Soc. Zoologists, Am. Fisheries Soc., Am. Elasmobranch Soc. Achievements include describing swimming performance and metabolism in species of shark, validating aging technique in shark species, producing triploid hybrid crappie. Office: U Miss Dept Biology University MS 38677

PARSONS, WILLIAM HUGH, chemist, researcher; b. Rochester, N.Y., Apr. 26, 1950; s. William Fraser and Lucy (Lynn) P.; m. Karen Grosser, Feb. 15, 1992. BS, U. Rochester, 1972; PhD, U. Vt., 1977. Postdoctoral fellow dept. chemistry U. Rochester, N.Y., 1977-80; sr. rsch. chemist Merck Rsch. Labs., Rahway, N.J., 1980-84, rsch. fellow, 1984-86, asst. dir., 1986—. Contbr. articles to Jour. Organic Chemistry, Tetrahedron, Biochemical and Biophysical Rsch. Communication, Jour. Med. Chemistry, Bioorganic and Medicinal Chemistry Letters, Tetrahedron Letters. Mem. Am. Chem. Soc., AAAS, N.Y. Acad. Scis. Baptist. Home: 303 Timber Oaks Rd Edison NJ 08820

PARSONSON, PETER STERLING, civil engineer, educator, consultant; b. Reading, Mass., Oct. 18, 1934; s. Alfred Horace and Elvera (Moran) P.; m.

Marilyn Shepherd, July 6, 1962 (div. Mar. 1984); children: Sheryl Elaine Parsonson Peeples, Ellen Marie Parsonson Milberger, Peter Shepherd; m. Sarah Irby, Oct. 6, 1990. BS, MIT, 1956, MS, 1959; PhD, N.C. State U., 1966. Registered profl. engr., Ga., Fla., Calif. Civil engr. Fay, Spofford and Thorndike, Boston, 1956-57, Ingenieria de Suelos, S.A., Caracas, Venezuela, 1959-61, Tippetts-Abbett-McCarthy-Stratton, N.Y.C., 1961-64; asst. prof. Coll. Engring. U. S.C., Columbia, 1966-69; assoc. prof. dept. civil engring. Ga. Inst. Tech., Atlanta, 1970-82; prof. dept. civil engring. Ga. Tech. Inst., Atlanta, 1982—; pres. Parsonson and Assocs., Inc., Atlanta, 1976—; cons. hwy.-engring. litigation; instr., researcher hwy. design, constrn., operation and maintenance. Author: Management of Traffic-Signal Maintenance, 1982, Signal Timing Improvement Practices, 1992; co-author: Traffic Detector Handbook, 1985, (chpt.) Computer Applications in Highway Engineering, 1987; contbr. articles to profl. jours. Scoutmaster Boy Scouts of Am., Atlanta, 1980-82. Fellow Inst. Transp. Engrs. (Marble J. Hensley Outstanding Individual Activity award 1974, Herman J. Hoose Disting. Svc. award, 1984, Karl A. Bevins Traffic Ops. award, 1992); mem. NSPE, Transp. Rsch. Bd., Nat. Acad. Forensic Engrs. Episcopalian. Achievements include development of training films and publications in traffic-responsive signalization. Home: 105 Mark Tr NW Atlanta GA 30328 Office: Ga Inst Tech Sch Civil Engring Atlanta GA 30332-0355

PARTHÉ, ERWIN, crystallographer, educator; b. Vienna, Austria, Mar. 29, 1928; s. Leopold and Stephanie (Brosch) P. PhD in Phys. Chemistry, U. Vienna, Austria, 1954; Dr.h.c. (hon.), U. Savoie, Chambery, France, 1980; Prof.h.c. (hon.), U. Vienna, 1990. Lectr. MIT, Cambridge, Mass., 1956-60; prof. materials sci. U. Pa., Phila., 1960-70; prof. structural crystallography U. Geneva, Switzerland, 1970—. Author: Crystalchemistry of Tetrahedral Structures, 1964, Crystallochimie des Structures Tetraedriques, 1972, Elements of Structural Chemistry, 1990; contbr. over 200 articles to profl. jours. Recipient William Hume Rothery award Am. Minerals, Metals and Materials Soc., 1991. Office: Univ of Geneva, 24 Quai Ernest Ansermet, Geneva Switzerland CH1211

PARTHENIADES, EMMANUEL, civil engineer, educator; b. Athens, Greece, Nov. 3, 1926; came to U.S., 1954; s. Michael and Alexandra (Vranika) P.; m. Dora Gutierrez-Aguero, Jan. 3, 1967. Diploma in civil engring., Tech. U. Athens, 1952; MSCE, U. Calif., Berkeley, 1954, PhD in Civil Engring., 1962. Design engr. various firms Athens, 1952-54; field engr., engring. analyst Dames and Moore, San Francisco, 1955-59; asst. prof. San Jose (Calif.) State U., 1962-63, MIT, Cambridge, 1963-66; assoc. prof. civil engring. SUNY, Buffalo, 1966-68; prof. engring. sci., coastal engring. U. Fla., Gainesville, 1968—; prof., chair hydraulic structures U. Thessaloniki, Greece, 1973-83; cons. engr. UN, India, 1973, 79, Argentina, 1987. Author: Introduction to Urban Hydraulics, Part A, Water Supply, 1980, Hydraulic Structures, Part 1, Open Channel Systems and Management of Natural Channels, 1982, Introduction to Maritime Hydraulics and Wave Mechanics, 1985; contbr. chpts. to books, articles and reports on cohesive sediments, stratified flow, thermo pollution, turbulence tides to profl. publs. Grantee EPA, NSF, U.S. Corps Engrs. Mem. ASCE. Achievements include pioneering work in cohesive sediment dynamics, advances in stratified flow dynamics.

PARTINEN, MARKKU MIKAEL, neurologist; b. Helsinki, Finland, Dec. 4, 1948; s. Väinö and Kerttu Elisabeth (Havunen) P.; divorced; children: Väinö Eemil, Evert Edvard J. MD, Faculty Medicine, Montpellier, France, 1975; DSc in Medicine, Epidemiology, Faculty Medicine, Helsinki, 1982; Docent Neurology, U. Helsinki, Leppävirta, Finland. Gen. practitioner Health Care Ctr., Leppavirta, Finland, 1975-76; asst. physician Clinic Neurophysiology and Medicine, Helsinki, 1975-78; resident in neurology U. Helsinki, 1978-82, asst. dept. pub. health sci., 1980-81, asst. prof. neurology, 1981-83, dir. sleep disorders unit, dept. neurology, 1983-84, staff neurologist, 1987—; dir. Ullanlinna Sleep Disorders Clinic and Research Ctr., Helsinki, 1984—, Vaajasalo Hosp., Kuopio, Finland, 1990-91; chief neurologist Kivelä Hosp., 1991—; sr. researcher epidemiology, Inst. Occupational Health, Helsinki, 1983-85; research fellow Sleep Disorders Ctr., Stanford, Calif., 1985-86; vis. lectr. Coll. Nurses, Helsinki, 1979-83; docent U. Helsinki, 1987. Spl. editor Annales Clin. Research (Sleep), 1985; editorial adv. bd. Jour. Sleep, 1986—; assoc. editor Jour. Sleep Rsch.; contbr. articles to profl. jours. Served to sub lt. medicine, Finnish Armed Forces, 1976-77. Sleep and Heart Found. fellow Cardiovascular Research Finland, 1980, internat. research fellow Fogarty Internat. Pub. Health Service-NIH, Stanford, Calif., 1985-86; grantee Paavo Nurmi Found., 1983-84, Miina Sillanpaa Found., 1983-87. Mem. Finnish Neurol. Soc., Finnish Brain Research Soc., Scandinavian Sleep Research Soc. (exec. bd. 1982—, pres. 1988—), European Sleep Research Soc. (sci. com. 1986-90, v.p. 1992—), Sleep Research Soc. U.S., Finnish Sleep Research Soc. (pres. 1988—). Evangelist Lutheran. Avocations: skiing, golf, fishing, painting, wines, tennis. Office: Kivelä Dept Neurology, Sibeliuksenkatu 12-14, SF 00260 Helsinki Finland

PARTINGTON, JOHN EDWIN, retired psychologist; b. Union Springs, N.Y., Nov. 13, 1907; s. Eliezer and Flora (Hobson) P.; m. Gwen L. Gray, Aug. 18, 1938. AB, Earlham Coll., 1929; MA in Psychology, U. Ky., 1938; postgrad., U. Chgo., 1946, Purdue U., 1959-62. Diplomate in counseling Am. Bd. Profl. Psychology; cert. psychologist, Ind. Tchr. Ky. Houses of Reform, Lexington, 1930-35; asst. to rsch. psychologist USPHS Hosp., Lexington, 1935-40; psychologist USES, Washington, 1940-42; counsellor VA, Roanoke, Va., 1946-50; psychologist US. Naval Exam. Ctr., Great Lakes, Ill., 1950-58; chief test devel. U.S. Army Enlisted Evaluation Ctr., Ft. Harrison, Ind., 1958-70; chief of rsch. U.S. Army Enlisted Evaluation Ctr., Ft. Harrison, 1970-72. Author: Leiter-Partington Adult Performance Scale, 1950, Helpful Hints for Better Living, 1990; contbr. articles to profl. jours. Chmn. adv. com. Ret. Sr. Vol. Program, 1973-88; chmn. ofcl. bd. Downtown Bapt. Ch., Lexington, Ky., 1989-91. Maj. AUS, 1942-46. Recipient Lt. Govs. Outstanding Kentuckian award, 1983. Mem. APA, Ind. Psychol. Assn. Address: 3458 Flintridge Dr Lexington KY 40517

PARTON, KATHLEEN, veterinarian; b. East Orange, N.J., Mar. 4, 1952; d. George Francis and Helen Marie (Tabor) Henry; m. Michael Conrad Parton, Dec. 31, 1990. BS, Kans. State U., 1976, DVM, 1978. Lic. vet. Vet./instr. Tsaile (Ariz.) Community Coll., 1978-79, Chinle (Ariz.) Pub. Schs., 1979-82; extension vet. specialist U. Ariz., Tucson, 1982-88, grad. assoc., 1988-89, vet. assoc., 1989-93, vet. assoc./grad. prog.-toxicology, 1990-93; sr. lectr. Vet. Clin. Sci. Massey U., New Zealand, 1993—; vet. cons. U.S. Army, Ft. Huachuca, Ariz., 1984. Author/editor: (video) Dairy Goat Care, 1988, co-author: (video) Horse Care, 1986. Bd. dirs., pres. Tucson chpt. Ariz. Right to Life, 1991-93; bd. dirs., sec. Pima County Mounted Sheriff's Posse, Tucson, 1988-90; coord. Pima County Adult Detention Ministry, Tucson, 1985-92; bd. dirs. Diocese of Tucson Pro-Life Commn., 1988-92; candidate Ariz. Senate, Dem. Party, Tucson, 1990. Mem. Am. Vet. Med. Assn., Kans. Vet. Med. Assn., Am. Assn. Lab. Animal Sci. Roman Catholic. Avocations: horse breeding and riding, photography, bowling, flying. Office: Massey U, Vet Clin Svcs, Palmerston North New Zealand

PARTOW-NAVID, PARVIZ, information systems educator; b. Iran, July 24, 1950; came to U.S., 1974; s. Mohammad and Parvin (Maleki) P.-N.; m. Mahroo Changizi, July 6, 1976; children: Puya, Rod. BBA, Tehran Bus. Coll., 1973; MBA, U. Tex., 1976, PhD, 1981. Asst. prof. Mich. Tech. U., Houghton, 1981-83; assoc. prof. Calif. State U., L.A., 1983-87, prof., 1987—; cons. Nichols Inst., San Juan Capistrano, Calif., 1987-89, Security Pacific Bank, L.A., 1990—; assoc. dir. Ctr. for Info. Resource Mgmt., L.A., 1987—; presenter in field. Author: Microcomputer Software Tools, 1987; contbr. articles, revs. to profl. publs. Mem. Inst. Mgmt. Sci., Soc. Info. Mgmt., Beta Gamma Sigma. Office: Calif State U LA Dept Info Svcs 5151 State University Dr Los Angeles CA 90032-4221

PARTRIDGE, NICOLA CHENNELL, physiology educator; b. Sydney, Australia, May 15, 1950; came to U.S., 1984; d. Malcolm Chennell and Ailsa Marion (Brockman) Smith. BS with honors, U. Western Australia, 1973, PhD, 1981. Biochemist Repatriation Gen. Hosp., Melbourne, Victoria, Australia, 1978-81; sr. rsch. officer U. Melbourne, 1981-84; rsch. assoc. Wash. U. Dental Sch., St. Louis, 1984-85; asst. rsch. prof. St. Louis U., 1985-89, asst. prof., 1989-91, assoc. prof., 1991—; ad hoc reviewer NIH, Washington, 1989—; pub. affairs Am. Soc. Bone and Mineral Rsch., Washington, 1989—. Contbr. over 50 articles to profl. jours. and book chpts. Recipient CJ Martin fellowship Nat. Health and Med. Rsch. Coun., 1984-86, Rsch. Project grant Shriners Hosps., 1986-88, NASA, 1987—, NIH, 1988—, Orthopaedic Rsch. Edn. Found., 1986-90. Office: St Louis U Med Sch 1402 S Grand Blvd Saint Louis MO 63104

PARVIN, PHILIP E., retired agricultural researcher and educator; b. Manatee, Fla., July 3, 1927; s. Clinton Fisk and Beatrice (Ward) P. MS, Miss. State U., Starkeville, 1950; PhD, Mich. State U., 1965. Asst. prof. U. Fla., Gainesville, 1952-55; extension specialist U. Calif., Davis, 1963-66; gen. mgr. Rod McLellan Co., San Francisco, 1966-68; horticulturist Maui Agrl. Rsch. Ctr. U. Hawaii, Kula, 1968-93. Contbr. over 100 articles to profl. jours. With U.S. Army, 1945-46. Fellow Am. Soc. for Hort. Sci.; mem. Am. Acad. Horticulture (hon.), South African Protea Producers and Exporters (hon. life), Internat. Protea Assn. (hon. life, chmn. rsch. com. 1989—), Protea Growers Assn. Hawaii (hon. life), Rotary (pres. Maui chpt. 1981-82). Republican. Methodist. Achievements include development of Hawaii protea industry. Home: 2395 Nuremberg Blvd Port Charlotte FL 33983

PARYANI, SHYAM BHOJRAJ, radiologist; b. Bhavnagar, Gujarati, India, July 18, 1956; came to U.S., 1966; s. Bhojraj Thakurdas and Sarswati (Shewarkanani) P.; m. Sharon Dale Goldman, May 12, 1979; children: Lisa Ann, Jason Bhojraj, Gregory Shyam. BSEE, U. Fla., 1975, MSEE, 1979, MD, 1979. Diplomate Am. Bd. of Radiology. Intern U. Tex., M.D. Anderson Hosp., Houston, 1979-80; resident Stanford (Calif.) U. Hosp., 1980-83, chief resident, 1983; dir. Williams Cancer Ctr., Bapt. Med. Ctr., Jacksonville, Fla., 1983—, Fla. Cancer Ctr., Jacksonville, Fla., 1985—; bd. dirs. Bapt. Med. Ctr., Jacksonville, 1986—, Meml. Med. Ctr., Jacksonville, 1987—, Meth. Hosp., Jacksonville, 1988—. Contbr. articles to profl. jours. Pres. Am. Cancer Soc., Jacksonville, 1992; bd. dirs. Jacksonville C. of C., 1991; adv. bd. Boy Scouts, 1990—. Mem. Am. Cancer Soc. (pres. 1992), Rotary Club. Republican. Hindu. Achievements include patent in Scott-Paryani Quick Implanter. Office: Fla Cancer Ctr #1500 3599 University Blvd S Jacksonville FL 32216

PASAMANICK, BENJAMIN, psychiatrist, educator; b. N.Y.C., N.Y., Oct. 14, 1914; s. Alex and Elizabeth (Moskalik) P.; m. Hilda Knobloch, May 1, 1942 (div. July 1982); m. Lidia Laba, Aug. 27, 1982. A.B., Cornell U., 1936; M.D., U. Md., 1941. Intern State Hosp., Bklyn., 1941, Harlem Hosp., N.Y.C., 1942; resident N.Y. State Psychiat. Inst., 1943; asst. Clinic Child Devel., Yale U., New Haven, 1944-46; chief children's in-patient and outpatient services Neuropsychiat. Inst., U. Mich. Hosp., Ann Arbor, 1946-47; chief children's psychiat. services Kings County Hosp., Bklyn., 1947-50; psychiatrist Phipps Clinic, Johns Hopkins Hosp., Balt., 1952-55, Harriet Lane Home, 1951-55; asst. prof. Johns Hopkins U. Sch. Hygiene and Pub. Health, 1950-52, asso. prof. pub. health adminstrn., 1953-55; prof. psychiatry Ohio State U., 1955-65, adj. prof. sociology and anthropology, 1963-65; clin. prof. psychiatry U. Ill. Coll. Medicine, Chgo., also Chgo. Med. Sch.; asso. dir. research Ill. Dept. Mental Health, 1965-67; pres. N.Y. Sch. Psychiatry, N.Y.C., 1968-72; Sir Aubrey and Lady Hilda Lewis prof. social psychiatry N.Y. Sch. Psychiatry, 1972-84; asso. commr. N.Y. State Dept. Mental Hygiene, 1968-76; adj. prof. psychology NYU, 1968-75, research prof. psychiatry, 1984—; adj. prof. epidemiology Sch. Pub. Health and Adminstrv. Medicine, Columbia, 1967-77; adj. prof. pediatrics Albany Med. Coll., 1972-77, research prof., 1978-80, prof. emeritus, 1980, research prof. med. library scis., 1978-84; clin. prof. preventive and community medicine N.Y. Med. Coll., 1976-77; Cutter lectr. preventive medicine Harvard, 1960; research prof. psychiatry and behavioral sci. Sch. Medicine, SUNY, Stony Brook, 1978-84. Author 17 books.; contbr. numerous articles to profl. jours.; mem. numerous editorial bds. of professional jours. Mem. Am. Psychopath. Assn. (Stratton award 1961, pres. 1967, Hamilton medal 1968), Am. Orthopsychiat. Assn. (pres. 1971), Am. Psychiat. Assn. (Hofheimer Rsch. award 1949, 67, Agnes Purcell McGavin award 1986), Am. Pub. Health Assn. (governing coun. 1956, Rema Lapouse Gold medal 1977), Am. Anthrop. Assn., Am. Sociol. Assn., Am. Coll. Psychiatrists, Am. Coll. Epidemiologists, Am. Coll. Neuropsychopharmacology, Psychonomic Soc., Soc. Biol. Psychiatry, Acad. Child Psychiatry, Am. Psychol. Assn. (pres. div. child, youth and family svcs. 1987), Theobald Smith Soc. (founder, pres. 1986-93, sec. 1993), History of Sci. Soc., Soc. for Study of Social Problems, others. Address: 12 N Ferry St Schenectady NY 12305

PASCALI, RARESH, mechanical engineering educator; b. Bucharest, Romania, Jan. 20, 1966; s. Dan D. and Cornelia (Luca) P. BS in Aerospace Engring., Poly. U., Bklyn., 1990; MS in Aeronautics Astronautics, Poly. U., 1993. Adj. prof. Poly. U., Bklyn., 1989-90, instr., 1990—; cons. GASL, Long Island, N.Y., 1991. Mem. AIAA, Tau Beta Pi, Sigma Gamma Tau. Democrat. Greek Orthodox. Home: 86-03 102nd St Apt 2L Jamaica NY 11418 Office: Poly U 333 Jay St Brooklyn NY 11201

PASCUCCI, JOHN JOSEPH, environmental engineer; b. Glen Cove, N.Y., Oct. 14, 1949; s. John Joseph and Dorothy Esther (Rhoades) P.; m. Joanne Palestri; children: Justin, Jonathan, Jesse. B in Engring., Stevens Inst. Tech., 1971; M in Engring., NYU, 1975. Profl. engr. N.Y.; diplomate Am. Acad. Environ. Engrs. Chief engr. Nassau County Dept. Pub. Works, Mineola, N.Y., 1971--. Mem. N.Y. Water Environment Assn. (com. chmn. 1976—). Republican. Roman Catholic. Office: Nassau County Dept Pub Work 1 West St Mineola NY 11501

PASETTI, LOUIS OSCAR, dentist; b. Tampa, Fla., Dec. 27, 1916; s. Joseph G. and Carmen (Gonzalez) P.; m. Mary Mendez, Jan. 11, 1942; children: Louis M., Arleen Pasetti Shearer. BS, U. Fla., 1937; DDS, Emory U., 1941; postgrad., U. Pa., 1978. Capt. U.S. Army, 1942-46; dentist pvt. practice Tampa, Fla., 1947—. Past. pres. Tampa Civitan Club, 1953; past lt. gov. Civitan Clubs of Tampa, 1962; past dep. gov. Civitan Internat., Tampa, 1964; fin. officer Am. Legion Post 248. Named Fla. Dentist of the Yr., Fla. Acad. Gen. Dentistry, 1983; recipient meritorious Svc. award Fla. Acad. Gen. Dentistry, 1989, Disting. Svc. award, 1985. Fellow Acad. Gen. Dentistry, Am. Coll. Dentists, Internat. Coll. Dentists, Acad. Dentistry Internat.; mem. Am. Dental Assn., Fla. Dental Assn., Fla. Acad. Gen. Dentistry (pres. 1981), Tampa Bay Acad. Gen. Dentistry (pres. 1977-78), Elks Club, Round Table of Civic Clubs of Tampa (sec. 1953), Palma Ceia Golf and Country Club.. Democrat. Roman Catholic. Avocations: photography, orchid culture. Home: 10023 Hampton Pl Tampa FL 33618-4227 Office: 220 E Madison St Ste 250 Tampa FL 33602-4826

PASKIN, ARTHUR, physicist, consultant; b. N.Y.C., Feb. 15, 1924; s. Max and Pauleen (Jacobs) P.; m. Charlotte R. Lipson, Nov. 1, 1953; children: Judith A., Carol J., Amy R. BS in Physics, S.D. Sch. Mines, 1948; PhD in Physics, Iowa State U., 1953. Engring. physicist Sylvania Electric Products, Mineola, N.Y., 1953-55; physicist U.S. Army Materials Rsch. Lab., Watertown, Mass., 1955-63; asst. head metallurgy Brookhaven Nat. Lab., Upton, N.Y., 1963-68; prof. physics CUNY, 1968-90; sr. ptnr. Forensic Scis., Port Jefferson, N.Y., 1990—; cons. Brookhaven Nat. Lab., Upton, 1968—, Time-Life Publs., N.Y.C., 1963-68. Editor Physics & Chemistry of Liquids jour., 1963-68. Mem. Bd. Edn., Port Jefferson, 1968-74. With Signal Corps, 1942-45, ETO. Recipient Sec. of Army fellowship U.S., 1960, Outstanding Rsch. award U.S. Dept. Energy, 1988. Fellow Am. Phys. Soc.; mem. Soc. Automotive Engrs., Sigma Xi. Achievements include first introduction of computer graphics of accident reconstruction into U.S. Court; 3-D film of solid (atomic) melting; computer simulations of fracture at atomic level. Home: 8 High Path Belle Terre NY 11777 Office: Forensic Scis PO Box 563 Port Jefferson NY 11777

PASQUALE, FRANK ANTHONY, engineering executive; b. Jersey City, Oct. 27, 1954; s. Frank F. and Josephine (Marano) P.; m. Elaine J. Rinaldi; children: Frank A. II, Phillip, Marielle. BSME, Fairleigh Dickinson U., 1976; postgrad., U. Mich., 1977. Registered profl. engr., N.J. Supr. chassis engr. Ford Co., Dearborn, Mich., 1976-80; engr. mgr. Bavarian Motor Works of N.Am., Montvale, N.J., 1980-82; engring. mgr. Purolator Products Inc, Rahway, N.J., 1982-85; dir. ops. Inverness Inc., Fairlawn, N.J., 1985-87; v.p. engring. Transworld Inc., East Rutherford, N.J., 1987-91; sr. v.p. engring. Henschel-Steinau, Inc., Englewood, N.J., 1991-92; pres. FPM Devel. Corp., Towaco, N.J., 1992—; cons. Marine-Tech. Corp., Wood-Ridge, N.J., 1985—; bd. dirs. Spin-Tech. Corp., Hoboken, N.J., Internat. Offshore Tackle, Inc., Montville, N.J. Patentee in field. Mem. ASME, Soc. Plastic Engrs., Soc. Automotive Engrs., Soc. Mfg. Engring. Avocations: sailing, woodworking, fishing, tennis, reading. Home: 9 Mulbrook Ln Towaco NJ 07082

PASQUEL, JOAQUIN, chemical engineer; b. Mexico City, Sept. 26, 1960; s. Roberto J. and Yolanda (Guerra) P.; m. Consuelo C. Cifuentes, Dec. 27, 1985; children: Joaquin, Alvaro. B.Chem.Engring., U. Iberoamericana, Mexico City, 1984; MSc in Chem. Engring., U. Simon Bolivar, Caracas, Venezuela, 1986. Rsch. staff U. Iberoamericana, Mexico City, 1983-85, U. Simon Bolivar, Caracas, 1985-86; project engr. Procter & Gamble Latin Am., Caracas, 1987-89, group mgr., 1989—. Contbr. articles to profl. jours. Office: Procter & Gamble PD M-108 PO Box 20010 Miami FL 33102-0010

PASSARELLA, LOUIS ANTHONY, systems engineer; b. Camden, N.J., Apr. 1, 1964; s. Louis Anthony and Ermina (Gattuso) P.; m. Kimberley Beth Rogers, Aug. 26, 1989. AS in Electronics, Lincoln Tech. Inst., 1984. Engring. technician Franklin Computers, Pennsauken, N.J., 1985-86; systems analyst Corp. Tech., Cherry Hill, N.J., 1986-90; sr. systems engr. Bloc Bus. Systems Advanced Tech. Group, Cherry Hill, 1990—; pres. Dycom Inc., Blackwood, N.J., 1990—; v.p. ISIS Group, Mt. Laurel, N.J., 1992—. Mem. Smithsonian Assocs. Republican. Roman Catholic. Home: 28 Blue Jay Dr Clementon NJ 08021 Office: ISIS Group Inc 3747 Church Rd Ste 100 Mount Laurel NJ 08054

PASSNER, ALBERT, physicist, researcher; b. N.Y.C., Aug. 30, 1938; s. Sam and Bertha (Katz) P.; m. Florence Edith Feldman, June 10, 1962; children: Jonathan M., Deborah S., Daniel B. BS, CUNY, 1960; MS, NYU, 1966. Engr. RCA, Heightstown, N.J., 1961-63; mem. staff Princeton (N.J.)-Penn Accelerator, 1963-69; mem. tech. staff AT&T Bell Labs., Murray Hill, N.J., 1969—; mem. adv. com. Queensborough Community Coll., N.Y.C., 1987—. Contbr. articles to IEEE Jour. Quantitative Electronics; contbr. articles to profl. jours. co-chair Jewish Community Rels. Coun., 1992—. Jewish. Achievements include research to produce transverse lasing in a semiconductor, optical bistability in a semiconductor, producing a positron plasma in the laboratory, building 65 Tesla pulsed magnet. Home: 3 Disbrow Ct East Brunswick NJ 08816-3569 Office: AT&T Bell Labs 600 Mountain Ave New Providence NJ 07974-2010

PASSWATER, RICHARD ALBERT, biochemist, author; b. Wilmington, Del., Oct. 13, 1937; s. Stanley Leroy and Mabel Rosetta (King) P. BS, U. Del., 1959; PhD, Bernadean U., 1976. Cert. firefighter; m. Barbara Sarah Gayhart, June 2, 1964; children: Richard Alan, Michael Eric. Supr. instrumental analysis lab. Allied Chem. Corp., Marcus Hook, Pa., 1959-64; tech. svcs. rep. F&M Sci. Corp., Avondale, Pa., 1965; dir. applications lab. Am. Instrument Co., Silver Spring, Md., 1965-77; dir. Am. Gen. Enterprises, Minn.; former daily broadcaster Sta. WMCA, N.Y.C., 1980-88; former daily broadcaster Sta. WRNG, Atlanta, 1982-85, Sta. WMCA; rsch. dir. Solgar Nutritional Rsch. Ctr., 1979—, corp. v.p. Solgar Co., Inc.; chmn. Worcester County Emergency Planning Com.; bd. dirs. Worcester Meml. Hosp., Atlantic Gen. Hosp., River Run Assn; pres. 1989-92, Subaqueous Exploration and Archeology Ltd. Author: Guide to Fluorescence Literature, vol. 1, 1967, vol. 2 1970, vol. 3, 1974, Supernutrition, 1975, Supernutrition For Healthy Hearts, 1977, Super Calorie, Carbohydrate Counter, 1978, Cancer and Its Nutritional Therapies, 1978, 83, 93, The Easy No-Flab Diet, 1979, Selenium As Food and Medicine, 1980, The Slendernow Diet, 1982, (with Dr. E. Cranton) Trace Elements, Hair Analysis and Nutrition, 1983, The New Supernutrition, 1991, The Longevity Factor, 1993; contbg. author: Fire Protection Guide to Hazardous Materials, 1991; editor Fluorescence News, 1966-77, Jour. Applied Health Scis., 1982-83; mem. editorial bd. Nutritional Perspectives, 1978-96, The Body Forum, 1979-80, Jour. Holistic Medicine, 1981-88, VIM Newsletter, 1979—; contbg. editor Firehouse Mag., 1988—, Jour Applied Nutrition; contbr. over 290 health articles to mags.; co-editor booklet series Your Good Health; sci. adv. and columnist Whole Foods mag.; patentee in field. bd. dirs. Sci. Documentation Ctr., Dunfermline, Eng.; Am. Found. Firefighter Health and Safety; chief Ocean Pines Vol. Fire Dept., 1984—; active Emergency Med. Tech.; adviser Nat. Inst. Nutrition Edn.; past adv. bd. Stephen Decatur High Sch., Worcester County Dept. Edn. Cubmaster, 1975-79. Named Citizen of Yr. Ocean Pines, Md., 1987, 5th Ann. Achievement award, 1989, VFW Cert. of Commendation, 1988, Industry award Nat. Inst. Nutritional Edn., 1991, . Fellow Internat. Acad. Preventive Medicine, Am. Inst. Chemists; mem. Am. Chem. Soc., Gerontology Soc., Am. Geriatric Soc., Am. Aging Assn., Soc. Applied Spectroscopy, Internat. Found. Preventive Medicine (v.p.), Internat. Union Pure and Applied Chemistry, Royal Soc. Chemistry (London), Internat. Acad. Holistic Health and Medicine, ASTM, Capital Chem. Soc., AAAS, Nutrition Today Soc., Am. Acad. Applied Health Scis. (pres., bd. dir.), Internat. Found. Preventive Medicine (v.p., dir.), Inst. Nutritional Rsch., Internat. Platform Assn., N.Y. Acad. Scis., Nat. Fire Protection Assn. (cert. firefighter level III, com. on properties of hazardous chemicals), Pi Kappa Alpha. Office: 11017 Manklin Meadows Ln Berlin MD 21811-9342

PASTAN, IRA HARRY, biomedical science researcher; b. Winthrop, Mass., June 1, 1931; s. Jacob and Miriam (Ceder) P.; m. Linda Olenik, June 14, 1953; children—Stephen, Peter, Rachel. B.S., Tufts U., 1953, M.D., 1957. Med. house officer Yale U. New Haven Hosp., 1957-59; clin. assoc. Nat. Inst. Arthritis and Metabolic Disease, NIH, Bethesda, Md., 1959-61, sr. investigator sect. on endocrine biochemistry, clin. endocrinology br., 1963-69; postdoctoral fellow Lab. of Cellular Physiology Nat Heart and Lung Inst. NIH, Bethesda, Md., 1961-62; head molecular biology sect. endocrinology br. Nat. Cancer Inst., NIH, Bethesda, Md., 1969-70, chief lab. molecular biology, 1970—. Author: An Atlas of Immunofluorescence, 1985; author, editor: Endocytosis, 1985; contbr. articles to profl. jours. Recipient Van Meter prize Am. Thyroid Assn., 1971, Superior Service award Dept. HEW and NIH, 1973, Meritorious Service medal USPHS, NIH, Nat. Cancer Inst., 1983, Disting. Service medal, 1985. Mem. Nat. Acad. Scis., Am. Soc. Biol. Chemists, Am. Soc. Clin. Investigation, Am. Soc. Cell Biology, Am. Soc. Microbiology, Peripatetic Club. Office: Nat Cancer Inst 900 Rockville Pike 37-4E16 Bethesda MD 20892

PASTERNACK, ROBERT FRANCIS, chemistry educator; b. N.Y.C., Sept. 20, 1936; 2 children. B.A., Cornell U., 1957, Ph.D. in Chemistry, 1962. Research assoc. in chemistry U. Ill., Champaign, 1962-63; from asst. to prof. chemistry Ithaca Coll., N.Y., 1963-76, Charles A. Dana Endowed prof. chemistry, 1976-82; Edmund Allen prof. chemistry Swarthmore Coll., Pa., 1984—; invited speaker seminars, colls., univs., nat., internat. meetings, confs. including Bioinorganic Chem., Italy, Portugal, Gordon Rsch. Confs., Spanish Royal Soc. Chem., many others; lectr. series Nankai U., China, U. Messina, Italy; mem. adv. com. Rsch. Corp.; mem. sci. & art com. Franklin Inst.; co-organizer, chmn. workshop on rsch. at undergrad. instn. NSF, mem. undergrad. curriculm chem.; vis. prof., vis. rschr. U Messina, U. Paris, Nakai, Rome, King's Coll., London, Fritz Haber Inst., Berlin; co-developer A Unified Lab. Program; initiator, chmn. C.P. Snow Lectr. Series. Author, co-author more than 90 sci. pubs. Mem. com. on sci. and the arts Franklin Inst., 1992—. Grantee NSF, 1965-68, 69-72, 77-78, 83-84, 86—, Petroleum Rsch. Fund, 1967-74, 86-88, NIH, 1971-89, Monsanto Corp., 1986-92, Rsch. Corp., 1974-75, 78-79, 84-85, Danforth Assocs., 1978-84, Camille and Henry Dreyfus Found., 1981, 85, NATO, 1979, 88-89; recipient Camille and Henry Dreyfus Teaching/Scholar award, 1987-89, NSF Manpower Improvement award, King's Coll., U. London, 1977-78; NSF sci. faculty fellow U. Rome, 1968-70. Mem. AAAS. Am. Chem. Soc., N.Y. Acad. Sci., Sigma Xi. Office: Swarthmore Coll Dept Chemistry Swarthmore PA 19081

PASTEUR, NICOLE, population geneticist; b. Terrasson, Dordogne, France, July 11, 1944; d. Andre and Luce (Brachet) Mercier; m. Georges Pasteur, Aug. 25, 1965; 1 child, Aude. PhD, Southwestern Med. Sch., Dallas, 1972; D of State, U. de Montpellier II, France, 1977. Libr. Inst. Sci. Cherifien, Rabat, Morocco, 1965-67; attaché de recherche CNRS, Montpellier, France, 1972-80; dir. de recherche CNRS, Montpellier, 1981—. Recipient Silver medal Centre Nat. de la Recherche Scientifique, France, 1987. Achievements include research in genetics of resistance to insecticides by insects; molecular and population studies of gene amplification. Office: U de Montellier II, E Bataillon, 34095 Montpellier France

PASTOR, RICHARD WALTER, research chemist; b. Oceanside, N.Y., June 21, 1951; s. William Henry and Alma Dolores (Strachy) P.; m. Dale Melanie Seecof, June 21, 1981; children: William Abraham, Joseph Mark. BA, Hamilton Coll., 1973; MS, Syracuse U., 1977; PhD, Harvard U., 1984. Sr. staff fellow Ctr. for Biology Evaluation & Rsch., FDA, Bethesda, Md., 1984-90, rsch. chemist, 1990—; adj. prof. Am. U., Washington, 1991—. Contbr. articles to profl. jours. Achievements include development of structural and dynamic microscopic model of lipid bilayers; development of stochastic dynamic computer simulation methods. Home: 8 Ingleside Ct Rockville MD 20850 Office: CBER/FDA Biophysics Lab 8800 Rockville Pike Bethesda MD 20892

PASTOR, STEPHEN DANIEL, chemistry educator, researcher; b. New Brunswick, N.J., Feb. 15, 1947; s. Stephen and Irene (Bors) P.; m. Joan Ordemann, Apr. 3, 1971 (div. 1979); 1 child, Melanie; m. Joanne Behrens, July 13, 1985 (div. 1990). BA in Chemistry, Rutgers U., 1969, MS in Chemistry, 1978, PhD in Chemistry, 1983. Chemist Nat. Starch and Chem. Corp., Bridgewater, N.J., 1972-79; rsch. group leader CIBA-Geigy Corp., Ardsley, N.Y., 1979-84, rsch. mgr., 1985-87; group leader Cen. Rsch. Labs. CIBA-GEIGY AG, Basel, Switzerland, 1987-89; rsch. fellow CIBA-GEIGY Corp., Ardsley, 1989-90, rsch. mgr., 1990—; asst. adj. prof. PACE U., Pleasantville, N.Y., 1984—, assoc. adj. prof., 1989—; mechanistic organic chemistry; conformation of large-membered heterocycles. Contbr. articles to profl. jours. 63 patentees in field. Served to 1st lt. U.S. Army, 1969-71, Vietnam. Mem. Am. Chem. Soc., Swiss Chem. Soc. Current work: Organophosphorus and organosulfur chemistry, organometallic chemistry, asymmetric synthesis, homogeneous catalysis. Home: 27 Crows Nest Ln Unit 4F Danbury CT 06810-2005

PASTUSZAK, WILLIAM THEODORE, hematopathologist; b. Stamford, Conn., Feb. 18, 1946; s. William and Ethel Louise (Dance) P.; m. Patricia Marie Nelson, June 14, 1969. BS, Tufts U., 1968; MD, U. Conn., 1972. Cert. anatomic pathology and hematology Am. Bd. Pathology. Asst. dir. hematology Hartford (Conn.) Hosp., 1976-81, dir. hematology, 1981—, asst. dir. pathology and lab. medicine, 1986—; pres. Hartford (Conn.) Pathology Assocs., P.C., 1992—; pres. Conn. Soc. Pathologists, 1983-84; mem. govt. affairs com. Coll. Am. Pathologists, Washington, 1986—, vice chmn., 1991—. Fellow Am. Soc. Clin. Pathology, Am. Soc. Hematology, Coll. Am. Pathologists. Office: Hartford Hosp Dept Pathology 80 Seymour St Hartford CT 06115

PATANKAR, SUHAS V., engineering educator. BE in Mech. Engring., Coll. Engring., Poona, India, 1962; M Tech. in Mech. Engring., Indian Inst. Tech., Bombay, 1964; PhD in Mech. Engring., Imperial Coll. Sci. and Tech., London, 1967. Various teaching and rsch. positions Indian Inst. Tech., Bombay, Imperial Coll., London, U. Waterloo, Can.; prof. mech. engring. U. Minn., Mpls., 1975—. Author, co-author 4 books; contbr. over 100 articles, papers to profl. publs. Recipient Alumni award Indian Inst. Tech., 1983. Mem. ASME (divsn. heat transfer, presenter various seminars, confs., Heat Transfer Meml. award 1991). Achievements include research in the calculation method for two-dimensional boundary layers, the development of computational techniques for fluid flow and heat transfer, and their application to practical problems. Office: Univ of Minnesota Dept or Mechanical Eng 111 Church St SE Minneapolis MN 55455*

PATCHETT, ARTHUR ALLAN, medicinal chemist, pharmaceutical executive; b. Middletown, N.Y., May 28, 1929; s. Arthur Allan and Anna Gertrude (Vossler) P.; m. Lois Rhoda Mc Neil, Aug. 18, 1962; Thomas John, Steven Edward. BA, Princeton U., 1951; PhD, Harvard U., 1955. Rsch. assoc. NIH, Bethesda, Md., 1955-57; rsch. chemist Merck Rsch. Labs., Rahway, N.J., 1957-62; dir. synthetic chem. rsch. Merck Rsch. Labs., Rahway, 1962-69, sr. dir. synthetic chem. rsch., 1969-71, sr. dir. new lead discovery, 1971-76, exec. dir. new lead discovery, 1976-88, v.p. exploratory chemistry, 1988—. Contbr. 125 papers to profl. jours., sci. confs. Elected N.J. Inventors Hall of Fame, N.J. Inst. Tech., 1990; recipient Discoverers award Pharm. Mfrs. Assn., 1992. Mem. AAAS, Am. Chem. Soc. (chmn. div. medicinal chemistry 1971, E.B. Hershberg Important Discoveries in Medicinally Active Substances award 1993). Achievements include 126 U.S. patents (co-holder); co-inventor antihypertensive drug Vasotec; key contbr. to discovery of cholesterol lowering drug Mevacor. Home: 1090 Minisink Way Westfield NJ 07090 Office: Merck Rsch Labs PO Box 2000 Rahway NJ 07065

PATE, FINDLAY MOYE, agriculture educator, university center director; b. Davisboro, Ga., Jan. 24, 1941; s. William Wayne and Valeria Moye P.; m. Vicky Lee Scruggs, Jan. 15, 1961; children: Julie, Celia, Joel, Craig. Student, Abraham Baldwin Agr., Tifton, Ga., 1961-63; BS, U. Ga., 1965, PhD, 1970; MS, Oreg. State U., 1967. From asst. to assoc. prof. Everglades rsch. ctr. U. Fla., Belle Glade, 1970-83; prof., dir. Ona rsch. ctr. U. Fla., Ona, 1983—. Methodist. Achievements include development of a feeding system of adding natural protein to a liquid molasses feed for cattle. Office: U Fla Ona Rsch Ctr Rt 1 Box 62 Ona FL 33865

PATE, SAMUEL RALPH, engineering corporation executive; b. Thorsby, Ala., Oct. 27, 1937; s. Ralph Elvin and Frances Roberta (Marcus) P.; children: Lisa, Sherri, Frances. BS in Aero. Engring., Auburn U., 1960; MS in Mech. Engring., U. Tenn., 1965, PhD, 1977. Registered profl. engr., Tenn. With Sverdrup Tech., Inc., 1960-69, project and rsch. engr. AEDC, 1969-74, mgr. rsch., engring., and testing projects br., 1975-78, dep. dir., Propulsion Wind Tunnel Facility, 1978-83; v.p. and gen. mgr. Tech. Group, Tullahoma, Tenn., 1983, sr. v.p. Tech. and Operational Groups, 1983—; cons., 1984-87; v.p., gen. mgr. AEDC Engring. and Ops. Group, 1987-90, Sverdrup Corp A & E Group, 1990—; pres. Sverdrup Tech., Inc.; adviser U.S.-German Aero. Data Exchange; guest lectr., U. Tenn. Contbr. articles, reports to profl. jours. Served with NG, 1955-59. Recipient Arch T. Colwell Merit award Soc. Automotive Engrs., 1983. Fellow AIAA (assoc., chmn. nat. ground testing com., Gen. H.H. Arnold award Tenn. sect. 1969). Office: 600 William Northern Blvd PO Box 884 Tullahoma TN 37388

PATEL, BHAVIN R., chemical engineer; b. Ahmedabad, Gujarat, India, June 8, 1963; came to U.S., 1983; s. Rambhai M. and Shardaben R. (Manibhai) P.; m. Trupti B. Patel, May 20, 1986; 1 child, Neil B. Diploma chem. engring., S.B.M. Polytech, Bombay, India, 1983; MSChemE, N.J. Inst. Tech., 1987. Process engr. J.M. Huber Corp., Edison, N.J., 1987-90; mgr. chem. mfg. Div. Harcros Chems., Hardman, Belleville, N.J., 1990—. Home: 25 New Brooklyn Rd Edison NJ 08817

PATEL, CHANDRA KUMAR NARANBHAI, communications company executive, researcher; b. Baramati, India, July 2, 1938; came to U.S., 1958, naturalized, 1970; s. Naranbhai Chaturbhai and Maniben P.; m. Shela Dixit, Aug. 20, 1961; children: Neela, Meena. B.Engring., Poona U., 1958; M.S., Stanford U., 1959, Ph.D., 1961. Mem. tech. staff Bell Telephone Labs., Murray Hill, N.J., 1961-93, head infrared physics and electronics rsch. dept., 1967-70, dir. electronics rsch. dept., 1970-76, dir. phys. rsch. lab., 1976-81, exec. dir. rsch. physics and acad. affairs div., 1981-87, exec. dir. rsch., materials sci., engring. and acad. affairs div., 1987-93; trustee Aerospace Corp., L.A., 1979-88; vice chancellor rsch. UCLA, 1993—; mem. governing bd. NRC, 1990-91; bd. dirs. Newport Corp., Fountain Valley, Calif., Cal Micro Devices Corp., Milpitas, Calif. Contbr. articles to tech. jours. Recipient Ballantine medal Franklin Inst., 1968, Coblentz award Am. Chem. Soc., 1974, Honor award Assn. Indians in Am., 1975, Founders prize Tex. Instruments Found., 1978, award N.Y. sect. Soc. Applied Spectroscopy, 1982, Schawlow medal Laser Inst. Am., 1984, Thomas Alva Edison Sci. award N.J. Gov., 1987. Fellow AAAS, IEEE (Lamme medal 1976, medal of honor 1989), Am. Acad. Arts and Scis., Am. Phys. Soc. (coun. 1987-91, exec. com. 1987-90, George E. Pake prize 1988), Optical Soc. Am. (Adolph Lomb medal 1966, Townes medal 1982, Ives medal 1989), Indian Nat. Sci. Acad. (fgn.); mem. NAS (coun. 1988-91, exec. com. 1989-91), NAE (Zworykin award 1976), Gynecol. Laser Surgery Soc. (hon.), Am. Soc. for Laser Medicine and Surgery (hon.), Third World Acad. Scis. (assoc.). Home: 1171 Roberto Ln Los Angeles CA 90077 Office: UCLA Office of Chancellor 405 Hilgard Ave Los Angeles CA 90024-1405

PATEL, JAYANT RAMANLAL, mechanical engineer administrator; b. Ahmedabad, India, July 15, 1946; came to the U.S., 1968; s. Ramanlal S. and Chanchalben R. Patel; m. Gita Jayant, May 14, 1972; children: Reepal, Reena. BS in Mech. Engring., Gujarat U., 1967; MS, N.D. State U., 1969; MBA, U. Minn., 1990. Registered profl. engr. Minn., Ill. Energy engr. Michaud, Cooley, Erickson Cons. Engr., Mpls., 1969-77; energy prin. engr. Honeywell Tech. Strategy Ctr., Mpls., 1977-81; energy cons. Honeywell Def. Group, Mpls., 1981-91; engring. mgr. hosp. bus. unit mktg. group Honeywell Inc., Mpls., 1991—. Contbr. articles to Energy Engring. Jour., Assn. Energy Engrs. Jour., ASHRAE JOur. Recipient 9th Ann. Energy award State of Minn., 1984, 85, Innovation in Energy Conservation award U.S. Dept. Energy, 1985. Mem. Assn. Energy Engrs. (bd. dirs. cert. energy contractor bd. 1985-88), ASHRAE. Home: 2229 Heritage Ln New Brighton MN 55112 Office: Honeywell Inc 2701 4th Ave S MN 27-7246 Minneapolis MN 55408

PATEL, KANTI SHAMJIBHAI, civil engineer; b. Munjiasar, Gujarat, India, June 28, 1938; came to U.S. 1968; s. Shamjibhai Premjibhai and Raniben (Dholaria) Ramani; m. Shanta, Mar. 4, 1967; children: Manish, Rajni. BSCE, Gujarat U., India, 1962. Registered profl. engr., Pa. Engr. Capitol Engring., Dillsburg, Pa., 1968-69, J.B. Ferguson, Hagerstown, Md., 1969-70, Buchart-Horn, Inc., York, Pa., 1970-73, Greenhorne-O'Mara Inc., Riverdale, Md., 1973-76; chief civil engr. Williams and Sheladia, Inc., Mt. Rainier, Md., 1976-78; engr. Ebasco, Inc., Atlanta, 1978-80; dept. head civil engring. Orba Corp., Mt. Lakes, N.J., 1980-88; project mgr. VEP Assocs., Inc., Parsippany, N.J., 1988—; cons. E. Harris Assocs., Fredericks, Md., 1976-78, A.K. Data Corp., Randolph, N.J., 1986-88. Home: 26 Delbrook Rd Morris Plains NJ 07950

PATEL, KIRITKUMAR NATWERBHAI, researcher, chemist; b. Eldoret, Kenya, Dec. 10, 1946; s. Natwerbhai Ishwerbhai and Savitaben Patel; m. Ranjanben K. Patel, Dec. 12, 1971; children: Prital, Jayna. BSc, S.P. U., V.V. Nagar, India, 1967. Chemist Kenya Tanning Exp. Co. Ltd., Thika, 1968-77; co. chemist E.A. Tanning Exp. Co. Ltd., Eldoret, 1978-80; rsch. officer, co-chemist Eatec Ltd., Eldoret, 1981—, div. head, 1991—. Bd. dirs. Lions Club Internat., Eldoret, 1984-87; chmn. Eld. Brotherhood, Eldoret, 197-89. Grantee The Agrl. Rsch. Fund, 1992. Mem. Am. Chem. Soc., Internat. Soc. of Mushroom Sci., Soc. of Leather Technologists and Chemists. Achievements include rsch. on brine preservation of mushrooms, mushroom munchies, production of liquid Ammonia from mushroom tunnels, adhesives from vegetable tanning extracts for plywood/chipboard, seed coating systems for cereal grains, medicinal capsules from shiitake mushrooms, mushroom spawn technology in Kenya. Home: Kenmosa Village, Eldoret Kenya Office: Eatec Ltd, PO Box 190 Kapgat Rd, Eldoret Kenya

PATEL, KISHOR MANUBHAI, nuclear medicine physicist; b. Tavdi, India, Aug. 22, 1953; came to U.S., 1986; s. Manubhai Morarji and Savitaben (Mangubhai) P.; m. Kusum Kishor, Nov. 15, 1976; children: Nisha, Silpa, Sheena. BS, Liverpool, 1975; MS, Birmingham, 1976; PhD, Surrey, 1981. Health physicist British Nuclear Fuels Ltd., Sellafield, England, 1981-83; sr. systems engr. GE, St. Albans, England, 1983-84; sr. physicist Leicester (England) Royal Infirmary, 1984-86; med. physicist Med. Coll. Va., Richmond, 1986-87; nuclear med. physicist Allegheny Gen. Hosp., Pitts., 1987-91, The Christ Hosp., Cin., 1991—. Mem. Am. Assn. Physicists Medicine, Soc. Nuclear Medicine. Office: The Christ Hosp 2139 Auburn Ave Cincinnati OH 45219

PATEL, MANU AMBALAL, environmental engineer; b. Navli, Gujarat, India, Aug. 30, 1943; came to U.S., 1969; s. Ambalal Ranchhodbhai and Shardaben A. (Shardaben) P.; m. Ranjan Patel, Apr. 29, 1967; children: Hamang, Rita. BSCEn, U. Baroda, India, 1965; MPH in Engring., U. Baroda, 1968; MS in Sanitary Engring., U. N.Mex., 1970. Registered profl. engr., Md., Pa. Civil engr. Narmada (India) Irrigation Project, 1965-66; grad. asst. U. N.Mex., Albuquerque, 1969-70; chief of facilities sect. div. wastewater engring. City of Balt. Dept. Pub. Works, 1970—. Contbr. tech. papers to profl. publs. Active temple, other Hindu religious orgns., Balt., Washington. Mem. ASCE, Water Environ. Fedn. Democrat. Hindu. Achievements include research in removing nitrogen and phosphorus from wastewater. Home: 6612 Hunters Wood Cir Baltimore MD 21228 Office: Balt Dept Pub Works 900 Abel Wolman Mcpl Bldg Baltimore MD 21202

PATEL, MUKUND RANCHHODLAL, electrical engineer, researcher; b. Bavla, India, Apr. 21, 1942; came to U.S., 1966; s. Ranchhodlal N. and Shakariben M. Patel; m. Sarla Shantilal, Nov. 4, 1967; children: Ketan, Bina, Vijal. BEng, Sardar U., Vidyanagar, India; MEng with honors, Gujarat U., Ahmedabad, India; PhD in Engring., Rensselaer Poly. Inst., 1972. Registered profl. engr., Pa.; chartered mech. engr., U.K. Lectr. elec. engring. Sardar U., Vidyanagar, India, 1965-66; sr. devel. engr. GE, Pittsfield, Mass., 1967-76; mgr. R & D, Bharat Bijlee (Siemens) Ltd., Bombay, 1976-80; fellow engr. Westinghouse R & D Ctr., Churchill, Pa., 1980-84, mem. senate, 1982-84; pres. Induction Gen., Inc., Pitts., 1984-86; prin. engr. rsch. and devel. Space Divsn. GE, Princeton, Pa., 1986—; cons. Nat. Productivity Coun., New Delhi, 1976-80. Assoc. editor IEEE Insulation Mag.; contbr. articles to nat. and internat. profl. jours. Fellow Instn. Mech. Engrs.; mem. IEEE (sr.), Am. Soc. Sci. Rsch. (hon.), Elfun Soc. Vols., Tau Beta Pi, Eta Kappa Nu, Omega Rho. Achievements include patents and invention awards on electromechanical design of superconducting generators; NASA award for research on space power systems; international authority in the area of electromechanical design of large power transformers. Home: 1199 Cobblestone Ct Yardley PA 19067-4751 Office: GE Astro Space PO Box 800 Princeton NJ 08543-0800

PATEL, SURESH, chemist, researcher; b. New Delhi, India, Aug. 16, 1953; came to U.S., 1975; s. Eviram Rafel and Skiekra Suva (Ravaramasam) P.; m. Kandra Levon, Mar. 16, 1978; children: Sidam, Selvaran. BA in Chemistry, Indian Inst. Tech., 1974; MS in Chemistry, MIT, 1977, PhD, 1979. Asst. rsch. scientist SUNY, Buffalo, 1979-82; rsch. chemist Chevron Techs., Saratoga, N.Y., 1982-87; rsch. chemist, asst. head chem. lab. Werik Lab., Buffalo, 1987-89, sr. rsch. chemist, assoc. head chem. lab., 1989-92, head chemist, dir. chem. rsch. tech., 1992—; lectr. SUNY, Buffalo, 1980-82, 90—; head program com. 13th Annual Conf. Chemists and Chem. Engring. Sci., Buffalo, 1989; cons. Chevron Techs., Saratoga, Lambert-Howe Industries, Syracuse, USFDA, Washington, Calvert Products, Middletown, N.Y., Ford. Motor Co. Assoc. editor Jour. Phys. Chemistry; reviewer Jour. Chem. Engring., Titration Review; contbr. over 115 articles to sci. publs. Rsch. grantee NAS, 1984, Ford Motor Co., 1987. Mem. AAAS, AICE (Outstanding Rsch. award 1989), Am. Chem. Soc., (bd. dirs., sec. N.Y. chtp. 1990—, Young Rsch. Chemist award 1981, Best Paper award 1992), Phys. Chemistry Soc. Am. (bd. dirs., sec. 1988-90, Howard Ezel Bakker award 1993), Nat. Assn. Rsch. Chemistry, Assn. Analytical Chemists, Soc. for Chemstry in Industry (Petroleum Chemistry award 1985), Indsl. Chem. Soc. (bd. dirs.), Petroleum Rsch. Assn., N.Y. Chem. Soc., N.Y. Soc. Chem. Rsch., N.Y. Acad. Scis., Buffalo C. of C., Sigma Xi. Achievements include co-discovery of new bonding process for inceased oil viscosity, new refinement techniques for industrial applications; research in bipolar molecual chemistry, chemical waste management safety. Office: Werik Lab 2289 Delaware Ave Buffalo NY 14216-2632

PATEL, TARUN R., pharmaceutical scientist; b. Borsad, Gujarat, India, Feb. 18, 1952; s. Ramesh C. and Savitaben R.; m. Nilima T. Patel, Feb. 17, 1981; children: Vishal, Shalini, Neha. BS in Pharmacy, Tex. So. U., 1975; PhD in Pharmaceutics, U. Iowa, 1980. Registered pharmacist Tex., Ill., Ind. Scientist Ortho Pharm. Corp., Raritan, N.J., 1980-82; sr. scientist Ortho Pharm. Corp., Raritan, 1982; group leader, mfg. engr. Bristol Labs., Syracuse, N.Y., 1982-83; mgr. process engring. Bristol Labs., Syracuse, 1983-84, dir. process engring., 1984-86; assoc. dir. process engring. Bristol Myers-USPNG, Evansville, Ind., 1986-90; dir. pharm. devel. Schering-Plough Corp., Memphis, 1990—; adj. prof. Univ. Tenn., Memphis, 1991—. Contbr. articles to profl. jours. Recipient Remington award Tex. So. Univ., 1975. Mem. Am. Assn. Pharm. Scientists, Am. Pharm. Assn. Office: Schering-Plough Corp 3030 Jackson Ave Memphis TN 38151-0001

PATELL, MAHESH, pharmacist, researcher; b. Ahmedabad, Gujarat, India, June 14, 1937; came to U.S., 1962.; s. Kantilal K. and Maniben K. Patell; m. Rajeshvari S. Amin, Sept. 8, 1967; children: Milan, Rupel. BS in Pharmacy, L.M. Coll. of Pharmacy, Ahmedabad, 1960; MS in Pharmacy, St Louis Coll., 1964. Rsch. pharmacist Rexall Drug Co., St. Louis, 1968-74; mgr. tech. info. svcs. Cord Labs., Bloomfield, Colo., 1974-77; rsch. investigator K.V. Pharms., St. Louis, 1977-79; section head-health care product devel. Bristol Myers-Squibb, Hillside, N.J., 1979—. Mem. Am. Assn. Pharmaceutical Scientists, Controlled Release Soc., Pa. Mfg. Confectioners Assn. Hindu. Achievements include patents for enteric coated tablet and process of making, uniquely designed capsule shaped tablets, taste masking pharmaceutical agents. Home: 4 Farrington St Edison NJ 08820 Office: Bristol Myers Squibb 1350 Liberty Ave Hillside NJ 07207-6050

PATERSON, EILEEN, radiation oncologist, educator; b. Bklyn., Oct. 16, 1939; d. John Alexander and Frances (Rabito) P.; m. Bruce Leroy Benedict, Jan. 2, 1981. BA, Wilson Coll., Chambersburg, Pa., 1961; MD, Woman's Med. Coll. Pa., 1965. Diplomate Am. Bd. Radiation Oncology, Am. Bd. Nuclear Medicine. Intern Highland Hosp., Rochester, N.Y., 1965-66; resident radiology (radiation therapy) U. Rochester, 1966-69; asst. prof. radiation oncology U. Rochester, N.Y., 1970-83, assoc. prof., 1983—; chief dept. radiation oncology Rochester Gen. Hosp., 1983—; cons. Arnot Ogden Hosp., Elmira, N.Y., 1970-74, Genesee Hosp., Rochester, 1983—. Contbr. articles to med. jours. Mem. Am. Coll. Radiology, Am. Soc. Therapeutic Radiology and Oncology. Avocations: raising and showing German shepherds and Missouri fox trotting horses. Office: Rochester Gen Hosp 1425 Portland Ave Rochester NY 14621-3001

PATHAK, DEV S., pharmaceutical administrator, marketing educator; b. Ahmedabad, India, Nov. 18, 1942; came to U.S., 1965, naturalized, 1975; m. Diane Lynn Pathak, Nov. 21, 1970; 1 child, Jay Ryan. MCom in Applied Stats., LLB, Gujarat (India) U., 1965; MS in Econs., So. Ill. U., 1966; MBA in Mktg., Mich. State U., 1969, D Bus. Adminstrn. in Mktg., 1972. Instr., asst. prof. Saginaw Valley Coll. Sch. Bus., 1970-73; asst. prof., dir. Inst. for Rsch. in Textile Mktg., Phila. Coll. Textiles and Sci. Sch. Bus., 1973-74; assoc. prof., assoc. dir. Bur. Econ. and Bus. Rsch., Appalachian State U., Boone, N.C., 1974-76; prof. pharm. adminstr. and data analysis in psychiatry U. Ill. Med. Ctr. Coll. Pharmacy, Chgo., 1978-80; assoc. prof. Coll. Pharmacy, Ohio State U., Columbus, 1976-78, prof. mktg., 1980-86, 88—, prof. div. pharm. adminstrn., 1980-86, Merrell Dow prof., 1988—, chmn. div. pharm. adminstrn., 1980-86, 88—, assoc. dean rsch. and grad. studies, 1986-88; cons. on devel. mktg. and mgmt. strategies, rsch. design, fin. analysis, and evaluation pharm. programs, including Comprehensive Pharmacy Network, Medina, Ohio, Upjohn, Kalamazoo, Peruvian Ministry Health, Lima, Walgreen Drug Stores, Inc., Chgo., Gilaxo Inc., Research Triangle Park, N.C.; over 100 presentations to profl. assns., speaker, seminar presenter in field. Author: (monograph) Downtown and Its Shoppers, 1973, Student Involvement Guide to Accompany Marketing Principles: The Management Process, 1977, Marketing Professional Pharmacy Services, 1980, Improving Health Care Delivery in Medicaid Programs, 1982, (with R. Lambert) Student Involvement Guide: Marketing Principles, 1980, (with A. Escovitz and S. Kucukarsian) Promotion of Pharmaceuticals: Issues, Trends, Options; former member editorial bd. various jours.; contbr. over 100 articles to Jour. Pharm. Mktg. and Mgmt., Jour. Health Care Mktg., Jour. Rsch. in Pharm. Econs., also others. Mem. Am. Assn. Colls. Pharmacy, Am. Soc. Hosp. Pharmacists, Am. Pharm. Assn., Am. Mktg. Assn., Assn. for Consumer Rsch., Acad. Mgmt., Am. Collegiate Retailing Assn., Soc. for Med. Decision Making. Home: 7739 Strathmoore Rd Dublin OH 43017 Office: Ohio State U 136A Lloyd M Parks Hall 500 W 12th Ave Columbus OH 43210

PATHAK, VIBHAV GAUTAM, electrical engineer; b. Bhavnagar, India, July 7, 1964; s. Gautam Yeshwantrai and Jyoti (Gautam) P. BSEE, Calif. State U., Long Beach, 1985, MSEE, 1987. Registered profl. engr., Calif. Mem. tech. staff Logicon, Inc., San Pedro, Calif., 1987-89; elec. systems engr. Rockwell Internat. Corp., Canoga Park, Calif., 1990—; founder Softouch, Inc., Norwalk, Calif., 1988—. Calif. State U. scholar, 1983. Mem. Cerrito's Music Circle (co-bd. dirs. 1989—), Workstation Users Group, Toastmasters Internat., Eta Kappa Nu. Avocations: drums, biking.

PATIENCE, JOHN FRANCIS, nutritionist; b. Thamesford, Ont., Nov. 13, 1951; s. Alwyn Francis Patience and Ellen Sophie Rosenberg; m. Ann Valerie German, Apr. 27, 1974; children: Emily, Matthew, Michael. BSc in Agr., U. Guelph, Ont., 1974, MSc, 1976; PhD, Cornell U., 1985. Cert. profl. agrologist. Extension swine specialist Sask. Agriculture, Regina, 1975-78; nutritionist Federated Co-opLtd., Saskatoon, Sask., 1978-82; grad rsch. asst. Cornell U., Ithaca, N.Y., 1982-85; rshc. assoc. U. Sask., Saskatoon, 1987-89, assoc. prof., 1989-91; pres. Prairie Swine Ctr. Inc., Saskatoon, 1991—; vis. fellow Animal Rsch. Ctr., Ottawa, Ont., 1985-87. Co-author: Swine Nutrition Guide, 1989. Mem. Am. Inst. Nutrition, Am. Soc. Animal Sci., Can. Soc. Animal Sci. (pres. 1993—). Office: Prairie Swine Centre Inc, 2105 8th St E PO Box 21057, Saskatoon, SK Canada S7H 5N9

PATIL, PRABHAKAR BAPUSAHEB, electrical and electronic research manager; b. Sankeshwar, Karnatak, India, Mar. 4, 1950; came to U.S., 1972; s. Bapusaheb Tatyaji and Shanta Hirachand (Shah) P.; m. Stephanie A. Olson, Apr. 26, 1975; children: Anand, Vijay. B in Tech., Indian Inst. of Tech., Bombay, India, 1972; MS, U. Mich., 1974, PhD, 1978. Sr. rsch. engr. Ford Motor Co., Dearborn, Mich., 1978-83, prin. rsch. engr. assoc., 1983-87, prin. staff engr., 1987-90, mgr. vehicle electronics, 1990—; sci. com. mem. Elec. Vehicle Symposium, 1986—. Contbr. over 30 publs. to profl. jours. Gen. sec. Aero. Students Assn., Bombay, India, 1971-72. Recipient Disting. Achievement award U. Mich., 1974, Henry Ford Tech. Achievement award Ford Motor Co., 1991. Mem. Soc. Automotive Engrs., Tau Beta Pi, Sigma Xi. Achievements include 10 patents in automotive electronics and electric vehicles. Home: 30225 Vernon Southfield MI 48076 Office: Ford Motor Co Rm S-2036 SRL Dearborn MI 48121-2053

PATINO, HUGO, food science research engineer; b. Monterey, Nuevo Leon, Mex., Oct. 1, 1952; came to U.S., 1982; s. Francisco De Paula and Aurora (Leal) P.; m. Leslie Ellen Nickels, May. 20, 1988; children: Erica, Laura Elizabeth. B of Engring., Monterrey Inst. Tech., 1984; PhD, U. Waterloo, 1989. Asst. prof. Monterrey (Mex.) Inst. Tech., 1975-76; mgr. brewing product engring. Cuauhtemoc Breweries, Monterey, 1979-82; asst. prof. U. Calif., Davis, 1982-84; dept. head rsch. Coors Brewing Co., Golden, Colo., 1984-91, dir. rsch. & devel., 1991—; bd. dirs. Colo. Inst. Rsch. in Biotech., Boudler. Co-author: Quality Control in Commercial Vegetable Processing, 1989; contbr. articles to profl. jours. Mem. Am. Soc. Brewing Chemists (edit. bd. mem. 1991—), Master Brewers Assn. Am. (tech. com. mem. 1991—, Presdl. award Best Paper 1990), Am. Inst. Chem. Engring., Inst. Food Technologists. Roman Catholic. Achievements include 1 patent using freeze concentration to prepare malt liquers. Office: Coors Brewing Co BC600 Golden CO 80401

PATRICK, JANET CLINE, medical society administrator; b. San Francisco, June 30, 1934; d. John Wesley and Edith Bertha (Corde) Cline; m. Robert John Patrick Jr., June 13, 1959 (div. 1988); children—John McKinnon, Stewart McLellan, William Robert. B.A., Stanford U., 1955; postgrad. U. Calif.-Berkeley, 1957, George Washington U., 1978-82. English tchr. George Washington High Sch., San Francisco, 1957, K.D. Burke Sch., San Francisco, 1957-59, Berkeley Inst., Bklyn., 1959-63; placement counselor Washington Sch. Secs., Washington, 1976-78, asst. dir. placement, 1978-81; mgr. med. personnel service Med. Soc. D.C., 1981-89, pres. Med. Pers. Svcs. Inc., 1989— . Chmn. area 2 planning com. Montgomery County Pub. Schs. (Md.), 1974-75; mem. vestry, corr. sec., Christ Ch., Kensington, Md., 1982-84, vestry, sr. warden, 1984-85, vestry, chmn. ann. giving com., 1986-89; chmn. long-range planning com., 1989-92, sec. 1992. Mem. Met. D.C. Med. Group Mgmt. Assn., Phi Beta Kappa. Republican. Episcopalian. Club: Jr. League (Washington). Home: 5206 Carlton St Bethesda MD 20816-2306 Office: Med Personnel Svcs Inc 1899 L St NW Ste 705 #705 Washington DC 20036

PATRICK, RUTH (MRS. CHARLES HODGE), limnologist, diatom taxonomist, educator; b. Topeka, Kans.; d. Frank and Myrtle (Jetmore) P.; m. Charles Hodge, IV, July 10, 1931; 1 son, Charles, V. BS, Coker Coll., 1929, LLD, 1971; MA, U. Va., 1931, PhD, 1934; DSc, Beaver Coll., 1970, PMC Colls., 1971, Phila. Coll. Pharmacy and Sci., 1973, Wilkes Coll., 1974, Cedar Crest Coll., 1974, U. New Haven, 1975, Hood Coll., 1975, Med. Coll. Pa., 1975, Drexel U., 1975, Swarthmore Coll., 1975, Bucknell U., 1976, Rensselaer Poly. Inst., 1976, St. Lawrence U., 1978; LHD, Chestnut Hill Coll, 1974; DSc, U. Mass., 1980, Princeton U., 190, Lehigh U., 1983, Temple U., 1985, Emory U., 1986, Wake Forst U., 1986, U. S.C., 1989, Clemson, 1989. Asso. curator microscopy dept. Acad. Natural Scis., Phila., 1939-47; curator Leidy Micros. Soc., 1937-47, curator limnology dept., 1947—, chmn. limnology dept., 1947-73; occupant Francis Boyer Research Chair, 1973—, chmn. bd. trustees, 1973-76, hon. chmn. bd. trustees, 1976—; lectr. botany U. Pa., 1950-70, adj. prof., 1970—; guest Fellow of Saybrook Yale, 1975; participant Am. Philos. Soc. limnology expdn. to, Mexico, 1947; leader Catherwood Found. expdn. to, Peru and Brazil, 1955; del. Gen. Assembly, Internat. Union Biol. Scis., Bergen, Norway, 1973; Dir. E.I. duPont, Pa.

Power and Light Co.; Chmn. algae com. Smithsonian Oceanographic Sorting Center, 1963-68; mem. panel on water blooms Pres.'s Sci. Adv. Com., 1966; mem. panel on water resources and water pollution Gov.'s Sci. Adv. Com., 1966; mem. nat. tech. adv. com. on water quality requirements for fish and other aquatic life and wildlife Dept. Interior, 1967-68; mem. citizen's adv. council Pa. Dept. Environ. Resources, 1971-73; mem. hazardous materials adv. com. EPA, 1971-74, exec. adv. com., 1974-79, chmn. com.'s panel on ecology, 1974-76; mem. Pa. Gov.'s Sci. Adv. Coun., 1972-78; mem. exec. adv. com. nat. power survey FPC., 1972-75; mem. coun. Smithsonian Instn., 1973—; mem. Phila. Adv. Council, 1973-76; mem. energy R & D adv. coun. Pres.'s, Energy Policy Office, 1973-74; mem. adv. coun. Renewable Natural Resources Found., 1973-76, Electric Power Rsch. Found., 1973-77; mem. adv. com. for rsrch. NSF., 1973-74; mem. gen. adv. com. ERDA, 1975-77; adv. bd. sec. energy, 1975-89; mem. com. on human resources NRC, 1975-76; mem. adv. council dept. biology Princeton, 1975—; mem. com. on sci. and arts Franklin Inst., 1978—; mem. univ. council com. Yale Sch. Forestry and Environ. Studies, 1978—; mem. sci. adv. council World Wildlife Fund-U.S., 1978—; trustee Aquarium Soc. Phila., 1951-58, Chesnut Hill Acad., Lacawac Sanctuary Found., Henry Found.; bd. dirs. Wissahickon Valley Watershed Assn.; mem. adv. council French and Pickering Creeks Conservation Trust; bd. govs. Nature Conservancy; bd. mgrs. Wistar Inst. Anatomy and Biology. Author: 4 books including (with Dr. C.W. Reimer) Diatoms of the United States, Vol. 1, 1966, Vol. II, Part 1, 1975; mem. editorial bd. (with Dr. C.W. Reimer) Science, 1974-76, American Naturalist; trustee Biological Abstracts, 1974-76; contbr. over 150 articles to profl. jours. Recipient Disting. Dau. of Pa. award, 1952, Richard Hopper Day Meml. medal Acad. Natural Scis., 1969, Gimbel Phila. award, 1969; Gold medal YWCA, 1970, Lewis L. Dollinger Pure Environment award Franklin Inst., 1970, Pa. award for excellence in sci. and tech., 1970, Eminent Ecologist award Ecol. Soc. Am., 1972, Phila. award, 1973, Gold medal Pa. State Fish and Game Protective Assn., 1974, Internat. John and Alice Tyler Ecology award, 1975, Gold medal Phila. Soc. for Promoting Agr., 1975, Pub. Service award Dept. Interior, 1975, Iben award Am. Water Resources Assn., 1976, Outstanding Alumni award Coker Coll., 1977, Francis K. Hutchinson medal Garden Club of Am., 1977, Golden medal Royal Zool. Soc. Antwerp, 1978, Green World award N.Y. Bot. Garden, 1979, Hugo Black award U. Ala., 1979, Sci. award Gov. Pa., 1988, Founders award Soc. Environ. Toxicology and Chemistry, 1982, Environ. Regeneration award Rene Du Bois Ctr., 1985, Disting. Citizen award Pa., 1989. Fellow AAAS (com. environ. alterations 1973-74); mem. Nat. Acad. Scis. (chmn. panel com. on pollution 1966, mem. com. sci. and pub. policy 1973-77, mem. environ. measurements panel of com. on remote sensing programs for earth resource surveys 1973-74, mem. nominating com. 1973-75), Nat. Acad. Engring. (com. environ. engring.'s ad hoc study on explicit criteria for decisions in power plant siting 1973), Am. Philos. Soc., Am. Acad. Arts and Scis., Bot. Soc. Am. (mem. Darbarker prize com. 1956, merit award 1971), Phycol. Soc. Am. (pres. 1954), Internat. Limnological Soc., Internat. Soc. Plant Taxonomists, Am. Soc. Plant Taxonomy, Am. Soc. Limnology and Oceanography, Water Pollution Control Fedn. (hon.), Soc. Study Evolution, Am. Soc. Naturalists (pres. 1975-76), Ecol. Soc. Am., Am. Inst. Biol. Scis., Internat. Phycol. Soc., Soc. Sigma Xi. Presbyterian. Office: Acad Natural Scis 19th at Benjamin Franklin Pkwy Philadelphia PA 19103

PATTEN, BERNARD MICHAEL, neurologist, educator; b. N.Y.C., Mar. 23, 1941; s. Bernard M. and Olga (Vaccaro) P.; m. Ethel Doudine, June 18, 1964; children: Allegra, Craig. AB summa cum laude, Columbia Coll., 1962; MD, Columbia U., 1966. Med. intern N.Y. Hosp. Cornell Med. Ctr., N.Y.C., 1966-67; resident neurologist Columbia Presbyn. Med. Ctr., N.Y.C., 1967-69, chief resident neurologist, 1969-70; assoc. prof. neurology Baylor Coll. Medicine, Houston, 1973—; asst. chief med. neurology NIH, Bethesda, Md., 1970-73; mem. med.bd. Nat. Myasthenia Gravis Found., 1973—, Nat. AmyoTrophic Lateral Sclerosis Found., 1982—. Contbr. over 100 articles to profl. jours. With USPHS, 1970-73. Rsch. grantee NIH, pvt. founds., nat. health orgns. Fellow ACP, Royal Coll. Physicians. Roman Catholic. Achievements include discoverer (with others) L-Dopa for Parkinson's disease; pioneered use of immune suppression for myasthenia gravis, diagnosis and treatment of medical and neurologcal complications of breast implants. Home: 1019 Baronridge Seabrook TX 77586 Office: Baylor Coll Medicine 1 Baylor Plz Houston TX 77030

PATTERSON, DAVID ANDREW, computer scientist, educator, consultant; b. Evergreen Park, Ill., Nov. 16, 1947; s. David Dwight and Lucie Jeanette (Ekstrom) P.; m. Linda Ann Crandall, Sept. 4, 1967; children: David Adam, Michael Andrew. BS in Math., UCLA, 1969, MS in Computer Sci., 1970, PhD, 1976. Mem. tech. staff Hughes Aircraft Co., L.A., 1972-76, Thinking Machines Corp., Cambridge, Mass., 1979; prof. computer sci. div. U. Calif., Berkeley, 1977—, chmn., 1990-93, Pardee chair, 1992—; cons. Sun Microsystems Inc., Mountain View, Calif., 1984—, Thinking Machines Corp., Cambridge, 1988—; mem. sci. adv. bd. Data Gen. Corp., Westborough, 1984—; com. to study scope and role of computer sci. NAS, Washington, 1990-92; chmn. program com. 4th Symposium on Archtl. Support for Oper. Systems and Computer Architecture, Santa Clara, Calif., 1991; mem. Mgmt. Ops. Working Group NASA, 1992—; co-chair program com. Hot Chips Symposium IV, Stanford, Calif., 1992. Author: A Taste of Smalltalk, 1986, Computing Unbound, 1989, Computer Architecture: A Quantative Approach, 1990, Computer Organization & Design: The Hardware/Software Interface, 1993. Recipient Disting. Teaching award U. Calif., Berkeley, 1982, corp. fellow Thinking Machines Corp., Cambridge, 1989, ACM Karl V. Karlstrom Outstanding Educator award, 1991. Fellow IEEE; mem. NAE, Computing Rsch. Assn. (bd. dirs. Washington 1991—), Spl. Interest Group on Computer Architecture of Assn. Computing Machinery (bd. dirs. 1987-90). Avocations: biking, football, soccer. Office: U Calif Computer Sci Div 571 Evans Hall Berkeley CA 94720

PATTERSON, DONALD EUGENE, research scientist; b. El Paso, Tex., Feb. 7, 1958; s. Donald M. Patterson and Beverly Lee (Viles) McElroy; m. Mary Jane Ingram, May 6, 1989. BS, U. Tex., 1982, MS, 1984; MA, Rice U., 1987, PhD, 1989. Rsch. scientist Rice U., Houston, 1989-91; sr. rsch. scientist Houston Advanced Rsch. Ctr., The Woodlands, Tex., 1989-93; scientist SI Diamond Tech. Inc., Houston, 1991--; sr. rsch. scientist TSA, Inc., The Woodlands, 1992--. Contrb. articles to profl. jours. Recipient Harry B. Wieser award Rice U., 1988; Rice U. Graduate fello, 1984; UTEP Grad. scholar, 1983, Davis and Bertha Green scholar, 1982, VFW Voice Democracy scholar, 1974. Mem. AAAS, Materials Rsch. Soc., Am. Chem. Soc., Phi Kappa Phi, Sigma Xi. Achievements include one patent. Home: 4045 Linkwood #119 Houston TX 77025 Office: SI Diamond Tech Inc 2435 North Blvd Houston TX 77098

PATTERSON, JAN EVANS, epidemiologist, educator; b. Ft. Worth, May 13, 1956; d. C. Wayne and Zona (Horn) Evans; m. Thomas F. Patterson, June 22, 1985. BA, Hardin-Simmons U., 1978; MD, U. Tex., 1982. Diplomate Am. Bd. Internal Medicine, Am. Bd. Infectious Diseases. Asst. prof. medicine and lab. medicine Yale U. Sch. Medicine, New Haven, 1988-92; assoc. prof. medicine and pathology Health Sci. Ctr. U. Tex., San Antonio, 1993—; hosp. epidemiologist Bexar County Hosp. Dist., San Antonio, 1993—, Audie L. Murphy Meml. Vets. Hosp., San Antonio, 1993—. Contbr. articles to profl. jours. Fellow ACP; mem. Infectious Disease Soc. Am., Am. Fedn. Clin. Rsch., Soc. Hosp. Epidemiologists (nominating com. 1992-93), Bur. on Issues in Infectious Mgmt., Alpha Omega Alpha. Office: Health Sci Ctr U Tex Divsn Infectious Diseases 7703 Floyd Curl Dr San Antonio TX 78284

PATTERSON, MANFORD K(ENNETH), JR., foundation administrator, researcher, scientist; b. Muskogee, Okla., Aug. 20, 1926; s. Manford Kenneth Sr. and Sara Lou (Patton) P.; m. Nancy Beverly Wilson, Aug. 21, 1953; 1 child, Shelley Lynn. BS, U. Okla., 1953, MS, 1954; PhD, Vanderbilt U., 1961. Rsch. chemist S. R. Noble Found., Ardmore, Okla., 1951-53, 54-57, sr. rsch. chemist, 1961-66, head nutrition sect., 1966—, v.p., dir. biomed. div., 1973—; cons. ICNND Nutrition Survey, Jordon, 1962; chair adv. bd. Snyder Found., Winfield, Kans., 1989. Co-author: Tissue Culture: Methods & Application, 1973; editor In Vitro, Cellular & Devel. Biology Jour., 1979-86. Vice chmn. Carter County Bd. of Health, Ardmore, 1976; pres. Cross Timbers Hospice, Ardmore, 1984-91. With USN, 1944-46. Fellow AAAS; mem. Am. Soc. Biochem. Molecular Biology, Okla. Acad. Sci. (Scientist of the Yr. 1990), Tissue Culture Assn. (treas. 1972-76). Achievements include patent for Method for Detecting Factor XIII in Plasma. Home: 2215 Hickory Dr Ardmore OK 73401-3433 Office: Samuel Roberts Noble Foundation Biomedical Division PO Box 2180 Ardmore OK 73402

PATTERSON, MILES LAWRENCE, psychology educator; b. Palatine, Ill., Jan. 17, 1942; s. James Albert and Rose Elizabeth (Corrigan) P.; m. Dianne Marie Ruder, July 21, 1972; 1 child, Kevin Thomas. BS, Loyola U., Chgo., 1964; MS, Northwestern U., 1966, PhD, 1968. Asst. prof. psychology Northwestern U., Evanston, Ill., 1968-69; asst. prof. psychology U. Mo., St. Louis, 1969-74, assoc. prof., 1974-79, prof., 1979—, dept. chairperson, 1982-85. Author: Nonverbal Behavior and Social Psychology, 1982, Nonverbal Behavior: A Functional Perspective, 1983; editor jour. Nonverbal Behavior, 1986-91; contbr. numerous articles and chpts. to profl. publs. Fellow APA; mem. Soc. Exptl. Social Psychology. Roman Catholic. Avocation: running. Office: U Mo Dept Psychology 8001 Natural Bridge Rd Saint Louis MO 63121-4499

PATTERSON, PATRICIA LYNN, applied mathematician, geophysicist, engineer; b. Kearny, N.J., Feb. 25, 1946; d. Clifford and Helen (Matthews) P. AA, St. Petersburg (Fla.) Jr. Coll., 1965; BA magna cum laude in Physics, U. South Fla., Tampa, 1966; MS in Geophys. Scis., Ga. Inst. Tech., Atlanta, 1976; PhD in Geophysics, Ga. Inst. Tech., 1980. Elec. engr. Burns & McDonnell Engring. Co., Miami, Fla., 1967-69; tchr. biology and physiology Orange County Bd. Edn., Orlando, Fla., 1969-70; acoustical cons. Bolt Beranek & Newman Inc., Downers Grove, Ill., 1971-72; geophysicist Exxon Co., U.S.A., New Orleans, 1976-77; comm. systems engr. E-Systems, St. Petersburg, Fla., 1980-85; pres. Solitonics (Rsch. & Cons.), St. Petersburg and Dallas, 1985—; image-processing engr. E-Systems, Garland, Tex., 1991-93. Contbr. articles to profl. jours.; patentee in field. Recipient Sigma Xi research awards, 1977, 81, others. Mem. IEEE (reviewer tech. papers), Soc. Indsl. and Applied Math., Am. Geophys. Union. Achievements include research in remote sensing and image processing, biomedical engineering, data compression, coding and information theory, numerical modeling and parallel processing.

PATTERSON, RICHARD GEORGE, retired aerospace engineer; b. Cascade, Iowa, May 26, 1928; s. George Samuel and Mary Anna (Heffernen) P.; m. Jeanne Shirley Lundberg, Oct. 10, 1953; children: Christa M., Courtney J. BSEE, Purdue U., 1949, MSEE, 1950. With engring. mgmt. Honeywell, Mpls., 1969-73, sect. head, 1973-74, with engring. mgmt. space shuttle engine control, 1974-78, chief engr., 1978-79, engring. mgr., 1979-81, chief engr. NAV systems, 1981-87, dir. quality, 1987-92; cons. in field Mpls., 1992—. Mem. Am. Soc. Quality Control, AIAA, Tau Beta Pi, Eta Kappa Nu. Roman Catholic. Achievements include patents for auto throttle, synchronizer. Home and Office: 4336 Avondale Rd Minneapolis MN 55416

PATTERSON, ROY, physician, educator; b. Ironwood, Mich., Apr. 26, 1926; s. Donald I. and Helmi (Lantta) P. M.D., U. Mich., 1953. Diplomate: Am. Bd. Internal Medicine, Am. Bd. Allergy and Immunology. Intern U. Mich. Hosp., Ann Arbor, 1953-54; med. asst. research U. Mich. Hosp., 1954-55, med. resident, 1955-57, instr. dept. medicine, 1957-59; attending physician VA Research Hosp., Chgo., Northwestern Meml. Hosp.; mem. faculty Northwestern U. Med. Sch., Chgo., 1959—; prof. medicine Northwestern U. Med. Sch., 1964—, Ernest S. Bazley prof. medicine, chief sect. allergy-immunology dept. medicine. Editor: Jour. Allergy and Clin. Immunology, 1973-78. Served with USNR, 1944-46. Fellow Am. Acad. Allergy (pres. 1976), A.C.P.; mem. Central Soc. for Clin. Research (pres. 1978-79). Office: Northwestern U Med Sch Dept Medicine 303 E Chicago Ave Chicago IL 60611-3008

PATTERSON, SCOTT PAUL, civil engineer, consultant; b. Coronado, Calif., Mar. 5, 1948; s. Warren Nelson and Mary Lemoyne (Herbert) P.; m. Jane Pocoroba, Apr. 23, 1972; children: Christopher, Timothy, Sara, Joanna. BSCE, U. Cin., 1971. Lic. profl. engr., 3 states; lic. gen. contractor, Fla. Asst. engr. Ebasco Svcs., N.Y.C., 1971-72; project engr. Conduit and Found. Corp., Elmwood Park, N.J., 1972-79, project mgr., 1979-85; v.p. Bellemead Devel. Corp., Roseland, N.J., 1985—; prin. Civil and Constrn. Engring. Svcs., Long Valley, N.J., 1990—; pres. Profl. Solar Consultants, Pompton Lakes, N.J., 1977-78; instr. County Coll. of Morris, Randolph, N.J., 1975-79. Coach twp. and recreation teams Pompton Lakes, Long Valley, 1972—; mem. St. Mark's Parish Life Coun., Long Valley, 1990-92. Mem. ASCE (younger mem. of yr. 1972, dir. N.J. sect. 1978-79), NSPE Profl. Engrs. in Constrn. (regional vice chmn. 1990-92, sec. 1993—), N.J. Profl. Engrs. in Constrn. (chmn. 1987-88, regional v.p. 1989-91), N.J. Soc. Profl. Engrs. (pres. 1992-93). Republican. Home: 78 Winay Terrace Long Valley NJ 07853 Office: Bellemead Devel Corp 4 Becker Farm Rd Roseland NJ 07068

PATTON, ALEXANDER JAMES, engineer, consultant; b. Westerly, R.I., June 25, 1945; s. Samuel James and Margaret (Thomson) P.; m. Linda L. Olcowick, Sept. 4, 1965 (div. June 1987); children: Daniel J., David A. BS, U. R.I., 1967, PhD, 1972; MS, U. Mich., 1968. Registered profl. engr., R.I. Mem. faculty Roger Williams Coll., Bristol, R.I., 1972-74; consulting engr. Safety Svcs., Providence, 1974-79; project engr. fluid systems divsn. G & W, Warwick, R.I., 1979-81; cons. Kingston, R.I., 1981—. Author: Fire Litigation Sourcebook, 1986; contbr. articles to profl. jours. Musician Westerly Band, 1988—, Wakefield Civic Band, Peacedale, R.I., 1988—. Dept. Interior fellow U. R.I., 1970, Nat. Def. Ednl. Account fellow U. Mich., 1968. Mem. ASME (pres. R.I. sect.), Nat. Fire Protection Agy., Am. Soc. Safety Engrs., R.I. Soc. Profl. Engrs. (pres., Young Engr. Yr.), Sigma Xi. Home and Office: 15 Upper College Rd Kingston RI 02881-1309

PATTON, FINIS S., JR., nuclear chemist. Dep. mgr. Martin Marietta Energy Systems, Piketon, Ohio. Recipient Robert E. Wilson Nuclear Chem. Engring. award Am. Inst. Chem. Engrs., 1992. Office: Martin Marietta Energy Systems POB 628 Piketon OH 45661*

PATTON, JOHN ANTHONY, chemical engineer; b. Birmingham, Ala., May 16, 1963; s. Earnest Howell Jr. and Catherine Ann (Bellanca) P.; m. Jeanne Elaine Patterson, June 5, 1982; 1 child, John Anthony Jr. B Chem. Engring., Auburn U., 1985. Registered profl. engr., Ala. Engr. So. Natural Gas Co., Birmingham, 1986-89, project engr., 1989-91; project engr. So. Natural Gas Co., Jackson, Miss., 1992-93, sr. project engr., 1993; transmission supt. So. Natural Gas Co., Pickens, Miss., 1993—. Mem. Am. Gas Assn. (corrosion com. 1987-91, welding supervisory com., pipeline rsch. com. 1987-93), So. Gas Assn. (chmn. corrosion roundtable 1991). Baptist. Home: 205 Hemlock Brandon MS 39042 Office: So Natural Gas Co PO Box 280 Pickens MS 39146

PATZELT, PAUL, nuclear chemist; b. Neu-Mohrau, Silesia, Germany, July 18, 1932; s. Josef and Gertrud (Lowack) P.; m. Brigitte Krautwurst, Aug. 2, 1963; children: Michael, Marc Christopher, Luc Stephen. Diploma Chemistry, U. Mainz, 1960, D Natural Scis. in Nuclear Chemistry, 1965. Rsch. asst. U. Mainz, 1961-67; rsch. assoc. CERN-European Orgn. Nuclear Rsch., Geneva, 1967-70; rsch. asst. U. Marburg, Germany, 1970-72, prof. nuclear chemistry, 1972—, dean Faculty Phys. Chemistry, 1975-76, 83-84, 89-90. Author papers in field. Mem. German Chem. Soc., Assn. German Natural Scientists and Physicians. Roman Catholic. Home: 17 Gabelsbergerstrasse, D-35037 Marburg Germany Office: Hans Meerwein-Strasse, Gebaeude J, D-35043 Marburg Germany

PAUL, ANTON DILO, chemical engineering consultant, researcher; b. Kandy, Sri Lanka, Aug. 3, 1951; came to U.S., 1981; s. William Reginald and Mary Evelyn (Snell) P.; m. Naomi K. Arasaratnam, June 4, 1987; children: Rachel Camala, Rebekah Evelyn. MS, Va. Poly. Inst. and State U., 1984, PhD, 1988. Mgr. R & D, Imperial Chem. Industries, Colombo, Sri Lanka, 1978-81; instr. dept. chem. engring. Va. Poly. Inst. and State U., Blacksburg, 1981-84, rsch. assoc. dept. mining and minerals engring., 1984-88; sr. rsch. engr. Eos Techs., Inc., Arlington, Va., 1988-92; sr. engr. SAIC (Sci. Applications Internat. Corp.), McLean, Va., 1992—; conf. presenter in field; tech. cons. on fossil energy systems to U.S. Dept. of Energy. Contbr. articles to profl. jours. Mem. AICE, Soc. for Mining, Metallurgy and Exploration. Office: SAIC MS 2-2-1 1710 Goodridge Dr Mc Lean VA 22102

PAUL, BIRAJA BILASH, engineering consulting company executive; b. Burdwan, West Bengal, India; Sept. 30, 1933; s. Devendra Nath and Shushila S. Paul; B.Sc., Calcutta U., 1952; A.A., Indian Inst. Sugar Tech., 1955; M.Chem. Engring., La. State U., 1958, Ph.D., 1960; m. Nita B., May 26, 1965; children: Baishali, Basbab B. Dep. gen. mgr. tech. Daurala Sugar Works, 1960-65; tech. supt. sugar and distillery div. K.C.T. & Bros. Pvt. Ltd., Calcutta, India, 1965-70; tech. dir. G.R. Engring. Works Pvt. Ltd., Bombay, India, 1970-74; tech. cons. Miwani Sugar & Distillery, Kenya, 1970-74; mng. dir. B.B. Cons. N Engring. Pvt. Ltd., Bombay, 1975—; dir. V.K. Engring. Works Pvt. Ltd., 1977—; cons. sugar distillery industries, 1974, Asian Devel Bank, 1977, IBRD, Washington, 1974; Commonwealth Secretariat, London, 1986. Contbr. over 135 articles to profl. jours.; holder 14 patents for process equipment and process tech. for sugar and distillation industry. Recipient Nat. award Pres. India, 1972; Noel Deer gold medal, 1967. Fellow Am. Inst. Chem. Engrs., India Inst. Engrs., Sugar Technologists Assn. India (v.p. 1975-76), Deccan Sugar Technologists Assn.; mem. Internat. Soc. Sugarcane Technologists, Otters, Lions. Home: Bandra, 5th Fl, Galaxy Apts, 239-A Byramjee Jeejibhoy Rd, Bombay India 400 050 Office: B B Cons N Engring Pvt Ltd, Chapsey Terr 3rd Fl, Bombay India 400026

PAUL, DONALD ROSS, chemical engineer, educator; b. Yeatesville, N.C., Mar. 20, 1939; s. Edgar R. and Mary E. (Cox) P.; m. Sally Annette Cochran, Mar. 28, 1964; children: Mark Allen, Ann Elizabeth. B.S., N.C. State Coll., 1961; M.S., U. Wis., 1963, Ph.D., 1965. Research chem. engr. E.I. DuPont de Nemours & Co., Richmond, Va., 1960-61; instr. chem. engring. dept. U. Wis., Madison, 1963-65; research chem. engr. Chemstrand Research Center, Durham, N.C., 1965-67; asst. prof. chem. engring. U. Tex., Austin, 1967-70; assoc. prof. U. Tex., 1970-73, prof., 1973—, T. Brockett Hudson prof., 1978-85, Melvin H. Gertz Regents chmn. chem. engring., 1985—, chmn. dept. chem. engring., 1977-85, dir. Center for Polymer Research, 1981—; Turner Alfrey vis. prof. Mich. Molecular Inst., 1990-91; cons. in field. Author: (with F.W. Harris) Controlled Release Polymeric Formulations, 1976, (with S. Newman) Polymer Blends, 2 vols, 1978; mem. editorial adv. bds. Jour. Membrane Sci, 1976—, Polymer Engring. and Sci, 1975—, Jour. Applied Polymer Sci, 1979—, Chemical Engring. Edn., 1986—, Polymer, 1987—, Jour. Polymer Sci., 1989—; editor: Indsl. and Engring. Chemistry Rsch.; contbr. articles to profl. jours. Recipient award Engring. News Record, 1975, Ednl. Service award Plastics Inst. Am., 1975, awards U. Tex. Student Engring. Council, 1972, 75, 76; award for engring. teaching Gen. Dynamics Corp., 1977; Joe J. King Profl. Engring. Achievement award, 1981. Mem. Am. Chem. Soc. (Doolittle award 1973, Phillips award in applied polymer sci. 1984), Am. Inst. Chem. Engrs. (South Tex. best fundamental paper award 1984, Materials Engring. and Scis. Div. award 1985), Soc. Plastics Engrs. (Outstanding Achievement in Rsch. award 1982, Internat. Edn. award 1989, Internat. award 1993), Fiber Soc., Nat. Acad. Engring., Nat. Materials Adv. Bd., Phi Eta Sigma, Tau Beta Pi, Phi Kappa Phi, Sigma Xi. Home: 7001 Valburn Dr Austin TX 78731-1818 Office: U Tex Dept Chem Engring Ctr Polymer Rsch EPS 206 Austin TX 78712

PAUL, FRANK WATERS, mechanical engineer, educator, consultant; b. Jersey Shore, Pa., Aug. 28, 1938. BSME, Pa. State U., 1960, MSME, 1964; PhD in Mechanical Engring., Lehigh U., 1968. Registered profl. control engr., Calif. Control engr. Hamilton Standard div. United Techs. Corp., 1961-64; instr. mechanical engring. Lehigh U., Bethlehem, Pa., 1964-68; asst. prof. mechanical engring. Carnegie-Mellon U., Pitts., 1968-73, asst. prof., 1973-77; assoc. prof. Clemson (S.C.) U., 1977-79, prof., 1979-83, McQueen Quattlebaum prof., 1983—; cons. numerous cos. including Westinghouse Electric, 1969, 82-83, Alcoa Rsch. Labs., 1976-80, State of N.J., Dept. Higher Edn., 1986, Dunlop Sports, Inc., 1988, BPM Tech., 1992—; hon. prof. engring. Hull U., Eng., 1990—, Dora Jones vis. prof. elec. engring., 1993; bd. dirs. Ctr. for Advanced Mfg.; lectr. to colls. and univs., U.S. and abroad. Author: (book, with others) Progress in Heat and Mass Transfer, Vol. 6, 1972, Metals: Processing and Fabrication, Encyclopedia of Materials Science and Engineering, 1986; contbr. articles to IEEE Control Systems mag., Jour. of Engring. for Industry (ASME), Jour. of Dynamic Systems Measurement and Control (ASME), and other scholarly publs. Sabbatical United Techs. Rsch. Ctr., 1985-86, Hull U., 1993. Mem. ASME (participant and paper reviewer Dynamic Systems and Control divsn. 1968—, chmn. panel on robotics 1985-87), Am. Soc. Engring. Educators, Soc. Mech. Engrs. (charter mem. Robotics Internat.), Pi Tau Sigma, Tau Beta Pi, Sigma Tau, Sigma Xi. Achievements include patents related to manufacturing automation. Office: Clemson U Coll Engring Riggs Hall Rm 300-D Clemson SC 29634-0921

PAUL, LEENDERT CORNELIS, medical educator; b. Jan. 31, 1946; married; 2 children. Student, Leiden State U. Med. Sch., The Netherlands, 1964-69; MD cum laude, State U. of Leiden, The Netherlands, 1969, PhD cum laude, 1979. Cert. gen. med. lic., The Netherlands; E.C.F.M.G.; cert. specialist in internal medicine, The Netherlands, cert. specialist in internal medicine, Can. Rotating intern U. Hosp. Leiden, 1969-72, resident in internal medicine, 1972-77, clin. fellow in nephrology, 1977-79, rsch. fellow in transplantation immunology, 1977-79, staff nephrologist, 1979-87, chef de policlinique Renal divsn., 1979, chef de clinique Renal Transplant unit, 1981-87; rsch. fellow in transplantation immunology & immunogenetics Renal divsn. Brigham and Women's Hosp./Harvard Med. Sch., Boston, 1979-81; from asst. prof. to assoc. prof. of medicine Leiden U., 1979-87; prof. of medicine U. Calgary, Can., 1987—, head divsn. nephrology, 1989—; staff nephrologist Foothills Hosp., Calgary, Can., 1987—, dir. Clinic for Organ Transplantation, 1987—, chief renal divsn., 1989—; prof. medicine U. Toronto, Can., 1993—; chief nephrology St. Michael's Hosp., Toronto, 1993—; guest prof. Free Univ. Brussels, 1987-90; med. dir. Human Orgn. Procurement and Exch. Program, South Alberta, 1991-92; head divsn. nephrology exec. com. dept. medicine, U. Calgary, 1989—, ad hoc reviewer conjoint med. ethics com., 1987—, chmn. immunol. scis. rsch. group, 1991-92, mem. rsch. com., 1992—, mem. search com. divsn. head rheumatology, clin. immunology and dermatology, 1992—, mem. search com. Jessie Bowden Lloyd Professorship in Immunology, 1990-92; grant reviewer Dutch Orgn. for Med. Rsch., Dutch Kidney Found., Med. Rsch. Coun. Can., Kidney Found. Can., Profl. Physicians Org., Searle USA, 1981—; adv. bd. Eurotransplant, 1984-87; sci. coun. Dutch Kidney Found., 1984-87; adv. bd. Dutch Soc. Cardiac Transplant Recipients, 1984-87; mem. panels Dutch Orgn. for Fundamental Med. Rsch., 1985, 1986-87; sec. Bd. Dutch Renal Failure Register, 1986-87; ext. adv. Hoffman-LaRoche, Basle, Switzerland, 1989-90; med. adv. bd., fellowship com. Kidney Found. Can., 1989—, sci. coun., 1990—, scholarship and fellowhip com., 1993—; chmn. Provincial Renal Programs Adv. Com., Alberta, Can., 1990—; mem. com. on immunology and transplantation Med. Rsch. Coun., 1991—; chmn. renal section Can. Transplantation Soc., 1991—; co-chair renal and pancreas subcom. Nat. Health Com. on Organ Transplantation, 1993—; lectr. Int. Soc. for Study of Hypertension in Pregnancy, Amsterdam, Indonesian Soc. of Immunology, Indonesian Soc. of Nephrology, Can. Soc. of Lab. Technologists Congress, Am. Soc. Transplant Physicians/Am. Soc. Transplant Surgeons, Houston, Banff Conf. Allograft Pathology, Alberta, also various other symposiums and confs. Editor: Management of Cyclosporin in Renal Transplantation, 1985; co-editor: Organ Transplantation: Long-Term Results, 1990-92, Integrins in Health and Disease, 1993, Recent Advances in Nephrology and Kidney Transplantation, 1987; editor-at-large Marcel Dekker, Inc., 1992—; mem. editorial bd. Transplantation, 1986—; manuscript reviewer for Transplantation, Clin. Immunology and Immunopathology, Lab. Investigation; contbr. numerous chpts. to books, also articles and abstracts to profl. jours. Recipient Anna Ida Overwater award for sci. contbns. in immunology of human kidney transplantation, 1983, Alta. Heritage Found. for Med. Rsch. Scientist award, 1988—, Herb Hughes Meml. Endowment award U. Calgary, 1990-91; USPHS Internat. Rsch. fellow, 1979-81; grantee Dutch Orgn. for Fundamental Med. Rsch., 1981-84, Dutch Kidney Found., 1985-87, 86-87, Alta. Heart and Stroke Found., 1988-89, U. Calgary, 1988-89, Searle Can./USA, 1989, 91-93, Heart and Stroke Found. Can., 1989-91, Kidney Found. Can., 1989-90, 90-91, 91-94, Schering, Berlin, 1991, Heart and Stroke Found. Alberta, 1992—, Med. Rsch. Coun. Can., 1993—. Fellow Royal Coll. Physicians and Surgeons of Can. (nephrology cert. 1989); memtion Soc., Transplantation Soc., Internat. Soc. of Nephrology, European Soc. for Organ Transplantation, Can. Soc. for Clin. Investigation, Am. Soc. Nephrology (mem. abstract rev. panel 1989), Can. Soc. of Immunology, Can. Soc. of Nephrology (councillor 1990-91, treas. 1992, pres.-elect 1993), Am. Soc. of Transplant Physicians, Internat. Soc. for Heart Transplantation. Achievements include research in organ transplantation and mechanisms of rejection, histocompatibility antigens, immunosuppressive drugs, mechanism of actions, side effects; mechanisms of progressive loss of renal function. Office: U Toronto/St Michaels Hosp, 30 Bond St, Toronto, ON Canada T2N 2T9

PAUL, STEVEN M., psychiatrist. Dir. intramural rsch. program NIMH NIH. Office: Nat Inst of Mental Health 5600 Fishers Ln Rockville MD 20857*

PAUL, VERA MAXINE, mathematics educator; b. Mansfield, La., Dec. 14, 1940; d. Clifton and Virginia (Smith) Hall; m. Alvin James Paul III, June 14, 1964; children: Alvin J., Calvin J., Douglas F. BS, So. U., 1962; MS, Roosevelt U., 1975. Tchr. Shreveport (La.) Bd. Edn., 1962-64; tchr. Chgo. Bd. Edn., 1964-81, asst. prin., 1981-86, 86—; tchr. South Bend (Ind.) Community Sch., 1967-68. Mem. Chgo. Bd. Edn., 1964-92. Recipient Disting. Vol. Leadership award March of Dimes, Chgo., 1982, Mayoral Tribute award City of Pontiac (Mich.), 1987, Disting. Svc. award City Coun. Detroit, 1988, Svc. award City Coun. Cleve., 1990, Disting. svc. award Zeta Phi Beta, 1992, State of Mich. Cert. of Merit Sen. Jackie Vaughn III, Great Lakes Svc. award, 1992, Svc. award Mich. Senate, 1992. Mem. NAACP, Am. Fedn. Tchrs., Chgo. Tchr. Union, Ill. Coun. Affective Reading Edn., Ill. Coun. Tchrs. Math., Nat. Coun. Tchrs. Math., Nat. Alliance Black Sch. Educators, Zeta Phi Beta (regional dir. 1986-90, Zeta of Yr. 1988). Lutheran. Avocations: reading, computer games, walking, piano.

PAUL, WILLIAM, physicist, educator; b. Deskford, Scotland, Mar. 31, 1926; came to U.S., 1952; s. William and Jean (Watson) P.; m. Barbara Anderson Forbes, Mar. 28, 1952; children—David, Fiona. M.A., Aberdeen U., Scotland, 1946; Ph.D., Aberdeen U., 1951; A.M. (hon.), Harvard U., 1960. Asst. lectr., then lectr. Aberdeen U., 1946-52; mem. faculty Harvard U., 1953—, Gordon McKay prof. applied physics, 1963-91, Mallinckrodt prof. applied physics, 1991—, prof. physics, 1980—; professeur associé U. Paris, 1966-67; cons. solid state physics, 1954—; Ripon prof., Calcutta, 1984. Author: Handbook on Semiconductors: Band Theory and Transport Properties, 1982; co-editor: Solids Under Pressure, 1963, Amorphous and Liquid Semiconductors, 1980. Carnegie fellow, 1952-53; Guggenheim fellow, 1959-60; Humboldt awardee, 1990; fellow Clare Hall Cambridge U., 1974-75. Fellow Am. Phys. Soc., Brit. Inst. Physics, N.Y. Acad. Scis., Royal Soc. Edinburgh; mem. AAAS, Sigma Xi. Home: 2 Eustis St Lexington MA 02173-5612 Office: Harvard U Pierce Hall Cambridge MA 02138

PAUL, WILLIAM ERWIN, immunologist, researcher; b. Bklyn., June 12, 1936; s. Jack and Sylvia (Gleicher) P.; m. Marilyn Heller, Dec. 25, 1958; children—Jonathan M., Matthew E. A.B. summa cum laude, Bklyn. Coll., 1956; M.D. cum laude, SUNY-Downstate Med. Ctr., 1960, DSc (hon.), 1991. Intern, asst. resident Mass. Meml. Hosp., Boston, 1960-62; clin. assoc. Nat. Cancer Inst. NIH, Bethesda, Md., 1962-64; postdoctoral fellow, instr. NYU Sch. Medicine, N.Y.C., 1964-68; sr. investigator Lab. Immunology Nat. Inst. Allergy and Infectious Diseases, NIH, Bethesda, Md., 1968-70, chief Lab. Immunology, 1970—; mem. bd. sci. advisors Jane Coffin Childs Meml. Fund for Med. Research, 1982-90; G. Burroughs Mider lectr. NIH, 1982; mem. sci. rev. bd. Howard Hughes Med. Inst., 1979-85, 87-91, mem. med. adv. bd., 1992—; mem. bd. sci. cons. Meml. Sloan-Kettering Cancer Ctr., N.Y.C., 1984-92; chmn. adv. com. Harold C. Simmons Arthritis Rsch. Ctr., 1984-90; bd. dirs. Fed. Am. Soc. Experimental Biology, 1985-88; mem. bd. basic biology Nat. Res. Council, 1986-89; mem. select com. Alfred P. Sloan Jr. Prize, Gen. Motors Cancer Res. Fedn., 1986-87; sci. adv. coun. Cancer Rsch. Inst., 1985—; mem. com. to visit div. med. sci., bd. overseers Harvard Coll., 1987-93; Carl Moore lectr. Sch. Medicine, Washington U., St. Louis, 1986; mem. adv. com. Pew Scholars Program in Biomed. Scis., 1988—; Richard Gershon lectr. Yale U. Sch. Medicine, 1986; Nelson Med. lectr., U. Calif., Davis, 1988; Disting. Alumnus lectr., Univ. Hosp., Boston, 1989; Anderson med. lectr. U. Va., 1990, La Jolla sci. lectr., 1991, Wellcome vis. prof. Wayne State U., 1991, mem. Adv. Com. dept. of Molecular Bio., Princeton U, 1993—, annual lectr., Dutch Soc. for Immunology, 1992, Yamamuro Meml. lectr. Osaka U., 1992, Kunkel lectr., Johns Hopkins U Sch. of Medicine, 1993; Welcome vis. prof. SUNY Stony Brook, 1993; Benacerrat lectr. Harvard Med. Sch., 1993. Editor: Fundamental Immunology, 1984, 2d edit., 1989, 3d edit., 1993, Annual Rev. of Immunology, Vols. 1-12, 1983—; adv. editor Jour. Exptl. Medicine, 1974—; assoc. editor Cell, 1985—; transmitting editor Internat. Immunology, 1989—; corr. editor Procs. of Royal Soc. Series B, 1989—; mem. editorial bd. Molecular Biology of the Cell, 1990-92; contbg. editor Proceedings of the Nat. Acad. Scis., U.S.A., 1992—; contbr. numerous articles to profl. jours. Served with USPHS, 1962-64, 75—. Recipient Founders' prize Tex. Instruments Found., Dallas, 1979, Alumni medallion SUNY-Downstate Med. Ctr., 1981, Disting. Svc. medal USPHS, 1985, 3M Life Scis. award, 1988, Tovi Cemet-Wallerstein prize CAIR Inst., Bar-Ilan U., 1992, 6th Ann. award for excellence in immunologic rsch. Duke U., 1993. Fellow Am. Acad. Arts and Scis.; mem. NAS, Inst. Medicine NAS, Am. Soc. Clin. Investigation (pres. 1980-81), Am. Assn. Immunologists (pres. 1986-87), Assn. Am. Physicians, Scandinavian Soc. Immunology (hon.). Office: NIH Bldg 10 Rm 11N311 9000 Rockville Pike Bethesda MD 20892

PAUL, WOLFGANG, physics educator; b. Lorenzkirch, Germany, Aug. 10, 1913; s. Theodor and Elisabeth (Ruppel) P.; m. Liselotte Hirsche, 1940 (dec. 1977), m. Doris Waloh, 1979; four children. Diplom, Techn. Hochschule, Berlin, 1937; Habilitation, U. Kiel/Göttingen, Germany, 1944; Dr. honoris causa, U. Uppsala, U. RWTH Aachen, Germany, U. Thessaloniki, Greece, U. Poznan, Poland; DCL (hon.), U. Kent, UK, 1992. Dozent U. Göttingen, Germany, 1944; ord. prof., dir. Inst. Physics U. Bonn, Germany, 1952-80, now prof. emeritus; pres. Alexander von Humboldt-Stiftung, Bonn, Germany, 1979-89; Dek. Mathemat.-Naturwissenschaft. Fak. U. Bonn, 1957-58, vis. scient. CERN (Centre Européen de Recherche Nucléaire, Genf) 1960-62, chrm. Vorstand der Kernforschungsanlage Jülich, 1963-64, chrm. Wissenshaftlicher Rat, 1963-64, dir. nuclear physics div. CERN, 1965-67, man. dir. DESY, 1971-73, chrm. Wissenschaftlicher Rat v. CERN, 1973-79. Contbr. numerous articles to profl. jours. Pres. Alexander von Humboldt-Stiftung, 1979-89. Recipient Nobel prize in physics, 1989, Robert W. Pohl prize. Mem. German Phys. Soc., European Phys. Soc., German Acad. for Natural Sci. (Leopoldina), Acad. Sci., Order 'Pour le mérite'. Achievements include co-founding of DESY, DORIS, PETRA. Address: Stationsweg 13, 5300 Bonn Germany Office: U Bonn, Nussallee 12, D-5300 Bonn Federal Republic of Germany

PAULIN, ANDREJ, metallurgical engineer, educator; b. Ljubljana, Yugoslavia, Sept. 22, 1939; s. Robert and Marija (Pintar) P.; diploma engr. U. Ljubljana, 1962; diploma Imperial Coll. Sci. and Tech., London, 1967, Ph.D., 1968, DSc in Metallurgy (honoris causa) Marguis Guiseppe Scieluna Internat. U. Found., Del., 1988, Ph.D. in Metallurgy (honoris causa) Albert Einstein Internat. Acad. Found., Del., 1990; m. Slavka Levec, Jan. 9, 1971; children: Maja, Irena. Research asst. Inst. Metallurgy, Ljubljana, 1962-63; tchr. asst. dept. mining and metallurgy U. Ljubljana, 1964-71, reader, 1971-77, assoc. prof., 1977-82, prof., 1982—, head metall. div., 1973-76, dep. head dept. geology, mining and metallurgy, 1983-85, mem. exec. council Faculty for Natural Scis. & Tech., 1973-76; mem. Internat Com. for Study of Bauxites, Alumina and Aluminum, 1977—; mem. exec. council Research Council of Slovenia for Geology, Mining and Metallurgy, 1978-80, mem. presidium, 1981-83; organizer 3d Yugoslav Symposium on Aluminum, Sibenik, 1978, 4th Yugoslav Internat. Symposium on Aluminum, Titograd, 1982, 5th symposium, Mostar, 1986, 6th symposium, T. Uzice, 1990; pres. com. for tech. terminology at Slovenian Acad. of Sci. and Arts, 1987—. Recipient Cross of Merit, Albert Einstein Internat. Acad. Found., 1992. Fellow Am. Biog. Inst. Rsch. Assn.; mem. Am. Biographical Inst. Rsch. Assn. (dep. gov. 1989, fellow 1992, medal of honor 1992), Instn. of Mining and Metallurgy (Great Britain, fellow 1988—), Union Slovenian Mining and Metall. Engrs. (v.p. 1973—, recipient award 1973, 83, exec. com.), Union Yugoslav Mining and Metall. Engrs. (exec. com. 1978-80, award 1979). Editor metallurgy Ency. Slovenia, 1975—, Tech. Ency. (Zagreb), 1979—; author numerous works in field of extractive metallurgy of lead, aluminum, copper, zinc, mercury, silver and gold. Home: 2 b V dolini, 61113 Ljubljana Slovenia Office: 12 Askerceva, 61000 Ljubljana Slovenia

PAULING, LINUS CARL, chemistry educator; b. Portland, Oreg., Feb. 28, 1901; s. Herman Henry William and Lucy Isabelle (Darling) P.; m. Ava Helen Miller, June 17, 1923 (dec. Dec. 7, 1981); children: Linus Carl, Peter Jeffress, Linda Helen, Edward Crellin. BS, Oreg. State U., Corvallis, 1922; ScD (hon.), Oreg. State Coll., Corvallis, 1933; PhD, Calif. Inst. Tech., 1925; ScD (hon.), U. Chgo., 1941, Princeton, 1946, U. Cambridge, U. London, Yale U., 1947, Oxford U., 1948, Bklyn. Poly. Inst., 1955, Humboldt U., 1959, U. Melbourne, 1964, U. Delhi, Adelphi U., 1967, Marquette U. Sch. Medicine, 1969; LHD, Tampa U., 1950; UJD, U. N.B., 1950; LLD, Reed Coll., 1959; Dr. h.c., Jagiellonian U., Montpellier (France), 1964; DFA, Chouinard Art Inst., 1958; also others. Teaching fellow Calif. Inst. Tech., 1922-25, research fellow, 1925-27, asst. prof., 1927-29, assoc. prof., 1929-31, prof. chemistry, 1931-64; chmn. div. chem. and chem. engring., dir. Calif. Inst. Tech. (Gates and Crellin Labs. of Chemistry), 1936-58, mem. exec. com., bd. trustees, 1945-48; research prof. (Center for Study Dem. Instns.), 1963-67; prof. chemistry U. Calif. at San Diego, 1967-69, Stanford, 1969-74; pres. Linus Pauling Inst. Sci. and Medicine, 1973-75, 78—, research prof., 1973—; George Eastman prof. Oxford U., 1948; lectr. chemistry several univs. Author several books, 1930—, including How to Live Longer and Feel Better, 1986; contbr. articles to profl. jours. Fellow Balliol Coll., 1948; Fellow NRC, 1925-26; Fellow John S. Guggenheim Meml. Found., 1926-27; Recipient numerous awards in field of chemistry, including; U.S. Presdl. Medal for Merit, 1948, Nobel prize in chemistry, 1954, Nobel Peace prize, 1962, Internat. Lenin Peace prize, 1972, U.S. Nat. Medal of Sci., 1974, Fermat medal, Paul Sabatier medal, Pasteur medal, medal with laurel wreath of Internat. Grotius Found., 1957, Lomonosov medal, 1978, U.S. Nat. Acad. Sci. medal in Chem. Scis., 1979, Priestley medal Am. Chem. Soc., 1984, Chem. Edn. award, 1987,Tolman medal, 1991, award for chemistry Arthur M. Sackler Found., 1984, Vannevar Bush award Nat. Sci. Bd., 1989. Hon., corr., fgn. mem. numerous assns. and orgns. Home: Salmon Creek 15 Big Sur CA 93920 Office: Linus Pauling Inst Sci and Medicine 440 Page Mill Rd Palo Alto CA 94306-2025

PAULL, WILLIAM BERNARD, standards engineer, biologist; b. Dayton, Ohio, Nov. 28, 1952; s. William John and Alice Marie (Brownfield) P.; m. Sharon Michelle Berk, July 30, 1976; children: Brooke Michelle, William Bradley. BS in Biology, Wright State U., 1977. Food technologist, standards engr. Hobart Corp., Troy, Ohio, 1977-90; mgr. product approvals Frymaster Corp., Shreveport, La., 1990—; mem. foodsvc. joint com. Nat. Sanitation Found., Ann Arbor, Mich., 1989—. Mem. ASTM (chair subcom. 1990—), Am. Soc. Quality Control, Gas Appliance Mfrs. Assn. (mem. govt. affairs com. 1992—). Republican. Roman Catholic. Office: Frymaster Corp 8700 Line Ave Shreveport LA 71106

PAULSON, JOHN DANIEL, toxicologist; b. McKeesport, Pa., Feb. 24, 1950; s. Evert D. and Lillian M. (O'Connell) P.; m. Patricia L. Kelly, May 7, 1987; children: Matthew J., Kelly M., Daniel A., Jonathan E. BS, U. Pitts., 1971, MS in Hygiene, 1974, PhD, 1977. Diplomate Am. Bd. Toxicology. Chief exptl. pathology Shadyside Hosp. Inst. Pathology, Pitts., 1975-80; group leader genetic toxicology McNeil Pharm. Co., Spring House, Pa., 1980-82, program dir. cancer therapy, 1982-83, dir. toxicology, 1983-87; dir. rsch. found. Ethicon, Inc., Somerville, N.J., 1987-91, dir. regulatory affairs, 1991—. Contbr. articles on genetic toxicology, histopathology of cancerous and precancerous lesions and carcinogenesis, 1975—. Recipient career devel. award NIH, U. Pitts., 1973-75. Mem. AAAS, Soc. Toxicology, Mid-Atlantic Soc. Toxicology (councilor, pres. 1990-92), Regulatory Affairs Profl. Soc. Achievements include participation in development of more than 20 medical devices and pharmaceutical products, AAMI and ISO expert committees on medical device biocompatibility. HIMA Silicone/medical device task force. Office: Ethicon Inc PO Box 151 Somerville NJ 08876

PAULSON, JOHN FREDERICK, research chemist; b. Providence, Oct. 29, 1929; s. Frederick Holroyd and Doris (Kerfoot) P.; m. Marjorie Mae Johnson, Sept. 24, 1955; children: David, Suzanne. AB, Haverford Coll., 1951; PhD, U. Rochester, 1958. Project assoc. U. Wis., Madison, 1958-59; rsch. chemist Air Force Geophysics Lab., Hanscom AFB, Mass., 1959—, group leader, 1962—; br. chief, 1992—. Contbr. articles to profl. jours., chpts. to books. Mem. Am. Chem. Soc., Am. Phys. Soc., Am. Geophys. Union. Home: 93 Carlisle Pines Carlisle MA 01741 Office: Ionospheric Effects Div Geophys Directorate Hanscom AFB MA 01731

PAULTRE, PATRICK, civil engineering educator. Prof. civil engring. U. Sherbrooke, Que., Can. Recipient Sir Casimir Gzowoski medal Can. Soc. Civil Engring., 1991. Office: Univ de Sherbrooke, Dept de Genie Civil, Sherbrooke, PQ Canada J1K 2R1*

PAUMGARTNER, GUSTAV, hepatologist, educator; b. Neumarkt, Styria, Austria, Nov. 23, 1933; s. Gustav and Grete (Egghart) P.; grad. Bundesrealgymnasium, Graz, Austria; student Princeton U.; MD, U. Vienna, 1960; m. Dagmar List, June 24, 1963. Fellow pharmacology U. Vienna, 1961-63; resident internal medicine U. Vienna Hosp., 1963-65, 66-71; fellow medicine N.J. Coll. Medicine, 1965-66; assoc. dir. dept. clin. pharmacology U. Berne (Switzerland), 1974-79; prof. medicine, chmn. dept. Medicine II, U. Munich (Germany), 1979—; sec. European Assn. Study of Liver, 1971-73; pres. European Assn. Study of Liver, 1989, German Soc. Gastroenterology, 1992. Author: The Liver, Quantitative Aspects of Structure and Function, 1973; also articles. Home: 13 Tassilostrasse, D8032 Graefelfing Federal Republic of Germany Office: Med Klinik II, Klinikum Grosshadern, D8000 Munich 70, Germany

PAUSTENBACH, DENNIS JAMES, environmental toxicologist; b. Pitts., Oct. 29, 1952; s. Albert Paustenbach and Patricia Jean (Iseman) Murray; m. Louise Dunning, Feb. 23, 1985; children: Mark Douglas, Anna Louise. BSChemE, Rose-Hulman Inst. Tech., 1974; MS in Indsl. Hygiene, U. Mich., 1977; MS in Indsl. Psychology, Ind. State U., 1978; PhD in Environ. Toxicology, Purdue U., 1982. Diplomate Am. Bd. Toxicology; cert. indsl. hygienist, safety profl., environ. assessor. Chem. process engr. Eli Lilly & Co., Clinton, Ind., 1974-76; indsl. hygiene engr. Eli Lilly & Co., Lafayette, Ind., 1977-80; prof. toxicology and indsl. hygiene Purdue U., West Lafayette, Ind., 1979-82; risk assessment scientist Stauffer Chem. Co., Westport, Conn., 1982-84; mgr. indsl. and environ. toxicology Syntex Corp., Palo Alto, Calif., 1984-87; v.p. McLaren/Hart Environ. Engring., Alameda, Calif., 1987-91, chief tech. officer, 1991—; cons. IBM, Maxus Energy, Kodak, Hercules, Weyerhauser, PPG Industries, 1980-82, RCA, Indpls., 1980-82, Hewlett-Packard, San Diego, Palo Alto, 1984-86, Semiconductor Indsl. Assn., San Jose, Calif., 1984-86, Maxus Energy Corp, Dallas, 1987-93, Hughes Aircraft, L.A., 1987-92, Weyerhauser, Seattle, 1992; mem. nat. coun. on radiol. protection and sci. adv. bd. U.S. EPA. Contbr. over 100 articles to profl. jours., 10 chpts. to books; author coll. textbook on environ. risk assessment. Recipient Kusnetz award in Indsl. Hygiene. Fellow Am. Acad. Toxicological Scis.; mem. AICE, Am. Indsl. Hygiene Assn., Soc. Toxicology, Soc. Risk Analysis, Soc. Environ. Toxicology and Chemistry, Soc. Exposure Assessment, Am. Conf. Govtl. Indsl. Hygienists, N.Y. Acad. Scis., Sigma Xi. Roman Catholic. Avocations: antique furniture, jogging, golf, baseball. Home: 65 Roan Pl Woodside CA 94062 Office: McLaren/Hart Environ Engrng 1135 Atlantic Ave Alameda CA 94501

PAVELIC, ZLATKO P., physician, pathologist; b. Slavonski Brod, Croatia, Aug. 14, 1943. Med. dr. degree, U. Zagreb, 1969, MS, 1969-71, D of Sci., 1974. Intern. Teaching Hosps. U. Zagreb, Gen. Hosp. Pakrac Surgical Dept., 1969-71; resident in pathologic anatomy U. Zagreb, 1971-74, med. faculty, 1974, asst. prof. dept. pathology, 1975, assoc. prof., 1975; assoc. cancer rsch. scientist Dept. Experimental Therapeutis Roswell Park Meml. Inst., Buffalo, 1976-83; asst. clin. rsch. prof., med. sch. dept. Pharmacology SUNY, Buffalo, 1978-81, asst. rsch. prof. dept. Pathology, 1978-89; assoc. cancer rsch. scientist V, Dir., Lab. for Pathology Roswell Park Meml. Inst., Buffalo, 1983-89; assoc. prof. Coll. Medicine U. Cin., 1989-91, assoc. prof. dept. otolaryngology head and neck surgery, 1991-92, prof. dept. otolaryngology head and neck surgery, 1992—; rsch. fellow Med. Sch. Oxford, Eng., 1975; cons. Ruder Boskovic Inst. Dept. Experimental Biology and Medicine, Zagreb, 1970-75; appointment and promotion com. Grace Cancer Drug Ctr. Roswell Park Meml. Inst., Buffalo, 1986-89; chmn. Tumor Procerement Com. Dept. Pathology and Lab. Medicine U. Cin., 1989-90; rsch. com. Dept. Otolaryngology Head and Neck Surgery U. Cin., 1991—; dir. divsn. Molecular Oncology Dept. Otolaryngology Head and Neck Surgery U. Cin., 1991—; cons. E-Z-EM inc., Westbury, N.Y., 1987-92, Molecular Oncology Inc., Gaithersburg, Md., 1992—, Med. Ctr. Info. and Comm. U. Cin., 1992—, Centocor Inc., Malvern, Pa., 1992—; speaker in field. Edit. bd.: Libri Oncologici, 1991—; contbr. 170 articles to profl. jours. NIH grantee 1984-93. Mem. Med. Assn. Croatia, Croatian Assn. Immunology, Croatian Assn. Pathology, Am. Assn. Cancer Rsch., Am. Soc. for Clin. Oncology, Am. Assn. for Pathology, The N.Y. Acad. Scis., Am. Assn. for the Advancement Sci., Sigma Xi. Home: 2401 Ingleside Ave Ste 10A Cincinnati OH 45206 Office: U Cin Coll Medicine 231 Bethesda Ave ML 528 Cincinnati OH 45267

PAVELKA, ELAINE BLANCHE, mathematics educator; b. Chgo.; d. Frank Joseph and Mildred Bohumila (Seidl) P.; B.A., M.S., Northwestern U.; Ph.D., U. Ill. With Northwestern U. Aerial Measurements Lab., Evanston, Ill.; tchr. Leyden Community High Sch., Franklin Park, Ill.; prof. math. Morton Coll., Cicero, Ill.; invited speaker 3d Internat. Congress Math. Edn., Karlsruhe, Germany, 1976. Recipient sci. talent award Westinghouse Elec. Co. Mem. Am. Edn. Research Assn., Am. Math. Assn. 2-Year Colls., Am. Math. Soc., Assn. Women in Math., Can. Soc. History and Philosophy of Math., Ill. Council Tchr. of Math., Ill. Math. Assn. Community Colls., Math. Assn. Am., Math. Action Group, Ga. Center Study and Teaching and Learning Math., Nat. Council Tchrs. of Math., Sch. Sci. and Math. Assn., Soc. Indsl. and Applied Math., Northwestern U. Alumni Assn., U. Ill. Alumni Assn., Am. Mensa Ltd., Intertel, Sigma Delta Epsilon, Pi Mu Epsilon. Home: PO Box 7312 Westchester IL 60154-7312 Office: Morton Coll 3801 S Central Ave Chicago IL 60650-4306

PAVIS, JESSE ANDREW, sociology educator; b. Washington, Sept. 15, 1919; s. Abraham and Ethel (Rein) P.; m. Mary Fleming Bennet, Mar. 9, 1942 (div. Sept. 1963); children: Amaranth, Athar, Arne, Andrea; m. Mary Margaret Monahan, May 9, 1964 (dec. Oct. 1992); children: Deidre, Shira. BA, George Washington U., 1942; MA, Howard U., 1947; PhD, NYU, 1969. Instr. Hofstra U., Hempstead, N.Y., 1948-49; prof. Borough of Manhattan Community Coll. (CUNY), N.Y.C., 1964—; mem. faculty New Sch. for Social Rsch., N.Y.C., 1973-79; researcher Bur. Applied Social Rsch., Columbia U., N.Y.C., 1952-53, Grad. Staff, 1953-54; mem. Cumberland Mental Health Community Ctr., Bklyn., 1977-78; reviewer Contemporary Sociology, Sociology of Sci., 1974; mem. panel Study of Nurse Edn. Needs in So. N.Y., 1964-65. Pres. Pub. Sch. #8 PTA, Bklyn., 1976-80; mem. Parent Coun. Sch. Dist. 13, Bklyn., 1986-90; mem. Bklyn. Child Safety Com., 1976-80. With U.S. Army, 1942-45. Recipient Founders Day award NYU, 1970. Mem. Am. Sociol. Soc., N.Y. Acad. Scis., Order of Artus, Pi Gamma Mu. Democrat. Achievements include patent for probability and educational game construction (U.S, Can.). Office: BMCC-CUNY 199 Chambers St New York NY 10007-1006

PAVLIK, THOMAS JOSEPH, quality control engineer, statistician; b. Pottstown, Pa., Aug. 15, 1951; s. John Michael and Helen Eleanor (Kopcho) P.; m. Patricia Antoinette Lamond, Feb. 17, 1973 (div. Nov. 1986); children: Erica, Gia, Thomas Jr., Gia, Mara; m. Stephanie Carolyn Longo, Dec. 27, 1987. BS in Math., St. Joseph's U., 1973. Sr. tech. engr. Firestone Tire and Rubber Co., Pottstown, Pa., 1974-80; quality control engr. Hatfield Wire and Cable Co., Linden, N.J., 1980-82; dir. quality mgmt. Marcal Paper Mills, Elmwood Park, N.J., 1982-90; dir. quality control Texwipe Co., Upper Saddle River, N.J., 1990—. Mem. IEEE, ASTM, Am. Soc. Quality Control. Home: 168 Mountain Way Rutherford NJ 07070 Office: Texwipe Co 650 E Crescent Ave Upper Saddle River NJ 07458

PAVLOVIC, MILIJA N., structural engineer, educator; b. Belgrade, Serbia, Yugoslavia, Jan. 13, 1950; s. Nikola M. and Tatjana M. (Vasiljevic) P.; m. Vera B. Timotic, Apr. 19, 1982; children: Nikola, Nada, Djordje, Miloš. B of Engring. with 1st class honors, U. Melbourne (Australia), 1971, M. Engring. Sci., 1974; PhD in Engring., U. Cambridge (Eng.), 1978; PhD in Medieval Spanish, U. London, 1984. Lectr. civil engring. Imperial Coll., U. London, 1978-87, reader structural engring., 1987—; cons. in field. Contbr. numerous articles to scholarly, sci. and profl. jours. and books. Recipient Baker medal Instn. of Civil Engrs. London, 1989. Orthodox. Avocations: history, literature, opera, football. Office: Imperial Coll, Dept Civil Engring, London SW7 2BU, England

PAVLOVICH, DONALD, technical writer; b. Euclid, Ohio, Mar. 25, 1957; s. Paul George and Carroll Rose (McDonald) P. BS, Cleve. State Univ., 1981. Tech. writer, instr. Cleve. (Ohio) Inst. Electronics, 1982-85; instr. Lorain (Ohio) County Community Coll., 1985-86; tech. writer Picker Internat.-Gov. systems, Highland Heights, Ohio, 1986-87; sr. tech. writer Picker Internat.-MRI, Highland Heights, Ohio, 1987—; reviewer IEEE Microprocessor and Microcomputer standards subcom., 1990—. Author (lesson books): ROMS, PROMs, and PLAs, 1983; co-author (lesson books): Karnaugh Maps, 1984, One-Shots, Astables and Schmitt Triggers, 1983; editor: Microprocessor Course, 1985; book reviewer. Vol. Cleve. Mus. of Art, 1991. Named Jr. Tech. Student of Yr. Cleve. State Engring. Alumni Assn., 1980. Mem. IEEE, Alpha Beta Kappa. Home: 19015 Van Aken Blvd # 404 Shaker Heights OH 44122-3551 Office: Picker Internat 5500 Avion Park Dr Highland Heights OH 44143

PAWELEC, GRAHAM PETER, immunologist; b. Cambridge, Eng., Nov. 23, 1951; s. Frank Robert and Hilary Irene (Lander) P.; m. Monika Rita Schneider, July 16, 1983; children: Michael Robert, Maria Elisabeth Sophie. MA, Cambridge (Eng.) U., 1973, PhD, 1982; Dozent, Tübirgen U., 1987. Rsch. asst. dept. surgery U. Cambridge, 1974-78; established scientist 2nd dept. internal medicine U. Tübingen, Germany, 1978—. Translator sci. texts; contbr. over 150 articles to profl. jours. Recipient Deutsche Forschungsgemeinschaft grant 1983—, Deutsche Krebshilfe grant, 1991—. Mem. AAAS, Brit. Soc. Immunology, Brit. Transplantation Soc., Am. Soc. for Histocompatibility and Immunogenetics, German Soc. Immunology, European Found. for Immunogenetics. Office: Medizinische Klinik, Abt Innere Medizin II, D-72076 Tübingen Germany

PAXTON, JOHN WESLEY, electronics company executive; b. Camden, N.J., Jan. 9, 1937; s. John Irving and Francis Rose (Jones) P.; m. Janet Rose Croteau, Nov. 12, 1975; children: John M., David R., William A., John Wesley Jr., Jacqueline R., Thomas W., Scott A. M of Bus. Mgmt., LaSalle, 1990; postgrad. in mfg. tech. U. Mich., 1976; postgrad. Rivier Coll., 1979. Registered profl. engr., Calif. With RCA, Camden, Hightstown, N.J. and Burlington, Mass., 1959-75, mgr. quality assurance, 1972-75; dir. mfg. Kollsman Instrument Co. div. Sun Chem. Corp., Merrimack, N.H., 1975-80; dir., plant mgr. Ocala ops. Martin Marietta Corp. (Fla.), 1980-83, dir. product ops., Orlando, Fla., 1983-84; gen. mgr. Midland Press, Urbana, Ohio, 1984-86; pres. Intermec Corp., Everett, Wash., 1986-88, pres., chief exec. officer, chmn. bd., 1988-92; sr. v.p., group exec. Litton Industries, 1992—. With USN, 1955-59. Mem. Nat. Soc. Profl. Engrs., Am. Soc. Mfg. Engrs. Republican. Presbyterian. Lodges: Lions, Masons.

PAXTON, R(ALPH) ROBERT, chemical engineer; b. Zion, Ill., Mar. 4, 1920; s. James Robert and Hazel Marie (Lawrence) P.; widowed; children: Nancy, Anne. BSChemE, U. Ill., 1943; ScD, MIT, 1949. Jr. engr. Amoco Corp. (formerly Standard Oil Co. of Ind.), Whiting, 1943-45; instr. chem. engring. U. Colo., Boulder, 1945-47; teaching asst. MIT, Cambridge, 1947-49; asst. prof. chem. engring. Stanford (Calif.) U., 1949-56; sr. engr. chem. engring. GE Co., Pittsfield, Mass., 1956-58; v.p. engring. and planning Pure Carbon Co., St. Mary, Pa., 1958-85; cons. Paxton Cons., St. Mary, 1986—. Author: Mechanical Carbon, 1979. Mem. St. Mary Area Sch. Bd., 1983-85. Fellow Am. Inst. Chemists, Soc. Tribologists and Lubrication Engrs.; mem. ASTM (sec. C-5 com. 1979-81), Am. Inst. Chem. Engrs. (assoc. editor Lubrication Engring. 1970—), Kiwanis (pres. 1980), St. Mary Country Club (sec. 1979-81). Achievements include 9 patents in field. Office: Pure Carbon Co 441 Hall Ave Saint Marys PA 15857-1497

PAYN, CLYDE FRANCIS, technology company executive, consultant; b. Auckland, New Zealand, Jan. 17, 1952; came to U.S., 1973; s. Phillip Francis and Ngaire Eunice P.; m. Betsy Ann Dannels, June 17, 1978; children: Tamara, Brittany, Erik. Cert., Auckland Inst. Tech., 1971; MBA, Vanderbilt U., 1980. Tech. mgr. Carborundum (N.Z.) Ltd., Auckland, 1968-73; mem. product application tech. staff Carborundum Co., Niagara Falls, N.Y., 1973-78; mgr. product mktg. Universal Abrasives, Phila., 1978-80; bus. mgr., catalyst advocate Johnson Matthey, Inc., Phila., 1980-84; pres. Catalyst Cons., Inc., Phila., 1984—; CEO Catalyst Group, Phila., 1988—. Pres. Hideaway Hill Civic Assn., Maple Glen, Pa., 1988, 89. Mem. AICE, Am. Chem. Soc., Catalysis Soc., Comml. Devel. Assn., Chem. Mktg. Rsch. Assn., Polymer Mfg. Engrs. Assn. Achievements include development of process technology, catalyst and product development for petroleum, petrochemical,

chemical and environmental industries. Office: The Catalyst Group Inc PO Box 637 Spring House PA 19477

PAYNE, ANITA HART, reproductive endocrinologist, researcher; b. Karlsruhe, Baden, Germany, Nov. 24, 1926; came to U.S., 1938; d. Frederick Michael and Erna Rose (Hirsch) Hart; widowed; children: Gregory Steven, Teresa Payne-Lyons. BA, U. Calif., Berkeley, 1949, PhD, 1952. Rsch. assoc. U. Mich., Ann Arbor, 1961-71, asst. prof., 1971-76, assoc. prof., 1976-81, prof., 1981—, assoc. dir. Ctr. for Study Reprodn., 1989—; vis. scholar Stanford U., 1987-88; mem. reproductive biology study sect. NIH, Bethesda, Md., 1978-79, biochem. endocrinology study sect., 1979-83, population rsch. com. Nat. Inst. Child Health and Human Devel., 1989-93. Assoc. editor Steroids, 1987-93; contbr. book chpts., articles to profl. jours. Recipient award for cancer rsch. Calif. Inst. Rsch. Cancer, 1953, Acad. Women's Caucus award U. Mich., 1986. Mem. Endocrine Soc. (chmn. awards com. 1983-84, mem. nominating com. 1985-87, coun. 1988-91), Am. Soc. Andrology (exec. coun. 1980-83), Soc. for Study of Reprodn. (bd. dirs. 1982-85, sec. 1986-89, pres. 1990-91). Office: U Mich Steroid Rsch Unit Lab L1225/ 0278 Womens Hosp Ann Arbor MI 48108-0278

PAYNE, ANTHONY GLEN, clinical nutritionist, naturopathic physician; b. Lubbock, Tex., Aug. 17, 1955; s. Walter Wade and Mary G. (Whittle) P.; m. Gretchen Karen Lejeune, Jul. 01, 1980. BS, Columbia Pacific Univ., 1982, MA with honors, 1987; NMD, Hearts of Jesus & Mary Coll., 1991; DSc, Notre Dame de Lafayette Univ., 1989. Cert. Wellness counselor; lic. Mercian practitioner. Product devel. dir. Human Scis. Rsch., Dallas, 1990—; staff clinician Steenblock Medical Clinic, El Toro, Calif., 1988-90; tech. adv. Global Vision, Inc., Livermore, Calif., 1991—; asst. clinical prof. San Francisco Sch. Chinese Medicine, 1989—; tech. cons. P.C. Teas Co., Burlingame, Calif., 1989—; bd. dirs. Barotech, Inc., Las Vegas, 1992; field faculty Notre Dame de Lafayette Univ., Aurora, Colo., 1992—. Author: (booklet) Naturopathic Medicine: A Primer for Laypersons, 1990; contbr. articles to profl. jours. Fellow Am. Nutritional Medical Assn., 1992; recipient Cert. Merit Am. Nutrimedical Assn., 1992. Mem. MENSA, Am. Nutritional Medical Assn. (profl.), British Guild of Drugless Practitioners, (lifetime), Am. Coll. Naturopathy (dir. 1987-88), Am. Nutritional Medical Assn. (state dir. 1988-89). Roman Catholic. Achievements include development of and empirically testing of 10 phytoceutical plant based formulas currently in production. Home: 103 High School Dr # 107 Grand Prairie TX 75050

PAYNE, CLAIRE MARGARET, molecular and cellular biologist; b. N.Y.C., Mar. 2, 1943; d. Frederick John Luscher and Florence Muriel (Seiler) Nothdurft; m. Thomas Bennett Payne, Apr. 19, 1969. BS, SUNY, Stony Brook, 1963; MS, Adelphi U., 1965; PhD, SUNY, Stony Brook, 1971. Biology tchr. North Babylon (N.Y.) High Sch., 1970-72; rsch. asst. Dept. Pathology U. Ariz., Tucson, 1972-73, lectr. Dept. Pathology, 1973-86, rsch. assoc. prof. Pathology, 1986-89, rsch. prof. Pathology, 1989-93; rsch. prof. Ariz. Rsch. Labs. and Dept. Microbiology & Immunology, Tucson, 1993—; supr. Clin. Electron Microscopy Lab., U. Ariz., Tucson, 1973-86, adminstrv. chief, 1986-93. Vol. Am. Cancer Soc., 1988—. Ariz. Disease Control Rsch Comm. grantee, 1990-93, grantee Mathers Found., 1992—. Mem. AAAS, Soc. for Diagnostic Ultrastructural Pathology (sec.-treas. 1991—), Ariz. Soc. for Electron Microscopy and Microbeam Analysis (pres. 1983-84), Ariz. Soc. Pathologists, Tucson Soc. of Pathologists, Am. Soc. Cell Biology, European Soc. for Cutaneous Ultrastructure Rsch., Biomedical Diagnostics & Rsch. (pres.). Avocations: camping, water-skiing, fishing, traveling, photography. Office: Dept Microbiology & Immunology 1501 N Campbell Ave Tucson AZ 85724-0001

PAYNE, DAVID N., optics scientist. Prof. optics U. Southhampton, Eng. Recipient John Tyndall award Optical Soc. Am., 1991. Office: U of Southhampton-Phototonics, Highfield, Southhampton SO9 5NH, England*

PAYNE, FRED J., physician; b. Grand Forks, N.D., Oct. 14, 1922; s. Fred J. and Olive (Johnson) P.; m. Dorothy J. Peck, Dec. 20, 1948; children: Chris Ann Payne Graebner, Roy S., William F., Thomas A. Student U. N.D., 1940-42; BS, U. Pitts., 1948, MD, 1949; MPH, U. Calif., Berkeley, 1958. Diplomate Am. Bd. Preventive Medicine. Intern, St. Joseph's Hosp., Pitts., 1949-50; resident Charity Hosp., New Orleans, 1952-53; med. epidemiologist Ctr. Disease Control, Atlanta, 1953-60; prof. tropical medicine La. State U. Med. Ctr., New Orleans, 1961-66; dir. La. State U. Internat. Ctr. for Med. Rsch. and Tng., San Jose, Costa Rica, 1963-66; exec. sec. 3d Nat. Conf. on Pub. Health Tng., Washington, 1966-67; epidemiologist Nat. Nutrition Survey, Bethesda, Md., 1967-68; chief pub. health professions br. NIH, Bethesda, 1971-74, med. officer, sr. rsch. epidemiologist Nat. Inst. Allergy and Infectious Diseases, 1974-78; asst. health dir. Fairfax County (Va.) Health Dept., 1978—, dir. HIV/AIDS case mgmt. program, 1988—; clin. prof. La. State U., 1966—; cons. NIH, 1979-81; leader WHO diarrheal disease adv. team, 1960. Contbr. articles to profl. jours. Served with AUS, 1942-46, 49-52. Decorated Combat Medic Badge. Fellow Am. Coll. Preventive Medicine, Am. Coll. Epidemiology; mem. AAAS, AMA, Am. Soc. Microbiology, Internat. Epidemiology Assn., Soc. Epidemiologic Rsch., USPHS Commd. Officers Assn., Sigma Xi. Home: 2945 Ft Lee St Herndon VA 22071-1813 Office: 10777 Main St Fairfax VA 22030

PAYNE, WINFIELD SCOTT, national security policy research executive; b. Denver, Jan. 20, 1917; s. Winfield Scott and Mildred (Hulse) P.; m. Barbara P. Reid, Nov. 18, 1945; children: Judith P. Beland, Patricia P. Dominiguez. AB. U. Colo., 1939, MA (grad. scholar), 1941; postgrad. (fellow) Syracuse U., 1942; MPA, Harvard U., 1948, PhD, 1955. Economist, Bur. Budget, Washington, 1945-46; staff Inter-Univ. Case Program, Washington, 1948-50; indsl. analyst Pres.'s Materials Policy Commn., Washington, 1950-52; project leader Ops. Research Office, Johns Hopkins U., Bethesda, Md., 1952-63; sr. research staff, panel dir. Inst. for Def. Analyses, Arlington, Va., 1963-72; asst. to pres. System Planning Corp., Arlington, 1972-86; cons. 1986-88; adj. rsch. staff Inst. Def. Analyses, 1989—; assoc. prof., lectr. George Washington U., Washington, 1963-65; cons. Def. Advanced Research Project Agy., 1972-76; guest lectr., various univs. Mem. Cabin John (Md.) Fire Bd., 1955-65. Served with USMC, 1942. Littauer fellow, 1946-48. Mem. AAAS, Am. Polit. Sci. Assn., Am. Econ. Assn., U.S. Strategic Inst., Cosmos Club, Phi Gamma Delta, Pi Gamma Mu. Contbr. articles to profl. jours.; contbr.: Public Administration and Policy Development: A Case Book, 1951. Home: 209 Providence Rd Annapolis MD 21401-6309

PAYYAPILLI, JOHN, engineer; b. Trichur, Kerala, India, June 2, 1947; came to U.S., 1985; s. George Ittira and Rose (Chacko) P.; m. Pushpamma Lukose, Nov. 1, 1976; children: Rose, George, Luke. BSc in Mech. Engring., Regional Engring. Coll., Calicut, India, 1969; MS in Naval, Arch. and Marine Engring., U. Mich., 1980, MS in Indsl. Ops. Engring., 1980. Registered profl. engr., N.Y.; first class engr., Ministry of Transport, India; chief engr., Liberia; merchant marine lic. From jr. engr. to chief engr. Shipping Corp. of India, Bombay, 1970-77; chief engr. Overseas Shipping Ltd., Singapore, 1977-78; 1st asst. to chief engr. Kapal Mgmt. Ltd., Singapore, 1981-82, Wallem Ship Mgmt., Hong Kong, 1982-85; chief engr. Pacific Internat., Singapore, 1983-85; plant mgr. Joann Internat., Bklyn., 1985-86; asst. mech. engr. N.Y.C. Dept. Gen. Svcs., 1986-87; assoc. staff analyst N.Y.C. Transi Authority, 1987-88; mgr. N.Y.C. Transit Authority, 1988—. Mem. ASME, ASHRAE, Soc. Naval Architecture and Marine Engrs., Assn. Energy Engrs. (bd. dirs., v.p. 1991—), Am. Soc. Indsl. Engrs., India Cath. Assn. Am. (life), Univ. Mich. Alumni Assn. (life), Holiday Inn Priority Club. Roman Catholic. Home: 662 E 24th St Brooklyn NY 11210 Office: NYC Transit Authority 130 Livingston St Brooklyn NY 11201

PAZ, NILS, physicist; b. La Paz, Bolivia, Aug. 7, 1950; came to U.S., 1955; s. Mario and Marina (Salinas) P. BS in PHysics, Miami (Fla.) Internat. U., 1975, BA in Psychology, 1975; MS in Physics, U. New Orleans, 1983. Geophysicst Naval Oceanographic Office, Bay St. Louis, Miss., 1977-80; mathematician Naval Oceanographic Office, North St. Louis, Miss., 1980-86; oceanographer, cons., advisor U.S. Sixth Fleet Office, Naples, Italy, 1986-89; acoustic analyst Systems Integrated, San Diego, 1990-92; rsch. physicist Universal Computing, San Diego, 1992—. Mem. AAAS, Am. Acoustical Soc. Achievements include research on the spread of spectrum processing for moving source and receiver. Office: Universal Computing 5055 View Ridge Ave San Diego CA 92123

PAZ-PUJALT, GUSTAVO ROBERTO, physical chemist; b. Arequipa, Peru, Aug. 9, 1954; came to U.S. 1973; s. Manuel Eduardo and Raquel Maria (Pujalt) Paz-Bishop; m. Ellen Frances Coey, Nov. 27, 1985; 1 child, Martin. BS in Chemistry, U. Wis., Eau Claire, 1977; PhD in Phys. Chemistry, U. Wis., Milw., 1985. Rsch. scientist Eastman Kodak Co., Rochester, N.Y., 1986-89, sr. scientist, 1989—; vis. scientist MIT, 1988, Kodak-Pathe, Chalon-sur-Saone, France, 1991. Co-editor Materials Rsch. Soc. Procs., fall, 1993; contbr. articles to profl. jours.; patentee in field. V.p. Spanish Action Coalition, Rochester, 1990-91; mem. allocations subcom. United Way Greater Rochester, grad. leadership devel. program, 1992; bd. dirs. Pyramid Art Ctr., 1993—. Mem. Am. Chem. Soc., Matls. Rsch. Soc. Roman Catholic. Home: 114 Shepard St Rochester NY 14620-1818 Office: Eastman Kodak Co 343 State St Rochester NY 14650-2011

PEACOCK, HUGH ANTHONY, agricultural research director, educator; b. Cairo, Ga., May 30, 1928; s. Leslie Hugh and Annie John (Aldrege) P.; m. Mary Helen Willis, Oct. 8, 1949; children: Ramon Anthony, Elizabeth Ann, Mary Evelyn. BS in Agronomy, U. Fla., 1952, MS in Agronomy, 1953; PhD, Iowa State U., 1956. Rsch. assoc. Iowa State U., Ames, 1955-56; asst. agronomist U. Fla., Leesburg, 1957-58; rsch. agronomist Ga. Exptl. Sta. USDA, Experiment, 1959-73; dir. Agrl. Rsch. & Edn. Ctr. U. Fla., Jay, 1973—. Contbr. articles to Agronomy Jour., Crop Sci., Plant Disease, Jour. Nematology. Director Hist. Soc. Santa Rosa County, Milton, Fla., 1979-80, Kiwanis Club of Milton, 1988-89, chmn. agrl. com., 1987-89. Cpl. U.S. ARmy, 1946-47, CBI. John D. Rockefeller Inst. scholar, 1953. Mem. Environ. Enhancement Assn. West Fla. (bd. dirs. Pensacola chpt. 1976—), Am. Soc. Agronomy (assoc. editor 1980-83), Am. Genetic Assn., Crop Sci. Soc. Am. (assoc. editor 1976-80), Alpha Zeta, Phi Kappa Phi, Sigma Xi, Gamma Sigma Delta. Methodist. Achievements include research in number of genes controlling defoliation vs. non-defoliation in cotton; effect of heterosis and combining ability on yield of cotton, effect of nitrogen fertilization on yield; effect of skip-row culture on fiber characteristics of cotton; yield response of cotton to spacing, effect of seed source on seedling vigor, yield and lint of upland cotton; a cone-type planter for experimental plots; subsurface sweep for applying herbicides; yield responses of soybean cultivars to Meloidogyne incognita. Office: U Fla Agr Rsch Ctr RR 3 Box 575 Jay FL 32565-9523

PEAK, DAVID, physicist, educator, researcher; b. Bklyn., Nov. 28, 1941; s. William Henry and Blanche Ethel (Seckendorff) P.; m. Terry L. Handwerger, June 3, 1983. BS, SUNY, New Paltz, 1965; PhD, SUNY, Albany, 1969. Physics instr. SUNY, Albany, 1971-75; asst. prof. physics Union Coll., Schenectady, N.Y., 1975-78; assoc. prof. physics Union Coll., Schenectady, 1978-85, prof. physics, 1985-87, Frank & Marie Bailey prof. physics, 1987—; E. Claiborne Robins Disting. Prof. Sci. U. Richmond, 1993—; vis. fellow Princeton (N.J.) U., 1978-79; vis. scientist Argonne (Ill.) Nat. Lab., 1983-84; mem. Union Coll. Bd. Trustees, Schenectady, 1985-88, Nat. Confs. on Undergrad. Rsch. Governing Bd., 1988—, chair 1992—; trustee Dudley Observatory, 1992—. Author: Order and Chaos: Art and Magic, 1991; contbr. 40 articles to profl. jours. Recipient Profl. Devel. award NSF, Washington, 1978, Meritorious Svc. award, Alumni Coun. Union Coll., Schenectady, 1990, New Liberal Arts Spl. Leave award, Sloan Found., 1990-91. Mem. Coun. on Undergrad. Rsch. (founding mem., sec. 1985-87), Am. Physical Soc. (mem. exec. com. N.Y. state), Am. Astron. Soc., Am. Assn. Physics Tchrs., Sigma Xi (com. on sci. edn.). Achievements include research on nature of reactions in condensed media, ion beam processing of thin films, mechanical properties of soft condensed matter. Office: Union Coll Physics Dept Schenectady NY 12308

PEAKE, THADDEUS ANDREW, III, environmental engineer; b. Jan. 6, 1948. BS in Environ. Engring., U. Louisville, 1978, MEngring, 1983; MBA, Kennesaw State Coll., 1989. Registered profl. engr., Ga., Ala., Ky., Tenn.; cert. indsl. engr.; diplomate Am. Acad. Environ. Engring., diplomate Nat. Acad. Forensic Engring. Air planning engr., water enforcement officer, field rsch. U.S. EPA, Atlanta, 1979-91; chief engr. Peake Engring., Inc., Marietta, Ga., 1988—. Ga. Water and Pollution Control Assn. (chmn. indsl. com.), Cons. Engrs. Coun. Ga. (chmn. environ. issues com.), Air and Waste Mgmt. Assn. (bd. dirs., chmn. Ga. chpt.). Achievements include indentification of health risk from industrial facilities; design and management installation of environmental monitoring systems, monitoring networks, emission inventory algorithms, development of EPA regulations. Home: 3111 Vandiver Dr NW Marietta GA 30066-4643 Office: 3111 Vanowen Dr Marietta GA 30066

PEALE, STANTON JERROLD, physics educator; b. Indpls., Jan. 23, 1937; s. Robert Frederick and Edith May (Murphy) P.; m. Priscilla Laing Cobb; June 25, 1960; children: Robert Edwin, Douglas Andrew. BSE, Purdue U., 1959; MS in Engring. Physics, Cornell U., 1962, PhD in Engring. Physics, 1965. Research asst. Cornell U., Ithaca, N.Y., 1962-64, research assoc., 1964-65; asst. research geophysicist, asst. prof. astronomy UCLA, 1965-68; asst. prof. physics U. Calif., Santa Barbara, 1968-70, assoc. prof., 1970-76, prof., 1976—; mem. com. lunar and planetary exploration NAS-NRC, Washington, 1980-84, lunar and planetary geosci. rev. panel, 1979-80, 86-89, Planetary Systems Sci. Working Group, 1988-93, Lunar and Planetary Sci. Coun., 1984-87; lunar sci. adv. group NASA-JPL, Pasadena, Calif., 1970-72. Assoc. editor: Jour. Geophys. Research, 1987; contbr. articles to profl. jours. Recipient Exceptional Scientific Achievement medal NASA, 1980, James Craig Watson award Nat. Acad. Scis., 1982; vis. fellowships U. Colo., Boulder, 1972-73, 1979-80. Fellow AAAS (Newcomb Cleveland prize 1979), Am. Geophys. Union; mem. Am. Astron. Soc. (mem. div. planet sci., div. dynamic astronomy, Dick Brouwer award 1992), Internat. Astron. Union. Avocation: gardening. Office: U Calif Dept Physics Santa Barbara CA 93106

PEARCE, ELI M., chemistry educator, administrator; b. Bklyn., May 1, 1929; s. Samuel and Sarah (Reitzen) Perlmutter; m. Maxine I. Horowitz, Feb. 21, 1951 (div. 1978); children:Russell Gane, Debra Nore; m. Judith Handler, May 29, 1980. B.S., Bklyn. Coll., 1949; M.S., NYU, 1951; Ph.D., Poly. Inst. Bklyn, 1958. Research chemist NYU-Bellevue Med. Ctr., N.Y.C., 1949-53, DuPont, Wilmington, Del., 1958-62; sec. mgr. J.T. Baker, Phillipsburg, N.J., 1962-68; tech. supr. Allied Corp., Morristown, N.J., 1968-72, research cons., 1972-73; dir. Dreyfus Lab. Research Triangle Inst., Research Triangle Park, N.C., 1973-74; prof. polymer chemistry and chem. engring. Poly. Inst. N.Y., Bklyn., 1974—, dir. Polymer Research Inst., 1981—, prof., 1990—, head dept. chemistry, 1976-82, dean arts and scis., 1982-90; cons. AMP Inc., Harrisburg, Pa., Arco, Newton Square, Pa., 1979—, Colgate, Piscataway, N.J., Dupont, Richmond, Va. Co-author: Laboratory Experiments in Polymer Synthesis and Characterization, 1982; contbr. scientific polymer articles to profl. jours.; editor: Macromolecular Synthesis, vol. 8/ 1982; co-editor: Fiber Chemistry, 1983, Contemporary Topics in Polymer Sci., vol. 2, 1977, Flame Retardance of Polymeric Materials, vols. 1, 2, 3; sect. editor: Ency. of Materials Sci., 1983. Bd. dirs. Petroleum Research Fund, 1982-84; bd. dirs. Nat. Materials Adv. Bd., 1975-77. Served with U.S. Army, 1953-55. Recipient Edn. Service award Plastics Inst. Am., 1973; recipient Disting. Faculty citation Poly. Inst. N.Y., 1980, Paul J. Flory Polymer Edn. award, 1992, Kaufman Lectr. award Ramapo Coll., 1992, Gold Medal award N.Y. Inst. Chemists, 1992, Reed-Lignin Lectr. award U. Wis., 1987. Fellow Am. Inst. Chemists, AAAS, N.Am. Thermal Analysis Soc., N.Y. Acad. Scis. (chmn.polymer sect. 1972-73), Soc. Plastics Engrs. (Internat. Edn. award, 1988); mem. Am. Chem. Soc. (councilor 1978-86, chmn. polymer div. 1980, coun. policy com., assoc. com. sci.), Sigma Xi. Home: 2 Fifth Ave New York NY 10011 Office: Poly U Polymer Rsch Inst 6 Metrotech Ctr Brooklyn NY 11201-2990

PEARCE-PERCY, HENRY THOMAS, physicist, electronics executive; b. Melbourne, Victoria, Australia, Sept. 7, 1947; came to U.S., 1970; s. Thomas Walker Pearce-Percy and Valda Marion (Mills) Woinarski; m. Virginia Kathleen Shattuck, Apr. 18, 1975; children—Patrick Walker, Nicole Kathleen. B.S., U. Melbourne, 1968, M.S., 1970; Ph.D., Ariz. State U., 1975. Guest scientist Inst. für Elektronenmikroskopie, Max-Planck-Gesellschaft, West Berlin, Fed. Republic Germany, 1974-76; mem. tech. staff Tex. Instruments, Inc., Dallas, 1977-84; mgr. KLA Instruments, Inc., Santa Clara, Calif., 1984-86; pvt. practice cons., Los Gatos, Calif., 1986-88; prin. engr. Etec Systems, Inc. (formerly Perkin-Elmer Corp.), Hayward, Calif., 1988—. Author numerous research articles. Pres. Richardson Noon Toastmasters, Tex., 1983. Mem. AAAS, Am. Phys. Soc., Electron Microscopy Soc. Am., Am. Vacuum Soc., Toastmasters Club. Office: Etec Systems Inc 26460 Corporate Ave Hayward CA 94545-3914

PEARL, WILLIAM RICHARD EMDEN, pediatric cardiologist; b. N.Y.C., Nov. 1, 1944; s. William Emden and Sara (Gilston) P.; m. Karlyn Katsumoto, July 9, 1978; children: Jeffrey, Kristine. BA, Queens Coll., 1966; MD, SUNY, Bklyn., 1970. Diplomate Am. Bd. Pediatrics, Am. Bd. Pediatric Cardiology. Intern Roosevelt Hosp., N.Y.C., 1970-71; resident N.Y. Hosp.-Cornell Med. Ctr., N.Y.C., 1971-72; fellow Albert Einstein Coll. Medicine, N.Y.C., 1972-74; asst. prof. U. Hawaii, Honolulu, 1974-76; asst. prof. Tex. Tech. Med. Sch., El Paso, 1976-82, assoc. prof., 1982-92; chief pediatric cardiology William Beaumont Army Med. Ctr., El Paso, 1976—; cons. Miami (Fla.) Children's Hosp., 1988, Driscol Children's Hosp., Corpus Christi, Tex., 1992, Thomason Hosp., El Paso, 1976-92. Contbr. articles to profl. jours. Col. USAR, 1974-92. N.Y. State Bd. Regents scholar, 1962-66, Fed. Health Careers scholar, 1967-70; NIH fellow, 1972-73; recipient Dept. of Army Commendation for outstanding sci. achievement, 1984. Fellow Am. Acad. Pediatrics, Am. Coll. Cardiology; mem. Am. Heart Assn. (coun. on cardiovascular disease in the young 1982). Office: William Beaumont Army MC 5000 Piedras St El Paso TX 79920

PEARLMAN, SETH LEONARD, civil engineer; b. Steubenville, Ohio, Aug. 6, 1956; s. Abraham and Rita Joy (Morov) P.; m. Pamela Diane Bretton, Mar. 29, 1987; children: Isaac Joseph, Julian Brett. BSCE, Carnegie Mellon U., 1978, MSCE, 1979. Registered profl. engr., Pa., Va. Sr. engr. GAI Cons., Pitts., 1979-82; v.p. mktg. Belot Concrete Industries, Tiltonsville, Ohio, 1982-86; chief design engr. Nicholson Constrn. Co., Bridgeville, Pa., 1986-93, regional mktg. mgr., 1993—; speaker, lectr. in field. Author conf. publs. Fundraiser United Jewish Fedn., Pitts., 1988—; bd. dirs. Beth El Congregation of the South Hills, Pitts., 1991—. Mem. ASCE (mem. geotech. sect. com. Pitts. chpt. 1990—), Am. Concrete Inst. (mem. fiber reinforced concrete com. 1982—, co-chair state of art report, mem. concrete piling com. 1989—, bd. dirs. Pitts. chpt. 1984-87), ADSC Internat. Assn. Found. Drilling (earth retention com.), Nat. Soc. Profl. Engrs. (pres. Wheeling, W.Va. chpt. 1986). Democrat. Home: 266 Twin Hills Dr Pittsburgh PA 15216-1108 Office: Nicholson Constrn PO Box 98 Bridgeville PA 15017

PEARLMUTTER, FLORENCE NICHOLS, psychologist, therapist; b. Bklyn., Mar. 17, 1914; d. William and Marie Elizabeth (Rugamer) Griebe; m. Wilbur Francis Nichols, Aug. 17, 1940 (dec. 1967); 1 child, Roger F.; m. F. Bernard Pearlmutter, June 27, 1969. BS, NYU, 1934, postgrad., 1965-75; MS, Yeshiva U., 1960. Psychologist P.P.P. Counseling Ctr., Northport, N.Y., 1967-69; hypno-therapist Robert E. Peck, M.D., Syosset, N.Y., 1969-75; therapist Arthur J. Gross, M.D., Hicksville, N.Y., 1975—; rsch. asst. and prof. rsch. in field. Mem. NEA, AAUW, Nassau County Psychol. Assn., N.Y. State Psychol. Assn. (assoc.), Kappa Delta Pi. Avocations: cooking, phototgraphy, travel, fishing.

PEARSALL, HARRY JAMES, dentist; b. Bay City, Mich., Apr. 12, 1916; s. Roy August and Gladys Agnes (Tierney) P.; m. Betty Almina Dahlke, Oct. 5, 1946 (dec. Nov. 1982); 1 child, Paul Roy. BS, Marquette U., 1937, DDS, 1939. Gen. practice dentistry Bay City, 1939—; cons. Delta Dental Ins., Lansing, Mich., 1975-86. Mem. Bay City chpt. Revision Com., 1965-66; bd. dirs. Downtown Bay City, 1962-73. Served to maj. U.S. Army, MC, 1940-46. Mem. ADA, Am. Coll. Dentists, Internat. Coll. Dentists, Mich. Dental Assn. (pres. 1972-73), Saginaw Valley Dental Soc. (pres. 1955-56), Bay County Dental Soc. (pres. 1950-51), Am. Legion. Lodge: Elks. Home: 1820 E Worfolk Dr Apt 1 Essexville MI 48732 Office: 404 Shearer Bldg Bay City MI 48708

PEARSALL, SAM HAFF, engineer, consultant; b. Guthrie, Ky., July 17, 1923; s. Samuel Haff and Susie Claire (Miller) P.; m. Isabelle Ikard, July 20, 1946; children: Sam III, Susan Claire Pearsall Morris, Sallie Pearsall Rogalle, Timothy C.H. B Engring., Vanderbilt U., 1948, MS, 1958. Registered profl. engr., Tenn. Engr. WSM, Inc., Nashville, 1947-56; assoc. prof. engring. Vanderbilt U., Nashville, 1957-62; rsch. scientist R.W. Benson and Assocs., 1962-71; v.p., gen. mgr. Cutters Electronics, 1971-81; ind. cons., 1981—; cons. E.I. duPont de Nemours, Old Hickory, Tenn., MEC, Nashville, So. Sales, Nashville. Grantee NSF, 1960. Mem. IEEE (life, sr., chpt. chair 1975), Elks, Sigma Xi (sr.). Republican. Achievements include patents in electronic instrumentation and control systems. Home and Office: 118 Spring Valley Rd Nashville TN 37214

PEARSE, GEORGE ANCELL, JR., chemistry educator, researcher; b. Stoneham, Mass., May 18, 1930; s. George Ancell and Marguerite Mae (Velmure) P.; m. Janet Rose Chaves, June 13, 1953; children: William, Kathleen, Nancy, Susan, James. BS, U. Mass., 1952; MS, Purdue U., 1956; PhD, U. Iowa, 1959. Rsch. chemist E.I. DuPont, Seaford, Del., 1959-60; asst. prof. chemistry Le Moyne Coll., Syracuse, N.Y., 1960-64, assoc. prof. chemistry, 1964-70, chmn. dept. chemistry, 1965-72, prof. chemistry, 1970—; vis. rsch. prof. U. Stockholm, Sweden, 1978-79, Cambridge (Eng.) U., 1985-86, James Cook U., Townsville, Australia, 1993. Mem. editorial bd. Microchemical Jour., 1984—. Mem. Am. Chem. Soc., Sigma Xi, Phi Lambda Upsilon, Alpha Chi Sigma. Republican. Roman Catholic. Office: Le Moyne Coll Dept Chemistry Syracuse NY 13214

PEARSE, JAMES NEWBURG, electrical engineer; b. LaCrosse, Wis., June 5, 1930; s. Richard Henry and Edith (Newburg) P.; m. Jeanette Kubiak, Dec. 27, 1951; children: James C., John P., Thomas G. BSEE, U. Wis., 1957. Registered profl. engr., Wis. Project engr. Allen-Bradley Co., Milw., 1957-67; v.p. product design and rsch. Appleton Electric Co., Chgo., 1967-75; group v.p. engring. Leviton Mfg. Co., Inc., Little Neck, N.Y., 1975—. Author: (with others) Standards Management: A Handbook for Profits, 1990, Tool and Manufacturing Engineers..Handbook, 1988; contbr. articles to profl. jours. With U.S. Army, 1951-53. Mem. IEEE, AAAS, ASTM, Intelligent Bldg. Inst., Illuminating Engring. Soc., Nat. Inst. Bldg. Scis., Nat. Lighting Bur. (bd. dirs. 1986—), Nat. Elec. Mfrs. Assn. (coms.), Nat. Fire Protection Assn. (elec. sect. 1969—, indsl. fire protection sect. 1969—, nat. elec. code com. 1984—), Newcomen Soc., Am. Nat. Standards Inst. (bd. dirs. 1988—, exec. com. 1988—, fin. com. 1988—, chmn. bd. dirs. 1989, 90, 91, others), Underwriters Labs. Industry Adv. Conf. (coms.), Eta Kappa Nu, Tau Beta Pi. Achievements include 23 patents, 2 patents pending, 50 foreign patents in electronics, relays, hazardous location enclosures, protective electrical products for personnel. Home: 12 Buckingham Dr Dix Hills NY 11746 Office: Leviton Mfg Co Inc 59-25 Little Neck Pkwy Little Neck NY 11362

PEARSON, HUGH STEPHEN, mechanical and metallurgical engineer; b. Nashville, June 8, 1931; s. Stephen Raymond and Amanda Orey (Wear) P.; m. Morceline Elizabeth Rowland, June 15, 1953; 1 child, Mary Elizabeth Pearson McLenden. B of Mech. Engring., Georgia Inst. Tech., 1953, M of Metallurgy, 1969. Engr. Lockheed Co., Marietta, Ga., 1953-57; salesman The Ellison Co., Atlanta, 1958-59; sr. engr. Lockheed Co., 1959-73; project engr. Pratt & Whitney Aircraft, West Palm Beach, Fla., 1974-75; v.p. Applied Tech. Svcs., Marietta, 1975-82; specialist engr. Lockheed Co., 1982-88; pres. Pearson Testing Labs., Marietta, 1984—. Contbr. chpts. to books and articles to profl. jours. Vol. probation officer Vol. Probation Program, Cobb County, Ga., 1991—. 1st lt.U.S. Army Corps Engrs., 1954-56. Mem. Am. Soc. Metals. Home and office: 2780 Pinestream Dr Marietta GA 30068

PEARSON, JEREMIAH W., III, military career officer, federal agency official; b. Indpls., Jan. 25, 1938; m. Patricia Van Gysel. B in Aeronautical Engring., Ga. Inst. Tech., 1960; MS in Systems Mgmt., U. So. Calif., 1973; grad., U.S. Naval Test Pilot Sch., 1966, Armed Forces Staff Coll., 1973, Air War Coll., 1982. Commd. 2d lt. USMC, 1960, advanced through grades to maj. gen., 1989; naval aviator 2d Marine Aircraft Wing USMC, Beaufort, S.C., 1961-64; flight instr. Naval Air Station USMC, Meridian, Miss., 1964-65; flight test divsn. U.S. Naval Test Pilot Sch. USMC, Patuxent River, Md., 1966-67; 1st Marine Aircraft Wing USMC, DaNang, Vietnam, 1968-69; project officer AWG-14 Radar, AIM-7F missile, Naval Missile Test Ctr., Flight Test USMC, Pt. Mugu, Calif., 1969-73; maintenance officer VMFA-232 USMC, Western Pacific, 1973-74; aviation plans officer Office Dep. Chief of Staff Aviation, HQ USMC, Washington, 1974-77; comd. HQ and Maintenance Squadron 24, 1st Marine Amphibious Brigade USMC, Marine Corps Air Station, Kaneohe Bay, Hawaii, 1977-78, comd. VMFA-235, 1978-

81; asst. chief staff logistics, chief staff 5th Marine Amphibious Brigade, 3d Marine Aircraft Wing USMC, El Toro, Calif., 1982-84, comd. Marine Aircraft Group 11, 1984-86; asst. wing comdr. 3d Marine Aircraft Wing, FMF. Pacific, Marine Corps. Air Station USMC, El Toro and Santa Ana, Calif., 1986; comdr. forward HQ Element/Inspector Gen. U.S Ctrl. Command USMC, MacDill AFB, Fla., 1986-88; asst. dep. chief staff aviation HQ USMC, 1988-89; commdg. gen. 4th Marine Aircraft Wing USMC, New Orleans, 1989-90; commdg. gen. Marine Corps. Rsch., Devel., Acquistion Command USMC, Washington, 1990; dep. comdr. Marine Ctrl. Command Riyadh, Saudi Arabia; assoc. adminstr. Office of Space Flight NASA, Washington, 1992. Decorated Defense Disting. Svc. medal, D.S.M., D.F.C., Bronze Star with Combat "V", Air medals (26) with gold star, Combat Action ribbon, Navy Unit Commendation, Nat. Def. Svc. medal, Vietnam Svc. medal, Southwest Asia Svc. medal, Rep. Vietnam Meritorious Unit Citation, Rep. Vietnam Campaign medal, Kuwait Liberation medal, others. Office: Nasa Space Flight Office Assoc Administr Washington DC 20546

PEARSON, JOHN, mechanical engineer; b. Leyburn, Yorkshire, U.K., Apr. 24, 1923; came to U.S., 1930, naturalized, 1944; s. William and Nellie Pearson; m. Ruth Ann Billhardt, July 10, 1944; children—John, Armin, Roger. B.S.M.E., Northwestern U., 1949, M.S., 1951. Registered profl. engr., Calif. Rsch. engr. Naval Ordnance Test Sta., China Lake, Calif., 1951-55, head warhead rsch. br., 1955-58, head solid dynamics br., 1958-59, head detonation physics group, 1959-67; head detonation physics div. Naval Weapons Ctr., China Lake, Calif., 1967-83, sr. rsch. scientist, 1983—; cons., lectr. in field; founding mem. adv. bd. Ctr. for High Energy Forming, U. Denver; mem. bd. examiners Sambalpur U., India, 1982-83. Author: Explosive Working of Metals, 1963; Behavior of Metals Under Impulsive Loads, 1954; contbr. articles to profl. publs; patentee impulsive loading, explosives applications. Charter mem. Sr. Exec. Svc. U.S., 1979. With C.E., U.S. Army, 1943-46, ETO. Recipient L.T.E. Thompson medal, 1965, William B. McLean medal, 1979, Superior Civilian Svc. medal USN, 1984, Haskell G. Wilson award, 1985, cert. of recognition Sec. Navy, 1975, merit award Dept. Navy, 1979, cert. of commendation Sec. Navy, 1981, Career Svcs. award Sec. Navy, 1988, John A. Ulrich award Am. Def. Preparedness Assn., 1991; 1st disting. fellow award Naval Weapons Ctr., 1989. Fellow ASME; mem. Am. Soc. Metals, Am. Phys. Soc., N.Y. Acad. Scis., AIME, NSPE, Fed. Exec. League, Sigma Xi, Tau Beta Pi, Pi Tau Sigma, Triangle. Home and Office: PO Box 1390 858 N Primavera Rd Ridgecrest CA 93556

PEARSON, JOHN MARK, civil engineer; b. Eugene, Oreg., June 15, 1950; s. Robert Lowell and Virginia Dale (Burt) P.; m. Donna Faye Bowman, June 19, 1976; children: David Andrew, Virginia Ruth, Elizabeth Rose, Sarah Lanelle, Abigail Michele, Daniel Joseph. BSCE, Oreg. State U., 1973. Registered profl. engr. Wash., Oreg., Idaho, MIch., Ala. Design engr. Chgo. Bridge and Iron Co., Houston, 1973-81; sr. design engr. Wyatt Industries, Inc., Houston, 1981-82; chief engr. GH Progressive Metals, Houston, 1982-83; v.p., gen. mgr. Security Concepts, Inc., Houston, 1983-87; civil/structural dept. mgr. Evergreen Engring., Inc., Eugene, 1987-91; project engr. Appel Engring. Svcs., Eugene, 1991; mgr. engring. Bergeson-Boese & Assocs., Inc., Eugene, 1991—; Registered profl. engr., Wash., Oreg., Mich., Ala., Idaho. Mem. NSPE, ASCE, ASME, Profl. Engrs. Oreg. (bd. dirs. 1992-94). Republican. Baptist. Office: Bergeson Boese & Assocs Inc 65 Centennial Loop Eugene OR 97401

PEARSON, MARK LANDELL, molecular biologist; b. Toronto, Ont., Can., June 2, 1940; s. William Holmes and Marguerite Rachel (Landell) P.; m. Katharine Anne Brown, Aug. 2, 1962; children: Scott Wallace, Jennifer Anne. BASc, U. Toronto, 1962, MA, 1964, PhD, 1966. Asst. to assoc. prof. U. Toronto, 1969-79; sect. head devel. genetics NCI-FCRF, Frederick, Md., 1979-83, dir. lab. molecular biology, 1982-83; asst. dir. life scis. E.I. DuPont de Nemours & Co., Wilmington, Del., 1983-84, dir. molecular biology, 1985-90; exec. dir. cancer and inflamatory diseases rsch. The DuPont Merck Pharm. Co., Wilmington, 1991-92; pres., CEO Darwin Molecular Corp., Kirkland, Wash., 1992—; adj. prof. U. Toronto, 1969-79, U. Md., 1982-83, 86—; sci. adv. bd. Genex Corp., Rockville, Md., 1978-83; bd. mem. Keystone Symposia on Molecular and Cellular Biology, 1986—, exec. com., 1990—; com. on sci. and the arts Franklin Inst., 1991—. Trustee Life Sci. Rsch. Found., 1988—, U. Del. Rsch. Found., 1990—; mem. program adv. com. human genome NIH, 1988—, Human Genome Orgn., 1989—; bd. dirs. Alliance Aging Rsch., 1989—, joint informatics task force NIH/DOE, 1990—. Josiah Macy Jr. fellow, 1977-78; Helen Hay Whitney postdoctoral fellow, 1966-69. Mem. AAAS, Am. Chem. Soc., Am. Soc. Microbiology, Am. Soc. Virology, Biophys. Soc., Am. Soc. Cell Biology, Am. Soc. Biochem. and Molecular Biology, Am. Soc. Human Genetics, Am. Assn. for Cancer Rsch., Can. Biochem. Soc., Can. Soc. Cell Biology, N.Y. Acad. Sci. Home: 5302-I Lake Washington Blvd NE Kirkland WA 98033 Office: Darwin Molecular Corp 2405 Carillon Point Kirkland WA 98033

PEARSON, RALPH GOTTFRID, chemistry educator; b. Chgo., Jan. 12, 1919; s. Gottfrid and Kerstin (Larson) P.; m. Lenore Olivia Johnson, June 15, 1941 (dec. June 1982); children—John Ralph, Barry Lee, Christie Ann. B.S., Lewis Inst., 1940; Ph.D., Northwestern U., 1943. Faculty Northwestern U., 1946-76, prof. chemistry, 1957-76; prof. chemistry U. Calif., Santa Barbara, 1976-89, prof. emeritus, 1989—; Cons. to industry and govt., 1951—. Co-author 5 books. Served to 1st lt. USAAF, 1944-46. Guggenheim fellow, 1951. Mem. Am. Chem. Soc. (Midwest award 1966, Inorganic Chemistry award 1969), Nat. Acad. Sci., Phi Beta Kappa, Sigma Xi, Phi Lambda Upsilon (hon.). Lutheran. Achievements include being originator prin. of hard and soft acids and bases.

PEARSON, ROGER ALAN, chemical engineer; b. Rockford, Ill., Apr. 2, 1956; s. Miles Addison and Pauline (Hammond) P. BA, Knox Coll., 1977, postgrad., U. Ill., 1977-80. R&D engr. Clinton Electronics Corp., Loves Park, Ill., 1980—. Contbr. articles to profl. publs. Mem. Am. Chem. Soc., Phi Beta Kappa. Achievements include development of measurement techniques and data associated with monochrome cathode ray phosphor persistance, numerous monochrome cathode ray tube phosphor blends, technique for measuring phosphor coulombic aging characteristics. Home: Unit B 124 2929 Sunnyside Dr Rockford IL 61114 Office: Clinton Electronics Corp 6701 Clinton Rd Loves Park IL 61131

PEARSON, SCOTT ROBERTS, economics educator; b. Madison, Wis., Mar. 13, 1938; s. Carlyle Roberts and Edith Hope (Smith) P.; m. Sandra Carol Anderson, Sept. 12, 1962; children—Sarah Roberts, Elizabeth Hovden. B.S., U. Wis.-Madison, 1961; M.A., Johns Hopkins U., 1965; Ph.D., Harvard U., 1969. Asst. prof. Stanford U., Calif., 1968-74, assoc. prof., 1974-80, assoc. dir. Food Research Inst., 1977-84, dir., 1992—, prof. food econs., 1980—. Cons. AID, World Bank, Washington, 1965—; staff economist Commn. Internat. Trade, Washington, 1970-71. Author: Petroleum and the Nigerian Economy, 1970; (with others) Commodity Exports and African Economic Development, 1974, (with others) Rice in West Africa, Policy and Economics, 1981, (with others) Food Policy Analysis, 1983, (with others) The Cassava Economy of Java, 1984, (with others) Portuguese Agriculture in Transition, 1987, (with Eric Monke) The Policy Analysis Matrix, 1989, (with others) Rice Policy in Indonesia, 1991. Mem. Am. Agrl. Econs. Assn., Am. Econ. Assn., Royal Econ. Soc. Home: 691 Mirada Ave Palo Alto CA 94305-8477 Office: Stanford U Food Rsch Inst Stanford CA 94305

PEARSON, THOMAS ARTHUR, epidemiologist, educator; b. Berlin, Wis., Oct. 21, 1950; married; 2 children. BA, Johns Hopkins U., 1973, MD, 1976, MPH, 1976, PhD in Epidemiology, 1983. Fellow cardiology Johns Hopkins Sch. Medicine, 1981-83, from asst. prof. to assoc. prof. medicine, epidemiology, 1983-88; prof. epidemiology Columbia U., 1988—; dir. Mary Imogene Bassett Rsch. Inst., 1988—; chmn. monitoring bd. CARDIA Project, Nat. Heart, Lung & Blood Inst., 1987—, rsch. com. Md. Heart Assn., 1986-88, nat. rsch. com. Am. Heart Assn., 1987-88; mem. Coun. Epidemiology, 1987—; commr. Md. Coun. Phys. Fitness, 1985-88; mem. clin. applications & prevention commn. NIH, 1987-91. Mem. Am. Fedn. Chem. Rsch., Am. Coll. Epidemiology, Am. Coll. Preventive Medicine, Assn. Tchrs. Preventive Medicine, Soc. Epidemiol. Rsch. (Rsch. prize 1978). Achievements include research in the etiology and pathogenesis of atherosclerosis. Office: Mary Imogene Bassett Med Rsch Inst 1 Atwell Rd Cooperstown NY 13326*

PEARSON, WILLIAM ROWLAND, nuclear engineer; b. New Bedford, Mass., Sept. 30, 1923; s. Rowland and Nellie (Hilton) P.; BS, Northeastern U., 1953; postgrad. U. Ohio, 1960; m. Arlene Cole Loveys, June 14, 1953; children: Denise, Robert, Rowland, Nancy. Engr., Goodyear Atomic Corp., Portsmouth, Ohio, 1953-63, Cabot Titania Corp., Ashtabula, Ohio, 1963-64; supr. United Nuclear, Wood River, R.I., 1964-72; sr. engr. Nuclear Materials and Equipment Co., Apollo, Pa., 1972-74; engr. U.S. Nuclear Regulatory Commn., Rockville, Md., 1974-90, ret., 1990. Served with USNR, 1942-45. Decorated Air medal. Mem. AAAS, Am. Nuclear Soc., Am. Inst. Chem. Engrs. (chmn. 1966-67). Republican. Baptist. Clubs: Masons, Elks. Home: 60 Meeting Hill Rd Hillsborough NH 03244-4854

PEARTON, STEPHEN JOHN, physicist; b. Hobart, Tasmania, Australia, Jan. 15, 1957; came to U.S., 1982; s. Dennis Gregory and Margaret Faye (Godfrey) P.; m. Cammy R. Abernathy, June 28, 1993. BS with honors, U. Tasmania, Australia, 1979, PhD, 1983. Exptl. officer Australian Atomic Energy Commn., Sydney, 1981-82; postdoctoral fellow U. Calif., Berkeley, 1982-83; mem. tech. staff AT&T Bell Labs., Murray Hill, N.J., 1984—. Author: Hydrogen in Crystalline Semiconductors, 1991; editor: (conf. proceedings) O.C.H&N in Crystalline Silicon, 1986, Defects in Electronic Materials, 1988, Ion Implanatation for Compound Semiconductors, 1990, Degredation Mechanisms in Compound Semiconductor Devices and Structures, 1990, advanced HI-V Compound Semiconductor Growth, Processing and Devices, 1992, III-V Electronic and Photonic Device Fabrication and Performance, 1993. Recipient scholarship Australian Inst. Nuclear Sci. and Engring., 1979-81. Mem. Am. Physical Soc. (life), Materials Rsch. Soc., Am. Vacuum Soc., IEEE, Electrochemical Soc. Republican. Achievements include several patents involving use of ion implantation and dry etching to fabricate semiconductor devices; contributions in areas of hydrogen in semiconductors, dry etching and ion implantation of semiconductors. Home: 294 Country Club Ln Scotch Plains NJ 07076 Office: AT&T Bell Labs 600 Mountain Ave New Providence NJ 07974-2010

PEASE, HOWARD FRANKLIN, structural engineer; b. Hyannis, Mass., Mar. 21, 1939; s. William Howard and Pauline Mary (Sabens) P.; m. June Brant, Jan. 27, 1962; children: Nancy Heather, Stephanie Brant, Andrew Scott. BS in Mech. Engring., Northeastern U., 1962. Registered profl. engr., Pa. Asst. engr. Woods Hole (Mass.) Oceanographic, 1957-62; structural analyst Avco, Wilmington, Mass., 1962-65; engr. Electric Boat, Groton, Conn., 1965-66; sr. project engr. Raymond Engring., Middletown, Conn., 1966-75; ind. cons. engr. Durham, Conn., 1975-76; engring. projects mgr. Precision Components Corp., York, Pa., 1976-90; pres. Pease Assocs., York, 1990—. Contbr. articles to Machine Design, SN Standardization News. Pres. Secular Franciscan Order, York, 1986-89; v.p. St. Augustine Province, Pitts., 1989-90; mem. bd. edn. Coginchaugh Regional Sch. Dist., Durham, 1973-76. Mem. Nat. Soc. Profl. Engrs., Pa. Soc. Profl. Engrs. (v.p. 1993, bd. dirs. 1978-81), Profl. Engrs. in Pvt. Practice (chairperson 1990—). Office: 1689 Randow Rd York PA 17403

PEASLEE, DAVID CHASE, physics educator; b. White Plains, N.Y., July 23, 1922; s. Arthur Frank and Anita Quigley (Clark) P.; m. Virginia Perry Close (dec. Aug. 1972); children: Anne Close, Frank David, Graham Frederick; m. Lillian Teresa Fuls, Mar. 3, 1973. AB, Princeton U., 1943; PhD, MIT, 1948. Asst. prof. physics Washington U., St. Louis, 1950-51; rsch. assoc. in physics Columbia U., N.Y.C., 1951-54; assoc. prof., prof. Purdue U., West Lafayette, Ind., 1954-59; prof. Australian Nat. U., Canberra, 1959-76; vis. prof. Brown U., Providence, 1976-77; project officer high energy div. U.S. Dept. Energy, Germantown, Md., 1977-81; vis. prof. U. Md., College Park, 1981—; cons. nuclear power div. Westinghouse, Pitts., 1955-59. Author: Elements of Atomic Physics, 1955; editor conf. proc. Hadron 91, 1992. Fulbright fellow, Australia, 1958. Fellow Am. Phys. Soc. Home: 11822 Goya Dr Potomac MD 20854 Office: U Md Physics Dept College Park MD 20742

PEASLEE, KENT DEAN, metallurgical engineer; b. Grand Junction, Colo., May 19, 1956; s. Don E. and Verla M. (Allen) P.; m. Mary Ann Hamilton, Aug. 21, 1976; children: Michael Andrew, Sarah Michelle, Matthew Kyle. BS in Metall. Engring., Colo. Sch. Mines, 1978; PhD candidate, U. Mo., Rolla, 1991—. Registered profl. engr. Metallurgist CF&I Steel, Pueblo, Colo., 1978-79; melt shop metallurgist Raritan River Steel, Perth Amboy, N.J., 1979-81; mgr. metallurgy and quality control Border Steel, El Paso, Tex., 1981-85; gen. mgr. tech. svcs. Bayou Steel, LaPlace, La., 1985-91; metall. engr. U.S. Bur. Mines, Rolla, 1991—. GAANN fellow U.S. Dept. Edn., 1991. Mem. AIME, Iron and Steel Soc., ASM Internat., Minerals, Metals and Materials Soc. Home: 16 Elmwood Dr Rolla MO 65401

PECK, A. WILLIAM, aerospace engineer; b. Greenfield, Mass., Sept. 15, 1938; s. Abner C. and Grace Nellie (Lyman) P.; m. Mary Elizabeth Pilkington, Dec. 27, 1968; children: Christopher A., Stephen W. BS in Aero. and Astro. Engring., MIT, 1961, MS, 1961. Flight test engr. Avco Systems Div., Wilmington, Mass., 1964-67, program mgr., 1967-77; dir. product devel. Timex Indsl. Products Group, Middlebury, Conn., 1977-79; dir. engring. Timex Clock Co., Middlebury, 1980-82; dir. prodn. programs Textron Def. Systems, Wilmington, 1982-87, program dir., 1987—. Founder Andover (Mass.) Sch. Montessori, 1974. 1st lt. USAF, 1961-64. Mem. AIAA (tech. com. mgmt.), Am. Def. Preparedness Assn., Assn. U.S. Army. Home: 1 Heritage Ln Andover MA 01810

PECK, DALLAS LYNN, geologist; b. Cheney, Wash., Mar. 28, 1929; s. Lynn Averill and Mary Hazel (Carlyle) P.; m. Tevis Sue Lewis, Mar. 28, 1951 (dec.); children: Ann, Stephen, Gerritt. B.S., Calif. Inst. Tech., 1951, M.S., 1953; Ph.D., Harvard U., 1960. With U.S. Geol. Survey, 1954—; asst. chief geologist, office of geochemistry and geophysics U.S. Geol. Survey, Washington, 1967-72; geologist, geologic div. U.S. Geol. Survey, 1972-77, chief geologist, 1977-81, dir., 1981-93, geologist, 1993—; mem. Lunar Sample Rev. Bd., 1970-71; chmn. earth scis. adv. com. NSF, 1970-72; vis. com. dept. geol. scis. Harvard U., 1972-78; mem. Earthscis. Adv. Bd., Stanford U., 1982—; chmn. com. earthscis., Fed. Coord. Coun. Sci., Engring and Tech., 1987-92; mem. sci., tech. com. UN Decade for Nat. Disaster Reduction. Recipient Meritorious Service award Dept. Interior, 1971, Disting. Service award, 1979; Presdl. Meritorious Exec. award, 1980, Disting. Alumni award Calif. Inst. Tech., 1985. Fellow AAAS, Geol. Soc. Am., Am. Geophys. Union (pres. sect. volcancology, geochemistry and petrology 1976-78). Home: 2524 Heathcliff Ln Reston VA 22091-4225 Office: US Geol Survey 12201 Sunrise Valley Dr Herndon VA 22092

PECK, DIANNE KAWECKI, architect; b. Jersey City, June 13, 1945; d. Thaddeus Walter and Harriet Ann (Zlotkowski) Kawecki; m. Gerald Paul Peck, Sept. 1, 1968; children: Samantha Gillian, Alexis Hilary. BArch, Carnegie-Mellon U., 1968. Architect, P.O.D. Research & Devel., 1968, Kohler-Daniels & Assos., Vienna, Va., 1969-71, Beery-Rio & Assocs., Annandale, Va., 1971-73; ptnr. Peck & Peck Architects, Occoquan, Va., 1973-74, Peck, Peck & Williams, Occoquan, 1974-81; corp. officer Peck Peck & Assos., Inc., Woodbridge, Va., 1981—; chief exec. officer, interior design group Peck Peck & Assoc., 1988—. Work pub. in Am. Architecture, 1985. Vice pres. Vocat. Edn. Found., 1976; chairwoman architects and engrs. United Way; mem. Health Systems Agy. of No. Va., commendations, 1977; mem. Washington Profl. Women's Coop.; chairwoman Indsl. Devel. Authority of Prince William, 1976, vice chair, 1977, mem., 1975-79; developer research project Architecture for Adolescents, 1987-88; mem. inaugural class Leadership Am., 1988, Leadership Greater Washington. Recipient commendation Prince William Bd. Suprs., 1976, State of Art award for Contel Hdqrs. design, 1985, Best Middle Sch. award Coun. of Ednl. Facilities Planners Internat., 1989, Creativity award Masonry Inst. Md., 1990, First award, 1990, Detailing award, 1990, Govt. Workplace award for renovations of Dept. of Labor Bldg., 1990, Creative Use of Materials award Inst. of Bus. Designers, 1991; named Best Instl. Project Nat. Comml. Builders Coun.; subject of PBS spl.: A Success in Howard Co. Mem. Soc. Am. Mil. Engrs., Prince William C. of C. (bd. dir.). Republican. Roman Catholic. Club: Soroptimist. Research on inner-city rehab., adolescents and the ednl. environ. Office: 1942 Davis Ford Rd Woodbridge VA 22192-2416

PECK, FRED NEIL, economist, educator; b. Bklyn., Oct. 17, 1945; s. Abraham Lincoln and Beatrice (Pikholtz) P.; m. Jean Claire Ginsberg, Aug. 14, 1971; children: Ron Evan, Jordan Shefer, Ethan David. BA, SUNY, Binghamton, 1966; MA, SUNY, Albany, 1969; PhM, NYU, 1984; PhD, Pacific Western U., 1984. Lectr. SUNY, Albany, 1969-70; research asst. N.Y. State Legislature, Albany, 1970; sales and research staff Pan Am. Trade Devel. Corp., N.Y.C., 1971; v.p., economist The First Boston Corp., N.Y.C., 1971-88; mng. dir. Sharpe's Capital Mkt. Assocs. Inc., N.Y.C., 1988-89; pres., chief economist Hillcrest Econs. Group, N.Y.C., 1989-93; dir. edn. The Ednl. Advantage, Inc., New City, N.Y., 1990—; adj. prof. Hofstra U., Hempstead, N.Y., 1975; lectr. NYU, 1982; mem. faculty New Sch. for Social Rsch., N.Y.C., 1974—; tchr. computer aided edn. N.Y.C. Bd. of Edn., 1990—. Author, editor: (biennial publ.) Handbook of Securities of U.S. Government, 1972-86. Mem. ASCD, Am. Econ. Assn., Ea. Econ. Assn., Econometric Soc., Nat. Assn. Bus. Economists, Am. Statis. Assn., Beta Gamma Sigma (hon. soc.), Phi Delta Kappa. Democrat. Jewish. Lodges: Knights Pythias, Knights Khorassan. Office: CS 66 1001 Jennings St Bronx NY 10460

PECK, GAILLARD RAY, JR., aerospace business and healthcare consultant; b. San Antonio, Oct. 31, 1940; s. Gaillard Ray and Lois (Manning) P.; m. Jean Adair Hilger, Dec. 23, 1962 (div. Oct. 1969); children: Gaillard III, Katherine Adair; m. Peggy Ann Lundt, July 3, 1975; children: Jennifer Caroline, Elizabeth Ann. BS, Air Force Acad., 1962; MA, Cen. Mich. U., 1976; MBA, U. Nev. Las Vegas, 1990. Lic. comml. pilot, flight instr. Commd. 2d lt. USAF, 1962, advanced through grades to col., 1983, ret., 1988, air force instr. pilot, fighter pilot, 1963-72; instr. Fighter Weapons Sch. USAF, Nellis AFB, 1972-75; fighter tactics officer Pentagon, Washington, 1975-78; aggressor pilot, comdr. 4477th Test & Evaluation Squadron, Nellis AFB, Nev., 1978-80; mil. advisor Royal Saudi Air Force, Saudi Arabia, 1980-82; student Nat. War Coll., Washington, 1982-83; dir. ops., vice comdr. Kadena Air Base, Japan, 1983-85; wing comdr. Zweibrucken Air Base, Germany, 1985-87; dep. dir. Aerospace Safety directorate USAF, Norton AFB, Calif., 1987-88; rsch. asst. U. Nev., Las Vegas, 1988-90; mktg. cons. Ctr. for Bus. & Econ. Rsch. U. Nev., Las Vegas, 1990; adminstr. Lung Ctr. of Nev., Las Vegas, 1991-93; cons. Las Vegas, 1993—. Author: The Enemy, 1973. Recipient Silver Star, Legion of Merit, DFC (3), Air medal. Mem. Phi Kappa Phi Nat. Honor Soc., Order of Daedalians, Red River Fighter Pilots Assn., Air Force Assn., Ky. Col., U. Nev. Las Vegas and Air Force Acad. Alumni Assn., The Ret. Officers Assn. Avocations: flying, auto restoration, computer sci., hiking, camping, family activities. Home: 1775 Sheree Cir Las Vegas NV 89119-2716

PECK, JOAN KAY, systems engineer; b. Cedar Rapids, Iowa, Sept. 22, 1959; d. Leonard Allen and Mildred Jane (Keller) P. BS in Indsl. Engring., Iowa State U., 1983; MS in Space Tech., Fla. Inst. Tech., 1986. Student intern Rockwell-Collins, Cedar Rapids, 1979; coop. student Amana (Iowa) Refrigeration, 1981; sr. engr. Harris Govt. Aerospace Systems, Palm Bay, Fla., 1983-88; sr. systems engr. McDonnell Douglas Space Systems Co., Kennedy Space Ctr., Fla., 1988—. V.p. programming Inst. Indsl. Engrs., Ames, 1982-83; victim advocate Sexual Assault Victims Svcs., Fla. State Attys. Office, Brevard County, 1991-92. Recipient Group Achievement award NASA, 1991. Home: PO Box 034119 Indialantic FL 32903-0119 Office: McDonnell Douglas Space Sys PO Box 21233 Kennedy Space Center FL 32815

PECK, MICHAEL DICKENS, burn surgeon; b. Iola, Kans., May 10, 1953; s. Dewey Alvin Peck and Delma Dickens Pleines; m. Celeste Brickel, May 9, 1981; children: Mark Andrew, Julie Michelle, David Alexander. BA, Harvard U., 1976; MD, U. Colo., Denver, 1981; ScD, U. Cin., 1990. Resident U. Ariz. Health Scis. Ctr., 1981-86; asst. prof. surgery U. Miami Sch. Medicine, Fla., 1990-93, asst. dir. Jackson Meml. Med. Ctr., 1990—, assoc. prof. surgery, 1993—. Contbr. articles to profl. jours. Fellow ACS (assoc.); mem. Am. Burn Assn., Am. Soc. Parenteral and Enteral Nutrition, Surg. Infection Soc. Office: Univ of Miami Dept Surgery R-310A PO Box 016310 Miami FL 33101

PECK, ROBERT MCCRACKEN, naturalist, science historian, writer; b. Phila., Dec. 15, 1952; s. Frederick William Gunster and Matilda (McCracken) P. BA in Art History, Princeton U., 1974; MA, U. Del., 1976. Dir. Pocono Lake (Pa.) Preserve Nature Ctr., 1971, 72; asst. to dir. Natural History Mus. Acad. Natural Scis., Phila., 1976-77; tech. dir. Bartram Heritage Study U.S. Dept. Interior and Bartram Trail Conf., Atlanta and Montgomery, Ala., 1977-78; spl. asst. to pres. Acad. Natural Scis., Phila., 1977-82, acting v.p. Nat. History Mus., 1982-83, fellow, 1983—; cons. BBC, Eng., 1987-92; bd. dirs. Phila. Conservationists, Natural Lands Trust, Phila., Libr. Co. of Phila., Phila. City Inst.; mng. editor Frontiers, 1979-82; lectr. in field. Author: A Celebration of Birds: The Life and Art of Louis Agassiz Fuertes, 1982, Headhunters and Hummingbirds: An Expedition Into Ecuador, 1987, Wild Birds of America: The Art of Basil Ede, 1991, Land of the Eagle: A Natural History of North America, 1991, German edit., 1992; author: (with others) William Bartram's Travels, 1980, John Cassin's Illustrations of the Birds of California, Texas, Oregon, British and Russian America, 1991; author: (foward) The Birds of America by John James Audubon, 1985; editor: Bartram Heritage Report, 1978; author (with others), editor: Philadelphia Wildfowl Exposition Catalog, 1979; contbr. chpts. to books, articles to profl. jours. Trustee Chestnut Hill Acad., Phila.; bd. dirs. RARE Ctr. Tropical Bird Conservation, Mus. Coun. of Phila. Recipient Richard Hopper Day Meml. award Acad. Natural Scis. of Phila., 1991. Fellow Royal Geographic Soc., Explorers Club (various coms. 1983—, Explorers award 1988); mem. Soc. History of Natural History, Sigma Xi. Achievements include naming of a new species of frog, Eleutherodactylus pecki; research on orthogptera indigenous to the Caribbean, status of invasive African Desert Locusts in the West Indies, preservation of the endangered Cuban Parrot, the Orinoco River and its tributaries, botanical, entomological, ichthyological, herpetological, and malacological specimens for the Smithsonian Instn. and the Acad. of Natural Sciencs; organization and participation in expeditions which discovered several new species of fish in Guayana Shield, Venezuela, penetrated the tribal lands of the Shuar (Jivaro) Indians in Ecuador, discovered several new species of amphibians and insects as well as two new races of birds in Ecuador, investigated the ecological, economic and political impact of instream-flow legislation on the Yellowstone River Basin. Office: Academy of Natural Sciences 1900 Benjamin Franklin Pkwy Philadelphia PA 19103

PECKHAM, P. HUNTER, biomedical engineer, educator; b. Elmira, N.Y., June 23, 1944; married; 2 children. BS, Clarkson Coll. Tech., 1966; MS, Case Western Res. U., 1968, PhD in Biomed. Engring., 1972. Rsch. assoc. Case Western Res. U., 1972-74, instr. divsn. orthapedic surgery, 1974-78; rsch. biomed. engr. Med. Ctr. Cleve. VA, 1976—; assoc. prof. orthopedics Case Western Res. U., 1978—, assoc. prof. biomed. engring., 1979—; mem. staff Cuyohoga Coun. Hosp., 1981—. Recipient Rsch. Career Devel. award NIH, 1978. Mem. IEEE, Rehab. Engring. Soc. NAm., Internat. Med. Soc. Paraplegia. Achievements include research in technology for rehabilitation of the severely disabled, restoration of movement of the arm and hand using functional neuromuscular stimulation. Office: Case We Reserve U Rehab Engring Ctr 3395 Scranton Rd Rm H-611 Cleveland OH 44109*

PECORA, LOUIS MICHAEL, physicist; b. Hazleton, Pa., July 31, 1947; s. Michael Angelo and Helen (Motway) P.; m. Judith Diane Miller, Aug. 28, 1976; children: Andrew, Daniel, Anna. BS in Physics, Wilkes Coll., 1969; PhD, Syracuse U., 1977. Teaching asst. Purdue U., West Lafayette, Ind., 1969-71; foreman Hazleton (Pa.) Weaving Co., 1971-72; rsch. asst. Syracuse (N.Y.) U., 1972-77; postdoctoral fellow Naval Rsch. Lab., Washington, 1977-79, rsch. physicist, 1979—. Author: Applied Chaos, 1993; editor: 1st Experimental Chaos Conference, 1992; contbr. articles to profl. jours. Recipient Physics award Wilkes Coll., 1969, NRL Tech. Achievement award Naval Rsch. Lab., 1989. Mem. AAAS, Am. Phys. Soc., Sigma Chi. Achievements include synchronization of chaotic systems; patents filed in field for Pseudoperiodic Driving, Cascading Synchronized Chaotic Systems and Tracking Unstable States. Office: Naval Rsch Lab Code 6341 Washington DC 20375

PECORINI, THOMAS JOSEPH, materials engineer; b. N.Y.C., July 8, 1962; s. Andrew and Ida (Anselmet) P.; m. Lisa Kay Bush, Jan. 30, 1993. BS, Worcester Poly., 1984; MS, Lehigh U., 1988, PhD, 1992. Devel. engr. Chandler Evans Inc. div. Colt Industries, West Hartford, Conn., 1984-86; rsch. asst. materials sci. dept. Lehigh U., Bethlehem, Pa., 1986-92; advanced materials engr. Eastman Chem. Co., Kingsport, Tenn., 1992—. Contbr. articles to Jour. Materials Sci., Polymer Composites, Polymer. Recipient Excellence in Polymer Sci. award Hoecst-Celanese, 1987. Mem. ASME, ASM Internat., Sigma Xi, Tau Beta Pi, Pi Tau Sigma. Democrat.

Roman Catholic. Achievements include research in fracture behavior of polymers and polymer composites. Office: Eastman Chem Co PO Box 1972 Kingsport TN 37662

PECORINO, LAUREN TERESA, biologist; b. Bronx, N.Y., June 17, 1962; d. Joseph Salvatore and Raffaela (Rapillo) P. BS in Biology, SUNY, Stony Brook, 1984, PhD, 1990. Postdoctoral fellow Ludwig Inst. for Cancer Rsch., London, 1991—. Contbr. articles to profl. jours. Postdoctoral fellow European Molecular Biology Orgn., 1991-93, NATO, 1993—. Mem. Brit. Sub-Aqua Club, Sigma Xi. Home: 340 Alfred St North Babylon NY 11703 Office: Ludwig Inst Cancer Rsch, 91 Riding House St, London England

PECTOR, SCOTT WALTER, telecommunications engineer; b. Chgo., Nov. 29, 1957; s. Samuel and Stella (Celmer) P.; m. Elizabeth A. Fichtner, June 13, 1981; 1 child, David R. BEE, Northwestern U., 1978, MEE, 1980. Mem. tech. staff AT&T Bell Labs., Naperville, Ill., 1979—. Mem. ACLU. Democrat. Achievements include patents for time div. system control arrangement and method, control info. communication arrangement for a distributed control switching system; rsch. in devel. estimation and estimation techniques. Office: AT&T Bell Labs 2000 N Naperville Rd Naperville IL 60566

PEDDICORD, KENNETH LEE, academic administrator; b. Ottawa, Ill., Apr. 5, 1943; s. Kenneth Charles and Elizabeth May (Hughes) P.; m. Patricia Ann Cullen, Aug. 2, 1969; children: Joseph, Clare. BSME, U. Notre Dame, 1965; MSNE, U. Ill., 1967, PhD, 1972. Registered profl. engr., Tex. Rsch. nuclear engr. Swiss Fed. Inst. for Reactor Rsch., Würenlingen, Switzerland, 1972-75; asst. prof. nuclear engring. Oreg. State U., Eugene, 1975-79, assoc. prof. nuclear engring., 1979-82; prof. nuclear engring. Tex. A&M U., College Station, 1983—, head dept. nuclear engring., 1985-88, asst. dir. rsch. Tex. Engring. Experiment Sta., 1988-91, dir. Tex. Experiment Sta., 1991—, assoc. dean coll. engring., 1990-91, interim dean coll. engring., 1991—, interim dep. chancellor engring., 1993—; vis. scientist Joint Rsch. Centre-Ispra Establishment, EURATOM, Ispra, Italy, 1981-82; cons. EG&G Idaho, Inc., Idaho Falls, Idaho, 1979, Portland (Oreg.) Gen. Electric Co., 1980, Battelle Human Affairs Rsch. Ctr., Bellevue, Va., 1984, Los Alamos (N.Mex.) Nat. Lab., 1984-89, Pa. Power & Light, Allentown, Pa., 1986, Argonne Nat. Lab., Idaho Falls, 1989, Univs. Space Rsch. Assn., Houston, 1989—, Ark. Dept. Higher Edn., Little Rock, 1990; speaker in field; mem. Nat. Rsch. Coun. Com. on Advanced Space Based High Power Techs., 1987-88, NASA OAET Aerospace Rsch. and Tech. subcom., 1987—, SP-100 Materials Sci. Rev. Com., DARPA/NASA/DOE, 1984-86, Nuclear Engring. Edn. for Disadvantaged, 1978—, sec., 1978-80, vice chair, 1985-88, chair, 1988-91; John and Muriel Landis Scholarship Com., 1979—, chair, 1985—; mem. adv. bd. for nuclear sci. and engring., 1984-86; mem. tech. program com. Internat. Conf. on Reliable Fuels for Liquid Metal Reactors, 1985-86; mem. Tex. Transp. Inst. Rsch. Coun., 1987—, Tex. A&M Rsch. Found. Users' Coun., 1989—. Contbr. articles to profl. jours. Recipient Best Paper award AIAA 9th Ann. Tech. Symposium Johnson Space Ctr., 1984. Mem. ASME, NSPE, Am. Nuclear Soc. (cert. governance 1983, 84, 85, Materials Sci. and Tech. Divsn. Chmn's. award 1989, mem. materials sci. and tech. disvsn. exec. com. 1983-86, sec.-treas. 1985-86, vice-chmn./chair-elect 1986-87, chair 1987-88, bd. dirs. 1988-91, mem. edn. divsn. exec. com. 1979-82, sec.-treas. 1982-83, vice-chmn./chair-elect 1983-84, chair 1984-85, mem. student activities com. 1976—, chair 1978-81, Oreg. sect. Edn. com. 1976, 79, 80, Oreg. sect. bd. dirs. 1977), Am. Soc. Engring. Edn., Tex. Soc. Profl. Engrs., Univ. Space Rsch. Assn. (mem. sci. coun. engring.), Pi Tau Sigma, Alpha Nu Sigma. Office: Tex A&M U Tex Engring Experiment Sta 301 Wisenbaker College Station TX 77843-3126

PEDERSEN, BENT CARL CHRISTIAN, cytogeneticist; b. Jutland, Denmark, Sept. 14, 1933; s. Jonathan and Lydia (Hansen) P.; m. Lene Reymann, July 3, 1965; children: Susanne, Kare. MB/BS, U. Copenhagen, 1960, MD, 1969; PhD, U. Cambridge, Eng., 1970. Assoc. researcher Inst. Human Genetics U. Copenhagen, 1961-67; sr. researcher Postgrad. Sch. of Medicine U. Cambridge, 1967-69; sr. researcher Cancer Rsch. Inst. Danish Cancer Soc., Aarhus, Denmark, 1970-78, assoc. rsch. dir., 1978-87, head dept. cytogenetics, 1987—. Author: Cytogenetic Evolution of Chronic Myelogenous Leukaemia, 1969, The Cytological Basis of Progression in Chronic Myeloid Leukaemia, 1970; contbr. articles to profl. jours. Recipient awards William Nielsens Found., 1971, 87, Generalkonsul Ernst Carlsens og Hustru Adolphine Carlsens Legat, 1989. Home: 94 Tingstedet, DK-8220 Brabrand Denmark Office: Danish Cancer Soc, Aarhus Amtssygehus, DK-8000 Aarhus C, Denmark

PEDERSON, THORU JUDD, biologist, research institute director; b. Syracuse, N.Y., Oct. 10, 1941; s. Thorvald and Ruth (Judd) P.; m. Judith Bennett, Mar. 26, 1966; children: David, Christopher. AB, Syracuse U., 1963, PhD, 1968. Rsch. assoc. Albert Einstein Coll. Med., N.Y.C., 1968-71; staff scientist Worcester Found. for Exptl. Biology, Shrewsbury, Mass., 1971-73, sr. scientist, 1973-83, prin. scientist and dir. Cancer Ctr., 1983—, pres., sci. dir., 1985—; trustee Mass. Biotech. Rsch. Inst.; dir. Hybridon, Inc.; mem. Med. Rsch. Com. Mass. div. Am. Cancer Soc., 1980-90; chmn. fellowships com. Marine Biol. Lab., Woods Hole, Mass.; sci. advisor Matritech, Inc. Mem. editorial bd. Jour. Cell Biology; contbr. numerous articles, papers to publs. in field. Trustee James Mountain Meml. Fund. Recipient Hudson Hoagland award Worcester Found., 1982; named Scholar Leukemia Soc. Am., 1972-77, Nat. lectr. Sigma Xi, 1984-86; grantee NIH, 1971—. Fellow AAAS; mem. Am. Soc. Cell Biology, Am. Soc. Biochemistry and Molecular Biology. Office: Worcester Found for Exptl Biology 222 Maple Ave Shrewsbury MA 01545-2732

PEDLEY, TIMOTHY ASBURY, IV, neurologist, educator, researcher; b. Phoenix, Aug. 31, 1943; s. Timothy Asbury Pedley III and Mary Adele (Newcomer) Melis; m. Barbara S. Koppel, Mar. 17, 1984. B.A., Pomona Coll., 1965; M.D., Yale U., 1969. Cert. neurology, 1975, electroencephalography, 1975. Intern Stanford U. Hosp., 1967-70, resident in neurology, 1970-73, post-doctoral fellow in neurophysiology, 1973-75, asst. prof. neurology, 1975-79; assoc. prof. neurology, 1979-83, prof., vice chmn. dept. neurology, 1983—; dir. clin. neurophysiology labs. Columbia-Presbyterian Med. Ctr., N.Y.C., 1983—. Contbr. articles to profl. jours. Bd. dirs. Epilepsy Found Am., 1984—, chmn. profl. adv. bd., 1985-87, pres., 1991-93, chmn. 1993—. mem. rev. com. NIH Nat. Inst. Neurol. and Chronic Diseases and Strokes, 1978-79; vis. fellow in exptl. neurology Inst. Psychiatry, London, 1978. Fellow Am. Acad. Neurology, Am. Electroencephalographic Soc. (pres. 1989-90, bd. dirs. 1981, 85); mem. Am. Neurol. Assn. (coun. 1992—), Am. Epilepsy Soc. (treas. 1980-83, pres. 1991-92), Soc. for Neurosci., Alpha Omega Alpha. Club: Yale (N.Y.C.). Office: The Neurological Inst 710 W 168th St New York NY 10032-2699

PEDREGAL, PABLO, mathematician, educator; b. Fuenllana, Ciudad Real Spain, Dec. 9, 1963; s. Jose and Petra (Tercero) P.; m. María José Pastor, Dec. 21, 1991. Lic., U. Complutense, Madrid, 1986; PhD, U. Minn., 1989, U. Complutense, Madrid, 1989. Prof. U. Complutense, Madrid, 1990—. Contbr. articles to profl. jours. Avocations: music, reading, hiking. Office: U Complutense, Ciudad Universitaria, 28040 Madrid Spain

PEEBLES, WILLIAM REGINALD, JR., mechanical engineer; b. Pontotoc, Miss., Jan. 18, 1945; s. William Reginald and Mable Alice (Palmer) P.; m. Sarah Ann Bradley, May 31, 1975; children: Gregory Morton, Stephen Bradley. BSME, Miss. State U., 1969, MSME, 1970, PhD in Mech. Engring., 1975. Registered profl. engr., Ill., S.C., Ark., Miss., La. Rsch. engr. Newport News Shipbuilding & Dry Dock, 1970-72; instr. Miss. State U., Mississippi State, 1972-75; supr. Sargent & Lundy, Chgo., 1975-88; staff rsch. engr. E.I. DuPont de Nemours & Co., Inc., Aiken, S.C., 1988-89; fellow Westinghouse Savannah River Co., Aiken, S.C., 1989-91; mgr. analysis Sargent & Lundy, Chgo., 1991—. Pres. Briarcliffe-West Townhome Owners Assn., Wheaton, Ill., 1983-86; v.p. Coun. for Energy Independence, Chgo., 1978-80. Mem. ASME, Am. Nuclear Soc. (pres. Chgo. sect. 1987-88, registration chair nat. reactor ops. div. mtg. 1987), Western Soc. Engrs., Ill. Engring. Coun. Achievements include design contribution to more than 35 nuclear power plants worldwide, and three production reactor plants; reactor containment design for the effects of safety-relief valve and loss-of-coolant accident phenomena. Office: Sargent & Lundy 31W64 55 E Monroe St Chicago IL 60603

PEEK, LEON ASHLEY, psychologist; b. DeLand, Fla., Mar. 31, 1945; s. Cecil McIntosch and Margaret Virginia (Taylor) P.; m. Roberta Harper Brent (div.); children: Jacob, Ashley, Margaret; m. Lori Luanne Rohloff. Student, Rancolph-Macon Coll., Ashland, Va., 1964-66; MS in Clin. Psychology, Va. Commonwealth U., 1973, PhD in Psychology, 1976. Lic. psychologist. Researcher Med. Coll. Va., Richmond, 1973-75, Va. Dept. Corrections, Richmond, 1974-76; asst. prof., assoc. prof. U. North Tex., Denton, 1975-91; dir. rsch. Wilmington Inst., Dallas, 1985—; cons. McCarron Dial Systems, Dallas, 1979—. Author: Individual Trial Analysis Program, 1985, Custody Quotient Test, 1987, 88, Jury Forecast Audit, 1992, Trial Science Reporter, 1992. Trustee Selwyn Sch., Denton, 1978-88. Mem. APA, Am. Psychol. Soc., Tex. Psychol. Assn., North Tex. Behavioral Sci. Found. (pres. 1989-91). Democrat. Episcopalian. Achievements include research in treatments effects of Delta 9-THC, prison factor profile, jury selection expert. Home: 2271 Scripture St Denton TX 76201 Office: Wilmington Inst Trial and Settlement Sci 4000 Spring Valley Ln Dallas TX 75244

PEERAN, SYED MUNEER, airline executive; b. Thanjavur, Tamil Nadu, India, Mar. 22, 1947; s. Syed Mahmoud and Mazharunnisa (Begum) P.; m. Lydia Rani Rathnam, Oct. 19, 1981; children: Yasmin, Leyla. BS, U. Madras, India, 1969. Self employed Thanjavur, India, 1969-73; instr. Ministry of Edn., Addis Ababa, Ethiopia, 1973-79; tech. instr. Saudi Arabian Airlines, Jeddah, 1979-83, sr. tech. instr., 1983-91, sect. mgr., 1991—. Mem. ASTD, ASME, IEEE, AIAA, ASTM, Soc. Automotive Engrs., Am. Vacuum Soc., Am. Soc. Metals, Instrument Soc. Am., Planetary Soc. Office: Saudi Arabian Airlines, P B 167 CC 815, Jeddah 21231, Saudi Arabia

PEERCE-LANDERS, PAMELA JANE, chemical researcher; b. Balt., Feb. 27, 1951; d. Edward Raymond and Eleanor Jane (Simkivicius) Peerce; m. Bruce Kneeland Landers, July 3, 1981. BS in Chemistry, Rensselaer Poly. Inst., 1973; PhD in Chemistry, Calif. Inst. Tech., 1978. Sr. mem. tech. staff Occidental Rsch. Corp., Irvine, Calif., 1979-83; project leader Betz Labs., Inc., The Woodlands, TEx., 1983-85; chem. rsch. mgr. Elf Atochem N.Am., King of Prussia, Pa., 1985—; lectr. U. Calfi., L.A., 1981-82. Contbr. articles to Jour. Electroanalytical Chemistry, Analytical Chemistry, Inorganic chemistry, Jour. of Electrochem. Soc. Mem. Am. Soc. Quality Control, Am. Chem. Soc., Electrochem. Soc. Achievements include patents in water treatment; in Breaking Emulsions of Methane Sulfonyl Chloride in Hydrochloric Acid. Office: Elf Atochem N Am PO Box 1536 900 1st Ave King Of Prussia PA 19406

PEERHOSSAINI, MOHAMMAD HASSAN, engineering and physics educator; b. Isfahan, Iran, Apr. 12, 1951; s. Kazem and Ezzat (Malek-Mohammad) P.; m. Farideh Banisadr, Nov. 15, 1951; children: Sina and Donia. MsME, Tehran (Iran) U., 1974; postgrad., Stanford U., 1979; DSc, U. Paris, 1987. Rsch. scientist Sch. Superieure Physics and Chemistry Paris, 1982-87; prof. Inst. Engring., U. Nantes, France, 1988—; guest scientist German Aerospace Establishment, Göttingen, Fed. Republic Germany, 1988; head thermo fluids rsch. group Laboratoire de Thermocinétique; coord. masters degree program on thermoscis. U. Nantes, officer Internat. Rels. Inst. Engrs.; dir. advanced rsch. workshop on Görtler Vortex Flows NATO; presenter to confs. in field. Author numerous papers in field; referee Jour. of Fluids, Internat. Jour. Heat and Mass Transfer, Exptl. Thermal and Fluid Scis., ASME Jour. Fluids Engring., Jour. de Physique. Recipient Merit award for the best paper in 1993 ann. meeting French Heat Transfer Soc. Mem. ASME, N.Y. Acad. Scis., Nat. Coun. Univs. France (conseil Nat. des Univs.), Am. Phys. Soc., French Phys. Soc., Assn. U. Mech., Fedn. European Assn. Nat. Engrs. Achievements include research on hydrodynamic instability, Görtler Vortex, Dean instability, chaotic advection, gas turbine blade cooling, direct resistence heating. Avocations: tennis, mountain climbing. Office: U Nantes, ISITEM La Chantrerie C P 3023, F 44087 Nantes 03, France

PEETERS, THEO LOUIS, biochemical engineering educator; b. Tienen, Belgium, Feb. 6, 1943; s. Henri and Josephine (Hendrickx) P.; m. Nadine Angèle Bonte, July 18, 1967; children: Bruno, Karen. Degree in biochem. engring., U. Leuven, 1965, PhD, 1968. Rsch. asst. Belgian Sci. Found., Leuven, Belgium, 1966-67, U. Ky., Lexington, 1968-69; rsch. asst. U. Leuven, 1970-79, lectr., 1983-93, chair dept. med. rsch., 1987-93; prof., 1993—; vis. prof. No. Ky. U., Highland Heights, 1980. Contbr. 100 articles to profl. jours., chpts. to books. Lt. Belgian Air Force, 1966. Office: Gut Hormone Lab, Gasthuisberg ON, B-3000 Leuven Belgium

PEGNA, JOSEPH, engineer, educator; b. Le Creusot, France, Oct. 17, 1956; m. Evelyne S. Vinas; children: Frederik, Leslie, Guillaume. MS in Mech. Engring., Ecole Normale Supericure, Cachan, 1980; diploma in engring., Institut Superieur Materiaux Construction Mecanique, Paris, 1982; PhD, Stanford U., 1988. Researcher Laboratoire Mecanique Technologie, Ecole Normale Superieure, Cachan, France, 1982-85; asst. prof. engring. U. Calif., Irvine, 1988-89, Rensselaer Poly. Inst., Troy, N.Y., 1990—. Editor: Product Modeling for Computer Aided Design and Manufacturing, 1991. Mem. N.Y. Acad. Sci. Office: Rensselaer Poly Inst Dept Engring JEC 5022 Troy NY 12180

PEINEMANN, MANFRED K.A., marketing executive; b. N.Y.C., Mar. 2, 1939; s. Hermann and Elisabeth (Lohrberg) P.; m. Marilyn Vicki Barlow, Aug. 24, 1963; children: Robert, Thomas. MS in Astronatics, Polytech. Inst. of N.Y., 1962; MBA, U. So. Calif., 1979. Rsch. engr. Grumman Aerospace Corp., Bethpage, N.Y., 1960-73, sr. aerodynamicist, 1970-73; sr. aerodynamicist space transp. systems div. Rockwell Internat., Downey, Calif., 1973-76, group leader, 1976-79; program devel. mgr. rocketdyne div. Rockwell Internat., Canoga Park, Calif., 1979-89, mktg. mgr., 1989—. Mem. AFA, AIAA, Am. Fin. Assn. Office: Rocketdyne 6633 Canoga Ave Canoga Park CA 91304

PEIRCE, PAMELA KAY, horticulturist, writer; b. Contra Costa County, Calif., Sept. 28, 1943; d. Sheldon James and Lynton (Wicks) P.; m. David Goldberg, Sept. 7, 1985. BA in Botany, Butler U., 1966; postgrad., U. Ill., 1966-67. Project coord. Mayor's Office Community Devel., San Francisco, 1980-81; lectr. Strybing Arboretum, Pier 39, San Francisco, 1980—; photographer, photo editor, freelance writer Ortho Books, San Ramon, Calif., 1983—; freelance writer, editor, photographer Rodale Press, Philips Interactive Media, Home Planners, Inc., 1983—; instr. City Coll. San Francisco, 1984—; founder and bd. pres. San Francisco League Urban Gardeners, 1983-87, bd. sec., 1987-90, bd. mem., 1990—. Author: Environmentally Friendly Gardening: Controlling Vegetable Pests, 1991, Golden Gate Gardening: A Guide to Year-Round Abundance in the San Francisco Bay Area and Coastal California. Mem. Garden Writers Assn. Am. (Cert. Merit 1992), Sigma Xi. Office: City Coll San Francisco OH Dept 50 Phelan Ave San Francisco CA 94112

PEIRO, JOSE MARIA, psychologist, educator; b. Torrent, Valencia, Spain, Mar. 5, 1950; s. Jose Maria and Amparo (Silla) P.; m. Otilia Alicia Salvador, Oct. 25, 1980; children: Teresa, Begoña. Lic. in philosophy, U. Valencia, Spain, 1975; lic. in psychology, U. Complutense, Madrid, 1976; PhD, U. Valencia, 1977. Asst. dept. psychology, assoc. prof. U. Valencia, 1976-81; prof. psychology U. Complutense, Madrid, 1981-82; prof. psychology U. Valencia, 1982-84, prof. social and orgnl. psychology, 1985—, dean psychology faculty; vis. prof. U. Sheffield, 1985, Argentina, 1989, U. Tillburg, 1992; mem. adv. bd. of the Work and Orgn. Rsch. Ctr. Univ. Tilburg. Author: Psicologia de la Organizaciòm. 1983, 91, Organizaciones Nuevas Perspectivas Psicosociológicas, 1990, Desencadenantes del estres laboral, 1993; co-author: Psicologia Contemporanea, 1981, Madurez Vocacional, 1986, Circulos de Calidad, 1993, Control del estres laboral, 1993; editor: Revista de Historia de la Psicologia, Revista de Psicologia Social Aplicada; co-editor: La socializacion laboral, 1987. Pres. Instituto Pro-Desarrollo Found., Torrent, 1982-87; v.p. Caja de Ahorros de Torrent, 1985-88. Mem. APA, European Network of Work and Orgnl. Psychologists, European Assn. Work and Orgnl. Psychology (exec. com. mem.), Spanish Soc. Psychology (pres. Valencia br.), Colegio Ofcl. de Psicologos of Spain (exec. com. mem.). Office: U Valencia Psychology Fac, Avda Blasco Ibanez 21, 46010 Valencia Spain

PELCZAR, OTTO, electrical engineer; b. Vienna, Austria, Aug. 9, 1934; moved to Australia, 1950; s. Joseph Franz and Brigitte (Von Witkowski) P.; m. Amy Margueritte Ludovici, May 9, 1959; children: Suzanne Patricia, Vicki Josephine, Michelle Amy, Paul Daniel. Diploma Elec. Engring., Perth Tech. Coll., W. Australia, 1967; B.Applied Sci., Inst. Tech., Perth, 1978; MBA, U. Western Australia, Perth, 1982. Labourer, welder Roads Bd., Maylands, 1950-56; chief draughtsman Westate Elec. Ind., Perth, 1956-70; dist. engr. Pub. Works Dept., 1970-72; comdg. officer Her Majesty's Australian Ship Acute, 1973-78; engr. Pub. Works Dept., Perth, 1972-85; lectr. Tech. Coll., Subiaco, 1979—; supervising engr. Dept. Marine and Harbours, 1985-90; sr. engr. State Energy Commn., 1990—; nominee W. Australian Parliament, 1983. Elected to Perth City Coun., 1990—. Served to comdr. Royal Australian Naval Res., 1956—. Decorated Res. Forces, Royal Naval Res.; Australian Commonwealth scholar, 1967; recipient Civic Medallion, 1993. Fellow Inst. Engrs.; mem. Royal United Svcs. Inst. (v.p. 1983—), Inst. Draughtsmen, Australian Inst. Internat. Affairs. Club: Old Austria (life mem.). Lodge: Rotary (program dir. 1984—). Avocations: chess, tennis, yachting. Home: 230 Oceanic Dr, City Beach 6015, Australia

PELCZER, ISTVÁN, spectroscopist; b. Esztergom, Hungary, Sept. 22, 1953; came to U.S., 1988; s. István and Katalin (Klinda) P.; children: Anna, Éva. MSc in Chemistry, Jòzsef Attila U., Szeged, Hungary, 1978, PhD in Chemistry summa cum laude, 1989. Rsch. nuclear magnetic resonance fellow dept. organic chemistry Jòzsef Attila U., 1978-80; rsch. NMR spectroscopist Inst. for Drug Rsch., Budapest, Hungary, 1980-82, head spectroscopy group, 1988-89; rsch. NMR spectroscopist EGIS Pharm., Budapest, 1982-88; NMR ops. mgr. chem. dept. Syracuse (N.Y.) U., 1989-91, rsch. asst. prof. chem. dept., 1991—; applications scientist New Methods Rsch., Inc., Syracuse, 1991—; vis. researcher Univ. Chem. Lab., Cambridge, Eng., 1984, chem. dept. Syracuse U., 1988, Lab. Chem. Physics, NIH, Bethesda, Md., 1989. Contbr. articles to profl. publs. Mem. AAAS, Am. Chem. Soc., Hungarian Chem. Soc., N.Y. Acad. of Scis., Smithsonian Inst. Office: Syracuse U Chemistry Dept NMR & Data Processing Lab CST Bldg Syracuse NY 13244-4100

PELED, ISRAEL, electrical engineer, consultant; b. Israel, Apr. 9, 1947; came to U.S., 1979; s. Zvi and Miryam (Klein) Shtul; m. Nurith Tarab, 1971 (div. Jan. 1988); children: Tamir, Limor; m. Sandra Faith Cashman, Jan. 14, 1989 (dec. Sept. 1991). Assoc. in Mech. Engring., Israel Def. Force, 1966; Assoc. in Elec. Engring., Yad Singalovsky, Israel, 1973; BEE, Bridgeport Engring. Inst., 1983. Plant instr. and controls engr. Monsanto, Ashdod, Israel, 1974-77; plant engr. Minn. Dakota Farmers Coop., Whapeton, N.D., 1978-80; sr. engr. John Brown Inc., Stamford, Conn., 1980-92; engring. supr. HR Internat., Edison, N.J., 1992—; cons. I.P. Computer Svcs., Westport, Conn., 1986—. With Israeli Navy, 1964-68. Mem. Instrument Soc. Am.

PELISSIER, EDOUARD-PIERRE, surgeon; b. Bastia, Corse, France, Sept. 1, 1937; s. Jean-Baptiste and Francoise (Geronimi) P.; m. Sylvie Geoffroy-Emmanuelli, Feb. 9, 1967; children: Emmanuelle, Pierre, Francois, Anne-Catherine. M.D., U. Paris, 1968. Ancien interne des Hopitaux de Paris, 1963-67; attache cons. Centre Hospitalier Universitaire de Besancon, France, 1975; surgeon Clinique St. Vincent, Besancon, 1968—; sec. Conseil de l'Ordre des Medecins, Doubs, France, 1980-83. Contbr. articles to med. revs. Served with French Marine Corps, 1961-63. Fellow Am. Coll. Surgeons; mem. Academie de Chirurgie de France (assoc.), Collegium Internationale Chirurgiae Digestivae, Société Nationale Francaise de Gastro-Enterologie, Assn. Française de Chirurgie, Internat. Soc. Surgery, Internat. Gastro-Surg. Club. Office: Clinique St Vincent, 40 Chemin des Tilleroyes, 25000 Besançon France

PELLE, EDWARD GERARD, biochemist; b. N.Y.C., Jan. 20, 1950; s. Enrico and Maria Donata (Cello) P.; m. Evangeline Solero, June 23, 1973; children: Edward G., John L., Gina M., Anthony C. BS, Fordham U., 1972; MS, NYU, 1978. Rsch. asst. Rockefeller U., N.Y.C., 1972-78, NYU Med. Ctr., N.Y.C., 1978-82; prin. scientist Estee Lauder Rsch. Lab., Melville, N.Y., 1982—; peer rev. com. Am. Inst. Biol. Scis., Washington, 1989—. Author: RNA: Biological Aspects, 1980, (with others) Antioxidant Protection Against Ultraviolet Light-Induced Skin Damage, 1993; contbr. articles to Archives Biochem. & Biophysics, Annals-N.Y. Acad. Scis., Cancer Rsch., Proceedings of the Nat. Acad. Scis. Recipient N.Y. State Regents scholarship N.Y. State Regent, 1968-72, Fordham U. scholarship, 1968-72. Mem. Soc. Investigative Dermatology, N.Y. Acad. Scis., AAAS. Achievements include patents in Cu-DIPS, an antioxidant used in sunscreen composition; in an L-Tocopherol derivative molecule with antioxidant properties; in a derivative molecule with UVA sunscreen ability. Office: Estee Lauder Rsch Labs 125 Pinelawn Rd Melville NY 11747-3145

PELLEGRINI, ROBERT J., psychology educator; b. Worcester, Mass., Oct. 21, 1941; s. Felix and Teresa (Di Muro) P.; 1 child, Robert Jerome. BA in Psychology, Clark U., 1963; MA in Psychology, U. Denver, 1966, PhD in Social Psychology, 1968. Prof. San Jose (Calif.) State U., 1967—; rsch. assoc. U. Calif., Santa Cruz, 1989-90; pres. Western Inst. for Human Devel., San Jose, 1985—. Author: Psychology for Correctional Education; contbr. articles to profl. jours. Mem. Phi Beta Kappa. Office: San Jose State U Dept Psychology San Jose CA 95192

PELLEGRINO, CHARLES ROBERT, author; b. N.Y.C., May 5, 1953; s. John and Jane (McAvinue) P.; m. Gloria Tam Pellegrino, July 17, 1988. BA, L.I. U., 1975, MS, 1977; PhD, Victoria U., Wellington, New Zealand, 1982. co-organizer AAAS Symposium on Interstellar Travel and Communication; tech. advisor to Arthur C. Clarke. Co-author: Darwins Universe: Origins and Crises in the History of Life, 1988, 2d edit., 1992, Chariots for Apollo, 1986, Chronic Fatigue Syndrome; author: Her Name Titanic, 1988, Bible Stories for Archeologists, 1993, Time Gate, 1983, Flying to Valhalla, 1993, Unearthing Atlantis: An Archaeological Odyssey, 1991; contbr. articles to profl. jours. Designer U.S./Soviet joint Mars Mission, 1983; organized AAAS Symposium on Internat. Coop. in Space; co-framer 1992 Internat. Space Yr. Fellow Brit. Interplanetary Soc.; mem. AAAS, N.Y. Acad. Scis., Nat. Heritage Found., The Challenger Ctr. (founding sponsor). Republican. Achievements include designing the Valkyrie rocket; development of the downblast theory, proposal that oceans may exist beneath the icy crusts of certain Jovian and Saturnian moons. Office: 170 W Broadway 3F Long Beach NY 11561

PELLEMANS, NICOLAS, patent agent, mechanical engineer; b. Montreal, Quebec, Can., Mar. 23, 1968; s. Wilhelm B. Pellemans and Suzanne Charbonneau. B of Engring., Ecole Poly., Montreal, 1990. Patent agt. ROBIC, Montreal, 1991—. Mem. Soc. Automotive Engring., Patent and Trademark Inst. Can. Office: ROBIC, 55 St Jacques, Montreal, PQ Canada H2Y 3X2

PELLERIN, ROY FRANCIS, civil engineer, educator; b. Everett, Wash., May 29, 1934; s. Clifford Francis and Vera (Johnson) P.; m. Patricia Ann Bateman, May 17, 1958; children: Karen, Douglas, Thomas. BS, Wash. State U., 1959; MS, U. Idaho, 1970. Prof. materials sci. and engring. Wash. State U., Pullman, 1959-85, prof. civil engring., 1985—; tech. adviser Am. Inst. Timber Constrn., Vancouver, Wash., 1980—; cons. FAO, Bejing, 1990, Westinghouse Co., Richland, Wash., 1990-92; vis. prof. Nanjing (Republic of China) Forestry U., 1990; chmn. symposia on nondestructive testing of wood. Author tech. publs. Mem. Forest Products Rsch. Soc. (sect. officer 1971-76, nat. com. 1982-83, L.J. Markward award engring. rsch. 1972, Woodworking and Furniture Digest award 1975), Soc. Wood Sci. and Tech., Internat. Union Forestry Rsch. Orgns., Sigma Xi (chpt. pres. 1986-87). Democrat. Lutheran. Achievements include patents in nondestructive method of grading wood materials, method and apparatus for nondestructive testing of beams, load testing machine, nondestructive evaluation methods. Home: NW 1215 Clifford St Pullman WA 99163 Office: Wood Materials/Engring Wash State Univ Pullman WA 99164

PELLERITO-BESSETTE, FRANCES, research biologist; b. Patchogue, N.Y., July 5, 1960; d. Fred J. and Marian (Taylor) P.; m. Peter Eric Bessette, June 10, 1990. BS in Biology, Siena Coll., 1982; postgrad., Rutgers U., 1992—. Sci. intern N.Y. State Dept. Pub. Health, Albany, 1982; sr. rsch. tech Albert Einstein Coll. of Medicine, N.Y.C., 1982-84; scientist Schering Plough Rsch., Bloomfield, N.J., 1984-91; med. rsch. assoc. Schering Plough Rsch. Inst., Kenilworth, N.J., 1991—. Contbr. articles to profl. jours. Named Pres.'a Award nominee Schering Plough Rsch. Inst., 1989, 88, 86. Mem. Internat. Soc. for Analytical Cytology, Assn. Clin. Pharmacology,

Drug Info. Assn., Healthcare Bus. Women's Assn. Roman Catholic. Office: Schering Plough Rsch Inst 2015 Galloping Hill Rd Kenilworth NJ 07033

PELLETIER, CLAUDE HENRI, biomedical engineer; b. Riviere-Ouelle, Can., Que., Dec. 15, 1941; s. Lucien Pelletier and Ernestine Michaud. Immatriculation sr., Coll. Universitaire U. Sherbrooke, 1961; B.Sc.A., U. Sherbrooke, 1966; M.Sc.A., Ecole Polytechnique, U. Montreal, 1972. Project engr. Alcan, Alma, Can., 1966-69; mgr. computer ctr. in physiology dept. faculty medicine U. Montreal, 1972-73; biomed. engr. Sacre-Coeur Hosp., Montreal, 1973-75; chief engr. biomed. engring. dept. Montreal Heart Inst., 1975—; lectr. faculty of medicine U. Montreal, 1972-74, research asst. faculty of medicine, 1973-75; cons. Montreal Heart Inst., 1975—. Contbr. articles to profl. jours. Mem. Order of Engrs. Que., IEEE, Assn. Advancement Med. Instrumentation, Assn. Des Physiciens Et Ingenieurs Biomedicaux Du Que. Roman Catholic. Avocations: swimming; tennis. Home: 5732 Plantagenet, Montreal, PQ Canada H3S 2K3

PELOSI, GIANCARLO, cardiologist; b. Milan, July 7, 1937; s. Antonio and Cesarina (Fausti) P.; m. Jolanda Sanna, May 20, 1964; children: Antonella, Alessandra, Ilaria. Degree in medicine, U. Milan, 1962, Libera Docenza in Pharmacology, 1969, Libera Docenza Systematic Med. Therapy, 1971. Asst. Ospedale Maggiore, Milan, 1968-70, vice-head physician, 1970-79; head physician Ospedale Viarana, Besana Brianza, Italy, 1979-80, Ospedale M. Melloni, Milan, 1980—; researcher CNR, Milan, 1965-70; coun. mem. Ospedale Maggiore, Milan, 1978-83, 91; pres. Unità Socio Sanitaria Locale 75/17, Milan, 1983-90. Patentee in field, Monitoraggio continuo di alcuni parametri ematochimici; contbr. over 80 articles to sci. publs. Rsch. grantee CNR, 1975-90; recipient Borsa di Studio D.Libre per mig/lior Tesi Laurea U. Milan, 1963. Assn. Nat. Medici Cardiologi Ospedalieri, Soc. Italy Medicina d'Urgenza, European Soc. Noninvasive Cardiovascular Dynamics. Roman Catholic. Avocations: sky, tennis, cycling. Home: via Pellegrine Rossi n 15, 20161 Milan Italy Office: Ospedale M Melloni, via Mecedonio Melloni 52, 20129 Milan Italy

PELTEKOF, STEPHAN, systems engineer; b. Plovdiv, Bulgaria, Feb. 11, 1929; came to U.S. 1956; s. Peter and Velika (Christova) P.; m. Colette Gringoire, July 6, 1957; 1 child, Brigitte. PhD, U. Santa Barbara, 1978. Cert. profl. mgr. Inst. Cert. Profl. Mgrs. Engr. Lockheed Missile & Space Co., Sunnyvale, Calif., 1967-69; systems analyst Martin-Marietta Aerospace Div., Vandenberg AFB, Calif., 1981-83; systems engr. Lockheed Space Ops. Co., Vandenberg AFB, Calif., 1983-86, Computer Software Analysts, Inc., Camarillo, Calif., 1987-89; sr. analytical specialist ITT Fed. Svcs. Corp., Vandenberg AFB, 1989-91; dir. edn. Cook's Inst. Engring., Jackson, Miss., 1991—. Author: The Conquest of Space, 1967, Commitment to Excellence, 1985, New Era Leadership, 1991, Forty Years in Exile, 1992. Recipient Outstanding Performance award Martin-Marietta Aerospace, 1982, Lockheed Space Ops. Co., 1985, Space Shuttle Directorate, 1986; named Employee of the Yr. Nat. Mgmt. Assn., 1986. Mem. Nat. Mgmt. Assn., Inst. Cert. Profl. Mgrs., N.Y. Acad. Scis. Eastern Orthodox. Achievements include development/implementation "Vandenberg Launch and Landing Site Excellence Program". Home: 1485 Calle Segunda Lompoc CA 93436

PELTZER, DOUGLAS LEA, semiconductor device manufacturing company executive; b. Clinton, Ia., July 2, 1938; s. Albert and Mary Ardelle (Messer) P.; m. Nancy Jane Strickler, Dec. 22, 1959; children: Katharine, Eric, Kimberly. BA, Knox Coll., 1960; MS, N.Mex. State U., 1964; MBA, U. Phoenix, 1990. Rsch. engr. Gen. Electric Co., Advanced Computer Lab., Sunnyvale, Calif., 1964-67; large scale integrated circuit engr. Fairchild Camera & Instrument, Rsch. & Devel. Lab., Palo Alto, Calif., 1967-70, supervisory engr. bipolar memory devel., Mountain View, Calif., 1970-73, process engring. mgr., bipolar memories div., 1973-83, tech. dir., 1977-83; v.p. tech. ops. Trilogy Systems Corp., Cupertino, Calif., 1983-85 ; pres. Tactical Fabs, Inc., 1985-89; v.p. process devel. Chips and Techs. Inc., 1989-92; pres, CEO CAMLAN, Inc., San Jose, Calif., 1992—. NSF fellow, 1962-63; recipient Sherman Fairchild award for tech. excellence, 1980, Semiconductor Equipment and Materials Inst. award, 1988; Inventor of Yr. award Peninsula Patent Law Assn., 1982. Mem. AAAS, IEEE, Sigma Pi Sigma. Inventor in field; patentee in field. Home: 10358 Bonny Dr Cupertino CA 95014-2908 Office: CAMLAN Inc 2381 Zanker Rd Ste 100 San Jose CA 95131-1122

PELZER, CHARLES FRANCIS, molecular geneticist, biology educator, researcher; b. Detroit, June 5, 1935; s. Francis Joseph and Edna Dorothy (Ladach) P.; m. Veronica Ann Killeen, July 7, 1972; 1 child, Mary Elizabeth. BS in Biology, U. Detroit, 1957; PhD in Human Genetics, U. Mich., 1965. Postdoctoral fellow Wabash Coll., Crawfordsville, Ind., 1965-66; instr. U. Detroit, 1966-68; asst. prof. Saginaw Valley State U., University Center, Mich., 1969-74, assoc. prof., 1974-79, prof., 1979—; rsch. assoc. Mich. State U., East Lansing, 1976-77; rsch. fellow Henry Ford Hosp., Detroit, 1982-83, 88-92; v.p. Saginaw Valley Retinitis Pigmentosa Found., Mich., 1979-81; vis. scientist Am. Inst. Biol. Scis., Washington, 1975-78; grant reviewer U.S. Dept. Edn., Washington, 1984-87, 91. Contbr. articles to profl. jours. Recipient Alumni award Saginaw Valley State U. Alumni Assn., 1971; grantee Fund for Ford Hosp., 1983, Mich. State U., 1977, Saginaw Valley State U. Found., 1979-82, 83-85, 86-89, Mich. Rsch. Excellence Fund, 1993, Kettering Found., 1965-66, Kellogg Found., 1961, NIH, 1961-64, Monsanto Co. rsch. grant, 1987, Dow Chem., 1988, 89, Dow Corning, 1988, 89. Fellow Human Biology Coun.; mem. Am. Soc. Human Genetics, Genetics Soc. Am., N.Y. Acad. Sci., Internat. Electrophorisis Soc., Nat. Assn. Biology Tchrs. (dir. for Mich. Outstanding Biology Tchrs. award), others. Home: 4900 Schneider St Saginaw MI 48603-4513 Office: Saginaw Valley State U Dept Biology SI53 University Center MI 48710

PENA, ELEUTERIO, utility executive; b. Brownsville, Tex., Sept. 12, 1944; s. Federico and Antonia (Alvarado) P.; m. Felicitas Mesa, July 12, 1970; children: Damaris, Eric, Melissa, Aaron. AA, Tex. Southmost Coll., 1991; BBA, U. Tex., Brownsville, 1992. Wastewater operator Pub. Utilities Bd., Brownsville, 1966, water treatment supt., 1968—. Active med. com., Brownsville, 1990—, retirement com., 1991—. With U.S. Army, 1966-68. Mem. Citrus Water and Wastewater Assn., Tex. Water Utilities Assn., Tex. Safety Assn., Am. Water Works Assn., Phi Theta Kappa. Baptist. Home: 403 Windcrest Dr Brownsville TX 78521

PENA, SERGIO DANILO JUNHO, physician; b. Belo Horizonte, MG, Brazil, Oct. 17, 1947; s. Danilo Drumond and Maria Aparecida (Brandao) P.; m. Betania Maria Andrade, Feb. 15, 1971; 1 child, Frederico Augusto Andrade. MD, U. Federal Minas Gerais, Belo Horizonte, Brazil, 1970; PhD, U. Manitoba, Winnepeg, Can., 1977. Vis. scientist Nat. Inst. Med. Rsch., London, 1977-78; asst. prof. McGill U., Montreal, Can., 1978-82; assoc. prof. U. Fed. Minas Gerais, Belo Horizonte, Brazil, 1982-85, prof., 1985—; pres. Nucleo de Genetica Medica de Minas Gerais, Belo Horizonte, 1982—; pres. Latin Am. Program for Human Genome, 1992—; pres. Brazilian Soc. Biochemistry and Molecular Biology, 1991-92. Recipient Postdoctoral fellowship MRC of Can., 1974-78, Scholarship, MRC of Can., 1978-82, Killan Found., 1978-82, Lafi award for Med. Rsch., Lafi Labs., 1984, "Best of 1990" in Sci. and Technology, 1991. Home: R Sao Paulo 2024/1002, 30170 Belo Horizonte Brazil Office: Gene/MG, Ave Afonso Pena 3111/9, 30130 Belo Horizonte Brazil

PENCE, THOMAS JAMES, mechanical engineer, educator, consultant; b. Milw., July 7, 1957; s. James Thomas and Norma Jean (Hannes) P.; m. Lora Casper, July 18, 1987. BS in Engring. Mechanics, Mich. State U., 1979; PhD in Applied Mechanics, Calif. Inst. Tech., 1983. Asst. prof. math. dept. U. Wis., Madison, 1983-86; asst., assoc. prof. dept. materials sci. and mechanics Mich. State U., East Lansing, 1986—; cons. Exxon Prodn. Rsch. Co., Houston, 1983, Johnson Controls Inc., Milw., 1989—. Contbr. articles to profl. jours. Named N.A.T.O. Postdoctoral fellow, 1984-85. Mem. ASME, Am. Soc. for Engring. Edn., Soc. of Indsl. and Applied Math. Achievements include rsch. in mathematical modeling of phase transformations; prediction of stress, buckling and failure in solids; analysis of smart materials and structures. Office: Mich State U Dept Materials Sci and Mechanics East Lansing MI 48824-1226

PENCZEK, STANISLAW, chemistry educator; b. Warsaw, Poland, Jan. 24, 1934; s. Marcin and Natalia (Libenbaum) P.; m. Irena Budzinska, Mar. 5, 1959; children: Wojciech, Alina. PhD, Leningrad. (USSR) Inst. Tech., 1963; DSc, Lodz (Poland) Poly. Inst., 1970. Head dept. polymerization Inst. Polymers, Warsaw, 1958-68; head dept. polymer chemistry Polish Acad. Sci., Lodz, 1972—, prof., 1975—; vis. prof. U. Ghent, U. Mainz, U. Kyoto, U. Paris; Titular prof. French Acad. Sci., 1992. Author: Catonic Ring Opening Polymerization, Vol.I, 1980, Vol. II, 1985, Models of Biopolymers, 1990; editor: Ring Opening Polymerization, 1975; mem. editorial bd. Jour. Polymer Sci., Makromoleculare Chemie, Polymer Internat., Jour. Bioactive Polymers. Mem. Polish Chem. Soc. (chmn. sect. kinetics 1982-89, chmn. sect. polymers 1989—). Achievements include establishment of several new mechanisms of polymerizations and synthesis of new models of biopolymers. Home: 10 Lutego 7A/11, 90 303 Lodz Poland Office: Polish Acad Sci, Sienkiewicza 112, 90 369 Lodz Poland

PENDER, NOLA J., community health nursing educator, researcher; b. Lansing, Mich., Aug. 16, 1941; d. Frank and Eileen (Schoenhals) Blunk; m. Albert R. Pender, Aug. 28, 1965; children: Andrea, Brent. BS, Mich. State U., 1964, MS, 1965; PhD, Northwestern U., Evanston, Ill., 1969. Prof. nursing No. Ill. U., DeKalb, dir. health promotion rsch. program; dir. Ctr. Nursing Rsch., Ann Arbor, Mich.; vis. prof. U. Mich. Sch. Nursing, researcher in field. Contbr. articles to med. jours. Fellow Am. Acad. Nursing; mem. ANA (chmn. rsch. cabinet 1982-84), Midwest Nursing Rsch. Soc. (pres. 1985-87), Am. Acad. Nursing (pres.-elect 1989—), NIH (nat. adv. coun. nursing rsch. Nat. Ctr. for Nursing Rsch. 1987-90, rsch. program grantee 1985-87). Office: U Mich Ctr Nursing Rsch 400 N Ingalls Rm 4236 Ann Arbor MI 48109*

PENDLETON, JOAN MARIE, microprocessor designer; b. Cleve., July 7, 1954; d. Alvin Dial and Alta Beatrice (Brown) P. BS in Physics, Elec. Engring., MIT, 1976; MSEE, Stanford U., 1978; PhDEE, U. Calif., Berkeley, 1985. Sr. design engr. Fairchild Semiconductor, Palo Alto, Calif., 1978-82; staff engr. Sun Microsystems, Mountain View, Calif., 1986-87; chief exec. officer Harvest VLSI Design Ctr. Inc., Palo Alto, 1988—; cons., designer computer sci. dept. U. Calif., Berkeley, 1988-90. Contbr. articles to profl. jours.; inventor, patnetee serpentine charge transfer device. Recipient several 1st, 2d and 3d place awards U.S. Rowing Assn., Fairchild Tech. Achievement award, 1982, 1st place A award Fed. Internat. Soc Aviron, 1991. Mem. IEEE, Assn. for Computing Machinery, Lake Merritt Rowing Club, Stanford Rowing Club, U.S. Rowing Assn. Avocations: rowing, skiing, backpacking. Home: 1950 Montecito Ave Apt 22 Mountain View CA 94043-4334

PENDLETON, VERNE H., JR., geologist; b. Medford, Oreg., Sept. 17, 1945; s. Verne H. and Ilene Clara Koepsell) P.; m. Paula Jean Obenshain, June 22, 1968. BS in Geology, Oreg. State U., 1973. Geologist/engring. inspector/tech. Soils Testing Lab., Inc., Medford, 1976-81; rsch. tech. Dept. Civil Engring., U. Idaho, Moscow, 1982-84; constrn. geologist, lab. mgr. Soils Testing Lab., Inc., Medford, 1984-88; quality control insp. LTM Inc., Medford, 1988-91; sr. materials specialist/coord. Hwy. div. Oreg. Dept. Transp., Bend, 1991-92; ATE region materials inspector hwy. div. Oreg. Dept. Transp., Roseburg, 1992—. With U.S. Army, 1966-69. Republican. Ch. of the Nazarene. Avocations: fly fishing, camping, hiking. Home: 1834 NE Todd St Roseburg OR 97470

PENECILLA, GERARD LEDESMA, botany educator; b. Iloilo, Philippines, Nov. 25, 1956; s. Felipe Sayco and Desposoria Grajo (Ledesma) P.; m. Teresa Villarete Maligad, May 5, 1988; children: Gerard Lorenz, Gerard Jr. BS, U. Philippines, Iloilo City, 1977; MS, U. Santo Tomas, Manila, The Philippines, 1980; EdD, West Visayas State U., 1992. Rsch. asst. U. of the Philippines, Quezon City, 1977-80; sr. researcher Nat. Rsch. Coun. of the Philippines, Manila, 1980-82; lectr. U. of the City of Manila, 1981-82; instr. Philippine Normal U., Manila, 1981-82, West Visayas State U., Iloilo, The Philippines, 1984-85; asst. prof. West Visayas State Coll., Iloilo, The Philippines, 1986-90, assoc. prof., 1991-93, prof., 1993—; professorial lectr. U. of the Philippines in Visayas, 1991—; rsch. cons. Philippines Nat. Oil Co., Manila, 1981-83; project leader Office of Rsch. West Visayas State U., Iloilo, 1985-87, Dept. Sci. & Tech., 1990—; rsch. study leader Philippines Coun. for Health Rsch., 1990—; postgrad. fellow in immunology U. Ga., Athens, 1989-90; instr. Japanese Ctrl. Philippine U., Iloilo City, 1992—; rsch. collaborator anti cancer screening dept. chem. Va. Inst. Tech., 1992—. Contbr. articles to profl. jours.; patentee in field. Extension svc. West Visayas State U., 1986-87, project planning, 1988-89; project developer Nat. Econ. & Devel. Authority, Iloilo, 1990. Grantee U.S. Agy. for Internat. Devel., 1989-90, Sci. Edn. Inst., 1993. Mem. AAAS, Nat. Rsch. Coun. Philippines, Philippine Assn. for Advancement Sci., Am. Cons. League, Systematic Biologists Assn., Philippine Biochem. Soc., Biology Tchr. Assn., Soc. Econ. Botany Recording Arts. Avocations: singing, songwriting, lawn tennis, bowling, badminton. Home: 451 H Lopez Jaena St Molo, Iloilo City 5000, The Philippines Office: W Visayas State U, Luna St, Iloilo 5000, The Philippines

PENG, LIANG-CHUAN, mechanical engineer; b. Taiwan, Feb. 6, 1936; came to U.S., 1965, naturalized, 1973; s. Mu-Sui and Wang-Su (Yang) P.; diploma Taipei Inst. Tech., 1960; M.S., Kans. State U., 1967; m. Wen-Fong Kao, Nov. 18, 1962; children—Tsen-Loong, Tsen-Hsin, Lina, Linda. Project engr. Taiwan Power Co., 1960-65; asst. engr. Carlson & Sweatt, N.Y.C., 1966-67; asst. engr. Pioneer Engrs., Chgo., 1967-68; mech. engr. Bechtel, San Francisco, 1969-71; sr. specialist Nuclear Services Co., San Jose, Calif., 1971-75; sr. engr. Brown & Root, Houston, 1975; stress engr. Foster Wheeler, Houston, 1976; staff engr. AAA Technologists, Houston, 1977; prin. engr. M.W. Kellogg, Houston, 1978-82; pres., owner Peng Engring., Houston, 1982—; instr. U. Houston; condr. piping tech. seminars. Chmn. South Bay Area Formosan Assn., 1974, No. Calif. Formosan Fedn., 1975. Registered profl. engr., Tex., Calif. Developer. (computer progarams) SIMFLEX; condr. seminars in field. Mem. ASME, Nat. Soc. Profl. Engrs. Confucian. Home: 3010 Manila Ln Houston TX 77043-1312

PENHALE, POLLY ANN, marine biologist, educator; b. St. Louis, Dec. 18, 1947. BA, Earlham Coll., 1970; MS, N.C. State U., 1972, PhD in Zoology, 1976. Mem. Am. Soc. Limnology and Oceanography (sec. 1985), Am. Geophys. Union, Phycological Soc. Am., Internat. Assn. Theoretical and Applied Limnology, Antarctican Soc. (pres.). Achievements include research in macrophyte-epiphyte productivity in seagrass communities, nutrient cycling, macrophyte-epiphyte interactions, seagrass ecosystems. Office: Coll William and Mary Va Inst Marine Sci Rt 1208 Gloucester Point VA 23062*

PENHUNE, JOHN PAUL, science company executive, electrical engineer; b. Flushing, N.Y., Feb. 13, 1936; s. Paul and Helene Marguerite (Beux) P.; m. Nancy Leigh Peabody, Sept. 6, 1958 (div. Apr. 1982); children: Virginia Burdet, James Peabody, Sarah Slipp; m. Marcellite Helen Porath, Feb. 15, 1986; 1 child, Marcellite Helen Broadhurst. BSEE, MIT, 1957, PHDEE, 1961; postgrad., U. Grenoble, France, 1959, Harvard U., 1973. Asst. prof. elec. engring. MIT, Cambridge, Mass., 1962-64; mem. tech. staff Lincoln Lab. MIT, Lexington, Mass., 1964-66; supr. radar group Bell Telephone Labs., Whippany, N.J., 1966-68; asst. dir. Advanced Ballistic Missile Def. Agy., Washington, 1968-69; pres. Concord Rsch. Corp., Burlington, Mass., 1969-73; pvt. practice sci. cons. Carlisle, Mass., 1973-79; bd. dirs. Phys. Dynamics, Inc., La Jolla, Calif., 1979-81; sr. v.p. for rsch. Sci. Applications Internat. Corp., San Diego, 1981—; indsl. adv. bd. Inst. for biomed. Engring., U. Calif., San Diego, 1992—, connect. steering com., 1993—. Author: Case Studies in Electromagnetism, 1960; patentee in field. Bd. dirs. La Jolla Chamber Music Soc., 1985-87. 1st lt. U.S. Army, 1961-62. Recipient Meritorious Civilian Svc. award Dept. Army, 1968. Mem. Cosmos Club (Washington), San Diego Yacht Club., Eta Kappa Nu, Tau Beta Pi, sigma Xi, Phi Kappa Sigma. Avocations: boating, classical music. Home: 6730 Muirlands Dr La Jolla CA 92037-6315 Office: Sci Applications Internat Corp 1241 Cave St La Jolla CA 92037

PENMAN, PAUL DUANE, nuclear power laboratory executive; b. Williston, N.D., Sept. 25, 1937; s. Robert Roy and Kathryn Erica (Hagstrom) P.; m. Cornelia Dennis, Jan. 9, 1960 (div. June 1986); children: Anne, Robert, Jill; m. Carrie B. Silverblatt, July 14, 1986. BS in Engring. Physics, U. Colo., 1959; MS in Physics, U. Louisville, 1965. Asst. prof. U. Louisville, 1962-65; engr. Bettis Atomic Power Lab., West Mifflin, Pa., 1965-71, mgr., 1971-77, in charge lab. ops., 1977-82, in charge nuclear core mfg., 1982-92; mgr. product assurance Westinghouse Electro-Mech. Div., Cheswick, Pa., 1992—. Leader Boy Scouts Am., Pitts., 1977-80. Lt. USN, 1959-64. Mem. U.S. Naval Inst., Gyro Internat. (bd. dirs., pres. 1988-90), U. Colo. Alumni Assn. (pres. Pitts. 1971-8o). Republican. Home: 105 Urick Ct Monroeville PA 15146-4919

PENN, SHERRY EVE, communication psychologist, educator; b. Jersey City, Nov. 25, 1941; d. Herman Joseph and Ida (Eventoff) P.; m. Donald Eugene Crawford, Aug. 15, 1987; stepchildren: Dan, Helen, David. BA in Psychology and Theatre, U. Louisville, 1963; MA in Theatre and Music History, U. Fla., 1967; PhD in Communication Psychology, Performing Arts, Union Inst., 1975. Dir. dance, assoc. prof. Miami (Fla.)-Dade Community Coll., 1967-78; assoc. dean baccalaureate program World U., Miami, 1978; pres. sec., dir. communication Jefferson County (Ky.) Judge Exec. A. Mitch McConnell, 1979-81; assoc. dean, asst. v.p. Union Inst. Grad. Sch., 1984-87, core faculty prof., 1984—; v.p. communication Penn-Crawford Assocs., 1987—; chair program policy bd., external masters degree program in psychology Lone Mountain Coll., Miami, 1975-76; cons. in pub. rels., comm., 1970—; vis. prof. U. Fla., 1969, Calif. State U., San Francisco, 1973, 74, Fla. Internat. U., 1975-76, U. Louisville, 1982-84, Webster U., 1983-84; presenter, speaker in field. Artistic dir., producer Miami Jazz Dance Ensemble, Miami Mime Artists, 1967-79; writer, producer pub. TV programs, 1980; exec. producer weekly pub. affairs series Consumer Corner, 1980-81; choreographer for various dance, mime and operatic prodns. Fellow Ford Found., 1961. Mem. ASTD, Pub. Rels. Soc. Am., Am. Women in Radio and TV, Women in Communication. Home: PO Box 46030 Eden Prairie MN 55344

PENNER, KAREN MARIE, civil engineer; b. Walla Walla, Wash., Aug. 3, 1972; d. Ted G. and Sharon K. (Lederer) P. BSCE, Gonzaga U., 1994. Coop. program U.S. Army Corps of Engrs., Walla Walla, Wash., 1991; transp. technician Wash. State Dept. of Transp., Spokane, 1992—. Mem. advisor Setons-Svc. Orgn., Spokane, 1991-93; mem. Gonzaga U. Ambassadors, Spokane, 1991-92. Recipient scholarship Wash. Soc. Profl. Engrs., scholarship Kaiser Engrs.; named Washington scholar Wash. State Ednl. Assn., 1990. Mem. Student Chtp. ASCE (v.p. 1990-92), Delta Tau Sigma.

PENNEY, ROBERT ALLAN, engineering executive; b. Mineola, N.Y., Mar. 18, 1959; s. Robert Leroy and Mary Elizabeth (Rumpf) P.; m. Mary Hidegard McGowan, Jan. 23, 1988; children: Rebecca Elspeth, Andrew Napier. BSME, Union Coll., Schenectady, N.Y., 1981. Cert. profl. mech. engr., Wash.; cert. energy mgr.; cert. demand-side mgmt. profl. Mech. engr. Rosser Fabrap Internat., Atlanta, 1982-86; dir. engring. svcs. Southface Energy Inst., Atlanta, 1987-91; v.p. Conservation Techs. Internat., Atlanta, 1990-91; clearinghouse lead engr. Wash. State Energy Office, Olympia, 1991—; computer applications com. ASHRAE, Atlanta, 1985-86; speakers bur. Union of Concerned Scientists, Cambridge, Mass., 1990—; sec. N.W. Power Quality Svcs. Ctr., Portland, Oreg., 1992—; co-facilitator Lighting Design Lab. Specifications Com., Seattle, 1992—; presenter at nat. and internat. confs. Contbr. tech. articles to profl. publs. Scoutmaster Boys Scouts Am., Tampa, Fla.; climbing com. The Mountaineers, Olympia, 1992. Recipient Patriotic Svc. award U.S. Dept. Treasury, 1992. Mem. Assn. Energy Engrs. (chpt. pres. 1990-91, sr. mem., demand-side mgmt. soc., Regional Energy Profl. of Yr. 1989, Chpt. Energy Engr. of Yr. 1989), Environ. Engrs. Mgrs. Inst., Assn. of Profl. Energy Mgrs. (Wash. state chpt. pres.), Southface Energy Inst., Sierra Club (energy com., chpt. membership chair, 1987), Rocky Mountain Inst. (assoc.). Democrat. Methodist. Achievements include development of computer-aided design and drafting standards; of lamp and ballast disposal potions; of energy information electronic bulletin boards; of power quality information transfer mechanisms; of minimization of the environmental effects of energy use. Home: 1415 11th Ave SW Olympia WA 98502-5787 Office: Electric Ideas Clearinghouse 925 Plum St SE PO Box 43171 Olympia WA 98504-3171

PENNICA, DIANE, molecular biologist; b. Fredonia, N.Y., July 25, 1951; d. Frank James and Mamie Marie (Tampio) P.; m. Frank Eugene Hagie, Mar. 16, 1990. BS, SUNY, Fredonia, 1973; PhD, U. R.I., 1977. Sr. rsch. scientist Genentech, Inc., South San Francisco, 1980—. Contbr. articles to profl. jours. including Nature, Virology, Biochemistry, others. Recipient Svc. to Humanity award Fredonia C. of C., 1988, Inventor of Yr. award Intellectual Property Owners Found., 1989; named One of Scientists involved in top 100 innovations of 1984-85, Sci. Digest Mag., 1984-85, One of 4 people for the articles of the people behind some of the bright ideas of 1987, N.Y. Times Newspaper, 1988. Mem. AAAS. Achievements include patents for cloning and expression of the cDNA for human tissue plasminogen activator (t-PA or Activase). Office: Genentech Inc 460 Point San Bruno Blvd South San Francisco CA 94080

PENNINGER, JOHANNES MATHIEU L., chemical engineer; b. Heerlen, Netherlands, Apr. 30, 1942; s. Andreas and Anna-Maria (Adriaens) P.; m. Gertrudis-Germaine Abels, June 11, 1966; children: Yasmin, Andrea. Degree in chem. tech., Tech. U., Eindhoven, Netherlands, 1965, PhD in chem. tech., 1968. Asst. prof. Middle East Tech. U., Ankara, Turkey, 1969-71; sr. lectr. U. Twente, Enschede, Netherlands, 1971-74; project mgr. bus. devel. AKZO Chems., Amersfoort, Netherlands, 1974-77; assoc. prof. chem. engring. U. Cin., 1977-78; sr. rsch. engr. Occidental Rsch. Corp., Irvine, Calif., 1978-81; head organic products rsch. Akzo Salt & Basic Chems., Hengelo, Netherlands, 1981-92, head chem. mfg. rsch., 1992—; prof. Tech. U., 1988—; bd. dirs. chemistry Netherland Cert. Com., 1991—; chair adv. bd. Enschede Polytechnics, 1987—. Author: Economic Principles of Chemical Process Design, 1989; editor: Supercritical Fluid Science and Technology, 1989; contbr. articles to Fuel, Internat. Jour. Chem. Kinetics, Jour. Catalysis. Named Vis. Scientist U. Calif., Berkeley, 1988, Coral Industries Vis. Scholar U. Hawaii, 1991; recipient PhD award Royal Dutch/Shell, 1965. Mem. Royal Netherlands Chem. Soc., AICE, Am. Chem. Soc., N.Y. Acad. Scis. Roman Catholic. Achievements include patent for process for preparation of polyarylene sulphide having a low alkali metal content, reactions of hydrocarbons in water at hig htemperature and high pressure, direct conversion of natural gas in hydrocarbons, zero-emission chemical manufacturing. Home: Boerhaavelaan 18, 7555 BC Hengelo The Netherlands

PENNINGTON, EDWARD CHARLES, medical physicist; b. McCook, Nebr., May 30, 1956; s. Duane Edward and Faye Arlene (Waterman) P.; m. Marneé Jean Wimberley, Nov. 11, 1953; children: Jill, Elizabeth. BS, U. Nebr., 1978; MS, U. Tex., Arlington, 1980. Med. physicist St. Paul Hosp., Dallas, 1980-85, U. Iowa, Iowa City, 1985—. Office: U Iowa Hosp & Clinics Radiation Oncology Iowa City IA 52242

PENNINGTON, RODNEY EDWARD, molecular biologist; b. Tulsa, Okla., Dec. 10, 1956; s. W.E. and Lucy Lee (Beattie) P.; m. Janice Kay Green, Dec. 17, 1982; children: Katherine Lee, Andrea Jean. BS, Okla. State U., 1980, PhD, 1991. Postdoctoral rsch. assoc. dept. Plant Pathology Okla. State U., Stillwater, 1991--. Co-author: Microbial Enhanced Oil Recovery, 1989. Mem. Am. Phytophthological Soc., Okla. Acad. Sci., Sigma Xi, Phi Lambda Upsilon. Presbyterian. Office: Okla State U 110 NRC Stillwater OK 74078

PENNINGTON, RODNEY LEE, engineer; b. Bloomsburg, Pa., Oct. 17, 1946; s. Ernest Eli and Ellen M. (Albertson) P.; m. Patricia Ann Bond, Sept. 4, 1965 (div. 1983); 1 child, Denise Rene; m. Linda Rae Petruna, Aug. 8, 1984. Engring. Tech. Cert., Williamsport Tech. Inst., Pa., 1965; AS in Engring., Williamsport Area Coll., 1972; BS in Engring. Sci., Pa. State U., 1974. Registered profl. engr. N.J. Design engr. Piper Aircraft, Lock Haven, Pa., 1965-66; staff engr. Armstrong World Ind., Lancaster, Pa., 1974-75; project mgr. REECO, Morris Plains, N.J., 1975-78, sales mgr., 1978-80, sales adminstr., 1980-81, engring. mgr., 1981-82, mktg. devel. mgr., 1982-84, v.p. engring. and R&D, 1982—; dir ARTCO, Inc., Morristown. Patentee in field; contbr. articles to profl. jours. With USAF, 1966-70. Recipient Alcan Engring. Sci. award, 1972. Mem. Nat. Coil Coaters Assn. (bd. dirs. 1991-94, chmn. environ. com. 1989-91). Avocations: computers, hunting, landscaping. Office: REECO PO Box 1500 Somerville NJ 08876-1251

PENNISTEN, JOHN WILLIAM, computer scientist, linguist, actuary; b. Buffalo, Jan. 25, 1939; s. George William and Lucy Josephine (Gates) P. AB in Math. and Chemistry with honors, Hamilton Coll., 1960; postgrad., Harvard U., 1960-61, U.S. Army Lang. Sch., 1962-63; MS in Computer Sci.

with honors, N.Y. Inst. Tech., 1987; cert. in taxation, NYU, 1982; cert. in profl. banking, Am. Inst. of Banking of Am. Bankers Assn., 1988.; cert. Asian Langs., NYU, 1992. Actuarial asst. New Eng. Mut. Life Ins. Co., Boston, 1965-66; asst. actuary Mass. Gen. Life Ins. Co., Boston, 1966-68; actuarial assoc. John Hancock Mut. Life Ins. Co., Boston, 1968-71; asst. actuary George B. Buck Cons. Actuaries, Inc., N.Y.C., 1971-75, Martin E. Segal Co., N.Y.C., 1975-80; actuary Laiken Siegel & Co., N.Y.C., 1980; cons. Bklyn., 1981—; timesharing and database analyst banklink corp. cash mgmt. div. Chem. Bank N.Y.C., 1983-85; programmer analyst Empire Blue Cross and Blue Shield, N.Y.C., 1986-88, Mt. Sinai Med. Ctr., N.Y.C., 1988-89, French Am. Banking Corp. (subs. Banque National de Paris), N.Y.C., 1989; sr. programmer analyst Dean Witter Reynolds, Inc., N.Y.C., 1989-92; computer specialist for software N.Y.C. Dept. Fin., 1992—; enrolled actuary U.S. Fed. Pension Legis. Bklyn., 1976—. Contbr. articles to profl. jours. With U.S. Army, 1961-64. Mem. AAAS, MLA, Soc. Actuaries (fellow), Practising Law Inst., Assn. Computing Machinery, IEEE Computer Soc., Am. Assn. Artificial Intelligence, Linguistic Soc. Am., Assn. Computational Linguistics, Am. Math. Soc., Math. Assn. Am., Nat. Model R.R. Assn. (life), Nat. Ry. Hist. Soc., Ry. and Locomotive Hist. Soc. (life), Bklyn. Heights Assn., Met. Opera Guild, Am. Friends of Covent Garden, Harvard Gra. Soc., Am. Legion, Phi Beta Kappa, others. Home: 135 Willow St Brooklyn NY 11201-2255

PENNOCK, DONALD WILLIAM, mechanical engineer; b. Ludlow, Ky., Aug. 8, 1915; s. Donald and Melvin (Evans) P.; B.S. in M.E., U. Ky., 1940, M.E., 1948; m. Vivian C. Kern, Aug. 11, 1951; 1 son, Douglas. Stationary engring., constrn. and maintenance Schenley Corp., 1935-39; mech. equipment design engr. mech. lab. U. of Ky., 1939; exptl. test engr. Wright Aero. Corp., Paterson, N.J., 1940, 1941, investigative and adv. engr. to personnel div., 1941-43; indsl. engr. Eastern Aircraft, div. Gen. Motors, Linden, N.J., 1943-45; factory engr. Carrier Corp., Syracuse, N.Y., 1945-58, sr. facilities engr., 1958-60, corporate material handling engr., 1960-63, mgr. facilities engring. dept., 1963-66, mgr. archtl. engring., 1966-68, mgr. facilities engring. dept., 1968-78. Staff, Indsl. Mgmt. Center, 1962, midwest work course U. Kan., 1959-67. Mem. munitions bd. SHIAC, 1950-52; trustee Primitive Hall Found., 1985—. Elected to Exec. and Profl. Hall of Fame, 1966. Registered profl. engr., Ky., N.J. Fellow Soc. Advancement Mgmt. (life mem., nat. v.p. material handling div. 1953-54); mem. ASME, NSPE, Am. Material Handling Soc. (dir. 1950-57, chmn. bd., pres. 1950-52), Am. Soc. Mil. Engrs., Am. Mgmt. Assn. (men. packaging council 1950-55, life mem. planning council), Nat. Material Handling Conf. (exec. com. 1951), Found. N.Am. Wild Sheep (life), Internat. Platform Assn., Tau Beta Pi. Protestant. Mng. editor Materials Handling Engring. (mag. sect.), 1949-50; mem. editorial adv. bd. Modern Materials Handling (mag.), 1949-52. Contbr. articles to tech. jours. Contbg., cons. editor: Materials Handling Handbook, 1958. Home: 24 Pebble Hill Rd Syracuse NY 13214-2406

PENNYPACKER, BARBARA WHITE, plant pathologist; b. Phila., Mar. 3, 1946; d. Jonathan Winborn Jr. and Rosalind (Christman) W.; m. Stanley Paul Pennypacker, Dec. 29, 1971; 1 child, Kathryn Rose. BS, Pa. State U., 1968, MS, 1971, PhD, 1991. Rsch. asst. Dept. Plant Pathology, Pa. State U., University Park, 1968-75, rsch. assoc., 1975-86, grad. rsch. asst., 1986-91; rsch. assoc. Dept. Agronomy, Pa. State U., University Park, 1991—. Author: The Role of Mineral Nutrition in the Control of Verticillium Wilt, 1989, Anatomical Changes Involved in the Pathogenesis of Plants by Fusarium, 1981; assoc. editor Phytopathology, 1993-96; contbr. articles to Phytopathology and Agronomy Jour., Canadian Jour. of Plant Pathology, Plant Disease. Vol. Spl. Olympics, University Park, 1992; lector Blessed Kateri Tekakwitha Cath. Ch., Spring Mills, Pa., 1983—, eucharistic min., 1987—, choir dir., 1983-86. Recipient Richard R. Hill Achievement award N.Am. Alfalfa Improvement Conf., 1992; postdoctoral fellowship grantee USDA Nat. Rsch. Initiative Competitive Grant Program, 1991, rsch. grantee Richard C. Storkan Found., 1990; recipient Outstanding Paper of Yr. award U.S. Reg. Pasture Rsch. Lab. USDA-ARS, North Atlantic Area, 1992. Mem. Crop Sci. Soc. Am., Am. Soc. Agronomy, Am. Phytopathol. Soc., Pa. Forage and Grassland Coun., Gamma Sigma Delta, Sigma Xi, Phi Epsilon Phi. Democrat. Achievements include discovery that resistant alfalfa can act as a symptomless carrier of verticillium wilt, mechanisms of pathogenesis of verticillium albo-atrum in alfalfa, resistance to verticillium albo-atrum in alfalfa is stable under drought. Office: Agronomy Dept Pa State U 245 Agrl Scis & Inds Bldg University Park PA 16802

PENRAAT, JAAP, architect; b. Amsterdam, Apr. 11, 1918; came to U.S. 1958; s. Gerrit Johannes and Maria (Leenslag) P.; m. Ottoline Henriette Gerarde Jongejans, Oct. 17, 1950; children: Marjolyn Renee, Mirjam Ruth, Noelle Denise. Grad., Acad. for Art, Amsterdam, 1932-37. Registered architect, Netherlands, N.Y. Founder Jaap Penraat Assocs., Leeds, N.Y.C., 1958—; lectr. in field. Contbr. numerous articles to profl. jours. throughout the world. Recipient Italian Carrara prize. Mem. Am. Inst. Graphic Arts, Am. Cotton Batting Inst., Package Design Coun. Achievements include 12 patents. Home: PO Box 395 Leeds NY 12451 Office: Jaap Penraat Assocs 315 Central Park W New York NY 10025

PENROSE, ROGER, mathematics educator; b. Colchester, Essex, England, Aug. 8, 1931; s. Lionel Sharples; m. Joan Isabel Wedge (marriage dissolved, 1981); three children: m. Vanessa Dee Thomas, Apr. 19, 1988. BS in Math., U. London; PhD, St. John's Coll., Cambridge, London, 1975. Asst. lectr. math. Bedford Coll., London, 1956-57; research fellow St. John's Coll., Cambridge, 1957-60; NATO research fellow Princeton U. and Syracuse U., 1959-61; research assoc. King's Coll., London, 1961-63; vis. assoc. prof. U. Tex., Austin, 1963-64, reader, 1964-66, prof. applied math., 1966-73; Lovett prof. Rice U., Houston, 1983-87; Rouse Ball prof. math. U. Oxford, 1987—; vis prof. Yeshiva U., Princeton U., Cornell U., 1966-67, 69. Author: Techniques of Differential Topology in Relativity, 1973; co-author: (with W. Rindler): Spinors and Space-time, vol. 1, 1984, vol. 2, 1986, The Emperor's New Mind, 1989; contbr. many articles to sci. jours. Recipient Adams prize Cambridge U., 1966-67, Dannie Heineman prize Am. Physics. Soc. and Am. Inst. Physics, 1971, Eddington medal Royal Acad. Sci., 1975, Royal medal Royal Soc., 1985, Wolf prize in Physics, 1988, Dirac medal and prize, IOP, 1989. Mem. London Math. Soc., Cambridge Philos. Soc., Inst. for Math. and its Applications, Internat. Soc. for Gen. Relativity and Gravitation. Avocations: reading sci. fiction, 3 dimensional puzzles, miniature stone carving, piano. Office: Oxford U, Math Inst, 24-29 St Giles, Oxford OX1 3LB, England

PENRY, DEBORAH L., biologist, educator; b. Fort Dix, N.J., Feb. 28, 1957. BA in Biol. Sci. with High Honors and Distinction, U. Del., 1979; MA in Marine Sci., Coll. William and Mary, 1982; PhD in Oceanography, U. Wash., 1988. Rsch. asst. dept. invertebrate ecology Va. Inst. Marine Sci., Gloucester Point, 1979-82; rsch. assoc. U.S. Dept. Energy program brine disposal monitoring McNeese State U., Lake Charles, La., 1982-83; lab. chemist Core Labs., Inc., Lake Charles, 1982-83; rsch. asst. Sch. Oceanography U. Wash., Seattle, 1983-88, postdoctoral rsch. assoc., 1988-90; rsch. assoc. Horn. Point Lab. U. Md., Cambridge, 1990-92; asst. prof. dept. integrative biology U. Calif., Berkeley, 1991—; adj. asst. prof. Coll. Marine Studies U. Del., Lewes, 1990—. Contbr. articles to profl. jours. Recipient Alan T. Waterman award NSF, 1993, Young Investigator award; fellow NSF; H. Rodney Sharp scholar U. Del., Whitson scholar U. Wash.; grantee U. Wash. Mem. AAAS, Am. Geophysical Union, Am. Soc. Limnologists and Oceanographers, Am. Soc. Naturalists, Oceanography Soc., Phi Beta Kappa, Phi Kappa Phi, Beta Beta Beta. Office: Univ of Calif Dept of Integrative Biology Berkeley CA 94720*

PENRY, JAMES KIFFIN, physician, neurology educator; b. Denton, N.C., Aug. 21, 1929; s. Robert Lee and Addie Cordelia (Leonard) P.; m. Sarah Doub, Mar. 20, 1955; children: Martin D., Denny K., Edith C. BS magna cum laude, Wake Forest U., 1951, MD, 1955. Diplomate Am. Bd. Psychiatry and Neurology. Commd. 2d lt. USAF, 1955, advanced through grades to maj., 1964; rotating intern Pa. Hosp., Phila., 1955-56; asst. resident in internal medicine and neurology N.C. Bapt. Hosp., Winston-Salem, 1956-58; asst. in neurology Bowman Gray Sch. Medicine, Wake Forest U., 1957-58; asst. resident, then resident in neurology Boston City Hosp., 1958-60; fellow in neurology Harvard Med. Sch., Cambridge, Mass., 1959-60; neurologist, chief neurology svc. USAF Hosp., Maxwell AFB, Ala., 1960-62; chief neurology svc. USAF Hosp., Tachikawa, Japan, 1962-65, Andrews AFB, Md., 1965-66; resigned, 1966; dir. USPHS, Bethesda, Md., 1966-79; dir. Neurol. Disorders Program, chief epilepsy br. Nat. Inst. Neurol. and Communicative Disorders and Stroke, NIH, 1975-79; prof. neurology, assoc. dean for rsch. devel. Bowman Gray Sch. Medicine, Wake Forest U., 1979-89, prof. neurology, sr. assoc. dean for rsch. devel., 1989—; pres. Epilepsy Inst. N.C.; hon. prof. U. Santo Domingo, Dominican Republic, 1987; dir. Comprehensive Epilepsy Ctr., N.C. Bapt. Hosp., Winston-Salem, 1989—; cons. FDA, 1975-80, Nat. Inst. Neurol. and Communicative Disorders and Stroke, 1979-90, Burroughs Wellcome Co., 1979—; lectr. in field. Author 30 books and monographs; co-author: (with J.R. Lacy) Infantile Spasms, 1976; (with M.E. Newmark) Photosensitivity and Epilepsy: A Review, 1979, Genetics of Epilepsy: A Review, 1980; editor: Epilepsy: The Eighth International Symposium, 1977; co-editor: (with others) Antiepileptic Drugs, 1972, 3d edit., 1989, Experimental Models of Epilepsy, 1972, Antiepileptic Drugs: Quantitative Analysis and Interpretation, 1978, Advances in Epileptology: (Xth-XIIth) Epilepsy International Symposium, 1980-82, Genetic Basis of the Epilepsies, 1982, Antiepileptic Drug Therapy in Pediatrics, 1983, Idiosyncratic Reactions to Valproate: Clinical Risk Patterns and Mechanism of Toxicity, 1992; editor Merritt-Putnam epilepsy clin. series, 1982-83; mem. editorial bd. Forefronts of Neurology, 1979—, Metabolic Brain Disease, 1986—; mem. internat. adv. bd. Advances in Neurology; contbr. numerous articles to profl. jours., book chpts. Decorated Meritorious Svc. medal; recipient Disting. Alumnus award Bowman Gray Sch. Medicine, 1975, 80, 90. Fellow Am. Acad. Neurology; mem. AMA, Am. Neurol. Assn., Am. Epilepsy Soc. (membership com. 1970-71, chmn. constn. com. 1971-72, coun. 1971-73, program com. 1973, nat. pres. 1974, mem. long range planning com. 1978-79, William G. Lennox award 1980), Am. Neurol. Assn., Assn. for Rsch. in Nervous and Mental Disease, Am. Electroencephalographic Soc., Ea. Assn. Electroencephalographers, Epilepsy Found. Am. (hon. life, Disting. Svc. award 1975, 25 Yrs. Svc. award 1993, Pearce Bailey award 1979, bd. dirs. 1989—), Phi Beta Kappa, Alpha Omega Alpha. Office: Wake Forest U Bowman Gray Sch Medicine Medical Center Blvd Winston Salem NC 27157-1023*

PENSE, ALAN WIGGINS, metallurgical engineer, academic administrator; b. Sharon, Conn., Feb. 3, 1934; s. Arthur Wilton and May Beatrice (Wiggins) P.; m. Muriel Drews Taylor, June 28, 1958; children—Daniel Alan, Steven Taylor, Christine Muriel. B.Metall. Engring., Cornell U., 1957; M.S., Lehigh U., 1959, Ph.D., 1962. Research asst. Lehigh U., Bethlehem, Pa., 1957-59, instr., 1960-62, asst. prof., 1962-65, assoc. prof., 1965-71, prof., 1971—, chmn. dept. metallurgy and materials engring., 1977-83, assoc. dean Coll. Engring. and Applied Scis., 1984-88, dean, 1988-90, v.p., provost, 1990—; assoc. dir. Ctr. Advanced Tech. for Large Structural Systems NSF, 1986-89; cons. adv. com. on reactor safeguards NRC, 1965-86. Author: (with R.M. Brick and R.B. Gordon) Structure and Properties of Engineering Materials, 4th edit, 1978; also articles. Recipient Robinson award Lehigh U., 1965, Stabler award, 1972; Danforth fellow, 1974-86. Fellow Am. Soc. metals, Am. Welding Soc. (William Spraragan award 1963, Adams membership award 1966, Jennings award 1970, Adams lectr. 1980, William Hobart medal 1982, hon. fellow 1991); mem. ASTM, Am. Soc. Engring. Edn. (Western Elec. award 1968), Internat. Inst. Welding, Nat. Acad. Engring. Republican. Evang. Congregationalist (pres. bd. trustees Evang. Sch. Theology). Home: 2227 West Blvd Bethlehem PA 18017-5025 Office: Lehigh U Alumni Bldg 27 Bethlehem PA 18015

PENTAS, HERODOTOS ANTREAS, civil engineer; b. Nicosia, Cyprus, June 8, 1955; came to U.S., 1982; s. Antreas John and Keti (Herodotou) P.; m. Maria Pyrgou, Dec. 28, 1975; children: Keti, Nicholas, Andry-Ann, Georgiana. BSCE cum laude, U. Ala., 1984, MSCE, 1985; PhD, La. State U., 1990. Registered profl. engr., La. Plant engr. J&P Civil Engrs. Ltd., Dubai, United Arab Emirates, 1978-82; rsch. assoc. U. Ala., Birmingham, 1985-86; civil engr. La. State Dept. Transp., Baton Rouge, 1990-93; sr. structural engr. Dames & Moore, Baton Rouge, 1993—. Bd. mem. Greek Orthodox Ch., Baton Rouge, 1992. Mem. ASCE, La. Soc. Profl. Engrs., Tau Beta Pi, Phi Kappa Phi. Achievements include research in design considerations for bridge. Home: 9989 Burbank Dr Apt F100 Baton Rouge LA 70810 Office: Dames & Moore 4949 Essen Ln Ste 900 Baton Rouge LA 70809

PENZIAS, ARNO ALLAN, astrophysicist, research scientist, information systems specialist; b. Munich, Germany, Apr. 26, 1933; came to U.S., 1940, naturalized, 1946; s. Karl and Justine (Eisenreich) P.; m. Anne Pearl Barras, Nov. 25, 1954; children: David Simon, Mindy Gail, Laurie Ruth. BS in Physics, CCNY, 1954; MA in Physics, Columbia U., 1958, PhD in Physics, 1962; Dr. honoris causa, Observatoire de Paris, 1976; ScD (hon.), Rutgers U., 1979, Wilkes Coll., 1979, CCNY, 1979, Yeshiva U., 1979, Bar Ilan U., 1983, Monmouth Coll., 1984, Technion-Israel Inst. Tech., 1986, U. Pitts., 1986, Ball State U., 1986, Kean Coll., 1986, U. Pa., 1992, Ohio State U., 1988, Iona Coll., 1988; Drew U., 1989; ScD (hon.), Lafayette Coll., 1990, Columbia U., 1990, George Washington U., 1992, Rensselaer Univ., 1992, U. Pa., 1992. Mem. tech. staff Bell Labs., Holmdel, N.J., 1961-72, head radiophysics rsch. dept., 1972-76; dir. radio research lab. Bell Labs., 1976-79, exec. dir. rsch., communications scis. div., 1979-81, v.p. rsch., 1981—, now exec. dir. rsch. physics divsn.; bd. dirs. A.D. Little; adj. prof. earth and scis. SUNY, Stony Brook, 1974-84, Univ. Disting. lectr., 1990; lectr. dept. astrophys. scis. Princeton U., 1967-72, vis. prof., 1972-85; rsch. assoc. Harvard Coll. Obs., 1968-80; Edison lectr. U.S. Naval Rsch. Lab., 1979; Kompfner lectr. Stanford U., 1979; Gamow lectr. U. Colo., 1980; Jansky lectr. Nat. Radio Astronomy Obs., 1983; Michelson Meml. lectr., 1985; Grace Adams Tanner lectr., 1987; Klopsteg lectr. Northwestern U., 1987; grad. faculties alumni Columbia U., 1987-89; Regents' lectr. U. Calif., Berkeley, 1990; Lee Kuan Yew Disting. visitor Nat. U. Singapore, 1991; mem. astronomy adv. panel NSF, 1978-79, mem. indsl. panel on sci. and tech., 1982—, disting. lectr., 1987, affiliate Max-Planck Inst. für Radioastronomie, 1978-85, chmn. Fachbeirat, 1981-83; researcher in astrophysics, info. tech., its applications and impacts. Author: Ideas and Information Managing in a High-Tech World, 1989 (pub. in 10 langs.); mem. editorial bd. Ann. Rev. Astronomy and Astrophysics, 1974-78; mem. editorial bd. AT&T Bell Labs. Tech. Jour., 1978-84, chmn. 1981-84; assoc. editor Astrophys. Jour., 1978-82; contbr. over 100 articles to tech. jours.; several patents in field. Trustee Trenton (N.J.) State Coll., 1977-79; mem. bd. overseers U. Pa. Sch. Engring. and Applied Sci., 1983-86; mem. vis. com. Calif. Inst. Tech., 1977-79; mem. Com. Concerned Scientists, 1975—, vice chmn., 1976—; mem. adv. bd. Union of Couns. for Soviet Jews, 1983—; bd. dirs. IMNET, 1986-91, Coun. on Competitiveness, 1989—. Served to lt. Signal Corps, U.S. Army, 1954-56. Named to N.J Lit, Hall of Fame, 1991; recipient Herschel medal Royal Astron. Soc, 1977, Nobel prize in Physics, 1978, Townsend Harris medal CCNY, 1979, Newman award, 1983, Joseph Handleman prize in the scis., 1983, Grad. Faculties Alumni award Columbia U., 1984, Achievement in Science award Big Brothers Inc., N.Y.C., 1985, Priestly award Dickinson Coll., 1989, Pender award U. Pa., 1992. Mem. NAE, NAS (Henry Draper medal 1977), AAAS, IEEE (hon.), Am. Astron. Soc., Am. Phys. Soc. (Pake prize 1990), Internat. Astron. Union, World Acad. Arts and Sci.

PEPE, TERI-ANNE, development chemist; b. Oakland, N.J., Sept. 9, 1967; d. David and Anne Marie Gardner. BS in Physics, Fairfield U., 1989; MS in Chemistry, Fairleigh Dickinson U., 1993, MBA in Fin., 1993. Cert. profl. tutor chemistry, physics, math., N.J., Conn., N.Y. Analytical chemist L&F Products, Inc., Montvale, N.J., 1989-91; sr. devel. chemist new products group Lever Bros. Co., Edgewater, N.J., 1991—; summer intern detergents lab L&F Products, Inc., Montvale, 1987, summer intern analytical lab., 1988, name change task force mem., 1990; judge Regional Sci. Fair, Hackensack, N.J., 1990-91. Advt. mgr. Mirror Newspaper, Fairfield, Conn., 1987-89. Recipient award for piano excellence Trinity Coll. of London, achievement award for outstanding acads. in physics, 1987. Mem. Am. Chemical Soc., Am. Oil Chemists Soc., Am. Inst. Physics, Liberty Sci. Ctr. (edn. com.), Sigma Pi Sigma, Pi Mu Epsilon. Achievements include development of explanatory video on performance review sessions for use in L&F Leadership Development Program. Home: 57 Calumet Ave Oakland NJ 07436-2902

PEPIN, JOHN NELSON, materials research and design engineer; b. Lowell, Mass., June 5, 1946; s. Nelson Andre and Leanne Florine (Boucher) P. BS in Mech. Engring., Northeastern U., 1968; MS in Aerospace Engring., MIT, 1970. Aero. engr. Bradway STOL Amphibian Ltd., Raymond, Maine, 1979; staff engr. Fiber Materials, Inc., Biddeford, Maine, 1979-84; pres. Pepin Assocs., Inc., Scarborough, Maine, 1984—; cons. Foster-Miller Engrs., Waltham, Mass., 1985—, Johnson & Johnson Orthopedic Div., Braintree, Mass., 1984-86, Allied Signal Aerospace, South Bend, Ind., 1985—, B.F. Goodrich, Akron, Ohio and Marlboro, Mass., 1986-87. Patentee in field. Capt. USAF, 1974-78. NSF grantee, 1982-84, U.S. Dept. Transp. grantee, 1989—, U.S. Dept. of Energy grantee, 1990—. Mem. Soc. for Advancement of Materials and Process Engring., Seaplane Pilots Assn., MIT Club of Maine (bd. dirs. Portland chpt. 1988). Democrat. Roman Catholic. Avocations: volleyball, cross-country skiing, tennis, canoeing, camping. Home: RR 1 Box 109N Alfred ME 04002-9801

PEPLINSKI, DANIEL RAYMOND, project engineer; b. Chgo., Sept. 23, 1951; s. Alex J. and Florence (Turkowski) P. BS in Chemistry, Loyola U., Chgo., 1973; MS in Chemistry, Ill. Inst. Tech., 1977. Rsch. asst. Rsch. Inst., ADA, Chgo., 1978; rsch. assoc. The Aerospace Corp., El Segundo, Calif., 1978-80, mem. tech. staff, 1980-90, project engr., 1990—. Contbr. articles to profl. jours. Mem. Am. Phys. Soc., Soc. Info. Display. Republican. Roman Catholic. Avocations: mountain biking, reading. Office: The Aerospace Corp Mail Sta M5/722 PO Box 92957 Los Angeles CA 90009-2957

PEPPER, DAVID J., electrical engineer; b. Washington, July 12, 1962; s. William H. and Marian Pepper. BSEE, Lehigh U., 1984; MSEE, Ga. Tech., 1985, PhD, 1990. Engr.-in-tng., Ga. MTS AT&T Bell Labs., Whippany, N.J., 1984-88; GRA Ga. Tech., Atlanta, 1987-90; mem. tech. staff Bellcore, Morristown, N.J., 1991—. Mem. disaster action team ARC, Atlanta, 1984, 87-91; instr. trainer FA/CPR, ARC, Madison, N.J., 1989—; mem. EMT, Madison Vol. Ambulance Corps, 1991—. Mem. Tau Beta Pi, Eta Kappa Nu. Home: 89 North St Apt 2 Madison NJ 07940 Office: Bellcore Rm 2E-261 445 South St Morristown NJ 07962

PEPPER, DOROTHY MAE, nurse; b. Merill, Maine, Oct. 16, 1932; d. Walter Edwin and Alva Lois (Leavitt) Stanley; m. Thomas Edward Pepper, July 1, 1960; children: Walter Frank, James Thomas. RN, Maine Med. Ctr. Sch. Nursing, Portland, 1954. RN, Calif. Pvt. duty nurse Lafayette, Calif.; staff nurse Maine Med. Ctr., Portland, 1954-56, Oakland (Calif.) VA Hosp., 1956-58; pvt. duty nurse, dir. RN's Alameda County, Oakland. Mem. Profl. Nurses Bur. Registry. Avocation: writing.

PEPPER, WILLIAM BURTON, JR., aeronautical engineer, consultant; b. Montrose, Colo.; s. William B. and Grace P. (McGinnis) P.; m. Doris Jernigan, June 11, 1971. BS in Aeronautical Engring., U. Minn., 1946; MS in Aeronautical Engring., U. Colo., 1957. Registered profl. engr. N.Mex. Aeronautical rsch. scientist NASA, Langley Field, Va., 1949-52; mem. tech. staff Sandia Nat. Lab., 1953-85; cons. Pepper Consulting Co., Albuquerque, 1985—; cons. Geo-Ctrs., Albuquerque, 1986, Pioneer Systems, Melborne, Fla., 1988; cons. to chief engr. Irvin Inds., Santa Ana, Calif., 1989-91. Contbr. over 120 tech. papers. Lt. (j.g.) USN. Fellow AIAA (assoc., sect. chmn. 1986). Achievements include patent on disappearing parachute. Home: 3809 Parsifal NE Albuquerque NM 87111

PERATT, ANTHONY LEE, electrical engineer, physicist; b. Belleville, Kans., Feb. 26, 1940; s. Galvin Ralph and Arlene Frances (Friesen) P.; m Glenda Delores White, Dec. 19, 1966; children: Sarah, Galvin, Mathias. BSEE, Calif. State Poly. U., 1963; MSEE, U. So. Calif., L.A., 1967, PhD, 1971. Staff The Aerospace Corp., El Segundo, Calif., 1971-72, Lawrence Livermore (Calif.) Nat. Lab., 1972-79; guest scientist Max Planck Inst. fur Plasmaphysik, Garching, Germany, 1975-77; sr. scientist Maxwell Labs., San Diego, 1979-81; staff Los Alamos (N.Mex.) Nat. Lab., 1981—; vis. scientist Royal Inst. Tech., Stockholm, 1985, 88; adv. bd. Mus. Sci. and Industry, Chgo., 1990—; mem. Excom, IEEE Nuclear and Plasma Sci. Soc., 1987, 88, 89, 90; gen. chmn. IEEE Internat. Conf. on Plasma Sci., 1994. Author: Physics of the Plasma Universe, 1992; editor IEEE Transactions on Plasma Sci. Jour., 1986, 89, 90, 92, Laser and Particle Beams, 1988. Recipient Award of Excellence, Dept. Energy, 1987. Mem. IEEE, Am. Phys. Soc., Am. Astron. Soc., Eta Kappa Nu. Achievements include coining of term "plasma universe"; research in modeling magnetic fields in galaxies with 3D particle-in-cell simulations and prediction of bisymmetric magnetic fields in spiral galaxies. Office: Los Alamos Nat Lab MS-D406 Los Alamos NM 87545

PERCHIK, BENJAMIN IVAN, operations research analyst; b. Passaic, N.J., May 3, 1941; s. Morris and Frances (Antman) P.; m. Ellen Mae Colwell, Aug. 25, 1963; children: Joel, Dawn. BA, Rutgers U., 1964; postgrad. N.Y. Inst. Tech., 1964-65. Quality control E.R. Squibb Corp., New Brunswick, N.J., 1964-67; edn. specialist Signal Sch., Ft. Monmouth, N.J., 1967-74; edn. specialist Armor Sch., Ft. Knox, Ky., 1974-75, ops. rsch. analyst, 1975-78; ops. rsch. analyst HQ TRADOC, Ft. Monroe, Va., 1978-80; ops. rsch. analyst Army Materiel Command, Alexandria, Va., 1980—, exec. officer USAREUR ORSA Cell, 1988-90; chmn. supervisory com. credit union, 1985-88, 91—; cons. Delta Force, Carlisle Barracks, Pa., 1982-84, Internat. Policy Inst., 1983-85, World Future Soc., 1982—; nat. coord. Mensa Investment SIG, 1983—; coord. econ. forecasting group Met. Washington Mensa, 1983—; chmn. security com. Watergate at Landmark, 1985-88. Author: ADP Program and Repair, 1972; writer, editor, pub. internat. newsletter Speculation and Investments, 1983—. Chmn. credit com. Darcom Fed. Credit Union, 1982-85. Mem. Inst. Mgmt. Scis., Internat. Platform Assn., Ops. Rsch. Soc. Am. Club: Old Dominion Boat. Office: 5001 Eisenhower Ave Alexandria VA 22333-0002

PERCUS, JEROME KENNETH, physicist, educator; b. N.Y.C., June 21, 1926; s. Philip M. and Gertrude B. (Schweiger) P.; m. Ora Engelberg, May 20, 1965; children: Orin, Allon. B.S.E.E., Columbia U., 1947, M.A., 1948, Ph.D., 1954. Instr. elec. engring. Columbia U., N.Y.C., 1952-54; asst. prof. Stevens Inst. Tech., Hoboken, N.J., 1955-58; assoc. prof. NYU, N.Y.C., 1958-65; prof. physics NYU, 1965—; dir. Nat. Biomed. Research Found. Author: Many-Body Problem, 1963, Combinatorial Methods, 1971, Combinatorial Methods in Developmental Biology, 1977, Mathematical Methods in Developmental Biology, 1978, Mathematical Methods in Enzymology, 1984, Lectures on the Mathematics of Immunology, 1986; editor: Pattern Recognition, Jour. Statis. Physics, Jour. Comp. Molecular Biology. With USN, 1944-46. Recipient Pregel Chemistry Physics award N.Y. Acad. Scis., 1975, Joel Henry Hildebrand award in the Theoretical and Exptl. Chemistry of Liquids, Am. Chem. Soc., 1993, Pattern Rec. Soc. award, 1992. Fellow AAAS, Am. Phys. Soc.; mem. Am. Math. Soc., Sigma Xi. Office: NYU 251 Mercer St New York NY 10012-1185

PERDANG, JEAN MARCEL, astrophysicist; b. Remerschen, Luxembourg, May 3, 1940; s. Léandre Eugène and Erica Marguerite (Schneider) P.; m. Esmée Marie-Jeanne Ewert, July 15, 1965; children: Pascale, Paul, Patrick. Licence in Physical Scis., U. Liège, Belgium, 1963, Cert. in Statis. Mechanics, 1968, Doctorat in Physical Scis., 1969. Asst. U. Liège, 1963-64; aspirant U. Liège, 1964-69, chargé de recherches, 1970-74, researcher, 1974—, maître de conférences, 1978—; chargé de cours extraordinaire U. Officielle du Burundi, Bujumbura, 1971; libero docente U. Padova, Padua, Italy, 1974, 75, 78; vis. prof. U. Fla., Gainesville, 1978, 81, 84, Internat. Sch. for Advanced Studies, Trieste, Italy, 1982; astronomer Observatoire de Paris à Meudon, 1989, 91; rschr. U. Montreal, 1971, Columbia U., 1974, Inst. for Theoretical Physics, U. Calif., 1990; mem. Com. Nat. Belge de Géophysique, 1964. Author: Stellar Oscillations: The Asymptotic Approach; co-editor: Chaos in Astrophysics, Applying Fractals in Astronomy; author numerous papers in field. Recipient postdoctoral fellow NSF, 1969, Leverhulme fellow in astronomy, 1972-73, European Exchange fellowships Royal Soc. Belgian Fonds National de la Recherche Scientifique, 1979-80, 83, 84, 85, 86, 87, 88, 89, 90, 91. Fellow Royal Astron. Soc.; mem. Internat. Astron. Union, Com. Belge d'Astronomie (assoc.), Inst. Grand Ducal Sect. des Scis. (hon.). Avocations: philosophy and history of science. Office: Inst d'Astrophysique, 5 Avenue de Cointe, B-4000 Liège Belgium

PERDUE, PAMELA PRICE, computer engineer; b. Richmond, Va., Mar. 24, 1962; d. Clyde C. and Lois W. Price; m. Paul S. Perdue, Sept. 7, 1992; 1 child, Brandon. BS, Va. Commonwealth U., 1984; BA, N.C. Wesleyan U., 1985. Sr. account mgr. Signet Data Systems, Midlothian, Va., 1984-85; quality assurance adminstr. Baker Equipment Engring., Richmond, 1985-87; computer systems engr. Dept. Corrections, Richmond, 1987-90; systems analyst Dept. Youth, Richmond, 1990-92, computer systems sr. engr., 1992—; cons. Dept. Corrections, Richmond, 1987-92. Mem. Office Automation Soc. Republican. Baptist. Office: Dept Youth Family Svcs PO Box 1110 Richmond VA 23208-1110

PEREGRINE, DAVID SEYMOUR, astronomer, consultant; b. Telluride, Colo., June 9, 1921; s. William David and Ella Bethea (Hanson) P. AB, UCLA, 1950; postgrad., U. Calif., Berkeley, 1956-59. Leadman N.Am. Aviation, Inglewood, Calif., 1940-44; sr. physicist N.Am. Aviation, Downey, Calif., 1960-66; photogrammetric cartographer U.S. Geol. Survey, Denver, 1950-56; exec. and sci. specialist space div., Chrysler Corp., New Orleans, 1966-68; cons. Denver, 1970—. Co-author: (environ. manuals) Moon, 1963, Mars, 1965, Venus, 1965. Served with U.S. Army, 1944-46, PTO. Mem. Am. Astron. Soc., Am. Soc. Photogrammetry, Sigma Xi. Home: 190 S Marion Pky Denver CO 80209-2526

PEREIRA, JOSE FRANCISCO, plant physiologist; b. San Jose, Costa Rica, May 29, 1935; came to Venezuela, 1964; s. Jose Maria and Ramona Clemencia (Marin) P.; m. Aura Felicia Ugarte, Dec. 14, 1963 (div. May 1976); 1 child, Jose Francisco; m. Ludmila Balevich. Ingeniero Agronomo, U. Costa Rica, San Jose, 1960; MS, U. Calif., Davis, 1961; PhD, U. Ill., 1970. Plant physiologist U. Costa Rica, San Jose, 1960-64; founder prof. U. Oriente, Monagas, Venezuela, 1964, prof., 1977—, dean grad. studies, 1977-87, cons. Cromoflora Co., Caripe, Monagas, Venezuela, 1988—; vis. prof. U. Ill., Urbana-Champaign, 1985; vis. scholar U. Calif., Davis, 1986. Mem. AAAS, Venezuelan Assn. Advancement Sci., N.Y. Acad. Sci., Sigma Xi. Avocations: agricultural economics, classical music, computers. Office: Cromoflora Co, La Placeta, Caripe 6210, Venezuela

PERERA, UDUWANAGE DAYARATNA, plant reliability specialist, engineer; b. Homagama, Sri Lanka, Jan. 13, 1945; arrived in Eng., 1970; s. Uduwanage Pabilis and Podinona (Jayatunga) P.; m. Seetha Wijetunga, May 11, 1978; children: Nirosha, Kapila. BSc with honors, Hatfield (Eng.) Poly., 1973; MSc, Aston U., Birmingham, Eng., 1974; PhD, Bradford (Eng.) U., 1983. Cert. engring. Technician Ceylon Steel Corp., Athurugiriya, Sri Lanka, 1966-69; quality engr. Alfred Herbert Machine Tools Ltd., Coventry, Eng., 1974-80; reliability engr. Brit. Gas Corp., London, 1980-85; plant reliability specialist IBM (UK) Ltd., Havant, Eng., 1985—. Contbr. papers to tech. lit. Mem. Instn. Prodn. Engrs., Soc. Environ. Engrs. (com. mem. product assurance group 1987—), IEEE (reliability soc.). Buddhist. Avocations: cricket, snooker, reading, gardening. Home: 9 Marigold Close, Fareham PO15 5HF, England Office: IBM UK Ltd, Customer Assurance, PO Box 6, Hampshire PO9 1SA, England

PEREZ, JEAN-YVES, engineering company executive; b. 1945. Ingenieur Civil Engring., Ecole Centrale des Arts et Manufactures, Paris, 1970; MS, U. Ill., 1971. With Soletanche Enterprise, 1972—; pres. Woodward-Clyde Cons., Denver. With French Air Force, 1971-72. Office: Woodward-Clyde Cons 4582 S Ulster St Ste 600 Denver CO 80237-2637*

PEREZ, JOHN CARLOS, biology educator; b. Park City, Utah, Apr. 29, 1941; s. John Cano and Elma May (Ivie) P.; m. Patsy May, Oct. 26, 1963; 9 children. B.S. in Molecular and Genetic Biology, U. Utah, 1967; M.A. in Zoology, Mankato State Coll., Minn., 1972; Ph.D. in Bacteriology, Utah State U., 1972. Research assoc. in bacteriology Utah State U., Logan, 1970-72; asst. prof. biology Tex. A&I U., Kingsville, 1972-75, a prof., dir. MBRS-NIH program, 1975-85, prof., 1981—. Contbr. articles to profl. jours. Served with U.S. Army, 1960-63. Named Outstanding Citizen of Kingsville, 1976; recipient Disting. Research award Tex. A&I U., 1979; recipient numerous grants for research and teaching, 1976—. Mem. Am. Soc. Microbiology, Internat. Soc. Toxinology, Tex. Assn. Coll. Tchrs., Sigma Xi. Mormon (stake pres.). Office: Tex A&I U Biology Dept Campus Box 158 A&I Kingsville TX 78363

PEREZ, JOSE LUIS, computer scientist, engineer, philosopher; b. Bogota, Colombia, Oct. 31, 1951; came to U.S., 1982; s. Jose Alberto and Elvira Paulina (Restrepo) P.; m. Liliana Maria Pelaez, Nov. 1, 1974 (div. 1981); 1 child, Jose David; m. Debra Helene Dambeck, Nov. 27, 1988; 1 child, Andrew Paul. BA in Philosophy and Lit., Rosario Coll., Bogota, 1968; MSCE, Columbia U., 1978; MS in Computer Sci., Maharishi Internat. U., Fairfield, Iowa, 1986. Registered profl. engr., Colombia. V.p. Jose Perez Ins. Co., Bogota, 1975-78; design engr. Acerias Paz del Rio, Belencito, Colombia, 1978-81; internat. financier Jose L. Perez & Co., L.A., 1981-83; dir. engring. lab. Indsl. Electronic Engrs., Van Nuys, Calif., 1983-84; software engr. Planning Rsch. Corp., Washington, 1986-87, Micom Digital Corp., Washington, 1987-88, Glazier Electronics Systems Group (name now Donatech Corp.), Fairfield, 1988—; cons. Maharishi Global Trading Group, Livingston Manor, N.Y., 1984—, Vision Link, N.Y.C., 1989—; founder Creative Capital Corp., 1990; mem. Radio Tech. Commn. for Aeronautics SC-147. Designer comm. and avionics instrument systems, traffic alert and collision avoidance system for airplanes, test access and control interface software for Boeing 777. Mem. Creating Coherence Program for World Peace, Fairfield, 1984—, Colombian Soc. Friends of Country, Bogota, 1969—; founder, pres. Drug Rehab. Found., Fairfield, 1989. Scholar U. Los Andes, Bogota, 1969, City of L.A., 1983. Mem. Radio Tech. Commn. for Aero. SC-147, U.S. Chess Fedn. (master level), Smithsonian Assocs., Toastmasters. Avocations: chess, bicycling, swimming, volleyball, meditation. Office: 116 W Burlington Fairfield IA 52556

PEREZ, LUZ LILLIAN, psychologist; b. Ponce, P.R., Aug. 7, 1946; d. Emiliano and Maria D. (Torres) P.; children: Vantroi, Maireni. BA, Herbert H. Lehman Coll., 1974; PhD, NYU, 1989. Lic. psychologist, N.Y. Staff psychologist Soundview Throgs Neck Community Mental Health Ctr., Bronx, 1980-88; coord. early childhood program Crotona Park COmmunity Mental Health Ctr., Bronx, 1988-91; cons. psychologist Highbridge Adv. Coun. Presch. Program, Bronx, 1991—, Coalition for Hispanic Family Svcs., Bklyn., N.Y., 1991—. Nat. Inst. Mental Health grantee, 1974-77. Mem. Am. Psychol. Assn., Assn. Hispanic Mental Health Profls. Avocations: Flamenco dancing, crochet, embroidery. Office: Highbridge Adv Coun 1181 Nelson Ave Bronx NY 10452

PEREZ, REINALDO JOSEPH, electrical engineer; b. Palm River, Cuba, July 25, 1957; came to U.S., 1975; s. Reinaldo I. and Palminia Ulloa (Rodriguez) P.; m. Madeline Kelly Reilly, Mar. 11, 1989; children: Alexander, Laura-Marie. BSc in Physics, U. Fla., 1979, MSc in Physics, 1981; MScEE, Fla. Atlantic U., 1983, PhD, 1989. Communications engr. Kennedy Space Ctr., NASA, Cape Canaveral, Fla., 1983-84; rsch. engr. jet propulsion lab. JPL, Calif. Inst. Tech., Pasadena, 1988—; instr. engring. UCLA, 1990—. Contbr. articles to profl. publs. Mem. AAAS, IEEE (book rev. editor 1990—), NSPE, Electromagnetic Compatibility Soc. (assoc. editor jour.), Am. Soc. Physics Tchrs., N.Y. Acad. Scis., Applied Computational Electromagnetic Soc. (assoc. editor jour.), Phi Kappa Phi. Republican. Baptist. Avocations: flying, skiing, fishing. Office: JPL Calif Inst Tech 4800 Oak Grove Dr MS: 301-460 Pasadena CA 91109

PEREZ, RONALD A., mechanical engineering educator; b. Jacksonville, Fla., Feb. 18, 1961; s. Mario A. and Lourdes M. Perez. MSME, Purdue U., 1985, PhD, 1990. Registered profl. engr., Wis. Grad. instr. Purdue U., West Lafayette, Ind., 1986-90; asst. prof. mech. engring. U. Wis., Milw., 1990—; cons. Minority High Sch. Apprenticeship Program, Milw., 1991—, Coll. For Kids, Milw., 1991. Contbr. articles to profl. jours. David Ross fellow, PurdueU., 1988-90, Proctor & Gamble fellow, 1986. Mem. AIAA (sr. mem.), IEEE, ASME (assoc. mem.), Sigma Xi. Republican. Roman Catholic. Home: Apt 202 9036 N 75th St Milwaukee WI 53223

PEREZ, VICTOR, medical technologist, laboratory director; b. Arecibo, P.R., Apr. 3, 1930; s. Perez Victor and Cintron Catalina; m. Iraida Quiñones, June 1, 1964; children: Victor Jaime, Victor Lucas, Victor Ramon, Maria Victoria. BS, CAAM, Mayaguez, P.R., 1952; grad. in Med. Tech., Sch. Medicine, San Juan, P.R., 1953. Cert. med. technologist. Med. technologist blood bank Sch. Medicine, San Juan, 1953-57, dir. exptl. surgery, 1957-71; lab. dir. Metro. Hosp. Clin. Lab., Rio Piedres, P.R., 1971—; pres. P.R. Bd. Med. Technologists, 1971-74, Coll. P.R. Med. Technologists, 1976-77, P.R. Bd. Examiners, mem., 1986—. Pres. Club Exchange-Cupey, Rio Piedres, 1960, Parent Tchr. Student Assn., San Jose Coll., Rio Piedres, 1972. Mem. Am. Soc. Med. Technicians, Coll. Med. Technologists (pres. 1976), Am. Assn. Med. Tech., N.Y. Acad. Scis., Am. Assn. Clin. Chemistry, P.R. Coll. Med. Technolgists, Nat. Cert. Agy., Omicron Sigma. Roman Catholic. Home: Sagrado Corazon Brigida # 1705 Rio Piedras PR 00926

PEREZ-RAMIREZ, BERNARDO, biochemist, researcher, educator; b. Valdivia, Chile, Aug. 20, 1960; s. Jose Esteban Perez and Rosenda Ramirez; m. Mariana Castells, Jan. 14, 1989; children: Alexander, Ariana. BS in Cell Biology, U. Austral, Valdivia, Chile, 1983; MS in Biochemistry, Med. Coll. Va., 1987; PhD in Biochemistry, U. Mo., Kansas City, 1990. Teaching asst. U. Mo., Kansas City, 1987-88, rsch. asst., 1988-89; rsch. assoc. Brandeis U., Waltham, Mass., 1990—. Contbr. articles and abstracts to profl. jours. Mem. AAAS, N.Y. Acad. Sci., The Protein Soc., Biophysical Soc. Achievements include development of methods to study the topology of transmembrane proteins; research on structure and function of proteins, membrane-bound receptors, ligand-protein interactions, signal transduction mechanisms, protein chemistry. Office: Brandeis U 415 South St Waltham MA 02154-2700

PERHAM, ROY GATES, III, industrial psychologist; b. Hackensack, N.J., Apr. 22, 1958; s. Roy Gates Jr. and Titania Joan (Robbitts) P. BA with honors, Bates Coll., 1980; MS, Stevens Inst. Tech., 1982, PhD, 1989. Intern Sen. Edmund S. Muskie, Washington, 1978; psychometrician Lab. Psychol. Studies Stevens Inst. Tech., Hoboken, N.J., 1981-83, instr., 1985, adj. asst. prof., 1990—; adj. asst. prof. Fairleigh Dickinson U., Rutherford, N.J., 1986; sr. assoc. AAI Orgnl. Performance Cons., Florham Park, N.J., 1990—; WordStar coord. N.Y. Computer Soc., N.Y.C., 1985-88. Chmn. Juvenile Conf. Com., Hasbrouck Heights and Wood-Ridge, N.J., 1985—; mem. N.J. State Juvenile Delinquency Commn., Trenton, N.J., 1988-91; county exec.'s rep. Bergen County Youth Svcs. Commn., 1990—. Named Citizen of Yr., Lions Club of Hasbrouck Heights, N.J., 1988. Mem. APA, Am. Psychol. Soc., Met. N.Y. Assn. for Applied Psychology, Soc. for Indsl./Orgnl. Psychology, Inc., Phi Beta Kappa, Psi Chi. Home: 269 Raymond St Hasbrouck Heights NJ 07604-1723 Office: AAI Orgnl Performance Cons 23 Vreeland Rd Florham Park NJ 07932

PERHEENTUPA, JAAKKO PENTTI, pediatrician; b. Eurajoki, Finland, Mar. 7, 1934; s. Antti and Suoma (Wikstroem) P.; m. Tuula Baeckman, June 19, 1964; children: Antti, Aaro, Laura, Ilkka, Inna. MD, U. Helsinki, Finland, 1959, D Med. Sci., 1967, Pediatrician, 1967, Pediatric Endocrinologist, 1979. Resident Children's Hosp./Univ. Helsinki, 1963-66, dir., 1984—; assoc. prof. Univ. Helsinki, 1973-84, prof., chmn. in pediatrics, 1984—; fellow in pediatrics The Johns Hopkins U. and Hosp., Balt., 1966-68; vis. assoc., prof. pediatrics Univ. Calif. Med. Sch., L.A., 1982-83. Contbr. articles to profl. jours. Mem. Endocrinol. Soc. Finland (hon.), Estonian Pediatric Assn. (hon.), European Soc. Pediatric Endocrinology (pres. 1981-82), Finnish Pediatric Assn. (pres. 1989-92), Deutsche Gesellschaft Kinderheilkunde (corres. mem.). Office: Childrens Hosp, Stenbackinkatu 11, 00290 Helsinki Finland

PERICH, MICHAEL JOSEPH, medical entomologist consultant; b. Omaha, June 17, 1957; s. Andrew John and Rita Theresa (Kazor) P.; m. Audrey Denise Hare, Aug. 7, 1982; 1 child, Sarah Virginia. BS, Iowa State U., 1979; MS, Okla. State U., 1982, PhD, 1985. Bd. cert. entomologist Med./Vet. Entomology. Chemist Omaha Pub. Power Dist., Fort Calhoun, Nebr., 1979; grad. rsch. asst. Okla. State U., Stillwater, 1979-85, rsch. assoc., 1985-86; rsch. med. entomologist U.S. Army Biomed. Rsch. & Devel. Lab., Frederick, Md., 1986-88; sr. rsch. med. entomologist U.S. Army Biomed. Rsch. & Devel. Lab., Frederick, 1988-92, Walter Reed Army Inst. Rsch., Washington, 1992—; med. entomology cons. Selema (Ala.) U., 1989-91, Va. State U., Petersburg, 1992—. Contbr. articles to Jour. Am. Mosquito Control, Jour. Med. Entomology, Med. Vet. Entomology. Com. mem. Frederick (Md.) County Sch. Sci. Rev. Com., 1991-92. Recipient Exceptional Performance awards U.S. Army, Ft. Detrick, 1989-92. Mem. Am. Soc. Tropical Medicine and Hygiene, Am. Mosquito Control Assn., Am. Soc. Animal Scientists, Entomol. Soc. Am., Sigma Xi (chpt. pres. 1989-91). Democrat. Roman Catholic. Achievements include design and evaluation of alternative vector control methologies for control of malaria, denque and leishmaniasis in the tropics; isolation and evaluation of selected botanical extract for disease vector supression. Home: 6993 Alabaster Ct Middletown MD 21769 Office: Walter Reed Army Inst Rsch Washington DC 21702

PERILSTEIN, FRED MICHAEL, consulting electrical engineer; b. Phila., Oct. 25, 1945; s. Paul Pincus and Adeline Sylvia (Schneyer) P.; m. Abigail Siff, June 13, 1971. BS in Econs., CCNY, 1968; BSEE, Newark Coll. Engring., 1972; MSEE Power, N.J. Inst. Tech., 1977. Registered profl. engr., N.J., Pa., N.Y., Calif. Applications engr. Fed. Pacific Electric Corp., Newark, 1972-78; project mgr. Vector Engring., Inc., Springfield, N.J., 1978-84; pres. Tramlec Corp., Cons. Engrs., Springfield, 1982—; seminar instr. Multi-Amp Corp., Springfield, 1982; mem. IEEE Consultant's Network, no. N.J., 1992—; lectr. IEEE Montech 86, Montreal, Can., 1986. Contbr. articles to IEEE Transactions, EC&M Mag. Regents scholar N.Y. Bd. Regents, 1963; recipient 3d prize trophy World Wide Inventor Expo '82, 1982. Mem. IEEE Power Engring. and Indsl. Application Socs., NSPE. Achievements include U.S. patent for Polyphase Variable Frequency Inverter. Office: Tramlec Corp 30 Benjamin Dr Springfield NJ 07081-3019

PERKINS, JAMES FRANCIS, physicist; b. Hillsdale, Tenn., Jan. 3, 1924; s. Jim D. and Laura Pervis (Goad) P.; A.B., Vanderbilt U., 1948, M.A., 1949; Ph.D., 1953; m. Ida Virginia Phillips, Nov. 23, 1949; 1 son, James F. Sr. engr. Convair, Fort Worth, Tex., 1953-54; scientist Lockheed Aircraft, Marietta, Ga., 1954-61; physicist Army Missile Command Redstone Arsenal, Huntsville, Ala., 1961-77; cons. physicist, 1977—. Served with USAAF, 1943-46. AEC fellow, 1951-52. Mem. Am. Phys. Soc., Sigma Xi. Contbr. articles to profl. jours. Home and Office: 102 Mountain Wood Dr SE Huntsville AL 35801-1809

PERKINS, RICHARD BURLE, II, environmental engineer, international consultant; b. Houston, May 25, 1960; s. Richard Burle I and Mariam (Jamail) P. BSChemE, U. Tex., Austin, 1983; postgrad., Ariz. State U., 1988-90. Engr. Magcobar group DiChem div. Dresser Industies, Houston, 1979-82; engr. Honeywell Satellite Systems Ops., Phoenix, 1984-92; sr. assoc. ICF Kaiser Internat. Inc., Washington, 1992—; chmn. electrostatic discharge control com. Honeywell, Inc., Satellite Systems Operation, Phoenix, 1985-92; mem. Honeywell Corp. CFC Reduction Task Force, 1990-92, chmn., 1991-92. Presenter papers at profl. confs. Rep. del., Austin, 1980, alt. del., 1982; chmn. City of Glendale Citizens Recycling Task Force, 1990-92; mem. Glendale Strategic Planning Com., 1991-92, Phoenix Hash House Harriers, 1990—; grad. Glendale Leadership Advancement and Devel., 1990. Recipient Excellence award Honeywell Mfg. Bd., 1991. Mem. Ariz. EOS/ESD Assn. (founding mem., sec. 1991), Ariz. Tex. Execs. (pres. 1984, 87, communications chmn. 1988-92), White House Harriers. Avocations: racquetball, southwestern contemporary art, cycling, cats. Home: 1665 A South Hayes St Arlington VA 22202-2713

PERKINS, ROBERT BENNETT, mechanical engineer; b. Woburn, Mass., May 4, 1932; s. George Burton and Gladys Beatrice (Jones) P.; m. Marjorie Ellen Kellerer, Jan. 25, 1979; children: Joyce Elaine Perkins Emery, Beverly Ann Perkins Poulin. Student, MIT, 1957; BS, U. Mass., 1961. Registered profl. engr., N.H. Mech. engr. USN, Kittery, 1961-68; gen. engr. PERA (SS), Kittery, Maine, 1968-88, SUBMEPP Naval Acty., Portsmouth, N.H., 1988—. Asst. editor: Portsmouth Shipyard History, 1979. Organizer Kittery Taxpayers Assn., 1988-92; activist Anti-War Organizer, Eastern Seaboard, 1991-92. Mem. ASME, Am. Sons Am. Revolution, Profl. Engrs. N.H., N.Y. Acad. Scis., Internat. Fedn. Profl. and Tech. Engrs. Democrat. Achievements include patents pending for new wrench, apple tree variety; research on experimental clam farm. Home: 192 Whipple Rd Kittery ME 03904 Office: PO Box 7002 CO SUBMEPP Attn 1845 6 Portsmouth NH 03801

PERKOWSKI, CASIMIR ANTHONY, biopharmaceutical executive, consultant; b. Derby, Conn., Apr. 12, 1941; s. Stanley Joseph and Stefanie (Lencewicz) P.; m. Janet Christine Kowalski, Aug. 18, 1962; children: Stefanie Christine, Paul Casimir, Leon Joseph. BS in Biology, Alliance Coll., 1962; MS in Biology, Emory U., 1964. Instr. Limestone Coll., Gaffney, S.C., 1964-65; asst. prof. biology Alliance Coll., Cambridge Springs, Pa., 1965-69; sect. head E.R. Squibb & Sons, New Brunswick, N.J., 1969-74; prodn. mgr. Hoffmann La-Roche, Belvidere, N.J., 1975-83; dir. ops. Cell Products Inc., New Brunswick, 1983-85; tech. mgr. H-R Internat., Inc., Edison, N.J., 1985-87; dir. biotech. Raytheon Engrs. & Constructors Inc., Phila., 1987-93; v.p. mfg. Promega Corp., Madison, Wis., 1993—; mem. editorial bd. BioPharm, Eugene, Oreg., 1987—. Mem. Roosevelt (N.J.) Bd. of Health, 1970-75; scoutmaster Boy Scouts Am., East Stroudsburg, Pa., 1976-81. NIH fellow, 1963. Mem. ASME (bioprocess equipment standards com. 1990—, chmn. subcom. gen. requirements), ASTM (E-48 main com. 1989—), N.Y. Acad. Scis., Soc. Indsl. Microbiologists, Am. Chem. Soc. Office: Promega Corp 2800 Woods Hollow Rd Madison WI 53711-5399

PERKUHN, GAYLEN LEE, civil/structural engineer; b. Grand Forks, N.D., Nov. 15, 1955; s. Gustav C. and Genevieve R. (Nygaard) P. BSCE summa cum laude, U. N.D., 1978. Registered profl. engr., Minn. Shareholder, dir. computer svcs., project structural engr. Van Doren-Hazard-Stallings, Inc., Mpls., 1978—. Mem. Home: 13965 36th Ave N Minneapolis MN 55447-5305

PERL, ANDRAS, immunologist, educator; b. Budapest, Hungary, Sept. 20, 1955; came to U.S., 1985; s. Miklos and Ibolya (Molnar) P.; m. Katalin Banki, July 23, 1983; children: Annamaria, Marcel Adam, Daniel Peter. MD, Semmelweis Med. Sch., Budapest, 1979, PhD, 1984. Resident, fellow Semmelweis U. Med. Sch., 1979-84, asst. prof., 1984-85; cancer rsch. fellow U. Rochester, N.Y., 1985-88; sr. instr. dept. medicine U. Rochester, 1988-89; asst. prof. dept. microbiology and immunology SUNY, Buffalo, 1989-92; cancer rsch. scientist Roswell Park Cancer Inst., Buffalo, 1989-92; assoc. prof. medicine, microbiology & immunology SUNY, Syracuse, 1992—; presenter, lectr. in field. Contbr. articles to profl. publs. Wilmot Cancer Rsch. fellow, 1985-88; Pardee Found. Cancer Rsch. grantee, 1991—; recipinet Arthritis Investigator award Arthritis Found., 1989—. Mem. Am. Assn. Immunologists, N.Y. Acad. Scis., Clin. Immunology Soc. Achievements include discovery of new human endogenous retrovial sequence first to encode protein, appearing to be involved in autoimmunity and tumorigenesis. Office: Roswell Park Meml Inst 666 Elm St Buffalo NY 14263-0001

PERLMUTTER, LAWRENCE DAVID, aerospace engineer; b. St. Louis, July 1, 1936; s. George and Sarah (Tofield) P.; m. Amy Schaffner, Mar. 19, 1960; children: Deborah Perlmutter Crahan, Donna. BSME, Washington U., St. Louis, 1958; MS in Instrumentation Engring., U. Mich., 1959. Engr. Project Mercury, McDonnell Douglas Corp., St. Louis, 1959, group engr. for advanced design, 1960-69, project dynamics engr., 1970-75, sect. chief for guidance and control, 1976-79, br. chief for guidance and control, 1979-84, McDonnell Douglas Corp fellow, staff mgr., 1984—. Contbr. articles to AIAA Jour., Nat. Aerospace Electronics Record, NATO Agard Record. Fellow AIAA (assoc., tech. com. guidance, navigation and control 1988-90, organizer, chmn. navigation and flight control sessions Guidance, Navigation and Control Conf. 1990, 91, 92); mem. Soc. Automotive Engrs. (aerospace guidance and control com. 1989—). Achievements include development of re-entry guidance techniques for both low lift-to-drag and high lift-to-drag vehicles (includes optimal guidance techniques); ascent abort guidance for Space Shuttle; thrust profile shaping system for spin-stabilized rocket-propelled space vehicles (technology disclosure to U.S. government); low-cost aided-inertial navigation systems; research in innovative alternate control methods for agile missles, strapdown inertial sensor requirements, unaided tactical guidance flight testing. Office: McDonnell Douglas Aerospace-East MC 3064025 PO Box 516 Saint Louis MO 63166-0516

PERLMUTTER, MILTON MANUEL, chemist, accountant, financial consultant, property manager, real estate appraiser, home building inspector; b. Montreal, Que., Can., July 21, 1956; s. Max and Edith (Liszauer) P. Student, Dawson Coll., Montreal, 1975; BSC, McGill U., Montreal, 1978; PhD, Queen's U., Kingston, Ont., 1984. Cert. profl. chemist. Lectr./tutor Queen's U., Kingston, 1984-85; postdoctoral scholar UCLA Sch. Medicine, 1986-89; sr. chemist P.G. & E. Co., L.A., 1989-92; pres., chief exec. officer MMP Chem. and Environ. Cons. Svcs. Co., L.A., 1992—; sr. rsch. scientist U. So. Calif. Sch. Dentistry, L.A., 1993—; cons. in field. Contbr. articles to profl. jours. Vol. UCLA Sch. Medicine, 1989—, Chabad House, L.A., 1986—. R. Samuel McLaughlin scholar, 1978-79, Natural Scis. and Engring. Rsch. Coun. scholar, 1979-82, 92—. Fellow Am. Inst. Chemists, Am. Biog. Inst., Internat. Biog. Inst.; mem. AAAS, Am. Chem. Soc., Internat. Union Pure and Applied Chemists, N.Y. Acad. Sci., Soc. Nuclear Medicine. Jewish. Avocations: singing (opera), cooking, reading, sports, jogging. Home: 741 Gayley Ave Apt 416 Los Angeles CA 90024-2410 Office: MMP Chem Environ Cons Svcs Co 1015 Gayley Ave Ste 1260 Los Angeles CA 90024-3424

PERLOFF, ROBERT, psychologist, educator; b. Phila., Feb. 3, 1921; s. Myer and Elizabeth (Sherman) P.; m. Evelyn Potechin, Sept. 22, 1946; children: Richard Mark, Linda Sue, Judith Kay. A.B., Temple U., 1949; M.A., Ohio State U., 1949, Ph.D., 1951; D.Sc. (hon.), Oreg. Grad. Sch. Profl. Psychology, 1984; D.Litt. (hon.), Calif. Sch. Profl. Psychology, 1985. Diplomate: Am. Bd. Profl. Psychology. Instr. edn. Antioch Coll., 1950-51; with personnel research br. Dept. Army, 1951-55, chief statis. research and cons. unit., 1953-55; dir. research and devel. Sci. Research Assos., Inc., Chgo., 1955-59; vis. lectr. Chgo. Tchrs. Coll., 1955-56; mem. faculty Purdue U., 1959-69, prof. psychology, 1964-69; field assessment officer univ. Peace Corps Chile III project, 1962; Disting. Svc. prof. bus. adminstrn. and psychology U. Pitts. Joseph M. Katz Grad. Sch. Bus., 1969-90, prof. emeritus, 1991—; dir. rsch. programs U. Pitts. Grad. Sch. Bus., 1969-77; dir. Consumer Panel, 1980-83; bd. dirs. Book Center.; cons. in field, 1959—; adv. com. assessment exptl. manpower research and devel. labs. Nat. Acad. Scis., 1972-74; mem. research rev. com. NIMH, 1976-80, Stress and Families research project, 1976-79. Contbr. articles to profl. jours.; editor: Indsl. Psychologist, 1963-65, Evaluator Intervention: Pros and Cons; book rev. editor: Personnel Psychology, 1952-55; co-editor: Values, Ethics and Standards sourebook, 1979, Improving Evaluations; bd. cons. editors: Jour. Applied Psychology; bd. advs.: Archives History Am. Psychology, Psychol. Service Pitts., Recorded Psychol. Jours.; guest editor: Am. Psychologist, May 1972, Education and Urban Society, 1977, Profl. Psychology, 1977. Bd. dirs., v.p. Sr. Citizens Svc. Corp., Calif. Sch. Profl. Psychology, 1985-91; chmn. nat. adv. com. Inst. Govt. and Pub. Affairs, U. Ill., 1986-89. Decorated Bronze Star; Robert Perloff Grad. Rsch. Assistantship in Inst. Govt. and Pub. Affairs, U. Ill., named in his honor, 1990; Robert Perloff Career Achievement award Knowledge Utilization Soc., named in his honor, 1991. Fellow AAAS, Am. Psychol. Assn. (mem.-at-large exec. com. div. consumer psychology 1964-67, 70-71, pres. div. 1967-68, mem. coun. reps. 1965-68, 72-74, chmn. sci. affairs com., div. consumer psychology 1968-69, edn. and tng. bd. 1969-72, chmn. finance com., treas., dir. 1974-82, chmn. investment com. 1977-82, pres., 1985, mem. adv. bd., author column Standard Deviations in jour.), Am. Soc. for Info. Scis., Ea. Psychol. Assn. (pres. 1980-81, dir. 1977-80), mem. Am. Psychol. Soc., Internat. Soc. Psychology, Internat. Assn. Applied Psychology, Pa. Psychol. Assn. (Disting. Svc. award 1985), Assn. for Consumer Rsch. (chmn. 1970-71), Am. Psychol. Found. (v.p. 1988-89, pres. 1990-92), Am. Evaluation Assn. (pres. 1977-78), Soc. Psychologists in Mgmt. (Disting. Contbn. to Psychology Mgmt. award 1989, pres. 1993—), Knowledge Utilization Soc. (pres. 1993—), Sigma Xi (pres. U. Pitts. chpt. 1989-91), Beta Gamma Sigma, Psi Chi. Home: 815 St James St Pittsburgh PA 15232-2112

PERNICONE, NICOLA, catalyst consultant; b. Florence, Italy, Mar. 4, 1935; s. Vincenzo and Nella (Bartoli) P.; m. Lucia Ersini, Oct. 15, 1966; 1 child, Massimo. Degree in Chemistry, U. Florence, 1958. Researcher Edison Chem., Milan, 1959-60, Montecatini, Novara, Italy, 1961-65; group leader Montedison, Novara, Italy, 1966-79, head dept., 1980-86; head R & D Ausimont, Novara, Italy, 1987-90, ret., 1991; pvt. cons. Novara, Italy, 1991—; vis. scientist Acad. Sci., USSR, 1981; sr. scientist Montedison, 1984; contract prof. U. Venice, Italy, 1988-89. Corr. news brief Applied Catalysis, 1981-83, mem. editorial bd., 1984-86; contbr. chpts. to books and articles to profl. jours. Chmn. # 12 Sci. Congress. Mem. ASTM (mem. com. D-32 1979-87), IUPAC (porous materials com. 1988—), CEE-BCR (mem. com. 4, 13 1976—). Avocations: bridge, tennis, skiing. Home: Via Pansa 7, 28100 Novara Italy

PERPER, JOSHUA ARTE, forensic pathologist; b. Bacau, Rumania, Dec. 17, 1932; came to U.S. 1967; s. Eli and Etty (Rachmuth) P.; m. Sheila Marcovici, Oct. 16, 1956; children: Edward, Harry, Irene. MD, Hebrew U.,

Jerusalem, 1961, LLB, 1966; MSc, Johns Hopkins U., 1969. Diplomate Am. Bd. Pathology. Intern Hadassah U. Hosp., Jerusalem, 1961-62; resident in pathology Tel-Hashomer Govt. Hosp., 1962-67; resident in forensic pathology Office of Chief Med. Examiner, Balt., 1967-68, assoc. med. examiner and sr. rsch. fellow, 1968-72; dir. labs. Mayview State Hosp., Pitts., 1973-80; assoc. pathologist Cen. Med. Pavilion Hosp., Pitts., 1973-80; chief forensic pathologist Allegheny County Coroner's Office, Pitts., 1972-80, acting coroner, 1980-81, coroner, 1982—; clin. prof. pathology U. Pitt., 1988—, clin. prof. epidemiology, 1985—, clin. prof. psychiatry, 1989—; adj. prof. pathology Duquesne U., Pitts., 1989—; cons. forensic medicine WHO, 1975-77; medico-legal cons. State Bd. Med. Edn. and Licensure, 1985—; guest lectr. Armed Forces Inst. Pathology, 1977-79; inspector Nat. Assn. Med. Examiners, 1990—, others; lectr. in field. Assoc. editor Am. Jour. Forensic Medicine and Pathology, 1990—; editor Forensic Horizons, 1988—; editorial bd. Am. Jour. Forensic Medicine and Pathology, 1979-90, feature editor, 1987; editor News and Views in Forensic Pathology, 1976-82, 85-87; mem. policy devel. and rev. Pa. State Bd. Medicine, 1987—, com. on health related profls., 1987—, newsletter com., 1988—, steering com., 1989—, chmn., 1990—; contbr. numerous articles to profl. jours. With Israeli Def. Forces, 1951-53. Recipient Spl. Recognition award Sudden Infant Death Found., 1986, Recognition award H.J. heinz Co. Found., 1988. Fellow Am. acad. Forensic Scis. (News and Views award in appreciation for svcs. as editor 1976-82), Am. Coll. Legal Medicine; mem. AMA (Physicians Award in Continuing Med. Edn. 1972-76, 84-87), Pa. Med. Soc., Allegheny County Med. Soc., Pitts. Soc. Pathology, Pa. Soc. Clin. Pathology, Internat. Acad. legal Medicine and Social Medicine, Am. Assn. Medico-Legal Cons., Am. profl. Practice Assn., AAAS, others. Democrat. Jewish. Office: Allegheny County Coroner 542 4th Ave Pittsburgh PA 15219

PERRAULT, GEORGES GABRIEL, chemical engineer; b. Vincennes, France, July 31, 1934; s. René and Georgette (Salin) P. Diploma in engring., Ecole Nat. Supérieure de Chimie de Strasbourg, France, 1958; D, U. Strasbourg, France, 1960, PhD, 1964. Rsch. scientist in phys. chemistry solids Nat. Ctr. for Scientific Rsch., Strasbourg, France, 1958-64; rsch. scientist electrochemistry Nat. Ctr. for Scientific Rsch., Meudon, France, 1964-67, Duke U., Durham, N.C., 1967-68, Meudon, 1968-88; rsch. scientist in biogeography and environment Nat. Ctr. for Sci. Rsch., Paris, 1988—. Editor: Nouvelle Revue Entomologie, 1984—; co-author: Encyclopedia of Electrochemistry Elements, 1978, Standard Potential of Aqueous Solutions, 1985, Biogeography of Carabidae of Mountains and Islands, 1991; contbr. articles to profl. jours. Sgt. French Marines, 1961-63. Mem. N.Y. Acad. Scis., Electrochem. Soc. (treas. European sect. 1981-84, chmn. 1984—), French Soc. Chemistry (electrochem. group com. mem. 1991-92), N.Y. Entomological Soc., French Entomological Soc., Am. Entomological Soc., French Soc. Astronomy, Camping Club. Roman Catholic. Avocations: travel, photography. Home: BP 21, 92290 Chatenay-Malabry France

PERRELLA, ANTHONY JOSEPH, electronics engineer; b. Boulder, Colo., Sept. 16, 1942; s. Anthony Vincent and Mary Domenica (Forte) P.; B.S., U. Wyo., 1964, postgrad., 1965; postgrad. U. Calif. at San Diego, 1966-67, U. Calif. at Irvine, 1968-70; m. Pamela Smith, July 19, 1980; 1 child, Kathleen. Flight engr. U.S. Naval Tng. Devices Center, San Diego, 1965-67; rsch. engr. Collins div. Rockwell Internat. (formerly Collins Radio Co.), Newport Beach, Calif., 1967-69, electromagnetic interference and TEMPEST group head, 1969-74, supr., 1974-75, mgr., 1975-77, mgr. systems integration, 1977, mgr. space communication systems, 1977-78; sr. mem. tech. staff ARGOSystems Inc., Sunnyvale, Calif., 1978—, program mgr., 1978-81, dep. dept. mgr. EW Systems, 1980-83, div. EW staff engr., 1983-84, dept. mgr., 1984-87, Sun Microsystems Inc., Mountain View, 1987-89; prin. A.J. Perrella-Cons., Cupertino, Calif., 1989—; v.p. rsch. and devel. Things Unlimited, Inc., Laramie, Wyo., 1965-72, pres., 1972-75; bd. dirs. Columbian Credit Union. Mem. IEEE, AAAS, Am. Mgmt. Assn., N.Y. Acad. Scis., Assn. Old Crows, KC, Tau Kappa Epsilon. Roman Catholic. Home: 931 Brookgrove Ln Cupertino CA 95014-4667 Office: 2550 Garcia Ave Mountain View CA 94043-1100

PERRENOUD, JEAN JACQUES, cardiologist, educator; b. Neuchatel, Switzerland, Apr. 14, 1947; s. Marcel and Germaine (Joho) P.; m. Francoise Molin, Dec. 13, 1972; 1 child, Jacques Francois. Diploma in medicine, U. Lausanne, Switzerland, 1972; MD, U. Geneva, Switzerland, 1976. Resident medicine and surgery U. Neuchatel Hosp., 1973-76; resident medicine U. Lausanne Hosp., Switzerland, 1976-77; resident cardiology ctr. U. Geneva, 1977-79, clinic chief cardiology ctr., 1979-84; pvt. practice cardiologist Geneva, 1984—; cons., educator Geriatric Hosp., U. Geneva, 1983—, asst.-lectr. Med. Faculty, 1990. Author: Echography and Cardiac Arrhythmias, 1989, (monograph) Ischaemic Cardiomyopathy, 1988, Sudden Death, 1989; contbr. articles to profl. jours. Mem. Cardiologic Soc. Switzerland, Echocardiographic Soc. Switzerland, Geneva Med. Soc. Avocations: music, painting. Home: 12 Chemin Thury, 1206 Geneva Switzerland Office: 22 chemin Beau-Soleil, 1206 Geneva Switzerland

PERRET, GERARD ANTHONY, JR., orthodontist; b. New Orleans, Feb. 13, 1959; s. Gerard A. and Marie M. (Gamino) P. BS in Chemistry, U. N.C., 1981; DDS, La. State U., 1986, cert. orthodontics, 1989. Clin. asst. prof. La. State U. Sch. Dentistry, New Orleans, 1986-87; pvt. practice dentistry Lakeside Dental Group, Metairie, La., 1986-87; pvt. practice orthodontics Jacksonville, Fla., 1989-91, Tampa, Fla., 1991—; founder, pres. Orthogap, Inc., Tampa, 1993—. Mem. ADA, Am. Assn. Orthodontists, Fla. Assn. Orthodontists, Hillsborough County Dental Soc., Hillsborough County Dental Rsch. Clinic, So. Assn. Orthodontists, Temple Terrace C. of C., Omicron Kappa Upsilon. Avocations: sailing, music, golf. Office: 14201 Bruce B Downs Blvd Ste 2 Tampa FL 33613-3913

PERRETEN, FRANK ARNOLD, surgeon, ophthalmologist; b. Boulder, Colo., Jan. 13, 1927; s. Arnold Ervin and Keene (Nichols) P.; B.A., U. Colo., 1948; M.D., U. Pa., 1952; m. Marilyn Ann Peterson, June 26, 1953; 1 son, Michael Peterson. Intern, St. Lukes Hosp., Denver, 1952, Denver Gen. Hosp., 1953; resident postgrad. Mass. Eye and Ear Infirmary of Harvard, 1957; practice medicine, specializing in ophthalmology, Winston-Salem, N.C., 1957-60, Denver, 1960—; mem. staff Children's, St. Joseph's, St. Luke's, Mercy, Presbyn., Luth. hosps.; cons. Brighton (Colo.) Community Hosp.; assoc. prof. U. Colo., 1961—; dir. N.C. Eye Bank, 1957-60. Bd. dirs. Collegiate Sch. of Denver, Goodwill Industries. Served with USNR, 1944-46. Diplomate Am. Bd. Ophthalmology. Fellow Am. Acad. Ophthalmology and Otolaryngology; mem. AMA, N.C., S.C. eye, ear, nose and throat socs., Colo. Ophthalmology Soc., Colo., Denver County med. socs., Colo. Soc. to Prevent Blindness (bd. dirs.), N.Y. Acad. Sci., Newcomen Soc., U. Colo. Alumni (bd. dirs.), Delta Tau Delta Alumni (pres.). Clubs: Denver, Cherry Hills Country (Denver). Lodge: Lions (bd. dirs.). Contbr. Pediatric Ophthalmology. Home: 60 S Birch St Denver CO 80222-1015 Office: 1801 High St Denver CO 80218

PERRIER, PIÉRRE CLAUDE, aeronautical engineer, researcher; b. Paris, June 30, 1935; s. Georges Marie and Marguerite Marie (Pellissier) P.; m. Anne Congnard, July 15, 1967; children: Emmanuel, Agnes, Claire. Baccalaureat, Stanislas Coll., Paris, 1953; Engenieur Civil, Ecole Nationale Superieure Aeronautique, Paris, 1958; Ingenieur Docteur, Ecole Nationale Superiour Aeronautique, Paris, 1959. Researcher Centre Nat. Res. Scientifique, Meudon, France, 1958-59; engr. Dassault Aviation, Paris, 1959—; Mem. com. applications l'Académé des Scis., Paris, 1987; pres., chmn. com.sci. and tech. CNRS, Paris, 1988, Brussels, 1989., vice chmn., founder European Rsch. Community Flow Turbulence and Combustion. Author: Karozoutha, Mshamshana; contbr. articles to profl. jours. Mem. coun. sci. def. Ministry of Armies, Paris, 1983-87. Recipient Médaile de l'Aéronautique award French Govt., Paris, 1980. Mem. AIAA, AAAF, French Acad. Sci. (corr.), Acad. Air and Space. Roman Catholic. Achievements include development of Computational Fluid Dynamics, Turbulence Modeling by Homogeneisation and basic principles of CAO; oral structures in thinking and sayings (application to gospel oral generation). Home: 16 Rue De Mouchy, 78000 Versailles France Office: Daussalt Aviation, 78 Quai Marcel Dassault, 92240 Saint-Cloud France

PERRON, PIERRE O., science administrator; b. Louiseville, Que., Can., Aug. 19, 1939; m. Rachel DesRosiers. BA in Scis. with honors, Laval (Can.) U., 1959, BASc in Metall. Engring., 1963; PhD in Metallurgy, U. Strathclyde, Glasgow, Scotland, 1966. Rsch. officer Chalk River Nuclear Labs. Atomic Energy of Can. Ltd., 1966-68; mgr. radiation protection dept. Hydro Que., 1968-71; dir. material scis. Que. Indsl. Rsch. Ctr., 1971-75, dir. R&D, 1975-82; assoc. dep. minister Dept. Energy and Resources Govt. of Que., Quebec City, 1982-85, Dept. Energy, Mines and Resources Can., Ottawa, 1985-89; pres. Nat. Rsch. Coun. Can., 1989—. Office: Nat Rsch Coun of Can, Montreal Rd, Ottawa, ON Canada K1A 0R6

PERROTTA, GIORGIO, engineering executive; b. Rome, Italy, May 8, 1940; s. Donato and Iole (Tamantini) P.; m. Maria Cecilia Magistri, Sept. 3, 1966; children: Tiziana, Francesca, Dario. Grad., Universita Studi, Rome, 1965. Registered profl. engr. Design engr. Selenia Industrie, S.p.A., Rome, Italy, 1965-70; system engr. Selenia Industrie, S.p.A., Rome, 1970-73, head space communications dept., 1973-78, project mgr., 1979-82; sect. head satellite elec. subsystems and payloads Selenia Spazio, S.p.A., Rome, 1983-87; studies & strategies mgr. Alenia Spazio, S.p.A., Rome, 1987-91; lectr. in field; mem. editorial bd. Space Communications & Broadcasting, 1986-90. Contbr. articles to profl. jours. Recipient Premio for Electronic Innovation, Selenia Industries, 1985, Gold medal for Outstanding Contbns. to Aero., Internat. Ctr. for Arts & Culture, 1986. Mem. AIAA. Achievements include 6 patents for aerospace and electronic techniques, space systems and applications. Home: Via Latina 293, 00179 Rome Italy Office: Alenia Spazio SpA, Via Saccomuro 24, 00131 Rome Italy

PERRY, DANIEL PATRICK, science association administrator; b. N.Y.C., Mar. 17, 1945; s. Kenneth Anderson Perry and Vida Beth (Daniels) Price; m. Carol Ann Fennema, June 30, 1967; children: Colleen Ann, Matthew Daniel. BS, U. Oreg., 1967. Staff writer The Bulletin, Bend, Oreg., 1968-69; reporter Evening News, Buffalo, 1969-72; spl. asst. U.S. Senator Alan Cranston, Washington, 1972-85; nat. affairs specialist March of Dimes Birth Defects Found., Washington, 1985-90; exec. dir. Alliance for Aging Rsch., Washington, 1986—; bd. dirs. Inst. for Advanced Studies in Immunology and Aging, Washington, 1988—; advisor Agy. for Health Care Planning and Rsch., Washington, 1990—, Nat. Inst. on Aging Elder Abuse Study Project, Washington, 1991—; mem. Nat. Leadership Coun. of Aging Orgns., 1991—; appointed Fed. Task Force Aging Rsch., 1992—. Author: (with others) A Good Old Age? The Paradox of Setting Limits, 1990; editor The Research Gap, 1990; editor, pub. Aging Research on the Threshold of Discovery, 1987. Dir. candidate activities Cranston for Pres., 1982-84; founder Alliance for Aging Rsch.; pres. Com. for Dem. Consensus, 1992—. Mem. Am. Geriatrics Soc., Gerontol. Soc. Am., Am. Soc. on Aging, Former Senate Aides Assn. Office: Alliance for Aging Rsch 2021 K St NW # 305 Washington DC 20006-1003

PERRY, DENNIS GORDON, computer scientist; b. Bakersfield, Calif., July 8, 1942; s. Cleo Hoot and Amanda Katherine (Johnson) P.; m. Linda Ellen Jones, June 27, 1964; children: Lynellen, Elizabeth. BA, Westmont Coll., Santa Barbara, Calif., 1964; PhD, U. Wash., 1970; MBA, U. N.Mex., 1983. Staff mem. Los Alamos (N.Mex.) Nat. Lab., 1972-85; program mgr. Def. Advanced Rsch. Project Agy., Arlington, Va., 1985-87; dir. tech. Paramax Systems Corp., McLean, Va., 1987—. Contbr. articles to profl. jours. Mem. AAAS, IEEE (sr.), Assn. Computing Machinery, Nat. Coun. System Engrs. Republican. Achievements include co-discover of 7 new isotopes. Home: 3389 Monarch Ln Annandale VA 22003 Office: Paramax Systems Corp 8201 Greensboro Dr Mc Lean VA 22102

PERRY, FREDERICK SAYWARD, JR., corporate executive; b. Kittery Point, Maine, Aug. 14, 1940; s. Frederick Sayward Sr. and Rita Alice (Contant) P.; m. Judith Ann Golden, June 21, 1963 (div. 1973); 1 child, Elizabeth; m. Sarah Winthrop Smith, Aug. 26, 1979; children: Mariah, Justus. BA in Maths., Harvard U., 1963. Applications engr. Block Engring. Inc., Cambridge, Mass., 1963-67, 69-71; sci. officer ACDA, Washington, 1967-69; sales mgr. Infrared Industries Inc., Waltham, Mass., 1971-75; mktg. rep. Honeywell Radiation Ctr., Lexington, Mass., 1975-77; pres. Boston Electronics Corp., Brookline, Mass., 1977—. Contbr. article to Laser Focus mag. Mem. Alexandria (Va.) Dem. Com., 1969-70, Brookline Town Meeting, 1988; bd. dirs. Brookline Green Space Alliance, 1987—. Mem. SPIE, Optical Soc. Am., Laser Inst. Am. Home: 32 Bowker St Brookline MA 02146-6955 Office: Boston Electronics Corp 72 Kent St Brookline MA 02146-7360

PERRY, GEORGE WILSON, oil and gas company executive; b. Pampa, Tex., July 18, 1929; s. Frank M. and Ruth (Ingersoll) P.; m. Patricia Carberry Bowen; children: Sally, Jett Perry Pemrick, Susan Jeanne Perry Bynder, Virginia Anne Perry Haynie, Tobe Jackson Perry. BS in Petroleum Engring., U. Tulsa, 1952. Registered profl. engr., Tex. Engr. Stanolind Oil & Gas Co., Oklahoma City, 1952-53, Parker Drilling Co., Tulsa, 1953-54; drilling engr. Holm Drilling Co., Tulsa, 1954-55; drilling mgr. Mobil Oil, Victoria, Tex., Lake Charles, La., Paris, France, Anaco, Venezuela, N.Y.C., Tehran, Iran, Stavanger, Norway, New Orleans, La., 1955-79; exec. v.p. Loffland Bros. Co., Tulsa, 1979-89; pres. Gas Well Properties, Dallas, 1989—, George Perry Farms, Tunica, Miss., 1989—. Mem. Delta Tau Delta.

PERRY, HARRY MONTFORD, computer engineering executive, retired officer; b. Newark, Jan. 22, 1943; s. Harry M. Sr. and Jeanette Jay (Decker) P.; m. Dolores Louise Guson, June 23, 1967 (div. 1985); children: Kimberly J., Eric M.; m. Lois Joy Barnes, Dec. 31, 1987; children: Philip C., Robert J. AA, AS, SUNY, Albany, 1978, BS, 1980. Enlisted USN, 1959, commd. lt. (j.g.), 1977, advanced through grades to lt. comdr., 1981, ret., 1983; program mgr. Advanced Tech. Inc., Virginia Beach, Va., 1983-86; dir. Computer Dynamics Inc., Virginia Beach, 1986-90; v.p. Info. Tech. Solutions Inc., Hampton, Va., 1990—; bd. dirs. L.J. Perry Ins., Inc., Virginia Beach, Visionetics Inc., Virginia Beach, Info. Tech. Solutions Inc. Contbr. articles to engring. publs. and govt. documents. Active Dem. City Com., Virginia Beach, 1990—; deacon, officer 1st Bapt. Ch., Norfolk. Mem. Am. Soc. Naval Engrs., U.S. Naval Inst., Broad Bay Country Club. Avocations: golf, religious studies, accounting, politics. Home: 2560 Landview Cir Virginia Beach VA 23454-1200 Office: Info Tech Solutions 850 Greenbrier Cir Ste I Chesapeake VA 23320-2644

PERRY, I. CHET, petroleum company executive; b. Phila., Jan. 18, 1943; s. Irving Chester Sr. and Erma Jackson (McNeil) P.; m. Dayanne Schurecht, Aug. 27, 1966 (div. 1983); 1 child, London Schade. BA in Psychology, Bus., Lake Forest Coll., 1965. Lic. real estate broker, Ill. Sr. mgmt. trainee British Overseas Airways Corp., London, Eng., 1968-69; owner Itec Internat. Ltd., Barrington, Ill., 1970—, Itec Refining & Mktg. Co., Ltd., Barrington, 1970—. Lt. U.S. Army, 1965-68, Vietnam. Mem. Am. Petroleum Inst., Barrington Bd. Realtors (bd. dirs. 1974-78) Forest Grove Club, Barrington Tennis Club. Republican. Quaker. Avocations: tennis, photography. Home: 444 W Russell St Barrington IL 60010-4123

PERRY, JAMES ALFRED, natural resources director, researcher, educator, consultant; b. Dallas, Sept. 27, 1945. BA in Fisheries, Colo. State U., 1968; MA, Western State Coll., 1973; PhD, Idaho State U., 1981. Sr. water quality specialist Idaho Div. Environ., Pocatello, 1974-82; area mgr. Centrac Assocs., Salt Lake City, 1982; prof. forest water quality U. Minn., St. Paul, 1982—, dir. natural resources policy and mgmt., 1985—, dir. grad. studies and water resources, 1988-92; vis. scholar Oxford U., Green College, Eng., 1990-91; internat. cons. in water quality. Charter mem. Leadershp Devel. Acad., Lakewood, Minn., 1988. Am. Inst. Indian Studies fellow, Delhi and Madras, 1985, Nat. Acad. Sci. fellow, Lodz, Poland, 1991-92. Mem. Minn. Acad. Scis. (bd. dirs. 1987-90), N.Am. Benthol Soc. (exec. bd. Albuquerque 1990-91), Soc. Am. Foresters, Sigma Xi, Xi Sigma Pi, Gamma Sigma Delta. Office: U Minn Dept Forest Resource 110 Green Hall 1530 Cleveland Ave N Saint Paul MN 55108-1027

PERRY, JOHN STEPHEN, meteorologist; b. Lynbrook, N.Y., Oct. 18, 1931; s. Stephen Augustine and Phyllis Mary (Brown) P.; m. Olive Barbara Jones, Sept. 20, 1953; children: Stephen K. Perry, Robert J. Perry. BS, Queens Coll., 1953; PhD, U. Washington, 1966. Commd. 2d lt. USAF, 1953, advanced through grades to col., 1973; weather officer, 1953-70; program mgr. Advanced Rsch. Projects Agy. USAF, Washington, 1971-74; retired USAF, 1974; exec. scientist Nat. Acad. Scis., Washington, 1974-79; staff dir. Bd. on Atmospheric Scis. and Climate, Washington, 1979-91; dir. Bd. on Global Change, Washington, 1991—; Trustee MENSA Ednl. and Rsch. Found., 1988-93; cons. World Meteorol. Orgn., Geneva, Switzerland, 1976-78, 91; cons. Internat. Inst. for Applied Systems Analysis, Luxembourg, Austria, 1991-92. Author: The U.S. Global Change Research Program, 1992; contbr. articles, revs. to profl. jours. Fellow AAAS, Am. Meteorol. Soc.; mem. Am. Geophysical Union, Sigma Xi. Democrat. Home: 6205 Tally Ho Ln Alexandria VA 22307

PERRY, LELAND CHARLES, quality engineer, quality assurance analyst; b. Saranac Lake, N.Y., Oct. 4, 1942; s. Leo H. and Stella D. P.; m. Karen A. Ware, Apr. 15, 1967; children: Jason M., Christiaan D., Vanessa L., Shane M., Rachel A., Jeffrey I. AAS, Paul Smith's Coll., 1962; BA, SUNY, Plattsburg, 1966; postgrad., U. Vt., 1968-70. Supr. Ayerst Pharm. Labs., Rouses Point, N.Y., 1967-68; sr. rsch. asst. U. Vt. Med. Sch., Burlington, 1968-70; tech. asst. to exec. dir. CIBA Geigy Pharm. Co., Summit, N.J., 1970-74; plant mgr. Norwich Pharm. Co., Greenville, S.C., 1974-77; supr. quality control lab. Abbott Labs., Spartanburg, S.C., 1977-85; testing mgr. Schmid Labs., Anderson, S.C., 1987-88; quality assurance mgr. Smith & Nephew Med. Co., Columbus, Ga., 1989-90; quality analyst Decora Mfg. Co., Ft. Edward, N.Y., 1990—. Author teaching materials. Mem. Coun. on Adoptable Children, Greenville, S.C., 1977; minister Evangel. Fellowship Internat., Spartanburg, S.C., 1992. Mem. TAPPI, Am. Soc. Quality Control (chair tech. program 1991-92). Republican. Achievements include research in chromosome mapping, mechanisms of stimulating interferon production and/or release, isolation of amino acid utilizing mutants from B-subtillis by using various mutogenic agents. Home: 3 Mohawk Trail Queensbury NY 12804

PERRY, LEWIS CHARLES, emergency medicine physician, osteopath; b. La Plata, Mo., Apr. 22, 1931; s. Lewis C. and Emily B. Perry; m. M. Sheryl Gupton, Oct. 30, 1953; children: David, Susan, Stephen, John. BS, U. Mo., 1958; postgrad., Louisville Presbyn. Sem., 1958-60; DO, Kirksville Coll. Osteo. Medicine, 1967. Intern Midcities Meml. Hosp., Arlington, Tex.; parish min. Presbyn. Bd. Nat. Missions, Canada, Ky., 1960-62; intern Mid Cities Meml. Hosp., Arlington; pvt. practice, Ingleside, Tex., 1968-72, Tucson, 1972-81; emergency physician Tucson Gen. Hosp., 1981-88, pres. med. staff, 1978-79, clin. instr., 1981-88; emergency physician Meml. Med. Ctr. East Tex., Lufkin, 1988—; clin. instr. Osteo. Coll. Pacific, Pomona, Calif., 1985-88. Pres. Helping Hands, Ingleside, 1969-72; bd. dirs., pres. Salvation Army, Tucson, 1978-81; commr. Cub Scouts Am., Tucson, 1975-76; bd. dirs. Unity of Tucson, Inc., 1986-88. 1st lt. USAF, 1952-56. Named Physician of Yr., Tucson Gen. Hosp., 1978. Mem. Am. Legion, Rotary, Masons, Scottish Rite. Avocations: cooking, gardening. Home: 1 Columbia Ct Lufkin TX 75901-7212

PERRY, ROBERT LEE, sociologist, ethnologist; b. Toledo, Dec. 6, 1932; s. Rudolph R. and Katherine (Bogan) P.; m. Dorothy LaRouth Smith, Aug. 23, 1969; children: Baye Kito, Kai Marlene, Ravi Kumar. BA, Bowling Green (Ohio) State U., 1959, MA, 1965; PhD, Wayne State U., 1978. Lic. counselor, Ohio; cert. prevention cons. Asst. prof. sociology Detroit Inst. Tech., 1967-70; asst. prof. sociology Bowling Green State U., 1970-79, assoc. prof. ethnic studies and sociology, 1979-91, prof., 1991—, prof., chmn. dept. ethnic studies, 1979—; cons. Northwest Ohio Regional Council on Alcoholism, Toledo, 1984-86. Contbr. articles to profl. jours. Records reviewer Lucas County Juvenile Ct., Toledo, 1979; mem. com. Govnrs. Coun. Recovery Svcs., Columbus, 1985-86; mem. Mayor City Toledo Ohio's police task force, 1990; mem. bd. Urban Minority Drug Alcohol Outreach Program. With USAF, 1953-57. Mem. Am. Sociol. Soc., N. Cen. Sociol. Assn., Nat. Assn. of Ethnic Studies, Nat. Assn. Black Studies, Am. Soc. Criminology, Kappa Alpha Psi, Sigma Delta Pi, Alpha Kappa Delta. Methodist. Avocations: running, swimming, biking. Office: Bowling Green State U Dept Ethnic Studies Bowling Green OH 43403-0216

PERRY, ROBERT RYAN, manufacturing engineer; b. Newark, N.J., July 7, 1956; s. Obbie and Bertha Lee (Clark) P. AS in Elec. Tech., N.J. Inst. Tech., 1980, BS in Engring. Tech., 1985. Ind. engr. technician Manville (N.J.) Corp., 1978-81; project engr. Boyle-Midway div. Am. Home, Cranford, N.J., 1981-84; corp. engr. Castrol-Inc., Wayne, N.J., 1984-88, corp. project engr., 1989-93; sr. project engr. Castrol-Inc., Wayne, 1993—. Mem. ASHRAE. Democrat. Baptist. Achievements include development of packaging testing procedure for new bottle design. Home: 1406 Coolidge St Plainfield NJ 07062 Office: Castrol Inc 1500 Valley Rd Wayne NJ 07470-2040

PERRY, SEYMOUR MONROE, physician; b. N.Y.C., May 26, 1921; s. Max and Manya (Rosenthal) P.; m. Judith Kaplan, Mar. 18, 1951; children: Grant Matthew, Anne Lisa, David Bennett. B.A. with honors, UCLA, 1943; M.D. with honors, U. So. Calif., 1947. Diplomate: Am. Bd. Internal Medicine. Intern L.A. County Hosp., 1946-48, resident, 1948-51, mem. staff outpatient dept., 1951; examining physician L.A. Pub. Schs., 1951-52; sr. asst. surgeon Phoenix Indian Gen. Hosp., USPHS, 1952; charge internal medicine USPHS Outpatient Clinic, Washington, 1952-54; fellow hematology UCLA, 1954-55, asst. rsch. physician atomic energy project, 1955-57; asst. prof. medicine, head Hematology Tng. Program, Med. Ctr., 1957-60; instr. medicine Coll. Med. Evangelists, 1951-57; attending specialist internal medicine Wadsworth VA Hosp., Los Angeles, 1958-61; sr. investigator, medicine br. Nat. Cancer Inst., 1961-65, chief medicine br., 1965-68, mem. clin. cancer tng. com., 1966-69, chief human tumor cell biology br., 1968-71, assoc. sci. director clin. trials, 1966-71, assoc. sci. dir. for program planning, div. cancer treatment, 1971-73, dep. dir., 1973-74, acting dir., 1974; spl. asst. to dir. NIH, 1974-78, assoc. dir., 1978-80, acting dep. asst. sec. health (tech.), 1978-79; acting dir. Nat. Ctr. Health Care Tech., OASH, 1978-80, dir., 1980-82; dep. dir. Inst. for Health Policy Analysis Georgetown U. Med. Ctr., Washington, 1983-89, prof. medicine, prof. community and family medicine, 1983—, interim chmn. dept., 1989-90, chmn., 1990—, dir. Inst. for Health Care Rsch. and Policy; mem. adv. com. on rsch. and on the therapy of cancer Am. Cancer Soc., 1966-70, adv. com. chemotherapy and hematology, 1975-77, chmn. epidemiology, diagnosis and therapy com., 1971, grantee, 1959-60; med. dir. USPHS, 1961-80; asst. surg. gen., 1980-82; mem. radiation com. NIH, 1963-70, co-chmn., 1971-73; pres. Nat. Blood Club, 1971; chmn. Interagy. Com. on New Therapies for Pain and Discomfort, 1978-80; mem. adv. panel on med. tech. and costs of medicare program Congress of U.S., 1982-84; chmn. criteria working group (bioseparation) NASA, 1984; cons. Nat. Ctr. Health Svcs. Rsch. and Health Care Tech., DHHS, 1985-90, Nat. Libr. of Medicine, 1985-89, Agy. for Health Care Policy and Rsch, DHHS, 1990—, Hosp. Assn. N.Y. State, 1990-91; mem. procedures rev. com. and profl. adv. panel Blue Cross/Blue Shield Nat. Capitol Area, 1987—; advisor WHO Programme on Tech. Devel., Assessment and Transfer. Assoc. editor Internat. Jour. Tech. Assessment in Health Care, 1984—; mem. editorial bd. Jour. Health Care Tech., 1984-87, Health Care Instrumentation, 1984-87; mem. editorial adv. bd. Health Tech.: Critical Issues for Decision Makers, 1987-90, Courts, Health and the Law, 1990-91. Decorated comendador Order of Merit, Peru; comendador Orden Hipólito Unanue, Peru; Pub. Health Service commendation, 1967; Meritorious Service medal USPHS, 1980. Fellow ACP (adv. com. to gov. Md. on coll. affairs 1969-76, gov. for USPHS and HHS 1980-82, subcom. on clin. efficacy assessment 1982-85, chmn. health and pub. policy com. D.C. met. area 1987—); mem. Inst. Medicine of NAS (mem. evaluation panel coun. health care tech. 1987-90, com. on evaluation med. techs. in clin. use 1981-84, chmn. rev. com. on Inst. Medicine Report on Hip Fracture 1990, mem. rev. com. on renal disease 1990, rev. com. on artificial heart 1991), Am. Pub. Health Assn., Assn. Health Svcs. Rsch. (health svcs. rsch. adv. com. 1990—), Assn. Acad. Health Ctrs., Internat. Soc. Tech. Assessment in Health Care (pres. 1985-87, bd. dirs. 1989—), Cosmos Club. Achievements include elucidation of leukocyte physiology; initiation of the consensus development process and the technology assessment forum method to resolve controversial issues in medical care. Office: Dept Community & Family Medicine Georgetown U Med Ctr Kober Cogan Bldg 3750 Reservoir Rd Washington DC 20007

PERSAUD, BISHNODAT, sustainable development educator; b. New Amsterdam, Berbice, Guyana, Sept. 22, 1933; arrived in U.K., 1974; s. Dhwarka and Dukhni (Surujbali) P.; m. Lakshmi Persaud, Aug. 1962; 3 children: Rajendra, Avinash, Sharda. BSc in Econs., Queen U., Belfast, U.K., 1960; postgrad. diploma agrl. econs., U. Reading, U.K., 1963, PhD in Econs., 1973. Jr. rsch. fellow Inst. of Social and Econ. Rsch., U. W.I., Mona, Jamaica, 1964-66, rsch. fellow, 1966-74; chief econs. officer Commodities div. Commonwealth Secretariat, London, 1974-76, asst. dir. econ.

affairs div., 1976-81, dir. and head, 1981-92; prof. sustainable devel. U. West Indies, Kingston, Jamaica, 1992—; bd. dirs. Commonwealth Equity Fund, London; mem. U. Guyana Commn., Georgetown, 1991—; assoc. fellow Ctr. for Caribbean Studies, Devel. Econs. Rsch. Ctr., U. Warwick, Coventry, 1991—; mem. panel of judges World Devel. Awards for Bus., 1991—. Co-author: Developing with Foreign Investment, 1987; contbr. articles to profl. jours. Fellow Royal Soc. of Arts, Royal Econ. Soc.; mem. RAC, Commonwealth Trust and the Royal Overseas League. Home and Office: U West Indies, 11 Long Mountain Rd, College Common Kingston Jamaica

PERSHE, EDWARD RICHARD, civil engineer; b. Omaha, July 30, 1924; s. Joseph Edward and Theresa Elizabeth (Mikich) P.; m. Clotilde Amalia (Sintes-Roscelli), Apr. 28, 1954; children: John Charles, Robert Andrew. BS in Civil Engring., U. Ill., 1949; MS in Sanitary Engring., MIT, 1950; PhD in Civil Engring., U. Ill., 1966. Registered profl. engr., Ohio, Fla. Design engr. Gannett Fleming Engrs., Harrisburg, Pa., 1953-56; project engr. Black & Veatch Engrs., Kansas City, Mo., 1956-58; asst. prof. Civil Engring. U. Nebr., Lincoln, 1958-62; assoc. prof. Northeastern U., Boston, 1966-70; dir. rsch. Whitman & Howard, Inc., Boston, 1970-84; asst. dist. drainage engr. Dept. Transp. State of Fla., Deland, 1987—. Contbr. articles to profl. jours. Pres. Cranes Roost Homeowners Assn., Altamonte Springs, Fla., 1988; pres. Altamonte Springs Garden Club, 1990-91. 1st. lt. USAF, 1950-53. Recipient NSF Sci. Faculty fellowship U. Ill., 1962, USPHS rsch. fellowship U. Ill., 1963, 64, 65, Ford Found. grant U. Ill., 1964, 65. Fellow ASCE (life mem., chmn. environ. engring. com. 1976-84). Republican. Home: 217 Mallard St Altamonte Springs FL 32701 Office: Fla Dept Transp 719 S Woodland Blvd Deland FL 32720

PERSON, VICTORIA BERNADETT, civil engineer, consultant; b. Garden City, Mich., Dec. 4, 1958; d. Miguel and Evelyn May (Trosin) Bernadett; m. Andrew Paul Person; children: Zoe Bernadette, Thor Anders. BSCE, U. Mich., 1982, MSCE, 1985. Registered profl. engr., Mich. Soils engr. Bechtel Power Corp., Ann Arbor, Mich., 1982-84; project engr. NTH Cons., Ltd., Farmington Hills, Mich., 1985-91; project mgr. Metcalf & Eddy, Inc., Detroit, 1992—. Mem. ASCE (geoenvironmental chmn. 1989--). Lutheran. Office: Metcalf & Eddy Inc One Detroit Ctr 500 Woodward Ave Ste 1510 Detroit MI 48226-3406

PERUMPRAL, JOHN VERGHESE, agricultural engineer, administrator, educator; b. Trivandrum, Kerala, India, Jan. 14, 1939; came to U.S., 1963; s. Verghese John and Sarah (Gevorghese) P.; m. Shalini Elizabeth Alexander, Dec. 27, 1965; children: Anita Sarah, Sunita Anna. BS in Agrl. Engring., Allahabad (UP India) U., 1962; MS in Agrl. Engring., Purdue U., 1965, PhD, 1969. Postdoctoral rsch. assoc. agrl. engring. dept. Purdue U., West Lafayette, Ind., 1969-70; asst. prof. agrl. engring. dept. Va. Poly. Inst. and State U., Blacksburg, 1970-78, assoc. prof., 1978-83, prof., 1983-86, Wm. S. Cross Jr. prof., head dept. agrl. engring., 1986—. Contbr. over 30 articles to scholarly and profl. jours. Mem. Am. Soc. Agrl. Engring. (outstanding faculty award student br. 1976, 81, cert. teaching excellence 1979, assoc. editor, transaction of ASAE 1985-86), Fluid Power Soc., Sigma Xi, Alpha Epsilon. Presbyterian. Office: Va Poly Inst and State U Agrl Engring Dept Blacksburg VA 24061

PERUTZ, MAX FERDINAND, molecular biologist; b. May 19, 1914; s. Hugo and Adele Perutz; m. Gisela Peiser, 1942; 1 son, 1 dau. Ed., U. Vienna; Ph.D., U. Cambridge, 1940. Dir. Med. Research Council Unit for Molecular Biology, 1947-62; chmn. European Molecular Biology Orgn., 1963-69; reader Davy Faraday Research Lab., 1954-68, Fullerian prof. physiology, 1973-79; chmn. Med. Research Council Lab. Molecular Biology, 1962-79. Author: Proteins and Nucleic Acids, Structure and Function, 1962, Is Science Necessary and Other Essays, 1989, Stereochemical Mechanisms of Cooperativity and allosteric Regulation in Proteins, 1990, Protein Structure: New Approaches to Disease and Therapy, 1992. Fellow Royal Soc. (Royal medal 1971, Copley medal, 1979); mem. Royal Soc. Edinburgh, Am. Acad. Arts and Scis. (hon.), Austrian Acad. Scis. (corr.), Am. Philos. Soc. (fgn.), Nat. Acad. Scis. (fgn. asso.), Royal Netherlands Acad. (fgn.), French Acad. Scis., Bavarian Acad. Scis., Nat. Acad. Scis. of Rome (fgn.), Accademia dei Lincei (Rome) (fgn.), Pontifical Acad. Scis. Office: MRC Lab Molecular Biology, Cambridge CB2 2QH, England

PESACRETA, GEORGE JOSEPH, optical metrology engineer; b. Beacon, N.Y., Apr. 24, 1960; s. Joseph and Marie Josephine (Noel) P.; m. Mary Kathleen Parrella, Dec. 2, 1989; 1 child, Mara Alyssia. BS, Stevens Inst. Tech., Hoboken, N.J., 1982, MS, 1982. Microlithography engr. IBM, Hopewell Junction, N.Y., 1982-89; optical metrology engr. Storage Tek, Louisville, Colo., 1989-92, Rocky Mountain Magnetics, Louisville, Colo., 19926. Mem. Internat. Soc. for Optical Engring. Office: Rocky Mountain Magnetics 2270 S 88th St Louisville CO 80028-8188

PESERIK, JAMES E., electrical engineer; b. Beloit, Wis., Sept. 30, 1945; s. Edward J. and G. Lucille Peserik; m. Elaine L. Peserik, May 6, 1972. BSEE, U. Wis., 1968; MS, St. Joseph's U., 1990. Registered profl. engr., registered profl. land surveyor; cert. fire and explosion investigator, cert. fire investigation instr. Development and instrumentation engr. Square D Co., Milw., 1968-71; product engr. I-T-E Imperial Corp., Ardmore, Pa., 1971-72; project engr. Harris-Intertype Corp., Easton, Pa., 1972-74; elec. engr. Day & Zimmerman, Inc., Phila., 1974-76; pvt. practice Coopersburg, Pa., 1976—; sr. elec. engr. S.T. Hudson Engrs., Inc., Phila., 1980-81; mem. adv. coun. Swenson Skills Ctr., Phila., 1990—. Treas. Salford-Fraconia Joint Parks Commn., Montgomery County, Pa., 1980-83. Mem. IEEE (sec. indsl. applications group Phila. chpt., 1980, chmn. 1981), Nat. Soc. Profl. Engrs., Pa. Soc. Profl. Engrs., Nat. Fire Protection Assn., Internat. Assn. Arson Investigators, Nat. Assn. Fire Investigators. Office: PO Box 181 Coopersburg PA 18036-0181

PESTANA-NASCIMENTO, JUAN M., civil, geotechnical and geoenvironmental engineer, consultant; b. Caracas, Venezuela, June 24, 1963; came to U.S., 1986; s. Domingos Pestana and Maria Cisaltina Nascimento; m. Sandra Mattar, Oct. 29, 1988; 1 child, Maria Teresa Pestana. BS summa cum laude, U. Catolica Andres Bello, Caracas, 1985; MS, MIT, 1988, DSc, 1993. Registered civil engr., Venezuela. Teaching asst. U. Catolica Andres Bello, Caracas, 1982-85; instr., lectr. U. Catolica Andres Bello, Cambridge, 1986; cons. engr. GEODEC, Geotech. Cons., Caracas, 1986-92; civil engr. CALTEC, Hydraulic Cons., Caracas, 1983-86, asst. engr., 1983-85; asst. engr. T.W. Lambe, Inc., Cambridge, Mass., 1990; cons. Portfolio Mgmt., Cambridge, 1990-92. Contbr. articles to profl. jours. Gran Mariscal de Ayacucho scholar, 1986-88, INTEVEP scholar, 1989-92. Mem. ASCE, Can. Soc. Geotech. Engrs., Internat. Soc. Soil Mechanics and Found. Engring., Sigma Xi. Achievements include development of a model to describe the behavior in compression of cohesionless soils, development of a generalized constitutive model to describe the mechanical behavior of sands and clays under drained and undrained conditions. Home: 51 Pette St # 27 Newton MA 02164 Office: MIT 77 Mass Ave MIT 1-371 Cambridge MA 02139

PESTANO, GARY ANTHONY, biologist; b. New Amsterdam, Guyana, Sept. 9, 1966; came to the U.S., 1985; s. Clive Patrick and Vivia (Madramootoo) P. BS, CCNY, 1990; postgrad., CUNY, 1991—. Health rsch. intern Bur. of Labs., N.Y.C., 1989; grad. rsch. asst. rsch. found. CUNY, 1990—. Contbr. articles to Jour. Cellular Biochemistry, Jour. AIDS, AIDS Rsch. and Human Retroviruses. CCNY scholar, 1992, 93. Mem. Sigma Xi (assoc.). Achievements include co-discovery of unique HIV-1 isolates in Ugandan population. Office: City Coll NY Dept Biology 138th and Convent Ave New York NY 10031

PESTRONK, ALAN, neurologist; b. Cambridge, Mass., Jan. 19, 1946; s. Seymour and Judith P. AB, Princeton U., 1966; MD, Johns Hopkins U., 1970. Diplomate Am. Bd. Psychiatry and Neurology. Resident in neurology Johns Hopkins U., Balt., 1971-74, neuromuscular fellow, 1974-77; asst./assoc. prof. neurology Johns Hopkins U., 1979-89; prof. neurology, dir. neuromuscular div. Washington U., St. Louis, 1989—. Mem. Am. Neurol. Assn., Am. Acad. Neurology. Office: Washington U Sch Medicine Box 8111 660 S Euclid Ave Saint Louis MO 63110

PETCHCLAI, BENCHA, physician, researcher, inventor; b. Nonthaburi, Thailand, July 14, 1937; s. Cherm and Rabieb (Shy Pung) P.; m. Phairin Petchclai, Spet. 2, 1982; children: Busaracum, Banjapuck. MD, Mahidol U., Bangkok, 1961. Intern Siriraj Hosp., Bangkok, 1961-62; instr. clin. pathology Mahidol U. Faculty Med. Tech., 1962-68; resident in immunology Albert Einstein Med. Ctr., Phila., 1964-66; dir. clin. immunology lab. faculty medicine Ramathibodi Hosp., Bangkok, 1968-91, chmn. dept. pathology, 1991—; dir. Ctr. for Rsch. and Indsl. Prodn. Diagnostics, Inst. Scis. and Tech., Mahidol, 1987—; researcher monitoring AIDS in Bangkok, Japanese Found. for AIDS Prevention, 1988—; temp. advisor on prodn. of immunodiagnostics WHO, 1986—. Inventor over 30 immunodiagnostics for infectious diseases. Recipient Prize for Tech. R&D, IBM Thailand, 1985, Awards for Useful Inventions, Nat. Rsch. Coun. Thailand, 1983-84. Mem. Thai Med. Assn., Soc. Ornamental Plant of Thailand (pub. rels. officer 1975—). Buddhist. Avocations: gardening, tennis, singing, playing piano and accordion, writing. Home: 72 Moo 2 Bangroey, 11130 Nonthaburi Thailand Office: Ramathibodi Hosp Medicine, Rama VI Rd, 10400 Bangkok Thailand

PETERING, DAVID HAROLD, chemistry educator; b. Peoria, Ill., Sept. 16, 1942; married, 1966; 2 children. BA, Wabash Coll., 1964; PhD in Biochemistry, U. Mich., 1969. From asst. prof. to assoc. prof. chemistry and biochemistry U. Wis., Milw., 1971-82, prof. chemistry, 1983—. Fellow Northwestern U., Am. Cancer Soc., 1969-71; sr. vis. fellow Nat. Inst. Environ. Health Sci., 1981—. Mem. Am. Chem. Soc., Am. Soc. Biol. Chemists, Sigma Xi. Achievements include research in the metabolism of essential transition metals zinc, iron, and toxic metals and their complexes, role of zinc in normal and tumor cell proliferation, of cadmium in biological toxicity and various metal complexes in cancer chemotherapy, biochemistry of metallothionein. Home: 7229 N Santa Monica Blvd Milwaukee WI 53217-3506 Office: U Wis-Marine & Freshwater Biomed Core Ctr PO Box 413 Milwaukee WI 53201-0413*

PETERLE, TONY JOHN, zoologist, educator; b. Cleve., July 7, 1925; s. Anton and Anna (Katic) P.; m. Thelma Josephine Coleman, July 30, 1949; children—Ann Faulkner, Tony Scott. BS, Utah State U., 1949; MS, U. Mich., 1950, PhD (univ. scholar), 1954; Fulbright scholar, U. Aberdeen, Scotland, 1954-55; postgrad., Oak Ridge Inst. Nuclear Studies, 1961. With Niederhauser Lumber Co., 1947-49, Macfarland Tree Svc., 1949-51; rsch. biologist Mich. Dept. Conservation, 1951-54; asst. dir. Rose Lake Expt. Sta., 1955-59; leader Ohio Coop. Wildlife Rsch. unit U.S. Fish and Wildlife Svc., Dept. Interior, 1959-63; asso. prof., then prof. zoology Ohio State U., Columbus, 1959-89, prof. emeritus, 1989; chmn. faculty population and environmental biology Ohio State U., 1968-69, chmn. dept. zoology, 1969-81, dir. program in environ. biology, 1970-71; liaison officer Internat. Union Game Biologists, 1965-93; co-organizer, chmn. internat. affairs com., mem. com. ecotoxicology XIII Internat. Congress Game Biology, 1979-80; pvt. cons., 1989—; mem. com. rev. EPA pesticide decision making Nat. Acad. Scis.-NRC; mem. vis. scientists program Am. Inst. Biol. Scis.-ERDA, 1971-77; mem. com. pesticides Nat. Acad. Scis., com. on emerging trends in agr. and effects on fish and wildlife; mem. ecology com. of sci. adv. council EPA, 1979-87; mem. research units coordinating com. Ohio Coop. Wildlife and Fisheries, 1963-89; vis. scientist EPA, Corvallis, 1987. Author: Wildlife Toxicology, 1991; editor: Jour. of Wildlife Mgmt., 1969-70, 84-85, 2020 Vision Meeting the Fish and Wildlife Conservation Challenges of the 21st Century, 1992. Served with AUS, 1943-46. Fellow AAAS, Am. Inst. Biol. Scis., Ohio Acad. Sci.; mem. Wildlife Disease Assn., Wildlife Soc. (regional rep. 1962-67, v.p. 1968, pres. 1972, Leopold award 1990, hon. mem. 1990), Nat. Audubon Soc. (bd. dirs. 1985-87), Ecol. Soc., INTECOL-NSF panel U.S.-Japan Program, Xi Sigma Pi, Phi Kappa Phi. Home: 4072 Klondike Rd Delaware OH 43015-9513 Office: Ohio State U Dept Zoology 1735 Neil Ave Columbus OH 43210-1293

PETERLIN, BORIS MATIJA, physician; b. Ljubljana, Slovenia, July 4, 1947; came to U.S., 1961; s. Anton and Leopoldina (Leskovic) P.; m. Anne Scheel-Larsen, July 21, 1984; children: Anton Alexander, Sebastian Bogomir. BS, Duke U., 1968; MD, Harvard U., 1973. Diplomate Am. Bd. Internal Medicine, Am. Bd. Rheumatology. Intern, resident Stanford (Calif.) Univ. Hosp., 1973-75, sr. resident, 1977-78; fellow in rheumatology, immunology U. Calif., 1978-81; asst. prof. U. Calif., San Francisco, 1981-88, assoc. prof., 1988—; asst. investigator HHMI, Bethesda, Md., 1984-89, assoc. investigator, 1989—. Contbr. articles to Nature, Cell, Genes and Development, others. Lt. commdr. USPHS, 1975-77. Rosalind Russell Arthritis scholar U. Calif., 1981. Fellow Am. Soc. for Clin. Investigation; mem. Am. Assn. Immunology, Am. Fedn. Clin. Rsch., Am. Soc. for Microbiology, Am. Coll. Rheumatology, Phi Beta Kappa, Phi Lambda Upsilon. Democrat. Roman Catholic. Achievements include diagnosis of bare lymphocyte syndrome; discovery of mechanism of action of HIV Tat protein; fundamental studies in replication of HIV. Home: 14 Hill Point San Francisco CA 94117-3603 Office: U Calif San Francisco-HHMI 3d and Parnassus San Francisco CA 94143-0724

PETERMANN, GOTZ EIKE, mathematics educator, researcher; b. Berlin, Aug. 24, 1941; came to Sweden, 1945; s. Erwin Karl Eduard and Selma Alwine Ida (Dunker) P.; m. Gunlog Birgitta Bjorkhem, May 17, 1970; children—Veronica, Ingemar, Waldemar, Ingmarie, Elisabeth, Johannes. Fil. Kand., U. Stockholm, 1963, Fil. Mag./Fil. Lic., 1970; Fil. Dr., Royal Inst. Tech., Stockholm, 1974. Asst., lectr. U. Stockholm, 1962-73; lectr. math. Royal Inst. Tech., Stockholm, 1973—. Author: Konvexitet och Optimering, 1973; Analytiska och Numeriska Metoder, 1981, 89. Contbr. articles to profl. jours. Mem. Swedish Math. Soc. (treas. 1975-78), Am. Math. Soc. Avocations: music, reading. Home: Satravagen 1A, S184 52 Osterskar Sweden Office: Royl Inst Tech, S100 44 Stockholm Sweden

PETERS, ALAN, anatomy educator; b. Nottingham, Eng., Dec. 6, 1929; came to U.S., 1966; s. Robert and Mabel (Woplington) P.; m. Verona Muriel Shipman, Sept. 30, 1955; children: Ann Verona, Sally Elizabeth, Susan Clare. BSc, Bristol (Eng.) U., 1951, PhD, 1954. Lectr. anatomy Edinburgh (Scotland) U., 1958-66; vis. lectr. Harvard, 1963-64, prof., chmn. dept. anatomy and neurobiology Boston U., 1966—; Mem. anatomy com. Nat. Bd. Med. Examiners, 1971-75; mem. neurology B study sect. NIH, 1975-79, chmn., 1978-79. Author: (with S.L. Palay and H. deF Webster) The Fine Structure of the Nervous System, 1970, 2d edit., 1976, 3rd edit., 1990, Myelination, 1970; contbr. (with A.N. Davison) articles profl. jours.; mem. editorial bd. Anat. Record, 1972-81, Jour. Comparative Neurology, Jour. Neurocytology, 1972-89, 93—, Jour. Cerebral Cortex, 1990—, Studies of Brain Function, Anat. and Embryology, 1989-92; editor book series: (with E.G. Jones) Cerebral Cortex, 1984—. Served to 2d lt. Royal Army Med. Corps, 1955-57. Mem. Anat. Soc. Gt. Britain and Ireland (Symington prize anatomy 1962, overseas mem. coun. 1969), Assn. Anatomy Chmn. (pres. 1976-77), Am. Anat. Assn. (exec. com. 1986-90, pres. 1992-93), Am. Soc. Cell Biology, Soc. Neuroscis., Internat. Primatological Soc., Cajal Club (Harman lectr. 1990, Cortical Discoverer award 1991). Home: 16 High Rock Cir Waltham MA 02154-2207 Office: Boston U Sch Medicine Dept Anatomy and Neurobiology 80 E Concord St Roxbury MA 02118-2394

PETERS, CAROL BEATTIE TAYLOR (MRS. FRANK ALBERT PETERS), mathematician; b. Washington, May 10, 1932; d. Edwin Lucius and Lois (Beattie) Taylor; B.S., U. Md., 1954, M.A., 1958; m. Frank Albert Peters, Feb. 26, 1955; children—Thomas, June, Erick, Victor. Group mgr. Tech. Operations, Inc., Arlington, Va., 1957-62, sr. staff scientist, 1964-66; supervisory analyst Datatrol Corp., Silver Spring, Md., 1962; project dir. Computer Concept, Inc., Silver Spring, 1963-64; mem. tech. staff, then mem. sr. staff Informatics Inc., Bethesda, Md., 1966-70, mgr. systems projects, 1970-71, tech. dir., 1971-76; sr. tech. dir. Ocean Data Systems, Inc., Rockville, Md., 1976-83; dir. Informatics Gen. Co., 1983-89; pres. Carol Peters Assocs., 1989—. Mem. Assn. Computing Machinery, IEEE Computer Group. Home and Office: 12311 Glen Mill Rd Potomac MD 20854-1928

PETERS, CHARLES MARTIN, research and development scientist, consultant; b. Brunssum, Netherlands, Aug. 31, 1955; s. Martin Eduard and Augustine (Daemen) P. Electronic Tech., LTS, Brunssum, 1972; electronic engr., MTS, Sittard, 1976; sci. tchr., Gelderseleerg, Nymegen, 1980; d in Physics, U. Nymegen, 1984, 87. Gen. mgmt. E.I.B. Peters, Brunssum, 1973-74; compiler implementor KUN-Nymegen, Nymegen, 1979-86; designer IBS-Computertechnik, Bielefeld, 1981; rsch. and devel. interim mgr. PéCèN BV, Nymegen, 1985-86; rsch. and devel. sci. NMi NV, Delft, 1988—; cons. Kath. Un Leuven, 1984; lectr. PATO, Eindhoven, 1990-91. Inventor: PCN System, 1985; Assymetrical top collisions in sudden approximation, 1983, Calcas System, 1989; discovery Semi Classical Q.H.E. Theory, 1986, Hall-Crystal Structure, 1986. Mem. Nerg, 1990. Mem. ISO/TAG4/WG3, NEN 400 69.03, ISO/TC69, Nederlandse Natuurkundige Vereniging. Avocations: philosophy, physics, animal welfare, nature conservation, art, dancing, painting, poetry. Home: Heyendaalseweg 245, Nymegen The Netherlands Office: Nederlands Meetinstituut, Van Swinden Laboratorium BV, Schoemakerstraar 97 NL-2600 Delft AR, The Netherlands

PETERS, ESTHER CAROLINE, aquatic toxicologist, pathobiologist, consultant; b. Greenville, S.C., May 9, 1952; d. Otto Emanuel and Winifred Ellen (Bahan) P.; m. Harry Brinton McCarty, Jr., May 27, 1984; children: Rachel Elizabeth, William Brinton. BS, Furman U., 1974; MS, U. South Fla., 1978; PhD, U. R.I., 1984. Rsch. asst. Environ. Rsch. Lab., U.S. EPA, Narragansett, R.I., 1980-81; grad. rsch. asst. U. R.I., Kingston, 1981-84; assoc. biologist JRB Assocs., Narragansett, 1984-85; postdoctoral fellow Dept. of Invertebrate Zoology, Nat. Mus. Natural History, Washington, 1985-86, resident rsch. assoc., 1986-89; rsch. fellow Registry Tumors in Lower Animals, Nat. Mus. of Natural History, Washington, 1987-91; sr. scientist Tetra Tech, Inc., Fairfax, Va., 1991—; sci. adv. panel Project Reef-keeper, Am. Littoral Soc., Miami, Fla., 1988—; courtesy asst. prof. Dept. Marine Sci., U. South Fla., St. Petersburg, 1987—; cons. The Nature Conservancy, Arlington, Va., 1991. Author: (with others) Pathobiology of Marine and Estuarine Organisms, 1993, Disease Processes of Marine Bivalve Molluscs, 1988; contbr. articles to profl. jours. Recipient Nat. Rsch. Svc. postdoctoral tng. fellowship NIH, Bethesda, Md., 1987-91. Mem. AAAS, Am. Fisheries Soc., N.Y. Acad. Scis., Soc. for Environ. Toxicology and Chemistry, Soc. Invertebrate Pathology, Sigma Xi. Office: Tetra Tech Inc 10306 Eaton Pl Ste 340 Fairfax VA 22030

PETERS, FRANK ALBERT, chemical engineer; b. Washington, June 3, 1931; s. Charles Albert and Dorothy Lynette (Paine) P.; m. Carol Beattie Taylor, Feb. 25, 1955; children: Thomas, June, Erick, Victor. BSChemE, U. Md., 1955. Devel. engr. Celanese Corp. Am., Cumberland, Md., 1955-58; chem. Engr. U.S. Bur. Mines, College Park, Md., 1958-66; project leader U.S. Bur. Mines, College Park, 1966-70, rsch. supr., 1970-77; chief process evaluation U.S. Bur. Mines, Washington, 1977—. Contbr. over 20 articles to profl. jours. Mem. Am. Inst. Chem. Engrs., Am. Assn. Cost Engrs. Avocations: photography, model railroading. Home: 12311 Glen Mill Rd Potomac MD 20854 Office: US Bur of Mines 810 7th St NW # 6202 Washington DC 20241

PETERS, JAMES EMPSON, mechanical and industrial engineering educator; b. Seymour, Ind., Sept. 11, 1954; s. Empson Geyer and Lela Evelyn (Durham) P.; m. Teresa Jane Wolf, June 7, 1980; children: Molly Elizabeth, Anna Katherine, Samuel James. BS, Purdue U., 1976, MS, 1978, PhD, 1981. Asst. prof. U. Ill., Urbana, 1981-86, assoc. prof., 1986-91, assoc. head, grad and rsch. studies, 1991--, prof., 1991--. Recipient Amoco award, 1983, Ralph R. Teetor award Soc. Automotive Engrs., 1984, Everitt award U. Ill., 1988, Sr. Zerox award, 1991; U. Ill. scholar, 1986. Fellow AIAA (assoc., Propellants and Combustion Best Paper award 1991); mem. The Combustion Inst., Am. Soc. for Engring. Edn., Am. Soc. Mechanical Engrs. Office: U Ill 1206 W Green St Urbana IL 61801

PETERS, LEO FRANCIS, environmental engineer; b. Melrose, Mass., Aug. 14, 1937; s. Joseph Leander and Mary Gertrude (Phalen) P.; m. Joan Catherine Anderson, May 20, 1961; children: Elizabeth M., Susan J., Carolyn A., Jennifer L. BS in Civil Engring., Northeastern U., Boston, 1960, MS in Civil Engring., 1966; postgrad., Harvard U., 1989. Registered profl. engr., Mass., N.H.; diplomate Am. Acad. Environ. Engrs. Jr. engr. N.Y. Dept. Transp., Albany, 1960-61; chief engr. John M. Cashman, Weymouth, Mass., 1961-62; project engr. Metcalf & Eddy, Inc., Boston, 1962-65; project engr. Weston & Sampson, Boston, 1965-67, assoc., 1967-70, ptnr., 1970-76; exec. v.p. Weston & Sampson Engrs., Inc., Boston, 1976-82; pres. Weston & Sampson Engrs., Inc., Wakefield and Peabody, Mass., 1982—; mem. Northeastern U. Corp., 1992. Clk., mem. Melrose (Mass.) Planning Bd., 1969-91. Named Young Engr. of Yr. Mass. Soc. Profl. Engrs. Fellow Am. Cons. Engrs. Coun.; mem. Am. Water Works Assn., Am. Pub. Works Assn., Water Environ. Fedn., Am. Cons. Engrs. Coun. New Eng. (pres. 1990-91), New Eng. Water Works Assn. (pres. 1989-90). Roman Catholic. Home: 187 E Emerson St Melrose MA 02176 Office: Weston & Sampson Engrs Inc 5 Centennial Dr Peabody MA 01960

PETERS, LEON, JR., electrical engineering educator, research administ; b. Columbus, Ohio, May 28, 1923; s. Leon P. and Ethel (Howland) Pierce; m. Mabel Marie Johnson, June 6, 1953; children: Amy T. Peters Thomas, Melinda A. Peters Todaro, Maria C., Patricia D., Lee A., Roberta J. Peters Cameruca, Karen E. Peters Ellingson. B.S.E.E., Ohio State U., 1950, M.S., 1954, Ph.D., 1959. Asst. prof. elec. engring. Ohio State U., Columbus, 1959-63; assoc. prof. Ohio State U., 1963-67, prof., 1967-93, prof. emeritus, 1993—; assoc. dept. chmn. for rsch. Ohio State U., Columbus, 1990-92, dir. electro sci. lab., 1983—. Contbr. articles to profl. jours. Served to 2d lt. U.S. Army, 1942-46, ETO. Fellow IEEE. Home: 1410 Lincoln Rd Columbus OH 43212-3208 Office: Ohio State U Electrosci Lab 1320 Kinnear Rd Columbus OH 43212-1191

PETERS, MERCEDES, psychoanalyst; b. N.Y.C. Student Columbia U., 1944-45; BS, L.I. U., 1945; MS, U. Conn., 1953; tng. in psychotherapy Am. Inst. Psychotherapy and Psychoanalysis, 1960-70; cert. in Psychoanalysis Postgrad. Ctr. For Mental Health, 1976; PhD in Psychoanalysis, Union Inst., 1989. Cert. psychoanalyst Am. Examining Bd. Psychoanalysis; cert. mental health cons. Social worker various agys., pub. instns., 1945-63; sr. psychotherapist Community Guidance Svc., 1960-75; staff affiliate Postgrad. Ctr. for Mental Health, 1974-76; pvt. practice psychoanalysis and psychotherapy, Bklyn., 1961—. Contbr. articles to profl. jours. Bd. dirs. Brookwood Child Care Assn. Fellow Am. Orthopsychiat. Assn.; mem. LWV, NAACP, NASW, Brooklyn Heights Mus. Soc., Postgrad. Ctr. Psychoanalytic Soc., Assn. For Psychoanalytic Self Psychology (program com.), N.Y. State Clin. Social Workers, Wednesday Club. Office: 142 Joralemon St Brooklyn NY 11201-4709

PETERS, RALPH IRWIN, JR., biology educator, researcher; b. Tulsa, June 30, 1947; s. Ralph I. and Margenelle M. (MacDowell) P.; m. Marsha A. Lerenberg; 1 child, Caitlin Louise. BS, U. Tulsa, 1969; PhD, Wash. State U., 1975. NIH pre-doctoral trainee Wash. State U., Pullman, 1971-75; rsch. assoc. Tex. A&M U., College Station, 1975-76; NIH post-doctoral fellow Wash. State Coll. of Vet. Medicine, Pullman, 1976-77; asst. prof. Bates Coll., Lewiston, Maine, 1977-80; from asst. to assoc. prof. Wichita (Kans.) State U., 1980-89; prof., chmn. Lynchburg (Va.) Coll.; reviewer West Pub. Co., 1983-84, Worth Pubs., 1985, NSF, 1987, Internat. Jour. of Comparative Psychology, 1988. Contbr. articles and abstracts to profl. jours.. With U.S. Army, 1969-71. Summer rsch. fellow USAF, 1987; rsch. grantee NSF, USAF, also others. Mem. AAAS, Sci. Rsch. Soc., Soc. for Neurosci., Internat. Brain Rsch. Orgn. Office: Lynchburg Coll 1501 Lakeside Dr Lynchburg VA 24501-3199

PETERS, RANDY ALAN, scientist; b. Glen Cove, N.Y., Apr. 23, 1953; s. Noralee Joyce (Miller) Askew; m. Angelia Kaye Fink, Jan. 17, 1973; 1 child, Tonia Michelle. BS, East Tex. State U., 1981; MS, U. Houston, Clear Lake, 1987, MBA, 1992. Cert. med. technologist. Chemist Midwest Rsch. Inst., Kansas City, Mo., 1982-84; rsch. analyst U. Tex. Med. Br., Galveston, Tex., 1984-86; rsch. scientist Krug, Clear Lake, Tex., 1986-89; sr. scientist Lockheed, Clear Lake, Tex., 1989—; cons. Krug Internat., 1989; presenter sci. meetings. With USAF, 1973-77. Charles B. Roher scholar East Tex. State U., 1980. Mem. Am. Chem. Soc., Am. Soc. Mass Spectrometry. Achievements include development of database of volatile organic compounds found in air samples in support of shuttle program, macro programs for GC-MS data systems, methods for material analysis by GC-MS; design of components of a miniature GC-MS prototype; research on pyrolysis sampling.

PETERS, RAYMOND EUGENE, computer systems company executive; b. New Haven, Aug. 24, 1933; s. Raymond and Doris Winthrop (Smith) P.; m. Mildred K. Mather, July 14, 1978 (div. Nov. 1983). Student, San Diego City Coll., 1956-61; cert., Lumbleau Real Estate Sch., 1973, Southwestern Coll., Chula Vista, Calif., 1980. Cert. quality assurance engr. Founder, pub.

Silhouette Pub. Co., San Diego, 1960-75; co-founder, news dir. Sta. XEGM, San Diego, 1964-68; news dir. Sta. XERB, Tijuana, Mex., 1973-74; founder, chief exec. officer New World Airways, Inc., San Diego, 1974-75; co-founder, exec. vice chmn. bd. San Cal Rail, Inc.-San Diego Trolley, San Diego, 1974-77; founder, pres. Ansonia Sta., micro systems, San Diego, 1986—; co-founder, dir. S.E. Community Theatre, San Diego, 1960-68; commr. New World Aviation Acad., Otay Mesa, Calif., 1971-77; co-founder New World Internat. Trade & Commerce Commn., Inc., 1991—. Author: Black Americans in Aviation, 1971, Profiles in Black American History, 1974, Eagles Don't Cry, 1988; founder, pub., editor Oceanside Lighthouse, 1958-60, San Diego Herald Dispatch, 1959-60. Co-founder, bd. dirs. San Diego County Econ. Opportunity Commn., 1964-67; co-founder Edn. Cultural Complex, San Diego, 1966-75; co-founder, exec. dir. S.E. Anti-Poverty Planning Coun., Inc., 1964-67; mem. U.S. Rep. Senatorial Inner Circle Com., Washington, 1990; mem. United Ch. Crist. With U.S. Army, 1950-53, Korea. Mem. Am. Soc. Quality Control, Nat. City C. of C., Internat. Biog. Soc., Afro-Am. Micro Systems Soc. (exec. dir. 1987—), Negro Airmen Internat. (Calif. pres. 1970-75, nat. v.p. 1975-77), Internat. Masonic Supreme Coun. (Belgium), Internat. Platform Assn., U.S. C. of C., Greater San Diego Minority C. of C. (bd. dirs. 1974—, past chmn. bd.), Masons (Most Worshipful Grand Master), Shriners (Disting. Community Svc. award 1975), Imperial Grand Potentate, Nubian Order. Republican. Avocations: creative writing, golf, world history. Home: 6538 Bell Bluff Ave San Diego CA 92119-1015

PETERS, ROBERT WILLIAM, speech and hearing sciences educator; b. Boyden, Iowa, Sept. 3, 1921; s. William Joseph and Janet Drucilla (Morris) P.; m. Helen Abramson, Sept. 3, 1949; children: Colin James, Sheila Mary. AB, U. Minn., 1948; PhD, Ohio State U., 1953. Rsch. assoc. Ohio State U. Rsch. Found., Columbus, 1953-55; asst. prof. and dir. speech and hearing clinic to prof. and chmn. dept. speech and hearing scis. U. So. Miss., Hattiesburg, 1955-68, dir. office of rsch. and projects, 1963-69; prof. and dir. inst. speech and hearing scis. U. N.C., Chapel Hill, 1969-80, prof. divsn. speech and hearing scis., dept. med. allied professions, sch. medicine, 1981-92, rsch. prof. speech, hearing scis. and psychology, 1992—; vis. rsch. scholar dept. exptl. psychology U. Cambridge, England, 1981-82, 83, 84, 86, 89, 91, 92; vis. scholar applied psychology unit Med. Rsch. Coun., Cambridge, 1981-82. Sgt. U.S. Army, 1943-46, NATOUSA. Fellow Am. Speech-Lang.-Hearing Assn., 1961; recipient Honors of the Assn., N.C. Speech, Hearing and Lang. Assn., 1976. Methodist. Achievements include research on auditory filters and aging, pitch discrimination and phase sensitivity and its relationship to frequency selectivity in young and elderly subjects, detection of temporal gaps in sinusoids by elderly subjects with and without hearing loss. Office: U NC Chapel Hill NC 27599

PETERS, THEODORE, JR., research biochemist, consultant; b. Chambersburg, Pa., May 12, 1922; s. Theodore and Miriam (Lenhardt) P.; m. Margaret Campbell, June 9, 1945; children: Theodore D., James C., Melissa Peters Barry, William L. BS in Chem. Engring., Lehigh U., 1943; PhD in Biol. Chemistry, Harvard U., 1950. Diplomate Am. Bd. Clin. Chemistry. Grad. asst. MIT, Cambridge, 1943-44; rsch. fellow Harvard Med. Sch., Boston, 1948-50; instr. U. Pa. Sch. Medicine, Phila., 1950-51; biochemist U.S. VA Hosp., Boston, 1953-55; rsch. biochemist Mary Imogene Bassett Hosp., Cooperstown, N.Y., 1955-88, rsch. scientist emeritus, 1988—; vis. scientist Carlsberg Laboratorium, Copenhagen, Denmark, 1958-59; guest worker NIH, Bethesda, Md., 1971-72; vis. rsch. prof. U. Western Australia, Perth, 1982; chmn. classification panel FDA, Washington, 1976-79; bd. dirs. Nat. Com. for Clin. Lab. Standards, Villanova, Pa., 1986-87. Editor: Plasma Protein Secretion by the Liver, 1983; chmn. bd. editors Clin. Chemistry, 1979-84; contbr. articles to profl. jours. Chmn. Sewer Bd., Cooperstown, 1975—; mem. Water Bd., Cooperstown, 1973—; chmn. lake com. Otsego County Conservation Assn., Cooperstown, 1972-78. Comdr. USNR, 1944-47, 51-53. Recipient Gold medal Biol. div. Electron Microscope Soc. Am., 1966. Fellow Am. Assn. Clin. Chemistry (pres. 1988, awards 1976, 77, 91); mem. Am. Chem. Soc., Am. Soc. Biol. Chem. Molecular Biology (emeritus), Am. Soc. for Cell Biology (emeritus), Protein Soc., Nat. Acad. for Clin. Biochemistry (diplomate), Acad. Clin. Lab. Physicians and Scientists, Phi Beta Kappa. Avocations: tennis, hiking, music. Home: 30 River St Cooperstown NY 13326-1317 Office: Mary Imogene Bassett Hosp Atwell Rd Cooperstown NY 13326-1302

PETERSDORF, ROBERT GEORGE, medical educator, association executive; b. Berlin, Feb. 14, 1926; s. Hans H. and Sonja P.; m. Patricia Horton Qua, June 2, 1951; children: Stephen Hans, John Eric. BA, Brown U., 1948, DMS (hon.), 1983; MD cum laude, Yale U., 1952; ScD (hon.), Albany Med. Coll., 1979; MA (hon.), Harvard U., 1980; DMS (hon.), Med. Coll. Pa., 1982, Brown U., 1983; DMS, Bowman-Gray Sch. Medicine, 1986; LHD (hon.), N.Y. Med. Coll., 1986; DSc (hon.), SUNY, Bklyn., 1987, Med. Coll. Ohio, 1987, Univ. Health Scis., The Chgo. Med. Sch., 1987, St. Louis U., 1988; LHD (hon.), Ea. Va. Med. Sch., 1988; DSc (hon.), Sch. Medicine, Georgetown U., 1991, Emory U., 1992, Tufts U., 1993, Mt. Sinai Sch. Medicine, 1993. Diplomate Am. Bd. Internal Medicine. Intern, asst. resident Yale U., New Haven, 1952-54; sr. asst. resident Peter Bent Brigham Hosp., Boston, 1954-55; fellow Johns Hopkins Hosp., Balt., 1955-59; chief resident, instr. medicine Yale U., 1957-58; asst. prof. medicine Johns Hopkins U., 1957-60, physician, 1958-60; assoc. prof. medicine U. Wash., Seattle, 1960-62, prof., 1962-79, chmn. dept. medicine, 1964-79; physician-in-chief U. Wash. Hosp., 1964-79; pres. Brigham and Women's Hosp., Boston, 1979-81; prof. medicine Harvard U. Med. Sch., Boston, 1979-81; dean, vice chancellor health scis. U. Calif.-San Diego Sch. Medicine, 1981-86; clin. prof. infectious diseases Sch. Medicine Georgetown U., 1986—; pres. Assn. Am. Med. Colls., Washington, 1986—; cons. to surgeon gen. USPHS, 1960-79; cons. USPHS Hosp., Seattle, 1962-79; mem. spl. med. adv. group VA, 1987—. Editor: Harrison's Priciples of Internal Medicine, 1968-90; contbr. numerous articles to profl. jours. Served with USAAF, 1944-46. Recipient Lilly medal Royal Coll. Physicians, London, 1978, Wiggers award Albany Med. Coll., 1979, Robert H. Williams award Assn. Profs. Medicine, 1983, Keen award Brown U., 1980, Disting. Svc. award Baylor Coll. Medicine, 1989, Scroll of Merit Nat. Med. Assn., 1990, 2d Ann. Founder's award Assn. Program Dirs. in Internal Medicine, 1991; named Disting. Internist of 1987, Am. Soc. Internal Medicine. Fellow AAAS, ACP (pres. 1975-76, Stengel award 1980, Disting. Tchr. award 1993), Am. Coll. Phys. Execs. (hon.); mem. Inst. Medicine of NAS (councillor 1977-80), Assn. Am. Physicians (pres. 1976-77), Cosmos Club. Home: 1827 Phelps Pl NW Washington DC 20008-1846

PETERSEN, PERRY MARVIN, agronomist; b. Bancroft, Iowa, Oct. 29, 1933; s. Jens Norgaard and Ella Marie (Saxton) P.; m. Jean Louise Bjaastad, Dec. 30, 1954; children: Patricia Lynn, Lisa Rene, Todd Perry, Timothy John. BS in Farm Ops., Iowa State U., 1959. Cert. profl. agronomist. Sales div. mgr. Walnut Grove Products, Atlantic, Iowa, 1960-65; area sales mgr., chief agronomist Amoco Oil Co., 1965-83; chief agronomist, trainer Cropmate Co., Omaha, 1983-85; agronomist, trainer Agri-Growth Rsch., Hollandale, Minn., 1986-89; mgr. tng. and devel. Terra Internat., Inc., Sioux City, Iowa, 1989—. Contbr. articles to profl. jours., various tng. matls. With U.S. Army, 1954-55. Mem. ASTM, Am. Soc. Agronomy, ARCPACS, Elks, Alpha Zeta. Office: Terra Internat Inc 600 4th St Sioux City IA 51101-1744

PETERSEN, RALPH ALLEN, chemist; b. Milw.; s. Ralph Allen Sr. and Mary Theresa P. BS in Chemistry, U. Wis., Milw., 1975, MS in Chemistry, 1978; PhD in Chemistry, U. Wis., Madison, 1983. Sr. chemist Johnson Controls Inc., Milw., 1983—. Mem. Elecrochem. Soc. Inc. (vice-chmn. So. Wis. sect. 1986-87, chmn. 1987-88, nat. sec. coun. local sects. 1992-93), Am. Chem. Soc. Achievements include 4 patents in the area of lead-acid battery processing and prodn., 2 patents pending. Office: Johnson Controls Inc PO Box 591 Milwaukee WI 53201

PETERSEN, RICHARD HERMAN, government executive, aeronautical engineer; b. Quincy, Ill., Oct. 9, 1934; s. Herman Hiese and Nancy (Getty) P.; m. Joandra Windsor Shenk, Sept. 15, 1959; children: Eric Norman, Kristin. BS in Aero. Engring., Purdue U., 1956, Dr. Engring. (hon.), 1986; MS in Aeronautics, Calif. Inst. Tech., 1957; D in Pub. Service (hon.), George Washington U., 1987; DSc (hon.), Coll. of William and Mary, 1992. Rsch. engr. NASA Ames Rsch. Ctr., Moffett Field, Calif., 1957-63, aerospace engr., 1963-65, 66-70, br. chief, 1970-73, div. chief, 1975-80; aerospace engr. NASA, Washington, 1965-66; exec. Nielsen Engring. & Rsch. Inc., Mountain View, Calif., 1973-75; dep. dir. NASA Langley Research Ctr., Hampton, Va., 1980-85, dir., 1985-91; assoc. adminstr. Aeronautics and Space Tech. NASA Hdqrs., Washington, 1991—. 1st lt. USAF, 1957-60. Recipient Disting. Alumnus award Purdue U., 1980, Meritorious Exec. award U.S. Pres., 1982, Disting. Exec. award U.S. Pres., 1989; Sloan exec. fellow Stanford U., 1973. Fellow AIAA (bd. dirs. 1984-90, Sylvanus A. Reed Aeronautics award 1991). Republican. Avocations: golf, skiing. Home: 6 Bray Wood Rd Williamsburg VA 23185-5504 Office: NASA Hdqrs Washington DC 20546

PETERSMEYER, JOHN CLINTON, architect; b. Regina, Sask., Can., Sept. 10, 1945; s. Karl Clifford and Dora Ileen (Bourne) P.; m. N. Jane Simpkins, May 7, 1966 (div. 1991); children—Brooke D., J. Croft. B.Arch., U. Man., Winnipeg, 1969. Registered architect. Design architect GBR Architects, Winnipeg, 1969-72, prin.-in-charge design mgmt., 1973-91, v.p., 1992—. Chmn. Man. Bd. Referees Unemployment Ins. Commn., Winnipeg, 1973-85; bd. dirs., chmn. ops. com. Man. Theatre Ctr., 1991—; chmn. Faculty of Arch., U. Manitoba (endowment com.), 1991—. Fellow Royal Archtl. Inst. Can.; mem. Man. Assn. Architects (pres. 1983), Ont. Assn. Architects. Liberal. Presbyterian. Avocation: graphic design. Home: 103 Fulham Ave, Winnipeg, MB Canada R3N 0G5 Office: GBR Architects, 1760 Ellice Ave, Winnipeg, MB Canada R3H 0B6

PETERSON, BRYAN CHARLES, biochemist; b. St. Croix Falls, Wis., Aug. 29, 1956; s. William Charles and Phyllis Ann (Sandstrom) P.; m. Julie Ann Baran, June 23, 1979; children: Kimberly Joy, Jennifer Lynn. BS magna cum laude, U. Minn., 1977; PhD, U. Wis., 1983. Postdoctoral fellow Northwestern U., Evanston, Ill., 1983-85; rsch. biochemist Baxter, Round Lake, Ill., 1985-86, Abbott Labs., Abbott Park, Ill., 1986—. Contbr. articles to jours., revs. to confs. Vol. PADS homeless shelter, Lake County, Ill., 1991-92, Habitat for Humanity, Lake County, 1991-92. Tng. grantee USPHS, 1985. Mem. Sigma Xi. Lutheran. Office: Abbott Labs Dept 93J-APIA 1 Abbott Park Rd Abbott Park IL 60064

PETERSON, CHARLES LOREN, agricultural engineer, educator; b. Emmett, Idaho, Dec. 27, 1938; s. Clarence James and Jane (Shelton) P.; m. Julianne Rekow, Sept. 7, 1962; children—Val, Karl, Marianne, Cheryl Ann, Charles Lauritz, Brent. B.S., U. Idaho, 1961, M.S., 1965; Ph.D. in Engring. Sci, Wash. State U., Pullman, 1973. Registered profl. engr., Idaho, Wash. Exptl. engr. Oliver Corp., Charles City, Iowa, 1961; farmer Emmett, 1962-65; instr. math. Emmett High Sch., 1962-63; instr. freshman engring., then extension agrl. engr. U. Idaho, Moscow, 1963-67; prof. agrl. engring. U. Idaho, 1973—; asst. prof. Wash. State U., 1968-73; cons. in field. Contbr. numerous articles profl. jours. Rep. precinct committeeman, 1972-75; sec. 5th legis. dist. Idaho, 1976-79; mem. Latah County Planning and Zoning Commn., 1980-90, 1st counselor Pullman Wash. Stake Presidency, 1989—. Grantee Wash. Potato, 1971-73, U & I, Inc., 1974, Amalgamated Sugar Co., 1974, 92-93, Beet Sugar Devel. Found., 1975-85, Phillips Chem. Co., 1976-80, USDA, 1976-91, Idaho Dept. Water Resources, Energy div., 1992-93, Star Found., 1978; Excellence in Rsch. award U. Idaho, 1992-93. Fellow Am. Soc. Agrl. Engrs. (chmn. Inland Empire chpt. 1978-79, chmn. nat. environ. stored products com. 1978-80, chmn. biomass energy com. 1985-86, chmn. Pacific N.W. region 1984-85, chmn. T-11 Energy com. 1989-90, dir. dist. 5, 1989-91, Engr. of Year award Inland Empire chpt., 1978, nat. Blue Ribbon award 1975, Engr. of Yr. award Pacific N.W. sect. 1990, Outstanding Paper award 1990); mem. Nat. Soc. Profl. Engrs., Am. Soc. Engring. Edn., Potato Assn. Am., Am. Soc. Sugarbeet Technologists, Idaho Soc. Profl. Engrs., Nat. Assn. Colls. and Tchrs. of Agriculture (Tchr. award 1990), Phi Kappa Phi, Sigma Xi, Gamma Sigma Delta (Outstanding Rsch. in Agriculture award 1989). Mem. LDS Ch. Office: U Idaho Agrl Engring AEL 81-B Moscow ID 83843

PETERSON, CHARLES MARQUIS, medical educator; b. N.Y.C., Mar. 8, 1943; s. Charles William and Elisabeth (Marquis) P.; m. Lois Jovanovic, Jan. 6, 1987; children: Kevin, Larisa, Boyce. BA in cum laude, Carleton Coll., 1965; MD, Columbia Coll., 1969. Intern Harlem Hosp., 1969-70, resident, 1970-73, chief resident, 1972-73; guest investigator, asst. physician The Rockefeller Univ., 1971-73, assoc. physician, 1973-78, asst. prof., 1973-78, assoc. prof., 1978-84; clin. prof. medicine U. So. Calif., 1985—; vis. clin. fellow Columbia Coll. Physicians and Surgeons, 1970-73; asst. vis. physician Harlem Hosp., 1973-84; cons. pediatrics Cornell U. Med. Ctr., 1975-84; assoc. attending medicine Beth Israel Med. Ctr., 1976-84; lectr. Mt. Sinai Sch. Medicine, 1977—; adj. assoc. prof. dept. medicine Cornell U. Med. Ctr., 1980-84; assoc. attending physician dept. medicine N.Y. Hosp., 1980-84; attending physician in medicine Cottage Hosp., Santa Barbara, Calif., 1984—; dir. rsch., med. dir. Sansum Med. Rsch. Found., 1984—; dir. diabetes Endocrine Clinic, Santa Barbara County, 1984—; dir. clin. lab. Sansum Med. CLinic, 1989—. Author: Self Monitoring of Blood Glucose: A Physician's Guide, 1981, Take Charge of Your Diabetes, 1982, Diabetes Management in the 80's, 1982; co-author: The Diabetes Self-Care Method, 1990, A Touch of Diabetes, 1991, Vivere con il Diabete, 1992, and many others; mem. editorial bd. Diabetes Care, 1980-84, Diabetes in the News, 1985—, Diabetes News Bureau, 1985—, Diabetes Profl., editor-in-chief, 1988-91, Diabetic Nephropathy/Jour. of Diabetic Complications, 1982-91; contbr. numerous articles to Prensa Medica, Jour. Lab. and Clin. Medicine, New England Jour. Medicine, Annals of Internal Medicine, Archives of Neurology, Blood, Jour. Nat. Med. Assn., Am. Jour. Obstetrics and Gynecology, many others. Mem. med. adv. bd. Cooley's Anemia Vols., 1975-84; bd. mem. Diabetes Control Found., 1980-88; dir. Diabetes Self Care Program, 1978-84, med. dir., 1981-84; bd. mem. Leake and Watts, 1978-84, Gifts for Life, 1986-89; bd. dirs. Sports Tng. Inst., 1984-86, others. Fellow ACP; mem. AAAS, Am. Chem. Soc., Am. Diabetes Assn., Am. Fedn. Clin. Rsch., Am. Med. Writers Assn., Am. Soc. Clin. Investigation, Am. Soc. Hematology, Am. Soc. Pharmacology and Experimental Therapeutics, Coun. Biology Editors, Diabetes and Pregnancy Study Group West (founder), N.Y. Acad. Scis., Rsch. Soc. Alcoholism, Soc. Experimental Medicine and Biology, Am. Med. Writers Assn., Am. Diabetes Assn. (founding bd. mem. Santa Barbara chpt. 1988—, pres. 1991-92), Sigma Xi. Home: 1075 San Antonio Creek Rd Santa Barbara CA 93111 Office: Sansum Med Rsch Found 2219 Bath St Santa Barbara CA 93105

PETERSON, DAVID MAURICE, plant physiologist, researcher; b. Woodward, Okla., July 3, 1940; s. Maurice Llewellyn and Katharine Anne (Jones) P.; m. Margaret Ingegerd Sundberg, June 18, 1965; children: Mark David, Elise Marie. BS, U. Calif., Davis, 1962; MS, U. Ill., 1964; PhD, Harvard U., 1968. Rsch. biologist Allied Chem. Corp., Morristown, N.J., 1970-71; plant physiologist U.S. Dept. Agr.-Agrl. Rsch. Svc., Madison, Wis., 1971—; from asst. to full prof. U. Wis., Madison, 1971—. Capt. U.S. Army, 1968-70. Fellow AAAS; mem. Am. Soc. Plant Physiologists (editorial bd. 1984-86), Am. Assn. Cereal Chemists (assoc. editor 1988-91), Crop Sci. Soc. Am. (assoc. editor 1975-78). Office: USDA Cereal Crops Rsch Unit 501 N Walnut St Madison WI 53705

PETERSON, DONN NEAL, forensic engineer; b. Northwood, N.D., Jan. 1, 1942; s. Emil H. and Dorothy (Neal) P.; m. Lorna Jean Kappedal, July 8, 1962 (div. July 1966); m. Donna Sue Butts Daiker, Aug. 26, 1967; children: Barbara Daiker, Elizabeth Wagner, Phoebe, Phaedra, Rosalind Peterson. BSME, U. N.D., 1963; MSME, U. Minn., 1972. Registered profl. engr. Advanced engring. courses student GE, Evendale, Ohio, 1963-66; systems engr. GE Aircraft Engine Group, Evendale, Ohio, 1963-70; prin. Donn N. Peterson & Assocs., Mpls., 1971-74; pres. Donn N. Peterson & Assocs., Inc., Mpls., 1974-85, Peterson Engring., Inc., Mpls., 1985—; instr. GE Edn. Program, 1968-69; seminar presenter State Bd. of Registration, Mpls., 1980; seminar leader Minn. Fedn. Engring. Socs., Mpls., 1990-91; speaker in field; expert witness 100 ct. trials and 100 depositions. Del. Minn. 6th Dist. Rep. Conv., Brooklyn Park, Minn., 1982. Fellow Am. Acad. of Forensic Scis. (sect. chmn. 1989-90, Founders award 1991), Nat. Acad. of Forensic Engrs.; mem. ASME (Young Engr. of Yr. 1978, state chmn. 1979-80), NSPE, Profl. Engrs. in Pvt. Practice (state pres. 1987-88, Svc. award 1988), Soc. of Automotive Engrs., Rotary Club of Brooklyn Park (sec. 1990-93, v.p. 1993—, svc. award 1992), C. of C. (city hwy. 610 corridor com. 1992—). Lutheran. Achievements include devel. of successulf math. models to simulate jet engine transient performance and wave dynamics in gas flow, computer simulations for vehicle and occupant dynamics during collisions. Home: 15720 15th Pl N Plymouth MN 55447 Office: Peterson Engring Inc 7601 Kentucky Ave N Brooklyn Park MN 55428

PETERSON, DOUGLAS ARTHUR, physician; b. Princeton, N.Y., Sept. 13, 1945; s. Arthur Roy William and Marie Hilma (Anderson) P.; m. Virginia Kay Eng., June 24, 1967; children: Rachel, Daniel, Rebecca. BA, St. Olaf, 1966; PhD, U. Minn., 1971, MD, 1975. Postdoctoral fellow U. Pitts., 1971-72; intern Hennepin County Med. Ctr., Mpls., 1975-76, resident in medicine, 1976-78; physician Bloomington Lake Clinic, Mpls., 1978-82; staff physician Mpls. VA Med. Ctr., Mpls., 1983-92, chief compensation & pension, 1992; asst. prof. U. Minn., 1985—. Bd. dirs. Rolling Acres Home, Victoria, Minn., 1985—. Maj. M.C., USAR. Mem. AAAS, Am. Assn. Pathologists, N.Y. Acad. Scis. Achievements include introduction of concept of reductive activation of receptors. Home: PO Box 24201 Minneapolis MN 55410 Office: VA Med Ctr One Veterans Dr Minneapolis MN 55417

PETERSON, DWIGHT MALCOLM, chemist; b. Mpls., July 11, 1957; s. Malcolm and Donna P. BA in Chemistry, Math., St. Olaf Coll., 1979; PhD in Chemistry, U. Minn., 1987. Computer analyst Control Data/Honeywell, Mpls., 1976-81; chemist Procter & Gamble, Cin., 1986-92; vis. prof. U. Cin., 1992—. Contbr. articles to profl. publs. Mem. AAAS, Am. Chem. Soc. Achievements include discovery of new mechanism of action for anti-cancer drug Mitomycin C, patent pending on protease inhibitors.

PETERSON, ERIC FOLLETT, engineering executive; b. San Francisco, June 2, 1960; s. Peter T. and Gail F. Peterson; m. Sylvia Oltion, Feb. 16, 1985; 1 child, Teresa. BSc in Petroleum Engring., Colo. Sch. Mines, 1982. Registered profl. engr. CAlif. Drilling engr. Sun Exploration and Production Co., Valencia, Calif., 1982-85, Oklahoma City, Okla., 1985-88; sr. engr. W.W. Irwin, Inc., Long Beach, Calif., 1988-90, mgr. ops., 1990—. Fund raiser YMCA, Long Beach, 1991, 92. Mem. Soc. Petroleum Engrs. Achievements include project engr. for first ever closure of site with groudwater contaminated by leaking fuel tank within Los Angeles basin.

PETERSON, ERIC SCOTT, physical chemist; b. Mpls., Sept. 27, 1962; s. Franklin Charles and Phyllis Audrey (White) P. BA in Chemistry, Gustavus Adolphus Coll., 1985; PhD in Chemistry, U. Calif., Berkeley, 1992. Rsch. assoc. Albert Einstein Coll. Medicine, Bronx, N.Y., 1992—. Mem. Phi Beta Kappa, Sigma Xi. Republican. Lutheran. Home: 1111 Midland Ave # 4M Bronxville NY 10708 Office: Albert Einstein Coll Med Dept Physiology & Biophys 1300 Morris Park Ave Bronx NY 10461

PETERSON, GARY, child psychiatrist; b. Lumberton, N.C., Apr. 22, 1945; s. Henry and Josephine Ruth (Grittner) P. M Engring., U. Fla., 1968; MD, U. South Fla., 1974. Diplomate Am. Bd. Med. Examiners, Am. Bd. Psychiatry and Neurology in Child Psychiatry and Gen. Psychiatry. Resident psychiatry U. Oreg. Health Sci. Ctr., Portland, 1975-77; child psychiatry fellow UCLA, 1977-79, asst. clin. prof., 1979-80; asst. prof. U. N.Mex., Albuquerque, 1979-80; rsch. fellow U. N.C., Chapel Hill, 1984-86; rsch. psychiatrist Dorothea Dix Clin. Rsch. Unit, Raleigh, N.C., 1986-88; clin. assoc. prof. U. N.C., Chapel Hill, 1984—; cons. Willie M, Dept. Human Resources, Raleigh, 1988, Southeast Inst., Chapel Hill, 1989—, Structure House, Durham, N.C., 1991—. Co-author: MPD Explained for Kids, 1991; contbr. articles to profl. jours. Bd. mem. Orange-Person-Chatham Mental Health Ctr., N.C., 1991—. Capt. USAF, 1967-71. Dept. Engring. scholar U. Fla., Gainesville, 1967-68, Coll. Medicine scholar U. South Fla., Tampa, 1971-72; named Competent Toastmaster Toastmasters Internat., 1991. Fellow Am. Psychiat. Assn., Am. Acad. Child and Adolescent Psychiatry, Orthopsychiatry; mem. Internat. Soc. for Study of Multiple Personality and Dissociation (chair child and adolescent dissociative disorders com., Disting. Achievement award 1992). Achievements include research which shows that intracisternal oxytocin can induct maternal-like behavior in developing rats; efforts towards awareness of childhood dissociative disorders in our society and has been instrumental in altering the psychiatric taxonomy to include the dimension of childhood dissociative disorders. Office: Southeast Inst 103 Edwards Ridge Chapel Hill NC 27514

PETERSON, GEORGE P., mechanical engineer, research and development firm executive; b. N.Y.C., Mar. 30, 1930; s. Peter and Evangeline (Soumakis) P.; m. Judith A. Moyer, May 1, 1965; children—Theodore S., Evangeline P., Antonia L. B.S.M.E., Columbia U., 1951. Devel. engr. materials lab. U.S. Air Force, Wright-Patterson AFB, Ohio, 1951-74, dir. materials lab., 1974-77, dep. dir. Wright Aero. Labs., 1977-80, dir. materials lab., 1980-85; pres. George Peterson Resources, Inc., Miamisburg, Ohio, 1985—; chmn. materials program NATO, Paris, 1977-79. Contbr., co-contbr. articles, papers on composite materials to profl. publs. Trustee, Engring. and Sci. Found. of Dayton, Ohio, 1983, v.p., 1985. Served to 1st lt. USAF, 1952-53. Recipient Structural Dynamics and Materials award AIAA, 1973; Decoration for Exceptional Civilian Service U.S. Air Force, 1980; Presdl. Meritorious Exec. award, 1982, Gustus L. Larson Meml. award ASME, 1992. Mem. NAE, Am. Soc. Metals (hon. life mem.), Soc. Aerospace Materials and Process Engrs. (hon. life mem.). Greek Orthodox. Avocation: tennis. Home: 9877 Washington Church Rd Miamisburg OH 45342-4511 Office: George P Peterson 9877 Washington Church Rd Miamisburg OH 45342-4511

PETERSON, GLENN STEPHEN, chemist; b. Bradford, Pa., May 24, 1952; s. Gordon Alan and Joan Elizabeth (Bredenberg) P.; m. Eleanor Mary Lennon, June 17, 1978; children: Christopher Sean, Gregory Brian, Kevin Michael. BS, Rensselaer Poly. Inst., 1974; MS, So. Conn. State U., 1982. Sr. chemist Uniroyal Chemical, Naugatuck, Conn., 1974-85; svc. engr. Nermag, Fairfield, N.J., 1985-86; chemist Yale U., New Haven, 1987-88; sr. scientist Envirite Analytical Svcs., Watertown, Conn., 1988-89; mgr. gas chromatography/mass spectrometry Adirondack Environ. Svcs., Albany, N.Y., 1989-91; sales rep. Finnigan MAT, San Jose, Calif., 1991—; sec. Conn. Mass Spectrometry Disc. Group, Wallingford, 1984-85, co-chmn., 1985-86. Author: Mass Spectrometry, vol. 193, 1989; contbr. articles to Biol. Mass Spectrometry, Pesticide Sci. Den leader troop 78 Cub Scouts, Loudonville, N.Y., 1990-91; asst. scoutmaster troop 78 Boy Scouts Am., Loudonville, 1992. Mem. Am. Soc. for Mass Spectrometry (contbr. papers to nat. meetings 1985, 88), Am. Chem. Soc. Achievements include 1 patent in field. Home: 110 Greenleaf Dr Latham NY 12110 Office: Finnigan MAT Ste 711 7 Regent St Livingston NJ 07039

PETERSON, HOLGER MARTIN, electrical engineer; b. Colman, S.D., Nov. 16, 1912; s. Peter and Karen Marie (Jensen) P.; m. Myrtle Berthine Teigen, Mar. 26, 1939; children: Robert Kent, Janice Marie (Peterson) Priddy. BS in Elec. Engring., S.D. State U., 1933. Chief draftsman bridge dept. S.D. Highway Commn., Pierre, 1935-39; draftsman U.S. Army Corps of Engrs., Tulsa, Okla., 1939-40; tech. instr. Army Air Corps, Rantoul, Ill., 1940-41; supr. Plans & Programming Mgr. USAF Tech. Tng. Command, Biloxi, Miss., 1941-50, Sheppard AFB, Tex., 1950-70; program mgr. Individual Devel. Ctr., Wichita Falls, Tex., 1970-72, cons., 1972-80; owner, mgr. Creative Leaded Glass Co., Wichita Falls, 1981-92; mem. Commn. on Disability, 1990—. Editor (tech. manuals, extension courses) Aircraft Maintenance, 1941-42, 1968-70. Recipient Exceptional Civilian Svc. award USAF, 1961. Mem. Elks. Democrat. Lutheran. Avocations: audio equipment, stained glass art, fishing. Home and Office: 4810 Marsha Ln Wichita Falls TX 76302-4006

PETERSON, JACK MILTON, retired physicist; b. Portland, Oreg., Apr. 25, 1920; s. Adolph Julius and Anna Sabina (Persson) P.; m. Beverly Lael Begole, Aug. 31, 1946; children: Laelanne Seville, Sharon, Diane. SB summa cum laude, Harvard U., 1942; PhD, U. Calif., Berkeley, 1950. Staff mem. radiation lab. MIT, Cambridge, Mass., 1942-46; asst. dir. vacuum tube devel. com. Columbia U., N.Y.C., 1943-46; div. leader Lawrence Livermore (Calif.) Nat. Lab., 1952-64; sr. staff scientist Lawrence Berkeley Lab., 1964-85; sr. scientist SSC Lab., Dallas, 1985-93, retired, 1993; lectr. physics dept. U. Calif., Berkeley, 1952-54; Fulbright fellow Bohr Inst., Copenhagen, 1960-61; vis. scientist Max Planck Inst., Munich, 1973; mem. neutron cross sect. adv. com. AEC, Washington, 1953-60; cons. in field. Author, editor numerous articles and book contbns. in field. Mem. Am. Phys. Soc., Phi Beta Kappa. Home: 350 Cordell Dr Danville CA 94526

PETERSON, JAMES ALGERT, geologist, educator; b. Baroda, Mich., Apr. 17, 1915; s. Djalma Hardaman and Mary Avis (McAnally) P.; m. Gladys Marie Pearson, Aug. 18, 1944; children—James D., Wendy A., Brian H. Student, Northwestern U., 1941-43, U. Wis., 1943; B.S. magna cum laude, St. Louis U., 1948; M.S. (Shell fellow), U. Minn., 1950, Ph.D., 1951. Mem. staff U.S. Geol. Survey, Spokane, Wash., 1949-51; instr. geology

Wash. State U., Pullman, 1951; geologist Shell Oil Co., 1952-65; geologist div. stratigrapher, 1958-63, sr. geologist, 1963-65; instr. geology N. Mex. State U., San Juan, (P.R.), br., 1959-65; prof. geology U. Mont., Missoula, 1965—; cons. U.S. Geol. Survey, 1976-82, rsch. geologist, 1982—. Editor: Geology of East Central Utah, 1956, Geometry of Sandstone Bodies, 1960, Rocky Mountain Sedimentary Basins, 1965, (with others) Pacific Geology, Paleotectonics and Sedimentation, 1986; Contbr. (with others) articles to profl. jours. Served to 1st lt. USAAF, 1943-46. Recipient Alumni Merit award St. Louis U., 1960. Fellow AAAS, Geol. Soc. Am.; mem. Am. Assn. Petroleum Geologists (pres. Rocky Mountain sect. 1964, Pres.'s award 1988, Disting. Svc. award 1992), Rocky Mountain Assn. Geologists (Outstanding Scientist award 1987), Four Corners Geol. Soc. (hon., pres. 1962), Am. Inst. Profl. Geologists (pres. Mont. sect. 1971), Soc. Econ. Paleontologists and Mineralogists (hon. 1985, sec.-treas. 1969-71, editor 1976-78, Disting. Pioneer Geologist award 1988), Mont. Geol. Soc. (hon. 1987), Utah Geol. Soc. Club: Explorers. Home: 301 Pattee Canyon Dr Missoula MT 59803-1624

PETERSON, LAUREN MICHAEL, physicist, educator; b. Minn., June 1943. BS, U. Minn., 1966; MS, Pa. State U., 1968, PhD, 1972. Physicist U.S. Bur. Mines, Mpls., 1965-67; rsch. asst. Pa. State U., University Park, 1966-72; rsch. physicist Environ. Rsch. Inst. Mich., Ann Arbor, 1972—; assoc. prof. U. Mich., Ann Arbor, 1980—. Mem. Optical Soc. Am. (local officer). Achievements include patents in field. Office: Environ Rsch Inst Mich Box 134001 Ann Arbor MI 48113-4001

PETERSON, PAUL MICHAEL, agrostologist; b. Pleasanton, Calif., May 2, 1954; s. Paul Gene and Darleen Roberta (Schmitt) P.; m. Carol Ruth Annable, Apr. 1, 1982. MS, U. Nev., Las Vegas, 1984; PhD, Wash. State U., 1988. Range technician Bur. Land Mgmt., Salmon, Idaho, 1979, U.S. Forest Svc., Mammoth Lakes, Calif., 1980-81; rsch. asst. Coop. Nat. Park Resources Studies Unit, Las Vegas, 1981-83; curatorial asst. Wash. State U., Pullman, 1984-86, teaching asst., 1987-88, instr., 1988; assoc. curator Smithsonian Instn., Washington, 1988—; co-editor New World Grass Flora Project, Washington, 1992—. Contbr. articles to Systematic Botany, Am. Jour. Botany, Wasmann Jour. Biology, others. Grantee Nat. Geog. Soc., 1991-92, NSF, Mex., 1986-88, Rsch. Opportunities, 1989-91, others. Mem. Am. Soc. Plant Taxonomists, Bot. Soc. Am., Calif. Bot. Soc., Sigma Xi. Achievements include using chloroplast-DNA restriction site variation as evidence for the monophyletic origin of the subtribe Muhlengerginae which includes Bealia, Blepharoneuron, Chaboissaea, Lycurus, Muhlenbergia, and Pereilema. Office: Smithsonian Instn NHB-166 10th and Constitution Ave Washington DC 20560

PETERSON, STEPHEN CARY, mechanical engineer; b. Roseville, Calif., Sept. 17, 1960; s. Victor Marsh and Mardelle Tressiemay (Sawyer) P.; m. Lorry Lawrence Peterson, Aug. 1, 1987; children: Hannah Rae, Haley Victoria. BSME, U. Calif. Berkeley, 1983. Registered profl. engr., Ariz. Engr. Garrett Turbine Engine Co., Phoenix, 1983-86; instr. Nat. U. for Def. Tech., Changsha, Hunan, China, 1986, Changsha Transp. Inst., Changsha, Hunan, China, 1987-89; factory supr. Fuzhou FuDa Electronics Ltd., Fujian, Fuvian, China, 1990; instr. Ctr. for Acad. Precocity, Ariz. State U., Tempe, 1990-91; engr. Brown & Caldwell, Phoenix, 1991, CRS Sirrine Engrs., Phoenix, 1991-93. Refugee sponsor Faith E.V. Free Ch., Ariz., 1984-86; refugee relief worker Youth with a Mission, Thialand, 1982; vol. Internat. Students Inc., Ariz., 1990-93. Named one of Outstanding Young Men of Am., 1985. Mem. ASME, Am. Soc. Heating, Refrigeration, and Air Conditioning Engrs. Achievements include solar crop dryer; instructional labs for Chinese Univs. Home: 1050 S Stanley #136 Tempe AZ 85281 Office: CRSS Inc 3200 E Camelback Ste #170 Phoenix AZ 85012

PETERSON, VICTOR LOWELL, aerospace engineer, research center administrator; b. Saskatoon, Sask., Can., June 11, 1934; came to U.S., 1937; s. Edwin Galladet and Ruth Mildred (McKeeby) P.; m. Jacqueline Dianne Hubbard, Dec. 21, 1955; children: Linda Kay Peterson Landrith, Janet Gale, Victor Craig. BS in Aero. Engring., Oreg. State U., 1956; MS in Aerospace Engring., Stanford U., 1964; MS in Mgmt., MIT, 1973. Rsch. scientist NASA-Ames Rsch. Ctr., Moffett Field, Calif., 1956-68, asst. chief hypersonic aerodyns., 1968-71, chief aerodyns. br., 1971-74, chief thermo and gas dynamics div., 1974-84, dir. aerophysics, 1984-90, dep. dir., 1990—; mem. nat. adv. bd. U. Tenn. Space Inst., Tullahoma, 1984—. Contbr. numerous articles to profl. jours. Treas. Woodland Acres Homeowners Assn., Los Altos, Calif., 1978—. Capt. USAF, 1957-60. Recipient medal for outstanding leadership NASA, 1982; Alfred P. Sloan fellow MIT, 1972-73. Fellow AIAA. Republican. Methodist. Achievements include development of numerical aerodynamic simulation system for aerospace, of method for reconstructing planetary atmosphere structure from accelerations of body entering atmosphere, of theory for motions of tumbling bodies entering planetary atmospheres. Home: 484 Aspen Way Los Altos CA 94024-7126 Office: NASA-Ames Rsch Ctr Mail Stop 200-2 Moffett Field CA 94035

PETERSON, WILLIAM FRANK, physician, administrator; b. Newark, Sept. 28, 1922; s. Edgar Charles and Margaret Benedict (Heyn) P.; m. Margaret Henderson Lee, June 28, 1946 (div. 1978); children: Margaret Lee, Edward Charles; m. 2d, Mary Ann Estelle McGrath, Nov. 29, 1980. Student, Cornell U., 1940-43; MD, N.Y. Med. Coll., 1946. Commd. lt. U.S. Air Force, 1946, advanced through grades to col., 1963; med. officer U.S. Air Force, 1946-70; chmn. dept. ob-gyn Washington Hosp. Ctr. 1970-93; dir. Women's Clinic. Washington, 1971—, Ob-Gyn Ultrasound Lab., Washington, 1974—. Contbr. articles to profl. jours. Chmn., Maternal Mortality Com., 1981—. Decorated Legion of Merit, 1960, 70; Cert. Achievement, Office Surgeon Gen., USAF, 1967. Fellow Am. Coll. Ob-Gyn, ACS, Nat. Bd. Med. Examiners (diplomate), Washington Gynecol. Soc. (exec. council 1980-85). Republican. Episcopalian. Home: 50 Stonegate Dr Silver Spring MD 20905-5701 Office: Washington Hosp Ctr 110 Irving St NW Washington DC 20010

PETHACHI, MUTHIAH CHIDAMBARAM, textile industry executive; b. Pallathur, India, Sept. 3, 1933; s. M.CT.M. Chidambaram Chettyar and CT. Vallimammai (Achi) Chettyar; m. Sivagami Pethachi, Jan. 26, 1958 (dec. 1990); children: M.CT.P. Chidambaram, M.CT.P. Muthiah. BS, Purdue U., 1954. Dir. Travancore Rayons Ltd., Rayonpuram, Kerala, India, 1968-86, mng. dir., 1986—, vice chmn. Mem. Am. Chem. Soc. Home: Bedford Villa 9, Leith Castle North St, Madras 600 028, India Office: 742 Anna Salai, PB 2730 Madras 600 002, India

PETIT, WILLIAM, chemist; b. Port of Spain, Trinidad, June 23, 1965; came to U.S. 1987.; s. William Granger Petit Sr. BS, U. Toronto, 1987; postgrad., Boston U., 1993—. Rsch. technician Inst. Marine Affairs, Chaguaramas, Trinidad, 1983-84; teaching fellow Boston U., 1987-90; tech. writer Bioinformation Assocs., Boston, 1989; lectr. Franklin Inst. Boston, 1990; writing fellow Boston U., 1990-91; chemist Polaroid Corp., Waltham, Mass., 1991--; instr. chemistry Engring. and Scientific Resources for Advancement, Boston, 1992--. Co-author: Biodegradable Homopolymers and Photodegradability of Commercial Polymers, 1989; editor: Caribbean Voice, 1992. Mem. Nat. Orgn. for Black Chemists and Chem. Engrs. (sec. 1991--), Phi Beta Sigma. Office: Polaroid Corp 1265 Main St W6-1A Waltham MA 02254

PETITPIERRE, EDUARD, genetics educator; b. Barcelona, Catalonia, Spain, Feb. 14, 1941; s. Eduard and Helena (Vall) P.; m. Carmelen Pedrerol, Sept. 8, 1976; children: Claudia, Pol. Grad. Biol. Scis., U. Barcelona, Spain, 1965; PhD in Biol. Scis., U. Barcelona, 1972. Asst. prof. genetics U. Barcelona, 1966-71, provisional assoc. prof., 1972-74, assoc. prof., 1975-80; prof. genetics U. Baleares, Palma de Mallorca, Spain, 1981—. Co-editor: Biology of Chrysomelidae, 1988. Mem. AAAS, Soc. Catalana Biology, Inst. Catalana Natural History, Assn. Española Entomologia, European Assn. Coleopterology. Home: Ramon Llull 8K, 07190 Esporles Spain Office: Univ Balearics, Crta Valldemossa km 7.5, 07071 Palma Mallo Spain

PETITTE, JAMES NICHOLAS, poultry science educator; b. Camden, N.J., July 25, 1957; s. Nicholas James and Barbara Agnes (Schmid) P.; m. Jane Marie Schlegel, Nov. 24, 1979; children: Jennifer, John. AB, Susquehanna U., 1979; MS, U. Maine, Orono, 1981; PhD, U. Guelph, Ont., Can., 1986. Asst. prof. dept. poultry sci. N.C. State U., Raleigh, 1990—; cons. Embrex, Inc., Research Triangle Park, N.C., 1991—. Contbr. articles to profl. jours. Grantee USDA, 1991, 92, N.C. Biotech. Ctr., 1991. Mem. AAAS, Soc. for Study of Reprodn., Poultry Sci. Assn., World's Poultry Sci. Assn. Roman Catholic. Office: NC State U Box 7608 Raleigh NC 27695-7608

PETRICH, MARK ANTON, chemical engineer, educator; b. Evergreen Park, Ill., May 10, 1961; s. George M. and Rosemary P.; m. Katherine McHugh, July 2, 1983; 1 child, Joseph William. BS, Washington U., St. Louis, 1982; PhD in Chem. Engring., U. Calif., Berkeley, 1987. Registered profl. engr., Ill. Asst. prof. chem. engring. Northwestern U., Evanston, Ill., 1987—; Morris E. Fine Jr. prof. materials and mfg. McCormick Sch. Engring., Northwestern U., 1990-93; cons. Advanced Fuel Rsch., East Hartford, Conn., 1989—, Ind. Environ. Svcs., Homewood, Ill., 1987—. Mem. Am. Inst. Chem. Engrs., Am. Chem. Soc., Electrochem. Soc., Am. Phys. Soc. Roman Catholic. Achievements include work on various solid waste resource recovery schemes, including manufacture of activated carbon from scrap tires. Office: Northwestern U Dept Chem Engring Evanston IL 60208

PETRICOLA, ANTHONY JOHN, chemical engineer; b. Timmins, Ont., Can., May 30, 1936; came to U.S. 1980; m. Denise Mary Maisonneuve, Dec. 16, 1937; children: Mario, Gina, Christine. BA, U. Toronto, 1959. Rsch. engr. Internat. Paper Co., Hawkesbury, Ont., 1959-66; mfg. supt. Hiram Walker, Windsor, Ont., 1966-80; chief process engr. Archer Daniels Midland, Decatur, Ill., 1980-84; sr. design engr. Brown-Forman, Louisville, 1984; plant mgr., corp. v.p. engring. Midwest Grain Products, Pekin, Ill., 1985—. Mem. Pekin Rotary (pres. 1992-93). Achievements include ethanol plant design; novel energy efficient mash cooking process; non-polluting, low energy feed drying process; waste heat evaporation processes. Home: 1918 St Clair Dr Pekin IL 61554 Office: Midwest Grain Products South Front St Pekin IL 61554

PETRIDES, GEORGE ATHAN, ecologist, educator; b. N.Y.C., Aug. 1, 1916; s. George Athan and Grace Emeline (Ladd) P.; m. Miriam Clarissa Pasma, Nov. 30, 1940; children: George H., Olivia L., Lisa B. B.S., George Washington U., 1938; M.S., Cornell U., 1940; Ph.D., Ohio State U., 1948; postdoctoral fellow, U. Ga., 1963-64. Naturalist Nat. Park Service, Washington and Yosemite, Calif., 1938-43, Glacier Nat. Park, Mont., 1947, Mt. McKinley Nat. Park, Alaska, 1959; game technician W.Va. Conservation Commn., Charleston, 1941; instr. Am. U., 1942-43, Ohio State U., 1946-48; leader Tex. Coop. Wildlife Unit; assoc. prof. wildlife mgmt. Tex. A. and M. Coll., 1948-50; assoc. prof. wildlife mgmt., zool. and African studies Mich. State U., 1950-58, prof., 1958—; research prof. U. Pretoria, S. Africa, 1965; vis. prof. U. Kiel, Germany, 1967; vis. prof. wildlife mgmt. Kanha Nat. Park, India, 1983; del. sci. confs. Warsaw, 1960, Nairobi and Salisbury, 1963, Sao Paulo, Aberdeen, 1965, Lucerne, 1966, Varanasi, India, Nairobi, 1967, Oxford, Eng., Paris, 1968, Durban, 1971, Mexico City, 1971, 73, Banff, 1972, Nairobi, Moscow, The Hague, 1974, Johannesburg, 1977, Sydney, 1978, Kuala Lumpur, 1979, Cairns, Australia, Mogadishu, Somalia, Peshawar, Pakistan, 1980; participant NSF Expdn., Antarctic, 1972, FAO mission to, Afghanistan, 1972, World Bank mission to, Malaysia, 1975. Author: Field Guide to Trees and Shrubs, 1958, 2d edit., 1972; Editor wildlife mgmt. terrestial sect.: Biol. Abstracts, 1947-72; Contbr. articles to biol. publs. Served to lt. USNR, 1943-46. Fulbright research awards in E. Africa Nat. Parks Kenya, 1953-54; Fulbright research awards in E. Africa Nat. Parks Kenya, Uganda, 1956-57; N.Y. Zool. Soc. grantee Ethiopia, Sudan, 1957; N.Y. Zool. Soc. grantee Thailand, 1977; Mich. State U. grantee Nigeria, 1962; Mich. State U. grantee Zambia, 1966; Mich. State U. grantee Kenya, 1969; Mich. State U. grantee Africa, 1970, 71, 73, 81; Mich. State U. grantee Greece, 1974, 83; Mich. State U. grantee Iran, 1974; Mich. State U. grantee Botswana, 1977; Mich. State U. grantee Papua New Guinea, Thailand, 1979; Iran Dept. Environment grantee, 1977; Smithsonian Instn. grantee India and Nepal, 1967, 68, 75, 77, 83, 85; World Wildlife Fund grantee W. Africa, 1968. Mem. Am. Ornithologists Union, Am. Soc. Mammalogists, Wildlife Soc. (exec. sec. 1953), Wilderness Soc., Am. Comm. Internat. Wildlife Protection, Ecol. Soc., Fauna Preservation Soc., E. African Wildlife Soc., Internat. Union Conservation Nature, Zool. Soc. So. Africa, Sigma Xi. Presbyterian. Home: 4895 Barton Rd Williamston MI 48895-9649 Office: Mich State U Dept Fisheries and Wildlife East Lansing MI 48824

PETRIDIS, PETROS ANTONIOS, electrical engineer; b. Veria, Macedonia, Greece, July 17, 1959; came to U.S., 1980; s. Antonios Petridis and Anthi Dimitriadou Petridou. Diploma in engring., Salford, Manchester, U.K., 1980; BS, No. Ill. U., 1983. Rsch. asst. No. Ill. U., Dekalb, 1983-85; quartz crystal engr. CTS, Sandwich, Ill., 1985-91, project mgr., 1991—. Contbr. papers in field. Mem. IEEE. Achievements include patent for preferential etching of a piezoelectric material. Office: CTS Corp 400 Reimann St Sandwich IL 60548-1866

PETRIE, GEORGE WHITEFIELD, III, retired mathematics educator; b. Pitts., May 6, 1912; s. George Whitefield Jr. and Mabel Margaret (O'Reiley) P.; m. Dorothy May Koeppen, May 28, 1944 (dec. 1965); children: George Whitefield IV, Charles Richard; m. Mildred Pearl McClary, Jan. 1, 1967. BS in Math., Carnegie Mellon U., 1933, MS in Math., 1936; PhD, Lehigh U., 1949. Assoc. prof. math. S.D. Sch. of Mines, Rapid City, 1936-41; piping engr. Pitts. Piping & Equipment Co., 1941-42; navigation instr. USNR Midshipmen's Sch., N.Y.C., 1943-45; spl. engr. Bethlehem (Pa.) Steel Co., 1945-47; asst. prof. math. and astronomy Lehigh U., Bethlehem, 1947-50; system engr., edn. exec. IBM Corp., 1950-66; prof. math. New Coll., Sarasota, Fla., 1966-68; prof., dept. chmn. U. Wis., Green Bay, 1968-76; ret. U. Wis.; cons. IBM Corp., Armonk, N.Y., 1967-77; tax cons. S.E. Bank Trust Co., Sarasota, Fla., 1977-78; supr. testing hydraulic intensifier Bethlehem Steel Co. Contbr. articles to profl. jours. and publs. Treas. Healthcare Resources, Inc., Sarasota, 1983—; tax cons. Am. Assn. Ret. Persons, Sarasota, 1989—; pres. Sarasota Inst. Lifetime Learning, 1984-86, Sarasota Community Concert Assn., 1985-87. Mem. Am. Math. Soc., Math. Assn. Am., Am. Guild Organists, Bird Key Yacht Club. Avocations: music, chess, bridge. Home: 4573 N Lake Dr Sarasota FL 34232-1938

PETROPOULOS, LABROS S., physicist, researcher; b. Katerini, Pierias, Greece, Aug. 5, 1962; came to U.S., 1986; s. Spiros L. and Elli (Theodosiadou) P. MSc, DePaul U., Chgo., 1989; PhD in Physics, Case We. Res. U., 1993. Cert. magnetic resonance imaging specialist. Teaching asst. dept. physics DePaul U., Chgo., 1987-89; rsch. asst. Case We. Res. U., Cleve., 1989—; cons. Picker Internat. Inc., Highland Heights, Ohio, 1989—. Contbr. articles to Rev. Sci. Instruments, Physics in Medicine and Biology, Jour. Magnetic Resonance, Measurement Sci. and Tech. Recipient Govt. award Greek Minister of Edn., 1980. Mem. Soc. Magnetic Resonance Imaging, Am. Phys. Soc. Achievements include patents for elliptical coil, asymmetric coil. Home: 42 Hepirou St, Athens Greece 15341

PETROVIC, ALEXANDRE GABRIEL, physician, physiology educator, medical research director; b. Belgrade, Yugoslavia, July 10, 1925; naturalized French citizen; s. Gabriel M. and Maria S. (Miskovic) P.; m. Suzanne Durry, Feb. 25, 1956; 1 child, Nicole Gasson. MD, Strasbourg Med. Sch., 1954; DSc, Strasbourg U., 1961; postgrad. McGill U. Med. Sch., 1961-62. Assoc. staff physician, asst. prof. Northwestern U. Med. Sch., Chgo., 1965-68; prof. U. Montreal Med. Sch., 1970-71; now dir. rsch. Nat. Inst. Health and Med. Research, Strasbourg, 1968—; prof. human physiology U. Louis Pasteur Med. Sch., Strasbourg, 1976-90, lectr. in biomed. rsch., methology, 1989—, also mem. sci. coun.; vis. rsch. scientist Center for Human Growth and Devel., U. Mich., Ann Arbor, 1976-78; vis. prof. La. State U. Med. Ctr., New Orleans, 1979—; van der Klaauw prof. U. Leiden, Netherlands, 1985; prof. U. Cattolica del Sacro Cuore, Rome, Italy, 1992—; prof. honoris causa U. Camilo Castelo Branco, São Paulo, Brazil; charge of French med. missions to USSR, 1969, Yugoslavia, 1969, 74, 76, 78, 81, Argentina, Peru, Brazil and Chile, 1974, U.S., 1977, 78, 82, Cuba, 1986. Recipient prize Vlès, Strasbourg Med. Sch., 1954; prize Laborde, Biol. Soc., 1961; E. Sheldon Friel award European Orthodontic Soc., 1976; Calvin Case award for orthodontic rsch., 1984. Mem. Soc. Cryobiology (charter), Acad. Medicine (Belgrade) (honor mem.), Club Internat. de Morphologie Faciale, Assn. des Physiologistes, European Tissue Culture Soc., etc. Contbg. author various books and sci. papers on a cybernetic theory of the mechanisms of craniofacial bone growth, on cytopathogenesis of craniostenosis and on philosophy of biomed. research; discovered feasibility of orthopedically stimulating the growth of the mandible; described new ways in orthodontic decision making; pioneer research studies on treatment of otospongiosis by sodium fluoride and disphosphonates, on theory of auto-immune origin of otospongiosis; new classification of bone tumors; discovered possibility of prefecondatory hereditary male contbribution by penetration of spermatozoary DNA into intraovarian ovocytes. Home: 2 rue de Rome, 67000 Strasbourg France Office: Inst Nat de la Sante et de la Recherche Medicale, 2 Rue de Rome, 67000 Strasbourg France

PETRUCCI, JANE MARGARET, medical technologist, laboratory director; b. Providence, R.I., May 26, 1955; d. James Francis and Dorothy Mary (Markarian) P. BA in Biology, R.I. Coll., 1979. Med. technologist Miriam Hosp., Providence, R.I., 1979-82; med. technologist R.I. Blood Ctr., Providence, 1981-83, lab. supr., 1983-89; lab. dir. R.I. Blood Ctr., 1989—. Mem. Vet's. Meml. Auditorium Preservation Assn., Providence, 1989—. Named to Nat. Honor Soc., 1974. Mem. R.I. Blood Banker's Soc. (bd. dirs. 1989—), Am. Assn. Clin. Pathologists (cert. blood bank technologist), Am. Assn. Blood Banks, Am. Assn. Bioanalysts. Avocations: travel, astronomy, philosophy. Office: Rhode Island Blood Ctr 405 Promenade St Providence RI 02908-4823

PETRY, HEYWOOD MEGSON, psychology educator; b. Hartford, Conn., Mar. 7, 1952; s. Henry Anthony and Madeline Heywood (Megson) P.; m. Michele Falzett, Sept. 3, 1983; 2 children. BA, Bates Coll., 1974; MA, Conn. Coll., 1976; PhD, Brown U., 1981. Postdoctoral fellow Vanderbilt U., Nashville, 1980-82, Brown U., Providence, 1982-84; asst. prof. SUNY, Stony Brook, 1984-91; assoc. ophthalmology and visual scis. dept. U. Louisville, 1991—, assoc. prof. psychology dept., 1991—. Reviewer profl. jours.; contbr. articles to profl. jours. including Brain Rsch., Nature, Jour. Comparative Neurology, Visual Neurosci., Vision Rsch., Progress in Brain Rsch., Psychophysiology, Am. Jour. Psychology. Grad. fellow Conn. Coll., 1974-76; rsch. grantee NIH, 1982-83, 87-93, Fight-for-Sight, Inc., 1992-93; recipient postdoctoral traineeship NIMH, 1980-82, NIH, 1982-84. Mem. AAAS, Assn. for Rsch. in Vision in Ophthalmology, Optical Soc. Am., Soc. for Neurosci., Sigma Xi. Office: U Louisville Dept Psychology Louisville KY 40292

PETTEE, DANIEL STARR, neurologist; b. N.Y.C., Feb. 15, 1925; s. Allen Danforth and Helen Marien (Starr) P.; m. Dimetra Marie Peters, June 24, 1961; children: William, Margaret, Allen. BA, Yale U., 1951; MD, Columbia U., N.Y.C., 1955. Cert. Am. Psychiatry and Neurology, 1965, Am. Bd. Clin. Neurophysiology, 1984. Rotating internship Strong Meml. Hosp. U. Rochester, N.Y., 1955-57, residency neurology, 1957-62; neurologist pvt. practice, Rochester, N.Y., 1962—; clinic dir. Rochester (N.Y.) Area Multiple Sclerosis Chpt., Rochester, N.Y., 1962-76; assoc. prof. neurology U. Rochester (N.Y.) Sch. Medicine, 1978—; clin. assoc. Dept Neurology Strong Meml. Hosp., Rochester, N.Y., 1978—; head neurology Div. Dept. Medicine The Genesee Hosp., Rochester, N.Y., 1972—; pres. Genesee Neurological Assocs., Rochester, 1974—; mem. bd. dirs. Rochester (N.Y.) Area Multiple Sclerosis Chpt., 1970-76. Contbr. articles to profl. jours. Mem. bd. dirs., singer Rochester (N.Y.) Oratorio Soc., 1960-61, 1955-78. Recipient Purple Heart, Bronze Star U.S. Army, 1944, Bronze Hope Chest for Svc. award Rochester (N.Y.) Area Multiple Sclerosis Chpt., 1976. Mem. N.Y. Acad. Sci., Rochester Acad. Sci. (astronomy sect. 1989—, bd. dirs., ibidem, 1993—). Home: 150 Summit Dr Rochester NY 14620 Office: Genesee Neurology Assocs 220 Alexander St Rochester NY 14607

PETTERSEN, BJØRN RAGNVALD, astronomer, researcher; b. Sandefjord, Vestfold, Norway, June 13, 1950; came to U.S., 1979; s. Ragnvald and Unni (Kristiansen) P.; m. Torill Marie Valum, July 18, 1975. Cand. real., U. Oslo, 1978; D in philosophy, U. Tromsø, 1986. Postdoctoral rsch. assoc. U. Tex., Austin, 1979-80; asst. prof. U. Tromsø, Norway, 1981-85, assoc. prof., 1985-86; prin. rsch. scientist Norwegian Rsch. Coun., Oslo, 1987-92; head Space Rsch. Geodetic Inst. Norwegian Mapping Authority, Honefoss, 1992—; Editor: Activity in Cool Star Envelopes, 1988, Flare Stars in Clusters, 1990, Stellar Flares, 1991, Astronomish Tidsskuft, 1992; contbr. numerous articles to profl. publs. Editor: Activity in Cool Star Envelopes, 1988, Flare Stars in Clusters, 1990, Stellar Flares, 1991; contbr. numerous articles to profl. publs. With Norwegian Army, 1970. Recipient Sr. Fulbright award Fulbright Found., 1983-84, Stranger Cultural award Rolf Stranger Found., 1987. Mem. Internat. Astron. Union, Am. Astron. Soc., Norwegian Astron. Soc., European Astron. Soc. Achievements include discovery of new flare stars. Office: Geodetic Inst, Norwegian Mapping Authority, N-3500 Honefoss Norway

PETTIS, FRANCIS JOSEPH, JR., electrical engineer; b. Portland, Maine, Oct. 2, 1930; s. Francis Joseph and Mida (Pedersen) P. BSEE, U. Maine, 1960. Electronic technician CAA, Burlington, Vt., 1957, CAA/FAA, New Bedford, Mass., 1958-59; electronic engr. FAA, Portland, 1960-65, Boston, 1965-68, N.Y.C., 1968-69; electronic technician FAA, Bangor, Maine, 1969-79; gen. engr. FAA, Washington, 1979—. Cpl. U.S. Army, 1953-54. Mem. IEEE, AAAS, AIAA. Achievements include numerous contributions to the the development and improvement of the National Airspace System (NAS). Office: FAA 800 Independence Ave SW Washington DC 20591-0001

PETTIT, GEORGE ROBERT, chemistry educator, cancer researcher; b. Long Branch, N.J., June 8, 1929; s. George Robert and Florence Elizabeth (Seymour) P.; m. Margaret Jean Benger, June 20, 1953; children: William Edward, Margaret Sharon, Robin Kathleen, Lynn Benger, George Robert III. B.S., Wash. State U., 1952; M.S. Wayne State U., 1954, Ph.D., 1956. Teaching asst. Wash. State U., 1950-52, lecture demonstrator, 1952; rsch. chemist E.I. duPont de Nemours and Co., 1953; grad. teaching asst. Wayne State U., 1952-53, rsch. fellow, 1954-56; sr. rsch. chemist Norwich Eaton Pharma., Inc., 1956-57; asst. prof. chemistry U. Maine, 1957-61, assoc. prof. chemistry, 1961-65, prof. chemistry, 1965; vis. prof. chemistry Stanford U., 1965; chmn. organic div. Ariz. State U., 1966-68, prof. chemistry, 1965—; vis. prof. So. African, Univs., 1978; dir. Cancer Rsch. Lab., 1974-75, Cancer Rsch. Inst., 1975—; lectr. various colls. and univs.; cons. in field. Contbr. articles to profl. jours. Mem. adv. bd. Wash. State U. Found., 1981-85. Served with USAFR, 1951-54. Recipient Disting. Rsch. Professorship award Ariz. State U., 1978-79, Alumni Achievement award Wash. State U., 1984; named Dalton Prof. Medicinal Chemistry and Cancer Rsch., 1986—, Regents Prof. Chemistry, 1990—. Fellow Am. Inst. Chemists (Pioneer award 1989); mem. Am. Chem. Soc. (awards com. 1968-71, 78-81), Chem. Soc. (London), Pharmacognosy Soc., Am. Assn. Cancer Rsch., Sigma Xi, Phi Lambda Upsilon. Office: Ariz State U Cancer Rsch Inst Tempe AZ 85287-1604

PETTIT, HORACE, allergist, consultant; b. Jan. 28, 1903; s. Horace and Katherine (Howell) P.; B.S., Harvard Coll., 1927; M.D., 1931; m. Millicent Lewis, Nov. 22, 1924 (dec.); children: Emily Connery, Horace (dec.), Deborah Myers, Norman; m. Jane Mann Hiatt, May 13, 1950; 1 adopted child, Barbara Mann Ralph. Intern, Bryn Mawr Hosp., 1933-34; asst. instr., instr., assoc. bacteriology U. Pa. Sch. Medicine, 1932-39, instr. medicine, 1939-53; pvt. practice allergy, 1940-42, 1947-75; cons. in allergy Bryn Mawr Hosp., 1963-79, emeritus staff mem., 1979—; cons. allergist Bryn Mawr Coll., 1963-75. Served from maj. to lt. col. AUS, 1942-46. Fellow Coll. Physicians of Phila.; mem. AMA, Am. Acad. Allergy, Am. Soc. Microbiology, Phila. County, Med. Soc., Pa. Med. Soc., Phila. Allergy Soc. (pres. 1958-59), United World Federalists (mem. nat. exec. coun. 10 years), St. Andrew's Soc. Phila. Unitarian. Clubs: Harvard-Radcliffe (Phila.); Merion Cricket (Haverford, Pa.); Camden (Maine) Yacht. Home and Office: 123 Kennedy Ln Bryn Mawr PA 19010-2808

PETTY, OLIVE SCOTT, geophysical engineer; b. Olive, Tex., Apr. 15, 1895; s. Van Alvin and Mary Cordelia (Dabney) P.; m. Mary Edwina Harris, July 19, 1921; 1 son, Scott. Student Ga. Inst. Tech., 1913-14; BS in Civil Engring., U. Tex., 1917, CE, 1920. Registered profl. engr., Tex. Adj. prof. civil engring. U. Tex., 1920-23; structural engr. R.O. Jameson, Dallas, 1923-25; pres. Petty Geophys. Engring. Co., Petty Labs., Inc., San Antonio, 1925-52, chmn. bd., 1952-73; chmn. bd. Petty Geophys. Engring. Co. de Mex. S.A. de C.V., 1950-73; ptnr. Petty Ranch Co., 1968—; ranching, oil, timber and investment interests, 1937—. Author: Seismic Reflections, Recollections of the Formative Years of the Geophysical Exploration Industry, 1976; A Journey to Pleasant Hill, The Civil War Letters of Capt. E. P. Petty, C.S.A., 1982; patentee geophys. methods, instruments, equipment, including electrostatic seismograph detector in op. on the moon and on Mars and now NASA's standard for space exploration. Benefactor, San Antonio Symphony

Soc., McNay Art Mus.; mem. exec. com., founding mem. chancellor's council U. Tex. System, Austin, also founding mem., hon. life mem. Geology Found.; mem. Inst. Texan Cultures, mem. devel and adv. bds; adv. council U. Tex., Austin. Served as lt. Engrs., U.S. Army, 1917-18; AEF in France. Hon. adm. Tex. Navy; recipient Disting. Grad. award U. Tex. Coll. Engring., Austin, 1962; Tex. Acad. Sci. (hon. life), N.Y. Acad. Sci.; mem. ASCE (hon. life), AIME (Legion of Honor), Am. Assn. Petroleum Geologists, AAAS (50-yr. mem.), Am. Petroleum Inst., Nat. Soc. Profl. Engrs. (life), Am. Geophys. Union, Houston Geophys. Soc., San Antonio Geophys. Soc., South Tex. Geol. Soc., Soc. Petroleum Engrs. (Legion of Honor), Am. Assn. Petroleum Geologists (life), Soc. Am. Mil. Engrs. (life), Soc. Exptl. Geophysicists (founding mem.; hon. life), Soc. 1st Inf. Div. (founding), Tex. Soc. Profl. Engrs., Explorers Club (life), Mil. Profl. Engrs. in Pvt. Practice, Wisdom Soc., Am. Geol. Inst. (Centurian Club), Am. Assn. Petroleum Geologists Trustee Assn. (life), Chi Epsilon (hon. life), Theta Xi, Tau Beta Pi. Baptist. Clubs: San Antonio Country, Argyle, St. Anthony, Giraud. Home: 101 E Kings Hwy San Antonio TX 78212-2993 Office: 711 Navarro St San Antonio TX 78205-1721

PETTY, RICHARD EDWARD, psychologist, educator, researcher; b. Garden City, N.Y., May 22, 1951; s. Edmund and Josephine (Serzo) P.; m. Virginia Lynn Oliver, Aug. 29, 1978. BA, U. Va., 1973; PhD, Ohio State U., 1977. Asst. prof. psychology U. Mo., Columbia, 1977-80, assoc. prof. psychology, 1981-83, Middlebush prof. psychology, 1984-85; vis. fellow Yale U., New Haven, Conn., 1986; prof. psychology Ohio State U., Columbus, 1987—; advisor com. on dietary guidelines implementation NAS, 1989-91. Author: Attitudes and Persuasion, 1981, Communication and Persuasion, 1986; editor Personality and Social Psychology Bull., 1988-92; contbr. articles to Jour. Personality and Social Psychology, Jour. Exptl. Social Psychology; mem. editorial bd. of 6 jours. Grantee NIMH, 1978-79, NSF, 1984-88, 90-94. Fellow AAAS, APA, Am. Psychol. Soc. Achievements include origination of the elaboration likelihood model of persuasion. Home: 2955 Scioto Pl Columbus OH 43221-4753 Office: Ohio State U 1885 Neil Ave Columbus OH 43210-1281

PETUCHOWSKI, SAMUEL JUDAH, physicist; b. Cin., May 2, 1952; s. Jakob Josef and Elizabeth Rita (Mayer) P.; m. Silvia Rita Chepal, July 4, 1974; children: Daniela, Ethan. BS, Hebrew U., 1974; PHD, U. Ill., 1979; postgrad., Georgetown U., 1993—. Rsch. assoc. U. Md., College Park, 1980-82; scientist US Naval Rsch. Lab., Washington, 1982-83; physicist NASA Goddard Space Flight Ctr., Greenbelt, Md., 1983—, legal intern, 1992—; mem. G.S. Rich Am. Inn. of Ct., Washington, 1992—. Contbr. articles to profl. jours. Mem. ABA, Am. Astron. Soc., Am. Phys. Soc., Sigma Xi. Jewish. Achievements include 2 patents on fiber optic sensors. Office: NASA Goddard Space Flight Ctr M/C 685 Greenbelt MD 20771

PETZ, THOMAS JOSEPH, internist; b. Detroit, Feb. 10, 1930; s. Arthur J. and Marie (McCarthy) P.; m. Catherine Crowe, June 13, 1959; children: Thomas Jr., William, David, John, Catherine. BS, U. Detroit, 1951; MD, Wayne State U., 1955. Diplomate Am. Bd. Internal Medicine and Pulmonary Disease. Intern Harper Hosp., Detroit, 1955-56, resident, 1958-59, 60-62; resident U. Calif., San Francisco, 1959-60; clin. instr. Wayne State U., Detroit, 1962-72, assoc. prof., 1972-76, clin. assoc. prof., 1976—; pvt. practice pulmonary disease and internal medicine Detroit, 1962-72, St. Clair Shores, Mich., 1977—; chief pulmonary Wayne State U., Detroit, 1974-76, Harper Hosp., Detroit, 1972-79; dir. med. intensive care unit Harper Hosp., Detroit, 1977-83; chmn. dept. medicine Bon Secours Hosp., Grosse Pointe, Mich., 1984-86. Bd. govs. Wayne State Sch. of Medicine Alumni Assn., Detroit, 1981-85. Fellow Detroit Acad. Medicine (pres. 1982-83), Am. Coll. Chest Physicians; mem. Am. Coll. Physicians, Detroit Med. Club. Republican. Roman Catholic. Avocations: golf, skiing. Office: 23201 Jefferson Ave Saint Clair Shores MI 48080-1980

PÉWÉ, TROY LEWIS, geologist, educator; b. Rock Island, Ill., June 28, 1918; s. Richard E. and Olga (Pomrank) P.; m. Mary Jean Hill, Dec. 21, 1944; children—David Lee, Richard Hill, Elizabeth Anne. AB in Geology, Augustana Coll., 1940; MS, State U. Iowa, 1942; PhD, Stanford U., 1952; DSc (hon.), U. Alaska, 1991. Head dept. geology Augustana Coll., 1942-46; civilian instr. USAAC, 1943-44; instr. geomorphology Stanford, 1946; geologist Alaskan br. U.S. Geol. Survey, 1946-93; chief glacial geologist U.S. Nat. Com. Internat. Geophys. Year, Antarctica, 1958; prof. geology, head dept. U. Alaska, 1958-65; prof. geology Ariz. State U., 1965-88, prof. emeritus, 1988—, chmn. dept., 1965-76; dir. Mus. Geology, 1976—; lectr. in field, 1942—; mem. organizing com. 1st Internat. Permafrost Conf. Nat. Acad. Sci., 1962-63, chmn. U.S. planning com. 2d Internat. Permafrost Conf., 1972-74, chmn. U.S. del. 3d Internat. Permafrost Conf., 1978, chmn. U.S. organizing com. 4th Internat. Permafrost Conf., 1979-83; com. to study Good Friday Alaska Earthquake Nat. Acad. Scis., 1964-70, mem. glaciological com. polar research bd., 1971-73, founding chmn. permafrost com., mem. polar research bd., 1975-81; organizing chmn. Internat. Assn. Quarternary Research Symposium and Internat. Field Trip Alaska, 1965; mem. Internat. Commn. Periglacial Morophology, 1964-71, 80-88 ; mem. polar research bd. NRC, 1975-78, late Cenozoic study group, sci. com. Antarctic research, 1977-80. Contbr. numerous papers to profl. lit. Recipient U.S. Antarctic Service medal, 1966; Outstanding Achievement award Augustana Coll., 1969; recipient Internat. Geophysics medal USSR Nat. Acad. Sci., 1985; named second hon. internat. fellow Chinese Soc. Glaciology and Geocryology, 1985. Fellow AAAS (pres. Alaska div. 1956, com. on arid lands 1972-79), Geol. Soc. Am. (editorial bd. 1975-82, chmn. cordilleran sect. 1979-80, chmn. geomorphology div. 1981-82), Arctic Inst. N.Am. (bd. govs. 1969-74, exec. bd. 1972-73), Iowa Acad. Sci., Ariz. Acad. Sci. (pres. 1982-83); mem. Assn. Geology Tchrs., Glaciological Soc., N.Z. Antarctic Soc., Am. Soc. Engring. Geologists, Internat. Permafrost Assn. (founding v.p. 1983, pres. 1988—), Am. Quaternary Assn. (pres. 1984-86), Internat. Geog. Union. Club: Cosmos. Home: 538 E Fairmont Dr Tempe AZ 85282-3723

PEYGHAMBARIAN, NASSER, optical science educator; b. Mar. 26, 1954. MS in Physics, Ind. U., 1979, PhD in Physics, 1982. Postdoctoral fellow Physics dept. U. Ind., 1981-82; postdoctoral fellow Optical Scis. U. Ariz., Tucson, 1982-83, rsch. asst. prof. Optical Sci. Ctr., 1983-85, asst. rsch. prof. Optical Sci. Ctr., 1985-88, asst. prof. Optical Sci. Ctr., 1988-91; assoc., prof., dir. Optical Sci. Ctr. Optical Circuitry Coop., Tucson, 1991—; cons. U.S. Army Res. and Devel. Command, 1986-87, Honeywell Corp., 1985—, Celenese Co., 1988-89. Author: Introduction to Semiconductor Optics, 1993; editor: Nonlinear Optical Materials and Devices for Photonic Switching IV, 1988, Nonlinear Photonics, 1990, Optical Bistability, 1988, (jour.) Optics Letters, 1992—; contbr. articles to prof. jours. Recipient TRW Young Faculty award, 1989-90, 3M Co.'s Young Faculty award, 1987-89. Fellow Optical Soc. Am.; mem. Am. Phys. Soc., Soc. Optical Engrs. Office: U Ariz Optical Scis Ctr Tucson AZ 85721

PEYTON, DAVID HAROLD, biochemistry educator; b. Biloxi, Miss., Oct. 31, 1956; s. Bob Brown and Ruth Anne (Toothman) P.; m. Julie Ann Hoard, July 3, 1978. MA in Chemistry, U. Calif., Santa Barbara, 1980, PhD in Chemistry, 1983. Asst. prof. Westmont Coll., Santa Barbara, 1982-83; postdoctoral researcher Cornell U. Med. Coll, N.Y.C., 1983-85, U. Calif., Davis, 1985-87; asst. prof. biochemistry Portland (Oreg.) State U., 1987-91, assoc. prof. biochemistry, 1991—; cons. VA Hosp., Portland, 1990-92, Stone & Webster Engring., Richland, Wash., 1992. Contbr. articles to profl. publs. Grantee Am. Heart Assn. of Oreg., 1989, Med. Rsch. Found. of Oreg., 1988, NSF, 1988. Mem. AAAS, Am. Chem. Soc. Achievements include development of nuclear magnetic resonance techniques for studying large and complex (protein) biological molecules. Office: Portland State U Dept Chemistry Portland OR 97207-0751

PEYTON, JAMES WILLIAM RODNEY, consultant surgeon; b. Belfast, Northern Ireland, Jan. 19, 1949; s. Richard Rozmond and Ella (Hogg) P.; m. Margaret Elizabeth Lynne Walker, Oct. 1, 1977; children: Christopher, Jonathan, Timothy. BSc with honors, Queens U., Belfast, 1970, MB, BChir, B of Art of Obstetrics, 1973, MD, 1982. Intern, resident Royal Victoria Hosp., Belfast, 1973-82; fellow surgery Med. Coll. Va., Richmond, 1980-81; cons. surgeon So. Health and Social Svcs. Bd. No. Ireland, 1982-86; sr. cons. surgeon South Tyrons Hosp., Dungannon, Tyrone, Ireland, 1986—; mem. com. Advanced Trauma Life Support, London, 1990—; coun. mem. U.K. Trauma Found., 1990—; mem. trauma working group Dept. Health No. Ireland, Belfast, 1990—. Contbr. articles to profl. jours. Served with U.K. Army Res., 1966—. Holder Territorial Decoration. Fellow Royal Coll. Physicians, Royal Coll. Surgeons (mem. com.), Royal Coll. Surgeons in Ireland. Episcopalian. Avocations: badminton, private pilot. Home: Beechlyn Ct, Ballynorthland Pk, Dungannon BT71 6BT, Northern Ireland Office: South Tyrons Hosp, Carland Rd, Dungannon BT71 4AU, Northern Ireland

PEZESHKI, SADRODIN REZA, plant ecophysiology educator; b. Khuzistan, Iran, Apr. 29, 1948; s. M. Sadegh and Ameneh (Adelzadeh) P.; children: Anna, David, Mona. BS, U. Tehran, Iran, 1971; MS, U. Wash., 1977, PhD, 1982. Agrl. engr. Safiabad Agrl. Exptl. Sta., Iran, 1971-75; rsch. asst. U. Wash., Seattle, 1976-82, postdoctoral researcher, 1982; postdoctoral researcher La. State U., Baton Rouge, 1983-86, asst. prof., 1987-89, assoc. prof. plant ecophysiology, 1990—, assoc. prof. forestry, 1987—, assoc. prof. horticulture, 1988—, mem. grad. faculty, 1989—. Contbr. articles to profl. publs. Grantee La. Edn. Quality Support Fund, 1987, 91, USDA, 1991, 92, NSF, USGS. Mem. Am. Soc. Plant Physiologists, Soc. Wetland Scientist, Soc. Am. Foresters, Sigma Xi. Democrat. Achievements include identification and cloning superior stress-tolerant coastal grasses for wetland restoration; interest in physiology of plants, rhizosphere-plant interactions, plant-environmental interactions. Office: La State U Westland Biogeochem Inst Baton Rouge LA 70803

PFALSER, IVAN LEWIS, civil engineer; b. Anthony, Kans., July 11, 1930; s. John Lewis and Gladys Evelyn (Murry) P.; m. Viola Florence Pfalser, Apr. 13, 1952; children: Ann, John, Jane, Jean. Student, U. Wichita, 1948-50; BS in Civil Engring., U. Kans., 1952. Registered profl. engr., Okla. Structural engr. Boeing Aircraft Co., Wichita, 1954-56; structural engr. Phillips Petroleum Co., Bartlesville, Okla., 1956-59, geotech. engr., 1959-65, material handling engr., 1965-70, civil/geotech./environ. prin. engr., 1970-92; civil and geotech. cons. Alyeska Pipeline Co., Houston, 1970-73, Bonney LNG, Ltd., Lagos, Nigeria, 1980-83; airport engr. City of Bartlesville, 1980-92. 1st lt. C.E. U.S. Army, 1952-54. Mem. ASCE, Okla. Profl. Engrs. Republican. Achievements include patent for groundwater sealing for cathopic protection well uranium mine deep shaft sealing and abandiment closure; devel. and testing of constrn. equipment and methods for Trans-Alaska pipeline, devel. of frozen pit LNG storage facilities; devel. and testing of construction equipment and methods for Trans-Alaska Pipeline.

PFAU, MICHEL ALEXANDRE, chemist, researcher, consultant; b. Geneva, Switzerland, June 3, 1931; s. Alexandre Stanislas and Emma Ferona (Fiffel) P. Licentiate chemistry, U. Geneva, 1958, diploma chemistry, 1958; D of Phys. Scis., U. Paris, 1963. Asst. U. Geneva, 1957-58; rsch. assoc. Princeton (N.J.) U., 1963-64; chargé mission Lehigh U., Bethlehem, Pa., 1969; apprentice researcher Nat. Ctr. Sci. Rsch., Paris, 1962, attaché rsch., 1963, chargé rsch., 1964-74, maitre rsch., 1974-84, dir. rsch., 1984—; surveys Nat. Ctr. Sci. Rsch., Paris, 1982; cons. Kirex Co., Geneva, 1984—. Contbr. articles to profl. jours.; patentee; discoverer new asymmetric synthesis. Rotary Club grantee, 1958. Mem. French Soc. Chemistry, Am. Chem. Soc. Avocations: collecting jazz records and films, skiing, swimming. Office: Lab Chim Organic ESPCI, 10 rue Vauquelin, 75231 Paris Cedex 05, France

PFAU, RICHARD OLIN, forensic chemist, forensic science educator; b. Youngstown, Ohio, Dec. 6, 1936; s. John Olin and Sydonah Homer (Buckley) P.; m. Carol Ann Tsaras, Sept. 24, 1960; children: Cynthia Jane Pfau Null, Constance Lynn Pfau Kendall, Christine Susan Pfau Davis. BS, Youngstown U., 1959; MS, Ohio State U., 1988, PhD, 1991. Clin. chemist Trumbull Meml. Hosp., Warren, Ohio, 1958-62; forensic chemist Ohio State Hwy. Patrol, Columbus, 1962-64; forensic chemist Columbus Police Dept., 1964-1970, lab. dir., 1970-92; adv. forensic sci. Franklin County Coroner's Office, Columbus, 1984—; adj. asst. prof. Sch. Allied Med. Professions, Columbus, 1984—; adj. lectr. Capital U., Bexley, Ohio, 1988—. Contbr. articles to profl. jours. Mem. Nat. Safety Coun., Chgo., 1969—; vice comdt. Ohio Naval Militia, 1985—, Columbus, 1985—. Capt. USAR, 1960-84. Fellow Am. Acad. Forensic Scis. (toxicology), Am. Inst. Chemists.; mem. Royal Soc. Chemistry (chartered chemist), Midwest Assn. Forensic Scientists, (charter mem., v.p. 1974), Nat. Sojourners (pres. chapt. #10 1990-91), Sigma Nu (v.p. Alumni House Corp. 1988-90). Republican. Episcopalian. Home: 93 S Remington Rd Columbus OH 43209-1855 Office: Columbus State C C c/o Law Enforcement Tech 550 E Spring St Columbus OH 43215-9999

PFEFFER, CYNTHIA ROBERTA, psychiatrist, educator; b. Newark, May 22, 1943; d. Edward I. and Ann Pfeffer. BA, Douglas Coll., 1964; MD, NYU, 1968. Assoc. dir. child psychiatry inpatient unit Albert Einstein Coll. Medicine, Bronx, N.Y., 1973-79; chief child psychiatry inpatient unit N.Y. Hosp. Cornell Med. Ctr., White Plains, N.Y., 1979—; assoc. prof. clin. psychiatry Cornell U. Med. Coll., N.Y.C., 1984—; prof. psychiatry Cornell U. Med. Coll., 1989—; pres. N.Y. Coun. on Child and Adolescent Psychiatry, N.Y.C., 1989—. Author: The Suicidal Child, 1986, Difficult Moments in Child Psychotherapy, 1988; editor: Youth Suicide: Perspectives on Risk and Prevention, 1989. Recipient Erwin Stengel award Internat. Assn. Suicide Prevention, 1987, Wilford Hulse award N.Y. Coun. on Child & Adolescent Psychiatry, 1989. Fellow Am. Psychiat. Assn., Am. Acad. Child and Adolescent Psychiatry (councillor-at-large 1989—, Norbert Rieger award 1988), Am. Psychopathological Assn.; mem. Am. Assn. Suicidology (pres. 1987, Young Contbrs. award 1981, 82). Office: NY Hosp Westchester Div 21 Bloomingdale Rd White Plains NY 10605-1596 also: 40 E 89 St New York NY 10016

PFEFFER, PHILIP ELLIOT, biophysicist; b. N.Y.C., Apr. 8, 1941; s. Charles and Della (Smith) P.; m. Judith Stadlen, Dec. 22, 1962; children: Charles, Ari, Shira. AB, Hunter Coll., 1962; MS, Rutgers U., 1964, PhD, 1966. Rsch. asst. dept. chemistry Rutgers U., New Brunswick, N.J., 1964-66; rsch. fellow dept. chemistry U. Chgo., 1966-68; rsch. scientist Ea. Regional Rsch. Ctr. USDA, Phila., 1968-88, rsch. leader Ea. Regional Rsch. Ctr., 1976-88, lead scientist Ea. Regional Rsch. Ctr., 1988—; editor-at-large Marcel Dekker, N.Y.C., 1990—. Editor: Nuclear Magnetic Resonance in Agriculture, 1989; author: (jours.) Plant Physiology, Carbohydrate Rsch., Biochemica Acta, Biophysica; mem. editorial bd. Jour. of Carbohydrate Chemistry, 1985—. Recipient Bond award Am. Oil Chemists Soc., 1976, Fed. Svcs. award Phila. Fed. Assn., 1979; fellow Orgn. for Econ. Cooperation and Devel., 1989; Agrl. Rsch. Svc. rsch. fellow, 1989; vis. scientist grantee Centre d'Etudes Nucleaires de Grenoble, 1986, Oxford U., 1989. Mem. AAAS, Internat. Soc. for Magnetic Resonance, Am. Chem. Soc. (Phila. Sect. Scientist of Yr. 1982), Soc. for Applied Spectroscopy. Achievements include patents and publs. concerning use of alpha-anions; discovery of deuterium isotope shift NMR method for determining carbohydrate structures; development of P-31 NMR in vivo methodology for studying metal ion transport and plant/microbe interactions in nitrogen fixing plant nodules. Office: USDA 600 E Mermaid Ln Philadelphia PA 19118

PFEFFER, RICHARD LAWRENCE, geophysics educator; b. Bklyn., Nov. 26, 1930; s. Lester Robert and Anna (Newman) P.; m. Roslyn Ziegler, Aug. 30, 1953; children—Bruce, Lloyd, Scott, Glenn. B.S. cum laude, CCNY, 1952; M.S., Mass. Inst. Tech., 1954, Ph.D., 1957. Research asst. MIT, 1952-55, guest lectr., 1956; atmospheric physicist Air Force Cambridge Research Center, Boston, 1955-59; sr. scientist Columbia U., 1959-61, lectr., 1961-62, asst. prof. geophysics, 1962-64; assoc. prof. meteorology Fla. State U., Tallahassee, 1964-67; prof., dir. Geophys. Fluid Dynamics Inst., 1967—; cons. NASA, 1961-64, N.W. Ayer & Son, Inc., 1962, Ednl. Testing Service, Princeton, N.J., 1963, Voice of Am., 1963, Grolier, Inc., 1963, Naval Research Labs., 1971-76; Mem. Internat. Commn. for Dynamical Meteorology, 1972-76. Editor: Dynamics of Climate, 1960; Contbr. articles to profl. jours. Bd. dirs. B'nai B'rith Anti-Defamation League; chmn. religious concern and social action com. Temple Israel, Tallahassee, 1971-72. Fellow Am. Meterol. Soc. (program chmn. ann. meeting 1963); mem. Am. Geophys. Union, N.Y. Acad. Scis. (chmn. planetary scis. sect. 1961-63), Sigma Xi, Chi Epsilon Pi, Sigma Alpha. Home: 926 Waverly Rd Tallahassee FL 32312-2813

PFEIFER, HOWARD MELFORD, mechanical engineer; b. St. Louis, Aug. 23, 1959; s. Howard William and Ruth Joyce (Norby) P. BS in Applied Sci. & Tech., Charter Oak State Coll., 1990; BSME, U. Hartford, 1991. Engr. in tng. Engr. asst. Pratt & Whitney, East Hartford, Conn., 1984-89; devel. engr. Chomalloy Rsch. and Tech. Divsn., Orangeburg, N.Y., 1991-93; process devel. engr. Howmet Corp., North Haven, Conn., 1993—; vice chmn. U. Hartford Engring. Alumni Adv. Bd., Bloomfield, Conn., 1992—; cons. in field. Mem. ASME, NSPE, Sigma Xi (assoc.). Republican. Achievements include research, design and construction of a human powered helicopter. Home: 96-1 Cosey Beach Ave East Haven CT 06512 Office: Howmet Corp 30 Corporate Dr North Haven CT 06473

PFEIFFER, JUERGEN WOLFGANG, biologist; b. Buchloe, Bavaria, Germany, Feb. 6, 1964; came to U.S. 1979; s. Wolfgang Karl and Inge (Kohlmus) P.; m. Paula Jane Drake, Aug. 27, 1988 (div. 1991). BS in Biology and German, Western Ky. U., 1986, MS in Biology with an emphasis in Biochemistry, 1988. Cons. EMS Togo, Hopkinsville, Ky., 1986; grad. asst. Western Ky. U., Bowling Green, 1986-88; biol. rsch. asst. 2 Letterman Army Inst. Rsch., San Francisco, 1988-92; assoc. scientist II XOMA, Berkeley, Calif., 1992—. Contbr. articles to profl. jours. With U.S. Army, 1988-92. Mem. Am. Chemical Soc., Sigma Xi (assoc.). Office: XOMA Corp 2910 7th St Berkeley CA 94710

PFENDT, HENRY GEORGE, retired information systems executive, management consultant; b. Frankfurt, Germany, Sept. 19, 1934; s. Georg and Elisabeth K. (Schuch) P.; m. Jane Ann Gussard, July 15, 1961; children: Katherine Ann, Henry G. Jr., Karen Jane. BS, U. Rochester, N.Y., 1972, postgrad., 1972; postgrad., U. Mich., 1986. Dir. No. info. ctr. Eastman Kodak Internat., Göteborg, Sweden, 1972-73; sr. project mgr. Eastman Kodak Internat., Stuttgart, Fed. Republic of Germany, 1973-75; dir. adminstrv. svcs. Kodak Australasia Party Ltd., Coburg, Australia, 1975-77; dir. customer svcs. div. Kodak Australasia Party Ltd., Coburg, 1977-81; dir. mktg. Kodak Australasia Party Ltd., Australia, 1981-84; dir. architecture devel. Eastman Kodak Info. Systems, Rochester, 1984-86, dir. corp. info. systems, 1986-93; ret., 1993, bus. and info. mgmt. cons., 1993—; bd. dirs. client adv. coun. Compu Ware, Detroit. Creator concepts and mgmt. processes in field. Charter mem. adv. bd. Rochester Inst. Tech. Sch. Computer Sci. and Tech., 1987; bd. dirs. YMCA of Maplewood, Rochester, 1989—. With USAF, 1955-59. Recipient Industry Visionary award of 25 Most Influential Communications Execs., 1991. Mem. Soc. for Info. Mgmt., Coun. of Logistics Mgmt., Ctr. for Info. Systems Rsch., Strategic Mgmt. Soc., Internat. Platform Assn., Interact Network (assoc.), C. of C. Republican. Lutheran. Avocations: reading, golf, gardening, jogging, travel. Home: 1229 East Ave Rochester NY 14610-1514

PFENNINGER, ARMIN, chemist, researcher; b. Zürich, Switzerland, Nov. 5, 1950; s. Fritz and Erna (Trachsel) P.; m. Elisabeth Stöckli, June 7, 1974; children: Lukas David, Matthias Samuel. Diploma in Chemistry, Eidgenössische Tech. Sch., Zürich, 1974; PhD, U. Bern, Switzerland, 1978. Post-doctoral fellow U. Cambridge, Eng., 1978-79; rsch. chemist CU Chemie Uetikon AG, Uetikon/ Zürich, 1979-84, prodn. mgr., 1984-87, head R&D, 1987—. Mem. parliament Reformed Ch., Zürich, 1983-91; v.p. Reformed Ch., Uetikon, 1990—. Mem. Am. Chem. Soc., Neue Schweizerische Chemische Gesellschaft. Office: CU Chemi Uetikon AG, Seestrasse 108, 8707 Uetikon Zfric, Switzerland

PFENNINGER, KARL H., cell biology and neuroscience educator; b. Stafa, Switzerland, Dec. 17, 1944; s. Hans Rudolf and Delie Maria (Zahn) P.; m. Marie-France Maylié, July 12, 1974; children—Jan Patrick, Alexandra Christina. M.D., U. Zurich, 1971. Research instr. dept. anatomy Washington U., St. Louis, 1971-73; research assoc. sect. cell biology Yale U., New Haven, 1973-76; assoc. prof. dept. anatomy and cell biology Columbia U., N.Y.C., 1976-81, prof., 1981-86; prof., chmn. dept. cellular and structural biology U. Colo. Sch. Medicine, Denver, 1986—; dir. interdeptmental program in cell and molecular biology Columbia U. Coll. Physicians and Surgeons, N.Y.C., 1980-85. Contbr. articles to profl. jours. Recipient C.J. Herrick award Am. Assn. Anatomists, 1977; I.T. Hirschl Career Scientist award, 1977; Javits neurosci. investigator awards NIH, 1984, 91. Mem. Am. Soc. for Cell Biology, AAAS, Harvey Soc., Soc. for Neurosci., Internat. Brain Research Orgn., Internat. Soc. for Neurochemistry, Am. Soc. Biol. Chemists. Office: U Colo Health Scis Ctr Dept Cellular and Structural Biology B-111 4200 E 9th Ave Denver CO 80262

PFISTER, DANIEL F., electrical engineer; b. Lakewood, Ohio, Feb. 18, 1963; s. Frederick and Greta (Flottmann) P. BSEE & Computer Sci., U. Colo., 1987. Design engr. CGH Med., Inc., Lakewood, Colo., 1988—. Named to U. Colo. Engring. Dean's List, 1984-87, Nat. Dean's List, 1984-87. Mem. IEEE, Eta Kappa Nu, Tau Beta Pi.

PFISTER, DONALD HENRY, biology educator; b. Kenton, Ohio, Feb. 17, 1945; s. William A. and Dorothy C. (Kurtz) P.; m. Cathleen C. Kennedy, July 1, 1971; children: Meghan, Brigid, Edith. AB, Miami U., Oxford, Ohio, 1967; PhD, Cornell U., 1971; AM (hon.), Harvard U., 1980. Asst. prof. biology U. P.R., Mayaguez, 1971-74; asst. prof. biology, asst. curator Farlow Herbarium Harvard U., Cambridge, Mass., 1974-77, assoc. prof. biology, assoc. curator Farlow Herbarium, 1977-80, prof. biology, curator Farlow Herbarium, 1980—, dir. univ. herbaria, 1983—; vis. mycologist U. Copenhagen, 1978; vis. prof. field station U. Minn., Itasca, 1979; master Kirkland House Harvard U., 1982—. Contbr. over 80 articles to profl. jours. Grantee NSF, 1973-75, 81-85, 85—, Am. Philos. Soc., 1975-76, Whiting Found., 1986. Fellow Linnean Soc. London; mem. Mycol. Soc. Am. (sec. 1988-91, v.p. 1993-94), Am. Phytopath. Soc., Am. Microbiol. Soc., New Eng. Bot. Club, Sigma Xi. Office: Harvard U Herbarium 22 Divinity Ave Cambridge MA 02138-2020

PFISTERER, FRITZ, physicist; b. Oehringen, Fed. Republic Germany, June 6, 1940; s. Ernst and Anna M.L. (Heinrich) P.; Heide Luckert, Nov. 25, 1966; 1 child, Bettina. Diploma in physics, Tech. U. Munich, 1967; D Engring., U. Stuttgart, Fed. Republic Germany, 1987. Mem. sci. staff Inst. Phys. Electronics, U. Stuttgart, 1968-82, acad. officer, 1982-89, sr. acad. officer, 1990—; head dept. Ctr. Solar Energy and Hydrogen Rsch., Stuttgart, 1989—. Contbr. articles to Jour. Applied Physics, Crystal Growth, other sci. publs. Mem. com. for rsch. and tech., com. for energy policy Free Dem. Party, Fed. Republic Germany, 1984-88. Mem. Internat. Solar Energy Soc., German Solar Energy Soc. (assoc.), Am. Solar Energy Soc., Verein Deutscher Ingenieure. Achievements include patents for methods of production of solar cells. Home: Charlottenstrasse 71, D 74349 Lauffen Germany Office: U Stuttgart Inst Phys Electronics, Pfaffenwaldring 47, D 70569 Stuttgart 80, Germany

PFITZMANN, ANDREAS, computer security educator; b. Berlin, Mar. 18, 1958; s. Ernst and Gisela (Muller) P.; m. Birgit Meyer, June 10, 1983. Diploma in informatics, U. Karlsruhe, Fed. Republic of Germany, 1982, D in Nat. Scis., 1989. Tutor U. Karlsruhe, 1979-82, researcher, 1983-89, sr. researcher, 1989-91; sr. researcher U. Hildesheim, Fed. Republic of Germany, 1991-93; prof. computer sci. U. Dresden, Germany, 1993—; cons. Commn. of the European Communities, Brussels, 1990—, German Info. Security Agy., Bonn, Fed. Republic of Germany, 1991—. Author: Dienststeigerierende Kommunikationsnetze mit Teilnehmeruberprufbarem Datenschutz, 1990; editor: VIS 91, Verlassliche Informationssysteme, Proceedings, 1991; mem. bd. editors Sci. Jour. Datenschutz und Datensicherung, 1990—. GI rep. large interdisciplinary project; Legal Control of Info. Tech., 1989-92; corr. Jour. Computer und Recht, Munich, 1986-91. Studienstiftung Deutschen Volkes grantee, 1977-82; recipient Scheffel Soc. award, 1977. Mem. IEEE, Internat. Assn. Cryptologic Rsch., German Assn. Data Protection, Assn. Computing Machinery, GI Spl. Interest Group (chmn. dependable computer systems 1991—), German Soc. Informatics. Avocations: badminton, soccer, biking, walking. Office: U Dresden Faculty Informatics, Mommsenstr 13 D-0 Theoretical Informatic, D-01062 Dresden Germany

PFLEIDERER, HANS MARKUS, physicist; b. Winnenden, Germany, Mar. 26, 1928; s. Erwin Gotthilf and Marie Elisabeth P.; m. Irmgard Auguste Mühlhäusser, Apr. 10, 1969; children: Roland, Julia. Diplom-Physiker, Technische Hochschule, Stuttgart, Germany, 1956, Dr. rer. nat., 1962. Telefunken, Backnang, Germany, 1957-58; Siemens Corp Rsch, Erlangen, Germany, 1962-71, München, Germany, 1971-93. Contbr. articles to profl. jours. Mem. Deutsche Physikalische Gesellschaft. Achievements include understanding of magnetodiode; ambipolar field-effect transistor; amorphous-silicon solar cell. Home: Gärtnerstrasse 36, D 80992 München

Germany Office: Siemens Corp Rsch, Otto-Hahn-Ring 6, D 81739 Munich Germany

PFLIEGER, KENNETH JOHN, architect; b. Washington, Feb. 20, 1952; s. Chester John Pflieger and Madeline (Maben) Wineke; m. Katherine Colleran Greeves, Oct. 1, 1977; children: Kristian, Kevin, Karissa. BA in Pre-Architecture, Clemson U., 1975, MArch, 1977. Registered architect. Intern architect Clark Assocs., Inc., Anderson, S.C., 1977-80; project architect Greene & Assocs., Architects, Inc., Greenville, S.C., 1980-83, exec. v.p., 1983-89, pres., 1989-90; pres. ID/A, Anderson, S.C., 1990—; adj. lectr. Coll. Architecture, Clemson U., 1982. Chmn. bd. dirs. Hospice of Anderson, 1991—; bd. mgmt. Anderson Family YMCA, 1992—. Mem. AIA (dir. S.C. 1990—, design-merit award 1986), Standard Bldg. Code Congress Internat., Interfaith Forum of Religion, Art and Architecture, Anderson Area C. of C., Kiwanis, Tau Sigma Delta. Avocations: parenting, reading, drawing, fishing. Office: ID/A 2702-B N Main St Anderson SC 29621

PFLUG, LEO JOSEPH, JR., civil engineer; b. Plainfield, N.J., July 2, 1952; s. Leo J. and Theresa Ann (Abbott) P.; m. Donna M. Tango, Nov. 9, 1980; children: Jennifer, Angela. BSCE, N.J. Inst. Tech., 1975. Registered profl. engr., N.J. Svc. engr. Rsch.-Cottrell Inc., Somerville, N.J., 1976-80, tech. svc. mgr., 1980-85; project mgr. LAD Constrn. and Engring., Old Bridge, N.J., 1985—. Mem. N.J. Soc. Profl. Engrs. (chpt. officer 1991-93), Prof. Engrs. in Constrn. (chmn. 1989-93), Am. Concrete Inst. Republican. Roman Catholic. Home: 64 Richland Dr Berkeley Heights NJ 07922 Office: LAD Constrn & Engring PO Box 561 Old Bridge NJ 08857

PFLUM, WILLIAM JOHN, physician; b. N.Y.C., July 30, 1924; s. Peter Arthur and Caroline (Schmidt) P.; BS, Georgetown U., 1947; MD, Loyola U., Chgo., 1951; m. Roseann Sarah Stubing, Oct. 13, 1956; children: Carol Jean, Jeannine, Suzanne, Denise, Peter. Intern, St. Vincent's Hosp., N.Y.C., 1951-52, resident in internal medicine, 1954-55; resident in internal medicine NYU div. Goldwater Meml. Hosp., N.Y.C., 1952-53; resident in allergy Inst. Allergy, Roosevelt Hosp., N.Y.C., 1956; attending internal medicine (allergy and immunology) Overlook Hosp., Summit, N.J., 1958—; assoc. attending Inst. Allergy, Immunology and Infectious Diseases, Roosevelt Hosp., N.Y.C., 1957-92; pvt. practice medicine, specializing in allergy and immunology, Summit, 1957-92; ret.; cons. in field. Served with USAAF, 1943-45; ETO. Decorated Purple Heart, Air medal with two clusters. Diplomate Am. Bd. Allergy and Immunology. Fellow Am. Acad. Allergy, Am. Coll. Allergists, Am. Assn. Clin. Immunology and Allergy; mem. Summit Med. Soc., Am. Assn. Clin. Immunology and Allergy (pres. Mid-Atlantic region 1975-76), Disabled Am. Vets., Mil. Order Purple Heart, Am. Ex-Prisoners of War, 8th Air Force Hist. Soc., World Marathon Runners Assn., Robert A. Cooke Allergy Alumni Assn. Roman Catholic. Home: PO Box 465 Rumson NJ 07760-0465

PFOHL, DAWN GERTRUDE, laboratory executive; b. Balt., June 7, 1944; d. Harry F. Coleman and Gertrude (Stahlin) Scharmer; m. Ronald J. Pfohl, June 18, 1967; 1 child, Shara Lynn. BS, Capital U., 1966; MS, Miami U., Oxford, Ohio, 1972; M Health Adminstrn., Xavier U., 1991. Grad. teaching asst. Mich. State U., East Lansing, 1966-67; rsch. asst. Istituto di Microbiologia, U. Palermo, Italy, 1967-69; grad. rsch. asst. Miami U., 1970-71; microbiology instr. Wilmington (Ohio) Coll., 1972; med. microbiologist McCullough-Hyde Hosp., Oxford, 1972-78, dir. clin. lab. svcs., 1978—. Treas., mem. ch. coun. Faith Luth. Ch., Oxford, 1990—; parent rep. S.W. Edn. Resource Reg. Coun., Cin., 1986-87; bd. dirs., editor newsletter, pres. Oxford Assn. for Children with Learning Disabilities, 1985-88. Mem. Am. Soc. Microbiologists, Am. Soc. Clin. Pathologists, South Ctrl. Assn. Clin. Microbiology, Clin. Lab. Mgmt. Assn. (scholar 1990), Oxford C. of C. Office: McCullough-Hyde Meml Hosp 110 N Poplar St Oxford OH 45056-1292

PFOUTS, RALPH WILLIAM, economist, consultant; b. Atchison, Kans., Sept. 9, 1920; s. Ralph Ulysses and Alice (Oldham) P.; m. Jane Hoyer, Jan. 31, 1945 (dec. Nov. 1982); children: James William, Susan Jane (Mrs. Osher Portman), Thomas Robert (dec.), Elizabeth Ann (Mrs. Charles Klenowski); m. Lois Bateson, Dec. 21, 1984. B.A., U. Kans., 1942, M.A., 1947; Ph.D., U. N.C., 1952. Rsch. asst., instr. econs. U. Kans., Lawrence, 1946-47; instr. U. N.C., Chapel Hill, 1947-50, lectr. econs., 1950-52, assoc. prof. econs., 1952-58, prof. econs., 1958-87, chmn. grad. studies dept. econs. Sch. Bus. Adminstrn., 1957-62, chmn. dept. econs. Sch. Bus. Adminstrn., 1962-68; cons. econs. Chapel Hill, 1987—; vis. prof. U. Leeds, 1983; vis. rsch. scholar Internat. Inst. for Applied Systems Analysis, Laxenberg, Austria, 1983; prof. Cen. European U., Prague, 1991. Author: Elementary Economics-A Mathematical Approach, 1972; editor: So. Econ. Jour, 1955-75; editor, contbr.: Techniques of Urban Economic Analysis, 1960, Essays in Economics and Econometrics, 1960; editorial bd.: Metroeconomica, 1961-80, Atlantic Econ. Jour, 1973—; contbr. articles to profl. jours. Served as deck officer USNR, 1943-46. Social Sci. Research Council fellow U. Cambridge, 1953-54; Ford Found. Faculty Research fellow, 1962-63. Mem. AAAS, Am. Statis. Assn., N.C. Statis. Assn. (past pres.), Am. Econ. Assn., So. Econ. Assn. (past pres.), Atlantic Econ. Soc. (v.p. 1973-76, pres. 1977-78), Population Assn. Am., Econometric Soc., Math. Assn. Am., Phi Beta Kappa, Pi Sigma Alpha, Alpha Kappa Psi, Omicron Delta Epsilon. Home and Office: 127 Summerlin Dr Chapel Hill NC 27514-1925

PFRANG, EDWARD OSCAR, association executive; b. New Haven, Aug. 9, 1929; s. Luitpold and Anna P.; m. Jacquelyn Marcia Montefalco, June 7, 1958; children: Lori Ann, Leslie Jean, Philip Edward. B.S., U. Conn., 1951; M.E., Yale U., 1952; Ph.D., U. Ill., 1961. Registered profl. engr., Md., N.Y., Calif. Sect. chief structures sect., bldg. research div. Nat. Bur. Standards, Washington, 1966-83; mgr. housing tech. program Nat. Bur. Standards, 1970-73, chief structures materials safety div., 1973-83; exec. dir. ASCE, N.Y.C., 1983—. Contbr. articles to profl. jours. Served with USNR, 1953-56. Fellow ASCE, Am. Concrete Inst.; mem. Earthquake Engring. Inst., Sigma Xi, Tau Beta Phi, Chi Epsilon, Sigma Tau. Office: ASCE 345 E 47th St New York NY 10017-2330*

PFUND, EDWARD THEODORE, JR., electronics company executive; b. Methuen, Mass., Dec. 10, 1923; s. Edward Theodore and Mary Elizabeth (Banning) P.; BS magna cum laude, Tufts Coll., 1950; postgrad U. So. Calif., 1950, Columbia U., 1953, U. Calif., L.A., 1956, 58; m. Marga Emmi Andre, Nov. 10, 1954 (div. 1978); children: Angela M., Gloria I., Edward Theodore III; m. Ann Lorenne Dille, Jan. 10, 1988 (div. 1990). Radio engr., WLAW, Lawrence-Boston, 1942-50; fgn. svc. staff officer Voice of Am., Tangier, Munich, 1950-54; project. engr. Crusade for Freedom, Munich, Ger., 1955; project mgr., materials specialist United Electrodynamics Inc., Pasadena, Calif., 1956-59; cons. H.I. Thompson Fiber Glass Co., L.A., Andrew Corp., Chgo., 1959, Satellite Broadcast Assocs., Encino, Calif., 1982; teaching staff Pasadena City Coll. (Calif.), 1959; dir. engring., chief engr. Electronics Specialty Co., L.A. and Thomaston, Conn., 1959-61; with Hughes Aircraft Co., various locations, 1955, 61-89, mgr. Middle East programs, also Far East, Latin Am. and African market devel., L.A., 1971-89, dir. internat. programs devel., Hughes Communications Internat., 1985-89; mng. dir. E.T. Satellite Assocs. Internat., Rolling Hills Estates, Calif., 1989—; dir. programs devel. Asia-Pacific TRW Space and Tech. Group, Redondo Beach, Calif., 1990-93. With AUS, 1942-46. Mem. AIAA, Phi Beta Kappa, Sigma Pi Sigma. Contbr. articles to profl. jours. Home: 25 Silver Saddle Ln Palos Verdes Peninsula CA 90274-2437

PHADKE, ARUN G., electrical engineering educator; b. Gwalior, M.P., India, Aug. 27, 1938; came to U.S., 1959; s. Gajanan G. and Indira G. Phadke; m. Kusum K. Joglekar, Sept. 14, 1964; 1 child, Ajit A. BS, Agra U., India, 1955; B in Tech. with honors, Indian Inst. Tech., 1959; MS, Ill. Inst. Tech., 1961; PhD, U. Wis., 1964. Systems engr. Allis Chalmers, Milw., 1963-67; asst. prof. elec. engr. dept. U. Wis., Madison, 1967-69; cons. engr. Am. Elec. Power Svc. Corp., N.Y.C., 1969-82; AEP prof. elec. engring. dept. Va. Poly Inst. & State U., Blacksburg, 1982—; cons. various electric utilities, equipment mfrs., 1980—. Co-author: Computer Relaying for Power Systems, 1988; patentee in field. Disting. Svc. citation, 1987, Centennial medal U. Wis., 1991. Fellow IEEE (chmn. working groups 1980—, outstanding educator Power Engring. Soc. 1991); mem. Edison Elec. Inst. (outstanding educator 1986), Conf. Internat. Grand Reseaux Electrique (chmn. working groups), Nat. Acad. Engring. Avocations: painting, tailoring. Office: Va Poly Inst & State U Elec Engring Dept 426 Whittemore Hall Blacksburg VA 24061-0111

PHALAN, ROBERT F., environmental scientist; b. Fairview, Okla., Oct. 18, 1940; married, 1966; 2 children. B in Physics, San Diego State U., 1964, M in Physics, 1966; PhD in Biophysics, U. Rochester, 1971. Engring. aide advanced space systems dept. Gen. Dynamics/Astronautics, San Diego, 1962-63; asst. to radiation safety officer, lab. teaching asst. San Diego State U., 1964-66, instr. physics dept., 1966; mem. summer faculty biology dept. Rochester (N.Y.) Inst. Tech., 1970-72; rsch. assoc. aerosol physics dept. Lovelace Found. for Med. Edn. and Rsch., Albuquerque, 1972-74; from adj. asst. prof. to assoc. prof. in residence dept. community and environ. medicine U. Calif., Irvine, 1974-84, prof. in residence, dir. Air Pollution Health Effects Lab., 1985—; reviewer Am. Review of Respiratory Disease, Applied Indsl. Hygiene, Bulletin of Math. Biology, Exptl. Lung Rsch., Jour. Toxicology and Environ. Health, Jour. Toxicology and Applied Pharmacology, Jour. Aerosol Sci., Sci.; reviewer, mem. editorial bd. Fundamental and Applied Toxicology, 1986-92, Inhalation Toxicology, Jour. Aerosol Medicine; mem. safety and occupational health study sect. NIH, 1988-90, mem. spl. study sects., 1980, 81, chmn. spl. study sects., 1982, 83, 84, 87, 88, 92, mem. site visit teams., 1980, 81, 82, 83, 84, 88; mem. expert panel on sulfur oxides EPA, mem. inhalation toxicology divsn. peer rev. panel, 1982, session chmn., 1983, participant workshop on non-oncogenic lung disease, 1984, mem. grants rsch. sci. rev. panel on health rsch., 1985-88; mem. task group on respiratory tract kinetic model Nat. Coun. Radiation Protection, 1978—; mem. adv. panel on asbestos Am. Pub. Health Assn., 1978; chmn. atmospheric sampling com. Am. Coun. Govtl. Indsl. Hygienists, 1982—; chmn. NIOSH spl. study sect., 1983; panelist workshop Nat. Heart, Lung and Blood Inst., 1982; sci. advisor Prentice Day Sch., 1986—. Author: Inhalation Studies: Foundations and Techniques, 1984; author: (with others Advances in Air Sampling, 1988, Concepts in Inhalation Toxicology, 1989; contbr. numerous articles to profl. jours. Am. Legion scholar. Mem. AAAS, Am. Assn. Aerosol Rsch. (charter, chmn. ann. meeting 1985), Am. Conf. Govtl. Indsl. Hygienists, Am. Indsl. Hygiene Assn. (jour. reviewer, chmn. ann. conf. 1981, 85, 86), Brit. Occupational Hygiene Soc., Fine Particle Soc., Soc. for Aerosol Rsch., Health Physics Soc., Soc. Toxicology (chmn. 20th ann. meeting 1981). Achievements include research in nasal, tracheobronchial and pulmonary transport of inhaled deposited particles and effects of pollutant exposure on transport kinetcs, laboratory simulation and characterization of airborne environmental pollutants, respiratory tract deposition and clearance models for inhaled particles, including species comparisons and body size effects, behavior of highly-concentrated aerosols with respect to deposition in the respiratory tract. Office: University of California Air Pollution Health Effects Lab Dept of Community & Environ Irvine CA 92717*

PHAN, CHUONG VAN, biotechnologist; b. Trungthanh, Vinhlong, Vietnam, July 6, 1942; came to U.S. 1988; s. Nhac Van and Lua Thi (Nguyen) P.; m. Bang Tam Tran; 1 child, Trongvan; m. Hitomi Shimono; children: Maichi, Milan, Mica. BSc in Agr., Nat. Ctr. Agr., Saigon, Vietnam, 1969; MSc in Agr., Kyushu U., Japan, 1976, PhD in Agr., 1980. Instr. U. Cantho, Vietnam, 1969-73; rsch. scientist U. Guelph, Can., 1982-87, Hoechst Can., Inc., Regina, 1987-88; sr. rsch. scientist Agrigenetics Co., Madison, Wis., 1988-89, Sungene Tech., San Jose, Calif., 1989-90, Petoseed Co., Inc., Woodland, Calif., 1990—. Contbr. articles to profl. jours. Mem. Internat. Assn. Plant Tissue Culture, In Vitro, Japanese Jour. Breeding. Buddhist. Achievements include patents pending for combination of cytoplasmic traits, atrazine resistance and cytoplasmic male sterility, production of rapeseed exhibiting an improved fatty acid profile; combination of cytoplasmically inherited traits by haploid protoplast fusion in oil rape; high production of haploids by means of microspore/anther/ovule culture in Brassica species, vegetables and ornamentals for application in plant breeding; increase of oleic acid content mutagenesis in Rapeseed. Home: 413 Grande Ave Davis CA 95616-0213 Office: Petoseed Co 37437 State Hwy 16 Woodland CA 95695-9353

PHAN, JUSTIN TRIQUANG, electrical project engineer; b. Viet Nam, May 15, 1964; s. Cu Quang Phan and Hung (Thi) Chau. BSEE, Fla. State U., 1989. Cert. continuing profl. devel. Elec. engr. Attapulgus operation Engelhard Corp, Fort Meade, Fla., 1989—. Adv. bd. United Way, Lakeland, 1990-91. Mem. NSPE, Fla. Soc. of Engrs., Assn. of Energy Engr. Office: Engelhard Corp 3225 State Rd 630W Attapulgus GA 33841

PHAN-TAN, TAI, scientist, researcher, educator; b. Kien-Giang, Vietnam, July 7, 1939; arrived in Fed. Republic Germany, 1960; s. Van-To Phan and Thi-Ngai Nguyen; m. Thanh-Van Do, Nov. 8, 1968; children: Thanh-Thao, Thanh-Thu, Thanh-Uy. Diploma in engring., Tech. Coll. Engring., Aachen, Fed. Republic Germany, 1966; postgrad., French Inst. Indsl. Freezing, Paris, 1966-68; D of Engring., U. Hannover, Fed. Republic Germany, 1982. Head of sect., Material Rsch. Inst. U. Hannover, 1984-86, lectr., 1992—. Contbr. articles to profl. publs. Mem. Verein Deutscher Eisenhuttenleute, Verein Deutscher Korosionfachleute. Office: Univ Hannover, Appelstrasse 11 A, 30167 Hannover 1, Germany

PHARES, LINDSEY JAMES, consultant, retired physicist and engineer; b. Onego, W.Va., May 31, 1916; s. Emerson P. and Luella (Smith) P.; m. Gloria Corriols, Apr. 16, 1944; children: Gloria C., Michael C. BS in Physics, Davis and Elkins Coll., 1936, PhD in Sci., 1982. Constrn. engr. various cos., 1936-41; constrn. supt. Raymond Internat. Inc., N.Y.C., 1941-46, gen. supt., 1946-52, constrn. mgr., 1952-57, v.p., 1957-79, v.p., cons., 1979-81; constrn. cons., 1981—. Lt. USN, 1942-45, PTO. Fellow ASCE (life). Republican. Episcopalian. Achievements include 18 patents in field. Home: 2630 Country Club Blvd Sugar Land TX 77478

PHARES, VICKY, psychology educator; b. San Diego, Oct. 13, 1960; d. William Dean and Rita Harriet (Yunker) P. BA in Psychology, UCLA, 1982; MS in Psychology, U. Wyo., 1984; PhD in Clin. Psychology, U. Vt., 1990. Lic. clin. psychologist. Asst. prof. dept. psychology U. Conn., Storrs, 1990-92, U. South Fla., Tampa, 1992—; presenter in field. Ad hoc reviewer Jour. Consulting & Clin. Psychology, 1990—, Profl. Psychology, 1990—, Jour. Abnormal Child Psychology, 1992—; contbr. articles to Am. Psychologist, Psychol. Bulletin, Jour. Consulting & Clin. Psychology, Child Devel., Devel. Psychology. Mem. APA (Dissertation Rsch. award 1989), Am. Psychol. Soc., Soc. for Rsch. in Child and Adolescent Psychopathology, Children's Def. Fund (supporter). Democrat. Achievements include research in the role of fathers in developmental psychopathology. Office: U South Fla BEH 339 Psychology Dept Tampa FL 33620

PHARR, GEORGE MATHEWS, materials science and engineering educator; b. Atlanta, May 20, 1953; s. George Mathews and Patricia (Steele) P.; m. Marilyn Walker, July 14, 1979; children: George Mathews, Adam Walker. BSME, Rice U., 1975; MS in Materials Sci., Stanford U., 1977, PhD in Materials Sci., 1979. Registered profl. engr., Tex. Postdoctoral fellow Cambridge (Eng.) U., 1979-80; asst. prof. materials sci. and engring. Rice U., Houston, 1980-85, assoc. prof., 1985-91, prof., 1991—; cons. Kalium Chems., Inc., Chgo., 1987—. Editor: Thin Films: Stresses and Mechanical Properties, 1990; also over 50 articles. Recipient faculty devel. award Exxon, 1982, IBM, 1984, Young Alumni Achievement award Rice U., 1985; fellow Hertz Found., 1977. Mem. Metall. Soc. of AIME, Am. Ceramic Soc. (assoc. editor Jour. 1989—), Materials Rsch. Soc., Am. Soc. for Metals (Bradley Stoughton award 1984),. Home: PO Box 2011 Houston TX 77252 Office: Rice U Dept Mech Engring PO Box 1892 Houston TX 77251

PHARR, PAMELA NORTHINGTON, physiology educator, researcher; b. Owensboro, Ky., Apr. 5, 1945; d. Leroy Wells and Nettie Clark (Grissom) Northington; m. Ronald Porter, Jan. 25, 1969 (div.); m. Walter Morgan Pharr, Mar. 5, 1983; 1 child, Edward James. BA, U. Ky., 1967; PhD, Duke U., 1972. Postdoctoral fellow U. Calif., San Diego, La Jolla, 1971-74, Duke U., Durham, N.C., 1974-76; rsch. assoc. U. Nebr., Lincoln, 1976-78; instr. Med. U. S.C., Charleston, 1978-80, asst. prof., 1980-84, assoc. prof., 1984—. mem. editorial bd. Experimental Hematology, 1980-84, Internat. Jour. of Cell Cloning, 1983-88; contbr. articles to profl. jours. Mem. standing com. Circular Congregational Ch., 1979-81. Predoctoral fellow NIH, 1967-71, postdoctoral fellow NIH, 1972-74. Mem. Internat. Soc. for Experimental Hematology, Am. Soc. for Cell Biology, Sigma Xi. Achievements include development of a stochastic model for mast cell proliferation, and research in mechanisms of hemopoietic cell differentiation in cells infected with a constitutively activated erythropoietin receptor. Home: 1545 Montclair St Charleston SC 29407 Office: Med Univ South Carolina 109 Bee St Charleston SC 29403

PHAUP, ARTHELIUS AUGUSTUS, III, structural engineer; b. Lawton, Okla., Oct. 13, 1965; s. Arthelius Augustus Jr. and Lorena Ann (Keech) P. BSCE, Va. Poly. Inst. and State U., 1987, MSCE, 1988. Registered profl. engr., Va. Structural engr. Reid and Cornwell, Ltd., Virginia Beach, Va., 1989-91, Ralph Whitehead and Assoc., Richmond, Va., 1991—.

PHELAN, FREDERICK ROSSITER, JR., chemical engineer; b. Phila., Mar. 2, 1960; s. Frederick Rossiter Sr. and Joan Marie (Gustave) P.; m. Sandra Jean Bognaski, Aug. 11, 1984. BSChemE summa cum laude, SUNY, Buffalo, 1983; PhD in Chem. Engring., U. Mass., 1989. Chem. engr. NIST/Polymers Div., Gaithersburg, Md., 1989—. Author: Advanced Composites Materials, New Development and Applications. Instr. religious edn. St. John the Evangelist Parish, Frederick, Md., 1990-91. Mem. Soc. Rheology, Tau Beta Pi. Roman Catholic. Achievements include rsch. interests in modeling of fluid mechanics and heat and mass transfer, flow in porous media, polymer composites, polymer processing. Office: NIST Bldg 224/A209 Gaithersburg MD 20899

PHELPS, ARTHUR VAN RENSSELAER, physicist, consultant; b. Dover, N.H., July 20, 1923; s. George Osborne and Helen (Ketchum) P.; m. Gertrude Kanzius, July 21, 1956; children: Wayne Edward, Joan Susan. ScD in Physics, MIT, 1951. Cons. physicist rsch. labs. Westinghouse Elec. Corp., Pitts., 1951-70; sr. rsch. scientist Nat. Bur. Standards, Boulder, Colo., 1970-88; fellow Joint Inst. Lab. Astrophysics U. Colo., Boulder, 1970-88, adjoint fellow, 1988—, chmn., 1979-81; chmn. Gordon Rsch. Conf., Plasma Chemistry, 1990. Recipient Silver Medal award Dept. Commerce, 1978. Fellow Am. Phys. Soc. (Will Allis prize 1990). Achievements include patent for Schulz-Phelps ionization gauge; research on electron and atomic collision processes involving low energy electrons, molecules, ions, metastable atoms and resonance radiation; on laser processes and modeling; on gaseous electronics. Home: 3405 Endicott Dr Boulder CO 80303-6908 Office: U Colo Joint Inst Lab Astrophysics Campus Box 440 Boulder CO 80309-0440

PHELPS, GLENN HOWARD, mechanical carbon process engineer; b. Addison, N.Y., May 23, 1943; s. Norman B. and Helen J. (Crain) P.; m. Lynn Jean Mauriello; children: Norman B., Nancy M. BS in Ceramic Engring., Alfred U., 1966. Process engr. Union Carbide Carbon Products, Niagara Falls, N.Y., 1966-70; devel. engr. Poco Graphite, Decatur, Tex., 1970-73, Pure Carbon, St. Marys, Pa., 1973-74; tech. dir. Metallized Carbon, Ossining, N.Y., 1974—. Coach Am. Youth Soccer Orgn., Croton on Hudson, N.Y., 1982—. Mem. ASTM, Soc. Tribologists and Lubrication Engrs. Achievements include development of commercially successful mech. carbon grades for mech. seal primary rings, bearings and pump vanes. Office: 19 S Water St Ossining NY 10562

PHELPS, JAMES EDWARD, electrical engineer, consultant; b. Knoxville, Tenn., Apr. 20, 1952; s. Edgar William and Gladys Jean (Price) P. BSEE, U. Tenn., 1977, MSEE, 1980. Researcher, instr. U. Tenn., Knoxville, 1977-80; sr. devel. engr., inventor Oak Ridge (Tenn.) Nat. Lab., 1980-87; cons. P.S. Help Engring. Cons., Knoxville, 1987-91; sr. systems engr., scientist QUALTEQ, Inc., Knoxville, 1990—; pres. P.S. Help Electronic Cons., Knoxville, 1987-91. Author: High Temperature Switching Regulating Power Supplies, 1980; contbr. articles to profl. jours. Nuclear activist Environ. Peace, Knoxville, 1987—; ethical adv. Oak Ridge Nat. Lab., 1987—; organizer Ch. St. United Meth. Ch., Knoxville, 1977—. Recipient Significant Event award Martin-Marietta Engring. Systems, 1986. Achievements include invention of USRADS system for automatic surveying of radiation contaminated area; design of PEARLS system for comml. application for lowest radiactivity measurement used by EPA, of NASA Servemanipulator Control System. Home: 1600 Buttercup Cir Knoxville TN 37921-4147 Office: Quality Inc 9041 Executive Pk Ste 500 Knoxville TN 32923

PHELPS, JAMES SOLOMON, III, astrodynamic engineer; b. Balt., Mar. 2, 1952; s. James S. Jr. and Rachel (Bruton) P.; m. Suzanne Rowell, July 24, 1975; 1 child, Caroline. BSNE, N.C. State U., 1974, MS, 1977. Nuclear engr. Combustion Engring., Windsor, Conn., 1978-79; sr. engr. Advanced Tech., Reston, Va., 1980-81, Nuclear Power Cons., Rockville, Md., 1981-83; chief engr. GE Space System div. MDSO, Springfield, Va., 1983-89; chief systems engr. Hughes STX, Vienna, Va., 1989—; cons. Submarine Maintenance Monitoring Support Office USN, Washington, 1988-90. Contbr. articles to profl. jours. Mem. Am. Nuclear Soc., Sigma Xi (assoc.). Achievements include devel. of a new approach to the use of process error in sequntial estimation.

PHILIP, A. G. DAVIS, astronomer, editor, educator; b. N.Y.C., Jan. 9, 1929; s. Van Ness and Lillian (Davis) P.; m. Kristina Drobavicius, Apr. 25, 1964; 1 dau., Kristina Elizabeth Elanor. B.S., Union Coll., 1951; M.S., N.Mex. State U., 1959; Ph.D., Case Inst. Tech., 1964. Tchr. physics, math. and chemistry Brooks Sch., 1954-59; instr. Case Inst. Tech., 1962-64; asst. prof. astronomy U. N.Mex., 1964-66; asst. prof. astronomy SUNY-Albany, 1966-67, assoc. prof., 1967-76, mem. exec. com. Arts and Scis. Council, 1975-76; prof. astronomy Union Coll., Schenectady, 1976—, astronomer Dudley Obs., 1967-81, Frank L. Fullam chair astronomy, 1980-81, editor Dudley Obs. Reports, 1977-81; astronomer Van Vleck Obs., Weslayan U., 1982—, editor contbns. of VVObs., 1982—; vis. prof. Yale U., 1972, 73, La. State U., 1973, 76, 86, Acad. Scis. Lithuania, USSR, 1973, 76, 79, 86, Stellar Data Ctr., Strasbourg, France, 1978, 79, 80, 82, 85, 86; bd. dirs., sec.-treas. N.Y. Astron. Corp.; pres., treas. L. Davis Press, Inc., 1982—, Inst. Space Observations, 1986—; trustee Fund Astrophys. Rsch., 1985—; bd. dirs. NE N.Y. IBM PC User Group, 1990—. Exhibited: 2d Ann. Photography Regional, Albany, 1980, author: (with M. Cullen and R.E. White) UBV Color - Magnitude Diagrams of Galactic Globular Clusters, 1976; (with A. Robucci, M. Frame, K.W. Philip) Mm, Fractal Series, Vol. 1, Midgets on the Spike, 1991; editor: The Evolution of Population II Stars, 1972, (with D.S. Hayes) Multicolor Photometry and the Theoretical HR Diagram, 1975, (with M.F. Mc Carthy) Galactic Structure in the Direction of the Galactic Polar Caps, 1977, (with D. H. DeVorkin In Memory of Henry Norris Russell, 1977, (with Hayes) The HR Diagram, 1978, Problems in Calibration of Multicolor Systems, 1979, (with M.F. McCarthy and G.V. Coyne) Spectral Classification of the Future, 1979, X-Ray Symposium, 1981, (with Hayes) Astrophysical Parameters for Globular Clusters, 1981, (with A.R. Upgren) The Nearby Stars and the Stellar Luminosity Function, 1983, (with Hayes and L. Pasinetti) Calibration of Fundamental Stellar Quantities, 1985, (with D.W. Latham) Stellar Radial Velocities, Horizontal-Branch and UV-Bright Stars, 1985, Spectroscopic and Photometric Classification of Population II Stars, 1986, (with J. Grindley) IAU Symposium No. 126, Globular Cluster Systems in Galaxies, 1987, (with Hayes and Liebert) IAU Colloquium No. 95, The Second Conference on Faint Blue Stars, (with Hayes and Adelman) New Directions in Spectrophotometry, 1988, Calibration of Stellar Ages, 1988, (with A.R. Upgren) Star Catalogues; A Centennial Tribute to A.N. Vyssotsky, 1989, (with P. Lu) The Gravitational Force Perpendicular to the Galactic Plane, 1989, (with D.S. Hayes and S.J. Adelman) CCDs in Astronomy. II. New Methods and Applications of CCD Technology, 1990, (with A.R. Upgren and K.A. Janes) Precision Photometry: Astrophysics of the Galaxy, 1991, (with Rebucci, Frame and Philip K.) Midgets on the Spike, vol. I, 1991, (with A.R. Upgren) Objective-Prism and Other Surveys, 1991, N.Y. State Astronomy, 1992, (with B. Hauck and A.R. Upgren) Workshop on Databases for Galactic Structure, 1993; lectr. tours (with K.W. Philip) An Introduction to the Mandelbrot Set, 1988-91; contbr. chpts. to books, articles to profl. jours. Served with AUS, 1951-53. Yale U. vis. fellow, 1976; rsch. grantee Rsch. Corp., NSF, NASA, Nat. Rsch. Lab., NAS, Am. Astron. Soc. Fellow AAAS, Royal Astron. Soc., Am. Phys. Soc.; mem. Am. Astron. Soc. (Harlow Shapley lectr. 1973-92, auditor 1977, 79-85), Am. Math. Soc., Internat. Astron. Union (chmn., sec. various coms. and commns., pres. commn. 30, 1982-85, chmn. working group on spectroscopic and photometric data 1985-91), N.Y. Acad. Scis., Astron. Soc. Pacific, Astron. Soc. N.Y. (sec.-treas. 1969—, editor newsletter 1974—), Sigma Xi. Achievements include being 1st U.S. observer Soviet 6M telescope, 1980. Home: 1125 Oxford Pl Schenectady NY 12308-2913 Office: Union Coll Physics Dept Schenectady NY 12308 also:n Vleck Obs Middletown CT 06457

PHILIPP, MANFRED HANS WILHELM, biochemist, educator; b. Rostock, Germany, Sept. 30, 1945; came to U.S., 1949; m. Gisela Nielsen; children: Ralf, Erik. BS in Chemistry, Mich. Technol. U., 1966; PhD in Biochemistry, Northwestern U., 1971. Vis. scientist Weizmann Inst. Sci., Rehovoth, Israel, 1974, U. Munich, 1984; vis. prof. U. Ulm, Germany, 1983; prof. Lehman Coll. CUNY, 1977—, prof. biochemistry Grad. Ctr., 1979—; dir. Project SEED (Summer Ednl. Experience for Disadvantaged Students), Lehman Coll., 1988—. Contbr. articles on nucleic acid chemistry and enzymology to sci. publs. Grantee Cottrell Rsch. Corp., 1984, NIH, 1986, NSF, 1987. Achievements include 3 patents on enzymology and nucleic acid chemistry. Office: Lehman Coll CUNY Dept Chemistry Bedford Park Blvd W Bronx NY 10468

PHILIPPI, EDMOND JEAN, airline executive; b. Alexandretta, Turkey, Oct. 27, 1936; s. Jean Christo Philippi and Alice (Michel) Yatros; m. Zoya W. Asad, Mar. 5,1 967; children: Lara, Priscilla, Faris. B.BP. in Bus., Am. U. Beirut, 1957, postgrad., 1957-59. Sales rep. T. Gargour & Fils, Beirut, 1959-61; sr. sales rep. Lufthansa German Airlines, Beirut, 1961-75; v.p. cargo Royal Jordanian Airlines, Amman, Jordan, 1975-89; pres. Z.W.A. Aviation, Amman, 1989—; cons. Arab Air Cargo, Amman, 1982-83. Mem. S.O.S. Children's Village, Amman, 1987. Mem. Airline Assn., IAPA (Airline Ins.), Internat. Automobil Club, Sheraton Club, Royal Yachting Club, Grand Orient Lodge (Beirut, knight 1961). Avocations: boxing, fishing, swimming, skin diving, reading. Office: ZWA Aviation, PO Box 12, Amman Jordan

PHILIPSON, LENNART CARL, microbiologist, science administrator; b. Stockholm, July 16, 1929; s. Carl and Greta (Svanström) P.; m. Malin Jondal, 1954; children: Niklas, Andreas, Tomas. MD, Uppsala (Sweden) U., 1957, D in Med. Scis., 1958, PhD (hon.), 1982, Dr. Med. (hon.), 1987. Rsch. asst. Inst. Bacteriology Uppsala U., 1953-57, rsch. asst. Inst. Biochemistry, 1957-58, asst. prof. Inst. Virology, 1958-59; Sophie Fricke fellow Swedish Royal Acad. Sci. Rockefeller Inst., N.Y.C., 1959-61; from asst. to assoc. prof. virology, Swedish Med. Rsch. Coun. Uppsala U., 1961-68, founder, dir., The Wallenberg Lab., 1967-76, prof. microbiology, dept. microbiology, 1968-82; dir. gen. European Molecular Biology Lab., 1982-93; dir. Skirball Inst. Molecular Medicine NYU Med. Ctr., N.Y.C., 1993—; vis. prof. Columbia U., 1971, MIT, 1977; fgn. assoc. U.S. Nat. Acad. Sci., Washington, 1992. Mem. editorial bd. EMBO Jour., 1982-86, Molecular Biology and Medicine, 1983-86, Jour. Virology, 1972-90, Nucleic Acid Rsch., 1983-91, BBA Revs. on Cancer, 1980—, Cancer Letters, 1985—, Genomics, 1986—; contbr. articles to profl. jours. Recipient Axel Hirsch prize Karolinska Inst., 1976, hon. prof. sci. faculty, Heidelberg U., Fed. Rep. Germany, 1985. Mem. Swedish Royal Acad. Sci., Royal Swedish Acad. Engring. Scis., European Molecular Biology Orgn., Heidelberg Acad. Scis., Am. Soc. for Microbiology, Soc. for Gen. Microbiology, N.Y. Acad. Scis., Swedish Soc. for Microbiology, Fedn. of European Biochem. Socs., Harvey Soc. (N.Y.). Avocations: golf, sailing. Office: NYU Med Ctr Skirball Inst Biomolecular Medicine New York NY 10016

PHILLIPS, ADRAN ABNER (ABE PHILLIPS), geologist, oil and gas exploration consultant; b. Sugden, Okla., Feb. 6, 1924; s. James M. and Jennie Elizabeth (Norman) P.; m. Carmel Darlene Pesterfield, Aug. 20, 1949 (div.); 1 son, John David. B.S. in Geology, U. Okla., 1949. With Exxon Corp. and affiliates, 1949-79, dist. geologist, Chico, Calif., 1959-64, ops. geologist, Sydney, Australia, 1964-67, exploration coordinator North Slope Alaska, Houston, 1968-70, div. geologist, Denver, 1970-71, exploration mgr. P.T. Inc., Stanvac, Jakarta, Indonesia, 1971-73, exploration mgr. ESSO exploration, Singapore, 1973-76; div. mgr. Exxon U.S.A., Denver, 1976-79; v.p. Coors Energy div., Golden, Colo., 1979-80, pres., 1980-92; oil and gas exploration cons., 1992—. Bd. dirs. Mountain States Legal Found., 1991—. Mem. Am. Assn. Petroleum Geologists, Ind. Petroleum Assn. Mountain States (past pres.), Ind. Petroleum Assn. Am. (dir.), Nat. Coal Council. Home and Office: 2194 Augusta Dr Evergreen CO 80439

PHILLIPS, ANTHONY, optical engineer. BS, mathematics, Carnegie Inst. Technology, Pittsburgh, Penn., 1966. Lens designer Owens-Illinois, 1966-73; chief lens designer GCA Tropel, Fairport, N.Y., 1973—. Recipient Engring. Excellence award Optical Soc. Am., 1992. Office: GCA Tropel Inc 60 O'Conner Rd Fairport NY 14450*

PHILLIPS, ARTHUR MORTON, III, botanist, consultant; b. Cortland, N.Y., Jan. 20, 1947; s. Arthur Morton Jr. and Ruth (Mason) P.; m. Diedre Weage, Sept. 3, 1988. BS, Cornell U., 1969; PhD, U. Ariz., 1977. Instr., dept. biol. sci. U. Ariz., Tucson, 1971-73, rsch. asst., dept. geosciences, 1973-76; rsch. botanist Mus. No. Ariz., Flagstaff, 1976-80, curator biology, 1980-89; environ. cons. Flagstaff, 1990—; mem. Ariz. plant recovery team U.S. Fish & Wildlife Svc. Endangered Species, Phoenix, 1981—, Natural Areas Adv. Coun. Ariz. State Pks., Phoenix, 1980-89, chmn. 1985-86; adj. prof. No. Ariz. U., Flagstaff, 1984—. Author: Grand Canyon Wildflowers, 2d edit., 1990; co-author: Checklist, Vascular Plants, Grand Canyon National Park, 1987, High Country Wildflowers, 1987, Expedition to San Francisco Peaks, 1989, 4 endangered plants recovery plans, 1984-87. Fellow Ariz.-Nev. Acad. Sci.; mem. Ecol. Soc. Am., Am. Quaternary Assn., Soc. for Conservation Biology, Flagstaff Rotary Club (sec. 1989-92, achievement 1990, pres. 1993-94). Achievements include research in Late Pleistocene climate and vegetation change in Grand Canyon, in status evaluation and ecological assessments of 110 species of rare and endangered plants in southwestern U.S. Home and Office: Bot & Environ Cons PO Box 201 Flagstaff AZ 86002-0201

PHILLIPS, ARTHUR WILLIAM, JR., biology educator; b. Claremont, N.H., Sept. 25, 1915; s. Arthur William and Jane Helen (Daley) P.; m. Mary Catherine Mich, Oct. 21, 1950; children: Marilynn, William (dec.). BS, U. Notre Dame, 1939, MS, 1941; DSc, MIT, 1947. Rsch. asst. Lobund lab. U. Notre Dame, Ind., 1937-41, rsch. scientist, 1943-45; rsch. assoc. MIT, 1947-49; rsch. assoc. prof., head div. bioengring. Lobund lab. U. Notre Dame, Ind., 1949-54; rsch. scientist dept. biology and bioengring. MIT, Cambridge, 1942-43, rsch. fellow dept. food tech., 1945-47, rsch. assoc. dept. food tech., 1947-49; rsch. assoc. prof. dept. bacteriology Syracuse (N.Y.) U., 1954-58, prof. microbiology, 1959-86, prof. emeritus, 1986—, founder, dir. biol. rsch. labs., 1955-65, head radiation and isotope lab., 1956-63, dir. germ-free life rsch. lab., 1956-84; mem. Internat. Congress on Nutrition, Washington, 1960, Internat. Congress for Microbiology, Montreal, Can., 1962, Moscow, 1966, Internat. Congress for Germ-Free Life Rsch., Nagoya, Japan, 1967; mem. com. on nutrition NAS-NRC, Washington, 1964-66; mem. Conf. on Germ-Free Life and Gnobiotics, Madison, Wis., 1986, Internat. Conf. on Gnotobiology, Versailles, France, 1987; cons. NSF, Washington, Cradle Soc. Inc., Evanston, Ill., GE, Syracuse, Am. Cyanamid, Pearl River, N.Y., Carnation Co., L.A., C.V. Mosby, St. Louis, Can. Dry Corp., Greenwich, Conn., Chocolate Mfrs. Assn., Washington, Continental Can Co., Syracuse. Contbr. articles to profl. jours., chpts. to books. Refrigeration Rsch. Found. fellow, 1945-47; NIH grantee, 1956-80. Mem. Am. Soc. for Microbiology (placement com. 1968-78), Gnotobiotics Assn., Soc. for Gen. Microbiology. Avocations: history, genealogy, hiking. Home: Clark Hollow Rd East Poultney VT 05741-0604 Office: Syracuse U Dept Biology 108 College Pl Syracuse NY 13244-1270

PHILLIPS, BESSIE GERTRUDE WRIGHT, school system administrator, museum trustee; b. Erie, Pa.; d. Charles Clayton and Mary Gertrude (Allen) Wright; m. Stephen Phillips, Oct. 2, 1942 (dec. Jan. 1971); children: Jane Appleton, Margaret Duncan (Mrs. Robert Cummings), Ann Willard (Mrs. Kevin Waters). AB, Fla. State Coll. Women, Tallahassee, 1930; MA, Mount Holyoke Coll., 1933. Sec. internat. hdqrs. World YMCA, Geneva, 1933-34; math. tchr. Washington Sem., Pa., 1930-32; acad. head, math tchr. Milw.-Downer Sem., Wis., 1934-37; trustee New Eng. Coll., Henniker, N.H., 1952-61, Orme Sch., Mayer, Ariz., 1962—, Penobscot Marine Mus., Searsport, Maine, 1978—; bd. advisers Coun. for Advancement Small Colls., Washington, 1959-69; trustee Peabody Mus., Salem, Mass., 1971—, vis. com. for edn., 1983—; ednl. cons. in field. Bd. dirs. Salem Female Charitable Soc., 1943—, Mack Indsl. Sch., Salem, 1943—. Mem. Nat. Soc. Colonial Dames, Bostonian Soc., Cum Laude Soc., Union Club, Chilton Club, Eastern Yacht Club. Republican. Presbyterian. Home: 30 Chestnut St Salem MA 01970-3129

PHILLIPS, DONALD LUNDAHL, research ecologist; b. Wilmington, Del., July 15, 1952; s. Donald Delaney and Mary Evelyn (Lundahl) P.; m. Margaret Cornelia Arentz, Aug. 5, 1978; children: Kevin, Paul. BS in Zoology, Mich. State U., 1974; MS, Utah State U., 1977, PhD in Biology, 1978. Cert. sr. ecologist. Asst. prof. Dept. Biology, Emory U., Atlanta, 1978-83; biostatistician CDC, Ctr. for Environ. Health, Atlanta, 1983-88; rsch. ecologist Environ. Rsch. Lab., U.S. EPA, Corvallis, Oreg., 1988—; cons. Haday, Inc., Salt Lake City, 1980, Oak Ridge Nat. Lab., 1982. Contbr. articles to profl. jours. Bd. trustees Highlands Biol. Found., 1980-88. Nat. Merit scholar, 1970-74; NSF rsch. grantee, 1980-84, U.S. Forest Svc. rsch. grantee, 1981-84; recipient Sec. of U.S. Dept. Health & Human Svcs. Recognition, 1987. Mem. Ecol. Soc. Am. Achievements include research in areas of plant ecology, statistical ecology and environ. health; examination of potential environmental effects of global climate change. Office: US EPA Environ Rsch Lab 200 SW 35th St Corvallis OR 97333

PHILLIPS, EARLE NORMAN, electro-optical engineer; b. Oneida, Tenn., Mar. 9, 1948; s. Robert and Fannie May (Marcum) P.; m. Susan Martha Appleton, Nov. 23, 1967; children: Norman Scott, Richard Alan, Jennifer Sue. AS in Engring. Tech., Polk C.C., 1969; postgrad., U. South Fla., 1969-71. Mech. engr. Monnfield Industries, Garland, Tex., 1971-72; quality assurance engr. Varo Inc., Garland, 1972-78; program mgr. Applied Devices Corp., Kissimmee, Fla., 1978-83; tech. group mgr. electro-optical divsn. ITT Def., Roanoke, Va., 1983—. Author: Day/Night Weaponsights for Small Arms, 1989, Gen III Night Vision Systems for Solic Missions, 1990. Scholar Superior Paving Corp., 1968-69. Mem. Am. Def. Preparedness Assn. Republican. Baptist. Achievements include patents for optical element output for an image intensifier device, universal image intensifier tube, telescopic sight for day/night viewing, collimator for binocular viewing system; 8 patents pending. Office: ITT Def Electro Optical Products 7635 Plantation Rd Roanoke VA 24019

PHILLIPS, GARY W., psychometrician; b. Charleston, W.Va., Dec. 30, 1947; s. William M. and Betty J. (Harler) P.; m. Joyce Ann Phillips, Apr. 2, 1973; children: Jimmy M., Christopher A. BA in Psychology, Philosophy, W.Va. State Coll., 1971, MA in Sch. Psychology, 1977; PhD in Applied Stats. and Psychometrics, U. Ky., 1983. Social worker Kanawha County Dept. Welfare, Charleston, 1971-73; supr. Charles Pack, Charleston, 1974; rsch. assoc. social work dept. grad. studies U. W.Va., Charleston, 1975; sch. psychology intern Regional Edn. Svcs. Agy., 1976-77; teaching rsch. asst. grad. studies U. W.Va., 1975-79; instr. U. Ky., Lexington, 1977-80; statistician, measurement specialist Md. State Dept. Edn., Balt., 1980-83, chief measurement and stats. evaluation sect., 1983-84, chief, rsch., evaluation and statis. svcs. br., 1985—; pres. Corp. Measurement and Stats., Balt., 1983—; adj. prof. W.Va. State Coll., 1976-79, U. W.Va., 1978, U. Md., 1982-83; cons., lectr. in field. Contbr. articles to The Ninth Mental Measurements Yearbook, 1984, Sch. Psychology, Applied Psychol. Measurement, Measurement and Evaluation in Guidance, Jour. Ednl. Measurement, Rev. Rsch. in Edn., Jour. Ednl. Stats., tech. papers in field; reviewer Jour. Ednl. Stats., Divsn. D, Ednl. Evaluation Policy Analysis, Ednl. Rschr. Mem. APA, Am. Ednl. Rsch. Assn., Nat. Coun. Measurement in Edn., Ea. Ednl. Rsch. Assn. Achievements include rsch. in path analysis, structural equations, LISREL models, meta-analysis, gen. linear models, item response theory models, application of item response theory models to criterion-referenced testing. Office: US Dept Edn 555 New Jersey Ave NW Washington DC 20208

PHILLIPS, JAMES CHARLES, physicist, educator; b. New Orleans, Mar. 9, 1933; s. William D. and Juanita (Hahn) P.; m. Joanna Vandenberg, July 4, 1993. B.A., U. Chgo., 1952, B.S., 1953, M.S., 1955, Ph.D., 1956. Mem. tech. staff Bell Labs., 1956-58; NSF fellow U. Calif. at Berkeley, 1958-59, Cambridge (Eng.) U., 1959-60; faculty U. Chgo., 1960-68, prof. physics, 1965-68; mem. tech. staff Bell Labs., 1968—. Sloan fellow, 1962-66; Guggenheim fellow, 1967. Fellow Am. Phys. Soc. (Buckley prize 1972), Minerals, Metals and Materials Soc. (William Hume-Rothery award 1992); mem. NAS. Home: 204 Springfield Ave Summit NJ 07901-3909

PHILLIPS, JAMES MACILDUFF, material handling company executive, engineering and manufacturing executive; b. Carrick, Pa., June 13, 1916; s. John MacFarlane and Harriet (Duff) P.; m., Majorie Watson, June 1940 (div. 1964); children: James M. Jr., William W.; m. Regina Leininger, Apr. 1964 (dec.); children: Jeffrey M., Molly M., Becky J., Thomas S. BSME, Carnegie Inst. of Tech., 1938; grad., Pitts. Inst. Aeronautics, 1939; ME refresher, Pa. State U. State Coll., 1960; grad., Internat. Corespondence Sch., Scranton, Pa., 1988. Profl. engr. Pa. Draftsmen, engr. Phillips Mine & Mill Supply Co., Pitts., 1933-40, v.p., 1941-77; v.p. engring. Salem Brosius, Inc., Carnegie, Pa., 1956-64; pres. Phillips Corp., Bridgeville, Pa., 1977-83, Phillips Jet Flight, Bridgeville, Pa., 1977-83, Phillips Mine & Mill Inc., Pitts., 1964—; also chmn. bd. Phillips Mine & Mill Inc. Inventor in field; contbr. articles to profl. mags. Bd. dirs. Brashear Assn., Pitts. Mem. Air Force Assn., Aero Club of Pitts., Pa. Pilots Coun. (pres.), Quiet Birdmen (pres.), Early Bird Pilots, Exptl. Aircraft Assn. Aircraft Owners and Pilots Assn. (founding), OX-5 Pioneer Airmen (pres. 1987), St. Clair County Club. Methodist. Avocations: airline transport pilot, flight instructor, golf, tennis. Office: Phillips Mine & Mill Inc 1738 N Highland Rd Pittsburgh PA 15241-1200

PHILLIPS, JOHN A(TLAS), III, geneticist, educator; b. Sanford, N.C., Jan. 24, 1944; s. John A. and Rachael (Sloan) P.; m. Gretchen Lynch, Aug. 1, 1965; children: Jennifer Allene, John Atlas IV, Charles Andrew, James William. Student, U. N.C., 1962-65; MD, Bowman Gray Coll. Medicine, 1969. Diplomate Am. Bd. Pediatrics, Am. Bd. Med. Genetics. Intern Children's Hosp. Med. Ctr., Boston, 1969-70, jr. resident, 1970-71, sr. resident, 1973-74, chief resident, 1974-75; asst. prof. Johns Hopkins U., Balt., 1978-82, assoc. prof., 1982-84; prof. pediatrics Vanderbilt U., Nashville, 1984—, prof. biochemistry, 1986—, David T. Karzon chair genetics, 1992—; bd. sci. counselors Nat. Inst. Child Health, Washington, 1984-88; counsilor Ctr. Study Polymorphisme Humain, Paris, 1988—; adv. com. Ctr. Reproductive Biology, Nashville, 1990—; bd. dirs. March of Dimes Birth Defects Found., Nashville, 1986; adv. bd. Nat. Neurofibromatosis Found., Tenn., 1990—; mem. Tenn. Genetics Adv. Com., Nashville, 1984—. Contbr. to profl. publs. Lt. comdr. USNR, 1971-73. Recipient Sidney Farber award Children's Hosp., Boston, 1975, E Mead Johnson award Mead Johnson Co., 1984; Pediatric Postdoctoral fellow Johns Hopkins U. Sch. Medicine, 1975-77. Mem. Am. Soc. Clin. Investigations, Soc. Pediatric Rsch., Phi Beta Kappa, Alpha Omega Alpha. Achievements include discovery of cause of hemoglobin H disease in Black Americans, chromosomal location of multiple genes in humans, improved diagnoses of cystic fibrosis, hemophilia, inborn metabolic errors, familial neurodegenerative diseases. Office: Vanderbilt U Dept Genetics DD 2205 Med Ctr N Nashville TN 37232-2578

PHILLIPS, JOHN BENTON, chemical engineer; b. Wheeling, W.Va., Aug. 17, 1959; s. James Edward and Alice Jean (Kuhar) P.; m. Catharine Marie Monroe, July 28, 1984; 1 child, Amy Catharine. BS, Fla. State U., 1982; M of Engring., Tulane U., 1989. Rsch. technician Monell Chem. Senses Ctr., Phila., 1981-82; rsch. chemist R & K Graphics, Bryn Mawr, Pa., 1982; assoc. engr. William S. Wood & Assocs., West Chester, Pa., 1982-84; chmn. dept. chemistry Northeastern Christian Coll., Villanova, Pa., 1982-84; project mgr. Universal Chem. Co., Tampa, Fla., 1982—; lectr. U. Pa., Phila., 1984; teaching asst. Tulane U. Sch. Engring., New Orleans, 1988-89, instr., 1989—; sr.process engr. Walk, Haydel and Assocs., New Orleans, 1991-92; sr. environ. process engr. Badger Design and Constructors, Cambridge, Mass., 1992-93, Raytheon Engrs., Phila., 1993—; cons. in field. Contbr. articles to Jour. Heterocyclic Chemistry, Jour. Phys. Chemistry, Applied Biochemist and Biotechnology, Biotechnology Progress. Pre. Ocean Ecology Club, Tallahassee, Fla., 1980; treas. Silvercrest PTA, New Orleans, 1990. Recipient Dept. scholarship Fla. State U., 1980, Weymouth-Campbell scholarship, 1985, Eastman Kodak Grad. fellowship, 1988, Outstanding Teaching award Omega Chi Epsilon, 1990, Rsch. Travel award Omega Chi Epsilon, 1990. Mem. Am. Inst. Chem. Engrs. (First Prize Regional Paper Contest award 1988), Omega Chi Epsilon (pres. 1989-90), Omicron Delta Kappa, Sigma Xi. Office: Tulane U Dept Chem Engring 6823 St Charles Ave New Orleans LA 70118-5665

PHILLIPS, JULIA MAE, physicist; b. Freeport, Ill., Aug. 17, 1954; d. Spencer Kleckner and Marjorie Ann (Figi) Phillips. BS, Coll. William and Mary, 1976; PhD, Yale U., 1981. Mem. tech. staff AT&T Bell Labs., Murray Hill, 1981-88, supr. thin film rsch. group, 1988—; program mgr. Consortium Superconducting Elecs., 1989-92. Editor: Heteroepitaxy on Silicon Technology, 1987; prin. editor Jour. Materials Rsch., 1990—; mem. editorial bd. Applied Physics Letters and Jour. of Applied Physics, 1992-94; contbr. articles to profl. jours. Mem. APS, Materials Rsch. Soc. (sec. 1987-89, councillor 1991-93, 2d v.p. 1993), Sigma Xi. Phi Beta Kappa. Office: AT&T Bell Labs 600 Mountain Ave NW New Providence NJ 07974

PHILLIPS, KEVIN JOHN, consulting engineer; b. N.Y.C., Feb. 3, 1948; s. William Charles and Estella (Kearney) P.; m. Suzanne B. Marciano, July 9, 1972; children: Alexander, Christopher. B of Engring., CCNY, 1970; MSCE, MIT, 1972, engrs. degree in environ. engring., 1973; PhD, Poly. U. of N.Y., 1978. Registered profl. engr., N.Y., N.J., Mass., Conn., Ala., Ga. Structural engr. Volmer Assocs., N.Y.C., 1970; rsch. asst., teaching asst. MIT, Cambridge, Mass., 1970-73; project engr., sr. project engr. Weston Environ. Cons. and Designers, Roslyn, N.Y., 1973-79; pres. Environ. Mgmt. and Engring., N.Y., 1979-80; prin. Fanning Phillips and Assocs., Plainview, N.Y., 1980-82, Fanning, Phillips and Molnar, Ronkonkoma, N.Y., 1982—; dir. Hydrotron, N.Y.C., 1984—. Contbr. articles to profl. jours. Recipient Engring. Excellence award Cons. Engrs. Coun. of Pa., 1978. Mem. ASCE, Nat. Waterwell Assn., Am. Water Works Assn., Water Pollution Control Fedn., Am. Acad. Environ. Engrs. (past diplomat), Sigma Xi, Tau Beta Pi. Home: 22 Norwood Rd Northport NY 11768 Office: Fanning Phillips Molnar 909 Marconi Ave Ronkonkoma NY 11779

PHILLIPS, LARRY H., II, neurologist; b. Clarksburg, W.Va., Dec. 30, 1947; m. Elayne K. Phillips, 1985; children: Joshua, Melanie. AB in Biology, Princeton U., 1970; MD, W.Va. U., 1974. Diplomate Am. Bd. Psychiatry and Neurology (test com. 1990—), Am. Bd. Electrodiagnostic Medicine; lic. Minn., Va. Intern U. Wis. Hosps., Madison, 1974-75; resident in neurology Mayo Clinic, Rochester, Minn., 1975-78, rsch. fellow in neurophysiology, 1978-79; instr. neurology Mayo med. sch. U. Minn., Rochester, 1979-80; asst. prof. U. Va. Med. Sch., Charlottesville, 1981-87, assoc. prof., 1987—; dir. electromyography lab. U. Va. Med. Ctr., Charlottesville, 1981—, dir. neuromuscular ctr, Muscular Dystrophy Assn. clinic, 1983—; cons. neurology, Mayo Clinic, Rochester, 1979-80, VA Hosp. Salem, Va., 1981—; cons. panel AMA Diagnostic and Therapeutic Tech. Assessment, 1989—, arbitrator panel, 1990—; expert panel mem. NIH, 1991; med. adv. com. Diabetes Rsch. and Training Ctr., U. Va., 1981-88; com. for respiratory intensive care unit U. Va. Hosp., 1988; various coms. Nat. Bd. Med. Examiners. Editorial reviewer Jour. Urology, 1986—, Western Jour. Medicine, 1990—, Jour. Applied Physiology, 1990—, Jour. Neurosurgery, 1990—, So. Med. Jour., 1993—; contbr. articles to profl. jours. Active med. adv. bd. Myasthenia Gravis Found., 1983—, bd. dirs. 1986—; bd. dirs. ARC Ctrl. Va. Chpt., 1985-91, exec. com. 1990-91. Recipient Young Investigator Travel award Internat. Congress Electromyography, 1979. Mem. Am. Neurological Assn., Am. Acad. Neurology (neuroepidemiology sect. 1978—, chmn. edn. com. 1992—, clin. neurophysiology sect. 1992—), Am. Assn. Electrodiagnostic Medicine (chmn. course com. 1992—, annual meeting coord. com. 1992—, and other coms.), Assn. Univ. Profs. Neurology (undergraduate edn. com.), Va. Neurological Soc. (pres. 1993—), Sigma Xi. Home: 1405 Wendover Dr Charlottesville VA 22901 Office: U Va Med Ctr Dept Neurology Box 394 Charlottesville VA 22908*

PHILLIPS, MARK DOUGLAS, aerospace company executive; b. N.Y.C., Oct. 23, 1953; s. Herbert and Dorinne (Borenstein) P.; m. Deanne Sue Pace, Dec. 28, 1979; children: Cheri, Brooke, Brittany. BA, SUNY, New Paltz, 1975; MA, Calif. State U., Northridge, 1982. Staff scientist Integrated Scis. Corp., Santa Monica, Calif., 1980-82; sr. engr. Martin Marietta Co., Denver, 1982-83; program mgr. CTA, Inc., Englewood, Colo., 1983-88, regional tech. officer, 1988-89; v.p., dep. dir. air traffic systems div. CTA, Inc., Rockville, Md., 1989-92; v.p., dir. air traffic systems div. CTA Inc., Rockville, Md., 1992—. Contbr. articles to profl. jours.; developer human factors engring. methods and tools for air traffic control. Mem. Air Traffic Control Assn. (Chmn.'s Citation of Merit 1984), Human Factors Soc., Requirements and Tech. Concepts for Aviation. Avocations: jazz guitar, tennis. Office: 6116 Executive Blvd Rockville MD 20852-4920

PHILLIPS, SHELLEY, psychologist, writer; b. Melbourne, Victoria, Australia, Mar. 18, 1934; children: Catherine, Odette. BA in Lit. and History with honors, U. Melbourne; PhD in Psychology, U. Sydney (Australia), 1973. Registered psychologist. Lectr. child devel. U. NSW, Sydney, 1972-77; guest lectr. Ga. State U., Atlanta, 1975; vis. fellow U. Birmingham (Eng.), 1975-76; vis. prof. U. Alta. (Can.), 1975-76; vis. scholar Inst. of Edn., London U., 1979, 82-83; sr. lectr. U. NSW, Sydney, 1977-85, dir. unit for child studies, 1979-85; dir. Found. for Child and Youth Studies, Sydney, 1985-90; pvt. practice psychol. cons. Sydney, 1985—; vis. scientist, program assoc. Tavistock Clinic, London, 1975, 78, 83; vis. scholar Inst. Early Childhood, Melbourne, 1982; chairperson children's programme com. Australian Broadcasting Tribunal, Sydney, 1984-87; lectr. Inst. of Psychiatry, Sydney, 1985-91; occasional lectr. staff tng. A.C.T. Schs.' Authority, Canberra, Australia, 1979-86. Author: Medieval Thinking, 1972, Young Australians, 1979, Self Concept and Sexism in Language, 1982, Relations with Children, 1986, Beyond the Myths, Mother Daughter Relations in Psychology, History, Literature and Everyday Life, 1991. Mem. Australian Psychol. Soc., Australian Soc. Authors, Australian Soc. Women Writers, Bd. Ednl. and Devel. Psychologists, Nat. Trust Australia, Jane Austen Soc. Office: 21/3 Plunkett St, Kirribilli NSW 2061, Australia

PHILLIPS, STEPHEN MARSHALL, electrical engineering educator; b. Radford, Va., Dec. 25, 1962; s. Winfred M. and Lynda (Bartlett) P. BSEE, Cornell U., 1984; PhD, Stanford U., 1988. Registered profl. engr., Ohio. Engr. Litton Guidance and Control, Woodland Hills, Calif., 1985; asst. prof. elec. engring. Case Western Res. U., Cleve., 1988—; faculty fellow NASA Lewis Rsch. Ctr., Cleve., 1990-92. Contbr. articles to Internat. Jour. Control, Analog Integrated Circuits and Signals Procs, Jour. Robotic Systems, Sensors and Actuators, IEEE Trans. on Electron Devices. Mem. IEEE, AIAA, Sigma Xi, Tau Beta Pi, Eta Kappa Nu. Office: Case Western Res U 10900 Euclid Ave Cleveland OH 44106-7221

PHILLIPS, THOMAS DEAN, mechanical engineer; b. Denver, Mar. 20, 1954; s. Walter J. and Grace (Towar) P.; m. Harriet Pickle, July 5, 1986; children: Brent Thomas, Brian James. BS in Archtl. Engring., Colo. U., 1978. Registered profl. engr., Tex. Field engr. Fluor E&C, Irvine, Calif., 1978-81; constrn. coord. Spantec, Inc., San Antonio, 1981-82; project mgr. Frey Mech., San Antonio, 1983-84; staff engr., project mgr. Mueller & Wilson, Inc., San Antonio, 1984—. Mem. NSPE, Airplane Owner and Pilot Assn., Am. Welding Soc. Office: Mueller & Wilson Inc 2107 Danbury San Antonio TX 78217

PHILLIPS, WALTER MILLS, III, psychologist, educator; b. N.Y.C., Sept. 29, 1947; s. Walter Mills and Grace Mary (Mullen) P. BS, Fordham U., 1970; MA, U. S.D., 1973, PhD, 1975. Lic. clin. psychologist, Conn.; m. Anne Marie Boyle, July 3, 1971; children: Jonathan, Elizabeth. Adolescent resident counselor Hawthorne (N.Y.) Cedar Knolls Sch., 1970-71; NIMH tng. fellow, 1971-75; clin. psychology intern Inst. of Living, Hartford, Conn., 1974-75, clin. staff psychologist, 1975-79, sr. staff psychologist, 1979-82, asst. dir. dept. clin. psychology, 1980-82, dir. clin. psychology tng., 1980-82; co-dir. outpatient psychiatry U. Conn., Farmington, 1982-88; asst. prof. psychiatry, dir. psychiatry evaluation svc. U. Conn. Health Ctr., 1982-88 ; pvt. practice psychotherapy, Hartford, 1976—, dir. Anxiey Rsch. and Treatment Ctr., 1985-88; dir. adolescent/young adult svc. Grandview Psychiat. Resource Ctr., Waterbury, Conn., 1988-90; dir. psychology Waterbury Hosp., 1990—; asst. clin. prof. psychiatry Sch. Medicine Yale U., New Haven, Conn., 1988—; mem. psychology exec. com. Sch. Medicine Yale U., New Haven. Mem. Am. Psychol. Assn., Conn. Psychol. Assn., Soc. Psychotherapy Rsch., Soc. Personality Assessment, Conn. Hosp. Assn. (chmn., dir. psychology conf.), Sigma Xi. Contbr. articles to profl. jours. Home: 70 Beverly Dr Avon CT 06001 Office: 60 Westwood Ave Ste 115 Waterbury CT 06708-2460

PHILLIPS, WILLIAM ROBERT, fluid dynamics engineering educator; b. Adelaide, Australia, Apr. 14, 1948; came to U.S., 1986; s. Robert Ray and Eileen Marjorie (Richter) P. BE with honours, Adelaide U., 1970; MEng, McGill U., Montreal, Que., Can., 1974; PhD, Cambridge (Eng.) U., 1978. Rsch. engr. Mt. Isa Mines Ltd., Queensland, Australia, 1971-73; rsch. assoc. McGill U., 1975; lectr. Nat. U. Singapore, 1979-81, sr. lectr., 1981-84; sr. rsch. fellow U. Melbourne, Australia, 1984-85; vis. scientist Cornell U.,

Ithaca, N.Y., 1986, assoc. vis. prof., 1987-89; assoc. prof. fluid dynamics Clarkson U., Potsdam, N.Y., 1989—. Contbr. numerous articles to sci. jours. Commonwealth U. scholar Govt. of Australia, 1966-70, scholar Nat. Coun. Can., 1974-75; Rolls Royce rsch. fellow Churchill Coll., Cambridge, 1975-78; grantee NSF, 1990-93. Mem. Soc. for Indsl. and Applies Math., Am. Phys. Soc., N.Y. Acad. Sci., Sigma Xi. Avocations: golf, skiing. Home: Riggs Dr Box 133 Hannawa Falls NY 13647 Office: Clarkson U Box 5725 Potsdam NY 13699

PHILLIPS, WILLIAM THOMAS, nuclear medicine physician, educator; b. Ulysses, Kans., June 29, 1952; s. Harold Wesley and Sheila (McGinnis) P.; m. Lauren Dorley, Sept. 20, 1981; children: Alison, Trevor, Caden. BA, Ea. N.Mex. U., 1975; MD, U. Tex., Galveston, 1980. Diplomate Am. Bd. Nuclear Medicine, Am. Bd. Family Practice. Resident in family practice Tex. Tech U., Lubbock, 1981-84; resident in nuclear medicine U. Tex., San Antonio, 1985-87, asst. prof. nuclear medicine, 1988-93; assoc. prof. nuclear medicine, 1993—. Mem. Soc. of Friends. Achievements include patents for new method of labeling liposomes with 99m Technetium, new method of treating diabetes by delaying gastric emptying; discovery that many diabetics have rapid gastric emptying. Office: U Tex Health Sci Ctr 7703 Floyd Curl Dr San Antonio TX 77284

PHILLIS, JOHN WHITFIELD, physiologist, educator; b. Port of Spain, Trinidad, Apr. 1, 1936; came to U.S., 1982; s. Ernest and Sarah Anne (Glover) P.; m. Pamela Julie Popple, 1958 (div. 1968); children: David, Simon, Susan; m. 2d Shane Beverly Wright, Jan. 24, 1969. B.V.Sc., Sydney U., Australia, 1958; D.V.Sc., Sydney U., (Australia), 1976; Ph.D., Australian Nat. U., Canberra, 1961; D.Sc., Monash U., Melbourne, Australia, 1970. Lectr./sr. Monash U., 1963-69; vis. prof. Ind. U.-Indpls., 1969; prof. physiology, assoc. dean research U. Man., Winnipeg, Can., 1970-73; prof., chmn. dept. physiology U. Sask., Saskatoon, Can., 1973-81, asst. dean research, 1973-75; prof. physiology, chmn. dept. physiology Wayne State U., Detroit, 1982—; wellcome vis. prof. Tulane U., 1986; mem. scholarship and grants com. Can. Med. Research Council, Ottawa, Ont., 1973-79; mem. sci. adv. bd. Dystonia Med. Research Found., Beverly Hills, Calif., 1980-85; sci. adv. panel World Soc. for Protection of Animals, 1982—. Author: Pharmacology of Synapses, 1970; editor: Veterinary Physiology, 1976, Physiology and Pharmacology of Adenosine Derivatives, 1983, Adensine and Adenine Nucleotides as Regulators of Cellular Function, 1991, The Regulation of Cerebral Blood Flow, 1993; editor Can. Jour. Physiology and Pharmacology, 1978-81, Progress in Neurobiology, 1973—. Mem. grants com. Am. Heart Assn. of Mich., 1985-90, mem. rsch. coun., 1991-92, mem. rsch. forum com. 1991, chair, 1992-93. Wellcome fellow London, 1961-62; Can. Med. Research Found. grantee, 1970-81; research prof., 1980; NIH grantee, 1983—. Mem. Brit. Pharmacol. Soc., Physiol. Soc., Can. Physiol. Soc. (pres. 1979-80), Am. Physiol. Soc., Soc. Neurosci., Internat. Brain Rsch. Orgn. Home: 25501 Circle Dr Southfield MI 48075-6127 Office: Wayne State U Dept Physiology 540 E Canfield St Detroit MI 48201-1998

PHILPOT, JOHN LEE, physics educator; b. Kansas City, Mo., May 7, 1935; s. John L. and Millie Ann (Johnson) P.; m. Nancy J. Tucker, Nov. 22, 1956 (div. Jan. 1978); children: John Michael, Patricia Ann; m. Sharon Carol Wuckowitsch, May 15, 1982; 1 child Mellissa Phillips. AB, William Jewell Coll., 1957; MS, U. Ark., 1962, PhD, 1965. Prof. physics U. Mo., Kansas City, 1963-68, William Jewell Coll., Kansas City, 1962—; thin film lab. Bendix Corp., Kansas City, 1963, 64. Contbr. articles to profl. jours. NDEA fellow, 1961-64. Mem. Am. Phys. Soc., Am Assn. Physics Tchrs., Mo. Acad. Sci. Baptist. Home: 349 Don Allen Rd Liberty MO 64068 Office: William Jewell Coll 500 College Hill Liberty MO 64068

PHINNEY, WILLIAM CHARLES, geologist; b. South Portland, Maine, Nov. 16, 1930; s. Clement Woodbridge and Margaret Florence (Foster) P.; m. Colleen Dorothy Murphy, May 31, 1953; children—Glenn, Duane, John, Marla. B.S., MIT, 1953, M.S., 1956, Ph.D., 1959. Faculty geology U. Minn., 1959-70; chief geology br. NASA Lyndon B. Johnson Space Center, Houston, 1970-82; chief planetology br. NASA Lyndon B. Johnson Space Center, 1982-89; NASA prin. investigator lunar samples. Contbr. articles to profl. jours. Served with C.E. AUS, 1953-55. Recipient NASA Exceptional Sci. Achievement medal, 1972; NASA Cert. of Commendation, 1987; NASA research grantee, 1972—; NSF research grantee, 1960-70. Mem. Am. Geophys. Union, AAAS, Mineral. Soc. Am., Geol. Soc. Am., Minn. Acad. Sci. (dir.), Sigma Xi. Home: 18523 Barbuda Ln Houston TX 77058-4005 Office: NASA Lyndon B Johnson Space Ctr SN 2 Houston TX 77058

PIAN, THEODORE HSUEH-HUANG, engineering educator, consultant; b. Shanghai, China, Jan. 18, 1919; came to U.S., 1943; s. Chao-Hsin Shu-Cheng and Chih-Chuan (Yen) P.; m. Rulan Chao, Oct. 3, 1945; 1 child, Canta Chao-Po. B in Engring., Tsing Hua U., Kunming, China, 1940; MS, MIT, 1944, DSc, 1948; DSc (hon.), Beijing U. Aeros. and Astronautics, 1990; PhD (hon.), Shanghai U. of Tech., 1991. Engr. Cen. Aircraft Mfg. Co., Loiwing, China, 1940-42, Chengtu Glider Mfg. Factory, 1942-43; stress analyst Curtis Aircraft Div., Buffalo, 1944-45; teaching asst. MIT, Cambridge, 1946-47, rsch. assoc., 1947-52, asst. prof., 1952-59, assoc. prof., 1959-66, prof., 1966-89, prof. emeritus, 1989—; vis. assoc. Calif. Inst. Tech., Pasadena, 1965-66; vis. prof. U. Tokyo, 1974, Tech. U., Berlin, 1975; vis. chair prof. Nat. Tsing Hua U., Hsin Chu, Taiwan, 1990, Nat. Ctrl. U., ChungLi, Taiwan, 1992; hon. prof. Beijing U. Aero. and Astronautics, Beijing Inst. Tech., Southwestern Jaiotong U., Dalian U. Tech., Huazhong U. Sci. and Tech., Changsha Rwy. U., Ctrl.-South U. Tech., Hohai U., Nanjing U. of Aero. and Astronautics. Recipient von Karman Meml. prize TRE Corp., Beverly Hills, Calif., 1974. Fellow AAAS, AIAA (assoc. editor jour. 1973-75, Structures, Structural Dynamics and Materials award 1975); mem. NAE, ASME (hon.), Am. Soc. for Engring. Edn., Internat. Assn. for Computational Mechanics (hon. assoc. gen. coun.). Home: 14 Brattle Cir Cambridge MA 02138-4625 Office: MIT Dept Aeronautics and Astronautics 77 Massachusetts Ave Cambridge MA 02139-4307

PIATETSKI-SHAPIRO, ILYA, mathematics educator; b. Moscow, Mar. 30, 1929; arrived in Israel, 1976; s. Joseph and Sheina (Gurevitch) P.-S.; m. Edith Piatetski-Shapiro, 1978; children: Gregory, Vera, Shlomit. PhD, Moscow U., 1951, D, 1958. Prof. math. Moscow U.,, 1964-72, Tel Aviv U.,, 1976—, Yale U., New Haven, 1977—. Recipient Laureate of Israel prize in math., Wolf Found. Math. prize, 1990. Avocation: hiking. Office: Yale U Dept Math 10 Hillhouse Ave PO Box 2155 New Haven CT 06520 also: Tel Aviv U Sch Math Scis, Ramat Aviv, Tel Aviv Israel

PIATKOWSKI, STEVEN MARK, environmental engineer; b. Jersey City, June 30, 1962; s. Arthur Edward and Marie Alice (Bataglio) P.; m. Dawn Hazel Seemon, June 15, 1991. BSChemE, Rutgers U., 1984; MS in Environ. Engring., Stevens Inst. Tech., 1993. Prin. environ. engr. N.J. Dept. Environ. Protection, Trenton, N.J., 1985-89; sr. cons. Consulting Svcs., Inc., Exton, Pa., 1989—. Mem. NSPE, Nat. Waterwell Assn., Am. Acad. Environ. Engrs. Achievements include environ. audits on over 200 indsl. facilities. Office: Consulting Svcs Inc 415 Eagleview Blvd Exton PA 19341

PIATTELLI-PALMARINI, MASSIMO, cognitive scientist; b. Rome, Apr. 29, 1942; came to U.S., 1985; s. Alberto Piattelli and Marcella (Bosco) Palmarini; m. Elisabeth Michahelles, Nov. 24, 1963 (div. 1972); 1 child, Simone. D Physics, U. Rome, 1969. Asst. prof. U. Rome, 1970-72; rsch. scientist Inst. Pasteur, Paris, 1972-74; lectr. Sch. Higher Studies, Paris, 1974-79; dir. Ctr. for Sci. of Man, Paris, 1974-79, Ctr. for Philosophy of Sci., Florence, Italy, 1980-85; prin. rsch. scientist Ctr. for Cognitive Sci., MIT, Cambridge, 1985—. Author: S Come Cultura, 1987, La Voglia di Studiare (The Will To Study, transl. into French, Spanish, Portugese and German), 1991, L'illusione Di Sapere, 1993; editor: Language and Learning, 1980; (books transl. into 6 langs.); editor-in-chief Kos—Monthly Jour. Sci., 1983-85; mem. editorial bd. Cognition, 1991—, Behavioral and Brain Sci., 1991—; also articles. Mem. Cognitive Sci. Soc., Soc. for Psychology and Philosophy, Italian Soc. for Logic nd Philosophy, Italian Coun. for Social Studies. Office: MIT 77 Massachusetts Ave Cambridge MA 02139

PICARDI, ANTHONY CHARLES, systems engineer, market research consultant; b. Boston, Apr. 23, 1948; s. E. Alfred and Mary Elizabeth (Long) P.; m. Shirley Mae Ives, Sept. 4, 1970. BS, MIT, 1971, MS, 1972, DSc, 1975. Head systems analysis Devel. Analysis Assn., Cambridge, Mass., 1975-80; sr. rsch. assoc. Charles River Assocs., Boston, 1980-82; dir. U.S. model Mgmt. Techs., Wellesley, Mass., 1982-83; program mgr. Higher Order Software, Cambridge, 1983-85; sr. product mgr. Cortex Corp., Waltham, Mass., 1985-89; v.p. mktg. and strategy Pilot Exec. Software, Boston, 1989; dir. software rsch. Internat. Data Corp., Framingham, Mass., 1990—. Contbr. 28 articles to profl. jours. Recipient James A. Peacock Quality in Market Rsch. award, 1992. Mem ASCE, Computer Soc. of IEEE, Soc. for Computing Machinery, Decus, Boston Yacht Club, Sigma Xi, Chi Epsilon. Home: 58 Washburn Ave Wellesley MA 02181-5224

PICCININO, LINDA JEANNE, statistician; b. Phila., July 3, 1956; d. Isaia and Genevieve H. Piccinino. BA, Cornell U., 1979, MPS, 1983; doctoral studies, U. N.C., 1993. Acquisitions editor, edit. asst. Plenum Publishing Co., N.Y.C., 1982-85; rsch. analyst Internat. Projects Assistance Svcs., Carrboro, N.C., 1988-91; population affairs officer UN, N.Y.C., 1990; social sci. analyst U.S. Bur. of the Census, Washington, 1991; statistician, demographer Nat. Ctr. for Health Stats., Hyattsville, Md., 1991--. Mem. Am. Pub. Health Assn., Population Assn. Am. Office: Nat Ctr for Health Statistc 6525 Belcrest Rd Rm 840 Hyattsville MD 20782

PICCIONE, NICOLAS ANTONIO, economist, educator; b. Buenos Aires, Sept. 18, 1925; s. Nicolás and Rosa Francisca (Aguggini) P.; m. Betty Raquel Saigg, Jan. 24, 1959; children: Lidia Cristina, Marcelo Nicolás, Juan Carlos. Degree in acctg., Buenos Aires U., 1953, econ. licenciate, 1963, DSc in Econ., 1968; degree in fin. adminstrn., Fin. Exec. Inst., Mexico City, 1968. Cont. Mfg. Tabacos Piccardo S.A., Buenos Aires, 1954-60; trustee, adminstrn. mgr. Ind. Autom. Sta. FE-DKW Auto Union, Buenos Aires, 1960-63; adminstrn. mgr., gen. acct. Kellogg Co. Argentina, Buenos Aires, 1968-71; gen. mgr. Inst. OBRA Social Ministerio Bienestar Social, Buenos Aires, 1972-74; econ. and fin. cons. Sindicatura Gen. Empresas Publ., Buenos Aires, 1975-90; titular prof. MBA program; trustee Bs. As. Catering S.A.; titular prof. fin. mgmt. Belgra U., Buenos Aires, 1977—; titular prof. master of bus. adminstrn. Salvador U., Buenos Aires, 1988—. Author: Economic/Financial Administration, 1968, 3d edit. 1988, Economic/Financial Administration of Enterprises, Added Value Tax, 1974; contbr. articles to profl. jours. Fellow Fin. and Devel. Internat. Monetary Fund; mem. Profl. Coun. Econ. Sci. Roman Catholic. Avocations: football, tennis, fishing. Home: Malvinas Argentinas N 84 A, 1406 Buenos Aires Capitol, Argentina Office: Professional Studio, Carlos Pellegrini 1175 8 A, 1009 Buenos Aires Argentina

PICHAL, HENRI THOMAS, electronics engineer, physicist, consultant; b. London, Feb. 14, 1923; came to the U.S., 1957; s. Henri and Mary (Conway) P.; m. Vida Eloise Collum Jones, Mar. 7, 1966; children: Chris C., Henri T. III, Thomas William Billingsley. PhD in Physics, London U., 1953, MSc in Engring., 1955. Registered profl. engr., Wash., Fla. Product engr. John Fluke Mfg. Corp., Everett, Wash., 1970-73; engring. specialist Harris Corp., Melbourne, Fla., 1973-75; pres., prin. Profl. Engring. Co., Inc., Kissimmee, Fla., 1975—. Contbr. articles to Electronics, Microwaves, and others. Named one of Two Thousand Men Achievement, 1972. Mem. Inst. Physics, Am. Phys. Soc., Fla. Engring. Soc. (sr.), Inst. Environ. Scis. (sr.), IEEE (past chmn. microwave theory and techniques communications systems), Aerospace/Navigational Electronics, Space Electronics and Telemetry, Mil. Electronics. Republican. Achievements include 69 patents in microwave RF, high frequence and high speed analog ultra linear technology, large dynamic range performance and intermodulation phenomena, noise, and congested area systems problems. Home: PO Box 969 Kingston WA 98346-0969

PICHETTE, CLAUDE, former banking executive, university rector, research executive; b. Sherbrooke, Que., Can., June 13, 1936; s. Donat and Juliette (Morin) P.; m. Renee Provencher, Sept. 5, 1959; children: Anne-Marie, Martin, Philippe. B.A., U. Sherbrooke, 1956; M.Sc.Soc. (Econ.), U. Laval, 1960; Doct. d'Etat es Sc. Econ. U. D'Aix-Marseille, France, 1970. Prof. U. Sherbrooke, 1960-70; civil servant Govt. Que., 1970-75; vice rector adminstrn. and fins. U. Que., Montreal, 1975-77; rector U. Que., 1977-86; pres., chief exec. officer La Financière prêts-épargne, 1986-90, La Financière Entraide-Cooperants (holding co.), 1987-90; pres. Que. Found. Econ. Edn., 1979-81; CEO Institut Armand-Frappier Rsch. Inst.; chmn. bd. La Financière courtage immobilier, 1989-90, La Financière crédit bail, 1989-90; bd. dirs. Shermag, Rona-Dismat. Author: Analyse micro-economique et cooperative, 1972. Can. Council grantee, 1958; Federation nationale des cooperatives de consommation de France grantee, 1973. Mem. Que. Assn. Econs. (pres. 1977-78). Club: St.-Denis (Montreal). Home: 745 Hartland, Outremont, PQ Canada H2V 2X5 Office: Armand-Frappier Institute, 531 Blvd des Prairies, Laval, PQ Canada H7N 4Z3

PICHLER, J(OHANN) HANNS, economics educator; b. Aspach, Austria, May 12, 1936; s. Franz and Berta (Mayr) P.; m. Hannelore M. Haslinger, Sept., 1963; children: Adelheid, Regine, Markus. M of Bus. and Econs., U. Econs. Vienna, Austria, 1958; D of Econs., U. Econs. Vienna, 1960, D habil., 1965; MSc in Econ. Theory, U. Ill., 1963. Mem. faculty U. Econs., Vienna, 1960-65; lectr. U. Ill., Urbana, 1962-63; sr. economist World Bank, Washington, 1965-71; res. rep. World Bank Group, Islamabad, Pakistan, 1971-74; prof. polit. economy and internat. devel. U. Econs., Vienna, 1974—; head faculty econs. U. Econs., 1975—; cons. World Bank, UNIDO; bd. dirs. Austrian Latin Am. Inst., Vienna, Afro-Asian Inst., Vienna, Schumpeter Soc., Vienna, Austrian Assn. Agr. and Forestry Policy, European Coun. Small Bus.; chmn. Inst. Small Bus. Rsch., Assn. for Devel. Coop. Austria, Ges. fuer Ganzheitsforschung, Vienna; vice chmn. Austrian Found. Devel.; mem. adv. bd. Sued-Ost Treuhand/Ernst & Young, Vienna; mem. Govt. Adv. Bd. on Devel. Author books, contbr. to profl. publs. Decorated Austrian Cross of Honor for Sci. and Arts 1st class; recipient Card. Innitzer Sci. award, 1966, Univ. Econs. prize, 1982, L. Kunesak prize, 1990; Schumpeter fellow Harvard U., 1990. Mem. European Acad. Scis. Arts, Am. Econ. Assn., Ges. Wirtschafts-i. Sozialwiss., List Ges., Rencontres de St. Gall, Internat. Coun. Mgmt. Population Programs, Am. Biog. Inst. Roman Catholic. Avocations: philosophy, astronomy, skiing, internat. travel. Office: U Econs, Augasse 2-6, A 1090 Vienna Austria

PICK, JAMES BLOCK, management and sociology educator; b. Chgo., July 29, 1943; s. Grant Julius and Helen (Block) P. BA, Northwestern U., 1966; MS in Edn., No. Ill. U., 1969; PhD, U. Calif., Irvine, 1974, C.D.P., 1980. C.S.P., 1985, C.C.P., 1986. Asst. rsch. statistician, lectr. Grad. Sch. Mgmt. U. Calif., Riverside, 1975-91, dir. computing, 1984-91; co-dir. U.S.-Mex. Database Project, 1988-91; assoc. prof. mgmt. and bus., dir. info. mgmt. program U. Redlands, 1991—; cons. U.S. Census Bur. Internat. Div., 1978: mem. Univ. Commons Bd., 1982-86; mem. bd. govs. PCCLAS, Assn. Borderlands Studies, 1989-92. Trustee Newport Harbor Art Mus., 1981-87, 88—, chmn. permanent collection com., 1987-91, v.p., 1991—. Recipient Thunderbird award Bus. Assn. Latin Am. Studies, 1993. Mem. AAAS, Assn. Computing Machinery, Assn. Systems Mgmt. (pres. Orange County chpt. 1978-79), Am. Statis. Assn., Population Assn. Am., Internat. Union for Sci. Study of Population, Soc. Info. Mgmt. Club, Standard (Chgo.). Author: Geothermal Energy Development, 1982, Computer Systems in Business, 1986, Atlas of Mexico, 1989; condr. research in info. systems, population, environ. studies; contbr. sci. articles to publs. in fields.

PICKANDS, JAMES, III, mathematical statistician, educator; b. Cleve., Sept. 4, 1931; s. James II and Sarah Cornelia (Martin) P.; m. Nancy Jane McCulloch, Aug. 19, 1961 (div. Aug. 1987); children: James IV (dec.), Holly; m. Donna Lee Morgan Smith. BA, Yale U., 1954; postgrad., Brown U., 1955-56; PhD, Columbia U., 1965; MA (hon.), U. Pa., 1973. Computer programmer Ballistics Rsch. Labs., Aberdeen, Mo., 1954-55; grad. asst. Columbia U., N.Y.C., 1960-64; instr. Rutgers U., New Brunswick, N.J., 1964-65; asst. prof. Va. Poly. Inst. and State U., Blacksburg, Va., 1965-69; assoc. prof. U. Pa., Phila., 1969-84, prof., 1984—. Contbr. articles to profl. publs. With U.S. Army, 1956-58. Mem. Am. Statis. Assn., Internat. Statis. Inst., Inst. Math. Stats., Sigma Xi. Achievements include contributions to the theory of inference for extreme values and rare events and to representation theory for multivariate extremes, others. Office: U Pa Dept Stats 3620 Locust Walk Philadelphia PA 19104-6302

PICKEN, HARRY BELFRAGE, aerospace engineer; b. Grimsby, Ont., Can., Jan. 8, 1916; s. John Belfrage and Leila Lucinda (Jarvis) P.; m. Florence Elizabeth Runciman, July 7, 1945; children: Roger Belfrage, Donald William, Wendy Elizabeth. BSc in Aero. Engring., U. Mich., 1940. Registered profl. engr., Ont., Can. Chief engr. White Can. Aircraft Ltd., Hamilton, Ont., 1940-45, Weston Aircraft Ltd., Oshawa, Ont., 1946-47, Field Aviation Ltd., Oshawa, 1947-51; pres., chief engr. Genaire Ltd. (Aerospace), St. Catharines, 1951-81; v.p., tech. dir. Ardrox Ltd. (Chems.), Niagara on the Lake, 1968-75, Avionics Ltd. (Electronics), Niagara on the Lake, 1953-67; v.p. Rotaire Ltd. (Helicopters), St. Catharines, Ont., 1958-63; design approval rep. acting on behalf of Dept. of Transport Can., Ottawa, 1948-78; mem. bd. govs. Niagara Coll., Welland, Ont., 1974-80. Editor, pub.: Early Architecture Town and Township of Niagara, 1968, architecture student edit., 1991, Map of the Colonial Town of Niagara-on-the-Lake, 1981; composer (music book) Calgary Song Suite, 1983, Chacun a son Goût, 1991;. Chmn. Planning Bd. of Niagara-on-the Lake, 1963-65; pres. Niagara-on-the-Lake C. of C., 1961-62; bd. dirs., v.p. Niagara Found., Niagara-on-the-Lake, 1963-80; mem. tech. adv. bd. Niagara Coll., 1966-74; vice chmn. bd. govs. Niagara Coll. Applied Arts and Tech., 1979-81; mem. Ont. Coun. of Regents, 1987-93. Named Citizen of Yr. Niagara-on-the-Lake C. of C., 1968; recipient Award of Merit, Mohawk Community Coll., Hamilton, Ont., 1990, Medal-Community Svc., Profl. Engrs. Ont., 1981, Citation for Outstanding and Meritorious Work, Transport Can. Civil Aviation Ont. Region, 1978, Caring and Sharing award Niagara Regional Govt., 1992, Citation from Premier Ont., 1993. Fellow Can. Aero. and Space Inst. (assoc.); mem. AIAA, Am. Helicopter Soc., Assn. Profl. Engrs. Ont. (lic. profl. engr.), Composers, Authors and Music Pubs. of Can. (assoc.), Am. Fedn. Musicians (bd. dirs. local 298 Niagara Falls, Ont.). Achievements include patent for developing an entirely new type of honeycomb primary structure and beams fabricated using staples and acrylic adhesives; research in thermal electric modules independently used in cooling and refrigeration techniques. Home: 494 Glenwood Dr, Ridgeway, ON Canada L05 1N0 Office: Genaire Ltd, Niagara Dist Airport Box 84, Saint Catharines, ON Canada L2R 6R4

PICKENS, JIMMY BURTON, earth and life science educator, military officer; b. Silver City, N.Mex., Oct. 7, 1935; s. Homer Calvin and Edna (Burton) P.; m. Joana Holterman, Oct. 7, 1955; children: Kathleen Jo Pickens Grace, Danette Lynn Pickens Fouch. BS, N.Mex. State U., 1957; MEd, U. Ariz., 1971; cert., Abilene Christian U., 1980. Cert. secondary sch. tchr., Tex. Commd. 2nd lt. USAF, 1957, advanced through grades to lt. col., 1972; flight instr. USAF, Harlingen, Tex., 1957-60; instr. USAF Acad., San Antonio, 1960-64; edn. cons. USAF, San Antonio, 1972-75; combat aircrew USAF, Vietnam, 1964-67, advisor, 1971-72; instr. USAF Acad., Colorado Springs, Colo., 1967-70; commdr. USAF, Sacramento, Calif., 1975-78; staff officer USAF, Abilene, Tex., 1978-80, ret., 1980; tchr. earth and life sci. Wylie Mid Sch., Abilene, 1980—; sci. textbook coord. Wylie Ind. Sch. Dist., Abilene, 1980-91, awards coord., 1982-91, mem. coordinating com. Campaign worker Tex. Rep. Party, Abilene, 1978-91; bd. dirs. Wylie Ind. Sch. Dist. United Way, 1983-87. Decorated Bronze Star, Air Medal with three oak leaf clusters, Cross of Gallantry with Palm (Vietnam); recipient Teaching Excellence award Tex. Edn. Agy., Austin, 1984. Mem. Assn. Tex. Profl. Educators (pres. 1982-83), Sci. Tchrs. Assn. Tex., Air Force Assn., Mil. Order of World Wars, Phi Delta Kappa. Republican. Baptist. Avocations: golf, hunting, fishing, gardening. Office: Wylie Mid Sch 3158 Beltway South Abilene TX 79606-5724

PICKERING, BARBARA ANN, pharmaceutical sales representative, nurse; b. Crawfordsville, Ind., Mar. 7, 1949; d. Francis Eugene and Mary Katherine (Smith) Weaver; m. James Bruce Pickering, June 13, 1970; children: Michelle Ann, Kara Rachelle. BS, Ind. Cen. Coll., 1969. RN, Ind. Nurse Meth. Hosp., Indpls, 1983-85; sales rep. Johnson & Johnson, Ft. Washington, Pa., 1984-87; restricted card specialist Blue Cross/Blue Shield, Govt., Indpls., 1987-88; med. sales rep. BASF/Knoll Pharm., Whippany, N.J., 1988—. Vol. Am. Cancer Soc., 1993—; active program bd. Project Leadership Svc., Butler U., Indpls., 1990. Recipient Creative Sales award Sweeney & Ptnrs., 1988. Mem. Indpls. Assn. Pharmacists, Indpls. Med. Reps. Soc. Avocation: travel. Home: 6310 Dahlia Dr Indianapolis IN 46217-3839 Office: BASF/Knoll Pharm 30 N Jefferson Rd Whippany NJ 07981-1045

PICKETT, JAMES MCPHERSON, speech scientist; b. Clyde, Ohio, Jan. 18, 1921; s. Royce M. and Mattie (Gail) P.; m. Betty H., Mar. 10, 1952. PhD, Brown U., 1951. Phonetics lab. asst. Oberlin (Ohio) Coll., 1943-47; asst. prof. U. Conn., Storrs, 1950-52; rsch. psychologist USAF, Cambridge, Mass., 1952-61; NIH fellow Royal Inst. Technology, Stockholm, 1961-62; scientist Melpar Inc., Falls Church, Va., 1962-63; prof. Gallaudet U., Washington, 1964-87, prof. emeritus, 1987—. Speech editor Jour. of Acoustical Soc. Am., 1978-81; author: (textbook) Sounds of Speech Communication, 1980; contbr. articles to profl. jours. With USN, 1945-46, WWII, PTO. Named Outstanding Phonetician of Yr., Internat. Soc. Phonetic Scis., 1988. Fellow Acoustical Soc. Am.; mem. Cosmos Club. Achievements include leadership in field of speech processing aids for handicapped. Home and Office: Windy Hill Lab PO Box 198 Surry ME 04684

PICKETT, JOLENE SUE, aerospace research scientist. BS in Astronomy/Physics with distinction, U. Iowa, 1982. Sr. rsch. asst. U. Iowa, Iowa City, 1983—. Contbr. articles to profl. jours. including Jour. of Geophys. Rsch., Jour. of Spacecraft and Rockets, IEEE Transactions on Nuclear Sci. and Advances in Space Rsch. Mem. AIAA. Achievements include discovery that space shuttle chem. releases such as water dumps and thruster firings, have various effects on the neutral and plasma environ. surrounding the shuttle. Office: U Iowa Dept Physics/Astronomy Iowa City IA 52242

PICKHARDT, CARL EMILE, III, psychologist; b. Cambridge, Mass., May 4, 1939; s. Carl Emile Pickhardt Jr. and Marjorie Sachs Wilson; m. Irene Linke, May 11, 1975; children: Amy, Daniel; children by previous marriages: Ravel, Nigel. BA, Harvard U., 1961, MEd, 1966; PhD, U. Tex., 1970. Lic. psychologist, Tex. Program dir. Teenage Employment Skills Tng., Cambridge, Mass., 1964-66; rsch. assoc. Tchr. Edn. R&D Ctr., U. Tex., Austin, 1967-69; trainer Deseg. Ctr., U. Tex., Austin, 1969-70; asst. prof. Mich. State U., East Lansing, 1970-71; cons. psychologist Edn. Svc. Ctr., Austin, 1971-81; psychologist in pvt. practice Austin, 1981—. Author: From Cell to Society, 1965, Parenting the Teenager, 1986; columnist (weekly col.) Parenting the Teenager, 1981-85, (mo. col.) Working It Out (Divorce/Single Parenting), 1981-90, (mo. col.) Organizational Life Management, 1982-83; contbr. articles to profl. jours. Mem. APA, Tex. Psychol. Assn., Capital Area Psychol. Assn. Home: 3311 Bryker Dr Austin TX 78703 Office: 3406 Glenview Austin TX 78703

PICKLE, LINDA WILLIAMS, biostatistician; b. Hampton, Va., July 19, 1948; d. Howard Taft and Kathryn Lee (Riggin) Williams; 1 child from previous marriage, Diane Marie; m. James B. Pearson, Jr., Oct. 14, 1984. BA, Johns Hopkins U., 1974, PhD in Biostats., 1977; postgrad., George Washington U., 1986-87. Computer programmer Comml. Credit Computer Corp., Balt., 1966-69; systems analyst, computer programmer Greater Balt. Med. Ctr., Balt., 1969-72; grad. teaching asst. biostats. Johns Hopkins U., Balt., 1974-77; adj. asst. prof. div. biostats. and epidemiology Georgetown U. Med. Sch., Washington, 1983-88, assoc. prof. div. biostats and epidemiology, 1988-91, dir. biostats. unit, V.T. Lombardi Cancer Rsch. Ctr., 1988-91; biostatistician Nat. Cancer Inst. NIH, Bethesda, Md., 1977-88; math. statistician office rsch. methodology Nat. Ctr. for Health Stats., Hyattsville, Md., 1991—. Author: Atlas of U.S. Cancer Mortality Among Whites: 1950-80, 1987, Atlas of U.S. Cancer Mortality Among Nonwhites: 1950-1980, 1990; contbr. articles to med. and statis. jours. Sr. troop leader Girl Scouts U.S., 1981-83; sci. fair judge, 1983—. Mem. The Biometric Soc., Am. Statis. Assn., Soc. Epidemiologic Research, Soc. Indsl. and Applied Math., Sigma Delta Epsilon (pres. Omicron chpt. 1984), Phi Beta Kappa. Achievements include research in statistical methods in epidemiology, mapping health statistics. Office: Nat Ctr for Health Stats ORM/STS Presdl Bldg Rm 915 Hyattsville MD 20782

PICKLE, WILLIAN NEEL, II, molecular biologist, researcher; b. Evansville, Ind., Feb. 10, 1961; s. William Neel and Barbara Francis (Moore) P. BS, Purdue U., 1983, MS, 1987. Lab. technician Purdue U., West Lafayette, Ind., 1983-87; assoc. rsch. scientist Molecular Diagnostics, West Haven, Conn., 1987—. Asst. scoutmaster Boy Scouts Am., 1988-92, com. chmn., 1992—. Mem. Purdue Alumni Assn. (life), Order of Arrow. Home: 90 Annawon Ave West Haven CT 06516 Office: Molecular Diagnostics 400 Morgan Ln West Haven CT 06516

PICKLES, JAMES OLIVER, physiologist. MA, Cambridge (Eng.) U., 1968; PhD, Birmingham (Eng.) U., 1972; DSc, Cambridge U., England,

1993. Reader U. Birmingham; Australian sr. rsch. fellow U. Queensland, Australia, 1991—. Author: Introduction to the Physiology of Hearing, 1988; contbr. over 100 articles to profl. jours. Fellow Acoustical Soc. Am.; mem. Physiol. Soc. (U.K.), Assn. for Rsch. in Otolaryncology. Office: U Queensland, Dept Physiology/Pharmacology, Queensland 4072, Australia

PICKNEY, CHARLES EDWARD, environmental engineer; b. Walnut Ridge, Ark., July 13, 1944; s. Leslie Exavier and Allie May (Manus) P.; m. Barbara Lea Mathews, Apr. 4, 1966; 1 child, James. BS, Arks. State U., 1972. Registered environ. mgr. Lab technician St. Louis Testing Labs., 1966-69; plating supr. Crane Co., Jonesboro, Ark., 1969-70; plant chemist Monroe Auto Equipment Co., Paragould, Ark., 1970-90, facilities, 1990--. Mem. Am. Electroplates Soc. (Memphis Mid-South br., pres. 1973-74), Ark. Environ. Fedn. (dir. 1991--). Office: Monroe Auto Equipment 1601 Highway 49-B Paragould AR 72450

PICKRELL, THOMAS RICHARD, retired oil company executive; b. Jermyn, Tex., Dec. 30, 1926; s. Mont Boll and Martha Alice (Dodson) P.; m. M. Earline Bowen, Sept. 9, 1950; children—Thomas Wayne, Michael Bowen, Kent Richard, Paul Keith. B.S., North Tex. State U., 1951, M.B.A., 1952; postgrad., Ohio State U., 1954-55; advanced mgmt. program, Harvard U., 1979. CPA, Tex. Auditor, acct. Conoco, Inc., Ponca City, Okla., 1955-62; mgr. acctg. Conoco, Inc., Houston, 1965-67; asst. controller Conoco, Inc., Ponca City, 1967-81; v.p., controller Conoco, Inc., Stamford, Conn., 1982-83, Wilmington, Del., 1983-85; asst. prof. Okla. State U., Stillwater, Okla., 1962-63; controller Douglas Oil Co., Los Angeles, 1963-65; mem. adv. bd. dept. acctg. North Tex. State U., Denton, 1978-85; mem. adv. bd. Coll. Bus., Kansas State U., Manhattan, 1979-81. Bd. dirs. YMCA, Ponca City, 1976-78, Kay Guidance Clinic, Ponca City, 1971-74, United Way, Ponca City, 1979-81; chmn. Charter Rev. Com., Ponca City, 1971-72. Served to sgt. U.S. Army, 1944-46; ETO. Mem. AICPA, Fin. Execs. Inst. (pres. Okla. chpt. 1972), Am. Petroleum Inst. (acctg. com., gen. com.), Ponca City Country Club (pres. 1980-81), Rotary (pres. Ponca City club 1973-74), Beta Gamma Sigma, Beta Alpha Psi. Republican. Presbyterian. Home: RR 4 Box 209B Santa Fe NM 87501-9804

PICOU, GARY LEE, technical publication editor; b. Bryn Mawr, Pa., June 10, 1957; s. John Harvey and Patricia Sue (Hoke) P.; m. Helen Margaret Randall, Oct. 20, 1978; children: Erin Margret, Kyle Zachary. AET, Spartan Sch. Aeronautics, 1976. Field engr. Narco Avionics, Vero Beach, Fla., 1978-80, King Radio Corp., Vero Beach, 1980-86; product specialist Bendix/King Co., Olathe, Kans., 1986-88, reliability engr., 1988, mgr. support program, 1988-89; exec. tech. editor Avionics Rev. Mag., Olathe, 1989-92, editor, 1992—. Contbg. editor Aviation Consumer Mag., Greenwich, Conn., 1987-92. Pres. Laurelwood Homeowners Assn., vero Beach, 1979. Mem. AIAA, Aircraft Electronics Assn., Aviation/Space Writers Assn., Wild Goose Assn. Libertarian. Office: Avionics Rev PO Box 2536 Olathe KS 66062

PICRAUX, SAMUEL THOMAS, physics researcher; b. St. Charles, Mo., Mar. 3, 1943; s. Samuel F. and Jeannette P.; m. Danice R. Kent, July 12, 1970; children—Jeanine, Laura, Daryl. B.S. in Elec. Engring., U. Mo., 1965; postgrad. Cambridge U., Eng., 1965-66; M.S. in Engring. Sci., Calif. Inst. Tech., 1967, Ph.D. in Engring. Scis. and Physics, 1969. Mem. tech. staff Sandia Nat. Labs., Albuquerque, 1969-72, div. supr., 1972-86, dept. mgr., 1986—; vis. scientist dept. physics Aarhus U., Denmark, 1975; NATO lectr., 1979, 81, 83, 86.; NSF lectr. 1976, 81. Author: Materials Analysis by Ion Channeling, 1982; editor: Applications of Ion Beams to Metals, 1974, Metastable Materials Formation by Ion Implantation, 1982, Nuclear Instruments and Methods International jour., 1983—, Surface Alloying by Ion Electon and Laser Beams, 1986, Beam-Solid Interactions and Transient Processes, 1987; contbr. numerous articles to profl. jours. Fulbright fellow, 1965-66. Recipient Ernest Orlando Lawrence Meml. award U.S. Dept. Energy, 1990. Fellow Am. Physics Soc.; mem. Materials Research Soc., IEEE, Electrochem. Soc., AIME. Democrat. Office: Sandia Nat Labs POB 969 Livermore CA 94550

PIEDRAHITA, RAUL HUMBERTO, aquacultural engineer; b. Medellin, Colombia, Mar. 29, 1954; came to U.S., 1980; s. Raul and Maria (Uribe) P.; m. Mariluz Holguin, May 7, 1977; children: Miguel A., Ricardo A. MASc, U. B.C., 1980; PhD, U. Calif., Davis, 1984. Lectr. U. Calif., Davis, 1985, asst. prof., 1985-91, assoc. prof. aquacultural engring., 1991—. Contbr. articles to profl. jours. Mem. Am. Soc. Agrl. Engring. (chair aquacultural engring. 1984-85), World Aquaculture Soc., Sigma Xi, Alpha Epsilon. Achievements include development of computer models of water quality in aquaculture ponds; co-development of equipment for automated monitoring of water quality; co-development of techniques to estimate carbon dioxide removal from aquaculture ponds. Office: U Calif Dept Biol and Agrl Engring Davis CA 95616

PIEN, SHYH-JYE JOHN, mechanical engineer; b. Kaohsiung, Taiwan, Republic of China, July 14, 1956; came to U.S., 1980; s. Ke-Lee and Sue-Jean (Shen) P.; m. Fong-Ling Yang, July 15, 1983; children: Irene J., Jennifer M. BSME, Nat. Taiwan U., 1978; MSME, U. Ill., 1983, PhD, 1985. Asst. prof. U. Notre Dame, South Bend, Ind., 1985-89; sr. engr. Alcoa Tech. Ctr., Alcoa Ctr., Pa., 1989-91, staff engr., 1991—; panelist on heat transfer in mfg. at profl. confs. Editor symposium procs.; contbr. articles to profl. publs. Pres. Chinese Inst. Engrs. in USA, U. Ill., Urbana, 1984; v.p. Orgn. Chinese Ams., Pitts., 1993. Recipient Rsch. Initiation award NSF, 1988. Mem. ASME (materials processing and manufacturing com. heat transfer div. 1990—), Sigma Xi, Phi Kappa Phi. Roman Catholic. Achievements include patent on continuous casting machine design, research on hysteresis effect in natural convection, thermal/fluid phenomena in continuous casting, electronic packaging, welding process control, extrusion flow modeling. Office: Aluminum Co Am Alcoa Tech Ctr Alcoa Center PA 15069

PIERCE, BENJAMIN ALLEN, biologist, educator; b. Birmingham, Ala., Nov. 15, 1953; s. John Rush and Amanda (Allen) P.; m. Marlene Francis Tyrrell, July 19, 1980; children: Sarah Elizabeth, Michael Stephen. BS in Biology, So. Meth. U., 1976; PhD, U. Colo., 1980. Rsch. asst. U. Colo., Boulder, 1979-80; asst. prof. biology Conn. Coll., New London, 1980-84; asst. prof. biology Baylor U., Waco, Tex., 1984-88, assoc. prof. biology, 1988—. Author: The Family Genetic Source Book, 1990; contbr. articles to profl. publs. Achievements include research in population genetics and evolution. Office: Baylor U Dept Biology PO Box 97388 Waco TX 76798-7388

PIERCE, CHARLES EARL, software engineer; b. Edenton, N.C., July 13, 1955; s. Charles William and Carrie (Rankins) P. BS in Math., L.I. U., 1977. Rsch. analyst Equitable Life, N.Y.C., 1977-80; systems analyst CTEK Software, N.Y.C., 1980-83; sr. systems cons. Bank N.Y., N.Y.C., 1987—; cons. Nibor Assocs., N.Y.C., 1983-85, Vital Cons., N.Y.C., 1985-87. Mem. IEEE, N.Y. Acad. Scis., Data Processing Mgmt. Assn., Assn. for Computing Machinery, Math. Assn. Am. Mem. Pentecostal Ch. Achievements include devel. of English text interpreter/command processing for mainframe at CTEK Software.

PIERCE, CHESTER MIDDLEBROOK, psychiatrist, educator; b. Glen Cove, N.Y., Mar. 4, 1927; s. Samuel Riley and Hettie Elenor (Armstrong) P.; m. Jocelyn Patricia Blanchet, June 15, 1949; children: Diane Blanchet, Deirdre Anona. A.B., Harvard U., 1948, M.D., 1952; Sc.D. (hon.), Westfield Coll., 1977, Tufts U., 1984. Instr. psychiatry U. Cin., 1957-60; asst. prof. psychiatry U. Okla., 1960-62, prof., 1965-69; prof. edn. and psychiatry Harvard U., 1969—; pres. Am. Bd. Psychiatry and Neurology, 1977-78; mem. Polar Research Bd.; cons. USAF. Author publs. on sleep disturbances, media, polar medicine, sports medicine, racism; mem. editorial bds. Advisor Children's TV Workshop; chmn. Child Devel. Assn. Consortium; bd. dirs. Action Children's TV, Student Conservation Assn. Served with M.C. USNR, 1953-55. Fellow Royal Australian and N.Z. Coll. Psychiatrists (hon.); mem. Inst. Medicine, Black Psychiatrists Am. (chmn.), Am. Orthopsychiat. Assn. (pres. 1983-84). Democrat. Home: 17 Prince St Jamaica Plain MA 02130-2725

PIERCE, FRANCIS CASIMIR, civil engineer; b. Warren, R.I., May 19, 1924; s. Frank J. and Eva (Soltys) Pierce; student U. Conn., 1943-44; B.S., U. R.I., 1948; M.S., Harvard U., 1950; postgrad. Northeastern U., 1951-52; m. Helen Lynette Steinouer, Apr. 24, 1954; children—Paul F., Kenneth J., Nancy L., Karen H., Charles E. Instr. civil engring. U. R.I., Kingston, 1948-49, U. Conn., Storrs, 1950-51; design engr. Praeger-Maguire & Ole Singstad, Boston, 1951-52; chief found. engr. C.A. Maguire & Assocs., Providence, 1952-59, assoc., 1959-69, v.p., 1969-72; sr. v.p. C.E. Maguire, Inc., 1972-76, officer-in-charge Honolulu office, 1976-78, exec. v.p., corp. dir. ops., 1975-87; dir. The Maguire Group, Inc., 1979—, gen. mgr. East Atlantic Casualty Co., Ltd., 1987-88; also dir.; pres. Magma, Inc., tech. ops. service co., 1986-88; lectr. found. engring. U. R.I., 1968-69, trustee, 1987—; mem. Coll. Engring. adv. council, 1986—, U.S. com. Internat. Commn. on Large Dams. Vice chmn. Planning Bd. East Providence, R.I., 1960-73; bd. dirs. R.I. Civic Chorale and Orch., 1986-90. Served with AUS, 1942-46. Recipient Chester H. Kirk Disting. Engr. award U. R.I. Coll. Engring., 1987. Mem. ASCE (chpt. past pres., dir.), R.I. Soc. Profl. Engrs. (nat. dir., engr. of year award 1973), Am. Soc. Engring. Edn., Soc. Am. Mil. Engrs., ASTM, Soc. Marine Engrs. and Naval Architects, Am. Soc. Planning Ofcls., Harvard Soc. Engrs., Scientists, Providence Engrs. Soc., R.I. Soc. Planning Agys. (past pres.). Contbr. articles to profl. jours. Recipient USCG Meritorious Pub. Service award, 1987. Home: 3830 St Girons Dr Punta Gorda FL 33950-7870 Office: 225 Foxborough Blvd Foxboro MA 02035

PIERCE, GEORGE FOSTER, JR., architect; b. Dallas, June 22, 1919; s. George Foster and Hallie Loise (Crutchfield) P.; m. Betty Jean Reistle, Oct. 17, 1942; children: Ann Louise Pierce Arnett, George Foster III, Nancy Reistle Pierce Brumback. Student, So. Meth. U., 1937-39; BA, Rice U., 1942, BS in Architecture, 1943; diplome d'architecture, Ecole des Beaux-Arts, Fontainebleau, 1958. Pvt. practice architecture, 1946—; founding ptnr. Pierce, Goodwin, Alexander & Linville (architects, engrs. and planners), 1947-86, of counsel, 1987—; design cons. Dept. Army, 1966-70; instr. archtl. design Rice U., 1945, preceptor dept. architecture, 1962-67; past trustee, mem. exec. com. Rice Ctr. for Community Design and Rsch.; bd. dirs. Billboards Ltd. Prin. works include projects 8 bldgs., Rice U. Campus, Exxon Brook Hollow Bldg. Complex, Houston Mus. Natural Sci. and Planetarium, Michael Debakey Center Biomed. Edn. and Research, Baylor Coll. Medicine; 4 terminal bldgs., master plan Houston Intercontinental Airport; S.W. Bell Telephone Co. office bldg.; U. Tex. Med. Sch. Hosp., Galveston, U. Houston Conrad Hilton Sch. Hotel & Club Mgmt., U. Houston Univ. Center; 40 story office bldg. for Tex. Eastern Transmission Corp.; outpatient facility U. Tex. M.D. Anderson Cancer Hosp.; 44 story Marathon Oil Tower, others; contbr. articles to profl. jours. Mem. exec. adv. bd. Sam Houston Area Coun. Boy Scouts Am.; past mem. grad. coun. Rice U.; adv. coun. Rice Sch. Architecture, chmn. Houston Mcpl. Sign Control Bd.; past chmn. bd. trustee Contemporary Arts Mus., Houston; past trustee Houston Mus. Natural Sci.; trustee emeritus, past pres. Tex. Archtl. Found. With USNR, WWII. Recipient numerous nat., state, local archtl. awards for design; named one of five Outstanding Young Texans Tex. Jr. C. of C., 1954; named Disting. R Man Rice U., 1992. Fellow AIA (past nat. chmn. com. on aesthetics, student affairs and chpt. affairs), La. Sociedad de Architectos Mexicanos (hon.); mem. Tex. Soc. Architects (past pres., L.W. Pitts award 1985), Houston C. of C. (past chmn. future studies), Rice U. Pres.'s Club (founding chmn. 1970-72), Rice U. Assocs., SAR, Tex. Golf Assn. (bd. dirs.), Kappa Alpha. Methodist. Home: 5555 Del Monte Apt 1103 Houston TX 77056 Office: 5555 San Felipe Ste 1000 Houston TX 77056

PIERCE, JEFFREY LEO, power systems engineer, consultant; b. Pitts., Nov. 27, 1951; s. Francis Leo and Hilda Elizabeth (Swartzer) P.; m. Vicki Lynn Freeman, Sept. 20, 1976 (div. Mar. 1985); 1 child, Rebecca Lynn; m. Samantha Elizabeth Cannon, Mar. 16, 1986; 1 child, Jonathan Leo. BSCE, U. Pitts., 1973, MSCE, 1977. Registered profl. engr. Asst. mgr. planning/studies Chester Engrs., Pitts., 1973-77; mgr. environ./indsl. engring. SE Techs., Inc., 1977-84; mgr. ind. power engr. SE Techs., Inc., L.A., 1984—. Contbr. articles to profl. jours. Mem. ASCE, ASME (chmn. western chpt. solid waste divsn. 1991-93), Assn. Energy Engrs. (charter), Water Pollution Control Fedn. Achievements include development, engineering, start-up of lowest air emissions coal-fired electric power plant in U.S., largest landfill gas-fired renewable energy electric power plant in U.S. Home: 4500 Spencer St Torrance CA 90503 Office: SE Techs Inc 98 Vanadium Rd Bridgeville PA 15017

PIERCE, JOHN ALVIN, physics researcher; b. Spokane, Dec. 11, 1907; s. John Alvin and Harriet Converse (Fernald) P.; m. Marion Caroline Rogers, Jan. 26, 1929 (div. Aug. 1938); 1 child, Martha Jane; m. Catherine Sophronia Stillman, Feb. 4, 1939; children: Robert Bancroft, Margaret Lovejoy. BA in Physics, U. Maine, 1937. Mem. staff Harvard U., Cambridge, Mass., 1935-39, supr., 1939-41, rsch. fellow, 1946-49, sr. rsch. fellow, 1949-74; staff mem. radiation lab. MIT, Cambridge, 1941-46; researcher in field. Editor: Loran; contbr. articles to tech. jours. Recipient cert. of merit Pres. of U.S., 1948, Morris Liebman award Inst. Radio Engring., 1953, Conrad award USN, 1975, Elmer A. Sperry award Five Engring. Socs., 1988, others. MEm. IEEE (medal for engring. excellence 1990), ASME, AIAA, Inst. Navigation (Thurlow award 1947, mem. various coms. 1947-66). Achievements include over 10 patents including development of loran and omega aids to navigation and phonographic reproduction and pickup design. Home and Office: 334 Flanders Memorial Rd Weare NH 03281

PIERCE, JOHN THOMAS, industrial hygienist, toxicologist; b. Coffeyville, Kans., Mar. 15, 1949; s. John Gordon and Mary Ellen (McGrath) P.; m. Janet D. Brousseau, Aug. 7, 1981. MPH, U. Okla., 1977, PhD, 1978. Assoc. prof., dir. environ. and occupational health U. North Ala., Florence, 1981-88, Va. Commonwealth U., Richmond, 1988-90; prof., chair dept. industrl. hygiene Cen. Mo. State U., Warrensburg, 1990-92; indsl. hygienist, rsch. asst. prof. U. Kans. Med. Ctr., Kansas City, 1992—. Author: Study Guide: Fundamentals of Industrial Hygiene, 1990; asst. editor Applied Occupational and Environ. Hygiene. Comdr. USNR, 1968—. Mem. Am. Acad. Indsl. Hygiene (local treas. 1993), Am. Acad. Clin. Toxicology, Am. Chem. Soc. (local treas. 1984-87), Am. Bd. Toxicology, Sigma Xi, Phi Kappa Phi. Democrat. Roman Catholic. Office: Kans U Med Ctr 39th and Rainbow Blvd Kansas City KS 66103

PIERCE, LAMBERT REID, architect; b. Evanston, Ill., Apr. 12, 1930; s. Ellsworth Reid and Jessie (Lambert) P.; m. Julia Ellen Sellers, Nov. 20, 1948; children: Kenneth Reid, Rebecca June, Wendy Lynn. BSCE, Northwestern u., 1953, postgrad., 1955. Registered architect, Ill., U.S. Virgin Islands; engr. in tng., Ill. Draftsman Rader & Co. Builders, Skokie, Ill., 1949-58; assoc. architect Richard E. Dobroth & Assoc., Deerfield & Glenview, Ill., 1958-67; v.p. Richard E. Dobroth & Assoc., Glenview, 1967-71; prin. Lambert R. Pierce Architect, Glenview, 1971-78, Lambert R. Pierce & Assocs. Architects, St. Croix, V.I., 1979—; bd. dirs. Virgin Island Properties, Inc., St. Croix. Prin. works include Pelican Cove Beach Club, St. Croix, Queens Quarter Hotel, St. Croix, Fin. Bldg., St. Croix, Red Cross Bldg, St. Croix, Chase Manhattan Bank, St. Croix. Del. 1st Internat. Congress on Religion Architecture and Visual Ats, N.Y.C. and Montreal, Can., 1967; pres., bd. dirs. Beacon Neighborhod House, Chgo., 1967; mem. program cabinet Prebytery of Chgo., 1972-73; pres. Glenview C. of C., 1978. Mem. AIA, St. Croix C. of C., Our Town Frederiksted. Presbyterian. Avocations: reading, woodworking, snorkeling. Home: PO Box 2389 Kingshill VI 00851-2389 Office: Lambert R Pierce & Assoc 5 Strand St Frederiksted Saint Croix VI 00840

PIERCE, NAOMI ELLEN, biology educator, researcher; b. Denver, Oct. 19, 1954; d. Arthur Preble and Ruiko (Ishizaka) P. BS, Yale U., 1976; PhD, Harvard U., k1983. Fulbright postdoctoral fellow Griffith U., Brisbane, Australia, 1983-84; rsch. lectr. Christ Ch., U. Oxford, Eng., 1984-86; asst. prof. Princeton U., N.J., 1986-91; Sydney A. and John H. Hessel prof. biology, curator lepidoptera Harvard U. and Harvard Mus. Comparative Zoology, Cambridge, Mass., 1991—. Contbr. articles to profl. jours. MacArthur Found. fellow, Chgo., 1988—. Office: Harvard U Museum Comparative Zoology Cambridge MA 02138

PIERCE, ROBERT RAYMOND, materials engineer, consultant; b. Helena, Mont., Feb. 17, 1914; s. Raymond Everett and Daisy Mae (Brown) P.; m. Stella Florence Kankos, June 12, 1938; children: Keith R., Patricia L., Diana L. BS in Chem. Engring., Oreg. State U., 1937. Process supr. Pennwalt Corp., Portland, Oreg., 1941-45, asst. tech. svc. mgr., Tacoma, 1945-47, gen. mgr., Phila., 1947-58, Natrona, Pa., 1958-65, tech. mgr., Phila., 1965-78, sr. tech. cons., Phila., 1978-80; self-employed cons., also Ohio State U., 1980—; pres. Pierce CorMat Svcs., Inc. Contbr. articles to prof. jours. Patentee in field. Vice chmn. Phila. Air Pollution Control Bd., Phila., 1969-79, chmn. Ad Hoc #1, 1974-79; Ky. Colonel, Louisville 1975—; mem. People to People del. on corrosion, People's Republic China, 1986. Recipient Phila. award City of Phila., 1973, Resolution award, City of Phila., 1979, Disting. leadership award Am. Biog. Inst., 1986, World Decoration of Excellence for Exceptional Contributions to World Communities award, 1980-90. Mem. AICE (Spl. Half-Century Membership and Contbrns. to the Advancement of Chem. Engring. award 1992), Nat. Assn. of Corrosion Engr., Internat. Com. for Industrial Chimneys (recipient best paper award Dusseldorf, Germany 1970), Am. Ceramic Soc., Nat. Inst. of Ceramic Engrs., Rotary (Paul Harris fellow 1988). Lutheran. Current work: Engineering of coatings; engineering of brickwork; stress analyses of coatings; stress analyses of brickwork; corrosion of metallic and non-metallic materials; design of polymeric concrete; design of chimney linings. Subspecialties: Corrosion; Materials.

PIERCE, SHELBY CRAWFORD, oil company executive; b. Port Arthur, Tex., May 26, 1932; s. William Shelby and Iris Mae (Smith) P.; B.S.E.E., Lamar U., Beaumont, Tex., 1956; student M.I.T. Program for Sr. Execs., 1980; m. Marguerite Ann Grado, Apr. 2, 1954; children—Cynthia Dawn, Melissa Carol. With Amoco Oil Co., 1956—, zone supr., gen. foreman, maintenance, 1961-67, operating supt., 1967-69, coordinator results mgmt., Texas City (Tex.) refinery, 1969-72, dir. results mgmt., corp. hdqrs., Chgo., 1972-75, ops. mgr. refinery, Whiting, Ind., 1975-77, asst. refinery mgr., 1977-79, dir. crude replacement program, Chgo., 1979-81, mgr. refining and transp. engring., 1981-92, gen. mgr. engring. & constrn., 1992, v.p. internat. bus. devel., 1993—. Fin. chmn. Bay Area council Boy Scouts Am., 1974; dir. JETS (Jr. Engring. Tech. Soc.); chmn. bd., chmn. fin. com. Methodist Ch., 1967-72; bd. dirs. Waste Tech. Svcs.; mem. steering com. Contractor Safety U.S. Dept. of Labor. Mem. Am. Inst. Chem. Engrs. (chmn. engring. constrn. contracting div.), Constrn. Industry Inst. (chmn. bus. roundtable coun., mem. strategic planning com.), The Bus. Roundtable (N.W. Ind. chpt., chmn. coun., chmn. exec. com., chmn. constrn. cost effectiveness task force), Flossmoor Country Club, Sigma Tau. Republican. Home: 18840 Loomis Ave Homewood IL 60430-4047 Office: 200 East Randolph Dr Chicago IL 60680-0707

PIERCE, WILLIAM SCHULER, cardiac surgeon, educator; b. Wilkes-Barre, Pa., Jan. 12, 1937; s. William Harold and Doris Louis (Schuler) P.; m. Peggy Jayne Stone, June 12, 1965; children: William Stone, Jonathan Drew. B.S., Lehigh U., 1958; M.D., U. Pa., 1962. Intern U. Pa., 1962-63; resident in surgery Hosp. U. Pa., 1963-70; asst. prof. M.S. Hershey Med. Ctr., Pa. State U. Coll. Medicine, Hershey, 1970-73, assoc. prof., 1973-77, prof. surgery, 1977—, chief div. cardiothoracic surgery, 1991—. Contbr. over 300 articles to profl. jours.; inventor cardiac valve, blood pump. Served with USPHS, 1965-67. Fellow ACS; mem. AMA, AAAS, Internat. Cardiovascular Soc., Am. Soc. Artificial Internal Organs, Soc. Vascular Surgery, Am. Heart Assn., Assn. Acad. Surgery, Inst. Medicine, So. Pa. Assn. Thoracic Surgery, Soc. Univ. Surgeons, Am. Surg. Assn., Soc. Clin. Surgery, Inst. of Medicine. Office: Milton S Hershey Med Ctr PO Box 850 Hershey PA 17033-0850

PIERRE, FRANÇOISE, physics educator. Prof. physics Cen-Seclay, Gif-sur-Yvette, France. Recipient W. K. H. Panofsky prize, Am. Physical Soc., 1991. Office: Cen-Seclay, Dept of Physics, 91191 Gif-sur-Yvette Cedex, France

PIERSOL, ALLAN GERALD, engineer; b. Pitts., June 2, 1930; s. Robert James and Irene Laticia (Dematty) P.; m. Gertrud Teresia Moller, June 8, 1958; children: Allan Gerald Jr., Marie Theresa, John Robert. BS in Engring. Physics, U. Ill., 1952; MS in Engring., U. Calif., 1961. Lic. profl. engr., Calif. Rsch. engr. Douglas Aircraft Co., Santa Monica, Calif., 1952-59; mem. tech. staff Ramo Wooldridge Corp., Canoga Park, Calif., 1959-63; v.p. Measurement Analysis Corp., Santa Monica, Calif., 1963-71; prin. scientist Bolt Beranek and Newman, Inc., Conoga Park, 1971-85; sr. scientist Astron Corp., Santa Monica, 1985-88; lectr. U. So. Calif., L.A., 1965--; sole proprietor Piersol Engring., Woodland Hills, Calif., 1988--. Co-author: Measurement and Analysis of Random Data, 1966, Random Data: Analysis and Measurement Procedures, 1971, 86, Engineering Application of Correlation and Spectral Analysis, 1980, 93, Noise and Vibration Control Engineering, 1992. Mem. Inst. Environ. Scis. (Irwin Vigness meml. award 1991), Am. Soc. Mech. Engrs., Acoustical Soc. Am. Achievements include a patent for a method and apparatus for determining terrain surface profiles. Home: 23021 Brenford St Woodland Hills CA 91364 Office: Piersol Engring Co 23021 Brenford St Woodland Hills CA 91364

PIERSON, STEVE DOUGLAS, aerospace engineer; b. St. Paul, Feb. 25, 1966; s. Robert Frederick and Mariann Theresa (Walz) P. BS, U. Minn., 1989. Engr. Irvin Industries Inc., Santa Ana, Calif., 1989—. Mem. AIAA. Home: 26371 Paloma # 14 Foothill Ranch CA 92610

PIERSON, WILLIAM R., chemist; b. Charleston, W.Va., Oct. 21, 1930; s. Roy H. Pierson and Gay Harris; m. Juliet T. Strong, May 20, 1961; children: Elizabeth T., Anne H. BSE, Princeton U., 1952; PhD, MIT, 1959. Rsch. assoc. Enrico Fermi Inst. for Nuclear Studies U. Chgo., 1959-62; rsch. scientist Ford Motor Co., Dearborn, Mich., 1962-87; exec. dir. Energy and Environ. Engring. Ctr. Desert Rsch. Inst., Reno, 1987—. Contbr. over 50 articles to profl. jours. Bd. dirs. Reno Chamber Orch., 1992—. Lt. USN, 1952-55. Mem. AAAS, Am. Phys. Soc., Am. Chem. Soc., Air and Waste Mgmt. Assn., Am. Assn. for Aerosol Rsch. (bd. dirs. 1982-85). Office: Desert Rsch Inst 5625 Fox Ave PO Box 60220 Reno NV 89506

PIETRA, FRANCESCO, chemist; b. Carrara, Italy, July 7, 1933; s. Mario and Margherita (Ratti) P.; m. Gudrun Burgstaller, Mar. 14, 1964; 1 child, Kristina. PhD in Chemistry, U. Padova, Italy, 1958; libero docente, Ministry Pub. Instrn., Rome, 1968. Asst. prof. U. Camerino, Italy, 1958-62, U. Padova, Italy, 1962-63, U. Perugia, Italy, 1963-64, U. Pisa, Italy, 1966-67; assoc. prof. U. Pisa, 1968-74; prof. U. Catania, 1975-78, U. Trento, 1976—. Author: A Secret World, Natural Products of Marine Life, 1990; contbr. reviews and articles to internat. profl. jours. Achievements include degradation of biomolecules by triplet oxygen under biological conditions; discovery of branched C15 acetogenins; rationalization of chemistry of troponoids in basic media. Office: Univ Trento Inst Chimica, Sommarive 14, 38050 Povo-Trento Italy

PIGNAL, PIERRE IVAN, computer scientist, educator; b. Saint Julien en Genevois, France, Mar. 2, 1961; s. Ivan Edgard and Yvette Marie Thérèse (Vernaz) P.; m. Claudine Julliard, Au g. 9, 1984; children: Julien, Clélia, Alexis. Engr. of communications, Ecole Nat. Superieure Telecommunications, Paris, 1985. Cons. IBM, La Gaude, France, 1985-87; chief designer IBM, La Gaude, 1988-89, sr. advisor, 1990—; cons. Aerospatiale, Paris, 1988—; prof. Ecole Nat. Superieure Telecommunications, Mines and Ecole Cent., Paris, 1987—; expert Dept. Industry, Paris, 1988—, Internat. Telegraph and Telephone Consultative Com., 1990—. Patentee in field; contbr. articles to profl. jours. Mem. IEEE (chmn. conf. 1989, author Standards, 1989), Internat. Assn. for Sci. and Tech. Devel. (chmn. internat. confs. 1988, 89), Assn. Computing Machinery, Assn. French Pour la Cybernetique Econs. and Techniques. Avocations: volleyball, skiing. Office: IBM, 06610 LaGaude France

PIGNATARO, LOUIS JAMES, engineering educator; b. Bklyn., Nov. 30, 1923; s. Joseph and Rose (Capi) P.; m. Edith Hoffmann, Sept. 12, 1954; 1 child, Thea. B.C.E., Poly. Inst. Bklyn., 1951; M.S., Columbia U., 1954; Dr. Tech. Sci., Tech. U. Graz, Austria, 1961. Registered profl. engr., N.Y., Calif., Fla. Faculty Poly. Inst. N.Y., 1951-85, prof. civil engring., 1965—, dir. div. transp. planning, 1967—, head dept. transp. planning and engring., 1970, dir. Transp. Tng. and Research Center, 1975; Kayser prof. transp. engring. City Coll., N.Y.C., 1985-88; assoc. dir. Inst. Transp. Systems City Coll., 1985-88; mem. faculty N.J. Inst. Tech., 1988—, disting. prof. transp. engring., 1988—, dir. ctr. transp. studies and research, 1988—; cons. govtl. agys., pvt. firms. Mem. Gov.'s Task Force Advisers on Transp. Problems, Gov.'s Task Force on Alcohol and Hwy. Safety; commr.'s council advisers N.Y. State Dept. Transp.; mem. adv. bd. freight services improvement conf. Port Authority N.Y. and N.J.; mem. adv. com. N.Y.C. Dept. Transp.; mem. rev. com. N.Y.C. Dept. City Planning; mem. Mayor's Transp. Commn., City of Newark. Sr. author: Traffic Engineering-Theory and Practice, 1973; contbr. over 70 papers to profl. jours. Served with AUS, 1943-46. Recipient Distinguished Tchr. citation Poly. Inst. Bklyn., 1965, Dedicated Alumnus award, 1971, Distinguished Alumnus award, 1972; citation for distinguished research Poly. chpt. Sigma Xi, 1975; named Engr. of Year N.Y. State Soc. Profl. Engrs., 1974. Fellow ASCE (dir.), Inst. Transp. Engrs. (Transp. Engr. of Yr. Met. sect. N.Y. and N.J. 1982); mem. Am. Rd. and Transp. Builders Assn. (div. dir.), Transp. Research Bd. (univ. liaison rep., Outstanding Paper award 1980 ann. meeting), Transp. Research Forum, Nat. Soc. Profl. Engrs., Sigma Xi, Chi Epsilon, Tau Beta Pi. Home: 230 Jay St Brooklyn NY 11201-1948 Office: NJ Inst Tech Ctr Transp Studies Rsch 323 Martin Luther King Blvd Newark NJ 07102

PIGOTT, MELISSA ANN, social psychologist; b. Ft. Myers, Fla., Jan. 28, 1958; d. Park Trammell and Leola Ann (Wright) P.; m. David H. Fauss, Jan. 1, 1988. BA in Psychology, Fla. Internat. U., Miami, 1979; MS in Social Psychology, Fla. State U., 1982, PhD in Social Psychology, 1984. Rsch. asst. Fla. Internat. U., 1978-79, Fla. State U., Tallahassee, 1980-84; dir. mktg. rsch. Bapt. Med. Ctr., Jacksonville, Fla., 1984-89; rsch. assoc. Litigation Scis., Inc., Atlanta, 1989-91; sr. litigation psychologist Trial Cons., Inc., Miami, 1991—; adj. prof. psychology U. North Fla., Jacksonville, 1985-89. Author: Social Psychology: Study Guide, 1990, Social Psychology: Instructors Manual, 1990; contbr. articles to profl. jours. Mem. Am. Psychol. Assn., Am. Civil Liberties Union, Am. Psychol. Law Soc., Amnesty Internat., Civitan Internat., Southeastern Psychol. Assn., Soc. for Psychol. Study of Social Issues, Soc. Personality and Social Psychology, Greenpeace, Psi Chi. Democrat. Avocations: concerts, playing piano, going to the beach.

PIH, NORMAN, chemical engineer; b. Johnson City, N.Y., July 19, 1959; s. Hui and Mabel Pih. BS in Chem. Engring., U. Tenn., 1982; MS in Chem. Engring., U. Del., 1984. Coop. engr. Oak Ridge (Tenn.) Nat. Lab., 1978-82; engr. DuPont Polymer Products Dept., Parkersburg, W.Va., 1984-86; engr. rsch. DuPont Engring. Tech. Lab., Wilmington, Del., 1986-87; area engr. DuPont Med. Products Dept., Glenolden, Pa., 1987-89; rsch. engr. DuPont Polymers Dept., Wilmington, Del., 1989-90; devel. engr. DuPont Ctrl. R&D, Wilmington, 1990—. Grad. fellow U. Del., 1982-84. Mem. Am. Inst. Chem. Engrs., Alpha Chi Sigma, Tau Beta Pi (engring. futures program instr. 1989—, engring. futures program bd. 1991-92, dist. dir. 1992—). Achievements include research in biotech. and bioengring. Home: 70 Helios Ct Newark DE 19711 Office: DuPont Expt Sta PO Box 80357 Wilmington DE 19880

PIHL, JAMES MELVIN, electrical engineer; b. Seattle, May 29, 1943; s. Melvin Charles and Carrie Josephine (Cummings) P.; m. Arlene Evette Housden, Jan. 29, 1966 (div. Dec. 1990); 1 child, Christopher James. AASEE, Seattle, 1971; postgrad., City Univ., Bellevue, Wash., 1982—. 1st class operators lic.; lic. in real estate sales. Journeyman machinist Svc. Exch. Corp., Seattle, 1964-67; design engr. P.M. Electronics, Seattle, 1970-73, Physio Control Corp., Redmond, Wash., 1973-79; project engr. SeaMed Corp., Redmond, 1979-83; sr. design engr. Internat. Submarine Tech., Redmond, 1983-85; engring. mgr. First Med. Devices, Bellevue, Wash., 1985-89; rsch. engr. Pentco Products, Bothell, Wash., 1989—. Inventor, patentee protection system for preventing defibrillation with incorrect or improperly connected electrodes, impedance measurement circuit. With U.S. Army,, 1961-64. Mem. N.Y. Acad. Scis. Avocations: dirt bike riding, high-performance automotive competition, target shooting, violin. Home: 14303 82d Ave NE Bothell WA 98011

PIHL, LAWRENCE EDWARD, electrical engineer; b. Quincy, Mass., Dec. 26, 1944; s. Wilfred Irving and Mary Barbara (Donahue) P.; 1 child, Stephanie; m. Cheryl Trenholme, June 19, 1982; children: Wesley, Lauryl. BSEE, Worcester Poly. Inst., 1966; MBA, Northeastern U., 1973. Test engr. Norden div. U.A.C.., Norwalk, Conn., 1966-67; engr. missile systems div. Raytheon, Bedford, Mass., 1967-72; applications engr. Sanders Assocs., Manchester, N.H., 1972-75; western sales mgr. Omni Spectra, Merrimack, N.H., 1975-79; sales rep. BBC, Chelmsford, Mass., 1979-84; pres. TMR Assocs., Amherst, N.H., 1984—; founder, pres. New Eng. Microwave Reps., 1988-90. Pres. Merrimack Jaycees, 1980. Recipient Charles Kulp Meml. award U.S. Jaycees, 1980. Achievements include development of data security system for small business which allows theft, electromagnetic and water damage protection. Office: TMR Assocs PO Box 1149 Amherst NH 03031-1149

PIKE, JOHN NAZARIAN, optical engineering consultant; b. Boston, Feb. 13, 1929; s. Arthur Thorndike and Sarah Lucy (Nazarian) P.; m. Margaretta May Horner, Dec. 28, 1957; children: Sally Katharine, Susan Horner. AB, Princeton U., 1951; PhD in Physics & Optics, U. Rochester, 1958. Staff scientist Parma (Ohio) Rsch. Ctr., Union Carbide Corp., 1956-63; sr. scientist Tarrytown (N.Y.) Tech. Ctr., Union Carbide Corp., 1963-85; pres. J.J. Pike & Co., Inc., Pleasantville, N.Y., 1986—. Contbr. numerous articles to profl. jours. Mem. Optical Soc. Am., Soc. Photo-Optical Instrument Engrs., Internat. Soc. for Optical Engring. Achievements include technical consulting in prototype instrument design and construction, exploratory research in applied optics, patentee in field. Home: 71 Cedar Ave Pleasantville NY 10570-1932 Office: JJ Pike & Co Inc PO Box 186 Pleasantville NY 10570-0186

PIKE, THOMAS HARRISON, plant chemist; b. West Palm Beach, Fla., Oct. 9, 1950; s. Rufus Draper and Dora Marie (Thomason) P.; m. Julie Lynn Simpson, Aug. 19, 1972; 1 child, Thomas Simpson. BS, Baylor U., 1972. Sci. instr. Valliant (Okla.) Pub. Sch., 1975-76; sch. adminstr. Swink (Okla.) Pub. Sch., 1976-81; plant chemist Western Farmers Electric Coop., Ft. Towson, Okla., 1981—; mem. adv. bd. Kiamichi Vo Tech Sch., Idabel, Okla., 1985-87. Charter mem. Valliant Youth Assn., 1987-91. Mem. ASME (co-chmn. task force 1988-90), ASTM, Nat. Assn. Corrosion Engrs. Mem. Assembly of God Ch. Achievements include rsch. in corrosion control of condensers, case history of turbine problems, improving boiler efficiency and preservation of turbines during extended outages. Home: Rt 1 Box 299 Garvin OK 74736 Office: Western Farmers Electric Coop Box 219 Fort Towson OK 74735

PILBEAM, DAVID ROGER, paleoanthropology educator; b. Brighton, Sussex, Eng., Nov. 21, 1940; came to U.S., 1968; s. Ernest Winton and Edith (Clack) P.; m. Maryellen Ruvolo, Dec. 18, 1982; 1 child, Katharine Alexandra. B.A., Cambridge U., 1962, M.A., 1966; Ph.D., Yale U., 1967; M.A. (hon.), Harvard U., 1982. Demonstrator in anthropology Cambridge U., Eng., 1965-68; asst. prof. anthropology Yale U., 1968-70, assoc. prof., 1970-74, prof., 1974-81, prof. anthropology, geology and geophysics, 1974-81, prof. anthropology, 1981-90; Henry Ford II prof. social sci. Harvard U., Cambridge, Mass., 1990—, dean undergrad. edn., 1987-92; dir. Peabody Mus., 1990—. Author: Evolution of Man, 1970, Ascent of Man, 1972; coauthor: Human Biology, 3d edit., 1988; co-editor: Cambridge Encyclopedia of Human Evolution, 1992. Fellow Am. Anthropol. Assn.; mem. Am. Acad. Arts and Scis., Nat. Acad. Scis. (fgn. assoc.). Office: Harvard U Peabody Mus 11 Divinity Ave Cambridge MA 02138-2096

PILGRIM, SIDNEY ALFRED LESLIE, soil science educator; b. Berlin, N.H., Dec. 24, 1933; s. Sidney Alfred and Vivian Gertrude (Larocque) P.; m. Faith Patricia Martin, June 9, 1956; children: Paul Garret, Kelly Lynn. BS, U. N.H., 1955. Cert. profl. soil scientist Am. Registry Cert. Profls. in Agronomy, Crops and Soils; cert. soil scientist, N.H. Soil scientist Soil Conservation Svc., USDA, Woodsville, N.H., 1955, 58-61, Concord, N.H., 1955-56, Plymouth, Ind., 1961-64; state soil scientist USDA, Durham, N.H., 1964-89; adj. assoc. prof. soil sci. U. N.H., Durham, 1979—. Contbr. articles to profl. jours. With U.S. Army, 1956-58. Recipient cert. of merit USDA, 1960, 68, 84, Silver Spade award N.E. Coop. Soil Survey, 1988, Disting. Svc. award U. N.H., 1989, Conservation award Strafford Rivers Conservancy, 1990. Mem. Am. Soc. Agronomy, Soc. Soil Scientists No. New Eng. (pres. 1975). Achievements include research on hydric soil determination using soil taxonomy, classification of New England soils. Home: 87 Mill Rd Durham NH 03824-2933

PILKINGTON, THEO CLYDE, biomedical and electrical engineering educator; b. Durham, N.C., June 23, 1935. BEE, N.C. State U., 1958; MS, Duke U., 1960, PhD, 1963. Instr. Duke U., Durham, N.C., 1958-63, asst. prof. elec. engring., 1963-66, assoc. prof., 1966-69, prof. biomed. engring. and elec. engring., 1969—; dir. engring. rsch. ctr. Emerging Cardiovascular Techs., Durham. Co-author: (with Robert Plonsey) Biophysical Principles of Electrocardiography, 1982; contbr. numerous articles to tech. jours. Mem. task force for cardiovascular tng. Nat. Heart-Lung Inst., 1981-82; bd. dirs. NSF Engring. Rsch. Ctr., 1987—. With U.S. Army, 1961. Mem. IEEE, Am. Soc. Engring. Edn., Ridge Woode Beagles Club (master beagler 1980—). Home: 2932 Ridge Rd Durham NC 27705-5528 Office: Duke U Engring Rsch Ctr 301 School Of Engring Durham NC 27706*

PILLERS, DE-ANN MARGARET, neonatologist; b. San Pedro, Calif., Aug. 1, 1957; d. Lauritz and June Pillers; m. Robert Nourse, July 25, 1981. BS in Chem. Engring., AB, Washington U., St. Louis, 1979; MD, Oreg. Health Sci. U., 1984, PhD, 1986. Resident pediatrics Oreg. Health Scis. U., Portland, 1986-89, chief resident pediatrics, 1988-89, fellow neonatal medicine, 1989-91, asst. prof. pediatrics, 1991—, asst. prof. molecular and med. genetics, 1991—. Contbr. articles to profl. jours. Recipient David Smith Pediatric Resident Rsch. award We. Soc. Pediatric Rsch., Carmel, Calif., 1991. Fellow Am. Acad. Pediatrics; mem. AMA, Am. Soc. Human Genetics, Am. Inst. Chem. Engrs. Achievements include description of unusual eye findings in boys with X chromosome deletions now called Oregon Eye Disease. Office: Oreg Health Scis U 3181 SW Sam Jackson Park Rd Portland OR 97201-3011

PILÓ-VELOSO, DORILA, chemistry educator, researcher; b. Belo Horizonte, Brazil, Oct. 4, 1944; d. Raimundo Veloso and Maria Idalina (Oliveira) Piló-Veloso; m. Waldo Silva; children: André, Iara. B in Chemistry, Inst. Scis. Exatas, 1966; PhD in Chemistry, U. Fed. Minas Gerais, Belo Horizonte, Brazil, 1973; D Scis. and Physics, Sci. and Med. U. Grenoble, France, 1978. Instr. U. Fed. Minas Gerais, Belo Horizonte, 1969-73, asst. prof., 1973-78; assoc. prof. Univ. Fed. de Minas Gerais, Belo Horizonte, 1979-90, prof., 1991—; stagiaire de recherche Grenoble (France) Ctr. for Nuclear Studies, 1973-78, post-doctoral fellow, 1978-79; post-doctoral fellow U. East Anglia, Norwich, E Anglia, Eng., 1980; adv. bd. Secretariat de Estado de Ciencis and Tech., Belo Horizonte, 1987, Commn. Rels. Internat. Ritoria de U. Fed. Minas Gerais, 1987-89; cons. Fundação Ezequiel Dias, Belo Horizonte, 1988-93. Contbr. articles to Jour. Chem. Rsch., Jour. Phys. Chemistry, Jour. Brazilian Chem. Soc., Holzforscung, others. Fellow Ministry Fgn. Affairs, France, 1973-77, Joliet-Curie Found., France, 1978; grantee Nat. Coun. Sci. Devel., Brasilia, 1979—, Found. Amparo Pesquisa de Minas Gerais, Belo Horizonte, 1983-93, Financiadora Etudes Projetos, Rio de Janeior, 1983-93. Mem. Regional Coun. of Chem. (Sci. Merit award 1992), Brazilian Soc. Chemistry, Am.Chemical Soc., Brazilian Soc. for the Advance of Scis. Achievements include research in organic and wood chemistry, natural products, organic synthesis. Office: U Fed de Minas Gerais Dept, Quimica ICEX UFMG, Cidade Univ Pampulha, 31270 Belo Horizonte MG, Brazil

PILSON, MICHAEL EDWARD QUINTON, oceanography educator; b. Ottawa, Ont., Can., Oct. 25, 1933; came to U.S., 1958; s. Edward Charles and Frances Amelia (Ferguson) P.; m. Joan Elaine Johnstone, July 6, 1957; children: Diana Jane, John Edward Quinton. BSc, Bishops U., Lennoxville, Que., Can.; MSc, McGill U., Montreal, Que., Can., 1958; PhD, U. Calif., San Diego, 1964. Chemist Windsor Mills (Can.) Paper Co., 1954-55; asst. chemist Macdonaldl Coll. of McGill U., 1955-58; biologist Zool. Soc. San Diego, 1963-66; asst. prof. U. R.I., Narragansett, 1966-71, assoc. prof., 1971-78, prof., 1978—; dir. Marine Ecosystems Rsch. Lab., Narragansett, 1976—. Contbr. articles to profl. and popular jours.; author chpts. for 5 books. Grantee NSF, NOAA, EPA, NIH. Mem. AAAS, AGU, ASLO, Oceanography Soc., Am. Soc. Mammalogists, Saunderstown Yacht Club (bd. govs. 1974-87, commodore 1985-87). Home: PO Box 27 Saunderstown RI 02874 Office: U RI Grad Sch Oceanography Narragansett RI 02882

PILZ, GÜNTER FRANZ, mathematics educator; b. Bad Hall, Austria, Mar. 19, 1945; s. Franz and Johanna (Zehetner) P.; m. Gerti A. Goldmann, Aug. 2, 1969; children: Michaela, Martin. PhD, U. Vienna, 1967. Asst. prof. U. Vienna, 1966-68, Tech. U. Vienna, 1968-69; rsch. assoc. U. Ariz., Tucson, 1969-70; asst. prof. U. Linz, Austria, 1970-72, lectr., 1972-74, prof., 1974—; vis. prof. U. Lafayette, La., Tex. A&M U., U. Tasmania, U. Tainan, Taiwan, U. Parma, Italy; dept. head math. U. Linz, 1980-83. Author: Nearrings, 1976, 83, Applied Abstract Aglebra, 1984, Einführung in die Mathematik, 1981 and others; contbr. articles to profl. jours. Fuibright Rsch. grantee, Vienna, 1970 and others, 1974-90; recipient Rsch. award Govt. Upper Austria, Linz, 1969. Mem. Am. Math. Soc., Edinburgh Math. Soc., Österreichische Math. Assn. Avocations: discus, weight lifting, acupuncture. Home: Hausleitnerweg 25, A-4020 Linz Austria Office: U Linz, Altenbergerstr 69, A-4040 Linz Austria

PIMENTA, GERVÁSIO MANUEL, chemist, researcher; b. Barreiro, Portugal, May 14, 1964; s. Albano Ferreira and Maria Celeste (Almeida) P.; m. Anabela de Almeida Campos, Jul. 30, 1992. Tech. chemistry, Faculty of Scis., 1987. Grantee Lab. Nac. Eng. Tecn. Industrial, Lisbon, Portugal, 1987-89, Electrochem. Ctr. Univ. Lisbon, 1989-90, Fed. Inst. Materials Rsch. & Testing, Berlin, 1990-92, Materials Dept. New Univ., Lisbon, 1992—. Contbr. articles to profl. jours. Fellow Portuguese Electrochemical Soc., Portuguese Chem. Soc. Home: Ave Bocage 9 11 andar Letra A, P-2830 Barreiro 2830, Portugal Office: Molecular Physics Ctr, Complexo 1, IST, P-1000 Lisbon Portugal

PINCUS, PHILIP A., chemical engineering educator. Prof., chemical engineering U. California, Santa Barbara, Calif. Recipient High-Polymer Physics Prize, Am. Chemical Soc., 1992. Office: U of California-Dept of Chem Eng 552 University Ave Santa Barbara CA 93106

PINCUS, THEODORE, microbiologist, educator. AB, Columbia U., 1961; MD, Harvard U., 1966. Assoc. Sloan-Kettering Inst., N.Y.C., 1973-75; asst. prof. medicine-immunology Stanford (Calif.) U., 1975-76; adj. assoc. prof. medicine-rheumatology U. Pa., Phila., 1976-80; prof. medicine & microbiology Vanderbilt U., Nashville, 1980—; prof. Wistar Inst., Phila., 1976-80; dir. clin. immunology lab. Stanford U. Hosp., 1975-76. Fellow ACP, Am. Rheumatism Assn., Am. Soc. Microbiology. Achievements include research in morbidity and mortility of rheumatoid arthritis, host variables in chronic diseases, host genetic control of virus infection, psychological and economic consequences of chronic disease, socioeconomic status and chronic disease. Office: Vanderbilt U Arthritis & Lupus Ctr T-3219 MCN Nashville TN 37232*

PINCZUK, ARON, physicist; b. San Martin, Argentina, Feb. 15, 1939; s. Faiwel and Ester (Wejeman) P.; m. Gladys Norma Teitelman, June 14, 1962; children: Ana Gabriela, Guillermo Fabian. Licenciado, U. Buenos Aires, Argentina, 1962; PhD, U. Pa., 1969. Staff mem. Nat. Rsch. Coun., Argentina, 1971-76; head physics dept. Faculty of Scis., U. Buenos Aires, Argentina, 1974; vis. scientist Max Planck Inst., Stuttgart, Germany, 1976, IBM Rsch., Yorktown Heights, N.Y., 1976-77; staff mem. AT&T Bell Labs., Murray Hill, N.J., 1978—; sec. Argentina Phys. Soc., Buenos Aires, 1972-75; editor Solid State Communications, 1989-92, assoc. editor in chief, 1992—. Contbr. over 180 articles to profl. jours. and numerous chpts. to books. Fellow Am. Phys. Soc.; mem. AAAS, Materials Rsch. Soc. Achievements include use and devel. novel optical methods in studies of structural phase transitions, semiconductor interfaces and interactions of free electrons in semiconductors; discovered novel phenomena in studies of quantum electron fluids. Office: AT&T Bell Labs 600 Mountain Ave Rm 1D-433 Murray Hill NJ 07974

PIND, JÖRGEN LEONHARD, computational linguist, psycholinguist; b. Reykjavik, Iceland, May 8, 1950; s. Kaj Leonhard Jensen and Anna Gudrun (Sand) P.; m. Aldis Unnur Gudmundsdóttir, June 24, 1972; children: Ólöf Hildur, Anna Gudrun, Finnur Kári. BA, U. Iceland, Reykjavik, 1973; MSc, U. Sussex, Eng., 1977, PhD, 1982. Tchr. Hamrahlid Coll., Reykjavik, Iceland, 1973-76, 77-79, 1980-83; rsch. assoc. Inst. Lexicography U. Iceland, Reykjavik, 1983-89, dir., 1990—; vis. rsch. scientist Rsch. Lab. Electronics MIT, 1993-94; dir. studies Hamrahlid Coll., Reykjavik, 1974-76; bd. dirs. Inst. Lexicography, Reykjavik. Co-author: Psychology (Icelandic), 1981, rev. edit. 1988; editor: Frequency Dictionary of Icelandic, 1991; contbr. numerous articles to profl. jours. Mem. Acoustical Soc. Am. Office: Inst Lexicography, Neshaga 16, 107 Reykjavik Iceland

PINE, DAVID JONAH, aerospace engineer; b. Bklyn., Oct. 23, 1942; s. Charles and Rose Pine; m. Randy Lynn Hurzwitz, June 27, 1971; children: Laura Elizabeth, Meredith Renee. BS in Aero. Engring., Bklyn. Poly., 1964; MS in Engring Adminstrn., George Washington U., 1975. Engr. NASA Hdqrs./Goddard Space Flight Ctr., 1966-88; dep. program mgr. Hubble space telescope NASA Hdqrs., 1988-90, asst. comptr. systems analysis, 1990—. Mem. AIAA, Internat. Soc. for Parametric Analysis. Office: NASA Hdqrs/Code B Washington DC 20546

PINEAU, DANIEL ROBERT, structural engineer; b. Fall River, Mass., Sept. 7, 1954; s. Robert Maurice and Alphonsine J. (Guillôme) P.; m. Sally Frances Thomas, Oct. 8, 1988. BS in Civil Engring., U. Mass., 1976; postgrad., Northeastern U., 1977-80. Registered profl. engr. Structural designer Badger Am., Cambridge, Mass., 1976-78; structural engr. Keyes Assocs., Waltham, Mass., 1978-84, project engr., 1984-91; structural engr. Badger Engrs., Cambridge, 1991-93. Mem. Am. Soc. Civil Engrs., U.S. Power Squadrons. Roman Catholic. Home: 43 Lakeside Blvd North Reading MA 01864 Office: Badger Engrs One Broadway Cambridge MA 02142

PINES, ALEXANDER, chemistry educator, researcher; b. Tel Aviv, June 22, 1945; came to U.S., 1968.; s. Michael and Neima (Ratner) P.; m. Ayala Malach, Aug. 31, 1967 (div. 1983); children: Itai, Shani; m. Ditsa Kafry, May 5, 1983; children: Noami, Jonathan, Talia. BS, Hebrew U., Jerusalem, 1967; PhD, MIT, 1972. Asst. prof. chemistry U. Calif., Berkeley, 1972-75, assoc. prof., 1975-80, prof., 1980—, Pres.'s chair, 1993; faculty sr. scientist materials scis. div. Lawrence Berkeley Lab., 1975—; cons. Mobil Oil Co., Princeton, N.J., 1980-84, Shell Oil Co., Houston, 1981—; chmn. Bytel Corp., Berkeley, Calif., 1981-85; vis. prof. Weizmann Inst. Sci., 1982; adv. prof. East China Normal U., Shanghai, People's Rep. of China, 1985; sci. dir. Nalorac, Martinez, Calif., 1986—; Joliot-Curie prof. Ecole Superieure de Physique et Chemie, Paris, 1987; Walter J. Chute Disting. lectr. Dalhousie U., 1989, Charles A. McDowell lectr. U. B.C., 1989, E. Leon Watkins lectr. Wichita State U., 1990; Hinshelwood lectr., U. Oxford, 1990, A.R. Gordon Disting. lectr. U. Toronto, 1990, Venable lectr. U. N.C., 1990, Max Born lectr. Hebrew U. of Jerusalem, 1990; William Draper Harkins lectr. U. Chgo., 1991, Kolthoff lectr. U. Minn., 1991; Md.-Grace lectr. U. Md., 1992; mem. adv bd. Nat. High Magnetic Field Lab., Inst. Theoretical Physics, U. Calif. Santa Barbara; mem. adv. panel chem. Nat. Sci. Found.; Randolph T. Major Disting. Lectr. U. Conn., 1992; Peter Smith lectr. Duke U., 1993, Arthur William Davidson lect. U. Kansas, 1992, Arthur Birch lect. Australian Nat. U., 1993. Editor Molecular Physics, 1987-91; mem. bd. editors Chem. Physics, Chem. Physics Letters, Nmr: Basic Principles and Progress, Advances in Magnetic Resonance; adv. editor Oxford U. Press; contbr. articles to profl. jours.; patentee in field. Recipient Strait award North Calif. Spectroscopy Soc., Outstanding Achievement award U.S. Dept. of Energy, 1983, 87, 89, R & D 100 awards, 1987, 89, Disting. Teaching award U. Calif., E.O. Lawrence award, 1988, Pitts. Spectroscopy award, 1989, Wolf Prize for chemistry, 1991, Donald Noyce Undergrad. Teaching award U. Calif., 1992, Robert Foster Cherry award for Great Tchrs. Baylor U., Pres.'s Chair for undergrad. edn. U. Calif., 1993; Guggenheim fellow, 1988, Christensen fellow St. Catherine's Coll., Oxford, 1990. Fellow Am. Phys. Soc. (chmn. div. chem. physics), Inst. Physics; mem. NAS (adv. panel chem.), Am. Chem. Soc. (mem. exec. com. div. phys. chemistry, Signature award, Baekeland medal, Harrison Howe award 1991), Royal Soc. Chemistry (Bourke lectr.), Internat. Soc. Magnetic Resonance (v.p.). Office: U Calif Chemistry Dept D 64 Hildebrand Hall Berkeley CA 94720

PINIEWSKI, ROBERT JAMES, geologist; b. Buffalo, Aug. 22, 1955; s. Brownie John and Patricia Ann (Petrella) P.; m. Christi Lynn Sha, May 10, 1986; children: Katelyn, Alex. AA in Civil Tech., Erie Coll., Amherst, N.Y., 1975; BS in Geology, SUNY, Buffalo, 1980. Well logging engr. NL Baroid, Anchorage, 1980-83, drilling engr., 1983-86; geologist Terra Vac, Tampa, Fla., 1987; project mgr. Terra Vac, Tampa, 1988-89; ops. mgr. Terra Vac, Rochester, N.Y., 1989; div. mgr. Terra Vac, Temperance, Mich., 1990—; lectr. in field. Contbr. articles to profl. jours. Mem. Assn. of Groundwater Scientists and Engrs. Office: Terra Vac 9030 Secor Rd Temperance MI 48182

PINILLA, ANA RITA, neuropsychologist, researcher; b. N.Y.C., May 20, 1957; d. Louis and Luz Maria (Diaz) P.; m. Jorge Rosado Rosado, Dec. 01, 1979; children: Jorge Javier, Juan Carlos, Ana Mari. BS magna cum laude, U. P.R., Rio Piedras, 1978; MS, Caribbean Ctr., San Juan, P.R., 1980, PhD, 1988. Lic. psychologist, P.R. Prof. psychology Inter-Am. U., San Juan, 1980-91; neuropsychologist Neuropsychol. Svcs. to Developmental Deficiencies Children, Bayamon, P.R., 1987-88; asst. dir. Gov.'s Prevention Program, San Juan, 1988-90; exec. dir. Learning Disability Ctr., San Juan, 1990—; external evaluator prevention program Roberto Clemente Sports City, Carolina, P.R., 1990—; cons. in field; adviser, evaluator drug prevention programs. Author: Analysis of Wisc-R, 1988; contbr. articles to profl. publs. Mem. Internat. Neuropsychol. Soc., Nat. Acad. Neuropsychology. Achievements include development of program of services to learning disabled children using neuropsychological approach, development of tests for measurement character traits. Home: Crisantemo C-8 Estancias de Bairoa Caguas PR 00725 Office: Ctr Learning Disabilities RR 3 Box 3120 Rio Piedras PR 00928

PINK, HARRY STUARD, petroleum chemist; b. Hartford, Conn., May 20, 1943; s. Wilson Vandervoort and Evelyn (Frankenfield) P.; m. Betsy Lynne Sadler, Dec. 18, 1971; children: Christopher Sadler, Jeffrey Wilson. BA, Rutgers U., 1965; PhD, Syracuse U., 1975. Instr. chemistry Onondaga C.C., Syracuse, N.Y., 1974-75; rsch. assoc. Empire State Paper Rsch. Inst., Syracuse, 1975-76; sr. rsch. chemist GAF Corp., Wayne, N.J., 1976-78; sr. chemist Exxon Rsch. and Engring. Co., Linden, N.J., 1978-81, staff chemist, 1981-84, sr. staff chemist, 1984—. Contbr. articles to profl. jours. including Proceedings of the Assn. Asphalt Paving Technologists, Chem. Physics, Jour. Phys. Chemistry. Recipient W.J. Emmons award Assn. Asphalt Paving Technologists, 1980. Mem. ASTM (sub-chmn. 1987-88), SAE, Nat. Lubricating Grease Inst., N.Y. Acad. Scis., Sigma Xi. Democrat. Achievements include patents on aluminum complex grease and on water resistant grease compositions. Office: Exxon Rsch and Engring Co PO Box 51 Linden NJ 07036

PINKER, STEVEN A., cognitive science educator; b. Montreal, Que., Can., Sept. 18, 1954; came to U.S., 1976; s. Harry and Roslyn (Wiesenfeld) P. BA, McGill U., Montreal, 1976; PhD, Harvard U., 1979. Asst. prof. Harvard U., Cambridge, Mass., 1980-82, Stanford U., Palo Alto, Calif., 1981-82; prof. cognitive sci. MIT, Cambridge, 1982—; cons. Xarox Palo Alto Rsch. Ctr., 1982—. Author: Language Learnability and Language Development, 1984, Learnability and Cognition, 1989; editor: Visual Cognition, 1985, Connections and Symbols, 1988; assoc. editor Cognition: Internat. Jour. Cognitive Sci., 1984—. Recipient grad. teaching award MIT, 1986. Fellow AAAS; mem. APA (Disting. Early Career award 1984, Boyd McCandless award 1986). Office: Dept Brain-Cognitive Scis MIT E10-016 Cambridge MA 02139

PINKERT, DOROTHY MINNA, chemist; b. N.Y.C., June 2, 1921; d. Harry and Frieda Dorothy (Pinkert) Klein. A.B., Bklyn. Coll., 1944; M.S., Bklyn. Poly. Inst., 1952. Creep lab. technician Am. Brakeshoe Co., Mahwah, N.J., 1942-44; rsch. and quality control chemist Reed and Carnrick, Jersey City, 1944-48; chief quality control chemist Gold Leaf Pharmacal Co., New Rochelle, N.Y., 1948-56; rsch. chemist Internat. Salt Co., Watkins Glen, N.Y. and N.Y.C., 1957-61; sr. assoc. drug regulatory affairs Hoffmann-LaRoche Inc., Nutley, N.J., 1962-83. Mem. AAAS, Am. Chem. Soc., Am. Inst. Chemists, Am. Soc. Quality Control, Poly. U. Alumni Assn. (life dir.), N.Y. Acad. Scis. Republican.

PINNINTI, KRISHNA RAO, economist; b. Chinabadam, India, May 15, 1950; came to U.S., 1985; s. Jagannaikulu and Janakamma Pinninti; m. Prema Kumari Priya, May 13, 1971; children: Uma, Usha. BS, Andhra Loyola Coll., Vijayawada, India, 1970; MS, Indian Inst. Tech., New Delhi, 1972, PhD, 1975. Economist, faculty Adminstrv. Staff Coll. of India, Hyderabad, India, 1974-81; post-doctoral fellow Harvard U., Cambridge, Mass., 1979; dir. Centre for Devel. Rsch., India, 1981-93, Ctr. Devel. Rsch., Plainsboro, N.J.; assoc. prof. Rutgers U., New Brunswick, N.J., 1985-87; chmn. Govt. of Andhra Pradesh Travel and Tourism Devel. Corp., Hyderabad, 1987-89; internat. cons. Devel. Alternatives, Inc., Washington, 1990-

92; with The Equitable, Edison, N.J., 1992—, Ctr. Devel. Rsch., Plainsboro, N.J., 1992—; vis. prof. Inst. Pub. Enterprise, Hyderbad, 1982; vis. faculty Rutgers U., New Brunswick, 1989; cons. UN Devel. Program, N.Y., 1978, Internat. Labor Office, Geneva, 1982-83, Japan Internat. Cooperation Agy., Tokyo, 1990-91, World Resources Inst., Washington, 1986; mem. econ. adv. coun. Chesapeake Bay Mgmt./USEPA, Annapolis, Md., 1993. Chief editor, founder (quar.) Developing India, 1984-86. Mem. del. for China, Peoples to People Internat., Washington, 1986. Nat. Merit scholar Govt. of India, 1967; Sr. Fulbright fellow Coun. for Internat. Exch. of Scholars, Harvard U., 1985. Mem. Am. Econ. Assn., Assn. Environ. and Resource Economists (internat. corr. India 1987-91), Internat. Assn. Polit. Risk Analysts (hon.). Office: Dir Ctr Devel Rsch 2312 Pheasant Hollow Dr Plainsboro NJ 08536

PINOLI, BURT ARTHUR, airline executive; b. Santa Rosa, Calif., Nov. 23, 1954; s. Norris L. and Grace G. (Williams) P.; m. So Yen, May 9, 1987; 1 child, Lucas. BS in Agri-Bus., Calif. State U., Fresno, 1979; M. Internat. Mgmt., Am. Grad. Sch. Internat. Mgmt., Glendale, Ariz., 1988. Loan officer, mgmt. trainee Lloyds Bank Calif., Sanger, 1979-81; mgr. sales/bus. devel. Transamerica Airlines, Oakland, Calif., 1981-86; credit analyst, comml. loan officer Farm Credit Bank System, Ukiah, Calif., 1986-87; city mgr. Northwest Airlines, Beijing, 1988-90, Shanghai, People's Rep. of China, 1991—. Del. to India, Internat. Youth Exch. Named Nat. 4-H Coun. Nat. winner health project, 1974; recipient Blue Key, 1978. Mem. Alpha Gamma Rho (pres. 1978-79), Alpha Zeta. Home: 1551 Boonville Rd Ukiah CA 95482-9303

PINSKER, WALTER, allergist, immunologist; b. Bay Shore, N.Y., Mar. 27, 1933; s. Albert and Irene (Kuchlick) P.; m. Tillene Giller, June 15, 1958; children: Neil, Andrew, Susann. BA, U. Rochester, 1954; MD, Chgo. Med. Sch. U. Health Sci., 1958. Diplomate Am. Bd. Allergy and Immunology. Intern L.I. Jewish Hosp., New Hyde Park, N.Y., 1958-59; resident internal medicine Bklyn. VA Hosp., 1959-60; resident internal medicine Long Beach (Calif.) VA Hosp., 1960-61, resident allergy and immunology, 1961-62; chief of allergy Letterman Army Hosp., San Francisco, 1962-64; pres. Bay Shore Allergy Group, 1964—; attending physician Mather Hosp., Port Jefferson, N.Y., St. Charles Hosp., Port Jefferson, 1981, Southside Hosp., Bay Shore, 1964—, chief of allergy, Good Samaritan Hosp., West Islip, N.Y., 1964—, chief of allergy; asst. clin. prof. medicine SUNY, Stony Brook, 1968—. Contbr. articles to profl. jours. Bd. visitors Pilgrim State Hosp., Brentwood, N.Y., 1974-77; pres. Suffolk Assn. Children with Learning Difficulties, N.Y., 1972-74; trustee Leeway Sch., Stony Brook, 1974-75, Bay Shore Jewish Ctr., 1974-84; com. for handicapped West Islip Schs., 1971—. Capt. U.S. Army, 1962-64. Recipient Physician's Recognition award AMA, 1969—. Fellow Am. Acad. Allergy and Immunology, Am. Coll. Allergy and Immunology, Am. Assn. Certified Allergists, Am. Coll. Chest Physicians, Am. Assn.-Study of Headaches, N.Y. Acad. Scis., Suffolk Acad. Medicine, Nassau-Suffolk Allergy Soc. (officer, bd. dirs. 1970—, pres. 1980-82). Avocations: golf, boating, photography. Office: Bay Shore Allergy Group P C 649 Montauk Hwy Bay Shore NY 11706

PINSKY, LEONARD, geneticist; b. Montreal, Que., July 2, 1935; married, 1960; 4 children. BSc, McGill U., 1956, MD, CM, 1960. Investigator somatic cell genetics, dir. divsn. Med. Genetics Lady Davis Inst. Med. Rsch. Jewish Gen. Hosp., Montreal, Que., 1967—; mem. oper. grants sect. Med. Rsch. Coun. Can., 1967—; asst. prof. McGill U., 1968—, assoc. prof., 1973—; prof. pediatrics Ctr. Human Genetics, 1979—. Fellow Royal Coll. Physicians and Surgeons (Can.); mem. Am. Soc. Human Genetics, Soc. Pediatric Rsch., Can. Coll. Med. Geneticists (pres. 1978-80). Achievements include research in Androgen resistance, disorders of sexual development and nosology of malformation syndromes. Office: McGill U Ctr Human Genetics, 1205 Dr. Penfield, Montreal, PQ Canada H3G 1B1*

PINSTRUP-ANDERSEN, PER, educational administrator; b. Bislev, Denmark, Apr. 7, 1939; came to U.S., 1965; s. Marinus and Alma (Pinstrup) Andersen; m. Birgit Lund, June 19, 1965; children: Charlotte, Tina. BS, Royal Vet. & Agrl. U., Copenhagen, 1965; MS, Okla. State U., 1967, PhD, 1969. Agrl. economist Centro Internacional de Agricultura Tropical, Cali, Colombia, 1969-72, head econ. unit, 1972-76; dir. agro.-econ. div. Internat. Fertilizer Devel. Ctr., Florence, Ala., 1976-77; sr. rsch. fellow, assoc. prof. Econ. Inst. Royal Vet. & Agr. U., 1977-80; rsch. fellow Internat. Food Policy Rsch. Inst., Washington, 1980, dir. food consumption and nutrition divsn., 1980-87, dir. gen., 1992—; dir. food and nutrition policy program, prof. food econs. Cornell U., Ithaca, N.Y., 1987-92; cons. The World Bank, Washington, 1978-92, Can. Internat. Devel. Agy., 1982-83, 86, UNICEF, N.Y.C.; cons. subcom. on nutrition UN, Rome, 1980-87; mem. bd. mng. editors Fertilizer Rsch., 1990—. Author: The World Food and Agricultural Situation, 1978, Agricultural Research and Economic Development, 1979, The Role of Fertilizer in World Food Supply, 1980, Agricultural Research and Technology in Economic Development, 1982; editor: (with Francis C. Byrnes) Methods for Allocating Resources in Applied Agricultural Research in Latin America, 1975, (with Margaret Biswas) Nutrition and Development, 1985, Food Subsidies in Developing Countries: Costs, Benefits, and Policy Options, 1988, Macroeconomic Policy Reforms, Poverty, and Nutrition: Analytical Methodologies, 1990, The Political Economy of Food and Nutrition Policies, 1993; contbr. numerous articles to profl. jours., chpts. to books. With Danish Army, 1958-59. Recipient Ford Internat. fellowship, 1965-66, People to People Cert. of Appreciation, 1967, Kellogg Travel fellowship, 1979, Competition prize Nordic Soc. Agrl. Rschrs. and Norsk Hydro, 1979, Cert. Merit, Gamma Sigma Delta, 1991, Disting. Alumnus award U. Colo., 1993. Mem. Am. Assn. Agrl. Econs. (PhD Thesis award 1970, Outstanding Jour. Article award 1977), Internat. Assn. Agrl. Econs., Ctr. for Internat. and Devel. Econs. Rsch., Internat. Ctr. Rsch. on Women (bd. dirs.), Columbian Nat. Orgn. Profls. in Agr. (hon.). Home: 1451 Highwood Dr Mc Lean VA 22101 Office: Internat Food Policy Rsch Inst 1200 17th St NW Washington DC 20041

PIOT, PETER, medical microbiologist, epidemiologist; b. Leuven, Belgium, Feb. 17, 1949; m. Greet Kimzeke; children: Bram, Sara. MD, U. Ghent, Belgium, 1974; PhD in Microbiology, U. Antwerp, Belgium, 1981. Sr. fellow infectious diseases U. Wash., Seattle, 1978-79; asst. Inst. Tropical Medicine, Antwerp, 1974-78; prof., head dept. microbiology Inst. Tropical Medicine, 1981-92; asst. prof. pub. health Free U., Brussels, 1989—; asst. dir. Global Program AIDS World Health Orgn., Geneva, Switzerland; dir. WHO Collaborating Ctr. on AIDS, Antwerp; bd. dirs. Project SIDA, Kinshasa, Zaire, STD/AIDS Project, Nairobi, Kenya; chair WHO Steering Com. on the Epidemiology of AIDS, 1989-92. Editor: (with others) Chlamydial Infection, 1982, (with J. Mann) AIDS and HIV Infection in the Tropics, 1988, (with P. Lamptey) The Handbook on AIDS Prevention in Africa, 1990, AIDS in Africa, 1991, (with F. Andre) Hepatitis B, and STD in Heterosexuals, 1991, (with others) Basic Laboratory Procedures in Clinical Bacteriology, 1991, (with others) Reproductive Tract Infections in Women, 1992, (with others) AIDS in Africa: A Handbook for Physicians, 1992; mem. editorial bd. sci. publs. Mem. expert coms. Recipient Health Rsch. award Brussels, 1989, AMICOM award for med. rsch., 1991, Pub. Health award Flemish Community, 1990, H. Breurs prize, 1992, A. Jaunioux prize, 1992, van Thiel award Leiden, 1993; NATO fellowm 1978. Mem. Royal Acad. Medicine (corr.), Internat. AIDS Soc. (pres.), profl. socs. in Europe, U.S.A. and Africa. Home: K Mercierlei 60, 2600 Antwerp Belgium Office: GPA World Health Orgn, 27 Geneva Switzerland

PIPER, JON KINGSBURY, ecologist, researcher; b. Keene, N.H., Sept. 6, 1957; s. Roy Kingsbury and Anne (Macechok) P.; m. Elizabeth Brown, Aug. 3, 1980; children: Joshua, Emily, Samuel. BS in Biology, Bates Coll., 1979; PhD in Botany, Wash. State U., 1984. Teaching asst., rsch. asst. Wash. State U., Pullman, 1980-84, technician, vis. faculty, 1985; rsch. assoc. The Land Inst., Salina, Kans., 1985—; adj. faculty Benedictine Coll., Salina, 1989; mem. Sand Prairie adv. bd. Bethel Coll., North Newton, Kans., 1989. Co-author: Farming in Nature's Image, 1991; contbr. articles to Ecology, Am. Jour. Botany, Oikos, Can. Jour. Botany, Am. Midland Naturalist. Grantee NSF, 1982, Eppley Found., 1988. Mem. Ecol. Soc. Am., Brit. Ecol. Soc., Soc. for Conservation Biology, Soil Sci. Soc. Am., Torrey Bot. Club. Achievements include research in natural ecosystem processes as models for sustainable agriculture. Office: The Land Inst 2440 E Water Well Rd Salina KS 67401-9051

PIPER, LLOYD LLEWELLYN, II, engineer, service industry executive; b. Wareham, Mass., Apr. 28, 1944; s. Lloyd Llewellyn and Mary Elizabeth (Brown) P.; BSEE, Tex. A&M U., 1966; MS in Indsl. Engring., U. Houston, 1973; m. Jane Melonie Scruggs, Apr. 30, 1965; 1 child, Michael Wayne. With Houston Lighting & Power Co., 1965-74; project mgr. Dow Chem. Engring. & Constrn. Svcs., Houston, 1974-78; project mgr. Ortloff Corp., Houston, 1978, mgr. engring., 1979-80, v.p., 1980-83; pres., chief exec. officer Plantech Engrs. & Constructors, Inc. subs. Dillingham Constrn. Corp., Houston, 1983-86; pres. The Delta Plantech Co., Houston, 1985-86; dir. on-site tech. devel. Chem. Waste Mgmt., Inc., Oak Brook, Ill., 1986-88; mgr. projects Chem. Waste Mgmt., Inc., Houston, 1988—; bd. dirs. Harris County Water Control and Improvement Dist., 1973-83, pres., 1973-83; Ponderosa Joint Powers Agy. Harris County, 1977-83, pres., 1977-83; pres. bus. and industry adv. coun. North Harris Montgomery C. C. Dist., 1991-92. Recipient Disting. Svc. award Engrs. Coun. Houston, 1970, Outstanding Svc. award Houston sect. IEEE, 1974; named Tex. Young Engr. of Yr., 1976, Nat. Young Engr. of Yr., 1976; registered profl. engr., Tex, diplomate hazardous waste mgmt. Am. Acad. Environ. Engrs. Mem. IEEE, Nat. Soc. Profl. Engrs. (chpt. pres. 1978, nat. chmn. engrs. in industry div. 1977, nat. v.p. 1977, chmn. nat. polit. action com. 1979-82, vice chmn. nat. engrs. week 1988-92, nat. trustee edn. found. 1988-90), Project Mgmt. Inst., Phi Kappa Phi, Tau Beta Pi. Contbr. articles to profl. jours. Home: 2214A Nantucket Dr Houston TX 77057-2908 Office: Chem Waste Mgmt Inc 100 Glenborough Dr Fl 14 Houston TX 77067-3600

PIPPIN, JOHN ELDON, electronics engineer, electronics company executive; b. Kinard, Fla., Oct. 7, 1927; s. Festus and Mary Elvie (Scott) P.; m. Barbara A. Pippin, June 15, 1952; children: Carol Jean Pippin Franklin, John F., Mary Christine Pippin Mobley. B.E.E., Ga. Inst. Tech., 1951, M.S.E.E., 1953; Ph.D. in Applied Physics, Harvard U., 1958. Research. engr. Ga. Inst. Tech. Expt. Sta., Atlanta, 1951-53; head research dept. Sperry Microwave Electronics Co., Clearwater, Fla., 1958-64; v.p., dir. research Scientific-Atlanta, Inc., 1964-68; pres. Electromagnetic Scis., Inc., Norcross, Ga., 1968—; adj. prof. U. Fla., 1962-64; cons. Cascade Research Corp., 1953-58. Contbr. articles to profl. jours. Served in USN, 1945-46. NSF fellow; Gen. Commns. fellow. Fellow IEEE (Outstanding Engr. Region III 1972, Engr. of Yr. 1972); mem. Briarean Soc., Am. Phys. Soc., Microwave Theory and Techniques Soc. (adminstrv. com.), Sigma Xi, Tau Beta Pi, Phi Kappa Phi, Eta Kappa Nu. Achievements include research in microwave physics; radar tracking problems. Home: 3760 River Dr Duluth GA 30136-6101 Office: Electromagnetic Sci Inc PO Box 7700 Norcross GA 30091-7700

PIQUETTE, GARY NORMAN, reproductive endocrinologist; b. Milton, Mass., Apr. 21, 1958; s. Herman Royal and Vera (Johnson) P. BS in Biology, U. Minn., St. Paul, 1981, BS in Animal Sci., 1982; PhD in Anatomy, U. S.D., 1988. Lab technician dept. surgery U. Minn., Mpls., 1977-82, sr. lab technician dept. neurosurgery, 1982-83; rsch./teaching asst. dept. anatomy U. S.D., Vermillion, 1983-88; postdoctoral fellow dept. reprodn. medicine U. Calif., San Diego, 1988-91; postdoctoral fellow dept. ob-gyn. Stanford (Calif.) U., 1991—. Recipient Nat. Rsch. Svc. award NIH, Bethesda, 1990. Mem. AAAS, Soc. for Study Reprodn., The Endocrine Soc. Achievements include research in morphological and histochem. studies of cultured rabbit ovarian surface epithelium, granulosa cells and peritonel mesothelium; hormonal regulation of the ovary and uterus in the human and rodent. Office: Stanford U Dept Ob-Gyn 300 Pasteur Dr Stanford CA 94305

PIQUETTE, JEAN CONRAD, physicist; b. St. Augustine, Fla., Feb. 15, 1950; d. Jean H. and Marie G. (Parks) P.; m. Mariceann Janow, Aug. 22, 1972; children: Renee M., Laura D. BA, Rutgers U., 1972; PhD, Stevens Inst. Tech., 1983. Rsch. physicist Naval Rsch. Lab., Orlando, Fla., 1981—; Contbr. articles to profl. jours. Lt. USN, 1977-81. Rutgers U. scholar, 1972; recipient Rsch. award Naval Rsch. Lab., 1986, 88, 92. Mem. Acoustical Soc. Am., Phi Beta Kappa. Achievements include invention of method for underwater acoustic transducer transient suppression; invention, development and implementation of method of symbolic special function integration on computer algebra systems; invention of ONION underwater acoustic panel measurement method. Office: Naval Rsch Lab Underwater Sound Reference PO Box 568337 Orlando FL 32856-8337

PIRKL, JAMES JOSEPH, industrial designer; b. Nyack, N.Y., Dec. 27, 1930; s. James and Ida Bertha (Gigrich) P.; m. Sarah B. W. Woolsey, June 8, 1974; children: Theo, James, Philip. Cert. Advt. Design, Pratt Inst., 1951, B of Indsl. Design cum laude, 1958. With design staff Gen. Motors Corp., Warren, Mich., 1958-65; sr. designer Gen. Motors Corp., 1961-64, asst. chief designer, 1964-65; instr. indsl. design Center for Creative Studies, Detroit, 1963-65; faculty dept. design Syracuse (N.Y.) U., 1965-92, assoc. prof., 1969-73, prof. indsl. design, 1974-92, prof. emeritus, 1992—, coord. indsl. design program, 1979-84, chmn. dept. design, 1985-91; exec. council chmn. Sch. Art, 1976-78, 80-81; sr. rsch. fellow All-U. Gerontology Ctr., 1990—; prin. James J. Pirkl/Design, 1965—; cons. Brownlie Design, Inc., 1972—, Rolland Co., 1993, Arthritis Found., 1993, Prince Corp., 1991, Ford Motor Design Ctr., 1992, Loretto Geriatric Ctr., Sage Marcom Inc., 1988-90, Hazard Mgmt. Co., 1985, Marcom Switches Inc., 1977-82, Cazenovia Abroad Ltd., 1973-81, Holistic Mgmt. Group, Inc., 1981, Pulos Design Assocs., 1972-80, Beck Assocs., 1976, Fed. Prison Industries, 1974, Gen. Electric Co., 1967-70, Genesee Labs., Inc., 1968, N.Y. State Council on Arts, 1968-69, Stettner-Trush, Inc., 1972-78, Strathmore Chem. Coatings, Inc., 1969, 72, Village of Cazenovia, 1979-93, Xerox Corp., 1975; chmn. accreditation council Design Found., 1982-84; interviewed on Nat. Pub. Radio. Author: Transgenerational Design: Products for an Aging Population, 1993; co-author: Guidelines and Strategies for Designing Transgenerational Products, 1988, Transgenerational Design, 1993; co-editor: State of the Art and Science of Design, 1971; co-designer: Gen. Motors Futurama Exhbn., N.Y. World's Fair, 1964-65; contbr. articles to profl. jours. including Design Mgmt. Jour., Jour. Indsl. Designers Soc. Am., Bus. Adminstrn. Jour., Design News, Design Perspectives, Indsl. Design. Mem. Everson Mus. Art, 1977-85; chmn. planning commn. Town of Cazenovia, N.Y., 1988-93; mem. senate Syracuse U., 1973-80; mem. adv. bd. SEARS Project, 1989-91; chmn. chancellor's citation com., 1988-92; mem. exhbns. com. Syracuse Cultural Resources Coun., 1992-93. With SeaBees USN, 1951-55. Fellow Indsl. Designers Soc. Am. (chmn. universal design com. 1991—, chmn. NASAD liaison com. 1984-88, mem. archives com. 1988-92, nat. bd. dirs. 1977-81, chmn. Central N.Y. chpt. 1977-78, v.p. Mid-East region 1978-80, dir., chmn. edn. com. 1980-81, U.S. rep., del. Internat. Congress Socs. Indsl. Design 1989); mem. AAUP, The Design Found. (chmn. accreditation coun. 1982), Nat. Assn. Schs. Art and Design (accreditation evaluator 1985—), nat. Ctr. for a Barrier Free Environment (adv. task force 1981), Human Factors Soc. (consumer products tech. group 1978, tech. group on aging 1987, forensics tech. group 1988), Internat. Congress Socs. Indsl. Design (edn. com.), Author's Guild. Achievements include patent for 4-way handle. Home: 66 Camino Barranca Placitas NM 87043

PIRRUNG, MICHAEL CRAIG, chemistry educator, consultant; b. Cin., July 31, 1955; s. Joey Matthew and Grace (Fielman) P. BA, U. Tex., 1975; PhD, U. Calif., Berkeley, 1980. NSF postdoctoral fellow Columbia U., N.Y.C., 1980-81; asst. prof. Stanford (Calif.) U., 1981-89; sr. scientist Affymax Rsch. Inst., Palo Alto, Calif., 1989; assoc. prof. Duke U., Durham, N.C., 1989—; cons. Am. Cyanamid, 1992—; sci. advisor Arrymax Rsch. Inst., 1991—, Genta, 1993—. Recipient Newcomb Cleveland prize AAAS, 1991; Sloan Found. fellow, 1986-88. Mem. Am. Chem. Soc., Am. Soc. Plant Phys. Achievements include invention of spatially directed light addressable parallel chemical synthesis; research on mechanism of ethylene biosynthesis. Office: Duke U Dept Chemistry PM Gross Lab PO Box 90346 Durham NC 27708

PITCHER, WAYNE HAROLD, JR., biotechnology company executive; b. St. Louis, Jan. 5, 1944; s. Wayne Harold Sr. and Ethel Pauline (Gehrke) P.; m. Julia Frances Liberace, Aug. 22, 1970; children: Wayne Harold III, Maria Beatrice. BS in Chem. Engring., Calif. Inst. Tech., 1966; SM in Chem. Engring., MIT, 1968, ScD in Chem. Engring., 1972. Sr. chem. engr. Corning (N.Y.) Glass Works, 1972-74, sr. research chem. engr., 1974-76, engring. supr., 1976-81, mgr. biotech. portfolio, 1981-83; v.p. devel. Genencor, South San Francisco, 1983-89; v.p. rsch. and devel., 1989-90; v.p. technology Genencor Internat., South San Francisco, 1990-92; sr. v.p. technology, 1992—. Editor: Immobilized Enzymes for Food Processing, 1980; contbr. articles to profl. jours.; contbr. chtps. to books; patentee in field. Mem. AAAS, Am. Inst. Chem. Engrs., N.Y. Acad. Sci., Am. Chem. Soc., Sigma Xi. Office: Genencor 180 Kimball Way South San Francisco CA 94080-6299

PITELKA, FRANK ALOIS, zoologist, educator; b. Chgo., Mar. 27, 1916; s. Frank Joseph and Frances (Laga) P.; m. Dorothy Getchell Riggs, Feb. 5, 1943; children: Louis Frank, Wenzel Karl, Vlasta Kazi Helen. B.S. with highest honors, U. Ill., 1939; summer student, U. Mich., 1938, U. Wash., 1940; Ph.D., U. Calif. at Berkeley, 1946. Mem. faculty U. Calif. at Berkeley, 1944—, prof. zoology, 1958—, chmn. dept., 1963-66, 69-71, Miller research prof., 1965-66; curator birds Mus. Vertebrate Zoology, 1945-63, assoc. dir., 1982—; exec. com. Miller Inst. Basic Rsch. in Sci., 1967-71, chmn., 1967-70; Panel environ. biology NSF, 1959-62, panel polar programs, 1978-80; mem. panel biol. and med. scis., com. polar research Nat. Acad. Scis., 1960-65; research assoc. Naval Arctic Research Lab., Barrow, Alaska, 1951-80; ecol. adv. com. AEC, 1956-1958; adv. coms. U.S. Internat. Biol. Program, 1965-69; mem. adv. com. U. Colo. Inst. Arctic and Alpine Research, 1968-73; mem. U.S. Tundra Biome Program, 1968-73, dir., 1968-69; vis. prof. U. Wash. Friday Harbor Labs, summer 1968; mem. U.S. Commn. for UNESCO, 1970-72. Contbr. research papers in field.; Editorial bd.: Ecology, 1949-51, 60-62, editor, 1962-64; mem. editorial bd. U. Calif. Press, 1953-62, chmn., 1959-62; mem. editorial bd. Pacific Coast Avifauna, 1947-60, Ecol. Monographs, 1957-60, Systematic Zoology, 1961-64, The Veliger, 1961-85, Studies in Ecology, 1972-84, Current Ornithology, 1980-85; asst. editor Condor, 1943-45, assoc. editor, 1945-62; mem. editorial bd. Studies in Avian Biology, 1979-84, editor, 1984-87. Guggenheim fellow, 1949-50; NSF sr. postdoctoral fellow Oxford (Eng.) U., 1957-58; Research fellow Center for Advanced Study in Behavioral Scis., Stanford, 1971; recipient Disting. Teaching award U. Calif.-Berkeley, 1984; The Berkeley citation, 1985. Fellow Arctic Inst. N.Am., Am. Ornithologists Union (Brewster award 1980), Calif. Acad. Scis., AAAS, Animal Behavior Soc.; mem. Ecol. Soc. Am. (Mercer award 1953, Eminent Ecologist award 1992), Soc. Study Evolution, Cooper Ornithol. Soc. (hon. mem.; pres. 1948-50), Brit. Ecol. Soc., Am. Soc. Mammalogists, Am. Soc. Naturalists, Am. Inst. Biol. Scis., Western Soc. Naturalists (pres. 1963-64), Nat. Audubon Soc., Sierra Club, Phi Beta Kappa, Sigma Xi. Home: PO Box 9278 Berkeley CA 94709-0278

PITMAN, FRANK ALBERT, aerospace engineer; b. Pitts., Oct. 28, 1964; s. Frank and Mary Pitman. BS in Aerospace Engring., U. Colo., 1987; MS in Aerospace Engring., Purdue U., 1988. Teaching asst. Purdue U., West Lafayette, Ind., 1987-88, U. Tenn., Knoxville, 1991—. Mem. AIAA, ASME, Planetary Soc. Home: 808 Sutters Mill Ln Knoxville TN 37909

PITOT, HENRY CLEMENT, III, physician, educator; b. N.Y.C., May 12, 1930; s. Henry Clement and Bertha (Lowe) P.; m. Julie S. Schutten, July 29, 1954; children: Bertha, Anita, Jeanne, Catherine, Henry, Michelle, Lisa, Patrice. BS in Chemistry, Va. Mil. Inst., 1951; MD, Tulane U., 1955, PhD in Biochemistry, 1959. Instr. pathology Tulane U. Med. Sch., 1955-59; postdoctoral fellow McArdle Lab., U. Wis., Madison, 1959-60; mem. faculty U. Wis. Med. Sch., Madison, 1960—, prof. pathology and oncology, 1966—, chmn. dept. pathology, 1968-71, acting dean, 1971-73, dir. McArdle Lab., 1973-91. Recipient Borden undergrad. rsch. award, 1955, Lederle Faculty award, 1962, Career Devel. award Nat. Cancer Inst., NIH, 1965, Parke-Davis award in exptl. pathology, 1968, Noble Found. Rsch. award, 1983, Esther Langer award U. Chgo., 1984, Disting. Svc. award Am. Cancer Soc., 1989, Hilldale award U. Wis., 1991, Chem. Industry Inst. of Toxicology Founders award, 1993. Recipient Borden undergrad. rsch. award, 1955, Lederle Faculty award, 1962, Career Devel. award Nat.Cancer Inst., NIH, 1965, Park-Davis award in exptl. pathology, 1968, Noble Found. Rsch. award, 1983, Esther Langer award U. Chgo., 1984, Disting. Svc. award Am. Cancer Soc., 1989, Hilldale award U. Wis., 1991, Founders award Chem. Industry Inst. Toxicology, 1993. Fellow N.Y. Acad. Scis.; mem. AAAS, Am. Soc. Cell Biology, Am. Assn. Cancer Rsch., Am. Soc. Biochemistry and Molecular Biology, Am. Chem. Soc., Am. Soc. Investigative Pathology (pres. 1976-77), Soc. Exptl. Biology and Medicine (pres. 1991-93), Soc. Surg. Oncology (Lucy J. Wortham award 1981), Am. Soc. Preventive Oncology, Soc. Toxicology, Soc. Toxicologic Pathologists, Japanes Cancer Soc. (hon.). Roman Catholic. Home: 1812 Van Hise Ave Madison WI 53705-4053 Office: U Wis McArdle Lab Cancer Rsch 1400 University Ave Madison WI 53706

PITT, BERTRAM, cardiologist, consultant; b. Kew Gardens, N.Y., Apr. 27, 1932; s. David and Shirley (Blum) P., m. Elaine Liberstein, Aug. 10, 1962; children—Geoffrey, Jessica, Jillian. BA, Cornell U., 1953; MD, U. Basel, Switzerland, 1959. Diplomate Am. Bd. Internal Medicine, Am. Bd. Cardiology. Intern Beth Israel Hosp., N.Y.C., 1959-60; resident Beth Israel Hosp., Boston, 1960-63; fellow in cardiology Johns Hopkins U., Balt., 1966-67; from instr. to prof. Johns Hopkins U., 1967-77; prof. medicine, dir. div. cardiology U. Mich., Ann Arbor, 1977-91, assoc. chmn. dept. medicine, 1991—. Author: Atlas of Cardiovascular Nuclear Medicine, 1977; editor: Cardiovascular Nuclear Medicine, 1974. Served to capt. U.S. Army, 1963-65. Mem. Am. Coll. Cardiology, Am. Soc. Clin. Investigation, Assn. Am. Physicians, Am. Physiol. Soc., Am. Heart Assn., Am. Coll. Physicians, Assn. Univ. Cardiology, Am. Coll. Chest Physicians, Royal Soc. Mich., Royal Soc. Medicine. Home: 24 Ridgeway St Ann Arbor MI 48104-1739 Office: U Mich Divsn Cardiology 1500 E Medical Ctr Dr Ann Arbor MI 48109

PITT, ROBERT ERVIN, environmental engineer, educator; b. San Francisco, Apr. 25, 1948; s. Wallace and Marjorie (Peterson) P.; m. Kathryn Jay, Mar. 18, 1967; children: Gwendolyn, Brady. BS in Engring. Sci., Humboldt State U., 1970; MSCE, San Jose State U., 1971; PhD in Civil and Environ. Engring., U. Wis., 1987. Registered profl. engr., Wis. Environ. engr. URS Rsch. Co., San Mateo, Calif., 1971-74; sr. engr. Woodward-Clyde Cons., San Francisco, 1974-79; cons. environ. engr. Blue Mounds, Wis., 1979-84; environ. engr. Wis. Dept. Natural Resources, Madison, 1984-87; assoc. prof. depts. civil engring. and environ. health scis. U. Ala., Birmingham, 1987—; mem. Resource Conservation and Devel. Coun., Jefferson County, Ala., 1992—; mem. com. on augmenting natural recharge of groundwater with reclaimed wastewater NRC, 1991—. Author: Small Storm Urban Flow and Particulate Washoff Contributions to Outfall Discharges, 1987; co-author: Manual for Evaluating Stormwater Runoff Effects in Receiving Waters, 1993; author software in field. Asst. scoutmaster Boy Scouts Am., Birmingham, 1988—. Recipient Fed. Water Pollution Control Adminstrn. fellowship, 1970-71, GE, 1984-86; recipient 1st Place Nat. award U.S. Soil Conservation Svc. Earth Team, 1989, award of recognition U.S. Dept. Agr., 1990, 1st Place Vol. award Take Pride in Am., 1991. Mem. ASCE, N.Am. Lake Mgmt. Soc. (profl. speakers award 1992), Water Environ. Fedn. (1st place nat. award 1992), Am. Water Resources Assn., Ala. Acad. Sci., Sigma Xi. Achievements include development of small storm urban hydrology prediction methods, toxicant control devices for stormwater source flows, methods to identify and correct inappropriate discharges to storm drain systems. Office: U Ala Birmingham Dept Civil Engring 1150 10th Ave S Birmingham AL 35294-4461

PITTACK, UWE JENS, engineer, physicist; b. Lübeck, Germany, Apr. 11, 1935; came to U.S., 1968; s. Moritz and Ella (Hoffman) P.; m. Ilse C. Heier, May 12, 1961; children: Christian, Catrin. MS in Physics, U. Kiel, Germany, 1962, PhD in Physics cum laude, 1964. Asst. prof. U. Kiel, 1965; mem. tech. staff Phillips Germany, 1965-68; sr. engr. Hughes Aircraft Co., Torrance, Calif., 1968-72; prin. engr. Raytheon Co., Waltham, Mass., 1972-85, Marlboro, Mass., 1985—; adv. bd. for critical components U.S. Undersec. of Def., Washington, 1984. Contbr. tech. papers in plasma physics, lasers, microwave tubes to sci. publs. Achievements include invention of dual mode traveling wave tube for electronic countermeasure systems, development of traveling wave tube for Intelsat satellite transmitter, verification of Eigenbroadening of helium spectral lines. Office: Raytheon Co 1001 Boston Post Rd Marlborough MA 01752

PITTARELLI, GEORGE WILLIAM, agronomist, researcher; b. Rome, Italy, Feb. 11, 1925; came to U.S., 1952; s. Ernest Maria and Irma (Bianchini) P.; m. Rachel Sbarra, July 12, 1952; children: Patricia, Loretta, Ernest. Grad., Coll. Agrl., Rome, Italy, 1949; postgrad., U. Rome, 1949-52, U. Md., 1963, '66. Asst. agronomist U. Md., College Park, 1952-55; agronomist USDA Agrl. Rsch. Svc. PS, Beltsville, Md., 1955-88, assoc. researcher, 1988—. Contbr. over 30 articles to profl. jours., including Jour.

Heredity, Plant Physiology, Crop Sci., Beitrage zur Tabakforschung Internat., Phytopathology, Phytochemistry, Plant Sci. Recipient Points of Light award, 1992. Mem. Am. Genetic Soc. Roman Catholic. Achievements include development of two new male sterile varieties of tobacco for genetic research; patent for discovery of biological insecticide from indigenous nicotiana tabacum. Home and Office: 7030 Hunter Ln Hyattsville MD 20782-1149

PITTELKOW, MARK ROBERT, dermatology educator, researcher; b. Milw., Dec. 16, 1952; s. Robert Bernard and Barbara Jean (Thomas) P.; m. Gail L. Gamble, Nov. 26, 1977; children: Thomas, Cameron, Robert. BA, Northwestern U., 1975; MD, Mayo Med. Sch., 1979. Intern then resident Mayo Grad. Sch., 1979-84; asst. prof. dermatology Mayo Med. Sch., Rochester, Minn., 1984-91, assoc. prof. biochemistry and molecular biology, 1991—, assoc. prof. biochemsitry and molecular biology, 1992; cons. Mayo Clinic/Found., Rochester, 1984—. Fellow Am. Acad. Dermatology; mem. AAAS, Soc. Investigative Dermatology, Am. Burn Assn., Am. Soc. Cell Biology, N.Y. Acad. Scis., Chi Psi. Home: 721 12th Ave SW Rochester MN 55902-2027 Office: Mayo Clinic 200 1st St SW Rochester MN 55905-0001

PITTENGER, GARY LYNN, biomedical educator; b. Canandaigua, N.Y.C., Sept. 4, 1949; s. Francis William and Bernice Ethel (Kingsley) P.; m. Vija Piziks, Sept. 12, 1976; children: Karina Boehm, Alija Kristina. MS, U. Mich., 1986, PhD, 1990. Rsch. assoc. dept. surgery U. Mich., Ann Arbor, 1977-84, regents fellow dept. anatomy and cell biology, 1984-87, NIH reproductive scis. fellow, 1987-89, rsch. assoc., 1989-90; dir. Protein Chemistry Lab. Diabetes Insts., Norfolk, Va., 1990—; asst. prof. anatomy and neurobiology, internal medicine Ea. Va. Med. Sch., Norfolk, 1990—. Contbr. articles to profl. jours. NIH fellow reproductive scis. NIH, 1987-89; regents fellow U. Mich., 1984-87. Mem. AAAS, Am. Soc. Cell Biology (student travel award 1988), Am. Diabetes Assn., N.Y. Acad. Scis. Achievements include originating patented device-hollow views tanometer; participation in studies demonstarting release and action of luminal hormones; identification of low molecular weight heat shock protein in rat testis; isolation of pancreatic B-cell growth factor, studying role of antibodies in diabetic neuropathy, and the role of the low molecular weight heatshock protein in platelet function. Office: The Diabetes Insts 855 W Brambleton Ave Norfolk VA 23510

PITTINGER, CHARLES BERNARD, JR., civil engineer; b. Balt., July 30, 1949; s. Charles Bernard Pittinger and Mary Helen (Dietz) Mezger; m. Beverly Ann Porter, July 22, 1972; children: Susan Marie, Charles Bernard III. BS in Aero. Engring., Embry-Riddle Aero. U., Daytona Beach, Fla., 1971. Registered profl. engr., profl. land surveyor, Md. Civil engr. Century Engring., Inc., Towson, Md., 1971-76; cons. Reisterstown, Md., 1976-80; civil engr. U.S. Army Directorate of Engring. and Housing, Ft. Meade, Md., 1980-88; exptl. facilities engr. NASA Hdqrs., Washington, 1988—; fed. constrn.-coun. coms. NRC/Nat. Acad. Sci. and Engring., Washington, 1989-93. Fellow ASCE; mem. NSPE, Md. Soc. Profl. Engrs. (pres. 1992-93). Democrat. Roman Catholic. Achievements include 46 U.S. patents and international patents and patents pending. Home: 1830 Hanniford Dr Finksburg MD 21048-1500 Office: NASA Hdqrs Code JXG 300 E St SW Washington DC 20546

PITTMAN, CONSTANCE SHEN, physician, educator; b. Nanking, China, Jan. 2, 1929; came to U.S., 1946; d. Leo F.-Z. and Pao Kong (Yang) Shen; m. James Allen Pittman, Jr., Feb. 19, 1955; children: James Clinton, John Merrill. AB in Chemistry, Wellesley Coll., 1951; MD, Harvard U., 1955. Diplomate Am. Bd. Internal Medicine, sub-bd. Endocrinology. Intern Baltimore City Hosp., 1955-56; resident U. Ala., Birmingham, 1956-57; instr. in medicine U. Ala. Med. Ctr., Birmingham, 1956-59, fellow dept. pharmacology, 1957-59, from asst. prof. to assoc. prof., 1959-70, prof., 1970—; prof. medicine Georgetown U., Washington, 1972-73; mem. diabetes and metabolism tng. com. NIH, Bethesda, Md., 1972-76, mem. nat. arthritis, metabolism and digestive disease coun., 1975-78, mem. gen. clin. rsch. ctrs. com., 1979-83, 87-90. Fellow ACP; mem. Assn. Am. Physicians, Am. Soc. for Clin. Investigation, Endocrine Soc. (coun., 1978-79, pres. women's caucus 1978-79), Am. Thyroid Assn. (pres. 1990-91). Achievements include research in activation and metabolism of thyroid hormone; kinetics of thyroxine conversion to triiodothyrine in healthy and disease states; control of iodine deficiency disorders. Office: U Ala Div Endocrinology/Metab Univ Sta Birmingham AL 35294

PITTMAN, JACQUELYN, mental health nurse, nursing educator; b. Pensacola, Fla., Dec. 22, 1932; d. Edward Corry Sr. and Hettie Oean (Wilson) P. BS in Nursing Edn., Fla. State U., 1958; MA, Columbia U., 1959, EdD, 1974. Physician asst. Med. Ctr. Clinic, Pensacola, 1953-55; clin. instr., asst. dir. nursing svc. Sacred Heart Hosp., Pensacola, 1955-56; instr.psychiatric nurse Fla. State Hosp., Chattahoochee, 1958, Pensacola Community Coll., 1959-60, 62-63; chmn. div. nursing Gulf Coast Community Coll., Panama City, Fla., 1963-66; asst. prof. U. Tex., Austin, 1970-72, assoc. prof., 1972-80; prof. nursing, coord. curriculum and teaching Grad. Program La. State U. Med. Ctr. Sch. Nursing, New Orleans, 1980—; curriculum cons. Nicholls State U., Thibodaux, La., 1982, Our Lady of the Lake Sch. Nursing, Baton Rouge, 1983; rsch. liaison So. Bapt. Hosp., New Orleans, 1987-89, Med. Ctr. La., 1992—; mem. Sci. Misconduct Inquiry comm. La. State U. Med. Ctr., 1992—. Tchr. Christian edn. program for mentally retarded St. Ignatius Martyr Ch., 1979-80; tchr. initiation team Rite of Christian Initiation of Adults, Our Lady of the Lake Cath. Ch., Mandeville, La., 1983-86; ethics com., bd. trustee Hotel Dieu Hosp., New Orleans, 1987-91; v.p., bd. dirs. St. Tammany Guidance Ctr., Inc., Mandeville, 1987-91; mem. Dem. Nat. Comm., Presdl. Task Force, 1992, Ctr. for Study of Presidency; judge Internat. Sci. and Engring. Fair Assn., 1990, 92; del. La. State Nurses' Assn. State Conv., 1992. Mem. ANA, LWV, N.Y. Acad. Sci., Acad. Polit. Sci., Nat. Trust for Historic Preservation, La. Endowment for Humanities, La. State Nurses Assn. (archivist 1987—, state task force com. to preserve hist. documents 1987—), So. Nursing Rsch. Soc., Nat. League Nursing, Boston U. Nursing Archives, Women's Inner Cir. of Achievement N.Am. Communities, Internat. Order Merit, World Found. Successful Women, Wilson Ctr. Assocs., Kappa Delta Pi, Sigma Theta Tau. Democrat. Roman Catholic. Avocations: swimming, golf, travel, reading, Louisiana history. Home: 204 Woodridge Blvd Mandeville LA 70448-2604 Office: La State U Med Ctr 1900 Gravier St New Orleans LA 70112-2262

PITTMAN, JAMES ALLEN, JR., endocrinologist, dean, educator; b. Orlando, Fla., Apr. 12, 1927; s. James Allen and Jean C. (Garretson) P.; m. Constance Ming-Chung Shen, Feb. 19, 1955; children—James Clinton, John Merrill. BS, Davidson Coll., 1948; MD, Harvard, 1952; DSc (hon.), Davidson Coll., 1980, U. Ala., Birmingham, 1984. Intern, asst. resident medicine Mass. Gen. Hosp., Boston, 1952-54; teaching fellow medicine Harvard U., 1953-54; clin. assoc. NIH, Bethesda, Md., 1954-56; instr. medicine George Washington U., 1955-56; chief resident U. Ala. Med. Ctr., Birmingham, 1956-58, instr. medicine, 1956-59, asst. prof., 1959-62, assoc. prof., 1962-64, prof. medicine, 1964—, dir. endocrinology and metabolism div., 1962-71, co-chmn. dept. medicine, 1969-71, also prof., physiology and biophysics, 1967—, prof. medicine, 1964—; dean U. Ala. Med. Ctr. (Sch. Medicine), 1973-92; asst. chief med. dir. rsch. and edn. in medicine U.S. VA, 1971-73; prof. medicine Georgetown U. Med. Sch., Washington, 1971-73; mem. endocrinology study sect. NIH, 1963-67; mem. pharmacology, endocrinology fellowships rev. coms., 1967-68; chmn. Liaison Com. Grad. Med. Edn., 1976; mem. Grad. Med. Edn. Adv. Com., 1976-78, HHS Coun. on Grad. Med. Edn., 1986; mem. nat. adv. rsch. resources coun. NIH, 1991—. Author: Diagnosis and Treatment of Thyroid Diseases, 1963; Contbr. articles in field to profl. jours. Fellow ACP (life); mem. Assn. Am. Physicians, Endocrine Soc., Am. Thyroid Assn., N.Y. Acad. Scis. (life), Soc. Nuclear Medicine, Am. Diabetes Assn., Am. Chem. Soc., Wilson Ornithol. Club (life), Am. Ornithologists Union, Am. Fedn. Clin. Research (pres. So. sect., mem. nat. council 1962-66), So. Soc. Clin. Investigation, Harvard U. Med. Alumni Assn. (pres. 1986-88, mem. com. on grad. med. evaluation 1987—), Phi Beta Kappa, Alpha Omega Alpha, Omicron Delta Kappa. Office: U Alabama-Sch of Med Disting Prof CAMS 75 SDB Birmingham AL 35294-0007

PITTMAN, VICTOR FRED, computer systems engineer; b. Atlanta, Feb. 9, 1958; s. William Fred and Faye Elizabeth (Futch) P.; m. Maria C. Fico, June 20, 1981 (div. May 1993); 1 child, Jenniferann. BS in Computer Sci., So. Coll. Tech., 1993. Transp. engr. Ga. Dept. Transp., Atlanta, 1980-83, Lee Wan and Assoc., Decatur, Ga., 1983-85, Giffels/Hart, Atlanta, 1985-88; systems mgr. Hill-Fister Engrs., Decatur, 1988-90; systems engr. Simons-Eastern Cons., Decatur, 1990—. Mem. Assn. for Computing Machinery (assoc.), Digital Equipment Computer Users Soc. Home: 304 Parkway Ave Smyrna GA 30080 Office: Simons-Eastern Cons 1 West Court Sq Decatur GA 30030-2509

PITTMAN, WILLIAM CLAUDE, electrical engineer; b. Pontotoc, Miss., Apr. 22, 1921; s. William Claude and Maude Ella (Bennett) P.; m. Eloise Savage, Apr. 20, 1952; children: Patricia A. Pittman Ready, William Claude III, Thomas Allen. BSEE, Miss. State Coll., 1951, MSEE, 1957. From electronic engr. to supr. elec. engring. dept. U.S. Army Labs., Redstone Arsenal, Ala., 1951-59; supr. electronic engr. to aero. engring. supr. NASA/Marshall Space Flight Ctr., 1960; electronic engr. Army Missile Labs., 1962-82; program mgr. Army Labs. and R&D Ctr., Redstone Arsenal, 1982—. Sgt. USMC, 1940-46, PTO. Recipient Medal of Honor DAR. Fellow AIAA (assoc.; chmn. Miss.-Ala. chpt. 1981-82, Martin Schilling award 1980); mem. IEEE (sr.), NSPE, First Marine Div. Assn., DAV, IRE (chmn. Huntsville sect. 1957-58), Madison Hist. Soc., SAR (pres. Tenn. Valley chpt. 1984-85, Ala. Soc. 1990-91, Cert. 1991, Patriot medal), Tau Beta Pi, Phi Kappa Phi, Kappa Mu Epsilon. Avocations: history, genealogy. Home: 704 Desoto Rd SE Huntsville AL 35801-2032 Office: US Army Missile Command Redstone Arsenal AL 38598-5253

PITTS, JOHN ROLAND, physicist; b. Dallas, Jan. 16, 1948; s. Jack Roland and Grace Ella (Allen) P.; m. Lise Ann LaCroix, May 19, 1979; children: Benjamin, Jessica, Emily. BS in Physics, N.Mex. Inst. Mining & Tech., 1970; MS in Physics, Oreg. Grad. Inst., 1972; PhD in Physics, U. Denver, 1985. Grad. teaching asst., rsch. asst. U. Colo., Boulder, 1975-78; assoc. physicist Nat. Renewable Energy Lab., Golden, Colo., 1978-80, staff physicist, 1980-84, sr. physicist, 1984—; mem. physics adv. com. U. Colo., Denver, 1987—. Contbr. articles to profl. jours. Lt. (j.g.) USN, 1972-75. Recipient Outstanding Performance award Nat. Renewable Energy Lab., 1992, Tech. Excellence award Dept. Energy, 1982; vis. fellow Max-Planck Gesellschaft, 1985-86. Mem. ASM Internat. (tech. transfer com. 1990-93), Am. Vacuum Soc. (bd. dirs. Rocky Mountain chpt. 1988—). Achievements include patents in field. Home: 2005 Applewood Dr Lakewood CO 80215-1003 Office: Nat Renewable Energy Lab 1617 Cole Blvd Lakewood CO 80401-3393

PITTS, MARCELLUS THEADORE, civil engineer, consultant; b. Frankfurt, Germany, Apr. 15, 1957; (parents Am. citizens); s. George Theadore and Nellie Ruth (Fowler) P. BSCE, S.C. State U., 1979; MS in Managerial Engring., N.J. Inst. Tech., 1982. Registered profl. engr., Ga. Design chief for civil engring. Mayes, Sudderth, Etheredge, Atlanta, 1979-80; mgr. civil design N.J. Dept. Transp., Newark, 1980-82; mgr. constrn. and civil engring. Fowler-Pitts Enterprises/Boyd Constrn., Mountville, S.C., 1982-83; environmentalist civil dept. Giffels-Harte, Inc., engrs., architects, Atlanta, 1983-84; civil design engr. Williams, Russell, Johnson Inc., engrs., architect, planners, Atlanta, 1984-85; section chief water engring. Fulton County Dept. Planning and Econ. Devel., Atlanta, 1985-87; div. chief infrastructure planning Fulton County Dept. Pub. Works, 1987—; prin. cons. Cole, Hinch, Jones, Pitts and Assocs., Atlanta, 1986—. Founding mem. Christmas in July, Inc., Atlanta, 1987; mem. adv. bd. Capital Area Ministries, Atlanta, 1986, Big Bros. and Big Sisters, Atlanta, 1989—; mem. sanctuary choir Ben Hill United Meth. Ch., Atlanta, 1983—, mem. adminstrv. v., 1986-89, 90—, mem. coun. on ministries, 1986-89, 90—. Recipient Outstanding Alumni award S.C. State U. Alumni Assn., 1987, Outstanding Svc. award Ben Hill United Meth. Ch., 1989, svc. award Fulton County Employees Assn., 1990. Mem. ASCE, NSPE, Am. Pub. Works Assn., Am. Backflow Prevention Assn., Masons (sec. 1982-83), Kiwanis (pres. S.W. Atlanta 1987-88, President's award 1988), 100 Black Men of South Metro Atlanta, Alpha Phi Alpha (life, Outstanding Svc. award 1987)), Kappa Kappa Psi (life). Avocations: step aerobics, tennis, skiing, mathematics tutoring. Office: Fulton County Dept Pub Work 141 Pryor St SW Ste 6001 Atlanta GA 30303-3468

PITTS, NATHANIEL GILBERT, science foundation director; b. Macon, Ga., Apr. 2, 1947; s. Raymond J. and Kathleen C. Pitts; m. Lisa V. Leonard, Feb. 14, 1992. BS, Whittier Coll., 1969; PhD, U. Calif., Davis, 1974. Postdoctoral fellow Rockefeller U., N.Y.C., 1974-77; health sci. adminstr. NIH, Bethesda, Md., 1985-86; congl. fellow Senate Labor Com., Washington, 1988-89; asst. program dir. NSF, Washington, 1977-79, assoc. program dir., 1979-85, program dir., 1985-88, dep. dir. direct bns., 1990-91, dir. Office Sci. and Tech., 1991—. Contbr. articles to sci. jours. Mem. Soc. for Neurosci., Inst. Brain Rsch. Organic. Office: NSF 1800 G St Washington DC 20550

PITZER, KENNETH SANBORN, chemist, educator; b. Pomona, Calif., Jan. 6, 1914; s. Russell K. and Flora (Sanborn) P.; m. Jean Mosher, July 1935; children—Ann, Russell, John. B.S., Calif. Inst. Tech., 1935; Ph.D., U. Calif., 1937; D.Sc., Wesleyan U., 1962; LL.D., U. Calif. at Berkeley, 1963, Mills Coll., 1969. Instr. chemistry U. Calif., 1937-39, asst. prof., 1939-42, asso. prof., 1942-45, prof., 1945-61, asst. dean letters and sci., 1947-48, dean coll. chemistry, 1951-60; pres., prof. chemistry Rice U., Houston, 1961-68, Stanford, Calif., 1968-70; prof. chemistry U. Calif. at Berkeley, 1971—; tech. dir. Md. Rsch. Lab. for OSRD, 1943-44; dir. research U.S. AEC, 1949-51, mem. gen. adv. com., 1958-65, chmn., 1960-62; Centenary lectr. Chem. Soc. Gt. Britain, 1978; mem. adv. bd. U.S. Naval Ordnance Test Sta., 1956-59, chmn., 1958-59; mem. commn. chem. thermo-dynamics Internat. Union Pure and Applied Chemistry, 1953-61; mem. Pres.'s Sci. Adv. Com., 1965-68; dir. Owens-Ill., Inc., 1967-86. Author: (with others) Selected Values of Properties of Hydrocarbons, 1947, Quantum Chemistry, 1953, (with L. Brewer) Thermodynamics, rev, 1961, Activity Coefficients in Electrolyte Solutions, 2d edit., 1992; editor: Prentice-Hall Chemistry series, 1955-61; contbr. articles to profl. jours. Trustee Pitzer Coll., 1966—; Mem. program com. for phys. scis. Sloan Found., 1955-60. Guggenheim fellow, 1951; recipient Precision Sci. Co. award in petroleum chemistry, 1950, Clayton prize Instn. Mech. Engrs., London, 1958, Priestley medal Am. Chm. Soc., 1969, Nat. medal for sci., 1975, Robert A Welch award, 1984, Clark Kerr award U. Calif., Berkeley, 1991; named 1 of 10 Outstanding Young Men U.S. Jaycees, 1950. Fellow Am. Nuclear Soc., Am. Inst. Chemists (Gold medal award 1976), Am. Acad. Arts and Scis., Am. Phys. Soc.; mem. NAS (councilor 1964-67, 73-76), AAAS, Am. Chem. Soc. (award pure chemistry 1943, Gilbert N. Lewis medal 1965, Williard Gibbs medal 1976), Faraday Soc., Geochem. Soc., Am. Philos. Soc., Chem. Soc. (London), Am. Coun. Edn. Clubs: Chemists (hon.), Bohemian; Cosmos (Washington). Home: 12 Eagle Hl Kensington CA 94707-1408 Office: U Calif Dept Chemistry Berkeley CA 94720

PIVER, M. STEVEN, gynecologic oncologist; b. Washington, Sept. 29, 1934; s. Harry Samuel and Sonia (Bard) P.; m. Susan Myers, June 25, 1958; children: Debra Ellen, Carolyn Jan, Kenneth Stuart. BS, Gettysburg Coll., 1957; MD, Temple U., 1961. Diplomate Am. Bd. Ob-Gyn, Am. Coll. Surgeons. Intern Nazareth Hosp., Phila., 1961-62; resident Johns Hopkins U. Hosp., Balt., 1962; resident ob-gyn. Pa. Hosp., U. Pa., Phila., 1965-68; fellow gynecologic oncology U. Tex., Hosp. and Tumor Inst., Houston, 1968-70; asst. prof. gynecologic oncology U. N.C. Sch. Medicine, 1970-71; assoc. chief gynecologic oncology Roswell Park Cancer Inst., Buffalo, 1972-83, founder, dir. Gilda Radner Familial Ovarian Cancer Registry, 1981—, chief gynecologic oncology, 1984—; clin. prof., dir. div. gynecologic oncology SUNY, Buffalo, 1986—. Cons. editor Year Book of Cancer, 1972-88; assoc. editor Nat. Cancer Inst., PDQ, 1984—; mem. editorial bd. The Female Patient, 1989—; author: Ovarian Malignancies: Clinical Care of Adults of Adolescents, 1983; editor: Ovarian Malignancies: Diagnostic and Therapeutic Advances, 1987, Manual of Gynecologic Oncology/Gynecology, 1989, Conversations About Cancer, 1990; contbr. 265 sci. articles to profl. jours. Bd. dirs. United Way of Buffalo and Erie County, 1986-91; trustee D'Youville Coll., Buffalo, 1989—; pres. Friends of the Night People (homeless shelter), Buffalo, 1988—. Capt. USAF, 1962-64. Hon. fellow Phi Beta Kappa, Gettysburg Coll., 1956, Tex. Assn. Obstetricians and Gynecologists, 1983; named Citizen of Yr., Buffalo News, 1989; recipient YMCA Leadership award Buffalo YMCA, 1990, Brotherhood/Sisterhood Award in Medicine (Western N.Y. Region), NCCJ, 1991, St. Marguerite D'Youville Coll. Community Svc. award, 1992. Fellow ACS, Am. Coll. Obstetricians and Gynecologists; mem. Am. Soc. Clin. Oncology, Soc. Gynecologic Oncologists, Soc. Surg. Oncology, Am. Radium Soc., Phi Beta Kappa. Achievements include documentation of hydroxyurea as a radiation sensitizer in cervix cancer that significantly improves cure rate and that ovarian cancer can be inherited. Home: 315 Lincoln Pky Buffalo NY 14216-3127 Office: Roswell Park Cancer Inst Elm and Carlton Sts Buffalo NY 14263

PIXLEY, JOHN SHERMAN, SR., research company executive; b. Detroit, Aug. 24, 1929; s. Rex Arthur and Louise (Sherman) P.; B.A., U. Va., 1951; postgrad. Pa. State U., 1958-59; m. Peggy Marie Payne, Oct. 16, 1949; children—John Sherman, Steven, Lou Ann. Asst. cashier Old Dominion Bank, Arlington, Va., 1953-56; tech. dir. John I. Thompson & Co., research and engring. firm, Bellefonte, Pa., 1956-65; co-founder, exec. v.p. Potomac Research Inc., Alexandria, Va., 1965-80; v.p. Gov. Sevs. Div. Electronic Data Systems, 1980-81; co-founder, pres. Potomac Research Inc. Inc., Alexandria, Va., 1981—. Owner Edgeworth Farm, Orlean, Va. Mem. Fairfax County Republican Com., Annandale, Va., 1964-72; mem. fin. com. for U.S. Rep. Joel T. Broyhill, Republican, Va., 1970-72. Served to 1st lt. AUS 1952-53; maj. Res. ret. Decorated Army Commendation medal. Mem. IEEE, Am. Radio Relay League, Sleepy Hollow Woods Civic Assn. (v.p., pres. 1969-71). Episcopalian. Club: Quantico (Va.) Flying (charter mem.). Home: 3711 Sleepy Hollow Rd Falls Church VA 22041-1021 Office: Potomac Rsch Inc 11320 Random Hills Rd Ste 300 Fairfax VA 22030

PIZARRO, ANTONIO CRISOSTOMO, agricultural educator, researcher; b. Pilar, Bataan, Philippines, Sept. 21, 1926; s. Marciano Venegas Pizarro and Rita (Reyes) Crisostomo; m. Celia Abelardo Villegas, Jan. 25, 1958; children: Peter, Rosario, Matheresa. BS in Agr., U. Philippines, 1951; MS in Plant Pathology, U. Wis., 1957, PhD in Virology, 1967; diploma in nematology, U. London, 1969. Jr. agronomist Bureau of Plant Industry, Manila, 1951-52, jr. plant pathologist, 1952-54, various positions, 1955-70; div. plant pathology Standard Fruit Corp., Davao City, Philippines, 1971-75; dean grad. sch., coll. agr. and forestry Gregorio Araneta U. Found., Manila, 1990—; cons. Republic Planters Bank, Manila, 1975-78, asst. mgr., 1984-86; Stanfilco trainee Limon, Costa Rica, 1971, chmn. bd. dirs., 1988—; mem. Ctrl. Luzon Ctr. Emergency Aid and Rehab., Inc. Author: (poster presentation) Little Leaf a New Disease of Papaya, 1992; contbr. articles to profl. jours. Recipient Govt. Pensionado award Republic of Philippines, Manila, 1955-56, Govt. scholar, 1962-66; Colombo Plan scholar Great Britian, London, 1969-70; recipient Appreciation diploma British Embasy, Manila, 1969. Roman Catholic. Achievements include recovery of rice plants from tungro virus infection, naturally en masse and permanently at any time of year; discovery of a new viral disease of papaya called little leaf, infected plants tend to recover during hot summer months; naming of new species of nematode, Panagrobelus petersi, from England. Home: 65 Tangali Sta Mesa Heights, Quezon City Philippines Office: G Araneta U Found, Victoneta Park, Malabon Manila The Philippines

PIZZO, JOE, physics educator; b. Houston, Oct. 30, 1939; s. Joseph Francis and Irene Emily (Thibodeaux) P.; m. Roberta Applegate, Mar. 13, 1991; children: Stephen, Charlotte, Paul Christopher, Francesca, Samuel. BA, U. St. Thomas, 1961; PhD, U. Fla., 1964. Rsch. participant Oak Ridge (Tenn.) Nat. Lab., 1965-68; prof. Lamar U., Beaumont, Tex., 1964-92; vis. prof. U.S. Mil. Acad., West Point, N.Y., 1992—. Editor (monthly column) The Physics Tchr. Jour., 1989-92. Mem. Am. Assn. of Physics Tchrs. (pres. Tex. sect. 1984-85, Disting. Svc. Citation, 1989), Phi Beta Kappa. Achievements include creation of a lending library of physics demonstrations available to pre-college physics and physical science teachers used as a national model by the NSF. Office: Lamar U Physics Dept Box 10046 Beaumont TX 77710

PIZZOLI, ELSA MARIA, chemistry educator; b. Bologna, Emilia, Italy, Dec. 11, 1930; d. Primo and Adriana (Paracchi) P.; m. Vito Mazzacane, Feb. 3, 1973 (dec. 1981). D in Chimica, U. Bologna, 1957. Asst. prof. U. Bologna, 1957-58; asst. prof. U. Bari, Italy, 1959-74, full prof., 1975—. Contbr. articles to jours. Applied Chemistry, Analiticla Chemistry, Reed Rsch. Mem. Soroptimist (pres. Bari club 1986). Roman Catholic. Home: M L King 45, 70124 Bari Italy Office: U Bari, Rosalba 53, 70124 Bari Italy

PLAA, GABRIEL LEON, toxicologist, educator; b. San Francisco, May 15, 1930; arrived in Can., 1968; s. Jean and Lucienne (Chalopin) P.; m. Colleen Neva Brasefield, May 19, 1951; children: Ernest (dec.), Steven, Kenneth, Gregory, Andrew, John, Denise, David. BS, U. Calif., Berkeley, 1952; PhD, U. Calif., San Francisco, 1958. Diplomate Am. Bd. Toxicology. Asst. toxicologist City/County San Francisco, 1954-58; asst. prof. Sch. Medicine Tulane U., New Orleans, 1958-61; assoc. prof. U. Iowa, Iowa City, 1961-68; prof. U. Montreal, Que., Can., 1968—, chmn. Dept. Pharmacology, 1968-80, vice-dean Faculty Medicine, 1982-89, dir. Interuniv. Ctr. Rsch. in Toxicology, 1991—; chmn. Can. Coun. Animal Care, Ottawa, Ont., Can., 1985-86. Editor Toxicology and Applied Pharmacology, 1972-80; contbr. over 200 articles to profl. jours. 1st lt. U.S. Army, 1952-53, Korea. Recipient Thienes award Am. Acad. Clin. Toxicology, 1977. Mem. Am. Soc. Pharmacology, Soc. Toxicology Can. (pres. 1981-83, Henderson award 1969), Pharm. Soc. Can. (pres. 1973-74), Soc. Toxicology (pres. 1983-84, Achievement award 1967, Lehman award 1977, Edn. award 1987). Roman Catholic. Home: 236 Meredith Ave, Dorval, PQ Canada H9S 2Y7 Office: U Montreal, Dept Pharmacology, PO Box 6128 Sta A, Montreal, PQ Canada H3C 3J7

PLAKAS, STEVEN MICHAEL, biological research scientist; b. Boston, Nov. 9, 1952; s. Michael J. and Martha (Koulouris) P. AS, U. R.I., 1973, MS, 1979, PhD, 1984; BS, U. Wash., 1976. Rsch. biologist FDA, Dauphin Island, Ala., 1984—. Contbr. 25 articles to profl. jours. W.F. Thompson scholarship U. Wash., 1976, Eppley scholarship U. R.I., 1973, Monbusho scholarship Kagoshima (Japan) U., 1978. Mem. N.Y. Acad. Sci., Am. Fisheries Soc., World Aquaculture Soc., Assn. Ofcl. Analytical Chemists. Office: US FDA PO Box 158 Dauphin Island AL 36528

PLATA-SALAMAN, CARLOS RAMON, neuroscientist; b. Guadalajara, Jalisco, Mex., Feb. 7, 1959; came to U.S., 1988; s. Manuel Plata and Eva Maria Salamán; m. Kyoko Matsumoto, Mar. 12, 1987. MD, U. Guadalajara, 1984; D of Med. Sci., Kyushu U., Fukuoka, Japan, 1988. Lectr. in pathophysiology, coord. rsch. sect. Dept. Pathophysiology Faculty of Medicine U. Guadalajara, 1979-82; coord. sci. activities Soc. of Boarding Med. Students, Civil Hosp. of Guadalajara, 1981-82; collaborator exptl. physiology teaching program for med. students Kyushu U., 1984-88; postdoctoral fellow Dept. Psychology U. Del., Newark, 1988-90, lectr. of neurobiology, neuroimmunology and physiology Sch. Life and Health Scis., 1990—, assoc. prof. Sch. Life and Health Scis., 1990—; referee Am. Jour. Physiology, Neuroscience Letters, Peptides, Physiology and Behavior, Brain, Behavior and Immunity, Brain Rsch. Bull. Contbr. articles to profl. jours. including Jour. Neurophysiology, Brain Rsch., Am. Jour. Physiology, Physiology and Behavior, Peptides, Life Scis., NeuroReport, Neurosci. Letters, Brain Rsch. Bull., Neurosci. Biobehavior Rev., Digestive Diseases, Environ. Jour. Pharmacology, Brain, Behavior and Immunity, Neurosci. Rsch. Comm. Mem. AAAS, N.Y. Acad. Scis., Soc. for Neurosci., Pavlovian Soc., Soc. for the Study of Ingestive Behavior, Human Anatomy and Physiology Soc. Achievements include research in communication between the immune and nervous systems, humoral and neural substrates in the control of feeding and drinking, effects of peptides in the nervous system; in cellular neurophysiology. Office: U Del Sch Life and Health Scis Newark DE 19716

PLATÉ, NICOLAI A., polymer chemist; b. Moscow, Nov. 4, 1934; s. Alfred and Raisa (Zelinskaya) P.; m. Natalia Shakhova, Mar. 30, 1956; children: Alexei, Ekatherina. MD, Lomonossov U., Moscow, 1956, PhD, 1960; D.Sci., Petrochem. Inst., Moscow, 1966. Rsch. fellow Lomonossov U., 1956-62, asst. prof., 1962-66, dir. lab., 1966-85, prof. chemistry, 1967—; dir. Inst. Petrochem. Synthesis, Moscow, 1985—; cons. Monsanto Chem. Co., Springfield, Mass., 1991—. Author: Macromolecular Reactions, 1977, Comb-Shaped Polymers and Liquid Crystals, 1983, Physiologically Active Polymers, 1987; author, editor: Liquid Crystal Polymers, 1993. Recipient Nat. prize for sci. USSR Govt., 1985, Golden Wolf prize Internat. Com. Sci. for Peace, 1988. Mem. Russian Acad. Scis. (vice chmn. div. gen. chemistry 1991—), USSR Acad. Scis. (Kargin award 1981), N.Y. Acad. Scis., Pugwash Internat. Com. Scientists. Achievements include discovery of relationships in properties of block and graft copolymers, novel polymeric liquid crystals, theory of macromolecular reactivity, biomedical polymer based on macromonomers, polymeric membranes for gas separation. Office: 29 Leninsky prosp, 117912 Moscow Russia

PLATIKA, DOROS, neurologist; b. Bucharest, Romania, Jan. 31, 1953; came to U.S., 1962; s. Emanuel and Liana (Catsica) P.; m. Patricia Anne Curran, Sept. 10, 1988; children: Christopher Adrian, Alexander Michael. BA in Biology and Psychology, Reed Coll., 1975; MD, SUNY, Stony Brook, 1980. Diplomate Am. Bd. Med. Examiners. Intern in medicine Mass. Gen. Hosp., Boston, 1980-81, resident in medicine, 1981-83, resident in neurology, 1983-86, chief resident in neurology, 1985-86; physician scientist Whitehead Inst. for Biomed. Rsch., Cambridge, Mass., 1986-89; neurology instr. Harvard U., Cambridge, Mass., 1986-91; asst. in neurology Mass. Gen. Hosp., Boston, 1986-91; asst. prof. neurology and neuroscis. Albert Einstein Coll. Medicine, N.Y.C., 1991-93; v.p. R & D Progenitor, Inc., Columbus, Ohio, 1993—; pres. Boston Med. Diagnostics, Inc., 1983-88;. Contbr. articles to Gastroenterology, Proc. Nat. Acad. Scis. USA, Trans. Assn. Am. Physicians, The New Biologist. Recipient 1st prize Rsch. award Boston Soc. Neurology and Psychiatry, 1986; Martin Luther King scholar, 1971, N.Y. State Regents scholar, 1971, Danforth scholar, 1972, John Hairgrove scholar, 1975; grantee NSF, 1974-76. Mem. AMA, AAAS, Am. Acad. Neurology, Soc. for Neurosci., Mass. Gen. Hosp. Soc. Fellows, U.S. Chess Fedn., N.Y. Acad. Scis., Mass. Med. Soc., Phi Beta Kappa. Achievements include patent (with others) for use of immortalized cells of neurons in scientific research; development (with others) of artificial intelligence paradigm for computer aided medical diagnosis, of new treatment for brain tumors; research on effects of 4,4-dichlorobiphenyl on calcium metabolism of Salmonoid fish, on temporal and somatotopic patterns of response in the Nucleus Interpositus of cats and rats to mechanical stimulation of forepaws and hind paws, on the Michigan polybrominated biphenyl contamination, on molecular mechanisms of neuronal development, regeneration and synapse formation, on the development of molecular vectors for gene transfer into neurons and on retroviral vectors for inducible selective ablation of targeted cells. Office: Progenitor Inc 132 N Woods Blvd Columbus OH 43235

PLATT, JUDITH ROBERTA, electrical engineer; b. Chgo., Feb. 8, 1939; d. Harry and Lena (Sedloff) Feldman; m. Marvin Platt, June 8, 1968; children: Jennifer, Allison. BSEE, Ill. Inst. Tech., 1959, MSEE, 1962, PhD in Elec. Engring., 1971. Rsch. engr. Ill. Inst. Tech. Rsch. Inst., Chgo., 1959-71; mem. tech. staff Gen. Rsch. Corp., Denville, N.J., 1972-73, David Sarnoff Rsch. Ctr., Princeton, N.J., 1973-76; sr. mem. engring. staff RCA Govt. Systems, Moorestown, N.J., 1977-80; mem. tech. staff Riverside Rsch. Inst., N.Y.C., 1980-82; sr. staff engr. Bendix Guidance Systems, Teterboro, N.J., 1982-84; prin. mem. tech. staff ITT Avionics, Nutley, N.J., 1984-86; staff engr. systems lab. Adv. Tech. Labs., GE, Moorestown, 1986-87; pres., cons. J.R. Platt Assocs., Inc., Princeton, 1988—; bd. dirs. Glendale Protective Techs., Inc. Contbr. articles to profl. jours. Founder, sec., bd. dirs. Solomon Schechter Day Sch., East Brunswick, 1981-82. Mem. IEEE (sr.), Armed Forces Comm. and Electronics Assn., Assn. Old Crows. Home: 12 Shady Ln East Brunswick NJ 08816-4127 Office: JR Platt Assocs Inc 5 Independence Way Princeton NJ 08540-6627

PLATTS, FRANCIS HOLBROOK, plastics engineer; b. Brunson, S.C., Sept. 15, 1939; s. Holbrook Trowbridge and Mildred Ruth (Thomar) P.; m. Martha Ann Price, July 1963; children: Martha Susan Platts Gilliam, David Holbrook. BS in Chem. Engring., U. S.C., 1962. Chem. engr. U.S. Naval Weapons Lab., Dahlgren, Va., 1962-64; engr. Westinghouse Electric Corp., Hampton, S.C., 1964-74, sr. engr., 1974-89, mgr. engring. and quality assurance, 1989-91, mgr. design and mktg. svcs., 1991-93, div. engr. mgr., 1993—; chmnm. NEMA DLATC Engring. Com., Washington, 1991—; mem. Color Marketing Group, Alexandria, Va., 1991—. Pres., mem. Hampton Jaycees, 1964-74; chmn., bd. dirs. Hampton County Watermelon Festival, 1965-76; mem., chmn. blood bank bd. South Atlantic region ARC, Savannah, Ga., 1974—; mem. Western Carolina Higher Edn. Com., Allendale, 1982—. Mem. ASTM (sr., pres. 1972—), Hampton Rotary (sr., pres. 1981-82, Oustanding Mem. award 1982), Hampton Gamecock Club (pres. 1966—). Methodist. Achievements include development of decorative high-pressure laminate specialty product, research on color and design trends for interior finish applications. Office: Westinghouse Electric Corp PO Box 248 Hampton SC 29924-0248

PLATZ, TERRANCE OSCAR, utilities company executive; b. Cadillac, Mich., Jan. 20, 1943; s. Jay and Gladys Pearl (Bigelow) P.; m. Nellie Mae Cross, Dec. 15, 1961 (div. Oct. 1967); children: Michael, Christopher, Michelle; m. Dorothy Fay Beasley, Aug. 4, 1984. AS in Electronics Engring. Tech., Pensacola Jr. Coll., 1972. Enlisted USN, 1962, resigned, 1981; adj. instr. in electronics Pensacola (Fla.) Jr. Coll., 1981—; instrument/elec. control technician Escambia County Utilities Authority, Pensacola, 1981-87, instrument/elec. control supr., 1987—; instrument controls advisor/cons. to various engring. firms, 1987—. Mem. Instrument Soc. Am. Democrat. Mem. Ch. of God. Achievements include design of cost saving variable speed drive systems, instrument control systems, SCADA systems. Home: 5045 Bankhead Dr Pensacola FL 32526-9413 Office: Escambia County Utilities Authority 401 W Government St Pensacola FL 32501

PLAUGER, P.J., science writer; b. Petersburg, W.Va., Jan. 13, 1944; s. James H. and Jessie Pearl (Mowry) P.; m. Tana Lee Eastman, Nov. 29, 1977; 1 child, Geoffrey James. AB, Princeton U., 1965; PhD, Mich. State U., 1969. Mem. tech. staff Bell Labs., Murray Hill, N.J., 1969-75; v.p. Yourdon Inc., N.Y.C., 1975-78; pres. Whitesmiths, Ltd., Westfield, Mass., 1978-88; chief scientist Intermetrics, Inc., Cambridge, Mass., 1988-89; pvt. practice Concord, Mass., 1989--; sec. ANSI/X3J11, Washington, 1983-92; convenor ISO WG14, Geneva, Switzerland, 1987—. Co-author: The Elements of Programming Style, 1974, Software Tools, 1976, Software Tools in Pascal, 1980, Standard C, 1990, ANSI and ISO Standard C, 1992; author: Programming on Purpose (3 vols.), The Standard C Library, 1992. Recipient John W. Cambpell award World Sci. Fiction Conv., 1975. Home and Office: 398 Main St Concord MA 01742

PLAUT, MARSHALL, physician, medical administrator, educator; b. Balt., Apr. 14, 1944; s. Howard H. and Victoria (Rosenfeld) P.; m. Rita Leah Kipper, Dec. 12, 1982; children: Jonathan, Ari, Eli. BA with honors, Johns Hopkins U., 1963, MD, 1967. Diplomate Am. Bd. Internal Medicine, Am. Bd. Allergy and Immunology. Intern U. Fla. Hosps., Gainesville, 1967-68, resident in internal medicine, 1968-69; med. officer streptococcus lab. clin. bacteriology unit ctr. for disease control USPHS, Altanta, 1969-71; postdoctoral fellow Johns Hopkins U., Balt., 1971-74, instr. in medicine divsn. clin. immunology, 1974-75, asst. prof. medicine sch. medicine, 1975-82, assoc. prof. sch. medicine, 1982-91, also mem. subdepartment of immunology, 1983-88; chief asthma and allergy br. divsn. allergy, immunology and transplantation Nat. Inst. Allergy and Infectious Disease/NIH, Bethesda, Md., 1991—. Editorial bd. Jour. Allergy & Clin. Immunology, 1987-92, Jour. Immunology, Bethesda, 1977-81, immunopathology editor, 1981-85; contbr. articles and abstracts to med. jours. Recipient Nat. Ints. Allergy Infectious Diseases Rsch. Career Devel. award NIH, 1977-82; grantee NIAID, Nat. Heart, Lung, Blood Inst., 1975-91. Fellow Am. Acad. Allergy and Immunology; mem. Am. Fedn. Clin. Rsch., Am. Assn. Immunologists, Phi Beta Kappa, Alpha Omega Alpha. Office: DAIT/NIAID NIH Solar Bldg Rm 4A23 900 Rockville Pike Bethesda MD 20892

PLEE, STEVEN LEONARD, mechanical engineer; b. St. Joseph, Mich., Aug. 17, 1951; s. Billy Edwin and Natalie (Eberhart) P.; m. Shelley Ann McCabe, Oct. 24, 1987; children: Michael, Amber, Tara. BSME, Purdue U., 1973, MSME, 1975, PhD, 1978. Staff rsch. engr. Gen. Motors Rsch. Labs., Warren, Mich., 1978-85; v.p., gen. mgr. Barrack Labs., Marlborough, Mass., 1985-90; engring. mgr. Motorola AIEG, Dearborn, Mich., 1991—; adj. prof. Oakland U., Rochester, Mich., 1980-85, Lawrence Inst. Tech., Southfield, Mich., 1979-85. Contbr. chpts to books, articles to Combustion and Flame, ASME, SAE Transactions, others. Mem. ASME, Combustion Inst., Soc. Automotive Engrs. (Horning Meml. award 1982), Sigma Xi. Lutheran. Achievements include patents covering sensors, diagnostics and control techniques for spark ignition and diesel engines. Home: 2226 Pine Hollow Brighton MI 48116 Office: Motorola AIEG 15201 Mercantile Dearborn MI 48120

PLISCHKE, LE MOYNE WILFRED, research chemist; b. Greensburg, Pa., Dec. 11, 1922; s. Fred and Ruth Naomi (Rumbaugh) P.; m. Joan Harper, Mar. 11, 1966. BS, Waynesburg Coll., 1948; MS, W.Va. U., 1952. Rsch. chemist U.S. Naval Ordinance Test Sta., China Lake, Calif., 1952-53; asst. prof. chemistry Commonwealth U., Richmond, Va., 1953-54; rsch. chemist E.I. du Pont, Gibbstown, N.J., 1955-57, Monsanto Chem. Co., Pensacola, Fla., 1957—. Mem. Am. Chem. Soc. Achievements include 15 U.S. patents and 48 fgn. patents in field. Home: 2100 Clubhouse Dr Lillian AL 36549-5402 Office: Monsanto Co The Chem Group PO Box 97 Gonzalez FL 32560-0097

PLISETSKAYA, ERIKA MICHAEL, biology and physiology educator; b. Leningrad, USSR, Dec. 8, 1929; came to U.S., 1980; d. Michael Israel and Amalia Zachary (Utevskaya) P. BS in Biology, State U., Leningrad, 1952, PhD in Physiology, 1958; DSc, Pavlov's Inst. Physiology Acad. Sci., USSR, 1972. Rsch. scientist Inst. Evolutionary Physiology Acad. Sci., Leningrad, 1958-79; rsch. assoc. dept. zoology U. Wash., Seattle, 1980-84, rsch. scientist dept. zoology, 1984-89, rsch. prof., 1989—. Author: Hormonal Regulation of Carbohydrate Metabolism in Lower Vertebrates, 1975; editor Evolution of Pancreatic Islets, 1977; assoc. editor Jour. Exptl. Zoology, 1990—, General Comparative Endocrinology, 1991—; contbr. articles to profl. jours. Rsch. grantee NSF, 1985—, Nato For Intern Collaboration, 1986—, USDA, 1991—. Fellow AAAS; mem. NSF (panel 1989), Am. Soc. Zoologists, Soc. Endocrinology, European Soc. Comparative Endocrinologists, Am. Fisheries Soc. Avocations: travel, music, flowers, animals. Office: U Wash Sch Fisheries HF-15 Seattle WA 98195

PLOETZ, LAWRENCE JEFFREY, ceramic engineer; b. North Hornell, N.Y., Aug. 10, 1946; s. George Lawrence Ploetz and Marie Mary Buenning Cramer; m. Denise Chapnick, July 13, 1969; children: Kristin Lisa, Jeffrey. BS in Ceramic Sci., Alfred U., 1968, MS in Ceramic Sci., 1975. Mfg. engr. Westinghouse Elec. Corp., Fairmont, W.Va., 1972-74, Zenith Radio Corp., Lansdale, Pa., 1975; chief engr. Wheaton Glass Co., Millville, N.J., 1975-85; sr. scientist Owens Corning, Granville, Ohio, 1985—; tech. dir. Makintech Corp., Cleve., 1987-92. Pres. Christ Luth. Ch., Heath, Ohio, 1990-91, choir mem., 1990—. Capt. U.S. Army, 1968-71. Decorated Army Commendation medal. Mem. Am. Chem. Soc. Republican. Achievements include development of pull control devices, oxygen burner; perfection of electric melting of neutral biosilicate tubing glass NSV. Home: 843 Jonathan Ln Newark OH 43055 Office: Owens-Corning Tech Ctr 2790 Columbus Rd Rt 16 Granville OH 43023-1200

PLOMMET, MICHEL GEORGES, microbiologist, researcher; b. St. Fons, Rhone, France, Sept. 24, 1927; s. Jacques Marie and Georgette Emelie (Andriot) P.; m. Anne-Marie Laurent, May ll, 1954. DVM, Maisons Alfort, Paris, 195l; Ingenieur in dairy tech., Nat. Inst. Agronomy, Paris, 1953; postgrad. in microbiology, Inst. Pasteur, Paris, 1954. Rsch. asst. Nat. Inst. Agronomy Rsch., Jouy, France, 1953-57, in-charge rsch., 1957-63; instr. rsch. Nat. Inst. Agronomy Rsch., Tours-Nouzilly, France, 1963-66, dir. rsch., 1966—, dir. lab., 1966-86, adminstr., 1968-72; chmn. microbiology com. Nat. Inst. Agronomy Rsch., Paris, 198l-84; expert on brucellosis WHO, Geneva, 1975—; chmn. sci. com. Fedn. Nat. Groupments Def. Sanitaire, Paris, 1979-87. Editor Prevention of Brucellosis in the Mediterranean Countries, 1992; co-editor: Brucella Melitensis, 1985; contbr. over 200 articles to profl. jours. Lt. Vet. Svc., French Army, 1950-5l. Mem. French Soc. Microbiology (adminstrv. bd. 1980-89), AAAS, N.Y. Acad. Scis. Roman Catholic. Home: 7 Rue de Bagatelle, 37540 Saint Cyr sur Loire France Office: Nat Inst Agronomy Rsch, 37380 Nouzilly France

PLOSSER, CHARLES IRVING, economics educator; b. Birmingham, Ala., Sept. 19, 1948; s. George Gray and Dorothy (Irving) P.; m. Janet Schwert, June 26, 1976; children: Matthew, Kevin, Allison. B.E. cum laude, Vanderbilt U., 1970; MBA, U. Chgo., 1972, PhD, 1976. Cons. Citicorp Realty Cons., N.Y.C., 1972-73; lectr. Grad. Sch. Bus., U. Chgo., 1975-76; asst. prof. Grad. Sch. Bus. Stanford (Calif.) U., 1976-78; asst. prof. econs. W.E. Simon Grad. Sch. Bus., U. Rochester (N.Y.), 1978-82, assoc. prof., 1982-86, prof., 1986-89; Fred H. Gowen prof. econs. U. Rochester, N.Y., 1989-92, John M. Olin Disting. prof. econs. and pub. policy, 1992—, acting dean W.E. Simon Grad. Sch. Bus., 1990-91, 92—. Editor, Jour. Monetary Econs., 1983—, Carnegie-Rochester Conference Series on Public Policy, 1989—; contbr. articles to profl. jours. 1st lt., U.S. Army, 1972-73. NSF research grantee, 1982, 84. Mem. Am. Econs. Assn., Econometrics Soc., Am. Fin. Assn., Am. Statis. Assn., Tau Beta Pi, Beta Gamma Sigma. Home: 95 Ambassador Dr Rochester NY 14610-3402 Office: U Rochester Dean of Simon Grad Sch Bus Rochester NY 14627

PLOTSKY, PAUL MITCHELL, neuroscientist, educator; b. Kansas City, Mo., Feb. 1, 1952; s. Herbert L. and Francis P. (Kern) P.; m. Andrea Gayle Halpern, Aug. 22, 1971; children: Melissa Michelle, Alyson Rose. BA, U. Kans., 1974; PhD, Emory U., 1981. Asst. prof. The Salk Inst., La Jolla, Calif., 1982-88; assoc. prof. The Salk Inst., La Jolla, 1988-92; prof. Emory U., Atlanta, 1992—; adv. bd. Integrated Rsch. & Info. Svcs., Atlanta, 1986—; ad hoc reviewer NIH, Bethesda, 1986—. Author: Neurobiology & Neuroendocrinology of Stress, 1991; editorial bd.: Endocrinology, Iowa City, 1986-89; contbr. articles to Biol. Psychiatry, Jour. Neuroendocrinology, Sci. Advisor neuroscience com. San Diego (Calif.) High Sch., 1992. Recipient Mellon Found. Faculty Scholars award Mellon Found./The Salk Inst., La Jolla, 1982-85, Martyn Jones Meml. Lectureship and medal Martyn Jones Trust, Bristol, England, 1989. Mem. Internat. Soc. Neuroendocrinology, The Endocrine Soc., Soc. for Neuroscience. Achievements include first direct demonstration of stimulus-specific and multifactor regulation of adrenocorticotropin secretion. Office: Emory Univ Dept Psychiatry 1639 Pierce Dr PO Drawer AF Atlanta GA 30322

PLOUFFE, LEO, JR., endocrinologist; b. Montreal, Que., Can., Nov. 26, 1957; s. Leo and Yvette (LaFontaine) P.; m. Evelyn Woodward White, Sept. 22, 1990. Diploma Collegiate Studies, Ahuntsic Coll., Montreal, 1975; MD, McGill U., Montreal, 1980. Diplomate gen. ob-gyn. and reproductive endocrinology. Resident in ob-gyn. McGill U., Montreal, 1980-85; fellow in reproductive endocrinology Med. Coll. Ga., Augusta, 1985-87, asst. prof., 1987-92, assoc. prof. Endocrinology, 1992—. Editor: (book chpt.) Comprehensive Management of Menopause, 1993; assoc. editor Jour. Pediatric & Adolescent Gynecology, 1987—. Fellow Am. Coll. of Ob-gyn., AAAS; mem. Am. Fertility Soc., Ga. Genetics Soc. (sec.-treas. 1992-93). Achievements include demonstration of homebox gene expression in ndometrium, techniques for screening for sexual problems. Office: Med Coll Ga Dept OB-Gyn Augusta GA 30912

PLOWMAN, R. DEAN, federal agriculture agency administrator; b. Smithfield, Utah, Aug. 25, 1928; s. Ronald O. and LaRue B. Plowman; m. Kathleen Simmons, 1950; children: Kenneth, Stephen, Maurine, Douglas, Anne. BS in Agr., Utah State U., 1951; MS, U. Minn., St. Paul, 1955, PhD, 1956. Rsch. asst. U. Minn., 1954-56; head dept. animal dairy and vet. scis. Utah State U., Logan, 1984-88; dairy rsch. scientist Agrl. Rsch. Svc., USDA, Beltsville, Md., 1957-63, investigations leader dairy rsch. bd., 1963-68, chief dairy cattle rsch. br., 1968-72; area dir. for Utah, Idaho and Mont. USDA, Logan, Utah, 1972-80, area dir. for Utah, Ariz., Nev. and N.Mex., 1980-83, area dir. for Utah, Colo., Wyo., Nev., Ariz. and N.Mex., 1983-84; adminstr. USDA, Washington, 1988-92, asst. sec. agrl. for sci. and edn., 1992—. Contbr. over 75 articles to sci. jours. Sgt. U.S. Army, 1946-47. Recipient Disting. Svc. award USDA, 1991, Alumnus of Yr. award Utah State U., 1993. Office: USDA Agrl Rsch Svc 12th & Independence Ave SW Washington DC 20250*

PLUCKNETT, DONALD LOVELLE, scientific advisor; b. DeWitt, Nebr., Sept. 9, 1931; s. William Donald and Phyllis Corrine (Barkey) P.; m. Ida Sue Richards, May 14, 1955; children—Karen, Roy, Duane. B.S., U. Nebr., 1953, M.S., 1957; Ph.D., U. Hawaii, 1961. Cert. profl. soil scientist, profl. agronomist. Grad. asst. U. Hawaii, Honolulu, 1958-60, instr., 1960-61, asst. prof., 1961-65, assoc. prof., 1965-70, prof., 1970-80; Ford Found. sr. agrl. cons. Egypt-Aswan Agrl. Devel., 1965; chief soil and water mgmt. div. Office Agr. AID, 1973-76; dep. exec. dir. Bd. Internat. Food and Agrl. Devel., Washington, 1978-79; chief agrl. and rural devel. div. Asia bur. AID, 1979-80, sci. advisor consultative group on internat. agrl. research World Bank, 1980—; cons. in field. Author: Common Weeds of the Philippines, 1969, The World's Worst Weeds, 1977, Farming Systems Research at the International Agricultural Research Centers, 1978, A Geographic Atlas of World Weeds, 1979, Managing Pastures and Cattle Under Coconuts, 1979, Azolla as a Green Manure, 1982, Gene Banks and the World's Food, 1987; (free verse) The Roof Only Leaked When It Rained, 1985; editor: Vegetable Farming Systems in China, 1981, Small Scale Processing and Storage of Tropical Root Crops, 1979; (series) Westview Tropical Agricultural, Detecting Mineral Nutrient Deficiencies in Tropical and Temperate Crops, 1989; contbr. articles to profl. jours. Served to 1st lt. U.S. Army, 1953-55. Recipient Superior Honor award AID, 1976; NSF fellow, 1960. Fellow AAAS, Am. Soc. Agronomy, Soil Sci. Soc. Am., Crop Sci. Soc. Am.; mem. Internat. Soc. Tropical Root Crops (hon. life, pres. 1976-79, 79-83), Asian Pacific Weed Sci. Soc. (sec. 1969-70), 73-81), Indian Soc. Root Crops, Soc. for Econ. Botany, Hawaiian Acad. Sci. Methodist. Home: 4205 St Jerome Dr Annandale VA 22003-3714 Office: 1818 H St NW Washington DC 20433-0002

PLUMMER, DIRK ARNOLD, electrical engineer; b. Stamford, Conn., Apr. 18, 1930; s. Charles Arnold Plummer and Edwina Woodling Johnson; m. Janis Susan Lowery Stuart, Feb. 18, 1967 (div. 1973); 1 child, Julie. B-SChEngr, MIT, 1952; BSEE, U. Calif., Berkeley, 1961; MSEE, Monmouth Coll., 1985. Cert. nondestructive test examiner of inspectors for radiography, magnetic particle, liquid penetrant and ultrasonic testing methods; cert. comml. pilot. Chem. engr. Foster Wheeler Corp., N.Y.C., 1952; engr. The M.W. Kellogg Co., N.Y.C., 1954; project engr. Am. Machine & Fdry. Co., Greenwich, Conn., 1955-56; devel. engr. Aerojet-Gen. Corp., Azusa, Sacramento, San Ramon, Calif., 1956-61; sr. mem. tech. staff Aerospace Communication & Controls Div. RCA, Burlington, Mass., 1961-62; engr. Elec. Boat Div. Gen. Dynamics Corp., Groton, Conn., 1963; electronics engr. U.S. Civil Svc., various locations, 1963-88; pvt. practice profl. engring. Sea Bright, N.J., 1988—. Contbr. articles to profl. jours. Archtl. control officer Sea Bright Village Assn., 1991. 1st lt. U.S. Army, 1952-54. Recipient Meritorious Svc. medal Pres. of U.S., 1982, Cert. for Commendable Svc. Def. Supply Agy., 1972. Mem. AAAS, AICE (profl. devel. officer 1990), IEEE (chmn. nuclear and plasma sci. chpt. 1990), Am. Phys. Soc., Math Assn. Am. Home and Office: 45 Village Rd Sea Bright NJ 07760

PLUMMER, ERNEST LOCKHART, industrial research chemist; b. Buffalo, Mar. 1, 1940; s. Ernest William and Katherine Swan (Lockhart) P.; m. Marjorie Joan Baxter, June 23, 1962; children: Marjorie Elizabeth, Karen Margaret Fisk. BA in Chemistry, U. Rochester, 1962; MS in Chemistry, U. Dayton, 1966; PhD in Chemistry, SUNY, Buffalo, 1974. Rsch. chemist FMC Corp., Princeton, N.J., 1974-76; rsch. chemist FMC Corp., Middleport, N.Y., 1976-82; rsch. assoc. FMC Corp., Princeton, 1982-84; rsch. fellow FMC Corp., 1984-92, dir. computational and analytical sci. dept., 1992—. Contbr. articles to profl. jours.; patentee in field. Capt. USAF, 1962-69. mem. Am. Chem. Soc. Office: FMC Corp PO Box 8 Princeton NJ 08543-0008

PLUMMER, GAYTHER L(YNN), climatologist, ecologist, researcher; b. Indpls., Jan. 27, 1925; s. Conley L. and Rowena H. (Huber) P.; m. H. Eileen Barr, June 3, 1950. BS, Butler U., 1948; MS, Kans. State U., 1950; PhD, Purdue U., 1954. Instr. biology Knox Coll., Galesburg, Ill., 1950-51; naturalist Ind. Dept. Conservation, various locations, 1947-52; asst. prof. biology Antioch Coll., Yellow Springs, Ohio, 1954-55; prof. botany U. Ga., Athens, 1955—, state climatologist, 1978—; rsch. fellow Oak Ridge (Tenn.) Inst. Nuclear Studies, 1958-62. Author: Georgia Weather Watchers, 1991, Georgia Temperatures, 1993; cartographer 160 vegetation maps of Ga., 1972-74; editor Ga. Jour. Sci., 1977-84; author over 200 rsch. reports. 2d lt. USAAF, 1943-46. Fellow AAAS; mem. Ecol. Soc. Am., Ind. Acad. Sci., Ga. Acad. Sci., Soil Sci. Soc. Am., Crop Sci. Soc., Agron. Soc. Am., Sigma Xi, Phi Kappa Phi. Achievements include research in droughts in southeast U.S. relating to astrogeophysical processes via geomagnetics, lightning history in Piedmont for over 70 million years etched in Stone Mountain granite. Office: Inst Natural Resources Univ Ga Athens GA 30602

PLUMMER, JAMES D., electrical engineering educator. Fluke profl elec. engring. Stanford (Calif.) U. Fellow IEEE. Office: Stanford U Dept of Elec Engring Stanford CA 94305

PLUMMER, JAMES WALTER, engineering company executive; b. Idaho Springs, Colo., Jan. 29, 1920. BS, U. Calif., Berkeley, 1942; MS, U. Md., 1953. V.p., asst. gen. mgr. space system Lockheed Missiles & Space Co., Inc., 1965-68, v.p., asst. gen. mgr. rsch. and develop., 1968-69, v.p., gen. mgr. space system, 1969-73, exec. pres., 1976-83; undersec. USAF, 1973-76; ret., 1983; trustee Aerospace Corp., El Segundo, Calif., 1983-85, chmn. bd. trustees, 1985—. Fellow AIAA (Robert H. Goddard Astronautics award 1992), Am. Astronautical Soc.; mem. Nat. Acad. Engring. (mem. space applied bd. 1978-83). Office: Aerospace Corp 2350 E El Segundo Blvd El Segundo CA 90245-4691*

PLUMSTEAD, WILLIAM CHARLES, testing engineer, consultant; b. Two Rivers, Wis., Nov. 2, 1938; m. Peggy Bass, July 19, 1959 (div. July 1968); children: Kevin, Keith, William Jr., Jennifer; m. Vicki Newton, June 27, 1981. Student, Temple U., 1966-72, Albright Coll., 1973-75; BSBA, Calif. Coast U., 1985, MBA, 1989. Registered profl. engr., Calif. V.p. U.S. Testing Co., Inc., Hoboken, N.J., 1963-76; div. mgr. Daniel Internat., Inc., Greenville, S.C., 1976-83; group mgr. Bechtel Group, Inc., San Francisco, 1983-89; prin. engr. Flour Daniel, Inc., Greenville, 1989—. Author: (with others) Code/Specification Syndrome, 1976, NDT Laboratories Update, 1991, NDT in Construction, 1991; contbr. articles to profl. jours. Fellow Am. Soc. for Nondestructive Testing (coun. chmn. 1985-88, nat. sec./treas. 1992-93, nat. v.p. 1993-94); mem. ASTM (sec. 1989-93), Toastmasters Internat. (pres. local chpt. 1990-91, Competent Toastmaster award 1986, Able Toastmaster award 1993). Avocations: sports, wine tasting. Home: 806 Botany Rd Greenville SC 29615-1608 Office: Flour Daniel Inc 100 Flour Daniel Dr Greenville SC 29607-2762

PLUNKETT, ROY J., retired chemical engineer. AB in Chemistry, Manchester Coll., 1932, DSc (hon.); PhD in Chemistry, Ohio State U., 1936, DSc; DSc, Washington Coll. With dept. organic chem. Jackson Lab. E. I. DuPont de Nemours & Co., Deepwater, N.J., 1936-39, chem. supr. chamber works, 1939-45, supt., 1945-49, supt. Ponsol Colors area, 1949-50, asst. mgr. works, 1950-52; mgr. chem. devel. section dept. organic chems. E. I. DuPont de Nemours & Co., Wilmington, Del., 1952-54, asst. mgr. tech. sect., 1954-57, mgr., 1957-60, dir. rsch. divsn. freon products, 1960-70, dir. ops., 1970-75. Recipient award City of Phila., Soc. Plastics Industry, Nat. Assn. Mfrs., Moissan medal Chem. Soc. France, Holley Medal ASME, 1990; named to Plastics Hall of Fame, 1973, Nat. Inventors Hall of Fame, 1985. Mem. AAAS, AICE, Am. Chem. Soc., Am. Inst. Chemists. Achievements include discovery of Teflon and research in fluorocarbon regrigerants. Home: 14113 Jackfish Ave Corpus Christi TX 78418*

PLUTH, JOSEPH JOHN, chemist, consultant; b. Cook, Minn., Feb. 18, 1943; s. Albert Matthew and Frances Rose (Mavetz) P.; m. Lucy Julie Corsetti, Aug. 25, 1973; children: Sara, Matthew, David. BA, Bemidji State Coll., 1965; PhD, U. Wash., 1971. X-ray crystallographer U. Chgo., 1972-81, sr. rsch. assoc., 1981—; ind. cons., Chgo., 1980—. Contbr. articles to sci. publs. Chmn. Richton Park (Ill.) Planning Commn., 1980-93. Mem. Am. Chem. Soc., Am. Crystallographic Soc., Sigma Xi. Achievements include research on structural properties of zeolites and other molecular sieves, structures using synchrotron radiation. Home: 22530 Lawndale Ave Richton Park IL 60471 Office: Univ Chgo 5734 S Ellis St Chicago IL 60637

PNIAKOWSKI, ANDREW FRANK, structural engineer; b. Grodno, Poland, Aug. 18, 1930; s. Josef Leon and Janina (Kodzynski) P.; Diploma Engr., Politechnika Warszawska, 1952; m. Margaret M. Czajkowski, Aug. 15, 1957; 1 dau., Mary. Bridge design and field engr. Govt. of Poland, Ministry of R.R., Warsaw, 1952-57; bridge design engr. Dept. Hwys., of Ont. (Can.), Toronto, 1958-66; sr. structural engr. Sverdrup & Parcel Assos. Inc., Boston, 1967-71; chief structural engr. Louis Berger & Assos. Inc., Waltham, Mass., 1972—; cons. engr. in transp., bridges, hwys., railroads, pub. bldgs., others. Registered profl. engr., Ont., Mass., Maine, N.H. Mem. Assn. Profl. Engrs. of Province Ont., NSPE. Roman Catholic.

POARCH, MARY HOPE EDMONDSON, science educator; b. Columbia, S.C., July 1, 1958; d. Homer Vincent and Janis (Bland) Edmondson; m. Robert Daniel Poarch, June 4, 1983; 1 child, Matthew Vincent. BS, Tex. A&M U., 1982; MEd, U. Tex., 1985. Sci. tchr. Evant (Tex.) Ind. Sch. Dist., 1982-83, Copperas Cove (Tex.) Ind. Sch. Dist., 1983-85, Northside Ind. Sch. Dist., San Antonio, 1985-90, Alamo Heights Ind. Sch. Dist., San Antonio, 1990—; mem. adv. bd. Merrill Pub. Co., 1989-90. Vol. San Antonio Zoo, 1991—. P.A.C.T. grantee Am. Chem. Soc., 1992. Mem. Sch. Sci. and Math. Assn., Nat. Sci. Tchrs. Assn. (Honors Tchr. award 1988), Coun. for Elem. Sci. Internat., Nat. Mid. Level Sci. Tchrs. Assn. Methodist. Home: 6114 Hopes Ferry San Antonio TX 78233 Office: Alamo Heights Jr Sch 7607 N New Braunfels San Antonio TX 78209

POBER, JORDAN S., pathologist, educator; b. N.Y.C., May 13, 1949; s. Irving and Ruth (Gardner) P.; m. Barbara Rose Herzog, June 6, 1971; children: Jeremy, Jonathan. BA, Haverford Coll., 1971; MD, Yale U., 1977, PhD, 1977. Diplomate Nat. Bd. Anatomic Pathology. Resident in pathology Yale-New Haven Hosp., 1977-78, Brigham and Women's Hosp., Boston, 1980-81; fellow in biochemistry Harvard U., Cambridge, Mass., 1978-80; asst. prof. pathology Med. Sch. Harvard U., Boston, 1981-86, assoc. prof. pathology Med. Sch., 1986-91; prof. pathology and immunobiology Yale U., New Haven, 1991—; dir. molecular cardiobiology Boyer Ctr. for Molecular Medicine, New Haven, 1991—; mem. pathology A study sect. NIH, Bethesda, Md., 1992—. Author textbook: Cellular and Molecular Immunology, 1991. Mem. Am. Assn. Pathologists (Warner-Lamber/Parke Davis award 1989). Office: Boyer Ctr Molecular Med 295 Congress Ave New Haven CT 06536

POCH, HERBERT EDWARD, pediatrician, educator; b. Elizabeth, N.J., Sept. 4, 1927; s. William and Min (Herman) P.; m. Leila Kosberg, Aug. 27, 1952; children: Bruce Jeffrey, Andrea Susan, Lesley Grace. AB, Columbia U., 1949, MD, 1953. Diplomate Am. Bd. Pediatrics. Intern Kings County Hosp. Ctr., Bklyn., 1953-54; resident Babies Hosp.-Columbia-Presbyn. Med. Ctr., N.Y.C., 1954-56; pvt. practice medicine specializing in pediatrics Elizabeth, 1956-92; chmn. dept. pediatrics, 1973-83; pres. med. staff, 1989, attending pediatrician Elizabeth Gen. Med. Ctr., 1973, sr. attending pediatrician, 1990; attending pediatrician St. Elizabeth Hosp., 1968, chmn. dept. pediatrics, 1971-81, attending pediatrician Monmouth Med. Ctr., 1991—, assoc. program dir. pediatrics; instr. pediatrics Columbia U., 1956-72, asst. clin. prof. pediatrics, 1972-91; honorary staff Elizabeth Gen. Med. Ctr., 1993—. With AUS, 1945-46. Fellow Am. Acad. Pediatrics; mem. N.J. Med. Soc. Address: 124 Chilton St Elizabeth NJ 07202

POCKER, YESHAYAU, chemistry, biochemistry educator; b. Kishinev, Romania, Oct. 10, 1928; came to U.S., 1961; naturalized, 1967.; s. Benzion Israel and Esther Sarah (Sudit) P.; m. Anna Goldenberg, Aug. 8, 1950; children: Rona, Elon I. MSc, Hebrew U., Jerusalem, 1949; PhD, Univ. Coll., London, Eng., 1953; DSc, U. London, 1960. Rsch. assoc. Weizmann Inst. Sci., Rehovot, Israel, 1949-50; humanitarian trust fellow Univ. Coll., 1951-52, asst. lectr., 1952-54, lectr., 1954-61; vis. assoc. prof. Ind. U., Bloomington, 1960-61; prof. U. Washington, Seattle, 1961—; bicentennial lectr. Mont. State U., Bozeman, 1976, U. N.C., Chapel Hill, 1977, guest lectr. U. Kyoto, Japan, 1984; Edward A. Doisy vis. prof. biochemistry St. Louis U. Med. Sch., 1990; plenary lectr. N.Y. Acad. Sci., 1983, Fast Reactions in Biol. Systems, Kyoto, Japan, 1984, NATO, 1989, Consiglio nat. delle Richerche, U. Bari, Italy, 1989, Sigma Tau, Spoleto, Italy, 1990; Internat. lectr. Purdue U., 1990; cons. NIH, 1984, 86, 88; Spl. Topic lectr. on photosynthesis, Leibniz House, Hanover, Fed. Republic Germany, 1991; enzymology, molecular biology lectr., Dublin, Ireland, 1992. Mem. editorial adv. bd. Inorganica Chimica Acta-Bioinorganic Chemistry, 1981-89; bd. reviewing editors Sci., 1985—; contbr. numerous articles to profl. jours.; pub. over 220 papers and 12 revs. Numerous awards worldwide, 1983-90. Fellow Royal Soc. Chemistry; mem. Am. Chem. Soc. (nat. speaker 1970, 74, 84, chmn. Pauling award com. 1978; plaque awards, 1970, 74, 84, outstanding svc. award 1979), Soc. Exptl. Biology, Am. Soc. Biol. Chemists, N.Y. Acad. Sci., Sigma Xi (nat. lectr. 1971). Avocations: Aramaic, etymology, history, philology, poetry. Office: U Wash Chemistry Dept BG-10 Seattle WA 98195

PODILA, GOPI KRISHNA, biochemistry and molecular biology educator; b. Paralakhemidi, Orissa, India, Sept. 14, 1957; came to U.S., 1981; s. Surya Prakasarao and Rama (Lakshmidevi) P.; m. Vani Ramadugu, Dec. 4, 1989. BSc, Nagarjuna U., India, 1974, MSc, 1978; MS, La. State U., 1983; PhD, Ind. State U., 1986. Grad. rsch. asst. La. State U., Baton Rouge, 1981-83; grad. teaching fellow Ind. State U., Terre Haute, 1983-86; postdoctoral rsch. assoc. Ohio State U. Biotech. Ctr., Columbus, 1987-90; asst. prof. biochemistry Mich. Tech. U., Houghton, 1990—; reviewer U.S. Dept. Agr. Nat. Rsch. Initiative Competitive Grants Program, Washington, 1991—. Bd. dirs. Keeweenaw Coop., Hancock, Mich., 1991—. Fellow Gamma Sigma Delta. Mem. AAAS, Am. Phytopath. Soc. (mem. biochem.-molecular biology com. 1992—), Internat. Soc. Plant-Microbe Interactions, Sigma Xi. Hindu. Achievements include first to develop an in vitro system to study fungal gene activation in isolated nuclei, antisense RNA technique to regulate lignin biosynthesis in woody plants; first to show involvement of tyrosine kinase in fungal gene regulation, pathogenicity genes in transgenic fungi. Office: Mich Tech U Dept Biol Sci 1400 Townsend Dr Houghton MI 49931

PODOLNY, WALTER, JR., structural engineer; b. Cleve., Oct. 23, 1929; s. Walter and Mary (Osowski) P.; m. Jean Marie Hoecker, Oct. 20, 1956; children: Michael Walter, Richard Albert, Laura Jean, John Joseph. BS in Structural Engring., Cleve. State U., 1952, BS in Civil Engring., 1953; MS in Civil Engring., Case Western Reserve U., 1960; PhD in Civil Engring., U. Pitts., 1971. Registered profl. engr., Ohio, Ind. Ky., Pa. Draftsman, designer Wilbur Watson Assocs., Cleve., 1952-53; jr. engr. Fischer & Assocs., Cleve., 1953-54; design engr. Fulton, DelaMotte, Larson, Nassau & Assocs., Cleve., 1956-59; chief engr. Geo. Rackle & Sons, Co., Cleve., 1959-61; chief engr. Cleve. precast concreted div. Cleve. Builder's Supply Co., 1961-63; ptnr. Levigne and Podolny Cons. Structural Engrs., Cleve., 1964-65; structural engr. Mktg. Tech. Svcs. U.S. Steel Corp., Pitts., Pa., 1965-71; structural engr. bridge div. office of engring. Fed. Hwy. Adminstrn., Washington, 1971—; Fed. Hwy. Adminstrn. advisor to Kuwait Ministry Pub. Works, 1979-87; mem. adv. panel Cornell U., NSF, to Nat. Arecibo Observatory in Puerto Rico.; mem. Gov.'s Bd. Inquiry Loma Prieta earthquake; advisor to Mayor Guatemala City, Guatemala, on Incenso bridge distress. Author: Construction and Design of Cable Stayed Bridges, 1976, rev. 2nd edit., 1986, Construction and Design of Prestressed Concrete Segmental Bridges, 1982; contbr. over 60 articles to tech. jours. Mem. Mayor's Urban Renewal Adv. Com., Cleve., 1965, Planning Bd., McCandless Twp., Allegheny County, Pa., 1968-71. Recipient Nat. Defense Fellowship award, 1970, Student award Am. Soc. for Testing and Materials, 1971, Outstanding Engring. Alumnus award, Cleve. State U., 1983, Silver medal, Sec. Transp., 1988; named Hwy. Adminstrn. Engr. of Yr., Nat. Soc. Profl. Engrs., 1983. Fellow ASCE (many offices including pres. Cleve. sect. 1956, Past Pres. award 1983, T.Y. Lin award 1986), Am. Concrete Inst. (fiber re-inforced concrete com. 1966-77); mem. Prestressed Concrete Inst. (mem. many coms. chmn. com. on segmental constrn. 1977-81), Reinforce Concrete Rsch. Coun. (exec. com. 1982-88), Transp.Rsch. Bd., Internat. Assn. for Bridge and Structural Engring., Post Tensioning Inst., Am. Segmental Bridge Inst., Fed. Internat. de La Precontrainte. Home: 9410 Raintree Rd Burke VA 22015-1946 Office: Fed Hwy Adminstrn Bridge Div HNG-32 400 7th St SW Washington DC 20590-0002

POEHLER, THEODORE OTTO, university provost, engineer, researcher; b. Balt., Oct. 20, 1935; s. Theodore O. and Marion E. (Rohde) P.; m. Anne Otter Evans, Dec. 30, 1961; children: Theodore, Jeffrey. BS, Johns Hopkins U., 1956, DEng, 1961. Mem. sr. staff Applied Physics Lab., Laurel, Md., 1963-68, prin. staff physicist, 1969-73, supr. Quantum Electrons Group, 1974-83; dir. Eisenhower Rsch. Ctr. Johns Hopkins U., Laurel, 1983-89; assoc. dean rsch. Sch. Engring. Johns Hopkins U., Balt., 1990-92, vice provost for rsch., 1992—; chmn. bd. dirs. Tech. Devel. Ctr., Balt., 1991—. Author: (with others) Detectors in Methods of Experimental Physics, 1990; contbr. more than 130 articles to profl. jours. including Phys. Rev., Applied Physics. Capt. U.S. Army, 1962-63. Recipient Nat. Capital award Coun. Engring. and Archtl. Socs. Mem. AAAS, IEEE, Am. Phys. Soc., Am. Chem. Soc., Materials Rsch. Soc. Achievements include 8 patents in optical information storage field. Office: Johns Hopkins U 34th and Charles Sts Baltimore MD 21218

POEHLMANN, CARL JOHN, agronomist, researcher; b. Calif., Mo., Jan. 29, 1950; s. Edwin William and Lucille Albina (Neu) P.; m. Linda Kay Garner, Dec. 29, 1973; children: Anthony, Kimberly. BS, U. Mo., 1972, MS, 1978. Farmer Jamestown, Mo., 1972-73; vocat. agrl. tchr. Linn (Mo.) Pub. Schs., 1973-75, Columbia (Mo.) Pub. Schs., 1975-78; dir., mgr. agronomy rsch. ctr. U. Mo., Columbia, 1978—. Mem. Am. Soc. Agronomy (div. A-7 chair 1985-86, bd. mem. 1991-94, cert. crop advisor 1993), Crop Sci. Soc. Am., Soil Sci. Soc. Am. Mem. Christian Ch. (Disciples of Christ). Office: U Mo 4968 S Rangeline Rd Columbia MO 65201-8973

POEHLMANN, GERHARD MANFRED, cartography educator; b. Gotha, Thuringia, Germany, Nov. 28, 1924; s. Karl Gustav and Elsa (Huschke) P.; m. Ilse Dora Schmidt, Dec. 2, 1945; children: Christine, Angelika. Degree in Engring., Acad. for Architecture, Berlin, 1952; postgrad., ETH & U. Zürich, 1959; doctors degree, Free U. Berlin, 1974. Cert. cartography prof. Map editor, prodn. mgr. Justus Perthes Geographical Pub., Gotha, Darmstadt, Fed. Republic Germany, 1947-56; asst. head cartographer Prof. Imhof Cartographic Inst., Zürich, 1956-59; editor-in-chief Mair's Geographical Pub. House, Stuttgart, Fed. Republic Germany, 1959-62; lectr., prof. Technische Fachhochscule Berlin, 1962-88, dean faculty, 1981-84; dir. Lab. Map Compilation & Design, Berlin, 1970-88; scientist, adviser various gov. positions, 1965—; dir. DGFK Working Group Thematic Cartography, Berlin, 1979-87, Spl. Rsch. Project 69 Subproject D1, Berlin, 1981-87; consulting Nat. Remote Sensing Agy., Hyderabad, India, 1982-85; project mgr. Gen. Petroleum Co., Cairo, 1982-88; guest lectr. on cartography T.U. Zagreb Geodetski Fakultet, Yugoslavia, 1965, 87; advisor desert rsch. Academia Sinica China, 1986; adviser Environ. Protection Xinjiang Inst., China, 1987; adviser mapping in the 1990s Kaduna Poly., Nigeria, 1989; Asian devel. bank cons. in cartography Geol. Survey Pakistan, 1991-92. Editor: Berliner Geowiss Abhandlungen, 1980-88; author numerous maps including Geol. Map Egypt, 1982; inventor Modular Map Prodn. System, 1982. Lt. German Army, 1942-45, WWII. Mem. Deutsche Gesellschaft für Kartographie, Gesellschaft für Erdkunde zu Berlin, Geographische Gesellschaft Zürich, Schweizerische Gesellschaft für Kartographie. Home and Office: Cimbernstrasse 11/i, D-14129 Berlin Germany

POETTMANN, FREDERICK HEINZ, retired petroleum engineering educator; b. Germany, Dec. 20, 1919; s. Fritz and Kate (Hussen) P.; m. Anna Bell Hall, May 29, 1952; children—Susan Trudy, Phillip Mark. B.S., Case Western Res. U., 1942; M.S., U. Mich., 1944, Sc.D., 1946; grad., Advanced Mgmt. Program, Harvard U., 1966; PhD in Mining Scis. (hon.), Mining U. Leoben, Austria, 1992. Registered profl. engr., Colo., Okla. Research chemist Lubrizol Corp., Wickliffe, Ohio, 1942-43; mgr. production research Phillips Petroleum Co., Bartlesville, Okla., 1946-55; asso. research dir. Marathon Oil Co., 1955-72; mgr. comml. devel. Marathon Oil Co., Littleton, Colo., 1972-83; prof. petroleum engineering Colo. Sch. Mines, 1983-90; retired; cons. in field. Contbr. articles to numerous publs.; co-author, editor 9 books in field; patentee in field. Chmn. S. Suburban Met. Recreation and Park Dist., 1966-71; chmn. Littleton Press Council, 1967-71; bd. dirs. Hancock Recreation Center, Findlay, Ohio, 1973-77. Recipient Charles F. Rand Meml. Gold medal of AIME, 1992, Katz medal Gas Processors Assn., 1993. Mem. Nat. Acad. Engring., Soc. Petroleum Engrs.(DeGolyer Disting. Svc. medal, 1990), Am. Inst. Chem. Engring., Am. Chem. Soc., Am. Petroleum Inst., Sigma Xi, Tau Beta Pi, Alpha Chi Sigma, Phi Kappa Phi, Pi Epsilon Tau. Republican. Home: 47 S Eagle Dr Littleton CO 80123-6644 Office: Colo Sch Mines Dept Petroleum Engring Golden CO 80401

POGGIO, TOMASO ARMANDO, physicist, educator, computer scientist, researcher; b. Genoa, Italy, Sept. 11, 1947; came to U.S., 1981; s. Angelo and Maria Adele (Moro) P.; m. Barbara Venturini-Guerrini, July 29, 1972; children: Martino, Allegra. D in Physics summa cum laude, U. Genoa, 1971. Researcher Max Planck Inst., Federal Republic of Germany, 1971-82; assoc. prof. to prof. MIT, Cambridge, 1982-88, Uncas and Helen Whitaker prof., dept. brain and cognitive scis., 1988—; corp. fellow Thinking Machines Corp., Cambridge, 1985—; adv. bd. MIT/Bradford Press series on Computational Models of Cognition and Perception, 1984, VNY Sci. Press series of monographs in neuroinformatics and robotics, 1984. Contbr. over 100 sci. papers to profl. publs.; assoc. editor Systems and Control Letters, 1984 mem. editorial bd. Synapse, 1986, Visual Neurosci., 1987, Neural Computation, 1988, Network, 1989, others; adv. bd. Jour. Math. Biology, 1977, Lecture Notes in Biomathematics, 1979. Recipient Cassa di Risparmio of Genoa award, 1966, Otto-Hahn-Medaille award Max-Planck Soc., 1979, Columbus prize Instituto Internazionale delle Communicazioni Genoa, 1982, Premio Luigi Carlo Rossi award (with V. Torre) Elsag Elettronica, Italy, 1984, Max-Planck-Forschungs Preis Alexander von Humboldt-Stiftung, 1992; Angelo delle Riccia grad. fellow, 1969-70, fellow CNR Lab. Biophysics and Cybernetics, Italy, 1971. Mem. IEEE, Am. Math. Soc., Optical Soc. Am. Office: MIT Artificial Intelligence Lab Cambridge MA 02139

POGLAZOV, BORIS FEDOROVICH, biochemist, researcher, administrator; b. Verkholensk, Irkutsk, Russia, Mar. 23, 1930; s. Fedor Ivanovich Poglazov and Evdokia Ivanovna Zotina; m. Margarita Nikolaevna Shalnova, July 13, 1956 (div. June 1991); children: Alexander B. Poglazov, Tatjana B. Panjukova; m. Irina Yurievna Suvorova, May 13, 1993. Rsch. student, Moscow State U., 1949-54; MS, A.N. Bach Inst. Biochemistry, Moscow, 1958; PhD, Moscow State U., 1966. Jr. rschr. Inst. Molecular Biology, Moscow, 1958-65; head dept. Moscow State U., 1965-77; head of lab., dir. Inst. Biochemistry A.N. Bach Inst. Biochemistry-Russian Acad. Acis., Moscow, 1977—, dir., 1988—; lab. head Moscow State U., 1975—. Author 8 books including Structure and Functions of Contractile Proteins, 1966; contbr. over 200 publs. to profl. jours.; chief editor jours. Advances in Biol. Chemistry, 1981—, Applied Biochemistry and Microbiology, 1993—. Recipient A.N. Bach prize in biochemistry Presidium of USSR, Moscow, 1978. Mem. Russian Acad. Sci. (corr.), Russian Acad. Natural Scis. (academician). Achievements include discovery of proteins actin and myosin in non-muscle cells and tissues; established principles of the T4 phage particles assembly. Office: Bakh Inst Biochemistry, Leninskiy Prospekt 33, 117071 Moscow Russia

POGORELSKY, IGOR VLADISLAV, laser physicist; b. Moscow, May 22, 1946; came to U.S., 1989; s. Vladislav V. and Katia A. (Kotlar) P.; m. Bella I. Khazanov, Feb. 5, 1970; children: Jan, Ilya. PhD, Lebedev Physics Inst., Moscow, 1979. Rsch. physicist Lebedev Physics Inst., 1970-79; sr. scientist Cen. Inst. Engring. Materials, Moscow, 1980-88, STI Optronics, Seattle, 1990-93; sr. assoc. Brookhaven Nat. Lab., Upton, N.Y., 1990—; presenter workshops on advanced accelerator concepts. Mem. Internat. Soc. Optical Engring., Optical Soc. Am. Achievements inlcude development of first xenon-oxide laser with explosive optical pumping, 4 patents on laser technology, research on E-Beam laser acceleration. Home: 1 Sorrent Ct Miller Place NY 11764 Office: Brookhaven Nat Lab Upton NY 11764

POGUE, WILLIAM REID, former astronaut, foundation executive, business and aerospace consultant; b. Okemah, Okla., Jan. 23, 1930; s. Alex W. and Margaret (McDow) P.; m. Jean Ann Pogue; children: William Richard, Layna Sue, Thomas Reid. B.S. in Secondary Edn., Okla. Bapt. U., 1951, D.Sc. (hon.), 1974; M.S. in Math., Okla. State U., 1960. Commd. 2d lt. USAF, 1952, advanced through grades to col., 1973; combat fighter pilot Korea, 1953; gunnery instr. Luke AFB, Ariz., 1954; mem. acrobatic team USAF Thunderbirds, Luke AFB and Nellis AFB, Nev., 1955-57; asst. prof. math. USAF Acad., 1960-63; exchange test pilot Brit. Royal Aircraft Establishment, Ministry Aviation, Farnborough, Eng., 1964-65; instr. USAF Aerospace Research Pilots Sch., Edwards AFB, Calif., 1965-66; astronaut NASA Manned Spacecraft Center, Houston, 1966-75; pilot 3d manned visit to Skylab space sta.; now with Vutara Services of Springdale, Ark. Decorated Air medal with oak leaf cluster, Air Force Commendation medal, D.S.M. USAF; named to Five Civilized Tribes Hall of Fame, Choctaw descent; recipient Distinguished Service medal NASA, Collier trophy Nat. Aero. Assn.; Robert H. Goddard medal Nat. Space Club; Gen. Thomas D. White USAF Space Trophy Nat. Geog. Soc.; Halley Astronautics award, 1975; de la Vaalx medal Fedn. Aeronautique Internat., 1974; V.M. Komarov diploma, 1974. Fellow Acad. Arts and Scis. of Okla. State U., Am. Astron. Soc.; mem. Soc. Exptl. Test Pilots, Explorers Club, Sigma Xi, Pi Mu Epsilon. Baptist (deacon). Office: Vutara Services 1101 S Old Missouri Rd Ste 30 Springdale AR 72764

POHL, JOHN JOSEPH, JR., retired mechanical engineer; b. Newport News, Dec. 4, 1927; s. John Joseph and Christine (Gentile) P.; m. Irene Lewis, June 30, 1951 (dec.); 1 child, John Joseph. BSME, Va. Poly. Inst & State U., 1951. Staff trainee Newport News Shipbuilding Co., 1951-55, nuclear engr., 1955-59, group leader, 1959-88, prog. security officer in research/devel., 1988-92. Contbr. articles to profl. jours. Official Spl. Olympics, 1985—. 1st lt. U.S. Army, 1946-47, 51-53. Named Engr. of the Yr., Peninsula Engrs. Council, 1988. Mem. ASME, Soc. Naval Architects and Marine Engrs., Propeller Club of U.S., Am. Legion, K.C. Home: 7 Stanton Ct Newport News VA 23606-2925 Office: Newport News Shipbuilding 4101 Washington Ave Newport News VA 23607-2734

POHL, ROBERT OTTO, physics educator; b. Gottingen, Fed. Republic of Germany, Dec. 17, 1929; came to U.S., 1958; s. Robert Wichard and Auguste Elenore (Madelung) P.; m. Karin Ursula Koehler, May 6, 1961; children—Helen M., Robert S., Otto C. Vordiplom, U. Freiburg, Fed. Rep. Germany, 1951; diploma, U. Erlangen, Fed. Rep. Germany, 1955, Dr. rer. nat., 1957. Asst. U. Erlangen, 1957-58; research assoc. Cornell U., Ithaca, NY, 1958-60, asst. prof., 1960-63, assoc. prof., 1963-68, prof., 1968—; vis. prof. Tech. Hochschule Stuttgart, 1966-67, Tech. U. Munchen, 1973-74, Konstanz U., Regensburg U., 1987-88, all Fed. Republic Germany; vis. scientist Nuclear Research Ctr., Juelich, Fed. Rep. Germany, 1980-81. Contbr. articles on solid state physics to profl. jours. Recipient Sr. Scientist award Alexander von Humboldt Found., 1980; Guggenheim Found. fellow, 1973, Erskine fellow U. Canterbury, New Zealand, 1988. Fellow Am. Inst. Physics (O.E. Buckley award 1985), AAAS. Office: Cornell U Physics Dept Ithaca NY 14853-2501

POHLAND, FREDERICK GEORGE, environmental engineering educator, researcher; b. Oconomowoc, Wis., May 3, 1931; s. Arnold Ernest and Eda Karoline (Petermann) P.; m. Virginia Ruth Simmons, Sept. 10, 1966; 1 child, Elizabeth Eda. BS in Civil Engring., Valparaiso U., 1953; MS in Civil Engring., Purdue U., 1958, PhD, 1961. Profl. engr.; diplomate Am. Acad. Environ. Engrs. Civil engr. Erie Railroad Co., Huntington, Ind., 1953; preventive medicine specialist U.S. Army, Ft. Bragg, N.C., 1953-56; grad. rsch. asst. Purdue U., West Lafayette, Ind., 1956-61; asst. prof. Ga. Inst. Tech., Atlanta, 1961-64, assoc. prof., 1964-71, prof., 1971-88; Weidlein prof. U. Pitts., 1989—; vis. scholar U. Mich., Ann Arbor, 1967-68; guest prof. Delft Univ. Tech., The Netherlands, 1976-77; mem. EPA Sci. Adv. Bd., Washington, 1989—; mem. sci. adv. com. Gulf Coast Hazardous Substance Rsch. Ctr., Beaumont, Tex., 1989—; mem. EPRI Sci. Adv. Com., Palo Alto, Calif., 1990—; mem. adv. coun. Purdue U., West Lafayette, 1990—. Author: Emerging Technologies in Hazardous Waste Management, 1990, 91, 93, Design of Anaerobic Processes for the Treatment of Industrial and Municipal Waste, 1992; editor (jour.) Water Rsch., 1983—; author over 100 publs. in field. Served with U.S. Army, 1953-56. Recipient Harrison Prescott Eddy medal Water Pollution Control Fedn., 1964, Charles Alvin Emerson medal, 1983, Gordon Maskew Fair medal, 1989; recipient Rsch. award Water Pollution Control Assn. Pa., 1991. Fellow Am. Soc. Civil Engrs.; mem. Am. Acad. Environ. Engrs. (diplomate; pres. 1992-93), Assn. Environ. Engring. Profs. (sec.-treas. 1970-71, Disting. lectr. 1992), Solid Waste Assn. N.Am. (Lawrence lectr. 1992), Am. Water Works Assn. (life mem.), Nat. Acad. Engring., Am. Chem. Soc., Am. Inst. Chem. Engrs., Am. Soc. for Microbiology, NSPE, Ga. Soc. Profl. Engrs., Internat. Water Quality Assn., Pa. Soc. Profl. Engrs., Pa. Water & Pollution Control Assn., Sigma Xi, Tau Beta Pi, Chi Epsilon, others. Achievements include major contributions to phase separation in anaerobic treatment processes; originated concept of leachate recirculation for accelerated stabilization in landfill bioreactors. Home: 118 Millstone Ln Pittsburgh PA 15238 Office: Univ of Pittsburgh Environmental Engineering Pittsburgh PA 15260

POHOST, GERALD M., cardiologist, medical educator; b. Washington, Oct. 27, 1941; married; 3 children. BS, George Washington U., 1963; MD, U. Md., 1967. Diplomate Am. Bd. Internal Medicine, Am. Bd. Cardiovascular Disease, Am. Bd. Nuclear Medicine. Intern Montefiore Hosp. & Med. Ctr., Bronx, N.Y., 1967-68, asst. resident, 1968-69; sr. resident Jacobi Hosp. Albert Einstein Coll. Medicine, Bronx, 1969-70; cardiology resident Montefiore Hosp. & Med. Ctr.; clin. & rsch. fellow in medicine Mass. Gen. Hosp., Boston, 1971-73; rsch. fellow in medicine Harvard Med. Sch., Boston, 1971-73; instr. medicine Harvard Med. Sch., 1974-77, from asst. prof. to assoc. prof. medicine, 1977-83; prof. medicine, radiology U. Ala., Birmingham, 1983—, Mary Gertrude Waters chair cardiovascular medicine divsn. cardiovascular disease, 1991—; cons. nuclear medicine radiology dept. Mass. Gen. Hosp., 1977-83, asst. in medicine gen. med. svc., 1978-82; dir. divsn. cardiovascular disease U. Ala. Hosps. U. Ala., 1983—, dir. ctr. NMR R&D, 1986—. Editor (with others): Noninvasive Cardiac Imaging, 1983, New Concepts in Cardiac Imaging, 1985, 86, 87, 88, 89, The Principles and Practice of Cardiovascular Imaging, 1991; contbr. more than 300 articles, abstracts, reviews, book chpts., editorials to profl. jours.; speaker in field; mem. edit. bds. J. Am. Coll. Cardiology, Circulation, J. Magnetic Resonance in Medicine, Internat. Jour. Cardiology, NMR in Medicine, Coronary Artery Disease, others; rsch interests in radionuclide and nuclear magnetic resonance studies of the heart, myocardial metabolism, cardiac pathophysiology. Recipient project grants NIH, 1990—, Dept. Energy, 1992—, Nat. Ctr. Rsch. Resources, 1992—; tng. grant NIH, 1992—. Fellow Am. College Cardiology (chmn. cardiac imaging com. 1982-88, current procedural terminology com. 1988—, gov. rels. com. 1989—); mem. AMA (chmn. panel nuclear magnetic resonance imaging 1985-88), Am. Fedn. Clin. Rsch., Am. Heart Assn. (fellow coun. clin. cardiology, 1975 , Mass. affiliate 1975-83, established investigator 1979-84, Richard and Hinda Rosenthal award for excellence in clin. investigation, 1985, chmn. advanced cardiac technology com. of coun. on clin. cardiology 1981-86, mem. exec. com. coun. clin. cardiology 1981—, Ala. affiliate 1983—, long range planning com. 1986-89, vice chmn. exec. com. coun. clin. cardiology 1988—, rsch. com. fellowship subgroup A, 1988-91, chmn. long range planning com. coun. clin. cardiology 1989-91, nom. com. coun. clin. cardiology 1989-91, chmn. budget com. coun. clin. cardiology 1989-91, chmn. exec. com. coun. clin. cardiology 1991-93), Soc. Nuclear Medicine (coun. nuclear energy 1990—), Soc. Magnetic Resonance in Medicine (exec. com. 1982-88, pres. 1986-87, chmn. exec. com. 1987, sci. program com. 1988-89), Nat. Heart, Lung and Blood Inst. (program project review com. A 1984-88, cardiovascular and renal study sect. 1991—), So. Med. Assn., Assn. Am. Physicians, NIH Reviewers Reserve, U.S. Nuclear Regulatory Commn. (adv. com. 1984—), Am. Soc. Clin. Investigators, Assn. Univ. Cardiologists, Assn. Profs. Cardiology, Sigma Xi. Home: 4301 Kennesaw Dr Birmingham AL 35213 Office: U Ala Specialized Ctr Rsch U Sta Birmingham AL 35294*

POIESZ, BERNARD JOSEPH, oncologist; b. Phila., Sept. 4, 1948; m. Elvira Celia Poiesz; children: Michael, Erica. BA, LaSalle U., 1970; MD, U. Pa., 1974; LHD (hon.), Lemoyne Coll., 1988. Rsch. asst. Albert Einstein Med. Ctr., Phila., 1966-70, Inst. for Cancer Rsch., Fox Chase, Pa., 1970-74; resident in internal medicine SUNY Health Sci. Ctr., Syracuse, 1974-77, prof., 1980—; from clin. assoc. to rsch. assoc. NCI NIH, Bethesda, Md., 1977-80; rsch. investigator Barbara Kopp Rsch. Ctr., Auburn, N.Y., 1980-84; immunology com. Cancer Acute Leukemia Group B, 1980; rsch. adv. bd. Upstate Med. Ctr., Syracuse, 1980, 89—; peer rev. com. Nat. Inst. for Allergy and Infectious Disease, 1986; virology com. AIDS Treatment Evaluations Unit, 1987; sci. adv. com. NYU, 1988; HIV adv. com. SUNY Health Sci. Ctr., 1990, departmental exec. com. dept. medicine, 1990, departmental appointments and promotions com. dept. medicine, 1991, governing bd. reps. dept. medicine, 1992—; Retrovinology Rsch. adv. com. Dept. VA Med. Ctr., Balt., 1992—. Contbr. articles to Jour. clin. Microbiology, Med. Virology, Cancer. Sci. affairs subcom. on neoplasia Am. Soc. Hematology, 1988; steering com. Internat. Retrovirology Assn., Pasteur Inst., Paris, 1992—. With USPHS, 1977-80. SUNY scholar, 1990; Jr. Clin. Faculty fellow Am. Cancer Soc., 1981-84; recipient Bausch & Lomb award for med. rsch. U. Pa., 1974, Clin. Investigator award VA Med. Ctr., Syracuse, 1984-87, Pres. award for rsch. SUNY Health Sci. Ctr., 1991. Mem. AAAS, Am. Soc. Clin. Oncology, Am. Soc. for Microbiology, Am. Assn. for Cancer Rsch., Am. Soc. Hematology, Am. Soc. Clin. Investigation, Am. Soc. Virology, N.Y. Acad. Sci. Achievements include discovery of human T-cell lymphoma/leukemia virus type I. Office: SUNY Health Sci Ctr 750 E Adams St Syracuse NY 13210

POINAR, GEORGE ORLO, JR., insect pathologist and paleontologist, educator; b. Spokane, Wash., Apr. 25, 1936; s. George O. and Helen Louise (Ladd) P.; m. Eva I. Hecht; children: Hendrik, Maya; m. 2d, Roberta Theresa Heil; 1 child, Gregory. BS, Cornell U., 1958, MS, 1960, PhD, 1962. Prof. dept. entomology U. Calif., Berkeley, 1964—. Author: 5 books on nematodes and insect pathology, 1 book on life in amber, 1975-92; editor: Nematode Pathology, 2 vols., 1988. Grantee NATO, NSF, NIH, 1962-72. Avocations: photography, classical piano, tennis. Office: U Calif Dept Entomology Berkeley CA 94720

POINDEXTER, KIM M., electrical engineer, consultant; b. Long Beach, Calif., Oct. 22, 1955; s. Marshall Edwin and Bettie Jane (Kraft) P.; m. Deborah Davidson, Apr. 25, 1981; 1 child, Eric Calen. BSEE, DeVry Inst. Tech., Phoenix, 1978; MSECE, U. Calif., Santa Barbara, 1981. Engr. Burroughs, Santa Barbara, Calif., 1978-81; staff engr. IBM, Tucson, 1981-83; engr. Mentor Graphics, Beaverton, Oreg., 1983-87; cons. Mentor Graphics, San Diego, 1988-92; mktg. engr. Mentor Graphics, Wilsonville, Oreg., 1992—. Mem. Soc. Concurrent Engring. Republican. Achievements include patent for network shareability of Hardware Modeling Devices, 1987. Office: Mentor Graphics 8005 SW Boeckman Rd Wilsonville OR 97070

POIROT, JAMES WESLEY, engineering company executive; b. Douglas, Wyo., 1931; m. Raeda Poirot. BCE, Oreg. State U., 1953. With various constrn. firms, Alaska and Oreg.; with CH2M Hill Inc., 1955, v.p., Seattle and Atlanta, from 1967; chmn. bd. CH2M Hill Inc., Englewood, Colo., 1983-93; former chmn. Western Regional Coun., Design Profls. Coalition, Accreditation Bd. Engring. and Tech., Indsl. Adv. Coun.; mem. Oreg. Joint Grad. Schs. Engring., Engring. Coun. Named ENR Constrn. Man of Yr., 1988. Fellow ASCE (chmn. steering com. quality manual from 1985, pres. 1993-94), Am. Cons. Engrs. Coun. (pres. 1989-90), Nat. Acad. Engring. Office: CHZM Hill Inc PO Box 22508 Denver CO 80222-0508

POITOUT, DOMINIQUE GILBERT M., orthopedic surgeon, educator; b. Paris, Dec. 1, 1946; s. Pierre Augustin M. and Helene Marie J. (Baudrais) P.; m. Anne Marie S. Gros, Apr. 4, 1972; children: Pierre-Brice R., Jean-Roch D., Marie-Elodie A. D Medicine, U. Paris, 1973; M Human Biology, U. Marseille, 1976; M in Biomechanics, 1976, M in Anthropology, 1976, M in Neuroanatomy, 1977, M in Gen. Anatomy, 1977, DEA in Geology and Anthropology, 1978. Extern hosp. Paris, 1967-70, intern hosp., 1970-71; intern hosp. Marseille, France, 1971-76; asst. in anatomy, 1973-76, chief of clinic, 1976-82; prof. orthopedic surgery Univ. Hosp. Ctr. of U. Aix-Marseilles, 1982—, chief of svc., chmn. orthopedic dept., 1986—; dir. surg. orthopedic rsch. lab. Fac-Medicine Marseille North, 1991—; nat. dir. D.E.A. for orthopedic biomechanics and redaction and medical communication, 1984—; dir. Nat. Diploma of Surg. Scis. for Biomechanics and Biomaterials, 1986—; organizer profl. confs.; orthopedic and traumatologic expert Aix and Marseille Ct. of Law, 1983; dir. surg. orthopedic rsch. lab.-Fac-Medicine Marseille North, 1991—; dir. diploma orthopedic oncology, 1991—. Author: Locomotor System Allografts, 1986, Orthopedic Biomechanic, 1987, Atlas of Orthopedic and Traumatologic Surgery of the Knee, 1992; editor Orthopedic and Traumatologic Letter; contbr. articles, papers to numerous profl. jours. Councillor of the Townhall of Marseille, 1983; v.p. of the commn. of Social Action of the Townhall of Marseille, 1983; pres. Perspectives for Health Futures in France; mem. Council for France Future, Paris. Recipient High French Com. for Civil Def. Mem. Med. Expert Assn. (pres. ct. law aix and narseille), French Institut, European Soc. Biomechanics, European Soc. Biomaterials, French Soc. Orthopedic Surgery, Nat. Acad. Surgery, French Coll. Orthopedic Surgeons, Internat. Soc. Orthopedic Surgery, SICOT-SIROT Rsch. Commn. (pres. 1993), Rsch. Assn. on Intraosseous Circulation, French Tissue Assn. (treas.), Rsch. Group Biomaterials and Grafts (pres.), Freedom and Family Club Assn. (pres.), Tastevin Club (commdr.), Lions (past pres.), Knight of St. Jean of Jerusalem, Rhodos and Malta. Achievements include research in allografts of bone, cartilage and ligaments and orthopedic oncology. Home: 120 Rue du Cdt Rolland, Les Jardins de Thalassa, 13008 Marseilles France Office: Centre Hosp U, Nord Chemin des Bourrely, 13015 Marseilles France

POKLEMBA, RONALD STEVEN, engineer; b. Bridgeport, Conn., Sept. 3, 1947; s. Frank Anthony and Hele Victoria (Macieski) P.; m. Dale Anne Selman, jane 23, 1971 (div. Feb. 1991); children: Allison Brie, Marla Jean. BSME, U. Conn., 1970; MBA, U. Hartford, 1978. Product devel. Brand Rex Co., Willimantic, Conn., 1970-80, Phalo Corp., Shrewsbury, Mass., 1980-86; engring. mgr. Hitachi Cable Manchester (N.H.), Inc., 1986–. Contbr. articles to profl. jours. Am. Screw Found fellow. Mem. Wire Assn., Soc. Plastics Engrs. Home: 53 Eagle Nest Way Manchester NH 03104 Office: Hitachi Cable Manchester 900 Holt Ave Manchester NH 03109

POKORNY, ALEX DANIEL, psychiatrist; b. Taylor, Tex., Oct. 18, 1918; s. John Robert and Olga Frances (Susen) P.; m. Jeanice Brooke Allen, Mar. 13, 1948; children: Martha, Ross, Ellen, Sally. BA, U. Tex., 1939; MD, U. Tex., Galveston, 1942. Diplomate Am. Bd. Psychiatry and Neurology. Psychiatrist VA Hosp., Houston, 1949-55, chief psychiatry and neurology svc., 1955-73; from instr. to prof. psychiatry Baylor Coll. Medicine, Houston, 1949-89, acting chmn. dept. psychiatry, 1968-72, vice chmn. dept. psychiatry, 1972-89, ret. Editor (with others) 7 books, including Phenomenology and Treatment of Anxiety, 1979, Phenomenology and Treatment of Alcoholism, 1980, Phenomenology and Treatment of Psychosexual Disorders, 1983, Phenomenology and Treatment of Psychiatric Emergencies, 1984; editor numerous publs.; contbr. 100 articles to profl. jours. Capt. U.S. Army, 1943-46. Recipient Amersa award for Excellence in Med. Edu. Assn. Med. Edn. & Rsch. Substance Abuse, 1989, Dublin award Am. Assn. Suicidology, 1992. Fellow AAAS, Am. Psychiat. Assn. (life), Am. Coll. Psychiatrists (life); mem. Soc. Psychophysiological Rsch. Democrat. Unitarian. Home: 813 Atwell St Bellaire TX 77401-4718

POLAKOFF, PEDRO PAUL, II, neurosurgeon; b. Phila., Mar. 13, 1926; s. Pedro Paul and Ruth (Mann) P.; m. Thelma Naome Krasney, Apr. 7, 1951; 1 child, Pedro Paul III. BS, U. Pa., 1949; MD, Hahnemann Med. Coll., 1953. Diplomate Am. Bd. Neurosurgery. Intern Hahnemann Hosp., Phila., 1953-54, gen. surg. resident, 1954-55; neurosurg. resident Grad. Hosp. of U. Pa., Phila., 1955-58; pvt. practice neurosurgery Phila., 1958—. Contbr. articles to profl. jours. Fellow ACS; mem. AMA, Congress of Neurol. Surgeons, Am. Assn. Neurologic Surgery, Pa. Med. Soc., Phila. County Med. Soc., Gerontol. Soc., Geriatric Soc., Phila. Neurol. Soc., Am. Acad. Neurology, Am. Soc. Neuroimaging, AAAS, Phi Beta Kappa, Alpha Omega Alpha. Republican. Jewish. Home: 783 Jenkintown Rd Elkins Park PA 19117-1646 Office: Polakoff & Ferrara Neuro As 1010 Fox Chase Rd Rockledge PA 19046

POLANYI, JOHN CHARLES, chemist, educator; b. Jan. 23, 1929; m. Anne Ferrar Davidson, 1958; 2 children. BSc, Manchester (Eng.) U., 1949, MSc, 1950, PhD, 1952, DSc, 1964; DSc (hon.), U. Waterloo, 1970, Meml. U., 1976, McMaster U., 1977, Carleton U., 1981, Harvard U., 1982, Rensselaer U., Brock U., 1984, Lethbridge U., Sherbrooke U., Laval U., Victoria U., Ottawa U., 1987, Manchester U. and York U., England, 1988, U. Montreal, Acadia U., 1989, Weizmann Inst., Israel, 1989, U. Bari, Italy, 1990, U. B.C., 1990, Concordia U., 1990, McGill U., 1990; LLD (hon.), Trent U., 1977, Dalhousie U., 1983, St. Francis-Xavier U., 1984. Mem. faculty dept. chemistry U. Toronto, Ont., Can., 1956—; prof. U. Toronto, 1962—, Univ. prof., from 1974; William D. Harkins lectr. U. Chgo., 1970; Reilly lectr. U. Notre Dame, 1970; Purves lectr. McGill U., 1971; F.J. Toole lectr. U. N.B., 1974; Philips lectr. Haverford Coll., 1974; Kistiakowsky lectr. Harvard U., 1975; Camille and Henry Dreyfus lectr. U. Kans., 1975; J.W.T. Spinks lectr. U. Sask., Can., 1976; Laird lectr. U. Western Ont., 1976; CIL Disting. lectr. Simon Fraser U., 1977; Gucker lectr. Ind. U., 1977; Jacob Bronowski meml. lectr. U. Toronto, 1978; Hutchinson lectr. U. Rochester, N.Y., 1979; Priestley lectr. Pa. State U., 1980; Barré lectr. U. Montreal, 1982; Sherman Fairchild disting. scholar Calif. Inst. Tech., 1982; Chute lectr. Dalhousie U., 1983; Redman lectr. McMaster U., 1983; Wiegand lectr. U. Toronto, 1984; Edward U. Condon lectr. U. Colo., 1984; John A. Allan lectr. U. Alta., 1984; John E. Willard lectr. U. Wis.; 1984, Owen Holmes lectr. U. Lethbridge, 1985; Walker-Ames prof. U. Wash., 1986, John W. Cowper disting. vis. lectr. U. Buffalo, SUNY, 1986; vis. prof. chemistry Tex. A&M U., 1986; Disting. vis. speaker U. Calgary, 1987; Morino lectr. U. Japan, 1987; J.T. Wilson lectr. Ontario Sci. Ctr., 1987; Welsh lectr. U. Toronto, 1987; Spiers Meml. lectr. Faraday div. Royal Soc. Chemistry, 1987; Polanyi lectr. Internat. Union Pure & Applied Chemistry, 1988; W.N. Leis lectr. Atomic Energy of Can. Ltd., 1988; Consol. Bathurst vis. lectr. Concordia U., 1988; Priestman lectr. U. N.B.; 1988, Killam lectr. U. Windsor, 1988; Herzberg lectr. Carleton U., 1988; Falconbridge lectr. Laurentian U., 1988; DuPont lectr. Ind. U., 1989; C.R. Mueller lectr. Purdue U., 1989; mem. sci. adv. bd. Max Plank Inst. for Quantum Optics, Fed. Republic Germany, 1982; mem. Nat. Adv. Bd. on Sci. and Tech., 1987; hon. cons. Inst. Molecular Sci., Okazaki, Japan, 1989-91; founding mem. Can. Com. on Sci. and Scholars. Co-editor: (with F.G. Griffiths) The Dangers of Nuclear War, 1979; contbr. articles to jours., mags., newspapers; producer: film Concepts in Reaction Dynamics, 1970. Bd. dirs. Can. Ctr. for Arms Control and Environment; founding mem. Can. Pugwash Com., 1960. Decorated officer Order of Can., companion Order of Can.; recipient Marlow medal Faraday Soc., 1962; Centenary medal Chem. Soc. Gt. Brit., 1965; with N. Bartlett Steacie prize, 1965; Noranda award Chem. Inst. Can., 1967; award Brit. Chem. Soc., 1971; Mack award and lectureship Ohio State U., 1969; medal Chem. Inst. Can., 1976; Henry Marshall Tory medal Royal Soc. Can., 1977; Remsen award and lectureship Am. Chem. Soc., 1978, Nobel Prize in chemistry, 1986, Isaac Walton Killam Meml. prize, 1988; co-recipient Wolf Prize in Chemistry, 1982; Sloan Found. fellow, 1959-63; Guggenheim fellow, 1979-80. Fellow Royal Soc. Can. (founding mem. com. on scholarly freedom), Royal Soc. London (Royal medal 1989), Royal Soc. Edinburgh; mem. Nat. Acad. Scis. U.S. (fgn.), Am. Acad. Arts and Sci. (hon. fgn., mem. com. on internat. security studies), Pontifical Acad. Scis., Rome. Office: U Toronto Dept Chemistry, 80 St George St, Toronto, ON Canada M5S 1A1

POLASEK, EDWARD JOHN, electrical engineer, consultant; b. Cudahy, Wis., Oct. 12, 1927; s. John Vincent and Mary Ann (Totka) P.; m. Alice S. Nee (Harnecki), Aug. 18, 1948. BSEE, Marquette U., 1948. Registered profl. engr., Wis., Fla. Cons. engr. Eau Claire, Wis., Gainesville, Fla., 1955-60, various countries, Korea, Vietnam, Nicaragua, 1960-72; v.p., dir. Finley Engring. Co., Eau Claire, 1972-78; pres. Chippewa Devel. Co., Eau Claire, 1978-82; planning engr. Harza Engring. Co. in Cairo, Egypt and Dominican Rep., 1982-86; cons. engr. Gainesville, 1986—; cons. Lake Altoona Rehab. Dist., Eau Claire, 1974. Author: Planning Methods, 1982, Feasibility Study, 1984; editor: Field Engineer's Handbook, 1982. Chmn. Eau Claire chpt. Am. Cancer Soc.; master gardner U. Fla. Ext. Svc., Gainesville, 1990. With USN, 1944-46, PTO. Mem. Nat. Soc. Profl. Engrs. (pres. 1956), IEEE, Audobon Soc., Tau Beta Pi, Eta Kappa Nu. Avocations: mycology, fishing, arts.

POLIACOF, MICHAEL MIRCEA, electrical engineer; b. Galati, Rumania, Nov. 19, 1932; came to U.S. 1968; s. Emmanuel and Rachel (Aronovici) P.; m. Michele Abecassis, Jan. 26, 1969 (div. 1990); children: Charles E., Ann. MSEE, Poly. Inst. Bucharest, 1958. Registered profl. engr., N.J. Elec. engr. Electrotechnica, Bucharest, 1958-63, AAPAVE, Mulhouse, France, 1964-68, Lockwood Gen. Engring., N.Y.C., 1968-70; asst. chief elec. engr. Port Authority of N.Y. and N.J., N.Y.C., 1970—; lectr. CCNY, 1981—. Co-author: IEEE/Gray Book, 1990. Achievements include patent for ground fault detector. Home: 84 Casino St Freeport NY 11520 Office: Port Authority of NY/NJ 1 World Trade Ctr #72S New York NY 10048

POLICH, JOHN MICHAEL, experimental psychologist; b. Phila., Sept. 2, 1947; s. John Joseph and Jessie Pauline (Rafa) P.; m. Deborah Jean Dawson, Aug. 15, 1973. Student, Conception (Mo.) Sem., 1965-67; BS with honors, U. Iowa, 1969; MA, Wayne State U., 1972; PhD, Dartmouth Coll., 1977. Asst. prof. psychology, behavioral sci. div. U. Wis.-Parkside, Kenosha, 1977-78; postdoctoral fellow dept. psychology Cognitive Psychophysiology Lab. U. Ill., Champaign, 1978-80; asst. rsch. neuropsychologist dept. neurology Calif. Coll. Medicine, U. Calif., Irvine, 1980-82; rsch. assoc. Neuropsychology Lab. The Salk Inst., San Diego, 1982-84; postdoctoral scholar dept. psychiatry U. Calif., San Diego, 1983-84; lectr. dept. psychology U. Calif.-San Diego, San Diego, 1983-85; sr. rsch. assoc. dept. neuropharmacology The Scripps Rsch. Inst., La Jolla, 1984-86, asst. mem., 1986-91, assoc. mem., 1991—, rsch. coord. div. med. psychology, 1989—; vis. asst. prof. psychology Dartmouth Coll., Hanover, N.H., 1976-77; adj. faculty dept. psychology San Diego State U., 1983-87; asst. adj. prof. dept. psychology U. Calif.-San Diego, La Jolla, 1985-89, assoc. adj. prof., 1989-93, adj. prof., 1993—; referee NSF, NIH, Dept. Vet. Affairs, NRC Can. Mem. editorial bd. Internat. Jour. Psychophysiology, 1992—; reviewer various jours.; contbr. articles to profl. jours., chpts. to books. With U.S. Army, 1969-71. USPHS fellow, 1972-73, Dartmouth grad. fellow, 1973-75. Mem. Psychonomic Soc., Soc. Psychophysiol. Rsch. (program com. 1988, co-chair mem. com. 1992—), Am. Psychol. Soc., Internat. Orgn. Psychophysiology. Achievements include research in cognitive neuroscience-use of event-related brain potentials to examine human info. processing; clin. electrophysiology-devel. of event-related potentials for the assessment and diagnosis of neurol. disorders of cognitive function; hemisphereic specialisation: analysis of the fundamental processing distinctions underlying hemispheric differences. Office: The Scripps Rsch Inst 10666 N Torrey Pines Rd La Jolla CA 92037

POLINSKY, JOSEPH THOMAS, purchasing manager; b. Kingston, Pa., Mar. 10, 1947; s. Joseph Patrick and Margaret Ceclia (Matej) P.; m. Donna Lee Miles, Dec. 28, 1968 (div. Nov. 1990); children: Jon Douglas, Jennifer Susan, Jeffrey David. BSBA, King's Coll., 1968; MBA in Mgmt., Fairleigh Dickinson U., 1977. Fin. svcs. specialist Bell Labs., Murray Hill, N.J., 1968-74; adminstrv. asst. Bell Labs., Whippany, N.J., 1974-76, supr. adminstrn. svcs., 1977; tech. employment rep. Bell Labs., Holmdel, N.J., 1977-80; sr. systems analyst Bell Labs., Short Hills, N.J., 1981-82; mgr. tech. employment Bellcore, Piscataway, N.J., 1983-85; mgr. logistics Bell Labs., Piscataway, N.J., 1986-92; mgr. purchasing Bellcore, Piscataway, N.J., 1992—. Mem. indsl. com. United Way, Morris County, N.J., 1977, allocation com., Monmouth County, N.J., 1985; cub master Boy Scouts Am., Raritan, N.J., 1983-86; mem. Bd. Adjustment, Raritan, 1988-89. With U.S. Army, 1969-70. Roman Catholic. Avocations: fishing, philately, gardening, boating, cooking. Home: 14 Normandie Ln Raritan NJ 08869 Office: Bellcore 8 Corporate Pl Piscataway NJ 08854

POLITES, MICHAEL EDWARD, aerospace engineer; b. Belleville, Ill., Mar. 19, 1944; s. Matthew Charles and Edith Louise (Schwarz) P. BS in Systems & Automatic Controls, Washington U., St. Louis, 1967; MSEE, U. Ala., 1971; PhD in Elec. Engring., Vanderbilt U., 1986. Aerospace engr. guidance, navigation & control NASA/Marshall Space Flight Ctr. Structures & Dynamics Lab., Huntsville, Ala., 1967—. Contbr. 29 articles to profl. jours.; referee various jours. & confs. Fellow AIAA (assoc., guidance navigation & control tech. com. 1990-93); mem. IEEE (sr.), ASME, Am. Astronautical Soc., Mensa. Achievements include patent for Rotating Unbalanced-Mass Devices for Scanning. Office: NASA/Marshall Space Flight Ctr Structures & Dynamics Lab Control Systems Divsn Huntsville AL 35812

POLKOSNIK, WALTER, physicist; b. L.I., Feb. 3, 1968; s. Edward and Maria (Bera) P. BA, Queens Coll., 1991. Grad. asst. physics dept. Queens Coll., Flushing, N.Y., 1991-92. Contbr. articles to profl. jours. Mem. Am. Inst. Physics, Optical Soc. Am., Sigma Xi.

POLL, DAVID IAN ALISTAIR, aerospace engineering educator, consultant; b. Mirfield, Yorkshire, Eng., Oct. 1, 1950; s. Ralph Angus and Mary (Hall) P.; m. Elizabeth Mary Read, May 31, 1975; children: Edward John, Robert Frederick, Helen Elizabeth. BSc in Engring. with honors, Imperial Coll., London, 1972; PhD, Cranfield Inst. Tech., 1978. Chartered engr. Engr. future projects Hawker-Siddeley Aviation, Kingston-upon-Thames, 1972-75; lectr. aerodynamics Cranfield Inst. Tech., U.K., 1978-85, sr. lectr., 1985-87; prof. aero. engring., dir. Goldstein Rsch. Lab. U. Manchester, Eng., 1987—, head engring. dept., 1991—; mng. dir. Flow Sci., 1990—; cons. Brit. Aerospace-Mil. Aircraft, 1988-90, Brit. Aerospace-Civil Aircraft, 1990—; mem. Agard Fluid Dynamics Panel NATO, 1990—; speaker in field. Contbr. over 100 articles to confs. and profl. jours. Recipient ACGI, City and Guilds Inst. London, Fletcher Trophy, London U. Air Squadron. Fellow Royal Aero. Soc.; mem. AIAA, Engring. Coun. (chartered engr.). Achievements include research on the physics of laminar to turbulent transition in flows over aircraft and re-entry vehicle wings, convective heat transfer in high speed flows, vortex induced forces on inclined bodies and surface drag reduction by active and passive means. Home: 4 Beech Ct Macclesfield Rd, Cheshire SK9 2AW, England Office: U Manchester Dept Engring, Simon Bldg Oxford Rd, Manchester M13 9PL, England

POLLACK, GERALD LESLIE, physicist, educator; b. Bklyn., July 8, 1933; s. Herman and Jennie (Tenenbaum) P.; m. Antoinette Amparo Velasquez, Dec. 22, 1958; children—Harvey Anton, Samuela Juliet, Margolita Mia, Violet Amata. B.S., Bklyn. Coll., 1954; Fulbright scholar, U. Gottingen, 1954-55; M.S., Calif. Inst. Tech., 1957, Ph.D., 1962. Physics student trainee Nat. Bur. Standards, Washington, 1954-58; solid state physicist Nat. Bur. Standards, 1961-65; cons. Nat. Bur. Standards, Boulder, Colo., 1965-70; assoc. prof. dept. physics Mich. State U., East Lansing, 1965-69; prof. Mich. State U., 1969—; cons. NRC, Ill. Dept. Nuclear Safety; physicist Naval Med. Research Inst., Bethesda, Md., summer 1979; physicist USAF Sch. Aerospace Medicine, San Antonio, Tex., summer 1987. Contbr. articles to profl. jours. Fellow Am. Phys. Soc.; mem. AAAS, Biophys. Soc., Am. Assn. Physics Tchrs. Office: Mich State U Dept Physics and Astronomy East Lansing MI 48824-1116

POLLACK, HENRY NATHAN, geophysics educator; b. Omaha, July 13, 1936; s. Harold Myron and Sylvia (Chait) P.; m. Lana Beth Schoenberger, Jan. 29, 1963; children—Sara Beth (dec.), John David. A.B., Cornell U., 1958; M.S., U. Nebr., 1960; Ph.D., U. Mich., 1963. Lectr. U. Mich., 1962, asst. prof., assoc. prof., prof. geophysics, 1964—, assoc. dean for research, 1982-85, chmn. dept. geol. scis., 1988-91; rsch. fellow Harvard U., 1963-64; sr. lectr. U. Zambia, 1970-71; vis. scientist U. Durham, U. Newcastle-on-Tyne, Eng., 1977-78, U. Western Ont., 1985-86; chmn. Internat. Heat Flow Commn., 1991-95. Fellow AAAS, Geol. Soc. Am.; mem. Am. Geophys. Union. Achievements include research on thermal evolution of the earth, recent climate change. Office: U Mich Dept Geol Scis Ann Arbor MI 48109

POLLACK, HOWARD MARTIN, radiologist, teacher, author, researcher; b. Phila., June 13, 1928; s. Max and Esther (Schwartz) P.; m. Shanlee Miriam Kirsh, Dec. 24, 1950; children: Andrew, Matthew, Stuart. BA, Temple Univ., 1947, MD, 1951. Diplomate Am. Bd. Urology, Am. Bd. Radiology. Resident urology Episcopal Hosp. Temple Univ., Phila., 1952-55; chief urologist 1100th USAF Hosp., Washington, 1955-58; pvt. practice Phila., 1958-63; resident radiology Temple Univ., 1963-66; pvt. practice radiology Rolling Hill Hosp., Phila., 1966-68; chief radiologist Episcopal Hosp., Phila., 1968-77; prof. radiology Univ. Penna Hosp. & Medical Ctr., Phila., 1977—. Author: Clinical Urography, 1990, Radiological Exam Urinary Tract, 1970; contbr. over 200 articles to profl. jours. Capt. USAF, 1955-58. Recipient Disting. Svc. award Am. Urol. Assn., 1991, Alumni Achievement award Temple Univ., 1992, Honorary AOA, 1992, Gold medal Pa. Radiol. Soc., 1994; recipient rsch. grants in field. Fellow Am. Coll. Radiology; mem. Soc. Uroradiology (pres.), Phila Roentgen Ray Soc. (pres.). Achievements include development of equipment for ultrasound localization for biopsy, interventional uroradiology; co-developer lithotropsy surface coils for MRI of pelvis. Home: 531 Ashmead Rd Cheltenham PA 19012 Office: Univ Penna Hosp 3400 Spruce St Philadelphia PA 19104

POLLACK, MURRAY MICHAEL, physician; b. Bklyn., Nov. 1, 1947; s. Louis R. and Shirley (Schilling) P.; m. Mona Michaels, Dec. 3, 1973; children: Seth, Haley. BA in Biology, U. Rochester, 1970; MD, Albert Einstein Sch. Medicine, 1974. Diplomate Am. Bd. Pediatrics, Am. Bd. Pediatric Critical Care. Intern, then resident in pediatrics Children's Nat. Med. Ctr., Washington, 1974-77, intensivist, 1978—, dir. health svcs. and clin. rsch., 1990—; prof. anesthesiology and pediatrics George Washington U. Med. Sch., 1988—. Editorial bd.: Resusitation, Critical Care Medicine; contbr. articles to profl. jours. Grantee PHHS, 1989—, Robert Wood Johnson Found., 1986-89. Fellow Coll. Critical Care Medicine; mem. Soc. for Pediatric Rsch., Am. Acad. Pediatrics, Soc. Critical Care Medicine (faculty, reviewer, moderator 1987—), Nat. Assn. Children's Hosps. (quality com. 1991—), Am. Bd. Pediatrics (sub-bd. critical care 1991—). Achievements include research in quantifying the relationship between physiologic instability and mortality risk, reduced risk of death associated with pediatric intensive care, creation of pediatric risk of mortality score. Office: Childrens Nat Med Ctr 111 Michigan Ave NW Washington DC 20010

POLLACK, ROBERT ELLIOT, biological sciences educator, writer, scientist; b. Bklyn., Sept. 2, 1940; s. Hyman Ephraim and Molly (Pollack) P.; m. Amy Louise Steinberg, Dec. 23, 1961; 1 child, Marya. B.A. in Physics, Columbia U., 1961; Ph.D. in Biology, Brandeis U., 1966. Asst. prof. pathology Med. Sch. NYU, N.Y.C., 1969-70; sr. scientist Cold Spring Harbor Lab., N.Y., 1971-75; prof. microbiology Med. Sch., SUNY-Stony Brook, 1975-78; prof. biol. sci. Columbia U., N.Y.C., 1978—; dean Columbia Coll., N.Y.C., 1982-89; instr. Pratt Archtl. Sch., Bklyn., 1970; vis. prof. pharmacology Albert Einstein Coll. Medicine, Bronx, N.Y., 1977-92; lectr. Rosenthal Colloquium, March of Dimes, 1989; McGregory lectr. Colgate U., 1979; du Vigneaud lectr. Med. Sch., Cornell U., 1983. Co-editor: Readings in Mammalian Cell Culture, 1973, 3d rev. edit., 1981, Signs of Life, 1994; mng. editor BBA Revs. on Cancer 1980-86; contbr. numerous rsch. articles on molecular cell biology to profl. jours. Trustee N.Y. Found., Brandeis U. Recipient Rsch. Career Devel. award NIH, 1974; NIH spl. fellow Weizmann Inst., Rehovot, Israel, 1970-71; grantee Nat. Cancer Inst., NIH, 1968—, Am. Cancer Soc., 1985—; John Simon Guggenheim fellow, 1993. Fellow AAAS; mem. N.Y. Acad. Scis., Am. Soc. Microbiology, Assn. Am. Colls., Sigma Xi. Jewish. Office: Columbia U Dept Biol Scis 813 Fairchild Hall New York NY 10027

POLLACK, ROBERT WILLIAM, psychiatrist; b. N.Y.C., May 22, 1947; s. George and Esther P.; m. Pam Gregory, Sept. 15, 1984; 1 child, Jessie. BS in Biology, Yale U., 1969; MD, SUNY Downstate Med. Ctr., Bklyn., 1973. Diplomate Am. Bd. Psychiatry and Neurology. Tng. resident U. Fla., 1973-76, chief resident Dept. of Psychiatry, 1975-76; asst. prof. Dept. of Psychiatry U. Fla., Gainesville, 1976-77; clin. asst. prof. dept. psychiatry Shands Hosp., Gainesville, 1977—; chief dept. psychiatry Fla. Hosp., Orlando, 1983, 84; clin. dir. assessment and evaluation team West Lake Hosp., Longwood, Fla., 1984-87, clin. dir. intensive evaluation unit, 1987-89; med. dir. Fla. Psychiat. Assocs., Fla. Psychiat. Mgmt., Winter Park, 1989—; med. dir. consultation, liaison svc., and spl. med. unit Winter Park Meml. Hosp., 1992. Contbr. 4 articles to profl. jours.; author sci. reports. Chmn. Retinitis Pigmentosa Casino Night, Orlando, 1988-92; vice-chmn. nat. championship com. U.S. Blind Golfers Assn., 1991-92, chair 48th ann. championship com., 1992-93; bd. dirs. Tennis with a Different Swing, Orlando, 1988-92. Achievements include introduction of use of computerized topographical brain mapping as a diagnostic tool in central Florida. Office: Fla Psychiat Assocs 1276 Minnesota Ave Winter Park FL 32789-4864

POLLARD, THOMAS DEAN, biologist, educator; b. Pasadena, Calif., July 7, 1942; s. Dean Randall and Florence Alma (Dierker) P.; m. Patricia Elizabeth Snowden, Feb. 7, 1964; children: Katherine, Daniel. BA, Pomona Coll., Claremont, Calif., 1964; MD, Harvard U., 1968. Intern Mass. Gen. Hosp., Boston, 1968-69; staff assoc. NIH, Bethesda, Md., 1969-72; from asst. prof. to assoc. prof. Harvard Med. Sch., Boston, 1972-78; prof., dir. dept. cell biology and anatomy Johns Hopkins Sch. Medicine, Balt., 1977—; chmn. Commn. on Life Sci., Nat. Rsch. Coun., 1993—. Guggenheim fellow, 1984. Fellow Am. Acad. Arts and Scis.; mem. NAS, Am. Soc. Cell Biology (pres. 1987-88, K.R. Porter lectr. 1989), Nat. Rsch. Coun. (commn. on life scis. 1993-96), Biophys. Soc. (pres. 1992-93), Marine Biol. Lab. (trustee 1991-97). Office: Johns Hopkins Med Sch Dept Cell Biology-Anatomy 725 N Wolfe St Baltimore MD 21205-2105

POLLEY, RICHARD DONALD, microbiologist, polymer chemist; b. Bklyn., Feb. 23, 1937; s. George Weston and Evelyn (Tuttle) P.; m. Linda R. Radford, Sept. 21, 1991; children from previous marriage: Gordon MacHeath, Jennifer Elizabeth, Tabitha Isabelle, Sean Sullivan; m. Linda R. Radford, 1991. Student, Trinity Coll., 1954-57; BS, BA, Hofstra U., 1960. Asst. advt. mgr. tech. Permatex Chem. Corp., Huntington, N.Y., 1960-61, Sun Chem. Corp., 1961-63; advt. mgr. Celanese Plastics Co., Newark, 1963-67; account dir. McCann Indsl. Tech. Sci. Mktg., N.Y.C., 1967-68, v.p., gen. mgr., Miami, 1968-70; pres. Intercapital Belgium S. Am., Brussels, 1970-72; pres., tech. dir. Iodinamics Corp., Lancaster, Pa., El Paso, Tex., 1972-76; tech. dir. Hydrodine Corp., Miami, 1976—, chmn., CEO, 1986—, also bd. dirs.; chmn., CEO, COO Polymorphic Polymers Corp., Miami, 1978-90; COO Omnidine Corp., Miami, 1980—, bd. dirs.; pres. Skin Care Labs., Inc., Miami, 1979-90; bd. dirs., microbiology dir. Pure H2O

Techs., Inc., Boca Raton, Fla., 1992—; chief exec. officer Polllabs, São Paulo, Brazil, 1990—; tech. dir. Internat. Tech. Corp., Bangkok, Thailand, 1992—, Grupos Gomez y Jalscienses, Guadalajara, 1992—; chmn. Peer Group Influencers, Ltd., London, also Miami, Fla., 1988—; bd. dirs., v.p. Internat. Airlines, Long Beach, Calif., 1984—; overseas dir. Field Iodine Goiter Med. Demonstration Projects, Beth Israel Hosp.-Harvard Med. Sch., 1977—; tech. dir. and chief scientist Tremos, Ltd., Turks and Caicos, B.W.I., 1993—; cons. water microbiology and disinfection control Pan Am. Health Orgn., others. Mem. editorial adv. bd. Chem. Week, 1988; contbr. articles to profl. jours.; patentee in field. Mem. AAAS, Am. Concrete Inst., Water Quality Assn., N.Y. Acad. Scis., Internat. Iodine Inst. (chmn. bd.), Associaçao de Ciencia e Tecnologia Ambiental (bd. dirs. São Paulo 1993—). Republican. Office: Hydrodine Corp 9264 Bay Dr Surfside FL 33154

POLLEY, WILLIAM ALPHONSE, power systems engineer; b. Milw., Dec. 1, 1942; s. William O. and Florence V. (Bruckbauer) P.; m. Connie A. Pippert, Aug. 28, 1965; children: Christopher, Karen, Craig, Carl. BSME, Marquette U., 1966. Engr. co-op Allis Chalmers, Milw., 1962-66; sales engr. Westinghouse, Duluth, Minn., 1966-79; applications engr. Westinghouse, Appleton, Wis., 1979-88, systems engr., 1988-89; power systems engr. Kimberly-Clark, Neenah, Wis., 1989—. Chmn. St. Michael's Parish Coun., Duluth, 1977-79; scoutmaster Boy Scouts Am., Appleton, 1989-92, scout commr., 1992—; exec. couple Nat. Marriage Encounter, Appleton, 1985-87. Mem. IEEE, Instrument Soc. of Am. Avocations: computers, gardening. Home: 1021 E Shady Ln Neenah WI 54556-1225 Office: Kimberly Clark Box 999 Neenah WI 54957-0999

POLLMER, JOST UDO, food chemist; b. Himmelpforten, Germany, May 14, 1954; s. Johannes and Ruth (Krugel) P. Staatsprufung fur, Lebensmittelchemiker U., Munich, 1981. Self-employed food chemist, cons., lectr. Germersheim, Fed. Republic of Germany. Author: IB und stirb Chemie in unserer Nahrung, 1982; contbr. articles to profl. jours. Mem. AAAS, Internat. Epidemiol. Assn., Japan Soc. Biosci., Biotech. and Agrochem., Behavior Toxicology Soc., Inst. Food Technologists, Am. Chem. Soc., Swiss Soc. Food Environ. Chemistry, Nutrition Soc. (U.K.), Assn. Ofcl. Analytical Chemists, N.Y. Acad. Scis. Home and Office: Posthiusstr 6, D-76726 Germersheim Germany

POLLMER, WOLFGANG GERHARD, agronomy educator; b. Wolkenstein, Fed. Republic Germany, Aug. 21, 1926; s. Willy and Wally Natalie (Bauer) P.; m. Emmi Maria Lukas, July 22, 1953; children: Beate, Bernd Michael, Evemarie, Christiane. Diploma in agronomy, U. Halle, Halle, Fed. Republic Germany, 1951; D. in Agronomy, U. Munich, 1956; D. Habilitation, U. Hohenheim, Stuttgart, Fed. Republic Germany, 1963. Plant breeder Seed Breeding Industry, Fed. Republic Germany, 1951-59; from asst. prof. to assoc. prof. U. Hohenheim, Stuttgart, 1959-69; maize breeder U. Hohenheim, 1960-92, prof. plant breeding, 1969-92, ordinarius prof. plant breeding, 1981-92, emeritus, 1992—; head maize breeding sta. U. Hohenheim, Eckartsweier, Fed. Republic Germany, 1960-92; bd. trustees Internat. Maize and Wheat Improvement Ctr., Mex., 1978—; cons. Fed. Ministry Econ. Coop., Bonn, Fed. Republic Germany, 1974-92; mem. Sci. Bd. Agronomy, Paris, 1981-89. Editor: Improvement of Quality Traits of Maize for Grain and Silage Use, 1980; editorial bd. MAYDICA, Milan, 1980-88, Indian Soc. Crop Improvement, New Delhi, 1979; contbr. articles to profl. jours. Mem. Am. Soc. Agronomy, Crop Sci. Soc. Am., EUCARPIA, German, Assn. Plant Scis., German Maize Com., German Coun. for Tropical and Subtropical Agrl. Rsch. Avocations: tennis, windsurfing, sailing, writing (composing). Home: Egilolfstrasse 25, D 70599 Stuttgart Germany

POLLOCK, ADRIAN ANTHONY, physicist; b. Stafford, Eng., Sept. 2, 1943; came to U.S., 1975, naturalized, 1985; s. Erskine Reginald Seton and Barbara (Winsome) Bostock; m. Jane Ellen Coffin, Apr. 19, 1986; children: Andrew James, Cameron James. BA, Cambridge U., Eng., 1964, MA, 1965; PhD, Imperial Coll., London, 1970. Sr. physicist Cambridge Cons., 1970-73; mgr. tech. svcs. Endevco Corp., Royston, Eng., 1974-75; dir. rsch. and application Dunegan/Endevco, San Juan Capistrano, Calif., 1975-81; ptnr. Acoustic Emission Assocs., Laguna Niguel, Calif., 1982-84; dir. tng. Phys. Acoustics Corp., Princeton, N.J., 1985—. Mem. adv. editorial bd. Jour. Ultrasonics, Guildford, Eng., 1970-76; producer videotape Varzin: A Fragment of the Life of Arthur Stanley Eddington, 1982; author approximately 70 publs. on acoustic emission. Recipient Gold medal award Acoustic Emission Working Group, 1985. Mem. ASTM, Am. Soc. Nondestructive Testing (bd. dirs. 1991—, sec. edn. and qualification coun. 1991-92, mem. nat. cert. bd. 1990—). Achievements include patents and inventions in acoustic emission including zone location and the dBae scale. Home: 202 Elm Ave Morrisville PA 19067 Office: Phys Acoustics Corp PO Box 3135 Princeton NJ 08543

POLLOCK, DONALD KERR, anthropologist; b. Huntsville, Ala., Mar. 19, 1950; s. Donald Kerr and Harriett (Donnelly) P.; m. Kathleen Lindberg, Aug. 23, 1972 (div.); m. Dorothy M. Taylor, Jan. 10, 1981. BA, U. Minn., 1972; MA, U. Chgo., 1974; PhD, U. Rochester, 1985. Asst. prof. Boston U., 1986-89; rsch. fellow Med. Sch. Harvard U., Boston, 1986—; asst. prof. SUNY, Buffalo, 1989—. Assoc. editor L.Am. Anthropology Rev., 1990—; contbr. articles to Social Sci. and Medicine, Am. Anthropologist, JAMA, NEJM, Disease and Ethnicity. Bd. dirs. Primary Care Coun. of Buffalo, 1992. Fellow U. Rochester, 1976-80; Grantee NIH, 1989, 90, MacArthur Found., 1992, NEH, 1993, Kaiser Found., 1986-89; named Woodbridge scholar U. Minn., 1971-72. Fellow Am. Anthrop. Assn., Royal Anthrop. Inst. Great Britain; mem. Soc. for Med. Anthropology, Soc. for L.Am. Anthropology, Phi Beta Kappa. Achievements include major research among Kulina Indians of Western Amazonia, culture of teriary care medicine in U.S., cultural shaping of depression, and health consequences of environmental change in developing world. Office: SUNY Dept Anthropolgy Buffalo NY 14261

POLLOCK, JACK PADEN, biology and dental educator, consultant, treelance writer, retired army officer; b. Columbus, Miss., May 12, 1920; s. Samuel Lafayette and Pauline Elizabeth (Pollock) O'Neal; m. Anne Olamae Silbernagel, Aug. 25, 1945; children: Poli A., Elizabeth D. Student, Gulf Coast Mil. Acad., 1936-38, Tulane U., 1938-41; BS, Southeastern La. U., 1942; DDS cum laude, Loyola U., New Orleans, 1945; diploma, Army War Coll., 1965, Indsl. Coll. Armed Forces, 1966; hon. degree, Baylor U., 1974. Asst. prof., Loyola U., New Orleans, 1945-46, prof., 1977-86, U.S.A. Command and Gen. Staff Coll., 1959-60, prof., 1977-86; commd. 1st lt. U.S. Army, 1946, advanced through the grades to brigadier gen., 1972; career officer U.S. Army, 1946-77; dental advisor to Sec. of Defense, 1970-73; cons. Surgeon Gen. U.S. Army, Washington, 1981-84; U. rank and tenure com. Loyola U., New Orleans, 1982-86; prof. emeritus Loyola U., 1986—. Author: International Communism: Its Future Prospects, 1965. Contbr. articles to profl. jours. Active nat. coun. Boy Scouts Am., 1973-92, Nat. Exploring Com., 1973-92. USAR, 1942-46. Decorated D.S.M., Legion of Merit with 2 oak leaf clusters, Bronze Star with one oak leaf cluster, Master Parachutist badges (U.S. and foreign); recipient Silver Beaver award, Boy Scouts of Am., numerous other military, civilian awards and honors. Recipient Disting. Alumnus award Southeastern La. U., 1974. Fellow Am. Coll. Dentistry (life), Internat. Coll. Dentistry (life); mem. ADA (life), La. Dental Assn. (life), New Orleans Dental Assn. (life), Am. Coll. Dentists (life), Internat. Coll. Dentists (life), Fedn. Dentaire Internationale, Assn. Mil. Surgeons of U.S. (life), Pierre Fauchard Acad., C. Victor Vignes Odontological Soc. (life), Am. Assn. Dental Schs., NRA (life), Assoc. U.S. Army (life), Military Order of World Wars (life), Retired Officers Assoc. (life), Disabled Am. Vets. (life), Am. Legion, Delta Sigma Delta(life), Kappa Sigma, Phi Kappa, Omicron Kappa Upsilon. Clubs: Army and Navy (Washington), Univ. (San Antonio). Avocations: writing, travel, conservation. Home: 118 Bayberry Dr Covington LA 70433-4702 Office: PO Box 1423 Mandeville LA 70470-1423

POLLOCK, KENNETH HUGH, statistics educator; b. Quirindi, Australia, Jan. 10, 1948; came to U.S., 1969; s. Ewen Nelson and Alma Jane (Clarke) P.; m. Mary Martha Bechtold, Nov. 10, 1973 (div. 1981); 1 child, Pamela Marie; m. Mary Watson Nooe, Apr. 25, 1987. BSc in Agrl., U. Sydney, Australia, 1969; MS, Cornell U., 1972, PhD, 1974. Lectr. U. Reading, Eng., 1974-78; vis. asst. prof. U. Calif., Davis, 1978-80; assoc. prof. N.C. State U., Raleigh, 1980-86, prof., 1986—. Assoc. editor Biometrics, 1984-89, Jour. of Wildlife Mgmt., 1989-91; contbr. articles to profl. jours. Recipient G.W. Snedecor award for best publ. in biometry, 1991. Mem. Wildlife Soc. (bd. dirs. N.C. chpt. 1992—, mem. awards com. southeastern sect. 1991—, Biometrics Soc. (mem. coun. 1992—, regional com. Ea. N.Am. region 1987-90). Democrat. Unitarian. Office: NC State U Dept Statistics Box 8203 Raleigh NC 27695

POLLOCK, NEAL JAY, electronics executive; b. Phila., Feb. 4, 1947; s. Sol J. and Shirley (Buchsbaum) P. BA in Physics, U. Pa., 1968; MS in Engring. Sci., Pa. State U., 1972; MBA, Temple U., 1975; postdoctoral, George Washington U., 1978-82. Student trainee Naval Air Devel. Ctr., Warminster, Pa., 1964-68, physicist, 1968-69, electronics engr., 1969-75, plans and programs asst., 1975-76; asst. for interface Naval Air Systems Command, Washington, 1976-78, budget and fin. mgr., 1978-79, asst. program mgr. for acoustic sensors, 1979-84; project engr. Naval Sea Systems Command, Washington, 1984-87; br. head, supr. electronics engring. Space and Naval Warfare Systems Command, Washington, 1987-90, div. head, 1990—; EEO counselor Naval Sea Systems Command, Washington, 1986, total quality leadership mgmt. facilitator, team leader, 1991—, instr. prevention of sexual harrassment, 1992. Co-author: Extended Radiometer Analysis-The Point Target; contr.: Organizations in a Changing Society, 1977. Unit commr. Boy Scouts Am., 1975-76; vol. income tax asst. Ayuda and Spanish Catholico, Washington, 1976-77; active, life mem. Save the Redwoods League, San Francisco, 1987, Nature Conservancy, Charlottesville, Va., 1988, Archeological Conservancy, Nat. Parks and Conservation Assn., 1990—, Suriname termites and Hawaii dolphins expeditions Earthwatch; active mem., patron sponsor Pearl Buck Found., Prevention of Blindness Soc., Internat. Rescue Com., others; officer of elections Arlington County. USN Student Engring Devel. scholar, 1964, Phila. Mayor's scholar, 1964, Nat. Sci. Found. scholar Stevens Inst. Tech., 1963; recipient Combined Fed. Campaign Eagle award 1990, 91, 92, Sustained Superior Performance award, 1983, Supervisory Excellence award, 1991, Outstanding Performance award PMRS, 1987, 89, 90, 92, 93. Mem. Soc. Naval Architects and Marine Engrs., Assn. Scientists and Engrs. (life, keyman 1987), U.S. Holocaust Mus. (charter), Alaska Natural History Assn. (life), Pa. State Alumni Assn. (life), Centurion Club, Clipper Club (life), Admirals Club (life), Ionosphere Club (life), Worldclub (life), Amb's. Club (life), Red Carpet Club (life), U.S. Air Club (life), Friends of the Arlington Libr. (life, vol.), Delta Crown Club (life), C.G. Jung Inst. Chgo. (life), Northeast High Sch. Alumni Assn., Pa. State U. Alumni Assn., Thomas Jefferson Pronaos Ancient Mystical Order Rosae Crucis (master 1980-82, Atlantis Lodge sec. 1984-85, treas. 1985-88, chmn. conv. 1980), Masons (32 degree), Beta Gamma Sigma. Republican. Avocations: visiting national parks, collecting oriental carpets, reading, psychology. Home: 2500 S Fern St Arlington VA 22202-2538 Office: SPAWARSYSCOM 2451 Crystal Dr PMW 163-22 Arlington VA 22202-5200

POLLOCK, RAPHAEL ETOMAR, surgeon, educator; b. Chgo., Dec. 25, 1950; s. George Howard and Beverly (Ufit) P.; m. Lesley Kittredge Newton, Aug. 27, 1984; children: Jessica L., Samuel W. BA, Oberlin Coll., 1972; MD, St. Louis U., 1977; PhD, U. Tex. Health Sci. Ctr., 1990. Intern U. Chgo. Hosps. and Clinics, Houston, 1977-78, resident, 1978-79; fellow in Surg. Oncology U. Tex. M.D. Anderson Hosp. and Tumor Inst., Houston, 1982-84, faculty assoc., 1984-85; asst. prof. U. Tex. M.D. Anderson Hosp. and Tumor Inst., 1985-90, assoc. prof., 1990—, dept. chmn. for rsch., 1991—; asst. prof. surgery U. Tex. Med. Sch., Houston, 1987-90, assoc. prof., 1990—; mem. exptl. therapy II study sect. Nat. Cancer Inst., Bethesda, Md., 1991—. Contbr. articles to Cancer Rsch., Jour. Immunology; assoc. editor: Surg. Oncology, 1991—. Recipient Clinician Investigator award Nat. Cancer Inst., 1984-89, FIRST award, 1989—. Fellow ACS (commr. 1991—); mem. Am. Assn. for Cancer Rsch., Soc. Surg. Oncology (program chmn. 1991-92), Soc. Univ. Surgeons. Home: 2826 Tangley Houston TX 77005 Office: U Tex MD Anderson Cancer Ct PO Box 106 1515 Holcombe Blvd Houston TX 77030

POLLOCK, ROBERT ELWOOD, nuclear physicist; b. Regina, Sask., Can., Mar. 2, 1936; s. Elwood Thomas and Harriet Lillian (Rooney) P.; m. Jean Elizabeth Virtue, Sept. 12, 1959; children—Bryan Thomas, Heather Lynn, Jeffrey Parker, Jennifer Lee. B.Sc. (Hons.), U. Man., Can., 1957; M.A., Princeton U., 1959, Ph.D., 1963. Instr. Princeton U., 1961-63; Nat. Research Council Can. postdoctoral fellow Harwell, Eng., 1963-64; asst. prof. Princeton U., 1964-69, research physicist, 1969-70; asso. prof. Ind. U., 1970-73, prof., 1973-84, disting. prof., 1984—, dir. Cyclotron Facility, 1973-79, mem. Nuclear Sci. Adv. Com., 1977-80; Recipient Alexander von Humboldt sr. U.S. scientist award, 1985-87. Fellow Am. Phys. Soc. (Bonner prize 1992); mem. Can. Assn. Physicists. Home: 1261 Winfield Rd Bloomington IN 47401 Office: West Ind U Swain Hall Dept of Physics Bloomington IN 47405

POLMANN, DONALD JEFFREY, water resources engineer, environmental consultant; b. Hackensack, N.J., May 15, 1957; s. John James and Joan Ethel (Fink) P.; m. Lynn Ellen Marshall, Mar. 23, 1980. BS, Rensselaer Poly. Inst., 1979; M of Engring., U. Fla., 1983; PhD in Hydrology, MIT, 1990. Registered profl. engr., Ga. Engring. aide, drafter John A. Grant Jr., Engrs./Surveyors, Boca Raton, Fla., 1978; environ. engr. Camp Dresser & McKee, Cons. Engrs., Ft. Lauderdale, Fla., 1979; grad. rsch. asst., environ. engring. sci. dept. U. Fla., Gainesville, 1979-81; hydrologist, water resources engr. Law Engring. Testing Co., Inc., Tampa, Fla., 1981-83; water resources engr. Environ. Sci. & Engring., Inc., Tampa, 1983-84; grad. rsch. asst. civil engring. dept. MIT, Cambridge, Mass., 1984-89; sr. water resources engr. Law Environ., Inc., Kennesaw, Ga., 1989—. Contbr. articles to Water Resources Rsch., others. Mem. ASCE, Am. Geophys. Union, Assn. Ground-Water Scientists and Engrs., Sigma Xi, Tau Beta Pi. Christian. Achievements include extensions to theory of transient moisture and solute transfer through natural unsaturated soils; development of new set of generalized soil hydraulic property models and use of pertubation and spectral techniques to develop predictive models for large-scale moisture and solute transfer; development of analytical expressions for multi-dimensional mean flow, effective moisture content, and macrodispersion. Office: Law Environ Inc 112 Townpark Dr Kennesaw GA 30144

POLSKY, MICHAEL PETER, mechanical engineer; b. Kiev, Ukraine, Aug. 5, 1949; s. Peter and Basheva P.; m. Maya, June 28, 1975; children: Alan, Gabriel. BSME, Kiev Poly. Inst., 1973; MBA, U. Chgo., 1987. Registered profl. engr., Ill., Mich. Sr. devel. engr. Indsl. Power Corp., Kiev, Ukraine, 1973-76; mech. engr. Bechtel Power Corp., Ann Arbor, Mich., 1976-78; sr. application engr. Brown Boveri Corp., St. Cloud, Minn., 1978-80; product mgr. congeneration Fluor/Daniel, Chgo., 1980-85; pres. Indeck Energy Svcs., Wheeling, Ill., 1985-90, Polsky Energy Corp., Northbrook, Ill., 1990—; bd. dirs. Ind. Power Producers of N.Y., Albany, 1988-89. Author: Public Utilities Fortnightly, 1985, Power, 1984, 83, Hydrocarbon Processing, 1981, 82; author: (book chpt.) Handbook of Power Plant Engineering, 1991. Mem. ASME, Soc. Energy Engrs. Office: Polsky Energy Corp 650 Dundee Rd Ste 170 Northbrook IL 60062

POLUNIN, NICHOLAS, environmentalist, author, editor; b. Checkendon, Oxfordshire, Eng.; s. Vladimir and Elizabeth Violet (Hart) P.; m. Helen Lovat Fraser, 1939 (dec.); 1 child, Michael; m. Helen Eugenie Campbell, Jan. 3, 1948; children: April Xenia, Nicholas V. C., Douglas H. H. Open scholar, Christ Ch., 1928-32; BA (1st class honors), U. Oxford, 1932, MA, 1935, DPhil, 1935, DSc, 1942; MS, Yale U., 1934. Participant or leader numerous sci. expdns., 1930-65, primarily in arctic regions, including Spitsbergen, Greenland, Alaska, Can. western, ea. Arctic, North Pole; curator, tutor, demonstrator, lectr. various instns., especially Oxford U., 1933-47; vis. prof. botany McGill U., 1946-47, Macdonald prof. botany, 1947-52; Guggenheim fellow, rsch. assoc. Harvard U., 1950-53; earlier fgn. research asso., USAF botanical Ice-island research project dir., lectr. plant sci. Yale, also biology Brandeis U., 1953-55; prof. plant ecology and taxonomy, head dept. botany, dir. U. Herbarium and Botanic Garden, Baghdad, Iraq, 1955-58; guest prof. U. Geneva, 1959-61, 75-76; adviser establishment, founding prof. botany, dean faculty sci. U. Ife, Nigeria, 1962-66; founding editor Plant Sci. Monographs and World Crops Books, 1954-78, Biol. Conservation, 1967-74, Environ. Conservation, 1974—; chmn. internat. steering com., organizer, editor procs. Internat. Conf. Environ. Future, Finland, 1971; chmn. internat. steering com., sec. gen., editor proc. 2d Internat. Conf. Environ. Future, Reykjavik, Iceland, 1977, 3d Internat. Conf. Environ. Future, Edinburgh, Scotland, 1987; sec.-gen., joint editor procs. 4th Internat. Conf. Environ. Future, Budapest, Hungary, 1990; pres. Found. for Environ. Conservation, 1975—; participant Internat. Bot. Congresses, Stockholm, 1950, Paris, 1954, Edinburgh, 1964, Seattle, 1969, Leningrad, 1975, Sydney, 1981; councillor (pres.) World Council For The Biosphere, 1984—. Author: Russian Waters, 1931, The Isle of Auks, 1932, Botany of the Canadian Eastern Arctic, 3 vols., 1940-48, Arctic Unfolding, 1949, Circumpolar Arctic Flora, 1959, Introduction to Plant Geography, 1960 (various fgn. edits), Eléments de Géographie botanique, 1967; editor: The Environmental Future, 1972, Growth Without Ecodisasters?, 1980, Ecosystem Theory and Application, 1986, (with Sir John Burnett) Maintainance of The Biosphere, 1990, Surviving with the Biosphere, 1993, Environ. Monographs and Symposia, 1979—, (with Muhamad Nagim) Environmental Challenges, 1993; founding chmn. edit. bd. Cambridge Studies in Environmental Policy, 1984—; contbr. to various jours. Decorated comdr. Order Brit. Empire, 1975; recipient undergrad., grad. student scholarships, fellowships, rsch. associateships Yale U., 1933-34, Harvard U., 1936-37; Rolleston Meml. prize, 1938; D.S.I.R. spl. investigator, 1938; Leverhulme Rsch. award, 1941; from sr. scholar to sr. rsch. fellow New Coll., Oxford; Arctic Inst. N.A. rsch. fellow, 1946-47; Guggenheim fellow, 1950-52; recipient Ford Found. award Scandinavia, USSR, 1966-67, Marie-Victorin medal Can., 1957, Indian Ramdeo medal, 1986, Internat. Sasakawa Environ. prize, 1987, USSR Vernadsky commemoration, 1988, Chinese Academia Sinica medal, 1988, Vernadsky medal USSR Acad. Scis., 1988, 89, Founder's (Zéchenyi) medal Hungarian Acad. of Scis., 1990; named to Netherlands Order of the Golden Ark, 1990 (officer), UN Environ. Programme Global 500 Roll of Honour, 1991. Fellow Royal Geog. Soc., Royal Hort. Soc., Linnean Soc. London, AAAS, Arctic Inst. N.A., INSONA (v.p.), NECA (India); mem. Internat. Soc. for Environ. Edn. (life), Internat. Acad. Environment (mem. conseil de fondation), Torrey Bot. Club (life), Bot. Soc. Am. (life), N.Am. Assn. for Environ. Edn. (life), Asian Soc. Environ. Protection (life), various fgn. and nat. profl. and sci. socs. Clubs: Harvard (N.Y.C.) (life); Canadian Field Naturalists (Ottawa, life); Reform (London) (life). Achievements include confirming existence of Spicer Islands in Foxe Basin and made world's last major land discovery, Can. Arctic, 1946; past rsch. plant life and ecology of arctic, subarctic and high-altitude regions; present occupation environ. conservation at the global level; initiator of ann. worldwide Biosphere Day, 1991, Biosphere Fund and Prizes, 1992, Biosphere Clubs, 1993. Address: Found Environ Conservation, 7 Ch Taverey 7th & 8th Fls, Geneva Switzerland

POLUZZI, AMLETO, chemical company consultant; b. Milan, Italy, Oct. 1, 1919; s. Nemore and Santina (Triva) P.; m. Elena Cattaneo, June 27, 1945; 1 child, Claudio. D in Indsl. Chemistry, State U., Milan, 1943; degree in industry mgmt., Poly. U., Milan, 1951. Chief gen. affaris div. Italian Lombardia region labor office, liaison sec., interpreter Allied Mil. Govt. Lombardia Region Labor Office, Milan, 1945; cons., interpreter Italian Regional Labor Office, Milan; tech. mgr. R&D R&D Alcrea (now Ashland Chem. Italiana), Milan, 1946-77; cons. Ashland Chem. Co., Columbus, Ohio, 1978-91. Contbr. articles to profl. jours. Mem. Assn. Italiana Tecnici Industrie Vernici and Affini (bd. dirs. 1950—), Fedn. Assns. Techniciens des Industries des Peintures del Europe Continentale (bd. dirs. 1984—, mem. hon. com.), Union des Assns. de Techniciens de Culture Mediterraneénne (bd. dirs. 1992—), Internat. Conf. Organic Coatings Sci. and Tech. (adv. bd. 1983—). Avocations: travel, photography, foreign languages, reading, scientific articles. Home: Boite Postale 529, 83616 Frejus Cedex France Office: Via Taranto 4, 20142 Milan Italy

POLYDORIS, NICHOLAS GEORGE, electronics executive; b. Evanston, Ill., July 7, 1930; s. George and Annetta (Karas) P.; m. Gloria Ann Lucas, Dec. 28, 1952; children: Steven, Janet, Lynn, Susan, Nancy. BSEE, Northwestern U., 1954. Trainee Fairbanks Morris Co., 1954-55; dist. mgr. Fasco Industries, Rochester, N.Y., 1955-57; founder, pres. ENM Co., Chgo., 1957—, also bd. dirs.; bd. dirs. Universal Clay Products, Inc., Sandusky, Ohio, Gladston-Norwood Bank, Chgo., Grt. Hellenic Found., Chgo., North Shore Mental Health Assn., Chgo. Patentee in field. Recipient Svc. award Northwestern U., Evanston, Ill., 1965. Mem. Soc. Automotive Engrs., Aircraft Owners and Pilots Assn., Ill. Soc. Profl. Engrs., Mich. Shores Club, Old Willow Club, Kenilworth Club, John Evans Club (pres.), Tau Beta Pi. Republican. Avocations: private pilot. Home: 1630 Sheridan Rd Wilmette IL 60091-1855 Office: ENM Co 5617 N Northwest Hwy Chicago IL 60646-6135

POMERANTZ, JAMES ROBERT, psychology educator, academic admininstrator; b. N.Y.C., Aug. 21, 1946; s. Mihiel Charles and Elizabeth (Solheim) P.; m. Sandra Elaine Jablonski, Oct. 1, 1945; children: Andrew Emil, William James. BA, U. Mich., 1968; PhD, Yale U., 1974. Prize teaching fellow Yale U., New Haven, 1973-74; asst. prof. psychology Johns Hopkins U., Balt., 1974-77; assoc. prof. SUNY, Buffalo, 1977-83, prof., 1983-88, chmn. dept. psychology, 1986-88, assoc. dean, 1983-86; dean social scis., Elma W. Schneider prof. psychology Rice U., Houston, 1988—; adj. prof. Baylor Coll. Medicine, 1992—. Editor: Perceptual Organization, 1981, The Perception of Structure, 1991. Fellow APA, Am. Psychol. Soc.; mem. Psychonomic Soc. Office: Rice U Sch Social Scis Houston TX 77251-1892

POMEROY, KENT LYTLE, physical medicine and rehabilitation physician; b. Phoenix, Apr. 21, 1935; s. Benjamin Kent and LaVerne (Hamblin) P.; m. Karen Jodelle Thomas (dec. Dec. 1962); 1 child, Charlotte Ann; m. Margo Delilah Tuttle, Mar. 27, 1964 (div. Jan. 1990); children: Benjamin Kent II, Janel Elise, Jonathan Barrett, Kimberly Eve; m. Brenda Pauline North, Sept. 1, 1990. BS in Phys. Sci., Ariz. State U., 1960; MD, U. Utah, 1963. Diplomate Am. Bd. Phys. Medicine and Rehab. Rotating intern Good Samaritan Hosp., Phoenix, 1963-64; resident in phys. medicine and rehab. Good Samaritan Hosp., 1966-69, asst. tng. dir. Inst. Rehab. Medicine, 1970-74, dir. residency tng., 1974-76, asst. med. dir., 1973-76; dir. Phoenix Phys. Medicine Ctr., 1980-85, Ariz. Found. on Study Pain, Phoenix, 1980-85; pvt. practice, Scottsdale, Ariz., 1988—; lectr. in field. Contbr. articles to med. jours. Leader Theodore Roosevelt coun. Boy Scouts Am.; mem. exec. posse Maricopa County Sheriff's Office, Phoenix, 1981—, posse comdr., 1992—; mem. med. adv. bd. Grand Canyon-Saguaro chpt. Nat. Found. March of Dimes, 1970-78. Capt. M.C., U.S. Army, 1964-66. Recipient Scouter's Tng. award Theodore Roosevelt coun. Boy Scouts Am., 1984, Scouter's Woodbadge, 1985. Mem. AMA, Am. Acad. Phys. Medicine and Rehab., Internat. Rehab. Medicine Assn., Am. Assn. Orthopaedic Medicine (co-founder, sec.-treas. 1982-88, pres. 1988-90), Pan Am. Med. Assn. (diplomate), Prolotherapy Assn. (pres. 1981-83), Am. Pain Soc., Western Pain Soc., Am. Assn. for Study Headache, Am. Thermographic Soc. (charter), Am. Soc. Addiction Medicine (sec. Ariz. chpt.), Am. Acad. Pain Medicine, Wilderness Medicine Soc., Acad. Clin. Neurophysiology, Ariz. Soc. Phys. Medicine (pres. 1977-78), Ariz. Med. Assn., Maricopa County Med. Soc., Mil. Officers World Wars (1st vice comdr. Phoenix chpt. 1993—), others. Mem. LDS Ch. Avocations: camping, drawing, painting, writing poetry, music. Office: Royal Orthopedic & Pain Rehab Assocs 9755 N 90th St Ste A-205 Scottsdale AZ 85258 also: 2536 N 3rd St Ste 3 Phoenix AZ 85004

PON-BROWN, KAY MIGYOKU, information systems specialist; b. Ft. Lewis, Wash., Mar. 15, 1956; d. Gin Ung and Toyo (China) Pon; m. John Joseph Brown, July 28, 1979; 1 child, J. Jason. BS in Chemistry, U. Idaho, 1978, BA in Zoology, 1978; BS in Math., Boise State U., 1984. Chemist Century Labs., Inc., Boise, Idaho, 1982-84; from customer support specialist to tech. support specialist Learned-Mahn, Inc., Boise, 1984-87; from computer programmer to coord. Idaho Power, Boise, 1987-90, customer solution rep., 1990-92; tech. specialist Hewlett-Packard Co., Boise, 1992—; cons. Eclipse, Inc., Boise, 1991—, bd. dirs. 1991—, pres., 1992—, Discovery Ctr. Idaho, Boise, 1991—. Bd. dirs. Idaho Zool. Soc., Boise, 1986—, Ada County Divsn. Am. Heart Assn., Boise, 1992—, divsn. sec., 1992-93; mem. Jr. League Boise, 1989—; vol. Am. Cancer Soc., 1990—; vol. newsletter editor Idaho Soc. Profl. Engrs., 1992—. 1st Place husband & wife team Royal Victoria Marathon, 1992. Mem. Data Processing Mgmt. Assn. (bd. dirs. 1993, newletter editor 1993), Meridian Toastmasters (pres. 1986, dist. 15 best newsletter 1986), Idaho PC Users Group, Greater Boise Road Runners Club (bd. dirs. 1993—, charter pres. 1993). Avocations: running, music-flute, bowling. Home: 1053 Egurrola Pl Boise ID 83709 Office: Hewlett-Packard Co 11311 Chinden Blvd Boise ID 83714

PONGSIRI, NUTAVOOT, chemical engineer; b. Chachoengsao, Thailand, Oct. 1, 1961; s. Paiboon and Pairough (Jantarawat) P. B Chem. Engring., Chulalongkorn U., Bangkok, 1983; MSc in Chem. Engring., U. Mo., Columbia, 1989. Registered profl. engr., Thailand. Process engr. Signetics (Thailand), Ltd., Bangkok, 1983-86; prodn. engr. Unocal Thailand, Ltd.,

Bangkok, 1989-90, mgr. prodn. engring., 1991—; co-presenter at profl. confs. Contbr. articles to Indsl. Engring. Chem. Rsch., Engring. Jour. Chulalongkorn U., Thailand Engring. Jour., 5th Ascope Conf., 1993. Mem. Am. Inst. Chem. Engrs., Am. Chem. Soc., Am. Inst. Chemists, N.Y. Acad. Scis. Home: 21/5 Mu 7 Nuangket, 24000 Muang Chachoengsao Thailand Office: Unocal Thailand Ltd, 1693 Phaholyothin Rd, 10900 Bangkok Thailand

PONNAMPERUMA, CYRIL ANDREW, chemist; b. Galle, Sri Lanka, Oct. 16, 1923; came to U.S., 1959, naturalized; s. Andrew and Grace (Siriwardene) P.; m. Valli Pal, Mar. 19, 1955; 1 child, Roshini. BA, U. Madras, 1948; BSc, U. London, 1959; PhD, U. Calif., Berkeley, 1962; DSc, U. Sri Lanka, 1978. Research assoc. Lawrence Radiation Lab., U. Calif., 1960-62; research scientist Ames Research Ctr., NASA, Mountain View, Calif., 1962-70; prof. chemistry U. Md., College Park, 1971—; dir. Arthur C. Clarke Ctr., Sri Lanka, 1984-86; sci. advisor to pres., Sri Lanka, 1984—; dir. Inst. Fundamental Studies, Sri Lanka, 1984-91. Author: Origins of Life, 1972; Cosmic Evolution, 1978. Contbr. articles to profl. jours. Pres. Third World Found., 1991. Fellow Royal Inst. Chemistry, Third World Acad. Scis.; mem. Am. Chem. Soc., Astron. Assn., Am. Soc. Biol. Chemists, AAAS, Geochem. Soc., Radiation Research Soc. Office: U Md Lab Chem Evolution College Park MD 20742

PONOSOV, ARCADY VLADIMIROVITCH, mathematician; b. Perm, Ural, USSR, June 29, 1957; came to Germany, 1990; s. Vladimir Pavlovitch and Olga Moiseevna (Kosak) P.; m. Tatjana Lurje, Aug. 12, 1977; children: Olga, Marina. Student, Byelorussian State U., Minsk, USSR, 1975-77; MD with honors, Perm (USSR) State U., 1979; PhD, Ural State U., Ekaterinburg, USSR, 1983. Lectr. Perm State U., 1983-85, sr. lectr., 1985-86, assoc. prof., 1986-88; sr. rsch. assoc. Perm Poly. Inst., 1988-90; rsch. fellow Ruhr U., Bochum, Germany, 1991—; rsch. fellow Alexander von Humboldt Found., 1991-92, Deutsche Forschungsgemeinschaft, 1993—. Contbr. articles on math. to profl. jours. Recipient Best Young Scientist award Perm State U., 1983, Best Rsch. Work awards Perm Poly. Inst., 1989, 90. Avocations: foreign languages, travel, plays. Office: Ruhr U Bochum, Universitätstraße 150, 4630 Bochum Germany

PONTE, CHARLES DENNIS, pharmacist, educator; b. Waterbury, Conn., Jan. 17, 1953; s. Americo Joseph and Irene (Poirier) P. BSc in Pharmacy, U. Conn., 1975; D Pharmacy, U. Utah, 1980. Diplomate Am. Acad. Pain Mgmt.; cert. diabetes edn. Intern Woodbury (Conn.) Drug Co., 1975; hosp. pharmacy resident Yale-New Haven Hosp., 1975-76, ambulatory staff pharmacist, 1976-78; prof. clin. pharmacy, family medicine Robert C. Byrd Health Scis. Ctr. of W.Va. U., Morgantown, 1980—; adv. bd. ambulatory care and family practice DICP, Cin., 1985—; mem. practice interest adv. panel Am. Soc. Hosp. Pharmacists, Bethesda, 1990-92, adv. panel on family practice USP Conv., Inc., Rockville, Md., 1990—; vis. faculty UpJohn Co., Kalamazoo, 1986; coord. Sch. Pharmacy, Spencer State Tng. Ctr., 1984-88; participant Practical Aspects of Diabetes Care: A Conf. for Pharmacy Educators, 1989; chmn. Van Liere Rsch. Convocation for Med. Students, 1990. Contbr. to profl. publs. Grantee Robert Wood Johnson Found., 1981. Fellow Am. Soc. Hosp. Pharmacists; mem. Am. Coll. Clin. Pharmacy, Am. Coll. Clin. Pharmacology, Soc. Tchrs. Family Medicine, Am. Diabetes Assn. (pres. W.Va. affiliate 1985-86), Phi Kappa Phi, Phi Lambda Sigma. Roman Catholic. Office: W Va U Robert C Byrd Health Sci Ct Sch Pharmacy Morgantown WV 26506

PONTIROLI, ANTONIO ETTORE, medical educator; b. Milan, Italy, July 25, 1947; s. Luigi and Andreina (Bardelli) P.; m. Paola Giuliana Rietti, Dec. 2, 1972; children: Andrew, Francesca. Sci. diploma, Leonardo Da Vinci, Milan, 1965; medical diploma, State U., Milan, 1971. Vol. physician State U., Milan, 1971-72; staff physician Inst. San Raffaele, Milan, 1973-75; rsch. assoc. Wayne State U., Sinai Hosp., Detroit, 1976; established investigator State U., Milan, 1977-88, assoc. prof., 1988—; vice chmn. Inst. San Raffaele, Milan, 1981—, clin. investigations, 1988—; lectr. in field. Contbr. articles to profl. jours.; inventor in field. Recipient Lepetit prize, State U., Milano, 1972. Mem. Endocrine Soc. USA, European Soc. Clin. Investigation (councilor 1993—), European Assn. Study Diabetes, Soc. Italian Diabetes (scientific coun. 1986—), Soc. Italian Endocrinology, Sport Club Milan. Avocations: classical and opera music, painting, skiing, tennis, literature. Home: Residenza Cerchi-Milano 2, 20090 Segrate Italy Office: Inst San Raffaele, Via Olgettina 60, 20132 Milan Italy

POOLE, MARION RONALD, civil engineering executive; b. Thomasville, N.C., July 2, 1936; s. Everette Worth and Mabel Emma (Crotts) P.; m. Janice Rochelle Dickens, May 9, 1959; children: Anne Kathleen, Amy Kathryn, Marie Elizabeth. BCE, N.C. State U., 1958, MSCE in Transp., 1961, PhDCE in Transp., 1982. Registered profl. engr., N.C., registered land surveyor, N.C. Engr.-in-tng. N.C. State Hwy. Commn., Winston-Salem, 1958; hwy. planning engr. N.C. State Hwy. Commn., Raleigh, 1961-63; thoroughfare planning engr. N.C. State Hwy. Commn./N.C. Dept. Transp., Raleigh, 1963-84; asst. mgr. planning and environ. N.C. Dept. Transp., Raleigh, 1984-91, mgr. statewide planning br., 1991—; chmn. com. AIDO5 Transp. Rsch. Bd., NAS, 1992—, sec., 1986-92; mem. com. AIBO7 Transp. Rsch. Bd., NAS, 1988—, mem. com. AIDO4, Transp. Rsch. Bd., 1988—. Contbr. articles to Transp. Rsch. Record, Jour. Advanced Transp.; author reports. Club rep. to Tarheel Swim Assn., Glen Forest Swim Club, Raleigh, 1982-88, v.p., 1982-84, bd. dirs., 1975-78. 2nd lt. C.E., U.S. Army, 1958-59, capt. C.E., USAR, 1960-66. Mem. NSPE, Inst. Transp. Engrs., Am. Soc. Value Engrs. Baptist. Achievements include research on the benefits matrix model for transportation project evaluation, on procedure for synthesizing travel movements. Home: 4605 Woodridge Dr Raleigh NC 27612 Office: NC Dept Transp 1 S Wilmington St Raleigh NC 27611

POOLE, RICHARD WILLIAM, economics educator; b. Oklahoma City, Dec. 4, 1927; s. William Robert and Lois (Spicer) P.; m. Bertha Lynn Mehr, July 28, 1950; children: Richard William, Laura Lynne, Mark Stephen. B.S., U. Okla., 1951, M.B.A., 1952; postgrad., George Washington U., 1957-58; Ph.D., Okla. State U., 1960. Research analyst Okla. Gas & Electric Co., Oklahoma City, 1952- 54; mgr. sci. and mfg. devel. dept. Oklahoma City C. of C., 1954-57; mgr. Office of J.E. Webb, Washington, 1957-58; instr., asst. prof., assoc. prof., prof. econs. Okla. State U., Stillwater, 1960-65; prof. econs., dean Coll. Bus. Adminstrn. Okla. State U., 1965-72, v.p., prof. econs., 1972-88, Regents Disting. Service prof., prof. econs., 1988—; cons. to adminstr. NASA, Washington, 1961-69; adviser subcom. on govt. rsch. U.S. Senate, 1966-69; lectr. Intermediate Sch. Banking, Ops. Mgmt. Sch., Okla. Bankers Assn., 1968-89. Author: (with others) The Oklahoma Economy, 1963, County Building Block Data for Regional Analysis, 1965. Mem. Gov.'s Com. on Devel. Ark.-Verdigris Waterway, 1970-71; Past v.p., bd. dirs., mem. exec. com. Okla. Council on Econ. Edn.; past bd. dirs., past chmn. Mid-Continent Research and Devel. Council. Served to 2d lt., arty. U.S. Army, 1946-48. Recipient Delta Sigma Pi Gold Key award Coll. Bus. Adminstrn., U. Okla., 1951. Mem. Southwestern Econ. Assn. (past pres.), Am. Assembly Collegiate Schs. Bus. (past bd. dirs.), Nat. Assn. State Univs. and Land Grant Colls. (past chmn. commn. on edn. for bus. professions), Southwestern Bus. Adminstrn. Assn. (past pres.), Okla. C. of C. (past bd. dirs.), Stillwater C. of C. (past bd. dirs. and pres.), Beta Gamma Sigma (past bd.dirs.), Phi Kappa Phi, Phi Eta Sigma, Omicron Delta Kappa. Home: 124 W Georgia Ave Stillwater OK 74075-3706

POON, CHI-SANG, biomedical engineering researcher; b. Kowloon, Hong Kong, Nov. 12, 1952; came to U.S., 1977; s. Chee-Kong and Man-Hing (Wong) P.; m. Sau-Chun Ng, Aug. 12, 1978; 1 child, Ivan Yun-Wye. BS in Engring. with hons., U. Hong Kong, 1975; MPhil, Chinese U., Hong Kong, 1977; PhD, U. Calif., 1981. Asst. prof. N.D. State U., Fargo, 1981-85, assoc. prof., 1985-89; prin. rsch. scientist MIT, Cambridge, 1989--; cons. NIH, Bethesda, Md., 1989, 93, NSF, Washington, 1990. Contbr. articles to profl. jours. Recipient NIH Rsch. award, 1983, 84, 88, 92, Rsch. award Am. Heart Assn., 1983, Rsch. award NSF, 1985, 87, 92, Interant. Exchange award Nat. Sci. and Engring. Coun., 1986, Whitaker Found. Rsch. award, 1986. Mem. IEEE, Biomed. Engring. Soc. (Harold Lamport award 1983), Sigma Xi (Outstanding Rsch. award 1987). Achievements include patent for Respirator Triggering Mechanism and for a computationalluy efficient and physically implementabel multilayer perception architecture. Office: MIT 77 Massachusetts Ave Cambridge MA 02139

POON, WILLIAM WAI-LIK, program analyst; b. Beijing, Apr. 27, 1965; came to the U.S., 1984; s. Chun Yuan and Wing Kuen (Yeung) P. BS, U. Calif., Berkeley, 1988; MBA, So. Meth. U., 1992. Infosystem specialist Cen. Design Group Superconducting Super Collider, Lawrence Berkeley Lab., 1988-89; ops. rsch. analyst Superconducting Super Collider Lab., Dallas, 1989-92, program analyst, 1992—. Mem. NSPE, Inst. Indsl. Engrs., Inst. Cert. Mgmt. Accts. Office: SSC Lab MS1070 2550 Beckleymeade Ave Dallas TX 75237

POP, EMIL, research chemist; b. Tirgu Mures, Romania, Aug. 12, 1939; came to U.S., 1983; s. Victor and Rosalia (Graf) P.; m. Elena Petrina Petri, Apr. 28, 1964; 1 child, Andreea Christina. BS, Babes-Bolyai U., Cluj., Romania, 1961; PhD, Inst. Chemistry, Cluj. and Supreme Coun. for Sci. Titles, Dept. of Edn. Bucharest, Romania, 1973. Chemist Chem. Pharm. Research Inst., Cluj-Napoca, Romania, 1962-65, researcher, 1965-78, sr. researcher, group leader, compartment leader, 1978-83; researcher Rugjer Boskovic Inst., Zagreb, Yugoslavia, 1971-72; postdoctoral research assoc. U. Fla., Gainesville, 1983-86; sr. research scientist Pharmatec, Inc., Alachua, Fla., 1986-87, group leader, 1987-88; assoc. dir. chem. devel., 1988-91, dir. chemistry, 1992, Pharmos Corp., Alachua, Fla., 1992—. Contbr. articles to profl. jours.; inventor in field. Recipient N. Teclu award Romanian Acad. Sci., 1980. Fellow Am. Inst. Chemists; mem. Am. Chem. Soc., AAAS, Am. Assn. Pharm. Scientists, Internat. Union Pure and Applied Chemistry, N.Y. Acad. Scis., Internat. Soc. Quantum Biology and Pharmacology. Greek Catholic. Achievements include design and synthesis of pharmaceutical compounds in particular brain specific chemical drug delivery systems; M.O. calculations. Home: 810 SW 51st Way Gainesville FL 32607-3856

POPA, PETRU, federal agency administrator, engineering executive, educator; b. Vaslui, Romania, June 30, 1938; s. Stefan-Stefan and Maria (Mihail) P.; m. Ana Iorgulescu, Aug. 14, 1962; children: Claudia, Catalin-Stefan, Sorin-Petru. Grad., Poly. Inst., Bucharest, Romania, 1961; PhD in Engring., Inst. Atomic Physics, Bucharest, 1974. Sci. rschr. Inst. Atomic Physics, Bucharest, 1961-70; sr. insp. State Com. Nuclear Energy, Bucharest, 1970-89; sr. insp., gen. insp., dir. Nat. Commn. Control Nuclear Activities, Bucharest, 1990, chmn., 1990—. Contbr. over 20 sci. papers, articles and revs. ot profl. jours. Christian Orthodox. Office: Commn Control Nuclear Activities, Bd Libertastii 12, Bucharest Romania*

POPE, CAREY NAT, neurotoxicologist; b. Pasadena, Tex., Sept. 3, 1952; s. R.E. and Myra Mae (Parker) P.; m. Kerry Lee Nelson, Jan. 25, 1984; children: Eric Lee, Natalee Mae, Nelson Lynn. MS, Stephen F. Austin State U., 1979; PhD, U. Tex., Houston, 1985. NAS rsch. assoc. U.S. EPA, Research Triangle Park, N.C., 1986-89; asst. prof. neurotoxicology N.E. La. U., Monroe, 1989—; mem. sci. adv. panel U.S. EPA, Washington, 1992. Contbr. articles to profl. publs. U.S. EPA grantee, 1989, 92. Mem. AAAS, South Ctrl. Soc. Toxicology (councilor 1991-92, sec. 1994), Internat. Soc. Behavioral Neuroscis. Achievements include discovery of potentiation of organophosphate induced delayed neurotoxicity. Home: 181 Morgan Hare Monroe LA 71203 Office: Northeast La U Coll Pharmacy & Health Sci Monroe LA 71209-0470

POPE, HARRISON GRAHAM, JR., psychiatrist, educator; b. Lynn, Mass., Dec. 26, 1947; s. H. Graham and Alice (Rider) P.; m. Mary M. Quinn, June 7, 1974; children: Kimberly, Hilary, Courtney. AB summa cum laude, Harvard U., 1969, MPH, 1972, MD, 1974. Diplomate Am. Bd. Psychiatry and Neurology. Resident in psychiatry McLean Hosp., Belmont, Mass., 1974-77, clin. rsch. fellow Mailman Rsch. Ctr., 1977-79, asst. psychiatrist, 1979-84, assoc. psychiatrist, 1984-92, psychiatrist, 1992—, chief biol. psychiatry lab., 1984—; Dupont-Warren rsch. fellow Harvard Med. Sch., Boston, 1976-77; instr. psychiatry Harvard Med. Sch., Boston, 1977-82, asst. prof., 1982-85, assoc. prof., 1985—; staff psychiatrist Hampstead (N.H.) Hosp., 1976-80; vis. fellow The Maudsley Hosp., London, 1977, Hôp. Ste. Anne, Paris, 1977; mem. Am. Psychiat. Assn., 1976-80, adv. com. on schizophrenic, paranoid and affective disorders, 1979, adv. com. on preparation of DSM-III-R, 1984, task force on nomenclature and stats., 1979, 84. Author: Voices from the Drug Culture, 1971, The Road East, 1974, (with Hudson J.I.) New Hope for Binge Eaters: Advances in the Understanding and Treatment of Bulimia, 1984; co-editor: The Psychobiology of Bulimia, 1987, Use of Anticonvulsants in Psychiatry: Recent Advances, 1988; contbr. over 200 papers on biol. psychiatry, with particular emphasis on diagnosis of psychotic disorders, treatment of mood disorders and eating disorders, and substance abuse, particularly abuse of anabolic steroids by athletes; mem. editorial bd. European Psychiatry, Paris, 1984—, Internat. Jour. of Eating Disorders, 1984—. Named one of Outstanding Americans under 40 Esquire mag., 1984; fellow Scottish Rite Schizophrenia Program, No. Masonic Jurisdiction, 1977-81, Charles A. King Trust, Boston, 1977-79. Avocation: weightlifting. Office: McLean Hosp 115 Mill St Belmont MA 02178-1048

POPE, RICHARD M., rheumatologist; b. Chgo., Jan. 10, 1946. Student, Procopius Coll., 1963-65, U. Ill., 1965-66; MD, Loyola U., 1970. Diplomate Am. Bd. Internal Medicine. Intern in internal medicine Med. Ctr. Michael Reese Hosp., Chgo., 1970-71, resident in internal medicine, 1971-72; fellow in rheumatology U. Wash., Seattle, 1972-74; asst. clin. prof. medicine U. Hawaii, 1974-77; asst. prof. medicine U. Tex. Health Sci. Ctr., San Antonio, 1976-81, assoc. prof. medicine, 1981-85; assoc. prof. medicine Northwestern U. Med. Sch., 1985-88, prof. medicine, 1988—; attending physician Northwestern Meml. Hosp., Chgo., 1985—, VA Lakeside Med. Ctr., Chgo., 1985—, Rehab. Inst. Chgo., 1985—; chief divsn. rheumatology VA Lakeside Med. Ctr., 1985-91, divsn. arthritis-connective tissue diseases Northwestern Meml. Hosp., 1989—, Northwestern Med. Faculty Found., 1989—; mem. program com. Ctrl. Soc. Clin. Rsch., 1987, ctrl. region Am. Rheumatism Assn., 1987; mem. sci. com. Ill. chpt. Arthritis Found., 1988—, bd. dirs., 1990—, mem. chpt. review grants subcom., 1983-88, chmn. chpt. rsch. grant subcom., 1986-88, mem. rsch. com., 1986-88; mem. site visit teams NIH, 1986, 87, 89; cons. reviewer VA Merit Review Bd., 1984, 87, 91; cons. reviewer Arthritis Soc. Can., 1986, 87; mem. editorial adv. bd. Jour. Lab. and Clin. Medicine, 1991—. Author: (with others) The Science and Practice of Clinical Medicine, 1979, Proceedings of the University of South Florida International Symposium in the Biomedical Sciences, 1984, Concepts in Immunopathology, 1985, Biology Based Immunomodulators in the Therapy of Rheumatic Diseases, 1986, Primer on the Rheumatic Diseases, 1988; contbr. numerous articles to profl. jours. With U.S. Army, 1974-76. Anglo-Am. Rheumatology fellow, 1983. Mem. Am. Coll. Physicians, Am. Coll. Rheumatology (councillor ctrl. region coun. 1990—, program com. 1983-86, 91), Am. Assn. Immunologists, Am. Fedn. Clin. Rsch., Am. Soc. Clin. Investigation, Lupus Found. Ill. (med. adv. bd. 1990—), Chgo. Rheumatism Assn. (pres. 1991—), Ctrl. Soc. Clin. Investigation, Soc. Irish and Am. Rheumatologists (sec., treas. 1989—), Univ. Rheumatology Coun. Chgo., Alpha Omega Alpha. Achievements include research in pathophysiology of rheumatoid arthritis, T cell activation, T cell receptor, macrophage migration. Office: Northwestern U Multipurpose Arthritis Ctr 303 E Chicago Ave Chicago IL 60611-3008

POPE, WILLIAM DAVID, III, pipeline engineer; b. Jackson, Tenn., Apr. 22, 1952; s. William David Jr. and Betty Lawrence (Taylor) P.; m. Emily Louise Veillon, Aug. 7, 1978; children: W. David IV, Christopher Michael, Matthew Scott, Kevin Paul. BSME, U. Tenn., 1974. Registered profl. engr., La. Gas engr. Texaco, Inc., Morgan City, La., 1975-78; asst. mgr. Sugar Bowl Gas Corp., Thibodaux, La., 1978-83; ops. mgr. Acadian Gas Pipeline System, Thibodaux, 1984-88, v.p. engring. and ops., 1988—. Area coord. MATHCOUNTS, Bayou chpt., La., 1988-90; pres. Thibodaux Biddy Basketball Assn., 1992-93. Mem. ASME, NSPE (pres. Bayou chpt. 1987-88), La. Engring. Soc. (pres. Bayou chpt. 1987-88), Jaycees (Outstanding Young Man of Thibodaux 1992). Home: 205 Abigail Dr Thibodaux LA 70301 Office: Acadian Gas Pipeline System 514 Waverly Rd Thibodaux LA 70301

POPEK, EDWINA JANE, pathologist; b. Lexington, Mo., Sept. 8, 1952; d. Edward and Erma Lea (McMullin) P. BS, U. Mo., Kansas City, 1974; DO, Kansas City U. Health Scis., 1981. Diplomate Am. Bd. Pathology, Am. Bd. Anatomic/Clin. Pathology, Am. Bd. Pediatric Pathology. Commd. U.S. Army, 1981, advanced through grades to lt. col., 1993; intern. William Beaumont Army Med. Ctr., El Paso, Tex., 1981-82; resident Letterman Army Med. Ctr., San Francisco, 1982-85; fellow Children's Hosp., Denver, 1987-89; chmn. pediatric pathology Armed Forces Inst. Pathology, Washington, 1989-93; with Tex. Children's Hosp., Houston, 1993—. Fellow Soc. Pediatric Pathologists (practice com. 1991); mem. AMA, U.S. and Can. Acad., Pathology. Office: TX Childrens Hosp 6621 Fannin Houston TX 77030

POPKIN, CAROL LEDERHAUS, epidemiologist; b. N.Y.C., Nov. 19, 1944; d. Herman William and Bertha (Tamm) Lederhaus; m. Barry Matthew Popkin, June 17, 1981 (div.); 1 child, Matthew Lederhaus; m. Forrest Matthew Council, July 17, 1990. BA, U. Fla., 1966; MSPH, U. N.C., 1974. Statis. rsch. asst. Dept. Psychiatry, U. Chgo., 1969-70; rsch. assoc. Dept. Epidemiology, U. N.C., Chapel Hill, 1971-78, Hwy. Safety Rsch. Ctr., U. N.C., 1978-88; staff assoc. Hwy. Safety Rsch. Ctr., 1988-89; mgr. for driver studies Hwy. Safety Rsch. Ctr., U. N.C., 1990—; sec. Transp. Rsch. Bd. com. on alcohol, other drugs and transp., 1990—, mem. com. on accident records and analysis, 1987—, mem. com. on operator edn. and regulation, 1987—, mem. older driver task force, 1990—, mem. subcom. on accident stats.; mem. exec. com. Internat. Com. on Alcohol, Drugs and Traffic Safety, Internat. Workshop on Women Alcohol and Traffic; mem. N.C. DHR Older Driver Com., 1990—; mem. Nat. Hwy. Traffic Safety Adminstrn., expert panel on future directions of traffic safety programs, 1992; advisor hwy. safety com. N.C. Med. Soc.; cons. Gov.'s Inst. on Alcohol and Substance Abuse, 1992. Author: (with others) Alcohol, Accidents, and Injuries, 1986, Alcohol, Drugs and Traffic Safety-T86, Women, Alcohol, Drugs and Traffic; mem. editorial bd. Jour. of Traffic Medicine; contbr. articles to profl. jours. Mem. Nat. Coalition to Prevent Impaired Driving, 1992; mem. steering com. Decision '90, '91, '92; mem. Gov.'s Interagy. Team on Alcohol and Drug Abuse, 1992; bd. dirs. YMCA, Chapel Hill, 1977-82. Mem. APHA, Inst. for Rsch. in the Social Scis., Assn. for the Advancement of Automotive Medicine (reviewer transp. rsch. bd., Best Scientific Paper 1986), N.C. Passenger Safety Assn., N.C. DWI Coordinating Coun., Alpha Lambda Delta, Alpha Kappa Delta, Phi Kappa Phi. Achievements include rsch. on increasing involvement of women in drinking driving behavior, prolonged involvement of non-whites in drinking driving activity. Home: 2477 Foxwood Dr Chapel Hill NC 27514 Office: U NC Hwy Safety Rsch Ctr CB 3430 Chapel Hill NC 27599-3430

POPLE, JOHN ANTHONY, chemistry educator; b. Burnham, Somerset, Eng., Oct. 31, 1925; s. Herbert Keith and Mary Frances (Jones) P.; m. Joy Cynthia, Sept. 22, 1952; children: Hilary Jane, Adrian John, Mark Stephen, Andrew Keith. B.A. in Math., Cambridge U., Eng., 1946, M.A. in Math., 1950; Ph.D. in Math., Cambridge U., 1951. Research fellow Trinity Coll., Cambridge U., Eng., 1951-54, lectr. in math., 1954-58; Ford vis. prof. chemistry Carnegie Inst. Tech., Pitts., 1961-62; Carnegie prof. chem. physics Carnegie-Mellon U., Pitts., 1964-74, J.C. Warner prof., 1974-91; prof. Northwestern U., Evanston, Ill., 1986—. Recipient Wolf Found. prize in chemistry, 1992. Fellow Royal Soc. London, AAAS; fgn. mem. NAS. Office: Northwestern U Sheridan Rd Evanston IL 60208

POPOV, EGOR PAUL, engineering educator; b. Kiev, Russia, Feb. 19, 1913; s. Paul T. and Zoe (Derabin) P.; m. Irene Zofia Jozefowski, Feb. 18, 1939; children—Katherine, Alexander. B.S., U. Calif., 1933; M.S., Mass. Inst. Tech., 1934; Ph.D., Stanford, 1946. Registered civil, structural and mech. engr., Calif. Structural engr., bldg. designer Los Angeles, 1935-39; asst. prodn. engr. Southwestern Portland Cement Co., Los Angeles, 1939-42; machine designer Goodyear Tire & Rubber Co., Los Angeles, 1942-43; design engr. Aerojet Corp., Calif., 1943-45; asst. prof. civil engring. U. Calif. at Berkeley, 1946-48, assoc. prof., 1948-53, prof., 1953-83, prof. emeritus, 1983—, chmn. structural engring. and structural mechanics div., dir. structural engring. lab., 1956-60; Miller rsch. prof. Miller Inst. Basic Rsch. in Sci., 1968-69. Author: Mechanics of Materials, 1952, 2d edit., 1976, Introduction to Mechanics of Solids, 1968, Engineering Mechanics of Solids, 1990; Contbr. articles profl. jours. Recipient Disting. Tchr. award U. Calif.-Berkeley, 1976-77, Berkeley citation U. Calif.-Berkeley, 1983, Disting. Lectr. award Earthquake Engring. Rsch. Inst., 1993. Fellow AAAS (assoc.); mem. NAE, Am. Soc. Metals, Internat. Assn. Shell Structures (hon. mem.), ASCE (hon. mem., Ernst E. Howard award 1976, J. James R. Croes medal 1979, 82, Nathan M. Newmark medal 1981, Raymond C. Reese research prize 1986, Norman medal 1987, von Karman medal 1989), Soc. Exptl. Stress Analysis (Hetenyi award 1967, William M. Murray medallion 1986), Am. Soc. Engring. Edn. (Western Electric Fund award 1976-77, Disting. Educator award 1979), Am. Concrete Inst., Soc. Engring. Sci., Internat. Assn. Bridge and Structural Engring., Am. Inst. Steel Constrn. (adv. com. specifications, chmn. subcom. on seismic design), Sigma Xi, Chi Epsilon, Tau Beta Pi. Home: 2600 Virginia St Berkeley CA 94709-1031

POPOVICH, ROBERT P., biochemical engineer, educator; b. Sheboygan, Wis., Jan. 9, 1939. BS, U. Wis., 1963; MS, U. Wash., 1968, PhD in Chemical Engring., 1970. Chem. engr. Battelle Meml. Inst. Pacific Northwest Labs., 1963-65; teaching asst. chem. engring. U. Wash., 1965-70, asst. prof., 1970-72, from assoc. prof. to prof., 1970-83; prof. chem. and biochem. engring. U. Tex., 1983—; cons. chem. and biochem. engring. various pvt., pub. and govt. agys., 1972—; pres. Hemotherapy Instruments, Inc., 1974—; co-dir. Moncriet-Popovich Rsch. Inst. Inc., 1977—; biomed. engring. cons. Baxter-Travenol Labs. Inc., 1977—. Recipient Dialysis Pioneering award Nat. Kidney Found., 1983. Mem. AICE, Am. Soc. Artificial Internal Organs, Nat. Soc. Profl. Engrs., Am. Soc. Nephrol., European Dialysis & Transplant Assn., Internat. Soc. Peritoneal Dialysis. Achievements include research in industrial and biomedical applications of membrane systems, development of artificial internal organs, transport phenomena and biomedical instrumentation, continuous ambulatory peritoneal dialysis. Office: U Tex Austin Biomed Transport Lab ENS-610 Austin TX 78712*

POPOVICI, GALINA, physicist; b. Byhov, Mogilev, Bielorussia, Feb. 24, 1940; came to U.S., 1990; d. Nikolai M. and Valentina T. (Solonovich) Kornienko; m. Mihai P. Popovici, Nov. 6, 1959; children: Alexander-Mihai, Andrei-Dan. MS in Physics, Leningrad U., Leningrad, Russia, 1961; PhD in Physics, Inst. of Physics, Bucharest, Romania, 1970. Rsch. scientist Inst. of Physics of Romanian Acad. Scis., Bucharest, Romania, 1962-77; sr. rsch. scientist Inst. of Physics and Tech. of Materials, Bucharest, Romania, 1977-90; visitor Stanford (Calif.) U., 1990-91; rsch. assoc. prof. nuclear engring. dept. U. Mo.-Columbia, Columbia, Mo., 1991—. Contbr. over 60 papers to profl. publs. Mem. Am. Phys. Soc., Materials Rsch. Soc., Sigma Xi. Achievements include devel. of technologies to manufacture all-evaporat-edCu2S-CDS solar cells, electrolumniescent diodes and semiconductor lasers, wide band gap photovoltaic cells; investigation of recombination radiation of A3B5 and A2B6 semiconductors, CVD diamond. Home: 4406 Germantown Dr Columbia MO 65203 Office: U of Missouri Engring Bldg East Columbia MO 65211

POPP, CARL J., university administrator, chemistry educator; b. Chgo., Apr. 3, 1941; s. John and Blanche Virginia (Kozdron) P.; m. Barbara Ann Reiman, Sept. 4, 1965; children: Christine Marie, Diana Lyn, Julie Catherine. BS, Colo. State U., 1962; MA, So. Ill. U., 1965; PhD, U. Utah, 1968. Instr. So. Ill. U., Carbondale, 1965; postdoctoral fellow U. Utah, Salt Lake City, 1968; from asst. to assoc. prof. chemistry N.Mex. Tech., Socorro, 1969-73, prof., 1982—, v.p. acad. affairs, 1983—. Contbr. articles to profl. jours. Mem. AAAS, Am. Chem. Soc., Am. Geophys. Union, Sigma Xi. Office: N Mex Tech Campus Sta Socorro NM 87801

POPP, DALE D., orthopedic surgeon; b. Tama County, Iowa, July 6, 1923; s. Herbert John and Millie (Rayman) P.; m. Dorothy L. Higgins (div. July, 1970); children: Mark, Craig, Gordon, Brian, Nancy, James, Melissa; m. Carla Jean Drobny, Aug. 27, 1970; 1 child (stepson) Gary. BA, U. Iowa, 1944, MD, 1947. Am. Bd. Orthopedic Surgeons. Orthopedic surgeon Spokane (Wash.) Orthopedic Clinic, 1954-82, Inland Medic Evaluations, Spokane, 1986-87. Capt. USAF, 1951-53. Mem. Am. Acad. Orthopedic Surgeons, North Pacific Orthopedic Soc., Western Orthopedic Assn., Spokane Surg. Soc. Republican. Avocations: fishing, photography, computer, western and wildlife art.

POPPEN, ANDREW GERARD, environmental engineer; b. Ashland, Wis., Mar. 20, 1965; s. Lyle Gene and Dorothy May (Kicherer) P.; m. Mary Jane Murken, Feb. 14, 1992. AS, C.C. of Air Force, 1987; BSCE, U. Minn., 1991. Registered engr.-in-tng., Minn. Student engr. Mower County Hwy. Dept., Austin, Minn., 1991; environ. engr. HQ AMCCOM U.S. Army, Rock Island, Ill., 1992—. With USAF, 1984-87. Mem. ASCE. Office: US Army-HQ AMCCOM Bldg 108 AMSMC-EQE Rock Island IL 61299

POPPER, CHARLES WILLIAM, child and adolescent psychiatrist; b. Chgo., Mar. 30, 1946; s. Hans and Lina (B.) P. AB, Princeton U., 1967; MD, Harvard U., 1972. Cert. psychiatry, child psychiatry. Rsch. in psychiatry Mass. Gen. Hosp., Boston, 1968-72, resident in psychiatry, 1972-73; rsch. assoc. NIMH, Bethesda, Md., 1973-76; resident in child psychiatry McLean Hosp., Belmont, Mass., 1976-79, child psychiatrist, 1979—; cons. in psychiatry Children's Hosp., Boston, 1980—; clin. instr. Harvard Med. Sch., Boston, 1979—; chmn. sci. program New Eng. Coun. Child Psychiatry, Boston, 1980-90, continuing med. edn. Am. Acad. Child Psychiatry, Washington, 1986-91. Co-author: Concise Guide to Child and Adolescent Psychiatry, 1991; editor: Psychiatric Pharmacosciences of Children and Adolescents, 1987; editor Jour. Child and Adolescent Psychopharmacology, 1990—. Lt. comdr. USPHS, 1973-75. Fellow Am. Psychiat. Assn. (sci. program com. 1985-91), Am. Acad. Child and Adolescent Psychiatry (author quarterly column in newsletter 1988—). Office: McLean Hosp 115 Mill St Belmont MA 02178

POPPER, KARL (RAIMUND), author; b. Vienna, Austria, July 28, 1902; s. Simon Siegmund Carl and Jenny (Schiff) P.; m. Josefine Anna Henninger, Apr. 11, 1930 (dec. Nov. 1985). PhD, U. Vienna, 1928; DLitt (hon.), U. London, 1948; LLD (hon.), U. Chgo., 1962, U. Denver, 1966; LittD (hon.), U. Warwick, Eng., 1971, U. Canterbury, New Zealand, 1973; DLitt (hon.), City U. London, 1976, Salford (Eng.) U., 1976, Oxford, 1982; Dr. (hon.), U. Mannheim, Fed. Republic Germany, 1978; Dr. rer. nat. (hon.), U. Vienna, 1978; DLitt (hon.), U. Guelph, Can., 1978; Dr. rer. pol. (hon.), U. Frankfurt, Can., 1979; PhD (hon.), U. Salzburg, Austria, 1979; LittD (hon.), U. Cambridge, Eng., 1980, U. Oxford, 1982; DSc (hon.), Gustavus Adolphus Coll., St. Peter, Minn., 1982, U. London, 1986; PhD (hon.), U. Eichstatt, 1991, U. Madrid, 1991, U. Athens, 1992. Sr. lectr. U. N.Z., 1937-45; reader, then prof. logic and sci. method U. London, 1949-69, emeritus, 1969—; William James lectr. Harvard U., 1950; Compton Meml. lectr. Washington U., St. Louis, 1965; Henry Broadhead Meml. lectr. U. Christchurch, N.Z., 1973; Herbert Spencer lectr. Oxford U., 1961, 73; Shearman Meml. lectr. U. London, 1961; Romanes lectr. Oxford U., 1972; Darwin lectr., Cambridge U.; fellow Ctr. Advanced Studies, Stanford U., 1956-57, Inst. Advanced Studies, Canberra, Australia, 1962, Vienna, 1964; vis. prof. univs. in U.S., Australia; hon. fellow Darwin Coll., Cambridge, 1980; hon. research fellow dept. history and philosophy of sci. Chelsea Coll., U. London, 1982. Author: Logik der Forschung, 9th edit., 1989, The Open Society and Its Enemies, 14th edit., 1983, The Poverty of Historicism, 9th edit., 1976, The Logic of Scientific Discovery, 13th edit., 1987, Conjectures and Refutations, 10th edit., 1989, Objective Knowledge, 7th edit., 1983, Unended Quest: An Intellectual Autobiography, 7th edit., 1986, Realism and the Aim of Science, 1983, The Open Universe, 1982, Quantum Theory and the Schism in Physics, A World of Propensities, 1990; co-author: The Self and Its Brain, 1977, Auf der Suche nach einer besseren Welt, 4th edit., 1989, Die Zukunft ist offen, 2d edit., 1989. Created knight, 1965; decorated insignia Order of Companions of Honor, Grand Cross with star; recipient prize City of Vienna, 1965, Sonning prize U. Copenhagen, 1973, Lippincott award Am. Polit. Sci. Assn., 1976, Grand Decoration of Honour in gold, Austria, 1976, Dr. Karl Renner prize, Vienna, 1977, Dr. Leopold Lucas prize U. Tubingen, 1981, Internat. prize Catalonian Inst. of Mediterranean Studies, 1989, Kyoto prize Inamori Found., 1992. Fellow Royal Soc., Brit. Acad., London Sch. Econs. (hon.), Darwin Coll. Cambridge U. (hon.); mem. l'Inst. de France; mem. Am. Acad. Arts and Scis. (fgn. hon.), Internat. Acad. Philosophy Sci. (titulaer), Acad. Royale de Belgique (assoc.), Royal Soc. N.Z. (hon.), Acad. Internat. d'Histoire des Scis. (hon.), Deutsche Akademie für Sprache und Dichtung (hon.), Acad. Européene des Sciences, des Arts et des Lettres, Soc. Straniero dellé Accad. Nazionale dei Lincei, Austrian Acad. Sci. (hon.), Phi Beta Kappa (Harvard U. chpt.). Office: care London Sch of Econs, Houghton St, Aldwych, London WC2A 2AE, England

POPPERS, PAUL JULES, anesthesiologist, educator; b. Enschede, Netherlands, June 30, 1929; came to U.S., 1958; naturalized, 1963; s. Meyer and Minca (Ginsburg) P.; m. Ann Feinberg, June 3, 1969; children: David Matthew, Jeremy Samuel. MD, U. Amsterdam, 1955. Diplomate Am. Bd. Anesthesiology. Instr. anesthesiology Columbia U., N.Y.C., 1962-63, assoc., 1963-65, asst. prof. anesthesiology, 1965-71, assoc. prof. anesthesiology, 1971-74; prof., vice chmn. dept. anesthesiology NYU, 1974-79; prof., chmn. dept. anesthesiology SUNY, Stony Brook, 1979—; cons. Brookdale Med. Ctr., Bklyn., 1975—, VA Med. Ctr., Northport, N.Y., 1979—, Booth Meml. Hosp., Flushing, N.Y., 1979—, Am. Hosp. Paris, 1989-93; cons., lectr. USN Regional Med. Ctr., Portsmouth, Va., 1975-83. Author: Regional Anesthesia, 1977; editor: Beta Blockade and Anaesthesia, 1979; section editor Jour Clin. Anesthesia, 1990—; mem. editorial bd. Internat. Jour. Clin. Monitoring and Computing, 1990—; internat. bd. editors Anaesthesiology Digest, 1991—; contbr. numerous articles to profl. jours. NIH postdoctoral rsch. fellow, 1961; recipient medal Polish Acad. Scis., Poland, 1987, Univ. medal Jagiellonian U., Krakow, Poland, 1987, 1st Sci. award Post-grad. Assembly in Anesthesiology; named Hon. Prof. Anesthesiology, U. Leiden, The Netherlands, 1977. Fellow Am. Coll. Anesthesiology, Am. Coll. Ob-Gyns., Royal Soc. Medicine, Post-Grad. Assembly in Anesthesiology (hon. chmn. 1989—); mem. Am. Soc. Anesthesiologists, Assn. Univ. Anesthesiologists, Soc. Acad. Anesthesia Chmn., Internat. Anesthesia Rsch. Soc., Soc. Obstetric Anesthesia and Perinatology, Jerusalem Acad. Medicine, Am. Soc. Pharmacology and Exptl. Therapeutics, Fedn. Am. Soc. Exptl. Biology, Sigma Xi. Office: SUNY Sch Medicine Health Scis Ctr Stony Brook NY 11794-8480

POPPLESTONE, ROBIN JOHN, computer scientist, educator; b. Bristol, U.K., Dec. 9, 1938; came to U.S., 1986; s. Jack Raymond and Iris Lillian (Sanders) P.; m. Rose Elaine Cornock (div. 1969); children: Michael, Jennifer, Sally; m. Kristin Diane Morrison, Sept. 29, 1985. BSc, Queens U., Belfast, U.K., 1960. Lectr. Edinburgh U., Scotland, 1968-82, reader, 1982-86; prof. U. Mass., Amherst, 1986—. Mem. editorial bd. Internat. Jour. Robotics Rsch., 1984—. Fellow AAAI (founding). Home: 35 S Orchard Dr Amherst MA 01002 Office: U Mass Dept Computer Sci Amherst MA 01002

POPRICK, MARY ANN, psychologist; b. Chgo., June 25, 1939; d. Michael and Mary (Mihalcik) Poprick; B.A., De Paul U., 1960, M.A., 1964; Ph.D., Loyola U., Chgo., 1968. Intern in psychology Elgin (Ill.) State Hosp., 1961-62; staff psychologist, 1962; staff psychologist Ill. State Tng. Sch. for Girls, Geneva, 1962-63, Mt. Sinai Hosp., Chgo., 1963-64; lectr. psychology Loyola U. at Chgo., 1964-67; asst. prof. Lewis U., Lockport, 1967-70, assoc. prof., 1970-75, chmn. dept., 1968-72 (on leave 1972-73); postdoctoral intern in clin. psychology Ill. State Psychiat. Inst., Chgo., 1972-73; pvt. clin. practice David Psychiat. Clinic, Ltd., South Holland Ill., 1973-87; pvt. practice, South Holland, Ill., 1987—; assoc. sci. staff Riveredge Hosp., Forest Park, Ill., 1975-76; ltd. lic. practitioner dept. psychiatry Christ Hosp., Oak Lawn, Ill., 1983—. Co-chmn. commn. on personal growth and devel. Congregation of 3d Order St. Francis of Mary Immaculate, Joliet, 1970-71; clin. resource person Cath. Archdiocese of Chgo., 1977—. Mem. Am. Psychol. Assn. (rep. from Ill. 1985-88), Ill. (sec.-treas. acad. sect. 1975-77, mem. student devel. com. 1975-77, chmn. acad. sect. 1977-78, 78-79, mem. program com. 1977-78, sec. 1979-81, pres.-elect 1981-82, pres. 1982-83, past pres. 1983-84, chmn. program com. 1981-82, awards com. 1983-86, rep. Com. of ET and Minority Affairs 1988-89, rep. Cook County 1989-91), Anxiety Disorders Assn. Am., Midwestern Psychol. Assn. (Cook County rep. 1989-91), Soc. for Sci. Study Religion, AAAS, Chgo. Assn. Psychoanalytical Psychology (rsch. com. 1988), Kappa Gamma Pi, Psi Chi (sec. 1964-65, pres. 1965-66). Home: 547 Marquette Ave Calumet City IL 60409-3316 Office: 16284 Prince Dr South Holland IL 60473-3233

POPSTEFANIJA, IVAN, electrical engineer; b. Belgrade, Yugoslavia, Mar. 6, 1959; came to U.S., 1984; s. Petro and Irena (Sekosan) Pop-Stefanija; m. Marija Ololovska, Apr. 17, 1983; children: Igor, Aleksandar. Diploma in engring., U. Ljubljana, Slovenia, 1982; PhD, U. Mass., 1991. Rsch. assoc. Inst. Mihailo Pupin, Belgrade, 1983-84; grad. rsch. asst. U. Mass., Amherst, 1985-91; sr. rsch. fellow Microwave Remote Sensing Lab., U. Mass., Amherst, 1991—; sr. rsch. cons. Quadrant Engring., Inc., Amherst, 1991—. Mem. IEEE, Am. Geophysical Union. Office: Univ Mass Knowles Engring Bldg Amherst MA 01003

PORAN, CHAIM JEHUDA, civil and environmental engineer, educator; b. Haifa, Israel, Mar. 21, 1952; came to U.S., 1980; s. Menachem and Sarah Forschner. BSCE, Technion, Haifa, 1977; MS in Engring., U. Calif., Davis, 1983, PhD in Engring., 1985. Registered profl. engr., Calif., N.Y., N.C. Asst. lab. engr. Technion-Israel Inst. Tech., Haifa, 1974-77; project engr., then team leader Solel Boneh Internat., Inc., Nigeria, 1978-80; teacing assoc., then rsch. asst. U. Calif., Davis, 1980-84; asst. prof. civil engring. Poly. U., Bklyn., 1985-88, U. N.C., Charlotte, 1988—; cons. engring cos., 1985—. Contbr. articles to ASCE, Soils and Founds., Can. Geotech. Jour., other jours. Grantee NSF, 1987, 89, 91, Cray Rsch., Inc., 1991, 92, VIC, Ltd., 1990, 92—. Mem. ASCE (chair N.C. sect. geotech. group 1992-93, contbr. articles to jours.), Internat. Soc. Soil Mechanics and Found. Engring., Internat. Geosynthetics Soc., N.Am. Geosynthetics Soc. Achievements include development of analytical method and commercial software for computing shear wave velocity profiles in soils based on surface waves measurement; technique for construction quality assurance of dynamic compaction using telemetry. Office: GEI Cons Inc 7721 Six Forks Rd Ste 136 Raleigh NC 27615-5014

PORCHER, FRANK BRYAN, II, systems engineer; b. Ft. Knox, Ky., Apr. 10, 1953; s. Frank Bryan and Patsy Delores (Horton) P.; children: Frank Bryan III. AA, N.Mex. Mil. Inst., 1973; BBA, N.Mex. State U., 1977. Systems analyst Phys. Sci. Lab. N.Mex. State U., Las Cruces, 1974-79; system integration engr. Link div. Singer Corp., Houston, 1979-84; data processing cons. Sperry Corp., Arlington, Va., 1984-86; system applications specialist Planning Rsch. Corp., Houston, 1987-90; systems engr. Barrios Tech., Inc., Houston, 1990—. 1st lt. USAR, 1978-86. Mem. AIAA, Planetary Soc., Nat. Mgmt. Assn., Assn. Computing Machinery, Inst. Elec. and Electronics Engrs. Home: 628 Spring Breeze League City TX 77573 Office: NASA JSC DP41 2101 NASA Rd 1 Houston TX 77058

PORKERT, MANFRED (BRUNO), medical sciences educator, author; b. Decin, Czechoslovakia, Aug. 16, 1933; arrived in West Germany, 1945; s. Bruno and Elfriede (Walter) P.; m. Elisabeth Friederike Herrmann, 1974 (div. 1978); 1 child, Christine Franka. PhD, Universite de Paris, 1957. Rsch. fellow Centre Nat. de la Recehereche Sci., Paris, 1955-57, Deutsche Forschungsgemeinschaft, Munich and Bonn, Fed. Republic Germany, 1959-69; dozent Universitat Munich, 1970-75, prof., 1975-78, prof. extraordinary, 1978—. Editor, pub.: Acta medicinae sinensis, 1980-85; cons. editor Chinesische Medizin, 1986—; exec. editor-in-chief International Normative Dictionary of Chinese Medicine, 1989—; contbr. numerous articles to profl. jours. Mem. interdisciplinary lectures com. U. Munich, 1975-79, mem. univ. coun., 1977-79, sec. philos. faculty, 1975-77. Mem. Internat. Chinese Medicine Soc. (founder, pres. Munich chpt. 1978-85). Avocation: photography. Office: U Munich, Kaulbachstrasse 51a, 80539 Munich Germany

PORKOLAB, MIKLOS, physics educator, researcher; b. Budapest, Hungary, Mar. 24, 1939; came to U.S., 1963; s. Tivadar Ferenc and Piroska (Joachim) P. BASc, U. B.C., Vancouver, Can., 1963; PhD, Stanford U., 1967. Mem. rsch. staff Princeton (N.J.) Plasma Physics Lab., 1967-71, rsch. physicist, 1971-74; sr. rsch. physicist, lectr. Princeton Plasma Physics Lab. and U., 1974-75; prof. physics MIT, Cambridge, 1977—, assoc. dir. Plasma Fusion Ctr., 1991—; cons. Gen. Atomics, San Diego, 1987—, Lawrence Livermore (Calif.) Nat. Lab., 1978—; cons. numerous orgns., 1971-89. Editor jour. Physics Letters A, 1991—; contbr. numerous articles to profl. jours. Recipient U.S. Sr. Scientist award Humboldt Stiftung, Max Planck Inst., 1975; Woodrow Wilson fellow Stanford U., 1964. Fellow Am. Phys. Soc. (award for excellence for plasma rsch. 1984); mem. Internat. Unon of Pure and Applied Physics. Office: MIT NW17-286 175 Albany St Cambridge MA 02139

PORTAL, JEAN-CLAUDE, physicist; b. Villevayre, Aveyron, France, Apr. 2, 1941; s. Antonin and Gabrielle (Davy) P.; m. Rosine Marty, Dec. 17, 1966; children: Gilles, Beatrice, Jerome. Licence in Sci., U. P. Sabatier, Toulouse, France, 1966, PhD, 1969, D és-Scis., 1975. Asst. prof. solid state physics Inst. Nat. Sci. Appliquees, Toulouse, 1966-69, maitre, 1969-80, prof., 1980-87, prof. 1st class, 1987—; rsch. group leader Ctr. Nat. Rsch. Sci.-Inst. Nat. Scis. Appliquees, Toulouse, 1975, High Magnetic Field Lab., Ctr. Nat. Rsch. Sci., Grenoble, France, 1981; mem. several European rsch. programs. Co-editor: Optical Properties of Narrow Gap Low Dimensional Structures, 1987, Electronic Properties of Multi-layers and Low Dimensional Semiconductors Structures, 1990; mem. editorial bd. Semiconductor Sci. and Tech. Jour., 1990; contbr. book chpts. and over 300 sci. publs. to internat. jours. and confs. Mem. Am. Phys. Soc., Internat. Soc. for Optical Engring., Semiconductors and Advanced Microelectronics Sci. Instrumentation Engrs. Avocations: music, swimming, walking, reading. Office: Ctr Nat Rsch Sci/Sci Appliquees, 156 Ave Ranguell, 31077 Toulouse France also: Svc. Nat. Champs Intenses, 25 Ave Martyrs BP166, 38042 Grenoble France

PORTE, DANIEL, JR., physician, educator, health facility administrator; b. N.Y.C., Aug. 13, 1931; s. Daniel and Majorie Veronica (Clark) P.; m. Eunice Claire Ungerleider, Mar. 21, 1951; children—Jeffrey, Michael, Kenneth. B.A., Brown U., 1953; M.D., U. Chgo., 1957. Assoc. prof. medicine U. Wash., Seattle, 1969-73; prof. medicine U. Wash., 1973—; dir. U. Wash. Diabetes-Endocrinology Rsch. Ctr., 1977—; ACOS for research and devel. VA Med. Ctr., Seattle, 1971—; chief div. endocrinology and metabolism VA Med. Ctr., 1975—; chmn. med. sci. adv. bd. Juvenile Diabetes Found., N.Y.C., 1977-78; cons. Scripps Clinic and Research Found., San Diego, 1978-81; Connaught lectr. Can. Diabetes Assn., Toronto, Ont., 1984. Assoc. editor Am. Jour. Physiology, 1976-78, Jour. Diabetes, 1984—; contbr. numerous articles, chpts. to profl. publs., 1965—. Served to lt. comdr. USN, 1959-61. Recipient Career Devel. award NIH, 1968-70, David Rumbaugh award Juvenile Diabetes Found., 1984; Guggenheim fellow, 1985. Fellow ACP; mem. Assn. Am. Physicians, Am. Soc. Clin. Investigation, Am. Diabetes Assn. (pres. 1986-87, Eli Lilly award 1970), Am. Diabetes Assn. (Wash. affiliate) (pres. 1984-85), Western Assn. Physicians (councillor 1982-85). Avocation: skiing. Home: 2707 Boyer Ave E Seattle WA 98102-3931 Office: U Wash Diabetes-Endocrinology Rsch Ctr-VA Med Ctr 2B-21 1660 S Columbian Way Seattle WA 98108 1597*

PORTELA, ANTONIO GOUVEA, retired mechanical engineering researcher; b. Lisbon, Portugal, Jan. 26, 1918; s. Raul Lello and Esther Gouvea Portela; married, Sept. 29, 1965; 2 children. Engr. degree, Inst. Superior Tecnico, Lisbon, 1960; prof. degree, Inst. Sup. Tecn., Lisbon, 1958. Cert. mech. engring. Contbr. articles to profl. jours. Mem. AAAS, Am. Nuclear Soc., ASME, N.Y. Acad. Scis. Office: Inst Superior Tecnico, Av Rovisco Paes, Lisbon Portugal

PORTENIER, WALTER JAMES, aerospace engineer; b. Davenport, Iowa, Oct. 9, 1927; s. Walter Cleveland and Doris Lucile (Williams) P.; m. Martha L. Dallam, Aug. 26, 1950 (dec. April, 1986); children : Andrea Ellen, Renee Suzanne; m. Patty Grosskopf Caldwell, Oct. 3, 1992. B in Aero Engring., U. Minn., 1950; MS in Aero Engring., U. So. Calif., 1958, Engr. in Aerospace Engring., 1969. Sr. engr. aerodynamics N. Am. Aviation, L.A., 1951-61; MTS project engr., mgr. The Aerospace Corp., El Segundo, Calif., 1961-85; instr. U. So. Calif., L.A., 1979; cons. L-Systems, Inc., El Segundo, 1985-89. Pres., bd. dirs. First United Meth. Ch., Santa Monica, Calif., 1988-90; judge, range officer Internat. Shooting Union, 1989--. Recipient Bronze Medal Internat. Shooting Union, 1990. Fellow Am. Inst. Aeronautics and Astronautics (assoc.). Republican. Achievements include discovery of F-100 wing transonic buffet solution, blowing definition and test, design definition and test of area variation, F-108 mach 3 cruise canard and shock lift effectiveness, XB-70 transport wing definition for subsonic lift and supersonic cruise; re-entry systems analysis of vehicle design, payload, observables for systems procurement and technical direction; development of re-entry technology support for nosetip shape change, boundary layer transition, flow field codes, maneuvering technology; DoD-space transportation system support for space transportation system management (program definitions, manpower); launch on demand requirements; support of new booster systems performance options; reliability review of current systems; re-entry systems test in arms control environment; definition of space transportation system DoD reference missions and mission modeling for cost effectivemess, effective V/STOL aircraft implementation options. Home and Office: 2443 La Condesa Dr Los Angeles CA 90049

PORTER, ARTHUR T., oncologist, educator; b. June 11, 1956; m. Pamela Porter; 3 children. Student, U. Sierra Leone, 1974-75; BA in Anatomy, Cambridge U., 1978, M.B.B.Chir./M.D., 1980, MA, 1984; DMRT, Royal Coll. Radiologists, Eng., 1985; postgrad., U. Alta., 1984-86; FRCPC, Royal Coll. Physicians and Surgeons, Can., 1986; cert. for physicians mgr. program, U. Toronto, 1990; postgrad., LaSalle U. Lic., bd. cert., Mich., Can., Eng. House physician gen. medicine Norfolk and Norwich Hosp., Eng., 1981; house sugeon gen. surgery New Addenbrookes Hosp., Cambridge, Eng., 1981-82; sr. house officer clin. hematology No. Gen. Hosp., Sheffield, Eng., 1982; sr. house officer gen. medicine Huntington County Hosp., Hinchingbrooke Hosp., Eng., 1982-83; sr. house officer radiotherapy and oncology Norfolk and Norwich Hosp., Norwich, 1983-84; chief resident radiation oncology Cross Cancer Inst., Edmonton, Alta., Can., 1984-86, radiation oncologist, 1986-87, sr. radiation oncologist, 1987; asst. prof. faculty medicine U. Alta., Edmonton, 1987, assoc. clin. prof. dept. surgery faculty medicine, 1988; head divsn. radiation oncology U. Western Ont., London, Can., 1988; cons. radiation oncologist, chief dept. radiation oncology London Regional Cancer Ctr., 1988, program dir. radiation oncology, 1989-91; chmn. dept. oncology Victoria Hosp. Corp., London, 1990; assoc. prof. dept. oncology U. Western Ont., 1990; program dir. radiation oncology Wayne State U., Detroit, 1991-92, prof., chmn. dept. radiation oncology Sch. Medicine, 1991—; chief Gershenson Radiation Oncology Ctr. Harper Hosp., Detroit, 1991—; radiation oncologist-in-chief Detroit Med. Ctr., 1991—; pres., CEO Radiation Oncology R & D Ctr., Detroit, 1991—; dir. multidisciplinary svcs. Meyer L. Prentice Comprehensive Cancer Ctr., Detroit, 1992—; chmn. radiation oncology Grace Hosp., Detroit, 1993—; vis. prof. U. London, Eng., 1990, U. Mich., 1991, U. Ky., 1992, U. Rochester, 1992; cons. neutron therapy Dept. Health Govt. of U.K., 1990; mem. editorial bd. Endocurietherapy/Hyperthermia Oncology, 1991—, Cambridge Cancer Series, 1991—, Baxter Adminstrv. Manual, 1991—, Oncore, 1989-91, Internat. Monitor Oncology, 1992—; mem. genito-urinary com. Radiation Therapy Oncology Group, 1986—, new investigators com., 1986-87, bladder task force, 1986-88, time dose and fractionation, 1987—, large field working group, 1987—, full mem. com., 1987-91, exec. com., 1991; mem. radiation oncology com. Nat. Cancer Inst. Can., 1987-90, radiation quality assurance subcom., 1987-90, G.U. com., 1987-90; prin. investigator Radiation Therapy Oncology Group, U. Alta., 1987-88, U. Western Ont., 1989-91; mem. working group on bladder cancer Internat. Consensus, 1988, working group on prostate cancer, 1988; mem. Can. Uro-Oncology Group, 1990; chmn. brachytherapy subcom. Radiation Therapy Oncology Group, 1990-92, spl. populations com., 1991, systemic radionucleides com., 1991; mem. G.U. com. Southwest Oncology Group, 1991, selection com. Windsor Cancer Ctr., 1992; mem. cancer grant review conf. Nat. Cancer Inst., 1992; mem. NIH Sub-Saharan African Health Rsch. Initiative, 1993; chmn. South Western Ont. Uro-Oncology Group, 1988-91, Profl. Adv. Com. Radiation Oncology, 1990-91, Site Com. for Prostate Cancer, 1990-91; mem. Ont. Comm. Radiation Oncology, 1989-91, Cancer 2000 Com., 1990-91, PET com. Children's Hosp. Mich., 1991—, adv. com. dept. radiology Detroit Receiving Hosp., 1991—; dir. Univ. Physicians, Inc., 1991—; bd. dirs. MLPCCCMD, Binary Therapeutics, Vetrogen Corp., MedCyc, Med. Knowledge Systems, Am. Cancer Soc.; bd. trustees Fund Med. Rsch. Edn., 1991—, Meyer L. Prentice Comprehensive Cancer Ctr., 1992—; mem. exec. com. Am. Cancer Soc., Wayne County, 1992—; mem. Amersham Internat. Adv. Bd., 1992—; co-chmn. regional adv. bd. Am. Cancer Soc., 1992—, v.p., pres. elect, 1993—; mem. Medi-Pysics Adv. Bd., 1992—; pres. Biomide Corp. Bd., 1993—; lectr. in field. Author: (with others) Fundamental Problems in Breast CaProgress in Urological Cancers, 1988, Proceedings of the Consensus Meeting of the Treatment of Bladder Cancer-1987, 1988, Brachytherapy, 1989, High and Low Dose Rate Brachytherapy, 1989, Brachytherapy of Prostate Cancer, 1991; co-editor Treatment of Cancer, 1991—; assoc. editor Can. Jour. Oncology, 1990—, Antibody and Radiopharmaceuticals, 1992—; contbr. articles to profl. jours. Recipient Nat. award Sierra Leone, 1975-80, Commonwealth Found. scholarship, 1980, Best Doctor in Am. award, 1992, Testimonial Resolution, City of Detroit, 1993, Testimonial Resolution, County of Wayne, 1993. Fellow Am. Coll. Angiology, Detroit Acad. Medicine, Royal Soc. Medicine; mem. Am. Soc. Therapeutic Radiation Oncology, Am. Radium Soc., Am. Soc. Clin. Oncology, Am. Coll. Radiology, Am. Med. Assn. (Physicians Recognition award 1986), Am. Acad. Med. Adminstrs., Am. Acad. Oncology Adminstrs., Am. Endocurietherapy Soc. (sec. Can. chpt. 1988, sec.), Mich. State Med. Soc., Mich. Soc. Therapeutic Radiation Oncology, Mich. Radiol. Soc., Detroit Med. Soc. (Ann. award for Excellenc 1993), Wayne County Med. Soc., European Soc. Therapeutic Radiation Oncology, Brit. Inst. Radiology, Can. Oncology Soc., Can. Assn. Radiation Oncology, Royal Coll. Radiologists, Sierra Leone Med. and Dental Assn., Greater Detroit C. of C., Sigma Xi. Achievements include patent in a perineal applicator; research in novel methods in delivery dose, brachytherapy, intraoperative therapy, unsealed source therapy, verification and dosimetry, real time portal imaging, three-dimensional and planning, unsealed source dosimetry, the design of perineal applicators. Office: Radiation Oncology Rsch & Dev Ctr 4201 St Antoine Detroit MI 48201

PORTER, BRUCE JACKMAN, military engineer, computer software engineer; b. El Paso, Tex., Aug. 7, 1954; s. Covington Baskin and Carolyn Fee (Bruce) P.; m. Janette Anne Brown, Oct. 19, 1985; children: Laura, Holly, Travis. BS, US Mil. Acad., 1976; MS in Computer Sci., Stanford U., 1985, MS in Civil Engring., 1985. Engr. in tng., Pa. Commd. 1st lt. U.S. Army, 1979-80, advanced through grades to lt. col., 1993; co-commdr. 17th armored engr. bn. U.S. Army, Ft. Hood, Tex., 1977-80; constrn. engr. U.S. Army, Misawa, Japan, 1981-83; orgnl. evaluator U.S. Army, Ft. Leavenworth, Kans., 1989-90; ops. officer 5th engr., engr. combat bn. U.S. Army, Saudi Arabia and Iraq, 1990-91; assoc. prof. mathematics U.S. Mil. Acad., West Point, N.Y., 1985-88; chief concepts officer USA Engr. Sch., Ft. Leonard Wood, Mo., 1991—. Decorated Bronze Star, Legion of Merit. Home: 9 West Elm St Fort Smith AR 72956 Office: US Army Engr Sch Bldg 3200 Rm 166 Fort Leonard Wood MO 65473

PORTER, COLIN ANDREW, optics scientist; b. Melbourne, Victoria, Australia, Oct. 8, 1956; s. George Henry and Phyllis Violet (Fitzhenry) P. B of Applied Sci., Royal Melbourne Inst. Tech., 1979; grad. diploma in digital electronics, Swinburne Inst. Tech., Melbourne, 1983. Lab. tech. Methodists Ladies Coll., Kew, Victoria, 1978-80; rsch. officer Broken Hill Proprietry Co., Clayton, Victoria, 1980-84; sr. optics scientist Varian Australia Pty. Ltd., Mulgrave, Victoria, 1984—. Designer (optical design) Cary 4/5 Spectrophotometer, 1988. Mem. Australian Optical Soc. (founding mem.), Optical Soc. Am., Internat. Soc. Optical Engring., Planetry Soc., Australian Mensa Inc., Melbourne Cricket Club, Kelvin Club. Avocations: astrophotography, painting, computer programming, reading. Office: Varian Australia Propriety Ltd, 679 Springvale Rd, Mulgrave 3170 Victoria, Australia

PORTER, DALE WAYNE, biochemist; b. Pitts., Sept. 22, 1963; s. Richard Lee and Joan Janet (Jackson) P. BS in Biology, Allegheny Coll., 1986; MS in Agrl. Biochemistry, W.Va. U., 1988, PhD in Biochem. Genetics, 1992. Rsch./teaching asst. W.Va. U., Morgantown, 1986-92; NRC rsch. assoc. U.S. Army Med. Rsch. Inst. Chem. Def., Aberdeen, Md., 1992—; presenter at internat. sci. meetings. Contbr. articles to profl. jours., chpt. to book. Mem. Sigma Xi. Office: US Army Med Rsch Inst Chem Def Aberdeen Proving Ground MD 21010-5425

PORTER, LORD GEORGE, chemist, educator; b. Stainforth, Yorkshire, Eng., Dec. 6, 1920; s. John Smith and Alice Ann (Roebuck) P.; m. Stella Jean Brooke, Aug. 12, 1949; children: John B., Andrew C. G. B.Sc., U. Leeds, Eng., 1941; M.A., Ph.D., Cambridge (Eng.) U., 1949, Sc.D., 1959; D.Sc. (hon.), U. Utah, 1968, Sheffield U., 1968, U. East Anglia, U. Surrey, U. Durham, 1970, U. Leicester, U. Leeds, U. Heriot-Watt, City U., 1971, U. Manchester, U. St. Andrews, London U., 1972, Kent U., 1973, Oxford U., 1974, U. Hull, 1980, U. Rio de Janiero, 1980, Inst. Quimico de Sarria, Barcelona, 1984, U. Coimbra, Portugal, 1984, Open U., U. Pa., 1984, U. Philippines, 1985, U. Notre Dame, U. Bristol, U. Reading, 1986, U. Loughborough, 1987, U. Brunel, U. Bologna. Asst. research dir. phys. chemistry Cambridge U., 1952-54; asst. dir. Brit. Rayon Research, 1954-55; prof. phys. chemistry U. Sheffield, Eng., 1955-63, Firth prof., head dept. chemistry, 1963-66; dir., Fullerian prof. chemistry Royal Instn., 1966-85; pres. The Royal Soc., 1985-90; chmn. Ctr. for Photomolecular Scis. Imperial Coll., London, 1990—; prof. photochemistry Imperial Coll.; vis. prof. chemistry Univ. Coll. London; hon. fellow Emmanuel Coll., Cambridge; hon. prof. phys. chemistry U. Kent; Richard Dimbleby lectr., 1988, John P. McGovern lectr., 1988. Author: Chemistry for the Modern World, 1962; author BBC TV series Laws of Disorder, 1965; Time Machines, 1969-70; Natural History of a Sunbeam, 1976. Editor: Progress in Reaction Kinetics;

contbr. to profl. jours. Trustee Bristol Exploratory, 1986—; pres. London Internat. Youth Sci. Fortnight, 1987, 88. Served with Royal Navy, 1941-45. Recipient (with M. Eigen and R. G. W. Norrish) Nobel prize in chemistry, 1967; Kalinga prize, 1977; created knight, 1972; decorated mem. Order of Merit, 1989. Fellow Royal Soc. (Davy medal 1971, Rumford medal 1978, Michael Faraday medal 1992, Copley medal 1992, pres. 1985-90), Royal Scottish Soc. of Arts, Royal Soc. of Edinburgh; mem. NAS (fgn. assoc. Washington), Chem. Soc. (pres. 1970-72, pres. Faraday div. 1973-74, Faraday medal 1979, 80, Longstaff medal 1981), Sci. Research Council (Brit.) (council, sci. bd. 1976-80), Comite Internat. de. Photobiologie (pres. 1968-72), N.Y. (hon.), Göttingen (corr.), Pontifical acads. scis., La Real Academia de Ciencias (Madrid; fgn. corr.), Am. Acad. Arts and Scis. (fgn. hon.), Soviet Acad. Scis., Hungarian Acad. Sci., Indian Acad. Sci., Acad. Lincei (Rome), Nat. Assn. Gifted Children (pres. 1975-80). Achievements include research on fast chemical reactions, photochemistry, photosynthesis. Office: Ctr for Photomolecular Scis, Imperial Coll Dept Biology, London SW7 6BB, England

PORTER, IRENE RAE, civil engineer; b. Charlotte, Mich., July 21, 1961; d. Tonnis Willem and Bernadine Kay (Steele) Zomerman; m. Thomas Francis Porter, June 1, 1991. BSCE, Mich. Tech. U., 1983, postgrad., 1983-86. Registered profl. engr., Ill., N.D. Rsch. engr. U.S. Army Corps Engrs., Hanover, N.H., 1984-86; airports engr. FAA, Des Plaines, Ill., 1986-92; mgr. Bismark ADO FAA, Bismark, N.D., 1992—. Co-author: Freezing & Thawing of Soil-Water Systems, 1985, Cold Regions Engineering-4th International, 1986. Mem. NSPE, ASCE (assoc.), Ill. Soc. Profl. Engrs./Profl. Engrs. Govt. (sec. 1991, 92).

PORTER, JOHN LOUIS, aerospace engineer; b. Kansas City, Mo., July 19, 1959; s. Louis Edward and Dorothy May (Johnston) P.; m. Kara Jean Rogers, Oct. 10, 1964 (div. 1988); children: Sean Michael, Craig Daniel, Kara Elizabeth; m. Pamela Marie Cortner, Oct. 21, 1989 (div. 1992). BS, U. Kans., 1961; MS, Calif. Inst. Tech., 1962; DSc, Washington U., St. Louis, 1972. Mgr. propulsion simulation McDonnell Aircraft Co., Hazelwood, Mo., 1963-74; mgr. visual systems Redifon Electronics, Arlington, Tex., 1974-75; mgr. propulsion tech. Advanced Tech. Ctr., Grand Prairie, Tex., 1976-80; mgr. advanced missiles LTV, Dallas, 1980-86, mgr. R&D aero. and def. div., 1986-88; dep. gen. mgr., v.p. Sverdrup Tech., Inc., Arnold AFB, Tenn., 1988-90; tech. chief Sverdrup Tech., Inc., Eglin AFB, Fla., 1990—; chmn. airframe integration panel Joint Army-Navy-NASA-Air Force com., 1987—; mem. mil. critical tech. working group U.S. Dept. Def., Washington, 1991—; pvt. cons., Navarre, Fla., 1990—; presenter at profl. confs.; lectr. in field. Spl. editor Aero. Am., 1992; assoc. editor Jour. Aircraft, 1983-86; contbr. papers to tech. publs. Founding mem., CEO, co-chair Creating Ednl. Opportunities, Tullahoma, Tenn., 1989-90; bd. dirs. Jr. Achievement Tenn., 1990, Manchester (Tenn.) unit March of Dimes, 1989. Fellow AIAA (chmn. com. on computational fluid dynamics stds. 1992—, Wright Bros. award 1974); mem. Am. Def. Preparedness Assn., Soc. Am. Magicians, Air Force Assn., Internat. Brotherhood Magicians (pres. ring 284 1992—), Nat. Mgmt. Assn., Sigma Gamma Tau, Tau Beta Pi. Achievements include invention of modified Rutowski method of flight path optimizaiton with variable throttle and process III ejectors. Home: 2000 Wind Trace Rd S Navarre FL 32566 Office: Sverdrup Tech Inc Bldg 260 PO Box 1935 Eglin A F B FL 32542

PORTER, MAX L., engineering educator; b. Hamburg, Iowa, Oct. 25, 1942; s. Harry E. and Mary O. P.; m. Monica G. Munyer, Aug. 2, 1969; 1 child, Nathan A. BS in Civil Engring., Iowa State U., 1965, MS in Structural Engring., 1968, PhD in Structural Engring., 1974. Registered structural engr., Iowa. Party chief survey crew Iowa Dept. Transp. (formerly Iowa State Hwy. Commn.), Atlantic, Iowa, 1965; instr. Iowa State U., Ames, 1966-74, asst. prof., 1974-77, assoc. prof., 1977-81, prof., assoc. engr. Ames Lab., 1981—; civil engring. asst. bridge dept. County of L.A., 1966; tech. subcom. 5 Masonry Bldg. Seismic Safety Coun., 1989-92; mem. Iowa State Bd. Engring. Examiners Peer Review Com., 1991, 92; mem. tech. adv. bd. Espeland Mktg. and Rsch., 1992-93; cons. Pitney, Hardin, Kipp, Szuch, Interstate Power, Associated Engrs., Otto-Culver, Chuck Barnes, Am. Iron and Steel Inst., Computerized Designs, Inc., Teltech, Inc., Skinner, Beattie & Wilson, Wheeling Corrugating Co. Wheeling-Pittsburg Steel Corp, City of Dubuque, Iowa, Gen. Svcs. Adminstrn., Edwards Precast, Allison, McCormac & Nicholaus, White & Johnson, Nady Engring., Engring. Tech. Assistance Program, Teltech Svcs., Midwest Planners, David Koench, Architect, City of Ames, City of Spencer, Iowa Mut. Ins., ASC Pacific Corp, others; expert witness in field; presenter numerous seminars; participant numerous nat. confs. Co-author: Commentary on the Specifications for the Design and Construction of Composite Slabs, 1985, Specifications for the Design and Construction of Composite Slabs, 1985, Form and Style Manual for ASCE Codes and Standards, 1990, Commentary on Specifications for Masonry Structures, 1988, Commentary on Building Code Requirements for Masonry Structures, 1988, Specifications for Masonry Structures, 1988, Building Code Requirements for Masonry Structures, 1988, Introduction to the Codes and Standards Development, 1991, Commentary to Chapter 24, Masonry of the Uniform Building Code 1988 Edition, 1990, Commentary on Specifications for the Construction and Inspection of Composite Slabs, 1992, Commentary on Building Standard for Structural Design of Composite Slabs, 1992, Commentary on Chapter 24, Masonry of the Uniform Building Code 1991 Edition, 1992, Specifications for Masonry Structures, 1992, Building Code Requirements for Masonry Structures, 1992, State-of-the-Art-Report on Fiber Reinforced Concrete, 1993, Commentary on the Standard for the Structural Design of Composite Slabs, 1993, Commentary on the Standard Practice for the Construction and Inspection of Composite Slabs, 1993, Standard Practice for the Construction and Inspection of Composite Slabs, 1993, Standard for Structural Design of Composite Slabs, 1993; contbr. articles to profl. jours. Vice chmn. King-Am Neighborhood Assn., 1982-88, pres. 1991—; bd. dirs. Ames Econ. Devel. Commn., 1987-89; trustee Presbyn. Ch., 1989-91, deacon, 1992—. Fred Loy Meml. scholar in Civil Engring., 1964; recipient Citation of Merit Enerpac hydraulic divsn. Applied Power Industries, Inc., 1974; grantee Agy. Internat. Devel., Masonry Inst. Iowa, Masons Union Iowa, Civil Engring. Rsch. Assistantship, Engring. Project Devel. Funding, 1981-84, Wheeling Corrigating Co.divsn. Wheeling-Pittsburg Steel Co., 1985-90, 87, 89-92, Vulcraft divsn. Nucor Corp., 1985-88, 86-90, U.S. Nuclear Regulatory Commn., 1984, Carter-Waters, Monarch Cement Co., Mac-Fab Corp. St. Louis, 1985-87, Thermomass Tech. Inc., 1986-87, Amoco Foam Products Co., 1986-87, NSF, 1987-90, 92—, Iowa Tech. Coun., 1988, Walker divsn. Butler Mfg., 1989, Dow Chem. Co., Shafer Fiberglass Co., 1989-90, Ctr. Advanced Tech. and Devel., 1990-92, Iowa Hwy. Rsch. Bd., 1989-93, Hammer's Plastic Recycling Corp., Composite Tech. Corp., 1990-92, DUR-O-WAL, Inc., 1991-92, DEHA, Martin Marietta Energy Systems, Inc., 1992, Wallace Techs., 1992-93, Espeland Mktg. & Rsch., Inc., 1992—, others; named Most Valuable Player Sixth Internat. Specialty Conf. Cold Steel Structures, 1982. Mem. ASCE (png coms. 1978-79, moderator student conf. 1978, com. composite constrn. divsn. structures 1977-84, 86-90, chmn. 1979-82, exec. com. tech. coun. codes and standards 1986-89, vice-chmn. 1988-89, adminstrv. com. bldg. codes 1987-93, steel deck & concrete com. 1977-92, standards com. design loads for structures during constrn. 1989-92, chmn. subcom. devel. mgmt. group plan 1988-90, task group oper. procedures and ops. manual 1989-90, masonry limit states standards com. 1989-92, chmn. structural standards exec. com. 1989-93, chmn. design subcom. allowable stress and empirical design 1990-93, vice chair nat. masonry standards joint com. 1990-93, nat. earthquake hazard reduction program adv. com. 1990-93, mgmt. group F 1991—, chair subcom. design composite deck slabs 1992—), NSPE (budget com. 1982-84, parliamentarian nat. convs. 1982-84, 85-86, 89, 90), chmn. constn. & bylaws com. 1983-84), AAUP (Iowa State U. chpt., chmn. faculty appeals grievance com. 1988-91), Am. Concrete Inst. (Iowa-Minn. chpt. com. 530 new masonry bldg. code standards 1984-89, bd. dirs. 1985-88, chpt. v.p. 1988-89, chpt. pres. 1989-90, immediate past pres., bd. dirs., chair nominations 1990-91, vice chair com. 530 masonry standards 1990-92, chair design subcom. 1990-92, com. fiber reinforced plastic 1991-93, com. fiber reinforced concrete 1991-93, subcom. bond and devel. 1991-93), Am. Inst. Parliamentarians, Nat. Assn. Parliamentarians, Nat. Inst. Bldg. Scis. (reviewer gen. svcs. adminstrn. model codes implementation study project com. 1989), Nat. Inst. Engring. Ethics, Iowa State Assn. Parliamentarians, Assn. Internat. Cooperation and Rsch. Steel-Concrete Composite Structures, Iowa Engring. Soc. (Anson Marston chpt. bd. dirs. 1978-79, chmn. young engrs. com. 1977-78, chmn. profl. devel. com., membership com. 1979-81, activities com. 1979-80, v.p. 1980-82, chmn. nominations com. 1982-86. pres.-elect 1982-83, co-chair mathcounts 1983-85, pres. 1983-84), Bldg. Seismic Safety Coun., Earthquake Engring. Rsch. Inst., Structural Engrs. Assn. Ctrl. Iowa (bd. dirs. 1977-79), Concrete Reinforcing Steel Inst., Soc. Exptl. Stress Analysis, Internat. Conf. Bldg. Officials, N.Y. Acad. Scis., Masonry Soc. (rsch. com. 1986-92, bd. dirs. Ctrl. Region 1987-92, codes com. 1988-92, masonry limit states standards com. 1989-92, vice chair, chair design subcom. and nat. masonry standards joint com. 1990-92, assoc. editor The Masonry Soc. Jour. 1990-93, chair pubs. com. 1991-92, editor The Masonry Soc. News 1992-93), Precast/Prestressed Concrete Inst., Rotary (Paul Harris sustaining 1986—, chmn. fellowship and brotherhood com. 1987-89), Iowa State U. Track Ofcls. Club , Iowa State U. Alumni Assn. (life), Iowa State Meml. Union (life), Scabbard & Blade, Order Knoll, Sigma Xi, Phi Kappa Phi, Tau Beta Pi, Chi Epsilon, Phi Eta Sigma. Office: Iowa State U Dept Civil & Constrn Engring 416A Town Engring Bldg Ames IA 50011

PORTER, MICHAEL BLAIR, applied mathematics educator; b. Que., Can., Sept. 19, 1958; s. Ronald and Muriel (Jones) P. BS, Calif. Inst. Tech., 1979; PhD, Northwestern U., 1984. Scientist Naval Ocean Systems Ctr., San Diego, 1983-85; rsch. physicist Naval Rsch. Lab., Washington, 1983-88; sr. scientist SACLANT Undersea Rsch. Ctr. NATO, La Spezia, Italy, 1988-91; assoc. prof. Math. N.J. Inst. Tech., Newark, 1991—. Contbr. articles to profl. jours. Mem. Soc. Indsl. and Applied Maths., Acoustical Soc. Am. Office: NJ Inst Tech Univ Heights Newark NJ 07102

PORTER, ROBERT PHILIP, electrical engineering educator; b. Brockton, Mass., Sept. 21, 1942; s. George Porter and Frances Carol (Nagle) Parkinson; m. Janet Cohen, Dept. 8, 1963 (div. Apr. 1975); children: Melissa J., Timothy R.; m. Charlotte Rose Lin, Sept. 21, 1980. SBEE, MIT, 1965, SMEE, 1965, postgrad., 1966; PhD, Northeastern U., 1970. Assoc. scientist Woods Hole (Mass.) Oceanographic Inst., 1971-77; dir. mechanics Schlumberger Doll Rsch., Ridgefield, Conn., 1978-82, dir. sensor physics, 1984-85; v.p. Nippon Schlumberger K.K., Tokyo, 1982-84; chair elec. engring. U. Wash., Seattle, 1985-88, prof. elec. engring., chair ocean acoustics, 1989—; v.p. Fairchild-Weston, Syosset, N.Y., 1988-89; mem. Naval Rsch. Adv. Com., Arlington, Va., 1988—; cons. Sperry Corp., Blue Bell, Pa., 1986, Schlumberger Ltd., N.Y.C., 1986-87. Contbr. over 70 articles to profl. publs., chpts. to books. Fellow Acoustical Soc. Am., Optical Soc. Am.; mem. IEEE (assoc.), Soc. Exploration Geophysicists, Internat. House Japan, Eta Kappa Nu, Tau Beta Pi, Phi Kappa Phi. Achievements include patents for downhole seismic exploration device, instrument for measuring motion of moored systems, pulse doppler underwater acoustic navigation system, tracking system for underwater objects. Office: U Wash Applied Physics Lab 1013 NE 40th St Seattle WA 98105

PORTER, ROBERT WILLIAM, mechanical engineering educator; b. Pontiac, Mich., Jan. 19, 1938; s. William J. and Lillian J. (Brasch) P.; m. Lourden Guiao, July 15, 1984. BS, U. Ill., 1961; MS, Northwestern U., 1963, PhD, 1966. Registered profl. engr., Ill. Engr. Bendix Corp., Mishawaka, Ind., 1962-63; prof. mech. engring. Ill. Inst. Tech., Chgo., 1966—; cons. to elec. power industry. Assoc. editor ASME Jour. Engring. for Gas Turbins and Power, 1982—, ASME Jour. Turbomachinery, 1982—. Recpient Octave Chanute medal Western Soc. of Engrs., Chgo., 1974. Fellow ASME; mem. AAUP, Sigma Xi. Roman Catholic. Office: Ill Inst Tech 10 W 32nd St Chicago IL 60616

PORTER, WILLIAM HUDSON, clinical chemist; b. Wilson, N.C., Mar. 12, 1940; s. Frank L. and Mildred McCollum P.; m. Faye Hardee, Mar. 23, 1963; children: Tracy M., Mark W. MS, Med. U. S.C., 1967; PhD, Vanderbilt U., 1971. Diplomate Am. Bd. Clin. Chemistry (bd. dirs. 1985-91). Postdoctoral fellow Fla. State U., Tallahassee, 1970-72; asst. rsch. prof. U. Tenn. Meml. Rsch. Ctr., Knoxville, 1972-74; asst. prof. Med. U. S.C., Charleston, 1974-78; assoc. prof. clin. chemistry U. Ky. Med. Ctr., Lexington, 1978-86, prof., 1986—. Contbr. articles to profl. jours. Mem. Nat. Registry in Clin. Chemistry (bd. dirs. 1991-93), Am. Assn. for Clin. Chemistry, Assn. of Clin. Scientists, Am. Chem. Soc., Am. Soc. Biochemistry and Molecular Biology. Office: Univ of Ky Med Ctr HA616 Dept Pathology Lab Medicine Lexington KY 40536

PORTER, WILLIAM L., electrical engineer; b. Leeds, N.D., July 2, 1929; s. Ernest Cecil and Dena Grace (Thompson) P.; m. Mary Lynn Lindsey, Oct. 9, 1948; children: Belinda Joyce, William Harry, Terry Jane, Derek Lewis, Michael Ronald. AA, Springfield Coll., 1960; BSEE, U. Ill., 1963. Registered profl. engr., Ill., Minn., Iowa, N.D., S.D., Ohio, Mich., Wis., Ind., Nebr. Lineman City Water, Light and Power, Springfield, Ill., 1947-54, troubleshooter, 1954-62, gen. supt. elec. divsn., 1962-76; prin. engr. R.W. Beck and Assocs., Columbus, Nebr., 1976-77; engring. mgr. R.W. Beck and Assocs., Mpls., 1977-80, ptnr., mgr., 1980-90, sr. cons., 1990—; speaker on engring. and utilities; cons. to electric utilities. Author numerous engring. reports and engring. and utilities papers. Street light com. chair City of Springfield, 1964, mem. CATV com., 1966; mem. Planning Commn. Spring Park (Minn.), 1978-79; chair environ. quality com. region IV Ill. Soc. Profl. Engrs., chair ethics and practices com. Capital chpt.; mem. tech. adv. com. Fed. Power Commn's Nat. Power Survey; chair engring. and ops. com. Am. Pub. Power Assn., 1967-70, chair power supply planning com., 1973-74. Named Engr. of Yr., Capital chpt. Ill. Soc. Profl. Engrs., 1975. Mem. NSPE, IEEE (chmn. Cen. Ill. sect. 1974-75), Minn. Soc. Profl. Engrs., Cons. Engrs. Coun., Am. Bus. Club, Eta Kappa Nu. Republican. Home: 4349 Channel Rd Spring Park MN 55384 Office: R W Beck and Assocs 8300 Norman Center Rd # 860 Minneapolis MN 55437

PORTERFIELD, CRAIG ALLEN, psychologist, consultant; b. Geneva, N.Y., May 11, 1955; s. Paul Laverne and Elizabeth Louise (Mearns) P.; m. Alta Marie Herring, Aug. 1977; children: Aleine Michelle, Brian Matthew. Student, Sorbonne U., Paris, 1975-76; BA, St. John Fisher Coll., 1977; MA, U. Tex., Austin, 1982, PhD, 1985. Lic. psychologist, N.Y., Del.; cert. sch. psychologist, N.Y. Program evaluation intern Austin Ind. Sch. Dist., 1980, psychol. intern, 1982-83; program evaluator Austin Child Guidance Ctr., 1981-82; evaluation mgr. Child, Inc., Austin, 1981-82; staff therapist Psychotherapy Inst., Austin, 1984-85; consulting psychologist Albany (N.Y.) Psychol. Assocs., 1987-90; staff psychologist Berkshire Farm Ctr. and Svcs. for Youth, Canaan, N.Y., 1985-87, dir. rsch., 1987-90; psychologist Del. Psychiatry Svcs., Dover, 1990—; adj. asst. prof. SUNY, Albany, 1986-87, 89-91; psychologist privileges dept. psychiatry Kent Gen. Hosp., Dover, 1990—; mem. adv. com. life skills curriculum Lake Forest Sch. Dist., Harrington, Del., 1991; co-founder, advisor Children with Attention Deficit Disorders Kent County, Del., 1991—; mem. Children with Attention Deficit Disorders State Attention Deficit Disorder Coun. Del., 1993—. Grantee N.Y State Integrated Task Force on Substance Abuse Programs for Youth, 1988. Mem. APA, Del. Psychol. Assn., Nat. Trust Historic Preservation, Wild Quail Country Club. Avocations: Victorian house restoration, antiques, exercise, public speaking. Office: Del Psychiatry Svcs 1001 S Bradford St Dover DE 19901-4141

PORTERFIELD, WILLIAM WENDELL, chemist, educator; b. Winchester, Va., Aug. 24, 1936; s. Donald Kennedy and Adelyn (Miller) P.; m. Dorothy Elizabeth Dail, Aug. 24, 1957; children—Allan Kennedy, Douglas Hunter. B.S., U. N.C., 1957, Ph.D., 1962; M.S., Calif. Inst. Tech., 1960. Sr. research chemist Hercules, Inc., Cumberland, Md., 1962-64; asst. prof. chemistry Hampden-Sydney (Va.) Coll., 1964-65, assoc. prof., 1965-68, prof. chemistry, 1968—, Charles Scott Venable prof. chemistry, 1989—, chmn. natural sci. div., 1973-77, chmn. dept. chemistry, 1982-85; vis. fellow U. Durham (U.K.), 1984. Author: Concepts of Chemistry, 1972, Inorganic Chemistry, 1984, 2d edit., 1993; contbr. articles to profl. jours. Mem. Am. Chem. Soc., Royal Chem. Soc. (London, Eng.), Phi Beta Kappa. Home: PO Box 697 Hampden Sydney VA 23943

PORTNOY, STEPHEN LANE, statistician; b. Kankakee, Ill., Dec. 2, 1942; s. Samuel L. and Rosella (Goldman) P.; m. Esther R. Portnoy, June 19, 1965; children: Gerald L., A. Rachel. PhD, Stanford (Calif.) U., 1969. Asst. prof. Harvard U., Cambridge, Mass., 1969-74; prof. U. Ill., Urbana, 1974—. Contbr. articles to profl. jours. including Annals of Statistics, Jour. Am. Stat. Assn., Econometrica. Rsch. grant NSF, 1975—. Fellow Inst. of Math. Stats. (assoc. editor 1979-80); mem. ACLU (Champaign chpt. 1987), Am. Stats. Assn. (assoc. editor 1991—). Office: U Ill Dept Stats 725 S Wright Champaign IL 61820

POSADAS, MARTIN POSADAS, physician, educator, businessman; b. San Carlos, Pangasinan, Philippines, Nov. 11, 1921; m. Rosalina Quebral, Feb. 14, 1950; 1 child, Marili (Mrs. Angelo Juan). BA, U. Philippines; MD, U. Santo Tomas, Philippines; career exec. cert. Devel. Acad. Pogy. Past pres. Boy Scout Coun.; chmn. Posadas Clan Found.; founder-pres. Binalatongan Hist. and Cultural Soc., Western Civic League; past pres. San Carlos Parish Coun., archdiocese coun.; bd dirs. Assn. Pvt. Schs., Colls. and Univs., Philippine Tripartite Wage Bd.; past bd. dirs. Assn. Philippine Med. Colls. and Univs.; pres. Pang Tourism Coun. Named Most Outstanding Physician, Philippine Med. Assn., Most Outstanding Educator, San Carlos City Govt., Pioneer in Countryside Devel., San Carlos City Govt., Most Outstanding Alumnus U. Santo Tomas, Outstanding Educator of the Philippines, 1992; recipient Papal award Pro-Eclesia Pontificae, Vatican, Lay Christian Apostolate award. Mem. Pangasinan Med. Soc. (past pres.), Philippine Hosp. Assn. (past v.p.), San Carlos Medicare Council (past chmn.), Filipino Assn. Med. Educators (treas.), San Carlos City C. of C. and Industry, Posadas Clan of the Philippines (pres.), N.Y. Acad. Scis., Jaycees (past pres., v.p. Luzon), Archdiocesan Catholic Action Club (Pangasinan) (pres. 1980), KC (4th degree, faithful navigator, grand knight, dist. dep.), Rotary (past pres., past sec.). Avocation: golf. Home: Taloy Dist, San Carlos City, Pangasinan 0740, The Philippines Office: Virgen Milagrosa Complex, Taloy, Pangasinah San Carlos City 0740, The Philippines

POSAMENTIER, ALFRED STEVEN, mathematics educator, university administrator; b. N.Y.C., N.Y., Oct. 18, 1942; s. Ernest and Alice (Pisk) P.; children—Lisa Joan, David Richard. A.B., Hunter Coll., 1964; M.A., CCNY, 1966; postgrad., Yeshiva U., N.Y.C., 1967-69; Ph.D., Fordham U., 1973. Tchr. math Theodore Roosevelt High Sch., Bronx, 1964-70; asst. prof. math. edn. CCNY, N.Y.C., 1970-76, assoc. prof., 1977-80, prof., 1981—, dept. chmn. dept. secondary and continuing edn., 1974-80, chmn., 1980-86; assoc. dean Sch. Edn. CCNY, 1986—, dir. select program in sci. and engring., 1978—; dir. CCNY, U.K., iniatives program dir., 1983—; dir. Germany/CCNY Exch. Program CCNY, 1985—, dir. Austria/CCNY Exch. Program, 1987—, dir. Czechoslovakia/CCNY Exch. Program, 1989—, dir. sci. lectr. program, 1981—, dir. Ctr. for Sci. and Maths. Edn., 1986—; chmn. bd. dirs. Salvadori Ednl. Ctr. on Built Environ., 1988—; dir. Exxon sponsored early childhood math. specialist tng. program at City Coll., 1988—; supr. math. and sci. Mamaroneck High Sch., N.Y., 1976-79; project dir. Math Proficiency Workship, Ossining, N.Y., 1976-79, NSF math. devel. program for secondary sch. tchrs. math., 1978-82, N.Y.C., Profl. Preparation of Math. and Sci. Tchrs., 1978-79; project dir. numerous NSF sponsored math./sci. tchr. devel. insts., 1976—; cons. Inst. Ednl. Devel. N.Y.C., 1970-73, Croft Ednl. Services, New London, 1971, Design and Evaluation, 1973, N.Y.C. Bd. Edn., 1973-75, N.Y.C. Bd. Edn. Office of Evaluation, 1974-80, N.Y.C. Bd. Edn. Examiners, 1979—, Ossining Bd. Edn., 1975-83, numerous others; coordinator NSF N.E. Resource Ctr. Sci. and Engring., 1981—, lectr. various convs. and meetings; vis. prof. U. Vienna, Austria, 1985, 87, 88, 90, Tech. U., Berlin, 1989, Tech. U., Vienna, 1993, Pedagogical Inst., Vienna, 1993. Author: Geometric Constructions, 1973, Geometry, Its Elements and Structure, 1972, rev. edit., 1977, Challenging Problems in Geometry, 2 vols, 1970, Challenging Problems in Algebra, 2 vols., 1970, A Study Guide for the Scholastic Aptitude Test in Math., 1969, rev. edit., 1983, Excursions in Advanced Euclidean Geometry, 1980, 2d edit., 1984, Teaching Secondary School Mathematics: Techniques and Enrichment Units, 1981, 3d edit., 1990, Uncommon Problems for Common Topics in Algebra, 1981, Unusual Problems for Usual Topics in Algebra, 1981, Using Computers in Mathematics, 1983, 2d edit., 1986, Math Motivators: Investigations in Pre-Algebra, 1982, Math Motivators: Investigations in Geometry, 1982, Math Motivators: Investigations in Algebra, 1983, Using Computers: Programming and Problem Solving, 1984, 2d edit., 1989, Advanced Geometric Constructions, 1988, Challenging Problems in Algebra, 1988, Challenging Problems in Geometry, 1988; contbr. numerous articles to profl. jours. Trustee Demarest Bd. Edn., 1977-80. Fulbright scholar U. Vienna, Austria, 1990. Mem. Math. Assn. Am., Sch. Sci. and Math. Assn., Nat. Council Tchrs. Math., (reviewer new publs., referee articles Math. Tchr. Jour.), Assn. Tchrs. Math. N.Y.C. (exec. bd. 1966-67, referee articles assn. jour.), Assn. Tchrs. of Math. of N.Y. State, Assn. Tchrs. Math. N.J. (editorial bd. N.J. Math. Tchr. Jour. 1981-84), Nat. Council of Suprs. of Maths. Home: 634 Caruso Ln River Vale NJ 07675-6210 Office: CCNY New York NY 10031

POSCHMANN, ANDREW WILLIAM, information systems and management consultant; b. N.Y.C., June 24, 1939; m. Anne Florence Fugarini, July 14, 1962; children: Stephen, Robert. BBA in Mgmt. and fin., Baruch Coll., N.Y.C., 1968; MBA in Fin., Marist Coll., Poughkeepsie, N.Y., 1981. Cert. info. systems auditor, mgmt. cons., systems profl. EDP applications engr. GE, N.Y.C., 1964-66; software cons. Western Union, Mahwah, N.J., 1966-69; cons. A.T. Kearney & Co., N.Y.C., 1969-73; systems mgr. Curtiss-Wright Corp., Wood-Ridge, N.J., 1973-77; cons. Ernst & Whinney, N.Y.C., 1977-83; applications mgr. NYU Med. Ctr., 1983-86; info. systems mgr. McKinsey & Co., N.Y.C., 1986-93; v.p. Advanced Mgmt. Inc., East Fishkill, N.Y., 1993—; mem. adv. bd. Advanced Mgmt., Inc., East Fishkill, N.Y., 1975—. Author: Standards and Procedures For Systems Documentation, 1984, Score-Company Review and Evaluation, 1985. Mem. Am. Arbitration Assn. (arbitrator), Nat. Railway Hist. Soc. Office: Advanced Mgmt Inc Kensington Dr East Fishkill NY 12533

POSEY, DANIEL EARL, analytical chemist; b. Corpus Christi, Tex., Apr. 9, 1947; s. Earl Lloyd and Mary Lucille (Williams) P.; m. Mary Jewell King, Dec. 7, 1968; children: Amanda America, Matthew Daniel. BS in Chemistry, U. Houston, 1970. Rsch. technician Getty Oil Co. Exploration & Prodn. Rsch. Labs., Houston, 1968-69; lab. mgr. Inst. for Rsch., Inc., Houston, 1969-79, Am. Convertors, El Paso, Tex., 1979-84; tech. dir. Inst. for Rsch.-Austin, 1984-86; cons. chemist Spectro Chem Inc., Austin, 1986-88; quality engring. supr. Advanced Micro Devices, Austin, 1988—; mem. Internat. Nonwoven & Disposables Assn., N.Y.C., 1981-83. Contbr. tech. papers to scholarly jours. Recipient Tech. Svc. award Am. Convertors R&D, 1981, Tech. Mgmt. award, 1982; recipient Cert. of Achievement, Am. Men and Women of Sci., 1986. Fellow Am. Inst. Chemists; mem. Am. Chem. Soc., Am. Soc. for Quality Control, Phi Eta Sigma. Republican. Achievements include patent for cleaning product for removal of mold and mildew composition and method of manufacture; invention of first lint particle generation test method for nonwoven fabrics; development of first antimicrobial surgical fabric. Office: Advanced Micro Devices 5204 E Ben White Blvd Mail Stop 519 Austin TX 78741

POSHNI, IQBAL AHMED, microbiologist; b. Calcutta, India, June 3, 1939; came to U.S., 1963; s. Ghulam Ahmed Poshni and Razia Begum; m. Azma Paul, Dec. 10, 1966; children: Kashif, Faiza. BS with hons., Karachi U., Pakistan, 1960, MS, 1961; PhD, U. Mo., 1968. Cert. clin. lab. dir., N.Y. Sr. rsch. scientist Pakistan Coun. Scientific and Indsl. Rsch., Karachi, 1968-70; prodn. in-charge Glaxo Labs. Ltd., Lahore, Pakistan, 1970-78; lab. mgr. Okla. Animal Disease Diagnostic Lab., Stillwater, Okla., 1978-79; chief, immunology and virology N.Y.C. Dept. Health, 1979—; adj. assoc. prof. CUNY, 1982-90; cons. Govt. Pakistan, 1984, N.Y.C. area hosps., 1982—. Pres. Pakistan-League of Am., N.Y., 1985-87. Recipient Fullbright Travel grant, 1963, scholarship N.Y. Inst. Higher Edn., 1965-66. Mem. AAAS, N.Y. Acad. Sci., Am. Soc. Microbiology. Moslem. Achievements include patents for hand lotion, domestic water filter (Pakistan). Home: 87-21 144th St Briarwood NY 11435-3121 Office: NYC Dept Health 455 1st Ave Rm 201 New York NY 10016-9102

POSLER, GERRY LYNN, agronomist, educator; b. Cainsville, Mo., July 24, 1942; s. Glen L. and Helen R. (Maroney) P.; m. O. Shirley Weeda, June 23, 1963; children: Mark L., Steven C., Brian D. BS, U. Mo., 1964, MS, 1966; PhD, Iowa State U., 1969. Asst. prof. Western (Macomb) Ill. U., 1969-74; assoc. prof. Kans. State U., Manhattan, 1974-80, prof., 1980—, asst. dept. head, 1982-90, dept. head, 1990—. Contbr. articles to profl. jours. and popular publs., abstracts, book reviews. Fellow Am. Soc. Agronomy, Crop Sci. Soc. Am.; mem. Am. Forage Grassland Coun., Crop Science Soc. Am. (C-3 div. chmn. 1991), Coun. Agrl. Science Tech. (Cornerstone club), Nat. Assn. Colls. Tchrs. Agr. (tchr. fellow award 1978, ensminger interstate dist. teaching award, 1987, north cen. region dir. 1989, v.p. 1990, pres. 1991; life mem.), Kans. Assn. Colls. Tchrs. Agr. (pres. 1983-85), Kans. Forage Grassland Coun. (bd. dirs. 1989-92), Gamma Sigma Delta (Outstanding Faculty award 1991, pres. 1987). Home: 3001 Montana Ct Manhattan KS 66502-2300 Office: Kans State U Dept Agronomy Throckmorton Hall Manhattan KS 66506

POSNER, GARY HERBERT, chemist, educator; b. N.Y.C., June 2, 1943; s. Joseph M. and Rose (Klein) P.; children: Joseph, Michael. BA, Brandeis U., 1965; MA, Harvard U., 1965, PhD, 1968. Asst. prof. Johns Hopkins U., Balt., 1969-74, assoc. prof., 1974-79, prof. dept. chemistry, 1979—, Scowe prof. chemistry, 1989—; prof. dept. environ. chemistry Johns Hopkins U., 1982—, chmn. dept. of chemistry, 1987-90; cons. Batelle Meml. Inst., Columbus, Ohio, 1983, S.W. Rsch. Inst., San Antonio, Nova Pharm. Co., Balt.; mem. Fulbright-Hays Adv. Screening Com. in Chemistry, 1978-81; Fulbright lectr. U. Paris, 1976; Michael vis. prof. Weizmann Inst. Sci., Rehovot, Israel, 1983; leader Round Table discussion Welch Found. Conf. Chem. Rsch., Houston, 1973, 83; Plenary lectr. Nobel Symposium on Asymmetric Synthesis, Sweden, 1984. Author: Introduction to Organic Synthesis Using Organocopper Reagents, 1980; mem. editorial bd. Organic Reactions, 1976-89. Named Chemist of Yr., State of Md., 1987; fellow Japan Soc. for Promotion Sci., 1991. Mem. AAAS, Am. Chem. Soc., AAUP, NIH (medicinal chemistry study sect. 1986-89), Phi Beta Kappa. Office: Johns Hopkins U Dept Chemistry 3300 N Charles St Baltimore MD 21218

POSNER, JEROME BEEBE, neurologist, educator; b. Cin., Mar. 20, 1932; s. Philip and Rose (Goldberg) P.; m. Gerta Grunen, Aug. 29, 1954; children: Roslyn, Joel, P.J. B.S., u. Wash., 1951; M.D., U. Wash., 1955. Intern King County Hosp., Seattle, 1955-56; asst. resident in neurology U. Wash. Affiliated Hosps., Seattle, 1956-59; fellow in neurology U. Wash. Affiliated Hosps., 1958-59; spl. fellow NIH, U. Wash., 1961-63; instr. medicine U. Louisville Sch. Medicine, 1959-61; attending neurologist King County Hosp., 1962-63; asst. prof. neurology Cornell U. Med. Coll., N.Y.C., 1963-67; assoc. prof. Cornell U. Med. Coll., 1967-70, prof., 1970—, vice chmn. dept. neurology, 1978-87; asst. attending neurologist N.Y. Hosp., 1963-67, asso. attending neurologist, 1967-70, attending neurologist, 1970—; asso. Cotzias Lab. of Neuro-Oncology, Sloan Kettering Inst. Cancer Research, N.Y.C., 1967-76; mem. Cotzias Lab. of Neuro-Oncology, Sloan Kettering Inst. Cancer Research, 1976—; chief neuropsychiat. service, attending physician dept. medicine Meml. Hosp. for Cancer and Allied Diseases, 1967-75, attending physician, 1975—, chmn. dept. neurology, 1975-87, 89—, Cotzias chair neuro-oncology, 1986; mem. med. adv. bd. Burke Rehab. Center, White Plains, N.Y., 1973—; adj. prof., vis. physician Rockefeller U. and Hosp., N.Y.C., 1973-75; vis. physician Rockefeller U. Hosp., 1980—; mem. med. adv. bd. Assn. for Brain Tumor Research, 1974—; mem. neurology B study sect. NIH, 1972-76. Author: (with F. Plum) Diagnosis of Stupor and Coma, 3d edit., 1980, (with H. Gilbert and L. Weiss) Brain Metastasis, 1980; mem. editorial bd. Archives of Neurology, 1971-76, Annals of Neurology, 1976-80, Am. Jour. Medicine, 1978-93; contbr. articles to med. jours. Served with M.C., U.S. Army, 1959-61. Served with M.C. U.S. Army, 1959-61. Mem. AAASA, AMA, Am. Acad. Neurology (Farber Brain Tumor award 1983), Am. Assn. Cancer Rsch., Am. Fedn. Clin. Rsch., Am. Neurol. Assn., Am. Physiol. Soc., Harvey Soc., Internat. Assn. Study of Pain, N.Y. Acad. Sci., Inst. of Medicine, Soc. Neuroscis, Can. Neurol. Soc. (hon.), Alpha Omega Alpha. Office: Meml Sloan-Kettering Cancer 1275 York Ave New York NY 10021-6094

POSNER, NORMAN AMES, medical educator; b. N.Y.C., July 28, 1933; s. Lewis Bernard and Valeria (Shapiro) P. BA, Washington and Jefferson U., 1953; MD, NYU, 1957. Diplomate Am. Bd. Ob-Gyn. Intern St. Vincent's Hosp. N.Y., Bklyn., 1957-58; resident Maimonides Med. Ctr., Bklyn., 1958-62; asst. clin. instr. ob-gyn. SUNY Sownstate Med. Ctr., Bklyn., 1961-62, instr., 1964-66, asst. prof., 1966-72, assoc. prof., 1972-84; lectr. SUNY Sownstate Med. Ctr., 1984—; assoc. prof. clin. ob-gyn. Albany Med. Coll., 1984-89, prof., 1989—; vis. prof. Albany Med. Coll., 1982, adj. assoc. prof., 1982-84; ab-gyn. adv. com. dept. health City of N.Y., 1975-84; diabetes detection com., N.Y., chmn. subcom. phys. soc. Maimonides Med. Ctr., 1967-68. Author: (with others) Carbohydrate Metabolism in Pregnancy, 1977, Circumcision, 1981, Postpartum Depression, 1985. Fellow N.Y. Obstet. Soc.; mem. AAAS, Am. Fertility Soc., Northeastern Ob-gyn. Soc., Am. Soc. Psychosomatic Obs/Gyn., Bklyn. Gynecological Soc. (historian 1971-73, sec. 1973-76, 2d v.p. 1978-79, pres. 1979-80, coun. 1980-82). Office: Albany Med Ctr Dept Ob-Gyn 47 New Scotland Ave Albany NY 12208

POSPISIL, LEOPOLD JAROSLAV, anthropology educator; b. Olomouc, Czechoslovakia, Apr. 26, 1923; came to U.S., 1949, naturalized, 1954; s. Leopold and Ludmila (Petrlak) P.; m. Zdenka Smyd, Jan. 31, 1945; children: Zdenka, Mira. J.U.C. in Law, Charles U., Prague, 1947, 91; BA in Sociology, Willamette U., Salem, Oreg., 1950; MA in Anthropology, U. Oreg., 1952; PhD (Ford Found. fellow, Sr. Sterling fellow), Yale U., 1956; ScD (hon.), Willamette U., 1969; JUDr, Charles U., Prague, 1991. Instr. Yale U., New Haven, 1956-57; asst. prof. Yale U., 1957-60; asst. curator Peabody Mus., 1956-60, assoc. prof., 1960-65, prof., curator, 1965—, dir. div. anthropology, 1966—. Author: Kapauku Papuans and Their Law, 1958, Kapauku Papuan Economy, 1963, Kapauku Papuans of West New Guinea, 1963, Anthropology of Law, 1971, Ethnology of Law, 1972, Anthropologie des Rechts, 1981; contbr. articles to profl. jours. Guggenheim fellow, 1962; NSF fellow, 1962, 64-65, 1967-71; SSRC grantee, 1966; NIMH fellow, 1973-79. Fellow AAAS, N.Y. Acad. Scis., Am. Anthrop. Assn.; mem. Explorers Club, Sigma Xi, Czechoslovakian Acad. Arts and Scis. (past pres.), Nat. Acad. Scis., Coun. Free Czechoslovakia, Assn. for Polit. and Legal Anthropology (pres.-at-large), Assn. for Social Anthropology in Oceania, Soc. for Econ. Anthropology. Home: 554 Orange St New Haven CT 06511-3819 Office: Yale U Dept Anthropology 51 Hillhouse Ave New Haven CT 06511-3703

POSSINGHAM, HUGH PHILIP, mathematical ecologist; b. Adelaide, Australia, July 21, 1962; m. Karen Anne Fiegert, July 21, 1985; children: Nicholas Lawrence, Alexandra Constance. BSc with honors, U. Adelaide, 1983; D. Phil., Oxford (Eng.) U., 1987. Postdoctoral rsch. scientist Stanford (Calif.) U., 1987-88; Queen Elizabeth II fellow Australian Nat. U., Canberra, Australia, 1989-90; lectr. (tenurable) U. Adelaide, 1991—, assoc. dean info. tech. maths faculty, 1992—; cons. Resource Assessment Commn., Canberra, 1990-91, Australian Nat. Pks. and Wildlife Svc., 1993, NSW Nat. Pks. and Wildlife Svc., 1993. Contbr. articles to Sci. (U.S.), Am. Naturalist, Ecology (U.S.). Rhodes scholar Rhodes Trustees, 1983, George Murray scholar Adelaide U., 1984. Mem. Ecol. Soc. Australia (v.p. 1992—). Home: 61 Salop St, Beulah Park 5067, Australia Office: U Adelaide Dept Applied Math, GPO Box 498, Adelaide 5001, Australia

POST, LAURA CYNTHIA, biologist; b. Rutland, Vt., June 28, 1969; d. Harold Robert and Evelyn Mary (Brabant) P. BS in Biology, Pa. State U., 1991. Biologist NCI, Frederick, Md., 1991—. Mem. Am. Assn. Advancement in Sci.

POSTMA, HERMAN, physicist, consultant; b. Wilmington, N.C., Mar. 29, 1933; s. Gilbert and Sophia Postma; m. Patricia Dunigan, Nov. 25, 1960; children: Peter, Pamela. BS summa cum laude, Duke U., 1955; MS, Harvard U., 1957, PhD, 1959. Registered profl. engr., Calif. Summer staff Oak Ridge Nat. Lab., 1954-57, physicist thermonuclear div., 1959-62, coleader DCX-1 group, 1962-66, asst. dir. thermonuclear div., 1966, asso. dir. div., 1967, dir. div., 1967-73, dir. nat. lab., from 1974; v.p. Martin Marietta, 1984-88, sr. v.p., 1988-91; cons., 1991—; vis. scientist FOM-Instituut voor Plasma-Fysica, The Netherlands, 1963; cons. Lab. Laser Energetics, U. Rochester; mem. energy research adv. bd. spl. panel Dept. Energy; bd. dirs. Fed. Res. Bank Atlanta, Nashville br., ICS Corp., PAI Corp. Mem. editorial bd. Nuclear Fusion, 1968-74; contbr. numerous articles to profl. jours. Bd. dirs. The Nucleus; chmn. bd. trustees Hosp. of Meth. Ch.; mem. adv. bd. Coll. Bus. Adminstrn., U. Tenn., 1976-84, Energy Inst., State of N.C.; bd. dirs., exec. com. Tenn. Tech. Found., 1982-88, Venture Capital Fund; vice chmn., commr. Tenn. Higher Edn. Commn., 1984-92; trustee Duke U., 1987—; chmn. Meth. Hosp. Found., 1990; mem. adv. bd. Inst. Pub. Policy Vanderbilt U., 1986-88, conf. chmn. 1987. Fellow Am. Phys. Soc. (exec. com. div. plasma physics), AAAS, Am. Nuclear Soc. (dir.); mem. C. of C. (v.p. 1981-83, chmn. 1987), Indsl. Rsch. Inst., Gas Rsch. Inst. (adv. bd. 1986-88), Phi Beta Kappa, Beta Gamma Sigma, Sigma Pi Sigma, Omicron Delta Kappa, Sigma Xi, Pi Mu Epsilon, Phi Eta Sigma. Home: 104 Berea Rd Oak Ridge TN 37830-7829

POSTOL, THEODORE A., physicist, educator. Prof. dept. physics MIT, Cambridge, Mass. Recipient Leo Szilard award Am. Phys. Soc., 1990. Office: MIT Dept of Physics 9 Cambridge Ctr Cambridge MA 02139*

POSTON, ANN GENEVIEVE, psychotherapist, nurse; b. Sioux City, Iowa, July 28, 1936; d. Frank Earl and Ella Marie (Stanton) Gales; m. Gerald Connell Poston, June 27, 1959; children: Gregory, Mary Ann, Susan. BSN, Briar Cliff Coll., 1958; MA, U. Mo., 1978; postgrad., Family Inst. of Kansas City, Inc., 1989-91. RN, Kans., Mo.; lic. counselor, Mo. Staff nurse, sr. team leader St. Joseph Mercy Hosp., Sioux City, 1958-59; head nurse St. Anthony's Hosp., Rock Island, Ill., 1960, charge nurse, 1966-69; charge nurse St. Mary's Hosp., Mpls., 1970-71, North Kansas City (Mo.) Hosp., 1972-73, Tri-County Mental Health Ctr., North Kansas City, 1973-79; psychotherapist VA Med. Ctr., Kansas City, 1979-84, Leavenworth, Kans., 1984-85; psychotherapist The Kans. Inst., Olathe, 1985—; cons. Synergy House, Parkville, Mo., 1974-75, North Kansas City Hosp., 1978-79, VA Hosps., Kansas City and Leavenworth, 1979-85, Cath. Charities, Kansas City, 1983-87, Olathe Med. Ctr., 1985—, Humana Med. Ctr. Overland Park, Kans., 1986—, St. Joseph Med. Ctr., Kansas City, Mo., 1990—. Author, presenter (video) Depression & Suicide, 1980. Third officer King's Daus., Moline, Ill., 1960-69; campaign worker Rep. Party, Moline, 1963-68; community asst. New Mark Community Affairs, Kansas City, 1972-76; nursing rep. Combined Fed. Campaign, Kansas City, 1982; coord. mental health program com. Midwest Health Congress, Kansas City, 1981. Mem. ACA, ANA (cert.), Internat. Assn. for Marriage and Family Counselors, Am. Assn. Marriage and Family Therapy (clinical), Nat. Bd. Cert. Counselors, Mo. Assn. Marriage and Family Therapy, Sigma Theta Tau. Roman Catholic. Avocations: theater, traveling, bridge. Office: The Kans Inst 20375 W 151st St Ste 206 Olathe KS 66061-5360

POSTON, JOHN MICHAEL, biochemist; b. Kalispell, Mont., Oct. 16, 1935; s. Howard Joseph and Mabel Lenore (Iverson) P.; m. Annette Marlane Dapp, July 15, 1967; children: Janice Marie, Susanne Marlene, Todd Russell. BS in Chemistry, Mont. State Coll., 1958; MS in Biochemistry, U. N.D., 1960, PhD in Biochemistry, 1970. Chemist Nat. Heart Inst. NIH, Bethesda, Md., 1961-69, rsch. biochemist Nat. Heart, Lung and Blood Inst., 1970—; mem. organizing com. Int Symposium on Cellular Regulation, Bethesda, 1984, Symposium on Cellular Regulation, New Orleans, 1990. Contbr. over 25 articles in biochemistry to profl. jours. Bd. dirs. Glenbrook Found., Bethesda, 1982-85, The Ivymount Sch., Rockville, Md., 1985—. NIH predoctoral fellow NIH, 1969-70; recipient travel grant Int. Symposium on Metabolism, Charleston, S.C., 1980. Mem. Am. Chem. Soc., Am. Soc. Biochemistry and Molecular Biology, Am. Soc. Microbiology, Sigma Xi. Lutheran. Home: 29 Orchard Way S Rockville MD 20854-6129 Office: NHLBI Lab of Biochemistry 3/216 NIH Bethesda MD 20892

POTENTE, EUGENE, JR., interior designer; b. Kenosha. Wis., July 24, 1921; s. Eugene and Suzanne Marie (Schmit) P.; Ph.B., Marquette U., 1943; postgrad. Stanford U., 1943, N.Y. Sch. Interior Design, 1947; m. Joan Ciofe, Jan. 29, 1946; children—Eugene J., Peter Michael, John Francis, Suzanne Marie. Founder, pres. Studios of Potente, Inc., Kenosha, Wis., 1949—; pres., founder Archtl. Services Assos., Kenosha, 1978—, Bus. Leasing Services of Wis. Inc., 1978—; past nat. pres. Inter-Faith Forum on Religion, Art and Architecture; vice chmn. State Capitol and Exec. Residence Bd., 1981—. Sec., Kenosha Symphony Assn., 1968-74. Bd. dirs. Ctr. for Religion and the Arts, Wesley Theol. Sem., Washington, 1983-84. Served with AUS, 1943-46. Mem. Am. Soc. Interior Designers (treas., pres. Wis. chpt. 1985—, chmn. nat. pub. service 1986), Illuminating Engring. Soc. N.Am., Inst. Bus. Designers, Sigma Delta Chi. Roman Catholic. Lodge: Elks. Home: 8609 2d Ave Kenosha WI 53143 Office: 914 60th St Kenosha WI 53143

POTHITT, KATHLEEN MARIE, physical oceanography researcher; b. Santa Monica, Calif., Jan. 11, 1964; d. John Andrew and Eileen Sharon (Melcher) Alexander; m. Richard Pothitt, Jan. 22, 1993; children from a pervious marriage: Richard Alexander Island, Timothy Demetrius Island. Student, UCLA, 1981-83; BS in Applied Math. summa cum laude, Calif. State U., Northridge, 1986. Rsch. analyst Arete Assocs., Sherman Oaks, Calif., 1985-91, mem. rsch. staff, 1991—; property mgr. Ronald Craig, Inc., Northridge, 1987-90; tutor math. Culver City 1986—. Recipient Acad. Achievement award, Calif. State U., 1986. Mem. Am. Geophys. Union, Am. Math. Assn., Soc. for Indsl. and Applied Math., Soc. Women Engrs., Women in Sci. and Engring., Nat. Parks Conservation Assn. Republican. Home: 3873 Girard Ave # 4 Culver City CA 90232 Office: Arete Assocs PO Box 6024 Sherman Oaks CA 91413-6024

POTHOS, EMMANUEL, neuroscientist; b. Athens, Greece, Jan. 1, 1965; s. Nikolaos and Zacharoula (Panayotopoulos) P. BA with distinction, Deree Coll., Athens, 1988, U. Athens, 1989; MA, Princeton U., 1990, PhD, 1993. Rsch. asst. Deree Coll., 1985-88, U. Cambridge, Eng., 1987. Contbr. articles to profl. publs. Mem. AAAS, APA, Am. Psychol. Soc., Soc. for Neurosci., Soc. for Study of Ingestive Behavior. Achievements include discovery of low levels of synaptic dopamine in the mesolimbic system of underweight rats, reduction of mesolimbic dopamine response to drugs or food intake in underweight rats. Office: Princeton U Program in Neurosci Dept Psychology Princeton NJ 08544-1010

POTTER, MICHAEL, genetics researcher, medical researcher; b. East Orange, N.J., Feb. 27, 1924. AB, Princeton U., 1945; MD, U. Va., 1949. Research asst. dept. microbiology U. Va. Med. Sch., 1952-54, biologist, 1954-70, head immunochemistry sect. Lab. Cell Biology, 1970-82; head genetics lab., cancer biology and diagnosis div. Nat. Cancer Inst., Bethesda, Md., 1982—. Recipient Paul-Ehrlick & Ludwig-Darmstaedter prize, 1983, Lasker award in basic med. research, 1984. Mem. Nat. Acad. Sci., Am. Assn. Cancer Research, Am. Assn. Immunologists. Office: Nat Cancer Inst 9000 Rockville Pike Bldg 31 Bethesda MD 20892-0001

POTTER, PAUL EDWIN, geologist, educator, consultant; b. Springfield, Ohio, Aug. 30, 1925; s. Edwin Forest and Mabel (Yanser) P. MS in Geology, U. Chgo., 1950, PhD in Geology, 1952; MS in Stats., U. Ill., 1959. Research assoc. Ill. Geol. Survey, Urbana, 1952-54, asst. geologist, 1954-61; assoc. prof. geology Ind. U., Bloomington, 1963-65, prof., 1965-71; prof. U. Cin., 1971-92, prof. titular, 1992—. Author: Atlas and Glossary of Sedimentry Structures, 1964, Sand and Sandstone, 2d edit., 1987, Paleocurrents and Basin Analysis, 2d edit., 1977, Introduction to Petrography of Fossils, 1971, Sedimentology of Shale, 1980. Served with U.S. Army, 1944-46. Sr. NSF fellow, 1958, Guggenheim fellow, 1961-62; recipient Francis J. Pettijohn Sedimentary medal, 1992. Republican. Office: Geociecias/UNESP, Caixa 178, Rio Claro 13506 CP, Brazil

POTTORFF, BEAU BACKUS, astronautical engineer; b. Ft. Pierce, Fla., Sept. 21, 1959; s. Robert Larry and Ann Claire (vanRavesteyn) Jinks; m. Marcia Ann Janssen, Sept. 28, 1985; 1 child, Amber Marie. BCS, Santa Clara (Calif.) U., 1983. Enrolled astron. engring., orbital mechanics. Commd. 2d lt. USAF, 1982, advanced through grades to capt., 1986; aerospace data analyst USAF, Aurora, Colo., 1983-86; space ops. officer interrange ops. USAF, Sunnyvale, Calif., 1986-88, dep. dir., 1988-90, dir., 1990-91; sr. field systems engr. Jet Propulsion Lab., Bendix Field Engring., Pasadena, Calif., 1991—; co. chmn. Radio Frequency Working Group, Sunnyvale, 1983-91; mem. Radio Frequency Spectrum Analysis Com., Washington, 1990-91; chmn. Satellite Requirements Group, Sunnyvale, 1990-91. Decorated Air Force Commendation medal with oak leaf cluster, Air Force Achievement medal with oak leaf cluster. Mem. AIAA, IEEE, Assn. Americans for Advancement of Sci., Planetary Soc., Profl. Assn. Diving Instrs. (divemaster 1989—), Nat. Space Soc. Republican. Lutheran. Achievements include discovery of radio frequency 2nd harmonic interference effects within Jet Propulsion Lab. Deep Space Network. Office: Allied Signal Tech Svcs Corp Jet Propulsion Lab 129 N Hill Ave M/S 507-215 Pasadena CA 91106

POTTS, ALBERT M., ophthalmologist; b. Balt., June 8, 1914; s. Isaac and Leah (Mintz) P.; m. Esther Topkis, June 14, 1938; children: William T., Leah Potts Fisher, Deborah. AB in Chemistry, Johns Hopkins U., 1934; PhD in Biochemistry, U. Chgo., 1938; MD, Western Res. U., 1948. Resident in ophthalmology Univ. Hosps., Cleve., 1948-51; instr. ophthalmology Western Res. U., Cleve., 1951-54, asst. prof. ophthalmic rsch., 1954-59; prof. ophthalmology U. Chgo., 1959-75; prof., chmn. dept. ophthalmology U. Louisville, 1975-84; acting head dept. ophthalmology U. Ariz., Tucson, 1984-85, clin. prof., 1984-88; v.p. Bahill Intelligent Computer Systems, Tucson, 1989—; cons. on ophthalmic toxicology, 1954—. Author: The World's Eye, 1982; editor: The Assessment of Visual Function, 1972; contbr. articles to profl. jours. Recipient Friedenwald award Assn. Rsch. in Vision and Ophthalmology, 1961, L.V. Johnson Lectureship Cleve. Ophthalmology Club, 1975, Sanford Gifford Meml. Lecture Chgo. Ophthalmol. Soc., 1977. Fellow AAAS, Am. Acad. Ophthalmology (life); mem. Assn. Rsch. in Ophthalmology (pres. 1974). Achievements include research in physiological pharmacology, the effects of drugs upon the eye and toxic responses of the eye, medical informatics, electronic imaging. Office: Bahill Intelligent Computer Systems 500 N Tucson Blvd Tucson AZ 85716-1602

POTTS, JOHN THOMAS, JR., physician, educator; b. Phila., Jan. 19, 1932; married; 3 children. B.A., LaSalle Coll, Phila., 1953; M.D., U. Pa., Phila., 1957. From intern to asst. resident in medicine Mass. Gen. Hosp., Boston, 1957-59; resident Nat. Heart Inst., 1959-60, research fellow in medicine, 1960-63, sr. research staff, 1963-66, head sect. polypeptide hormones, 1966-68; chief endocrine unit Mass. Gen. Hosp., Boston, 1968-81, chief gen med. svc., 1981—; from asst. to assoc. prof. medicine Harvard U. Med. Sch., Boston, 1968-75, prof., 1975-81, Jackson prof. clin. medicine, 1981—; chief endocrine unit Mass. Gen. Hosp., Boston, 1968-81, chief gen. med., 1981—. Recipient Ernest Oppenheimer award, Andre Lichwitz prize Endocrine Soc., 1968, Fred Conrad Koch award Endocrine Soc., 1991, William F. Neumann award Am. Soc. Bone and Mineral Rsch. Fellow AAA; mem. Am. Soc. Biol. Chemistry, Endocrine Soc. (pres. 1987), Assn. Am. Physicians, Am. Fedn. Clin. Research, Am. Soc. Clin. Investigation, Inst. Medicine. Office: Mass Gen Hosp Med Svcs Med Svcs Fruit St Boston MA 02114-2620

POUDYAL, SRI RAM, economics educator, consultant; b. Tuhurepasal, Tanahu, Nepal, Sept. 7, 1950; s. Ram Chandra and Ganga (Devi) P.; m. Pabitra Poudyal, 1972; children: Suraj, Supriya, Sumita. MA in Econs., Tribhuvan U., Kathmandu, Nepal, 1971; diploma in econ. devel., U. Manchester, Eng., 1974, MA in Econs., 1975; PhD in Econs., U. Delhi, India, 1987. Adviser Nepal Rastra Bank, Kathmandu, 1986-88; economist Tribhuvan U., 1978-79, lectr., 1979-81, reader, 1981-89, prof., 1990—; adviser Nat. Planning Commn., 1992; mem. Nepal Trade Promotion Bd., Kathmandu, 1990-91; cons. World Bank, Asian Devel. Bank, UNDP and other internat. agys. Author: Planned Development in Nepal, 1983, Foreign Trade, Aid and Development in Nepal, 1988; contbr. articles in nat. and internat. profl. jours. Avocation: hiking. Home: Sunar Gaon, Kalanki, GPO Box 5386, Kathmandu Nepal Office: Tribhuvan U Dept Econs, Kirtipur, Kathmandu Nepal

POULEUR, HUBERT GUSTAVE, cardiologist; b. Bouffioulx, Belgium, June 6, 1948; m. Michelle Leonet, July 7, 1973; children: Anne-Çatherine, Jean-Hubert. MD, U. Louvain, Belgium, 1973, PhD, 1980. Intern, resident, then fellow in internal medicine U. Louvain, Belgium, 1973-77; Pub. Health Service internat. research fellow U. Calif, San Diego, 1977-79; asst. prof. U. Louvain, Brussels, 1979-83, assoc. prof., 1983-91, prof., 1991—; disting. clin. scientist Syntex Clin. Rsch., Palo Alto, Calif., Maidenhead, U.K., 1988-93; assoc. dir. clin. rsch. Pfizer Inc., Groton, Conn., 1993—. Contbr. numerous sci. articles to profl. jours. Recipient Damman prize Damman Found., 1977, Bekales prize Bekales Found., 1986; Squibb Cardiovascular fellow Belgian Soc. Cardiology, 1982. Fellow Am. Coll. Cardiology, Coun. of Circulation; mem. Am. Heart Assn. (Coun. of Circulation fellow, Coun. Clin. Cardiology internat. fellow), Belgian Fly Fishing Club (Brussels). Avocation: fly fishing. Home: ave des Papillons 6, 1410 Waterloo Belgium Office: U Louvain, av Hippocrate 55 Hedy/5560, 1200 Brussels Belgium

POULIQUEN, MARCEL FRANÇOIS, space propulsion engineer, educator; b. Guimiliau, France, June 19, 1945; s. Jean Yves and Yvonne Marie (Cornily) P.; m. Renée Yannik Maza, June 27, 1969; 1 child, Elodie Marie. BCE, Sup'Aero, Paris, 1969. Design engr. Societe Europeenne de Propulsion, Paris, 1970-71; tech. mgr. Societe Europeenne de Propulsion, Villaroche, France, 1971-74; program mgr. Societe Europeenne de Propulsion, Vernon, France, 1974-83; R&D mgr. Societe Europeenne de Propulsion, Suresnes, France, 1983-89, chief engr., 1990-91, chief engr. advanced systems, 1991—; prof. various French and internat. univs., 1973—. Contbr. chpt. to book, numerous articles to profl. jours. Served with French Air Force, 1969-70. Mem. AIAA, Assn. Aeronautique et Astronautique de France, Internat. Astronautical Fedn., Internat. Acad. Astronautics. Roman Catholic. Achievements include patents for rocket engine components, contribution to the design of two European cryogenic rocket engines, combined engines for hypersonic flights-small launchers projects. Office: SEP, 24 Rue Salomon-de-Rothschild, 92150 Suresnes France

POULSEN, DENNIS ROBERT, environmentalist; b. Boston, Jan. 17, 1946; s. Stephen Dudley and Dorothy Hope (Davis) P.; m. Bonnie Lou Reed; children: David, Zachery, Patrick. AS in Forestry, U. Mass., Stockbridge-Amherst, 1965; AS in Indsl. Supervision, Chaffey Coll., Alta Loma, Calif., 1977; BS in Bus. Adminstrn., U. Redlands (Calif.), 1979; postgrad., U. Calif., Riverside, 1986, U. Calif., Davis, 1991-93; cert. program, U. Calif., Davis, 1991—. Cert. environ. profl., registered environ. profl., registered environ. assessor, Calif., cert. hazardous materials mgr., cert. lab. technolgoist; diplomate Inst. Hazardous Materials Mgmt. Water control technician Weyerhaeuser Co. Chem. Lab., Fitchburg, Mass., 1965-69; environ. rsch. technician Kaiser Steel Corp., Fontana, Calif., 1969-78, environ. rsch. engr., 1978-83, asst. environ. dir., 1983-87; mgr. environ. svcs. Calif. Steel Industries Inc., Fontana, 1987—; mem. adv. group Calif. EPA (CAL EPA), 1993—; originator AIEE Nat. Environ. Com., Pitts., papers chmn., 1993; mem. adv. group Calif. Environ. Protection Agy., 1993—. Editorial adv. bd. Indsl. Wastewater Mag., 1993—; contbr. articles and papers on environmental issues to profl. publs. Del. U.S. Environ. Delegation, Soviet Union, 1990; mem. U.S. Citizens Network of the UN Conf. on Environment and Devel. Mem. Nat. Assn. Environ. Profls. (cert. review bd., mem. internat. com. 1992—), Air and Water Mgmt. Assn., Nat. Environ. Health Assn., Environ. Info. Assn., Hazardous Materials Control Rsch. Inst., Water Environment Fedn. (groundwater com.), World Safety Orgn. (cert. hazardous materials supr.), Assn. Energy Engrs. (environ. engrs. mgrs. inst., environ. project of yr. award 1992), Chino Basin Water Dist. Watermaster Adv. Coun., Calif. Water Pollution Control Assn., Inst. Hazardous Materials Mgmt. (ethics com., publs. sub-com.), People to People Internat., U. Redlands Alumni Assn. (bd. mem., nominee Gordon Adkins award for profl. achievement). Avocation: travel. Home: 5005 Hedrick Ave Riverside CA 92505-1425 Office: Calif Steel Industries Inc 14000 San Bernardino Ave Fontana CA 92335-5259

POUND, ROBERT VIVIAN, physics educator; b. Ridgeway, Ont., Can., May 16, 1919; came to U.S., 1923, naturalized, 1932; s. Vivian Ellsworth and Gertrude C. (Prout) P.; m. Betty Yde Andersen, June 20, 1941; 1 son, John Andrew. B.A., U. Buffalo, 1941; A.M. (hon.), Harvard U., 1950. Research physicist Submarine Signal Co., 1941-42; staff mem. Radiation Lab. Mass. Inst. Tech., 1942-46; jr. fellow Soc. Fellows, Harvard U., 1945-48; asst. prof. physics Harvard U., 1948-50, assoc. prof., 1950-56, prof., 1956-68, Mallinckrodt prof. physics, 1968-89, emeritus, 1989—, chmn. dept. physics, 1968-72, dir. Physics Labs., 1975-83; Fulbright research scholar Oxford, 1951; Fulbright lectr., Paris, 1958; vis. prof. Coll. de France, 1973; vis. fellow Joint Inst. Lab. Astrophysics, U. Colo., 1979-80; vis. research fellow Merton Coll., Oxford U., 1980; Zernike vis. prof. U. Groningen (Netherlands), 1982; vis. sr. scientist Brookhaven Nat. Lab., 1986-87; vis. prof. U. Fla., 1987. Author, editor: Microwave Mixers, 1948; Contbr. articles to profl. jours. Trustee Associated Univs., Inc., 1976—. Recipient B. J. Thompson Meml. award Inst. Radio Engrs., 1948, Eddington medal Royal Astron. Soc., 1965, Nat. Medal Sci. NSF, 1990; John Simon Guggenheim fellow, 1957-58, 1972-73. Fellow Am. Phys. Soc., Am. Acad. Arts and Scis., AAAS; mem. Nat. Acad. Scis., Soc. Franc. de Physique (membre du conseil 1958-61), Acad. des Scis. (fgn. assoc.) (France), Phi Beta Kappa, Sigma Xi.

POWE, RALPH ELWARD, university administrator; b. Tylertown, Miss., July 27, 1944; s. Roy Elward and Virginia Alyne (Bradley) P.; m. Sharon Eve Sandifer, May 20, 1962; children: Deborah Lynn, Ryan Elward, Melanie Colleen. B.S. in Mech. Engring., Miss. State U., 1967, M.S. in Mech. Engring., 1968; Ph.D. in Mech. Engring., Mont. State U., 1970. Student trainee

NASA, 1962-65; research asst., lab. instr. Miss. State U., 1968, instr. dept. mech. engring., 1968; research asst., teaching asst. Mont. State U., Bozeman, 1968-70, asst. prof. dept. mech. engring., 1970-74; assoc. prof. Miss. State U., 1974-78, prof., 1979-80, assoc. dean engring., dir. engring. and indsl. research sta., 1979-80, assoc. v.p. research, 1980-86, v.p., 1986—; bd. dirs. Coalition of Experimental Program to Stimulate Competitive Rsch. States, Gulf Univs. Rsch. Consortium, Tenn.-Tombigbee Project Area Coun., Oktibbeha Devel. Coun.; rep. rsch. coun. Nat. Assn. State Univs. and Land Grant Colls., So. Growth Policies Bd., Miss. Mineral Resources Inst., Sci. and Tech. Coun. States. Disting. Engring. fellow Coll. Engring.; active Miss. Univ. Res. Authority, Coun. on Rsch. Policy, So. Growth Policies Bd.; rep. Miss. Mineral Resources Inst.; gov. rep. Sci. and Tech. Coun. of States; cons. energy conservation programs, coal fired power plants, torsional vibrations, accident analysis; dir. Miss. Energy Rsch. Ctr., 1979-81, Ctr. for Environ. Studies, 1980—; univ. rep. on lignite task force, rep. on bd. dirs. Miss.-Ala. Sea Grant Consortium, rep. to Council Oak Ridge Associated Univs., rep. to Tenn.-Tombigbee project area coun.; chmn. Miss. Rsch. Consortium; mem. S.E. Univs. Rsch. Assn. Named Outstanding Egr., Engring. Socs. Contbr. articles to profl. jours. Mem. Miss. Econ. Coun., univ. coord. United Way, 1983, 85; tchr. adult Sunday Sch. class 1st Papt. Ch. Recipient Ralph E. Teeter award Soc. Automotive Engrs.; named Outstanding Engr. in No. Miss. Joint Engr. Soc. Fellow ASME; mem. Nat. Assn. State Universities and Land Grant Colls., Starkville Community Theatre Wing. Edn., Wind Energy Soc. Am., Miss. Acad. Scis., Miss. Engring. Soc., Toastmasters, Starkville Quarterback Club, Rotary, Starkville C. of C., Blue Key, Sigma Xi (Miss. State U. research award), Tau Beta Pi, Kappa Mu Epsilon, Pi Tau Sigma, Phi Kappa Phi, Omicron Delta Kappa. Baptist. Lodge: Rotary. Avocations: hunting, fishing, gardening. Home: 110 Pinewood Rd Starkville MS 39759-4128 Office: Miss State U PO Box 6343 Mississippi State MS 39762-6343

POWELL, ALAN, mechanical engineer, scientist, educator; b. Buxton, Derbyshire, Eng., Feb. 17, 1928; came to U.S., 1956; s. Frank and Gwendolen Marie (Walker) P.; m. June Sinclair, Mar. 28, 1956. Student, Buxton Coll., 1939-45; diploma in aeros., Loughborough Coll., 1948; B.Sc. in Engring. with 1st class honors, London U., 1949; honours diploma 1st class, Loughborough Coll., 1949; D.Tech. (hon.), Loughborough U. Tech., 1980; Ph.D., U. Southampton, 1953. Chartered aero. engr., mech. engr. Engr. Percival Aircraft Co., Luton, Eng., 1949-51; research asst. U. Southampton, Eng., 1951-53; lectr. U. Southampton, 1953-56; research fellow Calif. Inst. Tech., Pasadena, 1956-57; engr. Douglas Aircraft Co., 1956; assoc. prof. UCLA, 1957-62, prof. engring., 1962-65, head Aerosonics lab., 1957-65; assoc. tech. dir., head acoustics and vibration lab. David Taylor Model Basin, Dept. Navy, Washington, 1965-66; tech. dir. David Taylor Model Basin, Dept. Navy, 1966-67, David Taylor Naval Ship Research & Devel. Center, Bethesda, Md., 1967-85; mem. Undersea Warfare Research & Devel. Council, 1966-76, chmn., 1971-72; mem. council on Fed. Labs., 1972-85; prof. mech. engring. U. Houston, 1985—, chmn., 1985-87; mem. com. on hearing bioacoustics and biomechs. NAS-NRC, 1961-85, advisor, 1985—, exec. council, 1963-65, chmn., 1965-66, mem. naval studies bd., 1990—; advisor Chinese U. Devel. Project, NAS, 1989-91; cons. Douglas Aircraft Co., various aerospace and acoustics cos., 1956-65; mem. adv. council Internat. Towing Tank Conf., 1981-85; mem. sci. com. Internat. Union Theoretical and Applied Mechanics, 1983-85; advisor U.S.-Japan Program Natural Resources, 1987—, mem., 1979-86; chmn. internat. conf. ComputerAided Design, Manufacture and Ops. in Marine and Offshore Industries, 1987-88; cons. Scientific Applications Internat., Inc., 1987-90. Contbr. articles to profl. jours. Recipient Navy Meritorious Civilian Service award, 1970; Brit. Empire scholar, 1945, Per Bruel gold medal, 1991; named Meritorious Exec. Pres. of U.S., 1982; Capt. Robert Dexter Conrad gold medal for sci. achievement Sec. Navy, 1984. Fellow AIAA (assoc. fellow, Aeroacoustics award 1980), Royal Aero. Soc. London (Baden-Powell prize 1948, Wilbur Wright prize 1953), Acoustical Soc. Am. (biennial award 1962, assoc. editor Jour. 1962-67, chmn. edn. com. 1964-66, exec. coun. 1966-69, chmn. medals and awards com. 1978-81, v.p. elect 1981-82, v.p. 1982-83, pres. elect 1989-90, pres. 1990-91, past pres. 1991-92, Silver medal in engring accoustics 1992), Instn. Mech. Engrs., Inst. Acoustics (U.K.); mem. ASME (Rayleigh lectr. 1988, Per Bruel Noise Control and Acoustics Gold medal 1991), IEEE (sr.), Inst. Noise Control Engrs. (initial mem., dir. 1974-77, Disting. lectr. 1975, 83, v.p. 1981-84), Acoustics, Speech and Signal Processing Soc. (sr., exec. com. 1969-72, awards com. 1971-73, bylaws com. chmn. 1973-75), Am. Soc. Naval Engrs. (life), Tau Beta Pi (hon. life mem.). Office: U Houston Dept Mech Engring Houston TX 77204-4792

POWELL, BRIAN HILL, software engineer; b. Austin, Tex., Dec. 5, 1962; s. Daniel Benjamin and Anna Marie (Lettermann) P.; m. Lynn Dees, Apr. 28, 1990. BA, U. Tex., Austin, 1984. With U. Tex., Dept. Computer Sci., Austin, 1984-87; software engr. Nat. Instruments Corp., Austin, 1988—. Mem. SAR. Office: Nat Instruments Corp 6504 Bridge Point Pky Austin TX 78730-5039

POWELL, CHRISTOPHER ROBERT, systems analyst; b. Summit, N.J., Feb. 2, 1963; s. Robin Powell and Nancy Mae (Spurling) Gould; m. Bonnie Jean Manning, June 10, 1989. BS in Math. and Computer Sci., Clarkson U., 1984; postgrad. in Computer Sci., Syracuse U., 1988; postgrad. in Philosophy, SUNY, Binghamton, 1990. Sr. assoc. program IBM Corp., Endicott, N.Y., 1984-90; sr. systems analyst/programmer Supercomputer Systems, Inc., Eau Claire, Wis., 1990-93; sr. systems programmer Network Systems Corp., Brooklyn Park, Minn., 1993—. Mem. Ass. for Computing Machinery, Nat. Systems Programmer's Assn. Democrat. Home: 8220 6th St NE Spring Lake Park MN 55432

POWELL, JUDITH CAROL, clinical psychologist; b. Tuscaloosa, Ala., May 7, 1949; d. Fred Bouch and Ella Louise (Harper) P. BS in Chemistry, Furman U., 1971; MA in Psychology, Bowling Green (Ohio) State U., 1985, PhD in Psychology, 1987. Lic. psychologist, N.C. Forensic chemist Ga. State Crime Lab., Atlanta, 1971-73; scientific glassblower Quality Glass Apparatus, Ann Arbor, Mich., 1973-79; clin. psychologist Wake County Mental Health Ctr., Raleigh, N.C., 1987-91; clin. supr. Wake County Mental Health Ctr., Raleigh, 1990-91; patient care advisor Psychol. Health Care, Chapel Hill, N.C., 1991—; pvt. practice Chapel Hill, 1991—. Bd. dirs. Our Own Place-Lesbian Community Ctr., Durham, N.C., 1989-90, N.C. Lesbian and Gay Health Project/The AIDS Svcs. Project, Durham, 1992-93. Mem. APA (divsns. 35 and 44), Assn. Women in Psychology, Nat. Assn. Lesbian and Gay Psychologists U.S. Home: 113 Simpson St Carrboro NC 27510-1235 Office: PO Box 4723 Chapel Hill NC 27515-4723

POWELL, MARGARET ANN SIMMONS, computer scientist; b. Gulfport, Miss., May 26, 1952; d. William Robert and Nancy Rita (Schloegel) Simmons; m. Mark Thomas Powell, Sept. 11, 1983. AS in Math., N.W. Miss. Jr. Coll., 1972; BS in Edn., Memphis State U., 1977; BS in Computer Sci., U. Md., 1988; MS in Computer Sci., Johns Hopkins U., 1991. Tchr. Sacred Heart Sch., Walls, Miss., 1973-80; office mgr. Hyman Builders Supply, Memphis, 1980-84; tech. instr. Bendix Field Engring. Corp., Greenbelt, Md., 1985-87; software engr. Assurance Technology Corp., Alexandria, Va., 1987-89, Naval Rsch. Lab., Washington, 1989-93; computer scientist Naval Info. Systems Mgmt. Ctr., Washington, 1993—. Bd. dirs. Greenbrook Village Homeowners Assn., 1992—. Named one of Outstanding Young Women Am., 1977. Mem. IEEE Computer Soc., Assn. for Computing Machinery, Phi Kappa Phi, Kappa Delta Pi, Phi Theta Kappa, Mu Alpha Theta. Roman Catholic. Avocations: needlework, tropical fish. Home: 7810 Somerset Ct Greenbelt MD 20770-3022 Office: NISMC Bldg 166 Washington Navy Yard Washington DC 20374

POWELL, MARTHA JANE, botany educator; b. Charlotte, N.C., Jan. 27, 1948; d. John James and Martha Lee (Martin) P.; m. Will Hoyle Blackwell, Dec. 23, 1977. BS, Western Carolina U., 1969; PhD, U. N.C., 1974. Postdoctoral researcher Purdue U., West Lafayette, Ind., 1974-76; prof. Miami U., Oxford, Ohio, 1976—. Contbr. articles to profl. jours. Recipient Arts and Scis. Disting. Educator award Miami U., 1990, Sigma Xi Researcher of Yr. award, 1985, Alexopoulos Rsch. award Mycological Soc. of Am., 1981, Coker fellowship U. N.C., 1972-73, Grad. Student Best Paper award, 1973. Mem. Mycological Soc. of Am. (pres. 1991-92, v.p. 1989-90, treas. 1986-89), Internat. Soc. for Evolutionary Protistology (sec., treas. 1979-83). Democrat. Methodist. Achievements include elucidation of roles of single-membrane bounded organelles in zoospores of fungi and protista. Home: 2890 Harris Rd Hamilton OH 45013 Office: Botany Dept Miami Univ Oxford OH 45056

POWELL, PATRICIA ANN, mathematics and business educator; b. Covington, Ga., Apr. 6, 1956; d. John Doyle Sr. and Pauline Josephine (Thompson) Dunn; m. Jackie Lee Powell, May 10, 1980; 1 child, Jackie Lee II. BS, Lee Coll., 1978; MEd in Adminstrn. and Supervision, U. Tenn., 1993. Br. loan officer Am. Nat. Bank and Trust, Chattanooga, 1979-81; instr. math., careers Hamilton County Schs., Chattanooga, 1983-85; customer svc. rep. First Union Nat. Bank, Atlanta, 1986-88; instr. tech. bus., typing DeKalb County Schs., Decatur, Ga., 1989; grad. asst. U. Tenn., Chattanooga, 1991-93; instr. math. Hamilton County Sch. System, 1993—; instr. English, bus. math. and bus. skills Urban League Bus. Skills Tng. Ctr., Chattanooga; adj. faculty Chattanooga State Tech. Community Coll., 1991-92. Co-author: Career Orientation-Grade 8, 1985 (monetary award 1984-85). Singer, Mayor's Office Performing Artists Against Drugs, Atlanta, 1990; vol. Chattanooga Community Kitchen (Cert. of Svc. award 1991), 1990—; tutor, coord. math., reading United Way's Adult Reading Program, Chattanooga, 1991—; instr. aerobics Am. Heart Assn., Chattanooga, 1991—; vol. Warner Park Zoo, Chattanooga; active Big Brother/Big Sister Program, Chattanooga, 1991—; treas. Looking to the Word Ministries, Inc., 1985—; v.p. parents group First Cumberland Child Devel. Ctr. Recipient Black Grad. fellowship U. Tenn., Chattanooga, 1992; named Woman of Yr. and Mrs. Congeniality, Mrs. Chattanooga-Am. Pageant, 1990; Endowment scholar, 1977-78. Mem. AAUW, Friends of the Zoo Preservation Group, Delta Sigma Theta, Kappa Delta Pi. Mem. Pentecostal Ch. Avocations: singing, aerobics, crafts, sewing, stamp collecting. Home: PO Box 24912 Chattanooga TN 37422-4912

POWELL, SAUL REUBEN, research scientist, surgical educator; b. Phila., Oct. 13, 1953; m. Pnina Korbashi, Sept. 7, 1989; 1 child, Keren. BS in Pharmacy, Phila. Coll. Pharmacy and Sci., 1976; PhD in Pharmacology, Med. Coll. Pa., 1983. Assoc. rsch. scientist The Okla. Med. Rsch. Found., Oklahoma City, 1983-85, sr. rsch. scientist, 1985-87; rsch. assoc. Hebre U. Jerusalem, Israel, 1987; rsch. scientist North Shore U. Hosp., Manhasset, N.Y., 1987—; asst. prof. surgery Med. Coll. Cornell U., N.Y.C., 1990—. Named NIH grantee, 1993. Mem. Am. Heart Assn. (grantee nat. ctr. 1989, grantee N.Y. affiliate 1992), Am. Chem. Soc., Am. Soc. Pharm. and Exptl. Therapists, The Oxygen Soc., N.Y. Acad. Scis. Achievements include rsch. in the role of oxidative injury in postischemic cardiac damage and the protective effects of antioxidants such as zinc. Office: North Shore Univ Hosp 350 Community Dr Manhasset NY 11030

POWER, WALTER ROBERT, geologist; b. Seattle, Nov. 7, 1924; s. Walter Robert and Marie (Madden) P.; m. Martha Ann Thompson, June 18, 1960; children: John Robert, Joseph Patrick. BS in Geology, U. Wash., 1949; PhD, Johns Hopkins U., 1959. Geologist U.S. Geol. Survey, Boston, 1950-54, U.S. Atomic Energy Commn., Olancha, Calif., 1956-57; asst. prof. U. Ga., Athens, 1957-61; ch. geologist Ga. Marble Co., Atlanta, 1961-67, mgr. mineral resources, 1987-89; prof. geology Ga. State U., Atlanta, 1967-87, prof. emeritus, 1987—; cons. Ga. Marble Co., Kennesaw, 1989—. Author: (with others) Dimension Stone. With U.S. Army, 1943-45. Fellow Geol. Soc. Am.; mem. AAAS, Soc. for Mining, Metalurgy and Exploration. Home: 792 7th Box 704 Frisco CO 80443

POWERS, CLIFFORD BLAKE, JR., communications researcher; b. Macon, Ga., July 15, 1960; s. Clifford Blake Sr. and Virginia (Davis) P. BA in Journalism with honors, Columbia Coll., 1983; MS in Communications, U. Tenn., 1989. Photographic intern Playboy, Chgo., 1983; corr.-at-large SpaceWorld Mag., 1983-86; sr. sci. writer Schneider Svcs. Internat., Arnold AFB, Tenn., 1985-87; grad. rsch. asst. U. Tenn., Knoxville, 1988-89; writer II Space & Def. div. Essex Corp., Huntsville, Ala., 1990—; project mgr. Spacelab J, USML-1, TSS-1 missions; instr. sci. Faulkner U., Huntsville, 1990-91. Author: The Soviet Watchers, 1990, (with others) Unlocking The Mysteries of the Universe: A Guide to Astro-1 Observations, 1990, Astro-1 Postmission Summary Report, 1991, Spacelab J: Microgravity and Life Sciences, 1992, The First United States Microgravity Laboratory, 1992, The First Mission of the Tethered Satellite System, 1992, USML-1 90-Day Science Report, 1992; contbr. articles to SpaceWorld Mag., Chgo. Sun-Times and others. Recipient Right Stuff award Space Acad., Huntsville, 1986, NASA Commendation for USML-1, 1992, Group Achievement award NASA, 1992, Disting. Tech. comm. award 1st Mission of Tethered Satallite System, 1993, Best of Show award Birmingham and Huntsville STC chpt., The First Mission of the Tethered Satellite System, 1993; Bickle scholar, 1988-89. Mem. Nat. Assn. Sci. Writers (assoc.), N.Y. Acad. Scis., Kappa Tau Alpha. Home: 3003 Flag Cir Apt 2514 Madison AL 35758-1980 Office: Essex Corp 690 Discovery Dr NW Huntsville AL 35806-2813

POWERS, DORIS HURT, engineering company executive; b. Indpls., Jan. 17, 1927; d. James Wallace Hurt Sr. and Mildred (Johnson) Devine; m. Patrick W. Powers, Nov. 12, 1950 (dec. 1989); children: Robert W. Powers, Jaye P., Laura S. Powers. Student, So. Meth. U., 1944-45; BS in Engring., Purdue U., 1949; postgrad., U. Tex., W. Tex., 1952-53, Ecole Normale Du Musique, Paris, 1965-68; grad., Harford County Leadership Acad., 1991. Flight instr. Red Leg Flying Club, El Paso, Lawton, Okla., 1951-57; check pilot Civil Air Patrol, El Paso, Lawton, Okla., 1952-57; ground instr. Civil Air Patrol, Washington, Tex., Okla, 1957-61; exec. v.p. T&E Internat., Inc., Bel Air, Md., 1979-88, pres., 1989-91; exec. v.p. T.E.I.S., Inc., Bel Air, 1979-88, pres., 1989-91; pres. Shielding Technologies, Inc., Bel Air, 1987—. Mem. Northeastern Md. Tech. Coun., 1991—; bd. dirs. Leadership Acad., 1991—. Recipient Svc. award U.S. Army, 1978, Cert. of Appreciation U.S. Army Test and Evaluation Command, 1988. Mem. CAP (lt. maj. 1951-58), Soc. of Women Engrs. (sr., v.p. 1977, treas. 1979, sec. rep. 1986-88, mentor 1986—, speaker 1978—, selected to Coll. of Fellows 1993), Engring. Soc. Balt. (speaker 1980—), 99's (pres. 1951-53), Am. Soc. Indsl. Security, Am. Def. Preparedness Assn., Hartford County Econ. Devel. Coun., Assn. of U.S. Army, Northeastern Md. Tech. Coun. Avocations: ice dancing, music. Home: 6 Mcgregor Way Bel Air MD 21014-5631

POWERS, ELDON NATHANIEL, data processing executive; b. Wichita, Kans., Feb. 14, 1932; s. Ernie Lee and Bessie Othella (Loomis) P.; m. Betty Jean Zeigler, Sept. 4, 1954; children: Rebekah Jean, Robert John, Samuel Tyler. Student, Friends U., 1950-51; BA in Missions, Central Bible Coll., 1954; BA in Modern Lang. Edn., Evangel Coll., 1963; MS in Math., Tulsa U., 1971. Pastor Assembly of God Ch., Hays, Kans., 1955-60; data processing technician Gospel Pub. House, Springfield, Mo., 1960-63; data processing analyst Amoco Prodn. Co., Oklahoma City, 1963-65; research scientist Amoco Prodn. Co., Tulsa, 1965-67, staff research scientist, 1968-81; sr. system analyst Electro Mech. Research, Bloomington, Minn., 1967-68; mgr. info. service Fox Drilling Co., Tulsa, 1981-82; pres. ENP Software, Inc., Sapulpa, Okla., 1982—; cons. in field. Contbr. articles to profl. jours.; author various computer programs. Adv. bd. Cen. Okla. Vocat. Tech. Sch., Sapulpa, 1986-89. Mem. Computer Oriented Geol. Soc., Internat. Assn. Math. Geology. Democrat. Methodist. Office: ENP Software Inc 1215 Ridgeoak Cir PO Box 370 Sapulpa OK 74067

POWERS, EVA AGOSTON, clinical psychologist; b. Budapest, Hungary, Mar. 30, 1938; came to U.S., 1940, naturalized, 1945; d. Tibor and Jeanne Iseult (Watson) Agoston; A.B., Smith Coll., 1960; M.A., Boston U., 1962; Ph.D., 1969; m. James F. Powers, July 4, 1960; children—Wayne, Glenn. Psychologist, Childrens' Hosp. Med. Center Boston, 1964-69, Newton (Mass.) Sch. System, 1969-71, Conway (N.H.) Sch. System, 1972-78; dir. child and youth services Seacoast Regional Counseling Center, Portsmouth, N.H., 1979-80; pvt. practice psychol. counseling, Portsmouth, 1980—; cons. to Maine Sch. System, 1978, Center of Hope, Conway, N.H., 1971-73; supr. tng. program N.H. Dept. Edn., Concord, 1972-74. NIMH grantee, 1961, 62; S. Burt Wolbach Research Fund grantee, 1968. Mem. Am. Psychol. Assn., Maine Psychol. Assn., N.H. Psychol. Orgn., Mass. Psychol. Assn., Nat. Assn. Psychologists, Sigma Xi, Phi Beta Kappa. Contbr. articles to profl. publs. Office: 127 High St Portsmouth NH 03801-3708

POWERS, KIM DEAN, optical engineer; b. Missoula, Mont., June 5, 1962; s. Vincent Ralph and Thelma Rose (Jorve) P. BS, Sonoma State U., 1984; MS, U. Ariz., 1987. Optical engr. Wyko Corp., Tucson, 1987-90; rsch. assoc. U. Ariz., Tucson, 1988-89; optical engr. Viratec Thin Films, Faribault, Minn., 1991—. Contbr. articles to profl. jours. Life mem. Calif. Scholarship Found., 1980—. Recipient achievement award in math. Bank of Am., 1980, Chemistry award Am. Chem. Soc., 1980. Mem. Am. Phys. Soc. Achievements include development of antireflection coatings applied directly to CRT's, of method of evaporating and sputtering (simultaneously) high temperature superconducting thin films. Office: Viratec Thin Films 2150 Airport Dr Faribault MN 55021

POWERS, NELSON ROGER, entomologist; b. San Diego, Sept. 23, 1946; s. Melvin Ardath and Joy Jacqueline (Nelson) P. BS in Zoology, San Diego State Coll., 1970; MS in Biology, San Diego State U., 1973; PhD in Entomology, U. Calif., Riverside, 1979; MPH, U. Tex., 1993. Entomologist Internat. Inst. Tropical Agriculture, Nigeria, 1980; commd. capt. U.S. Army, 1980, advanced through grades to maj., 1989; entomologist Letterman Army Inst. Rsch., Calif., 1980-83, Preventive Medicine Svc., Fort Hood, Tex., 1984-87, Acad. Health Scis., Ft. Sam Houston, Tex., 1987-88, 5th Preventive Medicine Unit, Korea, 1988-89, Preventive Medicine Svc., Fort Sam Houston, 1989-92; Preventive Medicine Svc., Panama, 1992—; entomologist Joint Task Force Bravo Med. Element, Honduras, 1985. Contbr. articles to Hilgardia, Mutation Rsch., Toxicology, Drosophila Info. Svc., Tex. Preventable Disease News, Jour. Med. Entomology, Clin. Infectious Diseases, Yonsei Reports on Tropical Medicine. Mem. Entomol. Soc. Am., Am. Soc. Tropical Medicine and Hygiene, Soc. Vector Ecologists, Sigma Xi. Republican. Methodist. Office: USA MEDDAC Attn: HSXU-PM APO AA 34004-5000

POWERS, ONIE H. See **ADAMS, ONIE H. POWERS**

POWERS, RUNAS, JR., rheumatologist; b. Jackson's Gap, Ala., Dec. 11, 1938; s. Runas and Geneva (Burton) P.; m. Mary Alice Shelton, Feb. 4, 1969; children: Tiffany, Trina, Runas Coley III. BS, Tenn. State U., 1961; MD, Meharry Med. Coll., Nashville, 1966. From intern to resident in internal medicine Hurley Hosp., Flint, Mich., 1966-67, 69-72; postdoctoral fellow Stanford (Calif.) Med. Ctr., 1972-76; pvt. practice Alexander City, Ala. Contbr. with others articles to Jour. of Rheumatology, Alcohol Myopathy and Myoglobinuric Nephrosis. Lt. comdr. MD, USN, 1967-69. Decorated Purple Heart, Bronze Star; named Man of Yr. Alexander City C. of C., 1991. Fellow Am. Coll. Rheumatology; mem. AMA, N.Y. Acad. Sci, Ala. State Med. Assn., Nat. Med. Assn., Am. Fedn. Clinical Rsch. Achievements include research in antigenic components of saline extracted nuclear antigen and their reactivity with SLE Sera, Metabolism of I-125 C3 in Systemic Lupus Erythematosus. Office: 1506 Hwy 280 Byp Ste 108 Alexander City AL 35010-2662

POWIS, ALFRED, natural resources company executive; b. Montreal, Que., Can., Sept. 16, 1930; s. Alfred and Sarah Champe (McCulloch) P.; m. Louise Finlayson, Nov. 1977; children: Timothy Alfred, Nancy Alison, Charles Robert. B in Commerce, McGill U., 1951. Mem. staff investment dept. Sun Life Assurance Co. Can., 1951-55, now dir.; with Noranda Inc. (formerly Noranda Mines Ltd.), Toronto, 1956—, asst. treas., 1958-62, asst. to pres., 1962-63, exec. asst. to pres., 1963-66, v.p., 1966-67, exec. v.p., 1967-68, pres., 1968-82, chmn., chief exec. officer, 1977—, chmn., 1990—, dir., 1964—; bd. dirs. MacMillan Bloedel Ltd., Noranda subsidiaries, Noranda Forest Inc., Can. Imperial Bank Commerce, Ford Motor Co. Can. Ltd., Gulf Can. Resources Ltd., Norcen Energy Resources Ltd., N. Can. Oils Ltd., Noranda Forest Inc., Sears Can. Inc., Kerr Addison Mines Ltd., Sun Life Assurance Co. Can., Brascan Ltd., Dal-Tile Group Inc.; mem. Brtish N. Am. Com.; dir. Canadian Inst. Internat. Affairs. Trustee Toronto Gen. Hosp. Recipient Inco medal Can. Inst. Mining and Metallurg., 1991. Mem. Conf. Bd., Conf. Bd. Can. (chmn.), Order of Can. (bd. dirs.), York Club, Toronto Club, Mount Royal Club (Montreal). Anglican. Office: Noranda Inc Box 755 BCE Pl, 181 Bay St Ste 4100, Toronto, ON Canada M5J 2T3

POWSNER, GARY, computer consultant; b. Providence, Sept. 28, 1952; s. Clement and Jeri (Rich) P.; m. Laurie Smith, May 25, 1980; 1 child, Jonah Louis Smith. BA, Franconia (N.H.) Coll., 1973; MA, Goddard-Cambridge Coll., 1977. Tech. draftsman Ekman Klarson, Warwick, R.I., 1975-77; mktg. dir. B&L, Concord, Mass., 1977-83; with MBS, Springfield, Mass., 1983-85; MIS staff NEI/NLP, Amherst, Mass., 1985—; ptnr. Blue Moon Studios, Conway, Mass., 1975—; tech. advisor FCC, Greenfield, Mass., 1989—; v.p., bd. dirs. Rowe Campand Conf. Ctr., Rowe, Mass., 1980-86; tech. adv. bd. Small Bus. Community Roundtable, Greenfield, 1990—. Office: Blue Moon Studios PO Box 130 Conway MA 01341-0130

POZZI, ANGELO, executive, civil engineer; b. Wattwil, Switzerland, Aug. 12, 1932; m. Verena Schubiger; children: Lucia, Monica, Martina. Grad. civil engring., Swiss Fed. Inst. Tech., Zürich, Switzerland, 1956, D. in Tech. Sci., 1970. Project mgr. different orgns. N.Am., Europe, 1956-67; rsch. fellow Swiss Fed. Inst. Tech., Zürich, 1968-70, prof., 1971-82; CEO Motor-Columbus Ltd., Baden, Switzerland, 1983, chmn., CEO, 1985-92, chmn., 1992-93; chmn. Pozzi & Ptnrs. Ltd., Baden, Switzerland, 1993—; chmn. Aare-Tessin Ltd., 1987—; bd. dirs. Holderbank Ltd., Alusuisse-Lonza Ltd. Home: Ländliweg 9a, CH 5400 Baden Switzerland Office: Pozzi & Ptnrs Ltd, Mellingerstrasse 1, CH 5400 Baden Switzerland

PRABHAKAR, ARATI, federal administration research director, electrical engineer; b. New Delhi, India, Feb. 2, 1959; came to U.S., 1962; d. Jagdish Chandra and Raj (Madan) P. BS in Electrical Engring., Tex. Tech U., 1979; MS in Electrical Engring., Calif. Inst. Tech., 1980, PhD in Applied Physics, 1984. Congl. fellow Office Tech. Assessment U.S. Cong., Washington, 1984-86; program mgr. electronic sci. divsn. DARPA, Arlington, Va., 1986-90, dep. dir. defense sci. office, 1990-91, dir. microelectroncs tech. office, 1991-93; dir. Nat. Inst. Standards & Tech., Gaithersburg, Md., 1993—. Contbr. articles to profl. jours. Rsch. fellow Calif. Inst. Tech., 1979-84; grad. rsch. program for women Bell Labs., 1979, 80. Mem. IEEE, Am. Physical Soc., Eta Kappa Nu, Tau Beta Pi. Office: Nat Inst of Stnds & Technology US Dept of Commerce Rte 270 Bldg 101 Gaithersburg MD 20899*

PRABHUDAS, MERCY RATNAVATHY, microbiologist, immunologist; b. Madras, India, Sept. 8, 1960; d. Roberts and Clara (Joseph) P. MS in Microbiology, Loma Linda U., 1985, PhD in Microbiology, 1991. Postdoctoral fellow Brigham & Women's Hosp., Boston, 1991—. Contbr. articles to profl. jours. Musician Univ. Christian Fellowship, U. Calif., Riverside, 1989-91. Mem. Am. Soc. Microbiology, Sigma Xi. Office: Brigham & Women's Hosp Ctr Neurologic Diseases 221 Longwood Ave LMRC Boston MA 02115

PRADO, NEILTON GONÇALVES, urologist; b. Tupaciguara, Brazil, July 3, 1940; d. Arthur Goncçalves and Leonina Prado (Campos) Rodrigues; m. Vera Pereira Santos Prado, May 9, 1946; children: Vanessa, Lisia, Lucas. Student, Puca, Sao Paulo, Brazil, 1961; MD, Sao Paulo U., 1967. Med. diplomate. Intern Sao Paulo U., 1967, resident, 1971; chief Maua (Brazil) Emergency Ctr., 1967-71; asst. prof. Botucatu (Brazil) Sch. Medicine, 1969; asst. prof. Uberlandia (Brazil) U., 1972-76, prof. medicine, 1976-85, chmn. urology, 1986—; councillor Internat. Soc. Urologic Endoscopy, 1982-85, Internat. Soc. Urologic Endoscopy, 1985-88; cons. nat. residence program, Brasilia, Brazil, 1979-86, Sch. Nurses, Uberlandia, 1978-85. Author: Medical Residency Manual, 1981, Surgery-Emergency, 1985, Endourology, 1985, Surgical Residency Manual, 1988. Pres. Tchr. Confederation, Uberlaândia, 1980-81. Fellow Urology Duke U.; mem. Am. Urological Assn., N.Y. Acad. Sci., Soc. Internat. D'Urologie, Soc. Basic Urological Rsch. Avocations: horses, photography. Home: Ave Liberdade 1175, 38400 Uberlandia Brazil Office: Urologic Clinic, Ave Getulio Vargas 295, 38400 Uberlandia MG, Brazil

PRADOS, JOHN WILLIAM, educational administrator; b. Spring Hill, Tenn., Oct. 12, 1929; s. Gustave Olivier and Elizabeth (Branham) P.; m. Ruth Lynn Baird, Sept. 2, 1951; children: Elizabeth Prados Bowman, Laura Lynn, Anne Prados Lynch. B.S. in Chem. Engring, U. Miss., 1951; M.S., U. Tenn., 1954, Ph.D., 1957. Registered profl. engr., Tenn. Asst. prof. U. Tenn., Knoxville, 1956-59, asso. prof., 1959-64, prof. chem. engring., 1964—, Univ. prof., 1989—, asso. dean engring., 1969-71, dean admissions and records, 1971-73, acting chancellor Knoxville campus, 1973, acting chancellor Martin campus, 1979; v.p. acad. affairs statewide U. Tenn. System, 1973-81, v.p. acad. affairs and rsch., 1981-88, v.p. emeritus, 1989—, head chem. engring. dept., 1990-93; cons. nuclear div. Union Carbide Corp.,

Oak Ridge, 1957-84, Martin Marietta Energy Systems, Inc., 1984-86. With USAF, 1951-53. Served with USAF, 1951-53. Recipient Outstanding Tchr. award U. Tenn., 1967, 92. Fellow Am. Inst. Chem. Engrs. (mem. council 1975-77, Knoxville-Oak Ridge Chem. Engr. of Yr. 1977), Am. Inst. Chemists; mem. Am. Chem. Soc., Am. Soc. Engring. Edn., Engring. Accreditation Commn. Accreditation Bd. Engring. and Tech., Inc. (vice chmn. 1981-84, chmn. 1984-85, bd. dirs. 1988—, sec. 1989-90, pres. 1991-92), So. Assn. Colls. and Schs. (commn. on colls. 1986-92, exec. coun. 1986-89), Sigma Xi (dir. at large, chmn. com. 1976—, pres. 1983-84, treas. 1990—), Torch Club, Tau Beta Pi (exec. coun. 1986-90), Alpha Tau Omega. Roman Catholic. Home: 7021 Stagecoach Trl Knoxville TN 37909-1112 Office: U Tenn Knoxville TN 37996-2200

PRADZYNSKI, ANDRZEJ HENRYK, chemist; b. Plock, Poland, Jan. 1, 1924; came to U.S., 1969; s. Maurycy and Frania (Goldkind) Nejman; m. Halina Romana Bromberger, Apr. 1, 1946; children: Richard E. Neuman, Zgibniew Jacek. BS, U. Wroclaw, Poland, 1949, MS, 1951. Asst. prof. crystallography, chmn. dept. U. Wroclaw, 1948-51; sect. mgr. materials testing Inst. Aviation, Warsaw, Poland, 1951-57; adj. prof. Polish Acad. Scis., Warsaw, 1957-68; dept. dir. Atomic Energy Commn. Poland, Warsaw, 1959-68; rsch. assoc. IV nuclear reactor U. Tex., Austin, 1969-80; exec. v.p. Halinco Skin Care Products, Inc., Austin, 1980—; cons. IAEA, Vienna, Austria, 1968-69. Author: Industrial Radiography (in Polish), 1957; also over 3o articles in IAEA Conf. Procs., Nukleonika, ISA Trans., also others. Mem. Am. Chem. Soc., Soc. Cosmetic Chemists, Air Pollution Control Assn. Achievements include patent for method and apparatus for collection and analysis of mercury in the atmosphere; developer method of photo-nuclear activation analysis of copper in ores and concentrates, synthetic standards for EDX-ray analysis, method of collection and analysis of mercury in air, method of nondestructive X-ray analysis of heavy metals in toys. Developed pre-concentration methods of trace elements in water for EDX-ray analysis. Office: Halinco Skin Care Products PO Box 9405 Austin TX 78766

PRAGANA, RILDO JOSÉ DA COSTA, information processing company executive; b. Recife, Brazil, Apr. 11, 1951; s. Rildo and Terezinha (Lucena) P.; m. Fernanda Wanderley, May 23, 1977; children: Julius, Rildo. BA in Physics, Fed. U. Pernambuco, Recife, 1974; MA in Physics, Fed. U. Rio de Janeiro, 1976. Asst. prof. biophysics UFPE, Recife, 1975-76, head digital systems lab. physics dept., 1979-86; v.p. TWR Informatica, Recife, 1987—; pres. Percomp Systems, Recife, 1986—; tech. cons. Inst. Tech. Estado Pernambuco, Recife, 1988-90. Inventor corisco microcomputer MT100 micro-terminal. Mem. IEEE, Soc. Usuaries Computadoels (sec. 1990-91, Honor medal 1990). Roman Catholic. Avocations: piano, chess, amateur radio, photography. Home: Estr Aldeia KM 10.5, PO Box 7440, 52010 Recife PE, Brazil

PRAGER, MICHAEL HASKELL, fishery population dynamicist; b. Providence, Nov. 21, 1948; s. Irving H. and Ruth (Barsky) P.; m. Juanita Madeleine Remien, Mar. 14, 1987. SB, MIT, 1971; PhD, U. R.I., 1984. Asst. prof. oceanography Old Dominion U., Norfolk, Va., 1984-90; ops. rsch. analyst Southwest Fisheries Sci. Ctr., NOAA, La Jolla, Calif., 1985-87; rsch. fishery biologist Southeast Fisheries Sci. Ctr., NOAA, Miami, Fla., 1990—; rev. panelist EPA, Washington, 1989-92, NOAA, Monterey, Calif., 1990-91. Co-author: Basic Fishery Science Programs, 1988; contbg. author: Self-Organizing Methods in Modeling, 1984; assoc. editor N.Am. Jour. Fisheries Mgmt., 1992—; contbr. articles to N.Am. Jour. Fishery Mgmt., Can. Jour. Fisheries and Aquatic Sci., other profl. publs. NSF fellow, 1980-83; faculty rsch. fellow Old Dominion U., 1984, 88; vis. fellow Coop. Inst. Marine and Atmospheric Studies, U. Miami, Fla., 1990. Mem. Am. Fisheries Soc., Biometric Soc., Resource Modeling Assn. Achievements include research in effects of contaminants on exploited fish populations, population dynamics of tunas, methods for fish stock assessment. Office: Southeast Fisheries Sci Ctr 75 Virginia Beach Dr Miami FL 33149

PRAKASH, LOUISE, biophysics educator; b. Lyon, France, Apr. 11, 1943; m. Satya, Dec. 4, 1965. BA, Bryn Mawr Coll., 1963; MA, Washington U., 1965; PhD, U. Chgo., 1970. Asst. prof. biophysics U. Rochester (N.Y.) Sch. of Medicine, 1972-86, assoc.a prof. biophysics, 1978-86, prof. biophysics, 1986-93; prof. human biol. chem., genetics and sr. scientist Sealy Ctr for Molecular Sci., Galveston, Tex., 1993—. Home: 45 Middlebrook Ln Rochester NY 14618 Office: Univ Tex Med Br Sealy Ctr Molecular Sci Med Rsch Bldg Rt J61 11th & Mechanic St Galveston TX 77555

PRAKASH, RAVI, scientific counselor, biomedical engineering educator; b. Varanasi, No., India, Mar. 30, 1950; came to U.S., 1992; s. Raj Kishore Lal and Bimla Devi Srivastava; m. Nirupama Prakash, Mar. 9, 1976; children: Gaurav, Saurabh. BS in Mech. Engring., Hindu U., Banaras, Varanasi, India, 1971; MS in Engring., U. Salford, Eng., 1973; PhD in Materials, Cranfield (Eng.) Inst. Tech., 1975. Lectr. in mech. engring. Inst. of Tech. Banaras Hindu U., Varanasi, India, 1975-79; reader in mech. engring. Inst. of Tech. Banaras Hindu U., Varanasi, 1980-85, prof. biomed. engring., 1985-92; post doctoral rsch. officer Bath (Eng.) U., 1979-80; sci. counselor Embassy of India, Washington, 1992—; Contbr. more than 50 articles and book chpts. to profl. pubs. Achievements include development of many non-destructive testing techniques for carbon-fibre reinforced composites, development of new technique for monitoring fracture healing process in bones. Office: Embassy of India 2536 Massachusetts Ave NW Washington DC 20008

PRASAD, B.H., utility company executive; b. Rudrur, India, Aug. 1, 1955; came to U.S. 1978; s. Krishna Murthy and Sudha (Murthy) Rao; m. Sheela Prasad, May 30, 1985; children: Neil B., Megan B. BSME, Osmania U., Hyderabad, India, 1977; MSIE, Okla. State U., 1980. Methods engr. in trng. Okla. Gas & Elec., Oklahoma City, 1980-82, methods engr., 1982-84, sr. engr. HVAC, 1984-87, mgr. mkt. svcs., 1987-89, mgr. load mgmt., 1989-92; mgr. Demand Side Mgmt., 1993—. Mem. ASHRAE (Profl. Award of Excellence 1988, Vol. of the Yr. 1991, pres. 1992-93, Golden Gavel award), Assn. of Demand Side Mgmt. Profls. (bd. dirs. 1989—, exec. v.p. 1991-92, pres., 1993—), Toastmasters. Office: Okla Gas and Elec 101 N Robinson Oklahoma City OK 73101

PRASAD, CHANDAN, neuroscientist; b. Jan. 1, 1941; m. Shail Gupta; children: Anand, Amit, Anoop. ISc, St. Andrews Coll., 1960; BS with honors, G.B. Pant U. Agriculture Tech., 1964, MS, 1966; PhD, La. State U., 1970. Teaching assoc. in food microbiology and biochemistry U. Alberta, Edmonton, Canada, 1966-67; rsch. asst. in microbial physiology La. State U., Baton Rouge, 1967-70, rsch. assoc. in mosquito genetics, 1970, chief lab. neuroscis. Pennington Biomed. Rsch. Ctr., 1989-93; vis. fellow lab. molecular biology NIH, Bethesda, Md., 1970-73, vis. assoc. lab. biochem. genetics, 1973-74, sr. staff fellow, 1974-78; asst. prof. medicine La. State U. Med. Ctr., New Orleans, 1978-82, assoc. prof., 1982-86, prof., 1986—; advisor tech. workshop Gulf Breeze (Fla.) Environ. Rsch. Lab., 1978; mem. VA R&D Com., 1985—; scientist Charity Hosp., New Orleans; ad hoc grant reviewer EPA, NSF, VA. Manuscript reviewer: Endocrinology, Life Scis., Jour. Nat. Cancer Inst., Jour. Clin. Investigation, Jour. Clin. Endocrinology and Metabolism, Peptides, Neuroendocrinolgoy, Molecular and Biochem. Parasitology, Brain Rsch., Jour. Pharmacology and Exptl. Therapeutics, Proceedings Soc. Exptl. Biology, Brain Rsch. Bull., Am. Jour. Physiology, Jour. Neurochemistry; contbr. articles to Applied Microbiology, Biochem. Jour., Jour. Bacteriology, Biol. Chem., Jour. Water Pollution Control Fedn., others. Mem. European Brain and Behavior Soc., Soc. for Neurosci., British Brain Rsch. Assn., So. Soc. for Clin. Investigation, Am. Soc. Biol. Chemists, Endocrine Soc., Internat. Soc. for Neurochemistry, N.Y. Acad. Scis., AAAS, Am. Fedn. for Clin. Rsch., Sigma Xi. Achievements include research in biochemistry, endocrinology, and pharmacology of brain peptides, neurochemistry of aging, neuroendocrinology and neurobiology of schizophrenia and tardive dyskinesia, nutrition and brain function. Office: La State U Med Ctr 1542 Tulane Ave New Orleans LA 70112

PRASAD, K. VENKATESH, electrical and computer engineer; b. Madras, India, Jan. 2, 1959; came to U.S., 1984; s. Vadenthadesikan and Jaya Krishnaswamy; m. Malini Raghavan, Jan. 17, 1988. BSEE, U. Madras, 1980; MSEE, Indian Inst. Tech., Madras, 1984, Wash. State U., 1987; PhD in Elec. and Computer Engring., Rutgers U., 1990. Project assoc. Indian Inst. Tech., Madras, 1980-83; rsch. asst. Wash. State U., Pullman, 1984-87; rsch. asst. Rutgers U., New Brunswick, N.J., 1987-90, postdoctoral fellow, 1990-92; vis. researcher Calif. Inst. Tech., Pasadena, 1991-92, mem. prof. staff, 1992—; Caltech affiliate Jet Propulsion Lab., Pasadena, 1992—. Co-author: Computational MEthods of Signal Recovery, 1992; reviewer Visual Communication and Image Representation jour., 1987—. Mem. IEEE, Optical Soc. Am., Sigma Xi. Achievements include formulation of new theory for the recovery of structure from blurred images. Home: 20875 Valley Green Dr # 60 Cupertino CA 95014-1718 Office: Calif Inst Tech Mail Stop 216 76 Pasadena CA 91125

PRASAD, MAREHALLI GOPALAN, mechanical engineering educator; b. Hassan, Karnataka, India, July 22, 1950; came to U.S., 1977; s. Marehalli S. Gopalan; m. Geetha Prasad; children: Tejasaji, Yashaswi. MS, Indian Inst. Tech., Madras, India, 1974; PhD, Purdue U., 1980. Cert. noise control engring. Lectr. mech. engring. BMS Coll. Engring., Bangalore, India, 1971-72; aero. engr. Hindustan Aeronautics, Bangalore, 1974-77; grad. instr. Purdue U., West Lafayette, Ind., 1978-80; asst. prof. mech. engring. Stevens Inst. Tech., Hoboken, N.J., 1980-84, assoc. prof. mech. engring., 1984-90, prof. mech. engring., 1990—; noise control expert UN (UNIDO), Vienna, Austria, 1989; tech. cons. IBM, Poughkeepsie, N.Y., 1988, AT&T, Murray Hill, N.J., 1986. Co-editor (conf. procs.) Noise-Con 91, 1991; contbr. articles to profl. jours. including Jour. Sound and Vibration, Jour. Acoustical Soc. Am. Tchr. Vidyapith (Cultural Ednl. Instn.), N.J., 1990—; co-chair edn. com. Hindu Temple and Cultural Soc., Bridgewater, N.J., 1993—; mem. Inst. Indian Culture, N.Y., 1988—. Recipient Acoustical Paper award (co-author) Nelson Industries, Wis., 1984; named for Best Paper, Inst. of Noise Control, Calif., 1989. Mem. ASME (assoc. editor Applied Mechanics Revs. 1986—), Inst. of Noise Control Engring. (cert., v.p. external affairs 1987-91). Achievements include research in the area of acoustical source characterization in ducts. Home: 1 Osborne Terr Maplewood NJ 07040 Office: Stevens Inst Tech Castle Point on the Hudson Hoboken NJ 07030

PRASAD, RAMJEE, telecommunications scientist, electrical engineer; b. Babhnaur, Bihar, India, July 1, 1946; s. Sabita and Chndakala (Devi) Nath; m. Jyoti Sinha Prasad, Mar. 7, 1969; children: Neeli Rashmi, Anand Raghawa, Rejeev Ranjan. BSc I., Gaya (India) Coll., 1963; BSc in Engring., Bihar Inst. Tech., Sindri, India, 1968; MSc in Engring., Birla Inst. Tech., Ranchi, India, 1970, PhD, 1979. Sr. rsch. fellow Birla Inst. Tech., Ranchi, 1970-71, tutor, 1971-72, asst. prof., 1972-79; sr. sci. officer small industries tng. and devel. orgn. Ranchi, 1979-80; assoc. prof. Birla Inst. Tech., Ranchi, 1980-83; assoc. prof. U. Dar es Salaam (Tanzania), 1983-86, prof., 1986-88; prof. Delft (the Netherlands) U. Tech., 1988—; assoc. prof. Govt. Women's Poly., Ranchi, 1976-83; head cons. instruments servicing cell BIT, Ranchi, 1975-83; project responsible satellite communications for rural zones (coop. between U. Dar es Salaam/U. Eindhoven), 1983-88; presenter tutorials on mobile and indoor radio communicaitons at various univs., tech. instns. and IEEE confs.; assoc. tech. editor IEEE Communications Mag.; organizer, chmn. IEEE Veh. Tech. Comm. Soc. (joint chpt.); working group European coop. in field of sci. and tech. rsch. (COST-231) project on mobile and personal communications as expert for The Netherlands; chmn. Tech. Prog. Com. PIMRC'94 Conf. IEEE, Netherlands. Editor-in-chief Internat. Jour. Wireless Personal Communications; contbr. over 150 sci. articles to profl. jours. Merit scholar (Indian) Edn. Dept., 1968. Fellow IEE (U.K.), Instn. Electronics and Telecommunications Engring.; mem. IEEE (sr., mem. program com. and internat. adv. com. several confs. and communications), Plasma Soc. India, N.Y. Acad. Scis. Avocations: yoga, football. Home: Issac da Costalaan 12, 2624 ZD Delft The Netherlands Office: Tech U Delft Elec Engring, PO Box 5031, 2600 GA Delft The Netherlands

PRASAD, RAVI, chemical engineer; b. Bareilly, India, Nov. 29, 1944; came to U.S. 1966; m. Aug. 30, 1966 (dec. 1991); children: Madhvi, Malini, Neil. BS in Chem. Engring., Indian Inst. Tech., 1966; MS in Chem. Engring., U. Rochester, 1968. Engr./group leader Gen. Foods Corp., Tarrytown, N.Y., 1968-78; lab. mgr. Maxwell House div. Gen. Foods, Hoboken, N.J., 1978-80, 81-86, Bremen, Germany, 1979-81; sr. engr. Philip Morris U.S.A., Richmond, Va., 1986—. Mem. Inst. Food Technologists, Am. Inst. Chem. Engrs. Achievements include patents in process for fixing volatile enchancers in sucrose; extraction of caffeine from coffee beans; extraction of nicotine from tobacco. Home: 10821 Hinshaw Dr Midlothian VA 23113 Office: Philip Morris USA PO Box 26583 Richmond VA 23261

PRASAD, SATISH C(HANDRA), physicist; b. Chapra, India, Apr. 1, 1944; came to the U.S., 1966; s. Shib Chandra and Sitapati (Devi) P.; m. Jayshri Sahay, May 2, 1970; children: Monica, Anita, Sunita. PhD, U. Mass., 1972; MS, U. Colo., 1976. Instr. Washington U. Med. Ctr., St. Louis, 1976-79, asst. prof., 1979-81; asst. prof. SUNY Health Sci. Ctr., Syracuse, 1981-84, assoc. prof., 1984-92, prof., 1992—. Contbr. articles to profl. jours. Mem. Am. Assn. Physicists in Medicine, Am. Phys. Soc., Am. Coll. Radiology, AAAS.

PRASAD, SURYA SATTIRAJU, environmental systems engineer; b. Hyderabad, India; came to U.S., 1970; s. Sattiraju Somayajulu and S. Bhanumati (Tallapragada) P.; m. Annapurna Sattiraju Chunduru; children: Swati, Ravi. MA, Sam Houston State U., Huntsville, Tex., 1971; PhD, Rutgers U., 1977. Cert. profl. soil scientist and agronomist. Rsch. intern Rutgers U.; teaching asst. Rutgers U. and Sam Houston State U.; environ. systems engr. Argonne Nat. Lab., Washington; chmn. session 2-environ. issues Third Mid-Atlantic Regional Biomed. Support Symposium, Charlotte, N.C., 1979; guest speaker internat. risk assessment workshop East-West Ctr., Honolulu, 1985; coord. protecting groundwater, pesticides and agrl. practices workshop EPA, Bethesda, 1988; participant U.S. Antarctic program NSF, 1990-91. Co-author: (EPA) Protecting Ground Water: Pesticides and Agricultural Practices, 1988; invited author: Trends In Analytical Chemistry, The Netherlands, 1993; author, co-author several publs. Pre-doctoral Scholarship grantee Watumull Found.; Internat. scholar Rotary Baccalaureate Program; recipient letter of commendation EPA, 1985, Cert. Merit with Hon. Mention, Asia Found., 1975, Outstanding Svc./Contribution award tech. com. Tana-World Telugu Conv., 1993. Mem. Am. Soc. Agronomy (presiding officer conf. session 1979, 86), Am. Water Resources Assn. (nat. capital sect.), Soil Sci. Soc. Am., Internat. Soc. of Soil Sci., Hazardous Material Control Resources Inst., Soc. Risk Analysis (nat. capital area chpt.), Beautification Com. City Gaithersburg, Montgomery County, Md., Sigma Xi, Beta Kappa Chi, Delta Tau Alpha. Office: Argonne Nat Lab 370 L'Enfant Promenade SW Rm 702 Washington DC 20024

PRASUHN, ALAN LEE, civil engineer, educator; b. Columbus, Ohio, Feb. 19, 1938; s. Oscar Warren and Martha Louise (Graetz) P.; m. Ramona Ernst, Dec. 19, 1959; children: Mona Leigh Prasuhn Mosier, Warren Ernst, Randall Alan. BCE, Ohio State U., 1961; MS, U. Iowa, 1963; PhD, U. Conn., 1968. Registered profl. engr., Ohio. From asst. prof. to prof. Calif. State U., Sacramento, 1968-78; prof. civil engring. S.D. State U., Brookings, 1978-90; prof., chair dept. civil engring. Lawrence Tech. U., Southfield, Mich., 1990—; vis. fellow U. Birmingham, Eng., 1977. Author: Fundamentals of Fluid Mechanics, 1980, Fundamentals of Hydraulic Engineering, 1987; author booklets in field; contbr. articles to profl. jours. Fellow ASCE (bd. dirs. 1985-88, chmn. history and heritage com. 1990—, and others); mem. Am. Soc. Engring. Edn. (several nat. coms.), Rotary. Home: 31450 Hunters Circle Farmington Hills MI 48334 Office: Lawrence Tech U Dept Civil Engring 21000 W Ten Mile Rd Southfield MI 48075

PRATHER, BRENDA JOYCE, librarian; b. Guam, June 20, 1956; d. Huston Garfield and Julia Victoria (Jones) P. BA in History, U. Ga., 1978; MLS, U. Denver, 1983. Reference libr., coord. on-line svcs. Ga. Inst. Tech., Atlanta, 1984-89; info. specialist BNR, Inc., Atlanta, 1989-90; rsch. chief Southpoint Mag., Atlanta, 1990-91; libr. Amoco Fabrics and Fibers Co., Atlanta, 1990—. Mem. Spl. Libr. Assn., Assn. Records Mgrs. and Adminstrs., Toastmasters Internat., Little People Am. Democrat. Office: Amoco Fabrics & Fibers Co 260 The Bluffs Austell GA 30001

PRATHER, DENZIL LEWIS, petroleum engineer; b. Elizabeth, W.Va., Mar. 18, 1921; s. Elias Hugh and Mary Deborah (Lewis) P.; m. Madeline Shimer, May 13, 1945; children: Denzil Jr., Kathy Taylor, Linda Holtgrewe, Mary Smith, Ann Scarbro, David. Student, Potomac State U., 1941-42; BS in Petroleum Engring., Marietta Coll., 1945. Registered profl. engr., W.Va. Supt. Weva Oil Co., Elizabeth, 1945-50; pvt. practice as cons. Midland, Tex., 1950-57; div. mgr. Southwestern Devel. Co., Parkersburg, W.Va., 1957-70; ptnr. Loper and Prather, Parkersburg, 1970-80; pres. Adena Petroleum, Inc., Parkersburg, 1980—. Pres. PTA, Belpre, Ohio, 1961-63; active Congl. Ch., Belpre, 1957-90. 1st lt. USAF, 1942-45. Mem. Soc. Petroleum Engrs. (sr. mem.), Appalachian Geol. Soc., Ohio Acad. Sci., Ohio Acad. History. Democrat. Congregationalist. Home: 814 Boulevard Dr Belpre OH 45714 Office: Adena Petroleum Inc PO Box 250 Mineral Wells WV 26150

PRATHER, RITA CATHERINE, psychology educator; b. Marietta, Ohio, Nov. 20, 1948; d. Lloyd R. Sr. and Rita C. (Alkazin) Peters; m. Robert E. Prather, Dec. 20, 1969. BA, U. Cen. Fla., 1983; MA, La. State U., 1985, PhD, 1989. Lic. clin. psychologist, Tex. Asst. prof. Tex. A&M U., College Station, 1988-89; psychologist Psychol. Assn. Tex., P.C., Houston, 1989-90; postdoctoral fellow U. Tex. Med. Sch. Houston, 1990-91, asst. prof. psychology, 1991—; speaker on pub. rels. U. Tex.-Houston, Judge Houston Sci. and Engring. Fair, 1993. Contbr. over 30 articles to profl. jours. and confs. U. Miss. Med. Ctr. rsch. grantee, 1988. Mem. APA (divsn. clin. and health psychology), Tex. Psychol. Assn., Houston Psychol. Assn. (speakers bur. 1990—), Soc. Behavioral Medicine, Assn. Advancement Behavior Therapy. Avocations: arts, wellness activities. Home: PO Box 920814 Houston TX 77292-0814 Office: U Tex Med Sch Psychiatry Dept 1300 Moursund Houston TX 77030-3497

PRATT, DAVID LEE, oil company executive; b. Graham, Tex., Feb. 7, 1957; s. Preston Persley Sr. and Rozella Gray (Parkinson) P. BBA, U. Tex., 1980. With Petroleum Svcs. Inc., Anchorage, 1980-81, Larry O. Hulsey & Co., Graham, Tex., 1981-82; mgr. Reeves Resources Inc., Sayre, Okla., 1982-83; v.p. Tex. Internat. Oil Co. Inc., Houston, 1983-85; v.p. corp. devel. Global Natural Resources Inc., Houston, 1985-91; pres., bd. dirs. Galvest Inc., Houston, 1988—; charter mem., bd. dirs. Tatex, Almetyevsk, Tatarstan, Russia. Home: 361 N Post Oak Ln #236 Houston TX 77024 Office: Galvest Inc 1000 Louisana Ste 4975 Houston TX 77002

PRATT, JEREMY, human ecologist, researcher; b. Winslow, Wash., June 10, 1954; s. Richard King Robertson and Drusilla Adele Pratt. BA, Evergreen State Coll., 1977; MS in Environ. Sci., Wash. State U., 1979. Environ. analyst Corff and Shapiro, Seattle, 1976; assoc. in rsch. Environ. Rsch. Ctr., Wash. State U., Pullman, 1977-79; ecologist Seattle City Light Office Environ. Affairs, Seattle, 1979-80; mem. rsch. staff energy and utilities coms. Wash. state Legislature, Olympia, 1979-81; prin. BioSystems Analysis, Inc., Tiburon, Calif., 1981—; exec. dir. Inst. Human Ecology, Sonoma, Calif., 1982—; commr., chmn. City of Pullman Environ. Quality Commn., 1977-79; mem. Wash. Environ. Policy Act Rev. Bd., Pullman, 1978-79; adv. com. Nature Conservancy Skagit River Bald Eagle Nat. Area, 1979-80. Author, editor: Human Ecology: Steps to the Future, 1990, Carrying Capacity: Concept and Application in Human Ecology, 1993; author monographs, papers, reports, articles for profl. jours. GRad. rsch. fellow Energy Rsch. and Devel. Adminstrn., Wash. State U., 1977-79. Fellow Inst. Human Ecology (exec. dir. 1982—), Pacific Energy and Resource Ctr.; mem. U.S. Soc. Human Ecology (bd. dirs. 1987-92, nat. exec. dir. 1988-90), Ecol. Soc. Am. (cert.), Seattle Audubon Soc. (bd. dirs. 1980-81), Western Wash. Solar Energy Assn. (bd. dirs. 1980-81). Achievements include leadership in development and application of human ecology with focus on carrying capacity, common problems and ecology of knowing. Office: BioSystems Analysis Inc 3152 Paradise Dr Bldg 39 Tiburon CA 94920

PRATT, MARK ERNEST, mechanical engineer; b. Jackhorn, Ky., Nov. 22, 1939; s. James Corbit and Anna Marie (Johnson) P.; m. Yvonne Rose, Sept. 13, 1958; children: Mark Ernest Jr., Mary Yvonne, Bobby Lee, James Paul. Student, I.C.S. Ctr., Scranton, Pa., 1983. Lic. 1st class engr. Engr. Mercy Hosp., Miami, Fla., 1965-70; chief engr. Palmetto Gen. Hosp., Hialeah, Fla., 1970-72, Osteo. Gen. Hosp., North Miami Beach, Fla., 1972-77; owner, operator N. Am. Van Lines, Ft. Wayne, Ind., 1977-78; dir. plant ops. Internat. Hosp., Miami, 1978-79, Cypress Hosp., Pompano Beach, Fla., 1979-81, Johnston Meml. Hosp., Abingdon, Va., 1981-87, North Ridge Med. Ctr., Ft. Lauderdale, Fla., 1988-90; adminstrv. dir. plant ops. St. Jude Med. Ctr., Kenner, La., 1990—; mechanic Bristol (Va.) Compressors, 1991—. With U.S. Army, 1955-56. Mem. Am. Soc. Hosp. Engrs., Nat. Fire Protection Assn., Fla. Hosp. Engrs. Assn., Nat. Assn. Power Engrs. (instr. 1980-81, chpt. pres. 1980-81). Baptist. Avocations: restoring classic cars, fishing, hunting, camping. Home: RR 8 Box 452 Abingdon VA 24210-8637 Office: St Jude Med Ctr 180 W Esplanade Ave Kenner LA 70065-2467 Also: Bristol Compressors 649 Industrial Park Rd Bristol VA 24210

PRATT, ROBERT GEORGE, electrical engineer; b. Detroit, Jan. 20, 1943; s. Glenn Beedzler and Lucille Evelyn (Davenport) P.; m. Karen Joyce Foytek, Sept. 11, 1965; 1 child, Robert Donald. BSEE, Wayne State U., 1965, MSEE, 1967. Registered profl. engr., Mich. Co-owner Millington Enterprises, Farmington Hills, Mich., 1975-80; instr. Lawrence Technol. U., Southfield, Mich., 1987; prin. engr. Detroit Edison Co., 1967—; mem. contractor rev. panel Electric Power Rsch. Inst., Palo Alto, 1980-82; photovoltaic tech. mgr. Utility Tech. Svcs., Detroit, 1991—. Contbr. articles to profl. jours. Planning commr. City of Farmington Hills, 1981-87. Mem. IEEE, Eta Kappa Nu. Achievements include patents for fiber optic displacement transducer and microprocessor electric vehicle charging and parking meter. Office: Detroit Edison Co 2000 Second Ave Detroit MI 48226

PRATTO, FELICIA, psychologist; b. Boulder, Colo., Apr. 29, 1961; d. David John and Marlene Rose (Massaro) P.; m. Thomas Anthony Wood, Sept. 6, 1987. BA, Carnegie Mellon U., 1983; MA, NYU, 1986, PhD, 1988. Postdoctoral fellow U. Calif., Berkeley, 1988-90; asst. prof. psychology Stanford (Calif.) U., 1990—; reviewer NSF, Washington, 1991—, Jour. Personality and Social Psychology, Washington, 1990—, Cognition and Emotion, Washington, 1992—. Contbr. article to profl. jour., chpt. to book. Student leader SANE/FREEZE, N.Y.C., 1986; co-chair Peace Alliance, Pitts., 1983; mem. Pax Christi, N.Y. and Calif., 1987—. Mem. APA, Am. Psychol. Soc., Psychometric Soc. Achievements include identification of a significant personality variable, social dominance orientation, which is a general predictor of prejudice, ideology, and attitudes related to intergroup relations; found experimental evidence for unconscious detection of the valence of stimuli. Office: Stanford U Psychology Dept Bldg 420 Stanford CA 94305-2130

PRAY, DONALD GEORGE, aerospace engineer; b. Troy, N.Y., Jan. 19, 1928; s. George Emerson and Jansje Cornelia (Ouwejan) P.; m. Betty Ann Williams, Oct. 1, 1950; children: Jennifer Loie, Jonathan Cornelius, Judy Karen, Jeffrey Donald. BA in Physics, Tex. Christian U., 1955; MS in Mech. Engring., So. Meth. U., 1979. Sr. structures engr. Gen. Dynamics Corp., Ft. Worth, 1955-62, 67-84; engring. specialist LTV Astronautics Corp., Dallas, 1962-65, sr. engring. specialist, 1989-91; aero. group engr. space div. Chrysler Corp., New Orleans, 1965-67; group engr. Bell Helicopter Textron, Ft. Worth, 1984-89; mgr. structural integrity program Oklahoma City Air Logistics Ctr., Tinker AFB, 1991—; prin. Donald G. Pray, Con., Ft. Worth, 1959-61. Contbr. articles to tech. publs. Gov., pres. Soc. Mayflower Descendants in State of Tex.; edn. com. Gen. Soc. Mayflower Descendants, Plymouth, Mass., chmn. scholarship program, Dallas; chmn. bd. trustees Cope Cemetery Assn., Johnson County, Tex. Mem. ASME, NSPE, NRA, Acoustical Soc. Am., Masons, Scottish Rite, Shriners, Sigma Pi Sigma, Pi Mu Epsilon. Baptist. Achievements include analytical engineering contributions to numerous aircraft and spacecraft programs including B-36, B-58, NX-2, Robot, Dynasoar, Scout, Apollo, F/FB-111, F-16, C-17, E-3 AWACS. Home: 3628 Wedgway Dr Fort Worth TX 76133-2135 Office: Oklahoma City Air Logistics OC-ALC/LAKRA Tinker AFB OK 73145

PRAY, LLOYD CHARLES, geologist, educator; b. Chgo., June 25, 1919; s. Allan Theron and Helen (Palmer) P.; m. Carrel Myers, Sept. 14, 1946; children: Lawrence Myers, John Allan, Kenneth Palmer, Douglas Carrel. B.A., Carleton Coll., 1941; M.S., Calif. Inst. Tech., 1943, Ph.D. (NRC fellow 1946-49), 1952. Geologist Magnolia Petroleum Co., summer 1942, U.S. Geol. Survey, 1943-44; hydrographic officer USN, 1944-46; Geologist U.S. Geol. Survey, 1946-56 part time; instr. to assoc. prof. geology Calif. Inst. Tech., 1949-56; sr. research geologist Denver Research Ctr., Marathon Oil Co., 1956-62, research assoc., 1962-68; prof. geology U. Wis., Madison, 1968-88; emeritus prof. geology, 1989—; short course vis. prof. U. Tex., 1964, U. Colo., 1967, U. Miami, 1971, U. Alta., 1969, Colo. Sch. Mines, 1985; vis. scientist Imperial Coll. Sci. and Tech., London, 1977, U. Calif. Santa Cruz, 1987. Author articles sedimentary carbonates, the Permian Reef complex, stratigraphy and structural geology So. N.M. and W. Tex., porosity of carbonate facies, Calif. rare earth mineral deposits. Pres. Colo. Diabetes Assn., 1963-67, v.p., 1968; mem. adv. panel earth scis. NSF, 1973-76. Served as hydrographic officer USNR, 1944-46. Named Layman of Year Am. Diabetes Assn., 1968; recipient Disting. Teaching award U. Wis. Madison, 1988; Disting. Achievement citation Carleton Coll., 1991. Fellow Geol. Soc. Am. (research grants com. 1965-67, com. on nominations 1973, com. Penrose medal 1979-81); mem. Am. Assn. Petroleum Geologists (research com. 1958-61, lectr. continuing edn. program 1966-69, Matson trophy 1967, continuing edn. com. 1978-80, disting. lectr. 1986-87, 87-88), Soc. Econ. Paleontologists and Mineralogists (hon. life mem. Permian Basin sect.; hon. mem. nat. assn.; sec.-treas. 1961-63, v.p. 1966-67, pres. 1969-70), Am. Geol. Inst. (edn. com. 1966-68, ho. bd. of dels. 1970-72), Phi Beta Kappa. Office: U Wis Madison WI 53706

PRAY, RALPH EMERSON, metallurgical engineer; b. Troy, N.Y., May 12, 1926; s. George Emerson and Jansje Cornelius (Owejan) P.; student N.Mex. Inst. of Mining and Tech., 1953-56, U. N.Mex., 1956; BSMetE, U. Alaska, 1961; DScMetE. (Ideal Cement fellowship, Rsch. grant), Colo. Sch. of Mines, 1966; m. Beverley Margaret Ramsey, May 10, 1959; children: Maxwell, Ross, Leslie, Marlene. Engr.-in-charge Dept. Mines and Minerals, Ketchikan, Alaska, 1957-61; asst. mgr. mfg. rsch. Universal Atlas Cement div. U.S. Steel Corp., Gary, Ind., 1965-66; rsch. metallurgist Inland Steel Co., Hammond, Ind., 1966-67; owner, dir. Mineral Rsch. Lab., Monrovia, Calif., 1968—; pres., Keystone Canyon Mining Co., Inc., Pasadena, Calif., 1972-79, U.S. Western Mines, 1973—, Silveroil Rsch. Inc., 1980-85; v.p. Mineral Drill Inc., 1981-90; pres., CEO Copper de Mex. S.A. de C.V.; prime contractor def. logistics agy. U.S. Dept. Def., 1989-92; owner Precision Plastics, 1973-82; bd. dirs. Bagdad-Chase Inc., 1972-75; ptnr. Mineral R&D Co., 1981-86; lectr., Purdue U., Hammond, Ind., 1966-67, Nat. Mining Seminar, Barstow (Calif.) Coll., 1969-70; guest lectr. Calif. State Poly U., 1977-81, Western Placer Mining Conf., Reno, Nev., 1983, Dredging and Placer Mining Conf., Reno, 1985, others; v.p., dir. Wilbur Foote Plastics, Pasadena, 1968-72; strategic minerals del. People to People, Republic of South Africa, 1983; vol. Monrovia Police Dept.; city coord. Neighborhood Watch, 1990—. With U.S. Army, 1950-52. Fellow Geol. Mining and Metall. Soc. India (life), Am. Inst. Chemists, South African Inst. Mining and Metallurgy; mem. Soc. Mining Engrs., Am. Chem. Soc., Am. Inst. Mining, Metall. and Petroleum Engrs., NSPE, Can. Inst. Mining and Metallurgy, Geol. Soc. South Africa, Sigma Xi, Sigma Mu. Achievements include research on recovery of metals from refractory ores, benefication plant design, construction and operation, underground and surface mine development and operation, mine and process plant management; syndication of natural resource assets with finance sources; contbr. articles to sci. jours.; guest editor Calif. Mining Jour., 1978—; patentee chem. processing and steel manufacture. Office: 805 S Shamrock Ave Monrovia CA 91016-3651

PREBLE, DARRELL W., systems analyst; b. Portsmouth, N.H., Dec. 8, 1946; s. Carlton B. and Ellen F. (Plummer) P.; m. Deborah Young, Dec. 19, 1970; children: Jennifer, Megan. BA in Physics, Vanderbilt U., 1969; MS in Systems Adminstrn., George Washington U., 1973; MA in Physics, Ga. State U., 1980. Sr. systems analyst So. Co. Svcs., Atlanta, 1984—; fellow Space Studies Inst., Princeton, N.J., 1982—; participant NASA Space Exploration Outreach Study, 1991. Elder Riverdale Presbyn. Ch. Mem. Atlanta AI Users Group (v.p. 1988, pres.). Office: So Co Svcs 64 Perimeter Ctr E Bin 081 Atlanta GA 30346

PREDESCU, VIOREL N., electrical engineer; b. Craiova, Dolj, Romania, Sept. 14, 1950; s. Nicolae I. and Constanta (Ciobanescu) P.; m. Rodica G. Apostoleanu, Sept. 14, 1974; 1 child, Dan Paul. MSEE, Poly. Inst., Bucharest, Romania, 1974. Registered profl. engr., Calif. Elec. engr. Romania, 1974-86; helper electrician CESSOP Electric Constrn., Tustin, Calif., 1986; elec. designer, drafter Sierra Pacific Tech. Svcs., Inc., Laguna Hills, Calif., 1986; asst. engr. Boyle Engring. Corp., Newport Beach, Calif., 1986-88; project engr. Hallis Engring., Inc., L.A., 1988-89; profl. engr. Elec. Bldg. Systems, Inc., North Hollywood, Calif., 1989-91; mgr. of elec. dept. William J. Yang Assocs., Inc., Burbank, Calif., 1991—. Mem. NSPE, IEEE. Christian Orthodox. Achievements include electrical design for large variety of projects: comml. Shanghai World Trade Ctr., So. Calif. Gas Co. Hdqrs., Torrance, Calif., indsl. pump stations, water and wastewater plants indsl. bldgs., military facilties. Home: 12426 Lemay St North Hollywood CA 91606

PREDPALL, DANIEL FRANCIS, environmental engineering executive, consultant; b. Rochester, N.Y., May 22, 1946; s. Daniel Francis and Elizabeth K. Predpall; m. Carolyn Jane Kokish, May 29, 1981; 1 child, Robert. BS in Physics, Stevens Inst. Tech., 1968; MS in Marine Sci., L.I. U., 1974; MBA, NYU, 1985. Registered profl. engr., Pa. Mgr. environmental projects Ebasco Services, N.Y.C., 1970-73, 1977-82; v.p., mgr. bus. unit, sr. prin. Woodward-Clyde Cons., Wayne, N.J., 1973-77, 1982—; mgr. environ. projects; speaker, tech. presenter on quality mgmt., risk assessment, site selection and waste mgmt. Contbr. numerous articles to profl. jours. Pres. bd. govs. Packanack Lake Community, 1987-92. Mem. Soc. Profl. Engrs., Air and Waste Mgmt. Assn. Republican. Lutheran. Avocations: stock market, tennis, golf, astronomy. Home: 68 High Point Rd Bloomingdale NJ 07403 Office: Woodward-Clyde Cons 201 Willowbrook Blvd Wayne NJ 07470-7025

PREECE, BETTY P., engineer, educator; b. Decatur, Ill.; d. George A. and Margaret (Stock) Peters; m. Raymond G. Preece; children: Eric, George. BSEE, U. Ky.; MS in Sci. Edn., Fla. Inst. Tech. Cert. Master Gardener, U. Fla. Engr. GE, various cities, 1947-50; project engr. Air Force Missle Test Ctr., Patrick AFB, Fla., 1951-54; faculty physics, math. Melbourne (Fla.) High Sch., 1972-90; exec. sec. Fla. Acad. Scis., Indialantic, 1991—; physics teaching resource agt., 1987—. Contbr. over 50 articles to profl. jours. Code bd. mem., chair Indialantic Code Enforcement Bd., 1987—; mem., chair bd. and local history editor So. Brevard Hist. Soc., 1975—. Fla. Nominee for Presdl. Award for Sci. Teaching; named Outstanding Sci. Tchr., Sigma Xi. Fellow Soc. Women Engrs. (nat. conv. program chair 1992, region D dir. 1992-94, local past pres., Corning award 1984-89); mem. AIAA, NSTA, IEEE, Assn. for Women i Sic., Am. Assn. of Physics Tchrs., Fla. Assn. Sci. Tchrs., Fla. Hist. Soc., Missile Space and Range Pioneers, Women's Engring. Soc. of the U.K., DAR, Daus. Am. Colonists, Duas. of War of 1812, Delta Kappa Gamma. Achievements include being U.S. delegate to Internat. Coun. on Physics Edn: Interamerican, 1987, U.S./China/Japan (trilateral), 1989, 91, 93. Home: 615 N Riverside Indialantic FL 32903 Office: Fla Acad Scis PO Box 033012 Indialantic FL 32903

PREECE, RAYMOND GEORGE, electrical engineer; b. Buffalo, Mar. 29, 1927; s. J. Eric and Marion (Black) P.; m. Betty Peters; children: J. Eric, B. George. BSEE, U. Ky., 1948. Field engr. Hughes Aircraft Co., various cities, 1954-57; radio guidance engr. Gen. Elec. Co., Cape Canaveral Air Force Sta., Fla., 1957-66; spacecraft checkout engr. Grummond Aerospace Corp., Kennedy Space Ctr., Fla., 1966-75; mem. tech. staff Rockwell Internat., Kennedy Space Center, Fla., 1976—. Maj. USAF. Recipient Cert. of Appreciation, NASA, 1990. Mem. IEEE, AIAA, Reserve Officers Assn, The Retired Officers Assn., Air Force Assn., Missile, Space and Range Pioneers. Home: 615 N Riverside Indialantic FL 32903 Office: Rockwell Internat Kennedy Space Center Cape Canaveral FL 32920

PREEDOM, BARRY MASON, physicist, educator; b. Stamford, Conn., Dec. 31, 1940; children: Bonnie Marie, Richard Lawrence. BS, Spring Hill Coll., 1962; MS, U. Tenn., 1964, PhD, 1967. Grad. fellow Oak Ridge (Tenn.) Nat. Lab., 1964-67; rsch. assoc. Mich. State U., East Lansing, 1967-70; asst. prof., then assoc. prof. U. S.C., Columbia, 1970-76, prof. physics, 1976—, Carolina rsch. prof., 1986—; vis. prof. Swiss Inst. Nuclear Rsch., Villigen, 1976; vis. staff Los Alamos (N.Mex.) Nat. Lab., 1972-91, tech. adv. panel, 1982-85; guest scientist Brookhaven Nat. Lab., Upton, N.Y., 1987—. Contbr. rsch. articles to sci. publs. Grantee Rsch. Corp., 1971, Office Naval Rsch., 1972-75, NSF, 1975—. Mem. Am. Phys. Soc., Sigma Xi. Achievements include research and study of nuclear reaction mechanisms and nuclear structure, reaction probes including gamma rays, mesons, protons, deuterons and light ions. Office: Univ SC Dept Physics Columbia SC 29208

PREEG, WILLIAM EDWARD, oil company executive; b. N.Y.C., Oct. 16, 1942; s. Ernest Winfield and Claudia Teresa (Casper) P. BE in Marine Engring., SUNY, 1964; MS in Nuclear Sci. and Engring., Columbia U., 1967, PhD in Nuclear Sci. and Engring., 1970. Project engr. U.S. AEC, N.Y.C., 1964-67; physics specialist Aerojet Nuclear Systems Co., Sacramento, Calif., 1970-71; group leader Los Alamos (N.Mex.) Sci. Lab., 1971-80; dir. fluid-mechanics-nuclear dept. Schlumberger-Doll Rsch., Ridgefield, Conn., 1980-85, v.p., dir. rsch., 1990—; mgr. nuclear dept. Schlumberger Well Svcs., Houston, 1985-87, v.p. engring., 1987-90; instr. mech. engring. CCNY, N.Y.C., 1968-70; cons. AEC, Gamma Process Co., Los Alamos Sci. Lab., Lawrence Livermore Lab. Mem. Am. Nuclear Soc., Am. Phys. Soc., Am. Inst. Physics (adv. com on corp. assocs.), Soc. Profl. Well Log Analysts, Soc. Petroleum Engrs. Home: 71 Oak St New Canaan CT 06840 Office: Schlumberger-Doll Rsch Old Quarry Rd Ridgefield CT 06877

PREISS, JACK, biochemistry educator; b. Bklyn., June 2, 1932; s. Erool and Gilda (Friedman) P.; m. Judith Weil Rosen, June 10, 1959; children: Jennifer Ellen, Jeremy Oscar, Jessica Michelle. BS in Chemistry, CCNY, 1953; PhD in Biochemistry, Duke U., 1957. Scientist NIH, Bethesda, 1960-62; asst. prof. dept. biochemistry, biophysics U. Calif., Davis, 1962-65, assoc. prof., 1965-68, prof., 1968-85, chair dept. biochemistry, 1971-74, 77-81; prof. dept. biochemistry Mich. State U., East Lansing, 1985—, chair dept., 1985-89; Mem. physiol. chemistry study panel NIH, Bethesda, 1967-71, study panel Metabolic Biology Div., NSF, 1972-75, study sect. biochemistry and molecular biology carcinogenesis Am. Cancer Soc., 1989-92. Editorial bd. mem. Jour. Bacteriology, 1969-74, Arch. Biochem. Biophysics, 1969—; editorial bd. Plant Physiology 1969-74, 77-80, assoc. editor July 1980—; editor Jour. Biol.Chemistry 1971-76, 78-83. Recipient Camille and Henry Dreyfus Disting. Scholar award Calif. State U., 1983, Alexander von Humboldt Stiftung Sr. U.S. Scientist award, 1984, Alsberg-Schoch Meml. Lecture award Am. Assn. Cereal Chemists, 1990, Nat. Sci. Coun. lectr. Republic of China, 1988, award of merit Japanese Soc. Starch Sci., 1992; Guggenheim Meml. fellow, 1969-70; grantee NIH, 1963—, NSF, 1978—; Japan Soc. for Promotion of Sci. fellow, 1992-93, award of merit of the Japanese Soc. of Starch Sci., 1992. Mem. AAAS, Am. Chem. Soc. (Charles Pfizer award in enzyme chemistry 1971), Biochem. Soc., Am. Soc. Biol. Chemists and Molecular Biology, Am. Soc. Chem. Microbiologists, N.Y. Acad. Scis., Am. Soc. Plant Physiologists, Soc. for Complex Carbohydrates, Protein Soc. Home: 1005 Prescott Dr East Lansing MI 48823-2445 Office: Mich State Univ Dept of Biochemistry East Lansing MI 48824

PREISSNER, EDGAR DARYL, engineering executive; b. Chgo., Nov. 2, 1938; m. Waldtraut Benz, Feb. 3, 1973; children: Paul Fredrick, Karl Matthew. BS in Civil Engring., Northwestern U., 1961; MS in Civil Engring., U. Wis., 1964; MBA in Fin. and Banking, U. Chgo., 1973. Registered profl. engr., Ill. V.p. ops. UOP-Allied Signal Corp., Des Plaines, Ill., 1970-80; corp. v.p. ops. Browning-Ferris Industries, Houston, 1980-82; dir. bus. devel. IT Corp., Houston, 1982-84; sr. mgr. products NUS/Brown & Root J.V., Houston, 1984-86; corp. prin. for hazardous waste group Sverdrup Corp., St. Louis, 1986-89; pres., CEO Harza Environ. Svcs., Chgo., 1989-91; pres. Riedel Ind. Waste Mgmt., St. Louis, 1991—; mem. Hazardous Waste Treatment Coun., Washington, 1980-81. pres. Homeowners Assn., Park Ridge, Ill., 1980; chmn. ch. fin. com., Houston, 1984-86; asst. scoutmaster troop 523 Boy Scouts Am., Ellisville, Mo., 1986-91. Commdr. USPHS, 1961-92. Mem. ASCE, NSPE, ACEC (sec., bd. dirs. 1987-91, Svc. award 1991), Mo. Soc. Profl. Engrs., Hazardous Waste Action Coalition. Achievements include patents for sewage sludge drying and processing, solids waste handling, low level radioactive waste storage and disposal, process drying of sludge moisture measurement. Home: 159 Highgrove Ct Clarkson Valley MO 63005-7115 Office: Riedel Indsl Waste Mgmt 18207 Edison Ave Chesterfield MO 63005

PREJEAN, J. DAVID, toxicologist; b. Pampa, Tex., Feb. 9, 1940; s. J.C. and Blanche (Groves) P.; m. Linda Hall (div. 1974); children: Michael David, K'Anne; m. Candice Dominick, Jan. 21, 1978. BS, Stephen F. Austin State Coll., 1963; MS, East Tex. State U., 1965; PhD, Tex. A&M U., 1969. Rsch. asst. Baylor U. Med. Ctr., Dallas, 1965, Tex. A&M U., College Station, 1965-69; sr. rsch. biologist Tex. A&M U., Birmingham, Ala., 1969-72; head chem. car. sect. So. Rsch. Inst., Birmingham, Ala., 1972-75, head chem. car. div., 1975-83, assoc. dir. chemotherapy rsch. dept., 1983-90, dir. toxicology dept., 1990-92; v.p. chemotherapy and toxicology rsch. So. Rsch. Inst., Birmingham, 1992—; adj. assoc. prof. toxicology U. Ala., Birmingham, 1989—; presenter in field. Contbr. articles, reports to profl. publs. Mem. Am. Cancer Soc. (bd. dirs. Jefferson County unit 1972-78), Am. Assn. Cancer Rsch., Am. Coll. Toxicology, N.Y. Acad. Sci., Sigma Xi. Home: 5231 Harvest Ridge Ln Birmingham AL 35242 Office: So Rsch Inst 2000 9th Ave S Birmingham AL 35205

PRELL, MARTIN, mechanical engineer; b. Chgo., Nov. 14, 1918; s. Benjamin and Anna Prelutsky; m. Ruth Dorothy Sosin, Mar. 14, 1943 (dec. 1975); children: Michael jack, Joel Lawrence; m. Marilyn Feldman, July 23, 1986. BSME, Tilden Tech., Chgo., 1941; M Nuclear Engring., U. Chgo., 1948. Registered profl. engr. Aircraft designer Consolidated Aircraft Corp., San Diego, 1941-42, Ft. Worth, Tex., 1942-45, Wayne, Mich., 1945-46; design engr. Argonne (Ill.) Nat. Lab., 1946-50; aircraft designer, R&D engr., mgr. exptl. design dept. Lockheed Aircraft Corp., Burbank, Calif., 1950-86; ret. Mem. AAAS, AIAA, Nat. Mgmt. Assn.

PRELOG, VLADIMIR, chemist; b. Sarajevo, Bosnia, Yugoslavia, July 23, 1906; s. Milan and Mara (Cettolo) P.; m. Kamila Vítek, Oct. 31, 1933; 1 child, Jan. Chem. Engring. degree, Inst. Tech. Sch. Chemistry, Prague, Czechoslovakia, 1928, D Chem. Engring., 1929; D honoris causa, U. Zagreb, Yugoslavia, 1954, U. Liverpool, Eng., U. Paris, 1963, Cambridge U., U. Brussels, 1969, U. Manchester, 1971, Inst. Quim. Sarria, Barcelona, 1978, Weizmann Inst., Rehovot, 1985; postgrad., U. Ljubljana, Yugoslavia, 1989, U. Osijek, Yugoslavia, 1989, U. Chem. Tech., Prague, 1989. Chemist Lab. G.J. Dríza, Prague, 1929-35; docent U. Zagreb, 1935-40, assoc. prof., 1940-41; mem. faculty Swiss Fed. Inst. Tech., Zurich, 1942—, prof. chemistry, 1950—, head Lab. Organic Chemistry,, 1957-65, ret., 1976; bd. dir. CIBA Geigy Ltd., Basel, Switzerland, 1963-78. Recipient Werner medal, 1945, Stas medal, 1962, medal of honour Rice U., 1962, Marcel Benoist award, 1965, A.W. Hofmann medal, 1968, Davy medal, 1968, Roger Adams prize, 1969, Nobel prize for chemistry, 1975, Paracelsus medal, 1976. Fellow Royal Soc., 1962; mem. Am. Acad. Arts and Scis. (hon.), Nat. Acad. Scis. (fgn. assoc.), Acad. dei Lincei (Rome, fgn.), Leopoldina, Halle/Saale, Acad. Scis. USSR (fgn.), Royal Irish Acad. (hon.), Royal Danish Acad. Scis. (hon.), Acad. Pharm. Scis. (hon.), Am. Philos. Soc., Acad. Scis. (Paris, fgn. mem.), Pontificia Acad. Sci. (Rome). Research, numerous publs. on constn. and stereochemistry alkaloids, antibiotics, enzymes, other natural compounds, alicyclic chemistry, chem. topology. Office: Eidgenossische Techn, Hochschule, Universitatsstrasse 16, 8092 Zurich Switzerland

PREMPREE, THONGBLIEW, oncology radiologist; b. Thailand, Feb. 1, 1935; s. Korn and Kam (T.) P.; m. Amporn Lohsuwand, Apr. 30, 1963. Pre-Med., Chulalongkorn U., Thailand, 1954; MD, Siriraj U. of Med. Sci., Thailand, 1958; PhD in Radiobiology, John Hopkins U., 1968. Diplomate Am. Bd. Radiology; Fellow in Am. Coll. Radiology. Instr. Dept. Radiology John Hopkins U., Balt., 1968-69; asst. prof. John Hopkins U. Dept. Radiol. Scis., Balt., 1969; staff mem. Dept. of Radiology Ramathibodi Hosp. Mahidol U., Bankok, Thailand, 1969-70; dir. radiobiology Dept. Therapeutic Radiology Tuft New England Med. Ctr., Boston, 1970; asst. prof. sch. of medicine John Hopkins U., Balt., 1971-74; assoc. prof. dept. radiology U. Md. Hosp., Balt., 1974, acting dir., assoc. prof., 1977-79, prof., chief, div. radiation oncology, 1979-83; prof., chmn. dept. radiation oncology U. Hosp., Jacksonville, Fla., 1983—. Recipient PhD Fellowship awards John Hopkins U., 1963-68. Mem. Am. Soc. Therapeutic Radiobiologists, Am. Coll. Radiology, AAAS, AMA, Radiation Research Soc., Fedn. Am. Scientists, Radiol. Soc. N. Am., The John Hopkins Med. & Surgical Assn. Home: 104 Lamplighter Island Ct Ponte Vedra Beach FL 32082-1940 Office: U Fla 655 W 8th St Jacksonville FL 32209-6595

PREMUS, ROBERT, economics educator; b. Connellsville, Pa., Nov. 20, 1939; s. Albert C. and Susan (Christian) P.; m. Dolores E. Ptak, Sept. 3, 1966; children: Deborah S., Robert F. BA in Edn., Bob Jones U., Greenville, S.C., 1963; MA in Econs., Ohio U., 1968; PhD in Econs., Lehigh U., Bethlehem, Pa., 1974. Tchr. Philo (Ohio) High Sch., 1963-66; instr. R.I. Coll., Providence, 1967-68; asst. prof. Elizabethtown (Pa.) Coll., 1970-72; asst. prof. econs. Wright State U., Dayton, Ohio, 1972-76, assoc. prof. econs., 1976-80, prof. econs., 1981—; mem. summer faculty Air Force Engring. Svcs., Panama City, Fla., 1978; staff economist Joint Econ. Com., Washington, 1981-84; cons. bus. and govt. throughout U.S. and Europe, 1972—; speaker bus. and govt. confs. throughout U.S. and Europe, 1972—; bd. dirs., fin. dir. Gatlings Auto Stores, Inc., Connellsville, Pa.; dir. M.S. program in social and applied econs. dept. econs. Wright State U., 1978-79. Author: New Opportunities in Venture Capital, 1986, The Climate for Entrepreneurship and Innovation in the United States, 1985, Venture Capital and Innovation, 1985, (with David Karns and Anthony Robinson) Socioeconomic Regulations and the Federal Procurement Market, 1985; Industrial Policy Movement in the United States: Is It the Answer?, 1984, Location of High Technology Firms and Regional Economic Development, 1982, Air Force Fiscal Impact Module, 1982; contbr. articles to profl. jours. Advisor Ohio Rep. Gubernatorial Campaign, Columbus, 1982, Ohio Senate Econ. Devel. Ctr., Columbus, 1988—; advisor, cons. Dayton Area Progress Coun., 1987—; mem. Taskforce on Handicapped Edn., Greene County, Ohio, 1988—; bd. trustees Dayton Area Tech. network, 1987—; bd. dirs. Greene County Mental Health and Retardation Bd., 1981; mem. com. econ. advisors Greene County Regional Planning and Coord. Commn., 1977-79; mem. econ. resources com. Miami Valley Regional Planning, 1976-79. NSF NDEA doctoral fellow, 1969-71; Ford Found. dissertation grantee, 1971-72; recipient Outstanding Svc. awards Wright State U. Coll. Bus., 1971-72, 88-89, Outstandibg Rsch. award Wright State U. Coll. Bus., 1987-88. Mem. Am. Econ. Assn., World Futures Soc., Pub. Choice Soc., Nat. Tax Assn., So. Econs. Assn., Western Regional Sci. Assn., Dayton Area Tech. Network, Dayton Area C. of C. (econ. adv. com. 1987—), Ohio Pub. Fin. Roundtable, Sertoma (pres. Beavercreek chpt. 1990-91). Baptist. Avocations: astronomy, horseback riding, dogsled racing, hiking, travel. Home: 3165 Indian Ripple Rd Dayton OH 45440-3608 Office: Wright State U Dept Econs Dayton OH 45435

PRENDERGAST, ROBERT A., pathologist educator; b. Bklyn., Nov. 6, 1931. BA, Columbia U., 1953; MD, Boston U., 1957. Intern Bellevue Hosp., 1957-58; resident Boston City Hosp., 1958-59, Meml. Sloan-Kettering Hosp., 1959-61; vis. physician Rockefeller U., 1963-65, asst. prof., 1965-70; Assoc. prof. opthamology and pathology, sch. medicine Johns Hopkins U., 1970—; prof. Rsch. Prevent Blindness, Inc., 1971—. Mem. Am. Assn. Immunology, Am. Soc. Exp. Pathology, Transplantation Soc., Reticuloendothelial Soc. Achievements include research in delayed hypersensitivity and cellular immunology, ontogeny of the immune response, transplantation immunology, immunology of neoplasia, viral immunopathology, immunopathology of ocular inflammatory diseases. Office: Johns Hopkins U Ocsular Immunology Lab Wilmer 600 N Broadway Rm 457 Baltimore MD 21205*

PRENDERGAST, WALTER GERARD, computer engineer; b. Dublin, Ireland, Jan. 3, 1965; s. John A. and Breege (McMahon) P. BA in Math., Trinity Coll., Dublin, 1986, BAI in Computer Engring., 1987, MA, 1990. Computer engr. GPA Group PLC, Shannon, Ireland, 1987—. Mem. IEEE, Assn. Computing Machinery, Internat. Neural Network Soc., MENSA. Roman Catholic. Avocations: accordion, guitar, rugby, track. Home: 8 Maple House Drumqeely, Shannon Ireland Office: GPA Group PLC, GPA House, Shannon Ireland

PRENDERGAST, WILLIAM JOHN, ophthalmologist; b. Portland, Oreg., June 12, 1942; s. William John and Marjorie (Scott) P.; m. Carolyn Grace Perkins, Aug. 17, 1963 (div. 1990); children: William John, Scott; m. Sherryl Irene Guenther, Aug. 25, 1991. BS, U. Oreg., Eugene, 1964; MD, U. Oreg., Portland, 1967. Diplomate Am. Bd. Ophthalmology. Resident in ophthalmology U. Oreg., Portland, 1970-73; pvt. practice specializing in ophthalmology Portland, 1973-82; physician, founder, ptnr. Oreg. Med. Eye Clinic, Portland, 1983—; founder, pres. Med. Eye Assocs., Inc. Ophthalmic Clinic Networking Venture, Portland, 1992—; clin. asst. prof. dept. ophthalmology Oreg. health Sci. Ctr., 1985—; pres. Med. Eye Assocs. Vol. surgeon N.W. Med. Teams, Oaxaca, Mexico, 1989, 90. With USPHS, 1968-70. Fellow Am. Acad. Ophthalmology; mem. Met. Bus. Assn., Multnomah Athletic Club, Mazamas Mountaineering Club, Portland Yacht Club, Phi Beta Kappa, Alpha Omega Alpha. Avocations: yacht racing, mountaineering. Office: Oregon Med Eye Clinic 1955 NW Northrup St Portland OR 97209-1689

PRENTICE, ANN ETHELYND, academic adminstrator; b. Grafton, Vt., July 19, 1933; d. Homer Orville and Helen (Cooke) Hurlbut; divorced; children: David, Melody, Holly, Wayne. AB, U. Rochester, 1954; MLS, SUNY, Albany, 1964; DLS, Columbia U., 1972; LittD (hon.), Keuka Coll., 1979. Lectr. sch. info. sci. and policy SUNY, Albany, 1971-72, asst. prof., 1972-78; prof., dir. grad. sch. library and info. sci. U. Tenn., Knoxville, 1978-88; assoc. v.p. info. resources U. South Fla., Tampa, 1988-93; dean Coll. of Libr. and Info. Svcs. U. Md., 1993—. Author: Strategies for Survival, Library Financial Management Today, 1979, The Library Trustee, 1973, Public Library Finance, 1977, Financial Planning for Libraries, 1983, Professional Ethics for Librarians, 1985; editor Pub. Library Quar., 1978-81; co-editor: Info. Sci. in its Disciplinary Context, 1990; assoc. editor Library and Info. Sci. Ann., 1987-90. Cons. long-range planning and pers. Knox County Libr. System, 1980, 85-86, Richland County S.C. Libr. System, 1981, Upper Hudson Libr. Fedn., N.Y., State Libr. Ohio, 1986; trustee Hyde Park (N.Y.) Free Libr., treas., 1973-75, pres., 1976; trustee Mid-Hudson Libr. System, Poughkeepsie, N.Y., 1975-78; trustee adv. bd. Hillsborough County Libr., 1991-93. Recipient Disting. Alumni award SUNY/Albany, 1987, Columbia U., 1991. Mem. ALA, Am. Soc. Info. Sci. (exec. bd. 1986-89, conf. chair 1989, pres. 1992-93), AAUP, Assn. Info. Mgrs., Assn. for Libr. and Info. Sci. Edn. (pres. 1986). Office: Univ Md Coll Libr and Info Svcs 4105 Hornbake Bldg College Park MD 20742

PRENTICE, ROBERT CRAIG, cardiologist; b. Chgo., Sept. 2, 1951; s. Robert Lee and Helen (Virginia) P.; m. Mary Ellen Toomey, Oct. 3, 1981; children: Ryan, Laura, Sarah. BA, Wabash Coll., Crawfordsville, Ind., 1973; MA, So. Ill. U., 1978; DO, Chgo. Coll. Osteo. Medicine, Chgo., 1982; PhD, U. Ill., 1982. Faculty asst. So. Ill. U., Carbondale, 1974-78; med. scientist Chgo. Osteo. Hosp., 1978-82; resident in medicine Hines (Ill.) VA Hosp., 1982-85; fellow in cardiology Loyola U., Maywood, Ill., 1985-87; staff asst. prof. medicine Loyola U., 1987-88; cardiologist Olympia Fields (Ill.) Med. Ctr., 1988-90; cardiologist, dir. cardiac lab. Michael Reese Hosp., Chgo., 1990-92; pvt. practice interventional cardiology Hazel Crest, Ill., 1992—; rsch. asst. U. Ill., 1980-82; dir. cardiac lab. Michael Reese Hosp., 1990-92. Grantee Chgo. Heart Assn., 1985. Fellow Am. Coll. Cadiology, ACP; mem. AMA, Chgo. Med. Soc. Avocations: skiing, running, tennis, golf. Office: 17850 S Kedzie Ave Hazel Crest IL 60423

PRESCOTT, JOHN HERNAGE, aquarium executive; b. Corona, Calif., Mar. 16, 1935; s. Arthur James and Henrietta (Hernage) P.; m. Sandra Baker, Sept. 26, 1985; children by previous marriage—Craig C., Blane R. B.A., UCLA, 1957; postgrad., U. So. Calif., Los Angeles, 1958-60; cert. advanced mgmt. program, Harvard U. Curator Marineland of the Pacific, Palos Verdes, Calif., 1957-70, v.p., 1966-70, gen. mgr., 1970-72; exec. dir., v.p. New Eng. Aquarium, Boston, 1972—; corporator Woods Hole Oceanographic Inst., Mass., 1976-90; chmn. mem. com. sci. advisers Marine Mammal Commn., Washington, 1977-80; dir. Mus. Inst. Teaching Sci., Boston, 1984-92; chmn. Humpback Whale Recovery Team NOAA, Washington, 1987—; mem. U.S delegation Internat. Whaling Commn., 1989-92. Author: Aquarium Fishes of the World, 1976. Editor: Georges Bank: Past, Present, Future, 1981, Right Whales: Past and Present Status, 1986. Bd. dirs. Boston Mcpl. Rsch. Bur., 1981—, Boston Am. Heart Assn., 1983-86, Nat. Oceanographic and Atmospheric Adminstrn., Washington, 1987-92; mem. Marine Fisheries Adv. Com.; trustee 100 Friends of Mass., 1991—. Recipient commendation for efforts to conserve whales U.S. Ho. of Reps., 1971, Ann. Sci. award for Conservation, Am. Cetacean Soc., 1969. Fellow Am. Assn. Zool. Parks and Aquariums (bd. dirs. 1985-92); mem. AAAS, Soc. Marine Mammalogy, Am. Assn. Mus., Sea Edn. Assn. (trustee 1986-92), Explorers Club (chmn. New Eng. sect. 1981-85). Office: New Eng Aquarium Corp Central Wharf Boston MA 02110

PRESCOTT, LAWRENCE MALCOLM, medical and health writer; b. Boston, July 31, 1934; s. Benjamin and Lillian (Stein) P. BA, Harvard U.,

1957; MSc, George Washington U., 1959, PhD, 1966; m. Ellen Gay Kober, Feb. 19, 1961 (dec. Sept. 1981); children: Jennifer Maya, Adam Barrett; m. Sharon Lynn Kirshen, May 16, 1982; children: Gary Leon Kirshen, Marc Paul Kirshen. Nat. Acad. Scis. postdoctoral fellow U.S. Army Rsch., Ft. Detrick, Md., 1965-66; microbiologist/scientist WHO, India, 1967-70, Indonesia, 1970-72, Thailand, 1972-78; with pub. rels. Ted Klein & Co., Hill & Knowlton, Interscience, , Smith, Kline, Beecham, others, 1984—; cons. health to internat. orgns., San Diego, 1978—; author manuals; contbr. articles in diarrheal diseases and lab. scis. to profl. jours., numerous articles, stories, poems to mags., newspapers, including Living in Thailand, Jack and Jill, Strawberry, Bangkok Times, Sprint, 1977-81; mng. editor Caduceus, 1981-82; pub., editor: Teenage Scene, 1982-83; pres. Prescott Pub. Co., 1982-83; med. writer numerous jours. including Modern Medicine, Dermatology Times, Internal Medicine World Report, Drugtherapy, P&T, Clinical Cancer Letter, Hospital Formulary, Female Patient, Australian Doctor, Inpharma Weekly, American Family Physician, Ophthalmology Times, Group Practice News, Newspaper of Cardiology, Paacnotes, Genetic Engineering News, Medical Week, Medical World News, Urology Times, Gastroenterology and Endoscopy News; author: Curry Every Sunday, 1984. Home and Office: 18264 Verano Dr San Diego CA 92128-1262

PRESLEY, ALICE RUTH WEISS, physicist, researcher; b. N.Y.C.; d. Joseph and Rose (Klein) Weiss; m. John David Presley; children: David Michael, Margaret Celia, Joseph William. BA in Physics and Math. with honors, Hunter Coll., 1960; MS in Physics, Stevens Inst., 1964; MS in Plant, Soil and Water, U. Nev., 1978; MS in Atmospheric Sci., Drexel U., 1980, PhD in Physics, 1990. Mem. rsch. faculty Desert Rsch. Inst., Reno, Nev., 1969-71; rsch. cons., data analyst U. Nev., Reno, 1974-78, programming cons., 1971-72, asst. acting state climatologist of Nevada, 1976-78; grad. student rsch. asst. Drexel U., Phila., 1978-91; tech. specialist Comm. Cable Co., Malvern, Pa., 1992—. Author: Synoptic Patterns and Precipitation in the Pinyon Juniper Zone of the Great Basin, 1978, Light Scattered from Interacting Finite Dielectric Scatterers, 1990; contbr. articles to Jour. Applied Meteorology, Jour. Applied Optics. Founder N.W. Reno chpt. Extension Svc. Homemakers, 1975; editor newsletter Del. Orchid Soc., 1991—; prin. Temple Emanual Sunday Sch., Reno, 1968-77; chmn. Summit Area Civic Assn., Malvern, Pa., 1988-89; mem. East Whiteland Environ. Adv. Coun., 1993—. Mem. Am. Meterol. Soc. sec., vice chmn., chmn. Reno chpt. 1973-76, editor Nev. Weather Watch Reno chpt. 1976-78), Am. Phys. Soc., Biophys. Soc., Sigma Xi, Pi Mu Epsilon, Sigma Pi Sigma. Achievements include development of a model for light scattered from different shape dielectrics which includes self-interactions, of a model to analyze climate and precipation patterns of the Great Basin region, and of a salt particle sizing and detection procedure; implementation of a computer-data bank of precipitation data collected in the Sierra Nevadas and its foothills.

PRESS, FRANK, geophysicist, educator; b. Bklyn., Dec. 4, 1924; s. Solomon and Dora (Steinholz) P.; m. Billie Kallick, June 9, 1946; children: William Henry, Paula Evelyn. BS, CCNY, 1944, LLD (hon.), 1972; MA, Columbia U., 1946, PhD, 1949; DSc (hon.), 23 univs. Rsch. assoc. Columbia U., 1946-49, instr. geology, 1949-51, asst. prof. geology, 1951-52, assoc. prof., 1952-55; prof. geophysics Calif. Inst. Tech., 1955-65, dir. seismol. lab., 1957-65; prof. geophysics, chmn. dept. earth and planetary scis. MIT, 1965-77; sci. advisor to pres., dir. Office Sci. and Tech. Policy, Washington, 1977-80; inst. prof. MIT, 1981; pres. Nat. Acad. Scis., 1981—; mem. Pres.'s Sci. Adv. Com., 1961-64; mem. Com. on Anticipated Advances in Sci. and Tech., 1974-76; mem. Nat. Sci. Bd., 1970—; mem. lunar and planetary missions bd. NASA; participant bilateral scis. agreement with Peoples Republic of China and USSR; mem. U.S. delegation to Nuclear Test Ban Negotiations, Geneva and Moscow. Author: (with M. Ewing, W.S. Jardetzky) Propagation of Elastic Waves in Layered Media, 1957, (with R. Siever) Earth, 1986; also over 160 publs.; co-editor: (with R. Siever) Physics and Chemistry of the Earth, 1957—. Decorated Cross of Merit (Germany); recipient Columbia medal for excellence, 1960, Pub. Svc. award U.S. Dept. Interior, 1972, Gold medal Royal Astron. Soc., 1972, Pub. Svc. medal NASA, 1973, Legion of Honor award, 1989, Japan prize Sci. and Tech. Found. Japan, 1993; named Most Influential Scientist in Am., U.S. News and World Report, 1982, 84, 85. Mem. Am. Acad. Arts and Scis., Geol. Soc. Am. (councilor), Am. Geophys. Union (pres. 1973), Soc. Exploration Geophysicists, Seismol. Soc. Am. (pres. 1963), AAUP, NAS, Am. Philos. Soc., French Acad. Scis., Royal Soc. (U.K.), Nat. Acad. Pub. Adminstrn., Acad. Scis. of USSR (fgn. mem.). Office: Carnegie Inst Washington Dept Terrestrial Magnetism 5241 Broad Branch Rd Rm 173 Washington DC 20015-4376

PRESS, WILLIAM HENRY, astrophysicist, computer scientist; b. N.Y.C., N.Y., May 23, 1948; s. Frank and Billie (Kallick) P.; m. Margaret Ann Lauritsen, 1969 (div. 1982); 1 dau., Sara Linda; m. Jeffrey Foden Howell, Apr. 19, 1991. A.B., Harvard Coll., 1969; M.S., Calif. Inst. Tech., 1971, Ph.D., 1972. Asst. prof. theoretical physics Calif. Inst. Tech., 1973-74; asst. prof. physics Princeton (N.J.) U., 1974-76; prof. astronomy and physics Harvard U., Cambridge, Mass., 1976—; chmn. dept. astronomy Harvard U., 1982-85; mem. numerous adv. coms. and panels NSF, NASA, NAS, NRC; vis. mem. Inst. Advanced Study, 1983—; mem. Def. Sci. Bd., 1985-89, sci. adv. com. Packard Found., 1988—, program com. Sloan Found., 1985-91; chmn. adv. bd. NSF Inst. Theoretical Physics, 1986-87; cons. MITRE Corp., Los Alamos Nat. Lab., Lawrence Livermore Nat. Lab.; trustee Inst. Def. Analysis, 1988—. Author: Numerical Recipes, 1986; assoc. editor Annals of Physics, 1984-91; contbr. articles to profl. jours. Sloan Found. research fellow, 1974-78. Fellow AAAS, Am. Phys. Soc.; mem. Am. Astron. Soc. (Helen B. Warner prize 1981), Internat. Astron. Union, Internat. Soc. Relativity and Gravitation, Assn. for Computing Machinery. Office: Harvard U 60 Garden St Cambridge MA 02138-1596

PRESTING, HARTMUT, physicist; b. Freiburg, Fed. Republic Germany, Sept. 29, 1956; s. Dieter and Jutta (Hilgenfeld) P. MS, U. Karlsruhe, Fed. Republic Germany, 1982, PhD, U. Stuttgart, Fed. Republic Germany, 1985. Mem. tech. staff AT&T Bell Lab., Holmdel, N.J., 1985-87, AEG Rsch. Lab., Ulm, Fed. Republic Germany, 1987-90; mem. tech. staff, project leader Daimler Benz Rsch. Ctr., Ulm, 1990—. Contbr. articles to profl. jours.; patentee in field. With German mil., 1975-76. Mem. Optical Soc. Am., German Phys. Soc. Avocations: tennis, skiing, biking, ham radio, computers. Office: Daimler Benz AG, Wilhlem Runge Str 11, D 89075 Ulm Germany

PRESTON, ANDREW JOSEPH, pharmacist, drug company executive; b. Bklyn., Apr. 19, 1922; s. Charles A. and Josephine (Rizzutto) Pumo; B.Sc., St. John U., 1943; m. Martha Jeanne Happ, Oct. 10, 1953; children: Andrew Joseph Jr., Charles Richard, Carolyn Louise, Frank Arthur, Joanne Marie, Barbara Jeanne. Cert. bus. intermediary. Mgr. Press Club, Bklyn. Nat. League Baseball Club, 1941-42; purchasing agt. Drug and Pharm. div. Intrassind, Inc., 1947; chief pharmacist Hendershot Pharmacy, Newton, N.J., 1949; agt. Bur. of Narcotics, U.S. Treasury Dept., 1948-49; owner Preston Drug & Surg. Co., Boonton, N.J., 1949-86; chief exec. officer Preston Pharmaceutics, Inc., Butler, N.J., 1970-80, chief exec. officer Preston Bus. Cons., Inc., Kinnelon, N.J., 1987—; commr. N.J. State Bd. Pharmacy, 1970-72, pres., 1973; organizer State of N.J. Drug Abuse Speakers Program, 1970-76; chmn. Morris County Drug Abuse Coun., 1969-70; lectr. drug abuse and narcotic addiction various community orgns., 1968-78; mem. adv. bd. Nat. Community Bank, Boonton, N.J., 1973. Chmn. bldg. fund com. Riverside Hosp., Boonton, 1963; mem. Morris County (N.J.) Rep. Fin. Com., 1972; pres. Ronald Reagan N.J. Re-Election Adv. Bd., 1984; mem. exec. com. Gov. Tom Kean Annual Ball, 1985-86; chmn. Pharmacists of N.J. for election of Pres. Ford, 1976, Pharmacists for Gov. Tom Kean, 1981-84, N.J. Pharmacists for Reagan/Bush '84; mem. exec. com. Morris County Overall Econ. Devel. Com., 1976-82; chmn. Pharmacists for Fenwick, 1982; v.p. Kinnelon Rep. Club, 1980, Rep. Com., Kinnelon, 1990. Served to lt. (j.g.), USNR, 1943-46. Recipient Bowl Hygeia award Robbins Co., 1969, E.R. Squibb President's award, 1968, N.J. Pharm. Square Club award, 1969. Mem. Am. Pharm. Assn., N.J. Pharm. Assn. (mem. econs. com. 1960-65, pres. 1967-68, Oscar Singer Meml. award 1987), Nat. Assn. of Retail Druggists, Internat. Narcotic Enforcement Officers Assn., N.J. Narcotic Enforcement Officers Assn., Nat. Assn. Realtors, N.J. Assn. Realtors, Morris County Bd. Realtors, Internat. Bus. Brokers Assn. (cert. bus. intermediary), Inst. Bus. Appraisers, Pharmacists Guild Am. (pres. N.Y. div. 1946-47), Pharmacists Guild of N.J., N.J. Public Health Assn., Morris County Pharm. Assn., Morris-Sussex Pharmacists Soc., Am. Legion, St. John's Alumni Assn. Roman Catholic. Clubs: Elks, K.C., Smoke Rise. Contbr. editorials to profl. jours. Home and Office: 568A Pepperidge Tree Ln Kinnelon NJ 07405

PRESTON, CHARLES BRIAN, orthodontist, school administrator; b. Johannesburg, South Africa, Nov. 19, 1937; s. David Charles and Mary (Meerkotter) P.; m. Joy Pretorius, Jan. 1, 1966; 1 child, Bridgette. B.D.S., Witwatersrand Sch. Dentistry, 1961, diploma orthodontics, 1973, M.Dent., 1974, Ph.D., 1988 Lectr. dentistry Sch. Dentistry, Johannesburg, 1967-73, lectr. orthodontics, 1973-77, acting head dept. orthodontics, 1977-79, prof., head dept. orthodontics, 1979-84, dep. dean, 1983, 84, 85, 86; dean and dir. Oral and Dental Teaching Hosp. U. of the Witwatersrand, 1987. Cons., referee Am. Jour. Orthodontics; contbr. articles to profl. jours. Active South African Council Alcoholism; life mem. Operation Wild Flower. Recipient Middleton-Shaw award Dental Assn. of South Africa, 1987; grantee Research, Edn. and Devel. Fund, 1979-80, Med. Research Council, 1979-82; Elida Gibbs research fellow Dental Assn. South Africa, 1979-80. Mem. South African Soc. Orthodontists (exec. bd.), European Orthodontic Soc., Am. Assn. Orthodontists, Internat. Assn., Dental Research, South African Coll. Medicine (assoc.), Fed. Coun. of the Dental Assn. of South Africa, Med. and Dental Coun. of South Africa, Aircraft Owners and Pilots Assn., Univ. Flying Club (chmn. 1980-81), Emmarentia Sailing Club, Lions, Alpha Omega. Office: U Witwatersrand, 1 Jan Smuts Ave, Johannesburg 2001, South Africa

PRESTON, DEAN LAVERNE, physicist; b. Oswego, N.Y., Feb. 4, 1953; s. Verne Elmer and Marjorie Alma (Connors) P.; m. Marsha Ann Weselak, Jan. 4, 1975; children: Melissa, Christopher; m. Sally Elaine Archuleta, Feb. 14, 1983; children: Daniel, Matthew. BS, Rensselaer Polytechnic Inst., Troy, N.Y., 1975; PhD, Princeton U., 1980. Cons. Los Alamos (N.Mex.) Nat. Lab., 1982-83; asst. prof. U. N.H., Durham, 1983-85; staff physicist Nuclear Applications Group - LANL, Los Alamos, 1985—; Contbr. articles to profl. jours. Mem. Am. Phys. Soc. Achievements include devel. of constitutive model for metals and alloys that encompasses homogeneous plastic flow, the shear modulus, and melting. It is becoming the prin. materials model for weapon design at Los Alamos. Office: Los Alamos Nat Lab X-4 MS-F664 Los Alamos NM 87545

PRESTON, R. KEVIN, software engineer; b. Danville, Ky., Dec. 13, 1959; s. Jopat and Dorthy Ann (Edwards) P.; m. Patricia Ann Adkins, May 24, 1986; children: Ryan Adkins, Ray Edward. BS, Eastern Ky. U., 1982; MS, U. Ala., Huntsville, 1992. Intern U.S. Army Intern Tng. Ctr., Red River, Tex., 1982-83; software quality engr. U.S. Army Missle Command, Redstone Arsenal, Ala., 1983-85; lead software quality engr. Gen. Electric Co., Huntsville, 1985-90; sr. software Hilton Systems, Inc., Huntsville, 1990--; instr. Eastern Ky. U., Richmond, Ky., 1981-82, So. Inst., Huntsville, 1984-85; assoc. mem. U. Ala., Huntsville, 1992--. Mem. bldg. com. property com. choir, Sunday sch. tchr. Hillwood Bapt. Ch., Huntsville, 1983--. Mem. IEEE. Home: 10001 Todd Mill Rd Huntsville AL 35803 Office: Hilton Systems Inc 2227 Drake Ave Ste 31 Huntsville AL 35805

PRESTON, ROBERT LESLIE, cell physiologist, educator; b. Stevens Point, Wis., Apr. 24, 1942; s. Roy Leslie and Lois (Pearson) P.; m. Joyce Welcome, June 27, 1965; children: Elisabeth, David, Catherine, Bryan. BA, U. Minn., 1966; PhD, U. Calif., Irvine, 1970. Postdoctoral fellow dept. physiology Yale U., New Haven, Conn., 1970-73, rsch. assoc., 1973-74; asst. prof. dept. biol. scis. Ill. State U., Normal, 1974-81, assoc. prof., 1981-88, prof., 1988—; vis. prof. physiol. lab. U. Cambridge, England, 1981-82, dept. zoology U. Hawaii, Honolulu, 1987; trustee Mount Desert Island Biol. Lab., Salsbury Cove, Maine, 1988—; mem. math/sci. task force Nat. Collegiate Honors Coun., 1991—. Sect. editor: Amino Acids, 1991—; author: Comparative Physiology, vol. 7: Comparative Aspects of Sodium Cotransport Systems, 1990; contbr. articles to Jour. Gen. Physiology, Biochimica et Biophysica Acta, Comparative Biochemistry and Physiology, Jour. Exptl. Zoology, Life Sci., Bull. Environ. Contamination and Toxicology. Grantee NSF, 1985-88, NIH, 1981-82, 87-90, Am. Heart Assn., 1984-85, 90-92, Nat. Insts. Environ. Health Scis., 1986-88, 89—, NATO, 1985-86, Nat. Sea Grant and Nat. Oceanic and Atmospheric Adminstrn., 1991—. Mem. Biochem. Soc., AAAS, Am. Soc. Zoologists, Soc. Gen. Physiologists. Office: Ill State U Dept Biol Scis Normal IL 61761

PREVOR, RUTH CLAIRE, psychologist; b. N.Y.C., June 20, 1944; d. Gustav and Greta (Dreifuss) Strauss; m. Sydney Joseph P., July 4, 1963; children: Joy, Grant, Jed. BA, U. P.R., 1966; PhD, Caribbean Ctr. of Postgrad., Studies, San Juan, 1988. Cert. forensic psychologist. Asst. dean Caribbean Ctr. of Postgrad. Studies, 1986-87; dir. prenatal edn. Ashford Meml. Hosp., San Juan, 1987; pvt. practice San Juan, 1984—; advisor, field faculty Vt. Coll., Norwich U., 1990-91; trustee Caribbean Ctr. for Advanced Studies, San Juan, Miami, Fla., 1990—. Trustee Jewish Community Ctr., Miramar, P.R., 1986—, bd. dirs. pre-sch., 1990—; pres. Home and Sch./St. John's Prep., San Juan, 1980-81, P.R. chpt. Hadassah Sch., 1972-74; presdl. adv. com., 1990-92. Mem. Am. Psychol. Assn., Assn. of Psychology of P.R. (hon. award 1984), Caribbean Counselors Assn., Caribe Hilton Club, Nat. Assn. Children with Learning Disabilities, Nat. Register Health Svc. Providers in Psychology. Jewish. Office: Ashford Med Ctr San Juan PR 00907

PREW, DIANE SCHMIDT, information systems executive; b. Orange, N.J., Jan. 21, 1945; d. Herman and Elfriede (Witt) Schmidt; m. Jonathan Prew, Jan. 27, 1968; 1 child, Heather Diane. BSBA, U. N.H., 1967. Cert. systems profl. Programmer analyst Eastman Kodak Co., Rochester, N.Y., 1967-70; program and system mgr. Nat. Acad. Scis., Washington, 1970-72; owner Active Info. Systems, Nashua, N.H., 1974-79; dir. info. svcs. City of Manchester, N.H., 1980—; bd. dirs. Manchester Mcpl. Employees Credit Union, v.p., 1993. Mem. Data Processing Mgmt. Assn. (sec. 1982-84, exec. v.p., 1984-85, pres. 1985-86, treas. 1986—, Bronze award 1988, Silver award 1991), Rotary Club. Avocations: gardening, swimming, hiking. Home: 50 Mack Hill Rd # 877 Amherst NH 03031-3223 Office: City of Manchester Info Systems Dept 100 Merrimack St Manchester NH 03101-2208

PREWITT, CHARLES THOMPSON, geochemist; b. Lexington, Ky., Mar. 3, 1933; s. John Burton and Margaret (Thompson) P.; m. Gretchen B. Hansen, Jan. 31, 1958; children: Daniel Hansen. SB, MIT, 1955, SM, 1960, PhD, 1962. Research scientist E.I. DuPont De Nemours & Co. Inc., Wilmington, Del., 1962-69; assoc. prof. SUNY, Stony Brook, 1969-71, prof., 1971-86, chmn. dept. earth and space scis., 1977-80; dir. Geophys. Lab., Carnegie Inst. of Washington, 1986—; sec.-treas. U.S. Nat. Com. for Crystallography, Washington, 1983-85; gen. chmn. 14th Meeting of Internat. Mineral. Assn., Stanford, Calif., 1986; chmn. NRC/Nat. Acad. Scis. com. on physics and chemistry of earth materials, 1985-87; mem. bd. govs. Consortium for Advanced Radiation Svcs. Editor: (jour.) Physics and Chemistry of Minerals, 1976-85; contbr. over 140 articles to profl. jours. Capt. USAR, 1956-65. NATO sr. postdoctoral fellowship, 1975, Churchill overseas fellowship, 1975, Japan Soc. for Promotion of Sci. fellowship, 1983; named Disting. Vis. Prof. Chemistry, Ariz. State U., 1983. Fellow Mineral. Soc. Am. (pres. 1983-84), Am. Geophys. Union; mem. Geol. Soc. Am., Am. Crystallographic Assn., Materials Research Soc., Mineral. Soc. Gt. Britain and Ireland. Home: 2728 Unicorn Ln NW Washington DC 20015-2234 Office: Carnegie Inst Geophys Lab 5251 Broad Branch Rd NW Washington DC 20015-1305

PREWITT, NATHAN COLEMAN, aerospace engineer; b. Batesville, Miss., Oct. 4, 1964; s. Maston Levi and China Ray (Yeates) P.; m. Tracye Wynne Malone, May 25, 1985. BS, Miss. State U., 1987, MS, 1988. With Naval Sea Systems Command, Alexandria, Va., 1984-85; teaching asst. dept. aero. engring. Miss. State U., Starkville, 1987-88; engr. Sverdrup Tech., Inc., Eglin AFB, Fla., 1988—. Author tech. paper. Mem. AIAA, Sigma Gamma Tau, Phi Kappa Phi, Tau Beta Pi. Methodist. Achievements include development of IGGy, Interactive Grid Generation System, for use in computational fluid dynamics applications. Home: 113 Golf Course Dr Crestview FL 32536 Office: Sverdrup Tech Inc PO Box 1935 Eglin A F B FL 32542

PRIBOR, HUGO CASIMER, physician; b. Detroit, June 12, 1928; s. Benjamin Harrison and Wanda Frances (Mioskowski) P.; m. Judith Elinor Smith, Dec. 22, 1955; children: Jeffrey D., Elizabeth F., Kathryn A. BS, St. Mary's Coll., 1949; M.S., St. Louis U., 1951, Ph.D., 1954, MD, 1955. Diplomate Am. Bd. Pathology. Intern Providence Hosp., Detroit, 1955-56; resident pathologic anatomy and clin. pathology NIH, Bethesda, Md., 1956-59; field investigator gastric cytology rsch. project Nat. Cancer Inst., Bowman-Gray Sch. Medicine, Winston-Salem, N.C., 1959-60; assoc. pathologist, dir. clin. lab. Bon Secours Hosp., Grosse Pointe, Mich., 1960-63; pathologist, dir. labs. Samaritan Hosp. Assn., East Side Gen. Hosp., Detroit, 1963-64, Anderson Meml. Hosp., Mt. Clemens, Mich., 1963-64; cons. pathologist Middlesex County Med. Examiners Office, New Brunswick, N.J., 1964-73; dir. dept. labs., chief pathologist, sr. attending physician Perth Amboy (N.J.) Gen. Hosp., 1964-73; chmn., chief exec. officer Ctr. Lab. Medicine, Inc., Metuchen, N.J., 1968-77; v.p. med. affairs Damon Corp., Med. Svcs. Group, 1977-78; exec. med. dir. MDS Health Group, Inc., Red Bank, N.J., 1978-80; med. dir. Internat. Clin. Labs., Inc., Nashville, 1981-88; med. dir. SmithKline Beechman Clin. Labs., Nashville, 1990—; physician, pathologist Assoc. Pathologists (P.C.), Nashville, 1981—; rsch. assoc. dept. pathology St. Louis U. Sch. Medicine, 1954-55; instr. pathology Bowman-Gray Sch. Medicine, Winston-Salem, N.C., 1959-60; asst. prof. chemistry U. Detroit, 1961-64; instr. pathology Wayne State U. Sch. Medicine, Detroit, 1961-64; clin. assoc. prof. dept. pathology Med. Sch. Rutgers U., The State U., New Brunswick, N.J., 1966-68; cons. Health Facilities Planning and Constrn. Svc., USPHS, HEW, Rockville, Md., 1970-71; prof. biomed. engring. Coll. Engring., Rutgers, The State U., New Brunswick, N.J., 1971-75, 80-82; chmn. bd. trustees St. Mary's Coll., Winona, Minn., 1972-74, chmn. fin. com., 1971-72; clin. prof. pathology Vanderbilt U. Sch. Medicine, Nashville, 1981—. Author: (with G. Morrell and G. H. Scherr) Drug Monitoring and Pharmacokinetic Data, 1980, The Laboratory Consultant, 1992; contbr. articles to profl. jours. Fellow Am. Soc. Clin. Pathologists (Silver award 1968); mem. AMA , Am. Assn. Exptl. Pathology, Coll. Am. Pathologists (chmn. subcom. 1974-78), Internat. Acad. Pathology, Pan Am. Med. Assn. (life), N.J. State Med. Soc., Acad. Medicine N.J. (chmn. clin. pathology sect. 1965-67), N.J. Soc. Pathologists (exec. com. 1965-67), Sigma Xi. Republican. Roman Catholic. Home: 200 Olive Branch Rd Nashville TN 37205-3220

PRIBRAM, KARL HARRY, psychology educator, researcher; b. Feb. 25, 1919. BS, U. Chgo., 1938, MD, 1941; PhD in Psychology (hon.), U. Montreal, Can., 1992. Diplomate Am. Bd. Med. Psychotherapists. Lectr. Yale U., Hartford, Conn., 1951-58; dir. Inst. of Living, Hartford, Conn., 1951-58; fellow Stanford (Va.) U., 1958-59, assoc. prof., 1959-62, rsch. career prof., 1962-89, prof. emeritus, 1989—; eminent scholar Radford (Va.) U., 1989—; vis. scholar, hon. lectr. MIT, 1954, Clark U., 1956, Harvard, 1956, Haverford Coll., 1961, U. So. Calif., 1961, U. Moscow, 1962, Beloit Coll., 1966-67, U. Alberta, Can., 1968, Ctr. for Study Dem. Insts., 1967-75, U. Coll., London, 1972, U. Chgo., 1973, Menninger Sch. Psychiatry, 1973-76, Ohio State U., 1975, Inst. for Higher Studies, 1975 and numerous others, 1954—. Author: Brain and Behavior, vol. 1-4, 1969, What Makes Man Human, 1971, Languages of the Brain: Experimental Paradoxes and Principles in Neuropsychology, 1971; The Neurosciences: Third Study Program, 1974, Brain and Perception: Holonomy and Structure in Figural Processing, 1991; editor, mem. consulting bd. Neuropsychologia, Jour. Math. Biology, Internat. Jour. Neurosci., Behavioral and Brain Scis., Jour. Mental Imagery, Jour. Human Movement Studies, Jour. Social and Biol. Structures, ReVision, STSM Quarterly, Indian Jour. Psychophysiology, Interam Jour. Psychology, Internat. Jour. Psychophysiology, Cognition and Brain Theory; contbr. over 170 articles to profl. jours. Recipient Lifetime Rsch. Career award in neurosci. NIH, 1962-89, Humanitarian award INTA, 1980, Realia honor Inst. Advanced Philosophic Rsch., 1986, Outstanding Contbns. award Am. Bd. Med. Psychotherapists. Fellow Am. Acad. Arts and Scis., N.Y. Acad. Scis. (hon. life); mem. AAUP, AMA, AAAS, APA (pres. div. physiol. and comparative psychology 1967-68, pres. div. theoretical and philos. psychology 1979-80), Internat. Neuropsychol. Soc. (founding pres. 1967-69), Internat. Assn. Study of Pain, Soc. Exptl. Psychologists, Am. Psychol. Soc., Am. Psychopathological Assn. (Paul Hoch award 1975), Am. Acad. Psychoanalysis, Soc. Biol. psychiatry (Manfred Sakel award 1976), Soc. Clin. and Exptl. Hypnosis (Henry Guze award 1991), Soc. Neurosci., Internat. Soc. Rsch. Agression, Profs. For World Peace (pres. 1982-85), Sigma Xi. Home: 102 Dogwood Ln Radford VA 24141 Office: Radford Univ Ctr Brain Rsch Box 6977 Radford VA 24142

PRICE, ALEXANDER, retired osteopathic physician; b. Stepanitz, Russia, Mar. 7, 1913; came to U.S., 1921; s. Abraham and Mariam (Cohen) P.; m. Frances Ahrens, 1944 (dec. Sept. 1987); children: John O., Carl A. BA, U. Pa., 1936; DO, Phila. Coll. Osteo. Medicine, 1941; postgrad., Columbia U., 1947-57. Diplomate Am. Osteo. Bd. Internal Medicine, Am. Osteo. Bd. Cardiology. Intern Warren Hosp., Phillipsburg, N.J., 1941-43; postgrad. in cardiology and internal medicine; pvt. practice Camden and Cherry Hill, N.J., 1943-44, 46-80, 82-87, Ft. Lauderdale, Fla., 1980-82; mem. med. staff Met. Hosp., Phila., 1948-63, chief staff, 1961-63; mem. med. staff Cherry Hill (N.J.) Hosp., 1960-87, chmn. dept. medicine, 1960-72, chmn. div. cardiology, 1960-80; ret., 1987; assoc. clin. prof. medicine Phila. Coll. Osteo. Medicine, 1965-80; clin. prof. medicine N.J. Sch. Osteo. Medicine, Rutgers U., Camden, 1970—; physician, cons. in cardiology Primary Health Care Clinic, Broward County, Fla., 1991—. Contbr. numerous articles to osteo. medicine jours. Vol. Medivan (home health care for elderly poor, indigent, disabled), Broward County, Fla., 1988—. With M.C., AUS, 1944-45, MTO. Fellow Am. Coll. Osteo. Internists (pres. 1974-75, Meml. lectr. ea. div. 1970, Meml. lectr. nat. div. 76, Disting. Svc. award 1977), Coll. Physicians Phila.; mem. Am. Osteo. Assn., N.J. Osteo. Assn., Masons. Home: 1071 NW 85th Ave Fort Lauderdale FL 33322-4622

PRICE, CLIFFORD WARREN, metallurgist, researcher; b. Denver, Apr. 22, 1935; s. Warren Wilson and Vivian Fredricka (Cady) P.; m. Carole Joyce Watermon, June 14, 1969; children: Carla Beth, Krista Lynn Kilton. MetE, Colo. Sch. Mines, 1957, PhD, 1973, MS, Ohio State U., 1970. Design engr. Sundstrand Aviation-Denver, 1957-60; materials specialist Denver Rsch. Inst., 1960-63; sr. metallurgist Rocky Flats div. Dow Chem. Co., Golden, Colo., 1963-66; staff metallurgist Battelle Columbus (Ohio) Labs., 1966-75; sr. scientist Owens-Corning Fiberglas, Granville, Ohio, 1975-80; metallurgist Lawrence Livermore (Calif.) Nat. Lab., 1980—. Contbr. articles to profl. jours. Battelle Columbus Labs. fellow, 1974-75. Mem. AAAS, Metall. Soc. of AIME, Electron Microscopy Soc. Am. (treas. Denver 1961-62), Am. Soc. for Metals. Republican. Achievements include research on electron, scanning probe and optical microscopy, secondary ion mass spectroscopy, deformation, fracture and recrystallization mechanisms in metals, recrystallization kinetics. Office: Lawrence Livermore Nat Lab PO Box 808 Livermore CA 94550

PRICE, DENNIS LEE, industrial engineer, educator; b. Taber, Alberta, Can., Oct. 24, 1930; s. Walter and Wilma Harlan (Nance) P.; m. Barbara Ann Shelton; children: Denice Lynn Price Thomas, Philip Walter. BA, Bob Jones U., 1952; BD, MA, Am. Bapt. Sem. of the West, Berkeley, Calif., 1955; MA, Calif. State U., Long Beach, 1967; PhD in Indsl. Engring., Tex. A&M U., 1974. Cert. product safety mgr., cert. hazard control mgr. Clergyman Am. Bapt. Conv., Calif., 1953-66; mem. tech. staff autonetics div. Rockwell Internat., Anaheim, Calif., 1966-69; sr. engr. Martin Marietta Aerospace, Orlando, Fla., 1969-72; rsch. assoc. Tex. A&M U., College Station, 1972-74; teaching asst. Calif. State U., Long Beach, 1963-66; asst. prof. dept. indsl. engring. and operations rsch. Va. Poly. Inst. and State U., Blacksburg, 1974-78, assoc. prof. dept. indsl. and systems engring., 1979-83, prof., 1984—; cons., expert witness in safety engring. Human Factors, 1978—; dir. safety projects office Va. Poly. Inst. and State U., Blacksburg, 1975; apptd. mem. U.S. Nuclear Waste Tech. Rev. Bd., 1989—, USA tech. adv. group Internat. Standards Tech. Com. 159 Ergonomics, 1987—; chmn. com. on transp. of hazardous materials Nat. Rsch. Coun., 1981-87, chmn. group 3 coun. emerging issues subcom. transp. rsch. bd., 1987-89; chmn. task force on pipeline safety Nat. Acad. Scis., 1986. Mem. editorial bd. Human Factors, Santa Monica, Calif., 1989—; author: (with K.B. Johns, J.W. Bain) Transportation of Hazardous Materials, 1983; contbr. chpts. to books, articles to profl. jours.; reviewer in field. Recipient Disting. Svc. award Nat. Rsch. Coun. NAS, 1987, 89, Outstanding Svc. commendation Transp. Rsch. Bd. NAS, 1981; grantee NIOSH, Va. Dept. Transp. and Safety, 1977-82, 86-87, IBM, 1981-84, USN Office of Naval Rsch., 1978-80, USN Naval Systems Weapons Command, 1978-79. Mem. Inst. Indsl. Engrs. (sr.), Am. Soc. Safety Engrs. (profl.), Human Factors Soc. (rep. to rev. panel Guideline for the Preparation of Material Safety Data Sheets), Systems Safety Soc. (Educator of Yr. 1993), Alpha Pi Mu. Avocation: flying. Home: 1011 Evergreen Way Blacksburg VA 24060-5366 Office: Va Poly Inst and State U Dept Indsl and Systems Engring 302 Whittemore Blacksburg VA 24061

PRICE, DONALD RAY, agricultural engineer, university administrator; b. Rockville, Ind., July 20, 1939; s. Ernest M. and Violet Noreen (Measel) P.; m. Joyce Ann Gerald, Sept. 14, 1963; children—John Allen, Karen Sue, Kimberly Ann, Daniel Lee. B.S. in Agrl. Engring., Purdue U., 1961, Ph.D. in Agrl. Engring., 1971; M.S. in Agrl. Engring., Cornell U., 1963. Registered profl. engr., Fla. From asst. prof. to prof. Cornell U., Ithaca, N.Y., 1962-80, dir. energy programs, 1975-77, 78-80; program mgr. Dept Energy, Washington, 1977-78, cons.; assoc. dean research U. Fla., Gainesville, 1980-83; dean Grad. Sch., U. Fla., Gainesville, 1983-84; v.p. research U. Fla., Gainesville, 1984—; pres. U. Fla. Research Found., Inc.; chmn. bd. dirs. Progress Research, Inc.; cons. to Pres. Carter, Washington, 1978; bd. dirs. Nat. Food and Engring. Council, Columbia, Mo., 1978-85, S.E. Healthcare Found., Gainesville, Fla., 1985. Contbr. numerous articles on engring. to profl. jours.; patentee mech. device. Mem. Ithaca Sch. Bd., N.Y., 1979-80; deacon Ch. of Christ, Gainesville, Fla., 1983—. Recipient citation Pres. Carter, 1979, Disting. Alumnus award Purdue U., 1990. Fellow Am. Soc. Agrl. Engrs. (dir. 1990, paper awards 1963, 77, 78, Young Engr. of Yr. award 1980); mem. Soc. Research Adminstrs., Nat. Assn. Univ. Research Adminstrs., S.E. Univ. Research Assn., Research Univs. Network. Democrat. Lodge: Rotary. Avocations: tennis, jogging, woodworking. Home: 8507 SW 5th Pl Gainesville FL 32607-1468 Office: U Fla 223 Grinter Hall Gainesville FL 32611

PRICE, DOUGLAS ARMSTRONG, chiropractor; b. Pitts., Feb. 17, 1950; s. Walter Coachman and Janet (Armstrong) P.; m. Ann Georgette Martino, Jan. 31, 1989; 2 children. BA, Brown U., 1972; D Chiropractic, Life Chiropractic Coll., Atlanta, 1983. Diplomate Am. Bd. Chiropractic Examiners; cert. rehab. doctor; life extension physician; independent medical examiner, Fla. Owner, chief exec. officer Athletic Attic-Westshore, Tampa, Fla., 1976-80, Applied Biomech. and Musculoskeletal Rehab., Tampa, 1989—, All Am. Chiropractic Clinic; pvt. practice Tampa, 1984—. Producer therapeutic exercise video for cervical and lumbar rehab.; contbr. articles to profl. jours. Magnetic Resonance Imaging fellow; named to Brown U. Athletic Hall of Fame; Southeastern Masters Champion Shotput, Discus, 1990-91. Fellow Am. Coll. Sports Medicine, Chiropractic Rehab. Assn., Am. Genotol. Assn.;mem. Am. Chiropractic Assn., Fla. Chiropractic Assn., Hillsborough County Chiropractic Soc. (bd. dirs. 1990-93, pres.1992-93). Democrat. Roman Catholic. Achievements include research in Russian stimulation applications in low back rehabilitation. Avocations: weightlifting, reading, coaching weight events in track and field. Home: 801 S Newport Tampa FL 33609 Office: 3421 W Saint Conrad St Tampa FL 33607

PRICE, ELIZABETH CAIN, environmental chemist; b. Seattle, Dec. 13, 1934; d. Russell Aaron and Miriam (Craig) Cain; m. Paul D. Price, Dec. 29 1956 (div. 1982); children: Catherine, Kenneth. BS in Chemistry, Duke U., 1956; MS in Environ. Engring., N.J. Inst. Tech., 1976. Chemist Pacquin Leeming Div. Chas. Pfizer, Parsippany, N.J., 1956-61; dir. Lakeland Labs., Wharton, N.J., 1971-78, Environ. Lab. CEE Dept. NJIT, Newark, 1978-84; adj. prof. N.J. Inst. Tech., Newark, 1979--; tech. dir. ICM Labs, RAndolph, N.J., 1984--; environ. cons. Price Engring., Lake Hopatcong, N.J., 1976-84; mem. Lake Hopatcong Regional Planning Bd., 1981--. Author: Biogas Production and Utilization, 1981; co-author: Pollution Control handbook, 1981, Biotechnology Handbook, 1984, Water and Sewage Works, 1980. Mem. Jefferson Twp. Environ. Com., Jefferson, N.J., 1987-90. Mem. Sigm Xi (assoc.). Home: 32 Old Fourth Dr Oak Ridge NJ 07438 Office: ICM Labs 1152 Rt 10 Randolph NJ 07869

PRICE, FREDERICK CLINTON, chemical engineer, editorial consultant; b. Ashland, Pa., Aug. 28, 1927; s. Edwin Warfield and Ruth Whilhemina (Strohmeier) P.; m. Dorothy Mae Williams, June 24, 1951 (div. Feb. 1978); children: Randall, Eric, Darren, Jason; m. Wilma Laura Cortines, May 6, 1978. BSChemE, Carnegie Inst. Tech., 1951. Chem. engr. Consolidation Coal Co., Library, Pa., 1951-60; mng. editor Chem. Engring. McGraw-Hill, N.Y.C., 1960-77; pub. rels. supr. KMG Internat., Houston, 1978; proposals mgr. King-Wilkinson Inc., Houston, 1978-80, CE Lummus Crest, Houston, 1980-86, ABB Lummus Crest, Houston, 1991—; cons. Price Cons. Svcs., The Woodlands, Tex., 1986-91. Editor: Report on Business and the Environment, 1972. With U.S. Army, 1945-46. Fellow Am. Inst. Chem. Engrs. Republican. Roman Catholic. Home: 96 W Rainbow Ridge The Woodlands TX 77381 Office: ABB Lummus Crest Inc 12241 Wickchester Houston TX 77079

PRICE, JAMES MELFORD, physician; b. Onalaska, Wis., Apr. 3, 1921; s. Carl Robert and Hazel (Halderson) P.; m. Ethelyn Doreen Lee, Oct. 23, 1943; children: Alta Lee, Jean Marie, Veda Michele. B.S. in Agr., U. Wis., 1943, M.S. in Biochemistry, 1944, Ph.D. in Physiology, 1949, M.D., 1951. Diplomate Am. Bd. Clin. Nutrition. Intern Cin. Gen. Hosp., 1951-52; mem. faculty U. Wis. Med. Sch., 1952—, prof. clin. oncology, 1959—, Am. Cancer Soc.-Charles S. Hayden Found. prof. surgery in cancer research, 1957—; on leave as dir. exptl. therapy Abbott Labs., 1967—, v.p. exptl. therapy, 1968, v.p. corp. research and exptl. therapy, 1971—, v.p. corp. sci. devel., 1976-78; v.p. med. affairs Norwich-Eaton Pharms., 1978—, v.p. internat. R&D, 1980-82; pres. RADAC Group, Inc., 1982-90, Biogest Products, Inc., 1984-88; mem. metabolism study sect. NIH 1959-62, pathology B study sect., 1964-68; sci. adv. com. PMA Found.; chmn. research adv. com. Ill. Dept. Mental Health; sci. com. Nat. Bladder Cancer program; mem. Drug Research Bd. Nat. Acad. Scis./NRC. Bd. dirs. Grandview Coll., Des Moines, 1977-78. Served with USNR, 1944-45. Fellow Am. Coll. Nutrition, Royal Soc. Medicine London; mem. Am. Soc. Pharmacology and Exptl. Therapeutics, Am. Assn. Cancer Research, Am. Cancer Soc. (com. etiology 1957-61), Pharm. Mfrs. Assn. (chmn. research and devel. sect. 1974-75), Am. Soc. Biol. Chemists, Am. Inst. Nutrition, Am. Soc. Clin. Nutrition, Research Dirs. Assn. Chgo., Soc. Exptl. Biology and Medicine, Soc. Toxicology. Spl. research trytophan metabolism, metabolism vitamin B complex, chem. carcinogenesis; research and devel. pharm., diagnostic and consumer products; licensing and bus. devel. Home: PO Box 211 Edmeston NY 13335-0211

PRICE, JEFFREY BRIAN, quality engineer; b. Waynesville, N.C., May 30, 1963; s. Jack and Mollie Lee (Parton) P.; m. Cora Lee Messer, Dec. 17, 1983; 1 child, Jennifer Michelle. BS, Western Carolina U., 1987. Quality engr. DAY Internat., Arden, N.C., 1988--. Mem. Am. Soc. for Quality Control, Am. Inst. for Aeronautics and Astronautics. Methodist. Home: 566 Camp Branch Rd Waynesville NC 28786 Office: Day Interant 95 Glenn Bridge Rd Arden NC 28786

PRICE, JONATHAN G., geologist; b. Danville, Pa., Feb. 1, 1950; s. A. Barney and Flora (Best) P.; m. Elisabeth McKinley, June 3, 1972; children: Alexander D., Argenta M. BA in Geology and German, Lehigh U., 1972; MA, U. Calif., Berkeley, 1975, PhD, 1977. Cert. profl. geologist. Geologist Anaconda Copper Co., Yerington, Nev., 1974-75; geologist U.S. Steel Corp., Salt Lake City, 1977, Corpus Christi, Tex., 1978-81; rsch. assoc. Bur. Econ. Geology, U. Tex., Austin, 1981-85, rsch. sci., 1984-88, program dir., 1987-88; dir. Tex. Mining & Mineral Resources Rsch. Inst., Austin, 1984-88; dir., state geologist Nev. Bur. Mines & Geology, U. Nev., Reno, 1988-92; staff dir. Bd. on Earth Scis. & Resources Nat. Rsch. Coun., Washington D.C., 1993—; asst. prof. Bucknell U., Lewisburg, 1977-78. Author, editor: Igneous Geology of Trans-Pecos Texas, 1986. Vol. instr. CPR and 1st aid ARC, 1983—, bd. dirs. U. Heidelberg, Sierra Nev. chpt., 1991-92. German Acad. Exchange Svc. fellow, 1972-73. Fellow Geol. Soc. Am., Soc. Econ. Geologists; mem. Am. Assn. Profl. Geologists (Nev. sect. pres. 1992), Mineral. Soc. Am., Phi Beta Kappa. Office: Nat Rsch Coun BESR-HA372 2001 Wisconsin Ave NW Washington DC 20007

PRICE, KAREN OVERSTREET, pharmacist, medical editor; b. South Boston, Va., Oct. 28, 1964; d. Alvin Keith and Catherine Coggin (Marshall) Overstreet; m. David McRoy Price, June 18, 1988. BS in Pharmacy, U. N.C., 1987; MS in Drug Info., L.I. U., 1990. Pharmacist Eckerd, Burlington, N.C., 1988; Lasdon Rsch. fellow Internat. Drug Info. Ctr., Bklyn., 1988-90; drug info. analyst Am. Soc. Hosp. Pharmacists, Bethesda, Md., 1990-91; med. editor Adverceutics, Inc., Laurel, Md., 1991-93; dir. Meniscus Ednl. Inst., Phila., 1993—. Co-author: Athletic Drug Reference, 1991; co-author, co-editor (computer program) Athletic Drug Reference, 1992. Recipient Upjohn award for excellence in rsch. Upjohn Pharms., 1990. Mem. Am. Pharm. Assn. (reviewer 1991—), Am. Med. Writers Assn. (manuscript editor 1991—), Am. Soc. Hosp. Pharmacists, Drug Info. Assn., Rho Chi Honor Soc. Office: Meniscus Ltd Ste 200 107 N 22d St Philadelphia PA 19103

PRICE, LAWRENCE H(OWARD), psychiatrist, researcher, educator; b. Detroit, July 26, 1952; s. Philip and Shirley Price; m. Ann Back. BS, U. Mich., 1974, MD, 1978. Intern Norwalk (Conn.) Hosp., 1978-79; resident Dept. Psychiatry, Yale U. Sch. Med., New Haven, 1979-82; instr. dept. psychiatry Yale U. Sch. Medicine, New Haven, 1982-83, asst. prof., 1983-89, assoc. prof., 1989—. Recipient Clin. Investigator award NIMH, 1985-88. Mem. Soc. for Neurosci., AAAS, Am. Coll. Neuropsychopharmacology. Office: Yale U Sch Medicine Dept Psychiatry 34 Park St New Haven CT 06519

PRICE, RICHARD GEORGE, electrical engineer; b. Chgo., Mar. 12, 1950; s. Edward Walter and Agnes (Grabske) P.; m. Diane Kay Price, Aug. 30, 1986. BSEE, Ill. Inst. Tech., 1971; MBA, U. Chgo., 1976. Elec. engr. Rockwell Internat., Cicero, Ill., 1971-74; sr. project engr. Johnson and Johnson Baby Prodn., Palos Heights, Ill., 1974-81; mgr. ride systems Walt Disney Imagineering, Bubank, Calif., 1981-82; mgr. elec. engring. Johnson and Johnson Baby Products, Skillman, N.J., 1982-85; sr. cons. engr. Automation Concepts Inc., Burr Ridge, Ill., 1985-88; pres., CEO Automation Horizons Inc., Des Plaines, Ill., 1988--; advisor NBS Map Com., Washington, 1984-85. Mem. IEEE, Instrument Soc. Am., Engring. Soc. Detroit, Robotics Internat. Office: Automtion Horizons Inc 2200 E Devon Des Plaines IL 60018

PRICE, ROBERT A., mechanical engineer; b. Chgo., July 12, 1924; s. Alfred F. and Marie G. (Coleman) P.; m. Jananne Sivill, Sept. 12, 1947; children: Denise, Todd, Alan. BSME, Northwestern U., 1948; MBA, Mich. State U., 1964. Registered profl. engr., Mich., Ind. Plant engr. U.S. Rubber Co., Mishawaka, Ind., 1948-50; design engr. Bendix Automotive, South Bend, Ind., 1950-51; engring. specialist Douglas Aircraft Co., Santa Monica, Calif., 1951-59; engring. supr. Bendix Aerospace, South Bend, 1959-62; engring. and sales specialist Bendix Energy Controls, South Bend, 1963-70; project engr. Sullair Corp., Michigan City, Ind., 1970-74; v.p. ops. Pure Aire Corp. subs. Sullair Corp., Charlotte, N.C., 1974-76; sr. engr. Sullair Corp., Michigan City, 1976-89; co. rep. CAGI Pneurop, Michigan City, 1987-90. Mem. Sch. Bd., River Valley, Mich., 1968; pres. Birchwood Assn. (Home Owner's Assn.)A Harbert, Mich., 1964-88. Mem. Elks, Beta Gamma Sigma, Phi Kappa Psi. Achievements include patents for helicopter rotor brake, for mono-rail brake; design of DC-8 Landing Gear-Ford T-Bird seperate unit power steering system many rothry screw compressor packages and compressed air dryer systems. Home: Box 324 Lakeside MI 49116

PRICE, ROBERT EDMUNDS, civil engineer; b. Lyndhurst, N.J., Jan. 8, 1926; s. William Evans and Charlotte Ann (Dyson) P.; B.S. in Civil Engring., Dartmouth Coll., 1946; M.S., Princeton U. 1947; m. Margaret Akerman Menard, June 28, 1947; children—Robert Edmunds, Alexander Menard. Mgr., P&S Standard Vacuum Oil Co., N.Y., London and Sumatra, 1947-55; project engr. Metcalf & Eddy, Cons. Engrs., Boston, 1956-59; structural engr. Lummis Co., Cons. Engrs., Newark, 1960-61; mgr. engring. materials Interpace Corp., Wharton, N.J., 1961-78; pres. Openaka Corp., Denville, N.J., 1979—; cons. cement and concrete design and constrn. Mem. Denville Bd. Health, 1963-66, chmn., 1966; mem. Denville Bd. Adjustment, 1966-69. Served with USNR, 1943-46. Registered profl. engr., N.J., Md. Fellow Am. Concrete Inst. (dir. 1981-84); mem. ASTM (chmn. subcom. spl. cements 1976-84, hon. mem. com. C-1), Nat. Assn. Corrosion Engrs. Episcopalian. Home: Lake Openaka Denville NJ 07834 Office: Openaka Corp 565 Openaki Rd Denville NJ 07834-9642

PRICE, RONALD FRANKLIN, air transportation executive; b. Quito, Ecuador, Oct. 10, 1948; came to U.S., 1951; s. Alexander and Ingeborg Teresa (Majewski) P.; m. Deborah Linda Mazen, July 16, 1970; children: Karen Rachel, Elisha Helen. BSME, Newark Coll. Engring., 1969; MBA, Rutgers, 1973. Registered profl. engr., N.J., Va., Conn., Fla. Engr. M. W. Kellogg Co., Plainfield, N.J., 1969-72; sr. assoc. Arnold Thompson Assocs., Inc., White Plains, N.Y., 1972-78; mgr. aviation projects CH2M Hill, Reston, Va., 1978-87; sr. v.p. Thompson Cons. Internat., Inc., Briarcliff Manor, N.Y., 1987-92, pres., 1992—. Bd. dirs. Com. for Dulles Internat. Airport, Washington, 1989-91. Mem. ASCE (exec. com. air transport divsn. 1984-88, chmn. 1988-89, Cert. Appreciation 1989), Airport Cons. Coun. (bd. dirs. 1990-92). Office: Thompson Cons Internat Inc 575 N State Rd Briarcliff Manor NY 10510

PRICE, THOMAS RANSONE, neurologist, educator; b. Hampton, Va., July 31, 1934; s. William Spencer and Virginia (Ransone) P.; m. Nancy Worrell Franklin, June 28, 1958; children: Franklin Ransone, Catherine Blair. BA in Psychology, U. Va., 1956, MD, 1960. Diplomate Am. Bd. Psychiatry and Neurology. Chief resident in neurology U. Va. Hosp., Charlottesville, 1965-66; instr. U. Va. Hosp., 1966-67, U. Md. Sch. Medicine, Balt., 1967-68; asst. prof. U. Md. Sch. Medicine, 1969-72, assoc. prof., 1972-78, prof., 1978—. Editor: Cerebro Vascular Diseases, 1979; contbr. articles to profl. jours. Mem. Govs. Task Force on Alzheimers Disease, Md., 1984-88. Capt. USAF, 1961-63. Fellow Am. Neurological Assn., Am. Acad. Neurology (chmn. bylaws com. 1978-79, 83-85), Am. Heart Assn. (chmn. stroke sub-com. 1986-87); mem. Md. Neurologic Soc. (pres. 1979-80). Office: U Md Hosp 22 S Green St Baltimore MD 21201

PRICE, WILLIAM ANTHONY, psychiatrist; b. Youngstown, Ohio, Aug. 15, 1959; s. Edward J. and Margaret (Krispli) P.; divorced; children: Matthew, Nicole; m. Pamela R. Gardner, Nov. 18, 1985; 1 child, Andrew A. BS, Kent State U., 1983; MD, Northeastern Ohio U., 1983. Diplomate Am. Bd. Psychiatry and Neurology, Nat. Bd. Med. Examiners; lic. psychiatrist, Ohio, Pa. Intern U. Health Ctr., Pitts., 1983-85; resident in psychiatry Northeastern Ohio U., 1985-86, chief resident, clin. assoc. prof., 1986-87; psychiatrist Specialty Care Psychiat. Svcs., Boardman, Ohio, 1987—; med. dir. PsyCare, Boardman; med. dir. Premenstrual Syndrome Program, Parkview Counseling Ctr., Youngstown, 1985-88, Child and Adolescent Diagnostic and Devel. Ctr., Youngstown, 1988—, Psycare; assoc. med. dir. psychiatry Windsor Hosp., Chagrin Falls, 1989—; clin. asst. prof., Northeastern Ohio U., 1987—; chief clin. officer, Mahoning County Mental Health Bd, 1990—; clerkship dir. psychiatry, St. Elizabeth Hosp. Med. Ctr., Youngstown, 1987—; Western Res. Care System, Youngstown, 1988—; reviewer Jour. Clin. Psychiatry, 1987—, Am. Jour. Psychiatry, 1987—, Psychosomatics, 1987—, lectr. various confs., forums, seminars, 1985—. Author: (with others) Opiate Addiction in the Biological Foundations of Clinical Psychiatry, 1986, Nootropics: Toward the Mind in the Biological Foundations of Clinical Psychiatry, 1986, (audiocassettes) Mitral Valve Prolapse and Bipolar Affective Disorder, 1985, Dealing with PCP, 1985; contbr. numerous articles to profl. publs., poems and 2 short stories to nat. mags. Recipient Founders Day award for Sci. Rsch., Ohio Psychiat. Assn. Edn. and Rsch. Found., 1986, hon. mention Lebenson award, Am. Assn. Gen. Hosp. Psychiatrists, 1987, Founder's award, Am. Assn. Psychiatrists in Alcoholism and Addictions, 1987, fellowship award, Assn. Acad. Psychiatrists, 1987; Laughlin fellow, Am. Coll. Psychiatrists. Mem. Am. Psychiat. Assn., Am. Acad. Clin. Psychiatry, Am. Soc. Clin. Pharmacology, Am. Assn. Psychiatrists in Alcoholism and Addiction, Am. Gen. Hosp. Psychiatrists, Am. Soc. Psychosomatic Ob-Gyn., Am. Coll. Psychiatrists, Internat. Soc. Psychosomatic Ob-Gyn., Nat. Fedn. Ind. Bus., Assn. Acad. Psychiatrists, Cen. Neuropsychiatric Assn., Soc. Neurosciences, Soc. Menstrual Cycle Rsch., Ohio State Med. Assn., Ohio Psychiat. Assn., Mahoning County Med. Assn., Mahoning County Mental Health Bd., Coun. Chiefs Psychiatry, Parents Supporting Parents (adv. bd. dirs.), Youngstown, Phi Delta Epsilon. Avocations: gourmet cooking, wine tasting, art collecting, skiing. Office: Psycare 843 Boardman Canfield Rd Boardman OH 44512-4235

PRICE, ZANE HERBERT, cell biologist, research microscopist; b. Dunbar, Nebr., June 17, 1922; s. Frank Robert and Carrie Helen (Benner) P.; m. Maxine May Fitzgerald, Sept. 26, 1943; 1 child, Peggy May. BA, Walla Walla Coll., 1950; postgrad., UCLA, 1951-54. Grad. researcher in microbiology UCLA Med. Sch., 1953-56, jr. rsch. microbiologist, 1956-59, asst. rsch. microbiologist, 1959-62, assoc. specialist, 1962-67, specialist, 1967-89, emeritus, 1989—; dir. Nina Anderton Lab. Electron Microscopy, UCLA Med. Sch., 1957-60; vis. scientist Inst. Poly. Nat., Mexico City, 1967-70. Contbr. numerous articles to profl. jours.; producer teaching films. Cpl. USMC, 1942-45. Recipient Electronic Eye award Biol. Photo Assn., 1959; named Alumnus of Yr. Walla Walla Coll., 1981. Fellow Royal Microscopical Soc. (Eng.), Los Angeles County Heart Assn.; mem. AAAS, Am. Microscopical Soc., So. Calif. Soc. Electron Microscopy (pres. 1971-72), N.Y. Acad. Sci., L.A. Microscopical Soc. Home: 12424 Everglade St Los Angeles CA 90066-1816

PRICHARD, ROBERT ALEXANDER, JR., telecommunications engineer; b. Paris, Tex., Feb. 23, 1953; s. Robert Alexander and Reba Marie (Fields) P.; m. Debra Ruth Holbrooks, Apr. 9, 1977; children: Robert Ross, Christopher Dean. BS, Trinity U., San Antonio, Tex., 1975; cert. communications engring., Capitol Radio Engring. Inst., Washington, 1978; MS in Telecommunications, Denver U., 1989. Owner Stamford (Tex.) Communications, 1979-85; sr. telecommunications engr. Martin-Marietta Data Systems, Denver, 1985-86, mgr. plans and analysis, 1986-87; supr. telecommunications engring. Pub. Svc. Co. Colo., Denver, 1987—; mgr. Telecom Engring., Denver, 1988, Telecom Network Planning, Denver, 1990-92; mgr. systems engring. Nova-Net Communications, Inc., Englewood, Colo., 1992—; gen. mgr. Jones Lighware of Denver, Inc., Jones Lightwave Ltd., 1992—; cons. broadcast engr. Sta. KDWT Radio, Stamford, 1980, 86; chief sound engr. U. hills Bapt. Ch., Denver, 1987; adj. prof. Denver U., Denver. Contbr. articles to profl. publs. Adjunct tchr. Denver Jr. Achievement, 1986. 1st lt. U.S. Army, 1975-79. Named one of Outstanding Young Men Am., 1985; recipient Internat. Leaders in Achievement award Internat. Biog. Ctr., Eng., Disting. Leadership award Internat. Dir. Am. Biog. Inst. Mem. IEEE, Nat. Assn. Radio and Telecommunications Engrs. (cert.), Internat. Platform Assn., Alpha Chi. Republican. Baptist. Avocations: home computers, martial arts. Home: 7506 E Long Ave Englewood CO 80112 Office: Jones Lightwave Denver Inc 9697 E Mineral Ave Englewood CO 80112

PRICHARD, ROGER KINGSLEY, university administrator, dean, educator; b. Melbourne, Victoria, Australia, Oct. 24, 1944; arrived in Can., 1984; s. Allan Joseph and Lorraine Anne (Larman) P.; m. Lynette Bacon, May 20, 1967; children: Rachael Simone, Julienne Belinda, Cressida Justine. BSc with honors, U. New South Wales, Sydney, Australia, 1966, PhD, 1969. Merck, Sharp & Dohme Postgrad. fellow U. New South Wales, 1966-69; from rsch. to sr. rsch. scientist Commonwealth Sci. & Indsl. Rsch. Orgn., Sydney, 1969-79; prin. rsch. scientist Commonwealth Sci. & Indsl. Rsch. Orgn., 1979-84, leader program parasite biochemistry and chemotherapy, 1980-84; prof. parasitology McGill U., Montreal, Que., Can., 1984—, assoc. prof. medicine, 1985—, prof. animal sci., 1985—, dir. inst. parasitology, 1984-90, dean grad. sch., 1990—, vice-prin. rsch., 1990—; vis. fellow inst. parasitology U. Zurich, Switzerland, 1980; assoc. dept. biochemistry U. Sydney, 1980-84; dir. Nat. Reference Ctr. Parasitology, Health and Welfare, Can., 1984-90; consulting specialist Montreal Gen. Hosp., 1985—; cons. Schering Plough, Kenkworth, N.J., 1986-91; vis. lectr. tropical medicine Nat. Def. Med. Ctr., Ottawa, 1986-88; vis. prof. chemotherapy sch. pharmacy U. Montreal, 1988; mem. Com. Advise Tropical Medicine and Travel, Health and Welfare, Can., 1990-91; mem. adv. bd. Montreal Neurol. Inst., 1990—; expert cons. food and agr. orgn. UN, 1991—; mem. task force tech. transfer Natural Scis. & Engring. Rsch. Coun. Can., 1992—; bd. govs. rsch. inst. Montreal Children's Hosp.; bd. dirs. Pulp & Paper Rsch. Inst. Can., CITEC Biotechnology, Montreal; grant reviewer various orgns.; mem. various coms. McGill U.; presenter in field. Mem. editorial bd. Molecular and Biochemical Parasitology, 1979-82, Internat. Jour. Parasitology, 1982-85; contbr. 14 chpts. to books and over 100 articles to profl. jours. Bd. dirs. Montreal Consortium Human Rights Advocacy Tng., 1992—. Grantee Natural Scis. & Engring. Rsch. Coun., 1984—, Fonds FCAR, 1984—, Med. Rsch. Coun., 1987-93; R. W. Griffith Meml. fellow U. Coll. Wales, 1973-75, Roche Found. fellow U. Zurich, 1980—; Wool Rsch. Trust Fund scholar U. New South Wales, 1962-66. Mem. Am. Soc. Parasitology, Am. Soc. Tropical Medicine and Hygiene, Am. Assn. Vet. Parasitologists (v.p. 1988-89, pres.-elect 1989-90, pres. 1990-91, past. pres. 1991-92), Can. Assn. Grad. Schs. (bd. dirs. 1992—), Can. Soc. Immunology, Can. Soc. Internat. Health (chmn. divsn. tropical medicine 1989-90, dir. 1989-90), Can. Soc. Zoologists, Can. Assn. Univ. Rsch. Adminstrs. (pres.-elect 1993—), Brit. Soc. Parasitology, N.Y. Acad. Sci., Assn. Adminstrs. Univ. Rsch. Que. (pres. 1992-93), World Assn. Advancement Vet. Parasitology (dir. 1989—, bd. dirs. 1989—). Roman Catholic. Achievements include patents for Potentiation of Anthelmintics, Characterization and Usage of Monoclonal Antibodies against Nematode Tubulin, Use of Peptides and Vaccine Composition; research in biochemical pharmacology of antiparasite drugs, parasite biochemistry, molecular biology and molecular immunology, elucidation of mode of action of benzimidazole anthelmintics and mechanism of drug-resistance to benzimidazole antiparasitic drug, anthelmintic pharmacokinetics, parasite control for animals and tropical diseases. Home: 597 Lakeshore, Beaconsfield, PQ Canada H9W 4K5 Office: McGill U, 853 Sherbrooke St W Rm 308, Montreal, PQ Canada H3A 2T6

PRICKETT, DAVID CLINTON, physician; b. Fairmont, W.Va., Nov. 26, 1918; s. Clinton Evert and Mary Anna (Gottschalk) P.; m. Mary Ellen Holt, June 29, 1940; children: David C., Rebecca Ellen, William Radcliffe, Mary Anne, James Thomas, Sara Elizabeth; m. Pamela S. Blackstone, Nov. 17, 1991. AB, W.Va. U., 1944; MD, U. Louisville, 1946; MPH, U. Pitts., 1955. Lab. asst., instr. in chemistry, W.Va. U., 1943; intern, Louisville Gen. Hosp., 1947; surg. resident St. Joseph's Hosp., Parkersburg, W.Va., 1948-49; gen. practice, 1949-50, 55-61; physician USAF, N.Mex., 1961-62, U.S. Army, Calif., 1963-64, San Luis Obispo County Hosp., 1965-66, So. Calif. Edison Co., 1981-84; assoc. physician indsl. and gen. practice Los Angeles County, Calif., 1967—; med. dir. S. Gate plant GM, 1969-71; physician staff City of L.A., 1971-76; relief med. practice Applachia summer seasons, 1977, 1986, 1988-93. Med. Officer USPHS, Navajo Indian Reservation, Tohatchi (N.Mex.) Health Ctr., 1953-55, surgeon, res. officer, 1957-59; pres. W.Va. Pub. Health Assn., 1951-52, health officer, 1951-53, sec. indsl. and pub. health sect. W.Va. Med. Assn., 1956. Author: The Newer Epidemiology, 1962, rev., 1990, Public Health, A Science Resolvable by Mathematics, 1965. Served to 2d lt. AUS, 1943-46. Dr. Thomas Parran fellow U. Pitts. Sch. Pub. Health, 1955; named to Hon. Order Ky. Cols. Mem. Am. Occupational Med. Assn., Western Occupational Med. Assn., Am. Med. Assn., Calif. Med. Assn., L.A. County Med. Assn., Am. Acad. Family Physicians, Phi Chi. Address: PO Box 4032 Whittier CA 90607

PRICKETT, GORDON ODIN, mining, mineral and energy engineer; b. Morris, Minn., Nov. 26, 1935; s. Glenn Irvin and Edna Margaret (Erickson) P.; m. Jean Carolyn Strobush, Oct. 8, 1958; children: Karen Joan Keating, Laura Jean, Glenn Thomas. B Mining Engring., U. Minn., 1958, MS in Mineral Engring. and Econs., 1965. Registered profl. engr., Mo., Ill. U.S. Steel fellow U. Minn., Mpls., 1963-65; rsch. mineral engr. Internat. Minerals & Chem. Corp., Skokie, Ill., 1965-68; mgmt. sci. cons. Computer Mgmt. Cons., Northfield, Ill., 1968-71; mgr. tech. systems Duval Corp., Tucson, Ariz., 1971-77; dir. mgmt. info. systems Arch Mineral Corp., St. Louis, 1977-78; supr. mine planning projects Peabody Coal Co., St. Louis, 1978-82; mgr. elec. tech. transfer, nuclear plant simulator, rsch. Union Electric Co., St. Louis, 1983—; presenter papers at industry confs. Contbr. articles to profl. jours. Co-founder, chmn. Lake Forest - Lake Bluff (Ill.) Com. for Equal Opportunity, 1968-71; com. Confluence St. Louis, 1987—; bd. dirs., officer ch. bds., polit. twp. orgn., Lake Forest, Tucson, St. Louis, 1968—. Lt. USN, 1958-63; to comdr. USNR, 1963-79. Mem. AIME (chair programs 1958—), Assn. Quality and Participation (chair programs 1986-90), Norwegian Soc. St. Louis, Engrs. Club St. Louis (chair affiliated socs. 1987-88). Avocations: running, photography, Norwegian Singing Club, skiing. Home: 125 W Cedar Ave Saint Louis MO 63119-2905 Office: Union Electric Co 1901 Chouteau Ave Saint Louis MO 63103-3085

PRIDHAM, THOMAS GRENVILLE, research microbiologist; b. Chgo., Oct. 10, 1920; s. Grenville and Gladys Etheral (Sloss) P.; m. Phyllis Sue Hokamp, July 1, 1943; children: Pamela Sue, Thomas Foster, Grenville Thomas, Rolf Thomas, Montgomery Thomas. BS in Chemistry, U. Ill., 1943, PhD in Bacteriology, 1949. Instr. bacteriology U. Ill., Champaign-Urbana, 1947; rsch. microbiologist No. Regional Rsch. Lab., USDA, Peoria, Ill., 1948-51, 53-65, U.S. Indsl. Chems., Balt., 1951-52; supr. tech. ops. Acme Vitamins, Inc., Joliet, Ill., 1952-53; sr. rsch. biologist U.S. Borax Rsch. Corp., Anaheim, Calif., 1965-67; supervisory rsch. microbiologist No. Regional Rsch. Ctr., No. Regional Rsch. Ctr., USDA, Peoria, 1967-81; head agrl. rsch. culture collection No. Regional Rsch. Lab. USDA, Peoria, 1967-81; ret., 1981; cons. Mycogen Corp., San Diego, 1985-87; U.S. sr. scientist Fed. Republic Germany, Darmstadt, 1977. Contbg. author: Actinomycetales: The Boundary Microorganisms, 1974, Bergey's Manual of Determinative Bacteriology, 1974, Synopsis and Classification of Living Organisms, 1982; mem. editorial bd. Jour. Antibiotics, 1969-81; contbr. articles to Jour. Bacteriology, Applied Microbiology, Phytopathology, Actinomycetes, Mycologia, Devel. Indsl. Microbiology, Jour. Antibiotics, Internat. Bull. Bacteriological Nomenclature Taxonomy, Antibiotics Ann., Antimicrobial Agts., Chemotherapy, also others. With USNR, 1943-45, with Rsch. Res., 1945-54, lt. ret. Fulbright scholar, Italy, 1952; grantee Soc. Am. Bacteriologists, 1957. Fellow Am. Acad. Microbiology (ASM state network 1991—); mem. Am. Soc. Microbiology (com. mem., workshop presenter), Soc. Indsl. Microbiology, Mycol. Soc. Am., U.S. Fedn. Culture Collections (v.p. 1981). Episcopalian. Achievements include patents in fermentative production of riboflavin and of antibiotics; research in microbial culture collection technology and management, systematics of streptomycetes, industrial microbiology, and air pollution. Home: 980 Looking Glass Ln Las Vegas NV 89110-2711

PRIEBE, STEFAN, psychiatrist; b. Berlin, Oct. 24, 1953; s. Karl Heinz and Barbara (Schippert) P.; 1 child, Roman. Abitur, Canisius-Kolleg, Berlin, 1973; Dipl.-Psychologist, U. Hamburg, 1978, MD, 1980; student, Free U. Berlin, 1991—. Rsch. fellow Middlesex Hosp., London, 1981, Neurol. Dept., U. Hamburg, 1981-82; asst. Albertinen Hosp., Hamburg, 1982-83; rsch. fellow dept clin. psychiatry Free U. Berlin, 1983-88, dep. head dept. social psychiatry, 1988-91, acting head dept. social psychiatry, 1991—; rschr. Med. Rsch. Inst., Palo Alto, Calif., 1983, Social Psychiat. U. Hosp., Berne, Switzerland, 1986, Med. Rsch. Coun., Social Psychiat. Unit, London, 1985. Editor and author several books; contbr. articles to profl. jours. Mem. Internat. Assn. for Emergency Psychiatry, Soc. for Psychotherapy Rsch., Deutsche Gesellschaft für Psychiat. und Nervenheilkunde, Deutscher Hfuer Fachverband Verhaltenstherapie, Platane 19 e.V (pres.), Dahlemer Tennis Club. Avocations: tennis, antiques, music (saxophone and clarinet). Office: Free U, Platanenallee 19, 14050 Berlin Germany

PRIEST, MELVILLE STANTON, consulting hydraulic engineer; b. Cassville, Mo., Oct. 16, 1912; s. William Tolliver and Mildred Alice (Messer) P.; m. Vivian Willingham, Mar. 22, 1941 (dec.); m. Virginia Young, Dec. 16, 1983. B.S., U. Mo., 1935; M.S., U. Colo., 1943; Ph.D., U. Mich., 1954. Registered profl. engr., Ala., La., Miss. Jr. engr. U.S. Engrs. Office, 1937-39; from jr. to asst. engr. Bur. Reclamation, 1939-41; from instr. to assoc. prof. civil engring. Cornell U., 1941-55; prof. hydraulics Auburn (Ala.) U., 1955-58, prof. civil engring., head dept., 1958-65; dir. Water Resources Research Inst., Miss. State U., 1965-77; UN adviser on hydraulics, Egypt, 1956, 57, 60; Mem. Ala. Bd. Registration Profl. Engrs., 1962-65. Contbr. articles to profl. jours. Fellow ASCE (pres. Ala. 1962, exec. com., pipeline div. 1971-74), Am. Water Resources Assn. (dir. 1973-75), Sigma Xi, Tau Beta Pi, Chi Epsilon, Pi Mu Epsilon. Address: PO Box 541 Starkville MS 39759

PRIESTER, GAYLE BOLLER, engineer, consultant; b. Mpls., July 1, 1912; s. George Charles and Lulu May (Boller) P.; m. Rachel Edith Miller, Sept. 3, 1938; children: Peggy Lu, George William, Phyllis Ann. B in Mech. Engring., U. Minn., 1933; MS, Harvard U., 1934; Mech. Engr., Case Inst. Tech., 1943. Registered profl. engr., Md. Industry rep. Mpls. Gas Light Co., 1934-35; application sales engr. Carrier Corp., N.Y.C., Chgo., Mpls., 1935-41; instr., asst. prof., assoc. prof. Case Inst. Tech., Cleveland, 1941-46; air conditioning engr. Balt. Gas & Electric Co., Balt., 1946-73, chief civil engr., 1973-77; pvt. practice cons. engr. Balt., 1977-89, Advance, N.C., 1989—; dir. Accreditation Bd. for Engring. and Technology, 1976-82. Coauthor: (textbook) Refrigeration & Air Conditioning, 1948, rev., 1956. Sec. Stoneleigh Community Assn., Balt. County, 1947-48, Boy Scout Troop Com., 1952-54; pres. Stoneleigh Elem. Sch. PTA, Balt. County, 1952-53; maj. United Appeal Campaign, Balt., 1953; bd. dirs. Alzheimer's Triad N.C. chpt., 1990—, treas., 1992—. Recipient Disting. Svc. award ASHRAE (Balt. chpt.), 1972. Fellow ASHRAE (chpt. pres. 1946, 50, 53, regional chmn. 1965-67, treas. 1970-71, Disting. Svc. award 1968, Disting. 50 Yr. Svc. award 1984); mem. Engring. Soc. Balt. (bd. dirs. 1951-52), Tau Beta Pi. Republican. Presbyterian. Club: Bermuda Run.

PRIESTLEY, G. T. ERIC, manufacturing company executive; b. Belfast, Northern Ireland, May 7, 1942; came to U.S., 1990; s. Thomas John McKee P.; m. Carol Elizabeth Gingles Nelson, June 8, 1966; children: Peter, Gaye, Simon. BS, Queens U., 1963; postgrad., Harvard Bus. Sch., 1989. Sales trainee Burroughs Machines Ltd., 1963-64; dealer, sales devel. Regent Oil Co., 1964-66; ops. mgr. RMC (Ulster) Ltd., 1967-70; distbn. mgr. Bass Charrington, Ireland, 1970-71; dir., gen. mgr. Farrans Ltd., 1971-80; dir., CEO Redland plc/British Fuels/Cawoods, 1980-88; dir. Bowater plc, London, 1988-90; pres., chief exec. officer Rexham Inc., Charlotte, N.C., 1990—; bd. dirs. Rexham Inc., Bowater plc, MiTek Inc., St. Louis, Interprint Formularios Ltd, Sao Paulo, Brazil. Mem. Moortown Golf Club, Aloha Golf Club, Royal Ulster Yacht Club, Quail Hollow Golf & Country Club. Home: 8520 Greencastle Dr Charlotte NC 28210 Office: Rexham Inc 4201 Congress St Ste 340 Charlotte NC 28209

PRIGOGINE, VICOMTE ILYA, physics educator; b. Moscow, Russia, Jan. 25, 1917; s. Roman and Julie (Wichmann) P.; m. Marina Prokopowicz, Feb. 25, 1961; children: Yves, Pascal. PhD, Free U. Brussels, 1942; hon. degree, U. Newcastle, Eng., 1966, U. Poitiers, France, 1966, U. Chgo., 1969, U. Bordeaux, France, 1972, U. de Liège, Belgium, 1977, U. Uppsala, Sweden, 1977, U de Droit, D'Economie et des Scis., d'Aix-Marseille, France, 1979, U. Georgetown, 1980, U. Cracovie, Poland, 1981, U. Rio de Janeiro, 1981, Stevens Inst. Tech., Hoboken, 1981, Heriot-Watt U., Scotland, 1985, Universidad Nacional de Educacion a Distancia, Madrid, 1985, U. Francois Rabelais de Tours, 1986, U. Peking, People's Republic of China, 1986, U. Buenos Aires, 1989, U. Cagliari, Sardinia, Italy, 1990, U. Sienne, Italy, 1990; DS (hon.), Gustavus Adolphus Coll., 1990; Membre d'Honneur, l'Academie Nationale d'Argenti, 1990, l'Academie des Sciences Nature, lles de Republique Federale de Russie, 1991; President d'Honneur, l'Academie Nationale des Sciences de Republique de San Marino, 1991; Membre d'Honneur, l'Academie Chilienne des Scis., 1991, de l'Université de Nice-Sophia-Antipolis, Nice, France, 1991, de l'Univ. Philippines System, Quezon City, 1991, del'Université de Santiago, Chile, del'Université de Tucumán, Argentine, 1991; Doctuer Monous, Caula de l' Université lo Monosov de Moscow, 1993; Docteur Honoris Causa de, L'Universite Lomonosov de Moscow, Russie, 1993. Prof. U. Brussels, 1947—; dir. Internat. Insts. Physics and Chemistry, Solvay, Belgium, 1959—; dir. Ilya Prigogine Ctr. for Studies in Statis. Mechanics, Thermodynamics and Complex Systems, U. Tex., Austin, 1967—; hon. prof. U. Nankin, People's Republic of China, 1986, Banaras Hindu U., Varasani, India, 1988; Ashbel Smith regental prof. U. Tex., Austin, 1984—. Author: (with R. Defay) Traite de Thermodynamique, conformement aux methodes de Gibbs et de De Donder, 1944, 50, Etude Thermodynamique des Phenomenes Irreversibles, 1947, Introduction to Thermodynamics of Irreversible Processes, 1954, 62, 67 (with A. Bellemans, V. Mathot) The Molecular Theory of Solutions, 1957, Statistical Mechanics of Irreversible Processes, 1954, 2nd edit. 1962, 3rd edit. 1967, (with others) Non Equilibrium Thermodynamics, Variational Techniques and Stability, 1966, (with R. Herman) Kinetic Theory of Vehicular Traffic, 1971, (with R. Glansdorff) Thermodynamic Theory of Structure, Stability and Fluctuations, 1971, (with G. Nicolis) Self-Organization in Nonequilibrium Systems, 1977, From Being to Becoming-Time and Complexity in Physical Sciences, 1979, French, German, Japanese, Russian, Chinese and Italian edits., (with I. Stengers), Order Out of Chaos, 1983, La Nouvelle Alliance, Les Métamorphoses de la Science, 1979, German, English, Italian, Spanish, Serbo-Croatian, Romanian, Swedish, Dutch, Danish, Russian, Japanese, Chinese, Portugese, Bulgarian, Korean and Polish edits., (with G. Nicolis) Die Erforschung des Komplexen, 1987, (with G. Nicolis) Exploring Complexity, 1989, Chinese, Russian, Italian, French and Spanish edits. également en langue espagnole, (with I. Stengers) Entre le temps et l'Eternité, 1988, Dutch edit., 1989, Italian edit., 1989, Portuguese and Spanish edit., 1990; mem. editorial bur. Ukrainian Phys. Jour., 1990. Mem. sci. adv. bd. Internat. Acad. for Biomed. Drug Rsch., 1990. Fellow RGK Found. Centennial, U. Tex. 1989-90; decorated comdr. Légion d'Honneur, France, comdr. de l'Ordre de Leopold, comdr. de l'Ordre Leopold II, Grande Croix de l'Ordre de Leopold II, Médaille Civique de Premiere Classe, comdr. de l'Ordre National du Mérite, France, comdr. de l'Ordre des Arts et des Lettres, France; named Mem. d'Honneur de l'Acad. Nat. d'Argentine, 1990, de l'Acad. des. Scis. Naturelles de la République Fédérale de Russie, 1991, de l'Acad. Nat. des. Scis. de la République de San Marino, 1991, de l'Acad. Chilienne des Scis., 1991; recipient Prix Franqui, 1955, Prix Solvay, 1965, Nobel prize in chemistry, 1977, Honda Prize, 1983, Rumford gold medal Royal Soc. London, 1976, Karcher medal Am. Crystallographic Assn., 1978, Descartes medal U. Paris, 1979, Prix Umberto Biancamano, 1987, award recipient Gravity Rsch. Found., 1988, Artificial Intelligence Sci. Achievement award Internat. Found. for Artificial Intelligence, 1990, others. Fellow Nat. Acad. Scis. India (hon.); mem. Royal Acad. Belgium, Am. Acad. Sci. (medal 1975), Royal Soc. Scis. Uppsala (Sweden), Nat. Acad. Scis. U.S.A. (fgn. assoc.), Soc. Royale des Scis. Liège Belgium (corr.), Acad. Gottingen Germany, Deutscher Acad. der Naturforscher Leopoldina (Cothenius medal 1975), Osterreichische Acad. der Wissenschaften (corr.), Chem. Soc. Poland (hon.), Internat. Soc. Gen. Systems Rsch. (pres.-elect 1988), Royal Soc. Chemistry Belgium (hon.), Nat. Acad. Agentina (hon.), Acad. Natural Scis. Fed. Rep. Russia (hon.), Nat. Acad. Scis. Rep. San Marino (hon. pres.), Chilian Acad. Scis. (hon.), World Inst. Sci., others. Address: 67 Ave Fond Roy, 1180 Brussels Belgium Office: U Tex Ilya Prigogine Ctr Studies Statis Mechanics Austin TX 78712

PRIHODA, JAMES SHELDON, endocrinologist; b. Chgo., Mar. 25, 1959; s. Donald Thomas and Millicent Lee (Clock) P.; m. Julia Elizabeth Tank, Jan. 15, 1989. BS, U. Ill., 1981, MD, 1985. Diplomate Am. Bd. Internal Medicine, subspecialty cert. in endocrinology; cert. Am. Bd. Clin. Nutrition. Intern, resident Oreg. Health Scis. U., Portland, 1985-88, fellow in endocrinology diabetes and nutrition, 1988-91, instr. endocrinology, 1991-92; instr. molecular genetics U. Tex. Southwestern Med. Ctr., Dallas, 1992—; mem. tech. adv. com. Multnomoh County Med. Soc., Portland, 1992—. Coauthor: The Reproductive Endocrinology of the Adrenal Gland in Reproductive Medicine and Surgery, 1993; contbr. articles to profl. jours. Oreg. Heart Assn. fellow, 1989-90, grantee 1990-92. Fellow Am. Heart Assn. Coun. on Arteriosclerosis; mem. Am. Heart Assn. (Clin. Scientist award 1990—), ACP, AAAS, Physicians for Social Responsibility. Episcopalian. Office: U Tex Southwestern Med Ctr 5323 Harry Hines Blvd Dallas TX 75235-9046

PRIMACK, ALICE LEFLER, librarian; b. Kent, Ohio, Feb. 14, 1939; d. Glenn Q. and Mary S. (Staley) Lefler; m. Robert B. Primack; children: Eric, Mary-Anne, Glenn. BS, Ea. Ill. U., 1961; MLS, U. Wis., 1962. Libr. intern Ohio State U., Columbus, 1962-63, reference, bot. and zool. librarian, 1963-64, head Pharmacy Libr., 1964-66; from asst. to assoc. librarian U. Fla., Gainesville, 1972-92, Univ. librarian, 1992—. Author: How to Find Out in Pharmacy, 1969, Finding Answers in Science and Technology, 1984, Journal Literature of the Physical Sciences, 1992. Mem. adv. bd. 4-H of Alachua County, Fla., 1989—; officer Unitarian-Universalist Fellowship of Gainesville, 1980—. Democrat. Office: U Fla Marston Sci Libr Gainesville FL 32611

PRIME, ROGER CARL, marketing professional; b. Driffield, Yorkshire, Eng., Nov. 1, 1938; came to U.S., 1988; s. Reginald and Joyce Eileen (Pring) P.; m. Margaret Colyer, June 1, 1961 (div. Oct. 1987); children: Jacquiline, Andrew, Gillian; m. Maggie Gallo, June 3, 1989; children: Kevin, Ron. Cert. in mech. engring., South East London Coll., 1953-59. Registered profl. engr., London. Sr. designer Standard Telephones & Cables Ltd., Sidcup, Kent, Eng., 1959-65; engring. mgr. Rediffusion Simulation Ltd., Crawley, Sussex, Eng., 1965-69, mgr. mech. engring., 1978-85; sr. design engr. Kerney & Trecker Ltd., Brighton, Sussex, Eng., 1969-72; sr. prodn. and design engr. Worcester Value Co. Ltd., Haywardsheath, Sussex, Eng., 1972-78; head mil. design Singer Link Miles Ltd., Lancing, Sussex, Eng., 1985-93; mktg. mgr. AAI/Microflite Simulation Internat., Binghamton, N.Y., 1988—; cons. Brighton Poly., 1983-85; robotics cons. With English Army, 1959-61. Fellow Instn. Engring. Designers; mem. AIAA (sr.), London Coun. Register of Engring. Achievements include research in cost effective flight training device. Office: AAI/Microflite Simulation Internat 33 Lewis Rd Binghamton NY 13902

PRINCE, GEORGE EDWARD, pediatrician; b. Erwin, N.C., Nov. 25, 1921; s. Hugh Williamson and Helen Herman (Hood) P.; m. Millie Elizabeth Mann, Nov. 26, 1944; children: Helen Elizabeth, Millie Mann, Susan Hood, Mary Lois. MD, Duke U., 1944. Diplomate Am. Bd. Pediatrics, Am. Bd. Med. Examiners. Intern Boston Children's Hosp. Harvard Svc., Boston, 1944-45; resident pediatrics Children's Hosp., Louisville, 1945-47; instr. pediatrics U. Louisville, 1947; founder Gastonia (N.C.) Children's Clinic, 1947, pediatrician, 1947-86; pub. health physician Gaston County Health Dept., Gastonia, N.C., 1986—; chmn. bd. dirs. Carolina State Bank; bd. dirs. So. Nat. Bank, Gastonia, 1979-93, Hospice, Gastonia, 1987-92; organizer, dir. AIDS Adv. Coun., Gaston County, N.C., 1988-93; coord. N.C. chpt. Pediatric Rsch. in Office Setting, 1986-92. Contbr. articles to profl. jours. Mem. Gaston County Human Rels. Com., Gastonia, 1966; mem. Sch. Health Adv. Coun., Gaston County, 1980-93. Maj. USAF, 1955-57. Recipient Balthis Heart Assn. award Gaston County, 1981, Good Ambassador award Health Dept., 1986. Fellow Am. Acad. Pediatrics (pres. N.C. chpt. 1984-86); mem. AMA, N.C. Pediatric Soc. (hon., pres. 1970), N.C. Med. Soc., Gaston County Med. Soc. (pres. 1966), Rotary (pres. 1984), County Club (bd. dirs. 1975-76). Democrat. Methodist. Avocations: golf, flying, skiing, sailing, bridge. Home: 2208 Cross Creek Dr Gastonia NC 28056-8808 Office: Gaston County Health Dept 991 Hudson Blvd Gastonia NC 28052-6430

PRINCE, STEPHEN, software developer, researcher; b. Nottingham, Eng., Dec. 30, 1959; s. Ronald Arthur and Margaret Lillian (Maltby) P.; m. Janis May Dodd, July 6, 1986 (div. Dec. 1990); children: Christopher Stephen, Chloe Lee. B in Engring., Swinburne Inst. Tech., Melbourne, Victoria, Australia, 1982; programming cert., Monash U., Melbourne, Victoria, Australia, 1983. Design engr. I&I Australia, Boronia, Victoria, 1981-82, '83-84, Info. Mechanics Inc., Melbourne, 1982-83; software cons. Micronics, Melbourne, 1984; systems programmer Labtam Ltd., Braeside, Victoria, 1984-90; software mgr. Chancery Ln. Computer Svcs., Melbourne, 1990—; speaker Uniforum Singapore Conf., 1990, FUAA Conf., 1993. Mem. IEEE, Assn. for Computing Machinery, USENIX, Australian Unix Users Group (chmn., mem. several coms. 1990-92, pres. Victorian chpt. 1991—, contbr. articles to jour.). Avocations: chess, bonsai, bushwalking. Home: 60 Peppermint Grove, Knoxfield Victoria 3180, Australia Office: Chancery Ln Computer Svcs, Level 25 385 Bourke St, Melbourne 3000, Australia

PRINCE, WARREN VICTOR, mechanical engineer; b. Kansas City, Mo., May 21, 1911; s. Charles William and Bertha (Lybarger) P.; m. Edna Skinner Scott, Aug. 31, 1975; children—Charlotte E. Prince Smith, Leslie Warren (dec.), Charles Allan, Charlene Diane Prince Tercovich. Student engring. Baker U., 1930-34. Registered mech. engr., Calif. Design engr. Hoover Co., North Canton, Ohio, 1934-39; tool and machine design Thompson Products, Inc., Cleve., 1939-41; devel. engr. The Acrotorque Co., 1941-42; asst. chief devel. engr. The Weatherhead Co., 1942-45; pres. Prince Indsl. Plastics Corp., 1945-46; cons. engr., mech., plastics and plant prodn. problems, Kansas City and Los Angeles, 1946-50; project engr. Aerojet Gen. Corp., 1950-63, chief engr. Deposilube Mfg. Co., 1963-65; cons., 1965-66; sr. mech. engr. Avery Label Co., 1966-68; sr. project engr. machine design projects AMF, Inc., 1969-72; mech. cons. engr. as machine and product design specialist, 1972-80; pres. Contour Spltys., Inc., 1980-85; v.p. mech. engring. HEP Inc., 1985-88, mechanical cons. engr., 1988—; evening instr. Mt. San Antonio Coll., 1954—. Recipient Soc. Plastics Engrs. 1948 Nat. award for establishing basic laws of plastic molding process, Commendable Recognition award First Internat. Conf. on Machine Tech., Hong Kong, 1991. Mem. Soc. Plastics Engrs., Am. Soc. for Testing Materials (D-2 com.), Soc. Mfg. Engrs., Kappa Sigma. Presbyterian. Lodges: Masons, Shriners, Rotary. Contbr. articles to profl. jours. Patentee in field. Achievements include an explanation of break through performance with a new principle of anti-friction. Office: 838 N West St Anaheim CA 92801-4302

PRINCIPE, JOSEPH VINCENT, JR., environmental engineer; b. N.Y.C., Feb. 5, 1946; s. Joseph Vincent and Ruth Marie (Horan) P.; m. Jeanne Gerber, May 9, 1970; children: Benjamin, Matthew. BS, East Stroudsburg U., 1982. Engring. asst. Metro. Edison Co., Reading, Pa., 1975-84; environ. project leader Gen. Battery Corp., Reading, Pa., 1985; power plant supr. Fairchild Republic Co., Farmindale, N.Y., 1985-87; environ. engr. Stone & Webster Engring. Corp., N.Y.C., 1987—. Chmn. Cub Pack 86, Boy Scouts Am., Stroudsburg, 1983-87. With USN, 1967-75. Mem. ASME. Republican. Home: 1827 Laurel St Stroudsburg PA 18360 Office: Stone & Webster Environ 250 W 34th St New York NY 10119

PRINGLE, RONALD SANDY ALEXANDER, seismic inspector; b. Oakland, Calif., Sept. 12, 1945; s. Ronald Alexander and Ruth Florian (Collins) P.; m. Elizabeth Ann, Feb. 17, 1990; children: Ronald III, Devin Justin, Tyler Morton, Molly Morton. Student, Nassau Coll., 1963. Cert. seismic spl. dep., inspector. Owner, pres. Sandy Pringle Masonry, L.A., 1974—; inspector, sr. dep. Sandy Pringle Inspection, Manhattan Beach, Calif., 1986—; owner, pres. Ace Repainting Svc., L.A., 1988—; mem. Earthquake Engring. Rsch. Inst. Author: Guideline for Drilled-In Concrete Anchors, 1992. Active Calif. Preservation Found. Recipient Cert. of Meritorious Achievement Seismic Spl./Dep./Inspectors Assn., 1990-91. Mem. Structural Engrs. Assn., Am. Constrn. Inspectors Assn. (v.p., L.A. chpt. spl. inspection liaison, seismic safety commn./Calif. bldg. ofcls. joint com.), Seismic Spl. Dep. Assn., Internat. Conf. Bldg. Ofcls., Calif. Office Emergency Svcs. Structural Engrs. Assn. of So. Calif. (chair inspection practices, com. mem. hazardous bldgs. com.),. Achievements include development of pressure grout injection process, and development of L.A. city guideline 3. Office: Sandy Pringle Inspection PO Box 635 Manhattan Beach CA 90266-0635

PRINJA, ANIL KANT, nuclear engineering educator; b. Mombasa, Kenya, Apr. 9, 1955; came to U.S., 1980; s. Kapil Dev and Kushal (Dharney) P.; m. Renu Mohan, Sept. 18, 1983; children: Vivek Kapil, Akash Prinja. BSc in Nuclear Engring. with 1st class honors, London U., 1976, PhD in Nuclear Engring., 1980. Asst. rsch. engr. UCLA, 1980-87; asst. prof. nuclear engring. U. N.Mex., Albuquerque, 1987-89, assoc. prof., 1989—; chmn., host Internat. Conf. Transport Theory, 1991, U.S. Edge Plasma Physics: Theory and Applications Workshop, 1993; cons. Sandia Nat. Labs., Albuquerque, 1987—, Sci. Applications Internat., Inc., Albuquerque, 1987—; vis. prof. dept. reactor physics Chalmers U. Tech., Goteborg, Sweden, 1993. Contbr. chpts. to books and articles to profl. jours. Recipient Outstanding Acad. Achievement award Instn. Nuclear Engrs., 1976; grantee Dept. of Energy, Sandia Nat. Lab., Los Alamos Nat. Lab., Culham Labs., U.K., KFA Julich, Germany, 1989—, others. Mem. Am. Phys. Soc., Am. Nuclear Soc., Materials Rsch. Soc., Brit. Nuclear Energy Soc., N.Y. Acad. Sci. Hindu. Avocations: travel, reading. Office: U New Mex 209 Farris Engring Ctr Albuquerque NM 87131

PRISCO, DOUGLAS LOUIS, physician; b. N.Y.C., Nov. 30, 1945; s. Frank James and Isabel (Gaetano) P.; AB, Georgetown U., 1967; postgrad. N.Y. U., 1967-68; MD, U. Rome, 1974; m. Marianne Paula Mangano, Jan. 8, 1972; children: Jennifer Leigh, Douglas Louis, Dana Lauren, Andrew Michael. Intern, Mt. Sinai Svcs., Elmhurst, N.Y., 1974-75, resident in medicine, 1975-77, pulmonary medicine fellow, 1977-79; practice medicine specializing in pulmonary medicine, N.Y.C., New Hyde Park, N.Y., 1979-81; clin. asst. in medicine Bklyn. Hosp., 1979-81; pulmonary cons. and admitting physician Booth Meml. Hosp.; chief pulmonary medicine Deepdale Gen. Hosp.; clin. asst. Mt. Sinai Sch. Medicine N.Y.C., 1977-79; physician adviser St. Barnabas Hosp., 1981-82; pres. Met. Pulmonary Assocs., P.C., 1980—, Met. Pulmonary, P.C., 1985—; physician adv. to Elmhurst Gen. Hosp., Queens County Profl. Standards Rev. Orgn., 1979-85; co-chmn. quality assurance com. downstate region Island Peer Rev. Orgn., 1990—,vice- chmn. pro-tem regional quality assurance com., N.Y., 1993—. Bd. dirs. Queens County Profl. Standards Rev. Orgn., 1984-85; v.p. med. affairs Little Neck Community Hosp. (formerly Deepdale Gen. Hosp.), 1993—; pres. Med. Staff Soc. 1992—, v.p. med. bd. 1993—; mem., cons. Queens div. Island Peer Rev. Orgn., 1985—. Diplomate Am. Bd. Internal Medicine, sub-bd. Pulmonary Diseases. Mem. Rep. Senatorial Inner Cir., 1990. Fellow Am. Coll. Chest Physicians; mem. ACP, Am. Lung Assn. Queens (bd. dirs. 1988—), U.S. Power Squadron, Nat. Assn. of Residents and Interns, N.Y. Acad. Scis., Queens County Med. Soc. Roman Catholic. Club: Port Washington Yacht (former chmn. jr. activities 1987-88, fleet surgeon 1991-93), Capitol Hill Club (Washington). Office: 1575 Hillside Ave New Hyde Park NY 11040-2501

PRITCHARD, DAVID EDWARD, physics educator; b. N.Y.C., Oct. 15, 1941; m. Andrea Hasler; children: Orion, Alexander. BS, Calif. Inst. Tech., 1962; PhD, Harvard U., 1968. Postdoctoral fellow MIT, Cambridge, Mass., 1968, instr., 1968-70, asst. prof., 1970-75, assoc. prof., 1975-80, prof., 1980—; vis. scientist Stanford Rsch. Inst., 1975; vis. prof. U. Pais Sud Orsay, 1983; disting. visitor Joint Inst for Lab. Astrophysics, 1989. Div. assoc. editor Phys. Rev. Letters, 1983-88; contbr. articles to profl. jours. Polaroid fellow Harvard U., 1962-63, NSF predoctoral fellow, 1963-68. Fellow AAAS, Am. Phys. Soc. (disting. traveling lectr. laser sci. topical group, 1992—, rep. steering com. laser sci. topical group, rep. to joint coun. on quantum electronics, Broida prize 1991), Riverton (R.I.) Yacht Club. Avocations: sailing, carpentry. Office: MIT Rm 26-237 Dept Physics Cambridge MA 02139

PRITCHARD, DAVID GRAHAM, research scientist, educator; b. Montreal, Que., Can., Oct. 2, 1945; came to U.S., 1967; s. Carson Angus and Hazel (Habkirk) P.; m. Marianne Louise Egan, Dec. 27, 1975; children: Barbara Lynn, Ross Arthur. BA, U. Sask., Regina, 1967; PhD in Biochemistry, UCLA, 1975. Jr. rsch. scientist City of Hope Med. Ctr., Duarte, Calif., 1973-75, asst. rsch. scientist, 1975-76; rsch. assoc. U. Ala., Birmingham, 1976-81, from rsch. asst. prof. to rsch. assoc. prof., 1981-92, prof., 1992—. Contbr. 60 articles to profl. jours. Office: U Ala Dept Microbiology Sch Medicine/Joint Depts Birmingham AL 35294

PRITCHARD, DONALD WILLIAM, oceanographer; b. Santa Ana, Calif., Oct. 20, 1922; s. Charles Lorenzo and Madeleine (Sievers) P.; m. Thelma Lydia Amling, Apr. 25, 1943; children—Marian Lydia, Jo Anne, Suzanne Louise, Donald William, Albert Charles. B.A., UCLA, 1942; M.S. Scripps Instn. Oceanography, La Jolla, 1948, Ph.D., 1951; D.Sc. (hon.), Coll. William and Mary, 1985. Research asst. Scripps Instn. Oceanography, 1946-47; oceanographer USN Electronics Lab., 1947-49; assoc. dir. Chesapeake Bay Inst., Johns Hopkins, Balt., 1949-51; dir. Chesapeake Bay Inst., Johns Hopkins, 1951-74, chief scientist, 1974-79, prof., 1958-79, chmn. dept. oceanography, 1951-68; assoc. dir. for research, prof. Marine Scis. Rsch. Ctr., SUNY at Stony Brook, 1979-86, acting dir., dean, 1986-87, assoc. dean, 1987-88, prof. emeritus, 1988—; cons. C.E., U.S. Army, USPHS, AEC, Internat. AEC, NSF, Adv. Panel Earth Scis.; mem. adv. bd. to Sec. Natural Resources, State of Md. Bd. editors: Jour. Marine Research, 1953-70, Bull. Bingham oceanographic Collection, 1960-70, Geophys. Monograph Bd., Am. Geophys. Union, 1959-70, Johns Hopkins Oceanographic Studies, 1962—; Author articles in field. Served from 2d lt. to capt. USAAF, 1943-46. Recipient Mathias Sci. medal The Chespeake Rsch. Consortium, 1990. Fellow Am. Geophys. Union (past pres. oceanography sect.); mem. AAAS, Nat. Acad. Engring., Nat. Acad. Engring., Am. Soc. Limnology and Oceanography (past v.p.), Am. Meteorol. Soc., Nat. Acad. Scis. (past mem. com. oceanography, past chmn. panel radioactivity in marine environment), Sigma Xi (past chpt. pres.). Office: SUNY Marine Scis Rsch Ctr Stony Brook NY 11794

PRITSKER, A. ALAN B., engineering executive, educator; b. Phila., Feb. 5, 1933; s. Robert and Gertrude (Liebowitz) P.; m. Anne Gruner, Sept. 22, 1956; children—Caryl Anne, Pamela Sue, Kenneth David, Jeffrey Richard. B.S.E.E., Columbia U., 1955, M.S. in Indsl. Engring., 1956; Ph.D., Ohio State U., 1961; DSc (hon.), Ariz. State U., 1992. Registered profl. engr., Tex., Ariz. Engr. Battelle Inst., 1956-62; Prof. engring. Ariz. State U., Tempe, 1962-69; prof. engring. Va. Poly. Inst., Blacksburg, 1969-70; prof. engring. Purdue U., West Lafayette, Ind., 1970-81, adj. prof. engring., 1981—; pres. Pritsker Corp., West Lafayette, 1973-86, 91—, chmn., 1973—; cons. Rand Corp., Gen. Electric Co., Gen. Motors Corp., Bethlehem Steel Co. Author: Simulation With Gasp IV, 1974, Modeling and Analysis Using Q-Gert Networks, 1979, Management Decision Making, 1984, Introduction to Simulation and SLAM II, 1986, TESS: The Extended Simulation Support System, 1987, SLAM II Network Models for Decision Support, 1989, Papers, Experiences, Perspectives, 1990. Recipient Gilbreth Indsl. Engring. award Am. Inst. Indsl. Engrs., 1991. Fellow Inst. Indsl. Engrs. (Rsch. and Innovation award 1978, Gilbreth award 1991); mem. NAE, Inst. Mgmt. Scis. (coll. on simulation, Disting. Svc. award 1991), Ops. Rsch. Am. Office: Pritsker Corp PO Box 2413 West Lafayette IN 47906-0413

PRITZKER, ANDREAS EUGEN MAX, physicist, administrator, author; b. Baden, Aargau, Switzerland, Dec. 4, 1945; s. Boris and Alice (Kamer) P.; m. Martha L. Ehrlich, Dec. 4, 1970. PhD, Swiss Fed. Inst. Tech., Zurich, Switzerland, 1974. Scientist ETH, Zurich, 1970-75, Alusuisse, Zurich, 1975-77; cons. Motor Columbus, Baden, 1977-80; scientist Swiss Inst. Nuclear

Rsch., Villigen, Switzerland, 1980-83; asst. to pres. Bd. ETH, Zurich, 1983-87; head adminstrn. Paul Scherrer Inst., Villigen, 1987—. Author: Filberts Verhaengnis, 1990, Das Ende der Taeuschung, 1993, also several short stories; contbr. articles to profl. jours. Mem. Swiss Phys. Soc., Am. Phys. Soc.

PROCHASKA, OTTO, engineer; b. Alexandria, La., Sept. 19, 1933; s. Otto and Bertha (Goldstein) P.; m. Elizabeth Sanford, June 2, 1956; children: Eric, Laura, Kurt. Registered profl. engr., La., Fla. Project engr. Esso Standard Oil Co., Baton Rouge, 1958-63, Am. Cyanamid, Milton, Fla., 1963-74; engr., supr. Am. Cyanamid, Milton, 1974-85; profl. engr., sect. head water program Fla. Dept. Environ. Regulation, Pensacola, 1986-88; dir. water and gas dept. Escambia County Utility Authority, Pensacola, 1989—. Served as lt. USAF, 1955-58, as capt. with Res., 1958-70. Mem. ASME, Assn. Energy Engrs. Episcopalian. Avocations: sailing, restoring fgn. cars, constructing model ships. Office: Escambia County Utility Authority PO Box 15311 Pensacola FL 32514-0311

PROCHÁZKA, KAREL, chemistry educator; b. Prague, Czechoslovakia, Sept. 5, 1947; Jaroslav and Vilemína (Slavíková) P.; m. Naděžda Paplhámová, Feb. 28, 1971; children: Andrea, Kristina. MSc, Charles U., Prague, 1970; RNDr, Charles U., 1970; PhD, Czechoslovak Acad. Scis., 1973. Asst. prof. Dept. of Phys. Chemistry, Charles U., 1973-90; assoc. prof. Dept. of Phys. and Macromolecular Chemistry, Charles U., 1990—; vis. prof. U. Quebec a Montreal, Can., 1985-86; external mem. scientific bd. Inst. of Macromolecular Chem., Prague, 1991—. Co-author: Methods of Chemical Research, Academia, Prague, 1987; contbr. articles to profl. jours. Mem. Am. Chem. Soc. Achievements include rsch. on theory of the size-exclusion chromatography of reversibly associating systems, dynamics of block co-polymer micelles formation and disociation. Home: Mechenice 127, 252 06 Prague West Czech Republic Office: Charles U Dept Phys Chemistry, Albertov 2030, 12840 Prague Czech Republic

PROCKOP, DARWIN JOHNSON, biochemist, physician; b. Palmerton, Pa., Aug. 31, 1929; s. John and Sophie (Gurski) P.; m. Elinor Sacks, Apr. 15, 1961; children: Susan Elizabeth, David John. AB, Haverford Coll., 1951; MA, Oxford U., 1953; MD, U. Pa., 1956; PhD, George Washington U., 1962; DSc (hon.), U. Oulu, 1983, U. So. Fla., 1993. Investigator NIH, 1957-61; assoc., asst. prof., asso. prof., prof. medicine and biochemistry U. Pa., Phila., 1961-72; prof., chmn. dept. biochemistry U. Medicine and Dentistry of N.J. (Rutgers Med. Sch.), Piscataway, N.J., 1972-86; prof., chmn. dept. biochemistry and molecular biology Jefferson Med. Coll., Phila., 1986—; dir. Jefferson Inst. Molecular Medicine. Contbr. articles to profl. jours.; research on collagen. Served with USPHS, 1958-61. Fulbright fellow Oxford U., 1951-53; NIH, grantee, 1961—; recipient Disting. Alumnus award George Wash. U., 1991, U. Pa., 1994. Mem. NAS, Inst. Medicine, Acad. Finland, Soc. Biol. Chemists, Am. Soc. Clin. Investigation, Am. Assn. Physicians, Phi Beta Kappa, Alpha Omega Alpha. Home: 291 Locust St Philadelphia PA 19106-3913 Office: Jefferson Med Coll 1025 Walnut St Philadelphia PA 19107

PROCTOR, CHARLES LAFAYETTE, II, mechanical engineer, educator, consultant; b. Crawfordsville, Ind., Nov. 21, 1954; s. Charles Lafayette and Marjorie E. (Purdue) P.; m. Dixie Lee Huffer, May 22, 1976; children: Christina Nicole, Courtney Alexandra. BSME, Purdue U., 1976, MSME, 1979, PhD, 1981. Registered profl. engr., Fla. Asst. prof. mech. engring. U. Fla., Gainesville, 1981-86, assoc. prof., 1986—, dir. Combustion Lab., 1981-89, mem. grad. studies faculty, 1982-87, mem. doctoral rsch. faculty, 1987—; presenter tech. meetings; mem. Fed. Emergency Mgmt. Agy. Emergency Mgmt. Inst. on Designing Bldg. Fire Safety; mem. hazardous waste combustion sampling and analysis workshop EPA, also mem. hazardous waste incenerator permit writers workshop; owner, mgr. Proctor Engring. Rsch. & Cons., Gainesville, 1984—; pres. ENVIRECO, Inc., Gainesville, 1986—. Contbr. articles to profl. jours., also ency. sects. Mem. ASTM (sec. 1984-86, chmn. com. on thermal processes 1986-91), ASME (regional oper. bd. 1986-88, Region XI Outstanding Faculty Advisor award 1986), AIAA, Am. Soc. Engring. Edn., Soc. Automotive Engrs. Combustion Inst. (oper. bd. ea. sect. 1985-91), Nat. Fire Protection Assn., Air and Waste Mgmt. Assn., Pi Tau Sigma. Methodist. Home: 6051 NW 19th Ln Gainesville FL 32605 Office: U Fla Dept Mech Engring Gainesville FL 32611

PROCTOR, RICHARD J., geologist, consultant; b. L.A., Aug. 2, 1931; s. George Arthur and Margaret Y. (Goodman) P.; m. Ena McLaren, Feb. 12, 1955; children: Mitchell, Jill, Randall. BA, Calif. State U., L.A., 1954; MA, UCLA, 1958. Engring. geologist, Calif.; cert. profl. geologist Am. Inst. Profl. Geologists. Chief geologist Met. Water Dist., L.A., 1958-80; pres., cons. geologist Richard J. Proctor, Inc., Arcadia, Calif., 1980—; vis. assoc. prof. Calif. Inst. Tech., Pasadena, 1975-78. Co-author: Citizens Guide to Geologic Hazards, 1993; editor: Professional Practice Guidelines, 1985, Engineering Geology Practice in Southern California, 1992. Pres., dir. Arcadia Hist. Soc., 1993. Fellow Geol. Soc. Am. (rep. 1989, Burwell Meml. award 1972); mem. Assn. Engring. Geologists (pres. 1979), Am. Inst. Profl. Geologists (pres. 1989, Van Couvering Meml. award 1990, hon. mem. 1992), Am. Geol. Inst. (sec.-treas. 1983). Home and Office: 327 Fairview Ave Arcadia CA 91007

PROCTOR, ROBERT NEEL, biologist, historian, educator; b. Corpus Christi, Tex., June 25, 1954; s. Norman N. and Eugenia (Milton) P.; children: Geoffrey Schiebinger, Jonathan. BS, Ind. U., 1976; PhD, Harvard U., 1984. Instr. biology Harvard U., Cambridge, Mass., 1978-84; vis. lectr. U. Calif., Berkeley, 1985; faculty mem. New Sch. for Social Rsch., N.Y.C., 1986-90; assoc. prof. Pa. State U., University Park, 1990-93, prof., 1993—; cons. Human Genome Project, Washington, 1990—. Author: Racial Hygiene: Medicine Under the Nazis, 1988, Value-Free Science? Purity and Power in Modern Knowledge, 1991. NSF fellow, 1976-79, Andrew Mellon fellow, 1984-86, fellow Ctr. for Advanced Study in Behavioral Scis., 1986, Davis Ctr. fellow Princeton U., 1992-93. Office: Pa State U Dept History University Park PA 16802

PRODAN, RICHARD STEPHEN, electrical engineer; b. Yonkers, N.Y., Oct. 26, 1952; s. Victor L. and Stella (Erbe) P.; m. Judith Mirabella, June 17, 1978; children: Adam L., Michael D. BEE, CCNY, 1976, MEE, 1979; PhD, Columbia U., 1981. Sr. mem. rsch. staff Philips Labs., Briarcliff Manor, N.Y., 1976-88; sr. engr. StellaCom, Inc., Arlington, Va., 1988-90; dir. Cable TV Labs., Inc., Boulder, Colo., 1990—. Mem. Eta Kappa Nu, Sigma Xi. Achievements include patents for Vertical pre-filter for picture-in-picture TV, Interlace to sequential display conversion using MLE. Office: Cable TV Labs Inc 1050 Walnut St # 500 Boulder CO 80302

PROEBSTING, EDWARD LOUIS, JR., research horticulturist; b. Woodland, Calif., Mar. 2, 1926; s. Edward Louis and Dorothy (Critzer) P.; m. Patricia Jean Connolly, June 28, 1947; children: William Martin, Patricia Louise, Thomas Alan. BS, U. Calif., Davis, 1948; PhD, Mich. State U., 1951. Asst. horticulturist Wash. State U., Prosser, 1951-52, horticulturist, 1963-93, superintendent irrigated agrl. rsch. and extension ctr., 1990-93; supt. irrigated agrl. rsch. and extension ctr., 1993; vis. prof. Cornell U., Ithaca, N.Y., 1966; vis. scientist Hokkaido U., Sapporo, Japan, 1978, Victoria Dept. Agr., Tatura, Australia, 1986—. Contbr. numerous articles to profl. jours. Scoutmaster Boy Scouts Am., Prosser, 1963-76, dist. chmn., 1976-78. Served to lt. USNR, 1943-46, 52-54. Recipient Silver Beaver award Boy Scouts Am.; fellow Japan Soc. Promotion Sci., Sapporo, 1978, Res. Bank. Australia, 1986. Fellow AAAS, Am. Soc. Hort. Sci. (pres. 1983-84, sci. editor jour. 1993—). Methodist. Avocations: backpacking, native plants. Home: 1929 Miller Ave Prosser WA 99350-1532 Office: Wash State U Irrigated Agrl Rsch Ctr and Ext RT 2 Box 2953A Prosser WA 99350-0030

PROFET, MARGIE, biomedical researcher; b. Berkeley, Calif., Aug. 7, 1958. BA in Polit. Philosophy, Harvard U., 1980; BA in Physics, U. Calif., Berkeley, 1985. Prof. dept. biology U. Calif., Berkeley; comupter programmer, Munich, Germany, 1980-81. Author: (with others) The Adapted Mind, 1992; contbr. articles to Quarterly Review of Biology. MacArthur fellow John D. and Katherine T. MacArthur Found., 1993. Achievements include research on evolutionary theory as a guide for research in human physiology and pathology, the function of allergy, immunological defense against toxins, pregnancy sickness as adaptation, a deterrent to maternal ingestion of teratogens, menstruation as a defense against pathogens transportede by sperm. Office: Univ of California Dept of Biology Berkeley CA 94720*

PROFETA, SALVATORE, JR., chemist; b. Phila., May 1, 1951; m. Catherine Mary Cherry, Sept. 20, 1980; children: Luisa, Theresa. BA, Temple U., 1973; PhD, U. Ga., 1978. Postdoctoral fellow chemistry dept. Fla. State U., Tallahassee, 1979-80; postdoctoral fellow pharm. chemistry dept. U. Calif., San Francisco, 1980-81, teaching fellow, 1981-82; instr. chemistry dept. La. State U., Baton Rouge, 1982-84; sr. scientist Allergan Pharms., Inc., Irvine, Calif., 1984-87; project mgr. computational chemistry Glaxo Rsch. Inst., Research Triangle Park, N.C., 1987-90, head chemistry systems, 1990-93; dir. MCNC Info. Techs. Rsch. Inst. (formerly N.C. Supercomputing Ctr. Rsch. Inst.), Research Triangle Park, 1993—; cons. CADD-CAMM Smith, Line & French, Phila., 1980-82, Squibb Rsch. Inst., Princeton, N.J., 1982-84; mem. allocation com. N.C. Supercomputing Ctr., 1989-93. Mem. editorial bd. Jour. Molecular Graphics, 1989—; contbg. editor Chem. Design Automation News, 1991—; contbr. articles to Jour. Am. Chem. Soc. NSF fellow, 1976-78; Petroleum Rsch. Found. grantee, 1984-88. Fellow N.Y. Acad. Scis.; mem. Am. Chem. Soc. Achievements include patents in anticancer drug design; co-author MM1, MM2, MM3 and AMBER molecular mechanical force fields. Office: IT Rsch Inst MCNC 3021 Cornwallis Rd Research Triangle Park NC 27709

PROFFITT, ALAN WAYNE, engineering educator; b. London, Feb. 23, 1956; s. Bobby Wayne and Mary (Dunne) P.; m. Linda Lee Dawes, Dec. 24, 1980; children: Amanda, Andrew. BS in Engring., Ark. Tech. U., 1977; MS in Instrumental Scis., U. Ark., 1984. Registered profl. elec. engr. Grad. asst. U. Ark., Little Rock, 1982-84; sr. elec. engr. Lockheed Aerospace Systems Corp., Marietta, Ga., 1984-88; asst. prof. Ark. Tech. U., Russellville, 1988—. 1st lt. U.S. Army, 1978-81, maj. U.S. Army Res., 1982—. Mem. IEEE, Am. Soc. Engring. Educators, Instrument Soc. Am., Res. Officer Assn., Army Engr. Assn., Mensa, Lambda Chi Alpha. Achievements include integrated extensive use of software in engineering curriculum. Home: 1507 N Louisville Russellville AR 72801

PROKHOROV, ALEKSANDR MIKHAILOVICH, radiophysicist; b. Atherton, Australia, June 28, 1916. Grad., Leningrad State U., 1939; postgrad., Physics Inst. USSR Acad. Scis., 1939-41, 44-46. Joined Communist Party Soviet Union, 1950; mem. staff Lebedev Physics Inst., USSR Acad. Scis., 1954-72, head of lab., 1972-83, vice dir., 1983—, also bd. dirs., academician-sec. gen physics and astronomy dept., 1971—; chmn. Nat. Commn. Soviet Physicists. Author standard works on laser. Served with Soviet Army, 1941-43. Decorated Hero of Socialist Labor; recipient Lenin prize, 1959; Nobel prize, 1964; Lomonosov medal Soviet Acad. Scis., 1987. Mem. Am. Acad. Sci. and Art., Russian Acad. Scis. Office: USSR Acad Scis Dept Gen Physics, Ulitsa Vavilova 38, 117942 Moscow Russia*

PROSTAK, ARNOLD S., chemical instrumentation engineer; b. Bklyn., Apr. 8, 1929; s. Jacob Louis and Rose S. Prostak; me. Meredith Ann Prostak, 1971; two children. BS, Coll. of William and Mary, 1950; MA, Johns Hopkins U., 1955; PhD, U. Mich., 1969. Chemist U.S. Army, Edgewood, Md., 1950-58; rsch. assoc. U. Mich., Ann Arbor, 1958-60; sr. rsch. and devel. engr. Bendix Corp., Ann Arbor, 1960-71; project engr. Gen. Motors Corp., Milford, Mich., 1972—. Mem. AAAS, Am. Chem. Soc., Soc. Automotive Engrs. Home: 1707 Harding Rd Ann Arbor MI 48104 Office: Gen Motors Corp 31 W-VEL Proving Ground Milford MI 48380

PROTHRO, EDWIN TERRY, psychologist educator; b. Robeline, La., Dec. 11, 1919; Edwin Thomas and Frances Lillian (Terry) P.; m. Dorothy Kenworthy, Apr. 26, 1943 (div. 1967); children: Martha Carol Wells, Edwin Terry Jr.; m. Najla Salman, July 31, 1968; 1 child, Gwendolyn. PhD, La. State U., 1942. Asst. prof. psychology La. State U., Baton Rouge, 1946-49; assoc. prof. psychology U. Tenn., Knoxville, 1949-51; prof. psychology Am. U. Beirut, Lebanon, 1951-85; fellow Ctr. for Middle East Studies Harvard U., Cambridge, Mass., 1960; dean, provost Am. U. Beirut, Lebanon, 1965-73; dep. dir. Edn. Abroad Program U. Calif., Sant Barbara, 1975-77; v.p. Hariri Found., Washington, 1986--; cons. Middle East Office Ford Found., Beirut, 1973-75; mem. bd. trustee Am. Community Sch., Beirut, 1970-73; mem. edit. bd. Jour. Social Psychology, 1955-70. Co-author: Psychology: A Biosocial Study of Behavior, 1950, 72; Changing Family Patterns in the Arab East, 1974; contbr. articles to profl. jours. Lt. USNR, 1943-46, WWII. Recipient Order of the Cedars Rep. Lebanon, 1969; Nat. Inst. Mental Health grantee, 1966-70, fellow 1963, 64. Mem. Am. Psychology Soc., La. Psychology Assn. (pres. 1949), Sigma Xi (pres. Beirut chpt. 1955). Office: Hariri Found 1020 19th St NW Washington DC 20036

PROUT, GEORGE RUSSELL, JR., medical educator, urologist; b. Boston, July 23, 1924; s. George Russell and Marion (Snow) P.; m. Loa Katherine Wheatley, Oct. 17, 1950; children: George Russell III, Elizabeth Louise. Student, Union Coll., 1943, DSc (hon.), 1990; MD, Albany Med. Coll., 1947; MA (hon.), Harvard U., 1969. Intern Grasslands Hosp., Valhalla, N.Y., 1947-48; resident N.Y. Hosp., N.Y.C., 1952-56; asst. attending physician Meml. Ctr. for Cancer and Allied Disease, N.Y.C., 1956-57; asst. clinician in surgery James Ewing Hosp., N.Y.C., 1956-57; assoc. prof., chmn. div. urology U. Miami, 1957-60; prof., chmn. div. urology Med. Coll. Va., 1960-69; chief urol. svc. Mass. Gen. Hosp., Boston, 1969-89; prof. surgery Harvard Med. Sch., 1969-89; emeritus prof. surgery Harvard Med. Sch., Boston, 1989—; hon. urologist Mass. Gen. Hosp., Boston, 1993—. chmn. Adjuvants in Surg. Treatment of Bladder Cancer; mem. adv. task force to Nat. Cancer Inst., 1968—, expert cons. div. surveillance, 1991—, Finland coop. ATBC study, 1991—; chmn. Nat. Bladder Coop. Group, 1973. With USNR, 1950-52. Fellow ACS, Acad. Medicine Toronto (corr.); mem. AMA, AAUP, Am. Urol. Assn., Can. Urol. Assn., Am. Cancer Soc., Soc. Pelvic Surgeons, Soc. Surg. Oncology, Soc. Univ. Urologists, Dallas. So. Clin. Soc. (hon.), Am. Assn. Genitourinary Surgeons, Soc. Pediatric Urology, Soc. Urol. Oncology, Soc. Internat. Urologists, Alpha Omega Alpha. Home: 27 Prospect Bay Dr W Grasonville MD 21638-9671

PRUCZ, JACKY CAROL, mechanical and aerospace engineer, educator; b. Bacau, Romania, Dec. 25, 1949; came to the U.S., 1981; s. Carol Haim and Rella (Kaltmann) P.; m. Migri Marlene Caner, Oct. 25, 1972; children: Shirley, Roni. BS, Israel Inst. Tech., 1972, MS, 1979; PhD, Ga. Inst. Tech., 1985. Project engr. Israeli Air Force, Tel Aviv, 1972-75, test engr., 1975-77, sr. rsch. engr., 1977-81; grad. rsch. asst. Ga. Inst. Tech., Atlanta, 1981-85; assoc. prof. mech. and aero. engring. W.Va. U., Morgantown, 1985—, asst. dir. concurrent engring. rsch. ctr., 1988—; dir. structures lab. W.Va. U., 1986-88, chmn. faculty search com., 1986-88; chmn. workshops at nat. confs., 1987—. Contbr. chpt. to book and articles to profl. jours. Captain Israeli Air Force, 1973-81. Mem. AIAA, ASME, Am. Soc. Engring. Edn., Soc. for Advancement Materials and Processing Engring., Am. Soc. Composites, Internat. Orgn. for Standardization, Mech.Vibration and Shock, Computer-Aided Acquisition and Logistics Support Initiative. Avocations: travel, exercise, biking, skiing. Home: 109 Greenwood Dr Morgantown WV 26505-2555 Office: WVa U Mech Aero Engring PO Box 6101 Morgantown WV 26506-6101

PRUESS, DAVID LOUIS, biochemist; b. Terre Haute, Ind., Sept. 2, 1938; s. Louis Martin and Cora (Olson) P.; m. Jacquelyn Olson, June 28, 1963 (div. June 1969); m. Ellen Susan Enhoffer, Oct. 11, 1972. BA, Cornell U., 1960; MS, U. Wis., 1964, PhD, 1967. Biochemist Hoffmann LaRoche, Nutley, N.J., 1967—. Contbr. 50 articles to profl. jours. and revs. Achievements include patents in field of new natural products. Home: 12 Brookside Terr North Caldwell NJ 07006-4115 Office: Hoffmann LaRoche 340 Kingsland St Nutley NJ 07110

PRUETT, CHARLES DAVID, aerospace research scientist, educator; b. Durham, N.C., June 28, 1948; s. Charles Danny and Edna Allen (Layne) P.; m. Suzanne Leone Fiederlein, Oct. 17, 1987. BSME, Va. Poly. Inst. & State U., 1970; MEd, U. Richmond, 1974; MS in Applied Math., U. Va., 1980; PhD in Applied Math., U. Ariz., 1986. High sch. math tchr. Henrico County Schs., Highland Springs, Va., 1973-76; design engr. Kentron HTC, Hampton, Va., 1976-78; engring. analyst Analytical Mechs. Assn., Hampton, 1979-81; asst. prof. math. scis. Va. Commonwealth U., Richmond, 1986-89; rsch. assoc. NRC-NASA Langley Rsch. Ctr., Hampton, 1989-91; rsch. scientist Analytical Svcs. & Materials, Inc., Hampton, 1991—. Contbr. articles to profl. jours. Mem. social concerns com. Williamsburg (Va.) Friends Meeting, 1990—. 1st lt. USAF, 1970-72. Recipient NAS Supercomputing grant Nat. Aerodynamics Simulator, NASA-Ames, Calif., 1991-92. Mem. AIAA, Soc. Indsl. & Applied Math., Sierra Club, Nature Conservancy. Mem. Soc. of Friends. Achievements include conducting the first direct numerical simulation of laminar breakdown in high-speed boundary-layer flows. Office: Analytical Svcs Materials Inc 107 Rsch Dr Hampton VA 23666

PRUETT, CLAYTON DUNKLIN, biotechnical company executive; b. Montgomery, Ala., June 16, 1935; s. William Rogers and Myra Eleanor (Ganey) P.; m. Barbara Clapp, Feb. 22, 1974; children: Christopher Blair, Tyler Michael. BSCE, Auburn U., 1956. Profl. engr., Ala. Civil engr. Ala. State Hwy Dept., Montgomery, 1956-57; lt. USAF, 1957; civil engr. USAF, Andrews AFB, 1959-64; staff assoc. Gen. Atomic div. Gen. Dynamics, La Jolla, Calif., 1964-68; exec. v.p. Enviro-Med Inc., La Jolla, 1968-73; chmn. bd. CDP Inc., Atlanta, 1973—; pres., CEO Advanced Cancer Techs., Inc., Atlanta, 1988—. Bd. dirs. Am. Cancer Soc., 1986-90, U. Calif., San Diego Cancer Ctr. Found, 1989—, Nat. Childhood Cancer Found., 1990—. Mem. San Diego Yacht Club. Home: PO Box 2304 Rancho Santa Fe CA 92067-2304 Office: Advanced Cancer Techs Inc 1050 Crown Point Pky Atlanta GA 30338-7700

PRUETT, KYLE DEAN, psychiatrist, writer, educator; b. Raton, N.Mex., Aug. 27, 1943; s. Ozie Douthitt Pruett and Velma Lorraine Smith; m. Leslie Ann Bloom, Aug. 14, 1965; children: Elizabeth Storr, Emily Farrar. BA in History, Yale U., 1965; D of Medicine, Tufts U., 1969. Intern Mt. Auburn Hosp.-Harvard U., 1969-70; resident in psychiat. medicine Tufts-New England Med. Ctr., Boston, 1970-72; child psychiatry fellow Child Study Ctr., Yale U., New Haven, 1972-74, asst. clin. prof. psychiatry, 1975-79, assoc. clin. prof., 1979-87, clin. prof., 1987—; dir. child devel. unit Yale U., 1982—; attending physician dept. child psychiatry Yale-New Haven Hosp., 1972—; cons. psychiatrist Guilford (Conn.) Pub. Schs., 1977—; vis. scholar Sch. Medicine, U. Vt., 1987, Sch. Medicine, U. N.Mex., 1988; mem. editorial bd. Med. Problems of Performing Artists jour., 1983—, Fathers mag., 1987—; bd. dirs. Zero to Three: Nat. Ctr. for Clin. Infant Programs, Washington; cons. CBS News, Lifetime. Author: The Nurturing Father, 1988 (Am. Health Book award 1988); contbr. numerous articles to profl. jours.; host biweekly TV series Your Child 6 to 12 with Dr. Kyle Pruett Lifetime TV, 1993—. Mem. med. adv. bd. Conn. Multiple Sclerosis Soc., Hartford, 1985—, Scholastic, Inc., 1988—, Yale U. Program for Humanities in Medicine, 1989—, World Assn. for Infant Mental Health, 1979—; cons. Neighborhood Music Sch., New Haven, 1986—, CBS TV Family time, 1989—, Wellesley Child Study Ctr.; prin. tenor Conn. Chancel Opera Co., New Haven, 1983—. Vis. fellow Anna Freud Clinic, London, 1975; recipient Mayoral Citation, City of Indpls., 1987. Mem. Am. Acad. Child and Adolescent Psychiatry, Am. Psychiat. Assn., Soc. for Rsch. in Family Therapy, Physicians for Social Responsibility, Nat. Assn. Physician Broadcasters (CBS news cons. 1989—), Yale U. Glee Club Alumni Assn. (pres. 1985-89). Avocations: rowing, sailing, skiing, vocal chamber music, running. Home: 10 Fernwood Dr Guilford CT 06437-2349 Office: Yale Child Study Ctr 333 Cedar St New Haven CT 06510

PRUITT, ALICE FAY, mathematician, engineer; b. Montgomery, Ala., Dec. 17, 1943; d. Virgil Edwin and Ocie Victoria (Mobley) Maye; m. Mickey Don Pruitt, Nov. 5, 1967; children: Derrell Gene, Christine Marie. BS in Math., U. Ala., Huntsville, 1977; postgrad. in engring., Calif. State U., Northridge, 1978-79. Instr. math. Antelope Valley Coll., Quartz Hill, Calif., 1977-78; space shuttle engr. Rockwell Internat., Palmdale, Calif., 1979-81; programmer, analyst Sci. Support Svcs. Combat Devel.-Experimentaton Ctr., Ft. Hunter-Liggett, Calif., 1982-85; sr. engring. specialist Loral Vought Systems Corp., Dallas, 1985-92; rsch. scientist Nichols Rsch. Corp., Huntsville, Ala., 1992—. Mem. DeSoto (Tex.) Coun. Cultural Arts, 1987-89. Mem. AAUW (sch. bd. rep. 1982, phone chmn. 1987-89, legal advocacy Fund Chairperson 1989-91), Toastmasters Internat., Phi Kappa Phi. Republican. Methodist. Avocations: dancing, gourmet cooking. Office: Nichols Rsch Corp PO Box 40002 4040 S Memorial Pkwy Huntsville AL 35815-1502

PRUS, JOSEPH STANLEY, psychology educator, consultant; b. Glen Cove, N.Y., Mar. 9, 1952; s. Joseph Stanley and Constance Mary (Lamp) P.; m. Audrey Kaye Mink, Apr. 19, 1980; children: Erin Marie, Elizabeth Lauren Scudder. Student, St. John Fisher Coll., Rochester, N.Y., 1970-72; BA, U. Ky., 1974, MA, 1975, PhD, 1979. Lic. sch. psychologist, S.C., Ky.; nat. cert. sch. psychologist. Coord. psychology human devel. program U. Ky., Lexington, 1978-79, assoc. dir. human devel. program, 1979-80; from asst. to assoc. prof. Winthrop U., Rock Hill, S.C., 1980-88, prof.; dir. Office of Assessment, Rock Hill, S.C., 1988—; cons. Cardinal Hill Hosp., Lexington, 1977-80, U. Ariz., Tucson, 1984-85, Lancaster (S.C.) Mental Retardation Bd., 1987—; numerous presentations in field. Author: Handbook of Certification for School Psychologists, 1989; also numerous articles. Bd. dirs. Wesley Found., 1985-87; chmn. bd. dirs. Rock Hill Girls' Home, 1987-91. Recipient Disting. Prof. award Winthrop U., 1989; numerous grants from state and nat. agys. Mem. APA, NASP (Presdl. citation 1993), S.C. Assn. Sch. Psychologists (Outstanding Contbns. to Sch. Psychology award 1991), Phi Beta Kappa, Psi Chi. Avocations: reading, sports, music. Home: 2430 Colebrook Dr Rock Hill SC 29732-9411 Office: Winthrop U Dept Psychology Rock Hill SC 29733

PRUSINER, STANLEY BEN, neurology and biochemistry educator, researcher; b. Des Moines, May 28, 1942; s. Lawrence Albert and Miriam (Spigel) P.; m. Sandra Lee Turk, Oct. 18, 1970; children: Helen Chloe, Leah Anne. AB cum laude, U. Pa., 1964, MD, 1968. Diplomate Am. Bd. Neurology. Intern in medicine U. Calif., San Francisco, 1968-69, resident in neurology, 1972-74, asst. prof. neurology, 1974-80, assoc. prof., 1980-84, prof., 1984—, prof. biochemistry, 1988—, acad. senate faculty rsch. lectr., 1989-90; prof. virology U. Calif., Berkeley, 1984—; mem. neurology rev. com. Nat. Inst. Neurol. Disease and Strokes, NIH, Bethesda, Md., 1982-86, 90-92; mem. sci. adv. bd. French Fedn., L.A., 1985—; mem. sci. rev. com. Alzheimer's Disease Diagnostic Ctr. & Rsch. Grant Program, State of Calif., 1985-89; chmn. sci. adv. bd. Am. Health Assistance Found., Rockville, Md., 1986—. Editor: The Enzymes of Glutamine Metabolism, 1973, Slow Transmissible Diseases of the Nervous System, 2 Vols., 1979, Prions - Novel Infectious Pathogens Causing Scraple and CJD, 1987, Prion Diseases of Humans and Animals, 1992, Molecular and Genetic Basis of Neurologic Disease, 1992; contbr. over 170 rsch. articles to profl. jours. Mem. adv. bd. Family Survival Project for Adults with Chronic Brain Disorders, San Francisco, 1982—, San Francisco chpt. Alzheimer's Disease and Related Disorder Assn., 1985—. Lt. comdr. USPHS, 1969-72. Alfred P. Sloan Rsch. fellow U. Calif., 1976-78; Med. Investigator grantee Howard Hughes Med. Inst., 1976-81; grantee for excellence in neurosci. Senator Jacob Javits Ctr., NIH, 1985-1990; recipient Leadership and Excellence for Alzheimer's Disease award NIH, 1990—, Potamkin prize for Alzheimer's Disease Rsch., 1991, Med. Rsch. award Met. Life Found., 1992, Christopher Columbus Discovery award NIH and Med. Soc. Genoa, Italy, 1992, Charles A. Dana award for pioneering achievements in health, 1992, Dickson prize for outstanding contbns. to medicine U. Pitts., 1992, Max Planck Rsch. award Alexander von Humboldt Found. and Max Planck Soc., 1992, Gairdner Found. Internat. award, 1993. Mem. NAS (Inst. Medicine, Richard Lounsbery award for extraordinary achievements in biology and medicine 1993), Internat. Soc. Neurochemistry, Am. Acad. Arts and Scis., Am. Acad. Neurology (George Cotzias award for outstanding rsch. 1987, Presdl. award 1993), Am. Assn. Physicians, Am. Soc. Microbiology, Am. Soc. Neurochemistry, Am. Soc. Virology, Am. Neurol. Assn., Concordia Argonaut Club.

PRYOR, DOUGLAS KEITH, clinical psychologist, consultant; b. Oakland, Calif., Aug. 12, 1944; s. Richard Smith and Clara Margaret (Ford) P.; m. Roxanne Rae, Oct. 14, 1990. BA, Calif. State U., Chico, 1967; MA, U. Pacific, 1971; PhD, U. So. Calif., 1976. Lic. psychologist, marriage, family and child counslor, ednl. psychologist, Calif. Social svc. worker Dept. Pub. Assistance, Stockton, Calif., 1967-70; asst. to program dir. Stockton State Hosp., 1970-72; sch. psychologist, vice prin. Mesrobian Elem. and High Sch., Pico Rivera, Calif., 1972-74; sch. psychologist Pasadena (Calif.) Unified Sch. Dist., 1974-83; pvt. practice clin. psychology, Santa Ana, Calif., 1980-87, Sacramento, 1987—; pres. Behavior Mgmt. Cons., Sacramento, 1982—; mgr. Calif. Behavior Modification Workshop, Stockton, 1972; clin. fellow Human

Factors Programs, Santa Ana, 1977-79, Milton Erikson Advanced Inst. Hypnotherapeutic and Psychotherapeutic Studies, Santa Ana, 1979-80; presenter, lectr. in field. Mem. APA, Calif. Assn. Sch. Psychologists (rep. region V, 1982-88, officer 1992), Calif. Psychol. Assn. (officer 1992), Sacramento Valley Psychology Assn. (bd. dirs. 1990—, treas. Div. 1, 1990, pres. Div. 1, 1992). Democrat. Buddhist. Avocations: photography, numismatics. Office: Behavior Mgmt Cons 650 Howe Ave Ste 560 Sacramento CA 95825-4732

PRYOR, WAYNE ROBERT, astronomer; b. Oakland, Calif., May 21, 1961; s. Arthur William and Lila Marie (Carlin) P. AB in Physics and Applied Math., U. Calif., Berkeley, 1983; PhD in Astrophys. Planetary & Atmos. Sci, U. Colo., 1989. Rsch. assoc. Lab. for Atmospheric and Space Physics, Boulder, Colo., 1989—; co-investigator Neptune data analysis program, NASA, 1990, Pioneer Venus guest investigator program, 1991, Galileo ultraviolet spectrometer team, 1992. Contbr. articles to profl. jours. Methodist. Office: Lab Atmospheric & Space Physics 1234 Innovation Dr Boulder CO 80303

PRYSCH, PETER, electrical engineer; b. Winterthur, Zurich, Switzerland, Aug. 4, 1962; came to U.S., 1991; m. Andrea Daniela Trutmann, Aug. 6, 1988. BSc in Engring., Technikum Winterthur, 1986; Fach-Ingenieur, Neu-Technikum Buchs, Switzerland, 1987. Rsch. engr. Alos, Ltd., Zurich, 1988-90; project leader Sprecher Energie, Ltd., Oberentfelden, Switzerland, 1990-91; dead tank products mgr. GEC Alsthom, Blauvelt, N.Y., 1991—; dir. Fremont Enterprises, Inc., Lander, Wyo. Contbr. articles to profl. jours. Achievements include devel. of formal specification method, datasharing by means of a "ctrl. repository" for integrated programming support environ. Home: RR 2 Box 116A Campbell Hall NY 10916 Office: GEC Alsthom T&D 200 Corporate Dr Blauvelt NY 10913

PRZELOZNY, ZBIGNIEW, physicist; b. Opole, Poland, Feb. 9, 1954; arrived in Australia, 1988; s. Edmund and Ludomira (Mazur) P.; m. Iwona Neugebauer, Sept. 14, 1956; children: Adrian, Jan. MS in Physics, U. Krakow, Poland, 1978, M in Math/Computer Sci., 1982, PhD in Physics, 1983. Lectr. physics U. Cracow, 1980-85; rschr. Deutsche Electronen Synchrotron, Hamburg, Germany, 1985-88, Monash U., Melbourne, Australia, 1988—. Contbr. articles to profl. jours. Fellowship Monash U., 1988. Achievements include research on existence of multiquantum Josephson vortex, general formulation of lattice dynamics in type II superconductors, prediction and experimental confirmation of different magnetic properties of high Tc superconductor of different grains size. Home: 43 Jacksons Rd, Melbourne 3174, Australia Office: Monash U, Wellington Rd, Melbourne 3168, Australia

PRZYBYCIEN, TODD MICHAEL, chemical engineering educator; b. Clarkson, N.Y., Aug. 19, 1962; s. Edward Francis and Diana (Graesel) P.; m. Valerie Patrick, Oct. 27, 1990. BA in Chemistry, Washington U., St. Louis, 1984, BSChemE, 1984; MSChemE, Calif. Inst. Tech., 1986, PhD in Chem. Engring., 1989. Sr. rsch. engr. Monsanto Agrl. Co., St. Louis, 1989-90; asst. prof. chem. engring. Rensselaer Poly. Inst., Troy, N.Y., 1991—, Isermann Jr. faculty chair, 1991—; cons. Monsanto Agrl. Co., 1988-89. Contbr. articles to profl. publs., chpt. to book. Bd. dirs. Howe Farm Assn., West Stockbridge, Mass., 1990—. NSF grantee, 1992; Langsdorf fellow, 1980-84, Hon. Mention Grad. fellow NSF, 1984, 85. Mem. AAAS, Am. Inst. Chem. Engrs. (local student sec. 1983-84), Am Chem. Soc. (local student pres. 1982-83), N.Y. Acad. Scis., Sigma Xi, Phi Beta Kappa, Tau Beta Pi, Omicron Delta Kappa. Republican. Achievements include development of predictive model for protein precipitation and process-scale refolding, studies of recombinant therapeutic proteins and interactions with drug delivery device materials, use of spectroscopy to characterize structure of protein aggregates. Office: Rensselaer Poly Inst Ricketts 127 Troy NY 12180-3590

PSALTAKIS, EMANUEL P., environmental engineer; b. Flushing, N.Y., Aug. 11, 1965; s. Pantelis N. and Bess (Sandalis) P.; m. Anna E. Kapsis, Jan. 13, 1991. BCE, The Cooper Union, 1987; M in Environ. Engring., Manhattan Coll., 1993. Prin. engr. Hazen and Sawyer, P.C., N.Y.C., 1987—. Mem. ASCE, Water Environment Fedn. Office: Hazen & Sawyer PC 730 Broadway New York NY 10003-9511

PSARIS, AMY CELIA, manufacturing engineer; b. Bklyn., June 18, 1963; d. Arnold S. and Ellen Marion (Wachtel) Levitt. BSME, U. Tex., 1985; MBA, U. Mich., 1991. Mfg. engr. BOC Power Train GM, Flint, Mich., 1985-87; balance engr. BOC Power Train GM, Flint, 1987-88; mfg. engr. BOC Power Train GM, Pontiac, Mich., 1988-91; sr. mgr. engr. Orbital Engine Co., Tecumseh, Mich., 1991-93; sr. project engr. Johnson & Johnson Corp., Austin, Tex., 1993—. Lit. vol. Williamson County Lit. Coun., 1992-93; mem. Beta Sigma Phi Svc. Sorority, 1992—, Austin Civic Wind Ensemble; tutor Vols. for Adult Literacy, 1987-88. Mem. ASME, Mich. Soc. Profl. Engrs. (bd. dirs. 1987-90, Young Engr. of Yr. 1988, MathCounts chmn. 1988), Soc. Mfg. Engrs., Saginaw Valley Engrs. Coun. (banquet publicity chmn. 1987-88), Tex. Exes Alumni Orgn. Avocations: piano, flute, skiing, golf. Home: 7508 Montaque Dr Austin TX 78729-9999

PSCHUNDER, WILLI, semiconductor engineer; b. Bietigheim, Germany, June 12, 1944; m. Hildegard Krause, June 4, 1971. Diploma engring., Fachhochschule, Heilbronn, Germany, 1968. Space and terrestrial engr. AEG-Telefunken, Heilbronn, 1968-79, solar cell devel. terrestrial engr., 1980-87; solar cell prodn. engr. Telefunken Electronic, Heilbronn, 1987-88; solar cell application engr. Telefunken Systemtechnik, Heilbronn, 1988-91; solar cell quality control engr. Telefunken Systemtechnik, Deutsch Aerospace, Heilbronn, 1992—. Home: Im Ring 43, 74360 Ilsfeld Germany Office: Deutsche Aerospace AG, Theresienstr 2, 74072 Heilbronn Germany

PSHTISSKY, YACOV, electrical engineer; b. Tel-Aviv, Jan. 5, 1952; came to the U.S., 1977, m. Debra Blonder, May 19, 1979. BEE, Poly. U., N.Y.C., 1979, M in Computer Sci., 1985. V.p. engring. Vicon Industries, Melville, N.Y., 1990—. Mem. IEEE, SMPTE. Achievements include patents for encoding/decoding video, bi-directional video amp, modular video control. Home: 1 Bay Club Dr Bayside NY 11360

PSUTY, NORBERT PHILLIP, marine sciences educator; b. Hamtramck, Mich., June 13, 1937; s. Phillip and Jessie (Proszykowski) P.; m. Sylvia Helen Zurinsky, June 13, 1959; children: Eric Anthony, Scott Patrick, Ross Phillip. BS, Wayne State U., 1959; MS, Miami U., Oxford, Ohio, 1960; PhD, La. State U., 1966. Rsch. assoc. Coastal Studies Inst., La. State U., Baton Rouge, 1962-64; instr. dept. geography and dept. geology U. Miami, Coral Gables, Fla., 1964-65; asst. prof. geography U. Wis., Madison, 1965-69; assoc. prof. geography and geol. scis. Rutgers U., New Brunswick, N.J., 1969-73, prof., 1973—, chmn. dept. marine and coastal scis., 1991—, dir. Marine Scis. Ctr., 1972-76, dir. Ctr. for Coastal and Environ. Studies, 1976-90; assoc. dir. Inst. Marine and Coastal Scis., New Brunswick, N.J., 1990—; mem. sci. com. Thalassas, Vigo, Spain, 1985—. Co-author: Living with the New Jersey Shore, 1986, Coastal Dunes, 1990; mem. editorial bd Coastal Mgmt., 1981—, Jour. Coastal Rsch., 1987—; contbr. numerous articles to scholarly jours., chpts. to books, monographs. Mem. Water Policy Bd., East Brunswick, N.J., 1981-83, N.J. Shoreline Adv. Bd., Trenton, 1984-86; chmn. N.J. Gov.'s Sea Level Rise Com., Trenton, 1987-90; referee U.S. Volleyball Assn. Recipient Disting. Pub. Svc. award Pres. of Rutgers U. 1988; numerous grants including NSF, Nat. Park Svc., EPA, Office Naval Rsch., Nat. Sea Grant Program, NOAA, 1961—. Mem. AAAS, Assn. Am. Geographers (Honors award 1993), Coastal Soc. (pres. 1980-82), Internat. Geog. Union (vice chair commn. on coastal environ. 1988-92, chmn. commn. on coastal systems, 1992-96, editor newsletter 1984-96), N.J. Acad. Sci. (pres. 1982). Avocations: gardening, reading. Office: Rutgers U Inst Marine & Coastal Scis Cook Campus New Brunswick NJ 08903

PUCK, THEODORE THOMAS, geneticist, biophysicist, educator; b. Chgo., Sept. 24, 1916; s. Joseph and Bessie (Shapiro) Puckowitz; m. Mary Hill, Apr. 17, 1946; children: Stirling, Jennifer, Laurel. B.S., U. Chgo., 1937, Ph.D., 1940. Mem. commn. airborne infections Office Surgeon Gen., Army Epidemiol. Bd., 1944-46; asst. prof. depts. medicine and biochemistry U. Chgo., 1945-47; sr. fellow Am. Cancer Soc., Calif. Inst. Tech., Pasadena, 1947-48; prof. biophysics U. Colo. Med. Sch., 1948—, chmn. dept., 1948-67, disting. prof., 1986—; dir. Eleanor Roosevelt Inst. Cancer Research, 1962—; Disting. research prof. Am. Cancer Soc., 1966—; nat. lectr. Sigma Xi, 1975-76. Author: The Mammalian Cell as a Microorganism: Genetic and Biochemical Studies in Vitro, 1972. Mem. Commn. on Physicians for the Future. Recipient Albert Lasker award, 1958, Borden award med. rsch., 1959, Louisa Gross Horwitz prize, 1973, Gordon Wilson medal Am. Clin. and Climatol. Assn., 1977, award Environ. Mutagen Soc., 1981, E.B. Wilson medal Am. Soc. Cell Biology, 1984, Bonfils-Stanton award in sci., 1984, U. Colo. Disting. Prof. award, 1987, Henry M. Porter medal, 1992; named to The Colo. 100, Historic Denver, 1992; Heritage Found. scholar, 1983; Phi Beta Kappa scholar, 1985. Fellow Am. Acad. Arts and Scis.; mem. Am. Chem. Soc., Soc. Exptl. Biology and Medicine, AAAS (Phi Beta Kappa award and lectr. 1983), Am. Assn. Immunologists, Radiation Research Soc., Biophys. Soc., Genetics Soc. Am., Nat. Acad. Sci., Tissue Culture Assn. (Hon. award 1987), Paideia Group, Santa Fe Inst. Sci. Bd., Phi Beta Kappa, Sigma Xi. Achievements include pioneering contributions to establishment of somatic cell approaches to mammalian cell genetics, to the identification and classification of the human chromosomes; measurement of mutation in mammalian cells; demonstration of the reverse transformation reaction and the genome exposure defect in cancer; development of quantitative approaches to mammalian cell radiobiology. Office: Eleanor Roosevelt Inst Cancer Rsch 1899 Gaylord St Denver CO 80206-1299

PUCKETT, ELBRIDGE GERRY, mathematician, educator; b. Boston, Dec. 1, 1956; s. Elbridge Smith and Gloria Hope (Yost) P.; m. Carolyn Louise Doyle, Sept. 10, 1988. BA in Math., Sonoma State U., 1981; PhD in Math., U. Calif., Berkeley, 1987. Postdoctoral fellow Lawrence Livermore (Calif.) Nat. Lab., 1987-90; asst. prof. math. U. Calif., Davis, 1990-93, assoc. prof., 1993—; cons. Lawrence Livermore Nat. Lab., 1992—. Contbr. articles to profl. jours. U.S. Dept. Energy applied math. scis. fellow, 1989-90. Mem. Am. Phys. Soc. Office: Univ of Calif Dept Math Davis CA 95616

PUCKETT, HOYLE BROOKS, agricultural engineer, research scientist, consultant; b. Jesup, Ga., Oct. 15, 1925; s. Lawrence Parham and Martha Elizabeth (Mizell) P.; m. Faye Eloise Bowden, June 22, 1945; children: Carol P. Keeley, Hoyle B. Jr., Kristina P. Berbaum. BS in Agrl. Engring., U. Ga., 1948; MS in Agrl. Engring., Mich. State U., 1949; student, Ga. Inst. Tech., 1964-65. reg. profl. engr. Ill. Res. engr. GS 7-9 USDA/ARS, Oxford, N.C., 1949-55; res. engr. GS 11-13 USDA/ARS, Urbana, Ill., 1955-68, res. ldr. GS 14-15, 1968-55; ret., 1985; engring. cons. pvt. practice, Champaign, Ill., 1985—. Fellow Am. Soc. Agrl. Engrs.; mem. Exchange Club of Urbana. Home: 407 W Univ Apt 104 Champaign IL 61820

PUDDY, DONALD RAY, mechanical engineer; b. Ponca City, Okla., May 31, 1937; s. Lester Andrew and Mildred Pearl (Olson) P.; m. Dana C. Timberlake, Sept. 8, 1956; children: Michael R., Douglas A., Glenn L. BSME, U. Okla., 1960; MBA, U. Houston, 1978; postgrad., Harvard U., 1991. Officer USAF, Fort Walton Beach, Fla., 1960-64; flight controller NASA Johnson Space Ctr., Houston, 1964-66, head module systems sec., 1966-69, asst. chief of lunar module systems br., 1969-72, asst. chief of space sci. and tech. br., 1972-74, chief of mission ops. br., 1974-75, flight dir. Apollo, Skylab, ASTP, Shuttle, lead flight dir. for shuttle orbiter programs, 1976-82, chief mission ops. systems div., 1982-85, asst. dir. for systems in mission ops., 1985-86, dir. flight crew ops., 1988-92; acting dep. dir. NASA's Ames Rsch. Ctr., Calif., 1986; asst. assoc. adminstr. spaceflight NASA Hdqs., Washington, 1986-87; asst. to NASA adminstr., 1991; spl. asst. joint U.S./Russian Programs NASA, Houston, 1992—. Pres. PTA, Ed White Elem. Sch.; past troop com. chmn. and asst. scoutmaster Boy Scouts Am. Mem. AIAA (mem. coun. 1983-84), Pi Tau Sigma, Sigma Tau. Office: NASA Johnson Space Ctr Joint US/Russian Office Houston TX 77058

PUETTMER, MARCUS ARMIN, metallurgical engineer; b. Stuttgart, BW, Germany, Aug. 28, 1964; s. Hermann and Erika (Hirluksch)P.; m. Sabine Bettig, Oct. 16, 1992. Abitur, Max-Born-Gymasium, Backnang, BW, Germany, 1984; diploma, Universitat Karlsruhe, Baden-Württemberg, BW, Germany, 1991. Engr. Gotz Gmbh Metall und Anlagenbau, Fellbach, BW, Germany, 1991—; mng. dir. P and K Energietechnik Gmbh, Fellbach, 1992—. Serg 1st l, Signal Bn., 1984-85. Roman Catholic. Achievements include patents held in absorbing-reflecting solar blinds, solar cut renewing facade system, solar cooling system. Office: Goetz Metall und Anlagenbau, Hoehen St 16, 7012 Fellbach Germany

PUETZ, WILLIAM CHARLES, engineering company executive; b. Chelsea, Mass., May 29, 1950; s. William Michael and Munita Laura (Marks) P.; m. Lynne Ellen Brewer, Jan. 9, 1971; children: Brittney Ellen, Katherine Lynne. BS in Engring. Mgmt., U. Mo., Rolla, 1976. Salesman Carp's, Rolla, 1971-73; asst. mgr. Mohr Value, Rolla, 1973-75; loss prevention cons. Factory Mut. Engring. Assn., St. Louis, 1976-78, adjuster, 1978-79, staff adjuster, 1979-82, sr. adjuster, 1982-86, dist. mgr., 1986-88, dist. mgr., gen. adjuster, 1988—. Home: 606 Jasmin Dr Saint Louis MO 63122-2550 Office: Factory Mutual Engring Assn 3300 S Rider Trl Ste 600 Earth City MO 63045-1310

PUGAY, JEFFREY IBANEZ, mechanical engineer; b. San Francisco, June 26, 1958; s. Herminio Salazar and Petronila (Ibanez) P. BSME, U. Calif., Berkeley, 1981, MSME, 1982; MBA, Pepperdine U., 1986, MS in Tech. Mgmt., 1991. Registered profl. engr., Calif. Engring. asst. Lawrence Berkeley Nat. Lab., 1978-80; assoc. tech. staff Aerospace Corp., L.A., 1981; mem. tech. staff Hughes Space & Comm. Co., El Segundo, Calif., 1982-85, project engr., 1985-88, tech. head, 1988-89, sr. staff engr., 1989-90, mgr. tech. ops. and strategic systems, 1991-92, ops. leader, info. tech., 1992—. White House Fellow regional finalist, 1991, 92. Mem. ASME, Soc. Competitor Intelligence Profls., Am. Mgmt. Assn., L.A. World Affairs Coun., Make A Wish Found., Pi Tau Sigma, Delta Mu Delta. Republican. Roman Catholic. Avocations: racquetball, scuba diving, sailing, backpacking, volleyball. Home: 8180 Manitoba St Unit 120 Playa Del Rey CA 90293-8651 Office: Hughes Space & Comm Co PO Box 92919 Los Angeles CA 90009-2919

PUGH, PAUL FRANKLIN, engineer consultant; b. Orrvile, Ohio, Aug. 1, 1922; s. Frank Kaggey and Hazel Hanora (Yenser) P.; m. Ruth Morgan, Sept. 22, 1951; 1 child, Paul F. Pugh, Jr. BEEE with honors, U. So. Calif., L.A., 1948; postgrad., Wayne State U., 1950-51, Case Inst., 1952-53. Registered profl. engr., Ohio, Calif., Wash., Ala. Cable engr. GE, Schenectady, N.Y., Detroit, Cleve., San Francisco, 1949-67; cons., inventor, mgr. utilities divsn. Beamer/Wilkinson & Assocs., 1968-75; cons., inventor Paul F. Pugh and Profl. Engrs., Oakland, Calif., 1975-88; engring. mgr., cons. Turnupseed Electric Svc., Tulare, Calif., 1989-92; cons., inventor Sprinville, Calif., 1992—. Contbr. over 17 tech. articles on cables, heat pipe, monitoring to profl. mags. Pres. Republican Club, North Royalton, Ohio, 1967. With U.S. Army, 1942-46. Named for Engring. Achievement, Specifying Engr. Mag., 1981. Mem. IEEE (Power, Indsl. Applications, Insulations Socs., voting mem. insulated conductor com. 1968—), Am. Legion, Eta Kappa Nu, Tau Beta Pi. Achievements include patents for Underground Cable System, for Electric Battery, for Wind Machine, for Heat Pipe, for Coreless Transformer, 5 others; research on method for using flue gas to prevent air pollution. Home and Office: 33112 Globe Dr Springville CA 93265

PUGH, TIM FRANCIS, II, software engineer, aeronautical engineer; b. Binghamton, N.Y., Apr. 17, 1963; s. Francis Leo and Mary Louis Pugh; m. Teresa J. Diaz, Dec. 21, 1991. BS in Aero. Engring., Calif. Poly. State U., San Luis Obispo, 1987, MS, 1993. Engring. technologist Jenike & Johanson, Inc., San Luis Obispo, 1985-86, rsch. engr., 1987-92; faculty rsch. asst., sci. programmer dept. oceanography Oreg. State U., Corvallis, 1992—; software engr. Visual Engring. Solutions, Salem, Oreg., 1992—; editor Mid-Willamette NEXT Users Group, Corvallis, 1992. Office: Visual Engring Solutions 5127 Vale Ct SE Salem OR 97306

PUGMIRE, GREGG THOMAS, optical engineer; b. Montpelier, Idaho, Sept. 23, 1963; s. Vaughn Rich and Yvonne (Thomas) P.; m. Linda Lee Harris, July 17, 1987; children: Lindsay, Stephanie. BS, MS, Brigham Young U., 1990. Summer rsch. engr. Nat. Security Agy., Ft. Meade, Md., 1988; rsch. asst. Brigham Young U., Provo, Utah, 1988-90; optical engr., scientist ESL Inc., Sunnyvale, Calif., 1990—; 1993 conf. presenter in field. Engring. scholar Brigham Young U., 1989-90. Mem. Optical Soc. Am., Golden Key, Eta Kappa Nu. Mem. LDS Ch. Achievements include novel system to monitor and regulate etch depth of D-type fibers; co-development of method to put diffraction gratings directly on a fiber optic cable, coupling guided light to free space; development of high proformance analog fiber optic links; research of high speed (GB/s) digital communications Fiber Optics Links. Home: 2184 Royal Dr Santa Clara CA 95050-3610 Office: ESL Inc 495 E Java Dr Sunnyvale CA 94088-3510

PUHK, HEINO, chemist, researcher; b. Sulustvere, Poltsamaa, Estonia, Dec. 15, 1923; came to U.S., 1950; s. Karl and Elviine (Rosenbaum) P., m. Marianne Elisabet Zettergren, May 19, 1955; children: Christina Elizabeth, Eva Maria. BS, Case Inst. Tech., 1954, MS, 1959. Rsch. chemist Tremco Mfg. Co., Cleve., 1954-55, Lubrizol Corp., Cleve., 1955-57; grad. asst. Case Inst. Tech., Cleve., 1957-59; sr. chemist Glidden Co., Cleve., 1960-75; scientist Glidden div. SCM Corp., Cleve., 1976-85, Imperial Chem. Industries, Cleve., 1986-88. Patentee in field. Pres. Estonian Luth. Ch., Cleve., 1977-78. Mem. Am. Chem. Soc., Estonian Philatelic Soc., Alpha Chi Sigma, Sigma Xi. Avocations: Estonian history, nature, photography. Home: 23902 Fairlawn Dr North Olmsted OH 44070-1507

PULIAFITO, CARMEN ANTHONY, ophthalmologist, laser researcher; b. Buffalo, Jan. 5, 1951; s. Dominic F. and Marie A. (Nigro) P.; m. Janet H. Pine, May 19, 1979. AB cum laude, Harvard Coll., 1973, MD magna cum laude, 1978. Diplomate Am. Bd. Ophthalmology. Intern Faulkner Hosp., Tufts U. Sch. Medicine, 1978-79; resident Mass. Eye and Ear Infirmary, Boston, 1979-82, retina fellow, 1982-83; instr. Harvard Med. Sch., Boston, 1983-85, asst. prof., 1985-89, assoc. prof., 1989-91; dir. divsn. continuing edn. dept. ophthalmology Harvard Med. Sch., 1989-91; vis. scientist MIT Regional Laser Ctr., Cambridge, 1982, asst. prof. health scis. and tech. program, 1987-89, assoc. prof., 1989—; mem. staff Mass. Eye and Ear Infirmary, Boston, 1983; bd. dirs. Morse Laser Ctr. Mass. Eye and Ear Infirmary, 1986-91; dir. New Eng. Eye Ctr., 1991—; prof., chmn. dept. ophthalmology Tufts U. Sch. Medicine, 1991—; prof. biomed. engring. Tufts U., 1991—. Author (with D. Albert) Foundations of Ophthalmic Pathology, 1979; (with R. Steinert) Principles and Practice of Ophthalmic YAG Laser Surgery, 1984; editor-in-chief jour. Lasers in Surgery and Medicine, 1987—; contbr. 60 sci. articles to profl. jours. Fellow Am. Acad. Ophthalmology. Roman Catholic. Home: 69 Pigeon Hill Rd Weston MA 02193 Office: New England Eye Ctr 750 Washington St Boston MA 02111

PULICH, WARREN MARK, JR., plant ecologist, coastal biologist; b. Tucson, Oct. 29, 1946; s. Warren Mark and Anne Marie (Doles) P.; m. Joyce Elaine Ilse, Dec. 29, 1968; children: Michelle, Mark, Jennifer. BS in Biol. Sci., Loyola U., New Orleans, 1967; PhD in Biology, Rice U., 1971. Rsch. biologist Marine Sci. Inst., U. Tex., Port Aransas, 1971-87; coastal ecologist Tex. Parks and Wildlife Dept., Austin, 1987—; environ. cons., coastal Tex., 1979-86; program reviewer Sea Grant Coll. Program, Silver Spring, Md, 1987—; mem. tech. adv. com. coastal ocean program NOAA, Silver Spring. Contbr. articles to Plant Physiol. Jour., Sci., Aquatic Botany Jour., Jour. Exptl. Marine Biology and Ecology, Jour. Coastal Rsch. Chmn. Port Aransas Planning Commn., 1979-82. Rsch. grantee NOAA, 1985-87. Mem. Estuarine Rsch. Fedn., Am. Soc. Limnology and Oceanography, Sigma Xi. Roman Catholic. Achievements include development of GIS techniques for assessing change in coastal wetland ecosystems and sensitivity to environmental stresses; documented effects of freshwater inflows from rivers on coastal Texas wetlands; research on coastal submerged vegetation in Gulf of Mexico, using physiological ecology and community dynamics techniques. Home: 2300 Westway Cir Austin TX 78704 Office: Tex Parks and Wildlife Dept 4200 Smith School Rd Austin TX 78744

PULIDO, CARLOS ORLANDO, electronics engineer, economist, consultant; b. Bogota, Colombia, Aug. 22, 1949; s. Antonio and A. Lucia Pulido; m. Rosario I., Aug. 5, 1978; children: Angelica, Andrea. BSEE, Dist. U. Bogota, 1972; MBA, Andes U., 1976; MS in Projects Econs., U. Javeriana, 1992. Equipment engr. Falconbridge Nickel, Dominican Republic, 1972-73; planing engr. Ecopetrol Oil Pipelines, Bogota, 1973-74; projects engr. Core Labs., Dallas, 1975-76; mktg. engr. Hervasquez Co., Colombia, Ecuador, 1977-78; mktg. mgr. Panamerican Engring., Colombia, 1979-83, pres., 1984—; mktg. cons. PBC Internat., Moscow, 1988—; cons. in engring., oil, gas, and mining, and mktg. engring.; advisor Soviet Republiques and Latin Am. countries. Author: Strategies for Marketing of High Technology Equipment for Western World, 1991, Development of Soviet Country Economics, 1992. Affiliate Aciem Engring. Assn., Bogota, 1973; mem. Interam. Mktg. Inst., Bogota, 1976. Fellow Colombian Engring. Assn.; mem. InterAm. Mktg. Inst. Roman Catholic. Office: Panamerican Engring World Trade Ctr Tower A, CRA 8 A 99-51 Ofc 701, Bogota 8, Colombia

PULLEN, MARGARET I., genetic physicist; b. Nebr., Sept. 13, 1950; d. Robert and Martha (Holtort) P. AA, Stephens Coll., 1971; BA in Internat. Rels., Econs., Bus. & Trade, U. Colo., 1975; BS in Physics, Northeastern U., 1983; MS in Physics, Tufts U., 1984; postgrad., U. Calif., 1984-86. Mathematician lawrence Livermore Lab., Calif., 1987; cons. Porterfield Enterprise, 1988; entrepreneur Evergreen Applied Rsch. Inc., 1988--; mem. Biotechnolgy Roundtable, Denver, 1987--. Vol. Nat. Sports Ctr. for the Physically Disabled, Winter Park, Colo., 1988-92. Tufts U. grantee, U. Colo. grantee; Nat. Sci. fellow; Stephens scholar, Perry Mansfield Ctr. for the Preforming Arts scholar. Mem. Am. Phys. Soc. Office: Evergreen Applied Rsch Inc PO Box 2870 Evergreen CO 80439

PULLIAINEN, ERKKI OSSI OLAVI, zoology educator, legislator; b. Varkaus, Finland, June 23, 1938; s. Ossian Ferdinand and Hilkka-Liisa (Sinervo) P.; m. Anneli Kuvaja, 1961 (div. 1973); children: Harri, Virpi; m. Riitta Inkeri Haaranen, 1974; children: Annariina, Annamiina, Rauli. Degree, U. Helsinki, Finland, 1960; MSc, U. Helsinki, 1964, PhD, 1966, M Agrl. and Forestry Sci., 1968. Scientist U. Helsinki, 1965-74; sr. scientist Acad. of Finland, Helsinki, 1975; prof. zoology U. Oulu, Finland, 1975—, dir. dept. zoology, 1978-87, dean of sci., 1980-87; dir. Värriö Rsch. Sta., U. Helsinki, Salla, 1967—. Author: Great Predators of Finland, 1974, Trade Union Policy of Civil Servants, 1975, Finland in the Year 2000, 1982, Predators and Man, 1984, 7 other books and over 400 scientific papers. Mem. City Coun. Oulu, 1985—; mem. Parliament of Finland, 1987—, group chmn., 1989-91, group vice-chmn., 1993—. Knight first class order of Finnish Lion. Mem. Mammalogical Soc. Finland (pres. 1987—), N.Y. Acad. Scis., Adminstrv. Coun. Vapo, Univ. Tchrs. and Employees Union (hon. pres. 1978). Mem. Green League. Evangelical Lutheran. Home: Rantakallontie 6, 90800 Oulu Finland Office: Parliament of Finland, 00102 Helsinki Finland

PULLIAM, CURTIS RICHARD, chemistry educator; b. Springfield, Ill., Oct. 16, 1957; s. Charles Richard P.; m. Joni L. Lawson, Nov. 24, 1979. BS in Chemistry, Western Ill. U., 1979; PhD in Inorganic Chemistry, U. Wis., 1986. Sr. rsch. staff mem. USG Corp., Libertyville, Ill., 1986-87; asst. prof. Utica (N.Y.) Coll. of Syracuse U., 1987-91, assoc. prof., 1991—. Recipient grant NSF, Utica Coll., 1988, 93. Mem. Am. Chem. Soc., Am. Crystallographic Assn., Coun. on Undergrad. Rsch. Office: Utica Coll Syracuse Univ 1600 Burrstone Rd Utica NY 13502

PULLIAM, HOWARD RONALD, ecology educator; b. Miami Beach, Fla., Sept. 7, 1945; s. Joe J. Jr. and Rachel Elizabeth (Miller) P.; m. Janice L. Crowder, June 20, 1969; children: Juliet, Tomlin. BS, U. Ga., 1968; PhD, Duke U., 1969. Asst. prof. U. Ariz., Tucson, 1971-78; rsch. biologist H. S. Colton Rsch. Ctr., Flagstaff, Ariz., 1978-80; assoc. prof. SUNY, Albany, 1980-84; prof. U. Ga., Athens, 1984—, dir. Inst. Ecology, 1987—. Author 2 books; contbr. over 60 articles to sci. jours. Bd. dirs. Ga. Conservancy, 1987—. Mem. Ecol. Soc. Am. (v.p. 1986-87, pres. 1991-92), Southeastern U. Rsch. Assn. (v.p. 1990-91). Office: U Ga Institute of Ecology 103 Ecology Bldg Athens GA 30602

PULLIAM, TERRY LESTER, chemical engineer; b. Garden City, Kans., Apr. 29, 1949; s. Lester Orville and Lucille Melisse (Chandler) P.; m. Page Ellen Pate, Oct. 7, 1989. BS in Chem. Engring., Oreg. State U., 1972. Project engr. ITT Rayonier Inc., Port Angeles, Wash., 1972-78; tech. dir. Boise Cascade Corp., Deridder, La., 1978-82; prodn. mgr. Finch Pruyn Paper Co., Glens Falls, N.Y., 1982-83; tech. supt. Great So. Paper Co., Cedar Springs, Ga., 1983-86; cons. Orion CEM Inc., Atlanta, 1986-87; sr. staff engr. BE & K Engring., Birmingham, Ala., 1987—. Loaned exec. United

Fund, Salem, Oreg., 1978. Staff sgt. USNG, 1968-74. Mem. TAPPI. Republican. Achievements include research in pulp and paper. Home: 4636 Lake Valley Dr Birmingham AL 35244 Office: BE& K Engring Co 2000 International Park Dr Birmingham AL 35243

PULLIN, JORGE ALFREDO, physics researcher; b. Buenos Aires, Feb. 26, 1963; came to U.S., 1989; s. Archie Ernest and Evangelina (Rostagno) P.; m. Gabriela Ines Gonzalez, Oct. 7, 1989. MSc in Physics, Inst. Balseiro, Bariloche, Argentina, 1986, PhD in Physics, 1988. Asst. prof. U. Cordoba, Argentina, 1988—, mgr. computer systems physics dept., 1988-89; rsch. assoc. Syracuse (N.Y.) U., 1989-91, U. Utah, 1991-93; asst. prof. Pa. State U., Collegeville, 1993—; grant reviewer ANEP, Govt. of Spain, 1991—. Reviewer Math. Revs., 1988—, Soc. for Indsl. & Applied Math. Rev., 1991—, Cambridge U. Press, 1991—; referee Physics Letters, 1990—, Phys. Rev., 1990—, Am. Jour. Physics, 1991—, Jour. Math. Physics, 1991—; contbr. articles to profl. jours. Mem. Internat. Soc. Gen. Relativity.

PUMARIEGA, JOANNE BUTTACAVOLI, mathematics educator; b. Coral Gables, Fla., May 27, 1952; d. Ciro Charles and Rosaria Frances (Calabrese) Buttacavoli; m. Andres Julio Pumariega, Dec. 26, 1975; children: Christina Marie, Nicole Marie. BA in Math. and Edn. magna cum laude, U. Miami, 1973, MA in Math., 1974; postgrad., U. Houston, 1991-92. Cert. secondary math. tchr., Tex., Fla., Tenn., N.C. Grad. teaching asst. U. Miami, Coral Gables, Fla., 1973-74; substitute tchr. Dade County Pub. Schs., Miami, 1975; math. instr. Miami Dade Community Coll., 1975-76; math. and G.E.D. instr. Durham (N.C.) Tech. Inst., 1976-77; math. instr. Durham High Sch., 1977-78, Durham Acad., 1978-80, Univ. Sch. of Nashville, 1980-83; pvt. practice math. instr. Houston, 1984-86; tutor Clear Lake Tutoring Svc., Houston, 1987-90; pvt. practice math. and S.A.T. instr. League City, Tex., 1990-92, Columbia, S.C., 1992—. Chair Bd. Edn. St. Mary Parish, League City, 1988-90, lector, 1990-92; treas. St. Thomas More Women's Club, Houston, 1985-86; v.p. Parent-Student Assn., League City, 1990—; v.p., then pres. housestaff med. wives Duke U., Durham, N.C., 1978-80. Mem. Newcomers of Greater Columbia (chair pub. rels. chpt. 1993—), Welcome Neighbors of Bay Area (v.p., program chmn. 1991-92), Tex. Med. Aux., Bay Area Med. Wives, Phi Kappa Phi, Kappa Delta Pi, Alpha Lambda Delta (Woman of Yr. 1972), Univ. of S.C. Faculty Women's Club, (v.p. 1993—). Roman Catholic. Avocations: reading, modeling, public speaking, traveling. Home and Office: 2121 Bee Ridge Rd Columbia SC 29223

PUNGOR, ERNO, chemist, educator; b. Vasszecseny, Hungary, Oct. 30, 1923; s. Jozsef and Franciska (Faller) P.; diploma of chemistry Pazmany Peter U., Budapest, 1948; Dr.h.c., Tech. U. Vienna, 1983; m. Elisabeth Lang, Oct. 26, 1950; children: Erno, Andras, Katalin; m. Tünde Horváth, Sept. 8, 1984. Asst. prof. Inst. Inorganic and Analytic Chemistry, Eotvos Lorand U., Budapest, 1948-51, reader, 1951-53, assoc. prof., 1953-62; prof. Inst. Analytical Chemistry, U. of Chem. Industry, Veszprem, 1962-70; prof. Inst. for Gen. and Analytical Chemistry, Tech. U. Budapest, 1970—; mem. nat. environ. com. Com. for Nat. Tech. Devel. of Hungary; redwood lectr. English Soc. for Analytic chemistry, 1979. Recipient 3d Robert Boyle gold medal in analytical chemistry, 1986, Talanta Gold medal, 1986. Mem. Internat. Union Pure and Applied Chemistry, Fedn. European Chem. Socs. (chmn. working group of European analysts), Hungarian Chem. Soc. (head analytical group), Hungarian Acad. Sci. (head analytical div.), Czechoslovakian Acad. Scis. (hon. mem. chemistry div.), Austrian Analytical and Microanalytical Soc. (hon.), Finnish Chem. Soc. (corr.), Japanese Analytical Chem. Soc. (hon.). Author: Oscillometry and Conductometry, 1965; Flame Photometry, Theory, 1967. Mem. editorial bd. Acta Chimica Hungarica, 1967—, Periodica Polytech., 1972, Mikrochimica Acta, 1964, Kemiai Kozlemenyek, 1970, Talanta, 1968, Analyst, 1970, Analitica Chimica Acta, 1966, Analytical Letters, 1967, Bull. des Soc. Chimiques Belges, 1974, Bunseki Kegaku, 1981. Mem. adv. bd. Analytical Chemistry, 1985-88. Contbr. over 300 articles to profl. jours. Home: 4 Meredek, 1112 Budapest Hungary Office: Orszagos Atomenergia Bizottsag, POB 565, 1374 Budapest Hungary

PUNNAPAYAK, HUNSA, microbiologist, educator; b. Bangkok, Thailand, Apr. 10, 1951; s. Prasit and Rabieb P. Punnapayak. BSc in Botany, Chulalongkorn U., Bangkok, 1974; MS in Microbiology, U. S.W. La., 1980; PhD of Microbiology, U. Ark., 1984. Teaching asst. U. S.W. La., Lafayette, 1978-80; grad. asst. U. Ark., Fayetteville, 1980-84; asst. rsch. scientist U. Ariz., Tucson, 1984-87; instr. Chulalongkorn U., Bangkok, 1987-90, asst. prof., 1990—; cons. Mushroom Rsch. Unit, Chula, Bangkok, 1987—; PhD examiner U. Calcutta, India, 1990—; prin. investigator Royal Thai Army, Bangkok, 1989—. Author (invention): Bioconversion of Grindelic, 1988, Xylose fermentation, 1986, Fermentation of Agave, 1990, Ganoderma liquid culture, 1990; contbr. articles to profl. jours. Grantee Japan Internat. Coop. Agy. 1990. Mem. Soc. for Indsl. Microbiology, Sci. Soc. Thailand. Buddhist. Home: 419-6 Ratchawithee Rd, Bangkok 10400, Thailand Office: Chulalongkorn U Dept Botany, Faculty of Sci, Bangkok 10330, Thailand

PURANDARE, YESHWANT K., chemistry educator, consultant; b. Poona, Purandare, India, Sept. 19, 1934; came to U.S., 1961; s. Kashinath Purandare and Indira Deshpande; m. Margarita Renella, Feb. 8, 1964; children: Sarita, Amar, Jasmine, Ravi. BS with honors, U. Bombay, 1956; MS, U. Poona, 1960, Fordham U., 1964; PhD, NYU, 1981; MD, Spartan Health Sci. U., St. Lucia, B.W.I., 1985. Cert. clin. lab. specialist; cert. lab. dir. Grad. teaching asst. Fergusson Coll., Poona, 1956-60; quality control chemist Sardesai Bros., Bilimora, India, 1960-61; part time clin. chemist Brunswick Hosp., Amityville, N.Y., 1965-81; instr. chemistry Coll. Tech., SUNY, Farmingdale, N.Y., 1964-67; asst. prof. Coll. Tech., SUNY, Farmingdale, 1967-69, assoc. prof., 1969-74, prof., 1974—, chmn. dept. chemistry, 1984—; cons. Boat-Life Industries, Old Bethpage, N.Y., 1986—. Author (manual) Experiments in Physiol. Chemistry, 1985. Leader Boy Scouts of Am., 1974—. Mem. Am. Chem. Soc., Am. Assn. Clin. Chemistry (cert.), Am. Soc. Clin. Pathologists (cert. chemistry specialist), Rotary (Farmingdale pres. 1989-90, Paul Harris fellow 1992). Office: SUNY Coll Tech Chemistry Dept Lupton Hall Farmingdale NY 11735

PURCELL, EDWARD MILLS, physics educator; b. Taylorville, Ill., Aug. 30, 1912; s. Edward A. and Mary Elizabeth (Mills) P.; m. Beth C. Busser, Jan. 22, 1937; children: Dennis W., Frank B. B.S. in Elec. Engring, Purdue U., 1933, D. Engring. (hon.), 1953; Internat. Exchange student, Technische Hochschule, Karlsruhe, Germany, 1933-34; A.M., Harvard U., 1935, Ph.D., 1938. Instr. physics Harvard U., 1938-40, asso. prof., 1946-49, prof. physics, 1949-58, Donner prof. sci., 1958-60, Gerhard Gade Univ. prof., 1960-80, emeritus, 1980—; sr. fellow Soc. of Fellows, 1949-71; group leader Radiation Lab., MIT, 1941-45. Contbg. author: Radiation Lab. series, 1949, Berkeley Physics Course, 1965; contbr. sci. papers on nuclear magnetism, radio astronomy, astrophysics, biophysics. Mem. Pres.'s Sci. Advisory Com., 1957-60, 62-65. Co-winner Nobel prize in Physics, 1952; recipient Oersted medal Am. Assn. Physics Tchrs., 1968, Nat. Medal of Sci., 1980, Harvard medal, 1986. Mem. Am. Philos. Soc., NAS, Phys. Soc., Am. Acad. Arts and Scis., Royal Soc. (fgn. mem.). Office: Harvard U Dept Physics Cambridge MA 02138

PURCELL, JERRY, chemist. With Budd Co., Troy, Mich. Recipient Clare E. Bacon Person of Yr. award Soc. Plastics Industry, 1991. Office: Budd Co 3155 W Big Beaver Dr Troy MI 48007*

PURCELL, ROBERT HARRY, virologist; b. Keokuk, Iowa, Dec. 19, 1935; s. Edward Harold and Elsie Thelma (Melzl) P.; m. Carol Joan Moody, June 11, 1961; children: David Edward, John Leslie. BA in Chemistry, Okla. State U., 1957; MS Biochemistry, Baylor U., 1960; MD, Duke U., 1962. Intern in pediatrics Duke U. Hosp., Durham, N.C., 1962-63; officer USPHS, 1963, advanced through grades to med. dir. (O-6), 1974; with Epidemic Intelligence Svc., Communicable Disease Ctr. Atlanta; assigned to vaccine br. Nat. Inst. Allergy and Infectious Diseases, Bethesda, Md., 1963-65; sr. surgeon Lab. Infectious Diseases, NIH, Bethesda, Md., 1965-69, med. officer, 1969-72, med. dir., 1972-74, head hepatitis viruses sect., 1974—; organizer, invited participant, speaker numerous nat. and internat. symposia, confs., workshops, meetings; temporary advisor WHO, 1967—; expert cons. in hepatitis U.S.—China, U.S.—Taiwan, U.S.—Japan, U.S.—Russia, U.S.—India, U.S.—Pakistan Bilateral Sci. Agreements; lectr. various virology classes. Mem. editorial bd. and/or reviewer Am. Jour. Epidemiology, Gastroenterology, Hepatology, Infection and Immunity, Jour. Clin. Microbiology, Jour. Infectious Diseases, Jour. Med. Virology, Jour. Nat. Cancer Inst., Nature, Science; contbr. over 400 articles to profl. jours., chpts. to books. Co-patentee in field. Decorated D.S.M.; recipient Superior Svc. award USPHS, 1972, Meritorious Svc. medal USPHS, 1974, Gorgas medal, 1977, Disting. Alumni award Duke U. Sch. Medicine, 1978, Eppinger prize 5th Internat. FALK Symposium on Virus and Liver, Switzerland, 1979, Medal of City of Turin, Italy, 1983, Gold medal Can. Liver Found., 1984, Inventor's Incentive award U.S. Commerce Dept., 1984; fellowships Baylor U., 1959-60, Duke U., 1960-62. Fellow Washington Acad. Scis.; mem. Am. Epidemiology Soc., Am. Soc. Microbiology, Am. Soc. Virology, Am. Acad. Microbiology, AAAS, Soc. Epidemiol. Rsch., Infectious Diseases Soc. Am. (Squibb award 1980), N.Y. Acad. Scis., Am. Soc. Clin. Investigation, Assn. Am. Physicians, Am. Coll. Epidemiology, Am. Assn. Study of Liver Diseases, Internat. Assn. Study and Prevention Virus Associated Cancers, Internat. Assn. Biol. Standardization, Internat. Assn. Study Liver, Soc. Exptl. Biology and Medicine (Disting. Scientist award 1986), Nat. Acad. Scis. Office: NIH Lab Infectious Diseases Rm 202 Bldg 7 Bethesda MD 20892

PURCELL, THOMAS OWEN, JR., chemical company executive; b. Louisville, Feb. 15, 1944; s. Thomas Owen and Mildred (Cox) P.; m. Janice Lee Riddle, Nov., 1962 (div. 1973); 1 child, Sasha Annette; m. Sandra Lou Brown, Oct. 19, 1973. BS in Chemistry and Math., Campbellsville Coll., 1965; MA in Chemistry, Western Ky. U., 1966; PhD in Chemistry, U. Louisville, 1970. Chemist Celanese Coatings Co., Louisville, 1966-68; group leader Borg-Warner Chemicals, Parkersburg, W.Va., 1970-72, sect. mgr., 1972-74, assoc. tech. dir., 1974—; tech. dir. Borg Warner/GE Plastics, Parkersburg, Pittsfield, Mass., 1985-90; assoc. dir. rsch. Ferro Corp., Cleve., 1990-91, v.p. R & D, 1991—; mem. indsl. adv. bd. chem. engring. dept. W.Va. U., Morgantown, 1983-89, polymer dept. U. So. Miss., Hattiesburg, 1984-90, Ctr. for Applied Polymer Rsch., Cleve., 1989—; bd. dirs. Tech. Transfer Confs., Inc., Nashville, Cleve. Area Rsch. Dirs., Edison Polymer Innovation Corp. Author: Solid State Polymerization of D.A.A., 1970, (with others) MBS Impact Modifiers, 1975; contbr. articles to sci. jours. Pres., bd. dirs. Sheltered Workshop of Wood County, Parkersburg, 1987-89; campaign chmn. GE Plastics United Way, Pittsfield, 1989-90. Recipient Rotary award, 1965. Mem. Soc. Plastics Engrs. (sr., pres. sect. vinyl bd. 1970-75), Licensing Execs. Soc. (editor, int. tech. dir. 1985-90). Republican. Methodist. Achievements include rsch. in intellectual property mgmt., univ.-industry collaboration, impact modification of polyvinyl chloride; development of hundreds of new products. Office: Ferro Corp Tech Ctr 7500 E Pleasant Valley Rd Independence OH 44131

PURCIFULL, DAN ELWOOD, plant virologist, educator; b. Woodland, Calif., July 1, 1935; s. Ernest Lee and Virginia (Margaroli) P.; m. Marcia Ann Weatherby, Sept. 7, 1966; children: Scott, Douglas. B.S., U. Calif., Davis, 1957, M.S., 1959, Ph.D., 1964. Asst. prof. plant pathology U. Fla., Gainesville, 1964-69; assoc. prof. U. Fla., 1969-75, prof., 1975—; dept. grad. coord., 1988-91; mem. plant virus subcom. Internat. Com. for Taxonomy of Viruses, Internat. Legume Virus Working Group, 1973. Assoc. editor Phytopathology, 1971-73, Plant Disease, 1987-89; contbr. articles to profl. jours. Mem. Morningside Nature Center Commn., City of Gainesville, 1978-81, treas., 1981. Served with U.S. Army, 1957. Fellow Am. Phytopathol. Soc. (Lee Hutchins award 1981, Ruth Allen award 1993); mem. AAAS, Fla. State Hort. Soc., N.Y. Acad. Sci., Am. Soc. Virology, Sigma Xi. Home: 3106 NW 1st Ave Gainesville FL 32607-2504

PURDEY, GARY RUSH, materials science and engineering educator, dean; b. Edmonton, Alta., Can., Oct. 8, 1936; s. Kent Edward and Bertha (McNaught) P.; m. Ruby Elinore Smith, June 3, 1961; children: Anne, Ruth, Jonathan, Daniel. BSc in Mining Engring., U. Alta., 1957, MSc in Metallurgy, 1959; PhD in Metallurgy, McMaster U., 1962. Cert. profl. engr., Ont. Rsch. assoc. U. Alta., 1959-60; postdoctoral fellow McMaster U., Hamilton, Ont., 1962-63, asst. prof., 1963-67, assoc. prof., 1967-71, prof., 1971—; vis. prof. Ctrl. Electricity Rsch. Labs., Eng., 1969-70, Royal Inst. Tech., Stockholm, 1976-77, Inst. Nat. Polytechnique de Grenoble, 1983-84; assoc. dean sch. grad. studies McMaster U., 1971-74, chair materials sci. and engring., 1978-81, 84-86, acting dean, 1986-87, dean, 1989—; lectr. in field. Editor: (with others) Solute-Defect Interactions, Theory and Experiment, 1986, Advances in Phase Transitions, 1988, Fundamentals and Applications of Ternary Diffusion, 1990; contbr. over 100 articles to profl. jours. Named C. D. Howe Meml. fellow, 1969, Hon. Prof., U. Sci. and Tech., Beijing, 1983, fellow Royal Soc. Can., 1991; recipient Can. Metal Physics medal, 1990. Office: McMaster University, Dean of Engineering, Hamilton, ON Canada L8S 4L8

PURDOM, PAUL WALTON, JR., computer scientist; b. Atlanta, Apr. 5, 1940; s. Paul Walton and Bettie (Miller) P.; m. Donna Armstrong; children—Barbara, Linda, Paul. B.S., Calif. Inst. Tech., 1961, M.S., 1962, Ph.D., 1966. Asst. prof. computer sci. U. Wis.-Madison, 1965-70, asst. prof., 1971-82; mem. tech. staff Bell Telephone Labs., Naperville, Ill., 1970-71; assoc. prof., chmn. computer sci. dept. Ind. U., Bloomington, 1977-82, prof. computer sci., 1982—. Author: (with Cynthia Brown) The Analysis of Algorithms; assoc. editor: Computer Surveys; contbr. articles to profl. jours. NSF grantee, 1979, 81, 83, 92. Mem. AAAS, Soc. for Indsl. and Applied Math., Assn. Computing Machinery, Sigma Xi. Democrat. Methodist. Home: 2212 S Belhaven Ct Bloomington IN 47401-6803 Office: Ind U 215 Lindley Hall Computer Sci Dept Bloomington IN 47405-4101

PURDY, DAVID LAWRENCE, biotechnical company executive; b. N.Y.C., Sept. 18, 1928; s. Earl and Mabel (Roberts) P.; m. Margaret Helen Rye, July 7, 1951; children: Susan Lee, John F. (dec.), Ross David, Thomas Griffith. BSME, Cornell U., 1951; degree in advanced & creative engring., GE, 1955, degree in profl. bus. mgmt., 1956. Devel. engr. GE, Valley Forge, Pa., 1953-64; mgr. energy conversion divsn. Nuclear Materials and Equipment Corp. (acquired by ARCO), Apollo, Pa., 1964-69, Atlantic Richfield Corp., Apollo, 1969-72; founder, pres., chmn. Biocontrol Tech., Inc., Indiana, 1972-93; chmn., treas. Diasense, Inc., Indiana, 1989—. Contbr. over 22 articles to profl. jours. 1st lt. USAF, 1961-63. Fellow ASME; mem. AAAS, Am. Diabetes Assn. Achievements include patents for generator of electrical energy by radioisotope thermoelectric conversion, for radioisotope powered cardiac pacemaker, for radioisotope powered artificial heart, for thermoelectric apparatus for high thermoelectric efficiency by cascading materials, for methodof metals joining and articles produced by such method - brazing cooper to tungsten, for thermoelectric apparatus for high thermoelectric efficiency by cascading materials, for radioisotope powered cardiac pacemaker, for generator of electrical energy by radioisotope thermoelectric conversion, for artificial pancreas, for glucose sensor. Office: Biocontrol Tech Inc 300 Indian Springs Rd Indiana PA 15701

PURDY, DONALD GILBERT, JR., soil scientist; b. St. Louis, Apr. 10, 1953; s. Donald Gilbert Purdy and Nelda Jean Bates Pierce; m. Janice Shirley Bogacki, Aug. 10, 1974; children: Erin Elizabeth, Justin Joseph. BS, U. Mo., 1975; MS, U. Nebr., 1981. Cert. profl. soil scientist/agronomist; registered environ. profl. Insp. USDA-Agrl. Mktg. Svc., St. Louis, 1975-76; staff rsch. asst. U. Nebr., Lincoln, 1976-78; soil scientist Mo. Dept. Natural Resources, Jefferson City, Mo., 1978-80; project mgr. N. Am. Coal Corp., Dallas, 1980-89; sr. project mgr. SCI Environ. Inc., Chesterfield, Mo., 1989-91; asst. dir. ops. Environ. Mgmt. Corp., Creve Coeur, Mo., 1991-92; pres. D. G. Purdy & Assocs., Inc., Manchester, Mo., 1992—. Contbr. articles to profl. jours. Mem. Am. Soc. Agronomy, Soil Sci. Soc. Am., Profl. Soil Scientist Assn. Tex., Mo. Assn. Profl. Soil Scientists. Home: 709 Boleyn Pl Manchester MO 63021-5327 Office: D G Purdy & Assocs Inc 709 Boleyn Pl Manchester MO 63021

PURETZ, DONALD HARRIS, educator; b. N.Y.C., Nov. 5, 1934; s. Henry Puretz; m. Anne Louise Botsford, Mar. 27, 1983; children: Henry Herman Bordman, Edward Abel Botsford. BS, Bklyn. Coll., 1960; MA, NYU, 1962, PhD, 1972; MPH, Columbia U., 1978. Jr. high sch. tchr. N.Y.C. Bd. Edn., 1960-63; faculty mem. R.I. Coll., Providence, 1963-69, SUNY-Dutchess C.C., Poughkeepsie, 1969—; cons. N.Y.C. Bd. Edn., 1975-76. Contbr. articles to profl. jours. With U.S. Army, 1954-56. Recipient Chancellor's award SUNY, 1975. Fellow Am. Sch. Health Assn.; mem. AAAS, Dutchess C.C. Am. Fedn. Tchrs. (pres. 1970-72). Achievements include research on school health and drug education. Home: 11 Garfield Pl Poughkeepsie NY 12601 Office: Dutchess CC Poughkeepsie NY 12601

PURGATHOFER, WERNER, computer science educator; b. Vienna, Austria, Oct. 27, 1955; s. Alois and Ingrid (Clar) P.; m. Brigitte Dorner, Sept. 24, 1982; children: Nina, Armin, Julia. Diploma in engring., Tech U. Vienna, 1980, D Tech., 1984. Rsch. asst. Tech. U. Vienna, 1980-88, prof. computer sci., 1988—; chmn. Eurographics '91 Conf., Vienna, 1991. Author: Graphische Datenverarbeitung, 1985; co-editor: Advances in Computer Graphics V, 1989, Austrographics '88, 1988, Austrographics '86, 1986. Mem. IEEE Computer Soc., Eurographics (3d best paper award 1981, best paper award 1986), N.Y. Acad. Scis., Assn. for Computing Machinery, Assn. for Computing Sci., Soc. for Informatiks, Computer Graphics Soc., Swiss Computer Graphics Assn. Office: Tech U Inst Computer, Graphics, Karlsplatz 13/186, A-1040 Vienna Austria

PURI, PUSHPINDER SINGH, chemical engineer; b. India, Feb. 3, 1946; came to U.S., 1976; m. Manjit Sethi, Jan. 19, 1975; children: Jasmine, Bandhana. BS, Panjab U., India, 1967; MS, Panjab U., 1969; PhD, Ottawa U., Can., 1975. Registered profl. engr., Fla. Prin. engr. CPC Internat., Union, N.J., 1976-80; res. mgr. C.D. Med., Miami, Fla., 1980-85; mgr. Air Products and Chem., Allentown, Pa., 1985-91; dir. R&D PERMEA, Inc., St. Louis, 1991—. Contbr. articles to profl. jours. Mem. AIChE, N.Am. Membrane Soc. (editor). Achievements include patents in field. Office: PERMEA Inc 11444 Lackland Rd Saint Louis MO 63146

PUROHIT, MILIND VASANT, physicist; b. New Delhi, Aug. 16, 1957; came to the U.S., 1978; s. Vasant Chintaman and Sudha Vasant (Indumati) P.; m. Uma Milind, Mar. 3, 1986; 1 child, Ashwin. MS, IIT Delhi, 1978; PhD, Calif. Inst. Tech., 1984. Grad. rsch. assoc. Calif. Inst. Tech., Pasadena, 1978-83; rsch. assoc. Fermilab, Batavia, Ill., 1983-86, Wilson fellow, 1986-88; asst. prof. Princeton (N.J.) U., 1988—. Contbr. articles to profl. jours. Named Outstanding Jr. Investigator Dept. of Energy, 1989. Mem. Am. Phys. Soc. Office: Princeton U Dept Physics PO Box 708 Princeton NJ 08544

PUROHIT, SHARAD CHANDRA, mathematician; b. Indore, India, Aug. 11, 1949; s. Balkrishna C. and Rukmini (Bagora) P.; m. Urvashi T. Bhatt, Feb. 24, 1973; children: Sudarshan, Viswas, Abhilasha. MS, B.I.T.S., 1969; PhD, IIT, Bombay, 1972. Scientist Vikram Sarabhai Space Ctr., Trivandrum, India, 1973-80; prin. investigator Air Force Office Scientific Rsch., Dayton, Ohio, 1981-83; sr. scientist Indian Space Rsch. Orgn., Trivandrum, 1984-88; program coord. Ctr. for Devel. of Advanced Computing, Pune, India, 1989—; vis. scientist U. Dayton Rsch. Inst., 1981-83, Nat. Aero. Lab., Bangalore, India, 1988; mem. dept. sci. and tech. Govt. of India, New Delhi, 1987-91, coord. dept. space, 1983-88; mem. Coun. Scientific and Indsl. Rsch., New Delhi, 1989—. Editor: Frontiers in Parallel Computing, 1990; author: Role of Computers in Science, 1987. Mem. AIAA, Indian Sci. Congress Assn., Nat. Math. Soc. Home: 1 Gulmohar Park Aundh, Pune 411 007, India Office: Ctr Devel Advanced Computing, Pune U Campus, Pune 411 007, India

PURPURA, DOMINICK P., neuroscientist, university dean; b. N.Y.C., Apr. 2, 1927; m. Florence Williams, 1948; children—Craig, Kent, Keith, Allyson. A.B., Columbia U., 1949; M.D., Harvard U., 1953. Intern Presbyn Hosp., N.Y.C., 1953-54; asst. resident in neurology Neurol. Inst., N.Y.C., 1954-55; Prof., chmn. dept. anatomy Albert Einstein Coll. Medicine, Yeshiva U., N.Y.C., 1967-74, sci. dir. Kennedy Ctr., 1969-72, dir. Kennedy Ctr., 1972-82, prof., chmn. dept. neurosci., 1974-82, dean, 1984—; dean Stanford U., Calif., 1982-84; mem. neurophysiol. panel Internat. Brain Rsch. Orgn., pres., 1986-92; v.p. med. affairs, 1987, UNESCO, 1961—; chmn. internat. congress com. Internat. Brain Rsch. Orgn./World Found. Neuroscientists, 1983—. Mem. editorial bd. Brain Rsch., 1965—, editor-in-chief, 1975—; editor-in-chief Brain Rsch. Revs., 1975—, Developmental Brain Rsch., 1981—, Molecular Brain Rsch., 1985—, Cognitive Brain Rsch., 1991—. Served with USAAF, 1945-47. Fellow N.Y. Acad. Scis.; mem. Inst. Medicine of Nat. Acad. Scis., Nat. Acad. Scis., Am. Acad. Neurology, Am. Assn. Anatomists, Am. Assn. Neurol. Surgeons, Am. Epilepsy Soc., Am. Physiol. Soc., Assn. Research in Nervous and Mental Disease, Soc. Neurosci., Sigma Xi. Office: Yeshiva U Albert Einstein Coll Medicine 1300 Morris Park Ave Bronx NY 10461-1924

PURVIS, GEORGE DEWEY, III, computational chemist; b. Memphis, May 10, 1947; s. George Dewey Jr. and Janet Louise (Mitchell) P.; m. Nancy Jean More; children: Kevin, John, George, Patrick. BS in Chemistry, Davidson Coll., 1969; PhD in Chemistry, U. Fla., 1973. Vis. lectr. U. Utah, Salt Lake City, 1976; rsch. scientist Battelle Columbus (Ohio) Labs., 1977-81; rsch. scientist, chemistry computer mgr. U. Fla., Gainesville, 1981-88; v.p. tech., chief scientist CAChe Sci., Inc., Beaverton, Oreg., 1988—. Contbr. articles to Jour. Chem. Physics, Jour. Am. Chem. Soc., Jour. Comp. Info. Sci. Capt. USAR 1971-86. Mem. AAAS, Am. Chem. Soc. (divsn. phys. chemistry, divsn. computer and info. sci.). Office: Cache Sci Inc PO Box 500 MS 13-400 Beaverton OR 97077

PUSZYNSKI, JAN ALOJZY, chemical engineer, educator; b. Zywiec, Poland, Oct. 8, 1950; came to U.S., 1982; s. Jozef and Irena (Mitus) P.; m. Malgorzata Maria Ratajczak, Sept. 23, 1972; children: Maciej, Lukasz. B-SChemE, MSChemE, Tech. U., Wroclaw, Poland, 1973; PhD in Chem. Engring., Inst. Chem. Tech., Prague, Czechoslovakia, 1980. Asst. prof. Tech. U., Wroclaw, 1980-82; vis. asst. prof. SUNY, Buffalo, 1982-86, rsch. prof., 1986-91; assoc. prof. S.D. Sch. Mines & Tech., Rapid City, 1991—; cons. INRAD, Northvalle, N.J., 1989-91, Materials & Electrochem. Rsch., Tucson, 1991—, Novel Techs., Inc., Cayuga, N.Y., 1991—. Co-author: Non-Oxide Ceramic Materials, 1993, Ullman Ency., 1992; contbr. articles to profl. jours. Recipient Sandvig Professorship S.D. Sch. Mines, 1993. Mem. AICE, Am. Ceramic Soc., Am. Assn. Combustion Synthesis, Combustion Soc., Sigma Xi. Roman Catholic. Achievements include development of new technique for the synthesis of aluminum nitride, silicon nitride, titanium and tantalum nitrides, sialons, and ceramic-ceramic composite materials. Home: 8420 Blackbird Ct Rapid City SD 57702 Office: SD Sch Mines & Tech 501 E Saint Joseph St Rapid City SD 57701

PUTA, DIANE FAY, medical staff services director; b. Manitowoc, Wis., Mar. 6, 1947; d. Ruben William and Gertrude Katherine (Novak) P. BSN, Alverno Coll., 1971; MS in Ednl. Adminstrn., U. Wis., Milw., 1979, PhD in Urban Edn., 1991. Staff nurse St. Mary's Hosp., Milw., 1971-72, St. Anthony's Hosp., Milw., 1972-74; nurse coord. Pvt. Initiative in PSRO, Wis., 1974-75; insvc. instr. Deaconess Hosp., Milw., 1975-77, insvc. coord., 1977-81; dir. nursing staff devel./quality assurance Good Samaritan Med. Ctr., Milw., 1981-84, dir. quality assurance, 1984-85, dir. utilization mgmt., 1985-88; mgr. quality mgmt. Sinai Samaritan Med. Ctr., Milw., 1988-89, dir. med. staff svcs. and quality mgmt., 1989—. Author: (with others) Interdisciplinary QA: Issues in Collaboration, 1991; author poem. Mem. Channel 10/36 Friends, Milw., Friends of Pub. Mus., Milw., 1991. Mem. ANA, Wis. Nurses Assn., Milw. Nurses Assn., Wis. Assn. for Healthcare Quality, Nat. Assn. for Healthcare Quality, Am. Soc. Quality Control, Alverno AlumnaeAssn., U. Wis. Alumnae, Delta Epsilon Sigma, Kappa Gamma Pi. Avocation: ballroom dancing. Home: 4050 W Rivers Edge Cir Apt 2 Milwaukee WI 53209-1129 Office: Sinai Samaritan Med Ctr PO Box 342 Milwaukee WI 53201-0342

PUTATUNDA, SUSIL KUMAR, metallurgy educator; b. Sanhpur, W. Bengal, India, Jan. 31, 1948; came to U.S., 1983; s. Provat Chandra and Santi Kana Putatunda; m. Ivy M. George, June 7, 1984; children: Sujata, Shibani. BS, Instn. Engrs., Calcutta, 1975; MS, U. Mysore (India), 1979; PhD, Indian Inst. Tech., Bombay, 1983. Metallurgist Hindustan Copper Ltd., Khetri, Rajesthan, 1973-77; grad. rsch. asst. U. Mysore, Mangalore, India, 1977-79; rsch. and devel. engr. Hindustan Electrographics, Bhopal, India, 1979-80; grad. rsch. asst. Indian Inst. Tech., Bombay, 1980-83; Fulbright scholar U. Ill., Urbana, 1983-84; assoc. prof. metallurgy Wayne State U., Detroit, 1985—. Recipient Scholarships, Govt. India, New Delhi, 1977, 80; Fulbright fellow USIA, Washington, 1982. Mem. Am. Soc. Metals, The Metall. Soc., ASTM (editor spl. tech. pub. on fractography 1989), Iron and Steel Soc., Engring. Soc. Detroit. Home: 5200 Anthony Wayne Dr Apt 1414 Detroit MI 48202-3977 Office: Wayne State U Coll Engring 5050 Anthony Wayne Dr Detroit MI 48202-3902

PUTCHAKAYALA, HARI BABU, chemical engineer; b. Maddirala, India, July 15, 1949; came to the U.S., 1978; s. Seshadri Chowdary and Sambrajyam (Penubothu) P.; m. Vijay Lakshmi, Aug. 9, 1976; children: Sashi Manohar, Gopi Krishna. BS in Chem. Engring., REC, Warangal, India, 1971; MS, BITS, Pilani, India, 1974; PhD, IIT, New Delhi, 1978. Registered profl. engr., Mich., Md., Calif., Pa. Trainee Fertilizer Corp., Bombay, 1971; environ. engr. Madison Madison Internat., Detroit, 1978-81, project coord., 1981-84, project mgr., 1984-89, v.p., 1990—, also bd. dirs.; bd. dirs. Spack Inc., Brownstown, Mich., 1990—. Contbr. articles to Canadian Jour. Chem. Engring. Rsch. fellow Univ. Grants Commn., 1974-78; recipient Cert. Boiler Efficiency Inst., 1981, U. Wis., 1986, 1992, Mich. State U., 1989, Ctr. for Hazardous Materials Rsch., 1990. Mem. AICE, NSPE. Achievements include development of design modifications for incineration plants, O&M manuals for numerous water and wastewater facilities; design of waste water treatment systems; research into energy and value engineering studies. Home: 654 Fox River Dr Bloomfield Hills MI 48304 Office: Madison Internat 1420 Washington Blvd Detroit MI 48226

PUTMAN, THOMAS HAROLD, electrical engineer, consultant; b. Pitts., Nov. 22, 1930; s. Henry Van de Vere and Ruth Marion (Williams) P.; m. Norma Alfreda Burns, June 30, 1956; children: Henry, Lesley, Harold. BSEE, Union Coll., 1952; MS, MIT, 1954, ScD, 1958. Registered profl. engr., N.Y. Instr. MIT, Cambridge, 1952-58; sr. rsch. engr. Westinghouse Rsch. Labs., Pitts., 1958-66, mgr. dept. electromechanics, 1966-74, cons. engr., 1974-92; ind. cons. Pitts., 1992—. Contbr. articles to profl. publs. Recipient Order of Merit, Corp. Signature Award of Excellence Westinghouse. Mem. IEEE, Soc. Naval Architects and Marine Engrs. Presbyterian. Achievements include 15 patents generally in the areas of power system and transportation vehicle control. Home: 354 Stoneledge Dr Pittsburgh PA 15235

PUTNAM, HUGH DYER, environmental engineer, educator, consultant; b. Carrington, N.D., Feb. 12, 1928; s. Hugh Rodney and Blanche Putman; m. Natalie Joy Knott, Dec. 23, 1950; children: Mark, Lynn, Charles. BA in Microbiology, U. Minn., 1953, MS in Microbiology, 1956, PhD in Pub. Health Biology/Environ. Health, 1963. Rsch. asst. U. Minn., 1952-54, teaching asst., 1954-55, instr., 1956-59, rsch. fellow, 1959-63; asst. prof. sanitary sci., asst. rsch. prof. U. Fla., Gainesville, 1963-66, assoc. prof. environ. sci., 1966-70, prof. environ. engring. studies, 1970-74, adj. prof. environ. engring. scis., 1974-85; founder, v.p. environ. scis. divsn. Environ. Sci. and Engring., Inc., Gainesville, 1965-74; v.p., prin. scientist Water and Air Rsch., Inc., Gainesville, 1974—; vis. prof. U. Minn., 1974-75; mem. ad hoc com. on environ. rsch. and toxicology NAS; mem. environ. biology review panel EPA; vice chmn. sect. 1000 15th Edit. Standard Methods for Water and Wastewater, chmn. sect. 1000 16th and 17th Edits.; coord. water and wastewater biol. examination sect. 17th Edit. Standard Methods; mem. site selection bd. U.S. Army Med. Rsch. abd Devel. Command Effect of Munition Wastes on Aqatic Life; mem. site vis. team for evaluation NERQ Corvalis Ecol. programs Inst. Ecology, Madisin, Wis.; guest lectr. NASA, U. Minn, Duluth, USAF Acad.; cons. EPA, TVA, Muscle Shoals, Ala., Met. Waste Control Commn., St. Paul, Dept. Waste and Power, L.A. Contbr. over 75 profl. reports and publs. to sci. jours. Mem. tech. awareness com. City of Tampa Water Reuse Com. Master sgt. U.S. Army, 1946-48, 50-52. Recipient Best Paper award TAPPI Environ. Conf. Planning Com., 1993. Mem. Am. Soc. Limnology and Oceanography, Am. Soc. Microbiology, Ecol. Soc. Am., Water Pollution Control Fedn., Internat. Soc. Theoretical and Applied Limnology, Sigma Xi. Office: Water & Air Rsch Inc 6821 SW Archer Rd Gainesville FL 32608

PUTNAM, ROBERT ERVIN, chemistry educator; b. Northampton, Mass., Oct. 18, 1927; s. Ervin Earl and Mary Gertrude (Connelly) P.; m. Caroline Wright, Aug. 23, 1952; children: David Earl, Mary Caroline, Robert Edward, Andrew Wright. BS in Chemistry, U. Mass., 1950; PhD in Organic Chemistry, U. Ill., 1953. Rsch. chemist E. I. du Pont de Nemours, Wilmington, Del., 1953-59; rsch. supr. E. I. du Pont de Nemours, Wilmington, Del., 1959-65, sr. rsch. supr., 1965-67; sr. rsch. supr. E. I. du Pont de Nemours, Parkersburg, W.Va., 1967-78, rsch. lab. supt., 1978-82, rsch. mgr., 1982-85; adj. mem. faculty Washington State C.C., Marietta, Ohio, 1985—; pvt. practice (cons.) Marietta, 1985—; alumni adv. coun. U. Mass. Chemistry Dept., Amherst, 1975-78; instr. chemistry Marietta Coll., 1982-89, mem. adv. coun., 1989—. Contbr. 20 articles and 2 chpts. to profl. jours. With USNR, 1945-46. NSF fellow, U. Ill., 1952-53. Fellow AAAS; mem. Am. Chem. Soc. (chmn. O.V. sec. 1976-78). Democrat. Mem. Unitarian Ch. Achievements include patents on fluorine containing polymers and monomers, ion exchange resins; research on industrial processes for nylon polyacetals, acrylics, rubber toughened plastics, fluorinated plastics. Home: 100 Alden Ave Marietta OH 45750-1138

PÜTSEP, PEETER ERVIN, electronics executive; b. Stockholm, Dec. 7, 1955; s. Ervin P. and Liidia (Voolaid) P.; m. Katrin Laan, Sept. 5, 1987; children: Liivia, Martin. MBA in Econs. and Bus. Adminstrn., Stockholm Sch. Econs. & Bus., 1978. Mktg. dir. Volvo-Beijer Fgn. Office, Stockholm, Tokyo, 1980-82; sr. cons. SAM, Brussells, 1982-86; pres. Bulten Internat., Stockholm, 1986-88; bus. group dir. AGFA-Gevaert, Stockholm, Sweden, 1988-92; CEO Swedish Telecombe, Stockholm, 1992—. Contbr. editorials to newspapers. Mem. Spl. Forces. Mem. Eesti Ulioppilaste Selts (vice chmn.), Swedish-Estonian Assn. (bd. mem.).

PUTT, JOHN WARD, propulsion engineer, retired, consultant; b. Huntingdon, Pa., Feb. 11, 1924; s. Donald Reed Putt and Mary Ruth (Lee) Benton; m. Florence Geary, Jan. 31, 1948 (div. Feb. 1967); 1 child, John G.; m. Joan Grant Kennedy, Feb. 4, 1967; stepchildren: Corinne, Constance. BS, Juniata Coll., 1945. Rsch. chemist Tidewater Assoc. Oil Co., Bayonne, N.J., 1945-51; chief chemist Walter Kidde & Co., Inc., Belleville, N.J., 1951-63; sect. head Hughes Aircraft Co., El Segundo, Calif., 1963-78, dept. mgr., 1978-82; product line mgr. Hughes Aircraft Co., Elsegundo, Calif., 1982-86; co-owner CompuType, L.A., 1986—; cons. Thiokol Corp., Ogden, Utah, 1986-91, Hitco (Brit. Petroleum), Gardena, Calif., 1986—. Fellow AIAA. Democrat. Achievements include patents in distillation equipment, catalytic decomposition of hydrogen peroxide, high temperature bearing lubricant, liquid vapor separator for zero G, liquid vapor separator. Home and Office: 8031 Bleriot Ave Los Angeles CA 90045

PUTTLITZ, KARL JOSEPH, SR., metallurgical engineer, consultant; b. Kingston, N.Y., Aug. 4, 1941; s. Adalbert and Elizabeth Anges (Barthel) P.; m. Dianne Elizabeth Markle, Sept. 16, 1967; children: Kirk, Christian, Karl Joseph Jr., Sara Ann. AAS in Chemistry, SUNY, Farmingdale, 1961; BS in Metall. Engring., Mich. State U., 1965, MS in Metall. Engring., 1967, PhD in Metallurgy and Material Sci. Engring., 1971. Assoc. metallurgist IBM, East Fishkill, N.Y., 1965-67, sr. assoc. metallurgist, 1967-71, staff metallurgist, 1971-72, adv. metallurgist, 1972-84, sr. engr., 1984—; pvt. practice cons., Wappingers Falls, 1975—. Contbr. over 7 articles to profl. jours. Team mgr. Little League, Wappingers Falls, 1982, Sr. League, Wappingers Falls, 1984-88; pres. Lake Oniad Lot Owners Assn., Wappingers Falls, 1974-76. IBM Corp. scholar Mich. State U., 1968-71. Mem. ASM (pres. Mich. State U. chpt. 1964-65), N.Y. Acad. Scis., Internat. Soc. Hybrid Microelectronics, Am. Welding Soc., Sigma Xi, Phi Lambda Tau. Republican. Roman Catholic. Achievements include over 30 published inventions including U.S. and foreign patents in microelectronic interconnection technology. Home: 21 Central Ave Wappingers Falls NY 12590-3803 Office: IBM B/330 81A Hopewell Junction NY 12533-0999

PUZDROWSKI, RICHARD LEO, neuroscientist; b. Detroit, Aug. 21, 1959; s. Richard Leo and Rose Mary (Drummond) P. MS in Biology, U. Mich., 1985, PhD, 1988. Teaching asst. U. Mich., Ann Arbor, 1982-86; staff rsch. assoc. U. Calif.-San Diego, La Jolla, 1987, 88; assoc. mem. Marine Biomed. Inst. U. Tex. Med. Br., Galveston, 1988—, asst. prof. anat. and neurosci., 1993—; postdoctoral councilor Galveston chpt. Soc. for Neurosci., 1992-93. Contbr. articles to jours. Fellow U. Mich., 1987, NIH, 1991. Mem. AAAS, Internat. Congress of Vertabrate Morphology, Am. Soc. Zoologists, Soc. for Neurosci., Sigma Xi. Office: U Tex Med Br Marine Biomed Inst Galveston TX 77555-0843

PUZELLA, ANGELO, microwave engineer; b. Summit, N.J., Aug. 19, 1961; m. Anna Rizzi, June 15, 1991. BA in Physics, Math., Ohio Wesleyan U., 1983; M Engring., U. Utah, 1984, D Engring., 1986. Engr. Raytheon Co., Wayland, Mass., 1983—. Mem. IEEE, Phi Beta Kappa. Home: 217 Wilson St Haverhill MA 01830 Office: Raytheon Co 430 Boston Post Rd Wayland MA 01778

PYE, LENWOOD DAVID, materials science educator, researcher, consultant; b. Little Falls, N.Y., May 16, 1937; s. Lenwood George and Elizabeth Marie (Murphy) P; m. Constance Lee Lanphere, Sept. 6, 1958; children: DeAnn, Lorie, Lisa, Brien. BS, Alfred U., 1959, PhD, 1968. Rsch. engr. PPG Industries, Pitts., 1959-60; rsch. scientist Bausch & Lomb, Rochester, N.Y., 1960-61, 62-64; prof. glass sci. N.Y. State Coll. Ceramics, Alfred U., 1968-92, dir. Inst. Glass Sci. and Engring., 1984—, dir. Industry-Univ. Ctr. Glass Rsch., 1986-92. 1st lt. U.S. Army, 1960-62. Mem. Am. Ceramics Soc. (trustee). Office: Alfred U NY State Coll Ceramics Ctr Glass Rsch New York NY 14802

PYER, JOHN CLAYTON, analytical chemist; b. Neenah, Wis., June 15, 1963; s. Ronald Lee and JoAnn (Brooks) P.; m. Dolores M. Buffington, July 1, 1983. Student, Tenn. Temple U., Chattanooga, 1981-84; BA in Chemistry, U. So. Ind., Evansville, 1989. Chemist Nat. Labs., Evansville, 1987-88, Core Labs. (now Standard Labs., Evansville, 1988-90; assoc. analytical chemist Eli Lilly & Co., Lafayette, Ind., 1990—; mem. chemistry alumni adv. bd. U. So. Ind. Baptist. Home: 909 S Wagon Wheel Trail Lafayette IN 47905

PYERITZ, REED EDWIN, medical educator, physician, consultant; b. Pitts., Nov. 2, 1947; s. Paul L. and Ida Mae (Meier) P.; m. Jane Ellen Tumpson, May 28, 1972; 2 children. S.B. in Chemistry, U. Del., 1968; A.M., Harvard U., 1971, Ph.D. in Biochemistry, 1972, M.D., 1975. Diplomate Am. Bd. Internal Medicine, Am. Bd. Med. Genetics. Intern Peter Bent Brigham Hosp., Boston, 1975-76, resident, 1976-77; resident Johns Hopkins Hosp., Balt., 1977-78; from instr. to assoc. prof. medicine and pediatrics Sch. Medicine, Johns Hopkins U. Balt., 1978-93, prof. medicine and pediatrics Med. Coll. Pa., Phila., 1993—; dir. Ctr. Human Genetics Allegheny Health Edn. and Rsch. Found., 1993—; chmn. dept. human genetics Allegheny Singer Rsch. Inst., Pitts., 1993—; chief physician Md. Athletic Commn., Balt., 1978-93; chmn. med. adv. bd. Nat. Marfan Found., N.Y.C., 1982-93. Mem. editorial bd. New Eng. Jour. Medicine; contbr. numerous articles to profl. publs. Lt. col. USMCR; NIH grantee. Fellow ACP; mem. Am. Heart Assn., AAAS, Hastings Ctr. Avocation: triathlon competition. Office: Allegheny Singer Rsch Inst 320 E North Ave Pittsburgh PA 15212-4772

PYKE, THOMAS NICHOLAS, JR., government science and engineering administrator; b. Washington, July 16, 1942; s. Thomas Nicholas and Pauline Marie (Pingitore) P.; m. Carol June Renville, June 22, 1968; children—Christopher Renville, Alexander Nicholas. B.S., Carnegie Inst. Tech., 1964; M.S. in Engring., U. Pa., 1965. Electronic engr. Nat. Bur. Standards, Gaithersburg, Md., 1964-69, chief computer networking sect., 1969-75, chief computer systems engring. div., 1975-79, dir. ctr. for computer systems engring., 1979-81, dir. ctr. programming sci. and tech., 1981-86; asst. adminstr. for satellite and info. services NOAA, Washington, 1986-92, dir. high performance computing and comm., 1992—; organizer profl. computer confs., 1970-86; mem. Presdl. Adv. Com. on Networking Structure and Function, 1980, Interagy. com. on Info. Resources Mgmt., 1983-84, bd. dirs., 1984-87, vice chmn. 1986-87 (Exec. Excellence award 1991), chmn. Interagy. Working Group on Data Mgmt. for Global Change, 1987-93; speaker in field. Editorial bd. Computer Networks Jour., 1976-86; contbr. articles to profl. jours. Bd. dirs. Glebe Commons Assn., Arlington, Va., 1976-79, v.p., 1977-79; chmn. Student Congress, Carnegie Inst. Tech., 1963-64; mem. Task Force on Computers in Schs., Arlington, 1982-85; pres. PTA, Arlington, 1983-84. Recipient Silver medal Dept. Commerce, 1973, award for exemplary achievement in pub. adminstrn. William A. Jump Found., 1975, 76, Presdl. Rank award of Meritorious Exec., 1988; Westinghouse scholar Carnegie Inst. Tech., Pitts., 1960-64; Ford Found. fellow U. Pa., Phila., 1964-66. Fellow Washington Acad. Scis. (Engring. Sci. award 1974); mem. Am. Fedn. Info. Processing Socs. (bd. dirs. 1974-76), IEEE (sr. mem.), Computer Soc. of IEEE (bd. govs. 1971-73, 75-77, vice chmn. tech. com. on personal computing 1982-86, chmn. 1986-87), AAAS, Assn. Computing Machinery, Sigma Xi, Eta Kappa Nu, Omicron Delta Kappa, Pi Kappa Alpha (chpt. v.p. 1963-64). Episcopalian. Office: NOAA Washington DC 20235

PYLE, THOMAS EDWARD, oceanographer, academic director; b. N.Y.C., Oct. 31, 1941; s. Thomas Edgar and Mary Bernadette (Keane) P.; m. Claudia Cecelia Ian, June 29, 1963; children: Laura, Holly. BS, Columbia Coll., N.Y.C., 1963; MS, Tex. A&M U., 1966, PhD, 1972. Geologist U.S. Geol. Survey, Corpus Christi, Tex., 1970-72; asst. to assoc. prof. Dept. Marine Sci., U. S. Fla., St. Petersburg, 1969-76; program dir. Office Naval Rsch., Bay St. Louis, Miss., 1976-80; dep. dir. NOAA, Rockville, Md., 1981-82, chief scientist, 1982-86; dir. ocean drilling Joint Oceanographic Insts., Inc., Washington, 1986—, v.p., 1988—; cons. Dillingham Corp., Ocean Optics, Inc., Geo-Marine, Inc.; bd. visitors Office Naval Rsch., Arlington, Va., 1990; mem. IAPSO Commn. on Tides and Mean Sea Level, 1984-85. Contbr. articles to profl. jours. Columbia Coll. Regents and L. Wright scholar, 1959-63; NASA fellow, 1965-68; rsch. grantee, NSF, Office Naval Rsch., 1969—. Mem. AAAS, Oceanographic Soc., Am. Geophys. Union. Office: JOI Inc 1755 Massachusetts Ave NW Washington DC 20036-2102

PYNN, ROGER, physicist; b. Maidstone, Kent, Eng., Feb. 15, 1945; s. Herbert John and Kathleen (Coleman) P. MA, Trinity Coll., Cambridge, Eng., 1966; PhD, U. Cambridge, 1969. Postdoctoral fellow AB Atomenergi, Studsvik, Sweden, 1970-71; rsch. fellow Inst. for Atomenergi, Kjeller, Norway, 1971-73; assoc. physicist Brookhaven Nat. Lab., Upton, N.Y., 1973-75; staff scientist Inst. Laue-Langevin, Grenoble, France, 1975-87; dir. Manuel Lujan Jr. Neutron Scatterin Ctr. Los Alamos (N.Mex.) Nat. Lab., 1987—. Mem. Norwegian Phys. Soc., Materials Rsch. Soc. Office: Los Alamos Nat Lab Manuel Lujan Neutron Scattering Ctr MS H805 Los Alamos NM 87545

PYTKO, STANISLAW JERZY, mechanical engineering educator; b. Pacanow, Poland, Oct. 19, 1929; s. Wiktor and Aniela (Badyl) P.; m. Krystyna Zasucha, July 2, 1955; children: Jolanta, Pawel. BSc in Engring., Tech. U. Mining & Metallurgy, Kraków, Poland, 1954, DSc in Engring., 1964, prof. Du Sc., 1978. Asst. Tech. U. Mining and Metallurgy, Kraków, 1952-58, adj., 1958-71, asst. prof., 1971-78, prof., 1978—; with Scientific Centre for Terotech., Radom, 1990—; v.p. Internat. Tribology Counsil, London, 1991—. Author: Principles of Tribology, 1984-89; author, editor: Problems of Contact Strength, 1982, Ultrasound Diagnostics of Iriction Joints, 1989; editor Exploitation Problem of Machines, 1956—; contbr. over 100 papers on trilology. Mem. Polish Tribology Soc. (chmn. 1990—). Roman Catholic. Home: Orlat 5, 31-518 Kraków Poland Office: Tech U Mining & Metallurgy, Al Mickiewicza 30, 30-059 Kraków Poland

PYTLINSKI, JERZY TEODOR, physicist, research administrator, educator; b. Warsaw, Poland, Apr. 1, 1938; s. Stanislaw and Natalia (Matuszewska) P.; m. Bonnie Laurie Bennett, Dec. 30, 1969; 1 child, Christine Barbara. MS, Tech. U. Warsaw, 1962; PhD in Plasma Physics with distinction, U. Paris, 1967. Program mgr., acting div. head N.Mex. State U., Las Cruces, 1977-80; sr. scientist, div. head U. P.R., Mayaguez, 1981-83; program dir., sr. scientist U. P.R., San Juan, 1983-86, sr. scientist, founding dir. Univ.-Industry Rsch. Ctr., 1986-89; pres. Univ.-Industry Rsch. Ctr., Tampa, Fla., 1989—; mem. adv. bd. on solar energy UNESCO, 1979-85; referee Am. Jour. Physics, 1980—, Solar Energy Jour., 1983-87, 38th Internat. Sci. and Engring. Fair, 1987; mem. U.S. tech. adv. group of ISD TC-180, 1981—. Editorial bd. Internat. Jour. of Energy, Environ., Econs., 1990—; co-editor procs. Internat. Conf. Energy for Americas, 1987; contbr. articles to profl. jours., procs. Grantee state and fed. agys., various founds.; Postdoctoral fellow U. Liverpool, England, 1968-69; recipient commendation State of Kans., 1977. Mem. Am. Phys. Soc., Nat. Coun. Univ. Rsch. Adminstrs., Soc. Rsch. Adminstrs., Internat. Solar Energy Soc., Internat. Energy Soc. (sci. coun. 1985—), Sigma Phi Sigma.

QADRI, SYED BURHANULLAH, physicist; b. Hyderabad, India, Aug. 20, 1949; came to U.S., 1974; s. Syed Waliullah and Tahera (Begum) Q.; m. Mahamooda Begum, May 2, 1983; children: Syed Amanullah, Syed Noorullah, Sana Ayesha. MA in Physics, Kent State U., 1975; MS in Math., Ohio State U., 1979; PhD in Physics, Ohi State U., 1979. Lectr. Ohio State U., Columbus, 1980; rsch. physicist Colo. State U., Ft. Collins, 1980-81, Sachs/Freeman Assocs. Inc., Bowie, Md., 1981-89, U.S. Naval Rsch. Lab., Washington, 1989—; cons. Bendix Aerospace Corp., Columbia, Md., 1985-86, Allied Chems., Morristown, N.J., 1986. Contbr. over 200 rsch. papers to profl. publs. Mem. Am. Phys. Soc., Am. Vacuum Soc. (Best Poster award 1984), User's Exec. Com.. Achievements include patents on method of making substantially single phase superconducting oxide ceramics having a Tc above 85 degrees, stabilized zirconia/cocraly high temperature coating. Office: US Naval Rsch Lab 4555 Overlook Ave SW Washington DC 20375

QAIM, SYED MUHAMMAD, nuclear chemist, researcher, educator; b. Fatehpur, Pakistan, Jan. 5, 1941; came to Germany, 1968, naturalized; s. Syed Raisul and Zahra Bibi (Akhtar) Hasan; m. Anneliese Josefa Grieger, Dec. 30, 1965; children: Yasmin, Matin. MS, Panjab U., Lahore, Pakistan, 1961; PhD, Liverpool (Eng.) U., 1964; DSc, Birmingham (Eng.) U., 1977; habilitation, Cologne (Germany) U., 1993. Rsch. assoc. Birmingham U., 1964-66; sr. scientist Atomic Energy Centre, Lahore, 1966-68; Humboldt fellow Mainz (Germany) U., 1968-70; scientist Rsch. Centre Jülich, Germany, 1970-75; group leader Rsch. Centre Jülich (Germany), 1975-85, div. leader, 1985—; cons. UN Tokten Scheme, Islamabad, Pakistan, 1983, 88; short term cons. Internat. Atomic Energy Agy. Author, editor: School Chemistry Book, 1968; mem. adv. bd. Radiochimica Acta, 1990—; contbr. rsch. articles to profl. jours.; chmn. Internat. Conf. Nuclear Data for Science and Technology, Jülich, 1991, editor Proceedings, 1992. Recipient Jari award medal Pergamon Press, Princeton, 1990, Roland Eötvös medal, Budapest, 1988. Fellow Royal Soc. Chemistry (London), Inst. Physics (London); mem. German Chem. Soc., Nuclear Energy Agy. (nuclear data com. sec. 1981-83, chmn. 1989-91, nuclear sci. com. 1991—), Hungarian Phys. Soc. (hon. mem.), Pakistan Acad. Scis. (fgn. mem). Islam. Avocations: reading, hiking. Home: Gutenberg Strasse 28, D-52425 Jülich Germany Office: Rsch Ctr Jülich, Inst Nuclear Chemistry, D-52425 Jülich Germany

QAZILBASH, IMTIAZ ALI, engineering company executive, consultant; b. Peshawar, North West Frontier, Pakistan, July 15, 1934; s. Nawazish Ali and Jahan Ara (Samdani Khan) Q.; m. Rubina Satti, Dec. 20, 1964; children: Zulfiqar Ali, Haider Ali, Zainab. Intermediate cert. Islamia Coll., Peshawar, Pakistan, 1951; gen. cert. edn. advanced level Coll. Tech., Northampton, U.K., 1952; diploma in French lang. Geneva U., 1953; B.Sc. in Engring., Imperial Coll. Sci. and Tech., London, 1957; Assoc., City and Guilds of London Inst., 1957. Registered profl. engr., Pakistan. Engr. N.Z. Power Co. Nord Sjaeland Elektricitet og Sporveje Aktieselskab, Copenhagen, 1957; telefoningenior Copenhagen Telephone Co., Kopenhavns Telefon Aktieselskab, Copenhagen, 1957-58; telecomm engr. Pakistan Indsl. Devel. Corp., Karachi, 1958-59; asst. dir. telecommunications Water and Power Devel. Authority, Lahore, Pakistan, 1959-64, dir. telecommunications, 1964-74; mng. dir., pres., CEO Engrs. Internat., Lahore, Peshawar, 1975—; expert to study com. on communications Internat. Conf. on Large High Voltage Electric Systems CIGRE, Paris, 1974—; expert roster UN, N.Y.C., 1970—; leader engrs. select com. West Pakistan Gov.'s Panel on Water and Power Devel. Authority Reorgn., Lahore, 1969-70; mem. Pakistan delegation Internat. Conf. on Large High Voltage Electric Systems, CIGRE, Paris, 1970, 74, 76; chmn. session Conf. on Implementation of Adminstrn. Reforms, Lahore, 1974-75; convenor com. on orgn. and adminstrn. Nat. Conf. on Acceleration of Devel. Process, Lahore, 1974; mem. energy panel Nat. Sci. Policy Group Islamabad, 1974-75; mem., organizer Nat. Seminar on Role of Hydroelectric Resources in Pakistan's Devel., Lahore, 1975; mem. selection com. U. Engring. and Tech., Lahore, 1969—; cons. Pakistan Adminstrv. Staff Coll., Lahore, 1972-74; mem. Energy Working Group Planning Commn., 1991—; speaker, author, nat. seminar on 8th Devel. Plan. Contbr. articles to profl. jours. Founder mem., central council mem. Fedn. Engring. Assns. Pakistan, Lahore, 1969—; v.p. Service of Elec. Engrs. Assn., Lahore, 1969-71; convenor ann. conv. Instn. Engrs., Dacca, 1970. Fellow Instn. Engrs. Pakistan (exec. council, vice-chmn. elec. sect. 1986-91, chmn. Peshawar ctr. 1991—); mem. Pakistan Engring. Congress (council 1970-72, 76), IEEE, Instn. Elec. Engrs. U.K., Clubs: Lahore Gymkhana; Peshawar; Punjab (Lahore); Golf (Peshawar). Avocations: books; music; ballet; golf; trout fishing; shooting; flying. Home and Office: 9 Mulberry Rd, University Town, Peshawar Pakistan

QIAN, DAHONG, laser development engineer; b. Shanghai, China, Oct. 29, 1965; came to U.S., 1989; s. Jinshen and Yang (Shen) Q.; m. Qian Chen Qian, May 31, 1991. BSE, Zhejiang U., Hangzhou, China, 1988; MSE, U. Tex., 1991. Design engr. Changzhou (China) Elec. Equipment Co., 1988-89; software engr. FoxJet, Inc., Arlington, Tex., 1992; laser devel. engr. Dallas Semiconductor Co., 1992—. Contbr. to profl. publs. Mem. IEEE. Achievements include development of high-precision motion measurement system. Office: Dallas Semiconductor Co 4401 S Beltwood Pky Dallas TX 75244

QIAN, YONGJIA, physicist; b. Shanghai, China, Nov. 10, 1939; came to U.S., 1989; m. Huali Li; 1 child, Chong. Grad., U. Sci. and Tech., Beijing, 1963. From asst. rsch. fellow to assoc. rsch. fellow Inst. Physics, Academia Sinica, Beijing, 1963-79; assoc. prof. Fudan U., Shanghai, 1984-86, 87-89; rsch. assoc. U. Wis., Milw., 1986-87, 1989-90; instrumentation specialist U. Cin., 1990—; vis. scholar Northwestern U., Evanston, Ill., 1979-81; vis. prof. U. Fla., Gainesville, 1990; mem. superconducting com. Shanghai Met., 1987-89. Mem. Am. Phys. Soc. Achievements include co-discovery of new superconducting phase of heavy fermion UPt3, of the weak link properties of high Tc superconductors; development of 77K RF SQUID model upon this weak link. Office: U Cin 400 Geology Physics Bldg ML11 Cincinnati OH 45221

QIN, NING, engineering educator; b. Nanjing, Jiangsu, China, Dec. 22, 1958; s. Song Tao and Shu Fan (Dai) Q.; m. Beining Chen, Sept. 16, 1987; 1 child, Gordon Philip. BSc in Maths., Nanjing Aero. Inst., 1982, MSc in Aerodynamics, 1984; PhD in Aerospace Engring., U. Glasgow, Scotland, 1987. Rsch. fellow U. Glasgow, 1987-92; sr. lectr. sch. engring. Coventry U., 1992—. Contbr. articles to Acta Aerodynamica Sinica, Lecture Notes in Physics, Notes on Numerical Fluid Mechs., Aero. Jour. Brit. Govt. scholar, 1984, U. Glasgow scholar, 1984; recipient 3 grants. Mem. AIAA. Achievements include research in numerical simulation of hypersonic flows around aerospace shapes, aerodynamic heating in hypersonic flows, fast solvers for computational fluid dynamics-sparse quasi-Newton method, parallel computing, numerical analysis. Office: Coventry U Sch Engring, Coventry CV1 5FB, England

QIU, SHEN LI, physicist; b. Guzhen, An Hui, China, May 1, 1947; came to U.S., 1980; s. Cong Liang and Dao Ying (Qin) Q.; m. De Huai Chen; children: Dong, Shuang. MA, CCNY, 1981, PhD, 1985. From rsch. assoc. to assoc. physicist Brookhaven (N.Y.) Nat. Lab., 1985-90; asst. prof. Fla. Atlantic U., Boca Raton, 1990—. Contbr. articles to Phys. Rev., Solid State Comm., Jour. Vacuum Sci. Tech. V.p. Alumni Fudan U. in U.S.A., N.Y., 1985—; advisor Chinese Assn. for Sci. and Tech., N.J., 1992. Grantee NSF, 1992—, Brookhaven Nat. Lab., 1991, Nat. Synchrotron Light Source, 1992, FAU Found., 1992. Mem. Am. Phys. Soc. Achievements include AC detection of secondary photoelectrons. Office: Florida Atlantic Univ Physics Dept 500 NW 20th St Boca Raton FL 33431

QUACK, MARTIN, physical chemistry educator; b. Darmstadt, Germany, July 22, 1948; m. Roswitha, May 24, 1977; children: Till, Niels, Manfred. Vordiplom, TH Darmstadt, Germany, 1969; Diplom, U. Göttingen, Germany, 1971; Dr ès sces techn, Ecole Poly. Federale, Lausanne, Switzerland, 1975. Privatdozent, prof. U. Göttingen, Germany, 1978-82; prof. U. Bonn, Germany, 1982-83; prof. ordinarius ETH Zürich, Switzerland, 1983—. Author: Molekulare Thermodynamik und Kinetik, 1986; editor Molecular physics, 1984-87; contbr. more than 150 sci. articles to profl. jours. Recipient Nernst-Haber-Bodenstein prize, 1982, Otto Klung prize Free U. Berlin, 1984, Otto Bayer prize, 1991. Fellow Am. Phys. Soc. Office: Lab Physikalische Chemie, ETH Zurich, CH-8092 Zurich Switzerland

QUAIL, JOHN WILSON, chemist, educator; b. Bklyn., Mar. 19, 1936; arrived in Canada, 1957; s. John Wilson and Mary Magdeline (Yousch) Q.; m. Florence Annie Nowik, May 2, 1959; children: Douglas Wilson, Kevin Steven, Eric William, Jacqueline Mary. BS in Chemistry with honors, U. British Columbia, 1959, MS, 1961; PhD, McMaster U., 1963. Postdoctoral fellow McMaster U., Hamilton, Ontario, Canada, 1963-64; asst. prof. dept. chemistry U. Saskatchewan, Saskatoon, Canada, 1964-69, assoc. prof., 1969-83, prof., 1983—; on sabbatical leave Cambridge Labs. Cambridge (England) U., 1972-73, dept. chemistry U. Alberta, Canada, 1981-82, dept. crystallography U. London, 1988-89, dept. chemistry and biochemistry Massey U., New Zealand. Contbr. articles to Advances in Space Rsch., Canadian Jour. Chemistry, Acta Crystallographia, Nature, Jour. Biol. Chemistry, Jour. Molecular Biology, Inorganic Chemistry. Fellow Chem. Inst. Canada; mem. Am. Crystallographic Assn. Office: U Saskatchewan, Dept Chemistry, Saskatoon, SK Canada S7N 0W0

QUALLS, CHARLES WAYNE, JR., veterinary pathology educator; b. Oklahoma City, Feb. 8, 1949; s. Charles Wayne and Mary Opal (Howard) Q.; m. Cheryl Lynn Lightfoot, Aug. 9, 1969; children: Kerry Lynn, Julie Elizabeth. BS, Okla. State U., 1971, DVM, 1973; PhD, U. Calif., Davis, 1980. Diplomate Am. Coll. Vet. Pathologists. Postdoctoral fellow U. Calif., Davis, 1973-77; asst. prof. La. State U., Baton Rouge, 1977-82; assoc. prof. Okla. State U., Stillwater, 1982-87, prof. vet. pathology, 1988—, acting head dept. vet. pathology, 1991-92. Contbr. articles to profl. jours., chpts. to books. Grantee Dept. of Def., U.S. Army Rsch. and Engring. Program, others. Mem Soc. Toxicologic Pathologists, Soc. Environ. Toxicology and Chemistry, Internat. Acad. Pathology, Am. Vet. Med. Assn. (student sponsor 1983—), Phi Kappa Phi, Phi Zeta. Home: 4 Liberty Cir Stillwater OK 74075-2014 Office: Okla State U Dept Vet Pathology Stillwater OK 74078

QUALLS, ROBERT GERALD, ecologist; b. Boone, N.C., May 20, 1952; s. Edward Spencer Qualls and Lynn Brown Heffner. MPH, U. N.C., 1981; PhD, U. Ga., 1989. Rsch. assoc. U. N.C., Chapel Hill, 1979-83; asst. rsch. prof. Duke U., Durham, N.C., 1989-92. Contbr. articles to profl. jours. NSF rsch. grantee, 1982, EPA acad. rsch. grantee, 1982; NSF Dissertation Improvement awardee, 1985; recipient Internat. Assn. on Water Pollution Founders award for best paper in Water Rsch., 1986. Mem. Soil Sci. Soc. Am., Ecol. Soc. Am. Presbyterian. Achievements include development and validation of mathematical model of disinfection process in ultraviolet light reactors; demonstration of the means by which soluble organic nutrients are conserved within a natural forest ecosystem, others. Home: 29F 311 S LaSalle St Durham NC 27705 Office: Duke Univ Sch of the Environment Durham NC 27706

QUAN, RALPH W., engineer, researcher; b. N.Y.C., Apr. 2, 1963. BS, Rensselaer Poly., 1984, MS, 1986; PhD, U. Colo., Boulder, 1991. Project asst. GE Rsch., Schenectady, N.Y., 1983; engr. Lawrence Livermore (Calif.) Nat. Lab., 1984, 85, Superior Electric Co., Bristol, Conn., 1986, NASA-Langley Rsch., Hampton, Va., 1989; rsch. assoc. USAF Acad., Nat. Rsch. Coun., Colorado Springs, Colo., 1991-93. Recipient Rensselaer fellowship Rensselaer Poly. Inst., 1984-85, Century XXI fellowship Dept. Edn., 1989-91. Mem. IEEE, AIAA. Home: Apt 289 4670 Anille Way Colorado Springs CO 80917 Office: US Air Force Acad FJSRL/NA U S A F Academy CO 80840

QUAN, XINA, polymer engineer; b. Gloucester, N.J., Dec. 23, 1957; m. Charles Roxlo, June 29, 1980; 1 child, Thomas. BS in Chem. Engring., MIT, 1980, MS in Chem. Engring., 1980; PhD, Princeton U., 1986. Assoc. mem. tech. staff AT&T Bell Labs., Murray Hill, N.J., 1980-82, mem. tech. staff, 1982-91, disting. mem. tech. staff, 1991—. Mem. Am. Phys. Soc., Am. Chem. Soc. Office: AT&T Bell Labs Rm 7F 212 PO Box 636 Murray Hill NJ 07874

QUANG, EIPING, nuclear engineer; b. Beijing, Jan. 2, 1957; came to U.S., 1980; s. Jinkuan and Peizhen (Chen) Kuang; m. Xiao Jian Huang, Dec. 25, 1985; children: Leigh Jian, Daniel Xin. BS cum laude, Poly. Inst. N.Y., 1983; MS, U. Mich., 1986, PhD, 1990. Rsch. asst. Phoenix Meml. Lab. U. Mich., Ann Arbor, 1983-89; nuclear engr. Nat. Inst. Standards and Tech., Gaithersburg, Md., 1990-92; sr. analyst EcoTek Lab. Svcs. Inc., Atlanta, 1992—. Contbr. articles to profl. publs. Mem. Am. Phys. Soc., Am. Nuclear Soc., Tau Beta Pi. Home: 6355 Memorial Dr K3-303 Stone Mountain GA 30083

QUATE, CALVIN FORREST, engineering educator; b. Baker, Nev., Dec. 7, 1923; s. Graham Shepard and Margie (Lake) Q.; m. Dorothy Marshall, June 28, 1945 (div. 1985); children: Robin, Claudia, Holly, Rhodalee; m. Arnice Streit, Jan., 1987. B.S. in Elec. Engring, U. Utah, 1944; Ph.D., Stanford U., 1950. Mem. tech. staff Bell Telephone Labs., Murray Hill, N.J., 1949-58; dir. research Sandia Corp., Albuquerque, 1959-60, v.p. research, 1960-61; prof. dept. applied physics and elec. engring. Stanford (Calif.) U., 1961—, chmn. applied physics, 1969-72, 78-81, Leland T. Edwards prof. engring., 1986—; assoc. dean Sch. Humanities and Scis., 1972-74, 82-83; sr. research fellow Xerox Research Ctr., Palo Alto, Calif., 1984—. Served as lt. (j.g.) USNR, 1944-46. Recipient Rank prize for Opto-electronics, 1982, Pres.'s Nat. medal of Sci. NSF, 1992. Fellow IEEE (medal of honor 1988), Am. Acad. Arts and Scis., Acoustical Soc.; mem. NAE, NAS, Am. Phys. Soc., Royal Microscop. Soc. (hon.), Sigma Xi, Tau Beta Pi. Office: Stanford U Dept Applied Physics Palo Alto CA 94305

QUAY, PAUL DOUGLAS, oceanography educator; b. N.Y.C., Oct. 10, 1949. BA, CUNY, 1971; PhD in Geology, Columbia U., 1977. Rsch. assoc. geology Quaternary Rsch. Ctr., 1977-80; from asst. prof. to prof. dept. Geology U. Wash., Seattle, 1980—. Recipient Newcomb Cleveland prize AAAS, 1991-92. Achievements include determination of ocean mixing rates by using radiocarbon distribution, modeling global carbon dioxide and radiocarbon distributions in nature studying the geochemistry of lakes, rivers and marine systems using naturally occurring radioisotopes, chemical oceanography. Office: Univ of Washington School of Oceanography Seattle WA 98195*

QUAY, WILBUR BROOKS, biologist; b. Cleve., Mar. 7, 1927; s. Paul Quilliam Quay and Katharine (Brooks) Pennebaker; m. Catherine Charlet Knox, Dec. 24, 1976; 1 child, Karen Worthington. AB, Harvard U., 1950; MS, PhD, U. Mich., 1952. Instr. U. Mich. Med. Sch., Ann Arbor, 1952-56; asst. prof. to prof. U. Calif., Berkeley, 1956-73, rsch. assoc., 1983-90; rsch. fellow Netherlands Cen. Inst. Brain Rsch., Amsterdam, 1966; prof. U. Wis., Madison, 1973-77, U. Tex. Med. Br., Galveston, 1977-83; sr. scientist Healthdyne Inc., Adv. Devel. Div., Napa, Calif., 1983-84; vis. prof. U. Calif., Berkeley, 1983-90; prin. investigator rsch. grants NIH, NSF, others, 1956-81; co-investigator and/or trainer rsch. and tng. grants, Ford Found., NIH, others, 1973-79. Author: Pineal Chemistry in Cellular and Physiological Mechanisms, 1974; co-founder, assoc. editor Jour. Pineal Rsch., 1984—; contbr. over 400 articles to profl. jours. With USN, 1945-46. Recipient Rsch. Professorship, Miller Inst. Basic Rsch. in Sci., U. Calif., 1964-65, 71-72, Jimmy Dickens Meml. award U. Tex. Med. Br., 1979, 80, 81. Fellow AAAS; mem. Am. Chem. Soc., Am. Ornithol. Union, Am. Physiol. Soc., Am. Soc. Neurochemistry, Internat. Brain Rsch. Orgn. Achievements include discovery and demonstration of circadian rhythmicity and contribution to control of some physiological rhythms by the pineal gland in mammals, use of sperm-release timing in wild and migrating birds in nature, rhythmic brain mechanisms in mammals based upon neurochemistry and cellular physiology of biogenic amines. Home: 1627 State Rd Y New Bloomfield MO 65063-9719 Office: Bio-Rsch Lab 1627 State Rd Y New Bloomfield MO 65063-9719

QUEEN, BRIAN CHARLES, chemist; b. Virginia Beach, Va., Feb. 22, 1964; s. Bobby Delmer and Esther Lee (Caruso) Q.; m. Jennifer Elizabeth Scott, Oct. 10, 1992. BS in Chemistry, We. Carolina U., 1986; postgrad., Mercer U., 1993—. Microscopist Clayton Environ. Cons., Inc., Kennesaw, Ga., 1987-88, McCrone Assocs., Inc., Atlanta, 1989-90; project mgr. EcoTek LSI, Atlanta, 1991—. Mem. Am. Chem. Soc., Internat. Union Pure and Applied Chemistry. Office: EcoTek LSI 3342 International Park Dr Atlanta GA 30316

QUEEN, DANIEL, acoustical engineer, consultant; b. Boston, Feb. 15, 1934; s. Simon and Ida (Droker) Q.; 1 child, Aaron Jacob. Student, U. Chgo., 1951-54. Quality control mgr. Magnacord, Inc., Chgo., 1955-57; project engr. Revere Camera Co., Chgo., 1957-62; dir. engring. for Amplivox products Perma Power Co., Chgo., 1962-70; prin. engr. Daniel Queen Assocs., Chgo., 1970—; pres. Daniel Queen Labs., Inc., Chgo., 1980—; chmn. Am. Nat. Standards Subcom. PH7-6, 1969-84, mem. com. PH-7; mem. standards com. P8-5 Electronic Industries Assn., 1967-82. Contbr. editor Sound and Communications, 1973—; patentee in field; contbr. papers to profl. jours., also articles to trade and popular jours.; editorial bd. Jour. Audio Engring. Soc., 1978—. Fellow Audio Engring. Soc. (standards mgr., chmn. tech. coun.); mem. IEEE (sr.), ASTM, AAAS, Am. Nat. Standards Inst. (sec. com. S4 on audio engring.), Acoustical Soc. Am. (chmn. Chgo. regional chpt. 1976-78, mem. engring. acoustics com.), Midwest Acoustics Conf. (pres. 1971-72), Chgo. Acoustical/Audio Group (pres. 1969-70), Assn. Ednl. Comms. and Tech., Soc. Motion Picture and TV Engrs. (audio rec./reprodn. com.), Am. Pub. Health Assn., Nat. Coun. Acoustical Cons., Inst. Noise Control Engring. Office: 239 W 23d St New York NY 10011

QUEENEY, DAVID, computer engineer; b. Balt., Mar. 24, 1957; s. Paul Joseph and Elizabeth Jane (Golden) Q. BSEE, Va. Poly. Inst. & State U., 1978, MSEE and Computer Sci., 1981. Rsch. asst. Spatial Data Analysis Lab., Blacksburg, Va., 1980-81; engr. Pfister GMBH, Augsburg, Germany, 1981-82; lab. head Ing. Büro Wild, Augsburg, Germany, 1983; engr. Thompson Brandt, Paris, 1984; chief engr. Thompson Micro GMBH, Mörfelden, Germany, 1985; cons. Logica GMBH, Darmstadt, Germany, 1986, Citibank, Paris, Zurich, Frankfurt, 1986-87, Alcatel SEL AG, Germany, 1987—; com. mem. Expert Systems Application Conf., Vienna, Austria, 1990, Berlin, 1991; tech. dir. Spomed AG, Zug, Switzerland, 1987-92. Patentee in field; author: The Calorymetric System, 1988; contbr. tech. papers to profl. publs. Mem. IEEE (software engring. tech. com. 1992—), N.Y. Acad. Scis. Home and Office: D-71640, Ludwigsburg Germany

QUENEAU, PAUL ETIENNE, metallurgical engineer, educator; b. Phila., Mar. 20, 1911; s. Augustin L. and Jean (Blaisdell) Q.; m. Joan Osgood Hodges, May 20, 1939; children: Paul Blaisdell, Josephine Downs (Mrs. George Stanley Patrick). B.A., Columbia U., 1931, B.Sc., 1932, E.M., 1933; postgrad. (Evans fellow), Cambridge (Eng.) U., 1934; D.Sc., Delft (Netherlands) U. Tech., 1971. With Corps of Engrs. USAR, 1937-62; With Internat. Nickel Co., 1934-69, dir. research, 1940-41, 46-48, v.p., 1958-69, tech. asst. to pres., 1960-66, asst. to chmn., 1967-69; vis. scientist Delft U. Tech., 1970-71; prof. engring. Dartmouth Coll., 1971-87, prof. emeritus, 1987—; cons. engr., 1972—; vis. prof. U. Minn., 1974-75, U. Utah, 1987-91; geographer Perry River Arctic Expdn., 1949; chmn. U.S. Navy Arctic Research Adv. Com., 1957; gov. Arctic Inst. N.Am., 1957-62; mem. engring. council Columbia U., 1965-70; mem. vis. com. MIT, 1967-70; mem. extractive metallurgy and mineral processing panels NAS; pres. Q-S Oxygen Processes Inc., 1974-79, dir., 1974—. Author: (with Hanson) Geography, Birds and Mammals of the Perry River Region, 1956; Cobalt and the Nickeliferous Limonites, 1971; editor: Extractive Metallurgy of Copper, Nickel and Cobalt, 1961; (with Anderson) Pyrometallurgical Processes in Nonferrous Metallurgy, 1965; The Winning of Nickel, 1967; contbr. articles to profl. jours.; patentee processes and apparatus employed in the pyrometallurgy, hydrometallurgy and vapometallurgy of nickel, copper, cobalt, lead, zinc and iron, extractive metallurgy oxygen tech. including INCO oxygen flash smelting, oxygen top-blown rotary converter, lateritic ore matte smelting, nickel high pressure carbonyl and iron ore recovery processes; co-inventor Lurgi QSL direct lead-making, QSOP direct coppermaking and nickelmaking reactors, Lurgi direct steelmaking reactors, and Dravo oxygen sprinkle copper smelting furnaces. Bd. dirs. Engring. Found., 1966-76, chmn., 1973-75. Served with U.S. Army, World War II, ETO; col. C.E. (ret.). Decorated Bronze Star, ETO medal with 5 battlestars, Commendation medal; recipient Egleston medal Columbia U., 1965, Fletcher award Dartmouth Coll., 1991. Fellow Metall. Soc. of AIME (dir. 1964, 68-71, pres. 1969, Extractive Metallurgy Lecture award 1977, Paul E. Queneau Internat. Symposium on Extractive Metallurgy of Copper, Nickel and Cobalt 1993); mem. AIME (Douglas Gold medal 1968, v.p. 1970, dir. 1968-71, Henry Krumb lectr. 1984, keynote lectr. ann. meeting 1990), NAE, NSPE, Can. Inst. Mining and Metallurgy, Inst. Mining and Metallurgy U.K. (overseas mem. council 1970-80, Gold medal 1980), Australasian Inst. Mining and Metallurgy, Sigma Xi, Tau Beta Pi. Office: Dartmouth Coll Thayer Sch Engring Hanover NH 03755

QUESADA, MARK ALEJANDRO, physical chemist; b. Munich, Bavaria, Germany, Aug. 3, 1955; came to U.S., 1958; s. Gonzalo James and Esperanza (Solorzano) Q.; m. Michelle Lynn Hilton, Nov. 22, 1989. BA in Chemistry, U. Calif., Berkeley, 1980; PhD in Phys. Chemistry, U. Calif., Santa Cruz, 1987. Postdoctoral rschr. U. Calif., L.A., 1987-90, Berkeley, 1990—; cons. Newport Rsch. Corp., Fountain Valley, Calif., 1992. Contbr. articles to profl. jours. Fellowship U. Calif., 1981-83. Mem. AAAS, Am. Chem. Soc. Achievements include patents pending in high speed, high throughout DNA sequencer using capillary array electrophoresis and fluorescence detection, multiple tag labeling method for DNA sequencing. Home: 1922 48th Ave San Francisco CA 94116 Office: Dept Chemistry U Calif Berkeley Berkeley CA 94720

QUICKSALL, CARL OWEN, chemist, researcher; b. Oak Park, Ill., June 14, 1941; s. Henry Carl and Lillie (Owen) Q.; m. Elizabeth Norman, July 31, 1965; children: Emily, Andrew. BSChemE, Northwestern U., 1964; MA in Chemistry, Princeton U., 1966, PhD in Chemistry, 1971. Postdoctoral rsch. fellow Ames (Iowa) Lab. Ames (Iowa) Lab., U.S. Atomic Energy Commn., 1969-71; asst. prof. chemistry Georgetown U., Washington, 1971-78; sr. rsch. chemist Foote Mineral Co., Exton, Pa., 1978-82; mgr. chem. and mineral tech. Internat. Minerals & Chems., Terre Haute, Ind., 1982-84, dir. chem. and engring. rsch., 1984-87; div. dir. R & D Mallinckrodt Specialty Chems. Co., St. Louis, 1987-90, v.p. sci. and tech., 1990—; mem. adv. bd. Rose Hulman Inst. Tech., Terre Haute, 1987-92, U. Mo., St. Louis, 1990-92. Contbr. articles to sci. jours. Mem. Am. Chem. Soc., Am. Crystallographic Assn., St. Louis Rsch. Coun. (pres. 1990-91), Sigma Xi. Office: Mallinckrodt Specialty Chems Co 3600 N Second St Saint Louis MO 63147

QUILÈS, PAUL, French federal official; b. Saint Denis du Sig, Algeria, Jan. 27, 1942; S. René Quilès and Odette Tyrode; m. Josephe-Marie Bureau, 1964; 3 children. Student, Lycée Lyautey, Paris, Casablanca, Paris, Lycée Chaptal et Louis le Grand, Paris. Engr. Shell Francaise, 1964-78; socialist dep. Nat. Assn., 1978-83, 86-88, min. town planning and housing, 1983-85, min. transport, 1984-85, min. def., 1985-86, min. posts, telecoms. and space, 1988-91, min. pub. works, housing, transp. and space rsch., 1991-92, min. the interior and pub. security, 1992-93; mem. econ. and social coun. Nat. Assn., 1974-75. Author: La Politique n'est pas ce que vous croyez, 1985, Nous vivons une époque intéressante, 1992. *

QUILLIN, PATRICK, nutritionist, writer; b. Ottawa, Ill., May 29, 1951; s. John Paul and Margaret Mary (Fagot) Q.; m. Noreen Marian Quinn, Jan. 7, 1978. Pre-medicine, Univ. Notre Dame, 1969-71; BA in Nutrition, San Diego State U., 1977; MS in Nutrition, No. Ill. U., 1979; PhD in Nutrition Edn., Kensington U., 1981. Registered dietitian, 1979; lic. dietitian. Dietitian Mercy Hosp., Aurora, Ill., 1978-79; lectr. Nutrition Dimension, San Diego, 1986-88, Mesa Coll., San Diego State Nat. U., 1979-89; co-editor Nutrition Times, San Diego, 1986-89; v.p. nutrition Cancer Treatment Ctrs. Am., Tulsa, 1990—; cons. LaCosta Spa, Carlsbad, Calif., 1983-89, Scripps Clinic, San Diego, 1981-83. Author: LaCosta Prescription for Longer Life, 1985, Healing Nutrients, 1987, LaCosta Book of Nutrition, 1988, Safe Eating, 1990; contbr.: (textbook) Blunt Suction Lipectomy, 1984, 90; author, prodr., dir. 3 videos; guest numerous radio and TV shows. Mem. Sierra Club, San Francisco, 1985—, Greenpeace, 1985—, Tree People, L.A., 1990—. Mem. Internat. Acad. Preventive Medicine, Am. Coll. Nutrition, Am. Inst. Cancer Rsch., N.Y. Acad. Scis. Achievements include organization of first international symposium on adjuvant nutrition in cancer treatment. Office: Cancer Treatment Ctr 8181 S Lewis St Tulsa OK 74137-1200

QUIMBY, FRED WILLIAM, pathology educator, veterinarian; b. Providence, Sept. 19, 1945; s. Edward Harold and Isabel (Barber) Q.; m. Cynthia Claire Connelly, Aug. 21, 1965; children—Kelly Ann, Cynthia Jane. V.M.D., U. Pa., 1970, Ph.D., 1974. Diplomate Am. Coll. Lab. Animal Medicine. Hematology fellow Tufts Med. Coll., Boston, 1974-75, instr. pathology, 1975-76, asst. prof., 1976-79; assoc. prof. pathology Cornell Med. Coll., N.Y.C., 1979—, N.Y. State Vet. Coll., Ithaca, 1979—; dir. lab. animal medicine Tufts-New Eng. Med. Ctr., Boston, 1975-79; dir. Ctr. Research Animal Resources, Cornell U., Ithaca, 1979—. Editor: Clinical Chemistry of Laboratory Animals, 1988, Animal Welfare, 1992, Lab. Animal Sci.; chmn. editorial bd. ILAr News, 1988-91; contbr. 100 sci. papers and abstracts. Greenfield Trust scholar, 1966-70; N.H. Rural Rehab. Corp. scholar, 1966-70; U. Pa. scholar, 1969-70. Mem. Am. Assn. Lab. Animal Sci. (pres. Northeast br. 1978-79; B. Trum award 1979), World Vet. Assn. (sec. exec. com. animal welfare 1990—). Episcopalian. Home: 115 Terraceview Dr Ithaca NY 14850 Office: NYSCVM Cornell U 221 VRT Ithaca NY 14853-6401

QUIMBY, GEORGE IRVING, anthropologist, former museum director; b. Grand Rapids, Mich., May 4, 1913; s. George Irving and Ethelwyn (Sweet) Q.; m. Helen M. Ziehm, Oct. 13, 1940; children: Sedna H., G. Edward, John E., Robert W. B.A., U. Mich., 1936, M.A., 1937, grad. fellow, 1937-38; postgrad., U. Chgo., 1938-39; LHD (hon.), Grand Valley State U., Mich., 1992. State supr. Fed. Archaeol. Project in La., 1939-41; dir. Muskegon (Mich.) Mus., 1941-42; asst. curator N.Am. archaeology and ethnology Field Mus. Natural History, 1942-43, curator exhibits, anthropology, 1943-54, curator N.Am. archeology and ethnology, 1954-65, research assoc. in N. Am. archaeology and ethnology, 1965—; curator anthropology Thomas Burke Meml. Wash. State Mus.; prof. anthropology U. Wash., 1965-83, emeritus prof., 1983—, mus. dir., 1968-83, emeritus dir., 1983—; lectr. U. Chgo., 1947-65, Northwestern U., 1949-53; Fulbright vis. prof., U. Oslo, Norway, 1952; archaeol. expdns. and field work, Mich., 1935, 37, 42, 56-63, Wis., 1936, Hudson's Bay, 1939, La., 1940-41, N.Mex., 1947, Lake Superior, 1956-61. Author: Aleutian Islanders, 1944, (with J. A. Ford) The Tchefuncte Culture, an Early Occupation of the Lower Mississippi Valley, 1945, (with P. S. Martin, D. Collier) Indians Before Columbus, 1947, Indian Life in the Upper Great Lakes, 1960, Indian Culture and European Trade Goods, 1966, A Thing of Sherds and Patches: The Autobiography of George Irving Quimby, American Antiquity, vol. 58, # 1, 1993; prodr. documentary film (with Bill Holm) In the Land of the War Canoes, 1973, Edward S. Curtis in the Land of the War Canoes: A Pioneer Cinematographer in the Pacific Northwest, 1980; Contbr. articles to profl. jours. Honored by festschrift U. Mich. Mus. Anthropology, 1983. Fellow AAAS, Am. Anthrop. Assn.; mem. Soc. Am. Archaeology (pres. 1958, 50th Anniversary award 1983, Disting. Svc. award 1989), Am. Soc. Ethnohistory, Wis. Archeol. Soc., Soc. Historical Archeology (council 1971-74, 75-78, J.C. Harrington medal 1986), Assn. Sci. Mus. Dirs. (pres. 1973-74), Arctic Inst. N.Am., Am. Assn. Museums (council 1971-74), Sigma Xi, Phi Sigma, Chi Gamma Phi, Zeta Psi. Home: 6001 52d Ave NE Seattle WA 98115 Office: U Washington Thomas Burke Meml Wash State Mus Seattle WA 98195

QUINLAN, KENNETH PAUL, chemist; b. Somerville, Mass., Jan. 13, 1930; s. Cornelius Francis and Helen A. (Tevanon) Q.; m. Margo Welch, Mar. 30, 1964; children: Ellen, Maureen, Kenneth, Joseph. MS, Tufts U., 1954; PhD, U. Notre Dame, 1960. Rsch. chemist Army Engrs., Fort Belvoir, Va., 1954-55, Am. Cyanamid Co., Winchester, Mass., 1955-56, USAF, Bedford, Mass., 1960—. Contbr. articles to profl. jours. Eucharistic min. Mass. Gen. Hosp., Boston, 1980—. Mem. Electrochem. Soc. Democrat. Roman Catholic. Avocation: painting. Home: 70 Grasmere St Newton MA 02158

QUINN, CHARLES LAYTON, JR., electrical engineer; b. Birmingham, Ala., Feb. 23, 1951; s. Charles L. and Jewell (Kilpatrick) Q.; m. Patricia S. Headley, Aug. 25, 1972; children: James S., Jason M., Jonathan R. BS in Engring., U. Ala., Birmingham, 1974. Registered profl. engr., Ala., Ga.; cert. mgr. Asst. engr. So. Co. Svcs., Inc., Birmingham, Ala., 1974-76, engr. II, 1976-78, engr. I, 1978-81, sr. engr., 1981-87, 90—; sr. engr. I Ga. Power Co., Atlanta, 1987-88, Birmingham, 1988-90; cons. Ga. Corp., Atlanta, 1988-90. Contbr. to profl. publs. Dir. Inverness Homeowner's Assn., Birmingham, 1991-93. Mem. IEEE (sr.), So. Co. Svcs. Leadership Devel. Assn. Republican. Baptist. Home: 2921 MacAlpine Circle Birmingham AL 35242

QUINN, EDWARD FRANCIS, III, orthopedic surgeon; b. Washington, Apr. 28, 1944; s. Edward F. Jr. and Louise Q.; m. Audrey Dickinson; 1 child, Edward Francis IV. BS, U. Md., 1968, MD, 1969. Diplomate Am. Bd. Orthopedic Surgery, Am. Bd. Neurol. and Orthopedic Surgery. Intern Ohio State Univ. Hosp., Columbus, 1969-70; resident USN Hosp., Bethesda, Md., 1971-74; staff Milford (Del.) Meml. Hosp., 1975—, pres. med. staff, 1991-93, chief surgery, 1993—; cons. staff Beebe Hosp., Lewes, Del., 1975—, Nanticoke Meml. Hosp., Seaford, Del., 1975-90, Kent Gen. Hosp., Dover, Del., 1981-90, attending staff, 1990—. Bd. advisors So. Campus of Goldey Beacon Coll., 1987-91. Lt. comdr. USN, 1970-75. Fellow ACS, Am. Acad. Neurol./Orthopeadic Surgery, Internat. Coll. Surgeons; mem. AMA, Med. Soc. Del., Sussex County Med. Soc., Am. Fracture Assn., So. Med. Assn., Del. Soc. Orthopaedic Surgeons, Ea. Orthopaedic Assn., So. Orthopaedic Assn., Acad. Neuro-Muscular Thermography, Chemonucleolysis Adv. Bd., Milford Rotary Club, Rotary Internat. (Paul Harris fellow 1986), Am. Acad. Disability Evaluation Rsch. Physicians, Am. Legion, So. Del. C. of C. (bd. dirs. 1983-90). Republican. Avocations: boating, skiing, firearms, hunting. Office: Dickinson Med Group 800 N Du Pont Hwy Milford DE 19963-1091

QUINN, JARUS WILLIAM, physicist, association executive; b. West Grove, Pa., Aug. 25, 1930; s. William G. and Ellen C. (DuRoss) Q.; m. Margaret M. McNerney, June 27, 1953; children: J. Kevin, Megan, Jennifer, Colin, Kristin. BS, St. Joseph's Coll., 1952; postgrad., Johns Hopkins U., 1952-55; PhD, Cath. U. Am., 1964. Rsch. assoc. physics Johns Hopkins U., 1954-55; staff scientist Rsch. Inst. Advanced Study, 1956-57; rsch. assoc. physics Cath. U. Am., 1958-60, instr., 1961-64, asst. prof., 1965-69; exec. dir. Optical Soc. Am., Washington, 1969-93; governing bd. Am. Inst. Physics, 1973—. Fellow Optical Soc. Am.; mem. Am. Phys. Soc., Am. Soc. Assn. Execs., Coun. Engring. and Sci. Soc. Execs., Am. Assn. Engring. Socs. (bd. govs. 1990-93). Office: Optical Soc of Am 2010 Massachusetts Ave NW Washington DC 20036-1023*

QUINN, JOHN ALBERT, chemical engineering educator; b. Springfield, Ill., Sept. 3, 1932; s. Edward Joseph and Marie (Von De Bur) Q.; m. Frances Wilkie Daly, June 22, 1957; children: Sarah D., Rebecca V., John E. B-SChemE, U. Ill., 1954; PhDChemE, Princeton U., 1959. Faculty mem. chem. engring. U. Ill., Urbana, 1958-70; prof. chem. engring. U. Pa., Phila., 1971—, Robert D. Bent prof. chem. engring., 1978—, chmn. dept., 1980-85; vis. prof. chem. engring. Imperial Coll. U. London, 1965-66; vis. sci. MIT, 1980; vis. prof. chem. U. Roma/La Sapienza, 1992; mem. sci. adv. bds. Sepracor, Inc., Marlborough, Mass., 1984—, The Whitaker Found., Mechanicsburg, Pa., 1987—; Mason lectr. Stanford U., 1981; Katz. lectr. U. Mich., 1985; Reilly lectr. U. Notre Dame, 1987. Contbr. articles to profl. publs.; editorial advisor Jour. Membrane Sci., 1975—, Indsl. and Chem. Engring. Rsch., 1987-88, Revs. in Chem. Engring., 1980—; pioneer researcher on mass transfer and interfacial phenomena. Recipient S. Reid Warren Jr. award for disting. teaching U. Pa., 1974; sr. postdoctoral fellow NSF, 1965-66; Sherman Fairchild scholar Calif. Inst. Tech., 1985. Fellow AAAS; mem. NAE, Am. Acad. Arts and Scis., Am. Chem. Soc., Am. Inst. Chem. Engrs. (Allan P. Colburn award 1966, Alpha Chi Sigma award 1978), Internat. Soc. Oxygen Transport to Tissue, Sigma Xi, Phi Lambda Upsilon, Tau Beta Pi. Home: 275 E Wynnewood Rd Merion Station PA 19066-1627 Office: Univ Pa Towne Bldg 220 S 33rd St Philadelphia PA 19104-6393

QUINN, JOHN MICHAEL, physicist, geophysicist; b. Denver, May 8, 1946; s. Leonard Simon and Winifred Ruth (Doolan) Q.; m. Pamela Dagmar Shield, May 28, 1983. BS in Physics, U. Va., 1968; MS in Physics, U. Colo., 1982. Physicist U.S. Naval Rsch. Lab., Washington, 1967-73; prin. engr. Singer Simulation Products, Silver Spring, Md., 1973-74; rsch. physicist U.S. Naval Rsch. Lab, Washington, 1979-80; geophysicist U.S. Naval Oceanographic Office, Stennis Space Ctr., Miss., 1974-79, 82-85, geophysicist, mathematician, 1985—; investigator Polar Orbiting Geomagnetic Survey Experiment, 1990—; prin. investigator Def. Meteorol. Satellite Program Polar Orbiting Geomagnetic Survey Ext., 1991—; chmn. com. on earth and planetary geomagnetic survey satellites Internat. Assn. Geomagnetism and Aeronomy, 1991—, mem. internat. geomagnetic ref. field com., 1989—. Author: Epoch World Geomagnetic Model, 1985, 1990. With U.S. Army, 1968-71. Mem. Am. Geophys. Union, Am. Math. Soc., European Geophys. Soc., Math. Assn. Am. Achievements include creation of official Department of Defense world magnetic models which are used by military and civilian agencies for navigational purposes and basic rsch. of the earth's magnetic field; project coord. USN Project MAGNET. Home: 107 Shore

Dr Long Beach MS 39560-3121 Office: US Naval Oceanographic Office Stennis Space Center Bay Saint Louis MS 39522-5001

QUINN, MARY ELLEN, science educator; b. Chgo., Sept. 24, 1923; d. Frank J. and Mary Grace (Harrison) Q. BS, St. Mary-of-the-Wood Coll., 1956; MS, DePaul U., 1964; MA, U. Tex., San Antonio, 1980; EdD, U. Pa., 1971. Cert. sch. adminstrn., supr., physics/math/ESL tchr., Tex., cert. physics and math tchr., Ind. Tchr. various middle schs., Ind., Ill., 1944-58, various high schs., Ind., Ill., 1958-68; instr. U. Pa. U., Phila., 1969-71; assoc. prof. Immaculata Coll. of Washington, D.C., 1973-75; dir. curriculum Edgewood Sch. Dist., San Antonio, 1975-80; secondary tchr. Alamo Heights Sch. Dist., San Antonio, 1980-88; vis. prof. math Our Lady of the Lake U., San Antonio, 1988—; cons. numerous sch. dists., Washington, Tex., Calif., N.Mex., N.Y., 1972—; cons., rschr. Ctr. for Applied Linguistics, Washington, 1986—. Author: Science for Language Learners, 1989; contbr. chpts. to Science Education and Cultural Enviroments in Americas, 1986, ESL and Science, 1987, Cooperative Language Learning, 1992. Coun. mem. Ind. Jr. Acad. Sci., Indpls., 1960-66; external assessor U. Malaysia, 1973; cons. Intercultural Devel. Rsch. Assoc., San Antonio, 1987—. Atomic Energy grantee NSF & Atomic Energy Commn., Oak Ridge Inst. Nuclear Studies, 1965, U. Calif., Berkeley grantee NSF, Lawrence Rad Labs., 1968; scholar, fellow U. Pa., 1969-71. Mem. AAAS, Nat. Assn. for Rsch. in Sci. Teaching (Outstanding Paper 1981), Am. Assn. Physics Tchrs., Nat. Sci. Tchrs. Assn. Home: 3123 Clearfield San Antonio TX 78230

QUINN, RICHARD KENDALL, environmental engineer; b. Cleve., July 15, 1957; s. William Jerome and Nancy Drysdale (Kendall) Q.; m. Ann Marie (Beebe), Jan. 28, 1984; children: Ryan K., Eric E., Margaret A., Emily E. BS in Civil Engring., U. Mich., 1979; BS in Geology, Colo. Sch. Mines, 1981. Registered profl. engr., Colo., Nebr., Iowa, Mich., Ill. Geophysic surveyor Shell Oil Co., Denver, 1980-84; profl. engr., environ. engr. Camp Dresser & McKee Engring., Denver, 1985-91; engr. Harco Tech. Corp., Schaumburg, Ill., 1991-92, Walker div. Chgo. Bridge & Iron, Aurora, Ill., 1992—. Cons. Iowa Community Devel. Block Grants, Des Moines, 1988—; active Big Bros. Assn. Grantee Iowa Engring. Soc., 1989. Mem. ASCE (chpt. sec. 1986-87), NSPE, Am. Pub. Works Assn. (chpt. sec. 1987—), Am. Water Works Assn., Water Pollution Control Fedn., Am. Soc. Mil. Engrs., Iowa Engring. Soc., Ill. Engring. Soc., Mich. Engring. Soc., Nat. Assn. Corrosion Engrs. Republican. Roman Catholic. Avocations: golf, hockey, flying. Home: 280 Westbrook Cr Naperville IL 60565 Office: Chgo Bridge & Iron Co Walker Divsn Ste 102 1245 Corporate Blvd Aurora IL 60504

QUINN, ROBERT WILLIAM, physician, educator; b. Eureka, Calif., July 22, 1912; s. William James and Norma Irene (McLean) Q.; student Stanford U., 1930-33; M.D., C.M., McGill U., 1938; m. Julia Rebecca Province, Jan. 21, 1942; children—Robert Sean Province, Judith D. Rotating intern Alameda County Hosp., Oakland, Calif., 1938-39; postgrad. tng. internal medicine U. Calif. Hosp., San Francisco, 1939-41, research fellow internal medicine, 1940-41; postgrad. tng. internal medicine Presbyn. Hosp., N.Y.C., 1941-42; research fellow Yale U., 1946- 47, instr. preventive medicine, 1947-49; assoc. prof. preventive medicine and student health U. Wis., 1949-52; prof., head dept. preventive medicine and public health Vanderbilt U., 1952-80, emeritus prof., 1980; dir. venereal disease and tuberculosis control Met. Nashville Health Dept. Bd. dir. Planned Parenthood Assn. Nashville; adv. com. Family Planning Tenn. and Nashville; vol. Vanderbilt U. Vets. Med. Ctr., 1992, Teen Clinic Nashville Planned Parenthood Assn., 1992-93. Served as capt. M.C., USN, 1942-46, USNR 1946-77. Diplomate Am. Bd. Preventive Medicine. Mem. Am. Acad. Preventive Medicine, Infectious Diseases Soc. Am., Middle Tenn. Heart Assn. (pres. 1957), Assn. Am. Med. Colls., Nashville Acad. Medicine, Assn. Tchrs. Preventive Medicine, Am. Public Health Assn., Am. Epidemiol. Soc., Am. Venereal Disease Assn., Physicians for Social Responsibility (pres. Greater Nashville chpt.). Club: Belle Meade Country. Author med. articles. Home: 508 Park Center Dr Nashville TN 37205-3430

QUINNAN, GERALD VINCENT, JR., medical educator; b. Boston, Sept. 7, 1947; s. Gerald Vincent and Mary (Lally) Q.; children: Kevin, Kylie, Kathleen, John; m. Leigh A. Sawyer. AB in Chemistry, Coll. Holy Cross, 1969; MD cum laude, St. Louis U., 1973. Diplomate Am. Bd. Internal Medicine. Intern, resident, fellow Boston U. Med. Ctr., 1973-77; med. officer Bur. Biologics, USPHS, Bethesda, Md., 1977; advanced through grades to asst. surgeon gen. USPHS, 1992; dir. herpes virus br., dep. dir. div. virology Bur. Biologics, Bethesda, 1980-81; dir. div. virology Ctr. for Drugs and Biologics, Bethesda, 1981-88; dep. dir. Ctr. Biologics Evaluation and Rsch., Bethesda, 1988—, acting dir., 1990-92; dep. dir. Ctr. Biologics Evaluation and Rsch., 1992-93, dir. off-blood rsch. and rev., 1993; prof. dept. preventive medicine Uniformed Svcs. U. Health Scis., Bethesda, 1993—; cons. World Health Orgn., Geneva, Switzerland, Pan Am. Health Orgn., liaison adv. coms. Contbr. chpts. to books, numerous articles to profl. jours. Fellow Infectious Diseases Soc. Am.; mem. Am. Assn. Immunology, Am. Soc. for Clin. Investigation, Sigma Xi, Alpha Omega Alpha. Roman Catholic. Office: Uniformed Svcs U Hlth Scis Div Tropical PH 3401 Jones Bridge Rd Bethesda MD

QUIÑONES, JOSE ANTONIO, structural engineer, consultant; b. Camuy, P.R., June 28, 1939; s. Antonio and Julia (Lopez) Q.; m. Dorisoraida Rivera Mathews, July 17, 1965; children: Antonio, Rafael, Doris Julia, Zoraida. BSCE magna cum laude, U. P.R., 1963; postgrad. in Marine Engring., UCLA, 1973-74. Registered profl. engr., P.R. Design engr. Barrett & Hale, Columbus, Ohio, 1963-65; assoc. design engr. R.A. Bermudez & Assocs./P.R. Testing Svcs., San Juan, 1965-69; prin. ptnr. Jose A Quiñnones, Jr. Assocs., San Juan, 1970—; v.p. Metro. Soils and Engring. Materials Lab., Inc. Contbr. articles to profl. jours. Mem. AAAS, ASCE (sec. P.R. chpt. 1969-70), NSPE, Am. Acad. Mechanics, Am. Concrete Inst., Am. Soc. Testing and Materials, Am. Welding Soc., Am. Water Works Soc., Am. Inst. Steel Constrn., Internat. Soc. Soil Mechanics and Foundations, Sociedad Española del Pretensado, Colegio de Ingenieros, Arquitectos y Agrimensores de P.R., Geotech. Soc. P.R. (sec. 1970-77), P.R. Soc. Profl. Engrs., P.R. Soc. Engrs., P.R. Planning Soc., P.R. Geotech. Soc., P.R. Inst. Architects, Prestressed Concrete Inst., Interam. Planning Soc., N.Y. Acad. Scis., London Concrete Soc., Inst. Eduardo Torroja, Calif. Alumni Assn. Roman Catholic. Achievements include construction of longest three hinges arech structure for Dorado Reef in the Caribbean, Sport Colosseum at Bayamon, largest prefabricated floating caisson structure, Palmas del Mar Marina, Humacao, P.R., unique three story loading and unloading structure floating in a pontoon, San Juan Bay, P.R., tallest microwave communication structure in the Caribbean, La Punta Mountain, Jayuya, P.R., largest regional shopping mall in the Caribbean, Plaza las Americas, hato Rey, P.R. Home: GPO PO Box 71387 San Juan PR 00936 Office: Jose Qui nones Jr Assocs Mayaguez Ave 41 San Juan PR 00917

QUINTANILLA, ANTONIO PAULET, physician, educator; b. Peru, Feb. 8, 1927; came to U.S., 1963, naturalized, 1974; s. Leandro Marino and Edel Paulet Q.; m. Mary Parker Rodriguez, May 2, 1958; children: Antonio Paulet, Angela, Francis, Cecilia, John. PhD, San Marcos U., 1948, MD, 1957. Assoc. prof. physiology U. Arequipa, Peru, 1960-63; assoc. in physiology Cornell U., N.Y., 1963-64; prof. physiology U. Arequipa, 1964-68; assoc. prof. medicine Northwestern U., 1969-80, prof., 1980—; chief renal sect. VA Lakeside Hosp., 1976-90; cons. nephrologist Northwestern Meml. Hosp., Evanston Hosp., 1990—; lectr. nat. Ctr. Advanced Med. Edn., Chgo.; mem. adv. bd. Kidney Found. Ill., Am. Fedn. Clin. Rssch. Fellow ACP; mem. Chgo. Heart Assn. (hypertension council), Central Soc. Clin. Rsch., Am. Soc. Clin. Pharmacology and Therapeutics, Am., Internat. socs. nephrology, Chgo. Soc. Internal Medicine, Am. Physiol. Soc. Contbr. articles on renal disease to med. jours.; author books, poetry, short stories. Home: 9352 Karlov Ave Skokie IL 60076-1415 Office: 2650 Ridge Ave Evanston IL 60201-1718

QUINTERO, HÉCTOR ENRIQUE, science educator; b. Mayaquez, P.R., Feb. 20, 1951; s. Hector E. and Carmen Elisa (Vilella) Q.; m. Irma H. Mendez, Dec. 28, 1974; children: Elisa, Mercedes, Raiza. BS in Biology, U. P.R., 1974, MS in Zoology, 1977; PhD in Ecology and Evolution, Fla. State U., 1983. Asst. researcher P.R. Nuclear Ctr., Mayaquez, 1975-76; inspector NOAA/Marine Fishery Svc., P.R., 1977-78; instr. U. P.R., Ponce, 1977-78; cons. sci. curriculum devel. resources ctr. sci., engring. U. P.R., Rio Piedras, 1989—; assoc. prof. Inter Am. U., San German, P.R., 1983-92, dean arts and scis., 1992—. Contbr. articles to profl. publs. Mem. Southwestern Ednl. Soc. (bd. dirs. 1992—), Sociedad Puertorriqueña de Conservación, Ecol. Soc. Am., Sociedad Ecologica Venezolana. Achievements include re-discovery of cave with water at Mona Island, P.R. Home: Box 6212 Mayaquez PR 00681 Office: Inter Am U Dept Biology San German PR 00683

QUINTO, P. FRANK, aerospace engineer; b. Antigua, Guatemala, Oct. 5, 1956; came to U.S., 1966; s. Felipe and Ana (Lee) Q. BS in Aero. Engring., Va. Tech., 1980. Aero-space technologist NASA Langley Rsch. Ctr., Hampton, Va., 1980-90, facility mgr., 1990—. Contbr. tech. papers to NASA. Sec. Chinese Community Assn. Hampton Rds., Virginia Beach, Va., 1991. Named Outstanding Vol. City of Norfolk, 1985. Mem. AIAA (sr. mem., co-author papers, jour. aircraft trimming adv. fighter 1983), Nat. Space Soc. Roman Catholic. Achievements include rsch. in thrust-induced effects on fighter aircraft, trimming high lift for STOL fighter aircraft. Office: NASA Langley Rsch Ctr Mail Stop 286 17 W Taylor St Hampton VA 23681-0001

QUIRK, FRANK JOSEPH, management consulting company executive; b. N.Y.C., Feb. 27, 1941; s. Frank J. and Madeline B. Quirk; B.A., Cornell U., 1962, M.B.A., 1964; m. Betty Josephine Mauldin, Jan. 7, 1967; children—Laura Josephine, Katherine Elizabeth. Assoc., Booz, Allen & Hamilton, Inc., Chgo. and Washington, 1967-72; exec. v.p. Macro Internat., Inc., Silver Spring, Md., 1972-79, pres., chief exec. officer, 1980—. Served to capt. U.S. Army, 1964-66. Club: Belle Haven Country. Home: 2110 Foresthill Rd Alexandria VA 22307-1128 Office: Macro Internat Inc 11785 Beltsville Dr Beltsville MD 20705

QUO, PHILLIP C., mechanical engineering educator; b. Fuchow, Fuchien, China, Oct. 4, 1930; came to U.S., 1955; m. Consuelo Perez, Oct. 8, 1959; children: Marcia, Geoffrey, Stacey, Brian. BSME, U. Kans., 1960, MSME, 1965; LLB, JD, U. Amoy, China, 1949. Project engr. Black & Veatch Engrs., Kansas City, Mo., 1960-64, head computer applications, 1964-66; dir. computer engring. A.M. Kinney, Inc. Engrs., Cin., 1966-70, acting dir. power engring., 1970-74, v.p., 1974-85; prof. U. Cin., 1986-88, prof., dir. mech. indsl. and nuclear engring., 1988—; speaker in field. Author: Introduction to Fortran Programming, 1967, CPM Programming and Network Scheduling, 1971, Computer Aided Piping Design and Stress Analysis, 1982, Just-In-Time Manufacturing Toward Factory Automation, 1983, Simultaneous Engineering in Product and Process Design, 1990; contbr. articles to profl. and refereed jours. Coach, Pleasant Run Say Soccer Team, Cin., 1972-74. Recipient Excelent Teachin of Yr. award U. Cin., 1975, 76, 90, Profl. Accomplishment of Yr. award Sci. Socs. Coun. Cin., 1978, Bicentennial medal ASME, 1980; named to Am. Computer Delegation to China, 1983. Fellow ASME (chmn. nat. bd. metrication 1991—, group chair tech. and indsl. affairs 1984-90, nat. disting. lectr. 1992—, Dedicated Svc. award 1991); mem. SME (sr.), Engring. Soc. Cin. (pres. 1984). Home: 12067 Deerhorn Dr Cincinnati OH 45240 Office: U Cin ML 72 Mech Indsl Nuclear Engring Cincinnati OH 45221-0072

QURESHI, IQBAL HUSSAIN, nuclear chemist; b. Ajmer, India, Sept. 27, 1936; arrived in Pakistan, 1950; s. Ashiq Hussain and Hasina Begum Q.; m. Zeenat, Sept. 21, 1965 (wid. July 1984); children: Adnan Iqbal, Imran Iqbal; m. Khurshid, June 9, 1987. BS, Govt. Coll., Hyderabad, Pakistan, 1956; MS, U. Sind, Hyderabad, Pakistan, 1958, U. Mich., 1962; PhD, Tokyo U., 1963. Lectr. Govt. Coll., 1956-60; officer on spl. tng. Pakistan Atomic Energy Commn., Karachi, 1960-63; sr. scientific officer Pakistan Atomic Energy Commn., Lahore, 1963-68; prin. scientific officer Pakistan Atomic Energy Comm., Lahore, 1969-76; rsch. chemist U.S. Nat. Bur. Standards, Washington, 1967-68; chief scientific officer Pakistan Atomic Energy Comm., Islamabad, 1976-88; chief scientist Pakistan Atomic Energy Commn., Islamabad, 1988—. vis. scientist AEC, Roskilde, Denmark, 1970-72; dir. Pakistan Inst. Nuclear Sci. and Tech., Islamabad, 1984-91; tech. mem. Pakistan AEC, Islamabad, 1991. Recipient Chancellor's Gold Medal for scis. U. Sind, 1958, Gold Medal for phys. scis. Pakistan Acad. Scis., 1988, Star of Distinction Sci. award Gov. Pakistan, 1992. Mem. Internat. Union Elementologists. Avocations: poetry, music. Home: House No 211 St No 18, F-10/2, Islamabad Pakistan Office: Pakistan Atomic Energy Comm, PO Box 1114, Islamabad Pakistan

QURESHI, SAJJAD ASLAM, biologist, researcher; b. Multan, Pakistan, June 16, 1954; s. Mohammad Aslam Qureshi and Mussarat Ashraf. BS, U. Punjab; MSc, U. Karachi, Pakistan, 1983; PhD, CUNY, 1991. Sales asst. Pakistan Airlines, Peshawar, 1975-78; flight attendant Pakistan Airlines, Karachi, 1978-85; rsch. asst. Hunter Coll., N.Y.C., 1986-91, grad. asst., 1988-91; postdoctoral assoc. Rockefellar U., N.Y.C., 1991—; adj. lectr. Hunter Coll., N.Y.C., 1986-88, pres. grad. student orgn., 1989-91; adv. com. Grad. Program Biology, CUNY, 1987-88. Contbr. rsch. articles to profl. jours. Cancer Rsch. Inst. fellow, 1992—, Beatric Konheim fellow, 1990; Mina S. Rees scholar, 1991. Mem. AAAS, N.Y. Acad. Scis. Muslim. Achievements include discovery of v-Src induced signaling; research in study of intracellular signaling to understand growth control and devise way to prevent cancer. Office: Rockefellar Univ 1230 York Ave Box 279 New York NY 10021

RAAB, HARRY FREDERICK, JR., physicist; b. Johnstown, Pa., May 9, 1926; s. Harry Frederick and Marjorie Eleanor (Stiff) R.; m. Phebe Ann Duerr, June 16, 1951; children: Constance Diane, Harry Frederick, Cynthia Ann Raab Morgenthaler. Student Navy Electronics Tech. Sch., 1944-45; SB and SM E.E., MIT, 1951; postgrad. Oak Ridge Sch. Reactor Tech., 1954-55. Reactor control engr. Bettis Atomic Power Lab. Westinghouse Electric Corp., West Mifflin, 1951-54, mgr. surface ship physics, 1955-62, mgr. light water breeder reactor physics, 1962-72, chief physicist Navy Nuclear Propulsion Directorate, Washington, 1972—. Patentee light water breeder reactor. Lay reader Episc. Ch. of the Good Shepherd, Burke, Va., 1957—, Sunday Sch. tchr., 1957-72, dir. liturgy, 1987—, healing Min., 1989—, stewardship chmn., 1979-82, 84, 92—; healing ministry, 1989—; sr. warden, 1983, 85, mem. stewardship com. Diocese of Va., 1983—; chaplain for mentally retarded No. Va. Tng. Ctr., 1983—. With USNR, 1944-46, PTO. Fellow Am. Nuclear Soc.; mem. Internat. Platform Assn., Sigma Xi, Tau Beta Pi, Eta Kappa Nu. Republican. Lodge: Masons. Home: 8202 Ector Ct Annandale VA 22003-1342 Office: Naval Sea Systems Command Code 08A Washington DC 20362-5101

RABB, GEORGE BERNARD, zoologist; b. Charleston, S.C., Jan. 2, 1930; s. Joseph and Teresa C. (Redmond) R.; m. Mary Sughrue, June 10, 1953. BS, Coll. Charleston., 1951; MA, U. Mich., 1952, PhD, 1957. Teaching fellow zoology U. Mich., 1954-56; curator, coord. rsch. Chgo. Zool. Park, Brookfield., Ill., 1956-64; assoc. dir. rsch. and edn. Chgo. Zool. Park, 1964-75, dep. dir., 1969-75, dir., 1976—; rsch. assoc. Field Mus. Natural History, 1965—; lectr. dept. biology U. Chgo., 1965-89; mem. Com. on Evolution Biology, 1969—; pres. Chgo. Zool. Soc., 1976—; mem. steering com. Species Survival Commn., Internat. Union Conservation of Nature, 1983—, vice chmn. for N.Am., 1986-88; dep. chmn., 1987-89, chmn., 1989—; chmn. policy adv. group Internat. Species Info. System, 1974-89, chmn. bd., 1989-92. Fellow AAAS; mem. Am. Soc. Ichthyologists and Herpetologists (pres. 1978), Herpetologists League, Soc. Systematic Zoology, Soc. Mammalogists, Soc. Study Evolution, Ecol. Soc. Am., Soc. Conservation Biology (council mem. 1986), Am. Soc. Zoologists, Soc. Study Animal Behavior, Am. Assn. Museums, Am. Soc. Naturalists, Am. Assn. Zool. Parks and Aquariums (dir. 1979-80), Internat. Union Dirs. Zool. Gardens, Am. Com. Internat. Conservation (chmn. 1987—), Chgo. Coun. Fgn. Relations (Chgo. com.), Sigma Xi. Club: Economic (Chgo.), Tavern. Office: Chgo Zool Park 3300 Golf Rd Brookfield IL 60513-1064

RABE, JÜRGEN P., chemical physicist; b. Neuss, Fed. Republic Germany, Nov. 20, 1955; s. Günter and Helga (Hoss) R. Diploma in physics, RWTH Aachen, Fed. Republic Germany, 1981; D. in Natural Scis., Tech. U. Munich, 1984; Habilitation, Joh. Gutenberg U., Mainz, Fed. Republic Germany, 1993. Researcher Tech. U. Munich, 1981-84; vis. scientist IBM Almaden Rsch. Ctr., San Jose, Calif., 1984-86; sr. researcher Max-Planck-Inst. für Polymerforschung, Mainz, 1986—; prof. Joh. Gutenberg U., Mainz, 1993—. Contbr. articles to profl. jours. Mem. German Phys. Soc., Am. Phys. Soc. Roman Catholic. Home: Walpodenstr 10, D-55116 Mainz Germany Office: Joh Gutenberg U Inst Phys Chemie, Jakob-Welder-Weg 11, D-55099 Mainz Germany

RABEN, DANIEL MAX, biochemist; b. St. Louis, Feb. 18, 1949; s. Hymen and Harriet (Goodman) R.; m. Marian Virginia Gillooly, May 31, 1975; children: Samuel G., Timothy G. BS, U. Mich., 1976; PhD, Washington U., St. Louis, 1981. Postdoctoral fellow U. Calif., Irvine, 1981-86; asst. prof. physiology Johns Hopkins Med. Sch., Balt., 1986-91, assoc. prof., 1991—; adv. bd. Biochem. Jour., Eng., 1992—. Contbr. articles to profl. jours. Grantee Am. Heart Assn., 1988, NIH, 1989—. Mem. Am. Chem. Soc., Am. Soc Cell Biology, Am. Soc. Biochemistry and Molecular Biology. Achievements include identification of phosphatidycholine as a source of diglycerides in mitogen-stimulated fibroblasts. Office: Johns Hopkins Med Sch Dept Physiology 725 N Wolfe St Baltimore MD 21205

RABIDEAU, PETER WAYNE, university dean, chemistry educator; b. Johnstown, Pa., Mar. 4, 1940; s. Peter Nelson and Monica (Smalley) R.; m. Therese Charlene Newquist, Sept. 1, 1962 (div.); children—Steven, Michael, Christine, Susan; m. Jennifer Lee Mooney, Nov. 15, 1986; children: Mark, Leah. B.S., Loyola U., Chgo., 1964; M.S., Case Inst. Tech., Cleve., 1967; Ph.D., Case Western Res U, Cleve., 1968. Postdoctoral asst. U. Chgo., 1968-69, instr., 1969-70; asst. prof. Ind. U.-Purdue U., Indpls., 1970-73, assoc. prof., 1973-76, prof., 1976-90, chmn. dept. chemistry, 1985-90; dean Coll. Basic Scis. La. State U., Baton Rouge, 1990—; program officer NSF, 1988-89. Contbr. numerous articles to profl. jours. Recipient research award Purdue Sch. Sci. at Indpls., 1982. Mem. Am. Chem. Soc. (chmn. Ind. sect. 1974, councilor 1981-90). Home: 15160 Old Oak Ave Baton Rouge LA 70810 Office: La State U Office of the Dean 338 Choppin Baton Rouge LA 70803

RABIN, AARON, neurologist; b. Bklyn., July 4, 1945; s. Mordecai and Rachel (Kopelowitz) R.; m. Abigail Teitz, Dec. 15, 1970; children: Aliza, Ariel, Moriah. BA summa cum laude, Yeshiva U., 1967; PhD, Rockefeller U., 1974; MD, Albert Einstein Coll. Medicine, 1976. Diplomate Am. Bd. Psychiatry and Neurology; cert. Am. Soc. Neurorehabilitation. Resident in medicine Brookdale Hosp., Bklyn., 1976-77; resident in neurology Albert Einstein Coll. Medicine, Bronx, 1977-80; attending physician Englewood (N.J.) Hosp., 1981—, chief sect neurology, 1989—; asst. clin. prof. neurology Columbia Presbyn. Med. Ctr., N.Y.C., 1984—. Author: The Vestibular System and Motor Control, 1974. Fellow Stroke Coun. Am. Heart Assn. Mem. Am. Acad. Clin. Neurophysiology, Am. Soc. Clin. Evoked Potentials, Am. Assn. Electromyography and Electrodiagnosis, Am. Electroencephalographic Soc., Am. Med. Electroencephalographic Assn., Am. Acad. Neurology, Am. Assn. Anatomists, Soc. Neurosci., British Brain Rsch. Assn. (hon.), European Brain and Behavior Soc. (hon.). Office: 177 N Dean St Englewood NJ 07631-2527

RABIN, MICHAEL O., computer scientist, mathematician; b. Breslau, Germany, Sept. 1, 1931; s. Israel A. and Else (Hess) R.; m. Ruth Scherzer, May 31, 1954; children: Tal, Sharon. M.S. in Math., Hebrew U., Jerusalem; Ph.D., Princeton U., 1956. Instr. Princeton (N.J.) U., 1956-57; mem. Inst. Advanced Study, Princeton, 1957-58; prof. computer sci. Harvard U., Cambridge, Mass., 1981—, Gordon McKay prof., now Thomas J. Watson sr. prof. computer sci.; vis. prof. U. Calif.-Berkeley, 1962, MIT, 1972-78, U. Paris, 1965, Yale U., 1967, NYU, 1970; Albert Einstein prof. Hebrew U., Jerusalem, 1965—; cons. computer industry. Inventor oblivious transfer algorithm. Recipient Rothschild prize in math., 1974, Turing award in computer sci., 1976, Harvey prize in sci. and tech., 1980. Mem. Israel Acad. Sci., NAS (fgn. assoc. mem.), Am. Acad. Arts and Scis. (fgn. hon. mem.), Am. Philos. Soc. Office: Harvard U Div Applied Scis Cambridge MA 02138

RABINOW, JACOB, electrical engineer, consultant; b. Kharkov, Russia, Jan. 8, 1910; came to U.S., 1921, naturalized, 1930; s. Aaron and Helen (Fleisher) Rabinovich; m. Gladys Lieder, Sept. 26, 1943; children: Jean Ellen, Clare Lynn. B.S. in Elec. Engring, Coll. City N.Y., 1933, E.E., 1934; D.H.L. (hon.), Towson State U., 1983. Radio serviceman N.Y.C., 1934-38; mech. engr. Nat. Bur. Standards, Washington, 1938-54; pres. Rabinow Engring. Co., Washington, 1954-64; v.p. Control Data Corp., Washington, 1964-72; research engr. Nat. Bur. Standards, 1972-89; cons. Inst. Standards and Tech., Gaithersburg, Md., 1989—; Regent's lectr. U. Calif.-Berkeley, 1972; lectr., cons. in field. Author. Recipient Pres.'s Certificate of Merit, 1948; certificate appreciation War Dept., 1949; Exceptional Service award Dept. Commerce, 1949; Edward Longstreth medal Franklin Inst., 1959; Jeffersom medal N.J. Patent Law Assn., 1973; named Scientist of Yr. Indsl. R&D mag., 1980. Fellow IEEE (Harry Diamond award 1977), AAAS; mem. Nat. Acad. Engring., Philos. Soc. Washington, Audio Engring. Soc., Sigma Xi. Club: Cosmos (Washington). Patentee in field. Home: 6920 Selkirk Dr Bethesda MD 20817-4750 Office: Inst Standards and Tech Gaithersburg MD 20899

RABINOWICZ, THÉODORE, neuropathology educator; b. Fribourg, Switzerland, May 8, 1919; s. Charles and Rose (Pinkwater) R.; m. Lucienne Umiglia-Foro, Apr. 9, 1969. Diploma medicine, U. Geneva, 1947, MD, 1955. Med. resident internal medicine, psychiatry, pathology Geneva U. Hosp., 1947-55; chief lab. neuropathology dept. pathology U. Hosp., Lausanne, Switzerland, 1955-62; fgn. resident Max Planck Inst. Hirnforschung, Munich, 1959; fgn. resident neurology and neuropathology Mass. Gen. Hosp., Harvard U., Boston, 1962; head assoc. prof. neuropathology Lausanne, 1962-84; vis. prof. Armed Forces Inst. Pathology, Washington, 1963-84, U. Calif.-Irvine, U. Calif.-Davis, U. So. Calif., 1963-84; vis. prof., lectr. neuropathology U. Calif.-Irvine, 1981; vis. prof., cons. dept. pathology U. Geneva, 1985—; cons. WHO, Geneva, 1979-80, Coun. Internat. Orgn. Med. Sci., Geneva, 1984-90; pres. Swiss Found. Rsch. Mental Retardation, Geneva, 1987—. Mem. editorial bd. Devel. Psychobiology Internat. Jour., 1968-78; contbr. articles to profl. jours. Active community confs. on health and mcpl. history. Rsch. grantee NIH, Bethesda, Md., 1959-68, Swiss NSF, Bern, Switzerland, 1969-76, Swiss Found. Rsch. on Mental Retardation, Geneva, 1978-93. Mem. Internat. Brain Rsch. Orgn., Am. Assn. Neuropathologists, Swiss Assn. Neuropathologists, Swiss Assn. Pathologists, French Soc. Neuropathologists. Avocations: music, amateur astronomy, gardening. Home: 30 Rte des Eaux-Belles, CH-1243 Cara/Presinge Geneva, Switzerland Office: U Med Ctr Dept Pathology, 1 rue Michel-Servet, CH-1206 Geneva Switzerland

RABINOWITZ, ARTHUR PHILIP, hematologist; b. N.Y.C., Oct. 8, 1957; s. Joel and Frances (Rothman) R.; m. Maria A. Ponsillo, June 21, 1987; children: Benjamin, Rebecca.. BS with honors, U. Ill., 1979; MD, Loyola U., 1984. Diplomate Nat. Bd. Med. Examiners, Am. Bd. Internal Medicine, Subspecialty in Hematology. Fellow hematology div. U. So. Calif., L.A., 1987-90; fellow autologous bone marrow transplant program U. So. Calif.-Norris Cancer Ctr., L.A., 1990-91; staff physician hematology sect. Lahey Clinic Med. Ctr., Burlington, Mass., 1991—. Mem. ACP, Am. Soc. Hematology, Mass. Med. Soc. Office: Lahey Clinic Hematology Sect 41 Mall Rd Burlington MA 01805

RABINOWITZ, SIMON S., physician, scientist, pediatric gastroenterologist; b. Bklyn., Apr. 8, 1953; s. Herman and Lola (Berman) R.; m. Lynne Susan Heckman, May 29, 1988; children: Brandon Paul, Jesse Mark, Jake Harris. BA, Vassar Coll., 1975; PhD, Univ. Wis., 1981; MD, Univ. Miami, 1983. Diplomate Am. Bd. Pediatrics, Am. Bd. Pediatric Gastroenterology. Rsch. asst. dept. physiol. chem. Univ. Wis., Madison, 1975-81; fellow pediatric gastroenterology dept. pediatrics Mt. Sinai Med., N.Y.C., 1985-87; chief pediatric gastroenterology Kings County Hosp. Ctr., Bklyn., 1987—; pres. Kings County Hosp. nutrition com., 1990—, Bklyn. Pediatric Assocs., 1992—; mem. med. edn. del. to Ea. Europe, People to People, Spokane, Wash., 1992; mem. Bklyn. gov. bd. Univ. Hosp. Contbr. articles to profl. jours. and books. Named Attending of Yr., Children's Med. Ctr., Bklyn., 1992. Mem. Am. Acad. Scis., Am. Acad. Pediatrics, Am. Coll. Gastroenterology, N.Am. Soc. Pediatric Gastroenterology, Crohns and Colitis Found., Met. Pediatric Gastroenterology Club, Bklyn. Gastroenterology Club, Sigma Xi. Jewish. Office: Downstate Med Ctr 450 Clarkson Ave Box 49 Brooklyn NY 11203

RABÓ, JULE ANTHONY, chemical research administrator, consultant; b. Budapest, Hungary; came to U.S., 1957; m. Sheelagh Ennis; children: Benedict, Sebastian. BSChemE, Poly. U., Budapest, 1946, DSc in Chemistry, 1949, D honoris causa, 1986. From asst. prof. to assoc. prof. Poly U., Budapest, 1946-54; assoc. dir. Hydrocarbon Rsch. Inst., Budapest,

1951-56; rsch. assoc. Union Carbide Corp., Buffalo, 1957-60; rsch. mgr. Union Carbide Corp., Tarrytown, N.Y., 1960-72, corp. fellow, 1969-82, sr. corp. fellow, 1982—; sr. corp. fellow UOP, Tarrytown, 1988—; cons. in chemistry and catalysis, Armonk, N.Y.; former mem. adv. bd. Ctr. for Advanced Materials, Lawrence-Berkeley Lab; mem. adv. bd. Lehigh U. Chemistry Lab. Author: Zeolite Chemisty and Catalysts; former mem. editorial bd. Jour. Catalysis, Applied Catalysis; contbr. articles to profl. jours.; patentee in field. Recipient Kossuth award Govt. of Hungary, 1953, Excellence in Catalysis award N.Y. Catalysis Soc., 1982, Humboldt award, Fed. Republic of Germany, 1990. Mem. Am. Chem. Soc. (E.V. Murphree award 1988), Am. Catalysis Soc. (Eugene J. Houdry award 1989), Hungarian Acad. Sci. (Varga medal 1991), Am. Inst. Chemists (Chem. Pioneer award 1993).

RABOLT, JOHN FRANCIS, optics scientist; b. N.Y.C., May 14, 1949; married, 1990; 1 child. BS, SUNY, Oneonta, 1970; PhD in Physics, Ill. U., Carbondale, 1974. Nat. Rsch. Coun., NAS assoc. Nat. Bur. Standards, 1976-77; scientist polymers rsch. staff rsch. lab IBM Corp., San Jose, Calif., 1978—. Recipient Coblentz award, 1985, Williams-Wright award, 1990, Ellis R. Lippincott award Optical Soc. Am., 1993. Achievements include research in the use of Fourier transform (FT) infared and FT and Conventional Raman spectoscropy to investigate crystal and molecular structure of long chain molecules and polymers, integrated optical techniques in conjunction with Raman Spectoscopy to investigate submicron polymer films and polymer surfaces, FTIR studies of self assembled and Langmoir-Blodgett films on metals and dielectrics, and co-development of FT Raman spectroscopy. Office: IBM Almaden Research Center K95/801 650 Harry Road San Jose CA 95120*

RABON, WILLIAM JAMES, JR., architect; b. Marion, S.C., Feb. 7, 1931; s. William James and Beatrice (Baker) R; m. Martha Ann Hibbitts, Mar. 7, 1987. BS in Arch., Clemson (S.C.) Coll., 1951; BA rch, N.C. State Coll., 1955; MA rch, MIT, 1956. Registered architect, Calif., Ky., N.C., Ohio, Pa., Ga. Designer archtl. firms in N.Y.C. and Birmingham, Mich., 1958-61; designer, assoc. John Carl Warnecke and Assocs., San Francisco, 1961-63, 64-66, Keyes, Lethbridge and Condon, Washington, 1966-68; prin. archtl. ptnr. A.M. Kinney Assocs. and William J. Rabon, Cin., 1968-85; v.p., dir. archtl. design A.M. Kinney, Inc., Cin., 1977-85; v.p., dir. programming svcs. Design Art Corp., 1977-85; sr. architect. John Portman & Assocs., Atlanta, 1985-88; dir. architectural design Robert and Co., Atlanta, 1988-89; studio dir., design prin. Carlson Assocs., Atlanta, 1990-93; prin. William Rabon Assocs., 1993—; lectr. U. Calif., Berkeley, 1963-65; asst. prof. archtl. design Cath. U. Am., 1967-68. Prin. works include Kaiser Tech. Ctr., Pleasanton, Calif. (Rsch. Devel. Lab. of Yr. award), 1970, Clermont Nat. Bank, Milford, Ohio, 1971, Pavilion bldg. Children's Hosp. Med. Ctr., Cin. (Cin. AIA design award), 1973, EG&G, Hydrospace, Inc., Rockville, Md. (Potomac Valley AIA design award), 1970, Mead Johnson Park, Evansville, Ind. (Rsch. Devel. Lab. of Yr. merit award), 1973, Hamilton County Vocat. Sch., Cin., 1972, hdqrs. lab. EPA, Cin., 1975, Arapahoe Chem. Co. Rsch. Ctr., Boulder, Colo. (Rsch. Devel. Lab. of Yr. award 1976, Concrete Reinforced Steel Inst. Nat. Design award, Regional AIA Design award), 1976, corp. hdqrs. Ohio River Co., Cin., 1977, Children's Hosp. Therapy Ctr., Cin. (Cin. AIA design award 1978, award of merit Am. Wood Council 1981), VA Hosp. addition, Cin. (Cin. ASHRAE Design award 1980), NALCO Chem. Co. Rsch. Ctr., Naperville, Ill. (Ohio and Cin. AIA design awards 1980, 81), 1980, Proctor & Gamble-Winton Hill Tunnel, Cin. (Ohio AIA design award), 1978, Toyota Regional Ctr., Blue Ash, Ohio (Ohio AIA and Ohio Masonry Council combined design award 1981), planning cons. Nat. Bur. Standards, Republic of China, 1982, East and West fleet hdqrs. and Data Ctr. Librs. of Royal Saudi Arabian Navy, 1985, corp. hdqrs. The Drackett Co., Cin., 1983, corp. hdqrs. Brown & Williamson, Louisville, 1984, Inst. Paper Sci. and Tech., Atlanta, 1989, 93. 1st lt. AUS, 1951-53, Korea. Decorated Silver Star, Bronze Star with V device, Bronze Star, Purple Heart with bronze cluster; MIT Grad. Sch. scholar, 1956; Fulbright scholar, Italy, 1957-58. Mem. AIA, Nat. Council Archtl. Registration Bds.

RABOSKY, JOSEPH GEORGE, engineering consulting company executive; b. Sewickley, Pa., May 20, 1944; s. Mary Helen (Mayer) Rabosky; m. Suzanne Lazzelle, Aug. 23, 1969. BS, Pa. State U., 1966; MS in Engring., W.Va. U., 1969, MSCE, 1973; PhD, U. Pitts., 1984. Registered profl. engr., Pa., Tenn., W.Va., Mo. Project engr. Chester Engrs., Coraopolis, Pa., 1969-70; project mgr. Calgon Corp., Pitts., 1970-73, sect. leader, 1979-85, mktg. mgr., 1985-86; sr. environ. specialist Mobay Chem. Corp., Pitts., 1975-79; project engr. Morris Knowles, Inc., Pitts., 1973-74; project mgr. Penn Environ. Cons., 1974-75; engring. mgr. Baker/TSA, Inc., Pitts., 1986-89; mgr. Chester Engrs., Pitts., 1989-92; prin. AquaTerra, Inc., Harmony, Pa., 1992—; adj. prof. U. Pitts., 1985-88, Pa. State U.-Beaver, McKeesport and New Kensington campuses, 1985—. Bd. dirs. Moon Twp. Mcpl. Authority, 1980-89. Mem. NSPE, ASCE, Pa. Soc. Profl. Engrs. (sec., bd. dirs. 1989-90), Am. Acad. Environ. Engrs. (diplomate, waste water com.), Water Pollution Control Fedn., WaterPollution Control Assn. Pa. (chmn. rsch. com. 1984-89, 91-92, mem. program com. 1984-89), Western Pa. Water Pollution Control Assn. (officer, pres. 1991-93), Internat. Water Conf. (mem. exec. bd. 1989—, gen. chmn. 1991-93). Home: 104 Wynview Rd Coraopolis PA 15108-1033

RABSON, ALAN SAUL, physician, educator; b. N.Y.C., July 1, 1926; s. Abraham and Florence (Shulman) R.; m. Ruth L. Kirschstein, June 11, 1950; 1 son, Arnold B. B.A., U. Rochester, N.Y., 1948; M.D., State U. N.Y. Downstate, 1950. Intern Mass. Meml. Hosp., Boston, 1951-52; resident in pathology N.Y. U. Hosp., 1952-54, USPHS Hosp., New Orleans, 1954-55; pathologist Nat. Cancer Inst., Bethesda, Md., 1955—; prof. pathology Georgetown U. Med. Sch., 1974—, Uniformed Services U. Health Scis., 1978—; professorial lectr. pathology George Washington U., 1978—. Asso. editor: Am. Jour. Pathology; Contbr. articles to med. jours. Mem. Am. Assn. Pathologists, Phi Beta Kappa, Sigma Xi, Alpha Omega Alpha. Address: NIH-National Cancer Institute Bldg 31-Cancer Biology 9000 Rockville Pike Bethesda MD 20892

RABSON, ROBERT, plant physiologist, administrator; b. Bklyn., Mar. 4, 1926; s. Samuel and Rose (Strauss) R.; m. Eileen K. Rabson, Aug. 27, 1950; children: Michael, Barbra, Laurel. BS, Cornell U., 1951, PhD, 1956. Rsch. assoc. biolog. div. Oak Ridge (Tenn.) Nat. Lab., 1956-58; asst. prof. U. Houston, 1958-62, assoc. prof., 1962-63; biochemist civ. biology and medicine AEC, Washington, 1963-67, asst. br. chief, 1967-73; first officer plant breeding and genetics sect. FAO/IAEA, Vienna, Austria, 1973-76; dir. div. energy bioscis. U.S. Dept. Energy, Washington, 1979—. Mem. plant scis. adv. bd. McKnight Found., Mpls., 1981—. With U.S. Army, 1944-45, PTO, ETO. Fellow AAAS; mem. Am. Soc. Plant Physiologists (chmn. publ. com. 1984-86, treas. 1988—, Adolph Gude award 1986). Office: US Dept Energy Basic Energy Scis ER-17 Washington DC 20545

RABSON, THOMAS AVELYN, electrical engineering educator, researcher; b. Houston, July 31, 1923; s. Charles Avelyn and Sara Kathleen (Drake) R.; m. Sylvia Mary Jenny, Aug. 22, 1957; children: William, Tamara, Robert. BA, Rice Inst. (now U.), 1954, BSEE, 1955, MA, 1957, PhD, 1959. Registered profl. engr., Tex. Asst. prof. elec. engring. Rice U., Houston, 1959-63, assoc. prof., 1963-70, prof., 1970—; tech. advisor Energy Rsch. and Edn. Found., Houston, 1974-79. Contbr. numerous articles to profl. jours. Fellow NSF, Basel, Switzerland, 1965, Rice Quantum Inst., 1987. Mem. IEEE, Am. Phys. Soc., Optical Soc. Am. Methodist. Achievements include 7 patents, construction of memory transistor with lithium niobate gate. Home: 4521 Ivanhoe St Houston TX 77027-4807 Office: Rice U Dept Elec Engring PO Box 1892 6100 South Main St Houston TX 77251

RACANIELLO, LORI KUCK, cellular and molecular biologist; b. Bklyn., May 22, 1961; d. Kenneth Herbert and June Carol (Johnson) Kuck; m. Vincent Joseph Racaniello, Aug. 3, 1985; children: Amanda Jeanne, Kristen Nicole. BS, SUNY, 1983; MS, Fla. Inst. Tech., 1989. Rsch. technician U. Mass. Med. Ctr., Worcester, 1984-85; rsch. scientist Fla. Inst. Tech., Melbourne, 1986-89, 1989-90; quality control technician Pharmafair Co., Inc., Hauppauge, N.Y., 1990; microbiologist Brookhaven Nat. Lab., Upton, N.Y., 1991—. Contbr. articles to profl. jours. Barbara Hammer Meml Fund scholar, 1979, Duffy Found. scholar, 1979, Glen Mohawk Found. scholar, 1979. Mem. Sigma Psi Beta, Sigma Xi. Home: 312 Starr Blvd Calverton NY 11933

RACANIELLO, VINCENT RAIMONDI, microbiologist, medical educator; b. Paterson, N.J., Jan. 2, 1953; s. Pasquale and Christine (Raimondi) R.; m. Doris Frances Cully, Oct. 10, 1982. BA, Cornell U., 1974; PhD, Mt. Sinai Sch. Medicine, 1979. Postdoctoral fellow MIT, Cambridge, 1979-82; asst. prof. Coll. Physicians and Surgeons, Columbia U., N.Y.C., 1982-88, assoc. prof., dir. grad. program dept. microbiology, 1988-90, prof. dept. microbiology, 1990—; cons. Lederle Labs., Pearl River, N.Y., 1988—; mem. virology study sect. NIH, Bethesda, Md., 1989—; Harvey Soc. lectr., 1991. NIH fellow, 1980; recipient Career Scientist award I.T. Hirschl Trust, 1983, Scholar award Searle Scholars Program, 1984, Eli Lilly and Co. Rsch. award Microbiology and Immunology Am. Soc. Microbiology, 1992. Mem. Am. Soc. Microbiology, Harvey Soc. Office: Columbia U 701 W 168th St New York NY 10032-2704

RACCA, GIUSEPPE DOMENICO, aerospace engineer; b. Marene, Cuneo, Italy, Nov. 22, 1957; arrived in The Netherlands, 1987; s. Antonio and Lucia (MOnasterolo) R.; m. Maria Grazia Bertola, Oct. 30, 1982; children: Matteo, Maurizio, Caterina. D in Nuclear Engring., Politecnico di Torino, 1982. Thermal analyst Aeritalia Space Systems Group, Torino, Italy, 1983-85, thermal systems engr., 1985-87; sr. thermal engr. European Space Agy., Noordwijk, The Netherlands, 1987-90, prin. mech. systems engr., 1990-93, study mgr. future sci. project office., Directorate of Sci. Programmes, 1993—. Contbr. articles to conf. proceedings. With Italian Mil., 1982. Mem. AIAA. Roman Catholic. Office: European Space Agy, Keplerlaan 1, 2200AG Noordwijk The Netherlands

RACE, LISA ANNE, environmental chemist; b. Casper, Wyo., Nov. 10, 1961; d. George L. and Sherry L. (Scheafermeyer) R. BS in Biochemistry, U. Wyo., 1984. Biol. aide High Plains Grassland Rsch. STa., USDA, Cheyenne, Wyo., summer 1981-84; chem. analyst technician divsn. labs. Wyo. Dept. Agr., Laramie, 1984-87; chemist Chem. Rsch. Labs., Santa Maria, Calif., 1987-88; chemist, environ. health and safety officer Enseco-Chem. Rsch. Labs., Santa Maria, 1988-90, acting lab. mgr., 1990, quality control and assurance officer, 1990; mobile lab. mgr. Coast to Coast Analytical Svcs., San Luis Obispo, Calif., 1991—. Avocations: softball, volleyball. Home: 269 E Waller Ln Santa Maria CA 93455-2061 Office: Coast to Coast Analytical Svcs 141 Suburban Rd San Luis Obispo CA 93401

RACHLIN, HOWARD, psychologist, educator; b. N.Y.C., Mar. 10, 1935; s. Irving and Gussie (Kugler) R.; m. Nahid, Feb. 1, 1961; 1 child, Leila. MA, New Sch. Social Rsch., 1962; PhD, Harvard U., 1965. Asst. prof. Harvard U., Cambridge, Mass., 1965-69; prof. SUNY, Stony Brook, 1969—. Author: Behavior and Mind: Two Psychologies, Judgment, Decision and Choice, 1990. Home: 501 E 87th St New York NY 10128 Office: SUNY Psychology Dept Stony Brook NY 11794

RACHLIN, STEPHEN LEONARD, psychiatrist; b. N.Y.C., Mar. 6, 1939; s. Murray and Sophie (Rodnitsky) R.; m. Florence Einsidler, Nov. 22, 1962; children: Michael Ira, Robert Alan. BA, NYU, 1959; MD, Albert Einstein Coll. Medicine, 1963. Diplomate Nat. Bd. Med. Examiners, Am. Bd. Forensic Psychiatry, Am. Bd. Psychiatry and Neurology. Internship UCLA, 1963-64; resident, chief resident in psychiatry Mt. Sinai Hosp., N.Y., 1964-67; staff psychiatrist Bronx Psychiat. Ctr., Bronx, N.Y., 1969-72; asst. chief svc. Bronx Psychiat. Ctr., 1970-72, chief svc., 1972-74; dep. dir. Meyer-Manhattan Psychiat. Ctr., N.Y.C., 1974-76; acting dir. Meyer-Manhattan Psychiat. Ctr., 1976-77; dep. dir. Manhattan Psychiat. Ctr., N.Y.C., 1977; clin. dir. dept. psychiatry & psychology Nassau County Med. Ctr., E. Meadow, N.Y., 1978-80; assoc. chmn. dept. psychiatry & psychology Nassau County Med. Ctr., 1979-80, chmn. dept. psychiatry & psychology, 1980—; assoc. prof. clin. psychiatry sch. medicine SUNY, Stony Brook, 1978-87, prof. clin. psychiatry, 1987—; spl. prof. law sch. law Hofstra U., Hempstead, N.Y., 1983—. Editor in chief Psychiat. Quar., 1990—; assoc. editor Bull. of the Am. Acad. os Psychiatry and the Law, 1989—; contbr. articles to profl. jours. Lt. comdr. USNR, 1967-69. Mem. Am. Psychiat. Assn. (chmn. com. adminstrv. psychiatry 1987-92, mem. assembly 1991—), N.Y. State Psychiat. Assn. (chmn. com. on pub. psychiatry 1986—), Am. Assn. Psychiat. Adminstrs. (pres. 1989-90), Am. Acad. Psychiatry and Law (pres. tri-state chpt. 1988-90), Am. Assn. Gen. Hosp. Psychiatrists (pres. 1993—), Am. Bd. Forensic Psychiatry (dir. 1990—, treas. 1992-93), Am. Hosp. Assn. (gov. coun. sect. psychiat. and substance abuse 1991-92), Hosp. Assn. N.Y. State (chmn. mental health 1992—). Office: Nassau County Med Ctr Dept Psychiatry and Psychology 2201 Hempstead Tpke East Meadow NY 11554-5400

RACUSEN, LORRAINE CLAIRE, pathologist, researcher; b. Burlington, Vt., July 1, 1947; d. Charles Edward and Lenore Evelyn (Ulrich) Parent; m. Richard Harry Racusen, Jan. 12, 1970; children: Christopher Charles, Darren David. BA, U. Vt., 1970, MD, 1975. Diplomate Am. Bd. Pathology, Nat. Bd. Med. Examiners. Postdoctoral fellow Yale-New Haven (Conn.) Med. Ctr., 1975-78; resident, fellow Sch. Medicine Johns Hopkins U., Balt., 1979-82, rsch. fellow Sch. Medicine, 1982-83, instr. Sch. Medicine, 1983-84, asst. prof. Sch. Medicine, 1984-92, assoc. prof. Sch. Medicine, 1992—. Author: (with others) Renal Pathology, 1993; editor: Acute Renal Failure: Diagnosis Treatment, Prevention, 1990, Kidney Transplant Rejection: Diagnosis/Rx, 1991; contbg. editor to Lab. Invest. Mem. Md. adv. bd. Nat. Kidney Found., Balt., 1987—; mem. kidney coun. Am. Heart Assn. Recipient Rsch. Svc. award USHPS, 1976-78, Mellon Clinician Sci. award Johns Hopkins Sch. Medicine, 1989-91; John Dewey fellow U. Vt., 1969, 70, Am. Heart Assn., 1982. Fellow Coll. Am. Pathologists; mem. AAAS, Am. Soc. Nephrology, Internat. Acad. Pathology, Internat. Soc. Nephrology, N.Y. Acad. Scis. Achievements include discovery of glomecular podocyte alterations in ischemic renal injury, altered tubular cell adhesion with ischemic renal tubular injury, direct hormonal effects of VIP somatostatin enkephaline on intestinal transport, development of a human renal proximal tubular cell line. Office: The Johns Hopkins U Med Sch 624 Ross 720 Rutland Ave Baltimore MD 21205

RACZKOWSKI, CYNTHIA LEA, chemist; b. Sutersville, Pa., Sept. 10, 1956; d. Stanley Michael and Sharon Lea (Ellis) R. BS in Secondary Edn. Chemistry, California (Pa.) State Coll., 1978. Physics tchr. Mon Valley Cath. High Sch., Chaleroi, Pa., 1978-79; rsch. chemist Gulf Oil Corp., Harmarville, Pa., 1979-81; chemist Warren Petroleum, Mont Belvieu, Tex., 1981—. EMT Tex. Dept. Health, 1984—. Achievements include supervision of first above ground bioremediation of hydrocarbon contaminated dirt in state of Tex. Home: Rt 3 Box 131-C Dayton TX 77535 Office: Warren Petroleum Co PO Box 10 Mont Belvieu TX 77580

RADEMACHER, JOHN MARTIN, sanitary engineer; b. Hammond, Ind., Sept. 21, 1924; s. Martin Edward and Anna F. (Rabe) R.; m. Phyllis Anderson, June 2, 1951; children: Lisa Ann, Karen Sue, Christian Edward. BS in Civil Engring., Purdue U., 1949; MS in Environ. Engring., Northwestern U., 1961; postgrad., Fed. exec. Inst., 1978. Registered profl. engr.; diplomate Am. Acad. Environ. Engring. Field engr. Ind. State Bd. of Health, Indpls., 1949-51; sales/svc. engr. Wallace & Tiernan Co., Indpls., 1951-56; dir. tech. svcs. Water Pollution Control Adminstrn., Washington, 1966-68; reg. VII adminstr. WPCA, WQ Adminstn., EPA, Kansas City, 1968-71; sr. environ. advisor State of Md., Annapolis, 1973-77; v.p. environ. health and regulatory affairs Velsicol Chem. Corp., Chgo., 1978-86; environ. mgmt. cons. PR & Assocs., Kingwood, Tex., 1986—; pres. Fed. Water Quality Assn., Washington, 1978-79; mem. Ind. Waste Coun., Ill. Inst. Tech., Chgo., 1984-86; cons. Indsl. Adv. Com. Westheimer Fin. Group, Houston, 1988—. Contbr. articles to profl. jours. Chmn. Village Water Commn., Park Forest, Ill., 1963-64; asst. commr. Fairfax (Va.) County Boy Scouts, 1972-78. Capt. USPHS, 1956-65. Mem. Zero Population Growth (Environ. Man of Yr. 1971), Water Environ. Fedn. (life), Chem. Mfrs. Assn. (environ. mgmt. com. 1983-86), Toastmasters (pres. 1963), Order of the Boar. Lutheran. Office: PR & Assocs PO Box 6136 Kingwood TX 77325-1136

RADER, CHARLES GEORGE, chemical company executive; b. Niagara Falls, N.Y., Apr. 9, 1946; s. Carl Franklin and Eileen (Adler) R.; m. Sheila Ann Dunlop, Oct. 30, 1971; children: Carla Beth, Kevin Alexander. B-SChemE, Rensselaer Polytech. Inst., 1968; MS, U. Rochester, 1970; PhD, SUNY, Buffalo, 1974. Sr. research engr. Occidental Chem. Corp., Grand Island, N.Y., 1974-77, group leader, 1977-78; tech. mgr. Occidental Chem. Corp., Niagara Falls, N.Y., 1978-81; tech. dir. Occidental Chem. Corp., Grand Island, 1981-84, dir. tech., 1984—; v.p. D.S. Ventures, Dallas, 1987—; bd. dirs. TreaTek, Grand Island. Patentee in field; contbr. articles to profl. jours. Industrial adv. bd. dept. chem. enring. SUNY, Buffalo. Mem. Am. Inst. Chem. Engrs., Am. Chem. Soc., Electrochem. Soc., Regional Tech. Strategy Com., Niagara Frontier Assn. Research and Devel. Dirs., Tau Beta Pi, Phi Lambda Upsilon, Sigma Xi, Delta Tau Delta. Office: Occidental Chem Corp Tech Ctr 2801 Long Rd Grand Island NY 14072-1244

RADER, ELLA JANE See ASHLEY, ELLA JANE

RADESPIEL, ROLF ERNST, aerospace engineer; b. Elmshorn, Germany, Feb. 16, 1957; s. Alfred Friedrich and Hella (Piening) R.; m. Bettina Piesch, Oct. 20, 1983; children: Steffen, Lisa. Diploma in engring. sci., Tech. U., Braunschweig, Germany, 1981, D of Engring. Sci., 1986. Rsch. scientist Inst. for Design Aerodynamics, Braunschweig, 1983-87, br. head aerothermodynamics, 1989—; mem. working group Adv. Group for Aerospace R & D, 1983, Group of Aero. Rsch. and Tech. in Europe, 1985—; vis. scientist NASA, Hampton, Va., 1988, 91, 93; lectr. in field. Contbr. articles to Jour. of Aircraft, also conf. papers and rsch. reports. Recipient Dr. Ernst Zimmermann Meml. award Motoren-und Turbinen Union, 1986, Wilhelm-Hoff/Johann-Maria-Boykow award Deutsche Forschungsanstalt für Luft- und Raumfahrt, 1988. Mem. AIAA (contbr. articles to jour.). Achievements include demonstration of feasibility of laminar-flow tech. for drag reduction of aircraft engine nacelles; of 1st efficient multigrid solutions of 3D Navier-Stokes equations for transonic flows; for efficient numerical flow solutions for hypersonic flows with demonstrated multigrid. Office: DLR-Inst for Design Aerodynamics, Am Flughafen, Braunschweig 38108, Germany

RADEV, IVAN STEFANOV, electronics engineer; b. Samokov, Sofia County, Bulgaria, Dec. 16, 1958; s. Stephan Radev and Simeonka Pavlova (Aneva) Todorov. Engr. in radioelectronics degree, computer sci., Tech. U., Sofia, Bulgaria, 1983; postgrad., Cen Inst. Computing Technique and Tech., Sofia, 1985-87, Inst. for Instrumentation and Computer Tech., Sofia, 1987-88. Computer design engr. Cen. Inst. Computing Technique and Tech., Sofia, 1983-85; rsch. assoc. Inst. for Instrumentation and Computer Tech., Sofia, 1990-91; part-time instr. Tech. U. Sofia, U. Sofia, 1992—. Contbr. articles to profl. jours. Mem. Bulgarian Soc. Cognitive Sci., Planetary Soc., Am. Math. Soc. Avocation: foreign languages. Home: PO Box 718, 1000 Sofia Bulgaria Office: Inst Instrmntn Comp Tech care Dr, K A-Blvd Tzarigradsko Chausse 7 KM, 1184 Sofia Bulgaria

RADFORD, LINDA ROBERTSON, psychologist; b. Winnipeg, Man., Can., Nov. 6, 1944; came to U.S., 1954; d. William and Edith Aileen (Wheatley) Robertson; 1 child, Drew Richard; m. Richard D. Polley, Sept. 21, 1991. BA, Seattle Pacific U., 1970; MEd, U. Wash., 1972, PhD, 1980. Lic. psychologist, Fla.; cert. hypnotherapist. Dir. support svcs. Highline-West Seattle Mental Health Clinic, 1973-75; rsch. asst. in human affairs Battelle, Seattle, 1976-80, rsch. scientist, 1982-87; sr. cons. Martin Simmonds Assoc., Seattle, 1980-82; pres., owner R.R. Assocs., Seattle and Miami, 1982—; pres. PGI Inc., Miami and London, 1989—; pvt. clin. psychologist Bay Harbor Island, Fla., 1991—, West Palm Beach, Fla., 1991—; vis. sr. assoc. Joint Ctr. for Environ. and Urban Problems, North Miami, Fla., 1986-88; cons. Health Ministry Govt. of Thailand, Bangkok, 1989—. Contbr. articles to profl. jours. Community Mental Health Ctr fellow, Seattle, 1972-73. Mem. Am. Psychol. Assn., Am. Soc. Clin. Hypnosis, N.Y. Acad. of Sci. Avocations: tennis, music, snorkeling, racquetball, fishing. Home: 9264 Bay Dr Surfside FL 33154 Office: 1160 Kane Concourse Ste 401 Bal Harbour FL 33154

RADKE, WILLIAM JOHN, biology educator; b. Mankato, Minn., June 8, 1947; s. Gerhard William and Ruth Ida (Stegeman) R.; m. Christine Maria Albasi, Aug. 11, 1984; children: Sarah Catherine, Julia Ruth. BS in Sci., Mankato State U., 1970, MS, 1972; PhD, U. Ariz., 1975. Asst. prof. biology U. Ctrl. Okla., Edmond, 1975-80, assoc. prof. biology, 1980-85, prof. biology, 1985—. Co-author: (book/manual) Laboratory Anatomy of the Perch, 1991, Laboratory Anatomy of the Vertebrates, 1992; author: (book/manual) Human Anatomy for Allied Health Students, 1991. Mem. Okla. City Audubon Soc., Oklahoma City, 1976—; past pres. Edmond Naturalists Club, Edmond, 1977. Mem. Okla. Soc. Physiologists (pres.-elect 1993-94), Okla. Ornithological Soc. Bull. (assoc. editor 1987-88), Sigma Xi (pres.-elect 1993-94). Achievements include research in the control of the thyrotropic hormone and aldosterone secretion in the fowl; research in SEM verification of pores in plasma membranes of fowl epidermis; development of a technique for collection of stomach contents in birds by emesis. Home: 2600 Meadowview Edmond OK 73013 Office: Univ of Ctrl Oklahoma Dept of Biology 100 University Dr Edmond OK 73034-0177

RADVANYI, PIERRE CHARLES, physicist; b. Berlin, Apr. 29, 1926; s. Ladislas and Netty (Reiling) R.; m. Odette Sabate; 1 child, Jean; m. Marie-France Bouvet, Dec. 20, 1956; children: Francois, Michel. Licence-es-sciences, U. Paris, 1948, DSc, 1954. With Centre National de la Recherche Scientifique, Paris, 1948, attache, charge, maitre de recherche, 1949-64, dir. de recherche, 1964—; dir. adjoint Laboratoire Nat. Saturne, Saclay, France, 1978-85; counsellor for internat. rels. Centre Nat. de la Recherche Scientifique, Paris, 1985—. Author: La Radioactivité Artificielle, 1984, Histoires d'Atomes, 1988; editor Bull. of French Phys. Soc., 1969—; contbr. articles to profl. jours. Fellow Inst. Physics (G.B.); mem. European Phys. Soc. (exec. com. 1976-81), Am. Phys. Soc., French Phys. Soc. (sec.-gen. 1975-80). Office: Laboratoire Nat Saturne, CE-Saclay, 91191 Gif sur Yvette France

RADWIN, ROBERT GERRY, science educator, researcher, consultant; b. Bklyn.. MS/MSE, U. Mich., 1979, PhD, 1986. Rsch. fellow U. Mich., Ann Arbor, 1986-87; asst. prof. indsl. engring. U. Wis., Madison, 1987-91, assoc. prof. indsl. engring., 1992—; mem. Am.Nat. Stds. Inst., Z.365 Nat. Stds. Com., 1981—. Contbr. articles to profl. jours. Recipient Presdl. Young Investigator award NSF, 1991, Spl. Emphasis Rsch. Career award Nat. Inst. for Occupational Safety and Health, 1991, Rsch. grant Whittaker Found., 1990. Mem. IEEE, IIE, Am. Soc. Biomechanics, Human Factors Soc., Am. Indsl. Hygiene Assn. (program chmn. 1992), Ergonomics Soc., Am. Nat. Stds. Inst. (Z.365 nat. stds. com., 1981—). Achievements include development of electronic instruments and analytical methods for measuring and assessing exposure to physical stress in the workplace; of ergonomics guidelines for design and use of hand-operated equipment and power tools; research in the causes and prevention of cumulative trauma disorders in manual work. Office: Univ Wis 1513 University Ave Madison WI 53706

RADYS, RAYMOND GEORGE, laser scientist; b. Kaunas, Lithuania, Aug. 30, 1940; came to U.S., 1949; s. Valerian Felix and Ella Lydia (Heinrich) R. BSEE, U. Ill., 1963. Engr. Hughes Aircraft, Culver City, Calif., 1963-76; scientist Hughes Aircraft, El Segundo, Calif., 1985—; sr. engr. Transaction Tech., Santa Monica, Calif., 1976-83; participant GM electric vehicle impact project, 1991-93. Recipient tech. award NASA, 1980, 89. Mem. U.S. Chess Found., Hughes Chess Club (champion 1971). Libertarian. Avocations: chess, computers, hiking, investments. Office: Hughes Aircraft Bldg E1 MS B122 PO Box 902 El Segundo CA 90245-0902

RAETHER, SCOTT EDWARD, mechanical engineer; b. Waukesha, Wis., Dec. 27, 1960; s. Edward William and Arlene Joyce R.; m. Linda Maria Hollar, Aug. 18, 1984; 1 child, Alyssa K. BSME, Purdue U., 1983. Registered profl. engr., Wis. Design engr. Pratt & Whitney Aircraft, West Palm Beach, Fla., 1983-84; project engr. TV products div. Owens-Ill., Columbus, Ohio, 1984-90; mgr. engring nuclear, internat. products Sentry Equipment Corp., Oconomowoc, Wis., 1990—. Mem. ASME, Soc. Profl. Engrs. Achievements include patent for automated mold changing mechanism for television glass pressing machine. Home: 922 Duchess Dr Oconomowoc WI 53066 Office: Sentry Equipment Corp 856 E Armour Rd Oconomowoc WI 53066

RAFEA, AHMED ABDELWAHED, computer science educator; b. Cairo, Egypt, Apr. 7, 1950; s. Rafea R.; m. Azza Rifaah, Aug. 20, 1978; children; Mohammed, Nora. BSc. in Electronics, Faculty Engring., Cairo, Egypt, 1973; diploma in computer sci., Inst. Stats., Cairo, Egypt, 1975; DEA in Computer Sci., Faculty of Sci., Toulouse, France, 1977, PhD in Computer

Sci., 1980. Engr. Cairo U., 1973-76; researcher U. Paul Sabatier, Toulouse, 1976-80; asst. prof. Cairo U., 1980-85, assoc. prof., 1986-87; assoc. prof. San Diego State U., 1985-86; assoc. prof. Am. U. in Cairo, 1987-91, prof., 1991—; project dir. Food and Agrl. Orgn., Cairo, 1989—; cons. Cen. Agy. for Pub. Mobilzation and Stats., Cairo, 1989—; regional chair 2d World Congress on Expert Systems, Portugal, 1994. Contbr. over 40 articles to profl. jours. Fellow IEEE (computer soc.); mem. Assn. for Computing Machinery, Am. Assn. for Artificial Intelligence, Assn. for Computational Linguistics. Moslem. Office: Am U in Cairo, 113 Kasr El-Aini St, Cairo Egypt

RAFETTO, JOHN, podiatrist; b. Phila., Mar. 29, 1950; s. Willard John and Anne (Brumbaugh) R.; m. Eleni Mallas, Dec. 5, 1987; children: Dominic Giovanni, Gianna Maria. BS, Pa. State U., 1974; postgrad., Ceux Sch. Medicine, Cuernavaca, Mex., 1977-79; D Podiatric Medicine, Ohio Coll. Podiatric Medicine, 1984. Diplomate Am. Bd. Podiatric Surgery. Resident Parkview Hosp., Phila., 1984-86; pvt. practice Paoli, Pa., 1986—. Fellow Am. Coll. Foot Surgeons; mem. Am. Podiatric Med. Assn., Am. Diabetic Assn. Avocations: golf, weight lifting, opera. Office: Orlando Foot & Ankle Clinic 1509 S Orange Ave Orlando FL 32806

RAFEY, LARRY DEAN, microbiologist; b. St. Louis, Mar. 30, 1948; s. Ernest George and Roslyn (Broida) R. BS, Am. U., 1981; postgrad., Georgetown U., 1981, Howard U., 1993—. Microbiologist First Medic, McLean, Va., 1973—; ops. dir. CCD Cons.-Forensic Sci. Svcs., Alexandria, Va., 1991—; mem. infectious diseases del. to China & Vietnam, 1993. Mem. Am. Soc. Tropical Medicine and Hygiene, Am. Soc. Clin. Pathology, N.Y. Acad. Sci. Home: Apt 1613 5021 Seminary Rd Alexandria VA 22311-1943

RAFFANIELO, ROBERT DONALD, research scientist, educator; b. Bklyn., Nov. 5, 1957; s. Thomas and Rose (Sansaverino) R.; m. Lori Ann Beranato, Nov. 30, 1986; 1 child, Jamie Ann. BS, Coll. Staten Island, 1980; MS, Long Island Univ., 1982; PhD, N.Y. Univ., 1988. Post doctoral fellow North Shore U. Hosp.-Cornell, 1988-89; rsch. scientist SUNY, Bklyn., 1989-93; rsch. asst. prof. SUNY, 1993—. Mem. N.Y. Acad. Scis., Soc. Experimental Biology & Medicine. Office: SUNY 450 Clarkson Ave Box 1196 Brooklyn NY 11203

RAFTERY, M. DANIEL, chemistry researcher; b. Berkeley, Calif., Jan. 18, 1962; s. Michael A. and Judith H. Raftery; m. Ana Maria Gomez-Bravo, 1992. AB, Harvard Coll., 1984; PhD, U. Calif., Berkeley, 1991. Technician C.E.R.N., Geneva, Switzerland, 1984-85; postdoctoral fellow U. Pa. Dept. Chemistry, Phila., 1991—. Postdoctoral fellow NSF, 1992—; Chateaubriand scholar, 1987-88, Harvard Coll. scholar, 1982-84. Mem. Am. Phys. Soc., Am. Chem. Soc., Sigma Xi. Democrat. Achievements include new method to study ultrafast solution phase reactions; new method to study surfaces using polarized xenon and nuclear magnetic resonance. Office: U Pa Dept Chemistry Philadelphia PA 19104

RAGAN, JAMES OTIS, engineer, consultant; b. Prineville, Oreg., Apr. 9, 1942; s. Robert Ray Ragan and Lois (Renfro) Williams. Student, Seattle U., 1961, Cerritos Coll., 1962-63. Profl. quality engr., cert. nondestrictive testing level III. Sr. engr. Combustion Engring., Inc., Chattanooga, 1973-78; quality control mgr. Brown & Root, Inc., Houston, 1978-84; nondestructive testing level III Texas Utilities Electric Co., Dallas, 1984—; cons. FMC Corp., Stephenville, Tex., 1990—, Bonded Inspection, Inc., Garland, Tex., 1992—. With U.S. Army, 1964-66. Recipient First Use award Electric Power Rsch. Inst. Comanche Peak Stream Electric Station, Glen Rose, Tex., 1991. Mem. ASME, Am. Soc. Nondestructive Testing. Republican. Avocation: computers. Home: RR 2 Box 184 Bluff Dale TX 76433-9749 Office: Tex Utilities Electric PO Box 1002 Glen Rose TX 76043-2300

RAGHUNATHAN, RAGHU SRINIVASAN, aeronautical engineer, educator, researcher; b. Anekal, Mycore, India, June 15, 1943; arrived in No. Ireland, 1970; s. Shrinivasan and Thangamani Chakravarthy; m. Suman Shanbag, Apr. 15, 1970; children: Bharath, Seema. PhD, Indian Inst. Tech., Bommbay, 1970; DSc, Queen's U., Belfast, No. Ireland, 1991. Chartered Engr. Lectr. Indian Inst. Tech., Bombay, 1967-70; sr. rsch. assoc. Loughborough (No. Ireland) U., 1970-74; lectr. Queens U., Belfast, 1976-83, sr. lectr., 1983-86, reader, 1986—. Contbr. 85 articles to profl. jours. Fellow Royal Aero. Soc. (vice chmn. Belfast 1986-90). Hindu. Achievements include patent for a buffet breather for aerofoils; design of wells turbine for wave power station at Islay, Scotland. Office: Queens U, Dept Aeronautical Engring, Stranmillis Rd, Belfast BT9, Northern Ireland

RAGUSA, PAUL CARMEN, mechanical and structural engineer; b. Utica, N.Y., Dec. 22, 1961; s. Cosmo and Dianne (Gigliotti) R.; m. Kelly M., Sept. 8, 1986; 1 child, Reagan. AS, Mohawk Valley Community Coll., Utica, N.Y., 1983; BS, Syracuse U., 1985. Layout engr. Utica Corp., Whitestown, N.Y., 1986-89, structural/mech. engr., asst. plant engr., 1989-92; mech. engr. Rome (N.Y.) Cable Corp.; milit. aviation photographer, writer Aviation Mag., 1987—. Contbr. articles to profl. jours. Mem. Nat. Rep. Congl. Com., Washington, 1989—. With USMC, 1986. Recipient Amelia Earhart award Civil Air Patrol, 1983, Asst. award Air. Force Recruiters, 1984. Mem. ASME, AIAA, SAE, Am. Inst. Plant Engrs., N.Y. Soc. Profl. Engrs. Republican. Home: PO Box 272 Marcy NY 13403 Office: Air to Ground Replicas 407 Van Roen Rd Utica NY 13502-2414

RAHIMIAN, AHMAD, structural engineer; b. Babol, Iran, Dec. 9, 1955; came to the U.S., 1979; s. Ghassem and Laya (Dadashpour) R. BS in Structural Engring., Arya-Mehr U. Tech., 1979; MCE, Poly. U., 1980, PhD, 1986. Registered profl. engr., N.Y., Calif. Cons. engr. Ahmad Rahimian Cons., N.Y.C., 1981-83; structural engr. Office of Irwin G. Canton, P.C., N.Y.C., 1983-85; dir. analysis dept. Office Irwin G. Cantor, N.Y.C., 1985-88, v.p., 1988—, ptnr., 1993—; asst. prof. structural engring. Poly. U., Bklyn., 1990—, lectr., 1987-90; vis. assoc. prof. Pratt Inst., Bklyn., 1990-91. Contbr. articles to profl. jours. Mem. Internat. Conf. Bldg. Ofcls., NSPE, ASCE (tall bldg. com. 1990), Sigma Xi. Achievements include research in reliability of structures under seismic activity, design of numerous tall buildings. Office: Cantor Seinuk Group 600 Madison Ave New York NY 10022

RAHMAN, AHMED ASSEM, project engineer, stress analyst; b. Cairo, May 26, 1940; came to U.S., 1967; s. Abdel Fathalla and Wagida (Shorbagy) R.; m. Zinab Mostafa, Oct. 5, 1975 (div. Jan. 1980); 1 child, Mona. B of Aeronautical Engring., U. Cairo, 1962; M of Aerospace Engring., U. Toronto, 1971, PhD, 1976. Registered profl. engr., Mich. Aerodynamicist, aeroelectrician Egyptian Aero-Orgn., Cairo, 1962-67; liaison engr. Douglas Aircraft Can., Ltd., Malton, Ont., 1967-71; prof. mech. engring. Oakland U., Rochester, Mich., 1977-79, U. Mich., Dearborn, 1979-81; project engr., stress analyst AM Gen. div. LTV Corp., Livonia, Mich., 1981—. Contbr. articles on heat transfer and energy conversions to profl. jours.; invented modification of rotary engine at Oakland U., 1977. NSF fellow, 1974-75. Mem. ASME, IEEE, Am. Soc. for Metals, Soc. Automotive Engrs. Democrat. Avocations: swimming, jogging, tennis, handball and other sports. Home: 12156 Cavell St Livonia MI 48150-2304 Office: AM Gen Divs LTV Corp 11900 Hubbard St Livonia MI 48150-1733

RAHMAN, ANWARUR, physicist; b. Hyderabad, India, Oct. 10, 1929; came to U.S., 1990; s. Fazlur and Saleha R.; m. Tayaba Anwarur, June 9, 1958; children: Samina, Aziz, Mahboob. MSc, Osmania U., Hyderabad, 1950, PhD, 1970. Lectr., reader, prof. physics dept. Osmania U., India, 1950-89; prof. U. Tabriz and Rezaieh, Iran, 1970-80; cons. Tech. Edn. Rsch. Ctr., Cambridge, Mass., 1992—. Mem. Am. Phys. Soc. Muslim. Achievements include research on the first visibility of the lunar crescent and the lunar calendar. Home: 35 Windsor Ct East Taunton MA 02718 Office: TERC 2067 Massachusetts Ave Cambridge MA 02140

RAHMAN, KHANDAKER MOHAMMAD ABDUR, engineering educator; b. Nabinagar, Bangladesh, Oct. 29, 1938; came to U.S., 1961; s. Ali Ahmed and Wazernessa Khandaker; m. Roushanara Rahman, Dec. 25, 1956; children: Nahar, Mahbub, Niru, Nipu. BSCE, Bangladesh Engring. U., Dhaka, 1960; MSCE, Tex. A&M U., 1963, PhD, 1974. Registered profl. engr., Tex. Asst. prof. Bangladesh Engr. U., Dhaka, 1964-70; design and rsch. engr. Metyko & Assocs., Inc., Houston, 1974-76; sr. devel. engr. CMC Specialists, Houston, 1977-78; sr. project engr. Metalic Braden Bldg. Co., Bellaire, Tex., 1978-83; head civil engring. Prairie View (Tex.) A&M U., 1985-91, assoc. prof., 1983-84, 91—; design engr. A.M. Ahmad Cons. Engr., Dhaka, 1963-64. Author: Lab Manual for Fluid Mechanics, 1984, Lab Manual for Soil Mechanics, 1984; editor (tech. report) Infiltration/Inflow Study, 1976; co-author (tech. report) Alternative Solutions for Water Resources Development, 1974. Pres. Bangladesh Assn., Houston, 1981-82, mem., 1978—; mem. Islamic Soc. Greater Houston, 1985—. Merit scholar Bangladesh Engring. U., 1956-60, scholar for higher studies, 1970-74; scholar US AID, 1961-63; recipient Excellence in Teaching award Gen. Dynamics, 1986-87. Fellow ASCE; mem. NSPE, Am. Soc. Engring. Edn., Tex. Soc. Profl. Engrs., Am. Concrete Inst., Am. Water Works Assn., Water Environ. Fedn., Tex. Water Environ Assn.

RAHMAN, MUHAMMAD ABDUR, mechanical engineer; b. Sylhet, Bengal, India, Mar. 1, 1930; came to U.S., 1950; s. Haji Sajjad Ali Khan and Momotaj Khanom. BSME, U. Toledo, 1953, MSME, 1968; PhD in Engring., Calif. Coast U., 1985. Registered profl. engr., Calif. Mech. design engr. various cons. firms, L.A., 1955-61; aerospace engr. Douglas Aircraft Co., Santa Monica, Calif., 1962-63, N.Am. Aviation, Inc., L.A., 1963-64, NASA Manned Spacecraft Ctr., Houston, 1964-70; safety engr. U.S. Dept. Labor, OSHA, Washington, 1975-86; invention researcher Arlington, Va., 1987—; Contbr. articles to profl. jours. Mem. N.Y. Acad. Scis. Democrat. Islam. Achievements include patent for solar energy collector, supersonic MHD generator system; copyrights for hypothesis on unified field theory and creation of the unverse, on the mechanism of superconductivity, others. Home and Office: 1805 Cyrstal Dr # 1013-S Arlington VA 22202-4407

RAHMAN, SAMI UR, environmental engineer; b. Karachi, Pakistan, Jan. 25, 1962; came to U.S., 1972; s. Mohammad H. and Anis F. (Hakim) R.; m. Sabahat Siddiqui, Aug. 10, 1991. AA in Engring., St. Louis C.C. Forest Park, 1986. Assoc. engr., environ. scientist D. W. Ryckmann & Assocs., Inc., St. Louis, 1986-88; site survey and monitoring sect. supr., health and safety specialist Versar, Inc., Alameda, Calif., 1988-90; sr. staff engr., project mgr. Converse Environ. West, San Francisco, 1990-91; sr. staff environ. specialist, project mgr., acting health and safety officer Environ. Geotech. Cons., Inc., Hayward, Calif., 1991; pres. SUR Environ., St. Louis, 1992—. Mem. Am. Indsl. Hygiene Assn., Pakistani Engrs. and Scientists Assn., Nat. Security Inst.

RAHMAN, YUEH-ERH, biologist; b. Kwangtung, China, June 10, 1928; came to U.S., 1960; d. Khon and Kwei-Phan (Chan) Li; m. Aneesur Rahman, Nov. 3, 1956; 1 dau., Aneesa. B.S., U. Paris, 1950; M.D. magna cum laude, U. Louvain, Belgium, 1956. Clin. and postdoctoral research fellow Louvain U., 1956-60; mem. staff Argonne (Ill.) Nat. Lab., 1960-72, biologist, 1972-81, sr. biologist, 1981-85; prof. pharmaceutics Coll. Pharmacy, U. Minn., Mpls., 1985—, dir. grad. studies, pharmaceutics, 1989-92, head dept. pharmaceutics, 1991—; vis. scientist State U. Utrecht, Netherlands, 1968-69; adj. prof. No. Ill. U., DeKalb, 1971-85; cons. NIH.; Mem. com. of rev. group, div. research grants NIH, 1979-83. Author. Recipient IR-100 award, 1976; grantee Nat. Cancer Inst., Nat. Inst. Arthritis, Metabolic and Digestive Diseases. Fellow Am. Assn. Pharm. Scientists; mem. AAAS, Am. Soc. Cell Biology, N.Y. Acad. Scis., Radiation Rsch. Soc., Assn. for Women in Sci. (1st pres. Chgo. area chpt. 1978-79). Unitarian. Patentee in field. Home: 902 Dartmouth Pl SE Minneapolis MN 55414-3158 Office: Coll Pharmacy U Minn Minneapolis MN 55455

RAHNAMAI, KOUROSH JONATHAN, electrical engineering educator; b. Teheran, Iran, Sept. 8, 1955; came to U.S. 1978; s. Mohammad and Maliheh Rahnamai; m. Christine Marie Hazen, May 26, 1984. MSEE, Wichita State U., 1981, PhD in Electrical Engring., 1985. Teaching asst. Wichita (Kans.) State U., 1979-81, instr. electrical engring. dept., 1983-85; sr. engr. Charles River Analytics, Inc., Cambridge, Mass., 1985-87; assoc. prof. Western N.Eng. Coll., Springfield, Mass., 1987--; cons. Charles River Analytics, Inc., Cambridge, 1987--. PhD fellow Wichita State U., 1981-85. Mem. IEEE. Home: 271 Brookwood Dr Longmeadow MA 01106 Office: Western New Eng Coll 1215 Wilbraham Rd Springfield MA 01119

RAICH, ABRAHAM LEONARD, rabbi, quality control professional; b. Pueblo, Colo., Oct. 25, 1922; s. Isaac and Anna (Silverstein) R.; m. Adelyn Wilma Strait, Mar. 16, 1947; children: David Gary, Kenneth Edward, Robert Aaron. BS in Chemistry, U. Denver, 1947; MS, U. Wash., 1949; degree in Rabbinical Studies, Yeshiva U., 1984. Ordained rabbi, 1984. Teaching fellow U. Wash., Seattle, 1947-48; SQC expert Tech. Assistance Sect. UN, Calcutta, 1956-57; quality control statistician CF&I Steel Corp., Pueblo, 1949-80; quality press referee Am. Soc. Quality Control, Milw., 1988—; rabbi B'nai Israel Congregation, Hampton, Va., 1984-88, Tree of Life Congregation, Clarksburg, W.Va., 1988-90, Temple Beth El, Santa Maria, Calif., 1990—; lectr. various religious orgns., Colo., Va. and Calif., 1957—. Contbr. numerous articles to profl. jours. With USN, 1943-46, 51-52, PTO. Fellow Am. Soc. Quality Control (founding officer Denver chpt. 1949-56, mem. ednl. adv. com. 1958-61), AAAS; mem. Rabbinical Assembly of Am., Interfaith Peace Alliance. Democrat. Jewish. Home: 1230 Jackie Ln Santa Maria CA 93454-5926 Office: Temple Beth El 1501 East Alvin Ave Santa Maria CA 93456-5217

RAIDER, LOUIS, physician, radiologist; b. Chattanooga, Sept. 7, 1913; s. Leah Reevin; m. Emma Silberstein, Oct. 19, 1940; children: Lynne Dianne, David Bernard, Paula Raider Olichney. BS, Bklyn. Coll., 1935; MD, Dalhousie U., 1941. Diplomate Am. Bd. Radiology. Intern Met. Hosp., N.Y.C., 1940-41, resident in radiology, 1941-42; resident in radiation therapy Bellevue Hosp., N.Y.C., 1942-43; fellow in cancer therapy NIH, N.Y.C., 1943-44; chief of radiology Vets. Hosp., New Orleans, 1947-50; radiologist, chief radiology Providence Hosp., Mobile, Ala., 1950-76; clin. prof. Med. Sch. U. South Ala., Mobile, 1987—. Contbr. articles to profl. jours. Maj. AUS, 1944-47. Fellow Am. Coll. Radiology, Am. Coll. Chest Physicians; mem. Radiol. Soc. N.Am., Am. Roentgen Ray Soc., AMA, Ala. Acad. Radiology (pres. 1970-71, Silver medal 1989), So. Med. Assn. (chmn. sect. radiology 1973-74), Soc. Thoracic Radiology, So. Radiol. Conf. Democrat. Jewish. Home: 1801 S Indian Creek Dr Mobile AL 36607-2309 Office: Hosp U South Ala 2451 Fillingim St Mobile AL 36617-2293

RAIFMAN, IRVING, clinical psychologist; b. Bklyn., Oct. 21, 1924; s. Samuel and Gussie (Feldberg) R.; m. Grace Schacht, Sept. 18, 1948; children: Lawrence J., Alan M. Gregory R. MA, Columbia U., 1949; PhD, NYU, 1952. Diplomate, Am. Bd. Profl. Psychology. Clin. psychologist VA, N.Y.C., 1947-51; chief clin. psychologist VA, Washington, 1952-53; tng. officer, clin. psychologist U.S. Naval Hosp., Bethesda, Md., 1953-63; dir. Consultation and Guidance Ctr., Silver Spring, Md., 1960-79; clin. psychologist pvt. practice, Chevy Chase, Md., 1979—; cons. psychologist Dept. Mental Health, Rockville, Md., 1954-74, Community Psychiat. Clinic, Bethesda, 1954-74; chmn. Bd. Examiners in Profl. Psychology for State of Md., 1967-69. Mem. editorial bd. Jour. Psychoanalysis, 1980-90; contbr. articles to profl. jours. Pres. Assn. Psychologists in Pvt. Practice, Montgomery County, Md., 1972-74. Ensign USNR, 1943-46, PTO. Fellow D.C. Psychol. Assn., Soc. Personality Assessment; mem. Am. Psychol. Assn. (div. psychoanalysis), Potomac Psychoanalytic Soc. (pres. 1982-86, treas. 1988-90). Home and Office: 3102 Woodhollow Dr Chevy Chase MD 20815

RAINA, RAJESH, computer engineer; b. New Delhi, Oct. 17, 1963; came to U.S., 1984; s. Niranjan Nath and Durga Raina; m. Karuna Sazawal, Mar. 6, 1993. BTech in Electronics/Elec. Comm. Engring, Indian Inst. Tech., Kharagpur, 1984; MSEE, Mich. Tech. U., 1986; PhDEE, Duke U., 1991. Sr. engr. Digital Equipment Corp., Hudson, Mass., 1991—. Referee, reviewer Internat. Test Conf., 1991—, IEEE Transactions on Cirs. and Systems, 1991—; contbr. articles to profl. jours. including Internat. Jour. of Computer and Elec. Engring., 1987, Internat. Test Conf., 1989, Internat. Symposium on Fault-Tolerant Computing, 1991. Tchr. Durham Edn. Vols., Duke U., Durham, N.C., 1986-87; mem. Habitat for Humanity, Hudson, 1991. Summer rsch. fellowship Duke U., 1987, 89, Grad. sch. conf. travel fellowship, 1989. Mem. Sigma Xi. Achievements include development of prototype of programmable fault-tolerant clock receiver, VLSI chip; use of non-linear feedback shift registers in signature analysis. Office: Digital Equipment Corp MS HL02-1/J12 77 Reed Rd Hudson MA 01749-2895

RAINE, WILLIAM ALEXIS, physicist; b. Neuilly/Seine, France, Feb. 24, 1959; came to U.S., 1970; s. Harry Gregg and Elaine Claire (de Bertier) R. MS, Georgetown U., 1989, PhD, 1991. Teaching asst. Georgetown U., Washington, 1986-91, postdoctoral fellow, 1991—. Recipient Henry M. Leslie award U.S. Textile Assn., 1980. Mem. Am. Phys. Soc., Sigma Xi. Office: Georgetown U 37th and O Sts NW Washington DC 20057

RAINEY, BARBARA ANN, sensory evaluation consultant; b. Fond du Lac, Wis., Nov. 11, 1949; d. Warren and Helen Eileen (Ginther) Bradley; m. Phillip Michael Rainey, Sept. 5, 1970; 1 child, Nicolette. BS, Kans. State U., 1975. Group leader Armour & Co. R&D Ctr., Scottsdale, Ariz., 1976-80; owner Barbara A. Rainey Cons., Manteca, Calif., 1980—. Contbr. articles to profl. jours. Kans. State Alumni fellow Kans. State U. Alumni Assn., 1990. Mem. Inst. Food Technologists (prof., sensory div., chmn. 1984-85, sec. 1980-82, Cen. Valley subsect. chmn. 1992-93, chmn.-elect/sec. 1991-92, treas. 1989-91, short course speaker 1979-81, 50th Anniversary SED chmn. 1989), ASTM, Phi Kappa Phi, Beta Sigma Phi, Delta Zeta Iota (sec. 1989-90, treas. 1984-85, named for Best Program 1989, 90, 92, chmn. 1993—). Avocations: cooking, recipe development. Office: PO Box 622 Manteca CA 95336-0622

RAINEY, LARRY BRUCE, systems engineer; b. Scottsbluff, Nebr., Mar. 21, 1951; s. Marshall Bruce and Florine Elloise (Huff) R.; m. Christine Louise Thurman, July 16, 1976; 1 child, Jeremy Bruce. BS, U. Colo., 1974; MS, Northrop U., 1975, Ohio State U., 1981; postgrad., Ohio State U., 1983; PhD, Union Inst., Cin., 1991; grad., Squadron Officer Sch, 1976, Air Command and Staff Coll., 1986. Cert. space shuttle guidance engr. Commd. USAF, 1975—, advanced through grades to maj.; engring. analyst Specialized Systems Program Office, Wright Patterson AFB, Ohio, 1975-77; ops. rsch. analyst Avionics Lab., Wright Patterson AFB, Ohio, 1977-80; sci. analyst HQ Air Force Logistics Command, Wright Patterson AFB, Ohio, 1980-84; chief guidance navigation and control br. 6595 Shuttle Test Group, Vandenberg AFB, Calif., 1984-87; chief engring. plans br. 3d Satellite Control Sqdn., Falcon AFB, Colo., 1987-90; chief space safety divsn. HQ Air Force Space Command, Peterson AFB, Colo., 1990-91, chief analytical support divsn., 1991-92, development engring. mgr., 1992—; adj. prof. U. Colo., Colorado Springs, 1992; founder grad. astronautical engring. program Vandenberg AFB, Calif., 1986. Editor: Space Launch Operations Textbook, 1992; editor: Space System Modeling and Simulation Textbook, 1993; contbr. articles to profl. jours. Recipient Most Significant Article award Air Force Journal Logistics, 1984. Mem. AIAA (sr. mem.), Air Force Assn. Achievements include expanding literature in application of systems science to command and control systems, management control systems, total quality management, and space operations; recommended improvements in inertial guidance for space shuttle after Challenger accident; executive editor for two first-ever textbooks in field of space operations; initiator and team leader to apply space modeling and simulation to Air Force operational units. Office: AFSPACECOM/CN 150 Vandenberg St Ste 1105 Peterson AFB CO 80914-4110

RAISBECK, JAMES DAVID, aircraft design executive; b. Milw., Sept. 29, 1936; s. Clifford Clinton and Minnie (Hommersand) R.; BS in Aero. Engring., Purdue U., 1961; m. Sherry Bylund; 1 child, Jennifer Lee Raisbeck Hunter. Aero. rsch. engr. Boeing Co., Seattle, 1961-69, rsch. aerodynamist, 1961-64, commercial aircraft preliminary design, 1965-66; liaison to U.S. Air Force, 1966-68, program mgr. comml. STOL programs, 1968-69; chmn. bd., chief exec. officer Robertson Aircraft Corp., Renton, Wash., 1969-73; v.p. tech. Am. Jet Industries, L.A., 1973-74; chmn. bd., chief exec. officer, founder Raisbeck Group, L.A., 1974-80; founder, chmn. bd., chief exec. officer Raisbeck Engring., Inc., 1981—; cons. Served with SAC, USAF, 1955-58. Recipient Disting. Engring. Alumnus in Aeronautics, Purdue U. Fellow AIAA (assoc.); mem. Soc. Automotive Engrs., Tau Beta Pi, Sigma Gamma Tau, Phi Eta Sigma. Patentee in field of wing design, propellors and aircraft systems; designs, builds aerodynamic cleanup packages for business and corp. turbine-powered airplanes. Address: 7536 Seward Park Ave S Seattle WA 98118

RAISMAN, ALLAN LESLIE, food products executive; b. Chgo., Jan. 29, 1952; s. Bernard and Marion Temra (Brown) R.; m. Mary Ann Roberto, Jan. 21, 1977; children: David Hadrian, Rebecca Marie. BA in Biology and Chemistry, North Park Coll., 1974; MS in Biology, Northeastern U., 1976; postgrad., Harvard U., 1976-77; MA in Mktg., Webster U., 1993. Hematology supr. VA Rsch. Hosp., Chgo., 1972-74, Beth Israel Hosp., Boston, 1974-76; clin. chemist Bioran Labs., Cambridge, Mass., 1976-77; dir. hematology Metpath Clin. Labs., Chgo., 1977-79; founder, owner Ben-Al Fisheries, Chgo., 1980-84; sales mgr. Vita Foods, Chgo., 1984—; biology instr., lectr. Northeastern U., Boston, 1974-77. Contbr. articles to profl. jours. Pres. Boston-Gainsboro Tennant Assn., 1975-77, Back-Bay Food Coop., Boston, 1976; room father Red Oak Grade Sch., Highland Park, Ill., 1989-90. Fellow Food Mktg. Inst., Chgo. Acad. Sci., N.Y. Acad. Sci., Chef de Cuisine Inst. (Outstanding Leadership award 1991); mem. Am. Soc. Clin. Pathologists, Am. Soc. Microbiologists, Midwest Frozen Food Assn. Jewish. Home: 473 Ridge Rd Highland Park IL 60035-4368 Office: Waxler Co 565 Lakeview Pky Vernon Hills IL 60061

RAIS-ROHANI, MASOUD, aerospace and engineering mechanics educator; b. Tehran, Iran, Mar. 8, 1960; came to U.S., 1979; s. Mohammad and Ehteram (Agha-Seyed-Mirza) R.; m. Grace Regino, June 5, 1989. BS magna cum laude, Miss. State U., 1983, MS, 1985; PhD, Va. Polytech Inst. & State U., 1991. Instr. Miss. State U., Starkville, 1984-87, asst. prof., 1991—; rsch. assoc. NASA Langley Rsch. Ctr., Hampton, Va., 1990. Contbr. articles to profl. jours. Va. Space Grant Consrotium co-investigator grantee, 1992. Mem. AIAA, Am. Soc. Engring. Edn., Ctr. for Composite Materials and Structures, Tau Beta Pi, Phi Kappa Phi, Sigma Gamma Tau, Gamma Beta Phi. Islam. Achievements include research in design optimization, aerospace structures with emphasis on composites and aeroelasticity. Office: Miss State U Aerospace Engring Dept Mississippi State MS 39762

RAJADURAI, PATHMANATHAN, pathologist, educator, consultant; b. Klang, Malaysia, Aug. 13, 1953; s. Rajadurai Arumugam and (Sivapackiadevi) Sangarapillai; m. Prema Pathmanathan, Mar. 11, 1985; children: Rohini Pathmanathan, Gayathiri Pathmanathan. MB BS, U. Malaya, Malaysia, 1978, M in Pathology, 1982; MRC in Pathology, Royal Coll. Pathologists, Eng., 1986. Cert. Malaysian Med. Coun., Gen. Med. Coun. U.K. Med. officer U. Hosp., Kuala Lumpur, Malaysia, 1978-79; trainee pathologist dept. pathology U. Malaya, 1979-82, lectr., clin. specialist, 1982-86, cons. pathologist, 1986—, assoc. prof. pathology, 1986-92, chmn. dept., 1991—, prof., dep. dean postgrads., 1992—; sr. cons. histopathologist dept. pathology Univ. Hosp., Kuala Lumpur, 1986—; cons. Tumorzentrum Regensburg, Germany, 1991—, dept. microbiology and tumor immunology U. Regensburg, 1991—. Contbr. over 100 articles to profl. jours. Recipient Pathology prize U. Malaya, 1974, 82. Fellow Acad. Medicine Malaysia, Internat. Coll. Tropical Medicine, Royal Australasian Coll. Pathologists; mem. Malaysian Med. Soc., Malaysian Soc. Pathologists, Royal Coll. Pathologists U.K., N.Y. Acad. Scis., Internat. Acad. Cytology. Hindu. Avocations: music, reading, painting, photography, mountaineering. Home: No 14 Lorong Siputih, Kuala Lumpur 58000, Malaysia Office: Univ Hosp Dept Pathology, Kuala Lumpur 59100, Malaysia

RAJAGOPAL, RANGASWAMY, geography and engineering educator; b. Bombay, July 20, 1944; m. Carol Ann Lauderbach, Aug. 25, 1973; 1 child, Julie Priya. BS in Math./Physics, U. Bombay, 1964, postgrad., 1967; MS in Ops. Rsch., U. Fla., 1969; PhD in Water Resources Mgmt., U. Mich., 1973. Prof. geography and civil and environ. engring. U. Iowa, Iowa City, 1987—; vis. scientist U.S. EPA Environ. Monitoring Systems Lab., Las Vegas, 1986-88. Author books and jour. articles in fields of ground water protection, environ. monitoring and info. systems; founding editor quar. jour. The Environ. Profl. Grantee Dept. Energy, Nat. Geog. Soc., U.S. EPA, U.S. Forest Svc., NSF, others. Mem. Nat. Assn. Environ. Profls. (Disting. Svc. award 1985). Office: U Iowa Dept Civil Environ Engring 308JH Iowa City IA 52242

RAJAKARUNA, LALITH ASOKA, civil engineer; b. Pannala, Srilanka, Aug. 25, 1958; came to U.S., 1985; s. Weerathilaka and Padma (Wickramanayaka) R.; m. Janitha Munaweera, Sept. 25, 1985; children: Reginald, Ashley, Keith. BS in Civil Engring., U. Moratuwa, 1984; MS in Civil Engring., Poly. U., N.Y., 1991. Lic. Profl. Engr., N.Y. State. Civil engr. Colombo Internat. Airport, Srilanka, 1984-85; assoc. of firm, chief engr. Ettlinger & Ettlinger, P.C., Staten Island, N.Y., 1985—; adjunct prof.

Coll. Staten Island, Staten Island, N.Y., 1992—; dir. profl. engring. rev. course Am. Soc. Civil Engrs., N.Y., 1992—. Mem. NSPE. Home: 32 Endor Ave Staten Island NY 10301 Office: Ettlinger & Ettlinger PC 125 Lake Ave Staten Island NY 10303

RAJAKUMAR, CHARLES, mechanical engineer; b. Coonoor, Tamilnadu, India, Apr. 12, 1947; came to U.S., 1981; s. Rajamony and Thangammal (Kumaradhas) Charles; m. Lalitha Theresa Daniel, June 12, 1975; children: Vinod, Anita. BEME, Annamalai U., Tamilnadu, India, 1969; MEME, Indian Inst. Sci., Bangalore, India, 1971; MSc in Mech. Engring., Cranfield Inst. Tech., Eng., 1974; PhD in Mech. Engring., Stevens Inst. Tech., 1985. Aero. engr. Hindustan Aeronautics, Ltd., Bangalore, 1971-73, dep. design engr., 1974-81; sr. rsch. engr. Swanson Analysis Systems, Inc., Houston, Pa., 1985—. Contbr. articles to profl. jours. Coun. mem. Our Redeemer Luth. Ch., McMurray, Pa., 1988-90; road to recovery vol. driver Am. Cancer Soc., Pitts., 1988—. Mem. ASME, Toastmasters Internat. (sec.-treas. 1989), Sigma Xi. Achievements include research in the areas of computational structural acoustics, fluid mechanics, rotordynamics, and eigenvalue analysis. Office: Swanson Analysis Systems In PO Box 65 Johnson Rd Houston PA 15342

RAJAPAKSA, YATENDRA RAMYAKANTHI PERERA, structural engineer; b. Colombo, Sri Lanka, Feb. 4, 1957; came to U.S., 1982; d. Lionel and Iranganie (Karunarathne) Perera; m. Dudley Piyasiri Rajapaksa, July 8, 1982. BSCE, U. Peradeniya, Sri Lanka, 1980; MSCE, U. Okla., 1985, PhD of Civil Engring., 1988. Registered profl. engr., Tex. Site engr. State Engring. Corp., Colombo, Sri Lanka, 1980-82; rsch., teaching asst. U. Okla., Norman, 1982-88; structural engr. Branham Industries, Conroe, Tex., 1988-91; engr. ABS Americas, Houston, 1991—. Co-author: (book chpt.) Finite Element Solution of Immiscible Fluid Flow, 1989; contbr. articles to profl. jours. Mem. ASCE, Soc. Petroleum Engrs., Chi Epsilon. Buddhist. Home: 58 N Duskwood Pl The Woodlands TX 77381 Office: ABS Americas 16855 Northchase Dr Houston TX 77060-6008

RAJARAVIVARMA, RATHIKA, electrical engineer; b. Pannaipuram, Tamilnadu, India, May 27, 1961; came to U.S., 1985; d. Kannaku and Kamala Murugesan; m. Veeramuthu Rajaravivarma, Aug. 21, 1985; 1 child, Omshri. MSEE, Tenn. Tech. U., 1987, PhD in Elec. Engring., 1992. Rsch. engring. trainee Electric Control Equipment, Madurai, India, 1983-84; sortware engr. trainee Madras (India) Refineries Ltd., 1984-85; grad. asst. Elec. Power Ctr. Tenn. Tech. U., Cookeville, 1986-87; grad. instr. Elec. Power Ctr. Tenn. Tech. U., Cookerville, 1987-88, grad. instr. Mfg. Ctr., 1988-91; asst. prof. computer info. systems Morehead (Ky.) State U., 1992—. Contbr. articles to profl. publs. Mem. IEEE, IEEE Circuits and Systems Soc., IEEE Signal Processing Soc., Sigma Xi, Phi Kappa Phi, Tau Beta Pi, Eta Kappa Nu, Alpha Mu Gamma. Home: 280 Heights Ave # 4 Morehead KY 40351 Office: Morehead State U UPO 1051 Morehead KY 40351

RAJENDRAN, NARASIMHAN, civil engineer; b. Madurai, Lakshmipuram, India, Dec. 25, 1955; came to U.S., 1982; s. Narasimhan Veerappan and Lingamal (Algarsamy) Narasimhan; m. Jayashree Ramakrishnan, Sept. 1, 1989; 1 child, Nithya Lashmi. BSCE, Madras U., Coimbatore, India, 1977; MTech, Regional Engring. Coll., Suthkal, Karnataka, India, 1980; MS, S.D. Sch. Mines and Tech., 1983; PhD, Drexel U., 1985. Resident engr. Renuga Textile Mills, Theni, Tamilnadu, India, 1979-81, Desman Assocs., N.Y.C., 1990-91; asst. engr. Madras (India) Port Trust, 1981-82; rsch. specialist Drexel U., Phila., 1983-85; sr./asst. engr. N.J. State Dept. Transp., Trenton, 1985-90; concrete specialist Bechtel Savannah River Inc., North Augusta, S.C., 1991—. Contbr. articles to profl. publs.; mem. transp. rsch. bd. Magnesium Phosphate Based Cement, 1987. Mem. Am. Concrete Inst. (mem. com. 1992—). Achievements include development of suitable test methods for rapid hardening cements. Home: 4127 Saddlehorn Dr Evans GA 30809 Office: Bechtel Savannah River Inc 802 E Martintown Rd North Augusta SC 29841

RAJLICH, VACLAV THOMAS, computer scientist, researcher, educator; b. Prague, Czech Republic, May 3, 1939; came to U.S., 1980; s. Vaclav and Marie (Janovska) R.; m. Ivana m. Bartova, Aug. 6, 1968; children: Vasik, Paul, John, Luke. MS, Czech Tech. U., Prague, 1962; PhD, Case Western Res. U., 1971. Rsch. engr. Rsch. Inst. for Math. Machines, Prague, 1963-67, scientist, 1971-75, mgr., 1975-79; vis. assoc. prof. computer sci. Calif. State U., Fullerton, 1980-81; assoc. prof. computer and communication sci. U. Mich., Ann Arbor, 1982-85; prof. Wayne State U., Detroit, 1985—, chair dept. computer sci., 1985-90; vis. scientist Carnegie-Mellon U., Pitts., 1987, Harvard U., Cambridge, Mass., 1988. Contbr. articles to profl. jours. Recipient Chrysler Challenge Fund, 1988. Mem. Computer Soc. of IEEE, Assn. for Computing Machinery. Roman Catholic. Achievements include development of tools for software maintenance, program comprehension, software design methods, and parallel grammars. Office: Wayne State U Dept Computer Sci Detroit MI 48202

RAJOTTE, RAY V., biomedical engineer, researcher; b. Wainwright, Alta., Can., Dec. 5, 1942; s. Sam and Bernadette (Tremblay) R.; m. Gloria A. Yackimetz, Aug. 20, 1966; children: Brian, Michael, Monique. RT, No. Alta. Inst. Tech., 1965; BSc in Elec. Engring., U. Alta., 1971, MSc in Elec. Engring., 1973, PhD in Biomed. Engring., 1975. Postdoctoral fellow U. Alta. Dept. Medicine, Edmonton, 1975-76, Oak Ridge (Tenn.) Nat. Lab., 1976-77, Washington U., St. Louis, 1977, UCLA, 1977; rsch. assoc. dept. medicine U. Alta., Edmonton, 1977-79, asst. prof. dept. medicine, 1979-82, asst. prof. dept. medicine & surgery, 1983-84, assoc. prof., 1984-88, prof., 1988—; dir. Islet Cel Transplant Lab. U. Alta., 1982—, Divsn. Exptl. Surgery U. Alta., 1988—; assoc. dir. Surg.-Med. Rsch. Inst. U. Alta., 1984-87, dir., 1987—; co-dir. Juvenile Diabetes Fund Diabetes Interdisciplinary Rsch. Program U. Alta., 1992—. Co-editor: The Immunology of Diabetes Mellitus, 1986; mem. adv. bd. Diabetologia jour., 1993—; author over 350 pubIs. in field; contbr. 18 chpts. to books. Mem. Cell Transplantation Soc. (founding mem., councillor 1991—), Internat. Pancreas and Islet Transplant Assn. (founding mem., treas. 1991—), Can. Transplantation Soc. (Western councillor 1987—), European Assn. for Study of Diabetes, Assn. Profl. Engrs., Geologists & Geophysicists Alta., Soc. Cryobiology, Am. Diabetes Assn., Can. Soc. Clin. Investigation, N.Y. Acad. Sci., Transplantation Soc., Can. Diabetes Assn., Acad. Surg. Rsch., Internat. Diabetes Fedn. Achievements include patent for glucose sensor. Office: U Alta, Surgical-Med Rsch Inst, Edmonton, AB Canada T6G 2N8

RAJU, KRISHNAM, chemist; b. Lankala Koderu, Andhra Pradesh, India, Nov. 1, 1954; s. Chitti Uddaraju and Chettemma Raju; m. Usha A.V. Raju, Aug. 9, 1979; children: Nitin U., Navin U. MSc, Osmania U., Hyderabad, India, 1976, PhD, 1980. Lectr. Osmania U., 1980-84; postdoctoral fellow U. Okla., Norman, 1984-87, rsch. scientist, 1987-90; lab. scientist Saudi Aramco, Dharan, Saudi Arabia, 1990—; cons. Bio-Med. Rsch. Labs., Oklahoma City, 1989-90. Mem. Am. Chem. Soc., Soc. Petroleum Engrs., Nat. Assn. of Corrosion Engrs. Hindu. Achievements include copyright for scale prediction program OKSCALE. Office: Saudi Aramco, PO Box 2815, Dhahran 31311, Saudi Arabia

RAJU, PERUMAL REDDY, chemist; b. Sabbiah-puram, Tamil Nadu, India, July 13, 1952; s. Perumal Reddy and Valli Ammal Raju; m. Subba Reddy Indra, July 8, 1982; children: Vasanth Babu, Prasanth Babu. BS in Agr., Agrl. Coll., Madurai, India, 1973; diploma, Inst. UN Studies, New Delhi, 1976; postgrad., Cen. Plant Protection Tng Inst, Hyderabad, India, 1979. Asst. rsch. officer State Dept. Agr., Tamilnadu, India, 1973—; activist for consumer orgn. concentrating on supply of quality pesticides, Madurai, India, 1985—. Editor jour. Tillers' Digest, 1992; contbr. articles to profl. publs. Fellow Indian Chem. Soc., United Writers' Assn. India; mem. Plant Protection Assn. (life), Am. Chem. Soc., Indian Sci. Congress, Arid Zone Rsch. Assn. Hindu. Achievements include research in study of sources and methods of adulteration of pesticides and evolving methods of identification, pesticides and environment. Home: GST Rd, 183 A Thiru Murugan Colony, Madurai 625 006, India Office: Pesticides Testing Lab, Gandhiji Rd Thirunagar, Madurai Tamil Nadu 625 006, India

RAKSIS, JOSEPH W., chemist; b. Wilkes Barre, Pa., Mar. 9, 1942; married, 1982; 3 children. BS, Wilkes Coll., 1963; PhD chemistry, U. Calif., Irvine, 1967. Rsch. chemist Dow Chem. Co., 1967-71, rsch. mgr., 1972-77, dir., 1977-82; v.p. rsch. divsn. W R Grace & Co., 1982—. Mem. Am. Chem. Soc., Indsl. Rsch. Inst. Achievements include research in design and synthesis of polymeric materials having specific functional properties. Office: W R Grace & Co Rsch Div Washington Rsch Ctr 7379 Route 32 Columbia MD 21044-4098*

RAKUTIS, RUTA, chemical economist; b. Marijampole, Lithuania, Aug. 16, 1939; came to U.S., 1949; s. Juozas Rakutis and Natalia Pavilcius. BS cum laude, U. Ill., 1961; PhD, U. Iowa, 1968; MBA, Northeastern U., 1974. Sr. scientist Jet Propulsion Lab, Pasadena, Calif., 1968-70; scientist color lab. Polaroid Corp., Cambridge, Mass., 1970-74; sales mgr. Am. Cyanamid Co., Wayne, N.J., 1974-79; mktg. mgr. Lignin Chems., Am. Can. Co., Greenwich, Conn., 1979-81; cons. SRI Internat., Menlo Park, Calif., 1981-88; mgr. tech. devel. Imperial West Chem. Co., Antioch, Calif., 1988-92; product mgr. polymers Garratt-Callahan Co., Millbrae, Calif., 1992—. Contbr. articles to profl. jours. Dunlop fellow, 1967-68. Mem. Chem. Mktg. Rsch. Assn. of Am. Chem. Soc., Lakeview Assocs., Sigma Xi. Avocations: horseback riding, skiing, tennis, Lithuanian-Am. community.

RALL, LLOYD LOUIS, civil engineer; b. Galesville, Wis., Dec. 7, 1916; s. Louis A. and Anna (Kienzle) R.; m. Mary Moller, July 12, 1952; children: Lauris, David, Christopher, Jonathan. BCE, U. Wis., 1940. Commd. 2d lt. U.S. Army, 1940, advanced through grades to col., 1972, ret., 1977; dir. Washington ops. Itek Optical Systems, Washington, 1977-91. Decorated Legion of Merit with oak leaf cluster, Bronze star. Home: 301 Cloverway Alexandria VA 22314

RALPH, JAMES R., physician; b. Lowell, Mass., Mar. 23, 1933; s. Richard Henry and Alice Claire (Walwood) R.; m. Edith Marguerite Aeschliman, June 7, 1958; children—James R., Lee P., Jon D., David G. BA Middlebury Coll., 1954; MD Yale U., 1959. Diplomate Nat. Bd. Med. Examiners, Am. Bd. Family Practice. Intern, Akron Gen. Med. Ctr., Ohio, 1959-60, resident Akron Gen. Hosp. and Akron Children's Hosp., 1960-61; staff physician Univ. Health Services, U. Mass., Amherst, 1963-72; team physician dept. intercollegiate athletics, U. Mass., 1965—, asst. med. dir. Univ. Health Services, 1971—, coordinator Sports Medicine U. Health Services, 1987—; assoc. family practice community medicine U. Mass. Med. Sch.-Ctr., Worcester, 1980—; attending physician U.S. VA, Florence, Mass., 1964-69; attending staff active Wing Meml. Hosp., Palmer, Mass., 1967-70. Contbr. articles to profl. jours. Bd. dirs. Inter-Faith Housing, Amherst, 1984—, Boston Med. Library, 1985—. Served to capt. USAF, 1961-67. U. Mass. Faculty research grantee, 1967-69; recipient Chancellors Citation award, 1991. Fellow Am. Acad. Family Physicians, Am. Med. Soc. Sports Medicine, 1991; mem. Mass. Acad. Family Physicians (bd. govs. 1980-89), Mass. Med. Soc. (exec. bd., bd. dirs. postgrad. med. inst.), Hampshire Dist. Med. Soc. (pres. elect 1984-86, pres. 1986-88), Am. Coll. Sports Medicine, Sigma Xi. Democrat. Roman Catholic. Avocations: tennis; hiking; camping; coin and stamp collecting. Home: 66 Hills Rd Amherst MA 01002-1839 Office: U Mass Univ Health Svcs Amherst MA 01003

RALSTON, JOHN PETER, theoretical physicist, educator; b. Reno, Nev., Aug. 8, 1951; s. Robert Lee and Katherine A. (Casserly) R.; m. Cherie Wagner, Sept. 1, 1973; children: Amy, John Charles, Joseph. PhD, U. Oreg., 1980. Design engr. Hamilton Co., Reno, 1974; postdoctoral fellow McGill U., Montreal, Can., 1980-82; rsch. assoc. Argonne (Ill.) Nat. Lab., 1982-84; asst. prof. U. Kans., Lawrence, 1984-88, assoc. prof., 1988-93; prof. U. Kans., 1993—; sci. assoc. CERN, Geneva, Switzerland, summer 1991, CEN Saclay, France, summer 1990, Stanford (Calif.) Linear Accelerator Ctr., summer 1989. Contbr. over 80 articles to profl. jours. Recipient Maitre de Recherche Ecole Polytechnique, France, 1990-91. Achievements include studies in quantum chromodynamics and high energy physics. Office: U Kans Dept Physics and Astronomy Lawrence KS 66045

RALSTON, PATRICIA A. STARK, engineering and computer science educator; b. Elizabethtown, Ky., Apr. 2, 1957; d. William Stanley and Anna Katherine (Higdon) Stark; m. Dale Lee Ralston, Mar. 19, 1983; children: Haylee Anne, Kendra Jean. M of Engring., U. Louisville, 1980, PhD in Chem. Engring., 1983. Systems engr. No. Petrochem., Morris, Ill., 1982; instr. U. Louisville, 1982-83, asst. prof., 1983-86, assoc. prof., 1986-93; prof., 1993—; bd. dirs. Ky. Sci. & Tech. Coun., Lexington, 1989—. Co-editor: Procs. Internat. Fuzzy Systems and Intelligent Control Conf., 1992, 93; contbr. articles to profl. jours. Bd. dirs. Walnut St. Bapt. Ch. Daycare Ctr., Louisville, 1991-93. Mem. AICE, Instrument Soc. Am., Sigma Xi, Tau Beta Pi. Office: Engring Math & Computer Sci U Louisville Louisville KY 40292

RALSTON, ROY B., petroleum consultant; b. Monmouth, Ill., June 7, 1917; s. Roy Crews and Helen Ruth (Boggs) R.; m. Catherine Elizabeth Thompson, Aug. 6, 1940; 1 child, John Richard. BA, Cornell Coll., 1939; student, Iowa U., 1938; postgrad., U. Ill., 1940-41. Pretroleum cons. Scottsdale, Ariz., 1977—; dist. mgr. exploration Skelly Oil Co., Evansville, Ind., 1941-46; div. mgr. exploration and prodn. Ashland Oil & Refinery Co., Henderson, Ky., 1946-50; mgr. exploration and prodn. Ashland Oil & Refinery Co., Ashland, Ky., 1950-54; div. mgr. exploration Phillips Petroleum Co., Evansville, 1955-58, Amarillo, Tex., 1958-65, Oklahoma City, 1965-69; v.p. exploration and prodn. Phillips Petroleum Can. Ltd., Calgary, Alta., Can., 1969-72; exploraton mgr. North Am. Phillips Petroleum Co., Bartlesville, Okla., 1973-75; regional mgr. exploration and prodn. Phillips Petroleum Co., Denver, 1975-77; petroleum cons. 1st Nat. Bank, Amarillo, Tex., 1977—, Valley Nat. Bank, Phoenix, 1977—, 1st Interstate Bank, Phoenix, 1977-86. Youth career dir. Oklahoma City C. of C., 1966-68. Recipient Disting. Svc. award Okla. Petroleum Coun., 1968, Svc. award Land Mgmt. Sch. Okla. U., 1978. Mem. Am. Assn. Petroleum Geologists (publicity dir. nat. conv. Oklahoma City 1968, cert. petroleum geologist), Am. Assn. Petroleum Landmen, Soc. Petroleum Engrs. AIME, Ariz. Geol. Soc., N.Mex. Geol. Soc., Rotary. Republican. Presbyterian. Avocations: hunting, horseback riding, water sports.

RAMACHANDRAN, NARAYANAN, aerospace scientist; b. Bombay, Jan. 6, 1958; came to U.S., 1980; m. Geetha Chandra. BE with honors, U. Madras, India, 1980; MSME, U. Mo., Rolla, 1983, PhD, 1987. Grad. rsch. asst. U. Mo., Rolla, 1980-87; rsch. assoc. Univs. Space Rsch. Assn., Huntsville, Ala., 1987-88, assoc. scientist, 1988—; prin. investigator NASA, 1987; co-investigator NASA Shuttle Flight, 1992. Contbr. 20 papers to sci. jours. Mem. AIAA, ASME, Sigma Xi.

RAMADHYANI, SATISH, mechanical engineering educator; b. Bangalore, Karnataka, India, Aug. 1, 1949; came to U.S., 1975; s. K. R. Keshavachar and Padma (Iyengar) R.; m. Rachel B. Sparrow, June 17, 1979. B of Tech., Indian Inst. of Tech., Madras, India, 1971; MS, U. Minn., 1977, PhD, 1979. Asst. engr. devel. Motor Industries Co., Bangalore, 1971-75; asst. prof. Dept. Mech. Engring., Tufts U., Medford, Mass., 1979-83; asst. prof. Sch. of Mech. Engring., Purdue U., West Lafayette, Ind., 1983-86, assoc. prof., 1986-91, prof., 1991—; expert cons. Teltech, Inc., Mpls., 1990—; expert reviewer numerous profl. jours. and funding agys., 1979—; book reviewer McGraw-Hill, John Wiley, Prentice-Hall, Cambridge, 1980—. Contbr. over 40 articles to profl. jours. Organizer Indiana Fiddlers' Gathering, Battle Ground, Ind., 1990-92; organizer, rschr. Tippecanoe Environ. Coun., Lafayette, 1988-89; actor/sound tech. Civic Theatre of Greater Lafayette, 1990-91. Recipient Pres. of India Gold Medal Govt. India, 1971. Mem. AIAA, ASME (mem. heat transfer div. K-6), Am. Soc. of Engring. Educators, Phi Kappa Phi, Tau Beta Pi, Sigma Xi. Achievements include development of new finite-element and finite-volume computational techniques for heat transfer and fluid flow problems, liquid immersion cooling technology for high-speed digital computers, mathematical models to predict performance of industrial furnaces. Office: Purdue U Sch Of Mech Engring West Lafayette IN 47907

RAMAKRISHNAN, RAMANI, acoustician; b. Madurai, Tamil Nadu, India, Sept. 22, 1948; came to Can. 1982; s. Sivaramakrishnan and Rajammal Meenakkshisundaram Ramakrishnan; m. Suman Murthy, July 15, 1985; 1 child, Jahnavi. MS, George Washington U., 1973, DSc, 1977. Registered profl. engr., Ont. Rsch. asst. Joint Inst. Acoustics and Flight Scis., NASA, Langley Rsch. Ctr., Va., 1971-77; sr. postdoctoral rsch. fellow Inst. Sound & Vibration Rsch., U. Southampton, Eng., 1977-79; assoc. scientist Lockheed-Ga. Co., Marietta, Ga., 1980-82; mgr. engring. svcs. div. Vibron Ltd., Misasissauga, Ont., 1982-85; faculty Ryerson Poly. Inst., Toronto, Ont., 1984—; project analyst noise assessment and systems support unit Ministry of the Environment, Province of Ont., 1985-91; tech. specialist noise control Ontario Hydro, Toronto, 1991—. Editor: Environmental Noise Assessment in Land Use Planning, 2d edit., 1989; contbr. to Introductory Environ. Noise Control Manual, 1988, Certificate Environ. Noise Course Manual, 1989; contbr. articles to profl. jours. Vice pres. Vedanta Soc. of Toronto, 1983, 90. Mem. Acoustical Soc. Am., Can. Acoustical Assn., Assn. profl. Engrs. Ont., Inst. Noise Control Engring., Can. Standards Assn. (subcom. on ground vibrations from transp. sources), Amnesty Internat. Hindu. Achievements include 1 patent on Air Handling Systems, 1991; developer a PC based computer software to evaluate the duct silencer performances, 1989-90. Home: 41 Watson Ave, Toronto, ON Canada M6S 4C9

RAMAKRISHNAN, VENKATASWAMY, civil engineer, educator; b. Coimbatore, India, Feb. 27, 1929; came to U.S., 1969, naturalized, 1981; s. Venkataswamy and Kondammal (Krishnaswamy) R.; m. Vijayalakshmi Unnava, Nov. 7, 1962; children: Aravind, Anand. B.Engring., U. Madras, 1952, D.S.S., 1953; D.I.C. in Hydropower and Concrete Tech, Imperial Coll., London, 1957; Ph.D., Univ. Coll., U. London, 1960. From lectr. to prof. civil engring., head dept. P.S.G. Coll. Tech., U. Madras, 1952-69; vis. prof. S.D. Sch. Mines and Tech., Rapid City, 1969-70; prof. civil engring. S.D. Sch. Mines and Tech., 1970—, dir. concrete tech. research, 1970-71, head grad. div. structural mechanic and concrete tech., 1971—, program coordinator materials engring. and sci. Ph.D. program, 1985-86. Author: Ultimate Strength Design for Structural Concrete, 1969; also over 200 articles. Colombo Plan fellow, 1955-60; recipient Outstanding Prof. award S.D. Sch. Mines and Tech., 1980. Mem. Internat. Assn. Bridge and Structural Engring., ASCE (vice chmn. constrn. div. publs. com. 1974), Am. Concrete Inst. (chmn. subcom. gen. considerations for founds., chmn. com. 214 on evaluation of strength test results, sec.-treas. Dakota chpt. 1974-79, v.p. 1980, pres. 1981), Instn. Hwy. Engrs., Transp. Rsch. Bd. (chmn. com. on admixtures and curing, chmn. com. on mech. properties concrete), Am. Soc. Engring. Edn., NSPE, Internat. Coun. Gap-Graded Concrete Rsch. and Application, Sigma Xi. Address: 1809 Sheridan Lake Rd Rapid City SD 57702

RAMALINGAM, MYSORE LOGANATHAN, mechanical engineer; b. Mysore, Karnataka, India, Dec. 12, 1954; came to U.S., 1982; s. Mysore Laxmanaswamy and Sundarabai (Arunachalam) Loganathan; m. Samyuktha Alalasundaram, June 22, 1981; children: Suraj Kumar, Shyma. PhD, Ariz. State U., 1986. Asst. devel. engr. Jyothi Pumps Ltd., Baroda, Gujarat, India, 1977; engr. Indian Space Rsch. Orgn., Bangalore, Karnataka, 1977-82; grad. rsch. asst. Ariz. State U., Tempe, 1982, rsch. assoc., 1984-86; sr. rsch. scientist Universal Energy Systems, Inc., Dayton, Ohio, 1986-92, prin. rsch. scientist, 1992—; tech. cons. Ariz. State U., Tempe, 1984-86; tech. specialist K-Tron Internat., inc., Phoenix, 1985; space rsch. specialist Vikram Sarabhai Space Ctr., Trivandrum, India, 1979. Contbr. articles to profl. jours. Transport coord. Tamilnadu Found. of Dayton, 1989; tournament organizer India Club of Greater Dayton, 1990, 91. Grantee USAF; recipient Recognition of Sci. Rsch. award ISRO, India. Fellow AIAA (assoc., Recognition of Svc. to Engring. Community award); mem. ASME, Am. Soc. Metals. Office: UES Inc 4401 Dayton-Xenia Rd Dayton OH 45432

RAMANI, NARAYAN, mechanical engineer; b. Thanjavur, Tamil Nadu, India, Dec. 9, 1942; s. Krishnan Narayan and Meenakshi Narayan; m. Chandra Ramani, May 26, 1971. MS, Indian Inst. Tech., Madras, 1984, PhD in Mech. Engring., 1990. Asst. devel. engr. Bajaj Tempo Ltd., Pune, India, 1964-67; tool designer Zahnradfabrik AG, Sch Gmund, Germany, 1967-69; works mgr. Pilot Engines Ltd., Madras, 1969-76; gen. mgr. R&D TVS Suzuki Ltd., Hosur, India, 1976—. Editor: Enjoy Rotary, 1992. Fellow Verein deutscher Ingenieure Germany, Rotary Club of Hosur (pres. 1992-93). Home: C17 TVS Nagar, Hosur India 635110 Office: TVS Suzuki Ltd, Harita, Hosur India 635109

RAMANI, RAJA VENKAT, mining engineering educator; b. Madras, India, Aug. 4, 1938; came to U.S., 1966; s. Natesa and Meenakshi (Srinivasan) Rajaraman; m. Geetha V. Chalam, July 9, 1972; children: Deepak, Gautam. BSc. with honors, Indian Sch. Mines, Dhanbad, Bihar, 1962; MS, Pa. State U., 1968, PhD, 1970. Registered profl. engr., Pa.; lic. first class mine mgr. Mining engr., mgr. Andrew Yule & Co., Asansol, West Bengal, India, 1962-66; grad. asst. Pa. State U., University Park, 1966-70, asst. prof., 1970-74, assoc. prof., 1974-78, prof. mining engring., 1978—, chmn. mineral engring. mgmt. sect., 1974—, head dept. mineral engring., 1987—; chmn. com. post-disaster survival/rescue NAS, Washington, 1979-81; cons. UN, UN Devel. Program, Dept. Econ. and Social Devel., N.Y.C., 1983; cons./expert panels U.S. Dept. Labor, 1979, 92, HHS, 1977, 92, U.S. Dept. State, 1986, 87, Dept. Environ. Resources, Commonwealth of Pa., 1990, 92; co-dir. Generic Mineral Tech. Ctr. on Respirable Dust, U.S. Bur. Mines, 1983—, Nat. Mines/Land Reclamation Ctr., 1988—, Std. Oil Ctr. of Excellence on Longwall Tech., 1983-89. Sect. editor, author: Computer Methods for the Eighties, 1979, SME Mining Engineering Handbook, 1992; editor: State-of-the-Art in Longwall-Shortwall Mining, 1981, Proceedings: 19th International APCOM Symposium, 1986. Recipient Disting. Alumni award Indian Sch. Mines, Dhanbad, India, 1978, Ednl. Excellence award Pitts. Coal Mining Inst., 1986, Environ. Conservation award AIME, N.Y.C., 1990, Howard N. Eavenson award SME/AIME, N.Y.C., 1991, Percy N. Nicholls award AIME/ASME Joint Soc. award, N.Y.C., 1992, Fulbright scholar award to Soviet Union Coun. Internat. Exch. of Scholars, Washington, 1989-90. Mem Internat. Coun. for Application of Computers in the Mineral Industry (Disting. Achievement award 1989), Soc. Mining, Metall. and Exploration (Disting. Mem. 1989), Mine Ventilation Soc. South Africa, Inst. Mgmt. Scis. Achievements include reserach in health, safety, environmental and productivity aspects in underground and surface mining engineering. Home: 285 Oakley Dr State College PA 16803 Office: Pa State U Dept Mineral Engring 104 Mineral Scis Bldg University Park PA 16802

RAMANUJA, TERALANDUR KRISHNASWAMY, structural engineer; b. Mysore, Mysore, India, June 23, 1941; came to U.S., 1967, naturalized, 1979; s. Teralandur R. and Padmammal Krishnaswamy; m. Jayalakshmi Ramanuja, Jan. 18, 1971; children: Srinivasan, Rekha. BSCE, U. Mysore, 1962; MS in Structural Engring., U. Notre Dame, 1969. Registered profl. engr., Ill., Mich., Ind., N.Y. Sub-divisional officer Mil. Engring. Svcs., Bangalore, India, 1962-67; structural engr. Clyde E. Williams and Assocs., South Bend, Ind., 1969-73; head structural engring. dept. Ayres, Lewis, Norris & May, Cons. Engrs., Ann Arbor, Mich., 1973-76; sr. project mgr. Johnson & Anderson Cons. Engrs., Pontiac, Mich., 1976-78; supr. Bechtel Power Corp., Ann Arbor, 1978-85; supr. Shoreham Nuclear Power Sta. Lilco, N.Y.C., 1985-89; supervising engr. Clinton (Ill.) Power Sta. Ill. Power Co., 1989—. Fellow ASCE; mem. Am. Concrete Inst.; mem. Chi Epsilon. Achievements include structural and foundation design of facilities for fossil and nuclear power plants, water/waste treatment plants, petrochemical plants, pulp and paper mills and for heavy equipment/machinery for these plants; seismic and dynamic analysis of structures, systems and components in nuclear power plants. Home: 2006 Hidden Lake Rd Bloomington IL 61704-7283 Office: Ill Power Co Clinton Power Sta PO Box 678 Clinton IL 61727

RAMAPRASAD, KACKADASAM RAGHAVACHAR, physical chemist; b. Bangalore, India, Dec. 8, 1938; came to U.S., 1965, permanent resident, 1971; s. Kackadasam Raghavachar and Saroja (Narasimhachar) R.; m. Rukmani Raghavachari, July 14, 1968; children: Saroja, Venkat. BS in Chemistry with honors, U. Mysore, Bangalore, 1958; MS in Phys. Chemistry, NYU, 1971, PhD, 1972. Trainee Bhabha Atomic Rsch. Ctr. Tng. Sch., Bombay, India, 1958-59, rsch. asst., jr. sci. officer chemistry div., 1959-65; teaching fellow N.Y.U., N.Y.C., 1965-71, duPont teaching asst., 1967-68; maitre-assistant dept. de chimie physique, U. Geneva, 1972-73; chemist Ecole Poly-technique Federale de Lausanne, Switzerland, 1974; rsch. assoc. dept. chemistry Princeton (N.J.) U., 1974-77, rsch. assoc., mem. profl. rsch. staff dept. chem. engring., 1977-79; sr. scientist Chronar Corp., Princeton, 1979-89; sr. scientist Electron Transfer Techs., Inc., Princeton, 1990-93; staff scientist TRI/Princeton, 1993—. Recipient Founder's Day award N.Y. U., 1972. Mem. Am. Chem. Soc., Sigma Xi. Contbr. articles to profl. publs. Office: TRI/Princeton PO Box 625 Princeton NJ 08542

RAMASAMY, RAVICHANDRAN, chemistry educator; b. Cuddalore, Tamil Nadu, India, Mar. 21, 1965; came to U.S., 1985; s. Ramasamy and Rajalakshmi (Chakravarthy) R. BS in Chemistry, Loyola Coll., Madras, India, 1985; PhD, Loyola U., 1989. Teaching asst. Loyola U., Chgo., 1985-

87, rsch. asst., 1987-89; rsch. assoc. in chemistry U. Tex. at Dallas, Richardson, 1989—; presenter in field. Contbr. articles to Jour. Am. Chem. Soc., Inorganic Chemistry, Investigative Radiology, Jour. Physiology, Clinica Chimica Acta, others, chpts. to books. Social worker Loyola chpt. Nat. Social Svc., Madras, 1982-83. Dissertation fellow Loyola U., Chgo., 1989. Mem. Am. Chem. Soc., Biophys. Soc. (Talbot award 1989), Soc. Magnetic Resonance in Medicine (Travel Stipend award 1988), Sigma Xi (Grad. Rsch. Forum first prize). Hindu. Achievements include patent in polyazamacrocyclic compounds for complexation of metal ions. Home: 7220 McCallum Blvd Apt 905 Dallas TX 75252-6166 Office: U Tex at Dallas 2601 N Floyd Rd Richardson TX 75080-1407

RAMASUBBU, SUNDER, aeronautical engineer, researcher; b. Poona, Maharashtra, India, Aug. 8, 1953; s. Ramasubbu Ramamurthy and Laxmi Ramasubbu; m. Kala Parthasarathy; 1 child, Shubha. M.Tech in Aeroengring., Inst. Civil Aviation, Kiev, USSR, 1975; PhD, Inst. Civil Aviation, 1978. Vis. scientist Wright Aero. Labs., Wright Patterson AFB, Ohio, 1986-88; scientist Nat. Aero. Lab., Bangalore, India, 1978-86; sr. scientist, head divsn. structural integrity Nat. Areo. Lab., Bangalore, 1988—; chmn. Com. Airframe Structural Integrity, 1990-91; reviewer UN Devel. Program Projects, India, 1990-91; mem. Aircraft Life Ext. Programme, 1992—. Contbr. articles to profl. jours. Recipient Tech. award Coun. Sci. and Indsl. Rsch., India, 1991; sr. rsch. associateship NRC and U.S. Acad. Scis., Ohio, 1985. Mem. ASTM, Materials Rsch. Soc. Hindu. Achievements include devel. of fractographic technique for crack closure measurement, of application software for fatigue testing, of microprocessor controller for servohydraulics in testing, link between short crack behavior and notch root fatigue phenomena. Office: Nat Aero Lab, Head, Structural Integrity Divsn, 560017 Bangalore India

RAMER, JAMES LEROY, civil engineer; b. Marshalltown, Iowa, Dec. 7, 1935; s. LeRoy Frederick and Irene (Wengert) R.; m. Jacqueline L. Orr, Dec. 15, 1957; children: Sarah T., Robert H., Eric A., Susan L. Student U. Iowa, 1953-57; MCE, Washington U., St. Louis, 1976, MA in Polit. Sci., 1978; postgrad. U. Mo., Columbia, 1984—. Registered profl. engr., land surveyor. Civil and constrn. engr. U.S. Army C.E., Tulsa, 1960-63; civil and relocations engr. U.S. State Dept., Del Rio, Tex., 1964; project engr. H. B. Zachry Co., San Antonio, 1965-66; civil and constrn. engr. U.S. Army C.E., St. Louis, 1967-76, tech. advisor for planning and nat. hydropower coordinator, 1976-78, project mgr. for EPA constrn. grants, Milw., 1978-80; chief architecture and engring. HUD, Indpls., 1980-81; civil design and pavements engr. Whiteman AFB, Mo., 1982-86, project mgr. maintenance, 1993—; soil and pavements engr. Hdqtrs. Mil. Airlift Command, Scott AFB, Ill., 1986-88; project manager AF-1 maintenance hangar; cattle and grain farmer, 1982—; pvt. practice civil-mech. engr., constrn. mgmt., estimating, cost analysis, cash flow, project scheduling, expert witness, Fortuna, Mo., 1988—, chief construction inspector divsn. Design and Construction, State of Mo., 1992-93; adj. faculty civil engring. Washington U., 1968-78, U. Wis., Milw., 1978-80. Ga. Mil. Coll., Whiteman AFB.; adj. research engr. U. Mo., Columbia, 1985—. Author tech. writing operation and maintenance manuals, fin. reports and environ. control plans; holder 25 U.S. patents in diverse art, 8 copyrights; developer solar waterstill, deep shaft hydropower concept. Mem. ASCE, NSPE, AAUP, Soc. Am. Mil. Engrs. Lutheran. Club: Optimists Internat. Home: Rt 1 PO Box 50-AA Fortuna MO 65034

RAMESH, KALAHASTI SUBRAHMANYAM, materials scientist; b. Tiruchi, Madras, India, Mar. 22, 1949; s. Subrahmanyam Veeraragaviah and Kuntala (Chinnaswami) Kalahasti; m. Atsumi Yoshida Ramesh, Jan. 30, 1990; 1 child, Siva. MS in Ceramic Engring., Benaras Hindu U., India, 1973; D in Engring., Tokyo Inst. Tech., 1986. Asst. rsch. mgr. Steel Authority India Ltd., Ranchi, Bihar, 1979-80; lectr. dept. ceramic engring. Benaras U., Varanasi, India, 1980-82; tech. mgr. ceramics div. TYK Corp., Tokyo, 1986-89; mgr. rsch. and devel. Mer Corp., Tuscon, Ariz., 1989; prin. scientist Battelle meml. Inst., Columbus, Ohio, 1989--; adv. Internat. Bus. Svc., Tokyo, 1988-89; cons. HTP Inc., Sharon, Pa., 1989--; mem. U.S. Dept. Energy Ceramics Adv. Com., Washington, 1991--. Mombusho Rsch. fellow Min. Edn. Japan, 1982-85. Fellow Indian Inst. Ceramics, Inst. Ceramics U.K.; mem. Japan India Assn., Found for Indsl. Rsch. (expert), Max Planck Soc. Hindu. Achievements include development of several ceramic and refractories for iron, steel and gas turbine/engine applications. Home: 100 Hillview Dr Richland WA 99352 Office: Pacific Northwest Labs K3-59 PO Box 999 Richland WA 99352

RAMESH, K(ALIAT) T(HAZHATHVEETIL), mechanical engineer, educator; b. Bangalore, India, Mar. 31, 1959; came to U.S., 1983; s. K.T.K. and K.T. (Sitalaxmi) Nambiar; m. Priti Nair, Aug. 22, 1986; 1 child, Rohan N. ScM in Engring., Brown U., 1985, ScM in Applied Math., 1987, PhD in Engring., 1987. Postdoctoral fellow U. Calif., San Diego, 1987-88; asst. prof. mech. engring. Johns Hopkins U., Balt., 1988-93; assoc. prof. mech. engring. Johns Hopkins U., 1993—; cons. Reisinger and Assocs., Balt., 1990-91. Contbr. articles on mechanics, materials sci. and tribology to tech. jours. Bd. dirs. Worthington Glen Homeowners' Assn., Owings Mills, Md., 1992. Grantee NSF, Army Rsch. Office, others; recipient Rsch. Initiation award NSF, 1989. Mem. ASME (exptl. mechanics com., recording sec., gen. com., organizer, chair conf. symposia, Best Paper award 1987), Soc. Engring. Sci., Minerals, Metals and Materials Soc., Soc. Rheology, Am. Acad. Mechanics. Achievements include development of new techniques for studying material properties of liquids. Office: Johns Hopkins U Dept Mech Engring Baltimore MD 21218

RAMESH, KRISHNAN, mechanical engineer; b. Nagercoil, Tamilnadu, India, Dec. 17, 1959; came to U.S., 1982; s. T. Krishna Iyer and S. Lakshmi Ammal; m. Alamelu Ramesh, Aug. 24, 1986; 1 child, Krishnan Jr. B Engring., Regional Engring. Coll., Trichy, India, 1982; MS, N.C. State U., 1984. Registered profl. engr., N.C. Grad. rsch. asst. dept. mech. and aerospace engring. N.C. State U., Raleigh, 1983-84; project engr. Sud Assocs., P.A., Durham, N.C., 1984-93; mech. engr. U. Wis. Madison Planning and Constrn., 1993—. Mem. Assn. Energy Engrs., Phi Kappa Phi. Hindu. Achievements include analytical study of suction boundary layer control in v/stol inlets. Home: 5314 Mathews Rd # 2 Madison WI 53562 Office: U Wis Planning and Constrn 915 WARF Bldg 610 Walnut St Madison WI 53705

RAMESH, SWAMINATHAN, chemist; b. Madras, India, June 10, 1949; came to the U.S., 1976; s. Sumaithangi and Kamalakshi (Venkata Subramanian) Swaminathan; m. Sumati Subbaratnam, Apr. 17, 1981; 1 child, Sheela L. BS in Chemistry, Madras U., 1969; PhD in Organic Chemistry, Indian Inst. Sci., 1976. Rsch. assoc. U. Ill., Chgo., 1976-80; mass spectroscopist Mar Inc., Chgo., 1980-81, Bionetics Inc., Chgo., 1981-83; rsch. scientist BASF Corp., Wyandotte, Mich., 1983-85, rsch. assoc., 1986-88, sr. rsch. assoc., 1989-90; sr. R&D assoc. BASF Corp., Southfield, Mich., 1991—. Contbr. articles to Jour. Agrl. Food Chemistry, Jour. Forensic Sci., others. Mem. Am. Chem. Soc., Am. Soc. Mass Spectrometry. Achievements include research in use of pyrolysis GC/MS for analyzing and characterizing polymers, new class of polymers for use in automotive coatings, use of supercritical fluids for removing volatiles in latex, soil and water. Office: BASF Corp Telegraph Rd Southfield MI 48086

RAMEY, CRAIG T., pschology educator. BAin Psychology, W.Va. U., 1965, MA in Psychology, 1967, PhD in Devel. Psychology, 1969; postdoctoral in Devel. Psychology, U. Calif., Berkeley, 1969. Asst. prof. psychology Wayne State U., 1969-71; assoc. prof., dept psychology U. N.C., Chapel Hill, 1971-78; sr. investigator, dir. infant rsch. Franf Porter Graham Child Devel. Ctr., U. N.C., Chapel Hill, 1971-74, dir. rsch., 1975-89, assoc. dir., 1978-89; prof. psychology U. N.C., Chapel Hill, 1979-90, prof. pediatrics, Sch. Medicine, 1984-90; dir. Civilan Internat. Rsch. Ctr., U. Ala., Birmingham, 1990—, Sparks Ctr. for Devel. and Learning Disorders, U. Ala., Birmingham, 1990—; prof., depts. psychology, pediatrics and pub. health U. Ala., Birmingham, 1990—. Contbr. to profl. jours. including Journal of the American Medical Association, American Journal of Public Health, Am. Psychologist, Educational Psychology. Office: University of Alabama Civilan Internat Rsch Ctr PO Box 313 UAB Sta Birmingham AL 35294*

RAMEY, HARMON HOBSON, JR., materials scientist; b. Augusta, Ark., Dec. 4, 1930; s. Harmon Hobson and Dorothy Lorene (O'Neal) R.; m. Jenell Evangeline Bostater, Aug. 15, 1954; children: Joseph William, Deborah Ann. BS, U. Ark., 1951, MS, 1952; PhD, N.C. State U., 1959. Geneticist, fiber scientist Nat. Cotton Coun. Am., Memphis, 1961-70; investigation leader Agrl. Rsch. Svc. USDA, Beltsville, Md., 1970-72; rsch. leader Agrl. Rsch. Svc. USDA, Knoxville, Tenn., 1972-80, New Orleans, 1980-84; br. chief Agrl. Mktg. Svc. USDA, Washington, Memphis, 1984—; convenor cotton fibers working group Internat. Standards Orgn., Geneva, Switzerland, 1965-80; chair Internat. Calibration Cotton Standards Com., Memphis, 1986—. Contbr. articles to profl. jours. Cpl. U.S. Army, 1952-54. Mem. Am. Soc. for Quality Control (sect. chair 1965-67, Featured Engr. 1967), Fiber Soc. Achievements include research in effects of variations in fiber properties on textile manufacturing and product quality, techniques for measuring fiber properties including materials for calibrating measurement instruments. Office: USDA 4841 Summer Ave Memphis TN 38122

RAMEY, HENRY JACKSON, JR., petroleum engineering educator; b. Pitts., Nov. 30, 1925; married, 1948; 4 children. B.S., Purdue U., 1949, Ph.D., 1952. Asst. chem. engr. unit ops. lab. Purdue U., 1949, asst. radiant heat transfer from gases, 1951-52; sr. research technologist petroleum prodn. research Magnolia Petroleum Co., Socony Mobil Oil Co., Inc., 1952-55; project engr. Gen. Petroleum Corp., 1955-60; staff reservoir engr. Mobil Oil Co. Div., 1960-63; mem. faculty petroleum engring. dept. Tex. A&M U., 1963-66; prof. petroleum engring. Stanford U., Calif., 1966—, chmn. dept., 1976-86; cons. Chinese Petroleum Corp., Taiwan, 1962-63, other cos. Contbr. papers, articles to profl. lit.; patentee in field. Served as officer USAAF, 1943-46. Recipient Purdue U. Disting. Engr. award, 1975, Mineral Edn. award AIME, 1987. Mem. Am. Inst. Chem. Engrs., AIME (Ferguson medal 1959), Nat. Acad. Engring. Office: Stanford U Dept Petroleum Engring Stanford CA 94305

RAMEY, JAMES MELTON, chemist; b. Waco, Tex., Sept. 1, 1928; s. Ernest Sylvester and Audrey Lee (McCasland) R.; m. Frankie Jo Montomery, Aug. 23, 1952; children: Marlana Ramey Valdez, James Monte Ramey, Douglas Dwain Ramey, Susan Elizabeth Ramey Attebery, Angela Dawn Ramey. BS, Southwestern U., 1949. Cert. product safety profl., profl. source for safety cons. Chemist Celanese Chem. Co., Bishop, Tex., 1954-57, group leader, spl. problems lab., 1957-60, dir. spl. problems lab., 1960-67; dir. quality assurance to dir. product standards Celanese Chem. Co., N.Y.C., 1967-76, dir. indsl. hygiene and toxicology, 1976-78; corp. dir. of product safety Celanese Corp., N.Y.C., 1978-83, dir. safety, 1983-85, dir. environ. health and safety audit program, 1985-87; dir. environ. health and safety audit programs Hoechst Celanese Corp., Somerville, N.J., 1987-88; cons. environ. health scis. Horseshoe Bay, Tex., 1988—; chmn. bd. Formaldehyde Inst., Washington, 1978-82, chmn. OSHA com. SOCMA, Washington, 1978-84; mem. environ. audit roundtable, Washington, 1985-88. Author: (with others) Encyclopedia of Chemical Process and Design, 1989; contbr. articles to profl. jours.; patents in field. Adv. Jesse Owens Found., N.Y., 1982-88; del. Tex. Dem. Conv., Tex., 1964-68; v.p. Internat. Amateur Athletic Assn., N.Y., 1982-88; bd. dirs. Property Owners Assn., Horseshoe Bay, 1991-92; mem. Horseshoe Bay Chapel Choir, 1992; mem. PLAN. Mem. Bishop (Tex.) C. of C. Methodist. Achievements include development of health surveillance and info. systems for Celanese Chem. Co., one of first comprehensive product safety programs in chemical industry and charter mem. of product safety forum, one of first comprehensive environ. health audit programs. Home and Office: PO Box 8634 Horseshoe Bay TX 78654

RAMEZAN, MASSOOD, mechanical engineer; b. May 9, 1956; m. Sara, Jan. 24, 1987. BS, W.Va. U., 1977, MS, 1979, PhD, 1984. Registered profl. engr., Pa., W.Va. Rsch. engr. W.Va. U., Mortantown, 1977-79, rsch. fellow, 1981-84, asst. prof., 1984-86; cons. engr. Hosp. Utility, Inc., Pitts., 1979-81; rsch. engr. METC, Morgantown, 1986-88; sr. principal engr. Burns & Roe, Pitts., 1988—. Contbr. articles to profl. jours. Fellow ASME. Office: Burns & Roe PO Box 18288 Pittsburgh PA 15236

RAMHARACK, ROOPRAM, research scientist; b. Berbice, Guyana, Aug. 26, 1952; m. Sita Ramdat, Dec. 30, 1982; children: Jeremy, Kenny, Sharon. BS, U. Guyana, 1976; PhD, Polytechnic U., 1984. Researcher Cornell Med. Ctr., N.Y.C., 1979-82; rsch. specialist 3M, St. Paul, Minn., 1984-87; rsch. scientist Polaroid, Boston, 1987-89; rsch. assoc. Nat. Starch, Bridgewater, N.J., 1989—. Contbr. articles to profl. jours. Fundraiser Com. for Democracy in Guyana, 1990-92. Mem. Am. Chem. Soc., Soc. Rheology, Am. Phys. Soc. Office: Nat Starch & Chem Co 10 Finderne Ave Bridgewater NJ 08807

RAMIREZ, JULIO JESUS, neuroscientist; b. Bridgeport, Conn., Dec. 25, 1955; s. Julio Pastor and Elia Rosa (Cortes) R. BS in Psychology magna cum laude, Fairfield U., 1977; MA, Clark U., 1980, PhD in Biopsychology, 1983; postgrad., MIT, 1985-86. Asst. prof. Coll. of St. Benedict, St. Joseph, Minn., 1981-85; vis. scientist MIT, Cambridge, Mass., 1985-86; asst. prof. Davidson (N.C.) Coll., 1986-89, assoc. prof. dept. psychology, 1989—; cons. dept. neurosci. U. Va., Charlottesville, Va., 1983-92; vis. scientist Centre Nationale de Recherche Scientifique, Strasbourg, France, summer 1982, Ludwig-Maximilians-Universitat, Munich, summer 1988, Yale U., New Haven, spring 1991, U. Va., Charlottesville, summers 1983-92; panelist NSF, 1991—, NIMH, 1992—; chmn., co-founder Faculty for Undergrad. Neurosci., 1981—. Contbr. articles to profl. publs. Mem. Habitat for Humanity, Davidson, 1989—, Union Concerned Scientist, Cambridge, 1987—; mem. adv. action com. Project Kaleidoscope, Washington, 1993—. Named N.C. Prof. of Yr., Nat. Gold Medalist Coun. for Advancement and Support of Edn., 1989, MacArthur Asst. Prof., 1986-89; recipient rsch. award Nat. Inst. Mental Health, 1992, NSF, 1991, Nat. Inst. Neurol. Disorders and Stroke, 1987, 93, N.C. Bd. Sci. & Tech., 1987. Mem. AAAS, APS, Coun. on Undergrad Rsch. (councillor 1992—), Soc. Neuroscience. Democrat. Achievements include demonstration that ganglioside administration facilitates recovery from central nervous system injury. Home: PO Box 26 Davidson NC 28036 Office: Davidson Coll Dept Psychology Davidson NC 28036

RAMIREZ CANCEL, CARLOS MANUEL, psychologist, educator; b. Mayaguez, P.R., Nov. 10, 1944; s. Carlos M. Ramirez-Forestier and Celia Cancel Ramirez; m. Ana Isobel Martinez, Dec. 20, 1969; children: Ana Isabel, Marie Claire, Carlos Javier. BA in Psychology, U. P.R., 1969; MA in Psychology, Inter-Am. U., 1971; PhD in Psychology, Tex. A&M U., 1975. Counselor U. P.R., Mayaguez, 1969-72; teaching asst. Tex. A&M U., College Station, 1972-75; psychologist, assoc. prof. U. P.R., Mayaguez, 1975-86; part-time prof. U. P.R. Med. Sci. Campus, Mayaguez, 1977—; psychologist Mental Health Ctr. Hosp., Rincon, P.R., 1976—; pvt. practice Mayaguez, 1976—; psychologist S.S. Disability Determination Program, San Juan, 1976—; evaluations psychologist P.R. Dept. Edn., San Juan, 1976—, Dept. Social Svcs., San Juan, 1976—. Active mem. Statehood for P.R., 1976—, Citizens for Animal Protection, 1976—; active PTA, Immaculate Conception Sch., 1989-91; bd. dirs. southwestern Sch., Mayaguez, 1976-78. Mem. APA, Alliance Francaise d'Mayaguez (pres. 1991—). Republican. Roman Catholic. Office: PO Box 426 Mayaguez PR 00681

RAMIREZ-GARCIA, EDUARDO AGUSTIN, biomedical engineer; b. Mexico City, Mex., July 14, 1950; arrived in Norway, 1980; s. Eduardo Ramirez Varela and Carmen (Garcia) Lopez. Mech. and Elec. Engring., U. Nacional Mex., Mexico City, 1975; M, Toronto U., Can., 1976; PhD, Strath Clyde U., Scotland, 1980. Rsch. asst. Inst. Engring. U. Nacional Mex., 1975-76, head instrumentation and svc. dept. Inst. Biomed. Rsch., 1977; supr. rsch. project Toronto U., 1976; postdoctoral rsch. fellow Sch. Medicine U. Trondheim, Norway, 1980-81; main egr. Haukeland Hosp., Bergen, Norway, 1982; mng. and chief engr. Med. Rsch. Ctr. U. Bergen, 1983—; cons. Bergen City Chambers, 1988—, Mexican Embassy in Oslo, Norway, 1991. Editor bull. Med-Net, 1990; coord. bull. Net-ARt, 1991. Del. Internat. sect. Norwegian Red Cross, Oslo, Norway, 1986—; vol. Red Cross Bergen, 1987—. Wisth Mexican mil., 1966-67. Mem. N.Y. Acad. Scis., The Planetary Soc., Norwegian Image Processing and Pattern Recognition, Internat. Soc. for Arts, Sci. and Tech., Verteft Computer Club for Youth (founder). Avocations: painting, drawing, photography, computer art. Office: U Bergen Med Rsch Ctr, Haukeland U Hosp, Bergen N0-5021, Norway

RAMÍREZ-RONDA, CARLOS HÉCTOR, physician; b. Mayaguez, P.R., Jan. 24, 1943; s. Carlos Manuel Ramirez-Silva and Flor de Maria Ronda-Davila; m. Crimilda Ramirez-Pérez, Aug. 17, 1963; children: Carlos, Ivan A. BSM, Northwestern U., 1964, MD, 1967. Intern Northwestern U. Med. Ctr., Chgo., 1967-68; resident and cardiology fellow U. P.R. Sch. Medicine, San Juan, 1968-73; dir. infectious diseases rsch. San Juan (P.R.) VA Med. Ctr., 1975—, assoc. chief of staff for R&D, 1975-89, chief dept. medicine, 1989—; dir. infectious disease U. P.R. Sch. Medicine, San Juan, 1975—, prof. medicine, 1985—; dir rsch. program San Juan AIDS Inst., 1989—; prof. microbiology U. Caribbean Med. Sch., Bayamon, P.R., 1991—; mem. adv. com. immunization practices, Atlanta, 1991—; cons. Harvard Inst. for Internat. Devel., Cambridge, 1989—. Contbr. over 150 sci. papers to profl. publs., chpts. to books; editorial bd. Infectious Disease in Clin. Practice jour., 1991—. Maj. USAF, 1969-71. NIH fellow, 1973-75; recipient award P.R. Soc. Microbiology, 1989. Fellow ACP (pres. P.R. chpt. 1980-82), Infectious Disease Soc. Am.; mem. P.r. Med. Assn. (chmn. sci. com., editor jour. 1977-83), So. Soc. Clin. Investigation (coun. 1976-79). Home: 81 Mirador St Paseo Alto Rio Piedras PR 00926 Office: San Juan VA Med Ctr Dept Medicine One Veterans Plaza III San Juan PR 00927-5800

RAMO, SIMON, engineering executive; b. Salt Lake City, May 7, 1913; s. Benjamin and Clara (Trestman) R.; m. Virginia Smith, July 25, 1937; children: James Brian, Alan Martin. BS, U. Utah, 1933, DSc (hon.), 1961; PhD, Calif. Inst. Tech., 1936; DEng (hon.), Case Western Res. U., 1960, U. Mich., 1966, Poly. Inst. N.Y., 1971; DSc (hon.), Union Coll., 1963, Worcester Polytechnic Inst., 1968, U. Akron, 1969, Cleve. State U., 1976; LLD (hon.), Carnegie-Mellon U., 1970, U. So. Calif., 1972, Gonzaga U., 1983, Occidental Coll., 1984, Claremont U., 1985. With Gen. Electric Co., 1936-46; v.p. ops. Hughes Aircraft Co., 1946-53; with Ramo-Woolridge Corp., 1953-58; dir. U.S. Intercontinental ballistic missile program, 1954-58; dir. TRW Inc., 1954-85, exec. v.p., 1958-61, vice chmn. bd., 1961-78, chmn. exec. com., 1969-78, cons., 1978—; pres. The Bunker-Ramo Corp., 1964-66; chmn. bd. TRW-Fujitsu Co., 1980-83; bd. dirs. Arco Power Techs.; vis. prof. mgmt. sci. Calif. Inst. Tech., 1978—; Regents lectr. UCLA, 1981-82, U. Calif. at Santa Cruz, 1978-79; chmn. Center for Study Am. Experience, U. So. Calif., 1978-80; Faculty fellow John F. Kennedy Sch. Govt., Harvard U., 1980-84; mem. White House Energy Research and Devel. Adv. Council, 1973-75; mem. adv. com. on sci. and fgn. affairs U.S. State Dept., 1973-75; chmn. Pres.'s Com. on Sci. and Tech., 1976-77; mem. adv. council to Sec. Commerce, 1976-77, Gen. Atomics Corp., 1988—, Aurora Capital Ptnrs., 1991—, Chartwell Investments, 1992—; co-chmn. Transition Task Force on Sci. and Tech. for Pres.-elect Reagan; mem. roster consultants to adminstr. ERDA, 1976-77; bd. advisors for sci. and tech. Republic of China, 1981-84; chmn. bd. Aetna, Jacobs & Ramo Venture Capital, 1987-90, Allenwood Ventures Inc., 1987—. Author: The Business of Science, 1988, other sci., engring. and mgmt. books. Bd. dirs. L.A. World Affairs Coun. 1973-85, Mus. Ctr. Found., L.A., L.A. Philharm. Assn., 1981-84; life trustee Calif. Inst. Tech., Nat. Symphony Orch. Assn., 1973-83; trustee emeritus Calif. State Univs.; bd. visitors UCLA Sch. Medicine, 1980—; bd. dirs. W.M. Keck Found., 1983—; bd. govs. Performing Arts Coun. Mus. Ctr. L.A., pres., 1976-77. Recipient award IAS, 1956; award Am. Inst. Elec. Engrs., 1959; award Arnold Air Soc., 1960; Am. Acad. Achievement award, 1964; award Am. Iron and Steel Inst., 1968; Disting. Svc. medal Armed Forces Communication and Electronics Assn., 1970; medal of achievement WEMA, 1970; awards U. So. Calif., 1971, 79; Kayan medal Columbia U., 1972; award Am. Cons. Engrs. Coun., 1974; medal Franklin Inst., 1978; award Harvard Bus. Sch. Assn., 1979; award Nat. Medal Sci., 1979; Disting. Alumnus award U. Utah, 1981; UCLA medal, 1982; Presdl. Medal of Freedom, 1983; named to Bus. Hall of Fame, 1984; recipient Aesculapian award UCLA, 1984, Durand medal AAIA, 1984, John Fritz medal, 1986, Henry Townley Heald award Ill. Inst. Tech., 1988, Nat. Engring. award Am. Assn. Engring. Socs., 1988, Franklin-Jefferson medal, 1988, Howard Hughes Meml. award, 1989. Fellow IEEE (Electronic Achievement award 1953, Golden Omega award 1975, Founders medal 1980, Centennial medal 1984), Am. Acad. Arts and Scis.; mem. N.Y. Acad. of Sci., Nat. Acad. Engring. (founder, coun. mem. Bueche award), Nat. Acad. Scis., Am. Phys. Soc., Am. Philos. Soc., Inst. Advancement Engring., Internat. Acad. Astronautics, Eta Kappa Mu (eminent mem. award 1966). Office: 9200 Sunset Blvd Ste 801 Los Angeles CA 90069-3506

RAMOS, ANGEL SALVADOR, veterinarian; b. Dingras, Philippines; s. Angel Sr. and Francisca (Salvador) R. DVM, U. of the Philippines, Quezon City, Philippines, 1965; MSc, Ontario Vet. Coll. U. Guelph, Guelph, Ontario, 1972; PhD, Tex. A&M U., 1975; MPH, Johns Hopkins U., 1988. Lic. med. technologist. Rsch. asst., instr. Tex. A&M U., College Station, 1972-75; rsch. fellow, instr. Harvard Med. Sch., Boston, 1975-77; rsch. fellow WHO Endocrine Rsch. Unit, Karolinska Inst., Stockholm, Sweden, 1977-79; asst. prof. vet. sci. U. Nebr., Lincoln, 1978-80; prof. histology King Fiasal U., Al-Hassa, Saudi Arabia, 1980-85; field vet. med. officer USDA, Miami, 1988-90; staff vet. med. officer USDA, Hyattsville, Md., 1990—. Contbr. articles to profl. jours. Named Most Outstanding Sr. Clinican, U. Philippines, 1965, Vet. Sci. scholar U. Philippines, 1963-65, Can. Med. Rsch. Coun. fellow, 1972-77, Rsch. fellow Harvard U., 1975, WHO Postdoctoral fellow, 1972-77. Mem. AAAS, Am. Vet. Med. Assn., Nat. Assn. Fed. Vets. Democrat. Roman Catholic. Home: 8306 Greg Marc St Laurel MD 20708

RAMOS-LEDON, LEANDRO JUAN, biologist; b. Havana, May 3, 1924; came to the U.S., 1962; s. Leopoldo and Elena (Ledon) R.; m. Josefina Moragues, Nov. 12, 1954; children: Josie C., Leandro J. Jr. BS in Chemistry, DSc in Botany, U. Havana, 1947; MS in Biology, U. Miami, 1970; PhD, Fla. Internat. U., 1987. Asst. prof. botany Agrl. Experiment Sta., Santiago de las Vegas, Cuba, 1948-59, assoc. prof., dir., 1959-60, prof., chmn. biochemistry, 1960-62; rsch. assoc. biology U. So. Calif., L.A., 1962-66; biologist U. Fla., Homestead, 1966-87, sr. scientist molecular biology, 1987—; adj. asst. prof. biology Miami-Dade (Fla.) Coll., 1975-81. Contbr. articles to Agrotecnia, Phytopathology, Plant Disease, Plant Pathology; contbr. chpts.: Cuban Agricultural Development, 1992. Active St. Vincent of Paul, Miami, 1992. Mem. Am. Soc. Plant Physiologists, AAAS, Cuban Assn. Engrs. and Scientists (bd. dirs. 1970-92, Julian Acuna Rsch. award 1987), Sigma Xi. Roman Catholic. Achievements include co-discovery of viral nature and vector Sogatodes Orizicola of Rice Hoja Blanca disease; developed new tomato cultivar Flora-dade for Florida, identified Botryosphaeria Ribis causing mango dieback; research in immunohistochemical localization of proteinases in Ipomoea Batatas, chloroplast DNA systematic of Cucurbita species; anatomy of kenaf (Hibiscus cannabinus). Home: 8220 SW 163rd St Miami FL 33157 Office: U Fla Tropical Rsch 18905 SW 280th St Homestead FL 33031-3314

RAMPINO, MICHAEL R., earth scientist; b. Bklyn., Feb. 8, 1948; s. Michael A. and Annette (Cohen) R. BA, Hunter Coll., 1968; PhD, Columbia U., 1978. Rsch. assoc. Goddard Inst. for Space Studies NASA, N.Y.C., 1978-85; asst. prof. dept. applied sci. NYU, 1985-92, assoc. prof. and chair dept. applied sci., 1992—, now chair dept. earth system sci.; cons. Goddard Inst. for Space Studies NASA, N.Y.C., 1985—. Editor: Climate: History, Periodicity, and Predictability, 1987; contbr. articles to profl. jours. Rsch. grantee NASA, Am. Philos. Soc., U.S. Dept. Energy, NSF. Mem. Geol. Soc. Am., Am. Geophys. Union, Am. Assn. Sedimentologists, Am. Assn. Geology Tchrs., N.Y. Acad. Scis. (chmn. geol. scis. sect. 1990-91). Achievements include research in galactic carrousel theory of comet showers and mass extinctions, effects of volcanic eruptions on climate, periodicity in the geologic record. Home: 25 Union Square W PHF New York NY 10003

RAMSAHOYE, LYTTLETON ESTIL, geophysicist, consultant, educator; b. Fellowship, Demerara, Guyana, Aug. 11, 1930; arrived in Barbados, 1978; s. Edward Jairad and Wilhelmina Alverna (Fenton) R.; m. Elizabeth Mariai Kerry, June 2, 1962; children: Debra, Carol. BSc in Physics with honours, King's Coll., London, 1954; PhD in Geophysics, Imperial Coll., London, 1957; diploma in pub. health engring., U. Minn., 1959. Geophysicist, hydrologist Geol. Survey Guyana, 1957-59; chief Pure Water Supply, Guyana, 1960-62; prof. physics U. Guyana, 1963-78, dean natural scis., dep. vice chancellor, 1964-78; rsch. dir. Caribbean Meteorol. Inst., Christchurch, Barbados, 1978-83; ret., 1983; cons. hydrologist Aubrey Barker Assocs., Guyana, 1963-78. Contbr. articles to profl. jours.; patentee for cement extender. Vol. Queen Elizabeth Hosp. Lab., Barbados, 1987—; mem. parliament Guyana, 1973-78. Scholar Govt. of Guyana, 1949; fellow Pan Am. Health Orgn., 1960. Home: Cardomar, Maxwell Main Rd, Christchurch Barbados

RAMSAY, DONALD ALLAN, physical chemist; b. London, July 11, 1922; s. Norman and Thirza Elizabeth (Beckley) R.; m. Nancy Brayshaw, June 8,

1946; children: Shirley Margaret, Wendy Kathleen, Catharine Jean, Linda Mary. BA, Cambridge (Eng.) U., 1943, MA, 1947, PhD, 1947, ScD, 1976; D honoris causa, U. Reims, France, 1969; Filosofie hedersdoktor, U. Stockholm, Sweden, 1982. With divs. chemistry Nat. Research Council Can., Ottawa, Ont., 1947-49; with divs. physics Nat. Research Council Can., 1949-75; with Herzberg Inst. Astrophysics, 1975—, sr. research officer, 1961-68, prin. research officer, 1968-87; vis. prof. U. Minn., 1964, U. Orsay, 1966, U. Stockholm, 1967, 71, 74, U. Calif., Irvine, 1970, U. Sao Paulo, 1972, 78, U. Bologna, 1973, U. Western Australia, 1976, Australian Nat. U., 1976, U. Canterbury, Christchurch, New Zealand, 1991, U. Ulm, Germany, 1992. Editor: (with J. Hinze) Selected Works of Robert S. Mulliken, 1975; contbr. numerous articles on molecular spectra and molecular structure to profl. jours. Recipient commemorative medal for 125th anniversary Confederation Can, 1992; decorated Queen Elizabeth Silver Jubilee medal. Fellow Royal Soc. London, Royal Soc. Can. (hon. treas. 1976-79, 88-91, Centennial medal 1982), Am. Phys. Soc., Chem. Inst. Can. (Chem. Inst. Can. medal 1992); mem. Can. Assn. Physicists. Mem. United Ch. of Canada (organist 1954—). Club: Leander (Henley-on-Thames, Eng.). Home: 1578 Drake Ave, Ottawa, ON Canada K1G 0L8 Office: Nat Research Council, 100 Sussex Dr, Ottawa, ON Canada K1A 0R6

RAMSBACHER, SCOTT BLANE, civil engineer; b. Collbarn, Colo., May 31, 1960; s. Richard Joseph and Marcella Ruth (Feist) R. AS, Victoria Coll., 1980; BSCE, Tex. A&M U., 1985. Project engr. P.A.W.A. Winkelmann Inc., Dallas, 1985-87, Campbell Taggart Inc., Dallas, 1987—. Mem. ASCE, ASME. Lutheran. Home: 909 Castleglen Dr Garland TX 75043 Office: Campbell Taggart Inc 6206 Peeler St Dallas TX 75235

RAMSDELL, RICHARD ADONIRAM, marine engineer; b. Hartford, Conn., Feb. 28, 1953; s. Robert Allen and Irene Ella (Lewis) R.; m. Vicki Lynn Pepin, July 1, 1978 (div. Mar. 1984); children: Eric Charles, Ryan Amber, Alexander Richard. BS in Marine Engring., Maine Maritime Acad., 1975. Plant operator Ga. Pacific, Woodland, Maine, 1975-77; 2d asst. engr. Farrell Lines, Inc., N.Y.C., 1977-83; steam plant foreman Jackson Lab., Bar Harbor, Maine, 1984-86; plant operator Babcock-Ultrapower Jonesboro, Maine, 1986-90; results, environ. engr. Maine Power Svcs., Bangor, 1990-92; plant engr. Babcock-Ultrapower West Enfield, Maine, 1992—. Mem. Assn. Energy Engrs. Office: Babcock Ultrapower West Enfield Rt 2 PO Box 317 West Enfield ME 04493

RAMSEY, BONNIE JEANNE, mental health facility administrator, psychiatrist; b. Tucson, Dec. 9, 1952; d. William Arnold Jr. and Doris Marie (Gaines) R. BS cum laude, U. S.C., 1971-75, MD, 1981. Diplomate Am. Bd. Psychiatry and Neurology; lic. child and adult psychiatrist S.C., N.C., Ga. Chief resident in psychiatry William S. Hall Psychiat. Inst., Columbia, S.C., 1983, unit dir. adolescent girls, 1986-89; chief child and adolescent in-patient program William S. Hall Psychiat. Inst., Columbia, 1989—, interim dir. child and adolescent div., 1989-92; interim chmn. child and adolescent div. dept. neuropsychiatry U. S.C., Columbia, 1989-92; instr. Sch. of Medicine U. S.C., Columbia, 1986-89, asst. prof. Sch. of Medicine, 1989—. Mem. choir Trinity Meth. Ch., West Columbia, 1981—, vice chmn. bd. trustees, 1989—, trustee, 1990—, mem. at large adminstrn. bd., 1993—; adv. coun. Habitat for Humanity. Named one of Outstanding Young Women of Am., 1985. Mem. AMA (del. residents physician sect. 1983, 84, 86, housing staff sect. 1988—), Am. Psychiat. Assn. (local sec.-treas., pres., 1981—), Am. Acad. Child Psychiatry, S.C. Med. Soc., Columbia Med. Soc., Palmetto Club. Methodist. Office: William S Hall Psychiat Inst PO Box 202 Columbia SC 29202-0202

RAMSEY, GLENN EUGENE, pathologist, blood bank physician; b. Newark, Ohio, July 17, 1953; s. Roy Eugene and Jean (Little) R.; m. Rosalind Goldman, June 29, 1975; children: Ethan G., Caitlin Rose. BA magna cum laude, Case Western Res. U., 1974, MD, 1978. Diplomate Am. Bd. Anatomic and Clin. Pathology, Am. Bd. Blood Banking. Intern U. Rochester, N.Y., 1978-79, resident in pathology, 1979-83; blood bank fellow U. Pitts., Cen. Blood Bank, 1983-84; dir. immunohematology Cen. Blood Bank, Pitts., 1984-91; asst. prof. pathology U. Pitts., 1985-91, Northwestern U., Chgo., 1991—; med. dir. United Blood Svcs., Chgo., 1991—; mem. med. staff Northwestern Meml. Hosp.; Va. Lakeside Med. Ctr.; invited lectr. Hong Kong Soc. Haematology, 1989, North London Blood Transfusion Ctr., 1992, Internat. Coll. Surgeons, Chgo., 1992. Author: Standards for Blood Banks and Transfusion Services, 1991, 93; contbr. chpt. to book, Basic Science Review for Surgeons, 1992; contbr. articles to New Eng. Jour. Medicine, Jour. AMA, others. Grantee NIH, 1985-88, Blood Systems Rsch. Found., 1991—. Mem. Am. Assn. Blood Banks (transfusion practice com. 1987-89, abstract selection com. 1989-93, stds. com. 1990—), Am. Soc. Clin. Pathology (coun. on transfusion medicine 1993—), Internat. Soc. Blood Transfusion, Am. Soc. Hematology, Coll. Am. Pathologists, Licking County (Ohio) Geneal. Soc., Nature Conservancy. Achievements include research in immunohematology and organ transplantation, pioneering descriptions of red cell antibodies arising from liver transplants, Rh antibodies from kidney transplants, study of red cell antibody problems. Home: 969 Spruce St Winnetka IL 60093 Office: Northwestern Meml Hosp 303 E Superior St Chicago IL 60611

RAMSEY, LELAND JAY, clinical electrical engineer; b. Shelbyville, Ind., June 25, 1956; s. Buford L. and June E. (Carman) R.; m. Martha H. Wilson, Aug. 30, 1986; children: Aaron C., Alan W. BSEE, Purdue U., 1978. Clin. engr. Cleve. Clinic Found., 1978—. Mem. North Coast Clin. Engring. Tech. Assn. Office: Cleve Clinic Found 9500 Euclid Ave Cleveland OH 44195

RAMSEY, NORMAN F., physicist, educator; b. Washington, Aug. 27, 1915; s. Norman F. and Minna (Bauer) R.; m. Elinor Jameson, June 3, 1940 (dec. Dec. 1983); children: Margaret, Patricia, Janet, Winifred; m. Ellie Welch, May 11, 1985. AB, Columbia U., 1935; BA, Cambridge (Eng.) U., 1937, MA, 1941, DSc, 1954; PhD, Columbia U., 1940; MA (hon.), Harvard U., 1947; DSc (hon.), Case Western Res. U., 1968, Middlebury Coll., 1969, Oxford (Eng.) U., 1973; DCL (hon.), Oxford (Eng.) U., 1990; DSc (hon.), Rockefeller U., 1986, U. Chgo., 1989, U. Sussex, 1990, U. Houston, 1991, Carleton Coll., 1991, Lake Forest Coll., 1992, U. Mich., 1993. Kellett fellow Columbia U., 1935-37, Tyndall fellow, 1938-39; Carnegie fellow Carnegie Inst. Washington, 1939-40; assoc. U. Ill., 1940-42; asst. prof. Columbia U., 1942-46; assoc. MIT Radiation Lab., 1940-43; cons. Nat. Def. Research Com., 1940-45; expert cons. sec. of war, 1942-45; group leader, asso. div. head Los Alamos Lab., 1943-45; assoc. prof. Columbia U., 1945-47; head physics dept. Brookhaven Nat. Lab. of AEC, 1946-47; assoc. prof. physics Harvard U., 1947-50, prof. physics, 1950-66, Higgins prof. physics, 1966—; sr. fellow Harvard Soc. of Fellows, 1970—; Eastman prof. Oxford U., 1973-74; Luce prof. cosmology Mt. Holyoke Coll., 1982-83; prof. U. Va., 1983-84; dir. Harvard Nuclear Lab., 1948-50, 52-53, Varlan Assos., 1963-66; mem. Air Forces Sci. Adv. Com., 1947-54; sci. adviser NATO, 1958-59; mem. Dept. Def. Panel Atomic Energy; exec. com. Cambridge Electron Accelerator and gen. adv. com. AEC. Author: Nuclear Moments and Statistics, 1953, Nuclear Two Body Problems, 1953, Molecular Beams, 1956, 85, Quick Calculus, 1965; contbr.: articles Phys. Rev.; other sci. jours. on nuclear physics, molecular beam experiments, radar, nuclear magnetic moments, radiofrequency spectroscopy, masers, nucleon scattering. Trustee Asso. Univs., Inc., Brookhaven Nat. Lab., Carnegie Endowment Internat. Peace, 1962-85, Rockefeller U., 1977-90; pres. Univs. Research Assocs., Inc., 1966-72, 73-81, pres. emeritus, 1981—. Recipient Presdl. Order of Merit for radar devel. work, 1947, E.O. Lawrence award AEC, 1960, Columbia award for excellence in sci., 1980, medal of honor IEEE, 1983, Rabi prize, 1985, Monie Ferst award, 1985, Compton medal, 1985, Rumford premium, 1985, Oersted medal, 1988, Nat. medal of sci., 1988, Nobel Prize for physics, 1989, Pupin medal Columbia Engring. Sch. Alumni Assn., 1992, Einstein award, 1993; Guggenheim fellow Oxford U., 1954-55. Fellow Am. Acad. Sci., Am. Phys. Soc. (coun. 1956-60, pres. 1978-79, Davisson-Germer prize 1974); mem. NAS, French Acad. Sci., Am. Philos. Assn., AAAS (chmn. physics sect. 1977), Am. Inst. Physics (chmn. bd. govs. 1980-87), Phi Beta Kappa (senator 1979—, v.p. 1982-85, pres. 1985-88), Sigma Xi. Home: 24 Monmouth Ct Brookline MA 02146-5634 Office: Harvard U Lyman Lab Cambridge MA 02138

RAMSEY, WILLIAM DALE, JR., petroleum company executive; b. Indpls., Apr. 14, 1936; s. William Dale and Laura Jane (Stout) R.; m. Mary Alice Ihnet, Aug. 9, 1969; children: Robin, Scott, Kimberly, Jennifer. AB in Econs., Bowdoin Coll., 1958. With Shell Oil Co., 1958—, salesman, Albany, N.Y., 1960, merchandising rep., Milton, N.Y., 1961-63, real estate and mktg. investments rep., Jacksonville, Fla., 1963-65, dist. sales supr., St. Paul, 1965-67, employee relations rep., Chgo., 1967-69, spl. assignment mktg. staff-adminstrn., N.Y.C., recruitment mgr., Chgo., 1970-72, sales mgr., Chgo., 1973-75, sales mgr., Detroit, 1975-79, dist. mgr. N.J. and Pa., Newark, 1979-84, Mid-Atlantic dist. mgr. (Md., D.C., Va.) 1984-87, econ. advisor head office, Houston, 1987-89; mgr. Mktg. Concepts head office, Houston, 1989—; dir. N.Am. Fin. Services, 1971-72; lectr., speaker on energy, radio, TV, appearances, 1972—; guest lectr. on bus. five univs., 1967-72; v.p., dir. Malibu East Corp., 1973-74; prin. Robotics Rsch. Consortium, 1991—; mem. Am. Right of Way Assn., 1963-65. James Bowdoin scholar Bowdoin Coll., 1958. Active Chgo. Urban League, 1971-75; mem. program com., bus. adv. council Nat. Rep. Congl. Com.; mem. Gov.'s Council on Tourism and Commerce, Minn., 1965-67; mem. Founders Soc., Detroit Inst. Arts, 1978-80; bd. dirs. N.J. Symphony Orch. Corp., 1981-85. Capt. U.S. Army, 1958-60. Mem. Soc. Environ. Econ. Devel., Internat. Svc. Robot Assn. N.J. Petroleum Council (exec. com. 1979-84 vice chmn. 1982-84), Midwest Coll. Placement Assn., Md. Petroleum Council (exec. com. 1984-87). Presbyterian. Clubs: Ponte Vedra (Fla.); Bowdoin Alumni (Houston); Morris County (N.J.) Golf; Kingwood (Tex.) Country; Bethesda (Md.) Country. Author: Corp. Recruitment and Employee Relations Organizational Effectiveness Study, 1969.

RAMSHAW, JOHN DAVID, chemical physicist, mechanical engineer; b. Salt Lake City, Mar. 20, 1944; s. William Edwin and Margaret Louise (Park) R.; m. Jean Tomer, Sept. 7, 1968 (div. 1988); children: Michael John, David Scott. BS, Coll. of Idaho, 1965; PhD, MIT, 1970. Postdoctoral assoc. U. Md., College Park, 1970-71; rsch. assoc., instr. U. Utah, Salt Lake City, 1971-72; mem. tech. staff Applied Theory, Inc., L.A., 1972-73; assoc. scientist Aerojet Nuclear Co., Idaho Falls, 1973-75; mem. staff Los Alamos (N.Mex.) Nat. Lab., 1975-86; sci. and engring. fellow Idaho Nat. Engring. Lab., Idaho Falls, 1986—. Contbr. over 60 articles to profl. jours. Fellow NSF, 1966-70, Air Force Office of Sci. Rsch./Nat. Rsch. Coun., 1970-71. Atomic ENergy Commn., 1965-66. Mem. Am. Phys. Soc., Soc. for Indsl. and Applied Math., Am. Nuclear Soc. (Idaho sect.). Achievements include original contributions to equilibrium and nonequilibrium statistical physics, esp. theory of dielectric fluids, and to analytical and computational fluid dynamics, esp. multicomponent flows with chemical reactions. Office: Idaho Nat Engring Lab PO Box 1625 Idaho Falls ID 83415

RAN, CHONGWEI, geotechnical engineer; b. Chongqing, Sichuan, China, Apr. 16, 1956; came to U.S., 1988; s. Binqing and Mingyu (Zhou) R.; m. Yutie Dong, Sept. 21, 1986; children; Young, Lydia. BS, Chongqing U., 1981; MS, U. Ariz., 1990, PhD, 1993. Registered profl. engr. Engr. Chem. Mines Design and Rsch. Inst. of China, Lianyungang, Jiangsu, China, 1982-88; rsch. assoc. U. Ariz., Tucson, 1988-92; cons. Ian Farmer & Ptnrs. Ltd., Tucson, 1991—. Contbr. tech. reports to profl. publs. Mem. ASCE, Chinese Soc. Rock Mechanics, Chinese Soc. Chem. Engrs. Achievements include development of Bentonite Grouting Technology, underground structural analysis software. Home: 806 E 10th St Tucson AZ 85719 Office: Ian Farmer & Ptnrs Inc PO Box 44302 Tucson AZ 85733

RAN, JOSEF, electronics executive, engineer; b. Linz, Austria, Israel, Oct. 22, 1947; s. Izak and Bronia (Wilf) Rand; m. Edith Ruth Buchwald; children: Sharon Hela, Eyal Salomon. BS in Indsl. Engring., Technion, Haifa, Israel. V.p. programs Elbit Corp., Israel, 1982-84, v.p., gen. mgr., 1984-86, v.p. ops., 1987-89; pres. Opgal Optronics, Karmiel, Israel, 1990—. Avocation: gardening. Home: Newe Oved 6, 30500 Binyamina Israel

RANCOURT, JAMES DANIEL, optical engineer. BA in Physics, Bowdoin Coll., 1963; MS in Physics, Carnegie Tech., 1965; PhD in Optical Scis., U. Ariz., 1974. Engr. Itek Corp., Lexington, Mass., 1965-69; rsch. assoc. U. Ariz., Tucson, 1969-74; engr. OCLI, Santa Rosa, Calif., 1970—. Author: Optical Thin Films Users Handbook, 1987; inventor 12 patents in field. Office: OCLI 2789 N Point Pkwy Santa Rosa CA 95407

RAND, STEPHEN COLBY, physicist; b. Seattle, Nov. 20, 1949; s. Charles Gordon and Margaret (Colby) R.; m. Paula Dian Fraser, Sept. 6; children: Spencer Fraser, Kevin Colby. BSc, McMaster U., 1972; MSc and PhD, U. Toronto, 1978. World trade postdoctoral fellow IBM, San Jose, Calif., 1978-80; rsch. assoc. dept. physics Stanford (Calif.) U., 1980-82; mem. tech. staff Hughes Rsch. Lab., Malibu, Calif., 1982-87; assoc. prof. div. applied physics U. Mich., Ann Arbor, 1987—; invited prof. Ctr. nat. de Recherche Scientifique, 1988. Contbr. over 65 articles to profl. jours. Mem. Am. Phys. Soc., Optical Soc. Am. (topical editor jour. 1992—), Materials Rsch. Soc. Achievements include patents for Solid State Laser Employing Diamond, Hybrid Optical Fibers, Polarization Preserving Optical Fiber, Continuous Wave Pair Pumped Laser, Continuous Wave Trio Pumped Laser. Home: 3725 Dines Ct Ann Arbor MI 48105 Office: Univ Mich Randall Lab Div Applied Physics Ann Arbor MI 48109

RANDALL, BARBARA ANN, computer design engineer; b. Mpls., Oct. 19, 1958; d. Brayton Dean and Phyllis Virginia (Soley) Naused; m. Kevin Courtney Randall, Nov. 19, 1988. BS in Applied Math., U. Wis., Menomonie, 1981; MSEE, U. Minn., 1988. Computer programmer Mayo Found., Rochester, Minn., 1981-83, computer architecture analyst, 1983-85, computer design engr., 1985-92; lead project engr., 1993—. Contbr. rsch. articles of high speed cirs. and systems to profl. jours. Mem. IEEE Computer Soc., Internat. Electronics Packaging Soc. Avocations: hiking, canoeing, billiards, softball, volleyball. Office: Mayo Found 200 1st St SW Rochester MN 55905-0001

RANDALL, DAVID JOHN, physiologist, zoologist, educator; b. London, Sept. 15, 1938. BSc, U. Southampton, 1960, PhD, 1963, FRSC, 1981. From asst. to assoc. prof. U. B.C., 1963-73, prof. zoology, 1973—, assoc. dean grad. studies, 1990—; vis. lectr. Bristol U., 1968-69; vis. sci. Marine Labs U. Tex., 1970, Zool. Sta., Naples, Italy, 1973; NATO vis. sci. Acadia U., 1975, Marine Lab U. Tex., 1977; chief sci. Alpha Helix Amazon Expedition, 1976; mem. adv. bd. J. Comp Physiology, 1977—, J. Exp. Biol., 1981-84; chmn. animal biol. comt. Nat. Res. Coun., Can., 1974. Assoc. editor: Marine Behavior Physiology. Guggenheim Found. fellow, 1968-69. Fellow Royal Soc. Can.; mem. Can. Soc. Zoologists (Fry medal 1993), Soc. Exp. Biologists. Office: U BC, Dept Zoology 6270 Unv Blvd, Vancouver, BC Canada V6T 124

RANDALL, JANET ANN, biology educator, researcher; b. Twin Falls, Idaho, July 3, 1943; d. William Franklin and Bertha Silvia (Kalousek) Orr; m. Bruce H. MacEvoy. BS, U. Idaho, 1965; MEd, U. Wash., 1969; PhD, Wash. State U., 1977. Postdoctoral fellow U. Texas, Austin, 1977-79; from asst. to assoc. prof. biology Ctrl. Mo. State U., Warrensburg, 1979-87; assoc. prof. biology San Francisco State U., 1987-92, prof., 1992—; vis. prof. Cornell U., Ithaca, N.Y., 1984-85. Contbr. 25 articles to profl. jours. Rsch. grantee Nat. Geog. Soc., 1982, 86, NSF, 1984, 87, 88-89, 89-91, 91-93, 93—. Fellow Calif. Acad. Sci.; mem. Animal Behavior Soc. (mem. at large 1986-89), Am. Soc. Zoologists (program officer), Am. Soc. Mammalogists, Internat. Soc. Behavioral Ecologists, Sigma Xi. Avocations: folk dance, backpacking, hiking. Home: 3137 Monterey St San Mateo CA 94403 Office: San Francisco State U Dept Biology San Francisco CA 94132

RANDALL, ROBERT L(EE), industrial economist; b. Aberdeen, S.D., Dec. 28, 1936; s. Harry Eugene and Juanita Alice (Barstow) R. MS in Phys. Chemistry, U. Chgo., 1960, MBA, 1963. Market devel. chemist E.I. du Pont de Nemours & Co., Inc., Wilmington, Del., 1963-65; chem. economist Battelle Meml. Inst., Columbus, Ohio, 1965-68; mgr. market and econ. rsch. Kennecott Copper Corp., N.Y.C., 1968-74, economist, 1974-79, dir. new bus. venture devel., 1979-81; pres., mng. dir. R.L. Randall Assocs., Inc., 1981—; economist U.S. Internat. Trade Commn., Washington, 1983—; founder, pres., exec. dir. The RainForest ReGeneration Inst., 1986—; indsl. panel policy rev. of effect of regulation on innovation and U.S.-internat. competition U.S. Dept. Commerce, 1980-81. Contbr. articles to profl. jours.; contbg. author: Computer Methods for the '80's. Mem. AAAS (organizer ann. meeting Tropical Forest Regeration Symposium), AIME (econs. coun., sec. mineral econ. subsect.), Am. Econ. Assn., Am. Statis. Assn., Am. Chem. Soc., Soc. Mining Engrs., Chemists Club N.Y.C., Metall. Soc., N.Y. Acad. Scis., Nat. Econs. Club Washington (sec., reporter), Assn. Environ. and Resource Economists, Internat. Soc. Ecol. Economists. Home: 1727 Massachusetts Ave NW Washington DC 20036-2153 Office: US Internat Trade Com 500 E Street SW Washington DC 20436

RANDISH, JOAN MARIE, dentist; b. Seattle, Oct. 2, 1954; d. Matthew John and Margaret Cecelia (Waham) R. DDS, U. Wash., 1981. Dentist S.E. Dental Clinic, Seattle, 1981-82, Indian Health Bd., Seattle, 1982-84, Joe Whiting Dental Clinic, Seattle, 1984-85, Georgetown Dental Clinic, Seattle, 1981-88, King County Pub. Health, Seattle, 1987—; pvt. practice dentist Bellevue, Wash., 1981—. Recipient Women of Yr. award Bellevue Bus. and Profl. Women, Seattle, 1987. Mem. ADA, Seattle King County Dental Soc., Soroptimist Club of Bellevue (del. 1983-91). Avocations: reading, water sports, movies, fgn. langs., traveling. Office: 25 102nd Ave NE Bellevue WA 98004-5622

RANDLE, RONALD EUGENE, structural engineer; b. Cheyenne, Wyo., Nov. 20, 1946; s. John Gentry and Elizabeth Joann (Page) R.; m. Patricia Ann Lord, Sept. 27, 1968; children: Michelle Leigh, Steven Ronald. BSCE, U. Wyo., 1968, MCE, 1970. Registered profl. engr., Minn., Pa. Bridge design engr. Wyo. State Hwy. Dept., Cheyenne, summer 1968, 69; design engr. Continental Pipe Line Co., Cheyenne, 1970; bridge design engr. Wyo. State Hwy. Dept., Cheyenne, 1970-72; design engr. Meyer Industries, Red Wing, Minn., 1972-74, sales engr., 1974-81, v.p., dir. engring., 1981—. Contbr. articles to profl. jours. and mags. Bd. dirs. YMCA, Red Wing, 1989-92; bd. dirs. Boy Scouts Am., Rochester, 1987-92, dist. chmn., 1987-92. Mem. ASCE (chmn. com. 1984-90, chmn. stds. com. 1990—), IEEE, Kiwanis (v.p. 1991-92, bd. dirs.). Office: Thomas & Betts 1555 Lynnfield Rd Memphis TN 38119

RANDOLPH, BRIAN WALTER, civil engineer, educator; b. Dayton, Ohio, Feb. 23, 1959; s. John Francis and Joan Mary(Botkin) R.; m. Clare Ellen Luddy, June 22, 1985; ž child, Brigid Luddy. BSCE, U. Cin., 1982, MS, 1983; PhD, Ohio State U., 1989. Registered profl. engr., Ohio. Engr. Woolpert Cons., Dayton, 1979-82; rsch. asst. U. Cin., 1982-83; rsch. assoc. Ohio State U., Columbus, 1983-87; instr. U. Toledo, 1987-89, asst. prof., 1989-93, assoc. prof. civil engring., 1993—, founding dir. Environ. Geotech. Lab., 1987—. Contbr. articles to profl. publs. including Jour. Geotech. Engring.; referee Jour. of Hazardous Materials, 1991—. Grantee Ohio Dept. Transp./FHWA, 1989, 92, GE, 1986, Sokkia Corp., 1992, Ohio Bd. Regents, 1989, NSF, 1993. Mem. ASCE, Am. Soc. Engring. Edn. (exec. bd. NEE com. 1990-91, Dow Outstanding Young Faculty award 1993), Toastmasters (edn. v.p. 1990—), Sigma Xi, Chi Epsilon, Pi Mu Epsilon. Roman Catholic. Achievements include research on reliability analysis of groundwater flow parameters, microwave aquametry of contaiminated soils and shear properties of soil and polymer interfaces. Office: Univ Toledo Dept Civil Engring 2801 W Bancroft St Toledo OH 43606-3390

RANDOLPH, LINDA JANE, mathematics educator; b. Ypsilanti, Mich., Feb. 25, 1942; d. Roy Lawrence and Sarah (Jefferson) Robinson; children: Deborah L. Bolton, Sandra A. Randolph. BS in Teaching and Math., Ea. Mich. U., 1983, M in Math., 1989, postgrad., 1989—. Math. tutor Ea. Mich. U., Ypsilanti, 1980-82, supr. adult edn., tchr., 1983-85, 83—, instr. math.; substitute tchr. Tecumseh (Mich.) Pub. Sch., 1983;, 1983-91; instr. math., program coord. UAW-FORD/EMU Milan Plastic Plant, 1991—; peer-advisor Acad. Svcs. Ctr., Ea. Mich. U., 1979-83; lecturer computer sci. dept. Ea, Mich. U., 1986-89, equity program, 1990-91. Mem. Am. Math. Assn., Mich. Math. Assn., Bus. and Profl. Women (v.p. chpt. 1991-92), Nat. Edn. Computing Conf., Mich. Assn. Computer Users in Learning, Ea. Mich. U. Alumni Assn. (bd. dirs.). Home: 1419 Gregory St Apt 20 Ypsilanti MI 48197-1672 Office: Ea Mich U 34 N Washington Ypsilanti MI 48197

RANDZA, JASON MICHAEL, aerospace engineer, consultant; b. Ellwood City, Pa., June 18, 1963; s. Frank Anthony and Jean Ann (Tracy) R. BS in Aerospace Engring., Pa. State U., 1985. Cert. private pilot. Engr. Atlantic Rsch. Corp., Gainesville, Va., 1985-90; cons. Design Integrated Tech., Warrenton, Va., 1991; control rm. operator Hadson Power #11, Franklin, Va., 1991; tng. coord. LG&E Westmoreland Energy, Franklin, Va., 1991—. Mem. AIAA, Nat. Assn. Rocketry, Nat. Arbor Day Found., Air and Space Smithsonian, Tripoli Rocketry Assn., Cousteau Soc., Am. Kitefliers Assn., Tau Beta Pi, Sigma Gamma Tau. Republican. Roman Catholic. Avocations: rocketry, hang gliding, kite flying, wright training, aerobics. Home: 504 Fairlane Blvd New Galilee PA 16141

RANELLONE, RICHARD FRANCIS, shipbuilding company executive; b. Yonkers, N.Y., Dec. 12, 1940; s. Frank Anthony and Catherine Anne (Guadian) R.; m. Thelma Ester Aguilar, Mar. 29, 1976; children: Ashley Colette, Keith Alexander. BS in Elec. Engring., Manhattan Coll., 1964; M Nuclear Engring., U. Va., 1968. Registered profl. engr., Va. Project leader Def. Intelligence Agy., Washington, 1974-75; sr. project engr. Westinghouse Electric Corp., Madison, Pa., 1975-77, mgr. tech. programs, 1977-82; product devel. mgr. Newport News (Va.) Indsl. subs. Newport News Shipbldg., 1982-85; mgr. spl. projects Newport News Shipbldg., 1985-88, mgr. rsch. and devel., 1988-90, mgr. advanced tech. applications, 1990—; indsl. adv. bd. Continuous Electron Beam Accelerator Facility, Newport News, 1989—. Contbr. over 25 articles to profl. jours. Mem. IEEE (chmn. Hampton Roads sect. 1990-91), Am. Nuclear Soc., Am. Soc. Naval Engrs., Cryogenic Soc. Am. (co. rep. 1989—). Achievements include rsch. in applications of superconductivity to Navy ships including magnetohydrodynamic propulsion; devel. of processes for recycling radioactive scrap material; initiation of cooperative rsch. programs with nat. labs. and univs. Home: 22 Kenilworth Dr Hampton VA 23666 Office: Newport News Shipbuilding Co 4101 Washington Ave Newport News VA 23607

RANEY, LELAND WAYNE, nuclear engineer; b. Chanute, Kans., May 4, 1943; s. Merle Irvin and Elizabeth Louise (Romine) R.; m. Judith Ellen Gibson, June 7, 1969; children: Helen Elizabeth, Jennifer Ellen. BS in Nuclear Engring., Kans. State U., 1966, MS in Nuclear Engring., 1971. Registered profl. engr., Ill. Sr. reactor engr. Idaho Nuclear Corp., Idaho Falls, 1966-69; grad. asst. Kans. State U., Manhattan, 1969-71; gen. engr. Commonwealth Edison Nuclear Engring. Dept., Chgo., 1971-76; sr. engr. Project Mgmt. Corp. Clinch River Breeder Reactor, Oak Ridge, Tenn., 1976-82; nuclear safety group supr. Commonwealth Edison La Salle County Sta., Seneca, Ill., 1982-86, Commonwealth Edison Braidwood Sta., Braceville, Ill., 1986-90; staff engr., probablistic risk assessment group Commonwealth Edison Nuclear Engring. Dept., Downers Grove, Ill., 1990—. Rep., pres. Neosho County 4-H Coun., Erie, Kans., 1960; dir., pres. Orchard Knob Homeowners Assn., Clinton, Tenn., 1980, Hatcher Woods Recreation Assn., Morris, Ill., 1984. Recipient Eagle award Boy Scouts Am., Chanute, 1958, Who's Who Key award 4-H Club, Chanute, 1961, Del. award Sunflower Boy's State, Kans., 1960. Mem. Am. Nuclear Soc. Achievements include being licensed reactor engineer on Engineering Test Reactor, reactor engineer on Advanced Test Reactor and sr. reactor engineer. Office: Commonwealth Edison Co P O Box 767 PO Box 767 Chicago IL 60690

RANGEL-ALDAO, RAFAEL, biochemist; b. Caracas, Venezuela, May 5, 1946; s. J. Rafael and Georgina (Aldao) Rangel; m. Doris Serrano de Rangel, Dec. 26, 1970; children: Dorella, Rafael Eugenio. MD, Ctrl. U., Caracas, 1963-69; PhD, Yeshiva U., N.Y., 1972-77. Instr. biochemistry U. Ctrl. de Venezuela, Caracas, 1969-72; rsch. assoc. Albert Einstein Coll. Medicine, N.Y.C., 1977-78; asst. prof. U. Carabobo, Venezuela, 1978-80, assoc. prof., 1980-83, prof., 1983-87; prof. U. Simon Bolivar, Caracas, Venezuela, 1987—; nat. mgr. biotechnology Empresas Polar, Caracas, Venezuela, 1987—; chmn. Iberoameriman Biotechnology, Madrid, 1987—; cons. World Health Orgn., Geneva, 1987; sr. advisor Inst. Estudios Superiors en Administracion, Caracas, 1992—. Author: Methods in Enzymology, 1988, Bioquimica A Biologia Molecular, 1988. Bd. dirs. Mus. Scis., Caracas, Venezuela, 1992. Recipient First prize Ea. Forum Am. Med. Assn., Miami, Fla., 1976, Interamerican Prize of Sci. and Tech., Orgn. Am. States, Washington, 1985, Nat. Prize on Sci. Journalism, 1993. Mem. Hilton Club. Roman Catholic. Achievements include patent in detection of human papiloma virus finding of signals that mediate differentiation of trypanosoma cruzi elidation of certain mechanisms of control of protein kinases in mammals. Office: Empresas Polar, Los Cortijos de Lourdes, Caracas 1010A, Venezuela

RANGNEKAR, VIVEK MANGESH, molecular biologist, researcher; b. Bombay, Dec. 17, 1955; s. Mangesh Vithal and Sanjivani (Dewoolkar) R.; m. Vidya Vivek Varsha Kulkarni, May 15, 1981; children: Vidyuta, Viraj. MSc, U. Bombay, 1979, PhD, 1983. Postdoctoral fellow U. Chgo., 1983-86; rsch. assoc. Rush Med. Ctr., Chgo., 1986-87; asst. prof. U. Chgo., 1988-91, U. Ky., Lexington, 1992—. Contbr. articles to profl. jours. including Jour. Biol. Chemistry, Nucleic Acids Rsch. Mem. AAAS, Am. Soc. Microbiology. Achievements include identification of role of early genes in growth control of human tumor cells. Office: U Ky 800 Rose St Lexington KY 40536

RANIERE, LAWRENCE CHARLES, environmental scientist; b. Wilmington, Del., Mar. 20, 1931; s. Charles J. and Ruth B. (Stierle) R.; m. Phillis Marie Laquaglia, July 10, 1954; 1 child, Debra Rose Marsh. BS, Utah State U., 1952; MS, U. Del., 1958; PhD, Rutgers the State U., 1964. Meteorologist USAF Air Weather Svc., Frankfurt, Fed. Republic Germany, 1953-56; plant pathologist USDA Agrl. Rsch. Svc., Beltsville, Md., 1958-61; agrl. meteorologist U.S. Weather Bur., Washington, 1962-70; phys. scientist U.S. EPA, Raleigh, N.C., 1970-73; environmental scientist U.S. EPA, Corvallis, Oreg., 1973-86; environmental scientist viticulture/nursery Kings Valley Vineyard, Corvallis, 1986—; pres. South Willamette Winegrowers Assn., Corvallis, 1984-85; chmn. Rsch. Devel. Oreg. Winegrowers Assn., Corvallis, 1985. Author: (with others) U.S. Agriculture Year Book, 1966; editor: E.P.A. Air Pollution Control Document, 1971. Chmn. United Fund, EPA Rsch. Lab., Corvallis, 1982. Lt. col. USAF, 1968-69. Mem. Rotary (pres. Philomath, Oreg. chpt. 1985-86, Paul Harris fellow 1988). Achievements include raised bed solar heating system. Home: 16150 NW Claremont Dr Portland OR 97229-7836

RANKER, TOM A., botanist, educator. Prof. dept. botany U. Colo. Recipient Edgar T. Wherry award Botanical Soc. Am., 1992. Office: U of Colorado Dept of Botany 914 Broadway University Of Colorado CO 80309*

RANKIN, JAMES EDWIN, computer systems consultant; b. Pitts., Sept. 9, 1943; s. Harvey Walter and Helen Elise (Engel) R. MS, U. N.C., 1968; PhD, Yeshiva U., 1971. Staff scientist Los Alamos (N.Mex.) Nat. Lab., 1978-80; pvt. practice computer cons. N.Y.C., 1980-89, 93—; systems mgr. Bantam Doubleday Dell Pub. Group, Inc., N.Y.C., 1990-93; mem. Com. of Concerned Scientists, Queens, N.Y., 1984—. Contbr. Classical and Quantum Gravity, Internat. Jour. Theor. Phys., IEEE Trans. Electron Devices. Mem. Am. Phys. Soc. (small coms. 1984—), N.Y. Acad. Scis., Sigma Xi (assoc. mem.). Democrat. Home: Apt 4J 228 W 71 St New York NY 10023 Office: 527 3d Ave Ste 298 New York NY 10016

RANNEY, CARLETON DAVID, plant pathology researcher, administrator; b. Jackson, Minn., Jan. 23, 1928; s. Carleton Oran and Ada Elizabeth (Harriman) R.; m. Mary Kathryn Ransleben, July 16, 1949; children: David Clayton, Mary Elizabeth. AA, Chaffey Jr. Coll., Ontario, Calif., 1952; BS, Tex. A&M U., 1954, MS, 1955, PhD, 1959. Plant pathologist Crops Rsch. Div. Agrl. Rsch. Svc. USDA, College Station, Tex., 1955-58, Stoneville, Miss., 1958-70; investigations leader Crops Rsch. Div. Agrl. Rsch. Svc. USDA, Beltsville, Md., 1970-72; area dir. Ala. No. Miss. area Agrl. Rsch. Svc. USDA, Starkville, Miss., 1973-78; area dir. Delta States agea Agrl. Rsch. Svc. USDA, Stoneville, Miss., 1978-84; area dir. Mid-South area Agrl. Rsch. Svc. USDA, Stoneville, 1984-87; asst. dir. Miss. Agrl. and Forestry Exptl. Stas., Stoneville, 1987—; adj. prof. agronomy Miss. State U., 1970—; sr. exec. svc. USDA, Stoneville, Miss., 1984-87; adv. bd. Belt Wide Meetings Nat. Cotton Coun., Memphis, Tenn. 1987—. Contbr. articles to profl. jours. Sect. advisor SE2 Order Arrow Boy Scouts Am., Miss. and W. Tenn., 1973-83; pres. Delta Area Coun. Boy Scouts Am., Clarksdale, Miss., 1990-91. Sgt US Air Force, 1946-49. Recipient Silver Beaver Boy Scouts Am., Stoneville, 1981, Disting. Svc. Order of Arrow, Stoneville, 1983. Mem. Am. Soc. Agronomy, Nat. Cotton Disease Coun. (chmn. 1961-63), Lions Club, Sigma Xi, Alpha Zeta, Phi Kappa Phi. Methodist. Achievements include development of fungicide control seedling diseases; definition of relationship of microclimate to boll rot of cotton; development of non-mercurial seed treatments. Office: Delta Rsch & Ext Ctr PO Box 197 Stoneville MS 38776-0197

RANNEY, MAURICE WILLIAM, chemical company executive; b. Buffalo, Jan. 13, 1934; s. Maurice Lynford and Helen Hart (Birdsall) R.; m. Theresa Ann Berthot, Oct. 24, 1953 (div. 1974); children: William, Linda, Laurel, James; m. Elisa Ramirez Villegas, Dec. 21, 1974; 1 stepchild, Elisa. BS in Chemistry, Niagara U., 1957, MS in Organic Chemistry, 1959; PhD in Phys. Organic Chemistry, Fordham U., 1967. Group leader, tech. mgr. Union Carbide Corp., Tarrytown, N.Y., 1957-75; gen. mgr. Union Carbide Japan KK, 1976-80; exec. v.p. Showa Union Gosei Co., Ltd., 1976-80; rep. dir. Union Carbide Svcs. Eastern Ltd., 1976-80; dir. Nippon Unicar Co., Ltd., 1976-80, Union Showa KK, 1976-80; pres. Union Carbide Formosa Co., Ltd., Hong Kong, Tokyo, 1980-82; bus. dir. Union Carbide Eastern, Inc., Tokyo, 1982-85; pres., rep. dir. Union Carbide Japan KK, 1986—; pres. Union Indsl. Gas Corp.; rep. dir. Oriental Union Chem. Corp., 1980-82; mng. dir. Nippon Unicar Co., Ltd., 1982-85; v.p. internat. Union Carbide Chems. and Plastics Co., 1986—; vice chmn., rep. dir. Nippom Unicar, 1986—; pres. Nihon Parylene, 1992; lectr. in field. Author: Flame Retardant Textiles, 1970, Power Coatings, 1971, Synthetic Lubricants, 1972, New Curing Techniques, 1976, Fuel Additives, 1976, Durable Press Fabrics, 1976, Silicones, Vols. I, II, 1977, Reinforced Plastics and Elastomers, 1978, Offshore Oil Technology, 1979, Oil Shale and Tar Sands, 1980, Primary Electrochemical Cell Technology, 1981; contbr. articles to profl. jours. Union Carbide fellow, Mellon Inst. Indsl. Rsch., 1958-60. Achievements include numerous patents in field. Home: Homat Governor 402, 5-17 Roppongi 1-chome, Minato-ku Tokyo Japan 105 Office: Union Carbide Japan KK, Hiroo Sk Bldg 36-13 Ebisu, 2 Chom Shibuya-Ku, Tokyo 150, Japan 105

RANPURIA, KISHOR PRAJARAM, wire and cable engineer; b. Jetpur, Gujarat, India, Sept. 15, 1941; came to U.S., 1968; s. Prajaram and Lilavati (Prajaram) R.; married; children: Anish, Reena. MSME, N.D. State U., 1969. Mfg. engr. Adirondack Wire and Cable Inc., Farmington, Conn., 1977-82; product engring. mgr. Monroe Wire and Cable Corp., Middletown, N.Y., 1982-84; product design engr. NEK Cable Inc., Ronkonkoma, N.Y., 1984-87; wire and cable engring. mgr. Am. Electric Cable, Holyoke, Mass., 1987-88; engring. mgr. Hitemp Wires Inc., Bohemia, N.Y., 1988-90; engring. mgr., cons. Ronkonkoma, 1990-93; process engr. Carol Cable Co. Inc., Woonsocket, R.I., 1993—; adv. mem. Underwriters Labs. Tech. Adv. Panel for Communication, Melville, N.Y., 1988-90. Mem. Am. Soc. for Quality Control (cert. quality auditor 1993), Nat. Elec. Mfg. Assn. (tech. com. 1984-90). Home: 738 Peconic St Ronkonkoma NY 11779-6558

RANSDELL, TOD ELLIOT, pharmaceutical, in vitro diagnostics biotechnologist; b. Imperial, Nebr., May 17, 1953; s. Merrill Guy and Rosalie E. (Nissen) R. BS in Botany, Mont. State U., Bozeman, 1977. Police officer Dillon (Mont.) Police Dept., 1979-80; dept. mgr. Woolco, Bozeman, Mont., 1980-83; lab. coord. Skyland Sci. Svcs., Inc., Bozeman, Mont., 1983-86; sales assoc. S&P Office Supply, Bozeman, Mont., 1986-91; validation specialist Skyland Sci. Svcs., Inc., Bozeman, Mont., 1991-92, sr. validation specialist, 1992; sr. validation specialist Genetic Systems Corp., Redmond, Wash., 1992—; cons. Skyland Sci. Svcs., Inc., Bozeman, Mont., 1987-92. Pres. Bozeman (Mont.) Jaycees, 1983, 85, Crime Stoppers, 1982; mem. Benevolent and Prtective Order of the Elks, Bozeman, Mont. 1989—. Mem. Am. Assn. Advancement of Sci., Union of Concerned Scientists, Parenteral Drug Assn., Bozeman Jr. C. of C., Internat. Soc. Pharm. Engring. Home: 18675 NE 62nd Ct # 3022 Redmond WA 98052 Office: Genetic Systems Corp Sanofi Diagnostics Pasteur 6565 185th Ave NE Redmond WA 98052

RANSOM, BRUCE ROBERT, neurologist, neurophysiologist, educator; b. Santa Fe, Aug. 5, 1945; s. Henry Robert and Alberta J. (Hoenig) R.; m. Sally Jean Lynum, Feb. 11, 1966 (div. July 1980); children: Rebecca, Christopher; m. Joann Grace Elmore, Feb. 7, 1989. BA, U. Minn., 1967; MD, Washington U., St. Louis, 1972, PhD, 1972. Diplomate Am. Bd. Psychiatry and Neurology. Intern Washington U., 1972-73; resident in neurology Stanford U., 1976-79; asst. prof. neurology Stanford U., Palo Alto, Calif., 1979-86; assoc. prof. neurology Yale U., New Haven, 1987—; mem. sci. adv. bd. Paralyzed Vets. Am., Washington, 1991—. Founder, editor-in-chief Jour. GLIA, 1988; contbr. numerous articles to profl. jours. Lt. comdr. USPHS, 1973-76. Recipient Rsch. Career Devel. award NIH, 1980-85, Javits Neurosci. Investigator award NIH, 1991—; NIH rsch. grantee, 1979—. Fellow Am. Neurology Assn.; mem. Am. Acad. Neurology, Neuroscis. Soc. Republican. Achievements include research in the role of glial cells in the brain in the control of neuronal microenvironment; pathophysiology of anoxic injury to central axons. Office: Yale Med Sch 333 Cedar St New Haven CT 06510

RANSOM, PERRY SYLVESTER, civil engineer; b. Atlanta, July 3, 1929; s. Perry Sylvester and Eva James (Smith) R.; m. Wilma Ruth Cone, June 1, 1951; children: Beverly Kay, Barbara Ann. BSCE, Auburn U., 1958. Registered profl. engr., La., Miss., Ala.; cert. land surveyor, La., Miss. Asst. timekeeper Swift & Co., Montgomery, Ala., 1947-51; trainman Atlantic Coast Line RR, Montgomery, 1951-58; lab. mgr. A.W. Williams Inspection Co., Mobile, Ala., 1958-60; owner, CEO Cons. Engrs., Inc., Biloxi, Miss., 1960—. Mem. Civitan Club, Mobile, 1959, Rotary Internat., Moss Point, Miss., 1965; pres. Gulf Coast chpt. Miss. Engring. Soc., Biloxi, 1965-66, chmn. Pepp sect., Jackson, 1967-68; bd. dirs. Miss. sect. ASCE, Jackson, 1972-74. With U.S. Army, 1951-53. Named Boss of Yr., Miss. Nat. Sec. Assn., 1975-76, for Outstanding Svc., Miss. Engring. Soc., 1966, Outstanding Supporter, Boys Clubs Am., Biloxi, 1991; recipient Cert. of Appreciation, Boys Clubs Am., Biloxi, 1990. Mem. Miss. Cons. Engrs. Coun., Aircraft Owners and Pilots Assn., VFW (Merit/Disting. Svc. 1989), Masons (life). Republican. Baptist. Home: 711 Twin Oaks Dr Ocean Springs MS 39564 Office: Cons Engrs Inc 430 Caillavet St Biloxi MS 39530

RAO, ATAMBIR SINGH, nuclear engineer. Principal engineer, BWR program GE Nuclear Energy, San Jose. Recipient George Westinghouse Silver Medal ASME, 1990. Office: GE Nuclear Energy 175 Curtner Ave M/C 754 San Jose CA 95125

RAO, GOPALA UTUKURU, radiological physicist, educator; b. Berhampur, Orissa, India, Nov. 15, 1930; came to U.S., 1962; s. Seshagiri Utukuru and Sakuntala (Utukuru) R.; m. Frances Rakestraw, Aug. 22, 1968. DSc, Johns Hopkins U., 1966; diploma in radiol. physics, Am. Coll. Radiology, 1971. Sci. officer Bhabha Atomic Ctr., Bombay, India, 1957-62; grad. asst. Johns Hopkins U., Balt., 1962-66, instr. in radiol. physics, 1966-68, asst. prof. radiol. physics, 1968-78, prof. radiol. physics, 1978—; mem. subcom. Nat. Coun. Radiation Protection, 1982-84; examiner Am. Bd. Radiology, 1983—; UN cons. in radiol. physics to India and Guyana, 1985-91. Author over 200 book chpts. and rsch. publs. in field. Pres. India Forum, Balt., 1975. Grantee NIH, USPHS, 1978-82; recipient awards for best sci. exhibits Am. Assn. Physicists in Medicine, Radiol. Soc. N.Am. Grantee NIH, USPHS, 1978-82. Achievements include work in medical imaging, hospital planning, radiation oncology physics; inventor radiation equipment and evaluation tools and methodology. Office: Johns Hopkins Hosp Dept Radiology Baltimore MD 21205

RAO, JOSEPH MICHAEL, chemist; b. Syracuse, N.Y., May 2, 1950; s. Samuel and Phyllis (Simiele) R.; m. Patricia M. Glynn, Aug. 2, 1974; children: Janel Margaret, Jordan Samuel. BS, Syracuse U., 1972, PhD, 1976. Rsch. chemist Allied Signal, Syracuse, 1976-79, group leader, 1979-85; sect. supr. Eastman Kodak Co., Rochester, N.Y., 1985-91, unit dir., 1991—; mem., rep. com. on analytical reagents Am. Chem. Soc., Washington, 1991—. Contbr. numerous articles to profl. jours. Mem. ASTM (chmn. com. E-15 indsl. chems. 1982—), Nat. Assn. Photographic Mfrs., Am. Nat. Standards Inst. (vice-chmn. com. Ph-4 photographic chems. 1991—). Achievements include rsch. on stabilization of high and low oxidation states of substituted Iron (II) Terpyridine complexes, electrochemical behavior of tridentate imine ligand complexes of Manganese (II). Office: Eastman Kodak Co Rsch Labs Bldg 65 Rochester NY 14650-1807

RAO, K. PRABHAKARA, aerospace engineering educator; b. Madanapalle, India, Oct. 14, 1939; s. K. Krishnamoorthi and K. Subbamma Rao; m. Keelapatla Geetha, Dec. 23, 1970; children: K. Krishna Chaitanya, K. Jyothi. ME, Indian Inst. Sci., Bangalore, India, 1964; PhD, Imperial Coll., London, 1969. Aeronautical engr. Hindustan Aeronautics Ltd., Bangalore, India, 1964-66; commonwealth scholar Imperial Coll., London, 1966-69; from asst. prof. to assoc. prof., prof. Indian Inst. Sci., Bangalore, 1969—; Alexander von Humboldt fellow Stuttgart (Germany) U., 1976-77; NRC sr. rsch. assoc. Wright Patterson AFB, Dayton, Ohio, 1986-88; cons. Nat. Aeronautical Lab., Bangalore, 1990, Indian Space Rsch. Orgn., Trivandrum, 1985, Aeronautical Devel. Agy., Bangalore, 1991. Co-author: Theory of Shells, 1980, Composites Design, 1987, International Guide Book on Structural Analysis Software, 1986; contbr. articles to profl. jours. Recipient Khosla Rsch. Gold medal Roorkee U., 1974. Fellow Instn. Engrs. (India); mem. Aero. Soc. India, AIAA. Achievements include contributions to the design of reinforcements around cut outs inclusions in shells; finite element analysis of composite beams plates shells; optimun design for buckling of fibre reinforced plastic shells. Office: Aerospace Engring Dept, Indian Inst Sci, Bangalore 560012, India

RAO, MING, chemical engineering and computer science educator; b. Gejiu, Yunnan, Peoples Republic of China, June 24, 1954; came to U.S., 1984; s. Jin Rao and Lie Zhang; m. Xiaomei Zheng, Oct. 28, 1981; children: Mai Rao, Diana Rao. BS, Kunming Inst. of Tech., 1976; MS, U. Ill., 1987; PhD, Rutgers U., 1989. Engr. Kaiyuan Chemicals Inc., Yunnan, 1976-78; lectr. Yunnan Inst. Tech., 1981-84; rsch. engr. FAA Tech. Ctr., Atlantic City, N.J., 1989; supr. Rutgers U., Piscataway, N.J., 1988-89, asst. prof., 1989-91, assoc. prof., 1991—; dir. Intelligence Engring. Lab. U. Alta., Edmonton, Can., 1989—. Author: Integrated System for Intelligent Control, 1991, Process Control Engineering, 1992, Integrated Distributed Intelligent Systems in Manufacturing, 1993, Integrated Distributed Intelligent Systems for Engineering Design, 1993, Advances in Modeling and Control for Paper Machines, 1993; guest editor: Intelligent Process Control, 1992; contbr. 150 articles to profl. jours. and internat. conf. procs. Recipient Doctoral Excellence fellowship Rutgers U., Piscataway, 1987-89. Mem. Am. Assn. Artificial Intelligence, IEEE, Can. Soc. Chem. Engrs., Can. Paper and Pulp Assn., Am. Inst. Chem. Engrs., Sigma Xi, Tau Beta Pi. Avocations: reading, travel, fishing. Office: U Alberta, Dept Chem Engring, Edmonton, AB Canada

RAO, VAMAN, economics educator; b. Gogi, Karnataka, India, Aug. 31, 1933; came to U.S., 1970; s. Venkatesh and Bhima (Bai) Anjutgi; m. Geetha Parikshitraj, Dec. 5, 1963; children: Kavita, Anita. BS in Chemistry, Botany and Zoology, Osmania U., India, 1952, BA in English Lit., 1955, BE in Advanced Psychology, 1957, MA in Econs., 1958; PhD in Econs., U. Mo., 1973. Lectr. edn. Govt. Coll. Edn. Osmania U., Hyderabad, India, 1961-65, sr. lectr. econs., 1965-70; research asst. Dept. Econs. U. Mo., Columbia, 1970-73; asst. prof. econs. U. Mo., Rolla, 1973-79; assoc. prof. Western Ill. U., Macomb, 1979-82, prof., 1982—, disting. prof., 1993-94, Disting. prof., 1993-94; ann. faculty lectr., 1991; joint dir. summer sch. for univ. econs. tchrs. Osmania U., Hyderabad, 1967; assoc. dir. Mo. Economy Study Project U. Mo., 1977; dir., prin. investigator state social services projects U. Mo., 1977-78, ethanol prodn. impact project Ill. Corn Bd., 1986-87; prin. investigator energy substitution prospects project Western Ill. U., Macomb, 1980-81; prin. investigator, assoc. dir. employment implications of gasohol prodn. project Dept. Commerce and Community Affairs, State of Ill., 1980-81; instr. rural devel. tng. project Office Internat. Coop. and Devel. USDA, Tanzania, 1982; vis. prof. econs. Soochow U., Taipei, Republic of China, 1982-83, Feng Chia U., Taichung, Republic of China, 1982-83, Nat. Chung Tshing U., Taichung, 1982-83; chmn. hon. doctorates com. Western Ill. U., 1985-86, council on fiscal mgmt., 1986-87, univ. pers. com. 1986-88, vice chmn. 1985-86, chmn. faculty excellence awards com., 1988-89, mem. presdl. merit awards com., 1984, faculty excellence task force, 1985-86, asst. dean search com., 1985, univ. budget council, 1986—, campus adv. com. for presdl. search, 1987—, faculty senate, 1990—, univ. computer com., 1987—, salary study task force Bd. Govs., Ill., 1985—, co-chair, 1989-91, chmn., mem. COB undergrad. com., 1989-91, mem. BGU Coun. Faculties, 1991-92, COB dean search com., 1989-90, COB MBA task force, 1991-92, vice chmn. univ. senate, 1992—. Editor Edn., 1957; editor and pub. Medhavi, 1960-62; mem. editorial bd. Jour. of Developing Areas; assoc. editor Internat. Jour. Indian Studies, 1991—; asst. editor The Jour of Econs., 1980—; book rev. editor The Jour. of Social Econ.; contbr. numerous articles to profl. jours. Recipient Best Tchr. award Coll. Edn. Osmania U. 1957, Excellent Tchr. award Western Ill. U. 1984, 89, 90, 91, 92, Best Svc. Activities award 1990, Best Rsch. Activities award 1991. Mem. Am. Econs. Assn., Assn. Indian Econs. Studies (editor Newsletter 1981—, mem. exec. com. 1979-87, pres. 1991—), Indian Econometric Soc. (life), Econometric Soc., Missouri Valley Econs. Assn., Midwest Econs. Assn., Univ. Profls. Ill. (treas. We. Ill. U. chpt. 1985—, mem. ho. of dels. 1985—, sec. local 4100, 1989-92), India Club of Macomb (pres. 1981). Avocations: polit. affairs, reading, writing, econs. research. Home: 130 Pam Ln Macomb IL 61455-3304 Office: Western Ill U Dept Econs 442 Stipes Hall Macomb IL 61455

RAO, VENIGALLA BASAVESWARA, biology educator; b. Donepudi, India, June 10, 1954; came to U.S. 1980; s. Venigalla Koteswara and Venigalla (Janaki) R.; m. Mangala Seetaramiah, Aug. 20, 1982; children: Prashant Arkalgud, Vishnu-Parkash Venigalla. MSc, Andhra U., Waltair, India, 1976; PhD, Indian Inst. Sci., Bangalore, India, 1980. Rsch. assoc. Med. Sch. U. Md., Balti., 1980-88, rsch. asst. prof. Med. Sch., 1988-89; asst. prof. Cath. U. Am., Washington, 1989--. Dept. Energy grantee, 1990-93. Mem. Am. Assn. for the Advancement Sci., Am. Soc. for Microbiology, Sigma Xi. Achievements include research publications in field. Home: 607 Hyde Rd Silver Spring MD 20902 Office: Cath U Am Dept Biology 620 Michigan Ave NE Washington DC 20064

RAO, VITTAL SRIRANGAM, electrical engineering educator; b. Inumpamula, India, June 8, 1944; came to U.S., 1981; s. Rangaiah Srirangam and Lakshmamma (Immadi) R.; m. Vijaya Morishetti, Feb. 28, 1965; children: Asha, Ajay. M of Tech., Indian Inst. Tech., 1972, PhD, 1975. Asst. prof. Indian Inst. of Tech., New Delhi, India, 1975-79; vis. prof. T.U., Halifax, N.S., Can., 1980-81; assoc. prof. U. Mo., Rolla, 1981-88, prof., 1988—; dir. Intelligent Systems Ctr., Rolla, 1991—; cons. Delco Remy, Anderson, Ind., 1985-87, Allison Gas Turbines, Indpls., 1986-87, U.S. Army Picatinny Arsenal, N.J., 1988-91. Contbr. articles to profl. jours. including Suboptimal/Near Optimal Control, Reduced Order Modeling Techniques, Robust Control, Large Space Structures, Smart Structures. Fellow AIAA (assoc.); mem. IEEE (sr., subsect. 1981-88, Centennial medal 1984). Achievements include devel. of reduced order modeling techniques for large space structures, interdisciplinary approach for control of smart structures. Home: 602 Fox Creek Rd Rolla MO 65401 Office: U Mo Intelligent Systems Ctr Rolla MO 65401

RAPAPORT, MARTIN BARUCH, safety engineer; b. N.Y.C., Nov. 9, 1963; s. Harold and Irene Nathalie (Chotner) R.; m. Carol Ann Cunningham, Dec. 17, 1988; 1 child, I. Molly Rapaport Egan. BS, Carnegie Mellon U., 1985; MS, Poly. U., Bklyn., 1990. Engr. aircraft div. Naval Air Warfare Ctr., Warminster, Pa., 1990—. Mem. ASME, Am. Helicopter Soc., Soc. Automotive Engrs., Survival and Flight Equipment. Home: 44 N Church St Doylestown PA 18901

RAPHAEL, GEORGE FARID, chemical engineer; b. Beirut, Lebanon, Aug. 26, 1962; s. Farid K. and Tamam (Hoayek) R. French Baccalaureate II, Notre-Dame, Jamhour, Lebanon, 1981; BSChemE, Calif. State Poly. U., 1989. Quality control technician Machlanburg Duncan Co., Industry, Calif., 1984-86; plant operator BKK Corp., West Covina, Calif., 1986-89; project engr. Chem. Waste Mgmt., Kettleman, Calif., 1989-90; mgr. tech. svcs. Greenfield Environ., Carlsbad, Calif., 1990-91; mgr. process engr. Greenfield Environ., Carlsbad, 1991—. Mem. NSPE, AICE, Hazardous Materials Control Resources Inst. Home: 1510 S Melrose Dr Apt 105 Vista CA 92083 Office: Greenfield Environ 5964 LaPlace Ct Ste 150 Carlsbad CA 92008

RAPIDIS, ALEXANDER DEMETRIUS, maxillofacial surgeon; b. Athens, Greece, Aug. 31, 1948; s. Demetrius Afentoulis and Calliope (Sbrinis) R.; m. Effie Stergiopoulos, May 8, 1978 (div. 1985); m. Iphigenia Thermidou, Jan. 22, 1986; 2 children. D.D.S. with honors, Athens U., 1973; Ph.D., Athens U., 1983; DSc (hon.) Marquis G. Scicluna Internat. U. Found., 1989. Clin. asst. dept. oral surgery Royal Dental Hosp., London, 1974; dept. oral and maxillofacial surgery London Hosp. Med. Coll., 1975; sr. house surgeon Whipp's Cross Hosp., London, 1976; research fellow dept. oral surgery Queen Mary's Hosp., London, 1977; attending maxillofacial surgeon St. Paul's Accident Hosp., Kifissia, Athens, 1978-81, Laikon Gen. Teaching Hosp. of Athens U., 1981-86; research assoc. dept. oral pathology U. Athens, 1978-86; hon. lectr., research assoc. dept. Maxillofacial Surgery King's Coll., Hosp. U. London, 1986—; hon. cons. maxillofacial surgeon U. Patras, Greece, 1985—; cons. maxillofacial surgeon dept. plastic surgery Nat. Health Hosp., Athens; cons., head dept. maxillofacial surgery Greek Anticancer Inst., Athens, 1990—. Contbr. articles to profl. jours. Recipient award Internat. Union Against Cancer, 1977. Mem. Greek Dental Assn., Royal Soc. Medicine London (fellow), Acad. Psychosomatic Medicine (fellow), Internat. Assn. Oral Surgeons (fellow), Internat. Assn. Maxillofacial Surgery. Home: 23 Asclipiou, 144 Athens Greece Office: 37 Chazitos St Kolonaki, Athens Greece

RAPIDIS, PETROS A., research physicist; b. Heraklion, Crete, Greece, Jan. 1, 1951; came to U.S., 1969; s. Panayotis A. and Athena Rapidis; m. Dorothea Poulos, 1984; 1 child, George. BS, MIT, 1973; MS, Stanford U., 1975, PhD, 1979. Rsch. assoc. Fermi Nat. Accelerator Lab., Batavia, N.Y., 1979-82, R.R. Wilson fellow, 1982-85, assoc. scientist, 1985-89, scientist, 1989—. Contbr. numerous articles to profl. jours. Mem. AAAS, Am. Phys. Soc., Sigma Xi. Achievements include research in particle and accelerator physics. Office: Fermi Nat Accelerator Lab PO Box 500 Batavia IL 60510

RAPIER, PASCAL MORAN, chemical engineer, physicist; b. Atlanta, Jan. 11, 1914; s. Paul Edward and Mary Claire (Moran) R.; m. Martha Elizabeth Doyle, May 19, 1945; children: Caroline Elizabeth, Paul Doyle, Mollie Claire, John Lawrence, James Andrew. BSChemE, Ga. Inst. Tech., 1939; MS in Theoretical Physics, U. Nev., 1959; postgrad., U. Calif., Berkeley, 1961. Registered profl. engr., Calif., N.J. Plant engr. Archer-Daniels-Midland, Pensacola, Fla., 1940-42; group supr. Dicalite div. Grefco, Los Angeles, 1943-54; process engr. Celatom div. Eagle Picher, Reno, Nev., 1955-57; project mgr., assoc. research engr. U. Calif. Field Sta., Richmond, 1959-62; project mgr. sea water conversion Bechtel Corp., San Francisco, 1962-66; sr. supervising chem. engr. Burns & Roe, Oradell, N.J., 1966-74; cons. engr. Kenite Corp., Scarsdale, N.Y., Rees Blowpipe, Berkeley, 1960-66; sr. cons. engr. Sanderson & Porter, N.Y.C., 1975-77; staff scientist III Lawrence Berkeley Lab., 1977-84; bd. dirs. Newtonian Sci. Found.; v.p. Calif. Rep. Assembly, 1964-65; discoverer phenomena faster than light, origin of cosmic rays and galactic red shifts. Contbr. articles to profl jours.; patentee agts. to render non-polar solvents electrically conductive, direct-contact geothermal energy recovery devices. Mem. Am. Inst. Chem. Engrs., Gideons Internat., Lions Internat., Corvallis, Sigma Pi Sigma. Presbyterian. Home and Office: 8015 NW Ridgewood Dr Corvallis OR 97330-3026

RAPIN, CHARLES RENÉ JULES, computer science educator; b. Lausanne, Switzerland, Jan. 10, 1935; s. René Louis Georges Octave and Mary-Coe (Reeves) R. BA ès lettres, Gymnases Cantonaux, Lausanne, Switzeralnd, 1954; physicist engr., Ecole Poly., Lausanne, 1959, PhD, 1964. Asst. Ecole Poly Univ., Lausanne, 1959-64, 1st asst., 1964-67; asst. prof. U. Ky., Lexington, 1967-68; sci. collaborator Ecole Fédérale, Lausanne, 1969-72, assoc. prof., 1972-78, prof., 1978—. Mem. Assn. Computing Machinery, Soc. Math. Suisse, Soc. Suisse des Informaticiens, Planetary Soc., Nat. Geographic Soc. Avocations: public transport, novels, classical music.

RAPKIN, JEROME, defense industry executive; b. Wilmington, Del., Aug. 1, 1929; s. Harry and Ida (Hermann) R.; B.S. in Marine Engring., U.S. Naval Acad., 1952; M.S. in E.E., U.S. Naval Postgrad. Sch., 1959; postgrad. Armed Forces Staff Coll., 1965, Catholic U. Am., 1978; m. Janet Vansant, Nov. 4, 1954; children—Keith, Leigh, Paige. Commd. ensign U.S. Navy, 1952, advanced through grades to capt., 1979; dir. Surface Warfare Systems Naval Sea Systems Command, Washington, 1974-75; comdr. Destroyer Squadron 26 Surface Force Atlantic, Norfolk, Va., 1975-78; head surface to surface warfare, chief naval ops., Washington, 1978; dir. programs and budget Chief Naval Material, Washington, 1979; v.p. engring. devel. Ocean Systems div. Gould, Inc., Cleve., 1979-83; head combat systems VSE Corp., 1983-85; v.p. def. systems Dynamac, 1985-88; sr. v.p., COO CASDE Corp., 1988-91; sr. v.p. Simms Industries, Inc., 1991—. Decorated Navy Meritorious Service medal, Navy Commendation medal with gold star. Mem. Navy League U.S. (nat. treas. 1987-92, nat. v.p. finance 1992—), U.S. Naval Inst., Am. Soc. Naval Engrs., Am. Def. Preparedness Assn., Surface Navy Assn. Home: 3139 Catrina Ln Annapolis MD 21403 Office: Simms Industries 9841 Broken Land Pkwy Ste 300 Columbia MD 21046

RAPOPORT, JUDITH, psychiatrist; b. N.Y.C., July 12, 1933; d. Louis and Minna (Enteen) Livant; m. Stanley Rapoport, June 25, 1961; children: Stuart, Erik. BA, Swarthmore Coll., 1955; MD, Harvard U., 1959. Lic. psychiatrist. Cons., child psychiatrist NIMH/St. Elizabeth's Hosp., Washington, 1969-72; clin. asst. prof. Georgetown U. Med. Sch., Washington, 1972-82, clin. assoc. prof., 1982-85, clin. prof. psychiat., 1985—; med. officer biol. psychiatry br. NIMH, Bethesda, Md., 1976-78, chief, child mental illness unit, biol. psychiat. br., 1979-82, chief, child psychiatry lab. of clin. scis., 1982-84, chief, child psychiatry div. intramural rsch. programs, 1984—; prof. psychiatry George Washington U. Sch. Med., Washington, 1979—; cons. in field. Author: (non-fiction) The Boy Who Couldn't Stop Washing, 1989 (best seller literary guild selection 1989), Childhood Obsessive Compulsive Disorder, 1989. Fellow Am. Psychiat. Assn., Am. Acad. Child Psychiat.; mem. Ea. Psychol. Assn., D.C. Psychiat. Assn., Am. Psychopath. Assn. Home: 3010 44th Pl NW Washington DC 20016-3557 Office: NIMH Bldg 10 Rm 6N240 Bethesda MD 20892

RAPOZA, NORBERT PACHECO, medical association administrator, virologist; b. New Bedford, Mass., Oct. 23, 1929; s. Ernest P. and Mary Dos Anjos (Silva) R.; m. Junia Bratter, Sept. 19, 1954; children: Darion, Julia Michelle. BA, Oberlin Coll., 1952; PhD, Ohio State U., 1960. Rsch. assoc. U. Pitts., 1960-61, instr., 1961-65, asst. prof., 1965-67; sr. scientist G.D. Searle and Co., Skokie, Ill., 1967-82; sr. scientist AMA, Chgo., 1982-89, dir. immunology and infectious diseases, 1989—; rsch. com. mem. Am. Cancer Soc., Chgo., 1977-79; bd. dirs. United Network for Organ Sharing, Richmond, Va. Author: (chpts.) Drug Evaluations, 1982—; editor: HIV Infection and Disease, 1990. Mem. AAAS, Am. Soc. Microbiology, Internat. Soc. Antiviral Rsch., Internat. AIDS Soc. Home: 530 Greenleaf Ave Wilmette IL 60091 Office: Am Med Assn 515 N State St Chicago IL 60610

RAPP, FRED, virologist; b. Fulda, Germany, Mar. 13, 1929; came to U.S., 1936, naturalized, 1945; s. Albert and Rita (Hain) R.; children: Stanley I., Richard J., Kenneth A.; m. Pamela A. Miles, Aug. 28, 1988. BS, Bklyn. Coll., 1951; MS, Albany Med. Coll., Union U., 1956; PhD, U. So. Calif., 1958. Jr. bacteriologist, div. labs. and research N.Y. State Dept. Health, 1952-55; from teaching asst. to instr. dept. med. microbiology Sch. Medicine U. So. Calif., 1956-59; cons. supervisory microbiologist Hosp. Spl. Surgery, N.Y.C., 1959-62; also virologist div. pathology Philip D. Wilson Research Found., N.Y.C.; asst. prof. microbiology and immunology Cornell U. Med. Coll., N.Y.C., 1961-62; assoc. prof. Baylor U. Sch. Medicine, Waco, Tex., 1962-66, prof., 1966-69; prof., chmn. dept. microbiology and immunology Pa. State U. Coll. Medicine, University Park, 1969-90, Evan Pugh prof. microbiology, 1978-90, prof. emeritus, 1990—, assoc. provost, dean health affairs, 1973-80, sr. mem. grad. faculty, assoc. dean acad. affairs, research and grad. studies, 1987-90, professor emeritus, 1990—; researh career prof. of virology Am. Cancer Soc., 1966-69, prof. virology, 1977-90; dir. Coll. Med. Pa. State U. (Specialized Cancer Research Ctr.), 1973-84; mem. del. on viral oncology, U.S./USSR Joint Com. Health Cooperation; chmn., Gordon Rsch. Conf. in Cancer, 1975; virology Task Force, 1976-79; chmn. Atlantic Coast Tumor Virology Group, Nat. Cancer Insts. Health, 1971-77; mem. council for projection and analysis Am. Cancer Soc., 1976-80; chmn. standards and exam. com. on virology Am. Bd. Med. Microbiology, 1977, 80; chmn. subsect. on virology program com. Am. Assn. Cancer Research, 1978-79; mem. adv. council virology div. Internat. Union Microbiol. Socs., 1978-84; referee Macy Faculty Scholar Award Program, 1979-81; mem. programme com. Fifth Internat. Congress for Virology, Strasbourg, France, 1981; mem. basic cancer rsch. group U.S.-France Agreement for Cooperation in Cancer Research, 1980-84; mem. organizing com. Internat. Workshop on Herpes viruses, Bologna, Italy, 1980-81, NATO Internat. Advanced Study Inst., Corfu Island, Greece, 1981; mem. Herpes viruses Study Group, 1981-84; mem. sci. adv. com. Wilmot Fellowship Program, U. Rochester Med. Ctr., 1981-90 ; mem. scientific rev. com. Hubert H. Humphrey Cancer Research Ctr., Boston U., 1981; mem. fin. com. Am. Soc. Virology, 1982-89, mem. council, 1984-88; mem. adv. com. persistent virus-host interactions research program R.J. Reynolds Scientific Bd./Wistar Inst., 1983—; mem. med. adv. bd. Herpes Resource Ctr., Am. Social Health Assn., 1983-90; bd. dirs. U.S.-Japan Found. Biomedicine, 1983-90; mem. council Soc. Exptl. Biology and Medicine, 1983-87; mem. scientific adv. com. Internat. Assn. Study and Prevention of Virus-Associated Cancers, 1983-90; mem. Basil O'Connor Starter Research Adv. Com., 1984-90; mem. council for research and clin. investigation awards Am. Cancer Soc., 1984—; mem. recombinant DNA adv. com. NIH, 1984-87; mem. outstanding investigator grant rev. com. Nat. Cancer Inst., 1984—; mem. organizing com. Fourth Symposium Sapporo Cancer Seminar, Japan, 1984, Second Internat. Conf. Immunobiology and Prophylaxis of Human Herpes virus Infections, Ft. Lauderdale, Fl., 1984-85, Internat. Congress of Virology, Sendai, Japan, 1984; mem. internat. sci. com. Internat. Meeting on Adv. in Virology, Catania, Italy, 1984-85; mem. adv. bd. Cancer Info., Dissemination and Analysis Ctr. Carcinogenesis and Cancer Biology, 1984-89; mem. internat. programme com. 7th Internat. Congress of Virology, Edmonton, Can., 1985-87; councilor div. DNA viruses Am. Soc. Microbiology, 1985-87; mem. adv. com. rsch. on etiology, diagnosis, natural history, prevention and therapy of multiple sclerosis Nat. Multiple Sclerosis Soc., 1985-89; mem. sci. adv. bd. Showa U. Research Inst. for Biomedicine in Fla., 1985-90; mem. sci. adv. coun. Pitts. Ctr. AIDS Rsch. U. Pitts., 1988. Sect. editor on oncology: Intervirology, 1972-84, assoc. editor, 1978-84, editor-in-chief, 1985-90; adv. bd. Archives Virology, 1976-81; editorial bd. Jour. Immunology, 1966-73, Jour. Virology, 1968-88; assoc. editor Cancer Research, 1972-79; editorial bd. Virology, 1979-83, editor, 1983-90. Recipient 1st CIBA-Geigy Drew award for biomed. research, 1977, Nat. award for teaching excellence in microbiology, U. Medicine and Dentistry N.J. Med. Sch., 1988; Wellcome vis. professorship in microbiology, 1989-90. Mem. AAAS, AAUP, Am. Acad. Microbiology (fellow, diplomate), Soc. Microbiology (mem. com. med. microbiology and immunology, bd. pub. sci. affairs 1979-88, chmn. DNA viruses div. 1981-82), Am. Soc. Virology (chmn. fin. com. 1987-88), , Am. Assn. Immunologists, The Harvey Soc., Soc. Exptl. Biology and Medicine, Am. Assn. Cancer Rsch., Assn. Med. Sch. Microbiology Chmn. (pres. 1980-81), Sigma Xi (Monie A. Ferst award 1990), Alpha Omega Alpha. Home: 68 Azalea Dr Hershey PA 17033-2602

RAPP, ROBERT ANTHONY, metallurgical engineering educator, consultant; b. Lafayette, Ind., Feb. 21, 1934; s. Frank J. and Goldie M. (Royer) R.; m. Heidi B. Sartorius, June 3, 1960; children: Kathleen Rapp Raynaud, Thomas, Stephen, Stephanie Rapp Surface. BSMetE, Purdue U., 1956; MSMetE, Carnegie Inst. Tech., 1959, PhDMetE, 1960. Asst. prof. metall. engring. Ohio State U., Columbus, 1963-66, assoc. prof., 1966-69, prof., 1969—, M.G. Fontana prof., 1988—, Univ. prof., 1989—; cons. DuPont, Lanxide, KB Alloys, IGT; vis. prof. Ecole Nat. Superior d'Electrochimie, Grenoble, France, 1972-73, U. Paris-Sud, Orsay, 1985-86, Ecole Nat. Superior de Chimie, Toulouse, France, 1985-86, U. New South Wales, Australia, 1987; Acta/Scripta Metallurgica lectr., 1991; rsch. metallurgist WPAFB, Ohio, 1960-63. Editor: Techniques of Metals Research, vol. IV, 1982, High Temperature Corrosion, 1984; translator Metallic Corrosion (Kaesche), 1986; bd. rev. jour. Oxid. Metals; contbr. numerous articles to profl. jours. Recipient W.R. Whitney award Nat. Assn. Corrosion Engring., 1986, NAE award, 1988, Disting. Engring. Alumnus award Purdue U., 1988, B.F. Goodrich Collegiate Inventor's award, 1991, 92, Ulrick Evans award Brit. Inst. Corrosion, 1992; Guggenheim fellow, 1972; Fulbright scholar Max Planck Inst. Phys. Chem., 1959-60. Fellow Am. Soc. Metals Internat., Mining, Metals and Materials Soc., Electrochem. Soc. (assoc. editor jour.), French Soc. Metals and Materials (hon.); mem. Nat. Assn. Corrosion Engrs. (bd. rev. jour. Corrosion 1984—, Oxid. Metals). Lutheran. Home: 1379 Southport Dr Columbus OH 43235-7649

RAPPAPORT, LAWRENCE, plant physiology and horticulture educator; b. N.Y.C., May 28, 1928; s. Aaron and Elsie R.; m. Norma, Nov. 21, 1953; children: Meryl, Debra Kramer, Craig. BS in Horticulture, U. Idaho, 1950; MS in Horticulture, Mich. State Coll., 1951; PhD in Horticulture, Mich. State U., 1956. Olericulturist Dept. Vegetable Crops U. Calif., Davis, 1956-58, asst. prof. Dept. Vegetable Crops, 1958-63, assoc. prof. Dept. Vegetable Crops, 1963-67, prof. Dept. Vegetable Crops, 1967-91, prof. emeritus, 1991—, dir. plant growth lab., 1975-78, chair Dept. Vegetable Crops, 1978-84; vis. scientist Calif. Inst. Tech., 1958; co-dir. Horticulture Subproject, Calif./Egypt project, 1978-82. Contbr. articles to profl. jours. 1st pres. Davis Human Rels. Coun., 1964-66; v.p. Jewish Fedn. Sacramento, 1969; pres. Jewish Fellowship, Davis, 1985-89. Sgt. maj. U.S. Army, 1952-53, Korea. Decorated Bronze star; Guggenheim Found. fellow, 1963, Fulbright fellow, 1964, USPHS Spl. fellow, 1970, Am. Soc. Horticulture Sci. fellow, 1987, Sir Frederick McMaster fellow, 1991. Achievements: discovery of evidence for gibberellin-binding protein in plants; evidence for the signal hypothesis operating in plants, positive evidence for phytochrome-mediated gibberellin metabolism and stem growth; isolation of somaclonal variants of celery bearing stable resistance to Fusarium oxysporum f. sp. apii. Home: 637 Elmwood Dr Davis CA 95616-3514 Office: U Calif Vegetable Crops Dept Davis CA 95616

RAPPAPORT, MARTIN PAUL, internist, nephrologist, educator; b. Bronx, N.Y., Apr. 25, 1935; s. Joseph and Anne (Kramer) R.; B.S., Tulane U., 1957, M.D., 1960; m. Bethany Ann Fitzgerald; children—Karen, Steven, Sheila. Intern, Charity Hosp. of La., New Orleans, 1960-61, resident in internal medicine, 1961-64; practice medicine specializing in internal medicine and nephrology, Seabrook, Tex., 1968-72, Webster, Tex., 1972—; mem. courtesy staff Mainland Ctr. Hosp. (formerly Galveston County (Tex.) Meml. Hosp.), 1968—, Bapt. Meml. System, 1969-72, 88—; mem. staff Clear Lake Regional Med. Ctr., 1972—; cons. staff St. Mary's Hosp., 1973-79; cons. nephrology St. John's Hosp., Nassau Bay, Tex.; fellow in nephrology Northwestern U. Med. Sch., Chgo., 1968; clin. asst. prof. in medicine and nephrology U. Tex., Galveston, 1969—; lectr. emergency med. technician course, 1974-76; adviser on respiratory therapy program Alvin (Tex.) Jr. Coll., 1976-82; cons. nephrology USPHS, 1979-80. Served to capt. M.C., U.S. Army, 1961-67. Diplomate Am. Bd. Internal Medicine, Nat. Bd. Med. Examiners. Fellow ACP, Am. Coll. Chest Physicians; mem. Internat., Am. Socs. Nephrology, So. Med. Assn., Tex. Med. Assn., Am. Soc. Artificial Internal Organs, Tex. Acad. Internal Medicine, Harris County Med. Soc., Am. Geriatrics Soc., Bay Area Heart Assn. (bd. govs. 1969-75), Clear Lake C. of C., Phi Delta Epsilon, Alpha Epsilon Pi, Tulane Alumni Assn. Lodge: Rotary. Home: 1818 Linfield Way Houston TX 77058-2324 Office: PO Box 57609 Webster TX 77598-7609

RAPPAPORT, THEODORE SCOTT, electrical engineering educator; b. Bklyn., Nov. 26, 1960; s. Eugene and Carol Ann (Cooper) R.; m. Brenda Marie Velasquez, May 30, 1981; children: Matthew B., Natalie M., Jennifer L. BSEE, Purdue U., 1982, MSEE, 1984, PhD, 1987. Registered profl. engr., Va. Engring. coop. Magnavox Govt./Ind. Elec. Co., Fort Wayne, Ind., 1980-81; engr. Harris Corp., Melbourne, Fla., 1983, systems engr., 1986; rsch., teaching asst. Purdue U., West Lafayette, Ind., 1982-87, postdoctoral rsch. assoc., 1987-91; asst. prof. Va. Poly. Inst. and State U., Blacksburg, 1992—, assoc. prof., 1992—; pres. TSR Tech., Inc.; cons. Ralph M. Parsons Co., Pasadena, Calif., 1992, Panasonic, Inc., Secaucus, N.J., 1992, Ericsson/GE Mobile Communications, Lynchburg, Va., 1992. Coauthor: Wireless Personal Communications, 1992; contbr. articles to profl. jours. V.p. Gilbert Linkous Elem. Sch. PTA, Blacksburg, 1990-91; exec. com., asst. den leader Boy Scouts of Am., Blacksburg, 1992. Named Marconi Young scientist IEEE, 1990, one of Young Men of Am., 1988, Pres. Faculty fellow NSF, 1992. Fellow Radio Club of Am.; mem. NSPE, IEEE (sr. mem.), Am. Radio Relay League, Am. Soc. for Engring. Edn.. Achievements include patents in Tunable Discone Antenna, computer-based bit error simulation for digital wireless communications; patents pending for real-time DSP receivers. Office: Va Tech MPRG-Bradley Dept Elec Engring 340 Whittemore Hall Blacksburg VA 24061-0111

RASHFORD, DAVID R., physicist, educator; b. Lancaster, Pa., July 23, 1947; s. Leonard Raymond and Emma Jean (Lehman) R.; m. Karen Louise Phelps, Dec. 12, 1975; children: Jonathan, Emily. BA in Physics, Stevens Inst. Tech., 1969; PhD, Worcester Poly. Inst., 1975. Postdoctoral rsch. fellow UCLA, 1976-78, rsch. scientist, 1978-83; asst. prof., researcher Worcester (Mass.) Poly. Inst., 1983-89, assoc. prof., 1989-92, assoc. head dept. physics, 1990-92; assoc. prof. Ga. Inst. Tech., Atlanta, 1992—; rsch. scientist Werik Rsch., Atlanta, 1992—; vis. assoc. prof. Boston U., 1989-90; cons. NASA, 1985—, Grumman Aerospace, 1987—. Co-author: Quantum Theory in Real Time Environments, 1989, Relativity Revisited, 1991; reviewer Jour. Quantum Mechanics, Jour. Motion Physics, Phys. Review, Physics Letters; contbr. over 75 articles to sci. and profl. publs. Active Atlanta chpt. Boy Scouts Am., 1992—, March of Dimes, Atlanta, 1992-93. Grantee NASA, 1985, 87-88, 90-92, Holt Found., 1989-90; Wendell McGee scholar Stevens Inst. Tech., 1969-71. Mem. AAAS, AAUP, ASTM, Am. Phys. Soc. (Edward Shaklee award 1990), Quantum Mechanics Soc. Am. (sec. 1980-85, pres. 1986-88, v.p. 1988-90, pres. 1990-92), Molecular Soc., Worcester Poly. Alumni Assn. (sec. 1976-77), Atlanta C. of C., Sigma Xi. Achievements include measurement of electron flux capacity in inert gasses; research in quantum mechanics theory and applications, polymer-bonded materials testing. Office: Weirk Rsch 3500 Piedmont Rd NE Ste 612 Atlanta GA 30305-1503

RASHID, KAMAL A., university international programs director, educator; b. Sulaimania, Kurdistan, Iraq, Sept. 11, 1944; came to U.S., 1972; s. Ahmad Rashid and Habiba M. Muhiedin; m. Afifa B. Sabir, May 23, 1970; children: Niaz K., Neian K., Suzanne K. BS, U. Baghdad, Iraq, 1965; MS, Pa. State U., 1974, PhD, 1978. Lab. instr. U. Baghdad, Iraq, 1966-72; mem. faculty U. Basrah, Iraq, 1978-80, U. Sulaimania, Iraq, 1980-83; sr. rsch. assoc., vis. prof. Pa. State U., University Park, 1983—, rsch. assoc. prof., 1992—; dir. Biotech. Workshop program Pa. State U., 1989—, dir. summer symposium molecular biology, 1991—, dir. Instrnl. and Internat. programs biotech. inst., 1993—; v.p. Cogenic Inc., State College, Pa., 1989-90; cons. in field; spkr. in field; developer workshops in field. Contbr. articles to profl. jours. Iraqi Ministry Higher Edn. scholar. Mem. Am. Chem. Soc., Environ. Mutagen Soc., Pa. Biotech. Assn. (mem. edn. com.), Rotary. Avocations: travel, swimming, reading. Home: 100 Berwick Dr Boalsburg PA 16827 Office: Pa State U 519 Wartik Lab University Park PA 16802

RASHKOVETSKY, LEONID, biochemist; b. Enakievo, Ukraine, Apr. 21, 1953; came to U.S., 1990; s. Grigori and Bella R.; m. Marianna Shulutko, Aug. 9, 1975; children: Arthur, Philip. MS, Moscow State U., 1975; PhD, Moscow Med. Inst., 1983. Sr. rsch. asst. Inst. Bio-Organic Chemistry, Moscow, 1974-75; sr. rsch. asst. Inst. Med. Enzymology, Moscow, 1975-79, scientist, 1979-87; sr. scientist Inst. Med. Enzymology (now Inst. Biol. and Med. Chemistry), Moscow, 1987-90; instr. Ctr. for Biochem./Biophys. Scis. and Medicine Harvard Med. Sch., Boston, 1990—. Contbr. articles to profl. jours. Recipient participant's award Nat. Achievements Exhbn., Moscow, 1989. Mem. AAAS, Am. Chem. Soc., Russian Nat. Biochem. Soc. Achievements include U.S.S.R. patent for method of determination of lactate dehydrogenase isoenzyme 1 in serum. Home: 1345 Center St Newton MA 02159 Office: CBBSM Harvard Med Sch SGM Bldg 250 Longwood Ave Boston MA 02115

RASINES, ISIDORO, chemist; b. Matanzas, Cuba, June 18, 1927; s. Isidoro and Perfecta (Linares) R. Licenciate in Chemistry, Facultad de Ciencias, Madrid, 1950, PhD in Chemistry, 1970. Prin. Colegio Gaztelueta, Lejona, Vizcaya, Spain, 1951-59; chancery assoc. U. Navarra, Pamplona, Spain, 1960-69, registrar, 1970-79; researcher Higher Coun. of Sci. Rsch., Madrid, 1980-88, dir. rsch., 1989—; councilman Nat. Coun. Edn., Madrid, 1964-67, European Community Chemistry Coun., London, 1990—. Author: Co,Ni,Cu,Zn Mixed Oxides, 1970; contbr. over 100 articles to profl. jours. Mem. UNESCO Spanish Com., Madrid, 1967-81. With Spanish Army, 1953-54. Fellow Chem. Soc. U.K.; mem. Am. Chem. Soc., Royal Spanish Soc. Chemistry (v.p. 1987-91), German Chem. Soc., Internat. Conf. Thermal Analysis, Materials Rsch. Soc., European Fedn. Chem. Socs. (councilman 1990-93). Roman Catholic. Avocations: running, climbing, swimming, music, reading. Office: Instituto Ciencia Materiales, CSIC Serrano 113, Madrid Spain 28006

RASK, MICHAEL RAYMOND, orthopaedist; b. Butte, Mont., Oct. 24, 1930; s. Barth John and Marguerite Sadie (Joseph) R.; m. Elizabeth Anne Shannon, May 21, 1948; children: Dagny Marguerite Rask-Regan, Badih John, Patrick Henry, Molly Michelle. BS, Oreg. State U., 1951; MD, Oreg. Health Scis. U., 1955; PhD, 1978, U. Humanistic Studies, 1986. Diplomate Am. Bd. Orthopaedic Surgery, Am. Bd. Neurological Orthopaedic Surgery, Am. Bd. Bloodless Surgery, Am. Bd. Medical-Legal Analysts, Am. Bd. Hand Surgery, Am. Bd. Sportsmedicine Surgery, Am. Bd. Spinal Surgery. Intern Kings County Hosp., Bklyn., 1955-56; orthopaedic resident U. Oreg. Med. Sch., Portland, 1959-63; with neurological orthopaedic surgery preceptorships Oreg. Emmanuel Hosp., Portland, 1962-76; pvt. practice in neurol. orthopedic surgery Las Vegas, 1976—; clin. instr. orthopaedics U. Oreg., 1964-71; prof. Am. Acad. Neurological and Orthopaedic Surgery, 1985—; editorial reviewer Clin. Orthopaedics & Related Rsch., 1978—, Am. Med. Reports, 1985—, Muscle & Nerve, 1987—, Am. Jour. Cranio-Mandibular Practice. Author: Seminoma, 1970, Orthopod, 1972; editor in-chief: Jour. Neurological Orthopedic Medicine & Surgery, 1976—; editorial rev. bd. Jour. Craniomandibular Practice; numerous lectures in field. Lectr. Arthritis Found., Las Vegas, 1976-78, cons. orthopaedist Easter Seal Ctr. for Crippled Children & Adults, Las Vegas, 1978-81, med. advisor so. Nev. chpt. Nat. Multiple Sclerosis Soc.; bd. dirs. Gov's. Com. on the Employment of the Handicapped, Nev., 1980-82. Capt. USAF, 1956-63. Fellow Cuban Soc. Orthopaedics Traumatology; mem. Am. Acad. Neurological Orthopaedic Surgeons (hon. 1979, course chmn. 1977-79, pres. 1978, chmn. bd. dirs. 1976—), Nev. State Pharmacy Assn., Am. Back Soc. (bd. dirs. 1983-88), Semmelweiss Sci. Soc. (pres. Nev. chpt. 1980—), Am. Fedn. Med. Accreditation (chmn. 1979—), Neurol. Orthopaedic Inst. (chmn. 1979—), Bd. Neurol. Orthopaedic Surgeons (chmn. 1977—), Sundry Primary Certifying Bds. (chmn. bd. dirs. 1976—), Silkworm Club, Caterpillar Club. Democrat. Avocations: painting, recording. Office: Am Acad Neurol Orthopaedic Surgeons 2320 N Rancho Dr Ste 108 Las Vegas NV 89102-4510

RASKIN, STEVEN ALLEN, academic program director; b. N.Y.C., July 6, 1953; s. Irving Robert and Adele (Yablon) R.; m. Martha Anne Koushel, June 8, 1975. BS, U. Houston, 1975; perfusion cert., Tex. Heart Inst., Houston, 1976; MBA, U. Houston, 1983. CPA Tex.; cert. perfusion. Nuclear physics technician M.D. Anderson Hosp., Houston, 1975-76; staff perfusionist Baylor Coll. Medicine, Houston, 1976-83, bus. mgr., 1983-87; ptnr. Willis & Co., CPA, Houston, 1986-88; program dir. Baylor Coll. Medicine, Houston, 1988—; Contbr. articles to Perfusion jour. Com. chmn. Brotherhood of Congregation Beth El, Missouri City, Tex., 1992; alternate del. state Republican Party, Houston, 1980; candidate Houston Ind. Sch. Dist., 1972. Mem. AICPA, Tex. Soc. CPA, Houston Chpt. CPA, Am. Soc. Extracorporeal Technologists, Am. Acad. CV Perfusion. Republican. Jewish. Home: 2114 Walnut Grove Ln Richmond TX 77469-6646 Office: Baylor Coll Medicine One Baylor Plz Houston TX 77030

RASMUSSEN, DENNIS ROBERT, behavioral ecologist; b. Glendale, Calif., July 14, 1949; s. Robert Harvey and Verla (Alvinia) R.; m. Carol Ann Buckey, Feb. 14, 1983; children: Blake, Mark, David. AB, Humboldt State U., 1971; MA, U. Calif., Riverside, 1972, PhD, 1978. Prin. investigator Calif. Primate Ctr., Davis, 1978-82; prin. investigator Wis. Primate Ctr., Madison, 1986-90, NIMH sr. rsch. assoc., 1986-89; founder, dir. Animal Behavior Rsch. Inst., Madison, 1982—; founder, dir. animal behavior rsch. unit Mikumi Nat. Park, Tanzania, 1974-76, Tamarin Res., Panama, 1981—: lectr. U. Calif., Davis, 1979-81, U. Calif., Berkeley, 1981, 82; asst. prof. Calif. State U., Sacramento. Contbr. articles to Animal Behaviour, Folia primatol, Psychol. Rev., Ecology and Behavior. Coach soccer team, Madison, 1990-91. Recipient Conservation award U. Western Ont.; fellow Darwin Coll. 1977, 78; grantee NSF, 1964, 66, 74-76, 88-89, Resources for the Future, 1974-75, Westinghouse scholarships, 1988. Mem. Am. Psychol. Soc., Am. Soc. Primatologists, Internat. Primatological Soc., Animal Behavior Soc., Behavioral and Brian Scis. Assoc., Internat. Soc. for Comparative Psychology. Achievements include field studies of yellow baboons, stumptail macaques and Panamanian tamarins, also, Japanese rhesus and bonnet macaques. Home and Office: Animal Behavior Rsch Inst 314 S Randall St Madison WI 53715

RASMUSSEN, GUNNAR, engineer; b. Esbjerg, Denmark, Nov. 23, 1925; s. Karl Sigurd and Frederikke Valentine (Gjeruiff) R.; m. Hanna Hertz, June 27, 1973; children: Jan, Lise, Per, Thue. Student, Aarhus Teknikum, Denmark, 1950. Mgr. quality control Brüel and Kjaer, Nerum, Denmark, 1950-54, with devel. div., 1955-69, with product planning div., 1969-74, with innovation div., 1975—; lectr. Danish Tech. U., Copenhagen, 1974-79, Med. Air Force Acad., Jegersborg, Denmark, 1978-79; examiner Danish Engring. Acad., Copenhagen, 1972—, Chalmers Tech. U., Gothenburg, Sweden, 1984-85. Editor: Intensity Measurements, 1989; inventor measurement microphones, accelerometers; contbr. articles to profl. jours. Chairman Audio Engring. Soc. Denmark, Cophenhagen, 1976. Recipient Danish Design prize for microphones, 1962, medal for contbn. to intensity techniques SETIM, 1990. Fellow Acoustical Soc. Am., Can. Acoustical Soc., Danish Medico Tech. Soc.; mem. Internat. Union Pure and Applied Physics (vice chmn. internat. commn. on accoustics), Danish Acoustical Soc. (bd. dirs.), Internat. Electronical Commn., Internat. Orgn. for Standarization. Home: Hojbjerggardsvej 15, 2840 Holte Denmark Office: Electronica, Høse Skodsborgues 22, 2942 Skodsborg Denmark

RASMUSSEN, JAMES MICHAEL, research mechanical engineer, inventor; b. Evanston, Ill., June 15, 1959; s. Walter Raymond and Nanci Jean (Litke) R.; m. Noreen J. Mucha, Sept. 30, 1978; children: Brian Joseph, Kevin Walter. Student, Wright Jr. Coll., Chgo., 1978-85. Draftsman Riddell Inc., Chgo., 1976-77; layout draftsman H.K. Porter Co., Chgo., 1977-78; design draftsman S & C Electric Co., Chgo., 1978-80; draftsman Cummins-Allison Corp., Chgo., 1980-83; jr. design engr. Cummins-Allison Corp., Mt. Prospect, Ill., 1983-86, design engr., 1986-88, rsch. engr., 1988—. Tech. illustrator svc. manuals; patent illustrator. Office: Cummins-Allison Corp 891 Feehanville Dr Mount Prospect IL 60056-6098

RASMUSSEN, NIELS LEE, geologist, environmental chemist; b. Douglas, Wyo., Dec. 26, 1956; s. Tom Charles and Iris JoAnn (Lee) R.; m. Wanda L. Stoumbaugh, Nov. 28, 1981; children: Niels Lee II, Katrina Rae-Ann. AA, Ea. Wyo. Coll., 1977; BS in Geology, U. Wyo., 1980. Logging geologist Exploration Logging, Houston, 1980-89; stack technician Western Environ. Svcs., Casper, Wyo., 1989-90, lab. supr., 1990—. Mem. Am. Water Works Assn. Home: 14 Lakota Ct Glenrock WY 82637 Office: Western Environ Svcs 6756 W Uranium Rd Casper WY 82604

RASMUSSEN, NORMAN CARL, nuclear engineer; b. Harrisburg, Pa., Nov. 12, 1927; s. Frederick and Faith (Elliott) R.; m. Thalia Tichenor, Aug. 23, 1952; children: Neil, Arlene. BA, Gettysburg (Pa.) Coll., 1950, ScD (hon.), 1968; PhD, MIT, 1956; Dr honoris causa, Cath. U. Leuven, Belgium, 1980. Mem. faculty MIT, 1958—, prof. nuclear engring., 1965—, McAfee prof. engring., 1983—, head dept., 1975-81; dir. reactor safety study AEC, 1972-75; mem. Def. Sci. Bd., 1974-78, sr. cons., 1978-84; mem. nat. sci bd. NSF, 1982-88; mem. Presdl. Adv. Group on Contbns. Tech. to Econs. Strength, 1975; dir. Bedford Engring. Corp.; adv. com. components tech. div. Argonne Univs. Assn., 1977-80; chmn. sci. rev. com. Idaho Nat. Engring. Lab., 1977-86 ; mem. adv. council Princeton Plasma Physics Lab., 1983-91; mem. safety adv. bd. Three Mile Island Unit-2 Program, 1980-90; trustee N.E. Utilities, 1977—. Co-author: Modern Physics for Engineers, 1966; contbr. articles to profl. jours. Chmn. Lincoln-Sudbury Sch. Com., 1971-72; bd. dirs. Atomic Indsl. Forum, 1980-84; mem. Utility Sci. Review Council for Nuclear Safety Analysis Center. Served with USNR, 1945-46. Recipient Disting. Achievement award Health Physics Soc., 1976; Disting. Service award NRC, 1976; Disting. Alumni award Gettysburg Coll., 1983; Enrico Fermi award Dept. Energy, 1985. Fellow Am. Nuclear Soc. (spl. award nuclear power reactor safety 1976, Theos J. Thompson award nuclear reactor safety div. 1981, bd. dirs. 1976-80); mem. Am. Acad. Arts and Scis., Nat. Acad. Engring., Nat. Acad. Sci., Soc. Risk Analysis, Health Physics Soc. Home: 80 Windsor Rd Sudbury MA 01776-2371 Office: MIT 77 Massachusetts Ave Cambridge MA 02139

RASMUSSON, DONALD C., agronomist, educator. Prof. U. Minn., St. Paul. Recipient Agronomic Achievement award Am. Soc. Agronomy, 1990. Office: Univ of Minnesota Saint Paul MN 55101*

RASMUSSON, EUGENE MARTIN, meteorology researcher; b. Lindsborg, Kans., Feb. 27, 1929; s. Martin Erick and Alma Sophia (Nelson) R.; m. Georgene Ruth Sachtleben, Aug. 7, 1960; children: Mary, Ruth, Elizabeth, Kristin. BS, Kans. State U., 1950; MS, St. Louis U., 1963; PhD, MIT, 1966. Forecaster Nat. Weather Service, St. Louis, 1956-64; rsch. meteorologist Geophysical Fluid Dynamics Lab, Princeton, N.J., 1964-70; chief, rsch. div. Ctr. for Experiment Design, NOAA, Washington, 1970-79; chief, diagnositc br. Climate Analysis Ctr., NOAA, Camp Spring, Md., 1979-86; sr. rsch. scientist U. Md., College Park, 1986—; mem. numerous coms. and panels U.S. Nat. Acad. Sci., World Climate Rsch. Program-UNESCO. Contbr. articles to profl. jours. 1st Lt. USAF, 1951-55. Recipient Silver medal, U.S.

Dept. Commerce, Washington, 1973, Adminstr. award, Nat. Oceanic and Atmospheric Adminstrn., Washington, 1983. Fellow Am. Meteorological Soc. (Jule Charney award, 1989); mem. Am. Geophysical Union, AAAS. Lutheran. Avocations: travel, history, gardening. Office: U Md Dept Meteorology College Park MD 20742

RASOR, ELIZABETH ANN, hydrogeologist, environmental scientist; b. Tuscaloosa, Ala., Aug. 30, 1962; d. Sam J. Rasor and Patricia Welling. BS, Coll. Charleston, 1984. Registered profl. geologist. Soil conservation technologist USDA Soil Conservation Svc., Charleston, S.C., 1984-86; geologist S.C. Dept. Health and Environ. Control, Columbia, 1986-88; compliance hydrogeologist HALLIBURTON NUS Environ. Corp., Aiken, S.C., 1988-92; waste mgmt. specialist HALLIBURTON NUS Environ. Corp., Aiken, 1989-92; hydrogeologist, regulatory specialist Martin Marietta Energy Systems, Inc., Oak Ridge, Tenn., 1992—; regulatory analyst HALLIBURTON NUS, Aiken, 1988—, project mgr., 1990—. Author: Use of Bentonite Grout, 1987. Vol. Habitat for Humanity, Aiken, 1992. Mem. Nat. Assn. Environ. Profls., Ctrl. Savannah River Area Geol. Soc. Avocations: horseback riding, volunteer elementary and middle school tutoring, reading. Office: Martin Marietta Energy Systems PO Box 2003 Oak Ridge TN 37831-7606

RASOULI, FIROOZ, chemical engineer, researcher; b. Tehran, Iran, Oct. 15, 1948; came to U.S., 1974; s. Khalil and Batool (Azami) R.; m. Mahnaz Allameh, July 4, 1978; children: Tannaz, Beeta. MSc, Ill. Inst. Tech., 1976, PhD, 1981. Prin. engr. Chamberlain GARD, Niles, Ill., 1981—. Contbr. articles to profl. jours. Fellow Am. Inst. Chemists; mem. Am. Inst. Chem. Engrs., Am. Chem. Soc., The Sci. Rsch. Soc. N.Am. Achievements include research in the development of a regenerative water recovery system for space applications (NASA); development of a catalytic process for detoxification of chlorinated hydrocarbons. Office: Chamberlain GARD 7449 N Natchez Ave Niles IL 60648-3801

RASSSEM, MOHAMMED HASSAN, sociology and cultural science educator; b. Munich, Apr. 27, 1922; arrived in Austria, 1968; s. Hassan and Elisabeth Fatima (Huber) R.; m. Theresia von Zumbusch, Oct. 31, 1957. Student, U. Munich, U. Vienna; MA, PhD, U. Basle, Switzerland, 1950. Contbr. Bavarian Radio, Munich, 1951-53; asst. U. Munich, 1954-59, lectr., 1959-63; prof. U. Saarbücken, Fed. Republic Germany, 1964-68, U. Salzburg, Austria, 1968-90; prof. emeritus U. Salzburg, 1990—; dean Philosophy Faculty U. Saarbrücken, 1966-67; mem. body tchrs. Hochschule Politik, Munich, 1975—. Author: Die Volkstumswissenschaften und der Etatismus, 2d edit., 1979, Stiftung und Leistung, Essais zur Kultursoziologie, 1979, Im Schatten der Apokalypse-Zur Deutschen Lage, 1984; editor: (with others) Zeitschrift für Politik, 1976—, Geschichte der Staatsbeschreibung, 1993; contbr. articles to profl. jours. Mem. Deutsche Gesellschaft Soziologie, Görresgesellschaft. Home: 6 Anton-Hochmuth-Str, A-5020 Salzburg Austria Office: Inst Kultursoziologie, 42 Rudolfskai, A-5020 Salzburg Austria

RATCHFORD, JOSEPH THOMAS, technical policy educator, consultant; b. Kingstree, S.C., Sept. 30, 1935; s. Raymond Howard and Elizabeth Arabella (Senn) R.; m. Joanne Walton Causey, June 18, 1960; children: Joseph Thomas, Laura Leigh, James Raymond, David Andrew. BS, Davidson (N.C.) Coll., 1957; MA, U. Va., 1959, PhD, 1961. Asst. prof. physics Washington and Lee U., Lexington, Va., 1961-64; physicist U.S. Air Force Office of Sci. Rsch., Arlington, Va., 1964-70; sci. cons. Com. on Sci. and Tech., U.S. Ho. of Reps., Washington, 1970-77; assoc. exec. officer AAAS, Washington, 1977-89; assoc. dir. Office of Sci. and Tech. Policy White House Office of Sci. and Tech. Policy, Washington, 1989-93; prof. internat. sci. and tech. policy George Mason U., Arlington, Va., 1993—; cons. internat. trade and tech., Alexandria, Va., 1993—, White House Office Sci. and Tech. Policy, 1993—; mem. tech. task force Coun. on Competitiveness, Washington, 1987-88; chmn. adv. com. for internat. programs NSF, Washington, 1984-87; chmn. adv. panel on energy from bio processes Office Tech. Assessment, 1979-80; chmn. Rsch. Coordination Panel, Gas Rsch. Inst., Chgo., 1976-79. Contbr. chpts. to books, numerous articles to profl. jours. Fellow AAAS; mem. Am. Phys. Soc., Coun. on Fgn. Fels., Va. Acad. Sci., Phi Beta Kappa, Sigma Xi. Achievements include development of legislation on a variety of science-based issues including establishment of Congl. Office of Technology Assessment; at the White House developed international science policy and math. and science edn. initiatives. Home: 8804 Fircrest Pl Alexandria VA 22308 Office: George Mason Univ Internat Inst 4001 N Fairfax Dr Ste 450 Arlington VA 22203

RATH, GUSTAVE JOSEPH, industrial engineering educator; b. N.Y.C., May 19, 1929; s. Gustavo Jules and Margaret Rose (Payor) R.; m. Clovia Pooch, July 23, 1953 (div. 1976); 1 child, Gustave Alexander; m. Karen S. Stoyanoff, Oct. 15, 1977. BS, MIT, 1952; PhD, Ohio State U., 1957. With IBM, 1957-59, Admiral Corp., Chgo., 1959-60, Raytheon Corp., Lexington, Mass., 1960-63; prof. indsl. engring. and mgmt. sci. Northwestern U., Evanston, Ill., 1963—; pres. Fundamental Systems, Inc., Evanston, 1967-70, Cassell, Rath & Stoyanoff, Ltd., Evanston, 1979-84; v.p. Resource Profls., Inc., Evanston, 1990—. Co-author: Marketing for Congregations, 1992; contbr. articles to profl. jours. 1st lt. USAF, 1952-54. Fellow AAAS, APA. Achievements include invention of computer assisted instrn. Home: 3555 Lyons St Evanston IL 60203 Office: Northwestern U Dept Indsl Engring Evanston IL 60208

RATHBUN, TED ALLAN, anthropologist, educator; b. Ellsworth, Kans., Apr. 11, 1942; s. Merle LeRoy and Wilma Louise (Hoth) R.; m. Babette Cowley, Nov. 24, 1964; 1 child, Joel Elliot. MA, U. Kans., 1966, PhD, 1971. Diplomate Am. Bd. Forensic Anthropology. Asst. prof. anthropology U. S.C., Columbia, 1971-74, assoc. prof., 1975-84, prof., 1984—; dep. state archaeologist, Columbia, 1985—; cons. pathology Med. U. S.C., Charleston, 1973—. Editor: Human Identification: Case Studies in Forensic Anthropology, 1984; contbr. articles to Am. Jour. Phys. Anthropology, Jour. Forensic Scis.; contbr. chpt. to Paleopathology of Transition to Agriculture, 1984; mem. editorial bd. Jour. Forensic Sci., 1987—. Vol. Peace Corps, Iran, 1966-68. Rsch. grantee NSF, 1986-87. Fellow Am. Acad. Forensic Sci. (chair phys. anthropology sect. 1991-92, sec. 1990-91); mem. Am. Bd. Forensic Anthropology (bd. dirs., treas. 1987-91). Achievements include research in forensic anthropology, human identification, African-American biohistory, health and disease, ancient Iran and Iraq health and disease. Office: U SC Dept Anthropology Columbia SC 29208

RATHI, MANOHAR, neonatologist, pediatrician; b. Beawar, Rajasthan, India, Dec. 25, 1933; came to U.S. 1969; s. Bagtawarmal and Sitadevi (Laddha) R.; m. Kamla Jajoo, Feb. 21, 1960; children: Sanjeev A., Rajeev. M.B.B.S., Rajasthan U., 1961. Diplomate Am. Bd. pediatrics; lic. physician, N.Y., Ill., Calif. Resident house physician internal medicine Meml. Hosp., Darlington, U.K., 1963-64; resident sr. house physician pediatrics Gen. Hosp., Oldham, U.K., 1964-65; dir. perinatal medicine, attending pediatrician Christ Hosp. Perinatal Ctr., Oak Lawn, Ill., 1974—; assoc. prof. pediatrics Rush Med. Coll., Chgo., 1979—; cons. obstetrician Christ Hosp., Oak Lawn, 1976—; cons. neonatologist Little Co. of Mary Hosp., Evergreen Park, Ill., 1972—, Palos Community Hosp., Palos Heights, Ill., 1978—, Silver Cross Hosp., Joliet, Ill., 1989—; cons./lectr. in field. Contbr. articles to profl. jours.; editor: Clinical Aspects of Perinatal Medicine, 1984, Vol. I, 1985, Vol. II, 1986, Current Perinatology, 1989, Vol. II, 1990; editor with others: Perinatal Medicine Vol. I, 1978, Vol. I, 1980, Vol. II, 1982. Hummell Found. grantee, 1976-77, WyethLab grantee, 1977-78; recipient Physicians Recognition award AMA, 1971-74, 91-92, Outstanding New Citizen's award State of Ill., 1978, Asian Human Svcs. of Chgo., 1988. Fellow Am. Acad. Pediatrics (perinatal sect.); mem. AMA, Chgo. Med. Soc., Ill. Med. Soc., Chgo. Pediatric Soc., Med. Soc. of County of Kings Bklyn., N.Y. Acad. Sci., Am. Thoracic Soc., Soc. Critical Care Medicine. Republican. Hindu. Office: Christ Hosp & Med Ctr Ste N232 4440 W 95th St Oak Lawn IL 60453

RATIU, MIRCEA DIMITRIE, mechanical engineer; b. Turda, Romania, Romania, Oct. 11, 1923; came to U.S., 1978; s. Augustin Stefan and Eugenia (Turcu) R.; m. Rodica P. Bucur; 1949; children: Tudor Stefan, Ion Mircea. M in engring., Polytech. Inst., Timisoara, Romania, 1948, D in Engring., 1967. Assoc. prof. Polytech. Inst., Timisoara, Romania, 1948-55, PhD adv., 1969-78; project mgr. The Machine Design Inst., Timisoara, Romania, 1950-54; division mgr. Nat. Inst. Metrology, Timisoara, Romania, 1954-67, Nat. Acad. Sci., Timisoara, Romania, 1967-70; rsch. mgr. The Inst. of Welding and Testing, Timisoara, Romania, 1970-78; lead sr. engr. EDS Nuclear, San Francisco, 1978-82; sr. tech. specialist Impell Corp., Walnut Creek, Calif., 1982-89, ASEA Brown Boveri Impell, San Ramon, Calif., 1989—; mem. working group Internat. Orgn. Legal Metrology, Paris, 1960-68, Internat. Orgn. Standardization, London, 1954-74, Internat. Inst. Welding, London, 1956-78. Author: Welding Residual Stresses, 1964, Reliability Concepts in Mechanical Engineering, 1973; co-author, editor; Testing and Analyses of Metals, 1964; co-author: Technique of Force Measurements, 1961, 2d edit., 1964; contbr. articles to profl. jours. Recipient Romanian Work Order, Romania, 1972, Romanian Scientific Order, Romania, 1967. Mem. ASTM. Republican. Roman Catholic. Achievements include founding of National Standards Laboratories for Measurements of Force, Hardness, and Stress/Strain Timisoara Romania with patented design of equipments validated by International Organization of Legal Metrology Paris, France; founder of Sclerometry as a part of Metrology dedicated to the Hardness Measurements. Home: 525 Mandana Blvd Apt 309 Oakland CA 94610 Office: ASEA Brown Boveri Impell Executive Pkwy 5000 San Ramon CA 94583

RATLIFF, HUGH DONALD, industrial engineering educator. Prof. engring. Ga. Inst. Tech., Atlanta. Recipient Baker Disting. Rschr. award Am. Inst. Indsl. Engrs., 1991. Office: Ga Inst Tech Detp Indsl Engring Atlanta GA 30332*

RATLIFF, ROBERT BARNS, JR., mechanical engineer; b. Narrows, Va., Oct. 24, 1950; s. Robert Barns and Rosemary (Simpson) R.; m. Marsha Meredith, Aug. 19, 1972; children: Lori Ann, Robert Barns III, Heather Michelle. BSME, Va. Tech, 1973. Registered profl. engr., N.C. Distbn. engr. Duke Power Co., Winston-Salem, N.C., 1973-76; distbn. svc. engr. Duke Power Co., Charlotte, N.C., 1976-77; supt. engring. Duke Power Co., Lenoir, N.C., 1977-79; gen. mgr. Floyd S. Pike Elec. Contractor, Inc., Mt. Airy, N.C., 1979-86, asst. v.p., 1986-90, v.p., 1990-91, exec. v.p., 1991—. Mem. exec. coun. Old Hickory Coun. of Boy Scouts, Winston-Salem, 1985; mem. C. of C., Mt. Airy, 1992; mem. adminstrv. bd. Ctrl. United Meth. Ch., Mt. Airy, 1992; bd. dirs. N.C. FREE. Mem. NSPE, Profl. Engr. of N.C., Internat. Soc. Arboriculture. Methodist. Office: Floyd S Pike Contractor Inc 351 Riverside Dr Mount Airy NC 27030

RATNER, BUDDY DENNIS, bioengineer, educator; b. Bklyn., Jan. 19, 1947; s. Philip and Ruth Ratner; m. Teri Ruth Stoller, July 7, 1968; 1 child, Daniel Martin. BS in Chemistry, Bklyn. Coll., 1967; PhD in Polymer Chemistry, Polytech. Inst. Bklyn., 1972. Postdoctoral fellow U. Wash., Seattle, 1972-73, from rsch. assoc. to assoc. prof., 1973-86, prof., 1986—; dir. Nat. ESCA and Surface Analysis Ctr., Seattle, 1984—. Editor: Surface Characterization of Biomaterials, 1988; mem. editoral bds. 8 jours. and book series; contbr. more than 200 articles to profl. jours. Recipient faculty achievement/outstanding rsch. award Burlington Resources Found., 1990, Perkin Elmer Phys. Electronics award for excellence in surface sci.; grantee NIH. Fellow Am. Inst. Med. Biological Engring. (founding), Am. Vacuum Soc.; mem. AAAS, Adhesion Soc., Am. Chem. Soc., Am. Inst. Chem. Engrs., Internat. Soc. Contact Lens Rsch. (coun. mem.), Materials Rsch. Soc., Soc. Applied Spectroscopy, Soc. for Biomaterials (pres. 1991, 1992, Clemson award 1989). Achievements include 10 patents in field. Office: U Wash Dept Chem Engring BF-10 Dept Chem Engring BF-10 Seattle WA 98195

RATNER, HAROLD, pediatrician, educator; b. Bklyn., June 19, 1927; s. George and Bertha (Silverman) R.; BS, Coll. City N.Y., 1948; MD, Chgo. Med. Sch., 1952; m. Lillian Gross, Feb. 4, 1961; children—Sanford Miles, Marcia Ellen. Intern, Jewish Hosp., Med. Center Bklyn., 1952-53, resident in pediatrics, 1953-55; practice medicine specializing in pediatrics, Bklyn.; clin. instr. pediatrics SUNY Downstate Med. Center, N.Y.C., 1955-67, clin. asst. prof., 1967-69, clin. assoc. prof., 1969-87; lectr. pediatrics, 1987—; chief of pediatrics Greenpoint Hosp., Bklyn., 1967-80, pres. med. staff, 1970-71, 74-80; dir. ambulatory services Woodhull Med. and Mental Health Center, Bklyn., 1980-83; clin. assoc. prof. pediatrics SUNY-Bklyn., 1983-87, lectr., 1987—; clin. assoc. prof. pediatrics, N.Y.U., 1987-90; med. specialist Nathan Kline Inst. for Psychiat. Research, Orangeburg, N.Y., Rockland Psychiat. Ctr., Orangeburg, N.Y., 1986-88, unit chief med. services, 1988-90; assoc. clin. dir. and dir. medicine Manhattan Psychiat. Ctr., N.Y.C., 1990—; mem. adv. council to pres. N.Y.C. Health and Hosp. Corp., 1970-71, 74-80, 81-83, sec., 1975, v.p., 1976-80; mem. med. bd., dir. Camp Sussex, camp for underprivileged children; bd. dirs. Kings County Health Care Rev. Orgn., Bklyn., 1976-84, past co-chmn. hosp. rev. com., continuing med. edn., med. care evaluation com. Trustee Village of Saddle Rock (N.Y.), 1980—. Served with AUS, 1945-47. Diplomate Nat. Bd. Med. Examiners, Am. Bd. Pediatrics. Fellow Am. Pediatric Soc., Am. Soc. Clin. Hypnosis, Bklyn. Pediatric Soc., Kings County Med. Soc., Royal Soc. Health; mem. AMA, Soc. Clin. and Exptl. Hypnosis, Am. Pub. Health Assn., Am. Soc. Clin. Hypnosis, N.Y. State Soc. Clin. Hypnosis, Kings County Med. Soc., Pan-Am. Med. Socs. Democrat. Jewish. Contbr. articles to med. jours. Home: 55 Blue Bird Dr Great Neck NY 11023-1001

RATNER, LILLIAN GROSS, psychiatrist; b. N.Y.C., Aug. 18, 1932; d. Herman and Sarah (Widelitz) Gross. BA, Barnard Coll., 1953; postgrad. U. Lausanne (Switzerland), 1954-56; MD, Duke U., 1959. Diplomate Bd. Pediatrics, Am. Bd. Psychiatry and Neurology, Am. Bd. Child Psychiatry; m. Harold Ratner, Feb. 4, 1961; children: Sanford Miles, Marcia Ellen. Intern Kings County Hosp., Bklyn., 1959-60, resident, 1967-70, fellow in child psychiatry, 1969-70, psychiatrist devel. evaluation clinic, 1970-72; resident Jewish Hosp. Bklyn., 1960-62, fellow in pediatric psychiatry, 1962-63; physician in charge pediatric psychiat. clinic Greenpoint (N.Y.) Hosp., 1964-67; pvt. practice psychiatry, Great Neck, N.Y., 1970—; clin. instr. psychiatry Downstate Med. Ctr., Bklyn., 1970-74, clin. asst. prof., 1974—; lectr. in psychiatry Columbia U., 1974—; psychiat. cons. N.Y.C. Bd. Edn., 1972-75, Queens Children's Hosp., 1975—; mem. med. bd. Camp Sussex (N.J.), 1963—, Saras Ctr., Great Neck, N.Y., 1977—. Fellow Am. Acad. Pediatrics, Am. Acad. Psychiatry, Am. Acad. Child Psychiatry, Am. Soc. Clin. and Experiential Hypnosis, N.Y. Soc. Clinical Hypnosis (past pres.); mem. AMA, Am. Psychiat. Assn., Nassau Psychiat. Assn., Bklyn. Psychiat. Assn., Bklyn. Pediatric Soc. (sr. mem.), Nassau Pediatric Socs., Soc. Adolescent Psychiatry, N.Y. Coun. Child Psychiatry, Soc. Clin. and Exptl. Hypnosis, Am. Med. Women's Assn. (past pres. Nassau), N.Y., Kings County med. socs., N.Y. Soc. Clin. Hypnosis (past pres.), Internat. Soc. for Study of Multiple Personality and Dissociation (founder, pres. L.I. component study group). Home and Office: 55 Blue Bird Dr Great Neck NY 11023-1001

RATNER, MARINA, mathemetician, educator, researcher; b. Moscow, Oct. 30. MA, Moscow State U., PhD. Asst. High Tech. Engring. Sch., Moscow, 1969-71; lectr. Hebrew U., Jerusalem, 1971-74, sr. tchr. pre-acad. sch., 1974-75; from acting asst. prof. to assoc. prof. U. Calif., Berkeley, 1975-82, prof., 1982—. Alfred P. Sloan rsch. fellow, 1977-79, Miller rsch. prof., 1985-86, John Simon Guggenheim fellow, 1987-88. Mem. NAS, AAAS. Office: Univ California Ctr Pure & Applied Math Berkeley CA 94720

RAU, ERIC, chemist; b. Weissenfels, Fed. Republic of Germany, Sept. 25, 1928; came to U.S., 1939; s. Simon and Beate (Schloss) R.; m. Anita Doris Goldrich, July 4, 1955; children: Allen Howard, Gerald Stuart. BA cum laude, NYU, 1951, PhD, 1955. Chemist U.S. Naval Air Rocket Test Sta., Dover, N.J., 1950-52; sr. engr. Bettis Atomic Power div. Westinghouse, Pitts., 1955-60; asst. dir. R&D indsl. chems. FMC Corp., Princeton, N.J., 1960-78; v.p. tech. CSI Inc., Horsham, Pa., 1979-84; dir. Spex Industries, Edison, N.J., 1985; v.p. E.A. Labs., Sparks, Md., 1986; asst. dir. N.J. Dept. Environ. Protection and Energy, Trenton, 1987—; mem. com. on nat. accreditation of environ. labs. EPA, Washington, 1991-92; gov.'s rep. to EPA/State focus group on nat. accreditation, 1993—. Contbr. articles to profl. jours. Chmn. edn. Jewish Fedn., Trenton, 1970-74, Bros. of Israel Synagogue, Trenton, 1965-72. Fellow Am. Inst. Chemists; mem. Assn. Rsch. Dirs. (bd. dirs.), Am. Chem. Soc. Achievements include 17 patents in field. Home: 17 Pine Knoll Dr Lawrenceville NJ 08648 Office: NJ Dept Environ Protection & Energy 380 Scotch Rd Trenton NJ 08625-0411

RAU, RALPH RONALD, physicist; b. Tacoma, Sept. 1, 1920; s. Ralph Campbell and Ida (Montgall) R.; m. Maryjane Uhrlaub, June 2, 1944; children: Whitney Leslie, Little Elise. B.S. in Physics, Coll. Puget Sound, 1941; M.S. in Physics, Calif. Inst. Tech., 1943, Ph.D. in Physics, 1948. Asst. prof. physics Princeton U., 1947-56; Fulbright research prof. physics Ecole Polytechnique, Paris, 1954-55; physicist Brookhaven Nat. Lab., Upton, N.Y., 1956-66; chmn. dept. physics Brookhaven Nat. Lab., 1966-70, assoc. dir. for high energy physics, 1970-81; adj. prof. U. Wyo.; vis. prof. MIT, 1984-88; staff scientist Desy Lab., Hamburg, Fed. Republic Germany, 1984-85. Trustee U. Puget Sound, 1978-84. Named Alumnus Cum Laude U. Puget Sound, 1968; recipient Alexander von Humboldt U.S. Sr. Scientist award 1988. Mem. Am. Phys. Soc., N.Y. Acad. Sci. Office: Brookhaven Nat Lab Upton NY 11973

RAUCH, ANDREW MARTIN, civil/structural engineer; b. Appleton, Wis., Mar. 1, 1961; s. Howard Franklin and Ruth Christine (Voss) R.; m. Donna Doris Thorston, May 7, 1988. BSCE, U. Minn., 1983. Registered profl. engr., Minn. Structural engr. BKBM Engrs., Inc., Mpls., 1983—, civil/structural group mgr., 1991—; mem. adv. com. Anoka (Minn.) Tech. Coll., 1991-93. Mem. ASCE, Cons. Engrs. Coun. Minn. (com. chair 1992). Lutheran. Office: BKBM Engrs 219 N 2nd St #200 Minneapolis MN 55401

RAUCH, LAWRENCE LEE, aerospace and electrical engineer, educator; b. L.A., May 1, 1919; s. James Lee and Mabel (Thompson) R.; m. Norma Ruth Cable, Dec. 15, 1961; children: Lauren, Maury Rauch. A.B., U. So. Calif., 1941; postgrad., Cornell U., 1941; A.M., Princeton U., 1948, Ph.D., 1949. Instr. math. Princeton U., 1943- 49; faculty U. Mich., 1949—, prof. aerospace engring., 1953-79, emeritus, 1979, chmn. instrumentation engring. program, 1952-63, chmn. computer, info. and control engring. program, 1971-76, asso. chmn. dept. elec. and computer engring., 1972-75; chief technologist telecommunication sci. and engring. div. NASA/Calif. Inst. Tech. Jet Propulsion Lab., 1979-85; vis. prof. Ecole Nationale Supérieure de L'Aéronautique et de l'Espace, Toulouse, France, 1970, Calif. Inst. Tech., Pasadena, 1977-85, U. Tokyo, 1978; cons. govt. and industry, 1946—; chmn. telemetering working group, panel test range instrumentation Research and Devel. Bd. Dept. Def., 1952-53; mem. exec. com. (Nat. Telemetering Conf.), 1959-64; Western Hemisphere program chmn. (1st Internat. Telemetering Conf.), London, 1963, program chmn., U.S.A., 1967; supr. air blast telemetering, Bikini, 1946; mem. project non-linear differential equations Office Naval Research, 1947-49; mem. research adv. com. on communications, instrumentation and data processing NASA, 1963-68. Author: Radio Telemetry, 1956; also numerous sci. articles and papers on radio telemetry. Recipient award for outstanding contbn. to WWII Army and Navy, 1947, annual award Nat. Telemetering Conf., 1960; Donald P. Eckman award for disting. achievement in edn. Instrument Soc. Am., 1966; Pioneer award Internat. Telemetering Conf./USA, 1985. Fellow IEEE (spl. award contbns. radio telemetry 1957, adminstrv. com. profl. group space electronics and telemetry 1958-64), AAAS, Explorers Club; mem. Am. Math. Soc., AIAA, U. Mich. Research Club, Phi Beta Kappa, Sigma Xi, Phi Eta Sigma, Phi Kappa Phi. Achievements include patent in field; development of first electronic time-division multiplex radio telemetering systems, of pre-detection recording; radio telemetry of first U.S. jet aircraft, of air blast over pressure for Operation Crossroads at Bikini Atoll; analysis of optimum demodulation of frequency-modulated signals. Address: 759 N Citrus Ave Los Angeles CA 90038-3401

RAUCHMILLER, ROBERT FRANK, JR., physicist; b. East Orange, N.J., July 17, 1959; s. Robert Frank Sr. and Joanne Mabel (Lick) R.; m. Lisa Charlene Roebuck, July 6, 1985; children: Melissa Ann, Geoffrey Cullen. BS, Ariz. State U., 1981; MS, U. Ariz., 1986. Geophysicist Gulf Oil Exploration & Prodn. Co., Houston, 1981-84; rsch. scientist ERIM, Ann Arbor, Mich., 1986-88; sr. engr. Gen. Dynamics Land Systems, Sterling Heights, Mich., 1988-89; rsch. engr. Indsl. Tech. Inst., Ann Arbor, 1989—; mem. proposal rev. com. Indsl. Tech. Inst., Ann Arbor, 1991-92; co-chair planning com. Intelligent Programmable Contr. '93 Conf., Detroit, 1992-93. Editor, writer newsletter Augusta News, 1991-92; contbr. articles to profl. journs. Co-chair Augusta Environ. Strategy Com., Whittaker, Mich., 1991-92; mem. Mich. Citizens Against Toxic Substances, Milan, Mich., 1992; organizer, founder Augusta Clean Air Classic 10Km foot race, Augusta Twp., Mich., 1992. Mem. Soc. Mfg. Engrs., Optical Soc. Am. Achievements include research in automatic target recoginition, image processing. Designer and builder of several unique machine vision systems for industrial inspection. Home: 6112 Bemis Rd Ypsilanti MI 48197 Office: Indsl Tech Inst 2901 Hubbard Rd Ann Arbor MI 48106

RAUF, ROBERT CHARLES, SR., systems specialist; b. N.Y.C., June 13, 1944; s. Charles Paul and Charlotte E. (Morten) R.; m. Susan E. Amadeo, June 6, 1964; children: Robert, Cheryl. AAS, Dutchess Community Coll., Poughkeepsie, N.Y., 1967; student, Marist Coll., Poughkeepsie, 1967-69. Program designer IBM, Poughkeepsie, 1964-70, mgr. system performance, 1970-75; mgr. job entry system design IBM, Gaithersburg, Md., 1976-77; mgr. office system and design IBM, Raleigh, N.C., 1978-82, mgr. office systems design, 1983-88; mgr. exec. support systems product IBM, Cary, N.C., 1988—. Home: 323 Lochside Dr Cary NC 27511 Office: IBM 8000 Regency Pky Cary NC 27511

RAUGHLEY, DEAN AARON, process engineer; b. Upland, Pa., Sept. 3, 1967; s. Ernest Ferguson and Alyce Leah (Knoll) R.; m. Susan Barbara Zgleszewski, Nov. 17, 1990. BS in Chem. Engring., Pa. State U., 1989. Process engr. Rohm and Haas Co., Bristol, Pa., 1989—. Mem. Am. Inst. Chem. Engrs. Office: Rohm and Haas Co Rt 413 and Old Rt 13 Bristol PA 19007

RAULS, WALTER MATTHIAS, engineering executive; b. Köln, Germany, July 29, 1930; s. Leopold and Klara (Neuburg) R.; m. Ingeborg Marlene Kerkow, July 29, 1963; children: Mathias, Bettina. Diploma in engring., Tech. U. Braunschweig, Germany, 1960; DEng, 1963. Asst. scientist Tech U. Braunschweig, Germany, 1963-65; head dept. Fried. Krupp Widia-Factory, Essen, Germany, 1965-73; sr. officer standards com. for testing materials in DIN, Berlin, 1973-79, mng. dir. standards com. for testing materials in DIN, Berlin, 1979—. Co-author: Introduction into the DIN-Standards, 1989. Mem. Assn. German Iron-Metallurgists, German Assn. Rsch. and Testing Materials (adv. coun. 1979), German Soc. Materials. Office: DIN Deutsches Inst Normung, Burggrafenstrasse 6, 10787 Berlin Germany

RAUSCH, JEFFREY LYNN, psychiatrist, psychopharmacologist; b. Butler, Pa., Jan. 10, 1953; s. John Kenneth and June Alice (Morrow) R.; m. Catherine Rebecca Montgomery, Aug. 24, 1974; children: Jeffrey David, Caroline Rebecca, Lauren Elizabeth. BS in Biology summa cum laude, Mercer U., Macon, Ga., 1974; MD, Med. Coll. Ga., 1978. Resident in psychiatry U. S.C., 1978-80; clin. psychopharmacology rsch. fellow U. Calif., San Diego, 1980-82; staff psychiatrist San Diego VA Med. Ctr., La Jolla, 1980-91; asst. prof. U. Calif., San Diego, 1982-89; assoc. prof. U. Calif., 1989-91, dir. psychoneuroendocrinology lab., 1990-91; acting assoc. chief psychiatry San Diego VA Med. Ctr., 1989-90; prof., vice chmn. dept. psychiatry Med. Coll. Ga., Augusta, 1991—; dir. lab. clin. neurosci. Augusta VA Med. ctr., 1991—; sci. cons. NIMH Study Sect. Contbr. articles to profl. jours. NIMH First award, 1987. Mem. AAAS, Am. Assn. Chairs Depts. Psychiatry, Nat. Alliance Mentally Ill, N.Y. Acad. Sci., Soc. Biol. Psychiatry, Psychiat. Rsch. Soc. Presbyterian. Avocations: surfing, skiing, biking, fishing, windsurfing. Office: Med Coll Ga Dept Psychiatry Health Behavior 1515 Pope St Augusta GA 30912-3800

RAUSCHER, FRANCES HELEN, psychologist; b. N.Y.C., May 4, 1957; d. Donald J. and Justine (Shir-Cliff) R. BA, Columbia U., 1984, MA, 1986, MPhil, 1988, PhD, 1992. Teaching asst. dept. psychology Columbia U., N.Y.C., 1984-89; mktg. rsch. assoc. DDB Needham, Worldwide, N.Y.C., 1989-90; rsch. assoc. dept. psychology Columbia U., N.Y.C., 1991-92; postdoctoral rsch. fellow Ctr. for Neurobiology of Learning and Memory, U. Calif., Irvine, 1992—; psychol. cons. Lintaas, N.Y., N.Y.C., 1992. Rschr.: Understanding the Language of Music, 1981; contbr. articles to profl. jours. Mem. APA, Am. Psychol. Soc., Calif. Psychol. Assn. Achievements include discovery that males and females accommodate the nonverbal components of their speech to that of their speaking partner, based on gender; that filled pauses are related to lexical word-choice options; gestures enhance lexical access; research in the hypothesis that early musical training enhances higher brain functions, i.e., abstract and spatial reasoning skills, listening to music facilitates spatial reasoning. Office: Univ Calif Ctr Neurobiology Learning Memory Bonney Ctr Irvine CA 92717

RAVAL, DILIP N., physical biochemist; b. Bombay, India, June 3, 1933; married, 1961. BS, U. Bombay, India, 1953; MS, U. Bombay, 1955; PhD Chemistry, U. Oreg., 1962. Fellow virus rsch. U. Calif., Berkeley, 1962-64; rsch scientist Palo Alto Medical Rsch. Inst., 1964-66; mgr. rsch. Varian Assocs., 1966-68; dir. clinicallabs. U. Calif. Med. Ctr., San Francisco, 1968-70; dir. rsch. U. Calif. Med. Ctr., 1970-72, gen. mgr. sci. and tech. divsn., 1972-73; v.p. rsch. and devel. divsn. Alcon Labs, 1973—. NIH fellow U. Oreg., 1961-62. Mem. Acad. Clinical Lab. Physicians and Scientist. Achievements include research in enzyme kinetics protein structure; medical instruments development; pharmaceutical drug development. Office: Alcon Labs Inc 6201 S Freeway Fort Worth TX 76134-2099*

RAVECHÉ, HAROLD JOSEPH, university administrator, physical chemist; b. N.Y.C., Mar. 18, 1943; s. Harold Edward Raveche and Helen Patricia (DeVincent) Gravino; m. Elizabeth Marie Scott, Jan. 26, 1974; children—John Vincent, Justin Blaise, Bernice Helen, Elizabeth Ann. B.A. in Chemistry, Hofstra U., 1963; Ph.D. in Phys. Chemistry, U. Calif.-San Diego, 1968. NRC postdoctoral assoc. Nat. Bur. Standards, Gaithersburg, Md., 1968-70, research chemist, 1970-78, chief thermophysics div., 1978-85; dean Sch. of Sci., prof. chemistry Rensselaer Poly. Inst., Troy, N.Y., 1985-88; pres. Stevens Inst. Tech., Hoboken, N.J., 1988—; bd. dirs. Nat. West N.J. and Bancorp, Atlantic Energy Inc.; commr. of sci. and tech., N.J. Editor: Perspectives in Statistical Physics, 1980; contbr. articles to profl. jours. Pres. Potomac Highlands Citizens Assn., Md., 1978-80. Recipient Disting. Young Scientist of Yr. award Md. Acad. Scis., 1975, U.S. Sr. Exec. Service award Nat. Bur. Standards, 1983, Equal Employment Opportunity award Nat. Bur. Standards, 1984. Mem. AAAS (commn. on sci. edn. 1972-75), Am. Phys. Soc. (adv. council 1975-78), Soc. for Indsl. and Applied Math. (adv. bd. conf. on large-scale computational problems 1984-88), Am. Chem. Soc., Sigma Xi. Roman Catholic. Avocations: hiking, swimming, skiing, music, theater. Office: Stevens Inst Tech Office of Pres Castle Point on Hudson Hoboken NJ 07030

RAVEN, BERTRAM H(ERBERT), psychology educator; b. Youngstown, Ohio, Sept. 26, 1926; s. Morris and Lillian (Greenfeld) R.; m. Celia Cutler, Jan. 21, 1961; children: Michelle G., Jonathan H. BA, Ohio State U., 1948, MA, 1949; PhD, U. Mich., 1953. Research assoc. Research Ctr. for Group Dynamics, Ann Arbor, Mich., 1952-54; lectr. psychology U. Mich., Ann Arbor, 1953-54; vis. prof. U. Nijmegen, U. Utrecht, Netherlands, 1954-55; psychologist RAND Corp., Santa Monica, Calif., 1955-56; prof., chmn. dept. psychology UCLA, 1956—; vis. prof. Hebrew U., Jerusalem, 1962-63, U. Wash., Seattle, U. Hawaii, Honolulu, 1968, London Sch. Econs. and Polit. Sci., London, 1969-70; external examiner U. of the West Indies, Trinidad and Jamaica, 1980—; participant Internat. Expert Conf. on Health Psychology, Tilburg, Netherlands, 1986; cons., expert witness in field, 1979—. co-dir. Tng. Program in Health Psychology, UCLA, 1979-88; cons. World Health Orgn., Manila, 1985-86; cons., expert witness various Calif. cts., 1978—. Author: (with others) People in Groups, 1976, Discovering Psychology, 1977, Social Psychology, 1983; editor: (with others) Contemporary Health Services, 1982; Policy Studies Review Annual, 1980; editor Jour. Social Issues, 1969-74; contbr. articles to profl. jours. Guggenheim fellow, Israel, 1962-63; Fulbright scholar The Netherlands, 1954-55, Israel, 1962-63, Britain, 1969-70; Citation from Los Angeles City Council, 1966, Rsch. on Soc. power by Calif. Sch. of profl. psychology, L.A., 1991; NATO sr. fellow, Italy, 1989. Fellow Am. Psychol. Assn. (chair bd. social and ethical responsibility 1978-82), Am. Psychol. Soc., Soc. for Psychol. Study of Social Issues (pres. 1973-74); mem. AAAS, Am. Sociol. Assn., Internat. Assn. Applied Psychology, Soc. Exptl. Social Psychology, Assn. Advancement of Psychology (founding, bd. dirs. 1974-81), Internat. Soc. Polit. Psychology, Interam. Psychol. Soc., Am. Psychology-Law Soc. Avocations: guitar, travel, international studies. Home: 2212 Camden Ave Los Angeles CA 90064-1906 Office: UCLA Dept Psychology Los Angeles CA 90024-1563

RAVENSCROFT, BRYAN DALE, alternate energy research company executive; b. Moscow, Idaho, Apr. 23, 1951; s. Vernon F. and Harriet B. Ravenscroft; m. Barbara Lewis, June 16, 1979; 1 child, Kara Lee. BS, U. Idaho, 1974; MS, Colo. State U., 1982. Mgr. R & D, Penta Post Co., Tuttle, Idaho, 1974-84, CEO, 1985—. Grantee Dept. Energy, 1980-82, 89-93. Memm. Sigma Xi. Republican. Achievements include research on biomass to diesel conversion, construction of biomass gasification demonstration plant. Office: Penta Post Co Interstate 84 Exit 147 Tuttle ID 83314

RAVI, VILUPANUR ALWAR, materials scientist; b. Madras, India, Aug. 21, 1960; came to U.S., 1983; s. Villupanur Sadagopa and Vembu (Srinivasan) Alwar; m. Radhu Rangachari, Feb. 22, 1990. BSc, U. Madras, 1980; B of Engring., Indian Inst. Sci., Bangalore, 1983; MS, Ohio State U., 1986, PhD, 1988. Grad. rsch. assoc. Ohio State U., Columbus, 1983-88; materials scientist Lanxide Corp., Newark, Del., 1988—. Co-editor: Processing and Fabrication of Advanced Materials for High Temperature Applications TMS, 1992; contbr. articles to profl. jours. and conf. proceedings. Recipient Alumni Rsch. award Ohio State U., 1986, Grad. Student award MRS, 1988, Outstanding Student Paper award TMS, 1985, 88. Mem. Minerals, Metals and Materials Soc. (mem. structural materials com. 1991—), Am. Foundrymen's Soc., Sigma Xi (assoc.). Achievements include patents pending in the field of advanced composites. Office: Lanxide Corp 1300 Marrows Rd Newark DE 19714-6077

RAVICHANDRAN, KURUMBAIL GOPALAKRISHNAN, chemist; b. Irinialakuda, Kerala, India, Dec. 22, 1960; s. Kurumbail S. Gopalakrishnan and Kurumbail (Ramakrishnan) Saraswathi; m. Ananthalakshmi Ramakrishnan, May 31, 1966. MS, Indian Inst. of Tech., Madras, India, 1983; PhD in Chemistry, Mich. State U., 1989. Rsch. assoc. Mich. State U., East Lansing, 1989-90; vis. rsch. assoc. Max-Planck Inst. Für Biochemistry, Martinsreid, Germany, 1989; rsch. assoc. Howard Hughes Med. Inst., Dallas, 1990—. Contbr. articles to profl. jours. Pres. India Club, East Lansing, 1984-87. Recipient Merit Level Teaching award Mich. State U., 1986, Merit cert. Royal Soc. of Chemistry, 1983. Mem. Am. Crystallographic Assn., Biophys. Soc. Hindu. Achievements include discovery of structure of hirudin thrombin complex which shows how the protein hirudin from the saliva of leeches is able to prevent blood clotting in its victims. Office: Howard Hughes Med Inst 5323 Harry Hines Blvd Dallas TX 75235-9050

RAVIKUMAR, THANJAVUR SUBRAMANIAM, surgical oncologist; b. Madras, India, Mar. 12, 1950; came to U.S., 1976; s. P. and Rajamani Subramaniam; m. Srikala Kandaswamy, Sept. 8, 1975; 1 child, Shruti. MS, Madras Med. Coll., 1976; MD, U. Edinburgh, Scotland, 1978. Diplomate Am. Bd. Surgery. Surg. resident Maimonides Med. Ctr., Maimonides Hosp., Bklyn., 1976-80; surg. oncology fellow U. Minn. Hosps., Mpls., 1980-82; rsch. fellow Harvard Med. SCh., Boston, 1982-84, asst. prof. surgery, 1986-89, assoc. prof. surgery, 1989-90; asst. prof. surgery SUNY Downstate Med. Ctr., Bklyn., 1984-85; assoc. prof., chief surg. oncology Yale U. Sch. Medicine, New Haven, 1990—; chmn. cancer com. Yale New Haven Hosp., 1991—; cancer program dir. Yale-China Assn., New Haven, 1992—. Contbr. articles to profl. jours. Mandelberg traveling fellow Maimonides Med. Ctr., Bklyn., 1978. Fellow Royal Coll. Surgeons, Am. Coll. Cryosurgery; mem. Soc. Surg. Oncology (clin. trials and govt. rels. com. 1991—, James Ewing Found. award 1983), Soc. Univ. Surgeons, Am. Assn. Cancer Rsch. Achievements include first human clin. trial of isolated hepatic perfusion chemotherapy using a novel double balloon catheter for treating liver cancers, new approaches to treat tumors spread into liver with cyrosurgery and laser surgery. Home: 204 Davis St Hamden CT 06517 Office: Yale U Sch Medicine 333 Cedar St PO Box 3333 New Haven CT 06510

RAVIOLA, ELIO, anatomist, neurobiologist; b. Asti, Italy, June 15, 1932; came to U.S., 1970; s. Giuseppe and Luigina (Carbone) R.; m. Trude Kleinschmidt; 1 child, Giuseppe. M.D. summa cum laude, U. Pavia, Italy, 1957, Ph.D. in Anatomy, 1963. Resident in neurology and psychiatry U. Pavia, asst. prof., 1958-70; asso. prof. anatomy Harvard U. Med. Sch., Boston, 1970-74; prof. human anatomy Harvard U. Med. Sch., 1974-89, Bullard prof. neurobiology, 1989—, prof. ophthalmology, 1989—. Editorial bd.: Jour. Cell Biology, 1978-80; asso. editor: Anat. Record, 1972—. Mem. Am. Soc. Cell Biology, Am. Assn. Anatomists, Soc. of Neurosci. Club: Harvard (Boston). Achievements include research on mechanism of vision. Office: Harvard U Med Sch Dept Neurobiology 220 Longwood Ave Boston MA 02115-6092*

RAVITZ, LEONARD J., JR., physician, scientist, consultant; b. Cuyahoga County, Ohio, Apr. 17; s. Leonard Robert and Esther Evelyn (Skerball) R. BS, Case Western Res. U., 1944; MD, Wayne State U., 1946; MS, Yale U., 1950. Diplomate Am. Bd. Psychiatry and Neurology, 1952. Rsch. asst. EEG to A.J. Derbyshire PhD Harper Hosp., Detroit, 1943-46; spl. trainee in hypnosis to Milton H. Erickson MD Wayne County Gen. Hosp., Eloise, Mich., 1945-46; rotating intern St. Elizabeth's Hosp., Washington, 1946-47; jr./sr. asst. resident in psychiatry Yale-New Haven Hosp.; asst. in psychiatry and mental hygiene Yale Med. Sch., 1947-49, rsch. fellow to Harold S. Burr PhD sect. neuro-anatomy, 1949-50; sr. resident in neuropsychiatry Richard S. Lyman svc. Duke Hosp., Durham, N.C., 1950-51; instr. Duke U. Med. Sch., Durham, 1950-51; assoc. to R. Burke Suitt MD Pvt. Diagnostic Clinic, Duke Hosp., Durham, 1951-53; assoc. Duke U. Med. Sch., 1951-53; vis. asst. prof. neuropsychiatry and asst. to vis. prof. Richard S. Lyman, MD Meharry Med. Ctr., Nashville, 1953; asst. dir. profl. edn. in charge tng. U. Wyo. Nursing Sch. affiliates; chief rsch. rehab. bldg. Downey VA Hosp. (now called VA Hosp.), North Chicago, Ill., 1953-54; assoc. psychiatry Sch. Medicine and Hosp., U. Pa., Phila., 1955-58; electromagnetic field measurement project Dep. Asst. Sec. Def. in Charge of Health and Med., E.H. Cushing, MD, Pentagon, 1958; dir. tng. and rsch. Ea. State Hosp., Williamsburg, Va., 1958-60; pvt. practice neuropsychiatry specializing in hypnosis Norfolk, Va., 1961—; psychiatrist, cons. Div. Alcohol Studies and Rehab., Va. Dept. Health (later Va. Dept. Mental Health and Mental Retardation), 1961-81; psychiatrist Greenpoint Clinic, Bklyn., 1983-87, 17th St. Clinic, N.Y.C., 1987-92; sec.-treas. Euclid-97th St. Clinic, Inc., Cleve., 1957-63, pres., 1963-69; spl. studies of epistemology, field physics and cybernetics with F.S.C. Northrop, PhD, 1973-80; cons. with Harold S. Burr, Lyme, Conn., 1973, Milton H. Erickson, 1980; clin. asst. prof. psychiatry, SUNY Health Sci. Ctr. Med. Sch., 1983—; psychiatrist Downstate Mental Hygiene Assocs., Bklyn., 1983—; pvt. cons., Cleve., 1961-69, Upper Montclair, N.J., 1982—; lectr. sociology Old Dominion U., Norfolk, 1961-62, cons. nutrition rsch. project, Old Dominion U. Research Found., 1978-90; spl. med. cons. Frederick Mil. Acad., Portsmouth, Va., 1963-71; cons. Tidewater Epilepsy Found., Chesapeake, Va., 1962-68, USPH Hosp. Alcohol Unit, Norfolk, 1980-81, Nat. Inst. Rehab. Therapy, Butler, N.J., 1982-83; participant 5th Internat. Congress for Hypnosis and Psychosomatic Medicine, Gutenburg U., Mainz, Fed. Republic Germany, 1970; organizer symposia on hypnosis in psychiatry and medicine, field theory as an integrator of knowledge, hypnosis in office practice, history of certain forensic and psychotherapeutic aspects of the study of man, Eastern State Hosp., Coll. William and Mary, Va. Soc. Clin. Hypnosis, Williamsburg, Va., 1959-60; founding pres. Va. Soc. Clin. Hypnosis, 1959-60, Found. for Study Electrodynamic Theory of Life, 1989—; rsch. on epistemology, field physics, neurocybernetics with F.S.C. Northrop, 1973-86. Asst. editor Jour. Am. Soc. Psychosomatic Dentistry and Medicine, 1980-83; mem. editorial bd. Internat. Jour. Psychosomatics, 1984—; contbr. sects. to books, articles, book revs., abstracts to profl. publs.; discoverer electromagnetic field correlates of hypnosis, emotions, psychiatric/med. disorders and aging and electrocyclic phenomena in humans which parallel those of other life forms, earth and atmosphere underwriting beginning short- and long-range predictions, such seemingly disparate phenomena united under a single regulating principle defined in terms of measurable field intensity and polarity. Sr. v.p. Willoughby Civic League, 1971-75. 1st lt. AUS, 1943-46. Lyman Rsch. Fund grantee, 1950-53. Fellow AAAS, Am. Psychiat. Assn., N.Y. Acad. Scis., Am. Soc. Clin. Hypnosis (charter), Royal Soc. Health (London); mem. Norfolk Acaicine, Soc. for Investigation of Recurring Events, Va. Med. Soc., Nu Sigma Nu, Sigma Xi. Office: SUNY Health Sci Ctr Med Sch Dept Psychiatry Box 1203, 450 Clarkson Ave Brooklyn NY 11203-2098 also: PO Box 9409 Norfolk VA 23505-0409

RAWSON, NANCY ELLEN, neurobiology researcher; b. New Haven, Conn., Sept. 15, 1956; d. Paul Olney and Lois (Zeruiah) R.; m. Louis Howard Braddock, Nov. 22, 1986. BS, Fairfield U., 1978; MS, U. Mass., 1982; PhD, U. Pa., 1993. Instr. Drexel U., Phila., 1982; rsch. nutritionist Campbell Soup Co., Camden, N.J., 1982-89; rsch. fellow Monell Chem. Senses Ctr., Phila., 1993—; chair nat. Nutrient Databank Conf.; mem. exec. adv. bd. Food Engring., 1987. Co-author: Fuel Metabolism and Appetite Control; contbr. articles to profl. jours. Preceptor Minority High Sch. Student Apprenticeship Program, Phila., 1992; big sister Big Bros./Big Sisters Am., Fairfield, Conn., 1976-79. Recipient Food Product of Yr. award Am. Health Found., 1988, Young Investigator award Soc. Study Ingestive Behavior, 1991, Grad Student Travel award Internat. Life Scis., Inst., 1992, Grad. Rsch. award Am. Inst. Nutrition, 1993; Biomed. Rsch. Support grantee 1990. Mem. Inst. Food Technologists, Soc. for Neuroscience, Soc. for the Study Ingestive Behavior Study (student rep. to bd. 1992—), Am. Physiol. Soc. Achievements include copyrighted computer database design for food and nutrition info., copyrighted data coding scheme for food and nutrition data; first to use NMR tech. to investigate mechanisms of appetite control. Office: Monell Chem Senses Ctr 3500 Market St Philadelphia PA 19104

RAY, GORDON THOMPSON, communications executive; b. N.Y.C., Jan. 31, 1928; s. John Henry and Hama Thompson (Potter) R. m. Ingrid Ray; children: Stuart, John, Lawrence, Carl. BEE, Rensselaer Poly. Inst., 1954; PhD, Midwest Coll. Engring., 1983. With The Bell System, various locations, 1954-83; successively engr., chief engr. N.Y. Tel., Albany, Utica, Jamaica, White Plains, N.Y., Bklyn., 1983; asst. v.p. long range planning N.Y. Tel., N.Y.C.; former mem. tech. staff Bell Tel. Labs., N.Y.C., Murray Hill and Holmdel, N.J.; engr. AT&T, N.Y.C.; dir. Bell System Computer Seminar; del. CCITT, Body Planning Internat. Direct Distance Dialing; v.p. planning NYNEX Materiel Enterprises Co. subs. NYNEX Corp., 1983-84, sr. v.p. NEC Am. Inc., Melville, N.Y., 1985-91, exec. v.p., 1992—; mem. bus. and industry ctr. SUNY, adv. bd. Marine Scis. Rsch. Ctr., Harriman Coll., Stonybrook, sci. & bus. coun. L.I. Rsch. Inst.; bd. dirs. NEC Am. Inc., NEC Rsch. Inst., Tel. Industry Assn., L.I. Assn., L.I. Forum for Tech.; exec. com. Computer and Communications Industry Assn., Nat. Comm. Forum. Trustee U.S. Coun. for Internat. Bus., N.Y. Hall of Sci.; mem. exec. com. Akron Golf Charities. With U.S. Army, 1946-49. Fellow Royal Soc. of Arts; mem. AAAS, NSPE (sr. mem.), IEEE (sr. mem.), IEEE Communications Soc. (past chmn. quality assurance mgmt. com., internat. communications field award com.), N.Y. Acad. Scis., Am. Mgmt. Assn., Internet Soc., Cedarbrook Country Club, Firestone Country Club (bd. govs.). Office: NEC Am Inc 8 Old Sod Farm Rd Melville NY 11747-3148

RAY, JAMES ALLEN, aerospace engineer; b. Memphis, Mar. 9, 1945; s. James Clark and Beverly Louise (Arnold) R.; m. Charlotte Ann Lebo, Aug. 26, 1967; children: Katherine Celene, Gregory Allen. BSME, U. Tenn., 1967; MSME, U. Mo., Rolla, 1970. Registered profl. engr., Mo., Ill. Aerospace engr. McDonnell Douglas Astronautics, St. Louis, 1967-69, U.S. Army Aviation and Troop Command, St. Louis, 1969-84; dep. dir. engring. U.S. Army Aviation Systems Command, St. Louis, 1984—; bd. dirs., v.p. Ener-Tech, Inc. Energy Conservation Cons. Engrs., St. Louis, 1977-88; instr. energy conservation St. Louis C.C., 1981-84; presenter in field. Author: Advanced Techniques in Failure Prevention, 1991; contbr. articles to Aviation Digest, FlightFax, Army Aviation, Vertiflight. Active St. Mark's United Meth. Ch., Florissant Mo., 1973—, various offices including chmn. social concerns, evangelism, vice-chmn. trustees, adminstrv. bd., rep. to St. Louis area Ch. and Mission Extension Soc., Sunday sch. tchr., chmn. adult Sunday sch.; dist. men's exec. bd., dist. scouting coord., conf. scouting coord. United Meth. Ch., St. Louis North Dist. and Mo. East Conf.; cubmaster, scoutmaster, com. chmn. troop and pack 528 Boy Scouts Am., 1981—, dist. commr. 1993. Recipient Cross and Flame award United Meth. Ch., 1988, Dist. award of Merit, Boy Scouts Am., 1989, Coun. Silver Beaver award 1991, Order of the Arrow Vigil Honor, 1991, HUD Cycles Passive Solar Design award, 1979. Mem. Am. Helicopter Soc., Internat. Solar Energy Soc. (past chmn. local chpt., past editor local newsletter), Army Assn. Am. (Exceptional Civilian Svc. award 1991, Achievement award 1991), Adv. Group for Aerospace R&D, ASHRAE, Soc. Automotive Engrs., ASME, Aerospace Industries Assn. (co-chair working group on flight safety parts), Am. Soc. Engring. Mgr., Royal Aeronautical Soc., Nat. Coun. on System Engrs., mech. Failures Prevention Group, Soc. Aircraft Weight Engrs. Office: US Army ATCOM 4300 Goodfellow Blvd Saint Louis MO 63120

RAY, JAMES HENRY, nuclear engineer; b. Cleve., Sept. 24, 1925; s. George Burrill and Louise (Hompe) R.; m. Barbara Westcott, June 10, 1950. BA in Physics, Harvard U., 1947; MS in Physics, U. Pa., 1954. Physicist Nuclear Devel. Assocs., White Plains, N.Y., 1955-60, United Nuclear Corp., Elmsford, N.Y., 1960-72, Gulf United Nuclear Fuels Corp., Elmsford, 1972-74; nuclear engr. Burns & Roe Inc., Paramus, N.J., 1974-83; nuclear reactor engr. N.Y. Power Authority, White Plains, 1984—. Mem. Irvington (N.Y.) Dem. Com., 1967—, chmn./sec., 1986-88, 92. Mem. Am. Nuclear Soc., Health Physics Soc. Home: 35 S Broadway Apt F-3 Irvington NY 10533

RAY, JOHN WALKER, otolaryngologist, educator; b. Columbus, Ohio, Jan. 12, 1936; s. Kenneth Clark and Hope (Walker) Ray; m. Susanne Gettings, July 15, 1961; children: Nancy Ann, Susan Christy. AB magna cum laude, Marietta Coll., 1956; MD cum laude, Ohio State U., 1960; postgrad. Temple U., 1964, Mt. Sinai Hosp. and Columbia U., 1964, 66, Northwestern U., 1967, 71, U. Ill., 1968, U. Ind., 1969, Tulane U., 1969. Intern, Ohio State U. Hosps., Columbus, 1960-61, clin. rsch. trainee NIH, 1963-65, resident dept. otolaryngology, 1963-65, 1966-67, resident dept. surgery 1965-66, instr. dept. otolaryngology, 1966-67, 70-75, clin. asst. prof., 1975-82, clin. assoc. prof., 1982-92, clin. prof., 1992—; active staff, past chief of staff Bethesda Hosp.; active staff, past chief of staff Good Samaritan Hosp., Zanesville, Ohio, 1967—; courtesy staff Ohio State U. Hosps., Columbus, 1970—, Meml. Hosp., Marietta, Ohio, 1992—; radio-TV health commentator, 1982—. Past pres. Muskingum chpt. Am. Cancer Soc.; trustee Ohio Med. Polit. Action Com. Capt. USAF, 1961-63. Recipient Barraquer Meml. award, 1965; named to Order of Ky. Col., 1966, Muskingum County Country Music Hall of Fame. Diplomate Am. Bd. Otolaryngology. Fellow ACS, Am. Soc. Otolaryn. Allergy, Am. Acad. Otolaryngology-Head and Neck Surgery (gov.), Am. Acad. Facial Plastic and Reconstructive Surgery; mem. Nat. Assn. Physician Broadcasters, Muskingum County Acad. Medicine (past pres.), AMA (del. hosp. med. Staff sect.), Ohio Med. Assn. (del.), Columbus Ophthalmol. and Otolaryngol. Soc. (past pres.), Ohio Soc. Otolaryngology (past pres.), Pan-Am. Assn. Otolaryngology and Bronchoesophagology, Pan-Am. Allergy Soc., Am. Acad. Invitro Allergy, Am. Auditory Soc., Am. Soc. Contemporary Medicine and Surgery, Acad. Radio and TV Health Communicators, Phi Beta Kappa, Alpha Tau Omega, Alpha Kappa Kappa, Alpha Omega Alpha, Beta Beta Beta. Presbyterian. Contbr. articles to sci., med. jours; collaborator, surg. motion picture Laryngectomy and Neck Dissection, 1964. Office: 2945 Maple Ave Zanesville OH 43701-1733

RAY, RICHARD EUGENE, osteopath, psychiatrist; b. Unionville, Mo., Oct. 7, 1950; s. Richard Raymond and Doris Maxine (Bunyard) R.; m. Elaine Marie Broda, Apr. 25, 1986. BS in Biology, N.E. Mo. State U., 1972; DO, Kirksville Coll. Osteopathy, 1978. Diplomate Nat. Bd. Examiners for Osteopathic Physicians & Surgeons; cert. ACLS, Am. Heart Assn. Intern Davenport (Iowa) Osteopathic Hosp., 1978-79; resident in pediatrics Kirksville (Mo.) Osteopathic Hosp., 1979; physician gen. practice Nat. Health Svc. Corps., Princeton, Mo., 1980-81; med. officer U.S. Bur. Prisons Fed. Correctional Inst., Alderson, W. Va., 1981-82; chief health programs U.S. Bur. Prisons Fed. Correctional Inst., Sandstone, Minn., 1982-83; resident physician in psychiatry Mich Osteopathic Med. Ctr., Detroit, 1983-86; staff psychiatrist Alaska Native Med. Ctr. Indian Health Svc., Anchorage, 1986-88, VA Med. Ctr., Sheridan, Wyo., 1988-89; chief psychiatrist in chem. dependency & psychogeriatrics Moose Lake (Minn.) Regional Med. Ctr., 1989-92; asst. prof. psychiatry, asst. dir. psychiat. med. edn., dir. consultation/liaison svcs. Chgo. Coll. Osteo. Medicine, 1992—; chmn. treatment review panel Moose Lake Regional Med. Ctr., 1989-92. Lt. comdr. USPHS, 1981-83. Mem. Am. Osteopathic Assn., Am. Acad. Osteopathy, Am Coll. Neuropsychiatrists, Am. Osteopathic Acad. of Addictionology (charter mem.), Kirksville Ostoepathic Med. Sch. Alumni Assn. (life mem.), N.E. Mo. State U. Alumni Assn. (life mem.). Republican. Baptist. Avocations: history, science, travel, art. Office: Chgo Coll Osteo Dept Psychiatry 5200 S Ellis Chicago IL 60615

RAY, ROBERT LANDON, nuclear physicist; b. Lubbock, Tex., Sept. 10, 1950; s. Cecil Armstrong and Joy Charlene (Andrews) R.; m. Mary Alice Gomez, Jan. 24, 1976; children: Jennifer Leigh, Michael Wayne. BS, U. Tex., Arlington, 1972; PhD, U. Tex., Austin, 1977. Mem. staff Los Alamos (N.Mex.) Nat. Lab., 1977-79, guest scientist, 1979-81; mem. staff U. Tex., Austin, 1979-82, rsch. scientist, vis. asst. prof., 1981—; cons. Los Alamos Nat. Lab., 1988-91. Contbr. articles to profl. publs. Fellow Am. Phys. Soc.; mem. Los Alamos Meson Physics Facility Users Group, Inc., Superconducting Super Collider Users Group, Solenoidal Tracker at the Relativistic Heavy Ion Collider Collaboration. Achievements include first to obtain quantitative information about neutron densities in atomic nuclei, demonstrate the importance of many body effects in proton-nucleus reactions at energies above pion production, study proton scattering from polarized atomic nuclei; co-developer new sub-field in nuclear physics which studies relativistic effects in nuclear reactions; research in relativistic and non-relativistic methods in medium energy proton-nucleus physics, calculations of medium effects, predictions for polarized nuclei, determination of neutron densities in nuclei. Office: U Tex Dept Physics Austin TX 78712

RAY, SANKAR, opto-electroic device scientist; b. Calcutta, India, Dec. 30, 1953; came to U.S., 1976; s. Asok Kumar and Sumitra (Dey) R.; m. Diana Konaszewski, June 8, 1985; children: Krishanu, Monisha. BSc in Physics with honors, U. Calcutta, 1974; MSc in Physics, Indian Inst. Tech., Kanpur, 1976; PhD in Physics, Brown U., 1981. Prin. rsch. scientist Honeywell Corp. Tech. ctr., Bloomington, Minn., 1981-87; lead rsch. scientist Boeing High Tech. Ctr., Bellevue, Wash., 1987-90, mgr. advanced components for fiber optic networks, 1990—; evaluator rsch. proposals Jet Propulsion Lab., Pasadena, Calif., 1990, Simon Fraser U., Vancouver, B.C., Can., 1990; chmn. organizing com. Boeing Photonics Symposium, Seattle, 1991; mem. Compound Semiconductor Adv. Com., Wash. Tech. Ctr., U. Wash., 1992. Contbr. articles to Phys. Rev. Letters, Applied Physics Letters, IEEE Electron devices; author, co-author over 25 jour. articles, conf. procs. Mem. IEEE, Am. Phys. Soc., Indian Physics Assn., Sigma Xi. Achievements include patent and patent pending; co-development of monolithic GaAs based laser transmitter chip at Honeywell; development of InP based monolithic fiber optic receiver chip and InGaAs/InAlAs based electronic chip at Boeing. Office: Boeing Def & Space Group PO Box 3999 Seattle WA 98124-2499

RAYBURN, RAY ARTHUR, audio engineer; b. Bklyn., Aug. 21, 1948; s. Ray B. and Anna E. (Shoremount) R.; m. Cornelia Bockwoldt, Nov. 4, 1972; children: Timothy, Christopher, Mark, James. ASET, N.Y. Inst. Tech., 1970. Rec. engr. RCA Records, N.Y.C., 1974-78; owner Rayburn Electronics, N.Y.C., 1978-82; engr. Broadway Video, N.Y.C., 1982-83; dir. engring. Broadcast Technologies Group, N.Y.C., 1983-84; dir. engring. svcs. Comcast Sound Comm., West Hartford, Conn., 1984-86; v.p. engring. Essential Telecomm. Corp., Glastonbury, Conn., 1986-91; assoc. Joiner Consulting Group, Arlington, Tex., 1991—. Mem. NRA (instr.), Audio Engring. Soc. (signal processing com. 1991—), Acoustical Soc. Am. Republican. Home: 4303 Solitude Ct Arlington TX 76017-1361 Office: Joiner Consulting Group 1184 A W Corporate Dr Arlington TX 76006

RAYMOND, ARTHUR EMMONS, aerospace engineer; b. Boston, Mar. 24, 1899. BS, Harvard U., 1920; MS, MIT, 1921; DSc (hon.), Poly. Inst. Bklyn., 1947. Engr. Douglass Aircraft Co., 1925-34, v.p. engring., 1934-60; cons. Rand Corp., 1960-85; ret., 1985; mem. Nat. Adv. Com. Aero., 1946-56; cons. NASA, 1962-68. Trustee Aerospace Corp., 1960-71, Rsch. Analysis Corp., 1965-71. Recipient Nat. Air and Space Mus. Trophy/Lifetime Achievement award Smithsonian Instn., 1991. Fellow AIAA (hon.); mem. NAS, Nat. Acad. Engring. Achievements include research in aeronautics and astronautics. Home: 73 Oakmont Dr Los Angeles CA 90049*

RAYMOND, DALE RODNEY, chemical engineer; b. Farmington, Maine, Mar. 8, 1949; s. Richard Frederick and Phyllis (Harnden) R.; m. Alysan Rae Baker, June 15, 1974; children: Dorrie, Donald, Alicia. BS, U. Maine, 1971, MS, 1973, PhD, 1975. Instr. U. Maine, Orono, 1973-75; from rsch. scientist to sect. leader R&D Union Camp Corp., Princeton, N.J., 1975-85; tech. dir. Kraft Div. Union Camp Corp., Prattville, Ala., 1985-92; tech. dir. Fine Paper Div. Union Camp Corp., Franklin, Va., 1992—. Mem. AICE, Tech.

Assn. of Pulp and Paper Industry, Sigma Xi. Baptist. Home: 148 Crescent Dr Franklin VA 23851

RAYMOND, EUGENE THOMAS, technical writer, consultant, retired aircraft engineer; b. Seattle, Apr. 17, 1923; s. Evan James and Katheryn Dorothy (Kranick) R.; m. Bette Mae Bergeson, Mar. 1, 1948; children: Joan Kay Hibbs, Patricia Lynn Adams, Robin Louise Flashman. BSME, U. Wash., 1944; postgrad., 1953-55; registered profl. engr., Tex. Rsch. engr. The Boeing Co., Seattle, 1946-59, sr. group engr., 1959-63, 66-71, sr. specialist engr., 1971-81, prin. engr. flight control tech., 1982-88; project design engr. Gen. Dynamics, Ft. Worth, 1963-66. Lt., USNR, 1943-46, 49-52; PTO. Recipient prize Hydraulics and Pneumatics mag., 1958. Mem. Soc. Automotive Engrs. (cert. of appreciation, chmn. adv. bd. com. A-6 nat. com. for aerospace fluid power and control tech. 1983-88, vice-chmn. com. 1986-88, cons.), Fluid Power Soc. (dir. northwest region 1973-74), Puget Sound Fluid Power Assn., AIAA, Beta Theta Pi, Meridian Valley Country Club, Masons, Shriners. Lutheran. Aircraft editorial adv. bd. Hydraulics and Pneumatics mag., 1960-70; achievements include 5 patents in Fluid Sealing Arrangements, Quasi-Open-Loop Hydraulic Ram Incremental Actuator with Power Conserving Properties, Rotary Digital Electrohydraulic Actuator, Two-Fluid Nonflammable Hydraulic System and Load-Adaptive Hydraulic Actuator System and Method for Actuating Control Surfaces; designed and developed mechanical systems for the XB-47 and B-52 jet bombers, 707 airliner and many other aircraft; contbr. over 20 technical papers and articles to profl. jours. Home and Office: 25301 144th Ave SE Kent WA 98042-3401

RAYMOND, JOHN CHARLES, physicist; b. Edgerton, Wis., Nov. 28, 1948; s. John Sayre and Marjorie (Price) R.; m. Bonnie Ecklund, May 30, 1975; children: Jennifer, Aaron. PhD, U. Wis., 1976. Rsch. assoc. Harvard Coll. Observatory, Cambridge, Mass., 1976-80; physicist Smithsonian Instn., Cambridge, 1980—. Mem. Am. Astron. Soc. Office: Ctr for Astrophysics 60 Garden St Cambridge MA 02138-1596

RAYMONT, WARWICK DEANE, chemical engineer, environmental engineer; b. Jamestown, Australia, June 14, 1941; s. Keith Mostyn and Florence Brenda Dewhirst (Langdon-Parsons) Ring; m. Sandra Kay Sullivan, Oct. 23, 1989; children: Jason, Jared, Jenelle, Aleksei. BS, diploma edn., U. New South Wales, 1962; PhD, Columbia U., 1970; Dip.T.Sec, U. South Australia, 1977; PhD, Pa. State U., 1972. Lic. organic chemistry, chem. engring. Tchr. sci. Cath. and Ind. Colls., South Australia, 1963-65; chemist Dairy Analytical Labs., Tasmania, 1965-68, chief chemist, 1968-70, dir., 1970-72; tchr. sci. Edn. Dept., South Australia, 1972-77; postdoctoral fellow WHO/Columbia U., N.Y.C., 1977-79; pvt. practice rsch. cons. South Australia and N.S.W., 1979—; del. WHO Codes Com., N.Y.C., 1970-71; mem. adv. com. Sr. Secondary Assessment Bd. South Australia, 1988-89; cons. South Australia Health Commn., 1989—. Leader Boy Scouts Assn., Australia, 1960—; chmn. Morella Community Assn., South Australia, 1985-87; exec. mem. South Australia Assn. State Sch. Orgns., 1973-78; chief instr. Air Tng. Corps, Royal Australian Air Force, South Australia, 1988—. Acting capt. Australian Army, 1961. Mem. Am. Chem. Soc., Australian Inst. Sci. Tech., Mentally Handicapped Children's Assn. (life), N.Y. Acad. Scis. Achievements include discovery of DDT in human breast milk. Home: PO Box 346, Golden Grove SA5125, Australia Office: Stolair Pty Ltd, 8 Aspiring Ct, Greenwith SA5125, Australia

RAYSON, GLENDON ENNES, internist, preventive medicine specialist, writer; b. Oak Park, Ill., Dec. 2, 1915; s. Ennes Charles and Beatrice Margaret (Rowland) R. AB, U. Rochester, 1939; MD, U. Ill., Chgo., 1948; MPH, Johns Hopkins U., 1965; MA, Northwestern U., 1965. Diplomate Am. Bd. Internal Medicine, Am. Bd. Preventive Medicine. Resident in internal medicine Presbyn.-St. Luke's Hosp., Chgo., 1953-56; physician-in-charge Contagious Disease Hosp., Chgo., 1956-58, asst. med. supt., 1958-64; rsch. assoc. Sch. Hygiene and Pub. Health Johns Hopkins U., Balt., 1966-71; internist Johns Hopkins Hosp., 1971-82, Columbia Free State Health Plan, Balt., 1984-91; pvt. practice Balt., 1984—; with Neurodiagnostics Assocs., 1990—; attending internist emergency rm. South Balt. Gen. Hosp., 1982-84; asst. prof. health sci. U. Ill., Chgo., 1958-64; fellow in gastroenterology and endocrinology Presbyn.-St. Luke's Hosp., 1956-58. Contbr. articles to med. jours., chpt. to book. Vol. physician, Vietnam, 1968, 71, 72, 73. Capt. M.C., USAF, 1951-53. Fellow Am. Coll. Preventive Medicine, Am Geriatrics Soc.; mem. AMA, Am. Pub. Health Assn. Avocations: writing poetry, short stories, composing songs. Home: 337 Poplar Point Rd Perryville MD 21903-1803 Office: 218 N Charles St #1407 Baltimore MD 21201-4021

RAZ, TZVI, industrial engineer; b. Montevideo, Uruguay, Dec. 18, 1951; came to U.S., 1980; m. Orna B. Witzthum, Mar. 11, 1974; children: Carmel Addie, Mical Orit. BS, Technion, Haifa, Israel, 1977; MS, U. Toronto, 1979; MSPH, U. Mo., 1982, PhD, 1982. Registered engr. Israel. Asst. prof. U. Iowa, Iowa City, 1987, assoc. prof., 1987-89; sr. lectr. Ben Gurion U., Beer Sheva, Israel, 1989-90; program mgr. IBM Corp., Roanoke, Tex., 1990—; mem. edit. bd. Inst. Indsl. Engring., 1985—. Editor: Expert Systems in Industrial Engineering, 1988, Design of Inspection Systems, 1990; contbr. over 60 articles to profl. jours. and confs. Achievements include models for inspection allocation; process analysis and re-engring.; methods and techniques for project mgmt. and scheduling. Office: IBM 5 W Kirkwood Roanoke TX 76799

RAZBOROV, ALEXANDER A., mathematician; b. Belovo, USSR, Feb. 16, 1963; s. Alexander A. and Ludmila A. (Odudenko) R. Degreea, Moscow State U., 1985; PhD, Steklov Math. Inst., Moscow, 1987. Fellow Steklov Math. Inst., Moscow, 1987-89, sr. fellow, 1989—. Contbr. articles to profl. jours. Recipient prize Moscow Math. Soc., 1986, Acad. Scis., USSR, 1987, Rolf Nevanlinna prize Internat. Math. Union, 1990. Office: Steklov Math Inst, Vavilova 42, 117966 Moscow Russia

RAZOUK, RASHAD ELIAS, retired chemistry educator; b. Dumiat, Egypt, Aug. 22, 1911; came to U.S., 1968; s. Elias A. and Martha A. (Israfil) R.; m. Emily S. Habib, Aug. 24, 1946 (dec. Dec. 1988); children: Reda R., Rami R.; m. Henrietta Doche, July 8, 1990. BSc with honors, Cairo U., 1933, MSc, 1936, PhD, 1939. Asst. prof. Cairo U., 1939-46, assoc. prof., 1946-50; prof. chemistry, chmn. dept. Ain Shams U., Cairo, 1950-66; prof. Am. U. Cairo, 1966-68; prof. Calif. State U., L.A., 1968-78, emeritus prof., 1978—; vice dean Faculty Sci. Ain Shams U., Cairo, 1954-60; acting dir. div. surface and coll. chem. Nat. Rsch. Ctr., Cairo, 1954-68; cons. Lockheed Aircraft Co., L.A., 1971-73. Contbr. articles on adsorption, active solids, wetting and wettability, solid reactions, surface tension, and contact angles to profl. jours. Fellow Am. Inst. Chemists (emeritus); mem. Am. Chem. Soc. (emeritus), Royal Soc. Chemistry (life). Democrat. Roman Catholic. Home: 1140 Keats St Manhattan Beach CA 90266-6810 Office: Calif State U 5151 State University Dr Los Angeles CA 90032-4221

RAZZAGHI, MAHMOUD, mechanical engineer; b. Shiraz, Iran, Apr. 30, 1951; came to the U.S., 1986; s. Ali and Sekineh (Torabi) R. BS in Mech. Engring., Pahlavi U., 1974; MS, Clarkson U., 1989. Mfg. engr., quality engr., design engr., cons. Iran, 1975-86; adj. prof., ind. researcher San Diego, 1989—. Contbr. articles to SPIE proceedings. Achievements include patent for compact high resolution high speed symetrical scanning optics using multiple deflections; patent pending for device for dynamic focusing of a laser beam and adaptive optical imaging. Home: 3740 Boyd Ave # 153 San Diego CA 92111

RE, RICHARD N., endocrinologist; b. Sept. 4, 1944; marrried; 3 children. AB summa cum laude, Harvard U., 1965, MD cum laude, 1969. Diplomate Am. Bd. Internal Medicine, Am. Bd. Endocrinology and Metabolism. Med. intern Mass. Gen. Hosp., Boston, 1969-70, med. resident, 1970-71, clin. and rsch. fellow in endocrinology, 1971-74, clin. asst. in medicine, 1974-76, chief Hypertension Clinic, 1975-79, asst. in medicine, 1976-79; rsch. fellow Harvard Med. Sch., Boston, 1971-74, instr. in medicine, 1974-76, asst. prof. medicine, 1977-79; mem. staff Ochsner Med. Instns., New Orleans, 1979—; assoc. clin. prof. medicine Tulane U. Sch. Medicine, New Orleans, 1980—; head sect. on hypertensive diseases Ochsner Clinic, New Orleans, 1981—; v.p., dir. rsch. Alton Ochsner Med. Found., New Orleans, 1985—; adj. prof. biology U. New Orleans, 1984—; chmn. clin. investigations com. Alton Ochsner Med. Found., 1980-86; mem. sci. rev. panel VA, 1988—; chmn. sci. resources com. Blood Rsch. Ctr., New Orleans, 1989—, trustee, 1989—; mem. adv. bd. Internat. Consortium for the Study of Tissue Renin Angiotensin Systems, 1990—; mem. rsch. com. Am. Heart Assn.-La., Inc., 1990-91, mem. sci. peer rev. com., 1990-91; mem. rsch. adv. com. Medicare Clinics Pilot Project, 1990-92. Author: Bioburst: The Impact of Modern Biology on the Affairs of Man, 1986; author: (with others) Clinical Pharmacy and Clinical Pharmacology, 1976, Biological Handbook II, 1977, Methods in Immunodiagnosis, 1981, Prostaglandins, Platelets, and Salicylates: Basic and Clinical Aspects, 1982, Systemic Disease in Dental Treatment, 1982, Kidney in Essential Hypertension, 1984, Current Clinical Practice, 1987, Current Advances in ACE Inhibition, 1989, Advances in Vascular Pathology, 1990; contbr. numerous articles to profl. jours. including Am. Jour. Hypertension, New Eng. Jour. Medicine, Jour. Inorganic Biochemistry, Current Opinion in Cardiology, Contemporary Internal Medicine, Lancet, Modern Medicine, Am. Jour. Cardiology. Fellow coun. on high blood pressure rsch. Am. Heart Assn. Grantee Nat. Heart and Lung Inst. NIH, 1981-85, La. Heart Assn., 1981-83, Indsl. Support, 1981-91. Fellow Am. Coll. Physicians; mem. AAAS, Am. Fedn. for Clin. Rsch., Assn. Am. Med. Colls. Group on Med. Edn., N.Y. Acad. Scis., Cen. Soc. for Clin. Investigation, So. Soc. for Clin. Investigation, So. Med. Assn., Internat. Soc. Hypertension, Endocrine Soc., Soc. for Exptl. Biology and Medicine, Phi Beta Kappa. Achievements include research on growth factors and cardiovascular structure, on angiotensin and regulation of cellular growth, and on hypertension. Office: Alton Ochsner Med Found 1516 Jefferson Hwy New Orleans LA 70121-2484*

READ, DOROTHY LOUISE, biology educator; b. Racine, Wis., July 18, 1938; d. Harlan Eugene and Margaret (Coxe) R.; m. David Michael Freifelder, Jan. 30, 1965 (div. Mar. 1981); children: Rachel, Joshua David Freifelder. BS, Antioch Coll., 1961; PhD, U. Calif., Berkeley, 1966. Postdoctoral fellow, rsch. assoc. Brandeis U., Waltham, Mass., 1966-67, 68-69; rsch. assoc. Brandeis U., 1974-78; lectr. in Biology Northeastern U., Boston, 1978; asst. prof. biology U. Mass, Dartmouth, 1978-88, assoc. prof. biology, 1988—; vis. prof. Stanford (Calif.) U., 1982-85, 90-91; cons. Stanford Rsch. Inst., Menlo Park, Calif., 1985. Contbr. articles to profl. jours. Recipient NIH Rsch. grant, 1988-91, NIH fellowship, 1963-66, Woodrow Wilson Found. fellowship, 1961-62. Mem. AAAS, Soc. for Indsl. Microbiology, Am. Soc. for Microbiology, Am. Soc. for Biochemistry and Molecular Biology. Office: U Mass Biology Dept North Dartmouth MA 02747

READ, LESLIE WEBSTER, electrical engineer; b. Alexandria, La., Sept. 5, 1937; s. Arthur Davis and Ethel Frances Mary (Woodgate) R. BA, Rhodes Coll., 1960; BEE, Ga. Tech. Inst., 1960, MSEE, 1962. Engr. Tex. Instruments, Dallas, 1963-71; ind. engr. Dallas, 1971-81; elec. engr. Sammons Communications, Dallas, 1981—. Mem. IEEE, Soc. Cable Television Engrs. (sr., bd. dirs. 1989-92). Presbyterian. Office: Sammons Communications Ste 800 3010 LBJ Fwy Dallas TX 75234

READER, ALEC HAROLD, physicist, researcher; b. Dartford, Kent, England, 1957; arrived in The Netherlands, 1983.; BS in Physics with honors, Birmingham U., Eng., 1979, PhD in Phys. Metallurgy, 1983. Postdoctoral rsch. fellow metallurgy dept. Delft U., Holland, The Netherlands, 1983-85; rsch. scientist Philips Rsch. Labs., Eindhoven, The Netherlands, 1985-90; rsch. project leader Philips Rsch. Labs., Eindhoven, 1990—; with SGS Thomson, Cedex, France. Author: (with others) Transmission Electron Microscopy, 1991, Microstructural Defects in IC-Materials, 1990, Silicides in Future IC's; contbr. articles to sci. jours.; patentee in field. Mem. London Inst. Physics (charter), London Inst. Metals and Materials, Materials Rsch. Soc. (Am.). Office: SGS Thomson Ctr de Crolles, 850 Rue de Jean Monnet, B016 38921 Crolles France

READER, GEORGE GORDON, physician, educator; b. Bklyn., Feb. 8, 1919; s. Houston Parker and Marion J. (Payne) R.; m. Helen C. Brown, May 23, 1942; children: Jonathan, David, Mark, Peter. BA, Cornell U., 1940, MD, 1943; DSc (hon.), Drew U., 1988. Diplomate Am. Bd. Internal Medicine. Intern N.Y. Hosp., N.Y.C., 1944; resident N.Y. Hosp., 1947-49, attending physician, 1962-92, dir. comprehensive care and teaching program, 1952-72, chief med. clinic, 1952-72; practice medicine specializing in internal medicine N.Y.C., 1949-93; chief div. ambulatory and community medicine N.Y. Hosp.-Cornell Med. Center, N.Y.C., 1969-72; prof. medicine Cornell U. Med. Sch., 1957-89, Livingston Farrand prof. Medicine and Pub. Health, 1972-89, prof. emeritus Pub. Health, Medicine, 1989—, chmn. dept. pub. health, 1972-92; Chmn. human ecology study sect. NIH, 1961-65; chmn. med. adv. com. Vis. Nurse Service N.Y., 1963—; mem. med. control bd. Health Ins. Plan Greater N.Y., 1964—; mem. Gov.'s Health Adv. Council, 1974-79. Author: (with R. Merton, P. Kendall) The Student Physician, 1957, (with Goss) Comprehensive Medical Care and Teaching, 1967, (with Goodrich and Olendzki) Welfare Medical Care: An Experiment, 1969; mem. editorial bd. Medical Care, 1969-70, Jour. Med. Edn, 1975-79; editor-in-chief Milbank Meml. Fund Quar.: Health and Society, 1972-76. Bd. dirs. N.Y.C. Vis. Nurse Svc., The Osborn, Health Ins. Plan Greater N.Y., 1983-93, Helen Keller Internat.; trustee Cornell U., 1982-87. Lt. USNR, 1944-46, PTO. Fellow ACP, Am. Coll. Preventive Medicine, Am. Public Health Assn. (governing council 1968-69); mem. AAAS, AMA, N.Y. Acad. Medicine (chmn. com. med. edn. 1968-71, v.p. 1978-80), Am. Sociol. Assn., Harvey Soc., Internat. Sociol. Assn., Internat. Epidemiological Assn., N.Y.C. Pub. Health Assn. (pres. 1956), Inst. Medicine Nat. Acad. Scis. (sr. mem.), Sigma Xi, Alpha Omega Alpha. Home: 155 Stuyvesant Ave Rye NY 10580-3112

READER, JOSEPH, physicist; b. Chgo., Dec. 1, 1934. BS, Purdue U., 1956, MS, 1957; PhD in Physics, U. Calif., 1962. Rsch. assoc. physics Argonne Nat. Lab., 1962-63; staff physicists Nat. Inst. Standards and Tech., Gaithersburg, Md., 1963—. Recipient Gold Medal Dept. Commerce. Fellow Am. Phys. Soc., Optical Soc. Am. (William F. Meggars 1992). Achievements includes research in experimental atomic physics, optical spectroscopy, hyperfine structure, electronic structure of highly ionized atoms, wave length standards, and ionization energies of atoms and ions. Office: Natl Inst Of Standards & Tech Gaithersburg MD 20899*

READING, HAROLD G., geology educator. Prof. geology U. Oxford, Eng. Recipient William H. Twenhofel medal Soc. Sedimentary Geology, 1994. Office: U Oxford Dept Earth Sci, Parks Rd, Oxford OX1 3PR, England*

REAL, THOMAS MICHAEL, civil engineer; b. Pittsfield, Mass., Dec. 18, 1962; s. Raymond Francis and Constance (Colby) R. BS in Civil and Environ. Engring. cum laude, Clarkson Coll. Tech., 1984. Registered profl. engr., Calif.; cert. competancy blasting ops., Mass., Alaska. Chainman survey F. Robert Bell & Assoc., Anchorage, 1983-84, instrument man survey, 1984-85, field engr., 1985; estimator, project coord. Rock & Waterscape Systems, Inc., Irvine, Calif., 1985-87, project mgr., 1987-89, sr. estimator, 1989-91, project engr., 1991—; engr. design and supervision Lost City Project Sun City, Bophuthatswana. Mem. NSPE, Calif. Soc. Profl. Engr., Tau Beta Pi, Chi Epsilon. Roman Catholic. Office: Rock Waterscape Systems Inc 11 Whatney Irvine CA 92718

REAMAN, GREGORY HAROLD, pediatric hematologist, oncologist; b. Akron, Ohio, Sept. 9, 1947; s. Harold J. and Margaret U. (D'Alfonso) R.; m. Susan J. Pristo, Sept. 7, 1974; children: Emily Margaret, Sarah Elizabeth. BS in Biology, U. Detroit, 1969; MD, Loyola U., Chgo., 1973. Diplomate Nat. Bd. Med. Examiners, Am. Bd. Pediatrics. Pediatric intern Loyola U. Med. Ctr., 1973-74; resident in pediatrics Montreal Children's Hosp., McGill U., 1974-76; clin. assoc. pediatric oncology br. Nat. Cancer Inst., NIH, Bethesda, Md., 1976-78, investigator pediatric oncology br., 1978-79; assoc. dept. hematology/oncology, attending physician Children's Nat. Med. Ctr., Washington, 1979-87, chmn. dept. hematology/oncology, 1987—; asst. prof. pediatrics Sch. Medicine and Health Scis., George Washington U., 1979-82, assoc. prof. pediatrics, 1982-87, prof. pediatrics, 1987—; mem. immunology devices panel FDA; assoc. chmn. Children's Cancer Group; bd. dirs., mem. med. affairs com., chmn., strategic planning com. Children's Oncology Scs. of Met. Washington. Mem. editorial bd. Cancer Physicians Data Query, Nat. Cancer Inst.; reviewer Cancer Treatment Resports, Blood, Jour. Clin. Oncology; contbr. articles to profl. publs. Trustee Nat. Childhood Cancer Found., Arcadia, Calif.; bd. dirs. Am. Cancer Soc., Atlanta; trustee, chmn. patient care and profl. edn. coms. Leukemia Soc. Am. Lt. comdr. USPHS, 1976-79, Res., 1979—. Folger Summer scholar Am. Cancer Soc.; recipient Spl. Fellowship Rsch. award Leukemia Soc. Am., 1980-82; grantee DHHS, Nat. Cancer Inst., 1987—. Mem. Soc. Pediatric Rsch., Am. Fedn. Clin. Rsch., Am. Soc. Clin. Oncology, Am. Assn. Cancer Rsch., Am. Soc. Pediatric Hematology/Oncology, Children's Cancer Group, Washington Blood Club, Alpha Omega Alpha. Democrat. Roman Catholic. Home: 7306 Brennon Ln Chevy Chase MD 20815 Office: Children's Nat Med Ctr 111 Michigan Ave NW Washington DC 20010

REAMES, SPENCER EUGENE, science educator; b. Bellefontaine, Ohio, Apr. 26, 1946; s. Lloyd Eugene and Evelyn Lucille (Williams) R.; m. Anne Marie Nicolosi, Aug. 28, 1972 (div. 1991); 1 child, Aaron Spencer. BS in Edn., Bowling Green State U., 1968; MS in Sci., Ball State U., 1978. Cert. tchr., Ohio. Tchr., sci. Benjamin Logan High Sch., Bellefontaine, 1968—, chmn. sci. dept., 1973—; sci. edn. cons., Bellefontaine, 1987—; co-chmn subcom. on partnerships USPHS, Washington, 1990-91; mem. tech. adv. com. State Supt. Pub. Instrn., Columbus, 1987-91, mem. state sci. curriculum adv. com., 1992—. Editor: Ohio Science Workbook: Biotechnology, 1993. Mem. Solid Waste Mgmt. Policy Com. Bellefontaine, 1989—, Logan Co. Litter Prevention and Recycling Com., 1989—. Recipient Presdl. award for excellence NSF, 1986, grantee Logan County Edn. Found., 1990-92; named Ohio Outstanding Biology Tchr. Nat. Assn. Biology Tchrs., 1984. Mem. AAAS, Assn. Presdl. Awardees in Sci. Teaching, Am. Soc. Microbiology, Soc. for Invertebrate Pathology, Ohio Acad. Sci. (Outstanding Tchr. award 1974, 77, 83, 86, v.p. sci. edn. sect. 1991-92, exec. com. 1985-89, asst. project dir. 1989—, Centennial Honoree). Office: Benjamin Logan High Sch 6609 State Route 47 E Bellefontaine OH 43311-9599

REAMES, THOMAS EUGENE, chemical engineer; b. Warren, Ark., May 10, 1940; s. Julian Thomas and Lucille (Greene) R.; Linda Lou Tomlinson, June 7, 1959; children: Timothy, Scott, Douglas. BSChemE, U. Ark., 1963. Jr. process engr. Gulf Oil Corp., Port Arthur, Tex., 1963-65, econ. anlayst, 1965-69; corp. econ. analyst Gulf Oil Corp., Pitts., 1969-74; project engr. Tosco, El Dorado, Ark., 1974-82, sr. process engr., 1982-84; plant engr. Ensco Inc., El Dorado, 1984-88, engring. mgr., 1989—. Cub and scout master Boy Scouts Am., El Dorado, 1974—. Mem. AICE, Constrn. Specifications Inst., Nat. Assn. Corrosion Engrs. Southern Baptist. Office: Ensco 309 American Rd El Dorado AR 71730

REARDON, FREDERICK HENRY, mechanical engineer, educator; b. Philadelphia, Oct. 22, 1932; s. John Nicholas and Gertrude (Setter) R.; m. Dorothy Hoeppner, June 16, 1956; children: Kenneth, Joy Moreton, Steven. BSME, U. Pa., 1954, MSME, 1956; PhD in Aeronaut. Engring., Princeton U., 1961. Sr. rsch. specialist Aerojet Liquid Rocket Co., Sacramento, 1961-66; asst., assoc. prof. mech. engr. Calif. State U., Sacramento, 1966-74, prof. mech. engring., 1974—, chair dept., 1978-83, assoc. dean Sch. Engring. and Computer Sci., 1983-88; cons. Aerojet-Gen. Corp., Sacramento, 1966-90, USAF, Wright-Patterson AFB, Ohio, 1976-86, United Technologies, San Jose, Calif., 1980-86. Editor: (with D.T. Harrje) Combustion Instability in Liquid Propellant Rockets, 1972. Fellow ASME, AIAA (assoc.); mem. Am. Soc. Engring. Edn., Combustion Inst., Sigma Xi, Tau Beta Pi. Democrat. Lutheran. Home: 5619 Haskell Ave Carmichael CA 95608 Office: Calif State Univ 6000 J St Sacramento CA 95819-6031

REARICK, JAMES ISAAC, biochemist, educator; b. Punxsutawney, Pa., Apr. 13, 1952; s. Albert Ross and Kathryn (Arble) R.; m. Daisy Whitesides, Mar. 15, 1980; children: Ross, William, Benjamin. BS, Pa. State U., 1974; PhD, Duke U., 1979. Postdoctoral fellow Washington U., St. Louis, 1979-82; sr. staff fellow Nat. Inst. Environ. Health Sci., Research Triangle Park, N.C., 1983-86; assoc. prof. Kirksville (Mo.) Coll. Osteo. Medicine, 1986—. Contbr. articles to Jour. Biol. Chemistry, Methods in Enzymology, Jour. Investigative Dermatology, others. James B. Duke fellow Duke U., 1974, postdoctoral fellow NIH, 1980; rsch. grantee Am. Cancer Soc., 1988. Mem. Am. Soc. for Biochemistry and Molecular Biology, The Sci. Rsch. Soc., Sigma Xi. Achievements include rsch. on biosynthesis of glycoproteins, regulation of differentiation in airway and skin, mechanism of action of vitamin A metabolites, regulation of cholesterol sulfation. Office: Kirksville Coll Osteo Medicine 800 W Jefferson Kirksville MO 63501

REARK, JOHN BENSON, consulting ecologist, landscape architect; b. Chgo., Aug. 1, 1923; s. Benson and Ruth (Garvey) R.; m. Muriel Frances Schafer, Sept. 9, 1950 (div. 1967); children: Michael Frederick, Eileen Frances; m. Barbara Hublitz, July 9, 1973. BS in Botany, U. Miami, 1950; MS in Tropical Forest Ecology, Centro Agronomico Tropico de Investigación y Ensaňza, Turrialba, Costa Rica, 1952; postgrad., U. Miami, 1958-60; PhD in Tropical Coastal Ecology, Clayton U., St. Louis, 1983. Lic. landscape architect, Fla. Landscape architect, ecologist, land use planner, South Miami, Fla., 1955—; instr. horticulture Dade Correctional Inst., Florida City, Fla., 1986-91; horticulturist Miami MetroZoo, 1987-91; mem. Dade County Landscape Ordinances Bd., Miami, 1964-65; mem. Pinewood Cemetery hist. site City of Coral Gables, Fla., 1985-89; newspaper columnist, 1979-83; architect major tourism projects, Bahamas, 1956-58, 67-68, Costa Rica, 1968-71, Honduras, 1975-76, Grand Cayman, 1970-71, Dominican Republic, 1974; town planner, Tegucigalpa, Honduras, 1977-78. Contbr. numerous articles to sci. publs.; innovator mist propagation of plants in full sun, 1951. Bd. dirs. Mus. Sci. and Planetarium, Miami, 1963-67. Capt. USAAF, 1942-46, ETO. Decorated Air medal. Mem. AAAS, Sigma Alpha Epsilon, Omicron Delta Kappa. Achievements include research into mangrove gall fungus (Cylindrocarpon didymum) in coastal monocultural plantings, low pressure greenhouse climate control, and polycultural algae species mixes suitable for tropical microphage large scale aquacultural production. Home and Office: 6870 SW 75th St Miami FL 33143-4425

REASONER, DONALD J., microbiologist; b. MacDonald, Kans., July 29, 1940; s. Arnold L. and Ruby Lucille (Graves) R.; m. Linda Lou Garver, Sept. 9, 1961; children: Jeffrey Scott, David Alan. BS, Colo. State U., 1966, PhD, 1971; MS, Wash. State U., 1969. Rsch. microbiologist drinking water rsch. divsn. Risk Reduction Engring. Labs. U.S. EPA, Cin., 1971-89, chief microbiologist drinking water rsch. divsn., 1989—. Contbr. articles to profl. publs., chpts. to book and ency. Scoutmaster Boy Scouts Am., Cin., 1979-85. With USAF, 1961-64. Mem. Am. Soc. for Microbiology (chmn. divsn. N 1987-88, lectr. Found. for Microbiology 1991-93), Am. Water Works Assn. (Best Paper award 1973), Sigma Xi, Phi Kappa Phi. Achievements include development of bacterial culture medium (R2A) widely used for drinking water analysis. Office: US EPA 26 W Martin Luther King Dr Cincinnati OH 45268

REBAGAY, TEOFILA VELASCO, chemist, chemical engineer; b. Pangasinan, Philippines, Feb. 5, 1928; came to U.S., 1965; d. Dionisio Opiniano and Antonia (Flora) Velasco; m. Guillermo Rabadam Rebagay, Apr. 4, 1956; children: Guillermo V., Teofilo V. BS in Chemistry, U. Philippines, Quezon City, 1951; BS in Chem. Engring., Nat. U., Manila, 1954; PhD in Chemistry, U. Ky., 1969. Postdoctoral fellow U. Ky., Lexington, 1969-71, U. Va., Charlottesville, 1971-72; rsch. assoc. U. Ky., Lexington, 1973-76; sr. chemist Ky. Ctr. Energy Rsch., Lexington, 1976-78; radiochemist Allied-Gen. Nuclear Svcs., Barnwell, S.C., 1978-83; sr. chemist Rockwell Hanford Ops., Richland, Wash., 1983-87; prin. scientist Westinghouse Hanford Co., Richland, 1987—. Contbr. articles to profl. publs. IAEA fellow UN, Tokyo, 1963; grantee Rockefeller Found., 1965. Mem. ASTM, Soc. Applied Spectroscopy. Office: Westinghouse Hanford Co Richland WA 99352

REBEIZ, CONSTANTIN ANIS, plant physiology educator; b. Beirut, July 11, 1936; came to U.S., 1969, naturalized, 1975; s. Anis C. and Valentine A. (Choueyri) R.; m. Carole Louise Conness, Aug. 18, 1962; children: Paul A., Natalie, Mark J. B.S., Am. U., Beirut, 1959; M.S., U. Calif. - Davis, 1960, Ph.D., 1965. Dir. dept. biol. scis. Agrl. Research Inst., Beirut, 1965-69; research assoc. biology U. Calif. - Davis, 1969-71; assoc. prof. plant physiology U. Ill., Urbana-Champaign, 1972-76, prof., 1976—. Contbr. articles to sci. publs. plant physiology and biochemistry. Recipient Beckman research awards, 1982, 85, Funk award, 1985, John P. Trebellas Research Endowment, 1986; named One of 100 Outstanding Innovators, Sci. Digest, 1984-85. Mem. Am. Soc. Plant Physiologists, Comite Internat. de Photobiologie, Am. Soc. Photobiology, AAAS, Lebanese Assn. Advancement Scis. (exec. com. 1967-69), Sigma Xi. Achievements include research on pathway of chlorophyll biosynthesis, chloroplast devel., bioengring. of photosynthetic reactors; pioneered biosynthesis of chlorophyll in vitro; duplication of greening process of plants in test tube, demonstration of operation of multibranched chlorophyll biosynthetic pathway in nature; formulation and design of laser herbicides, insecticides and cancer chemotherapeutic agents. Home: 301 W Pennsylvania Ave Urbana IL 61801-4918 Office: U Ill 240A Pabl Urbana IL 61801

REBELSKY, LEONID, physicist; b. Odessa, Ukraine, USSR, June 23, 1956; came to U.S., 1981; s. Israel Rebelsky and Adel Kumets; m. Katherine Pipko, Mar. 17, 1988. MS, Odessa State U., Ukraine, 1979; PhD, N.Y. U., 1987. Post doctoral Brookhaven Nat. Lab., Upton, N.Y., 1987-88; asst. physicist BNL, Upton, N.Y., 1989-91, assoc. physicist, 1991-92; pres. Advanced Materials & Technology, N.Y.C., 1992—. Mem. Am. Physical Soc., N.Y. Acad. Sci. Office: Advanced Materials & Tech 5 E 57th St New York NY 10022

REBENFELD, LUDWIG, chemist; b. Prague, Czechoslovakia, July 10, 1928; came to U.S., 1939, naturalized, 1946; s. Carl and Martha (Scheib) R.; m. Ellen Vogel, July 27, 1956. B.S., Lowell Tech. Inst., 1951; M.A., Princeton, 1954, Ph.D., 1955; D.Textile Sci. (hon.), Phila. Coll. Textiles and Sci. Rsch. fellow, sr. scientist Textile Rsch. Inst., Princeton, N.J., 1951-59, assoc. rsch. dir., also edn. program dir., 1960-66, v.p. edn. and rsch., 1966-70, pres., 1971-92, pres. emeritus and rsch. assoc., 1993—; vis. lectr., asso. prof., prof. chem. engring. Princeton, 1964—; mem. governing council, sec.-treas. Fiber Soc., 1962-84; mem. Nat. Council Textile Edn.; trustee Phila. College Textiles and Sci. (life). Recipient Distinguished Achievement award Fiber Soc., Harold DeWitt Smith Meml. medal ASTM. Fellow Am. Inst. Chemists, Brit. Textile Inst. (Inst. medal), Inst. Textile Sci. Can.; mem. Am. Chem. Soc., Am. Assn. Textile Chemists and Colorists (Olney medal 1987), AAAS, Sigma Xi, Phi Psi. Home: 49 Pardoe Rd Princeton NJ 08540-2617 Office: Textile Rsch Inst PO Box 625 601 Prospect Ave Princeton NJ 08540

REBER, RAYMOND ANDREW, chemical engineer; b. Bklyn., Apr. 16, 1942; s. Herbert and Dorothy Agnes (Schmidt) R.; m. Anita Jean Roe, June 22, 1963; children: Laura Jean Bucci, Paul Raymond, Jill Anita. BChemE, NYU, 1963, MChemE, 1966. Design engr. M.W. Kellogg, N.Y.C., 1964-66, process devel. engr., 1967-69; devel. engr., supr. Union Carbide, Tarrytown, N.Y., 1970-74, lic. bus. mgr., 1975-81; new bus. devel. mgr. Union Carbide, Danbury, Conn., 1982-84; tech. mgr. Union Carbide, Tarrytown, 1985-87; dir. of tech. UOP, Tarrytown, 1988-93; ind. cons., 1993—. Patentee in field. Commr. Montrose (N.Y.) Improvement Dist., 1970—; referee West County Approved Soccer Official Assn., N.Y. High Schs., 1977-91. Recipient Kirkpatrick award McGraw-Hill, 1967, award 1987. Mem. Am. Inst. Chem. Engrs., Nat. Soc. Profl. Engrs., Am. Water Works Assn. Episcopalian. Avocations: soccer, boating, table games, philatelist. Home and Office: 10 Bonnie Hollow Ln Montrose NY 10548-1314

REBOUL, JAQUES REGIS, computer company executive; b. Marseille, France, Jan. 6, 1947; s. Armand and Charlotte (Jambet) R.; m. Sylvie Bruley, Sept. 14, 1974. M Math. and Physics, U. Marseille, 1970; PhD in Physics, U. Paris, 1972. Dir. Burroughs Corp., Paris, 1974-77, Control Data Corp., Paris, 1977-86; v.p. Xidex Corp., Paris and Frankfurt, Fed. Republic Germany, 1986-88; group v.p. Euromagnetics Ltd., London, 1988-90; pres., dir. gen. Tandem Computers SA, Boulogne, 1990—. Home: 18 Ave Emile Deschanel, 75007 Paris France Office: Tandem Computers SA, 221 Bis Blvd Jean Jaures, 92514 Boulogne Billancourt, France

RECCHI, VINCENZO, manufacturing executive; b. Rome, July 24, 1945; s. Riccardo and Fernanda (Schiavoni) R.; m. Rosella Febo, Aug. 4, 1973; children: Chiara, Riccardo. Degree in Aero. Engring., Naples U., 1973. Rschr. Battelle Rsch. Ctr. Otb-Partecipazioni, Geneva, 1974-75; rsch. and devel. mgr. Otb-Partecipazioni, Bari, Italy, 1975-79, dir., 1980-86; gen. mgr. Tecnars, Bari, Italy, 1986-90, mng. dir., 1990—; Italian del. Internat. Energy Agy.-Advanced Heat Pumps; pres. Italian com. Internat. Flame Rsch. Found., Milan, Italy, 1983-88, mem. joint com., Ijmuiden, Holland, 1983-88. Contbr. articles to profl.jours. Mem. Italian Assn. Air Conditioning, Heating, and Refrigeration. Achievements include patents for Combustion Head, Low NOx Boilers. Home: Via Gen Dalla Chiesa 12/C, Bari Italy 70124 Office: Tecnars srl, Via Pionieri del Commercio Barese, Bari Italy 70123

RECHTIN, EBERHARDT, aerospace educator; b. East Orange, N.J., Jan. 16, 1926; s. Eberhardt Carl and Ida H. (Pfarrer) R.; m. Dorothy Diane Denebrink, June 10, 1951; children: Andrea C., Nina, Julie Anne, Erica, Mark. B.S., Calif. Inst. Tech., 1946, Ph.D. cum laude, 1950. Dir. Deep Space Network, 1949-67; asst. dir. Calif. Inst. Tech. Jet Propulsion Lab., 1949-67; dir. Advanced Research Projects Agy., Dept. Def., 1967-70, prin. dep. dir. def. research and engring., 1970-71, asst. sec. def. for telecommunications, 1972-73; chief engr. Hewlett-Packard Co., Palo Alto, Calif., 1973-77; pres., chief exec. officer Aerospace Corp., El Segundo, Calif., 1977-87; prof. U. So. Calif., 1988—. Served to lt. USNR, 1943-56. Recipient major awards NASA, Dept. Def., USN; Disting. Alumni award Calif. Inst. Tech., 1984. Fellow AIAA (Robert H. Goddard Astronautics award 1991), IEEE (Alexander Graham Bell award 1977); mem. NAE, Tau Beta Pi. Home: 1665 Cataluna Pl Palos Verdes Estates CA 90274 Office: U So Calif University Park Los Angeles CA 90089-2565

RECK, FRANCIS JAMES, electronics engineer; b. Shamokin, Pa., Mar. 14, 1955; s. Francis Clement and Elizabeth Delores (Coroniti) R.; m. Rosary Josephine Wobbe, June 4, 1983; children: Natalie, Anthony, Kathleen. BS in Electrical Engring., Lehigh U., 1977. Electronics engr. Naval Electronic Systems Engr., St. Inigoes, Md., 1977-82, head video systems br., 1982-84; electronics engr. Displays Systems Br. Defense Info. Systems Agy., Washington, 1984-87, chief installation and tech. SPT. br. Hardware Systems divsn., 1987-89, chief communication support div. Ctr. for Operational Svcs., 1989—. Mem. Armed Forces Communications and Electronics Assn. Roman Catholic. Office: The Pentagon Defense Info Systems Agy Washington DC 20301

RECKAMP, DOUGLAS E., military officer; b. Chgo., Mar. 31, 1968; s. Luke Leon and Celeste Marie (McClain) R.; m. Karen Frances Chester, July 21, 1990; 1 child, Katherine Jean. BS in Engring., U.S. Naval Acad., 1990; MS in Engring., U. Tex., 1992. Commd. lt. (j.g.) USN, 1986, submarine officer, 1990—. Mem. Acoustical Soc. Am., Sigma Xi, Pi Tau Sigma (v.p. 1989-90), Tau Beta Pi. Roman Catholic. Home: 12 Keklico Ct Goose Creek SC 29445

RECKASE, MARK DANIEL, psychometrician; b. Chgo., Aug. 31, 1944; s. John Daniel and Harriet (Stark) R.; m. Charlene Mae Repetny, Aug. 10, 1968; children: Erik Nathan, Debra Carol. MS, Syracuse U., 1971, PhD, 1972. Asst. prof. U. Mo., Columbia, 1972-75, assoc. prof., 1975-81; dir. Am. Coll. Testing, Iowa City, Iowa, 1981-84; asst. v.p. ACT, Iowa City, Iowa, 1984—; mem. Def. Adv. Com. on Mil. Pers. Testing, 1989—; mem. mgmt. com. Jour. Ednl. Stats., 1991-93. Contbr. articles to Jour. of Ednl. Measurementt, othrs. Capt. U.S. Army Res., 1966-75. Office of Naval Rsch. grantee, 1977-92. Fellow Am. Psychol. Soc.; mem. Psychometric Soc. (sec. 1986-91). Achievements include development of multidimensional item response theory. Office: ACT 2201 N Dodge St Iowa City IA 52243

REDALIEU, ELLIOT, pharmacokinetics executive; b. N.Y.C., Dec. 12, 1939; s. William and Anne (Kaplan) R.; m. Carmen Lydia Baretty, Dec. 20, 1962; children: David John, Daniel Jeffrey, Darryl Jason. BS in Pharmacy, Fordham U., 1961; MS, U. Mich., 1963, PhD, 1966. Postdoctoral fellow Stanford Rsch. Inst., Menlo Park, Calif., 1966-68; rsch. biochemist pharm. divsn. Clin. Pharmacokinetics and Disposition Ciba-Geigy Corp., Ardsley, N.Y., 1968-71, sr. scientist pharm. divsn., 1971-75, sr. scientist II pharm. divsn., 1975-80, mgr. pharm. divsn., 1980-87, asst. dir., 1987—. Author: (with others) Methods in Enzymology, 1971; contbr. 17 articles to profl. jours. Cubmaster Boy Scouts Am., Haverstraw, N.Y., 1973-76; coach Midget Football, Haverstraw, 1977-79; asst. coach Little League, Haverstraw, 1978; vol. United Fund Employees Drive, Ciba-Geigy Corp., 1988-90, co-chmn., 1989. Mem. AAAS, Am. Assn. Pharm. Scientists, Am. Chem. Soc., Internat. Soc. for the Study of Xenobiotics, N.Y. Acad. Sci., Phi Lambda Upsilon, Rho Chi. Office: Ciba-Geigy Corp 444 Saw Mill River Rd Ardsley NY 10502

REDD, RUDOLPH JAMES, chemical engineer, consultant; b. Charlotte, Va., Mar. 25, 1924; s. Hampton James and Leona (Lawson) R.; m. Mary Lou Camper, 1948; children: Rudolph J. Jr., Teresa M.; m. Noel Rebecca Willis, Sept. 3, 1985. BS, Morgan State U., 1948. Chemist Edgewood (Md.) Arsenal, 1948-70; chem. engr. Chem. Rsch. Labs., Aberdeen Proving Grounds, Md., 1970-75, Chem. Rsch. Dev and Engring. Ctr., Aberdeen Proving Grounds, Md., 1975-83; chem. engr., cons., electronic engr. Towson, Md., 1984—; pres., propr. Redd's Sounds Unltd., Inc., Towson, Md., 1984—; cons. Tel Tec, Mpls., 1983—. Mem. Charter Revision Commn., Balt. County, 1977, Cable Adv. Commn., Balt. City, 1989—. With U.S. Army, 1945-46. Mem. AICE, AAAS, Am. Chem. Soc., Am. Def. Preparedness Assn. Democrat. Methodist. Achievements include development of measurement techniques for use in evaluating protective masks; of new masks and protective clothing for the U.S. Army; research in methodology for measuring speech transmission for field use of protective masks. Home: 151E Versailles Cir Towson MD 21204-6936

REDDELL, DONALD LEE, agricultural engineer; b. Tulia, Tex., Sept. 28, 1937; s. Kimball Tuscola and Winonah (Claiborne) R.; m. Minnie Ellen Cox, Jan. 27, 1957; children: Revis Diane, Cheryl Reneé, Stephen Patrick. BS, Tex. Tech U., 1960; MS, Colo. State U., 1967, PhD, 1969. Registered profl. engr., Tex. Jr. engr. High Plains Underground Water Conservation Dist. No. 1, Lubbock, Tex., 1960-61, agrl. engr., 1961-63, engr., 1963-65; NSF trainee civil and agrl. engring. dept. Colo. State U., Ft. Collins, 1965-69; asst. prof. agrl. engring. dept. Tex. A&M U., College Station, 1969-72, assoc. prof. agrl. engring. dept., 1972-77, prof. agrl. engring. dept., 1977-89, prof. and head agrl. engring. dept., 1989-93, prof., 1993—. Author: Numerical Simulation of Dispersion in Groundwater Aquifers, 1970; assoc. editor: Jour. Environ. Quality, 1979-85, Energy in Agr. (The Netherlands), 1981-85; contbr. articles to profl. jours. Recipient Outstanding Jour. Paper award ASCE Jour. Irrigation and Drainage, 1989. Fellow Am. Soc. Agrl. Engrs. (pres. Tex. sect. 1979-80, ASAE Paper award 1977, 82, Agrl. Engr. of Yr. award Tex. sect. 1975, Disting. Young Agrl. Engr. of Yr. award SW region chpt. 1977); mem. Am. Geophys. Union, NSPE. Achievements include co-development of the Laplace Transform Finite Difference, Laplace Transform Finite Element, Laplace Transform Boundary Element, Laplace Transform Solute Transport techniques for modeling groundwater flow, eliminating the need for discretizing time in numerical simulations; development of Method of Characteristics used to describe solute transport in ground water, of automatic advance rate feedback furrow irrigation system. Home: 3808 Courtney Cir Bryan TX 77802-3407 Office: Tex A&M U Agrl Engring Dept College Station TX 77843-2117

REDDI, A. HARI, orthopaedics and biological chemistry educator; b. Madras, India, Oct. 20, 1942; came to U.S., 1968; s. Ramakrishna Reddi and Kausaly A. Reddy; m. Anu H. Reddy, June 4, 1972; children: Ajoy, Amit, Anand. MSc, Annamalai U., India, 1962; PhD, U. Delhi, India, 1966. Postdoctoral fellow Johns Hopkins Med. Sch., Balt., 1968-69; rsch. assoc. U. Chgo., 1969-72, asst. prof., 1972-76; rsch. biologist NIH, Bethesda, Md., 1976-78, chief bone cell biology sect., 1978-91; prof. dept. orthopaedics & biol. chemistry Johns Hopkins Med. Sch., Balt., 1991—; dir. Johns Hopkins U. Sch. Medicine Lab. Musculoskeletal Cell Biology, Balt., 1991—. Editor: Biochemistry of Extracellular Matrix, 1984, Extracellular Matrix, 1985. Recipient Dir. award NIH, Bethesda, 1990, Kappa Delta award Am. Acad. Orthopaedic Surgeons, 1991. Home: 3525 Raymoor Rd Kensington MD 20895 Office: Johns Hopkins University Ross Bldg 2nd Fl 720 Rutland Ave Baltimore MD 21205

REDDY, CHILECAMPALLI ADINARAYANA, microbiology educator; b. Nandimandalam, Cuddapah, India, July 1, 1941; came to U.S., 1965; s. C. Narayana and Balamma (Netapalle) R.; m. Sasikala C. Reddy, Oct. 9, 1972; 1 child, Sumabala C. DVM, Sri Venkateswara U., Tirupati, India, 1962; MS, U. Ill., 1967, PhD, 1970. Diplomate Am. Coll. Vet. Micrbiology. Vet. asst. surgeon Dept. Animal Husbandry, Proddatur, India, 1962-65; rsch. asst. U. Ill., Urbana-Champaign, 1965-70; rsch. assoc. U. Ga., Athens, 1970-72; asst. prof. microbiology Mich. State U., East Lansing, 1972-77, assoc. prof., 1977-85, prof., 1985—; cons. FAO of U.N., 1983, 89; vis. scientist Nat. Animal Disease Ctr., Ames, Iowa, 1979. Editor: Current perspectives on Microbial Ecology, 1984; contbr. over 100 articles to profl. jours. Dir. India Cultural Soc., Greater Lansing, Mich., 1984-86, U.S. Mich. State U./India SCOPE, East Lansing, 1987-89. Recipient award for rsch. excellence Smithkline Beecham Co., 1991, Rsch. Grants NIH, Dept. Energy, NSF, Rsch. Excellence Funds, State of Mich., other orgns. Fellow Am. Acad. Microbiology; mem. AAAS, Am. Soc. Microbiology, Am. Vet. Med. Assn., Soc. Indsl. Microbiology, Conf. Rsch. Workers in Animal Disease, Sigma Xi. Achievements include research in fermentative conversion of agroindsl. residues into nitrogen-rich feeds; defining the biochem. bases for interspecies hydrogen transfer in anaerobic ecosystems; first documentation on the importance of hydrogen peroxide in lignin degradation by fungi; first cloning of lignin peroxidase cDNAs; first isolation of nitrgen-deregulated mutants and lignin peroxidase-negative mutants; detailed investigations on taxonomy and physiology of animal pathogenic Coryne bacteria including the renaming of two bacterial species; rsch. on bioremediation. Office: Mich State U Dept Microbiology East Lansing MI 48824-1101

REDDY, GUVVALA NAGABHUSHANA, neurochemist; b. Madras, India, Dec. 12, 1959; s. Guvvala Narappa and Rangamma Reddy. MS, Madras U., 1982; PhD, Oreg. State U., 1987. Rsch. asst. Oreg. State U., Corvallis, 1983-88; rsch. assoc. U. Calif., Davis, 1988-89; staff scientist UCLA Sch. Medicine, 1989-90; scientific officer Swiss Nat. Inst. for Nuclear Rsch., Villigen, Switzerland, 1991—. Contbr. articles to Jour. Am. Chem. Soc., Jour. Nuclear Medicine. Office: Paul Scherrer Inst, Radiopharmacy PG-10 CH-5232, Villigen Switzerland

REDDY, J. NARASIMH, mechanical engineering educator. Prof., mechanical engineering Virginia Polytechnic Inst., Blacksburg, Vir. Recipient Worcester Reed Warner medal ASME, 1992. Office: Virginia Polytechnic Inst & State U Dept Engineering Science Blacksburg VA 24061

REDDY, NARAYANA MUNISWAMY, aerospace educator; b. Bangalore, Karnataka, India, Mar. 14, 1935; s. Narayana and Yellamma R.; m. Jayalakshmi, Jan. 5, 1967; children: Jalaja, Shailaja, Shilpa. BE, Mysore U., Bangalore, India, 1959; ME in Aerospace, Indian Inst. of Sci., Bangalore, India, 1961; PhD in Aerospace, Toronto U., Can., 1966. Sr. aerospace engr. Hindustan Aircraft Ltd., Bangalore, India, 1961-62; rsch. assoc. Toronto U., Can., 1962-66; NRC rsch. fellow NASA-Ames Rsch. Ctr., Calif., 1967-69; rsch. assoc. Wright-Patterson AFB, Dayton, Ohio, 1969-70; asst. prof. Indian Inst. of Sci., Bangalore, India, 1970-75, assoc. prof., 1975-80, prof., 1980—; mem. Aero. R & D Bd., New Delhi, India, 1980-90; internat. adv. com. mem. Internat. Symposium on Shock Waves and Shock Tubes, 1981; com. mem. Waves and Shock Tubes. Contbr. over 80 publs. to profl. jours. Recipient Burma Shell prize Aero. Soc. of India, 1976, Dr. V.M. Ghatge award, 1985, Indo-Can. Commonwealth fellowship, 1962. Fellow AIAA (assoc.), Instn. of Engrs. (India); mem. AAAS, Indian Laser Assn., Century Club. Achievements include significant rsch. and devel. contbns. in Aerospace Engring. in Hypersonic Aerodynamics. Home: 355 13th Main Rajmahavilas, Bangalore 560080, India Office: Indian Inst of Sci, Aerospace Engring Dept, Bangalore 560012, India

REDDY, RAJASEKARA L., mechanical engineer; b. Karedu, India, June 6, 1948; came to U.S., 1972; s. L. Ramana and Anna (Poorna) R.; m. Swarnalatha, Jan. 22, 1976; children: Aparna, Megha. BS, U. Jaipur, India, 1970; MS, U. Nev., 1975. Cert. profl. engr., Can. Project coord. Cool Air Systems, Toronto, Can., 1976-77; mech. engr. Fodor Engring., Toronto, 1978-79; design engr., project engr. Aldous Heating & Air Conditioning Co., Reno, Nev., 1979-88; project supt. Koller Mech., Sparks, Nev., 1988; project mgr. Ray Heating Products, Reno, 1989—. Mem. ASME (assoc.), ASHRAE (assoc.), Assn. Energy Engrs. (sr.). Home: 7269 Bold Venture Ct Reno NV 89502 Office: Ray Heating Products Inc 1008 E 4th St Reno NV 89512

REDDY, VENKAT NARSIMHA, ophthalmalogist, researcher; b. Hyderabad, India, Nov. 4, 1922; came to U.S., 1947; s. Malla and Manik (Devi) R.; m. Alvira M. DeMello, Dec. 10, 1955; children: Vinay Neville, Marlita Alvira. BSc, U. Madras, 1945; MS, PhD, Fordham U., 1952. Rsch. assoc. Coll. of Physicians and Surgeons Columbia U., N.Y.C., 1952-56, Banting and Best Inst., Toronto, Can., 1956; ass. and assoc. prof. ophthalmology Kresge Eye Inst. Wayne State U., Detroit, 1957-68; prof., biomed. scis., asst. dir. Eye Rsch. Inst. Oakland U., Rochester, Mich., 1968-75, prof., dir., 1975—; mem. study sect. NIH, Bethesda, 1966-70, nat. adv. eye coun., 1982-87, mem. bd. sci. counselors Nat. Eye Inst., 1977-81. Mem. editorial bd. Investigative Ophthalmology and Visual Scis., 1969-72, 78-88, Ophthalmic Research, 1978-90, Experimental Eye Research, 1985—; contbr. articles to profl. jours. Recipient Friendenwald award Assn. Research in Ophthalmology, 1979, Research Recognition award, Alcon Labs, 1984, Research award Cataract Research Found., 1987, Merit award Nat. Eye Inst., 1989; named Scientist of Yr. State of Mich., 1991. Mem. AAAS, Internat. Soc. Eye Rsch., The Biochem. Soc., Assn. Rsch. in Vision and Ophthalmology (pres. 1985), Am. Soc. for Biochemistry and Molecular Biology, Soc. Free Radicals, Oxygen Soc. Sigma Xi. Achievements include research on cataract etiology, intraocular fluids dynamics relating to glaucoma, cell biology of lens, ciliary body and retinal pigment epithelium, cell differentiation. Office: Oakland U Eye Rsch Inst 422 Dodge Hall Rochester MI 48309

REDEKER, ALLAN GRANT, physician, medical educator; b. Lincoln, Nebr., Sept. 10, 1924; s. Fred Julius and Fern Frances (Grant) R.; m. Andrea K. Siedschlag, June 16, 1979; children by previous marriage—Martha, James, Thomas. B.S., Northwestern U., Chgo., 1949, M.D., 1952. Intern Hollywood Presbyn. Hosp., Los Angeles, 1952-53; resident in internal medicine Hollywood Presbyn. Hosp., 1953-54; asst. prof. medicine U. So. Calif., 1959-62, assoc. prof., 1962-69, prof., 1969—; mem. Nat. Digestive Diseases Adv. Bd., 1985-88; mem. U.S.-Japan Med. Sci. Program, U.S. Dept. State, 1978—; bd. dirs. Am. Liver Found. Contbr. numerous articles to research jours. Served with AUS, 1943-46. Recipient Research Career Devel. award NIH, 1962-69; research fellow Giannini Found., 1956-58. Mem. Assn. Am. Physicians, Am. Soc. Clin. Investigation, Am. Gastroent. Assn. Home: 9323 Samoline Ave Downey CA 90240-2716 Office: 7601 Imperial Hwy Downey CA 90242-3496

REDER, LYNNE MARIE, cognitive science educator; b. San Francisco, Dec. 28, 1950; d. Melvin Warren and Edna B. (Oliker) R.; m. John Robert Anderson, Mar. 29, 1973; children: John Frank, Abraham Robert. BA, Stanford U., 1972; PhD, Mich. U., 1976. Postdoctoral fellow Yale U., New Haven, 1976-78; asst. prof. Carnegie Mellon U., Pitts., 1978—, assoc. prof., 1983-92, prof., 1992—; com. mem. NRC panel on techniques for enhancement of ultimate performance, 1991-93; mem. editorial bd. Jour. of Exptl. Psychology, 1983—, Memory & Cognition Jour., 1982-90. Office: Dept of Psychology Carnegie Mellon U Pittsburgh PA 15213

REDGATE, EDWARD STEWART, physiologist, educator; b. Yonkers, N.Y., Mar. 13, 1925; s. Carl Julius and Mildred (Lowndes) R.; divorced; children: Steven, Catherine, Daniel. BS, Bethany Coll., 1949; MS, U. Minn., 1951, PhD, 1954. Prof. U. Pitts., 1961—. Editor: Brain-Pituitary-Adrenal Gland Interrelationships, 1992. With U.S. Army, 1943-46. Mem. Am. Physiol. Soc., Endocrine Soc. Office: U Pitts Sch Medicine Pittsburgh PA 15261

REDHEFFER, RAYMOND MOOS, mathematician, educator; b. Chgo., Apr. 17, 1921; s. Raymond L. and Elizabeth (Moos) R.; m. Heddy Gross Stiefel, Aug. 25, 1951; 1 son, Peter Bernard. S.B., MIT, 1943, S.M., 1946, Ph.D., 1948; DSc (hon.), U. Karlsruhe, 1991. Rsch. assoc. MIT Radiation Lab., 1942-45, Rsch. Lab. of Electronics, 1946-48; instr. Harvard U., Radcliffe Coll., 1948-50; mem. faculty UCLA, 1950—, prof. math., 1960—; guest prof. Tech. U. Berlin, 1962, Inst. for Angewandte Math., Hamburg, 1966, Math. Inst. U. Karlsruhe, 1971-72, 81, 88, 91; U.S. sr. scientist Alexander von Humboldt Found., Karlsruhe, 1976, 85. Author: (with Ivan Sokolnikoff) Mathematics of Physics and Modern Engineering, 1958, (with Charles Eames) Men of Modern Mathematics, 1966, (with Norman Levinson) Complex Variables, 1970, Differential Equations, Theory and Applications, 1991, Introduction to Differential Equations, 1992; film author, animator, 1972-74; contbr. articles to profl. jours. Pierce fellow Harvard U., 1948-50; sr. postdoctoral fellow NSF, Göttingen, Germany, 1956; Fulbright rsch. scholar Vienna, 1957, Hamburg, 1961-62; recipient Disting. Teaching award UCLA Alumni Assn., 1969. Mem. Deutsche Akademie der Naturforscher (Leopoldina), Sigma Xi. Home: 176 N Kenter Ave Los Angeles CA 90049-2730 Office: UCLA Dept Mathematics 6224 Math Sci Bldg Los Angeles CA 90024

REDISH, EDWARD FREDERICK, physicist, educator; b. N.Y.C., Apr. 1, 1942; s. Jules and Sylvia (Coslow) R.; m. Janice Copen, June 18, 1967; children: A. David, Deborah. AB, Princeton U., 1963; PhD, MIT, 1968. CTP fellow U. Md., College Park, 1968-70, asst. prof., 1970-74, assoc. prof., 1974-79, prof., 1979—, chair dept. phys. astronomy, 1982-85; vis. prof. Ind. U., Bloomington, 1985-86, U. Washington, Seattle, 1992-93; vis. fgn. collaborator CEN, Saclay, France, 1973-74; co-dir. U. Md. Project in Physics and Ednl. Tech., 1983—, Comprehensive Unified Physics Learning Environment, 1989—; mem. Nuclear Sci. Adv. Com., Dept. of Energy/NSF, 1987-90; mem. program adv. com. Ind. U. Cyclotron Facility, 1985-89, chmn., 1986-89. Author: (software) Orbits, 1989; editor: Procs. Conf. on Computers in Physics Instrn., 1990; contbr. over 50 articles to profl. jours. Named Sr. Resident Rsch. Assoc., NAS-NRC, 1977-78; recipient Inst. medal Cen. Rsch. Inst. for Physics, 1979, Leo Schubert award Wash. Acad. Sci., 1988, Educator award Md. Assn. Higher Edn., 1989, Glover award Dickinson U., 1991. Fellow Am. Phys. Soc., AAAS, Wash. Acad. Sci.; mem. Am. Assn. Physics Tchrs. Office: U Md Dept Physics College Park MD 20742-4111

REDLICH, ROBERT WALTER, physicist, electrical engineer, consultant; b. Lima, Peru, Sept. 20, 1928; came to U.S., 1934; s. Richard and Barbara (Jacoby) R.; m. Joanne Delores Ashworth, June 14, 1960; children: Paul, Ruth. BS, Rensselaer Poly. Inst., 1950; MS, MIT, 1951; PhD in Physics, Rensselaer Poly. Inst., 1960. Assoc. prof. Clarkson Coll., Potsdam, N.Y., 1960-65; sr. fellow rsch. Sydney U., Australia, 1965-68; prof. Elec. Engring. Ohio U., Athens, 1968-85; ind. cons. pvt. practice, Athens, 1985—. Inventor: 20 patents, 1968—; contbr. articles to profl. jours. Sgt. U.S. Army, 1952-54. Recipient tuition scholarship, Rensselaer Poly., 1946, N.Y. State Scholarship, 1946, Tau Beta Pi Fellowship, 1950, NSF Fellowship, 1959. Mem. IEEE, N.Y. Acad. Scis.

REDLINGER, SAMUEL EDWARD, chemical engineering consultant; b. Pitts., July 30, 1949; s. William Albert and Teresa Marie (Dobbins) R.; m. Rita D. Riley, Aug. 8, 1970 (dec. 1975); children: Timothy J., Thomas M., Daniel P.; m. Linda Jane Barkley, Dec. 16, 1978; 1 child, Aaron J. BAChemE, U. Dayton, 1971, MS in Analytical Instrumentation, 1974; MA in Mgmt., Pacific Western U., 1984. Cert. firefighter, hazardous material response, Ohio. Lead designer U.S. Steel Corp., Pitts., 1967-68; instrumental analyst Miami Conservancy Dist., Dayton, Ohio, 1968-73; prodn. engr. Delco-Moraine GMC, Dayton, 1973-74; sr. project engr. A. M. Kinney, Inc., Cin., 1974-76; plant engr. Sherwin-Williams Chem., Cin., 1976-78; mgr. engring. Midwest Tech. Inc., Cin., 1978-82; engring. mgr. Hilton/Davis Chems., Cin., 1982-86; pres. RsCo Innovative Solutions, Cin., 1986—; cons. Butler Co. (Ohio) FCA, 1986—. Inventor reverse osmosis, solidification, anionic membranes. Bd. dirs. Presbyterians Pro-Life, Cin., 1989; com. chmn. Lakota Sch. Dist. Adv. Bd., Cin., 1986; vol. firefighter Union Twp. Fire Dept., 1985. Mem. Am. Inst. Chem. Engring., Gideons Internat., Masons (32d degree), K.T. Republican. Avocations: golfing, woodworking, reading. Home: 9978 Boxwood Ct Cincinnati OH 45241-1032 Office: RsCo Innovative Solutions 9978 Boxwood Ct Cincinnati OH 45241-1032

REDMAN, ROBERT SHELTON, pathologist, dentist; b. Fargo, N.D., Aug. 1, 1935; s. Kenneth and Elizabeth Francis (McMillan) R.; m. Barbara Darlien Klug, Sept. 14, 1958; 1 child, Melissa Darlien. Student, S.D. State U., 1953-55; BS, DDS, U. Minn., 1959, MSD, 1963; PhD, U. Wash., 1969. Cert. Am. Bd. Oral Pathology. Clin. asst. prof. sch. dentistry U. Minn., Mpls., 1963-64, assoc. prof., 1969-75; assoc. prof. sch. dentistry U. Colo., Denver, 1975-78; staff dentist, chief oral pathology rsch. lab. Dept. Vets. Affairs Med. Ctr., Washington, 1978—; Denver, 1975-78; clin. assoc. prof. Balt. Coll. Dental Surgery U. Md., 1989—; cons. Children's Orthopedic Hosp., Seattle, 1966-69; program specialist in oral biology Dept. Vets. Affairs, Washington, 1982-86. Contbr. 12 chpts. to books and over 60 articles to profl. jours. Capt. U.S. Army, 1959-61. Fellow Am. Acad. Oral

Pathology; mem. ADA, Am. Inst. Nutrition, Internat. Assn. Dental Rsch. (program chmn. salivary rsch. group 1982-86), Tissue Culture Assn., Omicron Kappa Upsilon. Presbyterian. Achievements include discovery and naming of an unique minor salivary gland in the rat; documentation of the relationship between weaning and maturation of salivary glands, of mitotic division of well-differentiated salivary gland cells of all types, including acinar, ductal and myoepithelial cells; determination of mode of inheritance of benign migratory glossitis. Office: Dept Vets Affairs Med Ctr Oral Pathology Rsch Lab 50 Irving St NW Washington DC 20422-0002

REDMAN, STEVEN PHILLIP, engineer; b. Edmonton, Alta., Can., Feb. 15, 1965. BS in Math., U. Wash., 1987. Asst. engr. Pacific Telecom Cable, Vancouver, Wash., 1990—. Mem. Am. Soc. Quality Control. Office: Pacific Telecom PO Box 9901 Vancouver WA 98668-8701

REDMOND, GAIL ELIZABETH, chemical company executive; b. Milw., July 28, 1946; d. George Foote and Doris Ruth (Roethke) R.; m. Stephen Thomas Koran, July 10, 1992. Student Coll. St. Catherine, 1964-66; BS magna cum laude, Utah State U., 1968. Tchr. pub. schs., Milw., 1968-70; staff coordinator Med. Personnel Pool, Milw., 1973-76; corp. manpower devel. mgr. Clark Oil & Refining Corp., Milw., 1976-80; sr. advisor communications, employee benefits Conoco, Inc., Ponca City, Okla., 1980-81, coord. profl. recruiting, 1981-82, coord. employee benefit communications, 1982-84, asst. mgr. Phase II Job Eval. Project, 1984-86, coord. job evaluation, 1986-87; dir. employee relations and job evaluation, 1987-90; sr. human resources cons. E.I. duPont de Nemours, Wilmington, Del., 1990—. With USNR, Persian Gulf. Recipient S.W. Asia svc. ribbon with with bronze star, Navy and Marine Corps overseas deployment ribbon, Nat. Def. ribbon. Mem. Naval Enlisted Res. Assn., Women in Mil. Svc. to Am. Meml. Found., Phi Kappa Phi.

REDMOND, ROBERT FRANCIS, nuclear engineering educator; b. Indpls., July 15, 1927; s. John Felix and Marguerite Catherine (Breinig) R.; m. Mary Catherine Cangany, Oct. 18, 1952 (dec. May 1988); children: Catherine, Robert, Kevin, Thomas, John. B.S. in Chem. Engring, Purdue U., 1950; M.S. in Math, U. Tenn., 1955; Ph.D. in Physics, Ohio State U., 1961. Engr. Oak Ridge Nat. Lab., 1950-53; scientist, adviser-cons. Battelle Meml. Inst., Columbus, Ohio, 1953-70; prof. nuclear engring. Ohio State U., Columbus, 1970-92; assoc. dean Coll. Engring. Ohio State U., dir. Engring. Experiment Sta., 1977-92, acting dean, 1990—. Contbr. articles to profl. jours. V.p. Argonne Univs. Assn., 1976-77, trustee, 1972-80; mem. Ohio Power Siting Commn., 1978-82; trustee Edison Welding Inst., 1988-92. With AUS, 1945-46. Mem. Am. Nuclear Soc. (chmn. Southwestern Ohio sect.), AAAS, Nat. Regulatory Rsch. Inst. (bd. dirs. 1988-92), Trans. Rsch. Ctr., Am. Soc. Engring. Edn., Sigma Xi, Tau Beta Pi. Home: 3112 Brandon Rd Columbus OH 43221-2210

REDMORE, DEREK, chemist; b. Horncastle, Lincs., Eng., Aug. 8, 1938; came to U.S., 1962; s. Hugh and Constance M. (Seymour) R.; m. Carolyn Ann Denison, June 12, 1965; children: Christopher Mark, Sandra Denise. BS, Nottingham (Eng.) U., 1959, PhD, 1962. Rsch. assoc. Washington U., St. Louis, 1962-65; chemist Petrolite Corp., St. Louis, 1965-70, rsch. mgr., 1970-89, dir. tech. support, 1989—. Author: (book) Ring Enlargement Reactions, 1968. Recipient Mo. Inventor of the Year, St. Louis Bar Assn. Mem. AAAS, Royal Soc. of Chemistry, Am. Chem. Soc. (St. Louis award 1983). Methodist. Achievements include 108 patents on corrosion inhibitors, scale inhibitors, organophosphorus compounds. Office: Petrolite Corp 369 Marshall Ave Saint Louis MO 63119

REDMOUNT, IAN H., physicist; b. N.Y.C., Sept. 25, 1956; s. Melvin Bernard and Florence Constance (Schweitzer) R. BS in Physics, Mich. State U., 1978; MS in Physics, Calif. Inst. Tech., 1981, PhD in Physics, 1984. Postdoctoral fellow Harvard Coll. Obs., Cambridge, Mass., 1983-86; vis. rsch. fellow Rsch. Inst. for Fundamental Physics, Kyoto U., Japan, 1984-85; rsch. assoc. physics Inst. Astronomy, U. Cambridge, Eng., 1986-89, Washington U., St. Louis, 1989-91, U. Wis., Milw., 1991-93; asst. prof. sci. and math. Parks Coll. of St. Louis U., Cahokia, Ill., 1993—; undergrad. rsch. asst. Mich. State U. Cyclotron Lab., East Lansing, 1976-78, Dept. Physics, 1977-78, undergrad. teaching asst.; grad. teaching asst. Calif. Inst. Tech., Pasadena, 1979-80; supr. cosmology course Emmanuel, Gonville and Caius and Queen's Colls., U. Cambridge, 1987-88; assoc. lectr. physics U. Wis., Milw., 1993. Co-author: Black Holes: The Membrane Paradigm, 1986; co-creator film: Evolution of Schwarzschild Space Through the Singularity, 1989; contbr. articles to profl. jours. Mich. State U. Alumni Disting. scholar, 1974-78; NSF predoctoral fellow, 1978-81, J.S. Fluor grad. fellow Calif. Inst. Tech., 1982, Japan Soc. for Promotion of Sci. rsch. fellow, 1984-85. Mem. Am. Phys. Soc., Am. Astron. Soc., N.Y. Acad. Scis., Japan Soc., Phi Beta Kappa, Sigma Xi, Phi Kappa Phi. Achievements include research contribution to the formulation of the "Membrane Paradigm" for black hole physics; co-discoverer of the DTR Relations for colliding lightlike shells in general relativity. Office: Parks Coll of St Louis U Dept Sci and Math Cahokia IL 62206

REDONDO, DIEGO RAMON, health, physical education and recreation educator; b. Chgo., Apr. 1, 1964; s. Diego and Susan Rae (Biebel) R.; m. Suzanne Barr Jones, Dec. 17, 1989. BA, So. Ill. U., 1987; MS, U. So. Miss., 1988, PhD, 1990. Asst. prof. dept. health, phys. edn. and recreation Old Dominion U., Norfolk, Va., 1990—, dir. lab. of kinesiological and biomechanical studies. Big bro. Big Bros./Sisters of Tidewater, Virginia Beach, Va., 1991—. Mem. Am. Coll. Sports Medicine, Sigma Xi. Office: Old Dominion U Dept Health Phys Edn and Recreation Norfolk VA 23529

REECE, ROBERT WILLIAM, zoological park administrator; b. Saginaw, Mich., Jan. 21, 1942; s. William Andrews and Mary Barbara (Murphy) R.; m. Jill Whetstone, Aug. 21, 1965; children: William Clayton, Gregory Scott, Mark Andrews. B.S., Mich. State U., 1964; postgrad., U. West Fla., 1969-71, U. South Fla., 1974-76. Dir. Northwest Fla. Zool. Gardens, Pensacola, Fla., 1970-72; zool. dir. Lion Country Ga., Stockbridge, 1972-73; asst. dir. Salisbury Zoo, Md., 1976-77; dir. zoology Wild Animal Habitat, Kings Island, Ohio, 1977—; exec. dir. The Wilds Internat. Ctr. for Preservation of Wild Animals, Columbus, Ohio, 1992—. Assoc. editor: Sci. Jour. Zoo Biology, 1982—. Lt. USN, 1964-69, Korea. Profl. fellow Am. Assn. Zool. Parks and Aqariums; mem. Cin. Wildlife Rsch. Fedn., Am. Soc. Mammalogists, Animal Behavior Soc., Captive Breeding Specialist Group, Species Survival Commn., Internat. Union for Conservation of Nature nd Natural Resources. Republican. Episcopalian. Home: 2772 Turnkey Ct Cincinnati OH 45244-3236 Office: The Wilds 85 E Gay St Ste 603 Columbus OH 43215

REED, BRUCE CAMERON, physics educator, astronomy researcher; b. Tornoto, Can., Oct. 25, 1954; came to U.S., 1992; s. Norman Bruce and Beryl Evelyn (Mackey) R.; m. Laureen Gaye Burgoyne, Apr. 24, 1985. BS, U. Waterloo, 1977, PhD, 1984. Assoc. prof. St. Mary's U., Halifax, Nova Scotia, Can., 1983-92, Alma (Mich.) Coll., 1992—. Author: Quantum Mechanics: A Frist Course, 1990; contbr. articles to profl. jours. Named Oper. grantee Nat. Sci. and Engring. Rsch. Coun. Can., 1984-92; recipient Cottrell Coll. Sci. award Rsch. Corp., 1993—. Mem. AAAS, Am. Phys. Soc., Am. Astron. Soc., Internat. Astron. Union, Sigma Xi. Office: Alma Coll Dept Physics Alma MI 48801

REED, CHARLES ELI, retired chemist, chemical engineer; b. Findlay, Ohio, Aug. 11, 1913. BS, Case Inst. Tech., 1934; ScD in Chem. Engring., MIT, 1937. Asst. prof. chem. engring. MIT, Cambridge, Mass., 1937-42; rsch. assoc. rsch. lab. GE, 1942-45, engring. mgr. chem. divsn., 1945-52, gen. mgr. silicone products dept., 1952-59, gen. mgr. metall. products dept., 1959-62, v.p., gen. mgr. chem. and metall. divsn., 1962-68, v.p., group exec. components and materials group, 1968-71, sr. v.p. corp. tech., 1971-79; ret., 1979. Recipient Nat. Medal of Technology, U.S. Dept. of Commerce Technology Admin., 191. Fellow AICE, Am. Inst. Chemist; mem. AAAS, Mat. Acad. Engring., Am. Chem. Soc. Achievements include research in colloid chemistry, high polymers, distillation. Home: 3200 Park Ave Bridgeport CT 06604*

REED, CHARLES KENNETH, physicist; b. Fairfield, Iowa, Jan. 28, 1917; s. Perrin Elmer and Grace Edna (Davies) R.; m. Elsie Steenberg; children: Lynda, Anita. BS, Parsons Coll., 1942; cert. meteorologist, U. Chgo., 1943; MS, U. Mich., 1949. Commd. 2d lt. U.S. Army, 1943, advanced through grades to lt. col., 1960; weather officer U.S. Army, St. Joseph, Mo., 1943-45; typhoon specialist U.S. Army, 1945-46; nuclear rsch. officer U.S. Army, Japan, U.S., 1949-57; asst. dir. Air Force Office of Scientific Rsch., Washington, 1960-65; ret. U.S. Army, 1965; spl. project officer Nat. Rsch. Coun., Washington, 1965-72, exec. dir. physics div., exec. dir. assembly of math. and phys. scis., 1978-82. Recipient Award for Disting. Svc. Nat. Acad. Scis., 1976. Mem. AAAS, Am. Phys. Soc., N.Y. Acad. Scis. Home: 7120 Leesville Blvd Springfield VA 22151

REED, CHRISTOPHER ROBERT, civil engineer; b. Charleston, W.Va., Feb. 12, 1948; s. Clarence Milton and Anne (Schaffner) R.; m. Mary Dandridge Kennedy, Mar. 4, 1983. Student W.Va. Inst. Tech., 1966-70, 76-77, Ga. State U., 1973-74. Designer, Sverdrup & Parcel, Charleston, 1970-72; asso. project engr. Mayes, Sudderth & Etheredge, Atlanta, 1973-76; project mgr. Sverdrup & Parcel, Washington, 1976-79; estimator Deleuw, Cather/Parsons, Washington, 1979-80; project mgr. Parsons Brinckerhoff, McLean, Va., 1980-85; assoc. Lolederman Assocs., Inc., Rockville, Md., 1985-86; assoc. Post Buckley Schuh and Jernigan Inc., Arlington, Va., 1986-89; assoc., dir. mcpl. engring. Lolederman Assocs., Inc., Rockville, 1989-90; mgr. CRS Donohue and Assocs. Inc., Fairfax, Va., 1990-92; asst. dist. location and design mgr. VDOT, Fairfax, 1992—. Mem. Constrn. Specifications Inst., Inst. Transp. Engrs., Am. Assn. Cost Engrs., Am. Ry. Engring. Assn., Soc. Am. Mil. Engrs., Am. Pub. Transit Assn., ASTM, Capital Yacht Club (sec. 1988-89, vice commodore 1990, commodore 1991), Corinthian Yacht Club (rear commodore). Home: 2334 Generation Dr Reston VA 22091-3029 Office: 3975 Fair Ridge Dr Fairfax VA 22033

REED, DIANE GRAY, business information service company executive; b. Trion, Ga., Sept. 5, 1945; d. Harold and Frances (Parker) Gray; m. Harry Reed, Oct. 2, 1982. Student, Jacksonville U., 1963-64, Augusta Coll., 1972-74; BS, Ga. State U., 1981. Various mgmt. positions Equifax Svcs., Inc., Atlanta, 1964-72, field rep., 1972-74, tech. rep., 1974-79, mgr. systems and programs, 1979-84, dir. tech., 1984-86, asst. v.p., 1986—; v.p. info. tech. sector Equifax Svcs., Inc., 1989—; presdl. adv. council Equifax Svcs., Inc., Atlanta, 1984—; cons. Ga. Computer Programmer Project, Atlanta, 1984-86, spkr. Oglethorpe U. Career Workshop, Atlanta, 1986. Bd. dirs. Atlanta Mental Health Assn., 1985-89; bd dirs., pres. Atlanta Women's Network, 1990; bd. dirs. United Way Bd. Bank, Atlanta, 1984-86; chairperson EquiFax United Way Campaign, 1988-89; vol. Cobb County Spl. Olympics, Marietta, Ga., 1984-87; mem. adv. coun. Coll. Bus. Adminstrn. Ga. State U. mgmt. info. systems industry adv. bd. U. Ga.; mem. Leadership Atlanta Class of '92, Girl Scouts Friendship Circle, Friends of Spelman Coll.; vol. coord. Atlanta Partnership Bus. and Edn.; co-chair salute to women of achievement YWCA, 1992. Named Woman of Achievement, Atlanta YWCA, 1987; recipient Decca award as one of top 10 bus. women in Atlanta, 1992. Mem. Women in Info. Processing, Inst. Computer Profls. (cert.), Soc. Info. Mgmt., Internat. Women's Alliance, Ga. State Alumni, LWV, Atlanta Yacht Club, Kiwanis Internat. (bd. dirs. Atlanta Buckhead chpt.). Avocations: water sports, golf. Office: Equifax Svcs Inc 1600 Peachtree St NW Atlanta GA 30309-2403

REED, DIANE MARIE, psychologist; b. Joplin, Mo., Jan. 11, 1934; d. William Marion and Olive Francis (Smith) Kinney; married; children: Wendy Robison, Douglas Funkhouser. Student, Art Ctr. Col., L.A., 1951-54; BS, U. Oreg., 1976, MS, 1977, PhD, 1981. Lic. psychologist. Illustrator J.L. Hudson Co., Detroit, 1954-56; designer, stylist N.Y.C., 1960-70; designer, owner Decor To You, Inc., Stamford, Conn., 1970-76; founder, exec. dir. Alcohol Counseling and Edn. Svcs., Inc., Eugene, Oreg., 1981-86, clin. supr., 1986; clin. supr. Christian Family Svcs., Eugene, 1986-87; pvt. practice Eugene, 1985—; co-founder, ptnr. ReGard Consulting, Eugene, Oreg., 1992—. Evaluator Vocat. Rehab. Div., Eugene, 1982—; alcohol and drug evaluator and commitment examiner Oreg. Mental Health Div., 1981-86; life mem. Rep. Presdle. Task Force. Mem. APA, Oreg. Psychol. Assn., Lane County Psychol. Assn. (pres. 1989-90), Lane Mental Health Providers Assn., C2 Investors (treas. 1987-88), Altair Ski and Sport, Oreg. Track. Avocations: photography, skiing, running, hiking, silversmithing. Office: 5 E 24th Ave Eugene OR 94705

REED, DONALD JAMES, biochemistry educator; b. Montrose, Kans., Sept. 26, 1930; married, 1949; 6 children. BS, Coll. of Idaho, 1953; MS, Oreg. State U., 1955, PhD in Chemistry, 1957. Asst. Oreg. State U., Corvallis, 1953-55, asst., then assoc. prof., 1962-72, prof. biochemistry, 1972—, dir. environ. health sci. ctr., 1981—; assoc. biochemist cereal investigations western regional rsch. lab., agr. rsch. svc. USDA, Calif., 1957-58; asst. prof. chemistry Mont. State U., Bozeman, 1958-62; mem. toxicology study sect. NIH, 1971-75, ad hoc rev. com. Nat. Cancer Inst., 1975-87, sabbatical scientist, 1984-85; mem. environ. sci. rev. panel health rsch. EPA, 1981, environ. health sci. rev. com. Nat. Inst. Environ. Health Sci., 1982-85; mem. Burroughs Wellcome Toxicology Scholar Award Selection Com., 1984-87; vis. prof. MRC Toxicology Unit, Carshalton, Eng., 1984; mem. task group health criteria, internat. program chem. safety WHO, 1986,biochem. & carcinogenesis rev. com. Am. Cancer Soc., 1984—; cons. U. Calif., San Francisco, 1985—. assoc. editor Jour. Toxicology & Environ. Health, 1980-84, Toxicology & Applied Pharmacology, 1981-84, Cancer Rsch., 1987—; editor Cell Biology & Toxicology. USPHS Spl. Rsch. fellow NIH, 1969-70, Eleanor Roosevelt Am. Cancer Soc. Internat. Cancer fellow Karolinska Inst., 1976-77; Burroughs Wellcome Travel grantee, 1984. Mem. Am. Soc. Biol. Chemistry, Am. Soc. Pharmacology and Exptl. Therapeutics, Am. Assn. Cancer Rsch., Soc. Toxicology, Sigma Xi. Achievements include research in biological oxidations, environmental toxicology, biochemical anticancer drugs, protective mechanisms of glutathione functions, vitamin E status. Office: Oreg State U Environ Health Scis Ctr 317 Weniger Hall Corvallis OR 97331-6504*

REED, DWAYNE MILTON, medical epidemiologist, educator; b. Kinsley, Kans., Dec. 10, 1933; s. John Milton and Margaret (Reger) R.; m. Leslie Smith, June 28, 1962 (div. 1968); children: Colin, Heather. BA, U. Calif., Berkeley, 1956; MD, U. Calif., San Franciso, 1960; MPH, U. Calif., 1962, PhD in Epidemiology, 1968. Dep. chief epidemiology Nat. Inst. Neurol. Disease NIH, Bethesda, Md., 1962-64; chief epidemiology Arctic Health Ctr. USPHS, 1964-66; assoc. prof. Sch. Pub. Health R. Tex., Houston, 1968-70; chief epidemiology NIH Neurology Field Ctr., Agana, Guam, 1970-72; epidemiologist South Pacific Commn., New Caledonia, 1972-75; chief epidemiology br. Nat. Inst. Child Health, NIH, Bethesda, 1975-78; dir. environ. epidemiology Calif. Dept. Health, Berkeley, 1978-79; dir. Honolulu health program Nat. Heart Lung Blood Inst. NIH, 1980—; adj. prof. epidemiology Sch. Pub. Health U. Calif., Berkeley, 1978-78, Sch. Pub. Health U. Honolulu, 1980—; cons. WHO-Pacific, Manila, 1983, 86-87, 89, Hawaii State Dept. Health, Honolulu, 1988—. Editor: The Epidemiology of Premature, 1977; contbr.. chpts. to books and over 100 sci. articles on med. epidemiology to profl. jours. Recipient Commendation medal USPHS, 1987. Fellow Am. Heart Assn.; mem. Am. Pub. Health Assn., Am. Coll. Epidemiology, Am. Epidemiology Soc., Soc. Epidemiologic Rsch., Delta Omega. Achievements include research in public health, epidemiology. Office: Buck Ctr for Rsch in Aging 505-H San Marin Dr Novato CA 94945*

REED, JOHN CHARLES, chemical engineer; b. Springfield, Ill., Dec. 3, 1930; s. Ralph Albert and Haldeen (Vandeveer) R.; m. Mary Helen Simpson, June 25, 1966; children: Katherine June Griffin, Kare Joy Griffin, Stephanie Lynn Reed. BSChemE, U. Ill., 1953; MSChemE, U. Del., 1954; PhD, U. Wis., 1961. Registered profl. engr., Ill., Okla. Chem. engr. Esso Rsch. and Engring., Linden, N.J., 1960-62; sr. engr. Pan Am. Petroleum Corp., Tulsa, 1962-69; assoc. prof. U. Tulsa, 1969-71; supr., tech. advisor Ill. EPA, Springfield, 1971—. With U.S. Army, 1954-56. Mem. AICE (Winston Churchill fellowship 1970), Am. Chem. soc., Am. Soc. of Cost Engrs. Home: 317 W Franklin St Edinburg IL 62531 Office: Ill EPA PO Box 19276 Springfield IL 62794

REED, LESTER JAMES, biochemist, educator; b. New Orleans, Jan. 3, 1925; s. John T. and Sophie (Pastor) R.; m. Janet Louise Gruschow, Aug. 7, 1948; children—Pamela, Sharon, Richard, Robert. B.S., Tulane U., 1943; D.Sc. (hon.), 1977; Ph.D., U. Ill., 1946. Research asst. NDRC, Urbana, Ill., 1944-46; research asso. biochemistry Cornell U. Med. Coll., 1946-48; faculty U. Tex., Austin, 1948—; prof. chemistry U. Tex., 1958—, Ashbel Smith prof., 1984—; research scientist Clayton Found. Biochem. Inst., 1949—, asso. dir., 1962-63, dir., 1963—. Contbr. articles profl. jours. Mem. Nat. Acad. Scis. U.S., Am. Acad. Arts and Scis., Am. Soc. for Biochemistry and Molecular Biology, Am. Chem. Soc. (Eli Lilly & Co. award in biol. chemistry 1958), Phi Beta Kappa, Sigma Xi. Home: 3502 Balcones Dr Austin TX 78731-5802 Office: U Tex Biochem Inst Experimental Sci Bldg 442 Austin TX 78712

REED, MICHAEL ALAN, scientist; b. Lima, Ohio, Oct. 11, 1953; s. Robert Dean and Janice Lou (Stumpp) R.; 1 child, Adam. BS, Ohio Univ., 1982, MS, 1985. Rsch. asst. Childrens Hosp., Columbus, Ohio, 1984-85; scientist Edison Animal Biotechnology Ctr., Athens, Ohio, 1985—. Contbr. articles to profl. jours. Mem. Tissue Culture Assn., Phi Beta Kappa, Phi Kappa Phi. Democrat. Achievements include patent on gene transfer using transformed, neodetermined, embryonic cells. Home: 170 E State St Athens OH 45701 Office: Ohio Univ Edison Ctr Wilson Hall W Green Athens OH 45701

REED, MICHAEL ROBERT, agricultural economist; b. Lawrence, Kans., July 11, 1953; s. Robert Stanley and Marian Lucille (Karr) R.; m. Patricia Gail Gurtler, Mar. 16, 1973; children: Laura Gail, Brian Michael. BS, Kans. State U., 1974; MS, Iowa State U., 1976, PhD, 1979. Asst. prof. U. Ky., Lexington, 1978-83, assoc. prof., 1983-89; prof., 1989—, exec. dir. Ctr. for Export Devel., 1988—; cons. U.S. Agy. for Internat. Devel., Washington, 1983-86. Editorial council So. Jour. of Agrl. Econs., 1983-86; contbr. articles to profl. jours. Recipient Outstanding Jour. Article award Soc. Farm Mgrs. and Rural Appraisers, 1986; grantee Farmer Coop. Svcs., 1982-84, 87-88, TVA, 1982-85, Fed. Crop Ins. Corp., 1985-87, USDA, 1986—. Mem. Am. Agrl. Econs. Assn., So. Agrl. Econs. Assn., Gamma Sigma Delta, Omicron Delta Epsilon. Home: 2216 Bonhaven Rd Lexington KY 40515-1150 Office: U Ky Dept Agrl Econs 300 Bradley Hall Lexington KY 40546-0215

REED, RAY PAUL, engineering mechanics measurement consultant; b. Abilene, Tex., May 26, 1927; s. Raymond Roseman and Gladys Daisy (Reddell) R.; m. Mary Antoinette Wied, Oct. 7, 1950; children: Mary Kathryn, Patricia Lynn. BSME, Tex. A&M U., 1950; MS in Engring. Mechanics, U. Tex., 1958, PhD, 1966. Registered profl. engr., N.Mex., Tex. Rsch. engr. S.W. Rsch. Inst., San Antonio, 1950-54; rsch. scientist U. Tex., Austin, 1954-56; mem. tech. staff Sandia Nat. Labs., Albuquerque, 1956-61, rsch. fellow, 1961-66, disting. mem. tech. staff, 1966—. Author: manual on the use of thermocouples; contbr. numerous reports and articles on shock measurement and thermometry to profl. jours. With USNR, 1945-46, PTO. NIH grantee U. Tex., 1962-66. Mem. ASTM (chmn. com. 1985—), ASME, ISA, Am. Phys. Soc., Sigma Xi. Avocations: photography, wood carving, cartooning, writing. Office: Sandia Nat Labs Ctr Applied Physics and Engring PO Box 5800 Albuquerque NM 87185-5800

REED, ROBERT MARSHALL, ecologist; b. Berea, Ohio, June 29, 1941; s. John Frederick and Mildred Gant (Stites) R.; m. Elizabeth Willoughby Newton, Sept. 3, 1966; children: Michael John, Eleanor Katrine. AB, Duke U., 1963; PhD, Wash. State U., 1964. Asst. prof. biology U. Ottawa, Ontario, Can., 1969-77; rsch. assoc. Oak Ridge (Tenn.) Nat. Lab., rsch. staff mem., group leader, sect. head, 1977—. Contbr. articles to profl. jours.; team leader, editor of environ. impact statements. Mem. Am. Inst. Biol. Scis., Nat. Assn. Environ. Profls., Ecol. Soc. Am., Sigma Xi. Office: Oak Ridge Nat Lab PO Box 2008 Oak Ridge TN 37831-6200

REED, SHARON LEE, infectious disease consultant; b. Austin, Tex., May 27, 1953; d. LEster James and Janet (Gruschow) R.; m. Howard Clemens Dittrich, July 18, 1981; children: Jessica Reed, Mark Edward. BS, Stanford U., 1974; MD, Harvard U., 1978; MS in Masters Clin. Tropical Medicine, London Sch. Hygiene, 1982. Bd. cert. in internal medicine and infectious diseases. Assoc. prof. pathology and medicine U. Calif. Med. Sch., San Diego, 1985-92, asst. prof. pathology and medicine, 1989-92, assoc. dir. microbiology lab., 1989—, assoc. prof. pathology and medicine, 1992—; infectious disease cons. U. Calif. Med. Ctr., San Diego, 1985—. Author: (with others) Harrison's Principles of Internal Medicine, 1992; contbr. articles to profl. jours. Infection and Immunity, Jour. Clin. Investigation, Nucleic Acids Rsch. Lucille P. Markey scholar Markey Charitable Trust, 1985-91. Mem. ACP (rsch. award 1978), Infectious Disease Soc., Am. Soc. Tropical Medicine and Hygiene, Royal Soc. Tropical Medicine and Hygiene. Office: U Calif Med Ctr 8416 200 W Arbor Dr San Diego CA 92103-8416

REED, WILLIAM PIPER, JR., surgeon, educator; b. Melrose, Mass., May 24, 1942; s. William Piper and Gertrude Harriet (Irons) R.; m. Martine Francoise Valentine Billet, Oct. 16, 1963; children: Antoinette Elsa, Christopher Llewellyn. AB, Harvard U., 1964, MD, 1968; diploma head, neck, cancer surgery, U. Paris (Villejuif), 1977. Diplomate Am. Bd. Surgery; diplomate Nat. Bd. Med. Examiners. Resident in surgery Stanford Med. Ctr., Stanford, Calif., 1976; dir. surgical ICU U. Md. Hosp., Balt., 1979-81, dir. tumor registry, 1981-86; asst. prof. surgery Sch. Medicine U. Md., Balt., 1978-83, assoc. prof. surgery, 1983-86; dir. surgical oncology Baystate Med. Ctr., Springfield, Mass., 1986—, dir. breast health ctr., 1992—; assoc. prof. surgery Tufts U., Boston, 1986-92, prof. surgery, 1992—; pres. Balt. County Unit Am. Cancer Soc., Balt., 1983-86; med. v.p. Springfield (Mass.) Unit Am. Cancer Soc., 1987—; bd. dirs. Mass. div. Am. Cancer Soc., Boston, 1989—. Contbr. articles to profl. jours. Mem. Conservation Commn. Longmeadow, Mass., 1989—. Capt. U.S. Army, 1970-72, Vietnam. Recipient Golden Apple award Baystate Med. Ctr., Springfield, 1988; grantee, NIH, 1973-75, Am. Cancer Soc. Md. Div., 1983-85, Margery Sadowski Found., Springfield, 1988-91. Fellow Am. Coll. Surgeons, Soc. Surgical Oncology (mem. issues com.); mem. Soc. for Surgery of Alimentary Tract, Soc. of Univ. Surgeons, Soc. Head and Neck Surgeons, New Eng. Surgical Soc., Soc. Am. Gastrointestinal Endoscopic Surgeons. Achievements include development and evaluation of new devices for vascular access in chemotherapy, dialysis and other treatments; development of microwave coagulation for hepatic resection and control of hemmorhage from other vascular organs; description and early characterization of macrophage derived tumor necrosis factor; research in use of pre-operative radiation in control of rectal cancer. Office: Baystate Med Ctr 759 Chestnut St Springfield MA 01199-0001

REEDER, MIKE FREDRICK, materials engineer, consultant; b. Woodriver, Ill., Mar. 10, 1955; s. Charles Fredrick and Bonnie Bell (Knight) R.; m. Julia A. Hodge, Jan. 15, 1977 (div. Sept. 1992); 1 child, Miranda E.; m. Terri Lynn Eckelbarger, Oct. 24, 1992; children: Shannon M. Crawford, Patric T. Crawford, Kristen E. Crawford. BA, Olivet U., 1977; postgrad., Ctrl. Mich. U., 1989-90. Assoc. engr. Carter Carburetor, St. Louis, 1977-80; pvt. practice Carlinville, Ill., 1981-85; mgr., materials and reliability engr. Walbro Engine Mgmt. Corp., Cass City, Mich., 1986—. Lectr. middle sch., Cass City, 1990, 92; organizer Native Am. Aid, Pine Ridge, S.D., 1992—. Mem. AAAS, Soc. Automotive Engrs. (tchr. Vision 2000, 1991—), Am. Soc. Materials. Achievements include patent for a flexible composite membrane; discovery of relationship of performance of fuel metering membranes; of cuase of corrosive action of ambient atmosphere in automotive fuel tanks on high pressure fuel pumps. Home: 4404 Oak St Cass City MI 48726 Office: Walbro Engine Mgmt Corp 6242 Garfield Cass City MI 48726

REEKE, GEORGE NORMAN, JR., neuroscientist, crystallographer, educator; b. Green Bay, Wis.; m. Gail E. Hunt. BS, Calif. Inst. Tech., Pasadena, 1964; PhD, Harvard U., 1969. Asst. prof. Rockefeller U., N.Y.C., 1970-76, assoc. prof., 1976—. Mem. Soc. Neurosci., Am. Soc. Biochemistry and Molecular Biology, Am. Crystallographic Assn. Office: Rockefeller Univ 1230 York Ave New York NY 10021

REEKER, LARRY H., computer scientist, educator; b. Spokane, Wash., Feb. 2, 1943; s. Walter M. and Frances (Miles) R.; m. Linda Karman (div. 1985); children: Philip, David, Christina; m. Gail Bleach, 1989; children: Greg, Seth. BA, Yale U., 1964; PhD, Carnegie-Mellon U., 1974. Asst. prof. Computer Sci. and Linguistics Ohio State U., Columbus, 1968-73; asst. prof. Computer Sci. U. Oreg., Eugene, 1973-75; assoc. prof. Computer Sci. U. Ariz., Tucson, 1975-78; reader and dept. head U. Queensland, Brisbane, Australia, 1978-82; prof. and dept. head Tulane U., New Orleans, 1982-85; v.p. and sr. prin. The BDM Corp., McLean, Va., 1985-89; rsch. staff mem. Inst. for Def. Analyses, Alexandria, Va., 1989—; vis. rsch. scientist Chem. Abstracts Corp., Columbus, Ohio, 1982, U.S. Naval Rsch. Lab., Washington, 1984-85; vis. scientist USAF Human Resources Lab., Denver, 1984; Nystal Vis. Fellow U. West Fla., Pensacola, 1989. Author: (book) Plan for

Revitalization of the American Machine Tool Industry, 1989; editor (book) Machine Learning of Natural Language and Ontology, 1991; editor (jour.) SIGACT News, 1970-78; contbr. more than 60 articles to profl. jours. Mem. Assn. for Computing Machinery (bd. dirs. 1986-89, Svc. award 1989, 90), IEEE, Am. Assn. for Artificial Intelligence. Home: 11001 Lockwood Dr Silver Spring MD 20901 Office: Inst for Defense Analyses 1801 N Beauregard St Alexandria VA 22311

REES, DAVID CHARLES, chemist, researcher; b. London, Apr. 10, 1958; s. Charles Wayne and Patricia Mary (Francis) R. BS, Southampton U., 1979; PhD, Cambridge U., 1982. Rsch. fellow Harvard U., Cambridge, Mass., 1982-84; head organic synthesis I Parke-Davis Neurosci. Rsch. Ctr., Cambridge, United Kingdom, 1985—. Contbr. articles to Jour. Medicinal Chemistry, Tetrahedron Letters, Bioorganic and Medicinal Chemistry Letters; author: Comprehensive Medicinal Chemistry, 1991. Mem. Royal Soc. Chemistry, Soc. Drug Rsch. Achievements include over 30 patents and publications in medicinal chemistry; discovery and development of novel kappa opioid agonists to treat pain or head injury, cholecystokinin antagonists, tachykinin antagonists. Office: Parke-Davis Neurosci Rsch, Addenbrookes Hospital Site, Cambridge CB2 2QB, England

REES, FRANK WILLIAM, JR., architect; b. Rochester, N.Y., June 5, 1943; s. Frank William and Elizabeth R. (Miller) R.; m. Joan Mary Keevers, Apr. 1, 1967; children: Michelle, Christopher. BS in Architecture, U. Okla., 1970; postgrad., Harvard U., Boston, 1979, 90. Registered architect, 36 states; cert. Nat. Coun. Archtl. Registration Bds. Sales mgr. Sta. KFOM, Oklahoma City, 1967-70; project architect Benham-Blair & Affiliates, Oklahoma City, 1970-75; pres., founder Rees Assocs., Inc., Oklahoma City, 1975—; pres., chmn. bd. Weatherscan Radio Network, Oklahoma City, 1973-78; chmn. bd. Weatherscan Internat., Oklahoma City, 1972-78; pres. Frontier Communications, Oklahoma City, 1980-84; chmn. architecture bd. U. Okla., Norman, 1988-91. Past pres. Lake Hefner Trails, Oklahoma City, Hosp. Hospitality House, Oklahoma City, Oklahoma City Beautiful; mem. Leadership Oklahoma City. Mem. AIA, Am. Assn. Hosp. Architects, Am. Healthcare Assn., Tex. Hosp. Assn., World Pres. Orgn., Nat. Assn. Sr. Living Industries. Home: 1104 Stone Gate Irving TX 75063 Office: 511 Carpenter Fwy #222 Irving TX 75062

REES, MARTIN JOHN, astronomy educator; b. York, Eng., June 23, 1942; s. Reginald and Joan (Bett) R. MA, PhD, Cambridge (Eng.) U., 1967; DSc (hon.), Sussex (Eng.) U., 1990, Leicester (Eng.) U., 1993. Rsch. fellow Calif. Tech. Inst., 1968; vis. rsch. fellow Inst. for Advanced Study, Princeton, N.J., 1969, 82; vis. scientist Harvard U., Cambridge, Mass., 1972, 87—; Regents fellow Smithsonian Instn., 1984-87; Plumian prof. astronomy Cambridge U., 1973-91; dir. Inst. Astronomy, Cambridge, 1977-91; rsch. prof. Royal Soc. Cambridge U., England, 1992—; vis. prof. Havard U., Princeton U., Calif. Tech. Recipient Heinemann prize Am. Inst. Physics, 1984, Gold medal Royal Astron. Soc., 1987, Balzan prize, 1989, Robinson prize, 1990, Bruce medal, 1993, Knight Bachelor, 1992, Officer of Order Arts and Letters, France; fellow King's Coll., Cambridge, Eng., 1969—. Fellow AAAS, Royal Soc. London, Indian Acad. Scis. (hon.), Swedish Acad. Sci. (Sci.), Am. Philosophy Soc.; mem. Nat. Acad. Scis. (fgn. assoc.), Pontifical Acad. Scis., Academia Europea, Inst. Phusics (Eng.) (Guthrie prize 1990). Anglican. Office: Inst Astronomy, Cambridge England CB3 0HA

REESE, COLIN BERNARD, chemistry educator; b. Plymouth, Eng., July 29, 1930; s. Joseph and Emily R.; m. Susanne Bird, June 29, 1968; children: Lucy, William Thomas. BA, Cambridge (Eng.) U., 1953, PhD, 1956, MA, 1957, ScD, 1972. Fellow Clare Coll., Cambridge, 1956-73; rsch. fellow Harvard U., Cambridge, Mass., 1957-58; demonstrator in chemistry Cambridge U., 1959-63, asst. dir. rsch., 1963-64, lectr., 1964-73; Daniell prof. chemistry King's Coll., London, 1973—. Contbr. articles to profl. jours. Fellow King's Coll., 1989. Fellow Royal Soc. (London), Royal Soc. Chemistry; mem. Am. Chem. Soc. Office: King's Coll London, London WC2R 2LS, England

REESE, HERSCHEL HENRY, mechanical engineer; b. Fairmont, W.Va., July 7, 1935; s. Clifford Carlton and Biddie Virginia (Harris) R.; m. Loraine Whetzel, Aug. 23, 1958; children: Herschel Henry Jr., Clifford, Allison Beth. BSME, W.Va. U., 1963. With Owens-Ill., 1963—; prodn. mgr. Owens-Ill., Charlotte, Mich., 1972-75; mgr. tech. support Owens-Ill., Geneva, Switzerland, 1975-80; mgr. tech. support internat. Owens-Ill., Toledo, 1982—. With USAF, 1954-58. Mem. ASMe. Avocations: golf, skiing, photography, hunting, fishing. Home: 4503 Sulgrave Dr Toledo OH 43623-2049 Office: Owens Brockway 1 Seagate Toledo OH 43604-1558

REESE, LYMON CLIFTON, civil engineering educator; b. Murfreesboro, Ark., Apr. 27, 1917; s. Samuel Wesley and Nancy Elizabeth (Daniels) R.; m. Eva Lee Jett, May 28, 1948; children: Sally Reese Melant, John, Nancy. BS, U. Tex. at Austin, 1949, MS, 1950; PhD, U. Calif. at Berkeley, 1955. Diplomate: Registered profl. engr., Tex., La. Internat. Boundary Commn. San Benito, Tex., 1939-41; surveyor U.S. Naval Constrn. Bns., U.S., Aleutian Islands, Okinawa, 1942-45; field engr. Assoc. Contractors & Engrs., Houston, 1945; draftsman Phillips Petroleum Co., Austin, 1946-48; research engr. U. Tex., Austin, 1948-50; asst. prof. civil engring. Miss. State Coll., 1950-51, 53-55; asst. prof. U. Tex., Austin, 1955-57; assoc. prof. U. Tex., 1957-64, prof., 1964—, chmn. dept., 1965-72, Taylor prof. engring., 1972-81, assoc. dean engring. for program planning, 1972-79, Nasser I. Al-Rashid Chair, 1981-84. Contbr. articles to profl. jours. Served with USNR, 1942-45. Recipient Thomas Middlebrooks award ASCE, 1958; Joe J. King Profl. Engring. Achievement award, 1977, Offshore Tech. Conf. Disting. Achievement award for Individuals, 1985, Disting. grad. Coll. of Engring., U. Tex., Austin., 1985. Mem. ASCE (Karl Terzaghi lectr. 1976 Terzaghi award, 1983, Tex. sect. award of Hon. 1985, hon. mem. 1984—), Nat. Acad. Engring., Nat. Soc. Profl. Engrs. Baptist (deacon). Office: U Tex Dept Civil Engring Austin TX 78712-1104

REESE, THOMAS SARGENT, neurobiology educator and researcher; b. Cleve., May 20, 1935; s. Thomas Sargent Reese and Jane Andrews; children: Andrea Cooney, Devin Andrews. BA magna cum laude, Harvard U., 1957; MD, Columbia U., 1962. Rsch. asst. psycho-acoustic lab. Harvard U., Cambridge, Mass., 1957-58; intern Boston City Hosp., 1962-63; rsch. assoc. NINDB NIH, Bethesda, Md., 1963-65, rsch. med. officer Lab. Neuropathology and Neuroanat. Scis. NiNCDS, 1966-70, head sect. on functional neuroanatomy, Lab. Neuropathology and Neuroanat. Scis., NINCCS, 1970-83, chief neurobiology lab., 1983—; rsch. fellow in anatomy Harvard Med. Sch., Boston, 1965-66; instr. neurobiology course Marine Biol. Lab. Woods Hole, 1972—, co-instr. in chief, 1980-84, trustee, exec. com. 1984; assoc. neurosci. rsch. program Rockefeller U., 1986—. Mem. editorial bd. Anat. Record, 1977—, Jour. Neurocytology, 1976-78, Neurosci., 1978—, Jour. Cell Biology, 1976-79, Cryoletters, 1979-83, Jour. Neurosci., 1980—, Jour. Cellular and Molecular Neurobiology, 1980—, Neuron, 1986-88, Jour. Electron Microscopy Technique, 1984—, Cell Regulation, 1989—; contbr. over 120 articles to profl. jours. With USPHS, 1964-66. Recipient Superior Svc.award USPHS, 1976, 1989, Mathilde Solowey award Found. for Advanced Edn. in the Scis., 1977, W. Alden Spencer award Columbia U., 1981, Joseph Mather Smith prize Coll. Physicians and Surgeons Columbia U., 1985, Grass Found. award Soc. for Neurosci., 1987, Fidia Neurosci. Found. Neurosci. award, 1988. Mem. NAS, Am. Soc. Cell Biology, Neurosci. Soc., Soc. Gen. Physiologists, Am. Assn. Anatomists (C. Judson Herrick award 1966). Avocations: tennis, bicycling, backpacking. Home: 3009 P St NW Washington DC 20007 Office: NIH Bldg 36 Rm 2A-29 Bethesda MD 20892

REESE, WILLIAM ALBERT, III, psychologist; b. Tabor, Iowa, Nov. 23, 1932; s. William Albert and Mary-Evelyn Hope (Lundeen) R.; B.A., U. Washington Reed Coll., 1955; M.Ed., U. Ariz., 1964, Ph.D., 1981; m. Barbara Diane Windermere, Dec. 22, 1954; children: Judy, Diane, William IV, Sandra-Siobhan, Debra-Anne, Robert-Gregory, Barbara-Joanne, Ronald. Clin. Psychology cons. Nogales Pub. Schs., Nogales-Tucson, Ariz., 1971-79; clin. psychologist Astra Found., N.Y.C., 1979-86, chief psychology svc., neuropsychiatry, 1980-89; chief psychologist Family Support Ctr. Community-Family Exceptional Mem. Svcs., Sonoita, Ariz., 1986-89, Psychol. Svc. Ctr., Mount Tabor, Iowa, 1989—; dir. religious Marriage and Family Life Wilderness Ctr., Berchtesgaden, W.Ger., summer 1981-82; exec. sec. Astra Ednl. Found., 1975-79, bd. dirs., 1979—, EEO officer, 1978—. Served with USAF, 1967-71: Vietnam. Decorated Bronze Star. Fellow in cons. psychology and holistic medicine Clin. Services Found., Ariz., 1979—. Fellow Am. Psychol. Soc.; mem. Calif. Psychol. Assn., Ariz. Psychol. Assn., Am. Counseling Assn., Iowa Psychol. Assn. Clubs: Los Padres Wilderness Center, Outdoor, Sierra, Skyline Estates Golf and Country (Tucson), K.C. Author: Developing a Scale of Human Values for Adults of Diverse Cultural Backgrounds, 1981, rev. edit., 1988. Office: Psychol Service Ctr RR 1 Box 139 Tabor IA 51653-9716

REEVE, LORRAINE ELLEN, biochemist, researcher; b. Cato, Wis., Aug. 12, 1951; d. Robert K. and Lila M. (Breneman) R.; m. Dennis L. Kiesling, July 21, 1990. BS, U. Wis., 1973, MS, 1978, PhD, 1981. Postdoctoral scholar U. Mich., Ann Arbor, 1981-86; project scientist Cleve. Clinic Found., 1986-88; sr. rsch. scientist R.P. Scherer Corp., Troy, Mich., 1988-89, Mediventures, Inc., Dearborn, Mich., 1989-92; prin. investigator Mediventures, Inc., Dearborn, 1992—. Contbr. articles to profl. jours. Mem. Founders Soc. Detroit Inst. Art, 1989—, Nat. Trust for Historic Preservation, 1991—. Mem. AAAS, N.Y. Acad. Sci. Achievements include patents in topical drug delivery, opthalmic drug delivery, drug delivery by injection and body cavity drug delivery all with thermo-irreversible gels. Home: PO Box 2962 Ann Arbor MI 48106 Office: Mediventures Inc 15250 Mercantile Dr Dearborn MI 48120

REEVES, ALVIN FREDERICK, II, genetics educator; b. Richmond, Ind., Jan. 21, 1941; s. Alvin Frederick and Ada Laurel (Wood) R.; m. Carol Ann Rundell, June 25, 1966; children: Susan Irene, Lawrence Andrew, Brian Douglas, Kenneth Scott. BS, Ind. U., 1963, MA, 1964; PhD, U. Calif., Davis, 1968. Asst. prof. U. Ark., Fayetteville, 1968-75; asst. prof. U. Maine, Orono, 1975-81, assoc. prof., 1981—; NE rep. Inter-Regional Potato Introduction Project (IR-1) Adv. Com., 1976-80. Contbr. articles to profl. jours. Treas. St. Anne's Epis. Ch., Mars Hill. Nat. Merit scholar IBM, 1959-63. Mem .Potato Assn. Am., Agronomy Soc. Am., Kiwanis, Sigma Xi. Achievements include participation in development and release of 12 potato varieties: Delta Gold, Allagash Russet, Yankee Chipper, Yankee Supreme, Islander, Campbell 14, Sunrise, Somerset, Prestile, MaineChip, Portage, St. John's. Office: Aroostook Farm Experiment Sta 59 Houlton Rd Presque Isle ME 04769-5207

REEVES, BARRY LUCAS, research engineer; b. St. Louis, Jan. 11, 1935; s. Raymond O. and Frances M. (Lucas) R.; m. Marilyn Alva Riester, May 8, 1954; children: Katherine, Michael, Janet. BS, Washington U., St. Louis, 1956, MS, 1958, PhD, 1960. Rsch. assoc. McDonnell Aircraft Corp., St. Louis, 1959-60; postdoctoral rsch. fellow Calif. Inst. Tech., Pasadena, 1960-64; staff scientist Avco Corp., Wilmington, Mass., 1964-68, sr. staff scientist, 1968-74, sr. cons. scientist, 1974-86; prin. scientist Textron Corp., Wilmington, Mass., 1986—; cons. Space Gen. Corp., L.A., 1962-63, Nat. Engring. and Sci. Co., Pasadena, 1963-64, Aerojet Corp., Azusa, Calif., 1964; contr. engring. symposiums; reviewer Jour. Fluid Mechanics, Physics of Fluids, Jour. Applied Mech., Internat. Jour. Heat and Mass Transfer, AIAA Jour. Contbr. over 30 articles to profl. jours. Mem. Snow Mountain Farms Assn., Vt., 1982—. Fellow Convair Aircraft Corp., 1958, NSF, 1959; postdoctoral fellow Air Force Office Sci. Rsch., 1960, 61, rsch. grantee, 1962, 63. Mem. Am. Inst. Physics, Smithsonian Assocs. (assoc.), Sigma Xi, Tau Beta Pi, Pi Tau Sigma (former v.p., treas.). Home: 10 Hillcrest Pky Winchester MA 01890-1427 Office: Textron Corp 201 Lowell St Wilmington MA 01887-2969

REEVES, BILLY DEAN, obstetrics/gynecology educator emeritus; b. Franklin Park, Ill., Jan. 17, 1927; s. Barney William and Martha Dorcus (Benbrook) R.; m. Phyllis Joan Faber, Aug. 25, 1951; children: Philip, Pamela, Tina, Brian, Timothy. BA, Elmhurst (Ill.) Coll., 1953; BS, U. Ill., Chgo., 1958, MD, 1960; post grad., UCLA, N.Mex. State U., 1953-54, 75-76. Diplomate Am. Bd. Ob-Gyn. Intern Evanston (Ill.) Hosp., 1960-61, resident ob-gyn., 1961-64; NIH fellow in reproductive endocrinology Karolinska Hosp. and Inst., Stockholm, 1968-69; pvt. practice Evanston, 1964-71, Las Cruces, N.Mex., 1972-77; from instr. to asst. prof. Dept. Ob-gyn. Northwestern U. Med. Sch., 1964-71; assoc. prof. Dept. Ob-gyn. Rush Med. Coll., Chgo., 1971-72; clin. assoc. in ob-gyn. U. N.Mex. Med. Sch., Albuquerque, 1972-77; clin. assoc. U. Ariz. Sch. Medicine, Tucson, 1975-78; from clin. prof. to prof. emeritus Tex. Tech. Med. Sch., El Paso, 1976-91. Contbr. 70 articles and chpts. to med., profl. jours., 1958-93. Adv. bd. Associated Home Health Svcs., Inc., Las Cruces; adv. com. N.Mex. State U. Nursing Sch.; community adv. com. Meml. Gen. Hosp., Las Cruces; tech. advisor on health edn. N.Mex. Health Systems Agy.; mem. N.Mex. U. Task Force 88; bd. dirs. Meml. Med. Ctr, Las Cruces, 1989-91, Parenthood Edn. Assn. of El Paso, Inc., Planned Parenthood of South Ctrl. N.Mex. (chmn. med. adv. com.). With USNR, 1945-46, 82-88, U.S. Army, 1946-47. Recipient Elmhurst Coll. Alumni Merit award, 1990, William W. Fry award for profl. excellence Tex. Tech. U. Sch. Medicine, 1979. Mem. AMA, ACS, AAAS, ACOG, Am. Coll. Physician Execs., Am. Assn. Advancement Humanities, Am. Fertility Soc., Assn. Profs. Ob-Gyn., North Am. Ob-Gyn. Soc., Ctrl. Assn. Ob-Gyn., Chgo. Gyn. Soc., Com. for Philosophy in Medicine, Dona Ana County Med. Soc. (assoc.), El Paso County Med. Soc., El Paso Surg. Soc., Endocrine Soc., Inst. Medicine in Chgo., Hasting Ctr., N.Mex. Med. Soc. (assoc.), N.Mex. Ob-Gyn. Soc., Soc. for Health and Human Values, Tex. Med. Assn., Tex. Assn. Ob-Gyn., U.S.-Mexico Border Health Assn. Home: 1620 Altura Ave Las Cruces NM 88001-1532

REEVES, RICHARD ALLEN, government aerospace program executive, lawyer; b. Chgo., July 10, 1944; s. Harry Mathew and Hazel Louise (Parker) R.; m. Patricia Dean McKenney, Dec. 30, 1967 (div. 1977); children: Garth Kirkman, Blakely Catherine, Janice Emily; m. Elizabeth Marie Litkowski, Nov. 22, 1980. BBA, Ga. State U., 1970; JD, U. Tenn., 1973. Bar: Tenn. 1974. Atty. NASA Marshall Space Flight Ctr., Huntsville, Ala., 1974-77, NASA Hdqrs., Washington, 1977-80; dep. chief counsel NASA Goddard Space Flight Ctr., Greenbelt, Md., 1980-82, dep. dir. mgmt., 1982-85; assoc. dir. NASA Ames Rsch. Ctr., Mountain View, Calif., 1985-88; dir. planning NASA Hdqrs., 1988-90; asst. dir. NASA Hdqrs. Space Exploration, Washington, 1990-91; dir. for instns. NASA Hdqrs., Washington, 1991—. Capt. U.S. Army, 1967-71, Vietnam. Mem. AIAA, Tenn. Bar Assn., Nat. Space Club, Women in Aerospace. Republican. Baptist. Avocations: genealogical research, photography, old house restoration. Home: 7811 Custer Rd Bethesda MD 20814

REEVES, ROBERT GRIER LEFEVRE, geology educator, scientist; b. York, Pa., May 30, 1920; s. Edward LeGrande and Helen (Baker) R.; m. Elizabeth Bodette Simmons, June 11, 1942; children: Dale Ann, Edward Boyd. Student, Yuba Coll., 1938-40; B.S., U. Nev., 1949; M.S., Stanford U., 1950, Ph.D., 1965. Registered profl. engr., Tex. Geophysicist, geologist U.S. Geol. Survey, 1949-69; with assignments as project chief iron ore deposits of Nev., tech. adviser Fgn. Aid Program; vis. prof. econ. geology U. Rio Grande do Sul, Brazil; staff geologist for research contracts and grants and profl. staffing Washington; prof. geology Colo. Sch. Mines, Golden, 1969-73; staff scientist Earth Resources Observation Systems Data Center, U.S. Geol. Survey, Sioux Falls, S.D., 1973-78; prof. geology U. Tex.-Permian Basin, Odessa, 1978-85, chmn. earth scis., 1978-82; dean Coll. Sci. and Engring. U. Tex. Permian Basin, 1979-84; cons. geologist, engr. ptnr. Orion, Ltd., Midland, Tex., 1984-87, cons. engr., geologist, 1987—; leader People to People Econ. Geology Trip to Brazil and Peru, 1985; co-leader Am. Geol. Inst. Internat. Field Inst. to Brazil, 1966; vis. geol. scientist Boston Coll., Boston U., 1967; sr. Fulbright lectr. U. Adelaide, Australia, 1969. Contbr. articles profl. jours. Served to maj. Signal Corps AUS, 1941-46, ETO. Recipient Outstanding Service award U.S. Geol. Survey, 1965. Fellow Geol. Soc. Am., Geol. Soc. Brazil, Soc. Econ. Geologists; mem. Am. Inst. Mining, Metall. and Petroleum Engrs. (sec. Washington 1964), Am. Soc. Photogrammetry (editor-in-chief Manual of Remote Sensing), Soc. Ind. Profl. Earth Scientists. Home and Office: 4025 Lakeside Dr Odessa TX 79762-7203

REFINETTI, ROBERTO, physiological psychologist; b. Sao Paulo, Brazil, Nov. 19, 1957; came to U.S., 1988; s. Renato and Maria Stella (Barroso) R.; m. Kathleen Diane Zylan, Mar. 5, 1988 (div. Aug. 1991); 1 child, Lauren Lynne. BA in Philosophy, Pontifical Cath. U., Sao Paulo, 1981; BS in Psychology, U. Sao Paulo, 1981, MA in Psychology, 1983; PhD in Psychology, U. Calif., Santa Barbara, 1987. Asst. prof. U. Sao Paulo, 1986-88; postdoctoral fellow U. Calif., Santa Barbara, 1988-89, U. Ill., Champaign, 1989-90, U. Va., Charlottesville, 1990-92; asst. prof. Coll. William and Mary, Williamsburg, Va., 1992—. Contbr. over 50 articles to profl. jours. Recipient Nat. Rsch. Svc. Individual award NIMH, 1991. Mem. Am. Physiol. Soc., Am. Psychol. Soc., Soc. Neuroscience, Soc. Rsch. on Biol. Rhythms. Office: Coll of William and Mary Dept Psychology Williamsburg VA 23187

REGALBUTO, MONICA CRISTINA, chemical engineer, research scientist; b. Monterrey, Mex., July 30, 1961; came to U.S., 1984; d. Horacio Gonzalez-Santos and Maria Concepcion (Banos) Gonzalez; m. John Robert Regalbuto, Dec. 30, 1983; children: Jose Ricardo, Maria Carolina, Jose Roberto. MS in Chem. Engring., U. Notre Dame, 1986, PhD, 1989. Teaching asst. dept. chem. engring. U. Notre Dame, Ind., 1984-85, rsch. asst. dept. chem. engring., 1984-88; chem. engr. rsch. scientist Chem. Tech. div. Argonne (Ill.) Nat. Lab., 1988—. Contbr. articles to Chem. Engring. Sci., Bull. Math. Biology, Separation Sci. and Tech., others. Computer instr. Our Lady of Perpetual Help Sch., Glenview, Ill., 1992-93; hispanic collection vol. Glenview Pub. Libr., 1993—. Recipient scholarship. Mem. Am. Inst. Chem. Engrs., Sigma Xi. Roman Catholic. Achievements include the use of applied mathematics in chemical engineering to simplify computer calculations. Office: Argonne Nat Lab 9700 S Cass Ave Argonne IL 60439-4837

REGAN, JOHN WARD, molecular pharmacologist; b. Washington, Oct. 27, 1950. MS in Nutrition, U. Calif., Berkeley, 1975; PhD in Pharmacology, U. Ariz., 1981. Rsch. fellow Duke U. Med. Ctr., Durham, N.C., 1981-84, rsch. assoc., 1984-89; asst. prof. U. Ariz., Tucson, 1989—; cons. Allergan, Inc., Irvine, Calif., 1990—, Syntex, Palo Alto, Calif., 1992, Abbott, Abbott Park, Ill., 1993. Contbr. articles to Sci., Jour. Biol. Chemistry, Proceedings Nat. Acad. Sci., Molecular Pharmacology, Biochemistry. Recipient Roche Lab. award in neurosci. Nat. Student Rsch. Forum, 1979, Nat. Rsch. Svc. award NIH, 1981-84, Pharmacology award Eli Lilly, 1991, Syntex prize in pharmacology Western Pharmacology Soc., 1992. Mem. Internat. Soc. Neurochemistry, Am. Heart Assn., AAAS. Achievements include purification and cloning of the human alpha-2 adrenergic receptors; co-discovery of alpha adrenergic receptor heretogeneity, G-protein receptor superfamily. Office: U Ariz Coll Pharmacy Dept Pharmacology Toxicology Tucson AZ 85721

REGDOS, SHANE LAWRENCE, nuclear chemist; b. Buffalo, Nov. 7, 1962; s. Casimer and Esther (Zebrasky) R. BS, Calif. Poly. State U., 1985; PhD, U. Md., 1992. Lab. asst. Chem. dept. Calif. Poly. State U., San Luis Obispo, 1981-85; chemist NIH Nuclear Medicine, Bethesda, Md., 1987—. Author: New Trends in Radiopharmaceutical Synthesis, Quality Assurance and Regulatory Control, 1991; contbr. articles to profl. jours. Mem. Soc. of Nuclear Medicine, Am. Chem. Soc., Holifield Heary Ion Rsch. Facility Users Group. Office: Nat Inst Health 9000 Rockville Pike Bethesda MD 20892

REGENER, VICTOR H., physicist; b. Berlin, Germany, Aug. 25, 1913; came to U.S., 1940; s. Erich and Victoria (Mincin) R.; m. Birgit Hamilton, Aug. 23, 1941; children: Eric, Vivian. D in Physics, T.H. Stuttgart, Germany, 1938. Rsch. fellow U. Padova, Italy, 1938-40; rsch. fellow U. Chgo., 1940-42, instr., 1942-46; from assoc. prof. to prof. U. New Nex., Albuquerque, 1946-79; owner VHR Systems Rsch. and Devel., Albuquerque, New Mex., 1979—; assoc. curator Mus. Sci and Industry, Chgo., 1945-46; chmn. Physics and Astronomy Dept. U. New Mex., Albuquerque, 1946-57, 1962-79. Grantee: NSF, USAF, U.S. Weather Bur., Rsch. Corp., Los Alamos Nat. Lab., 1958-79. Fellow Am. Phys. Soc., N.Y. Acad. Scis.; mem. Optical Soc. Am., Internat. Soc. Optical Engring., Am. Astron Soc. Office: VHR Systems 7200 Jefferson NE Albuquerque NM 87109

REGGIA, JAMES ALLEN, computer scientist, educator; b. Takoma Park, Md., Oct. 31, 1949; s. Frank and Betty Jo (Patterson) R. MD, U. Md., 1975, PhD, 1981. Diplomate Am. Bd. Psychiatry and Neurology. Asst. prof. U. Md., Balt., 1979-84; assoc. prof. U. Md., College Park, 1984-93, prof., 1993—. Author: Abductive Inference Methods for Diagnostic Problem-Solving, 1990; editor: Computer-Based Medical Decision Making, 1985; contbr. over 50 articles to profl. jours. Office: U Md Dept Computer Sci AVW Bldg College Park MD 20742

REGINO, THOMAS CHARLES, technology assessment, R&D benchmarking specialist; b. Dover, Oct. 21, 1950; s. Augustine A. and Frances L. (Bush) R.; m. Leslie Lynne Robinson, Oct. 6, 1990; 1 child, Lee Terrence. BS in Mechanical Engring., Newark Coll. Engring., 1973; MS in Environ. Engring., N.J. Inst. Tech., 1978. Staff engr. Buck, Seifert & Jost, Englewood Cliffs, N.J., 1973-75; project engr. Clinton Bogert Assocs., Fort Lee, N.J., 1975-77; sr. project engr. PQA, Inc./ Energy Resources Group, Wayne, N.J., 1977-79; dir. rsch. and tech. analysis Rain Hill Group, Inc., N.Y.C., 1979-84; global mgr. tech. assessment, R&D benchmarking Kline & Co., Inc., Fairfield, N.J., 1984—. Author: Principles and Applications of Solar Energy, 1978. Recipient Nat. Design award U.S. Dept. Housing and Urban Devel., 1978. Mem. Comml. Devel. Assn., The Planning Forum, The Quality in Rsch. and Devel. Network, The Licensing Execs. Soc. Office: Kline & Co Inc 165 Passaic Ave Fairfield NJ 07004

REGISTER, RICHARD ALAN, chemical engineering educator; b. Cheverly, Md., Sept. 6, 1963; s. Robert Leroy and Mary Nora (Berthold) R.; m. Jean Tom, Aug. 20, 1989. BS, MIT, 1984, MS, 1985; PhD, U. Wis., 1989. Asst. prof. Princeton (N.J.) U., 1990—. Contbr. articles to profl. jours.; author three book chpts. Recipient Young Investigator award NSF, 1992-97; Hertz fellow John and Fannie Hertz Found., U. Wis., 1987-89; Presdl. scholar Dodge Found., Md., 1980. Mem. Am. Chem. Soc. (PMSE mem.-at-large 1992-95), Am. Inst. Chem. Engrs., Am. Phys. Soc., Soc. Plastics Engrs. Office: Princeton Univ Dept Chem Engring Princeton NJ 08544

REH, JOHN W., engineer, consultant; b. Saline County, Kans., July 2, 1935; s. Leslie W. and Vera M. (Snyder) R.; m. Judith A. Kirkland, June 1, 1957; children: Elaine M. Edwards, Jeffrey K., Kirk W. BS in Agrl. Engring., Kans. State U., 1958; postgrad., U. Kans., 1964, Utah State U., 1973. Registered profl. engr., Kans., Mo. Hydraulic engr. USDA Soil Conservation Svc., Salina, Kans., 1958-61, 63-70; constrn. engr. USDA Soil Conservation Svc., Cheney, Kans., 1961-63; leader water resource planning staff USDA Soil Conservation Svc., Salina, 1970-84, asst. state conservationist, 1984-91; pres., owner REH & Assocs., Salina, 1992—; conservation and watershed advisor Kans. water plan Kans. Water Authority, Topeka, 1980-91. Recipient Disting. Svc. award State Assn. Kans. Watersheds, 1992. Mem. NSPE (v.p. 1966-93, engrs. in govt., chmn. various coms., rep. to First Nat. Water Symposium 1982, Leadership award 1985, award 1992), Soil and Water Conservation Soc., Kans. Engring. Soc. Republican. Roman Catholic. Achievements include development of hydrologic procedures. Office: Reh & Assocs Inc 909 E Wayne # 8 Salina KS 67401*

REH, THOMAS EDWARD, radiologist, educator; b. St. Louis, Sept. 12, 1943; s. Edward Paul and Ceil Anne (Golden) R.; m. Benedette Texada Gieselman, June 22, 1968; children: Matthew J., Benedette T., Elizabeth W. BA, St. Louis U., 1965, MD, 1969. Diplomate Am. Bd. Radiology, Nat. Bd. Med. Examiners. Intern St. John's Mercy Med. Ctr., St. Louis, 1969-70; resident St. Louis VA Hosp., 1970-73; fellow in vascular radiology Beth Israel Hosp., Boston, 1973-74; radiologist St. Mary's Health Ctr., St. Louis, 1974—, chmn. dept. radiology, 1986—; clin. asst. prof. radiology St. Louis U. Sch. Medicine, 1978—; clin. assoc. prof. radiology, 1989—. Mem. Am. Coll. Radiology, AMA, Radiol. Soc. N.Am., St. Louis Met. Med. Soc., Alpha Omega Alpha, Alpha Sigma Nu, Delta Sigma Phi. Republican. Roman Catholic. Clubs: St. Louis, Confrerie des Chevaliers du Tastevin. Home: 9850 Waterbury Dr Saint Louis MO 63124-1046 Office: Bellevue Radiology Inc 1699 S Hanley Rd Saint Louis MO 63144-2913

REHA, WILLIAM CHRISTOPHER, urologic surgeon; b. Bronx, N.Y., Dec. 17, 1954; s. William and Jeanette (Ford) R.; m. Lynda Marie Zeiders; children: David, Christine. BS in Biochemistry, SUNY, Binghamton, 1977; MD, N.Y. Med. Coll., Valhalla, 1981. Resident surgery/urology Georgetown U., Washington, 1981-87; attending surgeon Potomac Hosp., Woodbridge, Va., 1987—. Mem. AMA, Am. Assn. Clin. Urologists, Am. Urol. Assn., Washington Urol. Soc. (2nd pl resident essay competition 1987), Med. Soc. Va., Prince William County Med. Soc., Prince William County C. of C. Republican. Roman Catholic. Achievements include research in neonatal ascites and ureteral valves, rhabdomyolysis need for a high index of suspicion. Office: 2296 Opitz Blvd # 220 Woodbridge VA 22191

REHAK, JAMES RICHARD, orthodontist; b. Chgo., Jan. 2, 1938; s. James Joseph and Lydia Ann (Thomas) R.; BS, U. Ill., 1960, DDS cum laude, 1962, MS, 1967, cert. in orthodontics, 1965; m. Joann Marie Tabbert, Oct. 15, 1969; 1 child, Suzanne Therese. Practice dentistry, Chgo., 1962-63, practice orthodontics, Chgo., Arlington Heights, Ill., Cape Coral and Naples, Fla; asst. prof. U. Ill. Coll. Dentistry, 1966-68. Kellogg Found. fellow, 1958. Fellow Royal Soc. Health; mem. ADA, Ill. Dental Assn., Chgo. Dental Soc., Fla. Dental Assn., West Coast Dental Soc., Collier County Dental Assn., Am. Assn. Orthodontists, Am. Assn. Lingual Orthodontists, So. Soc. Orthodontists, Fedn. Dentaire Internationale, Psi Omega, Omicron Kappa Upsilon. Home: 859 Nelsons Walk Naples FL 33940-7870 Office: 785 Central Ave Naples FL 33940-6007

REHKUGLER, GERALD EDWIN, agricultural engineering educator, consultant; b. Lyons, N.Y., Apr. 11, 1935; s. Charles James and Minnie Sophie (Tange) R.; m. Nancy Carolyn Poole, Sept. 11, 1982. B.S., Cornell U., 1957, M.S., 1958; Ph.D., Iowa State U., 1966. Registered profl. engr., N.Y. Grad. asst. Cornell U., 1957-58, asst. prof. agrl. engring., 1958-64, assoc. prof., 1964-77, prof., 1977—, chmn. dept. agrl. and biol. engring., 1984-90, assoc. dean undergrad. programs, Coll. Engring., 1990—; vis. prof. Mich. State U., East Lansing, 1974; cons. in field. Patentee high capacity harvesting apparatus. Mem. Dryden Bd. Edn., N.Y., 1977-82, v.p., 1979-81, pres., 1981-82; mem. Bd. Coop. Ednl. Services, Ithaca, 1982-88. NSF sci faculty fellow, 1964-66. Fellow Am. Soc. Agrl. Engrs. (chmn. various coms. rsch. paper), Am. Inst. for Med. Biological Engring.; mem. Am. Soc. for Engring. Edn. Methodist. Home: 30 Rochester St Dryden NY 13053-9536 Office: Cornell U 223 Carpenter Hall Ithaca NY 14853

REHWALD, WALTHER R., physicist, researcher; b. Troppau, Czechoslovakia, June 4, 1930; came to Switzerland, 1962; s. Rudolf and Anna (Heinz) R.; m. Elisabeth Back, Dec. 27, 1960; children: Lotte, Ulrike, Anne, Uta. Dipl.-Phys., Techn. Hochschule, Darmstadt, Germany, 1957, Dr.-Ing., 1963; venia Legendi, U. Konstanz, Germany, 1988. Teaching asst. dept. elec. engring. Techn. Hochschule, Darmstadt, 1957-62; rsch. physicist RCA Labs., Zurich, 1962-87, Paul Scherrer Inst., Zurich, 1987-92; adj., rsch. physicist Paul Scherrer Inst., Wurenlingen, Switzerland, 1992—; privatdozent U. Konstanz, 1974—. Contbr. chpts. to books, articles on solid state physics and ultrasonics to profl. jours. Mem. German Phys. Soc., Swiss Phys. Soc. Roman Catholic. Achievements include patent applied for in ultrasonic detection of slip planes in silicon wafery. Office: Paul Scherrer Inst, Villigen Switzerland CH5232

REICH, RANDI RUTH NOVAK, software engineer; b. Chgo., July 10, 1954; d. Bernard Richard and Shirley Ann (Fiedorczyk) Novak; m. Barry Stephen Reich, Aug. 19, 1983; children: Rona Rachel, Bonnie Shaina. BS in Math., U. Calif., Santa Cruz, 1976, BA in Econs. with honors, 1976; postgrad., U. Rochester, 1976-78. Rsch. asst. U. Calif., Santa Cruz, 1974-76; Russian translator U. Chgo., 1977-78; intern economist Congl. Budget Office, Washington, 1977; engr. Lockheed MSC, Sunnyvale, Calif., 1978-82; software engr. contractor Silicon Valley Systems, Belmont, Calif., 1982, 83-84, Data Encore (subs. of Verbatim), Sunnyvale, 1982-83; systems programmer CompuPro/Viasyn Corp., Hayward, Calif., 1984-87; mem. tech. staff Network Equipment Techs., Redwood City, Calif., 1987-89; software engr. contractor Segue Setups, Burlingame, Calif., 1989-92, ptnr., 1992—; sr. mem. tech. staff NEC Am., San Jose, Calif., 1992—. Fellow Dept. Treasury, 1974-76, NSF, 1977-78, U. Rochester, Rush Rhees fellow. Mem. IEEE Computer Soc., Am. Math. Assn., Computer Profls. for Social Responsibility, Soc. for Computing and Info. Processing, Internat. Platform Assn., Calif. Scholarship Fedn. (life). Avocations: piano, oboe, music, photography, mathematics. Home: 75 Eastlake Ave Pacifica CA 94044-2835

REICH, ROBERT CLAUDE, metallurgist, physicist; b. Paris, France, Nov. 2, 1929; s. Felix and Nelly (Belestin) R. Engr., École Nationale Supérieure de Chimie Paris, 1953; Licence ès Scis. Physiques, U. Paris, 1954, Docteur ès Scis. Physiques, 1965; m. Francoise Thiébault, June 10, 1972; m. Michele Helene Brand'Huy, Dec. 29, 1981. Attaché de recherche Centre Nat. de La Recherche Scientifique, Centre d'Etudes de Chimie Métallurgique, Vitry sur Seine, France, 1953-54, 58-65; chargé de recherche Laboratoire de Physique des Solides, Orsay, France, 1965-71, maître de recherche, 1971-83, directeur de recherche, 1984—. Mem. Société Française de Physique, Société Française de Métallurgie, European Phys. Soc., Electrochem. Soc. (U.S.), Internat. Soc. Electrochemistry (past sec. French sect.), Am. Soc. Metal. Rsch. and publs. on purification of metals by zone-melting, elec. resistivity of metals versus purity (observed temperature squared term of ideal resistivity in nonmagnetic metals), deviations from Matthiessen's rule, supraconducting transition in tin, determination of characteristic Debye temperatures, ferminology studies in mercury, size-effect, ionic interactions in solutions, electrolyte glass transitions, anodic dissolution of metals, corrosion, superconducting oxides and quasicrystals. Home: 3 Allée des Mouille-Boeufs, 92290 Châtenay-Malabry France Office: Lab de Physique des Solides, Bâtiment 510/ U Paris-Sud, 91405 Orsay Cedex, France

REICHEK, NATHANIEL, cardiologist; b. N.Y.C., Mar. 25, 1941; s. Max and Bessie (Citron) R.; m. Carolyn Hirsh, Apr. 15, 1965 (div. 1986); children: Jennifer Laura, Matthew Solomon; m. Lesly Ann Armstrong, Sept. 19, 1986. BA summa cum laude, Columbia Coll., 1961; MD, Columbia U., 1965. Resident in internal medicine Albert Einstein Coll. Medicine, Bronx, N.Y., 1965-67, 69-70; fellow in cardiology Georgetown U., Washington, 1970-72; from asst. prof. medicine to prof. U. Pa., Phila., 1972—; prof. medicine Med. Coll. Pa., Pitts., 1992—; dir. cardiology Allegheny Gen. Hosp., Pitts., 1992—. Mem. editorial bd. Circulation, 1991-93, Jour. Am. Coll. Cardiology, 1988-93, Am. Jour. Cardiology, 1986—; contbr. articles to profl. jours. Pres. bd. Sch. in Rose Valley, Pa., 1976-77, mem. bd. Wallingford-Swarthmore (Pa.) Sch. Dist., 1981-85. Lt. comdr. USPHS, 1967-69. Fellow Am. Coll. Cardiology; mem. Am. Heart Assn. (bd. govs. S.E. Pa. 1977-80, 91-92), Am. Soc. Echocardiography (bd. govs. 1979-82, 87-91), Soc. Magnetic Resonance in Medicine. Achievements include development of methods for in vivo quantitation of weight of human left ventricular myocardium reliably; research in transdermal administration of nitrogylcerin; development and application of cardiac imaging methods using MRI. Office: Allegheny Gen Hosp 320 E North Ave Pittsburgh PA 15143

REICHEL, LEE ELMER, mechanical engineer; b. Cin., Aug. 30, 1944; s. Elmer William and Catherine Ann (Heidlemann) R.; m. Cheryl Christine Feagans, May 7, 1966; 1 child, Geoffrey William. A in Mech. Engring., Ohio Coll. Applied Sci., 1965; BS in Engring., U. Dayton, 1980. Registered mech. engr., Ohio. Project engr. Koehring, Master Div., Dayton, 1968-77, mgr. engring., 1978-81; project engr. Micro Devices, Dayton, 1977-78; mgr. mech. engring. dept. Xetron Corp., Cin., 1985; mgr. product engring. Dayton-Walther, Dayton, 1981-85, mgr. new product devel., 1985-91; projects mgr. Ethicon Inc., Cin., 1991—. Patentee in field. Mem. Alter High Sch. Booster Bd., Kettering, Ohio, 1984-86. Mem. NSPE, Soc. Automotive Engrs., Ohio Soc. Profl. Engrs. Republican. Roman Catholic. Avocations: guitar, racquetball, tennis. Home: 2325 Colony Way Dayton OH 45440-2510

REICHEN, JÜRG, pharmacology educator; b. Aarau, Switzerland, Jan. 23, 1946; s. Hans A. and Susi (Aeberhard) R.; m. Suzi Graden, may 29, 1970; children: Hansjakob, Annemarie, Katharina. BA, Gymnasium City, Burgdorf, Switzerland, 1964; MD, U. Bern, Switzerland, 1971. Diplomate medicine. Fellow dept. exptl. medicine Hofmann Laroche, Basel, Switzerland, 1972; fellow dept. clin. pharmacology U. Bern, 1973-76, prof. medicine, clin. pharmacologist, 1986—; intern, resident Georgetown U. and V.A. Med. Ctr., Washington, 1978-79; fellow div. GI UCHSC, Denver, 1980; asst. prof. div. GI and Clin. Pharmacology USHSC, Denver, 1980-84, assoc. prof., 1984-85; guest scientist Liver Unit, NIH, Bethesda, Md., 1976-78; mem. rsch. com. AASLD, Chgo., 1984-85; sec. scientific com. EASL, 1988-91. Editorial bd.: Hepatology, 1988—; assoc. editor: Jour. of Hepatology, 1989-93; contbr. articles to profl. jours. and books. Mem. Fed. Commn. for MDPhD, Bern, 1994; expert Fed. Commn. on Toxic Compounds, Bern, 1985-92; referee SNF, DFG, NIH and others. Recipient Price of the Faculty award U. Bern, 1976, Faculty Devel. award Pharm Mfg. Assn., Washington, 1981-83, Rsch. Career Devel. award NIH, 1983-88, Swiss Nat. Found. Scientific Rsch., Bern, 1986-91, Price award Swiss Soc. for Internal Medicine, 1987. Mem. European Assn. Study of the Liver, Am. Assn. Study of Liver Disease, Am. Gastroenterology Assn., Western Soc. Clin. Investigators, Internat. Assn. Study of the Liver, United European Gastroenterologists Week (counselor 1991-95), European Soc. Biochem. Pharms. Office: Univ Bern Dept Clin Pharmacology, Murtenstrasse 35, CH-3010 Bern Switzerland

REICHENBACH, THOMAS, veterinarian; b. N.Y.C., Jan. 6, 1947; s. Henry J. and Helen M. (Kelly) R.; m. Cleda L. Houmes, Nov. 23, 1984. BS in Chemistry, U. Notre Dame du Lac, 1968; MS in Chemistry, U. Calif., Davis, Calif., 1973; AA, Shasta Coll., 1975; DVM, U. Calif., Davis, Calif., 1981. Doctor of Veterinary Medicine, Diplomate Am. Bd. Veterinary Practitioner. Sentry dog handler U.S. Army, 1970-71; indsl. chemist Syntex, Palo Alto, Calif., 1973-75; gestation herd mgr. Llano Seco Rancho, Chico, Calif., 1975-76; pres. Veterinary Mgmt. Svcs., Salinas, Calif., 1981—; chief exec. officer Santa Barbara Vet. Emergency Group, 1991—; lectr. in field; adv. bd. Veterinary Post Grad. Inst., Santa Cruz, Calif., 1988. Contbr. articles to profl. jours.; author computer software Personal Wedding Planner, 1985, Veterinary Clinical Simulation, 1988. With U.S. Army, 1969-71. Mem. Am. Vet. Med. Assn., Calif. Vet. Med. Assn., Nat. Notre Dame Monogram Club. Republican. Roman Catholic. Avocations: artificial intelligence, computer-aided education, epidemiology. Office: Vet Mgmt Svcs 1887 Cherokee Dr # 1 Salinas CA 93906-2394

REICHERT, LEO EDMUND, JR., biochemist, endocrinologist; b. N.Y.C., Jan. 9, 1932; s. Leo and Anne (Holstein) R.; m. Gerda Sihler, July 20, 1957; children: Leo, Christine, Linda, Andrew. B.S., Manhattan Coll., N.Y.C., 1955; Ph.D., Loyola U., Chgo., 1960. Asst. prof. biochemistry Emory U. Med. Sch., Atlanta, 1960-66; assoc. prof. Emory U. Med. Sch., 1966-72, prof., 1972-79; prof., chmn. dept. biochemistry Albany (N.Y.) Med. Coll., 1979-88, prof. biochemistry, 1988—; dir. human and animal hormone isolation and distbn. program (NIH), Emory U. Med. Sch., 1960-75; mem. med. adv. bd. Nat. Pituitary Agy., 1971-74; com. on glycoprotein hormones Nat. Hormone and Pituitary Program, 1968-86; mem. reproductive biology study sect. NIH, 1971-75; mem. adv. panel on cellular physiology NSF, 1983-86, div. of integrative and neuro biology, 1992; mem. WHO Expert Adv. Panel on Biol. Standardization, 1984—, Nat. Bd. Med. Examiners, Part I, 1989-91. Mem. editorial bd. Endocrinology, 1967-75, Molecular and Cellular Endocrinology, 1977-83, 90—, Biology of Reproduction, 1968-70, 86-90, Andrology, 1983-86, Molecular Andrology, 1989—; contbr. over 250 articles to profl. jours.; U.S. patentee in field. Served with USMC, 1950-52. List among 75 endocrinologists, 1000 scientists most cited, 1965-78. Mem. AAAS, Am. Soc. Biol. Chemists, Endocrine Soc. (Ayerst award 1970), Andrology Soc. (coun. 1983-87), Soc. for Study of Reprodn. Home: 10 Laurel Dr Albany NY 12211-1618 Office: Albany Med Coll Dept Biochemistry Albany NY 12208

REICHLE, DAVID EDWARD, ecologist, biophysicist; b. Cin., Oct. 19, 1938; s. Edward John and Elsie (Nees) R.; m. Donna Rae Haubrich, Oct. 7, 1961; children: John, Deborah, Jennifer. BS, Muskingum Coll., 1960; MS, Northwestern U., 1961, PhD, 1964. Biophysicist Oak Ridge (Tenn.) Nat. Lab., 1966-70, program dir. terrestrial ecology, 1970-75, assoc. dir. environ. sci. div., 1975-86, dir. environ. sci. div., 1986-90, assoc. lab. dir. environ., life and social scis., 1990—; adj. prof. grad. program ecology U. Tenn., 1969—; state trustee, nat. bd. govs. Nature Conservancy, Washington, 1981—; bd. visitors Sch. Pub. and Environ. Affairs, Ind. U., Bloomington, 1986—; mem. ecology panels, sci. rsch. adv. com. NSF, Washington, 1970—; mem. environ. studies bd., bd. environ. toxicology and health Nat. Acad. Sci., Washington, 1974—. Editor advanced text: Springer Verlag, 1979—; editor books on forest ecology, global carbon cycles, 1970-85; contbr. articles to profl. jours. Mem. Oak Ridge Regional Planning Commn., 1961-72; mem., pres. Boys Club Oak Ridge, 1975—, Oak Ridge Rotary Club, 1981—; chmn. Tenn. area coun. Boys Club Am., 1989-91. Danforth Found. fellow, 1960; postdoctoral fellow U.S. AEC, Washington, 1964; recipient Sci. Achievement award Internat. Union Forest Rsch. Orgns., 1976. Fellow AAAS; mem. Am. Inst. Biol. Scis. (mem. coun., pub. affairs com.), Ecol. Soc. Am. (governing coun., chmn. pub. affairs, chmn. applied ecology sect.), Internat. Union Radioecologists (governing bd.), Assn. Women in Sci., Sigma Xi (pres. Oak Ridge chpt.). Achievements include pioneering in field of radionuclide cycles in forest ecosystems and systems ecology, quantified heterotrophic productivity in forest ecosystems, quantification of forest ecosystem metabolism and carbon exchanges. Home: Rte 4 Box 323 Kingston TN 37763 Office: Oak Ridge Nat Lab PO Box 2008 Oak Ridge TN 37831-6253

REICHMAN, GEORGE ALBERT, soil scientist, educator, consultant; b. Belgrade, Mont., Sept. 24, 1925; s. Lawrence Everett and Alice Lorania (Holdiman) R.; m. Virginia May Baum, Oct. 26, 1952; children: Lawrence William, Alice Marie. BS, Mont. State U., 1950, MS, 1954. Soils technician Bur. Reclamation, U.S. Dept. Interior, Miles City, Mont., 1948-49; soils asst. Mont. State U., Bozeman, 1950-54; soil scientist Agrl. Rsch. Svc. USDA, Mandan, N.D., 1955-91; adj. prof. N.D. State U., Fargo, 1979—. Contbr. articles to Health Physics, Trans. Am. Soc. Agrl. Engrs., Jour. Am. Soc. Sugarbeet Tech., Jour. Can. Agrl. Engring., Am. Soc. of Agronomy. 1st class petty officer USNR, 1943-46, PTO. Recipient Cert. of merit USDA, 1968, Cert. of Appreciation, 1991. Mem. Am. Soc. Agronomy, N.D. Acad. Sci., Orgn. Profl. Employees USDA (officer), N.D. R.R. Hist. Soc. (bd. dirs. 1987—), Rotary (pres. Mandan 1988-89, Paul Harris fellow 1989), Eagles, Elks. Presbyterian. Achievements include construction of large research lysimeters with undisturbed soil for irrigation crop studes; design and assembly of special water chamber for moisture probe calibration, feasibility of irrigation of layered till soils. Home: 306 6th Ave NW Mandan ND 58554-2512 Office: USDA No Gt Plains Rsch Lab S Hwy # 459 Mandan ND 58554

REICHMANN, PÉTER IVÁN, mathematician; b. Budapest, Hungary, Feb. 10, 1942; came to U.S., 1956; s. Rezső Rudolf and Margit (Grunberger) R. BSEE, Ill. Inst. Tech., 1967, MS in Math., 1973, PhD in Math., 1986. Elec. engr. Zenith Military and Motorola Comm. and Govt. divs., various cities, 1967-69; instr. math. and elec. engring. depts. Chgo. Tech. Coll., 1973-74; asst. prof. of math. Cath. U. Am., Washington, 1987-89. Grantee NASA, 1982. Mem. IEEE (congl. commendation 1989-92), Am. Math. Soc. Achievements include research on the introduction of a novel geometry for individual cell for negative Poisson's ratio foam and computing its volume. Home: 11800 Lisborough Rd Bowie MD 20720-3422

REICHSTEIN, TADEUS, botanist, scientist, educator; b. Wloclawek, Poland, July 20, 1897; s. Isidor and Gustava (Brokman) R.; m. Luise Henriette Quarles v. Ufford, July 21, 1927; children: Margrit, Ruth. Student, Industrieschule, Zurich, 1914-16; diploma in chem. engring., Eidg. Tech. Hochschule, Zurich, 1920, Dr. Ing.-Chem., 1922; D.Sc. (hon.), U. Sorbonne, Paris, 1947, U. Basel, 1951, U. Geneva, U. Abidjan, 1967, U. London, 1968, U. Leeds, 1970. Prof. Eidg. Techn. Hochschule, 1934-38; prof. U. Basel, Switzerland, 1938-67, prof. emeritus, 1967—; dir. Pharmacol. Inst., 1938-48, Inst. Organic Chemistry, 1946-67; research botanist on ferns, 1967—. Contbr. articles to profl. publs. Recipient Marcel-Benoist prize, 1948; corecipient Nobel prize in physiology or medicine, 1950, various other prizes. Fellow Royal Soc. London, Nat. Acad. Sci. (Washington), Royal Irish Acad., Chem. Soc. London (hon.), Swiss Med. Acad. (hon.), Indian Acad. Sci. (hon.); mem. Mus. Hist. Nat. Paris (corr.), Med. Faculty U. Basel (hon.), Linn. Soc. (London, hon.), Am. Fern Soc. (hon.), Acad. Royale Med. Belg. (hon.), Am. Rheumat. Assn. (hon.), Weizmann Inst. Rehovoth, Pharm. Soc. Japan (hon.), Deutsche Akad. Leopoldina, Am. Soc. Bd. Chem. (hon.). Home: 22 Weissensteinstrasse, CH-4059 Basel Switzerland Office: Inst fur Organische Chemie, St Johanns-Ring 19, CH-4056 Basel Switzerland

REID, JACK RICHARD, research executive; b. Youngstown, Ohio, Oct. 31, 1947; s. Carl Revere and Eva (Byrd) R.; m. Linda Newton, June 7, 1969; children: Lara Adrienne, Jonathan Scott. BS in Chemistry, Lebanon Valley Coll., 1969; MS, Lehigh U., 1972, PhD, 1973. Postdoctoral assoc. Sch. Pharmacy U. Kans., Lawrence, 1973-75; sr. rsch. chemist Lorillard Tobacco Co., Greensboro, N.C., 1975-87, mgr. organic chemistry, 1987-93, dir. rsch., 1993—. Contbr. articles to profl. publs. With U.S. Army, 1970-72. NDEA fellow, 1969. Mem. Am. Chem. Soc., Aircraft Owners and Pilots Assn., N.C. State Beekeepers Assn., Sigma Xi. Achievements include 2 patents in field. Office: Lorillard Tobacco Co 420 English St Greensboro NC 27405

REID, JANET WARNER, biologist consultant; b. Boston, Oct. 18, 1944; d. Clarence Steffens and Elizabeth Tyler (Lancaster) Warner; m. Willis Alton Reid Jr., Apr. 24, 1966; children: Blake Dietrich, Alexander Nathan. BS, Duke U., 1966; MS, N.C. State U., 1971, PhD, 1978. Prof. U. Brasilia, Brazil, 1981-82; pvt. practice biol. cons. Bethesda, Md., 1982—; sr. postdoctoral fellow Nat. Mus. Natural History, Smithsonian Instn., Washington, 1988-89, rsch. assoc. dept. invertebrate zoology, 1993—. Mem. editorial bd.: Acta Limnologica Brasiliensia; contbr. numerous articles to profl. jours. Mem. World Assn. Copepodologists (founding mem., mem. local organizing com. 1993 congress, gen. sec. 1993—), The Crustacean Soc., Biol. Soc. Washington (coun. 1990-92, pres.-elect 1992—), Am. Microscopial Soc., Internat. Assn. Meiobenthologists, N.Am. Benthological Soc., Sociedade Brasileira de Carcinologia, Sociedade Brasileira de Limnologia, Societas Internationalis Limnologiae. Home: 6210 Hollins Dr Bethesda MD 20817

REID, JOHN MITCHELL, biomedical engineer; b. Mpls., June 8, 1926; s. Robert Sherman and Meryl (Mitchell) R.; m. Virginia Montgomery, Dec. 31, 1949 (div.); children: Donald, Kathryn, Richard; m. Shadi Wang, June 30, 1983. BS, U. Minn., 1950, MS, 1957; PhD, U. Pa., 1965. Engring. assoc. U. Minn., Mpls., 1950-54; rsch. engr. St. Barnabas Hosp., Mpls., 1954-57; assoc. U. Pa., Phila., 1957-66; rsch. asst. prof. U. Wash., Seattle, 1966-72; rsch. engr. Providence Hosp., 1972-74; dir. bioengring. Inst. of Applied Physiology & Medicine, 1973-81; Calhoun prof. Drexel U., Phila., 1981—; adj. prof. radiology Thomas Jefferson Med. Sch., Phila., 1982—; cons. Inst. Applied Physiology & Medicine, Seattle. Contbr. over 100 articles to profl. jours.; 5 U.S. patents. Scoutmaster Boy Scouts Am., Mpls., 1955-57, Phila., 1960-65, cub and scoubmaster, Seattle, 1965-70. With USN, 1950-52, World War II. Grantee NIH. Fellow IEEE, Am. Inst. Ultrasound in Medicine (bd. govs. Pioneer award), Acoustical Soc. Am., Am. Inst. Med. and Biol. Engring. (Lifetime Achievement award 1993). Home: 722 Upper Gulph Rd Wayne PA 19087-2050 Office: Drexel U Inst Biomed Engring & Sci 32D Chestnut Philadelphia PA 19104

REID, JOSEPH LEE, physical oceanographer, educator; b. Franklin, Tex., Feb. 7, 1923; s. Joseph Lee and Ruby (Cranford) R.; m. Freda Mary Hunt, Apr. 7, 1953; children: Ian Joseph, Julian Richard. BA in Math., U. Tex., 1942; MS, Scripps Instn. Oceanography, 1950. Rsch. staff Scripps Instn. Oceanography, La Jolla, Calif., 1957-74; prof. oceanography Scripps Instn. Oceanography, La Jolla, 1974-91, ret., 1991; dir. Marine Life Rsch. Group, 1974-87; assoc. dir. Inst. Marine Resources, 1975-82; cons. Sandia Nat. Labs., Albuquerque, 1980-86. Author: On the Total Geostrophic Circulation of the South Pacific Ocean: Flow Patterns, Tracers and Transports, 1986, On the Total Geostrophic Circulation of the South Atlantic Ocean: Flow Patterns, Tracers and Transports, 1989; contbr. articles to profl. jours. Lt. USNR, 1942-46, ETO, PTO. Recipient award Nat. Oceanographic Data Ctr., Washington, 1984, Albatross award Am. Miscellaneous Soc., 1988, Alexander Agassiz medal NAS, 1992. Fellow AAAS, Am. Geophys. Union (pres. Ocean Scis. sect. 1972-74, 84-86); mem. Am. Meteorol. Soc., Oceanography Soc. Home: 1105 Cuchara Dr Del Mar CA 92014-2623

REID, MACGREGOR STEWART, aeronautics and microwave engineer; b. Johannesburg, South Africa, June 11, 1932; s. George Stewart and Nora Marion (Foote) R.; m. Anita Clair Menke, May 5, 1962; children: Stewart, Warwick, Matthew. PhD, U. Witwatersrand, Johannesburg, S. Africa, 1969. Sr. rsch. officer Coun. for Sci. and Indsl. Rsch., Johannesburg, 1960-69; tech. mgr. Jet Propulsion Lab., Pasadena, Calif., 1969-87; planning mgr. Jet Propulsion Lab., Pasadena, 1982-87, tech. exec., asst. to dir., 1987—; vis. researcher Jet Propulsion Lab., 1965-66; mem. adv. com. Dept. Energy's Solar Assessment Com., 1978-80, Blue Ribbon Rev. Panel Nat. Aeronautics & Space Adminstrn., Washington, 1990, tech. adv. group Am. Nat. Standards Inst., N.Y.C., 1990—. Contbr. over 90 articles to profl. jours. Fellow AIAA (chmn. standards exec. coun 1990—, chmn. standards tech. coun. 1988-90. v.p. 1991—), Internat. Astronautical Fedn. (dep. v.p. 1992—). Achievements include 1 U.S. patent for a conical scan tracking system, 1978, 5 NASA awards for the invention of new technol. Office: Jet Propulsion Lab 4800 Oak Grove Dr Pasadena CA 91109

REID, RALPH RALSTON, JR., electronics executive, engineer; b. Topeka, Nov. 19, 1934; s. Ralph Ralston Sr. and Else May (Whitebread) R.; m. Gloria Ann Cook, Feb. 3, 1957; children: Terri L., Jeffrey S. BS in Physics, Washburn U., 1956. Sr. v.p. engring. div. Loral Def. Systems Ariz., Litchfield, Ariz., 1963-89, sr. v.p., 1989—. Capt. USAF, 1957-60. Mem. Am. Electronics Assn., Am. Def. Preparedness Assn., Assn. U.S. Army. Republican. Avocations: skiing, running, hiking. Home: 2527 E Vogel St Phoenix AZ 85028 Office: Loral Def Systems Ariz PO Box 85 Litchfield Park AZ 85340

REID, ROBERT H., engineering consultant; b. Pitts., Oct. 8, 1960; s. Clarence George and Mary Kathryn (Haines) R.; m. Paula Ann Groetzinger, May 21, 1988; 1 child, Timothy. BS, U. Pitts., 1983; MS, Ill. Inst. Tech., 1985; postgrad., Carnegie Mellon U. Cert. profl. engrs. Engr. Sargert & Lundy Engrs., Chgo., 1983-85, Baker Engrs., Beaver, Pa., 1985-90; mgr. Sen. Engring., Pitts., 1990-92; owner, cons. Robert H. Reid P.E., Pitts., 1992—. Bd. dirs. Dear Valley YMCA Camp, Fort Hill, Pa., 1990--. Engring. Alumni Assn. fellow, 1983. Mem. Am. Soc. Civil Engrs., Tau Beta Pi, Chi Epsilon (editor 1982-83). Republican. Home and Office: 5207 Beeler St Pittsburgh PA 15217

REID, ROBERT LELON, college dean, mechanical engineer; b. Detroit, May 20, 1942; s. Lelon Reid and Verna Beulah (Custer) Cook; m. Judy Elaine Nestell, July 21, 1962; children: Robert James, Bonnie Kay, Matthew Lelon. ASE, Mott Community Coll., Flint, Mich., 1961; BChemE, U. Mich., 1963; MME, So. Meth. U., 1966, PhDME, 1969. Registered profl. engr., Tenn., Tex., Wis. Asst. rsch. engr. Atlantic Richfield Co., Dallas, 1964-65; assoc. staff engr. Linde Div., Union Carbide Corp., Tonawanda, N.Y., 1966-68; from asst. to assoc. prof. U. Tenn., Knoxville, 1969-75; assoc. prof. Cleve. State U., 1975-77; from assoc. to full prof. U. Tenn., Knoxville, 1977-82; prof., chmn. U. Tex., El Paso, 1982-87; dean Coll. Engring. Marquette U., Milw., 1987—; summer prof. NASA Marshall Space Ctr., Huntsville, Ala., 1970, EXXON Prodn. Rsch., Houston, 1972, 73, NASA Lewis Space Ctr., Cleve., 1986; cons. Oak Ridge Nat. Lab., 1974-75, TVA, 1978, 79, State of Calif., Sacramento, 1985, Tex. Higher Edn. Coordinating Bd., Austin, 1987. Contbr. 80 articles on heat transfer and solar energy to books, jours. and procs. Grantee NSF, DOE, TVA, NASA, DOI, 1976-87; named Engr. of Yr. Engring. Socs. El Paso, 1986. Fellow ASME (Centennial medallion, 1980, chmn. cryogenics com. 1977-81, chmn. solar energy div., 1983-84, chmn. Rio Grande sect., 1985-87); mem. ASHRAE, Engrs. and Scientists Milw. (bd. dirs. 1988—, v.p. 1989-90, pres. 1991-92). Lutheran. Avocations: travel, classic car restoration. Office: Marquette U Coll Engring 1515 W Wisconsin Ave Milwaukee WI 53233

REID, WILLIAM MICHAEL, mechanical engineer; b. Ames, Iowa, July 12, 1954; s. Richard James and Mary Lou (Moore) R. BS, Mich. State U., 1979. Registered profl. engr., Tex. Project engr. PRINTEK, St. Joseph, Mich., 1983-85; design engr. Tex. Instruments, Dallas, 1979-81, lead engr., 1981-83, project engr., 1985-88, proposal mgr., 1989, program mgr., 1989, 90—; cons., Dallas, 1987—. Mem. Texins. Republican. Roman Catholic. Avocations: numismatics, golf. Home: 617 E Ridge St Allen TX 75002-4724 Office: Tex Instruments Inc M/S 8458 PO Box 869305 Plano TX 75086

REIERSON, JAMES DUTTON, systems engineer, physicist; b. Seward, Nebr., Oct. 14, 1941; s. Vernon J. and Mary Louise (Dutton) R.; m. Pauline Lupo, June 7, 1975; children: Mary, Andrew. BS in Physics, U. Nebr., 1963; PhD in Physics, Iowa State U., 1969; MBA, Marymount U., Arlington, Va., 1984. Analyst Analytic Svcs., Inc., Alexandria, Va., 1969-73; vol. U. South Pacific, Peace Corps, Suva, Fiji, 1973-75; computer scientist Computer Sci. Corp., Alexandria, 1976-78; lead engr. Mitre Corp., McLean, Va., 1978—. Home: 3311 N George Mason Dr Arlington VA 22207

REIF, ARNOLD E., medical educator; b. Vienna, Austria, July 15, 1924; m. Jane C. Chess; children: Bertrand Paul, John Henry, Joseph Peter; (2nd marriage) Katherine E. Hume. BA, Cambridge U., Eng., 1945, MA, 1949; BSc, London U., 1946; MS, DSc, Carnegie-Mellon U., 1949, 50; M of Theol. Studies, Harvard U. Divinity Sch., 1993. Postdoctoral fellow U. Wis. Med. Sch., 1950-52, rsch. assoc., 1953; rsch. assoc. Lovelace Found. for Med. Edn. and Rsch., Albuquerque, 1953-57; asst. prof. surgery, biochemistry Tufts U. Sch. Medicine, 1957-68, asst. prof. surgery, oncology, 1968-69, assoc. prof. surgery, immunology, 1969-71, assoc. prof. surgery, 1971-75; rsch. pathologist chief, exptl. cancer immunotherapy Boston City Hosp., Mallory Inst. Pathology, 1973-89; rsch. prof. pathology Boston U. Sch. Medicine, 1975-90, prof. emeritus, 1990—; chmn. Nat. Cancer Inst. Conf. on Immunity to Cancer, 1984, mem. site vis. com. Nat. Cancer Inst., 1976, vice-chmn. Boston Cancer Rsch. Assn., 1976, chmn., 1990; numerous confs. in field.; vis. prof. U. Conn., Storrs, 1979. Editor: Immunity to Cancer, 1975, 1985, Thy-1, 1989. Numerous health related civic activities including chmn. Nat. Edn. Week, Interagy. on Smoking and Health, Greater Boston Area, 1970, Health Educators-Legislators Task Force on Sch. Health, 1974-76, vice chmn. 1976-77; mem. State Coord. Coun. on Mental Health, 1986; bd. dirs. Alliance for Mentally Ill, Mass., 1985-90; trustee Westborough State Hosp., 1989—; chmn. Morse's Pond Assn., Wellesley, 1981, 87; chmn. ski. and music coms. Appalachian Mountain Club, 1974-78, others. Mem. Am. Assn. Cancer Rsch., Am. Assn. Immunologists, N.Y. Acad. Scis., Health Physics Soc., Sigma Xi. Home: 39 College Rd Wellesley MA 02181-5703

REIFMAN, JAQUES, nuclear engineer, researcher; b. Rio de Janeiro, Nov. 28, 1957; s. Julio and Clara R. BS, Rio de Janeiro State U., 1980; BBA, Rio de Janeiro Fed. U., 1984; MS in Engring., U. Mich., 1985, PhD, 1989. Trainee engr. Portante Engenharia de Projetos, Rio de Janeiro, 1978-80; grad. student rsch. asst. U. Mich., Ann Arbor, 1986-89; postdoctoral appointee Argonne (Ill.) Nat. Lab., 1989-90, engr., 1990—. Big brother Jewish Children's Bur., Chgo., 1990—. Recipient Crada grant U.S. Dept. Energy, 1992. Mem. Am. Nuclear Soc. (Mark Mills award 1990, Alpha Nu Sigma honor 1986). Achievements include patent pending on expert system for process diagnosis through conservation equations and component classification, copyright pending on Prodiag program. Office: Argonne Nat Lab 9700 S Cass Ave RA 208 Argonne IL 60439

REIFSNIDER, KENNETH LEONARD, metallurgist, educator; b. Balt., Feb. 19, 1940; s. David Leonard and Daisy Pearl (Hess) R.; m. Loretta Lieb, June 15, 1963; children—Eric Scott, Jason Miles. BA, Western Md. Coll., 1963; BS in Engring., Johns Hopkins U., 1963, MS in Engring., 1965, PhD, 1968. Jr. instr. John Hopkins U., Balt., 1966-67; asst. prof. Va. Poly. Inst. and State U., Blacksburg, 1968-72, assoc. prof., 1972-75, prof., 1975-83, Reynolds Metals prof. engring. sci. and mechanics, 1983-90, Alexander Giacco prof., 1990—, also chmn. materials engring. sci. Ph.D. program, 1974-92, chmn. adminstrn. bd. Ctr. Composite Materials and Structures, 1984; dir. Va. Inst. for Material Systems, 1988—; engr. Lawrence Livermore Nat. Lab., 1981; cons. in materials sci. NATO, 1969, 75. Mem. troop 44 com. Boy Scouts Am., Blacksburg, Va. Recipient Va. Acad. Sci. J. Shelton Horsley award, 1978, Va. Poly. Inst. Alumni award, 1982, Disting. Rsch. award Am. Soc. Composites, 1992. Fellow ASTM (founder Jour. of Composites Tech. and Rsch., vice chmn. standing com. on publs., award of merit 1982); mem. ASME, Council on Engring. Editor, co-editor, author books, book chpts., articles for profl. publs.

REIK, RITA ANN FITZPATRICK, pathologist; b. Cleve., Mar. 9, 1951; d. Charles Robert Sr. and Rita Mae (Wilke) Fitzpatrick; m. Curtis A. Reik, Oct. 19, 1974. Nursing diploma, Luth. Med. Ctr., Cleve., 1974; BA in Chemistry, Fla. Internat. U., 1985; MD, U. Miami, 1989. Resident in pathology Jackson Meml. Hosp., Miami, Fla., 1989—; mem. faculty Dept. Pediatric Pathology U. Miami Sch. Medicine. Mem. AMA, NOW, U. Miami Med. Women (pres. 1988-89), Coll. Am. Pathologists, Am. Soc. Clin. Pathologists, Alpha Omega Alpha, Phi Kappa Phi. Avocations: painting, raising Japanese Koi, gardening. Office: U Miami Jackson Meml Hosp Dept Pathology 1600 NW 10th Ave Miami FL 33136-1090

REILLY, JEANETTE P., clinical psychologist; b. Denver, Oct. 19, 1908; d. George L. and Marie (Bloedorn) Parker; A.B., U. Colo., 1929; M.A., Columbia U., 1951, Ed.D., 1959; m. Peter C. Reilly, Sept. 15, 1932; children: Marie Reilly Heed, Sara Jean Reilly Wilhelm, Patricia Reilly Davis. Lectr. psychology Butler U., Indpls., 1957-58, 60-65; cons. child psychologist Mental Hygiene Clinic, Episcopal Community Services, Indpls., 1959-65; cons. clin. psychologist VA Hosp., Indpls., 1965-66; Christian Theol. Sem., 1968-70; pvt. practice clin. psychology, Indpls., 1967-89; cons. clin. psychologist St. Vincent's Hosp., 1973-86; adv. cons. middle mgmt. group Indpls. City Council, 1980-81. Mem. women's aux. council U. Notre Dame, 1953-65; trustee Hanover (Ind.) Coll., 1975-91; bd. dirs. Community Hosp. Found., Indpls., 1978-92, Regional Cancer Hosp. Bd., 1988-90, Indpls. Mus. Art, 1987-93; mem. Ind. Bd. Examiners in Psychology, 1969-73; mem. Com. for Future of Butler U., 1985-86. Mem. Am. Psychol. Assn., Am. Personnel and Guidance Assn., Am. Vocat. Assn., Ind. Psychol. Assn., Central Ind. Psychol. Assn., Ind. Personnel and Guidance Assn., Nat. Registry Psychologists in U.S.A. Office: 3777 Bay Rd North Dr Indianapolis IN 46240-2973

REILLY, JUDITH GLADDING, physics educator; b. Middletown, Conn., Nov. 28, 1935; d. George Roger and Emma Diantha (Rutty) Gladding; m. James Alfred Reilly, Sept. 15, 1958; 1 child, Thomas Edward. AB, Clark U., 1958, MA, 1960. Lectr. in physics Clark U., Worcester, Mass., 1963-75; tchr. of physics Assumption Preparatory Sch., Worcester, 1963; prof. of phys. sci., physics Quinsigamond C. C., Worcester, 1963—; cons. in field; adj. instr. in astronomy Worcester Pub. Schs. Magnetic Sch., 1980s. Author: (textbook) Physical Science, 1970, (workbook) 1971; contbr. articles to profl. jours. Mem. Assn. Am. Physics Tchrs. Achievements include ultrasonic artifacts produced by digital audio and the damage produced thereby. Home: 40 Loring St Auburn MA 01501 Office: Quinsigamond C C 670 West Boylston St Worcester MA 01606

REILLY, MARGARET ANNE, pharmacologist, educator; b. Port Chester, N.Y., Mar. 21, 1937; d. Thomas and Margaret (Byrnes) R. BA, Coll. New Rochelle, 1959; MS, NYU, 1978, PhD, 1981. Sr. lab. technician Sloan-Kettering Inst., Rye, N.Y., 1959-64; teaching asst. CUNY, 1965; adj. assoc. prof. Coll. New Rochelle, N.Y., 1978—; adj. asst. prof. Concordia Coll., Bronxville, N.Y., 1988—; rsch. scientist Nathan Kline Rsch. Inst., Orangeburg, N.Y., 1966—. Author book chpts., 1989, 92; book review editor Assn. for Women in Sci., 1985—; contbr. articles to profl. jours. Treas., bd. dirs. Port Chester/Rye Vol. Ambulance, 1981—; pres. Coll. New Rochelle Alumnae Assn., 1987-91. Recipient Angela Merici award Coll. New Rochelle, 1989. Mem. Histamine Rsch. Soc. N.Am. (sec./treas. 1980—), Am. Soc. Pharmacology and Exptl. Therapeutics, N.Y. Acad. Scis. Office: Nathan Kline Rsch Inst Orangeburg NY 10962

REILLY, MICHAEL THOMAS, biochemical engineer; b. Pitts., June 3, 1955; s. Thomas Paul and Doris Jane (Stoehr) R.; m. Laura Ann Bruening, May 13, 1989. BS in Biology, Muskingum Coll., 1977; MSChemE, Carnegie-Mellon U., 1980; PhD in Chem. Engring., Lehigh U., 1987. Sr. engr. Westvaco Corp., Laurel, Md., 1980-82; rsch. engr. Pfizer, Inc., Easton, Pa., 1982-83, DuPont Co., Wilmington, Del., 1986-92; rsch. scientist Upjohn Co., Kalamazoo, Mich., 1992—; cons. Assoc. Bioengrs. and Cons., Bethlehem, Pa., 1984-86. Author: Bioprocess Analysis, Monitoring and Control, 1992; contbr. article to profl. publ., chpt. to book. Fellow Carnegie-Mellon U., 1978, Benedum Found., 1979, Air Products and Chems., 1983. Mem. AAAS, Am. Inst. Chem. Engrs., Am. Chem. Soc., Inst. Food Technologists. Democrat. Roman Catholic. Home: 9181 Cotter's Ridge Rd Richland MI 49083 Office: Upjohn Co 1400-89-1 Bioprocess R & D Kalamazoo MI 49001

REILLY, PETER C., chemical company executive; b. Indpls., Jan. 19, 1907; s. Peter C. and Ineva (Gash) R.; AB, U. Colo., 1929; MBA, Harvard U., 1931; DSc (hon.), Butler U; m. Jeanette Parker, Sept. 15, 1932; children: Marie (Mrs. Jack H. Heed), Sara Jean (Mrs. Clarke Wilhelm), Patricia Ann (Mrs. Michael Davis). With accounting dept. Republic Creosoting Co., Indpls., 1931-32; sales dept. Reilly Tar & Chem. Corp. (became Reilly Industries, Inc. 1989), N.Y.C., 1932-36, v.p., Eastern mgr., 1936-52; v.p. sales, treas. both cos., Indpls., 1952-59, pres., 1959-73, chmn. bd., 1973-75, vice chmn., 1975-82, chmn., 1982-90, chmn. exec. com. 1990—; dir. Environ. Quality Control Inc.; past dir. Ind. Nat. Corp., Ind. Union Ry., Ind. Nat. Bank. Dir. Goodwill Industries Found.; bd. dirs. United Fund Greater Indpls., Indpls. Symphony Orch.; bd. govs. Jr. Achievement Indpls.; mem. adv. council U. Notre Dame Sch. Bus. Adminstrn., 1947—; mem. adv. council Winona Meml. Hosp.; life mem. Boy Scouts Am., Crossroads of Am. Coun. Recipient Sagamore of Wabash award; named Disting. Eagle Scout Boy Scouts Am. Mem. Chem. Spltys. Mfg. Assn. (life; treas. 1950-60, past dir.), Chem. Mfrs. Assn. (past dir.), Am. Chem. Soc., Soc. Chem. Industry (past dir. Am. sect.). Clubs: Union League, Harvard, Chemist (N.Y.C.); Larchmont (N.Y.) Yacht; Indianapolis Athletic, Pine Valley Golf (N.J.), Meridian Hills Country, Columbia (Indpls.); Rotary (Paul Harris award), One Hundred (past dir.); Crooked Stick Golf. Home: 3777 Bay Rd North Dr Indianapolis IN 46240-2973 Office: Reilly Industries Inc Market Sq Ctr 151 N Delaware St Ste 1510 Indianapolis IN 46204-2592

REIMSCHUESSEL, HERBERT KURT, chemist, consultant; b. Eherenberg, Germany, Sept. 7, 1921; came to U.S., 1957; s. Kurt E. and Martha Lindner R.; m. Annemarie C. Colditz, Mar. 29, 1947; children: Regina, Renate. PhD, Halle-Wittenberg U., Halle, Germany, 1952. Mgr. polyamide lab. Filmfabrik AGFA, Wolfen, Germany, 1949-54; head chemistry dept. H.J. Zimmer, Frankfurt, Germany, 1954-56; cons. Frederick Uhde, Offenbach, Germany, 1956-57; scientist Wright Air Devel. Ctr. Wright Patterson AFB, Dayton, Ohio, 1957-59; sr. rsch. assoc. Allied Signal, Inc., Morristown, N.J., 1959-88; cons. pvt. practice, Morristown, 1988—; cons. Allied-Signal, Morristown-Chgo., 1988—. Contbr. more than 70 articles to profl. jours., chpts. to books. Fellow N.Y. Acad. Scis.; mem. Am. Chem. Soc., Ges Deutscher Chemiker, Fibers Soc. Achievements include 31 patents in polymer science, synthetic fibers and organic chemistry related to novel lactam derivatives and their isomerization polymerization resulting in novel polyimides. Home: 20 Junard Dr Morristown NJ 07960

REINEKER, PETER, physics educator; b. Freudenstadt, Germany, Jan. 17, 1940; s. Paul and Hilde (Weiss) R.; m. Hilda Jacobi; children: Katja, Martina. Diploma in physics, U. Stuttgart, Fed. Republic Germany, 1966, PhD, 1971; Habilitation, U. Ulm, Fed. Republic Germany, 1974. Asst. U. Stuttgart, 1966-72; asst. U. Ulm, Fed. Republic Germany, 1972-75, sci. advisor, 1975-78, prof. physics, 1978—; chmn. conf. dept. physics German. Univs., 1993—. Author: Exciton Dynamics, 1981, Molecular Aggregates, 1983; also over 130 articles. Mem. Am. Phys. Soc., German Phys. Soc. (chmn. div. molecular physics 1990-92, div. chem. physics 1992-93, coun. 1992—). Office: U Ulm, Albert-Einstein-Allee 11, D-89081 Ulm Germany

REINERS, KARLHEINZ, neurologist, educator; b. Wegberg, Northrhine-Westfalia, Federal Republic of Germany, Nov. 16, 1950; s. Wilhelm and Gertrud (Bertrams) R. MD, U. Duesseldorf, Fed. Republic Germany, 1975. From registrar to oberarzt dept. neurology Heinrich-Heine U., Duesseldorf, Federal Republic of Germany, 1977-92, privat lectr., 1989-92; rsch. fellow Inst. Neurology, London, 1983-84; prof. neurology U. Wuerzburg, Germany, 1992—. Author (book) Neuropathie und Motorik, 1990; co-author (books) Charcot-Marie-Tooth Disorders, 1990, Therapie-Handbuch, 4th edit., 1992, Innere Medizin der Gegenwart, vols. 6, 7, 1990. Served with Federal Republic of Germany Air Force, 1975-76. Recipient Myopathie prize German Muscular Dystrophy Orgn., 1990; Rsch. grantee Ministerium for Wissenschaft and Forschung NRW, 1987-90. Fellow Royal Soc. Medicine; mem. Deutsche Assn. for Neurology, Deutsche EEG Assn., Deutsche Assn. for Neurol. Rehab., German Soc. for Study of Pain, European Neurosci. Assn., European Neurol. Soc. Roman Catholic. Avocations: painting, economy, promotion of medical education. Office: U Wuerzburg Dept Neurology, Josef-Schneider-Str 11, 97080 Wuerzburg Germany

REINERT, ERIK STEENFELDT, economist, researcher, administrator; b. Oslo, Feb. 15, 1949; s. Erik S. and Thora (Schultz) R.; m. Fernanda Aars, Oct. 18, 1974; children: Hugo, Sophus. BA, Hochschule, St. Gallen, Switzerland, 1973; MBA, Harvard U., 1976; PhD, Cornell U., 1980. Bd. sec. Pacific Internat. Trade Fair, Lima, Peru, 1969; researcher Latin Am. Inst., St. Gallen, Switzerland, 1971-72; expert Unctad/Gatt, S. Am., 1972-74; lectr. Hobart & William Smith Colls., Geneva, N.Y., 1977; rsch. assoc. Norwegian Rsch. Coun. for Social Scis. and the Humanities, Peru, U.S.A., 1978-79; cons. Telesis Inc., Paris, Dublin, 1980-81; pres. and chief exec. officer Matherson-Selig Group, Italy, 1981-91; rsch. scientist Norwegian Computing Ctr., Oslo, 1992—; pres. Matherson-Selig S.P.A., Presezzo, Bergamo, Italy, Matherson-Selig OY AB, Pargas, Finland. Author: International Trade and the Economic Mechanisms of Underdevelopment; contbr. articles to profl. jours. Mem. Am. Econ. Assn., Latin Am. Studies Assn. Home: N-3148, Hvasser Norway

REINERT, JAMES A., entomologist, educator; b. Enid, Okla., Jan. 26, 1944; s. Andrew J. and Emma Reinert; m. Anita Irwin; children: Travis J., Gina N., Mindy K., Melanie B., Gregory W., Teresa J. BS, Okla. State U., 1966; MS, Clemson U., 1968, PhD, 1970. Asst. state entomologist U. Md., College Park, 1970; asst. prof. entomology to prof. entomology Ft. Lauderdale Rsch. and Edn. Ctr., U. Fla., 1970-84; resident dir. and prof. entomology Rsch. and Ext. Ctr., Tex. A&M Univ. System, Dallas, 1984—. Contbr. over 265 articles to profl. jours. NDEA fellow, 1968; recipient Porter Henegar Meml. award., So. Nurserymen's Assn., 1982. Mem. Inter-Turfgrass Soc., Entomol. Soc. Am., Fla. Entomol. Soc. (pres. 1984, Entomologist of Yr. 1985), Fla. State Hort. Soc. (v.p. 1982), S.C. Entomol. Soc., Rsch. Ctr. Adminstrs. Soc. (state rep. 1991-92, sec. 1993), Dallas Agriculture Club (bd. dirs. 1989, v.p. 1990, pres. 1991). Roman Catholic. Home: 3805 Covinton Ln Plano TX 75023-7731 Office: Tex A&M Univ Rsch and Ext Ctr 17360 Coit Rd Dallas TX 75252-6599

REINES, FREDERICK, physicist, educator; b. Paterson, N.J., Mar. 16, 1918; s. Israel and Gussie (Cohen) R.; m. Sylvia Samuels, Aug. 30, 1940; children: Robert G., Alisa K. M.E., Stevens Inst. Tech., 1939, M.S., 1941; Ph.D., NYU, 1944; D.Sc. (hon.), U. Witwatersrand, 1966; D. Engring. (hon.), Stevens Inst. Tech., 1984. Mem. staff Los Alamos Sci. Lab., 1944-59; group leader Los Alamos Sci. Lab. (Theoretical div.), 1945-59; dir. (AEC expts. on Eniwetok Atoll), 1951; prof. physics, head dept. Case Inst. Tech., 1959-66; prof. physics U. Calif., Irvine, 1966-88, dean phys. scis., 1966-74, Disting. prof. physics, 1987-88, prof. emeritus, 1988—; Centennial lectr. U. Md., 1956; Disting. Faculty lectr. U. Calif., Irvine, 1979; L.I. Schiff Meml. lectr. Stanford U., 1988; Albert Einstein Meml. lectr. Israel Acad. Scis. and Humanities, Jerusalem, 1988; Goudschmidt Meml. lectr., 1990; co-discoverer elementary nuclear particles, free antineutrino, 1956. Contbr. numerous articles to profl. jours.; contbg. author: Effects of Atomic Weapons, 1950. Mem. Cleve. Symphony Chorus, 1959-62. Recipient J. Robert Oppenheimer Meml. prize, 1981, Nat. Medal Sci., 1983, medal U. Calif., Irvine, 1987, Michelson Morley award, 1990; co-recipient Rossi prize Am. Astron. Soc., 1990; Guggenheim fellow, 1958-59, Sloan fellow, 1959-63, Franklin medal Franklin Inst., 1992. Fellow Am. Phys. Soc. (W.K.H. Panofsky prize 1992), AAAS; mem. NAS, Am. Assn. Physics Tchrs., Argonne U. Assn. (trustee 1965-66), Am. Acad. Arts and Scis., Phi Beta Kappa, Sigma Xi, Tau Beta Pi. Office: U Calif Dept Physics Campus Dr Irvine CA 92717

REINHARDT, CHARLES FRANCIS, toxicologist; b. Spring Grove, Ind., Nov. 25, 1933; s. Charles H. and Frances N. R.; m. Linda Helen Ieler, Sept. 1, 1956; children: Amy Linn, Jeff Charles, Meg Susan, Jane Ellen. AB, Wabash Coll., 1955; MD, Ind. U., 1959; MSc, Ohio State U., 1965. Diplomate: Am. Bd. Preventive Medicine, Am. Bd. Toxicology (pres. 1983-84). Resident in occupational medicine Ohio State U.; plant physician Chambers Works, E.I. du Pont de Nemours & Co., 1964-66; physiologist Haskell Lab., Newark, Del., 1966-69; chief physiology Haskell Lab., 1969-70; research mgr. Environ. Scis. Group, 1970-71, asst. dir., 1971-74, asso. dir., 1974-76, dir., 1976—; mem. sci. adv. panel Chem. Industry Inst. Toxicology, 1980-83; mem. sci. adv. bd. EPA, 1985-89. Contbr. in field. Served to capt., M.C. USAF, 1960-62. Mem. AMA, Am. Coll. Occupational Medicine (past pres.), Soc. Toxicology, Am. Indsl. Hygiene Assn. Office: Haskell Laboratory El du Pont De Nemours and Co Elkton Rd Newark DE 19714

REINHARDT, KURT, retired nuclear physician; b. Limbach, Saar, Feb. 18, 1920; s. Friedrich and Elisabeth (Hock) R.; m. Maria Lefeber, Dec. 29, 1951. Student U. Berlin, 1939-40, U. Heidelberg, 1940-45; med. diploma, U. Innsbruck, 1945. Resident dept. radiology U. Homburg, 1951-58; head physician, dept. radiol. nuclear medicine Kreiskrankenhaus Volklingen, from 1958, habilitation, 1958, prof., from 1964, now ret. Decorated Cross of Merit 1st class Fed. Republic of Germany. Mem. Deutsche Roentgengesellschaft, Internat. Skeletal Soc. Author 10 monographs and books including Krankhafte Haltungs-änderungen Skoliosen und Kyphosen, Gedanken über Erinnerungen an dem RuBlandkrieg, 1991; contbr. 200 articles to med. jours. Home: 32 am Kirschenwäldchen, Volklingen Germany

REINHARDT, UWE ERNST, economist, educator; b. Osnabrueck, Germany, Sept. 24, 1937; came to U.S., 1964; s. Wilhelm and Edeltraut (Kehne) R.; m. Tsung-mei Cheng, May 25, 1968; children—Dirk, Kara, Mark. B.Comm. in Econs. with honors, U. Sask., Saskatoon, Can., 1964; M.A. in Econs., Yale U., 1965, M.Ph. in Econs., 1967, Ph.D., 1970; D.Sc. (hon.), Med. Coll. of Pa., 1987. Apprentice in export forwarding Schenker and Co., Cologne and Hamburg, Fed. Republic Germany, 1955-57; manifest clk. Saguenay Shipping Ltd. div. Aluminum Co. Can., Montreal, Que., 1957-60; asst. prof. econs. and pub. affairs Princeton (N.J.) U., 1968-74, assoc. prof., 1974-79, prof., 1979—, James Madison prof. polit. economy, 1984—; cons. Urban Inst., Washington, 1971-75, HEW, 1974—, HHS, Math., Inc., Princeton, 1970-80, AT&T, Basking Ridge, N.J., 1976-82, Nat. Westminster Bank USA, N.Y.C., 1979—, Phillips Petroleum Co., Bartlesville, Ohio, 1978—; trustee Tchrs. Ins. and Annuity Assn., 1978—; mem. Nat. Leadership Commn. Health Care, 1986—; mem. U.S. Physicians' Payment Rev. Commn., U.S. Congress, 1986—; pres. Assn. for Health Svcs. Rsch., 1989-90, Found. for Health Svcs. Rsch., 1990-91, bd. dirs. 1991—. Author: Physician Productivity and the Demand for Health Manpower, 1975; mem. editorial bd. Health Affairs, 1982—, New Eng. Jour. Medicine, 1989-92, Health Mgmt. Quar., Health Policy and Edn., Jour. AMA, 1991—; assoc. editor Jour. Health Econs., 1980—, mem. editorial bd., 1981-83; contbr. articles to profl. jours. Mem. Am. Econs. Assn., Am. Pub. Health Assn., Am. Fin. Assn., Inst. Medicine Nat. Acad. Scis. (gov. council 1979-82). Office: Princeton U Princeton NJ 08544

REINHART, MARY ANN, medical association administrator; b. Jackson, Mich., Aug. 14, 1942; d. Herbert Martin and Josephine Marie (Keyes) Conway; m. David Lee Reinhart, Dec. 28, 1963; children: Stephen Paul, Michael David. MA, Mich. State U., 1983, PhD, 1985. Rsch. asst. Mich. State U., East Lansing, 1979-82, 85, teaching asst. dept psychology, 1982-84, asst. prof. Office Med. Edn. R&D, Coll. Human Medicine, 1985-88; assoc. exec. dir. Am. Bd. Emergency Medicine, East Lansing, 1988—; cons. Am. Bd. Emergency Medicine, 1985-88; chairperson collegewide evaluation com. Coll. Human Medicine, Mich. State U., East Lansing, 1985-88; adj. prof. Office Med. Edn. Rsch. and Devel., Coll. Human Medicine, 1988—. Reviewer Annals of Emergency Medicine, 1987—. Bd. dirs. Neahtawanta Rsch. and Edn. Ctr., Traverse City, Mich., 1991—. Mem. APA (div. health psychology), AAAS, Phi Kappa Phi. Achievements include application of chart stimulated recall method of assessment in a nat. med. recert. exam.; devel. of nat. longitudinal study of emergency physicians, longitudinal study of factors predicting medical student success. Office: Am Bd Emergency Medicine 3000 Coolidge Rd East Lansing MI 48823

REINHART, ROBERT KARL, control engineer, consultant; b. Homestead, Pa., Dec. 1, 1962; s. Reinhart K. and Susan Irene (Laird) R.; m. Im Sook Chong, Sept. 14, 1991. BS in Chem. Engr., Univ. Pitts., 1985. Proposal engr. Bailey Controls, Wickliffe, Ohio, 1985-86, project engr., 1986-87, application engr., 1987-88; control system engr. Trimark Engrs., Port, Pa., 1988-90; pres., sr. process control engr. Controls Link, Carnegie, Pa., 1990—. Contbr. articles to monthly newsletter. Mem. AICE, Instruments Soc. Am. (program com. chmn. 1988-90, sec. 1990-91, treas. 1991-92, pres.-elect 1992—). Roman Catholic. Office: Controls Link Inc 150 East Mall Plaza Carnegie PA 15106

REINISCH, BODO WALTER, electrical engineering educator; b. Beuthen, Germany, Nov. 26, 1936; came to U.S., 1965; s. Kurt and Alice Ada (Walleiser) R.; m. Gerda Seidenschwand, June 1, 1963; children: Karin, Ulrike. MS, U. Freiburg, Germany, 1963; PhD, Lowell Tech. Inst., 1970. Rsch. asst. Ionosphären Inst., Breisach, Germany, 1961-63, physicist, 1963-65; physicist Lowell (Mass.) Tech. Inst., 1965-75; dir. Ctr. for Atmospheric Rsch. U. Mass., Lowell, 1975—, assoc. prof. elec. engring., 1980-83, prof., 1983—, dept. head, 1988—; cons. Royal Meteorol. Inst., Brussels, 1970-71; guest prof. physics U. Linz, Austria, 1978-79. Contbr. over 100 articles to Radio Sci., Jour. Geophys. Rsch., Advances of Space Rsch., others. Mem. sch. com. German Saturday Sch., Boston, 1979—. Grantee USAF, NASA, NSF, others; awarded Bundesverdienstkreuz by Pres. of Germany, 1987; Univ. Prof. U. Mass., Lowell, 1987-90; recipient Outstanding Achievement award U. Mass., Lowell, 1986, 87, 88. Sr. mem. IEEE; mem. Am. Geophys. Union, Internat. Union Radio Sci. (chmn. working group G3 on ionospheric informatics, exec. com. commn. G, mem. COSPAR/URSI Task Force on Internat. Reference Ionosphere), Sigma Pi Sigma. Achievements include development of global network of "digisonde" sounders; established HF Doppler observations for ionospheric drift studies; 2 patents. Office: U Lowell Ctr Atmosphere Rsch 450 Aiken St Lowell MA 01854-3692

REINISCH, LOU, medical physics researcher, educator; b. St. Louis, May 21, 1954; s. Louis Anthony and Kathleen Marie (Stief) R.; m. Sue Ann Bartchy, May 22, 1982; children: Peter Benjamin, Steven James. BS in Physics, U. Mo., Rolla, 1976; MS in Physics, U. Ill., 1978, PhD in Physics, 1982. Rsch. asst. physics U. Ill., Urbana, 1977-81, 82; rsch. assoc. Biophysical Inst. Biol. Rsch. Ctr., Szeged, Hungary, 1982; rsch. fellow Alexander von Humbolt Found., Muenster, Germany, 1982-84; asst. prof. physics Northeastern U., Boston, 1984-88; asst. prof. radiology Uniformed Svc. U., Bethesda, Md., 1988-91; asst. prof. otolaryngology Vanderbilt U., Nashville, 1991—; rsch. cons. Harvard Med. Sch. Wellman Labs of Photomedicine, Boston, 1988; future directions coun. Office of Naval Rsch. Molecular Biology, 1988; info. exch. and computer networking Nat. Gov.'s Assoc., 1988; faculty Senate rsch. com., 1989-90. Author: (with others) Hemoglobin and Oxygen Binding, 1982, Methods of Enzymology, Enzyme Structure, 1986, Experiments in Physics, 4th ed., 1987, Otolaryngology--Head and Neck Surgery, 2d edit., 1992; author: Er:YAG Laser as a Surgical Tool in Ostectomy, 1992, Laser Surgery on Otolaryngology, 1992; contbr. numerous articles to profl. jours.; presenter in field. Grantee Rsch. Corp., 1985, Rsch. and Scholarship Devel. Fund Northeastern U., 1985, Biomed. Grant Support Program Northeastern U., 1987, Office Naval Rsch., 1988-91, U.S. Army, 1990-91, Nat. Navy Med. Ctr., 1991. Fellow Am. Soc. Laser Medicine and Surgery; mem. Am. Physical Soc., Am. Soc. of Photobiology, Biophysical Soc. Achievements include pending patents for Surgical Laser System, 1991, Non-invasive Retina Diagnostics System, 1991. Home: 118 Sheffield Place Franklin TN 37064 Office: Vanderbilt U Otolaryngology S-2100 Medical Center N Nashville TN 37232

REINSCHKE, KURT JOHANNES, mathematician, educator; b. Zwickau, Saxony, Germany, Sept. 22, 1940; s. Eugen C.A. and Liselotte (Bahlig) R.; m. Gertrud G. Ullrich, July 27, 1967; children: Annegret, Johannes. Diploma math., Tech. U., Dresden, Germany, 1963, D of Engr., 1966; D for Natural Scis., Berg Acad., Freiberg, Germany, 1969; D Habilitation, Tech. U., Dresden, 1971. Rsch. asst. Tech. U., Dresden, 1963-65; researcher, dept. head indsl. enterprise RFT MeBelektronik, Dresden, 1965-78; rsch. fellow Acad. Wissenschaften der DDR, Berlin, 1978-87; prof. Engr. U., Cottbus, Germany, 1987-92, Tech. Univ., Dresden, Germany, 1992—. Author: Reliability of Systems, 1973, 74, Sensitivity and Reliability Modelling, 1979, Multivariable Control, 1988; co-author: C.A. Analysis of Linear Networks, 1976; translator, editor, and/or author of other books; contbr. articles to profl. jours. Bd. mem. Saxony Univ. Commn., 1991. Mem. IEEE, League Freedom of Sci., Internat. Linear Algebra Soc., Am. Math. Soc., Assn. Angewandte Math. and Mechanics. Avocations: music, reading, gardening. Home: Wachwitzer Bergstrasse 32, D-01326 Dresden Germany

REINSCHMIEDT, ANNE TIERNEY, nurse, health care executive; b. Washington, Mar. 6, 1932; d. Edward F. and Frances (Palmer) Tierney; m. Edwin Ruben Reinschmiedt, Sept. 20, 1959 (div. 1961); 1 child, Kathleen Frances Tierney. BS, Cen. State U., Edmond, Okla., 1975; JD, Oklahoma City U. Sch. Law, 1991. RN, Calif., Okla.; lic. residential care facility adminstr. Nurse San Jose (Calif.) Hosp., 1952-55; owner, operator Hominy Studio, 1960-62; dir. nurses, lab and x-ray, technician, adminstr. Hominy (Okla.) City Hosp., 1961-63; nurse Jackson County Dept. Health, Altus, Okla., 1963-65; adminstr. Propp's Inc., Oklahoma City, 1965-80; nursing homes cons. Propps & Self, Oklahoma City, 1965—; pres. Shamrock Health Care Ctr., Bethany, Okla., 1981—; operator Lakeview Lodging Residential Care Facility, 1981—; adult edn. instr., med. aide technicians East Central U., Ada, Okla., 1987-89; cons. residential care facilities, 1985—. Author: Recovery Room Procedures, 1958. Mem. Jackson County (Okla.) Draft Bd., 1965-70. Lt. USN, 1955-60. Mem. Am. Nurses Assn., Nat. Assn. Residential Care Facilities (sec., bd. dirs. 1983-85), Okla. Bar Assn., Okla. Assn. Residential Care Facilities (founding pres. 1981-87, bd. dirs. 1981—), Beta Sigma Phi, Phi Alpha Delta (vice justice, exec. bd. 1988-90), Iota Tau Tau. Republican. Roman Catholic. Avocations: reading, horses, astrology, ceramics, golf. Office: Shamrock Health Care PO Box 848 Bethany OK 73008-0848

REINSMA, HAROLD LAWRENCE, design consultant, engineer; b. Slayton, Minn., Sept. 6, 1928; s. Frank and Ida M. (Zabel) R.; m. Julia A. Tusek, Oct. 18, 1958; children: Frank, Michael, Diane. Student, Macalester Coll., 1948-50; BCE, U. Minn., 1953. Registered profl. engr., Ill. Cons. engr. GM Orr Engring. Co., Mpls., 1953-54; research test engr. Caterpillar Tractor Co., Peoria, Ill., 1955-58, research design engr., 1958-71, research project engr., 1971-73, research supervising engr., 1973-76, research staff engr., 1976-91; design cons. Dunlap, Ill., 1991—. Holder 42 patents including first viable sealed and lubricated track, sealed maintenance-free linkage for use in abrasive environments. With USMC, 1946-48. Avocations: skiing, cycling, hiking, gardening. Home and Office: 13600 Luscern Ct Dunlap IL 61525

REIS, DONALD JEFFERY, neurologist, neurobiologist, educator; b. N.Y.C., Sept. 9, 1931; s. Samuel H. and Alice (Kiesler) R.; m. Cornelia Langer Noland, Apr. 13, 1985. A.B., Cornell U., 1953, M.D., 1956. Intern N.Y. Hosp., N.Y.C., 1956; resident in neurology Boston City Hosp.-Harvard Med. Sch., 1957-59; Fulbright fellow, United Cerebral Palsy Found. fellow London and Stockholm, 1959-60; research asso. NIMH, Bethesda, Md., 1960-62; spl. fellow NIH, Nobel Neurophysiology Inst., Stockholm, 1962-63; asst. prof. neurology Cornell U. Med. Sch., N.Y.C., 1963-67; asso. prof. neurology and psychiatry Cornell U. Med. Sch., 1967-71, prof., 1971—, First George C. Cotzias Disting. prof. neurology, 1982—; mem. U.S.-Soviet Exch. Program; mem. adv. coun. NIH; bd. sci. advisers Merck, Sharpe & Dohm, Sterling Rsch. Group; cons. Eli Lilly, Servier Pharms.; bd. dirs. China Seas, Inc., Charles masterson Burke Rsch. Found. Contbr. articles to profl. jours.; mem. editorial bd. various profl. jours. Recipient CIBA Prize award Am. Heart Assn. Fellow AAAS, ACP; mem. Am. Physiol. Soc., Am. Neurol. Assn., Am. Pharmacol. Soc., Am. Assn. Physicians, Telluride Assn., Am. Soc. Clin. Investigation, Century Assn., Ellis Island Yacht Club (commadore), Phi Beta Kappa, Sigma Xi, Alpha Omega Alpha. Home: 190 E 72d St New York NY 10021 also: 73 Water St Stonington CT 06378 Office: 1300 York Ave New York NY 10021-4896

REIS, JOÃO CARLOS RIBEIRO, chemistry educator; b. Lisbon, Portugal, Mar. 18, 1945; s. Virgilio Ribeiro and Cecilia Macedo (Marques) R.; m. Leonor Corte-Real Silva, Aug. 18, 1970; children: Pedro, André. BSc, U. Lisbon, 1967; PhD, U. London, 1972; diploma, Imperial Coll., London, 1972; DSc, U. Lisbon, 1992. Aux. prof. faculty of scis. U. Lisbon, 1973-79, assoc. prof. faculty of scis., 1979—; rsch. group leader Ctr. for Electrochemistry and Kinetics, Lisbon, 1976—. Contbr. articles to profl. jours. Recipient Artur Malheiros prize Acad. of Scis. Lisbon, 1981. Mem. Soc. Portuguesa de Quimica, Royal Soc. Chemistry (assoc.). Roman Catholic. Avocations: reading, sleeping. Office: Faculty Scis Chemistry Dept, R Ernesto Vasconcelos C1, P 1700 Lisbon Portugal

REIS, VICTOR H., mechanical engineer, government official. PhD in Mech. Engring., Princeton U. Formerly sr. v.p. for strategic planning Sci. Applications Internat. Corp.; security adviser Office Sci. and Tech. Policy, Washington; spl. asst. to dir. Lincoln Labs MIT; dep. dir. Def. Advanced Rsch. Projects Agy., Arlington, Va., 1989-90, dir., 1990-93; asst. sec. Dept. Energy Def. Programs, Washington, 1993—. Office: Dept Energy Def Programs 1000 Independence Ave SW Washington DC 20585*

REISCHMAN, MICHAEL MACK, university official; b. Barnesville, Ohio, Sept. 26, 1942; s. Otto John and Mary Elizabeth (McWilliams) R.; m. Sue Ellen Medley, Sept. 7, 1963 (div. 1981); children: Stacy Dee, Todd Mack; m. Deborah Jean Freimiller, June 23, 1984; 1 child, Kristi Jean. Assoc. in Electronic Engring., Ohio Tech. Coll., Columbus, 1962; BS in Mech. Engring., N.Mex. State U., Las Cruces, 1967; MS in Mech. Engring., N.Mex. State U., 1969; PhD in Mech. Engring., Okla. State U., Stillwater, 1973. Electronic technician Los Alamos (N. Mex.) Scientific Lab., 1962-64; electronic technician (part time) Physical Scis. Lab., Las Cruces, 1964-67; rsch. assoc. Okla. State U., 1969-73; post doctoral, rsch. assoc. Nat. Rsch. Council, Washington, 1973-74; rsch. engr. Naval Ocean Systems Ctr., San Diego, 1974-83; program and div. dir. Office of Naval Rsch., Arlington, Va., 1983-90; assoc. dean grad. studies and rsch. Pa. State U., University Park, 1990—; mem. curriculum adv. com. mech. engring. dept. N.Mex. State U., 1989—; ABET evaluator, 1990—; adv. bd. Applied Rsch. Lab., Material Rsch. Lab., Pa. Transplantation Inst., Environ. Resources Rsch. Inst. Author articles in profl. jours; editor Conf. Proceedings. Trustees Pa. Dance Theatre; bd. dirs. Indsl. Modernization Ctr., 1990—. Named Outstanding Alumnus, Coll. of Engring., N. Mex. State U., 1988. Mem. AAAS, ASME (bd. engring. edn. 1989—, com. on awards and prizes 1993—, chmn. task force on NSF budget 1993—), Am. Phys. Soc., Am. Soc. for Engring. Edn. Democrat. Methodist. Office: Pa State U Coll Engring 101 Hammond Bldg University Park PA 16802

REISNER, ANDREW DOUGLAS, psychologist, chief clinical officer; b. Ithaca, N.Y., Dec. 28, 1955; s. Gerald Seymour and Estelle Ruth (Siegel) R.; m. Deborah Kay Dermen, Aug. 1, 1981; children: David Aaron, Alyssa Danielle. BA, Allegheny Coll., 1977; MA, Edinboro (Pa.) State U., 1978; D of Psychology, Baylor U., 1987. Lic. psychologist, Ohio. Psychology asst. Tiffin (Ohio) Devel. Mental Health Ctr., 1979-80; psychology asst. Community Counseling Svcs., Galion, Ohio, 1980-83, chief clin. officer, 1990-93; psychologist, 1990—; intern in clin. psychology Mich. State U., East Lansing, 1986-87; postdoctoral fellow in clin. psychology Harding Hosp., Worthington, Ohio, 1987-88; psychologist Ctr. for Individual Family Svcs., Mansfield, Ohio, 1988-90; chief clin. officer Alcohol, Drug Addictions and Mental Health Svcs. Bd., Crawford County, Ohio, 1993—; psychology cons. Crestline (Ohio) Meml. Hosp., 1989—. Contbr. articles to Phobia Practice and Rsch. Jour., Pastoral Psychology, Psychoanalytic Psychology. Mem. APA, N.Y. Acad. Scis., Ohio Psychol. Assn. Office: Community Counseling Svcs PO Box 954 Galion OH 44833-0954

REISNER, JOHN HENRY, JR., electronic research physicist; b. Nanking, China, Oct. 14, 1917; s. John Henry and Birtha (Betts) R.; m. Caroline Hawke Pennypacker, Apr. 27, 1946; children: John III, Caroline, Anne, James. BS, Davidson Coll., 1939; MS, U. Va., 1941, PhD, 1943. Engr. RCA Corp., Camden, N.J., 1943-50, project leader, 1950-70; mem. tech. staff RCA Labs., Princeton, N.J., 1970-79, fellow tech. staff, 1979-83; ret., 1983. Author: History of the Electron Microscope, 1989. Mem., pres. bd. edn. Haddonfield, N.J., 1960-71; mem. Franklin Inst. Medal Com., Phila., 1958—; hon. mem. Rotary Internat., 1960—. Fellow AAAS, Am. Phys. Soc.; mem. Electron Microscopy Soc. Am. (bd. dirs. 1956, 78, pres. 1959). Democrat. Presbyterian. Achievements include development of permanent magnet electron microscope, 19 U.S. patents. Home: 671 Euclid Ave Haddonfield NJ 08033

REISNER, MILTON, psychiatrist; b. N.Y.C., Jan. 30, 1934; s. Maximillian and Dora Reisner; m. Linda Ellis, Mar. 3, 1959 (div. 1975); children: Margaret Ann, Amanda Lee. BA, NYU, 1954; MD, Downstate Med. Ctr., 1958. Diplomate Nat. Bd. Med. Examiners, N.Y. State Bd. Psychiat. Examiners. Resident in psychiatry Kings County Hosp., Bklyn., 1959-62; sr. psychiatrist Manhattan VA Hosp., N.Y.C., 1962-66; assoc. dir. psychiatry Westchester Community Mental Health Bd., White Plains, N.Y., 1966-69; dir. psychiatry Westchester Mental Health Bd., White Plains, 1969-74; pvt. practice N.Y.C., 1976—; cons. Cath. Charities, N.Y.C., 1965-66, H.I.P., N.Y.C., 1973-74, NYU Med. Ctr., 1963-68. Contbr. articles to profl. jours. Lt. j.g. USPHS, 1958-59. Fellow Am. Soc. Psychoanalytic Physicians; mem. Am. Assn. Psychoanalytic Physicians (pres. 1985-86, 87-88, Plaque 1988), Nat. Arts Club, Phi Beta Kappa. Achievements include research in mirroring as a technique for treating delusions. Office: 200 E 84th St New York NY 10028

REISS, BETTI, biological and medical researcher, medical writer; b. Denver, May 23, 1944; d. Louis A. and Edna Eda (Emroch) R. BS, CUNY, 1965; MS, NYU, 1971, PhD, 1974. Rsch. assoc. Pub. Health Rsch. Inst., N.Y.C., 1974-75; assoc. Am. Health Found., Valhalla, N.Y., 1976-87, 91-92, Westchester Med. Ctr., Valhalla, 1988-90; med. writer, sci. comms. Purdue Frederick Co., Norwalk, Conn., 1992—. Author: Experimental Colon Carcinogenesis, 1982; author, editor: Advances in Modern Environmental Toxicology, 1985; contbr. articles to profl. jours. NSF fellow 1967. Achievements include development of organ culture system for study of carcinogenesis, cell culture methodology for study of asbestos toxicology and carcinogenicity. Home: 70 Mitchell Rd Somers NY 10589-1801 Office: Purdue Frederick Co 100 Connecticut Ave Norwalk CT 06850-3590

REISS, DONALD ANDREW, physicist; b. Newark, June 20, 1945; s. Wilhelm and Julia Reiss. BS in Physics, Drexel U., 1967; MS, U. Ala., Huntsville, 1976, PhD, 1985. Physicist NASA/Marshall Space Flight Ctr., Huntsville, 1967—. Contbr. articles to profl. jours. Mem. Am. Phys. Soc., Sigma Xi. Achievements include patents for method andapparatus for shaping and enhancing acoustical forces, crystal growths in a microgravity environment. Office: NASA Marshall Space Flight Ctr ES 76 Huntsville AL 35812

REISS, HOWARD, chemistry educator; b. N.Y.C., Apr. 5, 1922; s. Isidor and Jean (Goldstein) R.; m. Phyllis Kohn, July 25, 1945; children: Gloria, Steven. A.B. in Chemistry, NYU, 1943; Ph.D. in Chemistry, Columbia U., 1949. With Manhattan Project, 1944-46; instr., then asst. prof. chemistry Boston U., 1949-51; with Central Research Lab., Celanese Corp. Am., 1951-52, Edgar C. Bain Lab. Fundamental Research, U.S. Steel Corp., 1957, Bell Telephone Labs., 1952-60; asso. dir., then dir. research div. Atomics Internat., div. N.Am. Aviation, Inc., 1960-62; dir. N.Am. Aviation Sci. Center, 1962-67, v.p. co., 1963-67; v.p. research aerospace systems group N.Am. Rockwell Corp., 1967-68; vis. lectr. chemistry U. Calif. at Berkeley, summer 1957; vis. prof. chemistry UCLA, 1961, 62, 64, 67, prof., 1968-91, prof. emeritus, 1991—; vis. prof. U. Louis Pasteur, Strasbourg, France, 1986, U. Pa., 1989; vis. fellow Victoria U., Wellington, New Zealand, 1989; cons. to chem.-physics program Air Force Cambridge Rsch. Cambridge Rsch. Labs., 1950-52; chmn. editor proc. Internat. Conf. Nucleation and Interfacial Phenomena, Boston; mem. Air Force Office Sci. Rsch. Physics and Chemistry Rsch. Evaluation Groups, 1966—, Oak Ridge Nat. Lab. Reactor Chemistry Adv. Com., 1966-68; adv. com. math. and phys. scis. NSF, 1970-72, ARPA Materials Rsch. Coun., 1968—; chmn. site rev. com. NRC Associateships Program, Naval Rsch. Lab., 1989. Author: The Methods of Thermodynamics, 1965; author articles; editor in field.; editor: Progress in Solid State Chemistry, 1962-71, Jour. Statis. Physics, 1968-75, Jour. Colloid Interface Sci; mem. editorial adv. bd. Internat. Jour. Physics and Chemistry of Solids, 1955, Progress in Solid State Chemistry, 1962-73, Jour. Solid State Chemistry, 1969, Jour. Phys. Chemistry, 1970-73, Ency. of Solid State, 1970, Jour. Nonmetals, 1971—, Jour. Colloid and Interface Sci., 1976-79, Langmuir, 1985—. Guggenheim Meml. fellow, 1978. Fellow AAAS, Am. Phys. Soc. (exec. com. div. chem. physics 1966-69); mem. NAS, Am. Chem. Soc. (chmn. phys. chemistry sect. N.J. sect. 1957, Richard C. Tolman medal 1973, Kendall award in colloid and surface chemistry 1980, J.H. Hildebrand award in theoretical and exptl. phys. chemistry of liquids 1991), Phi Beta Kappa, Sigma Xi, Phi Lambda Upsilon. Office: U Calif Dept Chemistry Los Angeles CA 90024

REISSNER, ERIC (MAX ERICH REISSNER), applied mechanics educator; b. Aachen, Germany, Jan. 5, 1913; came to U.S., 1936, naturalized, 1945; s. Hans and Josephine R.; m. Johanna Siegel, Apr. 19, 1938; children—John E., Eva M. Dipl. Ing., Technische Hochschule, Berlin, 1935, Dr. Ing., 1936; Ph.D., MIT, 1938; Dr. Ing. (hon.), U. Hannover, Germany, 1964. Mem. faculty MIT, Cambridge, 1939-69, prof. math., 1949-66; prof. applied math. MIT, 1966-69; prof. applied mechanics U. Calif., San Diego, 1970-78, prof. emeritus, 1978—; aero. research scientist NACA, Langley Field, 1948, 51; vis. prof. U. Mich., Ann Arbor, 1949. Cons. editor Addison-Wesley Pub. Co., 1949-60; mng. editor Jour. Math. and Physics, 1945-67; assoc. editor Quar. Applied Math., 1946—, Studies Applied Math., 1970—, Internat. Jour. Solids and Structures, 1983—; contbr. chpts. to books, articles to profl. jours. Recipient Clemens Herschel award Boston Soc. Civil Engrs., 1956, Theodore von Karman medal ASCE, 1964; Guggenheim fellow, 1962. Fellow ASME (Timoshenko medal 1973, ASME medal 1988, hon. mem. 1991), AIAA (structures and materials award 1984), Am. Acad. Arts and Scis., Am. Acad. Mechanics; mem. NAE, Internat. Acad. Astronautics, Gesellschaft fur Angewandte Mathematik und Mechanik (hon.). Office: U Calif San Diego Dept Applied Mechs & Engring Scis La Jolla CA 92093-0411

REITER, STANLEY, economist, educator; b. N.Y.C., Apr. 26, 1925; s. Frank and Fanny (Rosenberg) R.; m. Nina Sarah Breger, June 13, 1944; children: Carla Frances, Frank Joseph. A.B., Queens Coll., 1947; M.A., U. Chgo., 1950, Ph.D., 1955. Rsch. assoc. Cowles Commn., U. Chgo., 1949-50; mem. faculty Stanford U., 1950-54, Purdue U., 1954-67; prof. econs. and math. Northwestern U., 1967—, now Morrison prof. econs. and math. Coll. Arts and Scis., Morrison prof. managerial econs. and decision scis. Kellogg Grad. Sch. Mgmt.; cons. in field. Trustee Roycemore Sch., Evanston, Ill., 1969-71, treas., 1970-71. Served with inf. AUS, 1943-45. Decorated Purple Heart. Fellow Econometric Soc., AAAS; mem. Soc. Indsl. and Applied Math., Inst. Mgmt. Scis., Ops. Research Soc. Am., Am. Math. Soc., Math. Assn. Am. Home: 2138 Orrington Ave Evanston IL 60201-2914 Office: Northwestern U Ctr for Math Studies 2001 Sheridan Rd Evanston IL 60201-2962

REITER, WILLIAM MARTIN, chemical engineer; b. Phila., Sept. 23, 1925; s. William Henry and Marie Catherine (Farrell) R.; m. Helen C. Fuchs, May 31, 1947; children—William L., Ann C. B.Sc. in Chem. Engring., Drexel U., 1949. Chem. engr. Allied Chem. Co., Claymont, Del., 1949-52; process design engr. Catalytic Constrn. Co., Phila., 1952-53; engring. group leader Allied Chem. Co., Phila., 1953-65; mgr. research and devel., vinyl products Allied Chem. Co., Painesville, Ohio, 1965-72, asst. corp. dir. air and water pollution control, Morristown, N.J., 1972-77, corp. dir. pollution control, 1977-86; pres. Cape Environ. Assocs., Ocean City, N.J., 1986-91; ind. cons. chem. engr., 1991—; mem. nat. air pollution control techniques adv. com. EPA, 1979-82, 85-89; mem. N.J. Hazardous Waste Adv. Commn., 1980-81; commr., chmn. Ocean City Utility Commn., 1990-92; mem. Cape May County Indsl. Pollution Control Fin. Authority, 1992-93; adj. asst. prof. chem. engring. Drexel U., Phila., 1960-65; pres. Springview Farms (Pa.), 1963-65. Contbr. articles to profl. jours. Served with AUS, 1943-46. Registered profl. engr., Pa., N.J. Mem. Am. Inst. Chem. Engrs., Water Pollution Control Fedn., Air Pollution Control Assn. (dir. sect. 1962-65, dir. Mid-Atlantic States sect.), Tau Beta Pi, Phi Kappa Phi. Home: 6 Coral Ln Ocean City NJ 08226-2638

REITZ, RICHARD ELMER, physician; b. Buffalo, Sept. 18, 1938; s. Elmer Valentine and Edna Anna (Guenther) R.; m. Gail Pounds, 1960 (div. 1990); children: Richard Allen, Mark David; m. Myrnna Mecenario, 1991. BS, Heidelberg Coll., 1960; MD, SUNY-Buffalo, 1964. Intern Hartford (Conn.) Hosp., 1964-65, resident in medicine, 1966-67; asst. resident in medicine Yale U., 1965-66; vis. rsch. assoc. NIH, Bethesda, Md., 1967-68; rsch. fellow in medicine Harvard Med. Sch., Mass. Gen. Hosp., Boston, 1967-69; dir. clin. investigation ctr. Naval Regional Med. Ctr., 1969-71; dir. Endocrine Metabolic Ctr., Oakland, Calif., 1973-92; med. dir. Nichols Inst. ; asst. prof. medicine U. Calif.-San Francisco, 1971-76; chmn. dept. medicine John Muir MEd. Ctr., 1975-77; assoc. clin. prof. medicine U. Calif.-Davis, 1976-86; clin. prof. med. 1986—; chief endocrinology Providence Hosp., Oakland, Calif., 1972-92. Contbr. articles to profl. jours., chpt. to book. Mem. scholarship com., Bank of Am., San Francisco, 1983. Served to comdr. USNR, 1969-71. Mem. Endocrine Soc., Am. Soc. Bone and Mineral Rsch., Am. Fedn. Clin. Rsch., Am. Fertility Soc., Am. Soc. Internal Medicine, AAAS, Am. Assn. Clin. Endocrinologists, Am. Coll. Endocrinology. Democrat. Office: Nichols Inst 3100 Summit St Oakland CA 94609-3410

REITZE, DAVID HOWARD, physicist; b. Washington, Pa., Jan. 6, 1961; s. Philip Howard and Marylou (McWade) R. BA, Northwestern U., 1983; PhD, U. Tex., 1990. Postdoctoral mem. tech. staff Bellcore, Red Bank, N.J., 1990-92; physicist Lawrence Livermore Nat. Lab., Livermore, Calif., 1992—. Contbr. articles to Phys. Rev., Applied Physics Letters, Optics Letters. Mem. Optical Soc. Am., Am. Phys. Soc., Phi Beta Kappa. Achievements include research in femtosecond spectroscopy of materials, ultrafast optics, femtosecond lasers; first determination of electrical and optical properties of liquid carbon using femtosecond spectroscopy. Office: Lawrence Livermore Nat Lab PO Box 808 L-251 Livermore CA 94550

REIZENSTEIN, PETER GEORG, hematologist; b. Feb. 25, 1928; s. Max L. and Bertha R.; m. Anna Lisa Hall; children: Johan A., Elisabet, Maja. MD, Karolinska, Stockholm, 1954, PhD, 1959. Rsch. assoc. AEC Brookhaven Nat. Lab., N.Y.C., 1959-61; intern, resident Karolinska Hosp., 1954-59, asst. chief medicine, 1962-67, chief med. orgn., 1968-71, chief hematology, 1972-85, chmn. hematology collaborative group, 1981-93, hematology cons., 1986—; prof. U. Paris VII, 1983-84; chief biomed. div. Nat. Inst. Radiation Protection, Stockholm, 1986—; prof., 1989—; pres. Swedish Soc. Hematology, 1972-75; bd. dirs. Swedish Soc. Nuclear Medicine, European Soc. Med. Oncology. Contbr., editor and editorial bd. mem. profl. jours. UN advisor Govt. Argentina, 1974, Govt. Tanzania, 1976; chmn. various local com. Recipient Boris Rajewsky award European Soc. Radiology, 1972. Mem. Swedish Assn. patients with Blood Diseases (bd. dirs. 1982—, founder), Lundblad Rsch. Found. (dir. 1990-92), Internat. Soc. Quality Assurance in Health Care (pres., hon.). Home: Villav 9, Stocksund 18275, Sweden Office: Nat Inst Radiation Protection, Karolinska Hosp, Stockholm 10401, Sweden

REKKAS, CHRISTOS MICHAIL, mechanical engineer; b. Athens, Greece, Jan. 21, 1962; s. Michail Dimitrios and Pinelopi Christos (Konstas) R. Diploma, Nat. Tech. U. Athens, 1984, PhD in Aeronautics, 1990. Rsch. asst. Nat. Tech. U. Athens, 1985-90; with Eurocontrol Agy., Brussels, 1985-90; postdoctoral researcher Nat. Tech. U. Athens, 1991—; cons. R&D Group, Athens, 1988-90. Contbr. articles to profl. jours. With Greek Mil., 1990-91. Grantee Greek Ministry Edn., 1986-90, Greek Secretariat for Rsch., 1986-87. Mem. IEEE, AIAA, Royal Inst. Navigation, Tech. Chambers Greece. Home: Nestoros 13, Holargos 15562, Greece Office: Tech U Athens, Dept Mech Engring, Patission 42, Athens 10682, Greece

RELLE, FERENC MATYAS, chemist; b. Gyor, Hungary, June 13, 1922; came to U.S., 1951, naturalized, 1956; s. Ferenc and Elizabeth (Netratics) R.; m. Gertrud B. Tubach, Oct. 9, 1946; children: Ferenc, Ava, Attila. B-SChemE, Jozsef Nador Poly. U., Budapest, 1944, MS, 1944. Lab. mgr. Karl Kohn Ltd. Co., Landshut, W.Ger., 1947-48; resettlement officer IRO, Munich, 1948-51; chemist Farm Bur. Coop. Assn., Columbus, Ohio, 1951-56; indsl. engr. N.Am. Aviation, Inc., Columbus, 1956-57; rsch.chemist Keever Starch Co., Columbus, 1957-65; rsch. chemist Ross Labs. div. Abbott Labs., Columbus, 1965-70, rsch. scientist, 1970-89; cons. in field. Chmn. Columbus and Central Ohio UN Week, 1963; pres. Berwick Manor Civic Assn., 1968; trustee Stelios Stelson Found., 1968-69; deacon Brookwood Presbyn. Ch., 1963-65, 92—, trustee, 1990-91. Mem. Am. Chem. Soc. (alt. councilor 1973, chmn. long range planning com. Columbus sect. 1972-76, 78-80), Am. Assn. Cereal Chemists (chmn. Cin. sect. 1974-75), Ohio Acad. Sci., Internat. Tech. Inst. (adv. dir. 1977-82), Nat. Intercollegiate Soccer Ofcls. Assn., Am. Hungarian Assn., Hungarian Cultural Assn. (pres. 1978-81), Ohio Soccer Ofcls. Assn., Columbus Männerchor, Germania Singing and Sport Soc., Civitan (gov. Ohio dist. 1970-71, dist. treas. 1982-83, pres. Eastern Columbus 1963-64, 72-73, gen. sec. for Hungary 1991-92, established 1st Civitan club in Hungary 1991, Internat. Gov. of Yr. award 1971, Internat. Honor Key 1992, master club builder award 1992, various other awards), World Fedn. of Hungarian Engrs. Home and Office: 3487 Roswell Dr Columbus OH 43227-3560

RELLER, L. BARTH, microbiologist, educator. Prof. med. ctr. Duke U., Durham, N.C. Recipient Becton Dickinson and Company award in Clin. Microbiology, Am. Soc. Microbiology, 1991. Office: Duke Univ Med Ct Box 3938 Durham NC 27710*

RELWANI, NIRMALKUMAR MURLIDHAR (NICK RELWANI), mechanical engineer; b. Aug. 9, 1954; married. BS in Mech. Engring., U. Baroda, 1976; student, U. Nebr., 1977-78; MS in Mech. Engring., U. Wis., Milw., 1980. Registered profl engr., Wis., Ill. Rsch. asst. dept. mech. engring. U. Nebr., Lincoln, 1978; design engr. Allis Chalmers Corp., Milw., 1978-80; engring. cons. Bombay, 1980-86; assoc. engr. IIT Rsch. Inst., Chgo., 1986; mech. engr. Gen. Energy Corp., Oak Park, Ill., 1987-89, Arrowhead Environ. Control, Chgo., 1989-90; environ. engr. Ill. Dept. Pub. Health, Bellwood, 1990-92; environ. protection engr. field ops. sect. bur. air Ill. EPA, Maywood, 1992—. Mem. ASME, ASHRAE (energy conservation award 1991), Assn. Energy Engrs. (sr.). Home: 413 S Home Ave Apt # 2B Oak Park IL 60302 Office: Ill EPA 1701 S First Ave Ste 600 Maywood IL 60153

REMBAR, JAMES CARLSON, psychologist; b. N.Y.C., May 4, 1949; s. Charles Isaiah and Billie Ann (Olsson) R.; m. Jill Bailin, June 4, 1988; 1 child, Lilianna. BA, Sarah Lawrence Coll., 1972; MA, U. Mich., 1976, PhD, 1978. Cert. Psychoanalyst, N.Y., psychologist. Clin. psychologist U. Mich. Med. Ctr., Ann Arbor, 1978-80; instr. psychology in psychiatry N.Y. Hosp. Cornell U. Med. Coll., White Plains, 1980-84, clin. asst. prof., 1984—, coord. child and adolescent psychology Westchester div., 1982-87; pvt. practice clin. psychologist Irvington, White Plains, N.Y., 1981—; faculty Westchester Ctr. for Study of Psychoanalysis and Psychotherapy, 1989—, dir. continuing edn., 1992—; cons. Andrus Children's Home, Yonkers, N.Y., 1987—. Contbr. articles to profl. jours., chpt. in book. Mem. Am. Psychol. Assn., N.Y. State Psychol. Assn., Westchester County Psychol. Assn., Psychoanalytic Assn. Westchester Ctr. Avocations: tennis, music. Home and Office: 9 Sunnyside Pl Irvington NY 10533-1300 Office: 510 N Broadway White Plains NY 10603-3242

REMILLARD, RICHARD LOUIS, automotive engineer; b. L.A., June 19, 1954; s. Richard Leo Remillard and Helen Lois (Dittberner) Freeman; m. Kyu-Kyu April Leong, Feb. 24, 1990; 1 child, Ian Louis. BA in Geology, U. Calif., Berkeley, 1978; BS in Mech. Engring., San Jose State U., 1987. Rsch. assoc. Lawrence Berkeley Lab., 1979-81; staff geophysicist Woodward Clyde Cons., Oakland, Calif., 1981-82; county geologists asst. Santa Clara County, San Jose, Calif., 1983-88; air resources engr. Calif. Air Resources Bd., El Monte, Calif., 1988-90; sr. cons. Booz, Allen, and Hamilton, L.A., 1990—. Mem. Soc. Automotive Engrs. (assoc.). Republican. Achievements include development of Locomotive Emission study which produced an accurate inventory of locomotive emissions in Calif. and evaluated techniques for reducing locomotive emissions, methodology for comparing emissions associated with electric vehicles against those of various i.c. passenger vehicles; comparison of emission rates between passenger cars and passenger trains. Office: Booz Allen and Hamilton 523 W Sixth St Ste 616 Los Angeles CA 90014

REMINE, WILLIAM HERVEY, JR., surgeon; b. Richmond, Va., Oct. 11, 1918; s. William Hervey and Mabel Inez (Walthall) ReM.; m. Doris Irene Grumbacher, June 9, 1943; children—William H., Stephen Gordon, Walter James, Gary Craig. B.S. in Biology, U. Richmond, 1940, D.Sc. (hon.), 1965; M.D., Med. Coll. Va., Richmond, 1943; M.S. in Surgery, U. Minn., Mpls., 1952. Diplomate Am. Bd. Surgery. Intern Doctor's Hosp., Washington, 1944; fellow in surgery Mayo Clinic, Rochester, Minn., 1944-45, 47-52; instr. surgery Mayo Grad. Sch. Medicine, Rochester, Minn., 1954-59, asst. prof. surgery, 1959-65, assoc. prof. surgery, 1965-70, prof. surgery, 1970-83, prof. surgery emeritus, 1983—; surg. cons. to surgeon gen. U.S. Army, 1965-75; surg. lectr., USSR, 1987, 89, Japan, 1988, 90, Egypt, 1990; lectr. Soviet-Am. seminars, USSR, 1987, 89. Sr. author: Cancer of the Stomach, 1964, Manual of Upper Gastro-intestinal Surgery, 1985; editor: Problems in General Surgery, Surgery of the Biliary Tract, 1986; mem. editorial bd. Rev. Surgery, 1965-75, Jour. Lancet, 1968-77; contbr. 200 articles to profl. jours. Served to capt. U.S. Army, 1945-47. Recipient St. Francis surg. award St. Francis Hosp., Pitts., 1976, disting. service award Alumni Council, U. Richmond, 1976. Mem. ACS, AAAS, Am. Assn. History of Medicine, AMA, Am. Med. Writers Assn., Am. Soc. Colon and Rectal Surgeons, Soc. Surgery Alimentary Tract (v.p. 1983-84), Am. Surg. Assn., Assn. Mil. Surgeons U.S., Internat. Soc. Surgery, Digestive Disease Found., Priestley Soc. (pres. 1968-69), Central Assn. Physicians and Dentists (pres. 1972-73), Central Surg. Assn., Soc. Med. Cons. Armed Forces, Mayo Clinic Surg. Soc. (chmn. 1964-66), Soc. Head and Neck Surgeons, Soc. Surg. Oncology, So. Surg. Assn., Western Surg. Assn. (pres. 1979-80), Minn. State Med. Assn., Minn. Surg. Soc. (pres. 1966-67), Zumbro Valley Med. Soc., Sigma Xi; hon. mem. Colombian Coll. Surgeons, St. Paul Surg. Soc., Flint Surg. Soc., Venezuelan Surg. Soc., Colombian Soc. Gastroenterology, Dallas So. Clin. Soc., Ga. Surg. Soc., Soc. Postgrad. Surgeons Los Angeles County, Japanese Surg. Soc., Argentine Surg. Digestive Soc., Bassanese Surg. Assn. (Italy), Tex. Surg. Soc., Omicron Delta Kappa, Alpha Omega Alpha, Beta Beta Beta. Methodist. Avocations: hunting, fishing, golf, photography, boating, music. Home: 129 Island Dr Ponte Vedra Beach FL 32082

REMINGTON, DELWIN WOOLLEY, soil conservationist; b. Vernal, Utah, May 10, 1950; s. Lyle H. and Muriel (Woolley) R.; m. Sylvia Bendixsen, July 20, 1973 (div. Nov. 1990); children: Holly, Reed Lyle, Roger Delwin, Kevin Bendixsen, Kollin Scott, Heidi; m. Marcia C. Cebull, June 1991. BS, Utah State U., 1975. Nurseryman Millcreek Gardens, Salt Lake City, 1975-79; yard foreman Brown Floral, Salt Lake City, 1979-82; agrl. salesman Steve Regan Co., Vernal, 1982-85; plant protection and quarantine aid APHIS USDA, Salt Lake City, 1986; soil conservationist USDA Soil Conservation Svc., Vernal, 1986—. Active Take Pride in Am., Vernal, 1989, Op. Desert Shield, Saudi Arabia, 1990. With Air NG, 1971-92. Mem. Utah Cert. Nurseryman, Soil & Water Conservation Soc. (fin. com. Utah chpt. 1989-90). Mem. LDS Ch. Avocations: genealogy, hiking. Home: 860 N 1500 E Apt A Vernal UT 84078-4623 Office: USDA Soil Conservation Svc 475 W 100 N Vernal UT 84078-2093

REMINGTON, SCOTT ALAN, laser engineer; b. Gt. Falls, Mont., Dec. 31, 1964; s. Derlin Raymond and Marjorie Barbara (Lopach) R. BS in Physics, Mont. State U., 1987. Laser engr. Big Sky Laser, Bozeman, Mont., 1990—. Office: Big Sky Laser Co PO Box 8100 Bozeman MT 59715

REMMER, HARRY THOMAS, JR., obstetrician/gynecologist; b. Utica, N.Y., Apr. 1, 1920; s. Harry Thomas and Ida Kathleen (Woodcock) R.; m. Constance Mary Conlan, Nov. 26, 1949; children: Constance Therese, Robert Joseph, Elizabeth Conlan, Randall Conlan. BA, Toronto (Can.) U., 1941; MD, U. Pa., 1944. Diplomate Am. Bd. Ob-Gyn. Intern Syracuse (N.Y.) U. Med. Ctr. Hosp., 1944-45, asst. resident gen. surgery, 1945-46, 48-49; resident in ob-gyn. Woman's Hosp. N.Y., N.Y.C., 1949-53; pvt. practice Utica, N.Y., 1953-68, South Miami, Fla., 1968—. Lt. (j.g.) USN, 1947-48. Fellow Am. Coll. Ob-Gyn.; mem. AMA, Fla. Med. Assn., Dade County Med. Assn., So. Med. Assn., Fla. Obstet. Soc., Kiwanis, Deering Bay Yacht and Country Club. Democrat. Roman Catholic. Avocations: fishing, golf, tennis. Office: Harry T Remmer MD PA Ste 205 6701 Sunset Dr South Miami FL 33143

REMSEN, CHARLES CORNELL, III, microbiologist, research administrator, educator; b. Newark, N.J., May 16, 1937; s. Charles Cornell Jr. and Elizabeth Havens (Atwood) Remsen; children: David Pratt, Linda Remsen Brandenburg, Stephen Dwyer, Andrew Walker; m. Margaret Ellis Fairchild, June 19, 1976; stepchildren: Elizabeth Hoffman Herzog, Jennifer Hoffman Jonas. BS in Food Chemistry and Microbiology, Delaware Valley Coll. Sci. and Agr., 1960; MS in Microbiology, Syracuse U., 1963, PhD in Microbiology, 1965. Rsch. asst. Schering Pharm., Bloomfield, N.J., 1959-60; rsch. asst. dept. preventive medicine Upstate Med. Ctr., Syracuse, N.Y., 1961-63; grad. teaching asst. Syracuse U., 1962-63, grad. rsch. asst., 1963-65; NIH post-doctoral fellow dept. gen. biology Swiss Fed. Inst. Tech., Zurich, Switzerland, 1965-67; asst. scientist Woods Hole (Mass.) Oceanographic Inst., 1967-71, assoc. scientist, 1971-75; rsch. assoc. in microbial ecology Marine Biol. Labs., Woods Hole, 1973-74; assoc. prof. dept. zoology/microbiology, assoc. scientist Ctr. Great Lakes Studies U. Wis., Milw., 1975-83, prof., sr. scientist, 1983—, coord. zoology/microbiology, 1976-84, acting dir. Ctr. Great Lakes Studies, Great Lakes Rsch. Facility, 1987-89, interim dir., 1989-92, dir., 1992—; mem. editorial bd. Jour. of Bacteriology, 1969-77; external examiner McGill U. Grad. Sch., 1971-73; chmn. joint com. for biol. oceanography MIT-WHOI PhD Program, 1971-74; mem. internat. adv. bd. ScienceQuest. Author: (with others) The Encyclopedia of Microscopy and Microtechnique, 1973, Effect of the Ocean Environment on Microbial Activities, 1974, The Phytosynthetic Bacteria, 1978, Responses of Marine Organisms to Pollutants, 1984, Structure of Photosynthetic Prokaryotes, 1991; contbr. articles to profl. jours. Del. Coun. on Ocean Affairs. Recipient NIH Foreign Postdoctoral fellowship, 1965-67, Disting. Svc. award Jour. of Bacteriology, 1977. Mem. Nat. Assn. Marine Labs. (exec. com.), Am. Geophys. Union, Am. Soc. for Microbiology, Internat. Assn. for Great Lakes Rsch., Coun. Great Lakes Rsch. Mgrs. (internat. joint commn.), N.E. Assn. Marine Labs. (co-v.p.), N.E. Assn. Marine and Great Lakes Labs., Nature Conservancy (bd. trustees Wis. chpt. 1987, exec. com. 1989—, sec. 1989—), Sigma Xi. Achievements include research in relating the structure of chemolithotrophs to their ecological niche, and to their response to the environment, methane-oxidizing bacteria and how these microorganisms fit into the overall carbon cycle in the Great Lakes, the study of sediment samples from Lakes Superior, Michigan and Huron in order to determine the extent and rate of organic matter diagenesis in sediments, the exchange of gases (CO2, CH4, O2) with overlying waters and ultimately the atmosphere, the ecological role and significance of chemosynthesis and photosynthesis in sublacustrine hydrothermal vents and gas fumaroles. Office: University of Wisconsin - Mil Ctr for Great Lakes Studies Milwaukee WI 53201

REMSEN, JAMES VANDERBEEK, JR., biologist, museum curator; b. Newark, Sept. 21, 1949; s. James V. and Elizabeth (Willox) R.; m. Catherine Cummins, Nov. 12, 1988. BA, Stanford U., 1971, MA, 1971; PhD, U. Calif., Berkeley, 1978. Dir., curator birds, prof. La. State U. Mus. Natural Sci., Baton Rouge, 1978—. Author: Annotated List of the Birds of Bolivia, 1989, (monograph) Community Ecology of Neotropical Kingfishers, 1990; contbr. over 50 articles to sci. jours. Fellow Am. Ornithologists Union; mem. AAAS, Am. Soc. Naturalists, Ecol. Soc. Am. Home: 545 Pecan Dr Saint Gabriel LA 70776-5513 Office: La State U Mus Natural Sci Baton Rouge LA 70803

REN, CHUNG-LI, engineer; b. Chefoo, China, June 1, 1931; came to U.S., 1955; s. Shantsai and Fooching (Wang) R.; m. Rosalie Fen Lo, Aug. 4, 1962; children: Eric W., Caroline W. BSEE, Taiwan Coll. Engring., 1953; MSEE, U. Notre Dame, 1957; PhD in Electro Physics, Polytech Inst. Bklyn., 1964. Teaching asst. U. Notre Dame, South Bend, Ind., 1956-57; grad. asst., sr. lectr. Polytech Inst. Bklyn., Microwave Rsch. Inst., 1959-65; disting. mem. tech. staff AT&T Bell Labs., North Andover, Mass., 1965-90; lead engr. Mitre Corp., Advanced Satellite Terminals and Tech., Bedford, Mass., 1990—; speaker and panelist in field. Patentee in field; contbr. articles to profl. jours. Rsch. fellow Polytech Inst. Bklyn., 1957-59. Mem. IEEE (sr., mem. review bd. 1982—), Sigma Xi. Avocations: tennis, soccer, classical music, landscape design. Office: The Mitre Corp Burlington Rd Bedford MA 01730

RENDINA, GEORGE, chemistry educator; b. N.Y.C., July 1, 1923; s. Gaetano and Giovannina (Barbero) R.; m. Irma Civia Esner, Sept. 28, 1948; children—Alan Ralph, Steven Jeremy, David Nathan, Frederick Thomas. A.B., NYU, 1949; M.A., Kans. U., 1953, Ph.D., 1955. Postdoctoral fellow U. Mich., Ann Arbor, 1955-57, Henry Ford Hosp., Detroit, 1958; instr. Johns Hopkins Med. Sch., Balt., 1959-61; chief biochem. research Tng. Sch., Vineland, N.J., 1961-65, Mendota State Sch., Madison, Wis., 1966; prof. chemistry Bowling Green State U., Ohio, 1967-84.

RENDU, JEAN-MICHEL MARIE, mining executive; b. Tunis, Tunisia, Feb. 25, 1944; s. Paul C. and Solange M. (Krebs) R.; m. Karla M. Meyer, Aug. 18, 1973; children: Yannick P., Mikaël P. Ingénieur des Mines, Ecole des Mines St. Etienne, France, 1966; MS, Columbia U., 1968, D. Engring. Sci., 1971. Mgr. ops. rsch. Anglovaal, Johannesburg, Republic of South Africa, 1972-76; assoc. prof. U. Wis., Madison, 1976-79; assoc. Golder Assocs., Denver, 1979-84; dir. tech. and sci. systems Newmont Mining Corp., Danbury, Conn., 1984-89; v.p. Newmont Gold Co., Denver, 1989-93, Newmont Mining Corp., Denver, 1993—. Author: An Introduction to Geostatistical Methods of Mineral Evaluation, 1978, 81; contbr. tech. papers to profl. jours. Fellow South African Inst. of Mining and Metallurgy (corr. mem. of coun.); mem. N.Y. Acad. Sci., Internat. Assn. for Math. Geology, Soc. Mining Engrs., Sigma Xi. Roman Catholic. Office: Newmont Gold Co 1700 Lincoln St Denver CO 80203-4501

RENEAU, DANIEL D., university administrator. Prof., head dept. biomed. engring. La. Tech. U., Ruston, 1973-80, v.p. acad. affairs, 1980-87, pres., 1987—. Office: La Tech U Tech Station PO Box 3168 Ruston LA 71272*

RENGARAJAN, SEMBIAM RAJAGOPAL, electrical engineering educator, researcher, consultant; b. Mannargudi, Tamil Nadu, India, Dec. 12, 1948; came to U.S., 1980; s. Srinivasan and Rajalakshmi (Renganathan) Rajagopalan; m. Kalyani Srinivasan, June 24, 1982; children: Michelle, Sophie. BE with honors, U. Madras, India, 1971; MTech, Indian Inst. Tech., Kharagpur, 1974; PhD in Elec. Engring., U. N.B., Fredericton, Can., 1980. Mem. tech. staff Jet Propulsion Lab., Pasadena, Calif., 1983-84; asst. prof. elec. engring. Calif. State U., Northridge, 1980-83, assoc. prof., 1984-87, prof., 1987—; vis. prof. UCLA, 1987-88, vis. researcher, 1984—; cons. Hughes Aircraft Co., Canoga Park, Calif., 1982-87, NASA/Jet Propulsion Lab., Pasadena, 1987-90, 92, Ericsson Radar Electronics, Sweden, 1990-92; guest researcher Chalmers U., Sweden, 1990, UN Devel. Program, 1993. Contbr. sci. papers to profl. publs. Recipient Outstanding Faculty award Calif. State U., Northridge, 1985, Meritorious Performance and Profl. Promise award, 1986, 88, Merit award San Fernando Valley Engrs. Coun., 1989, cert. of recognition NASA, 1991, 92; Nat. Merit scholar Govt. India, 1965-71. Fellow Inst. Advancement Engrs.; mem. IEEE (sr. L.A. chpt. sec., treas. antennas and propagation soc. 1981-82, vice chmn. 1982-83, chmn. 1983-84), Internat. Union Radio Sci. (U.S. nat. com.), N.Y. Acad. Scis., Sigma Xi. Avocations: swimming, camping, jogging, tennis. Office: Calif State U 18111 Nordhoff St Northridge CA 91330-0001

RENKA, ROBERT JOSEPH, computer science educator, consultant; b. Summit, N.J., Dec. 28, 1947; s. John and Elizabeth (Pierce) R. BA in Computer Sci., BS in Math., U. Tex., 1976, MA in Math., 1979, PhD in Computer Sci., 1981. Numerical analyst Oak Ridge (Tenn.) Nat. Lab., 1981-84; asst. prof. computer sci. U. North Tex., Denton, 1984-89, assoc. prof. computer sci., 1989—; cons. in scientific computing. Contbr. articles to profl. jours. With USN, 1967-69, Vietnam. Rsch. grantee U. North Tex., 1984-89, NSF, 1990—. Mem. Assn. for Computing Machinery (algorithms editor 1988—, editor-in-chief 1989—), Soc. Indsl. and Applied Math. Avocations: rock climbing, canoeing. Home: 1700 Kendolph Dr Denton TX 76205-6940 Office: U North Tex PO Box 13886 Dept Computer Scis Denton TX 76203

RENKEN, KEVIN JAMES, mechanical engineering educator; b. Oak Lawn, Ill., June 29, 1961; s. James Michael and Jean Marie (Sielepkowski) R. BSME, U. Ill., Chgo., 1983, MSME, 1985, PhD in Mech. Engring., 1987. Registered profl. engr.-in-tng. Engring. technician Navy Pub. Works Ctr., Great Lakes, Ill., 1982; student rsch. participant, engring. asst. Argonne (Ill.) Nat. Lab., 1983-84, resident student assoc., 1984; teaching asst., rsch. asst. U. Ill., Chgo., 1983-87, rsch. assoc., 1987; faculty rsch. participant Argonne Nat. Lab., 1988; asst. prof. U. Wis., Milw., 1987-93, assoc. prof., 1993—; reviewer for jours. in field. Contbr. articles to profl. jours. Grantee NSF, 1992-95, EPA, 1992-93, 91-92; recipient Rsch. Initiation award NSF, 1989-92, Outstanding Young Men of Am. award, 1987. Mem. ASME, AIAA, ASHRAE, Am. Assn. Radon Scientists and Technologists, Tau Beta Pi, Sigma Xi. Roman Catholic. Achievements include recognition for fundamental studies related to single phase and multiphase transport in porous media; contributions to modelling and analysis of thin porous-layer condensation heat transfer enhancements; pioneering work in radon mitigation and radon transport processes in concrete. Office: U of Wisconsin-Milw Mech Engring Dept PO Box 784 Milwaukee WI 53201-0784

RENNER, MICHAEL JOHN, psychologist, biologist, educator; b. Battle Creek, Mich., Aug. 7, 1957; s. John Wilson and Carol Jean (Fennel) R.; m. Catherine Ann Hackett, Mar. 11, 1989; 1 child, Moriah Jeannette. BA, Boise (Idaho) State U., 1977; MS, U. Okla., 1979; PhD, U. Calif., Berkeley, 1984. Cert. tchr., Okla. Tchr., curriculum developer Norman (Okla.) High Sch., 1978-79; vis. asst. prof. U. Wyo., Laramie, 1984-86; asst. prof. U. Wis., Oshkosh, 1986-88, Memphis State U., 1988-92, West Chester (Pa.) U., 1992—. Author (with M. Rosenzweig) Enriched and Impoverished Environments, 1987; contbr. articles to profl. jours. Mem. Am. Psychol. Soc., APA (Donald O. Hebb award, mem. Young Psychologists Del. to Internat. Congress of Psychology, Sydney, Australia), Psychonomic Soc., Animal Behavior Soc., Soc. for Neuroscience, Sigma Xi. Democrat. Roman Catholic. Achievements include experimentation on spontaneous expression of curiosity results in acquisition of functionally useful information, 1988, (studies) Investigation of environment has sequential structure of behavior that resembles liguistic grammar, 1993, Prosimians display sustained curiosity in investigation of nonfood objects, 1992. Office: Dept Psychology West Chester University PA 19383

RENOUX, GERARD EUGENE, immunologist; b. Paris, Mar. 23, 1915; s. Joseph Jean and Andrée (Bechofer) R.; Dr.Med., U. Montpellier (France), 1939; Lic.Sciences, U. Marseille, 1939; Agregation microbiologie, U. Paris, 1949; m. Micheline Dehais, Feb. 19, 1962; children: Jean, Catherine, Marianne, Frank. Staff, U. Montpellier, 1935-52, prof. microbiology, 1960-67; lab. chief Institut Pasteur, Tunis, Tunisia, 1952-55, dir., 1955-60; prof. immunology U. Tours (France), 1967—; chief Brucellosis Lab., Nat. Inst. Agronomic Researches, Nouzilly, France, 1967-73; asso. scientist Sloan Kettering Cancer Center, N.Y.C., 1973-82; expert in Brucellosis, WHO/FAO, 1950; Brucellosis sr. cons. FAO, 1952; expert trachoma WHO, 1955; cons. Public Health Labs., 1964-66. Decorated Légion d'Honneur, Croix de Guerre, medaille des Epidemies, Palmes Académiques, Laureat Faculté de Medecine, Montpellier; prix Galien, 1978. Mem. French Acad. Medicine, Italian Acad. Medicine, Argentine, French, Am. socs. microbiologists, Am. Soc. Immunologists, N.Y. Acad. Scis., AAAS, Soc. Française Pharmacologie, Soc. Française d'Immunologie, Soc. Royale Belge Med. Tropicale. Research and numerous publs. on Brucellosis including discovery of vaccine for livestock, new techniques of study, new species, new diagnostic methods; research on immunology: immunostimulation with thio-compounds, mechanisms, application to human diseases and to cancer; studies on antibodies and antigens in human diseases, lab. procedures for diagnostics. Home: 1 rue des Ursulines,, F 37000 Tours France

RENSHAW, AMANDA FRANCES, nuclear engineer; b. Wheelwright, Ky., Dec. 10, 1934; d. Taft and Mamie Nell (Russell) Wilson; divorced; children: Linda, Michael, Billy. BS in Physics, Antioch Coll., 1972; MS in Physics, U. Tenn., 1982, MS in Nuclear Engring., 1991. Rsch. asst. U. Mich., Ann Arbor, 1970-71; teaching asst. Antioch Coll., Yellow Springs, Ohio, 1971-72; physicist GE, Schenectady, N.Y., 1972-74, Union Carbide Corp., Oak Ridge, Tenn., 1974-79; rsch. assoc. Oak Ridge Nat. Lab., 1979-91, mgr. strategic planning, 1991—; asst. to counselor for sci. and tech. Am. Embassy, Moscow, 1990. Contbr. articles to profl. jours. Mem. AAUW, Am. Assn. Artificial Intelligence, Am. Nuclear Soc. (Oak Ridge chpt.), Soc. Women Engrs., Soc. Black Physicists. Avocations: reading, travelling. Home: 1850 Cherokee Bluff Dr Knoxville TN 37920 Office: Oak Ridge Nat Lab PO Box 2008 Bldg 1505 MS 6038 Oak Ridge TN 37831

RENSIMER, EDWARD R., internist, educator; b. Phila., Sept. 2, 1949; s. Edward H. and Florence (Shuda) R.; m. Jane Pendergast; children: Andrew James, John Edward, Connor Alexander. BA in Biology cum laude, Temple U., 1971, MD, 1975. Diplomate Am. Bd. Internal Medicine, Infectious Diseases. Asst. dir. Employee Health Svc. Northwestern Meml. Hosp., Chgo., 1978-79; adjunct prof. of Medicine Northwestern U. Sch. of Medicine, Chgo., 1978-79; clin. rsch. fellow infectious diseases U. Tex. Sch. of Medicine, Houston, 1979-81; pvt. practice infectious diseases Houston, 1981—; asst. clin. prof. of Medicine U. Tex., Houston, 1981—; chmn. Dept. Internal Medicine, Meml. City Med. Ctr., Houston, 1989-90, dir., 1990—; founder Internat. Medicine Ctr., Houston, 1990. Contbr. articles to profl. jours. Mem. Pastoral Coun. St Cecilia's Roman Cath. Ch., Houston, 1988-92, bible study leader, 1988-92. Recipient scholarship Sears Roebuck Found., Phila., 1967. Fellow Am. Coll. Physicians; mem. Infectious Diseases Soc. of Am., Harris County Med. Soc. (rep. mem. Harris County Ryan White HIV Health Svcs. Planning Coun., 1993). Avocations: photography, reading, bible study, computer software devel., entrepreneurial activities. Office: Edward R Rensimer MD FACP 920 Frostwood Dr # 670 Houston TX 77024-2302

RENSON, JEAN FELIX, psychiatry educator; b. Liège, Belgium, Nov. 9, 1930; came to U.S., 1960; s. Louis and Laurence (Crahai) R.; m. Gisèle Bouillenne, Sept. 8, 1956; children: Marc, Dominique, Jean-Luc. MD, U. Liege, 1959; PhD in Biochemistry, George Washington U., 1971. Diplomate Am. Bd. Psychiatry. Asst. prof. U. Liège, 1957-60; rsch. fellow U. Liege, 1966-72; clin. assoc. prof. dept. psychiatry U. Calif., San Francisco, 1978—; pvt. practice Lakeview Psychotherapy Ctr., Stockton, Calif., 1986—; vis. asst. prof. Stanford U., Palo Alto, Calif., 1972-77. Assoc. editor: Fundamentals of Biochemical Pharmacology, 1971. NIH fellow, 1960-66. Democrat. Avocations: neurosciences, music. Office: Lake View Psychotherapy Ctr # 12 2389 W March Ln Stockton CA 95207-5239

RENTHAL, ROBERT DAVID, biochemist, educator; b. Chgo., Oct. 29, 1945; s. Sidney and Helen (Greenwald) R.; m. Ann R. Lomax, July 25, 1977; children: William, Katherine. AB, Princeton U., 1967; PhD, Columbia U., 1972. NIH fellow Yale U., New Haven, 1972-74; NRC fellow NASA Ames Rsch. Ctr., Moffett Field, Calif., 1974-75; asst. prof. U. Tex., San Antonio, 1975-80, assoc. prof., 1980-87, prof. div. earth and phys. scis., 1987—; mem. faculty U. Tex. Health Sci. Ctr., 1977—. Contbr. articles to profl. publs. Office: U Tex Div Earth and Phys Scis 7000 N Anderson Loop San Antonio TX 78249

RENTZEPIS, PETER M., chemistry educator; b. Kalamata, Greece, Dec. 11, 1934; m. Alma Elizabeth Keenan; children—Michael, John. B.S., Denison U., 1957, D.Sc. (hon.), 1981; M.S., Syracuse U., 1959, Ph.D. (hon.), 1980; Ph.D., Cambridge U., 1963; D.Sc. (hon.), Carnegie-Mellon U., 1983. Mem. tech. staff, rsch. labs. Gen. Electric Co., Schenectady, 1960-62; mem. tech. staff AT&T Bell Labs., Murray Hill, N.J., 1963-73, head phys. and inorganic chemistry rsch. dept., 1973-85; Presdl. prof. chemistry U. Calif., Irvine, 1986—, Presdl. chair, 1985—, regent lectr., 1984; vis. prof. Rockefeller U., N.Y.C., 1971, MIT, Cambridge, to 1975; vis. prof. chemistry U. Tel Aviv; adj. prof. U. Pa., Phila.; with Ctr. Biol. Studies, SUNY-Albany, 1979—; adj. prof. chemistry and biophysics Yale U., New Haven, 1980—; mem. numerous adv. bodies; lectr. Robert A. Welch Found., 1975; faculty lectr. Rensselaer Poly. Inst., Troy, N.Y., 1978; IBM lectr. Williams Coll., 1979; lectr. disting. lecture series U. Utah, 1980; Xerox lectr. N.C. State U., 1980; Frank C. Whitmore lectr. in chemistry Pa. State U., 1981; Dreyfus disting. scholar lectr., 1982; regent lectr. U. Calif., 1982, UCLA, 1985; Harry S. Ganning disting. lectr. U. Alta., Can., 1984; mem. IUPAC Commn. on Molecular Structure and Spectroscopy; chmn. 1981 Internat. Conf. on Photochemistry and Photobiology; bd. dirs. KRIKOS Sci. and Tech. Resources for Greece; mem. com. on kinetics Nat. Acad. Scis., NRC; chmn. fast reaction chemistry U.S. Fgn. Applied Sci. Assessment Ctr.; with NATO Advanced Study Insts., 1984—; dir. Quanex Corp. Assoc. editor Chem. Physics, Jour. Lasers and Chemistry, Jour. Biochem. and Biophys. Methods; editorial bd. Biophys. Jour., Jour. Chem. Intermediates; contbr. articles, papers to profl. publs.; patentee in field. Recipient Scientist of Yr. award, 1977, award for significant contbns. to field of biochem. instrumentation ISCO, 1979, award for leadership in sci. and edn. AHEPA, Disting. Alumni award SUNY, 1982, H.S. Ganning award U. Alta., 1984; Camille and Henry Dreyfus disting. scholar Williams Coll., 1982; AAAS fellow, 1985; alumni scholar Denison U., 1978. Fellow N.Y. Acad. Scis. (A Cressy Morrison award 1978), Am. Phys. Soc. (chmn. chem. physics div. 1979-80, exec. com. 1980-82, chmn. nominating com. 1981, Irving Langmuir award 1973); mem. Nat. Acad. Scis., Nat. Acad. Greece, Am. Chem. Soc. (exec. com. div. phys. chemistry to 1978, Peter Debye award phsy. chemistry 1982), Inter-Am. Photochem. Soc. (nominating coun., chmn. phys. div., Laser Conf. prize, 1989). Office: U Calif Dept Chemistry Irvine CA 92717

REPINE, JOHN E., pediatrician, educator; b. Rock Island, Ill., Dec. 26, 1944; married, 1969, 88; 4 children. BS, U. Wis., 1967; MD, U. Minn., 1971. Instr., then assoc. prof. internal medicine U. Minn., Mpls., 1974-79; asst. dir divsn. exptl. medicine Webb-Waring Lung Inst., Denver, 1979-89, prof. medicine, dir., 1989—; assoc. prof. medicine U. Colo., Denver, 1979—, assoc. prof. pediatrics, 1981—; mem. rsch. com., site visitor Am. Lung Assn., NIH; co-chmn. steering com. Aspen Lung Conf., 1980, chmn., 1981. Young Pulmonary Investigator grantee Nat. Heart & Lung Inst., 1974-75; recipient Basil O'Connor Starter Rsch. award Nat. Found. March of Dimes, 1975-77. Mem. AAAS, Am. Assn. Immunologists, Am. Fedn. Clin. Rsch., Am. Heart Assn. (established investigator award 1976-81), Am. Thoracic Soc., Am. Soc. Clin. Investigators, Assn. Am. Physicians. Achievements include research in role of phagocytes and oxygen radicals in lung injury and host defense. Office: Webb-Waring Lung Inst 4200 E 9th Ave Denver CO 80262-0001*

REPPERGER, DANIEL WILLIAM, electrical engineer; b. Charleston, S.C., Nov. 24, 1942; s. Daniel William and Mary (Schurer) R.; m. Frances Sullivan, Jan. 2, 1988; children: Lisa A. Repperger Cornwell, Daniel William III. BSEE, Rensselaer Poly. Inst., 1967, MSEE, 1968; PhD in Elec. Engring., Purdue U., 1973. Registered profl. engr., Ohio. Instr. Purdue U., West Lafayette, Ind., 1968-71, David Ross rsch. fellow, 1971-73; postdoctoral fellow NRC, Washington, 1973-75; electronics engr. Armstrong Lab., Dayton, Ohio, 1975—; adj. prof. Wright State U., Dayton, Ohio, 1984—; Contbr. articles to profl. jours. Author 2 book chpts.; contbr. over 30 articles to profl. jours. Recipient Comdr.s Disting. Paper award Wright Patterson AFB, 1987; named Rsch. Scientist of Yr., Affiliates Coun., 1991. Mem. IEEE (sr. assoc. editor Transactions on Control Systems Tech. 1992—, chmn. Conf. on Control Applications 1992, H. Schuck award 1980, Biomed. Engring. award 1990), Dayton IEEE (chmn. Control Systems Soc. 1987, chmn. Engring. Medicine and Biology Soc. 1988-93), Sigma Xi. Achievements include 7 patents; 12 Air Force inventions; development of joint Department of Defense-VA program to transfer technology from the military uses to help handicapped people. Home: 833 Blossom Heath Rd Dayton OH 45419-1102

REPPUCCI, NICHOLAS DICKON, psychologist, educator; b. Boston, May 1, 1941; s. Nicholas Ralph and Bertha Elizabeth (Williams) R.; m. Christine Marlow Onufrock, Sept. 10, 1967; children: Nicholas Jason, Jonathan Dickon, Anna Jin Marlow. BA with honors, U. N.C., 1962; MA, Harvard U., 1964, PhD, 1968. Lectr., rsch. assoc. Harvard U., Cambridge, Mass., 1967-68; asst. prof. Yale U., New Haven, 1968-73, assoc. prof., 1973-76; prof. psychology U. Va., Charlottesville, 1976—, dir. grad. studies in psychology, 1986—; originator biennial conf. on community rsch. and action, 1986. Assoc. editor Law and Human Behavior, 1986—; edit. bd. Am. Jour. Community Psychology, 1974-88, 88-91; author: (with J. Haugaard) Sexual Abuse and Children, 1988; editor: (with J. Haugaard) Prevention in Community Mental Health, (with L. Weithorn, E. Mulvey and J. Monahan) Mental Health, Law and Children, 1984; contbr. over 90 articles to profl. jours., chpts. to books. Adv. bd. on prevention Va. Dept. Mental Health, Mental Retardation and Substance Abuse Svcs., Richmond, 1986-92. Disting. scholar in psychology Va. Assn. Social Scis., 1991. Fellow APA (chair task force on pub. policy 1980-84), Am. Psychol. Soc., Soc. for Community Rsch. and Action (pres. 1986), Phi Beta Kappa. Office: Univ of Va Dept of Psychology Charlottesville VA 22903

RESCH, HELMUTH, environmental science educator; b. Vienna, Austria, May 22, 1933; came to U.S., 1956, 1960; Diploma in Engring., U. Agr. and Forestry, Vienna, 1956, PhD, 1960; MS, Utah State U., 1957. Rsch. asst. U.S. Forest Svc., J.P. Neils Lumber Co., Mont., 1957-58; acad. asst. Univ. Agr. & Forestry, Vienna, Austria, 1958-60; asst. prof. wood tech. Utah State U., Logan, 1960-62; assoc. and asst. prof. wood sci. Univ. Calif., Berkeley, 1962-70; prof. and head Forest Products Dept. Oregon State U., Corvallis, 1970-87; rsch. dean SUNY, Coll. Environ. Sci. & Forestry, Syracuse, 1987-92; prof. Universität für Bodenkultur, Vienna, Austria, 1992—; dir. Austrian Forest Products Lab., 1992—. Office: Universität Bodenkultur A-1180, Gregor Mendelstr 33-35, Vienna Austria

RESCH, LAWRENCE RANDEL, mechanical engineer; b. Balt., July 12, 1961; s. Ronald Herman and Joan Carol (Waxter) R.; m. Cheryl Lynn Sullins, June 6, 1986; 1 child, Christine Elizabeth. BSME, U. Md., 1984, MS, 1988. Student trainee Naval Surface Warfare Ctr., Silver Spring, Md., 1981-84; mech. engr. hydroballistic facilities Naval Surface Warfare Ctr., Silver Spring, 1984-90, advanced projects group leader hypervelocity wind tunnel, 1990—; cons. Obert Assocs., Inc., University Park, Md., 1988-89. Contbr. articles to profl. publs. Mem. AIAA, Instrument Soc. Am., Pi Tau Sigma (treas. 1983), Tau Beta Pi. Office: Naval Surface Warfare Ctr 10901 New Hampshire Ave Silver Spring MD 20903-5000

RESCORLA, ROBERT ARTHUR, psychology educator; b. Pitts, May 9, 1940; s. Arthur R. and Mildred J. (Jenkins) R.; children: Eric, Michael. BA, Swarthmore Coll., 1962; PhD, U. Pa., 1966; MA, Yale U., 1974. Successively asst. prof., assoc. prof., prof. Yale U., New Haven, 1966-80; prof. psychology U. Pa., Phila., 1981—, James Skinner prof. sci., 1986—. Author: Pavlovian Second-Order Conditioning, 1980. Contbr. articles to profl. jours. Mem. APA (pres. div. 3 1985, Disting. Sci. Contbn. award 1986), Am. Psychol. Soc. (William James fellow 1988), NAS, AAAS (pres. sect. J, psychology 1988-89), Soc. Exptl. Psychologists (Warren medal 1991), Psychonomic Soc. (mem. governing bd. 1979-85, chmn. publ. bd. 1985-86),

Ea. Psychol. Assn. (bd. dirs. 1983-86, pres. 1986-87). Office: U Pa Dept Psychology 3815 Walnut St Philadelphia PA 19104-6196

RESDEN, RONALD EVERETTE, medical devices product development engineer; b. Littleton, N.H., Oct. 27, 1944; s. Lawerence A. and Rita Mae (Bowen) R.; m. Dee Kronenburg, Apr. 20, 1974; children: Philip, Alison. Cons. Franklin Mfg. Co., Norwood, Mass., 1984—, Boston Sci. Co., Watertown, Mass., 1985—, Via Med, Easton, Mass., 1986—, White Marsh Labs., Balt., 1989—, Spectraphos Malmo Sweden, 1991—, Vision Scis. Inc., Natick, Mass., 1991—; Cordis Corp., Miami, Fla., 1993—. Author: Hologram Control Transfer, 1984; inventor, patentee in field. Mem. NRA (life), Soc. Plastics Engrs., Soc. Mfg. Engrs., Nat. Geographic Soc., Mass. Chiefs of Police Assn., Citizens for Ltd. Taxation. Home and Office: Resden Rsch Inc 44 Arrowhead Rd Weston MA 02193

RESHOTKO, ELI, aerospace engineer, educator; b. N.Y.C., Nov. 18, 1930; s. Max and Sarah (Kalisky) R.; m. Adina Venit, June 7, 1953; children: Deborah, Naomi, Miriam Ruth. B.S., Cooper Union, 1950; M.S., Cornell U., 1951; Ph.D., Calif. Inst. Tech., 1960. Aero. research engr. NASA-Lewis Flight Propulsion Lab., Cleve., 1951-56; head fluid mechanics sect. NASA-Lewis Flight Propulsion Lab., 1956-57; head high temperature plasma sect. NASA-Lewis Research Center, 1960-61, chief plasma physics br., 1961-64; asso. prof. engring. Case Inst. Tech., Cleve., 1964-66, dean, 1986-87; prof. engring. Case Western Res. U., 1966-88, chmn. dept. fluid thermal and aerospace scis., 1970-76, chmn. dept. mech. and aerospace engring., 1976-79, Kent H. Smith prof. engring., 1989—; Susman vis. prof. dept. aero. engring. Technion-Israel Inst. Tech., Haifa, Israel, 1969-70; cons. United Technologies Research Ctr., United Research Corp., Dynamics Tech. Inc., Arvin/Calspan Inc., Martin-Marietta Corp., Rockwell Internat.; mem. adv. com. fluid dynamics NASA, 1961-64; mem. aero. adv. com. NASA, 1980-87, chmn. adv. subcom. on aerodynamics, 1983-85; chmn. U.S. Boundary Layer Transition Study Group, NASA/USAF, 1970—; U.S. mem. fluid dynamics panel AGARD-NATO, 1981-88; chmn. steering com. Symposium on Engring. Aspects Magneto-hydro-dynamics, 1966, Case-NASA Inst. for Computational Mechanics in Propulsion, 1985-92, USRA/NASA ICASE Sci. Coun., 1992; Joseph Wunsch lectr. Technion-Israel Inst. Tech., 1990. Contbr. articles to tech. jours. Chmn. bd. govs. Cleve. Coll. Jewish Studies, 1981-84. Guggenheim fellow Calif. Inst. Tech., 1957-59. Fellow ASME, AAAS, AIAA (Fluid and Plasma Dynamics award 1980, Dryden Lectureship in Rsch. 1994), Am. Phys. Soc., Am. Acad. Mechanics (pres. 1986-87); mem. NAE, AAUP, Ohio Sci. and Engring. Roundtable, Sigma Xi, Tau Beta Pi, Pi Tau Sigma. Office: Case Western Res Univ University Circle Cleveland OH 44106

RESMINI, RICHARD, electrical engineer; b. Somerville, Mass., June 15, 1958; s. Charles and Rosemary (Ricci) R.; m. Kathleen Mary Duggan, Oct. 24, 1981; children: Kathleen Mary, Jennifer Lynn. BS in Elec. Engring., Tufts U., 1980; MS, Northeastern U., 1985. Engr. Raytheon Missile Systems Div., Bedford, Mass., 1980-82; project engr. Textron Def. Systems, Wilmington, Mass., 1982-93. Mem. IEEE. Republican. Roman Catholic. Home: 95 Woods Ave Somerville MA 02144

RESNATI, GIUSEPPE PAOLO, chemistry researcher, chemistry educator; b. Monza, Milan, Italy, Aug. 26, 1955; s. Ambrogio and Ester (Motta) R.; m. MariaAntonia Civati, May 25, 1983; children: Chiara Maria, Claudia Maria. Grad. cum laude, U. Milan, 1979; PhD with Carlo Scolastico, Poly. Milan, 1986. Rsch. chemist Farmitalia, Milan, 1980-81, Pierrel, Milan, 1981-82; sr. rsch. chemist Lab. Prodotti Biol., Milan, 1982-83; asst. prof. U. Padua, Italy, 1987-90; sr. rsch. chemist Italian Nat. Rsch. Coun., Rome, 1983—; vis. prof. U. Paris XI, 1993. Author: B-Acylanions, 1982, Muscarinic Compounds, 1989; contbr. over 70 articles to profl. jours. Recipient Corrado Fuortes award Inst. Lombardo Sci. Lettere, 1986, fellowship NATO, 1990-91. Mem. Am. Chem. Soc., Royal Soc. Chemistry, Italian Chem. Soc. Avocation: climbing. Home: Caronni 9, Monza 20052, Italy Office: Poly Dept Chimica, Pza Leonardo Da Vinci 32, Milan 20133, Italy

RESNICK, HENRY ROY, pharmacist; b. N.Y.C., Dec. 12, 1952; s. Samuel and Miriam (Jacobson) R.; m. Mary Lee Monroe. Sept. 13, 1981; children: Jacob Monroe, Aaron Leo. BS in Pharmacy, L.I. U., 1975, MS in Drug Info. & Communication, 1978. Pharmacist Sagamore Childrens Ctr., Melville, N.Y., 1976-77; staff pharmacist Montefiore Hosp. and Med. Ctr., N.Y.C., 1977-80; mgr. clin. pharmacy svcs. Beth Israel Med. Ctr., N.Y.C., 1980-82, asst. dir. pharmacy, 1982-87, supervising pharmacist methadone maintenance treatment program, 1982-90, sr. assoc. dir., 1987-90; dir. pharmacy New Rochelle (N.Y.) Hosp. Med. Ctr., 1990—; adj. clin. instr. Arnold and Marie Schwartz Coll. Pharmacy Health Scis., N.Y.C., 1980-82. Mem. Am. Soc. Hosp. Pharmacists, N.Y. State Coun. Hosp. Pharmacists (monitoring com. on legislation 1989-90, joint com. with industry 1989-90, govt. affairs com. 1991—), Westchester Soc. Hosp. Pharmacists (chmn. legis., constn. and bylaws com. 1991—, exec. com. 1991—, pres.-elect 1992-93, pres. 1993-94, del. N.Y. State coun. hosp. pharmacists 1992-93). Avocations: fishing, photography. Office: New Rochelle Hosp Med Ctr 16 Guion Pl New Rochelle NY 10801-5503

RESNICK, PHILLIP JACOB, psychiatrist; b. Cleve., Feb. 27, 1938; s. Zelick and Betty (Piccus) R.; children: Heather, Robert, Kimberly. BA in Psychology, Case Western Res. U., 1959, MD, 1963. Cert. Am. Bd. Psychiatry and Neurology, Am. Bd. Forensic Psychiatry. Resident psychiatry Case Western Res. U., Cleve., 1966-69, prof. psychiatry, 1969—. Contbg. author: (chpts.) Modern Perspectives in Psycho-Obstetrics, 1972, Psychiatry, 1986, Clinical Assessment of Malingering and Deception, 1988 and numerous others; contbr. over 30 articles to profl. jours. Capt. U.S. Army, 1963-66. Recipient Cert. of Commendation, Ohio Senate, 1984. Fellow Am. Psychiat. Assn.; mem. Am. Acad. Psychiatry and the Law (pres. 1984-85, Seymour Pollack Disting. Achievement award 1991), Group for the Advancement of Psychiatry (law and psychiatry 1985—), Alpha Omega Alpha. Office: Case Western Reserve Univ 2040 Abington Rd Cleveland OH 44106

RESTER, ALFRED CARL, JR., physicist; b. New Orleans, July 11, 1940; s. Alfred Carl and Willietta (Voth) R.; m. Blanche Sue Bing, June 20, 1964 (div. Jan. 20, 1985); children: Andrea Dawn, Karen Alane; m. Sherry Alice Warren, Dec. 12, 1985. BS, Miss. Coll., 1962; MS, U. N.Mex., 1964; PhD, Vanderbilt U., 1969. Postdoctoral fellow U. Delft, Netherlands, 1969-70, scientist first class, 1970-71; guest prof. U. Bonn (Germany) Dept. Physics, 1972-74; asst. prof. Emory U., Dept. Physics, Atlanta, 1975-76; assoc. prof. Tenn. Tech. U., Dept. Physics, Cookeville, Tenn., 1976-77; vis. assoc. prof. U. Fla., Dept. Physics, Gainesville, 1978-81; assoc. rsch. scientist U. Fla. Space Astronomy Lab., Gainesville, 1981-88; dir. U. Fla. Inst. for Astrophysics and Planetary Exploration, Alachua, 1988-93; pres., CEO Constellation Tech. Corp., St. Petersburg, Fla., 1992—; cons. Lockheed Ga., Marietta, 1976, Oak Ridge (Tenn.) Nat. Lab., 1976-77. Editor: High Energy Radiation Background in Space, 1989; contbr. articles to profl. jours. Recipient Antarctic Svc. medal NSF-USN, 1988; grantee various fed. agys. Mem. AAAS, Am. Phys. Soc., Am. Astron. Soc., Sigma Xi Sci. Rsch. Soc. Avocations: writing fiction, fishing, nature walks. Office: Constellation Tech Corp One Progress Blvd Box 33 PO Box 14405 Saint Petersburg FL 33733-4405

RESTORFF, JAMES BRIAN, physicist; b. Wytheville, Va., June 22, 1949; s. William Poulson and Lula Avery (Wallace) R.; m. Kathleen Ann Leonard, June 2, 1973; 1 child, Cheryl. BS, U. Md., 1971, PhD, 1976. Rsch. physicist Naval Surface Warfare Ctr., Silver Spring, Md., 1976—. Mem. Am. Phys. Soc. Achievements include research on soft magnetic materials. Office: Naval Surface Warfare Ctr 10901 New Hampshire Ave Silver Spring MD 20903-5640

RETSAS, SPYROS, oncologist; b. Thessaloniki, Greece, Dec. 4, 1942; arrived in U.K., 1969; s. Stylianos and Panaghiota (Alexandri) R.; m. Diana Gillian Rees, July 8, 1972; 1 child, Philip-Alexander. MD, Aristotle U., Thessaloniki, 1967, U. Athens, 1978. Accredited specialist. Sr. house officer Royal Marsden Hosp., Surrey, Eng., 1971-72; registrar in medicine Whipps Cross Hosp., London, 1973-75; sr. lectr. Westminster Hosp., London, 1978-85; cons. med. oncologist Chelsea & Westminster Med. Sch., Charing Cross Hosps., London, 1985—; tchr. U. London, 1979—; chmn. adv. com. on cancer svcs. Riverside Health Authority, 1990; vis. prof. U. Ioannina (Greece) Med. Sch., 1991-92; lectr. cancer med. instns. in Europe, N.Am. S.Am. and China. Editor: Palaeooncology-The Antiquity of Cancer, 1986. Contbr. articles to profl. jours. Founder, pres. Hellenic Med. Soc. Gt. Britain, 1987. 2d lt. M.C. Greek Army, 1967-69. Fellow Royal Soc. Medicine; mem. Royal Coll. Physicians, British Assn. Cancer Physicians, British Assn. Cancer Rsch., European Soc. Med. Oncology, Am. Soc. Clin. Oncology. Avocations: skiing, horseback riding, travel, history of medicine. Home: Parnassus Park Hill, Essex, Loughton 1G10 4ES, England Office: Med Oncology Unit, Charing Cross Hosp, London W6 8RF, England

RETTERER, BERNARD LEE, electronic engineering consultant; b. Waldo, Ohio, Sept. 23, 1930; s. Calvin C. and Gertrude S. (Kries) R.; m. Mary Susan Gaster, Dec. 22, 1951; children: John, Jeffrey, Laura. BSEE, Ohio No. U., 1952; MS in System Mgmt., George Washington U., 1972. Registered profl. engr., Ohio. Program mgr. RCA, Cherry Hill, N.J., 1953-65; v.p. engring. ARINC Rsch., Annapolis, Md., 1965-90; Del. to Internat. Electrotechnical Commn., 1975-85; cons. Inst. for Def. Analyses, Alexandria, Va., 1990—. Contbr. 39 tech. articles on maintainability to profl. jours. Mem. adv. bd. Embry-Riddle U., Daytona Beach, Fla., 1988-90, Anne Arundel Coll., Arnold, Md., 1987-90. Mem. IEEE (v.p. tech-ops. 1982), Operation Rsch. Soc., Armed Forces Prep. Assn., Armed Forces Comms. Assn. Achievements include development (with others) of maintainability prediction technique for electronic equipment; research into the use of information content of failure symptoms as a predictor of diagnostic time. Home: 37 Whittier Pkwy Severna Park MD 21146

RETTERSTOL, NILS, psychiatrist; b. Oslo, Norway, Oct. 3, 1924; s. Kittel and Kathrine (Steen) R.; M.D., U. Oslo, 1950, Dr.med., 1966; m. Kirsten Christensen, Aug. 16, 1958; children: Trine Lise, Kjetil, Lars Jorgen. Med. officer Dikemark Hosp., Ulleval Hosp., Oslo, 1952-56, Runwell Hosp., Eng., 1956-57; resident in psychiatry Ullevål Hosp., Oslo, 1957-58; asso. prof., dep. dir. Univ. Psychiat. Clinic, Oslo, 1959-68; prof. psychiatry U. Bergen, head dir. Neevengarden Hosp., Bergen, 1969-73; prof. psychiatry U. Oslo, 1973—; head dir. Gaustad Hosp., Oslo, 1973—; head Norwegian Info. Bank for Narcotic problems; chmn. Norwegian Commn. for Forensic Pschiatry, 1983; cons. in field. Served to capt. Norwegian Army. Recipient gold medal for psychiat. research, H.M. The King of Norway; prize for research Norwegian Council Humanities and Sci., 1978; comdr. The Royal St. Olav Order, 1984. Mem. Norwegian Acad. Sci. (hon.), Swedish Psychiat. Assn., German. Assn. Neurology and Psychiatry, Finnish Assn. Psychiatry, Norwegian Med. Soc., Norwegian Psychiat. Soc., French Assn. Psychiatry, Assn. European Psychiatrists, Internat. Assn. Suicide Prevention and Crisis Intervention (pres.). Author: 30 books including Paranoid and Paranoiac Psychoses, 1966; Drug Dependence, 1967, Suicide, 1970, 5th edit., 1990, Prognosis in Paranoid Psychoses, 1970 (with L. Eitinger and A. A. Dahl) Crisis and Neuroses, 5th edit., 1990, (with L. Eitinger) Forensic Psychiatry, 4th edit., 1990, (with Eitinger and U. Malt) Psychoses, 4th edit., 1984, Suicide A European Perspective, 1993; editor Scand. Med. Yearbook, 1972, Eur. Archives of Phychiatry and Neurological Scis., 1975—; European editor Jour. Drug Issues, 1980; co-editor Psychopathology, 1983. Home: Nordseterveien 20 A, Oslo 11, Norway Office: Gaustad Hosp, Boks 24 Gaustad, 0320 Oslo 3, Norway

RETTIG, TERRY, veterinarian, wildlife consultant; b. Houston, Jan. 30, 1947; s. William E. and Rose (Munves) R.; m. Debra Holmes, Apr. 4, 1992; children from previous marriage: Michael Thomas, Jennifer Suzanne. B.S. in Zoology, Duke U., 1969, M.A.T. in Sci., 1970; D.V.M., U. Ga., 1975. Resident veterinarian, mgr. animal health The Wildlife Preserve, Largo, Md., 1975-76; wildlife veterinarian Dept. Environ. Conservation State of N.Y., Delmar, 1976-77; owner Atlanta Animal Hosp., 1976—; sec., dir. Atlanta Pet Supply, Inc., 1983-89; cons. Six Flags Over Ga., Yellow River Game Ranch, Stone Mountain Park Animal Forest, Atlanta Zoo. Author: (with Murray Fowler) Zoo and Wild Animal Medicine (Aardvark award 1978), 1978, 2d edit., 1986 (Order of Kukukifuku award 1986); contbr. articles to profl. jours. Del. Dekalb County Republican Conv., 1983; mem. Roswell United Meth. Ch., Boy Scouts Am., 1954—, mem. troop coun., asst. scoutmaster, scout master, Philmont expedition leader, 1988, 89. Spl. scholar Cambridge U. Coll. Vet. Medicine, 1973-74. Mem. AVMA, Ga. Vet. Med. Assn., Greater Atlanta Vet. Med. Assn., Dekalb Vet. Soc., Acad. Vet. Medicine, Am. Assn. Zoo Veterinarians, Am. Assn. Zool. Parks and Aquaria, Nat. Wildlife Health Found., Nat. Wildlife Assn., Atlanta Zool. Soc., Am. Fedn. Aviculturists, Cousteau Soc., Am. Assn. Avian Veterinarians, Am. Animal Hosp. Assn., Internat. Wildlife Assn., Soc. Aquatic Veterinary Medicine, Am. Buffalo Assn. Methodist. Home: 5035 Kimball Bridge Rd Alpharetta GA 30202-4543 Office: Atlanta Animal Hosp 5005 Kimball Bridge Rd Alpharetta GA 30202-4543

RETTURA, GUY, civil engineer; b. Nicastro, Catanzaro, Italy, Nov. 22, 1945; s. Giuseppe and Giuseppina (Caruso) R.; m. Cecilia Marie Rizzardi, Nov. 17, 1973; children: Christina Marie, Brian Joseph. AS, Point Park Coll., 1966; B. Engring., Youngstown State U., 1972. Registered profl. engr., Pa., Ohio, Md.; cert. profl. surveyor, Pa. Engr. Rust Internat. Co., Pitts., 1972-75, sr. design engr., 1975-85, structural, archtl., civil depts. mgr., 1985-90, chief engr., 1990—. Author: (with Jack D. Bakos, Jr.) Structural Analysis for Engineering Technology-Solutions Manual, 1973. Staff sgt. U.S. Army, 1968-72. Mem. ASCE, NSPE, Am. Concrete Inst., Am. Inst. Steel Constrn. Home: 2461 Summit St Bethel Park PA 15102-2074 Office: Rust Internat Co 441 Smithfield St Pittsburgh PA 15222-2219

RE VELLE, JACK B(OYER), consulting statistician; b. Rochester, N.Y., Aug. 2, 1935; s. Mark A. and Myril (Bubes) Re V.; m. Brenda Lorraine Newcombe, Aug. 2, 1968; 1 child, Karen Alyssa. BS in Chem. Engring., Purdue U., 1957; MS in Indsl. Engring. and Mgmt., Okla. State U., 1965, PhD in Indsl. Engring. and Mgmt., 1970. Commd. 2d lt. USAF, 1957, advanced through grades to major, 1968, resigned, 1968; adminstrv. asst. Gen. Dynamics, Ft. Worth, 1970-71; cons. engr. Denver, 1971-72; chmn. decision scis. U. Nebr., Omaha, 1972-77; dean Sch. Bus. and Mgmt. Chapman U., Orange, Calif., 1977-79; sr. staff engr. McDonnell Douglas Space Systems, Huntington Beach, Calif., 1979-81; head mfg. tng. and devel. Hughes Aircraft Co., Fullerton, Calif., 1981-82, sr. statistician, 1982-86; corp. mgr. R & D Hughes Aircraft Co., L.A., 1986-88, corp. chief statistician, 1988—; mem. bd. examiners Malcolm Baldrige Nat. Quality award Nat. Inst. for Standards and Tech., U.S. Dept. Commerce, Washington, 1990, 93; cons. to various pub. and pvt. orgns.; presenter in field; lectr. at seminars. Author: Safety Training Methods, 1980, The Two-Day Statistician, 1986, The New Quality Technology, 1988, (with others) Quest for Quality, 1986, Mechanical Engineers Handbook, 1986, Production Handbook, 1987, Handbook of Occupational Safety and Health, 1987, A Quality Revolution in Manufacturing, 1989, Quality Engineering Handbook, 1991; co-author: Quanititative Methods for Managerial Decisions, 1978. Bd. dirs. Assn. for Quality and Participation, Cin., 1985-86. Fellow Inst. for the Advancement Engring., 1986; recipient Disting. Econs. Devel. award Soc. Mfg. Engrs., 1990. Fellow Am. Soc. for Quality Control (co-chair total quality mgmt. com. 1990-92, nat. accreditation project dir. 1978-80), Am. Soc. Safety Engrs., Inst. Indsl. Engrs. (treas. 1992—, regional v.p. 1982-84). Office: Hughes Aircraft Co 7200 Hughes Ter Los Angeles CA 90045

REVELLE, WILLIAM ROGER, psychology educator; b. Washington, Nov. 22, 1944; s. Roger Randall and Ellen Virginia (Clark) R.; m. Eleanor McNown, June 20, 1965; children: Daniel James, David Robert. BA, Pomona Coll., 1965; PhD, U. Mich., 1973. Vol. U.S. Peace Corps, Sarawak, Malaysia, 1965-67; rsch. asst. U. Mich., Ann Arbor, 1968-73; asst. prof. psychology Northwestern U., Evanston, Ill., 1973-79, assoc. prof., 1979-84, prof. psychology, 1984—, chmn. dept. psychology, 1987—; mem. Cognitive Emotion Personality com. NIMH-IRG, 1986-90. Contbr. articles to profl. jours. NIH Fogarty Sr. Internat. fellow, Oxford U., 1981-82. Mem. APA, Am. Psychol. Soc., Internat. Soc. for Study of Individual Differences (dir. 1989-93), Soc. of Multivariate Exptl. Psychology (pres. 1984-85). Home: 2302 Orrington Evanston IL 60201 Office: Northwestern Univ Dept Psychology 2029 Sheridan Rd Evanston IL 60208

REVELS, MIA RENEA, science educator; b. DeQueen, Ark., Sept. 18, 1966; d. Richard Paul Revels and Novadean (Davis) Shelton. BS in Edn., Henderson State U., Arkadelphia, Ark., 1987, MS in Edn., 1990. Rsch. asst. Henderson State U., Arkadelphia, 1988-89, instr., 1989-90; instr. Bapt. Med. Systems, Little Rock, 1990-91; teaching asst. U. Ark., Fayetteville, 1991—. Vol. Spl. Olympics, Arkadelphia, 1988-91; judge Sci. Fair, Arkadelphia, 1988-91. Recipient Outstanding Achievement award U.S. Acad. Scis., 1987. Mem. Cooper Ornithol. Soc., Sigma Xi, Phi Kappa Phi, Beta Beta Beta. Home: 1007 W Lawson Fayetteville AR 72703 Office: U Ark 632 Sci-Engring Fayetteville AR 72701

REVICZKY, JANOS, mathematician, researcher; b. Budapest, Hungary, Mar. 17, 1954; s. Janos and Kornelia (Bakonyi) R.; m. Katalin Parkany, Sept. 9, 1978; children: Agnes, Katalin, Adam. MSc, Eötvös, Budapest, 1978, Kerteszeti, Budapest, 1981. Rsch. fellow Computer and Automation Inst. Compter and Automation Inst. Hungarian Acad. Scis., Budapest, 1978-85; mem. R&D staff Insotec Cons. GmbH, Munich, 1986-88, GMI Graphische Systeme (formerly Digidata GmbH), Munich, 1989—; cons. Internat. Standard Orgn., Deutsches Institut für Normung, Berlin., Germany, 1986—. Author: Area Filling Methods in Computer Graphics, 1985, Eurographics '85, 1985, Eurographics '87, 1987, GKS Theory and Practice, 1987, Proceedings of the GKS Review, 1987; contbr. articles to profl. jours. Community helper Schönstatt movement, Munich, 1988—. Sgt. Hungarian mil., 1972-73. Recipient Best Paper award Eurographics Assn., 1985, 2d Best Paper award, 1987, Schweitzer Miklos Math. award, 1973, award of young scis. Hungarian Acad. Scis., 1984. Mem. Bolyai Janos Math. Soc., Eurographics Assn., Hungarian Bridge Soc., Hungarian Mycology Soc. Roman Catholic. Avocations: music, bridge, mushroom, travelling, coin collecting. Home: Albert Schweitzer Str 56, 8000 Munich 83, Germany Office: GMI Graphische Systeme, Goethestr 17, 8000 Munich 2, Germany

REVOL, JEAN-PIERRE CHARLES, physicist; b. Neuville-les-Dames, Ain, France, Aug. 15, 1948; came to U.S., 1974; s. Joseph and Andrée (Ruet) R. Degree in engring., ENSAM, Paris, 1972; lic. de math., Paris VI U., 1973; PhD, MIT, 1981. Fellow CERN, Geneva, Switzerland, 1982-84, mem. sr. staff, 1989—; asst. prof. MIT, Cambridge, Mass., 1984-89, assoc. prof., 1989—; advisor to dir. gen. CERN, Switzerland, 1991—. Sgt. French Army, 1975-76. Recipient Lettre de Felicitations Ministre des Armees, France, 1976. Mem. Am. Phys. Soc., Sigma Xi. Achievements include co-discovery of W and Z Bosons. Office: MIT 77 Massachusetts Ave Cambridge MA 02139-4307

REW, WILLIAM EDMUND, civil engineer; b. Corning, N.Y., Nov. 24, 1923; s. Robert James and Clara (Neal) R.; m. Jean Ella Ohls, Aug. 16, 1947 (dec.); children: Virginia Ann, Robert James, John Edward. BE, Yale U., 1954, M in Engring., 1955. Registered profl. engr., N.Y., Fla., Calif., Ill. Project engr. Texaco & Affiliate, USA and Saudi Arabia, 1955-62; sr. engr. Martin-Marietta Corp., Cape Kennedy, Fla., 1962-63, Chrysler Corp, Cape Kennedy, 1963-65, The Boeing Co., Cape Kennedy, 1965-70; project mgr. Brevard Engring. Co., Cape Canaveral, Fla., 1970-74; city engr. City of Vero Beach (Fla.), 1974-77; resident engr. Post, Buckley, Schuh & Jernigan, Miami, Fla., 1977-85; mgr. Keith & Schnars, P.A., West Palm Beach, Fla., 1985-90; pvt. practice consulting Lake Placid, Fla., 1990—. Active Dem. Party of Brevard County. 1st lt. U.S. Army, 1942-46, ATO. Scholar of 2d rank Yale U., 1953, grad. scholar, 1955. Fellow ASCE (chmn. Fla. ann conv. 1971, Engr. of Yr. 1974); mem. NSPE, Soc. Am. Mil. Engrs. (bd. dirs. 1982-83), Fla. Engring Soc. (chpt. pres. 1976), Yale Club, Browning Assn. Club. Episcopalian. Avocations: woodworking, reading. Home: 1425 S Washington Blvd NW Lake Placid FL 33852-4031

REYES, MARCIA STYGLES, medical technologist; b. Winchester, Mass., July 15, 1950; d. Bernard Francis and Eleanore Cecilia (Nicgorska) Stygles; B.S. in Med. Tech., Merrimack Coll., North Andover, Mass., 1972; M.S. in Health Scis. (Kellogg Found. grantee), SUNY, Buffalo, 1977; m. Carlos Reyes, Aug. 5, 1978. Sr. med. technologist Symmes Hosp., Arlington, Mass., 1970-73; sr. microbiologist and serologist Mt. Auburn Hosp., Cambridge, Mass., 1973-75; asst. prof., clin. coordinator Quinnipiac Coll., Hamden, Conn., 1976-81; lab. supr. Canberra Clin. Labs., Meriden, Conn., 1981-86; lab. supr. Hill Health Ctr, New Haven, Conn., 1984—; cons. in med. tech. mgmt., allied health edn. Mem. Am. Soc. Clin. Pathologists, Am. Soc. Med. Tech., Conn. Soc. Med. Tech. (Speaker awards), Am. Soc. Microbiology, Am. Soc. Allied Health Profls. Home: 199 Dover St New Haven CT 06513-4818

REYES-GUERRA, ANTONIO, dental surgeon; b. San Salvador, El Salvador, Aug. 4, 1916; s. Antonio and Linda Elizabeth (Hardesty) Reyes-G.; m. Olive Isabella Scott, Nov. 27, 1954; children: Richard Bruce, Alan Scott. BA, BS, St. Joseph's Coll., 1938; DDS, U. El Salvador, 1943, U. Pa., 1946. Dir. dentistry pub. health Pub. Health Dept., El Salvador, 1944; intern, then resident in oral surgery Rosales Hosp., El Salvador, 1943-44; oral surgeon Polyclinic Hosp., N.Y.C., 1956-60, Lincoln Hosp., N.Y.C., 1955-62; editor Jour. Am. Soc. Advancement of Anesthesia in Dentistry, 1973-88. Mem. Am. Soc. Advancement Anesthesia in Dentistry (pres. 1970-71, exec. sec. 1971-88, dir. 1980—), ADA, N.Y. Dental Soc., Am. Dental Soc. Anesthesia, Internat. Dental Fedn. Anesthesists, N.Y. Diplomate Oral Surgery, Eastchester Dental Soc. Avocations: history, fencing, golf, colecting toy soldiers. Home: 50 Winter Hill Rd Tuckahoe NY 10707-4315 Office: Am Soc Advancement Anesthesia Dentistry 475 White Plains Rd Tuckahoe NY 10707-1517*

REYNICK, ROBERT J., materials scientist; b. Bayonne, N.J., Dec. 25, 1932; s. Mary Reynik; m. Georgiana M. Walker, Apr. 12, 1959; children: Michael, Christopher, Jonathan, Katherine, Steven, Kevin. BS in Math. and Physics, U. Detroit, 1956; MS in Elec. Engring., U. Cin., 1960, PhD in Phys. Chemistry, 1963. Assoc. dir. engring. materials program engring. divsn. NSF, Washington, 1970-71, dir. engring. materials program Divsn. Materials Rsch., 1971-74, dir. metallurgy program Divsn. Materials Rsch., 1974-83, acting head metallurgy, polymers, ceramics and electronic material sect. Divsn. Materials Rsch., 1985-90, head metallurgy, polymers, ceramics and electronic materials sect. Divsn. Materials Rsch., 1985-90, head office spl. programs in materials Divsn. Materials Rsch., 1990—; rsch. assoc. Sch. Metallurgical Engring. U. Pa., 1963-64; asst. prof. dept. materials engring. Drexel U., 1964-67, assoc. prof., 1967-70; vis. prof. dept. materials sci. and engring. U. Pa., 1982-83; cons. Litton Industries, 1966-67, RCA Corp., 1970, Univ. City Sci. Ctr., 1968; vis. scientist Am. Inst. Physics, Cin. High Schs., 1960-61. Recipient Univ. scholarships and Rsch. fellowships U. Cin., 1958-63, Post Doctoral fellowship U. Pa., 1963, ASM fellowship, 1993. Mem. AAAS, Am. Soc. Materials Internat. (coun. mem. at large materials sci. divsn. 1988—, chmn. com. nonmetallic materials 1984-85, internat. coun. alloy phase diagrams 1978-85), Am. Chem. Soc., Am. Phys. Soc., N.Y. Acad. Scis., Minerals Metals and Materials Soc. (mem., chmn. various coms.), Sigma Xi (past chpt. pres.), Tau Beta Pi. Office: Nat Sci Found Divsn Materials Rsch 1800 G St NW Washington DC 20550

REYNOLDS, BRADFORD CHARLES, industrial engineer, management consultant; b. Ft. Smith, Ark., May 15, 1948; s. Charles Francis and Beverly (Berry) R.; m. Kathleen Woods, June 19, 1971; 1 child, Karissa Elizabeth. BS in Engring., Tulane U., 1970, MBA in Fin., 1971. Registered profl. engr., Va. Prodn. supr. Philip Morris Inc., Richmond, Va., 1971, project engr., 1972-74, sr. indsl. engr., 1974-75, supr. dept. indsl. engring., 1975-78, supr. facilities planning, 1978-79, supr. dept. materials handling, 1980, mfg. mgr., 1980-83, mgr. mfg. coordination, 1983—; lectr. total quality mgmt. and fin. Program coord. Jr. Achievement, Richmond, 1976, lectr., tchr. project bus., 1984; bd. dirs. Offender Aid and Restoration. Texaco scholar Tulane U., 1970. Mem. Inst. Indsl. Engrs. (sr., v.p. 1979-80), TAPPI, Instrument Soc. Am., Am. Mgmt. Assn., Rotary. Episcopalian. Avocations: golf, boating, swimming, music. Home: 9514 Chatterleigh Dr Richmond VA 23233-4405 Office: Philip Morris Inc PO Box 26603 Richmond VA 23261-6603

REYNOLDS, C(LAUDE) LEWIS, JR., materials scientist, researcher; b. Roanoke, Va., Dec. 16, 1948; s. Claude Lewis and Lois Anne (Warren) R.; m. Sherryl Ann Allbright, June 27, 1970 (div. Apr. 1989); children: Karen, Brian, Kristin; m. Mary Elizabeth Derr, Aug. 29, 1989; 1 child, Jeff. BS in Physics, Va. Mil. Inst., 1970; MS in Materials Sci., U. Va., 1972, PhD in Materials Sci., 1974. Sr. scientist dept materials sci. U. Va., Charlottesville, 1974-75; rsch. assoc. dept. physics U. Ill., Urbana, 1975-77; sr. project engr. Union Carbide Corp., Indpls., 1977-80; disting. mem. tech. staff AT&T Bell Labs., Reading, Pa., 1980-92, Breiningsville, Pa., 1992—. Contbr. over 80 articles to profl. jours. Mem. IEEE, AAAS, Am. Phys. Soc., Materials Rsch. Soc., Am. Assn. Physics Tchrs., Sigma Xi. Methodist. Achievements include patents for LPE apparatus with improved thermal geometry, slefaligned rib-waveguide high power laser, fabrication of GaAs integrated circuits. Office: AT&T Bell Labs 9999 Hamilton Blvd Breinigsville PA 18031

REYNOLDS, DON WILLIAM, geologist; b. Centerburg, Ohio, Apr. 6, 1926; s. Loren William and Charlotte Lonas (Hunt) R.; m. Betty Jeannette Spears, Sept. 4, 1953; children: Don William, Jr., Richard Allen (dec.), Brenda Gay. BS, Ohio State U., 1952. Registered profl. geologist, Calif. Mgr., Geochem. Engring. Inc., Midland, Tex., 1950-52; geologist Union Oil Co. Calif., Midland, 1953-66, dist. exploration geologist, Anchorage, 1966-68, area geologist, Bakersfield, Calif., 1968-76, dist. devel. geologist, Ventura, Calif., 1976-86; dist. devel. geologist mid-continent dist., Oklahoma City, 1986-89, regional mgr. mid-continent devel., 1989-90, advisor geology, 1990-92; chmn. bd. Future Petroleum Corp., 1992—; gen. ptnr. Reynolds Farms, 1979—; sec. I.F.P., Inc., 1989—. Pres. Park Stockdale Civic Assn., Bakersfield, 1970, Clearpoint Home Owners Assn., Ventura, 1980-86; chmn. Kern County Freeway Com. Bakersfield, 1970-73. Served wtih USAF, 1944-45. Mem. Am. Assn. Petroleum Geologists, West Tex. Geol. Soc. (sec. 1965-66), Kans.-Okla. Oil and Gas Assn. (nomenclature com. 1987-92), San Joaquin Geol. Soc. (treas. 1974-75), Am. Assn. Petroleum Geologists (sec. Pacific sect. 1975-76). Republican. Methodist. Office: PO Box 1327 Bowie TX 76230

REYNOLDS, ELLIS W., chemist; b. Owensboro, Ky., May 16, 1932; s. Joseph A. and Hallie P. (Bevins) R.; m. Joanna, Jan. 16, 1957 (div. 1965); children: Jennifer, Lee Anne; m. Anetta D. Wonsettler, Apr. 19, 1970; stepchildren: Sandra, Joan, Kathryn. BS in Chemistry, Ky. Wesleyan U., 1958; Indsl. Tech., U. North Fla., 1980. Rsch. chemist Glidden-Durkee, Jacksonville, Fla., 1959-62, prodn. supt., 1962-69, project engr., 1969-73, orig. equip. mfg., 1973-76; forest rsch. mgr. Organic Chem. div. SCM, Jacksonville, Fla., 1976-86; tech. svc. dir. Glidco Organics, Jacksonville, Fla., 1986-90; cons., by-product recovery tech. svcs. and recovery systems Ellis Reynolds & Assocs., Jacksonville, Fla., 1990—; adj. prof. div. tech. U. North Fla., Jacksonville, 1982-83. Contbr. articles to profl. jours. Vol. Duval County Rd. Patrol, Jacksonville, 1966-70; vol. various polit. campaigns. With USMC, 1949-52. Mem. TAPPI, Pulp Chems. Assn. (adv. com. 1968-90, recovery com. 1968-90), Jacksonville Hist. Soc., Univ. Club, Optimist Club (pres.). Republican. Achievements include 6 patents. Home: 2506 Acadie Dr Jacksonville FL 32217 Office: Ellis Reynolds & Assocs PO Box 24072 Jacksonville FL 32241

REYNOLDS, MICHAEL EVERETT, architect; b. Peewee Valley, Ky., Aug. 9, 1945; s. David William and Marjorie Ellen (Mount) R.; m. Virginia Vetter, June 12, 1980 (div. 1987); stepchildren: Georgia, John, Brady Coleman; m. Christine Marie Palmeri, May 1, 1988; 1 child, Jonah; stepchildren: Justin, Tanya Simpson. BA, U. Cin., 1969. Registered architect N.Mex., Colo. Pvt. practice arch. Taos, N.Mex., 1969—; lectr in field; participant UN Conf. on Low Cost Housing for the Third World, 1974; cons. Venezuelan Govt. on use of recycled materials in housing for barrios in Caracas, 1976. Author: A Coming of Wizards, 1989, Earthship, Vol. I, 1990, Vol. II, 1991, Earthship, Vol. 3; contbr. numerous articles to profl. jours., newspapers; featured in: The Wizard's Eye, 1978, Home Sweet Dome (Germany), 1978, Design for a Limited Planet, People Who Live in Solar Houses, Fantastic Architecture, 1980, Building for Tomorrow, 1982. Recipient Burlington House award for interiors, 1975, Terra Alpha Tech. 1990 acknowledgement award, 1990; Nat. Endowment Arts Design Div. grantee, 1981. Achievements include research and development of self-sufficient housing made from recycled materials. Home and Office: PO Box 1041 Taos NM 87571

REYNOLDS, MICHAEL FLOYD, civil engineer, educator; b. Watertown, Wis., Oct. 17, 1953; s. Richard Lee and Shirley P. Reynolds; m. Debbie Carroll, June 2, 1977 (dec. Apr. 1992); children: Richard, Heidi, Christina. BSCE, USAF Acad., 1977; MS in Indsl. Engring., Purdue U., 1980; DEng, Tex. A&M U., 1990. Registered profl. engr., Colo. Commd. USAF, 1977, advanced through grades to lt. col. selectee, 1992; asst. chief ops. USAF, Peterson AFB, Colo., 1977-79; chief constrn. mgmt. USAF, Minot AFB, N.D., 1981-82, chief resources and requirements, 1982; constrn. engring. mgr. USAF, Riyadh, Saudi Arabia, 1982-83; asst. prof. dept. civil engring. USAF Acad., Colo., 1983-86; chief ops. and maintenance USAF, Peterson AFB, Colo., 1986-88; assoc. prof. civil engring., dept. civil engring. USAF Acad., Colo., 1990—. Decorated Meritorious Svc. medal with 1 oak leaf cluster, Air Force Commendation medal with 2 oak leaf clusters. Fellow ASCE, NSPE, Am. Soc. Engring. Edn., Soc. Am. Mil. Engrs., Assn. Grads. USAF Acad. Home: 11130 Teachout Rd Colorado Springs CO 80908 Office: Hdqrs USAF/Dept. Civil Engring USAF Acad Colorado Springs CO 80840

REYNOLDS, ROBERT JOEL, economist, consultant; b. Indpls., May 13, 1944; s. Joel Burr and Betty (Schimpf) R.; m. Lucinda Margaret Lewis, May 27, 1979; children: Joel, Sarah. BSBA in Fin., Northwestern U., 1965, PhD in Econs., 1970. Asst. prof. econs. U. Idaho, Moscow, 1969-73, assoc. prof., 1973-75; asst. dir, sr. economist econ. policy office Dept. Justice, Washington, 1973-81; sr. economist, v.p. ICF Inc., Washington, 1981-87, sr. v.p., 1987-91; exec. v.p., prin. Econsult Corp., Washington, 1991—; vis. assoc. prof. U. Calif., Berkeley, 1976-77, Cornell U., Ithaca, N.Y., 1981. Reviewer: NSF, Rand Jour. of Econs., Internat. Econ. Rev., Internat. Jour. Indsl. Orgn., Jour. Indsl. Econs., Am. Econ. Rev.; mem. editorial bd. Managerial and Decision Econs.; contbr. numerous papers to profl. jours. Recipient Dow Jones award Wall St. Jour., 1965; AT&T grantee, 1971-72, Brookings Instl. grantee, 1968-69; NDEA fellow, 1965-69. Mem. AAAS, Am. Econ. Assoc., Econometric Soc., Royal Econ. Soc., Am. Statis. Assn., European Assn. for Rsch. in Indsl. Econs., Soc. for the Promotion of Econ. Theory, Math. Assn. Am. Congregationalist. Home: 9228 Farnsworth Dr Potomac MD 20854-4503 Office: Econsult Corp Ste 370 901 15th St NW Washington DC 20005

REYNOLDS, TERRY SCOTT, social science educator; b. Sioux Falls, S.D., Jan. 15, 1946; s. Ira Ebenezer and Therasea Anne (Janzen) R.; m. Linda Gail Rainwater, June 4, 1967; children: Trent Aaron, Dane Adrian, Brandon Vincent, Derek Vinson. BS, So. State Coll., 1966; MA, U. Kans., 1968, PhD, 1973. Asst. prof. U. Wis., Madison, 1973-79, assoc. prof., 1979-83; dir. program in sci., tech. and soc. Mich. Technol. U., Houghton, 1983-87, prof., 1987—, head dept. social scis., 1990—; adv. editor History of Technology, Encyclopedia Americana, 1980-92; adv. bd. St. Martin's Press Great Engrs. Project, 1980-84. Author: Sault Ste Marie: The Hydroelectric Plant's History, 1982, Stronger Than a Hundred Men, 1983, History of American Institute of Chemical Engineers, 1983, The Engineer in America, 1991. Mem. Soc. for the History Tech. (chair awards com. 1988-89), Soc. for Indsl. Archeology (Norton prize 1985). Methodist. Office: Mich Technol Univ Dept Social Scis 1400 Townsend Dr Houghton MI 49931

REYNOLDS, THOMAS ROBERT, scientist, biotechnology company executive; b. Pitts., Aug. 15, 1962; s. Thomas Noll and Carol Ruth (Grimes) R.; m. Julie Yvone Synder, June 16, 1984; children: Nicholas Richard, Alexander Sammeul. BS in Sci./Biology, Pa. State U., 1984; postgrad., Va. Commonwealth U., 1988—. Rsch. asst. Carnegie-Mellon U., Pitts., 1984-86; lab. mgr. dept. microbiology and immunology Med. Coll. Va., Richmond, 1986—; v.p., founding mem. Commonwealth Biotechnologies Inc., Richmond, 1992—. Contbr. articles to profl. jours. including Genetics, Jour. Am. Human Genetics, Molecular Micro Biology, Biotechniques, Jour. Biol. Chemistry, Sci. Innovation '92. Mem. AAAS, Genetics Soc. Am. Democrat. Roman Catholic. Achievements include devel. of new strategies and techniques in synthetic DNA (oligonucleotide) synthesis and purification as well as PCR sequencing and automated fluroescent DNA sequencing; rsch. in applied molecular genetics. Office: Va Commonwealth U Box 678 MCV Sta Richmond VA 23298

REYNOLDS, WILLIAM CRAIG, mechanical engineer, educator; b. Berkeley, Calif., Mar. 16, 1933; s. Merrill and Patricia Pope (Galt) R.; m. Janice Erma, Sept. 18, 1953; children—Russell, Peter, Margery. B.S. in Mech. Engring., Stanford U., 1954, M.S. in Mech. Engring., 1955, Ph.D. in Mech. Engring., 1957. Faculty mech. engring. Stanford U., 1957—, chmn. dept. mech. engring., 1972-82, 89—, Donald Whittier prof. mech. engring., 1986—, chmn.Inst. for Energy Studies, 1974-81; staff scientist NASA/Ames Rsch. Ctr., 1987—. Author: books, including Energy Thermodynamics, 2d edit, 1976; contbr. numerous articles to profl. jours. NSF sr. scientist fellow Eng., 1964, Otto Laporte awd., Am. Physical Soc., 1992. Fellow ASME, Am. Phys. Soc.; mem. AAUP, AIAA, Nat. Acad. Engring. Stanford Integrated Mfg. Assn. (co-chmn. 1990—), Sigma Xi, Tau Beta Pi. Achievements include research in fluid mechanics and applied thermodynamics. Office: Stanford U Dept Mech Engring Bldg 500 Stanford CA 94305

REZACHEK, DAVID ALLEN, energy and environmental engineer; b. Manitowoc, Wis., Jan. 5, 1950; s. Paul Frank and Corrie (Iverson) R. BS in Chemistry, U. Minn., 1972; BS in Environ. Tech. and Urban Systems, Fla. Internat. U., 1976; MSME, U. Hawaii at Manoa, Honolulu, 1980, PhD in Ocean Engring., 1991. Registered profl. engr., Hawaii. Project engr. Advanced Reactors div. Westinghouse, Madison, Pa., 1974; quality assurance engr. Fla. Power & Light Co., Miami, Fla., 1974-76; grad. rsch. asst. oceanography U. Hawaii, Honolulu, 1976-77, jr. rschr. Ctr. Engring. Rsch., 1978-79; grad. rsch. asst. Hawaii Natural Energy Inst., Honolulu, 1979-80; asst. mech. engr. Hawaiian Sugar Planters Assn., Aiea, Hawaii, 1980-87; alternate energy specialist Energy div. State of Hawaii, Honolulu, 1987—; prin., owner Rezachek & Assocs., Honolulu, 1993—. Contbr. numerous articles and reports to profl. jours. Chief judge jr. display div. Hawaii State Sci. and Engring. Fair, Honolulu, 1990—. Lt. (j.g.) USN, 1972-74. Recipient Best Energy Edn. Promotion Program award U.S. Energy Program Mgrs., 1989. Mem. ASHRAE, ASME, Am. Solar Energy Soc., Elec. Vehicle Assn. Hawaii (pres. 1992—), Assn. Energy Engrs., Internat. Solar Energy Soc., Lunalilo Tower Apartment Owners Assn. (pres. bd. 1990—). Achievements include research in development of a solar pond system design computer model, application of heat pumps to residential water heating. Home: 710 Lunalilo St # 1107 Honolulu HI 96813

REZAIYAN, A. JOHN, chemical engineer, consultant; b. Tehran, Iran, Apr. 15, 1955; came to U.S., 1974; s. Satar and Sedigheh (Noori) R.; m. Stephanie Glover, Aug. 18, 1979; children: Sherene, Tina. BS in chem. engr., Univ. Md., 1979. Chem. engr. Benmol Corp., Alexandria, Va., 1980-81; lead engr. HRI, Princeton, N.J., 1981-84; project mgr. SHELADIA Assocs., Inc., Rockville, Md., 1984-88; program mgr. SAIC, McLean, Va., 1988-92; pres. Columbia Engring. and Tech. Group, Inc., Columbia, Md., 1992—; Presenter Coal Targets of Opportunity Workshop, 1988, Coke Oven Emission Control R & D Workshop, 1991. Bd. dirs. Fgn. Info. Refferal Network, 1989-90. Mem. AICE, Am. Chem. Soc., Columbia Rotary (bd. dirs.). Home: 8523 Moon Glass Ct Columbia MD 21045 Office: Columbia Engring & Tech 8525 Moon Glass Ct Columbia MD 21045

REZAK, RICHARD, geology and oceanography educator; b. Syracuse, N.Y., Apr. 26, 1920; s. Habib and Radia (Khoury) R.; m. Hifa Hider, July 1, 1944 (div. Mar. 1965); 1 child, Christine Sara; m. Anna Lucile Nesselrode, Mar. 12, 1965. MA, Washington U., St. Louis, 1949; PhD, Syracuse U., 1957. Geologist U.S. Geol. Survey, Denver, 1952-58; rsch. assoc. Shell Devel. Co., Houston, 1958-67; assoc. prof. oceanography Tex. A&M U., College Station, Tex., 1967-71; prof. Tex. A&M U., College Station, 1971-91, prof. emeritus, 1991—; mem. edit. bd. Geo-MArine Letters, N.Y.C., 1981—; coun. SEPM, Tulsa, Okla., 1968-69; mem. govs. adv. panel Offshore Oil & Chem. Spill Response, Austin, Tex., 1984-85. Co-author: Reefs and Banks of the Northwest Gulf of Mexico, 1985; co-editor: Contributions on the Geological/Oceanography of the Gulf of Mexico, 1972, Carbonate Microfabrics, 1993; contbr. articles to profl. jours. Comdr. USNR, 1942-64. Rsch. grantee various fed. agys., 1968-90. Mem. Lions. Epsicopalian. Home: 3600 Stillmeadow Dr Bryan TX 77802-3324 Office: Tex A&M U Dept Oceanography College Station TX 77843-3146

REZNICEK, ANTON ALBERT, plant systematist; b. Plöchingen, Germany, June 11, 1950; s. Joseph and Klara (Rose) R. MSc, U. Toronto, 1973, PhD, 1978. Asst. curator U. Mich., Ann Arbor, 1978-88, assoc. curator, 1988-92; curator, 1992—; dir. Matthaei Botanical Gardens, Ann Arbor, 1988-90; chmn. Endangered Species Tech. Com. for Plants, Lansing, Mich., 1990—. Editor, author: Evolution in Sedges, 1990. Mem. Am. Soc. Plant Taxonomists, Internat. Assn. for Plant Taxonomy, Botanical Soc. Am. Achievements include development of new hypotheses for evolution in sedges. Office: U Mich Herbarium N Univ Bldg Ann Arbor MI 48109-1057

REZNICEK, BERNARD WILLIAM, power company executive; b. Dodge, Nebr., Dec. 7, 1936; s. William Bernard and Elizabeth (Svoboda) R.; m. Mary Leona Gallagher; children—Stephen B., Michael J., Charles W., Mary E., Bernard J., James G. BSBA, Creighton U., 1958; MBA, U. Nebr., 1979. With Omaha Public Power Dist., pres., chief exec. officer, from 1981; exec. Sierra Pacific Power, Reno, 1979-80; now pres., chief operating officer Boston Edison Co.; dir. Inst. Nuclear Power Ops., Atlanta, Atomic Indsl. Forum, Guarantee Mut. Life Ins. Co. Trustee Father Flanagans Boys' Home, 1984. Mem. C. of C. (bd. dirs. 1981, treas. 1986-87), Beta Gamma Sigma. Republican. Roman Catholic. Club: Oak Hills Country (sec. 1985, v.p. 1986). Lodge: Rotary (bd. dirs. 1984-86), Ak-Sar-Ben. Office: Boston Edison Co 800 Boylston St Boston MA 02199-8001

RHA, CHOKYUN, biomaterials scientist and engineer, researcher, educator, inventor; b. Seoul, Republic of Korea, Oct. 5, 1933; came to U.S., 1956; d. Sea Zin and Young Soon (Choi) R.; m. Anthony John Sinskey, Aug. 22, 1969; children: Tae Minn Song, Tong Ik Lee Sinskey. BS in Life Scis., MIT, 1962, MS in Food Tech., 1964, MSChemE, 1966, ScD in Food Sci., 1967. Sr. rsch. engr. Anheuser-Busch, Inc., St. Louis, 1967-69; asst. prof. food and biol. process engring. U. Mass., Amherst, 1969-73; assoc. prof. food process engring. MIT, Cambridge, 1974-82, assoc. prof. biomaterials sci. and engring., 1982-90, prof. biomaterials sci. and engring., 1990—; prin. BioInfo. Assocs., Boston, 1980—; mem. sci. adv. bd. Genzyme, Cambridge, 1985—; pres. Rha-Sinskey Assocs., Boston, 1972—, XGen, Inc., Boston, 1992—; bd. dirs. Am. Flavor Inc., Boston. Mem. editorial bd. Carbohydrate Polymers, Food Hydrocolloids; author, editor: Theory, Determination and Control of Physical Properties of Food Materials, 1975; contbr. over 120 articles and sci. papers to sci. jours. Mem. AICE, Inst. Food Technologists, Soc. of Rheology, Sigma Xi. Achievements include patents for Encapsulated Active Material System, for Process for Encapsulation and Encapsulated Active Material System, for Hydrocarbon and Non-Polar Solvents Gelled with a Lipophilic Polymer Carbohydrate Derivative, for process of making powdered cellulose laurate, for method of making soybean beverages, for chewing gum, for method of utilizing an exocellular polysaccharide isolated zoogloea ramigera, for glucan compositions and process for preparation thereof, for liquid-liquid extractions, many others. Office: MIT 77 Massachusetts Ave # 56-137 Cambridge MA 02139-4307

RHEE, HYUN-KU, chemical engineering educator; b. Kyungki-do, Korea, Jan. 15, 1939; s. Chong-Hoon and Hee-Won (Kang) R.; m. Junhie Han, Oct. 7, 1973; children: Seungyoon, Sangwoo. BS, Seoul Nat. U., 1962; PhD, U. Minn., 1968. Asst. prof. U. Minn., Mpls., 1968-73; vis. prof. U. La Plata, Argentina, 1972, U. Waterloo, Can., 1976; lectr. NATO-Advanced Study Inst., Oporto, Portugal, 1978; vis. prof. U. Houston, 1980-81, 84-85; dean acad. affairs Seoul Nat. U., 1987-91, prof., 1973—. Author: First-order Partial Differential Equations, Vol. !, 1986, Vol. II, 1989; contbr. articles to Philos. Transactions of Royal Soc. London, Chem. Engring. Sci., AIChE Jour. Cpl. mil., 1962-64. Named to Nat. Order of Pomegranate Govt. of Korea., 1983. Mem. Korean Inst. Chem. Engrs. (jour. editor 1975-77, dir. planning 1985-87, Profl. Progress award 1977), Am. Chem. Soc., Am. Inst. Chem. Engrs. Home: 2-4 Jamwon-dong Shinbanpo 24-cha, Apt 342-1002 Seocho-ku, Seoul Korea 137-030 Office: Seoul Nat U Dept Chem Engring, San 56-1 Shillim-dong Kwanak-ku, Seoul Republic of Korea 151-742

RHEINSTEIN, PETER HOWARD, government official, physician, lawyer; b. Cleve., Sept. 7, 1943; s. Franz Joseph Rheinstein and Hede Henrietta (Neheimer) Rheinstein Lerner; m. Miriam Ruth Weissman, Feb. 22, 1969; 1 child, Jason Edward. B.A. with high honors, Mich. State U., 1963, M.S., 1964; M.D., Johns Hopkins U., 1967; J.D., U. Md., 1973. Bar: Md., D.C.; diplomate Am. Bd. Family Practice 1977, recert. 1983, 89. Intern USPHS Hosp., San Francisco, 1967-68; resident in internal medicine USPHS Hosp., Balt., 1968-70; practice medicine specializing in internal medicine Balt., 1970—; instr. medicine U. Md., Balt., 1970-73; med. dir. extended care facilities CHC Corp., Balt., 1972-74; dir. drug advt. and labeling div. FDA, Rockville, Md., 1974-82, acting dep. dir. Office Drugs, 1982-83, acting dir. Office Drugs, 1983-84, dir. Office Drug Standards, 1984-90; dir. medicine staff, Office Health Affairs FDA, 1990—; chmn. Com. on Advanced Sci. Edn., 1978-86, Rsch. in Human Subjects Com., 1990-92; adj. prof. forensic medicine George Washington U., 1974-76; WHO cons. on drug regulation Nat. Inst. for Control Pharm. and Biol. Products, People's Republic of China, 1981—; advisor on essential drugs WHO, 1985—; FDA del. to U.S. Pharmacopeial Conv., 1985-90. Co-author: (with others) Human Organ Transplantation, 1987; spl. editorial advisor Good Housekeeping Guide to Medicine and Drugs, 1977—; mem. editorial bd. Legal Aspects Med. Practice, 1981-89, Drug Info. Jour., 1982-86, 91—; contbr. articles to profl. jours. Recipient Commendable Svc. award FDA, 1981, group award of merit, 1983, 88, group commendable svc. award, 1989, 92, 93. Fellow Am. Coll. Legal Medicine (bd. govs. 1983—, treas., chmn. fin. com. 1985-88, 90-91, chmn. publs. com. 1988-93, chmn. jud. coun. 1993—, Pres.'s awards 1985, 86, 89, 90, 91, 93), Am. Acad. Family Physicians; mem. AMA, ABA, Drug Info. Assn. (bd. dirs. 1982-90, pres. 1984-85, 88-89, v.p. 1986-87, chmn. ann. meeting 1991, 94, N.Am. steering com. 1991—, Outstanding Svc. award 1990), Fed. Bar Assn. (chmn. food and drug com. 1976-79, Disting. Svc. award 1977), Med. and Chirurgical Faculty Md., Balt. City Med. Soc., Johns Hopkins Med. and Surg. Assn., Am. Pub. Health Assn., Md. Bar Assn., Math. Assn. Am., Soc. Indsl and Applied Math., Mensa (life), U.S. Power Squadrons, Mich. State U. Alumni Assn. (life), U. Md. Alumni Assn. (life), Johns Hopkins U. Alumni Assn., Fed. Exec. Inst. Alumni Assn. (life), Chartwell Golf and Country Club, Annapolis Yacht Club, Johns Hopkins Club, Delta Theta Phi. Avocations: boating, electronics, physical fitness, real estate investments. Home: 621 Holly Ridge Rd Severna Park MD 21146-3520 Office: FDA Office of Health Affairs Dir Medicine Staff 5600 Fishers Ln Rockville MD 20857-0001

RHIEW, FRANCIS CHANGNAM, physician; b. Korea, Dec. 3, 1938; came to U.S., 1967, naturalized, 1977; s. Byung Kyun and In Sil (Lee) R.; m. Kay Kyungja Chang, June 11, 1967; children: Richard C., Elizabeth. BS, Seoul Nat. U., 1960, MD, 1964. Intern, St. Mary's Hosp., Waterbury, Conn., 1967-68; resident in radiology and nuclear medicine L.I.U.-Queens Hosp. Ctr., N.Y., 1968-71; instr. radiology W. Va. U. Sch. Medicine, Morgantown, 1971-73; mem. staff Mercy Hosp. and Moses Taylor Hosp., Scranton, Pa., 1973—, also dir. nuclear medicine; clin. instr., Temple U., 1987—; pres. Radiol. Consultants, Inc., 1984—. Served with M.C., Korean Army, 1964-67. Recipient Minister of Health and Welfare award, 1963; certified Am. Bd. Nuclear Medicine. Mem. AMA, Soc. Nuclear Medicine, Radiol. Soc. N.Am., Am. Coll. Nuclear Medicine, Am. Coll. Radiology, Am. Inst. Ultra Sound, Country Club Scranton, Pres.'s Club U. Scranton, Elks. Home: 14 Lakeside Dr Clarks Summit PA 18411 Office: 746 Jefferson Ave Scranton PA 18501

RHIMES, RICHARD DAVID, civil engineer; b. Dromana, Victoria, Australia, June 2, 1947; s. Ernest Howard and Betty Grace (Hill) R.; m. Heather Marie Harcourt, Oct. 16, 1971; children: Timothy Harcourt, Andrew David, Anthony Jon. Diploma in Civil Engring., Caulfield Inst. Tech., 1969; B in Engring., U. Tamania, Australia, 1980; Diploma in Indsl. mgmt., Swinburne Coll. Tech., Burwood, Australia, 1974. Cert. profl. engr. 1975, Mcpl. engr., 1980, mcpl. bldg. surveyor, 1979, safety profl., 1992. With Mobil Oil Australia, 1973—; environ. health and safety advisor Mobil Oil Corp., Fairfax, Va., 1990-91; risk assessment advisor Mobil Oil Corp., Princeton, N.J., 1991—; mem. telecommunications com. Inst. Traffic Engrs., Australia, 1975; chmn. safety com. Australia Inst. Petroleum, Australia, 1989, chmn. security com. 1989. Recipient Safety Auditor award Internat. Loss Control Inst., 1988. Fellow Australian Inst. Mgmt. (assoc.); mem. Am. Soc. Civil Engrs., Australian Inst. Risk Mgrs., Australian Inst. Engrs. Office: Mobil Oil Corp 202 Carnegie Ctr Princeton NJ 08550

RHOADES, DOUGLAS DUANE, chemical engineer; b. Panama City, Fla., Jan. 23, 1960; s. Robert L. Rhoades and Georgia E. (O'Neill) Merklein; m. nancy J. Quarles, Oct. 14, 1989. BS in Chem. Engring., U. Mo., 1983. Project engr. Clean Air Engring. Inc., Palatine, Ill., 1983—. mem. Air and Water Mgmt. Assn., Am. Inst. Chem. Engrs. Office: Clean Air Engring Inc 500 W Wood St Palatine IL 60067

RHOADS, JONATHAN EVANS, surgeon; b. Phila., May 9, 1907; s. Edward G. and Margaret (Ely Paxson) R.; m. Teresa Folin, July 4, 1936 (dec. 1987); children: Margaret Rhoads Kendon, Jonathan Evans Jr., George Grant, Edward Otto Folin, Philip Garrett, Charles James; m. Katharine Evans Goddard, Oct. 13, 1990. BA, Haverford Coll., 1928, DSc (hon.), 1962; MD, Johns Hopkins U., 1932; D. Med. Sci., U. Pa., 1940, LLD (hon.), 1960; DSc (hon.), Swarthmore Coll., 1969, Hahnemann Med. Coll., 1978, Duke U., 1979, Med. Coll. Ohio, 1985; DSc (Med.) (hon.), Med. Coll. Pa., 1974, Georgetown U., 1981, Yale U., 1990; LittD (hon.), Thomas Jefferson U., 1979. Intern Hosp. of U. Pa., 1932-34, fellow, instr. surgery, 1934-39; asso. surgery, surg. research U. Pa. Med. Sch., Grad. Sch. Medicine, 1939-47, asst. prof. surg. research, 1944-47, asst. prof. surgery, 1946-47, assoc. prof., 1947-49; J. William White prof. surg. research U. Pa., 1949-51; prof. surgery Grad. Sch. Medicine, U. Pa., 1950—; prof. surgery and surg. research U. Pa. Sch. Med., 1951-57, prof. surgery, 1957-59; provost U. Pa., 1956-59, provost emeritus, 1977—, John Rhea Barton prof. surgery, chmn. dept. surgery, 1959-72, prof. surgery, 1972—, asst. dir. Harrison dept. surg. research, 1946-59, dir., 1959-72; chief surgery Hosp. U. Pa., 1959-72, chmn. med. bd., 1959-61; dir. surgery Pa. Hosp., 1972-74; surg. cons. Pa. Hosp., Germantown (Pa.); mem. staff Hosp. of U. Pa.; dir. J. E. Rhoads & Sons, Inc.; mem. bd. pub. edn., City of Phila., 1965-71; co-chmn. Phila. Mayor's Commn. on Health Aspects of Trash to Steam Plant, 1986, chief justice Pa. Com. on Phila. Traffic Ct.; former mem. bd. mgrs. Haverford Coll., chmn., 1963-72, pres. corp., 1963-78, emeritus bd. mgrs. 1989—; bd. mgrs. Friends Hosp. of Phila.; trustee Coriell Inst. Med. Rsch., 1957-90, v.p. sci. affairs, 1964-76, trustee, 1990—; trustee GM Cancer Rsch. Found.; chmn. bd. trustees Measey Found.; trustee emeritus Bryn Mawr Coll.; mem. com. in charge Westtown Sch.; treas. Germantown Friends Sch.; cons. Bur. State Services, VA, 1963; nat. adv. gen. medical scis. council USPHS, 1963; cons. to div. of medical scis. NIH, 1962-63; adv. council Life Ins. Med. Research Fund., 1961-66; Pres. Phila. div., 1955-56; chmn. adv. commn. on research on pathogenesis of cancer Am. Cancer Soc., 1956-57, del., 1956-61, dir. at large, 1965—, pres., 1969-70, past officer dir., 1970-77, hon. life mem., 1977—; chmn. surgery adv. com. Food and Drug Adminstrn., 1972-74; chmn. Nat. Cancer Adv. Bd., 1972-79; Mem. Am. Bd. Surgery, 1963-69, sr. mem., 1969—. Author, co-editor: Surgery: Principles and Practice, 1957, 61, 65, 70; author: (with J.M. Howard) The Chemistry of Trauma; mem. editorial bd. Jour. Surg. Rsch., 1960-71, Oncology Times, 1979—; co-editor: Accomplishments in Cancer Research, 1979-92; editor: Jour. Cancer, 1972-91, editor emeritus, 1991—; mem. editorial bd. Annals of Surgery, 1947-77, emeritus, 1977—, chmn., 1971-73; mem. editorial adv. bd. Guthrie Bull., 1986—; contbr. articles to med. jours., chpts. to books. Trustee John Rhea Barton Surg. Found. Recipient Roswell Park medal, 1973, Papanicolaou award, 1977, Phila. award, 1976, Swanberg award, 1987, Benjamin Franklin medal Am. Philos. Soc., Medal of the Surgeon Gen. of U.S., Disting. Alumnus award U. Pa., 1993; hon. Benjamin Franklin fellow Royal Soc. Arts. Fellow Am. Med. Writers Assn., Am. Philos. Soc. (sec. 1963-66, pres. 1977-84), ACS (regent, chmn. bd. regents 1967-69, pres. 1971-72), Royal Coll. Surgeons (Eng.) (hon.), Royal Coll. Surgeons Edinburgh (hon.), Deutsches Gesellschaft für Chirurgie (corr.), Assn. Surgeons India (hon.), Royal Coll. Physicians and Surgeons Can. (hon.), Coll. Medicine South Africa (hon.), Polish Assn. Surgeons (hon.), Royal Coll. Surgeons in Ireland (hon.), AAAS (sec. med. sci. sect. 1980-86); mem. Hollandsche Maatschappij der Wetenschappen (fgn.), Am. Public Health Assn., Assn. Am. Med. Colls. (chmn. council acad. socs. 1968-69, disting. service mem. 1974—), Fedn. Am. Socs. Exptl. Biology, Am. Assn. Surgery Trauma, Am. Soc. Clin. Nutrition, Am. Trauma Soc. (founding mem., v.p., chmn. bd. dirs. 1986—), AMA (co-recipient Goldberger award 1970, Dr. Rodman and Thomas G. Sheen award 1980), Pa. Med. Soc. (Disting. service award 1975), mem. of Judicial Coun., 1991—, Phila. County Med. Soc. (pres. 1970, Strittmater award 1968), Coll. Physicians Phila. (v.p. 1954-57, pres. 1958-60, Disting. Service award 1987), Phila. Acad. Surgery (pres. 1964-66), Phila. Physiol. Soc. (v.p. 1945-46), Am. Surg. Assn. (pres. 1972-73, Disting. Service medal, trustee found., vice chairman, 1992-93), Pan Pacific Surg. Assn. (v.p. 1975-77), So. Surg. Assn., The Internat. Surg. Group (pres. 1958), Internat. Fedn. Surg. Colls. (v.p. 1972-78, pres. 1978-81, hon. pres. 1987—), Fellows of Am. Studies, Soc. of U. Surgeons, Soc. Clin. Surgery (pres. 1966-68), Am. Assn. for Cancer Research, Am. Chem. Soc., Am. Physiol. Soc., Coun. Biology Editors, Internat. Soc. Surgery (hon.) N.Y. Acad. Scis., Surg. Infection Soc. (pres. 1984-85), Surgeons Travel Club (pres. 1976, hon. mem.), Am. Inst. Nutrition, World Med. Assn., Am. Acad. Arts and Scis., Inst. of Medicine (sr.), Soc. for Surgery Alimentary Tract (pres. 1967-68), Southeastern Surg. Congress, Soc. Surg. Chmn. (pres. 1966-68), Buckingham Mountain Found.

(sec., treas.), James IV Soc. (hon.), Phi Beta Kappa, Alpha Omega Alpha, Sigma Xi. Clubs: Rittenhouse, Union League, Philadelphia; Cosmos (D.C.). Achievements include demonstration that protein malnutrition could retard callus formation in experimental fractures; positive nitrogen balance could be induced in protein deficient patients who could not take things by mouth. Office: 3400 Spruce St Philadelphia PA 19104-4220

RHOADS, KEVIN GEORGE, consulting engineer; b. Northampton, Pa., Apr. 20, 1951; s. William George and Grace Arlene (Heiland) R.; m. Leonore Irene Katz, Aug. 22, 1983; 1 child, Thomas William Leo. MS in Engring., MIT, 1983, PhD, 1989. Pres. Kevin G. Rhoads Engring., Inc., Andover, Mass., 1989—; lectr. Concourse Program, MIT, Cambridge, 1991—. Contbr. articles to profl. jours. Ho. tutor MIT Dormitory System. Fellowship MIT, 1981, 86. Mem. Sigma Xi, Tau Beta Pi. Achievements include patents for power oscillator circuit and its application to veterinary, medical and self-protection electrical stimulation devices. Office: Kevin G Rhoads Engring Inc 113 Bellevue Rd Andover MA 01810-5318

RHODES, CHUCK WILLIAM, electrical engineer; b. Ruston, La., Nov. 4, 1954; s. Raymond William and Katie Eugenia (Grigsby) R.; m. Ruthie Mae Hilton, Jan. 26, 1986; children: Kate, Julie. BSEE, La. State U., 1979. Profl. engr., Colo. Distbn. engr. Ark. Power & Light, Little Rock, 1979-82; planning engr. Pub. Svc. Co. of Colo., Denver, 1982-88, sr. planning engr., 1988—; participant Photovoltaics for Utilities, Washington, 1992, Utility Renewable Energy Forum, Denver, 1992. Composer music Changing Scene Theatre, Denver, 1988—. Recipient Best Sound Design award Denver Drama Critics Circle, 1991. Achievements include proposal of first solar "village power system" in the continental U.S., for the town of White Pine, Colorado. Home: 808 S High St Denver CO 80209 Office: Pub Svc Co of Colo 5909 E 38th Ave Denver CO 80207

RHODES, DOUG, geologist. Recipient William J. Stephenson Outstanding Svc. award Nat. Speleological Soc., 1991. Office: PO Box 12334 Albuquerque NM 87195*

RHODES, JAMES RICHARD, economics educator; b. Omaha, Nov. 28, 1945; s. Jack Richard and Mary Elizabeth (Doherty) R.; m. Kimie Hyoto, Dec. 27, 1980; children: John Makoto, Emi Alicia. BA, U. Wash., Seattle, 1969, MA, 1973, PhD, 1981. Teaching asst. U. Wash., Seattle, 1974-75, teaching assoc., 1975-76, 77-79; asst. prof. Western Wash. U., Bellingham, 1976-77, Wash. State U., Pullman, 1979-80, Kans. State U., Manhattan, 1980-88; vis. prof. Internat. U. Japan, Urasa, Niigata, Japan, 1987-88, lectr., 1988—; assoc. prof. econs. Grad. Sch. Policy Sci., Saitama U., Urawa, Japan, 1988-91, prof., 1991—; bd. dirs. Children's Rsch. Ctr., Saitama U., 1990—; lectr. Rikkyo U., Tokyo, 1992—. Reviewer Wall St. Rev. Books, South Salem, N.Y., 1982-87, profl. jours., 1984—; regional editor (Japan) Bus. Libr. Rev., N.Y.C., 1990—; contbr. articles to profl. jours., news media. 1st lt U.S. Army, 1969-71, Vietnam; lt. col. USAR, 1987—. Decorated Bronze Star with two oak leaf clusters, Meritorious Svc. medal, Army commendation medal with one oak leaf cluster; Internat. Edn. Found. grantee, 1988; Inst. Humane Studies F. Leroy Hill fellow, 1984, Kellogg Found. grantee, 1984. Mem. Am. Econ. Assn., Civil Affairs Assn. (life), Res. Officers Assn. (life, membership chmn., sec.-treas. Far East chpt. 1990-93, treas. 1993—), Assn. for Asia Studies, Nat. Eagle Scout Assn. (life), Phi Alpha Theta, Beta Gamma Sigma, Alpha Sigma Nu. Avocations: swimming, jogging, lit., history, languages. Home: 635 Shimo Okubo, Urawa 338 Saitama, Japan Office: Grad Sch Policy Sci, Saitama U, Urawa 338 Saitama, Japan

RHODES, ROBERT LEROY, systems engineer; b. Longview, Wash., Jan. 22, 1953; s. Edward LeRoy and Jeanne Irene (Rhinehart) R. BS in Chemistry, U. Wash., 1975. Programmer N.Am. Controls, Clackamas, Oreg., 1977-79, Tera Indsl. Controls, Portland, Oreg., 1979-81; systems analyst Quantitative Tech. Corp., Tigard, Oreg., 1981-82; control systems engr. Wacker Siltronic Corp., Portland, 1982—. Contbr. article to book Objects in Action, 1993. Recipient Best Cost-saving award Object World, 1992. Mem. Assn. Computing Machinery. Home: 21160 NW Rock Creek Blvd Portland OR 97229 Office: Wacker Siltronic Corp 7200 NW Front Ave Portland OR 97210

RHODES, ROBERTA ANN, dietitian; b. Red Bank, N.J., Apr. 11; d. Franklin Galloway and Frances (Kieswetter) DuBuy; m. Albert Lewis Rhodes, Feb. 10, 1978. BS, Fla. State U., 1977, MS, 1988. Registered dietitian, Fla. Clin. dietitian Archbold Hosp., Thomasville, Ga., 1988-90; nutritionist Women, Infant and Children program, Tallahassee, 1990-91; sr. mgmt. clin. and adminstrv. dietitian Sunrise Community, Inc., Tallahassee, 1991-93; clin. svcs. specialist Heritage Health Care Ctr., Tallahassee, 1993—. Mem. Am. Dietetic Assn., Fla. Dietetic Assn., Sigma Xi, Omicron Nu. Home: 4112 Alpine Way Tallahassee FL 32303-2244 Office: Heritage Health Care Ctr 1815 Ginger Dr Tallahassee FL 32308

RHODES, RONDELL HORACE, biology educator; b. Abbeville, S.C., May 25, 1918; s. Leslie Franklin and Pearl Lee (Clinkscales) R.; B.S., Benedict Coll., Columbia, S.C., 1940; M.S., U. Mich., 1950; Ph.D., N.Y.U., 1960. Instr. biology Lincoln U., Jefferson City, Mo., 1947-49; asst. prof. Tuskegee (Ala.) Inst., 1950-55; teaching fellow N.Y.U., 1955-61; mem. faculty Fairleigh Dickinson U., Teaneck, N.Y., 1961—, prof. biol. scis., 1968-88, prof. emeritus, 1988—; chmn. dept., 1966-70, 73-76, 79-82. Served with AUS, 1942-46. Mem. AAAS, Am. Inst. Biol. Scis., Am. Soc. Zoologists, AAUP, Nat. Assn. Biology Tchrs., N.Y. Acad. Scis., Sigma Xi. Democrat. Episcopalian. Home: 122 Ashland Pl Apt 5H Brooklyn NY 11201-3910 Office: Fairleigh Dickinson U Teaneck NJ 07666

RHODES, WAYNE ROBERT, ergonomist, consultant; b. Toronto, Ont., Can., Feb. 10, 1951; s. James Albert and Dorothy May (Butler) R.; m. Jane Barbara Macneill, June 24, 1972; children: Jorden James, Lindsay Ruth. BA, York U., Downsview, Ont., 1975; MA, U. Toronto, 1977, PhD, 1982. Lectr., tutorial asst. York U., Toronto, 1977-78; lectr., lab. asst. U. Toronto, 1975-82; freelance cons. Don Mills, Ont., 1979-82; prin. cons. Rhodes & Assocs., Willowdale, Ont., 1982-86; pres. Rhodes & Assocs., Inc., Willowdale, 1986—; mem. robot safety com. Can. Standards Assn., Rexdale, Ont., 1991—; exec. mem. Can. Aeronatuics and Space Inst., Toronto, 1985—. Mem. Can. Aeronautics and Space Inst. (chmn. 1985—, svc. award 1987), Human Factors Assn. Can., Human Factors Soc., Soc. Logistics Engrs., Internat. Design Extreme Environments, Bd. Certification for Profl. Ergonomists (cert.). Achievements include research in ergonomics of building design, aerospace and military systems, nuclear facilities and protective clothing. Home and Office: Rhodes & Assocs Inc, 177 Jenny Wrenway, Willowdale, ON Canada M2H 2Z3

RHODES, WILLIAM GEORGE, III, physicist; b. Mt. Kisco, N.Y., Apr. 9, 1956; s. William George Jr. and Beverly Jean (Harrison) R.; m. Susan Kay Marshall, July 2, 1977 (div. 1979); m. Jacqueline Loudis, Aug. 22, 1981; children: William Peter, Andrew Bernard, David Saxton. BA, Wittenberg U., 1978; MS, U. Lowell, 1982. Radiologic tech. U. Tex. M.D. Anderson Hosp., Houston, 1978-80; grad. asst. U. Lowell, 1980-82; sr. scientist Gen. Physics Corp., Columbia, Md., 1982-83; lead engr. Gen. Electric Co., Schenectady, N.Y., 1983-90; mgr. Arthur D. Little, Inc., Cambridge, Mass., 1990—. Author: Radiation Protection Management, 1992. Pres. Grand Blvd. Fire Co., Niskayuna, N.Y., 1989, v.p., 1988. Mem. Health Physics Soc. (cert., pres. northeastern N.Y. chpt. 1988-89, pres.-elect 1987-88), Am. Nuclear Soc. Office: Arthur D Little Inc Acorn Park Cambridge MA 02140

RHODIN, ANDERS G.J., orthopaedic surgeon, chelonian researcher and herpetologist; b. Stockholm, Dec. 11, 1949; came to U.S., 1958; s. Johannes A.G. and Gunvor (Thorstenson) R.; m. Susan DeSanctis, Aug. 27, 1972; 1 child, Michael H.J. BA in Geology, Dartmouth Coll., 1971; MD, U. Mich., 1977. Diplomate Am. Bd. Orthopaedic Surgeons. Surg. resident Yale U., New Haven, 1977-82; prin. Orthopaedic Assocs. P.C., Fitchburg, Mass., 1982-90, Wachusett Orthopaedic Surgery, Leominster, Mass., 1990—; founder, dir. Chelonian Rsch. Found., Lunenburg, Mass., 1992—; assoc. herpetology Mus. Comparative Zoology, Harvard U., Cambridge, Mass., 1979—; dep. chmn. IUCN/SSC Tortoise and Freshwater Turtle specialist group World Conservation Union-Species Survival Commn., 1991—. Author, editor: (book) Advances in Herpetology and Evolutionary Biology, 1983, Conservation Biology of Freshwater Turtles, 1993, Chelonian Conservation and Biology, 1993; contbr. articles to profl. jours. Penrose Fund Rsch. grant Am. Philos. Soc., 1979. Fellow Am. Acad. Orthopaedic Surgeons; mem. Am. Soc. Ichthyologists and Herpetologists, Soc. Study of Amphibians and Reptiles, Herpetologists' League. Achievements include descriptions of several new species of turtles, mainly side-necked chelid turtles from South America and New Guinea region; unique bone growth and cartilaginous vascularization of the leatherback turtle, a mammalian-like reptile. Office: Chelonian Rsch Found 168 Goodrich St Lunenburg MA 01462

RIACH, PETER ANDREW, economist; b. Melbourne, Victoria, Australia, May 15, 1937; arrived in Eng., 1989; s. Andrew Brown and Doris Hope (Stanley) R.; m. Lorraine Margaret Trew, Jan. 16, 1960; 1 child, Emma Simone. B of Commerce, Melbourne U., 1958; PhD in Econs., London U., 1965. Asst. lectr. Melbourne U., 1960-61; lectr., reader Monash U., Australia, 1965-88; head econs. dept. De Montfort U., 1989—. Contbr. articles to profl. jours. Office: De Montfort U, The Gateway, Leicester LE1 9BH, England

RIBARIC, MARIJAN, physicist, mathematician; b. Ljubljana, Slovenia, Mar. 12, 1932; s. Miho and Joza (Kramer) R.; m. Pavla Peternelj, Dec. 12, 1959; children: Samo, Peter. Diploma in Physics, U. Ljubljana, 1954; PhD, 1959. Research fellow Inst. J. Stefan, Ljubljana, 1954-62; sr. sci. Inst. J. Stefan, 1964-71, head applied math., 1969-92, sci. counselor, 1971—, habilitation for asst. prof. math analysis, 1971, assoc. prof., 1979-86, prof., 1986—; resident research assoc. Argonne (Ill.) Nat. Lab., 1962-64. Author: Functional-Analytic Concepts and Structures of Neutron Transport Theory, 1973, Thermodynamics of Linear Transport Processes, 1975, Computational Methods for Parsimonious Data Fitting, 1984, Conservation Laws and Open Questions of Classical Electrodynamics, 1990; contbr. articles to profl. jours. Recipient Kidric Found. Praise for Sci. Achievement, 1966, 85. Mem. Gesellschaft fuer Angewandte Math. und Mechanik, European Phys. Soc. Home: 51 Trzaska, 61000 Ljubljana Slovenia Office: Inst J Stefan, 39 Jamova, 61000 Ljubljana Slovenia

RIBBLE, RONALD GEORGE, psychologist, educator, writer; b. West Reading, Pa., May 7, 1937; s. Jeremiah George and Mildred Sarah (Folk) R.; m. Catalina Valenzuela (Torres), Sept. 30, 1961; children: Christina, Timothy, Kenneth. BSEE, U. Mo., 1968, MSEE, 1969, MA, 1985, PhD, 1986. Cert. psychologist, Tex. Enlisted man USAF, 1956-60, advance through grades to lt. col., 1976; rsch. dir. Coping Resources, Inc., Columbia, Mo., 1986; referral devel. Laughlin Pavilion Psychiat. Hosp., Kirksville, Mo., 1987; program dir. Psychiat. Insts. of Am., Iowa Falls, Iowa, 1987-88; lead psychotherapist Gasconade County Counseling Ctr., Hermann, Mo., 1988; lectr. U. Tex., San Antonio, 1989—; assessment clinician Afton Oaks Psychiat. Hosp., San Antonio, 1989-91; psychologist Olmos Psychol. Svcs., Inc., San Antonio, 1991—; vol. assessor Holmgreen Children's Shelter, San Antonio, 1991-92. Contbr. essays to psychol. reference books and poetry to anthologies periodicals; columnist Feelings, 1993—. Del. Boone County (Mo.) Dem. Conv., 1984; vol. announcer Pub. Radio Sta., Columbia, 1983; vol. counselor Cath. Family and Children's Svc., San Antonio, 1989-91; chpt. advisor Rational Recovery Program for Alcoholics, San Antonio, 1991-92. Mem. APA, AAUP, NEA, ACLU, Internat. Soc. for Study of Individual Differences, Internat. Platform Assn., Bexar County Psychol. Assn., Air Force Assn., Ret. Officers Assn. Roman Catholic. Avocations: running and fitness, poetry, singing, pub. speaking. Home: 14023 N Hills Village Dr San Antonio TX 78249-2531 Office: U Tex Div Cultural and Behavioral Scis San Antonio TX 78249 also: Olmos Psychol Svcs Inc 14607 San Pedro Ste 205 San Antonio TX 78232

RICARDO-CAMPBELL, RITA, economist, educator; b. Boston, Mar. 16, 1920; d. David and Elizabeth (Jones) Ricardo; m. Wesley Glenn Campbell, Sept. 15, 1946; children: Barbara Lee, Diane Rita, Nancy Elizabeth. BS, Simmons Coll., 1941; MA, Harvard U., 1945, PhD, 1946. Instr. Harvard U., Cambridge, Mass., 1946-48; asst. prof. Tufts U., Medford, Mass., 1948-51; labor economist U.S. Wage Stabilization Bd., 1951-53; economist Ways and Means Com. U.S. Ho. of Reps., 1953; cons. economist, 1957-60; vis. prof. San Jose State Coll., 1960-61; sr. fellow Hoover Instn. on War, Revolution, and Peace, Stanford, Calif., 1968—; lectr. health svc. adminstrn. Stanford U. Med. Sch., 1973-78; bd. dirs. Watkins-Johnson Co., Palo Alto, Calif., Gillette Co., Boston; mgmt. bd. Samaritan Med. Ctr., San Jose, Calif. Author: Voluntary Health Insurance in the U.S., 1960, Economics of Health and Public Policy, 1971, Food Safety Regulation: Use and Limitations of Cost-Benefit Analysis, 1974, Drug Lag: Federal Government Decision Making, 1976, Social Security: Promise and Reality, 1977, The Economics and Politics of Health, 1982, 2d edit., 1985; co-editor: Below-Replacement Fertility in Industrial Societies, 1987, Issues in Contemporary Retirement, 1988; contbr. articles to profl. jours. Commr. Western Interstate Commn. for Higher Edn. Calif., 1967-75, chmn., 1970-71; mem. Pres. Nixon's Adv. Coun. on Status Women, 1969-76; mem. task force on taxation Pres.'s Coun. on Environ. Quality, 1970-72; mem. Pres.'s Com. Health Services Industry, 1971-73, FDA Nat. Adv. Drug Com., 1972-75; mem. Econ. Policy Adv. Bd., 1981-90, Pres. Reagan's Nat. Coun. on Humanities, 1982-89, Pres. Nat. Medal of Sci. com., 1988—; bd. dirs. Ind. Colls. No. Calif., 1971-87, Mt. Pelerin Soc., 1988—; mem. com. assessment of safety, benefits, risks Citizens Commn. Sci., Law and Food Supply, Rockefeller U., 1973-75; mem. adv. com. Ctr. Health Policy Rsch., Am. Enterprise Inst. Pub. Policy Rsch., Washington, 1974-80; mem. adv. coun. on social security Social Security Adminstrn., 1974-75; bd. dirs. Simmons Coll. Corp., Boston, 1975-80; mem. adv. coun. bd. assocs. Stanford Librs., 1975-78; mem. coun. SRI Internat., Menlo Park, Calif., 1977-90. Mem. Am. Econ. Assn., Mont Pelerin Soc. (bd. dirs. 1988—, v.p. 1992—), Harvard Grad. Soc. (coun. 1991), Phi Beta Kappa. Home: 26915 Alejandro Dr Los Altos Hills CA 94022 Office: Stanford U Hoover Instn Stanford CA 94305-6010

RICCIARDI, ANTONIO, prosthodontist, educator; b. Jersey City, June 5, 1922; s. Frank and Eugenia (Izzo) R.; student Upsala Coll., 1941-42, B.A. in Chemistry, 1951; D.D.S., Temple U., 1958; Diplomate Am. Bd. Oral Implantology, Am. Bd. Implant Dentistry; m. Lucy DePalma, June 18, 1945; children: Eugenia, Lynda. Purchasing agt. Dade Bros., Newark, Airport, 1951-52; asst. work mgr. Cooper Alloy Steel Co., Hillside, N.J., 1954; chemist White's Pharm. Co., Union, N.J., 1954; practice gen. dentistry, Westfield, N.J., 1958—; dentist Westfield Public Schs., 1958-60; mem. staff Mountainside Hosp., Montclair, N.J., St. Elizabeth's Hosp., Elizabeth, N.J.; implant staff John F. Kennedy Hosp., Edison, N.J., chief of prosthetics, 1980—; clin. chmn. implant study Columbia U. Sch. Oral Surgery and Dentistry; implant cons. Columbia Presbyn. Sch. Oral Surgery and Dentistry, N.Y.C.; cons. Implants Internat., N.Y.C., 1971—; pres. Universal Dental Implements, Inc., 1979—. Pres. Nat. Gymnastics Clinic, Sarasota, Fla., 1968—; v.p. rebound tumbling center Welmarick Inc., Plainfield, 1958—. Gymnastics ofcl. Eastern Coll. Conf., 1954—. Served to lt. col. USMCR, 1942-48, 52-54; Korea. Fellow Acad. Gen. Dentistry, Royal Soc. Health (Eng.), Internat. Coll. Oral Implantology (founding mem.), Am. Acad. Gen. Dentistry, Am. Acad. Implant Dentistry (program chmn. nat. conv. 1974, sec. 1976, pres. N.E. sect. 1978, chmn. ethics com. 1980, ethics chmn. 1981, credentialling mem. 1985—), Acad. Dentistry Internat., Am. Acad. Implant Dentistry, Fedn. Dentistry Internat.; hon fellow Italian and German implant socs.; mem. Inst. Endosseous Implants, Inst. for Advance Dental Research, ADA, Middlesex County Dental Assn., Union County and Plainfield Dental Soc. Fedn. Prosthodontics Orgns., Internat. Research Com. on Oral Implantology (pres. U.S. chpt.), Am. Acad. Oral Implantology, Nat. Gymnastics Judges Assn. (pres. Eastern div.; named to Hall of Fame 1978), Delta Sigma Delta. Writer, lectr. on implantology. Address: 1450 Fernwood Rd Mountainside NJ 07092

RICCO, RAYMOND JOSEPH, JR., computer systems engineer; b. Tullahoma, Tenn., Aug. 7, 1948; s. Raymond Joseph and Betty Jean (Collins) R.; m. Susan Rae Frey, Mar. 30, 1985. BS, Mid. Tenn. State U., 1971; MS, U. Tenn., Tullahoma, 1976. Rsch. asst. U. Tenn. Space Inst., Tullahoma, 1972-76; prin. analyst Teledyne Brown Engring., Huntsville, Ala., 1976-78; sr. analyst Sci. Applications Internat., Huntsville, 1978-82; sr. mem. tech. staff Mitre Corp., Colorado Springs, Colo., 1982-83; sr. engr. analyst Sci. Applications Internat., Huntsville, 1983-84; project mgr. Systems Devel. Corp., Dayton, Ohio, 1984-85; bus. assoc. Booz Allen & Hamilton Inc., Dayton, 1985-87; project dir. Bell Tech. Ops., Sierra Vista, Ariz., 1987-90; ptnr. Ricco-Thompson Cons. Engrs., Sierra Vista, 1990-92, Gazelle Affiliates, Sierra Vista, 1992—; sr. engr., scientist SAIC, Sierra Vista, 1992—; mem. adv. bd. Am. Security Coun., Arlington, Va., 1981—. Contbr. articles to profl. jours. Mem. NRA, IEEE, IEEE Computer Soc., Am. Def. Preparedness Assn., Armed Forces Comm. and Electronics Assn., Air Force Assn. Avocations: outdoor sports, philately, reading. Home: PO Box 3672 Sierra Vista AZ 85636-3672 Office: SAIC 333 W Wilcox Ste 200 Sierra Vista AZ 85635 also: Gazelle Affiliates 3323 E Willow Dr Sierra Vista AZ 85635

RICCOBONO, JUANITA RAE, solar energy engineer; b. N.Y.C., Sept. 20, 1963; d. Salvatore and Mary Louise (Keller) R. BA, Northwestern U., 1985; MS, U. Mass., 1993. Lab. aide Northwestern U., Evanston, Ill., 1983-84, rsch. assoc., 1983-85; chemistry instr. Kendall Coll., Evanston, 1984-85; subs. tchr. Ogdensburg (N.Y.) City Schs., 1985-86; vol. Peace Corps., Togo, West Africa, 1986-88; rsch., teaching asst. U. Mass., Lowell, 1989-91; sr. scientist Northeast Photosciences, Inc., Hollis, N.H., 1991--. Contbr. articles to profl. jours. Vol. U. Mass. Lowell Office Community Svc., 1991, Peace Corps, Washington, 1991-92; com. mem. U. Mass. Lowell Peace Project, 1991. Recipient Beyond the War award 1987, Letter of Appreciation, Pres. U.S. and dir. Peace Corps, 1988. Mem. Energy Engring Assn. (pres. 1990-92), Sigma Xi. Office: Northeast Photosciences 18 Flagg Rd Hollis NH 03049

RICE, CHARLES LANE, surgical educator; b. Atlanta, May 22, 1945; s. Marion Jennings and Molly Black (Moore) R.; m. Lynn Carol Inscoe, Dec. 27, 1968 (div. 1976); m. Judith Josephine Bousha, July 9, 1977; children: Aaron Nicholas, Patrick Marion. AB, U. Ga., 1964; MD, Med. Coll. Ga., 1968. Commd. ensign USN, 1966, advanced through grades to comdr., 1976, ret., 1977; intern Bowman Gray Sch. Medicine, Winston-Salem, N.C., 1968-69; resident Nat. Naval Med. Ctr., Bethesda, Md., 1969-73; asst. prof. surgery U. Chgo., 1977-80, assoc. prof. surgery, 1980-84; dir. intensive care unit Michael Reese Hosp., Chgo., 1977-84; prof., vice chmn. dept. surgery U. Wash., Seattle, 1985-92; surgeon-in-chief Harborview Med. Ctr., Seattle, 1985-92; Dr. Lee Hudson- Robert R. Penn prof., chmn., divsn. gen. surgery U. Tex. Southwestern Med. Ctr., Dallas, 1992—; Robert Wood Johnson Health Policy fellow, 1991-92; legis. asst. to U.S. senator Tom Daschle, 1991-92; vice chmn. Com. Trauma, Am. Coll. Surgeons, Chgo., 1992-93; ann. lectr. U. Wash., Soc. U. Surgeons. Assoc. editor Jour. of Surg. Rsch., 1983-90; contbr. articles to profl. jours. Fellow ACS (gov. 1992—. vice chmn. com. on trauma 1992-93), Am. Surg. Assn., Am. Assn. for Surgery of Trauma (com. chair 1989-91); mem. Soc. Univ. Surgeons, Am. Physiol. Soc., Shock Soc. (pres. 1991-92). Democrat. Episcopalian. Office: U Tex Southwestern Med Ctr 5323 Harry Hines Blvd Dallas TX 75235-9031

RICE, ERIC EDWARD, technologies executive; b. Fremont, Ohio. BS, U. Wis., 1967; PhD, Ohio State U., 1972. From rschr. to assoc. sect. mgr. Battelle Labs., Columbus, Ohio, 1972-84; dir. Astronautics Tech. Ctr., Madison, Wis., 1984-88; pres., CEO, chmn. Orbital Tech. Corp., Madison, 1988—; chmn. Wis. Space Inst., Madison, 1991—. Fellow AIAA (assoc., pres. Wis. sect. 1987-89).

RICE, JOHN ANDREW, statistician; b. N.Y.C., June 14, 1944; s. Frank A. and Ann C. (Sutton) R.; m. Diane E. LaMarche, Aug. 9, 1969; children: Andrew L., Marcelle A. BA, U. N.C., 1966; PhD, U. Calif., Berkeley, 1973. Prof. dept. math. U. Calif.-San Diego, La Jolla, 1973-91; prof. dept. stats. U. Calif., Berkeley, 1991—; mem. com. on applied and theoretical stats. Nat. Rsch. Coun., Washington, 1986-89, mem. doctoral and postdoctoral math. study group, 1990-91. Author: Mathematical Statistics and Data Analysis, 1988; contbr. articles to profl. jours. Fellow Inst. Math. Stats.; mem. AAAS, Am. Statis. Assn. Achievements include research on nonparametric statistics and on stochastic models of ion channels. Home: 3084 Buena Vista Way Berkeley CA 94708 Office: U Calif Berkeley Dept Stats Evans Hall Berkeley CA 94720

RICE, MARY ESTHER, biologist; b. Washington, Aug. 3, 1926; d. Daniel Gibbons and Florence Catharine (Pyles) R. AB, Drew U., 1947; MA, Oberlin Coll., 1949; PhD, U. Wash., 1966. Instr. biology Drew U., Madison, N.J., 1949-50; rsch. assoc. Columbia U., N.Y.C., 1950-53; rsch. asst. NIH, Bethesda, Md., 1953-61; curator invertebrate zoology and dir. Smithsonian Marine Sta., Smithsonian Instn., Washington, 1966—; mem. adv. panel on systematic biology NSF, Washington, 1977-78; mem. com. on marine invertebrates Nat. Acad. Sci., 1976-81; mem. overseers com. on biology Harvard U., Cambridge, Mass., 1982-88. Assoc. editor Jour. Morphology, Ann Arbor, Mich., 1985-91; editor: (with M. Todorovic) Biology of Sipuncula and Echiura, 1975, 2nd vol., 1976; (with F.S. Chia) Settlement and Metamorphosis of Marine Invertebrate Larvae, 1978; contbr. articles to profl. jours. Recipient Drew U. Alumni Achievement award in sci., 1980. Fellow AAAS; mem. Am. Soc. Zoologists (pres. 1979), Phi Beta Kappa. Office: Smithsonian Marine Sta 5612 Old Dixie Hwy Fort Pierce FL 34946-7303

RICE, PETER ALAN, physician, scientist; b. Orange, N.J., May 25, 1942; s. Alan Biddlecome and Josephine (Berkley) R.; m. Nancy Heyl Royster, Aug. 17, 1965; 1 child, Nicole Randolph. B in English, Yale U., 1964; MD, U. Pa., 1969. Diplomate Am. Bd. Internal Medicine, Am. Bd. Infectious Diseases. Intern, resident Yale-New Haven (Conn.) Hosp., 1969-71; epidemic intelligence officer Ctrs. for Disease Control, Atlanta, 1971-73; resident Peter Bent Brigham Hosp., Boston, 1973-74; postdoctoral fellow Boston City Hops.-Harvard Med. Sch., 1974-76, instr. medicine, 1976-77; from asst. prof. to prof. Sch. Medicine Boston U., 1977—; dir. Maxwell Finland Lab. for Infectious Diseases, Boston, 1990—; chief infectious diseases Boston City Hosp., Boston U. Sch. Medicine, 1990—; mem. sci. adv. bd. Hygeia Scis., Newton, Mass., 1980-90; bacteriology, mycology study sect. NIAID, NIH, Bethesda, Md., 1985-89, microbiology and infectious disease rsch. com. NIAID, NIH, Bethesda, 1991—. Editor: Proceedings of the 8th Internat. Pathogenic Neisseria Conf., 1992; contbr. articles to Jour. Clin. Invest. Sr. Surgeon USPHS, 1971-73. Recipient Roche award U. Pa., 1969; named Ranking scholar Yale U., 1964, grantee NIAID, NIH, 1970—. Fellow ACP, Infectious Disease Soc. Am.; mem. Mass. Infectious Disease Soc. (counselor 1992—). Achievements include patents awarded and pending in diagnostic assay for rapid identification of gonococcal infection and an anti-idiotype vaccine candidate for prevention of gonorrhea. Office: Boston City Hosp/ Maxwell Finland Lab Infectious Dis 774 Albany St Boston MA 02118

RICE, RAMONA GAIL, physiologist, phycologist, educator, consultant; b. Texarkana, Tex., Feb. 15, 1950; d. Raymond Lester and Jessie Gail (Hubbard) R.; m. Carl H. Rosen. BS, Ouachita U., 1972; MS, U. Ark., 1975, PhD, 1978; postgrad. Utah State U., 1978-80. Undergrad. asst. Ouachita U., Arkadelphia, Ark., 1970-72; grad. teaching asst. U. Ark., Fayetteville, 1972, 77-78, grad. rsch. asst., 1973-77; asst. rsch. scholar, scientist Fla. Internat. U., Miami, 1980-85; rsch. coord. in biology, Pratt Community Coll., Kans., 1985-87, faculty, 1985—; adj. instr. Miami Dade Community Coll., 1984-85, Wichita (Kans.) State U., 1986-88. Contbr. articles to profl. jours. Judge Pratt County Sci. Fairs, Dade County Sci. Fair, Fla., 1981-85, Barber County Sci. Fairs; tchr. Sunday Sch. First Baptist Ch., South Miami, Fla., 1982-85, leader girls in action, 1982-83, youth chaperone, 1982-85; patron Pratt Community Concert Series. Grantee NSF, 1981-83, Fla. Dept. Environ., 1981-83, EPA, 1983-85, So. Fla. Rsch. Ctr., Everglades Nat. Park, 1983-86; recipient Am. Biog. Inst. Disting. Leadership award, 1987. Mem. AAAS, AMA (Ninescah Valley Med. Soc. Aux.), Pratt Higher Edn. Assn. (sec. 1987-88), Fla. Acad. Scis., Phycological Soc. Am., Soc. Limnology and Oceanography, Epsilon Sigma Alpha (Epsilon Pi chpt. pres. 1991-92, counselor 1992-93, historian, 1993—, rec. sec. 1989-90, publicity com. 1988-89, philanthropic com. 1988-90, zone publicity co-chmn. 1989-90, chpt. v.p. 1990-91, zone 12 auditor 1990-91, zone co-chair 1993—, Kans. state coun. 1990—, zone 12 awards com. chair 1991-93.), Kans. Roadrunner dir. 1992-93, edn. com. chmn. 1993—, zone 12 outstanding sister 1990-91, Kans. State outstanding vol. roadrunner 1990-91, outstanding sister zone 12 1990-91), Delta Kappa Gamma, Sigma Xi. Republican. Avocations: pianist, crochet, needlework, photography, reading. Office: Pratt Community Coll Dept Biol Scis Pratt KS 67124

RICE, ROY WARREN, ceramic engineer; b. Seattle, Aug. 31, 1934; s. Warren Francis and Edith (Koch) R.; m. Doris Elfriede Gutzeit; children: Colleen Susan, Craig Ronald. BS in Physics, U. Wash., 1957, MS in Physics, 1962. Engr. The Boeing Co., Seattle, 1957-68; sect. head U.S. Naval Rsch. Lab., Washington, 1968-74, br. head, 1974-84; dir. materials rsch.

W.R. Grace and Co., Columbia, Md., 1984—; mem. rev. com. U. Ill. Ceramics Dept., Champaign, 1977-80, Los Alamos (N.Mex.) Materials Sci. and Tech. Div., 1981-87. Contbr. over 200 articles on ceramics to profl. jours. Recipient Navy Meritorious Civilian Svc. award U.S. Naval Rsch. Lab., 1978. Fellow Am. Ceramic Soc. (v.p. 1972); mem. AAAS, Am. Phys. Soc., Am. Soc. for Materials. Achievements include devel. of dense MgO CaO, Al_2O_3 and $BaTiO_3$ using LiF; ceramic welding; understanding of microstrutural dependence of ceramic mech. properties; devel. of ceramic composites and hot pressing for ceramic electronic packages. Home: 5411 Hopark Dr Alexandria VA 22310 Office: W R Grace 7379 Rte 32 Columbia MD 21044

RICE, STEVEN DALE, electronics educator; b. Valparaiso, Ind., Aug. 11, 1947; s. Lloyd Dale and Mary Helen (Breen) R.; m. Reyanna Danti, Mar. 4, 1972; children: Joshua, Breanna. AAS, Valparaiso Tech. Inst., 1969; BS Health Sci., Ball State U., 1973; BSEE, Valparaiso Tech. Inst., 1973; MS in Vocat. Edn., No. Mont. Coll., 1991. Electronics technician Heavy Mil. Electronic Systems GE, Syracuse, N.Y., 1969-70; electronics technician Ball State U., Muncie, Ind., 1974-75; with electronic sales Tandy Corp., Valparaiso, 1976-77; electronics technician Missoula (Mont.) Community Hosp., 1977-84; instr. electronics Missoula Vocat. Tech. Ctr., 1984-88, instr., chmn. dept., 1988—. Book reviewer Merrill Pub., 1988—. Bd. dirs. Victor (Mont.) Sch. Bd., 1989—, chmn. bd., 1992—. Mem. IEEE, Instrument Soc. Am., Mont. Fedn. Tchrs. Office: Missoula Vocat Tech Ctr 909 South Ave W Missoula MT 59801-7910

RICE, STUART ALAN, chemist, educator; b. N.Y.C., Jan. 6, 1932; s. Harry L. and Helen (Rayfield) R.; m. Marian Ruth Coopersmith, June 1, 1952; children—Barbara, Janet. B.S., Bklyn. Coll., 1952; M.A., Harvard, 1954, Ph.D., 1955. Jr. fellow Harvard, 1955-57; faculty U. Chgo., 1957—, prof. chemistry, 1960-69, Louis Block prof. phys. scis., 1969—, chmn. dept. chemistry, 1971-76, Frank P. Hixon disting. service prof., 1977—, dean phys. scis. div., 1981, dir. Inst. Study Metals, 1962-68; Mem. Nat. Sci. Bd., 1980-86. Author: Polyelectrolyte Solutions, 1961, Statistical Mechanics of Simple Liquids, 1965, Physical Chemistry, 1980; bd. dirs.: also numerous articles. Bull. Atomic Scientists. Guggenheim fellow, 1960-61; Falk-Plautt lectr. Columbia U., 1964; Riley lectr. Notre Dame U., 1964; NSF sr. postdoctoral fellow, 1965-66; USPHS spl. postdoctoral fellow U. Copenhagen, 1970-71; Univ. lectr. chemistry U. Western Ont., 1970; Seaver lectr. U. So. Calif., 1972; Noyes lectr. U. Tex., Austin, 1975; Foster lectr. SUNY, Buffalo, 1976; Frank T. Gucker lectr. Ind. U., 1976; Fairchild lectr. Calif. Inst. Tech., 1979; Baker lectr. Cornell U., 1985-86; Centenary lectr. Royal Soc. Chemistry, 1986-87. Fellow Am. Philos. Soc.; mem. Am. Chem. Soc. (award Pure Chemistry 1963, Leo Hendrik Baekland award 1971, Peter Debye award 1985, Hildebrand award 1987), Nat. Acad. Sci., Am. Acad. Sci., Am. Phys. Soc., AAAS, Faraday Soc. (Marlowe medal 1963), N.Y. Acad. Scis. (A. Cressy Morrison prize 1955), Danish Acad. Sci. and Letters (fgn.).

RICE, TREVA KAY, genetic epidemiologist; b. Poteau, Okla., Feb. 1, 1951; d. George Jr. and Martha L. (Baldwin) R. BS, U. Tex., Arlington, 1981; PhD, U. Colo., 1987. Grad. rsch. asst. Inst. Behavioral Genetics, U. Colo., Boulder, 1981-87; rsch. instr. div. biostats. Washington U. Sch. Medicine, St. Louis, 1987-89, rsch. asst. prof., 1989—. Mem. Am. Soc. Human Genetics, Internat. Genetic Epidemiology Soc., Am. Psychol. Soc., Behavior Genetics Assn. Achievements include research in genetic and environmental influences on intelligence and on cardiovascular disease risk factors such as lipids and obesity. Office: Washington U Sch Medicine Div Biostats 660 S Euclid Ave Box 8067 Saint Louis MO 63110

RICH, CHARLES ANTHONY, hydrogeologist, consultant; b. London, Eng., Nov. 5, 1951; came to U.S., 1955; s. Eric Hebert and Ilse (Renard) R.; m. Linda Christine Johnson, June 23, 1984; 1 child, Oliver Sandor. BS in Geology, Utica Coll. of Syracuse U., 1973; MA in Geology, Queens Coll., CUNY, 1975; MBA, Pepperdine U., 1989, cert. in dispute resolution, 1993. Cert. profl. geologist. Hydrologic technician U.S. Geol. Survey, Mineola, N.Y., 1973-75; hydrogeologist Holzmacher, McLendon & Murrel P.C., Melville, N.Y., 1975-76, Geraghty & Miller, Inc., Port Washington, N.Y., 1976-79; prin.-in-charge Dames and Moore, Cranford, N.J., 1979-82; pres. CA Rich Cons., Inc., Sea Cliff, N.Y., 1982—; cons. to various indsl. and govtl. orgns., comml. developers, 1975—; expert witness Nat. Forensic Ctr., 1989; speaker Innovative Remedial Techs., 1992, water summit forum, 1993. Mem. ASTM (standards com., environ. audits for real property transfer com., environ. monitoring com.), Am. Inst. Profl. Geologists (Northeast sect. pres. 1981-92, nat. govt. affairs com. 1983—), Am. Water Resources Assn., Am. Water Works Assn., Nat. Water Well Assn. (cons. com., keynote speaker 3d nat. USEPA/NWWA symposium 1983), N.Y. Water Pollution Control Assn., Am. Geol. Inst./Geol. Soc. Am., Nat. Forensic Ctr. (expert writer), Conn. Bus. and Industry Assn., Conn. Ground Water Assn. Home: 168 Baldwin Ave Locust Valley NY 11560-1920 Office: CA Rich Cons 404 Glen Cove Ave Sea Cliff NY 11579-2107

RICH, DANIEL HULBERT, chemist; b. Fairmont, Minn., Dec. 12, 1942; married, 1964; 2 children. BS, U. Minn., 1964; PhD in Organic Chemistry, Cornell U., 1968. Rsch. assoc. organic chemist Cornell U., 1968; rsch. chemist Dow Chem. Co., 1968-69; asst. prof. pharm. chemistry U. Wis., Madison, 1970-75, assoc. prof., 1975-81, prof. dept. medical chemistry, 1981—; cons. biorganic natural product study sect., NIH, 1980—, mem., 1981—. Recipient H.I. Romnes award, 1980; fellow NIH, 1968, Stanford U., 1969-70. Mem. AAAS, Am. Chem. Soc. (Ralph F. Hirschmann award in Peptide Chemistry 1993), Am. Pharm. Assn. Achievements include research in synthesis in peptides and hormones, inhibition of peptide receptors and proteases, characterization, synthesis and mechanisms of action of peptide natural products. Office: Univ of Wisconsin Dept of Medical Chemistry 500 Lincoln Dr Madison WI 53706*

RICH, HARRY LOUIS, physicist, marine engineering consultant; b. N.Y.C., Apr. 30, 1917; s. Meyer and Annie (Nemser) R.; m. Irene Silverman, July 3, 1941; children: Michelle, Margo. BA, Bklyn. Coll., 1939; postgrad., George Washington U., 1941-50. Physicist David Taylor Model Basin, USN, Carderock, Md., 1941-65; prin. physicist David Taylor Ship R & D Ctr., Carderock, 1965-74; pvt. practice cons., marine engring. Bethesda, Md., 1974—; U.S. rep. Internat. Electro Tech. Com., N.Y.C., 1964-74, Internat. Standards Com., N.Y.C., 1964—; sci. advisor USN and Korean Navy, 1971-72. Contbr. articles to Shock and Vibration Symposium, Inst. for Environ. Scis. Recipient Superior Civilian Svc. award USN, Washington, 1966, Civilian Svc. award USN, Seoul, 1972. Fellow Acoustical Soc. Am., Inst. for Environ. Scis.; mem. Cosmos Club (Washington). Achievements include patent for shock spectrum recorder. Home and Office: 6765 Brigadoon Dr Bethesda MD 20817-5418

RICH, RAPHAEL Z., mechanical engineer, researcher; b. Bobruisk, USSR, Aug. 19, 1929; came to U.S., 1979; s. Zalmon and Sima (Bibelnick) Rakhmilevich; widowed; 1 child, Natella Vaidman. BA in Viola, Violin, Conservatory, Moscow, 1944; MME, Bauman Inst. Tech., Moscow, 1950; PhD, Gubkin Petrochem. & Gas Inst., Moscow, 1955, Postdoctorate Thesis, 1973; sr. rsch. assoc., USSR Supreme Attestation Bd., 1967. Registered profl. engr. Sr. mech. engr., head of rsch. lab., vessel engr. State R & D Inst. of Petroleum and Chem. Equipment, Moscow, 1950-78; assoc., asst. prof. Gubkin Petrochem. and Gas Inst., Moscow, 1967-73; mech. design engr. Allstate Design & Devel. Co., Inc., Phila., 1979-80; pressure vessel design engr. The Lumus Co., Bloomfield, N.J., 1980-83; sr. scientist European-Am. Engring. and Trade Co., Ft. Lee, N.J., 1983—; pres. B'nai-Zion Soviet-Am. Scientists Div., N.Y.C., 1988—. Author: Heat Exchanger Design, 1979, Heat Exchanger Standardization, 1964, Heat Exchanger Design for Enlarged and Combined Process Plant, 1964, (with others) Heat Transfer Research and Heat Exchanger Design, 1963, Heat Exchangers for High Capacity Refineries, Petrochemical and Chemical Plants, 1966, (with others) Pressure Vessel Design, 1968; contbr. more than 100 articles to profl. jours. Mem. AAAS (hon.), ASME, ASCE, NSPE, Am. Inst. Chem. Engrs., N.Y. Acad. Scis., Am. Chem. Soc. Achievements include 5 inventions in the field of petrochemical equipment.

RICH, ROBERT F., political sciences educator, science administrator; married; 3 children. BA in Govt. with high honors, Oberlin Coll., 1971; student, Free U. of Berlin, 1971-72; MA in Polit. Scis., U. Chgo., 1973, PhD in Polit. Scis., 1975. Project dir., asst. rsch. scientist Ctr. for Rsch. on Utilization Sci. Knowledge, Inst. Social Rsch., U. Mich., lectr. dept. polit. sci., 1975-76; asst. prof. politics and pub. affairs Princeton U., 1976-82, coord. domestic and urban policy field Woodrow Wilson Sch., 1979-81; assoc. prof. polit. sci., pub. policy and mgmt. Sch. Urban and Pub. Affairs, Carnegie-Mellon U., 1982-86; prof. polit. sci., health resources mgmt., medical humanities and social svcs., community health, prof. Inst. Environ. Studies U. Ill., Urbana, 1986—, dir. Inst. Govt. and Publ. Affairs; acting head med. humanities and social scis. program U. Ill., Urbana-Champaign, 1988-90, prof. Inst. for Environ. Studies; cons. Carnegie-Mellon U., 1986—, MacArthur Found., NIMH, 1988-89, Food, Drug and Law Inst., HHS, 1989, others. Author: Social Science Information and Public Policy Making: The Interaction Between Bureaucratic Politics and the Use of Survey Data, 1981; co-author: Government Information Management: A Counter-Report of the Commission on Federal Paperwork, 1980; editor: Translating Evaluation into Policy, 1979, The Knowledge Cycle, 1981, Knowledge, Creation, Diffusion, Utilization, 1979-88, 88-91; co-editor: Competitive Approaches to Health Policy Reform, 1993; assoc. editor Society, 1984-88, Evaluation Rev., 1985-89; mem. editorial adv. rev. bd. Policy Studies Rev. Series, 1980-83; mem. editorial bd. Evaluation and Change, 1979-82; mem. editorial adv. bd. Law and Human Behavior, 1983-87; contbr. numerous articles to profl. jours., book chpts. Recipient Emil Limbach Teaching award Carnegie-Mellon U., Sch. Urban and Pub. Affairs, 1985; fellow German Acad. Exch. Program, Fed. Republic Germany, 1971-72, Nat. Opinion Rsch. Ctr. fellow, 1972-73, German Govt. fellow, 1974, Russel Sage Found. Rsch. fellow, 1974-75; vis. scholar Hastings Ctr. for Society, Ethics and Life Scis., 1982. Mem. Am. Psychol. Assn. (task force on victims of crime and violence 1982-84), Soc. for Traumatic Stress Studies (bd. dirs. 1980—), World Fedn. for Mental Health (chmn. com. on mental health needs of victims 1985—, vice chmn. 1981-83), Robert F. Rich rsch. ann. award established in his honor, sci. com. on mental health needs of victims 1983), Howard R. Davis Soc. for Knowledge Utilization and Planned Change (pres. 1986-89), Polit. Sci. 400, Phi Beta Kappa, Sigma Xi, Phi Kappa Phi. Home: 3014 E Oaks Rd Urbana IL 61801-9802 Office: U Ill Inst Govt and Pub Affairs 1007 W Nevada St #204 Urbana IL 61801-3889 also: 921 W Van Buren M/C 191 Chicago IL 60607

RICH, THOMAS HEWITT, curator; b. Evanston, Ill., U.S., May 30, 1941; arrived in Australia, 1973, naturalized, 1981; s. Albert Dyckman and Albert Ross (Woelfel) R.; m. Patricia Arlene Vickers, Sept. 3, 1966; children: Leaellyn, Timothy. AB, U. Calif., Berkeley, 1964, MA, 1967; PhD, Columbia U., 1973. Computer programmer U. Calif., Berkeley, Calif., 1964-67; curator Mus. of Victoria, Melbourne, Victoria, Australia, 1974—. Co-author: The Fossil Book, 1989. Mem. Earthwatch, Dinosaur Cove, Victoria, Australia, 1986-89. With U.S. Navy, 1959-65. Mem. Soc. Vertebrate Paleontology, Australian Mammal Soc., Geological Soc. Australia, Royal Soc. Victoria, Nat. Geographic Soc. (U.S.), Australian Rsch Coun., Sigma Xi. Avocation: collecting slide rules. Home: 24 Sunnyside Terr, 3782 Emerald Victoria, Australia Office: Mus of Victoria, 285-321 Russell St, 3000 Melbourne Victoria, Australia

RICHARD, SYLVAN JOSEPH, electrical engineer; b. Ville Platte, La., Feb. 14, 1934; s. Loyce and Armance (Pitre) R.; m. Mary Jane Fontenot, July 1, 1956; children: Rachel Ann, Janet Louise, Jeffery Paul. BS, U. Southwestern La., 1960. Registered professional engineer, La. Engr. SLEMCO, Lafayette, La., 1960-73; dir. utilities system City of Lafayette, 1973-80; gen. mgr. La. Energy and Power Authority, Lafayette, 1980—. Editor (newsletter) The Power Line, 1980—. Mem. del. to Europe and Russia, People to People, Spokane, Wash., 1974, mem. del. to China on renewable resources, 1980; mem. USL power adv. com. U. Southwestern La., Lafayette, 1990—. Mem. NSPE, IEEE (chpt. pres.), La. Engring. Soc. (chpt. pres., Officer of Yr. 1965-66), Lions Club Internat. (chmn. club telethon Lafayette 1981-82, Lion of Yr. 1981-82). Democrat. Roman Catholic. Office: La Energy and Power Authority 315 Johnston St Lafayette LA 70501-8021

RICHARDS, DIANA LYN, psychologist; b. Baton Rouge, Dec. 8, 1944; d. William Allen Richards and Julia Viola (Hamilton) Richards Hamilton. AA, Stephens Coll., 1964; BA, U. Colo., 1966; MA, Miami U., Oxford, Ohio, 1969, PhD, 1974. Lic. psychologist, Mo. Dir. community psychol. svcs. Malcolm Bliss Mental Health Ctr., St. Louis, 1975-77; mem. staff Women's Counseling Ctr., St. Louis, 1976-78; mem. faculty Gestalt Inst., St. Louis, 1977-80; instr. Washington U., St. Louis, 1977—; dir. psychology Lindenwood Coll. for Individualized Edn., St. Louis, 1978-80, core faculty in psychology, 1984—; psychologist in pvt. practice St. Louis, 1977—; mem. Psychoanalytic Study Group, St. Louis, 1980—. Contbr. articles to profl. jours. Mem. Operation Food Search, Defenders of Wildlife, Humane Soc., People for the Ethical Treatment of Animals, Arts and Edn. Fund, Mo. Bot. Garden, Digit Fund, Earth Island Inst.; founding mem. The Pleiades. Mem. Am. Psychol. Assn., Mo. Psychol. Assn., St. Louis Psychol. Assn. (program chair 1988-89), Network of Women Psychologists (program chair 1986), St. Louis Psychoanalytic Inst., World Wildlife Fund, Nature Conservancy, Audubon Soc., Humane Farming Assn. Democrat. Avocations: sailing, backpacking, gardening, literature cross-country skiing. Home: 2014 S Mason Rd Saint Louis MO 63131-1619 Office: 7396 Pershing Ave Saint Louis MO 63130-4206

RICHARDS, ERNEST WILLIAM, clinical nutritionist; b. New Brunswick, N.J., June 27, 1960; s. William Weldon and Dorothy Marie (Miller) R. AS, Middlesex Community Coll., 1980; BA, Rutgers U., 1982, PhD, 1988. Rsch. assoc. Bapt. Med. Ctrs., Birmingham, Ala., 1988-90; asst. rsch. scientist Bapt Med. Ctrs., Birmingham, 1990-92; adj. asst. prof. U. Ala., Birmingham, 1991-92; clin. rsch. assoc. Ross Labs., Columbus, Ohio, 1992—; cons. Marlboro (N.J.) State Psychiat. Hosp., 1987-88; nutrition counselor Open Door Rehab. Project, New Brunswick, 1987-88. Contbr. articles to Nutrition, Metabolism, Am. Jour. Clin. Nutrition. Speaker Healthtalks Speakers Bur., Birmingham, 1990-92; active Biomed. Rsch. Support Grant Rev. Com., Birmingham, 1991-92. Mem. Am. Soc. for Clin. Nutrition, Am. Inst. Nutrition, Am. Soc. for Parenteral and Enteral Nutrition, N.Y. Acad. Scis. Achievements include development of nutritional support products for AIDS and cancer patients; research in nutritional protection of the GI tract from therapeutic-induced injury. Office: Ross Labs 625 Cleveland Ave Columbus OH 43215-1724

RICHARDS, HUGH TAYLOR, physics educator; b. Baca County, Colo., Nov. 7, 1918; s. Dean Willard and Kate Bell (Taylor) R.; m. Mildred Elizabeth Paddock, Feb. 11, 1944; children: David Taylor, Thomas Martin, John Willard, Margaret Paddock, Elizabeth Nicholls, Robert Dean. BA, Park Coll., 1939; MA, Rice U., 1940, PhD, 1942. Research assoc. Rice U., Houston, 1942; scientist U. Minn., Mpls., 1942-43, U. Calif. Sci. Labs., Los Alamos, N.Mex., 1943-46; research assoc. U. Wis., Madison, 1946-47, mem. faculty, 1947-52, prof., 1952-88, prof. emeritus, 1988—, physics dept. chairperson, 1960-63, 66-69, 85-88; assoc. dean Coll. Letters and Sci., U. Wis, 1963-66. Author: Through Los Alamos 1945: Memoirs of a Nuclear Physicist; contbr. articles to profl. jours. Fellow Am. Phys. Soc.; mem. Am. Assn. Physics Tchrs. Unitarian-Universalist. Achievements include neutron measurements first A-Bomb test; fission neutron (and other) spectra by new photo-emulsion techniques; mock fission neutron source; spherical electrostatic analyzer for precise reaction energy measurements; negative ion sources for accelerators (He ALPHATROSS, SNICS); accurate proton, deuteron, and alpha particle scattering and reaction cross sections; systematics mirror nuclei; isospin violations in nuclear reactions. Home: 1902 Arlington Pl Madison WI 53705-4002 Office: Univ of Wis Dept of Physics Madison WI 53706

RICHARDS, IRA STEVEN, toxicology educator; b. Bklyn., May 6, 1948; s. Oscar Seymour and Jean (Fafarman) R.; m. Barbara Lynn Mildworm, June 6, 1976; children: Charles Michael, Jeffrey Adam, Elizabeth Pamela. BS, L.I. U., 1970; MS, U. Mass., Dartmouth, 1972; PhD, NYU, 1976. Asst. prof. physiology Rutgers U., Camden, N.J., 1976-80; rsch. prof. medicine U. Cin., 1981-86; assoc. prof. toxicology South Fla. Coll. Pub. Health, Tampa, Fla., 1986—; tech. advisor State of Fla. Toxic Substances Adv. Bd., Tallahassee, Fla., 1992-96; cons. in field. Contbr. articles to profl. jours. Recipient Rsch. award Am. Lung Assn., 1987. Mem. Soc. Toxicology, Internat. Soc. Toxicology, Am. Assn. Clin. Toxicology, Sigma Xi. Achievements include pioneering the field of the Electromechanical Properties of Human Airway Smooth Muscle, thus providing information on Mechanisms of Asthma, Bronchoconstriction, and the devel. of new Bronchodilatory drugs. Home: PO Box 1625 Lutz FL 33549 Office: U So Fla Coll of Pub Health MDC Box 56 13201 Bruce B Downs Blvd Tampa FL 33612-3805

RICHARDS, J. SCOTT, rehabilitation medicine professional. BA in Psychology cum laude, Oberlin Coll., 1968; Cert. in Elem. Edn., Wayne State U., 1969; MS in Resource Ecology, U. Mich., 1973; PhD in Psychology, Kent State U., 1977. Elem. tchr. Detroit Pub. Schs., 1968-71; rsch. asst. Inst. Fisheries Rsch., 1971-73; teaching asst. Kent State U., 1973-74; psychology intern, 1973-77; dir. psychology and instr. dept. rehab. medicine, co-dir. pain control program SRC, with depts. psychiatry and psychology U. Ala., Birmingham, 1977—, from asst. to assoc. prof. dept. rehab. medicine, 1980-90, prof., 1990—, dir. tng. med. rehab. rsch. tng. ctr., 1985-87, dir. rsch., 1987—, co-dir. UAB-SCI Care System and Nat. SCI Statis. Ctr., 1989—; cons. Ctrs. Disease Control, Nat. Inst. Disability & Rehab. Rsch. Contbr. articles to profl. jours. Fellow APA. Office: University of Alabama Med Rehab Rsch/Training UAB Sta Birmingham AL 35294*

RICHARDS, JOHN LEWIS, civil engineer, educator; b. Staten Island, N.Y., Feb. 27, 1943; s. John Rose Richards and Elaine (Martin) Brady; m. Rosa Lee Unger, June 10, 1972; children: John P., Catherine L., James E. BS, U.S. Mil. Acad., 1964; MS, U. Ill., 1967, PhD, 1973; MBA, U. Pitts., 1986. Registered profl. engr., Ill., Pa. Commd. 2d lt. U.S. Army, 1964, advanced through grades to lt. col., 1980, area engr. Chgo. dist., 1968-71, asst. prof. West Point (N.Y.) U.S. Mil. Acad., 1973-76, staff engr. various locations, 1976-81, dist. engr. Pitts. dist., 1981-84, ret., 1984; adj. prof. U. Pitts., 1984—; ind. engring. project mgr. Pitts., 1984—; cons. in field; constrn. arbitrator Am. Arbitration Assn., Pitts., 1984—. Contbr. articles to profl. publs. Com. chmn. United Way Southwestern Pa., Pitts., 1983; bd. dirs. Washington County unit ARC, 1986-88; coun. mem. Peters Twp., McMurray, Pa., 1986-88. Decorated Legion of Merit, 2 Bronze Star medals. Fellow Soc. Am. Mil. Engrs. (past pres. 1983-84); mem. ASCE, Tri-State Constrn. Users Coun., Phi Kappa Phi, Chi Epsilon. Achievements include development of a mathematical model for optimal project scheduling, mathematical model for resolving financial disputes. Home: 106 Shawnee Trail Venetia PA 15367 Office: Ste 205 505 Valleybrook Rd McMurray PA 15317

RICHARDS, RANDAL WILLIAM, chemist, educator; b. Wednesbury, United Kingdom, Aug. 30, 1948; s. William James and Anne Florence (Fletcher) R.; m. Patricia Margaret Magee, Sept. 7, 1972; children: Robert James, Gregory Peter. BS, U. Salford, 1971, PhD, 1974. Rsch. fellow Deutsches Kunststoff Ins., Darmstadt, Germany, 1974-75; rsch. assoc. Imperial Coll., London, 1975-77; lectr., sr. lectr. U. Strathclyde, Glasgow, Scotland, 1977-89; reader U. Durham, United Kingdom, 1989—; chmn. Neutron Beam Rsch. Com., U.K., 1991—; mem. sci. bd. Sc. and Engring. Rsch. Coun., U.K., 1991-93. Editor: Polymer; contbr. over 60 articles to profl. jours. Fellow Royal Soc. Chemistry (Courtaulds prize 1990), Am. Chem. Soc. Home: 25 Telford Close, High Shincliffe DH1 2YJ, England Office: U Durham, Dept Chemistry, Durham DH1 3LE, England

RICHARDS-KORTUM, REBECCA RAE, biomedical engineering educator; b. Grand Island, Nebr., Apr. 14, 1964; d. Larry Alan and Linda Mae (Hohnstein) Richards; m. Philip Ted Kortum, May 12, 1985; 1 child, Alexander Scott. BS, U. Nebr., 1985; MS, MIT, 1987, PhD, 1990. Asst. prof. biomed. engring. U. Tex., Austin, 1990—. Named Presdl. Young Investigator NSF, Washington, 1991; NSF presdl. faculty fellow, Washington, 1992; recipient Career Achievement award Assn. Advancement Med. Instrumentation, 1992, Dow Outstanding Young Faculty awd., Am. Soc. for Engineering Education, 1992. Mem. AAAS, Am. Soc. Engring. Edn. (Outstanding Young Faculty award 1992), Optical Soc. Am., Am. Soc. Photobiology. Achievements include research in photochemistry, photobiology, applied optics and bioengring. Office: U Tex Dept Elec & Computer Engring Austin TX 78712

RICHARDSON, BRYAN KEVIN, mathematics educator, naval aviator; b. Long Beach, Calif., Dec. 7, 1959; s. Wilbur R. and Virginia Mae (Kraft) R.; m. Janet Marie Vomero, May 25, 1984; children: Aaron Micheal, Daniel James, Matthew David. BA in Math., Calif. State U., Long Beach, 1986. Tchr.'s aide Long Beach Unified Sch. Dist., 1978-84; constrn. contractor Richardson Enterprises, Long Beach, 1980-84; communications technician Kenny's Auto Svc., Bellflower, Calif., 1984-86; elem. sch. tchr. St. Pancratins Elem. Sch., Lakewood, Calif., 1992; math. educator St. John Bosco High Sch., Bellflower, 1992—; mktg. cons. Aviation Week and Space Tech., Brunswick, Maine, 1989-91; col. Confederate Air Force, Midland, Tex., 1991—; mem. Planes of Fame Air Mus., Chino, Calif., 1989—. Author (daily publ.) VP26 Sports Corner, 1991. Lay eucharistic min. Brunswick Naval Air Sta. Chapel, 1990-91; bd. dirs. "A Time for US" Marriage Retreat Ministries, Lakewood, 1992—. Lt. USN, 1987-91. Mem. St. Pancratius Holy Name Soc. Republican. Roman Catholic. Office: St John Bosco High Sch 13640 S Bellflower Blvd Bellflower CA 90706

RICHARDSON, CAROL JOAN, pediatrician; b. San Angelo, Tex., July 6, 1944; d. Giles Otto and Noreen (Bailey) R. BA, U. Tex., 1966; MD, U. Tex., Galveston, 1970. Diplomate Am. Bd. Pediatrics, sub-bd. Neonatal-Perinatal Medicine. Intern U. Tex. Med. Br. at Galveston, 1970-71, resident in pediatrics, 1971-72; asst. prof. pediatrics Med. Br. U. Tex., Galveston, 1974-78, assoc. prof., 1978-83, prof. pediatrics, ob-gyn., 1983—. Contbr. numerous articles to profl. publs., 1972—. U. Calif. San Diego fellow, 1972-74. Mem. Am. Acad. Sci., So. Perinatal Assn., Tex. Med. Assn., Tex. Pediatric Soc., Tex. Perinatal Assn. (pres. 1978-79), Alpha Omega Alpha. Democrat. Methodist. Avocations: sailing, tennis. Home: 514 16th St Galveston TX 77550-4802 Office: U Tex Med Br Dept Pediatrics Galveston TX 77550

RICHARDSON, DEBORAH RUTH, psychology educator; b. Hampton, Va., July 5, 1950; d. Thomas Thornton and Sophia Ruth (South) R.; m. Bibb Latané, May 29, 1989. BS in Psychology, Va. Commonwealth U., 1972; MA in Psychology, Coll. of William & Mary, Williamsburg, Va., 1974; PhD in Psychology, Kent State U., 1978. Asst. prof. U. Ga., Athens, 1977-83, assoc. prof. psychology, 1983-89; assoc. prof. psychology Fla. Atlantic U., Boca Raton, 1990-93, prof. psychology, 1993—; instr. Asian div. U. Md., Yokota, Japan, 1982-83; co-dir. Nags Head Conf. Ctr., 1987—; vis. assoc. prof. U. N.C., Chapel Hill, 1988-89. Co-author chapts. Acquaintance Rape: The Hidden Crime, 1991, Social Psychology, 1991; contbr. over 40 articles to profl. jours. Office: Fla Atlantic U Dept Psychology Boca Raton FL 33431

RICHARDSON, DON ORLAND, agricultural educator; b. Auglaize County, Ohio, May 12, 1934; s. Dana Orland and Mary Isabell (Bowersock) R.; m. Shirley Ann Richardson (div. 1982); children: Daniel, Bradley, Eric, Laura. BS, Ohio State U., 1956, MS, 1957, PhD, 1961. Asst. prof. to prof. U. Tenn., Knoxville, 1963—, head Animal Sci. Dept., 1982-88, dean Agrl. Exptl. Sta., 1988—. Mem. Am. Dairy Sci. Assn., Am. Soc. Animal Sci., Coun. Agrl. Sci. and Tech., Holstein Assn. Am., Rotary Club Knoxville. Office: Tenn Agrl Exptl Sta 103 Morgan Hall PO Box 1071 Knoxville TN 37901-1071

RICHARDSON, DONALD CHARLES, engineer, consultant; b. Glendale, Calif., June 6, 1937; s. George Robert and Margaret Josephine (Buchholz) R.; m. Helen Mary Boyd, Aug. 9, 1984. BA in Sci., Calif. State U., 1965; MS in Engring., Queens U., 1981, M.Ed., 1983; PhD, Clarkson U., 1988. Sr. engr. Control Data Corp., Toronto, Ont., Can., 1972-75; tchr. Algonquin Coll., Kingston, Ont., Can., 1975-79; assoc. prof. Royal Mil. Coll. of Can., Kingston, 1982-84; instr. Clarkson U., Potsdam, N.Y., 1988; supercomputer cons., 1989; mgr. tech. support Multiflow Computer, Inc. Author of engring. papers. Served to lt. comdr. USN, 1964-69, Vietnam. Electrochem. Soc. fellow, 1984. Mem. Am. Soc. Engring. Edn., Mensa, Sigma Xi, Fraternal Order of Seals.

RICHARDSON, EVERETT VERN, hydraulic engineer, educator, administrator; b. Scottsbluff, Nebr., Jan. 5, 1924; s. Thomas Otis and Jean Marie (Everett) R.; m. Billie Ann Kleckner, June 23, 1948; children—Gail Lee, Thomas Everett, Jerry Ray. B.S., Colo. State U., 1949, M.S., 1960, Ph.D., 1965. Registered profl. engr., Colo. Hydraulic engr. U.S. Geol. Survey, Wyo., 1949-52; hydraulic engr. U.S. Geol. Survey, Iowa, 1953-66; rsch.

hydraulic engr. U.S. Geol. Survey, Ft. Collins, Colo., 1956-63, project chief, 1963-68; prof. civil engring., adminstr. engring. rsch. ctr. Colo. State U., Ft. Collins, 1968-82, prof. in charge of hydraulic program, 1982-88, prof. civil engring., 1988—, dir. hydraulic lab. engring. rsch. ctr., 1982-88, dir. Egypt water use project, 1977-84, dir. Egypt irrigation improvement project, 1985-90, dir. Egypt Water Rsch. Ctr., 1988-89; sr. assoc. Resource Consultants and Engrs., Inc., Ft. Collins, Colo., 1989—; dir. Consortium for Internat. Devel., Tucson, Ariz., 1972-87; cons. in field. Editor: Highways in the River Environment, U.S. Bur. Pub. Rds., 1975, 90; contbr. articles to profl. jours., chpts. to books. Mem. Ft. Collins Water Bd., 1969-84. With AUS, 1943-45. Decorated Bronze Star, Purple Heart; Combat Infantry Badge, U.S. Govt. fellow MIT, 1962-63. Fellow ASCE (J.D. Stevens award 1961); mem. Internat. Congress for Irrigation and Drainage (bd. dirs.), Sigma Xi, Chi Epsilon, Sigma Tau. Home: 824 Gregory Rd Fort Collins CO 80524-1504 Office: Resource Cons and Engrs PO Box 270460 Fort Collins CO 80527

RICHARDSON, JASPER EDGAR, nuclear physicist; b. Memphis, Nov. 8, 1922; s. Jasper Edgar and Katherine Cecil (Copp) R.; m. Nellie Carolyn Harwell, May 30, 1947; children: Ann Helen, Janet Katherine, Susan Carolyn, Patricia Lynn, Ellen Claire. BS in Physics, Yale U., 1944; MA In Physics, Rice U., 1948, PhD in Physics, 1950. Instr. physics U. Miss., Oxford, 1946-47; asst. prof. Auburn (Ala.) U., 1950-51; A.E.C. fellow Oak Ridge (Tenn.) Inst. Nuclear Studies, 1951-53; physicist U. Tex. M. D. Anderson Hosp., Houston, 1953-55; rsch. physicist Shell Bellaire Rsch. Ctr., Houston, 1955-69; sr. engr. Shell Oil Co., Midland, Tex., 1969-74; staff engr. Shell Oil Co., Houston, 1974-86; ret. Patentee in electronics, oil discovery, measurement; contbr. articles to profl. jours. With USN, 1944-46, Guam. Mem. Am. Phys. Soc., Soc. Petroleum Engrs. Episcopalian.

RICHARDSON, JOSEPH HILL, physician, educator; b. Rensselaer, Ind., June 16, 1928; s. William Clark and Vera (Hill) R.; M.S. in Medicine, Northwestern U., 1950, M.D., 1953; m. Joan Grace Meininger, July 8, 1950; children: Lois N., Ellen M., James K. Intern, U.S. Naval Hosp., Great Lakes, Ill., 1953-54; fellow in medicine Cleve. Clinic, 1956-59; individual practice medicine specializing in internal medicine and hematology, Marion, Ind., 1959-67, Ft. Wayne, Ind., 1967—; assoc. clin. prof. medicine, Ind. U. Sch. Medicine. Served to lt. MC USNR, 1954-56. Diplomate Am. Bd. Internal Medicine. Fellow ACP, AAAS; mem. Am. Fedn. for Clin. Research, AMA, Masons. Contbr. articles to med. jours. Home and Office: 8726 Fortuna Way Fort Wayne IN 46815-5725

RICHARDSON, KEVIN WILLIAM, civil and environmental engineer; b. Louisville, Apr. 29, 1954; s. William Henry and Mary (Pembroke) R.; m. Debra Ann Fisher, Sept. 27, 1980; children: Katie, Billy, Timmy. BSCE, U. Wis., 1978; MBA, Keller Grad. Sch., 1989. With Alvord, Burdick & Howson, Chgo., 1978-93, Foth & Van Dyke, Madison, Wis., 1993—. Recipient Charles Ellet award Western Soc. of Engrs., 1989. Mem. ASCE, Water Environment Fedn., Western Soc. Engrs., Am. Water Works Assn. (dir. chmn. 1990—). Achievements include design of first Zerba Mussel Control System for water treatment Plant's Intakes in the state of Wisconsin. Office: Foth & Van Dyke 406 Science Dr Madison WI 53711

RICHARDSON, ROBERT COLEMAN, physics educator, researcher; b. Washington, June 26, 1937; s. Robert Franklin and Lois (Price) R.; m. Betty Marilyn McCarthy, Sept. 2, 1962; children: Jennifer, Pamela. BS in Physics, Va. Poly. Inst. and State U., 1958, MS, 1960; PhD in Physics, Duke U., 1966. Research assoc. Cornell U., Ithaca, N.Y., 1966-67, asst. prof., 1968-71, assoc. prof., 1972-74, prof., 1975—; chmn. Internat. Union Pure and Applied Physics Commn. (C-5), 1981-84; mem. bd. assessment Nat. Bur. Standards, 1983—. Mem. editorial bd. Jour. of Low Temperature Physics, 1984—. Served to 2d lt. U.S. Army, 1959-60. Guggenheim fellow 1975, 83; recipient Simon Meml. prize Brit. Phys. Soc., 1976. Fellow AAAS, Am. Phys. Soc. (Oliver E. Buckley prize 1981); mem. Nat. Acad. Scis. Avocations: photography, gardening. Office: Cornell Univ Dept Physics Clark Hall Ithaca NY 14853

RICHARDSON, THOMAS HAMPTON, design consulting engineer; b. St. Louis, Nov. 25, 1941; s. Claude Hampton and Pearl Lily (Burks) R.; m. Lois Louise Atteberry June 8, 1963; children: Shelley Ann, David Hampton, Stephanie Lynn. BTEE, Wash. U., St. Louis, 1974. Registered profl. engr., Mo., Ill., Ind., Kans., Iowa, Fla., Ky., Miss. Elec. project designer Fruco Engrs. Inc., St. Louis, 1967-68; mgr., mech./elec. engr. MBA Engrs. Inc., St. Louis, 1968-74, Kenneth Balk and Assoc., St. Louis, 1974-76; instr. elec. engring. Wash. U., St. Louis, 1976; v.p., chief engr. John F. Steffen Assoc., St. Louis, 1976-79; prin. ptnr. Keeler, Webb and Richardson, St. Louis, 1979—; pres./owner The Richardson Engring. Group, St. Louis, 1979—. Contbr. articles to profl. jours. Recipient Internat. Lgt. Des. award Illuminating Engr. Soc. St. Louis 1985, Edwin F. Guth award of Merit Illuminating Engr. Soc. N.Am. 1986. Mem. NSPE, ASHRAE, Am. Cons. Engr. Coun., Illuminating Engring. Soc. Past pres.), Soc. for Mkt. Profl. Svcs. (v.p.), Profl. Svcs. Mgmt. Assn. (bd. dirs.), Mo. Soc. Profl. Engrs. (govt. rels. com.), Nat. Fire Protection Assn., Green Turtle Bay Yacht Club, Grand Lake Yacht Club, Ky. Lake Club. Avocations: sailing, flying, horses, photography. Office: The Richardson Engring 1700 Gilsinn Dr Fenton MO 63026-2004

RICHART, DOUGLAS STEPHEN, chemist; b. Harrisburg, Pa., June 6, 1931; s. Howard Winans and Muriel Matilda (Long) R.; B.S. in Chemistry, Franklin and Marshall Coll., Lancaster, Pa., 1954; m. Joan J. Lombardo, Apr. 19, 1986; children—Deborah, Sandra, Stephen, Catherine. Research chemist Union Carbide Corp., Bound Brook, N.J., 1954-60; group leader research and devel. Polymer Corp., Reading, Pa., 1960-65, mgr. research and devel. coatings, 1965-86; mgr. research and devel. Chem. div. Morton Internat. Powder Coating, Reading, 1986-89, sr. scientist, 1989—. Mem. Am. Chem. Soc., Soc. Plastics Engrs., Nat. Assn. Corrosion Engrs., AAAS. Republican. Episcopalian. Author in field; patentee powder coatings. Home: 6 Golfview Ln Reading PA 19606 9597 Office: Morton Internat Flying Hills Corp Ctr 3 Commerce Dr Reading PA 19607-9700

RICHES, KENNETH WILLIAM, nuclear regulatory engineer; b. Long Beach, Calif., Oct. 23, 1962; s. William Murray Riches and Carlene Katherine (Simmons) Anderson; m. Susan Ruth Flagg, Aug. 11, 1990; 1 child, Benjamin William Bancroft Riches. BSEE, U. Ill., 1984; MS in Engring. Mgmt., Santa Clara U., 1989. Registered profl. engr., Calif. Engr. Pacific Gas & Electric Co., San Luis Obispo, Calif., 1984-88, elec. engr., 1988—; prin. K.W. Riches & Assocs., Arroyo Grande, Calif., 1988—; owner The Peaberry Coffee Pub, Arroyo Grande, Calif., 1991-92. Mem. Rep. Nat. Com., 1986—; active Corp. Action in Pub. Schs., San Francisco, 1987, 88, World Wildlife Fund. Univs. Rsch. Assn. scholar, 1980. Mem. NSPE, IEEE (chpt. chmn. 1986-87, sect. dir. 1988-90), Power Engring. Soc. of IEEE (mem. nat. chpts. coun. 1988-92), Pacific Coast Engring. Assn., Nature Conservancy, Order of DeMolay (master counselor Paul Revere chpt. 1979). Methodist. Avocations: golf, skiing, reading, travel. Home: 775 Ridgemont Way Arroyo Grande CA 93420 Office: PG&E Diablo Canyon Power Plant PO Box 117 M/S 104/5/21A Avila Beach CA 93424

RICHEY, CLARENCE BENTLEY, agricultural engineering educator; b. Winnipeg, Manitoba, Can., Dec. 28, 1910; s. Raus Spears and Emily Cornelia (Bentley) R.; m. Marguerite Anne Jannusch, Dec. 27, 1936; children: David Volkman, Stephen Bentley. BS in Agrl. Engring., Iowa State U., 1933; BS in Mech. Engring., Purdue U., 1939. Registered agrl. engr., Calif. Instr. agrl. engring. Purdue U., West Lafayette, Ind., 1936-41; asst. prof. dept. agrl. engring. Ohio State U., Columbus, 1941-43; head devel., engr. Electric Wheel Co., Quincy, Ill., 1943-46; project engr. Harry Ferguson, Inc., Detroit, 1946-47; sr. project engr. Dearborn Motors Corp., Detroit, 1947-54; supt., chief rsch. engr. Ford Tractor Divsn., Birmingham, Mich., 1954-62; chief engr. Fowler (Calif.) divsn. Massey-Ferguson Ltd., 1964-69; product mgmt. engr. Massey-Ferguson Ltd., Toronto, Ont., Can., 1970-71; assoc. prof. agrl. engring. Purdue U., West Lafayette, 1971-76, prof. emeritus, 1976—; farm equipment cons. Ford Found., Allahabad, India, 1963. Author: (autobiography) Fifty Years of Engineering Farm Equipment, 1989; editor-in-chief: Agricultural Engineer's Handbook, 1961; contbr. bulls. and articles to profl. jours. Fellow Am. Soc. Agrl. Engrs. (Cyrus Hall McCormick Gold medal 1977); mem. Lafayette Kiwanis. Achievements include patent for farm equipment; holder or co-holder of 79 patents. Home: 2217 Delaware Dr West Lafayette IN 47906-1917

RICHMAN, JOHN EMMETT, architect; b. East Liverpool, Ohio, June 8, 1951; s. Ethel M. (Thompson) R.; m. Susan E. Nusser, May 29, 1971; children: Stephen T., Sarah J. BArch, BS, Kent State U., 1982. Registered architect Ohio, W.Va., Pa., Iowa, Ind.; ordained elder Evang. Presbyn. Ch., 1978. Draftsman Fairfield Machine Co., Columbiana, Ohio; archtl. draftsman Robert F. Beatty, Architect, East Liverpool, 1973-80; archtl. draftsman Smiths & and Assocs., Architects, Columbiana, 1981-84, architect, 1984-85, architect, v.p., 1985-89, pres., 1989—. Bd. dirs. Beaver Local High Sch. Alumni Assn., 1990—; pres. Beaver Local High Sch. Alumni Band, 1991—; chmn. Youth Camp com. First Evang. Presbyn. Ch., 1983-93, asst. dir. Youth Camp, 1982-88, chmn. worship com., 1985-86, supt. Sunday sch., 1986-89, mem. choir; co-founder Tri-State Teen World, East Liverpool, 1978, chmn. bd. dirs., 1981—, Bible quiz dir., 1982-90, coach all-star quiz team, 1985, 878. With U.S. Army, 1971-73. Mem. AIA (bd. dirs. Ea. Ohio chpt. 1985-90, sec. 1986-87, v.p. 1988, pres. 1989), Architects Soc. Ohio (alt. dir. 1986-87, trustee polit. action com. 1988), Nat. Inst. Bldg. Scis. (mem. Ohio consultive coun.), Youth Evangelism Assn. (mem. Bible quiz com. 1985—, bd. dirs. 1989—), Columbiana C. of C. (bd. dirs. 1987-89), Rotary (sec. 1988-90, v.p. 1992-93, pres. 1993—). Avocations: sports, music, travel, reading. Home: 47995 Calcutta Smith Ferry Rd East Liverpool OH 43920-9647 Office: Smith and Associated Architects 330 N Main St Columbiana OH 44408

RICHMOND, ANN WHITE, cell and molecular biologist, educator; b. Lake Village, Ark., Jan. 1, 1946; d. Carl Lehman White and Nora Dell (Ritchie) Carey; children: Mechelle Renee, Aimee Lorraine. BS, N.E. La. U., 1966; MSN, La. State U., 1972; PhD, Emory U., 1979. Post-doctoral fellow Emory Univ., Atlanta, 1979-82, asst. prof., 1982-87, assoc. prof., 1988-89; assoc. prof. Vanderbilt U., Nashville, 1989—; mem. CBY-2 study sect. NIH, Washington, 1990-94; ad hoc mem. ACS Study Sect., Atlanta, 1990; program com. AACR, Phila., 1992. Contbr. articles to Oncogene, European Molecular Biology Orgn. Jour., Jour. CellPhysiology, Cancer Rsch. Mem. PTO, Nashville, 1989—; judge Internat. Sci. & Engring. Fair, Nashville, 1992. Recipient Cokesbury award United Meth. Ch., 1976-77. Mem. Am. Soc. for Cell Biology, Am. Assn. for Cancer Rsch., Soc. for Developmental Biology, Sigma Xi (pres. 1988-89). Methodist. Office: Vanderbilt Univ Dept Cell Biology 1161 21st Ave South Nashville TN 37232-2175

RICHMOND, JAMES ARTHUR, entomologist; b. Orange County, N.C., July 22, 1940; s. Walter William and Viola Bell (Love) R.; m. Shirley Mae Brown; children: Mesina Shirlette, James Arthur Richmond, Jr., Valerie Jean, Monica Denise. BS, N.C. Agrl. and Tech. State U., Greensboro, N.C., 1963; BA, Shaw U., 1975; MS, N.C. State U., 1972, PhD, 1988. Lab. tech. USDA Forest Svc., Research Triangle Park, N.C., 1968-69; rsch. biologist USDA Forest Svc., Research Triangle Park, 1969-75, rsch. entomologist, 1975-92; asst. prof. N.C. State U., Raleigh, 1989—; adj. prof. Shaw U., Raleigh, 1991—; presenter in field. Author numerous scientific publs. With USAF, 1963-66, Turkey, Thailand. Mem. Entoml. Soc. Am., Ga. Entoml. Soc., Soc. Am. Foresters, Sigma Xi, Alplha Phi Alpha. Home: 602 Saddle Ridge Ave Durham NC 27704 Office: USDA Forest Svc 3041 Cornwallis Rd Research Triangle Park NC 27709

RICHMOND, JULIUS BENJAMIN, retired physician, health policy educator emeritus; b. Chgo., Sept. 26, 1916; s. Jacob and Anna (Dayno) R.; m. Rhee Chidekel, June 3, 1937 (dec. Oct. 9, 1985); children: Barry J., Charles Allen, Dale Keith (dec.); m. Jean Rabow, Jan. 11, 1987. BS, U. Ill., 1937, MS, MD, 1939; DSc (hon.), Ind. U., 1978, Rush-Presbyn.-St. Luke Med. Ctr., 1978, U. Ill., 1979, Georgetown U., 1980, Syracuse U., 1986, U. Ariz., 1991; D. Med. Sci. (hon.), Med. Coll. Pa., 1980; D. Pub. Service (hon.), Nat. Coll. Edn., Evanston, Ill., 1980; LHD (hon.), Tufts U., 1986. Intern Cook County Hosp., Chgo., 1939-41, resident, 1941-42, 46; resident Municipal Contagious Disease Hosp., Chgo., 1941; mem. faculty U. Ill. Med. Sch., Chgo., 1946-53, prof. pediatrics, 1950-53, dir. Inst. Juvenile Research, 1952-53; prof., chmn. dept. pediatrics Coll. Medicine, SUNY at Syracuse, 1953-65, dean med. faculty, chmn. dept. pediatrics, 1965-70; prof. child psychiatry and human devel., prof., chmn. dept. preventive and social medicine Harvard Med. Sch., 1971-77, prof. health policy, 1981-88, dir. div. health policy research and edn., John D. Mac Arthur prof. health policy and mgmt., 1983-88, prof. health policy emeritus, 1988—; also faculty Harvard Sch. Pub. Health; psychiatrist-in-chief Children's Hosp. Med. Center, Boston, 1971-77, adv. on child health policy, 1981—; dir. Judge Baker Guidance Center, Boston, 1971-77; asst. sec. health and surgeon gen. HHS, 1977-81; mem. Pres.'s Commn. on Mental Health, 1977. Author: Pediatric Diagnosis, 1962, Currents in American Medicine, 1969. Nat. dir. Project Head Start; dir. Office Health Affairs OEO, 1965-66. Served as flight surgeon USAAF, 1942-46. Recipient Agnes Bruce Greig Sch. award, 1966, Parents Mag. award, 1966, Disting. Service award Office Econ. Opportunity, 1967, Family Health Mag. award, 1977, Myrdal award Assn. For Evaluation Rsch., 1977, award for disting. sci. contbn. Soc. for Research in Child Devel., 1979, Dolly Madison award Inst. on Clin. Infants Programs, 1979, Public Health Disting. Service award HEW, 1980, Illini Achievement award U. Ill. Alumni Assn., 1982, Community Service award Health Planning Council Greater Boston, 1985, Lemuel Shattuck award Mass. Pub. Health Assn., 1985, 1st Ann. Ronald McDonald Children's Charities award for Outstanding Contbns. to Child Health and Welfare, 1986, Sedgwick award APHA, 1992. Fellow Am. Orthopsychiat. Assn.; Disting. fellow Am. Psychiat. Assn.; hon. mem. Am. Acad. Child Psychiatry; assoc. mem. New Eng. Coun. Child Psychiatry; mem. Inst. Medicine of NAS (1st Ann. Gustav O. Lienhard award 1986), AMA (AMA-ERF award in health edn. 1988), Am. Pediatric Soc. (John Howland award 1990), Am. Acad. Pediatrics (C. Anderson Aldrich award 1966, ann. award sect. on community pediatrics 1977, Outstanding Contbn. award sect. community pediatrics 1978), Soc. Pediatric Rsch., Am. Psychosomatic Soc., Am. Public Health Assn. (Martha May Eliot award 1970, Sedgwick Medal 1992), Sigma Xi, Alpha Omega Alpha, Phi Eta Sigma.

RICHMOND, RAYMOND DEAN, aerospace engineer; b. Cin., Dec. 30, 1958; s. Raymond Charles and Charmaine (LeDuke) R.; m. Maria Louise Walsh, Oct. 30, 1989. BS in Aerospace Engring., U. Cin., 1983. Sales engr. Parkway Aerospace, Erlanger, Ky., 1984—. Mem. AIAA. Republican. Office: Parkway Aerospace 1400 Jamike Ave Erlanger KY 41018

RICHMOND, RONALD LEROY, aerospace engineer; b. L.A., Aug. 16, 1931; s. William Paul and Martha Emelia (Anderson) R.; m. Mary Louise Gates, Jan. 2, 1955; children: Pandora Deanne Richmond Perry, Steven Lee. BSME, U. Calif., Berkeley, 1952; MS in Aero. Engring., Calif. Inst. Tech., 1953, PhD in Aero. Engring., 1957. Aerodynamicist Lockheed Aircraft Co., Burbank, Calif., 1952-54; teaching/rsch. asst. Calif. Inst. Tech., Pasadena, 1952-57; asst. group leader aero. performance Douglas Aircraft Co., Long Beach, Calif., 1957-59; chief engr. adv. devel. Ford Aerospace, Newport Beach, Calif., 1959-87; adj. assoc. prof. Sch. Engring., U. Calif., Irvine, 1987-88; dir. engring. Brunswick Def., Costa Mesa, Calif., 1988—; aerodynamics cons. Douglas Aircraft, 1956-57, Shelby-Am. (Ford) Auto., L.A., 1960-62; subgroup leader NATO Indsl. Adv. Group #16, Brussels, Belgium, 1984-86. Res. dep. Orange County Sheriff's Dept., 1976—. Calif. Inst. Tech. Rsch. assistantship, 1953, 54, 55, 56, 57, teaching asst., 1955, 56, 57, grantee, 1955, 56, 57. Assoc. fellow AIAA (Orange County sect. chmn. 1989-90); mem. Western States Assn. Sheriff's Air squadrons (comdr. 1987-88), Skylarks of So. Calif. (pres.,, chmn. bd. 1987-88). Republican. Achievements include experimentally proved that skin friction forces on long, slender cylinders was several times that on flat plates, at mach 5.8, for both laminar and turbulent boundary layers. Home: 1307 Seacrest Dr Corona Del Mar CA 92625 Office: Brunswick Def 3333 Harbor Blvd Costa Mesa CA 92628-2009

RICHTER, BURTON, physicist, educator; b. N.Y.C., Mar. 22, 1931; s. Abraham and Fanny (Pollack) R.; m. Laurose Becker, July 1, 1960; children: Elizabeth, Matthew. B.S., MIT, 1952, Ph.D., 1956. Research assoc. Stanford U., 1956-60, asst. prof. physics, 1960-63, assoc. prof., 1963-67, prof., 1967—, Paul Pigott prof. phys. sci., 1980—, tech. dir. Linear Accelerator Ctr., 1982-84, dir. Linear Accelerator Ctr., 1984—; cons. NSF, Dept. Energy; dir. Varian Corp., Litel Instruments; Loeb lectr. Harvard U., 1974; DeShalit lectr. Weizmann Inst., 1975. Contbr. over 300 articles to profl. publs. Recipient E.O. Lawrence medal Dept. Energy, 1975; Nobel prize in physics, 1976. Fellow Am. Phys. Soc. (pres.-elect 1993), AAAS; mem. NAS, Am. Acad. Arts and Scis. Achievements include research in elementary particle physics. Office: SLAC PO Box 4349 Palo Alto CA 94309

RICHTER, EDWIN WILLIAM, physicist, editor, consultant; b. N.Y.C., Nov. 26, 1922; s. Edward Oswald and Gertrude Anna (Muether) R.; m. Margaret Carrick Wright, Nov. 27, 1948; children: John, Stephen, Brian, Paul, Margaret, Dorothy. BS in Physics, CUNY, 1943; postgrad., NYU, 1946-48, Boston U., 1954-55. Registered profl. engr., Mass. Rsch. physicist Columbia U. Radiation Lab., N.Y.C., 1943-51; sr. project engr. Sperry Electron Tube Div., Gt. Neck, N.Y., 1951-54; sect. head missile systems div. Raytheon Co., Bedford, Mass., 1954-58; chief microwave engr. Ewen Knight Corp., Natick, Mass., 1958-62; sr. engring. scientist RCA Automated Systems, Burlington, 1962-85; cons. radio systems GTE Govt. Systems Div., Taunton, Mass., 1988; editor Transactions IEEE, N.Y.C., 1988-91; cons. editor Butterworth-Heinemann, Woburn, Mass., 1991—. Mem. sch. com., Acton, Mass., 1961-69, Acton Bd. Health, 1969-78; mem. adv. com. Wentworth Inst., Boston, 1975-85. Fellow IEEE (chmn. microwave chpt. Boston 1965-66, chmn. instrumentation and measurement chpt. Boston 1987-88), Sigma Xi. Achievements include patent on temperature compensated resonant cavities, on variable bandwidth tunable directional filter; magnetron design, design of microwave and millimeter wave instrumentation to 300 GHz, automated test system for military use. Home and Office: 32 Brewster Ln Acton MA 01720-4254

RICHTER, JEFFREY ALAN, chemical engineer; b. Paterson, N.J., Sept. 24, 1960; s. Richard Roger and Mary Emma (Glatcz) R. BSChemE, Lehigh U., 1982; MBA, Fairleigh Dickinson U., 1993. Registered profl. engr., N.J. Simulation engr. Exxon Rsch. and Engring. Co., Florham Park, N.J., 1982-84, team leader simulation, 1988-89, team leader-crude oil characterization, 1991—; lead simulation engr. Exxon Rsch. and Engring. Co., Rotterdam, The Netherlands, 1985; process control engr. Exxon Rsch. and Engring. Co., Baytown, Tex., 1986-87; lead process control engr. Exxon Rsch. and Engring. Co., Augusta, Italy, 1990. Trustee Hills Highlands Master Assn., Bedminster, N.J., 1993; runner N.Y.C. Marathon, 1988, 91, Marine Corp Marathon, Washington, 1989, Wine Glass Marathon, N.Y., 1989. Mem. AIChE, Madison Ave. Tiger Toastmasters (Madison, N.J.) (ednl. v.p. 1989, sec.-treas. 1991—, named Competent Toastmaster 1989). Roman Catholic. Home: 49 Encampment Dr Bedminster NJ 07921 Office: Exxon Rsch and Engring Co 180 Park Ave Florham Park NJ 07932

RICK, CHARLES MADEIRA, JR., geneticist, educator; b. Reading, Pa., Apr. 30, 1915; s. Charles Madeira and Miriam Charlotte (Yeager) R.; m. Martha Elizabeth Overholts, Sept. 3, 1938 (dec.); children: Susan Charlotte Rick Baldi, John Winfield. B.S., Pa. State U., 1937; AM, Harvard U., 1938, Ph.D., 1940. Asst. plant breeder W. Atlee Burpee Co., Lompoc, Calif., 1936, 37; instr., jr. geneticist U. Calif., Davis, 1940-44; asst. prof., asst. geneticist U. Calif., 1944-49, asso. prof., asso. geneticist, 1949-55, prof., geneticist, 1955—; chmn. coordinating com. Tomato Genetics Coop., 1950-82; dir. CMR Tomato Genetics Resource Ctr., 1975—; mem. genetics study sect. NIH, 1958-62; mem. Galapagos Internat. Sci. Project, 1964; mem. genetic biology panel NSF, 1971-72; mem. nat. plant genetics resources bd. Dept. Agr., 1975-82; Gen. Edn. Bd. vis. lectr. N.C. State U., 1956; Faculty Research lectr. U. Calif., 1961; Carnegie vis. prof. U. Hawaii, 1963; vis. prof. Universidade São Paulo, Brazil, 1965; vis. scientist U. P.R., 1968; centennial lectr. Ont. Agr. Coll. U. Guelph, Ont., Can., 1974; adj. prof. Univ. de Rosario, Argentina, 1980; univ. lectr. Cornell U., 1987; mem. Plant Breeding Research Forum, 1982-84. Contbr. numerous articles in field to books and sci. jours. Recipient award of distinction Coll. Agr. and Environ. Scis., U. Calif., Davis, 1991, Disting. Svc. award Calif. League Food Processors, 1993; Guggenheim fellow, 1948, 50; grantee NSF, USPHS/NIH, Rockefeller Found., 1953-83; Pa. State U. alumni fellow, 1991; C.M. Rick Tomato Genetics Resource Ctr. at U. Calif., Davis, named in his honor, 1990. Fellow Calif. Acad. Sci., AAAS (Campbell award 1959), Indian Soc. Genetics and Plant Breeding (hon.), Am. Soc. Horticultural Sci.; mem. Nat. Acad. Scis., Bot. Soc. Am. (Merit award 1976), Am. Soc. Hort. Sci. (M.A. Blake award 1974, Vaughan Research award 1946), Mass. Hort. Soc. (Thomas Roland medal 1983), Soc. Econ. Botany (named Disting. Econ. Botanist 1987), Nat. Council Comml. Plant Breeders (Genetic and Plant Breeding award 1987), Am. Genetics Assn. (Frank N. Meyer medal 1982). Office: U Calif Davis CA 95616

RICKABAUGH, JANET FRALEY, environmental chemistry educator; b. Chillicothe, Ohio, Oct. 17, 1939; d. John James and Gusta Marie (Tackett) F.; m. Charles Roger Rickabaugh, Nov. 30, 1958; children: Roger Steven, Robin, Tracey R. BS in Chemistry, U. Cin., 1974, MS in Environ. Engring., 1977, PhD in Environ. Sci., 1990. Instr. Coll. Engring. U. Cin., 1975-77, rsch. assoc., 1978-90, assoc. prof., 1991—. Grantee US EPA, 1980—. Mem. Am. Chem. Soc., Assn. Ofcl. Analytical Chemists, Sigma Xi. Achievements include rsch. in photochem. destruction of hazardous chems.; gaseous emissions from landfills; environ. instrumentation. Office: U Cin Dept Civil Engring ML-071 Cincinnati OH 45221

RICKARDS, MICHAEL ANTHONY, aerospace engineer; b. Manila, Aug. 6, 1930; came to U.S., 1949; s. Anthony Francis and Carmen (Velasco) R.; m. Georgette Equerme, May 17, 1957 (div. 1981); children: Gregory, Mark, (twins), Ronald, Paul; m. Natalie Marie Pistorio, Jan. 9, 1984; 1 stepchild, Kathy. BS, MIT, 1953; MS, U. S.C., 1955, PhD, 1976. Aerodynamicist Weber Aircraft, Burbank, Calif., 1955-63, chief scientist, 1963-71; systems engring. mgr. Litton Industries, L.A., Calif., 1971-73; mgr. seals devel. Rohr Marine, San Diego, 1973-81; pvt. practice San Diego, 1981-85; sect. mgr. aeroscis. E-Systems, Greenville, Tex., 1985-90; mgr. tech. support Weber Aircraft, Gainesville, Tex., 1990-92; pvt. practice aerospace cons. Rockwall, Tex., 1992—. Patentee aero. and marine related systems; inventor escape system; song writer. Republican. Roman Catholic. Avocations: music, sports. Home and Office: 202 St Mary St Rockwall TX 75087-4016

RICKE, P. SCOTT, obstetrician/gynecologist; b. Indpls., June 28, 1948; s. Joseph and Betty (Rae) R.; divorced; 1 child, Alaina Michelle. BA, Ind. U., 1970; MD, Ind. U. Sch. of Medicine, 1974. Bd. cert. ob-gyn., 1981. Intern St. Lukes Hosp., Denver, 1975; resident U. Calif. at Irvine, Orange, 1977-79; pvt. practice Ob-Gyn Tucson, 1981—. Inventor (med. instrument) Vaginal Retractor, 1989. Bd. dirs. City of Hope, Tucson, 1981-85, Am. Cancer Soc., Tucson, 1981-83. Fellow Am. Bd. Ob-Gyn. Avocations: golfing, swimming, photography. Home: 3755 N Tanuri Dr Tucson AZ 85715-1939 Office: 5501 N Oracle Rd Ste D Tucson AZ 85704-3850

RICKEL, ANNETTE URSO, psychology educator; b. Phila.; d. Ralph Francis and Marguerite (Calcaterra) Urso; m. Peter Rupert Fink, July 21, 1989; 1 child, John Ralph. BA, Mich. State U., 1963; MA, U. Mich., 1965, PhD, 1972. Lic. psychologist, Mich. Faculty early childhood edn. Merrill-Palmer Inst., Detroit, 1967-69; adj. faculty U. Mich., Ann Arbor, 1969-75; asst. dir. N.E. Guidance Ctr., Detroit, 1972-75; asst. prof. psychology Wayne State U., Detroit, 1975-81; vis. assoc. prof. Columbia U., N.Y.C., 1982-83; assoc. prof. psychology Wayne State U., 1981-87, asst. provost, 1989-91, prof. psychology, 1987—; Am. Coun. on Edn. fellow Princeton and Rutgers Univs., 1990-91; AAAS and APA Congl. Sci. fellow on Senate Fin. Subcom. on Health and Pres. Nat. Health Care Reform Task Force, 1992-93. Cons. editor Am. Jour. of Community Psychology; co-author: Social and Psychological Problems of Women, 1984, Preventing Maladjustment..., 1987; author: Teenage Pregnancy and Parenting, 1989; contbr. articles to profl. jours. Mem. Pres.'s Task Force on Nat. Health Care Reform, 1993; bd. dirs. Children's Ctr. of Wayne County, Mich., The Epilepsy Ctr. of Mich., Planned Parenthood League, Inc. Grantee NIMH, 1976-86, Eloise and Richard Webber Found., 1977-80, McGregor Fund, 1977-78, 82, David M. Whitney Fund, 1982, Katherine Tuck Fund, 1985-90; recipient Career Devel. Chair award, 1985-86; Congl. Sci. fellow AAAS, 1992-93. Fellow APA (div. pres. 1984-85); mem. Midwestern Psychol. Assn., Mich. Psychol. Assn., Soc. for Rsch. in Child Devel., Soc. for Rsch. in Child and Adolescent Psychopathology, Internat. Assn. of Applied Psychologists, Sigma Xi, Psi Chi. Roman Catholic. Office: Wayne State U Dept Psychology 71 W Warren Ave Detroit MI 48201-1305

RICKER, ALISON SCOTT, science librarian; b. Canton, Ohio, June 13, 1953; d. Robert Lane and Janet Elizabeth (Scott) Ricker; m. Edward J. Chesney Jr., Aug. 5, 1978 (div. 1983); m. Raymond Alexander English, Oct. 19, 1985; children: John Alexander, Michael Scott. BS, Alma Coll., 1975;

MLS, U. R.I., 1977. Libr. Skidway Inst. Oceanography, Savannah, Ga., 1977-83; sci. libr. Oberlin (Ohio) Coll., 1983—. Mem. ALA, Acad. Libr. Assn. Ohio (pres. 1992-93), Assn. Coll. and Rsch. Librs. (co-chair 1990-93, discussion group for coll. sci. librs. sci. and tech. sect.), Ohio Acad. Sci., Internat. Assn. Marine and Aquatic Sci. Librs., Info. Ctrs. Democrat. Episcopalian. Achievements include surveys of liberal arts college libraries to study funding for science collections, provision of library service and facilities for science collections, staffing and instruction activities in science libraries. Office: Oberlin Coll Kettering Hall Libr 130 W Lorain St Oberlin OH 44074-1083

RICKER, RICHARD EDMOND, metallurgical scientist; b. Newport News, Va., Feb. 26, 1952; s. Harry Hamlin and Edith Elizabeth (Slayton) R.; m. Winifred Lou Vinson, June 20, 1975; children: Carrie Elizabeth, Jacob Edmond. BS in Materials Engring., N.C. State U., 1975; PhD in Materials Engring., Rensselaer Poly. Inst., 1983. Engr. trainee NASA Marshal Space Flight Ctr., Huntsville, Ala., 1973-75; sr. engr. The Babcock and Wilcox Co. Inc., Lynchburg, Va., 1977-79; asst. prof. U. Notre Dame, Ind., 1984-86; metallurgist Nat. Inst. Standards and Tech., Gaithersburg, Md., 1986-90, group leader, 1990—. Co-editor: Environmental Effects on Advanced Materials, 1991; bd. review Jour. Metall. Transactions; contbr. articles to profl. jours. Recipient Bronze medal U.S. Dept. Commerce, 1991. Mem. AAAS, ASM Internat. (treas.-chmn. local chpt. 1988-93), The Metallurg. Soc. AIME, Electrochem. Soc., Nat. Assn. Corrosion Engrs., Sigma Xi, Alpha Sigma Mu (trustee 1987-89). Achievements include research in the mechanism of accelerated fatigue failure of aluminum aircraft alloys, the effect of ordering on the electrochemical behavior of intermetallics; discovery of intermetallic alloys susceptible to stress corrosion cracking, a brittle film's ability to induce cleavage in a ductile substrate. Home: 16549 Sioux Ln Gaithersburg MD 20878 Office: Nat Inst Standards & Tech Corrosion Group Materials Room B254 Gaithersburg MD 20899

RICKER, WILLIAM EDWIN, biologist; b. Waterdown, Ont., Can., Aug. 11, 1908; s. Harry Edwin and Rebecca Helena (Rouse) R.; m. Marion Torrance Cardwell, Mar. 30, 1935; children—Karl Edwin, John Fraser, Eric William, Angus Clemens. B.A., U. Toronto, 1930, M.A., 1931, Ph.D., 1936; D.Sc. (hon.), U. Man., 1970; LL.D., Dalhousie U., 1972. Sci. asst. Fisheries Research Bd. Can., Nanaimo, B.C., 1931-38; editor publs. Fisheries Research Bd. Can., Nanaimo, 1950-62, biol. cons. to chmn. and staff, 1962-63; acting chmn. Fisheries Research Bd. Can., Ottawa, 1963-64; chief scientist Fisheries Research Bd. Can., Nanaimo, 1964-73; jr. scientist Internat. Pacific Salmon Fisheries Commn., New Westminster, B.C., 1938-39; asst. prof., assoc. prof., prof. zoology Ind. U., 1939-50; dir. Ind. Lake and Stream Survey Ind. Dept. Conservation, 1939-50; vol., contract investigator Pacific Biol. Sta., Nanaimo, 1973-93. Contbr. articles to profl. jours. Decorated officer Order of Can., 1986; named Eminent Ecologist Ecol. Soc. Am., 1990. Fellow Royal Soc. Can. (Flavelle medal 1970), AAAS; mem. Wildlife Soc. (awards 1956, 59), Profl. Inst. Pub. Service Can. (gold medal 1966), Am. Fisheries Soc. (award of excellence 1969), Can. Soc. Zoologists (F.E.J. Fry medal 1983), Am. Soc. Limnology and Oceanography (pres. 1959), Arctic Inst. N.Am., Can. Soc. Wildlife and Fishery Biologists, Entomol. Soc. B.C., Internat. Assn. Limnology, Marine Biol. Assn. India, Ottawa Field-Naturalists Club, Wilson Ornithol. Club, Explorers Club, Sigma Xi. Home: 3052 Hammond Bay Rd, Nanaimo, BC Canada V9T 1E2

RICKERBY, DAVID GEORGE, physicist, materials scientist; b. Chorley, Eng., Apr. 8, 1952; arrived in Italy, 1980; s. Thomas and Gertrude Sylvia (Turner) R.; m. Elisabetta Maria Emma Golin, Sept. 16, 1978; children: Andrew, Mark. BSc, Leicester (Eng.) U., 1973; PhD, Cambridge (Eng.) U., 1977. Chartered physicist. Rsch. asst. Cavendish Lab., Cambridge, 1977; rsch. assoc. Pa. State U., University Park, 1977-80; vis. scientist Joint Rsch. Ctr., Ispra, Italy, 1980-81, sci. officer, 1982—, head electron microscopy, 1989—; lectr. Pa. State U., University Park, 1985; cons. Scott Paper Co., Phila., 1979; tech. reviewer Am. Soc. for Metals, 1985, Inst. Metals, London, 1987. Contbr. numerous articles to sci. jours.; patentee semiconductor electron detector. Recipient medal Internat. Metallographic Soc., 1989. EEC grantee, 1980. Fellow Royal Micros. Soc., Cambridge Philos. Soc. (rsch. scholar 1976); mem. Am. Soc. for Metals, Inst. Physics, European Phys. Soc., Italian Soc. Electron Microscopy, Corpus Assn., Oxford and Cambridge Soc. Avocations: skiing, philately. Home: Via Roveda 10, 21100 Varese Italy Office: Joint Rsch Ctr Inst Adv Mat, 21020 Ispra Italy

RICORDI, CAMILLO, transplant surgeon, diabetes researcher, educator; b. N.Y.C., Apr. 1, 1957; m. Valerie A. Grace, Aug. 8, 1986; children: M. Caterina, Eliana G., Carlo A. MD, U. Milan (Italy) Sch. Medicine, 1982. Intern dept. internal medicine San Raffaele Inst., Milan, 1978-82, grad. clin. trainee in gen. surgery, 1982-85; rsch. assoc. dept. pathology Washington U. Sch. Medicine, St. Louis, 1985-88; attending surgeon dept. surgery San Raffaele Inst., Milan, 1988-91; asst. prof. surgery dept. surgery div. transplantation U. Pitts., Pa., 1989-91; assoc. prof. surgery dept. surgery U. Pitts., 1991-93; prof. surgery, chief divsn. cellular transplantation U. Miami, Fla., 1993—, co-dir. Diabetes Rsch. Inst., 1993—; reviewer of applications for grants Can. Diabetes Assn., European Econ. Community, Juvenile Diabetes Found., Med. Rsch. Coun., London, NIH, Nora Eccles Treadwell Found., Stanley Thomas Johnson Found.; chmn. First Internat. Congress of the Cell Transplant Soc., Pitts., 1992; others; mem. editorial bd. Transplantation Sci., Cell Transplantation, Transplantation Procs. Editor: Pancreatic Islet Cell Transplantation, 1992; contbr. chpts. to books and articles to Jour. Immunology, Hepatology, Diabetes, Transplantation, Endocrinology, Procs. Nat. Acad. Sci. USA, Am. Jour. Physiology, Wolrd Jour. Surgery, Surgery, Lancet, Pancreas, and many others. Recipinet Juvenile Diabetes Found. Internat. Rsch. Grant award, 1988-90; grantee EEC/Nova Nordisk Industries, Denmark, 1989-91, Fondazione Centro San Romanello Del Monte Tabor, Italy, 1991, Am. Diabetes Assn., 1991—, Nat. Kidney Found., 1992-93, NIH-RO1, 1993—, Juvenile Diabetes Found., 1993—; recipient NIH trainee award, 1986. Mem. AAAS, The Cell Transplant Soc. (founder, pres. 1992—), Internat. Assn. for Pancreas and Islet Transplantation (co-founder, mem. steering com.), The Transplantation Soc., Am. Diabetes Assn., Am. Fedn. Clin. Rsch. Achievements include patent for Automated Method for Cell Separation. Home: 72 S Hibiscus Dr Hisbiscus Island Miami Beach FL 33139 Office: U Miami Diabetes Rsch Inst 1450 NW 10 Ave Miami FL 33136

RIDDICK, DOUGLAS SMITH, horticultural industrialist, industrial designer; b. High Point, N.C., Sept. 28, 1942; s. Delmar Smith and Irene Douglas (Sparks) R.; m. Marcia Ann, Feb. 24, 1968; children: Eric Smith, Adrea Anne. Student, Columbus (Ohio) Coll. Art and Design, 1961-65, U. Bridgeport, 1965-66; BFA, U. Del., 1978. Indsl. designer Harper Landell & Assocs., Phila., 1967-70, ILC Industries, Dover, Del., 1970-75, Leeds Travelwear, Clayton, Del., 1975-76, DuPont Co., Glasgow, Del., 1976-79, Consumer Electronics Div. RCA, Indpls., 1979-80; designer Brayton Internat. Coll., High Point, N.C., 1981; owner Riddick Landscape Nursery, Archdale, N.C., 1980—; mgr. Riddick Greenhouses & Nursery, Archdale, N.C., 1980—. Patentee; designer indsl. instruments in field. Mem. Dover Bicentennial Com., 1975-76, Gov. DuPont's Com. Promotion of Solar Energy and subcom. Consumer Protection, 1978-79. Recipient Best in Packaging award Print Casebooks III, Washington, 1977, Excellence in Advt. award Am. Assn. Nurserymen, 1981. Mem. AAAS, Nat. Space Soc., Planetary Soc. Baptist. Club: Bicycle Club Del. (coordinator 1974-78, pres. 1979). Avocation: bicycling. Home and Office: 7125 Turnpike Rd Archdale NC 27263

RIDDICK, FRANK ADAMS, JR., physician, health care facility administrator; b. Memphis, June 14, 1929; s. Frank Adams and Falba (Crawford) R.; m. Mary Belle Alston, June 15, 1952; children: Laura Elizabeth Dufresne, Frank Adams III, John Alston. BA cum laude, Vanderbilt U., 1951, MD, 1954. Diplomate: Am. Bd. Internal Medicine (bd. govs. 1973-80). Intern Barnes Hosp., St. Louis, 1954-55, resident in medicine, 1957-60; fellow in metabolic diseases Washington U., St. Louis, 1960-61; staff Ochsner Clinic (Ochsner Found. Hosp.), New Orleans, 1961—; head sect. endocrinology and metabolic disease Ochsner Clinic (Ochsner Found. Hosp.), 1976—, asst. med. dir., 1968-72, assoc. med. dir., 1972-75, med. dir., 1975-92; clin. prof. medicine Tulane U., New Orleans, 1977—; trustee Alton Ochsner Med. Found., 1973—, CEO, 1991—; chmn. bd. Ochsner Health Plan, 1983-92; pres. Orleans Svc. Corp., 1976-80, South La. Med. Assocs., New Orleans, 1978—; dir. Brent House Corp., New Orleans, 1980—; pres. Am. Group Practice Assn., 1992-94. Trustee St. Martin's Protestant Epis. Sch., Metairie, La., 1970-84; bd. govs. Isidore Newman Sch., New Orleans, 1987-93. Recipient Disting. Alumnus award Castle Heights Mil. Acad., 1979; recipient teaching award Alton Ochsner Med. Found., 1969, Physician Exec. award Am. Coll. Med. Group Adminstrs., 1984, Disting. Alumnus award Vanderbilt U. Sch. Med., 1988. Fellow ACP, Am. Coll. Physician Execs. (pres. 1987-88); mem. Am. Soc. Internal Medicine (trustee 1970-76, disting. internist award), AMA (ho. of dels. 1971, chmn. coun. on med. edn. 1983-85), Endocrine Soc., Am. Diabetes Assn., Nat. Acad. Scis. Inst. Medicine, Soc. Med. Adminstrs. (v.p. 1993—), Accreditation Coun. on Grad. Med. Edn. (chmn. 1986-87), Nat. Resident Matching Program (v.p. 1986-90, accreditation coun. on med. edn. 1988-90), Boston Club, New Orleans Country Club, Cosmos Club, Intern. Home: 1923 Octavia St New Orleans LA 70115-5651 Office: Ochsner Clinic 1514 Jefferson Hwy New Orleans LA 70121-2483

RIDE, SALLY KRISTEN, physics educator, scientist, former astronaut; b. Los Angeles, May 26, 1951; d. Dale Burdell and Carol Joyce (Anderson) R.; m. Steven Alan Hawley, July 26, 1982 (div.). B.A. in English, Stanford U., 1973, B.S. in Physics, 1973, Ph.D. in Physics, 1978. Teaching asst. Stanford U., Palo Alto, Calif.; researcher dept. physics Stanford U.; astronaut candidate, trainee NASA, 1978-79, astronaut, 1979-87; on-orbit capsule communicator STS-2 mission Johnson Space Ctr. NASA, Houston; on-orbit capsule communicator STS-3 mission NASA, mission specialist STS-7, 1983, mission specialist STS-41G, 1984; sci. fellow Stanford (Calif.) U., 1987-89; dir. Calif. Space Inst. of U. Calif. San Diego, La Jolla, 1989—; prof. Physics U. Calif. San Diego, La Jolla, 1989—; mem. Presdl. Commn. on Space Shuttle, 1986. Author: (with Susan Okie) To Space and Back, 1986, (with T. O'Shaughnessy) Voyager: An Adventure to the Edge of the Solar System, 1992. Office: Calif Space Inst A-021 U Calif San Diego La Jolla CA 92093

RIDENOUR, MARCELLA V., motor development educator; b. New Martinsville, W.Va., Nov. 29, 1945; married, 1973; 2 children. BS, Miami U., 1967; MS, Purdue U., 1968, PhD in Motor Devel., 1972. Prof. motor devel. Temple U., Phila., 1974—. Mem. Am. Soc. Testing and Materials. Office: Temple University Motor Development Lab Pearson Hall Philadelphia PA 19122*

RIDGWAY, MARCELLA DAVIES, veterinarian; b. Sewickley, Pa., Dec. 24, 1957; d. Willis Eugene and Martha Ann (Davies) R. BS, Pa. State U., 1979; VMD, U. Pa., 1983. Intern Univ. Ill., Urbana, 1983-84, resident in small animal internal medicine, 1984-87; small animal vet. Vet. Cons. Svcs., Savoy, Ill., 1987—. Contbr. articles to profl. jours. Mem. Am. Vet. Med. Assn., Am. Animal Hosp. Assn., Acad. Vet. Clinicians, Ednl. Resources in Environ. Sci. (bd. dirs.), Savoy Prairie Soc. (pres. 1989—), Grand Prairie Friends. Avocations: prairie conservation activities, hiking, horseback riding, long distance running, sketching. Home and Office: Vet Cons Svcs 194 Paddock Dr E Savoy IL 61874-9663

RIEBE, SUSAN JANE, environmental engineer; b. Laramie, Wyo., July 15, 1955; d. Leonard Adam and Lela Alexa (Quandt) R.; m. Kevin S. Moody. BSN, U. No. Colo., 1977; BSChemE, Colo. Sch. of Mines, 1986. Registered engr.-in-tng., Colo. Med./surg. RN St. Anthony's Hosp., Denver, 1977-79; obstets. RN Riverton (Wyo.) Meml. Hosp., 1979-80; med./surg. RN Hillcrest Meml. Hosp., Tulsa, 1980-82; pvt. practice RN Golden, Colo., 1982-86; ops. engr. Mobil Oil Corp., Bakersfield, Calif., 1986-88, environ. engr., 1989-92; sr. environ. engr. Mobil Oil Corp., Midland, Tex., 1992—. Vol. ARC, Tulsa, 1980-82, first aid instr. Colo. and Wyo., 1975-80; mem. Nat. Ski Patrol Systems, Colo., 1976-79, asst. leader Front Range Patrol, 1978-79; mem. LWV, Bakersfield, 1990-91; vol. activity organizer singles against cancer, Am. Cancer Soc., Bakersfield, 1987-90, Ptnrs. in Excellence, 1990-92; vol. tutor and mentor Ptnrs. in Excellence, 1990-91, Ptnrs. in Edn., 1992—. Mem. Soc. Petroleum Engrs., Soc. Women Engrs. (v.p. 1987-88, pres. 1989-91). Home: 1714 Holloway Midland TX 79701 Office: Mobil Oil Corp PO Box 633 Midland TX 79702

RIEDINGER, ALAN BLAIR, chemical engineer, consultant; b. Maquoketa, Iowa, Oct. 27, 1926; s. Clem Adolphus and Opal Lorraine (Griffin) R.; m. Arlene Janet Swiedom, Aug. 1, 1953; children: Jodi, Bruce. BS in Chem. Engring., Iowa State U., 1948; MS in Chemistry, Union Coll., Schenectady, N.Y., 1958. Engr. Gen. Electric Co., Schenectady, 1948-54, engr. Knolls Atomic Power Lab., 1954-58; rsch. staff medm. Gen. Atomic div. Gen. Dynamics, San Diego, 1958-67, Gulf Gen. Atomic, San Diego, 1967-74; sr. engr. Fluid Systems div. UOP, Inc., San Diego, 1975-86; cons. chem. engr. San Diego, 1986-92; ret. Fluid Systems div. UOP, Inc., San Diego, 1992. Contbr. chpts. to books. Mem. Sigma Xi, Tau Beta Pi, Alpha Chi Sigma. Lutheran. Achievements include a patent on method and equipment for treating demineralized water, a computer program for predicting the performance of reverse osmosis systems, design of a 12,000 cubic meter/day seawater reverse osmosis unit for Saudi Arabia. Home: 5925 Sagebrush Rd La Jolla CA 92037

RIEDNER, WERNER LUDWIG FRITZ, retired chemicals executive, industrial consultant; b. Mannheim, Fed. Republic Germany, Oct. 19, 1924; s. Georg Michael and Helene Barbara (Sanzen) R.; m. Heidi Inge Erika Kaethe Kohl, Apr. 30. 1954; children: Michael, Axel, Nicola. Diploma, Wirtschaftshochschule, Mannheim, 1950, Dr. Polit. Sci., 1952. Mgr. Metallgesellschaft AG, Frankfurt, Main, Fed. Republic Germany, 1952-59; joint mng. dir. Pigment Chemie GmbH, Duisburg, Fed. Republic Germany, 1959-64; joint mng. dir. DuPont Nemours Deutschland GmbH, Düsseldorf, Frankfurt, Fed. Republic Germany, 1964—, from asst. to mng. dir., 1965, from dep. mng. dir. to mng. dir. photo products, 1966-72, dir. photo products Europe, 1972; group mng. dir. DuPont Europe, Düsseldorf, Frankfurt, Fed. Republic Germany, 1982-87; cons. Fed. Republic Germany; mem. supervisory bd. DuPont Deutschland, Bad Homburg, Fed. Republic Germany, 1988-89; chmn. supervisory bd. Zanders Feinpapiere A.G., Bergisch Gladbach, Fed. Republic Germany, 1990—. Author: Die Widerstande gegen staatliche Preismassnahmen, 1953. Mem. Rotary.

RIEGEL, BYRON WILLIAM, ophthalmologist; b. Evanston, Ill., Jan. 19, 1938; s. Byron and Belle Mae (Huot) R.; B.S., Stanford U., 1960; M.D., Cornell U., 1964; m. Marilyn Hills, May 18, 1968; children—Marc William, Ryan Marie, Andrea Elizabeth. Intern, King County Hosp., Seattle, 1964-65; asst. resident in surgery U. Wash., Seattle, 1965; resident in ophthalmology U. Fla., 1968-71; pvt. practice medicine specializing in ophthalmology, Sierra Eye Med. Group, Inc., Visalia, Calif., 1972—; mem. staff Kaweah Delta Dist. Hosp., chief of staff, 1978-79; mem. staff Visalia Community Hosp. Bd. dirs., asst. sec. Kaweah Delta Dist. Hosp., 1983-90. Served as flight surgeon USN, 1966-68. Co-recipient Fight-for-Sight citation for research in retinal dystrophy, 1970. Diplomate Am. Bd. Ophthalmology, Nat. Bd. Med. Examiners. Fellow A.C.S., Am. Acad. Ophthalmology; mem. Calif. Med. Assn. (del. 1978-79), Med. Assn., Tulare County Med. Assns., Calif. Assn. Ophthalmology, Am. Soc. Cataract and Refractive Surgery, Internat. Phacoemulsification and Cataract Methodology Soc. Roman Catholic. Club: Rotary (Visalia). Home: 3027 W Keogh Ct Visalia CA 93291-4228 Office: 2830 W Main St Visalia CA 93291-4300

RIEGEL, GREGG MASON, ecologist, researcher; b. Sacramento, Calif., June 8, 1953; s. Mason Dudley and Patricia Marie (Armstrong) R.; m. Janice Leda Gauthier, Oct. 9, 1982 (div. 1990); m. Gail Adeline Burton, Aug. 29, 1992. BS, U. Calif., Davis, 1976; MS, Humboldt State U., 1982; PhD, Oreg. State U., 1989. Rsch. range sci. USDA Agrl. Rsch. Svc., Reno, 1989-91; ecologist USDA Forest Svc Silviculture Lab, Bend, Oreg., 1991—; cons. ecologist Linda Nelson Botanical, Reno, 1991—; courtesy faculty asst. prof. Oreg. State U., 1992—. Contbr. articles to profl. jours. Mem. Ecology Soc. Am., Soc. Range Mgmt., Soc. Wetland Scis., N.W. Sci. Assn., Calif. Botanical Soc. Home: 19816 Connarn Rd Bend OR 97701 Office: USDA Forest Svc Silviculture Lab 1027 NW Trenton Ave Bend OR 97701

RIEGEL, KURT WETHERHOLD, environmental protection, occupational safety and health; b. Lexington, Va., Feb. 28, 1939; s. Oscar Wetherhold and Jane Cordelia (Butterworth) R.; m. Lenore R. Engelmann, Nov. 15, 1974; children: Tatiana Suzanne, Samuel Brent Oscar, Eden Sonja Jane. BA, Johns Hopkins U., 1961; PhD, U. Md., 1966; PMD, Harvard U., 1977. Asst. prof. astronomy UCLA, 1966-74; prof. astronomy U. Calif. Extension, Los Angeles, 1968-74; mgr. energy conservation program Fed. Energy Adminstrn., Washington, 1974-75; chief tech. and consumer products energy conservation Dept. Energy, Washington, 1975-78, dir. consumer products div., conservation and solar energy, 1978-79; assoc. dir. environ. engring. and tech. EPA, 1979-82; head Astronomy Ctrs. NSF, 1982-89; dir. Environ. Protection Office USN, 1989—; cons. Aerospace Corp., El Segundo, Calif., 1967-70, Rand Corp., Santa Monica, Calif., 1973-74; vis. fellow U. Leiden, Netherlands, 1972-73; Mem. Casualty Council Underwriters Labs., Nat. Radio Astron. Observatory Users Com., 1968-74. Contbr. articles to profl. jours. Mem. AAAS, Am. Phys. Soc., Sierra Club, Audubon Soc., Internat. Radio Sci. Union, Am. Astron. Soc., Internat. Astron. Union, Assn. of Scientists and Engrs. Home: 3019 N Oakland St Arlington VA 22207-5320

RIEGER, PHILIP HENRI, chemistry educator, researcher; b. Portland, Oreg., June 24, 1935; s. Otto Harry and Carla (Oertli) R.; m. Anne Bioren Lloyd, June 18, 1957; 1 child, Christine Lloyd. B.A., Reed Coll., 1956; Ph.D., Columbia U., 1962. Prof. chemistry Brown U., Providence, 1962—. Contbr. articles to profl. jours. Mem. Am. Chem. Soc. (chmn. R.I. sect. 1978), Royal Soc. Chemistry, New Eng. Assn. Chemistry Tchrs. Episcopalian. Home: 119 Congdon St Providence RI 02906-1462 Office: Brown U Dept Chemistry Box H Providence RI 02912

RIEGER, PHILLIP WARREN, aquatic ecology educator, researcher; b. Ponca City, Okla., Sept. 22, 1948; s. Eldon Ivan and Bonita Delores (Rhodes) R.; m. Laura Jean Powell, Jan. 21, 1991. BS, N.W. Okla. State U., 1974; MS, Okla. State U., 1976; postgrad., Iowa State U., 1991—. Rsch. assoc. Okla. State U., Stillwater, 1974-76, Aquatic Systems, Inc., San Diego, 1990-91; ecologist U.S. Army C.E., L.A., Waltham, Mass., 1976-85; aquatic cons., L.A., 1985-89; instr. dept. animal ecology Iowa State U., Ames, 1991—. Contbr. articles to profl. publs. With USN, 1969-72. Recipient commendation U.S. Army C.E., 1979, 80, 85; fish and game fellow Okla. State U., 1974, Boehm fellow Am. Fishing Tackle Assn., 1992. Mem. Am. Fisheries Soc., Sigma Xi. Achievements include first to provide photographic evidence for mechanism of initial gasbladder inflation in striped bass; designer first fish ladder for rainbow smelt, constructed Quincy-Hingham Bay, Massachusetts. Home: 116 Cherry Ames IA 50010 Office: Dept Animal Ecology Iowa State U Ames IA 50011

RIEHLE, JAMES RONALD, volcanologist; b. Duluth, Minn., July 7, 1943; s. Lloyd Frank and Allie Alice (Dalbacka) R.; m. Kaaren Ruth Johnson, Jan. 14, 1978 (div. May 1987); m. Diedra Bohn, June 30, 1991; 1 child, Kirsti Maija. BSc, U. Minn., 1966; MSc, Northwestern U., Evanston, Ill., 1969, PhD, 1970. Asst. prof. SUNY, Binghamton, 1970-74; geol. engr. Alaska Div. Geol. Surveys, Anchorage, 1976-80; geologist U.S. Geol. Survey, Anchorage, 1980-91; dep. for volcanoes U.S. Geol. Survey, Reston, Va., 1991—; tchr. geology Chapman Coll., Ft. Richardson, Alaska, 1976-77; chmn. Fed. Com. Aviation and Ash, 1991—. Lobbyist Nordic Ski Club, Anchorage, 1990-91. NDEA fellow Northwestern U., Evanston, 1966-69; trustees rsch. grantee SUNY, Binghamton, 1971. Mem. Alaska Geol. Soc. (bd. dirs. 1977-78, jour. editor 1979-80). Achievements include first systematic database of Aleutian volcanic tephra deposits; team leader first systematic geologic investigation of Katmai Nat. Park; first model of compaction in ash-flow sheets. Office: US Geol Survey Sunrise Valley Dr Reston VA 22092

RIEHS, JOHN DARYL, state agency administrator; b. LaGrange, Tex., Oct. 27, 1953; s. Leroy William and Lucille (Byler) R.; m. Katherie Kubesch, Mar. 25, 1972; children: Jason, John. BS in Edn., Tex. A&M U., 1976. Cert. tchr., Tex.; cert. fire svc. instr., Tex. Sci. tchr. Brenham (Tex.) High Sch., 1976-77; power plant operator Lower Colo. River Authority, LaGrange, Tex., 1978—, supr., 1978—, tech. trainer, 1978—. Author training manuals for pvt. industry. Alderman City of LaGrange, 1990—; sec. LaGrange Vol. Fire Dept., 1980—. Mem. S.W. Utilities Tng. Group, Tex. Assn. Indsl. Trainers, Tex. Firemen's and Fire Marshal's Assn. Office: Lower Colo River Authority PO Box 519 LaGrange TX 78945

RIEKEN, DANNY MICHAEL, naval officer; b. Hastings, Minn., Aug. 31, 1967; s. Oscar Rieke and Dorothy Ruth (O'Toole) R.; m. Lisa Kae Bruer, June 8, 1991. B Aerospace Engring. and Mechanics, U. Minn., 1991. Enlisted USN, 1984, commd. ensign, 1991; advanced through grades to lt. (j.g.) USN, Ill., 1986-88; stationed at USN, Great Lakes, 1985-86; stationed at USNR, Mnpls., 1988, Pensacola, 1991—; co-engr./designer U. Minn./USRA/NASA, 1990-91; pub. affairs officer Res. Officers Tng. Ctr., U. Minn., Mpls., 1989-90, yearbook editor, 1989-90, adminstrv. asst., staff officer ROTC, 1991-92; NROTC profl. devel. officer Chief of Naval Edn. and Tng., Pensacola, 1992; pvt. pilot single engine land, 1990—. Co-author: Mars Integrated Transportation System, 1990-91; editor: Winds of Change, 1990. Pres Centennial Hall House 11, U. Minn., 1987-88, mem. Centennial Hall coun., 1986-89. USN scholar, 1988-91. Mem. AIAA, Aircraft Owners and Pilots Assn., U. Minn. ROTC Alumni Assn. Democrat. Luthern. Achievements include design in cooperation with classmates and NASA through USRA of Mars Integrated Transportation System, of new computer system MIRS to process over 1 million applicants into all branches of armed forces through 68 offices of U.S. Military Entrance Processing Command. Home: 1823 Country Dr Apt # 302 Grayslake IL 60030

RIEL, GORDON KIENZIE, research physicist; b. Columbus, Ohio, Oct. 26, 1934; s. Gordon Wilson and Emma Leota (Kienzie) R.; m. Ane Lee Rutledge, Aug. 21, 1954; children: Valerie Lee, Gordon Wilson II, Cynthia Grace. BChemE, U. Fla., 1956; MS, U. Md., 1961, PhD, 1967. Registered profl. engr.; cert. health physicist. Rsch. physicist Naval Surface Warefare Ctr., Silver Springs, Md., 1956—; cons. engr. Mayo, Md., 1968—; mem. numerous coms. Editor: (with others) The Dumand Project, 1977; contbr. over 50 articles to profl. jours. Mem., instr. vessel operation USCG Aux. Flotilla 71, Wheaton, Md. Fellow Washington Acad. of Sci.; mem. NSPE, Am. Bd. Health Physics, Internat. Radiation Protection Assn., Am. Phys. Soc., Health Physics Soc., Sigma Xi. Methodist. Achievements include 5 patents; development of underwater radiation spectometers, underwater selective extraction analysis in chemical oceanography, underwater neutrino detector. Office: Naval Surface Warefare Ctr R36 Silver Spring MD 20903-5640

RIENHOFF, OTTO, physician, medical informatics educator; b. Dortmund, Fed. Republic of Germany, Nov. 9, 1949; s. Otto and Lotte (Wiechers) R. Ed., Wilhelms U., Münster, Federal Republic of Germany, 1973, Hannover Med. Sch., Federal Republic of Germany, Y. Intern Hannover Med. Sch., 1974, rsch. asst. Inst. Med. Informatics, 1975-82, dep. dir. Inst. Med. Informatics, 1980-83, assoc. prof. med. informatics, 1982-85; prof., dir. Inst. Med. Informatics Philipps U., Marburg, Federal Republic of Germany, 1985—; lectr. Braunschweig U., 1978-83, Poly. Hannover, 1980-82, Tech. Acad. Wuppertal, 1980-82; coord. Quality Assurance program in perinatology, Lower-Saxony, 1979—; cons. med. documentation, edn., tng. and systems devel., 1977—; mem. editorial bd. various profl. jours. and publs. Editor books on med. informatics and edn.; contbr. numerous articles to profl. jours. Mem. Internat. Med. Informatics Assn. (chmn. working group 9, 1983-91), German Soc. Med. Informetics, Biometry and Epidemiology (nat. rep. 1988—, pres. 1993-95), South African Med. Informatics Group (corr.), Brazilian Soc. Informatics in Health (corr.), others. Office: Inst Med Informatics, Bunsenstr 3, D-3550 Marburg Germany

RIENNE, DOZIE IGNATIUS, technologist; b. Chickasha, Okla., July 22, 1954; s. James O. and Joy I. Rienne; m. Charlotte Roberts, Feb. 6, 1982; children: Tonnia, Chovia, Brittany. Student, Okla. State U., 1988. Project mgr. DF Young Constrn. Co., Dallas, 1981-84; constrn. mgr. VB Cons. Group, Oklahoma City, 1984-88; programs dir. Riennes Corp., Chickasha, Okla., 1989—. Editor: The Role of the Construction Managers, 1992, The APA Code Plus Built Homes, 1992. Mem. Am. Plywood Assn. Office: Riennes Constrn Co 1524 S First St Chickasha OK 73018-5908

RIES, EDWARD RICHARD, petroleum geologist, consultant; b. Freeman, S.D., Sept. 18, 1918; s. August and Mary F. (Graber) R.; student Freeman Jr. Coll., 1937-39; A.B. magna cum laude, U. S.D., 1941; M.S., U. Okla., 1943, Ph.D. (Warden-Humble fellow), 1951; postgrad. Harvard, 1946-47; m. Amelia D. Capshaw, Jan. 24, 1949 (div. Oct. 16, 1956); children: Rosemary Melinda, Victoria Elise; m. Maria Wipfler, June 12, 1964. Asst. geologist Geol. Survey S.D., White River area, 1941; geophys. interpreter Robert Ray Inc., Western Okla., 1942; jr. geologist Carter Oil Co., Mont., Wyo., 1943-

44, geologist Mont., Wyo., Colo., 1944-49; sr. geologist Standard Vacuum Oil Co., Assam, Tripura and Bangladesh, India, 1951-53, sr. regional geologist N.V. Standard Vacuum Petroleum, Maatschappij, Indonesia, 1953-59, geol. adviser for Far East and Africa, White Plains, N.Y., 1959-62; geol. adviser Far East, Africa, Oceania, Mobil Petroleum Co., N.Y.C., N.Y., 1962-65; geol. adviser for Europe, Far East, Mobil Oil Corp., N.Y.C., 1965-71, sr. regional explorationist Far East, Australia, New Zealand, Dallas, 1971-73, Asia-Pacific, Dallas, 1973-76, sr. geol. adviser Rsch. Geology, 1976-79, assoc. geol. advisor Geology-Geophysics, Dallas, 1979-82, sr. geol. cons., 1982-83; ind. internat. petroleum geol. cons. Europe, Sino-Soviet and S.E. Asia, 1986—. Grad. asst., teaching fellow U. Okla., 1941-43, Harvard, 1946-47. Served with AUS, 1944-46. Mem. N.Y. Acad. Scis., Am. Assn. Petroleum Geologists (asso. editor 1976-83, 50 Yr. Mem. Svc. award 1993). Geol. Soc. Am., Am. Hort. Soc., Internat. Platform Assn., Am. Geol. Inst., A.A.A.S., Nat. Audubon Soc., Nat. Wildlife Fedn., Soc. Exploration Geophysicists, Wilderness Soc., Am. Legion, Phi Beta Kappa, Sigma Xi, Phi Sigma, Sigma Gamma Epsilon. Republican. Mennonite. Club: Harvard (Dallas). Author numerous domestic and internat. proprietary and pub. hydrocarbon generation and reserve evaluations, reports and profl. papers. Home and Office: 6009A Royal Crest Dr Dallas TX 75230-3434

RIESE, ARTHUR CARL, environmental engineering company executive, consultant; b. St. Albans, N.Y., Jan. 2, 1955; s. Walter Herman and Katherine Ellen (Moore) R. BS in Geology, N.Mex. Inst. Mining and Tech., 1976, MS in Chemistry, 1978; PhD in Geochemistry, Colo. Sch. Mines, 1982. Lic. geologist, N.C.; registered profl. geologist, N.C., S.C., Ark., Fla., Tenn., Wyo. Asst. petroleum geologist N.Mex. Bur. Mines and Mineral Resources, Socorro, 1973-76; geologist Nord Resources, Inc., Albuquerque, 1975; rsch. asst. N.Mex. Inst. Mining and Tech., Socorro, 1976-78; vis. faculty Colo. Sch. Mines, 1978-81; rsch. geochemist Gulf R & D Co., Houston, 1982-84; sr. planning analyst/mgr. tech. planning Atlantic Richfield Co., L.A., 1984-87; sr. v.p. Harding Lawson Assocs., Denver, 1987—; mem. affiliate faculty U. Tex., Austin, 1983—; speaker, conf. chmn. in field. Numerous patents in field. Panel participant N.Mex. First, Gallup, 1990. Mem. Am. Inst. Hydrology (cert. profl. hydrogeologist 1988), Am. Inst. Profl. Geologists (cert. geol. scientist 1988). Office: Harding Lawson Assocs 2400 Arco Tower 707 Seventeenth St Denver CO 80202

RIESENBERGER, JOHN RICHARD, strategic marketing company executive; b. N.Y.C., Sept. 25, 1948; s. Richard Raymond and Marie Teresa (Long) R.; m. Patricia Ann Casey, Nov. 23, 1974; children: Christine, Jennifer. BS in Econs. and Bus., Hofstra U., 1970, MBA in Mgmt., 1975; cert. internat. sr. mgmt. program, Harvard U., 1989. Customer svc. supr. Chase Manhattan Bank, 1970-72; gen. sales rep. various regions Upjohn Co., Bklyn., 1972-75; sales rep., sales mgr. Upjohn Co., various locations, N.Y., 1976-81; profl. tng. and devel. officer Upjohn Co., Kalamazoo, Mich., 1981-83; dir. Chgo. sales area Upjohn Co., 1983-87, with, 1972-89; v.p., group mgr. Upjohn Co. of Can., Toronto, Ont., 1987-89; exec. dir. Worldwide Med. Scis. Liaison Upjohn Co., Kalamazoo, 1989-92; exec. dir. Worldwide Strategic Marketing, Kalamazoo, 1992—. Mem. mktg. com. United Way. Mem. Pharm. Advt. Coun., Pharm. Mfrs. Assn. (chmn. mktg. practices com.), Harvard Bus. Sch. Club. Avocations: golf, tennis. Home: 7398 Oak Shore Dr Portage MI 49002-7858

RIESENHUBER, HEINZ FRIEDRICH, German minister for research and technology; b. Frankfurt, Germany, Dec. 1, 1935; s. Karl and Elisabeth (née Birkner) Riesenhuber; m. Beatrix Walter, 1968; 4 children. Student, Gymnasium in Frankfurt, Univs. Frankfurt, Univs. Munich. With Erzgesellschaft mbH, c/o Metallgesellschaft, Frankfurt, 1966-71; technical mgr. Synthomer-Chemie GmbH, Frankfurt, 1971-82; fed. min. rsch. and tech., 1982-93; with CDU, 1961, chair Frankfurt br., 1973-78, mem. Bundestag, 1976—. Contbr. articles to profl. jours. Office: Bundesministerium fur Forchung, Und Technologie, Heinemanstr 2, PF 200240, 5300 Bonn 2, Germany*

RIESS, HENRI GERARD, chemist, educator; b. Haguenau, France, Mar. 7, 1932; s. Henri and Salome (Eschenbrenner) R.; m. Aimee Graf, July 30, 1955; children: Jean, Isabelle. Degree in engring., U. Strasbourg, 1957; PhD, U. Mulhouse, France, 1980. Asst. prof. Ecole Nat. Superieure Chimie, Mulhouse, France, 1959-68, prof., 1968—; vis. prof. Mich. Mol. Inst., Midland, 1979. Author: Copolymers, 1985; co-editor: Makromol. Chemie, New Polymeric Materials. Sgt. French Air Force, 1957-59, Germany. Recipient Acad. Palms Ministry of Edn., 1980, Paul Neumann prize Hoechst, 1985. Mem. French Chem. Soc., Am. Chem. Soc., French Polymer Group. Achievements include 25 patents in polymer science. Home: Rue Meunier 31, 68200 Mulhouse France Office: ENSCMu, 3 Rue Werner, 68200 Mulhouse France

RIETH, PETER ALLAN, business executive; b. Buffalo, Oct. 4, 1941; s. Hermann and Erna (Jordans) Father; m. Jo Helen Wilson, Dec. 21, 1970 (div. Mar. 1977); children: Peter Allan Jr., Timothy Paul, Derek Jordan, Heather Anne. AAS in Elec. Tech., Rochester (N.Y.) Inst. Tech., 1963; BS in Gen. Studies, Portland State U., 1976. Mktg. rep. IBM, Portland, 1963-73; application engr. Measurex Systems, Inc., Portland, 1973-74; account rep. Inforex, Inc., Portland, 1974-76; mktg. rep. Applied Theory, Inc., Corvallis, Oreg., 1976-80; owner Tusk Digital Controls, Corvallis, 1980-82; pres. The Drawbridge Corp., Corvallis, 1982—; adv. bd. Internat. Conf. on Scanning Tech. in Sawmilling, San Francisco, 1985-89. Mem. Mall Arts Commn., Eugene, 1971, citizens adv. bd. City Coun., Eugene, 1971; adv. bd. to Planning Commn., Portland, 1973. Named State Spoke of the Yr. Jaycees, 1970, Project of Yr., 1970. Republican. Achievements include development of automation products for sawmills and secondary wood products. Office: The Drawbridge Corp PO Box 761 Corvallis OR 97339

RIFE, JACK CLARK, physicist; b. Omaha, Oct. 21, 1945; s. Clark Augustine and Alice Virginia (Adams) R.; m. Mary Lou Ference, Nov. 19, 1983; children: Katherine Elizabeth, Luke Daniel. BS, U. Chgo., 1968; PhD, U. Wis., 1976. Rsch. physicist Nat. Inst. Standards and Tech., Gaithersburg, Md., 1977-80, Naval Rsch. Lab., Washington, 1980—; Contbr. chpt. to Electro-Optics Handbook, 1993. Mem. Am. Phys. Soc., Sigma Xi. Achievements include patent on vacuum ultraviolet grating/crystal monochromator, patent on solid state vacuum ultraviolet light source; optimal design and fabrication of vacuum ultraviolet multilayer-coated gratings. Office: Naval Rsch Lab Code 6686 Washington DC 20375

RIGALI, JOSEPH LEO, quality assurance professional; b. Frankfurt, Germany, Oct. 31, 1948; s. Joseph Leo Jr. and Theresa Virginia (Amato) R.; m. Linda Joy Owens, Dec. 6, 1969; children: Tiffany Lynn, Julie Lyn, Joseph Leo IV. BS in Polit. Sci., Kans. State U., 1977. Commd. 2d lt. U.S. Army, 1971, advanced through grades to capt., resigned, 1979; lead supr. Kaiser Aluminum, Ravenswood, W.Va., 1979-82; quality control engr. Warner Electric, South Beloit, Ill., 1982-84; engring. mgr. quality assurance Chamberlain Electronics Div., Nogales, Ariz., 1984-85; mgr. quality assurance Bohn Aluminum & Brass, Adrian, Mich., 1985-87, St. Clair Metal Products, Port Huron, Mich., 1987; gen. mgr. quality assurance Sanden Internat. (USA), Inc., Wylie, Tex., 1987—; instr. consortium supplier tng. Dallas County C.C. Dist.; bd. dirs. Ctr. for Quality and Prodn., U. North Tex., 1990—, cons., 1990—; mem. adv. bd. BCIS U. North Tex.; speaker in field. Author: TQM - Entering A New Era, 1990; contbr. articles to profl. publs. Recipient cert. of appreciation Stateline Area Schs., 1983, GE, 1985. Mem. Am. Soc. Quality Control (sr., chmn. edn. Toledo sect. 1986, chmn. Bluewater sect. 1987, chmn. edn. Dallas sect. 1989-91). Democrat. Roman Catholic. Home: 915 Edgewood Dr Richardson TX 75081 Office: Sanden Internat (USA) Inc 601 S Sanden Blvd Wylie TX 75098

RIGATTI, BRIAN WALTER, psychiatric researcher; b. Natrona Heights, Pa., May 1, 1968; s. Joseph Walter and Dorothy Helen (Klingensmith) R. BS in Chemistry, U. Pitts., 1989, BS in Behavioral Neurosci., 1991. Autopsy technician Children's Hosp. of Pitts., 1987-92; rsch. assoc. Western Psychiatric Inst. and Clinic, Pitts., 1990-91, rsch. specialist, 1991—. Mem. AAAS, Pitts. Neurosci. Soc., N.Y. Acad. Sci., Soc. for Neurosci. (student). Methodist. Office: Western Psychiatric Inst 3811 OHara St W1641 BST Pittsburgh PA 15213

RIGGENBACH, DUANE LEE, maintenance engineer; b. Paulding, Ohio, Oct. 15, 1956; s. Ray Henry and Dorothy Dean (Smith) R.; m. Brenda Harper, Sept. 19, 1984. BS in Mech. Tech., Purdue U., Fort Wayne, Ind., 1979. Tech. asst., plant engr. Gen. Portland Inc., Paulding, Ohio, 1979-85; maintenance engr. Lafarge Corp., Paulding, 1985—. Home and Office: 14956 US 127 Paulding OH 45879

RIGGS, PENNY KAYE, cytogeneticist; b. Laredo, Tex., Aug. 24, 1965; d. Michael Gene and Karen Ann (Redic) R. BS in Biology, Purdue U., 1987, MS in Animal Cytogenetics, 1991. Rsch. assoc. Purdue U. Cytogenetics Lab., West Lafayette, Ind., 1987-91; Regents' fellow Tex. A&M U., College Station, 1991—, pres. Veterinary Med. Student Coun. Author: (with others) Advances in Veterinary Science, 1990. Recipient Grad. Travel award Purdue U. Dept. Animal Sci., 1990. Mem. Am. Genetic Assn., Am. Assn. for Lab. Animal Sci., Tissue Culture Assn., Gamma Sigma Delta. Achievements include first to describe chromosomal fragile sites in the domestic pig. Office: Tex A&M U Dept Veterinary Pathobiology College Station TX 77843

RIGGSBY, ERNEST DUWARD, science educator; b. Nashville, June 12, 1925; s. James Thomas and Anna Pearl (Turner) R.; m. Dutchie Sellers, Aug. 25, 1964; 1 child, Lyn-Dee. BS, Tenn. Polytech. Inst., 1948; BA, George Peabody Coll. Tchrs., 1952, George Peabody Coll. Tchrs., 1953; MA, George Peabody Coll. Tchrs., 1956, EdS, 1961, EdD, 1964. Vis. grad. prof. U. P.R., Rio Piedras, George Peabody Coll., 1963-64; prof. Auburn (Ala.) U., Troy (Ala.) State U., Columbus (Ga.) Coll.; pres. Ednl. Developers, Inc., Columbus, Ga.; vis. grad. prof. George Peabody Coll., 1963-64. Contbr. articles to profl. jours. Col., USAF, 1944-85. Fellow AAAS; mem. Nat. Sci. Tchrs. Assn., World Aerospace Edn. Assn. (v.p. for the Ams.). Home: Columbus Coll Columbus GA 31993-2399

RIGNEY, E. DOUGLAS, biomedical engineer; b. Auburn, Ala., Jan. 13, 1958; s. E. Douglas Sr. and Eva (Armstrong) R.; m. Renita Stevens, June 2, 1979; children: Erin, Cole, Kathleen. BSE, U. Ala., 1980, MS, 1985, PhD, 1989. Jr. engr. Ala. Power Co., Mobile, 1981-83, generating plant engr., 1983; biomaterials technician U. Ala. Sch. Dentistry, Birmingham, 1987-89; asst. prof. U. Ala., Birmingham, 1989—. Contbr. articles to profl. jours. including Jour. of Am. Ceramic Soc., Biomaterials, Jour. of Materials Science: Materials Medicine, Ceramics in Substitutive and Reconstructive Surgery, Jour. of Biomed. Materials Rsch. Pres. Engring. Alumni Assn., Birmingham, 1992, v.p., 1991; advisor Cahaba Heights Community Schs., Birmingham, 1991-92, S.E. Consortium for Minorities in Engring., 1990-93. Mem. Biomed. Engring. Soc., Soc. for Biomaterials, Nat. Assn. of Corrosion Engrs., Am. Assn. of Dental Rsch., Phi Kappa Phi, Sigma Xi, Tau Beta Pi. Home: 3845 River Run Trail Birmingham AL 35243

RIHA, WILLIAM EDWIN, beverage company executive; b. New Brunswick, N.J., Sept. 15, 1943; s. William Edwin and Grace Blue (McDowell) R.; m. Joan Ann Murphy, June 25, 1967; children: William Edwin III, Jennifer Dawn. BS, Rutgers U., 1965, MS, 1969, PhD, 1972. Mgr. product devel. Hunt-Wesson Foods, Inc., Fullerton, Calif., 1972-76; dir. tech. svcs. Cadbury N.Am., Hazleton, Pa., 1976-78; mgr. food tech. Peter Paul Cadbury, Inc., Naugatuck, Conn., 1978-80; group mgr. U.S. product devel. PepsiCo, Valhalla, N.Y., 1980-84, dir. internat. product devel., 1984-89; v.p. R&D J. E. Seagram & Sons, Ltd., White Plains, N.Y., 1989—. Capt. USAR, 1965-73. Mem. Inst. Food Technologists (grad. fellowship Nestle 1968), Indsl. Rsch. Inst. (membership com. 1990-92), Sigma Xi. Avocations: golf, tennis, baseball memorobilia. Home: 231 Mimosa Cir Ridgefield CT 06877-2539 Office: JE Seagram & Sons Ltd 3 S-Corp Park Dr White Plains NY 10604

RIHANI, SARMAD ALBERT (SAM RIHANI), civil engineer; b. Beirut, Lebanon, Feb. 22, 1954; s. John Albert and Laureen Salim (Schoucair) R.; m. Ina Lee Hand, July 12, 1975; children: Cedar, Paul, Michael. BSCE, Oreg. State U., 1977. Registered profl. engr., D.C., Va., Mo., Oreg., Calif. Designer Butler Mfg. Co., Kansas City, Mo., 1977-79; applications analyst United Computing Systems, Overland Park, Kans., 1979-80; mgmt. info. systems supr. Zamil Steel Bldgs. Co., Saudi Arabia, 1980-81, sr. structural engr., 1981-82, design mgr., 1982-84, engring. mgr., 1984-87, bldg. products mgr., 1987-89; gen. mgr. multistory bldg. system Butler Mfg. Co., Kansas City, Mo., 1989-91; v.p. project mgmt. and engring. Beaman Corp., Greensboro, N.C., 1991-92; bd. dirs. Beaman Corp., Greensboro; divsn. svc. mgr. Varco-Pruden Bldgs., Little Rock, 1992—; bd. dirs. Engrs. Coun., Saudi Arabia, 1988-89. Am. Field Svc. scholar, 1970. Mem. ASCE (pres. Saudi Arabia 1988, bd. dirs. 1989, appreciation award 1989), Nat. Soc. Profl. Engrs., Am. Lebanese League (bd. dirs. 1990), Tau Beta Pi. Republican. Roman Catholic. Avocations: personal computers, reading, tennis, skiing.

RIKOSKI, RICHARD ANTHONY, engineering executive, electrical engineer; b. Kingston, Pa., Aug. 13, 1941; s. Stanley George and Nellie (Gober) R.; m. Giannina Batchelor Petrullo, Dec. 18, 1971 (div. 1979); children: Richard James, Jennifer Anne. BEE, U. Detroit, 1964; MSEE, Carnegie Inst. Tech., 1965; PhD, Carnegie-Mellon U., 1968; postdoctoral student, Case-Western Res. U./NASA, 1971. Registered profl. engr., Ill., Mass., Pa. Engr. 1st communication satellite systems Internat. Tel. & Tel., Nutley, N.J., 1961-64; engr. Titan II ICBM program Gen. Motors, Milw., 1964; trainee NASA, 1964-67; instr. Carnegie-Mellon U., Pitts., 1966-68; asst. prof. U. Pa., Phila., 1968-74; assoc. prof., dir. hybrid microelectronics lab., chmn. ednl. TV com. IIT, Chgo., 1974-80, chmn. ednl. TV com., 1974-80; rsch. engr. nuclear effects ITT Rsch. Inst., Chgo., 1974-75; pres. Tech. Analysis Corp., Chgo., 1980—; engr. color TV colorimetry Hazeltine Rsch., Chgo., 1969; engr. Metroliner rail car/roadbed ride quality dynamics analysis U.S. Dept. Transp., ENSCO, Inc., Springfield, Va., 1970; pres. Tech. Analysis Corp., Chgo., 1978-91; contractor analysis of color TV receiver safety hazards U.S. Consumer Product Safety Commn., 1977, analysis heating effect in aluminum wire Beverly Hills Supper Club Fire, Covington, Ky., 1978; engr. GFCI patent infringement study 3M Corp., St. Paul, 1979-81; elec. systems analyst Coca-Cola Corp., Atlanta, 1983-91; fire investigator McDonald's Corp., Oak Brook, Ill., 1987-90; engring. analyst telephone switching ctrs. ATT, Chgo., 1990-91, expert witness numerous other govtl. and corp. procs. Author: Hybrid Microelectronic Circuits, 1973; editor: Hybrid Microelectronic Technology, 1973; contbr. articles to profl. jours. Officer Planning Commn., Beverly Shores, Ind., 1987-93, trustee town coun., 1992—, police liason 1993—; mem. Chgo. Coun. Fgn. Rels., USAF SAC Comdrs. Disting. Vis. Program; adv. coun. Nat. Park Svc. Ind. Dunes Nat. Lake Shore, 1993—. NASA fellow, 1964-67, 70. Mem. IEEE (sr. ednl. activities bd. N.Y.C. 1970-74, USAB career devel. com. 1972-74, editor Soundings 1973-75, Cassette Colloquia 1973-74, del. Popov Soc. Tech. Exch. USSR, mgr. Dial Access Tech. Edn. program 1972), Assn. for Media Based Continuing Engring. Edn. (bd. dirs.), Nat. Fire Protection Assn., Sigma Xi, Tau Beta Pi, Eta Kappa Nu. Republican. Avocations: sailing, travel. Home: One E Lakefront Dr Beverly Shores IN 46301-0444 Office: Tech Analysis Corp 3600 N Lake Shore Dr Chicago IL 60613-4656

RILEY, CARROLL LAVERN, anthropology educator; b. Summersville, Mo., Apr. 18, 1923; s. Benjamin F. and Minnie B. (Smith) R.; m. Brent Robinson Locke, Mar. 25, 1948; children: Benjamin Locke, Victoria Smith Evans, Cynthia Winningham. A.B., U. N.Mex., 1948, Ph.D., 1952; M.A., UCLA, 1950. Instr. U. Colo., Boulder, 1953-54; asst. prof. U. N.C., Chapel Hill, 1954-55; asst. prof. So. Ill. U., Carbondale, 1955-60, assoc. prof., 1960-67, prof., 1967-86, Disting. prof., 1986-87, Disting. prof. emeritus, 1987—, chmn. dept., 1979-82, dir. mus., 1972-74; rsch. assoc. lab. anthropology Mus. N.Mex., 1987-90, sr. rsch. assoc., 1990—; rsch. collaborator Smithsonian Instn., 1988—; adj. prof. N.Mex. Highlands U., 1989—. Author: The Origins of Civilization, 1969, The Frontier People, 1982, expanded edit., 1987; editor: Man Across the Sea, 1971, Southwestern Journals of Adolph F. Bandelier, 4 vols., 1966, 70, 75, 84, Across the Chichimec Sea, 1978; others; contbr. numerous articles to profl. jours. Served in USAAF, 1942-45. Decorated 4 battle stars; grantee Social Sci. Research Council, NIH, Am. Philos. Soc., Am. Council Learned Socs., NEH, others. Home: 1106 6th St Las Vegas NM 87701-4311

RILEY, DOUGLAS SCOTT, quality assurance specialist, biochemist; b. Detroit, Jan. 22, 1958; s. Richard H. and Sally Ann (Deckert) R.; m. Janet Combs, May 19, 1988; 1 child, Angela Lynne. BS, Alma Coll., 1980. Quality control technician Van Diest Supply Co., Webster City, Iowa, 1983-85; quality control chemist Quality Plus Products, Ft. Dodge, Iowa, 1985-87; prodn. technician Monsanto Co., St. Louis, 1987-90; prodn. chemist Sigma Chem. Co., St. Louis, 1990-91, Centocor, Inc., St. Louis, 1991-92; quality assurance specialist Parke Davis Co., Rochester, Mich., 1992—. Office: Parke Davis Sterile Products Op 870 Parkdale Rd Rochester MI 48307

RILEY, FRANCENA, nurse, military officer; b. New Smyrna Beach, Fla., May 5, 1957; d. Willard Harrell and Jacqueline Delores (Griffen) R.; 1 child, Daniel Albert Cross. AA, U. Md., Heidelberg, Fed. Republic Germany, 1987. Enlisted U.S. Army, 1980, advanced through grades to sgt. 1st class, 1991, expert field med. badge, parachutist; practical nurse emergency room Keller Army Hosp., West Point, N.Y., 1981; bn. tng. noncommd. officer 34th Med. Bn., Ft. Benning, Ga., 1988-89, practical nurse 2d Mobile Army Surg. Hosp., 1989-91; wardmaster intensive care unit #1 2d MASH, 1990-91; practical nurse pediatric ward Walter Reed Army Med. Ctr., Washington, 1982-84; practical nurse, then nursing supr. 913th Med. Detachment, Kaiserslautern, Fed. Republic Germany, 1984-86; wardmaster surgery clinic Army Regional Med. Ctr., Landstuhl, Fed. Republic Germany, 1987; with 2D MASH 44th med. brigade operation desert shield U.S. Army, Saudi Arabia, 1990-91; ops. non-commd. officer 2d MASH, 1991-92; wardmaster newborn nursery USAMEDDAC, Ft. Polk, La., 1992—. Recipient med. badge U.S. Army, 1991. Baptist. Avocations: bowling, roller skating, bicycling, plate collecting, visiting the zoo. Home: PO Box 3057 Fort Polk LA 71459-6000 Office: USA MEDDAC Ward 4C Newborn Nursery Fort Polk LA 71459-0057

RILEY, JAMES ALVIN, civil engineer; b. Deventer, Mo., Sept. 11, 1943; s. James Clifton and Earlene (Nolen) R.; m. Loretta Ann Sagers, Dec. 11, 1970 (dec. Aug. 1976); children: Erik, Jared, April; m. Pauline Armstrong, Aug. 9, 1978; children: Scott Armstrong, Paige Armstrong, Cam Armstrong. BS in Civil Engring., U. Mo., Rolla, 1966; M of Civil Engring., Brigham Young U., 1972; PhD in Civil Engring., Colo. State U., 1982. Registered profl. engr., Utah. Planning team leader U.S. Bur. Reclamation, Provo, Utah, 1982-86, Phoenix, 1990-91; chief plan formulation U.S. Bur. Reclamation, Provo, 1986-88; advisor to govt. of Egypt U.S. Agy. for Internat. Devel., Cairo, 1988-90; prin. engr. Bookman-Edmonston Engring., Orem, Utah, 1991—. Contbr. articles to sci. jours. Varsity leader Boy Scouts Am., Provo, 1987-88, mem. advancement com., 1992. With U.S. Army, 1966-68. Mem. ASCE, Internat. Water Resources Assn. Mormon. Achievements include rsch. in salinity investigations, advancing understanding of salinity mechanisms contributing to salt loading of Utah Lake and the Sevier River Basin, rsch. in salinity reduction programs, water mgmt. in arid environments. Home: 690 E 3950 North Provo UT 84604

RILEY, PAUL EUGENE, electronics engineer; b. Syracuse, N.Y., Nov. 18, 1941; s. George Charles and Mary (Tobin) R. BS, U. Toronto, Ont., 1963; PhD, U. Hawaii, 1974. Postdoctoral fellow U. Tex., Austin, 1974-78, dept. crystallographer, 1978-80; postdoctoral fellow U. Calif., Berkeley, 1980-82; with Fairchild Camera Rsch. Ctr., Palo Alto, Calif., 1982-87, Westinghouse Advanced Tech. Lab., Balt., 1987-88, Digital Equip. Co., Hudson, Mass., 1988-90; electronics engr. Hewlett-Packard Labs., Palo Alto, 1990—. Contbr. over 75 articles to profl. jours. Robert Welch fellow, 1974-78, U. Hawaii Dept. Rsch. fellow, 1973. Mem. Am. Chem. Soc., Am. Vacuum Soc., Electrochem. Soc., Sigma Xi. Achievements include patents in silicon processing. Home: 6104 Elmbridge Dr San Jose CA 95129 Office: Hewlett-Packard Co 3500 Deer Creek Rd Palo Alto CA 94303

RILEY, WILLIAM FRANKLIN, mechanical engineering educator; b. Allenport, Pa., Mar. 1, 1925; s. William Andrew and Margaret (James) R.; m. Helen Elizabeth Chilzer, Nov. 5, 1945; children—Carol Ann, William Franklin. B.S. in Mech. Engring., Carnegie Inst. Tech., 1951; M.S. in Mechanics, Ill. Inst. Tech., 1958. Mech. engr. Mesta Machine Co., West Homestead, Pa., 1951-54; research engr. Armour Research Found., Chgo., 1954-61; sect. mgr. IIT Research Inst., Chgo., 1961-64, sci. adviser, 1964-66; prof. Iowa State U., Ames, 1966-78, Disting. prof. engring., 1978-88, prof. emeritus, 1989—; ednl. cons. Bihar Inst. Tech., Sindri, India, 1966, Indian Inst. Tech., Kanpur, summer 1970. Author: (with A.J. Durelli) Introduction to Photomechanics, 1965; (with J. W. Dally) Experimental Stress Analysis, 1991; (with D. Young, K. McConnell and T. Rogge) Essentials of Mechanics, 1974; (with A. Higdon, E. Ohlsen, W. Stiles and J. Weese) Mechanics of Materials, 4th edit., 1985; (with J. Dally and K. McConnell) Instrumentation for Engineering Measurements, 1993; (with L.W. Zachary) Introduction to Mechanics of Materials, 1989, (with L.D. Sturges) Engineering Mechanics- Statics and Dynamics, 1993; also numerous articles and tech. papers. Served to lt. col. USAAF, 1943-46. Fellow Soc. for Exptl. Mechanics (hon. mem.); mem. Soc. for Exptl. Stress Analysis (hon., M.M. Frocht award 1977). Home: 1518 Meadowlane Ave Ames IA 50010-5547

ŘIMAN, JOSEF, biology educator; b. Horní Suchá, Karviná, Czechoslovakia, Jan. 30, 1925; s. Alois and Hilda (Glaserová)Ř R.; m. Věra Tomková, July 16, 1950. MD, Charles U., Prague, Czechoslovakia, 1950; PhD, Czechoslovak Acad Sci., Prague, 1955, DSc in Chemistry, 1966; DSc in Biology (hon.), J.E. Purkyně U., Brno, Czechoslovakia, 1987. Rsch. physician 1st Clinic Pediatrics Charles U., Prague, 1950-51, prof. med. faculty, 1967-72; sr. scientist Inst. Organic Chemistry Czechoslovak Acad. Sci., Prague, 1951-74, dir. Inst. Molecular Genetics, 1974—, sci. sec., 1978-81, v.p., 1981-86, pres., 1986-90; acad. rep. UNESCO, 1980-86; Czechoslovak nat. rep. Internat. Coun. Scientific Unions, 1982-84; dep. Ho. Nations Fed. Assembly, Prague, 1986-89; chmn. commn. INTERKOSMOS, 1986-90. Mem. editorial bd. Neoplasma Slovak Acad. Scis., 1967, Acta Virologica, 1970, Biologica, 1982, Cancer Biochemistry and Biophysics, 1985; chmn., chief editor Czechoslovak Encyclopaedia, 1986-90. Recipient State Prizes Govt. of Czechoslovakia, 1968, 78, J.E. Purkyne medal Govt. of Czechoslovakia, 1979, Order of Labor Govt. of Czechoslovakia, 1983, Gold medal Slovak Acad. Sci., 1989, State Prize of USSR, 1979, J. Dimitrov medal Govt. of Bulgaria, 1986, Gold medal Nagoya U. Med. Sch., 1990. Fellow Indian Nat. Sci. Acad. (fgn.); mem. Acad. Sci. USSR (fgn., M.L. Lomonosov Gold medal 1987), Czech Acad. Sci. (chief editor Folia Biologica 1975, G.J. Mendel plaque 1975, numerous others, Medal of Honor 1989), German Acad. Scis. (fgn.), Hungarian Acad. Sci. (hon.), Slovak Soc. Biochemistry (hon.), Czechoslovak Soc. Immunology (hon.), Bulgarian Acad. Sci. (fgn.). Avocation: history of science. Office: Czechoslovak Acad Sci & Inst Mol, Genetics Flemingovo n2, 166 37 Prague 6, Czech Republic

RIMBEY, PETER RAYMOND, physicist, mathematical engineer; b. LaGrande, Oreg., Aug. 27, 1947; s. Raymond Lee and Margaret Mary (McGinty) R. BA, Ea. Oreg. State Coll., 1969; PhD, U. Oreg., 1974. Postdoctoral rsch. asst. Ind. U., Bloomington, 1973-74, Iowa State U., Ames, 1975-78; postdoctoral rsch. U. Wis., Milw., 1979-80; specialist engr. Boeing Aerospace Co., Seattle, 1980-83, sr. specialist engr., 1990—; vis. assoc. prof. Ea. Oreg. State Coll., LaGrande, 1983-84; asst. prof. physics Seattle U., 1984-90; organizing com. Symposium on Electrons and Atoms in Solids, U. Oreg., Eugene, 1983; cons. Vektor, Inc., Bellevue, Wash., 1985-88, RHO, Inc., Seattle, 1988-90; referee physics review bd. Physics Review Letters. Contbr. articles to profl. jours; pub. indsl. documents. Nat. Defense Edn. Act fellow U.S. Govt., 1969-71. Mem. Am. Phys. Soc., Am. Assn. Physics Tchrs., Mid-Am. States Univs. Assn. (chairperson Theoretical Physics Conf. 1977), Soc. Physics Students (student advisor 1984—), Sigma Xi, Sigma Pi Sigma. Home: 4129 39th Ave SW Seattle WA 98116 Office: Boeing Comml Airline Group PO Box 3707 MIS 19 AH Seattle WA 98124

RIMILLER, RONALD WAYNE, podiatrist; b. Rome, N.Y., July 21, 1949; s. Harold Henry and Ida (Scherz) R.; m. Janice Ann Bombara, June 30, 1973; children: Joseph Harold, Lori Ann. BA, Colgate U., 1971; D Podiatric Medicine, Pa. Coll. Podiatric Medicine, 1979. Diplomate Am. Bd. Podiatric Orthopedics. Pvt. practice, Elmwood, Conn., 1979—; cons. Hartford (Conn.) Hosp., 1982—. Lt., EMT, Tunxis Hose Co. No. 1, Unionville, Conn., 1990—. Recipient achievement award Tunxis Hose Co. No. 1, 1989. Fellow Am. Coll. Foot Orthopedists; mem. APHA, Internat. Coll. Podiatric Laser Surgery (assoc.), Acad. Ambulatory Foot Surgery (assoc.), Am. Acad. Podiatric Sports Medicine (assoc.), Am. Acad. Sports Medicine, Am. Podiatric Med. Assn. Roman Catholic. Avocations: gardening, sports, reading. Home: 65 Sylvan Ave Unionville CT 06085-1170 Office: 1123A New Britain Ave West Hartford CT 06110-2412

RIMOIN, DAVID LAWRENCE, physician, geneticist; b. Montreal, Que., Can., Nov. 9, 1936; s. Michael and Fay (Lecker) R.; m. Mary Ann Singleton, 1962 (div. 1979); 1 child, Anne; m. Ann Piilani Garber, July 27, 1980; children: Michael, Lauren. BSc, McGill U., Montreal, 1957, MSc, MD,

CM, 1961; PhD, Johns Hopkins U., 1967. Asst. prof. medicine, pediatrics Washington U., St. Louis, 1967-70; assoc. prof. medicine, pediatrics UCLA, 1970-73, prof., 1973—, chief med. genetics, Harbor-UCLA Med. Ctr., 1970-86; dir. dept. pediatrics, dir. Med. Genetics and Birth Defects Ctr., 1986—; Steven Spielberg chmn. pediatrics Cedars-Sinai Med. Ctr., L.A., 1989—; chmn. coun. Med. Genetics Orgn., 1993. Co-author: Principles and Practice of Medical Genetics, 1983, 90; contbr. articles to profl. jours., chpts. to books. Recipient Ross Outstanding Young Investigator award Western Soc. Pediatric Research, 1976, E. Mead Johnson award Am. Acad. Pediatrics, 1976. Fellow ACP; mem. Am. Fedn. Clin. Rsch. (sec.-treas. 1972-75), Western Soc. Clin. Rsch. (pres. 1978), Am. Bd. Med. Genetics (pres. 1979-83), Am. Coll. Med. Genetics (pres. 1991—), Am. Soc. Human Genetics (pres. 1984), Am. Pediatric Soc., Soc. Pediatric Rsch., Am. Soc. Clin. Investigator, Assn. Am. Physicians, Johns Hopkins Soc. Scholars, Inst. Medicine. Office: Cedars-Sinai Med Ctr 8700 Beverly Blvd Los Angeles CA 90048

RIMSON, IRA JAY, forensic engineer; b. Cleve., Mar. 1, 1935; s. Oscar and Goldie (Wachs) R.; m. Arlene Friedman, Oct. 14, 1959. AB, Harvard Coll., 1956; MS, U. So. Calif., 1969. Registered profl. engr., Calif. Commd. ensign USN, 1956, advanced through grades to comdr., 1977, aviator, 1956-77, dir. safety U.S. Naval Air Systems Command Hdqrs., 1975-77, ret., 1977; pres. System Safety Assocs., Ltd., Springfield, Va., 1977—; dir. Aviation Resources Group, Inc., Alexandria, Va., 1981—; bd. advisors Nat. Forensic Ctr., Princeton, N.J., 1986—. Contbr. articles to profl. and trade publs. Fellow Am. Acad. Forensic Scis., Internat. Soc. Air Safety Investigators (treas. 1989—, editor jour. 1976-84, 91-92); mem. ASTM (sec. com.), Helicopter Assn. Internat., Va. Aero. Hist. Soc., Mid-Atlantic Helicopter Assn., System Safety Soc. (sr.), Assn. Aviation Psychologists, Assn. Naval Aviation, Royal Aero Club (U.K., life), Harvard Club Washington, Phi Kappa Phi. Achievements include research in improving quality of forensic science investigations. Office: Aviation Resources Group Inc 205 S Whiting St # 405 Alexandria VA 22304-3632

RIN, ZENGI, economic history educator; b. Yuanlin, Taiwan, China, Sept. 15, 1935; arrived in Japan, 1962; s. Sankei and Sango (Ko) R.; m. Shien Gi, Dec. 28, 1969; children: Kotatsu, Jobun, Joan. BA, Zhong Xing U., Taiwan, 1960; MA, Kyoto U., 1965, postgrad., 1965-68. From asst. prof. to prof. economic history Nagoya Gakuin U., Seto, 1968—; vis. scholar, Harvard U., Cambridge, 1979-80, Peking U., 1985. Author: Asian Economic History, 1974, Introduction to Economic History, 1981, General Economic History, 1984, Lectures on Asian Economic History, 1987, The Current in the Economic History, 1992. Mem. Japan Assn. for Asian Polit. and Econ. Studies, Socio-Econ. History Soc. Japan, Bus. History Soc. Japan. Avocations: tennis, travel. Home: 1263-4 Shimokirido Arai, Owariasahi Aichi 488, Japan Office: Nagoya Gakuin U, 1350 Kamishinano-cho, Seto Aichi 480-12, Japan

RINDE, JOHN JACQUES, internist; b. Przemysl, Poland, Jan. 3, 1935; came to U.S., 1952; s. Maurice and Stella (Klein) R.; m. Toni Igel, June 21, 1959; children: Debbie Ann, Barbara Gail. BS, MIT, 1957, MS, 1958, MME, 1959; MSEE, Poly. Inst. Bklyn., 1965; MD, U. Ark., 1975. Cert. profl. engr.; diplomate Am. Bd. Internal Medicine. Sr. engr. Sperry Gyroscope Co., Great Neck, N.Y., 1959-67, 70-71; v.p. Olson Assocs. Inc., Huntington, N.Y., 1967-69; sr. engr. Hydrosystems, Inc., Farmingdale, N.Y., 1969-70; physician Clearwater, Fla., 1978—. NSF fellow, 1958. Mem. ACP, Am. Soc. Internal Medicine, Fla. Med. Assn., Pinellas County Med. Soc. (bd. govs. 1989-92). Avocations: tennis, skiing, photography, swimming. Office: 1305 S Ft Harrison Ave Clearwater FL 34616-3301

RINES, ROBERT HARVEY, lawyer, inventor, law center executive, educator; b. Boston, Aug. 30, 1922; s. David and Lucy (Sandberg) R.; m. Carol Williamson, Dec. 29, 1972; 1 son, Justice Christopher; children by previous marriage: Robert Louis, Suzi Kay Ann. BS in Physics, MIT, 1942; JD, Georgetown U., 1947; PhD, Nat. Chiao Tung U., 1972; DJ, New Eng. Coll. Law, 1974. Bar: Mass. 1947, D.C. 1947, N.H. 1974, Va. 1983, U.S. Supreme Ct., FCC, Tax Ct., U.S. and Can. patent offices; Registered profl. engr., Mass. Asst. examiner U.S. Patent Office, 1946; partner Rines & Rines, Boston, 1947—; pres., prof. law, chmn. Franklin Pierce Law Center, 1973—; bd. dirs. Megapulse, Inc., Astro Dynamics, Inc., PTC Research Found., Lord Corp., New England Fish Farming Enterprises Inc., Acad. Applied Sci. Project Orbis Bangladesh and Singapore Opthamology Programs, Sportsmans Handbook, KR Assos., Beltronics Inc., Stylus Innovation, Inc., Albavision Ltd., Promotion of Am. Chinese Tech., Bank of N.H., Ctr. Broadcasting Corp. of N.H.; Gordon McKay lectr. patent law Harvard, 1956-58; lectr. inventions and innovation Mass. Inst. Tech., 1962—; Mem. commerce tech. adv. bd. Dept. Commerce, 1963-67, mem. nat. inventors council, 1963-67, 81—; mem. N.H. Gov.'s Crime Study Com., 1976—. Author: A Study of Current World-Wide Sources of Electronic and Other Invention and Innovation, Computer Jurisprudence: Create or Perish--The Case for Patents and Inventions; patentee in field of radar and sonar. Campaign chmn. United Fund, Belmont Mass., 1960; com. mem. Nat. Inventors Hall of Fame, 1974—; adv. bd. Harvard-Mass. Inst. Tech. Biomed. Engring. Center, 1976—; bd. dirs. Allor Found., Inst. Applied Pharmacology. 2d lt. to capt. AUS, 1942-46. Recipient Civilian Disting. Svc. award U.S. Army, 1976, Inventions citation Pres. Carter and U.S. Dept. Commerce, 1980, N.H. High Tech. Entrepreneur award, 1989, Beyond Peace award, 1989, Bangladesh (N.Am.) Disting. Svc. award, 1990. Mem. IEEE (sr.), AAAS, ABA, Acad. Applied Sci. (pres., Medal of Honor, 1989), Am. Patent Law Assn., Sci. Research Soc. Am., Aircraft Owners and Pilots Assn., Nat. Acad. Engring. (patent com. 1969—, cons. to exec. officer 1979-80), Explorers Club, Sigma Xi. Unitarian (trustee 1954-56). Clubs: Harvard (Boston); Torquay Co. Theatrical Productions (N.Y.C.); Chemists (N.Y.C.); Mass. Institute Technology Faculty (Cambridge); Nat. Lawyers; Capitol Hill (Washington); Highland (Scotland); Commonwealth (England); Am. (London). Home: 13 Spaulding St Concord NH 03301-2571

RINGLER, DANIEL HOWARD, lab animal medicine educator; b. Oberlin, Ohio, Aug. 19, 1941. DVM, Ohio State U., 1965; MS, U. Mich., 1969; diploma, Am. Coll. Lab. Animal Medicine, 1971. From instr. to assoc. prof. lab. animal medicine U. Mich., Ann Arbor, 1969-79, prof., 1979—; cons. animal resources br. divsn. rsch. resources NIH, 1973—; mem. adv. coun. inst. lab. animal resource Nat. Rsch. Coun.-NAS, 1975-78. Mem. Am. Assn. Lab. Animal Sci. (Griffin award), Am. Vet. Medicine Assn., Am. Assn. Accreditation Lab. Animal Care. Achievements include research in spontaneous diseases of laboratory animals, use of animals in biomedical research, diseases and husbandry of amphibians, pathogenic bacteriology. Office: U Mich Unit Lab Animal Medicine Animal Rsch Facility Ann Arbor MI 48109*

RINGOIR, SEVERIN MARIA GHISLENUS, medical educator, physician; b. Aalst, Belgium, June 17, 1931; s. Benoni and Mariette (Vlasschaert) R.; children: Marc, Yves. MD, U. Gent, Belgium, 1956, PhD, 1967. Resident Med. Clinic U. Gent, 1958-61, instr., 1961-71, assoc. prof., 1971-75, prof. nephrology, 1975—, chief renal div., 1971—, chmn. medicine, 1981-84; chmn. biomed engring. program U. Gent, 1991—; chmn. 1st Internat. Symposium Single Needle Hemodialysis, Tampa, Fla., 1984; co-founder Internat. Faculty Artificial Organs. Inventor pressure-pressure single needle hemodialysis; contbr. sci. articles to profl. jours. Maj. Mil. Health Svc., 1956-58, Res., 1958-89. Decorated comdr. Order of Leopold, officer Order of Crown (Belgium); recipient J. Lemaire prize, 1970, Internat. Disting. medal Nat. Kidney Found., 1991. Fellow Royal Soc. Medicine; mem. Royal Acad. Medicine Belgium, Am. Soc. Artificial Internal Organs, Swiss, German, French and Dutch Soc. Nephrology, European Dialysis and Transplant Assn. (coun. 1981-84, pres. 1985 Congress), European Soc. for Artificial Organs (gov. 1988-92, pres. Congress 1989), Internat. Soc. Artificial Organs (v.p. 1981-85, gen. sec. 1984-90). Home: Vaderlandstraat 44, B9000 Ghent Belgium Office: Univ Gent, De Pintelaan 185, B9000 Ghent Belgium

RINK, CHRISTOPHER LEE, information technology consultant, photographer; b. Fullerton, Calif., May 20, 1952; s. Wesley Winfred and Doreen (Warman) R.; m. Donna Marie Wootton, Feb. 25, 1989; 1 child, Christopher Lee Jr. BS in Acctg., No. Ill. U., 1974, MBA, 1976. Teaching asst. No. Ill. U., Dekalb, 1974-76; region acct. Hewlett-Packard Co., Chgo., 1976-77, systems programmer, 1977-80; pres. Amity Systems Assocs., Elmhurst, Ill., 1978-81; programming mgr. Hewlett-Packard Co., Chgo., 1980-81, systems engr., 1981-84; response ctr. engr. Hewlett-Packard Co., Atlanta, 1984-86, info. tech. mgr., 1986-90, info. tech. cons., 1990—; established response ctr. info. tech. dept. Hewlett-Packard, Atlanta, 1986, tech. dir., 1990, established info. tech. resource mgmt. group. Author (manual) Stock Market Analysis System, 1976; designer Hewlett-Packard centralized call mgmt. systems, 1990; author computer programs. Coord. Ptnrs. in Edn. Cobb County, Atlanta, 1987-89. Mem. Inforum INner Cir., Interex, Computer Measurement Group. Avocations: personal computers, photography, tennis. Home: 2175 Deep Woods Way Marietta GA 30062-2586 Office: Hewlett Packard Co 2015 S Park Pl NW Atlanta GA 30339-2014

RINKENBERGER, RICHARD KRUG, physical scientist; b. Gridley, Ill., May 15, 1933; s. Burl E. and Olive J. (Krug) R.; divorced; children: Janice L., Ginger R., Rebekah P.; m. Ida Lee Vaughn, Mar. 22, 1985; children: Douglas W., Angela D. BA in Geology, U. Colo., 1959. Dir. prospecting Grubstake Assn., Sask., Can., 1958-59; engr. Martin-Marietta Aerospace Co., Denver, 1960-75; geologist U.S. Geol. Survey, Denver, 1975; geologist remote sensing U.S. Mine Safety and Health Adminstrn., Denver, 1975-79; pres., exploration geologist Banner Set, Ltd., Denver, 1980-84; pres., cons. geologist R.K. Rinkenberger & Assocs., Aurora, Colo., 1979-87; phys. scientist U.S. Dept. Energy, Germantown, Md., 1987—; pres., cons. geologist Earth Audit, Ltd., Rockville, Md., 1993—; educator prospecting Denver Sch. Prospecting, 1968-71, U. Colo., Denver, Boulder, 1970-75; rsch. geochemist Heritage Chem. Co., Englewood, 1984-86; prospecting researcher R.K. Rinkenberger & Assocs., Aurora, 1965—. Contbr. articles to profl. pubIs. Mem. parent adv. bd, supt. of schs. Westminster, Colo., 1982-83. Recipient High Quality Performance award U.S. Mine Safety and Health Dept., 1977; grantee Saskatchewan (Can.) Dept. Mineral Resources, 1958, 59, U.S. Geol. Survey, 1978. Mem. hon. geol. soc., Sigma Gamma Epsilon. Mem. Ch. of the Nazarene. Avocations: cross country skiing, geol. experiment and theory research, writing. Office: PO Box 5523 Rockville MD 20855-0523

RIORDAN, JOHN RICHARD, chemist; b. St. Stephen, N.B., Can., Sept. 2, 1943; married, 1970. BSc, U. Toronto, 1966, PhD in Biochemistry, 1970. Investigator biochemistry rsch. inst. Hosp. Sick Children, Toronto, Ont., Can., 1973—; prof. depts. biochemistry and clin. biochemistry U. Toronto, 1974—. Recipient Boehringer-Mannheim Can. prize Can. Biochem. Soc., 1991; fellow Max Planck Inst. Biophysics, 1970-73, Can. Cystic Fibrosis Found., 1970, grantee, 1982—; grantee Med. Rsch. Coun. Can., 1981—. Mem. Can. Fedn. Biol. Socs., Am. Soc. Biol. Chemistry. Achievements include studies of mammalian cell plasma membrane glycoproteins, particularly normal structure, function and aberrations thereof in genetic diseaseincluding cystic fibrosis, disorders of myclimation and cancer. Office: Research Institute, Hospital for Sick Children, Toronto, ON Canada M5S 1A1*

RIORDAN, WILLIAM JOHN, manufacturing process designer, consultant; b. Mishawaka, Ind., Sept. 27, 1955; s. Robert Emit and Angela Dolores (Malloy) R.; m. Debbi Jean Laird, Aug. 9, 1980; 1 child, Robert Alan. BS, Purdue U., 1976; MBA, U. Calif., Irvine, 1984. Coop. engr. Aluminum Co. Am., Newburgh, Ind., 1974-76; engr. Rockwell Internat., Newport Beach, Calif., 1977-78; engr. mgr. Printronix, Irvine, 1978-81, Ford Aerospace Co., Newport Beach, 1981-86; project engr. Northrop Co., Anaheim, Calif., 1986-89; ops. mgr. GDE Systems, Inc., San Diego, 1989—. Contbr. articles to profl. jours. Speaker, advisor Youth Motivation Task Force, Anaheim, 1985; coach Rancho Bernardo Youth Soccer League, San Diego, 1991. Mem. San Diego Deming User Group, Greater Orgn. Alliance Lawrence/QPC, Toastmasters.

RIORDON, JOHN ARTHUR, electrical engineer; b. Washington, June 12, 1941; s. Robert Callahan and Beda Tilda (Arey) R.; m. Elizabeth Esthermay Cowan, Dec. 15, 1962; children: David, James, Karen, Barbara, Jennifer, Stephen. BS in Engring., Ariz. State U., 1967; MS, Rensselaer Poly. Inst., 1969. Registered profl. engr., Tex. Commd. 2d lt. USAF, 1966, advanced through grades to maj., 1980; engr. Polaroid Corp., New Bedford, Mass., 1973-74, McDonnell Douglas, Houston, 1974-78, Martin Marietta, Denver, 1978-79, ORI, Inc., Silver Spring, Md., 1979-80; officer USAF, 1980-90; dept. mgr. ARC Profl. Svcs. Group, Ft. Huachuca, Ariz., 1990—. Recipient NASA Adminstr. award, 1977. Fellow AIAA (assoc.), NASA Alumni League. Republican. Roman Catholic. Achievements include development of lunar mass concentration simulations for Apollo program; flight procedures and crew training for space shuttle approach landing tests; research in optimization studies for space shuttle flight control system. Home: 4774 Green Oak Ln Hereford AZ 85615 Office: ARC Profl Svcs Group PO Box 719 Fort Huachuca AZ 85613-0719

RIPINSKY-NAXON, MICHAEL, archaeologist, art historian, ethnologist; b. Kutaisi, USSR, Mar. 23, 1944; s. Pinkus and Maria (Kokielov) R.; m. Agata Dutkiewicz; 1 child, Tariel. AB in Anthropology with honors, U. Calif.-Berkeley, 1966, PhD in Archeology and Art History, 1979. Rsch. asst. Am. Mus. Natural History, N.Y.C., 1964, U. Calif.-Berkeley, 1964-66; mem. faculty dept. anthropology and geography of Near East, Calif. State U.-Hayward, 1966-67; asst. prof. Calif. State U.-Northridge, 1974-75; researcher, assoc. UCLA, 1974-75, sr. rsch. anthropologist Hebrew U., Hadassah Med. Sch., Jerusalem, 1970-71; curator Anthropos Gallery of Ancient Art, Beverly Hills, Calif., 1976-78; chief rsch. scientist Archaeometric Data Labs., Beverly Hills, 1976-78; dir. Ancient Artworld Corp., Beverly Hills, 1979-82; dir. prehistoric studies Mediterranean Rsch. Ctr., Athens, 1989-91; prof., chairperson, Dept. Cultural Studies, Pedagogical U., Kielce, Poland; conducted excavations Israel, Egypt, Jordan, Mesopotamia, Mexico, Cen. Am; specialist in the development of early religions and shamanism, phenomenon of origins of domestication and camel ancestry; expert on art works from French Impressionists to ancient Egypt and classical world. Author: The Nature of Shamanism, 1993; contbr. articles to sci. and scholarly jours. dir. Cen. Am. Inst. Prehistoric and Traditional Cultures, Belize; chmn. bd. Am. Found. for Cultural Studies. Recipient Cert. of Merit for Sci. Endeavour, Dictionary of Internat. Biography, 1974. Mem. Archaeol. Inst. Am. (life), Soc. for Am. Archaeology, Am. Anthropol. Assn., Royal Anthropol. Inst., Am. Ethnol. Soc., History of Sci. Soc., Am. Chem. Soc., Assn. for Transpersonal Psychology, Soc. Ethnobiology, Soc. Archeol. Scis. (life), New England Appraisers Assns. Home: PO Box 2088 Cathedral City CA 92235-2088 Office: Pedagogical U, ul Żeromskiego 5, 25-369 Kielce Poland

RIPP, BRYAN JEROME, geological engineer; b. Tucson, Dec. 22, 1959; s. Jerome Peter and Helen Marie (Bussmuss) R.; m. Susan Sorensen, Nov. 7, 1987; 1 child, Aaron. BS in Geol. Engring., S.D. Sch. Mines, 1982; MS in Geol. Engring., U. Mo., Rolla, 1984. Registered profl. engr., Ill., Mo., profl. geologist, Ark., Am. Inst. Profl. Geologists. Roustabout Shell Oil Co., Yorba Linda, Calif., 1980; geol. engr. Tenneco Oil Co., Lafayette, La., 1981; staff engr. Shannon and Wilson, Inc., St. Louis, 1984-88; sr. engr. Geotechnology, Inc., St. Louis, 1988—; cons. Consolidation Coal Co., St. Louis, 1986—, Union Pacific RR, Omaha, 1985-88, Mallinckrodt Specialty Chems., St. Louis, 1992—. Author: Underground Storage Tank Closure Manual, 1992; author, editor: Hydrocarbon Assessment Manual, 1992; reviewer WASTECH '93 monographs; author reports. Mem. ASCE, Soc. Mining Engrs. (chmn. 1992-93), Assn. Engring. Geologists (treas. 1990-94), Ill. Soc. Profl. Engrs., Order of the Engr., Sigma Xi. Home: 26 Glenhaven Dr Glendale MO 63122 Office: Geotechnology Inc 2258 Grissom Dr Saint Louis MO 63146

RIPS, RICHARD MAURICE, chemist, pharmacologist; b. Neuilly Sur Seine, France, Oct. 7, 1930; s. Maurice E.M. and Lucie A.M. (Agier) R.; m. Ginette H. Lapote, Sept. 10, 1957; children: Florence, Claire. PhD, U. Paris, 1959. Attache de recherche Inst. Nat. de la Sante et de la Recherche Medicale, Paris, 1960-63, charge de recherche, 1963-66, maitre de recherche, 1967-69, dir. rsch., 1969-82, dir. rsch. de classe exceptionelle, 1982—; dir. Unite de Pharmacologie Chimique, Paris, 1971-87. Contbr. over 160 articles to profl. jours.; patentee in field. Pres. of The Sci. Commn. for rsch. in pharmacology, pharmacochemistry, biochemistry and nutrition, INSERM, France, 1987-91. Mem. Am. Chem. Soc., Soc. Chim. Therap. Assn. des Pharmacologistes. Home: 41 Rue Buffon, 75005 Paris France

RISBERG, ROBERT LAWRENCE, SR., electronics executive, consultant, engineer; b. Milw., Apr. 11, 1935; s. Theodore Carl and Maria K. (Kappes) R.; m. Ruth Elizabeth Brenner, Aug. 13, 1956; children: Robert L. Jr., Gustav Lars, Eric Leif. BSEE, Marquette U., 1962, MSEE, 1973; MS in Physics, U. Wis., 1981, PhD in Elec. Engring., 1985. Registered profl. engr., Wis. Sr. engr. Culter-Hammer Inc., Milw., 1959-70; tech. cons. Louis Allis Co., Milw., 1970-72; pres., cons. engr. Risberg Power Electronics Inc., New Berlin, Wis., 1972—; cons. ASEA Corp., Västerås, Sweden, 1986-88, USN, 1988—, many others. Inventee in field. With USAF, 1952-56. Mem. IEEE. Republican. Lutheran. Home: 16915 Judith Ln Brookfield WI 53005-6317 Office: Risberg Power Electronics 1810 S Calhoun Rd New Berlin WI 53151-1300

RISELEY, MARTHA SUZANNAH HEATER (MRS. CHARLES RISELEY), psychologist, educator; b. Middletown, Ohio, Apr. 25, 1916; d. Elsor and Mary (Henderson) Heater; BEd, U. Toledo, 1943, MA, 1958; PhD, Toledo Bible Coll., 1977; student Columbia U., summers 1943, 57; m. Lester Seiple, Aug. 27, 1944 (div. Feb. 1953); 1 child, L. Rolland, III; m. Charles Riseley, July 30, 1960. Tchr. kindergarten Maumee Valley Country Day Sch., Maumee, Ohio, 1942-44; dir. recreation Toledo Soc. for Crippled Children, 1950-51; tchr. trainable children Lott Day Sch., Toledo, 1951-57; psychologist, asst. dir. Sheltered Workshop Found., Lucas County, Ohio, 1957-62; psychologist Lucas County Child Welfare Bd., Toledo, 1956-62; tchr. educable retarded, head dept. spl. edn. Maumee City Schs., 1962-69; pvt. practice clin. psychology, 1956—; instr. spl. edn. Bowling Green State U., 1962-65; instr. Owens Tech. Coll., 1973-78; interim dir. rehab. services Toledo Goodwill Industries, summer 1967, clin. psychologist Rehab. Center, 1967—; staff psychologist Toledo Mental Health Center, 1979-84. Dir. camping activities for retarded girls and women Camp Libbey, Defiance, Ohio, summers 1951-62; group worker for retarded women Toledo YWCA, 1957-62; guest lectr. Ohio State U., 1957. Health care profl. mem. Nat. Osteoporosis Found., 1988—. Mem. Ohio Assn. Tchrs. Trainable Youth (pres. 1956-57), NW Ohio Rehab. Assn. (pres. 1961-62), Toledo Council for Exceptional Children (pres. 1965), Greater Toledo Assn. Mental Health, Nat. Assn. for Retarded Children, Ohio Assn. Tchrs. Slow Learners, Am. Assn. Mental Deficiency, Am. Soc. Psychologists in Marital and Family Counseling, Psychology and Law Soc., Am. (asso.), Ohio, NW Ohio (sec.-treas. 1974-77, pres. 1978-79), Am. Theater Orgn. Soc., Ohio Psychol. Assn. (continuing edn. com. 1978—), NEA, AAUW, Am. Soc. Psychologists in Pvt. Practice (nat. dir. 1976—), State Assn. Psychologists and Psychol. Assts., Bus. and Profl. Women's Club, (pres. 1970-72), Ohio Fedn. Bus. and Profl. Women's Clubs (dist. sec. 1970-71, dist. legis. chmn. 1972-74), Toledo Art Mus., Women's Aux. Toledo Bar Assn., League Women Voters (pres. Toledo Lucas County 1991-93), Y Matrons (pres. 1993—), Toledo Area Theater Orgn. Soc. (sec. 1991), Zonta Internat. (local pres. 1973-74, 78-79, area dir. 1976-78, Maumee River Valley Woman of Yr. for svc. to community and Zonta, 1992), Maumee Valley Hist. Soc., MBLS PEO (chpt. pres. 1950-51), Toledo Council on World Affairs, Internat. Platform Assn. Baptist. Home: 2816 Wicklow Toledo OH 43606 Office: 940 S Detroit Ave Toledo OH 43614-2701

RISER, BRUCE L., research pathologist; b. Marshall, Mich., Mar. 5, 1950; s. Erman W. and Betty J. (Amsler) R.; m. Jan E. Piper, Feb. 2, 1973; children: Melissa L., Rebecca E., Sarah C. AB, Albion Coll., 1973; MS, U. Mich., 1980, PhD, 1986. Head clin. microbiology Regional Med. Labs., Battle Creek, Mich., 1977-80; instr. microbiology, cell biology U. Mich. Sch. Pub. Health, Ann Arbor, Mich., 1980-82; postdoctoral fellow dept. pathology U. Mich. Med. Sch., Ann Arbor, 1985-88; sr. rsch. scientist Wellcome Rsch. Labs., Rsch. Triangle Park, N.C., 1988-90; sr. staff scientist Henry Ford Hosp., Detroit, 1990—. Contbr. articles to profl. jours. Recipient Young Investigator award Internat. Soc. Nephrology, Rsch. award Juvenile Diabetes Found. Internat., 1992; Horace H. Rackham grantee U. Mich., 1980-85, Henry Ford Found. grantee, 1992; Francis Payne predoctoral fellow U. Mich., 1982-84, fellow Am. Cancer Soc., 1985-88. Mem. AAAS, Nat. Kidney Found., Am. Assn. Cancer Rsch., Soc. Leukocyte Biology, N.Y. Acad. Scis. Episcopalian. Achievements include first to demonstrate productive replication of influenza virus in lung macrophages, and to suggest a relationship to virus strain virulence. First to show a role for the molecule thrombospondin in epithelial and squamous carcinoma cell recognition and adhesion to the substrate and to macrophages. First to show that mechanical stretching force as result of intraglomerular hypertension enhances extrcellular matrix production as a mechanism of progressive renal sclerosis. Office: Henry Ford Hosp Divsn Nephrology 2799 Grand Blvd Detroit MI 48202

RISLEY, ALLYN W(AYNE), petroleum engineer, manager; b. Great Bend, Kans., Sept. 22, 1950; s. Albert L. and Hazel M. (Hull) R.; m. Marit P. Gimre, Sept. 16, 1977; children: Jessica, Michael. BS in Petroleum Engring., U. Kans., 1972. Registered profl. engr., Kans. Engring. dir. Phillips Far East, Singapore, 1981-83; supt. Phillips U.K., Great Yarm Dist., Greatyarm Dist., Yarmouth, 1984-85; mgr., product engr. Phillips Europe/Africa, London, 1986; mgr. drilling and prodn. Phillips U.K., Woking, 1987-88; mgr. liquified natural gas sales Phillips 66 Natural Gas Co., Bartlesville, Okla., 1989-90; mgr. corp planning Phillips Petroleum Co., Bartlesville, Okla., 1991, v.p. drilling and prodn., 1992—. Active Boy Scouts of Am. Planning Commn., 1991-92. Mem. NSPE, soc. Petroleum Engrs. (chmn. Singapore 1983-84), Okla. Soc. Profl. Engrs., Rotary. Home: 2616 Camelot Ct Bartlesville OK 74006

RIST, HAROLD ERNEST, consulting engineer; b. Newcomb, N.Y., Aug. 6, 1919; s. Ernest DeVerne and Iva Cardine (Braley) R.; m. Vera Leona Basuk, July 30, 1942 (div. June 1980); children: Cherry Diana Rist Chapman, Harold Ernest II, Byron Basuk; m. Ruth Ann Mahony, Aug. 16, 1980. BCE, Rensselaer Poly. Inst., 1950, MCE, 1952. Registered profl. engr., N.Y., N.J., N.H., Vt., Mass., Pa., Md., Ky. Project mgr. Seelye, Stevenson, Value & Knecht, N.Y.C., 1952-58; prin. Harold E. Rist, Assocs., Glens Falls, N.Y., 1958-60; ptnr., chief exec. officer Rist, Bright & Frost, Glens Falls, 1960-64, Rist, Frost & Assocs., Glens Falls, 1964-79; pres., chief exec. officer Rist-Frost Assocs., P.C., Glens Falls, 1979-84, chmn. bd., 1984-89, chmn. emeritus, 1989—; ptnr., chief exec. officer Twenty-One Bay Partnership, 1970—, Smith Flats Partnership, 1976—; pres., chief exec. officer Mech.-Elec. Systems, Inc., 1983—, Glens Falls communications Corp., 1985—. Commr. Hudson River Valley Commn., Tarrytown, N.Y., 1970-78. Served with U.S. Army, 1942-45. Mem. NSPE, Cons. Engrs. Coun. N.Y. (pres. 1966-68), Am. Cons. Engrs. Coun. (v.p. 1970-72), Nat. Profl. Svcs. Coun. (bd. dirs. 1972-85), N.Y. State Assn. of the Professions (charter), Adirondack Regional C. of C. (v.p. 1984-89), Adirondack North Country Assn. (bd. dirs. 1959-86, vice chmn. 1985-92), Masons, Safari Club Internat., Surfside Vacation Assn. (Kihea, Maui, Hawaii, pres. 1987—). Republican. Mem. Meth. Episcopal Ch. Home: Lake Shore Dr Box A-1 Diamond Point NY 12824-9723 Office: Harold E Rist Assocs Cons Engrs 21 Bay St Glens Falls NY 12801-3049

RISTICH, MIODRAG, psychiatrist; b. Belgrade, Yugoslavia, July 19, 1938; came to U.S., 1967; s. Teodosije and Gordana (Isailovic) Ristic; m. Yvonne Muriel Cunliffe, May 6, 1967; children: Katharine Alexandra, Elizabeth Victoria. MD, U. Belgrade, 1962. Diplomate Am. Bd. Psychiatry and Neurology. Psychiatric resident Manhattan Psychiatric Ctr., NYU, 1980-83; Med. dir. Cambridge State Hosp., Cambridge, Minn., 1967-72; dir. Willowbrook State Sch., Staten Island, N.Y., 1972-74; med. dir. DeWitt Nursing Home, N.Y.C., 1976—; pvt. practice psychiatry, N.Y.C., 1973—. Treas. St. Bartholomew's Episcopal Ch., Hohokus, N.J. 1987. Mem. AMA, Am. Psychiat. Assn., Am. Assn. for Geriatric Psychiatry, Royal Coll. Psychiatrists. Republican. Avocation: tennis. Home: 37 Sunrise Ln Saddle River NJ 07458-1631 Office: DeWitt Nursing Home 211 E 79th St New York NY 10021-0891

RITCHIE, ROBERT OLIVER, materials science educator; b. Plymouth, Devon, U.K., Jan. 2, 1948; came to U.S., 1974; s. Kenneth Ian and Kathleen Joyce (Sims) R.; m. Connie Olesen (div. 1978); 1 child, James Oliver. BA with honors, U. Cambridge, Eng., 1969, MA, PhD, 1973, ScD, 1990. Cert. engr. Goldsmith's rsch. fellow Churchill Coll. U. Cambridge, 1972-74; Miller fellow in basic rsch. sci. U. Calif., Berkeley, 1974-76; assoc. prof. mech. engring. MIT, Cambridge, 1977-81; prof. U. Calif., Berkeley, 1981—; dir. Ctr. for Advanced Materials Lawrence Berkeley Lab., Cambridge, 1987—; dep. dir. materials sci. div. Lawrence Berkeley Lab., Berkeley, 1990—; cons. Alcan, Boeing, Chevron, Exxon, GE, Grumman, Instron, Northrop, Rockwell, Westinghouse, Baxter, Carbonmedics, Carbon Implants, Med. Inc., Shiley, St. Jude Med. Editor six books; contbr. over 200 articles to profl. jours. Recipient Curtis W. McGraw Rsch. award Am. Soc. Engring. Educators, 1987, Mathewson Gold medal Minerals, Materials, Metals, Soc., 1985; named one of Top 100 Scientists, Sci. Digest mag., 1984. Fellow Inst. Materials (London), Am. Soc. Metals Internat., Internat. Congress on Fracture (hon., v.p.); mem. Minerals, Materials and Metals Soc., Am. Orchid Soc. Avocations: skiing, orchids, tennis. Home: 590 Grizzly Peak Blvd Berkeley CA 94708-1238 Office: Lawrence Berkeley Lab MS-66-247 One Cyclotron Rd Berkeley CA 94720

RITSCHEL, JAMES ALLAN, computer research specialist; b. St. Paul, June 25, 1930; s. Florian Peter and Doris (Miller) R.; m. Dorothy Kvapil, Apr. 15, 1952. BA, Globe Coll., 1962. Cert. data processor, systems and office automation profl. Jr. underwriter St. Paul Cos., 1948-51; acct. Burlington North, St. Paul, 1951-66; bus. systems specialist 3M, Maplewood, Minn., 1966-85, micro rsch. specialist, 1985-92; cons. Mgmt. Assistance Project, Mpls., 1983—; co-founder Minn. Emerging Techs. Coun., 1988, 3M rep., 1989-92, advisor, 1993—; mem. professionalism adv. bd. Minn. Computer/Industry Coalition, 1990. Mem. Office Automation Soc. Internat. (exec. officer 1984-86, session chmn. and conf. speaker 1985), Assn. Inst. Cert. Computer Profls. (regional bd. dirs. 1987-88, internat. bd. dirs. 1989-92), Minn. Soc. Cert. Computer Profls. Avocations: genealogy, European history, computers. Home: 6059 48th St N Saint Paul MN 55128-1935

RITTER, DALE F., geologist, research association administrator; b. Allentown, Pa., Nov. 13, 1932; s. C. Century and Elizabeth (Bowden) R.; m. Jacqueline Leh, Aug. 15, 1953 (dec. Jan. 1961); children: Duane, Darryl, Glen; m. Esta Virginia Lewis, Nov. 23, 1962; 1 child, Lisa Diane. BA in Edn., Franklin and Marshall Coll., 1955, BS in Geology, 1959; MS in Geology, Princeton U., 1963, PhD in Geology, 1964. From asst. to assoc. prof. geology Franklin and Marshall Coll., Lancaster, Pa., 1964-72; prof. geology So. Ill. U., Carbondale, 1972-90; exec. dir. Quaternary Sci. Ctr. Desert Rsch. Inst., Reno, Nev., 1990—. Author: Process Geomorphology, 2d edit., 1986. Fellow NSF, 1968-69; recipient Lindback award Disting. Teaching, Lindback Found., 1970, Outstanding Teaching award Amoco Found., 1979. Fellow Geol. Soc. Am. (chmn. quaternary geol./geomorphology divsn. 1988-89); mem. Am. Quaternary Assn., Assn. Am. Geographers, Yellowstone-Big Horn Rsch. Assn. (pres. 1983-85). Office: Quaternary Sci Ctr Desert Rsch Inst 7010 Dandini Blvd Reno NV 89512

RITTER, DONALD LAWRENCE, congressman, scientist; b. N.Y.C., Oct. 21, 1940; s. Frank and Ruth R.; m. Edith Duerksen; children: Jason, Kristina. B.S. in Metall. Engring., Lehigh U., 1961; M.S. in Phys. Metallurgy, MIT, 1963, Sc.D., 1966. Mem. faculty Calif. State Poly. U., also contract cons. Gen. Dynamics Co., 1968-69; mem. faculty dept. metallurgy and materials scis., asst. to v.p. for research Lehigh U., 1969-76; mgr. research program devel., 1976-79; mem. 96th-102d congresses from 15th Pa. dist., 1979-93; chmn., pres. Nat. Environ. Policy Inst., Washington, 1993—; mem. energy and commerce com. and subcoms. telecommunications and fin.; ranking minority mem. transp. and hazardous materials; mem. sci., space and tech. com. and subcoms. environment and tech. and competitiveness; chmn. house Rep. task force on tech. and policy; co-chair Cngl. High Tech. Caucus; ranking minority mem. house Commn. on Security and Cooperation in Europe (Helsinki Commn.), mem. since 1980—; co-chmn. ad hoc com. on Baltic states and Ukraine; treas. Congl. steel caucus; mem. Congl. textile and apparel caucus; mem. environ. and energy study conf.; sci. exchange fellow U.S. Nat. Acad. Scis.-Soviet Acad. Sci., Baikov Inst., Moscow, 1967-68. Contbr. articles to sci., engring. and quality jours. Bd. dirs. Nat. Metric Coun., Baum Art Sch. of Allentown Art Mus., Bach Choir Bethlehem, Pa.; ex officio mem. bd. assocs. Muhlenberg Coll., Allentown; mem. polit. sci. vis. com. MIT, 1982-85. Recipient award for disting. pub. svc. IEEE, 1990. Fellow Am. Inst. Chemists (honor scroll award); mem. NSPE, Am. Soc. for Metals (disting. life), Sigma Xi, Tau Beta Pi, Pi Mu Epsilon. Unitarian. Home: 2746 Forest Dr Coopersburg PA 18036 Office: Nat Environ Policy Inst 1101 16th St NW Ste 502 Washington DC 20036

RITTER, GERALD LEE, chemical engineer; b. Kellogg, Idaho, Aug. 13, 1939; s. Max Edward and Margaret Ruth (Vergin) R.; m. Maureen Joyce McAllister, Aug. 19, 1961; children: Carolyn, Brian, Glenn, Michelle. BA in Chemistry, Pacific Luth. U., 1961; BS in Chem. Engring., U. Washington, 1962; MS in Chem. Engring., U. Calif., Berkeley, 1964. Mgr. separation process engring. Atlantic Richfield Hanford Co., Richland, Wash., 1969-71; mgr. process engring. Exxon Nuclear Co., Richland, 1972-76, mgr., reprocessing engr., 1976-79; v.p. tech. dept. Exxon Nuclear Idaho Co., Idaho Falls, 1979-81; mgr., process and equipment devel. Exxon Nuclear Co., Richland, 1981-85; mgr., fuel engring. and tech. svc. Advanced Nuclear Fuels Corp., Richland, 1985-90; mgr., rsch. and product devel. Siemens Power Corp., Richland, 1990-. Contbr. articles to profl. jours. Pres., treas. Good Shepherd Luth. Ch., Richland, 1966-76, worship com. chmn., 1983-85. Engring. scholar Rainer Pulp and Paper, 1961. Mem. Am. Nuclear Soc., Am. Inst. for Chem. Engring. (dir. nuclear div. 1985-86), Sigma Xi. Achievements include patents for solvent extraction process for purifying AM and CM and advanced boiling water reactor fuel assembly design. Office: Siemens Power Corp 2101 Horn Rapids Rd Richland WA 99352-0130

RITTER, GRANT L., water operations supervisor; b. San Jose, Calif., Mar. 1, 1955; s. Grant K. and Frances L. (Stephen) R.; m. Cynthia L. Schultz, Mar. 7, 1975 (div. July 1980); 1 child, Chiska L. Cert. water distbn. and water treatment profl., Alaska. Meter reader City of Juneau (Alaska) Water Dept., 1975-77, water distbn. operator I, 1977-80, water distbn. operator II, 1980-85, water systems leadman, 1985-89; water ops. supr. City of Juneau, 1989—, telemetry system integrator, 1990-91; system info. guide Juneau Sch. Dist., 1985—. Pace setter United Way campaign, Juneau, 1990. Recipient Civic Accomodation award City Assembly, 1989. Mem. Alaska Water Mgmt. Assn. (sec., treas. 1988-92). Roman Catholic. Office: City Borough Juneau Util 5433 Shaune Dr Juneau AK 99801

RITTER, JAMES ANTHONY, chemical engineer, educator, consultant; b. Syracuse, N.Y., July 21, 1960; s. James Archer and Rosalind Mary (Grosso) R.; m. Mary Margaret Snyder, Aug. 9, 1986; children: Daniel James, Holly Marie, Matthew James. AA, Onondaga C.C.; BS, SUNY, Buffalo, 1983, MS, 1985, PhD, 1989. Teaching asst. Onondaga C.C., Syracuse, 1980; teaching asst. SUNY, Buffalo, 1983-86, rsch. asst., 1983-89; faculty mem. Consortium of the Niagara Frontier, Buffalo, 1986-89; sr. engr. Westinghouse Savannah River Tech. Ctr., Aiken, S.C., 1989-93; asst. prof. chem. engring. U. S.C., Columbia, 1993—; conf. speaker, 1989—. Contbr. articles to sci. jours. and books. Vol. instr. Nat. Engrs. Week, Augusta, Ga., 1992. Mem. AICE (com. mem., chmn. Separations divsn. 1990—, dir. Savannah River sect. 1992-93), AAAS, Internat. Absorption Soc., Am. Chem. Soc., Tau Beta Pi, Sigma Xi. Achievements include research in separation processes, waste management and environmental restoration. Office: U SC Swearingen Engring Cr Dept Chem Engring Columbia SC 29208

RITTER, TERRY LEE, electrical engineer, educator; b. St. Paul, Mar. 22, 1952; s. William Henry and Lorraine B. (Jensen) Cole; m. Shamim Siddig, Apr. 21, 1990. BSEE, U. Minn., 1984; MSEE, Stanford U., 1987. Engring. mgr. Intel Inc., Santa Clara, Calif., 1984-88; researcher Apple Computer/Stanford U., Cupertino, Calif., 1988-89; cons., educator various high tech. firms Santa Clara, 1989-90; cons. nCHIP, Fremont, Calif., 1991; researcher Tessera Inc./Sematech, Santa Clara, 1992; cons., v.p. Exec. Financing, Inc. Co-author: Thin Film MCMs, 1992; contbr. articles to tech. publs. With USN, 1969-77. Mem. IEEE, Internat. Soc. Hybrid Mfrs. Home: 43904 S Moray St Fremont CA 94539

RITZEN, JOZEF MARIA MATHIAS, economist; b. Heerlen, Limburg, The Netherlands, Oct. 3, 1945. Student, U. Delft. Prof. edn. econs. Nijmegen U., The Netherlands, 1981-83; prof. pub. sector econs. Erasmus U., Rotterdam, The Netherlands, 1983-89; min. Ministry Edn. & Sci., Zoetermeer, The Netherlands, 1989—; advisor Min. Social Affairs. Office: Ministry of Education & Science, Europaweg 4, POB 25000, 2700 LZ Zoetermeer The Netherlands*

RIVERA, ARMANDO REMONTE, utilities engineer; b. Oas, Albay, Philippines, Jan. 27, 1940; came to U.S., 1969; s. Venanclo Rey Rivera and Eugenia (Raneses) Remonte; m. Carmelita Lim Chan, Dec. 11, 1971. BSME, U. of the Philippines, Quezon City, 1962; MBA, U. of the Philippines, Manila, 1967; postgrad., U. So. Calif., L.A., 1972. Registered profl. engr., Calif. Engr. Fluor Corp., L.A., 1969-71; energy svcs. engr. So. Calif. Edison Co., Rosemead, Calif., 1972-76; project engr. Aramco Overseas Co., Hague, Netherlands, 1976-82, London, 1982-84; sr. project engr. electric systems Anaheim (Calif.) Pub. Utilities, 1984—. EPA grantee U. So. Calif., 1972. Mem. ASME, Am. Assn. Cost Engrs., Nat. Soc. Profl. Engrs. Democrat. Roman Catholic. Avocations: antique, religious art, Japanese prints. Home: 1523 N Pacific Ave Glendale CA 91202-1213 Office: Anaheim Pub Utilities 201 S Anaheim Blvd Anaheim CA 92805-9999

RIVERA, LUIS RUBEN, electrical engineer; b. Canal Zone, Panama, Apr. 1, 1956; s. Ruben Rivera and Nancy Macchi Roman. BS in Physics, Herbert H. Lehman Coll., Bronx, 1981; BSEE, CCNY, 1984. System engr. Northrop Corp., Rolling Meadows, Ill., 1984-85, Grumman Corp., Bethpage, N.Y., 1985-88; electro-optics engr. Hughes Danbury Optical Systems, Danbury, Conn., 1988—. Recipient Gillet Award in Physics Herbert H. Lehman Coll., 1984. Republican. Achievements include development in digital imaging laser radar; target recognition in images. Office: Hughes Danbury Optical Sys 100 Wooster Heights Rd Danbury CT 06810

RIVERS, HORACE KEVIN, aerospace engineer, researcher, thermal-structural analyst; b. Chesterfield, S.C., Oct. 20, 1968; s. Horace William and Betty Faye (Raffaldt) R.; m. Sara Melissa Brown, Aug. 4, 1990; 1 child, Joshua William. BSME, Miss. State U., 1991. Rsch. asst. Miss. State U., Starkville, 1989-91; with NASA Langley Rsch. Ctr., Hampton, Va., 1987-90, aerospace engr., 1991—. Recipient Coleman Design award, Miss. State U., 1991; Pres.'s scholar Miss. State U., 1990-91. Mem. AIAA, ASME, Pi Tau Sigma, Phi Kappa Phi, Tau Beta Pi. Baptist. Office: NASA Langley Rsch Ctr MS 396 Hampton VA 23665

RIVERS, LEE WALTER, chemical engineer; b. N.Y.C., Apr. 28, 1929; s. Lee Walter and Vivian Katherine Rivers; m. Ada Lou Galigher, Feb. 7, 1948; children: Lynne Oliver, Lee Walter, Lawrence William. BS in Chem. Engring., NYU, 1950; MS in Chem. Engring., U. Del., 1955. With Allied Signal, Inc., Morristown, N.J., 1950-87, v.p. R&D, 1979-83, v.p. tech., 1983-84, corp. dir. planning, 1984-85; fellow, cons. on loan Indsl. Rsch. Inst., 1985-87; dir. tech. transfer U. S.C., Columbia, 1990-91; exec. dir. Nat. Tech. Transfer Ctr., Wheeling, W.Va., 1991—; chmn. rsch.-on-rsch. com. Indsl. Rsch. Inst., 1983-84, dir., 1985-88; cons. Pres. Reagan's Sci. Advisor, Washington, 1985-87. Mem. Bd. of Adjustment, Morris Plains, 1969-70, Planning Bd., 1971; v.p. United Way of Morris County, 1976-78. Fellow AAAS; mem. Am. Inst. Chem. Engrs., Am. Mgmt. Assn., Comml. Devel. Assn., Indsl. Rsch. Inst., Tech. Transfer Soc., Am. Chem. Soc., The Planning Forum, Assn. Univ. Tech. Mgrs., Assn. Fed. Tech. Transfer Mgrs. Office: Nat Tech Transfer Ctr 316 Washington Ave Wheeling WV 26003

RIVERS, ROBERT ALLEN, research pilot, aerospace engineer; b. Jacksonville, Fla., Oct. 2, 1951; s. Clarence Manning and Sara Clair (Williams) R. BS in Math., U. N.C., 1973; M in Aerospace Engring., U. Va., 1985. Pilot Ea. Airlines, 1979-86; rsch. pilot Johnson Space Ctr. NASA, Houston, 1986-90; rsch. pilot Langley Rsch. Ctr. NASA, Hampton, Va., 1990—. Contbr. articles to Jour. AIAA, Soc Exptl. Test Pilots. Comdr. USNR, 1973-79. Mem. AIAA, Soc. Exptl. Test Pilots, Tail Hook Assn. Achievements include research in low lift drag vehicle flight characteristics, agile A/C fighter tactical employment, in-flight determination of handling qualities, advanced transport operation system. Office: NASA Langley Rsch Ctr Mail Stop 255A Hampton VA 23681

RIVLIN, RICHARD SAUL, physician, educator; b. Forest Hills, N.Y., May 15, 1934; s. Harry Nathaniel and Eugenie (Graciany) R.; m. Barbara Melinda Pogul, Aug. 28, 1960 (div.); children: Kenneth Stewart, Claire Phyllis; m. Rita Klausner, Feb. 29, 1976; children: Michelle Elizabeth, Daniel Elliott. A.B. cum laude in Biochem. Scis., Harvard U., 1955; M.D. cum laude in Biochem. Scis., 1959. Diplomate Am. Bd. Internat Medicine. Intern Bellevue Hosp., N.Y.C., 1959-60; asst. resident in medicine Johns Hopkins U. Hosp., Balt., 1960-61; asst. resident Johns Hopkins U. Hosp., 1963-64; clin. assoc. endocrinology br. Nat. Cancer Inst., NIH, Bethesda, Md., 1961-63; fellow dept. physiol. chemistry, medicine Johns Hopkins U. Sch. Medicine, Balt., 1964-66; lectr. clin. medicine Johns Hopkins U. Sch. Medicine, 1965-66; attending physician med. service Balt. City Hosps., 1964-66; assoc. in medicine Columbia U. Coll. Physicians and Surgeons, N.Y.C., 1966-67; asst. prof. medicine Columbia U. Coll. Physicians and Surgeons, 1967-71, assoc. prof. medicine, 1971-79; mem. Inst. Human Nutrition, 1972-79; chief endocrinology, asst. physician Francis Delafield Hosp., N.Y.C., 1966-75; asst. physician Presbyterian Hosp., N.Y.C., 1966-73; assoc. attending physician Presbyterian Hosp., 1973-79; chief nutrition service Meml. Sloan-Kettering Cancer Ctr., N.Y.C., 1979—; prof. medicine Cornell U. Med. Coll., 1979—; chief div. nutrition dept. medicine N.Y. Hosp. -Cornell Med. Center, 1979—; NSF grant reviewer, 1970—; vis. prof. Creighton U., 1974, U. Guadalajara (Mexico), 1974; vis. prof. N.J. Coll. Medicine and East Orange VA Hosp., Newark Med. Sch., N.J., 1974, 1976, 1983; Upjohn vis. prof. in nutrition Med. Coll. Ga., 1976; vis. prof. Syracuse U., 1980; Nat. Dairy Council vis. prof. in nutrition U. Mich., Ann Arbor, 1982; vis. prof. Washington U.-Jewish Hosp., St. Louis, 1983; external examiner in physiology Calcutta U., India; vis. physician Rockefeller U., N.Y.C., 1979—; prin. investigator coop. core labs. amd clin. nutririon rsch. unit Meml. Sloan-Kettering Cancer Ctr., N.Y.C. Editor: Riboflavin, 1975, Contemporary Issues in Clinical Nutrition (series), 1981—; profl. conf. procs., referee numerous profl. jours.; contbr. articles to profl. jours. Served with USPHS, 1961-63. Recipient Grace A. Goldsmith Lectre award Am. Coll. Nutrition, 1981. Fellow ACP; mem. Am. Fedn. Clin. Research, Endocrine Soc., AAAS, Harvey Soc., Am. Thyroid Assn., Am. Physiol. Soc., Am. Soc. Clin. Investigation, Am. Inst. Nutrition, Soc. Exptl. Biology and Medicine. Home: 21 Woodcut Ln New Rochelle NY 10804-3417*

RIZKIN, ALEXANDER, photonics researcher, educator; b. Vitebsk, USSR, May 20, 1936; came to U.S., 1988; s. Abel Rizkin and Elizabeta Kulik; m. Sofia Volkovich, Mar. 2, 1957; 1 child, Mikhail. MSEE, Inst. Communic. Engring., Leningrad, USSR, 1958; PhD in Elec. Engring., Inst. Comm. Engring., Leningrad, USSR, 1968; MS in Edn., Advanced Edn. U., Moscow, 1982. Engr., group leader State Radio Equip. Rsch. Corp., Leningrad, 1959-65; rschr. Inst. Communic. Engring., Leningrad, 1965-68, prof. system dept., 1968-88, head advanced edn. tech. dept., 1979-88; cons. Holographics, Inc., Burlington, Mass., 1989-90; mgr. holographic mfg., rsch. scientist Phys. Optics Corp., Torrance, Calif., 1990—; cons. State Ministry of Communication, Moscow, 1980-85. Recipient Silver medal Nat. Exhbn. of Tech. Achievements, Moscow, 1982, Grand prize Nat. Soc. Def. Tech. Contbns., Moscow, 1983. Mem. Internat. Soc. Optical Engring. Achievements include patents in field; transmitting first known three-dimensional images over telecommunication channel, using holography. Office: Phys Optics Corp 20600 Gramercy Pl slo3 Torrance CA 90501

RIZVI, JAVED, facilities engineer; b. Karachi, Sindh, Pakistan, May 16, 1955; came to U.S., 1981; s. S. Ahmed and Bilquis (Ahmed) R.; m. Iram Rizvi, Dec. 27, 1962; children: Mehreen, Sherish. BS, Nadirshah Eduijee Dinshah U., Karachi, 1978; MS, U. Cin., 1982. Registered profl. engr., Calif.; cert. plant engr. Teaching asst. U. Cin., 1981; indsl. engr. Sierracin Sylmar, Calif., 1982-84; sr. mfg./indsl. engr. Transtech Corp., Sylmar, 1984-86; prin. facilities engr. Micropolis Corp., Chatsworth, Calif., 1986—. Named Dawood Edn. Ctr. scholar, 1971, Karachi Bd. Edn. scholar, 1970, U. Cin. scholar, 1981. Mem. Energy Mgrs. Muslim. Achievements include research in lighting design, HUAC design, clean room design, and operating efficiency. Home: 8015-2 Canby Ave Reseda CA 91335 Office: Micropolis Corp 21329 Nordhoff St Chatsworth CA 91311

RIZZO, THOMAS ANTHONY, psychologist; b. Rochester, N.Y., Mar. 26, 1958; s. Antonio and Mary (Mustardo) R.; m. Jean Rita Acosta, Nov. 4, 1983; children: Alicia Nicole, Michael Thomas. BA in Psychology, Ind. U., 1980, PhD in Devel. Psychology, 1986; MS in Exptl. Psychology, SUNY, Binghamton, 1982. Asst. prof. Med. Sch. Northwestern U., Chgo., 1986—; mem. Ctr. for Endocrinology, Metabolism and Nutrition, Northwestern U. Med. Sch., 1986—. Author: Friendship Development Among Children in School, 1989; contbr. articles to profl. jours. Achievements include demonstrating association between metabolic control of maternal diabetes during pregnancy and intellectual devel. of progeny in early childhood; demonstrating interplay among social-ecology, friendship processes in sch. children. Office: Northwestern Univ Searl 10-526 303 E Chicago Ave Chicago IL 60611

RIZZOLATTI, GIACOMO, neuroscientist; b. Kiev, USSR, Apr. 28, 1937; s. Pietro and Valentina (Fedorkova) R.; m. Leni Bronzin, Mar. 19, 1964; children: Pietro, Beatrice. Liceo Classico, Stellini, Udine, Italy, 1955; MD, U. Padua, Italy, 1961; Degree Neurology, U. Padua (Italy), 1964; PhD in Physiology, U. Rome, 1969. Asst. neurology U. Padua, 1961-64; asst. physiology U. Pisa, Italy, 1964-67; asst. prof. physiology U. Parma, Italy, 1967-69, assoc. prof. human physiology, 1969-75, prof., 1975-80, 81—, dir. Inst. Human Physiology, 1990—; vis. scientist McMaster U., Hamilton, Ont., Can., 1970-71; vis. prof. U. Pa., Phila., 1980-81; chmn. internat. sci. com. neurosci. program, Srasbourg, 1988-91; coun. mem. Academia Europaea, London, 1990-94. Contbr. articles on motor control visuomotor integration, and attention to profl. jours. Recipient Golgi award Accademia dei Lincei, Rome, 1982, 1st prize Italian Neurophysiol. Soc., 1970. Mem. Italian Neuropsychol. Soc. (pres. 1982-84), European Brain and Behavior Soc. (pres. 1984-86), European Neurosci. Assn. Home: via Salnitrara 3, 43100 Parma Italy Office: Inst di Fisiologia Umana, via Gramsci 14, 43100 Parma Italy

ROALDSET, ELEN, geologist; b. Oslo, Feb. 17, 1944; d. Peter and Kristine (Meltvedt) Solberg; m. Asbjørn Roaldset, Aug. 10, 1968; children: Hege, Eli, Ashild. Cand. real., U. Oslo, 1970, PhD, 1978. Geol. asst. Mining Co./Geol. Survey, Norway, 1966-69; rsch. fellow Norwegian Rsch. Coun., Oslo, 1970-73; lectr. U. Oslo, 1973-77, assoc. prof., 1978-81, 87-89; mgr. rsch. and devel. Norsk Hydro A/S, Bergen, Norway, 1981-85; chief geologist Norsk Hydro A/S, Oslo, 1985-87; spl. advisor Norwegian Rsch. Coun., Oslo, 1986-87; prof. Norwegian Inst. Tech., Trondheim, 1989—; cons. Norwegian Coun. for Sci. and Indsl. Rsch., Oslo, 1987—; bd. dirs. Norwave Tech. A/S, Oslo; organizer numerous nat. and internat. sci. meetings. Mem. editorial bd. Applied Clay Sci., 1985—. Chmn. Local Parent Orgn., Oslo, 1985-86; mem. Olso Local City Dist. Coun., 1988—. Recipient Norsk Varekrigsforsikrings award Norwegian Acad. Scis., 1979. Mem. Norwegian Geol. Soc. (pres. 1982-83), Nordic Soc. for Clay Rsch. (pres. 1981-83), Assn. Internat. for Study of Clays (v.p. 1993—), Clay Minerals Soc., Internat. Assn. Sedimetologists, Mineralogical Soc., Petroleum Exploration Soc. Great Britain. Home: Holtveien 5, 1177 Oslo Norway Office: Norwegian Inst Tech Geology, and Mineral Rsch Engring, Hogskoleringen 6, 7034 Trondheim Norway

ROAN, VERNON PARKER, JR., mechanical engineer, educator; b. Ft. Myers, Fla., Nov. 19, 1935; s. Vernon Parker and Mary Evelyn (Wall) R.; m. Paula Gail Phillippi, Sept. 4, 1955; children: Shannon Kay, Ronald Parker. BS, U. Fla., 1958, MS, 1959; PhD, U. Ill., 1966. Registered profl. engr., Fla., Ill. Sr. design engr. Pratt & Whitney Aircraft, West Palm Beach, Fla., 1959-64; instr. U. Ill. Aero. Engring. Dept., Urbana, 1964-66; prof. U. Fla., Gainesville, 1966—; dir. U. Fla. Ctr. for Advanced Studies in Engring., Palm Beach Gardens, 1989—; cons. DuPont Corp., Wilmington, Del., Pratt & Whitney Aircraft, West Palm Beach,Fla., 1985—; vis. prof. Brunel Univ., Uxbridge, Eng., 1979-80. Contbr. articles to profl. jours. Pres. PTA, Gainesville, Fla., 1970-72, Fla. Engring. Soc. North-Ctrl. Fla. Sect., 1972-76. Recipient 100 Top Researchers award U. Fla., 1989-90, 90-91, 91-92, Disting. Alumnus award U. Ill., 1990, Sci. Achievement award Jet Propulsion Lab., Pasadena, Calif.,1983, Blue Key Leadership award U. Fla., 1979. Mem. ASME, Soc. Automotive Engrs., Tau Beta Pi (advisor 1969-79), Pi Tau Sigma, Sigma Gamma Tau, Sigma Xi. Achievements include air enhanced turbo ramjet engine (patent pending). Office: U Fla Mech Engring Dept Gainesville FL 32611

ROBB, JEFFERY MICHAEL, forensic toxicologist; b. Santa Barbara, Calif., Mar. 1, 1949; s. Alan Russell Robb and Mary Louise (Jeffery) Page; m. Catherine Louise HIggins, June 15, 1974. BS, U. N.Mex., 1971. Grad. asst. chemistry dept. U. N.Mex., Albuquerque, 1973-76; chemist Am. Mgmt. and Exploration, Albuquerque, 1977; lab. scientist sci. lab. div. State of N.Mex., Albuquerque, 1977-81, prin. scientist sci. lab. div., 1981—; pvt. practice cons., Albuquerque, 1986—; mem. com. on alcohol and other drugs Nat. Safety Coun., 1986—. Contbr. articles to profl. publs. Fellow Am. Acad. Forensic Sciences (presenter); mem. Soc. Forensic Toxicologists. Achievements include research regarding effects of toluene on breath alcohol testers. Office: State of NM Sci Lab Div PO Box 4700 700 Camino de Salud NE Albuquerque NM 87196

ROBB, RICHARD ARLIN, biophysics educator, scientist; b. Price, Utah, Dec. 2, 1942; s. Max Arlin R.; m. Shanna Lee Woodruff, June 10, 1965; children: Michael, Rachelle, Jason, Jeffrey. AS, Carbon Coll., 1963; BA, U. Utah, 1965, MS, 1968, PhD, 1971. Assoc. in sci. Coll. Ea. Utah, 1963; rsch. assoc. U. Utah, Salt Lake City, 1971-72; NIH postdoctoral fellow Mayo Grad. Sch. Medicine, Rochester, Minn., 1972-75, rsch. assoc., 1973-75, instr. physiology, 1974-76, prof. biophysics, 1985—; cons. radiology Mayo Found., Rochester, 1976-85, cons. physiology, 1976—, bd. dirs. Mayo Biomed. Imaging Rsch., 1985—; bd. dirs. Mayo Biomed. Imaging Resource, Mayo Found., 1985—. Editor: Three-Dimensional Biomedical Imaging, 1985; author: (software system) ANALYZE, 1986-89; patentee in field. Grantee NIH, 1976-89. Mem. IEEE, AAAS, Am. Physiol. Soc., Biomed. Engring. Soc., Soc. Photo-Optical Instrumentation Engrs. Office: Mayo Found Grad Sch Medicine 200 First St SW Rochester MN 55904*

ROBB, WALTER LEE, retired electric company executive, management company executive; b. Harrisburg, Pa., Apr. 25, 1928; s. George A. and Ruth (Scantlin) R.; m. Anne Gruver, Feb. 27, 1954; children: Richard, Steven, Lindsey. B.S., Pa. State U., 1948; M.S., U. Ill., 1950, Ph.D., 1951; DEng (hon.), Worcester Poly. Inst. With Gen. Electric Co., 1951-93; mgr. research/devel. Silicone Products Dept. Gen. Electric Co., Waterford, N.Y., 1966-68; venture mgr. Med. Devel. Ops. Gen. Electric Co., Schenectady, 1968-71; sr. v.p., group exec. Med. Systems Group Gen. Electric Co., Milw., 1973-86; sr. v.p., Corp. Research and Devel. Gen. Electric Co., Schenectady, 1986-93; cons. Gen. Electric Co., 1993—; pres. Vantage Mgmt., Schenectady, N.Y., 1993—; bd. dir. Celgene Corp., Marquette Electronics, Inc., Cree Rsch., Neopath. Recipient Nat. medal of Tech., NSF, 1993. Mem. Nat. Acad. Engring. Acheivements include patentee in field of membranes and gas separation; research in permeable membranes, diagnostic imaging equipment. Home: 1358 Ruffner Rd Niskayuna NY 12309-2500 Office: Vantage Mgnt 1222 Troy-Schenetady Rd Schenectady NY 12309

ROBBINS, ALLEN BISHOP, physics educator; b. New Brunswick, N.J., Mar. 31, 1930; s. William Rei and Helen Grace (Bishop) R.; m. Shirley Mae Gernert, June 14, 1952 (div. 1978); children: Catherine Jean, Marilyn Elizabeth, Carol Ann, Melanie Barbara; m. Alice Harriet Ayars, Jan. 1, 1979. Student, Oberlin Coll., 1948-49; B.S., Rutgers U., 1952; M.S., Yale U., 1953, Ph.D., 1956. Research fellow U. Birmingham (Eng.), 1957-58, lectr., 1960-61; instr. physics Rutgers U., New Brunswick, N.J., 1956-57; asst. prof. physics Rutgers U., 1957-60, assoc. prof., 1960-68, prof., 1968—, chmn. dept. physics and astronomy, 1979—. Contbr. articles on nuclear physics to profl. jours. Recipient Lindbach Christian and Mary F. Lindbach Found., Rutgers U., 1975. Fellow Am. Phys. Soc.; mem. Am. Assn. Physics Tchrs., AAAS, Phi Beta Kappa, Sigma Xi. Office: Rutgers U Dept Physics & Astronomy PO Box 849 Piscataway NJ 08855-0849

ROBBINS, CHARLES MICHAEL, teratologist, research developmental biologist; b. Kadena, Okinawa, Japan, Nov. 1, 1966; (father Am. citizen); s. Charles Clayton Jr. and Mitsuko (Higa) R.; m. Wendy Nannette Saint, Oct. 18, 1992. BS in Biol. Sci. and English cum laude, East Tenn. State U., 1988; postgrad., East Tenn State U., 1988—. Rsch. asst. dept. English, East Tenn. State U., Johnson City, 1989-90, rsch. asst. dept. anatomy, 1990—; presenter at ann. meetings Teratology Soc., 1991-93. Contbr. article and abstracts to profl. jours. Mem. AAAS, Am. Assn. for Lab. Animal Sci., Grad. Student Assn. (pres.), Sigma Xi, Beta Beta Beta, Sigma Tau Delta, Gamma Beta Phi, Omicron Delta Kappa (v.p.). Achievements include research on developmental effects of purine nucleosides. Home: 1616 Seminole Dr # 220 Johnson City TN 37604 Office: East Tenn State U Anatomy Dept Box 70582 Johnson City TN 37614

ROBBINS, CONRAD W., naval architect; b. N.Y.C., Oct. 11, 1921; s. Girard David and Ethyl Rae (Bergman) R.; m. Danae Gray McCartney, Jan.

8, 1923 (dec. Jan. 1971); children: Lorraine, Linton, Jennifer; m. Melissa Jahn, Apr. 15, 1971 (dec. Mar. 1992). BSE, U. Mich., 1942. Estimator Pacific Electric Co., Seattle, 1946-47; pres. Straus-Dupanquet, Lyons, Alpha, Albert Pick, N.Y.C. and Chgo., 1947-67, C.W. Robbins, Inc., Carefree, Ariz., 1967—; cons. in field. Capt. floating drydock USN, 1942-46. Avocations: travel, gardening, gourmet cooking. Home: PO Box 2208 Carefree AZ 85377-2208 Office: CW Robbins Inc 7500 Stevens Rd Carefree AZ 85377

ROBBINS, FREDERICK CHAPMAN, physician, medical school dean emeritus; b. Auburn, Ala., Aug. 25, 1916; s. William J. and Christine (Chapman) R.; m. Alice Havemeyer Northrop, June 19, 1948; children: Alice, Louise. AB, U. Mo., 1936, BS, 1938; MD, Harvard U., 1940; DSc (hon.), John Carroll U., 1955, U. Mo., 1958, U. N.C., 1979, Tufts U., 1983, Med. Coll. Ohio, 1983; LLD, U. N.Mex., 1968. Diplomate Am. Bd. Pediatrics. Sr. fellow virus disease NRC, 1948-50; staff rsch. div. infectious diseases Children's Hosp., Boston, 1948-50, assoc. physician, assoc. dir. isolation svc., asso. rsch. div. infectious diseases, 1950-52; instr., assoc. in pediatrics Harvard Med. Sch., 1950-52; dir. dept. pediatrics and contagious diseases Cleve. Met. Gen. Hosp., 1952-66; prof. pediatrics Case Western Res. U., 1952-80, dean Sch. Medicine, 1966-80, univ. prof., dean emeritus, 1980—, univ. prof. emeritus, 1987—; pres. Inst. Medicine, NAS, 1980-85; vis. scientist Donner Lab., U. Calif., 1963-64. Served as maj. AUS, 1942-46; chief virus and rickettsial disease sect. 15th Med. Gen. Lab. investigations infectious hepatitis, typhus fever and Q fever. Decorated Bronze Star, 1945; recipient 1st Mead Johnson prize application tissue culture methods to study of viral infections, 1953; co-recipient Nobel prize in physiology and medicine, 1954; Med. Mut. Honor Award for, 1969; Ohio Gov.'s award, 1971. Mem. Assn. Am. Med. Colls. (Abraham Flexner award 1987), Nat. Acad. Scis., Am. Acad. Arts and Scis., Am. Soc. Clin. Investigation (emeritus mem.), Am. Acad. Pediatrics, Soc. Pediatric Research (pres. 1961-62, emeritus mem.), Am. Pediatric Soc., Am. Philos. Soc., Phi Beta Kappa, Sigma Xi, Phi Gamma Delta. Office: Case Western Res U Sch of Medicine 10900 Euclid Ave Cleveland OH 44106-4901

ROBBINS, JACKIE WAYNE DARMON, agricultural and irrigation engineer; b. Spartanburg, S.C., Feb. 6, 1940; s. Jack Dennis and Laura Christina (Champion) R.; m. Betty Jo Wright, June 17, 1963; children: Jackie Wayne Darmon II, Robin C.D. BS, Clemson U., 1961, MS, 1965; PhD, N.C. State U., Raleigh, 1970. Registered profl. engr., La., Tex.; cert. irrigation designer. Asst. prof. engring. La. State U., Baton Rouge, 1963-65; engring. rsch. assoc. N.C. State U., 1965-70; assoc. prof. engring. U. Mo., Columbia, 1970-71; prof. engring. La. Tech. U., Ruston, 1971-88; pres., engr. Irrigition Mart, Inc., Ruston, 1978—; cons. irrigation systems design and engring.; cons. agtl. waste gmt. processes and facilities; cons. crop prodn. techniques. Contbr. articles to Agrl. Engring., Jour. Water Pollution Control Fedn., others. E.C. McArthur fellow, 1962-63; NSF Hydrology Inst. fellow, 1965; NSF trainee, 1968; NSF grantee, 1977-78, others. Mem. NSPE, Am. Soc. Agrl. Engrs. (chair rural water supplies, vie chair land use group, vice chair S.W. Region, chair La. sect., numerous others), Irrigation Assn. (bd. govs. 1988-92, chair 1990-91), La. Engring. Soc. (chair Monroe chpt.), Am. Soc. Engring. Edn., La. Agrl. Waste Mgmt. Com., Sigma Xi, Tau Beta Pi, Alpha Epsilon. Achievements include patent for drip irrigation hose. Home: Rte 6 Box 1241 Ruston LA 71270

ROBBINS, JESSIE EARL, metallurgist; b. York, Ala., Feb. 17, 1944; s. Elbert Jessie and Ella Lee (Hurst) R.; m. Dolly Marie Welch, Apr. 7, 1977; children: Angela Michelle, Amanda Leigh. BS in Engring., U. Ala., 1974. Registered profl. engr. Tex. Welding engr. Chicago Bridge & Iron Co., Brimingham, Ala., 1974-77; metallurgist AMF Tuboscope, Houston, 1977-78; plant metallurgist Tubular Finishing Works, Navajoto, Tex., 1978-80; welding engr. Daniel Inds., Houston, 1980-82; welding engr., metallurgist TRW-Mission, Houston, 1982-85; quality assurance engr. Vincotte USA, Houston, 1986-87; welding engr. LTV Missiles & Electronics, Camden, Ark., 1987-90; mgr. tech. svcs. Mavrick Tube Corp., Conroe, Tex., 1990—. With USN, 1961-65. Mem. Am. Petroleum Inst. (com. mem. 1992), Am. Soc. Testing and Materials (com. mem. 1992), Nat. Assn. Corrosion Engrs. (com. mem. 1992), Soc. Petroleum Engring., Grangerland Lions Club (charter). Independent. Home: Rt 2 Box 679BB Conroe TX 77303

ROBBINS, KEITH CRANSTON, research scientist; b. Berwyn, Ill., Feb. 3, 1944; s. Ollice Cranston and Edith Eileen Robbins; m. Barbara Anne Byrne, June 17, 1967; children: Christine Anne, Nancy Edith, Elizabeth Erin. BA, Ind. U., 1967; MS, Am. U., 1972; PhD, George Washington U., 1977. Staff scientist Hazleton Labs., Vienna, Va., 1972-80; cancer expert Nat. Cancer Inst., Bethesda, Md., 1980-85, sect. chief, 1985-88; sect. chief Nat. Inst. Dental Rsch., Bethesda, 1988—, lab. chief, 1992—; invited lectr. Australian Biochem. Soc., Am. Heart Assn., Nat. Geriatric Soc., Am. Soc. Biochemistry & Molecular Biology, Argentine Congress on Cancer, Sci. Frontiers in Clin. Dentistry, Am. Soc. Microbiology, Mich. Cancer Found., S.Am. Symposium on Signal Transduction, Cellw Growth and Cancer, Upjohn Pharms., Pfizer Pharms., Sterling Drug, Nova Pharms., Cor Therapeutics, Pfizer Inc., also numerous univs.; ad hoc mem. various sci. study sects., NIH, NRSA, NCI. Contbr. over 100 articles to profl. jours. Vestryman United Christian Parish, Reston, Va., 1993. Mem. AAAS. Achievements include discovery of gene specifying platelet-derived growth factor; discovered a number of oncogenes; 2 patents. Office: Natl Inst of Dental Health Bldg 30 9000 Rockville Pike Bethesda MD 20892

ROBBINS, KENNETH CARL, biochemist; b. Chgo., Sept. 1, 1917; s. Samuel and Mary (Silberbrandt) R.; m. Pearl Podorowsky, Mar. 31, 1946; children: Paula Lange, Shelley R. BS, U. Ill., 1939, MS, 1940, PhD, 1944. Asst. prof. pathology Western Res. U. Sch. Medicine, Cleve., 1947-51; head protein sec. biochemistry rsch. The Armour Labs., Chgo., 1951-58; dir. biochemistry rsch., scientific dir. Michael Reese Rsch. Found., Chgo., 1958-84; prof. medicine and pathology Pritzker Sch. Med./Univ. Chgo., 1970-87, prof. emeritus, 1987—; dir. exptl. pathology Michael Reese Hosp. and Med. Ctr., Chgo., 1984-86; rsch. scientist, prof. hematology and oncology medicine Northwestern Univ. Sch. Medicine, Chgo., 1989—; mem. hematol. study sect. NIH, Bethesda, Md., 1971-75, 76-80, blood diseases & resources adv. com. Nat. Heart, Lung, and Blood Inst., NIH, 1976-80; chmn. Gordon Conf. Hemostasis, N.H., 1975; mem. Internat. Com. on Thrombosis and Haemostasis, 1980-86, chmn. subcom. on Fibrinolysis, 1980-82; lectr. in field. Mem. editorial bd. Jour. Biol. Chemistry, 1975-80; contbr. articles to profl. jours. Recipient fourth Elwood A. Sharp award Wayne State U. Sch. Medicine, Detroit, 1971, Prix Servier Medal and Prize, Fifth Intern. Congress Fibrinolysis, Malmo, Sweden, 1980; grantee NIH, Bethesda, 1960—. Mem. Am. Assn. Immunologists, Am. Soc. Biochemistry and Molecular Biology, Am. Soc. Hematology, Soc. Exptl. Biology and Medicine. Achievements include 10 patents in field; discovery of fibrin stabilizing factor, pancreatic elastase zymogen-proelastase, mammalian enzymatic omega oxidation of fatty acids system; development of oral thrombolytic therapy, hybrid pasminogen activators. Home: Apt 36C 6101 N Sheridan Rd E Chicago IL 60660-2804

ROBBINS, MARION LE RON, agricultural research executive; b. Inman, S.C., Aug. 18, 1941; s. Jack Dennis and Christina (Champion) R.; m. Margaret Elanor Wilson, Sept. 25, 1965; children: Jack, Rona, Jeff, Kyle. BS, Clemson U., 1964; MS, La. State U., 1966; PhD, U. Md., 1968. Asst. prof. Iowa State U., Ames, 1968-72; rsch. scientist Clemson U., Charleston, S.C., 1972-83; resident dir. Sweet Potato Rsch. Sta., Chase, La., 1984-88, Calhoun (La.) Rsch. Sta., 1988—; advisor Farm Bureau, Monroe, La., 1988—, Agribus. Coun., Monroe, 1988—. Editor: Jour. of Vegetable Crop Prodn., 1992—; assoc. editor: (jour.) Crop Prodn., 1972—; contbr. over 200 articles, abstracts, rsch. papers and reviews to profl. jours. Delegation leader People to People Internat., Spokane, Wash., 1985—. Mem. Am. Soc. for Horticultural Sci. (dir., pres. so. region 1982-83), Rsch. Ctr. Adminstrs. Soc. (dir. 1985), Rotary Club (pres. 1987-88), Exchange Club (pres. 1976-77). Presbyterian. Achievements include development of 22 varieties and genetic lines of crop plants, including 2 All-American winners and an All-Am. designate. Home and Office: Calhoun Rsch Sta 321 Hwy 80 Calhoun LA 71225

ROBBINS, PHILLIPS WESLEY, biology educator; b. Barre, Mass., Aug. 10, 1930; married, 1953; 2 children. A.B., Depauw U., 1952; Ph.D., U. Ill., 1955. Research assoc. Mass. Gen. Hosp., Boston, 1955-57; asst. prof. Rockefeller Inst., N.Y.C., 1957-59; mem. faculty MIT, Cambridge, 1959—; prof. biochemistry MIT, 1965—, now Am. Cancer Soc. prof. biology. Recipeint Eli Lilly award in biol. chemistry, 1966. Mem. Am. Soc. Biol. Chemistry, Am. Chem. Soc., Am. Acad. Arts and Scis. Office: Mass Inst Tech Ctr for Cancer Research E 17-233 Cambridge MA 02139-4325

ROBBINS, THOMAS OWEN, pathologist, educator; b. Mpls., Nov. 29, 1937; s. Owen Francis and Marjorie (Lumsden) R.; m. Mary Ann Butler, July 25, 1959; children: Frederick, Patrick, James, John, William. BS, U. Minn., 1960, MD, 1962. Diplomate Am. Bd. Pathology, Clin and Anatomic Pathology. Intern St. Mary's Hosp., Duluth, Minn., 1962-63; resident U.S. Naval Hosp., San Diego, 1964-68; physician Naval Recruiting Sta., Mpls., 1963-64; lectr. in hematology San Diego State Coll., 1966-68; chief lab. svcs. U.S. Naval Hosp., Republic of Vietnam, 1968-69; asst. head anat. pathology U.S. Naval Hosp., Bethesda, Md., 1969-71; assoc. attending pathologist William Beaumont Hosp., Royal Oak, Mich., 1971-72, attending pathologist, 1972-73, assoc. dir., chief div. surg. pathology, 1973-89, chief pathology svcs., chmn. anatomic pathology, 1990—, assoc. med. dir., 1991—; dir. S.E. Mich. Dermapathology Group, Royal Oak, 1986—; mem. profl. adv. com. S.E. MIch. Hospice, Detroit, 1979-81. Contbr. articles to profl. jours., chpts. to books. Trustee, v.p., pres. Troy (Mich.) Sch. Dist., 1975-79. Comdr. USN, 1970-71. Decorated Bronze Star. Fellow Am. Assn. Clin. Pathologists, Coll. Am. Pathologists; mem. AMA, Internat. Acad. Pathology, ASCP (mem. cancer task force 1989-91, dir. workshops 1977-79), Am. Urological Assn. (faculty GU pathology 1990—). Office: William Beaumont Hosp 3601 W 13 Mile Rd Royal Oak MI 48073-6769

ROBERGE, FERNAND ADRIEN, biomedical researcher; b. Thetford Mines, Que., Can., June 11, 1935. BAS, Engr., Poly. Sch. Montreal, Can., 1959, MScA, 1960; PhD in Control Engring., Biomedical Engring., McGill U., 1964. Devel. engr. numerical control Sperry Gyroscope Co., Montreal, 1960-61; from asst. prof. to prof. physiology faculty medicine U. Montreal, 1965-78, prof. biomedical engring., 1978—, dir. biomedical engring. inst. ecole poly., 1978-88, dir. rsch. group biomedical modeling, 1988—; mem. rsch. group neurol. sci. Med. Rsch. Coun. Can. U. Montreal, 1967-75, mem. grant com. biomedical engring., 1971-76; mem. Sci. Coun. Can., 1971-74; mem. sci. com. Can. Heart Found., 1974-77; mem. Killiam Program Can. Coun., 1974-77; mem. elec. engring. com. Nat. Sci. Engring. Rsch. Coun., Can., 1981-83, chmn., 1985-88. Recipient D. W. Ambridge award, 1964, Rousseau award Assn. Can.-France Advancement Sci., 1986, Leon Lortie award, 1987. Mem. IEEE, Can. Physiol. Soc., Can. Med. and Biol. Engring. Soc. (v.p. 1974-76), Internat. Fedn. Med. Electronics and Biol. Engring., Biomedical Engring. Soc. Achievements include research in membrane biophysics, cardiovascular regulation and control, cardiac arrhythmias; assessment of medical technologies. Office: U of Montreal-Inst of BioMed Eng, CP 6128 Succursale A, Montreal, PQ Canada H3C 3J7*

ROBERGE, LAWRENCE FRANCIS, neuroscientist, biotechnology consultant; b. Springfield, Mass., Mar. 16, 1959; s. Donald Richard and Cornelia Marie (Daly) R. BS in Zoology and Psychology cum laude, U. Mass., 1985; MS in Biomed. Sci., U. Mass., Worcester, 1989. Cert. radiation safety and protection. Sr. tech. Mass. Gen. Hosp., Boston, 1988; nuclear chemist Interstate Nuclear Svc., Springfield, 1989; tech. specialist NERAC, Tolland, Conn., 1989—; instr. Assumption Coll., Worcester, 1988-91, Quinsigamond Community Coll., Worcester, 1989; tchr. Price Lifeskill Inst., Leicester, Mass., 1988-89. Precinct mem. Ludlow, Mass., 1978-81. Mem. Internat. Fedn. Advancement of Genetic Engring. and Biotech., N.Y. Acad. Sci., Mortar Bd., Psi Chi. Achievements include research on reversible vasopressin control of hamster aggression, the role of steroids to control vasopressin expression in hamsters. Home: 62 Hubbard St Apt 13 Ludlow MA 01056-2701 Office: NERAC Inc 1 Technology Dr Tolland CT 06084-3900

ROBERT, DEBRA ANN, chemical engineer; b. Holyoke, Mass., Dec. 19, 1969; d. Gerald Roger and Linda Rose (Labonte) R. BS in Chem. Engring., U. Mass., 1992. Asst. engring. dept. U. Mass., Amherst, 1989-92; engring. intern, coop. rsch. engr. Am. Cyanamid Co., Stamford, Conn., 1990-91; R&D chemist Polycast Tech. Corp., Stamford, 1992—. Mem., sec. Hampshire County 4-H Internat. Exch. Club, 1983-88; pres., co-leader Granby 4-H Bike Club, 1986-87. Mem. AIChE, Order of Engrs. Roman Catholic. Home: 27 Northill St Unit 2P Stamford CT 06907 Office: Polycast Tech Corp 69 Southfield Ave Stamford CT 06902

ROBERT, LESLIE LADISLAS, research center administrator, consultant; b. Budapest, Hungary, Oct. 24, 1924; s. Louis and Elizabeth (Bardos) R.; m. Barbara Klinger, Nov. 19, 1949 (dec.); children: Marianne, Catherine, Elisabeth; m. Jacqueline Labat, Dec. 20, 1976. Student, U. Szeged, Budapest, Hungary, 1944-48; MD, U. Paris, 1953; PhD, U. Lille, France, 1977; Dr. Causa (hon.), Med. U. Budapest, 1991. Mem. med. faculty dept. biochemistry U. Paris, 1950-59; postdoctoral rsch. fellow dept. biochemistry Sch. Medicine U. Ill., 1959-60; postdoctoral rsch. assoc., spl. fellow Columbia U., N.Y.C., 1960-61, 67; dir. biochemistry lab. Inst. for Immunobiology INSERM/CNRS, Broussais Hosp., Paris, 1962-66; founder 1st rsch. ctr. on connective tissue biochemistry CNRS, U. Paris XII, Créteil, France, 1966—; rsch. dir. French Nat. Rsch. Ctr., Paris, 1974—; founder rsch. ctr. for clin. and biol. rsch. on aging Charles Foix-Jean Rostand Hosp., Ivry, France, 1993—; cons. several pharm. firms; mem. Sci. Coun. Arteriosclerosis Rsch. Inst., U. Munster, Fed. Republic of Germany, 1970—. Author 3 books on biology of aging, monograph series: Frontier of Matr's Biology, 11 vols.; mem. editorial bd. several sci. jours.; contbr. over 700 articles on connective tissues, biochemistry and pathology to sci. jours. Recipient Spl. Sci. prize Sci. Writer, 1966, Reiss prize in Opth† olmology, 1970. Mem. French Atherosclerosis Soc. (pres. 1993—). Home: 7 Rue Lully, 94440 Santeny France Office: Univ Paris Med Sch, 8 Rue de General Sarrail, 94010 Creteil France

ROBERTS, CHARLES S., software engineer; b. Newark, Sept. 25, 1937; s. Ben and Sara (Fasten) R.; m. Wendy Shadlen, June 8, 1959; children: Lauren Roberts Gold, Tamara G. Roberts. BS in Chemistry, Carnegie-Mellon U., 1959; PhD in Physics, MIT, 1963. MTS, radiation physics rsch. AT&T Bell Labs., Murray Hill, N.J., 1963-68, head info. processing rsch., 1968-73, head interactive computer systems rsch., 1973-82; head, advanced systems dept. AT&T Bell Labs., Denver, 1982-87; head software architecture planning dept. AT&T Bell Labs., Holmdel, N.J., 1987-88; R&D mgr., system architecture lab. Hewlett-Packard Co., Cupertino, Calif., 1988-90, R&D mgr. univ. rsch. grants, 1990-92; prin. lab. scientist Hewlett-Packard Labs., Palo Alto, Calif., 1992—. Contbr. articles to profl. jours. Westinghouse scholar Carnegie Mellon U., 1955-59; NSF fellow MIT, 1959-63. Mem. IEEE, Assn. for Computing Machinery, Am. Phys. Soc., Sigma Xi, Tau Beta Pi, Phi Kappa Phi. Achievements include 2 patents on associative information retrieval and dithered display system; development of early UNIX operating system for 32-bit computers; research on theory to explain electron loss in Van Allen Belts, on superimposed code techniques for associative information retrieval. Home: 210 Manresa Ct Los Altos CA 94022-4646 Office: Hewlett-Packard Labs 3500 Deer Creek Rd PO Box 10350 Palo Alto CA 94303

ROBERTS, EARL JOHN, carbohydrate chemist; b. Magee, Miss., May 14, 1913; s. William J. and Mary E. (Kennedy) R.; m. Mary Kirk Lilly, Oct. 28, 1944; children: Mary Karyl, John William. BA, Miss. Coll., 1939; MS, La. State U., 1942. Cert. chemist. Jr. chemist U.S. Dept. Agrl., New Orleans, 1942-44, prin. chemist, 1944-47, sr. chemist, 1948-72; sr. rsch. chemist Sugar Processing Rsch. Inst., New Orleans, 1972—. Contbr. articles to profl. jours. Recipient Sugar Crystal award. Achievements include patents for preparation of glue from peanut protein, recovery of aconitic acid from molasses, preparation of derivatives of cotton cellulose.

ROBERTS, EDWARD BAER, technology management educator; b. Chelsea, Mass., Nov. 18, 1935; s. Nathan and Edna (Podradchik) R.; m. Nancy Helen Rosenthal, June 14, 1959; children: Valerie Jo Roberts Friedman, Mitchell Jonathan, Andrea Lynne. BS and MS in Elec. Engring., MIT, 1958, MS in Mgmt., 1960, PhD in Econs., 1962. Founding mem. system dynamics program MIT, 1958-84, instr., 1959-61, asst. prof., 1961-65, assoc. prof., 1965-70, prof., 1970—, David Sarnoff prof. mgmt. of tech., 1974—, assoc. dir. research program on mgmt. of sci. and tech., 1963-73, chmn. tech. and health mgmt. group, 1973-88, chmn. mgmt. of tech. and innovation, 1988—, chmn. ctr. for entrepreneurship, 1992—, co-dir. internat. ctr. rsch. mgmt. tech., 1993; co-founder, pres. Pugh-Roberts Assocs., Inc., Cambridge, Mass., 1963-89, chmn., 1989—; co-founder, dir. Med. Info. Tech., Inc., Westwood, Mass., 1969—; dir. MIT-Boston VA Joint Ctr. on Health Care Mgmt., 1976-80; dir. MIT Mgmt. of Tech. Program, 1980-89, co-chmn., 1989—; co-founder, gen. ptnr. Zero Stage Capital Group, 1981—; bd. dirs. Advanced Magnetics, Inc., Cambridge, Laser Scis., Inc., Newton, Digital Producs Inc., Waltham, Mass., Selfcare Inc., Waltham. Author: The Dynamics of Research and Development, 1964, Systems Simulation for Regional Analysis, 1969, The Persistent Poppy, 1975, The Dynamics of Human Service Delivery, 1976, Entrepreneurs in High Technology, 1991; prin. author, editor: Managerial Applications of System Dynamics, 1978; editor (with others) Biomedical Innovation, 1981; editor: Generating Technological Innovation, 1987; mem. editorial bd. IEEE Trans. on Engring. Mgmt., Internat. Jour. Tech. Mgmt., Indsl. Mktg. Mgmt., Jour. Engring. and Tech. Mgmt., Jour. Product Innovation Mgmt., Tech. Forecasting and Social Change. Mem. IEEE, Inst. Mgmt. Sci., Sigma Xi, Tau Beta Pi, Eta Kappa Nu, Tau Kappa Alpha. Home: 300 Boylston St Boston MA 02116 Office: MIT 50 Memorial Dr Cambridge MA 02139

ROBERTS, ERIC STENIUS, computer science educator; b. Durham, N.C., June 8, 1952; s. James Stenius and Anne (Estep) R. AB, Harvard U., 1973, SM, 1974, PhD, 1980. Asst. prof. computer sci. Wellesley (Mass.) Coll., 1980-84; rsch. scientist DEC/SRC, Palo Alto, Calif., 1985-90; assoc. prof., assoc. chmn. computer sci. dept. Stanford (Calif.) U., 1990—; vis. lectr. Harvard U., Cambridge, Mass., 1984-85. Author: Thinking Recursively, 1986; contbg. author: Computers in Battle, 1987; editor column, Abacus jour., 1987-88; contbr. articles to profl. jours. Fellow NSF, 1973-76. Mem. Assn. Computing Machinery, Dem. Socialists of Am., Computer Profls. for Social Responsibility (nat. sec. 1987-90, pres. 1990—). Mem. Soc. of Friends. Office: Stanford U Dept Computer Sci Stanford CA 94305

ROBERTS, GODWIN, medical products manager, consultant; b. Colombo, Sri Lanka, Mar. 11, 1939; arrived in Australia, 1984; s. Victor and Anna Roberts; m. Sunita Niles, Sept. 27, 1970; 1 child, Gehan. BSc, U. Ceylon, Colombo, 1963; PhD, U. London, 1969. Head biochemistry dept. Tea Rsch. Inst., Sri Lanka, 1971-80; prof. chemistry U. Colombo, 1980-83; head quality assurance dept. Cordis Bio Synthetics, Melbourne, Australia, 1984-85; regulatory affairs mgr. Bio Nova Neo Technics, Melbourne, 1985-87; gen. mgr. Bio Nova Internat., Melbourne, 1987—, also bd. dirs.; cons. in field. Patentee in field; contbr. articles to profl. jours. Mem. Royal Australian Chem. Inst., Regulatory Affairs Profls. Soc. Home: 40 Kerrie Road, Glen Waverley 3150, Australia Office: Bio Nova Internat, 36 Munster Terr, North Melbourne 3051, Australia

ROBERTS, HAROLD ROSS, medical educator, hematologist; b. Four Oaks, N.C., Jan. 4, 1930; s. Walter Lee and Matilda Alicia (Daughtry) R.; m. Marilyn Claassen; children—Eric Michael, John Claassen. B.S., U. N.C., 1952, M.D., 1955. Research assoc. U. N.C., Chapel Hill, 1961-62, instr. medicine, 1962-64, asst. prof., 1964-67, assoc. prof., 1967-70, prof., 1970—, chief hematology, 1968-77, dir. Hemophilia Treatment Ctr., 1977-80, dir. Ctr. for Thrombosis and Hemostasis, 1978—, co-chief hematology and oncology, 1979-81, chief hematology, 1981—; vis. prof. U. Aarhus, Denmark, 1973-74; dir. clin. coagulation lab. N.C. Meml. Hosp., Chapel Hill, 1977—. Assoc. editor Thrombosis & Hemostasis, 1975-81; editor Hemostasis, 1975-83; mem. editorial bd. Blood, 1976-82, assoc. editor, 1983—; contbr. articles to profl. jours. Chmn. Orange County Bd. Adjustment, N.C. Recipient Disting. Career award Temple U. Health Sci. Ctr., Stockholm, 1983. Fellow ACP; mem. Assn. Am. Physicians, Am. Soc. for Clin. Investigation, AMA, Internat. Soc. on Thrombosis and Hemostasis. Anglican. Avocations: arborist; ornithology; philosophy. Home: 2502 Jones Ferry Rd Chapel Hill NC 27516 Office: U NC Chapel Hill Sch Medicine Ctr Thrombosis & Hemostasis Campus Box 7015 Chapel Hill NC 27514*

ROBERTS, HOWARD RICHARD, food scientist, association administrator; b. Eldred, Pa., July 6, 1932; s. Edward Euclid and Irene Victoria (Bills) R.; m. Marylyn Ann Morrissey, Dec. 28, 1957; children: Cynthia Anne, Mark Edward, Mary Beth, John Michael. BS, George Washington U., 1955, MS, 1957, PhD, 1962. Cert. quality engr., D.C. Instr. George Washington U., Washington, 1958-59; ops. analyst Johns Hopkins U., Bethesda, Md., 1958-59; rsch. dir. Booz, Allen & Hamilton, Washington, 1959-67; v.p. Booz, Allen & Hamilton, Kansas City, Mo., 1967-72; dir. bur. foods FDA, Washington, 1972-78; dir. sci. affairs Nat. Soft Drink Assn., Washington, 1978-85, sr. v.p., 1986—; cons. George Washington U. Med. Sch., Washington, 1958-59; adv. panel AMA, Chgo., 1974-80; food expert panel FDA, Washington, 1990-91. Author: Food Safety, 1981; co-author: Human Consumption of Caffeine, 1984, Agricultural & Food Chemistry, 1978; contbr. articles to Food Tech. Jour., Food Drug Cosmetic Law Jour. Coach, official Vienna (Va.) Youth Inc., 1971-74. With USNR, 1950-54. George Washington U. fellow, 1957; recipient FDA award of Merit, 1978. Fellow Am. Soc. Quality Control (chmn. 1968-70, Svc. award 1971); mem. AAAS, Am. Coll. Toxicology (coun. mem. 1985-87), Inst. Food Technologists (profl. mem.), Omicron Delta Kappa, Sigma Xi. Republican. Achievements include rsch. in risk assessment, consumption estimation. Office: Nat Soft Drink Assn 1101 16th St NW Washington DC 20036-4803

ROBERTS, HYMAN JACOB, internist, researcher, author, publisher; b. Boston, May 29, 1924; s. Benjamin and Eva (Sherman) R.; m. Carol Antonia Klein, Aug. 9, 1953; children: David, Jonathan, Mark, Stephen, Scott, Pamela. M.D. cum laude, Tufts U., 1947. Diplomate Am. Bd. Internal Medicine. Intern, resident Boston City Hosp., 1947-49; resident Mcpl. Hosp., Washington, 1949-50; research fellow, instr. med. Tufts Med. Sch., Boston, 1948-49, Georgetown Med. Sch., Washington, 1949-50; fellow in medicine Lahey Clinic, Boston, 1950-51; mem. active staff Good Samaritan and St. Mary's Hosps., West Palm Beach, Fla., 1955—; dir. Palm Beach Inst. Med. Research, West Palm Beach, 1964—, pres., Sunshine Sentinel Press, Inc. lectr. two day seminar on "The New Frontiers in Legal Medicine." U.S. rep. Council of Europe for Driving Standards, 1972. Author: Difficult Diagnosis, Spanish and Italian edits, 1958; The Causes, Ecology and Prevention of Traffic Accidents, 1971, Is Vasectomy Safe?, 1979, Aspartame (NutraSweet): Is It Safe?, 1989, Sweet'ner Dearest, 1992, Is Vasectomy Worth the Risk?, 1993, Mega Vitamin E: Is It Safe 2?, 1993(play) My Wife, The Politician; assoc. editor: Tufts Med. Alumni Bull, Boston, 1978-87; contbr. sci. and med. articles to profl. and theol. jours. Pres. Jewish Community Day Sch., West Palm Beach, Fla., 1975-76; disting. mem. pres. council U. Fla., Gainesville, 1974—; founder, dir. Jewish Fedn. Palm Beach County, West Palm Beach, 1960-72. Served to lt. USNR, 1951-54. Named Fla. Outstanding Young Man Jr. C. of C. Fla., 1958; hon. Ky. col.; recipient Gold Share cert. and silver certs. Inst. Agr. and Food Scis., U. Fla., 1974-78; Paul Harris fellow Rotary Found., 1980. Fellow Am. Coll. Chest Physicians, Am. Coll. Nutrition, Stroke Council; mem. AMA, ACP, Am. Soc. Internal Medicine, Am. Acad. Neurology, Endocrine Soc., Am. Diabetes Assn., Am. Heart Assn., Am. Fedn. Clin. Research, Am. Coll. Angiology (gov. 1981), Am. Coll. Legal Medicine, Pan Am. Med. Assn. (chmn. endocrinology 1982), So. Med. Assn., N.Y. Acad. Scis., Am. Physicians Fellowship of Israel Med. Assn., Confrerie de la Chaine des Rotisseurs, Alpha Omega Alpha, Sigma Xi. Club: Governors of West Palm Beach (a founder), Executive (founder). Lodges: Rotary; B'nai B'rith, Order St. George (knight of magistral grace 1992). Research in med. diagnosis, diabetes, hypoglycemia, postvasectomy state, Vitamin E metabolism, pentachlorophenol, heavy metal toxicity, narcolepsy, traffic accidents, thrombophlebitis, aspartame, Alzheimer's disease, brain tumors, nutrition and bioethics. Home: 6708 Pamela Ln West Palm Beach FL 33405-4175 Office: Palm Beach Inst Med Rsch 300 27th St West Palm Beach FL 33407-5299 also: Sunshine Sentinel Press Inc PO Box 8697 West Palm Beach FL 33407

ROBERTS, JOHN D., chemist, educator; b. Los Angeles, June 8, 1918; s. Allen Andrew and Flora (Dombrowski) R.; m. Edith Mary Johnson, July 11, 1942; children: Anne Christine, Donald William, John Paul, Allen Walter. A.B., UCLA, 1941, Ph.D., 1944; Dr. rer. nat. h.c., U. Munich, 1962; D.Sc. (hon.), Temple U., 1964, Notre Dame U., 1993, U. Wales, 1993. Instr. chemistry U. Calif. at Los Angeles, 1944-45; NRC fellow chemistry Harvard, 1945-46, instr. chemistry, 1946; instr. chemistry Mass. Inst. Tech., 1946, asst. prof., 1947-50, asso. prof., 1950-52; vis. prof. Ohio State U., 1952, Stanford U., 1973-74; prof. organic chemistry Calif. Inst. Tech., 1953-72, Inst. prof. chemistry, 1972-88, Inst. prof. chemistry emeritus, 1988—, dean of faculty, v.p., provost, 1980-83, lectr., 1988—, chmn. div.

chemistry and chem. engring., 1963-68, acting chmn., 1972-73; Foster lectr. U. Buffalo, 1956; Mack Meml. lectr. Ohio State U., 1957; Falk-Plaut lectr. Columbia U., 1957; Reynaud Found. lectr. Mich. State U., 1958; Bachmann Meml. lectr. U. Mich., 1958; vis. prof. Harvard, 1958-59, M. Tishler lectr., 1965; Reilly lectr. Notre Dame U., 1960; Am.-Swiss Found. lectr., 1960; O.M. Smith lectr. Okla. State U., 1962; M.S. Kharasch Meml. lectr. U. Chgo., 1962; K. Folkers lectr. U. Ill., 1962; Phillips lectr. Haverford Coll., 1963; vis. prof. U. Munich, 1962; Sloan lectr. U. Alaska, 1967; Disting. vis. prof. U. Iowa, 1967; Sprague lectr. U. Wis., 1967; Kilpatrick lectr. Ill. Inst. Tech., 1969; Pacific Northwest lectr., 1969; E.F. Smith lectr. U. Pa., 1970; vis. prof. chemistry Stanford U., 1973-74; S.C. Lind lectr. U. Tenn.; Arapahoe lectr. U. Colo., 1976; Mary E. Kapp lectr. Va. Commonwealth U., 1976; R.T. Major lectr. U. Conn., 1977; Nebr. lectr. Am. Chem. Soc., 1977; Leermakers lectr. Wesleyan U., 1980; Iddles Meml. lectr. U. N.H., 1981; Arapahoe lectr. Colo. State U., 1981; Winstein lectr. UCLA, 1981; Gilman lectr. Iowa State U., 1982; Marvel lectr. U. Ill., 1982; vis. lectr. Inst. Photog. Chemistry, Beijing, People's Republic of China, 1983, King lectr. Kans. State U., 1984, Lanzhou U., People's Republic of China, 1985, Davis lectr. U. New Orleans, 1986, Du Pont lectr. Harvey Mudd Coll., 1987, 3M vis. lectr. St. Olaf Coll., 1987, Swift lectr. Calif. Inst. Tech., 1987, Berliner lectr. Bryn Mawr Coll., 1988; Friend E. Clark lectr. W. Va. U., 1990; George H. Büchi lectr. MIT, 1991; Henry Kuivila lectr. SUNY Albany, 1991, Fuson lect. U. Nev., 1992; dir., cons. editor W.A. Benjamin, Inc., 1961-67; cons. E.I. du Pont Co., 1950—; mem. adv. panel chemistry NSF, 1958-60, chmn., 1959-60, chmn. divisional com. math., phys. engring. scis., 1962-64, mem. math. and phys. sci. div. com., 1964-66; chemistry adv. panel Air Force Office Sci. Research, 1959-61; chmn. chemistry sect. Nat. Acad. Scis., 1968-71; chmn. Nat. Acad. Scis. (Class I), 1976-78, councillor, 1980-83, nominating com., 1992; dir. Organic Syntheses, Inc. Author: Basic Organic Chemistry, Part I, 1955, Nuclear Magnetic Resonance, 1958, Spin-Spin Splitting in High-Resolution Nuclear Magnetic Resonance Spectra, 1961, Molecular Orbital Calculations, 1961, (with M.C. Caserio) Basic Principles of Organic Chemistry, 1964, 2d edit., 1977, Modern Organic Chemistry, 1967, (with R. Stewart and M.C. Caserio) Organic Chemistry-Methane To Macromolecules, 1971; (autobiography) At The Right Place at the Right Time, 1990; cons. editor: McGraw-Hill Series in Advanced Chemistry, 1957-60; editor in chief Organic Syntheses, vol. 41; mem. editorial bd. Spectroscopy, Organic Magnetic Resonance in Chemistry, Asymmetry, Tetrahedron Computer Methodology. Trustee L.S.B. Leakey Found., 1983-92; bd. dirs. Huntington Med. Rsch. Insts., Organic Syntheses Inc., Coleman Chamber Music Assn.; mem. Calif. Competitive Tech. adv. com., 1989—. Recipient Alumni Profl. Achievement award UCLA, 1967; Guggenheim fellow, 1952-53, 55-56; recipient Am. Chem. Soc. award pure chemistry, 1954; Harrison Howe award, 1957, Roger Adams award in organic chemistry, 967, Alumni Achievement award UCLA, 1967, Nichols medal, 1972, Tolman medal, 1975, Michelson-Morley award, 1976, Norris award, 1978, Pauling award, 1980, Theodore Wm. Richards medal, 1982, Willard Gibbs Gold medal, 1983, Golden Plate award Am. Acad. Achievement, 1983, Priestley medal, 1987, Madison Marshall award, 1989, (with W. v.E. Doering) Robert A. Welch award, 1990, Nat. Medal Sci. NSF, 1990, Glenn T. Seaborg medal, 1991, Award in nuclear magnetic resource, 1991, Svc. to Chemistry award, 1991; named hon. alumnus Calif. Inst. Tech., 1990, SURF dedicatee, 1992. Mem. NAS (com. sci. and engring. pub. policy 1983-87), AAAS (councillor 1992—), Am. Chem. Soc. (chmn. organic chemistry div. 1956-57, exec. com. organic div. 1953-57), Am. Philos. Soc. (coun. 1983-86), Am. Acad. Arts and Scis., Am. Assn. Adv. Sci. (councillor 1992—), Sigma Xi, Phi Lambda Upsilon, Alpha Chi Sigma. Office: Calif Inst Tech Div of Chem 164-30 CR Pasadena CA 91125

ROBERTS, LEIGH MILTON, psychiatrist; b. Jacksonville, Ill., June 9, 1925; s. Victor Harold and Ruby Harriet (Kelsey) R.; m. Marilyn Edith Kadow, Sept. 6, 1946; children: David, Carol Roberts Mayer, Paul, Nancy Mills. B.S., U. Ill., 1945, M.D., 1947. Diplomate: Am. Bd. Psychiatry and Neurology. Intern St. Francis Hosp., Peoria, Ill., 1947-48; gen. practice medicine Macomb, Ill., 1948-50; resident in psychiatry U. Wis. Hosps., Madison, 1953-56; staff psychiatrist Mendota (Wis.) State Hosp., 1956-58; mem. faculty U. Wis. Med. Sch., Madison, 1959-89; prof. psychiatry U. Wis. Med. Sch., 1971-89, acting chmn. dept., 1972-75; mem. spl. rev. bd. Wis. Parole Bd. Sex Crimes Law, 1962-88, forensic cons., 1988—; mem. Dane County Devel. Disabilities Bd., 1962-66, Wis. Planning Com. Mental Health, 1963-65, Wis. Planning Com. Health, 1969-71, Wis. Planning Com. Vocat. Rehab., 1966-68, Wis. Planning Com. Health Centers, 1967-71, Wis. Mental Health Adv. Com., 1973-78; bd. dirs. Methodist Hosp., Madison, Dane County Rehab. House, Dane County Assn. Mental Health; cons. in field. Editor: Community Psychiatry, 1966, Comprehensive Mental Health, 1968; contbr. articles profl. jours. Pres. Wis. Council Chs., 1976-78; bd. dirs. Madison Campus Ministry, St. Benedict Center; trustee N.Central Coll., Naperville, Ill. Served with USNR, 1943-45, 50-53. Decorated Bronze Star, Purple Heart. Fellow Am. Psychiat. Assn. (bd. trustees 1981-84), Wis. Psychiat. Assn. (pres. 1967). Methodist. Home and Office: 722A Sauk Ridge Trl # Tr Madison WI 53705-1157

ROBERTS, LIONA RUSSELL, JR., electronics engineer, executive; b. Sheffield, Ala., Apr. 9, 1928; s. Liona Russell Sr. and Julia Phillipia (Harrison) R.; m. Norma Jean Walker, Mar. 15, 1952 (div. 1972); children: Laura Lee, Boyd Harrison, John King, Jenna Lynne; m. Carole Jeanne Hedges, 1973. BS in Physics, U. Miss., 1958; MS in Electronics, Navy Postgrad. Sch., Monterey, Calif., 1961; PhD in Mech. Engring., Cath. U. Am., 1977. Cert. amateur radio oper. Commd. ensign USN, 1945, advanced through grades to capt., 1967, ret., 1970; chief scientist Interstate Electronics Corp., Anaheim, Calif., 1970-83; v.p. Enigmatics, Inc., LaHabra, Calif., 1983—. Author: Signal Processing Techniques, 1977; patentee in field, 1987. Mem. IEEE (sr.), Rsch. Soc. Am., Sigma Xi. Achievements include patent on detection of gas in drilling fluids. Home: 1885 Kashlan Rd La Habra CA 90631-8423

ROBERTS, LORIN WATSON, botanist, educator; b. Clarksdale, Mo., June 28, 1923, s. Lorin Cornelius and Irene (Watson) R.; m. Florence Ruth Greathouse, July 10, 1967; children: Michael Hamlin, Daniel Hamlin, Margaret Susan. B.A., U. Mo.-Columbia, 1948, M.A., 1950, Ph.D. in Botany, 1952. Asst. prof., then assoc. prof. botany Agnes Scott Coll., Decatur, Ga., 1952-57; vis. asst. prof. Emory U., 1952-55; mem. faculty U. Idaho, 1957—, prof. botany, 1967-91, prof. botany emeritus, 1991—; Fulbright research prof. Kyoto (Japan) U., 1967-68; research fellow U. Bari, Italy, 1968; Cabot fellow Harvard, 1974; Fulbright teaching fellow North-Eastern Hill U., Shillong, Meghalaya, India, 1977; Fulbright sr. scholar and fellow Australian Nat. U., Canberra, 1980; sr. researcher U. London, 1984; pres. botany sect. 1st Internat. Congress Histochemistry and Cytochemistry, Paris, 1960; Alexander von Humboldt vis. fellow Australian Nat. U., 1992. Author: Cytodifferentiation in Plants, 1976, (with J.H. Dodds) Experiments in Plant Tissue Culture, 1982, 2d edit., 1985, (with P.B. Gahan and R. Aloni) Vascular Differentiation and Plant Growth Regulators, 1988; contbr. articles to profl. jours. Served with USAAF, 1943-46. Alexander van Humboldt fellow, 1992; Decorated chevalier de l'Ordre du Merite Agricole France, 1961. Fellow AAAS; mem. N.W. Sci. Assn. (pres. 1970-71), Bot. Soc. Am., Am. Soc. Plant Physiologists, Internat. Assn. Plant Tissue Culture, Internat. Soc. Plant Morphologists, Am. Inst. Biol. Scis., Idaho Acad. Scis., Sigma Xi, Phi Kappa Phi, Phi Sigma. Home: 920 Mabelle Ave Moscow ID 83843-3834

ROBERTS, LOUIS DOUGLAS, physics educator, researcher; b. Charleston, S.C., Jan. 27, 1918; s. Louis Wigfall and Evelyn (Douglas) R.; m. Marjorie Violette Staveley-Lawson, Aug. 29, 1942; 1 child, Joyce Carol. AB with honors, Howard Coll., 1938; postgrad., John Hopkins U., 1938-39; PhD, Columbia U., 1941. Rockefeller Found. fellow Cornell U., Ithaca, N.Y., 1941-42; rsch. physicist GE, Schenectady, N.Y., 1942-46, U. Calif. at Berkeley, 1944-45; prin. physicist Oak Ridge (Tenn.) Nat. Lab., 1946-68; Ford Found. prof. U. Tenn., Knoxville, 1963-68; prof. physics U. N.C., Chapel Hill, 1968—, Alumni Disting. prof., 1980—. Contbr. articles on physics to profl. jours.; holder numerous patents in semiconductor devices, magnetron design, nuclear power, metals and alloys, etc. Recipient Tanner teaching award U. N.C., 1977; Fulbright fellow Oxford U., 1958-59, Guggenheim Found. fellow, 1958-59. Fellow Am. Phys. Soc. (mem. Southeastern sect., 1948—, vice chmn. 1954-55, chmn. 1955-56). Republican. Avocations: reading, music, travel, garden, photography. Home: 1116 Sourwood Cir Chapel Hill NC 27514-4912 Office: Univ NC Dept Physics and Astronomy Chapel Hill NC 27599-3255

ROBERTS, MALCOLM JOHN, steel company executive; b. Widnes, U.K., Nov. 8, 1942; s. Thomas and Cicely (Prescott) R.; m. Margaret Irene Turner, Dec. 5, 1964; children: Kathryn, Christopher, Jillian. B Engring., U. Liverpool, Eng., 1963, PhD, 1966. Research assoc. Cornell U., Ithaca, N.Y., 1966-67; research engr. Bethlehem (Pa.) Steel Corp., 1967-71, supr., 1971-79, sect. mgr., 1979-82, div. mgr., 1982-84, assoc. dir. research, 1984-85, dir. research, 1985—. Mem. AIME, Am. Soc. Metals, Am. Iron and Steel Inst., Indsl. Research Inst. (rep.). Office: Bethlehem Steel Corp 8th & Eaton Ave Bethlehem PA 18016*

ROBERTS, MARIE DYER, computer systems specialist; b. Statesboro, Ga., Feb. 19, 1943; d. Byron and Martha (Evans) Dyer; BS, U. Ga., 1966; student Am. U., 1972; cert. systems profl., cert. in data processing; m. Hugh V. Roberts, Jr., Oct. 6, 1973. Mathematician, computer specialist U.S. Naval Oceanographic Office, Washington, 1966-73; systems analyst, programmer Sperry Microwave Electronics, Clearwater, Fla., 1973-75; data processing mgr., asst. bus. mgr. Trenam, Simmons, Kemker et al, Tampa, Fla., 1975-77; mathematician, computer specialist U.S. Army C.E., Savannah, Ga., 1977-81, 83-85, Frankfurt, W. Ger., 1981-83; ops. rsch. analyst U.S. Army Contrn. Rsch. Lab., Champaign, Ill., 1985-87; data base adminstr., computer systems programmer, chief info. integration and implementation div. U.S. Army Corps of Engrs., South Pacific div., San Francisco, 1987—; instr. computer scis. City Coll. of Chgo. in Franfurt, 1982-83. Recipient Sustained Superior Performance award Dept. Army, 1983. Mem. Am. Soc. Hist. Preservation, Data Processing Mgmt. Assn., Assn. of Inst. for Cert. Computer Profls., Assn. Women in Computing, Assn. Women in Sci., NAFE, Am. Film Inst., U. Ga. Alumni Assn., Sigma Kappa, Soc. Am. Mil. Engrs. Author: Harris Computer Users Manual, 1983.

ROBERTS, MARY LOU, school psychologist; b. Green Bay, Wis., Sept. 28, 1950; d. Elmer David and Leona Theodora (Puyleart) DeGrand. BA in Elem. Edn. and English, U. Wis., Oshkosh, 1972, student, 1977; MS in Counseling Psychology, U. Cen. Tex., 1989; student, U. Hawaii, 1988. Lic. sch. counselor, Tex.; lic. practicing counselor, Tex.; lic. marriage and family therapist, Tex.; cert. tchr., Tex., Wis., Hawaii. Kindergarten tchr. Howard-Suamico ISD, Green Bay, 1972-78; elem. tchr. St. Anthony's Sch., Kalihi, Hawaii, 1978-80, Ave. E. Copperas Cove (Tex.) ISD, 1983-85; elem. tchr. Killeen (Tex.) ISD, 1985-89, sch. psychologist, 1989—; parent teaching cons. Chpt. I Killeen ISD, 1989—, tchr. inservice training, 1989—, child psychologist, 1989—, cons. family and community counselor, 1989—, community speaker, 1988—. Author: (handbook) Counseling Sessions for Small Groups, 1990; editor, reviewer: Let's Learn More About Responsibility, 1991. Vol. Families In Crisis Inc., Killeen, 1987-89; presenter Family Fair Killeen Ind. Sch. Dist., 1989; mentor for 2 at risk Children in Killeen. Mem. AACD, Am. Bus. Women of Am. (scholarship 1985, ways and means com. chmn. Globe chpt. 1986), Tex. Fedn. Tchrs., U. Cen. Tex. Alumni Assn., Tex. Assn. Counseling Devel., Tex. Sch. Counselor Assn., Mid-Tex. Assn. for Counseling Devl. (sec. 1990—). Democrat. Roman Catholic. Avocations: reading, presenting workshops, travel, counseling children. Office: Chpt 1 12th and Rancier Killeen TX 76543

ROBERTS, MELVILLE PARKER, JR., neurosurgeon, educator; b. Phila., Oct. 15, 1931; s. Melville Parker and Marguerite Louise (Reimann) R.; m. Sigrid Marianne Magnusson, Mar. 27, 1954; children: Melville Parker III, Julia Pell, Erik Emerson. B.S., Washington and Lee U., 1953; M.D. (James Hudson Brown research fellow), Yale U., 1957. Diplomate: Am. Bd. Neurol. Surgery. Intern Yale Med. Ctr., 1957, neurosurg. resident, 1958-60, 62-64, Am. Cancer Soc. fellow in neurosurgery, 1962-64, instr., 1964; asst. prof. surgery Sch. Medicine U. Va., Charlottesville, 1965-69; practice medicine specializing in neurol. surgery Hartford, Conn., 1970—; mem. sr. staff Hartford Hosp., John Dempsey Hosp.; asst. prof. surgery Sch. Medicine U. Conn., Farmington, 1970-71; assoc. prof. U. Conn., 1972-75, assoc. prof. neurology, 1974-77, chmn. div. neurosurgery, 1971-84, prof. surgery, 1975—, acting chmn. dept. neurology, 1973-77, acting chmn. dept. surgery, 1974-77, William Beecher Scoville prof. neurosurgery, 1976—. Author: Atlas of the Human Brain in Section, 1970, 2d edition, 1987; mem. editorial bd.: Conn. Medicine, 1973—; contbr. articles to profl. jours. Served as capt. M.C. U.S. Army, 1960-61. Fellow ACS; mem. AAUP, Am. Assn. Neurol. Surgeons, Soc. Neurol. Surgeons, Congress Neuol. Surgeons (bd. dirs. joint spinal sect. with Am. Assn. Neurol. Surgeons, chmn. ann. meeting 1987, sci. program chmn. ann. meeting 1988), Assn. for Rsch. in Nervous and Mental Diseases, Am. Assn. Anatomists, New Eng. Neurosurg. Soc. (bd. dirs. 1976-79, pres. 1989-91), Soc. Brit. Neurol. Surgeons, Royal Soc. Medicine (London), Rsch. Soc. Neurol. Surgeons, Soc. Rsch. into Hydrocephalus and Spina Bifida, Vereingung Schweizer-Neurochirugen, Mory's Assn., Graduates Club (New Haven), Sloane Club (London), Naval Club (London), Farmington Country Club, Sigma Xi. Episcopalian. Office: 85 Seymour St Hartford CT 06106-5501

ROBERTS, PETER CHRISTOPHER TUDOR, engineering executive; b. Georgetown, Demerara, Brit. Guiana, Oct. 12, 1945; came to U.S., 1979; s. Albert Edward and Dorothy Jean (Innis) R.; m. Julia Elizabeth Warner, Nov. 10, 1984; children: Kirsta Anne, Serena Amanda, Angelee Julia, Zephanie Elizabeth, Fiona Ann, Emrys Tudor, Peter Christopher Tudor Roberts II. BSc with honors, Southampton (Eng.) U., 1969, PhD in Microelectronics, 1975. Rsch. fellow dept. electronics Southampton U., 1974-77; prof. microcircuit dept. electronics INAOE, Tonantzintla, Mexico, 1977-79; staff scientist Honeywell Systems & Rsch. Ctr., Mpls., 1979-84; dir. advanced tech. Q-Dot Inc. R&D, Colorado Springs, Colo., 1984-86; program mgr. Honeywell Opto-Electronics, Richardson, Tex., 1986; vis. prof. U. N.Mex. CHTM, Albuquerque, 1987; supr. engring. Loral Inc. (formerly Honeywell), Lexington, Mass., 1988-90; mgr. engring. Litton Systems Inc., Tempe, Ariz., 1990—, cons. engr. Q-Dot, Inc. R&D, Colorado Springs, 1982—, pvt. stockholder, 1984—. Author: (with P.C.T. Roberts) Charge-Coupled Devices and Their Applications, 1980; contbr. articles to Boletin del INAOE, IEEE Transactions on Electron Devices, Procs. of the IEEE (U.K.), Procs. of the INTERNEPCON, Internat. Jour. Electronics, IEEE Electron Device Letters, Electronics Letters, Solid State and Electron Devices, IEEE Jour. Solid State Circuits, others. Republican. Christian. Achievements include patent for VHSIC bipolar ROM and RAM circuits; patents pending for GaAs High Speed Interface Circuit; for Fixed Algorithm Image Processing CCD Focal Plane; for Cryogenic A-to-D Converter using SCCDs; for GaAs 2 GHz by 16-Bit Digital Active Backplate, others. Home: 1017 W Peninsula Dr Gilbert AZ 85234-8903 Office: Litton Systems 1215 S 52d St Tempe AZ 85281

ROBERTS, RANDOLPH WILSON, health science educator; b. Scranton, Pa., Oct. 8, 1946; s. S. Tracy and Alecia Francis (Sullivan) R.; m. Martha Jeanne Burnite, July 12, 1969 (div. Dec. 1985); children: Gwendolyn Suzanne, Ryan Weylin; m. Ava Elaine Brown, June 17, 1989. AB in Biology, Franklin & Marshall Coll., 1968, MA in Geoscis., 1974; MS in Sci. Teaching, Am. U., 1977; MS in Counseling, Western Md. Coll., 1990; postgrad., U. Md., Towson State U., Union Inst., Cin., Johns Hopkins U. Cert. tchr., counselor, tax cons., health educator. Tchr. sci. Woodlawn Jr. High Sch., Balt., 1968-73, Deer Park Jr. High/Mid. Sch., Randallstown, Md., 1973-87, Franklin Mid. Sch., Reisterstown, Md., 1987-89; chmn. and counselor dept. health/sci. Balt. County Home & Hosp. Sch., 1989—; math. and sci. tchr. Loyola High Sch., Towson, Md., summers 1981-86, Talmudical Acad., Pikesville, Md., 1983-86, 93—; ednl. cons. Scott & Fetzer Co., Chgo., 1981-86; founder and pres. Tax Assistance, Ltd., Owings Mills, Md., 1981—; curriculum cons. Balt. County Bd. Edn., Towson, 1977, 78, 93. Author: Earth Sciences Workbook, 1979. Mem. Glyndon (Md.) Meth. Ch., 1993—, scholarship and fin. com., handbell choir mem.; treas. Boy Scouts Am. Pack 315, Reis, Md., 1986-90, Webelos Den leader, 1987-90, advancement chmn., com. mem. Troop 315, 1990—. Mem. NEA, ACA, Balt. Rd. Runners, Nature Conservancy, Chesapeake Bay Found., Phi Delta Kappa, Mu Epsilon Sigma, Chi Sigma Iota, Eta Sigma Gamma. Avocations: scouting, traveling, gardening, running, parenting. Home: 9 Indian Pony Ct Owings Mills MD 21117-1210 Office: Home and Hosp Sch 6229 Falls Rd Baltimore MD 21209-2199

ROBERTS, RICHARD, mechanical engineering educator; b. Atlantic City, N.J., Feb. 16, 1938; s. Harold and Marion (Hofman) R.; m. Rochelle S. Perelman, Oct. 2, 1960; children: Lori, Lisa, Scott. BSME, Drexel U., 1961; MSME, Lehigh U., 1962, PhD in Mech. Engring., 1964. Asst. prof. mech. engring. Lehigh U., Bethlehem, Pa., 1964-68, assoc. prof., 1968-75, prof., 1975—. Editor: Proceedings of the Thirteenth Nat. Symposium on Fracture Mechanics, 1980, ASME PVP Division's Design Handbook, Materials and Fabrication, Vol. III. Recipient W. Sparagen award Am. Welding Soc., 1972, Adams Meml. award, 1981. Home: 317 Bierys Bridge Rd Bethlehem PA 18017-1142 Office: Lehigh Univ MSE/200 W Packer Bethlehem PA 18015

ROBERTS, RICHARD JOHN, molecular biologist, consultant; b. Derby, Eng., Sept. 6, 1943; came to U.S., 1969; s. John Walter and Edna Wilhelmina (Allsop) R.; m. Elizabeth Dyson, Aug. 21, 1965 (dec.); children: Alison, Andrew; m. Jean E. Michaelis, Feb. 14, 1986; 1 child, Christopher. BS, Sheffield (Eng.) U., 1965, PhD, 1968. Research fellow Harvard U., Cambridge, Mass., 1969-70, research assoc., 1971-72; sr. staff investigator Cold Spring Harbor Lab., N.Y., 1972-87, asst. dir., 1987—; cons. New Eng. Biolabs, Beverly, Mass., 1974—; sci. adv. bd. Genex, Rockville, Md., 1977-85. Contbr. articles to profl. jours. Nobel Prize in Medicine, Nobel Foundation, 1993. John Simon Guggenheim Found. fellow, 1979. Mem. Am. Soc. Microbiology, Am. Soc. Biol. Chemists. Office: Cold Spring Harbor Lab 35 Pine Dr PO Box 100 Cold Spring Harbor NY 11724

ROBERTS, ROBERT CHADWICK, ecologist, environmental scientist, consultant; b. Yakima, Wash., Jan. 6, 1947. BA in Zoology, Humboldt State Coll., 1969; PhD in Ecology, U. Calif., Davis, 1976. Instr. U. Calif., Davis, 1971-73, 76-77; asst. prof. Western Mich. U., Kalamazoo, 1978-79; ind. cons. Eureka, Calif., 1979-80; instr. Humboldt State U., Arcata, Calif., 1982-83; dir. environ. svcs. Oscar Larson & Assocs., Eureka, 1980—; cons. Nat. Audubon Soc., Sacramento, 1984—, Calif. Native Plant Soc., Sacramento, 1987-89, U.S. Forest Svc., Eureka and Sacramento, 1988—; mem. Outer Continental Shelf Adv. Com., County of Humboldt, 1987-91, Creeks/Wetlands Adv. Com., City of Arcata, 1989-93; expert witness in 6 legal actions, 2 legis. hearings. Contbr. articles to profl. jours., chpts. to books; author, editor numerous tech. reports and environ. documents. Cons. Ballot Proposition 130 Steering Com., Sacramento, 1990. Grantee F.M. Chapman Fund., 1974; sci. trainee NSF, 1970-75. Mem. Ecol. Soc. Am., Cooper Ornithol. Soc., Wildlife Soc., Soc. Conservation Biology. Office: Oscar Larson & Assocs 317 3d St Eureka CA 95501

ROBERTS, ROBERT CLARK, sales engineer; b. Fostoria, Ohio, Dec. 26, 1946; s. F. Clark and Leatha May Roberts; m. Deborah Kay Edinger, June 4, 1972; children: Amy, Anell. BSEd, Ohio No. U., Ada, 1970; MA, Chapman Coll., Orange, Calif., 1974; MBA, West Coast U., Orange, 1976. Tchr. Grand Blanc (Mich.) City Schs., 1970-71; ops. mgr. The Optical Shops, San Bernardino, Calif., 1973-74; nat. sales mgr. Davidson Optronics, Inc., West Covina, Calif., 1974-76; eastern U. S. and European ops. mgr. R. Howard Strasbaugh, Inc., Woodbury, Conn., 1976—. Contbg. editor: Optical Mgmt., White Plains, N.Y., 1980-83; contbr. articles to Semicondr. Internat., Optical Mgmt., Photonics Spectra, Photonics Handbook, others. Served with U.S. Army, 1971-73. Mem. Optical Soc. Am., Soc. Photo-Optical Instrumentation Engrs., Electrochem. Soc. Achievements include initial development of first successful chemical-mechanical planarization of multi-layer interlevel dielectrics and multiple level metallized semiconductor wafers used in 16 Megabit to 1 Gigabit memory and logic chips. Office: RH Strasbaugh Inc 426 Main St N Woodbury CT 06798-2128

ROBERTS, THOMAS CARROL, management consultant; b. Manhattan, Kans., July 24, 1947; s. Thomas Charles and Dixie Darlene (Werner) R.; m. Karen Louise McDaniel, Dec. 21, 1968; children: Gregory Charles, Chad Daniel. BS in Nuclear Engring., Kans. State U., 1970, MS in Nuclear Engring., 1972. Registered profl. engr., Mo.; cert. mgmt. cons. Project officer U.S. Army Chem. Corps, Ft. McClellan, Ala., 1972-74; engr., adminstrv. coord. Black & Veatch Engrs.-Architects, Kansas City, Mo., 1974-80, mgr. coll. rels., 1979-83, dir. human resource devel., 1980-90; instr. U. Kans., Lawrence, 1988—; sales mgr. Kornfeld Thorp Electric Co., Kansas City, 1990-91; pres. Upward Consulting, Olathe, Kans., 1991—; mem. chmn. Inst. Mgmt. Cons., Kansas City, 1992—; TQM cons. Johnson County Community Coll., 1991—; engr. career counselor Kansas State U., Manhattan, 1983—; presenter in field. Co-author: (videotape) So You Want to be an Engineer?, 1985, (symposium) Energy Symposium: A Means of Educating Teachers, 1978. Dep. dist. commr. Boy Scouts of Am., Johnson County, Kans., 1992—; grand praetor Sigma Chi, Kans., Nebr., 1981—; mem. Order of Constantine, 1990. Named Outstanding Scoutmaster, 1989, Disting. Toastmaster, Toastmasters Internat., 1984; recipient Meritorious Sect. award Am. Nuclear Soc., 1979. Mem. Am. Soc. Engr. Edn. (chmn. SIGLTD 1987—), Mid-Am. Engr. Guidance Coun. (pres. 1984), Mo./Kans. Sect. Am. Nuclear Soc., Heart of Am. Archtl. Engring. Legis. Coun. (chmn. 1983-84), Sigma Chi. Republican. Methodist. Office: Upward Consulting 2015 Pierre Manhattan KS 66502

ROBERTS, THOMAS GEORGE, retired physicist; b. Ft. Smith, Ark., Apr. 27, 1929; s. Thomas Lawrence and Emma Lee (Stanley) R.; m. Alice Anne Harbin, Nov. 14, 1958; children: Lawrence Dewey, Regina Anne; foster child, Marcia Roberts Dale. AA, Armstrong Coll., 1953; BS, U. Ga., 1956, MS, 1957; PhD, N.C. State U., 1967. Research physicist U.S. Army Missile Command, Huntsville, Ala., 1958-85; cons. industry and govt. agys., 1970—. Contbr. articles to profl. jours. Patentee in field. Served to sgt. USAF, 1948-52. Fellow Am. Optical Soc.; mem. Am. Phys. Soc., IEEE, Huntsville Optical Soc. Am. (pres. 1980, 92). Episcopalian. Club: Toastmaster Internat. (pres. 1963). Current work: Laser physics, optics, particle beams and instrumentation; diagnostic devices and techniques development. Subspecialties: Laser physics; Plasma physics. Office: Technoco PO Box 4723 Huntsville AL 35815

ROBERTS, WALTER ARTHUR, computer systems scientist; b. Detroit, July 5, 1955; s. Walter Wesley and Gwendolyn (Gillespie) R.; m. Sandra Diane Felton, Sept. 3, 1988 (dec. June 1990). BS in Chem./Cell Biology, U. Mich., 1977; postgrad., Wayne State U., 1992—. Mgr. computer systems U. Mich., Ann Arbor, 1977-88, KMS Fusion, Inc., Ann Arbor, 1988-92; dir. tech. svcs. Strategic Innovations Corp., Dexter, Mich., 1992—. Contbr. articles to profl. jours. Treas. U. Luth. Chapel Endowment Fund, Ann Arbor, 1986-89; asst. med. vol. Haiti Luth. Missionary Soc., Port-du-Prince, 1991; vol. Sci. in Action High Sch. Program, Oakland County, Mich., 1992; founder SDF Roberts Meml. Nursing Endowed Scholarship, Valparriso U. Mem. AAAS, Am. Assn. of Physicists in Medicine. Lutheran. Achievements include pioneered integrated medical computer networks in clinical services; developed novel computer software in enhanced quality improvement in plant asset management. Home: 7780 4th St Dexter MI 48130

ROBERTS, WARREN HOYLE, JR., chemist; b. Charlotte, N.C., Nov. 16, 1955; s. Warren Hoyle and Adelaide (Weaver) R.; m. Robin Lynn Mauney, Dec. 20, 1981; 1 child, Laura Elizabeth. BS in Chemistry, Belmont (N.C.) Abbey, 1978; MS in Chemistry, E. Tenn. State U., 1982. Lab. mgr. United Merchants & Mfg., Langley, S.C., 1982-85, Am. Hoechst, Mt. Holly, N.C., 1985-87, Hoechst Celanese Corp., Mt. Holly, N.C., 1987—. Author: Lactone Ring Opening Kinetics, 1982. Tenn. Eastman Corp. summer intern rsch. grantee, 1981. Mem. Am. Chem. Soc., Assn. of Ofcl. Analytical Chemists. Home: 2611 Wilson Dr Dallas NC 28034-9411 Office: Hoechst Celanese E Catawba Ave Mount Holly NC 28120-2152

ROBERTS, WILBUR EUGENE, dental educator, research scientist; b. Lubbock, Tex., Nov. 16, 1942; s. Wilbur Eugene Roberts and Elva Etna (Chance) Turnwall; m. Cheryl Ann Jones, June 6, 1967; children: Jeffery Alan, Carrie Jean. DDS, Creighton U., 1967; PhD in Anatomy, U. Utah, 1969; cert. in orthodontics, U. Conn., 1974. Diplomate Am. Bd. Orthodontics. Rsch. fellow U. Utah, Salt Lake City, 1967-69; postdoctoral fellow U. Conn., Farmington, 1971-74; from asst. prof. to prof. dentistry U. Pacific, San Francisco, 1974-88; prof. chmn. dept. orthodontics Ind. U., Indpls., 1988-93, acting chmn. dept. of oral facial devel., 1993—, prof. physiology and biophysics Sch. Medicine, 1988—; mem. steering com. Biomechs. and Biomaterials Rsch. Ctr. Ind. U.-Purdue U., Indpls., Calif., 1980—; NRC sr. rsch. assoc. NASA Ames Rsch. Ctr., Moffett Field, Calif., 1982-83; dir. Bone Rsch. Lab., U. Pacific, 1980-88, Oral Devel. Clinic, 1980-86; rsch. cons. Neodontics Corp., Laguna Nigel, Calif., 1982-85, Denar Corp., Anaheim, Calif., 1985-87, Nobelpharma AG, Goteborg, Sweden, 1988, Dental Implant Clin. Rsch. Group, Ann Arbor, Mich., 1991, Oral Medicine and Biology Study sect. NIH, 1992, Rsch. Coun., ADA, 1992; adj. prof. mech. engring. Purdue U., Indpls., 1990—; assoc. prof. implantology and maxillofacial reconstructive surgery U. Lille, France, 1987—. Contbr. sci. articles to profl. jours. Rep. campaign worker, Contra Costa County, Calif., 1980-82; ch. sch. supt. San Ramon Valley Meth. Ch., Alamo, Calif., 1979-81; adult ministries council San Ramon Valley Meth. Ch., Danville, Calif., 1984-86; sci. cons. St. Isadore Sch. and San Ramon Valley High Sch., Danville, 1978-86; chmn. bldg. com. Sunrise at Geist United Meth. Ch., Indpls. Served to lt. comdr. USN, 1969-71, Vietnam. Recipient Cosmos Achievement award NASA, 1981, 88, 92; medal recipient City of Paris, 1989, City of Rouen, France, 1991; Dr. George Grieve Meml. lectr. Can. Dental Assn., 1993. Fellow Internat. Coll. Dentists, Am. Coll. Dentists; mem. Med. Dental Guild Calif. (pres. 1982-83, Gold Key award 1985), Am. Assn. Dental Rsch., Pacific Dental Rsch. Found. (pres. 1978-80), Omicron Kappa Upsilon, CCGVN (pres. Indpls. chpt. 1992—). Avocations: fishing, hunting, backpacking. Home: 8260 Skipjack Dr Indianapolis IN 46236-8429 Office: Ind U Sch Dentistry Orthodontics Dept 1121 W Michigan St Indianapolis IN 46202-5186

ROBERTS, WILLIAM JAMES CYNFAB, physician; b. Liverpool, Lancashire, Eng., May 31, 1938; s. Cynfab and Ellen James (Pritchard) R.; m. Beryl Ann Davies, July 18, 1962; children: Robert Sion Cynfab, William Owain Marc Cynfab. BA in Nat. Scis., Cambridge (Eng.) U., 1960, BChir, 1963, MA, MB, 1964. Diplomate Royal Coll. Obstetricians and Gynecologists. Intern St. Marys Hosp. Med. Sch., Paddington; resident Royal Hosp., Sheffield Yorks; mng. ptnr. Roberts, Roberts, Jones and Penry, Abgrystwyth Dyfed, Wales, 1967—; chmn. com. edn. Gen. Med. Practitioners Wales; mem. Coun. for Postgrad. Edn. of Drs. and Dentists, Wales; hon. med. adviser Astra Pharm. Ltd.; vis. prof. King Saud U., Riyhad, Saudi Arabia, summer 1992. Contbr. articles to profl. jours. Mem. Court of Nat. Libr. Wales, 1984—, Ct. and Coun. U. Coll. Wales, 1985—; trustee Pantyfedn. Charitable Found. Fellow Royal Coll. Gen. Practitioners; mem. Brits. Med. Assn. (chmn. West Wales div.), U.K. Council Royal Coll. Gen. Practitioners; vis. fellow King Saud Univ. Methodist. Avocations: reading history, hill walking, canoeing, playing bridge and chess. Office: Roberts Roberts Jones, and Penry, 26 N Parade, Aberystwyth Dyfe SY23 2NF, Wales

ROBERTS, 2AMES LEWIS, medical sciences educator; b. Lima, Peru, Oct. 23, 1951; U.S. citizen; s. David and Mary (Fuller) R.; m. Mariann Blum, Mar. 7, 1985. BS, Colo. State U., 1973; PhD, U. Oreg., 1977. Fellow U. Calif., San Francisco, 1977-79; asst. prof. Columbia U., N.Y.C., 1979-86, assoc. prof., 1986; dir., prof. Mt. Sinai Sch. Medicine, N.Y.C., 1986-90, prof., 1990—; cons. Calif. Biotech., Mountain View, Calif., 1986-88, NIH, Bethesda, Md., 1979—. Recipient Golden Lamport award, Excellence Basic Sci.; NIH rsch. grantee, 1979—, NSF rsch. grantee, 1981-84, Mellon Found. rsch. grantee, 1980-84. Mem. AAAS, Am. Chem. Soc., Soc. for Neurosci., Endocrine Soc., Internat. Endocrine Soc., N.Y. Acad. Scis., Am. Soc. Biochemists and Molecular Biologists. Achievements include research in biosynthesis and regulation of the adrenocorticotropin-endorphin precursor, recombinant DNA cloning of pituitary and brain adrenocorticotropin-endorphin, glucocorticoid and thyroid hormone regulation of gene expression, gene structure. Office: Mt Sinai Sch Medicine One Gustave Levy Pl New York NY 10029*

ROBERTSON, ABEL L., JR., pathologist; b. St. Andrews, Argentina, July 21, 1926; came to U.S., 1952, naturalized, 1957; s. Abel Alfred Lazzarini and Margaret Theresa (Anderson) R.; m. Irene Kirmayr Mauch, Dec. 26, 1958; children: Margaret Anne, Abel Martin, Andrew Duncan, Malcolm Alexander. BS, Coll. D.F. Sarmiento, Buenos Aires, Argentina, 1946; MD suma cum laude, U. Buenos Aires, 1951; PhD, Cornell U., 1959. Fellow tissue culture div. Inst. Histoloty and Embryology, Sch. Medicine Inst. Histology and Embryology, 1947-49; surg. intern Hosp. Ramos Mejia, Buenos Aires, 1948-50; fellow in tissue culture research Ministry of Health, Buenos Aires, 1950-51; resident Hosp. Nacional de Clinicas, Buenos Aires, 1950-51; head blood vessel bank and organ transplants Research Ctr. Ministry of Health, Buenos Aires, 1951-53; fellow dept. surgery and pathology Sch. Medicine Cornell U., N.Y.C., 1953-55; asst. vis. surgery U. Hosp. N.Y., N.Y.C., 1955-60; asst. prof. research surgery Postgrad. Med. Sch. NYU, N.Y.C., 1955-56; asst. vis. surgeon Bellevue Hosp., N.Y.C., 1955-60; assoc. prof. research surgery NYU, 1956-60, assoc. prof. pathology Sch. Medicine and Postgrad Med. Sch., 1960-63; staff mem. div. research Cleve. Clinic Found., 1963-73, prof. research, 1972-73; assoc. clin. prof. pathology Case Western Res. U. Sch. Medicine, Cleve., 1968-72, prof. pathology, 1973-82, dir. interdisciplinary cardiovascular research, 1975-82; exec. head dept. pathology Coll. Medicine, U. Ill., Chgo., 1982-88; prof. pathology Coll. Medicine, U. Ill., 1982-93, prof. emeritus pathology, 1993—; research fellow N.Y. Soc. Cardiovascular Surgery, 1957-58; mem. research study subcom. of heart com. N.E. Ohio Regional Med. Program, 1969—. Mem. internat. editorial bd.: Atherosclerosis, Jour. Exptl. and Molecular Pathology, 1964—, Lab. Investigation, 1989—, Acta Pathologica Japonica, 1991—; contbr. articles to profl. jours. Recipient Research Devel. award NIH, 1961-63. Fellow Am. Coll. Cardiology, Am. Coll. Clin. Pharmacology, Am. Heart Assn. (established investigator 1956-61, nominating com. council on arteriosclerosis 1972), Royal Microscopical Soc., Royal Soc. Promotion Health (Gt. Britain), Am. Geriatrics Soc., N.Y. Acad. Scis., Cleve. Med. Library Assn.; mem. AMA, AAAS, AAUP, Am. Soc. for Investigative Pathology, Am. Inst. Biol. Scis., Am. Judicature Soc., Am. Soc. Cell Biology, Am. Soc. Pathologists, Am. Soc. Nephrology, Assn. Am. Physicians and Surgeons, Assn. Computing Machinery, Electron Microscopy Soc. Am., Assn. Pathology Chmn., Internat. Acad. Pathology, Soc. Cardiovascular Pathology, Internat. Cardiovascular Soc., Internat. Soc. Cardiology (sci. council on arteriosclerosis and ischemic heart disease), Internat. Fed. on Genetic Engring. and Biotechnology, Internat. Soc. for Heart Rsch., Internat. Soc. Nephrology, Internat. Soc. Stereology, Pan Am. Med. Assn. (life, councillor in angiology 1966), Ill. Registry Anatomical Pathology (treas. 1985-87), Chgo. Pathology Soc., Reticuloendothelial Soc. Leucocyte Biology, Soc. Cryobiology, Tissue Culture Assn., Ohio Soc. Pathologists, Electron Microscopy Soc. Northeastern Ohio (pres., trustee 1966-68), Heart Assn. Northeastern Ohio, N.Y. Soc. Cardiovascular Surgery, N.Y. Soc. Electron Microscopists, Cuyahoga County Med. Soc., Cleve. Soc. Pathologists, The Oxygen Soc., Sigma Xi. Home: 947 S Adams St Hinsdale IL 60521-4315

ROBERTSON, CLIFFORD HOUSTON, physician; b. Owensboro, Ky., Aug. 29, 1912; s. Oscar Clifford and Golda Belle (Whitaker) R.; m. Eleanor Louise Harden, Oct. 15, 1932; children: James Clifford, Fred Houston, Caroline, Bruce, William Alan. DO, KCOM, 1935. Owner Robertson Clinic, Owensboro, 1935-80; founder Cliff Robertson Found., Owensboro, 1980, Acad. Health Profls., Whitesville, Ky., 1982; v.p. Nat. Food Assocs., Atlanta, Tex., 1981-82. Author: The Health Explosion, 1982, The Time is Ripe!. 1st lt. U.S. Army, 1927-29. Baptist. Avocations: organic farming, natural food stores. Home and office: 3766 Herbert Rd Whitesville KY 42378-9417

ROBERTSON, DAVID, clinical pharmacologist, physician, educator; b. Sylvia, Tenn., May 23, 1947; s. David Herlie and Lucille Luther (Bowen) R.; m. Rose Marie Stevens, Oct. 30, 1976; 1 child, Rose. B.A., Vanderbilt U., 1969, M.D., 1973. Diplomate Am. Bd. Internal Medicine, Am. Bd. Clin. Pharmacology. Intern, Johns Hopkins U., Balt., 1973-74, asst. resident, 1974-75, asst. chief service in medicine, 1977-78; fellow in clin. pharmacology Vanderbilt U., Nashville, 1975-77, asst. prof. medicine and pharmacology, 1978-82, assoc. prof., 1982-86, prof., 1986—, prof. neurology, 1991—; dir. clin. research ctr., 1987—; dir. Ctr. for Space Physiology and Medicine, 1989—; dir. Med. Sci. Tng. Program, 1993—; pvt. practice medicine specializing in disorders of blood pressure regulation, Nashville, 1978—; mem. staff Vanderbilt Hosp., Burroughs Wellcome scholar in clin. pharmacology, 1985-91. Author: (with B.M. Greene and G.J. Taylor) Problems in Internal Medicine, 1980, (with C.R. Smith) Manual of Clinical Pharmacology, 1981, (with Italo Biaggioni) Disorders of the Autonomic Nervous System, 1993; editor-in-chief Drug Therapy, 1991—; editorial bds. Jour. Autonomic Nervous System, Clin. Pharm. and Therapeutics, Clin. Autonomic Rsch. Recipient Research Career Devel. award NIH, 1981, Grant W. Liddle award for leadership in rsch., 1991; Adolph-Morsbach grantee Bonn, Germany, 1968; Logan Clendening fellow Reykjavik, Iceland, 1969. Fellow Am. Heart Assn. Council Hypertension and Circulation; mem. ACP (teaching and research scholar 1978-81), Am. Autonomic Soc. (pres. 1992—), Am. Acad. Neurology, Soc. Neurosci., Am. Inst. Aeronautics and Astronautics, U.S. Pharmacopeial Conv., Nat. Bd. Med. Examiners, Aerospace Med. Assn. (space station sci. and applications com.), FDA Consortium Rare Disorders, Am. Fedn. for Clin. Research, Am. Soc. Clin. Investigation, Brit. Pharmacological Soc., So. Soc. for Clin. Investigation, Am. Soc. for Clin. Pharmacology and Therapeutics, Phi Beta Kappa, Alpha Omega Alpha (hon.). Baptist. Home: 4003 Newman Pl Nashville TN 37204-4308 Office: Vanderbilt Hosp Clin Rsch Ctr 21st Ave S Nashville TN 37232-2195

ROBERTSON, ERLE SHERVINTON, virologist, molecular biologist; b. St. John's, Grenada, W.I., Oct. 31, 1962; came to U.S., 1983; s. Milton Nicholas and Merlyn Robertson. BSc, Howard U., 1987; PhD, Wayne State U., 1992. Researcher Wayne State U., Detroit, 1987-89, lab. instr., 1989-91; rsch. fellow Harvard Med. Sch., Boston, 1992—; instr. Howard Hughes Med. Inst., Wayne State U., 1991. Contbr. articles to profl. jours. Mem. AAAS, Am. Soc. Virology, Am. Soc. Microbiology (William T. Brown award 1991), Am. Soc. Genetics, N.Y. Acad. Scis., Phi Beta Kappa, Sigma Xi. Achievements include discovery of phosphorylation of Escherichia Coli translation factors on expression of T7 virus protein kinase expression in vivo; genetic analysis of Epstein Barr Virus DNA involved in transformation of B lymphocytes. Home: 114 Evans Rd Apt 6 Brighton MA 02135 Office: Harvard Med Sch 75 Francis St Boston MA 02115

ROBERTSON, IVAN DENZIL, III, chemical engineer; b. Beaumont, Tex., May 27, 1951; s. Ivan Denzil Jr. and Dolores Sharon (Mabry) R.; m. Nancy Lynn Hayes, Nov. 23, 1977 (div. Nov. 1982); 1 child, Annie Kay Robertson; m. Mary Ellen McBride, May 5, 1990. Student, Southwestern U., 1970-73; BSChemE, Lamar U., 1977. Registered profl. engr., Tex. Sr. process engr. Dow Chemical U.S.A., Freeport, Tex., 1977-86, Ausimont U.S.A., Inc., Orange, Tex., 1986-89; staff engr. Petrocon Engring., Inc., Beaumont, 1989—; vol. Beaumont Ind. Sch. Dist., 1992-93. Mem. AICE, Instrument Soc. Am. (sr.). Methodist. Achievements include optimization of hydrogen cyanide reactor; development of recycled wastewater system from Chloralkali plant back to brine treating area; renovation and installation of digital control system on caustic filter system; design and installation of first twin screw extrusion process for Halar plastic production; designed instrumentation and digital control system for ethylene furnace. Home: 195 W Circuit Dr Beaumont TX 77706-6427 Office: Petrocon Engring Inc 3105 Executive Blvd Beaumont TX 77705

ROBERTSON, JAMES MUELLER, civil engineer, educator; b. Champaign, Ill., Apr. 18, 1916; s. William Spence and Gertrude (Mueller) R.; m. Margaret Dillinger, Oct. 23, 1943; children: Bruce D., Alan S. BSCE, U. Ill., 1938; MS, U. Iowa, 1940, PhD, 1941. Asst. physicist U.S. Navy Dept., Taylor Model Basin, 1941-42; mem. engring. faculty, dir. Water Tunnel Pa. State U., 1942-54; rsch. engr. Douglas Aircraft, Santa Monica, Calif., 1944-45; prof. theoretical and applied mechanics U. Ill., Urbana, 1954-82, acting head dept., 1982, prof. emeritus, 1982—; cons. U.S. Army, various indsl. orgns., 1957-73; vis. lectr. hydraulic engring. Kans. State U., 1967; course instr. TAPPI, 1969; lectr. U. Tenn. Space Inst., 1973; vis. prof. civil engring. Colo. State U., 1974; adj. prof. mech. engring. Naval Postgrad. Sch., 1984; bd. dirs. Internat. Mgmt. and Engring. Ltd., Colo., Terabyte, Inc., Boulder, Colo. Author: Hydrodynamics in Theory and Application, 1965; contbr. numerous articles to profl. pubs. Sec., v.p., pres. Summit County Sr. Citizens, Frisco, Colo., 1983—; sec. Skyline Six Area Agy. on Aging, Frisco, 1986—; bd. dirs. Breckenridge (Colo.) Music Inst., 1984—, sec., 1989—. Fellow ASCE (life, Hilgard prize 1955), ASME (life); mem. Sigma Xi, Phi Eta Sigma, Chi Epsilon, Tau Beta Pi, Phi Kappa Phi. Achievements include co-design of hydrodynamics research water tunnel at Pennsylvania State University. Home: PO Box 1097 Silverthorne CO 80498

ROBERTSON, JERRY LEWIS, chemical engineer; b. Tulsa, Okla., Oct. 25, 1933; s. Jerry D. and Allene (Lewis) R.; m. Audrey Hansen, Oct. 12, 1956; children: Craig E., Carla J. BSChemE, Okla. State U., 1955; PhDChemE, Northwestern U., 1962. Rsch. engr. Exxon Rsch. and Engr. Co., Florham Park, N.J., 1962-69, sect. head, 1971-74, sr. engring. assoc., 1978-84; engring. advisor Exxon Rsch. and Engr. Co., Florham Park, 1984-93; start up leader Esso Italia, La Spezia, Italy, 1969-71; mgr. process systems Exxon Nuclear Co. Inc., Saratoga Springs, N.Y., 1974-78; with Energy Efficient Refinery Rsch. and Applications, 1978-93, ret., 1993; sr. cons. Baskings Ridge, N.J., 1993—; indsl. advisor Carnegie Mellon U., EDRC, Pitts., 1989-92; gen. chmn. Nat. Heat Transfer Conf., Mpls., 1991. Editorial adv. bd. Chem. Engring. Progress, N.Y., 1991—; contbr. articles to profl. jours. 2d lt. U.S. Army, 1957. Recipient citation for svc. Am. Petroleum Inst., 1993. Fellow AICE (div. chmn. 1986); mem. Am. Chem. Soc. (Editorial adv. bd. 1984-87), Am. Math. Soc. Achievements include research and implementation of energy efficiency and process systems improvements.

ROBERTSON, JOHN HARVEY, microbiologist; b. Cheyenne, Wyo., Dec. 6, 1941; s. Harvey John and Harriet (Outsen) R.; m. Kiara Blanche Lunn, Aug. 26, 1966; children: Denice Nicole, Andrea Lynn. BS, U. Wyo., 1968. Microbiologist The Upjohn Co., Kalamazoo, 1968-91, tech. cons., 1991—. Contbr. articles to profl. jours. Mem. Am. Chem. Soc., Inst. Environ. Sci. Republican. Methodist. Home: 7802 Pickering Portage MI 49002 Office: The Upjohn Co 7171 Portage Rd Kalamazoo MI 49001

ROBERTSON, KENNETH MCLEOD, mechanical engineer; b. Dundee, Angus, Scotland, Feb. 5, 1957; came to U.S., 1962; s. George Foster and Isabelle (McLeod) R.; m. Patricia Ann McGuire, Nov. 2, 1984; children: Heather Elizabeth, Kenneth McLeod II. Student, U. Colo., 1975-78. Registered comm. distbn. designer. Field engr. Henkels & McCoy, Blue Bell, Pa., 1978-83; engr. Siemons-Telplus, Somerset, N.J., 1983-88; mgr. engring. Electronic Mcht. Systems, King of Prussia, Pa., 1988-90; sr. project engr. Ray Comm., Inc., Norristown, Pa., 1990—. Mem. IEEE Computer Soc. Republican. Achievements include design and engring. of local and wide area (worldwide) data comm. system for U.S. govt. Office: Ray Comm Inc Ste 101 3 Bala Plaza E Bala Cynwyd PA 19004

ROBERTSON, LARRY WAYNE, toxicologist, educator; b. Lynchburg, Va., May 22, 1947; s. John C. and Clara J. (Dalton) R. BA in Chemistry, Stetson U., 1969; MS in Microbiology, U. Fla., 1971; MPH, U. Mich., 1972, PhD in Environ. Health Scis., 1981. Rsch. assoc. Tex. A&M U., College Station, 1981-82; project leader U. Mainz, Fed. Republic Germany, 1983-86; assoc. prof. toxicology U. Ky., Lexington, 1986—. Contbr. articles on polyhalogenated aromatic hydrocarbon toxicity to sci. publs. Grantee NATO, EPA, NIH. Mem. Soc. Toxicology (pres. Ohio Valley chpt. 1991-92), others. Office: Univ Ky Grad Ctr Toxicology 204 Funkhouser Bldg Lexington KY 40506-0054

ROBERTSON, LAURIE LUISSA, computer scientist; b. Dallas, Feb. 19, 1960; d. Jack Weldon and Gretchen Luissa (Roeschlaub) R. BA, Rice U., 1982; MLS, U. Pitts., 1983; cert. procurement and contracting, U. Va., 1993. Programmer/statistician Galveston (Tex.) Dist. Atty., 1980-82; programmer/analyst Def. Systems Inc., McLean, Va., 1983-85; computer scientist Computer Scis. Corp., Falls Church, Va., 1985—. Mem. Computer Soc. of IEEE, Assn. for Computing Machinery, Beta Phi Mu. Office: Computer Scis Corp 3160 Fairview Park Dr Falls Church VA 22042-4501

ROBERTSON, SAMUEL HARRY, III, transportation safety research engineer, educator; b. Phoenix, Oct. 2, 1934; s. Samuel Harry and Doris Byrle (Duffield) R.; m. Nancy Jean Bradford, Aug. 20, 1954; children: David Lyle, Pamela Louise. BS, Ariz. State U., 1956; D in Aviation Tech. (hon.), Embry-Riddle Aero. U., 1972. Registered profl. engr. Chief hazards div. Aviation Safety Engring. and Research, Phoenix, 1960-70; pres. Robertson Research Engrs., 1960-70; research prof., dir. Safety Ctr. Coll. Engring. and Applied Scis., Ariz State U., Tempe, 1970-79; pres. Robertson Research Inc., 1970-86, Robertson Aviation Inc., 1977-86, Internat. Ctr. for Safety Edn., 1982-86; pres., chief exec. officer Robertson Research Group, Inc., Tempe, 1986—; airplane design and accident investigator, 1961—; instr. aircrash investigation Internat. Ctr. Safety Edn., 1960—, inst. aerospace safety U. So. Calif., 1962-70, Armed Forces Inst. Pathology, 1970—, Dept. Transp. Safety Inst., 1970-89; pres. Flying R Ranches, 1976—; mem. adv. bd. Rio Salado Bank, Tempe, 1985—; mem. adv. coun. Ctr. Aerospace Safety Edn., Embry-Riddle Aero. U., Daytona Beach, Fla., 1986—, bd. trustees, 1992—. Contbr. 65 articles to profl. jours. and pubs.; patentee applying plastic to paper, fuel system safety check valves, crash resitant fuel system, safety aircraft seats; holder FAA STC's various fuel systems, fuel system components; designer, developer, mfr. crash resistant fuel systems for airplanes, helicopters, championship racing cars. Served as pilot USAF, 1956-60, Ariz. Army NG 1960-61, 70-74, Ariz. Air NG, 1961-69. Recipient Contbns. Automotive Racing Safety award CNA, 1976, Adm. Luis De Florez Internat. Flying Safety award, 1969, Cert. Commendation Nat. Safety Coun., 1969, Gen. W. Spruance award for safety edn., SAFE Soc., 1982; holder Nat. Speed Record for one class of drag racing car, 1955-62, 5 nat. records for flying model aircraft, 1950-56. Mem. AIAA, Internat. Soc. Air Safety Investigators (Jerome Lederer Aircraft Accident Investigation award, 1981), Aerospace Med. Assn., Exptl. Aircraft Assn., Soc. Automotive Engrs., Am. Helicopter Soc., Nat. Fire Protection Assn., Aircraft Owners and Pilots Assn., U.S. Automobile Club (mem. tech. com.). Office: 1024 E Vista Del Cerro Dr Tempe AZ 85281-5709

ROBERTSON, WILLIAM BELL, JR., environmental scientist; b. Atlanta, Nov. 15, 1949; s. William Bell and Beatrice (Yoder) R.; m. Elizabeth Mary Frances Breslin, Oct. 20, 1978; 1 child, Nancy Ann. AA, Marion Inst., 1970; BS, U. Ala., 1987. Technician P.E. Labreaux and Assocs., Tuscaloosa, Ala., 1978-85; environ. scientist Soil and Material Engrs., Raleigh, N.C., 1985-88; sr. environ. scientist Westinghouse Environ. Svcs., Cary, N.C., 1988-92, SEC-Donohue, Cary, 1992—. With USAF, 1970-75. Mem. Groundwater Profls. N.C. Office: SEC Donohue 3500 B Regency Pky Cary NC 27512

ROBERTSON, WILLIAM OSBORNE, physician; b. N.Y.C., Nov. 24, 1925; s. William Osborne and Barbara Konvalinka (Bennett) R.; m. Barbara Foster Simpson, Feb. 23, 1952; children—Kathy, Lynn, Kerry, Douglas, Andrew. B.A., U. Rochester, 1946, M.D., 1949. Intern Strong Meml. Hosp., Rochester, N.Y., 1949-51; resident Strong Meml. Hosp., 1951-52, Grace New Haven Hosp., 1954-56; acting med. dir. Ross Labs., Columbus, Ohio, 1956-59; mem. faculty Ohio State Coll. Medicine, 1956-63, assoc. prof. pediatrics, 1961-63; mem. faculty dept. pediatrics U. Wash., Seattle, 1963—; prof. U. Wash., 1972—, assoc. dean, 1967-72, med. dir., 1963-67, acting chmn. dept. pediatrics, 1972-73, 80-84, head div. ambulatory pediatrics, 1975-77, 78-79; dir. med. edn. div. Children's Orthopedic Hosp., 1971-90, dir. poison control center, 1971—; mem. staffs Children's, Harborview, Univ. hosps.; mem. advisory com., chmn. Wash. Alaska Regional Med. Program; bd. dirs. Wash. Med. Edn. and Research Found., 1968-73. Contbr. articles to profl. publs. Served with USNR, 1952-54. Mem. Am. Acad. Pediatrics (chmn. edn. com. 1971-73, med. liability com. 1985-90, chmn. task force on quality assurance), Am. Acad. Toxicology, King County Med. Soc. (pres. 1971-72), Wash. State Med. Assn. (pres. 1975-76), Am. Assn. Poison Control Ctrs. (pres. 1988-90), Phi Beta Kappa, Alpha Omega Alpha. Home: 18724 40th Pl NE Seattle WA 98155-2806 Office: U Wash Sch Medicine PO Box 5371C Seattle WA 98195-0001

ROBIE, DONNA JEAN, chemist; b. Toledo, July 12, 1966; d. Gerald Daniel and Judith Emelia (Borowiak) R. AA, Seminole C.C., 1985; BS, U. Cen. Fla., 1989; PhD, U. Fla., 1993. Lab. mgr. Seminole C.C., Sanford, Fla., 1986-89; grad. asst. U. Fla., Gainesville, 1989—. Recipient travel award Fedn. Analytical Chemistry and Spectroscopy Socs., 1992. Mem. Am. Chem. Soc., Soc. Applied Spectroscopy, Phi Kappa Phi, Phi Theta Kappa. Democrat. Roman Catholic. Office: Kellogg Co 235 Porter St Battle Creek MI 49016

ROBINOWITZ, CAROLYN BAUER, psychiatrist, educator; b. Bklyn., July 15, 1938; d. Milton Leonard and Marcia (Wexler) Bauer; m. Max Robinowitz, June 10, 1962; children—Mark, David. A.B., Wellesley Coll., 1959; M.D., Washington U., 1964. Diplomate Am. Bd. Psychiatry and Neurology. Chief physician tng. NIMH, Bethesda, Md., 1968-70; dir. pediatric liaison U. Miami Sch. Medicine, Fla, 1972-74, dir. child psychiatry tng., 1971-72; dir. edn. George Washington U. Sch. Medicine, Washington, 1972-74; project dir. Psychiatrist as Tchr., Washington, 1973-75; dep. med. dir. Am. Psychiat. Assn., Washington, 1976-86, dir. Office Edn., 1976-87, chief operating officer, 1986—; prof. psychiatry and pediatrics George Washington U. Sch. Medicine, 1982—; dir. Am. Bd. Psychiatry and Neurology, Evanston, Ill., 1979-86, sec. 1984, v.p., 1985, pres. 1986; professorial lectr. Georgetown U. Sch. Medicine, 1982—; clin. prof. psychiatry and behavioral scis., child health and devel. George Washington U., 1984—. Editor: Women in Context, 1976; contbr. articles to jours., chpts. to books. Admissions com. Wellesley Coll. Club, Washington, 1983-84; active Boy Scouts Am. Served with USPHS, 1966-69. Recipient NIMH Mental Health Career Devel. award, 1966-70; NIMH grantee, 1974—. Fellow Am. Psychiat. Assn. (Disting. Svc. award 1991), Am. Coll. Psychiatrists (bd. dirs. 1993); mem. Group for Advancement Psychiatry (dir. 1982-84, pres.-elect 1987-89, pres. 1989), Coun. of Med. Specialty Socs. (dir. 1977-82, pres. 1981-82). Home: 7204 Helmsdale Rd Bethesda MD 20817-4624 Office: Am Psychiat Assn 1400 K St NW Washington DC 20005-2403

ROBINS, ELI, psychiatrist, biochemist, educator; b. Houston, Feb. 22, 1921; m. Lee Nelken, Feb. 22, 1946; children: Paul, James, Thomas, Nicholas. BA, Rice U., 1940; MD, Harvard U., 1943; DSc (hon.), Wash. U., St. Louis, 1984. Intern Mt. Sinai Hosp., N.Y.C., 1944; resident Mass. Gen. Hosp., Boston, 1944-45, McLean Hosp., Waverley, Mass., 1945-46, Pratt Diagnostic Hosp., Boston, 1948-49; instr. neuropsychiatry Washington U. Sch. Medicine, St. Louis, 1951-53, asst. prof. psychiatry, 1953-56, assoc. prof., 1956-58, prof., 1958-66, Wallace Renard prof., 1966-91, prof. emeritus, 1991—, head dept. psychiatry, 1963-75. Author: (with M.T. Saghir) Male and Female Homosexuality: A Comprehensive Study, 1973; editor: (with others) Ultrastructure and Metabolism of the Nervous System, 1962, (with K. Leonhard) Classification of Endogenous Psychoses; The Final Months, A Study of the Lives of 134 Persons Who Committed Suicide, 1981; contbr. to profl. jours. Fellow Am. Psychiat. Assn. (life, Disting. Svc. award 1992), Am. Coll. Neuropsychopharmacology (hon.), Royal Coll. Psychiatrists (hon.); mem. Am. Soc. Clin. Investigation, Am. Soc. Biol. Chemists, Psychiat. Rsch. Soc., Am. Psychopath. Assn. Home: 1 Forest Ridge Pl Saint Louis MO 63105-3007 Office: 4940 Audubon Ave Saint Louis MO 63110-1081

ROBINSON, ALEXANDER JACOB, clinical psychologist; b. St. John, Kans., Nov. 7, 1920; s. Oscar Frank and Lydia May (Beitler) R.; m. Elsie Louise Riggs, July 29, 1942; children: Madelyn K., Alicia A., David J., Charles A., Paul S., Marietta J., Stephen N. BA in Psychology, Ft. Hays (Kans.) State U., 1942, MS in Clin. Psychology, 1942; postgrad., U. Ill., 1942-44. Cert. psychologist, sch. psychologist. Chief psychologist Larned (Kans.) State Hosp., 1948-53, with employee selection, outpatient services, 1953-55; sch. psychologist County Schs., Modesto, Calif., 1955-61, Pratt (Kans.) Jr. Coll., 1961-66; fed. grantee, writer assoc. dir. Exemplary Federally Funded Program for Spl. Edn., Pratt, 1966-70; dir. spl. edn., researcher Stafford County Schs., St. John, 1970-81, ret., 1981; supr. testing and data Incidence of Exceptional Children in Kansas, Kans. State U., Ft. Hays, 1946; writer, asst. dir. Best Exemplary Federally Funded Program on Spl. Edn., Pratt, 1966-70; fed. grantee, researcher, writer, study dir. Edn. for the High-Performance Child, St. John, 1970—, Psychogenesis of the Sociopathic Personality, a longitudinal study. Minister, The Ch. of Jesus Christ. Served to 2d lt. U.S. Army, 1944-46, PTO. Lodge: Lions (program chmn. St. John 1974-76). Avocations: history, ethnology, cultural anthropology, music, literature. Home and Office: RR 1 Box 121A Saint John KS 67576-9801

ROBINSON, ALFRED G., petroleum chemist; b. Thomasville, Ga., Feb. 19, 1928. BA, Emory U., 1949, MS, 1951, PhD in Chemistry, 1955. Chemist Hercules Powder Co., 1951-52, Tenn. Eastman Co Divsn., 1955-58; chemist Tex. Eastman Co. Divsn., 1958-73, develop. assoc., 1973-74, head develop. divsn., 1974-77, dir. rsch. divsn., 1977-84, dir. R&D, 1984-89, v.p. R&D, 1989—; v.p. R&D Eastman Kodak Co., 1989—. Mem. Am. Chem. Soc. Achievements include research in chemical properties of aliphatic carbonyl compounds, synthesis of polymers by condensation polymerization. Office: Tex Eastman Co-Research & Devel Lab Hwy 149 PO Box 7444 Longview TX 75607*

ROBINSON, BRUCE BUTLER, physicist; b. Chester, Pa., Oct. 13, 1933; s. George Senior and Dorothy Conerly (Butler) R.; m. Dorothy Ross, June 4, 1960; children: Douglas Ross, Christopher Scott. BS in Physics, Drexel U., 1956; PhD in Physics, Princeton U., 1961; MBA, Rider U., 1977. Rsch. assoc. U. Calif., San Diego, 1961-63; rsch. scientist RCA David Sarnoff Lab., RCA, Princeton, N.J., 1963-73; exec. dir., mem. commerce tech. adv. bd. U.S. Dept. Commerce, Washington, 1973-75; dir. policy integration, dir. coal and synfuels policy U.S. Dept. Energy, Washington, 1975-81; sr. science advisor to v.p. rsch. Exxon Rsch. and Engring. Co., Linden, N.J., 1981-84;

dep. dir. Office Naval Rsch., Arlington, 1984-87, dir. rsch., 1987—; prin. author nat. energy policy plan U.S. Dept. Energy, 1981; U.S. rep. to internat. energy agy., govt. expert group on tech., Paris, 1979-81; mem. internat. team to rev. R&D programs Dutch Ministry Econs. and Fin., The Hague, The Netherlands, 1979; presenter sci. lectures. Contbr. articles to sci. jours. NSF fellow Princeton U., 1956-58, NSF internat. summer fellow, Varenna, Italy, 1962; recipient Meritorious Presdl. Rank award Pres. of U.S., 1989. Mem. IEEE, Am. Phys. Soc., The Oceanography Soc. (founding).

ROBINSON, CHRISTINE MARIE, mathematics educator; b. Savannah, Ga.; d. Aaron Sr. and Lucille (Jones) Williams; m. Amos Robinson, Aug. 2, 1953; children: Michael Anthony, Pamela Michele. BS in Math. magna cum laude, Savannah State Coll., 1951; MA, U. Mich., 1965. Instr. in math. Chatham County Bd. of Instruction, Savannah, 1951-64; instr. in math. Duval County Bd. of Instruction, Jacksonville, Fla., 1964-71, master and resource tchr., 1971-76; prof. math. Fla. C.C., Jacksonville, 1976—; mem. faculty task force Fla. Dept. Edn./Fla. Assn. Community Colls., Tallahassee, 1979-81; on-site coord. Fla. Devel. Edn. Assn. Conv., Jacksonville, 1986; chmn. Fla. Community Coll. EA/EO Com., Jacksonville, 1988, 89. Mem. YWCA, Jacksonville, 1989—; vol. driver Wheels for Cancer-AKA Sorority, Jacksonville, 1986; chmn. United Way, Jacksonville, 1987. Recipient Outstanding Faculty Mem. award Fla. Community Coll., 1987, Teaching Excellence award U. Tex., 1988; scholar U. Mich., U. Ill.; grantee NSF. Mem. Am. Math. Assn. Two-Yr. Colls., Fla. Devel. Edn. Assn. (bd. dirs. Jacksonville chpt. 1983-86), Fla. Assn. Community Colls., Math. Assn. Am., So. Assn. Colls. and Schs. (Fla. com.), LWV, Alpha Kappa Alpha. Democrat. Roman Catholic. Avocations: reading, piano, dancing, bicycling. Home: 7426 Simms Dr Jacksonville FL 32209-1023 Office: Fla Community Coll 3939 Roosevelt Blvd Jacksonville FL 32205-8945

ROBINSON, DAVID ALLEN, computer engineer; b. Buxton, Eng., Feb. 23, 1956; came to U.S., 1984; s. Allen Sawkill and Nancy (Howes) R.; m. Janet Alison Groom, Apr. 12, 1975; 1 child, Matthew. AA in Telecomm., B.C.A.T., Eng., 1976; BSc in Mgmt., Golden Gate U., 1992. Technician Brit. Telecomm. Sta. KDGMXG, Bradford, Eng., 1972-77; engr. ICC, Leeds, Eng., 1977-79; engr. Hewlett Packard, Leeds, 1979-82, Richmond, B.C., Can., 1982-84; mktg. engr., software engr. Hewlett Packard, Cupertino, Calif., 1984-88; mgr. Hewlett Packard, Roseville, Calif., 1988—; producer Channel 8 Community Access TV, Roseville, 1992.

ROBINSON, DAVID ASHLEY, family physician; b. N.Y.C., Sept. 30, 1954; s. Richard Foster Robinson and Jane Bunnell Salmon Staley; m. Mary Clair Persina, July 27, 1985; 1 child, Kerstin Anna. BA, U. Pa., 1977; MD, Karolinska Inst., Stockholm, 1987. Diplomate Am. Bd. Family Practice. Rsch. asst. Dept. Medicine, NYU, N.Y.C., 1977-79, Karolinska Inst., Stockholm, 1979-83; chief resident family practice Albany (N.Y.) Med. Ctr., 1990-91; med. dir. stuent health Siena Coll., Loudenville, N.Y., 1991-92, Albany Med. Coll., 1992—; med. dir. family practice group Albany Med. Ctr., 1992—; instr. family practice Albany Med. Coll., 1991-92, asst. prof., 1992—. Contbr. articles to profl. jours. P. Weiterviks Fund/Karolinska Inst. grantee, 1981, 82. Mem. N.Y. Acad. Scis. Office: Family Practice Group Albany Med Ctr 1 Clara Barton Dr Albany NY 12208

ROBINSON, DAVID MASON, cell physiologist; b. Barton, Eng., July 7, 1932; came to U.S., 1969, naturalized, 1979; s. Thomas Leon Mason and Mabel (Orr) R.; B.Sc. with 1st class honours, U. Durham, 1955, Ph.D. (Philip Buckle Meml. scholar), 1958; m. Jean Marcia Smith, Sept. 10, 1965; children—Jane Leonie Mason, Simon Henry Mason. Mem. sci. staff Namulonge Research Sta., Kampala, Uganda, 1959-61; research officer, tutor Hope Dept. Zoology, Oxford (Eng.) U., 1961-63; mem. sci. staff, biophysics group Med. Research Council, Radiobiol. Research Unit, Harwell, Eng., 1963-66; prin. sci. officer, head cell biology Microbiol. Research Establishment, Porton, Eng., 1966-69; asst. research dir., head cell biology ARC Blood Research Lab., Bethesda, Md., 1969-73; prof. biology, assoc. mem. Vincent Lombardi Cancer Research Ctr., Georgetown U., Washington, 1974-80, adj. prof. anatomy and cell biology, Sch. of Medicine, 1982-90; professorial lectr. in liberal studies Georgetown U., 1980—; assoc. dir. for sci. programs, div. heart and vascular disease Nat. Heart, Lung and Blood Inst., NIH, Bethesda, Md., 1993—, acting dir., 1993—; mem. faculty biology and genetics NIH Grad. Sch., 1981-86. Capt. 1st Royal Green Jackets, 43d and 52d, Brit. Ter. Army, 1962-65. Empire Cotton Growing Corp. postgrad. scholar, 1957. Recipient Vicennial medal Georgetown U., 1992. Mem. Biophys. Soc., Soc. Complex Carbohydrates, Soc. Cryobiology (sec. 1975), Am. Soc. Cell Biology, Sigma Xi (pres. Georgetown chpt. 1978), Alpha Sigma Nu (hon.). Democrat. Episcopalian. Club: Royal Green Jackets Officers. Author: (with G.A. Jamieson) Mammalian Cell Membranes, 5 vols., 1973-76; contbr. articles to profl. jours. Home: Stoneleigh Cottage PO Box 2164 Shepherdstown WV 25443-2164 Office: NIH Fed Bldg Rm 416 Bethesda MD 20892

ROBINSON, DONALD EDWARD, electrical power industry executive; b. Worcester, Mass., May 8, 1947; s. Rollins Flagg and Juliette (Dietrich) R.; m. Margaret Anne Murphy, Nov. 17, 1974; children: Michael Paul, John Vincent. BS, Worcester Poly. Inst., 1969, MBA, 1983. Indsl. engr. ITT/Surprenant, Clinton, Mass., 1969-70; consumer svcs. rep. Mass. Electric, Marlboro, Mass., 1972-79; asst. dist. mgr. Mass. Electric, Hopedale, Mass., 1979; mgr. energy conservation New Eng. Power Svc. Co., Westborough, Mass., 1980-86, coord. conservation and load mgmt., 1987—; pres., chmn. Mass Save, Waltham, Mass., 1986-91. Contbr. articles to profl. jours. and publs. Coach Shrewbury (Mass.) Youth Soccer, 1992. 1st lt. U.S. Army, 1970-71, Vietnam. Recipient Award for Energy Innovation, U.S. Dept. Energy, 1984, 85, 86, 7th Annual Edison Electric Inst. Comml./Indsl. award, 1987. Republican. Congregationalist. Achievements include development of programs In-House Energy Conservation, Energy Efficient Homes, Energy Van, Enterprise Zone, Perfomance Contracting, Energy Initiative, Design 2000 Performance Engineering, and Complementary Demand-Side Programs. Home: 24 Rawson Hill Dr Shrewsbury MA 01545 Office: New Eng Power Svc Co 25 Research Dr Westborough MA 01582

ROBINSON, EDWARD NORWOOD, JR., physician, educator; b. Winston-Salem, N.C., Aug. 20, 1953; s. Edward Norwood and Pauline (Gray) R.; m. Pamela Martin Pittman, Apr. 22, 1978; children: Patrick Edward, Alexander Wood. BA, Duke U., 1975; MD, Bowman Gray Sch., Winston-Salem, 1979. Diplomate Am. Bd. Internal Medicine. Intern East Carolina U. Sch. Medicine, Greenville, N.C., 1979-80, resident, 1980-82; postdoctoral fellow U. Utah Sch. Medicine, Salt Lake City, 1982-85, asst. prof. medicine, 1985-86; asst. prof. medicine U. Louisville, 1986-88; chief infectious diseases Louisville VA Hosp., 1987-88; clin. assoc. prof. medicine U. N.C., Greensboro, 1988—; cons. infection control High Point (N.C.) Regional Hosp., 1990—; epidemiologist Greensboro Women's Hosp., 1990—. Contbr. articles to profl. jours. Fellow ACP; mem. Am. Soc. Microbiology, Infectious Disease Soc. Am., Am. Med. Infomatics Assn., Cousteau Soc. (founding mem.), Alpha Omega Alpha. Democrat. Methodist. Office: Moses H Cone Meml Hosp 1200 N Elm St 1200 N Elm St Greensboro NC 27401-1020

ROBINSON, EMILY WORTH, computer systems analyst, mathematician; b. Englewood, N.J., Jan. 3, 1931; d. John Browning and Grace (Kellsey) Worth; m. Richard Carleton Robinson, Nov. 5, 1949; children: Barbara, Ginger, Richard, Deborah. BS, U. Md., 1970. Math. tutor Kensington, Md., 1970-83; office mgr., technician Andrulis Rsch., Inc., Bethesda, Md., 1973-76; computer programmer FTC, Washington, 1976-80; programmer, analyst U.S. Nuclear Regulatory Commn., Washington, 1980-90, sr. computer system analyst, 1990—; co-chair federated Los Alamos Vulnerability and Risk Assessment of Washington user group, 1991—. Tchr. Sunday sch., St. Paul's United Meth. Ch., Kensington, Md., 1968—; Christian counselor, Kensington, 1975—. Named Competent Toastmaster Toastmasters Internat., Bethesda, 1982. Mem. DAR. Republican. Home: 4013 Cleveland St Kensington MD 20895 Office: US Nuclear Regulatory Commn Mail Stop 3C12 Washington DC 20555

ROBINSON, ENDERS ANTHONY, geophysics educator, writer; b. Boston, Mar. 18, 1930; s. Edward Arthur and Doris Gertrude (Goodale) R.; m. Eva Arborelius, Sept. 9, 1962 (div. 1973); children: Anna, Erik Arthur, Karin. BS in Math., MIT, 1950, MS in Econs., 1952, PhD in Geophysics, 1954. Dir. geophys. analysis group MIT, Cambridge, Mass., 1952-54; geophysicist Gulf Oil Corp., Pitts., 1954-55; instr. math. MIT, Cambridge, Mass., 1955-56; petroleum economist Standard Oil Co. N.J., N.Y.C., 1956-57; asst. prof. stats. Mich. State U., East Lansing, 1958; asst. prof. math. U. Wis., Madison, 1958-61, assoc. prof. math. (with tenure), 1961-62; dep. prof. stats. Uppsala (Sweden) U., 1960-64; v.p., dir. Digicon, Inc., Houston, 1965-70; pres. Robinson Rsch. Inc., Houston, 1970-82; vis. prof. theoretical and applied mechanics Cornell U., Ithaca, N.Y., 1981-82; McMan prof. geophysics U. Tulsa, 1983—. Author 25 books on sci. and tech. including Einstein's Relativity in Metaphor and Mathematics, 1990; editor: Internat. Jour. of Imaging Systems & Tech., 1988—; assoc. editor: Jour. of Time Series Analysis, 1984—; editorial bd. Multidimensional Systems and Signal Processing, An Internat. Jour., 1990—. 2d lt. U.S. Army, 1950-51. Recipient Conrad Schlumberger award European Assn. of Exploration Geophysicists, 1969, Donald G. Fink Prize award IEEE, 1984. Mem. NAE, Nat. Rsch. Coun. (com. on undiscovered oil and gas resources), Soc. Exploration Geophysicists (hon., Best Paper award, 1964, medal 1969). Home: 100 Autumn Ln Lincoln MA 01773-2407 Office: U of Tulsa Dept of Geophysics Tulsa OK 74104

ROBINSON, GARY DALE, aerospace company executive; b. Colcord, W.Va., Sept. 9, 1938; s. Samuel Claytor and Madge (Fraley) R. Jr.; m. Lorelei Mary Christl, June 25, 1967; children: John Claytor, Kirk Dean. BA in Latin Am. Econ. History, So. Ill. U., 1964; PhD in Orgn. Behavior, Case Western Res. U., 1977. Program tng. chief The Peace Corps, Washington, 1969-71; cons. self-employed Ohio, 1971-76; health planning advisor USAID, San Salvador, El Salvador, 1976; project dir. Cen. Am. Inst. for Pub. Adminstrn. and Ministry of Health, San Jose, Costa Rica, 1977-78; mgmt. advisor Agy. for Internat. Devel., Santo Domingo, Dominican Republic, 1978-79; indsl. rels. mgr. Boeing Comml. Airplane Co., Everett, Wash., 1979-83; human resource mgr. Boeing Marine Systems, Renton, Wash., 1983-85; indsl. rels. mgr. The Boeing Co., Seattle, 1985-86, internal audit mgr., asst. to v.p. controller, 1986-90, 90—; cons. in field; adj. prof. Cen. Wash. U., Ellensburg, 1984—; mem. adv. bd. Drake, Beam & Moran, Seattle, 1991—; mem. adv. bd. and faculty Sch. of Advanced Studies in Orgnl. Cons., Santiago, chile, 1992—. Contbg. author: International Organizational Behavior, 1986. chmn. Metrocenter YMCA, Seattle, 1990-91; mem. Peace Corps Nat. Adv. Coun., Wash., 1988-89; founding mem. Pacific N.W. Orgn. Devel. Network, Seattle, 1982-86; bd. advisors Nat. Found. for Study Religion & Edn., Greensboro, N.C., 1984-87; mem. edn. com. World Affairs Coun., Seattle, 1987-88; sec., treas. The Edmonds Inst., Lynnwood, Wash., 1989-90; mem. Internat. Rels. Com. Named Paul Harris fellow Rotary Internat., 1989. Mem. AIAA, The Wash. Ctr. for Mgmt. and Leadership (founder, bd. dirs.), Inst. for Internal Auditors (co-editor Pistas newsletter 1991—), Nat. Orgnl. Devel. Network, Acad. of Mgmt., Earth Svcs. Corps (adv. bd.). Avocations: running, walking, reading. Office: The Boeing Co PO Box 3707 M/S 11-KA Seattle WA 98124-2207

ROBINSON, GLENN HUGH, soil scientist; b. Rosedale, Ind., May 20, 1912; s. William Albert and Margaret Lorene (Brown) R.; m. Anna Mary Howald, Oct. 12, 1940 (div. 1976); children: G. Phillip, Joan E., Albert L.; m. Joan Yeager, Sept. 14, 1992. BS, MS, Purdue U., 1940; PhD, U. Wis., 1950. Cert. soil correlator Am. Registry of Cert. Profl. Soil Scientist. Asst. prof. agronomy Purdue U., Lafayette, Ind., 1940-42; with Guayula Rubber Project U.S. Forest Svc., Tex., 1943-44; prof. soils, soil scientist USDA Soil Conservation Svc., Madison, Wis., 1946-51; party chief, author soil surveys and reports Barron and Richland Counties, Wis. and Avery County, N.C., 1955-58; prof., sr. soil correlator USDA Soil Conservation Svc., Blacksburg, Va., 1953-61; sr. soil scientist Food & Agr. Orgn., Tanganika, Guiana, 1961-64; project mgr. Food & Agr. Orgn., Indonesia, Thailand, Sudan, 1964-75; sr. soil scientist Saudi Arabia/U.S. State Dept., Riyadh, 1976-80; farmer, soil cons. Carbon, Ind., 1980—; sr. soil scientist for evaluation soils for use in septic tank sewage disposal systems Ind. State Dept. Health, 1991-93; soil cons. Ind. State Dept. Health, 1991-93; mem. Internat. Seminar on Shifting Cultivation, Chiang Mai, Thailand, 1970, Symposium on Cotton Growth in Sudan, WadMedania, Sudan, 1969; cons. Abaca project USDA, Costa Rica, 1951, UN, Amman, Jordan, 1983, UN FAO, Malawi, 1965. Author: A Proposed Land Capability Appraisal System for Agricultural Use in Indonesia, 1975, A Land Capability Appraisal of Indonesia, 1974; contbr. articles to Soil Sci., Geoderma, Soil Sci. Soc. Am. Procs., Jour. Soil Sci., African Soils. Pres. PTO, Oreg., Wis., 1949, Oreg. Sch. Bd., 1948-50. Mem. Soil Sci. Soc. Am., Am. Soc. Agronomy, Internat. Soc. Soil Sci., Coun. for Agrl. Sci. & Tech., Rotary, Sigma Xi. Methodist. Achievements include development of methods and procedures for soil studies in developing countries including soil fertility studies, mineralogy, geology, kinds of crops, crop rotation and map production utilizing aerial photos and satellite images, plus studies of soil characteristics and land classification for the use and management of soil areas; research on rate and degree of development of various kinds of soil and the production of food crops for use in various developing countries. Home and office: RR 1 Box 210A Carbon IN 47837-9634

ROBINSON, HAROLD WENDELL, JR., systems engineer; b. Boston, Jan. 2, 1937; s. Harold Wendell and Mazie Edith (Bissett) R.; m. Barbara Jane Denney, Aug. 23,1958 (div. June 1967); children: Jeffrey, Diane Robinson Harris; m. Joan Reeves Lightcap, June 17, 1967; children: Kristin Reeves DeNoyelles, Cindy Reeves Hagy,Scott. BAE in Aero. Engring., Rensselaer Poly. Inst., Troy, N.Y., 1958; MSME, Rensselaer Poly. Inst., East Windsor Hill, Conn., 1963. Sr. analytic engr. Pratt & Whitney Aircraft, East Hartford, Conn., 1960-65; mgr. range instrumentation The Aerospace Corp., San Bernadino, Calif., 1965-72; program mgr., bus. mgr., mgr. mktg. Honeywell Electro Optics Div., Lexington, Mass., 1972-90; div. dir. Nichols Rsch. Corp., Wakefield, Mass., 1990—. Comdr. U.S. Power Squadron, Acton, Mass., 1986. Lt. USN, 1958-60. Mem. AIAA, Air Force Assn., Assn. of Old Crows, Am. Def. Preparedness Assn., AFCEA. Office: Nichols Rsch Corp 251 Edgewater Dr Wakefield MA 01880

ROBINSON, HURLEY, surgeon; b. L.A., Feb. 25, 1925; s. Edgar Ray and Nina Madge (Hurley) R.; m. Mary Anne Rusche, Mar. 14, 1953; children: Kathleen Ann Robinson Petschke, Mary Elizabeth, Lynda Jean Robinson Lamb, William Hurley, Patricia Kay Robinson Hardy, Paul Edgar. Student, U. Calif., Berkeley, 1943, U. Calif., Santa Barbara, 1946-48; BS, Northwestern U., 1950, MD, 1952. Diplomate Am. Bd. Surgery. Intern Wesley Meml. Hosp., Chgo., 1952-53; resident Milw. County Hosp., 1953-57; surgeon Abbott Med. Group, Ontario, Calif., 1957-59; pvt. practice Upland, Calif., 1959-64; ptnr. Robinson & Schechter Surg. Med. Group, Upland, 1964-92; sr. surg. staff San Antonio Community Hosp., Upland, 1958—, trustee, 1979-81, pres. med. staff, 1980; mem. staff Pomona (Calif.) Valley Med. Ctr., Dr.'s Hosp. Montclair, Calif., Ontatio Community Hosp.; exec. com. San Bernardino (Calif.) County Med. Ctr., 1974, adv. bd., 1974; clin. asst. vascular surgery London Hosp., Eng., 1973; cons. in field. Co-contbr. articles to Wis. Med. Jour. Chmn. troop com., camp dr. Boy Scouts Am., Upland, 1970-72. With U.S. Army, 1943-46. Fellow ACS, Am. Coll. Chest Physicians, Am. Coll. Angiology; mem. AMA, Am. Med. Soc. Vienna, Calif. Med. Assn., N.Y. Acad. Scis., San Bernardino County Med. Soc., Tri-County Surg. Soc. So. Calif. (pres.), Pan-Pacific Surgical Assn., Soc. Clin. Vascular Surgery, Royal Soc. Medicine. Republican. Presbyterian. Office: 415 W 16th St Upland CA 91786

ROBINSON, JAMES LEROY, architect, educator; b. Longview, Tex., July 12, 1940; s. Willie LeRoy and Ruby Nell R.; B.Arch., So. U., 1964; M.C.P. (Martin Luther King fellow, 1972), Pratt Inst., 1972; m. Annabell Hilton; children: James LeRoy II, Kerstin Gunilla, Maria Theresa Narvaez, Jasmin Marisol, Ruby Nell, Kenneth Arne. Architect, Port of N.Y. Authority, 1964; architect, store planner W.T. Grant, 1964; with Herbst & Rusciano, AIA, 1965; architect Carson, Lundin & Shaw, N.Y.C., 1966, Kennerly, Slomanson & Smith, N.Y.C., 1967-69, architect-on-bus., 1969; pres. Robinson Architects, P.C., N.Y.C., 1969—; pres. NAROB Devel. Corp.; vis. prof. CUNY; adj. prof. Pratt Inst. Bd. dirs. Boys Club Am. Served with U.S. Army, 1966. Decorated knight Order of St. John, Knights of Malta; recipient AIA design award, 1976. Mem. Am. Arbitration Assn. (arbitrator), N.Y. Council Black Architects. Democrat. Works include: Stuyvesant Heights Christian Church, David Chavis House, Fulton Ct. Houses, Sinclair Houses, Hamilton Heights Terr., Eliot Graham Houses, Sojourner Truth Houses, Nehemiah Plan, Casas Theresa, N.Y.C. Postal Data Ctr., Mt. Carmel Bapt. Ch., Consol. Edison Collection Center, Casas Theresa, Jasmin Houses, CityHomes CD&E. Home: 67 Murray St New York NY 10007-2126 Office: 5 Beekman St New York NY 10038

ROBINSON, JEFFERY HERBERT, modular building company executive; b. Atlanta, Nov. 30, 1956; s. Herbert W. and Annie Hue (Maxey) R.; children: Angela Marie, David Clifton, John Wiliam; m. Cynthia Moss Geeslin, Sept. 28, 1991; adopted children: William Damon Geeslin, Taylor Lauren Geeslin. AA, DeKalb Coll., Clarkston, Ga., 1979; BBA in Mktg., Ga. State U., 1980. Data comm. specialist GE, Atlanta, 1980-81; account exec. Burlington No. Air Freight, Atlanta, 1981-82; br. mgr. Spacemaster Internat., Charlotte, N.C., 1982-84; regional sales mgr. Profit Freight Systems, Atlanta, 1984-87; v.p. sales Eden Air Freight, Inc., Atlanta, 1987-93; pres. Modular Design & Bldg., Inc., Atlanta, 1993—. With U.S. Army, 1974-77. Lewis Gordon Meml. scholar Sales and Mktg. Execs. Atlanta, 1980. Baptist. Office: Modular Design & Bldg Inc 2584 Chestnut Dr N Atlanta GA 30360

ROBINSON, JOHN ABBOTT, mechanical engineer; b. Bridgeton, N.J., Sept. 14, 1939; s. Norman Trickett and Rose (McGowan) R.; m. Elizabeth Ann Jarman, Apr. 3, 1965; children: John A. II, Carol A. BS in Indsl. Engring., Lafayette Coll., 1961. Mech. engr. U.S. Army Combat Systems Test Activity, Aberdeen Proving Ground, Md., 1964-65, mech. engr., applied mechanics, 1965-76, sr. mech. engr., applied mechanics, 1976-78, supervisory mech. engr., 1978—; mem. adv. bd. Nat. Shock and Vibration Tech. Adv. Group, Springfield, Va., 1986—; mem. adv. bd., cons. U.S. Army Test and Evaluation Command, Internat. Standardization Group, Aberdeen Proving Ground, 1982—; mem. U.S. Army Test and Evaluation Command, Shock and Vibration Tech. Com., 1978—; tech. agt. for Shock and Vibration Field for Command, 1991—; mem. adv. bd. Mil. Standard 810 D/F Test Method Com., Dayton, Ohio, 1985—. Contbr. articles to profl. jours. Mem. Daretown (N.J.) Vol. Fire Co., 1969—, chief, 1982-84; mem., officer Elmer Grange and N.J. State Grange, 1953—; Sunday sch. supt. Pittsgrove Bapt. Ch., Daretown, 1968-70. 1st lt. U.S. Army, 1962-64. Mem. Inst. Environ. Scis. (asst. tech. program co-chair 1990, tech. program co-chair 1991). Achievements include development of major vibration test facility at U.S. Army Combat Systems Test Activity; derived the shock and vibration environments on all types of military ground logistic and combat vehicles; standardization of vibration test techniques for all proving grounds of U.S. Army Test and Evaluation Command. Home: RR 3 Box 354 Elmer NJ 08318-9720

ROBINSON, LAWRENCE WISWALL, project engineer; b. Keene, N.H., Sept. 16, 1948; s. Frank Taytor and Cherolyn Cora (Wyman) R.; m. JoAnn Ruth Davis, Oct. 7, 1972; children: Emily Ruth, Martha Alice. BS in Indsl. Engr., Keene State Coll., 1972. Tchr. Conant High Sch., Jaffrey, N.H., 1973-78; assembly tech. Kingsbury Machine Tool Co., Keene, 1978-81; quality assurance mgr. Pneumo Precision, Keene, 1981-86; plant mgr. Optical Filter Corp., Keene, 1986-90; project engr. Teleflex Corp., Jaffrey, 1990—. Chmn. Supr. of Checklist, Marlborough, N.H., 1982—. Republican. Mem. United Ch. of Christ. Achievements include development of combine twisted PTFE teflon computer tubing (multi-color) twisted and cured twisting machine. Home: 16 Laurel St Marlborough NH 03455 Office: TFX Med Tall Pines Park Jaffrey NH 03452

ROBINSON, MARK LOUIS, metallurgical engineer; b. Phila., May 24, 1950; s. E. Warren and Vera A. (Gebhardt) R.; m. Noreen M. Jennings, Mar. 1, 1975; children: James K., Kevin D. BS, Drexel U., 1972, MS, 1974, PhD, 1977. Registered profl. engr., Pa. Engr. R & D Westinghouse R & D Labs., Pitts., 1976-77; sr. metallurgist Inco Alloy Products R & D, Suffern, N.Y., 1977-84; supr. R & D Carpenter Tech. Corp., Reading, Pa., 1984-88; dir. R & D SPS Techs., Jenkintown, Pa., 1988—; adv. bd. NSF Steel Rsch. Ctr., Golden, Colo., 1985-88, Ben Franklin Tech. Ctr., Phila., 1988—. Asst. tech. editor ASME jour., 1987-93; contbr. 10 articles to profl. jours. Mem. ASTM, AIME, Am. Materials Soc. Internat., Metallurgical Soc., Indsl. Rsch. Inst. (alt. rep.). Achievements include patents in field. Office: SPS Techs Highland Ave Jenkintown PA 19046

ROBINSON, MELVYN ROLAND, urologist; b. Wolverhampton, Staffs, Eng., May 17, 1933; s. Noah and Hilda Emily (Griffiths) R.; m. Rosalind Robinson, Nov. 5, 1972; children: Emma Claire, Matthew James. MB, BS, Charing Cross Hosp., London, 1957. House officer Charing Cross Hosp., London, 1958-59; med. officer No. Nigerian Govt., 1960-62; orthopaedic registrar Charing Cross Hosp., London, 1963; surgical registrar Mount Vernon Hosp., London, 1964-67, Balham, London, 1968-69; sr. registrar Inst. Urology, London, 1969-72; urologist Pontefract, West Yorkshire, 1972—; hon. sr. clin. lectr. Inst. of Urology, London, 1986—. Contbr. articles to profl. jours. Fellow Royal Coll. Surgeons (Edinburgh and London), Royal Soc. Medicine; mem. Brit. Assn. of Urol. Surgeons, Rotary (pres. Hemsworth Club 1983-84). Mem. Conservative Party. Anglican. Avocations: walking, gardening, tennis. Home: 25 Went Hill Close, Pontefract WF7 7LP, England Office: Pontefract Gen Infirmary, Friars Wood, Pontefract WF8 1PL, England

ROBINSON, ROBERT ALAN, physicist; b. Chelmsford, Essex, Eng., Mar. 28, 1955; came to U.S., 1982; s. Leonard and Eva Jean (Webber) R.; m. Trude Schwarzacher, Dec. 27, 1984 (div. 1989). MA, Cambridge (Eng.) U., 1976, PhD, 1982. Tchr. Alsager (Eng.) Comprehensive Sch., 1977-79; A.S.O. AERE, Harwell, Eng., summer 1979; postdoctoral fellow Los Alamos (N.Mex.) Nat. Lab., 1982-84, mem. staff, 1985—; vis. scientist Institut Laue Langevin, Grenoble, France, winter 1984, KEK Nat. Lab. for Energy Physics, Tsukuba, Japan, spring 1992. Contbr. articles to profl. publs.; editor conf. procs. in field. Mem. Am. Phys. Soc. Office: Los Alamos Nat Lab Mail Stop H805 Los Alamos NM 87545

ROBINSON, RUDYARD LIVINGSTONE, economist, financial analyst; b. St. Elizabeth, Jamaica, June 6, 1951; s. Vivian and Elvira (Green) R.; m. Yvonne Evadney Nairne, Mar. 12, 1983; 1 child, Reneé. BA, McCormick Coll., Chgo., 1974; MPhil, U. Chgo., 1976, M in Social Sci., 1977, PhD, 1984. Prof. Govs. State U., Park Forest, Ill., 1978-79; lectr. Coll. Arts, Sci. and Tech., Kingston, Jamaica, 1981-82, U. W.I., Kingston, 1984, econ. advisor U.S. AID, Kingston, 1984-86; dir. for econ. rsch. and planning WTO, UN Devel. Program, Bridgetown, Barbados, 1986-88; chief economist Govt. of Cayman Islands, George Town, 1988—. Contbr. articles to profl. jours. Fellow McCormick Coll., 1973-74, Inter-Am. Found., 1981-83. Fellow Fin. Analysts Fedn.; mem. Am. Econ. Assn., Am. Fin. Assn., Assn. Investment Mgmt. and Rsch., Internat. Soc. Fin. Analysts, Soc. Quantitative Analysts. Presbyterian. Avocations: fishing, cycling, travel.

ROBINSON, SUSAN ESTES, pharmacology educator; b. Radford, Va., Apr. 26, 1950; d. Cecil Bennett and Helen Elizabeth (Buckner) Estes; m. Robert McMurdo Gillespie, Nov. 7, 1981; children: William Bennett, James Campbell. BA in Chemistry, Vanderbilt U., 1972, PhD in Pharmacology, 1976. Staff fellow Lab. Preclin. Pharmacology St. Elizabeth's Hosp., NIMH, Washington, 1976-79; asst. prof. dept. med. pharmacology and toxicology Coll. Medicine, Tex. A&M U., College Station, 1979-81; asst. prof. dept. pharmacology and toxicology Med. Coll. Va., Va. Commonwealth U., Richmond, 1981-87, assoc. prof., 1987-93; assoc. prof. dept. pathology Va. Commonwealth U., Richmond, 1991—; mem. study sect. Internat. and Coop. Projects Div. Rsch. Grants, NIH, Bethesda, Md., 1990-93. Contbr. articles to profl. jours. Recipient Lyndon Baines Johnson Rsch. award Am. Heart Assn., Tex. affiliate, 1980; rsch. grantee Nat. Inst. Drug Abuse, 1989, 90, NIH, 1989. Mem. Am. Soc. Neurochemistry, Am. Soc. Pharmacology and Exptl. Therapeutics (exec. com. div. for neuropharmacology 1992—), Soc. Neurosci. (councillor Cen. Va. chpt. 1989-90, pres. 1991-92), Neurotrauma Soc., Internat. Soc. Psychoneuroendocrinology. Office: Med Coll Va Va Commonwealth U Dept Pharmacology Toxicology Box 613 MCV Sta Richmond VA 23298-0613

ROBISON, CLARENCE, JR., surgeon; b. Tecumseh, Okla., Dec. 9, 1924; s. Clarence Sr. and Margaret Irene (Buzzard) R.; m. Patricia Antoinette Hagee, May 27, 1951; children: Timothy D., Paul D., John D., Rebecca A. AS, Stanford U., 1943; MD, U. Okla., 1948. Intern Good Samaritan Hosp., Portland, Oreg., 1948-49; fellow pathology and oncology U. Okla., 1949-51; pathologist USAF Hosp., Cheyenne, Wyo., 1951-53; resident in surgery Okla. U. Health Scis.-Va. Svc., Oklahoma City, 1953-56; mem. faculty surgery dept. Okla. U. Health Scis., Oklahoma City, 1956-57; clin. prof. surgery Okla. U. Health Scis-Va. Svc., Oklahoma City, 1957—; mem. bd. advisors Mercy Health Ctr., Oklahoma City, 1974-81, sec. of staff, 1974-84, chief surgery, 1992—; bd. dirs. Okla. Found. for Peer Rev., Oklahoma City.

Active Commn. on Mission Indian Nations Presbytery, 1980-91; bd. dirs. Found. Sr. Citizens, 1964—; elder Presbyn. Ch. Capt. USAF, 1951-53. Fellow ACS, Am. Cancer Soc. (past pres. Okla. divsn., exec. com., bd. dirs., nat. del. dir.); mem. AMA (del. hosp. med. staff Oklahoma City chpt. 1989—, Okla. alt. del. 1991-93), SAR, Oklahoma County State Med. Soc. (bd. dirs. Oklahoma County chpt. 1989-93), Okla. State Med. Assn. (alt. trustee Okla. 1989-92, trustee 1993—), Okla. Surg. Assn. (sec., treas. 1966-68), Oklahoma City Surg. Soc. (pres. 1967-69), Beacon, Oak Tree, Men's Dinner Club, Masons, Kiwanis, Shriners (presdl. elector 1960). Democrat. Presbyterian. Office: 4200 W Memorial Rd Oklahoma City OK 73120-8305

ROBISON, NORMAN GLENN, tropical research director; b. Littlefield, Tex., Oct. 5, 1938; s. J. Clifton and Opal Lee (McCain) R.; m. Mary Ruth Correl, Aug. 10, 1963; children: Molly Anne, Susan Wray, Andrew Glenn. BS, Tex. Tech. U., 1961; PhD, U. Nebr., 1967. Plant breeder DeKalb Plant Genetics, Hastings, Nebr., 1965-85; rsch. dir. DeKalb (Ill.) Plant Genetics, 1985—. Mem. Crop Sci. Soc. Am. Republican. Episcopalian. Achievements include development of commercial grain sorghum hybrids adapted to northern U.S., Mexico and France. Home: 110 Thornbrook DeKalb IL 60115 Office: De Kalb Plant Genetics 3100 Sycamore Rd De Kalb IL 60115

ROBISON, PETER DONALD, biochemist; b. Dallas, Dec. 4, 1950; s. Donald Elon and Margaret Ann (Waters) R.; m. Mary Ellen Stafford, Aug. 27, 1972; children: Megan, Brett, Caitlin. AB in Chemistry, Cornell U., 1972; PhD in Biochemistry, Syracuse U., 1978. Rsch. assoc. microbiology U. Tex., Austin, 1978-81; biochemist R&D Dept., Texaco, Inc., Beacon, N.Y., 1981-90, group leader analytical, 1990—. Recipient Gourevitch Grad. Student award Syracuse U., 1978; Nat. Inst. of Gen. Med. Scis. postdoctoral fellow, 1978-81. Mem. ASTM (com. on petroleum products and lubricants), Am. Chem. Soc. Office: Texaco Inc PO Box 509 Beacon NY 12508

ROBISON, WILBUR GERALD, JR., research biologist; b. Cheyenne, Wyo., Dec. 27, 1933; s. Wilbur Gerald and Irene (Decker) R.; m. Lucia Maria Panuncio, Sept. 20, 1957; children: Sylvia Lee, Stanley Jay, Nancy Kay, Lydia Joy. BA, Brigham Young U., 1958, MA, 1960; PhD, U. Calif., Berkeley, 1965; postgrad., Harvard U., 1966. Postdoctoral rsch. fellow Harvard Med. Sch., Boston, 1965-66; asst. prof. biology U. Va., Charlottesville, Va., 1966-72; sr. staff fellow Nat. Eye Inst., NIH, Bethesda, Md., 1972-76, geneticist, cell biologist, 1976-83, head exptl. anatomy, 1983-85, head pathophysiology, 1985—, acting head pathology, 1988—. Contbr. articles to profl. publs., chpts. to books. Mem. AAAS, Am. Assn. Cell Biology, Assn. for Rsch. in Vision and Ophtalmology, Sigma Xi. Mem. LDS Ch. Achievements include development of rat model for diabetic retinopathy and demonstration of prevention with aldose reductase inhibitors. Home: 1306 Gresham Rd Silver Spring MD 20904 Office: Nat Eye Inst NIH Bldg 6 Rm 316 Bethesda MD 20892

ROBLEK, BRANKO, science educator; b. Slovenia, Yugoslavia, Jan. 9, 1934; s. Viktor and Marija (Kern) R. Diploma in Physics, Univ. FNT, Ljubljana, Slovenia, 1959, diploma in math., 1962, prof. math., 1964. Tchr. primary sch., Zg. Gorje, Slovenia, 1960-62, secondary sch., Skofja Loka, 1962-69, 73-81; insp., cons. Inst. Edn., Ljubljana, 1969-73; tchr. math., physics and computer sci., Edn. Ctr., Skofja Loka, 1981-92; cons. Soc. Computer Sci., Ljubljana, 1973-81; mem. physics edn. faculty Inst. Edn., Ljubljana, 1981—. Mem. The Planetary Soc., Pasadena. Author: (with others) AAAIII Zbirka vaj, 1969; Racunalnistvo ZN, 1980. Chmn. Syndicate of Civilizing Workers, Skofja Loka, 1966. Fellow Soc. Mathematicians; mem. Physicists and Astronomers of R Slovenia. Home: Partizanska 46, 64220 Škofja Loka PO 15, Slovenia

ROBOHM, RICHARD ARTHUR, microbiologist, researcher; b. Jackson, Wyo., Feb. 29, 1936; s. Arthur John and Florence Elizabeth (Johnson) R.; m. Dorothy Ann Henry, June 6, 1958 (div. 1975); children: Erick, Kurt, Kim; m. Peggy Adler Walsh, Dec. 24, 1976 (div. 1993). BS in Bacteriology, Idaho State Coll., 1958; PhD in Microbiology and Immunology, U. Mich., 1970. Lab. instr. microbiology dept. U. Mich., Ann Arbor, 1962-63, rsch. asst. microbiology dept., 1964; microbiologist tech. lab. Nat. Marine Fisheries Svc., Ann Arbor, 1965-71; rsch. microbiologist Exptl. Biol. br. NOAA, Nat. Marine Fisheries Svc., Milford, Conn., 1971-89, chief microbiology investigation, 1989—; thesis advisor L.I. U., Greenvale, N.Y., 1973-74. Author: (with others) Proceedings First U.S.-Japan Conference on Toxic Microorganisms, 1970, Comparative Pathobiology, vol. 6, 1984; mem. editorial bd. Bull. of Environ. Contamination and Toxicology, 1991—; contbr. articles to jours. Applied Microbiology, Applied & Environ. Microbiology, Vet. Immunology and Immunopathology. Chmn. com. Children With Social and Learning Difficulties, Ann Arbor, 1969-71; pres. Parent and Tchrs. in Unity, Hamden, Conn., 1974; judge Conn. Sci. Fair Assn., Hartford, Conn., 1986—; v.p. SARAH Seneca, Guilford, Conn., 1992-93. 1st lt. U.S. Army, 1958-62. Recipient USPHS traineeship, 1962-63; USDA grantee, 1992. Mem. AAAS, Am. Soc. Microbiology, Internat. Soc. Devel. and Comparative Immunology. Achievements include research on effects of environmental stress on immune mechanisms in fish and mollusks. Office: NOAA Nat Marine Fisheries Svc-NE Fisheries Sci Ctr 212 Rogers Ave Milford CT 06460

ROBSON, ANTHONY EMERSON, plasma physicist; b. London, Mar. 29, 1932. BA, Oxford U., 1952, MA, DPhil, 1956. Scientific officer U.K. Atomic Energy Authority, Harwell, 1956-65, Culham, 1965-66; rsch. scientist U. Tex., Austin, 1966-72; head experimental plasma physics br. Naval Rsch. Lab., Washington, 1972-89, sr. scientist, 1989—. Fellow Am. Phys. Soc. Home: 2683 Centennial Ct Alexandria VA 22311

ROBSON, GEOFFREY ROBERT, geologist, seismologist, consultant; b. Stockton on Tees, Eng., Jan. 24, 1929; s. Robert and Mary (Darbyshire) R.; m. Dorothy Elizabeth Jewitt, Aug. 26, 1953; children: James Alexander, Charles Robert. BSc, U. Durham (Eng.), 1949, PhD, 1956. Registered geologist, mining engr. Sci. officer Brit. Ea. Caribbean area Colonial Rsch. Svc., 1952-60; head seismic rsch. U. W.I., Trinidad, 1960-68; econ. affairs officer energy resources br. UN Secretariat, N.Y.C., 1968-74, chief mineral resources br., 1974-86; hon. fellow dept. geol. scis. U. Durham, 1987—; cons. UN, N.Y.C., 1986—. Editor: Economics of Mineral Engineering, 1976, The Development Potential of Precambrian Mineral Deposits, 1982, Legal and Institutional Arrangements in Minerals Development, 1982, Mineral Processing in Developing Countries, 1982, Jour. Volcanology and Geothermal Rsch., 1970-92, Natural Resources Forum, 1988-91; author: Catalogue of Eastern Caribbean Earthquakes, 1964, Catalogue of Active Volcanoes of the West Indies, 1966; contbr. articles on earthquakes, volcanic activity and mineral resources to profl. jours. Fellow Geol. Soc. London, Instn. Mining and Metallurgy; mem. AIME, Royal Commonwealth Soc. Anglican. Avocation: gardening. Home: Little Leake, Nether Silton, Thirsk Y07 4BL, England Office: U Durham Dept Geol Scis, South Rd, Durham DH1 3LE, England

ROBY, CHRISTINA YEN, data processing specialist, educator; b. Shanghai, China; came to U.S., 1980; d. Hai Zhou and Yun Qui (Zhang) Yen; m. Ronald L. Roby; 1 child, Colin H. BS, Jiao-Tung U., Shanghai, 1957; MS, U. Balt., 1986. Lic. engr., Peoples Republic of China. Chief mech. engr. Shenyang Valve Rsch. Inst., China, 1958-1980; computer system operator U. Balt., 1984, rsch. asst., 1984-86; sales assoc. V. F. Assocs., Inc., Balt., 1986-88; system analyst Computer Data Systems, Inc., Rockville, Md., 1988-89; data processing specialist Dept. of Health and Mental Hygiene, Balt., 1989—; instr. Community Coll. of Balt., 1986, 88; cons. Nat. Ins. Agency, Balt., 1988. Author: Guide to Using MS-DOS, 1988; contbr. author Japanese-Chinese Electrical Mechanical Industry Dictionary, 1980; transl., editor Analysis of Gas, Impurities and Carbide in Steel, 1961; contbr. articles to profl. jours. Vol. tutor U. Balt., 1983; vol. tchr. Chinese Lang. Sch., Balt., 1985-86, 90—; lectr. Internat. Festival Exhbn., 1986. Recipient cert. of appreciation Chinese Lang. Sch., 1986. Mem. NAFE, Sci. and Tech. Assn., Beta Gamma Sigma, Delta Mu Delta. Avocations: calligraphy, meditation, Tai-Chi, foreign languages.

ROBY, RICHARD JOSEPH, research engineering executive, educator; b. Balt., Oct. 21, 1954; s. Robert Edelen Sr. and Lorraine Teresa (McDonough) R.; m. Jeanne Celia Busch, Aug. 2, 1986. AB in Chemistry, BS in Chem. Engring., Cornell U., 1977, MS in Mech. Engring., 1980; PhD in Mech. Engring., Stanford U., 1988. Registered profl. engr., Calif. Teaching asst. chemistry dept. Cornell U., Ithaca, N.Y., 1975-77, teaching asst. govt. dept., 1977-79, rsch. asst. mech. and aerospace engring. dept., 1977-79; rsch. engr. fuels and lubricants dept. Ford Motor Co., Dearborn, Mich., 1979-83; rsch. asst. mech. engring. dept. Stanford U., Palo Alto, Calif., 1983-86; from asst. prof. to assoc. prof. mech. engring. dept. Va. Poly. Inst. & State U., Blacksburg, 1987-92; dir. combustion rsch. Hughes Assocs., Inc., Columbia, Md., 1992—; mem. program sub-com. Internat. Symposium Combustion, 1982-84, 86-88, 88-90, 90-92; session chmn. Ea. states sect. Combustion Inst. Meeting, Orlando, Fla., 1990; adv. bd. chem. and thermal systems divsn. NSF, 1991; adj. prof. Va. Poly. Inst. and State U., 1992—. Reviewer, contbr. articles to Fire Tech., Combustion and Flame, Jour. Propulsion and Power; contbr. articles to Jour. Fire Protection Engring., Fire Engring. Jour., Thermochimica Acta; presenter papers at various symposia and confs. Firefighter Ithaca Vol. Fire Dept., 1973-79, dept. instr., 1976-79, lt. Neriton Fire Co., 1976-79; firefighter Belleville (Mich.) Vol. Fire Dept., 1980-81; mem. Calif. Dems. for New Leadership, L.A. and San Francisco, 1984-86; del. Montgomery County (Va.) Dem. Com., 1987-92. Mem. AIAA, ASME, Internat. Assn. for Fire Safety Sci., Soc. Fire Protection Engrs., Soc. Automotive Engrs. (faculty advisor Va. Tech. student chpt. 1987-92), The Combustion Inst., Planetary Soc., Sigma Xi (sec. Ford Motor Co. chpt. 1981-82). Achievements include patent pending for plasma torch for ignition and stabilization of supersonic combustion; for dual frequency microwave technique for measuring solid propellant burn rates; for flame quality detector for gas turbine engines; research includes pollutant formation and control, alternate fuels, light off detector for gas turbine engines, cold start assist device for alcohol fueled vehicles, single station smoke detector, principles of fire protection chemistry, carbon monoxide, smoke yields, and window breakage from compartment fires, and dynamic turbine blade temperature measurement. Office: Hughes Assocs Inc Ste 125 6770 Oak Hall Ln Columbia MD 21045

ROCABOY, FRANÇOISE MARIE JEANNE, acoustical engineer; b. Lyon, Rhone, France, July 18, 1962; d. Francis V. and Marie Th. (Prost) R. MS in Acoustics, U. London, Chelsea Coll., 1985. Tech. cons. Société de Recherches Réalisations Electriques Mécaniques/KARL Deutsch, La Flèche, France, 1990-91; devel. engr. Non Destructive Testing Systems, Parthenay, France, 1991, Direction Constructions Navales Indret/Centre Etude Structures et Matériaux Navals/Controles Non Destructifs, Nantes, France, 1992—. Contbr. articles to profl. publs. Recipient scholarship JAUW, 1987-88. Mem. Acoustical Soc. Am., Catgut Acoustical Soc. Home: Creviac, 44170 Nozay France Office: DCN INDRET/CESMAN/CND, 44620 La Montagne France

ROCHAIX, JEAN-DAVID, molecular biologist educator. Prof. dept. molecular biology U. Geneva, Switzerland. Recipient Gilbert Morgan Smith medal Nat. Acad. Scis., 1991. office: University of Geneva, Dept of Molecular Biology, Geneva Switzerland*

ROCHE, ALAIN ANDRE, research chemist; b. St. Etienne, Loire, France, Apr. 17, 1948; s. Rene Francisque and Marie Francine (Faurie) R.; m. Marie Jose De Abreu, Sept. 4, 1976 (div. Oct. 1984); children: Barbara, Gael; m. Royer Dominique Lise Claude, July 19, 1986; children: Xavier, Gregory. BA, Lycée Etienne Mimard, St. Etienne, 1967; MA in Chemistry and lic. en scis., U. Lyon, France, 1976, DEA, 1977, MA in Physics, 1978, PhD, 1983. Allocataire de recherche DGRST U. Claude Bernard, Lyon, 1977-79; postdoctoral fellow AFML/Wright Patterson AFB, Dayton, Ohio, 1979-80; attaché de recherche Ctr. Nat. Rsch. Sci./URA 417, Lyon, 1980-83, chargé de recherche, 1983-92, directeur de recherche, 1992—; cons. Techmetal-Promotion, Maizieres, France, 1985, ERIM, Lyon, 1986, LEFLON, St. Laurent, Chamousset, France, 1990-91, Vitus-Bador, St. Chamond, 1990—. Mem. editorial adv. bd. Jour. Adhesion Sci. and Tech., 1990—; contbr. numerous articles to profl. jours. Pres. Children Care Assn., La Doua, France, 1983-88; mcpl. councillor Town Hall, St. Jean de Niost, France, 1989—. Avocations: skying, swimming, bicycling. Home: Route du Port Neuf, 01800 Saint Jean De Niost France Office: U Claude Bernard Lyon, 1, CNRS, URA 417, 48, Bd du 11 Novembre 1918, F-69622 Villeurbanne France

ROCHE, JAMES RICHARD, pediatric dentist, university dean; b. Fortville, Ind., July 17, 1924; s. George Joseph and Nelle (Kinnaman) R.; m. Viola Marie Morris, May 15, 1949; 1 child, Ann Marie Roche Potter. DDS, Ind. U., 1947, MS in Dentistry, 1983. Diplomate Am. Bd. Pediatric Dentistry (exec. sec.-treas. 1982—). Prof. emeritus Ind. U. Sch. Dentistry, Indpls., 1968—, chmn. div. grad. pediatric dentistry, 1969-76, asst. dean faculty devel., 1976-80, assoc. dean faculty devel., 1980-87, assoc. dean for acad. affairs, 1987-88; cons. Council Dental Edn., Hosp. Dental Service and Commn. Accreditation, Chgo., 1977-83. Served to capt. U.S. Army, 1952-54. Fellow Internat. Coll. Dentists, Am. Coll. Dentists, Am. Acad. Pediatric Dentistry (bd. dirs. 1967-70), Pierre Fauchard Acad.; mem. ADA (cons. Bur. Dental Health Edn. 1977), Ind. Dental Assn. (v.p. 1973-74, chmn. legis. com. 1968-77, lobbyist 1970-77), Indpls. Dist. Dental Assn. (pres. 1967-68), Omicron Kappa Upsilon (hon. dental frat. 1959), Masons, Scottish Rite. Home and Office: 1193 Woodgate Dr Carmel IN 46033

ROCHE, KERRY LEE, microbiologist; b. Lynn, Mass., Apr. 6, 1964. BS, U. N.H., 1986; postgrad., Lesley Coll. Rsch. scientist Millipore Corp., Bedford, Mass., 1987-92, sr. rsch. scientist, 1991—. Contbr. articles to BioPharm, Jour. Am. Soc. Brewing Chemists, Jour. Pharm. and Biomed. Analysis. Recipient Achievement awrd ARC, 1992, Tech. Inovation award Millipore Corp., 1992. Mem. Parenteral Drug Assn., Am. Soc. Microbiology, Sigma Xi. Achievements include research in bacterial, mycoplasma, pseudomonas and viral removal from aseptically processed fluid streams. Office: Millipore Corp 80 Ashby Rd Bedford MA 01730

ROCHE, (EAMONN) KEVIN, architect; b. Dublin, Ireland, June 14, 1922; came to U.S., 1948, naturalized, 1964; s. Eamon and Alice (Harding) R.; m. Jane Tuohy, June 10, 1963; children: Eamon, Paud, Mary, Anne, Alice. B.Arch., Nat. U. Ireland, 1945; D.Sc. (hon.), Nat. U. Ireland, 1977; postgrad., Ill. Inst. Tech., 1948; D.F.A. (hon.), Wesleyan U., 1981. With Eero Saarinen and Assocs., Hamden, Conn., 1950-66; partner Kevin Roche John Dinkeloo and Assocs., Hamden, from 1966. Prin. works include Ford Found. Hdqrs., 1967, Oakland (Calif.) Mus, 1968, Met. Mus. Art, N.Y.C., Creative Arts Ctr., Wesleyan U., Middletown, Conn., 1971, Fine Arts Ctr., U. Mass., 1971, Union Carbide Corp. World Hdqrs., Conn., Gen. Foods Corp. Hdqrs., Rye, N.Y., 1977, 1978, Conoco Inc. Hdqrs., Houston, 1979, Central Pk. Zoo, N.Y.C., 1980, DeWitt Wallace Mus. Fine Arts, Williamsburg, Va., 1980, Bouygues World Hdqrs., Paris, 1983, J.P. Morgan and Co. Hdqrs., N.Y.C., 1983, UNICEF Hdqrs., N.Y.C., 1984, Leo Burnett Co. Hdqrs., Chgo., 1985, Corning (N.Y.) Glass Works Corp. Hdqrs., 1986, Merck & Co. Inc. Corp. Office Facility, N.J., 1987, Dai Ichi Hdqrs./Norinchukin Bank Hdqrs., Tokyo, 1989, NationsBank Hdqrs., Atlanta, 1989, N.Y.C. Exchs. Hdqrs., 1989, Pontiac Marina Pvt. Ltd., Singapore, 1990, Metropolitano, Madrid, 1990, Borland Internat., Inc., Scotts Valley, Calif., 1990, Internat. Trade Ctr., Dusseldorf, Germany, 1991, Eczacibasi Group Hdqrs., Istanbul, 1991. Mem. Fine Arts Commn., Washington; trustee Am. Acad. in Rome, 1968-71, Woodrow Wilson Center for Scholars in Smithsonian Instn. Recipient Creative Arts award Brandeis U., 1967, medal of honor N.Y. chpt. AIA, 1968, A.S. Bard award City Club N.Y., 1968, 77, 79, award Gov. Calif., 1968, N.Y. State award Citizens Union N.Y., 1968, total design award Am. Soc. Interior Design., Pritzker Arch. prize, 1982, Albert S. Bard award, 1990, Gold medal AIA, 1993. Mem. NAD (academician), Am. Acad. and Inst. Arts and Letters (Brunner award 1965, Gold medal 1990), Académie d'Architecture (Grand Gold medal 1977), Mcpl. Art Soc. N.Y. (Brendan Gill prize 1989), Acad. di San Luca. Office: Kevin Roche John Dinkeloo & Assoc PO Box 6127 20 Davis St Hamden CT 06517

ROCHELLE, WILLIAM CURSON, aerospace engineer; b. Houston, May 15, 1937; s. Wilbur Tillet and Lillian Verna (Curson) R.; m. Patricia Jane Perry, June 8, 1963; children: Jeremy Dee, William Shane, Derek Ernest. BS in Aerospace Engring., U. Tex., 1960, postgrad., 1961-63; MSME, Calif. Inst. Tech., 1961; grad., U.S. Army Guided Missile Sch., 1964. Registered profl. engr., Tex. Rsch. engr. Defense Rsch. Lab., Austin, Tex., 1961-63; mem. profl. staff TRW Systems, Houston, 1967-71; engr./scientist TRACOR, Inc., Austin, 1971-73; prin. engr. Lockheed Electronics Co., Houston, 1973-76; mem. tech. staff V Rockwell Internat., Houston, 1976-88; sr. advanced systems specialist Lockheed Engring. & Scis. Co., Houston, 1988—; instr. solar energy Coll. of Mainland, Texas City, Tex., 1976; cons. TRACOR, Inc., Austin, 1975; owner, operator William C. Rochelle Tax Svcs., 1993. Author: Entry Heating and Thermal Protection, 1980, Entry Vehicle Heating and Thermal Protection Systems, 1983; contbr. articles to profl. jours. Baseball coach NASA Area Little League and Pony League, Houston, 1970-79; football coach Tex. Jr. League, Houston, 1972-73; weight-lifting instr. Clear Lake City Recreation Ctr., Houston, 1974-77; Sun. sch. tchr. Univ. Bapt. Ch., Houston, 1974-84. Capt. U.S. Army, 1964-67. Grad. fellow Tau Beta Pi, 1960; recipient Group Achievement awards NASA/Johnson Space Ctr., Houston, 1981, 82, Astronaut's Silver Snoopy award NASA, Houston, 1990; named Weightlifting Champion, Tex. and So. U.S., 1954-62. Mem. AIAA (assoc. fellow, mem. thermophysics tech. com. 1993, ground test chmn. 1987-89, Best Paper award 1984-87, reviewer tech. jours. 1971—), Nat. Mgmt. Assn. Republican. Baptist. Home: 16231 Brookford Houston TX 77059

ROCHER, EDOUARD YVES, computer engineer; b. La Tronche, Xi, France, Aug. 22, 1932; came to U.S. 1968; s. Jean E. and Marguerite (Favre) R.; m. Francoise M. Kreitmann, Dec. 28, 1963; children: Eric A., Anne-Laure. Ingenieur Diplome, Ecole Centrale, Paris, 1956; Dr.Ing., Tech. Hochschule, Darmstadt, Germany, 1963. Devel. engr. IBM-France, La Gaude, 1960-67; rsch. staff IBM, Yorktown, 1968-77; pres. Telematic Inc., Millwood, N.Y., 1978-79; product mgr. AT&T, Basking Ridge, N.J., 1980-82; sr. cons. Aetna Telecomm. Cons., Centerville, Mass., 1983-84; pres. ULAN Corp., Centerville, Mass., 1984—. Contbr. articles to profl. jours. Lt. French Army, 1956-59. Mem. IEEE (sr.), ACM, Am. Phys. Soc. Democrat. Roman Catholic. Achievements include patents on electroplating, electronic devices, network protocols, Standards IEEE 802, and fiber optics (FDDI); fundamental paper on theoretical physics. Home: 77 Old Post Rd Centerville MA 02637-2920 Office: ULAN Corp PO Box 1087 Centerville MA 02632-1087

ROCK, IRVIN, research scientist; b. N.Y.C., July 7, 1922; s. Daniel Joel and Lilian (Weinberger) R.; m. Romola Hardy, Oct. 26, 1945 (div. July 1963); children: Peter, Alice; m. Sylvia Shilling, July 24, 1963; children: Lisa, David. BS, CCNY, 1947, MA, 1948; PhD summa cum laude, New Sch. for Social Rsch., N.Y.C., 1952. From rsch. asst. to assoc. prof. New Sch. for Social Rsch., N.Y.C., 1947-59; assoc. prof. Yeshiva U., N.Y.C., 1959-67; prof. Inst. for Cognitive Studies Rutgers U., Newark, 1967-81; prof. Program in Cognition Rutgers U., New Brunswick, N.J., 1981-87; prof. emeritus Rutgers U., New Brunswick, 1987—; fellow to instr. CCNY, 1947-49; adj. prof. U. Calif., Berkeley, 1987—; adv. bd. mem. Perception Jour., Bristol, Eng., 1986—. Author: The Nature of Perceptual Adaptation, 1966, An Introduction to Perception, 1975, Orientation and Form, 1973, The Logic of Perception, 1983, Perception, 1984; peer reviewer various psychology jours. With U.S. Army, 1943-45, ETO. Fellow AAAS, APA, Soc. Exptl. Psychologists. Achievements include discovery of one trial learning, the effect of perceived orientation on phenomenal shape; co-investigated the moon illusion; co-discovered the dominance of vision over touch; co-inventor new method for studying perception without attention; research includes theoretical analysis of intelligence of perception. Office: Dept Psychology Univ Calif Berkeley CA 94720

ROCKART, JOHN FRALICK, information systems reseacher; b. N.Y.C., June 20, 1931; s. John Rachac and Janet (Ross) R.; m. Elise Jean Feldmann, Sept. 16, 1961; children: Elise B. Liesl, Scott F. AB, Princeton U., 1953; MBA, Harvard U., 1958; PhD, MIT, 1968. Sales rep. IBM, 1958-61, dist. med. rep., 1961-62, fellow in Africa, 1962-64; instr. MIT, Cambridge, Mass., 1966-67; asst. prof. IBM, Cambridge, Mass., 1967-70, assoc. prof., 1970-74, sr. lectr., 1974—; dir. MIT, Cambridge, 1976—; bd. dirs. Keane, Inc., Boston, Comshare, Inc., Ann Arbor, Mich., Transition Systems, Inc., Boston, Multiplex, St. Louis. Co-author: Computers & Learning Process, 1974, Rise of Managerial Computing, 1986, Executive Support Systems, 1988 (Computer Press Assn. 1989); contbr. articles to profl. jours. Trustee New Eng. Med. Ctr., Boston; mem. Mass. Gov. Adv. Coun. on Info. Tech., Boston. Lt. USN, 1953-56. Mem. Assn. for Computing Machinery, Inst. for Mgmt. Sci., Ops. Rsch. Soc. Am., Soc. for Info. Mgmt. (bd. dirs. mem. at large 1989-93), New Eng. Med. Ctr. Audit Com., Weston (Mass.) Golf Club, Lake Sunapee Country Club (New London, N.H.). Republican. Unitarian. Home: 150 Cherry Brook Rd Weston MA 02193-1308 Office: CISR MIT Sloan Sch Mgmt E40-187 77 Massachusetts Ave Cambridge MA 02139-4307

ROCKEFELLER, JOHN DAVISON, IV (JAY ROCKEFELLER), senator, former governor; b. N.Y.C., NY, June 18, 1937; s. John Davison III and Blanchette Ferry (Hooker) R.; m. Sharon Percy, Apr. 1, 1967; children: Jamie, Valerie, Charles, Justin. B.A., Harvard U., 1961; student, Japanese lang. Internat. Christian U., Tokyo, 1957-60; postgrad. in Chinese, Yale U. Inst. Far Eastern Langs., 1961-62. Apptd. mem. nat. adv. council Peace Corps, 1961, spl. asst. to dir. corps, 1962, ops. officer in charge work in Philippines, until 1963; desk officer for Indonesian affairs Bur. Far Eastern Affairs, U.S. State Dept., 1963; later asst. to asst. sec. state for Far Eastern affairs; cons. Pres.'s Commn. on Juvenile Delinquency and Youth Crime, 1964; field worker Action for Appalachian Youth program, from 1964; mem. W.Va. Ho. of Dels., 1966-68; sec. of state W.Va., 1968-72; pres. W.Va. Wesleyan Coll., Buckhannon, 1973-75; gov. State of W.Va., 1976-84; U.S. senator from W.Va., 1985—, mem. vets. affairs com., fin. com., commerce, sci. and transp. com., chmn. Sen. steel caucus, Bipartisan Com. on Comprehensive Health Care; chmn. Nat. Commn. on Children, natural resources and environ. com. Nat. Govs. Assn., 1981-84. Contbr. articles to mags. including N.Y. Times Sunday mag. Trustee U. Chgo., 1967—; chmn. White House Conf. Balanced Nat. Growth and Econ. Devel., 1978, Pres.'s Commn. on Coal, 1978-80, White House Adv. Com. on Coal., 1980; active Commerce, Sci., and Transp. Com., Fin. Com.; chmn. Vet. Affairs Com. Office: US Senate 109 Hart Senate Bldg Washington DC 20510*

ROCKENSIES, JOHN WILLIAM, mechanical engineer; b. N.Y.C., May 30, 1932; s. John William and Wilma (Mercz) R.; m. Marion Pauline Peachman, Sept. 16, 1956; children: Kenneth John, Karen Martha Rockensies Steinbeck. B of Mech. Engring., CCNY, 1954, M of Mech. Engring., 1960; postgrad., Bklyn. Polytechnic Inst., 1955, Columbia U., 1956. Registered prof. engr., N.Y. Jet engine performance and compressor devel. Curtiss Wright Corp., Woodridge, N.J., 1954-56; product devel. engr. Sperry Gyroscope Corp., Lake Success, N.Y., 1956-60; sr. exptl. test engr. Pratt & Whitney Corp., East Hartford, Conn., 1960-62; project engr. Stratos Corp., Bayshore, N.Y., 1962; prin. propulsion engr. Republic Aviation Corp., Farmingdale, N.Y., 1963-64; power plant design engr., group and project leader, project engr., engr. specialist and mgr. Grumman Aerospace Corp., Bethpage, N.Y., 1964—; mem. SAE E-32 Engine Condition Monitoring com., 1983; instr. navigation Smithtown Bay Power Squadron. Author tech. papers in field. Deacon, trustee, elder First Presbyn Ch. of Smithtown. Recipient Apollo Achievement award NASA, Washington, 1970. Assoc. fellow AIAA; mem. NSPE, ASME, U.S. Power Squadrons (sr.). Avocations: sailing/boating, jogging, tennis, camping, model aircraft, travel. Home: 65 Parnell Dr Smithtown NY 11787-2428 Office: Grumman Aircraft Systems MS B69-001 Bethpage NY 11713-5820

ROCKETT, ANGUS ALEXANDER, materials science educator; b. Boston, Sept. 14, 1957; s. John Alexander and Abby (Burgess) R. BS, Brown U., Providence, 1980; PhD, U. Ill., 1987. Asst. prof. materials sci. U. Ill., Urbana, 1987—. Mem. Am. Vacuum Soc. (chpt. chair 1990-91, div. exec. com. 1991-93, student issues co-chair 1991-93). Achievements include research in TEM study of crystallography, influence of surface structure on growth, simulating diffusion on Si (100) 2x1 surfaces, $CuInSe_2$ for photovoltaic applications. Office: U Ill 1101 W Springfield Ave Urbana IL 61801

ROCKOWER, EDWARD B., physicist, operations researcher, consultant; b. Phila., May 29, 1943; s. Jacob R. and Helen (Grodman) R.; m. JoAnne Manetti, June 17, 1982; children: Joseph, David. BS in Physics, UCLA, 1964; MA, PhD in Physics, Brandeis U., 1975; postgrad., So. Meth. U., 1975-76, Temple U., 1973-74. Rsch. assoc. LTV Rsch. Ctr., Anaheim, Calif., 1964-65; v.p. Univ. Home Svcs., Inc., Phila. and, N.J., 1971-74; sr. ops. analyst F-16 program Gen. Dynamics, Ft. Worth, 1975-77; physicist Laser Isotope Separation program Ketron, Inc., Berwyn, Pa., 1977-79; physicist

Lawrence Livermore (Calif.) Nat. Lab., 1979-82; instr. U. Md., Japan, Germany, 1982-84; assoc. prof. ops. rsch. Naval Postgrad. Sch., Monterey, Calif., 1984-89; pres. Edward Brandt, Inc., Monterey, 1989—; cons. U.S. Army, Wuerzburg, Germany, 1984, RAM, Inc., Las Cruces, N.Mex., 1986, CTB McGraw-Hill, Monterey, 1988, Tetrex Internat., 1993; adj. ass. ptof. LaSalle Coll., Phila., 1977-78; invited scholar Harvard U., 1989; presenter, guest speaker in field. Contbr. numerous articles to refereed jours. Vol. Buddy Program, Monterey, 1992—. Recipient Commendation letter Merit Scholarship Program, 1961; NSF fellow, 1965-70. Mem. Am. Phys. Soc., Ops. Rsch. Soc. Am. (treas. Phila area, mem. SIG/Tech. sect. com.), Optical Soc. Am., Inst. Mgmt. Sci., Assn. Old Crows. Achievements include research in theoretical understanding of evolution and dynamics of quantum statistics of light; patent pending for Optical Image Translator. Home: 49 Alta Mesa Cir Monterey CA 93940-4601 Office: Edward Brandt Inc PO Box 2265 Monterey CA 93942

ROCKWELL, BENJAMIN ALLEN, physicist; b. Fort Smith, Ark, Aug. 23, 1964; s. John Allen and Carolyn (Kidd) R.; m. Cheryl Ann Minton, June 10, 1989. BS in Physics, Cent. Mo. State U., 1986; PhD in Physics, U. Mo., Columbia, 1991. Rsch. biophysicist, nat. rsch. coun. postdoctoral assoc. Armstrong Lab. USAF, Brooks AFB, Tex., 1991--. Contbr. articles to profl. jours. O.M. Stewart fellow, 1990. Mem. Am. Phys. Soc., Soc. Photo-Optical Instrumentation Engrs. Baptist. Achievements include the discovery that substrate heterostructures was found to have a dramatic effect on the pressure dependence of lattice mismatch strains in several technologically important materials; measurement of nonlinear optical constants of ocular media using ultra short pulsed lasers.

ROCKWELL, NED M., chemical engineer; b. Valparaiso, Ind., Nov. 29, 1956; s. Richard Vernon and Marcia Anne (Taylor) R.; m. Amy L. Hartzell, June 25, 1983; children: Ben, John. BS in Chemistry, Purdue U., 1979; MChemE, Ill. Inst. Tech., 1986; M in Mgmt., Northwestern-Kellogg U., 1990. Chemist Stepan Co., Northfield, Ill., 1979-83, process devel. chemist, 1983-85, rsch. engr., 1985-87, group leader, 1987-88, process devel. mgr., 1988-93, sr. mgr., 1993—. Contbr. articles to profl. jours. V.p., treas. Alliance for Excellence Lake Bluff, Ill., 1991, 92. Mem. AICE, AAAS, Am. Chem. Soc., Am. Oil Chemsits Soc. Achievements include patents for Phenyl Esters via Novel Catalyst, Novel Surfactant Compounds. Office: Stepan Co 22 W Frontage Rd Northfield IL 60093

ROCO, MIHAIL CONSTANTIN, mechanical engineer, educator; b. Bucharest, Romania, Nov. 2, 1947; came to U.S., 1980; s. Constantin M. and Armande Ch.-Ad. (Cantacuzino) R.; m. Ecaterina (Cathy) Roco, July 24, 1986; children: Constance-Armanda M., Charles Michael. PhD, Polytechnic Inst., Bucharest, 1976. Prof. U. Ky., Lexington, 1981—; program dir. engring. NSF, Washington, 1990—; part-time prof. Johns Hopkins U., Balt., 1993; vis. prof. U. Paderborn, Fed. Republic Germany, 1979, U. Sask., Can., 1990, Tohoku U., Sendai, Japan, 1988-89, Calif. Inst. Tech., Pasadena, 1988-89; cons. to industry in U.S., Can., Europe and Australia, 1981-92; cons. to U.S. govt. agys., 1983-89; lectr. postgrad. engring. Japan, Chile, Ga., Fla., 1982-91; trainee strategic planning and exec. leadership U.S. OPM, Washington, 1992. Author: (with others) Principles and Practice of Slurry Flow, 1991, Particulate Two-Phase Flow, 1992; assoc. tech. editor Jour. Fluids Engring., 1985-89; contbr. over 100 articles to sci. and engring. jours and several articles symbolist poetry in literary jours. Grantee NSF, Rsch. Founds. USA, Can., Fed. Republic Germany, OAS, 1979-91; recipient Carl Duisberg award Carl Duisberg Soc., Fed. Republic Germany, 1979, Outstanding Rsch. Professorship award U. Ky., 1986-87. Mem. ASME (editor internat. symposium series 1984-91), AIAA, Soc. Rheology, Acad. Mechanics, N.Y. Acad. Scis. Achievements include formulation of innovative numerical methods for fluid and particulate flows (finite volume, probabilistic, marching), computer-aided-engineering for centrifugal slurry pumps, hydrotransport of solids through pipelines, and electro-imaging (fast laser printers); 13 inventions of fluid machineries, wear resistant equipment, methods to increase pipe transport capacity, viscosimeter; research on multiphase flow modeling, laser Doppler anemometry, flow visualization, power stations, pumps, and wear mechanisms. Office: NSF Engring Directorate Rm 11-15 Washington DC 20550

RODBARD, DAVID, endocrinologist, biophysicist; b. Chgo., July 6, 1941. BA, U. Buffalo, 1960; MD, We. Reserve U., 1964. Intern in medicine King County Hosp., Seattle, 1964-65; resident Hahnemann Hosp., Phila., 1965-66; clin. assoc. med. rsch. NIH, Bethesda, Md., 1966-69, sr. investigator, 1969-78, head sects. med. rsch., biophysics and endocrinology, Endocrinology and Reproduction Rsch. Bur., Nat. Inst. Child Health and Human Devel., 1979—; cons. Internat. Atomic Energy Agy., 1970—, WHO, 1974—. Assoc. editor Am. Jour. Physiology, 1976—. Recipient Young Investigator award Clin. Radioassay Soc., 1979. Mem. Am. Physiol. Soc., Am. Soc. Biol. Chemistr, Endocrine Soc. (Ayerst award 1981), Am. Soc. Clin. Investigation. Achievements include research in endocrinology, physiology, biomathematics, biochemistry, radio immunoassay, physical-chemistry of proteins, neurotransmitters. Office: National Institutes of Health Computer Rsch & Technology 9000 Rockville Pike Bldg 12A Bethesda MD 20892*

RODDA, LUCA, computer company executive, researcher; b. Milan, June 23, 1960; s. Alessandro and Iole (Radaelli) R.; m. Daniela Togni, Oct. 4, 1990; 1 child, Martina Astrid. Grad. in physics, U. Studi, Milan, 1989. Guest researcher Nat. Rsch. Coun., Milan, 1982-84; cons. European Computer Trading SRL, Milan, 1984-85, owner, mgr., 1985—; lectr. U. Studi, 1983-85; cons. Pirelli Informatica Spa, Milan, 1985. Contbr. articles to profl. jours. Mem. IEEE, Assn for Computing Machinery. Avocation: collecting playing cards. Office: ECT-European Computer Trading SRL, Via Imperia 23, 20142 Milan Italy

RODDIS, WINIFRED MARY KIM, structural engineering educator; b. Arlington, Va., Nov. 5, 1955; d. Louis H. and Alice (Stets) R.; m. Lindsey L. Spratt, Dec. 24, 1979; 1 child, Hillary Roddis Spratt. MS, MIT, 1987, PhD, 1989. Registered profl. engr., Mass., Kans. Structural engr. Stone & Webster Engring. Corp., Boston, 1977-81, Souza & True Cons. Engrs., Watertown, Mass., 1981-84, AG Lichtenstein Engrs., Framingham, Mass., 1986; teaching asst. MIT, Cambridge, Mass., 1984, rsch. asst., 1985-86, rsch. fellow, 1986-88; asst. prof. civil engring. U. Kans., Lawrence, 1988—. Contbr. articles to profl. jours. Counselor MIT Edl. Coun., Cambridge, 1986—; v.p. Assn. MIT Alumnae, Cambridge, 1986-88. Student Rsch. fellow Am. Inst. Steel Constrn., 1990, 92; Hertz Found. fellow, 1986-88; recipient Rsch. Invitation award NSF, 1989. Mem. ASCE (mem. com., rsch. fellow 1986), Soc. Women Engrs. (nat. career guidance chair, 1989), Am. Soc. Engring. Edn., Am. Concrete Inst. (mem. coms.). Office: U Kans Dept Civil Engring Lawrence KS 66045

RODENHUIS, DAVID ROY, meteorologist, educator; b. Michigan City, Ind., Oct. 5, 1936; married; 2 children. BS, U. Calif., Berkeley, 1959, Pa. State U., 1960; PhD in Atmospheric Sci., U. Wash., 1967. From asst. prof. to assoc. prof. fluid dynamics and applied math. U. Md., College Park, 1972-76, assoc. prof. meteorology, 1976—; exec. scientist U.S. com. global atmospheric rsch. program NAS, 1972; sic. officer World Meteorology Orgn., 1975—; U.S.-U.S.S.R. exchange scientist, 1980. Mem. Am. Geophys. Union, Am. Meteorol. Soc. Achievements include research in tropical meteorology, convection models, dynamic climate models. Office: Dept of Commerce-Nat Weather Svc Natl Meterological Center 5200 Auth Rd Washington DC 20233*

RODER, HEINRICH, biophysicist, educator; b. Belp, Berne, Switzerland, Feb. 21, 1952; came to U.S., 1981; s. Martin and Ruth (Horst) R.; m. Fereshteh Allahyari, Apr. 23, 1976; children: Sanam C., Iman M., Navid A., Samyra K. MS, Fed. Inst. Tech., 1978, PhD, 1981. Rsch. assoc. U. Ill., Urbana, 1981-84; rsch. asst. prof. U. Pa., Phila., 1984-87, asst. prof., 1987-90, assoc. prof., 1990-91, adj. assoc. prof., 1991—; mem. Inst. for Cancer Rsch., Fox Chase Cancer Ctr., Phila., 1991—. Contbr. articles to jours. Nature, Sci., PNAS, Biochemistry. NIH grantee, 1986, 90, 91. Mem. AAAS, Biophys. Soc., Protein Soc. Achievements include research in mechanism of protein folding, dynamic aspects of protein structure, molecular recognition, nuclear magnetic resonance (NMR) spectroscopy; development of NMR and hydrogen exchange techniques for study of protein folding intermediates and protein-ligan interactions. Office: Fox Chase Cancer Ctr 7701 Burholme Ave Philadelphia PA 19111

RODGERS, IMOGENE SEVIN, toxicologist; b. Rochester, Pa., Nov. 13, 1945; d. Irvin Edward and Hester Pearl (Barto) Sevin; m. John W. Horm (div. 1974); m. James Earl Rodgers, July 4, 1982; 1 child, Kimberly. BS, U. Pitts., 1967; PhD, Duquesne U., 1975. Rsch. asst. U. Pitts., 1968-71; postdoctoral assoc. Allegheny Gen. Hosp., Pitts., 1975-76; chemist Dept. Health Human Svcs. Nat. Inst. Occupational Safety and Health, Rockville, Md., 1976-80; health scientist U.S. Dept. Labor/OSHA, Washington, 1980-89; sci. coord. EPA, Washington, 1989-93; cons. occupational and environ. health scis., 1993—; guest lectr. OSHA Tng. Inst., Chgo., 1989. Author: (with others) Handbook of Radiation Measurement and Protection, 1979, Alpha-2u-Globulin: Association with Chemically Induced Renal Toxicity and Neoplasia in the Male Rat, 1991; contbr. articles to profl. jours. Mem. APHA, Soc. for Risk Analysis, N.Y. Acad. Sci., Rho Chi. Achievements include research on occupational and environmental hazard assessment, on occupational exposure to formaldehyde, and on cancer risk assessment. Home and Office: 2302 Eagle Rock Pl Silver Spring MD 20906-3248

RODGERS, JAMES EARL, physicist, educator; b. L.A., Aug. 19, 1943; s. Royal E. and Doris Velma (Clawson) R.; m. Vera Margaret Rodgers, July 19, 1963 (div. 1978); 1 child, Brenda; m. Imogene Sevin, July 4, 1983; 1 child, Kimberly. BS in Physics and Math., Calif. State U., Long Beach, 1966; PhD in Physics, U. Calif., Riverside, 1972. Diplomate Am. Bd. Radiology. Rsch. physicist Naval Weapons Ctr., Corona, Calif., 1967-68; rsch. assoc. theoretical physics dept. physics SUNY, Albany, 1972-76; NCI rsch. fellow Tufts New Eng. Ctr., Boston, 1976-78, asst. prof. spec and sci. staff dept. therapeutic radiology, 1978-80; dir. radiation physics dept. Georgetown U. Sch. Medicine, Washington, 1980—, assoc. prof. radiation medicine, 1987—, dir. radiation sci. dept., 1989—; cons. in malpractice cases, 1984—; cons. U.S. NRC, Rockville, Md., 1986-92. Contbr. articles to profl. publs. Mem. Am. Assn. Physics in Medicine (pres. Mid Atlantic chpg. 1989-91), Am. Phys. Soc., Am. Coll. Med. Physics, Am. Coll. Radiology/Health Physics Soc. Office: Georgetown U Hosp/Rad Med 3800 Reservoir Rd Washington DC 20007

RODGERS, JAMES FOSTER, association executive, economist; b. Columbus, Ga., Jan. 15, 1951; s. Laban Jackson and Martha (Jackson) R.; m. Cynthia Lynne Bathurst, Aug. 20, 1975. B.A., U. Ala., Tuscaloosa, 1973; Ph.D., U. Iowa, 1980. Fed. intern Office Rsch. and Stats., Social Security Adminstrn., Washington, 1976-77; rsch. assoc. Ctr. Health Policy Rsch., AMA, Chgo., 1979-80, rsch. dir., 1980-82, asst. to dep. exec. v.p. AMA, 1982-85, dir., 1985—. Contbr. articles on health econs. to profl. jours. Pharm. Mfrs. Assn. grantee, 1978; NSF grantee, 1978; Hohenberg fellow, 1969-70. Mem. Am. Econ. Assn., Am. Soc. Assn. Exec., Am. Statis. Assn., So. Econ. Assn., Western Econ. Assn. Home: 2233 N Orchard Chicago IL 60614 Office: AMA Ctr for Health Policy Rsch 515 N State St Chicago IL 60610

RODGERS, RHONDA LEE, health facility administrator; b. Anawalt, W.Va., July 9, 1939; d. Joseph Charlie and Dorothy Lois (Jones) Music; m. Robert Allen Rodgers, June 5, 1960; children: Tammy, Tina, Toni, Terry. Diploma, Laird Meml. Sch. Nursing, Montgomery, W.Va. Nursing supr. W.Va. State Hosp., Spencer, 1960-62; nurse Dr. Robert Smith, Centreville, Md., 1962-63; nurse supr. Del. State Hosp., New Castle, 1964-66; asst. office mgr., bookkeeper, adminstr. Neurology Assocs., Wilmington, Del., 1966-92. Avocations: bowling, creative needlework, stamp collecting. Home: 29 Dempsey Dr Newark DE 19713-1930 Office: Neurology Assocs 1228 N Scott St Wilmington DE 19806-4060 *Died Nov. 4, 1992.*

RODRIGUEZ, AGUSTIN ANTONIO, surgeon; b. Hato Rey, P.R., Aug. 20, 1961; s. Agustin and Esther Rodriguez (Gonzalez) R.; m. Liana Esther Lopez, 1993. AB in Biology, Harvard Coll., 1982; MD, U. P.R., San Juan, 1986. Diplomate Nat. Bd. Med. Examiners. Surgical intern Boston U. Med. Ctr., 1986-87, surgical resident, 1988-93, acad. trainee surgery; vascular fellow Tufts U., N. Eng. Med. Ctr., 1993—. Contbr. articles to Jour. Cardiovascular Surgery, Jour. Vascular Surgery, Archives of Surgery. Mem. AMA, Am. Numismatic Soc., Am. Numismatic Assn., Mass. Med. Soc., N.Y. Acad. Scis., European Soc. Vascular Surger (assoc.), Alpha Omega Alpha (Psi chpt.). Republican. Home: 373 Commonwealth Ave Apt 502 Boston MA 02115-1806 Office: New Eng Med Ctr Dept Surgery 750 Washington Ave Boston MA 02111

RODRIGUEZ, ARMANDO ANTONIO, electrical engineering educator; b. N.Y.C., July 15, 1961; s. Armando and Neida (Pardo) R. EE, MIT, 1987, PhDEE, 1990. Asst. prof. IBM, Poughkeepsie, N.Y., 1981-83; engr. AT&T Bell Labs., Holmdel, N.J., 1983-87; design engr. Raytheon Missile Systems, Bedford, Mass., 1987-89; instr. MIT, Cambridge, 1989-90; prof. elec. engring., minority recruiter, mentor Ariz. State U., Tempe, 1990—; rsch. Aerospace Rsch. Ctr., Tempe, 1990—, Ctr. Systems Sci. and Engring., Tempe, 1990—; AFOSR rsch. assoc. Eglin AFB, Fla., 1992. Contbr. articles to profl. jours. Whitney M. Young Meml. scholar, 1979-80, Joseph Tauber scholar Poly. Inst. N.Y.,1979-80, IBM scholar, 1979-83; doctoral fellow AT&T Bell Labs., 1983-90; Undergrad. Ctrl. Systems Design Lab. grantee NSF, 1992-93, Faculty Rsch. Initiation award grantee AFOSR, 1993-94; recipient CEAS Young Faculty Teaching award, 1993. Mem. IEEE, AIAA, Tau Beta Pi, Eta Kappa Nu, Sigma Chi. Achievements include development of a theory for designing finite-dimensional controllers for distributed parameter systems, systematic procedures for designing controllers for flexible systems with hard nonlinearities, virtual instrumentation-based control systems laboratory. Home: 9618 S Ash Ave Tempe AZ 85284 Office: Ariz State U Ctr for System Sci and Engring Tempe AZ 85287-7606

RODRIGUEZ, FABIO ENRIQUE, electrical engineer; b. Bogota, Colombia, Mar. 22, 1959; came to U.S., 1988; s. Fabio Jose and Maria Dolores (Rodriguez) R.; m. Evelin Valentin, July 12, 1991. BSEE, Nat. U. Colombia, 1985; MSE, ME in Mgmt., Dartmouth Coll., 1991. Applications engr. Inst. Nuclear Affairs, Bogota, 1986-88; rsch. engr. Dartmouth Coll., Hanover, N.H., 1990-91; mgr. tech. support Gigatec (USA), Inc., Portsmouth, N.H., 1991—; referee IEEE Computer Mag., Hanover, 1991—. Contbr. to profl. publs. Vol. Am. Med. Resources Found., Wentworth, Mass., 1991. Mem. IEEE, Instrument Soc. Am., Sigma Xi. Democrat. Roman Catholic. Home: 265 Bank St Lebanon NH 03766 Office: Gigatec USA Inc 871 Islington St Portsmouth NH 03801

RODRÍGUEZ, FERDINAND, chemical engineer, educator; b. Cleve., July 8, 1928; s. José and Concha (Luís) R.; m. Ethel V. Koster, July 28, 1951; children: Holly Edith, Lida Concha. B.S., Case Western Res. U., 1950, M.S., 1954; Ph.D., Cornell U., 1958. Devel. engr. Ferro Corp., Bedford, Ohio, 1950-54; asst. prof. chem. engring. Cornell U., 1958-61, asso. prof., 1961-71, prof., 1971—; on sabbatic leave at Union Carbide Corp., 1964-65, Imperial Chem. Industries, Ltd., 1971, Eastman Kodak Co., 1978-79; cons. to industry. Author: Principles of Polymer Systems, 3d edit, 1989; contbr. numerous articles to profl. jours.; songwriter. Served with U.S. Army, 1954-56. Recipient Excellence in Teaching award Cornell Soc. Engrs., 1966, Edn. Achievement award Hispanic Engr. Mag., 1991. Fellow Am. Inst. Chem. Engrs.; mem. Am. Chem. Soc., Soc. Hispanic Profl. Engrs., Soc. Plastics Engrs. Lutheran. Home: 107 Randolph Rd Ithaca NY 14850-1720 Office: 230 Olin Hall Cornell U Ithaca NY 14853

RODRIGUEZ, J. LOUIS, civil engineer, land surveyor; b. N.Y.C., Sept. 8, 1920; s. Cesar and Carmen (Quintero) R.; m. Rita Victoria Fradera, Sept. 4, 1948; children: Carmen Brana, Christina McCarthy, Robert. BCE, City Coll. N.Y., 1942. Lic. profl. engr., N.Y., Conn., Ohio, Miss., Mass., Maine, Ontario, Alberta; lic. land surveyor, Conn. Office engr. Arthur A. Johnson Co., L.I., N.Y., 1945-48; project engr. Merritt Chapman & Scott Corp., N.Y.C., 1948-54; chief engr. Merritt Chapman & Scott Corp. of the Dominican Republic, Santo Domingo, Dominican Republic, 1954-55; estimator Hoggson Bros./F. H. McGraw, N.Y.C., 1956-59; mgr. engring. Diamond Internat. Corp., N.Y.C., 1959-86; owner North Stamford (Conn.) Surveyors, 1986—; constrn. adminstr. Holy Spirit Parish, Stamford, 1986-88; bldg. planning coord. Drug Liberation Program, Stamford, 1989-91. Stamford rep. S.W. Regional Planning Agy, Norwalk, Conn., 1973-75; bd. dirs. Drug Liberation Program, Inc., Stamford, 1987-91; keynote speaker Vets. Day Program, Stamford, 1982. 1st lt. USAF, 1942-45. Decorated Air medal with oak leaf cluster. Fellow ASCE (life); mem. NSPE (life), Profl. Engrs. Ontario, Conn. Soc. Profl. Engrs. (chpt. pres., state dir. 1963), Conn. Assn. Land Surveyors, North Stamford Exch. Club (bd. dirs., past pres., Man of Yr. award 1989), Eighth Air Force Hist. Soc., Air Force Escape and Evasion Soc., Tau Beta Pi. Republican. Roman Catholic. Achievements include designing structures for Molded Pulp Plant, Natchez, Miss., Tissue Mill, Old Town, Maine, 600 Ton Recovery Boiler, Old Town. Home and Office: 237 Russet Rd Stamford CT 06903-1823

RODRIGUEZ, LORRAINE DITZLER, biologist, consultant; b. Ava, Ill., July 4, 1920; d. Peter Emil and Marie Antoinette (Mileur) D.; m. Juan G. Rodriguez, Apr. 17, 1948; children: Carmen, Teresa, Carla, Rosa, Andrea. BEd, So. Ill. U., 1943; MS, Ohio State U., 1944; PhD, U. Ky., 1973. Asst. nutritionist OARDC, Wooster, Ohio, 1944-49; postdoctoral fellow U. Ky., Lexington, 1973-74, pesticide edn. specialist, 1978-89; pvt. cons. Lexington, 1974-79, 89—. Author ext. publs. in field; co-author rsch. publs. and book chpts. in field. Leader 4-H, Lexington, 1962-68. Named Outstanding 4-H Alumni Woman, Ky., 1969. Mem. Vegetation Mgmt. Assn. Ky. (chmn. adv. bd. 1985-89), Am. Chem. Soc. Democrat. Roman Catholic. Home: 1550 Beacon Hill Rd Lexington KY 40504-2304

RODRIGUEZ, MOISES-ENRIQUE, industrial engineer; b. Bogota, Colombia, June 6, 1962; arrived in Switzerland, 1987; s. Moises and Graciela (Gomez) R. BSc in Indsl. Engring., U. Hertfordshire, U.K., 1986; MSc in Computer Sci., Birmingham U., U.K., 1987. Indsl. engr. Portescap S.A., La Chaux-de-Fonds, Switzerland, 1987-90; grad. asst. in computer sci. U. Neuchatel, Switzerland, 1990-91, lectr. in computer aided prodn. mgmt., 1991-93, team leader QMIS project, 1991-93; logistics planning officer Serono Labs., Aubonne, Switzerland, 1993—; European corr. El Siglo Newspaper, Bogota, 1980—; liaison and student tng. supr. U. Hertfordshire, Switzerland. Contbr. articles to profl. jours. Mem. Instn. Mfg. Engrs. (grad. mem.), Fgn. Press Assn. (assoc.), Brit. Computer Soc. (assoc.), Inst. Indsl. Engrs. Roman Catholic. Avocations: squash, swimming, jogging, reading, collecting model soldiers. Home: Ave de la Gare 40, 1003 Lausanne Switzerland Office: Serono Labs, 1170 Aubonne Switzerland

RODRÍGUEZ-ARIAS, JORGE H., retired agricultural engineering educator; b. Ponce, P.R., Apr. 24, 1915; s. Herminio Rodríriguez Colón and Rosa María Arias Ríos; m. Carmen Teresa Quiñones Sepúlveda, May 9, 1948; children: Jorge H., Jaime Osvaldo, Nelson Rafael. BS in Agriculture, U. P.R., Mayaguez, 1936; BS in Agrl. Engring., Tex. A&M U., 1945; MS in Agrl. Engring., Kans. State U., 1947; PhD in Agrl. Engring., Mich. State U., 1956; D (hon.), U. P.R., Mayaguez, 1986. Instr. vocat. agriculture P.R. Reconstruction Adminstrn., Aibonito, 1936-37; instr. horticulture U. P.R., Mayaguez, 1937-43, from asst. to assoc. to prof. agrl. engring., 1947-77, dir. agrl. engring. dept., 1948-77; panelist UN Program for Devel., Lima, Peru, 1959. With USN, 1945-46. Fellow Am. Soc. Agrl. Engrs. (life), Instn. of Agrl. Engrs.; mem. Am. Soc. for Engring. Edn. (life), Inst. of Food Technologists (profl.). Home and Office: U PR Faculty Residences 3-B Mayaguez PR 00681-5158

RODRIGUEZ ARROYO, JESUS, gynecologic oncologist; b. Arecibo, P.R., Jan. 11, 1948; s. Jesus Rodriguez and Blanca Arroyo; m. Annie Arsuaga, June 3, 1972; children: Ivan, Patricia. BS, U. P.R., San Juan, 1968, MD, 1972, postgrad., 1976. Diplomate Am. Bd. Ob-Gyn. Assoc. dir. gynecologic oncology Oncology Hosp., Rio Piedras, P.R., 1978-83; assoc. prof. ob-gyn, dir. gynecologic oncology U. Hosp. Sch. Medicine, Rio Piedras, 1978-85; gynecologic oncologist Met. Hosp., Rio Piedras, 1979-88; obstetrician, gynecologist Auxilio Mutuo Hosp., P.R., 1981-91, San Pablo Hosp., Bayamon, P.R., 1981-91, Ashford Meml. Community Hosp., 1983-91; cons. gynecologic oncology Tchrs. Hosp., Hato Rey, P.R., 1979-85, Hermanos Melendez Hosp., Bayamon, 1979-91; instr. ob-gyn. U. P.R. Sch. Medicine, 1976-78, asst. prof., 1978-83, assoc. prof., 1984, dir. gynecologic oncology sect., 1978-84. Contbr. articles to med. jours. Mem. Citizen Ambassador Cancer Mgmt. Del. to USSR, 1990. Mem. AAAS, Am. Coll. Surgeons, P.R. Med. Assn. (jud. ethical coun. 1990-91), Internat. Gynecologic Cancer Soc., Soc. Gynecologic Oncologists, Dorado Beach Hotel, Caparra Country Club. Home: 1910 Pasionaria St Urban Santa Maria Rio Piedras PR 00927 Office: Caribbean Oncology & Ob-Gyn Assn PO Box 194557 Hato Rey San Juan PR 00919-4557

RODRIGUEZ-DEL VALLE, NURI, microbiology educator; b. San Juan, P.R., May 25, 1945; d. Paulino and Lucila (del Valle) Rodriguez-Rolan; m. Juan C. Perez-Otero, June 1, 1968; children: Juan Carlos, Claudia Rosalia. BS, U. P.R., 1967, MS, 1970, PhD, 1978. Rsch. asst. Coll. Pharmacy U. P.R., Rio Piedras, 1966-67, instr. biochemistry Med. Sch., 1970-78, asst. prof. microbiology, 1978-82, assoc. prof., 1982-92, prof., 1992—, ad honorem asst. prof. biology, 1984—; prin. investigator MBRS-NIH Grant, 1981—. Tchr. adult chatechism, 1992. Named Disting. Scientist Mobil Oil Co., San Juan, 1981, Disting. Prof. Student Rsch. Forum, San Juan, 1989. Mem. AAAS (coun. mem. Caribbean div. 1990-92, pres.-elect 1993—), Am. Soc. Microbiology (coun. mem. 1981-89, sec., pres. elect, pres. P.R. br. 1982-89, Dr. Arturo Carrion lectr. 1983), Soc. Exptl. Biology and Medicine, N.Y. Acad. Medicine, Sigma Xi. Roman Catholic. Office: U PR dept microbiology Med Sci Campus PO Box 365067 San Juan PR 00936-5067

RODRIGUEZ GARCIA, JOSE A., agronomist, investigator; b. Toa Baja, PR, Nov. 1, 1946; s. Pedro R. Rodriguez-Gónzalez and Maria A. Garcia-Fóntanez. BSA, Coll. Agriculture and Mechanical Arts, Mayagûez, PR, 1967; MSA, U. Mayagûez, 1979. Tchr. agr. Dept. Pub. Instrn. of P.R., 1967-68; asst. investigations Sta. of Agrl. Expts., 1968-76, asst. agronomics, 1976-83, assoc. agronomics, 1983-89, agronomics investigator, 1989—, adminstr., 1985—. Contbr. numerous articles to profl. jours. Mem. Little League of P.R., Recreation Assn. of Toa Baja, Red Cross of P.R.; bd. dirs. Baseball Team of Toa Baja. Mem. Coll. Agronomics of P.R., P.R. Soc. Agrl. Scis., Am. Soc. Agronomics, Caribbean Food Crops Soc., Am. Soc. Tropical Horticulture. Home: Apt 128 Toa Baja PR 00951 Office: U P R HC02 Box 10322 Bo Padilla Corozal PR 00643

RODTS, GERALD EDWARD, computer scientist; b. Paterson, N.J., Mar. 8, 1934; s. Peter J. and Josephine (Anderson) R.; m. Lila Henrietta Sellmann, Feb. 10, 1960; children: Gerald E. Jr., Constance E., Avery S., Shannon M. AB magna cum laude, Princeton U., 1956. Dir. industry mktg., dir. office systems IBM Corp., White Plains, N.Y., 1958-80; exec. v.p. Dyatron Corp., Birmingham, Ala., 1980-82; v.p. sales support Prime Computer, Natick, Mass., 1982-84; v.p. sales and mktg. Gould Computer Systems Div., Ft. Lauderdale, Fla., 1984-88, Micro Palm Computers, Clearwater, Fla., 1988-90; v.p., gen. mgr. DAP Technologies Corp., Tampa, Fla., 1990—. 1st lt. U.S. Army, 1956-58. Republican. Roman Catholic. Home: 10368 Carrollwood Ln # 234 Tampa FL 33618 Office: DAP Technologies Corp 1408 N Westshore Blvd #610 Tampa FL 33607

ROE, BYRON PAUL, physics educator; b. St. Louis, Apr. 4, 1934; s. Sam S. and Gertrude Harriet (Claris) R.; m. Alice Susan Krauss, Aug. 27, 1961; children: Kenneth David, Diana Carol. B.A., Washington U., St. Louis, 1954; Ph.D., Cornell U., 1959. Instr. physics U. Mich., Ann Arbor, 1959-61, asst. prof., 1961-64, assoc. prof., 1964-69, prof., 1969—; guest physicist SSC Lab., 1991. Author: Probability and Statistics in Experimental Physics, 1992. CERN vis. scientist Geneva, 1967, 89; Brit. Sci. Rsch. Coun. fellow, Oxford, 1979; recipient inventor's prize CDC Worldtech, Edina, Minn., 1982, 83. Fellow Am. Phys. Soc. Home: 3610 Charter Pl Ann Arbor MI 48105-2825 Office: U Mich Physics Dept 500 E University Ave Ann Arbor MI 48109-1120

ROE, GEORGEANNE THOMAS, information brokerage executive; b. Washington, Apr. 1, 1945; d. George Albert and Lois Rose (Baker) Haun; m. Frank S. Weidner, Feb. 6, 1966 (div. Apr. 1969); m. John Steadman Roe, Apr. 11, 1969. BA in English, Simmons Coll., 1971, MS in LS, 1972; MBA, Babson Coll., 1980. Dir. Holbrook (Mass.) Pub. Libr., 1972-79; adminstrv. asst. Comprehensive Group Resources, Newton, Mass., 1980-84; bus. svcs. libr. Southeastern Mass. U., Dartmouth, 1984-85; ptnr., cons. Perry, Roe & Assocs., Millis, Mass., 1985—; asst. to dir. New Eng. Wild Flower Soc., Framingham, Mass., 1989—. Trustee Millis Pub. Libr., 1985-91. Mem. ALA, Assn. Ind. Info. Profls., Mass. Libr. Trustees Assn., New Eng. Libr. Assn., Mass. Libr. Assn. (chmn. intellectual freedom com. 1976-78), Women

Entrepreneurs Homebased (exec. sec. 1988—), P.E.O. (guard chpt. AM-MA 1990—, corr. sec. 1992—). Avocations: miniatures, needlework, story-telling. Home: 111 Acorn St Millis MA 02054-1410 Office: New Eng Wild Flower Soc 180 Hemenway Rd Framingham MA 01701-2636

ROE, MICHAEL DEAN, psychologist; b. Pasadena, Calif., Apr. 8, 1951; s. Robert Harrison and Marylou (Miller) R.; m. Janice Ann Beck, June 14, 1975; children: Sean Stephen, Shannon Noel, Megan Seon Jin, Corey Woo Seong. BA, U. Calif., La Jolla, 1973; MEd, U. Wash., 1975, PhD, 1981. Cert. spl. edn. tchr. Animal trainer Sea World of San Diego, Calif., 1970-73; ednl. researcher Child Devel./Mental Retardation Ctr. U. Wash., Seattle, 1973-75, clin. asst. neonatal intensive care Follow-Up Clinic, 1975-76, rsch. assoc. Med. Sch., 1977; instr. to assoc. prof. psychology Bethel Coll., St. Paul, Minn., 1977-88; scholar-in-residence Ctr. for Global Edn./Augsburg Coll., Mpls., 1986; prof. and chmn., psychology dept. Seattle Pacific U., 1988—; cons., researcher Cowlitz Indian Tribe, Longview, Wash., 1990—, Iona Coll., Jamaica, 1986. Contbr. articles to profl. jours. Recipient faculty rsch. grants Seattle Pacific U., 1989, 91, 92, Weter award for scholarship, 1990, internat. study grant PEW Trust, Philippines, 1988, Disting. Faculty award Bethel Coll., 1987. Mem. AAAS, Am. Psychol. Assn., Internat. Assn. Cross Cultural Psychology, Calif. Scholarship Fedn. (life), Amnesty Internat., Fedn. Am. Scientists. Democrat. Mennonite. Office: Dept Psychology Seattle Pacific Univ Seattle WA 98119

ROEHRENBECK, PAUL WILLIAM, marketing professional; b. Jersey City, Apr. 15, 1945; s. William Joseph and Jean Cleary (Flanagan) R.; m. Joanmarie Colette McMahon, Sept. 12, 1976; 1 child, Jean. BS in Physics, Holy Cross Coll., 1967; postgrad., Drexel U., 1968-69, Pace U., 1978-79. Applications physicist RCA Astro-Electronics, Princeton, N.J., 1967-73; from product to group mgr. EG&G Princeton Applied Rsch., 1973-82; mktg. mgr. LeCroy Corp., Chestnut Ridge, N.Y., 1982-85, Hamamatsu Photonic Systems, Bridgewater, N.J., 1986—; adv. bd. mem. Physics Today Buyers Guide, N.Y.C., 1982—. Mem. Optical Soc. Am., Am. Inst. Physics, Internat. Soc. Optical Engring., Inc. (conf. chair 1992-93). Home: 24 Autumn Hill Rd Princeton NJ 08540 Office: Hamamatsu Photonic Systems 360 Foothill Rd Bridgewater NJ 08540

ROELOFS, WENDELL LEE, biochemistry educator, consultant; b. Orange City, Iowa, July 26, 1938; s. Edward and Edith (Beyers) R.; m. Donna R. Gray, Dec. 23, 1989; children by previous marriage: Brenda Jo, Caryn Jean, Jeffrey Lee, Kevin Jon. BA, Central Coll., Pella, Iowa, 1960, DSc (hon.), 1985; PhD, Ind. U., 1964, DSc (hon.), 1986; DSc (hon.), Hobart and William Smith Colls., 1988, U. of Lund, Sweden, 1989, Free U. Brussels, 1989. Asst. prof. Cornell U., Geneva, N.Y., 1965-69, assoc. prof., 1969-76, prof., 1976—, Liberty Hyde Bailey prof. insect biochemistry, 1978—, chmn. dept., 1991—. Contbr. over 250 articles to sci. jours. Recipient Alexander von Humboldt award in Agr., 1977, Outstanding Alumni award Central Coll., 1978, Wolf prize for agr., 1982, Disting. Alumnus award Ind. U., 1983, Nat. Medal of Sci., 1983, Disting. Svc. award USDA, 1986, Silver medal Internat. Soc. Chem. Ecology, 1990; postdoctoral fellow MIT, 1965. Fellow AAAS, Entomol. Soc. Am. (J. Everett Bussart Meml. award 1973, Founder's Meml. award 1980, Disting. Achievement award Eastern br. 1983); mem. NAS, Am. Chem. Soc., Am. Acad. Arts and Sci., Sigma Xi. Republican. Presbyterian. Patentee in field (10). Home: 4 Crescence Dr Geneva NY 14456-1302 Office: Cornell U Geneva NY 14456

ROEMER, EDWARD PIER, neurologist; b. Milw., Feb. 10, 1908; s. John Henry and Caroline Hamilton (Pier) R.; m. Helen Ann Fraser, Mar. 28, 1935 (dec.); children: Kate Pier, Caroline Pier; m. Marion Clare Zimmer, May 24, 1980. BA, U. Wis., 1930; MD, Cornell U., 1934. Diplomate Am. Bd. Neurology. Intern Yale-New Haven Hosp., 1934-36; resident internal medicine N.Y. Hosp., 1936; resident neurology Bellevue Hosp., N.Y.C., 1936-38; instr. Med. Sch. Yale U., New Haven, 1935-36; asst. prof. neurology Cornell U., N.Y.C., 1936-41; prof. neurology U. Wis., Madison, 1946-64; chief of neurology Huntington Meml. Hosp., Pasadena, Calif., 1964-78; pvt. practice Capistrano Beach, Calif., 1978—; founder, dir. Wis. Neurol. FDN, Madison, 1946-64; dir. Wis. Multiple Sclerosis Clinic, Madison, 1946-64; adv. bd. Inst. Antiquities and Christianity, Claremont Grad. Sch., 1970—; dir. found. Univ Good Hope, S.Africa. Contbr. rsch. articles on multiple sclerosis, neuropathies to profl. jours. Lt. col. med. corps U.S. Army, 1941-46, ETO. Fellow ACP, Royal Coll. Medicine, L.S.B. Leakey Found.; mem. Rotary Internat., Annandale Golf Club, El Niguel Country Club, Nu Sigma Nu, Phi Delta Theta. Republican. Achievements include significant findings in field of anthropology and archaeology in Egypt and southwest U. S. relative to prehistory and PreColumbian European influences. Home: 35651 Beach Rd Capo Beach CA 92624-1710

ROEMER, ELIZABETH, astronomer, educator; b. Oakland, Calif., Sept. 4, 1929; d. Richard Quirin and Elsie (Barlow) R. B.A. with honors (Bertha Dolbeer scholar), U. Calif., Berkeley, 1950, Ph.D. (Lick Obs. fellow), 1955. Tchr. adult class Oakland pub. schs., 1950-52; lab technician U. Calif. at Mt. Hamilton, 1954-55; grad. research astronomer U. Calif. at Berkeley, 1955-56; research asso. Yerkes Obs. U. Chgo., 1956; astronomer U.S. Naval Obs., Flagstaff, Ariz., 1957-66; asso. prof. dept. astronomy, also in lunar and planetary lab. U. Ariz., Tucson, 1966-69; prof. U. Ariz., 1969—; astronomer Steward Obs., 1980—; Chmn. working group on orbits and ephemerides of comets commn. 20 Internat. Astron. Union, 1964-79, 85-88, v.p. comm. 20, 1979-82, pres., 1982-85, v.p. commn. 6, 1973-76, 85-88, pres., 1976-79, 88-91; mem. adv. panels Office Naval Research, Nat. Acad. Scis.-NRC, NASA; researcher and author numerous publs. on astrometry and astrophysics of comets and minor planets including 79 recoveries of returning periodic comets, visual and spectroscopic binary stars, computation of orbits of comets and minor planets. Recipient Dorothea Klumpke Roberts prize U. Calif. at Berkeley, 1950, Mademoiselle Merit award, 1959; asteroid (1657) named Roemera, 1965; Benjamin Apthorp Gould prize Nat. Acad. Scis., 1971; NASA Spl. award, 1986. Fellow AAAS (council 1966-69, 72-73), Royal Astron. Soc. (London); mem. Am. Astron. Soc. (program vis. profs. astronomy 1960-75, council 1967-70, chmn. div. dynamical astronomy 1974), Astron. Soc. Pacific (publs. com. 1962-73, Comet medal com. 1968-74, Donohoe lectr. 1962), Internat. Astron. Union, Am. Geophys. Union, Brit. Astron. Assn., Phi Beta Kappa, Sigma Xi. Office: U Ariz Lunar and Planetary Lab Tucson AZ 85721

ROEMMELE, BRIAN KARL, electronics, publishing, financial and real estate executive; b. Newark, Oct. 4, 1961; s. Bernard Joseph and Paula M. Roemmele. Grad. high sch., Flemington, N.J. Registered profl. engr., N.J. Design engr. BKR Techs., Flemington, N.J., 1980-81; acoustical engr. Open Reel Studios, Flemington, 1980-82; pres. Ariel Corp., Flemington, 1983-84, Ariel Computer Corp., Flemington, 1984-89; pres., chief exec. officer Ariel Fin. Devel. Corp., N.Y.C., 1987-91; pres., CEO Avalon Am. Corp., Temecula, Calif., 1990—; CEO United ATM Credit Acceptance, Beverly Hills, 1992—, United ATM Card Acceptance Corp., Beverly Hills, 1992—; pres., CEO Coupon Book Ltd., 1987-89, Value Hunter Mags., Ltd., AEON Cons. Group, Beverly Hills, Calif.; bd. dirs. Waterman Internat., Whitehouse Station, N.J., United Credit Card Acceptance Corp., Temecula; electronic design and software cons., L.A., 1980—. Pub., editor-in-chief: Computer Importer News, 1987—. Organizer Internat. Space Week or Day, 1978-83; lectr. Trenton State Mus., N.J., 1983; chmn. Safe Water Internat., Paris. Mem. AAAS, AIAA, ABA, IEEE, Am. Bankers Assn., Bankcard Svcs. Assn., Boston Computer Soc., Ford/Hall Forum, Am. Soc. Notaries, Planetary Soc. Avocations: musician, surfing, cycling, reading, numismatics. Office: Avalon Am Corp PO Box 1615 Temecula CA 92593-1615

ROEPE, PAUL DAVID, biophysical chemist; b. Suffern, N.Y., June 20, 1960; s. David George and Hazelann (Kerr) R.; m. Sarah Wallace Hamilton, Oct. 8, 1989. BA, Boston U., 1982, MA, PhD, 1987. Grad. rsch. fellow Dept Physics Boston U., 1982-87; postdoctoral fellow Roche Inst. for Molecular Biology, Nutley, N.J., 1987-89; rsch. assoc. Howard Hughes Inst. UCLA, L.A., 1989-90; asst. mem. Sloan Kettering Cancer Ctr., N.Y.C., 1990–; asst. prof. Cornell U. Med. Coll., N.Y.C., 1990–. Contbr. articles to profl. jours. Recipient Sackler Biomed. Scholar award Raymond and Beverly Sackler Meml. Fund, 1991—; postdoctoral fellow Jane Coffin Childs Fund, 1987-91. Mem. Biophysical Soc., N.Y. Assn. Acad. Scis., Am. Assn. for the Advancement Sci. Office: Sloan Kettering Inst 1275 York Ave New York NY 10021

ROESER, ROSS JOSEPH, audiologist, educator; b. Louisville, Nov. 14, 1942; s. Carl Henry and Yvonne Marie (Phillips) R.; m. Sharon Lynn Hill, June 9, 1962; children: Wnedy Ann, Elizabeth Marie, Jennifer Yvonne. BS, Western Ill. U., 1966; MA, No. Ill. U., 1967; PhD, Fla. State U., 1972. Audiologist Anna (Ill.) State Hosp., 1967-69; chief of audiology Callier Ctr., Dallas, 1972-88, dir., 1988—; prof. audiology U. Tex., Dallas, 1975—; clin. assoc. prof. dept. otolaryngology UTSWMC, 1975—. Author: Auditory Disorders, 1982; founder Ear and Hearing jour., 1979. Recipient Alumni Achievement award Western Ill. U., 1978, Fla. State U., 1990; recipient Joel Wernick award Acad. Dispensing Audiologists, 1990. Home: 1921 Marydale Rd Dallas TX 75208-3034 Office: Callier Ctr 1966 Inwood Rd Dallas TX 75235-7298*

ROESKY, HERBERT WALTER, chemistry educator; b. Laukischken, Germany, Nov. 6, 1935; s. Otto and Lina (Hublitz) R.; m. Christel Glemser, July 24, 1964; children: Rainer, Peter. Diploma, U. Göttingen, Fed. Republic Germany, 1961, PhD, 1963; prof. (hon.), Nankai U., Teinjin, China, 1990; PhD (hon.), U. Bielefeld, Germany, 1992. Postdoctoral fellow DuPont Co., Wilmington, Del., 1965-66; docent U. Göttingen, 1968-71, dir. Inst. Inorganic Chemistry, 1980—; prof. inorganic chemistry U. Frankfurt, Fed. Republic Germany, 1971-80; vis. prof. Tokyo Inst. Tech., 1987, Kyoto U. Elsevier: Rings, Clusters and Polymers of Main Group and Transition Elements, 1990; contbr. over 600 articles to profl. publs. Recipient Dozenten prize Fonds der Chemischen Industrie Frankfurt, 1970, Alfred-Stock-Gedächtnis prize Gesellschaft Deutscher Chemiker, 1990. Mem. Acad. Scis., Akademie der Naturforscher Leopoldina, Austrian Acad. Scis. (corr.). Achievements include 12 patents in field. Home: Emil-Nolde-Weg 23, 37085 Göttingen Germany Office: Inst Anorganische Chemie, Tammannstrasse 4, 37077 Göttingen Germany

ROETH, FREDERICK WARREN, agronomy educator; b. Houston, Ohio, Aug. 21, 1941; s. Roy O. and Gertrude (Durand) R.; m. Carol Logemann, June 9, 1968; children: Alicia, Bradley, Cari. BS, Ohio State U., 1964; MS, U. Nebr., 1967, PhD, 1970. Asst. prof. Purdue U., West Lafayette, Ind., 1969-75; asst. prof. U. Nebr., Lincoln, 1975-76, assoc. prof., 1976-83, prof., 1983—. Assoc. editor: (jour.) Weed Technology. Mem. Clay Center (Nebr.) Sch. Bd., Clay Center Ambulance Assn. Recipient Disting. Svc. award Nebr. Coop. Extension, 1987. Mem. AAAS, Weed Sci. Soc. Am., Am. Soc. Agronomy, Coun. for Agr. and Tech. Office: U Nebr Box 66 Clay Center NE 68933

ROGALLO, FRANCIS MELVIN, mechanical, aeronautical engineer; b. Sanger, Calif., Jan. 26, 1912; s. Mathieu and Marie Rogallo; m. Gertrude Sugden, Sept. 14, 1939; children: Marie, Robert, Carol, Frances. Degree in Mech. and Aero. Engring., Stanford U., 1935. With NACA (name changed to NASA), Hampton, Va., 1936-70, Kitty Hawk Kites, Nags Head, S.C.; lectr. on high lift devices, lateral control, flexible wings and flow systems. Contbr. articles to profl. jours. Active Rogallo Found. Recipient NASA award, 1963, Nat. Air and Space Mus. trophy, Lifetime Achievement award Smithsonian Instn., 1992; named to N.C. Sports Hall of Fame, 1987. Mem. Am. Kite Assn., U.S. Hang Gliding Assn. Episcopalian. Achievements include patents for Flexible Wing, Corner Kite, 23 others; research in aerodynamics, wind tunnels. Home: 91 Osprey Ln Kitty Hawk NC 27949 Office: Kitty Hawk Kites PO Box 1839 Nags Head NC 27959*

ROGALSKI, JERZY MARIAN, biochemist, educator; b. Przemysl, Poland, Feb. 10, 1956; s. Bronislaw and Ludmila (Klisowska) R.; m. Jolanta Barbara Wojtas, Aug. 20, 1983. MSc, Marie Curie Sklodowska U., Lublin, Poland, 1979, PhD, 1986, habil., 1992. Jr. asst. Marie Curie Sklodowska U., Lublin, 1979-80, asst., 1980-83, sr. asst., 1983-87, lectr., 1987—; vis. researcher Tech. U. Vien, 1986, U. Helsinki, 1987-90; mem. biotech. team Polish Agriculture Ministry, 1993—. Patentee in field; co-author: chpts. in books; contbr. over 100 articles to profl. jours. Lt. Polish Chem. Div., 1985—. Recipinet award Polish Acad. Scis. 1989, Polish Ministry Edn., 1989, 93. Mem. AAAS, Polish Biochem. Soc., Polish Chem. Soc., Polish Genetic Soc., Polish Microbiology Soc., N.Y. Acad. Scis., Polish Hunting Corp., Polish Red Cross, Polish Fishing Corp., Polish Ecology Corp. Roman Catholic. Avocations: hunting, fishing, photography. Home: Hanki Sawickiej St 15, 37-700 Przemysl Poland Office: M Curie Sklodowska U, Biochemistry Dept, M C Sklodowska Pl No 3, 20-031 Lublin Poland

ROGERS, CHARLIE ELLIC, entomologist; b. Booneville, Ark., Aug. 13, 1938; s. Robert Wesley and Parthenia Fern (Mahoney) R.; m. Donna Carol Ray, Jan. 29, 1971; children: Christian Edward, Cheryl Elaine. PhD, Okla. State U., 1970. Cert. entomologist. Tchr. Dysart Pub. Sch., Glendale, Ariz., 1964-65; grad. rsch. asst. U. Ky., Lexington, 1965-67; grad. rsch. asst. Okla. State U., Stillwater, Okla., 1967-70, postdoctoral rsch. assoc., 1970-71; asst. prof. Tex. A&M U., College Station, Tex., 1971-74; rsch. entomologist USDA Agrl. Rsch. Svc., Bushland, Tex., 1974-83; supervisory rsch. entomologist USDA Agrl. Rsch. Svc., Tifton, Ga., 1983—; editor Biol. Control Acad. Press, San Diego, 1990—; rsch. com. Sunflower Assn. Am., Fargo, N.D., 1981-83; dir. elect Bd. Cert. Entomologist Entomol. Soc. Am., Lantham, Md., 1992. Author: Sunflower Species of the United States, 1982; co-editor: The Entomology of Indigenous and Naturalized Systems in Agriculture, 1988; contbr. articles to profl. publs. Sch. bd. mem. Bushland Ind. Sch. Dist., 1981-83; tchr. First Bapt. Ch., Tifton, 1986-91; mem. Rotary Club, Vernon, Tex., 1974-76. With U.S. Army, 1958-61. Recipient Profl. Excellence award Am. Agrl. Econ. Assn., 1979, Leadership award USDA, 1987, Profl. Svc. award Am. Registered Profl. Entomologists, 1987, Outstanding Performance award USDA, 1991, Disting. Alumni award, 1992. Mem. Entomol. Soc. Am. (com. 1987-92), Ga. Entomol. Soc., Entomol. Soc. S.C.,Sigma Xi, Tifton Club (pres. 1992-93). Republican. Baptist. Achievements include study of biology, ecology, and control of sheep nose bot, oestrus ovis; biology and augmentation of Propylea 14-punctata against the green bus; control of guar midge Conterina texana; control of insect pests of sunflowers; bionomics of a parasitic nematode Noctuidonema guyanense. Home: 1711 Sarah Dr Tifton GA 31793-0748 Office: USDA/ARS-Insect Biology & Population Mgmt Rsch Lab PO Box 748 Tifton GA 31793-0748

ROGERS, COLONEL HOYT, agricultural consultant; b. Mullins, S.C., Jan. 6, 1906; s. Colonel Cross and Mary (Page) R.; B.S., Clemson Coll., 1926; M.S. (Research fellow), U. Ky., 1927; Ph.D. in Plant Physiology (Teaching research fellow), Rutgers U., 1930; m. Justine Frances Harris, Sept. 27, 1927; children—James H., Richard L. Instr. biology Ark. State Coll., Jonesboro, 1927-28; instr. botany Rutgers U., New Brunswick, N.J., 1928-29, asst. research plant physiology, 1929-31; plant pathologist Tex. A&M U., Temple, 1931-42; plant pathologist tobacco and cotton research Coker's Pedigreed Seed Co., Hartsville, S.C., 1942-60, v.p., 1960-72, cons. U.S. and fgn. agr., 1972—; mem. bd. rev. tobacco N.C. State U. Mem. faculty adv. com. plant pathology and physiology and bd. visitors Clemson U., 1972-74. Named Man of Year, Progressive Farmer, 1969, Man of Year, Farmer Coops., 1972, Man of Year, Mullins (S.C.) C. of C., 1977, Man of Year in Agr., N.C. State U., 1976; recipient Disting. Service award S.C. Tobacco Warehouse Assn., 1969, N.C. State U., 1976; Disting. Alumnus award Clemson U., 1982. Mem. Am. Soc. Agronomy, Crop Sci. Soc. Am., Bot. Soc. Am., Am. Phytopathol. Soc., Am. Farm Bur., S.C. Farm Bur., Sigma Xi. Contbr. sci. and popular articles to profl. jours. Developer 21 varieties tobacco; leader tobacco breeding program in Italy. Address: Route 4 Box 532 Mullins SC 29574

ROGERS, CRAIG ALAN, mechanical engineering educator, university program director; b. Barre, Vt., Dec. 1, 1959; s. Alan Craig and Beverly Ann (Perkins) R.; m. Kathleen Alecia Tucker, May 24, 1980; children: Jennifer Kathleen, David Craig. Assoc in Mech. Engring., Vt. Tech. Coll., 1979; BSME, N.J. Inst. Tech., 1982; MSME, Va. Poly. Inst. and State U., 1983, Phd in Mech. Engring., 1987. Phys. design engr. Bell Telephone Labs., Holmdel, N.J., 1979-81; instr. mech. engring. Va. Poly. Inst. and State U., Blacksburg, 1983-87, vis. asst. prof., 1987-88, asst. prof., 1988-90, assoc. prof., 1990-93, prof., 1993—, dir. Ctr. for Intelligent Material Systems and Structures, 1990—; pres. Paradigm, Inc., Blacksburg, 1989—; cons. numerous materials and aerospace cos., 1987—; lectr. in Japan, Belgium, U.K., Sweden, Australia, Fed. Reoublic Germany, Switzerland, Russia, Republic of China; mem. internat. intelligent materials forum Japanese Ministry Sci. and Tech., 1990—. Editor: (monograph) Smart Material Systems and Structures, 1989; editor: U.S. Japan Workshop on Smart/Intelligent Materials and Systems, 1990, Recent Advances in Active Vibration and Acustic Control, 1991, Advances in Adaptive and Sensory Materials, 1992; editor-in-chief Jour. Intelligent Material Systems and Structures, 1990; contbr. articles to tech. jours. Recipient Young Investigator award Office Naval Rsch., 1988, Presdl. Young Investigator award NSF, 1991. Mem. ASME (nat. tech. com. on reliability stress analysis and failure prevention 1987—, nat. tech. com. on aerospace materials and structures 1990—, chair nat. tech. com. adaptive structures and material systems, Adaptive Structures and Material Systems prize 1993). Avocations: hang gliding, golf, travel. Home: 1509 Hoyt St Blacksburg VA 24060-2517 Office: Va Tech Ctr Intelligent Material Systems Blacksburg VA 24061-0261

ROGERS, DALE ARTHUR, electrical engineer; b. Denver, June 30, 1954; s. Gordon Wayne and Florence (Ruckus) R.; m. Jill Kathleen Gislason, Sept. 21, 1989; children: Abby, Lydia. BSEE, Colo. State U., 1983; MSEE, U. Colo., 1993. Engr. Martin Marietta Astronautics, Denver, 1983—. With USAF, 1972-78. Home: PO Box 621193 Littleton CO 80162 Office: Martin Marietta Astronautis PO Box 179 MS L5420 Denver CO 80201

ROGERS, DAVID ANTHONY, electrical engineer, educator, researcher; b. San Francisco, Dec. 21, 1939; s. Justin Anthony and Alice Jane (Vessey) R.; m. Darlene Olive Hicks, Feb. 20, 1965; 1 child, Stephen Arthur. BSEE cum laude, U. Wash., 1961, PhD in Elec. Engring., 1971; MSEE, Ill. Inst. Tech., 1964; MDiv cum laude, Trinity Evangelical Div. Sch., Deerfield, Ill., 1966. Registered profl. engr., Wash. Assoc. engr. Ford Aero., Newport Beach, Calif., 1961; tech. asst. IIT Rsch. Inst., Chgo., 1963, grad. fellow, 1963-64; predoctoral lectr. U. Wash., Seattle, 1964-65, 66-71, acting asst. prof., 1971-72; asst. prof. State U. of Campinas, Brazil, 1972-77, assoc. prof., 1977-80; assoc. prof. elec. engring. N.D. State U., Fargo, 1980-86; prof. 1986—; external MS thesis examiner Poly. Sch. U. Sao Paulo, Brazil, 1974, external Ph.D Thesis examiner Inst. Tech., Banaras Hindu U., India, 1989, 91; researcher microwaves, fiber optics, electromagnetics, profl. and rsch. ethics, engring. edn. Mem. editorial bd. IEEE Transactions Microwave Theory and Techniques, 1987—. Mem. rev. panel NSF, Quantum Electronics Waves and Beams program, 1989; mem. tech. program com. Internat. Symposium on Recent Advances in Microwave Tech., China, 1989, Reno, 1991, India, 1993; judge N.D. Sci. Olympiad, 1987—, S.E. N.D. Regional Sci. and Engring. Fair, 1993. 2d lt. Signal Corps, U.S. Army, 1961-62. NSF Summer fellow, 1965; grantee Ford Found., 1969-70, TELEBRÁS (Brazil), 1973-80, N.D. State U.-Bush Found., 1981-88, 91—; engr. and arch. rep. N.D. State Univ. Faculty Seminar (interdisciplinary, multi-cultural and internat. studies, 1991—). Mem. IEEE, Am. Soc. Engring. Edn. (internat. and other divs., grantee summer 1984), Vols. in Tech. Assistance, NSPE, N.D. Acad. Sci., Am. Geophysical Union, Applied Computational Electromagnetics Soc., Fargo-Moorhead Engrs. Club, Evangelicals for Social Action, Nat. Inst. Engring. Ethics, Am. Sci. Affiliation, Am. Radio Relay League (life), Order of Engr., Sigma Xi, Tau Beta Pi, Eta Kappa Nu. Evangelical. Co-author: Fiber Optics, 1984. Contbr. articles to profl. pubs. incl. IEEE Transactions on Antennas and Propagation, Transactions on Edn., Transactions on Microwave Theory and Techniques, Jour. of Quantum Electronics, Electronics Letters, Radio Sci., Engring. Edn., Computers in Edn. Jour. Office: ND State U Elec Engring Dept Fargo ND 58105

ROGERS, JAMES EDWIN, geology and hydrology consultant; b. Waco, Tex., Feb. 24, 1929; s. Charles Watson and Jimmie (Harp) R.; m. Margaret Anna Louise Bruchmann, Oct. 10, 1957; 1 child, James Fredrick. Student, Rice U., 1947-49, Baylor U., 1953; BS, U. Tex., 1955, MA, 1961. Geologist U.S. Geol. Survey, St. Paul, 1956-59; geologist U.S. Geol. Survey, Alexandria, La., 1959-63, supervisory hydrologist, 1963-85; ind. cons. Alexandria, 1985—; cons. geol. survey for map State of La., Baton Rouge, 1982-85. Author: Water Resources of Kisatchie Well-Field Area Near Alexandria, Louisiana, 1981, Preconstruction and Simulated Postconstruction Ground-Water Levels at Urban Centers in the Red River Navigation Project Area, Louisiana, 1983, Red River Waterway Project - Summary of Ground-Water Studies by the U.S. Geological Survey, 1962-85, 1988; co-author: Water Resources of Vernon Parish, Louisiana, 1965, Water Resources of Ouachita Parish, Louisana, 1972, Water Resources of the Little River Basin, Louisana, 1973. Scoutmaster Boy Scouts Am., Alexandria, 1971, 72. Sgt. U.S. Army, 1950-52, Japan. Mem. Assn. Ground Water Scientists and Engrs., Geol. Soc. Am., Gem Mineral and Lapidary Soc. Cen. La. (pres. 1972, 86, 87), Baton Rouge Geol. Soc., Phi Beta Kappa. Presbyterian. Avocations: numismatics, minerals, genealogy, travel, history. Home and Office: 4008 Innis Dr Alexandria LA 71303-4738

ROGERS, JAMES VIRGIL, JR., radiologist, educator; b. Johnson City, Tenn., Oct. 7, 1922; s. James Virgil and Mary Ruth (Collins) R.; m. Mildred Vandivere, June 9, 1945 (div. 1985); children: Rebecca Jean, James V. III, Janet Marie, Susan Margaret; m. Mary Lujean Craven, Mar. 18, 1989. BS, Emory U., 1943, MD, 1945. Intern Kings County Hosp., N.Y.C., 1945-46; resident in radiology Grady Meml. Hosp., Altanta, 1947-48, Emory U. Hosp., Altanta, 1948-50; instr. Emory U. Sch. Medicine, Atlanta, 1950-51, assoc., 1950-53, asst. prof., 1954-60, assoc. prof., 1960-64, prof., 1965-93; ret.; chief radiology svc. Emory U. Hosp., Atlanta, 1971-80, vice chmn. dept. radiology 1973-78, acting chmn., 1978-80; dep. sect. head Emory Clinic, Atlanta, 1971-78, sect. head, 1978-80, chief radiology svc., 1982-91; cons. radiology 3d Army Hdqrs., Atlanta; examiner Am. Bd. Radiology, 1970-79, site inspector for residency programs, 1972-83. Contbr. articles to profl. jours. Pres. PHRC Ga., Atlanta, 1990. 1st lt. U.S. Army, 1946-47. Fellow Am. Coll. Radiology (councilor); mem. Radiol. Soc. North Am., Am. Roentgen Ray Soc., So. Radiologic Conf. (past pres.), Ga. Radiologic Soc. (past pres.), Atlanta Radiologic Soc. (past. pres.), AMA, So. Med. Assn. (past pres. radiology sect.), Druid Hills Golf Club, Am. Legion. Republican. Methodist. Avocations: golf, fishing. Home: 1715 Silver Hill Rd Stone Mountain GA 30087-2212

ROGERS, JOHN JAMES WILLIAM, geology educator; b. Chgo., June 27, 1930; s. Edward James and Josephine (Dickey) R.; m. Barbara Bongard, Nov. 30, 1956; children: Peter, Timothy. BS, Calif. Inst. Tech., 1952, PhD, 1955; MS, U. Minn., 1952. Lic. geologist, N.C. From instr. to prof. Rice U., Houston, 1954-75, master Brown Coll., 1966-71, chmn. geol. dept., 1971-74; W.R. Kenan Jr. prof. geology U. N.C., Chapel Hill, 1975—. Co-author: Fundamentals of Geology, 1966, Precambrian Geology of India, 1987; co-editor: Holocene Geology of Galveston Bay, 1969, Precambrian of South India, 1983, Basalts, 1984, African Rifting, 1989, A History of the Earth, 1993; regional editor Jour. African Earth Scis., 1982-93; contbr. articles to profl. jours. Fellow Geol. Soc. Am., Geol. Soc. India, Geol. Soc. Africa (hon.); mem. Mineral. Soc. Am., Am. Assn. Petroleum Geologists, Soc. Econ. Paleontologists and Mineralogists. Home: 1816 Rolling Rd Chapel Hill NC 27514-7502 Office: U of NC Dept Geology CB #3315 Chapel Hill NC 27599-3315

ROGERS, JOHN RUSSELL, engineer; b. St. Louis, May 12, 1929; s. John Flint and Faye (Russell) R.; m. Lorraine Esther Klockenbrink, Sept. 15, 1951; children: John Oliver, Gail Joanne. AB in Econs., Washington U., St. Louis, 1951. Registered profl. engr., Mo., Ill. Mfg. engr. Day Brite Lighting Inc., St. Louis, 1957-59; plant supt. Day Brite Lighting Inc., Tuptlo, Miss., 1959-64; plant mgr. White Rodgers Ltd., Markham, Ont., Can., 1964-66; ops. mgr. Metal Goods Corp., St. Louis, 1966-71; v.p., prin. Ross & Baruzzini, Inc., Cons. Engrs., St. Louis, 1971-84; pres. John R. Rogers Assocs., Inc., Cons. Engrs., St. Louis, 1984—. Bd. dirs. Grace Hill Settlement House, St. Louis, Grace Hill Child Devel. Bd.; pres. Thompson Ctr., St. Louis, 1984. Capt. U.S. Army, 1951-54. Mem. NSPE, Inst. Indsl. Engrs. (sr.), Am. Cons. Engrs. Coun., Assn. Profl. Materials Handling Cons. (pres. 1986-88, bd. dirs.), Materials Handling & Mgmt. Soc. (v.p. 1983—, bd. dirs.), Soc. Mfg. Engrs. (sr.), Rotary. Avocations: tennis, amateur radio. Home and Office: John Rogers Assocs Inc 10332 Richview Dr Saint Louis MO 63127-1433

ROGERS, KATE ELLEN, educator; b. Nashville, Dec. 13, 1920; d. Raymond Lewis and Louise (Gruver) R.; MA in Fine Arts, George Peabody Coll., 1947; EdD in Fine Arts and Fine Arts Edn., Columbia U., 1956. Instr., Tex. Tech. Coll., Lubbock, 1947-53; co-owner, v.p. Design Today, Inc., Lubbock, 1951-54; student asst. Am. House, N.Y.C., 1953-54; asst. prof. housing and interior design U. Mo., Columbia, 1954-56, assoc. prof., 1956-66, prof., 1966-85, emeritus, 1985—, chmn. dept. housing and interior design, 1973-85; mem. accreditation com. Found. for Interior Design Edn. Research, 1975-76, chmn. standards com., 1976-82, chmn. research, 1982-85.

Mem. 1st Bapt. Ch., Columbia, Mo.; bd. dirs. Meals on Wheels, 1989-91. Nat. Endowment for Arts research grantee, 1981-82. Fellow Interior Design Educators Council (pres. 1971-73, chmn. bd. 1974-76, chmn. research com. 1977-78); mem. Am. Soc. Interior Designers, (hon., medal of honor 1975), Am. Home Econs. Assn., Columbia Art League (adv. bd. 1988-93), Pi Lambda Theta, Kappa Delta Pi, Phi Kappa Phi (hon.), Gamma Sigma Delta, Delta Delta Delta (Phi Eta chpt.), Phi Upsilon Omicron, Omicron Nu (hon.). Democrat. Author: The Modern House, USA, 1962; editor Jour. Interior Design Edn. and Research, 1975-78.

ROGERS, MCKINLEY BRADFORD, mechanical engineer; b. Sevierville, Tenn., June 6, 1955; s. Kenneth Bradford and Betty Jo (Jenkins) R.; m. Kimberly Lane Mason, Jan. 1, 1984. BSME, Larmar U., 1978. Registered profl. engr., Tex. Generations planning engr. Gulf States Utilities, Beaumont, Tex., 1978-86; energy analyst CSA, Inc., Arlington, Va., 1986-89, Exeter Assocs., Silver Spring, Md., 1989-90; dir. Meridian Corp., Alexandria, Va., 1990—. Contbr. articles to profl. jours. Mem. ASME, Assn. of Energy Engrs., Tau Beta Pi, Pi Tau Sigma. Office: Meridian Corp 4300 King St Alexandria VA 22302

ROGERS, ROBERT WAYNE, electronics engineer; b. Grovertown, Ind., Feb. 25, 1935; s. William Howard and Enid E. (Ecker) R.; m. Gail Aleda Ritenour, Aug. 29, 1954; children: Michael Lee, Allen Jay. BSEE, Ind. Tech. Coll., 1957. Engr. Bendix Corp., South Bend, Ind., 1957-59, Miles, Inc., Elkhart, Ind., 1959—; adv. bd. Elkhart Area Career Ctr., 1982—. Contbr. articles to Analytical Chemistry, other tech. publs. Mem. IEEE. Mem. Ch. of Brethren. Achievements include patent on differential conductivity measuring apparatus. Office: Miles Inc PO Box 70 Elkhart IN 46515

ROGERS, RODDY, geotechnical engineer; b. Springfield, Mo.. BSChemE cum laude, U. Mo., 1981, MSChemE, 1983, MS in Engring. Mgmt., 1990. Registered profl. engr., Mo. Asst. and staff engr. Dames and Moore Consulting Firm, Phoenix, Ariz., 1983-85; project mgr. City Utilities, Springfield, 1985-90, sr. engr. civil engring. sect. system engring., 1990—; teaching asst. soil mechanics U Mo., 1981-83, rsch. asst. soil mechanics lab., 1982-83; apptd. mem. Mo. Dam and Reservoir Safety Coun.; presented numerous papers. Contbr. articles to profl. jours. mem. First Bapt. Ch., Springfield, 1959—, sponsor for youth group, 1986, 87, coach ch. softball and basketball teams, 1986-88, class dir., 1986-89, usher adv. com. 1987-89, activities evaluation com. 1988-90, dir. prayer group com., 1989-90, missions com. 1989—; mem., associational basketball tournament com. Greene County, 1985-89, associational track meet com., 1986-89; judge sci. fair Springfield Pub. Schs., mem. task force to aid high sch. students in devel. reading, 1988-89; active Boy Scouts Am., 1990; dir. Jr. Achievement, 1986-87, presenter 1988, 89. Mem. NSPE (judge chpt. sci. fair com. 1987, chmn. 1988, chpt. treas. 1988-89, chpt. sec., mem. various coms., chpt. historian 1990-91, chpt. pres.-elect 1990-91, Young Engr. of Yr. Ozark chpt. 1990, Young Engr. of Yr. 1991, Edmund Friedmund Young Engr. award for Svc. to global community 1991, presenter numerous confs.), ASCE (Phoenix br. 1983-85, mid-Mo. br.), Am. Water Works Assn., Nat. Water Well Assn., Assn. State Dam Safety Ofcls. (award of excellence in dam safety), Mo. Soc. Profl. Engrs. (state publs. com. 1987-88, 88-89, Young Engr. of Yr. 1990, presenter numerous confs.), Extra Mile Resolution), Mid-Mo. Soc. Civil Engrs. (jr. dir. 1987-88, 88-89, 3d v.p. 1989-90, 2d v.p. 1990-91), U. Mo.-Rolla Civil Engrs. Alumni Adv. Coun., Tau Beta Pi, Chi Epsilon. Home: 3426 W Camelot Springfield MO 65807 Office: City Utilities Springfield PO Box 551 Springfield MO 65801

ROGERS, ROSS FREDERICK, III, nuclear engineer; b. Shreveport, La., Aug. 27, 1944; s. Ross Frederick and Jane Ann (Gleason) R.; m. Georganna M. Howe, Aug. 19, 1966 (div. 1974); children: Charles, Randolph; m. Ollie Delane Fields, July 31, 1974; children: Laura, David, Kristen. BS in Engring., U.S. Naval Acad., 1966; MEd in Math., Ga. So. U., 1982. Cert. chief naval nuclear engr., sta. nuclear engr.; licensed sr. reactor operator. Reactor inspector U.S. Nuclear Regulatory Commn., Atlanta, 1974-81, asst. dir., 1982; mgr. ops. Ga. Power Co./Hatch, Baxley, 1983; asst. plant mgr. Miss. Power & Light, Vicksburg, 1984, mgr. projects, 1985-89; with TVA, Chattanooga, 1989—; bd. dirs. EPRI Tech. Ctr., Syracuse, N.Y. Active Tenn. Consumer Adv. Coun., Nashville, 1990-92. Lt. USN, 1966-74. Decorated Commendation medal. Mem. Am. Nuclear Soc. (local chmn. 1991-92), Chattanooga C. of C. (nuclear power rep. 1989-92). Republican. Home: 2 Leith Circle Signal Mountain TN 37377 Office: TVA Nuclear Projects 1101 S Market St LP 5B Chattanooga TN 37402

ROGERS, VERN CHILD, engineering company executive; b. Salt Lake City, Aug. 28, 1941; s. Vern S. and Ruth (Child) R.; m. Patricia Powell, Dec. 14, 1962. BS, U. Utah, 1965, MS, 1965; PhD, MIT, 1969. Registered profl. engr. Assoc. prof. Brigham Young U., Provo, Utah, 1969-73; vis. assoc. prof. Lowell Tech. Inst., 1970-71; mgr. IRT Corp., San Diego, 1973-76; v.p. Ford, Bacon & Davis, Salt Lake City, 1976-80; pres. Rogers & Assocs. Engring. Corp., Salt Lake City, 1980—. Contbr. articles to profl. jours. Mem. Health Physics Soc., Am. Soc. Profl. Engrs., Am. Nuclear Soc., Am. Chem. Soc. Mormon. Office: Rogers & Assocs Engring Corp PO Box 330 Salt Lake City UT 84110-0330

ROGERS, VERNA AILEEN, mechanical and biomedical engineer, researcher; b. Talmo, Kans., Jan. 15, 1930; d. John Orville and Fleeta Fern (Lowell) Cory; m. George Jerry Rogers, Aug. 15, 1951; 1 child, Hedy Elaine Rogers Dick. BA cum laude, Rutgers U., 1957; BS, Drexel U., 1963, MS, 1967, PhD, 1970. NIH spl. fellow Drexel U., Phila., 1965-70; mgr. biomed. engring. U. City Sci. Ctr., Phila., 1970-74; project mgr. Computer Horizons, Cherry Hill, N.J., 1975-76; asst. prof. Drexel U., Phila., 1977-83; NASA rsch. assoc. Nat. Rsch. Coun., Hampton, Va., 1981-83; engring. specialist PRC Kentron at NASA, Hampton, Va., 1983-89; multi media pub. Williamsburg, Va. Contbr. articles to profl. jours. Bd. dirs. LWV, Moorestown, N.J., 1955. Recipient Faculty fellowship NASA, 1980, 81. Mem. IEEE (v.p. engring. in medicine and biology gourp 1976, pres. 1977, chmn. edn. 1980), N.Y. Acad. Sci., Phi Kappa Phi, Sigma Pi Sigma. Home: 650 Counselors' Way Williamsburg VA 23185

ROGGERO, ARNALDO, polymer chemistry executive; b. Novi Ligure, Italy, Aug. 11, 1934; s. Carlo and Olimpia Laura (Sciutti) R.; m. Nada Mossi, Oct. 9, 1966. D Indsl. Chemistry, U. Milan, 1960. Researcher Montedison, Milan, 1960-63; sr. researcher Snamprogetti Lab., San Donato Milanese, Italy, 1963-69, head of lab., 1970-78; head of lab. Assoreni, San Donato Milanese, Italy, 1978-84; mgr. Eniricerche, San Donato Milanese, Italy, 1985—, sr. scientist, 1991—; chmn. internat. congresses, Europe, 1988—. Contbr. chpt. to book, numerous articles to profl. jours. Brite-Euram Project grantee CEC, Brussels, 1992. Achievements include 94 patents for polymeric materials. Home: Via Libertà 72, 20097 San Donato Milanese, Italy Office: Eniricerche, Via Maritano 26, 20097 San Donato Milanese, Italy

ROGISTER, ANDRE LAMBERT, physicist; b. Melen, Liege, Belgium, Dec. 3, 1940; s. Joseph Adam and Francoise Jeanne (Pellens) R.; m. Marie Madeleine Jacobs, Nov. 16, 1963; children: Yves-André, Fabien. Ing. Physicien, U. de Liege, Belgium, 1963; PhD, Princeton U., 1968. Postdoctoral fellow European Space Rsch. Inst., Frascati, Italy, 1968-69; rsch. physicist European Space Rsch. Orgn., Frascati, Italy, 1969-72; sci. officer Euroatom, 1973-89, prin. sci. officer, 1989—. Contbr. over 50 articles to profl. jours. Belgian Am. Ednl. Found. fellow, 1964-65, European Space Rsch. Orgn., NASA fellow, 1965-67. Mem. Am. Phys. Soc., European Phys. Soc., N.Y. Acad. Scis. Home: Steinroth 50, 4700 Eupen Belgium Office: Inst Plasmaphysik des, Forschungszentrum Julich GmbH, D-52425 Jülich Germany

ROGLER, CHARLES EDWARD, medical educator; b. Jersey City, N.J., Dec. 31, 1946; s. Paul Vincent and Edith Alida (Meisinger) R.; m. Leslie E. Toth, Sept. 24, 1988; children: Christopher, Patrick, Kimberly. BS, Rutgers U., 1968, MS, 1970; PhD, U. Calif., Davis, 1974. Asst. prof. U. Md., Balt., 1977-79; rsch. assoc. Inst. Cancer Rsch., Phila., 1979-83; mem. faculty Albert Einstein Coll. Medicine, Bronx, N.Y., 1983—, prof. medicine, microbiology and immunology, 1992—. Contbr. chpts. to books, articles to Jour. Virology, Carcinogenesis, other profl. publs. Mem. Emmanuel Luth. Ch., Pleasantville, N.Y. Capt. U.S. Army, 1973. Grantee NIH, 1992—. Mem. N.Y. Acad. Sci., Harvey Soc. Achievements include pioneering in cloning of Hepatitis virus integrated DNA from liver cancer, integrated Hepatitis DNA from normal liver, discovery of Novel forms of Hepatitis virus DNA, and mechanism of viral DNA integration. Office: Albert Einstein Coll Med 1300 Morris Park Ave Bronx NY 10532

ROGNONI, PAULINA AMELIA, cardiologist; b. Panama City, Panama, Mar. 21, 1947; d. Mario Carlos-Enrique and Isabel Maria (Rodriguez) R.; B.S. in Chemistry (Tulane scholar), Tulane U., New Orleans, 1969, M.D., 1973. Rotating intern Gorgas Hosp., C.Z., Panama, 1973-74; intern internal medicine Touro Infirmary, New Orleans, 1974-75; resident in internal medicine Charity Hosp., New Orleans, 1975-77; fellow in cardiology Charity Hosp., 1977-79, VA Hosp., New Orleans, 1979-80; compulsory rural intern Panamanian Govt., Hosp. Amador Guerrero, Panam, 1980-81; dir. intensive care unit Gorgas Army Hosp., Meddac, Panama; cardiology cons. Clinica San Fernando. Mem. ACP, Med. Assn. Panama Canal Area, Am. Coll. Cardiology (asso.), Am. Med. Women's Assn., Am. Heart Assn., Mussor-Burch Soc., AMA. Assn. Mil. Surgeons U.S., Sociedad Panamena de Cardiologia, Sociedad Panamena de Mecidina Interna, Chi Beta, Beta Beta Beta, Alpha Epsilon Delta. Roman Catholic. Club: Panama Soroptomists. Home: U South Am Meddac, Panama Gorgas Army Hosp dept Medicine, 34004-5000 Panama Panama

ROGO, KATHLEEN, safety engineer; b. Carrollton, Ohio, Sept. 28, 1952; d. Silvio and Mary (Siragusano) R. Grad. high sch., Carrollton; PhD in Med. Sci. (hon.), Ohio Valley Pathologists, Inc., 1992. Cert. histotechnologist, emergency med. technologist, safety engr. Rsch. pathology trainee Aultman Hosp., Canton, Ohio, 1970-75, supr. anatomic pathology, 1974-75; lab. mgr. W. Morgan Lab., Canton, 1973-74; supr. anatomic pathology Dr.'s Hosp., Massillon, Ohio, 1975-78; emergency med. technician Canton Fire Dept., 1976-81; safety engr. Ashland Oil Co., Canton, 1980-82; rsch. pathologist assoc., med. cons, v.p. Ohio Valley Pathologists, Inc., Wheeling, W.Va., 1990—. Mem. Am. Soc. Clin. Pathology (cert. histotechnician), Am. Soc. Safety Engrs. (cert.), Am. Soc. Emergency Med. Technicians (cert.), Ohio State Med. Soc. Democrat. Roman Catholic. Avocations: professional model, dancer and musician. Home: 1 Cloverfields Wheeling WV 26003

ROGOWAY, LAWRENCE PAUL, civil engineer, consultant; b. Portland, Oreg., Dec. 9, 1932; s. Phillip F. and Shirley G. (Goldstein) R.; children: Cathy L. Hucke, Karen G. Callow. BS in Civil Engring., Oreg. State U., 1954. Registered profl. engr. Ariz., Calif., Colo., Fla., Nev., Okla., Oreg., Va., Wash. Assoc. engr. Lockheed Aircraft Co., Burbank, Calif., 1956-57; civil engring. asst. Dept. Pub. Works, Bur. Engring., L.A., 1957-59; civil engr. G. Shuirman & Assocs., L.A., 1959-66; v.p. Shuirman, Rogoway & Assocs., L.A., 1966-79; pres. RBA Ptnrs., Inc., L.A., 1979—. Chmn. Citizens Adv. Bd. for Com. Planning, L.A., 1970-72. Lt. (j.g.) USNR, 1954-56, Korea. Fellow aSCE; mem. NSPE, Am. Cons. Engrs. Coun. (various coms. 1989—, vice chmn. small firms coalition 1991—), Am. Water Works Assn., Cons. Engrs. and Land Surveyors Assn. Calif. Office: RBA Ptnrs Inc 3470 Wilshire Blvd Ste 900 Los Angeles CA 90010

ROHACK, JOHN JAMES, cardiologist; b. Rochester, N.Y., Aug. 22, 1954; s. John Joseph and Margaret Elizabeth (McLaughlin) R.; m. Charlotte McCown, Dec. 7, 1980; 1 child, Elisha Monique Feigle. BS, U. Tex., El Paso, 1976; MD, U. Tex., Galveston, 1980. Diplomate Am. Bd. Internal Medicine. Intern internal medicine U. Tex. Med. Br. Hosps., Galveston, 1980-81, resident internal medicine, 1981-83, chief resident internal medicine, 1983-84, fellow cardiology, 1984-86; instr. medicine U. Tex. Med. Br., Galveston, 1983-86; asst. prof. medicine Tex. A&M Coll. Medicine, College Station, 1986—; sr. staff cardiologist Scott and White Clinic, College Station, 1988—; med. dir. Fitlife Ctr., Tex. A&M U., College Station, 1990—. Mem. College Station Morning Lions Club, 1986-88, Nat. Bd. Med. Examiners, Phila., 1983-86; bd. dirs. Am. Heart Assn., Brazos Valley College Station, 1987-94, Am. Heart Assn. Tex. Affiliate, Austin, 1991-95. Mem. Tex. Med. Assn. (exec. coun. med. student sect. 1979-80, chmn. resident physician sect. 1981-82, house of dels. 1982-93), AMA (alternate del. house of dels. 1993-94), U. Tex. Med. Br. Alumni Assn. (bd. trustees 1989-94). Avocations: golf, gardening, salt water aquarium, reading. Office: Scott and White Clinic 1600 University Dr E College Station TX 77840-2199

ROHDE, JAMES VINCENT, software systems company executive; b. O'Neill, Nebr., Jan. 25, 1939; s. Ambrose Vincent and Loretta Cecilia R.; children: Maria, Sonja, Daniele. BCS, Seattle U., 1962. Chmn. bd. dirs., pres., Applied Telephone Tech., Oakland, 1974; v.p. sales and mktg. Automation Electronics Corp., Oakland, 1975-82; pres., chief exec. officer, chmn. bd. dirs. Am. Telecorp, Inc., 1982—; bd. dirs. Enerlogica, Inc., 1989-91. Chmn. exec. com., chmn. emeritus Pres.'s Coun. Heritage Coll., Toppenish, Wash., 1985—; chmn. No. Calif. chpt. Coun. of Growing Cos., 1990-93; bd. dirs. Ind. Colls. No. Calif., 1991—. Mem. Am. Electronics Assn. (bd. dirs. 1992—, vice chmn. No. Calif. coun. 1993-94, chmn. 1993—). Republican. Roman Catholic. Office: Am Telecorp Inc 100 Marine Parkway Redwood City CA 94065

ROHEIM, PAUL SAMUEL, physiology educator; b. Kiskunhalas, Hungary, July 11, 1925; s. Joseph Roheim and Aranka Schon; m. Judy Roheim; 1 child, John G. BS, Gymnasium Kiskunhalas, 1943; MD, Med. Sch. of Budapest. Intern Univ. Clinics, Budapest, 1951-52; asst. prof. physiology Med. Sch. of Budapest, 1952-56; rsch. assoc. Dept. Physiology, Hahnemann Med. Coll., Pa., 1957-58; instr. medicine Albert Einstein Coll. Medicine, N.Y.C., 1958-62, from NIH postodctoral fellow to assoc. prof. physiology, 1958-72, prof. physiology, 1972-76; prof. medicine La. State U. Med. Ctr., New Orleans, 1974-76, prof. physiology, pathology and medicine, 1976—, dir. div. lipoprotein metabolism and pathophysiology, 1991—; vis. prof. Dept. Medicine Hadassah Med. Sch., Jerusalem, 1970-71. Mem. edit. bd. Jour. Lipid Rsch., 1968-71, 73-75, 77—, Jour. Exptl. and Molecular Pathology, 1983—, Am. Jour. Physiology and Applied Physiology, 1969-72; assoc. editor Jour. Lipid Rsch. 1971-73, 76-77; contbr. numerous articles to profl. jours. Fellow Am. Heart Assn. (coun. on arteriosclerosis, mem. nominating com. 1980-82, chmn. nominating com. 1983-87, mem. exec. com. 1983-87, chmn. awards com. 1989-91); mem. NIH, Am. Physiol. Soc., Harvey Soc. Office: La State U Med Ctr Dept Physiology 1542 Tulane Ave New Orleans LA 70112-2865

ROHLF, F. JAMES, biometrist, educator; b. Blythe, Calif., Oct. 24, 1936. BS, San Diego State Coll., 1958; PhD in Entomology, U. Kans., 1962. Asst. prof. biology U. Calif., Santa Barbara, 1962-65; assoc. prof. statis. biology U. Kans., 1965-69; assoc. prof. biology SUNY, Stony Brook, 1969-72, prof., 1972—, chmn. dept. ecology and evolution, 1975-80, 90-91; statis. cons. N.Y. Pub. Svc. Commn., 1975—, IBM, 1977—, U.S. EPA, 1978-80, Applied Biomath Inc., 1984—; vis. scientist IBM, Yorktown Heights, N.Y., 1976-77, 80-81. Mem. Biometric Soc., Assn. Computing Machinery, Soc. Systematic Zoology, Classification Soc. Achievements include research in applications of multivariate analysis, cluster and factor analysis to systematics morphometrics, and population biology. Office: State Univ of NY at Stony Brook Ecology Laboratory Stony Brook NY 11794*

ROHN, ROBERT JONES, internist, educator; b. Lima, Ohio, June 16, 1918; s. Earl Crandall and Dorothy (Jones) R.; m. Annie Janet Smith, Oct. 21, 1942; children: Megan Marie, David Riis, Daniel Keith, Matthew Lee. AB, DePauw U., 1940; MD, Ohio State U., 1943. Diplomate Am. Bd. Internal Medicine. Intern Ohio State U. Starling Loving Hosp., Columbus, 1944; asst. resident medicine Ohio State U. Starling Loving Hosp., 1944-45, chief resident medicine, 1945-46, asst. resident hematology, 1948-49, chief resident, 1950; mem. faculty Sch. Medicine Ind. U., Indpls., 1950—, prof. medicine, 1961-72, B.K. Wiseman disting. prof. medicine, 1972-85, cancer coord., 1955-85, disting. emeritus prof. medicine, 1985—; adv. bd. sr. clin. cancer trainees USPHS. Contbr. articles, abstracts to profl. publs. Capt. U.S. Army, 1946-48. Named hon. Ky. Col., Sagamore of the Wabash; recipient Hadley award Soc. Ret. Execs., 1990. Fellow ACP; mem. AMA, N.Y. Acad. Sci., Am. Assn. Cancer Rsch. Republican. Achievements include demonstration of formation of Lupus Erythematosus cell. Home: 7334 A King George Dr Indianapolis IN 46260

ROHRBACH, JAY WILLIAM, biotechnology company supervisor; b. Reading, Pa., July 1, 1957; s. Walter Allen and Jean Catherine (Williams) R.; m. Laura Beth Schentes, Nov. 28, 1981; children: Kyle, Chelsea. BA, Calif. State U., 1981; MS, Calif. State Poly., 1985. Analyst CLMG, Inc., L.A., 1981-83; instr. Calif. State Poly., Pomona, 1983-85; rsch. technician UCLA Med. Ctr., Torrance, 1985; quality assurance technician Am. Bently, Irvine, Calif., 1985-87; lab. leader U. So. Calif. Med. Ctr., L.A., 1987; instr. anatomy Chaffey Community Coll., Alta Loma, Calif., 1987; mfg. supr. Amgen, Inc., Thousand Oaks, Calif., 1987—. Grad. rsch. grantee Calif. State Poly., 1984, 85. Mem. Sigma Xi (rsch. grantee 1984, 85). Office: Amgen Inc Amgen Ctr Thousand Oaks CA 91320-1789

ROHRBACH, ROGER PHILLIP, agricultural engineer, educator; b. Canton, Ohio, Oct. 12, 1942; s. Clarence A. and Beatrice E. (Burens) R.; m. M. Jeanette Weishner, June 12, 1965; children: Sharon E., Gregory A., Sara L. BS in Agrl. Engring., Ohio State U., 1965, PhD, 1968. From asst. prof. to assoc. prof. N.C. State U., Raleigh, 1968-78, prof., 1978—. Author: Design in Agricultural Engineering, 1986. Fellow Am. Soc. Agrl. Engrs. (Young Designer award 1981). Office: NC State U Biol and Agrl Engring Dept PO Box 7625 Raleigh NC 27695-7625

ROHRER, HEINRICH, physicist; b. Buchs, Switzerland, June 6, 1933. Diploma in Physics, Swiss Inst. Tech., Zurich, 1955, PhD in Physics, 1960; D. Sci. (hon.), Rutgers U., 1987, Marseille U., 1988, Madrid U., 1988. Rsch. asst. Swiss Inst. Tech., Zurich, 1960-61; post-doc. Rutgers U., New Brunswick, N.J., 1961-63; with IBM Rsch. Lab., Zurich, 1963—; vis. scholar U. Calif., Santa Barbara, 1974-75. Co-recipient King Faisal Internat. prize for sci., 1984, Hewlett Packard Europhysics prize, 1984, Nobel prize for Physics, 1986, Cresson medal Franklin Inst., Phila., 1987; Magnun Seal U. Bologna, Italy, 1988; IBM fellow, 1986. Fellow Royal Microscopical Soc. (hon. 1988); mem. NAS (fgn. assoc.), Swiss Acad. Tech. Scis., Swiss Phys. Soc. (hon. 1990), Swiss Assn. Engring. and Architecture (hon.), Zurich Phys. Soc. (hon. 1992). Office: IBM Rsch Divsn, Zurich Rsch Lab, Saeumerstrasse 4, CH-8803 Rueschlikon Switzerland

ROHRER, RICHARD JOSEPH, nuclear engineer; b. Lake Forest, Ill., May 14, 1960; s. William Joseph and Suzanne Marie (McGill) R. BS in Nuclear Engring., U. Ill., 1982; MS in Nuclear Engring., U. Wis., 1983; MS in Mgmt., Cardinal Stritch Coll., Milw., 1993. Registered profl. engr., Minn. Intern Savannah River Lab., Aiken, S.C., 1981; rsch. fellow U. Wis., Madison, 1982-83; nuclear engr. Commonwealth Edison, Seneca, Ill., 1984-89, No. States Power, Monticello, Minn., 1989—. Contbr. articles to profl. jours. Dir. Camp Sunrise, Mpls., 1992—. Named Free Thinker of the Yr. Freedom from Religion Found., 1990. Mem. Minn. Atheists, Triangle Fraternity, Tau Beta Pi, Phi Kappa Phi, Alpha Nu Sigma.

ROHRER, RONALD ALAN, electrical and computer engineering educator, consultant; b. Oakland, Calif., Aug. 19, 1939; m. Catherine (Casey) Jones. BSEE, MIT, 1960; MSEE, U. Calif., Berkeley, 1961, PhDEE, 1963. Asst. prof. elec. engring. U. Ill., Urbana, 1963-65; asst. prof. applied analysis SUNY, Stony Brook, 1965-66; asst. prof. elec. engring. and computer sci. U. Calif., Berkeley, 1966-68, assoc. prof., 1968-72; prof. elec. engring. Carnegie Mellon U., Pitts., 1974-75; vis. lectr. U. Colo., Boulder, 1976-77; prof., chmn. dept. elec. engring. U. Maine, Orono, 1977-79, So. Meth. U., Dallas, 1979-80; prof., chmn. dept. elec. engring and computer sci. U. Colo., Colorado Springs, 1980-81; prof. elec. and computer engring. Carnegie Mellon U., Pitts., 1985-89, Howard M. Wilkoff Univ. prof., 1989—; mem. tech. staff, sect. mgr. Fairchild R&D, Palo Alto, Calif., 1968-70; mgr. elec. engring. analysis programs SofTech Inc., Waltham, Mass., 1971-72; staff scientist GE, 1981, program mgr. Integrated Circuit Design Automation, 1982, dir. elec. products mktg., Calma Divsn., 1982-83; gen. mgr. MEDS divsn. scientific calculations, 1983-85; cons. Synopsys, Mountain View, Calif., 1987-89, Mentor Graphics, Beaverton, Oreg., 1987-91, Irwin Pub., Boston, 1989-91; bd. dirs. ISS, Inc., Research Triangle Park, N.C., (chmn.) Performance Signal Integrity, Inc., Pitts. Author: Circuit Theory: An Introduction to State Variable Approach, 1970; co-author: Theory of Linear Active Networks, 1967, Introduction to System Theory, 1972. Recipient Rsch. Paper award Nat. Elec. Conf., 1964, Best Paper award IEEE CAD Transactions, 1991, Best Paper award ACM/IEEE Design Automation Conf.; recipient Inventor Recognition award Semicondr. Rsch. Corp., 1990, Tech. Excellence award, 1991. Fellow IEEE (Browder J. Thompson award 1967); mem. Cirs. and Systems Soc. of IEEE (pres. 1987, Guillemin-Cauer award 1970, founding editor Transactions on Computer Aided Design 1980-84, IEEE Edn. medal 1993), NAE, Am. Soc. Engring. Edn. (Frederick Emmons Terman award elec. engring. div. 1978). Office: Carnegie-Mellon U Dept Elec & Computer Engr 5000 Forbes Ave Pittsburgh PA 15213-3890

ROHRICH, RODNEY JAMES, plastic surgeon, educator; b. Eureka, S.D., Aug. 5, 1953; s. Claude and Katie (Schumacher) R.; m. Diane Louise Gibby, July 3, 1990. BA summa cum laude, N.D. State U., 1975; MD with honors, Baylor Coll., 1979. Diplomate Am. Bd. Plastic Surgery, Nat. Bd. Med. Examiners. Instr. surgery Harvard Med. Sch. Mass. Gen. Hosp., Boston, 1985-86; asst. prof. U. Tex. Southwestern Med. Ctr., Dallas, 1986-89, assoc. prof., 1989—; chief plastic surgery Parkland Meml. Hosp., Dallas, 1989-91; prof., chmn. dir. plastic surgery U. Tex. Southwestern Med. Ctr., Dallas, 1991—; pres., faculty senate U. Tex. Southwestern Med. Ctr., Dallas. Mem. editorial bd. Selected Readings in Plastic Surgery, The Cleft Palate and Craniofacial Jour.; contbr. articles to med. jours. Bd. dirs. Save-the-Children Found., Dallas, March of Dimes, Dallas; class mem. Leadership Dallas, 1989-90; mem. Adopt-A-Sch., Dallas Summer Mus. Guild, Dallas Mus. Art, Dallas Symphony Assn., Tex. Health Found., Youth Leadership Dallas. Grantee Urban Rsch. Fund, 1982, United Kingdom Ltd. Ednl. Rsch. Fund, 1983, Oxford Cleft Palate Found., 1983, Am. Assn. Plastic Surgeons, 1985, Plastic Surgery Ednl. Found., 1985, 89, 90, U. Tex. Health Sci. Ctr. Dept. Surgery, 1986, Howmedica, 1989, ConvaTec-Squibb, 1989, 91, ConvaTec, 1991. Mem. AAAS, AMA (Thomas Cronin award 1988, 90, Clifford C. Snyder award 1990), Am. Assn. Hand Surgery, Am. Burn Assn., Am. Cleft Palate Assn., Am. Coll. Surgeons, Am. Soc. Law and Medicine, Am. Soc. Maxillofacial Surgeons, Am. Soc. for Surgery the Hand, Am. Soc. Plastic and Reconstructive Surgeons, Am. Trauma Soc., British Med. Assn., Nat. Vascular Malformations Found. Inc. (med. and sci. adv. bd.) Tex. Med. Assn., Tex. Soc. Plastic Surgeons, Mass. Gen. Hosp. Hand Club, Dallas County Med. Soc., Assn. Acad. Chmn. Plastic Surgery, Dallas Soc. Plastic Surgeons, Harvard Med. Sch. Alumni Assn., Inst. for Study of Profl. Risk, Plastic Surgery Rsch. Coun., Reed O. Dingman Soc. Plastic Surgeons, So. Med. Assn. Republican. Roman Catholic. Office: U Tex Southwestern Med Ctr 5323 Harry Hines Blvd Dallas TX 75235-7200

ROHRIG, TIMOTHY PATRICK, toxicologist, educator; b. May 30, 1956. BS, Rockhurst Coll., 1978; PhD, U. Mo. Kansas City, 1984. Diplomate Am. Bd. Forensic Toxicology. Asst. chemist, supr. analytical support group Midwest Rsch. Inst., Kansas City, Mo., 1978-80; lectr. Rockhurst Coll., Kansas City, 1980-83; forensic toxicologist Kans. Bur. Investigation, Topeka, 1983-85, chief forensic toxicologist, 1986-87; toxicologist Office Chief Med. Examiner, S. Charleston, W. Va., 1985-86; chief forensic toxicologist Office of Chief Med. Examiner, Okla. City, Okla., 1987—; lab. inspector Nat. Inst. on Drug Abuse, Nat. Lab. Certification Program, 1990—; adjunct asst. prof. pharmacy, U. Okla., Okla. City, 1989—; presenter at many profl. seminars, workshops, confs. Contbr. many articles to Jour. Analytical Toxicology, Medicolegal Gram, Rsch. Communications in Chem. Pathology and Pharmacology. Fellow Am. Acad. Forensic Scis. (toxicology sect., Gen. Sect. award 1989-90); mem. Nat. Safety Coun. (com. on alcohol and other drugs), Soc. Forensic Toxicologists, Midwestern Assn. of Forensic Scientists (toxicology sect. coord. 1987), Southwestern Assn. Toxicologists (counselor 1990-91, pres.-elect 1991-92, pres. 1992-93, bd. dirs. 1993-94), Rho Chi. Office: 901 W Stonewall Oklahoma City OK 73117

ROHRLICH, GEORGE FRIEDRICH, social economist; b. Vienna, Austria, Jan. 6, 1914; came to the U.S., 1938; s. Egon Ephraim and Rosa (Tenzer) R.; m. Laura Ticho, Feb. 3, 1946; children: Susannah Ticho Feldman, David Ephraim, Daniel Mosheh. D in Legal Scis., U. Vienna, Austria, 1937, Gold Dr.'s Diploma Law, 1937; PhD (univ. refugee scholar), Harvard U., 1943. Diplomate Consular Acad. Vienna, 1938. Social economist sect. public health and welfare Supreme Comdr. Allied Powers, Tokyo, 1947-50; socioecon. program analyst and developer U.S. Govtl. Policies, 1950-59; sr. staff mem. social security div. ILO, Geneva, 1959-64; vis. prof. social econs. and policy U. Chgo., 1964-67; prof. econs. and social policy Temple U., 1967-81, prof. emeritus, 1981—; dir. econs. and bus. programs Temple U. Japan, Tokyo, 1987-88; founder, past bd. dirs. Inst.

Social Econs. and Policy Rsch.; sr. lectr. Sch. Social Work, Columbia U., 1968-69; dir. rsch. P.R. Commn. Integral Social Security System, San Juan, 1975-76; cons. in field; lectr. USIA, Brazil, 1984; co-dir., Keynoter Nat. Conf. Community Dimensions of Econ. Enterprise, 1984; ILO cons. Govt. Mauritius, 1985. Author: Social Economics—Concepts and Perspectives, 1974; others; editor books, the most recent being: Checks and Balances in Social Security, 1986, Environmental Management: Economic and Social Dimensions, 1976; contbr. articles to profl. publs.; assoc. editor Rev. of Social Economy; editorial adv. bd. Internat. Jour. Social Econs., U.K. Former mem. bd. dirs. Health and Welfare Coun. Greater Phila. Recipient festschrifts Internat. Jour. Social Econs., vol. 10, no. 6/7, 1983, vol. 11, nos. 1/2 and 3/4, 1984; Brookings rsch. tng. fellow, 1941-42; Ford Found. travel grantee, 1966; Fulbright rsch. scholar N.Z., 1980. Mem. AAAS, Assn. Social Econs. (pres. 1978-79, disting. mem. 1983, disting. scholar-Divine award 1989), Am. Econ. Assn., Indsl. Rels. Rsch. Assn. (charter), Internat. Soc. Labor Law and Social Security, Am. Risk and Ins. Assn., Nat. Acad. Social Ins. (elected), Harvard Club of Phila. Democrat. Jewish. Avocation: music. Home: 7913 Jenkintown Rd Cheltenham PA 19012-1106 Office: Temple U Sch of Bus and Mgmt Philadelphia PA 19122

ROITBERG, BERNARD DAVID, biology educator; b. Windsor, Ont., Can., June 24, 1953; s. Harry and Jeanette Roitberg; m. Carol Ann Hubbard, July 8, 1977; 1 child, Gabriela. BS, Simon Fraser U., 1975; MS, U. B.C., 1977; PhD, U. Mass., 1982. Rsch. assoc. U. Mass., Amherst, 1982; from asst. prof. to assoc. prof. Simon Fraser U., Burnaby, Can., 1982-93, prof., 1993—; guest prof. German Rsch. Coun., U. Kiel, Germany, 1993. Editor: Chemical Ecology of Insects, 1992; assoc. editor: (jour.) Can. Entomologist, 1989-93. Mem. Entomol. Soc. Can. (dir. 1989-91, C. Gordon Hewitt award 1990), Entomol. Soc. B.C. (pres. 1988), Brit. Ecol. Soc., Animal Behavior Soc. Achievements include demonstration of pheromone learning in insects, a new mimicry form, suicide in insects; development and confirmation of new theory on life expectancy and reproduction. Office: Simon Fraser U, Dept Bioscis, Burnaby, BC Canada V5A 1S6

ROJHANTALAB, HOSSEIN MOHAMMAD, chemical engineer, researcher; b. Tehran, Iran, Sept. 26, 1944; came to U.S., 1984; s. Mohammad Rojhantalab and Sakineh (Fakhri) Nasser-Ghandi; m. Nastaran Danesh, July 14, 1979; 1 child, Ayda. MS, Calif. State U., Hayward, 1972; PhD, Oreg. State U., 1976. Asst. prof. Ahwaz (Iran) U., 1976-77, Shiraz (Iran) U., 1977-82; cons., chemist Water-Con Co., Tehran, 1982-84; rsch. assoc. U. Oreg., Eugene, 1985-88; lithography engr. Intel Corp., Hillsboro, Oreg., 1988-91; thin film engr. Intel Corp., Aloha, Oreg., 1991—; transl. Popular Sci. Pub. Co., Tehran, 1980-83, UNESCO workshop, 1984; editor, CEO, DNA Pub. Co., Tehran, 1982-84; vis. prof. chemistry Ore. State U., 1985. Editor, translator 4 books on genetic code, controlled nuclear fusion to Farsi, 1981-84; contbr. articles to sci. jours. Scholar Calif. State U., 1971-72; grantee Oreg. State U., 1975-76, CENTO, 1978-79. Mem. Electrochem. Soc. Am. Achievements include patent pending on single pass graded NSG/BPSG glass for deep trench fill; development of thin BPSG film for defect detection in 0.2 - 10 microns. Home: PO Box 6652 Aloha OR 97007

ROKACH, ABRAHAM JACOB, structural engineering and computer software consultant; b. N.Y.C., Nov. 14, 1948; s. David and Sara (Biro) R.; m. Pninah Abigail Kacev, June 19, 1977; children: David, Aaron Zvi, Moshe Mordecai, Aryeh Raphael Pesach, Chaya Esther, Malka Rachel. B of Engring., CCNY, 1969; MSCE, MIT, 1970. Registered structural engr., Ill., profl. engr., N.Y. Pres. Rokach Engring. P.C., Chgo., 1984—, Hypermedia Systems Inc., Chgo., 1989—; adj. prof. structural engring. U. Ill., Chgo., 1984-87; consulting editor McGraw-Hill, 1990—. Author: Reliability of Expert Systems for Computer-Aided Structural Design, 1986, Guide to Load and Resistance Factor Design of Structural Steel Buildings, 1986, Schaum's Outline of Structural Steel Design, 1991. Fellow, grantee NSF; recipient Cert. Honor, Structural Engrs. Assn. Ill., 1985. Mem. ASCE, Am. Soc. for Engring. Edn., Am. Inst. Steel Constrn. Office: 6754 N Whipple St Chicago IL 60645-4123

ROKHLIN, STANISLAV IOSEF, engineering educator; b. Leningrad, USSR, Oct. 21, 1944; came to U.S., 1984, then naturalized; MS, Leningrad Elec. Engring. U., 1967, PhD, 1972; postgrad. diploma, Leningrad State U., 1969. Rsch. assoc., teaching assoc. Leningrad Elec. Engring. U., 1969-72; sr. rsch. scientist Nat. Sci. Inst., Leningrad, 1973-74, group leader, 1974-76; sr. lectr. Ben Gurion U., Beer-Sheva, Israel, 1977-83, assoc. prof., 1983-85; assoc. prof. Ohio State U., Columbus, 1985-89, prof. dept. welding engring., 1990—; cons. Edison Welding Inst., NASA, USAF, various industries; chmn. organizing com. various confs. on materials and nondestructive evaluation. Assoc. tech. editor: Materials Evaluation, 1990—; editorial bd.: Jour. Adhesion Sci. and Tech., 1987—; contbr. over 150 rsch. papers to internat. jours. on phys. and gen. acoustics, nondestructive evaluation of materials, materials sci. and mech. engring. Recipient Alcoa Found. award, 1988, 89. Fellow Acoustical Soc. Am. (tech. com. phys. acoustics); mem. Am. Soc. Nondestructive Testing (rsch. and edn. coun., Fellowship award 1991, bd. dirs.), Am. Welding Soc. (Charles H. Jennings Meml. award 1986, A.F. Davis silver medal 1991). Achievements include 2 patents in field; pioneered application of guided ultrasonic waves, especially Lamb and interface waves, for nondestructive evaluation of materials; discovery, with experimental verification, of the solution for Lamb wave diffraction on cracks; development of several techniques for nondestructive reconstruction of elastic properties of composite materials; innovation of X-ray imaging application for arc welding sensing and control, and physical study of plasma liquid metal interaction. Home: 4085 Edgehill Dr Columbus OH 43220-4508 Office: Ohio State Univ 190 W 19th Ave Columbus OH 43210-1182

ROKOSZ, SUSAN MARIE, environmental engineer; b. Detroit, Nov. 12, 1957; d. Frank Peter and Bernice Jane (Krzeski) Winkler; m. Michael John Rokosz, June 28, 1986. BS in Chemistry, Wayne State U., 1979, MSChE, 1980. Registered profl. engr., Mich. Rsch. asst. Wayne State U., Detroit, 1979-80; process control engr. Arco Chem. Co., Channelview, Tex., 1980-81; rsch. engr. Ford Motor Co., Dearborn, Mich., 1981-88; environ. engr. Ford Motor Co., Dearborn, 1988—. Trustee Nat. Multiple Sclerosis Soc., Southfield, Mich., 1990-92. Mem. Engring. Soc. Detroit (prof. engr. rev. instr. 1985-92, Outstanding Leadership award 1984, Outstanding Young Engr. 1989, Disting. Svc. award 1993), Phi Beta Kappa, Sigma Xi, Tau Beta Pi. Office: Ford Motor Co 15201 Century Dr Ste 602 Dearborn MI 48120

ROKSTAD, ODD ARNE, chemical engineer; b. Kristiansund, Norway, May 27, 1935; s. Johannes Konrad and Elisabeth (By) R.; m. Randi Godske, Apr. 10, 1965; children: Grete, Elisabeth, Anne, Håkon. Degree in engring., Norwegian Inst. Tech., 1962, D in Engring., 1975. Fellow Norwegian Inst. Tech., Trondheim, 1962-67; rsch. scientist SINTEF, Trondheim, 1968-80, sr. scientist, 1982—; vis. scientist Exxon Rsch. and Engring., Linden, N.J., 1981-82; prof. Norwegian Inst. Tech., 1975—. Contbr. articles to Acta Chem. Scand., Indsl. Engring. Chem. Process Design Devel; co-author: Novel Production Methods, 1992. Fellow Norsk Hydro, 1962, 63, U. Norway, 1964, 65, Govt. of Norway, 1966, 67. Mem. Norwegian Chem. Soc., Norwegian Soc. Chartered Engrs., Am. Chem. Soc. (petroleum chemistry divsn.), N.Y. Acad. Scis. Lutheran. Home: Jonsvannsveien 55, 7017 Trondheim Norway Office: Sintef Applied Chemistry, N-7034 Trondheim Norway

ROLDAN, LUIS GONZALEZ, materials scientist; b. Garafia, Spain, Aug. 8, 1925; s. Jose and Donatila Dolores (Gonzalez-Cabrera) R.; m. Carmen Noguera-Jimenez, Jan. 25, 1961; children: Jose, Luis, Mary Carmen, Carlos. Degree in scis., U. Sevilla, Spain, 1950, DSc, 1957. Adj. prof. U. Sevilla, 1953-57; sr. physicist British Rayon Rsch. Assn., Manchester, United Kingdom, 1957-61; sr. scientist Allied Chem. Corp., Morristown, N.J., 1961-68; rsch. assoc. J.P. Stevens and Co., Inc., Garfield, N.J., 1968-81; dept. mgr. J.P. Stevens and Co., Inc., Greenville, S.C., 1981-86; dir. LGR Micro Rsch., Greer, S.C., 1986—; vis. prof. Univ. da Beira Interior, Covilha, Portugal, 1987; tech. expert Nat. Inst. Standards and Tech., Gaithersburg, Md., 1988—; adj. prof. N.C. State U., Raleigh, 1988—. Contbr. articles to profl. jours. Lt. Spanish Army, 1946-51. Mem. Fiber Soc., Am. Crystallographic Assn. Home and Office: LGR Micro Rsch 124 Becky Don Dr Greer SC 29651-1213

ROLF, HOWARD LEROY, mathematician, educator; b. Laverne, Okla., Nov. 25, 1928; s. James Walter and Edith (Yoho) R.; m. Anita Jane Ward, June 24, 1961; children—James Scott, Jennifer Jane, Stephanie Kaye, Rhonda Mary. B.S., Okla. Baptist U., 1951; M.A., Vanderbilt U., 1953, Ph.D., 1956. Instr. math. Vanderbilt U., 1954-56, asst. prof., dir. computer center, 1959-64; asst. prof. Baylor U., 1956-57, prof., 1964—, dir. acad. computing, 1968-70, chmn. dept. math., 1971—; asso. prof. Georgetown (Ky.) Coll., 1957-59; cons. in field. Author: (with William C. Brown) Mathematics, 1982, Finite Mathematics, 1988, 91, Mathematics for Management, Social and Life Sciences, 1991. Mem. Math. Assn. Am. (chmn. Tex. sect. 1977), Am. Math. Soc., Sigma Xi, Pi Mu Epsilon. Baptist. Home: RR 11 Box 155 Waco TX 76712

ROLFS, KIRK ALAN, agronomist; b. Ellsworth, Kans., Mar. 18, 1962; s. Merle Dean Rolfs and Arlene Marie (Watts) Phillips; m. Amy Marie Richey, Apr. 14, 1984; children: Bailey, Brooke, Spencer. BS in Agriculture, Colo. State U., 1987. Rsch. tech. Argl. Rsch. Svc., USDA, Fort Collins, Colo., 1982-87, Pioneer Hi-Bred, Connell, Wash., 1987-88; prodn. agronomist Pioneer Hi-Bred, Fresno, Calif., 1988—. Mem. Am. Soc. Agronomy, Crop Sci. Soc. Am., Fresno C. of C. Office: Pioneer Hi-Bred 4762 W Jennifer # 101 Fresno CA 93722

ROLLE, F. ROBERT, health care consultant; b. Jamaica, N.Y., May 21, 1939; s. Fred and Ruth D. (Lowes) R.; m. L. Norene Mahoney, Sept. 23, 1974; 1 child, Craig D. BS, Pratt Inst., 1961; MS, Purdue U., 1965, PhD, 1966. Rsch. fellow U. London, Egham, Eng., 1966-67; dir. Johnson & Johnson, New Brunswick, N.J., 1967-84; assoc. dir. Chicopee Divsn. Johnson & Johnson, Dayton, N.J., 1984-89; rsch. fellow corp. office Johnson & Johnson, New Brunswick, 1989-90; cons. Rolle and Assocs., Princeton, N.J., 1990—; cons. Johnson & Johnson, New Brunswick, 1990—. Contbr. articles to profl. jours. Mem. Am. Chem. Soc., Royal Soc. Chemistry. Achievements include formulating and implementing a world-wide effort to resolve dioxin in pulp issue, coordinating the evaluation of the claim "environmentally friendly" for Johnson & Johnson world-wide. Home: 1 Lafayette Rd W Princeton NJ 08540-2428

ROLLENCE, MICHELE LYNETTE, molecular biologist; b. Takoma Park, Md., Nov. 23, 1955; d. John Francis and Martha Jo (Jackson) R.; m. David H. Specht, June 3, 1978 (div. Sept. 1982). AA, Montgomery Coll., 1976; BS, U. Md., 1978. Lab. technician Dairy and Food Labs., San Francisco, 1979-81; rsch. asst. Genex Corp., Gaithersburg, Md., 1981-82, rsch. assoc., 1982-86, sr. rsch. assoc., 1986-88, rsch. scientist, 1989-93; rsch. assoc. Genetic Therapy, Inc., Gaithersburg, Md., 1993—. Contbr. articles to profl. publs.; patentee in field. Pres. Explorer Post div. Boy Scouts Am., Gaithersburg, 1973; youth advisor Neelsville Presbyn. Ch., Germantown, Md., 1990. Recipient Nat. Exploration award TRW/Explorers Club, 1973. Mem. AAAS, Am. Soc. Microbiology, DAR, Pleasant Plains of Damascus. Republican. Presbyterian. Avocations: bell choir, guitar, dance, hiking. Office: Genetic Therapy Inc 19 Firstfield Rd Gaithersburg MD 20878

ROLLER, RICHARD ALLEN, marine invertebrate physiologist; b. Etain, France, June 29, 1956; s. Oscar C. and Marble Marie (Jones) R.; m. Tina Louise Fisher, July 21, 1984; children: Mitchell Wade, Traci Lynn. BS in Biology, U. Ark., 1980; MS in Zoology and Physiology, La. State U., 1983, PhD in Zoology, Physiology, 1987. Rsch. asst. U. Ark., Little Rock, 1980; rsch. asst. La. State U., Baton Rouge, 1985-86, rsch. asst. III, 1986-87; asst. prof. U. Wis., Stevens Point, 1987-91, Lamar U., Beaumont, Tex., 1991--; mem. grant panel rev. NSF, Washington, 1991; speaker, host Electron Microscopy Soc. Am., Milw., 1988. Contbr. articles to profl. jours. Mem. Zoning Bd. Nelsonville, Wis., 1990-91, Grad. Coun. U. Wis., Stevens Point, 1987-90. NSF Instrumentation grantee, 1990. Mem. AAAS, Malacological Union, Am. Soc. Zoologists, Gulf Estuarine Rsch. Soc. Office: Lamar U Dept Biology PO Box 10037 Beaumont TX 77710

ROLLINS, ALBERT WILLIAMSON, civil engineer, consultant; b. Dallas, July 31, 1930; s. Andrew Peach and Mary (Williamson) R.; B.S. in Civil Engring., Tex. A. and M. U., 1951, M.S. in Civil Engring., 1956; m. Martha Ann James, Dec. 28, 1954; children—Elizabeth Ann, Mark Martin. Engring. asst. Tex. Hwy. Dept., Dallas, 1953-55; dir. pub. works City of Arlington (Tex.), 1956-63, city mgr., 1963-67; partner Schrickel, Rollins & Assos., land planners-engrs., Arlington, 1967—. Mem. Gov.'s Energy Adv. Council; chmn. Tex. Mass Transp. Commn.; bd. dirs. Tex. Turnpike Authority. Served as 1st lt. AUS, 1951-53. Registered profl. engr., Tex., La., Okla. Mem. Internat. City Mgmt. Assn., Nat. Soc. Profl. Engrs., ASCE, Am. Water Works Assn., Water Pollution Control Fedn., Sigma Xi, Phi Eta Sigma, Tau Beta Pi, Phi Kappa Phi, Chi Epsilon. Contbr. articles to profl. jours. Home: 3004 Yellowstone Dr Arlington TX 76013-1166 Office: 1161 Corporate Dr W Suite 200 Arlington TX 76006

ROLLINS, SCOTT FRANKLIN, chemist; b. Beverly, Mass., Jan. 2, 1959; s. Harold Franklin Jr. and Dorithy I. (Rowell) R.; m. Michelle Anne Heise, Oct. 10, 1986; children: Corey William, Ashley Elizabeth. AS in Law Enforcement, North Shore Community Coll., Beverly, Mass., 1980; BS in Chemistry, Salem State Coll., 1984. Chemist, rsch. and devel. Stahl Finish, Div. of Stahl USA, Peabody, Mass., 1985-90; sr. chemist, rsch. and devel. Paule Chem., Div. of Stahl USA, Peabody, Mass., 1990-91; chemist R&D K.J. Quinn Shoe Div., Malden, Mass., 1991—; mem. quality coun. Stahl USA, Peabody, 1991—. Mem. IUPAC, Am. Chem. Soc., Am. Leather Chemists Assn., New Eng. Soc. for Coatings Tech., Soc. Plastics Engrs. Office: KJ Quinn Shoe Div 209 Canal St Malden MA 02148-6701

ROLLS, BARBARA JEAN, biobehavioral health educator, laboratory director; b. Washington, Jan. 5, 1945; d. Howard Julian and Patricia Jane (Pratt) Simons; m. Edmund Thomson Rolls, Sept. 6, 1969 (div. Jan. 1983); children: Melissa May, Juliet Helen. BA, U. Pa., 1966; PhD, Cambridge (Eng.) U., 1970; MA (hon.), Oxford (Eng.) U., 1970. Mary Somerville rsch. fellow Oxford U., 1969-72, IBM rsch. fellow, 1972-74, jr. rsch. fellow Wolfson Coll., 1974-75, E.P. Abraham rsch. fellow Green Coll., 1979-82, fellow in nutrition, 1983-84; assoc. prof. psychiatry Johns Hopkins U. Sch. Medicine, Balt., 1984-91, prof. psychiatry, 1991-92, dir. Lab. for Study Human Ingestive Behavior, 1984—; Jean Phillips Shibley prof. biobehavioral health Pa. State U., 1992—; cons. to numerous large corps., 1983—. Author: Thirst, 1982; mem. editorial adv. bd. Jour. Appetite, 1981—; mem. editorial bd. Am. Jour. Physiology, 1985—, Trends in Food Sci. and Tech., 1991-93, Am. Jour. Clin. Nutrition, 1992—, Obesity Rsch., 1992-93, Nutrition Rev., 1993—; contbr. numerous articles to profl. jours. Recipient Rolleston Meml. prize, Oxford U., 1974; Thouron scholar Cambridge U., 1966-69; Med. Rsch. Coun. (U.K.) grantee, 1969-84, NIH grantee, 1987—. Mem. Am. Physiol. Soc., Soc. for Study Ingestive Behavior (bd. dirs. 1986-90, pres.-elect 1990-91, pres. 1991-92), N.Am. Assn. for Study Obesity (coun. 1991—), Am. Inst. Nutrition, Am. Soc. Clin. Nutrition. Office: Pa State U 104 Benedict House University Park PA 16802

ROLLWAGEN, JOHN A., federal official; b. 1940; married. BSEE, MIT, 1962; MBA, Harvard U., 1964. Mktg. rep. Control Data Corp., 1964-66; prodn. mgr. Monsanto Corp., 1966-68; v.p. Internat. Timesharing Corp., 1968-75; v.p. fin. Cray Rsch. Inc., 1975-76, v.p. mktg., 1976-77, pres., 1977-80, pres., chief exec. officer, 1980-88, chmn., chief exec. officer, 1981-1993, also bd. dirs.; dep. sec. Dept. of Commerce, Washington, 1993—; bd. dirs. Dayton Hudson Corp., Minn.; chmn. bd. dirs. Fed. Reserve Bank Minn. Office: Dept of Commerce 14th & Constitution Ave NW Washington DC 20230 also: Cray Rsch Inc 608 2nd Ave S Ste 154 Minneapolis MN 55402-1910 also: Cray Rsch Inc 655-ALone Oak Dr Eagan MN 55121

ROLOFF, THOMAS PAUL, combustion engineer; b. Vienna, Austria, June 30, 1965; came to U.S., 1973; s. Herbert Helmut and Gertrud (Heckmanns) R. BSME, Lafayette Coll., 1987; MSME, Va. Poly. Inst. and State U., 1988; postgrad., MIT, 1993—. Rsch. asst. Va. Poly. Inst. and State U., Blacksburg, 1987-88; combustion engr. GE Co., Schenectady, N.Y., 1989—. Mem. ASME, Pi Tau Sigma. Avocations: travel, investing. Home: 129 Franklin St # 141 Cambridge MA 02139 Office: GE Co 1 River Rd Bldg 53332 Schenectady NY 12345-6001

ROLSTON, HOLMES, III, theologian, educator, philosopher; b. Staunton, Va., Nov. 19, 1932; s. Holmes and Mary Winifred (Long) R.; m. Jane Irving Wilson, June 1, 1956; children: Shonny Hunter, Giles Campbell. BS, Davidson Coll., 1953; BD, Union Theol. Sem., Richmond, Va., 1956; MA in Philosophy of Sci., U. Pitts., 1968; PhD in Theology, U. Edinburgh, Scotland, 1958. Ordained to ministry Presbyn. Ch. (USA), 1956. Asst. prof. philosophy Colo. State U., Ft. Collins, 1968-71, assoc. prof., 1971-76, prof., 1976—; vis. scholar Ctr. Study of World Religions, Harvard U., 1974-75; lectr. Yale U., Vanderbilt U., others; official observer UNCED, Rio de Janiero, 1992. Author: Religious Inquiry: Participation and Detachment, 1985, Philosophy Gone Wild, 1986, Science and Religion: A Critical Survey, 1987, Environmental Ethics, 1988; assoc. editor Environ. Ethics, 1979—; mem. editorial bd. Oxford Series in Environ. Philosophy and Pub. Policy, Zygon: Jour. of Religion and Sci.; contbr. chpts. to books, articles to profl. jours. Recipient Oliver P. Penock Disting. Svc. award Colo. State U., 1983, Coll. Award for Excellence, 1991., Univ. Disting. Prof., 1992; Disting. Russell fellow Grad. Theol. Union, 1991, Disting. Lectr. Chinese Acad. of Social Scis., 1991, Disting. Lectr. Nobel Conf. XXVII. Mem. AAAS, Am. Acad. Religion, Soc. Bibl. Lit. (pres. Rocky Mountain-Gt. Plains region), Am. Philos. Assn., Internat. Soc. for Environ. Ethics (pres. 1989—), Phi Beta Kappa. Avocation: bryology. Home: 1712 Concord Dr Fort Collins CO 80526-1602 Office: Colo State U Dept Philosophy Fort Collins CO 80523

ROMAN, CECELIA FLORENCE, cardiologist; b. Phila., June 12, 1956; d. Stanley Jeremiah and Doris (Manus) Romanowski. BA magna cum laude, Boston U., 1977; DO, Phila. Coll. Osteo. Medicine, 1981. Intern, internal medicine resident Del. Valley Med. Ctr., Langhorne, Pa., 1981-84; cardiology fellow Deborah Heart and Lung Ctr., Browns Mills, N.J., 1984-86, cardiology attending dir. med. intensive care unit, 1986-90; clin. instr. dept. medicine Robert Wood Johnson Med. Sch. U. Medicine and Dentistry of N.J., Phila., 1987-90; pvt. practice Bristol, Pa., 1990-93; cardiology attending Clin. Cardiology Group, Langhorne, 1990-93; staff cardiologist Albert Einstein Med. Ctr., Med. Coll. Pa., Del. Valley Med. Ctr., Lower Bucks Hosp. Author med. videos; lectr. in field; contbr. articles to profl. jours. Recipient Physicians Recognition award, 1987—. Fellow Am. Coll. Angiology, Am. Coll. Osteo. Internists; mem. Am. Osteo. Assn., Pa. Osteo. Med. Assn., Am. Coll. Osteo. Internists-cardiology and geriatric divs. Avocations: sailing, flying, hot air ballooning, travel. Home: 12 Duffield Dr West Trenton NJ 08628

ROMAN, STANFORD AUGUSTUS, JR., medical educator, dean; b. N.Y.C.; s. Stanford Augustas and Ivy L. (White) R.; children: Mawiyah Lythcott, Jane E. Roman-Brown. AB, Dartmouth Coll., 1964; MD, Columbia U., 1968; MPH, U. Mich., 1975. Diplomate Nat. Bd. of Med. Examiners. Intern in medicine Columbia U.-Harlem Hosp. Ctr., 1966-69, resident in medicine, 1969-71, chief resident in medicine, 1971-73; assoc. dir. ambulatory care Columbia U. Harlem Hosp., N.Y.C., 1972-73; instr. in medicine Columbia U., N.Y.C., 1972-73; asst. physician Presbyn. Hosp., 1972-73; clin. dir. Healthco, Inc., Soul City, N.C., 1973-74; dir. ambulatory care, asst. prof. medicine/sociomed. scis. Boston City Hosp., 1974-78; asst. prof. medicine U. N.C., Chapel Hill, 1973-74; asst. dean Boston U. Sch. Medicine, 1974-78; med. dir. D.C. Gen. Hosp., Washington, 1978-81; assoc. dean acad. affairs Dartmouth Med. Sch., Hanover, N.H., 1981-86, assoc. prof., 1981-87, dep. dean, 1986-87; dean, v.p., prof. medicine Morehouse Sch. Med., Atlanta, 1987-89; sr. v.p., med. and profl. affairs Health and Hosps. Corp., N.Y.C., 1989-90; dean med. sch. CUNY, 1990—; dir. Boston Comprehensive Sickle Cell Ctr., 1975-78; bd. dirs. Nat. Bd. Med. Examiners, Phila., 1988-92. Contbr. chpt. to books and articles to profl. jours. Trustee Dartmouth Coll, Hanover, N.H., Dartmouth-Hitchcock Med. Ctr., Hanover. Mem. AMA, APHA, Nat. Med. Assn., N.Y. State Coun. Grad. Med. Education, N.Y. State Dept. Edn. Bd. Medicine. Democrat. Episcopalian. Avocations: photography, travel, music. Office: CUNY Med Sch J 909 Convent Ave and 138th St New York NY 10031

ROMANIUK, RYSZARD STANISLAW, electrical engineering educator, consultant; b. Kozmin, Poland, May 8, 1952; s. Stanislaw and Albina (Bielawna) R.; m. Irena Urszula Rutkowska, Feb. 18, 1978; 1 child, Adonis Laurentius. MSc, Warsaw U. Tech., 1975, PhD, 1980. Cert. 1st and 2d degree in electronics and telecom. and in high-tech. and tech. adminstrn. Fedn. Polish Tech. Assns.; cert. registered expert in electronics and telecom. Assn. of Polish Elec. Engrs. (SEP). Asst. prof. elec. engring. Warsaw U. Tech., 1980-82, assoc. prof. elec. engring., 1982-89, prof. elec. engring., 1991—; dir. Coun. Ministers and Ministry Nat. Edn., Warsaw, 1989-90; cons. Nat. Glass/Ceramics/Electronics Industry, Warsaw, 1980—; govtl. cons. Ministry of Communications, 1980—; adv. bd. Internat. Press; editorial bd. Photonics Spectra jour., N.J., Internat. Jour. Optoelectronics, London. Author, editor internationally pub. books and manuals; co-editor domestic and internat. elec. engring. proc. (Recognition award Polish Acad. Scis., 1986); contbr. numerous articles to profl. jours., conf. presentations, and speeches. Exec. dir. Polish Commn. for French Polish Found., Paris, Warsaw, 1989-90; chmn. Polish-Swedish Tng./Human Resources Commn., Stockholm, Warsaw, 1989-90, Polish-German Tng./Human Resources Commn., Bonn, Warsaw, 1989-90; Polish del. to Europeand Community/Task Force on Tng. and Human Resources, Brussels, Warsaw, 1989-90. Grantee Eisenhower Found., 1991, USIA, 1991; Eisenhower fellow, 1991. Fellow Internat. Soc. for Optical Engring. (SPIE)-(founder Polish chpt., hon. award 1986); mem. Assn. of Polish Elec. Engrs. (SEP) (mem. chambers of experts, internat. bd. and industrial bd., hon. award 1983, 1989), Polish Acad. Scis. (sec. electronics com., exec. sec. 1982, 86, 90), Polish Com. Optoelectronics (founding mem.), Polish Found. Electronics (v.p. 1990—), Polish Phys. Soc., Polish Cybernetical Soc., Polish Soc. of Theoretical and Applied Electronics, IEEE (sr. mem.), Optical Soc. Am., Soc. Optical Engring., European Phys. Soc., European Optical Soc., Laser Inst. of America, NY Acad. of Scis. Roman Catholic. Avocations: American culture, sports. Home: Mickiewicza 74/21, PL-01650 Warsaw Poland Office: Warsaw Univ Tech, Nowowiejska 15/19, PL-00665 Warsaw Poland

ROMANKIW, LUBOMYR TARAS, materials engineer; b. Zhowkwa, Ukraine, Apr. 17, 1931. BSc, U. Alta., 1955; MSc and PhD in Metallurgy, MIT, 1962. Mem. rsch. staff materials and processes, Thomas J. Watson Rsch. Ctr. IBM Corp, Yorktown Heights, N.Y., 1962-63, mgr. magnetic components divsn., 1965-68, mgr. magnetic material and devices, 1968-78, mgr. material and process studies, 1981—, dep. mgr.; instr. MIT, 1959-61; cons. East Fishkill Devel. Lab. & Mfg. IBM Corp, 1978-80. Recipient Perkin medal Am. Chem. Soc., 1993. Mem. IEEE, Electrochem. Soc. (sec.-treas. 1979-80), Am. Electroplaters Soc., Sigma Xi. Achievements include research in magnetic thin films, deposition of thin films, dielectrics, magnetic device design, material selection and fabrication, electrodeposition, magnetic materials, electronic and magnetic device fabrication, chemical engineering, and metallurgy. Office: IBM-Thomas J Watson Research Ctr POB 218 Yorktown Heights NY 10598*

ROMANO, LOUIS JAMES, chemist, educator; b. Orange, N.J., Nov. 19, 1950; s. Louis Frank and Josephine (Masciocchi) R.; m. Sheila Ann Laffey, Oct. 10, 1970; children: Eric, Carissa. BA, PhD., Rutgers U., 1972. Rsch. fellow Harvard Med. Sch., Boston, 1976-79, rsch. assoc., 1979-80; prof. chemistry Wayne State U., Detroit, 1980—. Recipient Career Devel. award NIH, 1985-90. Mem. Am. Chem. Soc., AAAS, Am. Assn. for Cancer Rsch., Am. Soc. for Biochemistry and Molecular Biology, Am. Cancer Soc. (carcinogenesis adv. com. 1991—). Achievements include patent for method for analyzing organic samples. Home: 2105 Burns Detroit MI 48214 Office: Wayne State U Dept Chemistry Detroit MI 48202

ROMANOWSKI, THOMAS ANDREW, physics educator; b. Warsaw, Poland, Apr. 17, 1925; came to U.S., 1946, naturalized, 1949; s. Bohdan and Alina (Sumowski) R.; m. Carmen des Rochers, Nov. 15, 1952; children—Alina, Dominique. B.S., Mass. Inst. Tech., 1952; M.S., Case Inst. Tech., 1956, Ph.D., 1957. Rsch. assoc. physics Carnegie Inst. Tech., 1956-60; asst. physicist high energy physics Argonne Nat. Lab., Ill., 1960-63; assoc. physicist Argonne Nat. Lab., 1963-72, physicist, 1972-78; prof. physics Ohio State U., Columbus, 1964—. Contbr. articles to profl. jours. and, papers to sci. meetings, seminars and workshops. Served with C.E. AUS, 1946-47. Fellow Am. Phys. Soc., AAAS; mem. Lambda Chi Alpha. Achievements include research in nuclear and high energy physics. Home: 1380 Kersey Ln Rockville MD 20854-6117 Office: Dept Energy Div High Energy Physics Washington DC 20585

ROMBERGER, JOHN ALBERT, scientist; b. near Klingerstown, Pa., Dec. 25, 1925; s. Ralph T. and Carrie (Bahner) R.; student Hershey Jr. Coll., 1947-49; B.A., Swarthmore Coll., 1951; M.S., Pa. State U., 1954; Ph.D., U.

Mich., 1957; post doctoral, Calif. Inst. Tech., 1957-60; m. Margery Janet Davis, June 17, 1951; children: Ann I., Daniel D. Plant physiologist, Forest Physiology Lab., U.S. Forest Service, U.S. Dept. Agr., Beltsville, Md., 1961-82; vis. scientist Swedish U. Agrl. Scis., Alnarp, 1983, Inst. Agrl. Scis., Zamosc, Poland, 1985, Agrl. U., Warsaw, 1988. Lay leader Unitarian Ch. Served with AUS, 1945-46. Fellow Poland-U.S. Interacad. Exchange Program, U. Silesia, Katowice, 1981, 83. Fellow AAAS; mem. Am. Soc. Plant Physiologists, Bot. Soc. Am., Am. Inst. Biol. Scis., Soc. for History Tech., Sigma Xi. Author: Meristems, Growth, and Development in Woody Plants, 1963, (with Z. Hejnowicz and J.F. Hill) Plant Structure: Function and Development, 1993; editor: Internat. Rev. Forestry Research, 1963-70, Beltsville Symposia in Agrl. Research, 1976-78. Contbr. articles on devel. and theoretical biology to profl. jours. Home: 2005 Forest Hill Dr Silver Spring MD 20903-1533

ROMER, ROBERT HORTON, physicist, educator; b. Chgo., Apr. 15, 1931; s. Alfred Sherwood and Ruth (Hibbard) R.; m. Diana Haynes, June 12, 1953; children: Evan James, David Hibbard, Theodore Haynes. B.A., Amherst Coll., 1952; Ph.D. in Physics, Princeton U., 1955. Faculty Amherst (Mass.), Coll., 1955—, prof. physics, 1966—, chmn. dept., 1966—; Research asso. Duke, 1958-59; guest physicist Brookhaven Nat. Lab., 1963—; vis. prof. physics Voorhees Coll., 1969-70. Author: Energy—An Introduction to Physics, 1976, Energy Facts and Figures, 1984. NSF fellow U. Grenoble, France, 1964-65. Fellow AAAS, Am. Phys. Soc.; mem. Am. Assn. Physics Tchrs. (asso. editor jour. 1968, book rev. editor 1982-88, editor 1988—), Phi Beta Kappa, Sigma Xi. Research low temperature physics, solar energy, electromagnetic theory. Home: 104 Spring St Amherst MA 01002-2332

ROMERO, EMILIO FELIPE, psychiatry educator, psychotherapist, hospital administrator; b. Havana, Cuba, Nov. 12, 1946; came to U.S., 1960; s. Emilio Jose and Isela Maria (Correoso) R. BA cum laude, U. Miami, Coral Gables, Fla., 1966; MD, U. Zaragoza (Spain), 1972. Cert. Am. Bd. Psychiatry and Neurology. Resident in psychiatry U. Tex. Health Sci. Ctr., San Antonio, 1976; assoc. dir. Therapeutic Community for Schizophrenics Audie L. Murphy VA Hosp., San Antonio, 1976-78, staff psychiatrist outpatient psychiatry svcs., 1978-81, dir. outpatient psychiatry svcs., 1981-89, acting chief psychiatry svc., 1989-90, chief psychiatry svc., 1990—; asst. prof. psychiatry U. Tex. Health Sci. Ctr., San Antonio, 1976-87, assoc. prof. psychiatry, 1987-93, prof., 1993—; examiner Am. Bd. Psychiatry and Neurology, Deerfield, Ill., 1987—; psycho-analytical cons. Med. Ctr. Hosp., San Antonio, 1981-87; lectr. on dream interpretation, U.S.A., Can., Europe, South Am., 1977—. Contbr. articles to nat. and internat. profl. jours.; judge Internat. Film Festival on Culture and Psychiatry, 1976, 78; cons. to Donald Sutherland on film Lost Angels, 1988. Active local TV and radio stas., San Antonio Express News & Light, San Antonio and Diario del las Americas, Miami, Fla., 1973—; mem. art com. San Antonio Mus. Assn., 1988—, mayor's Com. to Receive the King of Spain, San Antonio, 1987; mem. Mex. Rels. Com. to Receive His Holiness Pope John Paul II, San Antonio, 1987. Named knight Sovereign Order of Malta, Rome, 1987; recipient Silver plaque U. Salamanca (Spain), 1982; recognized with proclamation of day in his honor, Dade County, Fla., 1992. Fellow Am. Psychiat. Assn., Am. Assn. Social Psychiatry; mem. Tex. Soc. Psychiat. Physicians (chmn. internat. grads. 1985—), Assn. Mil. Surgeons U.S. (internat. com. 1986—), Am. Coll. Psychiatrists, Bexar County Psychiat. Soc. (pres. 1987-88, Leadership award 1988), Sociedad Espanola De Psiquiatria (Mem. of Honor 1988, Disting. Guest and Citizen City of Salmanaca Spain 1990). Republican. Roman Catholic. Avocations: painting, writing, travelling, international relations, reading. Home: 141 Twinleaf Ln San Antonio TX 78213-2516 Office: Audie Murphy VA Hosp 7400 Merton Minter Blvd San Antonio TX 78284

ROMERO, JORGE ANTONIO, neurologist, educator; b. Bayamon, P.R., Apr. 15, 1948; s. Calixto Antonio Romero-Barcelo and Antonia (de Juan) R.; m. Helen Mella, June 20, 1970 (div. 1983); children: Sofia, Jorge, Alfredo, Isabel. SB, MIT, 1968; MD, Harvard U., 1972. Diplomate Am. Bd. Psychiatry and Neurology. Intern U. Chgo. Hosp. and Clinics, 1972-73; resident Mass. Gen. Hosp., Boston, 1975-78; rsch. fellow in pharmacology NIMH, Bethesda, Md., 1973-75; asst. prof. neurology Harvard Med. Sch., Boston, 1979-92; mem. staff VA Med. Ctr., Brockton, Mass., 1979-92; assoc. physician Brigham and Women's Hosp., Boston, 1980-92; chmn. dept. neurology Ochsner Clin. Baton Rouge, 1993—; cons. Mass. Mental Health Ctr., Boston, 1987-92. With USPHS, 1973-75. Recipient Career Devel. award VA, 1979. Mem. Am. Acad. Neurology. Office: Ochsner Clin Baton Rouge 16777 Medical Center Dr Baton Rouge LA 70816

ROMERO, MIGUEL A., animal nutrition director; b. Tonalá, Mexico, Sept. 29, 1925; s. Elias Aurelio and Luisa (Sánchez) R.; m. Margarita Betha, Feb. 2, 1958; children: Miguel, Luisa, Alejandro. MA, Harvard U., 1953, PhD, 1955. Postdoctoral Imperial Coll., London, 1955-56; rsch. fellow U. Mexico, Mexico City, 1956-57; rsch. coord. G.D. Searle, Chgo., 1958-62; gen. dir. Grupo Romero, Tehuacán, Mex., 1963—; pres. acadmeic bd. U. de Las Americas, Puebla, Mex., 1990—; founder, owner Romero Mineral. Mus. Congressman Fed. Congress, Mexico City, 1985-88. Sgt. Mexican Army, 1944. Recipient Carnegie award, 1991; named to Hall of Fame, Latin Am. Poultry Orgn., 1989. Mem. Am. Chem. Soc., Mineral. Soc. Am., Mexican Mineral. Soc. (pres. 1984—, founder 1984), Poultry Sci. Assn., others. Achievements include finding of largest and most advanced integrated poultry operation known; discovery of two new mineral species: Mapimite and ojuelaite. Office: 7 Norte # 356, 75700 Tehuacán Mexico

ROMERO, RALPH, physicist; b. N.Y.C., Oct. 11, 1953; s. Ralph and Patria (Aromi) R.; m. Ana Iris Fernandez, Oct. 4, 1992; 1 child, Lil. BS in Physics, Havana U., 1975, MS, 1977, PhD, 1984. Instr. Havana (Cuba) U., 1975-83, rsch. scientist, 1983-84; dir. hybrid integrated circuits COPEXTEL, Havana, 1984-91; sr. scientist Advanced Photovoltaic Systems Inc., Princeton, N.J., 1991-92, dir. tech. ops., 1992—; exch. scientist AF Ioffe Physics Tech. Inst., Leningrad, USSR, 1979-82,Inst. Microelectronics, Sofia, Bulgaria, 1984. Contbr. articles to profl. jours. Recipient Medal of Honor 1st Soviet-Cuban Space Flight Acad. Scis. Cuba, 1980. Achievements include research in the disruption of solid-liquid interface in Germanium crystals; development of Gallium arsenide solar cells. Office: Advanced Photovoltaic System 195 Clarksville Rd Lawrenceville NJ 08648

ROMEY, WILLIAM DOWDEN, geologist, educator; b. Richmond, Ind., Oct. 26, 1930; s. William Minter and Grace Warring (Dowden) R.; m. Lucretia Alice Leonard, July 16, 1955; children—Catherine Louise, Gretchen Elizabeth, William Leonard. A.B. with highest honors, Ind. U., 1952; student, U. Paris, 1950-51, 52-53; Ph.D., U. Calif. at Berkeley, 1962. Asst. prof. geology and sci. edn. Syracuse U., 1962-66, assoc. prof., 1966-69; exec. dir. earth sci. ednl. program Am. Geol. Inst., 1969-72; prof., chmn. dept. geology St. Lawrence U., Canton, N.Y., 1971-76; prof. St. Lawrence U., 1976—, prof., chmn. dept. geography, 1983—; ednl. cons., 1962—; Nat. Acad. Sci. visitor USSR Acad. Sci., 1967; vis. geoscientist Am. Geol. Inst., 1964-66, 71; earth sci. cons. Compton's Ency., 1970-71; adj. prof. Union Grad. Sch., 1974—; mem. bd. research advisers and readers Walden U., 1981—. Author: (with others) Investigating the Earth, 1967, (with J. Kramer, E. Muller, J. Lewis) Investigations in Geology, 1967, Inquiry Techniques for Teaching Science, 1968, Risk-Trust-Love, 1972, Consciousness and Creativity, 1975, Confluent Education in Science, 1976; co-editor: Geochemical Prospecting for Petroleum, 1959; assoc. editor: Jour. Coll. Sci. Teaching, 1972-74, Geol. Soc. Am. Bull, 1979-84, Jour. Geol. Edn, 1980—; editor-in-chief: Ash Lad Press, 1975—; contbr. articles on geology, geography and edn. to profl. publs. Bd. dirs. Onondaga Nature Centers, Inc., 1966-69. Served to lt. USNR, 1953-57; lt. comdr. Res. Woodrow Wilson Found. fellow, 1959-60, 61-62; NSF sci. faculty fellow U. Oslo, 1967-68. Fellow Geol. Soc. Am., AAAS; mem. Nat. Assn. Geology Tchrs. (v.p. 1971-72), N.Y. Acad. Scis., Nat. Assn. Geology Tchrs. (pres. 1972-73), Assn. Am. Geographers, Am. Geophys. Union, Geol. Soc. Norway, Assn. Educating Tchrs. of Sci., Can. Assn. Geographers, Assn. for Can. Studies in U.S., Phi Beta Kappa, Sigma Xi, Phi Delta Kappa. Home: PO Box 294 East Orleans MA 02643 Office: St Lawrence U Dept Geography Canton NY 13617

ROMIG, ALTON DALE, JR., metallurgist, educator; b. Bethlehem, Pa., Oct. 6, 1953; s. Alton Dale and Christine (Groh) R.; m. Julie H. Romig. BS, Lehigh U., 1975, MS, 1977, PhD, 1979. Metallurgist, mem. tech. staff Sandia Nat. Labs., Albuquerque, 1979-87, supr. physical metallurgy, 1987-90, mgr. metallurgy, 1990-92, dir. materials and process scis., 1992—; part time full prof. N.Mex. Inst. Mining and Tech., Socorro, 1981—; Acta/Scripta Metallurgica Lectr., 1993; Fellow Am. Soc. for Metals (trustee 1992—, Outstanding Rsch. award 1992); mem. TMS, Electron Microscopy Soc. Am. (Burton Outstanding Young Sci. medal 1988), Microbeam Analysis Soc. (pres. 1990, Heinrich award for Outstanding Young Sci. 1991), Materials Rsch. Soc., Sigma Xi, Tau Beta Pi. Author: Principles of Analytical Electron Mecroscopy, 1986, Scanning Electron Microscopy, X-ray Microanalysis and Analytical Electron Microscopy, 1991, Scanning Electron Microscopy and Microanalysis, 1992; editor numerous procs. in phys. metallurgy and electron microscopy; contbr. over 140 articles to sci. jours. Home: 4923 Calle De Luna NE Albuquerque NM 87111-2916 Office: Sandia Nat Labs Ctr 1800 Albuquerque NM 87185

ROMINE, THOMAS BEESON, JR., consulting engineering executive; b. Billings, Mont., Nov. 16, 1925; s. Thomas Beeson and Elizabeth Marjorie (Tschudy) R.; m. Rosemary Pearl Melancon, Aug. 14, 1948; children—Thomas Beeson III, Richard Alexander, Robert Harold. Student, Rice Inst., 1943-44; B.S. in Mech. Engring, U. Tex., Austin, 1948. Registered profl. engr., Tex., Okla., La., Ga. Jr. engr. Gen. Engring. Co., Ft. Worth, 1948-50; design engr. Wyatt C. Hedrick (architect/engr.), Ft. Worth, 1950-54; chief mech. engr. Wyatt C. Hedrick (architect/engr.), 1954-56; chmn., chief mech. engr. Thomas B. Romine, Jr. (now Romine Romine & Burgess, Inc. cons. engrs.), Ft. Worth, 1956—; mem. heating, ventilating, and air conditioning controls com. NRC, 1986-88. Author numerous computer programs in energy analysis and heating and air conditioning field; contbr. articles to profl. jours. Mem. Plan Commn., City of Ft. Worth, 1958-62; mem. Supervisory Bd. Plumbers, City Ft. Worth, 1963-71, chmn., 1970-71; chmn. Plumbing Code Rev. Com., 1968-69; mem. Mech. Bd., City Ft. Worth, 1974—, chmn., 1976—; chmn. plumbing code bd. North Central Tex. Council Govts., Ft. Worth, 1971-75; Bd. mgrs. Tex. Christian U.-South Side YMCA, 1969-74; trustee Ft. Worth Symphony Orch., 1968—, Orch. Hall, 1975—. Served with USNR, 1943-45. Disting. fellow ASHRAE (pres. Ft. Worth chpt. 1958, nat. committeeman 1974—); fellow Am. Cons. Engrs. Coun., Automated Procedures for Engring. Cons. (trustee 1970-71, 75, 1st v.p. 1972-73, internat. pres. 1974); mem. NSPE, Tex. Soc. Profl. Engrs. (bd. dirs. 1956, treas. 1967), Cons. Engrs. Coun. Tex. (pres. North Tex. chpt., also v.p. state orgn. 1965, dir. state orgn. 1967), Starfish Class Assn. (nat. pres. 1970-73, nat. champion 1976), Delta Tau Delta (v.p. West div. 1980—), Pi Tau Sigma. Episcopalian (vestryman). Clubs: Colonial Country, Rotary. Home: 3232 Preston Hollow Rd Fort Worth TX 76109-2051 Office: Romine Romine & Burgess 300 Greenleaf St Fort Worth TX 76107-2392

RONDEAU, JACQUES ANTOINE, marketing specialist, chemical engineer; b. Roubaix, Nord, France, Feb. 2, 1947; s. Antoine J. and Renée J. (Vincent) R.; m. Francoise N. Regard, Aug. 31, 1974; children: Elizabeth, Xavier. MS, Nat. Sch. Indsl. Chemistry, Nancy, France, 1970; DSc, Inst. Nat. Poly. Lorraine, Nancy, France, 1976; MBA, Inst. Adminstrn. Enterprises, Paris, 1979. Jr. scientist Nat. Rsch. Coun., Nancy, 1970-75; project leader Kleber Tyres and Rubber, Paris, 1976-81; bus. devel. mgr. Elf Aquitaine Atochem, Paris, 1982-89; mktg. mgmt. Produits Chimiques Auxilliaires et de Synthese, Paris, 1989-93; mktg. and sales mgr. SEAC, Paris, 1993—. Mem. Am. Chem. Soc., French Chem. Soc. Home: 28 Genevieve Couturier, 92500 Rueil Malmaison France

RONEY, LYNN KAROL, psychologist; b. Stillwater, Okla., Oct. 22, 1946; d. Maurice William and Ruth Evangelene (Hobart) R.; m. Norman C. Lawson, Mar. 12, 1988; children: Emily, Rachael, Abigail. BS, Okla. State U., 1968, MA, 1970; PhD, U. Tex., 1979. Lic. psychologist, Tex. Staff psychologist Counseling Ctr., U. Iowa, Iowa City, 1979-81; dir. Counseling Ctr., Ithaca (N.Y.) Coll., 1981-85; pvt. practice Houston, 1985—. Contbr. articles to profl. jours. Fundraiser Dem. Party, Ann Richards campaign, 1990. Mem. APA, Am. Coll. Personnel Assn. (dir. commn. on counseling 1984), Tex. Psychol. Assn., Houston Psychol. Assn. (chmn. ethics com. 1987-89, officer for orgnl. affairs 1991—). Office: 4615 Post Oak Pl # 200 Houston TX 77027 also: 15835 Park Ten Pl Ste 106 Houston TX 77084

RONN, AVIGDOR MEIR, chemical physics educator, consultant, researcher; b. Tel Aviv, Nov. 17, 1938; came to U.S., 1959; m. Linda Ann Tenney, Aug. 25, 1963; children: David A., Karin J. BS in Chemistry, U. Calif., Berkeley, 1963; AM in Phys. Chemistry, Harvard U., 1964, PhD in Phys. Chemistry, 1966. Rsch. asst. Nat. Bur. Standards, 1966-68; from asst. prof. to assoc. prof. chemistry Poly. Inst. Bklyn., 1968-73; prof. chemistry Bklyn. Coll. CUNY, 1973—, Broeklundian prof. Bklyn. Coll., 1987-90, dir. Laser Inst., 1987—; v.p. Leron Assocs., Great Neck, N.Y., 1984—; sr. rsch. fellow Long Island Jewish Med. ctr, 1992—, Albert Enstein Med. Coll., 1992—; exec. officer PhD program in chemistry CUNY, Manhattan, 1984-90, exec. dir. Applied Sci. Inst., Bklyn., 1987-90; vis. prof. U. Tel Aviv, 1971-72; Fulbright Sr. scholar U. Sao Paulo, Brazil, 1983-84; v.p., gen. mgr. Lic Industries, Inc., Suffern, N.Y., 1979-80; cons. Eastman Kodak, Rochester, N.Y., 1981-85. Author: (with others) Advances in Chemical Physics, 1980, Techniques of Chemistry, 1981. Pres. Towne House 27, Inc., Great Neck, 1984—. 1st sgt. Israeli Army, 1956-58. Fellow Alfred P. Sloan Found., 1971-73, OAS, Sao Paulo, 1973. Mem. Israel Chem. Soc., Am. Phys. Soc., Am. Chem. Soc., Phi Beta Kappa. Achievements include 5 patents for Laser Initiated Chain Reactions for Producing a Sintered Product, Method for Forming Patents on Substrate or Support, Production of Chain Reaction by Laser Chemistry, Preparation of Metal Containing Polymeric Materials via Laser Chemistry, Method of Molecular Species Alteration by Nonresonant Laser Induced Dielectric Breakdown. Office: Bklyn Coll Bedford Ave # H Brooklyn NY 11222-3102

ROOD, ROBERT EUGENE, construction engineering executive, consultant; b. Syracuse, N.Y., Feb. 21, 1951; s. Robert Lloyd and Ruth Merribel (Perry) R.; m. Debra Janice Stahl, Aug. 31, 1974; children: Nathan Josiah, Ryan Kendell. BS in forestry, Syracuse Univ., 1973; BS in building construction, Coll. Environ. Sci., 1973. Registered profl. engr. Miss. Civil engr. Pyramid Construction Co., Liverpool, N.Y., 1973; quality control engr. United Engrs. and Construction, Salem, N.J., 1973-74; civil construction engr. Bechtel Power Corp., Gaithersburg, Md., 1974-86; lead civil engr. Bechtel Construction Inc., Gilberton, Pa., 1986-88; resident pipe/civil engr. North Am. Power Corp., Spring City, Tenn., 1988-89; area construction engr. mgr. Bechtel Savannah River, Inc., Aiken, S.C., 1989—; cons. Testing labs., Pigeon Forge, 1986-87. Active Miss. Youth Recreation Assn., Vicksburg, 1974-82, North Augusta Youth Recreation Assn., 1990-92. Recipient Total Quality award Westinghouse, 1992. Mem. Am. Concrete Inst., Am. Soc. Civil Engrs. Home: 1158 Ridgemont Aiken SC 29803

ROOKE, ALLEN DRISCOLL, JR., civil engineer; b. San Antonio, Oct. 5, 1924; s. Allen Driscoll and Jean Edna (Lackner) R.; m. Betty Ruth Whitson, Oct. 17, 1949; children: Victoria Lynn Lewis, Cornelia Ruth. BSCE, Tex. A&M U., 1957; MSCE, Miss. State U., 1980. Registered profl. engr., Miss. Enlisted U.S. Army, 1942, advanced through grades to brig. gen., ret., 1984; rsch. civil engr. U.S. Army Corps Engrs., Vicksburg, Miss., 1958-83; ptnr. F.B. Rooke & Sons, Woodsboro, Tex., 1964—; sr. engr. Sci & Tech. Corp., Vicksburg, Miss., 1984—; bd. dirs. First Nat. Bank, Woodsboro, 1985—. Author/co-author numerous tech. publs. Mem. Res. Officers Assn. U.S. (dept. pres. 1980-82, svc. award 1980, 84), Assn. of U.S. Army, Ret. Officer's Assn. (chpt. v.p. 1985-86), Soc. Am. Mil. Engrs. (post v.p. 1979). Episcopalian. Club: Army and Navy Vicksburg (pres. 1980, 82). Avocation: chess. Home: PO Box 732 Woodsboro TX 78393-0732

ROOMSBURG, JUDY DENNIS, industrial/organizational psychologist; b. Seoul, Korea, June 2, 1954; came to U.S., 1957; children: Michael Scott, Christopher Thomas. MA, Chapman Coll., Orange, Calif., 1983, U. Tex., 1988; PhD, U. Tex., 1989. Enlisted USAF, 1973, commd., 1979, advanced through grades to maj., 1990; human factors engr. Aero. Systems Div., Wright-Patterson AFB, Ohio, 1984-85; behavioral scientist Armstrong Lab., Brooks AFB, Tex., 1985-86; chief behavioral analysis Hdqrs. USAF, Washington, 1989-93; mgr. organizational devel. and tng. Allied Signal Inc., Columbia, Md., 1993—. Mem. APA, Am. Psychol. Soc., Soc. Indsl./Organizational Psychology, Acad. Mgmt., Am. Statis. Assn., Soc. Human Resource Mgmt. Achievements include research in utility as a function of BR and SR, biographical predictors in mil. aviation, tech. assessment. Office: Allied Signal Inc One Bendix Rd Columbia MD 21045-1897

ROONEY, JOHN CONNELL, consulting civil engineer; b. Racine, Wis., Feb. 4, 1967; s. James Francis and Nancy Lee (Schulz) R. BSCE, U. Wis., 1990. Registered profl. engr., Wis. Cons. civil engr. Nielsen, Madsen & Barber, S.C., Racine, Wis., 1990—. Mem. NSPE, Wis. Soc. Profl. Engrs. (Rock River Valley Chpt.). Democrat. Roman Catholic. Home: 1500 Michigan Blvd Racine WI 53402 Office: Nielsen Madsen & Barber SC 1339 Washington Ave Racine WI 53403

ROOP, JOSEPH MCLEOD, economist; b. Montgomery, Ala., Sept. 29, 1941; s. Joseph Ezra and Mae Elizabeth (McLeod) R.; B.S., Central Mo. State U., Warrensburg, 1963; Ph.D., Wash. State U., Pullman, 1973; m. Betty Jane Reed, Sept. 4, 1965; 1 dau., Elizabeth Rachael. Economist, Econ. Research Service, U.S. Dept. Agr., Washington, 1975-79; sr. economist Evans Econs., Inc., Washington, 1979-81; sr. research economist Battelle Pacific N.W. Labs., Richland, Wash., 1981—; instr. dept. econs. Wash. State U., 1969-71; with Internat. Energy Agy., Paris, 1990-91. Contbr. tech. articles to profl. jours. Served with U.S. Army, 1966-68. Dept. Agr. Coop. State Research Service research grantee, 1971-73. Mem. Am. Econ. Assn., Econometric Soc., Internat. Assn. Energy Economics. Home: 715 S Taft Kennewick WA 99336 Office: PO Box 999 Richland WA 99352

ROOP, MARK EDWARD, process control applications engineer; b. Pitts., Oct. 31, 1958; s. Edward J. and Helen Marie (Cox) R.; m. Jennifer Lea Rowe, Nov. 17, 1990; children: Devon, Ian, Adrienne. BS in Chem. Engring., W.Va. U., 1981. Sales engr. Bailey Controls, Houston and New Orleans, 1982-86; ops. mgr. Bailey Controls, Chgo., 1986-87; applications engr. Fisher Controls, Austin, Tex., 1987-89, corp. account mgr., 1989-91; internat. sales engr. Honeywell, Phoenix, 1991—. Mem. TAPPI, Am. Inst. Chem. Engrs., Paper Industry Mgmt. Assn. Office: Honeywell Inc 16404 N Black Canyon Hwy Phoenix AZ 85023

ROOP, ROBERT DICKINSON, biologist; b. Plainfield, N.J., Sept. 23, 1949; s. Robert Wendell and Katherine (Booth) R.; m. Edna Southerland, Jan. 25, 1981; children: Jay Lee, Sarah Maria. BA, Hiram Coll., 1971; MA, SUNY, Stony Brook, 1975. Staff ecologist The Inst. Ecology, Washington, 1974-76; rsch. assoc. Oak Ridge (Tenn.) Nat. Lab., 1976-89; project dir. Labat-Anderson, Inc., Oak Ridge, 1989—; chmn. ecology com. Water Environment Fedn., Alexandria, Va., 1991—. Mem. Ecol. Soc. Am. Unitarian. Office: Labat-Anderson Inc 575 Oak Ridge Tpk Oak Ridge TN 37830

ROOT, M. BELINDA, chemist; b. Port Authur, Tex., May 2, 1957; d. Robert A. and Charlene (Whitehead) Lee; m. Miles J. Root, Nov. 8, 1980; children: Jason Matthew, Ashley Erin. BS in Biology, Lamar U., 1979. Asst. chemist Merichem Co., Houston, 1979-81, project chemist, 1982-84, instrument chemist, 1984-85, quality assurance coord., 1986-89, product lab. supr., 1989-91; quality control supr. mfg. Welchem Inc. subs. Amoco, 1991—; mgr. Quality Control Petrolite Corp., 1993—. Editor (newsletter) Merichemer, 1989-91. Mem. MADD, 1989—, PTA, 1988—. Recipient Gulf Shore Regional award Cat Fanciers Assn., 1981, Disting. Merit award, 1990. Mem. NAFE, Am. Soc. for Quality Control (cert. quality auditor, quality engr.), United Silver Fancier (sec. 1980-82), Lamar U. Alumni Assn., Am. Chem. Soc., Beta Beta Beta (sec. 1978-79). Avocations: campaigning show cats, camping, gardening.

ROPCHAN, JIM R., research chemist, administrator; b. Leamington, Ont., Can., Apr. 14, 1950; s. William George and Katie (Rudyka) R. Degree in chem. tech., St. Clair Coll. Applied Arts and Tech., Ont., Can., 1971; BS with honors, Detroit Inst. Tech., 1972; PhD, U. Detroit, 1981. Quality control chemist Ford Motor Co., Windsor, Ont., 1973-76; postdoctoral scholar UCLA, 1981-85, assoc. investigator div. nuclear medicine and biophysics, 1983-85; dir., chief chemist chemistry sect., positron emission tomography facility VA Med. Ctr., L.A., 1986—, sec. radioactive drug rsch. com., 1987—, instr. radiopharmacy, 1987—; dir. radiopharm. chemistry rsch., Positron Emmission Tomography Facility, 1987—, dir. cyclotron targetry devel., 1989—, dir. eyclotron facility, 1992—; supr. radiopharmacy/nuclear medicine dept., VA Med. Ctr., L.A., 1987—, mem. Radiation Safety com., 1986—; part-time tchr. high sch. math. and sci., Windsor, Ont., 1977; part-time instr. organic chemistry Detroit Inst. Tech., 1977; lectr. on radiopharm. chemistry/cyclotron rsch., Italy, 1984, Can., 1989, Switzerland, 1991. Contbr. articles to profl. jours.; inventor lab. accessories; designer, fabricator new cyclotron targetry/chemistry systems, 1991—. Recipient Outstanding Scholar award, Detroit Inst. Tech., 1972, Supr. Performance award, VA Med. Ctr., L.A., 1989, 91, 92, Assoc. Investigator award UCLA, grantee, 1983; named ABI Man of Yr., 1991. Fellow IBC (Man of Yr. 1992), Am. Inst. Chemists; mem. AAAS, N.Y. Acad. Scis., Am. Chem. Soc. (div. Carbohydrate, Organic Medicinal Chemistry), Am. Heart Assn., Am. Mgmt. Assn., The Cousteau Soc., The Planetary Soc., Keepers Club, Diamond Club. Achievements include design of new targetry and chemistry processing systems, design of new radiopharmaceuticals and organic molecules. Office: VA Med Ctr-Wadsworth Bldg 500 Rm 91 Div Nuclear Medicine UltraSound Sawtelle and Wilshire Blvds Los Angeles CA 90073

ROPE, BARRY STUART, packaging engineer, consultant; b. Detroit, Aug. 12, 1942; s. Sanford Julian and Toby (Freedman) R.; m. Rosalyn Sarnoff, Dec. 13, 1965; children: Daniel Jay, Todd Loren. BS in Packaging, Mich. State U., 1969. Packaging engr. Eastman Kodak, Kingsport, Tenn., 1969-70; sr. packaging engr. Parke-Davis & Co., Detroit, 1970-74; packaging quality control supr. Faygo Beverages, Detroit, 1974-78; sr. packaging engr. Computer Peripherals, Rochester, Mich., 1978-83; mgr. corp. packaging engring. Unisys (Burroughs), Detroit, 1983-90; cons. Rope & Assocs., Farmington Hills, Mich., 1991—; mem. Teltech Resource Network, 1983—; editorial adv. bd. Jour. Packaging Tech., 1986—; mem. packaging coun. AMA, 1984—; judge Ameristar Packaging Competition. Author: Physical Distribution Handbook, 1984; producer videos on packaging; contbr. articles to profl. publs. With U.S. Army, 1960-63. Recipient Golden Cassette award Internat. TV Assn., 1986, Monitor award Internat. Monitor, 1987. Mem. ASTM (packaging com. 1983—), Inst. Packaging Profls. (chmn. com. protective packaging of computers and components 1986-87). Achievements include patents for cushioning materials in packaging, design of dispenser package, hosiery package, break-away pallet. Home and Office: Ste A 27631 Westcott Farmington Hills MI 48334

ROPP, RICHARD CLAUDE, chemist; b. Detroit, Mar. 26, 1927; s. Claude V. and Claire (Beusschemin) R.; m. Francisca Margarita Ropp, Aug. 20, 1952; children: Richard J., Michael C., Thomas A., Daniel F. AB, Franklin Coll., 1950; MS, Purdue U., 1952; PhD, Rutgers U., 1971. Staff scientist Allied Chem., Morristown, N.J., 1973-77; dir. rsch. Merco, Westwood, N.J., 1977-78; with Petrex Corp., Summit, N.J., 1979-83; cons. chemist, 1983-86; cons. Ramirez Assocs., Far Hills, N.J., 1986-88; mgr. quality control lab. CPG, Inc., West Caldwell, N.J., 1988-89; with rsch. lab. Metrigen, Inc., Piscataway, N.J., 1989; v.p. Internat. Superconductor Corp., Warren, N.J., 1990-91; tech. assistance for attys., Blue Bell, Pa., 1984—, Warren N.J., 1989. Author: Luminescence and the Solid State, 1991, Inorganic Polymeric Glasses, 1992, The Chemistry of Artificial Lighting Devices-Lamps, Phosphors and Cathode Ray Tubes, 1993; contbr. articles to profl. jours. Fellow Royal Soc. Chemists; mem. N.J. Inst. Chemists (pres. 1978-80), Am. Inst. Chemists (pres. 1979). Achievements include 56 patents and 4 patents pending. Home and Office: Lumi Tech 138 Mountain Ave Warren NJ 07059

ROSAN, BURTON, microbiology educator; b. N.Y.C., Aug. 18, 1928; s. Harry and Rae (Halpern) R.; m. Helen Mescon, Jan. 14, 1951; children: Rhea, Felice, Jonathan. BS, CCNY, 1950; DDS, U. Pa., Phila., 1957, MSc, 1962. Rsch. assoc. microbiology U. Pa. Dental Medicine, Phila., 1959-63, asst. prof., 1963-67, assoc. prof., 1967-75, prof., 1975—; vis. prof. Inst. Dental Rsch., Sydney, 1980, Royal Coll. Surgeons, Downe, Eng., 1985; mem. oral biology and med. study sect. NIH, Bethesda, Md., 1975-79; dental devices cons. FDA, Rockville, Md., 1990—. Editor: Molecular Basis Oral Microbial Adhesions, 1985; contbr. articles to profl. jours., chpts. to books. 1st lt. U.S. Army, 1957-60. Multiple grantee Nat. Inst. Dental Rsch., U. Pa., 1959—. Fellow Am. Acad. Microbiology; mem. Am. Soc. Microbiology, Internat. Assn. Dental Rsch. (pres. micro-immuno group 1979-80, chair fellowship com. 1980-81), Omicron Kappa Upsilon. Jewish. Achievements include research in serology or oral streptococci, how oral streptococci stick to teeth, mechanisms of dental plaque formation. Home:

113 Cambridge Rd Broomall PA 19008 Office: U Pa Sch Dental Medicine 4001 Spruce St Philadelphia PA 19104

ROSANDER, ARLYN CUSTER, mathematical statistician, management consultant; b. Mason County, Mich., Oct. 7, 1903; s. John Carl and Nellie May (Palmer) R.; m. Beatrice White, Aug. 26, 1933 (div.); children: Nancy Rosander Peck, Robert Richard Roger (dec.); m. Margaret Ruth Guest, Aug. 15, 1964. BS, U. Mich. 1925; MA, U. Wis., 1928; PhD, U. Chgo., 1933; postgrad. Dept. Agr., 1937-39. Rsch. asst. U. Chgo., 1933-34; rsch. fellow Gen. Edn. Bd. Tech. dir. Am. Youth Commn., Balt. and Washington, 1935-37; chief statistician urban study U.S. Bur. Labor Stats., Washington, 1937-39; sect. and br. chief War Prodn. Bd., Washington, 1940-45; chief statistician IRS, Washington, 1945-61; chief math. and stats. sect. ICC, Washington, 1961-69; cons. Pres.'s Commn. on Fed. Stats., Washington, 1970-71; cons., Loveland, Colo.; lectr. stats. George Washington U., 1946-52. Recipient Civilian War Service award War Prodn. Bd., 1945; Spl. Performance award Dept. Treasury, 1961; A.C. Rosander awardcreated by Service Industries divsn. CSCC, 1991. Fellow AAAS, Am. Soc. Quality Control (25 yr. honor award 1980, Howard Jones Meml. award 1984, chmn. emeritus svc. industries divsn. 1991); mem. Am. Statis. Assn. Author: Elementary Principles of Statistics, 1951; Statistical Quality Control in Tax Operations, IRS, 1958; Case Studies in Sample Design, 1977; Application of Quality Control to Service Industries, 1985, Washington Story 1985, The Quest for Quality in Services, 1989, Deming's 14 Points Applied to Services, 1991. Home and Office: 4330 Franklin Ave Loveland CO 80538-1715

ROSARIO, MYRA ODETTE, molecular biologist, pharmacist, educator; b. Ciales, P.R., Oct. 11, 1960; d. Joaquin Antonio and Nereida (Barrio) R.; BS in Pharmacy cum laude, U. P.R., 1982; MS in Med. Microbiology and Immunology, U. Okla., Oklahoma City, 1984, PhD in Med. Microbiology and Immunology, 1987. Lic. pharmacist, Tex. Rsch. asst. U. P.R. Med. Scis. Campus, San Juan, 1980-82; grad. dellow U. Okla. Health Scis. Ctr., 1983-87; postdoctoral fellow Lab. Molecular Genetics, Nat. Inst. Environ. Health Scis., NIH, Research Triangle Park, N.C., 1987-90, sr. staff fellow Lab. Reproductive and Devel. Toxicology, 1990-93, chmn. EEO adv. com., 1990-92; presenter and lectr. in field, 1980—; guest speaker Bennett Coll., Greensboro, N.C., 1992; rep. Hispanic Am. adv. com. NIH, 1990-92. Contbr. articles and abstracts to Devel. Biology, Molecular and Gen. Genetics, Arthritis and Rheumatism, Jour. Clin. Investigations. Catechist Holy Infant Cath. Ch., Durham, N.C., 1988-92; sponsor Elder Advocacy, Durham, 1989-92; vol. Handicap Encounter with Christ, Smithville, N.C., 1991-92. Fellow U. Okla., 1982-85, Gina Finzi Meml. summer fellow Lupus Found. Am., 1984, Okla. Med. Rsch. Found., 1985-87. Mem. N.C. Pharm. Assn. (continuing edn. com. 1992), Sigma Xi. Democrat. Achievements include identification of gene for developmentally expressed 70 kDa heat-shock protein of mouse spermatogenic cells.

ROSARIO-GUARDIOLA, REINALDO, dermatologist; b. Santurce, P.R., Sept. 17, 1948; s. Tomas and Aurea (Guardiola) Rosario; m. Fe Milagros Rivera, Aug. 19, 1972; children: Amarillis, Reinaldo, Gadiel. BS, U. P.R., 1968, MD, 1972. Rsch. fellow photobiology Harvard Med. Sch., Boston, 1976-77; asst. prof. U. P.R. Sch. of Medicine, Rio Piedras, 1979—; chief, dermatology sect. San Juan VA Hosp., Rio Piedras, 1979—. Bd. dirs. Wesleyan Acad. Guaynabo, P.R., 1990—. Grantee Dermatology Found., 1978. Fellow Am. Acad. Dermatology; mem. P.R. Dermatol. Soc. (pres. scientific com. 1978-79), Harvard-MGH House Officers Club, Alpha Omega Alpha. Office: El Monte Mall Ste 33A Hato Rey PR 00918

ROSATI, MARIO, mathematician, educator; b. Rome, Jan. 5, 1928; s. Aristide and Maria (Gabrielli) R.; m. Maria Luisa Marziale, Aug. 3, 1968; children: Francesca, Nicoletta, Giulio. Laurea in Math., U. Rome, 1950. Asst. prof. U. Rome, 1952-66; prof. math. U. Padua, Italy, 1966—; dir. Applied Math. Inst., Padua U., 1978-86, Dept. Pure Applied Math., 1987-92. Co-editor: (with G. Tedone) Collana di Informazione Scientifica, 1965-78; contbr. books and articles in field. Fellow U. Goettingen 1955. Mem. Italian Math. Union, Am. Math. Soc., Nat. Rsch. Ctr. Roman Catholic. Home: 43 G Leopardi, 35126 Padua Italy Office: Dept Pure Applied Math, 7 GB Belzoni, 35131 Padua Italy

ROSE, GILBERT JACOB, psychiatrist, writer, psychoanalyst; b. Malden, Mass., May 9, 1923; s. M. Edward and Sara (Freedman) R.; m. Anne Kaufman, Mar. 10, 1946; children: Renee Rose Shield, Daniel Asa, Cecily Rose Itkoff, Aron Dana. AB, Harvard U., 1944; MD, Boston U., 1947. Diplomate Am. Bd. Psychiatry and Neurology. Asst. clin. prof. psychiatry Med. Sch. Yale U., New Haven, 1961-67; assoc. clin. prof. psychiatry Yale U. Med. Sch., New Haven, 1967-83, lectr. in psychiatry Med. Sch., 1983-87; instr. Western New Eng. Psychoanalytic Inst., New Haven, 1970-76; pvt. practice Rowayton, Conn., 1955—. Author: Power of Form: A Psychoanalytic Approach to Aesthetic Form, 1980, expanded edit., 1992, Trauma & Mastery in Life & Art, 1987. Capt. USAF, 1953-55. Fellow Am. Psychiat. Assn. (life), Am. Coll. of Psychoanalysts; mem. Am. Psychoanalytic Assn., Yale U. Muriel Gardiner Program for Psychoanalysis and the Humanities. Avocations: traveling, saling, art, music. Home and Office: PO Box 215 Norwalk CT 06853-0215

ROSE, HUGH, management consultant; b. Evanston, Ill., Sept. 10, 1926; s. HOward Gray and Catherine (Wilcox) R.; m. Mary Moore Austin, Oct. 25, 1952; children: Susan, Nancy, Gregory, Matthew, Mary. BS in Physics, U. Mich., 1951, MS in Geophysics, 1952; MBA, Pepperdine U., 1982. Mgr. Caterpillar, Inc., Peoria, Ill., 1952-66; v.p., mktg. mgr. Cummins Engine Co., Columbus, Ind., 1966-69; pres., chief exec. officer Cummins Northeastern, Inc., Boston, 1969-77; pres. Power Systems Assocs., L.A., 1980-83, C.D. High Tech., Inc., Austin, Tex., 1984-87; mgmt. cons. Rose and Assocs., Tucson, 1984, 87—. Contbr. paleontol. articles to various publs. Bd. dirs. Raymond Alf Mus., Claremont, Calif., 1979—, Comstock Found., Tucson, 1988, Environ. Edn. Exch., 1991, Heart Ctr. U. Ariz., Tuscon, 1992. With USAAF, World War II. Fellow AAAS; mem. Soc. Vertebrate Paleontology, Beacon Soc. Boston (pres.), Algonquin Club Boston (v.p., bd. dirs. 1974-80), Duxbury Yacht Club, Longwood Cricket Club, Racquet Club Tucson, Sigma Gamma Epsilon, Beta Beta Beta. Republican. Presbyterian. Office: Rose & Assocs 5320 N Camino Sumo Tucson AZ 85718-5132

ROSE, JAMES TURNER, aerospace consultant; b. Louisburg, N.C., Sept. 21, 1935; s. Frank Rogers and Mary Burt (Turner) R.; m. Daniele Raymond, Sept. 15, 1984; children by previous marriage—James Turner, Katharine S. B.S. with high honors, N.C. State U., 1957. Aero. research engr. NASA, Langley Field, Va., 1957-59; project engr. NASA (Mercury and Gemini), Langley Field, Va. and Houston, 1959-64; program systems mgr. McDonnell Douglas Astronautics Co (MDAC), St. Louis, 1964-69; mgr. shuttle ops. and implementation (MDAC) McDonnell Douglas Astronautics Co., St. Louis, 1969-72; mgr. shuttle support (MDAC) St. Louis, 1972-74; dir. space shuttle engrng. NASA, Washington, 1974-76; mgr. space processing programs McDonnell Douglas Astronautics Co., St. Louis, 1976-83; dir. electrophoresis ops. in space McDonnell Douglas Astronautics Co (MDAC), St. Louis, 1983-86; asst. adminstr. comml. programs NASA, Washington, 1987-91; aerospace cons., 1992—. Recipient Lindberg award for mgmt. leadership AIAA, 1983, Presdl. Meritorious Rank award, 1989, NASA Exceptional Svc. medal, 1990, Laurels award Aviation Week, 1990, Aerospace Contribution to Soc. awd. AIAA, 1993. Mem. Phi Kappa Phi. Episcopalian.

ROSE, LUCY MCCOMBS, chemist; b. Birmingham, Ala., July 25, 1940; d. Carolyn Ivy Johnson; m. Jeremy Douglas Rose, Dec. 15, 1962; children: Robert, Amy, Anne. BS in Chemistry, Birmingham So. Coll., 1962; MA in Chem. Edn., U. Ala., Birmingham, 1986. Asst. chemist So. Rsch. Inst., Birmingham, 1964-69; assoc. chemist So. Rsch. Inst., 1969-77, rsch. chemist, 1977-89, staff chemist, 1989—, group leader, 1990—. Contbr. more than 30 articles to profl. jours. Active Met. Opera Guild, N.Y.C. Mem. Am. Soc. Mass Spectrometry, Sigma Xi (chpt. pres. 1991-92, state coord. 1990-91), Kappa Delta Pi, Kappa Delta Epsilon, Omicron Delta Kappa. Office: So Rsch Inst 2000 9th Ave S Birmingham AL 35205-2708

ROSE, MARIAN HENRIETTA, physics educator; b. Brussels, Belgium; (parents Am. citizens); m. Simon Rose, Oct. 20, 1948 (dec. Jan. 1981); children: Ann, James, David, Simon. BA, Barnard Coll., 1942; MA, Columbia U., 1944; PhD, Harvard U., 1947. Teaching fellow Harvard U., Cambridge, Mass., 1945-46; adj. asst. prof. Courant Inst., N.Y.C., 1947-48, rsch. assoc., 1951-65, sr. rsch. scientist, 1965-75; vis. fellow Yale U., New Haven, Conn., 1981—; bd. dirs. Minna-James-Heineman Stiftung, Essen, Fed. Republic of Germany, Jay Heritage Ctr., Rye, N.Y. Contbr. articles to profl. jours. Mem. Wetlands Control Commn., Bedford, N.Y., 1991—, Conservation Bd., Bedford, 1989-93. Mem. Sierra Club (conservation chair Atlantic chpt.), Phi Beta Kappa, Sigma Xi. Avocations: skiing, hiking. Office: Yale U Dept Physics 9 Hillhouse Ave New Haven CT 06511-6815

ROSE, NOEL RICHARD, immunologist, microbiologist, educator; b. Stamford, Conn., Dec. 3, 1927; s. Samuel Allison and Helen (Richard) R.; m. Deborah S. Harber, June 14, 1951; children: Alison, David, Bethany, Jonathan. BS, Yale U., 1948; MA, U. Pa., 1949, PhD, 1951; MD, SUNY, Buffalo, 1964; MD (hon.), U. Cagliari, Italy, 1990; ScD (hon.), U. Sassari, Italy, 1992. From instr. to prof. microbiology SUNY Sch. Medicine, Buffalo, 1951-73; dir. Center for Immunology SUNY Sch. Medicine, 1970-73, dir. Erie County Labs., 1964-70; dir. WHO Collaborating Center for Autoimmune Disorders, 1968—; prof. immunology and microbiology, chmn. dept. immunology and microbiology Wayne State U. Sch. Medicine, 1973—82; prof., chmn. dept. immunology and infectious diseases Johns Hokins U. Sch. Hygiene and Pub. Health, 1982—; joint appt. dept. medicine and environ. health scis. Johns Hopkins U. Sch. Medicine, 1982—; cons. in field. Editor: (with others) International Convocation on Immunology, 1969, Methods in Immunodiagnosis, 1973, 3d rev. edit., 1986, The Autoimmune Diseases, 1986, 2d edit., 1992, Microbiology, Basic Principles and Clinical Applications, 1983 Principles of Immunology, 1973, 2d rev. edit., 1979, Specific Receptors of Antibodies, Antigens and Cells, 1973, Manual of Clinical Laboratory Immunology, 1976, 2d rev. edit., 1980, 4d edit. 1992, Genetic Control of Autoimmune Disease, 1978, Recent Advances in Clinical Immunology, 1983, Clinical Immunotoxicology, 1992; editor in chief Clin. Immunology and Immunopathology, 1988—; contbr. articles to profl. jours. Recipient award Sigma Xi, 1952, award Alpha Omega Alpha, 1976, Lamp award, 1975, Faculty Recognition award Wayne State U. Bd. Govs., 1979, Pres.'s award for excellence in teaching, 1979, Disting. Service award Wayne State U. Sch. Medicine, 1982, U. Pisa medal, 1986; named to Acad. Scholars Wayne State U., 1981; Josiah Macy fellow, 1979. Fellow Am. Public Health Assn., Am. Acad. Allergy and Immunology, Am. Acad. Microbiology, mem. Acad. Clin. Lab. Physicians and Scientists, AAAS, Am. Assn. Immunologists, Am. Soc. Investigative Pathology, Am. Soc. Clin. Pathologists, Am. Soc. Microbiology (Abbott Lab. Clinical and Diagnostic Immunology award 1993), Brit. Soc. Immunology, Coll. Am. Pathologists, Société Française d'Immunologie, Can. Soc. Immunology, Soc. Exptl. Biology and Medicine Council, Clinical Immunology Soc. (sec., treas., pres. 1993), Sigma Xi (pres. Johns Hopkins U. chpt. 1988), Alpha Omega Alpha, Delta Omega. Office: Johns Hopkins U Sch Hygiene and Pub Health Dept Immunology 615 N Wolf St Baltimore MD 21205

ROSE, RAYMOND ALLEN, computer scientist; b. Alhambra, Calif., Aug. 20, 1951; s. David Bernard and Gerri (Swiryn) R.; m. Elue Maria Adams, Aug. 18, 1975 (div. 1981); children: James Michael, Brian Seth, Joshua Aaron, Jeffrey Steven, Kyle Christopher. AS in Computer Sci., Grossmont Coll., El Cajon, Calif., 1975; BS in Computer Engrng., San Diego State U., 1974, BS in Computer Sci. Programming, 1975; M in Computer Engrng., Stanford U., 1977. Chief program design W.P.S., Houston, 1977-79, v.p. program design, 1980-83; v.p. software design Computer Tape, Houston, 1983-88; pres., chief exec. officer C.A.R.D., San Diego, 1988—; Netware LANs adminstr. Grossmont Coll. Dist., 1991—; software devel. cons. U.S. Govt., Houston, 1983-88; hiring bd. adv. San Diego State U., 1990-91; instr. computer sci. Grossmont Coll. Contbr. articles to profl. jours. Lt. col. U.S. Army, 1968-75. NASA software devel. grantee, 1984; named Coach of the Yr., El Cajon Little League Assn., 1989. Office: CARD 294 Orlando St Apt 9 El Cajon CA 92021-7011

ROSE, THOMA HADLEY, environmental consultant; b. Williamston, N.C., Aug. 17, 1942; s. Howard Thomas and Bergie Glen (Bailey) R.; m. Patrein Jane Sullivan, Sept. 9, 1967; children: Margaret Jane Blount, Thomas Patrick. BS in Sci., Va. Poly. Inst., 1970. Air pollution control specialist Va. Air. Control Bd., Richmond, 1970-72; enforcement officer U.S. EPA, Athens, Ga., 1972-79; pres. Eastern Tech. Assn., Raleigh, N.C., 1979—. Author: (manuals) Sect. 3.12 EPA Quad Manual, 1984, Visible Emmission Field Guide, Classroom Manual Visible Emmissons, NASN Field Manual, 1975; author, programer: OPACICALC, 1984-92. Mem. Triangle Jr. Coun. Gov., 1986. Recipient Gov.'s award N.C., 1986, 87, First Flight award, 1990, Technologim Devel. award NC TAD, 1986. Mem. Source Evaluation Soc., Air and Waste Mgmt. Assn., Soc. Profl. Optical Engrs. Democrat. Achievements include patent for Instack Transmissometer and development of long path transmissometer and vinyl chloride measurement method. Office: ETA Box 58495 Raleigh NC 27658

ROSE, WILLIAM CUDEBEC, electrophysiologist; b. Balt., Mar. 13, 1959; s. Edward Andrews and Joy Louise (Waldron) R.; m. Karen Elaine Evans, May 25, 1985; children: Michelle Rebecca, William Robert, Allison deForest. AB in Physics cum laude, Harvard U., 1981; PhD in Biomed. Engring., Johns Hopkins U., 1989. Engr. Harvard U., Cambridge, Mass., 1981, Amco Sci., Tokyo, 1982; postdoctoral fellow Johns Hopkins U., Balt., 1989-92; vis. scientist DuPont, Wilmington, Del., 1992—. Reviewer Circulation jour. of Am. Heart Assn., Am. Jour. Physiology; author rsch. articles in field; contbr. articles to profl. jours. Mem. IEEE, Biomed. Engring. Soc., Tau Beta Pi. Achievements include determination of kinetic properties of calcium channels under realistic physiological conditions, measurement of mechanical properties of blood veins using random noise.

ROSE-HANSEN, JOHN, geologist; b. Copenhagen, May 2, 1937; s. Helmer and Laura (Nielsen) R.-H.; m. Elsemarie Berg Olesen, Feb. 4, 1961; children: Inge Berg, Vibeke Berg. Geologist, U. Copenhagen, 1963. Lectr. U. Copenhagen, 1963-69, assoc. prof. geology, 1969—; vis. prof. sci. U.S.A, France, Norway, Egypt, China. Author books and articles in mineralogy, petrology, econ. and environ. geology. Recipient silver medal Royal Danish Acad. Scis. and Letters, Copenhagen, 1985; Majorgen. J. F. Classens grantee, Copenhagen, 1985. Mem. AAAS, Internat. Assn. Geochemistry and Cosmochemistry, Am. Geophys. Union, Mineral. Soc. Am., Geochem. Soc., N.Y. Acad. Scis., Lions Internat. Avocation: classical music. Home: Landsevej 14A, 2840 Holte Denmark Office: U Copenhagen Geol Inst, Øster Voldgade 10, 1350 Copenhagen Denmark

ROSEIG, ESTHER MARIAN, veterinary researcher; b. Bklyn., July 23, 1917; d. Chone and Rebecca (Kaplan) Fogel; m. Seymour Roseig, Jan. 21, 1967. Cert., Med. Assts. Sch., N.Y.C., 1967; student, Orange County Community Coll., Middletown, N.Y., 1967-68. Cert. clin. lab. technician, N.Y. Gen. lab. technician Arden Hill Hosp., Goshen, N.Y., 1967-68; tech. rsch. asst. Lamont-Doherty Geol. Obs., 1968-70. Democrat. Achievements include research on the organism saccharomyces cerevisiae in its inactive dry state as brewers yeast or bakers yeast, and its ability to repel the parasites, fleas and ticks from domestic pets through a biochemical process of metabolism in conjunction with meat protein: the end product as CO(NH2)2" in solution in sweat; a coincidental process of coat pigment losses in both dogs and cats fed the initial Yeast was resolved by adjusting the B, A, D Vitamins and Calcium.

ROSELL, SHARON LYNN, physics and chemistry educator, researcher; b. Wichita, Kans., Jan. 6, 1948; d. John E. and Mildred C. (Binder) R. BA, Loretto Heights Coll., 1970; postgrad., Marshall U., 1973; MS in Edn., Ind. U., 1977; MS, U. Wash., 1988. Cert. profl. educator, Wash. Assoc. instr. Ind. U., Bloomington, 1973-74; instr. Pierce Coll. (name formerly Ft. Steilacoom (Wash.) Community Coll.), 1976-79, 82, Olympic Coll., Bremerton, Wash., 1977-78; instr. physics, math. and chemistry Tacoma (Wash.) Community Coll., 1979-89; instr. physics and chemistry Green River Community Coll., Auburn, Wash., 1983-86; researcher Nuclear Physics Lab., U. Wash., Seattle, 1986-88; asst. prof. physics Cen. Wash. U., Ellensburg, 1989—. Mem. Math. Assn. Am., Am. Assn. Physics Tchrs. (rep. com. on physics for 2-yr. colls. Wash. chpt. 1986-87, v.p. 1987-88, pres. 1988-89), Am. Chem. Soc., Internat. Union Pure and Applied Chemistry (affiliate). Democrat. Roman Catholic. Avocations: leading scripture discussion groups, reading, writing poetry, needlework. Home: RR 5 Box 880 Ellensburg WA 98926-9379 Office: Cen Wash U Physics Dept Ellensburg WA 98926

ROSELLA, JOHN DANIEL, clinical psychologist, educator; b. Phila., Sept. 12, 1938; s. Orazio and Angela Theresa (Cardone) R.; B.S. in Psychology, Villanova U., 1961; cert. in Edn., St. Joseph's U., 1963; M.Ed., Temple U., 1966, postgrad., 1969-72; Ph.D., Walden U., 1981; cert. hynotherapist; m. Rose Mary Theresa Malloy, Nov. 14, 1964; children—Anne-Marie, John Daniel. Tchr., counselor Father Judge High Sch., Phila., 1962-67; counselor Bristol Twp. Sch. Dist., Bucks County, Pa., 1967-69; prof. Bucks County Community Coll., Newtown, Pa., 1968—, founder coll. reading and study skills program, 1968-70, founding chmn. dept. basic studies, 1970-76; dir. psychol. services Fairless Hills (Pa.) Med. Center, 1978-89, dir. clin. svcs., 1989—; asst. clin. prof. Widener U., 1990; psychol. cons. Eugenia Hosp., 1980—, Bur. Disability Determination, 1982—, Human Growth Center, Inc., 1987—, Crestview North Nursing Home and Rehab. Ctr., 1990—; clin. assoc. prof. Dept. Mental Health Scis. Hahnemann U., 1982—; grad. clin. supr. Trenton State Coll., 1985-86; grad. counseling intern supr. Rider Coll., 1988-91; co-founder Employee Assistance Programs, 1985; participant 1st Internat. Colloquium on Family Health, Sri Lanka, 1983, Australia, 1988, ednl. profl. travel, Italy and Switzerland, 1991. Bd. dirs. Valley Day Sch., 1978-81, Bucks County Community Centers, 1980—; co-founder Newtown Twp. Dem. Party, 1978, 1st vice chmn., 1979-80, Dem. committeeman, 1989—; active Right to Read Task Force, 1972-73; mem. 8th Congressional Dist. Adv. Council on Health Care, 1981-83; project dir. Fairless Hills Psychiat. Hosp. bldg. program, 1982-83; pres. bd. trustees Friends of the Library Found., Bucks County Community Coll., 1984—. Recipient Man of Yr. award Assn. to Advance Ethical Hypnosis, 1976, Disting. Teaching recognition Phi Theta Kappa, 1981, 83, Faculty Svc. award, 1989, Profl. Achievement award Bucks County Community Coll. Alumni Assn., 1991. Lic. psychologist, Pa. Fellow Internat. Council for Sex Edn. and Parenthood of Am. U.; mem. Am. Psychol. Assn., Am. Assn. Marriage and Family Therapy (clin.), Pa. Psychol. Assn., Pa. Assn. Marriage and Family Therapy. Roman Catholic. Clubs: KC, Sons of Italy. Author: Reading and Study Skills: A Counseling Approach, 1970; Effects of the Basic Studies Program on the Academic Achievement of High Risk Students, 1973-74; The Professor and the Law, 1975; Research in Hypnosis for Students, 1976; Marriage and Family Therapy: Its Evolution from Revolution, 1980; others; (audio-tapes) Developing Successful Study Skills, Guided Imagery Exercises; also articles. Office: Offices at Oxford Crossing 333 Oxford Valley Rd Ste 202 Fairless Hills PA 19030-2618

ROSELLE, RICHARD DONALDSON, industrial, marine and interior designer; b. Garwood, N.J., Nov. 20, 1916; s. Ernest North and Mary Elizabeth (Donaldson) R.; m. Eunice Calpin, June 28, 1947 (div. Oct. 1981); children: Sheryn, Christina, Gail; m. Judith Marie Bishop, Nov. 13, 1982. Student, St. John's Mil. Acad., 1935, Aurora (Ill.) Coll., 1935-37, Bucknell U., 1937-39. Exec. trainee J.J. Newberry Co., N.Y.C., 1939-41, G. Fox & Co., Hartford, Conn., 1941-42; materials expeditor Pratt & Whitney Aircraft, Hartford, 1942-43; exec. trainee, indsl. engr. TWA Airline, N.Y.C. and Kansas City, 1943-47; with employee rels. staff R.H. Macy, N.Y.C., 1947-49; asst. tng. dir. J.C. Penney Co., N.Y.C., 1949-50; owner Roselle Tile Mfr., Seattle, 1950-56; sr. indsl. designer Walter Dorwin Teague Assocs., Seattle, 1956-63; owner Roselle Design Internat., Inc., Seattle, 1963—; dir. Roselle Design Tours Internat., Seattle, 1967-93; internat. bus. developer via confs., 1993—. Mem. Am. Soc. Interior Design (nat. edn. chair Seattle chpt. 1972-74), Indsl. Designers Soc. Am. (charter), Internat. Inst. Profl. Designers, Master Resources Coun. Internat. (founder, pres.), Rotary. Republican. Episcopalian. Achievements include design of airline coupon ticket form accept as transportation industry standard; white on fuselage of aircraft to accentuate graphics, standard airline office ticket counters, aircraft interiors, megayacht styling and interiors, passenger train interior and exterior graphics; packages for oysters and caviar; concrete "Flothaus" and trimaran/hydrofoil cruise ship, panelized building systems. Office: Roselle Design Internat Camelot Atelier 3854 140th Ave NE Bellevue WA 98005-1451

ROSEMBERG, EUGENIA, physician, scientist, educator, medical research administrator; b. Buenos Aires, Argentina, Apr. 25, 1918; came to U.S., 1948, naturalized, 1956; d. Pedro and Fanny (Hestrin) R. BS, Liceo Nacional de Senoritas, Buenos Aires, 1936; MD, U. Buenos Aires, 1944. Intern Hosp. Pirovano, Buenos Aires, 1940-41; resident Hosp. Nacional de Clinicas, U. Hosp., U. Buenos Aires, 1941-44, assoc. in pediatrics, 1943-48; instr. in anatomy Hosp. Nacional de Clinicas, U. Hosp., U. Buenos Aires (Med. Sch.), 1940-46, instr. pediatrics, 1946-48; practice medicine specializing in pediatrics, 1946-48; research in endocrinology Balt., 1948-51, Worcester, Mass., 1955—; Mead Johnson fellow dept. endocrinology Johns Hopkins Med. Sch., Balt., 1948-49; vis. scientist Med. Sch., U. Montevideo, Uruguay, 1950; research fellow NIH, Bethesda, Md., 1951-53, Nat. Inst. Arthritis and Metabolic Diseases, 1951-53, Med. Research Inst. and Hosp., Oklahoma City, 1953; mem. staff Worcester Found. Exptl. Biology, Shrewsbury, Mass., 1953-62; research dir. Med. Research Inst. of Worcester, Inc., 1962—; cons. Center for Population Research, Nat. Inst. Child Health and Human Devel., NIH, 1969-70, chief contraceptive devel. br., 1970-71; prof. pediatrics U. Md. Hosp., Balt., 1970-73; prof. medicine U. Mass. Med. Sch., Worcester, 1972—; mem. staff Worcester City Hosp., 1955-85, sec. human experimentation com., 1965-83, chmn., 1984-85, dir. clin. research, 1972-85; Sec. subcom. on gonadotropins Nat. Hormone and Pituitary Program, Nat. Inst. Arthritis, Diabetes, Digestive and Kidney Diseases, 1965-69, chmn., 1969-85, mem. med. adv. bd., 1969-72, 73-85, sec. subcom. on standards endocrinology study sect., 1968. Author: Gonadotropins, 1968, (with C.A. Paulsen) The Human Testis, 1970, Gonadotropin Therapy in Female Infertility, 1973, (with C. Gual) Hypothalamic Hypophysiotropic Hormones—Physiological and Clinical Studies, 1973; Mem. editorial bd.: Giner, 1970—, Procs. 1st Ann. Meeting Am. Soc. Andrology, supplement, Vol. 8, 1976, Andrologia, 1978—, Jour. Andrology, 1979-82, Internat. Jour. Andrology, 1978—; assoc. editor: Reproduccion, 1970—, Andrologia jour, 1974-77; Contbr. articles and book chpts. on research in endocrinology to med. texts and jours.; Translator: from Spanish Diagnosis and Treatment of Endocrine Disorders in Childhood and Adolescence (L. Wilkins). Patentee in field, U.S., Can., Europe. Fellow AAAS; mem. Am. Med. Women's Assn., Endocrine Soc. U.S. (mem. com. pub. affairs 1971, v.p. 1975-76), Soc. for Research in Biology of Reproduction, Soc. for Study of Reproduction, Am. Fertility Soc., Peru Fertility Soc. (fgn. corr.), N.Y. Acad. Scis., New Eng. Cardiovascular Soc., Am., Mass. heart assns., Argentine Endocrine Soc., Argentine Pediatric Soc., Sociedad Argentine Para El Estudio de la Esterilidad., Pan Am. Med. Women's Alliance, Am. Soc. Andrology (program chmn. 1975-76, exec. council 1976-78, chmn. publ. com. 1975-80, Disting. Andrologist award 1982), Internat. Com. for Study Andrology (exec. council 1976-79).

ROSEN, ALEXANDER CARL, psychologist, consultant; b. L.A., Feb. 2, 1923; s. Benjamin and Pauline (Katz) R.; m. Florence Friedman, Mar. 18, 1951 (div. Nov. 1973); children: Diane, Judith; m. Susan Margaret Gersbacher, Nov. 4, 1973; 1 child, Rebecca. AA, U. Calif., L.A., 1943; AB, U. Calif., Berkeley, 1946, PhD, 1953. Diplomate clin. psychology Am. Bd. Profl. Psychology; lic. psychologist. Psychologist Contra Costa County, Martinez, Calif., 1953-56; asst. rsch. psychologist Office Naval Rsch. and San Francisco State Coll., 1953-56, UCLA-Neuropsychiat. Inst., L.A., 1956-57; asst. prof. to prof. psychiatry and behavioral sci. UCLA Sch. Medicine, L.A., 1956-89; chief psychology UCLA Neuropsychiatric Inst., L.A., 1958-89; prof. emeritus UCLA Sch. Medicine, L.A., 1989—; pvt. practice psychology cons. L.A., 1973—; instr. San Francisco State Coll., 1955; instr. psychology Calif. Inst. Tech., Pasadena, 1969; staff assoc. Nat. Tng. Lab. Inst. Applied Behavioral Sci., 1962—; cons. tng. U.S. Veteran's Assn., Sepulveda (Calif.) Hosp., 1966—; bd. mem. L.A. Group Psychotherapy Tng. Inst., 1972-75; bd. mem., trustee Calif. Sch. Profl. Psychology, 1974-76, 78; nat. bd., regional bd. Cert. Cons. Internat. Cons. Cons. editor: Jour. Genetic Psychology and Genetic Psychology Monograph, 1984—; contbr. articles to profl. jours. Mem. gov. bd. Hillel Coun., So. Calif.; cons. San Fernando Valley Counseling Ctr., 1991-92, Pacific Ctr. for AIDS, L.A., 1991—. Fellow APA, AAAS; mem. Calif. State Psychology Assn. (pres. 1977-78), Western Psychol. Assn. Avocations: photography, music, drama. Home: 3625 Beverly Ridge Dr Sherman Oaks CA 91423 Office: Ste 408 4419 Van Nuys Blvd Sherman Oaks CA 91403

ROSEN, BARRY HOWARD, museum director, history educator; b. Phila., June 26, 1942; s. Robert R. and Sylvia (Chanin) R.; m. Ann Adair Gould, Feb. 14, 1970; 1 son, David Joshua. B.S., Temple U., 1963, M.A., 1966; Ph.D., U. S.C., 1974. Asst. to provost U. S.C., Columbia, 1973-74, asst. to pres., 1974-77, dir. mus., univ. archivist, dir. mus. mgmt. programs, 1975-82;

exec. dir. Kansas City Mus., Mo., 1982-86; pres. N.J. Hist. Soc., Newark, 1986-88; pres., CEO Milw. Pub. Mus., 1988—; field reader Inst. Mus. Svcs., Washington, 1981-84. Bd. dirs., pres. Westown Assn., 1988—. Mem. Am. Assn. State and Local History, Am. Assn Mus. (accrediting com. 1982—), Assn. Sci. Mus. Dirs. (chmn. future planning com. 1990—), Natural History Museums 2000 (steering com., 1989—, Southeastern Wis. Amniotrophic Lateral Sclerosis Soc. (bd. dirs. 1990—). Home: 5043 N Lake Dr Milwaukee WI 53217-5749 Office: Milw Pub Mus 800 W Wells St Milwaukee WI 53233-1478

ROSEN, COLEMAN WILLIAM, radiological physicist; b. Phila., June 5, 1957; s. Alvin and Rosalie Edith (Rosengarten) R.; m. Robin Lynn Carpenter, Apr. 29, 1984; children: Justin Ellis Rosen, Adon Franklin Grant Rosen. BS in Zoology, George Washington U., 1980; MS in Radiation Sci., Rutgers U., 1981. Diplomate Am. Bd. Radiology, Am. Bd. Med. Physics. Med. physicist Georgetown U. Hosp., Washington, 1981-84; rsch. physicist Bowman-Gray Sch. Medicine, Winston-Salem, N.C., 1984-86; med. physicist St. Barnabas Med. Ctr., Livingston, N.J., 1986-89; radiol. physicist Fairfax Hosp., Falls Church, Va., 1989—; adv. com. St. Barnabas Therapy Tech Sch., Livingston, 1988-89. Co-author: Thought Mapping: A Better Way to Learn, 1985; ad hoc rev. com. Physics in Medicine and Biology, London, 1987—, Med. Physics, N.Y.C., 1991—; contbr. to books and articles to profl. jours. Co-chmn. for image mgmt. com. INOVA, Fairfax, 1989-90. Grant U.S. EPA, 1992. Mem. APHA (counselor 1992—), Mid-Atlantic Am. Assn. of Physicists in Medicine (pres. 1991-93), Health Physics Soc., Am. Assn. Physicists in Medicine (task group for magnetic resonance imaging acceptance testing protocol, 1992). Democrat. Office: Fairfax Hosp 3300 Gallows Rd Falls Church VA 22046

ROSEN, DAVID LAWRENCE, physicist, researcher; b. Bklyn., Aug. 17, 1954; s. Lester and Marilyn (Ziskind) R.; m. Janet Weissman, June 28, 1986. BS, Bklyn. Coll. N.Y., 1976; PhD, CUNY, 1985. Rsch. physicist Sachs Freeman Assocs., Landover, Md., 1985-87; asst. prof. N.Mex. State U., Las Cruces, 1987-89; physicist Atmospheric Sci. Lab., White Sands Missile Range, N.Mex., 1989-92, Battlefield Environ. Directorate, White Sands Missile Range, N.Mex., 1992—. Contbr. articles to profl. jours. in spectroscopy, lasers, and lidar. Mem. Friends of Paleontology in Albuquerque, Toastmasters (sgt. of arms 1992—). Home: 435 Townsend Terr Las Cruces NM 88005 Office: Battlefield Environ Directorate AMSRL-BE-W White Sands Missle Range NM 88002-5501

ROSEN, HERMAN, civil engineer; b. N.Y.C., Aug. 8, 1935; s. Benjamin and Bertha (Colten) R.; m. Terry Miller, June 17, 1961 (div. 1981); children: Michelle, Wendy, Neal; m. Lenore E. Dani, Sept. 11, 1988. BCE, CCNY, 1960. Registered profl. engr., N.Y., Pa. Design engr. Hydrotechnic Corp., N.Y.C., 1960-62; water supply engr. N.Y.C. Bur. Water Supply, 1962-91; chief of distbn., reservoirs and pumping stas. N.Y.C. Bur. Water Supply, Pa., 1962-91. Contbr. articles to Am. Water Works Jour. Pres. Reform Temple, Monsey, N.Y., 1975; coach, organizer Little League, Monsey, 1976; active Jewish Community Ctr. Harrisburg, 1991—. Fellow Am. Water Works Assn. Achievements include design of improved suspender clasps, faster and more reliable method for hydrant flow testing. Home: 2133 N 2d St Harrisburg PA 17110

ROSEN, JOSEPH DAVID, chemist, educator; b. N.Y.C., Feb. 26, 1935; s. Rubin and Tobie (Greenspan) R.; m. Doris Stieber, Sept. 8, 1962; children: Todd, Amy, Mark, Dayan. BS in Chemistry, CCNY, 1956; PhD in Organic Chemistry, Rutgers U., 1963. Rsch. chemist E.I. DuPont de Nemours & Co., Parlin, N.J., 1963-65; from asst. rsch. prof. to rsch. prof. food science Rutgers U., New Brunswick, N.J., 1965—; sci. adviser FDA Regional Lab., Bklyn., 1974-78, 81-87; co-editor Jour. Food Safety, 1977-89; vis. prof. U. Calif., Berkeley, 1978-79. Editor: Applications of New Mass Spectrometric Techniques in Pesticide Chemistry, 1987. Office: Rutgers U Food Sci Bldg New Brunswick NJ 08903

ROSEN, LOUIS, physicist; b. N.Y.C., June 10, 1918; s. Jacob and Rose (Lipionski) R.; m. Mary Terry, Sept. 4, 1941; 1 son, Terry Leon. BA, U. Ala., 1939, MS, 1941; PhD, Pa. State U., 1944; DSc (hon.), U. N.Mex., 1973, U. Colo., 1989. Instr. physics U. Ala., 1940-41, Pa. State U., 1943-44; mem. staff Los Alamos Sci. Lab., 1944-90, group leader nuclear plate lab., 1949-65, alt. div. leader exptl. physics div., 1962-65, dir. meson physics facility, 1965-85, div. leader medium energy physics div., 1965-86, sr. lab. fellow, 1985-90, sr. fellow emeritus, 1990—; Sesquicentennial hon. prof. U. Ala., 1981; mem. panel on future of nuclear sci., chmn. subpanel on accelerators NRC of NAS, 1976, mem. panel on instnl. arrangements for orbiting space telescope, 1976; mem. U.S.A.-USSR Coordinating Com. on Fundamental Properties of Matter, 1971-90. Author papers in nuclear sci. and applications of particle accelerators.; bd. editors: Applications of Nuclear Physics; co-editor Climate Change and Energy Policy, 1992. Mem. Los Alamos Town Planning Bd., 1962-64; mem. Gov.'s Com. on Tech. Excellence in N.Mex.; mem. N.Mex. Cancer Control Bd., 1976-80, v.p., 1979-81; co-chmn. Los Alamos Vols. for Stevenson, 1956; Dem. candidate for county commr., 1962; bd. dirs. Los Alamos Med. Ctr., 1977-83, chmn., 1983; bd. govs. Tel Aviv U., 1986. Recipient E.O. Lawrence award AEC, 1963; Golden Plate award Am. Acad. Achievement, 1964; N.Mex. Disting. Public Service award, 1978; named Citizen of Year, N.Mex. Realtors Assn., 1973; Guggenheim fellow, 1959-60; alumni fellow Pa. State U., 1978; Louis Rosen prize established in his honor by bd. dirs. Meson Facility Users Group, 1984. Fellow AAAS (coun. 1989), Am. Phys. Soc. (coun. 1975-78, chmn. panel on pub. affairs 1980, div. nuclear physics 1985, mem. subcom. on internat. sci. affairs 1988). Home: 1170 41st St Los Alamos NM 87544-1913 Office: Los Alamos Sci Lab PO Box 1663 Los Alamos NM 87545-0001

ROSEN, PETER, health facility administrator, emergency physician, educator; b. Bklyn., Aug. 3, 1935; s. Isadore Theodore and Jessie Olga (Solomon) R.; m. Ann Helen Rosen, May 16, 1959; children: Henry, Monte, Curt, Ted. BA, U. Chgo., 1955; MD, Washington U., St. Louis, 1960. Diplomate Am. Bd. Surgery, Nat. Bd. Med. Examiners, Am. Bd. Emergency Medicine; cert. Advanced Cardiac Life Support Instr., Advanced Trauma Life Support Provider. Intern U. Chgo. Hosps. & Clinics, 1960-61; resident Highland County Hosp., Oakland, Calif., 1961-65; assoc. prof. divsn. emergency medicine U. Chgo. Hosps. & Clinics, 1971-73, prof. divsn. emergency medicine, 1973-77; dir. divsn. emergency medicine Denver City Health & Hosps., 1977-86, 87-89; asst. dir. dept. emergency medicine U. Calif., San Diego Med. Ctr., 1989—, dir. edn. dept. emergency medicine, 1989—, dir. emergency medicine residency program, 1991—; attending physician Hot Springs Meml. Hosp., Thermopolis, Wyo., Worland (Wyo.) County Hosp., Basin-Graybull Hosp., Basin, Wyo., 1968-71, U. Chgo. Hosps. & Clinics, 1971-77; dir. emergency medicine residency program, divsn. emergency medicine U. Chgo. Hosps. & Clinics, 1971-77; emergency medicine med. advisor State of Colo., 1977-85; dir. emergency medicine residency program Denver Gen. Hosp., St. Anthony Hosp. Systems, St. Joseph Hosp., 1977-88; clin. prof. divsn. emergency medicine Oreg. Health Scis. U., Portland, 1978-89; prof. sect. emergency medicine, dept. surgery U. Colo. Health Scis. Ctr., 1984-89; dep. mgr. med. affairs Denver Dept. Health & Hosps., 1986-87; med. dir. life flight air med. svc. U. Calif., San Diego Med. Ctr., 1989-91; mem. hosp. staff U. Calif., San Diego Med. Ctr., Tri-City Med. Ctr., Oceanside, Calif., 1989—; base hosp. physician, adj. prof. medicine & surgery U. Calif., San Diego Med. Ctr., 1989—; chair med. ethics com., mem. ethics consult team U. Calif., San Diego Med. Ctr., 1990—, mem. recruitment and admissions com., 1992—; lectr. in field; cons. in field. Author: (with others) Case Reports in Emergency Medicine: 1974-76, 1977, Encyclopedia Brittannica, 1978, 85, Principles and Practice of Emergency Medicine, 1978, 86, Protocols for Prehospital Emergency Care, 1980, 84, Cardiopulmonary Resuscitation, 1982, An Atlas of Emergency Medicine Procedures, 1984, Critical Decisions in Trauma, 1984, Emergency Pediatrics, 1984, 86, 90, Controversies in Trauma Management, 1985, Standardized Nursing Care Plans for Emergency Department, 1986, Emergency Medicine: Concepts and Clinical Practice, 1988, 92, The Clinical Practice of Emergency Medicine, 1991, Essentials of Emergency Medicine, 1991, Current Practice of Emergency Medicine, 1991, Care of the Surgical Patient, 1991, Diagnostic Radiology in Emergency Medicine, 1992, Pediatric Emergency Care Systems: Planning and Management, 1992, The Airway: Emergency Management, 1992; contbg. editor, editor abstracts sect. Jour. Am. Coll. Emergency Physicians, Annals of Emergency Medicine, 1976-83; mem. editorial bd. Topics in Emergency Medicine, 1979-82, ER Reports, 1981-83; consulting editor Emergindex Microindex, 1980—; editor in chief Jour. Emergency Medicine, 1983—; contbr. articles to profl. jours. Capt. USMC, 1965-68, lt. col. Res. inactive. Recipient AMA award, 1970, Am. Hosp. Assn. award, 1973. Fellow Am. Coll. Surgeons, Am. Burn Assn., Am. Coll. Emergency Physicians (chmn. edn. com. 1977-79, bd. dirs. Colo. chpt. 1977-80, pres. Colo. chpt. 1981-82, N.C. chpt. award 1976, Outstanding Contbns. and Leadership in Emergency Medicine award 1977, Silver Tongue Debater award 1980, John. D. Mills Outstanding Contbn. to Emergency Medicine award 1984); mem. Am. Trauma Soc. (founding), Soc. Acad. Emergency Medicine (Leadership award 1990), Alpha Omega Alpha Honor Med. Soc. (grad.), Coun. Emergency Medicine Dirs. Office: U of California-San Diego 200 W Arbor Dr. San Diego CA 92103-8676

ROSEN, ROBERT THOMAS, analytical and food chemist; b. Concord, N.H., Nov. 5, 1941; s. Maurice J. and Miriam M. (Miller) R.; m. Sharon Lynne Beres, Apr. 23, 1972. BA (cum laude), Nasson Coll.; PhD, Rutgers U. Sr. rsch. scientist Chem. Rsch. and Devel. Ctr., FMC Corp., Princeton, N.J., 1966-84; program dir. analytical support facilities Ctr. for Advanced Food Technology, Rutgers U., New Brunswick, N.J., 1984—; chmn. North Jersey ACS Mass Spectrometry Topical Group, 1987-88. Assoc. editor The Mass Spec Source, 1988-90; contbr. articles and book reviews to profl. jours. Fellow Am. Inst. Chemists; mem. Am. Soc. for Mass Spectrometry, Am. Chem. Soc., N.Y. Acad. Scis., North Am. Native Fishes Assn., Inst. Food Technologists, Phi Lambda Epsilon (hon.). Achievements include research in gas and liquid chromatography, free and glycosidically bound organic compounds in fruits and vegetables, determination of non-volatile and thermally labile pesticides and phytochemicals in food, related materials and the environment by liquid chromatography and mass spectrometry. Home: Keats Rd # 293 Pottersville NJ 07979-9999 Office: Ctr for Advanced Food Tech Cook Coll Rutgers U New Brunswick NJ 08903

ROSEN, STEVEN TERRY, oncologist, hematologist; b. Bklyn., Feb. 18, 1952; married, 1976; 2 children. MB, Northwestern U., 1972, MD, 1976. Genevieve Teuton prof., med. sch. Northwestern U., 1989—, dir. cancer ctr., 1989—; dir. clin. programs Northwestern Meml. Hosp., 1989—. Editor-in-Chief Jour. Northwestern U. Cancer Center, 1989—, Contemporary Oncology, 1990—. Mem. AAAS, ACP, AMA, Am. Soc. Hematology, Am. Soc. Clin. Oncology, Ctrl. Soc. Clin. Rsch. Achievements include research in cutaneous T-cell lymphomas, biology of lung cancer, biologic therapies, and hormone receptors. Office: NW Univ Cancer Ct 303 E Chicago Ave Chicago IL 60611*

ROSENBAUM, JAMES TODD, rheumatologst, educator; b. Portland, Oreg., Sept. 29, 1949; s. Edward E. and David Carol (Naftalin) R.; m. Sandra Jean Lewis, June 27, 1970; children: Lisa Susanne, Jennifer Lewis. AB, Harvard U., 1971; MD, Yale U., 1975. Diplomate Am. Bd. Internal Medicine, Am. Bd. Rheumatology. Intern, resident Stanford (Calif.) Med. Ctr., 1975-78, postdoctoral fellow, 1978-81; from instr. to asst. prof. U. Calif., San Francisco, 1981-83; sr. scientist Kuzell Inst., San Francisco, 1983-85; from asst. prof. to assoc. prof. Oreg. Health Scis. U., Portland, 1985-91, prof., asst. dean rsch., 1991—; bd. dirs. Fund for Arthritis/Infectious Disease Rsch., San Francisco, Oreg. Arthritis Found., Portland, Portland VA Rsch. Found.; mem. spl. rev. panel Nat. Eye Inst., Bethesda, Md., 1992. Dolly Green scholar Rsch. to Prevent Blindness, 1986. Mem. Am. Assn. Immunologists, Am. Coll. Rheumatology, Am. Uveitis Soc. Office: Casey Eye Inst 3375 SW Terwilliger Blvd Portland OR 97221

ROSENBERG, ALBERTO, pharmaceutical company executive, physician; b. Tucuman, Argentina, Aug. 13, 1937; came to U.S., 1963; s. Pedro and Fanny (Stisman) R.; m. Irene Zachari, Feb. 7, 1964; children: Diego, Marsha. BS, Nt. Coll. Mariano Moreno, Buenos Aires, 1955; MD, U. Buenos Aires, 1962; MS in Physiology, U. So. Calif., L.A., 1973. Asst. in rsch. Inst. of Physiology U. Buenos Aires, 1957-59, teaching asst., dept. physiology, 1957-59; intern Hosp. Tigre, Buenos Aires, 1961, Hosp. Fiorito, Buenos Aires, 1962; pvt. practice Buenos Aires, 1962-63; pharmacologist Riker Labs., Northridge, Calif., 1963-67; rsch. fellow Orange County Med. Ctr., Santa Ana, Calif., 1967-69; rsch. assoc. VA Ctr., L.A., 1969-70; assoc. dir. clin. rsch. dept. Stuart Pharms. div. ICI Ams. Inc., Wilminton, Del., 1970-78, dir. investigational drugs, clin. rsch. dept., 1978-81, v.p. clin. rsch. dept., 1981-85; exec. dir. clin. rsch. dept. Wyeth Labs., Radnor, Pa., 1985-86; med. reviewer cardio-renal div. FDA, Rockville, Md., 1986-87; v.p. clin. rsch. Wallace Labs Div. Carter-Wallace Inc., Cranbury, N.J., 1987—; rsch. assoc. gastrointestinal lab. Mt. Sinai Hosp., L.A., 1966-68; cons. in gastroenterology Riker Labs., Northridge, Calif., 1967-69. Contbr. articles to profl. jours. Rsch. fellow Am. Heart Assn., L.A., 1967-69, advanced rsch. fellow, 1969-70. Mem. AMA, Am. Pharm. Assn., Am. Coll. CDlin. Pharmacology, Am. Soc. for Clin. Pharmacology and Therapeutics, Am. Heart Assn., Am. Fedn. for Clin. Rsch., Am. Physiol. Soc., Am. Gastroent. Assn., Microcirculatory Soc., N.Y. Acad. Scis., Am. Soc. Hypertension Inc., Epilepsy Found. Am., Drug Info. Assn. Office: Carter-Wallace Inc Half-acre Rd # 1001 Cranbury NJ 08512*

ROSENBERG, CHARLES HARVEY, otorhinolaryngologist; b. N.Y.C., June 10, 1919; s. Morris and Bessie (Greditor) R.; m. Florence Rich, Dec. 27, 1943; children: Kenneth, Ina Garten. BA cum laude, Alfred U., 1941; MD, U. Buffalo, 1944. Intern Jewish Hosp. Bklyn., 1944-45; resident otolaryngology Mt. Sinai Hosp., N.Y.C., 1945-46, 48-50; teaching faculty, sr. clin. asst. Mt. Sinai Hosp. and Med. Sch., N.Y.C., 1950-72; attending surgeon Stamford (Conn.) Hosp., St. Joseph's Hosp., 1953—; dir. dept. otolaryngology Stamford Hosp. and St. Joseph's Hosp., 1973-79. Campaign chmn. United Jewish Fedn., Stamford, 1978-81, pres., 1981-83, exec. com., 1978—; mem. pres.'s coun. Alfred (N.Y.) U., 1990—. Capt. U.S. Army, 1945-46. Fellow Am. Coll. Surgeons; mem. AMA, Stamford Med. Soc., Fairfield Med. Soc., Conn. State Med. Soc., Am. Bd. Otolaryngology, Am. Acad. Ophthalmology and Otolaryngology. Democrat. Jewish. Home: 304 Erskine Rd Stamford CT 06903 Office: 810 Bedford St Stamford CT 06901

ROSENBERG, DALE NORMAN, psychology educator; b. St. Ansgar, Iowa, Dec. 12, 1928; s. Eddie Herman and Ella (Kirchgatter) R.; BS, Mankato State Coll., 1956; M.Ed., U. S.D., 1959; postgrad. Ball State Tchrs. Coll., 1962, U. Nebr., 1961, Colo. State Coll., 1963-67; D.Arts, U. Central Ariz., 1978; m. Delrose Ann Hermanson, Sept. 10, 1950; children—Jean Marie, James Norman, Julie Ann, Lisa Jo. Tchr. public schs., Holstein, Iowa, 1956-60; prin., guidance dir., Crystal Lake, Iowa, 1960-62; prin. Grafton (Iowa) Jr. High Sch., 1962-66; psychol. tester Dept. Rehab., State of Iowa, 1960-66; prof. psychology North Iowa Area Community Coll., Mason City, 1966—; vis. lectr. Buena Vista Coll., Storm Lake, Iowa, 1984; invited speaker Inst. Advanced Philosophic Research, 1984-85. Served with USAF, 1949-53. Mem. NEA, Iowa Edn. Assn., Kappa Delta Pi, Phi Delta Kappa. Lutheran. Author multi-media curriculum for teaching disadvantaged introductory welding; author textbook-workbook, 1985. Home: RR 3 Box 295 12373 Thrush Ave Mason City IA 50401-9828 Office: N Iowa Area Community Coll Mason City IA 50401

ROSENBERG, DAN YALE, retired plant pathologist; b. Stockton, Calif., Jan. 8, 1922; s. Meyer and Bertha (Naliboff) R.; A.A., Stockton Jr. Coll., 1942; A.B., Coll. Pacific, 1949; M.S., U. Calif. at Davis, 1952; m. Marilyn Kohn, Dec. 5, 1954; 1 son, Morton Karl. Jr. plant pathologist Calif. Dept. Agr., Riverside, 1952-55, asst. plant pathologist, 1955-59, assoc. plant pathologist, 1959-60, pathologist IV, 1960-63, program supr., 1963-71, chief exclusion and detection, div. plant industry, 1971-76, chief nursery and seed services div. plant industry, 1976-82, spl. asst. div. plant industry, 1982-87; pres. Health, Inc., 1972-73; agrl. cons., 1988—; mem. Gov.'s Interagy. Task Force on Biotech., 1986—; bd. dirs. Health Inc., Sacramento, 1967, pres., 1971-72, 79-81, 81-83. Served with AUS, 1942-46; ETO. Mem. Am. Phytopath. Soc. (fgn. and regulatory com. 1975—, grape diseases sect. 1977-79, grape pests sect. 1979-84), Calif. State Employees Assn. (pres. 1967-69). Contbr. articles to profl. jours. Home and Office: 2328 Swarthmore Dr Sacramento CA 95825-6867

ROSENBERG, IRWIN HAROLD, physician, educator; b. Madison, Wis., Jan. 6, 1935; s. Abraham Joseph and Celia (Mazursky) R.; m. Civia Muffs, May 24, 1964; 1 child, Ilana. BS, U. Wis., 1956; MD, Harvard U., 1959. Diplomate Am. Bd. Internal Medicine. Intern Mass. Gen. Hosp., Boston, 1959-60, resident, 1960-61; instr. medicine Harvard Med. Sch., Boston, 1965-66, assoc. in medicine, 1966-68, asst. prof., 1968-70; assoc. prof. medicine U. Chgo., 1970-75, prof., 1975-86, Sarah and Harold Lincoln Thomson prof. medicine, 1983-86; prof., dir. USDA Human Nutrition Rsch. Ctr. on Aging Tufts U., Boston, 1986—; cons. FDA, NIH, AID, 1970—; mem. food and nutrition bd. Nat. Acad. Scis., 1971-83, chmn., 1981-83. Co-chair local br. Med. Com. on Human Rights, Boston, 1967; mem. adv. bd. Hebrew Coll., Boston, 1987, 91; chmn. bd. dirs. Hillel Found., U. Chgo. With USPHS, 1961-64. Recipient Josiah Macy Faculty award Macy Found., 1974, Goldsmith award Am. Coll. Nutrition, 1984. Fellow AAAS; mem. Am. Soc. for Clin. Nutrition (pres. 1983-84, Herman award 1989), Internat. Life Sci. Inst. (editor nutrition revs. 1989—). Jewish. Avocations: sports, music, Judaica. Office: Tufts U USDA Human Nutrition Rsch Ctr 711 Washington St Boston MA 02111-1524

ROSENBERG, JACOB JOSEPH, orthodontist; b. N.Y.C., July 15, 1947; s. Louis and Pearl (Flaster) R.; m. Marylynn Borteck; children: Jonathan, Carolyn, Hilary. BA, U. Vt., 1968; MS, Colo. State U., 1970; DDS with honors, SUNY, Buffalo, 1975; cert. in Orthodontics, Columbia U., 1977. Diplomate Am. Bd. Orthodontics. Practice dentistry specializing in orthodontics Bethesda, Md., 1977—; alumni admission rep. U. Vt. Mem. ADA, Md. State Soc. Orthodontists (pres. 1986-87), Am. Assn. Orthodontists, Am. Bd. Orthodontics (mem. Coll. Diplomates), Orthodontic Edn. of Research Found., Alpha Omega. Avocations: golf, skiing, reading, photography, travel. Office: 4405 E West Hwy Bethesda MD 20814-4522

ROSENBERG, JOEL BARRY, government economist; b. Bronx, N.Y., Aug. 14, 1942; s. Benjamin and Miriam Dorothy (Yellin) R.; B.A., Queens Coll., 1964, M.A., 1966; Ph.D., Brown U., 1972; m. Judith Lynne Jackler, Aug. 26, 1965; children: Jeffrey Alan, Marc David. Cons., Commonwealth Svcs., Washington, 1970-71; asst. prof. econs. SUNY, Geneseo, 1971-75, Case Western Res. U., Cleve., 1975-76; mgr., industry economist IRS, Washington, 1976—. NDEA fellow, Brown U., 1966-69. Mem. Am. Econ. Assn., Nat. Assn. Bus. Economists, Am. Statis. Assn. Contbr. articles to profl. jours. Home: 16 Flameleaf Ct Gaithersburg MD 20878-1885 Office: IRS 500 Capitol St NW Washington DC 20221-0001

ROSENBERG, ROBERT ALLEN, psychologist, educator, optometrist; b. Phila., July 31, 1935; s. Theodore Samuel and Dorothy (Bailes) R.; m. Geraldine Bella Tishler, Sept. 3, 1961; children: Lawrence David, Ronald Joseph. BA, Temple U., 1957, MA, 1964; BS, Pa. Coll. Optometry, 1960, OD, 1961. Lic. optometrist, psychologist, Pa. Instr. Pa. Coll. Optometry, Phila., 1962-65, asst. prof., 1965-67; asst. prof. psychology Community Coll. Phila., 1967-76, assoc. prof., 1976—; pvt. practice optometry, Roslyn, Pa., 1965—. Contbr. articles to profl. jours. Named Humanitarian Chapel of Four Chaplains Bapt. Temple, 1980. Fellow Am. Acad. Optometry; mem. Am. Optometric Assn., Pa. Optometric Assn., Bucks-Montgomery Optometric Assn., Alumni Assn. Pa. Coll. Optometry (v.p. 1991—, sec. 1992—). Avocations: singing, acting, photography, writing, public speaking. Home: 970 Corn Crib Dr Huntingdon Valley PA 19006-3304 Office: Community Coll Phila 1700 Spring Garden St Philadelphia PA 19130-3991

ROSENBERG, SAUL ALLEN, oncologist, educator; b. Cleve., Aug. 2, 1927. BS, Western Res. U., 1948, MD, 1953. Diplomate Am. Bd. Internal Medicine. Intern Univ. Hosp., Cleve., 1953-54; resident in internal medicine Peter Bent Brigham Hosp., Boston, 1954-61; research asst. toxicology AEC Med. Research Project, Western Res. U., 1948-53; asst. prof. medicine and radiology Stanford (Calif.) U., 1961-65, assoc. prof., 1965-79, chief div. oncology, 1965-93, prof., 1970—, Am. Cancer Soc. prof., 1983-89, assoc. dean, 1989-92; chmn. bd. No. Calif. Cancer Program, 1974-80. Contbr. articles to profl. jours. Served to lt. M.C. USNR, 1954-56. Master ACP; mem. Am. Assn. Cancer Research, Inst. Medicine Nat. Acad. Sci., Am. Assn. Cancer Edn., Am. Fedn. Clin. Research, Am. Soc. Clin. Oncology (pres. 1982-83), Assn. Am. Physicians, Calif. Acad. Medicine, Radiation Research Soc., Western Soc. Clin. Research, Western Assn. Physicians. Office: Stanford U Sch Medicine Div Oncology Stanford CA 94305

ROSENBERG, SHELDON, psychologist, educator; b. Bklyn., Jan. 6, 1930; s. George and Antoinette (Mannarino) R.; m. Irma Blumstein, Jan. 27, 1951; children: Eric Mark, Jason Stuart. BA, Bklyn. Coll., 1954; MA, U. Minn., 1956, PhD, 1958. From instr. to asst. prof. Ind. U., Jeffersonville, Ind., 1957-59; rsch. scientist Tng. Sch. at Vineland, N.J., 1959-61; from asst. to assoc. prof. Peabody Coll., Nashville, 1961-66; rsch. scientist U. Mich., Ann Arbor, 1966-69; prof. psychology U. Ill., Chgo., 1969—. Editor: Applied Psycholinguistics, 1979-83, Topics in Applied Psycholinguistics, 1987—, six books in psycholinguistics; contbr. chpts. to books and articles to profl. jours. NIH postdoctoral fellow, 1966, rsch. fellow Ill. Inst. for Study of Devel. Disabilities, 1981-82. Mem. Psychonomic Soc., Am. Psychol. Soc., Midwestern Psychol. Assn. Achievements include developments in field of applied psycholinguistics. Office: U Ill Dept Psychology M/C285 Chicago IL 60607

ROSENBERG, SHERMAN, program director; b. Chgo., Nov. 7, 1933; s. Abe and Rae Emma (Greenberg) R.; m. Lillie L. Stanton, Nov. 14, 1954; children: Lisa Renee, Tani Dail. BS, Roosevelt U., 1956; MS, Mich. State U., 1957. Cert. tchr., N.C. Head sci. dept. Queens Coll., Yaba, Nigeria, 1959-67, Dacca (East Pakistan) Am. Soc. Sch., 1967-71, Air Am. Sch., Udorn, Thailand, 1971-73; guidance counselor Belair Sch., Mandeville, Jamaica, 1973-81, Cairo Am. Coll., Ma'adi, Egypt, 1981-84, Union Sch., Port-au-Prince, Haiti, 1984-89; sr. sch. counselor Overseas Childrens Sch., Battaramulla, Sri Lanka, 1989-91; dir. coll. guidance Koç özel Lisesi, Pendik-Istanbul, Turkey, 1991—; cons. Harvard Grad. Sch. Edn., Cambridge, Mass., 1962, Internat. Schs. Svcs., Princeton, N.J., 1974. Contbr. articles to Sch. Sci. Rev. Mem. AAAS, ACA, Nat. Sci. Tchrs. Assn. (life), Fedn. Am. Scientists (supporting), Am. Chem. Soc., Nat. Assn. Coll. Admission Counselors, Assn. Internat. Educators, Assn. Sci. Edn. (Eng.). Baha'i. Home: PK 38, Pendik-Istanbul Turkey Office: Koç özel Lisesi, PK 38, Pendik-Istanbul Turkey

ROSENBERG, STEVEN AARON, surgeon, medical researcher; b. N.Y.C., Aug. 2, 1940; s. Abraham and Harriet (Wendroff) R.; m. Alice Ruth O'Connell, Sept. 15, 1968; children—Beth, Rachel, Naomi. B.A., Johns Hopkins U., 1960, M.D., 1963; Ph.D., Harvard U., 1968. Resident in surgery Peter Bent Brigham Hosp., Boston, 1963-64, 68-69, 72-74; resident fellow in immunology Harvard U. Med. Sch., Boston, 1969-70; clin. assoc. immunology br. Nat. Cancer Inst., Bethesda, Md., 1970-72; chief surgery Nat. Cancer Inst., 1974—, assoc. editor Jour., 1974—; mem. U.S.-USSR Coop. Immunotherapy Program, 1974—, U.S.-Japan Coop. Immunotherapy Program, 1975—; clin. assoc. prof. surgery George Washington U. Med. Ctr., 1976—; prof. surgery Uniformed Services U. Health Scis. Contbr. articles to profl. jours. Author: The Transformed Cell: Unlocking the Mysteries of Cancer, 1992. Served with USPHS, 1970-72. Recipient Meritorious Service medal Pub. Health Service, 1981; co-recipient Armand Hammer Cancer prize, 1985; named 1990 Scientist of the Yr., R&D magazine. Mem. Soc. Univ. Surgeons, Am. Surg. Assn., Soc. Surg. Oncology, Surg. Biology Club II, Halsted Soc., Transplantation Soc., Am. Assn. Immunologists, Am. Assn. Cancer Research, Phi Beta Kappa, Alpha Omega Alpha. Office: Nat Cancer Inst 9000 Rockville Pike Bethesda MD 20892-0001

ROSENBERGER, DAVID A., research scientist, cooperative extension specialist; b. Quakertown, Pa., Sept. 14, 1947; s. Henry and Ada C. (Geissinger) R.; m. Carol J. Freeman, July 29, 1973; children: Sara, Matthew, Nathan. BS in Biology, Goshen Coll., 1969; PhD in Plant Pathology, Mich. State U., 1977. Asst. prof. Hudson Valley lab. Cornell U., Highland, N.Y., 1977-84, assoc. prof., 1984—; supt. Cornell U., Highland, 1990—. Mem. AAAS, Am. Phytopathological Soc., Coun. Agrl. Sci. and Tech. Avocations: religious studies, gardening, hiking, jogging. Office: Cornell U Hudson Valley Lab PO Box 727 Highland NY 12528

ROSENBERGER, FRANZ ERNST, physics educator; b. Salzburg, Austria, May 31, 1933; came to U.S., 1966; s. Franz and Hertha (Sompek) R.; m. Renate Hildegard Suessenbach; children: Uta, Bernd, Till. BS in Physics, U. Stuttgart, 1960, diploma in physics, 1964; PhD in Physics, U. Utah, 1970. Asst. prof. physics U. Utah, Salt Lake City, 1970-77, assoc. prof., 1977-81, prof., 1981-86; prof. physics U. Ala., Huntsville, 1986—, dir. Ctr for Microgravity and Materials Rsch., 1986—. Author: Fundamentals of Crystal Growth, 1979; editor Jour. Crystal Growth, 1981—; contbr. articles to profl. jours. Mem. Am. Phys. Soc., Am. Assn. for Crystal Growth,

Materials Rsch. Soc. Avocations: windsurfing, skiing, photography, home improvements. Home: 171 Stoneway Trail Madison AL 35758-8543

ROSENBLATT, GERD MATTHEW, chemist; b. Leipzig, Germany, July 6, 1933; came to U.S., 1935, naturalized, 1940; s. Edgar Fritz and Herta (Fisher) R.; m. Nancy Ann Kaltreider, June 29, 1957 (dec. Jan. 1982); children: Rachel, Paul; m. Susan Frances Barnett, Nov. 23, 1990. BA, Swarthmore Coll., 1955; PhD, Princeton U., 1960; Doctorate in Physics (hon.), Vrije Universiteit Brussel, 1989. Chemist Lawrence Radiation Lab., Univ. Calif., 1960-63, cons., guest scientist, 1968-84; from asst. to assoc. prof. chemistry Pa. State U., University Park, 1963-70, prof., 1970-81; assoc. div. leader Los Alamos (N.Mex.) Nat. Lab., 1981-82, chemistry div. leader, 1982-85; dep. dir. Lawrence Berkeley (Calif.) Lab., 1985-89, sr. chemist, 1985—; lectr. U. Calif., Berkeley, 1962-63; vis. prof. Vrije U. Brussels, 1973; vis. fellow Southampton U., 1980, King's Coll., Cambridge, 1980; adj. prof. chemistry U. N.Mex., 1981-85; cons. Aerospace Corp., 1979-85, Solar Energy Rsch. Inst., 1980-81, Xerox Corp., 1977-78, Hooker Chem. Co., 1976-78, Los Alamos Nat. Lab, 1978, mem. external adv. com. Ctr. for Materials Sci., 1985—, mem. rev. com. chemistry div., 1985; mem. rev. com. for chem. engring. div. Argonne Univ. Assn., 1974-80, chmn. 1977-78; mem. rev. com. for chem. sci. Lawrence Berkeley Lab., 1984; chmn. rev. com. for chem. and material sci. Lawrence Livermore Nat. Lab., 1984—; mem. bd. advs. Combustion Rsch. Facility, Sandia Nat. Lab., 1985-89; mem. bd. advs. rsch. and devel. div. Lockheed Missiles & Space Co., 1985-87; mem. U.S. Nat. Com., Com. on Data for Sci. and Tech., 1986—, Internat. Union of Pure and Applied Chemistry, 1986—; mem. basic scis. lab. program panel Dept. Energy, 1985-89; sec. IUPAC Commn. on High Temperature and Solid State Chemistry, 1992—. Editor: (jour.) Progress in Solid State Chemistry, 1977—; mem. editorial bd. High Temperature Sci., 1979—; contbr. articles to profl. jours. Du Pont grad. fellow, Princeton U., 1957-58; fellow Solvay Inst., 1973, U.K. Rsch. Coun., 1980. Fellow AAAS; mem. Am. Chem. Soc., Am. Phys. Soc., Electrochem. Soc., Nat. Rsch. Coun. (chmn. high temperature sci. and tech. com. 1977-79, 84-85, panel on exploration of materials sci. and tech. for nat. welfare 1986-88, sci. and tech. info. bd. 1990-91, chmn. numerical data adv. bd. 1986-90, solit state scis. com. 1988-91, chmn. western regional materials sci. and engring. meeting 1990). Achievements include first use of imaging detectors to obtain Raman compositional profiles and two-dimensional maps of chemical compositions, of rotational Raman scattering as a temperature and state-population probe in high temperature and combustion systems; elucidation of role of crystal defects and molecular structure in the evaporation of solid materials; first experimental determination of how molecular polarizability anisotrophies change with internuclear distance; estimation of thermodynamic properties and molecular structures for gaseous molecules. Home: 1177 Miller Ave Berkeley CA 94708-1754 Office: Lawrence Berkeley Lab Cyclotron Berkeley CA 94720-0001

ROSENBLATT, JAY SETH, psychobiologist, educator; b. N.Y.C., Nov. 18, 1923; s. Louis and Minnie (Epstein) R.; m. Gilda Rosen, Feb. 18, 1948; children: Daniel Lawrence, Nina Lara. MA, NYU, 1950, PhD, 1953; PhD (hon.), Gothenborg U., 1987. Lic. psychologist, N.Y. Lectr. psychology dept. CCNY, N.Y.C., 1949-57; assoc. rsch. specialist Inst. Animal Behavior, Rutgers U., Newark, 1959-65; assoc. prof. psychology dept. Rutgers U., Newark, 1965-67, prof. I psychology dept., 1967-73, prof. II psychology dept., 1973-87; D.S. Lehrman prof. psychobiology Inst. Animal Behavior, Rutgers U., Newark, 1987—, dir., 1972-89; rsch. assoc. dept. animal behavior Am. Mus. Natural History, 1948-81. Editor: (book series) Advances in the Study of Behavior, Vols. 5-19, 1972-88; contbr. articles to profl. jours. With U.S. Army, 1943-45, ETO. Recipient USPHS rsch. grants NIMH, 1959-92, USPHS tng. grants NIMH, 1972-87, Lindback award for rsch. Rutgers U., 1977, rsch. grant NSF, 1988-91. Fellow APA, Animal Behavior Soc.; mem. Internat. Soc. for Devel. Psychobiology (pres. 1984-85). Democrat. Jewish. Achievements include research in early learning among mammals, hormonal basis of maternal behavior among mammals, sexual behavior in male cats. Home: 126 Sherman Ave Teaneck NJ 07666 Office: Inst of Animal Behavior 101 Warren St Newark NJ 07102

ROSENBLATT, JOAN RAUP, mathematical statistician; b. N.Y.C., Apr. 15, 1926; d. Robert Bruce and Clara (Eliot) Raup; m. David Rosenblatt, June 10, 1950. AB, Barnard Coll., 1946; PhD, U. N.C., 1956. Intern Nat. Inst. Pub. Affairs, Washington, 1946-47; statis. analyst U.S. Bur. of Budget, 1947-48; rsch. asst. U. N.C., 1953-54; mathematician Nat. Inst. Standards and Tech. (formerly Nat. Bur. Standards), Washington, 1955—, asst. chief statis. engring., 1963-68, chief statis. engring. lab., 1969-78, dep. dir. Ctr. for Applied Math., 1978-88; dep. dir. Computing and Applied Math. Lab., Gaithersburg, 1988-93, dir., 1993—; mem. com. on indsl. rels. Dept. Stats. Ohio State U.; mem. adv. com. in math. and stats. USDA Grad. Sch., 1971—; mem. Com. Applied and Theoretical Stats., Nat. Rsch. Coun., 1985-88. Mem. editorial bd. Communications in Stats., 1971-79, Jour. Soc. for Indsl. and Applied Math., 1965-75, Nat. Inst. Stds. and Tech. Jour. Rsch., 1991-93; contbr. articles to profl. jours. Chmn. Com. on Women in Sci., Joint Bd. on Sci. Edn., 1963-64. Rice fellow, 1946, Gen. Edn. Bd. fellow, 1948-50; recipient Fed. Woman's award, 1971, Gold medal Dept. Commerce, 1976, Presdl. Meritorious Exec. Rank award, 1982. Fellow AAAS (chmn. stats. sect. 1982, sec. 1987-91), Inst. Math. Stats. (coun. 1975-77), Am. Statis. Assn. (v.p. 1981-83, dir. 1979-80, Founders award 1991), Washington Acad. Scis. (achievement award math. 1965); mem. AAUW, IEEE Reliability Soc., Am. Math. Soc., Royal Statis. Soc. London, Philos. Soc. Washington, Internat. Statis. Inst., Bernouilli Soc. Probability and Math. Stats., Caucus Women Stats. (pres. 1976), Assn. Women Math., Exec. Women Govt., Phi Beta Kappa, Sigma Xi (treas. Nat. Bur. Standards chpt. 1982-84). Home: 2939 Van Ness St NW Apt 702 Washington DC 20008-4628 Office: Nat Inst Stds and Tech Bldg 225 Rm B118 Gaithersburg MD 20899-0001

ROSENBLATT, JOSEPH DAVID, hematologist, oncologist; b. Bklyn., Oct. 20, 1953; s. Max and Frederika (Gluck) R.; m. Lily Teichman, July 26, 1977; children: Aliza, Eliana, Joshua. BSc in Chemistry, UCLA, 1976, MD, 1980. Diplomate Am. Bd. Internal Medicine, subspecialty Hematology and Med. Oncology. Intern, then resident in internal medicine UCLA, 1981-83, fellow in hematology-oncology, 1983-85, instr. of medicine, 1986-87, asst. prof. medicine, 1987-93, assoc. prof., 1993—, assoc. dir. AIDS Inst., 1992—. Contbr. articles to profl. publs. Recipient Physician-Scientist award NIH, 1986-91, award Calif. Inst. for Cancer Rsch., 1988-89, award Leukemia Rsch. Found., 1989-90, Shannon award NIH, 1991; U. Task Force on AIDS grantee, 1987-89; Fulbright scholar, 1989-90. Mem. AAAS, Am. Soc. Hematology, U.S. Fulbright Assn., Am. Israel Med. Soc., Am. Soc. Microbiology. Jewish. Office: UCLA Dept Medicine 10833 Le Conte Ave Los Angeles CA 90024-1678

ROSENBLATT, MURRAY, mathematics educator; b. N.Y.C., Sept. 7, 1926; s. Hyman and Esther R.; m. Adylin Lipson, 1949; children—Karin, Daniel. B.S., CCNY, 1946; M.S., Cornell U., 1947, Ph.D. in Math., 1949. Asst. prof. statistics U. Chgo., 1950-55; assoc. prof. math. Ind. U., 1956-59; prof. probability and statistics Brown U., 1959-64; prof. math. U. Calif., San Diego, 1964—; vis. fellow U. Stockholm, 1953; vis. asst. prof. Columbia U., 1955; guest scientist Brookhaven Nat. Lab., 1959; vis. fellow U. Coll., London, 1965-66, Imperial Coll. and Univ. Coll., London, 1972-73, Australian Nat. U., 1976, 79; overseas fellow Churchill Coll., Cambridge U., Eng., 1979; Wald lectr., 1970; vis. scholar Stanford U., 1982. Author: (with U. Grenander) Statistical Analysis of Stationary Time Series, 1957, Random Processes, 1962, (2d edit), 1974, Markov Processes, Structure and Asymptotic Behavior, 1971, Studies in Probability Theory, 1978, Stationary Sequences and Random Fields, 1985, Stochastic Curve Estimation, 1991; editor: The North Holland Series in Probability and Statistics, 1980; mem. editorial bd. Jour. Theoretical Probability. Recipient Bronze medal U. Helsinki, 1978; Guggenheim fellow, 1965-66, 71-72. Fellow Inst. Math Statistics, AAAS; mem. Internat. Statis. Inst., Nat. Acad. Scis. Office: U Calif Dept Math La Jolla CA 92093

ROSENBLUETH, EMILIO, structural engineer; b. Mexico City, Mex., Apr. 8, 1926; s. Emilio and Charlotte (Deutsch) R.; m. Alicia Laguette, Feb. 20, 1954; children: David, Javier, Pablo, Monica. CE, Nat. Autonomous U. Mex., 1948, PhD (hon.), 1985; MS in Civil Engring, U. Ill., 1949, PhD, 1951; PhD (hon.), U. Waterloo, Ont., Can., 1983; postgrad., Nat. Autonomous U. Mex., 1985; PhD (hon.), Carnegie Mellon U., 1989. Surveyor and structural engr., 1945-47; soil mechanics asst. Ministry Hydraulic Resources, also U. Ill., 1947-50; structural engr. Fed. Electricity Commn., also Ministry Navy, 1951-55; prof. engring. Nat. Autonomous U. Mex., 1956-87; prof. emeritus Nat. Autonomous U. Mexico, 1987—; regent Nat. Autonomous U. Mex., 1972-81; pres. DIRAC Group Cons., 1970-77; vice-minister Ministry Edn., 1977-82; pres. Réunion Internationale des Laboratoires d'Essais des Materiaux (RILEM), 1965-66. Co-author: Fundamentals of Earthquake Engineering, 1971; Co-editor: Seismic Risk and Engineering Decisions, 1976; Contbr. to profl. publs. Trustee Autonomous Metropolitana U., 1974-77; mem. working group engring. seismology UNESCO, 1965, UN ad hoc com. Experts Internat. Decade Natural Disaster Reduction, 1987-88, U.S. com. for Decade Natural Disaster Reduction, 1989-90. Recipient M. Hidalgo medal, Mex., 1985, Prince of Asturias prize for sci. and tech., Spain, 1985, Luis Elizondo prize, 1974, Disting. Svc. in Engring. award U. Ill., 1976, Bernardo A. Houssay prize in tech. OAS, 1988, Univ. award for sci. Rsch., 1986; prof. honoris causa Nat. U. Engring., Peru, 1964. Mem. NAS, NAE, Mexican Acad. Sci. Rsch. (pres. 1964-65 Sci. award 1963), Mexican Soc. Earthquake Engring., Mex. Soc. Soil Mechanics (trustee), Internat. Assn. Earthquake Engring. (pres. 1973-77), Mexican Assn. Civil Engrs. (M.A. Urquijo research prize 1977, N. Carrillo rsch. award 1984), N.Z. Soc. Earthquake Engring., Am. Concrete Inst. (hon.), ASCE (W.L. Huber Rsch. prize 1965, Moisseiff award 1966, Alfred M. Freudenthal medal 1976, Nathan M. Newmark medal 1987), Internat. Assn. Earthquake Engring. (pres. 1972-76), Nat. Acad. Arts and Scis. (fgn. assoc.), 3d World Acad. Scis. (bd. dirs.), Sigma Xi. Office: Inst de Ingenieria, Ciudad Univ, Mexico City 04510, Mexico

ROSENBLUM, JUDITH BARBARA, psychologist; b. Bklyn., July 18, 1951; d. Barnett and Bette (Bromberg) R.; m. Alan Scott Goldberg, Dec. 27, 1986. BA magna cum laude, Adelphi U., 1973; PhD, Yeshiva U., 1983; cert. analytic psychotherapy, Advanced Inst. Analytic Psychotherapy, Jamaica, N.Y., 1991. Cert. sch. psychologist. Psychotherapist Advanced Ctr. for Psychotherapy, Jamaica, 1977-86; police psychologist N.Y.C. Police Dept., Rego Park, N.Y., 1986-87; sch. psychologist N.Y.C. Bd. Edn., Bklyn., 1987-88, Glen Cove (N.Y.) City Schs., 1988—. Mem. APA, Nat. Assn. Sch. Psychologists, Nassau County Psychol. Assn., Suffolk County Psychol. Assn. Avocations: craft fairs, reading. Home: 28 W 11th St Deer Park NY 11729-4010

ROSENBLUM, MARVIN, mathematics educator; b. Bklyn., June 30, 1926; s. Isidore and Celia (Mendelsohn) Rosenblum; m. Frances E. Parker, May 30, 1959; children: Isidore, Mendel, Jessie, Rebecca, Sarah. B.S., U. Calif.-Berkeley, 1949, M.A., 1951, Ph.D., 1955. Instr. math. U. Calif.-Berkeley, 1954-55; asst. prof. U. Va., Charlottesville, 1955-59, assoc. prof., 1960-65, prof., 1965—, now Commonwealth prof.; mem. Inst. Advanced Study, 1959-60. Served with USNR, 1944-46. Mem. Am. Math. Soc., Am. Math. Assn., Soc. Indsl. and Applied Math. Jewish. Office: Dept Math Univ Va Math-Astro Bldg Charlottesville VA 22903

ROSENBLUM, ROBERT, computer graphics software developer. BS in Computer Sci., U. Ill., 1987; MS in Computer & Info. Sci., Ohio State U., 1990. Software researcher MetroLight Studios, L.A., 1990—. Office: MetroLight Studios Ste 400 5724 W 3rd St Los Angeles CA 90036-3078

ROSENDAHL, BRUCE RAY, dean, geophysicist, educator; b. Jamestown, N.Y., Dec. 28, 1946; s. Raymond Leslie and Marjorie (Anderson) R.; m. Susan Andrews, June 14, 1969; children: Tana, Andrew. BS, U. Hawaii, 1970, MS, 1972; PhD in Oceanography, U. Calif. San Diego, La Jolla, 1976. Rsch. asst. U. Hawaii, Honolulu, 1970-72; rsch. asst. Scripps Instn. Oceanography, San Diego, 1972-76, postdoctoral fellow, 1976; from asst. prof. to assoc. prof. Duke U., Durham, N.C., 1976-88, prof., 1988-89; dean, Lewis Weeks chair in Marine Geology & Geophysics U. Miami, Fla., 1989—; v.p., chief exec. officer Internat. Oceanographic Found., Miami, 1989—; bd. govs. Joint Oceanographic Instns., Washington, 1989—, mem. exec. com. deep earth sampling, 1989—; bd. dirs. Marine Coun., Miami, 1990; bd. advisors Maritime Sci. and Tech. Acad., Miami, 1990; mem. Coun. on Ocean Affairs, 1989—. Author, editor: Initial Reports Deepsea Drilling, Leg 54, 1980, African Rifting, 1989; tech. editor Sea Frontiers Mag., 1989—; contbr. articles to profl. jours. Trustee Mus. Sci., Miami, 1990. Mem. Am. Geophys. Union, Am. Assn. Petroleum Geologists. Home: Apt 802 1581 Brickell Ave Miami FL 33129-1236 Office: U Miami 4600 Rickenbacker Cswy Miami FL 33149-1031

ROSENDORFF, CLIVE, cardiologist; b. Bloemfontein, South Africa, Mar. 28, 1938; s. Karl and Rachel (Elkon) R.; m. Daphne Avigail Lynn, Dec. 30, 1962; children: Bryan Peter, Nicola, Adam. BSc with honors, U. Witwatersrand, Johannesburg, South Africa, 1958, MBBCh, 1962, MD, 1977, DSc Medicine, 1984. Med. cert. South Africa, U.K., N.Y. Lectr., cons. medicine St. Thomas Hosp., London, 1965-69; prof., chmn. physiology U. Witwatersrand Med. Sch., Johannesburg, 1970-91, dean, 1987-90; prof., assoc. chmn. medicine Mt. Sinai Sch. Medicine, N.Y.C., 1991—; chief medicine VA Med. Ctr., Bronx, N.Y., 1991—; vis. prof. Yale U., 1969-70, U. Calif., San Francisco, 1977, Hosp. Lariboisiere, Paris, 1991; rsch. fellow Am. Heart Assn., Yale U., 1969-70. Author: Clincial Cardiovascular and Pulmonary Physiology, 1988; author over 160 rsch. papers. Fellow Royal Soc. South Africa, Royal Coll. Physicians, Am. Coll. Cardiology. Achievements include research in cardiovascular disease. Office: Mt Sinai Sch Medicine Dept Medicine 1 Gustave L Levy Pl New York NY 10029

ROSENFELD, AZRIEL, computer science educator, consultant; b. N.Y.C., Feb. 19, 1931; s. Abraham Hirsh and Ida B. (Chadaby) R.; m. Eve Hertzberg, Mar. 1, 1959; children—Elie, David, Tova. B.A., Yeshiva U., 1950, M.H.L., 1953, M.S., 1954, D.H.L., 1955; M.A., Columbia U., 1951, Ph.D., 1957; D.Tech. (hon.), Linkoping U., Sweden, 1980. Ordained rabbi, 1952. Physicist Fairchild Controls Corp., N.Y.C., 1954-56; engr. Ford Instrument Co., Long Island City, N.Y., 1956-59; mgr. research electronics div. Budd Co., Long Island City and McLean, Va., 1959-64; prof., computer sci. dir. Ctr. for Automation Research U. Md., College Park, 1964—; vis. asst. prof. Yeshiva U., N.Y.C., 1957-63; pres. ImTech, Inc., Silver Spring, Md., 1975-92. Author, editor numerous books; editor numerous jours. Fellow IEEE (Emanuel R. Piore award 1985), Washington Acad. Scis. (Sci. Achievement award 1988), Am. Assn. for Artificial Intelligence (founding); mem. Assn. Computing Machinery, Math. Assn. Am., Machine Vision Assn. (bd. dirs. 1984-88, Pres.'s award 1987), Internat. Assn. Pattern Recognition (pres. 1980-82, K.S. Fu award 1988), Assn. Orthodox Jewish Scientists (pres. 1963-65), Nat. Acad. Engring. of Mex. (corr.). Home: 847 Loxford Ter Silver Spring MD 20901-1132 Office: U Md Ctr Automation Rsch Computer Vision Lab College Park MD 20742-3275

ROSENFELD, JOEL, ophthalmologist; b. Jan. 27, 1957; s. Jacques Maurice and Mazal (Attia) R. BS with high honors, U. Mich., 1976, MD, 1980; JD cum laude, U. Detroit, 1993. Diplomate Nat. Bd. Med. Examiners, Am. Bd. Ophthalmology. Intern Baylor Coll. Medicine, Houston, 1980-81; resident in ophthalmology Kresge Eye Inst./Wayne State U., Detroit, 1981-84; fellow in ultrasound U. Iowa, Iowa City, 1984; chief of ophthalmology Wheelock Hosp., Goodrich, Mich., 1984—, Huron Meml. Hosp., Bad Axe, Mich., 1988—. Author rsch. reports. Supporting mem. Boys and Girls Club Am., Pontiac, Mich., 1984—, Pontiac Rescue Mission, 1988—. Recipient Man of Yr. award Boys and Girls Clubs Am., 1988; named Hon. Citizen, Father Flanagan Boys Home, Boys Town, Nebr., 1991, Ptnr. of Conscience, Amnesty Internat., N.Y.C., 1991. Fellow AMA, Am. Acad. Ophthalmology; mem. Mich. State Med. Soc. Jewish. Avocations: reading, travel, exercise, water skiing, boating. Office: 1060 S Van Dyke Rd Bad Axe MI 48413-9712

ROSENFELD, LOUIS, biochemist, educator; b. Phila., Oct. 21, 1947; s. David and Anne (Nemirofsky) R.; m. Maria Elena Aguero, June 24, 1983; children: Andrew, Gregory. BS, San Jose State U., 1969; PhD, U. Calif., Berkeley, 1974. Postdoctoral fellow Johns Hopkins U., Balt., 1974-76, Temple U. Health Sci. Ctr., Phila., 1976-78, Columbia U. Coll. Physicians and Surgeons, N.Y.C., 1978-81; asst. prof. N.Y. Med. Coll., Valhalla, 1981-87, 89—; rsch. asst. prof. Hosp. of U. of Pa., Phila., 1987-89. Contbr. articles to profl. publs. Recipient Nat. Rsch. Svc. award Nat. Eye Inst., 1978-81, Am. Heart Assn. grant-in-aid, 1991-94. Mem. AAAS, Am. Soc. for Biochemistry and Molecular Biology, Internat. Thrombosis Soc., Am. Chem. Soc. Achievements include discovery of role of carbohydrate moieties of glycoproteins, role of lens lipids in cataractogenesis, study of structure-function of heparin and related glycosaminoglycans on regulation of thrombosis and endothelial cells. Office: NY Med Coll Vosburgh Pavilion Valhalla NY 10595

ROSENFELD, LOUIS, biochemist; b. Bklyn., Apr. 8, 1925; s. Joseph and Esther (Achtenberg) R. BS, CCNY, 1946; MS, Ohio State U., 1948, PhD, 1952. Rsch. assoc. U. Va., Charlottesville, 1952-53; clin. chemist Wayne County Gen. Hosp., Eloise, Mich., 1954-56; assoc. biochemist Beth-El Hosp., Bklyn., 1956-61; asst. prof. NYU Med. Ctr., N.Y.C., 1961-67, assoc. prof., 1967—; adj. instr. dept. sci. Rutgers U., Newark, 1957-63. Author: Origins of Clinical Chemistry, The Evolution of Protein Analysis, 1982, Thomas Hodgkin: Morbid Anatomist and Social Activist, 1993; contbr. articles to profl. jours. Rsch. grantee Pub. Health Svc., 1965-67. Fellow AAAS, Assn. Clin. Scientists; mem. Am. Assn. for Clin. Chemistry, N.Y. Acad. Scis., Sigma Xi. Office: NYU Med Ctr 530 1st Ave New York NY 10016

ROSENFELD, RON GERSHON, pediatrics educator; b. N.Y.C., June 22, 1946; s. Stanley I. and Deborah (Levin) R.; m. Valerie Rae Spitz, June 16, 1968; children: Amy, Jeffrey. BA, Columbia U., 1968; MD, Stanford U., 1973. Intern Stanford (Calif.) U. Med. Ctr., 1973-74, resident in pediatrics, 1974-75, chief resident pediatrics, 1975-76; pvt. practice Santa Barbara, Calif., 1976-77; postdoctoral fellow Stanford U. Sch. Medicine, 1977-80, from asst. to assoc. prof. pediatrics, 1980-89, prof. pediatrics, 1989-93; chmn., prof. pediatrics Oreg. Health Scis. U., 1993—; physician-in-chief Doernbecher Children's Hosp., 1993—; cons. Genentech Inc., South San Francisco, 1980—, Kabi Pharmacia, Inc., Stockholm, 1990—, Novo Nordisk, Inc., Copenhagen, 1991—, Diagnostic Systems Labs., Webster, Tex., 1991—. Editor: Growth Abnormalities, 1985, Turner Syndrome, 1987, Turner Syndrome: Growth, 1990, Growth Regulation; editorial bd.: Jour. Clin. Endocrinology and Metabolism, Growth Factors, Clin. Pediatric Endocrinology, Growth and Growth Factors, Growth Regulation. Recipient Ross Rsch. award Ross Laboratories, 1985. Mem. Endocrine Soc., Soc. for Pediatric Rsch., Lawson Wilkins Pediatric Endocrine Soc., European Soc. for Pediatric Endocrinology, Diabetes Soc. Office: Oreg Health Scis Univ Dept Pediatrics 3181 SW Sam Jackson Park Rd Portland OR 97201-3098

ROSENFIELD, RICHARD ERNEST, emeritus medical educator; b. Pitts., Apr. 7, 1915; s. Abe E. and Ernestine (Lowenthal) R.; m. Olive da Costa-Levy, Apr. 5, 1944; children: Richard Ernest, Allan Oliver, Phyllis Ann Rosenfield Steele. BS, U. Pitts., 1936, MD, 1940. Rsch. and clin. asst. Mt. Sinai Hosp., N.Y.C., 1948-53, asst. hematologist, 1953-57, assoc. hematologist, 1957-71, dir. blood bank, 1957-80; prof. medicine Mt. Sinai Sch. Medicine, N.Y.C., 1966-67, 79-85, prof. pathology, 1972-85, emeritus prof. medicine, 1985—; hematologist N.Y.C. Health Dept., 1948-72; mem. sci. adv. com. N.Y. Blood Ctr., N.Y.C., 1964-78; mem. nat. blood rsch. coun. Nat. Acad. Sci., Washington, 1952-62. Contbr. more than 200 articles to profl. jours. Recipient Humanitarian award Nat. Hemophilia Found., 1981, Landsteiner award Am. Assn. Blood Banks, 1972; NIH grantee, 1958-78, 62-77. Mem. AAAS, AMA (emeritus), Am. Acad. Forensic Scis. (emeritus), Am. Assn. Immunologists (emeritus), Am. Soc. Hematology (emeritus), Am. Pub. Health Assn. (emeritus), Am. Soc. Human Genetics (emeritus), Internat. Soc. Blood Transfusion, Internat. Soc. Hematology (emeritus), Am. Soc. Clin. Pathologists (emeritus, Philip Levine award 1975). Home: 4418 Waldo Ave New York NY 10471 Office: Mt Sinai Med Ctr 1 Gustave L Levy Plz New York NY 10029

ROSENKILDE, CARL EDWARD, physicist; b. Yakima, Wash., Mar. 16, 1937; s. Elmer Edward and Doris Edith (Fitzgerald) R.; m. Bernadine Doris Blumenstine, June 22, 1963 (div. Apr. 1991); children: Karen Louise, Paul Eric; m. Wendy Maureen Ellison, May 24, 1992. BS in Physics, Wash. State Coll., 1959; MS in Physics, U. Chgo., 1960, PhD in Physics, 1966. Postdoctoral fellow Argonne (Ill.) Nat. Lab., 1966-68; asst. prof. math. NYU, 1968-70; asst. prof. physics Kans. State U., Manhattan, 1970-76, assoc. prof., 1976-79; physicist Lawrence Livermore (Calif.) Nat. Lab., 1979—, cons., 1974-79. Contbr. articles on physics to profl. jours. Woodrow Wilson fellow, 1959, 60. Mem. Am. Phys. Soc., Am. Astron. Soc., Soc. for Indsl. and Applied Math., Am. Geophys. Union, Acoustical Soc. Am., Phi Beta Kappa, Phi Kappa Phi, Phi Eta Sigma, Sigma Xi. Republican. Presbyterian. Club: Tubists Universal Brotherhood Assn. (TUBA). Current Work: Nonlinear wave propagation in complex media. Subspecialties: Theoretical physics; Fluid dynamics.

ROSENKOETTER, GERALD EDWIN, engineering and construction company executive; b. St. Louis, Mar. 16, 1927; s. Herbert Charles and Edna Mary (Englege) R.; m. Ruth June Beekman, Sept. 10, 1949; children: Claudia Ruth, Carole Lee. BSCE, Washington U., St. Louis, 1951; MSCE, Sever Inst. Tech., St. Louis, 1957. Registered profl. engr. Colo., Del., D.C., Fla., Ga., Idaho, Kans., Mass., Mich., Mo., N.C., N.J., Ohio, Pa., Tex., Utah, Wis. Sr. structural engr. Sverdrup & Parcel, Inc., St. Louis, 1951-56, project engr., 1956-60; engring. mgr. Sverdrup & Parcel, Inc., Denver, 1960-62; project mgr. Sverdrup & Parcel & Assocs., St. Louis, 1962-69, chief engr., 1969-74, v.p., 1974-80; pres. SPCM, Inc., St. Louis, 1980-85; exec. v.p. Sverdrup Corp., St. Louis, 1985-88, vice-chmn., 1988—; pres. Sverdrup Hydro, Inc., 1988—; asst. prof. Washington U., 1955-60; bd. dirs. Sverdrup Corp. and 18 subsidiaries, 1976—; ptnr. 3 Sverdrup Partnerships, 1977—; expert witness Sverdrup & Parcel & Assocs., 1970-75. Councilman City of Berkeley, Mo., 1956-58, councilman-at-large, 1958-60, chmn. city planning and zoning com., 1963-65; dir. Conservatory and Sch. Arts, St. Louis, 1989-92. Sgt. U.S. Army, 1945-46. Engrs. Club of St. Louis scholar, 1950. Mem. ASCE (chmn. continuing edn. 1965-66, named Outstanding Sr. Engring. Student 1951), Forest Hills Country Club (St. Louis), Bent Tree Country Club (Sarasota, Fla.). Lutheran. Avocations: golfing, sailing.

ROSENMAN, JULIAN GARY, radiation oncologist; b. Cleve., July 24, 1945; s. Jacques and Ida R.; m. Mary Anne Schaub, Apr. 29, 1970; children: Daniel, James, Alexander. BS in Math., Kent State U., 1966; MS in Physics., U. Ill., 1968; PhD in Physics, U. Tex., 1971; MD, U. Tex. Health Sci. Ctr., 1977. Diplomate Am. Bd. of Radiology; lic. MD, N.C. Postdoctoral fellow dept. of physics U. Tex., Austin, 1972; math and physics tchr. DelValle High Sch., Austin, 1972-73; surgical intern The Waltham (Mass.) Hosp., 1977-78; resident radiation medicine Mass. Gen. Hosp., Boston, 1978-81; instr. div. radiation therapy U. N.C., Chapel Hill, 1981-82, acting dir. div. radiation therapy sch. of medicine, 1983-84, asst. prof. dept. radiation oncology sch. of medicine, 1982-88, adj. assoc. prof. computer sci., 1988—, assoc. prof. dept. radiation oncology sch. of medicine, 1988—; respiratory core mem. Cancer and Leukemia Group B, Nat. Coop. Cancer Group, 1990—; dir. med. oncology Radiation Oncology Conf., 1986—; mem grad. faculty of coll. of arts and scis. U. N.C.; dir. Head and Neck Oncology Conf., 1983-90, Pediatric Oncology Conf., 1986-88; NIH grant reviewer at large, 1991—. Contbr. over 50 articles to profl. jours. Grantee, Whitaker Found., 1986-89, NCI, 1985-88, 87-88, 89-94, Siemens Corp. Rsch., 1988—, NIH, 1989-91, 1991-95, NSF, 1990-92; recipient Shannon award, 1992-93. Mem. Am. Assn. for Cancer Edn., Am. Coll. Radiology, Am. Soc. Therapeutic Radiology. Jewish. Achievements include devel. of computer generated graphical techniques for 3-dimensional radiation therapy treatment planning. Office: U of North Carolina Sch of Medicine Manning Dr Chapel Hill NC 27514

ROSENMANN, DANIEL, physicist, educator; b. Lima, Peru, Sept. 6, 1959; came to U.S., 1991; s. Lothar and Eva (Roiter) R.; m. Patricia Edith Alvarado, Jan. 21, 1989. BS in Physics, U. Nac. Mayor de San Marcos, Lima, Peru, 1986; postgrad., No. Ill. U., 1991-93. Instr. U. Nacional Mayor de San Marcos, Lima, 1982-91; tchr. Coll. Leon Pinelo, Lima, 1986-91; teaching asst. No. Ill. U., DeKalb, 1991-93, grad. rsch. asst., 1993; lab. grad. participantship Argonne Nat. Lab., 1993—. Author: Lab. guide book, 1988, 89. Argonne Nat. Lab. scholar 1993—. Mem. AAAS, Am. Phys. Soc., N.Y. Acad. Scis., Nat. Geographic Soc., Sigma Pi Sigma. Home: 501 N Annie Glidden Rd D-12 De Kalb IL 60115

ROSENOF, HOWARD PAUL, electrical engineer; b. Newark, Dec. 26, 1948; s. Abraham and Zelda (Ginsberg) R.; m. Jane Emily Rosengarten, Mar. 3, 1990. BS, Cornell U., 1970; MSEE, Northeastern U., Boston, 1973. Registered profl. engr., Mass., N.Y. Engr. controls Stone & Webster Engring. Corp., Boston, 1972-78; from project engr. to mgr. The Foxboro (Mass.) Co., 1978-88; from sr. cons. to dir. consulting Gensym Corp., Cambridge,

Mass., 1988-91; mktg. mgr. Gensym Corp., Cambridge, 1991—. Co-author: Batch Process Automation Theory and Practice, 1987; contbr. articles to profl. jours. Mem. Instrument Soc. Am. (sr., pres. Boston sect. 1985-86, com. SP88 1990—). Achievements include producing a general software structure and engineering method for the automation of batch processing. Office: Gensym Corp 125 Cambridge Park Dr Cambridge MA 02140

ROSENSCHEIN, GUY RAOUL, pediatric and visceral surgeon, airline pilot; b. Paris, July 28, 1953; s. Maurice and Caroline (Meller) R. M.D., Lariboisiere-St. Louis, Paris, 1977. Qualified profl. transport pilot, 1992, flight instr., 1993—. Intern, Hôpital Saint-Louis, 1973-74, Hôpital Lariboisière, 1975-76; resident Hôpitaux de Paris, 1977-80, Hôpital Bretonneau, 1977-78, Hôpital Lariboisière, 1979-80; resident Hôpital de Monaco, Monte Carlo, 1980-81; resident Hôpital St. Vincent de Paul, Paris, 1981-82, attache, 1982-84, asst., 1984-86; chef de clinique U. Paris, 1984-86; attache Hôpital de Villeneuve St Georges, 1987—; attache C.H.S. Saine Anne (Paris, 1992—; maitre de stage hospitalier Faculté de Médecine de Creteil, 1987-92; pilot 1991; profl. transport. instr. Author: Pancreatite non traumatique et non infectieuse de l'enfant, 1982. Capt., M.C., French Armed Forces, 1977. Mem. Conseil Nat. de L'ordre des Medecins. Jewish. Club: A.C. Renault (Chavenay, France), C.A.P.V (Melun, France). Home: 61 rue de Picpus, 75012 Paris France Office: Ctr Raymond Garcin Gen Surgery, 2 bis Rue d' Alesia, 75014 Paris France

ROSENSTEEL, GEORGE T., physics educator, nuclear physicist; b. Balt., Sept. 30, 1947; s. Walter St. George and Marie Emily (White) R. BSc, U. Toronto, Can., 1973, PhD, 1975. Can. fellow NRC, 1976-78; prof. physics Tulane U., New Orleans, 1978—, chmn. dept., 1985-91; vis. fellow Brit. Sci. and Engring. Coun., U. Sussex, Eng., 1986; vis. prof. Nat. Inst. Nuclear Theory, U. Washington, 1992. Contbr. 85 articles to profl. jours. Delivered grad. sch. commencement address Tulane U., 1987; recipient 7 grants NSF, 1979—. Mem. Am. Phys. Soc., Am. Math. Soc., Sigma Xi (young scientist award 1987). Office: Tulane U Dept of Physics New Orleans LA 70118

ROSENSTEIN, MARVIN, public health association administrator; b. Sept. 5, 1939. BSChemE, U. Md., 1961, PhD in Nuclear Engring., 1971; MS in Environ. Engring., Renssselaer Poly. Inst., 1966. Rschr. U.S. Bur. Mines/College Park (Md.) Metall. Rsch. Sta., 1961; commd. ensign Commd. Corps Pub. Health Svc., 1962, advanced through grades to capt., 1983; with N.E. Radiological Health Lab., Winchester, Mass., 1962, program coord. analytical quality control svc. divsn. radiological health, 1962; with data collation and analysis sect. radioation surveillance ctr. Divsn. Radiological Health, Washington, 1966; chief radiation exposure intelligence sect. standards and intelligence br. Nat. Ctr. for Radiological Health, Rockville, Md., 1967; dep. chief radiation measurements and calibration br. divsn. electronic products Bur. Radiological Health, Rockville, 1971, spl. asst. to dir. divsn. electronic products, 1972, dep. dir. divsn. electonic products, 1973, sr. sci. advisor, 1979; dep. assoc. commr. for policy coordination office policy coordination FDA, Rockville, 1978; dir. office health physics Ctr. for Devices and Radiological Health, Rockville, 1982—; mem. USASI Standards Com. N101, 1968-69; guest worker Ctr. for Radiation Rsch., Nat. Bur. Standards, 1969-74; faculty rsch. assoc. lab. for polymer and radiation sci. dept. clin. engring. U. Md., 1971-74; asst. clin. prof. radiology sch. medicine and scis. George Washington U., 1977-90. Contbr. over 70 publs. to profl. and sci. jours. Recipient Fed. Engr. of Yr. award NSPE/Dept. Health and Human Svcs., 1987. Mem. Nat. Coun. on Radiation Protection and Measurements (coun. 1988—, com. SC44 1976—, com. SC62 1980-85, chmn. com. SC46-12 1992—), Health Physics Soc. (publs. com. 1967-77, del. to 4th Internat. Congress Internat. Radiation Protection Assn. 1977, contbg. editor newsletter 1982—) Com. on Interagency Radiation Rsch. and Poliy Coordination (alt. HHS policy panel 1984—, vice chmn. sci. panel 1985—, exec. com. 1985—, chmn. subpanel on use BEIR V and UNSCEAR 1988 in risk assessment 1989—), Internat. Commn. on Radiological Protection (com. 3 on radiological protection in medicine 1985—, corresp. mem. task group on revision publ. 21, 1979-88), Commd. Officers Assn., Sigma Xi. Achievements include patent for radiation dosimeter; research in absorbed dose from medical X rays, radiation risk estimates, dosimetry for epidemiological studies, absorbed dose to the public from radiation emergencies, electron depth-dose and dosimetry, radiochemistry and environmental health, radioactivity in food, general radiological health. Office: FDA Ctr Devices & Radiological Hlth 2094 Gaither Rd Rockville MD 20850

ROSENSTRAUS, MAURICE JAY, biologist; b. Bklyn., Mar. 13, 1951; s. Herman Rosenstraus and Clarice Zimmer; m. Paula Selkow, May 22, 1977; 1 child, David. BS in Physics, Rensselaer Poly. Inst., 1972; PhD in Biol. Sci., Columbia U., 1977. Postdoctoral fellow Princeton (N.J.) U., 1976-78; asst. prof. Rutgers U., New Brunswick, N.J., 1978-85; investigator Enzo Biochem, Inc., N.Y.C., 1985-89; prin. rsch. scientist Cytogen Corp., Princeton, 1989-93; sect. mgr. Roche Molecular Systems, Branchburg, N.J., 1993—. Contbr. articles to Cancer Rsch., Devel. Biology, Genetics, Biotechniques, others. Recipient faculty fellowship Columbia U., N.Y.C., 1972, Herbert C. MacGregor prize Columbia U., N.Y.C., 1974, postdoctoral fellowship NIH, 1976, rsch. grant Nat. Cancer Inst., 1979, rsch. grant Am. Cancer Soc., 1983. Mem. AAAS, Genetics Soc. Am., Soc. for Devel. Biology, Sigma Xi. Democrat. Jewish. Home: 40 Smith Rd Somerset NJ 08873

ROSENTHAL, ALAN IRWIN, geophysicist; b. Bklyn., Mar. 11, 1947; s. Arthur Fred and Ethel (Rosenberg) R.; m. Shelley Sherman, June, 1969 (div. Mar. 1981); children: Habeeba, Elisabeth; m. Jean Grant, Nov., 1984 div. Dec. 1985). BA, Cornell U., 1968; PhD, Harvard U., 1974. Sr. rsch. physicist Shell Devel. Corp., Houston, 1974—. Treas., v.p. Young Israel of Houston, 1987—. Mem. Am. Phys. Soc., Soc. of Exploration Geophysicists (assoc.). Jewish. Home: 6331 Dawnridge Houston TX 77035 Office: Shell Devel Co PO Box 481 Houston TX 77001

ROSENTHAL, JOHN THOMAS, surgeon, transplantation surgeon; b. Richmond, Va., Feb. 22, 1949; s. John L. and June L. (Smith) R.; m. Susan Moore, May 11, 1984; children: Abigail, Sam. BS, Johns Hopkins U., 1970; MD, Duke U., 1974. Intern then resident in surgery U. Va., Charlottesville, 1974-76; resident in urology Lahey Clinic Found., Boston, 1977-80; asst. prof. surgery, urology U. Pitts., 1980-86; assoc. prof. surgery, urology UCLA, 1986—; chief renal transplantation UCLA Med. Ctr., 1986—; exec. vice chair dept. surgery UCLA, 1992—. Bd. dirs. ACLU West Pa., Pitts., 1982-86. Office: UCLA Medical Center 66-121 CHS Los Angeles CA 90024

ROSENTHAL, ROBERT, psychology educator; b. Giessen, Germany, Mar. 2, 1933; came to U.S., 1940, naturalized, 1946; s. Julius and Hermine (Kahn) R.; m. Mary Lu Clayton, Apr. 20, 1951; children: Roberta, David C., Virginia. A.B., UCLA, 1953, Ph.D., 1956. Diplomate: clin. psychology Am. Bd. Examiners Profl. Psychology. Clin. psychology trainee Los Angeles Area VA, 1954-57; lectr. U. So. Calif., 1956-57; acting instr. UCLA, 1957; from asst. to assoc. prof., coordinator clin. tng. U. N.D., 1957-62; vis. assoc. prof. Ohio State U., 1960-61; lectr. Boston U., 1965-66; lectr. clin. psychology Harvard U., Cambridge, Mass., 1962-67, prof. social psychology, 1967—, chmn. dept. psychology, 1992—. Author: Experimenter Effects in Behavioral Research, 1966, enlarged edit., 1976, (with Lenore Jacobson) Pygmalion in the Classroom, expanded edit., 1992, Meta-analytic Procedures for Social Research, 1984, rev. edit., 1991, Judgement Studies, 1987; (with others) New Directions in Psychology 4, 1970, Sensitivity to Nonverbal Communication: The Pons Test, 1979; (with Ralph L. Rosnow) The Volunteer Subject, 1975, Primer of Methods for the Behavioral Sciences, 1975, Essentials of Behavioral Research, 1984, 2d edit., 1991, Understanding Behavioral Science, 1984, Contrast Analysis, 1985, Beginning Behavioral Research, 1993; (with Brian Mullen) BASIC Meta-analysis, 1985; editor: (with Ralph L. Rosnow) Artifact in Behavioral Research, 1969, Skill in Nonverbal Communication, 1979, Quantitative Assessment of Research Domains, 1980, (with Thomas A. Sebeok) The Clever Hans Phenomenon: Communication with Horses, Whales, Apes and People, 1981, (with Blanck and Buck) Nonverbal Communication in the Clinical Context, 1986, (with Gheorghiu, Netter, and Eysenck) Suggestion and Suggestibility: Theory and Research, 1989. Recipient Donald Campbell award Soc. for Personality and Social Psychology, 1988; Guggenheim fellow, 1973-74, fellow Ctr. for Advanced Study in Behavioral Scis., 1988-89; sr. Fulbright scholar, 1972. Fellow AAAS (co-recipient Sociopsychol. prize 1960), Am. Psychol. Soc., Am. Psychol. Assn. (co-recipient Cattell Fund award 1967); mem. Soc. Exptl. Social Psychology, Eastern Psychol. Assn. (Disting. lectr. 1989), Midwestern Psychol. Assn., Mass. Psychol. Assn. (Disting. Career Contbn. award 1979), Soc. Projective Techniques (past treas.), Phi Beta Kappa, Sigma Xi. Home: 12 Phinney Rd Lexington MA 02173-7717 Office: Harvard U 33 Kirkland St Cambridge MA 02138-2044

ROSENTHAL, SUSAN LESLIE, psychologist; b. Washington, Sept. 27, 1956; d. Alan Sayre and Helen (Miller) R. BA, Wellesley Coll., 1978; PhD, U. N.C., 1986. Postdoctoral fellow Yale Child Study Ctr., New Haven, 1986-88; asst. prof. clin. pediatrics U. Cin., 1988-93; assoc. prof. clin. pediatrics Yale Child Study Ctr., New Haven, 1993—; dir. psychology div. adolescent medicine U. Cin., 1988—, assoc. prof. clin. pediatrics, 1993—; adj. faculty dept. psychology Miami U., Oxford, Ohio, 1992—. Contbr. articles to profl. jours. Mem. APA (program chair div. 37 1992), Cin. Soc. Clin. Child Psychologists, Ohio Psychol. Assn., Soc. Behavioral Pediatrics, Soc. Rsch. on Adolescence. Office: Children's Hosp Med Ctr Div Adolescent Medicine Elland & Bethesda Aves Cincinnati OH 45229

ROSENZWEIG, MARK RICHARD, psychology educator; b. Rochester, N.Y., Sept. 12, 1922; s. Jacob and Pearl (Grossman) R.; m. Janine S.A. Chappat, Aug. 1, 1947; children: Anne Janine, Suzanne Jacqueline, Philip Mark. B.A., U. Rochester, 1943, M.A., 1944; Ph.D., Harvard U., 1949; hon. doctorate, U. René Descartes, Sorbonne, 1980. Postdoctoral research fellow Harvard U., 1949-51; asst. prof. U. Calif., Berkeley, 1951-56; assoc. prof. U. Calif., 1956-60, prof. psychology, 1960-91, assoc. research prof., 1958-59, research prof., 1965-66, prof. emeritus, 1991—; vis. prof. biology Sorbonne, Paris, 1973-74; mem. exec. com. Internat. Union Psychol. Sci., 1972—, v.p., 1980-84, pres., 1988-92; chmn. U.S. Nat. Com. for Internat. Union Psychol. Sci., NRC-Nat. Acad. Sci., 1985-88, mem., 1988—. Author: Biologie de la Mémoire, 1976, (with A.L. Leiman) Physiological Psychology, 1982, 2d edit., 1989, (with D. Sinha) La Recherche en Psychologie Scientifique; editor: (with P. Mussen) Psychology: An Introduction, 1973, 2d edit., 1977, (with E.L. Bennett) Neural Mechanisms of Learning and Memory, 1976, International Psychological Science: Progress, Problems, and Prospects, 1992; co-editor (with L. Porter) Ann. Rev. of Psychology, 1968—, (with M.J. Renner) Enriched and Impoverished Environments: Effects on Brain and Behavior, 1987; contbr. articles to profl. jours. Served with USN, 1944-46. Recipient Disting. Alumnus award U. Rochester; Fulbright research fellow; faculty research fellow Social Sci. Research Council, 1960-61; research grantee NSF, USPHS, Easter Seal Found., Nat. Inst. Drug Abuse. Fellow AAAS, APA (Disting. Sci. Contbn. award 1982), Am. Psychol. Soc.; mem. NAS, NAACP (life), Am. Physiol. Soc., Am. Psychol. Soc., Internat. Brain Rsch. Orgn., Soc. Exptl. Psychologists, Soc. for Neuroscience, Société Française de Psychologie, Sierra Club (life), Common Cause, Phi Beta Kappa, Sigma Xi. Office: U Calif Dept Psychology 3210 Tolman Hall Berkeley CA 94720*

ROSENZWEIG, MICHAEL LEO, ecology educator; b. Phila., June 25, 1941; s. Max and Phyllis (Fine) R.; m. Carole Ruth Citron, June 4, 1961; children: Abby Judith Rosenzweig Daniel, Juli Ellen, Ephron Solomon. AB in Zoology, U. Pa., 1962, PhD in Zoology, 1966. Asst. prof. biology Bucknell U., Lewisburg, Pa., 1965-69, SUNY, Albany, 1969-71; assoc. prof. U. N.Mex., 1971-75; prof. ecology and evolutionary biology U. Ariz., Tucson, 1975—; vis. asst. prof. Cranberry Lake Biol. Sta., SUNY, Albany, 1969; vis. prof. zoology U. Wis., Madison, 1990-91; vis. prof. biology Ben-Gurion U., Israel, 1981-82; cons. Susquehanna Econ. Devel. Assn., Pa., 1970-71, EPA, 1979, Inst. Ecology, 1979, Dept. of Energy, 1989, U. Minn. Dept. Ecology and Evolutionary Biology, 1990; cons. U.S. Congress, 1974, mem. sci. adv. panel to the com. on pub. works, mem. task force on population distbn. and carrying capacity, mem. task force on growth policy. Author: And Replenish the Earth: The Evolution, Consequences and Prevention of Overpopulation, 1974, (with others) The Science of Life, 1977; assoc. editor Paleobiology, 1983-86; founder, editor-in-chief Evolutionary Ecology, 1986—; contbr. articles to Am. Naturalist, Jour. Mammals, Sci., Ecology and Pollution, Ecology, Oecologia, Nature, Evolutionary Ecology, numerous others. Coop. grad. fellow NSF, 1962-65, Rudi Lemberg traveling fellow Australian Acad. Sci., 1989, Jock Marshall fellow Monash U., Melbourne, Australia, 1989, Brittingham fellow U. Wis., 1990-91; Jacob Blaudstein scholar Ben-Gurion U. of the Negev, Israel, 1992. Mem. Ecol. Soc. Am., Am. Soc. Naturalists, Am. Soc. Zoologists, Soc. for the Study of Evolution, Soc. Population Ecology (Japan) Brit. Ecol. Soc., Sigma Xi. Office: U Ariz Dept Ecology Fremont and Lowell Sts Tucson AZ 85721

ROSENZWEIG, NORMAN, psychiatry educator; b. N.Y.C., Feb. 28, 1924; s. Jacob Arthur and Edna (Braman) R.; m. Carol Treleaven, Sept. 20, 1945; 1 child, Elizabeth Ann. MB, Chgo. Med. Sch., 1947, MD, 1948; MS, U. Mich., 1954. Diplomate Am. Bd. Psychiatry and Neurology. Asst. prof. psychiatry U. Mich., Ann Arbor, 1957-61, asst. prof., 1963-67, assoc. prof., 1967-73; prof. Wayne State U., Detroit, 1973—; chmn. dept. psychiat. Sch. Med. Wayne State U., Detroit, 1987-90, Sinai Hosp., Detroit, 1961-90; spl. cons., profl. advisor Oakland County Community Mental Health Services Bd., 1964-65; mem. protem med. adv. panel Herman Kiefer Hosp., Detroit, 1970, psychiat. task force N.W. Quadrangle Hosps., Detroit, 1971-78, planning com. mental health adv. council Dept. Mental Health State of Mich., Lansing, 1984-90, tech. adv.rsch. com., 1978-82; psychiat. bed need task force Office Health and Med. Affairs State of Mich., 1980-84; bd. dirs. Alliance for Mental Health, Farmington Hills, Mich.; speaker in field. Author: Community Mental Health Programs in England: An American View, 1975; co-editor: Psychopharmacology and Psychotherapy-Synthesis or Antithesis?, 1978, Sex Education for the Health Professional: A Curriculum Guide, 1978; contbr. articles to profl. jours. and chpts. to books. Mem. profl. adv. bd. The Orchards, Livonia, Mich., 1963. Served as capt. USAF, 1955-57. Recipient Appreciation and Merit cert. Mich. Soc. Psychiatry and Neurology, 1970-71. Fellow Am. Coll. Mental Health Adminstrn., Am. Coll. Psychiatrists (hon. membership com., com. on regional ednl. programs, liaison officer to The Royal Australian and New Zealand Coll. Psychiatrists 1984-88), Am. Psychiat. Assn. (life fellow, council on internat. affairs 1970-79, chmn. 1973-76, assembly liaison to council on internat. affairs 1979-80, 82-84, reference com. 1973-76, nominating com. 1978-79, internat. affairs survey team 1973-74, assoc. representing Am. Psychiat. Assn. to Inter-Am. Council Psychiat. Assns. 1973-75, others, Rush Gold Medal award 1974, cert. Commendation, 1973-76, 78-80, Warren Williams award 1986); mem. AAUP, AMA (Physician's Recognition award 1971, 74, 77, 80-81, 84, 87, 90, 92), Am. Assn. Dirs. Psychiat. Residency Tng. (nominating com. 1972-74, task force on core curriculum 1972-74), Am. Assn. Gen. Hosp. Psychiatry, Puerto Rico Med. Assn. (hon.), Am. Hosp. Assn. (governing council psychiat. services sect. 1977-79, ad hoc com. on uniform mental health definitions, chmn. task force on psychiat. coverage under Nat. Health Ins. 1977-79, others), Brit. Soc. Clin. Psychiatrists (task force on gen. hosp. psychiatry 1969-74), Can. Psychiat. Assn., Mich. Assn. Professions, Mich. Hosp. Assn. (psychiat. and mental health services com. 1979-81), Mich. Psychiat. Soc. (com. on ins. 1965-69, chmn. com. on community mental health services 1967-68, chmn. com. on nominations of fellows 1972-73, mem. com. on budget 1973-74, task force on pornography 1973-74, chmn. commn. on health professions and groups 1974-75, pres. elect 1974-75, pres. 1975-76, chmn. com. on liaison with hosp. assns. 1979-81, chmn. subcom. on liaison with Am. Hosp. Assn. 1979-81, numerous others, Past Pres. plaque, 1978, cert. Recognition, 1980, Disting. Service award 1986), Mich. State Med. Soc. (vice chmn. sect. psychiatry 1972-73, chmn. sect. psychiatry 1974-75, mem. com. to improve membership 1977-78, alt. del for Mich. Psychiat. Soc. to House of Dels. 1978-79, del. from Wayne County Med. Soc. to Mich. Med. Soc. House of Dels. 1982-88), N.Y. Acad. Scis., Pan Am. Med. Assn., Wayne County Med. Soc. (com. on hosp. and prof. rels., 1983-84, com. on child health advocacy 1983-87, med. edn. com. 1983-87, mental health com. 1983-87), Royal Australian and New Zealand Coll. Psychiatrists (hon.), Indian Psychiat. Soc. (hon. corr.), World Psychiat. Assn., Sect. Gen. Hosp. Psychiat. Avocations: music, film, reading. Home: 1234 Cedarholm Ln Bloomfield Hills MI 48302-0902 Office: 26211 Central Park Blvd Ste 602 Southfield MI 4876-4164

ROSKO, JOHN JAMES, biology educator; b. Yonkers, N.Y., Oct. 17, 1947; s. Joseph Vincent and Louise Carolyn (Krajeski) R.; m. Dolores Catherine Donaldson, Nov. 20, 1976; children: Jeffrey, Gregory, Lauren. BS, Fordham U., 1969; MA, Lehman Coll., 1976; PhD, Fordham U., 1982. Instr. Misercordia Sch. Nursing, Bronx, N.Y., 1973-76, Mercy Coll., Dobbs Ferry, N.Y., 1972-75, Lehman Coll., Bronx, 1972-76, Westchester Community Coll., Valhalla, N.Y., 1976-79, Iona Coll., New Rochelle, N.Y., 1979-81; assoc. prof. biology St. Thomas Aquinas Coll., Sparkill, N.Y., 1981—; mem. adv. bd. Madam Curie Sci. Ctr., Sparkill, 1991—. Advisor Cornell Coop. Ext., Thiells, N.Y., 1991--. Mem. AAAS. Office: St Thomas Aquinas Coll Rt 340 Sparkill NY 10976

ROSNER, RONALD ALAN, mechanical engineer; b. Florissant, Mo., Jan. 14, 1967; s. Harold Richard and Ellen Mary (Moser) R. BS in Aerospace Engring., U. Mo., 1989; MSME, U. Ill., 1991. Flight test engr. McDonnell Aircraft Co., St. Louis, 1989-91; rsch. asst. U. Ill. Dept. Mechanical and Indsl. Engring., Urbana, 1990-91; mechanical engr. B&V Waste Sci. and Tech. Corp., Kansas City, Mo., 1991—. Sci. fair mentor, sponsor Mid. Sch. Bingham, Kansas City, 1991, 92, 93. Recipient A.P. Green medal U. Mo., 1989. Mem. AIAA (Abe M. Zarem award 1993), Air and Waste Mgmt. Assn. Achievements include application of particle image velocimetry to high speed flows; air quality and pollution prevention. Office: B&V Waste Sci & Tech Corp 4717 Grand Ave Ste 500 Kansas City MO 64112

ROSOCHA, LOUIS ANDREW, physicist; b. Harrison, Ark., Feb. 7, 1950; s. Stanley and Lottie Cecilia (Zarkowski) R.; m. Emma Louise Riemer, June 16, 1979. BS, U. Ark., 1972; PhD, U. Wis., 1979. Scientist Nat. Rsch. Group, Madison, Wis., 1978-81; staff scientist Los Alamos (N.Mex.) Nat. Lab., 1981-83, project mgr., 1983-89, dep. group leader, 1987—, mem. mid. mgmt. coun., 1992—; chair profl. confs., 1985—; referee profl. jours. Contbr. to profl. publs. Mem. Am. Phys. Soc., Internat. Ozone Assn., Phi Beta Kappa, Sigma Pi Sigma. Democrat. Unitarian. Achievements include patent on dielectric surface electrical discharge device, participation in development of world's largest krypton fluoride, inertial fusion laser system; established 2 laboratories and several projects for the application fo plasmas and electron beams to destruction of environmental pollutants. Office: Los Alamos Nat Lab Mail Stop J564 PO Box 1663 Los Alamos NM 87545

ROSS, AMY ANN, experimental pathologist; b. Glendale, Calif., Apr. 28, 1953; d. William F. Ross and Joyce V. (Stuart) Ruygrok. BA, Calif. State U., Northridge, 1981; PhD, U. So. Calif., 1986. Assoc. dir. rsch. Inst. Cancer and Blood Rsch., Beverly Hills, Calif., 1986-89; dir. R & D Biologic and Immunologic Sci. Labs., Reseda, Calif., 1989—; rsch. asst. Sch. Medicine U. So. Calif., L.A., 1975-80, rsch. assoc., 1982-86. Mem. AAAS, Am. Soc. Clin. Pathology, Assn. of Women in Sci. (pres. L.A. chpt. 1979-80). Achievements include development of post-embedding ultrastructural immuno-cytochemical technique for analyzing archival pathology specimens. Office: BIS Labs 19231 Victory Blvd Ste 12 Reseda CA 91335-6308

ROSS, BRUCE MITCHELL, psychology educator; b. Ames, Iowa, June 24, 1925; s. Earle Dudley and Ethel Eileen (Newbecker) R. BA, U. Wis., 1949, MA, 1950, PhD, 1953. Instr. Brown U., Providence, R.I., 1955-57; asst. prof. Rutgers U., New Brunswick, N.J., 1957-63; prof. Cath. U., Washington, 1964--. Co-author: Developmental Memory Theories: Baldwin and Piaget, 1978, Children's Concepts of Chance and Probability, 1982, (encyclopedia article) Les Fonctions des Stockage, 1987; author: Remembering the Personal Past, 1991. With U.S Army, 1943-45, ETO. Office: Life Cycle Inst Cath U Washington DC 20064

ROSS, CHARLES ALEXANDER, geologist; b. Champaign, Ill., Apr. 16, 1933; s. Herbert Holdsworth and Jean (Alexander) R.; m. June Rosa Pitt Phillips, June 27, 1959. BA, U. Colo., 1954; MS, Yale U., 1958, PhD, 1959. Rsch. assoc. Peabody Mus., Yale U., New Haven, 1959-60; asst., assoc. geologist Ill. State Geol. Survey, Urbana, 1960-64; asst., assoc., full prof. dept. geology Western Wash. U., Bellingham, 1964-82, chair dept. geology, 1977-82; sr. staff geologist Gulf Oil Co., Tech. Exploration Ctr., Houston, 1982-83, dir., mgr., 1983-85; biostratigrapher Chevron USA, Houston, 1985-92, GeoBioStrat, Bellingham, Wash., 1992—; rsch. assoc. dept. geology Western Wash. U., 1992—; sec., treas. Soc. Econ. Paleontologists and Mineralogists, Tulsa, 1982-84; pres. Cushman Found. for Foraminiferal Rsch., 1983-84, 90-91. Author, co-author and editor of 8 books; contbr. over 150 articles to profl. jours. 1st lt. U.S. Army, 1954-56. Recipient Best Paper award 1967 Jour. Palontology, 1968. Fellow Geol. Soc. Am., Cushman Found. for Foraminiferal Rsch.; mem. AAAS, Am. Assn. Petroleum Geologists, Soc. Sedimentary Geology (Permian Basin sect.). Achievements include study of paleobiogeography of late Paleozoic faunas and use of results to recognize the far traveled nature of accreted terranes around Pacific margin; devel. detailed sea-level fluctuation curve for late Paleozoic. Office: GeoBioStrat 600 Highland Dr Bellingham WA 98225-6410

ROSS, DONALD EDWARD, engineering company executive; b. N.Y.C., May 2, 1930; m. Jeanne Ellen McKessy, Apr. 4, 1954; children: Susan, Christopher, Carolyn. BA, Columbia U., 1952, BS in Mech. Engring., 1953; MBA, NYU, 1960. Registered profl. engr., N.Y., 14 other states. Engr. Carrier Corp., N.Y.C., 1955-70; v.p. Dynadata, 1970-71; with Jaros, Baum & Bolles, N.Y.C., 1971—, ptnr., 1977—. Vice chmn. adv. coun. Columbia U. Sch. Engring. and Applied Sci. Lt. (j.g.) USN, 1953-55. Fellow ASHRAE; mem. ASME, NSPE, Nat. Acad. Engrs., Am. Cons. Engrs. Coun., Nat. Bur. Engring. Registration, N.Y. Assn. Cons. Engrs. (pres. 1984-86), Coun. on Tall Bldgs. and Urban Habitat (vice chmn. N.Am., mem. steering group), Univ. Club (N.Y.C.), Nassau Country Club. Office: Jaros Baum & Bolles 345 Park Ave New York NY 10154-0004

ROSS, EDWARD, cardiologist; b. Fairfield, Ala., Oct. 10, 1937; s. Horace and Carrie Lee (Griggs) R.; BS, Clark Coll., 1959; MD, Ind. U., 1963; m. Catherine I. Webster, Jan. 19, 1974; children: Edward, Ronald, Cheryl, Anthony. Intern, Marion County Gen. Hosp., Indpls., 1963; resident in internal medicine Ind. U., 1964-66, 68, cardiology rsch. fellowship, 1968-70, clin. asst. prof. medicine, 1970; cardiologist Capitol Med. Assn., Indpls., 1970-74; pvt. practice medicine, specializing in cardiology, Indpls., 1974—; staff cardiologist Winona Meml. Hosp. (now Midwest Med. Ctr.), Indpls.; staff Meth. Hosp., Indpls., dir. cardiovascular patient care programs, chmn. cardiovascular sect., chmn., dir. Cardiovascular Ctr., 1990—; dir. cardiovascular ctr. Meth. Hosp., 1990-92; bd. dirs. Meth. Hosp. Heart-Lung Ctr. Mem. Cen. Ind. Health Planning Coun., 1972-73; bd. dir. Ind. chpt. Am. Heart Assn., 1973-74, multiphasic screening East Side Clinic, Flanner House of Indpls., 1968-71; med. dir. Nat. Ctr. for Health Service Rsch. and Devel., HEW, 1970; consumer rep. radiologic device panel health, FDA, 1988-92; dir. hypertensive screening State of Ind., 1974. Assoc. editor Angiology, Jour. Vascular Disease. Capt., MC, USAF, 1966-68. Woodrow Wilson fellow, 1959; Nat. Found. Health scholar, 1955, Gorgas Found. scholar, 1955. Diplomate Am. Bd. Internal Medicine. Fellow Royal Soc. Promotion of Health (Eng.), Am. Coll. Angiology (v.p. fgn. affairs), Internat. Coll. of Angiology, Am. Coll. Cardiology; mem. AMA, Am. Soc. Contemporary Medicine and Surgery, Nat. Med. Assn. (council sci. assembly 1985-89), Ind. Med. Soc., Marion County Med. Soc., Am. Soc. Internal Medicine, Am. Heart Assn., Ind. Soc. Internal Medicine (pres. 1987-89), Ind. State Med. Assn. (chmn. internal medicine sect. 1987-89), Aesculapean Med. Soc., Hoosier State Med. Assn. (pres. 1980-85, 90—), NAACP, Urban League, Ind. Med. Soc., Alpha Omega Alpha, Alpha Kappa Mu, Beta Kappa Chi, Omega Psi Phi. Baptist. Sr. editor Jour. Vascular Medicine, 1983—. Office: 3737 N Meridian St Ste 400 Indianapolis IN 46208-4348

ROSS, ERIC ALAN, civil engineer; b. Mineola, N.Y., Sept. 11, 1961; s. Howard Edward and Marjorie Jean (Sheldon) R.; m. Lauren Elizabeth O'Connell, May 31, 1986. BA in Math., Hope Coll., 1983; BE, Hofstra U., 1985. Registered profl. engr., Mich. Asst. civil engr. N.Y.C. Dept. Environ. Protection, 1985-86; project mgr., estimator Angelo Iafrate Constrn., Warren, Mich., 1986-90, purchasing agt., 1990-91; civil engr. McNeely & Lincoln Assocs., Inc., Northville, Mich., 1991—. Vol. ARC, Oakland, Mich., 1988—; rep. Northfield Hills Homeowners Assn., Troy, Mich., 1988—, Coun. Troy Homeowners Assn. Mem. ASCE, Nat. Soc. Profl. Engrs., Mich. Soc. Planning Ofcls. Republican. Methodist. Home: 1860 Fordham Dr Troy MI 48098-2542

ROSS, EUAN MACDONALD, pediatrician, educator; b. Welwyn, Herts, Eng., Dec. 13, 1937; s. James Stirling and Frances (Blaze) R.; Jean Mary Palmer, June 11, 1966; children: Matthew, James. MB, ChB, U. Bristol, Eng., 1962, MD, 1975. Registrar in pediatrics Dundee (Scotland) Teaching Hosp., 1964-69; lectr. pediatrics U. Bristol, 1969-74; sr. lectr. pediatrics Middlesex and St. Mary's Med. Schs., London, 1974-84; cons. pediatrician Riverside Health Authority, London, 1984-89; prof. community pediatrics King's Coll. London, 1989-93; chair Child Health Computing Com.,

London, 1989—, Brit. Pediatric Surveillance com., London, 1991—, external examiner Trinity Coll., Dublin, Ireland, U. Aberdeen (Scotland), 1991—; advisor child health project, WHO, Estonia, 1993—. Author books; contbr. articles to profl. jours. Fellow Royal Coll. Physicians (Diploma of Child Health 1965); mem. F.P.H.M., Brit. Pediatric Assn., Brit. Pediatric Neurology Assn., Athenaeum Club. Avocations: art, photography, writing, travel. Home: Linklater House, Mount Park Rd, Harrow HA1 3JZ, England Office: Kings Coll SW Hosp, Pucross Rd, London SW9 9NU, England

ROSS, JOHN, JR., physician, educator; b. N.Y.C., Dec. 1, 1928; s. John and Janet (Moulder) R.; children—Sydnie, John, Duncan; m. Lola Romanucci, Aug. 26, 1972; children: Adan, Deborah Lee. A.B., Dartmouth Coll., 1951; M.D., Cornell U., 1955. Intern Johns Hopkins Hosp., 1955-56; resident Columbia-Presbyn. Med. Center, N.Y.C., 1960-61, N.Y. Hosp.-Cornell U. Med. center, 1961-62; chief sect. cardiovascular diagnosis cardiology br. Nat. Heart Inst., Bethesda, Md., 1962-68; prof. medicine U. Calif., San Diego, 1968—, also dir. cardiovascular div., 1968—, prof. cardiovascular research, 1985—; mem. cardiology adv. com. Nat. Heart, Lung and Blood Inst., 1975-78, task force on arteriosclerosis, 1978-80, adv. council, 1980-84; bd. dirs. San Diego Heart Assn.; vis. prof. Brit. Heart Assn., 1990. Author: Mechanisms of Contraction of the Normal and Failing Heart, 1968, 76, Understanding the Heart and Its Diseases, 1976; mem. editorial bd. Circulation, 1967-75, 80-88, editor in chief 1988-93, Circulation Research, 1971-75, Am. Jour. Physiology, 1968-73, Annals of Internal Medicine, 1974-78, Am. Jour. Cardiology, 1974-79, 83-88; cons. editor Jour. Clin. Investigating, 1992—; contbr. chpts. to books, sci. articles to profl. jours. Served as surgeon USPHS, 1956-63. Recipient Enzo Ferrari prize Organizing Com. for Enzo Ferrari Prize, Modena, Italy, 1989, James B. Herrick award Am. Heart Assn., 1990. Fellow Am. Coll. Cardiology (v.p. trustee, pres. 1986-87, Disting. Scientist award 1990), ACP; mem. Am. Soc. Clin. Investigation (councillor), Am. Physiol. Soc., Assn. Am. Physicians, Cardiac Muscle Soc., Am. Soc. Pharmacology and Exptl. Therapeutics, Assn. Univ. Cardiologists, Assn. West Physicians (councillor), Interam. Soc. Cardiology (research com.), Council Clin. Cardiology, Am. Heart Assn. (Herrick Award in Clin. Cardiology, 1990). Home: 8599 Prestwick Dr La Jolla CA 92037-2025 Office: Univ California Dept Med M-013B San Diego CA 92037

ROSS, JOHN, physical chemist, educator; b. Vienna, Austria, Oct. 2, 1926; came to U.S., 1940; s. Mark and Anna (Krecmar) R.; m. Virginia Franklin (div.); children: Elizabeth A., Robert K.; m. Eva Madarasz. BS, Queens Coll., 1948; PhD, MIT, 1951; D (hon.), Weizmann Inst. Sci., Rehovot, Israel, 1984, Queens Coll., SUNY, 1987, U. Bordeaux, France, 1987. Prof. chemistry Brown U., Providence, 1953-66; prof. chemistry MIT, Cambridge, 1966-80, chmn. dept., 1966-71; chmn. faculty of Inst. MIT, 1975-77; prof. Stanford (Calif.) U., 1980—, chmn. dept., 1983-89; cons. to industries, 1979—; mem. bd. govs. Weizmann Inst., 1971—. Author: Physical Chemistry, 1980; editor Molecular Beams, 1966; contbr. articles to profl. jours. Served as 2d lt. U.S. Army, 1944-46. Recipient medal Coll. de France, Paris. Fellow AAAS, Am. Phys. Soc.; mem. NAS, Am. Acad. Arts and Scis., Am. Chem. Soc. (Irving Langmuir Chem. Physics prize 1992, Dean's award for Disting. Teaching 1992-93). Home: 738 Mayfield Ave Palo Alto CA 94305-1044 Office: Stanford U Dept Chemistry Stanford CA 94305-2060

ROSS, JOHN R., III, aerospace engineer; b. Fort Carson, Colo., Oct. 22, 1955; s. John R. Jr. and Alice Ross. BSME, U. Md., 1980; postgrad., George Washington U. Registered Engr.-in-tng., Md. Mech. engr. Rosenblatt & Sons, Arlington, Va., 1981, Naval Surface Weapons Ctr., Silver Spring, Md., 1981-82, Esystems Melpar, Fairfax, Va., 1982-84, DSI, McLean, Va., 1984-86; staff engr. Ardak, Inc., McLean, 1986-88; sr. R&D engr. Loral/Ford Aerospace, Reston, Va., 1988-91; pres. Westwind Corp., Vienna, Va., 1991—. Mem. AIAA (sr.), IEEE Computer Soc., Assn. for Computing Machinery. Achievements include research on advancement of small satellite technology, weapon fuzing technology, parallel processor systems, solar energy. Home: 1609 Sereno Ct Vienna VA 22182

ROSS, JOSEPH COMER, physician, educator, academic administrator; b. Tompkinsville, Ky., June 16, 1927; s. Joseph M. and Annie (Pinckley) R.; m. Isabelle Nevins, June 15, 1952; children: Laura Ann, Sharon Lynn, Jennifer Jo, Mary Martha, Jefferson Arthur. BS, U. Ky., 1950; MD, Vanderbilt U., 1954. Diplomate Am. Bd. Internal Medicine (bd. govs. 1975-81), Am. Bd. Pulmonary Disease. Intern Vanderbilt U. Hosp., Nashville, 1954-55; resident Duke U. Hosp., Durham, N.C., 1955-57, rsch. fellow, 1957-58; instr. medicine Ind. U. Sch. Medicine, Indpls., 1958-60, asst. prof., 1960-62, assoc. prof., 1962-66, prof., 1966-70; prof., chmn. dept. medicine Med. U. of S.C., Charleston, 1970-80; vis. prof. Vanderbilt U. Sch. Medicine, Nashville, 1979-80, prof. medicine, 1981—, assoc. vice chancellor for health affairs, 1982—; mem. cardiovascular study sect. NIH, 1966-70, program project com., 1971-75; mem. adv. coun. Nat. Heart, Lung and Blood Inst., 1982-86; mem. ad hoc coms. NAS, 1966, 67; mem. Pres.'s Nat. Adv. Panel on Heart Disease, 1972; mem. merit rev. bd. in respiration VA Rsch. Svc., 1972-76, chmn., 1974-76. Mem. editorial bd. Jour. Lab. and Clin. Medicine, 1964-70, Chest, 1968-73, Jour. Applied Physiology,1968-73, Archives of Internal Medicine, 1976-82, Heart and Lung, 1977-86; contbr. articles to profl. jours. With U.S. Army, 1945-47. Fellow ACP, Am. Coll. Chest Physicians (gov. S.C. 1970-76, vice chmn. bd. govs. 1974-75, chmn. bd. govs. 1975-76, exec. council 1974-80, pres.-elect 1976-77, pres. 1978-79, chmn. sci. program com. 1973), Am. Coll. Cardiology; mem. AMA (sect. on med. schs.), Am. Fedn. Clin. Rsch. (chmn. Midwest sect.), Am. Physiol. Soc., Am. Soc. Clin. Investigation, Assn. Am. Physicians, Assn. Profs. Medicine, Cen. Soc. Clin. Rsch., S.C. Med. Soc., Am. Thoracic Soc. (nat. councillor 1972-76), So. Soc. Clin. Rsch., S.C. Lung Assn. (v.p. 1974-75), Am. Soc. Internal Medicine, Phi Beta Kappa, Alpha Omega Alpha. Mem. Ch. of Christ (elder). Office: Vanderbilt University D 3300 Medical Ctr N Nashville TN 37232

ROSS, LAWRENCE JOHN, federal agency administrator; b. N.Y.C., Jan. 17, 1942; s. William Harvey and Marion (Hayes) R.; m. Carol Marie Wood, Oct. 30, 1965; children: Catherine M., Sharon M., Patricia A., James L. BSEE, Manhattan Coll., Riverdale, N.Y., 1963. Elec. engr. NASA Lewis Rsch. Ctr., Cleve., 1963-78, dir. launch vehicles, 1978-80, dir. space flight systems, 1980-87, dep. dir., 1987-90, dir., 1990—. Trustee Cleve. Nat. Air Show, 1990; bd. dirs. Greater Cleve. Growth Assn., 1991; alumnus Leadership Cleve., 1988. Mem. Nat. Space Club (bd. govs. 1990), AIAA, Smithsonian Assocs. (lifetime). Roman Catholic. Avocations: computers, reading. Office: NASA Lewis Rsch Ctr 21000 Brookpark Rd Cleveland OH 44135*

ROSS, LESA MOORE, quality assurance engineer; b. New Orleans, Jan. 25, 1959; d. William Frank and Carolyn West Moore; m. Mark Neal Ross, Nov. 30, 1985; children: Sarah Ann, Jacquelyne Caroline. BS in Engring., U. N.C., Charlotte, 1981; MBA in Quality and Reliability Mgmt., U. North Tex., 1991. Seismic qualification engr. Duke Power Co., Charlotte, N.C., 1981-82; quality assurance engr. Tex. Instruments Inc., Lewisville, Tex., 1982-91; compliance mgr. Am. Med. Electronics, Inc., 1992—. Recipient Nat. Sci. Found. Rsch. Grant, U. N.C., Charlotte, 1980. Mem. Am. Soc. for Quality Control (cert. quality engr., quality auditor, reliability engr.), Zeta Tau Alpha (pres. 1984-85). Avocations: crafts, cross-stitching, reading, travel. Home: 4925 Wolf Creek Trail Lewisville TX 75028-1955

ROSS, MARY HARVEY, entomology educator, researcher; b. Albany, N.Y., Apr. 1, 1925; d. Roy Newman and Myrtle Adele (King) Harvey; m. Robert Donald Ross, Dec. 20, 1947 (dec. Nov. 1983); children: Mary Jane, Robert Douglas, Nancy Ross Angel. BA, Cornell U., 1946, MA, 1947, PhD, 1950. Biologist Oak Ridge (Tenn.) Nat. Lab., 1950-51; instr. Va. Poly. Inst. & State U., Blacksburg, 1959-70, asst. prof., 1970-74, assoc. prof., 1974-80, prof., 1980—. Contbr. over 100 articles to profl. jours., 4 chpts. to books. Grantee NSF, 1971-74, 80-83, Office of Naval Rsch., 1977-85, Whitmire Rsch. Labs., Inc., 1986-91. Fellow Royal Entomol. Soc. (U.K.); mem. Am. Genetic Assn. (pres. 1983), Entomol. Soc. Am., Genetic Soc. Am., Genetics Soc. Can., Entomol. Soc. Washington, Sigma Xi, Phi Beta Kappa, Gamma Sigma Delta. Achievements include establishment of a formal genetics for the German cockroach, Blattella germanica. Office: Entomology Dept Va Poly Inst & State U Blacksburg VA 24061

ROSS, MURIEL DOROTHY, research scientist; b. Grand Rapids, Mich., Jan. 22, 1927; d. Theophilus Joseph and Marie Rose (Bonk) Karp; m. Bernard Alfred Ross, Mar. 31, 1951; children: Mary Katherine, Carol Anne, Patricia Lynn, Sharon Marie. BA in Biology, Aquinas Coll., 1948; MS in Anatomy, U. Mich., 1950, PhD in Anatomy, 1953. Instr. biology Wayne State U., Detroit, 1951-53; instr. anatomy U. Mich., Ann Arbor, 1966-67, assoc. prof., 1971-79, prof., 1979-86; chief space biology NASA Ames Rsch. Ctr., Moffett Field, Calif., 1986-88; dir., sr. rsch. scientist NASA Biocomputation Ctr., Moffett Field, Calif., 1988—; vis. prof. UCLA, 1987—. Author: (with others) Auditory Physiology, 1981, rev. edit. 1987, Basic and Applied Aspects of Vestibular Function, 1988; contbr. articles to profl. jours. Troop leader Girl Scouts U.S., Dearborn, Mich., 1962-64; chmn. Acad. Women's Caucus, U. Mich., Ann Arbor, 1982-84. Recipient Disting. Svc. award Women's Acad. Caucus, 1986, Excellence in Rsch. award NASA, Washington, 1990. Mem. IEEE, Barany Soc. (Sweden, rsch. award 1987), Am. Soc. Gravitational and Space Biology (adv. group, bd. dirs. 1986-89). Achievements include contributions toward understanding inner ear balance organs by the development of computer-assisted method for producing 3-D images of their neural organization, demonstrating that they are organized for weighted parallel processing of information, producing computer simulations of their functioning, showing that their neural network is adaptable to environmental change. Office: NASA Ames Rsch Ctr MS 261-2 Moffett Field CA 94035

ROSS, PATTI JAYNE, obstetrics and gynecology educator; b. Nov. 17, 1946; d. James J. and Mary N. Ross; B.S., DePauw U., 1968; M.D., Tulane, U., 1972; m. Allan Robert Katz, May 23, 1976. Asst. prof. U. Tex. Med. Sch., Houston, 1976-82, assoc. prof., 1982—, dir. adolescent ob-gyn., 1976—, also dir. phys. diagnosis, dir. devel. dept. ob-gyn.; speaker in field. Bd. dirs. Am. Diabetes Assn., 1982—; mem. Rape Coun. Diplomate Am. Bd. Ob-Gyn. Mem. Tex. Med. Assn., Harris County Med. Soc. So. Perinatal Assn., Houston Ob-Gyn. Soc., Assn. Profs. Ob-Gyn., Soc. Adolescent Medicine, AAAS, Am. Women's Med. Assn., Orgn. Women in Sci., Sigma Xi. Roman Catholic. Clubs: River Oak Breakfast, Profl. Women Execs. Contbr. articles to profl. jours. Office: 6431 Fannin St Houston TX 77030

ROSS, REUBEN JAMES, JR., paleontologist; b. N.Y.C., July 1, 1918; married, 1942; 4 children. AB, Princeton U., 1940; MS, Yale U., 1946, PhD in Geology, 1948. Field asst. Newfoundland Geol. Survey, 1938; asst. prof. geology Wesleyan U., 1948-52; geologist paleontology and strategic br. U.S. Geol. Survey, 1952-80; chmn. subcom. Ordovician Stratigraphy Internat. Union Geol. Scis, 1976-82; adj. prof. geology Colo. Sch. Mines, 1980—. Recipient Raymond C. Moore medal for Paleontology, 1993. Mem. Paleontology Soc., Geol. Soc. Am., Am. Assn. Petroleum Geologists, Brit. Palaeontograph Soc., Soc. Econ. Paleontology and Mineral. Achievements include research in invertebrate paleontology and Ordovician stratigraphy of Basin Ranges. Home: 5255 Ridge Trail Littleton CO 80123*

ROSS, ROBERT NATHAN, medical writer, consultant; b. N.Y.C., Feb. 24, 1941; s. William and Edith (Newburgh) R.; married; 3 children. BA, Williams Coll., 1963; PhD, Cornell U., 1969. Lectr. U. Calif., Irvine, 1974-75; rsch. scientist sch. medicine Boston U., 1975-79, dir. grad. programs sci. communication, 1981-82; vis. scientist MIT, Cambridge, 1978-86; cons. Denver Mus. Natural History, 1988-91, Ctr. for Ednl. Devel. in Health and Kellogg Found., 1990—. Author: Care and Punishment, 1988, Clinical Psychobiology, 1982; contbr. articles to profl. jours. Mem. Phi Beta Kappa. Home and Office: 16 Windsor Rd Brookline MA 02146

ROSS, RUSSELL, pathologist, educator; b. St. Augustine, Fla., May 25, 1929; s. Samuel and Minnie (DuBoff) R.; m. Jean Long Teller, Feb. 22, 1956; children: Valerie Regina, Douglas Teller. A.B., Cornell U., 1951; D.D.S., Columbia U., 1955; Ph.D., U. Wash., 1962; DSc (hon.P, Med. Coll. of Pa., 1987. Intern Columbia-Presbyn. Med. Ctr., 1955-56, USPHS Hosp., Seattle, 1956-58; spl. research fellow pathology sch. medicine U. Wash., Seattle, 1958-62, asst. prof. pathology and oral biology sch. medicine and denistry, 1962-65, assoc. prof. pathology Sch. Medicine, 1965-69, prof. Sch. Medicine, 1969—, adj. prof. biochemistry Sch. Medicine, 1978—, assoc. dean for sci. affairs sch. medicine, 1971-78, chmn. dept. pathology sch. medicine, 1982—; vis. scientist Strangeways Research Lab., Cambridge, Eng.; mem. research com. Am. Heart Assn.; mem. adv. bd. Found. Cardiologique Princess Liliane, Brussels, Belgium; life fellow Clare Hall, Cambridge U.; mem. adv. council Nat. Heart, Lung and Blood Inst., NIH, 1978-81; vis. prof. Royal Soc. Medicine, U.K., 1987. Editorial bd. Procs. Exptl. Biology and Medicine, 1971-86, Jour. Cell Biology, 1972-74, Exptl. Cell Rsch., 1982-92, Jour. Exptl. Medicine, Growth Factors, Am. Jour. Pathology, Internat. Cell Biology Jour.; assoc. editor Arteriosclerosis, 1982-92, Jour. Cellular Physiology, Jour. Cellular Biochemistry; reviewing editorial bd. Sci. mag., 1987-90; contbr. articles to profl. jours. Trustee Seattle Symphony Orch. Recipient Birnberg Research award Columbia U., 1975, Nat. Rsch. Achievement award Am. Heart Assn., 1990; Gordon Wilson medal Am. Clin. and Climatol. Assn., 1981; named to Inst. Medicine, Nat. Acad. Scis.; Japan Soc. Promotion of Sci. fellow, 1985, Guggenheim fellow, 1966-67. Fellow AAAS, Am. Acad. Arts and Scis.; mem. Am. Soc. Cell Biology, Tissue Culture Assn., Am. Assn. Pathologists (Rous-Whipple award 1992), Internat. Soc. Cell Biology, Electron Microscope Soc. Am., Am. Heart Assn. (fellow Coun. on Arteriosclerosis, Nat. Rsch. Achievement award 1990), Royal Micros. Soc., Harvey Soc. (hon.), Am. Soc. Biochemistry and Molecular Biology, Royal Belgian Acad. Scis. (fgn. corr. mem.), Sigma Xi. Home: 4811 NE 42d St Seattle WA 98105 Office: U Wash Sch Medicine 1959 NE Pacific St Seattle WA 98195-0001

ROSSAVIK, IVAR KRISTIAN, obstetrician/gynecologist; b. Stavanger, Rogaland, Norway, Nov. 3, 1936; came to U.S., 1982; s. Andreas and Bergit (Berge) R.; divorced; children: Line, Anne Britt, Kirsten, Solveig; m. Claudia Lagos, May 23, 1987; children: Claudia Kristina, Eevar Benjamin. MD, U. Oslo, 1962, PhD, 1982. Pvt. practice, medicine Stavanger, 1974; asst. chief, acting chmn. U. Tromsoe, Norway, 1974-76; clin. fellow Nat. Hosp. of Norway, Oslo, 1976-81, Norwegian Radium Hosp./U. Oslo, 1981-82; pvt. practice Oslo, 1977-82; rsch. asst. prof. Baylor Coll. Medicine, Houston, 1983-86; assoc. prof. U. Okla., Oklahoma City, 1987-93, prof., 1993—; dir. Ultrasound Svcs., Dept. Ob/Gyn., U. Okla. Inventor Rossavik Growth Equation, 1980; author: (textbook) Practical Obstetrical Ultrasound: With and Without A Computer, 1991. Lt. Royal Norwegian Navy, 1964-65. Mem. AMA, Am. Fertility Soc., Am. Inst. Ultrasound in Medicine, Okla. State Med. Assn., Irish and Am. Paediatric Soc., Internat. Perinatal Doppler Soc., So. Med. Assn., AAAS. Lutheran. Avocations: ultrasonography technology, computer technology, fetal growth studies. Office: Univ Oklahoma Dept Ob/Gyn PO Box 26901 Oklahoma City OK 73190-0001

ROSSBACH, PHILIP EDWARD, civil engineer; b. Omaha, Nebr., Oct. 6, 1959; s. Joseph James and Mary Carolyn (Clauser) R.; m. Therese Ann La Croix, July 31, 1981; children: Diane, Dan, Lauren. BS, U. Nebr., 1981. Registered profl. engr. Nebr., Kans.; registered structural engr. Mass. Structural engr. Gibbs-Hill, Inc., Omaha, 1981-82, Henningson, Durham and Richardson, Omaha, 1982-83; bridge engr. HDR Engring., Inc., Omaha, 1983-88, sect. mgr., bridges, 1988—; ASCE rep. Roundtable Profl. Engring. Socs., Omaha, 1991-93. Judge Met. Engring. and Sci. Fair, Omaha, 1987-93. named Young Engr. of Yr. Nebr. Soc. Profl. Engr., 1990. Mem. Am. Soc. Civil Engrs. (pres. Nebr. sect. 1991, v.p. Nebr. sect. 1989, dir. Nebrs. sect. 1988, treas. Nebr. sect. 1987), Nebr. Chpt. of Am. Concrete Inst., Nat. Soc. Profl. Engrs. Office: HDR Engring Inc 8404 Indian Hills Dr Omaha NE 68114

ROSSI, GUIDO A(NTONIO), mathematics educator, researcher; b. Moretta, Cuneo, Italy, Jan. 17, 1944; s. Giulio Cesare and Anna Maria (Ferraris di Celle) R.; m. Maria Emilia Zucchi, Mar. 27, 1978. Dr. in Math., Universita di Torino, Italy, 1967. Asst. Universitá di Torino, 1969-82, asst. prof. Facoltá di Economia e Commercio, 1969-82, assoc. prof. math., 1982-86, Prof. math. for social scis. and econs., 1986; full prof. 1986—, dir. Istituto di Matematica Finanziaria, 1974-81, 83-85, 1992—, nat. coord. rsch. projects, 1986—; prof. Scuola di Applicazione Italian Army, 1992—; mem. sci. bd. 3d A.F.I.R. Colloquium, 1993. Contbr. articles to profl. and sci. jours. Served to lt. Italian Army, 1967-68. Decorated Cavaliere dell'Ordine al Merito Civile di Savoia, 1990. Mem. Unione Matematica Italiana, Associazione per la Matematica Applicata alle Scienze Economiche Sociali (adminstr. 1990—), Associazione Museo Ferroviario Piemontese (pres. 1986—), Am. Mathematical Soc. Roman Catholic. Club: I Neoteri (pres. 1990-92). Achievements include contributions to the foundations of probability, decision, theory and financial decisions. Office: Istituto di Matematica Finanziaria, Piassa Arbarello 8, I-10122 Torino Italy

ROSSI, JOSEPH STEPHEN, research psychologist; b. Providence, Feb. 22, 1951; s. Joseph B. and Nicolina M. (Calise) R.; m. Susan R. Finkle, June 13, 1981. BA, R.I. Coll., 1975; PhD, U. R.I., 1984. Instr. U. R.I., Kingston, 1978-81, R.I. Coll., Providence, 1978-82; postdoctoral fellow U. R.I., Kingston, 1984-85, asst. prof., 1985-90, assoc. prof., 1990—; dir. rsch. Cancer Prevention Rsch. Ctr., Kingston, 1992—; cons. NIH, Bethesda, Md., 1989—, CDC, Atlanta, 1991—; lectr. various sci. confs. Reviewer numerous sci. jours. and profl. socs.; contbr. over 100 sci. articles to profl. jours. Recipient Faculty Merit awards U. R.I., 1987-89; sci. rsch. grantee Nat. Cancer Inst., 1985—, CDC, 1990-92, Sigma Xi, 1980-81. Mem. APA, AAAS, Am. Psychol. Soc., Psychometric Soc., Soc. Behavioral Med. Achievements include research on new ways to help prevent cancer through behavior change in the areas of smoking cessation, sun exposure and sunscreen use, weight control, diet, alcohol and cocaine use, radon gas exposure, HIV risk reduction, exercise adoption. Office: Univ of RI Cancer Prevention Rsch Ctr Kingston RI 02881

ROSSI, MIRIAM, chemistry educator, researcher; b. Asti, Italy, Mar. 8, 1952; came to U.S., 1956; d. Antonio and Aldegonda (Zanni) R. BA, Hunter Coll., 1974; PhD, Johns Hopkins U., 1979. Postdoctoral fellow Fox Chase Cancer Ctr., Phila., 1979-82; asst. prof. chemistry Vassar Coll., Poughkeepsie, N.Y., 1982-89, assoc. prof., 1989—, chair dept. chemistry, 1990—. Editor: Patterson and Pattersons, 1988. Office: Vassar Coll Box 484 Poughkeepsie NY 12601

ROSSIGNOL, ROGER JOHN, coatings company executive; b. N.Y.C., Jan. 19, 1941; s. Willard and Claire Rossignol; m. Gloria Jean Rossignol, Sept. 19, 1981; children: Vincent, Michael, David, Tricia. Salesman Bates Fabrics Inc., N.Y.C., 1964-75, Bates Mfg. Co., Lewiston, Maine, 1964-75; in mktg. Ethan Allen Co., Danbury, Conn., 1975-77; v.p. sales Internat. Flooring, Long Branch, N.J., 1977-80; sales mgr. Asbestos Corp., Long Branch, 1980-82; pres. Encapco Corp., Point Pleasant, N.J., 1982-84, Internat. Protective Coatings Corp., Ocean Twp., N.J., 1984—. Office: Internat Protective Coating 725 Carole Ave Oakhurst NJ 07755-1202

ROSSING, THOMAS D., physics educator; b. Madison, S.D., Mar. 27, 1929; s. Torstein H. and Luella E. Rossing; children: Karen, Barbara, Erik, Jane, Mary. BA, Luther Coll., 1950; MS, Iowa State U., 1952, PhD, 1954. Rsch. physicist Univac div. Sperry Rand, 1954-57; prof. physics St. Olaf Coll., 1957-71, chmn. physics dept., 1963-69; prof. physics No. Ill. U., DeKalb, 1971—, Disting. Rsch. prof., chmn. dept., 1971-73; rschr. Microwave Lab., Stanford (Calif.) U., 1961-62, Lincoln Lab., MIT, Cambridge, Mass., summer 1963, Clarendon Lab., Oxford (Eng.) U., 1966-67; rsch. assoc. Argonne (Ill.) Nat Lab., 1974-76, scientist-in-residence, 1990—; vis. lectr. U. New Eng., Armidale, Australia, 1980-81; vis. exch. scholar People's Republic of China, 1988; guest rschr. Physikalische-Technische Bundesanstalt, Braunschweig, Fed. Rep. Germany, 1988-89. Author 10 books in field; contbr. more than 200 articles to profl. publs. Fellow Acoustical Soc. Am. (Silver medal in mus. acoustics 1992); mem. Am. Phys. Soc., Am. Assn. Physics Tchrs. (pres. 1991), Catgut Acoustical Soc., IEEE Percussive Arts Soc., Sigma Xi (nat. lectr. 1984-87), Sigma Pi Sigma. Achievements include research in musical acoustics, psychoacoustics, speech and singing, vibration analysis, magnetic levitation, environmental noise conrol, surface effects in fusion reactors, spin waves in metals, physics education. Office: No Ill U Physics Dept De Kalb IL 60115

ROSSKOTHEN, HEINZ DIETER, engineer; b. Duisburg, Fed. Republic of Germany, Sept. 19, 1936; came to U.S., 1959; naturalized, 1965; s. Bernhard and Hedwig Rosskothen; m. Ilse Meyer, Oct. 26, 1963; children: Norman, Susan. Diploma tool and die maker, Bloomfield (N.J.) Tech., 1964. Elec. engr. Niederheinische Hutte, Duisburg, 1957-59; instrument designer Buders Tool Co., Caldwell, N.J., 1959-60; mgr. Monach Tool & Mfg., Coldwell, N.J., 1960-64; engr. Columbia U., N.Y.C., 1964-72, sr. st. assoc., 1972—, dir. instrumentation, 1974—; cons. Kreske Eye Inst., Detroit, 1983—. Elder Presbyn. Ch., 1991—. Mem. SME (sr.), MTA/SME (charter mem.). Achievements include development of wide field specular microscope; design of keratoprostheses (artificial cornea), special infrared pupil camera for mass screening clinical use; development of first Argon Laser Photocoagulation for eye use; development of technology for cornea tissue transplantation, of special antennae and phantoms for MRI; measurement and micropositioning system for cellular vibration and motility in the organ of corti; detection chambers for tissue fluid transport. Home: 293 Carlton Ter Teaneck NJ 07666-3403 Office: Columbia U 635 W 165th St New York NY 10032-3701

ROSSMANN, CHARLES BORIS, obstetrician/gynecologist; b. Brno, Moravia, Czechoslovakia, Nov. 19, 1945; came to U.S., 1988; s. Milos and Vlasta Boudna (Cernochova) Lota; m. Tatiana Elenka Hajossy, Oct. 19, 1979; children: Nathalie Nissa Cora, Nadine Nicole. MD, Purkyne U., Brno, Czechoslovakia, 1969. Bd. cert. ob-gyn., ACLS. Ob-gyn. specialist Skalica Gen. Hosp., Czechoslovakia, 1973-74; gen. practice Fed. Republic Germany, 1974-80; med. officer Fogo Island Hosp., Can., 1983-84, Baragwanath Hosp., Johannesburg, South Africa, 1984-85, Edendale Hosp., Pietermaritzburg, South Africa, 1985-86; ob-gyn. specialist Western Mem. Regional Hosp., Corner Brook, Can., 1986-88; ob-gyn. gen. practice Deuel County Mem. Hosp., Clear Lake, S.D., 1988-89; ob-gyn. specialist Bullock County Hosp., Union Springs, Ala., 1989, St. Joseph's Hosp., Huntingburg, Ind., 1989—; intern, resident Valtice Gen. Hosp., Czechoslovakia, 1969-73; resident pathology, Sunnybrook and Toronto Gen. Hosp., 1981-82, resident ob-gyn., Grace and St. Clare Gen. Hosp., St. John's, Can., 1982-83. Contbr. articles to profl. jours.; inventor in field. Mem. Ind. Med. Assn., Am. Assn. Gynecol. Laparoscopists, Am. Coll. Internat. Physicians (Ky. chpt.). Avocations: rsch., computers, reading. Office: Med Arts Plz Huntingburg IN 47542

ROSSMILLER, GEORGE EDDIE, agricultural economist; b. Gt. Falls, Mont., June 8, 1935; s. Albert E. and Romaine (Hennford) R.; m. Betty Ann Rinio, Dec. 20, 1955 (dec. Mar. 1990); children: David W., Diane J.; m. Norma Lee Adams, Apr. 18, 1992. B.S., Mont. State U., 1956, M.S., 1962; Ph.D., Mich. State U., 1965. Rsch. assoc. Mich. State U., East Lansing, 1965-66, asst. prof., 1967-71, assoc. prof., 1972-76, prof. agrl. econs., 1977-80; agrl. attache to OECD, Fgn. Agrl. Service, USDA, Paris, 1978-79; asst. adminstr. internat. trade policy Fgn. Agrl. Svc., USDA, Washington, 1979-81, dir. planning and analysis, 1981-85; sr. fellow and dir. Nat. Ctr. Food and Agr. Policy, Resources for the Future, 1986-92; also exec. dir. Internat. Policy Council on Agr. and Trade, 1988-92; chief situation and policy studies svc. Food and Agr. Orgn. of UN, Rome, 1992—. Author: The Grain-Livestock Economy of West Germany with Projections to 1970 and 1975, 1968, (with others) Korean Agricultural Sector Analysis and Recommended Development Strategies, 1971-1985, 1972; editor: (with others) Agricultural Sector Planning: A General System Simulation Approach, 1978. Served with U.S. Army, 1956-59. Recipient service citation Korean Ministry of Agrl. and Fisheries, 1973, service citation Office of Prime Minister of Korea, 1977, Superior Service award U.S. Dept. Agr., 1983, Fgn. Agrl. Service merit award, 1984. Mem. Am. Agrl. Econs. Assn. (Disting. Policy Contbn. award 1992), Internat. Assn. Agrl. Economists. Presbyterian. Home: Via Illiria 18 Apt B-10, 00183 Rome Italy Office: Food and Agrl Orgn UN, viale delle Terme di Caracalla, 00100 Rome Italy

ROSSOTTI, CHARLES OSSOLA, computer software company executive; b. N.Y.C., Jan. 17, 1941; s. Charles C. and V. Elizabeth (Ossola) R.; m. Barbara Jill Margulies, June 9, 1963; children: Allegra Jill, Edward Charles. AB magna cum laude, Georgetown U., 1962; MBA with high distinction, Harvard U., 1974. Mgmt. cons. Boston Cons. Group, 1964-65; prin. dep. asst. sec. Office of Systems Analysis, Dept. Def., Washington, 1965-70; prin. dep. asst. sec. of Def. Office of Systems Analysis, Dept. Def., 1969-70; pres. Am. Mgmt. Systems, Inc., Arlington, Va., 1970—, chief exec. officer, 1981—, chmn. bd., 1989—; bd. dirs. Index Tech., Inc., Sovran Bank, Caterair Internat., Nations Bank of Va. Bd. dirs. Georgetown U., 1969-77, 92—; trustee Nat. Cathedral Sch., Washington, 1987—, Woodstock Theol. Ctr., 1990—; chmn. Corp. Against Drug Abuse, 1993—. Mem. Coun. Fgn. Rels. Office: Am Mgmt Systems Inc 1777 N Kent St Arlington VA 22209-2110

ROSTOHAR, RAYMOND, chemist, chromatographer; b. Ilion, N.Y., Mar. 29, 1961; s. Brandt and Marlene Mildred (Kubecka) R. BS in Biology, U. Ala., Huntsville, 1989. Asst. chemist J&A Enterprises, Huntsville, 1984-85; chemist Tech. Micronics Control Inc., Huntsville, 1986-88; chromatography supr. South Eastern Analytical Svcs., Inc., Huntsville, 1988-92; project scientist Roy F. Weston, Inc., Auburn, Ala., 1992—. Office: Roy F Weston Inc Svcs Inc Ste 1 Auburn AL 36830

ROSTOKER, GORDON, physicist, educator; b. Toronto, Ont., Can., July 15, 1940; s. Louis and Fanny (Silbert) R.; m. Gillian Patricia Farr, June 29, 1966; children: Gary David, Susan Birgitta, Daniel Mark. BSc in Physics, U. Toronto, 1962, MA in Physics, 1963; PhD in Geophysics, U. B.C., Can., 1966. Postdoctoral fellow Royal Inst. Tech., Stockholm, 1966-68; asst. prof. physics U. Alta., B.C., 1968-73, assoc. prof., 1973-79, prof., 1979—, McCalla Rsch. Prof., 1983-84, ann. Killam Prof., 1991-92, dir. Inst. Earth and Planetary Physics, 1985-91; assoc. chmn. dept. physics U. Alta., 1976-79, univ. rep. to bd. dirs. Can. Network for Space Rsch., 1992—, mem. univ. rsch. policy com., 1987-91; cons. TRW Systems Group, 1973, Dome Petroleum Ltd., 1981, U. Western Ont., 1983, York U., 1986; contract researcher Energy, Mines and Resources, Can.; mem. assoc. com. space rsch. Nat. Rsch. Coun. Can., 1975-80, mem. com. on internat. sci. exchanges, 1977-79, others; mem. physics and astronomy com. Natural Scis. and Engring. Rsch. Coun., 1979-82, mem. spl. ad hoc com. on physics and astronomy, 1987-91, mem. grant selection com. for sci. publs., 1988-92; prin. investigator CANOPUS, 1989—; chmn. div. III Internat. Assn. Geomagnetism and Aeronomy, 1979-83; chmn. working group on data analysis phase of internat. magnetospheric study Sci. Com. on Solar Terrestrial Physics of Internat. Coun. Sci. Unions, 1980-86, co-chmn. steering com. for solar-terrestrial energy program, 1987-89, chmn., 1989—;. Editor Can. Jour. Physics, 1980-86, mem. editorial adv. bd., 1986—; contbr. over 250 articles to profl. publs. Mem. pub. adv. com. Govtl. Environ. Conservation Authority of Province of Alta., Edmonton, 1973-74. Recipient Steacie prize EWR Steacie Meml. Fund, 1979, Geophys. Centenary medal Acad. Scis. USSR, 1984. Mem. Am. Geophys. Union (assoc. editor Jour. Geophys. Rsch. 1976-79, 92—, assoc. editor Jour. Geomag. Geoelec. 1993—), Can. Assn. Physicists (sec.-treas. Can. Geophys. Union 1973-74, chmn. divsn. aeronomy and space phsyics 1977-78, chmn. publs. com. 1980-86). Achievements include use of ground magnetometer arrays to discover stepwise evolution of electric current systems which flow in the ionosphere and magnetosphere during episodes of strong auroral disturbance. Office: U Alta, Dept Physics, Edmonton, AB Canada T6G 2J1

ROTENBERG, DON HARRIS, chemist; b. Portland, Oreg., Mar. 31, 1934; s. Morris Hyman and Helen (Harris) R.; m. Barbara Ress, June 29, 1958; children: Laura, Debra. BA, U. Oreg., 1955; AM, Harvard U., 1956; PhD, Cornell U., 1960. Rsch. chemist Enjay Chem. Lab. Exxon Rsch. & Engring. Co., Linden, N.J., 1960-67, sr. rsch. chemist Enjay Polymer Lab., 1967-71; mgr. polymer sci. and engring. Am. Optical Corp., Southbridge, Mass., 1971-75, dir. materials and process lab., 1975-80, v.p. R&D, 1980-85; v.p., gen. mgr. precision products bus. Am. Optical Corp., Southridge, Mass., 1985-88; tech. dir. Coburn Optical Industries, Tulsa, 1988-92; cons. Plastics Tech. Assocs., Tulsa, 1992—. Contbr. articles to Advances in Chem. Series, Jour. Macromolecular Sci.-Chem. Todd Rsch. fellow Cornell U., Ithaca, N.Y., 1956-59. Mem. AAAS, Am. Chem. Soc. (plastics, polymer and rubber divs., contbr. to jour.), Soc. Plastics Engrs., Phi Beta Kappa, Sigma Xi. Achievements include patents in field; development of first mass-produced coated polycarbonate safety lenses, of first factory-produced coated plastic prescription lenses, of first photochromic plastic prescription lenses. Home: 4507 E 108th St Tulsa OK 74137-6850 Office: Plastics Tech Assocs. Ste 207 8086 S Yale Tulsa OK 74136-9060

ROTENBERG, MANUEL, physics educator; b. Toronto, Ont., Can., Mar. 12, 1930; came to U.S., 1946; s. Peter and Rose (Plonzker) R.; m. Paula Weissbrod, June 23, 1952; children: Joel, Victor. BS, MIT, 1952, PhD, 1956. Mem. staff Los Alamos (N.Mex.) Nat. Lab., 1955-58; instr. physics Princeton (N.J.) U., 1958-59; asst. prof. U. Chgo., 1959-61; prof. applied physics U. Calif., San Diego, 1961—, dean grad. studies and research, 1975-84, chair dept. elect. engring. and computer engring., 1988-93. Author: The 3-j and 6-j Symbols, 1959; founding editor: Methods of Computational Physics, 1963, Jour. of Computational Physics, 1962; editor: Biomathematics and Cell Kinetics, 1981. Fellow Am. Phys. Soc.; mem. AAAS, Sigma Xi. Office: U Calif San Diego La Jolla CA 92093-0407

ROTERT, KELLY EUGENE, engineer, consultant; b. Des Moines, Jan. 21, 1962; s. Leroy John and Joan (Johnson) R.; m. Pamela Lynn Ford, June 8, 1985; 1 child, Daniel Thomas. BSCE, Iowa State U., 1986. Registered profl. engr. Iowa. Engr., then mgr. constrn. material dept. Patzig Testing Labs., Inc., Des Moines, 1985-91, pres., 1991—. Mem. NSPE, Iowa Engring. Soc., Cyclone Corvette Club. Republican. Roman Catholic. Home: 6671 NW 48th St Johnston IA 50131 Office: Patzig Testing Labs Co Inc 3922 Delaware Ave Des Moines IA 50313

ROTH, ANNEMARIE, conservationalist; came to U.S., 1957.; Pediatrician Calif., 1957-63; rschr. Ctr. de la Nature du Mont Saint-Hillaire, Que., 1970—, Centre de Recherches, MacDonald, Que., Owl Rsch. and Rehab. Found., Ont., Vt. Inst. Nat. Scis.; founder Ctr. de Réhabilitation des Oiseaux Blessés, Montérégie. Recipient Snowy Owl Conservation award Québec Zoological Gardens, 1991. Mem. Nat. Wildlife Rehabilitation Assn., INternat. Wildlife Rehabilitation Coun., Soc. Preservation Birds. Office: Ctr rehabil osieaux Monteregie, 1357 Ozias Leduc, Otterburn Park, PQ Canada J3K 6B6*

ROTH, JAMES, engineering company executive; b. 1936. With Goodyear Aerospace, 1959-69, KMS Tech. Ctr., 1969-74; with Gen. Rsch. Corp., Vienna, Va., 1974—, pres., CEO, 1992—; pres., CEO GRC Internat., Vienna, Va., 1992—. Office: GRC International Inc General Research Corporation 1900 Gallows Rd Vienna VA 22182*

ROTH, KARL SEBASTIAN, pediatrician; b. N.Y.C., Mar. 3, 1941; s. Victor and Ruth Leila (Fisher) R.; m. Beverly Rochelle, Apr. 16, 1967 (div. July 1984); children: Christopher Geordie, Marcus Amadeus; m. Lydia Carole Noland, July 28, 1984; 1 child, Alexander Kristof Parham. BA, U. Rochester, N.Y., 1963; MD, Wake Forest U., 1969. Pediatric intern Med. Coll. Pa., Phila., 1969-70; pediatric resident Thomas Jefferson U., Phila., 1970-72; fellow genetics/metabolism U. Pa., Phila., 1972-75, asst. prof. pediatrics, 1975-82; prof. pediatrics, biochemistry and molecular biophysics Med. Coll. Va., Richmond, 1982—; cons. state newborn screening program Commonwealth Va., Richmond, 1982—; mem. gov.'s genetics adv. bd. Commonwealth Va., Richmond, 1982—; mem. bd. Clin. Pediatrics, Cleve., 1988—; vis. prof. pediatrics U. Zurich, Switzerland, 1993. Author: Metabolic Disease: A Guide to Early Recognition, 1983. Bd. mem. Agoraphobics Bldg. Ind. Lives, Richmond, 1987—, Gov.'s Coun. on Child Mental Health, Richmond, 1989—; participant Children's Miracle Network Telethon, Richmond, 1989—. Recipient Rsch. Career Devel. award NIH, 1978-83, Rsch. Grant, 1985-90; Daland fellow Am. Philos. Soc., 1976-78. Mem. Am. Pediatric Soc., Soc. Pediatric Rsch., Fedn. Am. Socs. Exptl. Biology, Soc. Inherited Metabolic Disorders, Sigma Xi. Lutheran. Achievements include devel. of prenatal therapy of an inborn error of metabolism for the second disease in med. history (Holocarboxylase Synthase Deficiency); establishment and characterization (biochemically and physiologically) the first animal model for the human renal faconi syndrome based upon use of a physiological produced compound; discovered and characterized a unique inborn error of metabolism. Office: Med Coll Va Box 239 MCV Sta Richmond VA 23298

ROTH, PETER HANS, chemical engineer; b. Vienna, June 16, 1935; came to the U.S., 1947; s. Joseph and Elizabeth (Tomor) R.; m. Joanne Hull, June 30, 1962; children: Patricia Elizabeth, Laura Anne, Douglas John. BSChE, CCNY, 1958; MS in Chemistry and Physics, Northeastern U., 1966. Chem. engr. Hooker Electrochem. Co., Niagara Falls, N.Y., 1958-59; emulsion scientist Polaroid Corp., Cambridge, Mass., 1959-67, mgr. film lab., 1967-70, mgr. dye lab., 1970-75; project mgr. Polaroid Corp., Waltham, Mass., 1975-82, sr. project mgr., 1982-90, sr. rsch. fellow, 1990—; vis. scientist Whitehead Inst., Cambridge, 1989-91; cons. expert image preservation Internat. Standards Orgn., Geneva, 1991—. Contbr. articles to profl. jours. Commr. Needham (Mass.) Conservation Commn., 1985-93; biosafety officer Needham0Damon Biotech. Corp., 1987-93; mem. land use commn. Needham Planning Bd., 1989-92. 1st lt. U.S. Army Corps Engrs., 1958-64. Mem. Image Sci. and Tech. Assn., Needham Photo Club. Republican. Achievements include 14 U.S. patents and 6 foreign patents on photo systems, film base, new photo products, graphic arts color proofing system and adhesion polymers, very high sensitivity energy recording systems. Home: 93 Garden St Needham MA 02192 Office: Polaroid Corp 1265 Main St Waltham MA 02254

ROTH, STEPHEN, biochemist, educator; b. Bklyn., Sept. 3, 1942; s. Irving Maurice and Beatrice (Wior) R.; m. Hallam Hurt, July 14, 1981; children: Adam, Rachel, Hallam. BA, Johns Hopkins U., 1964; PhD, Case Western Res., 1968; MA, U. Pa., 1981. Asst. prof. biology Johns Hopkins U., Balt., 1970-74; assoc. prof. biology Johns Hopkins U., 1974-80; prof. biology U. Pa., Phila., 1980-92, chmn. biol. dept., 1982-87, prof.-on-leave, 1992—; chief sci. officer Neose Pharms., Inc., Horsham, PA, 1992—; mem. editorial bd. Quar. Rev. Biology, Jour. Molecular Recognition. Contbr. more than 75 articles on carbohydrate chemistry and embryology to profl. jours. Recipient Rsch. Career Devel. award NIH, 1977-82, numerous grants, 1970—, Rsch. award Toyko (Japan) Med. Soc., 1988. Mem. AAAS, Soc. for Devel. Biology, Am. Soc. for Cell Biology, Soc. for Complex Carbohydrates. Achievements include first measurement of intercellular recognition, first demonstration of cell surface glycosyltransferases and retino-tectal recognition in embryonic cells; patent for saccharide compositions, methods and apparatus for their synthesis. Home: 1105 Rose Glen Rd Gladwyne PA 19035 Office: Neose Pharms Inc 102 Witmer Rd Horsham PA 19044

ROTH, THOMAS, psychiatry educator; b. Czechoslovakia, July 17, 1942; U.S. citizen; BA, CUNY, 1965; postgrad., Howard U., 1965-67; MA, U. Cin., 1969, PhD, 1970. Teaching asst. introductory and exptl. psychology, rsch. asst. dept. pharmacology Coll. Medicine Howard U., 1965-67; rsch. assoc. dept. community planning U. Cin., 1967-68, teaching asst. exptl. psychology, 1967-69; rsch. asst. Sleep and Dream Lab. dept. psychiatry VA Hosp., Cin., 1969-70; rsch. assoc. dept. psychiatry Coll. Medicine U. Cin., 1970-72, asst. prof. psychology dept. psychiatry Coll. Medicine, 1972-76; co-dir. Sleep Rsch. Lab. VA Hosp., Cin., 1972-78; assoc. prof. psychology Xavier U., Cin., 1976-78; assoc. prof. psychology dept. psychiatry Coll. Medicine U. Cin., 1976-78; co-dir. Sleep Disorders Ctr. Cin. Gen. Hosp., 1977-78; clin. prof. dept. psychiatry Sch. Medicine U. Mich., Ann Arbor, 1979—; divsn. head dept. psychiatry Henry Ford Hosp. Sleep Disorders and Rsch. Ctr., Detroit, 1978—; adj. assoc. prof. psychology U. Cin., 1976-78; mem. rsch. com. dept. psychiatry Coll. Medicine U. Cin., 1972-77, faculty com. on human rsch., 1975-77; chmn. rsch. com. Xavier U., 1972-77, rsch. com. dept. psychiatry Henry Ford Hosp., Detroit, 1978-88; mem. rsch. and edn. com. VA Hosp., Cin., 1976-78; mem. residency tng. com. dept. psychiatry Henry Ford Hosp., 1979-82, gen. clin. rsch. ctr. com., 1979-82, exec. com. dept. psychiatry, 1978—, rsch. com., 1985-89; bd. dirs. Assn. Profl. Sleep Socs., 1986-92, chmn. sci. program com., 1986—; mem. governing bd. World Fedn. Sleep Rsch. Socs., 1990—; pres. Nat. Sleep Found., 1990—; mem. joint coordinating coun. Project Sleep, U.S. Dept. Health, Edn. and Welfare, 1979-81, physicians syllabus subcom., 1979-81; active sleep disorders tng. com. Am. Thoracic Soc./Am. Sleep Disorders Assn., 1990—; bd. dirs. Am. Sleep Apnea Assn., 1990—; chmn. fellowship trainee grant com. World Fedn. Sleep Rsch. Socs., 1990; mem. editorial bd. Advances in Therapy, Sleep, Sleep Reviews, Stress Medicine, Human Psychopharmacology; chmn. hypnotics task force World Psychiatric Assn., 1992, spl. study sect. NIH, 1985, 86, 88, 89; reviewer various jours; cons. in field. Contbr. 174 articles to profl. jours, 59 chpts. to books. Mem. ethical review com. State of Ohio, 1976-78. Recipient Rush Bronze award Am. Psychiatric Assn., 1977. Mem. AAAS, Am. Sleep Disorders Assn. (exec. sec. treas. 1976-78, chmn. polysomnography accreditation com. 1976-78, pres. 1987-88, 88-89, exec. com. 1984-89, edn. com. 1976-81, 88-92, chmn. midwest region accreditation com. 1983-85, fin. com. 1985-88, chmn. devel. com. 1982-85, diagnostic classification steering com. 1987-90, grant review com. 1986-92, Nathaniel Kleitman award 1990), Am. Psychol. Assn., Can. Sleep Soc., N.Y. Acad. Scis., Sleep Rsch. Soc. (exec. com. 1975-77, 86—, pres. 1978-81, chmn. psychopharmacology com. 1976—), Psychonomic Soc., Clin. Sleep Soc. Office: Henry Ford Hosp Sleep Disorders Ctr 2921 W Grand Blvd Detroit MI 48202-2689*

ROTH, THOMAS J., physicist; b. Berwyn, Ill., Apr. 2, 1955; s. Raymond Edward and Elizabeth Ann (Robbins) R. BS, U. Ill., 1977, MS, 1981, PhD, 1986. Mem. tech. staff TRW ElectroOptics Rsch. Ctr., El Segundo, Calif., 1985-88; scientist TRW Group Rsch. Staff, Redondo Beach, Calif., 1988-90, TRW Rsch. Ctr., Redondo Beach, 1990—. Author: GaInAsP Alloy Semiconductors, 1982; contbr. to profl. publs. Mem. IEEE, Am. Physics Soc., Ave. A Athletic Assn. (sgt. at arms 1989-90), So. Calif. Crystal Growers (v.p., program chair 1988-92, pres. 1992—), Sigma Xi. Achievements include patents for phase-locked array of semiconductor lasers using closely-spaced antiguides, vertical-cavity, surface-emitting diode laser. Office: TRW D1/2519 1 Space Park Redondo Beach CA 90278

ROTHENBERG, ALBERT, psychiatrist, educator; b. N.Y.C., June 2, 1930; s. Gabriel and Rose (Goldberg) R.; m. Julia C. Johnson, June 28, 1970; children: Michael, Mora, Rina. A.B., Harvard U., 1952; M.D., Tufts U., 1956. Diplomate: Am. Bd. Psychiatry and Neurology. Intern Pa. Hosp., Phila., 1956-57; resident in psychiatry Yale U., West Haven (Conn.) VA Hosp., 1957-58, Grace-New Haven Hosp., 1958-59; resident in psychiatry Yale Psychiat. Inst., New Haven, 1959-60, chief resident, 1960-61; practice medicine specializing in psychiatry New Haven, 1960-61, 1963-75; chief neuropsychiatry Rodriguez U.S. Army Hosp., San Juan, P.R., 1961-63; practice medicine specializing in psychiatry Farmington, Conn., 1975-79, Stockbridge, Mass., 1979—, Canaan, N.Y., 1992—; dir. rsch. Austen Riggs Center, Stockbridge, Mass., 1979—; asst. dir. Yale Psychiat. Inst., 1963-64, sr. staff mem., 1964-83; mem. staff Yale-New Haven Med. Ctr., West Haven VA Hosp., U. Conn. Health Ctr., Farmington; cons., mem. editorial bd. various jours. in psychiatry and psychology; instr. dept. psychiatry Yale U. Sch. Medicine, 1960-61, 63-64, asst. prof., 1964-68, assoc. prof., 1968-71, clin. prof., 1974-84; prof. psychiatry U. Conn. Sch. Medicine, Farmington, 1975-79, dir. residency tng., 1976-78, dir. clin. svcs., 1975-78; prin. investigator Studies in the Creative Process, 1964—; vis. prof. Pa. State U., 1971, adj. prof., 1971-78; vis. prof. dept. Am. studies Yale U., 1974-76; lectr. dept. psychiatry Harvard U. Med. Sch., 1982-86, clin. prof., 1986—; researcher in psychotherapy. Author: (with B. Greenberg) Index of Scientific Writings on Creativity: Creative Men and Women, 1974, Index of Scientific Writings on Creativity: General 1566-1974, 1976; (with C.R. Hausman) The Creativity Question, 1976; The Emerging Goddess: The Creative Process in Art, Science and Other Fields, 1979; The Creative Process of Psychotherapy, 1988; Adolescence: Psychopathology, Normality, and Creativity, 1990; Creativity and Madness: New Findings and Old Stereotypes, 1990; contbr. numerous articles on the creative process, schizophrenia, anorexia nervosa, and psychotherapy to profl. and popular jours. Researcher on creativity in the arts, sci. and tech. Served with M.C. U.S. Army, 1961-63. Recipient Tufts Med. Alumni award 1956, Rsch. Scientist Career Devel. award NIMH 1964, 69, Golestan Found. award 1991, 92; Guggenheim Meml. fellow 1974-75, Ctr. Adv. Study in Behavioral Studies fellow 1986-87, Netherlands Inst. for Adv. Study in Humanities and Social Scis. fellow, 1992-93. Fellow Am. Psychiat. Assn. (life), Royal Soc. Health, Am. Coll. Psychoanalysts; mem. AAAS, Mass. Psychiat. Soc., Am. Soc. Aesthetics, N.Y. Psychiat. Soc., Sigma Xi. Democrat. Home: 4 Pine Ridge Rd Canaan NY 12029 Office: Austen Riggs Ctr Stockbridge MA 01262

ROTHENBERG, ELLEN, biologist; b. Northhampton, Mass., Apr. 22, 1952; d. Jerome and Winifred (Barr) R. AB summa cum laude, Harvard U., 1972; PhD, MIT, 1977. Postdoctoral fellow Meml. Sloan-Kettering Cancer Ctr., N.Y.C., 1977-79; asst. rsch. prof. The Salk Inst. for Biol. Studies, LaJolla, Calif., 1979-82; asst. prof. Calif. Inst. of Tech., Pasadena, 1982-88, assoc. prof., 1988—; Co-editor: Mechanisms of Lymphocyte Activation and Immune Regulation III, 1991; contbr. articles to profl. jours. Assoc. editor Jour. of Molecular and Cellular Immunology, 1984—, Jour. of Immunology, 1986-91, Molecular Reproduction and Development, 1987—. Scientific adv. bd. Hereditary Disease Found., 1991-94; mem. immunol. scis. study sect., div. of rsch. grants, NIH, Pub. Health Svc., 1988-92; rev. com. Calif. div., postdoctoral fellowship program, Am. Cancer Soc., 1982-85, 92—. Jane Coffin Childs Meml. fellowship, 1977-79. Mem. AAAS, Am. Assn. of Immunologists, Am. Soc. for Microbiology, Phi Beta Kappa. Achievements include first in vitro synthesis of biologically active retroviral genomic DNA; identification of distinct patterns of macromolecular synthesis at successive stages in T-lymphocyte differentiation; analysis of molecular basis for changes in functional reactivity in T-lymphocyte development; interleukin-2 gene regulation; single-cell molecular analysis of functional responses in developing T cells. Office: Calif Inst Tech Biology Divsn 1201 E California Blvd Pasadena CA 91125-0001

ROTHENBERG, ROBERT EDWARD, physician, surgeon, author; b. Bklyn., Sept. 27, 1908; s. Simon and Caroline A. (Baer) R.; m. Lillian Lustig, 1933 (dec. 1977); m. Eileen Fein, 1977 (dec. 1987); children: Robert Philip, Lynn Barbara Rothenberg Kay; m. Florence Richman, 1989. A.B., Cornell U., 1929, M.D., 1932. Diplomate Am. Bd. Surgery. Intern Jewish Hosp., Bklyn., 1932-34; attending surgeon Jewish Hosp., 1955-82; postgrad. study Royal Infirmary, Edinburgh, 1934-35; civilian cons. U.S. Army Hosp., Ft. Jay, N.Y., 1960-66; attending surgeon French Polyclinic Med. Sch. and Health Center, N.Y.C., 1964-76; pres., 1973-76, trustee, 1972-76; attending surgeon Cabrini Health Care Center, 1976-86; cons. surgeon Cabrini Med. Ctr., 1986—, dir. surg. research, 1981—; clin. asst. prof. environ. medicine and community health State U. Coll. Medicine, N.Y.C., 1950-60; clin. prof. surgery N.Y. Med. Coll., 1981-86, prof. emeritus, 1986—; pvt. practice, 1935-86; chmn. Med. Group Coun. Health Ins. Plan of Greater N.Y., 1947-64; cons. Office and Profl. Employees Internat. Union (Local 153) Health Plan, 1960-82, United Automobile Workers (Local 259) Health Plan, 1960-86, Sanitationmen's Security Benefit Fund, 1964-83; dir. Surgery Internat. Ladies Garment Workers Union, 1970-85; med. adv. bd. Hotel Assn. and Hotel Workers Health Plan, 1950-60, Hosp. Workers Health Plan, 1970-76; past bd. dirs. Health Ins. Plan of Greater N.Y. Author and/or editor: Group Medicine and Health Insurance in Action, 1949, Understanding Surgery, 1955, New Illustrated Med. Ency., 4 vols., 1959, New Am. Med. Dictionary and Health Manual, 1962, Reoperative Surgery, 1964, Health in Later Years, 1964, Child Care Ency., 12 vols., 1966, Doctor's Premarital Medical Adviser, 1969, The Fast Diet Book, 1970, The Unabridged Medical Encyclopedia, 20 vols., 1973, Our Family Medical Record Book, 1973, The Complete Surgical Guide, 1973, What Every Patient Wants to Know, 1975, The Complete Book of Breast Care, 1975, Disney's Growing Up Healthy, 4 vols., 1975, First Aid—What to Do in an Emergency, 1976, The Plain Language Law Dictionary, 1980; contbr. articles to med. jours. Served to lt. col. M.C., AUS, 1942- 45. Recipient Cabrini Gold medal, 1986. Fellow ACS; mem. AMA, Bklyn. Surg. Assn., N.Y. County Med. Soc., Alpha Omega Alpha. Home: 35 Sutton Pl New York NY 10022-2464 also: Monterosso, Camaiore Italy

ROTHFELD, LEONARD BENJAMIN, chemical/environmental engineer; b. N.Y.C., Feb. 20, 1933; s. Aaron and Sylvia R.; m. Mary Ellen Young, Apr. 1, 1967; 1 child, Rebecca Rose. BChemE, Cornell U., 1955; MS, U. Wis., 1956, PhD, 1961; postgrad., U. Ariz., 1987-88. Registered engr., Colo., Ariz., Calif. Various positions Shell Oil Co., 1961-73; dir. process design Arco Coal Co., L.A., 1973-74; mgr. process engring. Arco Coal Co., L.A., Denver, 1974-81; mgr. engring. and constrn. Arco Coal Co., Denver, 1981-82; assoc. rsch. dir. Anaconda Minerals Co., Denver, 1982; mgr. coal rsch. Anaconda Minerals Co., Tucson, 1982-85; environ. engr. U.S. Bur. Mines, Denver, 1988—; cons. Rothfeld Engring., Inc., Arco Coal Co., Pacific Coal Pty, Australia, 1985-89. Contbr. articles to profl. jours. Mem. Air and Waste Mgmt. Assn., Soc. for Mining, Metallurgy and Exploration, Nat. Assn. of Environ. Profls., Rocky Mountain Assn. Environ. Profls. (bd. dirs.), Extractive Metallurgy Chpt. of Denver, Denver Mining Club.

ROTHFIELD, NAOMI FOX, physician; b. Bklyn., Apr. 5, 1929; d. Morris and Violet (Bloomgarden) Fox; m. Lawrence Rothfield, Sept. 18, 1954; children—Susan, Lawrence, John, Jane. B.A., Bard Coll., 1950; M.D., N.Y. U., 1955. Intern Lenox Hill Hosp., N.Y.C., 1955-56; instr. N.Y. U. Sch. Medicine, 1956-62, asst. prof., 1962-68; assoc. prof. U. Conn. Sch. Medicine, Farmington, 1968-72; prof., chief div. rheumatic diseases U. Conn. Sch. Medicine, 1972—. Contbr. chpts. to books; contbr. articles to med. jours. Mem. Am. Soc. Clin. Investigation, Am. Rheumatism Assn., Assn. Am. Physics. Jewish. Home: 540 Deercliff Rd Avon CT 06001-2859 Office: U Conn Sch Medicine Div of Rheumatic Diseases Farmington CT 06030

ROTHMAN, ALAN BERNARD, consultant, materials and components technologist; b. Pitts., July 5, 1927; s. Harry and Lillian (Wolfe) R.; m. Leora Rose Lubovsky, June 15, 1963; 1 child, Robert Sim. BS in Chemistry, U. Pitts., 1949; PhD, Carnegie Inst. Tech., 1954. Sr. scientist Westinghouse Atomic Power Lab., Pitts., 1953-54; rsch. chemist Pitts. Plate Glass Co., Creighton, Pa., 1954-57; assoc. chemist Argonne (Ill.) Nat. Lab., 1957-60; adv. engr., mgr. NERVA flight safety Westinghouse Electric Corp., Pitts., 1960-64; adv. engr. Westinghouse Bettis Atomic Power Lab., Pitts., 1964-65; rsch scientist Space div. Chrysler Corp., New Orleans, 1965-68; chemist Argonne Nat. Lab., 1968-91, mgr. TREAT experiments sect., 1974-79; cons. MAC Tech. Svcs., Argonne, 1992—. Contbr. articles to profl. jours. Bd. dirs. Lake Hinsdale Village Homeowners Assn., Willowbrook, Ill., 1983-87, 92, pres., 1988. With USN, 1945-46. AEC fellow, 1952; Carnegie Inst. Tech. fellow, 1953. Mem. Am. Nuclear Soc. (emeritus), Phi Tau Phi, Phi Eta Sigma, Phi Lambda Upsilon. Achievements include invention and development of a byproduct recovery process for the chemical polishing of glass; invention of an electrode for measurement of the pH of solutions corrosive to glass. Home: 301 Lake Hinsdale Dr # 403 Willowbrook IL 60514-2239

ROTHMAN, DAVID J., history and medical educator; b. N.Y.C., Apr. 30, 1937; s. Murray and Anne (Beier) R.; m. Sheila Miller, June 26, 1960; children: Matthew, Micol. B.A., Columbia U., 1958; M.A., Harvard U., 1959, Ph.D., 1964. Asst. prof. history Columbia U., N.Y.C., 1964-67; assoc. prof. Columbia U., 1967-71, prof., 1971—, Bernard Schoenberg prof. social medicine, dir. Ctr. for Study of Society and Medicine; Fulbright-Hayes prof. Hebrew U., Jerusalem, 1968-69, India, 1982; vis. Pinkerton Prof. Sch. Criminal Justice, State U. N.Y., at Albany, 1973-74; Samuel Paley lectr. Hebrew U., Jerusalem, 1977; Mem. Com. for Study of Incarceration, 1971-74; co-dir. Project on Community Alternatives, 1978-82; chmn. adv. bd. on criminal justice Clark Found., 1978-82. Author: Politics and Power, 1966, The Discovery of the Asylum, 1971; co-author: Doing Good, 1978, Conscience and Convenience: The Asylum and its Alternatives in Progressive America, 1980, (with Sheila M. Rothman), The Willowbrook Wars, 1984, Strangers at the Bedside, 1991; Editor: The World of the Adams Chronicles, 1976, (with Sheila M. Rothman) On Their Own: The Poor in Modern America, 1972, The Sources of American Social Tradition, 1975, (with Stanton Wheeler) Social History and Social Policy, 1981. Bd. dirs. Mental Health Law Project, 1973-80, 82—. Recipient Albert J. Beveridge prize Am. Hist. Assn., 1971; NIMH fellow, 1974-75, 78-81; Law Enforcement Assistance Adminstrn. fellow, 1975-76. Mem. Am. Hist. Assn., Orgn. Am. Historians, N.Y. State Acad. Medicine, Phi Beta Kappa. Office: Ctr Study Society and Medicine Columbia U Coll Physicians and Surgeons 630 W 168th St New York NY 10032

ROTHMAN, JAMES EDWARD, cell biologist, educator; b. Haverhill, Mass., Nov. 3, 1950. BA, Yale U., 1971; PhD in Biochemistry, Harvard U., 1976. Fellow MIT, Cambridge, Mass., 1976-78, asst. prof., 1978-81, assoc. prof., 1981-84; prof. Stanford (Calif.) U., 1984-88; prof. dept. molecular biology Princeton (N.J.) U., 1988-91; Paul Marks chair, chmn. program in cellular biochemistry and biophysics Sloan-Kettering Inst., N.Y.C., 1991—. Office: Meml Sloan Kettering Cancer Ctr Rockefeller Rsch Lab 1275 York Ave Box 251 New York NY 10021-6007

ROTHSCHILD, NAN ASKIN, archaeologist; b. N.Y.C., May 19, 1937; d. Joseph S. and Alma (Durst) Askin; m. Edmund O. Rothschild, Aug. 30, 1958 (div. 1992); children: Oliver, Emily, Joshua, Adam. BA, Vassar Coll., 1959; PhD, NYU, 1975. Asst. prof. Lehman Coll. CUNY, 1976-78, asst. prof. Hunter Coll., 1978-80; asst. prof. NYU, 1980-81; asst. prof. Barnard Coll. Columbia U., N.Y.C., 1981-91, assoc. prof., 1991—, chair, 1993—; curator W.D. Strong Mus. Anthropology Columbia U., 1984—; cons. 100 Broad St. Project, Assay Site, N.Y.C., 1983-84, African Burial Ground video prodn., 1992-93; mem. mus. panel N.Y. State Coun. on Arts, N.Y.C., 1990; mem. com. on predictive models N.Y. State Bd. for Hist. Preservation, Albany, 1987. Author: New York City Neighborhoods, The 18th Century, 1990, Prehistoric Dimensions of Status, 1991; editor: The Research Potential of Anthropological Museum Collections, 1981. Trustee Little Red Sch. House, N.Y.C., 1980-91. Recipient Cert. of Excellence Mcpl. Arts Soc., N.Y.C., 1982; NSF grantee, 1983, 84. Fellow Am. Anthrop. Assn.; mem.

Soc. for Am. Antiquity (mem. exec. bd. dirs. 1987-89), N.Y. Archaeol. Coun. (pres. 1979-80), Profl. Archaeologists N.Y.C. (pres. 1991-92). Achievements include directing excavation of first two large archeological sites in lower Manhattan (the Stadt Huys Block and Seven Hanover Square Block); recovery of architectural remains of early New Amsterdam/New York and hundreds of thousands of objects. Office: Columbia U Barnard Coll Dept Anthropology New York NY 10027

ROTHSTEIN, ARNOLD JOEL, marine engineer, mechanical engineer; b. N.Y.C., July 31, 1928; s. Michael and Beatrice (Marks) R.; m. Naomi Goldberg, Jan. 10, 1953; children: Sheri, Marcus, Lynn, Susan Emily, Elliot. BS in Mech. Engring., N.Y. State Maritime Coll., Ft. Schuyler, 1949; MS, MIT, 1951. Marine engr., cert. energy mgr., lic. gen. contractor. Asst. prof. U.S. Naval Acad., Annapolis, 1953-55; project mgr. GE Aircraft Nuclear Propulsion, Cin., 1955-61; mgmt. staff GE Space Div., Valley Forge, Pa., 1962-68; v.p. program mgmt. Deepsea Ventures, Inc., Gloucester Point, Va., 1968-78; systems mgr. Pratt & Whitney Govt. Engring., West Palm Beach, Fla., 1978-81; v.p. Perry Oceanographics, Riviera Beach, Fla., 1981-85; pres. Aqua/Agri Energy Corp., North Palm Beach, Fla., 1985—; editorial bd. Marine Mining Jour., Belleview, Wash., 1975-78; chmn. energy tech. com. Pan Am. Assn. Engring. Socs., Caracas, Venezuela, 1984-90. Editor: (book) History of U.S. Aircraft Nuclear Propulsion, 1962, Aircraft Nuclear Propulsion Technology, 1964; author: Ocean Mining for Deep Ocean Minerals, 1972. Chmn. Solid Waste Mgmt. Com., Newport News, Va., 1974-77; mem. Dept. Def. System Adv. Panel, Washington, 1965-68, bd. dirs. World Trade Coun., Palm Beach County, 1986—. Named Engr. of Yr. Peninsula Engrs. Assn., Hampton Roads, Va., 1976; recipient Disting. Svc. award ASME, 1991; co-chair Internat. Conf. on Engring. Mgmt., 1986-90. Mem. ASME (sr., chmn. mgmt. div. 1979-83, Internat. Affairs Bd. 1987—), Assn. Energy Engrs. (sr.), MIT Alumni Assn. (pres. 1993-94), Rotary. Jewish. Office: AAE/The Facilities Svcs Co 420 U S Hwy One North Palm Beach FL 33408

ROTHSTEIN, HOWARD, biology educator; b. Bklyn., Aug. 25, 1935; s. Morris Arnold and Esther (Rosen) R.; m. Nurith Fish, June 24, 1956; children: Arlene Beth, Caren Jean, Jeffrey Lewis. BA, Johns Hopkins Univ., 1956; PhD, Univ. Pa., 1960. Asst. assoc. prof. U. Vt., Burlington, 1962-77; assoc. prof. ophthalmology Wayne State Sch. Medicine, Detroit, 1977-81; chmn. biology sci., prof. Fordham U., Bronx, N.Y., 1981-86, prof. biology, 1986—. Author: General Physiology: The Cellular and Molecular Basis, 1971; contbg. editor: Biology: The Behavioral Review, 1973; contbr. articles to profl. jours. Home: 100-403 High Point Dr Hartsdale NY 10530 Office: Fordham U Bronx NY 10458

ROTHWARF, ALLEN, electrical engineering educator; b. Phila., Oct. 1, 1935; s. Max and Bessie (Dichter) R.; m. Bernice Cecelia Golansky, June 16, 1957; children: Richard, Jeanne, David. BA in Physics, Temple U., 1957; MS in Physics, U. Pa., 1960, PhD in Physics, 1964. Instr. Rutgers, The State U., Camden, N.J., 1960-62; mem. tech. staff RCA Labs., David Sarnoff Rsch. Ctr., Princeton, N.J., 1964-72; postdoctoral fellow U. Pa., Phila., 1972-73; sr. scientist Inst. Energy Conversion, U. Del., Newark, 1973-79; prof. elec. engring. dept. Drexel U., Phila., 1979—; dir. Ben Franklin Superconductivity Ctr., Phila., 1989—; cons. RCA Labs., Princeton, 1979-83, Solarex Thin Film Div., Newtown, Pa., 1983—. Contbr. articles to profl. jours.; patentee in field. Recipient Rsch. award Drexel U., 1989. Fellow IEEE; mem. Am. Phys. Soc. Office: Drexel U Elec and Computer Engring Dept Philadelphia PA 19104

ROTHWELL, TIMOTHY GORDON, pharmaceutical company executive; b. London, Jan. 8, 1951; came to us., 1966; s. Kenneth Gordon Rothwell and Jean Mary (Stedman) Davey; m. Joanne Claire Fleming; children: Tiffany, Heather. BA, Drew U., 1972; JD, Seton Hall U., 1976; LLM, NYU, 1979, MBA, 1983. With Sandoz Pharms., East Hanover, N.J., 1972—, patent atty., 1974-77, patent and trademark counsel, 1980-82, mng. ops. planning and adminstrn., 1982-84, dir. mktg. ops., 1984-85, exec. dir. field ops., 1985-86, v.p. field ops., 1986-87, pres. profl. bus. ops., 1987-88, corp. v.p., chief oper. officer, 1988-89; sr. v.p. sales Squibb Pharm. Group, Princeton, N.J., 1989—. Patentee if field. Mem. N.J. State Bar Assn., N.Y. State Bar Assn., Am. Soc. for Pharmacy Law, Nat. Health Care Quality Coun., Am. Found. for Pharm. Exec. (bd. dirs.), N.J. Patent Law Assn. (pres. 1986). Republican. Episcopalian. Avocations: philately, coaching youth soccer, golf, tennis. Office: Squibb Pharm Group US PO Box 4500 Princeton NJ 08543-4500

ROTITHOR, HEMANT GOVIND, electrical engineering educator; b. Patan, India, July 5, 1958; came to U.S., 1985; s. Govind Hari and Suman (Govind) R.; m. Shubhada Hemant, Sept. 1, 1987; 1 child, Sagar. PhD in Elec. Engring., U. Ky., 1989. Devel. engr. ORG Systems, Baroda, India, 1981-82, Philips India, 1982-85; asst. prof. Worcester (Mass.) Poly. Inst., 1990—; mem. steering com. 3d Internat. Symposium on personal, indoor, mobile, radio comm., 1992. Contbr. articles to Internat. Jour. Mini and Micro Comuters, IEEE. Grantee Worcester Poly. Inst., 1990, NSF, 1992. Mem. IEEE, Assn. for Computing Machinery, Sigma Xi, Tau Beta Pi, Eta Kappa Nu. Achievements include research in instrumentation measurement, computer architecture and distributed computing. Home: 10 Hackfeld Rd Worcester MA 01609 Office: Worcester Poly Inst 100 Institute Rd Worcester MA 01609

ROTOLO, VILMA STOLFI, immunology researcher; b. Villa Diego, Argentina, Jan. 27, 1930; d. Eduardo and Manuela (Gallego) Stolfi; m. Jose Jaime Rotolo, Dec. 6, 1957; children: Gloria Claudia, Alejandro Claudio. Degree in pharmacology, U. Litoral Med. Sch., Rosario, Argentina, 1953, Lic. in Biochem. Sci., 1956, PhD in Biochem. Sci., 1966. Asst. dir., rsch. mem. oncology lab. Oncology Inst., Nat. Ministry Pub. Health, Rosario, 1956-67; rsch. mem. dept. pathology-tissue culture div. U. Rosario Med. Sch., 1967-72, chief Exptl. Immunopathology Inst., 1972-89, dir. immunotoxicology program, 1989-91; sr. rsch. scientist Prodinar Co., Rosario, 1991—; pres. internat. symposium Assn. Allergy and Immunopathology, Argentina, 1981; rapporteur internat. seminar Commn. European Communities and IPCS, WHO, ILO, UNEP, EPA, NIEHS, Luxembourg, 1984; cons. Fed. Justice, Argentina, 1989. Guest editor: African Jour. Clin. and Exptl. Immunology, 1982; contbr. paper to Triduo Sci. Ann. (hon. diploma award 1965). Active For the Life Found., Neuquén, Argentina, 1987-90; pres. Found. Nature and Sci., 1990—. Recipient rsch. visitor travel award WHO, 1983, expert immunotoxicology travel award Commn. European Communities and UNEP, ILO, IPCS, WHO, EPA, NIEHS, Luxembourg, 1984; WHO travel award fellow, 1973, 78; Nat. Coun. Sci. Investigation grantee, 1974-77, 79-81, 82-83. Mem. Internat. Soc. Immunopharmacology (mem. internat. coun.), Argentine Assn. Immunopharmacology and Immunotoxicology (pres. 1987), Argentine Environ. Protection Assn., Latin Am. Assn. Immunology, Argentine Soc. Ecology, Sigma Delta Epsilon. Avocations: drawing, painting, horticulture. Home: Mendoza 2435, Santa Fe, 2000 Rosario Argentina

ROTONDO, GAETANO MARIO, aerospace medicine physician, retired military officer; b. Taranto, Puglie, Italy, Jan. 30, 1926; s. Francesco and Clotilde (Guida) R.; m. Vittoria Maria Beltrami, June 23, 1957. MD, U. Bari, Italy, 1949; Physiology Docent Aerospace Medicine, Dept. Pub. Instn., Italy, 1969. Cert. physiologist, physiology docent aerospace medicine Dept. Pub. Inst., Italy; diplomate Italian Bd. Forensic Medicine, Italian Bd. Sports Medicine. Col., chief Italian Air Force Aeromed. Inst., Milan, 1972-74; brig. gen., chief Mil. Sch. Aviation Med., Rome, 1974-75; maj. gen., chief Italian Air Force Med. Svc., Rome, 1975-82; lt. gen., surgeon gen. Italian Air Force Med. Corp., Rome, 1980-86; prof. forensic medicine and med. bioclimatology U. Milan, 1973-93; prof. aerospace medicine U. Rome, 1975-93; cons. Dept. Transp., Rome, 1985-93, Dept. Pub. Health, Rome, 1984-93; nat. mem. Aerospace Med. Panel AGARD-NATO, Paris, 1976-92. Editor: Mil. Forensic Medicine, 1983, Forensic Aerospace Medicine, 1988, Aviation Medicine: Physiol. and Medico-legal Aspects, 1990 (Nat. Acad. History Med. Art award 1991), Medical Bioclimatology, 1993; co-editor: Minerva Aerospaziale, 1980-93; mem. editorial bd. U.S. Mil. Medicine, 1982-93; contbr. over 250 articles to profl. jours. and publs. Lt. Gen. Italian Air Force Med. Corps, 1950-91, ret. as gen. inspector. Decorated Golden Medal of merit for pub. health, Great Officer Cross Order of Merit, Golden Medal for Sr. Mil. Svc.; Hon. License and Wings Flight Surgeon (Germany); recipient Forensic Med. award Italian Inst. Social Medicine, 1979, Theodore C. Lyster award Aerospace Med. Assn., Washington, 1985. Fellow Am. Astronautics Soc., Aerospace Med. Assn., Royal Aero. Soc., British Interplanetary Soc., Am. Inst. Aeronautics and Astronautics (assoc.); mem. Nat. Soc. Mil. Medicine (v.p.), Italian Soc. Aerospace Medicine (v.p.), Italian Fedn. Sports Medicine, Aeroclub of Italy (flight security com.), Assn. U.S. Mil. Surgeons (hon.), Nat. Acad. History Health Art (academecian), Lancisiana Acad. Rome, Roman Acad. Med. and Biol. Scis., Internat. Acads. Astronautics and Aviation Space Medicine (academician). Roman Catholic. Achievements include expanding aeromedical aspects of National Air Search and Rescue Services, improving aeromedical evacuation in case of public disasters, organization/direction for medical selection of ESA Spacelab Payload Specialist national candidates. Home: Via Conca d'Oro 348, 00141 Rome Italy

ROTT, NICHOLAS, fluid mechanics educator; b. Budapest, Hungary, Oct. 6, 1917; came to U.S. 1951; s. Alexander and Margaret (Pollak) R.; m. Rosanna Saredi, Sept. 30, 1944; children: Paul, Kathy. Diploma in Mechanical engring., Swiss Fed. Inst. Tech., Zurich, 1940; PhD, ETH, Zurich, 1944. Rsch. asst., pvt. dozent Aerodynamics Inst., Zurich, 1944-51; prof. Grad. Sch. Aeronautical Engring. Cornell U., Ithaca, N.Y., 1951-60; prof. UCLA, 1960-67, ETH, Zurich, 1967-83; vis. prof. Stanford (Calif.) U., 1983—. Fellow AIAA, Am. Phys. Soc.; mem. NAE, Acoustical Soc. Am. Home: 1865 Bryant St Palo Alto CA 94301 Office: Stanford U Aero Astro Dept Stanford CA 94305

ROTTIERS, DONALD VICTOR, physiologist; b. Saginaw, Mich., Oct. 20, 1938; s. Victor Bernard Rottiers and Mildred Lavira (Willis) Leaym; m. Karla Emily Hildebrandt, Jan. 6, 1962; children: Victor Bernard, Jamie Lee. BS, U. Mich., 1963, MS, 1965; PhD, Pa. State U., 1990. Cert. fishery scientist. Fishery aid U.S. Fish & Wildlife Svc., Ann Arbor, Mich., 1962-65; fishery biologist U.S. Fish & Wildlife Svc., Marquette, Mich., 1965-68, Ann Arbor, 1968-79; physiologist U.S. Fish & Wildlife Svc., Wellsboro, Pa., 1979—; instr. Mansfield (Pa.) U., 1992. Mem. Am. Fishery Soc. Achievements include research on the effects of diet, photoperiod, temperature, salinity, season and disease on growth, survival, smoltification and migration of Atlantic salmon as part of a program to reestablish the species in the Merrimack River. Home: RD 6 Box 359 Wellsboro PA 16901 Office: RD 4 Box 63 Wellsboro PA 16901

ROTUNNO, RICHARD, meteorologist. BE in Engring. Sci., SUNY, Stony Brook, 1971, MS in Mechanics, 1972; MA in Geophysical Fluid Dynamics, Princeton U., 1974, PhD in Geophysical Fluid Dynamics, 1976. Postdoctoral fellow advanced study program Nat. Ctr. Atmospheric Rsch., Boulder, Colo., 1976-77, scientist II, 1980-83, scientist III, 1983-89, sr. scientist, 1989—; vis. fellow coop. inst. rsch. environ. scis. U. Colo., 1977-78, rsch. assoc., 1978-79; guest lectr. dept. math. Monash U., Australia, 1981, 84; mem. adv. com. Internat. Conf. Computational Methods and Exptl. Measurements, 1981-82; vis. assoc. prof. MIT, 1985; mem. adv. com. coop. inst. mesoscale meteorol. studies U. Okla., 1985—; chmn. panel coastal meteorology NAS, 1991. Co-chief editor Monthly Weather Rev., 1986-91. Recipient Ted Fujita award, 1983. Mem. Am. Meteorol. Soc. (mem. com. atmospheric and oceanic waves and stability 1980-83, com. severe local storms 1983-86, publs. commn. 1986-91, com. mesoscale processes 1990-93, Banner I. Miller award 1992). Office: Nat Ctr for Atmospheric Research PO Box 3000 Boulder CO 80307*

ROUAN, GREGORY W., internal medicine physician, educator; b. Cin., Sept. 8, 1954; s. Leo Wayne and Donna Lou (Umbel) R. BS, U. Dayton, 1976; MD, U. Cin., 1980. Intern and resident U. Cin., 1980-83, chief resident, 1983-84, clin. instr. dept. internal medicine, 1983-84, asst. prof. clin. medicine, 1984-89, assoc. prof. clin. medicine, 1989—, dir. grad. med. edn. program, 1990—; rsch. fellow Harvard Med. Sch./Brigham & Women's Hosp., 1986-87. Bd. dirs. U. Cin. Med. Ctr. Fund, 1991—. Kellogg fellow, 1986-87. Fellow ACP (Tchr. and Rsch. Scholars award 1986-87; mem. AAAS, Soc. Gen. Internal Medicine, Cin. Soc. Internal Medicine (bd. dirs. 1990—), Cen. Soc. for Clin. Rsch., Am. Fedn. Clin. Rsch., Univ. Club. Home: 8 Hampton Ct Cincinnati OH 45208 Office: 231 Bethesda Ave Cincinnati OH 45267-0557

ROUAYHEB, GEORGE MICHAEL, scientific research council advisor; b. Kobba, Batroun, Lebanon, Sept. 7, 1933; came to U.S., 1952; s. Michael George and Naimeh (Fattouh) R.; m. Leila Isaac Koussa, July 5, 1964; children: Michel, Marwan. BS, La. State U., 1956, MS, 1958; postgrad., Okla. State U., 1961. Rsch. scientist Conoco R&D, Ponca City, Okla., 1958-61; process engr. Medreco Refinery, Sidon, Lebanon, 1962-65; prodn. mgr. Esso Fertilizer Co., Beirut, 1965-70; dir. tech. div. Idcas Arab League, Cairo, 1970-73; dir. applied scis. div. Nat. Coun. Sci. Rsch., Beirut, 1973—; profl cons. ECWA UN, Beirut, 1978; regional advisor UNIDO, Vienna, Austria, 1982-83. Mem. Order of Engrs. Lebanon. Greek Orthodox. Achievements include patent on the use of tritium in gas chromatography ionization detector. Home: Kobba, Batroun Lebanon Office: Nat Coun Sci Rsch, PO Box 11-8281, Beirut Lebanon

ROUBAL, WILLIAM THEODORE, biophysicist, educator; b. Eugene, Oreg., Dec. 20, 1930; s. Frank J. and Irene I. (Ellenberger) R.; m. Carol Jean, Sept. 6, 1953; children: Diane Jeanette Roubal Daniel, Linda Ann Roubal Myrick, Sandra Mae Roubal, Cathy Roubal Hoover. BS, Oreg. State U., 1954, MS, 1959; PhD, U. Calif.-Davis, 1965. Chemist dept. entomology Oreg. State U., Corvallis, 1958-60; rsch. chemist Pioneer Rsch. Lab., Dept. Interior, Seattle, 1960-70; rsch. scientist Dept. Commerce, Seattle, 1970-75; sr. scientist NOAA, NMFS, Seattle, 1976—; biophysicist Environ. Conservation div. Nat. Marine Fisheries Svc., Seattle, 1968-75; assoc. prof. Sch. Fisheries, U. Wash., Seattle, 1976—; ednl. specialist Seattle Central Community Coll., 1974-85; instr., lectr. in field. Head usher Haller Lake United Methodist Ch., Seattle, 1974-89; workshop dir. Seattle YMCA Summer Family Camp, Seabeck, Wash., 1979—. Served as 1st lt. U.S. Army, 1954-56. Recipient award Seattle Central Community Coll., 1980. Mem. Am. Chem. Soc., Am. Oil Chemists' Soc. (A.E. MacGee award 1964), Am. Sci. Glassblowers Soc. (abstracts chmn. 1964-65), Claire Hammonds Guild, Sigma Xi. Home: 17840 Wayne Ave N Seattle WA 98133-5142 Office: NOAA NMFS EC Div 2725 Montlake Blvd E Seattle WA 98112-2097

ROUDYBUSH, FRANKLIN, diplomat, educator; b. Washington, Sept. 17, 1906; s. Rumsey Franklin and Frances (Mahon) R.; student U. Vienna, 1925, Ecole National des Langues Orintales Vivantes, Paris, 1926, U. Paris, 1926-28, U. Madrid, 1928, Academie Julian, Paris, 1967; B.Fgn. Svc., Georgetown U., 1930; postgrad. Harvard U., 1931; MA, George Washington U., 1944; PhD, U. Strasbourg (France), 1953; m. Alexandra Brown, May 22, 1941. Dean Roudybush Fgn. Svc. Schs., Washington, L.A., Phila., N.Y.C., 1932—. Prof. internat. econ. rels. Southeastern U., Washington, 1938-42; dir. Pan Am. Inst., Washington, 1934, London Econ. Conf., 1934; editor Affairs, 1934-45; v.p. France Libre, Washington; censor Diplomatic Pouch World War II; commodity economist, statistician Dept. State, 1945; with Fgn. Svc. Inst., Dept. State, 1945-48, Council of Europe, Strasbourg, 1948-54, Am. Embassy, Paris, 1954, Pakistan, 1955, Dublin, 1956; consular Acad. Vienna, 1925; mem. Punjab U., Lahore, Pakistan, 1954. Recipient prize Julian painting, Paris. Mem. Am. Soc. Internat. Law, Brit. Inst. Internat. and Comparative Law (London), Delta Phi Epsilon. Clubs: Assns. des Amis du Salon d'Automne (Paris); France Amerique; English Speaking Union (London); Nat. Press (Washington); Harvard (Paris), Royal Aberdeen Golf; Miramar Golf (Oporto, Portugal), Yacht (Angiers, France); Pormarnock Golf (Dublin); Les Societe des Artistes Independants Grand Palais (Paris). Author: The Twentieth Century; The Battle of Cultures; Diplomatic Language; Twentieth Century Diplomacy; The Present State of Western Capitalism, 1959; Diplomacy and Art, French Educational System, 1971; The Techniques of International Negotiation, 1979; The Diplomacy of the Cardinal, Duke de Richilieu, 1980, Talleyrand - The Diplomat, 1989, The French Government Political Science School, 1989, History of a Family During the XXth Century, 1990, The Flying Dutchman, 1990, Monsieur Fedeaux, 1990, Death in Darjeeling, 1991, The Alsatians, 1991, The Oriental Express to Constantinople and On To Teheran, 1992, The Elegant Facade of Macaó, 1992, Café Royal, 1993, The Roman Holiday, 1992, The Burlington Sisters, 1993, The Strange Fate of Madame Tarleton, 1980, Rendezvous in Basle, 1989, From Naples to Buenos Aires, 1977, Drawn Blinds, 1972. Home: Villa St Honoré, Moledo do Minho, Minho Portugal Office: 15 Ave du Pres Wilson, Paris 16, France also: Sauveterre de Rouerque, 12800 Aveyron France

ROUF, MOHAMMED ABDUR, microbiology educator; b. Dacca, Bangladesh, May 2, 1933; came to U.S., 1957; s. Kafiludden and Nadia Sarder; m. Reena Sarwar, Sept. 30, 1973; children: Arman, Alia. BS in Biology with honors, U. Dacca, 1954, MS in Botany, 1955; MA in Microbiology, U. Calif., Davis, 1959; PhD in Bacteriology and Biochemistry, Wash. State U., 1963. Lectr. botany Govt. Coll. Sylhet, Pakistan, 1955-57; rsch. asst. in bacteriology U. Calif., 1957-59; teaching and rsch. asst. Wash. State U., Pullman, 1959-63, asst. bacteriologist, 1963-64; asst. prof. microbiology U. Wis., Oshkosh, 1964-66, assoc. prof., 1966-68, prof., 1968—, chmn. dept. biology and microbiology, 1970-76, 82—; dir. survey medicinal plants U. Dacca, 1955-56; NSF rsch. participant Ill. Inst. Tech., Chgo., 1966; guest investigator Marine Biol. Lab., Woods Hole, Mass., 19*6; vis. prof. U. Wis., Madison, 1977-78; dir. rsch. Biomedica, Inc., Neenah, Wis., 1981—; exec. dir., sec., bd. dirs. Prompt & Precise Labs., Inc., Oshkosh, 1982-85; cons. to numerous cos., including Am. Can Co., Procter and Gamble, Rockwell Internat. Corp.; numerous presentations in field. Author: Laboratory Exercises in Microbiology, 1987; editor Sci. of Biology Jour., 1975-77; contbr. articles to Jour. Bacteriology, Bacterial Procs., Applied Microbiology, Sci. Biol., Current Microbiology, Applied and Environ. Microbiology, Can. Procs. Microbiology,. Pres. Aurelian Soc., Oshkosh, 1973, Georgia Garden Condominium, Inc., Oshkosh, 1975-77, 80-85, Fox Valley Islamic Soc., Neenah, 1990—. Recipient James F. Duncan rsch. award U. Wis.; scholar Wash. State U., 1959-61; fellow NSF, 1976-78; grantee Wis. State U., 1969-70, NSF, 1970-72, 76-78, 80-81, U. Wis., 1979-80. Fellow Am. Acad. Microbiology, Am. Inst. Chemists (accredited profl. chemist, councilor 1971-74, chmn. interprofl. rels. com. 1974-78); mem. Am. Soc. for Microbiology (pres. North Cen. br. 1975-76, 91-92, ednl. rep. 1982—, councilor 1978-81), Am. Chem. Soc., Wis. Acad. Arts and Sci., Sigma Xi (pres. U. Wis.-Oshkosh chpt. 1964-66, chmn. nat. meeting 1974). Achievements include research on microbiology, food and industrial microbiology, biotechnology. Home: 70 E Waukau Ave Oshkosh WI 54901 Office: Dept Biology & Microbiology U Wis Oshkosh WI 54901

ROULSTON, DAVID JOHN, engineering educator; b. London, Nov. 3, 1936; divorced; children: Christine, Helene, Philippe. BSc, Queens U., Belfast, Northern Ireland, 1957; PhD., Diploma, Imperial Coll., London, 1962. Rsch. engr. Compagnie Generale de Telegraphie sans Fil, Puteaux, Seine, France, 1962-66; assoc. prof. U. Waterloo, Ont., Can., 1966-71, prof., 1971—; cons. Thomson-CSF, France, 1976-87, UN Indsl. Devel. Orgn., India, 1985, 86; vis. fellow Wolfson Coll., Oxford, 1988, 89. Author: Bipolar Semiconductor Devices, 1990; inventor 6 patents; author over 100 tech. papers. Office: U Waterloo, Elect & Computer Engring Dept, Waterloo, ON Canada N2L 3G1

ROUND, FRANK E., botanist, educator. Prof. dept. botany U. Bristol, U.K. Recipient Gerald W. Prescott award Phycological Soc. Am., 1991. Office: Univ of Bristol, Dept of Botany, Bristol 8, England*

ROURK, CHRISTOPHER JOHN, electrical engineer; b. Orlando, Fla., Apr. 22, 1962; s. John Wescott and Rosemary (Vevera) R. BSEE, U. Fla., 1985; M in Engring., Rensselaer Poly. Inst., 1987. Registered profl. engr., Fla. Elec. engr. Bechtel Power Corp., Norwalk, Calif., 1985-86, Westinghouse Elec. Corp., Orlando, Fla., 1987-91; task mgr. U.S. Nuclear Regulatory Commn., Bethesda, Md., 1991—. Mem. IEEE. Achievements include patent in Structure and Method for Distributing Failure-Induced Transient Currents in a Multiphased Electrical Machine. Office: US Nuclear Regulatory Commn Washington DC 20555

ROUS, STEPHEN NORMAN, urologist, educator, editor; b. N.Y.C., Nov. 1, 1931; s. David H. and Luba (Margulies) R.; m. Margot Woolfolk, Nov. 12, 1966; children: Benjamin, David. A.B., Amherst Coll., 1952; M.D., N.Y. Med. Coll., 1956; M.S., U. Minn., 1963. Diplomate: Am. Bd. Urology. Intern Phila. Gen. Hosp., 1956-57, resident, 1959-60; resident Flower-Fifth Ave. and Met. Hosp., N.Y.C., 1957-59, Mayo Clinic, Rochester, Minn., 1960-63; practice medicine specializing in urology San Francisco, 1963-68; assoc. prof. urology N.Y. Med. Coll., N.Y.C., 1968-72; assoc. dean N.Y. Med. Coll., 1970-72; prof. surgery, chief div. urology Mich. State U., East Lansing, 1972-75; prof., chmn. dept. urology Med. U. S.C., Charleston, 1975-88; urologist-in-chief Med. U. S.C. and County hosps., Charleston, 1975-88; editorial dir. Norton Med. Books div. W.W. Norton and Co., 1988—; adj. prof. urology Med. U. S.C., 1988—, adj. prof. surgery Dartmouth Med. Sch., 1988-91, prof. surgery (urology), 1991—; staff urologist Dartmouth-Hitchcock Med. Ctr., 1991—; cons. urologist Saginaw VA Hosp., 1971-75, Charleston VA Hosp., 1975-88; hon. cons. St. Peter's Hosp., London, 1981-82; sr. vis. fellow Inst. Urology, London, 1981-82; mil. cons. in urology USAF Surgeon Gen., 1982-85; chmn. alumni devel. com. Mayo Clinic, 1979-82; hon. staff The Exeter Hosp., N.H., 1988—; mem. nat. bd. visitors N.Y. Med. Coll., 1988—; chief urology VA Med. Ctr., White River Junction, Vt., 1991—. Author: Understanding Urology, 1973, Urology in Primary Care, 1976, Spanish edit., 1978, Russian edit., 1979, Urology: A Core Textbook, 1985, The Prostate Book, 1988, rev. edit., 1992, 93, (with Judge Hiller B. Zobel) Doctors and the Law: Defendants and Expert Witnesses, 1993; editor Urology Ann., 1987—; Stone Disease: Diagnosis and Management, 1987; mem. editorial bd. Mil. Medicine mo. jour., 1984—; contbr. articles to profl. jours. Mem. East Lansing Planning Commn., 1974-75; vestryman, jr. warden All Saints Episcopal Ch., E. Lansing, Mich., 1973-75, lay reader, mem. diocesan com. on continuing edn., 1975-86; vestryman St. Michael's Episcopal Ch., 1979-82, Charleston, S.C., chmn. every mem. canvas, 1979-80, chmn. lay readers, 1983-86; mem. fin. com., lay reader Christ Episcopal Ch., Exeter, N.H., 1989-91; vestryman, stewardship chairman, lector St. Thomas Episcopal Ch., Hanover, N.H., 1991—, vestryman, 1992—; mem. selectman's alt. Hampton Falls Planning Bd., 1989-91. Col. USAFR, 1981-85; col. USAR, 1985—. Recipient "A" designator in urology, U.S. Army Surgeon Gen., 1986. Fellow ACS, Am. Acad. Pediatrics; mem. AMA, Soc. Univ. Urologists, Internat. Soc. Urology, Am. Urol. Assn., Nat. Urologic Forum, Soc. Pediatric Urology, Brit. Assn. Urol. Surgeons, German Urol. Assn. (hon.), Mayo Alumni Assn. (v.p. 1979-81, pres. 1983-85), Army and Navy Club (Washington), Lotos Club (N.Y.C.), Dartmouth Club (N.Y.C.), Sigma Xi, Alpha Omega Alpha. (hon.). Republican. Home: Main St Box 296 Norwich VT 05055 Office: Dartmouth Hitchcock Med Ctr Sect Urology Lebanon NH 03756

ROUSE, IRVING, anthropologist, emeritus educator; b. Rochester, N.Y., Aug. 29, 1913; s. Benjamin Irving and Louise Gillespie (Bohachek) R.; m. Mary Uta Mikami, June 24, 1939; children: Peter, David. BS, Yale U., 1934, PhD, 1938; D in Philosophy and Letters (hon.), Centro de Estudios Avanzados de Puerto Rico y el Caribe, 1990. Asst. anthropology Yale Peabody Museum, 1934-38, asst. curator, 1938-47, assoc. curator, 1947-54, research assoc., 1954-62, curator, 1977-85, emeritus curator, 1985—; instr. anthropology Yale U., 1939-43, asst. prof., 1943-48; assoc. prof. Yale, 1948-54; prof. Yale U., 1954-69, Charles J. MacCurdy prof. anthropology, 1969-84, prof. emeritus, 1984—. Author monographs on archaeology of Fla., Cuba, Haiti, P.R., Venezuela. Recipient Medalla Commemorativa del Vuelo Panamericano pro Faro de Colon Govt. Cuba, 1945, A. Cressy Morrison prize in natural sci. N.Y. Acad. Sci., 1951, Viking fund medal Wenner-Gren Found., 1960, Wilbur Cross medal Yale U., 1992; Guggenheim fellow, 1963-64. Mem. Am. Anthrop. Assn. (pres. 1967-68), Eastern States Archeol. Fedn. (pres. 1946-50), Assn. Field Archaeology (pres. 1977-78), Soc. Am. Archaeology (editor 1946-50, pres. 1952-53), Nat. Acad. Scis., Am. Acad. Arts and Scis., Soc. Antiquaries (London). Office: Box 208277 Yale Sta New Haven CT 06520-8277

ROUSE, JOHN WILSON, JR., research institute administrator; b. Kansas City, Mo., Dec. 7, 1937; s. John Wilson and Gail Agnes (Palmer) R.; m. Susan Jane Davis, May 3, 1981; 1 son, Jeffrey Scott. A.S., Kansas City Jr. Coll., 1957; B.S., Purdue U., 1959; M.S., U. Kans., 1965, Ph.D., 1968. Registered profl. engr., Mo., Tex. Engr. Bendix Corp., Kansas City, Mo., 1959-64; rsch. coord. Ctr. for Rsch., U. Kans., Lawrence, 1964-68; prof. elec. engring., dir. remote sensing ctr. Tex. A&M U., College Station, 1968-78; Logan prof. engr., chmn. elec. engring. U. Mo., Columbia, 1978-81; dean engring. U. Tex., Arlington, 1981-87; pres. So. Rsch. Inst., Birmingham, Ala., 1987—; mgr. microwave program NASA Hdqrs., Washington, 1975-77; bd. dirs. Protective Life Corp., Ala. Power Co.; chmn. bd. So. Rsch.

Techs. Inc. Contbr. articles to profl. jours. Recipient Outstanding Tchr. award Tex. A&M U., 1971; Outstanding Prof. award U. Mo., 1980; Engr. of Yr. Tex. Soc. Profl. Engrs., 1983. Mem. IEEE, Nat. Soc. Profl. Engrs., Am. Soc. Engring. Edn., Internat. Bus. Fellows, Internat. Union Radio Sci., Sigma Xi, Eta Kappa Nu., Tau Beta Pi. Home: 2004 Bridgelake Dr Birmingham AL 35244-1421 Office: Southern Research Institute PO Box 55305 2000 9th Ave S Birmingham AL 35205

ROUSE, ROBERT MOOREFIELD, mathematician, educator; b. Auburn, N.Y., Aug. 1, 1936; s. Lester Mallory and Margaret (Moore) R.; m. Mary Josephine Sellers, Aug. 3, 1968; 1 child, Meredeth Elizabeth. BEE, Clarkson U., 1958, M in Engring. Sci., 1972; MS, Syracuse U., 1962. Registered profl. engr., N.Y. Envr. Gulf Oil Corp., Phila, Port Arthur, Tex., 1958-61; prof. SUNY, Morrisville, 1964—. 1st. lt. U.S. Army, 1962-63. Mem. IEEE, ASCE, Am. Soc. for Engring. Edn. Republican. Presbyterian. Home: Hart Rd PO Box 159 Morrisville NY 13408-0159 Office: SUNY Morrisville NY 13408

ROUTBORT, JULES LAZAR, physicist, editor; b. San Francisco, May 15, 1937; s. Jules Lazar and Elaine Marian (Lipman) R.; m. Agnes Johanna Eickhorst, July 9, 1966; children: Julia C., Mark J. BS, U. Calif., Berkeley, 1960; PhD, Cornell U., 1964. Postdoctoral fellow Cavendish Lab., Cambridge, U.K., 1964-66; postdoctoral fellow dept. physics RPI, Troy, N.Y., 1966-68; physicist Argonne (Ill.) Nat. Lab., 1968—; fellow Inst. Transuranium Elements, Karlsruhe, Fed. Republic of Germany, 1973-74, 81-82; program mgr. U.S. Dept. of Energy, Washington, 1987-88. Author: Point Defects and Defect Related Properties, 1991; assoc. editor Applied Physics Letters, 1991—; N.Am. editor Jour. of Hard Materials, 1990—; contbr. over 150 articles to profl. jours. Recipient Alexander von Humboldt fellowship A.V. Humboldt Found., Fed. Republic of Germany, 1973. Fellow Am. Ceramic Soc. Office: Argonne Nat Lab Materials Sci Div Argonne IL 60439

ROUTH, JOSEPH ISAAC, biochemist; b. Logansport, Ind., May 8, 1910; s. William Arthur and Ethel Marie (Etnire) R.; m. Dorothy Francis Hayes, Sept. 4, 1937 (widowed May 1972); children: Joseph Hayes, John Michael; m. Elizabeth Marie Hayes, Dec. 4, 1976 (widowed Feb. 1990). BSChemE, Purdue U., 1933, MS, 1934; PhD, Michigan U., Ann Arbor, 1937. Diplomate Am. Bd. Clin. Chem. Instr. Biochemistry Dept. U. Iowa, Iowa City, 1937-42, asst. prof., 1942-46, assoc. prof., 1946-51, prof., 1951-78; dir. clin. biochemistry Lab. Univ. Hospitals, 1952-64; prof. pathology U. Iowa, 1970-78, dir. spl. clin. chem. lab., 1970-78; cons. VA Hosp., Iowa City, 1952-78; pres. Am. Assn. Clin. Chemists, Washington, 1957-58, Am. Bd. Clin. Chemistry, Washington, 1959-73. Author, co-author: 20th Century Chemistry, 1953-63, Essentials of General Organic and Biochemistry, 1969-77; contbr. articles to profl. jours. Recipient Outstanding Efforts in Edn. and Tng. of Clin. Chemists Am. Assn. Clin. Chemists, Washington, 1973. Fellow Am. Inst. Chemists, 1959—; mem. Chem. Abstracts (sect. editor), Clin. Chemistry. Republican. Roman Catholic. Achievements include research in first chemical and nutritional studies on powdered keratins. Home: Box 712 Cherokee Village AR 72525 Office: Dept Biochemistry Univ Iowa Iowa City IA 52242

ROUTTI, JORMA TAPIO, federal agency administrator, engineering educator; b. Jyväskylä, Finland, Dec. 17, 1938; s. Olli and Rauha (Kuusalo) R.; m. Irmeli Saurama, May 24, 1965; children: Laura, Heli. Diploma in engring., Helsinki (Finland) U., 1964; MSc, U. Calif., Berkeley, 1966, PhD, 1969. Scientist U. Calif., Berkeley, 1965-70; vis. scientist CERN, Geneva, 1970-73; prof. Helsinki U. of Technology, Espoo, Finland, 1973—; pres. Finnish Nat. Fund for R & D SITRA, Helsinki, 1985—; now chair atomic energy commn. Finnish Fed. Govt., Helsinki. Contbr. articles to profl. jours. Eisenhower Exch. fellow, 1978, Fulbright fellow, 1964-69. Mem. Finnish Acad. Tech. Scis. (pres. 1987—), Finnish Inst. Mgmt. (pres. 1987—). Home: Lutherinkatu 2 B 18, 00100 Helsinki Finland Office: Atomic Energy Commn, Aleksanterinkatu 10, 00170 Helsinki Finland*

ROUVILLOIS, PHILIPPE ANDRÉ MARIE, science administrator; b. Saumur, France, Jan. 29, 1935; s. Jean Rouvillois and Suzanne Hulot; m. Madeleine Brigol, 1960; 4 children. Student, Lycée Fustel-de-Coulanges, Strasbourg, Lycée Louis-le-Grand, Paris; Faculté de Droit, Inst. d'Etudes Politiques, Paris. Insp. of fin., 1959; with Office of Revenue, 1964; adviser Min. Econs. & Fin., 1966-68; dep. dir. Office of Revenue, Ministry Econs. & Fin., 1967, head svc., 1969, dep. dir.- gen. revenue, 1973, dir.- gen., 1976, insp.- gen. fin., 1981; dep. dir.- gen. Societe Nationale des Chemins de Fer Francais, 1983-87, dir.- gen., 1987-88, pres. adminstrn. bd., 1988; gen. mgr. Atomic Energy Commn., Paris, 1989—; pres. Atomic Energy Commn.- Industrie, Paris, 1989-92. Commdr. Légion d'honneur, Croix de Valeur militaire (France). Office: Commissariat a l'Energie, Atomique: 31-33 rue de la, Federacion, 75752 Paris Cedex 15, France*

ROVISON, JOHN MICHAEL, JR., chemical engineer; b. North Tonawanda, N.Y., June 15, 1959; s. John Michael and Veronica Marie (Donat) R.; m. Beverly Jean Farinet, Sept. 6, 1986 (div. Oct. 1989); m. Janet Marie Konieczny, Apr. 27, 1991; 1 child, Kevin Michael (dec.). BA in Biology, BSChemE, Washington U., 1982; MS in Cancer Biology, Niagara U., 1986. Assoc. process engr. Ag Chem. Group FMC Corp., Middleport, N.Y., 1982-83, process engr. Ag Chem. Group, 1983-84, sr. process engr. Ag Chem. Group, 1986-90; sr. process engr. divsn. peroxygen chem. FMC Corp., Buffalo, 1990-91, process group leader divsn. peroxygen chem., 1992-93, prod. area supr. divsn. peroxygen chem., 1993—; physics tchr. North Tonawanda High Sch., 1985; mem. new products evaluation bd. Chem. Engring. McGraw Hill, 1983-84; tech. cons. Ag Chem. Group FMC Corp., Middleport, 1985. Mem. Resolve through Sharing Parents Group, Williamsville, N.Y., 1992. Mem. Am. Inst. Chem. Engrs., Am. Chem. Soc. Roman Catholic. Achievements include redesigning Furadan Milling Plant to Reduce N2 usage, persulfate caking issues and development of mineral process development; originated mathematical system to study S1 endonuclease activity on plasmids in alcohol environments using hyperchromic shifts.. Home: 1394 Saybrook Ave North Tonawanda NY 14120 Office: FMC Corp Sawyer Ave and River Rd Tonawanda NY 14150

ROWAN, ANDREW NICHOLAS, biologist, educator; b. Bulawayo, Zimbabwe, May 14, 1946; came to U.S., 1978; s. Albertus Nicholas and Mary Katherine (Skaife) R.; m. Kathryn Julia Swallow, Apr. 7, 1974 (div. 1979); m. Kathleen Bridget Zuroski, Aug. 29, 1981; children: Jennifer, Andrea, Nicholas. BSc, Cape Town (S.Africa) U., 1968; BA, Oxford (Eng.) U., 1971, DPhil, 1975. Editor Pergamon Press, Oxford, 1975-76; sci. adminstr. FRAME, London, 1976-78; assoc. dir. study of animal problems Humane Soc. of U.S., Washington, 1978-82; asst. prof. Sch. Vet. Medicine Tufts U., Boston, 1983-87; assoc. prof. Tufts U., North Grafton, Mass., 1987-93, dir. Ctr. for Animals and Pub. Policy, 1983—; prof. Tufts U., North Gafton, Mass., 1993—; adj. asst. prof. dept. animal biology U. Pa., 1979-82; bd. advisers Johns Hopkins Ctr. Alternatives to Animal Testing, Balt., 1981—, adv. panel office of tech., 1984-85; bd. dirs. Pub. Responsibility in Medicine and Rsch., Boston, 1987—; working groups on ethics of animal rsch., Hastings Ctr., Briarcliff, N.Y., 1988-90, 91-92; sci. adv. panel Am. Humane Assn., Denver, 1988—; lab. animal resources com. Nat. Rsch. Coun., Washington, 1988-91; presenter workshops on animal tech. Author: Of Mice, Models and Men, 1984; editor: The Use of Alternatives in Drug Research, 1980, Animals and People Sharing the World, 1988; founding editor Instn. Jour. Study Animal Problems, 1980-82, Anthrozoos, 1987—; contbr. articles to profl. jours. and mags. Sec. Oxford S. African Scholarship Assn., 1971-74, Washington Cricket League, 1980-81, U.S. Cricket Assn., Phila., 1982. Scarbrow Trust bursary, 1973-75; Rhodes scholar, Oxford, 1968-71; recipient Felix Wankel prize Wankel Trust, Munich, 1980. Fellow Royal Soc. Medicine; mem. AAAS, Mass. Vet. Med. Assn. (chmn. centennial com. hist. rsch. 1985-87), Tissue Culture Assn., Lab. Animal Sci. Assn., Internat. Assn. Biol. Standardization, Coun. Biology Editors, Scientists Ctr. Animal Welfare (bd. dirs. 1987-92), Delta Soc. (bd. dirs. 1987—), Applied Rsch. Ethics Nat. Assn. Episcopalian. Office: Ctr Animals Pub Policy Tufts U 200 Westboro Rd North Grafton MA 01536

ROWE, ALVIN GEORGE, structural engineer, consultant; b. Dubuque, Iowa, May 21, 1933; s. Eldon Mathew Rowe and Eldoris Margaret (Hanson) Burkett; m. Donna Louise Curtis, Apr. 29, 1967; children: John, Richard. BSCE, U. Iowa, 1956; MS in Structural Engring., Iowa State U., 1963. Registered profl. engr., Ga., Ala., Miss., Iowa, Alaska. Commd. 2d lt. U.S. Army, 1956, advanced through grades to col., 1975, with C.E.,, 1956-83; ret., 1983; sr. scientist Lockheed Corp., Marietta, Ga., 1983-88; sr. cons. Surety and Constrn. Cons. (formerly CMA Cons. Group), Marietta, 1988—. Mem. Cobb County (Ga.) Rep. Com., 1983—, precinct chmn., 1988—; campaign mgr. Bob Barr for U.S. Senate, 1992. Decorated Legion of Merit with oak leaf cluster, Bronze Star with "V", Purple Heart; Honor medal 1st class (Vietnam). Fellow ASCE, Soc. Am. Mil. Engrs. (bd. dirs. Atlanta chpt. 1986-88, 91-93, pres. 1988-89); mem. Assn. U.S. Army, Tau Beta Pi, Phi Kappa Phi, Chi Epsilon. Methodist. Achievements include research on shear deformation in prismatic and non-prismatic members. Home: 45 Stonington Pl Marietta GA 30068-3770 Office: Surety and Constrn Cons 1355 Terrell Mill Rd Marietta GA 30067-5441

ROWE, GILBERT THOMAS, oceanography educator; b. Ames, Iowa, Feb. 7, 1942; s. Charles Gilbert and Catherine (Corkery) R.; m. Judith Lee Ingram, Nov. 27, 1962; 1 child, Atticus Ingram. BS in Zoology, Texas A&M U., 1964, MS in Oceanography, 1966; PhD in Zoology, Duke U., 1968. From asst. to assoc. scientist Woods Hole (Mass.) Oceanographic Inst., 1968-79; oceanographer Brookhaven Nat. Lab., Upton, N.Y., 1979-87; prof., head dept. oceanography Tex. A&M U., College Station, 1987—, co-dir. Inst. Marine Life Scis., 1993. Editor: (books) The Sea, 1983, Deep-Sea Food Chains, 1992; U.S. editor: Jour. Marine Systems. Fellow AAAS. Office: Texas A&M U Dept Oceanography College Station TX 77843

ROWE, HARRISON EDWARD, electrical engineer; b. Chgo., Jan. 29, 1927; s. Edward and Joan (Golden) R.; m. Alicia Jane Steeves, Feb. 10, 1951; children—Amy Rogers, Elizabeth Joanne, Edward Steeves, Alison Pickard. B.S. in Elec. Engring, Mass. Inst. Tech., 1948, M.S., 1950, Sc.D., 1952; M of Engring. (hon.), Stevens Inst. Tech., 1988. Mem. tech. staff Radio Research Lab., Bell Labs., Holmdel, N.J., 1952-84; Anson Wood Burchard prof. elec. engring. Stevens Inst. Tech., Hoboken, N.J., 1984—; vis. lectr. U. Calif., Berkeley, 1963, Imperial Coll., U. London, 1968; mem. Def. Sci. Bd. Task Force, 1972-74. Author: Signals and Noise in Communication Systems, 1965; asso. editor: IEEE Trans. on Communication, 1974-76; contbr. articles to profl. jours. Served with USN, 1945-46. Co-recipient Microwave prize, 1972, David Sarnoff award, 1977, AT&T Bell Labs Disting. Tech. Staff award, 1982. Fellow IEEE; mem. Internat. Union Radio Sci., Monmouth Symphony Soc., Monmouth Conservatory Opera Soc., Sigma Xi, Tau Beta Pi, Eta Kappa Nu. Unitarian. Clubs: Shrewsbury Sailing and Yacht, Appalachian Mountain. Patentee in field. Home: 9 Buttonwood Ln Rumson NJ 07760-1045

ROWE, JAY E., JR., research and development director; b. Tacoma, Wash., Jan. 10, 1947; s. Jay E. and Betty V. Rowe; m. Lisha L. Mohn, Aug. 5, 1991; children: Coby, Kistin, Felicity, Robbi, Daniel. BS in Chemistry, Bucknell U., 1968; PhD in Organic Chemistry, Lehigh U., 1973; MBA, St. Joseph's U., Phila., 1987. Rsch. chemist Crompton and Knowles Corp., Reading, Pa., 1974-79, group leader, 1979-91, dir. R & D, 1991—. Office: Crompton & Knowles Corp Dyes & Chemicals Div PO Box 341 Reading PA 19603

ROWE, JOSEPH EVERETT, electrical engineering educator, administrator; b. Highland Park, Mich., June 4, 1927; s. Joseph and Lillian May (Osbourne) R.; m. Margaret Anne Prine, Sept. 1, 1950; children: Jonathan Dale, Carol Kay. B.S. in Engring, U. Mich., 1951, B.S. Engring. in Math, 1951, M.S. in Engring, 1952, Ph.D., 1955. Mem. faculty U. Mich., Ann Arbor, 1953-74; prof. elec. engring. U. Mich., 1960-74, dir. electron physics lab., 1958-68, chmn. dept. elec. and computer engring., 1968-74; vice provost, dean engring. Case Western Res. U., Cleve., 1974-76; provost Case Inst. Tech., 1976-78; v.p. tech. Harris Corp., Melbourne, Fla., 1978-81; v.p., gen. mgr. Harris Corp. (Controls div.), 1981-82; exec. v.p. research and def. Gould Inc., 1982, vice chmn., chief tech. officer, 1983-87; sr. v.p., chief technologist Inst. Research Ill. Inst. Tech., Chgo., 1987; v.p. and chief scientist PPG Industries, Inc., Pitts., 1987-92; v.p., dir. Rsch. Inst., U. Dayton, Ohio, 1992—; cons. to industry.; mem. adv. group electron devices Dept. Def., 1966-78; bd. govs. Research Inst. of Ill. Inst. Tech.; chmn. Coalition for Advancement of Indsl. Tech., U. Ill.; mem. indsl. adv. bd. U. Ill. at Chgo.; mem. Army Sci. Bd., 1985-91. Author: Nonlinear Electron-Wave Interaction Phenomena, 1965, also articles. Recipient Disting. Faculty Achievement award U. Mich., 1970. Fellow AAAS, IEEE (chmn. adminstrv. com. group electron devices 1968-69, editor proc. 1971-73); mem. NAE, Am. Phys. Soc., Am. Soc. Engring. Edn. (Curtis McGraw rsch. award 1964), Am. Mgmt. Assn. (research and devel.), Sigma Xi, Phi Kappa Phi, Tau Beta Pi, Eta Kappa Nu. Office: U Dayton 300 College Pk Dayton OH 45469-0101

ROWE, STEPHEN COOPER, venture capitalist, entrepreneur; b. Glen Ridge, N.J., Dec. 24, 1951; s. Malcolm James and Audrey Ruth (Christian); m. Anne Mary Maddock, June 7, 1986; children: Lauren Elizabeth, Christopher Malcolm. BA, Harvard Coll., 1975; MBA, U. Calif., Berkeley, 1986. Mgr. Genentech, Inc., San Francisco, 1978-80, mktg. mgr., 1980-82, product mgr., 1982-86; dir. mktg. DAC, Foster City, Calif., 1987-88; founder Cell Genesys, Foster City, 1988-90, Cytotherapeutics, Providence, 1989-91; prin. Mayfield Fund, Menlo Park, Calif., 1989-92, Med Impact Ventures, Alameda, Calif., 1986-93; dir. bus. devel., founder Focal, Inc., Cambridge, Mass., 1993—. Home: 1 Chatham Circle Wellesley MA 02181 Office: Focal Inc 1 Kendall Sq Bldg 600 Cambridge MA 02139

ROWELL, JOHN THOMAS, psychologist, consultant; b. Lloyd, Fla., Mar. 21, 1920; s. Irvin Caleb and Esther Estelle (Rowden) R.; m. Mabel Zelma Hartwell Mason, Aug. 15, 1942; children: James Roger, Douglas Hugh, Martin Allen. BA, U. Mich., 1949; MA, Fla. State U., 1950, PhD, 1958. RN, Mass. Chief psychologist Milledgeville (Ga.) State Hosp., 1951-57; human factors scientist RAND Corp., Santa Monica, Calif., 1957, System Devel. Corp., Santa Monica, 1957-58, branch head System Devel. Corp., Ft. Lee, Vt., 1958-61; group leader System Devel. Corp., Santa Monica, 1961-63; br. mgr. System Devel. Corp., Falls Church, Va., 1963-69; pres. N.C. Leadership Inst., Greensboro, N.C., 1969-72; v.p. Essex Corp., Alexandria, Va., 1972; pres. John Rowell Assocs., Greensboro, 1972-87, New Bern, N.C., 1987-93; cons. indsl. orgns. in southeastern U.S.; bd. dirs. Hospice of Pamlico County, Bayboro, N.C., 1990-91. Co-author: National Document Handling Systems for Science and Technology, 1976; contbr. articles to profl. jours. Bd. dirs. Goodwill Industries, Greensboro, 1973-87, Greensboro Mental Health Assn., Greensboro, 1976-81, Epilepsy Assn. N.C., Greensboro, 1977-81. With U.S. Army, 1943-46. Recipient commendation White House, 1975. Mem. Am. Psychol. Assn., N.C. Psychol. Assn., S.E. Psychol. Assn., Guilford County Psychol. Assn. (pres. 1974, Citation 1975), Rotary (chmn. com.). Republican. Avocations: boating, fishing, hunting, genealogy. Home: 1107 Link Ln Box 748 Oriental NC 28571

ROWLAND, FRANK SHERWOOD, chemistry educator; b. Delaware, Ohio, June 28, 1927; m. Joan Lundberg, 1952; children: Ingrid Drake, Jeffrey Sherwood. AB, Ohio Wesleyan U., 1948; MS, U. Chgo., 1951, PhD, 1952, DSc (hon.), 1989; DSc (hon.), Duke U., 1989, Whittier Coll., 1989, Princeton U., 1990, Haverford Coll., 1992; LLD (hon.), Ohio Wesleyan U., 1989, Simon Fraser U., 1991. Instr. chemistry Princeton (N.J.) U., 1952-56; asst. prof. chemistry U. Kans., 1956-58, assoc. prof. chemistry, 1958-63, prof. chemistry, 1963-64; prof. chemistry U. Calif., Irvine, 1964—, dept. chmn., 1964-70, Aldrich prof. chemistry, 1985-89, Bren prof. chemistry, 1989—; Humboldt sr. scientist, Fed. Republic of Germany, 1981; chmn. Dahlem (Fed. Republic of Germany) Conf. on Changing Atmosphere, 1987; vis. scientist Japan Soc. for Promotion Sci., 1980; co-dir. western region Nat. Inst. of Globe Environ. Change; lectr., cons. in field. Contbr. numerous articles to profl. jours. Mem. ozone commn. Internat. Assn. Meteorology and Atmospheric Physics, 1980-88, mem. commn. on atmospheric chemistry and global pollution, 1979-91; mem. acid rain peer rev. panel U.S. Office of Sci. and Tech., Exec. Office of White House, 1982-84; mem. vis. com. Max Planck Insts., Heidelberg and Mainz, Fed. Republic Germany; ozone trends panel mem. NASA, 1986-88; chmn. Gordon Conf. Environ. Scis.-Air, 1987; mem. Calif. Coun. Sci. Tech., 1989—. Recipient numerous awards including John Wiley Jones award Rochester Inst. of Tech., 1975, Disting. Faculty Rsch. award U. Calif., Irvine, 1976, Profl. Achievement award U. Chgo., 1977, Billard award N.Y. Acad. Sci., 1977, Tyler World Prize in Environment Achievement, 1983, Global 500 Roll of Honor for Environ. Achievement UN Environment Program, 1988, Dana award for Pioneering Achievements in Health, 1987, Silver medal Royal Inst. Chemistry, U.K., 1989, Wadsworth award N.Y. State Dept. Health, 1989, medal U. Calif.-Irvine, 1989, Japan prize in Environ. Sci., 1989, Dickson prize Carnegie-Mellon U., 1991; Guggenheim fellow, 1962, 74. Fellow AAAS (pres. elect 1991, pres. 1992, chmn. bd. dirs. 1993), Am. Phys. Soc. (Leo Szilard award for Physics in Pub. Interest 1979), Am. Geophys. Union; mem. NAS (bd. environ. studies and toxicology 1986-91, com. on atmospheric chemistry 1987-89, com. atmospheric scis., solar-terrestial com. 1979-83, co-DATA com. 1977-82, sci. com. on problems environment 1986-89, Infinite Voyage film com. 1988-92, Robertson Meml. lectr. 1993, chmn. com. on internat. orgns. and programs 1993—), Am. Acad. Arts and Scis., Am. Chem. Soc. (chmn. div. nuclear sci. and tech. 1973-74, chmn. div. phys. chemistry 1974-75, Tolman medal 1976, Zimmermann award 1980, E.F. Smith lectureship 1980, Environ. Sci. and Tech. award 1983, Esselen award 1987, Peter Debye Physical Chem. award 1993). Home: 4807 Dorchester Rd Corona Del Mar CA 92625-2718 Office: U Calif Irvine Dept of Chemistry 571 PS1 Irvine CA 92717

ROWLAND, HELEN, geographer. Recipient Svc. to Profession of Geography award Can. Soc. Geographers, 1991. Office: 554 Victoria Ave, Montreal, PQ Canada H3Y 2R6*

ROWLAND, ROBERT CHARLES, clinical psychotherapist, writer, researcher; b. Columbus, Ohio, Jan. 18, 1946; s. Charles Albert and Lorene Bernadine (Friedlinghaus) R.; m. Saundra Marie Gardner, Dec. 21, 1968 (div. Mar. 1987); children: Carrie Ann, Marcus Jules Harrad, Heather Renée. BS in Physiol. Psychology, Ohio State U., 1971, MSW, 1981. Cert. marital and family therapist; cert. in drug and alcohol treatment; cert. sex therapist; cert. hypnotist. Respiratory therapist Mt. Carmel Med. Ctr., Columbus, Ohio, 1965-68; adj. prof. Columbus Ctr. Sci. and Industry, Columbus, Ohio, 1968-71; researcher in tetrahydrocannabinol/learning experiments Ohio State U. Rsch. Ctr., 1970-71; secondary tchr. Columbus (Ohio) Pub. Schs., 1971-73; case cons. Bur. Disability Determination, Columbus, 1973-80; clin. social worker Clarke County Out-Patient Mental Health Ctr., Springfield, Ohio, 1979-80; clin. social worker, Upham Hall Ohio (Columbus) State U. Hosps., 1980-81; clin. psychotherapist Psychol. Systems, Inc., Columbus, 1981-84; psychotherapist, cons. Columbus, 1974-87, Delray Beach, Fla., 1987—; dir. social svc., community rels. Apple Creek (Ohio) Devel. Ctr., 1981-82; pres., rsch. dir. Neurosocial Scis. Inst., Delray Beach, 1987—; former profl. tennis player. Author: Brain Wars-The End of the Drug Game, 1991; contbr. articles to profl. jours. Adv. Neighbor to Neighbor, Delray Beach, 1991-93. Recipient scholarship grant, Ohio State U. Coll. of Social Work, 1980-81. Mem. AAAS, NASW (chmn. Ohio Pace chpt., lobbyist 1980-81, Excellence award 1981, mem. Fla. chpt.), Fla. Freelance Writer's Assn., Union of Concerned Scientists, Palm Beach County Scis. Jour. Club, Alpha Delta Mu. Avocations: rockhounding, scuba diving, tennis, music, chess. Home: 15812 Philodendron Cir Delray Beach FL 33484

ROWLANDS, GEORGE, physics educator; b. Chirk, Wales, Nov. 17, 1932; s. Edmund and Elizabeth Ellen (Edwards) R.; m. Gwenda Dora Pearson, Sept. 17, 1956 (div. 1986); children: Gwilym David, Mathew Philip; m. Inga Starke Harland, Nov. 9, 1990; children: Robert Philip Harland, Janet Louise Pearce. BSc with honors, Leeds (Eng.) U., 1953, PhD, 1956. Postdoctoral fellow NRC, Ottawa, Ont., Can., 1956-57; fellow Atomic Energy Establishment, Harwell, Eng., 1957-60, sr. sci. officer, 1960-62; sr. sci. officer Atomic Energy Establishment, Culham, Eng., 1962-66; sr. lectr. physics U. Warwick, Coventry, Eng., 1966-72, reader, 1972-92; prof. U. Warwick, Coventry, 1992—. Author: Non-Linear Waves, Solitons and Chaos, 1990, Introduction to Non-Linear Phemomena, 1990. Fellow Inst. Physics. Avocations: tennis, walking. Home: 111 Bridge End, Warwick CV34 6PD, England Office: U Warwick Dept Physics, Coventry CV4 7AL, England

ROWND, ROBERT HARVEY, biochemistry and molecular biology educator; b. Chgo., July 4, 1937; s. Walter Lemuel and Marie Francis (Joyce) R.; m. Rosalie Anne Lowery, June 13, 1959; children: Jennifer Rose, Robert Harvey, David Matthew. BS in Chemistry, St. Louis U., 1959; MA in Med. Scis, Harvard U., 1961, PhD in Biophysics, 1963. Postdoctoral fellow Med. Rsch. Coun., NIH, Cambridge, Eng., 1963-65; postdoctoral fellow Nat. Acad. Scis.-NRC, Institut Pasteur, Paris, 1965-66; prof., chmn. molecular biology and biochemistry U. Wis., Madison, 1966-81; John G. Searle prof., chmn. molecular biology and biochemistry Med. and Dental Schs., Northwestern U., Chgo., 1981-90; leader cancer molecular biology program Cancer Ctr.Northwestern U., Chgo., 1982-89; prof. biochemistry, dir. Ctr. for Molecular Biology Wayne State U., Detroit, 1990—, interim chair dept. molecular biology and genetics, 1993—; vice chmn. Gordon Rsch. Conf. Extrachromosomal Elements, 1984, chmn., 1986; hon. rsch. prof. Biotech. Rsch. Ctr., Chines Acad. Agrl. Scis., Beijing, 1987—. Contbr. numerous articles to sci. jours. and books; mem. editorial bd. Jour. of Bacteriology, 1975-81, editor, 1981-90; assoc. editor Plasmid, 1977-87. Mem. troop com., treas. Four Lakes coun. Boy Scouts Am., Madison, 1973-77, mem. People to People Program del. of microbiologists to China, 1983; mem. Nat. Acad. Scis./Nat/ Rsch. Coun. Com. on Human Health Effects of Subtherapeutic Antibiotic Use in Animal Feeds, 1979-81; sr. tech. adv. recruitment cons. United Nations Devel. Program in China, 1987. NSF fellow, NIH fellow, 1959-66, rsch. grantee, 1966—, tng. grantee, 1970-79, 83-91; USPHS Rsch. Career Devel. awardee, 1968-73; recipient Alumni Merit award St. Louis U., 1984. Mem. NIH (microbial genetics study sect. 1978-82, dir. med. scientist tng. program 1982-90, adv. panel Nat. Rsch. Coun. 1974-77, chmn. 1976-77, adv. panel for devel. biology 1968-71), Am. Soc. Microbiology, Assn. Harvard Chemists, Am. Soc. Biol. Chemists, Am. Acad. Microbiology, N.Y. Acad. Scis. Home: 14010 Harbor Place Dr Saint Clair Shores MI 48080-1528

ROXEY, TIMOTHY ERROL, nuclear engineer, biomedical consultant; b. Diana, Fla., Sept. 10, 1950; m. Anne E. Todd, Aug. 16, 1971; children: Debbie, Donnalyn. BS, U. Fla., 1979, MS, 1981. Engr. Fla. Power and Light, Miami, 1981-83; sr. rsch. scientist R.T.S. Labs., Gainesville, Fla., 1983-85; owner, pres. Eclectic Technologies, Inc., various locations, 1985—; rsch.engr. U. Fla., Gainesville, 1985-87; engr. Am. Electric Power Svc. Co., Columbus, Ohio, 1987-91, sr. engr. Baltimore Gas and Electric, Lusby, Md., 1991—. Contbr. articles to Applied Optics, Circulation, other profl. jours. With U.S. Army, 1969-72. Fellow Am. Soc. Laser Medicine and Surgery. Achievements include patent for intra-ocular lens design. Office: Calvert Cliffs Nuclear Power Plant PO Box 1536 Lusby MD 20657

ROY, BIMALENDU NARAYAN, ceramic engineering educator; b. Calcutta, India; s. Sachindra Nath and Prative (Dhar) R.; m. Manju Gupta, Sept. 17, 1979; 1 child, Niladri. BSc, Calcutta U., 1961, BA, 1963, LLB, 1966; postgrad. diploma, Leeds (Eng.) U., 1968; PhD, London U., 1972. Lectr. U. Sains Malaysia, Penang, 1973-76; sci. officer Cen. Glass and Ceramic Rsch. Inst., Calcutta, 1977-78; sr. lectr. U. Sci. and Tech., Kumasi, Ghana, 1978-80; sr. rsch. fellow Newcastle (Eng.) Poly., 1984-85; chief rsch. officer U. Essex (Eng.), 1986-89; sr. lectr. Sheffield (Eng.) Hallam U., 1990-92, U. Brunei Darussalam, 1993—; freelance abstractor Derwent Pub., London, 1969-71, PRM Sci. and Tech. Agy., London, 1972-73, Inst. Metals, London, 1985-88; vis. prof. Mont. State U., 1981-82, Ariz. State U., 1983-84. Author: Crystal Growth from Melts: Applications to Growth of Groups 1 and 2 Metals, 1991, Thermodynamics, 1993; contbr. numerous articles to profl. jours. Gen. sec. community orgn., London, 1985-86, v.p., 1988-92. Rsch. grantee Lady Mountbatten Trust, Eng., Sidney Perry Found., Eng., Curzon Wylie Found., Eng., 1969-71, Ministry Edn., Malaysia, 1974, U. Sci. and Tech., Ghana, 1979, Royal Soc., London, 1991; recipient First Literary prize Belur Sch., India, 1951, 52, 53, 54, Rotary Internat., India, 1955, Nehru Birth Day Celebration Com., India, 1955, Ctrl. Calcutta Coll., 1960, 61, Second prize Bally Cultural and Lit. Soc., India, 1961, Youth Festival Com., India, 1964. Mem. AAAS, N.Y. Acad. Scis. Avocations: writing, travelling, photography, scouting, games. Home: 47 Arlesford Rd, London SW9 9JS, England Office: Univ Brunei Darussalam, Dept Physics, Gadong 3186 Bandai Seri Begawan Brunei

ROY, CHUNILAL, psychiatrist; b. Digboi, India, Jan. 1, 1935; came to Can., 1967, naturalized, 1975; s. Atikay Bandhu and Nirupama (Devi) R.; m. Elizabeth Ainscow, Apr. 15, 1967; children: Nicholas, Phillip, Charles. MB, BS, Calcutta Med. Coll., India, 1959; diploma in psychol. medicine, Kings Coll., Newcastle-upon-Tyne, Eng., 1963. Intern Middlesborough Gen. Hosp., Eng., 1960-61; jr. hosp. officer St. Luke's Hosp., Middlesborough, Eng., 1961-64, sr. registrar, 1964; sr. hosp. med. officer Parkside Hosp., Macclesfield, Eng., 1964-66; sr. registrar Moorehaven Hosp., Ivybridge,

Eng., 1966; reader, head dept. psychiatry Maulana Azad Med. Coll., New Delhi, 1966; sr. med. officer Republic of Ireland, County Louth, 1966; sr. psychiatrist Sask. Dept. Psychiat. Services, Can., 1967-68; regional dir. Swift Current, Can., 1968-71; practice medicine specializing in psychiatry Regina, Sask., Can., 1971-72; founding dir., med. dir. Regional Psychiat. Ctr., Abbotsford, B.C., Can., 1972-82; with dept. psychiatry Vancouver Gen. Hosp., 1987—; cons. to prison adminstrs.; hon. lectr. psychology and clin. prof. dept. psychiatry U. B.C.; ex-officio mem. Nat. Adv. Com. on Health Care of Prisoners in Can.; cons. psychiatrist Vancouver Hosp.; advisor Asian chpt. Psychosomatic Medicine, World Congress of Law and Medicine, New Delhi, 1985. Author: (with D.J. West and F.L. Nichols) Understanding Sexual Attacks, 1978; co-author: Oath of Athens, 1979; mem. editorial rev. bd. Evaluation, 1977—; assoc. editor Internat. Jour. Offender Therapy and Comparative Criminology, 1978—; mem. Bd. Internat. Law Medicine, 1979—; field editor Jour. of Medicine and Law; contbr. articles to profl. jours. Recipient merit award Dept. Health, Republic of Ireland, 1966, merit award Can. Penitentiary Svc., 1974, merit award Correctional Svcs. Can., 1983, citation by pres. U. B.C., 1983; knighted by Order of St. John Ecumenical Found., 1993; Hon. Consul designate Burkina Faso. Fellow Royal Coll. Psychiatry (Can.), Royal Coll. Psychiatry (Eng.), Pacific Rim Coll. Psychiatrists (a founder); mem. World Psychiat. Assn. (sec. sect. forensic psychiatry 1983), World Fedn. Mental Health, Internat. Coun. Prison Med. Services (founding sec.-gen. 1977), Can. Med. Assn., Can. Psychiat. Assn., Amnesty Internat., Internat. Acad. Legal Medicine and Social Medicine, Indian Psychiat. Assn. (life), Can. Assn. Profl. Treatment Offenders (founding dir. 1975), Assn. Physicians & Surgeons Who Work in Can. Prisons (founding pres. 1974), Internat. Found. for Tng. in Penitentiary Medicine and Forensic Psychiatry (founding pres. 1980, vice-chmn., sec.), World Psychiatry Assn., Australian Acad. Forensic Sci. (corr.), Can. Physicians Interested in South Asia (v.p. 1989, pres. 1990), Internat. Coll. Psychosomatic Med. (adv. Asian Chapt.), Internat. Conf. of Health, Culture, & Contemporary Soc. (chief advisor Bombay, India 1989, v.p. 1989, pres. 1990), Internat. Coun. Peniteniary Medicine (founding sec., bd. dirs.), World Psychiat. Assn. (vice chmn. forensic psychiat. sect. 1989), Order of St. John (knight 1992—), mem. bd. dirs., Vancouver Multi Cultural Soc. Home: 2439 Trinity St, Vancouver, BC Canada V5K 1C9 Office: 1417-750 W Broadway, Vancouver, BC Canada V5Z 1J4

ROY, FRANCIS CHARLES, electrical engineer; b. Iota, La., Nov. 28, 1926; s. fernan A. and Gussie Marie (Matte) R.; m. Pauline Bertha Dischler, June 4, 1949; children: Mary Monica, Michael Anthony, Richard Regan (dec.). BSEE, La. State U., 1949; MSEE, U. Tex., 1958. Registered profl. engr., La. Engr. Allis Chalmers Mfg. Co., Milw., 1949-52; instr. Jeff Davis Vocat.-Tech., Jennings, La., 1952-55; assoc. prof. engring. La. Tech. U., Ruston, 1955-65; v.p. Indsl. Equipment and Engring., Lake Charles, La., 1965—; adj. prof. elec. engring. McNeese State U., Lake Charles, 1984—; mem. La. State Bd. Registry for Profl. Engrs. and Land Surveyors, New Orleans, 1980-89, chmn. 1985-86; v.p., bd. dirs. Nat. Coun. Examiners for Engring. and Surveying, 1987-89, mem. exam. for profl. engrs. com., 1983—; cons. United Gas Rsch. Lab., Shreveport, 1965; speaker in field. Mem. edn. coun. La. Assn. Bus. and Industry, 1978-84; mem. constrn. design svcs. com. La. Legis.,1 981, econ. devel. com., 1980; mem. Task Force for tchr. Recruitment and Retention, 1983. Recipient Leo M. Odom award, 1982. Mem. IEEE (sr., ad hoc visitor for Engring. Accreditation Commn. of Accreditation Bd. Engring. and Tech. 1984-89), NSPE, La. Engring. Soc. (state pres. 1979-80, , Professionalism award 1980). Republican. Roman Catholic. Home: 1435 Watkins St Lake Charles LA 70601 Office: Indsl Equipment & Engring PO Box 1663 Lake Charles LA 70602

ROY, GABRIEL DELVIS, scientific researcher, research manager; b. Neyyoor, Tamilnadu, India, July 7, 1939; came to U.S., 1971; s. Arulappa and Lizzie (Jabez) Gabriel; m. Vimala Johncibel Baylis, May 25, 1961; children: Suchitra, Sitara, Navaroop, Sreeroop. BS, U. Kerala, Trivandrum, India, 1960, MS, 1965; PhD, U. Tenn., 1977. Lectr. Coll. Engring., Trivandrum, 1960-65, asst. prof., 1965-71; rsch. asst. U. Tenn. Space Inst., Tullahoma, 1971-76, sr. rsch. engr., 1976-78, group leader, 1978-81; program mgr. TRW Inc., Redondo Beach, Calif., 1981-87; scientific officer and program mgr. Office of Naval Rsch., Arlington, Va., 1987—; vis. prof. Calif. State U., Long Beach, 1983-87; program reviewer Dept. of Energy, Pitts., 1988—, Strategic Def. Initiative Orgn., Washington, 1988—; Indepent Rsch. and Devel. reviewer various industries in U.S., 1988—; tech. adv. com. Army Rsch. Office, 1993—. Contbr. articles to profl. jours. Patentee carbonaceous slurry injector, slagging combustor. Sec. Trivandrum Art Group, 1969-71; com. chmn. Jr. Chamber Internat., Trivandrum, 1986-70; judge local schs., Washington, 1989—; v.p. Mar Thoma Ch. of Greater Washington, Langley Park, Md., 1990—. Recipient Duthie Meml. award Scott Christian Coll., Nagercoil, India, 1955, 56, Arch. Design award Sri Ram Indsls., Coimbatore, India, 1970, John F. Louis award ASME Energy Systems, 1991, Roll of Honor award TRW, 1983. Fellow AIAA (assoc., mem. terrestrial energy systems 1993—, plasma dynamics and lasers com. 1987-90, vice chmn. propellants and compustion 1990-92, assoc. editor Jour. Propulsion and Power 1989—); mem. ASME, Sigma Xi. Avocations: painting, sculpture, photography, poetry, fiction. Home: 9944 Great Oaks Way Fairfax VA 22030 Office: Office Naval Rsch 800 N Quincy St Arlington VA 22030

ROY, HAROLD EDWARD, research chemist; b. Stratford, Conn., June 2, 1921; s. Ludger Homer and Meta (Jepsen) R.; B.A., Duke U., 1950; m. Joyce E. Enslin, Oct. 9, 1946 (div. 1975); children—Glenn E., Barbara Anne, Suzanne Elizabeth; m. Gail LaVer Jensen, Feb. 11, 1983. Chemist research div. Lockheed Propulsion Co., Redlands, Calif., 1957-61; sec., treas. The Halgene Corp., Riverside, Calif., 1961-63; self-employed chemist, Glendora, Calif., 1963-64; chief engr. propellant devel. Rocket Power, Inc., Mesa, Ariz., 1964-65; cons., Glendora, 1965-66; engring. specialist Northrop Corp., Anaheim, Calif., 1966-69; pres. Argus Tech., Beverly Hills, 1969-70, dir. Harold E. Roy & Assos., Glendora, 1969—. Served to lt. (j.g.) USNR, 1943-46. Mem. Exptl. Aircraft Assn., Am. Ordnance Assn., Am. Inst. Aeros. and Astronautics, Internat. Platform Assn., Acad. Parapsychology and Medicine, Calif. Profl. Hypnotists Assn., World Future Soc. Republican. Home: 143 Warren Rd PO Box 414 Selma OR 97538

ROY, KAREN MARY, limnologist, state government regulator; b. Ft. Kent, Maine, Dec. 7, 1953; d. Hercules Rosaire and Maureen Pricilla (Daigle) R.; m. Steven Carl Engelhart, June 1, 1985; children: Samuel Carl Engelhart, Noah Cyr Engelhart. BA in Human Ecology, Coll. of Atlantic, 1977; MS in Water Resources, U. Vt., 1984. Engring. aide Army C.E. Lab., Lyme, N.H., 1974; aquaculture rsch. asst. Coastal Resources Ctr., Bar Harbor, Maine, 1975-77; editorial rsch. mgr. WoodenBoat Mag., Brooklin, Maine, 1978-82; grad. rsch. asst. U. Vt., Burlington, 1982-84; project analyst Adirondack Pk. Agy., Ray Brook, N.Y., 1984—; bd. dirs. Lake Champlain Com., Burlington; cons. Davis Assocs., Wadhams, N.Y., 1990—. Named Outstanding Alumna Lectr., U. Vt., 1986-87. Mem. Assn. State Wetland Mgrs., Am. Water Resources Assn., New Eng. Assn. of Freshwater Biologists, N.Am. Lake Mgrs. Soc. Democrat. Unitarian Universalist. Office: Adirondack Pk Agy Rt 86 PO Box 99 Ray Brook NY 12977

ROY, KENNETH RUSSELL, educator; b. Hartford, Conn., Mar. 29, 1946; s. Russell George and Irene Mary (Birkowski) R.; BS, Central Conn. State Coll., New Britain, 1968, MS, 1974; 6th yr. degree in profl. edn. U. Conn., 1981, Ph.D., 1985; m. Marisa Anne Russo, Jan. 27, 1968; children: Lisa Marie, Louise Irene. Tchr. sci. Rocky Hill (Conn.) High Sch., 1968-73, N.W. Cath. High Sch., West Hartford, Conn., 1973-74; sci. and math. coord. Bolton (Conn.) High Sch., 1974-78; chmn. scis. Bacon Acad., Colchester, Conn., 1978-81; K-12 dir. sci. Glastonbury (Conn.) Pub. Schs., 1981—; mem. adj. faculty Manchester C.C., 1976—, Tunxis C.C., 1975—; instr. U. Conn. Coop. Program, 1974-78; cons./adv. Project Rise, 1978-81; lectr., sci. curriculum cons. various Conn. sch. dists.; nat. dir. Nat. Sci. Suprs. Assn., 1988-91, exec. dir. Leadership Inst. Cen. Conn. State U., New Britain, 1989-91; exec. dir. Nat. Sci. Suprs. Assn., 1992—. Co-editor Conn. Jour. Sci. Edn., 1984-88; editor Sci. Leadership Trend Notes, 1989-91; contbr. articles to profl. jours. Mem. St. Christopher Sch. Bd., 1982-83. Recipient Disting. Educator's and Conn. Educator's awards Milken Family Found., 1989; named Tchr. of Yr., Colchester, 1980; NSF grantee, 1968, staff devel. grantee, 1979, 80, Nat. Sci. Supr. Leadership Conf. grantee, 1980. Mem. ASCD, AAAS, Nat. Sci. Tchrs. Assn., Nat. Sci. Suprs. Assn. (pres.-elect 1986-87, pres. 1987-88), Conn. Sci. Tchrs. Assn., Conn. Sci. Suprs. Assn. (pres. 1985-86), Conn. Assn. Profl. Devel., Conn. Assn. Supervision and Curriculum Devel., Glastonbury Adminstrs. and Suprs. Assn., Nat. Ctr. Improvement Sci. Teaching and Learning (mem. adv. bd. 1988-91), Internat. Council Assns. Sci. Edn. (nat. rep. 1987-88, N.Am. region rep. and exec. com. mem. 1989—), Phi Delta Kappa. Roman Catholic. Office: Glastonbury Pub Schs Glastonbury CT 06033

ROY, ROB J., biomedical engineer, anesthesiologist; b. Bklyn., Jan. 2, 1933; m. Carole Ann Roy, Aug. 1, 1959; children: Robert Bruce, David John, Bruce Glenn. BSEE, Cooper Union, N.Y.C., 1954; MSEE, Columbia U., 1956; DEngSc, Rensselaer Poly. Inst., 1962; MD, Albany (N.Y.) Med. Coll., 1976. Profl. engr., N.Y.; diplomate Am. Bd. Anesthesiology. Prof. elec. engrin. dept. Rensselaer Poly. Inst., Troy, N.Y., 1962—, prof. biomed. engring. dept., 1980—, head biomed. engring. dept., 1986—; cons. numerous cos. Author: State Variables for Engineers, 1965; author 150 papers in field. Sr. mem. IEEE; mem. Am. Soc. Anesthesiologists, Sigma Xi. Home: 10 Sevilla Dr Clifton Park NY 12065 Office: Rensselaer Polytech Inst Dept Biomedical Engring Jonsson Engring Ctr Troy NY 12180-3590

ROY, ROBIN K., government official; b. Bellfonte, Pa., Apr. 13, 1959; s. R.R. and M.L. (Kelley) R.; m. Cathy R. Zoi, Aug. 16, 1987; children: Wyatt R., Susha B. BSEE, Stanford U., 1981, MS in Engring. Econ. Systems, 1981, PhD in Civil Engring., 1987. Energy economist Pacific Gas & Elec. Co., San Francisco, 1982-87; project dir., fellow Office of Tech. Assessment, Washington, 1987—. Contbg. author: Annual Review of Energy and Environment, 1992. Office of Tech. Assessment fellow, 1987, ARCS Found. fellow, 1985; recipient Blue Pencil award Nat. Assn. Govt. Comms., 1992. Home: Arlington VA 22201 Office: Office of Tech Assessment 600 Pennsylvania Ave SE Washington DC 20003

ROY, RUSTUM, interdisciplinary materials researcher, educator; b. Ranchi, India, July 3, 1924; came to U.S., 1945, naturalized, 1961; s. Narendra Kumar and Rajkumari (Mukherjee) R.; m. Della M. Martin, June 8, 1948; children: Neill, Ronnen, Jeremy. BS, Patna (India) U., 1942; MS, Patna (India) U., India, 1944; Ph.D., Pa. State U., 1948; DSc (hon.), Tokyo Inst. Tech., 1987, Alfred U, 1993. Research asst. Pa. State U., 1948-49, mem. faculty, 1950—, prof. geochemistry, 1957—, prof. solid state, 1968—, chmn. solid state tech. program, 1960-67, chmn. sci. tech. and soc. program, 1977-84, dir., 1984—, dir. materials research lab., 1962-85, Evan Pugh prof., 1981—; sr. sci. officer Nat. Ceramic Lab., India, 1950; mem. com. mineral sci. tech. Nat. Acad. Scis., 1967-69, com. survey materials sci. tech., 1970-74; exec. com. chem. div. NRC, 1967-70, nat. materials adv. bd., 1970-77, mem. com. radioactive waste mgmt., 1974-80, chmn. panel waste solidification, 1976-80; chmn. com. NRC, USSR and Eastern Europe, 1976-81; mem. com. material sci. and engring. NRC, 1986-89; mem. Pa. Gov.'s Sci. Adv. Com.; chmn. materials adv. panel Gov.'s Sci. Adv. Com., 1965; mem. adv. com. on engring. NSF, 1968-72, adv. com. to ethical and human value implications sci. and tech., 1974-76, adv. com. div. materials rsch., 1974-77; Hibbert lectr. U. London, 1979; bd. dirs. Kirkridge, Inc., Bangor, Pa.; cons. to industry; mem. adv. com. Coll. Engring., Stanford U., 1984-86. Author: Honest Sex, 1968, Crystal Chemistry of Non-metallic Materials, 1974, Experimenting with Truth, 1981, Radioactive Waste Disposal, Vol. 1, the Waste Package, 1983, Lost at the Frontier, 1985; also articles.; editor-in-chief: Materials Research Bull, 1966—, Bull. Sci. Tech. and Soc, 1981—. Chmn. bd. Dag Hammarskjold Coll., 1973-75; chmn. ad hoc com. sci., tech. and ch. Nat. Council Chs., 1966-68. Sci. policy fellow Brookings Instn., 1982-83. Fellow Indian Acad. Scis. (hon.); mem. AAAS (chmn. chemistry sect. 1985), NAE, Nat. Res. Coun. (internat. sci. lectr., 1991-92), Royal Swedish Acad. Engring. Scis. (fgn.), Indian Nat. Acad. Sci. (fgn.), Engring. Acad. Japan, Fedn. Materials Socs. (Nat. Materials Advancement award 1991), Ceramic Soc. Japan (Centennial award, 1991, hon. mem. 1991), Mineral. Soc. Am. (award 1957), Fine Ceramics Assn. Japan (Internat. award), Am. Chem. Soc. (Petroleum Rsch. Fund award 1960, Dupont award for Chem. of Materials, 1993), Am. Ceramic Soc. (Sosman lectr. 1975, Orton lectr. 1984, disting. life mem. 1993, Educator of Yr. 1993), Am. Soc. Engring. Educators (Centennial medal 1993, named to Hall of Fame, 1993), Nat. Assn. Sci., Tech. and Soc. (pres. 1988—), Materials Rsch. Soc. (founder, pres. 1976). Home: 528 S Pugh St State College PA 16801-5312 Office: 102 Materials Rsch Lab University Park PA 16802

ROY, TUHIN KUMAR, engineering company executive; b. Monghyr, India, Aug. 1, 1923; s. Rakhal Raj and Bijoyini (Gupta) R.; m. Silva Mardiste, Jan. 1, 1951; children: Dipak, Rupak, Indrek. BSc, Calcutta U., 1943, MSc, 1945; MS, MIT, 1949, ScD, 1951. Head metals rsch. Chem. Constrn. Corp., N.Y.C., 1951-54; prof., head chem. engring. dept. Jadavpur U., 1954-56, 58-60; cons. Freeport Minerals Co., New Orleans, 1956-58; mng. dir. Indsl. Cons. Bur., New Delhi, 1960-63; sr. exec. Sci. Design Co., N.Y.C., 1963-65; mng. dir. Chem. and Metall. Design Co. Pvt. Ltd., New Delhi, 1966-89; chmn. CMDC Design Pvt. Ltd., New Delhi, 1990—; v.p. SLBR Internat. Ltd., Hamilton, Ont., Can.; dir. Haldia Petrochem. Corp., Consultancy Consortium P. Ltd., Chemcrown (India) Ltd. Contbr. articles to profl. jours.; patentee in hydrometallurgy and chemical technology. Mng. trustee B. Jagtiani Charitable Trust. Fellow Indian Acad. Scis., Indian Nat. Acad. Engring.; mem. Am. Inst. Chem. Engrs., Indian Inst. Chem. Engrs. (Chem. Engr. of Yr. award 1983), Nat. Assn. Cons. Engrs. New Delhi (past pres.). Hindu. Home: C 6/3 Safdarjung Dev Area, New Delhi 110016, India Office: A-89, Malviya Nagar, New Delhi 110 017, India

ROYCHOUDHURI, CHANDRASEKHAR, physicist; b. Barisal, Bengal, India, Apr. 7, 1942; s. Hiralal and Amiyabala (Sengupta) R.; children: Asim, Onnesha. BS in Physics, Jadavpur U., India, 1963; MS in Physics, Jadavpur U., 1965; PhD, U. Rochester, 1973. Asst. prof. U. Kalyani, West Bengal, India, 1965-68; sr. scientist Nat. Inst. Astrophysics, Puebla, Mex., 1974-78; sr. staff scientist TRW Inc., Redondo Beach, 1978-86; mgr. laser systems Perkin-Elmer, Danbury, Conn., 1986-89; chief scientist Optics & Applied Tech. Lab., UTOS, West Palm Beach, Fla., 1990-91; dir. Photonics Rsch. Ctr. U. Conn., Storrs, 1991—. Author: chpt. Optical Shoptesting, 1978; contbr. articles to profl. jours. Fulbright scholar U. Vt., 1968. Mem. IEEE, Optical Soc. Am., Soc. Photo-optical Instrumentation, Am. Phys. Soc. (life). Avocations: hiking, spl. edn. programs for children.

ROYDS, ROBERT BRUCE, physician; b. Harrogate, England, Oct. 3, 1944; came to U.S., 1974; s. John Edmund and Ailsa Dorothea (Williams) R.; m. Marilyn Maria Valerio, Apr. 28, 1948; children: Elizabeth Caroline, Leslie Alexandra. M.B., B.S., U. London, 1967, M.R.C.P., 1970. Sr. house officer Royal Northern Hosp., London, 1968; sr. house officer Luton and Dunstable Hosp., Beds, England, 1968-69; registrar St. Albans City Hosp., Herts, England, 1969-70; research fellow clin. pharmacology dept. St. Bartholomew's Hosp., U. London, London, 1970-72; chief asst., sr. registrar med. professorial unit St. Bartholomew's Hosp., U. London, 1972-74; assoc. dir. Merck, Sharp & Dohme, Inc., Rahway, N.J., 1974-75; sr. research physician Hoffmann-La Roche Inc., Nutley, N.J., 1976-78; v.p. Besselaar Assocs., Princeton, N.J., 1979-82; pres. Theradex Systems, Inc., Princeton, 1982—; cons. Ctr. for Rsch. Mothers/Infants Nat. Inst. Child Health & Human Devel., Washington, 1983. Bd. trustees Chapin Sch., Princeton, 1984-89, pres. bd. trustees, 1986-89; pres. Riverside Condominium Assn., Cranford, N.J., 1978-79. Fellow Royal Soc. Medicine; mem. Royal Coll. Physicians, Am. Coll. Clin. Pharmacology Therapeutics, Am. Soc. for Clin. Research (sr. mem.), Am. Soc. Microbiology. Home: 5 Quick Ln Plainsboro NJ 08536-1424

ROYER, RONALD ALAN, entomologist, educator; b. Des Moines, Feb. 27, 1945; s. Charles Melvin and Sylva Edith (Noe) R.; children from previous marriages: Jon Michael, Lara Lanai, Jesse Alan, Adam Joseph; m. Margaret Ruth Tavis, Nov. 27, 1982; children: Daniel Grant, Emily Ann, Noah Michael. BS, Iowa State U., 1972; PhD, U. N.D., Grand Forks, 1984. Rsch. asst. Iowa State U., Ames, 1972-73; teaching asst. Bemidji (Minn.) State U., 1973-74; instrn. coord. Long Lake Conservation Ctr., Palisade, Minn., 1977-82; asst. prof. sci. edn. U. Minn., Morris, 1984-85; assoc. prof. sci. Minot (N.D.) State U., 1985—, dir. honors program, 1989—. Artist (cover illustration) Jour. of the Lepidopterists' Soc., 1988; author (book) Butterflies of North Dakota: an Atlas and Guide, 1988; contbr. articles to profl. jours. Mem. The Lepidopterists' Soc. (Great Plains coord., 1988—), Soc. for Rsch. on the Lepidoptera, The Xerces Soc., N.D. Natural Sci. Soc., N.D. Acad. Sci. (mem. exec. bd.). Presbyterian. Achievements include discovery of numerous range extensions and distribution records for N.Am. butterflies. Office: Minot State U 500 University Ave W Minot ND 58707

ROYSAM, BADRINATH, computer scientist, educator; b. T. Narasipur, India, Dec. 22, 1961; s. Nagaraj and Vimala Roysam. BTech, Indian Inst. Tech., Madras, 1984; MS, Washington U., St. Louis, 1987, PhD, 1989. Asst. prof. elec., computer and systems engring. Rensselaer Poly. Inst., Troy, N.Y., 1989—; cons. Inst. for Biomed. Computing, St. Louis, 1989—. Contbr. articles to Procs. of NAS, Digital Signal Processing, Jour. of Microscopy Rsch. and Tech., IEEE Transactions on Neural Networks, others. NSF grantee, 1991. Mem. IEEE (local arrangements chair, Prize Paper award), Electron Microscopy Soc. Am. (Presdl. Scholar award), Sigma Xi. Hindu. Achievements include a computer algorithm for low-count EM autoradiography; a method for computationally visualizing exit signs through smoke; a 3-D microscope and algorithms for automated pap smear analysis, neuron tracing, cell population analysis. Home: 55 Southbury Rd Clifton Park NY 12065 Office: Rensselaer Poly Inst Troy NY 12180

ROYSTON, IVOR, scientific director; b. Belford, Eng., Apr. 29, 1945; m. Colette Carson. BS in Human Biology, John Hopkins U., 1967, MD, 1970, postgrad., 1970. Diplomate Am. Bd. Internal Medicine, Am. Bd. Med. Oncology. Intern in internal medicine Stanford (Calif.) U. Hosp., 1970-71, resident in internal medicine, 1971-72; staff assoc. div. virology Bur. of Biologics (formerly div. Biologic Stds. at NIH), Bethesda, Md., 1972-73, chief viral oncology sect., div. virology, 1973-75; postdoctoral fellow Div. Oncology, Dept. Medicine, Stanford U. Med. Ctr., 1975-77; asst. prof. medicine Div. Hematology/Oncology, U. Calif. Sch. Medicine, San Diego, 1977-82; staff physician San Diego Vets. Adminstrn. Med. Ctr., 1977-78, clin. investigator, 1978-81; dir. Cell Surface Marker Lab. Cancer Ctr., U. Calif., San Diego, 1981-90; chief oncology sect. med. svc. San Diego VA Med. Ctr., 1982-84; assoc. prof. medicine hematology/oncology divsn. U. Calif., 1982-90, dir. clinical immunology program cancer ctr., 1984-90, prof. medicine, 1990-91; pres., sci. dir. San Diego Reg. Cancer Ctr., 1990—; bd. dirs. UniSyn Techs., Inc., San Diego, 1991—, Somatix Therapy Corp., Alameda, Calif.; adj. prof. medicine U. Calif. San Diego, Sch. Medicine, 1990—; founder, dir., cons. Hybritech, Inc., La Jolla, 1978-86, IDEC Pharms., Inc., LaJolla, 1985-92; immunology com. Cancer and Leukemia Group B., 1980-91; vice chmn. immunology com., 1981-91; cons. mem. biol. response modifiers program decision, network com., Div. of Cancer Treatment, Nat. Cancer Inst., 1982-85; mem. Clin. Cancer Program Project Rev. Com., Nat. Cancer Inst., 1983-88, Long Range Planning com. U. Calif. Cancer Ctr.; mem. adv. com. U. Calif. San Diego Biotechnology Transfer Faculty; co-dir. Internat. Conf. Monoclonal Antibody Immunoconjugates for Cancer, 1986—, Internat. Conf. Gene Therapy of Cancer, 1992—; mem. merit rev. bd. oncology Dept. Vets. Affairs, 1992—. Editorial bd. Jour. of Biol. Response Modifiers, Hybridoma; assoc. editor Jour. of clin. Lab. Analysis, Antibody, Immunoconjugates and Radio Pharmaceuticals, Molecular Biology of Cancer; frequent reviewer Cancer Research, Blood, Jour. of Clin. Oncology; adv. bd. CRC Critical Rev. in Oncology/Hematology. Bd. trustees La Jolla Mus. of Contemporary Art, 1985-86, La Jollar Playhouse, 1986-91, Francis Parker Sch., San Diego, 1989—; bd. dirs. Am. Cancer Soc. With USPHS, 1972-74. Ford Found fellowship, 1969-70; recipient Johns Hopkins Med. Soc. award 1970, Clin. Investigator award VA, 1978-81; named Bus. Leader of Yr., San Diego Venture Group, 1990, Man of Yr., 1991. Fellow ACP; mem. AAAS, Am. Soc. of Microbiology, Am. Fedn. for Clin. Rsch., Am. Assn. for Cancer Rsch., Am. Soc. for Clin. Oncology, Internat. Assn. for Comparative Rsch. on Leukemia, Am. Soc. of Hematology, Transplantation Soc., Am. Assn. of Immunologists, Am. Soc Clin. Investigation. Achievements include patents for monoclonal antibody compositions specific for single antigens in antigen aggregates; immunoglobulin secreting human hybridomas from a cultured lymphoblastoid cell line. Office: San Diego Regional Cancer Ctr Ste 200 2099 Science Park Rd San Diego CA 92121

ROZE, ULDIS, biologist, author; b. Riga, Latvia, Jan. 3, 1938; came to U.S., 1950; s. Ernests and Lucia R.; married, June 24, 1966; 1 child, Rachel. BS, U. Chgo., 1959; PhD, Washington U., St. Louis, 1964. Instr. Queens Coll., Flushing, N.Y., 1964-67, asst. prof., 1967-76, assoc. prof., 1976-90, prof. biology, 1990—. Author: The Living Earth, 1976, The North American Porcupine, 1989; contbr. articles to sci. publs. Mem. AAAS, Am. Soc. Mammalogists, Ecol. Soc. Am., Wilderness Soc. Office: Dept Biology Queens Coll Flushing NY 11367

ROZENBERGS, JOHN, electronics engineer; b. Leeds, Yorkshire, Eng., May 6, 1948; came to U.S., 1978; s. Peter Arvid and Austra (Alers) R. BS, Lakehead U., 1972, MS, 1974; PhD, Johannes Kepler U., Linz, Austria, 1977. Sr. scientist Internat. Telephone and Telegraph, Roanoke, Va., 1978-85; staff scientist Ortel Corp., L.A., 1986-87; product mgr. Applied Solar Energy Corp., L.A., 1987-89; sr. mem. tech. staff TRW, L.A., 1989—. Contbr. articles to profl. jours. Mem. IEEE, Am. Phys. Soc. Republican. Lutheran. Home: 4315 Massachusetts St Long Beach CA 90814-2942

ROZENFELD, GREGORY, petroleum engineer; b. Leningrad, Russia, Apr. 20, 1946; came to U.S., 1979; s. Haim I. and Geniy G. (Manulis) R.; m. Dina Goldberg, Dec. 31, 1972; children: Alexander, Michael. Mech. Engr., Leningrad Maintenance Coll., 1965; BS in Civil Engring., Leningrad Civil Engr. Inst., 1971, MS in Civil Engring., 1981. Registered profl. engr., N.Y., Calif. Engr., sr. engr. Main Constrn. TRust, Leningrad, 1965-70, head dept., chief engr., 1972-78; petroleum engr. Williams Bros. Engring. Co., Bakersfield, Calif., 1979-80; area spl. project engr. Texaco Inc.-U.S.A., Ventura, Calif., 1981-87; sr. spl. project engr. Texaco-Caltex, Indonesia, 1988-89; mgr. spl. projects Texaco Inc.-Europe, White Plains, N.Y., 1989—; v.p. Texaco Internat. Ops. Inc., White Plains, N.Y., 1992—. Recipient Bronze medal All Union Exhbn., Moscow, 1969, Citations for various tech. innovations and applications. Mem. ASCE, Soc. Petroleum Engrs. Republican. Jewish. Office: Texaco Internat Ops Inc 2000 Westchester Ave White Plains NY 10650

RUAN, JU-AI, physics and tribology researcher; b. Yugan, Jiang Xi, China, Nov. 2, 1959; came to U.S., 1984; s. Cai-Mian and Huo-Xiu (Li) R.; m. Ming Wu, Feb. 15, 1988; children: Wenly, Winston. BS, Shandong U., 1983; MS, U. Pitts., 1986, PhD, 1991. Grad. rschr. U. Pitts., 1984-91, rsch. assoc. in physics, 1991-92; rsch. assoc. in mech. engring. Ohio State U., Columbus, 1992—. Contbr. articles to profl. jours. Andrew Mellon fellow U. Pitts., 1990. Mem. Am. Physics Soc. Achievements include discovery that oxygen will reduce the probability of boron incorporation in chemical vapor deposition of diamond, that it is silicon impurity, not neutral vacancy, that is responsible for the luminenescence peak at 1.68eV in diamond; that the broad luminescence band -"Band A"- in diamond is not due to donor-acceptor pairs, this has ended a controversy which has lasted for over a quarter of a century; the variation of friction force between a sharp tip and a surface is proportional to the slope of the surface roughness. Office: Ohio State U 206 W 18th Ave Columbus OH 43210

RUBANO, RICHARD FRANK, civil engineer; b. N.Y.C., Mar. 10, 1946; s. Emil and Bessie Ann (Goldberg) R.; m. Jane M. Biamonte, June 5, 1975 (dec. 1987); children: David C., Kimberly A.; m. Claudine T. Wyrobnik, Mar. 27, 1988; children: David S. Sharbani, Jonathan E. Sharbani. BSE, CUNY, 1969; MSCE, Columbia U., 1976. Lic. profl. engr. N.Y., Tex., Fla., Ill., Mass. Structural engr. Burns & Roe Inc., Paramus, N.J., 1976-81; mech. engr. Stone & Webster Inc., N.Y.C., 1981-83; nuclear engr. cons. N.Y.C., 1983-86; auditor constrn. N.Y.C. Dept. Constrn., 1986-88; sr. engr., resident engr. Ammann & Whitney Inc., N.Y.C., 1988-92; sr. resident engr. Goodkind & O'Dea, Rutherford, N.J., 1993—; cons. Columbia U. Alumni Assn., N.Y.C., 1976—; guest speaker Deans Day Columbia, 1977.; resident engr. rehab. Triborough Bridge, N.Y.C. Contbr. articles to profl. jours. Jazz workshop Local Evening Program, Roslyn, N.Y., 1991; performer Community Big Band. Fellow ASCE; mem. Nat. Soc. Profl. Engrs., N.Y. Soc. Profl. Engrs. (sec. L.I. chpt. 1988—). Achievements include work on Olmstead Terrace Restoration of U.S. Capitol. Office: Goodkind & ODea Inc PO Box 1708 60 Feronia Way Rutherford NJ 07070

RUBBERT, PAUL EDWARD, engineering executive; b. Mpls., Feb. 18, 1937; s. Adolf Christian and Esther Ruth Rubbert; m. Mary Parpart, Oct. 6, 1958 (div. 1985); children: Mark, David, Stephen; m. Rita Monica Saiia, Oct. 7, 1989. BS with high distinction, U. Minn., 1958, MS in Aero. Engring., 1960; PhD in Aerodyn., MIT, 1965. Rsch. engr. The Boeing Co., Seattle, 1960-62, 65-72, unit chief aeordyns. rsch., comml. airplane group, 1972—; cons. NASA, 1989—, aeronautics adv. com., aerospace rsch. and tech. subcoms.; corp. vis. com. MIT, 1990—; served on various coms. Nat. Rsch. Coun. Panel; aerodyns. cons. GM; speaker in field. Contbr. articles to profl. jours. Recipient Arch T. Colwell Merit award Soc. Automotive Engrs. Fellow AIAA (Outstanding Tech. Mgmt. award PNW section, disting. lectr., assoc. editor jour., past mem. fellow selection com., dir., chmn. various workshops and coms.); mem. NAE. Achievements include three patents in field. Office: Boeing Comm Airplane Group MS 7H-91 PO Box 3707 Seattle WA 98124-2207

RUBBIA, CARLO, physicist; b. Gorizia, Italy, Mar. 31, 1934; s. Silvio and Bice (Liceni) R.; m. Marisa Rome, June 27, 1960; children—Laura, Andrea. Diploma, Scuola Normale, Pisa, 1958; A.M. (hon.), Harvard U., 1970; hon. degree, U. Geneva, 1983, Carnegie Mellon U., 1985, U. Genoa (Italy), 1985, U. Udine (Italy), 1985. Research fellow Columbia U., N.Y.C., 1960-61; asst. prof. physics U. Rome, 1961-62; prof. physics Harvard U., Cambridge, Mass., 1960-89; sr. rsch. physicist European Orgn. for Nuclear Rsch., Geneva, 1960—, dir. gen., 1989-93. Recipient Nobel prize in physics, 1984. Mem. European Physics Soc., Am. Physics Soc., NAS (fgn. assoc.), Pontifical Acad. Scis., Royal Soc. (U.K., fgn. mem.), USSR Acad. Scis. (fgn. mem.). Office: CERN European Lab, Particles Physics, 1211 Geneva 23, Switzerland

RUBEL, EDWIN W, neurobiologist; b. Chgo., May 8, 1942; s. Nathan Webster and Ruth (Mayer) R.; m. Wendy Rubel, Dec. 14, 1963; children: Trevor, Lisa. BS, Mich. State U., 1964, MS, 1967, PhD, 1969. Postdoctoral fellow dept. psychobiology U. Calif., Irvine, 1970-; asst. prof., then assoc. prof. dept. psychology Yale U., New Haven, 1971-77; assoc. prof., then prof. dept. otolaryngology U. Va., Charlottesville, 1977-86; prof. dept. otolaryngology U. Wash., Seattle, 1986—; mem. study sect. on hearing scis. NIH, Bethesda, Md., 1981-85; adv. coun. Nat. Inst. Deafness, Bethesda, 1991—. Editor: Developmental Psychoacoustics, 1992, Surgical Anatomy of the Temporal Bone, 1992; editorial bd. Hearing Rsch., Jour. Comparative Neurology, 1985—; contbr. articles to profl. jours. Recipient Career Devel. award NIH, 1977-82, named Jacob Javits Neurosci. Investigator, 1985-92; recipient Internat. prize for brain dysfunction rsch. Oasi Inst. Mental Retardation and Aging, Italy, 1991. Fellow AAAS; mem. Am. Assn. Anatomists, Soc. Neurosci., Assn. Rsch. in Otolaryngology, Collegium Oto-rhino-laryngologicum Amicitae Sacrum. Office: Univ Wash Hearing Devel Labs RL 30 Health Sci Bldg Rm BB1136 Seattle WA 98195

RUBEN, LAURENS NORMAN, biology educator; b. N.Y.C., May 14, 1927; s. Samuel and Rena Sylvia (Koch) R.; m. Judith Marion Starr, Aug. 29, 1950; children: Bruce L., Ellen P., Barbara J. AB, U. Mich., 1949, MS, 1950; PhD, Columbia U., 1954. Postdoctorate Nat. Cancer Inst. at Princeton, Princeton, N.J., 1954-55; faculty mem. Reed Coll., Portland, Oreg., 1955-83; Wm. R. Kenan Jr. prof. biology, 1988-93, prof. emeritus, 1992—; vis. faculty U. Geneva, Switzerland, 1962-63, U. Calif. San Diego, La Jolla, Calif., 1971-72, The Walter and Eliza Hall Inst. Med. Rsch., Melbourne, Australia, 1977, 82-83, Queen's Med. Ctr., Nothingham, U.K., 1988, 90, 92; chair Devel. Comp. Immunol. Am. Soc. Zoologists, 1973-74; pres. Internat. Soc. Devel. Comp. Immunol., 1989-91; mem. editorial bd. Devel. Comp. Immunol., Pergamon Press., 1970-80. Contbr. over 115 articles to profl. rsch. publs. Mem. Soc. Devel. Biology, Internat. Soc. Devel. and Comp. Immunol. Democrat. Jewish. Home: 3108 SE Crystal Spring Blvd Portland OR 97202 Office: Reed Coll 3203 SE Woodstock Portland OR 97202

RUBEN, ROBERT JOEL, physician, educator; b. N.Y.C., Aug. 2, 1933; s. Julian Carl and Sadie (Weiss) R.; children—Ann, Emily, Karin, Arthur. A.B., Princeton U., 1955; M.D., Johns Hopkins U., 1959. Intern Johns Hopkins Hosp., Balt., 1959-60; resident Johns Hopkins Hosp., 1960-64, dir. neurophysiology lab., div. otolaryngology, 1958-64; practice medicine specializing in otorhinolaryngology N.Y.C., 1964—; asst. prof. otorhinolaryngology N.Y. U. Sch. Medicine, 1966-68; mem. staffs hosps. Bronx Mcpl. Hosp. and North Ctrl. Bronx Hosp; prof. otorhinolaryngology Albert Einstein Coll. Medicine, Bronx, N.Y., 1979—; prof. pediatrics Albert Einstein Coll. Medicine, Bronx, 1983—, mem. dept. history medicine, 1973; assoc. prof. otolaryngology, dir. Albert Einstein Coll. Medicine, N.Y.C., 1968-70, prof., chmn. dept. otolaryngology, 1970—; prof. pediatrics Albert Einstein Coll. Medicine, 1983—, mem. dept. history medicine, 1973—; chmn. Nat. Com. for Research and Neurol. and Communicative Disorders, pres., 1982-84. Editor-in-chief: Internat. Jour. Pediatric Otorhinolaryngology, 1979—. Bd. dirs. N.Y. League Hard of Hearing, 1969-75, 76-85. Served to surgeon USPHS, 1964-66. Recipient Rsch. award Am. Acad. Opthalmology and Otoaryngology, 1962, Edmund Prince Fowler award, 1973, Gold medal Best Didactic Film XI World Congrl., 1977. Fellow ACS, N.Y. Acad. Medicine; mem. AMA, Am. Assn. Anatomists, Audiology Study Group N.Y. (pres. 1964-66), Acoustical Soc. Am., Am. Acad. Ophthalmology and Otolaryngology, Soc. Univ. Otolaryngologists, Am. Otol. Soc. (sec.-treas. rsch. fund 1979—), Soc. for Ear, Nose and Throat Advances in Children (pres. 1973), Assn. for Rsch. in Otolaryngology (pres. 1985-86), Am. Acad. Pediatrics (chmn. otol. bronchoesphology 1983-85), Am. Soc. Pediatric Otolaryngology (historian 1986—), Am. Soc. Ped. Otolaryngologists (historian 1986-93, pres.-elect 1993—), Johns Hopkins U. Soc. of Scholars. Home: 1025 Fifth Ave Apt 12C S New York NY 10028 Office: Montefiore Med Ctr 111 E 210th St Bronx NY 10467-2490

RUBENDALL, RICHARD ARTHUR, civil engineer; b. Pierre, S.D., Sept. 24, 1957; s. Quentin Theodore and Doris (Noe) R.; m. Sandra Mae Kovacich, June 8, 1985. BSCE, S.D. Sch. Mines & Tech., 1979; MSCE, U. N.Mex., 1993. Registered profl. engr., Mont., Ariz. Field engr. USPHS/Indian Health Svc., Lame Deer, Mont., 1979-86; sr. field engr. USPHS/Indian Health Svc., Many Farms, Ariz., 1986-89; sr. field engr. USPHS/Indian Health Svc., Sells, Ariz., 1989-90, dist. engr., 1990—. Comdr. USPHS, 1979—. Recipient USPHS Isolated Hardship award, 1980, 84, 88, 91, USPHS Hazardous Duty award 1984, USPHS Citation with plaque, 1990, USPHS Achievement medal, 1985; named Indian Health Svc. Engr. of Yr., Tucson area, 1990. Mem. ASCE, Am. Water Works Assn., Water Environment Fedn., USPHS Commd. Officer Assn. (pres. Tucson chpt. 1993—), Res. Officers Assn., Assn. Mil. Surgeons of the U.S. Home: 6944 E 42d St Tucson AZ 85730 Office: USPHS PO Box 548 Sells AZ 85634-0548

RUBENS, ROBERT DAVID, physician, educator; b. Woking, Eng., June 11, 1943; s. Joel and Dinah (Hasseck) R.; m. Margaret Chamberlin, Oct. 30, 1970; children: Abigail, Carolyn. BSc, King's Coll., London, 1964; MBBS, St. George's Hosp., London, 1967; MD, U. London, 1974. House physician St. George's Hosp., London, 1968, house surgeon, 1968-69; house physician Brompton Hosp., London, 1969, Hammersmith Hosp., London, 1969-70; med. registrar Royal Marsden Hosp., London, 1970, St. George's Hosp., London, 1970-72; rsch. fellow Imperial Cancer Rsch. Fund, London, 1972-74; cons. physician Guy's Hosp., London, 1975—; prof. clinical oncology United Med. and Dental Sch., London, 1985—; dir. ICRF Unit Guy's Hosp., London, 1985—; chmn. EORTC Breast Cancer Coop. Group, 1991—; examiner Royal Coll. of Physicians, London, 1987—; chief med. officer Mercantile and Gen. Reins Co. Ltd., London, 1977—. Author: (textbook) Clinical Oncology, 1980; editor-in-chief Cancer Treatment Revs., 1992; contbr. articles to profl. jours. Dir. Lee House, Wimbeldon, Eng., 1983; mem. Regional Cancer Orgn., Southeast Eng., 1983—. Fellow Royal Coll. Physicians; mem. Soc. Apothecaries, Assurance Med. Soc. (coun. 1982), Brit. Med. Assn., Am. Soc. Clinical Oncology, Am. Assn. for Cancer Rsch., Athenaeum (London), Royal Wimbledon Golf Club. Avocations: golf, music, reading. Office: Guy's Hosp, St Thomas St, London SE1 9RT, England

RUBENSTEIN, ARTHUR HAROLD, physician, educator; b. Johannesburg, South Africa, Dec. 28, 1937; came to U.S., 1967; s. Montague and Isabel (Nathanson) R.; m. Denise Hack, Aug. 19, 1962; children: Jeffrey Lawrence, Errol Charles. MB, BCh, U. Witwatersrand, 1960. Diplomate Am. Bd. Internal Medicine. Intern, then resident Johannesburg (South Africa) Gen. Hosp., 1961, 63-65, 66-67; fellow in endocrinology Postgrad. Med. Sch., London, 1965-66; fellow in medicine U. Chgo., 1967-68, asst. prof., 1968-70, assoc. prof., 1970-74, prof., 1974—, Lowell T. Coggeshall prof. med. sci., 1981—, assoc. chmn. dept. medicine, 1975-81, chmn., 1981—, dir. Diabetes Rsch. and Tng. Ctr., 1986-91; attending physician Mitchell Hosp., U. Chgo., 1968—; mem. study sect. NIH, 1973-77, Hadassah Med. Adv. Bd., 1986—, adv. council Nat. Inst. Arthritis, Metabolism and Digestive Diseases, 1978-80; chmn. Nat. Diabetes Adv. Bd., 1982, mem., 1981-83. Mem. editorial bd. Diabetes, 1973-77, Endocrinology, 1973-77, Jour. Clin. Investigation, 1976-81, Am. Jour. Medicine, 1978-81, Diabetologia, 1982-86, Diabetes Medicine, 1987-91, Annals of Internal Medicine, 1991—, Medicine, 1992—; contbr. articles to profl. jours. Mem. Gov.'s Sci. Adv. Coun. State of Ill., 1989—. Recipient David Rumbough Meml. award Juvenile Diabetes Found., 1978. Master ACP; fellow South African Coll. Physicians, Royal Coll. Physicians London; mem. Am. Soc. for Clin. Investigation, Am. Diabetes Assn. (Eli Lilly award 1973, Banting medal 1983, Solomon Berson Meml. lectr. 1985), British Diabetes Assn. (Banting lectr. 1987), Endocrine Soc., Am. Fedn. Clin. Rsch., Cen. Soc. Clin. Rsch. (v.p. 1988, pres. 1989), Assn. Am. Physicians (treas. 1984-89, councillor 1989—), Am. Bd. Internal Medicine (bd. govs. 1985-93, exec. com. 1989-93, chmn. 1992-93), Am. Acad. Arts amd Scis., Inst. Medicine (coun. 1991—), Assn. Profs. Medicine (councillor 1991—). Home: 5517 S Kimbark Ave Chicago IL 60637-1618 Office: U Chgo Dept of Medicine 5841 S Maryland Ave Chicago IL 60637-1470

RUBIK, BEVERLY ANNE, university administrator; b. Chgo., Jan. 12, 1951; d. Frank D. and Marian A. (LaPage) R. BS in Chemistry, Ill. Inst. Tech., 1972; PhD in Biophysics, U. Calif., 1979. Lectr. San Francisco State U., 1979-81, 83-88, Holy Names Coll., Oakland, Calif., 1985-88; sr. scientist Miles Labs., Berkeley, 1986-87; dir. Ctr. for Frontier Scis., Temple U., Phila., 1988—; mem. adv. bd. Office of Alternative Medicine, NIH, Bethesda, 1992—; bd. dirs. Soc. for Scientific Exploration; adv. bd. Internat. Centre for Earth Renewal, Vancouver, Ctr. for Functional Rsch., Inverness, Calif.; mem. adv. coun. Calif. Inst. Integral Studies, Ctr. for Sci./Spirituality, San Francisco, 1989-90; vis. scientist Am. Inst. Physics, Bucknell U., Pa., 1991, Gettysburg Coll., Pa., 1991; lectr. Tacoma City Club, Mensa, Inst. Noetic Scis., Italian Homeopathic Soc., Assn. Women in Sci. Editor: The Interrelationship Between Mind and Matter, 1992; editor Frontier Perspectives jour., 1990—; contbr. articles to profl. jours. Grantee John E. Fetzer Inst., 1988-89, Holmes Rsch. Found. for Holistic Health, 1981. Mem. AAAS, Bioelectromagnetics Soc., Internat. Soc. for Study of Subtle Energies and Energy Medicine, Soc. for Sci. Exploration, Sci. and Med. Network. Panentheist. Achievements include development of university-based international center to facilitate scientific research, education, and networking among scholars in frontier areas of science not yet mainstream. Office: Ctr for Frontier Scis Temple U Ritter 003-00 Philadelphia PA 19122

RUBIN, ALLAN MAIER, physician, surgeon; b. Bavaria, Germany, Aug. 4, 1947; s. Benjamin Rubin and Ida Spiegle; children: Alanna T., Marissa D., Sarina D.; m. Jean Tellander, Mar. 5, 1989. BS, McGill U., Montreal, Que., Can., 1968, MS, 1970; PhD, MD, U. Toronto, Ont., Can., 1979. Diplomate Am. Bd. Otolaryngology. Demonstrator neuroanatomy U. Toronto, 1971-73, resident, 1979-84; investigator Toronto Gen. Hosp., 1976-78; fellow otolaryngology Toronto East Gen. Hosp., 1985; asst. prof. dept. physiology Creighton U., Omaha, 1986-87; assoc. prof. dept. surgery Med. Coll. Ohio, Toledo, 1987-88, chmn., prof. dept. otolaryngology, 1988—; mem. subcom. on equilibrium Am. Acad. Otolaryngology-Head & Neck Surgery, Washington, 1988—, residnet edn. com. Soc. Univ. Otolaryngologists, 1992—, Blue Cross N.W. Ohio, Toledo, 1992-93, HMO/Toledo Health Plan, 1989-93; pres. Acad. Senate Med. Coll. Ohio, Toledo, 1991-92; chmn. search for urology chair Med. Coll. Ohio, Toledo, 1991-92, presdl. search com., 1991-93. Mem. internat. editorial adv. bd. Jour. Otolaryngology, 1991—; contbr. articles to profl. jours. Rsch. grantee Biomed. Rsch. Support Grant, 1984, NIH, 1986, 87. Fellow ACS, Royal Soc. Medicine, Am. Neurotology Soc., Am. Acad. Otolaryngology, Am. Acad. Otolaryngology-Head and Neck Surgery (subcom. on equilibrium 1988—); mem. Soc. Univ. Otolaryngologists (resident edn. com. 1992—). Achievements include management and treatment of vestibular dysfunction and dizziness in children; correlation and transcranial doppler (TCD) and brain single photon emission computed tomography (SPECT) in patients with dizziness. Office: Med Coll Ohio 3000 Arlington Ave Toledo OH 43614

RUBIN, ANDREW LAWRENCE, toxicologist, cell biologist; b. Oakland, Calif., Oct. 9, 1952; s. Harry and Dorothy Margaret (Schuster) R.; m. Morissa Ruth Miller, Mar. 31, 1985; children: Meira Navit, Ilan Naftali. BA, U. Calif.-San Diego, LaJolla, 1974; PhD, U. Wash., 1982. Postdoctoral scientist Lab. Toxicology, Harvard Sch. Pub. Health, Boston, 1983-88; asst. rsch. cell biologist Virus Lab., U. Calif., Berkeley, 1988-92; staff toxicologist med. toxicology br. Dept. Pesticide Regulation, Calif. EPA, Sacramento, 1992—. Contbr. 20 articles to profl. jours. Mem. AAAS, Genetic and Environ. Toxicology Assn., Soc. for Risk Analysis. Jewish. Achievements include exptl. support for the theory of progressive state selection to account for spontaneous and chemically-induced neoplastic transformation in cell populations. Home: 3839 Berrendo Dr Sacramento CA 95864 Office: Med Toxicology Br Dept Pesticide Regulation 1220 N St Sacramento CA 95814

RUBIN, BETSY CLAIRE, clinical psychologist, researcher; b. Long Beach, N.Y., Dec. 4, 1957; d. Fred and Sylvia R.; m. Charles Stephan Reichardt, Mar. 21, 1992. PhD, U. Tenn., 1988. Postdoctoral fellow U. Colo. Health Scis. Ctr., Denver, 1990; clin. assoc. U. Denver Child Clin. Doctoral Program, 1990—; clin. affiliate U. Denver Sch. Profl. Psychology, 1991—; asst. clin. prof. U. Colo. Health Scis. Ctr., Denver, 1991—. Office: Betsy Rubin PhD 469 S Cherry St Denver CO 80222

RUBIN, BRADLEY CRAIG, astrophysicist; b. Balt., Dec. 12, 1960; s. Lanny Rubin and Ruth Denise (Bogart) Gans. MS, Columbia U., 1983, PhD, 1990. Rsch. scientist Univ. Space Rsch. Assn., Huntsville, Ala., 1990—. Co-author articles for profl. jours. Mem. Am. Phys. Soc., Phi Beta Kappa, Sigma Pi Sigma. Office: NASA-Marshall Space Flt Ctr ES-64 Space Science Lab Huntsville AL 35812

RUBIN, BRUCE JOEL, chemical engineer, electrical engineer; b. Bklyn., Nov. 24, 1942; s. Irving and Isabel (Silverstein) R.; m. Roslyn Roberta Schwartz, Oct. 18, 1964; children: Holly, Michael. BChE, CCNY, 1964; MChE, Poly. Inst. N.Y., 1965; MBA, U. Rochester, 1973; MEE, Rochester Inst. Tech., 1988. Registered profl. engr., N.Y. Chemist Eastman Kodak, Rochester, 1965-68, rsch. chemist, 1968-73, sr. chemist, 1981-86, rsch. assoc., 1986-92, process engr., 1992—. Contbr. articles to profl. jours. Mem. Imaging Sci. & Tech. Assn. Achievements include 8 patents. Office: Eastman Kodak B205 3rd Fl KO ML03084 Rochester NY 14650-3084

RUBIN, GERALD MAYER, molecular biologist, biochemistry educator; b. Boston, Mar. 31, 1950; s. Benjamin H. and Edith (Weisberg) R.; m. Lynn S. Mastalir, May 7, 1978; 1 child, Alan F. B.S., MIT, 1971; Ph.D., Cambridge U., Eng., 1974. Helen Hay Whitney Found. fellow Stanford U. Sch. Medicine, Calif., 1974-76; asst. prof. biol. chemistry Sidney Farber Cancer Inst.-Harvard U. Med. Sch., Boston, 1977-80; staff mem. Carnegie Instn. of Washington, Balt., 1980-83; John D. MacArthur prof. genetics U. Calif., Berkeley, 1983—; investigator Howard Hughes Med. Inst., 1987—. Recipient Young Scientist award Passano Found., 1983, U.S. Steel Found. award Nat. Acad. Scis., 1985, Eli Lilly award in biochemistry Am. Chem. Soc., 1985, Genetics Soc. Am. medal, 1986. Mem. Nat. Acad. Scis. Office: U Calif Dept MCB 539 LSA Bldg Berkeley CA 94720

RUBIN, GUSTAV, orthopedic surgeon, consultant, researcher; b. N.Y.C., May 19, 1913; s. William and Rose (Strongin) R.; m. Mildred Synthia Holtzer, July 4, 1946 (dec. Dec. 1964); m. Esther Rosenberg Partnow, July 23, 1965; 1 stepchild, Michael Partnow. B.S., NYU, 1934; M.D., SUNY-Downstate Med. Ctr., 1939. Diplomate Am. Bd. Orthopedic Surgery. Intern Maimonides Hosp., Bklyn., 1939-41; resident in orthopedics Hosp. for Joint Diseases, N.Y.C., 1941-42, 1946; practice medicine specializing in orthopedics Bklyn., 1947-56; from orthopedic surgeon to dir. clinic VA Clinic, Bklyn., 1956-70; chief Spl. Prosthetic Clinic VA Prosthetics Ctr., N.Y.C., 1970-85, dir. spl. team for amputations, mobility, prosthetics/orthotics, 1985-87, mem. chief med. dir. adv. group on prosthetics services, rehab. research and devel., 1985-87, orthopedic cons., 1970-87, ret., 1987; pvt. practice N.Y.C., 1987—; med. advisor prosthetic research com. N.Y. State DAV, 1970—; lectr. prosthetics NYU, 1972-89; clin. prof. orthopedics N.Y. Coll. Podiatric Medicine, 1980—. Contbr. book chpts., articles to profl. jours. Served to capt. U.S. Army, 1942-46. Recipient Nat. Comdrs. award DAV, 1968, Amvets award for outstanding service, 1969, award for Service to Veterans Allied Veterans Meml. Com., 1970, Eastern Paralyzed Veterans Assn. award, 1977, award for Service to Israeli Wounded Israeli Govt. Dept. Rehab., 1981, Cert. of Merit, Nat. Amputation Found., 1972, Olin E. Teague award VA, 1984, Physician of Yr. award Pres.'s Commn. on Employment of People with Disabilities, 1984. Fellow Am. Acad. Orthopedic Surgeons, ACS, Am. Acad. Neurol. and Orthopedic Surgeons; mem. Alumni Assn. Hosp. Joint Disease, Sigma Xi. Jewish. Avocations: sculpting; oil painting. Home: 15 Circle Dr PO Box 572 Moorestown NJ 08057 Office: 41 E 29th St New York NY 10016-9101

RUBIN, JOSEPH WILLIAM, surgeon, educator; b. Montreal, Que., Can., Jan. 16, 1941; m. Edith Leah Hirsch, Aug. 13, 1961; children: Karen Elena, Heidi Lisette, Daphna Suzanne, Judah Simcha. BS in Physiology with honors, McGill U., 1961, MD, 1965. Diplomate Nat. Bd. Med. Examiners, Am. Bd. Surgery, Am. Bd. Thoracic Surgery, others. Intern Royal Victoria Hosp., Montreal, 1965-66; resident, chief resident surgeon, rsch. fellow in surgery Harvard Med. Sch./Boston City Hosp., 1966-71; resident, chief resident surgeon, teaching fellow Med. U. S.C., Charleston, 1971-73; chief divsn. cardiothoracic surgery VA Med. Ctr., Augusta, Ga., 1973-87; asst. prof. surgery Med. Coll. Ga., Augusta, 1973-76, assoc. prof. surgery, 1976-82, prof. surgery, 1982—, interim chmn. sect. thoracic & cardiac surgery, 1987-88, chief sect. thoracic & cardiac surgery, 1988—, also numerous coms.; cons. Dwight D. Eisenhower Army Med. Ctr., Ft. Gordon, Ga., 1979—, Ga. Med. Care Found., 1990—; vis. prof. U. Ky., Lexington, 1975, 76, Medtronic, Inc./Mpls. Pacemaker Symposium, 1982, Hershey Med. Ctr., 1983; numerous confs., symposia, seminars.; physician advisor MEDIPRO Med. Rev. Bd. VA Med. Dist. 9, 1985-87. Author: Chylothorax Complicating Intrathoracic Operations, 1979, Complications of Esophageal Diagnostic Procedures, 1979; contbr. over 50 publs. to profl. jours. Bd. dirs. Adas Yeshurun Synagogue, 1975-78; bd. gov. Augusta Sailing Club, 1986-89. Fellow ACS, Am. Coll. Cardiology, Am. Coll. Chest Physicians (com. on cardiovascular surgery 1976—, sec. So. chpt. 1976-80, pres.-elect 1980-81, pres. So. chpt. 1981-82); mem. ACS, AMA, Am. Assn. Thoracic Surgery, Am. Thoracic Soc., Soc. Thoracic Surgeons, Undersea Med. Soc., Soc. Critical Care Medicine, Am. Heart Assn., N.Y. Acad. Scis., Ga. Heart Assn., Ga. Thoracic Soc., Med. Assn. Ga., Richmond County Med. Soc., So. Thoracic Surg. Assn., So. Med. Assn., Alpha Omega Alpha. Office: Sect Thoracic/Cardiac Surg Med Coll Ga 15th St Augusta GA 30912-4040

RUBIN, KARL COOPER, mathematician; b. Urbana, Ill., Jan. 27, 1956; s. Robert J. and Vera (Cooper) R. AB, Princeton U., 1976; MA, Harvard U., 1977, PhD, 1981. Instr. Princeton (N.J.) U., 1982-83; mem. Inst. Advanced Study, Princeton, 1983-84; asst. prof. Ohio State U., Columbus, 1984-87; prof. Columbia U., N.Y.C., 1988-89, Ohio State U., 1987—. Contbr. articles to Inventiones Math. Postdoctoral fellow, NSF, 1981, Sloan fellow 1985; recipient Presdl. Young Investigator award, NSF, 1988. Mem. Am. Math. Soc. (Frank Nelson Cole Number Theory prize 1992), Phi Beta Kappa. Achievements include rsch. on elliptic curves, Tate-Shafarevich groups, Birch and Swinnerton-Dyer conjecture, Iwasawa theory and p-adic L-functions. Office: Ohio State U Dept Math 231 W 18th Ave Columbus OH 43210-1174

RUBIN, MARK STEPHEN, ophthalmic surgeon; b. Syracuse, N.Y., Dec. 22, 1946; s. Max Leon and Ruth (Dworski) R.; divorced; 1 child, Jonathan C. BA, SUNY, Buffalo, 1968; MD summa cum laude, U. Bologna, Italy, 1974. Diplomate Am. Bd. Ophthalmology. Intern Deaconess Hosp., Buffalo, 1976; resident in ophthalmology Wettlauffer Eye Clinic, Buffalo, 1979; chief ophthalmology Augsburg (Fed. Republic Germany) Army Hosp., 1979-80; pvt. practice Modena, Italy, 1980-88; head dept. ophthalmology Fla. Health Care Plan, Daytona Beach, 1988-90; pvt. practice Daytona Beach, Fla., 1990—; cons. USAF, Aviano Air Base, Italy, 1982-88; cons. surgeon Hesperia Hosp., Modena, 1980-88. Author: (with others) Extracapsular Cataract Surgery, 1988; translator: Lasers and Microsurgery, 1986, Ophthalmic Lasers, 1986. Bd. dirs. Ctr. for Visually Impaired, Daytona Beach; cons. Volusia County Health Dept. Fellow Am. Coll. Internat. Physicians, Am. Coll. Surgeons; mem. Italian Order Physicians and Surgeons, Internat. Assn. Ocular Surgeons, Am. Acad. Ophthalmology, Fla. Med. Soc., Eruopean Soc. Refractive Surgery, Italian Ophthalmologic Soc., Italian Soc. Profl. Ophthalmologists, Volusia County Med. Soc., Cen. Fla. Soc. Ophthalmology. Jewish. Avocations: scuba diving, ultralight flying, horseback riding, cooking, enology. Home: 891 N Beach St Ormond Beach FL 32174-4002 Office: 402 N Halifax Ave Daytona Beach FL 32118-4016

RUBIN, MELVIN LYNNE, ophthalmologist, educator; b. San Francisco, May 10, 1932; s. Morris and May (Gelman) R.; m. Lorna Isen, June 21, 1953; children: Gabrielle, Daniel, Michael. A.A., U. Calif., Berkeley, 1951, B.S., 1953; M.D., U. Calif., San Francisco, 1957; M.S., State U. Iowa, 1961. Diplomate Am. Bd. Ophthalmology (bd. dirs. 1977-83, chmn. 1984). Intern U. Calif. Hosp., San Francisco, 1957-58; resident in ophthalmology State U. Iowa, 1958-61; attending surgeon Georgetown U., Washington, 1961-63; asst. prof. surgery U. Fla. Med. Sch., Gainesville, 1963-66; assoc. prof. ophthalmology U. Fla. Med. Sch., 1966-67, prof. ophthalmology, 1967—, chmn. dept. ophthalmology, 1978—; eminent scholar U. Fla. Med. Sch., Gainesville, 1989; research cons. Dawson Corp.; ophthalmology cons. VA Hosp., Gainesville. Author: Studies in Physiological Optics, 1965, Fundamentals of Visual Science, 1969, Optics for Clinicians, 1971, 2d edit., 1974, The Fine Art of Prescribing Glasses, 1978, 2d edit., 1991; editor: Dictionary of Eye Terminology, 1984, 2d edit., 1990, Eye Care Notes, 1989; mem. editorial bd. Survey Ophthalmology; contbr. over 100 articles to med. jours. Co-founder Gainesville Assn. Creative Arts; co-founder Citizens for Public Schs., Inc.; co-founder ProArteMusica of Gainesville, Inc., 1969, pres., 1971-73; mem. Thomas Center Adv. Bd. for the Arts, 1978-84, nat. sci. adv. bd. Helen Keller Eye Rsch. Found., 1989—; bd. dirs. Hippodrome State Theater, 1981-87. Served with USPHS, 1961-63. Recipient Best Med. Book for 1978 award Am. Med. Writers Assn., 1979. Fellow ACS, Am. Acad. Ophthalmology (sec., dir. 1970-92, pres. 1988, Sr. Honor award 1987, Guest of Honor 1992), Found. Am. Acad. Ophthalmology (bd. trustees, 1988-94, chmn., 1992-94), Joint Commn. on Allied Health Pers. in Ophthalmology (Statesman of Yr. award 1987); mem. Assn. Rsch. in Vision and Ophthalmology (trustee 1973-78, pres. 1979), Retina Soc., Macula Soc., Club Jules Gonin, N.Y. Acad. Sci., Fla. Ophthal. Soc., Am. Ophthal. Soc., Pan Am. Soc. Ophthalmology, Ophthalmic Photographers Soc., Alachua County Med. Soc., Fla. Med. Assn., AMA (editorial bd. Archives of Ophthalmology 1975-85), N.Fla. Eye Bank (dir.), Sigma Xi, Alpha Omega Alpha., Phi Kappa Phi. Office: U Fla Med Ctr Box 100284 Gainesville FL 32610

RUBIN, STEPHEN CURTIS, gynecologic oncologist, educator; b. Phila., May 24, 1951; s. Alan and Helen (Metz) R.; m. Anne Loughran, May 30, 1985; children: Michael, Elisabeth. BS, Franklin & Marshall U., 1972; MD, U. Pa., 1976. Diplomate Am. Bd. Ob-Gyn., Nat. Bd. Med. Examiners. Intern in ob.-gyn. Hosp. of Univ. of Pa., Phila., 1976-77, residency in ob.-gyn., 1977-80, fellow in gynecology, 1980-82; asst. prof. of ob-gyn Med. Coll. of Pa., Phila., 1982-85, dir. surg. gynecology, 1982-85, chief gynecol. oncology, 1984-85; asst. mem. gynecol. staff Meml. Sloan-Kettering Hosp., N.Y.C., 1985-90, assoc. mem., 1990-93; asst. prof. ob-gyn Cornell U. Med. Coll., N.Y.C., 1985-90, assoc. prof., 1990-93; prof., dir. gynecologic oncology U. Pa., Phila., 1993—. Contbr. articles to profl. publs. Recipient Career Devel. award Am. Cancer Soc., 1987; Nat. Cancer Inst. grantee, 1991. Mem. ACS, Am. Coll. Obstetricians and Gynecologists, Am. Soc. Clin. Oncology, Soc. Gynecol. Oncologists, Soc. Gynecol. Investigation. Office: U Pa Med Ctr 3400 Spruce St Philadelphia PA 19104

RUBIN, STUART HARVEY, computer science educator, researcher; b. N.Y.C., Mar. 18, 1954; s. Jack and Rhoda Rochelle (Lentz) R. BS, U. R.I., 1975; MS in Indsl. and Systems Engring., Ohio U., 1977; MS, Rutgers U., 1980; PhD, Lehigh U., 1988. Lectr. U. Cin., 1977-78; electronic engr. U.S. Army Rsch. Labs., Ft. Monmouth, N.J., 1980-83; asst. prof. computer sci. Cen. Mich. U., Mt. Pleasant, 1988—, founder, dir. Ctr. for Intelligent Systems, 1990—; tech. cons. RCA, Princeton, N.J., 1982-83, Babcock and Wilcox Corp., Alliance, Ohio, 1990, Booz-Allen and Hamilton, Inc., San Diego, 1990-91, Adept Tech., San Jose, Calif., 1990-91; mem. rsch. coun. Scripps Clin. Contbr. articles to profl. jours.; inventor in field. Agt. United Fund Isabella County, Mt. Pleasant, 1988; supporting coach Mich. Spl. Olympics, Mt. Pleasant, 1990; event capt. San Diego Regional Sci. Olympic Competition, 1990, 92; judge 37th, 38th and 39th Ann. Greater San Diego Sci. and Engring. Fair, 1991-93. Recipient Am. Chem. Soc. award, 1972, U.S. Govt. Cert. of Merit, Washington, 1987, Letter of Appreciation, Gen. Charles C. McDonald, 1990; grantee NSF, Office Naval Tech., State of Mich., others, 1988—. Mem. IEEE, Am. Assn. Artificial Intelligence, Am.

Soc. Engring. Edn. (ONT postdoctoral fellow 1990-93), N.Y. Acad. Scis., Internat. Assn. Knowledge Engrs., Assn. for Computer Machinery. Avocations: boating, skiing, hiking and nature. Home: 1604 Canterbury Trail Apt E Mount Pleasant MI 48858 Office: Ctrl Mich U Dept Computer Sci Pearce Hall Mount Pleasant MI 48859

RUBIN, VERA COOPER, research astronomer; b. Phila., July 23, 1928; d. Philip and Rose (Applebaum) Cooper; m. Robert J. Rubin, June 25, 1948; children: David M., Judith S. Young, Karl C., Allan M. BA, Vassar Coll., 1948; MA, Cornell U., 1951; PhD, Georgetown U., 1954; DSc (hon.), Creighton U., 1978, Harvard U., 1988, Yale U., 1990. Research assoc. to asst. prof. Georgetown U., Washington, 1955-65; physicist U. Calif.-LaJolla, 1963-64; astronomer Carnegie Inst., Washington, 1965—; Chancellor's Disting. prof. U. Calif., Berkeley, 1981; vis. com. Harvard Coll. Obs., Cambridge, Mass., 1976-82, 92—, Space Telescope Sci. Inst., 1990-92; Beatrice Tinsley vis. prof. U. Tex., 1988; Commonwealth lectr. U. Mass. 1991, Yunker Lectr. Oregon state U., 1991, Bernhard vis. fellow Williams Coll., 1993, numerous lectures US, Chile, Russia, Armenia, India, Japan, China, Europe; trustee Associated Univs., Inc., 1993—. Assoc. editor: Astrophys. Jour. Letters, 1977-82; editorial bd.: Sci. Mag., 1979-87; contbr. numerous articles sci. jours.; assoc. editor: Astron. Jour., 1972-77. Pres.'s Disting. Visitor, Vassar Coll., 1987. Mem. Smithsonian Instn. Council, 1979-85; Phi Beta Kappa scholar, 1982-83; Nat. Medal of Sci., Nat. Sci. Found., 1993. Mem. Am. Astron. Soc. (coun. 1977-80), Internat. Astron. Union (pres. Commn. on Galaxies 1982-85), Assn. Univs. Rsch. in Astronomy (dir. 1973-76), Nat. Acad. Scis. (Space Sci. Bd. 1974-77, chmn. sect. on astronomy 1992—), Astron. Soc. Pacific (bd. dirs. 1991—), Am. Acad. Arts and Scis., Phi Beta Kappa. Democrat. Jewish.

RUBINOFF, IRA, biologist, research administrator, conservationist; b. N.Y.C., Dec. 21, 1938; s. Jacob and Bessie (Rose) R.; m. Roberta Wolff, Mar. 19, 1961; 1 son, Jason; m. Anabella Guardia, Feb. 10, 1978; children: Andres, Ana. B.S., Queens Coll., 1959; A.M., Harvard U., 1960, Ph.D., 1963. Biologist, asst. dir. marine biology Smithsonian Tropical Research Inst., Balboa, Republic of Panama, 1964-70; asst. dir. sci. Smithsonian Tropical Research Inst., 1970-73, dir., 1973—; assoc. in ichthyology Harvard U., 1965—; courtesy prof. Fla. State U., Tallahassee, 1976—; mem. sci. adv. bd. Gorgas Meml. Inst., 1964-88; trustee Rare Animal Relief Effort, 1976-85; bd. dirs. Charles Darwin Found. for Galapagos Islands, 1977—; chmn. bd. fellowships and grants Smithsonian Instn., 1978-79; vis. fellow Wolfson Coll., Oxford (Eng.) U., 1980-81; vis. scientist Mus. Comparative Biology-Harvard U., 1987-88. Author Strategy for Preservation of Moist Tropical Forests; Contbr. articles to profl. jours. Vice chmn. bd. dirs. Panama Canal Coll., 1989-93; bd. dirs. Internat. Sch. Panama, 1983-85, 90—, Fundacion Para La Conservacion de la Naturaleza, Found. for the Conservation of the Natural Resources; hon. dir. Instituto Latino Americano de Estudios Avanzados. Awarded Order of Vasco Nunez de Balboa of Republic of Panama. Fellow Linnean Soc. (London), AAAS; mem. Am. Soc. Naturalists, Soc. Study of Evolution, N.Y. Acad. Scis., Nat. Assn. Conservation of Nature Panama (bd. dirs.). Club: Cosmos (Washington). Home: Box 2072, Balboa Panama Office: Smithsonian Tropical Rsch Inst 27-6022 Unit 0948 APO AA 34002-0948

RUBINSTEIN, ARYE, pediatrician, microbiology and immunology educator; b. Tel Aviv, Oct. 2; came to U.S., 1971; s. Reuven and Kathe (Samson) R.; m. Orna Eisenstein, Dec. 7, 1965 (div. 1982); children: Ran, Yair, Avner, Noam; m. Charline Nezri, Dec. 27, 1983; children: Reuven, Rena, Rachel. MD, U. Berne (Switzerland), 1962. Bd. cert. Pediatrics, Israel, Switzerland, U.S.A.; allergy and immunology. Intern, pediatrics resident, fellow U. Tel Aviv, 1962-67; rsch. assoc., div. immunology Harvard Med. Sch., 1971-73; dir. div. immunology and bone marrow transplantation U. Berne, 1969-71; asst. prof. cell biology Albert Einstein Coll. Medicine, Bronx, 1973-80, asst. prof. pediatrics, 1973-77, assoc. prof. pedicatrics, 1977-82, assoc. prof. microbiology and immunology, 1981-85, prof. pediatrics, 1982—, prof. microbiology and immunology, 1985—; dir. div. clin. allergy and immunology Albert Einstein Coll. Med., Montefiore Med. Ctr.; dir. tng. program for allergy and immunology Albert Einstein Coll. Medicine; attending pediatrics Bronx Mcpl. Med. Ctr., Hosp. Albert Einstein Coll. Medicine; attending med. Hosp. Albert Einstein Coll. Medicine; mem. NIH Study Sect. AIDS Rsch. Editorial bd. mem. (jours.): AIDS, Internat. Jour. Pediatric Otorhinolaryngology, Annals of Allergy; reviewer: New England Jour. Medicine, Jour. for Clin. Investigation, Jour. of Pediatrice, Am. Journ. of Diseases of Children; contbr. over 100 articles to profl. jours. Lt. armed svcs., Israel, 1955-57. Recipient Lifetime award in Immunology, Humanitarian award DIFFA, Birch Svcs. for Children, Annual award U.S. Asst. Sec. of Health for excellence in AIDS rsch. and treatment, 1990; AIDS Rsch. Program grantee NIH, Bronx. Mem. Am. Acad. Allergy and Immunology, N.Y. Acad. Scis., Soc. Pediatric Rsch., The Harvey Soc., Am. Coll. Allergy, Clin. Immunology Soc., Clin. Immunology Soc. Office: Albert Einstein Coll of Medicine 1300 Morris Park Ave Bronx NY 10461

RUBINSTEIN, JACK HERBERT, health center administrator, pediatrics educator; b. N.Y.C., Aug. 4, 1925; s. Saul David and Anna (Gordon) R.; m. Thelma Regenstreif, Nov. 22, 1952 (dec. June 1988); m. Marlene Florence Tibbs, Sept. 1, 1990. AB, Columbia U., 1947; MD, Harvard U., 1952. Diplomate Am. Bd. Pediatrics. Intern in pediatrics Beth Israel Hosp., Boston, 1952-53; intern in pediatrics Mass. Gen. Hosp., Boston, 1953-54, sr. asst. resident, 1954-55; sr. asst. resident Children's Hosp. Med. Ctr., Cin., 1955-56, asst. med. dir., fellow pediatrics outpatient dept., 1956-57, attending pediatrician, 1957—; instr., asst. prof., assoc. prof. U. Cin., 1956-70, prof., 1970—; dir. Univ. Affiliated Clin. Program for Mentally Retarded, Cin., 1967-74, Univ. Affiliated Cin. Ctr. for Devel. Disorders, Cin., 1974—; dir. Hamilton County Diagnostic Clinic for Mentally Retarded, Cin., 1957-74, Children's Neuromuscular Diagnostic Clinic, Cin, 1962-74. Contbg. author: Medical Aspect of Mental Retardation, 1965, 2d edit., 1978; also articles. Mem. Ohio Devel. Disabilities Planning Coun., Columbus, Ohio Maternal and Child Health Block Grant Cons. Group, Columbus. With USAAF, 1944-46, ETO. Recipient longterm svc. award Children's Hosp. Med. Ctr., 1989, Founder's award Cin. Pediatric Soc., 1989. Fellow Am. Acad. Pediatrics; mem. Am. Pediatric Soc., Cin. Pediatric Soc., Teratology Soc., Am. Assn. on Mental Retardation, Am. Assn. Univ. Affiliated Programs for Persons with Devel. Disabilities (pres. 1978-79), Phi Beta Kappa, Alpha Omega Alpha. Achievements include reporting Rubinstein-Taybi syndrome, 1963. Home: 541 Ludlow Ave Cincinnati OH 45220 Office: Univ Affiliated Cin Ctr for Devel Disorders 3300 Elland Ave Cincinnati OH 45229*

RUBIO, PEDRO A., cardiovascular surgeon; b. Mexico City, Dec. 17, 1944; came to U.S., 1970; s. Isaac and Esther; m. Debra Rubio; children: Sandra, Edward, MD, U. Nacional Autónoma de Méx., 1968; MS in Surg. Tech., Pacific Western U., 1981, PhD in Biomed. Tech., 1982. Diplomate Am. Bd. Surgery, Am. Bd. Abdominal Surgery, Am. Bd. Laser Surgery, Am. Bd. Quality Assurance and Utilization Rev. Physicians, Am. Acad. Pain Mgmt.; profl. cert. law enforcement sci., Nat. Com. Profl. Law Enforcement Standards, 1972. Prof. neurology Escuela Normal de Especialización, Secretaria de Educación Publica, Mexico City, 1968-69; asst. instr. dept. surgery Baylor Coll. Medicine, Houston, 1971-76; clin. instr. dept. surgery U. Tex. Med. Sch. Houston, 1978-91; clin. assoc. prof. surgery dept. surgery U. Tex. Med. Sch., Houston, 1991—, clin. supr. psychiatry residency tng. program Tex. Research Inst. Mental Scis., Houston, 1979-85; surgeon, dir. Cardiovascular Surg. Ctr., Houston, 1976-85, Houston Cardiovascular Inst., 1985-89, Laser Gallbladder Surgery Ctr. of Houston, 1990—; course dir. Am. Laser Inst., 1991—, Houston Laser Inst., 1989-91; chmn. surgery dept. HCA Med. Ctr. Hosp., Houston, 1978—; research projects with FDA, NCI, HEW, VA ; pres. exec. com. Houston Chamber Singers, 1982-83. Decorated Palms Honor Cross (hon.), Mex. Army; recipient Recognition diploma bachelor's class Universidad Nacional Autonoma de Mex., 1961, Facultad de Medicina, 1966; named Outstanding Surg. Intern, Baylor Coll. Medicine, 1970-71. Fellow Academia de Ciencias Medicas del Instituto Mexicana de Cultura, Academia Mexicana de Cirugia, ACS (Best Paper award South Tex. chpt. 1976), Am. Coll. Angiology, Am. Coll. Chest Physicians, Houston Acad. Medicine. Internat. Coll. Angiology, Internat. Coll. Surgeons (N.Am. fedn. sec. 1991-92, pres. U.S. sect. 1988-89, pres. Tex. div. 1983-85, world pres.-elect 1993-94, historian 1985—, chmn. membership com. U.S. sect. 1984-86, 3d pl. sci. motion picture 1980), Israel Med. Assn. USA, Royal Soc. Medicine, Am. Heart Assn. (stroke council), Am. Geriatrics Soc., AMA (Recognition award 1971, 73—), Am. Trauma Soc., Denton A. Cooley Cardiovascular Soc., Harris County (Tex.) Med. Soc., Houston Surg. Soc. (1st pl. essay 1973, 75), Internat. Assn. Study Lung Cancer, Internat. Cardiovascular Soc., Internat. Soc. Laser Medicine and Surgery, Sociedad Mexicana de Angiologia (1st pl. nat. essay contest 1974), Soc. Internat. Chirurgie, Soc. Am. Gastrontestinal Endoscopic Surgeons, Soc. Laparoendoscopic Surgeons, Soc. for Minimally Invasive Surgery, Southwestern Surg. Congress, Tex. Med. Assn., World Med. Assn. Lodge: Rosicrucian. Author: (with E.M. Farrell) Atlas of Angioaccess Surgery, 1983; Atlas of Stapling Techniques, 1986; contbr. 245 sci. articles to publs.; patentee med. instrumentation. Office: Ste 1200 7400 Fannin Houston TX 77054

RUCH, WAYNE EUGENE, microlithography engineer; b. Sewickly, Pa., Nov. 24, 1946; s. Eugene Herbert and Marian Adelle (Moreth) R. BS in Chemistry, Carnegie Mellon U., Pitts., 1968; MS in Chemistry, U. Fla., 1975. Sr. engr. Harris Corp., Palm Bay, Fla., 1975-80, lead engr., 1980-85, staff engr., 1985-91; microlithography inspection staff engr., cons. KLA Instruments, San Jose, Calif., 1991—; adj. prof. chemistry Fla. Inst. Tech., Melbourne, Fla., 1979-85. Contbr. articles to profl. jours. With U.S. Army, 1969-71. Decorated Bronze Star with oak leaf cluster. Mem. N.Y. Acad. Scis., Am. Chem. Soc. Achievements include design of linear dimension submicron electron beam lithography process. Home: 15100 Fern Ave Boulder Creek CA 95006-9776

RUCKENSTEIN, ELI, chemical engineering educator; b. Botosani, Romania, Aug. 13, 1925; came to U.S., 1969; m. Velina Rothstein, May 15, 1948; children: Andrei, Lelia. BSChemE, Poly. Inst., Bucharest, Romania, 1949, PhD in Chem. Engring., 1967. Prof. Poly. Inst., Bucharest, 1949-69; vis. prof. U. London, 1969; NSF sr. scientist Clarkson Coll. Tech., Potsdam, N.Y., 1969-70; prof. U. Del., Newark, 1970-73; prof. SUNY, Buffalo, 1973-81, disting. prof., 1981—; vis. Humbolt prof. Bayreuth U., Fed. Republic Germany, 1986; Gulf vis. prof. Carnegie Mellon U., Pitts., 1988-89; disting. lectr. U. Waterloo, 1985, U. Mo., 1983; Fair Meml. lectr. U. Okla., 1987, Colburn Symposium lectr. U. Del., 1988, Van Winkle lectr. U. Tex., 1989. Contbr. articles, papers to profl. jours. Recipient Nat. award Romanian Dept. Edn., 1958, 64, Teaching award, 1961, George Spacu award Romanian Acad. Sci., 1963, Sr. Humbolt award Alexander von Humbolt Found., 1985, Creativity award NSF, 1985. Mem. NAE, Am. Inst. Chem. Engrs. (Alpha Chi Sigma award 1977, Walker award 1988), Am. Chem. Soc. (Kendall award 1986, Jacob F. Schoellkopf medal 1991), Materials Rsch. Soc. Office: SUNY Dept Chem Engring Buffalo NY 14260

RUCKER, JOSHUA E., environmental engineer; b. Lynchburg, Va., June 18, 1942; s. Joshua Eldon and Bess (Thompson) R.; m. Louisa Winfree, Sept. 4, 1965; children: Joshua, Jonathan, Mollie. BSCE, Va. Tech., 1965, MS in Environ. Engring., 1966. Engr. Union Carbide, South Charleston, W.Va., 1966-72; sr. engr. Mobil Oil, Princeton, N.J., 1972-75; regulatory analyst, dep. dir. Am. Petroleum Inst., Washington, 1975—; lectr. numerous seminars and confs. Contbr. numerous articles to profl. jours. Mem. NSPE, Water Pollution Control Assn., Air and Waste Mgmt. Assn. Republican. Episcopalian. Office: Am Petroleum Inst 1220 L St Washington DC 20005

RUCKMAN, MARK WARREN, physicist; b. Rolla, Mo., Dec. 26, 1954; s. Homer Leslie and Audrey (Warren) R. BS in Physics, Pa. State U., 1977; PhD in Physics, Rensselaer Polytechnic Inst., 1984. Asst. physicist Brookhaven Nat. Lab., Upton, N.Y., 1985-87, assoc. physicist, 1987-91, physicist, 1991—. Contbr. articles to profl. jours. Mem. Am. Phys. Soc., Am. Vacuum Soc., Am. Chem. Soc., Materials Rsch. Soc., Phi Beta Kappa, Phi Kappa Phi. Republican. Baptist. Office: Brookhaven Nat Lab 20 Pa Ave Upton NY 11973

RUCKTERSTUHL, RUSSELL M(ILTON), mechanical engineer, consulting engineer; b. Slingerlands, N.Y., Aug. 5, 1969; s. Robert and Meryl (Swanson) R. BSME summa cum laude, N.C. State U., 1991. With Rist Frost Assocs., Glens Falls, N.Y., 1991-92; consulting engr. Southerland Assocs., Charlotte, N.C., 1992—. Mem. ASME, NSPE, Phi Kappa Phi, Tau Beta Pi, Phi Eta Sigma. Home: 2123 A Dartmouth Pl Charlotte NC 28207

RUDBACH, JON ANTHONY, biotechnical company executive; b. Long Beach, Calif., Sept. 23, 1937; s. John Alexander and Lola (Whitcomb) R.; m. Inge Clye Steincke, July 4, 1959; children: Lucy Trine, Karl Kristian. BA, U. Calif., Berkeley, 1959; MS, U. Mich., 1961, PhD, 1964; MBA, Lake Forest Coll., 1986. Rsch. scientist Rocky Mountain Lab., Hamilton, Mont., 1964-70; prof. microbiology U. Mont., Missoula, 1970-77; mgr. exploratory rsch. Abbott Labs., North Chicago, Ill., 1977-79; dir. Stella Duncan Meml. Rsch. Inst., Missoula, 1979-82; head infectious disease rsch. Abbott Labs., Missoula, 1982-85; v.p. rsch. and devel. Ribi Immunochem Rsch., Inc., Hamilton, 1985—. Contbr. articles to profl. jours.; author 2 books; patentee in field. Mem. Bitteroot Community Bd., Hamilton, 1988—. USPHS grantee, 1970-79, fellow, 1964-66. Mem. Am. Assn. Immunologists, Am. Soc. Microbiology, Soc. for Exptl. Biology and Medicine, Soc. for Biol. Therapy, Lions (local v.p. 1989-90, dir. 1987—). Republican. Congregationalist. Avocations: skiing, fishing, hunting, hiking, band. Home: 243 Hilltop Dr Hamilton MT 59840-9317 Office: Ribi ImmunoChem Rsch 553 Old Corvallis Rd Hamilton MT 59840-1409

RUDD, LEO SLATON, psychology educator, minister; b. Hereford, Tex., Feb. 20, 1924; s. Charles Ival and Susan Leola (Horton) R.; m. Virginia Mae Daniel, Nov. 17, 1943; children: Virginia Kaye, Leo Jr., Bobbie Ann. BA, William Jewell Coll., Liberty, Mo., 1947; MRE, Cen. Bapt. Sem., Kansas City, Kans., 1948; MS, E. Tex. State U., Commerce, 1957; PhD, N. Tex. State U., Denton, 1959. Ordained to ministry, Southern Bapt. Conv. Bapt. student dir./instr. Smith County Bapt., Tyler, Tex.; psychology instr. Tyler (Tex.) Jr. Coll.; dir. missions, Linn County, Mo. Author: Syllabus for New Testament Studies, Syllabus for Old Testament Studies. With U.S. Army, 1942-43. Named Tchr. of the Yr., Tyler Jr. Coll., 1987, 1986 Best Tchr. Alumni award. Mem. Tex. Jr. Coll. Tchrs. Assn., Southwestern Bible Tchrs. Assn., E. Tex. Counselors Assn., DAV. Home: 1913 E 5th St Tyler TX 75701-3593

RUDDELL, JAMES THOMAS, civil engineer, consultant; b. Charlotte, N.C., Nov. 12, 1954; s. Joseph Preston and Patricia Ann (Conaway) R.; m. Martha Ann Wolford; 1 child, Hillary Steele W. Ruddell. BSCE, Stanford U., 1977. Profl. Engr. Calif., Va., Md., 1980, 88, 90. Constrn. engr. Fluor Arabia Ltd., Dhaharn, Saudi Arabia, 1977-80; sr. constrn. engr. Bechtel Petroleum, San Francisco, 1981-83; project supt. Bechtel Constructors, San Francisco, 1984-86; mgr. constrn. Dewberry & Davis, Fairfax, Va., 1987-92; constrn. mgr. Parsons Brinckerhoff, Herndon, Va., 1992—; mem. ASCE, 1976—, civil works subcommittee Va. Rd. and Transp. Builders Assn., Richmond, Va., 1991—. Contbr. articles to profl. jours. Vol. tutor Loudoun County Youth Shelter, Leesburg, Va., 1989-91; vol. Lovettsville (Va.) Pub. Schs., 1992; instr. Dewberry & Davis Inst., Fairfax, Va., 1991-92; vol. carpenter Christmas in April, Purcellville, Va., 1992-93. Home: Rt 2 Box 280A Lovettsville VA 22080 Office: Parsons Brinckerhoff Spring Pk Tech Ctr Herndon VA 22070

RUDERT, CYNTHIA SUE, gastroenterologist; b. Cin., Mar. 17, 1955; d. John Wayne and Hilda Wanda (Loftus) R. B.S. with honors, U. Ky., 1975; M.D., U. Louisville, 1979. Diplomate Am. Bd. Internal Medicine, Am. Bd. Gastroenterology. Intern internal medicine Emory U., Atlanta, 1979-80, resident, 1980-82, fellow gastroenterology, 1982-84, asst. prof. medicine, Emory U., Atlanta, 1984-91; guest speaker Alcoholism Conf., Kanasawa, Japan, 1987; nat. and internat. speaker in gastroenterology; author: Medicine for the Practicing Physician, 3rd rev. edition, 1991, (chpts.) Acute Pancreatitis, Chronic Pancreatitis, Ischaemic Hepatitis, Rudert, C.S. Alcohol Related Symptons. Recipient Newburg award U. Louisville, 1979. Mem. AMA, Am. Med. Women's Assn., Am. Assn. for Study Liver Disease, ACP, Am. Gastroent. Assn., Am. Assn. for Study Liver Diseases, So. Med. Assn., Am. Liver Found., Am. Acad. Scis., Ga. Gastroent. Soc., Med. Assn. Ga., Med. Assn. Atlanta.

RUDGE, WILLIAM EDWIN, IV, computational physicist; b. New Haven, June 14, 1939; s. William Edwin III and Abigail (Hazen) R.; m. Georgiana Ludmila Kopal, June 24, 1962; children: Julia R. Gulbransen, Marian K., William E. V. BS in Physics, Yale U., 1960; PhD in Physics, MIT, 1968. Physicist IBM Corp., Poughkeepsie, N.Y., 1960-63; rsch. staff mem. IBM Corp., San Jose, Calif., 1968—. Contbr. to Phys. Rev., Phys. Rev. Letters, Nature. Mem. Am. Phys. Soc., Sigma Xi. Home: 1187 Washoe Dr San Jose CA 95120-5542 Office: IBM Almaden Rsch Ctr 650 Harry Rd San Jose CA 95120-6099

RUDIN, ALFRED, chemistry educator emeritus; b. Alta., Can., Feb. 5, 1924; m. Pearl Rudin; children: Jonathan, Jeremy, Joel. BSc, U. Alta., 1949; PhD, Northwestern U., 1952. Profl. engr., Ont. Rsch. chemist CIL Cen. Rsch. Lab., 1952-60, rsch. group leader, 1960-64; lab. mgr. CIL Plastics, 1964-67; assoc. prof. chemistry U. Waterloo, Ont., 1967-69, prof. chemistry, 1969-73, prof. chemistry & chem. engring., 1973-89, prof. emeritus, 1990—; vis. prof. chem. engring. Technion, Israel, 1975-76, 81; vis. scientist IBM, San Jose, Calif., 1980; adj. prof. chemistry & chem. engring. U. Waterloo, 1989; dir. Inst. for Polymer Rsch., Waterloo, 1990—. Author: Elements of Polymer Science & Engineering, 1982; contbr. over 250 articles to profl. jours. Served with Royal Canadian Signals, 1942-45. Fellow Royal Soc. of Can., Chem. Inst. of Can.; mem. Chem. Soc. of Can. (Polysar Lecture award 1989, Protective Coatings award 1983). Achievements include 19 patents. Office: Institute for Polymer Research, U of Waterloo-Dept of Chemistry, Waterloo, ON Canada N2L 3G1

RUDNYK, MARIAN E., planetary photogeologist; b. Long Island, N.Y., June 14, 1960; s. Augustin J. and Romana O. (Ludkewycz) R. BS in Earth Sci., Calif. State Poly., 1983. Photogeologist cons. Jet Propulsion Lab. NASA, Pasadena, Calif., 1983—, astronomer Jet Propulsion Lab., 1985-87, data mgr., planetary photogeologist Jet Propulsion Lab., 1988—. Artist: (comic strip) Lunar Loonies, 1979—, copywrighted in 1986; contbr. articles to profl. jours.; author numerous poems. Mem. AIAA, Soc. Ukrainian Engrs., L.A. Astron. Soc. (cons. 1988—), Marslinks, Rose Float Club, Jet Propulsion Lab. Writers Club. Republican. Achievements include discovery of asteroid #4601 Ludkewycz, and nearly 200 other asteroids. Home: 732 W Hillcrest Blvd Monrovia CA 91016

RUDOLF, PHILIP REINHOLD, chemist, crystallographer; b. Heckmondwike, United Kingdom, Sept. 8, 1955; came to the U.S., 1977; s. Reinhold and Patricia (Moore) R.; m. Christine Anne Polansky, May 23, 1986. BSc with honors, Imperial Coll., 1977; PhD, Tex. A&M U., 1983. Postdoctoral fellow Tex. A&M U., College Station, 1983-84, sr. scientist, 1984-88; rsch. leader Dow Chem. Co., Midland, Mich., 1988—; cons. Molecular Structure Corp., Woodlands, Tex., 1983-88, A.E. Staley Mfg. Corp., Decatur, Ill., 1986-88, Shell Devel. Co., Westhollow, Tex., 1986-88, Amoco Corp., Naperville, Ill., 1985-88. Contbr. articles to profl. jours. Grantee Getty Oil Co., 1982-83. Mem. Am. Crystallographic Assn. Achievements include co-development of MSC/AFC diffractometer control software and TEXSAN structure analysis software; research in X-ray and neutron diffraction and scattering, crystallographic software development, image analysis, inorganic materials science. Office: Dow Chem Co Analytical Scis 1897 G Bldg Midland MI 48667

RUDOLPH, ANDREW HENRY, dermatologist, educator; b. Detroit, Jan. 30, 1943; s. John J. and Mary M. (Mizesko) R.; children: Kristen Ann, Kevin Andrew. MD cum laude, U. Mich., 1966. Diplomate Am. Bd. Dermatology. Intern, Univ. Hosp., U. Mich. Med. Center, Ann Arbor, 1966-67, resident dept. dermatology, 1967-70; practice medicine specializing in dermatology, 1972—; asst. prof. dermatology Baylor Coll. Medicine, Houston, 1972-75, assoc. prof., 1975-83, clin. prof., 1983—; chief dermatology svc. VA Hosp., Houston, 1977-82; mem. staff Meth. Hosp., Ben Taub Gen. Hosp., Tex. Children's Hosp., St. Luke's Episcopal Hosp., Hermann Hosp. Served as surgeon USPHS, 1970-72. Regent's scholar U. Mich., 1966. Fellow Am. Acad. Dermatology; mem. Am. Dermatol. Assn., AMA, So. Med. Assn., Tex. Med. Assn., Harris County Med. Soc., Houston Dermatol. Soc. (past pres.), Tex. Dermatol. Soc., Assn. Mil. Dermatologists, Internat. Soc. Tropical Dermatology, Royal Soc. Health, Royal Soc. Medicine, Dermatology Found., Skin Cancer Found., Am. Venereal Disease Assn. (past pres.), Assn. Mil. Surgeons U.S., Am. Soc. for Dermatol. Surgery, Soc. for Investigative Dermatology, S. Central Dermatologic Congress, Mich. Alumni Assn. (life), Alpha Omega Alpha, Phi Kappa Phi, Phi Rho Sigma, Theta Xi. Mem. editorial bd. Jour. of Sexually Transmitted Diseases, 1977-85. Contbr. to med. jours., periodicals and textbooks. Office: Surlock Tower 6560 Fannin St Ste 724 Houston TX 77030-2725

RUDOLPH, BRIAN ALBERT, computer science educator; b. Sandusky, Ohio, June 22, 1961; s. Louis John and Melvina Katherine (Martin) R. BS in Computer Sci. cum laude, Bowling Green State U., 1982, MS in Computer Sci., 1984. Instr. computer sci. Bowling Green (Ohio) State U., 1983-85, 86-87; asst. prof. computer sci. Ashland (Ohio) Coll., 1986, U. Wis., Platteville, 1987-91, Shawnee State U., Portsmouth, Ohio, 1991—; lectr., cons. in field. Commentator, lector St. Augustine's Univ. Parish, Platteville, 1988-90, trustee, 1989. Grantee NSF, 1989. Mem. IEEE, Internat. Brotherhood Magicians, Assn. Computing Machinery (referee tech. symposium on computer sci. edn. 1988—, dir. North Cen. Regional Programming Contest, 1989-91, dir. East Regional Programming Contest 1992, contest finals judge 1991, chair regional contests com. 1991, chair contest dirs. workshop 1991—, dir. regional contests 1991—, self-assessment com. 1991—), Phi Beta Kappa. Roman Catholic. Achievements include research in computer science education, concurrency, operating systems, parallel and distributed systems, algorithms and data structures, software engineering, and systems programming. Office: Shawnee State U Coll Engring Techs Vern Riffe Advanced Tech Portsmouth OH 45662

RUDOLPH, FREDERICK BYRON, biochemistry educator; b. St. Joseph, Mo., Oct. 17, 1944; s. John Max and Maxine Leah (Wood) R.; m. Glenda M. Myers, June 18, 1971; children: Anna Dorine, William K. BS in Chemistry, U. Mo., Rolla, 1966; PhD in Biochemistry, Iowa State U., 1971. Prof. biochemistry Rice U., Houston, 1972—, dir. Lab. for Biochem. and Genetic Engring., 1986—, exec. dir. Inst. Bioscience and Bioengineering; cons. World Book, Chgo., 1972—; mem. biochemistry study sect. NIH, Bethesda, Md., 1983-87; bd. dirs. S.W. Assn. Biotech. Cos., Houston, 1990-93. Contbr. over 150 articles to profl. jours. including Jour. Biol. Chemistry, Biochemistry, Transplantation, Exptl. Hematology, Jour. Parenteral and Enteral Nutrition, Jour. Molecular Biology, Applied and Environ. Microbiology, Life Scis., Archives Biochem. Biophysics, Critical Care Medicine, Archives Surgery, Sci.; also chpts. to books. Recipient Disting. Alumnus award Iowa State U., 1980. Mem. Am. Chem. Soc., Am. Soc. for Biochemistry and Molecular Biology. Achievements include research on dietary requirements for immune function, new techniques for protein purification, new methods for kinetic analysis of enzymes, structure and function of various enzymes. Office: Rice U Dept Biochemistry PO Box 1892 Houston TX 77251-1892

RUDOLPH, JEFFREY STEWART, pharmacist, chemist; b. Chgo., Oct. 30, 1942; married, 1967. 2 children. BS, U. Ill., 1966; MS, Purdue U., 1969, PhD in Pharmacy, 1970. Sr. rsch. pharmacist Ciba-Geigy Pham. Corp., 1970-72; sr. scientist McNeil Labs Divsn., Johnson & Johnson, 1972-75; group leader pharm. McNeil Labs Divsn., Johnson & Johnson Pilot Plant, 1975-76; asst. dir. McNeil Labs Divsn., Johnson & Johnson, 1977-80; dir. pharm. devel. Stuart Pharm Divsn., ICI Am. Inc., 1980-87; v.p. pharm. rsch. and devel. ICI Pharm., 1988—. Mem. Acad. Pharm. Sci., Am. Pharm. Assn., Am. Assn. Pharm. Scientists. Achievements include research in optimization of drug delivery systems; development of new dosage forms with emphasis on optimum bioavailability; evaluation of pharmaceutical processing equipment. Office: ICI Pharms Rsch & Devel Concord Pike Rd Wilmington DE 19897-0001*

RUDZKI, EUGENIUSZ MACIEJ, chemical engineer, consultant; b. Warsaw, Poland, Feb. 24, 1914; came to U.S., 1955.; s. Aleksander and Wanda (Lukaszewicz) R.; m. Fiorina Maria Di Vito, Feb. 23, 1952; children: Robert Alexander, Marcella Wanda Rudzki Meddick. Diploma with honors, Warsaw (Poland) Poly. Inst., 1937; Chem. Engr., The Polish U. Coll., London, 1951. Project devel. and field engr. Chance Bros. Ltd., Eng., 1951-54; head instr. chem. engring. dept. U. Toronto, Can., 1954-55; rsch. engr. T.C. Wheaton Co., N.J., 1955-56; rsch. engr. Bethlehem (Pa.) Steel Corp., 1956-61, supr. rsch. dept., 1961-82, ret., 1982; cons. Am. Flame Rsch. com., 1982—. Patentee in field, U.S., Eng., Can., Belgium; contbr. articles to profl. jours. Active The Polish Inst. Arts and Scis., N.Y.C., 1975, The Kosciuszko Found., N.Y.C., 1977; polit. prisoner Gulag Abis forced labor camp, Peczora, USSR, 1940-41. Maj. 2d Polish Army Corp. Gen. Anders, 1942-46. Decorated the Virtuti Militari Order, The Cross of Valor with bar, The Silver Order of Merit with Swords. Fellow The Inst. of Energy London, The Coun. Engring. Insts. London (chartered engr.); mem. AIME, Am. Inst. Chem. Engrs., The Combustion Inst., The Polish Vets. of World War II, Assn. Vets. of 2d Polish Army Corp. Roman Catholic. Avocations: gardening, reading, travel.

RUEDENBERG, KLAUS, theoretical chemist, educator; b. Bielefeld, Germany, Aug. 25, 1920; came to U.S., 1948, naturalized, 1955; s. Otto and Meta (Wertheimer) R.; m. Veronika Kutter, Apr. 8, 1948; children: Lucia Meta, Ursula Hedwig, Annette Veronika, Emanuel Klaus. Student, Montana Coll., Zugerberg, Switzerland, 1938-39; licence es Scis., U. Fribourg, Switzerland, 1944; postgrad., U. Chgo., 1948-50; PhD, U. Zurich, Switzerland, 1950; PhD (hon.), U. Basel, Switzerland, 1975, U. Bielefeld, Germany, 1991. Research assoc. physics U. Chgo., 1950-55; asst. prof. chemistry, physics Iowa State U., Ames, 1955-60; assoc. prof. Iowa State U., 1960-62, prof., 1964-78, disting. prof. in sci. and humanities, 1978—, sr. chemist Ames Lab., U.S. Dept. Energy, 1964—; prof. chemistry Johns Hopkins, Balt., 1962-64; vis. prof. U. Naples, Italy, 1961, Fed. Inst. Tech., Zurich, 1966, 67, Wash. State U. at Pullman, 1970, U. Calif. at Santa Cruz, 1973, U. Bonn, (Germany), 1974, Monash U. and CSIRO, Clayton, Victoria, Australia, 1982, U. Kaiserlautern, Fed. Republic Germany, 1987; lectr. univs., research instns. and sci. symposia, 1953—. Author articles in field; assoc. editor: Jour. Chem. Physics, 1964-67, Internat. Jour. Quantum Chemistry; Chem. Physics Letters, 1967-81, Lecture Notes in Chemistry, 1976—, Advances in Quantum Chemistry, 1987—; editor-in-chief Theoretica Chimica Acta, 1985—. Co-founder Octagon Center for the Arts, Ames, 1966, treas., 1966-71, also bd. dirs. Guggenheim fellow, 1966-67; Fulbright sr. scholar, 1982. Fellow AAAS, Am. Phys. Soc., Am. Inst. Chemists; elected Internat. Acad. for Quantum Molecular Scis.; mem. Am. Chem. Soc. (Midwest award 1982), AAUP, Sigma Xi, Phi Lambda Upsilon. Home: 2834 Ross Rd Ames IA 50010-4030

RUEGG, STEPHEN LAWRENCE, quality engineer, chemist; b. Harvard, Ill., May 12, 1959; s. Lawrence Roy and Carol Marie (Sayland) R.; m. Candace Lynn Davidson, Apr. 14, 1984; children: Melissa Ashley, Matthew Stephen, Mitchell Tyler. BA in Chemistry, Carthage Coll., 1981. Quality control technician Libby, McNeill & Libby, Inc., Darien, Ill., 1975-81; analytical chemist Gen. Mills, Inc., Mpls., 1981-84; quality control mgr. Pioneer Products, Inc. subs. Gen. Mills, Inc., Ocala, Fla., 1984-87; quality control contract field supr. Pioneer Products, Inc. subs. Gen. Mills, Inc., Mpls., 1987-91; quality engr. Gen. Mills, Inc., Lodi, Calif., 1988-91; mgr. quality control Gen. Mills., Inc., Woodland, Calif., 1991-93; mgr. quality assurance, tech. svcs. Am. Italian Pasta Co., Excelsior Springs, Mo., 1993—; statis. process control trainer Gen. Mills, Inc., Lodi, 1990-92. Mem. Am. Soc. Quality Control (cert. quality engr.). Democrat. Lutheran. Avocations: ice hockey, music, travel, home brewing, movies.

RUELLE, DAVID PIERRE, mathematical physicist, educator; b. Ghent, Belgium, Aug. 20, 1935; came to France, 1964; s. Pierre Paul Jules and Margareta Constantia (DeJonge) R.; m. Janine Rosa Armandine Lardinois, Oct. 8, 1960; children—Nicolas Elie, Anne Yvonne, Denise Armande. Licence in Physics, Universite Libre, Brussels, Belgium, 1957, Ph.D. in Physics, 1959; Hon. degree, Fed. Polytech. Sch., Lausanne, Switzerland, 1977. Asst. Free U., Brussels, 1957-59; research asst. and privat dozent Fed. Polytech. Sch., Zurich, Switzerland, 1960-62; mem. Inst. for Advanced Study, Princeton, N.J., 1962-64, 70-71; prof. math. Institut Des Hautes Etudes Scientifiques, Bures-Sur-Yvette, France, 1964—. Author: Statistical Mechanics, 1969; Thermodynamic Formalism, 1978, Chance and Chaos, 1991. Mem. Acad. Scis. Paris, Am. Acad. Arts and Scis. (fgn. hon.), Am. Phys. Soc. (Dannie Heineman prize 1985), Societe Mathematique de France, Am. Math. Soc., Internat. Assn. Math. Physicists. Office: Inst des Hautes Etudes, Scientifiques, 91440 Bures-sur-Yvette France

RÜETSCHI, PAUL, electrochemist; b. Regensdorf, Switzerland, Sept. 3, 1925; s. Charles and Louise (Baumann) R.; m. Yone Oberhánsli, Jan. 2, 1960 (dec. 1966); children: Gesima, Mathias, Johannes; m. Elisabeth Plüss, July 22, 1969; 1 child, David. PhD, Fed. Inst. Tech., 1952. Head electrochemistry div. Electric Storage Battery Co., Phila., 1955-64; dir. Leclanché S.A., Yverdon, Switzerland, 1964—. Adv. bd. Jour. Power Sources, Jour. Applied Electrochem.; contbr. numerous articles to profl. jours. Recipient Silver medal Fed. Inst. of Tech., 1952. Mem. U.S. Electrochem. Soc. (Young Author award 1957, Internat. Frank Booth award 1980, Internat. Planté medal 1993), Am. Chem. Soc. Achievements include 40 patents in the field of power sources.

RUF, JACOB FREDERICK, chemical engineer; b. Kansas City, Mo., Dec. 30, 1936; s. Paul William and Amalia (Maier) R.; m. Sondra Sue Ramsey, Aug. 30, 1957; children: Kurtis, Brian, Eric, Jake II, Sondra Sue II. AS in Engring., Kansas City Met. Jr. Coll., 1957; BSChE, U. Kans., 1959, MSChE, 1967. Systems mgr. Black Sivalls & Bryson, Kansas City, Mo., 1964-67; pres. Mid Continent Computing, Kansas City, Mo., 1967-69; dir. data systems M.A.R.C., Kansas City, Mo., 1967-69; exec. v.p. I.S.D. Inc., Kansas City, Mo., 1969-76; v.p. N.L.T.C.S. Inc., Kansas City, Mo., 1976-77; pres. Ruf Corp., Olathe, Kans., 1976—; mem. adv. bd. engring. dept. U. Kans., 1993—; mem. Kans. Tech. Enterprise Corp., 1991—, County Econ. Rsch. Inst., Johnson County, Kans., 1989—; adj. faculty U. Kans., Lawrence, 1987—; vice chmn. Johnson County Transp. Coun., 1985-91; presenter in field. Contbr. tech. papers to pubs. Mayor City of Olathe, 1991-93; mem. Olathe Libr. Bd., 1991-93, Pub. Bldg. Commn., Olathe, 1991-93. With USN, 1953-61. Mem. AICE, Assn. Computing Machinery, Univ. Kans. Engring. Adv. Bd., Urban & Regional Info. Systems Assn. Achievements include development of first fourth-generation relational data base language, development of a statistical system for a highly accurate marketing system, numerous other statistical and chemical engineering programs for information decision making. Home: 13700 Pflumm Olathe KS 66062 Office: Ruf Corp 1533 E Spruce Olathe KS 66061

RUFF, ROBERT LOUIS, neurologist, physiology researcher; b. Bklyn., Dec. 16, 1950; s. John Joseph and Rhoda (Alpert) R.; m. Louise Seymour Acheson, Apr. 26, 1980. BS summa cum laude, Cooper Union, 1971; MD summa cum laude, U. Wash., 1976, PhD in Physiology, 1976. Diplomate Am. Bd. Neurology and Psychiatry. Asst. neurologist N.Y. Hosp., Cornell Med. Sch., N.Y.C., 1977-80; asst. prof. physiology and medicine U. Wash., Seattle, 1980-84; assoc. prof. neurology Case Western Res. Med. Sch., Cleve., 1984-92, prof. neurology and neuroscis., 1993—; chief dept. neurology Cleve. VA Med. Ctr., 1984—; adv. Child Devel. and Mental Retardation Ctr., Seattle, 1980-84, Burien Devel. Disability Ctr., Wash., 1982-84; mem. med. adv. bd. Muscular Dystrophy Assn., Seattle, 1984, NE Ohio chpt. Multiple Sclerosis Soc., 1986—; mem. adv. bd. for Neurology Dept. Vets. Affairs, 1989—; chmn. med. adv. bd. N.E. Ohio chpt. Myasthenia Gravis Found., 1987—, nat. med. adv. bd., 1988—, grant and fellowship com., 1990—. Ad Hoc reviewer various profl. and scientific jours.; contbr. articles to profl. jours. and chpts. to books. Recipient Tchr. Investigator award NIH; NSF fellow, 1971; NIH grantee, Muscular Dystrophy Assn. grantee, Dept. Vets. Affair grantee; N.Y. State Regents med. scholar, 1971. Fellow Am. Heart Assn. (stroke coun.), Am. Acad. Neurology (scientific issues com., legis. action com.); mem. AMA, Am. Physics Soc., Neurosci. Soc., Biophys. Soc., Am. Neurol. Assn., N.Y. Acad. Sci., Am. Geriatrics Soc., Sigma Pi Sigma (v.p. 1970-71), Alpha Omega Alpha (v.p. 1975-76). Home: 2572 Stratford Rd Cleveland OH 44118-4063 Office: VA Med Ctr 10701 East Blvd Ste 127W Cleveland OH 44106-1702

RUGGE, HENRY FERDINAND, medical products executive; b. South San Francisco, Oct. 28, 1936; s. Hugo Heinrich and Marie Mathilde (Breiholz) R.; m. Sue Callow, Dec. 29, 1967. BS in Physics, U. Calif., Berkeley, 1958, PhD in Physics, 1963. Sr. physicist Physics Internat. Co., San Leandro, 1963-68; dir. adminstrn. and fin. Arkon Sci. Labs., Berkeley, Calif., 1969-71; v.p. Norse Systems, Inc., Hayward, Calif., 1972-74; v.p. Rasor Assocs., Inc., Sunnyvale, Calif., 1974-81, v.p., gen. mgr., 1983-87, exec. v.p. fin., 1988-89, pres., chief exec. officer, 1990—; pres. Berlinscan, Inc., Sunnyvale, 1981-82; cons. The Rugge Group, Berkeley, 1987—; bd. dirs. Rasor Assocs., Inc., Space Power Ind., Analatom, Inc. Patentee in area med. devices. U. Calif. scholar, 1954-58. Mem. Am. Heart Assn., Berkeley Bicycle (treas. 1983-84), Phi Beta Kappa. Avocations: bicycle racing, wine, food. Home: 46 Hiller Dr Oakland CA 94618 Office: Rasor Assocs Inc 253 Humboldt Ct Sunnyvale CA 94089-1300

RUGGE, HUGO ROBERT, physicist; b. South San Francisco, Calif., Nov. 7, 1935; s. Hugo Heinrich and Marie (Breiholz) R.; m. Coral Loy Irish, Dec. 28, 1969; children—Leslie Anne, Robert David. A.B., U. Calif.-Berkeley, 1957, Ph.D., 1962. Research physicist Lawrence Berkeley Lab., 1961-62; mem. tech. staff Aerospace Corp., Los Angeles, 1962-68, dept. head, 1968-79, prin. dir., 1979-81, lab. dir., 1981-89, v.p. lab. ops., 1989-91, v.p. tech. ops., 1991—. mem. avionic panel AGARD/NATO. Contbr. numerous articles on space sci. and astrophysics to profl. jours. Fellow Am. Phys. Soc.; mem. Am. Astron. Soc., Am. Geophys. Union, Internat. Astron. Union, Phi Beta Kappa, Sigma Xi.

RUGGERA, PAUL STEPHEN, biomedical engineer; b. Rock Springs, Wyo., Aug. 30, 1944; s. David Joe and Anna Eva (Ribovich) R.; m. Doris Ann Hankins, July 10, 1971. BSEE, U. Wyo., 1966, MS in Bioengring., 1968. Registered profl. engr., Va. Commd. officer USPHS, 1968, advanced through grades to capt., 1986; staff engr. Nat. Ctr. for Radiol. Health, Rockville, Md., 1968-72; spl. med. products engr. Bur. Radiol. Health, Rockville, 1972-82; spl. product engr. Ctr. for Devices and Radiol. Health, Rockville, 1982—; mem. C-63 com. on Electromagnetic Compatibility, Am. Nat. Standards Inst., N.Y.C., 1984—. Contbr. articles to IEEE EMC Soc. Proc., IEEE Trans. on Biomed. Engring., Internat. Jour. Hyperthermia, Cryobiology. Recipient commendation medal USPHS, 1980, 91, Outstanding Unit citation, 1991, PHS citation, 1991, 92. Mem. IEEE (sr.), Assn. for Advancement Med. Instrumentation (high frequency therapeutic devel. com. 1986—), Soc. Automotive Engrs. (AE-4 com. on electromagnetic compatibility 1972—), Eagles. Democrat. Roman Catholic. Achievements include patents for Helical Coil for Diathermy Apparatus, Diathermy Coil; patent pending for Method and Apparatus for Heating of Cryogenically Stored Organs; research on rapid warming of vitrified organs for transplant, electromagnetic susceptibility of medical devices to the ambient environment, and RF radiation bioeffects. Home: 1609 Woodmoor Ln Mc Lean VA 22101-5160 Office: Ctr for Devices and Radiol Health HFZ-133 12721 Twinbrook Pky Rockville MD 20857-0001

RUGGLES, HARVEY RICHARD, civil engineer; b. Newton, Mass., June 21, 1927; s. Clarence Victor and Cora Judson (Whitman) R.; m. Marjorie Joan Zimmer, Apr. 1, 1950; children: Harvey R. III, Robert W., Bradford P., Douglas A. BSCE, Tufts U., 1950. Registered profl. engr., Ohio; lic. surveyor, Ohio. Chief field surveys U.S. Geol. Survey, Arlington, Va., 1950-55; design engr. Dodson, Kinney & Lindblom Consulting Engrs., Columbus, Ohio, 1955-63; regional mgr. Prestressing Industries, Inc., San Antonio, 1963-65; exec. v.p. Conesco Midcontinent Inc., Hindsdale, Ill., 1965-77; regional mgr. The Prescon Corp., San Antonio, 1977-81, Dywidag Systems Internat., Inc., Lemont, Ill., 1981—; bd. dirs. first precast segmental bridge com. Prestressed Concrete Inst., Chgo., 1974-75. Scoutmaster Boy Scouts Am., Columbus, 1960-65. With USN, 1945-46, PTO. Mem. ASCE, Post-Tensioning Inst., Concrete Reinforcing Steel Inst. Baptist. Achievements include being first to introduce the technology, acceptance and use of post-tensioning concepts for prestressing jobsite-cast concrete building structures in the midwestern U.S. Home: 395 N Murray Hill Rd Columbus OH 43228-1352 Office: Dywidag Systems Internat Inc 1474 Grandview Ave Ste 1 Columbus OH 43212-2808

RUH, EDWIN, ceramic engineer, consultant, researcher; b. Westfield, N.J., Apr. 22, 1924; s. Harry John and Martha A. (Grasing) R.; m. Elizabeth J. Mundy, June 14, 1952; children: Edwin Jr., Elizabeth Jeanne. BS in Ceramic Engring. with honors, Rutgers U., 1949, MS in Ceramic Engring., 1953, PhD in Ceramics, 1954. Registered profl. engr., Pa. Rsch. engr. Harbison Walker Refractories Co., Pitts., 1954-57; asst dir. rsch. Harbison Walker Refractories Co., 1957-70; dir. rsch. Harbison Walker Refractories Div. Dresser Ind., Pitts., 1970-73; dir. advanced tech. Harbison Walker Refractories Div. Dresser Ind., 1973-74; v.p. rsch. Vesuvius Crucible Co., Pitts., 1974-76; adj. prof. Carnegie Mellon U., Pitts., 1976-84; rsch. prof. Rutgers U., New Brunswick, N.J., 1984—; pres. Ruh Internat., Inc., Pitts., 1976—. Editor Metallurgical Transactions, 1979-84; author chpts. to books; contbr. articles to profl. jours. With U.S. Army, 1943-46, ETO. Recipient ann. award Ceramic Assn. N.J., 1988. Fellow Am. Ceramic Soc. (pres. 1985-86, founders award Phila. sect. 1989, Bleininger award Pitts. sect. 1990), Inst. Ceramics; mem. Am. Soc. for Testing Materials, AAAS, AIME, Minerals, Metals and Materials Soc., ISS, Acad. Ceramics (profl.), Australasian Ceramic Soc., Keramos (nat. pres. 1970-72, Greaves-Walker Roll of Honor 1976). Republican. Presbyterian. Avocations: antiques, antique autos. Home: 892 Old Hickory Rd Pittsburgh PA 15243-1112 Office: Rutgers U Ctr for Ceramic Rsch Campus Busch Piscataway NJ 08855-0909

RÜHLMANN, ANDREAS CARL-ERICH CONRAD, biochemist; b. Hamburg, Germany, Feb. 1, 1961; came to U.S., 1990; s. Hans-Joachim Franz Friedrich and Christiane Ella Martha (Schuntermann) R.; m. Susanne Oerke, Aug. 8, 1992. BSc, U. Hamburg, 1984; MSc, Georg-August U., Göttingen, 1987; PhD, Carolo-Wilhelmina, Braunschweig, Fed. Republic Germany, 1990. Grad. fellowship Inst. for Organic Chemistry, Göttingen, 1985-87; post-grad. fellowship Max-Plank Inst. for Exptl. Medicine, Göttingen, 1987-90, Dana-Farber Cancer Inst./HMS, Boston, 1990-91, Beth Israel Hosp./HMS, Boston, 1991—. Contbr. articles to profl. jours. Doctoral fellowship Fonds of the Chem. Industry, Germany, 1987-89, Liebig fellowship, 1990-92. Mem. AAAS, Gesellschaft deutscher Chemiker, Gesellschaft deutscher Biochemiker. Achievements include patents for Fusionsprotein-vektoren (German), Vektor und seine Verwendung (German). Home: 103 Revere St #1 Boston MA 02114-3307 Office: Beth Israel Hosp Dept Medicine RE202 330 Brookline Ave Boston MA 02215

RUKOVENA, FRANK, JR., chemical engineer; b. Youngstown, Ohio, June 3, 1941; s. Frank and Julia (Husak) R.; m. Karen Lee Kelly, Sept. 10, 1976; children: Philip A., Kelly N., Lisa A. BChemE, Youngstown U., 1964; MBA, Kent State U., 1989. Process engr. Koppers, Pitts., 1964-68; design/project engr. Gen. Tire & Rubber, Akron, Ohio, 1968-74; dir. rsch. and devel. Norton Chem. Process Prodn. Corp., Stow, Ohio, 1974—; bd. dirs. Fractionation Rsch., Inc., Tulsa; rep. Separation Rsch. Project, Austin, Tex., 1985—. Mem. Am. Inst. Chem. Engrs. Achievements include patents on process improvement for Wellman-Lord Flue Gas Desulfurization Process, Mass Transfer Metal Structured Packing, Low Pressure Drop Packed Tower Internals. Office: Norton Chem Process Product Corp PO Box 350 Akron OH 44309-0350

RUMACK, BARRY H., physician, toxicologist, pediatrician; b. Chgo., Nov. 1, 1942; s. Alvin Eugene and Shirley (Kazan) R.; m. Carol Masters, June 10, 1964; children—Becky, Marc. B.S., in Microbiology, U. Chgo., 1964; M.D., U. Wis., 1968. Diplomate Nat. Bd. Med. Examiners, Am. Bd. Pediatrics, Am. Bd. Med. Toxicology (v.p.). Intern U. Colo., 1968-69, resident in pediatrics, 1971-72; fellow U. Colo. Med. Ctr., 1972-73; clin. assoc. Regional Poisoning Treatment Ctr., Royal Infirmary of Edinburgh, Scotland, 1973; assoc. prof. pediatrics U. Colo. Sch. Medicine, 1978-86, prof., 1986-92, dir. drug assay lab., 1975-77, Rocky Mountain Poison Ctr., 1974-92; pres., CEO Micromedex Inc., 1974—; chmn. pharmacy and therapeutics com., 1976-80; cons. Nat. Clearinghouse Poison Control Ctrs., 1975-91. Editor: (with A.R. Temple) Management of the Acutely Poisoned Patients, 1977; (with E. Salzman) Mushroom Poisoning, 1978; (with M.J. Bayer) Poisoning and Overdose, 1982; (with M.J. Bayer and L. Wanke) Toxicologic Emergencies: A Manual of Diagnosis and Management, 1984; contbr. chpts. to books, articles to profl. jours.; author abstracts. Active hazardous substances com. State of Colo., 1974-88; mem. adv. panel on toxicology U.S. Pharmacopeia, 1975-80; mem. Gov.'s Tech. Rev. com. on Rocky Mountain Arsenal, 1975; mem. adv. com. toxicology info. Nat. Library of Medicine, 1976-79; active Colo. State Bd. of Health, 1977-81; chmn. Internat. Congress of Clin. Toxicology, Colo., 1982; com. to advise Red Cross, 1984-91. Fellow Am. Acad. Clin. Toxicology (mem. edn. com. 1976-79, bd. dirs. 1978-81), Am. Acad. Pediatrics (com. on accidents and poisoning 1977-82, com. on drugs 1983-91); mem. Am. Assn. Poison Control Ctrs. (com. on standards 1974-79, bd. dirs. 1975-79, pres. 1982-84, v.p. 1980-82), Western Soc. Pediatric Research, Soc. Pediatric Research, Soc. Toxicology, Am. Coll. Emergency Medicine, N. Am. Mycological Assn. (toxicology com. 1975-79). Quaker. Office: Micromedex Inc Ste 600 600 Grant St Denver CO 80203-3527

RUMBAUGH, DUANE M., psychology educator; b. Maynard, Iowa, July 4, 1929; s. Arthur F. and Ida (Traudt) R.; m. E. Sue Savage, Dec. 29, 1976; 1 child, Joan Rubin. AB, U. Dubuque, 1950; MA, Kent State U., 1951; PhD, U. Colo., 1955. Instr. to prof. dept. psychology San Diego (Calif.) State Coll., 1954-69, asst. div. chmn. life sci. div., 1963-69; assoc. dir. Yerkes Primate Ctr., Emory U., Atlanta, 1969-71; chmn., prof. dept. psychology Ga. State U., Atlanta, 1971-89, Regent's prof. dept. psychology, 1984—; dir. Lang. Rsch. Ctr., Ga. State U., Atlanta, 1985—; mem. Nat. Inst. of Child Health & Human Devels. Maternal and Child Health Rsch. Com., 1983-87; sci. con. NASA Life Scis. Div. Rhesus Project, 1986—; mem. com. on well-being of nonhuman primates Inst. of Lab. Animal Resources, Washington, 1992—. Editor: Language Learning by a Chimpanzee: The LANA Project, 1977. Recipient grants NSF, 1961-71, Nat. Inst. of Child Health and Human Devel., Lang. Rsch. Ctr., 1971—, NASA, Lang. Rsch. Ctr., 1987—. Mem. APA (G. Stanley Hall lectr. in comparative psychology 1985, pres. Div. 6 1988), Ga. State U. Alumni Assn. (disting. prof. 1987), Kent State U. Alumni Assn. (Metro Atlanta chpt. Outstanding Alumnus award 1992), Phi Kappa Phi. Home: 3424 Riverhills Dr Ellenwood GA 30049 Office: Dept Psychology Ga State Univ Atlanta GA 30303

RUMMEL, DON, agronomist. With agrl. rsch. & ext. ctr. Tex. A&M U., Lubbock. Recipient CIBA-GEIGY Agronomy award Am. Soc. Agronomy, 1990. Office: Texas A&M University Ag research & Extension Ct Rt3 Box 219 Lubbock TX 79401*

RUMMERFIELD, PHILIP SHERIDAN, medical physicist; b. Raton, N. Mex., Feb. 27, 1922; s. Lawrence Lewis and Helen Antoinette (Roper) R.; m. Mary Evelyn Kubick, Dec. 29, 1979; children: Casey Regan, Dana Jay. BSME, Healds Coll., 1954; MSc, U. Cin., 1964, DSc, 1965. Registered profl. engr., safety, nuclear, Calif. Piping engr. Morrison Knudsen Co., Surabaja, E. Java, 1956-57; civil engr. State of Calif., San Francisco, 1957-59, constn. and radiation engr., 1959-63; hosp. physicist and radiation safety officer U. Calif., San Diego, 1966-73; prin. Applied Radiation Protection Svc., Encinitas, Calif., 1973—. Contbr. articles to Science, Bull. Atomic Scientists, Occupational Health Nursing, Health Physics Jour., Internat. Jour. Applied Radiation & Isotopes. Candidate for City Coun., Carlsbad, Calif., 1984. Grantee Teaching grant NSF, 1969-71. Mem. Am. Nuclear Soc., Calif. Soc. Profl. Engrs., Am. Indsl. Hygiene Assn., Am. Assn. Physicists in Medicine (pres. So. Calif. chpt. 1971-72), Calif. Soc. Profl. Engrs., Health Physics Soc. (pres. So. Calif. chpt. 1973-74). Democrat. Home: 3303 Dorado Pl Carlsbad CA 92009-7706 Office: Applied Radiation Protective Svcs 700 2nd St Ste C Encinitas CA 92024-4459

RUMMERY, TERRANCE EDWARD, nuclear engineering executive, researcher; b. Brockville, Ont., Can., 1937; s. Albert Edward and Evelyn Maud (Hayter) R.; m. Margaret Dianne Walker, Mar. 1967; children: Tara, Marcus. BSc with honours Engring. Chem., Queen's U., Kingston, PhD in Phys. Chemistry; student, Nat. Defence Coll., Can., 1984-85. Engr. Brockville Chems.; rsch. scientist Ont. Rsch. Found., Am. Labs. Airco Speer Carbon and Graphite; with Rsch. Chemistry br. AECL Whiteshell Labs., Pinawa, Man., Can., 1971; head Rsch Chemistry br. Whiteshell Labs., Pinawa, Manitoba, Can., 1976; dir. Nuclear Fuel Waste Mgmt. divsn. Atomic Energy Can. Ltd., Pinawa, Manitoba, Can., 1979-84; team leader reactor mktg. activities Atomic Energy Can. Ltd., The Netherlands, Yugoslavia, 1986-88; mgr. adv. unit to Dept. Nat. Defence Atomic Energy Can. Ltd., Ottowa, Can., 1988; acting pres. Rsch. Atomic Energy Can. Ltd., Ottawa, Ont., 1989; pres. Atomic Energy Can. Ltd., Ottowa, Can., 1990, acting pres., CEO, 1993; past mem. Queen's Univ. adv. coun. Engring.; adv. bd. U. Toronto Centre Nuclear Engring., Environ. Engring. program Carleton U. Contbr. numerous publications to profl. jours. Crystallography rsch. fellow University Coll. Fellow Chem. Inst. Can.; mem. Assn. Profl. Engrs. Ont., Canadian Nuclear Soc., Am. Nuclear Soc. Office: Atomic Energy Can Ltd, 344 Slater St, Ottawa, ON Canada K1A 0S4

RUMORE, MARTHA MARY, pharmacist, educator; b. N.Y.C., Feb. 29, 1956; d. Barney B. and Frieda A. (Sinacore) R. BS in Pharmacy, St. John's U., Jamaica, N.Y., 1978, PharmD, 1980; JD, Thomas Jefferson Coll., L.A., 1986; MS in Drug Info., Arnold & Marie Schwartz Coll., 1990. Registered pharmacist, N.Y., Fla., Conn. Lab. asst. St. John's U., 1973-77; pharmacy intern Queens Hosp. Ctr., Jamaica, 1977-78; sr. info. scientist Richardson-Vicks, Inc., Shelton, Conn., 1981-84; assoc. dir. profl. svcs. Sterling Drug Inc., N.Y.C., 1984-90; adminstr. Arnold & Marie Schwartz Coll. Pharmacy, Bklyn., 1988—, assoc. prof. pharmacy adminstrn., 1990—; clin. pharmacist Lenox Hill Hosp., N.Y.C., 1990-93. Contbr. over 60 articles to jours. Pharmacoepidemiology, Drug Info., Pharmacotherapeutics. Recipient Hosp. Pharmacy Achievement award L.I. Soc. Hosp. Pharmacists, 1978, Vis. Scientist award Pharm. Mfrs. Assn., 1990, 91; named Outstanding Young Women Am., 1988. Fellow Am. Pharm. Assn., Acad. Pharm. Practice and Mgmt., Am. Soc. Pharmacy Law; mem. Am. Pharm. Assn. (trustee, ho. of dels. 1988—, vice chmn. publs. 1988-89, polit. action com. 1990—, policy com. on sci. affairs), Drug Info. Assn. Republican. Roman Catholic. Office: Arnold & Marie Schwartz Coll Pharmacy 75 Dekalb Ave Brooklyn NY 11201-5497

RUNDE, DOUGLAS EDWARD, wildlife biologist; b. Chgo., July 29, 1954; s. Robert Mill and E. Shirley (Rittenhouse) R.; m. Laura Edith Campbell (div. 1979); m. Patricia Arlene MacLaren, July 30, 1983. MS, U. Vt., 1980; PhD, U. Wyo., 1987. Rsch. asst. Sch. Natural Resources, Burlington, Vt., 1978-80, Wyo. Coop. Fish and Wildlife Rsch. Unit, Laramie, 1981-86; wildlife biologist Fla. Game and Fresh Water Fish Commn., Quincy, 1987-93; wildlife ecologist Wyerhaeuser Timberlands Co., Tacoma, 1993—. Author: Florida Atlas of Breeding Sites for Herons and Their Allies: Update 1987-89, 1991; co-author: Wildlife Monograph No. III, 1990, contbr. articles to profl. jours. Student Conservation fellow Nat. Wildlife Fedn., 1983-85. Mem. The Wildlife Soc., Soc. for Conservation Biology, Ecol. Soc. Am., Wilson and Cooper Ornithological Socs., Sigma Xi. Democrat. Methodist. Achievements include participation in international wildlife conservation programs, ecological management of protected areas, monitoring conservation status of wildlife communities, censuses for rare and cryptic wildlife species. Home: 8913 Autumn Line Loop SE Lacey WA 98513-4783 Office: Weyerhaeuser Co WTC 1A5 Tacoma WA 98477

RUNDHAUG, JOYCE ELIZABETH, biochemist; b. Seattle, Sept. 8, 1952; d. Robert Norman and Elsie Elizabeth (Ohm) Ball; m. William George Rundhaug, July 16, 1977. BSN, U. Md., Balt., 1974; PhD, U. Hawaii, 1989. Teaching asst. S.W. Tex. State U., San Marcos, 1978-80; teaching asst. U. Hawaii, Honolulu, 1980-81, rsch. asst., 1982-89; staff fellow Nat. Inst. Environ. Health Sci., Research Triangle Park, N.C., 1989-92; postdoctoral fellow M.D. Anderson Cancer Ctr. Sci. Park, U. Tex., Smithville, 1992—. Contbr. articles to jours. Cancer Rsch., Carcinogenesis, Jour. Cellular Physiology. Capt. U.S. Army, 1970-77. Walter Reed Army Inst. Nursing scholar, 1970-74, Achievement Rewards for Coll. Scientists scholar, 1987. Mem. AAAS, Am. Assn. Cancer Rsch., N.Y. Acad. Scis., Phi Kappa Phi. Office: MD Anderson Cancer Ctr Sci Park-Rsch Divsn PO Box 389 Smithville TX 78957

RUNGE, ERICH KARL RAINER, physicist; b. Wiesbaden, Hassia, Germany, Apr. 21, 1959; came to U.S., 1990; s. Rudger Hanns and Margareta E. (Moldaenke) R. MS, U. Frankfurt, Germany, 1984; PhD, Max Planck Inst., Stuttgart, Germany, 1990. Researcher for solid state rsch. Max Planck Inst., Stuttgart, 1985-89; researcher Tech. U., Darmstadt, Germany, 1989-90; postdoctoral fellow Harvard U., Cambridge, Mass., 1990—; trainer Internat. Math. Olympiade, Germany, 1983-89. Author: Vielteilchen Theorie, 1986, Many-Particle Theory, 1991; contbr. articles to Phys. Review, Phys. Rev. Letters, Annals Physics. Studienstiftung fellow, Bonn, Germany, 1977-85. Mem. Am. Phys. Soc., German Phys. Soc. Achievements include discovery of existence theorem of time dependent density functional theory. Office: Harvard Univ Oxford St Cambridge MA 02138

RUNKA, GARY G., agricultural company executive. Pres. G.G. Runka Land Sense, Inc., Burnaby, B.C., Can. Recipient Fellowship award Agrl. Inst. Can., 1990. Office: G G Runka Land Sense Ltd, Box 80356, Burnaby, BC Canada V5H 3X6*

RUNKLE, ROBERT SCOTT, environmental company executive; b. Washington, Mar. 9, 1936; s. Lloyd Manor and Louise (Armstrong) R.; m. Betsy

Grater, Mar. 26, 1960 (div. July 1983); children: Beth R. Mackey, Brynn A.; stepchildren: Lori Anne Thompson, Jay M. Thompson; m. Joan Lewis, Aug. 6, 1983 (dec. Nov. 1987); m. Mary Beth Jorgensen, July 12, 1992; stepchildren: Elizabeth Jorgensen Feild, David Jorgensen Feild. BS in Bldg. Constrn., Ga. Inst. Tech ., 1960. Draftsman Ted Englehardt AIA, Silver Spring, Md., 1960-62; engr. Research Facilties Planning BD. div. Research Service NIH, Bethesda, Md., 1962-64; vice chmn. biohazards sect. Nat. Cancer Inst., NIH, Bethesda, 1964-67; research contracts mgr. Becton Dickinson & Co., Rutherford, N.J., 1967-69; adminstrn. mgr. Becton Dickinson Research Ctr., Raleigh, N.C., 1969-73; dir. adminstrn. Huntington (Eng.) Research Ctr., 1974-75; dir. rsch. liaison Becton Dickinson Co, Rutherford, N.J., 1976-78; v.p. ops. BBL microbiology systems div., Becton Dickinson Co., Balt., 1978-85; pres., chief exec. officer, chmn. bd. Pharmplastics Closures Inc., Balt., 1985-88; v.p. EA Labs, EA Engring., Sci. and Tech., Inc., Balt., 1989-91; v.p. bus. devel., sr. office leader EA Labs, EA Engring., Sci. and Tech., Inc., Chgo., 1992—; cons. Am. Inst. Biological Scis., Bethesda, 1963-67, ind., Balt., 1982—. Author: Microbial Contamination Control Facilities, 1969, Biomedical Applications Laminar Airflow, 1973; contbr. articles to profl. jours. Mem. Assn. for Corp. Growth, Am. Chem. Soc., Bldg. Futures Coun., Ga. Tech. Nat. Alumni Assn., Raleigh (N.C.) C. of C. Democrat. Episcopalian. Avocations: photography, racquetball, swimming. Home: 3440 Lake St Evanston IL 60203 Office: EA Midwest 444 Lake Cook Rd Ste 18 Deerfield IL 60015

RUOFF, HEINZ PETER, chemistry educator; b. Kirchheim unter Teck, Germany, May 15, 1953; s. Heinz and Hilde (Ruzicka) R.; m. Cathrine Lillo, July 26, 1974; children: Martin Ruoff-Lillo, Astrid Lillo-Ruoff. PhD, Oslo U., 1987. Rsch. asst. Oslo U., 1980-84; prof. chemistry Rogaland U. Ctr., Stavanger, Norway, 1984—. Contbr. articles to profl. jours. Odd Hassel fellowship Norwegian Rsch. Coun., 1988-90. Mem. Am. Chem. Soc., Norwegian Chem. Soc., Sigma Xi. Achievements include rsch. on the excitability in the Belousov-Zhabotinsky reaction, the exptl. finding of excitability in the Belousov-Zhabotinsky reaction after theoretical prediction by Richard J. Field and Richard M. Noyes. Office: Rogaland U Ctr, Stavanger 4004, Norway

RUPNIK, KRESIMIR, physicist, researcher; b. Cakovec, Croatia, Nov. 14, 1951; came to U.S., 1984; s. Milan and Ljubica Rupnik; m. Dubravka Rudar, Mar. 20, 1976; children: Ivan, Ana. PhD in Physics, U. Zagreb, Croatia, 1983. Rsch. assoc. Inst. Ruder Boskovic, Zagreb, 1981-84; postdoctoral rsch. assoc. chemistry dept. U.N.C., Chapel Hill, 1984-85; rsch. assoc. chemistry dept. La. State U., Baton Rouge, 1985—. Contbr. articles to Phys. Rev. Letters, Jour. Chem. Physics, Jour. Phys. Chemistry, Theoretica Chimia Acta, Jour. Molecular Structure, Nouvo Jour. Chemie, Jour. Electron Spectroscopy and Related Phenomena. Mem. IEEE, Optical Soc. Am., Am. Assn. Artificial Intelligence, Internat. Neural Network Soc., Internat. Soc. Theoretical Chem. Physics (U.S. rep. 1991). Achievements include research on radiation-matter interactions, on atomic and molecular spectroscopy (X-, UV, MCD, Raman), on nucleic acid bases and other biologically important molecules, on application of AI and adaptive neural nets in physics, and on radiation signatures. Home: 4735 Alvin Dark Ave Apt 3 Baton Rouge LA 70820-8902 Office: La State U Dept Chemistry Baton Rouge LA 70803

RUPPEL, EDWARD THOMPSON, geologist; b. Ft. Morgan, Colo., Oct. 26, 1925; s. Henry George and Gladys Myrtle (Thompson) R.; m. Phyllis Beale Tanner, June 17, 1956; children: Lisa, David, Douglas, Kristin. BA, U. Mont., 1948; MA, U. Wyo., 1950; PhD, Yale U., 1958. Cert. profl. geologist. Geologist U.S. Geol. Survey, Washington and Denver, 1948-86; chief cen. regional br. U.S. Geol. Survey, Denver, 1971-75, geologist, 1975-86; dir., state geologist Mont. Bur. Mines and Geology, Butte, 1986—. Contbr. about 40 articles to profl. jours. With USNR, 1943-46. Fellow Geol. Soc. Am., Soc. Econ. Geologists; mem. Am. Inst. Profl. Geologists, Mont. Geol. Soc., Assn. Am. State Geologists, Tobacco Root Geol. Soc., Rotary. Office: Mont Bur Mines & Geology West Park St Butte MT 59701

RUSCELLO, ANTHONY, aerospace engineer; b. Youngstown, Ohio, Jan. 27, 1965; s. Eugene Mario and Ruth M. (Joseph) R.; m. Julie Ann Wohlfrom, May 28, 1989; 1 child, Nicole Marina. BS in Aerospace Engring., U. Cin., 1988. Aerospace engr. Aero. Systems Ctr. Devel. Planning Directorate Studies & Analysis Divsn. Design Branch, Wright Patterson AFB, Ohio, 1988—. Mem. AIAA. Roman Catholic. Home: 130 Allspice Ct Springboro OH 45066 Office: ASC/XRED Aerospace Engring 1970 Third St Ste 7 Wright Patterson AFB OH 45433-7214

RUSCONI, LOUIS JOSEPH, marine engineer; b. San Diego, Calif., Oct. 10, 1926; s. Louis Edward and Laura Ethelyn (Salazar) R.; m. Virginia Caroline Bruce, Jan. 1, 1972. BA in Engring. Tech., Pacific Western U., 1981, MA in Marine Engring. Tech., 1982; PhD in Marine Engring. Mgmt., Clayton U., 1986. Cert. nuclear ship propulsion plant operator, surface and submarine. Enlisted USN, 1944, electrician's mate chief, 1944-65, retired, 1965; marine electrician planner U.S. Naval Shipyard, Vallejo, Calif., 1965-72; marine elec. technician Imperial Iranian Navy, Bandar Abbas, Iran, 1974-79; marine shipyard planner Royal Saudi Navy, Al-Jubail, Saudi Arabia, 1980-86; cons. in marine engring., 1986—. Author: Shipyards Operations manual, 1980, poetry (Golden Poet award 1989, Silver Poet award 1990). Mem. Rep. Presdl. Task Force, Washington, 1989-90, trustee, 1991. Mem. IEEE, U.S. Naval Inst., Soc. of Naval Architects and Marine Engrs. (assoc. mem.), Fleet Res., Nat. Geographic Soc. Avocations: creative writing, poetry, martial arts. Home: 949 Myra Ave Chula Vista CA 91911-2315

RUSSELL, ALAN JAMES, chemical engineering and biotechnology educator; b. Salford, Lancashire, Eng., Aug. 8, 1962; came to U.S., 1987; s. Francis Anthony and Yvonne (Heilbrunn) R.; m. Janice Elaine Quoresimo, Sept. 19, 1987; children: Hannah Justine Serena, Vincent Anthony Abscander. BSc with honors, U. Manchester (U.K.), 1984; PhD, Imperial Coll., London, 1987. NATO rsch. fellow MIT, Cambridge, 1987-89; Fulton C. Noss assoc. prof. dept. chem. engring. U. Pitts., 1989—; cons. Chem. and Pharm. Industries, 1988—. Contbr. articles to profl. jours. NATO fellow, 1988; recipient Presdl. Young Investigator award NSF, 1990. Mem. Am. Chem. Soc. (session chmn. 1990, 91, awards 1989, 93), Biochemistry Soc., Am. Inst. Chem. Engrs. Republican. Lutheran. Achievements include pioneering use of protein engineering to alter rationally the pH dependence of enzymes; discovery of the phenomenon of enzyme memory in organic solvents. Office: U Pitts 1235 Benedum Hall Pittsburgh PA 15261-2212

RUSSELL, CHARLES ROBERTS, chemical engineer; b. Spokane, Wash., July 13, 1914; s. Marvin Alvin and Dessie Corselia (Price) R.; m. Dolores Kopriva, May 17, 1943; children—Ann E., John C., David F., Thomas R. B.S. in Chem. Engring, Wash. State U., 1936; Ph.D. in Chem. Engring. (Procter and Gamble Co. fellow 1940-41), U. Wis., 1941. Egr. div. reactor devel. AEC, Washington, 1950-56; engr. Gen. Motors Tech. Center, Warren, Mich. and Santa Barbara, Calif., 1956-68; assoc. dean engring. Calif. Poly. State U., San Luis Obispo, 1968-73; prof. mech. engring. Calif. Poly. State U., 1973-80; mem. nuclear standards bd. Am. Nat. Standards Inst., 1956-78; cons., 1980—; sec. adv. com. reactor safeguards AEC, 1950-55. Author: Reactor Safeguards, 1962, Elements of Energy Conversion, 1967, Energy Sources, Ency. Britannica. Served with USNR, 1944-46. Mem. Am. Chem. Soc. Republican. Roman Catholic. Club: Channel City (Santa Barbara). Home and Office: 3071 Marilyn Way Santa Barbara CA 93105-2040

RUSSELL, CRAIG JOHN, management educator; b. Mason City, Iowa, Sept. 10, 1954; s. Ned Elmer and Carolyn Ann (Noelke) R.; m. Marla Joan Newton, Aug. 16, 1975; children: Johanna Marie, Ian Benjamin, Alec Myles. BS, U. Iowa, 1976, PhD, 1982. Asst. prof. U. Pitts., Pa., 1982-86, Inst. Mgmt. and Labor Rels., Rutgers U., New Brunswick, N.J., 1986-91; assoc. prof. dept. mgmt. Coll. Bus. Adminstrn., La. State U., Baton Rouge, 1991—; faculty rsch. appt. Navy Pers. Rsch. and Devel. Ctr., San Diego, summer 1985, 87; vis. prof. Purdue U. Kranner Grad. Sch. Mgmt., West Lafayette, Ind., 1990-91; cons. reviewer So. Mgmt. Conf., 1987—, Nat. Acad. Mgmt. meetings, 1988—; cons. L.A. Unified Sch. Dist., 1992—, Marriott Corp., Washington, 1991—, Goodyear Tire and Rubber, Akron, Ohio, 1988-92, others; presenter in field. Co-editor: The Training and Development Sourcebook, 1992.; mem. editoral bd.: The Indsl.-Orgnl. Psychologist, 1989—; cons. reviewer Acad. Mgmt. Jour., Acad. Mgmt. Rev., Human Resources Mgmt. Rev., Jour. Applied Psychology, Jour. Mgmt., Orgnl. Behavior and Human Decision Processes, Pers. Psychology, Psychol. Bulletin; contbr. articles to Jour. Applied Psychology, Acad. Mgmt. Jour., Applied Psychol. Measurement, Jour. Mgmt., Pers. Psychology. Mem. APA, Acad. Mgmt. (membership com. P/HR div. 1990-91, awards com. human resources div. 1992, awards com. rsch. methods div. 1992), Soc. Indsl. and Organized Psychology (awards com. 1989—, program com. 1988-91, Edwin E. Ghiselli award for rsch. design 1986), Beta Gamma Sigma. Office: Dept Mgmt Coll Bus La State Univ Baton Rouge LA 70803-6312

RUSSELL, DAVID EMERSON, consulting mechanical engineer; b. Jacksonville, Fla., Dec. 20, 1922; s. David Herbert and Wilhelmina (Ash) R.; B.Mech. Engring., U. Fla., 1948; postgrad. Oxford (Eng.) U. Mech. engr. United Fruit Co., N.Y.C., 1948-50, U.S. Army C.E., Jacksonville, 1950-54, Aramco, Saudi Arabia, 1954-55; v.p. Beiswenger Hoch and Assocs., Inc., Jacksonville, 1955-57; owner, operator David E. Russell and Assos., cons. engrs., Jacksonville, 1957—. Chmn. Jacksonville Water Quality Control Bd., 1969-73; bd. dirs. Jacksonville Hist. Soc., 1981-82; mem. Jacksonville Bicentennial Commn., 1973-79. Served to 2d lt. AUS, 1943-46. Recipient Outstanding Service award City of Jacksonville, 1974. Registered profl. engr., Fla., Ga. Mem. ASME (chmn. N.E. Fla. 1967-68), Nat. Soc. Profl. Engrs., ASHRAE, Soc. Am. Inventors, Fla. Engring. Soc. Episcopalian. Club: University (Jacksonville). Contbr. articles to profl. jours.; holder of 5 U.S. patents. Home: 4720 Timuquana Rd Jacksonville FL 32210-8231 Office: 110 Riverside Ave Jacksonville FL 32202-4995

RUSSELL, ELBERT WINSLOW, neuropsychologist; b. Las Vegas, N.Mex., June 4, 1929; s. Josiah Cox and Ruth Annice (Winslow) R.; children from previous marriage: Gwendolyn Marie Harvey, Franklin Winslow, Kirsten Nash, Jonathan Nash; m. Sally Lynn Kolitz, Apr. 2, 1989. BA, Earlham Coll., Richmond, Ind., 1951; MA, U. Ill., 1953; MS, Pa. State U., 1958; PhD, U. Kans., 1968. Clin. psychologist Warnersville (Pa.) State Hosp., 1959-61; clin. neuropsychologist VA Med. Ctr., Cin., 1968-71; dir. neuropsychology lab. VA Med. Ctr., Miami, Fla., 1971-89, rsch. psychologist, 1989—; adj. prof. Nova U., Ft. Lauderdale, 1980-87, U. Miami Med. Sch., 1980—, U. Miami, 1979—. Author: (with C. Neuringer and G. Goldstein) Assessment of Brain Damage, 1970; contbr. articles to profl. jours. Fellow Am. Psychol. Assn., Am. Psychol. Soc.; mem. Nat. Acad. Neuropsychology, Sigma Xi. Democrat. Soc. of Friends. Home: 6091 SW 79th St Miami FL 33143-5030 Office: VA Med Ctr 1201 NW 16th St Miami FL 33125-1624

RUSSELL, HENRY GEORGE, structural engineer; b. Tewkesbury, Eng., June 12, 1941; came to U.S., 1968. BE, Sheffield U., Eng., 1962, PhD, 1965. Registered structural engr., Ill.; registered profl. engr., Wash., Minn. Rsch. fellow Bldg. Rsch., Eng., 1965-68; structural engr. Constrn. Tech. Labs., Inc. (formerly Portland Cement Assn.), Skokie, Ill., 1968-74, mgr., 1974-79, dir., 1979-88, pres., 1989-91, v.p. 1991—. Contbr. articles on reinforced and prestressed concrete to profl. jours. Named one Those Who Made Marks in 1992, Engring. News Record. Fellow Am. Concrete Inst. (Delmar L. Bloem award 1986, Wason medal 1992); mem. Prestressed Concrete Inst. (Martin P. Korn award 1980). Office: Constrn Tech Labs Inc 5420 Old Orchard Rd Skokie IL 60077-1030

RUSSELL, JEFFREY SCOTT, civil engineering educator; b. Alliance, Ohio, June 14, 1962; s. Ronald Francis Russell and Georgia Ann (Charleston) Holmes; m. Vicki Carolina Radford, Aug. 17, 1985; children: Nicole Lynne, Jacob Thomas, Matthew David. BS, U. Cin., 1985; MS, Purdue U., 1986, PhD, 1988. Grad. teaching asst. Purdue U., West Lafayette, Ind., 1985-87, grad. rsch. asst., 1987-88, postdoctoral rsch. assoc., 1988-89; asst. prof. civil engring. U. Wis., Madison, 1989—; lectr. Tex. A&M U., College Station, 1988—, U. Tex., Austin, 1992—, U. Wis., Madison Ext., 1990—. Mem. Wis. Right to Life, Madison, 1989—; project coord. U. Wis. Coll. Testament Distbn., Gideon's Internat., Madison West Camp, 1990—; dir. evangelistic ministries First Ch. of Nazarene, Madison, 1992-93. Recipient Presdl. Young Investigator award NSF, 1990, Edmund Friedman Young Engr. award for profl. achievement, 1993. Mem. ASCE (assoc., sec. constrn. div. 1989-92, Collingwood Prize 1991, Outstanding Profl. Civil and Environ. Engring. 1991), Am. Assn. Cost Engrs. (assoc.), Am. Soc. Engring. Edn., Constrn. Mgmt. Assn. Am., Sigma Xi, Tau Beta Pi. Achievements include identification of causes of constrn. contractor failure; devel. of analytical models to assist in predicting contractor failure prior to contract award. Office: U Wis-Madison 2258 Engring Hall 1415 Johnson Dr Madison WI 53706

RUSSELL, JOHN BLAIR, chemistry educator; b. Rochester, N.Y., Dec. 13, 1929; s. John and Ruth (Fulton) R.; m. Barbara Ruth Woods, Sept. 12, 1955; 1 child, Deborah Ruth. AB, Oberlin (Ohio) Coll., 1951; PhD, Cornell U., 1956. Instr. Cornell U., Ithaca, N.Y., 1955-56; asst. prof. Humboldt State U., Arcata, Calif., 1956-61, assoc. prof., 1961-67, prof. chemistry, 1967-92, prof. emeritus, 1992—. Author: General Chemistry, 1980, 2nd edit., 1992. Mem. Am. Chem. Soc., Sigma Xi, Phi Beta Kappa, Alpha Chi Sigma. Achievements include first to directely measure the reduction potential of the electron electrode in liquid ammonia. Home: 6749 Fickle Hill Rd Arcata CA 95521 Office: Humboldt State U Arcata CA 95521

RUSSELL, JOSETTE RENEE, industrial engineer; b. Defiance, Ohio, June 14, 1964; d. Eugene Alvin and Carole Josette (Galusha) R. BS in Indsl. Engring., GMI Engring. and Mgmt. Inst., 1988. Electronics engr. aero. systems div. USAF, Wright Patterson AFB, Ohio, 1985-90; process engr. Masland Industries, Carlisle, Pa., 1990-91, prodn. supr., 1991-92, prodn. tech. dir., 1992-93; owner Auto Buy Line, 1992-93; mgr. quality McCord Winn Textron, Cookville, Tenn., 1993—. Appeared on cover of Woman Engr. Mag., 1988. Mem. Mensa, Sweet Adelines Internat. Republican. Avocations: Mickey Mouse memorabilia, composing, arranging, reading. Home: 120 B Westgate Cir Cookeville TN 38501 Office: McCord Winn Textron 815 Delman Dr Cookeville TN 38501

RUSSELL, KENNETH CALVIN, metallurgical engineer, educator; b. Greeley, Colo., Feb. 4, 1936; s. Doyle James and Jennie Frances (Smith) R.; m. Charlotte Louise Wolf, Apr. 13, 1963 (div. 1978); children: David Allan, Doyle John. Met.E., Colo. Sch. Mines, 1959; Ph.D., Carnegie Inst. Tec., 1963. Engr. Westinghouse Research and Devel. Center, 1959-61; NSF postdoctoral fellow Physics Inst., U. Oslo, 1963-64; asst. prof. metallurgy M.I.T., Cambridge, 1964-69; assoc. prof. M.I.T., 1969-78, prof. metallurgy, 1978—, prof. nuclear engring., 1979—. Contbr. articles to prof. publs. Served as 2d lt. U.S. Army, 1959-60. DuPont fellow, 1961-62; NSF grad. fellow, 1962-63. Mem. ASTM, AIME, Am. Phys. Soc., Am. Nuclear Soc., Am. Soc. for Metals. Office: MIT Rm 8-411 Cambridge MA 02139

RUSSELL, KENNETH WILLIAM, chemical engineer; b. Belleville, N.J., Nov. 11, 1963; s. Fred William and Joanne Philomena (Tomaselli) R.; m. Kathy Lynn LeBlanc, May 26, 1990; children: Kyleigh Danae, Kamron William. B Chem. Engring., Rutgers U., 1986; M Engring. Mgmt., Lamar U., 1992. Engring. aide Plastics Recycling Inst., Piscataway, N.J., 1985-86; rsch. technician Comet Design, Warren, N.J., 1986-88; process engr. J.M. Huber Polymer Svcs., Orange, Tex., 1988-90; pilot plant engr. Chevron Chem. Co., Orange, 1990-92, product devel. engr., 1992—; cons. Gateway Svcs., Orange, 1992—. Mem. AICE, Soc. Plastics Engrs. (newsletter editor 1989-91, bd. dirs. 1991-93), Houston Material Handling Soc. Republican. Roman Catholic. Achievements include development of reactor control scheme for increased quality products, design of strand die for improved product quality and reduced down-time. Home: Rt 10 PO Box 1366 Orange TX 77630 Office: Chevron Chem Co FM 1006 Orange TX 77631-7400

RUSSELL, RAY LAMAR, project manager; b. Maryville, Tenn., Oct. 13, 1944; s. Homer Ray and Edna (Moore) R.; m. Mary Elizabeth Knoll, Dec. 17, 1965; children: Mary Ruth, Matthew Hastings. BSEE, U. Tenn., 1966; MBA, Stetson U., 1971. Digital systems engr. Guidance & Control Div., Kennedy Space Center, Fla.; spacelab avionics engr. Elec. Systems Div. Am. Embassy, The Hague, Netherlands, 1974-80; shuttle engr. Elec. Systems Div., Kennedy Space Ctr., 1981-83; shuttle avionics engr. Cite & Software Br., Kennedy Space Ctr., 1981-83; payload integration engr. Payload Ops. & Checkout Office, Kennedy Space Ctr., 1983-87; sect. chief Elec. & Networks Systems, Kennedy Space Ctr., 1987-91; project mgr. Payload Mgmt. & Ops., Kennedy Space Ctr., 1991—; space lab. avionics engring. European Space Agy., The Hague, 1976-77. Chmn. Planning & Zoning Bd., Cape Canaveral, Fla., 1982-92. Office: Info Systems Br Mail Code CS-GSD-3 Kennedy Space Center FL 32899

RUSSELL, ROBERT PRITCHARD, ophthalmologist; b. Columbia, S.C., Apr. 30, 1945; s. Austin Henderson and Ruby Mae (Pritchard) R.; m. Olivia Louise Walker, Jan. 22, 1972; children: Robert Pritchard Jr., Denise Olivia. BA with distinction, U. Miss., 1967; MD, U. Miss., Jackson, 1971. Intern Miss. Bapt. Hosp., Jackson, 1971-72; resident U. Med. Ctr., Jackson, 1974-77; pvt. practice ophthalmologist Jackson, 1977—; mem. surg. staff St. Dominic Health Svcs., Jackson, 1977—; cons. Miss. Bapt. Med. Ctr., Jackson, 1977—, River Oaks Hosp., Flowood, Miss., 1981—; attending teaching staff mem. U. Med. Ctr., Jackson, 1977—. Lt. USNR, 1972-74. Mem. Am. Assn. Ophthalmology, Am. Soc. Contemporary Ophthalmology, Internat. Glaucoma Congress, Am. Intraocular Implant Soc., Contact Lens Assn. Ophthalmologists, Am. Acad. Ophthalmology, Phi Kappa Phi, Alpha Epsilon Delta, Phi Eta Sigma, Phi Chi, Sigma Chi. Republican. Avocations: outdoor sports, philately. Home: 139 Royal Lytham Jackson MS 39211-2516 Office: Watkins Med Bldg 1421 N State St Ste 501 Jackson MS 39202-1677

RUSSELL, ROGER WOLCOTT, psychobiologist, educator, researcher; b. Worcester, Mass., Aug. 30, 1914; s. Leonard Walker and Sadie (Stanhope) R.; m. Kathleen Sherman Fortescue, 1945; 2 children. BA, Clark U., 1935, MA, 1936; PhD, U. Va., 1939; DSc, U. London, 1954; DSc (hon.), Newcastle, 1978, Flinders, 1979. Instr. psychology U. Nebr., 1939-41, Mich. State Coll., 1941; rsch. psychologist USAF Sch. Aviation Medicine, 1941-42; from asst. prof. to assoc. prof. psychology U. Pitts., 1946-49; rsch. fellow neurophysiology Western Psychiat. Inst., 1947-49; Fulbright advanced rsch. scholar, dir. animal rsch. lab., Inst. Psychiat. U. London, 1949-50; prof. psychology, head dept. psychology U. Coll., London, 1950-57; from prof., chmn. dept. psychology to dean advanced studies Indiana U., 1959-67; vice chancellor acad. affairs, prof. psychol-biology, clin. pharmacology, therapeutics U. Calif., Irvine, 1967-72; vice chancellor, prof. psychobiology Flinders U. of So. Australia, 1972-79; prof. emeritus U. Calif., Irvine, 1980—, rsch. psychobiologist Sch. Biol. Scis., 1990—; vis. prof. dept. psychology U. Reading, 1977, U. Stockholm, 1977, U. Sydney, 1965-66; vis. prof. pharmacology Sch. Medicine Brain Rsch. Inst. UCLA, 1976-77, 1980—; vis. Erskine fellow U. Canterbury, New Zealand, 1966. Editor: Frontiers in Psychology, 1964, Frontiers in Physiological Psychology, 1966, Matthew Flinders: The Ifs of History, 1979, Behavioral Measures of Neurotoxicity, 1990. Sec. gen. Internat. Union of Psychological Sci., 1960-66, v.p. 1966-69, pres. 1969-72, exec. com. 1972-80; mem. Army Sci. adv. panel, 1958-66; active Nat. Rsch. Coun., 1958-61, 63-65, 67-71; mem. adv. com. psychopharm. USPHS, 1957-63, 67-70, 81-85; active Commonwealth Edn. R&D com., 1974-79; bd. dirs. Australian-Am. Edn. Found., 1972-79. With USAF, 1942-46. Decorated Bronze Star; Livermore scholar, Clark fellow in psychology, Clark U.; DuPont Rsch. fellow U. Va.; Payne scholar Vanderbilt U. Mem. APA (CEO 1956-59, bd. dirs. 1963-65, pres. div. 1 1968-69). Achievements include rsch. papers on neurochem. bases of behaviour, exptl. psycho-pathology, physiological, child and social psychology, psychopharmacology. Office: Univ Calif Ctr Neurobiol Learning and Memory Bonney Ctr Irvine CA 92717

RUSSELL, TED MCKINNIES, electronics technician; b. Marion, Va., Feb. 25, 1943; s. Frank and Virginia (Pollit) R.; m. Patricia Ann Gilroy, Feb. 25, 1979. B in Elec. Tech., Grantham Sch. Electronics, 1966. Sr. electronic technician DPL Svcs., 1965-76; ind. gen. contractor, 1976-78; electronics technician Teltronic Indsl. Systems, Silver Spring, 1978-80, Washington Met. Transit Authority, 1980—; pres. TMR Design, 1989—. Recipient Cert. Achievement Motorola Communications and Electronics, 1978. Home: 1414 17th St NW Apt 814 Washington DC 20036-6415 Office: Washington Met Transit Authority 600 5th St NW Washington DC 20001-2651

RUSSELL, THOMAS WILLIAM FRASER, chemical engineering educator; b. Moose Jaw, Saskatchewan, Can., Aug. 5, 1934; s. Thomas D. and Evelyn May (Fraser) R.; m. Shirley A. Aldrich, Aug. 1956; children: Bruce, Brian, Carey. BS, U. Alberta, Edmonton, Can., 1956, MS, 1958; PhD, U. Del., 1964. Registered profl. engr., Del. Asst. prof. U. Del., 1964-67, assoc. prof., 1967-70, prof., 1970-81, assoc. dean, 1974-77, acting dean, 1978-79, dir. Inst. of Energy Conversion, 1979—, Allan P. Colburn prof., 1981—, chmn. chem. engring. dept., 1986-91; cons. E.I. duPont de Nemours & Co., Inc., Wilmington, Del., 1968—. Author 2 books; contbr. articles to profl. jours., patentee in field. Mem. NAE, Am. Inst. Chem. Engring. (Thomas H. Chilton award 1988, Chem. Engring. Practice award 1987), Am. Chem. Soc., Am. Soc. Engring. Edn. (3M Leadership award chem. engring. div. 1984, Lecture award, 1984). Avocations: wind surfing, hiking, skiing. Home: 46 Darien Rd Newark DE 19711-2024 Office: U Del Inst Energy Conversion Wyoming Rd Newark DE 19716-0001

RUSSELL, WILLIAM ALEXANDER, JR., environmental scientist; b. Hovre de Grace, Md., Nov. 12, 1946; s. William Alexander Sr. and Margaret Adams Webster (Scott) R.; m. Nancy Dion Stacey, Jan. 4, 1965 (div. June 1971); 1 child, Angela Dion; m. Lynne Allison Ertle, July 10, 1971; children: Sara Lynne, Brent William. AA, Harford Community Coll., 1973; BS, Towson State U., 1983, MA, 1991; grad., Army Mgmt. Staff Coll., 1991. Cert. EMT level III firefighter. Environ. coord. U.S. Army Aberdeen Proving Ground (Md.), 1976-81; environ. protection specialist Hdqrs. Dept. Army NAt. Guard Bur., Washington, 1981-85, U.S. Army Environ. Hygiene Agy., Aberdeen Proving Ground, 1985—. Contbr. tech. papers. Bd. dirs. Md. Ornithological Soc., Balt., 1982—, Harford Glen Found., Bel Air, Md., 1989—; chmn. Harford County Environ. Adv. Bd., Bel Air, 1985—; vol., asst. chief, dir., others Aberdeen (Md.) Fire Dept., 1962—; asst. scout master Boy Scouts Am., 1989—. Mem. Nat. Assn. Environ. Profls., Nat. Wildlife Fedn. (life), Nat. Audubon Soc., Md. Conservation Fedn. (charter), Raptor Rsch. Found., Nature Conservancy, Internat. Geographical Honor Soc. Democrat. Avocations: birding, hiking, environ. conservation. Home: PO Box 823 Edgewood MD 21040

RUSSELL, WILLIAM EVANS, pediatric endocrinologist; b. Deroit, Nov. 29, 1949; s. Theodore William and Georgia (Bageris) R.; m. Barbara Kane, Sept. 7, 1987. BS, U. Mich., 1972; MD, Harvard U., 1976. Resident pediatrics Mass. Gen. Hosp., Boston, 1976-79; asst. prof. pediatrics Harvard Med. Sch., Boston, 1984-90; assoc. prof. pediatrics, cell biology Vanderbilt U., NAshville, 1990—. Pediatric Endocrinology fellow Mass. Gen. Hosp., Harvard Med. Sch., 1979-82, U. N.C., 1982-84. Office: Vanderbilt Med Ctr Div Pediatric Endocrinology Nashville TN 37232-2579

RUSSELL-HUNTER, W(ILLIAM) D(EVIGNE), zoology educator, research biologist, writer; b. Rutherglen, Scotland, May 3, 1926; came to U.S., 1963, naturalized, 1968; s. Robert R. and Gwladys (Dew) R-H.; m. Myra Porter Chapman, Mar. 22, 1951 (dec. 1989); 1 child, Peregrine D. B.Sc. with honors, U. Glasgow, 1946, Ph.D., 1953, D.Sc., 1961. Sci. officer Bisra, Brit. Admiralty, Millport, Scotland, 1946-48; asst. lectr U. Glasgow, Scotland, 1948-51, univ. lectr. in zoology, 1951-63; examiner in biology Pharm. Soc. Gt. Britain, Edinburgh, 1957-63; chmn. dept. invertebrate zoology Marine Biol. Lab., Woods Hole, Mass., 1964-68, trustee, 1967-75, 77-87, trustee emeritus, 1989—; prof. zoology Syracuse U., N.Y., 1963-90, emeritus prof., 1990—, continuing rsch. fellow in biology, 1990—; cons. editor McGraw-Hill Encys., 1977—; bd. dirs. Upstate Freshwater Inst., Syracuse, 1981—. Author: Biology of Lower Invertebrates, 1967, Biology of Higher Invertebrates, 1968, Aquatic Productivity, 1970, A Life of Invertebrates, 1979, The Mollusca: Ecology, 1983; mng. editor: Biol. Bull. Woods Hole, Mass., 1968-80; contbr. over 120 articles to sci. jours. William Wasserstrom award Syracuse U., 1988; Carnegie and Brown fellow, 1954; rsch. grantee NIH, 1964-70, NSF, 1971-81, U.S. Army C.E., 1985-87; confirmed Scottish armiger, 1967. Fellow Linnean Soc. London, Royal Soc. Edinburgh, Inst. Biology U.K., AAAS; mem. Ecol. Soc. Am., AAUP. Avocations: book collecting, small boat sailing, model railroading, painting (oils and acrylics). Office: Syracuse U 029L Lyman Hall Syracuse NY 13244

RUSSO, ALVIN LEON, obstetrician/gynecologist; b. Buffalo, Dec. 2, 1924; s. Anthony Joseph and Sarah (Leone) R.; m. Mary Rose Hehir, Sept. 19, 1953; children: Mary B., Sally A. Silvestri, Daniel J., Jeanne Wotherspoon, Margaret Battaile, Terri A., Anthony A. Student, Baylor U., Waco, Tex., 1943-44, U. Iowa, 1944; MD, U. Kans., Kansas City, 1949. Diplomate Am.

Bd. Obstetrics and Gynecology. Intern, then resident E. J. Meyer Meml. Hosp., Buffalo, 1949-55; Fellow in gynocological oncology Roswell Park Meml. Inst., Buffalo, 1955; pvt. practice ob/gyn. San Bernardino, Calif., 1955-89; med. dir. San Bernardino Community Hosp., 1989-92; ret., 1992; bd. dirs. San Bernardino Community Hosp., 1982-89, chmn. bd., 1982-85. Pres. San Bernardino unit Am. Cancer Soc., 1961-62; bd. dirs. More Attractive Community Found. Capt. USAF, 1951-53. Knight, St. John of Jerusalem, 1986--; recipient Distinguished Member award Boy Scouts Am., 1987. Fellow Am. Coll. Ob-Gyns.; mem. AMA, Calif. Med. Assn., N.Y. Acad. Scis., S.W. Ob-Gyn. Soc. (coun. mem. 1989—, v.p. 1992), San Bernardino-Riverside Ob-Gyn. Soc. (pres. 1966), Lions Internat. (dep. dist. gov. 1962-63), Serra Club (pres. San Bernardino chpt.), Arrowhead Country Club (bd. dirs., pres.). Republican. Roman Catholic. Avocations: golf, travel, gardening. Home: 3070 Pepper Tree Ln San Bernardino CA 92404-2313

RUSSO, GILBERTO, engineering educator; b. Rome, Aug. 23, 1948; s. Guido and Maria (Mazzoni) R. Laurea, Poly. Inst. Turin, Italy, 1975; PhD, MIT, 1980. Pres. Studio Russo, Inc. Engring. Cons., Turin, 1970; asst. prof. Poly. Inst. Turin, 1975-80; lectr. MIT, Cambridge, Mass., 1985; dr. dept. plastic and reconstructive surgery U. Chgo., 1992—; mem. designer selection bd. State of Mass., Boston, 1989. Contbr. articles to profl. publs., chpts. to books. Pres. Dante Alisheri Soc., Cambridge, 1986-88; treas. MIT/Poly. Alumni Assn., Turin, 1970. Fulbright fellow, 1978. Fellow Nat. Coun. Engring. Examiners; mem. Mass. Soc. Profl. Engrs. (v.p. 1991—), Tau Beta Pi (chpt. advisor 1985, Eminent Engr. 1985). Achievements include patents in solar energy collectors, development of computer aided therodynamics, computer methods for engineering, optimization of non-steady-state systems, compressible fluid flow with heat transfer. Office: U Chgo Dept Plastic-Reconstrv Surg Chicago IL 60637

RUST, LYNN EUGENE, geologist; b. York, Nebr., Apr. 1, 1952; s. Dale E. and Wilma J. (Wetzel) R.; m. C. Olander, Aug. 1974 (div. May 1979); m. Kristie L. Graham, Sept. 19, 1987. BA in Geology and History, U. Colo., 1975, postgrad., 1976-77; postgrad., Casper Coll., 1979-80. Registered Profl. Geologist, Wyo. Solar observer NOAA/Environ. Rsch. Labs./Space Environ. Svcs. Ctr., Boulder, Colo., 1970-75; phys. sci. tech. Geol. div. U.S. Geol. Survey, Lakewood, Colo., 1976-77; dist. environ. scientist Conservation div. U.S. Geol. Survey, Billings, Mont., 1977-78; regional environ. scientist Conservaton div. U.S. Geol. Survey Minerals Mgmt. Svc., Casper, Wyo., 1978-82; asst. dist. supr. Minerals Mgmt. Svc., Casper, 1982-83; geologist/fluids specialist U.S. Bur. Land Mgmt., Cheyenne, Wyo., 1983-90; chief, br. solid minerals U.S. Bur. Land Mgmt., Cheyenne, 1990—; lectr., seminar presenter; panel leader, mem., moderator on numerous panels for fed. and industry mineral workshops and confs.; mem. 5 nat. task forces on fed. mineral regulation, 1982—. Mem. Cheyenne (Wyo.) Symphony Orch., 1989—, patron, 1991—. Recipient Individual Skylab award NASA, 1976, Unit citation NOAA, 1976. Mem. AAAS, Am. Geol. Inst., Geol. Soc. Am., Am. Mensa (Wyo. Proctor coord. 1988—). Methodist. Home: 2417 E 16th St Cheyenne WY 82001 Office: US Bur of Land Mgmt 2515 Warren Ave Cheyenne WY 82001-3198

RUTAN, ELBERT L. (BURT RUTAN), aircraft designer; b. 1943; s. George and Irene R.; m. BS in Aero. Engring., Calif. State Polytech. U., 1965. Flight test project engr. Air Force Flight Test Ctr., Edwards AFB, Calif., 1965-72; dir. Bede Test Ctr., Kans., 1972-74; pres. Rutan Aircraft Factory, Mojave, Calif., 1974—; founder Scaled Composites, Mojave, Calif., 1982—; v.p. Beech Aircraft, 1985—. Designer more than 100 aircraft including VariViggen, Solitaire, Defiant, and other kits; designer Voyager aircraft, first to fly around-the-world without stopping, refueling. Recip. Spirit of St. Louis Medal, Am. Soc. Mech. Engrs., 1987, Best Design Award, Exptl. Aircraft Assn.; Air medal, 1970, Stan Szik design contbn. trophy, 1972, EAA Outstanding New Design, 1975-76, 78; Dr. August Raspet Meml. award, 1976, ABC World News Tonight Person of the Week, 1986, Presdl. Citizens medal, 1986, FAI gold medal for Voyager Constrn., 1987, medal for the City of Paris, 1987, NASA langley Rsch. Ctr. dirs. award, 1987, Soc. of NASA Flight Surgeons, W. Randolph Lovelace award 1987, Collier trophy for ingenious design and devel., 1987, medal of Outstanding Achievement and disting. leadership, 1987, Soc. of exptl. test pilots, 1987, Aviation Man of the Yr, 1987, Lindbergh Eagle award, 1987, USAF 40th anniversary award, 1987, The City of Genoa, Italy, Christopher Columbus Internat. Communications medal, 1987, British gold medal, 1987, Outstanding Engring. achievement awards, 1988, Franklin medal, 1987, Disting. inventor award 1988, Meritorious Svc. award, 1988, Internat. Aerospace Hall of Fame Honoree, 1988, medal of achievement, 1989, Crystal Eagle award, 1989, Meritorious Civilian Svc. medal, 1989, Leroy Randle Grumman medal, 1989, Structures, Structural Dynamics and Materials award AIAA, 1992. Mem. NAE. Office: Scaled Composites Inc Mojave Airport Hangar 78 Mojave CA 93501

RUTAN, SARAH COOPER, chemistry educator; b. Rochester, N.Y., July 31, 1959; d. Edward Charles and Carol (Gaylord) R. BS, Bates Coll., 1980; MS, Washington State U., 1983, PhD, 1984. Asst. prof. Va. Commonwealth U., Richmond, 1984-90, assoc. prof., 1990--; mem. edit. bd. Jour. Chemometrics, N.Y.C., 1985-92-; sec. div. Analytical Chemistry, Washington, 1992--. Assoc. editor Analytica Chimica Acta. Mem. Am. Chem. Soc. Office: Va Commonwealth U Box 2006 Dept Chemistry Richmond VA 23284

RÜTER, INGO, research chemist, toxicology consultant; b. Neumünster, Germany, Aug. 30, 1962; s. Hans-Dieter and Gerda (Ledig) R. Diploma in chemistry, U. Kiel, Germany, 1988, DSc, 1990. Rsch. chemist U. Kiel, 1988-91; toxicology cons., hazardous waste mgr. Dezernènt, Bezirk sregierung, Münster, 1991—. Contbr. articles to sci. jours. Avocations: astronomy, tennis. Home: Hopfenhorst 7, 24232 Schönkirchen Germany Office: Regierungspräsident Münster, Domplatz 1-3, 48143 Münster Germany

RUTGER, J. NEIL, agronomy research administrator; b. Noble, Ill., Mar. 3, 1934; s. Frank Russell and Jennie Marie (Pearce) R.; m. P. J. Kuyoth, Feb. 15, 1958; children: Ann Michele, Robyn Jean. BS, U. Ill., 1960; MS, U. Calif., Davis, 1962, PhD, 1964. Asst., then assoc. prof. Cornell U., Ithaca, N.Y., 1964-70; rsch. geneticist ARS-USDA, Davis, 1970-88; assoc. dir. Mid-South area ARS-USDA, Stoneville, Miss., 1989—; bd. dirs. Nat. Plant Genetics Resources Bd., Washington, 1988-90. Contbr. numerous articles to sci. and profl. jours. Sgt. U.S. Army, 1954-56. Fellow Am. Soc. Agronomy, Crop Sci. Soc. Am., AAAS. Office: USDA Agrl Rsch Svc PO Box 225 Stoneville MS 38776

RUTH, STAN M., chemical engineer; b. Tulsa, Okla., Jan. 12, 1956; s. Ralph Gustav and Eleanor Grace (Basinger) R.; m. Diane Marie Crain, May 22, 1982; children: Anna Leigh, Katherine Michelle. BSChE, U. Tulsa, 1978. Licensed profl. engr., Tex. Process engr. Unocal Corp., Nederland, Tex., 1978-80, ops. engr., 1980-82, sr. process engr., 1982-84, sr. ops. engr., 1984-87, supr. process control, 1987-90; sr. process automation engr. Michelin/Ameripol Synpol Co., Port Neches, Tex., 1990—. Mem. Nat. Soc. Profl. Engrs., Instrument Soc. Am., Am. Inst. Chem. Engrs. (bd. dirs. S.E. Tex. sect. 1984-87). Presbyterian. Home: 730 Monterrey Beaumont TX 77706 Office: Michelin/Ameripol Synpol Co PO Box 667 Port Neches TX 77651

RUTHERFORD, JOHN STEWART, chemistry educator; b. Johnstone, U.K., June 25, 1938; arrived in Republic of South Africa, 1982; s. Thomas and Elisabeth Gibb (Stewart) R.; m. Wendy Margaret Seale, June 20, 1964 (div. 1978); children: Shona Dawn, Julie Fiona; m. Sasha Bond, Aug. 11, 1982. BSc in Chemistry, Glasgow (Scotland) U., 1959; PhD in Chemistry, McMaster U., Hamilton, Ont., Can., 1967. Cert. tchr. Rsch. assoc. U. Pitts., 1967-68; rsch. fellow U. Essex, Colchester, Eng., 1968-69; programmer-analyst No. Electric Co., Bramalea, Ont., Can., 1970-71; rsch. assoc. U. Regina (Sask., Can.), 1972-75; tchr. Strathclyde (U.K.) Region, 1976-79; lectr. U. Swaziland, Kwaluseni, 1979-82; prof. U. Transkei, Umtata, South Africa, 1982—; dean faculty of sci. U. Transkei, Umtata, 1987-92. Contbr. articles to profl. jours. including Acta Crystallographica. Bd. dirs. Tsolo (Republic of South Africa) Agrl. Coll., 1988—. Mem. Am. Chem. Soc., Am. Inst. Physics, Am. Crystallographic Assn., South Africa Chem. Inst. Achievements include research in crystal structure analysis, in discrete mathematics in solid state chemistry. Office: U Transkei, Private Bag XI Unitra, Umtata Transkei, South Africa

RUTLEDGE, JANET CAROLINE, electrical engineer, educator; b. Detroit, May 26, 1961; d. Philip James and Violet (Eklund) R. BEE, Rensselaer Poly. Inst., 1983; PhD, Ga. Inst. Tech., 1990. Asst. prof. Northwestern U., Evanston, Ill., 1990—; mem. rsch. adv. coun. Nelson Industries, Stoughton, Wis., 1991—. Bd. trustees Rensselaer Poly. Inst., 1993—. Named Outstanding Tech. Nat. Soc. Black Engrs., 1991, one of "Top Forty under 40" Crain's Chgo. Bus., 1992; recipient Alumni Key Rensselaer Poly. Inst., 1992. Mem. IEEE, Acoustical Soc. Am., Rensselaer Alumni Club Chgo. (pres. 1990), Sigma Xi. Achievements include patent for apparatus and method for modifying a speech waveform to compensate for recruitment of loudness. Office: Northwestern U EECS Dept 2145 Sheridan Rd Evanston IL 60208-3118

RUTLEDGE, KATHLEEN PILLSBURY, sensory scientist, researcher; b. Monterey, Calif., May 7, 1952; d. Arthur Niles and Dawn Harwood Hull; m. Richard Keith Rutledge, May 16, 1975. BA in English, Pacific U., 1975; AS in Radiography, City Coll. San Francisco, 1977. Cert. in radiography. Radiographer French Hosp., San Francisco, 1978-80; test officer U.S. Army Tropic Test Ctr., Corozal, Panama, 1980-82; sensory analyst Provesta Corp., Bartlesville, Okla., 1989—. Viola player Bartlesville Symphony, 1991—. Mem. ASTM (E18 sensory divsn.), Am. Soc. Cereal Chemists, Inst. Food Technologists. Achievements include flavor research on theory of flavor enhancers and flavors using highly trained sensory descriptive flavor panelists. Home: 519 Lee Dr Bartlesville OK 74006 Office: Provesta Corp 501 SE 4th Bartlesville OK 74004

RUTLEDGE, SHARON KAY, research engineer; b. Fairview Park, Ohio, July 13, 1961; d. Harry E. and Wilma E. (Mills) R. B Chem. Engring., Cleve. State U., 1984; M Materials Sci. Engring., Case Western Res. U., 1988. Registered profl. engr., Ohio. Rsch. engr. NASA Lewis Rsch. Ctr., Cleve., 1984—; cons. on atomic oxygen reaction with power materials issues Space Sta. Freedom; presenter sci. demonstrations local schs., 1990—. Contbr. chjpt. to Ency. of Composites, 1990; contbr. articles to Jour AIAA, Jour. of Surfaces and Coatings Tech. Mem. AICE, Am. Soc. Materials Internat., Soc. Advancement of Material and Process Engring., Tau Beta Pi. Achievements include patent for ion beam sputter etching; development of thermal emittance enhanced surfaces. Office: NASA Lewis Rsch Ctr MS 302-1 21000 Brookpark Rd Cleveland OH 44135

RUTLEDGE, WYMAN CY, research physicist; b. Abrahamsville, Pa., Dec. 15, 1924; s. Coe Sanford and Bernice I. (Gregg) R.; Mary Louise Jones, June 6, 1945; Paul, Nancy, Mark. AB in Physics, Math. and Chemistry, Hiram Coll., 1944; MS in Physics, U. Mich., 1948, PhD in Physics, 1952; PhD in Sci. (hon.), Miami U., 1990. Jr. physicist Woods Hole (Mass.) Oceanographic Inst., 1946-47; rsch. assoc. Aero Rsch. Lab. U. Mich., Ann Arbor, 1947-48, rsch. assoc. Nuclear Spectros, 1948-50; rsch. assoc. Argonne Nat. Lab., Lemont, Ill., 1950-52; rsch. physicist Philips Labs., Irvington, N.Y., 1952-56; prin. scientist Mead Cent. Rsch. Labs, Chillicothe, Ohio, 1956-91; cons. Chillicothe, 1991--; cons. measurements Am. Inst. Paper, N.Y.C., 1965—; cons. advanced sensors Dept. Energy, Washington, 1980—; adv. bd. mem. Ohio U., Chillicothe, 1976—, chmn., 1984; trustee Hiram Coll., 1980-86. Contbr. 60 articles to profl. jours. Pres., v.p., treas., bd. dirs. Med. Ctr., Ross County, Ohio, 1968-74; mem., pres. Ross County Hosp. Commn., 1982-92, mem., 1974—; bd. dirs., pres. 4-Coun. Boy Scout Coun., Chillicothe, 1970-71, mem., 1970-92; v.p. Mid-Ohio Health Planning Fedn., Columbus, Ohio, 1973-75. With U.S. Army, 1944-46. Recipient Phoenix award U. Mich., 1950-51, Garfield Soc. award Hiram Coll., 1980; named Tech. Man of Yr. Inter Soc. Coun., 1975, 77, fellow, 1975, Layman of Yr. Kiwanis, 1973. Fellow Tech. Assn. Pulp and Paper, Instrument Soc. Am. (nat. v.p. 1974-78, Disting. Svc. award 1982, Golden Achievement award 1988, chmn. admissions com. 1989-91), Symposiarchs Am. (nat. pres. 1970-71, 91, Symposiarch of Yr. 1952), Sigma Xi. Republican. Achievements include patents infield. Home: 704 Ashley Dr Chillicothe OH 45601

RUTOLO, JAMES DANIEL, mechanical engineer; b. Stratford, Conn., Apr. 21, 1965; s. James Thomas and Carla Marie (Conti) R. BS in Aerospace Engring., Pa. State U., 1989. Carpenter's helper Century Tech. Builders, Feasterville, Pa., 1986; delinquent acct. auditor TSO Fin. Corp., Horsham, Pa., 1987; aeronautical engr. Pa. State U., University Park, 1988; math. proctor Pa. State U., Fogelsville, 1989; asst. mgr. Tuerkes, North Whales, Pa., 1988-90; mechanical engr. Tech. Products Inc., Hatfield, Pa., 1990-93; mech. engr. Hale Fire Pump Co., Conshohocken, Pa., 1993—; tutor in English Pa. State U., Altoona, 1983. Mem. AIAA, United Steelworkers Am. Republican. Roman Catholic. Home: 2126A Ted-Jim Dr Warrington PA 18976 Office: Hale Fire Pump Co 700 Spring Mill Ave Conshohocken PA 19428

RUTSTROM, DANTE JOSEPH, chemist; b. Beverly, Mass., Nov. 10, 1958; s. Eric and Laura (Grimaldi) R.; m. Melanie Sue Bragdon, June 25, 1983; children: Stephanie Lynne, Kristen Ashley, Daniel Joseph. BA, Gordon Coll., Wenham, Mass., 1980; PhD, Tufts U., 1985. Chemist Behlehem (Pa.) Steel Corp., 1985-86; analytical chemist Eastman Chem. Co., Kingsport, Tenn., 1986—. Co-author: Analytical Chemistry, vol. 57; contbr. articles to Jour. Electroanalytical Chemistry, Advances in Lab. Automation. Mem. Product Devel. Mgmt. Assn. Achievements include research in laboratory automation and robotics.

RUTTER, WILLIAM J., biochemist, educator; b. Malad City, Idaho, Aug. 28, 1928; s. William H. and Cecilia (Dredge) R.; m. Jacqueline Waddoups, Aug. 31, 1951 (div. Nov. 1969); children: William Henry II, Cynthia Susan; m. Virginia Alice Bourke, 1972 (div. 1978). BA, Harvard U., 1949; MA, U. Utah, 1950; PhD, U. Ill., 1952. USPHS postdoctoral fellow U. Wis., Madison, 1952-54, Nobel Inst., Stockholm, 1954-55; from asst. prof. to prof. biochemistry, dept. chemistry U. Ill., 1955-65; prof. biochemistry U. Wash., 1965-69; Hertzstein prof. biochemistry U. Calif., San Francisco, chmn. dept. biochemistry and biophysics, 1969-82, dir. Hormone Research Inst., 1983-89; chmn. bd. dirs. Chiron Corp., 1981—; mem. USPHS Biochemistry and Nutrition Fellowship Panel, 1963-66; cons. physiol. chemistry study sect. NIH, 1967-71; mem. basic sci. adv. exec. com. Nat. Cystic Fibrosis Research Found., 1969-74, chmn., 1972-74, pres.'s adv. council, 1974-75; exec. com. div. biology and agr. NRC, 1969-72; mem. developmental biology panel NSF, 1971-73; mem. biomed. adv. com. Los Alamos Sci. Lab., 1972-75; pres. Pacific Slope Biochem. Conf., 1972-73; mem. bd. sci. counselors Nat. Inst. Environ. Health Scis., 1976—; mem. adv. com. biology div. Oak Ridge Nat. Lab., 1976-80; adv. bd. Oak Ridge Nat. Lab. and Martin-Marietta Energy Systems, 1984-87; basic research adv. com. Nat. Found., 1976—; bd. dirs. Keystone Life Sci. Study Ctr., Meridian Instruments, 1982-88, Hana Biologies, 1980-83; sci. adv. bd. German Ctr. Molecular Biology, U. Heidelberg, 1983—, AMGEN, 1979-81; panel sci. advisors Internat. Ctr. Genetic Engring. and Biotechnology, 1984—; mem. bd. overseers Harvard U., 1992-96, special commn. NSF, 1992, sci. adv. bd. FDA, 1993. Asso. editor: Jour. Exptl. Zoology, 1968-72; editor: PAABS Revista, 1971-76, Jour. Cell Biology, 1976-78, Archives Biochemistry and Biophysics, 1978—, Developmental Genetics, 1979—; editorial bd. various jours. Served with USNR, 1945. Guggenheim fellow, 1962-63. Mem. Am. Soc. Biol. Chemists (treas. 1970-76, mem. editorial bd. jour. 1970-75), Am Soc. Cell Biology, Am. Chem. Soc. (Pfizer award enzyme chemistry 1967), Am. Soc. Developmental Biology (pres. 1975-76), Nat. Acad. Scis. Home: 80 Everson St San Francisco CA 94131-2659 Office: U Calif Hormone Rsch Inst San Francisco CA 94143-0534 also: Chiron Corp 4560 Horton St Emeryville CA 94608

RUUD, CLAYTON OLAF, engineering educator; b. Glasgow, Mont., July 31, 1934; s. Asle and Myrtle (Bleken) R.; children: Kelley Astrid, Kirsten Anne; m. Paula Kay Mannino, Feb. 24, 1990. BS in Metallurgy, Wash. State U., 1957; MS in Matl. Sci., San Jose State U., 1967; PhD in Materials Sci., U. Denver, 1970. Registered profl. engr., Calif., Colo. Asst. remelt metallurgist Kaiser Aluminum & Chem. Corp., Trentwood, Wash., 1957-58; devel. engr. Boeing Airplane Co., Seattle, 1958-60; mfg. rsch. engr. Lockheed Missiles & Space Corp., Sunnyvale, Calif., 1960-63; rsch. engr. FMC Corp., San Jose, 1963-67; sr. rsch. scientist U. Denver, 1967-79; prof. indsl. engring. Pa. State U., University Park, 1979—; cons. in field; bd. dirs. Denver X-Ray Inst. Inc., Altoona, Pa. Editor series of books: Advances in X-Ray Analysis, Vol. 12-22, 1970-80, Nondestructive Character of Materials, Vol. 1-6, 1983—; editorial com. Nondestructive Testing and Evaluation, 1976-84; contbr. over 100 sci. papers to profl. jours. Mem., chmn. Nat. Acad. Sci. Safe Drinking Water Com., Washington, 1976-78. Recipient IR 100 award, 1983, Gov.'s New Product Award, Pa. Soc. Profl. Engrs., 1988. Mem. ASM (interna. chmn. Resid. Stress Conf. 1989-91), Internat. Ctr. for Diffraction Data, Soc. Mfg. Engrs., Metall. Soc. of AIME. Achievements include patent on Method for Determining Internal Stresses in Polycrystalline Solids; patent on Stress-Unstressed Standard for X-Ray Stress Analysis; invention of a Fiber Optic Based Position Sensitive Scintillation X-Ray Detector; invention of an instrument for simultaneous stress and phase composition measurement; development of an X-ray diffraction instrument for manufacturing process quality control. Office: Pa State U 159 MRL State College PA 16801-6239

RUVKUN, GARY B., molecular geneticist; b. Berkeley, Calif., Mar. 26, 1952; s. Sam and Dora R.; m. Natasha Staller. AB, U. Calif., Berkeley, 1973; PhD, Harvard U., 1981. Postdoctoral fellow MIT, Cambridge, Mass., 1982-85; jr. fellow Soc. of Fellows Harvard U., Cambridge, 1982-85; asst. prof. genetics Harvard U., Cambridge, 1985-90, assoc. prof. genetics, 1990—. Recipient Faculty Rsch. award Am. Cancer Soc., 1989—. Achievements include findings regarding molecular basis of temporal pattern formation, molecular basis of cell lineage asymmetry, scores of homeobox genes in C. elegans. Office: Dept Molecular Biology Mass Gen Hosp Boston MA 02114

RUWWE, WILLIAM OTTO, automotive engineer; b. Cuba, Mo., July 25, 1930; s. Otto Albert and Maude May (Hines) R.; m. Helen Leona Haynes, Jan. 1, 1958; children: Teresa Lynn, Nancy Jean. BS, Cen. Mo. State U., 1959. Engring. clk. Wagner Brake div. Cooper Industries, St. Louis, 1959-64, engr., 1964-67, quality control chemist, 1967-68, mfg. mgr., 1968-82, plant mgr., 1982-93; ret., 1993. Inventor electroless nickle plating process for cast iron, 1964, dissolution of crystal formation in brake fluid, 1971. With U.S. Army, 1951-53. Mem. Soc. Automotive Engrs. (cert., product bus. com.1985-90), St. Louis Geneal. Soc. Avocations: genealogy, history. Home: 415 Meramec Way Saint Charles MO 63303-8447

RYALL, A(LBERT) LLOYD, horticulturist, refrigeration engineer; b. Phoenix, June 25, 1904; s. Lloyd Oliver and Kate Florence (Southam) R.; m. Mary Elizabeth Newton, Dec. 26, 1928; children: Patricia June, Philip Lloyd, Pamela Kate, Peter Newton. BS, N.D. Agrl. Coll., 1926; MS, Oreg. Agrl. Coll., 1928. Jr. pomologist USDA, Yakima, Wash., 1928-41; assoc. horticulturist USDA, Harlingen, Tex., 1942-49; horticulturist USDA, Fresno, Calif., 1949-57; sr. horticulturist USDA, Beltsville, Md., 1957-62; prin. horticulturist, br. chief USDA, Washington, 1962-69; asst. prof. N.Mex. State U., Las Cruces, 1971-75; hort. cons. various produce and transp. cos., 1970-82. Author: (with others) Handling, Transportation and Storage, Vol. 1, Vegetables and Melons, 1981, Handling, Transporatation and Storage, Vol. 2, Fruits and Tree Nuts, 1983; contbr. over 90 articles to profl. jours. Fellow AAAS, Am. Soc. Hort. Sci. (chmn. Pacific chpt. 1956); mem. ASHRAE, Am. Inst. Biol. Sci., Alpha Zeta, Phi Kappa Phi, Lambda Kappa Delta. Achievements include development of advanced applications in precooling, special storage requirements, packaging and transportation services for fresh fruits and vegetables. Home: 6101 Camelot Dr Harlingen TX 78550-8420

RYAN, CLARENCE AUGUSTINE, JR., biochemistry educator; b. Butte, Mont., Sept. 29, 1931; s. Clarence A. Sr. and Agnes L. (Duckham) R.; m. Patricia Louise Meunier, Feb. 8, 1936; children: Jamie Arlette, Steven Michael (dec.), Janice Marie, Joseph Patrick (dec.). BA in Chemistry, Carroll Coll., 1953; MS in Chemistry, Mont. State U., 1956, PhD in Chemistry, 1959. Postdoctoral fellow in biochemistry Oreg. State U., Corvallis, 1959-61, U.S. Western Regional Lab., Albany, Calif., 1961-63; chemist U.S. Western Regional Lab., Berkeley, Calif., 1963-64; asst. prof. biochemistry Wash. State U., Pullman, 1964-68, assoc. prof., 1968-72, prof., 1972—, Charlotte Y. Martin disting. prof., 1991—, chmn. dept. agrl. chemistry, 1977-80, fellow Inst. Biol. Chemistry, 1980—; faculty athletics rep. to PAC-10 & NCAA Wash. State U., 1991—; vis. scientist dept. biochemistry U. Wash., 1981, Harvard U. Med. Sch., 1982; cons. Kemin Industries, Des Moines, 1981—, Plant Genetics, Davis, Calif., 1987-89; research adv. bd. Frito-Lay, Inc., Dallas, 1982, Plant Genetic Engring. Lab., N.M. State U., Las Cruces, 1986-89; mem. adv. rev. bd. Plant Gene Exptl. Ctr., Albany, Calif., 1990-93; mgr. biol. stress program USDA Competitive Grants Program, Washington, 1983-84; former mem. adv. panels for H. McKnight Found., Internat. Potato Ctr., Lima, Peru, Internat. Ctr. Genetic Engring. and Biotech., New Delhi, Internat. Ctr. Tropical Agr., Cali, Columbia, Internat. Tropical Agr., Ibandan, Africa; mem. grant rev. panels NSF, USDA, DOE, NIH; co-organizer Internat. Telecommunications Symposium on Plant Biotech. Mem. edit. bd. several biochem. and plant physiology jours.; contbr. articles to profl. publs., chpts. to books; co-editor 2 books. Grantee USDA, NSF, NIH, Rockefeller Found., McKnight Foun.; recipient Merch award for grad. research Mont. State U., 1959, career devel. awards NIH, 1964-74, Alumni Achievement award Carroll Coll., 1986, Pres.'s Faculty Excellence award in research Wash. State U., 1986; named to Carroll Coll. Alumni Hall Fame, 1981; Carroll Coll. Basketball Hall Fame, 1982. Mem. AAAS, Nat. Acad. Scis. (elected 1986), Am. Chem. Soc. (Kenneth A. Spencer award 1992), Am. Soc. Plant Physiologists (Steven Hales Prize 1992), Am. Soc. Exptl.Biology, Biochem. Soc., Internat. Soc. Chem. Ecology, Internat. Soc. Plant Molecular Biology (bd. dirs.), Phytochem. Soc. N.Am., Nat. U. Continuing Assn. (Creative Programming award 1991), Phi Kappa Phi (Recognition award 1976, selected 1 of 100 centennial disting. alumni Mont. State U. 1993). Democrat. Avocations: fishing, basketball, golf. Office: Wash State Univ Inst Biol Chemistry Pullman WA 99164

RYAN, HUGH WILLIAM, engineer, consultant; b. San Diego, Oct. 25, 1946; s. William Warren and Frances Edith (O'Connor) R.; m. Carolyn Olsen, Nov. 23, 1969; children: Catherine, Sarah, William. BS in Aero. Engring., Okla. U., 1969; MSME, N.Mex. State U., 1971. Rsch. engr. Clinton P. Anderson Phys. Scis. Lab., Las Cruces, N.Mex., 1969-71; mem. staff Andersen Cons., Chgo., 1971-76, mgr., 1976-82, ptnr., 1982-86, 88—; tech. dir. Dept. Social Svcs., London, 1985-86, asst. sec., 1986-88. Editorial bd. Jour. Info. Systems, Chgo., 1985—, Corp. Computing, Chgo., 1991—. Achievements include development of architecture-driven development for systems development, automated systems design approach, integration of structured techniques for systems development methodology and early applications of client/server computing. Office: Andersen Cons 100 S Wacker Dr Chicago IL 60606

RYAN, JAMES WALTER, physician, medical researcher; b. Amarillo, Tex., June 8, 1935; s. Lee W. and Emma E. (Haddox) R.; children: James P.A., Alexandra L.E., Amy J.S. A.B. in Polit. Sci., Dartmouth Coll., 1957; M.D., Cornell U., 1961; D.Phil., Oxford U. (Eng.), 1967. Diplomate: Nat. Bd. Med. Examiners. Intern, Montreal (Que.) Gen. Hosp., McGill U., Can., 1961-62; asst. resident in medicine Montreal (Que.) Gen. Hosp., McGill U., 1962-63; USPHS research asso. NIMH, NIH, 1963-65; guest investigator Rockefeller U., N.Y.C., 1967-68; asst. prof. biochemistry Rockefeller U., 1968; asso. prof. medicine U. Miami (Fla.) Sch. Medicine, 1968-79, prof. medicine, 1979—; sr. scientist Papanicolaou Cancer Research Inst., Miami, 1972-77; hon. med. officer to Regius prof. medicine Oxford U., 1965-67; vis. prof. Clin. Research Inst. Montreal, 1974; mem. vis. faculty thoracic disease div., dept. internal medicine Mayo Clinic, 1974; cons. Ventrex Labs., Inc., Chugai Pharm. Co., Ltd., Tokyo. Contbr. numerous articles on biochem. research and pathology to sci. jours.; patentee in field. Rockefeller Found. travel awardee, 1962; William Waldorf Astor travelling fellow, 1966; USPHS spl. fellow, 1967-68; Pfizer travelling fellow, 1972; recipient Louis and Artur Luciano award for research of circulatory diseases McGill U., 1984-85. Fellow Am. Inst. Chemists; mem. Am. Chem. Soc., Biochem. Soc., Am. Soc. Biol. Chemists, Am. Heart Assn. (mem. council cardiopulmonary diseases 1972—, Council for High Blood Pressure Research 1976—), Microcirculatory Soc., So. Soc. Clin. Investigation, AAAS, N.Y. Acad. Scis., Sigma Xi. Baptist. Club: United Oxford and Cambridge U. (London); The Fisher Island (Miami). Home: 3420 Poinciana Ave Miami FL 33133-6525 Office: 1399 NW 17th Ave Miami FL 33125-2349

RYAN, KENNETH JOHN, physician, educator; b. N.Y.C., Aug. 26, 1926; s. Joseph M. R.; m. Marion Elizabeth Kinney, June 8, 1948; children: Alison Leigh, Kenneth John, Christopher Elliot. Student, Northwestern U., 1946-48; M.D., Harvard U., 1952. Diplomate: Am. Bd. Obstetrics and

Gynecology. Intern, then resident internal medicine Mass. Gen. Hosp., Boston, also Columbia-Presbyn. Med. Center, N.Y.C., 1952-54, 56-57; resident obstetrics and gynecology Boston Lying-in Hosp., also Free Hosp. for Women, Brookline, Mass., 1957-60; instr. obstetrics and gynecology Harvard U.; also dir. Harvard (Fearing Research Lab.), 1960-61; prof. obstetrics and gynecology, dir. dept. Western Res. U. Med. Sch., 1961- 70; prof. reproductive biology, dept. obstetrics and gynecology U. Calif.-San Diego, La Jolla, 1970-73; Kate Macy Ladd prof., chmn. dept. obstetrics and gynecology Harvard U. Med. Sch., 1973-93, dir. Lab. Human Reprodn. and Reproductive Biology, 1974-93, Disting. Kate Macy Ladd prof., 1993—; chief staff Boston Hosp. for Women, 1973-80; chmn. dept. obstetrics and gynecology Brigham's Women's Hosp., Boston. Chmn. Nat. Commn. for Protection Human Subjects Biomed. and Behavioral Research, 1974-78. Recipient Schering award Harvard Med. Sch., 1951, Soma Weis award, 1952, Bordon award, 1952; Ernst Oppenheimer award, 1964; Max Weinstein award, 1970; fellow Mass. Gen. Hosp., 1954-56. Fellow Am. Cancer Soc.; mem. Am. Coll. Obstetricians and Gynecologists, Am. Soc. Biol. Chemists, Endocrine Soc., Soc. Gynecol. Investigation, Mass. Med. Soc., Am. Soc. Clin. Investigation, Am. Gynecol. Soc., Alpha Omega Alpha. Office: Brigham & Woman's Hosp 75 Francis St Boston MA 02115-6195

RYAN, STEPHEN JOSEPH, JR., ophthalmology educator, university dean; b. Honolulu, Mar. 20, 1940; s. S.J. and Mildred Elizabeth (Farrer) F.; m. Anne Christine Mullady, Sept. 25, 1965; 1 dau., Patricia Anne. A.B., Providence Coll., 1961; M.D., Johns Hopkins U., 1965. Intern Bellevue Hosp., N.Y.C., 1965-66; resident Wilmer Inst. Ophthalmology, Johns Hopkins Hosp., Balt., 1966-69, chief resident, 1969-70; fellow Armed Force Inst. Pathology, Washington, 1970-71; instr. ophthalmology Johns Hopkins U., Balt., 1970-71, asst. prof., 1971-72, assoc. prof., 1972-74; prof., chmn. dept. ophthalmology Los Angeles County-U. So. Calif. Med. Ctr., L.A., 1974—; acting head ophthalmology div., dept. surgery Children's Hosp., L.A., 1975-77; med. dir. Doheny Eye Inst. (formerly Estelle Doheny Eye Found.), L.A., 1977—; chief of staff Doheny Eye Hosp., L.A., 1985-88; dean U. So. Calif. Sch. Medicine, L.A., 1993—; mem. advisory panel Calif. Med. Assn., 1974—. Editor: (with M.D. Andrews) A Survey of Ophthalmology—Manual for Medical Students, 1970, (with R.E. Smith) Selected Topics in the Eye in Systemic Disease, 1974, (with Dawson and Little) Retinal Diseases, 1985, (with others) Retina, 1989; assoc. editor: Ophthalmol. Surgery, 1974—; mem. editorial bd.: Am. Jour. Ophthalmology, 1981—, EYESAT, 1981—, Internat. Ophthalmology, 1982—, Retina, 1983—, Graefes Archives, 1984—; contbr. articles to med. jours. Recipient cert. of merit AMA, 1971; Louis B. Mayer Scholar award Research to Prevent Blindness, 1973; Rear Adm. William Campbell Chambliss USN award, 1982. Mem. Wilmer Ophthal. Inst. Residents Assn., Am. Acad. Ophthalmology and Otolaryngology (award of Merit 1975), Am. Ophthal. Soc., Pan-Am. Assn. Ophthalmology, Assn. Univ. Profs. of Ophthalmology, Los Angeles Soc. Ophthalmology, AMA, Calif. Med. Soc., Los Angeles County Med. Assn., Pacific Coast Oto-Ophthal. Soc., Los Angeles County Acad. Medicine, Pan Am. Assn. Microsurgery, Macula Soc., Retina Soc., Nat. Eye Care Project, Research Study Club, Jules Gonin Club, Soc. of Scholars of Johns Hopkins U. (life). Office: Doheny Eye Inst 1450 San Pablo St Los Angeles CA 90033-4681

RYAN, SUZANNE IRENE, nursing educator; b. Yonkers, N.Y., Mar. 13, 1939; d. Edward Vincent and Winifred E. (Goemann) R. BA in Biology, Mt. St. Agnes Coll., Balt., 1962; BSN, Columbia U., 1967, MA in Nursing Svc., 1973, MEd in Nursing Edn., 1975; MS in Oncology, San Jose (Calif.) State Coll. U., 1982. RN, N.Y. Prof. nursing Molloy Coll., Rockville Centre, N.Y., 1970—, co-dir. health svcs., dir. ednl. programs, 1987—, co-dir. mobile health van, adminstr. health edn., 1992—; pres., CEO SIR Enterprises, Inc., 1986; photographer Molloy Coll. Pubs., 1991—; mem. N.Y. State AIDS Coun., 1987—, L.I. Alcohol Consortium, 1987—; educator Nassau County Dept. Sr. Citizens Health, 1991—; photography dir. Bali-Art, Center Harbor, N.Y., Loon Conservancy, Lee's Mills, N.H. Group show exhbns. include permanent displays in photographic galleries in Carmel, Calif., Laconia and Wolfboro, N.H., 1963—; photographer 8 books on Monterey Peninsula, Carmel, Big Sur, San Francisco, New Hampshire, and New England. Health educator Nassau County Dept. of Sr. Citizens Outreach Program, Molloy Coll.; adminstr., chief AIDS counselor Interaction AIDS Counseling, Babylon, N.Y., 1992. USPHS fellow, 1962, Nat. Cancer Inst. fellow, 1981-82. Mem. AAUP, AAUW, Nat. Congress Oncology Nurses, N.Y. State Fedn. Health Educators, Inc., Nurses Assn. Counties L.I. Dist. 14, N.Y. State Nurses Assn., World Wildlife Orgn., Audubon Soc., Nature Conservancy, Sierra Club, Century Club of St. Labre, Indian Sch. Edn. Assn., Internat. Ctr. for Photography, Sigma Theta Tau (Epsilon Kappa chpt., rsch. grantee 1985, 87). Roman Catholic. Avocation: writing, photography. Home: 16 Walker St Malverne NY 11565

RYAN, TIMOTHY WILLIAM, analytical chemist; b. Wilmington, Del., Jan. 18, 1962; s. Joseph Harry and Ruthann (Sutton) R.; m. Susan Lynn Moloney, Oct. 19, 1991. BA, U. Del., 1985. Application chemist Analytical Instrument Devel., Avondale, Pa., 1984-85; metabolism chemist E.I. duPont de Nemours, Wilmington, 1985-88; prin. scientist Computer Chem. Systems, Avondale, 1988-90; rsch. assoc. Bristol-Myers Squibb, New Brunswick, N.J., 1990; analytical chemist Ganes Chemicals, Pennsville, N.J., 1991—; instrument design/mktg. cons. CDS Analytical, Oxford, Pa., 1992—. Contbr. articles to Jour. Chromatography, Intelligent Instruments and Computers, Jour. Liquid Chromatgraphy, Biopharm. Mem. Am. Chem. Soc., Am. Soc. Mass Spectroscopy, Nat. Geographic Soc. Achievements include rsch. in HPLC-UV photodiode array detection as a purity determining technique in bulk pharm. mfg., coupling of this technique with mass spectrometry; supercritical fluid extraction/supercritical fluid chromatography for determining additive levels in polyethylene; development of electrospray, EI/CI mass spectroscopy as a routine qualitative and quantitative technique in QC. Office: Ganes Chems 33 Industrial Park Rd Pennsville NJ 08070

RYAN, WILLIAM B. F., geologist; b. Troy, N.Y., Sept. 1, 1939; married, 1962; 2 children. BA, Williams Coll., 1961; PhD in Geology, Columbia U., 1971. Rsch. asst. oceanography Woods Hole Oceanographal Inst., 1961-62; rsch. asst. geologist Lamont-Doherty Geol. Observatory Columbia U., 1962-74, sr. rsch. assoc. geologist, 1974—. Recipient Francis P. Shepard medal, Soc. Sedimentary Geology, 1993. Am. Geophysics Union, Sigma Xi. Achievements include research in marine geology and geophysics. Home: 12 Clinton Ave Nyack NY 10960*

RYANS, JAMES LEE, chemical engineer; b. Kingsport, Tenn., Sept. 3, 1948; s. Homer Lee and Ollie Hazel (Gilliam) R.; m. Juanella Mae Robinson, Mar. 18, 1968 (div. June 1977); children: Karina Joy, Jeffrey Eric. BS in Math., E. Tenn. State, 1973; BSChemE, U. Tenn., 1973. Chem. engr. Eastman Chem., Kingsport, Tenn., 1973-79; chem. engr. Filter Products Div. Eastman Chem., Kingsport, 1979-86, sr. chem. engr. Cellulose Esters Divsn., 1986-89, sr. chem. engr. Engring. and Constrn. Divsn., 1989—; cons. Am. Chem. Soc., Washington, 1981-82, Environ. Protection Agy., Cin., 1982-83, Ctr. for Profl. Advancement, East Brunswick, N.J., 1983-84. Author: Process Vacuum System Design and Operation, 1986; contbr. articles to profl. jours. Capt. USAR, 1977-83. Mem. AICE, Am. Vacuum Soc., Civitan Internat. Office: Eastman Chem Co PO Box 511 Kingsport TN 37662-5054

RYDBECK, BRUCE VERNON, civil engineer; b. Schenectady, N.Y., Aug. 20, 1948; s. Vernon Arvid and Eleanor Delores (Johnson) R.; m. Cherith Corrine Brabon, June 9, 1973; children: Joel, Caleb, Lydia. BSCE, Clarkson U., 1970; MSCE, Northeastern U., 1975. Registered profl. engr., Mass., N.H., Maine. Engr. Stone & Webster Engr. Corp., Boston, 1970-75; engr. assoc. Haley and Ward Inc., Waltham, Maine, 1975-80; engr. World Radio Missionary Fellowship, Inc., Colorado Springs, 1980—. Mem. ASCE, Am. Concrete Inst., Coll. Civil Engrs. Ecuador. Evangelical. Home: Casilla 17 17 671, Quito Ecuador Office: World Radio Missionary Inc PO Box 39300 Colorado Springs CO 80949

RYDELL, EARL EVERETT, electrical engineer; b. Breckenridge, Minn., Jan. 29, 1944; s. Edwin Fredrick and Phoebe Theodora (Nordly) R.; m. Penny Lee Peterson, Dec. 19, 1970; children: Nicole Jean, Joseph Blaine, Heather Lynn. BSEE, N.D. State U., 1975, postgrad., 1977-78. Systems engr. USN, Port Hueneme, Calif., 1975-78; project engr. No. Ordnance Div. No. Ordnance Div. FMC Corp., Mpls., 1978-86; sr. systems engr. Rockwell-Collins, Cedar Rapids, Iowa, 1986—. Coun. mem. St. John's Luth. Ch., Ely, Iowa, 1988—, pres. 1992—. With USAF, 1966-70. Mem. Soc. Automotive Engrs. Avionics Systems Div. (working tasks groups). Republican. Achievements include patent for Token ring technology. Home: 2897 Skyview Ln NE Swisher IA 52338 Office: Rockwell-Collins 400 Collins Rd NE Cedar Rapids IA 52498

RYDER, OLIVER ALLISON, geneticist, conservation biologist, educator; b. Alexandria, Va., Dec. 27, 1946; s. Oliver A. Ryder and Elizabeth R. (Semans) Paine; m. Cynthia Ryan, Dec. 5, 1970; children: Kerry, Ryan. BA, U. Calif., Riverside, 1968; PhD, U. Calif., San Diego, 1975. Postdoctoral fellow U. Calif., San Diego, 1975-78; geneticist San Diego Zoo, 1978-86, Kleberg genetics chair, 1986—; assoc. adj. prof. biology U. Calif., San Diego, 1988—. Editor: One Medicine, 1984, Jour. Heredity, 1989—, Conservation Biology, 1993—; contbr. over 130 articles on genetics and endangered species conservation to profl. jours. Species coord. Asian Wild Horse Species Surival Plan, San Diego, 1979. Med. rsch. fellow Bank Am.-Giannini Found., 1976; grantee Pew Charitable Trusts, 1989—, NIH, 1976-88; named 91 San Diegans to Watch, 1991. Fellow N.Y. Zool. Soc. (sci.), San Diego Soc. for Natural History; mem. Am. Assn. Zool. Parks, Am. Soc. Mammalogists, Am. Genetics Assn. (coun. 1990—), Soc. for Conservation Biology (founding), Soc. for Systematic Zoology, Soc. for Study Evolution, Am. Soc. for Microbiology. Office: San Diego Zoo Ctr for Reprodn Endangered Species PO Box 551 San Diego CA 92112-0551

RYDER, ROBERT WINSOR, medical epidemiologist; b. N.Y.C., Aug. 6, 1946; s. James F. and Nancy N. (Nickerson) R.; children: Hilary, Abby. BSc, Middlebury Coll., 1968; MD, Columbia U., 1972; MSc, London U., 1981. Diplomate Am. Bd. Internal Medicine. Intern Boston City Hosp., 1972-73, resident in medicine, 1973-74; staff mem. epidemic intelligence svc. Ctrs. for Disease Control, Atlanta, 1974-77; fellow in infectious diseases Harvard U. Med. Sch., Boston, 1977-78; dir. AIDS project Ctrs. for Disease Cointrol, Kinshasa, Zaire, 1986-90; dir. div. infectious disease Dept. Epidemiology and Pub. Health, Sch. Medicine Yale U., New Haven; asst. prof. Sch. Medicine Johns Hopkins U., Balt., 1978-80, Tufts U., Boston, 1980-85; assoc. prof. Sch. Medicine Boston U., 1985-86; prof. epidemiology Mt. Sinai Sch. Medicine, N.Y.C., 1991-93; John Rodman Paul prof. of epidemiology Sch. Medicine Yale U., 1993—; cons. WHO, Geneva, 1975-90. Contbr. articles on HIV transmission to profl. publs. Lt. comdr. USPHS, 1974-77. Milbank clin. scholar Milbank Found., Sch. Medicine Tufts U., 1980-85. Achievements include design and set up of world's largest cancer vaccine trial using hepatitis B virus vaccine to prevent liver cancer in West Africa; dir. of largest HIV/AIDS research project in Africa. Office: Dept Edpidem and Pub Health Yale U Sch Medicine 60 College St PO Box 208034 New Haven CT 06520

RYERSON, SUNNY ANN, entomologist; b. Sweetwater, Tex., Jan. 9, 1949; d. Samuel Hyatt and Barbara Jo (Bennett) Estes; m. James D. Ryerson III, July 4, 1968 (div. Apr. 1976); children: Shannon Kelly, Erin Caroline. AA, Odessa Coll., 1974; BS, Ariz. State U., 1975; MS, U. Ariz., 1977; postgrad., Tex. Tech U., 1983-88, Walden U., 1991—. Cert. entomologist. Rsch. assoc. Tex. A&M U., El Paso, 1977-79; rsch. biolgist FMC Corp., Lubbock, Tex., 1979-86; owner Sun Danz Landscape and Cons., Lubbock, 1986-87; dir. R & D Waterbury Cos., Inc., Independence, La., 1988—. Contbr. articles to profl. publs. Fundraiser Battered Women's Shelter, Lubbock, 1981-86; co-chmn. NOW, Lubbock, 1980-85. Named Semifinalist Woman of Yr., Glamour mag., 1986. Mem. Southwestern Entomol. Soc. (sec.-treas. 1984-86), Am. Women in Sci. (mem. com. 1983), Entomol. Soc. Am., Chem. Specialty Mfr. Assn. (mem. com. 1988—). Democrat. Roman Catholic. Home: 918 Fletcher Isle Rd Panchatoula LA 70454 Office: Waterbury Cos Inc 100 Calhoun St PO Box 640 Independence LA 70443

RYLANDER, HENRY GRADY, JR., mechanical engineering educator; b. Pearsall, Tex., Aug. 23, 1921; married; 4 children. B.S., U. Tex., 1943, M.S., 1952; Ph.D. in Mech. Engring., Ga. Inst. Tech., 1965. Design engr. Steam Div., Aviation Gas Turbine Div., Westinghouse Elec. Corp., 1943-47; from asst. to assoc. prof. mech. engring. U. Tex., Austin, 1947-68, research scientist, 1950, prof. mech. engring., 1968—, Joe J. King prof. engring., 1980—; cons. engr. TRACOR, Inc., 1964-69; founding dir. Ctr. for Electromechanics, U. Tex., 1977-85, chmn., mech. engring. dept., 1976-86. Named Disting. Grad. Coll. Engring., U. Tex., Austin, 1989. Fellow ASME (Leonardo da Vinci award 1985); mem. ASME. Office: U Tex Coll Engring Austin TX 78712

RYMER, WILLIAM ZEV, research scientist, administrator; b. Melbourne, Victoria, Australia, June 3, 1939; came to U.S., 1971; s. Jacob and Luba Rymer; m. Helena Bardas, Apr. 10, 1961 (div. 1975); children: Michael Morris, Melissa Anne; m. Linda Marie Faller, Sept. 5, 1977; 1 child, Daniel Jacob. MBBS, Melbourne U., 1962; PhD, Monash U., Victoria, 1971. Resident med. officer dept. medicine Monash U., Victoria, 1964-66; Fogarty internat. fellow NIH, Bethesda, Md., 1971-74; rsch. assoc. Johns Hopkins U. Med. Sch., Balt., 1975-76; asst. prof. SUNY, Syracuse, 1976-78; asst. prof. Northwestern U., Chgo., 1978-81, assoc. prof., 1981-87, prof., 1987—; rsch. dir. Rehab. Inst. Chgo., 1989—. Contbr. articles to profl. jours. Grantee NIH, VA, Dept. of Def., Nat. Inst. Disability Rehab. Rsch., pvt. founds. Fellow Royal Australian Coll. Physicians; mem. Soc. Neurosci., Am. Soc. Biomechanics. Democrat. Avocations: tennis, racquetball. Office: Rehab Inst Chgo 345 E Superior St Chicago IL 60611-4496

RYMON, LARRY MARING, environmental scientist, educator; b. Portland, Pa., Nov. 16, 1934; m. Barbara Persons, June 29, 1962; children: Tyler, Holly. BS, East Stroudsburg (Pa.) U., 1958, MEd, 1964; PhD, Oreg. State U., 1968. Grad. asst., instr. Oreg. State U., Corvallis, 1964-68; prof. biology, coord. environ. studies East Stroudsburg U., 1968—. Contbr. chpt. to book, articles to profl. jours. Supr. Upper Mt. Bethel Twp., Mt. Bethel, Pa., 1976-82. Lt. USNR, 1958-63. Danforth assoc., 1970. Mem. Raptor Rsch. Found., African Wildlife Found., Sigma Xi, Phi Sigma, Phi Kappa Phi. Democrat. Office: East Stroudsburg U Environ Studies Program Dept Biol Scis East Stroudsburg PA 18301

RYNARD, HUGH C., engrineer, engineering executive. BASc in Civil Engring., U. Toronto, 1951. Design engr. Acres Internat. Ltd., Toronto, Ont., Can., 1951-53; liaison engr. Bersimis 1 Project, Acres Internat. Ltd., Toronto, Ont., Can., 1953-56; asst. resident engr. Chute des Passes Program, Acres Internat. Ltd., Toronto, Ont., Can., 1957-59; acting resident engr. Kelsey Project, Acres Internat. Ltd., Toronto, Ont., Can., 1960; head constrn. dept. Acres Internat. Ltd., Toronto, Ont., Can., 1960-64, v.p., dir., 1962; mgr. Acres Internat. Ltd., Montreal, Can., 1964-67; mem. policy bd. Acres Canadian Bechtel, Can., 1964; pres. Acres Internat. Ltd., Toronto, Ont., Can., 1971, Acres Internat. Corp., Toronto, Ont., Can., 1986; chmn. Acres Internat. Ltd., Toronto, Ont., Can., 1987—. Recipient Julian C. Smith medal Engring Inst. Can., 1991. Mem. Assn. Consulting Engrs. Can., Engring. Inst. Can., Assn. Profl. Engrs. Ont. Office: Acres Internat Ltd, 480 University Ave 13th Fl, Toronto, ON Canada M5G 1V2*

RYNIKER, BRUCE WALTER DURLAND, industrial designer, manufacturing executive; b. Billings, Mont., Mar. 23, 1940; s. Walter Henry and Alice Margaret (Durland) R.; B. Profl. Arts in Transp. Design (Ford scholar), Art Ctr. Coll. Design, Los Angeles, 1963; grad. specialized tech. engring. program Gen. Motors Inst., 1964; m. Marilee Ann Vincent, July 8, 1961; children: Kevin Walter, Steven Durland. Automotive designer Gen. Motors Corp., Warren, Mich., 1963-66; mgmt. staff automotive designer Chrysler Corp., Highland Park, Mich., 1966-72; pres., dir. design Transform Corp., Birmingham, Mich., 1969-72; indsl. designer, art dir. James R. Powers and Assocs., Los Angeles, 1972-75; sr. design products mgr. Mattel Inc., El Segundo, Calif., 1975—; dir. design and devel. Microword Industries, Inc., Los Angeles, 1977-80, also dir.; exec. mem. Modern Plastics Adv. Council, 1976-80; elegance judge LeCercle Concours D'Elegance, 1976-77; mem. nat. adv. bd. Am. Security Council, 1980; cons. automotive design, 1972—. Served with USMC, 1957-60. Mem. Soc. Art Ctr. Alumni (life), Mattel Mgmt. Assn., Second Amendment Found., Am. Def. Preparedness Assn., Nat. Rifle Assn. Designer numerous exptl. automobiles, electric powered vehicles, sports and racing cars, also med. equipment, electronic teaching machines, ride-on toys. Home: 21329 Marjorie Ave Torrance CA 90503-5443 Office: 333 Continental Blvd El Segundo CA 90245-5012

RYNTZ, ROSE ANN, chemist; b. Detroit, July 23, 1957; d. Raymond Leonard and Rose Marie (Schabel) R. BS, Wayne State U., 1979; PhD, U. Detroit, 1983. Rsch. chemist Dow Chem. Co., Midland, Mich., 1983-85; sr. rsch. chemist Ford Motor Co., Mt. Clemens, Mich., 1985-86; group leader DuPont, Mt. Clemens, 1986-88; coatings specialist Dow Corning Corp., Midland, 1988-89; lab. dir. Akzo Coatings, Inc., Troy, Mich., 1989; tech. dir. Akzo Coatings, Inc., Troy, 1990-92; tech. specialist Ford Motor Co., Dearborn, Mich., 1992—; adj. prof. U. Detroit, 1985—; cons. Maro Communications, Fla., 1990—. Patentee in field; contbr. articles to profl. jours. Tchr. Polymer Edn., Detroit, 1981. Mem. Am. Chem. Soc. (chem. chmn. 1988-90), Fed. Soc. Coating Tech. (tech. chmn. 1985—, profl. devel. com. 1990—, editorial rev. bd., 1989—, tech. adv. bd. 1992—), Soc. Automotive Engrs. Avocations: music, gardening, bicycling. Office: Ford Motor Co 24300 Glendale Detroit MI 48239

RYPIEN, DAVID VINCENT, materials research engineer; b. Youngstown, Ohio, July 31, 1956; s. Vincent Mark and Anna Marie (Carr) R. BS, Walsh Coll., 1983; MS, Ohio State U., 1985, PhD, 1990. Instr. Stark Tech. Coll., Canton, 1978-79; analyst Timken Co., Canton, 1979-83; rschr. Ohio State U., Columbus, 1983-85, instr., 1985-90; scientist Shell Oil Co., Houston, 1990—; judge Ohio Acad. Sci., Columbus, 1983-90. Author: (with others) Nondestructive Characteristics, 1987; contbr. articles to profl. jours. Bd. dirs. AGA-NDE Supervisory, Washington, 1990-92, CNDE Iowa State U., Ames, 1990-92; scientist Houston Children Mus., 1992. With USN, 1991—. Acoutstics scholarship NATO, 1985. Mem. Am. Nondestructive (lectr. 1987), Am. Welding Soc. (lectr. 1989), Sigma Xi (rsch. mem. 1986). Roman Catholic. Achievements include research in ultrasonic evaluation of cast aluminum, nonlinear acoustics evaluation of adhesive bonded joints, electromagnetic induced ultrasonic velocity changes to evaluate ferromagnetic materials, failure analysis and nondestructive evaluation of large metallic structures. Office: Shell Devel Co 3333 Highway 6 South Houston TX 77082

RYSKAMP, CARROLL JOSEPH, chemical engineer; b. Grand Rapids, Mich., Dec. 25, 1930; s. Henry C. and Edna E. (Robinson) R.; m. Joanne Ruth Winter, Nov. 17, 1951; children: Jan C., John M., Julie K., Jay A. BS in Chem. Engring., Wayne State U., 1953. Registered profl. control systems engr. Chem. engr. Reichhold Chem. Co., Ferndale, Mich., 1953-55; process supv. and specialist Marathon Oil Co., Detroit, 1955-65; process control coordinator Marathon Oil Co., Findlay, Ohio, 1965-70; control cons. Foxboro (Mass.) Co., 1970-85; owner Process Performance Co., Foxboro, 1986. Contbr. articles to profl. jours.; patentee in field. Bristol fellow, The Foxboro Co., 1985. Sr. mem. Instrument Soc. Am. (Philip T. Sprague award, 1981). Republican. Avocations: electronics, travel. Home and Office: 48 Prospect St Foxboro MA 02035-1724

RYSZYTIWSKYJ, WILLIAM PAUL, mechanical engineer; b. N.Y.C., Dec. 14, 1952; s. Semko and Eva R. BSME, Ga. Inst. Tech., 1974, MSME, 1976. Reliability engr. Allied Chem. Corp., Hopewell, Va., 1976-77; mech. engr. Corning (N.Y.) Glass Works, 1977-85; sr. rsch. engr. Corning, Inc., 1985—. Mem. NSPE, ASME, Fine Particle Soc., Soc. for Info. Display, Tau Beta Pi, Pi Tau Sigma. Achievements include patents for Glass Article with Defect-free Surface, Pultrusion Fiber-Reinforced Composites. Office: Corning Inc Sullivan Park Process Rsch Ctr Corning NY 14831

RYU, DEWEY DOO YOUNG, biochemical engineering educator; b. Seoul, Korea, Oct. 27, 1936; came to U.S., 1955; s. Hansang and Sonam (Kim) R.; children: Mina L., Regina P. BSChemE, MIT, 1961, PhD in Biochemical Engring., 1967. Sr. rsch. engr. Sqibb Inst. for Med. Rsch., New Brunswick, N.J., 1967-72; adj. prof. Rutgers U., New Brunswick, 1968-72; prof., chmn. Korea Advanced Inst. Sci., Seoul, 1973-81; prof., dir. biochemical engring. program U. Calif., Davis, 1982—, SUNY, Buffalo, 1991-92; vis. prof. MIT, Cambridge, 1972-73; com. mem. UN Indsl. Devel. Orgn. & Food and Agr. Orgn./UN, N.Y.C., 1981-86, NAS-NRC/Bioprocess Engring., Washington, 1991-92. Contbr. over 170 sci. articles to profl. jours. Recipient Nat. Order Civil Merit/Presdl. Medal Hon., Pres. of Korea, Korea, 1981. Fellow Am. Inst. Med. and Biological Engring. Socs.; mem. AAAS, AICE, Am. Chem. Soc., Am. Soc. for Microbiology. Achievements include 18 patents in the areas of bioprocess engineering and biotechnology. Home: 658 Portsmouth Ave Davis CA 95616 Office: U Calif Dept Chem Engring Davis CA 95616

RYU, KYOO-HAI LEE, physiologist; b. Seoul, Republic of Korea, Sept. 5, 1948; came to U.S., 1972; d. Hee Soon and Jung Ock Lee; m. David Tai-Hyung Ryu, May 13, 1978; children: Eugenia, Christina, John. BS, Yonsei U., Seoul, 1971; PhD, U. Minn., 1981. Postdoctoral fellow U. Minn., Mpls., 1980-81, staff scientist, 1981-82; sr. rsch. assoc. Wright State U., Dayton, Ohio, 1985-91; adminstr. Ohio Ctr. of Cosmetic Surgery, Bellefontaine, Ohio, 1991—. Mem. Am. Physiol. Soc., Biophys. Soc., Soc. Gen. Physiologists. Home: 15 Bexley Ave Springfield OH 45503-1103

RYZLAK, MARIA TERESA, biochemist, educator; b. Augustow, Poland, Feb. 27, 1938; came to the U.S., 1959; d. Stefan-Fabian and Wladyslawa (Debska) Kozlowski; m. Jan Ryzlak, Apr. 2, 1959; children: Danuta Alexandra Anita, Maria Valentine. BS in Chemistry and Physics, U. Manchester, 1966; MS in Engring. Sci., N.J. Inst. Tech., 1970; PhD in Biochemistry, Rutgers U., U. Medicine-Dentistry N.J., 1986. Rsch. chemist Troy Chem. Co., Newark, 1964-68; rsch. asst. Worcestern Found. for Exptl. Biology, Shrewsbury, Mass., 1969-74; rsch. assoc. Temple U. Med. Sch., Phila., 1974-78, McNeil Pharm., Fort Washington, Pa., 1978-81; Busch fellow Waksman Inst. Rutgers U., Piscataway, N.J., 1986-89, rsch. assoc., 1989-91; adj. asst. prof. U. Med. and Dentistry N.J., Newark, 1992—. Contbr. articles to Archives Biochemistry and Biophysics, Biochemica Biophysica Acta, Alcohol Clin. Exptl. Rsch. Active World Affairs Coun., Phila., 1980—, Polish Am. Congress, Phila., 1989. Mem. Soc. Exptl. Biology and Medicine, Rsch. Soc. on Alcoholism, Sigma Xi. Democrat. Achievements include discovery and isolation of glyceraldehyde-3-phosphate dehydrogenase isozymes in human brain, purification of human brain aldehyde dehydrogenase, endogenous inhibitor of aldehyde dehydrogenase from alcoholic liver. Office: U Med Dentistry NJ 138 S Orange Ave Newark NJ 08034

RZEPECKI, EDWARD LOUIS, packaging management educator; b. Chgo., Oct. 23, 1921; s. Louis and Emilia (Sawicki) R.; m. Dolores Helen Modeen, Sept. 21, 1946; children: Michael, JEffrey, Steven. BSChemE with honors, U. Ill., 1943. Product devloper 3M Co., St. Paul, 1946-50, supr. package tape devel., 1950-52, tech. mgr. packaging tapes, 1954-73, tech. dir. package system div., 1973-82; adj. prof. packaging U. Wis.-Stout, Menomonie, 1983-92; adj. prof. packaging mgmt. U. St. Thomas, St. Paul, 1989-92, 3M fellow, disting. vis. prof., 1991—; cons. SIAT Corp., Chiasso, Switzerland, 1983, Evergreen Soln Co., Minnetonka, Minn., 1991, Lectec Corp., Minnetonka, 1988-91. Author/editor: Packaging and Environmental Issues, 1991. Fund raiser ARC, St. Paul, 1962; precinct vice-chair Rep. Party, Mpls., 1967; active United Way, St. Paul, 1966. Lt. col. U.S. Army Corp. of Engrs., 1943-46. Mem. Inst. of Packaging Profls. Republican. Lutheran. Office: U St Thomas Mail Stop 108 2115 Summit Ave Saint Paul MN 55105-1096

SA, LUIZ AUGUSTO DISCHER, physicist; b. Lages, Brazil, Sept. 28, 1944; came to U.S., 1983; s. Catulo J.C. and Maria (Discher) S. MSc in Physics, Carnegie Mellon U., 1969; PhD in Elec. Engring., Stanford U., 1989. Asst. prof. Cath. U. of Rio, Rio de Janeiro, 1969-72, Fed. U. of Rio, Rio de Janeiro, 1973-83; post-doctoral scholar Stanford (Calif.) U., 1990-91, rsch. scientist, 1991—; mission ops. scientist for MDI instrument to be flown in SOHO spacecraft, NASA, 1992—. Contbr. articles to Jour. of Applied Physics, Jour. of Geophys. Rsch. Mem. Sigma Xi. Roman Catholic. Achievements include rsch. in non-linear plasma physics, naturally occurring plasmas. Office: CSSA ERL 328 Stanford CA 94305

SAARI, ALBIN TOIVO, electronics engineer; b. Rochester, Wash., Mar. 16, 1930; s. Toivo Nickoli and Gertrude Johanna (Hill) S.; m. Patricia Ramona Rudig, Feb. 1, 1958; children: Kenneth, Katherine, Steven, Marlene, Bruce. Student, Centralia Community Coll., Wash., 1950-51; AS in Electronic Tech., Wash. Tech. Inst., Seattle, 1958; BA in Communications, Evergreen State Coll., Olympia, Wash., 1977. Electronic technician Boeing Co., Seattle, 1956-59; field engr. RCA, Van Nuys, Calif., 1959-61; tv engr. Gen. Dynamics, San Diego, 1961-65, Boeing Co., Seattle, 1965-70; chief media engr. Evergreen State Coll., Olympia, Wash., 1970—; adv. bd. KAOS-

FM Radio, Olympia, 1979-82, New Mkt. Vocat. Skills Ctr., Tumwater, Wash., 1985—. Soccer coach King County Boys Club, Federal Way, Wash., 1968-70, Thurston County Youth Soccer, Olympia, 1973-78. With USAF, 1951-55. Recipient Merit award for electronic systems design Evergreen State Coll., 1978. Mem. Soc. Broadcast Engrs. (chmn. 1975-77), Soc. of Motion Picture and TV Engrs., IEEE, Audio Engring. Soc., Tele-Communications Assn., Assoc. Pub. Safety Communications Officers. Lutheran. Avocations: amateur radio, swimming. Home: 6617 Husky Way SE Olympia WA 98503-1433 Office: Evergreen State Coll Media Engring L1309 Olympia WA 98505

SAAVEDRA, JUAN ORTEGA, chemical engineering educator; b. Las Palmas, Spain, Feb. 26, 1951; s. Juan Ortega Garcia and Dolores (Saavedra) Gonzalez; m. Elena Velasco Sendino, July 28, 1990; 1 child, Elena Ortega Velasco. BS in Chem. Engring., ETS Ingenieros Indsl., Las Palmas, Spain, 1976; PhD in Chem. Engring., U. La Laguna, Tenerife, Spain, 1979. Instr. chem. engring. U. La Laguna, Tenerife, 1976-78, thermodynamic educator, 1978-82; researcher U. Las Palmas, Spain, 1982-86, prof., 1986—; head Thermodynamic group Las Palmas U., 1983, high sch. indsl. engring, Las Palmas U., 1986. Author, editor: Exercises in Thermodynamics, 1985, Thermodynamic Tables, 1986; contbr. articles to Jour. Chem. and Engring. Data, Fluid Phase Equilibria. Recipient Alfonso X el Sabio Cross Spanish Govt., 1986; grantee Esso Espanol, 1990. Mem. Am. Chem. Soc., Am Inst. Chem. Engrs., Chem. Inst. Canada, Chem. Soc. Japan. Achievements include design of VLE equipment, standard pattern substances for calibration of densimeters, smoothing equation for thermodynamic properties. Home: C/ Perchel 12, 35009 Las Palmas de Gran Canaria, Spain Office: U Las Palmas Gran Canaria, Tafira Baja (Las Palmas), 35071 Las Palmas Spain

SABANAS-WELLS, ALVINA OLGA, orthopedic surgeon; b. Riga, Latvia, Lithuania, July 30, 1914; d. Adomas and Olga (Dagilyte) Pipyne; m. Juozas Sabanas, Aug. 20, 1939 (dec. Mar. 1968); 1 child, Algis (dec.); m. Alfonse F. Wells, Dec. 31, 1977 (dec. 1990). MD, U. Vytautas The Great, Kaunas, Lithuania, 1939; MS in Orthopaedic Surgery, U. Minn., 1955. Diplomate Am. Bd. Orthopaedic Surgery. Intern Univ. Clinics, Kaunas, 1939-40; resident orthopaedic surgery and trauma Red Cross Trauma Hosp., Kaunas, 1940-44; orthopaedic and trauma fellow Unfall Krankenhous, Vienna, Austria, 1943-44; intern Jackson Park Hosp., Chgo., 1947-48; fellow in orthopaedic surgery Mayo Clinic, Rochester, Minn., 1952-55; assoc. orthopaedic surgery Northwestern U., 1956-72; asst. prof. orthopaedic surgery Rush Med. Sch., 1973-76; pvt. practice orthopaedic surgery Sun City, Ariz., 1976-89; pres. cattle ranch corp. Contbr. articles to profl. jours. Fellow ACS; mem. Am. Acad. Orthopaedic Surgery, Physicians Club Sun City, Mayo Alumni Assn. Republican. Mem. Evang. Reformed Ch. Avocations: art, antiques, environment. Home: 13443 N 107th Dr Sun City AZ 85351-2625

SABATA, ASHOK, materials engineer; b. Berhampur, India, Apr. 28, 1964; came to U.S., 1985; s. Balakrishna and Bishnu Priya (Panigrahi) S.; m. Shanta Sabata, July 9, 1990. BTech, Indian Inst. Tech., Bombay, 1985; MS, Va. Poly. Inst. and State U., 1986; PhD, Colo. Sch. Mines, 1989. Vis. scientistr IRSID, St. Germaine-en-Laye, France, 1986-87; rsch. engr. Armco Rsch. and Tech., Middletown, Ohio, 1989—. Contbr. articles to profl. publs. Mem. Am. Iron and Steel Inst. Achievements include patents in steel sheet with enhanced corrosion resistance with a silane-treated silicate coating. Home: 1068 Park Ln Apt D Middletown OH 45042-3420 Office: Armco Rsch & Tech 705 Curtis St Middletown OH 45043

SABATINO, DAVID MATTHEW, chemical engineer; b. Salem, Ohio, May 28, 1959; s. Cardini Angelo and Margaret Marie (Meisner) S.; m. Kathleen Marie Kurtz, Oct. 19, 1985; children: Matthew Thomas, Stephen David, Thomas Paul. BEChE, Youngstown State U., 1981. Registered profl. engr. S.C., Fla. Devel. engr. Dupont Co., Aiken, S.C., 1981-89; sr. process engr. Monsanto Co., Augusta, Ga., 1989-91, Internat. Minerals & Chem. Fertilizer Group, Inc., Mulberry, Fla., 1991—. Author: (publ.) Off-gas Treatment System for the DWPF, 1986, Process Safety Analysis in a Phosphate Plant, 1993. Mem. NSPE (treas. local sect. 1987), Am. Inst. Chem. Engrs. (treas. local sect. 1984-87), Fla. Engring. Soc. (Ridge chpt. Young Engr. of the Yr. 1993). Office: IMC Fertilizer Highway 640 Mulberry FL 33860

SABEL, BERNHARD AUGUST MARIA, research neuroscientist, psychologist; b. Trier, Fed. Republic Germany, Nov. 7, 1957; s. Hans and Ursula (Berekoven) S.; m. Elizabeth Ho-Jeong Kang, May 25, 1982; children: Torsten, Daniela. BA, U. Trier, 1978; MA, U. Düsseldorf, 1982; PhD, Clark U., 1984; Dr. med. habil., priv.-dozent, U. Munich, 1988. Postdoctoral assoc. and fellow MIT, Cambridge, 1984-86; rsch. scientist Inst. Med. Psychology U. Munich, 1986-92; prof., chmn. med. psychology U. Magdeburg (Germany) Med. Sch., 1992—; founder, chmn. Neurel, Inc., Boston, 1987-90; founder, chmn. sci. adv. bd. Polykinetix, Inc., N.Y., 1991—; vis. neuroscientist Mass. Gen. Hosp., Harvard Med. Sch., 1991. Co-editor: Pharmacological Approaches to the Treatment of Brain and Spinal Cord Injury, 1988; mem. editorial bd. Restorative Neurology and Neuroscience, 1989—, Zeitschrift für Medizinische Psychologie; contbr. articles on neurology and neurosci. to profl. jours.; spl. cons. for Germany to Sci. Mag., 1993—; patentee (U.S.) on controlled drug delivery system for treatment neural disorders, and drug delivery system for small, water soluble molecules, (German) device to convert sound of pianos. Mem. Cen. Mass. Symphony Orch., Worcester, 1978-86, Revere String Quartet, Boston, 1984-86. Fulbright scholar 1978, German Acad. Exchange Svc. 1982. Mem. Soc. for Neurosci., Internat. Brain Rsch. Orgn., Internat. Neurochem. Soc., European Neurosci. Assn., Controlled Release Soc., German Soc. Med. Psychology, European Brain and Behavior Soc. (Fulbright assoc.), Internat. Bus. Club (bd. dirs. 1988-89). Office: U Magdeburg Med Sch, Leipzigerstr 44, 39120 Magdeburg Germany

SABIDO, ALMEDA ALICE, mental health facility administrator; b. Blairsville, Pa., Sept. 24, 1928; d. George Jackson and Dora Irene (Byrd) McClellen; m. Frederick Lionel Harrison, Feb. 1, 1963; children: Frederick L.II., Derek M. BS in Secondary Edn., Indiana U. of Pa., 1950; MSW cum laude, U. Pitts., 1958. Staff psychiat. social worker S.I. Mental Health Soc., 1958-63, supr. psychiat. social worker, 1963-66, asst. dir. psychiat. social work, 1967-69, dir. psychiat. social work, 1969-81, acting dir. Children's Community Mental Health Ctr., 1981, dir. Children's Community Mental Health Ctr., 1982—. Mem. NAACP, NASW, N.Y. Urban League, Nat. Coun. Negro Women, S.I. Com. on Child and Adolescent Mental Health (pres. 1984-86), S.I. Mental Health Coun. (sec. 1982-84), S.I. Mental Health Soc. (Richard M. Silberstein award 1991). Presbyterian. Avocations: writing, reading, sports, music. Home: 142 Benedict Ave Staten Island NY 10314-2315 Office: SI Mental Health Soc 669 Castleton Ave Staten Island NY 10301-2028

SABINE, NEIL B., ecology educator; b. Syracuse, N.Y., Sept. 28, 1952; s. Miles Edward and Elizabeth (Prieto) S.; m. Kathleen Miller, Aug. 21, 1976; children: Matthew, Megan, Emily. MA, So. Ill. U., 1982; PhD, Brigham Young U., 1987. Prof. Indian Hills Community Coll., Ottumwa, Iowa, 1985-91; asst. prof. Indian U. East, Richmond, Ind., 1991—. Recipient Faculty fellowship Indian U. East, 1992, Faculty Achievement award Indian Hills Community Coll., 1988, Julia Greenwell award Brigham Young U., 1985, D. Eldon Beck award, 1982. Mem. Am. Ornithologists Union, Cooper Ornithol. Soc., Am. Mus. of Natural History. Office: Indian U East Div Natural Sci and Math Richmond IN 47374

SABISTON, DAVID COSTON, JR., educator, surgeon; b. Onslow County, N.C., Oct. 4, 1924; s. David Coston and Marie (Jackson) S.; m. Agnes Foy Barden, Sept. 24, 1955; children: Anne Sabiston Leggett, Agnes Sabiston Butler, Sarah Coston. BS, U. N.C., 1943; MD, Johns Hopkins U., 1947. Diplomate: Am. Bd. Surgery (chmn. 1971-72). Successively intern, asst. resident, chief resident surgery Johns Hopkins Hosp., 1947-53; successively asst. prof., assoc. prof., prof. surgery Johns Hopkins Med. Sch., 1955-64, Howard Hughes investigator, 1955-61; Fulbright research scholar U. Oxford, Eng., 1960; research assoc. Hosp. Sick Children, U. London, Eng., 1961; James B. Duke prof. surgery, chmn. dept. Duke Med. Sch., 1964—; chmn. Accreditation Council for Grad. Med. Edn., 1985-86. Editor: Textbook of Surgery, Essentials of Surgery, Atlas of General Surgery, Atlas of Cardiothoracic Surgery, A Review of Surgery; co-editor: Gibbon's Surgery of the Chest, Companion Handbook to Textbook of Surgery; chmn. editorial bd. Annals Clin. Rsch., ISI Atlas of Surgery: The Classics of Surgery Library, Surgery, Gynecology and Obstetrics, Jour. Applied Cardiology, Jour. Cardiac Surgery, World Jour. Surgery. Served to capt., M.C. AUS, 1953-55. Recipient Career Rsch. award NIH, 1962-64, N.C. award in Sci., 1978, Disting. Achievement award Am. Heart Assn. Sci. Coun., 1983 Michael E. DeBakey award for Outstanding Achievement, 1984, Significant Sigma Chi award, 1987, Coll. medalist Am. Coll. Chest Physicians, 1987, Disting. Tchr. award Alpha Omega Alpha, 1992. Mem. ACS (chmn. bd. govs. 1974-75, regent 1975-82, chmn. bd. regents 1982-84, pres. 1985-86), NAS Inst. Medicine, Am. Surg. Assn. (pres. 1977-78), So. Surg. Assn. (sec. 1969-73, pres. 1973-74), Am. Assn. Thoracic Surgery (pres. 1984-85), Soc. Clin. Surgery, Internat. Soc. Cardiovascular Surgery, Soc. Vascular Surgery (v.p. 1967-68), Soc. Univ. Surgeons (pres. 1968-69), Halsted Soc., Surg. Biology Club II, Soc. Thoracic Surgery, Soc. Surgery Alimentary Tract, Johns Hopkins U. Soc. Scholars, Soc. Surg. Chairmen (pres. 1974-76), Soc. Thoracic Surgeons Great Britain and Ireland, Soc. Internat. De Chirurgie, James IV Assn. Surgeons (bd. dirs. U.S. chpt.), Ill. Surg. Soc. (hon.), Phila. Acad. Surgery (hon.), Royal Coll. Surgeons Edinburgh (hon.), Royal Coll. Surgeons Eng. (hon.), Asociación de Cirugía del Litoral (Argentina) (hon.), Royal Coll. Physicians and Surgeons Can. (hon.), Royal Coll. Surgeons Ireland (hon.), Royal Australasian Coll. Surgeons (hon.), German Surgical Soc. (hon.), Colombian Surg. Soc. (hon.), Brazilian Coll. Surgeons (hon.), Japanese Coll. Surgeons (hon.), French Surg. Assn. (hon.), Surg. Congress Assn. Espanola de Cirujanos (hon.), Philippine Coll. Surgeons (hon.), Phi Beta Kappa, Alpha Omega Alpha. Clubs: Cosmos (Washington), Hope Valley Country Club (Durham), Treyburn City Club (Durham). Home: 1528 Pinecrest Rd Durham NC 27705-5817 also: Duke U Med Ctr Box 3704 Durham NC 27710

SABLIK, MARTIN JOHN, research physicist; b. Bklyn., Oct. 21, 1939; s. Martin C. and Elsie M. (Fuzia) S.; m. Beverly Ann Shively, Nov. 26, 1965; children: Jeanne, Karen, Marjorie, Larry. BA in Physics, Cornell U., 1960; MS in Physics, U. Ky., 1965; PhD, Fordham U., 1972. Jr. engr. The Martin Co., Orlando, Fla., 1962-63; half-time instr. U. Ky., Lexington, 1963-65; rsch. assoc. Fairleigh Dickinson U., Teaneck, N.J., 1965-67, instr. physics, 1967-1972, asst. prof., 1972-76, assoc. prof., 1976-80; sr. rsch. scientist Southwest Rsch. Inst., San Antonio, 1980-87, staff scientist, 1987—. Mem. editorial bd. Nondestructive Testing and Evaluation, 1989—; mem. adv. bd. Conf. on Properties and Applications of Magnetic Materials, 1990—, Workshop on Advances in Measurement Techniques and Instrumentation for Magnetic Properties Determination, 1993—; contbr. articles to profl. jours.; referee, patentee in field. Recipient Imagineer award Mind Sci. Found., 1989. Mem. Am. Phys. Soc., Am. Geophys. Union, Am. Soc. Nondestructive Testing (chmn. So. Tex. sect. 1983-84), IEEE, Am. Assn. Physics Tchrs. Roman Catholic. Office: SW Rsch Inst PO Box 28510 San Antonio TX 78228-0510

SABLOFF, JEREMY ARAC, archaeologist; b. N.Y.C., Apr. 16, 1944; s. Louis and Helen (Arac) S.; m. Paula Lynne Weinberg, May 26, 1968; children—Joshua, Saralinda. A.B., U. Pa., 1964; M.A., Ph.D., Harvard U., 1969. Asst. prof., asso. prof. Harvard U., Cambridge, Mass., 1969-76; asso. prof. anthropology U. Utah, Salt Lake City, 1976-77; curator anthropology Utah Mus. Natural History, Salt Lake City, 1976-77; prof. anthropology U. N.Mex., Albuquerque, 1978-86; chmn. dept. U. N.Mex., 1980-83; Univ. prof. anthropology and the history and philosphy of sci. U. Pitts., 1986—, chmn. dept. anthropology, 1990-92; rsch. assn. Carnegie Mus. of Natural History, 1987—; fellow Ctr. for the Philosophy of Sci., U. Pitts., 1987—; sr. fellow for Pre-Columbian Studies, Dumbarton Oaks, 1986-92, chmn. 1989-92. Author: (with G.R. Willey) A History of American Archaeology, 1974, 2d edit., 1980, 3d edit., 1993, Excavations at Seibal: Ceramics, 1975, (with C.C. Lamberg-Karlovsky) Ancient Civilizations: The Near East and Mesoamerica, 1979, (with D. A. Freidel) Cozumel: Late Maya Settlement Patterns, 1984, The Cities of Ancient Mexico, 1989, The New Archaeology and the Ancient Maya, 1990, (with G. Tourtellot) The Ancient Maya City of Sayil: The Mapping of a Puuc Region Center, 1991; editor: (with C.C. Lamberg-Karlovsky) The Rise and Fall of Civilizations, 1974; Editor: (with C.C. Lamberg-Karlovsky) Ancient Civilization and Trade, 1975, (with W.L. Rathje) A Study of Changing Pre-Columbian Commercial Systems, 1975, American Antiquity, 1977-81, (with G.R. Willey) Scientific American Readings in Pre-Columbian Archaeology, 1980, Simulations in Archaeology, 1981, Supplement to the Handbook of Middle American Indians: Archaeology, 1981, Archaeology: Myth and Reality: A Scientific American Reader, 1982, Analyses of Fine Paste Ceramics, 1982, (with D. Meltzer and D. Fowler) American Archaeology: Past and Future, 1986, (with E.W. Andrews V) Late Lowland Maya Civilization: Classic to Postclassic, 1986, (with J.S. Henderson) Lowland Maya Civilization in the Eighth Century A.D., 1993. Nat. Geog. Soc. grantee, 1972-74; NSF grantee, 1983-88; NEH grantee, 1990-91. Fellow Am. Anthrop. Assn., AAAS (sec. H, chair elect 1993—), Royal Anthrop. Inst., Soc. Antiquaries London; mem. Soc. Am. Archaeology (pres. 1989-91), Prehist. Soc., Internat. Soc. Comparative Study of Civilizations, Sigma Xi. Office: U Pitts Dept Anthropology Pittsburgh PA 15260

SABOTA, CATHERINE MARIE, horticulturist, educator; b. Bridgeton, N.J., Sept. 9, 1949; d. John Robert Sabota and Colleen Catherine Moran Schultz. BS, Tex. Tech. U., 1973, MS, 1975; PhD, U. Ill., 1983. Rsch. asst. Tex. Tech. U., Lubbock, 1973-75, rsch. assoc., 1975-76; asst. horticulturist U. Ill., Dixon Springs, 1978-80; rsch. assoc. U. Ill., Champaign, 1980-83; horticulturist, asst. prof. Ala. A&M, Normal, 1983-88, horticulturist, assoc. prof., 1988—; advisor Ala-Tenn Fruit & Vegetable Assn., Elora, 1985-90. Contbr. articles to profl. jours. Tex. State scholar, 1971-73; grantee CSRS-USDA, 1986, Soil Conservation Svc., 1986, Ala. U./TVA Consortium, 1987; recipient Award of Excellence Coop. Extension Program, Ala. A&M U., 1989. Mem. Am. Soc. for Hort. Sci. (chmn. awards com. 1990), Ala. Fruit & Vegetable Growers, So. Region Soc. for Hort. Sci. Achievements include research of shiitake mushrooms. Office: Ala A&M Univ PO Box 69 Normal AL 35762-0069

SABSHIN, MELVIN, psychiatrist, educator, medical association administrator; b. N.Y.C., Oct. 28, 1925; s. Zalman and Sonia (Barnhard) S.; m. Edith Goldfarb, June 12, 1955; 1 child, James K. B.S., U. Fla., Gainesville, 1944; M.D., Tulane U., New Orleans, 1948. Diplomate Am. Bd. Psychiatry and Neurology. Assoc. dir. Michael Reese Hosp. Psychosomatic and Psychiat. Inst., Chgo., 1953-61; prof., head dept. psychiatry U. Ill. Coll. Medicine, Chgo., 1961-74; med. dir. Am. Psychiat. Assn., Washington, 1974—. Author Depression, 1960, Psychiatric Ideology, 1961; Normality, 1978; Normality and Life Cycle, 1984. Served with U.S. Army, 1944. Recipient Bowen award Am. Coll. Psychiatrists, 1978, Disting. Psychiatrist award, 1985. Mem. Am. Coll. Psychiatrists (pres. 1974-75). Home: 2801 New Mexico Ave NW Washington DC 20007-3921 Office: Am Psychiat Assn 1400 K St NW Washington DC 20005-2403

SACHAR, DAVID BERNARD, gastroenterologist, medical educator; b. Urbana, Ill., Mar. 2, 1940; s. Abram Leon and Thelma (Horwitz) S.; m. Joanna Maud Belford Silver, Aug. 29,1 961; children: Mark Benson, Kenneth Hulbert Belford. AB magna cum laude, Harvard U., 1959, MD cum laude, 1963. Diplomate Bd. Gastroenterology Am. Bd. Internal Medicine. Intern medicine Beth Israel Hosp., Boston, 1963-65, resident, 1967-68; asst. chief clin. rsch. Pakistan-SEATO Cholera Rsch. Lab., Dhaka, Bangladesh, 1965-67; resident in gastroenterology Mt. Sinai Hosp., N.Y.C., 1968-70; from instr. to prof. medicine Mt. Sinai Sch. Medicine, CUNY, N.Y.C., 1970-92, 1st Burrill B. Crohn prof. medicine, 1992—; dir. div. gastroenterology Mt. Sinai Hosp., N.Y.C., 1983—, vice chmn. dept. medicine; co-chmn. work group on inflammatory bowel disease NIH, 1973-75; expert adv. panel on gastroenterology and nutrition U.S. Pharmacopeial Conv., 1980-85; chmn. rsch. devel. com. Nat. Found. for Ileitis and Colitis, 1984-89; co-founder, sec.-treas. Burrill B. Crohn Rsch. Foun., N.Y.C., 1984—; K.H. Koster meml. lectr. Danish Soc. of Gastroenterology, 1992. Author over 130 articles and chpts. on natural history and treatment of inflammatory bowel disease; editor 7 books and monographs on gastroenterology. Trustee Bangladesh Coun. of the Asia Soc., N.Y.C., 1972-75, Bd. Edn., Englewood Cliffs, N.J., 1973-75. Sr. surgeon, comdr. USPHS,1965-67. Fellow ACP, Am. Coll. Gastroenterology (program dirs. com. 1991—, Henry Baker Presdl. lectr. 1989); mem. Am. Gastroent. Assn. (chmn. subcom. on certification 1987, 1st chmn. clin. teaching project 1984-90), Crohn's and Colitis Found. Am. (grants rev. com. and coun. 1990—, Disting. Svc. award 1991, N.Y. Govs. medal, 1992), Internat. Orgn. for Study of Inflammatory Bowel Disease (1st Am. elected chmn. 1989-92), Phi Beta Kappa, Alpha Omega Alpha. Achievements include co-development of oral rehydration therapy for diarrhea; development of resources and standards for clinical teaching in gastroenterology. Office: Mt Sinai Med Ctr One Gustave L Levy Pl New York NY 10029

SACHAROW, STANLEY, chemist, consultant, writer; b. N.Y.C., Oct. 8, 1935; s. Max and Fannie (Rosenberg) S.; m. Beverly Lynn Levy, June 18, 1961; children—Scott Hunter, Brian Evan. A.B., Hunter Coll., 1957, M.A., 1965. Engr. Standard Packaging Corp., Clifton, N.J., 1960-65; sales engr. Archer Aluminum, Winston-Salem, N.C., 1965-67; tech. service mgr. Reynolds Metals Co., Richmond, Va., 1967-84; exec. dir. The Packaging Group Inc., Milltown, N.J., 1984—; cons. world wide basis, The Packaging Group Inc., 1984—. Author: Food Packaging, 1970; Principles of Packaging Development, 1972; A Packaging Primer, 1979; Packaging Regulations, 1979. Contbr. articles to profl. jours. Recipient Golden Keys award Club Printing N.Y. 1969, Best Tech. Article award Chilton Press 1974. Mem. Packaging Inst., Am. Chem. Soc., Coblentz Soc., Inst. Dirs. (U.K.), Inst. Packaging (U.K.). Republican. Clubs: Napoleonic Soc. (Clearwater, Fla.), Victorian Soc. (Phila.). Avocations: antiques; writing; Napoleonic battles. Home: 70 Valley Forge Dr East Brunswick NJ 08816-3278 Office: Packaging Group Inc PO Box 345 Milltown NJ 08850-0345

SACHAU, DANIEL ARTHUR, psychology educator; b. Harvey, Ill., June 30, 1959; s. Robert G. and Virginia (Wiley) S.; m. Beth Ann DeBeer, June 16, 1984. MS in Econs., U. Utah, 1983, MS in Psychology, 1986, PhD in Psychology, 1990. Assoc. prof. dept. psychology Mankato (Minn.) State U., 1989—, dir. Ilo psychology grad. program, 1990—. Contbr. articles to profl. jours. Mem. APA, Acad. Mgmt., Soc. Indsl./Orgnl. Psychology. Office: Mankato State U Dept Psychology Mankato MN 56002

SACHDEV, RAJ KUMAR, biochemical engineer; b. Kanpur, India, Sept. 13, 1951; came to U.S., 1981; s. Daya Ram and Kamla Devi (Bhagwanti) S.; m. Shakuntaia, June 17, 1981; children: Shweta, Samir. PhD, Indian Inst. Tech., 1981. Lab. head Amgen Inc., Thousand Oaks, Calif., 1984—; cons. Indian Inst. Tech., 1990. Contbr. articles to profl. jours. Danida fellow Tech. U. Denmark, 1980-81, Vis. fellow NIH, 1981, Postdoctoral fellow Drexel U., 1982-84. Mem. Am. Inst. Chem. Engrs., Am. Chem. Soc. Home: 5952 Palomar Cir Camarillo CA 93012 OFfice: Amgen Inc Amgen Ctr Thousand Oaks CA 91320

SACHDEV, VED PARKASH, neurosurgeon; b. Mitranwali, India, Feb. 22, 1932; came to U.S., 1968; s. Girdhari Lal and Amar Kaur Sachdev; m. Ranjit Kaur Sachdev, Apr. 17, 1970; children: Ulka, Rivka. MB BS, Govt. Med. Coll., Amritsar, Panjab, India, 1955. Diplomate Am. Bd. Neurosurgery. Asst. prof. neurosurgery Med. Inst., Chandigarh, India, 1964-69; intern St. Josephs Hosp., Lorain, Ohio, 1969-70; resident in neurosurgery Mt. Sinai Med. Ctr., N.Y.C., 1970-73, from asst. to assoc. clin. prof. dept. neurosurgery, 1974-88, clin. prof. dept. neurosurgery, 1988—; vice chmn. dept. neurosurgery Mt. Sinai Med. Ctr., N.Y.C., 1988-92. Author chpts. in 7 med. books. Surgeon lt., Indian Navy, 1957-60. Fellow ACS, Royal Coll. Surgeons Eng. (diplomate laryngology and otology). Avocation: music. Home: 128 Moorland Dr Scarsdale NY 10583 Office: Mt Sinai Med Ctr Dept Neurosurgery 1148 Fifth Ave New York NY 10028

SACHS, CLIFFORD JAY, research scientist; b. Bronx, Mar. 19, 1960; s. Norman and Irma (Finkelstein) S.; m. Nancy Lynn Hersh, Mar. 31, 1985; children: Steven, David. BS in Chemistry, SUNY, Binghamton, 1982; MBA in Pharm. Studies, Fairleigh Dickinson U., 1987. Sr. rsch. asst. Bristol-Myers Squibb Co., New Brunswick, N.J., 1982-85, rsch. assoc., 1985-88, asst. rsch. investigator, 1988-90, rsch. scientist I, 1990-92, rsch. scientist II, 1992—. Contbr. articles to profl. jours. Dir. Wyckoff's Mill Condo Assn., Hightstown, N.J., 1987-90. Mem. Am. Chem. Soc., Am. Assn. Pharm. Scientists. Office: Bristol-Myers Squibb 1 Squibb Dr Box 191 New Brunswick NJ 08903

SACHS, HARVEY M., policy analyst; b. Atlanta, Dec. 10, 1944; m. Susan E. Slaughter, June 1967; 1 child, Gregory. AB, Rice U., 1967; PhD, Brown U., 1973. Asst. prof. Case Western Res. U., Cleve., 1974-76, Princeton U., 1977-82; dir. environ. scis. ECRI, Phila., 1982-84; mem. tech. staff AT&T Bell Labs., Holmdel, N.J., 1984-87; asst. commr. cons. N.J. Dept. Commerce, Newark & Trenton, N.J., 1987-89; dir. policy rsch. U. Md., College Park, 1991—. Mem. Cranbury (N.J.) Environ. Com., 1979-83; v.p. Cranbury Bd. Edn., 1983-87. Home: 20 Wynnewood Dr Cranbury NJ 08512 Office: U Md Ctr for Global Change 7100 Baltimore Ave Ste 401 College Park MD 20740

SACHS, MARTIN WILLIAM, computer scientist; b. New Haven, Conn., Sept. 30, 1937; s. Benjamin and Lillian (Moskowitz) S.; m. Jane Stein Sugarman, June 30, 1968. AB, Harvard U., 1959; MS, Yale U., 1961, PhD, 1964. Postdoctoral fellow Weizmann Inst. Sci., Rehovoth, Israel, 1964-65, rsch. asst., 1966; rsch. assoc. Yale U., New Haven, Conn., 1967-72, sr. rsch. assoc., 1972-76; rsch. staff mem. IBM Watson Rsch. Ctr., Yorktown Heights, N.Y., 1976--; mem. ad-hoc panel on line computers in nuclear rsch. Nat. Rsch. Coun., Washington, 1968-70; mem. program com. Optical Fiber Communication Conf., Washington, 1990-92. Contbr. articles to profl. jours. NATO Postdoctoral fellow, 1965. Mem. IEEE, Assn. for Computing Machinery, Am. Phys. Soc., Sigma Xi. Achievements include patents in the field. Home: 28 Warnock Dr Westport CT 06880 Office: IBM PO Box 704 Yorktown Heights NY 10598

SACHS, THOMAS DUDLEY, biomedical engineering scientist; b. St. Louis, Jan. 29, 1923; s. Ernest and Mary Parmley (Koues) S.; m. Margaret Cripps, Jan. 1952 (div. 1960); children: Martin Thomas, Zoltan Naszay; m. Elizabeth Bennet Burroughs, May 29, 1988. BA in Chemistry, U. Calif., Berkeley, 1950; PhD in Physics/Math., U. Innsbruck, Austria, 1960. Owner Sachs Electronics, Berkeley, 1946-52; postdoctoral rsch. investigator Western Res. U., Cleve., 1960-62; assoc. prof. physics U. Vt., Burlington, 1962—; pres. Electronic Educator Inc., Burlington, 1969-77; rsch. dir. Wellen Assocs. Inc., Stowe, Vt., 1988—; assoc. prof. med. engring. Tokyo U., 1988-89; vis. prof. neurosurgery Kyorin U. Tokyo, 1989; cons. Ladd Rsch., IBM, Varian, 1962-80; mem. Vt. Regional Cancer Ctr., Burlington, 1980—, Cell and Molecular Biology Group, Burlington, 1984—. Contbr. articles to profl. jours. Presenter Beyond War, Burlington, 1984—; bd. dirs. Bt. Voice of Energy, Burlington, 1970-80; mem. Zero Population Growth, Burlington, 1970-80. Rsch. grantee NSF, 1961-70, Sloan Found. Neurosci., U. Vt., 1980, Technicon Instruments, Tast Assocs., Vt., 1982, Dept. Surgery, U. Vt. Med. Sch., 1989—. Achievements include 8 patents, 3 patents pending; invented thermo-acoustic sensing technique microbubble sensing system, perburbed acoustic propogation parameter measuring system, electronic educational system, deep-diving system, cerebro-spinal shunt flow measuring system, medical warning system; measurements of magneto-acoustic absorption relaxation, inter-molecular force. Home: 21 Grandview Ave Essex Junction VT 05452 Office: U Vt Physics Dept Cook Sci Hall Burlington VT 05405

SACHSE, GUENTHER, health facility administrator, medical educator; b. Zeitz, Sachsen, Fed. Republic of Germany, Feb. 21, 1949; s. Gerhard and Johanna (Waitz) S.; m. Regina Haas, Mar. 5, 1982; 1 child, Juliane Elisabeth. Medizinisches Staatsexamen, Georg-August U., Göttingen, Fed. Republic of Germany, 1974; MD, U. Göttingen, 1975. Leitender arzt in diabetes and medicine German Diagnostic Clinic, Wiesbaden, Fed. Republic of Germany, 1987—, med. dir., 1991—; chmn. cons. group Wyeth Pharms, Münster, Fed. Republic of Germany, 1991—. Author: Diabetologié, 1989. Mem. Internat. Diabetes Fedn., European Assn. Study of Diabetes, German DiabetesAssn. Avocations: classical music, mountain climbing, wine tasting. Office: German Diagnostic Clinic, Aukammallee 33, 6200 Wiesbaden Germany

SACHTLER, WOLFGANG MAX HUGO, chemistry educator; b. Delitzsch, Germany, Nov. 8, 1924; came to U.S., 1983; s. Gottfried Hugo and Johanna Elisabeth (Bollmann) S.; m. Anne-Lore Luise Adrian, Dec. 9, 1953; children: Johann Wolfgang Adriaan, Heike Kathleen Julia, Yvonne Rhea Valeska. Diplomchemiker, Tech. U., Braunschweig, Ger., 1949; Dr.rer.nat. (Ph.D), 1952. Research chemist Kon-Shell Lab., Amsterdam, Netherlands, 1952-71, dept. head, 1972-83; extraordinary prof. chemistry U. Leiden, Netherlands, 1963-83; V.N. Ipatieff prof. Northwestern U., Evanston, Ill.,

1983—; chmn. Gordon Research Conf. Catalysis, N.H., 1985; Rideal lectr. Faraday div. Royal Soc. Chemistry, 1981; F. Gault lectr., 1991. Mem. editorial bd. Jour. Catalysis, 1976-88, Applied Catalysis, 1983-87, Catalysis Letters, 1987—, Advances in Catalysis, 1987—; contbr. more than 280 articles to sci. jours. Mem. Royal Netherlands Acad. Scis., Internat. Congress Catalysis (pres. coun.), Royal Dutch Chem. Soc., Am. Chem. Soc. (E.V. Murphree award 1987, petroleum chemistry award 1992), Catalysis Soc. N.Am. (Robert L. Burwell award 1985, E. Houdry award 1993). Home: 2141 Ridge Ave Apt 2D Evanston IL 60201-2788 Office: Northwestern U Sheridan Rd Evanston IL 60208-0002

SACKHEIM, ROBERT LEWIS, aerospace engineer, educator; b. N.Y.C., N.Y., May 16, 1937; s. A. Frederick and Lillian L. (Emmer) S.; m. Babette Freund, Jan. 12, 1964; children: Karen Holly, Andrew Frederick. BSChemE, U. Va., 1959; MSChemE, Columbia U., 1961; postgrad., UCLA, 1966-72. Project engr. Comsat Corp., El Segundo, Calif., 1969-72; project mgr. TRW, Redondo Beach, Calif., 1964-69, sect. head, 1972-76, dept. mgr., 1976-81, mgr. new bus., 1981-86, lab. mgr., 1986-90, dep. ctr. dir., 1990—; mem. adv. bds. NASA, Washington, 1989—; mem. peer rev. bd. various univs. and govtl. agys., 1990—; guest lectr. various univs. and AIAA short courses. Author: Space Mission Analysis and Design, 1991; contbr. over 60 papers to profl. jours., confs. Mem. adv. bd. L.A. Bd. Edn., 1990-92; fund raiser March of Dimes, L.A., 1970-90, YMCA, San Pedro, Calif., 1974-86. Capt. USAF, 1960-63. Recipient Group Achievement award NASA, 1970, 78, 86. Mem. AIAA (chmn. com. 1980-83, J.H. Wyld Propulsion award 1992, Shuttle Flag award 1984). Achievements include 4 patents for spacecraft propulsion systems, devices and components. Office: TRW Space and Engring Group Bldg 01/RM 2010 1 Space Park Redondo Beach CA 90278

SACKMAN, GEORGE LAWRENCE, educator; b. Baxley, Ga., Mar. 15, 1933; m. Nancy Lee Davis, 1963; children: David, Anne. BSME, U. Fla., 1954, BSEE, 1957, MSEE, 1959; PhD, Stanford (Calif.) U., 1964. Registered profl. engr., N.Y. Rsch. engr. Litton Electron Tube div., San Carlos, Calif., 1963-64; prof. elec. engring. Naval Postgrad. Sch., Monterey, Calif., 1964-84; prof., chmn. elec. engring. SUNY, Binghamton, 1984-89, prof. elec. engring., 1990—. Author: (with others) Peace: Meanings, Politics, Strategies, 1989; contbr. articles to profl. jours. Sgt. U.S. Army, 1954-56. Fulbright scholar Coun. for Internat. Exch. of Schs., U. Malta, 1990-91. Mem. IEEE (sr.), Am. Soc. for Engring. Edn. (chmn. grad. studies div. 1991-92). Achievements include patent for ultrasonic camera tube, contour map underwater acoustic image system. Office: SUNY Watson Sch Binghamton NY 13902-6000

SACKS, COLIN HAMILTON, psychologist educator; b. Evanston, Ill., Sept. 7, 1957; s. Sheldon and Marjorie (Hamilton) S. BA in Psychology, Grinnell Coll., 1979; MA in Exptl. Psychology, U. Calif., Santa Barbara, 1982, PhD in Exptl. Psychology, 1986. Asst. prof. Grinnell (Iowa) Coll., 1988-90; rsch. assoc. Beryl Buck Inst. for Edn., Novato, Calif., 1990—. Reviewer Elem. Sch. Jour., 1991—, Am. Ednl. Rsch. Assn., 1992—; contbr. articles to profl. jours. Vol. tutor Contra Costa County Youth Homes, Concord, Calif., 1991—. Mem. Am. Psychol. Soc., Audio Engring. Soc., Am. Ednl. Rsch. Assn. Democrat. Achievements include research for empirical support of the diathesis-stress model of learned helplessness in adults and children; recent research concerns teacher cognition and behavior. Office: Beryl Buck Inst for Edn 18 Commercial Blvd Novato CA 94949

SACKS, HENRY S., medical researcher, infectious disease physician; b. N.Y.C., Apr. 27, 1942; s. Louis Robert and Anne (Rothaus) S.; m. Gillian S. Wachs, June 13, 1965. BA, Williams Coll., 1962; PhD, Albany Med. Coll., 1971, MD, 1975. Diplomate Am. Bd. Internal Medicine, sub-bds. Infectious Diseases and Geriatric Medicine. Instr. U. Conn. Sch. Medicine, Farmington, 1978-79; instr. Mt. Sinai Sch. Medicine, N.Y.C., 1978-82, asst. prof. medicine and biomath. sci., 1982-87, assoc. prof. medicine, 1989—, dir. clin. trials unit, 1983—, assoc. prof. community medicine and pediatrics, 1992-93, prof. community medicine and biomath. scis., 1993—. Contbr. more than 100 articles to profl. jours. NIH grantee. Fellow ACP, Infectious Disease Soc. Am.; mem. Soc. for Clin. Trials (programs com.), Am. Fedn. for Clin. Rsch. Achievements include development of criteria for meta-analysis. Office: Mt Sinai Med Ctr 1 Gustave L Levy Pl New York NY 10029

SADASIVAN, MAHAVIJAYAN, chemical engineering manager; b. Singapore, Singapore, Aug. 2, 1959; came to U.S., 1980; s. Kandasami and Kamala (Vasudaven) S.; m. Ruth Ann Reutter, Aug. 30, 1986; children: Nathan, Daniel, Evan. BSChemE, U. Mich., 1984; MS in Math., U. Mo., Kansas City, 1986. Lt. Singapore Armed Forces, Singapore, 1977-80; grad. teaching asst. U. Mo., Kansas City, 1984-86; instr. Bowling Green (Ohio) State U., 1987; resource mgmt. specialist Betz Labs, Detroit, 1987-92; chem. mgmt. systems mgr. Quaker Chem., Detroit, 1992—. Recipient Chmn.'s award Betz Labs, 1990. Mem. AICE, Soc. of Tribologists and Lubrication Engrs. Office: Quaker Chem 26677 W 12 Mile Rd Southfield MI 48034

SADEGHI, FARSHID, engineering educator; b. Masjed Solaiman, Iran, Dec. 23, 1956; came to U.S., 1975; s. Abbas Ali and Forogh (Kochakian) S.; m. Brenda Stevens, Mar. 10, 1984; Nina Michelle, Sara Ashley. MS, U. Tenn., 1981; PhD, N.C. State U., 1985. Rsch. asst. U. Tenn., Chattanooga, 1979-81; teachng asst. N.C. State U., Raleigh, 1981-83, rsch. asst., 1983-85; asst. prof. Purdue U., West Lafayette, Ind., 1986-91, assoc. prof., 1991—. Mem. ASME (planning com 1991—, Burt L. Newkirk award 1991), Soc. Automotive Engrs., Tau Beta Pi. Achievements include discovery that temperature effects in lubricated contacts are significant and cannot be neglected. Home: 2907 Browning St West Lafayette IN 47906 Office: Purdue U Sch Mechanical Engring West Lafayette IN 47907

SADI, MARCUS VINICIUS, urologist; b. Sao Paulo, Brazil, Aug. 18, 1956; s. Afiz and Leila (Maluli) S.; m. Angela Chofhi Atala, May 12, 1983; children: Amanda and Rodrigo. MD, Escola Paulista de Medicina, Sao Paulo, 1979; hosp. adminstr. degree, Faculdades Sao Camilo, Sao Paulo, 1992. Resident Escola Paulista Medicina, Sao Paulo, 1979-82; fellow Harvard Med. Sch., Boston, 1983-85; asst. prof. Escola Paulista Medicina, Sao Paulo, 1986-88; fellow Johns Hopkins Sch. Medicine, Balt., 1989-90; assoc. prof. Escola Paulista Medicina, Sao Paulo, 1991—, chief urologic oncology surgeon, 1986—. Mem. editorial bd. Urology Ency.; contbr. 22 book chpts., 34 articles to profl. jours. Mem. Internat. Soc. Urology, Am. Urological Assn., Am. Soc. Andrology, Am. Inst. Ultrasound in Medicine, Brazilian Med. Assn. (jour. editorial bd. mem. 1989—), Brazilian Coll. Surgeons, Brazilian Urological Assn. (bd. dirs. 1988-89), Sao Paulo State Med. Assn. (sec. 1988-89), N.Y. Acad. Scis. Avocations: computers, professional photo developing. Home: Rua Honduras 1108, 01428-001 São Paulo Brazil Office: Escola Paulista Medicina, Rua Napoleão de Barros 715, 04063 São Paulo Brazil

SADLER, JAMES BERTRAM, psychologist, clergyman; b. Albuquerque, Mar. 29, 1911; s. James Monroe and Mary Agnes (English) S.; m. Vera Ellen Ahrendt, Apr. 10, 1938. AB, U. N.Mex., 1938; BD, Crozer Theol. Sem., 1941, ThM, 1948; MA, U. Pa., 1941, EdD, 1959. Lic. psychologist, S.D.; ordained to ministry Baptist Ch., 1941. Pastor First Bapt. Ch., Mt. Union, Pa., 1941-42; chaplain USAF, 1943-48; pastor Hatboro (Pa.) Bapt. Ch., 1948-61; chmn. dept. psychology Sioux Falls (S.D.) Coll., 1961-75; pvt. practice psychology, Sioux Falls, 1975—; cons. in psychology and religion. Contbr. articles to profl. jours. Mem. ministers coun. Am. Bapt. Conv. Mem. APA, ACA, Soc. for Sci. Study Religion, Masons, Rotary (pres. 1960). Home: 4312 Glenview Rd Sioux Falls SD 57103-4935

SADLER, JOHN PETER, mechanical engineering educator; b. Rochester, N.Y., May 29, 1946; s. Joseph Elmer and Eleanor Loretta (Meisenzahl) S.; m. Sharon Anne Canham, Nov. 20, 1965; children: Deborah Lynn, Jennifer Anne, Katherine Jo. BS cum laude, Rensselaer Poly. Inst., 1968, MEngring., 1969, PhD in Engring., 1972. Registered profl. engr., N.D. Asst. prof. SUNY, Buffalo, 1972-75; asst. prof. U. N.D., Grand Forks, 1976-78, assoc. prof., 1978-84, prof. engring., 1984-86; assoc. prof. mech. engring. U. Ky., Lexington, 1986—. Editor for dymanics Jour. Mechanism and Machine Theory, 1989—, assoc. editor, 1982-89; assoc. editor Jour. Mech. Design, 1976-78; author: (with C.E. Wilson and W.J. Michels) Kinematics and Dynamics of Machinery, 1983, (with C.E. Wilson) Kinematics and Dynamics of Machinery, 2d edit., 1993). Fellow NDEA, 1968, Cluett-Peabody Found., 1971. Mem. ASME (various offices), Am. Soc. Engring. Edn. (Dow Outstanding Young Faculty award 1978, N.D. Young Engr. award 1981), Nat. Soc. Profl. Engrs., Soc. Mfg. Engrs., Sigma Xi, Tau Beta Pi, Pi Tau Sigma. Roman Catholic. Office: U Ky Mech Engring Dept Lexington KY 40506

SADOULET, BERNARD, astrophysicist, educator; b. Nice, France, Apr. 23, 1944; s. Maurice and Genevieve (Berard) S.; m. Elisabeth M.L. Chaine, Apr. 27, 1967; children: Loic, Helene, Samuel. Lic. in Physics., U. Paris, 1965; diploma, Ecole Polytechnique, Paris, 1965; Diploma in Theoretical Physics, U. Orsay, France, 1966, PhD in Phys. Scis., 1971. Fellow CERN, Berkeley, Calif., 1966-73; physicist, then sr. physicist CERN, France, 1976-84; postdoctoral fellow Lawrence Berkeley (Calif.) Lab., 1973-76; mem. faculty U. Calif., Berkeley, 1984—, prof. physics, 1985—, dir. Ctr. Particle Astrophysics, 1988—; mem. commn. astrophysics Internat. Union Pure and Applied Physics, 1991—; vis. com. Max Planck Inst., Heidelberg, Germany, 1991—, Fermilab, 1992—, Lawrence Livermore Nat. Lab., 1992—; mem. program initiation com. NAS, 1992; internat. adv. com. various profl. confs. Contbr. to Sky and Telescope, The Early Universe Observable from Diffuse Backgrounds, other profl. publs. Fellow Am. Phys. Soc., U.S. Nat. Res. Coun. (com. on astronomy & astrophysics 1992). Achievements include work on the problem of the dark matter which constitutes more than 90% of the mass of the universe; devel. of a high-pressure gas scintillation drift chamber; search for WIMPs using cryogenic detectors of phonons and ionization. Office: Univ Calif Berkeley Ctr Particle Astrophysics 301 Le Conte Hall Berkeley CA 94720

SADOW, HARVEY S., health care company executive; b. N.Y.C., Oct. 6, 1922; s. Nat and Frances Donna (Saveth) S.; m. Sylvia June Riber, Dec. 22, 1944 (div. 1966); children: Harvey Jr., Suzanne Gail, Todd Forrest, Gay Summer; m. Jacqueline Lucille Clavel, Jan. 24, 1969 (div. 1993); 1 adopted child, Daniel Jean Marie. BS, Va. Mil. Inst., Lexington, 1947; MS, U. Kans., 1949; PhD, U. Conn., 1953. Intelligence officer CIA, Washington, 1951-53; assoc. dir. rsch. Lakeside Labs., Inc., Milw., 1953-56; med. rsch. cons. Milw., 1956; dir. clin. rsch. U.S. Vitamin & Pharm. Corp., N.Y.C., 1957-64, v.p. rsch. and devel., 1964-68; sr. v.p. scientific affairs USV Pharm./Revlon Corp., N.Y.C., 1969-71; pres., chief exec. officer Boehringer Ingelheim, Ltd. (named changed to Boehringer Ingelheim Pharms., Inc. 1984), Ridgefield, Conn., 1971-88; pres., chief exec. officer Boehringer Ingelheim, Ridgefield, 1984-88, chmn. bd., 1988-90; chmn. bd. Roxane Labs., Inc., Columbus, Ohio, 1981-88, Boehringer Ingelheim Animal Health, Inc., St. Joseph, Mo., 1981-88, Henley Co., N.Y.C., 1986-88, U. Conn. Rsch. and Devel. Corp., Storrs, 1984-87; bd. dirs. Cortex Pharms., Inc., Irvine, Calif., 1989—, chmn. bd., 1991—; bd. dirs. Triton Biosciences, Inc., Alameda, Calif., 1989, Telios Pharms., Inc., La Jolla, Calif., Penederm Corp., Foster City, Calif., TargeTech, Inc., Meriden, Conn., 1992, Microgenesys Corp., Meriden, Conn., Delta Health Products Devel. Corp., Mill Valley, Calif., Conn. Innovations Inc., Rocky Hill, Neocrin Corp., Irvine; bd. dirs. Cholestech Corp., Hayward, Calif., chmn. bd., 1992—; mem. adv. bd. Salk Inst. Biotechnology/Indsl. Assocs., Inc., La Jolla, 1988-90. Co-author: Oral Treatment of Diabetes, 1967; author, co-author 23 papers on intermediary metabolism, diabetes, obesity and cardiovascular disease., 1963-72. Bd. dirs. Pharm. Mfrs. Assn., 1983-90; chmn. Pharm. Mfrs. Assn. Found., 1988-90; bd. dirs. Conn. Bd. Higher Edn., Hartford, 1977-83, Govs. Tech. Adv. Bd., Hartford, 1984-87; mem. Conn. Commn. on Bus. Opportunity, Def. Diversification and Indsl. Policy, 1991—; mem. bd. visitors Va. Mil. Inst., Lexington, 1987—, bd. pres., 1991—; chmn. bd. Conn. Law Enforcement Found., Hartford, 1981-86, 92—, Pharm. Mfrs. Assn. Found., Washington, 1988-90, U. Conn. Found., Storrs, 1984-87; chmn., pres.' coun. Am. Lung Assn., N.Y.C., 1986-87; York Sch., Monterey, Calif., 1988-89; trustee Conn. Coll., Groton, 1991—. Capt. U.S. Army, 1943-53, ETO, Korea. Decorated Disting. Svc. Cross, Fed. Republic of Germany, 1987; recipient Univ. medal U. Conn., 1987, Recognition award Nat. Hypertension Assn., 1990, Humanitarian award Am. Lung Assn. Conn., 1993. Mem. Am. Soc. for Clin. Pharmacology and Therapeutics, Am. Fedn. for Clin. Rsch., Am. Diabetes Assn., Danbury C. of C. (Abraham Ribicoff Community Svc. award City of Danbury 1987, bd. dirs. 1978-81), Union League (N.Y.C.), Landmark Club (Stamford, Conn.), Masons, Sigma Xi, Sigma Pi Sigma, Phi Lambda Upsilon. Avocations: art collecting, photography, music, writing, golfing. Home and Office: 120-36 Prospect St Ridgefield CT 06877-4648

SADUN, ALBERTO CARLO, astrophysicist, physics educator; b. Atlanta, Apr. 28, 1955; s. Elvio Herbert and Lina (Ottolenghi) S.; m. Erica Liebman. BS in Physics, Mass. Inst. Tech., 1977; PhD in Physics, MIT, 1984. Asst. prof. Agnes Scott Coll., Decatur, Ga., 1984-90, assoc. prof., 1990—, dir. Bradley Obs., 1984—; adj. prof. Ga. State U., Atlanta, 1986—; rsch. affiliate NASA/Caltech Jet Propulsion Lab., Pasadena, Calif., 1988-90, summer faculty fellow, 1987, 88. Contbr. articles to Nature, Astrophys. Jour., Astron. Jour., Publ. Astron. Soc. of the Pacific, Astrophys. Letters and Communications. Mem. Am. Jewish Com., Atlanta, 1984—. Fellow Royal Astron. Soc.; mem. Internat. Astron. Union, Am. Astron. Soc., N.Y. Acad. Scis. Democrat. Achievements include relocation of Agnes Scott College's telescope to Hard Labor Creek Observatory. Home: 112 Hampshire Ct Avondale Estates GA 30002-1558 Office: Agnes Scott Coll 141 E College Ave Decatur GA 30030-3797

SADUSKY, MARIA CHRISTINE, environmental scientist; b. Wilmington, Del., June 20, 1963; d. Joseph Anthony and Concetta Marie (Simeone) S. BS in Plant Sci. with distinction, U. Del., 1985, MS in Plant Sci., 1987. Scientist III, group supr. Geo-Ctrs., Inc., Ft. Washington, Md., 1987—. Contbr. articles to Soil Sci. Soc. Am. Jour., Environ. Toxicology and Chem. Recipient Potash and Phosphate Inst. Fellowship award, 1986. Mem. ASTM (coms. D-18 on soil and rock and E-47 on biol. effects and environ. fate), Soc. Environ. Toxicology and Chem., Am. Soc. Agronomy, Soil Sci. Soc. Am., Assn. Women Soil Scientists, Alpha Zeta (pres. 1983-85). Republican. Roman Catholic. Achievements include research in developing environmental test systems to obtain toxicological fate and effects data on chemicals and materials and their impact on the terrestrial ecosystem.

SAEGUSA, TAKEO, polymer scientist; b. Mukden, China, Oct. 18, 1927; s. Isamu and Choko Saegusa; m. Ayako Saegusa, May 6, 1956; children: Yumiko, Mamiko. Bachelor degree, Kyoto (Japan) U., 1950, Doctorate, 1956. From asst. to assoc. prof. Kyoto U., 1957-65, prof., 1965-91, prof. emeritus, 1991—; exec. v.p. Kansai Rsch. Inst., Kyoto, 1991—; pres. macromolecular divsn. Internat. Pure and Applied Sci., Oxford, Eng., 1985-89; pres. Pacific Polymer Fedn., Tokyo, 1991-92. Contbr. over 500 papers on polymer sci. to jours. including Macromolecules, Jour. Polymer Sci., Makromolekulare Chemie, Polymer Bull. Recipient award Chem. Soc. Japan, 1978, Wilhelm Exner medal Österreichischer Gewerbeverein, Vienna, 1990, H.F. Mark medal Austrian Inst. of Sci. and Rsch., Vienna, Purple Ribbon medal Japanese Govt., 1992, ACS award in Polymer Chemistry Am. Chem. Soc., Washington, 1993. Mem. Soc. of Polymer Sci. (hon.). Achievements include research in organic-inorganic polymers hybrids, no catalyst copolymerization via zwitterion Intermediate, new catalysts for the ring-opening polymerization reactions; creation of novel non-ionic hydrogels; invention of new types of non-ionic surfactants. Home: Kyoto Univ, 8-22 Toji-in Kitamachi, Kita-ku Kyoto Japan 603 Office: Kansai Rsch Inst, Kyoto Research Park, Shimogyo Kyoto Japan 600

SAENGER, WOLFRAM HEINRICH EDMUND, crystallography educator; b. Frankfurt, Hessen, Fed. Republic of Germany, Apr. 23, 1939; s. Hans-Heinrich and Else (Hemming) S.; m. Barbara Fey, Oct. 17, 1964; children: Nicole, Jörg. Diploma in chemistry, U. Darmstadt, Fed. Republic of Germany, 1964, DEng, 1965; postdoctoral, Harvard U., 1965-67; habilitation, U. Göttingen, Fed. Republic of Germany, 1972. Leader rsch. group Max Planck Inst. Exptl. Medicine, Göttingen, 1967-81; prof. Free U. Berlin, 1981—; vis. prof. Japanese Soc. for Promotion of Sci., Osaka, 1979. Author: Principles of Nucleic Acid Structure, 1983; (with G.A. Jeffrey) Hydrogen Bonding in Biological Structures, 1991; editor: Landolt-Bornstein Vol. I Biophysics, 1990; co-editor: Acta Crystallographica, 1990—; contbr. 300 articles to profl. jours. Recipient Leibniz award German Rsch. Found., 1987, Humboldt award Alexander von Humboldt Found., 1988. Mem. European Molecular Biology Orgn., Deutsche Gesellschaft für Kristallographie, Am. Crystallographic Assn., Gesellschaft Deutsche Chemiker. Achievements include research in three-demensional structures of snake toxin cobratoxin, enzymes proteinase K, DNA methylase and ribonuclease T1, DNA-binding proteins FIS and tetracyclin repressor, photosynthetic complex photosystem I. Home: Türksteinweg 39, 1000 Berlin 37, Germany Office: Free U Inst Crystallography, Takustrase 6, 1000 Berlin 33, Germany

SAETHER, OLA MAGNE, geochemist; b. Schleswig, Germany, Dec. 13, 1949; arrived in Norway, 1953; s. Edvin Andreas and Hjoerdis (Dahlsveen) S.; m. Allison Sargent, May 22, 1977 (div. 1986); 1 child, Erik Andreas. Student, U. Bergen, 1974; MS, U. Okla., 1976; PhD, U. Colo., 1980. Cons. geologist Sci. Applications Inc., Boulder, Colo., 1977; geologist Exxon Prodn. Rsch., Houston, 1978; postdoctoral fellow U. Oslo, 1980; geochemist Norwegian Geol. Survey, Trondheim, 1981—. Contbr. articles to Chem. Geology. Mem. Bondeungdoml Nidaros, Trondheim, 1985—. Lt. Norwegian Med. Corps, 1968-71. Mem. Am. Chem. Soc., Am. Geophys. Union, Soc. Exploration Geochemists, Norwegian Geol. Soc. (editor jour. 1985-92). Social Democrat. Lutheran. Office: Norwegian Geol Survey, Leiv Eiriksson Vei 39, N-7040 Trondheim Norway

SÁEZ, ALBERTO M., physics educator; b. Pamplona, Navarra, Spain, July 24, 1922; s. Mariano J. Sáez and María de los Angeles Fernández de Toro; m. María Dolores Soloaga, Jan. 2, 1954; children: María Eugenia, María de los Angeles, José Alberto. B. in Math., Cen. U., Madrid, 1946, B. in Physics, 1949, D. in Physics, 1952; postgrad., MIT, 1958. Asst. prof. Cen. U., Madrid, 1946-48; physicist Spanish Inst. Oceanography, Madrid, 1948-53; physicist researcher Optical Inst. "Daza Valdés", Madrid, 1949-53, prof. microscopy, 1952; head physics dept. U. Zulia, Maracaibo, Venezuela, 1953-61, dir. computing inst., 1960-61; head physics dept. U. Los Andes, Mérida, Venezuela, 1961-63; prof. physics Cen. U. Scis., Caracas, Venezuela, 1963-69, Cen. U. Engring., Caracas, 1969—; cons. oceanography UNESCO, Paris, 1953, cons. Open Nat. U., Caracas, 1980-81; cons.-prof. Met. U., Caracas, 1980-86. Author: Instituto Optica. Apodización-Difraccion, 1961, Contraste de Fase, 1952; contbr. articles to profl. jours. Pres. Zulia Assn. for Advancement Scis., Maracaibo, 1955-56, Condominum of Residencias Crillón, Caracas, 1987-88; treas. Puerto Azul Sailing Club Assn., Naiguata, Venezuela, 1970; mem. Italian Club, Caracas, 1968-85. Rsch. travel gratnee Sci. Rsch. Coun., Madrid, 1952-53, Italian Govt., 1952, grantee Ford Found. and Cen. U., Caracas, 1965. Mem. Optical Soc. Am. (Emeritus award 1986), Colegio de Ingenieros Venezuela (Emeritus award 1987), U.S. Chess Fedn. (sr.), Club Puerto Azul, Catalan Club. Roman Catholic. Avocations: philosophy, history, chess, bowling, racing-sailing. Home: Residencias Crillón 9-A,, Ave 4, Calle Ciega, Palos Grandes Caracas 1060, Venezuela Office: U Ctrl Ingenieria, Ciudad Universitaria, Caracas Venezuela

SAFAAI-JAZI, AHMAD, electrical engineering educator, researcher; b. Isfahan, Iran, Nov. 18, 1948; came to U.S., 1986; s. Ali and Talaat (Niroomand) S.-J.; m. Zohreh Azargoshasb, Mar. 1, 1983; children: Rokhsana, Amir-Arsalan. BSc, Sharif U. Tech., Tehran, Iran, 1971; MASc, U. B.C., Vancouver, Can., 1974; PhD, McGill U., Montreal, Que., Can., 1978. Asst. prof. dept. elec. and computer engring. Isfahan U. Tech., 1978-84; rsch. assoc. dept. elec. engring. McGill U., 1984-86; assoc. prof. Va. Poly. Inst. and State U., Blacksburg, 1986—. Contbr. articles to IEEE Trans. on Microwave Theory and Tech., IEEE Trans. on Ulstrsonics, Ferroelectrics and Frequency Control, Optical Soc. Am. Jour., Jour. Lightwave Tech., IEEE Trans. Antennas and Propagation, Radio Sci., Electronic Letters, Optics Letters, Acoustic Soc. Am. Jour., Applied Optics. Mem. IEEE (sr., treas.-sec. Va. Mountain sect. 1989-90, vice chmn. 1990-91, chmn. 1991-92), Optical Soc. Am. Achievements include patent for narrowband fiberoptic spectral filter formed from fibers having a refractive index with a W-profile and a step profile, longitudinal mode fiber acoustic waveguide with solid core and solid cladding, birefringent single-mode acoustic fiber, also others. Office: Va Poly Inst and State U Bradley Dept Elec Engring Blacksburg VA 24061

SAFAR, MICHAL, information scientist. BA in English-History, Butler U., 1972; MLS in Info. Sci., Rosary Coll., 1982. Credit reporter Dun & Bradstreet, Indpls., 1972-73, supr. svc., 1974-75; supr. telephone order dept. Dun & Bradstreet, Chgo., 1976, sales rep., 1977, adminstrv. asst., 1978; supr. Chgo. Blue Cross and Blue Shield, 1979-81; asst. libr. East-West U., Chgo., 1983-85; mgr. rsch. mfg. dept. IIT Rsch. Inst., Chgo., 1985—, info. specialist mfg. tech. info. analysis ctr., 1985-87, tech. coordinator, 1987-90, dir., 1990—. Mng. editor Mfg. Competitiveness Frontiers. Mem. ALA, Am. Soc. Info. Sci., Am. Def. Preparedness Assn. (treas. Chgo. chpt.), Nat. Contract Mgmt. Assn., Ill. Librs. Assn., Spl. Librs. Assn. Office: IIT Rsch Inst Mfg Productivity Ctr 10 W 35th St Chicago IL 60616-3799

SAFAR, PETER, emergency health care facility administrator, educator; b. Vienna, Austria, Apr. 12, 1924; came to U.S., 1949; s. Karl and Vinca (Landauer) S.; m. Eva Kyzivat, July 6, 1950; children: Elizabeth, Philip, Paul. MD, U. Vienna, 1948; Dr. Hon. Causa, Gutenberg U., Mainz, Fed. Republic Germany, 1972. Resident in surgery U. Vienna and Yale U., 1948-50; resident in anesthesiology U. Pa., Phila., 1950-52; chief anesthesiologist Nat. Cancer Hosp., Lima, Peru, 1953, Balt. City Hosp., 1955-61; anesthesiologist Johns Hopkins Hosp., Balt., 1954; asst. prof. anesthesiology Johns Hopkins U., Balt., 1955-61; prof., chmn. dept. anesthesiology/CCM U. Pitts. Med. Ctr., 1961-78; dir. Internat. Resuscitation Rsch. Ctr. U. Pitts., 1978—, disting. svc. prof. resuscitation medicine, 1978—; co-initiator cardiopulmonary-cerebral resuscitation; researcher in field; mem. emergency med. svcs. and resuscitation coms. NRC/NAS, Washington, 1950-70; mem. Emergency Med. Svcs. Interagy. Com., Washington, 1974-76. Co-author nat. and internat. guidelines for modern resuscitation. Considered Father of Modern Resuscitation. Mem. Soc. Critical Care Medicine (pres., co-initiator 1972-73), Am. Coll. Emergency Physicians (hon.), Nat. Assn. Emergency Med. Svcs. Physicians (hon.), Univ. Assn. Emergency Medicine (hon.), German Acad. Natural Scis. Leopoldina (hon., corr.), Am. Acad. Scis. (hon., corr.), Czechoslovak Med. Soc. J.E. Purkinje (hon., corr.). Avocations: snow skiing, mountaineering, water sports, music, piano. Office: U Pitts Internat Resuscitation Rsch 3434 5th Ave Pittsburgh PA 15260-0001

SAFARS, BERTA See FISZER-SZAFARZ, BERTA

SAFF, EDWARD BARRY, mathematics educator; b. N.Y.C., Jan. 2, 1944; s. Irving H. and Rose (Koslow) S.; m. Loretta Singer, July 3, 1966; children: Lisa Jill, Tracy Karen, Alison Michelle. BS with highest honors, Ga. Inst. Tech., 1964; PhD, U. Md., 1968. Asst. prof. U. Md., 1968; post-doctoral researcher Imperial Coll., London, 1968-69; asst. prof. math. U. South Fla., 1969-71, assoc. prof., 1971-76, prof., 1976-86, disting. rsch. prof., 1986—, dir. Ctr. for Math. Svcs., 1978-83, dir. Inst. for Constructive Math., 1985—; sr. vis. fellow Oxford U., 1978. Author: (with A.D. Snider) Fundamentals of Complex Analysis, 1976; (with A.W. Goodman) Calculus, Concepts and Calculations, 1981; (with A. Edrei and R.S. Varga) Zeros of Sections of Power Series; editor (with R.S. Varga) Pade and Rational Approximation: Theory and Applications, 1977; (with R.K. Nagle) Fundamentals of Differential Equations; (with D.S. Lubinsky) Strong Asymptotics for Extremal Polynomials Associated with Weights on R., 1988; editor in chief Constructive Approximation Jour., 1983—; editor Jour. Approximation Theory, 1990—. Fulbright fellow, 1968-69, Guggenheim fellow, 1978; NSF grantee, 1970-72, 80—; Hon. prof. Zhejiang Normal U. Mem. Am. Math. Soc., Math. Assn. Am., Sigma Xi. Home: 11738 Lipsey Rd Tampa FL 33618-3620 Office: U South Fla Dept Math Tampa FL 33620

SAFFER, LINDA DIANE, biology researcher; b. Rochester, N.Y., Feb. 23, 1941; d. Eric Howard and Dolores Jeanne (Laurini) Lewis; m. Jerry Benjamin Saffer, June 20, 1964; children: Amy Kathleen, Marnie Haiya. BS, U. Rochester, 1963; MS, Northwestern U., Evanston, Ill., 1964; PhD, U. Va., 1985. Instr. Evanston Hosp. Sch. Nursing, 1964-68; postdoctoral fellow U. Va., Charlottesville, 1988-89, rsch. assoc., 1989-92; postdoctoral fellow Johns Hopkins Med. Sch., Balt., 1992—. Contbr. articles to profl. jours. including Molecular and Cellular Biology, Am. Jour. Tropical Medicine, Jour. of Protozoology, Exptl. Protozoology and Infection and Immunity. Mem. AAAS, Am. Assn. Tropical Medicine and Hygiene. Achievements include research in cell biology, parasites, visualization of transcription and replication in chromation. Home: Apt H 3004 Fallstaff Manor Ct Baltimore MD 21209 Office: John Hopkins Med Sch Monument and Wolfe Sts Baltimore MD 21209

SAFFIOTTI, UMBERTO, pathologist; b. Milan, Jan. 22, 1928; came to U.S., 1960, naturalized, 1966; s. Francesco Umberto and Maddalena (Valenzano) S.; m. Paola Amman, June 21, 1958; children: Luisa M., Maria

Francesca. MD cum laude, U. Milan, 1951, splty. diploma occupational medicine cum laude, 1957. Intern Inst. Pathol. Anatomy U. Milan, 1951-52, asst. to chmn. occupational medicine, chief lab. pathology, Inst. Occupational Medicine, 1956-60, fellow Inst. Gen. Pathology, 1957-60; rsch. asst. oncology, rsch. assoc. Chgo. Med. Sch., 1952-55, from asst. prof. to prof. oncology, 1960-68; mem. staff Nat. Cancer Inst., NIH, Bethesda, Md., 1968—, assoc. dir. carcinogenesis, 1968-76, chief lab. exptl. pathology, 1974—, acting head Registry of Exptl. Cancers, 1988—; mem. pathology B study sect., NIH, 1964-68; mem. various adv. coms. govt. agys.; mem. cancer prevention com. Internat. Union Against Cancer, 1959-66, panel on carcinogenicity, 1963-66; chmn. ad hoc com. evaluation low levels environ. carcinogens HEW, 1969-70. Co-editor books; contbr. articles to profl. jours. Bd. dirs. Rachel Carson Trust, 1976-79. Recipient Career Devel. award NIH, 1965-68, Superior Svc. Honor award HEW, 1971, Pub. Interest Sci. award Environ. Def. Fund, 1977, Spl. Recognition award USPHS, 1980. Fellow NYAS; mem. AAAS, Am. Assn. Cancer Rsch. (pres. Chgo. chpt. 1966-67), Am. Soc. Investigative Pathology, Internat. Commn. Occupational Health, Soc. Occupational and Environ. Health (councillor 1972-76, v.p. 1976-78, pres. 1978-82), Soc. Toxicology, Sigma Xi. Democrat. Home: 5114 Wissioming Rd Bethesda MD 20816-2259 Office: NIH Nat Cancer Inst Lab Exptl Pathology Bldg 41 Bethesda MD 20892

SAFFIR, HERBERT SEYMOUR, structural engineer, consultant; b. N.Y.C., Mar. 29, 1917; s. A.L. and Gertrude (Samuels) S.; m. Sarah Young, May 9, 1941; children: Richard Young, Barbara Joan. BS in Civil Engring. cum laude, Ga. Inst. Tech., 1940. Registered profl. engr., Fla., N.Y., Tex., P.R., Miss. Civil engr. TVA, Chattanooga, 1940, NACA, Langley Field, Va., 1940-41; structural engr. Ebasco Services, N.Y.C., 1941-43, York & Sawyer & Fred Severud, N.Y.C., 1945; engr. Waddell & Hardesty, Cons. Engrs., N.Y.C., 1945-47; asst. county engr. Dade County, Miami, Fla., 1947-59; cons. engr. Herbert S. Saffir, Coral Gables, Fla., 1959—; adj. lectr. civil engring. Coll. Engring., U. Miami, 1964—; adviser civil engring. Fla. Internat. U., 1975-80; cons. on bldg. codes Govt. Bahamas; cons. on engring. in housing to UN; mem., chmn. Met. Dade County Unsafe Structures Bd., 1977-92; mem. Bldg. Code Evaluation Task Force after Hurricane Andrew; mem. Am. Nat. Stds. Inst. Commn. Bldg. Design Loads, Nat. Adv. Group on Glass Design, Dade County Bldg. Code Com. 1993—; cons. to govt. and industry, condr. seminars, Australia; reviewer for NSF. Author: Housing Construction in Hurricane Prone Areas, 1971, Nature and Extent of Damage by Hurricane Camille, 1972; contbg. author: Wind Effects on Structures, 1976; editor Wind Engr., 1986-92; contbr. articles to profl. jours.; designer Saffir/Simpson hurricane scale. With 23d Regiment, N.Y. Guard, 1942-43, AUS, 1943-44. Recipient Outstanding Service award Fla. Profl. Engrs., 1954, Pub. Service award Nat. Weather Service, 1975, Disting. Service award Nat. Hurricane Conf., 1987; named Miami Engr. of Year, 1978, Gov.'s Design award, 1986, Gov. Gilchrist award for Profl. Excellence, 1988, Wind Engring. Svc. award Wind Engring. Rsch. Coun., 1990, Albert H. Friedman Community Svc. award, 1992. Fellow ASCE (past pres., sec., aerodynamics com. 1983—, mem. mitigation of wind damage com. 1985—, chmn. com. on damage investigation 1989—), Fla. Engring. Soc. (award for outstanding tech. achievement 1973, Community Service award 1980); mem. Soc. Am. Mil. Engrs., Am. Concrete Inst., ASTM (mem. com. performance bldg. constrn.), Prestressed Concrete Inst., Internat. Assn. for Bridge and Structural Engring., Colegio de Ingenieros P.R., Am. Meteorol. Soc., Am. Arbitration Assn., Wind Engring. Research Coun. (past bd. dirs., Svc. award 1990), Coral Gables C. of C. (bd. dirs., past pres., past chmn.), Tau Beta Pi, Chi Epsilon (hon.). Club: Country of Coral Gables. Home: 4818 Alhambra Cir Coral Gables FL 33146-1615 Office: 255 University Dr Ste 211 Coral Gables FL 33134-6733

SAFIER, LENORE BERYL, research chemist; b. Bklyn., Mar. 15, 1932; d. Irwin and Syd (Blaustein) S. BA, Vassar Coll., 1952; MS, NYU, 1954. Technician NYU-Bellevue Med. Ctr., N.Y.C., 1954-57; rsch. asst. NYU Med. Ctr., 1957-62; rsch. chemist VA Med. Ctr., N.Y.C., 1962—. Contbr. articles to profl. jours. Mem. AAAS, N.Y. Acad. Scis. Democrat. Jewish. Achievements include contribution to research on human platelet lipid composition and transcellular metabolism, including identification of 2 new eicosanoids formed by neutrophils from platelet 12-HETE. Office: VA Med Ctr 423 E 23d St New York NY 10010

SAFO, MARTIN KWASI, chemist; b. Kumasi, Ghana, May 16, 1958; came to U.S., 1986; s. Kofi Aduboahen and Agnes (Mensah) S.; m. Lydia Nkansah, July 22, 1988. BS, U. Cape Coast, 1985; PhD, U. Notre Dame, 1991. Postdoctoral U. Notre Dame, Ind., 1991, Med. Coll. of Va., Richmond, 1991—. Contbr. articles to profl. jours. Mem. Am. Crystallographic Assn. Office: Med Coll of Va PO Box 540 Richmond VA 23298

SAGAN, CARL EDWARD, astronomer, educator, author; b. N.Y.C., Nov. 9, 1934; s. Samuel and Rachel (Gruber) S.; m. Ann Druyan; children: Alexandra, Sam; children by previous marriages: Dorion Solomon, Jeremy Ethan, Nicholas. AB with gen. and spl. honors, U. Chgo., 1954, BS, 1955, MS, 1956, PhD, 1960; ScD (hon.), Rensselaer Poly. Inst., 1975, Denison U., 1976, Clarkson Coll. Tech., 1977, Whittier Coll., 1978, Clark U., 1978, Am. U., 1980, U. S.C., 1984, Hofstra U., 1985, L.I. U., 1987, Tuskegee U., 1988; DHL (hon.), Skidmore Coll., 1976, Lewis and Clark Coll., 1980, Bklyn. Coll., CUNY, 1982; LLD (hon.), U. Wyo., 1978, Drexel U., 1986. Miller research fellow U. Calif.-Berkeley, 1960-62; vis. asst. prof. genetics Stanford Med. Sch., 1962-63; astrophysicist Smithsonian Astrophys. Obs., Cambridge, Mass., 1962-68; asst. prof. Harvard U., 1962-67; mem. faculty Cornell U., 1968—, prof. astronomy and space scis., 1970—, David Duncan prof., 1976—, dir. Lab. Planetary Studies, 1968—, assoc. dir. Center for Radiophysics and Space Research, 1972-81, Johnson Disting. lectr. Johnson Grad. Sch. Mgmt., 1985; pres. Carl Sagan Prodns. (Cosmos TV series), 1977—; nonresident fellow Robotics Inst., Carnegie-Mellon U., 1982—; NSF-Am. Astron. Soc. vis. prof. various colls., 1963-67, Condon lectr., Oreg., 1967-68; Holiday lectr. AAAS, 1970; Vanuxem lectr. Princeton U., 1973; Smith lectr. Dartmouth Coll., 1974, 77; Wagner lectr. U. Pa., 1975, Bronowski lectr. U. Toronto, 1975; Philips lectr. Haverford Coll., 1975; Disting. scholar Am. U., 1976; Danz lectr. U. Wash., 1976; Clark Meml. lectr. U. Tex., 1976; Stahl lectr. Bowdoin Coll., 1977; Christmas lectr. Royal Instn., London, 1977; Menninger Meml. lectr. Am. Psychiat. Assn., 1978, Adolf Meyer lectr., 1984; Carver Meml. lectr. Tuskegee Inst., 1981; Feinstone lectr. U.S. Mil. Acad., 1981; Pal lectr. Motion Picture Acad. Arts and Scis., 1982; Dodge lectr. U. Ariz., 1982; Disting. lectr. USAF Acad., 1983; Lowell lectr. Harvard U., 1984; Poynter fellow, Schultz lectr. Yale U., 1984; Disting. lectr. Fla. State U., 1984; Jack Disting. Am. lectr., Ind. U., Pa., 1984; Keystone lectr. Nat. War Coll., Nat. Def. U., Washington, 1984-86; Marshall lectr. Nat. Resources Def. Coun., Washington, 1985; Gifford lectr. in natural theology U. Glasgow, 1985; Lilenthal lectr. Calif. Acad. Sci., 1986;, Dolan lectr. Am. Pub. Health Assn., 1986; von Braun lectr. U. Ala., Huntsville, 1987; Gilbert Grosvenor Centennial lectr. Nat. Geog. Soc., Washington, 1988; Murata lectr., Kyoto, Japan, 1989; Bart Bok Centennial lectr. Astron. Soc. of the Pacific, 1989, James R. Thompson Leadership lectr. Ill. Math. and Sci. Acad., 1991, Nehru Meml. lectr. New Delhi, 1991; other hon. lectureships; mem. various adv. groups NASA and Nat. Acad. Scis., 1959—; mem. council Smithsonian Instn., 1975—; vice chmn. working group moon and planets, space orgn. Internat. Council Sci. Unions, 1968-74; lectr. Apollo flight crews NASA, 1969-72; chmn. U.S. del. joint conf. U.S. Nat. and Soviet Acads. Sci. on Communication with Extraterrestrial Intelligence, 1971; responsible for Pioneer 10 and 11 and Voyager 1 and 2 interstellar messages; mem. Voyager Imaging Sci. Team; judge Nat. Book Awards, 1975; mem. fellowship panel Guggenheim Found., 1976—; disting. vis. scientist Jet Propulsion Lab., Calif. Inst. Tech., 1986—; researcher physics and chemistry of planetary atmospheres and surfaces, origin of life, exobiology, Mariner, Viking and Voyager spacecraft observations of planets, nuclear winter. Author: Atmospheres of Mars and Venus, 1961, Planets, 1966, Intelligent Life in the Universe, 1966, Planetary Exploration, 1970, Mars and the Mind of Man, 1973, The Cosmic Connection, 1973, Other Worlds, 1975, The Dragons of Eden, 1977, Murmurs of Earth: The Voyager Interstellar Record, 1978, Broca's Brain, 1979, Cosmos, 1980, (novel) Contact, 1985, Comet, 1985, (with Richard Turco) Path Where No Man Thought: Nuclear Winter and the End of the Arms Race; also numerous articles; editor: Icarus: Internat. Jour. Solar System Studies, 1968-79, Planetary Atmospheres, 1971, Space Research, 1971, UFO's: A Scientific Debate, 1972, Communication with Extraterrestrial Intelligence, 1973; editorial bd.: Origins of Life, 1974—, Icarus, 1962—, Climatic Change, 1976—, Science 80, 1979-82. Mem. bd. advisors Children's Health Fund, N.Y.C., 1988—. Recipient Smith prize Harvard U., 1964; NASA medal for exceptional sci. achievement, 1972; Prix Galabert, 1973; John W. Campbell Meml. award, 1974; Klumpke-Roberts prize, 1974; Priestley award, 1975; NASA medal for disting. pub. service, 1977, 81; Pulitzer prize for lit., 1978; Washburn medal, 1978; Rittenhouse medal, 1980; Peabody award, 1981; Hugo award, 1981; Seaborg prize, 1981; Roe medal, 1981; Enivironment Programme medal UN, 1984; SANE Nat. Peace award, 1984; Regents medal Bd. Regents Univ. of State N.Y., 1984; Ann. award Physicians for Social Responsibility, 1985; Disting. Svc. award World Peace Film Festival, 1985; Honda prize Honda Found., 1985; Nahum Goldmann medal World Jewish Congress, 1986; Ann. award of merit Am. Cons. Engrs. Coun., 1986; Maurice Eisendrath award Cen. Conf. Am. Rabbis and Union Am. Hebrew Congregations, 1987; In Praise of Reason award Com. for Sci. Investigation of Claims of the Paranormal, 1987; Konstantin Tsiolkovsky medal Soviet Cosmonautics Fedn., 1987; George F. Kennan Peace award SANE/Freeze, 1988; Oersted medal Am. Assn. Physics Tchrs., 1990, Ann. award for Outstanding TV Script Writers Guild Am., 1991, UCLA medal UCLA, 1991; NSF fellow, 1955-60; Sloan research fellow, 1963-67. Fellow AAAS (chmn. astronomy sect. 1975), Am. Acad. Arts and Scis., AIAA, Am. Geophys. Union (pres. planetology sect. 1980-82), Am. Astronautical Soc. (council 1976-81, Kennedy award 1984), Brit. Interplanetary Soc., Explorers Club (75th Anniversary award 1980); mem. Am. Phys. Soc. (Leo Szilard award 1985), Am. Astron. Soc. (councillor, chmn. div. for planetary scis. 1975-76, Mazursky award 1991), Fedn. Am. Scientists (council 1977-81, bd. sponsors 1988—, Ann. award 1985), Am. Com. on East-West Accord, Soc. Study of Evolution, Genetics Soc. Am., Internat. Astron. Union, Internat. Acad. Astronautics, Internat. Soc. Study Origin of Life (council 1980—), Planetary Soc. (pres. 1979—), Authors Guild, Am. Com. on U.S.-Soviet Rels., Phi Beta Kappa, Sigma Xi. Office: Cornell U Space Sci Bldg 302 CRSR Ithaca NY 14853

SAGDEEV, ROALD ZINNUROVI, physicist educator. BSc, physics, Moscow State U., USSR, 1955; Ph.D., theoretical physics, Moscow Inst. for Physical Problems, 1960. Head Plasma Theory Lab., Inst. of Nuclear Physics, Novosibirsk, USSR; mem. Inst. of High-Temperature Physics, Moscow, USSR, 1971-73; dir. Inst. of Space Reseach, Moscow, USSR, 1973-90; Distinguished Prof. of Physics U. Maryland, College Park, Md., 1990—; dir. East-West Space Science Ctr, 1990—. Recipient John T. Tate Internat. award Am. Inst. Physics, 1992. Office: U of Maryland Dept of Physics College Park MD 20742*

SAGE, ANDREW PATRICK, JR., systems information and software engineering educator; b. Charleston, S.C., Aug. 27, 1933; s. Andrew Patrick and Pearl Louise (Britt) S.; m. LaVerne Galhouse, Mar. 3, 1962; children: Theresa Annette, Karen Margaret, Philip Andrew. BS in Elec. Enging, The Citadel, 1955; SM, MIT, 1956; PhD, Purdue U., 1960; DEng (hon.), U. Waterloo, Can., 1987. Registered profl. engr., Tex. Instr. elec. engring. Purdue U., 1956-60; assoc. prof. U. Ariz., 1960-63; mem. tech. staff Aerospace Corp., Los Angeles, 1963-64; prof. elec. engring. and nuclear engring. scis. U. Fla., 1964-67; prof., dir. Info. and Control Scis. Center, So. Methodist U., Dallas, 1967-74; head elec. engring. dept. Info. and Control Scis. Center, So. Methodist U., 1973-74; Quarles prof. engring. sci. and systems U. Va., Charlottesville, 1974-84; chmn. dept. chem. engring. U. Va., 1974-75, chmn. dept. engring. sci. and systems, 1977-84, assoc. dean, 1974-80; First Am. Bank prof. info. tech. George Mason U., Fairfax, Va., 1984—, assoc. v.p. for acad. affairs, 1984-85; dean Sch. Info. Tech. and Engring. George Mason U., 1985—; cons. Martin Marietta, Collins Radio, Atlantic Richfield, Tex. Instruments, LTV Aerospace, Battelle Meml. Inst., TRW Syustems, NSF, Inst. Def. Analyses, Planning Rsch. Corp., MITRE, Engring. Rsch. Assocs., Software Productivity Consortium; gen. chmn. Internat. Conf. on Systems, Man and Cybernetics, 1974, 87; mem. spl. program panel on system sci. NATO, 1981-82. Author: Optimum Systems Control, 1968, 2d edit., 1977, Estimation Theory with Applications to Communications and Control, 1971, System Identification, 1971, An Introduction to Probability and Stochastic Processes, 1973, Methodology for Large Scale Systems, 1977, Systems Engineering: Methodology and Applications, 1977, Linear Systems Control, 1978, Economic Systems Analysis, 1983, System Design for Human Interaction, 1987, Information Processing in Systems and Organizations, 1990, Introduction to Computer Systems Analysis, Design, and Applications, 1989, Software Systems Engineering, 1990, Decision Support Systems Engineering, 1991, Systems Engineering, 1992; assoc. editor: IEEE Transactions on Systems Sci. and Cybernetics, 1968-72; editor: IEEE Transactions on Systems, Man and Cybernetics, 1972—; assoc. editor: Automatica, 1968-81; editor, 1981—; mem. editorial bd.: Systems Engring, 1968-72, IEEE Spectrum, 1972-73, Computers and Electrical Engineering, 1972—, Jour. Interdisciplinary Modeling and Simulation, 1976-80, Internat. Jour. Intelligent Systems, 1986—, Orgn. Sci., 1990; editor Elsevier North Holland textbook series in system sci. and engring., John Wiley textbook series on systems engring., 1989—; co-editor-in-chief: Jour. Large Scale Systems: Theory and Applications, 1978-88, Information and Decision Technologies, 1988—; contbr. articles on computer sci. and systems engring. to profl. jours. Bd. trustees Ctr. for Naval Analysis, 1990—. Recipient Frederick Emmonds Terman award Am. Soc. for Engring. Edn., 1970, M. Barry Carlton award IEEE, 1970, Norbert Wiener award, 1980, Joseph G. Wohl career award, 1991; Case Centennial scholar, 1980. Fellow IEEE (Centennial medal 1984, Outstanding Contbn. award 1986), AAAS (chmn. sect. M 1990), IEEE Systems; mem. Man and Cybernetics Soc. (pres. 1984-85), Inst. Mgmt. Scis., Internat. Fedn. Automatic Control (Outstanding Svc. award), Am. Soc. for Engring. Edn. (Centennial cert. for exceptional contbn. 1993), Ops. Rsch. Soc. Am., Sigma Xi, Eta Kappa Nu, Tau Beta Pi. Home: 8011 Woodland Hills Ln Fairfax VA 22039-2433

SAGE, JAMES TIMOTHY, physicist, educator; b. Wilkinsburg, Pa., July 29, 1957; s. Milton James and Mary Elizabeth (Wilson) S. BS, Carnegie-Mellon U., 1979; PhD, U. Ill., 1986. Rsch. asst. U. Ill., Urbana, 1980-86; rsch. assoc. Northeastern U., Boston, 1986-91, sr. rsch. scientist, 1991—. Contbr. articles to profl. jours. Achievements include research in characterization of protein folding intermediates at low pH, comparison of protein structure in crystal and solution; relaxation and ligand migration in heme proteins. Office: Northeastern U Dept Physics 360 Huntington Ave Boston MA 02115

SAGER, RUTH, geneticist; b. Chgo., Feb. 7, 1918; married, 1973. BS, U. Chgo., 1938; MS, Rutgers U., 1944; PhD, Columbia U., 1948. Merck fellow Nat. Research Council, 1949-51; asst. in biochemistry Rockefeller Inst., 1951-55; research assoc. in zoology Columbia U., N.Y.C., 1955-60, sr. research assoc. in zoology, 1961-65; prof. biology Hunter Coll., CUNY, 1966-75; prof. cellular genetics Harvard Med. Sch., 1975-88, prof. emeritus, 1988—; chief cancer genetics div. Dana-Farber Cancer Inst., from 1975; mem. sci. adv. bd. Friedrich Miescher Inst., Basle, 1990—; mem. coun. Nat. Inst. Aging NIH, 1993—. Author: (with F.J. Ryan) Cell Heredity, 1961, Cytoplasmic Genes and Organelles, 1972. Recipient Gilbert Morgan Smith medal Nat. Acad. Scis., 1988; Guggenheim fellow, 1972-73. Fellow AAAS; mem. Am. Acad. Arts and Sci., Nat. Acad. Scis., Inst. of Medicine, Am. Soc. Cell Biologists, Genetics Soc. Am., Am. Assn. Cancer Rsch., Am. Soc. Biol. Chem., Sigma Xi, Phi Beta Kappa. Office: Dana-Farber Cancer Inst 44 Binney St Boston MA 02115-6084

SAGUE, JOHN E(LMER), mechanical engineer; b. Phila., July 23, 1933; s. John Taylor and Katherine (Flynn) S.; m. Florence Mary Aguero; children: Deborah, John, Donna, Allyson. BME, Villanova U., 1974. Product designer SKF Industries, Inc., Phila., 1951-58; dir. engring. Messinger Bearings Inc., Phila., 1958-77; v.p., dir. engring. Formmet Corp., Avon, Ohio, 1977-80; head mech. engring. Franklin Rsch. Ctr., Phila., 1980-86; prin. assoc., dir. Sague & Assocs., Phila., 1986—; instr. Am. Soc. Tribologists and Lubrication Engrs., 1990. Contbr. articles to profl. jours. Mem. ASME (chmn. Slewing Ring Bearing com. 1990—), Am. Soc. Metals, Am. Arbitration Assn. Achievements include patents for In-Situ Replaceable Bearing, Inspectable Slewing Ring Bearing, Anti-Friction Bearings with Hardened Race Inserts. Office: Sague & Assocs 702 Princeton Ave Philadelphia PA 19111

SAHA, DHANONJOY CHANDRA, biomedical research scientist; b. Bhola, Bangladesh, June 5, 1957; came to U.S., 1982; s. Gopi Ballav and Bina Pani Saha. DVM, Bangladesh Agrl. U., Mymensingh, Bangladesh, 1981; MS, Rutgers U., 1987. Assoc. rsch. scientist N.J. Agrl. Exptl. Sta., New Brunswick, N.J., 1984-85, rsch. scientist, 1986-88; postdoctoral fellow UMDNJ-N.J. Med. Sch., Newark, 1988; rsch. specialist Winthrop Univ. Hosp., Mineola, N.Y., 1988-89; rsch. mgr. St. Vincent's Hosp., N.Y.C., 1989—; sci. cons. Rutgers U., Piscataway, N.J., 1987-88. Contbr. articles to profl. jours. Recipient NST fellowship Govt. of Bangladesh, 1982; named Peter Selmer Loft scholar Loft Trust, 1986. Mem. AAAS, Am. Assn. Lab. Animal Sci., Can. Assn. Lab. Animal Medicine, N.Y. Acad. Sci. Achievements include improvement of turf grass performance; discovery of fungus in several grass species and its utilization. Office: Saint Vincent's Hosp 153 W 11 St New York NY 10011

SAHA, MURARI MOHAN, electrical engineer; b. Chaumuhani, Naokhali, Bangladesh, Jan. 10, 1947; arrived in Sweden, 1975; s. Debendra Kumar and Milan Bala (Bhuiya) S.; m. Rekha Chaudhary, Dec. 26, 1982; 1 child, Robin Mohan Rohin; 1 child from previous marriage, Patrick Bappi. BSEE, Bangladesh U. Engring. & Tech., Dacca, 1968, MScEE, 1970; MSEE, Warsaw (Poland) Tech. U., 1972, PhD, 1975. Chartered engr., Eng., European engr. Lectr. electrical engring. dept. Bangladesh U. Engring. & Tech., Dacca, 1969-71; postgrad. scholar Warsaw Tech. U., 1971-75; devel. engr. ASEA Relays AB, Västerås, Sweden, 1975-86, devel. engr. gen. engring. office, 1986-88; sr. devel. engr. rsch. group ABB Relays AB, Västerås, 1988-90, sr. devel. rsch. engr., rsch. dept., 1991—. Patentee fault location method, directional detection, range limitation. Mem. IEEE (sr. U.S.), Inst. Elec. Engrs. (U.K., hon. sec., Swedish br.), Eur. Ing., European Fedn. Nat. Eng. Assn. (Paris). Avocations: music, tennis, cards, reading, travelling. Home: Värmlandsvägen 11, S-722 44 Västerås Sweden Office: ABB Relays AB, Rsch Dept, S-721 71 Västerås Sweden

SAHAI, HARDEO, medical educator; b. Bahraich, India, Jan. 10, 1942; m. Lillian Sahai, Dec. 28, 1973; 3 children. BS in Math., Stats. and Physics, Lucknow U., India, 1962; MS in Math., Banaras U., Varanasi, India, 1964; MS in Math. Stats., U. Chgo., 1968; PhD in Stats., U. Ky., 1971. Lectr. in math. and stats. Banaras U., Varanasi, India, 1964-65; asst. stats. officer Durgapur Steel Plant, Durgapur West Bengal, India, 1965; statistician Rsch. and Planning div. Blue Cross Assn., Chgo., 1966; statis. programmer Cleft Palate Ctr., U. Ill., 1967, Chgo. Health Rsch. Found., 1968; mgmt. scientist Mgmt. Systems Devel. Dept. Burroughs Corp., Detroit, 1971-72; from asst. prof. to prof. dept. math. U. P.R., Mayaguez, 1972-82; vis. research prof. Dept. Stats. and Applied Math. Fed. U. of Ceara, Brazil, 1978-79; sr. research statistician Travenol Labs., Inc., Round Lake, Ill., 1982-83; chief statistician U.S. Army Hqrs., Ft. Sheridan, Ill., 1983-84; sr. math. statistician U.S. Bur. of Census Dept. of Commerce, Washington, 1984-85; sr. ops. rsch. analyst Def. Logistics Agy. Dept. Def., Chgo., 1985-86; prof. Dept. Biostats. and Epidemiology U. P.R. Med. Scis., San Juan, 1986—; cons. P.R. Univ. Cons., P.R. Driving Safety Evaluation project, Water Resources Rsch. Inst., Travenol Labs., Campo Rico, P.R., U.S. Bur. Census. Washington, Lawrence Livermore Nat. Lab., Calif., others; lectr. in field. Author: Statistics and Probability: Learning Module, 1984; author: (with Jose Berrios) A Dictionary of Statistical Scientific and Technical Terms: English-Spanish and Spanish-English, 1981, (with Anwer Khurshid) Statistical Methods in Epidemiologic Data Analysis, (with Wilfredo Martinez) Statisical Tables and Formulas for the Social, Physical and Biological Sciences, 1993; mem. editorial bd. Sociedad Colombiana de Matematicas, P.R. Health Scis Jour.; contbg. editor Current Index to Stats; reviewer: Collegiate Microcomputer, Communications in Statistics, Indian Jour. Statistics; contbr. more than 80 articles and papers to profl. and sci. jours., numerous articles to tech. mags. Active Dept. Consumer Affairs Svcs. Commonwealth of P.R., San Juan, Dept. Anti-Addiction Svcs., Commonwealth of P.R., San Juan., Inst. of AIDS, Municipality of San Juan, VA Med. Ctr. of San Juan. Recipient Dept. Army Cert. Achievement award, 1984, U. Ky. Outstanding Alumnus award, 1993; fellow Coun. Sci. and Indsl. Rsch., 1964-65, U. Chgo., 1965-68, Harvard U., 1979, Fulbright Found., 1982; U.P. Bd. Merit scholar, 1957-59, Govt. India Merit scholar, 1959-64; grantee NSF, 1974-77, NIMH, 1987-90, 91—, NIDA, 1991—. Fellow AAAS, Inst. Statisticians (charter statistician), N.Y. Acad Scis., Royal Statis. Soc.; mem. Internat. Statis. Inst.Internat. Assn. for Teaching Stats., Soc. Epidemol. Rsch., Inst. Math. Stats., Bernoulli Soc. for Math. Stats. and Probability, Biometrics Soc. Eastern N.Am. Region, Am. Soc. for Quality Control, Am. Stats. Assn., Japan Statis. Soc., Am. Pub. Health Assn., Can. Statis. Soc., Inter-Am. Statis. Inst., Internat. Assn. Statis. Computing, Sch. Sci. and Math. Assn., Sigma Xi. Avocations: religious studies, philosophy, reading, gardening. Home: Hill Mansions Calle 14 K5B Mayaguez PR 00680 Office: U PR Grad Sch Pub Health Med Scis Campus Dept Biostats & Epidemiology GPO Box 365067 San Juan PR 00936-5067

SAHATJIAN, RONALD ALEXANDER, science foundation executive; b. Cambridge, Mass., Oct. 1, 1942; s. Vartan and Roxy (Abrahamian) S.; m. Jean Khachadoorian, July 15, 1966; 1 child, Jennifer. BS in Chemistry, Tufts U., 1964; MS in Chemistry, U. Mass., 1968, PhD in Chemistry, 1969. Scientist color photographic rsch. lab. Polaroid Corp., Cambridge, 1971-73, sr. scientist color photograhic rsch. lab., 1973-75, sr. rsch. group leader photographic/optical materials, 1976-79, program mgr. polacolor transparency projects, 1979-81, mgr. applications rsch. lab., 1980-84; dir. R & D Chem. Fabrics Corp., Merrimack, N.H., 1984-87; v.p. corp. tech. Boston Sci. Corp., Watertown, Mass., 1987—; mem. adv. bd. Franklin Inst., Boston, 1989—. Contbr. articles to Jour. Polymer Sci., Macromolecules, Radiology. Fellow Am. Inst. Chemists; mem. ASTM, Radiol. Soc. N.Am., Watertown C. of C. (bd. dirs. 1991—). Achievements include patents on new medical devices, superelastic tennis string. Home: 29 Saddle Club Rd Lexington MA 02173-2121 Office: Boston Sci Corp 480 Pleasant St Watertown MA 02172-2414

SAHU, RANAJIT, engineer; b. Cuttack, Orissa, India, Dec. 6, 1961; came to U.S., 1983; s. Kailas Chandra and Bimala (Prusty) S.; m. Catherine Therese Sheahan, Aug. 28, 1988; children: Joseph, Daniel, Margaret. MS, Calif. Inst. Tech., 1984, PhD, 1988. Rsch. engr. Heat Transfer Rsch., Inc., College Station, Tex., 1988-89; devel. engr. Kinetics Tech. Internat., Monrovia, Calif., 1989-90; supervising engr. Engring. Sci., Pasadena, Calif., 1990-92, prin. engr., 1992—; mem. staff Grand Canyon Visibility Transport Commn., Denver, 1992—; instr. U. Calif., Riverside, Extension, 1992—, Loyola Marymount U., L.A., 1992—; presenter at profl. confs. Contbr. to profl. publs. Mem. ASME, Air and Waste Mgmt. Assn. Achievements include research in combustion, heat transfer, fluid mechanics, accoustics and air pollution engineering. Home: 311 N Story Pl Alhambra CA 91801 Office: Engring Sci 199 S Los Robles Ave Pasadena CA 91101

SAIF, MEHRDAD, electrical engineering educator; b. Tehran, Iran, Dec. 7, 1960; came to U.S. 1978; s. Jahangir and Shahrebanou (Jamshidi) S.; m. Maria De los Angeles Alvarez, May 20, 1989; 1 child, Cyrus Anthony. MSEE, Cleve. State U., 1984, DEng, 1987. Rsch. asst. Cleve. State U./NASA Lewis, 1984; rsch. assoc. Cleve. Adv. Mfg. Program (CAMP)/Adv. Mfg. Ctr. (AMC), 1985-87; instr. Cleve. State U., 1987; asst. prof. elec. engring. Simon Fraser U., Burnaby, B.C., 1987-93, assoc. prof. elec. engring., 1993—; assoc. fellow Inst. for Computational and Applied Math., Burnaby, 1990—; internat. com. 4th Internat. Symposium on Robotics and Mfg., Albuquerque, 1991. Co-editor: Robotics and Manufacturing, 1991; contbr. articles to profl. jours. Grantee, NSERC of Can., 1988-90, 91—; Adv. Systems Inst. grantee, 1990. Mem. IEEE, AIAA, Assn. Profl. Engrs. and Geoscientists B.C. (registered), Internat. Fedn. Automatic Control, Eta Kappa Nu. Office: Simon Fraser Univ, Sch Engring Sci, Burnaby, BC Canada V5A 1S6

SAIFER, MARK GARY PIERCE, pharmaceutical executive; b. Phila., Sept. 16, 1938; s. Albert and Sylvia (Jolles) S.; m. Phyllis Lynne Trommer, Jan. 28, 1961; children: Scott David, Alandria Gail. AB, U. Pa., 1960; PhD, U. Calif., Berkeley, 1966. Acting asst. prof. zoology U. Calif., Berkeley, 1966, postdoctoral fellow, 1967-68; sr. cancer research scientist Roswell Park Meml. Inst., Buffalo, 1968-70; lab. dir. Diagnostic Data Inc., Palo Alto, Calif., 1970-78; v.p. DDI Pharms., Inc., Mountain View, Calif., 1978—. Patentee in field. Mem. AAAS (life), Am. Assn. Pharm. Scientists, Parenteral Drug Assn. Office: DDI Pharms Inc 518 Logue Ave Mountain View CA 94043-4096

SAILLON, ALFRED, psychiatrist; b. Haïfa, Israel, Apr. 13, 1944; arrived in France, 1963; s. Joseph and Andree (Gemayel) Sahyoun; m. Anne Crouzat, Mar. 19, 1965; children: Helene, Marc, Antoine. Student, Am. U., Beirut, 1963; MD, Paris Med. Sch., 1971, psychiatry, 1974. Intern various hosps., Paris, 1971-74; pvt. practice specializing in psychiatry Fontainebleau, France, 1974—; asst. Hosp. de Coulommiers, 1976; pres. Les Jardins D'Eleusis Nursing Homes for Alzheimer Disease, France, 1986. Contbr. articles to profl. jours. Mem. Assn. Pour la Prevention des Comportements Toxico Maniaques (pres. 1985-87). Home and Office: 3 Bd Thiers, 77300 Fontainebleau France

SAIMA, ATSUSHI, science and engineering educator; b. Tokyo, Dec. 14, 1923; s. Hitoshi and Aiko Saima; m. Yasuko Saima, Oct. 21, 1953; children: Junko, Keiko. B. of Engring., Nihon U., Tokyo, 1946; D. of Engring., Tokyo U., 1964. Asst. prof. Nihon U., 1953-61, assoc. prof., 1961-68, prof., 1968—, dir. rsch. Inst. Sci. and Tech., 1988—. Home: 1-19-18 Jiyugaoka, Meguro-ku Tokyo 152, Japan Office: Nihon U, 1-8 Kanda Surugadai, Chiyoda-ku Tokyo 101, Japan

SAIN, MICHAEL KENT, electrical engineering educator; b. St. Louis, Mar. 22, 1937; s. Charles George and Marie Estelle (Ritch) S.; m. Frances Elizabeth Bettin, Aug. 24, 1963; children: Patrick, Mary, John, Barbara, Elizabeth. BSEE, St. Louis U., 1959, MSEE, 1962; PhD, U. Ill., 1965. Engr. Sandia Corp., Albuquerque, 1958-61, Vickers Electric Corp., St. Louis, 1962; instr. U. Ill., Urbana, 1962-63; asst. prof. U. Notre Dame (Ind.), 1965-68, assoc. prof., 1968-72, prof., 1972-82, Frank M. Freimann prof. elec. engring., 1982—; vis. scientist U. Toronto, Ont., Can., 1972-73; disting. vis. prof. Ohio State U., Columbus, 1987; cons. Allied-Bendix Aerospace, South Bend, Ind., 1976—, Deere & Co., Moline, Ill., 1981, 82, Garrett Corp., Phoenix, 1984, GM, Warren, Mich., 1984—; plenary speaker IEEE Conf. on Decision and Control, 1990. Author: Introduction to Algebraic System Theory, 1981; editor: Alternatives for Linear Multivariable Control, 1978; hon. editor: Ency. of Systems and Control, 1987; editor jour. IEEE Trans. on Automatic Control, 1979-83; contbr. 230 articles to profl. jours., books, and refereed proceedings. Grantee Army Rsch. Office, NSF, Ames Rsch. Ctr. and Lewis Rsch. Ctr. NASA, Office Naval Rsch., Air Force Office Sci. Rsch., Law Enforcement Assistance Adminstrn. Fellow IEEE; mem. Control Systems Soc. of IEEE (bd. govs. 1978-84, Disting. Mem. award 1983, Centennial medal 1984, Axelby prize 1991-93, awards com. 1993), Circuits and Systems Soc. of IEEE (co-chair internat. symposium on circuits and systems 1990, newsletter 1990-93, v.p. adminstrn. 1992-93), Soc. Indsl. and Applied Math. Republican. Roman Catholic. Avocations: photography, swimming, jogging. Office: U Notre Dame Dept Elec Engring Notre Dame IN 46556

SAINATH, RAMAIYER, electrical engineer; b. Madras, India, Oct. 21, 1949; came to U.S., 1976; s. Veeramoney and Thangam V.; m. Ramaa Ganesan, Feb. 27, 1980; 1 child, Kamalesh. BSEE, U. Madras, 1971; MSEE, U. W.Va., 1978. Registered profl. engr. Calif. Sr. engr. Instrumentation Ltd., Kota, India, 1971-76; rsch. asst. U. W.Va., Morgantown, 1976-78; control systems engr. Black & Yeath, Kansas City, Mo., 1978-79; faculty mem. DeVry Inc., Kansas City, Mo., 1979-81; control systems engr. Bechtel Group Inc., L.A., 1981-83; cons., sr. faculty mem. DeVry Inc., L.A., 1983-89; test engring. cons. Lucas Aerospace, Brea, Calif., 1989—; cons. Kamaltek Svcs., Ontario, Calif., 1988—; dir. Loom Specialities, Ontario, 1986—, Sishya Sch., Hosur, India, 1989—. Mem. IEEE. Office: Kamaltek Svcs 1806 E 5th St Ontario CA 91764

SAINES, MARVIN, hydrogeologist; b. Bronx, N.Y., Feb. 14, 1942. BS, Bklyn. Coll., 1963; MS, Miami U., Oxford, Ohio, 1966; PhD, U. Mass., 1973. Registered profl. engring. geologist, Oreg., hydrogeologist Am. Inst. Hydrology; cert. environ. mgr., Nev. Hydrogeologist Roy F. Weston Inc., West Chester, Pa., 1969-71, Harza Engring. Co., Chgo., 1971-80; sr. hydrogeologist Harza Engring. Co., Las Vegas, Nev., 1989-92, Woodard-Clyde, Chgo., 1980-81, Tetra Tech Internat., Arlington, Va., 1981-86; pres. Land & Water Co., Chgo., 1986-87; sr. hydrogeologist Donohue and Assocs., Chgo., 1987-89, Kleinfelder, Las Vegas, 1992; pvt. practice cons. Las Vegas, 1993—. Contbr. numerous articles to profl. jours. Mem. Assn. Engring. Geologists (founder Las Vegas chpt., chmn. 1990-93), Geol. Soc. Am., Nat. Groundwater Assn., Internat. Assn. Hydrology. Avocation: hiking.

SAINI, VASANT DURGADAS, computer software company executive; b. Bombay, Jan. 31, 1952; came to U.S., 1974; s. Durgadas D. and Pushpa (Sethi) S.; m. Sonia Juneja, May 20, 1983; children: Isha, Kaasha. B Tech. Electronics, Ind. Inst. Tech., 1974; MSEE, U. Rochester, 1975, PhD in Elec. Engring., 1979. Asst. prof. elec. engring. U. Rochester (N.Y.), 1980-88; pres., chief exec. officer Advanced Computer Innovations, Inc., Pittsford, N.Y., 1988—; cons. All-Pro Printers, Rochester, 1986, W. Main Ultrasound Group, Rochester, 1986; software developer Dantec Electronics, Denmark, 1987-89, Brother Industries Ltd., Japan, 1992-93. Co-author: Doppler Echocardiography, 1985, 2d edit., 1992; also articles. Mae Stone Goode Found. grantee, 1979-81. Avocations: Indo-jazz music, mathematics of music, squash. Home: 30 Burncoat Way Pittsford NY 14534-2216

ST. AMAND, PIERRE, geophysicist; b. Tacoma, Wash., Feb. 4, 1920; s. Cyrias Z. and Mable (Berg) St. A.; m. Marie Pöss, Dec. 5, 1945; children: Gene, Barbara, Denali, David. BS in Physics, U. Alaska, 1948; MS in Geophysics, Calif. Inst. Tech., 1951, PhD in Geophysics and Geology, 1953; Dr. honoris causa, U. De Los Altos, Tepatitlan, Mex., 1992. Asst. dir. Geophys. Lab., U. Alaska, also head ionospheric and seismologic investigations, 1946-49; physicist U.S. Naval Ordnance Test Sta., China Lake, Calif., 1950-54; head optics br. U.S. Ordnance Test Sta., 1955-58, head earth and planetary sci. div., 1961-88, now cons. to tech. dir., head spl. projects office; fgn. service with ICA as prof. geol. and geophys. Sch. Earth Scis., U. Chile, 1958-60; originator theory rotational displacement Pacific Ocean Basin; pres. Saint-Amand Sci. Services; adj. prof. McKay Sch. Mines, U. Nev., U. N.D.; v.p., dir. Covillea Corp.; v.p., dir. Mutual Corp.; cons. World Bank, Calif. Div. Water Resources, Am. Potash & Chem. Co., OAS; mem. U.S. Army airways communications system, Alaska and Can., 1942-46; cons. Mexican, Chilean, Argentine, Philipines, Can. govts.; mem. Calif. Gov.'s Com. Geol. Hazards; mem. com. magnetic instruments Internat. Union Geodesy and Geophys., 1954-59, Disaster Preparation Commn. for Los Angeles; charter mem. Sr. Exec. Service. Adv. bd. GeoScience News; contbr. 100 articles to scientific jours. Chmn. bd. dirs. Ridgecrest Community Hosp.; chmn. bd. dirs. Indian Wells Valley Airport Dist.; v.p. bd. dirs. Kern County Acad. Decathlon. Decorated knight Mark Twain, Mark Twain Jour.; recipient cert. of merit OSRD, 1945, cert. of merit USAAF, 1946, letter of commendation USAAF, 1948, Spl. award Philippine Air Force, 1969, Diploma de Honor Sociedad Geologica de Chile, Disting. Civilian Svc. medal USN, 1968, L.T.E. Thompson medal, 1973, Thunderbird award Weather Modification Assn., 1974, Disting. Pub. Svc. award Fed. Exec. Inst., 1976, Meritorious Svc. medal USN, 1988, Disting. Alumnus award U. Alaska, 1990; Fulbright rsch. fellow France, 1954-55. Fellow AAAS, Geol. Soc. Am., Eathquake Engr. Rsch. Inst.; mem. Am. Geophys. Union, Weather Modification Assn., Am. Seismol. Soc., Sister Cities (Ridgecrest-Tepatitlan) Assn. (pres.), Rotary (past pres., Paul Harris fellow), Footprinters Internat. (mem. grand bd., pres.), Sigma Xi. Achievements include patents in photometric instrument, weather and ordnance devices. Home: 1748 W Las Flores Ave Ridgecrest CA 93555-9672

SAINT-COME, CLAUDE MARC, science educator, consultant; b. Saint Marc, Haiti, May 23, 1949; came to U.S., 1968; s. Ernest and Marie Therese (Viau) Saint-Come; m. Cicely Patricia Leacock, Oct. 19, 1985. BS, Fordham Coll., 1975; MS, L.I. U., 1978; PhD, NYU, 1985. Adj., lectr. Medgar Evers Coll., Bklyn., 1975-78; lab. coord. NYU, N.Y.C., 1980-84; rsch. assoc. U. Mass. Med. Sch., Worcester, 1985-87; asst. prof. Bates Coll., Lewiston, Maine, 1987-88, So. Univ., Baton Rouge, 1988-89, U. S.C., Sumter, 1989-91, DeKalb Coll., Clarkston, Ga., 1991—; cons. NYU, 1990-91; advisor ednl. programs N.Y. Acad. Scis., N.Y.C., 1983-84. Co-author: Laboratory Manual, 1984, Teacher's Manual, 1983; contbr. articles to profl. jours. Asst. to coord. Community Ednl. Tng. Program, Medgar Evers Coll., Bklyn., 1977. Recipient Rsch. Devel. grant Biomeasure Inc., 1989, Conf. fellowship Multiple Sclerosis Soc., 1987, Postdoctoral fellowship Porter Devel., 1985-87, Dissertation fellowship NYU, 1983. Mem. AAAS, Human Anatomy and Physiology, Sigma Xi (assoc.). Achievements include research in the release of atrial natriuretic peptides from the right atrium of the rat is modulated by pressure in the pulmonary trunk; in administration of modified ACTH peptides during the early phases of nerve regeneration prevents degenerative changes in associated muscles. Office: DeKalb Coll 555 N Indian Creek Dr Clarkston GA 30021

ST. CYR, JOHN ALBERT, II, cardiovascular and thoracic surgeon; b. Mpls., Nov. 26, 1949; s. John Albert and Myrtle Lavira (Jensen) St. C.; m. Mary Helen Malinoski, Oct. 29, 1977. BA summa cum laude, U. Minn., 1973, BS, MS, 1975, 77, MD, 1980, PhD, 1988. Teaching asst. dept. biochemistry U. Minn., Mpls., 1973, rsch. asst. dept. surgery, 1977-78, intern surgery dept. surgery, 1980-81, resident surgery, 1981-88, cardiovascular rsch. fellow dept. surgery, 1983-86, with dept. surgery, 1991-92; rsch. assoc. fellow Cardiovascular Pathology, United Hosp., St. Paul, 1987-88; cardiovascular surg. resident U. Colo., Dept. Cardiovascular Surgery, Denver, 1988-91; med. advisor Organetics, Ltd., Mpls., 1992, med. dir., 1992—; med. advisor Aor Tech, Inc., St. Paul. 1992—; bd. dirs. Minn. Acad. Sci. Contbr. over 40 articles to profl. jours. and chpts. to books. Recipient NIH Rsch./Fellowship award, 1983-86, Grant in Aid Rsch. award Minn. Heart Assn., 1983-85, Med. Student Rsch. award Minn. Med. Found., 1980, Acad. Excellence award Merck Found., 1980. Mem. AAAS, ACS, AMA, Assn. Acad. Surgeons, Soc. Thoracic Surgeons, Am. Physiol. Soc., Am. Fedn. and Clin. Rsch., N.Y. Acad. Scis., Phi Kappa Phi. Republican. Achievements include patent in field with subsequent clin. studies.

ST. DENNIS, BRUCE JOHN, computer engineer, software researcher; b. Eugene, Oreg., Sept. 19, 1956; s. John Clarke and Aileen Louella (Steen) St. D.; m. Lynda Jean Myers, Feb. 21, 1982 (div. 1987); children: Honey Aileen, Haley Shea. BSEE, Oreg. State U., 1980; MEE, U. Utah, 1992. Design engr. Sperry Corp., Salt Lake City, 1980-87; sr. system engr. Unisys Corp., Salt Lake City, 1987-90; project engr. Evans and Sutherland, Salt Lake City, 1990—. Mem. Eta Kappa Nu. Republican. Achievements include development of 1750A processor board. Home: 1806 E Glider Ln Sandy UT 84093 Office: Evans and Sutherland Computer Corp 600 Komas Dr Salt Lake City UT 84108

ST. HILAIRE, CATHERINE LILLIAN, food company executive; b. Meyersdale, Pa., Oct. 15, 1949; d. Earl Wilson and Ann (Smith) Weimer; m. William E. Reed, Jr., Nov. 9, 1968 (div. 1977); m. James V. St. Hilaire, July 15, 1979; children: Devon, Robyn. BS in Sci. Edn., W.Va. U., 1971; PhD in Microbiology, Pa. State U., 1977. Legis. aide U.S. Ho. of Reps., Washington, 1977-78; postdoctoral fellow U. Calif., Berkeley, 1979-80; assoc. scientist Nat. Acad. Scis., Washington, 1980-83; project mgr. ENVIRON Corp., Washington, 1983-86; prin. ENVIRON Corp., Emeryville, Calif., 1988-90; dir. ILSI Risk Sci. Inst., Washington, 1986-88; dir. regulatory affairs Hershey Foods Corp., Hershey, Pa., 1990—. Editor Jour. of Am. Coll. Toxicology, 1988; contbr. articles to profl. jours. Am. Cancer Soc. rsch. fellow, 1977, Am. Soc. for Microbiology Congl. Sci. fellow, 1977. Mem. AAAS, Am. Coll. Toxicology, Soc. Risk Analysis (adv. com. on EPA Risk Assessment 1992), Soc. of Toxicology. Achievements include research in risk assessment under auspices of the Risk Science Institute. Office: Hershey Foods Corp 1025 Reese Ave Hershey PA 17033

ST. JOHN, CHARLES VIRGIL, retired pharmaceutical company executive; b. Bryan, Ohio, Dec. 18, 1922; s. Clyde W. and Elsie (Kintner) St. J.; m. Ruth Ilene Wilson, Oct. 27, 1946; children: Janet Sue St. John Amy, Debra Ann St. John Mishler. AB, Manchester Coll., 1943; MS, Purdue U., 1946. Asst. gen. mgr., dir. ops. Eli Lilly and Co., Clinton, Ind., 1971-75; gen. mgr. Eli Lilly and Co., Lafayette, Ind., 1975-77, v.p. prodn. ops. div., 1977-89; bd. dirs. Bank One of Lafayette, Lafayette Life Ins. Co., Bioanalytical Systems, Inc., West Lafayette, Ind. Pres. bd. dirs. United Way of Greater Lafayette and Tippecanoe County; bd. trustees Lafayette Symphony Found.; past chmn. lay adv. coun. St. Elizabeth Hosp.; mem. pres.'s coun. Purdue U.; trustee Manchester (Ind.) Coll. Recipient Elizabethan award, St. Elizabeth Hosp., Lafayette, 1985. Mem. Am. Chem. Soc., Purdue Rsch. Found., Greater Lafayette C. of C. (past bd. dirs.), Lafayette Country Club, Rotary. Republican. Methodist. Home: 321 Overlook Dr West Lafayette IN 47906-1249

ST. JOHN, MARGARET KAY, research coordinator; b. Clifton Forge, Va., Apr. 20, 1953; d. Clarence Robinson Jr. and Betty Jean (Miller) St.J. BS in Life Scis., Worcester Poly. Inst., 1975. Electron microscopy asst. St. Vincent Hosp., Worcester, Mass., 1974-80, med. and research technologist I, 1980-81; researcher U. Nebr. Med. Ctr., Omaha, 1981-85, research coordinator, 1985—. Contbr. articles to sci. jours. Counselor Personal Crisis Ctr., 1982-83; sec. Citizens Media Adv. Council, Omaha, 1983-85; mem. Episcopal Ch. Women, 1984—; sci. coach in biology, chemistry, physics NAACP-Afro-Am. Cultural Technol. Sci. Olympics Competition, 1985—; mem. Urban League, Omaha, 1987—. Mem. AAAS, New Eng. Soc. for Electron Microscopy, Electron Microscopy Soc. Am., Am. Assn. Profl. and Exec. Women, N.Y. Acad. Scis., Am. Mgmt. Assn. Democrat. Home: 423 N 40th St Apt 3 Omaha NE 68131-2346 Office: U Nebr Med Ctr Dept Pathology 600 S 42d St Omaha NE 68198-3135

ST. LOUIS, ROBERT VINCENT, chemist, educator; b. L.A., Dec. 15, 1932; s. Vincent Theodore and Eileen Alena (Mattson) St. L.; m. Nadine Margaret Small, July 30, 1960; 1 child, Leigh R. nee Paula Kay. BS, UCLA, 1954; PhD, U. Minn., 1962. Rsch. assoc. Johns Hopkins U., Balt., 1962-63; rsch. chemist U.S. Borax Rsch. Corp., Anaheim, Calif., 1963-66; rsch. assoc. U. So. Calif., L.A., 1966-68; prof. chemistry U. Wis., Eau Claire, 1968—; rsch. fellow ZIF-U. Bielefeld, Germany, 1985-86. Contbr. articles on infrared spectroscopy to Jour. Chem. Physics. Dow fellow U. Minn., 1957. Mem. Am. Chem. Soc. (newsletter editor 1989—), Sierra Club (sec. John Muir chpt. 1992—, legis. liaison 1986-92), Indianhead Track Club (membership chair 1986—). Achievements include patent on method for producing boric oxide. Office: U Wis Dept Chemistry Eau Claire WI 54701

ST. PIERRE, GEORGE ROLAND, JR., materials science and engineering administrator, educator; b. Cambridge, Mass., June 2, 1930; s. George Rol and Rose Ann (Levesque) St. P.; m. Roberta Ann Hansen, July 20, 1956; children: Anne Renee, Jeanne Louise, John David, Thomas George; m. Mary Elizabeth Adams, Dec. 11, 1976. BS, MIT, 1951, ScD, 1954. Rsch. metallurgist Inland Steel Co., 1954-56; mem. faculty Ohio State U., 1956—, prof. metall. engring., 1957-88, assoc. dean Grad. Sch., 1964-66, chmn. Metall. Engring., 1984-88, chmn. mining engring., 1985—; dir. Ohio Mineral Rsch. Inst., 1984—, prof., chmn. material sci. and engring., 1988—, Presdl. prof., 1988—; cons. in field, 1957—; vis. prof. U. Newcastle, New South Wales, Australia, 1975. Editor: Physical Chemistry of Process Metallurgy, vols. 7 and 8, 1961, Advances in Transport Processes in Metallurgical Systems, 1992; contbr. articles to profl. jours. Bd. dirs. Edward Orton Jr. Ceramic Found., 1990—. With USAF, 1956-57. Recipient Milton (Mass.) Sci. prize, 1947; MacQuigg award, 1971; Alumni Disting. Tchr. award, 1978; named Disting. scholar Ohio State U., 1988, Presdl. prof. Ohio State U., 1988. Fellow The Metallurgical Soc. AIME (bd. dirs. 1988—), Am. Soc. Materials Internat. (Bradley Stoughton Outstanding Tchr. award 1961, Gold medal 1987), Am. Soc. Engring. Edn., Am. Inst. Mining Metallurgical and Petroleum Engrs. (Mineral Industry educator award 1987), Am. Contract Bridge League (silver life master), Faculty Club (pres. 1990-92), Sigma Xi. Home: 1250 E Cooke Rd Columbus OH 43224-2058 Office: Ohio State U Dept Math Sci & Engring 2041 N College Rd Columbus OH 43210

ST-YVES, ANGÈLE, agricultural engineer; b. Charette, Que., Can., Apr. 30, 1940; d. Léonard St. Yves and Lucille Déziel; m. Denis Désilets, July 1, 1961; children: Luc, Marie-Christine, Valérie. BA in Teaching, Laval U., Ste.-Foy, Que., 1973, BS in Agrl. Engring., 1977, MS in Agrl. Engring., 1983. Tchr. Commn. Scolaire Shawinigan-Sud, Que., 1958-61, pvt. practice, Que., N.S. and Mich., 1961-70; corr. Farmer's Union, Que., 1970-72; engr. SNC, Morocco, 1977, Roche & Assocs., Que., 1979, Commn. de Protection des Terres Agricoles, Que., 1979-82; engr. Ministère de l'Environment du Que., 1982-85, chief divsn. mgmt. of agrl. nonpoint source pollution, 1985-89, chief divsn. agrl. sector, 1989-90, dir. agrl. sector and pesticide mgmt., 1990-91; dir. rsch. station Agriculture Can., Que., 1991—; sci. coord. for R & D on Sludge Utilization in Agriculture Consortium; mem. selection bds. for employment of profls. and mgrs.; assessor sci. papers; expert in agrl. vs. environ. issues; lectr. on vision of future, soil and water, environment, recycling residue or tech., rsch. and devel. issues in agriculture. Co-author: La Recherche du Matrimoine, 1991 (Laura Jameison award 1992); contbr. numerous tech. papers to profl. jours. and confs. V.p. fundraising campaign Laval U., Ste.-Foy, 1991-92. Recipient Comdr. of Order of Agrl. Merit award Que. Govt., 1989, Spl. Merit award Environment Que., 1990; named Personality of Week, La Presse Montreal, 1989, named Woman of Yr. in Agriculture, Salon de la Femme de Montreal, 1990. Mem. Am. Soc. Agrl. Engrs. (NABEC disting. lectr. 1990), Que. Order of Agronomists (bd. dirs. 1986-91, exec. com. 1986-91, v.p. 1987-89, pres. 1989-91, keynote speaker 1993), Que. Order Engrs. (examinor's com. 1986—, named one of Personalities of Yr. 1989), Que. Soc. of Agrl. Engrs. (pres. 1985-87), Can. Soc. Agrl. Engring. (in charge of environ. engring. 1992—), Agrl. Inst. Can. (land utilization com. 1984-87), Soil Conservation Soc. Am., Que. Soc. Water Works. Home: Agriculture Canada Research Stn, 1291 rue de la Visitation, Sainte-Foy, PQ Canada G1W 3K5 Office: Agriculture Can, Rsch Sta, 2560 Blvd Hochelaga, Sainte-Foy, PQ Canada G1V 2J3

SAITO, ISAO, chemist; b. Fukushima, Japan, Mar. 9, 1941; s. Kinjiro and Shie Saito; m. Chiyo, Jan. 16, 1970; children: Yumi, Kiyoshi, Mitsue. BS, Kyoto U., Japan, 1963, MS, 1965, PhD, 1968. Asst. prof. Kyoto U., Japan, 1968-86, assoc. prof., 1987-90, prof., 1991—, dept. chmn. synthetic chemistry, faculty engring., 1991—; internat. com. IUPAC Symposium Phochemistry, Coventry, U.K., 1989; com. U.S.-Japan Coop. Rsch. Program, Okazaki, Japan, 1989—. Author: Bioorganic Photochemistry, 1990, Chemistry of Peroxides, 1985; contbr. articles and revs. to profl. jours. and publs. Recipient Japan Chem. Soc. award, 1973, First Japan Photochemistry Assn. award, 1987, Asahi Glass Found. award, 1987. Mem. AAAS, Am. Chem. Soc., Am. Soc. for Photobiology, Japan Chem. Soc., Japan Photochemistry Assn. (exec. mem. 1980—). Avocation: golf. Home: 1-21 Shibayama Kanshuji, Yamashinaku, Kyoto 607, Japan Office: Kyoto U, Dept Synthetic Chemistry, Kyoto 606, Japan

SAITO, MITSURU, civil engineer, educator; b. Fujimi, Japan, July 7, 1952; came to the U.S., 1978; s. Otoji and Yone (Shiozawa) S.; m. Yasuko Mizuno, July 22, 1983; children: Selena Yuka, Rex Riki. Diploma, Gunma (Japan) Tech. Coll., 1973; BCE, Brigham Young U., 1981; MCE, U. Va., 1983; PhD, Purdue U., 1988. Asst. bridge engr. Nippon Koei Cons. Engrs., Tokyo, 1973-78; grad. rsch. asst. Va. Transp. Rsch. Coun. U. Va., Charlottesville, 1981-83; grad. rsch. asst. Purdue U., Lafayette, Ind., 1983-88, grad. teaching asst., 1985-88; asst. prof. Inst. for Transp. Systems CUNY, 1988-92, assoc. prof., 1993—. Referee Jour. Transp. Engrs., 1988—, Transp. Rsch. Bd. Records, 1988—; contbr. articles to profl. jours. Active Japanese Am. Citizens League, N.Y.C., 1990—; ward mission leader Ch. LDS, Caldwell, N.J., 1990—. Mem. ASCE (assoc.), Inst. Transp. Engrs. (assoc.), Transp. Rsch. Bd. (assoc.), Sigma Xi. Achievements include research in traffic engineering, intelligent vehicle highway systems, application of geographic information systems to transportation, traffic congestion mitigation, computer applications to civil engineering curriculum. Office: Inst Transp Systems CUNY Convent Ave at 138th St New York NY 10031

SAITO, SHUZO, electrical engineering educator; b. Nagoya, Aichi, Japan, Jan. 12, 1924; s. Sukesaburo and Masa Saito; m. Yoko Nakane, Mar. 26, 1953; children: Jun'ichiro, Ken'jiro. BSEE, Nagoya U., 1948, MSEE, 1953, PhD, 1962. Mem. tech. staff Elec. Community Lab. N.T.&T., Tokyo, 1953-64, chief rsch. sect., 1964-75, dir. rsch. dept., 1975-79; prof. speech sci. U. Tokyo, 1979-84; prof. elec. engring. Kogakuin U., 1984-92; prof. info. sci. Hokkaido Info. U., 1992—; mem. tech. staff Japanese Patent Agy., Tokyo, 1963; tech. specialist Japanese Ministry Transp., Tokyo, 1982. Author: Fundamental Speech Signal Processing, 1979; contbr. articles to profl. publs.; inventor PARCOR speech synthesis. Recipient promotion award Asahi Newspaper Co., 1981. Fellow Acoustical Soc. Am.; mem. IEEE (sr., chmn. acoustics, speech and signal processing Tokyo chpt. 1986-88, chmn. tech. program com. internat. conf. on acoustics, speech and signal processing 1986), Audio and Visual Rsch. Group (hon., pres. 1985-88), Inst. Elec. and Communication Engrs. Japan (adviser speech rsch. com. and pattern recognition com. 1983, paper award 1970, 79, achievement award 1973), Acoustical Soc. Japan (exec. coun. 1969-83, Sato paper award 1972). Avocations: golf, photography. Home: 1-1-3-38-704 Atsubetsu Chuo, Atsubetsu-ku, Sapporo 004, Japan Office: Hokkaido Info U, Nishinopporo 59-2, Ebetsu, Hokkaido 069, Japan

SAITO, YOSHITAKA, biochemistry educator; b. Nishinomiya, Hyogo-ken, Japan, Sept. 11, 1926; s. Harutaka and Kazue (Okuno) S.; m. Fusako Nishikawa, Feb. 11, 1956; children: Mitsuko, Ayako, Masaki. MD, Osaka (Japan) U., 1949; PhD, Osaka City U., 1959. Asst. Osaka City U. Med. Sch., 1950-55, lectr., 1955-58, asst. prof., 1958-72; prof. biochemistry Hyogo Med. U., Nishinomiya, 1972—; rsch. fellow NIH, U.S., 1956-58. Author: Advances in Enzymology, 1970. Fellow Japanese Biochem. Soc. Avocations: golf, rugby, classical music. Office: Hyogo Med U Dept Biochemist, Mukogawa-cho 1-1, Nishinomiya 663, Japan

SAITOH, TADASHI, electrical engineering educator; b. Otaru, Hokkaido, Japan, Apr. 10, 1940; s. Ryota and Miwa Saitoh; m. Tomoko Watanabe, Apr. 22, 1967; children: Atsushi, Mika, Hiroshi. BSc, Hokkaido (Japan) U., 1962, MS, 1964; D Engring., Osaka (Japan) U., 1978. Researcher Ctrl. Rsch. Lab., Hitachi Ltd., Kokubunji, Tokyo, 1964-71, sr. researcher, 1971-89; assoc. prof. Tokyo U. Agr. and Tech., Koganei, 1989-92, prof. elec. engring., 1992—. Cons. editor Progress in Photovoltaics, 1992. Recipient Tech. award OHM Co., Tokyo, 1987. Sr. mem. IEEE. Office: Tokyo U Agr and Tech, 2-24-16 Nakamachi, Koganei Tokyo 184, Japan

SAITOH, TAMOTSU, pharmacology educator; b. Tokyo, May 29, 1938; s. Jiro and Tayo Saitoh; m. Masako Hayashida, June 4, 1967. PhD, U. Tokyo, 1969. Postdoctoral fellow chemistry dept. UCLA, 1969-71; instr. U. Tokyo, 1968-76; asst. prof. Showa U., Tokyo, 1976-78; prof. Teikyo U., Tokyo, 1978—. Author: Natural Products Chemistry, Vol. 2, 1975, Pharmacognosy, 1983. Avocations: travel, photography, computer programming, painting. Home: 7-13-19 Tsukimino, Yamato 242, Japan Office: Teikyo U, Sagamiko-machi, Tsukui Kanagawa 199-01, Japan

SAIZ, BERNADETTE LOUISE, morphology technician; b. Albuquerque, Feb. 2, 1964; d. Dimas and Annette Felicitas (Armijo) S. BS, U. N.Mex., 1987, MS, 1992. Rsch. asst. U. N.Mex., Albuquerque, 1983-92, registration asst., 1988-89, morphology technician Office Med. Investigation, 1989—. Mem. Am. Soc. Microbiology, Sigma Xi. Roman Catholic. Home: 860 E Hwy 66 Tijeras NM 87059

SAIZ, ENRIQUE, chemist, educator; b. Mira, Spain, Mar. 10, 1950; s. Jose and Miguela Rosario (Garcia) S.; m. Maria Pilar Tarazona, Dec. 30, 1972; children: Ana, Pablo. M in Chemistry, U. Complutense, Madrid, 1972, PhD, 1975. Asst. prof. U. Complutense, 1972-75; postdoctoral fellow Stanford (Calif.) U., 1975-77; assoc. prof. U. de Extremadura, Badajoz, Spain, 1977-81; foreing scientist IBM Rsch. Lab., San Jose, Calif., 1980; assoc. prof. U. de Alcalá de Henares, Spain, 1981-86, head dept. phys. chemistry, 1981-91, prof., 1986—. Author: Dipole Moments and Birefringence of Polymers, 1992; contbr. 80 articles to profl. jours. Mem. Real Soc. Española de Química, Grupo Especializado de Polímeros, Am. Chem. Soc., Soc. Chilena de Química. Roman Catholic. Office: U de Alcala de Henares, Dept Phys Chemistry, 28871 Alcala de Henares Spain

SAJO, ERNO, nuclear engineer, educator, physicist; b. Budapest, Hungary; s. Erno L. F. and Eva (Kemenes) S.; m. Agatha Nicolette Schiller. BS, MS, Tech. U. Budapest, 1983; PhD, U. Lowell, Mass., 1989. Design engr. Eroterv, Budapest, 1983-85; rsch. asst. U. Lowell, 1985-89; asst. prof. La. State U., Baton Rouge, 1990—. Co-author: English-Hungarian Technical Dictionary, 1991; contbr. articles to profl. jours. Mem. Am. Nuclear Soc. (faculty advisor student chpt. 1990—), Sigma Xi, Alpha Nu Sigma. Achievements include discovery of new theorems concerning higher transcendental functions, research in atmospheric dispersion involving transient releases, radiation damage in materials, particle transport and diffusion, mathematical physics. Office: La State U Nuclear Sci Ctr Baton Rouge LA 70803

SAKAGUCHI, GENJI, food microbiologist, educator; b. Kobe, Hyogo, Japan, Sept. 17, 1927; s. Kazumasa and Yuko (Fujita) S.; m. Sumiko Igarashi, Nov. 25, 1951; children: Akiko, Ei, Fumiko, Hide, Kei. BSc, Hokkaido U., 1950, PhD, 1960; postgrad., U. Chgo., 1956. Tech. ofcl. Nat. Inst. Health, Tokyo, 1950-70, chief investigator, 1965-70; prof. U. Osaka Prefecture, Sakai, Japan, 1970-91; prof. emeritus U. Osaka, Sakai, Japan,

1991—; sci. adviser Japan Food Rsch. Labs., Suita, Osaka, 1991—; cons. WHO, Geneva, 1967-86, Chiba Serum Inst., 1970—; advisor Ministry of Health and Welfare, Tokyo, 1974—. Editor Internat. Jour. Food Microbiology, 1990—, assoc. editor-in-chief, 1993—; author books and articles in field. Grantee USPHS, 1962-67, WHO, 1967-70, Fulbright Found., 1955-56. Mem. Japanese Soc. Bacteriology (bd. dirs. 1991—, Asakawa-sho award 1988), Soc. Antibacterial and Antifungal Agts. (bd. dirs. 1991—), Japanese Soc. Food Microbiology (pres. 1990-93). Avocations: photography, classical music, driving, swimming, cooking. Home: 404 Oaza Koda 63-1, Tondabayashi Osaka 584, Japan Office: Japan Food Rsch Labs, 3-1 Toyotsu-cho, Suita-shi Osaka 564, Japan

SAKAI, KATSUO, electrophotographic engineer; b. Matsumoto, Nagano, Japan, Apr. 9, 1942; s. Mototeru and Fumi (Iida) S.; m. Toshiko Hagiwara, Feb. 7, 1976; children: Asako, Akiharu, Hirohiko. B in Applied Physics, Waseda U., Tokyo, 1967. Engr. Ricoh Co. Ltd., Tokyo, 1967—; lectr. in field. Contbr. articles to profl. jours.; inventor, patentee two-color electrophotography, 1980. Mem. Soc. Electrophotography of Japan, Soc. for Imaging Sci. and Tech. Club: Minami Fuji Country. Avocations: golf, reading. Home: 25-64 Moegino Midori-ku, Yokohama 227, Japan Office: Ricoh Co Ltd, 1-3-6 Nakamagome, Ohta-ku, Tokyo 143, Japan

SAKAI, KUNIKAZU, chemistry researcher; b. Taira, Fukushima, Japan, Mar. 9, 1941; s. Takiya and Naka (Abe) S.; m. Asako Aimi, July 1, 1972; children: Kazuo, Hiroyuki, Michie. BS, Tohoku U., Sendai, Japan, 1963, MS, 1965, DSc, 1968. Researcher Sagami Chem. Rsch. Ctr., Sagamihara, Japan, 1969-79, rsch. fellow, 1979-84, sr. rsch. fellow, 1984—. Mem. Internat. Union Pure and Applied Chemistry, Am. Chem. Soc., Japanese Cancer Assn., Chem. Soc. Japan, Soc. Synthetic Organic Chemistry Japan. Office: Sagami Chem Rsch Ctr, Nishi Ohnuma 4-4-1, Sagamihara 229, Japan

SAKAI, SHOGO, theoretical chemist; b. Itami, Hyogo, Japan, Oct. 15, 1948; s. Takeshi and Ikue S.; m. Misato Nakane, MAr. 30, 1984; children: Chika, Tamaki. B, Kansai U., Suita, Japan, 1972, M, 1974, PhD, 1978. Postdoctoral fellow Inst. for Molecular Sci., Okazaki, Japan, 1978-80, 82-84; rsch. assoc. U. Pitts., 1980-82, N.D. State U., Fargo, 1984-85; lectr. Osaka Gakuin, Suita, 1985-90; assoc. prof. Osaka Sangyo U., Daito, Japan, 1989—; lectr. Kansai U., Suita, 1987—; com. mem. Osaka (Japan) Sci. and Tech. Ctr., 1990-92. Grantee Ministry of Edn. Japan, 1987-90, 89-92, 92—. Mem. Chem. Soc. Japan, Am. Chem. Soc., Am. Phys. Soc., N.Y. Acad. Scis. Office: Osaka Sangyo U, 3-1 Nakakaito, Osaka Daito 574, Japan

SAKAI, TOSHIHIKO, engineer; b. Nagoya, Aichi, Japan, July 17, 1943; s. Saichi and Yoshie (Taguchi) S.; m. Junko Fukuoka, Oct. 7, 1970; children: Eiji, Kenji; m. Natsuko Kamei, Nov. 25, 1989; children: Rei, Rui. B.Sc., Nagoya Inst. Tech., Japan, 1967; M.Sc., Ga. Inst. Tech., 1969; PhD, U.M.I.S.T., U.K., 1974. Assoc. engr. Toyota Cen. R. & D. Labs., Inc., Aichi, Japan, 1967-75; researcher Toyota Cen. R. & D. Labs., Inc., 1975-84, sr. researcher, 1984—; staff for CEO, 1991—; mgr. secretariat Toyota Cen. R&D Labs., Inc., Aichi, 1982—. Contbr. articles to profl. jours.; inventor and patentee in field. Submit referee Textile Machinery Soc. Japan. Fellow Textile Inst. (UK); mem. Soc. Fiber Sci. and Tech. Japan, Soc. Automotive Engr., Nagoya Jr. Chamber (exec. v.p. 1982). Avocations: travel, driving, golf. Home: City Corp B1003, Ueda 3-1501, Tempaku Nagoya 468, Japan Office: Toyota Cen R&D Labs Inc, Yokomichi 41 1 Nagakute, Aichi 48011, Japan

SAKAMOTO, ICHITARO, oceanologist, consultant; b. Kobe, Hyogo, Japan, Feb. 22, 1926; s. Ikutaro and Kisa (Kawai) S.; m. Teruko Konaka, Aug. 28, 1950; children: Akane (Matsubara), Akira. B.Agr., Kyushu U., Fukuoka, 1948; Dr.Agr., Tohoku U., Sendai, 1962. Asst. Faculty of Agr., Tohoku U., Sendai/Miyagi, 1948-52; lectr. Faculty of Fisheries, Miye Prefectural U., Tsu/Miye, 1952-65, asst. prof., 1965-66, assoc. prof., 1966-68, prof., 1968-73; prof. Faculty of Fisheries, Miye U., Tsu/Miye, 1973-89, Faculty of Bioresources, 1987-89; tech. advisor Shin-Nippon Metocean Cons. Co. Ltd., Osaka, Japan, 1989—. Contbr. articles to profl. jours. spl. conferee Hydraulic Model Conf. of Ise Bay, 1972—. Named Person of Disting. Svcs. for Water Quality Preservation, Environ. Agy. Japan, 1986. Mem. Oceanographical Soc. Japan, La Sociētē Franco-Japonaise d'Océanographie, Japanese Soc. Sci. Fisheries, Japanese Soc. Fisheries Oceanography, The Plankton Soc. Japan. Hokke-shu Buddhism. Achievements include research on the Action Mechanism of Osmotic Balance upon the movement of the pelagic fish school, Load Control from the view point of Pisciculture, and the use of Respiration in the Sandy Beach or on the Tidal Flat. Home: 1700-19 Takajayakomori-cho, Tsu Japan 514 Office: Osaka Br Shin Nippon Metocean Cons Co Ltd, 2-23 Edobori 3 Nishi-Ku, Osaka 550, Japan

SAKAMOTO, MUNENORI, engineer educator, researcher, chemist; b. Kita-Kyuushu, Japan, 1936; s. Masanori and Ai (Sasaki) S.; m. Kyoko Sawano, Dec. 8, 1967; children: Shigenori, Hironori. B. Engring., Tokyo Inst. Tech., 1958, MSc, 1960, DSc, 1963. Asst. Tokyo Inst. Tech., 1963-70; rsch. assoc. U. Ariz., Tucson, 1964-66; assoc. prof. Tokyo Inst. Tech., 1970-84, prof., 1984—; mem. sci. and tech. com. Tokyo Textile Rsch. Inst., 1991-92. Editor Proceedings 7th Internat. Wool Rsch. Conf., 1985, Sen'i Gakkaishi, 1989-91; translator: Introdution to Reaction Injection Molding, 1983. Mem. Soc. Fiber Sci. Tech. Japan (chmn. rsch. com. on textile finishing, recipient award 1980), Chem. Soc. Japan, Soc. Polymer Sci. Japan, Am. Chem. Soc., Am. Assn. Textile Chemists Colorists. Buddhist. Home: Tsukushino 2-11-9, Machida-shi Tokyo 194, Japan Office: Tokyo Inst Tech, O-okayama 2-12-1, Meguro-ku Tokyo 152, Japan

SAKANASHI, MATAO, pharmacology educator; b. Kumamoto, Japan, Oct. 16, 1943; s. Hidefumi and Mine Sakanashi; m. Yukiko Sakanashi, Nov. 30, 1970; children: Mayuko, Makiko. B in Medicine and MD, Kumamoto U., 1968, postgrad., 1972. Asst. prof. Sch. of Medicine Kumamoto U., Kumamoto-City, 1969-70, instr. Sch. of Medicine, 1970-82, prof. faculty of medicine Sch. of Medicine U. Ryukyus, Nishihara-cho, Okinawa, Japan, 1982—. Author: Peripheral Dopaminergic Receptors, 1979, Vascular Neuroeffector Mechanisms, 1983, Progress in Hypertension, 1988, Cardiovascular Disease in Diabetes, 1992. Ministry of Edn. grantee, 1979-80, 89-90; recipient Kanae Fund awards, 1983. Mem. Japanese Circulation Soc., Japanese Pharmacological Soc., Japanese Coll. of Angiology, Japanese Soc. Circulation Rsch., Japanese Soc. Clin. Pharmacology and Therapeutics, Internat. Soc. Toxinology, Japanese Soc. Pharmacoanesthesiology. Home: 2-96-1-2-302 Shuri-ishimine-cho, Naha Okinawa 903, Japan Office: U Ryukyus Sch Medicine, 207 Uehara, Nishihara-cho, Okinawa 903-01, Japan

SAKATA, KIMIO, aerospace engineer, researcher; b. Tokyo, Japan, Jan. 6, 1947; s. Kazuyoshi and Ayako S.; m. Yoko Sakata, Oct. 26, 1972; children: Ritsuko, Kohtaro, Yoshiko. BSME, Sophia U., Tokyo, 1969; ME, Sophia U., 1972. Researcher Aeroengine div. Nat. Aerospace Lab., Tokyo, 1972-80, sr. researcher, 1980-89, head engine aerodynamics, 1989—; visitor of Mech. Engring. Dept., Stanford (Calif.) U., 1980-81; tech. officer Agy. of Indsl. Sci. and Tech. of Ministry of Internat. Trade and Industry, Tokyo, 1982-83; dep. dir. Space Planning div. Sci. and Tech. Agy., Tokyo, 1984-86. Contbr. articles to profl. jours. Mem. AIAA, Japan Soc. Mech. Engring., Gas Turbine Soc. Japan, Japan Soc. for Aero. and Space Scis. Achievements include patent for turbine cooling. Home: 5 30 2 Jindaijikita, Chofu Tokyo 182, Japan Office: Nat Aerospace Lab, 7 44 1 Jindaijihigashi, Chofu 182 Tokyo Japan

SAKELLAROPOULOS, GEORGE PANAYOTIS, chemical engineering educator; b. Kalamata, Greece, Apr. 10, 1944; s. Panayotis and Eudokia (Renieris) S.; m. Sofia A. Korili, Jan. 7, 1991. Diploma in chem. engring., Nat. Tech. U. Athens (Greece), 1966; PhD in Chem. Engring., U. Wis., 1974. Cert. chem. engring. Instr. Greek Naval Acad., Piraeus, Greece, 1966-69; process engr. Gen. Cement Co., Volos, Greece, 1969-70; vis. prof., lectr. chem. engring. U. Wis., Madison, 1975; asst. prof. chem. engring. Rensselaer Poly. Inst., Troy, N.Y., 1976-79; prof. chem. engring. Aristotle U. Thessaloniki, Greece, 1979—. Mem. Am. Inst. Chem. Engrs., Am. Chem. Soc., Electrochem. Soc., Tech. Chamber of Greece, Sigma Xi, Phi Kappa Phi. Orthodox. Home: 66 Kallidou St, 55131 Thessaloniki Greece Office: Aristotle U Thessaloniki, PO Box 1520, 54006 Thessaloniki Greece

SAKMANN, BERT, physician, cell physiologist. Postdoctoral fellow with Bernard Katz London; asst. prof. Max Planck Inst., Gottingen, Fed. Republic of Germany, from 1974; instr. Marine Biol. Lab. summer courses, Woods Hole, Mass., during 1980's; prof. Max Planck Inst. Für medizinische Forschung, Heidelberg, Fed. Republic of Germany. Co-recipient Nobel Prize in physiology or medicine, 1991. Office: Max Planck Inst, Pf 103820, 69028 Heidelberg Germany

SAKURADA, YUTAKA, chemist; b. Kyoto, Japan, Jan. 1, 1933; s. Ichiro and Chiyoko (Okumura) S.; m. Keiko Sugimoto, May 10, 1960; children: Kazuhiro, Akihiro. BS, Kyoto U., 1956, MS, 1958, PhD, 1966. Rsch. fellow Cen. Rsch. Lab. Kuraray Co. Ltd., Kurashiki, Japan, 1958-62, 64-66; internat. fellow Stanford Rsch. Inst., Menlo Pk., Calif., 1962-64; tech. rep. N.Y. Office Kuraray Co. Ltd., N.Y.C., 1968-71; mgr. Med. Bus. Devel. Div. Kuraray, Osaka, Japan, 1974-77; gen. mgr. Med. Products Div. Kuraray, Osaka, 1977-88, gen. mgr. Corp. Rand D Div., 1988-89; mng. dir. Kuraray Plastics Co. Ltd., Osaka, 1988—; bd. dirs. Kuraray Co. Ltd., Osaka, 1989-91, Kuraray Plastics Co., Ltd., Osaka, 1989-91, Haemonetics Corp., USA, 1991—; vice chmn. Japanese Soc. for Biomaterials, Tokyo, 1987—; pres. Haemonetics, Japan, 1991—. Recipient Technology award The Soc. Polymers, 1984, Japanese Chem. Soc., 1985. Achievements include development of ethylene vinyl alcohol copolymer hollow fiber for hemodialyzer; development of dental adhesives. Home: 14-615 Kawaracho, Yoshida Sakyo-ku, Kyoto 606, Japan Office: Haemonetics Japan Co Ltd, Shin-Kojimachi Bldg 1-2F 4-3-3, Chiyoda-ku Tokyo 102, Japan

SAKURAI, AKIRA, nuclear engineer. Prof. Inst. Atomic Energy Kyoto (Japan) U. Recipient Melville medal ASME, 1991. Office: Inst of Atomic Energy, Kyoto Univ, Gokasho Uji Kyoto 611, Japan*

SAKURAI, KIYOSHI, economics educator; b. Tokyo, Aug. 1, 1934; m. Noriko Ogawa, Oct. 22, 1966; children: Misako, Minako. D of Commerce, Meiji U., Tokyo, 1971. Prof. econs. Wako U., Tokyo, 1974—, chmn. dept. econs., 1980-82, dean dept. econs., 1990-93; lectr. Ministry of Fin., Tokyo, 1982-86; dir. Ednl. Institution Wako Gakuen, Tokyo, 1990-93; pres. Japanese Calligraphy Soc., Tokyo, 1970—; curator of the Library of Wako U., Tokyo, 1993—. Author: Historical Studies of the Staple System of England, 1974, Friction Between Japan and the U.S. in the Automobile Industry (Prewar Period), 1987; co-author: Adam Smith and His Age, 1977, Malthus, Ricardo and Their Age, 1981, Mill, Marx and Their Age, 1986, A. Marshall and His Age, 1991. Mem. Soc. Socio-Econ. History, Soc. Finance Theory, Econ. History Assn. Home: 3177-9 Honmachida, Machida-shi, Tokyo 194, Japan Office: 2160 Kanai, Machida-shi, Tokyo 195, Japan

SAKURAI, TAKEO, surgery educator; b. Wakayama, Japan, June 26, 1938; s. Tadao and Tsuyako Sakurai; m. Makiko Sakurai, Feb. 11, 1957; children: Teruhisa, Keiko, Mie, Ikuko. Grad., Wakayama (Japan) Med. Coll., 1963. Asst. dept. surgery Wakayama (Japan) Med. Coll., 1964-73; lectr. dept. thoracic surgery Wakayama Med. Coll., 1973-77, assoc. prof. dept. thoracic surgery, 1977-86, prof. dept. thoracic surgery, 1986—; prof. dept. surgery Kikoku -Bunin Hosp., Itogun Katserangi cho, 1986—, dir., 1992—. Home: 279-2 Mikazura, Wakayama 641, Japan Office: Kihoku-Bunin Hosp, 219 Myoji Katsuragi-cho, Itogun 649-71, Japan

SAKUTA, MANABU, neurologist, educator; b. Ichikawa, Japan, Oct. 31, 1947; s. Jun and Shizuko (Tsuji) S.; m. Yuko Fukushi, June 17, 1973; children: Akiko, Junko, Ken-Ichi. MD, U. Tokyo, 1973, PhD, 1978; MS in Neurology, U. Minn., 1981. Med. diplomate. Diplomat Japanese Bd. Neurology, Japanese Bd. Internal Medicine. Asst. Dept. Neurology U. Tokyo, Japan, 1980; rsch. fellow Dept. Neurology U. Minn., Mpls., 1980-81, asst. prof., 1981-82; head Dept. Neurology Japanese Red Cross Med. Ctr., Tokyo, 1982—; prof. Japanese Red Cross Women's Coll. Sch. Nursing, Tokyo, 1983-85, instr. 1986-88; lectr. dept. neurology, U. Tokyo, 1984—, dept. medicine U. Kobe, 1990—; cons. Nakayama Hosp., Ichikawa, Japan, 1980—. Contbr. articles to profl. jours. Fellow Royal Soc. Medicine (London); mem. Japanese Soc. Internal Medicine (pres. Kanto br. 1992), Japanese Soc. Neurology (mem. coun., 1985—, mem. coun. Kanto Br., 1984—, pres. Kanto Br. 1984, mem. editorial bd. 1988—), Japanese Soc. Diabetology, Japanese Soc. Electroencephalography and Electromyography, Japanese Soc. Autonomic Nervous System (mem. coun.), U. Minn. Alumni Club, Tetsumon Club. Liberal. Buddhist. Office: Japanese Red Cross Med Ctr, 4-1-22 Hiroo Shibuya-ku, Tokyo 150, Japan

SALA, MARTIN ANDREW, biophysicist; b. Buffalo, N.Y., Sept. 6, 1957; s. Paul and Adrian (Williams) Zahm; m. Erie Anne Wagner-Sala, Nov. 23, 1986; 1 child, Rebecca. BA in Biophysics, SUNY, Buffalo, 1981. Dir. clin. engring. Buffalo Columbus Hosp., 1982-85; lab. inst. designer Roswell Park Cancer Inst., Buffalo, 1985-89; v.p. for R&D MBS Foundry, Brook's Grove, N.Y., 1989—; cons. Lotus Link Found., Buffalo, 1990—, West N.Y. Clin. Engring. Assn., Buffalo, 1989—. Author: Theory & Design of Core Memory, 1979; editor various periodicals, 1970—. With USN, 1976-81. Grantee NIH, 1990. Mem. Am. Inst. Physics, Instrument Soc. Am., AAAS, Soc. for Advancement Med. Instrument Design. Mem. Anglican Ch. Achievements include patents pending for new surgical measuring tool, facsimile design, canine surgical tool; invention of various novel scientific instruments. Office: MBS Foundry Brooks' Grove 8161 Rte 408 Mount Morris NY 14510

SALABOUNIS, MANUEL, computer information scientist, mathematician, scientist; b. Salonica, Greece, Apr. 15, 1935; came to U.S., 1954; s. Anastasios and Marietta (Mytonidis) Tsalabounis; children from previous marriage: Stacy, Mary E., John; m. Baerbel Thekla Steinbach-Rushford, July 2, 1988. Cert., Anatolia Coll., Salonica, Greece, 1954; student, Morris Harvey Coll., 1954-56; BS in Engring. Sci., Cleve. State U., 1960; MS in Math., Akron U., 1964. Master Univ. Sch., Shaker Heights, Ohio, 1960-62; mathematician Babcock and Wilcox, Alliance, Ohio, 1962-66; dir. computer ctr. John Carroll U., University Heights, Ohio, 1966-68, pres. Electronic Service Assocs. Corp., Euclid, Ohio, 1968-73; v.p. North Am. Co., Chgo., 1974-79; sr. project leader Hibernia Bank, New Orleans, 1979-83; project mgr. Compuware Corp., Farmington Hills, Mich., 1984-90, mgr. spl. projects, 1990—. Accomplishments include work in internat. banking applications, 3 dim. thermostress analysis, generic tool definitions and design, law office info. systems, software delivery system, and other. Vol. Sts. Constantine and Helen Ch., Cleve., 1960-74, St. Nicholas, Detroit, 1986—. Avocations: classical music, opera, golf, fishing, cooking. Home: 28691 Bristol Ct Farmington MI 48334-2914 Office: Compuware Corp 31440 Northwestern Hwy Farmington MI 48334-2564

SALAKHITDINOV, MAKHMUD, mathematics educator; b. Namangan, Uzbekistan, Nov. 23, 1933; s. Salokhitdin and Zukhra Shamsutdin; m. Mukharram Rasulova, Aug. 3, 1955; children: Rustam, Bakhrom, Aghzam, Gulbakhor. Student, Ctrl. Asian State U., Toshkent, Uzbekistan, 1950-55, grad. student, 1955-58. Assoc. instr. Toshkent State U., 1958-59; jr. sci. fellow Inst. Math., Uzbekistan Acad. of Scis., Toshkent, 1959-60, sr. sci. fellow, 1960-64, dept. chief, 1964-66, dep. dir., 1966-67, dir., 1967-85, chief dept. differencial equations, 1974—; v.p. Uzbekistan Acad. Scis., Toshkent, 1984-85, minister higher edn., 1985-88, pres., 1988—. Contbr. over 140 articles on differencial equations to sci. jours. People's dep. Republic of Uzbekistan, Toshkent, 1990—; active People's Dem. Party of Uzbekistan, Toshkent, 1990—. Home: Uzbek Academy of Sciences, G Lopatina Str Apt 70 Fl 64, Toshkent 70, Uzbekistan 700031 Office: Acad of Scis Rep Uzbekistan, Gogol str 70, Toshkent Uzbekistan 70000

SALAM, ABDUS, physicist, educator; b. Jhang, Pakistan, Jan. 29, 1926. Student, Govt. Coll., Lahore, Pakistan, 1938-46, St. John's Coll., Cambridge (Eng.) U., 1946-49; BA, Cambridge U., 1949, PhD, 1952; 20 DSc hon. degrees including, Panjab U., 1957; 45 DSc hon. degrees including, U. Edinburgh, 1971, Hindu U., U. Chittagong, U. Bristol, U. Maiduguri, 1981, U. Khartoum, U. Complutense de Madrid, 1983; U. Cambridge, U. Glasgow, U. Exeter, U. Gent, 1985-88, U Ghana, U. Dakar, U. Tucumen, U. Lagos, U. S.C., U. West Indies, U. St. Petersburg, 1990-93. Prof. Govt. Coll., Lahore, 1951-54; prof., head math. dept. U. Panjab, Lahore, 1951-54; fellow St. John's Coll. Cambridge, 1951-56; prof. theoretical physics Imperial Coll., London, 1957—; founder, dir. Internat. Ctr. for Theoretical Physics, Trieste, Italy, 1964—; mem. AEC Pakistan, 1958-74, Pakistan Nat. Sci. Coun., 1963-75, South Commn., 1987; hon. sci. adviser to Pres. of Pakistan, 1961-74; mem. sci. and tech. adv. com. UN, 1964-75; gov. Internat. Atomic Energy Agy., Vienna, 1962-63; developer new physics ctrs. and schs. Pakistan, Peru, Jordan, Sudan, Colombia; hon. life fellow St. John's Coll., Cambridge, 1971—; hon. prof. Beijing Univ., 1987. Author: Symmetry Concepts in Modern Physics, 1965; Aspects of Quantum Mechanics, 1972; (with E. Sezgin) Supergravity in Diverse Dimensions, Vols. I and II, 1988. Mem. sci. coun. Stockholm Internat. Peace Rsch. Inst., 1970—; mem. coun. Univ. for Peace, Costa Rica, 1981-86. Recipient Hopkins prize Cambridge U., 1957, Adams prize, 1958; Maxwell medal and prize London Phys. Soc., 1961, Atoms for Peace prize, 1968, Oppenheimer prize and medal, 1971, Guthrie medal and prize Inst. Physics, London, 1976, Sir D. Sarvadhikary Gold medal Calcutta U., 1977, Matteucci medal Acad. Nat. dei XL, Rome, 1978, John Torrence Tate medal Am. Inst. Physics, 1978, Nobel Prize in Physics, 1979, Einstein medal UNESCO, 1979, Shri R.D. Birla award Indian Physics Assn., 1979, Josef Stefan Inst. medal, Ljubljiana, Yugoslavia, 1980, Peace medal Charles U., Prague, 1981, Gold medal for outstanding contbns. to physics Czechoslovak Acad. Scis., 1981, Lomonosov gold medal USSR Acad. Scis., 1983, Dayemi Internat. Peace award, Bangladesh, 1986, Premio Umberto Biancamano, Italy, 1986, 1st Edinburgh medal and prize, 1988, Internat. Devel. of Peoples prize, Genoa, Italy, 1988, Catalunya Internat. prize, 1990, Medal of 260th Anniversary of Havana, Cuba, 1991, Gold medal Slovak Acad. Scis., 1992, Mazhar-Ali Applied Sci. medal Pakistan League of Am., 1992; named Hon. Knight Comdr. Order of Brit. Empire, 1989; also numerous decorations. Fellow Royal Soc. London (Hughes medal 1964, medal 1978, Copley medal 1990), Royal Swedish Acad. Scis. (bd. dirs. Beijir Inst. 1986—), Pakistan Acad. Scis., Bangladesh Acad. Scis.; mem. U.S. Nat. Acad. Scis.(fgn. assoc.), Am. Acad. Arts and Scis. (fgn.), Acad. dei Lincei (Rome, fgn. assoc.), European Acad. Sci., Arts and Humanities, Acad. Scis. USSR, Internat. Union Pure and Applied Physics (v.p.), Third World Acad. Scis. (founding, pres. 1983—), Third World Network Sci. Orgns. (founding pres. 1988—), Club of Rome, also hon. mem. or hon. fellow numerous other worldwide sci. orgns. Office: Dir ICTP, PO Box 586, 34100 Trieste Italy also: Internat Ctr Theoretical, Physics PO Box 586, I-34100 Trieste Italy

SALAMON, MIKLOS DEZSO GYORGY, mining educator; b. Balkany, Hungary, May 20, 1933; came to U.S., 1986; s. Miklos and Sarolta (Obetko) S.; m. Agota Maria Meszaros, July 11, 1953; children: Miklos, Gabor. Diploma in Engring., Polytech U., Sopron, Hungary, 1956; PhD, U. Durham, Newcastle, England, 1962; doctorem honoris causa, U. Miskolc, Hungary, 1990. Research asst. dept. mining engring. U. Durham, 1959-63; dir. research Coal Mining Research Controlling Council, Johannesburg, South Africa, 1963-66; dir. collieries research lab Chamber of Mines of South Africa, Johannesburg, 1966-74, dir. gen. research org., 1974-86; disting. prof., dir. Ctr. for Advanced Mining Systems, Colo. Sch. Mines, Golden, 1986—, head dept. mining engring., 1986-90, dir. Colo. Mining and Mineral Resources Rsch. Inst., 1990—; 22d Sir Julius Wernher Meml. lectr., 1988; hon. prof. U. Witwatersrand, Johannesburg, 1979-86; vis. prof. U. Minn., Mpls., 1981, U. Tex., Austin, 1982, U. NSW, Sydney, Australia, 1990, 91—. Co-author: Rock Mechanics Applied to the Study of Rockbursts, 1966, Rock Mechanics in Coal Mining, 1976; contbr. articles to profl. jours. Mem. Pres.'s Sci. Adv. Council, Cape Town, South Africa, 1984-86, Nat. Sci. Priorities Com., Pretoria, South Africa, 1984-86. Recipient Nat. award Assn. Scis. and Tech. Socs., South Africa, 1971. Fellow South African Inst. Mining and Metallurgy (hon. life, v.p. 1974-76, pres. 1976-77, gold medal 1964, 85, Stokes award 1986, silver medal 1991), Inst. Mining and Metallurgy (London); mem. AIME, Internat. Soc. Rock Mechanics. Roman Catholic. Office: Colo Sch of Mines Dept of Mining Engring Golden CO 80401

SALAZAR-CARRILLO, JORGE, economics educator; b. Havana, Cuba, Jan. 17, 1938; came to U.S., 1960; s. Jose Salazar and Ana Maria Carrillo; m. Maria Eugenia Winthrop, Aug. 30, 1959; children: Jorge, Manning, Mario, Maria Eugenia. BBA, U. Miami, 1958; MA in Econs., U. Calif., Berkeley, 1964, cert. in econ. planning, 1964, PhD in Econs., 1967. Sr. fellow, non-resident staff mem. Brookings Instn., Washington, 1965—; dir., mission chief UN, Rio de Janeiro, Brazil, 1974-80; prof. econs. Fla. Internat. U., Miami, 1980—, chmn. dept. econs., 1980-89; dir. Ctr. Econ. Rsch. & Edn.; mem. coun. econ. advisors State of Fla.; advisor U.S. Info. Agy., advisor, contbg. editor Library of Congress, Washington, 1972—; chmn. program com. Hispanic Profs. of Econs. and Bus.; cons. econs. Agy. for Internat. Devel., Washington, 1979—; council mem. Internat. Assn. Housing, Vienna, 1981—; exec. bd. Cuban Am. Nat. Council, Miami, 1982—; bd. dirs., pres. Fla. chpt. Insts. of Econ. and Social Rsch. of Caribbean Basin, Dominican Republic and Costa Rica, 1983—, U.S.-Chile Council, Miami, 1984—, Fla.-Brazil Inst. Co-author: Trade, Debt and Growth in Latin America, 1984; Prices for Estimation in Cuba, 1985; The Foreign Debt and Latin America, 1983; External Debt and Strategy of Development in Latin America, 1985; The Brazilian Economy in the Eighties, 1987, Foreign Investment, Debt and Growth in Latin America, 1988; World Comparisons of Incomes, Prices, and Product, 1988, Comparisons of Prices and Real Products in Latin America, 1990, The Latin American Debt, 1992; author: Wage Structure in Latin America, 1982. Fellow Brit. Council, London, 1960, Georgetown U., Washington, 1961-62, OAS, Washington, 1962-64, Brookings Instn., Washington, 1964-65. Mem. Am. Econ. Assn., Internat. Assn. Research in Income and Wealth, Econometric Soc. Latin Am., N.Am. Econs. and Fin. Assn., Nat. Assn. Cuban Am. Educators (treas. exec. com.), Internat. Assn. Energy Economists (pres. Fla. chpt.), Nat. Assn. Forensic Economists, Assn. for Study Cuban Economy (exec. com.), Latin Am. Studies Assn., Knights of Malta. Roman Catholic. Home: 1105 Almeria Ave Miami FL 33134-5503 Office: Fla Internat U Tamiami Campus DM 347 Miami FL 33199

SALDICH, ROBERT JOSEPH, electronics company executive; b. N.Y.C., June 7, 1933; s. Alexander and Bertha (Kasakove) S.; m. Anne Rawley, July 21, 1963 (div. Nov. 1979); 1 child, Alan; m. Virginia Vaughan, Sept. 4, 1983; stepchildren: Tad Thomas, Stan Thomas, Melinda Thomas, Margaret Thomas. BSChemE, Rice U., 1956; MBA, Harvard U., 1961. Mfg. mgr. Procter & Gamble Mfg. Co., Dallas, Kansas City, Kans., 1956-59; rsch. asst. Harvard Bus. Sch., Boston, 1961-62; asst. to pres. Kaiser Aluminum & Chem. Corp., Oakland, Calif., 1962-64; mgr. fin. and pers., then gen. mgr. various divs. Raychem Corp., Menlo Park, Calif., 1964-83, with office of pres., 1983-87, sr. v.p. telecommunications and tech., 1988-90, pres., chief exec. officer, 1990—; pres. Raynet Corp. subs. Raychem Corp., 1987-88; chair mfg. com. of adv. bd. Leavy Sch. Bus. and Adminstrn., Santa Clara U. Chair mfg. com. adv. bd. Leavy Sch. Bus. and Administrn., Santa Clara U. Mem. Calif. Roundtable (dir. Bay Area Coun.), San Francisco Com. on Fgn. Rels. Jewish. Avocations: sailing, skiing. Office: Raychem Corp Mailstop 120/7815 300 Constitution Dr Menlo Park CA 94025*

SALE, PETER FRANCIS, biology educator, marine ecologist; b. Jerusalem, Jan. 12, 1941; came to U.S., 1988; s. Peter A. and Lily M. (Colby) S.; m. Donna M. Hindelang, Apr. 25, 1970; 1 child, Darian R.P. BS in Zoology, U. Toronto, Ont., Can., 1963, MA in Zoology, 1964; PhD, U. Hawaii, 1968. From lectr. to assoc. prof. U. Sydney, Australia, 1968-87; prof. zoology U. N.H., Durham, 1988-93, chmn. dept., 1988-92, dir. Ctr. Marine Biology, 1990-93; prof. biology, dept. head U. Windsor, Ont., Can., 1994—; mem. Australian Marine Sci. and Tech. Adv. Coun., Canberra, 1984-87; also numerous other editorial, rsch. adv. and other nat. coms., Australia, U.S.; Glaser vis. prof. Fla. Internat. U., 1993. Editor: The Ecology of Fishes on Coral Reefs, 1991; also over 85 articles. Mem. AAAS, Australian Coral Reef Soc. (life, pres. 1984-86), internat. Soc. for Reef Studies (pres. 1989-91), Am. Soc. Naturalists, Ecol. Soc. Am. Achievements include research on ecology of coral reefs, especially reef fishes. Office: U Windsor, Dept Biology, Windsor, ON Canada N9B 3P4

SALEM, IBRAHIM AHMED, electrical engineer, consultant, educator; b. Alexandria, Egypt, May 1, 1932; s. Ahmed Mohamed and Monira (Mahmoud) S.; m. Zahira Abdelrahman Haroun, Aug. 21, 1966; 1 child, Yasmin. BS, Faculty of Engring., Alexandria, 1953; PhD, Acad. of Sci., Brno, Czechoslovakia, 1967. Registered profl. engr., Egypt. Prof. microwaves Mil. Tech. Coll., Cairo, 1968—; chmn. radar and guidance dept. Ein-Shams U., Cairo, 1975-82, dean, 1982-84, prof. microwaves, 1975-82; cons. Amcon Group of Egypt, Cairo, 1986—; cons. Rel-Internat., Boynton Beach, Fla., 1986-89; bus. devel. rep. Hercules Def. Electronic Systems, Clearwater, Fla., 1989—. Author: Waveguides & Cavity Resonators, 1970, Magnetrons & Kylstrons, 1973; contbr. several articles to profl. jours. Mem. Heliopolis Sporting Club, Cairo, 1962, Gezira Sporting Club, Giza, 1968, Greek-Egyptian Friendship Club, Garden City Cairo, 1984. Recipient Medal of Tng., Pres. of Egypt, 1971, Honor medal, 1985. Mem. IEEE (sr.), AIAA, Internat. Radio Sci. Union, Assn. Old Crows, N.Y. Acad. Sci. Home: 17 Elqouba St, 11341 Cairo Arab Republic of Egypt Office: Amcon Group Egypt, 34 Syria-Elmohandseen, Giza Egypt

SALEM, KENNETH GEORGE, theoretical physicist; b. Johnstown, Pa., Dec. 5, 1928; s. George John and Sadie (Abraham) S.; m. Jean Mae Leone, Sept. 25, 1954. Grad. high sch., Johnstown. Machinist Thompson Products, Cleve., 1950; defense plant inspector Stevens Mfg., Ebensburg, Pa., Ireco, Eugene, Oreg.; steel inspector, indsl. engr. Bethlehem Steel Corp., Johnstown; indsl. sales rep. M. Glosser & Sons, Inc., Johnstown, 1964-86; theoretical physicist pvt. practice Johnstown, 1986—. Author: 2.8 Angstroms, 1990; contbr. articles to profl. jours. including Jour. Brit.-Am. Sci. Rsch. Assn., Bull. Pure and Applied Sci. Reporting. Pres. Franklin High Sch. Alumni Assn., 1965-66. With U.S. Navy, 1946-49. Mem. Am. Physical Soc., Am. Astron. Soc., Am. Assn. Physics Tchrs. Achievements include patents relating to discoveries that sub-micron particles are detectable with naked eye, that turbulence in fluid flows is primarily due to acoustical radiation, and for improving flashlight to emit diffuse (shadow free) light; research on universal force and a universal constant of acceleration directly related to Newton's "G", both of which involve a new definition of gravitational mass. Home and Office: PO Box 908 Johnstown PA 15907-0908

SALERNO, PHILIP ADAMS, infosystems specialist; b. Harrisburg, Pa., Oct. 25, 1953; s. Lewis Gabriel S. and Barbara Ellen (Garlinger) Hardisty. AAS, Baylor U., 1975; BS in Med. Tech., Our Lady of the Lake U., 1979. Cert. med. technologist, lab. technologist. Lab. technician U.S. Army, Stuttgart, West Germany, 1971-74; instr., Acad. Health Scis. U.S. Army, Ft. Sam Houston, 1976-80; asst. instr. Baylor U., 1976-78; staff med. technologist M.D. Anderson Hosp., Houston, 1980; lab. supv. Twelve Oaks Hosp., Houston, 1980-83; clin. lab. instr., Acad. Health Scis. U.S. Army, Ft. Sam Houston, 1982-90; dir. current product engring. Community Health Computing, Houston, 1983--; owner Salerno Systems Group, Houston, 1985--. Co-author: Basic Med. Parasitology, 1977. Del. Rep. State Conv., 1986, 88, 90; precinct chmn. Rep. Party of Harris County, Houston, 1988-92; vol. fireman, Kentland, Md., 1969-73. Named Soldier of the Year, U.S. Army Baden Wurtemburg Support Dist., Federal Republic of Germany, 1974. Mem. Soc. Armed Forces Med. Lab. Scientists; assoc. mem. Am. Soc. Clin. Pathologists. Lutheran. Avocations: stamp collecting, animals. Home: 8510 Ariel St Houston TX 77074 Office: Community Health Computing 5 Greenway Pla Ste 2000 Houston TX 77046

SALESS, FATHIEH MOLAPARAST, biochemist; b. Shiraz, Iran, Dec. 22, 1948; came to U.S., 1972; d. Mehdi and Sedigheh (Milani) Molaparast; m. Shahin Shahid Saless, Nov. 7, 1971; 1 child, Neema. BS, Nutritional Sci. Inst., Tehran, Iran, 1971; MS, U. Akron, 1973; MSc, U. London, 1974; PhD, U. Wis., 1978. Rsch. asst. U. Akron, Ohio, 1972-73; rsch. asst. U. Wis., Madison, 1974-78, project assoc., 1978-79, rsch. scientist, 1984-86; lab. mgr., asst. prof. Taleghani Med. Sch., Tehran, 1979-83; supr. Baxter Healthcare Corp., L.A., 1988-92; mgr. Baxter Healthcare Corp., Irwindale, Calif., 1992—. Mem. ASTM, AAAS, Health Industry Mfrs. Assn., Am. Assn. Clin. Chemistry, N.Y. Acad. Sci. Achievements include research on the effect of drugs on metabolism and mech. performance of heart, secondary carnitine deficiency, glyceroneogenis in adopose tissue. Home: 1980 El Arbolita Dr Glendale CA 91208 Office: Baxter Healthcare Corp 4401 Foxdale Ave Irwindale CA 91706

SALI, VLAD NAIM, chemical engineer; b. Bucharest, Romania, Apr. 22, 1962; came to U.S., 1989; s. Schender and Lolia (Logan) S. MS, Nat. Chemistry Inst., Bucharest, Romania, 1987. Chem. engr. Synthetic Fiber Plant, Corabia, Romania, 1986-89, Elec. Industries, Murray Hill, N.J., 1989—. Author: Titanium Activity in Mineral Acids, 1986. Roman Catholic. Home: 110 Hawthorne Dr New Providence NJ 07974 Office: Elec Industries 691 Central Ave Murray Hill NJ 07974

SALIBA, ANIS KHALIL, surgeon; b. Karaoun, Bekaa, Lebanon, Jan. 14, 1933; s. Khalil and Rosa (Sallum) S.; m. Siham Saliba, Nov. 8, 1959; children: Khalil Saliba, Nada Saliba. BA, Am. U., Lebanon, 1952; MD with Distinction, Damascus U., 1960. Diplomate Am. Bd. Surgery, Am. Bd. Thoracic and Cardiovascular Surgery, Am. Bd. Surg. Critical Care. Resident Coney Island, Bklyn., 1960-65, St. Vincent Charity, Cleve., 1965-67; cardiovascular surgeon Beebe Med. Ctr., Lewes, Del., 1980—. Trustee Beebe Med. Ctr., 1989—. Fellow Am. Coll. Surgeons; mem. AMA. Home: 68 Sassafrass SL Lewes DE 19958 Office: 431 Savannah N Lewes DE 19958

SALIBA, JOSEPH ELIAS, civil engineering educator; b. Btegrine, Lebanon, May 2, 1955; came to U.S., 1977; s. Elias Diab and Emilie Bechara (Sawaya) S.; m. Dorothea Louise Liem, Dec. 24, 1980; children: Elias Joseph, Maria Louise, David Joseph. BSCE, U. Dayton, 1979, MSCE, 1980, PhD in Engring., 1983. Registered profl. engr., Ohio. Asst. prof. U. Dayton, Ohio, 1980-87, assoc. prof., 1987—; cons. in field, 1980—; lectr. Affit, Dayton, 1985—; vis. scientist Material Lab. at Wright-Patterson AFB, 1990. Author: Composite Design and Micromechanics, 1990; reviewer (books) Numerical Methods for Engineers, 1988, Computers for Engineers, 1990; contbr. articles to profl. jours. Chair sidewalk com. City of Huber Heights, Ohio, 1987—. Recipient several grants. Mem. ASCE (faculty advisor), Am. Soc. Engring. Educators, Am. Concrete Insst., Masonry Soc. Home: 6131 Corsica Dr Huber Heights OH 45424 Office: U Dayton 300 College Park Dayton OH 45464-0243

SALISBURY, JOHN WILLIAM, research geophysicist, consultant; b. West Palm Beach, Fla., Feb. 6, 1933; s. John William Salisbury and Mary (Bates) Massey; m. Lynne Marie Trowbridge, Oct. 25, 1956; children: John William III, Matthew Trowbridge. BA, Amherst Coll., 1955; MS, Yale U., 1957, PhD, 1959. Br. chief Air Force Cambridge Rsch. Labs., Bedford, Mass., 1959-76; div. dir. U.S. Dept. Energy, Washington, 1976-81; scientist U.S. Geol. Survey, Reston, Va., 1981-89; rsch. prof. geophysics Johns Hopkins U., Balt., 1989—; cons. NASA, Washington, 1965—, CBS, N.Y.C., 1969-72, Earth Satellite Corp., Rockville, Md., 1992—. Author: Mid-Infrared Spectra of Minerals, 1992; editor: The Lunar Surface Layer, 1965; contbr. over 100 articles to profl. publs. 1st lt. USAF, 1959-62. Fellow Geol. Soc. Am.; mem. AAAS, Am. Geophys. Union, Sigma Xi, Phi Beta Kappa. Achievements include pioneering measurement of spectral signatures of minerals, rocks and soils for satellite remote sensing of composition on the Earth and planets. Home: 5529 Coltsfoot Ct Columbia MD 21045 Office: Johns Hopkins U Dept Earth/Planetary Sci Baltimore MD 21218

SALISBURY, TAMARA PAULA, foundation executive; b. N.Y.C., Dec. 14, 1927; d. Paul Terrance and Nadine (Korolkova) Voloshin; m. Franklin Cary Salisbury, Jan. 22, 1955; children: Franklin Jr., John, Elizabeth, Elaine, Claire. BA, Coll. Notre Dame, 1948; postgrad., Am. U., George Washington U. Chemist depts. pathology and chemotherapy NIH Cancer Inst., Bethesda, Md., 1946-52; asst. to chief of Chemistry Br. Office of Naval Rsch., Bethesda, 1953-55; v.p., COO Nat. Found. Cancer Rsch., Bethesda, 1973—. Bd. mem. ARC; mem. Krebsforschung Internat., Assn. Internat. Cancer Rsch. Recipient Outstanding Contbns. award Internat. Soc. Quantum Biology, 1983. Mem. Am. Chemical Soc., N.Y. Acad. Sci., Am. Assn. Advancement of Sci., Inst. Phys. and Chem. Biology (mem. assn. etrangers), Nat. Liberal Club, Order of Leopold (officer 1985). Home: 10811 Alloway Dr Potomac MD 20854 Office: Nat Found Cancer Rsch Ste 500 W 7315 Wisconsin Ave Bethesda MD 20814

SALIU, ION, software developer, computer programmer; b. Gemeni, Mehedinti, Romania, Mar. 9, 1950; came to U.S., 1985; s. Marin and Elena (Chiser) S.; m. Ofelia Foltean, June 30, 1979 (div. 1989); 1 child, Amy Ofelia. Diploma in Econs., Acad. Econ. Studies, Bucharest, Romania, 1976. Economist Silk Co., Deva, Romania, 1978-84; technician Inacomp Computer Ctr., Troy, Mich., 1987; worker Boyer Nursery, Biglerville, Pa., 1985; software developer Lotwon, Biglerville, Pa., 1989—. Author (software) on probability applied to lottery and computer programming, 1989-92; contbr. articles to profl. jours. Mem. N.Y. Acad. Scis. Achievements include discovery of repeat cycles in probabilistic events, incorporated in a computer software algorithm. Home and Office: Lotwon 3587 Mummasburg Rd Biglerville PA 17307

SALK, JONAS EDWARD, physician, scientist; b. N.Y.C., Oct. 28, 1914; s. Daniel B. and Dora (Press) S.; m. Donna Lindsay, June 8, 1939; children: Peter Lindsay, Darrell John, Jonathan Daniel; m. Francoise Gilot, June 29, 1970. BS, CCNY, 1934, LLD (hon.), 1955; MD, NYU, 1939, ScD (hon.), 1955; LLD (hon.), U. Pitts., 1955; PhD (hon.), Hebrew U., 1959; LLD (hon.), Roosevelt U., 1955; ScD (hon.), Turin U., 1957, U. Leeds, 1959, Hahnemann Med. Coll., 1959, Franklin and Marshall U., 1960; DHL (hon.), Yeshiva U., 1959; LLD (hon.), Tuskegee Inst., 1964. Fellow in chemistry NYU, 1935-37, fellow in exptl. surgery, 1937-38, fellow in bacteriology, 1939-40; Intern Mt. Sinai Hosp., N.Y.C., 1940-42; NRC fellow Sch. Pub. Health, U. Mich., 1942-43, research fellow epidemiology, 1943-44, research asso., 1944-46, asst. prof. epidemiology, 1946-47; asso. research prof. bacteriology Sch. Medicine, U. Pitts., 1947-49, dir. virus research lab., 1947-63, research prof. bacteriology, 1949-55, Commonwealth prof. preventive medicine, 1955-57, Commonwealth prof. exptl. medicine, 1957-63; dir. Salk Inst. Biol. Studies, 1963-75, resident fellow, 1963-84, founding dir., 1976—, disting. prof. internat. health scis., 1984—; developed vaccine, preventive of poliomyelitis, 1955, cons. epidemic diseases sec. war, 1944-47, sec. army, 1947-54; mem. commn. on influenza Army Epidemiol. Bd., 1944-54, acting dir. commn. on influenza, 1944; mem. expert adv. panel on virus diseases WHO; adj. prof. health scis., depts. psychiatry, community medicine and medicine U. Calif., San Diego, 1970—. Author: Man Unfolding, 1972, The Survival of the Wisest, 1973, (with Jonathan Salk) World Population and Human Values: A New Reality, 1981, Anatomy of Reality, 1983; Contbr. sci. articles to jours. Decorated chevalier Legion of Honor France, 1955, officer, 1976; recipient Criss award, 1955, Lasker award, 1956, Gold medal of Congress and presdl. citation, 1955, Howard Ricketts award, 1957, Robert Koch medal, 1963, Mellon Inst. award, 1969, Presdl. medal of Freedom, 1977, Jawaharlal Nehru award for internat. understanding, 1976. Fellow AAAS, Am. Pub. Health Assn., Am. Acad. Pediatrics (hon., assoc.); mem. Am. Coll. Preventive Medicine, Am. Acad. Neurology, Assn. Am. Physicians., Soc. Exptl. Biology and Medicine, Inst. Medicine (sr.), Phi Beta Kappa, Alpha Omega Alpha, Delta Omega. Office: Salk Inst Biol Studies PO Box 85800 San Diego CA 92186-5800

SALKIND, ALVIN J., electrochemical engineer, educator; b. N.Y.C., June 12, 1927; s. Samuel M. and Florence (Zins) S.; m. Marion Ruth Koenig, Nov. 7, 1965; children: Susanne, James. B.Ch.E., Poly. Inst. N.Y., 1949, M.Ch.E., 1952, D.Ch.E., 1958; postgrad. and mgmt. courses, Pa. State U., 1965, Harvard U., 1976. Registered profl. engr., N.Y., N.J. Chem. engr. U.S. Electric Mfg. Co., N.Y.C., 1952-54; sr. scientist Sonotone Corp., Elmsford, N.Y., 1954-56; research assoc. Poly. Inst. N.Y., 1956-58, adj. prof. chem. engring., 1960-70; with ESB-Ray OVAC Co., Yardley, Pa., 1958-79; dir. tech. ESB-Ray OVAC Co., 1971-72, v.p. tech., 1972-79; pres. ESB Tech. Co., 1978-79; prof., chief bioengring. div., dept. surgery UMDNJ-Robert Wood Johnson Med. Sch., Piscataway, N.J., 1970—, vis. prof. chem. engring., 1979-85; prof. biomed. engring. and chem. and biochem. engring Rutgers U., Piscataway, 1985—, dir. Bur. Engring. Rsch., assoc. dean Coll. Engring., 1989—; vis. prof. and exec. officer Case Ctr. for Electrochem. Sci., 1981-82; bd. dirs., cons. various cos., rsch. instns. and govt. orgns. Author: (with S.U. Falk) Alkaline Storage Batteries, 1969, (with Herbert T. Silverman and Irving F. Miller) Electrochemical Bioscience and Bioengineering, 1973; editor: (with E. Yeager) Techniques of Electrochemistry, 1971, vol. 2, 1973, vol. 3, 1978, History of Battery Technology, 1987; contbr. articles to profl. jours. Served with USNR, 1945-46. Recipient Alumnus citation Poly. Inst. N.Y., 1975, award Internat. Tech. Exch. Soc., 1992; Case Centennial scholar Case-Western Res. U., 1980. Fellow Acad. Medicine of N.J., Am. Coll. Cardiology, AAAS; mem. Electrochem. Soc. (past chmn. new tech. com., past chmn. battery div.), Assn. Advancement Med. Instrumentation, Indsl. Research Inst. (emeritus 1979), Internat. Soc. Electrochemistry, N.Y. Acad. Scis., Sigma Xi, Phi Lambda Upsilon. Home: 51 Adams Dr Princeton NJ 08540-5401 Office: Rutgers U Bur Engring Rsch Busch Campus PO Box 909 Piscataway NJ 08855

SALMIRS, SEYMOUR, aeronautical engineer, educator; b. N.Y.C., Sept. 5, 1928; s. Meyer and Sylvia (Halpern) S.; m. Myra Cohen, June 14, 1952; children: Diane, Steven Alan, Roberta. BS in Aero. Engring., Ga. Inst. Tech., 1951, MS in Aero. Engring., 1952. Engr. NACA, Hampton, Va., 1955-58, NASA, Hampton, Va., 1958-79; assoc. prof. Ariz. State U., Tempe, 1980-91, prof. Emeritus, 1992—. Author aero. papers. Lt. (j.g.) USN, 1952-55, Korea. Mem. AIAA (ground test tech. com. 1990—). Achievements include patents in radiation direction detection; in spacecraft separation systems; in acoustic personal intercoms; discovery of relation between control power and damping for helicopters; prediction method for turbulence reduction screens in wind tunnels; aerodynamic drag reduction front end shape for trucks; development of NASA TCV program which introduced pilot CRT displays and digital computer control systems in comml. aviation. Office: Ariz State Univ Dept Aero Tech Tempe AZ 85287

SALMON, DAVID CHARLES, mechanical engineer; b. Kingston, Ont., Can., Sept. 2, 1962; s. Ross and Lillian (Gazeley) S.; m. Fay Tian, Oct. 27, 1990. BASc in Engring. Sci., U. Toronto, Ont., 1984; MSME, U. Utah, 1988, PhD in Mech. Engring., 1992. Lic. profl. engr., Ont., Tenn. Lab. mgr. Quality and Integrity Design Engring. Ctr. U. Utah, Salt Lake City, 1987-89, instr. dept. mech. engring., 1989-92; short course instr. ASTM, FAA, U. Utah, 1987-92; project mgr., Robotics, Sarcos Rsch. Corp., Salt Lake City, 1993—. Contbr. article to profl. jour. Audio tape ministry coord., liturgist First Presbyn. Ch., Salt Lake City, 1986-92. Can. Coun. Profl. Engrs. scholar, 1991; Brown and Sharpe, Inc. grantee, 1989, 90; Rolls Royce rsch. fellow, 1986-87. Mem. Am. Soc. Metals, ASTM, Am. Cermaic Soc. (treas. U. Utah chpt. 1991-92), ASME (satellite programs chair, Nashville sect. 1993-94), Sigma Xi, Phi Kappa Phi. Achievements include development of apparatus for viewing fatigue crack growth in ceramic materials under scanning electron microscope, identification of discontinuous mechanism of fatigue crack propagation in silicon nitride ceramics. Home: 1840A Stargazer Dr Cookeville TN 38501 Office: Ste 44 390 Wakara Way Salt Lake City UT 84108

SALMON, FAY TIAN, mechanical engineering educator; b. Xian, Shaanxi, China, July 3, 1959; came to U.S. 1984; d. Runmin and Fengji (Chen) Tian; m. David Charles Salmon, Oct. 27, 1990. BS in Engring. Mechanics, Northwestern Polytech. U., Xian, China, 1981, MS in Engring. Mechanics, 1984; PhD in Mech. Engring., U. Utah, 1991. Rsch. asst. Northwestern Poly. U., Xian, 1982-84; rsch./teaching asst. U. Utah, Salt Lake City, 1985-91; asst. prof. mech. engring. Tenn. Tech. U., Cookeville, 1991—; cons. U. Utah, Salt Lake City, 1986-91. Contbr. articles to profl. jours. Mem. Chinese Student Assn., U. Utah, 1985-91. Rsch. grantee Textron Aerostructures, 1992-93, Tenn. Tech. U. Mem. Am. Soc. Metals, ASME (exec. com. Nashville sect. 1993-94), Sigma Xi, Phi Kappa Phi. Achievements include research on using three-dimensional techniques for modelling post-impact behavior of composite materials. Home: 1840A Stargazer Dr Cookeville TN 38501 Office: Tenn Tech Univ Dept Mech Engring Cookeville TN 38505

SALMON, WILLIAM COOPER, mechanical engineer, engineering academy executive; b. N.Y.C., Sept. 3, 1935; s. Chenery and Mary (Cooper) S.; m. Josephine Stone, Sept. 16, 1967; children—William Cooper, Mary Bradford, Pauline Alexandra. S.B. in Mech. Engring., MIT, 1957, S.M. in Mech. Engring., 1958, Mech. Engr., 1959, S.M. in Mgmt. Sci., 1969. Registered profl. engr., Mass. Research and teaching asst. MIT, Cambridge, 1957-59; sr. engr. Microtech, Cambridge, 1959-60; asst. sci. advisor U.S. Dept. State, Washington, 1961-74, sr. advisor for sci. and tech., 1978-86; counselor for sci. and tech. Am. embassy, Paris, 1974-78; exec. officer Nat. Acad. Engring., Washington, 1986—. Recipient Superior Honor award Dept. State, 1964; Meritorious Service award Pres. U.S., 1984; Sloan fellow MIT, 1968. Mem. ASME, Nat. Soc. Profl. Engrs. Episcopalian. Club: Cosmos. Lodge: Masons. Office: Nat Acad Engring 2101 Constitution Ave NW Washington DC 20418

SALOM, SCOTT MICHAEL, forest entomologist; b. Queens, N.Y., Sept. 14, 1959; s. Herbert D. and Phylis (Falick) S.; m. Siti Marwiah Hasim, May 29, 1987. BS, Iowa State U., 1981; MS, U. Ark., 1985; PhD, U. B.C., Vancouver, Can., 1989. Rsch. assoc. dept. entomology U. Wis., Madison, 1989; rsch. assoc. dept. entomology Va. Poly. Inst., Blacksburg, 1989-92, rsch. scientist, 1992-93; asst. prof. Va. Poly Inst. and State U., Blacksburg, 1993—; tchr. in field of undergrad. students. Contbr. articles to profl. jours. Vol. pen pal scientist Sci.-by-Mail Program, Richmond, Va., 1992—. Recipient 1st Place award Entomol. Soc. Can., 1987. Mem. AAAS, Soc. Am. Foresters, Entomol. Soc. Am. (Pres. award 1988), Sigma Xi, Xi Sigma Pi, Phi Kappa Phi. Office: Va Poly Inst Dept Entomolgy 216 Price Hall Blacksburg VA 24061-0319

SALOP, ARNOLD, internist; b. N.Y.C., Oct. 19, 1923; s. Alexander and Anna (Lefrak) S.; m. Maryellen Kolt, June 27, 1979; children: Andrea, Holly, Evan Arnold. AB, Oberlin Coll., 1943; MB, Northwestern U., 1949, MD, 1950. Intern, resident in internal medicine Beth Israel Hosp., 1949-52; resident in internal medicine Goldwater Meml. Hosp., 1950-51, Kingsbridge VA Hosp., 1952-53; pvt. practice medicine specializing in cardiology and internal medicine, Ossining, N.Y., 1957-90; pres. med. staff affairs, sr. attending Phelps Meml. Hosp., North Tarrytown, N.Y, 1988-90; sr. v.p., med. dir. Phelps Meml. Hosp, North Tarrytown, 1991—. Served with AUS, 1943-45, 1st lt. USAF, 1953-54. Fellow ACP, Am. Coll. Cardiology; mem. Am. Heart Assn., Am. Geriatrics Assn., Am. Rheumatism Assn., Alpha Omega Alpha. Office: care Phelps Meml Hosp Ctr 701 N Broadway Tarrytown NY 10591-1096

SALOVEY, PETER, psychology educator; b. Cambridge, Mass., Feb. 21, 1958; s. Ronald and Elaine Y. (Gross) S.; m. Marta Elisa Moret, June 15, 1986. BA in Psychology, Stanford U., 1980; PhD in Psychology, Yale U., 1986. Lic. psychologist, Conn. Asst. prof. Yale U., New Haven, Conn., 1986-90; assoc. prof. Yale U., New Haven; cons. psychologist West Haven (Conn.) VA Med. Ctr., 1986—. Author: Peer Counseling, 1983, The Remembered Self, 1993, Psychology, 1993; editor: Judgement and Inference in Clin. Psychology, 1988, The Psychology of Jealousy and Envy, 1991; contbr. numerous articles to profl. jours. Named Presidential Young Investigator, NSF, Washington, 1990. Mem. Am. Psychol. Assn., Am. Psychol. Soc., Internat. Soc. for Rsch. on Emotion (treas. 1992—), Phi Beta Kappa, Sigma Xi. Democrat. Jewish. Achievements include rsch. on psychological consequences of the arousal of mood and emotion. Office: Yale U Dept Psychology 2 Hillhouse Ave New Haven CT 06520

SALPETER, EDWIN ERNEST, physical sciences educator; b. Vienna, Austria, Dec. 3, 1924; came to U.S., 1949, naturalized, 1953; s. Jakob L. and Frieder (Horn) S.; m. Miriam Mark, June 11, 1950; children—Judy Gail, Shelley Ruth. M.S., Sydney U., 1946; Ph.D., Birmingham (Eng.) U., 1948; D.Sc., U. Chgo., 1969, Case-Western Reserve U., 1970. Research fellow Birmingham U., 1948-49; faculty Cornell U., Ithaca, N.Y., 1949—; now J.G. White prof. phys. scis. Cornell U.; mem. U.S. Nat. Sci. Bd., 1979-85. Author: Quantum Mechanics, 1957, 77; mem. editorial bd. Astrophys. Jour, 1966-69; assoc. editor Rev. Modern Physics, 1971—; contbr. articles to profl. jours. Mem. AURA bd., 1970-72. Recipient Gold medal Royal Astron. Soc., 1973, J.R. Oppenheimer Meml. prize U. Miami, 1974, C. Bruce medal Astron. Soc. Pacific, 1987, A. Devaucouleurs medal, 1992. Mem. NAS, Am. Astron. Soc. (v.p. 1971-73), Am. Philos. Soc., Am. Acad. Arts and Scis., Australian Acad. Sci., Deutsche Akademie Leopoldina. Home: 116 Westbourne Ln Ithaca NY 14850-2414 Office: Cornell U Newman Lab Ithaca NY 14853

SALTERS, RICHARD STEWART, engineering company executive; b. St. Johns, Mich., Apr. 4, 1951; s. Stewart Arthur and Mary Ann (Eiseler) S.; m. Patricia Lynn Shumsky, Oct. 23, 1971 (div. Mar. 1982); children: Tiffani, Destiny. BS in Engring., Purdue U., 1974. Field engr. Henkels & McCoy, Inc.,, Blue Bell, Pa., 1972-77; area mgr. engring. dept. Harris McBurney Co., Inc., Jackson, Mich., 1977-81; project engr. Lambic Telcom, Inc., Ridgewood, N.J., 1981-82; pres. S & H Assocs., Inc., Lafayette, La., 1982—. Mem. Engring. Soc. of Detroit, City Club of Lafayette. Roman Catholic. Avocations: raising, racing thoroughbred horses, skiing, golf, tennis. Office: S & H Assocs Inc PO Box 52721 Lafayette LA 70505-2721

SALTHOUSE, THOMAS NEWTON, cell biologist, biomaterial researcher; b. Fleetwood, U.K., Mar. 8, 1916; came. to U.S., 1960; s. William and Edith Alice (Croft) S.; m. Mary Reynolds, Apr. 30, 1942; children: Andrew John, Robert William. Fellow Inst. Med. Lab. Sci., Inst. Med. Lab. Sci., London, 1942, Fellow Royal Photographic Soc., 1951. Diplomate Royal Microscopical Soc. Sr. technologist U. Coll. E. Africa, Uganda, 1947-56; rsch. assoc. Atomic Energy Can., 1957-60; asst. pathologist E.I. DuPont, 1960-62; sr. scientist E.R. Squibb & Sons, 1962-68; sect. mgr. Ethicon Rsch. Found., 1968-81; vis. prof. bioengineering Clemson U., 1982-88. Author: (with others) Enzyme Histochemistry in Fundamentals of Biocompatibility, 1981, Biocompatibility of Sutures in Biocompatibility in Clinical Practice, 1982, Implant Shape and Surface in Biomaterials in Reconstructive Surgery, 1983; co-editor Soft Tissue Histology in Handbook of Biomaterials, 1986; editorial bd. Jour. Biomed. Material Rsch., 1978-84; contbr. articles to profl. jours. Recipient Phillip B. Hoffman award, 1974. Fellow N.Y. Acad. Scis.; mem. Soc. for Biomaterials (sec., treas. 1976-78, pres. 1980-81, Clemson award 1978), Inst. Med. Lab. Scis. (ret.). Achievements include research on cellular mechanisms involved in tissue response to surgical implantion of various devices. Home: 714 Teakwood Ct West Columbia SC 29169-4914

SALTYKOV, BORIS GEORGIEVICH, economist; b. Moscow, Dec. 27, 1940. Student, Moscow Inst. Physics & Tech. Rschr., head lab., divsn. ctrl. inst. econs. & math. USSR Acad. Scis., Moscow, 1967-86, head divsn. inst. econs. of prognosis for nat. economy, 1986-91, dep. dir. analytical ctr., 1991; minister of sci. Higher Sch. & Tech. Policy of Russian Fedn., Moscow, 1991-92; dep. prime minister Russia Moscow, 1992—. Office: Russian Parliament, Staraya pl 4, Moscow Russia*

SALTZMAN, BARRY, meteorologist, educator; b. N.Y.C., Feb. 26, 1931; s. Benjamin and Bertha (Burmil) S.; m. Sheila Eisenberg, June 10, 1962; children—Matthew David, Jennifer Ann. B.S., CCNY, 1952; S.M., Mass. Inst. Tech., 1954, Ph.D., 1957; M.A. (hon.), Yale, 1968. Research staff meteorologist MIT, 1957-61; sr. research scientist Travelers Research Center, Inc., Hartford, Conn., 1961-66, research fellow, 1966-68; prof. geophysics Yale U., 1968—, chmn. dept. geology and geophysics, 1988-91. Editor: Selected Papers on the Theory of Thermal Convection, 1962, Advances in Geophysics, 1977—; asso. editor Jour. Geophys. Research, 1971-74; mem. editorial bd. Climate Dynamics, 1986—, ATMOSFERA, 1987—; co-editor Milankovitch and Climate, 1984; contbr. articles to profl. publs. Fellow AAAS, Am. Meteorol. Soc.; mem. Conn. Acad. Sci. and Engring., Am. Geophys. Union, Acad. Scis. Lisbon (hon. fgn.), European Geophys. Soc., Phi Beta Kappa, Sigma Xi. Home: 9 Forest Glen Dr Woodbridge CT 06525-1420 Office: Yale U Dept Geology and Geophysics PO Box 6666 New Haven CT 06511-8101

SALVADOR, ARMINDO JOSE ALVES SILVA, biochemist; b. Pontes de Monfalim, Lisbon, Portugal, Jan. 10, 1965; s. Armindo Ventura da Silva and Maria Alves (Silva) S. Lic., U. Lisbon, 1988. Biochemist Inst. Bento da Rocha Cabral, Lisbon; researcher Instituto Bento da Rocha Cabral, 1987—. Recipient PhD grant Nat. Junta Sci. & Tech. Investigation, 1990. Mem. AAAS, Portuguese Soc. Chemistry, Portuguese Soc. Biochemistry, Portuguese Soc. Physics, Biochem. Soc., N.Y. Acad. Scis., Soc. Free Radical Rsch., Rotary. Avocation: photography. Home: Rua Carlos de Oliveira 3 7B, 1600 Lisbon Portugal Office: Instituto Bento da Rocha, Cabral, Calcada Bento da Rocha 14, 1200 Lisbon Portugal

SALVADOR, MARK Z., system safety engineer, aerospace engineer; b. N.Y.C., June 10, 1968; s. Oscar L. and Concepcion Juan (Zarate) S. BS in Aero. Engring., Syracuse U., 1990. System safety engr. Naval Weapons Ctr. Code 3687, China Lake, Calif., 1990-91, Naval Air Systems Command AIR-516C4, Washington, 1991-92, Naval Air Warfare Ctr., Patuxent River, Md., 1992—. Mem. AIAA. Home: 1503 Hope Cir Waldorf MD 20601

SALVADORI, MARIO, mathematical engineer; b. Rome, Italy, Mar. 19, 1907; came to U.S., 1939; s. Riccardo and Ermelinda (Alatri) S.; m. Giuseppina Tagliacozzo, July 30, 1935 (div. June 1975); 1 child, Vieri R.; m. Carol B. Salvadori, Apr. 5, 1975. Dr. Civil Engring., U. Rome, 1930, Dr. Math. Physics, 1933; DSc, Columbia U., 1977; Dr. Fine Letters, New Sch. for Social Rsch., 1990. Prof. U. Rome (Italy) Sch. Engring., 1932-38, Columbia U., N.Y.C., 1940-90; hon. chmn. Weidlinger Assocs., N.Y.C., 1957-90, 1991—; founder, chmn. Salvadori Ednl. Ctr. on Brief Environment, 1975-91, hon. chmn., 1993—. Author of 17 books; contbr. articles to profl. jours.

Recipient over 20 awards from univs., engring. and archtl. socs. and ednl. assns., 1970-92. Fellow ASME, Am. Concrete Inst.; mem. ASCE (hon.), AIA (hon.). Democrat. Achievements include research in applied mathematics and engineering structures; 27 new routes and 3 virgin peaks climbed in the Eastern Alps. Home: 2 Beekman Pl New York NY 10022-8058 Office: Weidlinger Assocs 333 Seventh Ave New York NY 10001

SALVAGGIO, JOHN EDMOND, physician, educator; b. New Orleans, May 19, 1933; s. Louis and Zenobia Ann (Engman-Riley) S.; m. Anne Poillon, Apr. 19, 1958; children: John, Garry, Wayne, Peggy. B.S., Loyola U., New Orleans, 1954; M.D., La. State U., 1957. Diplomate: Am. Bd. Internal Medicine (bd. govs. 1973-81), Am. Bd. Allergy and Immunology (pres. 1975-78). Intern Charity Hosp., New Orleans, 1957-58; resident Charity Hosp., 1958-60; fellow, instr. Mass. Gen. Hosp. and Harvard U., 1961-63; prof. medicine La. State U., 1964-74; Henderson prof. medicine Tulane U., 1974—, chmn. dept. medicine, 1982-87; vice chancellor for rsch. Tulane Med. Ctr., 1988—; vis. prof. U. Colo., 1972-73; mem. pulmonary diseases adv. com. and immunologic scis. study sect. NIH. Contbr. over 400 articles to profl. jours. Served with USAR, 1959-65. Mem. AAAS, ACP, Am. Fedn. Clin. Rsch., Am. Bd. Allergy and Immunology (pres. 1975-78), Am. Acad. Allergy and Immunology (pres. 1985-86), Am. Bd. Internal Med. (bd. govs. 1975-80), Am. Assn. Immunologists, Am. Soc. Clin. Investigation, Am. Thoracic Soc. (governing council 1977-79), Assn. Am. Physicians, Soc. Exptl. Biology and Medicine, Collegium Internat. Allergy, Royal Soc. Medicine. Home: 5726 St Charles Ave New Orleans LA 70115-5052 Office: 1430 Tulane Ave New Orleans LA 70112-2699

SALVATORE, SCOTT RICHARD, ecologist; b. Lewistown, Pa., Apr. 30, 1960; s. Richard Ebert and Helen Louise S. BS, Juniata Coll., 1982; MS, Miami U., Oxford, Ohio, 1984. Rsch. technician Pa. Fish Commn., Thompsontown, 1981-82; teaching asst. Miami U., 1982-84; dist. exec. Boy Scouts Am., West Chester, Pa., 1985-88; aquatic ecologist Triegel & Assocs., Inc., Berwyn, Pa., 1988—; resource profl. Pa. Forest Stewardship Program, 1991—. Leader Cub Scout Pak 60, Scout Troop 53 Boy Scouts Am., Kennett Square, Pa., 1992. Recipient Citizenship award B'nai Brith, Lewistown, 1978. Mem. Kennett Area Jaycees (mgmt. v.p. 1990-91, outstanding officer 1991), Univ. Rifle Club. Methodist. Achievements include work on the Am. Shad Restoration Project designed to restore shad to the Susquehanna River basin; rsch. was used to force power cos. on the lower river to put fish passage devices at each of their dams. Office: Triegel & Assocs Inc 1235 Westlakes Dr Ste 320 Berwyn PA 19312-2414

SALVINI, GIORGIO, physicist, educator; b. Milan, Italy, Apr. 24, 1920; s. Ascanio and Maria (Sardella) S.; m. Costanza Catenacci, Apr. 24, 1951; children: Paola, Francesco, Stefano, Giovanna, Pietro. Physics degree summa cum laude, U. Milan, 1942; engring. degree (hon.), U. L'Aquila, 1991. Assoc. prof. Superior Physics, Milan, 1945-48; vis. researcher U. Princeton, N.J., 1949; instr. U. Cagliari, Italy, 1951-52; project dir. U. Pisa, Italy, 1953-55; faculty gen. physics Physics Inst., Rome, 1955-65; researcher Nat. Labs., Frascati, Italy, 1966-74, Centro Europeo de Ricerche Nucleari, Rome, 1975-89; instr., researcher physics dept. U. Rome, 1990—; instr. physics U. La Sapienza, 1959-89. Collaborator, coord. various scientific encys.; collaborator Dictionary of Phys. Scis.; contbr. articles to profl. jours. Pres. del Comitato Sci. esatte e Naturali, UNESCO, Rome, 1989—; pres. com. Internat. Security and Arms Control, Rome, 1990—, Def. of Human Rights, Rome, 1990—. Mem. INFN (pres. 1966-69), ECFA (pres. 1971-73), Accademia Nazionale dei Lincei (pres. 1990—), Accademia Nazionale delle Sci. Avocation: painting. Home: Via Senafe 19, 00199 Rome Italy Office: Acad Nat dei Lincei, Via della Lungara 10, 00165 Rome Italy

SAMANTA ROY, ROBIE ISAAC, aerospace engineer; b. Calcutta, India, Sept. 24, 1968; s. R.C. and J.I. Samanta Roy. SB, MIT, 1989, SM, 1991. Teaching asst. MIT, Cambridge, Mass., 1990, rsch. asst., 1991—. Recipient NSF fellowship, 1991. Mem. AIAA, Sigma Xi. Office: MIT Rm 37-471 77 Massachusetts Ave Cambridge MA 02139

SAMARANAYAKE, GAMINI SARATCHANDRA, chemist, researcher; b. Galle, Sri Lanka, Mar. 20, 1955; came to the U.S., 1985; s. Tulet and K. Samaranayake; m. Deepani Padmanayana Kotalawala, Aug. 22, 1985. MS, U. Peradeniya, Sri Lanka, 1985; PHD, PhD, Va. Tech., 1990. Rsch. assoc. Va. Tech., Blacksburg, 1990—. Mem. Am. Chem. Soc. Office: Va Tech Sci and Forest Products 210 Cheatham Wood Blacksburg VA 24061

SAMARAWEERA, UPASIRI, research chemist; b. Matara, Sri Lanka, Jan. 19, 1951; came to U.S. 1982; s. Harischandra and Soma (Yapa) S.; m. Indrani Savithri Goonewardene, Dec. 8, 1979; children: Ravinda, Hasanga. BS, U. Sri Lanka, 1977, MPhil, 1982; PhD in Chemistry, N.D. State U., 1989. Asst. govt. analyst Govt. Analyst Dept., Colombo, Sri Lanka, 1981-82; postdoctoral fellow polymers and coatings dept. N.D. State U., Fargo, 1989-90; rsch. chemist Am. Crystal Sugar Co., Moorhead, Minn., 1990—. Contbr. articles to profl. jours. Recipient Roon award Coatings Found., 1991. Mem. Am. Chem. Soc., Assn. Sugar Beet Tech. Home: 3113 9-1/2 St N Fargo ND 58102 Office: Am Crystal Sugar Co 1700 N 11th St Moorhead MN 56560

SAMELSON, FRANZ, psychologist; b. Breslau, Germany, Sept. 23, 1923; came to U.S., 1952; s. Siegfried and Irmgard (Engel) S.; m. Phoebe Ellen Jones, June 10, 1955; 1 child, Karen Ann. Diploma in Psychology, U. Munich, Germany, 1952; PhD in Psychology, U. Mich., 1956. Investigator U.S. Mil. Govt. for Germany, Munich, 1945-48; rsch. assoc. U. Mich., Ann Arbor, 1955-57; from asst. prof. to prof. psychology Kans. State U., Manhattan, 1957-89, prof. emeritus, 1990—, vis. prof. Giessen (Germany) U., 1967-68. Contbr. chpts. to books, articles to profl. jous. Mem. precinct com. Dem. Party, Manhattan, 1988-90. Social Sci. rsch. tng. fellow, 1954-55; Fulbright travel grantee to Germany, 1967-68; NSF rsch. grantee, 1978-79, 85-86. Fellow APA; mem. Internat. Soc. for History of Social and Behavioral Scis. (chair rev. bd. 1984-85). Home: 2078 College Hts Rd Manhattan KS 66502 Office: Kans State U Dept Psychology Manhattan KS 66502

SAMII, ABDOL HOSSEIN, physician, educator; b. Rasht, Iran, June 20, 1930; came to U.S., 1947; s. Mehdi Ebtehaj and Zahra (Mojdehi-Akbar) S.; m. Shahla Khosrowshahi; children: Ali, Golnaz. Student, Stanford U., 1947-49; BA, UCLA, 1950, MA, 1952; MD, Cornell U., 1956. Intern N.Y. Hosp., N.Y.C., 1956, asst. in medicine, 1956-58; asst. in physiology Cornell U. Med. Sch., N.Y.C., 1958-59; resident and sr. resident N.Y. Hosp., Peter Bent Brigham Hosp. and Mass. Gen. Hosp., Boston, 1959-61; adj. prof. medicine Cornell U. Med. Sch., 1973-79, prof. clin. medicine, 1979—; rsch. fellow Harvard U., Boston, 1959-60; prof. medicine Nat. Univ. Iran, Tehran, 1963-68; med. dir. Pars Hosp., Tehran, 1968-73; dir. div. medicine N.Y. Hosp.-Cornell Med. Coll., White Plains, 1979—; chancellor Reza Shah Kabir Grad. Univ., Tehran, 1973-78; cons. med. rsch. WHO, Geneva, 1973-79; v.p. Imperial Acad. Sci., Tehran, 1974-78. Gen. editor: International Textbook of Medicine, 1981; author, editor: Medical Clinics of North America, 1983, Textbook of Diagnostic Medicine, 1987. Dep. minister, Ministry of Health, Tehran, 1963-65; minister, Ministry Sci. and Higher Edn., Tehran, 1973-75. Fellow Rockefeller Found., Helen Hay Whitney Found. Fellow Royal Soc. of Medicine; mem. N.Y. Acad. Medicine, Harvey Soc., Internat. Soc. Nephrology, Am. Fed. Clin. Rsch. Avocations: music, antiques. Office: NY Hosp CMC WD 21 Bloomingdale Rd White Plains NY 10605-1596 also: 449 E 68th St New York NY 10021

SAMIOS, NICHOLAS PETER, physicist; b. N.Y.C., Mar. 15, 1932; s. Peter and Niki (Vatick) S.; m. Mary Linakis, Jan. 12, 1958; children: Peter, Gregory, Alexandra. A.B., Columbia U., 1953, Ph.D., 1957. Instr. physics Columbia U., N.Y.C., 1956-59; asst. physicist Brookhaven Nat. Lab., Upton, N.Y., 1959-62; asso. physicist Brookhaven Nat. Lab., 1962-64, physicist, 1964-68, sr. physicist, 1968—, group leader, 1965-75, chmn. dept. physics, 1975-81, dep. dir. for high energy and nuclear physics, 1981, dir., 1982—; adj. prof. Stevens Inst. Tech., 1969-75, Columbia U., 1970—. Contbr. articles in field to profl. jours. Recipient E.O. Lawrence Meml. award, 1980, award in physics and math. scis. N.Y. Acad. Scis., 1980; named AUI Disting. Scientist, 1992, W.K.H. Panofsky prize, 1993. Fellow Am. Phys. Soc., Am. Acad. Arts and Scis.; mem. NAS, Phi Beta Kappa, Sigma Xi. Achievements include being an expert in field of high energy particle and nuclear physics. Office: Brookhaven Nat Lab Office of Dir Upton NY 11973

SAMMAD, MOHAMED ABDEL, cardiovascular surgeon, consultant; b. Misurata, Libya, Nov. 30, 1953; arrived in Austria, 1985; s. Makhzoum A. Sammad and Mabrouka Ahmed; m. Moufeda A. Fituri, Feb. 5, 1981; children: Manal M., Sara M. MD, Alfateh Med. Sch., Tripoli, Libya, 1980. House officer surgery Tripoli (Libya) U. Hosp., 1980-81, sr. house officer, 1981-83; sr. resident Tajour U. Hosp., Tajoura, Libya, 1983-85; resident in cardiovascular surgery U. Hosp. Vienna (Austria), 1985-87, sr. resident in cardiovascular surgery, 1987-89, fellow in cardiovascular surgery, 1989-91. Author: Hypertension and Related Illness Risk Factors and Prevention, 1989. Fellow Austrian Bd. Surgery; mem. Internat. Soc. Heart Transplantation, N.Y. Acad. Scis., Am. Assn. for Advancement Sci. Avocations: music (violinist), soccer player. Home: 19 Silbergasse 11/15, 1190 Vienna Austria Office: Vienna Univ Hosp 2d Surg, Spitalgasse 23, 1090 Vienna Austria

SAMOREK, ALEXANDER HENRY, electrical engineer, mathematics and technology educator; b. Detroit, Feb. 14, 1922; s. Walter and Gladys (Kurys) S.; m. Deloris Gehrig 1944 (dec. Mar. 1948); 1 child, David A.; m. Matilda Louise Dusincki, May 10, 1952. Student, U. Detroit, 1946-49; BSEE, Detroit Inst. Tech., 1961. Electronics instr. Radio Electronic and TV Sch., Detroit, 1946-49; electronics inspector USAF Procurement Office, Detroit, 1950-53; chief technician Wayne Engring. and Rsch. Inst., Wayne State U., Detroit, 1954-57; elec. engr. Control Engring. Co., Detroit, 1957-60; chief engr., engring. mgr. Weltronic Co. subs. Ransburg Corp., Clare, Mich., 1960-84; electronic instr. Redford High Sch., Redford Twp., Mich., 1966; instr. math. and elec./electronics Mid. Mich. Community Coll., Clare, 1984—; cons. Welsam Cons., Clare, 1984—. With USAAF, 1942-46. Mem. (life) IEEE, Soc. Automotive Engrs. Home: 323 Markley St Clare MI 48617-1848

SAMOUR, CARLOS M., chemist. BA in Chemistry, Am. U. Beirut, 1942, MA, 1944; MS in Organic Chemistry, MIT, 1947; PhD, Boston U., 1950. Postdoctoral rsch. fellow Boston U., 1950-52; rsch. chemist The Kendall Co., 1952-57, dir. Theodore Clark Lab., 1957-73, dir. Lexington Rsch. Lab., 1973-81; pres. Samour Assocs., 1981-84; chmn., scientific dir. MacroChem Corp., Lexington, Mass., 1982—; section chmn. Internat. Union Pure and Applied Chemistry, U. Mass., Amherst, 1982; session chmn. Biomaterials, Sardinia, Italy, 1988, internat. conf. MIT, 1982, tech. advisor; pres., chmn. Augusta Epilepsy Rsch. Found., Washington, 1989—; advisor univs. and med. ctrs. Contbr. numerous articles to profl. jours. Mem. Am. Chem. Soc. (cert. merit 1981, adminstr. Kendell award 1964-83), Am. Assn. Pharm. Scis., Controlled Release Soc. Achievements include over 50 U.S. patents and over 200 foreign patents; research in the fields of polymer chemistry, biomaterials, pharmaceuticals and dential materials. Office: Macrochem Corp 110 Hartwell Ave Lexington MA 02173

SAMPATH, KRISHNASWAMY, reservoir engineer; b. Madurai, India, Apr. 21, 1955; came to U.S., 1976; s. S. Krishnaswamy and Jeyamma (Srinivasan) Rawal; m. Harini Raghavan, June 16, 1980; children: Yashoda, Dhruvan. B Tech., Univ. Madras, 1976; MS, Ill. Inst. of Tech., 1978. Asst. chem. engr. Inst. of Gas Tech., Chgo., 1979-81; chem. engr. Mobile R&D Corp., Dallas, 1981-88, assoc., 1988-90; reservoir engr. adv. Mobil Exploration and Prod. US, Bakersfield, Calif., 1990-92; engring. adv. Mobil Exploration and Prod. Tech. Ctr., Dallas, 1992—. Editor: Soc. Of Petroleum Engrs., Dallas, 1990—; contbr. articles to profl. jours. Achievements include 10 patents on coreanalysis, enhanced oil recovery. Office: Mobil E&P Tech Ctr 13777 Midway Rd Dallas TX 75244

SAMPLES, JERRY WAYNE, army officer; b. Staunton, Va., July 18, 1947; s. Wilmer Clark and Nellie Virginia (Price) S.; m. Kathleen Miller, Nov. 2, 1969; children: Christopher John, Steven Wayne. BS, Clarkson Coll., 1969; MS, Okla. State U., 1979, PhD, 1983. Registered profl. engr., Va. Lab. asst. Columbia Ribbon & Carbon, Glen Cove, N.Y., 1969-70; commd. 2d lt. U.S. Army, 1969, advanced through grades to col., 1991; asst. prof. mech. engring. U.S. Mil. Acad., West Point, N.Y., 1979-82; with Air Command and Staff Coll., Maxwell AFB, Ala., 1983; exec. officer 10th Engr. Bn., 1983-85, bn. comdr. 10th engr. bn. 3d inf. div., Fed. Republic Germany, 1987-89; assoc. prof. mech. engring. U.S. Mil. Acad., West Point, 1985-87, assoc. prof. dept. civil and mech. engr., 1989—. Author (with others) Fundamentals of Engineering Examination, 1991. Decorated Army Commendation medal, Meritorious Svc. medal. Mem. AIAA, ASME, ASEE, Phi Kappa Phi. Home: 17 Wilson Rd # A West Point NY 10996-1706 Office: US Mil Acad Dept Civil and Mech Engring West Point NY 10996

SAMPSON, ROBERT NEIL, association executive; b. Spokane, Wash., Nov. 29, 1938; s. Robert Jay and Juanita Cleone (Hickman) S.; m. Jeanne Louise Stokes, June 7, 1960; children—Robert W., Eric S., Christopher B., Heidi L. B.S. in Agr, U. Idaho, 1960; M.Public Adminstrn., Harvard U., 1974. Soil conservationist Soil Conservation Service, Burley, Idaho, 1960-61; work unit conservationist Soil Conservation Service, Orofino, Idaho, 1962-65; agronomist Soil Conservation Service, Idaho Falls, Idaho, 1967-68; info. specialist Soil Conservation Service, Boise, 1968-70, area conservationist, 1970-72; land use specialist Soil Conservation Service, Washington, 1974-77; dir. environ. services div. Soil Conservation Service, 1977; land use program mgr. Idaho Planning and Community Affairs Agy., Boise, 1972-73; exec. v.p. Nat. Assn. Conservation Dists., Washington, 1978-84, Am. Forestry Assn., Washington, 1984—; instr. soils and land use Boise State U., 1972; dir. Am. Land Forum, Washington, 1978-88. Author: Farmland or Wasteland: A Time To Choose, 1981, For Love of the Land, 1985; contbr. articles to profl. and popular publs. Pres. Orofino Golf Assn., 1966, Clearwater County Search and Rescue Unit, 1966-67. Recipient President's citation Soil Conservation Soc. Am., 1978; named Boise Fed. Civil Servant of Year Boise Fed. Bus. Assn., 1972. Fellow Soil and Water Conservation Soc. (Hugh Hammond Bennett award 1992); mem. Soc. Am. Foresters, Am. Soc. Assn. Execs. Presbyterian. Home: 5209 York Rd Alexandria VA 22310-1126 Office: Am Forestry Assn 1516 P St NW Washington DC 20005-1932

SAMSON, FREDERICK EUGENE, JR., neuroscientist, educator; b. Medford, Mass., Aug. 16, 1918; s. Frederick Eugene and Annie Bell (Pratt) S.; m. Camila Albert; children Cecile Samson Folkerts, Julie Samson Thompson, Renée. DO, Mass. Coll. Osteopathy, 1940; PhD, U. Chgo., 1952. Asst. prof. U. Kans., Lawrence, 1952-57, prof. physiology, 1962-73, chmn., prof. dept. physiology and cell biology, 1968-73; prof. physiology U. Kans. Med. Ctr., Kansas City, 1973-89, prof. emeritus, 1989—; dir. Ralph L. Smith Rsch. Ctr. U. Kans., Kansas City, 1973-89; staff scientist neuroscis. rsch. program MIT, Cambridge, Mass., 1968-82, cons., 1982-91; vis. prof. neurobiology U. Catolica de Chile, Santiago, 1972; prof. Inst. de Investigaciones Citologicas, Valenica, Spain, 1981-89; hon. lectr. Mid-Am. State Univs. Assn., 1987. Editor: (with George Adelman) The Neurosciences: Paths of Discovery, II, (with Merrill Tarr) Oxygen Free Radicals in Tissue Damage; contbr. articles to profl. publs. Scientist, U.S.A., Spain Friendship Treaty, Madrid and Valencia, 1981. Staff sgt. U.S. Army, 1941-45, PTO. Recipient Rsch. Recognition award U. Kans. Med. Ctr., Kansas City, 1984; Van Liere fellow U. Chgo., 1948; Rawson fellow U. Chgo., 1949-51; USPHS fellow MIT, 1965. Fellow AAAS; mem. Am. Soc. Neurochemistry (chmn. program com. 1980), Am. Soc. Cell Biology (local host com. 1984), Am. Physiol. Soc. (emeritus 1990), Soc. Neurosci. (program com. 1972-73), The Oxygen Soc., N.Y. Acad. Sci., U. Chgo. Kansas City Club (chmn. alumni fund bd. 1975-82, pres. 1979-81), Sigma Xi (regional lectr. 1974-75, pres. Kansas City chpt. 1977-78, pres. neurosci. chpt. 1978). Avocation: hand balancing. Home: 171 Lake Shore S Lake Quivira KS 66106-9516 Office: U Kans Med Ctr Ralph L Smith Rsch Ctr Bldg 37 Kansas City KS 66160-7336

SAMSON, JOHN ROSCOE, JR., electrical engineer; b. Chgo., Aug. 27, 1948; s. John Roscoe Sr. and Seraphine Ann (Brunetti) S.; m. Kathleen Ann Kennedy, Aug. 21, 1971; children: John Roscoe III, Michael Vincent. BSEE, Ill. Inst. Tech., Chgo., 1970; SMEE, MIT, 1972, diploma in elec. engr., 1973; PhD, U. S. Fla., 1992. Recitation instr. MIT, Cambridge, 1972-73; staff mem. Lincoln Lab. MIT, Lexington, Mass., 1972-78; part-time instr. Grad. Sch. Engring. Northeastern U., 1978-81; prin. engr. Raytheon Co. Equipment Div., Sudbury, Mass., 1978-84; prin. engring. fellow Honeywell Inc., Clearwater, Fla., 1984—; mem. Nat. Rsch. Coun., Nat. Acad. Scis. Naval Studies Bd., Advanced Radar Tech. Panel, Washington, 1990-91. Fellow AIAA (assoc.); mem. IEEE, Sigma Xi. Achievements include research in fault tolerance, distributed processing, optimizing real-time fault tolerance design in VLSI and wafer scale architectures for real-time parallel processing; tech. papers on onboard processing for space and airborne applications.

SAMUDLO, JEFFREY BRYAN, architect, educator, planner, business owner; b. San Gabriel, Calif., Oct. 3, 1966; s. Lazaro and Grace (Alvarez) S. BArch, U. So. Calif., 1990. Ptnr. Design AID, Architects & Planners, L.A., 1987—; v.p. N.E. Design & Devel., Glendale, Calif., 1989-90; instr. L.A. Trade Tech. Coll., 1989—; asst. dir. program in historic preservation U. So. Calif. Sch. Architecture. Bd. dirs. Community Planning Adv. Bd., L.A., 1987-92; founding mem., bd. dirs. The Eagle Rock Assn., Inc., 1987—; co-chair Highland Park Main St. Urban Revitalization Com.; asst. dir. Frank Lloyd Wright-Freeman House. Mem. AIA (assoc.), vice chair preservation advocacy com.), Am. Planning Assn. (assoc.), Calif. Preservation Found., Arroyo Arts Collective, L.A. Conservancy, Soc. Archtl. Historians (life, bd. dirs. So. Calif. chpt.), U. Soc. Calif. Archtl. Guild, L.A. City Hist. Soc. (life). Republican. Avocations: historic preservation, flying. Office: Design AID 2320 Langdale Ave Los Angeles CA 90041-2912

SAMUELSON, ANDREW LIEF, civil engineer; b. N.Y.C., Feb. 17, 1938; s. Samuel Ben and Sylvia (Schulman) S. BA, Dartmouth Coll., 1959; BSCE, CCNY, 1965; MSCE, Colo. State U., 1973. Registered profl. engr., Pa., N.J., N.Y., Conn., Colo. Staff engr. Dames & Moore, N.Y.C., 1965-69; project engr. Pacific Architects and Engrs., Saigon, Republic of Viet Nam, 1969-71; div. engr. Parsons Brinkerhoff, N.Y.C., 1973-76; project mgr. Cahn Engrs. Inc., Wallingford, Conn., 1976-80; project engr. Carroll Engring. Corp., Warrington, Pa., 1982-85, dept. mgr., 1985-88, v.p., 1988—. Lt. (j.g.) USN, 1959-61. Fellow ASCE. Republican. Jewish. Home: PO Box 1311 Doylestown PA 18901-0117

SAMUELSON, ROBERT DONALD, retired combat aircraft executive; b. Wabasha, Minn., June 20, 1929; s. Fern Roberta (Price) S.; m. Aspacia LaCreta Jackson, July 13, 1950; children: Rebecca Susanne, Charles Ross, Donald Robert. Aero. engr., Aeronautical U., 1954; BS, U. Mo., St. Louis, 1989. From flight test engr. to program mgr. McDonnell Aircraft Co., St. Louis, 1954-90; v.p., ops. McDonnell Douglas Spain Ltd., Madrid, 1984-87. Recipient Kelly Johnson award Soc. Flight Test Engrs., 1978. Mem. AIAA (chmn. St. Louis sect. 1982-83, Outstanding Sect. award 1984). Republican. Baptist. Avocations: hunting, fishing, golf, woodworking, gardening. Home: 2 Chantilly Ct Lake Saint Louis MO 63367-1630

SAMUELSSON, BENGT INGEMAR, medical chemist; b. Halmstad, Sweden, May 21, 1934; s. Anders and Stina (Nilsson) S.; m. Inga Karin Bergstein, Aug. 19, 1958; children: Bo, Elisabet, Astrid. DMS, Karolinska Inst., Stockholm, 1960, MD, 1961; DSc (hon.), U. Chgo., 1978, U. Ill., 1983. Asst. prof. Karolinska Inst., 1961-66, prof. med. and physiol. chemistry, 1972—, chmn. physiol. chemistry dept., 1973-83, dean Med. Faculty, 1978-83, pres., 1983—; rsch. fellow Harvard U., 1961-62; prof. med. chemistry Royal Vet. Coll., Stockholm, 1967-72; Harvey lectr., N.Y.C., 1979; mem. Nobel Com. Physiology and Medicine, 1984—, chmn. com., 1987-89; mem. rsch. adv. bd. Swedish Govt., 1985-88; mem. Nat. Commn. Health Policy, 1987-90. Contbr articles to profl. jours. Recipient A. Jahres award Oslo U., 1970, Louisa Gross Horwitz award Columbia U., 1975, Albert Lasker basic med. research award, 1977, Ciba-Geigy Drew award in biomed. research, 1980, Lewis S. Rosenstiel award in basic med. research Brandeis U., 1981, Gairdner Found. award, 1981, Heinrich Wieland prize, 1981, Nobel prize in physiology or medicine, 1982, award medicinal chemistry div. Am. Chem. Soc., 1982, Waterford Bio-Med. Sci. award, 1982, Internat. Assn. Allergology and Clin. Immunology award, 1982, Abraham White sci. achievement award, 1984, Gregory Pincus Meml. award, 1984, Charles E. Culpepper award, 1985, Supelco award Am. Oil Chemists Soc., 1985, Chilton lectureship award, 1986, Abraham White Disting. Sci. award, 1991, City of Medicine award, 1992. Mem. AAAS (hon.), Royal Swedish Acad. Scis., Mediterranean Acad. Sci., Acad. Europaea (founding mem.), French Acad. Scis., Assn. Am. Physicians, Swedish Med. Assn., Am. Soc. Biol. Chemists, Italian Pharm. Soc., Acad. Nat. Medicina de Buenos Aires, Internat. Soc. Hematology, Fgn. Assn. U.S. Nat. Acad. Scis., Royal Soc. London (fgn. mem.), Spanish Soc. Allergology and Clin. Immunology, Royal Nat. Acad. Medicine Spain (hon.), Internat. Acad. Sci. (hon.). Office: Dept Physiological Chemistry, Karolinska Inst, S-104 01 Stockholm Sweden

SAN, NGUYEN DUY, psychiatrist; b. Langson, Vietnam, Sept. 25, 1932; s. Nguyen Duy and Tran Tuyet (Trang) Quyen; came to Can., 1971, naturalized, 1977; M.D., U. Saigon, 1960; postgrad. U. Mich., 1970; m. Eddie Jean Ciesielski, Aug. 24, 1971; children: Megan Thuloan, Muriel Mylinh, Claire Kimlan, Robin Xuanlan, Baodan Edward. Intern, Cho Ray Hosp., Saigon, 1957-58; resident Univ. Hosp., Ann Arbor, Mich., 1968-70, Lafayette Clinic, Detroit, 1970-71, Clarke Inst. Psychiatry, Toronto, Ont., Can., 1971-72; chief of psychiatry South Vietnamese Army, 1964-68; sr. psychiatrist Queen St. Mental Health Ctr., Toronto, 1972-74; unit dir. Homewood San., Guelph, Ont., 1974-80; cons. psychiatrist Guelph Gen. Hosp., St. Joseph's Hosp., Guelph; practice medicine specializing in psychiatry, Guelph, 1974-80; unit dir. inpatient service Royal Ottawa (Ont., Can.) Hosp., 1980-84, dir. psychiat. rehab. program, 1985-87; asst. prof. psychiatry U. Ottawa Med. Sch., 1980-85, assoc. prof. psychiatry, 1985-87; bd. dirs. Hong Fook Mental Health Service, Toronto, 1987—, dir. East-West Mental Health Ctr., Toronto, 1987—; chmn., bd. dirs. Access Alliance Multicultural Health Ctr., Toronto, 1988—; cons. UN High Commr. for Refugees, 1987—. Served with Army Republic of Vietnam, 1953-68. Mem. Can. Med. Assn., Can., Am. psychiat. assns., Am. Soc. Clin. Hypnosis, Internat. Soc. Hypnosis, N.Y. Acad. Scis. Buddhist. Author: Etude du Tetanos au Vietnam, 1960; (with others) The Psychology and Physiology of Stress, 1969, Psychosomatic Medicine, theoretical, clinical, and transcultural aspects, 1983, Uprooting, Loss and Adaptation, 1984, 87, Southeast Asian Mental Health, 1985, Ten Years Later: Indochinese Communities in Canada, 1988, Refugee Resettlement and Well-Being, 1989. Office: 2238 Dundas St W Ste 306, Toronto, ON Canada M6R 3A9

SANANMAN, MICHAEL LAWRENCE, neurologist; b. Bklyn., Oct. 11, 1939; s. Jack and Sarey (Bykofsky) S.; m. Elisa Joan Freeman, Apr. 12, 1964; children: Amy, Peter. AB, Swarthmore Coll., 1960; MD, Columbia U., 1964. Diplomate Am. Bd. Psychiatry and Neurology. Intern, Univ. Hosp., San Francisco, 1964-65; resident in neurology N.Y. Neurol. Inst., N.Y.C., 1966-69; practice medicine specializing in neurology, Elizabeth, N.J., 1972—; cons. neurologist St. Elizabeth's Hosp., Elizabeth Gen. Hosp., Rahway (N.J.) Hosp.; instr. neurology Columbia U., N.Y.C., 1971-75; asso. clin. prof. neurology U. Medicine and Dentistry N.J., Newark, 1975—; mem. adv. coun. N.J. chpt. Multiple Sclerosis Soc. Served to lt. comdr. M.C., USNR, 1969-71. Mem. AMA, Am. Acad. Neurology, Am. Epilepsy Soc. (adv. coun. N.J. chpt.), N.J. Acad. Medicine (chmn. neurology sect.), Am., Eastern EEG socs., Am. Assn. EMG and Electrodiagnosis. Office: 700 N Broad St Elizabeth NJ 07208

SAN BIAGIO, PIER LUIGI, physicist; b. Palermo, Italy, May 20, 1952; s. Carmelo and Liliana (Tovagliari) San B.; m. Daniela Giacomazza, Sept. 18, 1979; children: Marco, Livio. Maturitā Classica, Inst. G. Meli, Palermo, Italy, 1970; Laureate in Fisica, U. Palermo, Italy, 1975. Researcher physics dept. U. Palermo, 1980—; vis. assoc. Nat. Inst. Health, Bethesda, Md., 1986-87; vis. assoc. prof. physics dept. Union Coll., Schenectady, N.Y., 1990. Contbr. articles to profl. jours. Mem. Am. Phys. Soc., Biophys. Soc. Avocations: tennis, books, computers. Office: U Palermo Dept Physics, Via Archirafi 36, Palermo 90123, Italy

SANBORN, CHARLES EVAN, retired chemical engineer; b. Mankato, Minn., July 11, 1919; s. Walter A. and Gertrude Egryn (Evans) S.; m. Jane Martin McClanahan, June 24, 1941; children: Charles Evan, James Martin, Jane Ann Sanborn Russell, Rachel Elizabeth; m. Norma L. Gary, Mar. 15, 1980. B in Chem. Engring., U. Minn., 1941, PhD, 1949. Engr. Shell Devel., Emeryville, Calif., 1949-72, Shell Rsch. NV, Amsterdam, Netherlands, 1972-73; staff rsch. engr. Shell Devel., Houston, 1973-82; mem. faculty U. Calif. Engring. Extension, Berkeley, 1950-55. Contbr. articles to profl. jours. Capt. U.S. Army, 1941-45. Allied Chem. fellow U. Minn., 1948. Mem. Am. Inst. Chem. Engrs., Sigma Xi. Republican. Presbyterian. Achievements include patents in iodinative hydrogenation and iodine recovery, dehydration of alpha-methylbenzyl alcohol and others. Home: 205 Castle Hill Ranch Walnut Creek CA 94595

SANCAKTAR, EROL, engineering educator; b. Ankara, Turkey, July 13, 1952; came to U.S., 1974; s. Mehmet Ali and Ulker Mualla (Elveren) S.; m. Teresa Sue Sancaktar, Feb. 16, 1979; children: Orhan Ali, Errol Alan. BS in Mech. Engring., Robert Coll., Istanbul, Turkey, 1974; MS in Mech. Engring., Va. Poly. Inst. and State U., 1975, PhD, 1979. Teaching asst. Robert Coll., Istanbul, 1972-74; instr. Va. Poly. Inst. and State U., Blacksburg, Va., 1977-78; visiting scholar Kendall Co., Boston, 1985-86; assoc. prof. Clarkson U., Potsdam, N.Y., 1984—; cons. to the UN Devel. Programme, 1987, ALCOA, 1990-91, U.S. Army Benet Labs., 1991. Mem. editorial adv. bd. Jour. Adhesion Sci. Tech., 1993—; contbr. articles profl. jours.; rsch. on mech. behavior of polymers, adhesives and composite materials; patentee in field. Recipient various rsch. grants awarded by NSF, NASA, U.S. Army, N.Y., Grumman Corp., Kendall Co., GE, IBM. Mem. ASME.

SANCHEZ, DOROTHEA YIALAMAS, neuroscientist; b. Allentown, Pa., July 16, 1960; d. George Constantine and Agatha (Caramuchos) Yialamas; m. James Michael Sanchez, June 9, 1985. BS in Biology, Muhlenberg Coll., 1982; PhD in Zoology, U. Tex., 1988. Postdoctoral fellow U. Tex. Southwestern Med. Ctr., Dallas, 1988—. Contbr. articles to Jour. Gen. Physiology, Jour. Comparative Neurology, Jour. for Neurosci., Jour. Comparative Physiology. Mem. Biophys. Soc., Soc. for Neurosci., Philoptochos Soc. (recording sec. 1991—), Sierra Club. Office: U Tex Southwestern Med Ctr Dept Physiology Dallas TX 75235-9040

SANCHEZ, JAVIER ALBERTO, industrial engineer; b. San Cristobal, Tachira, Venezuela, Apr. 13, 1960; came to U.S., 1977; s. Leonidas and Ana Mireya (Albornoz) S. AA, Butler County C.C., El Dorado, Kans., 1979; BS in Indls. Engring., Wichita State U., 1982, MS in Engring. Mgmt., 1985. Indsl. cons. Ferronikel, C.A., Caracas, Venezuela, 1977-83; project mgr. Trabajos Viales, C.A., Caracas, Venezuela, 1980; applications engr. Major, Inc., Wichita, Kans., 1983; mfg. engr. L.S. Industries, Inc., Wichita, 1983-86; plant mgr. World Wide Mfg., Inc., Miami, Fla., 1986-88; prodn. mgr. Capitol Hardware Mfg. Co., Chgo., 1988-91; product mgr. Ready Metal Mfg.Co., Chgo., 1991-92; engring. tech. resources mgr. Taurus Internat. Mfg. Inc., Miami, 1992—; sr. cons. Ferronikel, C.A., Caracas, 1983—, mgr. internat. ops., 1986—; mem. adv. bd. Plastidrica, C.A., Caracas, 1988—. Recipient scholarship award Venezuelan Govt., 1977, Mariscal Ayacucho award, 1980. Mem. Am. Soc. for Metals, Soc. Mfg. Engrs., Inst. Indsl. Engrs. (sr.), Alpha Pi Mu. Roman Catholic. Achievements include development of applications of world class manufacturing techniques in tube fabricating, sheet metal operations and firearms production; cost estimating techniques for metal fabricators; design and manufacturing of retail store fixtures and racks; worldwide engineering and technical resources for firearms manufacturing and for international businesses. Home: 8357 W Flagler St Ste 308 Miami FL 33144 Office: Taurus Internat Mfg Inc 16175 W 49th Ave Miami FL 33140

SÁNCHEZ, LUIS RUBEN, environmental engineer; b. Mexico City, Mex., Sept. 18, 1963; s. Jorge and Dachka (Catano) S.; m. Lucrecia Infante, Aug. 30, 1991; 1 child, Dunai. Degree, U. Autonoma Metroplitana, Mexico City, 1988; postgrad., Boston U., 1992—. Asst. prof. U. Autonoma Metropolitana, Mexico City, 1986-89; program mgr. Dept. Distrito Fed., Mexico City, 1989-92; project mgr. Ingenieria y Gestion Ambiental, S.A., Mexico City, 1992; project mgr., Mexico City, 1988-89. Author: Mexico City Air Pollution, 1992; contbr. article to profl. jours. Fulbright fellow U.S.-Mex. Commn. on Cultural and Acad. Exch., 1992—, Ford-MacArthur fellow Ford Found. and MacArthur Found., 1992—. Mem. Air and Waste Mgmt. Assn., Sociedad Mexicanade Ingenieria Sanitoariay Aubiental AC. Home: Tuxpan 95-10 Col Romasur, 06760 Mexico DF Mexico Office: Ingenieria y Gestion Ambiental, Manzanillo 154 Altos, 06760 Mexico City Mexico

SANCHEZ, MIGUEL RAMON, dermatologist, educator; b. Havana, Cuba, May 5, 1950; came to the U.S., 1962; s. Rodolfo and Maria Sanchez. BS, CCNY, 1971; MD, Albert Einstein Coll. Medicine, 1974. Instr. Montefiore Dept. Family Medicine, Bronx, N.Y., 1978-79; sr. med. specialist Kingsborough Psychiat. Ctr., Bklyn., 1979-80; med. dir. Ten Communities Health Ctr., Tulare, Calif., 1980-82; teaching asst. dept. dermatology NYU, N.Y.C., 1982-83, asst. prof., 1983—; attending-in-chief dept. dermatology Bellevue Hosp., N.Y.C., 1983—; mem. Tulare County Mental Health Bd., 1980-81; mem. med. bd. Bellevue Hosp., 1990—. Contbr. articles to profl. jours. and chpts. to books; editor: (software) Derm-Rx, 1986-90, (book) Dermatology Educational Review Manual, 1993. Bd. dirs. Community Health Project, N.Y.C., 1993; mem. patient care com. community bd. Bellevue Hosp., 1990-93; founder Assn. Latino Faculty and Students. Recipient Testimonial of Appreciation So. Tulare County, 1981, 1st Place award Scientific Forum N.Y. Acad. Dermatology, 1985. Mem. Am. Acad. Dermatology, Acad. for Advancement Sci., Dermatologic Found. Democrat. Roman Catholic. Achievements include development of clinics for tropical dermatology, HIV skin disease, disorders of keratinization, connective tissue disease, and phototherapy; research in infectious diseases, dermatopharmacology and cutaneous manifestation of HIV infection. Home: 1 Washington Square Vlg New York NY 10012-1611 Office: NYU Dept Dermatology 562 1st Ave New York NY 10016-6402

SANCHEZ, PEDRO ANTONIO, JR., soil scientist, administrator; b. Havana, Cuba, Oct. 7, 1940; s. Pedro Antonio Sr. and Georgina (San Martin) S.; m. Cheryl Palm, 1990; children: Jennifer, Evan, Juliana. BS in Agronomy, Cornell U., 1962, MS in Soil Sci., 1964, PhD in Soil Sci., 1968. Grad. asst. U. Philippines/Cornell Grad. Edn. Program, Los Baños, Philippines, 1965-68; asst. prof. soil sci. N.C. State U., Raleigh, 1968-73, leader tropic soils program dept. soil sci., 1971-76, assoc. prof. soil sci., 1973-79, prof. soil sci., 1979-91, coord. tropic soils program, 1979-82, 84-91, prof. emeritus soil sci. and forestry, 1991—; co-leader Nat. Rice Program Peru N.C. State U. Agrl. Mission to Peru, Lambayeque, Peru, 1968-71; chief N.C. State U. Agrl. Mission to Peru, Lima, Peru, 1982-83; coord. beef-tropical pastures program Ctr. Internat. Agrl. Tropical, Cali, Colombia, 1977-79; tech. chief Inst. Nat. Investigation and Promotion Agropercuaria, 1982-83; dir. Ctr. for World Environment and Sustainable Devel. Duke U./N.C. State U./U. N.C. Chapel Hill, 1990-91; dir. gen. Internat. Ctr. for Rsch. in Agroforestry, Nairobi, Kenya, 1991—; adj. prof. tropical conversation Duke U., 1990; chmn. exec. com. Univ. Consortium on Soils of Tropics; lead analyst land and water sect. World Food and Nutrition Study, NAS, Washington, com. on selected biological problems of humid tropics, chmn. com. on sustainable and environment in humid tropics; conselho assessor do Centro de Pesquisa Agropecuarias dos Cerrados, AMBRAPA, Brasilis, Brazil; consejo directivo del Centro Nacional de Investigationes Agropecuarias de Carimagua, ICA-CIAT, Colombia; mem. tech. adv. bd. Commissao do Plano da Lavoura Cacauiera, Itabuna, Bahia, Brazil; coord. Soil Mgmt. Collaborative Rsch. Support Program Planning Grant, USAID/BIFAD, Washington; vice-chmn. internat. steering com. Red de Investigacion Agroecologica para la Amazonia-REDINAA, Amazon Rsch. Network, leader soil project; mem. steering com. Formation Internat. Bd. on Soils Rsch. and Mgmt., chmn. acid tropical soils network coord. com.; mem. tech. com. Soil Mgmt. CRSP; mem. com. on tropical deforestation Office of Tech. Assessment, U.S. Congress; mem. Internat. Com. on Land Clearing and Devel.; mem. coord. com. Tropical Soil Biology and Fertility Program; chmn. bd. mgmt. Tropical Soil Biology and Fertility Program; cons. Ford Found., USAID, Inst. Interamericano Scis. Agrl., Rockefeller Found., NAS, TVA, Venezuelan Soc. Soil Sci., Internat. Ctr. for Rsch. in Agroforestry, World Bank, New Zealand Soc. Soil Sci., Consultative Group on Internat. Agrl. Rsch., Empresa Brasileira de Pesquisa Agropecuaria, FAO, Royal Swedish Acad. Scis., Ecosystems Ctr. Woods Hole Oceanographic Inst., IBSRAM, Consejo Nat. Sci. and Tech. Peru, UNESCO-Main in Biosphere Program, U.S. EPA, CIAT, WWF, Rainforest Alliance, U.S. Congress. Author: Properties and Management of Soils in the Tropics, 1981; co-author: Suelos Acidos: Estrategia para su Manejo con Bajos Insumos en America Tropical, 1983; editor: A Review of Soils Research in Tropical Latin America, 1973; co-editor: Curso de Capacitacion sobre el Cultivo del Arroz, 1969, Multiple Cropping, 1976, Pasture Production in Acid Soils of the Tropics, 1979, Amazonia: Agriculture and Land Use Research, 1982, Land Clearing and Development in the Tropics, 1986, Management of Acid Tropical Soil for Sustainable Agriculture, 1987, Myths and Science of Soils of the Tropics, 1992; mem. editorial adv. bd. Field Crops Rsch., Tropical Agriculture, Agroforestry Systems, Geoderma; contbr. articles to sci. and profl. jours. Recipient Agronomy Achievement award Nat. Plant Food Inst., 1960, Diploma Merit Peru Min. Agriculture, 1971, Diploma Honor, Colombian Inst. Agropecuario, 1979, INIPA, 1985, Order Agrl. Merit, Govt. Peru, 1984; named hon. prof. U. Nat. Amazonia Peruana, 1987. Fellow Am. Soc. Agronomy (bd. dirs., chmn. divsn. internat. agronomy, Internat. Svc. in Agronomy award 1993), Soil Sci. Soc. Am. (Internat. Soil Sci. award 1993); mem. AAAS, Internat. Soc. Soil Sci. (vice chmn. commn. VI 1986), Latin Am. Assn., Agrl. Scis., Latin Am. Assn. Scis. Pecuarias Peru, Latin Am. Soc. Sci. Suelo, Assn. Investigators Agrl. and Pecuarias Peru, Soc. Colombia Sci. Suelo (hon., bd. dirs.), Soc. Peru. Sci. Suelo (hon.), Soil Sci. Soc. N.C., Sigma Xi, Sigma Iota Rho. Office: Internat Ctr Rsch in Agroforestry, PO Box 30677, Nairobi Kenya

SANCHEZ DE LA PEÑA, SALVADOR ALFONSO, biomedical chronobiologist; b. Mexico City, Nov. 14, 1948; came to U.S., 1979; s. Salvador and Maria Estela (de la Peña) S.; m. Maria Elizabeth Castro, Oct. 26, 1979; children: Salvador Ricardo, M. Amaranda. MD, U. Mex., 1973; MSc, U. Minn., 1985; PhD, Cinvestav-Mex., 1986. MD cert. Mex. Asst. prof. U. Mex., Mexico City, 1979; postdoctoral fellow U. Minn., Mpls., 1979-85, postdoctoral rschr., 1986-89; rsch. asst. prof. Albany (N.Y.) Med. Ctr., 1989-92; rsch. scientist VA Med. Ctr., Albany, 1991—; gen. dir. Mexican Chronooncology Rsch. Found., 1992—; coord., cons. on biotech. Falmex Group, Mex., 1987—. Contbr. articles to profl. jours. Grantee in field. Mem. Internat. Soc. for Chronobiology (award 1981), N.Y. Acad. Sci., Sigma Xi. Roman Catholic. Achievements include discovery of the feed-side wardphenomenon in the field of chronobiology circaiaan and circaseptan (7days), rhythm of murine death from malaria, circadian rhythm of renin I activity in the rat pineal gland, circadian-infradian intermodulation of melatonin content of rat pineal gland, chronomodulation of murine tumor growth by human recombinant interleukin 2 and melatonin. Home: Oxford Heights Hathaway #7 Albany NY 12203 Office: Samuel Stratton Med Ctr VA 113 Holland Ave C111-C Albany NY 12208

SANCHEZ MUÑOZ, CARLOS EDUARDO, physicist; b. Bogota, Colombia, May 9, 1957; s. Carlos and Ana (Muñoz) S.; 1 child, Carlos Javier Sanchez Plazas. Degree in physics, U. Nacional, Bogota, 1979, postgrad., 1982-85. Asst. tchr. Univ. Nacional, 1980; physicist Inst. de Asuntos Nucleares, Bogota, 1982-86, with, 1986-90, div. boss, 1990—; tchr. U. Jorge Tadeo Lozano, Bogota, 1988. Author/Editor: The Atlas of Solar Radiation of Columbia, 1992; Contbr. articles to profl. jours. Mem. Assn. Nacional de Fisicos, Soc. Colombiana de Energia Solar y no Convencionales. Achievements include design and installation of solar desalination plant.

SÁNCHEZ-RAMOS, JUAN RAMON, neurologist, researcher; b. Cabimas, Zulia, Venezuela, July 16, 1946; came to U.S., 1950; s. Juan R. and Carmen F. (Ramos) Sánchez; m. Catherine O'Neill, Aug. 19, 1984; children: Zachary, Zoe, Sofia. BS, U. Chgo., 1967, PhD, 1976; MD, U. Ill., Chgo., 1981. Diplomate Am. Bd. Neurology. Resident in neurology U. Chgo., 1982-85; movement disorders fellow U. Miami, Fla., 1985-87, asst. prof., 1987-92, assoc. prof., 1992—; cons. Nat. Parkinson Found., Miami, 1988—. Contbr. articles to profl. jours. Recipient Clin. Investigator Devel. award NIH, 1988-91, Prix Creatif Technologique for hologram Concourse des Technologie, Paris, 1991. Mem. AAAS, Am. Assn. Neurology, Movement Disorders Soc. Office: Univ Miami Dept Neurology 1501 NW 9th Ave Miami FL 33136

SANCHEZ SUDON, FERNANDO, renewable energy company executive; b. Merida, Badajoz, Spain, Nov. 20, 1949; s. Jose and Santa (Sudon) Sanchez; m. Tamara Kisielewska, Dec. 10, 1976; children: Olga, Mario. Lic. telecomm. engr. Control engr. Empresa Nacional de Ingenieria y Tecnologia, Spain, 1977-84; dir. Plataforma Solar Almeria, Spain, 1984-86; head solar div. Instututo de Energias Renovables Centro de Investigacio Nes Energeticas Medioambientales y Tecnologicas, Madrid, 1986-89, also dir. Instutuo De Energias Renovables, 1989—; chmn. internat. confs. on renewable energies; mem. sci. and organising coms. of confs. Co-author: Guide Lines for Economic Analysis; mem. editorial bd. Internat. Jour. Solar Energy. Mem. Jouce Com., Renewable Energy Working Party. Achievements include builder of first prototype of solar thermal power plants in Spain. Home: Lorca 16, 28230 Las Rozas Madrid, Spain Office: CIEMAT-IER, Avda Complutense 22, 28040 Madrid Spain

SAND, MICHAEL, industrial designer; b. Bklyn., Aug. 9, 1940; s. Joseph H. and Ethel (Lichtenstein) S.; m. Margaret Emma Schmidt, Aug. 8, 1969; children: Zoe, Jessica. BFA in Indsl. Design, R.I. Sch. Design, 1963. Design dir. Boston Children's Mus., 1963-64; pres. Michael Sand, Inc., Brookline, 1964—; acting dir. Richmond (Va.) Children's Mus., 1980-81, Muncie (Ind.) Children's Mus., 1977-78; vis. critic R.I. Sch. Design, Providence, 1967—; vis. instr. Ball State U., Muncie, 1977-78, Harvard U. Grad. Sch. Design, Cambridge, Mass., 1981-83; design dir. Edn. Devel. Ctr., Cambridge, 1969-77, Boston Zool. Soc., 1970-76; planner Nat. Mus. Boy Scouts Am., Murray, Ky., 1986-88. Achievements include design of Lowell (Mass.) Nat. Park, 1986, The Wave, Boston, 1977, Mus. of Sci. Big Dig Exhibit, Boston, 1992, Boston Bicentennial Exhibit, Health Edn. Ctr. Wis. Active Brookline Arts Coun., 1986; founder, dir. Brookline Found., 1986; bd. dirs. Coolidge Corner Theatre Found., 1993—. Mem. Am. Assn. Mus., Boston Computer Soc. (bd. dirs. 1992—), Am. Assn. Youth Mus., Internat. Coun. Mus. Home and Office: 157 Aspinwall Ave Brookline MA 02146

SANDAGE, ALLAN REX, astronomer; b. Iowa City, June 18, 1926; s. Charles Harold and Dorothy (Briggs) S.; m. Mary Lois Connelley, June 8, 1959; children: David Allan, John Howard. AB, U. Ill., 1948, DSc (hon.), 1967; PhD, Calif. Inst. Tech., 1953; DSc (hon.), Yale U., 1966, U. Chgo., 1967, Miami U., Oxford, Ohio, 1974, Graceland Coll., Iowa, 1985; LLD (hon.), U. So. Calif., 1971; D Honoris Causa, U. Chile, 1992. Astronomer Mt. Wilson Obs., Palomar Obs., Carnegie Instn., Washington, 1952—; Peyton postdoctoral fellow Princeton U., 1952; asst. astronomer Hale Obs., Pasadena, Calif., 1952-56; astronomer Obs. Carnegie Instn., Pasadena, Calif., 1956—; sr. rsch. astronomer Space Telescope Sci. Inst. NASA, Balt., 1986—; Homewood Prof. of Physics Johns Hopkins U., Balt., 1987-89; vis. lectr. Harvard U., 1957; mem. astron. expdn. to South Africa, 1958; cons. NSF, 1961-64; Sigma Xi nat. lectr., 1966; vis. prof. Mt. Stromlo Obs., Australian Nat. U., 1968-69, ; vis. rsch. astronomer U. Basel, 1985-92; rsch. astronomer U. Calif., San Diego, 1985-86; vis. astronomer U. Hawaii, 1986; Lindsey lectr. NASA Goddard Space Flight Ctr., 1989; Jansky lectr. Nat. Radio Astron. Obs., 1991; Grubb-Parsons lectr. U. Durham, Eng., 1992. With USNR, 1944-45. Recipient Pope Pius XI gold medal Pontifical Acad. Sci., 1966, Rittenhouse medal, 1968, Presdl. Nat. Medal of Sci., 1971, Adon medal Obs. Nice, 1988, Crafoord prize Swedish Royal Acad. Scis., 1991, Tomalla Gravity prize Swiss Phys. Soc., 1992; Fulbright-Hays scholar, 1972. Mem. Lincei Nat. Acad. (Rome), Am. Astron. Soc. (Helen Warner prize 1960, Russell prize 1973), Royal Astron. Soc. (Eddington medal 1963, Gold medal 1967), Astron. Soc. Pacific (Gold medal 1975), Royal Astron. Soc. Can., Franklin Inst. (Elliott Cresson medal 1973), Phi Beta Kappa, Sigma Xi. Home: 8319 Josard Rd San Gabriel CA 91775-1003 Office: 813 Santa Barbara St Pasadena CA 91101-1232

SANDERGAARD, THEODORE JORGENSEN, information technology director; b. Chgo., Sept. 27, 1946; s. Theodore Jorgensen Sandergaard and Hilda (Roberts) Stec; m. Laraine Ann Spina, May 29, 1965 (div. 1970); 1 child, Michael James; m. Ebba Mary Janet Groth, July 12, 1980. Diploma, USAF, 1964. Data processing mgr. Leo Pharm. Products., Ballerup, Denmark, 1969-73, sr. cons., 1973-80; data processing mgr. Leo Labs. Ltd., Princes Risborough, Eng., 1980-88; dir. info. tech. Dewe Rogerson Ltd., London, 1988—; cons. in field. Author: FLX Programming Language, 1979. Mem. Jr. C. of C., Ballerup, 1971. Mem. IEEE, IEEE Computer Soc., Assn. Computing Machinery, Brit. Computer Soc., Inst. Data Processing Mgmt., Inst. Dirs. Republican. Lutheran. Avocations: history, geopolitics, reading, philately, travel. Home: 12 Gainsborough Pl, Aylesbury HP19 3SF, England Office: Dewe Rogerson Ltd, 3 1/2 London Wall Bldgs, London EC2M 5SY, England

SANDERS, GARY GLENN, electronics engineer, consultant; b. Gettysburg, Pa., Dec. 21, 1944; s. James Glenn Sanders and Martha Maybelle (Fleming) Ehlert; m. Elizabeth Marie Rega, Sept. 9, 1977 (div. Sept. 1981). Cert. med. technologist, Chgo. Inst. Tech., 1970; AA, Mayfair Coll., 1972; BS in Electronic Engring., Cooks Inst., Jackson, Miss., 1982. Registered Internat. Med. Techs. Cons. engr. Electronics Design Services, Chgo., 1977-79; applications engr. Nationwide Electronics Systems, Streamwood, Ill., 1979-80; mng. engr. Electronics Design Ctr. Case Western Res. U., Cleve., 1980-82; sr. project mgr. Scott Fetzer Co., Cleve., 1982-89; v.p. engring. Penberthy, Inc., 1990—; comml. pilot; electronic transduction cons. Teltech Inc., Mpls., 1989—; mem. adv. bd. Electronics Search Group, Indpls., 1991—. Contbr. articles on electronics in medicine and biology to profl. confs. and publs.; patentee in biomed. electronics and indsl. instrumentation, inventor, 1985—. Served with U.S. Army, 1962-68, Vietnam. Decorated DFC, Bronze Star, Air medal, Purple Heart. Fellow Internat. Coll. Med. Technologists; mem. IEEE, AAAS, NRA, DAV, VFW, Instrument Soc. Am. (sr. mem.), Internat. Soc. Hybrid Microelectronics, N.Y. Acad. Scis., Ohio Acad. Sci., Nat. Fire Protection Assn., Am. Legion, Am. Soc. Materials Internat., Boy Scouts Am. Alumni, Nat. Eagle Scout Assn. Republican. Avocations: marquetry, archery. Home: 3104 Prophetstown Rd Rock Falls IL 61071-2556 Office: Penberthy Inc 320 Locust St Prophetstown IL 61277-1147

SANDERS, GEORGIA ELIZABETH, science and mathematics educator; b. Holmwood, La., July 14, 1933; d. Frederick Rudolph and Susie W. (Hackett) S. Student, La. Coll., 1951-53, La. State U., 1959-60; BS, then MS in Microbiology, U. Southwestern La., 1970; MS in Math., U. So. Miss., 1983. Instr. dept. biology U. New Orleans, 1976-79, instr. dept. math., 1983-86; tchr. East Baton Rouge Parish Schs., 1988—; tchr. math. St. Tammany Parish, 1990. Mem. NEA, ASCD, Am. Math. Soc., Math. Assn. Am., Nat. Coun. Tchrs. Math. Home: PO Box 968 Slidell LA 70459-0968

SANDERS, GILBERT OTIS, educational and research psychologist, addictions treatment therapist, consultant, educator; b. Oklahoma City, Aug. 7, 1945; s. Richard Allen and Evelyn Wilmoth (Barker) S.; m. Lidia Julia Grados-Ventura, Aug. 31, 1984. A.S., Murray State Coll, 1965; B.A., Okla. State U., 1967, U. State of N.Y.; M.S., Troy State U., 1970; Ed.D., U. Tulsa, 1974; postdoctoral studies St. Louis U., Am. Tech. U.; grad. U.S. Army Command and Gen. Staff Coll., Ft. Leavenworth, Kas., 1974. Dir. edn. Am. Humane Edn. Soc., Boston, 1975; chmn. dept. computer sci., dir. Individual Learning and Counseling Ctr., asst. prof. pschology and law enforcement Calumet Coll., Whiting, Ind., 1975-78; rsch. psychologist U.S. Army Rsch. Inst., Ft. Hood, Tex., 1978-79; pvt. practice counseling, Killen, Tex., 1978-79; engring. psychologist U.S. Army Tng. and Doctrine Command Systems Analysis Activity, White Sands Missile Range, N.Mex., 1979-80; project dir./rsch. psychologist Applied Sci. Assocs., Ft. Sill, Okla., 1980-81; pvt. practice counseling, Lake St. Louis, 1981-83; assoc. prof. Pittsburg State U., Kans., 1983-85; pres. Applied Behaviroal Rsch. Assocs. (fromerly Southwestern Behavioral Rsch.), Oklahoma City, 1985-92—; pvt. practice counseling Christian Family Counseling Ctr., Lawton, Okla., 1986-87; psychologist, systems analyst U.S Army Field Artillery Sch.-Directorate of Combat Devels., Ft. Sill, Okla., 1987; psychologist U.S. Army Operational Test and Evaluation Agy., Washington, 1988-89;psychologist, drug abuse program dir. Fed. Bur. of Prisons-Fed. Correctional Inst. El Reno, Okla., 1989-91; psychologist, clin. dir. drug abuse program U.S. Penitentiary, Leavenworth, Kans., 1991—; pvt. practice psychologist and profl. counselor Northwest Counseling Ctr., Oklahoma City, 1992—; adj. prof. bus. and psychology Columbia Coll.-Buder Campus, St. Louis, 1982-84; adj. prof. U.S. Army Command Staff and Gen. Coll., 1983-89, Columbia Pacific U., 1984—, Greenwich U., 1990—. Editor: Evaluation for a Manual Backup System for TACFIRE (ARI), 1978, Training/Humane Factors Implications--Copperhaed Operational Test II Livefire Phase, 1979, TRADOC Training Effectiveness Analysis Handbook, 1980, Cost and Training Effectiveness Analysis/TEA 8-80/Patroit Air Defense Missile System, 1980, Cost and Training Analysis/Infantry Fighting Vehicle (Bradley), 1980, Human Factors Implications for the Howitzer Improvement Program, 1989; author research reports. Hon. col. Okla. Gov.'s Staff, Oklahoma City, 1972; hon. ambassador Gov. Okla., 1974. Recipient Kavanough Found. Community Builder award, 1967; named Hon. Col. Okla. Gov. Staff, 1972, Hon. Amb., 1974. Mem. APA, ACA, Am. Assn. Marriage and Family Therapists, Am. Mental Health Counselors Assn., Res. Officers Assn., Commd. Officers Assn., U.S. Pub. Health Svc., Pi Kappa Phi, Alpha Phi Omega. Home: 5404 NW 65th St Oklahoma City OK 73132 Mailing Address: 3436 Tudor Dr Leavenworth KS 66048

SANDERS, JAMES GRADY, biogeochemist; b. Norfolk, Va., June 10, 1951; s. Allen Buford and Maple Seretha (Myers) S.; m. Carmen Lee Nance, Aug. 19, 1972. BS in Zoology, Duke U., 1973; MS in Marine Scis., U. N.C., 1975, PhD in Marine Scis., 1978. Postdoctoral investigator Woods Hole (Mass.) Oceanographic Instn., 1978-80; vis. scientist Chesapeake Biol. Lab. U. Md., Solomons, 1980-81; asst. curator Benedict (Md.) Estuarine Rsch. Lab. Acad. Natural Scis., 1981-85, assoc. curator Benedict (Md.) Estuarine Rsch. Lab., 1985-89, curator Benedict (Md.) Estuarine Rsch. Lab., 1989—, dir. Benedict (Md.) Estuarine Rsch. Lab., 1983—; cons. EPA Sweden, Stockholm, 1985-90; mem. Md. Sea Grant Adv. Com., College Park, 1983-90, Environ. Commn., Calvert City, Md., 1981-88; mem. enbiron. biology panel Office R & D EPA, Washington, 1986-93; regional rep. Coastal Resources Adv. Commn., Md., 1983-86. Contbr. over 50 articles to sci. jours. Grantee NOAA, 1981—, EPA, 1983—. Mem. AAAS, Am. Soc. Limnology and Oceanography, Oceanography Soc., Soc. for Environ. Toxicology and Chemistry, Estuarine Rsch. Fedn. Achievements include first identification of relationships between algal growth and chemical transformations of arsenic in aquatic systems. Office: Acad Natural Scis Benedict Estuarine Rsch Lab Benedict MD 20612

SANDERS, JOHN LYELL, JR., educator, researcher; b. Highland, Wis., Sept. 11, 1924; s. J. Lyell and Grace Ann (Waters) S.; m. Mary Jane Wade, Apr. 23, 1960; children: Alice Elizabeth, William Lyell, Jeanne Frances. BS, Purdue U., 1945; MS, MIT, 1950; PhD, Brown U., 1954; AM, Harvard U., 1960. Section head Nat. Adv. Com. for Aeronautics, Langley Field, Va., 1954-58; lectr. Harvard U., Cambridge, Mass., 1958-60, prof., 1960—; bd. of editors jour. of Math. and Physics, 1963-69; editor's bd. S.I.A.M. Jour. of Applied Math., 1969-79; adv. bd. Acta Mechanica Sinica, Beijing, 1990—. Fellow AAAS, ASME, Am. Acad. Mechanics; mem. Sigma Xi. Democrat. Roman Catholic. Office: Harvard Univ 312 Pierce Hall Cambridge MA 02138

SANDERS, MARC ANDREW, computer technical consultant; b. Chgo., Apr. 21, 1947; s. Edward and Elizabeth Sanders. BA, Roosevelt U., 1973; MAS, Fla. Atlantic U., 1987. Computer programmer Market Facts, Inc., Chgo., 1973-76, N.E. Ill. Planning Commn., Chgo., 1977; salesman Radio Shack, Tamarac, Fla., 1982-83; sr. analyst/tech. cons. Birch/Scarborough Rsch., Coral Springs, Fla., 1984—. Mem. IEEE (computer soc.), Phi Kappa Phi, Beta Gamma Sigma, Upsilon Pi Epsilon. Democrat. Jewish. Avocations: golf, bicycling, walking, writing, science fiction. Office: Birch/Scarborough Rsch PO Box 9742 Coral Springs FL 33075-9742

SANDERS, MICHAEL KEVIN, hypertension researcher; b. Birmingham, Ala., Apr. 9, 1970; s. Donald Hubert and Mary Jane (Philpott) S. BS, Birmingham So. Coll., 1992; postgrad., U. South Ala. Coll. Medicine, 1993—. Rsch. asst. neuropsychiatry dept. U. Ala. Sch. Medicine, Birmingham, 1988-92; rsch. asst. biology dept. Birmingham (Ala.) So. Coll., 1991-92; rsch. asst. hypertension dept. U. Ala. Sch. Medicine, Birmingham, 1992-93; rsch. asst. Paris, 1992. Mem. Tri Beta, Phi Sigma Iota, Sigma Xi. Baptist. Home: 3060 Overton Rd Birmingham AL 35223

SANDERS, ROBERT B., biochemistry educator; b. Augusta, Ga., Dec. 9, 1938; s. Robert and Lois (Jones) S.; m. Gladys Nealous, Dec. 23, 1961; children: Sylvia Lynne, William Nealous. BS, Paine Coll., 1959; MS, U. Mich., 1961, PhD, 1964; postdoctoral, U. Wis., 1964-66. Vis. scientist Battelle Meml. Inst., Richland, Wash., 1970-71; vis. assoc. prof. U. Tex. Med. Sch., Houston, 1974-75; program dir. NSF, Washington, 1978-79; from asst prof. to prof. U. Kans., Lawrence, 1966-86, prof. biochemistry, 1986—, assoc. dean The Grad. Sch., 1987—, assoc. vice chancellor, 1989—; cons. NSF, Washington, 1983-92; cons. Dept. Edn., Washington, 1983, NIH, Washington, 1982, Interx Rsch. Corp., Lawrence, 1972-80, Nat. Rsch. Coun., Washington, 1973—. Contbr. over 50 articles to profl. jours. Bd. dirs. United Child Devel. Ctr., Lawrence, Kans., 1968-93; mem. bd. high edn. United Meth. Ch., 1976-80. With USAR, 1955-62. Paine Coll. Alumni Assn. scholar, 1958; U. Mich. grad. fellow, 1959-64; NIH postdoctoral fellow, 1974-75, Am. Cancer Soc. fellow, 1964-66, Battelle Meml. Inst. fellow, 1970-71; recipient numerous rsch. grants. Mem. AAUP, Am. Soc. Biochemistry and Molecular Biology, Am. Soc. for Pharmacology and Exptl. Therapeutics, Sigma Xi. Office: U Kans Dept Biochemistry Lawrence KS 66045

SANDERS, ROBERT WALTER, ecologist, researcher; b. Washington, Feb. 25, 1952; s. George Bradley and Ida Mary (Harmon) S.; m. Elizabeth Hunter Davies, Dec. 28, 1976; children: Katherine Hunter, Nicholas Edward. BA in Biology, U. Va., 1979; MS in Zoology, U. Maine, 1982; PhD, U. Ga., 1988. Postdoctoral assoc. U. Ga., Athens, 1988-89; asst. curator Acad. Natural Scis. Phila., 1989—; peer reviewer NSF, NATO Sci. Affairs, Sea Grant Program, 1988—, Sci. Marine Biology, Marine Ecology Progress Series, Limnology & Oceanography, Archiv fur Hydrobiologie, Bull. Marine Sci., Can. Jour. Fisheries and Aquatic Sci., Jour. Great Lakes Rsch., Marine Microbial Food Webs, editorial bd., 1993—; adj. asst. prof. biology U. Pa., 1993—. Contbr. chpts: The Biology of Heterotrophic Flagellates, 1991, Ecology and Classification of North American Freshwater Invertebrates, 1991, Advances in Microbial Ecology, 1988; contbr. articles to Marine Ecology Progress Series, Limnology and Oceanography, Applied and Environ. Microbiology, Microbial Ecology. Grantee NSF, 1988, 90, Nat. Sea Grant program, 1991, 93. Mem. Am. Soc. for Microbiology, Am. Soc. Limnology and Oceanography, Soc. Protozoologists, Sigma Xi, Phi Kappa Phi. Achievements include research in ecology of marine and freshwater algae, protozoa, bacteria, invertebrates, fate and effects of toxic materials, bioassay, eutrophication of coastal and fresh waters. Office: Acad Natural Scis 1900 Bejamin Franklin Pkwy Philadelphia PA 19103

SANDERSON, GLEN CHARLES, science director; b. Wayne County, Mo., Jan. 21, 1923; married; 2 children. BS, U. Mo., 1947, MA, 1949; PhD, U. Ill., 1961. Game biologist Iowa State Conservation Commn., 1949-55, Ill. Dept. Conservation, 1955-60; game biologist Ill. Nat. History Survey, Champaign, 1955-60, assoc. wildlife specialist, 1960-63, wildlife specialist, acting head wildlife rsch., 1963-64, head sect. wildlife rsch., 1964-90, prin. scientist emeritus, dir., 1990—; prof. U. Ill., 1965—; adj. rsch. prof. So. Ill. U., 1964, adj. prof. 1964-84. Editor Jour. Wildlife Mgmt., 1971-72. Recipient Oak Leaf award Nature Conservancy, 1975. Mem. AAAS, Am. Soc. Mammal, Am. Inst. Biol. Sci., Wildlife Soc. (Aldo Leopold Meml. award 1992). Achievements include research in population dynamics of wild animals, especially furbearers, physiological factors of reproductive and survival rates, and lead poisoning in waterfowl. Office: Ill Natural History Survey Ctr Wildlife Ecology 711 S State St Champaign IL 61820*

SANDHAM, WILLIAM ALLAN, electronics and electrical engineer, educator; b. Glasgow, Scotland, July 17, 1952; s. Herbert Gladstone and Annie McDonald (Johnston) S. BSc in Electronic and Elec. Engring. with honors, U. Glasgow, 1974; PhD in Electronic, Elec. Engring., U. Birmingham, Eng., 1981. Registered engr.; registered geophysicist. Med. physicist Greater Glasgow Health Bds., 1974-75; rsch. student U. Birmingham, 1975-78, rsch. fellow, 1978-80; sr. programmer British Nat. Oil Corp., Glasgow, 1980-84; rsch. seismic analyst Britoil, Glasgow, 1984-86; lectr. U. Strathclyde, Scotland, 1986-90, sr. lectr., 1990—; sec., originator West Scotland Digital Signal Processing Consortium, 1982-86; sr. cons. British Gas, Glasgow, 1989-90, BP Exploration, Glasgow, 1988-90. Publs. chmn. Internat. Conf. Acoustics, Speech and Signal Processing, Glasgow, 1989; publs. and tech. com. Transputer Applications, Glasgow, 1991; gen. chmn. Internat. Workshop Parallel Architectures Seismic Data Processing, Glasgow, 1991. Mem. IEEE, European Assn. Exploration Geophysicists, Soc. Exploration Geophysicists, Internat. Conf. European Assn. Signal Processing (chmn. publs. and spl. session), Applied Signal Processing (mem. adv. bd. internat. jour. 1993—). Avocations: marathon running, weight training, music. Office: U Strathclyde, 205 George St, Glasgow G1 1XW, Scotland

SANDHU, BACHITTAR SINGH, ophthalmologist; b. Lyallpur, Punjab, Pakistan, Sept. 15, 1935; s. Kartar Singh and Pritam Kaur (Uppal) S.; m. Amarjeet Kaur Narula; children: Manjeet, Triptjeet, Jagjeet, Amanjeet, Harinderjeet. MBBS, Med. Coll. Amritsar, Punjab, India, 1958; DO, Royal Coll. Surgeons, London, 1962. Fellow Royal Coll. Surgeons, Eng.; Fellow Coll. Ophthalmologists, Eng.; med. cons. eye surgeon. Resident house surgeon ophthalmology Irwin Hosp., New Delhi, 1959, resident house physician gen. medicine, 1959; resident house officer gen. surgery Bury (Eng.) Gen. Hosp., 1960; resident sr. house officer ophthalmology Newcastle Gen. Hosp., Newcastle-Upton-Tyne, Eng., 1961; registrar in ophthalmology Derbyshire Royal Infirmary, Derby, Eng., 1963, Lewisham Group of Hosps., London, 1964; sr. registrar in ophthalmology St. Paul's Eye Hosp., Liverpool, Eng., 1964; cons. ophthalmologist Bedford (Eng.) Gen. Hosp., 1965-88, Lister Hosp. Stevenage, Hertfordshire, Eng., 1965—, Queen Elizabeth II Hosp., Welwyn Garden City, Eng., 1988—; hon. sec. Ophthalmic Adv. Com. N.W. Met. Regional Hosp. Bd., London, 1972-74; mem. North Herts Health Authority, North Herts Dist., 1981-89; chmn. Dist. Hosp. Med. Com., Stevenage Herts, 1987-89. Mem. East Herts Div. British Med. Assn. (pres. 1977-78), Oxford Ophthal. Congress. Sikh. Avocations: music, poetry. Home: 3 Rectory Croft, Rectory Lane Stevenage, Herefordshire SG1 4BY, England

SANDMEIER, RUEDI BEAT, agricultural research executive; b. Basel, Switzerland, Apr. 25, 1945; came to U.S., 1985; s. Emil and Margrit (Bolli) S.; m. Susann R. Guggenbühl, July 10, 1970; children: Corinna D., Franziska C. BS, U. Basel, 1968, PhD in Chemistry, 1973. Postdoctoral fellow Syntex Corp., Palo Alto, Calif., 1973-74; sr. rsch. chemist Sandoz Ltd., Basel, 1974-85; dir. biosci. Northrup King Seeds, Stanton, Minn., 1985-87; v.p. rsch. Sandoz Crop Protection, Palo Alto, 1987—. Patentee fungicides, herbicides, prodn. process. Mem. Soc. Chem. Industry, Internat. Soc. Plant Molecular Biology, Am. Chem. Soc. Avocations: cooking, hiking, nature walks. Office: Sandoz Crop Protection 975 S California Ave Palo Alto CA 94304-1104

SANDOR, GEORGE NASON, mechanical engineer, educator; b. Budapest, Hungary, Feb. 24, 1912; came to U.S., 1938, naturalized, 1949; s. Alexander S. and Maria (Adler) S.; m. Magda Breiner, Dec. 5, 1964; stepchildren: Stephen Gergely, Judith Patricia Gergely (Mrs. J. Peter Vernon). Diploma in Mech. Engring, Poly. U. Budapest, 1934, D. Eng. Sci., Columbia U., 1959; D (hon.), Budapest Technol. U., 1986. Registered profl. engr., Fla., N.J., N.Y., N.C., cert. of qualification Nat. Council Engring. Examiners. Asst. to chief engr. Hungarian Rubber Co., Budapest, 1935-36; mfg. dept. head Hungarian Rubber Co., 1936-38; design engr. Babcock Printing Press Corp., New London, Conn., 1939-44; v.p., chief engr. H.W. Faeber Corp., N.Y.C., 1944-46; chief engr. Time Inc. Graphic Arts Research Labs., Springdale, Conn., 1946-61, Huck Co., Inc., N.Y.C., 1961; assoc. prof. mech. engring. Yale, 1961-66; prof. mech. engring. Rennselaer Poly. Inst., Troy, N.Y., 1966-67; Alcoa Found. prof. mech. design, chmn. machines and structures div. Rennselaer Poly. Inst., 1967-75; rsch. prof. mech. engring. U. Fla., Gainesville, 1976-89, rsch. prof. emeritus, 1989—; dir. mech. engring. design and rotordynamics labs. U. Fla., 1979-87; instr. engring. U. Conn. Extension, New London and Norwich, 1940-44; lectr. mech. engring. Columbia U., N.Y.C., 1961-62; dir. Huck Design Corp., Huck Co., Inc., Montvale, N.J., 1964-70; cons. engr., printing equipment and automatic machinery, mech. engring. design, 1961—; cons. NSF Departmental and Instl. Devel. Program, 1970-72; cons. nat. materials adv. bd. Nat. Acad. Scis., 1974; cons. Xerox Corp., Burroughs Corp., Govt. Products div. Pratt & Whitney Aircraft Co., Time Inc., also others, 1961-92; prin. investigator, co-investigator NSF, U.S. Army Research Office and NASA sponsored research at Yale U.; dir. and co-dir. NSF, U.S. Army Research Office and NASA sponsored research at Rennselaer Poly. Inst. and; U. Fla. at Gainesville; chief U.S.A. del. to Internat. Fedn. for Theory Machines and Mechanisms, 1969-75; cons. for materials conservation through design Office Tech. Assessment, Congress U.S., 1977. Author: (with others) Mechanical Design and Systems Handbook, 1964, 2d edit., 1985, Linkage Design Handbook, 1977, Mechanism Design-Analysis and Synthesis, vol. 1, 1984, 2d edit., 1991, Advanced Mechanism Design, Analysis and Synthesis, vol. 2, 1984; mem. editorial bd. Jour. Mechanism, 1966-72, Machine and Mechanism Theory, 1972—, Robotica, 1982—; contbr. articles to profl. jours. Recipient Outstanding Achievement awrd Northctrl. sect. Fla. Engring. Soc., 1983; Fla. Blue Key Leadership award for disting. faculty mem. U. Fla., 1985; elected hon. mem. Internat. Fedn. for Theory Machines and Mechanisms, 1987, Hungarian Acad. Scis., Budapest, 1993. Fellow ASME (life, Machine Design award 1975, mechanisms com. award 1980, hon. mem. 1991); mem. NSPE, Am. Soc. Engring. Edn. (Ralph Coats Roe award 1985), N.Y. Acad. Scis., Am. Acad. Mechanics, Hungarian Acad. Scis. (hon. mem. 1993), Flying Engrs. Internat., Sigma Xi, Tau Beta Pi, Pi Tau Sigma. Achievements include patent for rotary-linear actuator for robotic manipulators, and 5 others. Home: 136 Broadview Acres Highland NC 28741

SANDQUIST, GARY MARLIN, engineering educator, researcher, consultant, author; b. Salt Lake City, Apr. 19, 1936; s. Donald August Sandquist and Lillian (Evaline) Dunn; m. Kristine Powell, Jan. 17, 1992; children from previous marriage: Titia, Julia, Taunia, Cynthia, Carl; stepchildren: David, Michael, Scott, Diane, Jeff. BSME, U. Utah, 1960, PhD in Mech. Engring., 1964; MS in Engring. Sci., U. Calif., Berkeley, 1961; postdoctoral student MIT, 1969-70. Registered profl. engr., Utah, New York; cert. health physicist, cert. quality auditor. Staff mem. Los Alamos Sci. Lab. (N.Mex.), 1966; vis. scientist MIT, Cambridge, 1960-70; research prof. surgery Med. Sch., U. Utah, Salt Lake City, 1974—; prof., dir. nuclear engring., mech. engring. dept. U. Utah, Salt Lake City, 1975—, acting chmn. dept., 1984-85; expert in nuclear sci. IAEA, UN, 1980-82; chief scientist Rogers and Assocs. Engring. Corp., Salt Lake City, 1980—; advisor rocket design Hercules, Inc., Bachus, Utah, 1962; sr. nuclear engr. Idaho Nat. Engring. Lab., Idaho Falls, Idaho, 1963-65; cons. various cos.; cons. nuclear sci. State of Utah, 1982—. Author: Geothermal Energy, 1973; Introduction to System Science, 1984. Served to comdr. USNR, 1954-56, Korea. Recipient Glen Murphy award in nuclear engring. Am. Soc. Engring. Edn., 1984. Fellow ASME, Am. Nuclear Soc.; mem. Am. Soc. Quality Control, Am. Health Physics Soc., Alpha Nu Sigma of Am. Nuclear Soc., Sigma Xi, Tau Beta Pi, Pi Tau Sigma. Republican. Mormon. Home: 2564 Neffs Circle Salt Lake City UT 84109-4055 Office: U Utah 1205 Merrill Engring Bldg Salt Lake City UT 84112

SANDRY, KARLA KAY FOREMAN, industrial engineering educator; b. Davenport, Iowa, Apr. 2, 1961; d. Donald Glen and Greta Geniveve (VanderMaten) Foreman; m. William James Sandry, Oct. 12, 1985; 1 child, Zachary Quinn. BS in Indsl. Engring., Iowa State U., 1983; MBA, U. Iowa, 1992. Quality control supr., indsl. engr. Baxter Travenol Labs, Hays, Kans., 1983-84; indsl. engr. HQ Amccom, Rock Island, Ill., 1984-86; mgmt. engr. St. Lukes Hosp., Davenport, 1986-90; adj. instr. engring. St. Ambrose U., Davenport, 1990—; chair space allocations St. Luke's Hosp., Davenport, 1987-90; pres. employee rels. coun. HQ Amccom, Rock Island, 1986, chair savings bonds, 1985; speaker in field. Vol., past counselor Fellowship Christian Athletes Cen. High Sch., Davenport, 1984-87, vol., adult chpt., 1988—; counselor Explorer Scout Troop, Davenport, 1984-85; leader, counselor ch. youth group, 1985-89. Mem. Inst. Indsl. Engrs. (sr. mem.), Healthcare Info. & Mgmt. Systems Soc. (recognition & communications com. 1988), Soc. for Health Systems (founding mem.), Found. for Christian Living, Iowa State U. Alumni Assn., Positive Thinkers Club. Avocations: vocalist, tennis, golf, violinist, playing with son. Office: St Ambrose U 518 W Locust St Davenport IA 52803

SANDS, HOWARD, pharmacologist, biochemist, research scientist; b. N.Y.C., Aug. 20, 1942. BA, Rutgers U., 1964; PhD in Pharmacology, Case Western Reserve U., 1969. NIH tng. grant renal disease Northwestern U., 1969-71; pharmacologist Nat. Jewish Hosp. & Rsch. Ctr., 1971-81; group leader pharmacology, toxicology New Eng. Nuclear, 1981-89; prin. rsch. sci. Dupont-Merck Pharmaceutical Co., Wilmington, Del., 1989—; pharmacologist Vet. Administrn. Rsch. Hosp. 1970-71; asst. prof. Dept. Oral Biol. Sch. Dentistry, U. Colo., 1975-77. Mem. Am. Assn. Cancer Rsch. Achievements include rsch. in biochemical aspects of drug actions, radio pharm. and related immunology, pharmacokinetics and metabolism of macromolecular drugs. Address: 2417 Landon Dr Wilmington DE 19810

SANDSTRÖM, GUNNAR EMANUEL, microbiologist; b. Lycksele, Sweden, June 1, 1951; s. Lars Åke Emanuel and Maj Dovi Helena (Holmgren) S.; m. Elsa Lillemor Johansson, May 30, 1986; children: Jörgen, Sandra. D. in Med. Sci., U. Umeå (Sweden), 1988. Asst. U. Umeå, 1970-80; asst. Nat. Def. Rsch. Establishment, Umeå, 1980-85, sr. rsch. officer, 1985-89; postdoctoral fellow US Army Med. Rsch. Inst., Ft. Detrick, Frederick, Md., 1989-90; head div. microbiology Nat. Def. Rsch. Establishment, Umeå, 1990-91, assoc. prof., 1991—; sr. rsch. officer NRC, Frederick, 1989-90. Contbr. tularemia articles to profl. jours.; inventor-patentee rapid diagnosis of bacteria. Chmn. Local Community Coun., Umeå, 1979-88. Mem. Lion Club (Sävar, Sweden). Avocations: music, fishing, traveling. Home: Vallmovägen 33, S 90352 Umeå Sweden Office: Nat Def Rsch Establishment, Cementvägen 20, S 901 82 Umeå Sweden

SANDSTROM, ROBERT EDWARD, physician, pathologist; b. Hull, Yorkshire, Eng., Apr. 4, 1946; came to U.S., 1946; s. Edward Joseph and Ena Joyce (Rilatt) S.; m. Regina Lois Charlebois (dec. May 1987); children: Karin, Ingrid, Erica. BSc, McGill U., Montreal, 1968; MD, U. Wash., 1971. Diplomate Am. Bd. Pathology, Am. Bd. Dermatopathology. Internship Toronto (Can.) Gen. Hosp., 1971-72; resident pathologist Mass. Gen. Hosp., Boston, 1974-78; clin. fellow Harvard U. Med. Sch., Boston, 1976-78; cons. King Faisel Hosp., Riyadh, Saudi Arabia, 1978; pathologist, dir. labs. St. John's Med. Ctr., Longview, Wash., 1978—; v.p. Intersect Systems Inc.; chmn. bd. Cowlitz Med. Svc., Longview, 1988; participant congl. sponsored seminar on AIDS, Wash., 1987. Script writer movie Blood Donation in Saudi Arabia, 1978; contbr. articles to profl. jours. Surgeon USPHS, 1972-74. Fellow Coll. Am. Pathologists, Royal Coll. Physicians; mem. Cowlitz-Wahkiakum County Med. Soc. (past pres.). Roman Catholic. Avocations: sport fishing, mountain climbing, philately. Home: 49 View Ridge Ln Longview WA 98632-5556 Office: Lower Columbia Pathologists 1606 E Kessler Blvd Ste 100 Longview WA 98632-1841

SANDSTRUM, STEVE D., engineer, marketing manager; b. Ulysses, Kans., Dec. 8, 1953; s. Don Eugene and Alleene (Lawrence) S.; m. Nancy Heinzer, Aug. 28, 1976; 1 child, Andrew. BS in Zoology, Okla. State U., 1976, MS in Engring., 1980. Registered engr. in training. Devel. engr. Phillips Driscopipe, Richardson, Tex., 1984-85, sr. tech. engr., 1985-86, quality assurance specialist, 1986-88; sr. polymer engr. Solvay Polymers, Houston, Tex., 1988; tech. svc. mgr. Poly Pipe Industry, Gainesville, Tex., 1988; group leader Solvay Polymers, Houston, 1988-91, mktg. mgr. profl. engr., 1991—; exec. bd. Plastics Pipe Inst., Wayne, N.J., 1989-91, adv. bd., 1991—, hydrostatic bd., 1991—; tchr. Okla. State U., Stillwater, 1979-80. Author: (chpt.) Pipeline Rehabilitation, 1987, Above Ground Applications, 1989, editor: Handbook of Polyethylene Pipe, 1991. Bd. dirs. Sea Lion Swim Team, Kingwood, Tex., 1991. Recipient rsch. award Gen. Motors, 1981. Mem. NSPE, ASTM (Cert. of Appreciation 1993), Soc. Plastic Engrs., Am. Water Works Assn., Plastic Pipe and Fittings Assn., Am. Pub. Works Assn., Tex. Soc. Profl. Engrs., Alpha Pi Mu. Republican. Presbyterian. Achievements include patent pending (with others). Office: Solvay Polymers 3333 Richmond Ave Houston TX 77098-3007

SANDT, JOHN JOSEPH, psychiatrist; b. N.Y.C., June 29, 1925; s. John Jacob and Victoria Theodora Sandt; m. Mary Cummings Evans, Sept. 14, 1946; children: Christine, Karen, John K., Kurt, Colin, Carol; m. Mary W. Griswold, July 10, 1992. BA, Vanderbilt U., 1948; MA, Yale U., 1951; MD, Vanderbilt U., 1957. Instr. English Vanderbilt U., Nashville, 1951-52, Syracuse (N.Y.) U. Coll., 1960-61; intern SUNY Upstate Med. Ctr., Syracuse, 1957-58, resident, 1958-61; instr. psychiatry Southwestern Med. Sch., Dallas, 1961-63; chief psychiatry VA Med. Ctr., Dallas, 1961-63; chief outpatient clinic Dept. Mental Health, Springfield, Mass., 1963-66; asst. prof. psychiatry U. Rochester (N.Y.) Med. Sch., 1966-75, clin. assoc. prof. psychiatry, 1975—; chief psychiatry Clifton Springs (N.Y.), 1985-88, VA Med. Ctr., Bath, N.Y., 1988—; cons. psychiatry VA Med. Ctr., Northampton, Mass., 1965-66, Springfield Coll., 1964-66, Brockport (N.Y.) State Coll., 1966-75, Fairport (N.Y.) Bapt. Home, 1966-88; asst. dir. ind. study program U. Rochester Med. Sch., 1971-75. Author: Clinical Supervision of Psychiatric Resident, 1972; contbr. articles on psychiat. consultation and treatment to profl. jours. Vestryman All Saints Episcopal Ch., South Hadley, Mass., 1963-66. With USNR 1944-46, PTO. Nathaniel Currier fellow Yale Grad. Sch., 1948-49. Mem. Am. Psychiat. Assn., Am. Assn. Geriatric Psychiatry, AAAS, Med. Soc. N.Y. State. Office: VA Med Ctr Bath NY 14810

SANDU, BOGDAN MIHAI, aeronautical engineer; b. Bucharest, Romania, Jan. 15, 1968; came to U.S., 1984; s. Tanase and Maria (Neacsu) S. BS, RPI, 1991, MS, 1993. Rsch. asst. Rensselaer Poly. Inst., Troy, N.Y., 1990; mgr. Carmella's Cafe, Colonie, N.Y., 1989—; teaching asst. Rensselaer Poly. Inst., 1991—. Mem. AIAA, Am. Helicopter Soc., Nat. Tech. Assn. Greek Orthodox. Home: 204 Washington Ave Albany NY 12210

SANDU, CONSTANTINE, development engineer; b. Costesti, Arges, Romania, Nov. 9, 1943; came to U.S. 1979, naturalized 1984; s. Dumitru and Maria (Calinoiu) S. Eng., U. Galatz, Romania, 1966; PhD, U. Wis., 1989. Plant engr. Fruit and Vegetables Co., Riureni, Romania, 1967-68; prof.'s asst. U. Galatz, Romania, 1968-75; vis. scientist Fed. Rsch. Ctr. Nutrition, Karlsruhe, Ger., 1975-77; R & D engr. Soc. for Ind. Heating & Engring., Krefeld, Ger., 1978-79; rsch. asst. U. Wis., Madison, 1979-86; sr. devel. engr. The Quaker Oats Co., Barrington, Ill., 1986—; adj. prof. Purdue U., W. Lafayette, Ind. 1989—. Contbr. articles to profl. jours.; editor: Fouling and Cleaning in Food Processing, 1985. Mem. Inst. Food Technologists, Math. Assn. Am., Sigma Xi, Phi Tau Sigma. Avocations: philosophy, history, foreign languages, body building, tennis. Office: The Quaker Oats Co 617 W Main St Barrington IL 60010-4199

SANDY, EDWARD ALLEN, obstetrician/gynecologist; b. Harrisonburg, Va., Nov. 7, 1958; s. Edward Allen and Joy Elizabeth (Rhodes) S.; m. LuAnn Gerig, Mar. 6, 1982; children: Christian Edward, Kathryn Elizabeth. BS in Chemistry, Eastern Mennonite Coll., 1981; MD, Eastern Va. Med. Sch., 1984. Diplomate Nat. Bd. Med. Examiners, Am. Bd. Ob/Gyn. Internship Ohio State U. Hosps. and Clinics, Columbus, 1984-85; resident physician, clin. instr. dept. ob-gyn Ohio State U., Columbus, 1984-88; pvt. practice ob/gyn Cambridge, Ohio, 1988-90, Mt. Pleasant, Iowa, 1991—; chmn. laser com. Guernsey Meml. Hosp., Cambridge, 1988-90, chmn. surg. com., 1990; dir. maternal child health com. Henry County Health Ctr., Mt. Pleasant, 1991—. Contbr. articles to Jour. Ob/Gyn, Am. Jour. Perinatology, Region V Perinatal Newsletter, Jour. Gynecologic Surgery, Soc. Perinatol Obstetrics., Am. Jour. of Ob-Gyn, Ob-Gyn. Bd. dirs. S.E. Ohio Symphony Assn., Cambridge, 1989-90; bd. dirs. Muskingham coun. Boy Scouts Am., Zanesville, Ohio, 1990; bd. dirs. S.E. Iowa Symphony Orch., Mt. Pleasant, 1991—. Fellow Am. Coll. Ob-Gyn; mem. Am. Fertility Soc,. Am. Assn. Gynecol. Laparascipists, Gynecol. Laser Soc., Cen. Assn. Ob-Gyn. Office: Park Plz Med Bldg Ste 27 Mount Pleasant IA 52641

SANES, JOSHUA RICHARD, biologist, researcher; b. Buffalo, N.Y., Sept. 5, 1949; s. Irving and Carlyn (Mildred) S.; m. Susan Corcoran, Dec. 27, 1982; children: Jesse, Amelia. BA, Yale U., 1970; MA, PhD, Harvard U., 1976. Asst. prof. Washington U. Med. Sch., St. Louis, 1980-85, assoc. prof., 1985-89, prof., 1989—; mem. neurology study sect. NIH, Washington, 1988—. Contbr. over 100 articles to profl. jours. Fellow AAAS; mem. NIH (bd. sci. counselor 1993—), Soc. for Neurosci. (councilor 1990—), Muscular Dystrophy Assn. (scientific adv. bd. 1991—). Office: Washington U Med Ctr Dept Anatomy Saint Louis MO 63110

SANETO, RUSSELL PATRICK, neurobiologist; b. Burbank, Calif., Oct. 10, 1950; s. Arthur and Mitzi (Seddon) S. BS with honors, San Diego State U., 1972, MS, 1975; PhD, U. Tex. Med. Br., 1981. Teaching asst. San Diego State U. 1969-75; substitute tchr. Salt Lake City Sch. Dist., 1975; teaching and research asst. U. Tex. Med. Br., 1976-77, NIH predoctoral fellow, 1977-81, postdoctoral fellow, 1981; Jeanne B. Kempner postdoctoral fellow UCLA, 1981-82, NIH postdoctoral fellow, 1982-87; asst. prof. Oreg. Regional Primate Rsch. Ctr. div. Neurosci., Beaverton, 1987-89; asst. prof. dept. cell biology and anatomy Oreg. Health Scis. U., Portland, 1988-90, U. Osteo. Medicine & Surgery, 1991—; lectr. rsch. methods Grad. Sch., 1982; vis. scholar in ethics So. Baptist Theol. Sem., Louisville, 1981. Contbr. articles to profl. jours. Recipient Merit award Nat. March of Dimes, 1978; named one of Outstanding Young Men in Am., 1979, 81, Man of Significance, 1985. Mem. Bread for World, Save the Whales, Sierra Club, Am. Soc. Human Genetics, AAAS, Winter Confs. Brain Research, Neurosci. Study Program, N.Y. Acad. Scis., Am. Soc. Neurochem., Soc. Neurosci., Am. Soc. Neurochemistry, Soc. Neurosci. Democrat. Mem. Evangelical Free Ch. Club: World Runners.

SANFORD, THOMAS WILLIAM LOUIS, physicist; b. Seattle, Jan. 24, 1943; s. F. Bruce and Wilma Louise (Steinhauser) S.; m. Joanne Marie Cicala, Jan. 23, 1969. BS in Math., U. Wash., 1965, BS in Physics, 1965; MA in Physics, Columbia U., 1967, PhD in Physics, 1973. Rsch. assoc. Rutherford Lab., Harwell, Eng., 1972-74, MIT, Boston, 1974-75, NYU, N.Y.C., 1975-78, Max-Planck Inst. for Physics, Munich, Germany, 1978-81; staff mem. Los Alamos (N.Mex.) Nat. Lab., 1981-82; sr. mem. tech. staff Sandia Nat. Labs., Albuquerque, 1982-90, disting. mem. tech. staff, 1990—. Contbr. articles to profl. jours. Recipient Recognition for Excellence award U.S. Dept. Energy, 1990. Mem. IEEE, AAAS, Am. Phys. Soc. Achievements include devel. novel diodes to control world's most powerful 20 MeV electron beam enabling generation of useful radiation fields; showed that the beam could be transported with stability over long distances for efficient energy transport. Home: 918 Tramway Ln NE Albuquerque NM 87122 Office: Sandia Nat Labs Orgn 1231 PO Box 5800 Albuquerque NM 87185-5800

SANGER, FREDERICK, retired molecular biologist; b. Rendcomb, Gloucestershire, Eng., Aug. 13, 1918; s. Frederick and Cicely Sanger; m. Joan Howe, 1940; children: Robin, Peter Frederick, Sally Joan. B.A., St. John's Coll., Cambridge U., 1940, Ph.D., 1943; D.Sc. (hon.), Leicester U., 1968, Oxford U., 1970, Strasbourg U., 1970. Beit Meml. Med. Research fellow U. Cambridge, 1944-51, research scientist dept. biochemistry, 1944-61, research scientist, div. head Med. Research Council Lab. of Molecular Biology,, 1962-83. Contbr. articles in field to sci. jours. Recipient Nobel prize for chemistry, 1958, 80; Gairdner Found. ann. award, 1971, 79, William Bate Hardy prize Cambridge Philos. Soc., 1976, Copley medal Royal Soc., 1977; fellow King's Coll., Cambridge U., 1954. Mem. Am. Acad. Arts and Scis. (hon. fgn. mem.), Am. Soc. Biol. Chemists (hon.), Fgn. Assn., NAS. Home: Far Leys Fen Ln, Swaffham Bulbeck, Cambridge CB5 ONJ, England

SANGREY, DWIGHT A., civil engineer, educator; b. Lancaster, Pa., May 24, 1940; m. 1964; 5 children. B.S., Lafayette Coll., 1962; M.S., U. Mass., 1964; Ph.D. in Civil Engring., Cornell U., 1968. Engr. H.L. Griswold, Cons. Engrs., 1960-64; project engr. Shell Oil Co., Tex., 1964-65; asst. prof. civil engring. Queen's U., Ont., Can., 1967-70, assoc. prof., 1970-77; prof. civil and environ. engring. Cornell U., 1977-79; prof. civil and engring., dept. chmn. Carnegie-Mellon U., Pitts., 1979-85; dean Sch. Engring. Rensselaer Poly. Inst., Troy, N.Y., 1985-88; pres. Oreg. Grad. Inst. Sci. and Tech., Portland, 1988—. Recipient Rsch. award ASTM, 1969, ASCE, 1985, 90, Teaching award Tau Beta Pi, 1970, Chi Epsilon, 1971, 77, ASCE, 1985. also: Oreg Grad Inst PO Box 91000 20000 NW Walker Rd Portland OR 97291-1000

SANI, ROBERT LEROY, chemical engineering educator; b. Antioch, Calif., Apr. 20, 1935; m. Martha Jo Marr, May 28, 1966; children: Cynthia Kay, Elizabeth Ann, Jeffrey Paul. B.S., U. Calif.-Berkeley, 1958, M.S., 1960; Ph.D., U. Minn., 1963. Postdoctoral researcher dept. math Rensselaer Poly. Inst., Troy, N.Y., 1963-64; asst. prof. U. Ill., Urbana, 1964-70, assoc. prof., 1970-76; prof. chem. engring. U. Colo., Boulder, 1976—; co-dir. Ctr. for Low-g Fluid Mechanics and Transport Phenomena, U. Colo., Boulder, 1986-89, dir., 1989—; assoc. prof. French Ministry Edn., 1982, 84, 86, 92; cons. Lawrence Livermore Nat. Lab., Calif., 1974—. Contbr numerous chpts. to profl. publs.; mem. editorial bd. Internat. Jour. Numerical Methods in Fluids, 1981—, Revue Européenne des Éléments Finis, 1990—,. Guggenheim fellow, 1970. Mem. AICE, Soc. for Applied and Indsl. Math., World User Assn. in Applied Computational Fluid Dynamics (bd. dirs.). Democrat. Office: U Colo Dept Chem Engring Campus Box 424 Boulder CO 80309

SAN JUAN, GERMAN MORAL, structural engineer, international consultant; b. Pasig, Rizal, Philippines, Feb. 14, 1940; s. Victoriano Mijares and Paz Antonio (Moral) S.J.; m. Sudsanguansri Sirisantana, Jan. 10, 1969; children: Rosellen, Rosalie, Rusukon, German Jr. AS in Geodetic Engring., Nat. U., 1958; BS in Structural Engring., U. Ill., 1973; MCE, NorthwesternU., 1976; postgrad., Ill. Inst. Tech., 1977. Registered profl. engr., Ill., N.Y., Calif., Ohio, Pa., W.Va., Mass., N.J., Ind.; registered civil engr. Thailand, Malaysia, Philippines. Sr. surveyor Quedding Surveying Co. Ltd., Manila, 1958-62; sr. project engr. Techdata Co., Ltd., Vietnam, Cambodia, Laos, 1962-69; structural engr. Engrs. Collaborative, Chgo., 1973-76; sr. structural engr. Consoer & Townsend, Chgo., 1976-77; sr. project structural engr., project mgr. Lester B. Knight & Assocs., Chgo., 1977-80; bridge structure cons. Malaysian Govt., Kuala Lumpur, 1981-82; pres., head cons. San Juan & Assocs. Co., Ltd., Bangkok, 1980—; pres. San Juan Internat., Manila; adj. lectr. Thammasat U., Bangkok, 1984-86; invited lectr. Asian Inst. Tech., Bangkok, 1983. Contbr. articles to profl. jours. Recipient

Media Unity and Progress award Media Exponent of Philippines, 1991. Recipient Media Unity and Progress award Media Exponent of Philippines, 1991, Press Media award, 1992, Internat. Humanitarian Record award, 1992 Outstanding Internat. Engring. Cons., Nat. Honor award Ctr. Youth Devel. Rsch., 1993, Engr. Yr.award Community Leadership Found., 1993. Mem. Soc. Profl. Engrs. Thailand (pres. 1991-92), Ill. Soc. Profl. Engrs., Structural Engrs. Assn. Ill., Nat. Soc. Profl. Engrs., ASCE, Am. C. of C., Structural Engrs. Assn. Philippines, Rotary (bd. dirs. Bangkok chpt. 1987-89, hon. treas. 1989-90). Roman Catholic. Achievements include development of high strength concrete in S.E. Asia. Office: 75 Soi Siripot, Sukhumvit 81, Bangkok 10250, Thailand

SANKAR, SUBRAMANIAN VAIDYA, aerospace engineer; b. New Delhi, India, June 22, 1959; came to U.S. 1982; s. V.S.S. and Bala (Sankar) Narayanan; m. Asha Govindarajan, July 31, 1988; 1 child, Sitara Sankar. B.Tech., Indian Inst. Tech., Madras, 1982; MSAE, Ga. Inst. Tech., Atlanta, 1983; PhD, Ga. Inst. Tech., 1987. R & D dir. Aerometrics, Inc., Sunnyvale, Calif., 1987—. Contbr. articles to profl. jours. J.N. Tata scholar, India. Mem. AIAA, Nat. Geog. Soc., AAAS. Home: 34211 Petard Ter Fremont CA 94555-2611 Office: Aerometrics Inc 550 Del Rey Ave Unit A Sunnyvale CA 94086

SAN MARTIN, ROBERT L., federal official. MME, U. Fla., PhD ME. Prof. mech. engring. New Mexico State U.; dep. asst. sec. renewable energy, solar energy, field ops. Office Conservation and Renewable Energy, Washington, 1978—; dep. asst. sec. utility techs. Office Energy Efficiency and Renewable Energy (formerly Office Conservation and Renewable Energy), Washington, 1990—; acting asst. sec. Office Energy Efficiency and Renewable Energy, 1993—; past chair Renewable Energy Working Party Internat. Energy Agy.; past dir. New Mexico Solar Energy Inst., New Mexico Energy Inst. Mem. ASME (former chair solar divsn.), Am. Solar Energy Soc. (past bd. dirs.). Office: US Dept Energy Renewable Energy 1000 Independence Ave SE Washington DC 20585-0001

SANNER, GEORGE ELWOOD, electrical engineer; b. Rockwood, Pa., Aug. 30, 1929; s. Dennis Charles and Alverda (Growall) S.; m. Marjorie Mary Hohman, July 1, 1951; children: George Bradley, Marjorie Rosalie, Cathy Ann. BS, U. Pitts., 1951; postgrad., Johns Hopkins U., 1957-59. Registered profl. engr., Md.; cert. cost acctg. mgmt. Supervisory engr. Westinghouse Electric Corp., Balt., 1952-58, chief scientist, cons. def. and space ctr., 1964-72; chief engr., program mgr. radio div. Bendix Corp., Balt., 1958-64; engring. mgr. jet propulsion labs. Bendix Corp., Pasadena, Calif., 1980-81; pres., gen. mgr. Santron Corp., Balt., 1972-79; v.p. engring. M-Tron Industries div. Curtiss Wright Corp., Yankton, S.D., 1979-80; sr. engring. specialist engring. ctr. Litton Data Systems, New Orleans, 1981-83; cons. engring. mgmt. AIL div. Eaton Corp., Deer Park, N.Y., 1983-87; sr. prin. engr. Am. Electronics Labs, Inc., Lansdale, Pa., 1987-92; cons. Atlanta, 1992—; rep. People to People Tour, various countries, 1978. Patentee in field. Ch. vestry Immanuel Ch., Sparks-Glencoe, Md., 1969-70, dir. advg. pony show, 1969-70; trustee St. Paul's Sch. for Boys, Balt., 1965-67; bishop's secretariat Diocese L.I., Garden City, N.Y., 1985-87; exec. com. Scriptural Coalition, Diocese, Phila., 1990-92. A.K. Mellon Found. scholar, 1947-50, Carnegie Inst. Tech. scholar, 1947-51. Mem. IEEE, Assn. Old Crows, Quarter Century Wireless Assn. Episcopalian. Address: 2998 Clary Hill Ct Roswell GA 30075-5430

SANO, KEIJI, neurosurgeon; b. Shizuoka Prefecture, Japan, June 30, 1920; s. Takeo and Haru (Sase) S.; m. Yaeko Sano. M.D., U. Tokyo, 1945, DMS, 1951. Asst., U. Tokyo, 1945-56, lectr., chief out patient clinic, 1956-57, assoc. prof. neurosurgery, 1957-62, prof. neurosurgery, 1962-81; emeritus prof. U. Tokyo, 1981—; prof. neurosurgery Teikyo U., 1981—; dir. Fuji Brain Inst., 1986—; pres. 5th Internat. Congress Neurol. Surgery, 1973; pres. Internat. Conf. on Cerebral Vasospasm, 1990; chmn., dir Nat. Com. for Brain Rsch., Sci. Coun. of Japan, 1987-91. Mem. Japan Neurosurg. Soc. (pres. 1965), Japanese Assn. Rsch. in Stereo-ancephalotomy (pres. 1966), Asian and Australasian Soc. Neurol. Surgeons (pres. 1967-71, hon. life pres. 1971—), World Fedn. Neurosurg. Socs. (pres. 1969-73, hon. life pres. 1973—), Japanese Soc. CNS CT (pres. 1978—), Am. Assn. Neurol. Surgeons (hon.), Deutsche Gesellschaft für Neurochirurgie (hon.), Academia Eurasiana Neurochirurgica (pres. 1986); Soc. Neurol. Surgeons (hon.), Am. Acad. Neurol. Surgery (hon.), Congress Neurol. Surgeons (hon.), Scandinavian Neurosurg. Soc. (corr.), Am. Surg. Assn. (sr.), Am. Neurol. Assn. (corr.), ACS (hon.). Research on treatment of brain tumors, aneurisms, stereo-encephalotomy, vascular lesions. Home: 4-22-6 Den-en-chofu, Ota-ku Tokyo 145, Japan

SANTANA-GARCIA, MARIO A., plant physiologist; b. Pabellon, Aguascalientes, Mex., Nov. 16, 1956; came to U.S., 1985; s. Norberto and Maria de la Luz (Garcia) Santana; m. Irene Macias, Apr. 11, 1981; children:Kristian Mario, Iris Stephanie, Abril Ilusion, Gibram Moises. BS, U. Autonoma Chapingo, Mex., 1978; MS, N.Mex. State U., 1985, PhD, 1991. Researcher Nat. Inst. Agrl. Rsch., Mex., 1978-83; researcher, prof. Technol. Inst. Agr., Aguascalientes, 1983—; agrl. adviser Mex. Congress, Mexico City, 1991—. Author books in field. Roman Catholic. Office: Sec Pub Edn, Paseo de la Cruz 803, 20250 Aguascalientes Mexico

SANTANAM, SURESH, chemical engineer, environmental consultant; b. Tiruchirapalli, India, Jan. 25, 1959; came to U.S. 1983; s. Saranthan and Jayalakshmi (Srinivasan) S. BTech in Chem. Engring., Kakatiya U., India, 1981; MS, Oreg. Grad. Ctr., 1985; ScD, Harvard U., 1989. Environ. engr. Engrs. India Ltd., New Delhi, 1981-83; rsch. asst. Oreg. Grad. Ctr., Beaverton, 1983-85; staff scientist ERT Inc., Acton, Mass., 1986-87; rsch. asst. Harvard U., Boston, 1986-89; sr. projects mgr. Galson Tech. Svcs., Syracuse, N.Y., 1989-90; unit mgr. Galson Corp., Syracuse, N.Y., 1990-92, v.p., 1992—; adj. asst. prof. chem./civil engring. Syracuse U., 1990—. Group judge Onondaga County Sci. Fair, Syracuse, 1989—. Mem. Am. Inst. Chem. Engrs., Am. Geophys. Union, Air and Waste Mgmt. Assn., Sigma Xi. Achievements include research on exposure assessment and indoor air quality; pollution prevention - engineering solutions; batch chemical processes. Office: Galson Corp 6601 Kirkville Rd East Syracuse NY 13057

SANTANGELO, GEORGE MICHAEL, molecular geneticist; b. Bronx, N.Y., Nov. 14, 1956; s. Mario M. and Catherine T. (Campanelli) S.; m. Joanne S. Tornow, July 15, 1989. BA, U. Pa., 1978; PhD, Yale U., 1984. Postdoctoral rschr. U. Calif., Irvine, 1983-85; postdoctoral rschr. U. Calif., Santa Cruz, 1985, asst. rsch. biologist, 1985-87; asst. prof. biochemistry Oreg. Health Scis. U., Portaldn, 1987-88; adj. asst. prof. biology Portland State U., 1987-89; asst. prof. biol. scis. U. So. Miss., Hattiesburg, 1989—. Contbr. articles to profl. jours. Recipient NSF, 1990-91, Instrumentation Lab. Improvement award NSF, 1990-92, Rsch. award NIH, 1992—, Acad. Rsch. Infrastructure Program award NSF, 1992—. Mem. AAAS, Am. Soc. Microbiology, Genetics Soc. Am., Miss. Acad. Sci., Sigma Xi. Achievements include discovery of the interdependence of two global transcriptional activators in Saccharomyces cerevisiae; generated the first comprehensive library of eukaryotic promoters. Office: Univ So Miss Dept Biol Scis 55 Box 5018 Hattiesburg MS 39406-5018

SANTI, PETER ALAN, neuroanatomist, educator; b. Miami, Fla., June 5, 1947. BS, Fla. State U., 1969, PhD, 1975. Teaching asst. Fla. State U. Dept. Biol. Sci., Tallahassee, 1969-73; rsch. assoc. U. Minn. Dept. Otolaryngology, Mpls., 1975-77, asst. prof., 1977-84; assoc. prof. otolaryngology U. Minn. Dept., Mpls., 1984—; mem. edit. bd. Hearing Rsch., Pitts., 1989—; ad hoc grant reviewer NASA/AIBS NSF, VA, NIH, Washington; jour. reviewer Hearing Rsch. Jout. Contbr. chpts. to books and articles to profl. jours. Chmn. Mahtomedi (Minn.) Planning And Zoning Commn., 1985-87. Predoctoral felloe NIH, Fla. State U., Dept. Biol. Sci., 1973-75; Rsch. grantee NIH, 1977—, Deafness Rsch. Found., N.Y.C., 1988-91, Am. Otological Soc., Davis., Calif., 1992-93; recipient Fla. State U. Outstanding Achievement award, 1989; Apple Macintosh Software 1st prize award, 1991. Mem. AAAS, Acoustical Soc. Am., Assn. Rsch. Otolaryngology, Soc. Neurosci. Achievements include development of a quantative method of analysis for pathology of the stria vascularis; development of computer courseware for learning cochlear anatomy; discovery of fibronectin and gangliosides in the cochlea. Office: U Minn 2001 6th St SE Minneapolis MN 55455

SANTIAGO, JULIO V., medical educator, medical association administrator; b. San German, Puerto Rico, Jan. 13, 1942. BS, Manhattan Coll., 1963; MD, U. Puerto Rico, 1967. Diplomate Am. Bd. of Internal Medicine, 1975. Asst. resident in medicine, Ward Med. Svc. Washington U., St. Louis, 1970-72, fellow in Metabolism and Endocrinology, 1972-74; asst. dir. and chief resident, Gen. Med. Svc. Barnes Hosp., St. Louis, 1974-75, asst. in medicine, 1970-74; instr. in medicine Washington Univ. Sch. Medicine, St. Louis, 1975-80; asst. physician and pediatrician Barnes Hosp. and St. Louis Children's Hosp., St. Louis, 1975-77, co-dir. clin. rsch. and edn. facility, Diabetes and Endocrinology Rsch. Ctr., 1975-77, asst. dir., Clin. Rsch. Ctr., 1975—, assoc. dir. for biomedical rsch., Diabetes Rsch. and Tng. Ctr., 1977-87, assoc. prof. of pediatrics, 1980-83; assoc. physician and pediatrician Barnes Hosp. and St. Louis Children's Hosp., 1980-83; prof. of pediatrics Washington Univ. Sch. Medicine, 1983—, assoc. prof. of medicine, 1983-92, co-dir. divsn. of endocrinology and metabolism, Dept. of Pediatrics, 1984—, program dir., Diabetes Rsch. and Tng. Ctr., 1987—, prof. of medicine, 1992—; cons. NIH-DRR Rsch. Ctrs. for Minority Insts., 1984-86, mem. NIH NIDDK Study Sect. for Spl. Programs in Endocrinology and Metabolism, 1988—, mem. endocrine adv. com. FDA, 1990—; mem. editorial bd. Diabetes Care, 1980-83, The Diabetes Educator, 1980-88. assoc. editor (jour.) Diabetes, 1977-79, Diabetes, 1991—. Capt. U.S. Army, 1968-70. Mem. Am. Soc. for Clin. Investigation, Am. Fedn. Clin. Rsch., Soc. for Pediatric Rsch., Am. Diabetes Assn., St. Louis Metro. Med. Soc., Internat. Study Group of Diabetes, Endocrine Soc., Juvenile Diabetes Found. (mem. med. adv. bd. 1976—). Home: 4 Forest Pky Manchester MO 63011 Office: St Louis Hosp at Washington U Sch Medicine 1 Childrens Place Saint Louis MO 63110*

SANTILLANES, SIMON PAUL, analytical chemist, biotoxicologist; b. Albuquerque, Apr. 27, 1967. BS, U. N.Mex., 1991. Lab. asst. Meth. Hosp., Lubbock, Tex., 1987-88; analytical chemist City of Albuquerque, 1989-91, dir. biotoxicology lab., 1991—; cons. Santillanes and Assocs., Albuquerque, 1987—; cons. Environ. Health Cons., Albuquerque, 1988—. Author: Standard Operating Proceedures for Biotoxicology Laboratories, 1991. Adv., dist. com. Albuquerque area Boy Scouts Am., 1985—; responder Nat. Disaster Med. Team, Albuquerque, 1987—, treas. Christian Aviators N.Mex., 1992—. Mem. Soc. Applied Spectroscopy, Am. Chem. Soc., N.Mex. Waste Mgmt. Soc. Home: PO Box 40528 Albuquerque NM 87196

SANTO, HAROLD PAUL, engineer, educator, researcher, consultant, designer; b. Santos, São Paulo, Brazil, Feb. 14, 1947; s. Harold S. and Rosa (Jesus) S.; m. Amélia Maria Marques, June 25, 1988. BS in Engring., Tech. U. Lisbon, Portugal, 1975, D in Engring., 1984. Tech. sch. tchr. Ministry of Edn., Lisbon, 1973-76; teaching asst. Tech. U. Lisbon, 1976-78, lectr., 1978-80, sr. lectr, 1980-87, prof. engring., 1987—; rschr. Ctr. for Str. Mech. and Engring., T.U., Lisbon; project engr., head computer ctr. Proplano; dir. Compugraphics ann. internat. conf. Computational Graphics and Visualization Techniques, Edugraphics biennial conf. on graphics edn.; implementor, organizer various univ. courses, sci. meetings and confs.; lectr., rschrs. in field. Author: Computer Graphics-A Guided Introduction, 1985; contbr. papers, reports, and articles to profl. jours. and scientific Confs. Mem. ASCE, AAAS, Engring Soc. Portugal, IEEE Computer Soc. (affiliate) Assn. Computing Machinery (Portuguese ACM chpt. 1988—, Lisbon ACM SIGGRAPH 1986-90), Eurographics, Am. Soc. Engring. Edn., Portuguese Soc. Authors, Portuguese Informatics Assn. Avocations: musician, history researcher, student of langs., computer. Home: Av Azedo Gneco 49-R/C D, Massama, 2745 Queluz Portugal Office: Tech U Lisbon, Dept Civil Engring, Av Rovisco Pais, 1096 Lisbon Portugal

SANTORO, ALEX, infosystems specialist; b. Kansas City, Mo., July 28, 1936; s. Alexander Luke and Mara Louise (Ratkaj) S. BS in Chem., U. Mo., Rolla, 1957; postgrad., U. Kans., 1959; MA in Math., U. Mo., Kansas City, 1965. Dir. computing svc. U. Kans. Med. Ctr., 1966-68; communications programmer United Computing Svc., Kansas City, 1969-71; programmer Trans Tech., Inc., Kansas City, 1971-72; database adminstr. Fed. Res. Bank Kansas City, 1972-84; prin. tech. assoc. Mut. Benefit Life Ins. Co., Kansas City, 1984-91; private cons. Kansas City, 1991—; rschr./decoder b-lang. Shakespearean works. With U.S. Army, 1959-62. Mem. Assn. for Computing Machinery, AFTRA, Nat. Hon. Soc. Home: 710 E Armour Blvd Kansas City MO 64109-2323

SANTORO, FERRUCIO FONTES, microbiologist; b. Salvador, Bahia, Brazil, Apr. 15, 1952; came to France, 1974; s. Florido Vidal and Edite Santos (Fontes) S.; m. Maria-Olga Vidal, May 25, 1974; children: Frederic, Ferrucio Jr. Pharmacist, Pharmacy Coll., Salvador, 1973; D Pharmacy, Pharmacy Coll., Lille, France, 1977; PhD, Medicine Coll., Lille, 1980. Fellow Oswaldo Cruz Inst., Salvador, 1973; fellow Pasteur Inst., Lille, 1974-77, rsch. asst., 1977-80, 83-85, rsch. dir., 1985-87; assoc. researcher NYU Med. Ctr., N.Y.C., 1980-83; rsch. dir. Oswaldo Cruz Inst., Salvador, 1987-89, Faculty Medicine, Grenoble, France, 1989-91, Oswaldo Cruz Inst., 1991—; cons. WHO, Geneve, 1979-80, Acta Tropica, Basel, 1979-91. Author: Parasitic Diseases, 1982, Immunity, 1991; inventor Radioimmunoprecipitation-Polyethilene Glycol Assay, 1976, DNA probe-toxoplasma, 1990. Grantee Rockefeller Found., N.Y.C., 1983, WHO, Geneva, 1985, Inst. Nat. de la Santé et de la Recherche Med. (INSERM), Paris, 1986, EEC, Brussels., 1986. Mem. Brazilian Soc. Tropical Medicine, Am. Assn. Immunolgoists, French Soc. Parasitology. Roman Catholic. Avocations: ranching, soccer, skiing. Office: CIBP Pasteur Inst, 1 rue Prof A Calmette, 59000 Lille France

SANTSCHI, PETER HANS, marine sciences educator; b. Bern, Switzerland, Jan. 3, 1943; came to U.S., 1976; s. Hans and Gertrud (Joss) S.; m. Chana Hoida, Mar. 28, 1972; children: Rama Aviva, Ariel Tal. BS, Gymnasium, Bern, 1963; MS, U. Bern, 1971, PhD summa cum laude, 1975; Privatdozent, Swiss Fed. Inst. Tech., Zurich, Switzerland, 1984. Lectr. chemistry Humboltianum Gymnasium, Bern, 1968-70; teaching rsch. asst. U. Bern, 1970-75; rsch. scientist Lamont-Doherty Geol. Obs., Columbia U., Palisades, N.Y., 1976-77; rsch. assoc. Lamont-Doherty Geol. Obs. Columbia U., Palisades, N.Y., 1977-81; sr. rsch. scientist Lamont-Doherty Geol. Obs., Columbia U., Palisades, N.Y., 1981-82, Swiss Inst. Pollution Control, Zurich-Duebendorf, Switzerland, 1982-88; prof. oceanography Tex. A&M U., College Station, 1988—; prof. marine scis. Tex. A&M U., Galveston, Tex., 1988—; sect. head chem. oceanography dept. oceanography Tex. A&M U., College Station, 1990—; head isotope geochemistry and radiology sect. Swiss Inst. Water Resources and Water Pollution Control, Zurich, 1983-88; mem. rev. panel on chem. oceanography NSF, 1990-91. Contbr. articles to profl. jours. Cpl. Swiss Army, 1964-65. Mem. AAAS, Am. Chem. Soc., Am. Geophys. Union, Oceanography Soc., Am. Soc. Limnology and Oceanography. Avocation: swimming. Office: Tex A&M U Galveston TX 77553-1675

SANTTI, GARY ALLEN, hazardous waste engineer; b. Ishpeming, Mich., Oct. 21, 1953; s. Leonard Nels and Hilda Amanda (Korpi) S. BS in Geology, Mich. State U., 1977, BSCE, 1978. Registered profl. engr., Fla.; registered profl. geologist, Fla. Water resources engr. State of Colo. Dept. Natural Resources, Denver, 1980; project engr. Arabian Am. Oil Co., Dhahran, Saudi Arabia, 1980-83; commodities-stock trader Tucson, 1984-86; planning engr. Magma Copper Co., Tucson, 1987-89; project mgr. Nat. Seal Co., Galesburg, Ill., 1989; sr. civil engr. Newmont Gold Co., Elko, Nev., 1990-91; hazardous wast mgr. S.W. dist. Fla. Dept. Environ. Protection, Tampa, 1992—; cons. civil engr. Phys. Resource Engring., Tucson, 1989, 90. Mem. ASME, Nat. Soc. Profl. Engrs., Fla. Engring. Soc. Office: Fla Dept Environ Protection 3804 Coconut Palm Dr Tampa FL 33619

SANTUCCI, ANTHONY CHARLES, neuroscientist, educator; b. Bronx, N.Y., Oct. 20, 1959; s. Edward and Catherine (Stella)C.; m. Annmarie Homola, Jan. 31, 1982; children: Jennifer, Robert. BA, Iona Coll., 1981; PhD, Kent State U., 1987. Postdoctoral rsch. fellow Bronx VA Med. Ctr.-Mt. Sinai Sch. Medicine, N.Y.C., 1986-90; asst. prof. psychology Manhattanville Coll., Purchase, N.Y., 1990—. Contbr. chpt.: Alzheimer's Disease and Related Disorders, 1989, Pharmacological Interventions on Central Cholinergic Mechanisms in Senile Dementia, 1989, Progress in Brain Research, 1990, Perspectigves on Cognitive Neuroscience, 1991. Grantee Am. Fedn. Aging Rsch., 1988, Mt. Sinai Sch. Medicine, 1989. Mem. APS, AAAS, Soc. Neurosci. Office: Manhattanville Coll Dept Psychology Purchase NY 10577

SAPADIN, LINDA ALICE, psychologist, writer; b. N.Y.C., Mar. 20, 1940; d. Samuel Miles and Helen Leah (Bogen) Fink; m. Seymour Sapadin, Nov. 10, 1962 (div. 1980); children: Brian, Glenn, Daniel; m. Ronald J. Goodrich, May 15, 1983. BA, Bklyn. Coll., 1960; MA, Temple U., 1961, CUNY, 1986; PhD, CUNY, 1986. Lic. psychologist, N.Y. Sch. psychologist N.Y.C. Bd. Edn., 1962-66, rsch. cons., 1985-87; tchr. Hewlett-Woodmere (N.Y.) Adult Edn., 1975-84; devel. dir. Ctr. for Women and Achievement, Island Park, N.Y., 1984-89; dir. Biofeedback and Stress Reduction Ctr., Valley Stream, N.Y., 1990—; pvt. practice Valley Stream, 1987—; forum leader, adj. prof. Hofstra U., N.Y. Inst. Tech., Five Towns Coll., Nassau Community Coll., L.I., 1974-90; cons. Nassau County Town of Hempstead, N.Y., 1986; adj. prof. continuing edn. Hofstra U., Uniondale, N.Y., 1985—; talk show host Sta. WGBB Radio, Merrick, N.Y., 1987. Columnist Chanry Communications, 1987-90, Richner Publs., 1992; contbr. articles to profl. jours. Chmn. psychology com. Nassau County NOW, Uniondale, 1983; speaker L.I. Assn. Planned Parenthood, Econ. Opportunities Coun., L.I. Libr. System, B'nai Brith, Women's Forum, Nat. Coun. Jewish Women, 1984—. Recipient Outstanding Community Svc. award State Senator Carol Berman, 1984. Mem. APA (media div., psychology of women div.), Nassau County Psychol. Assn. (women's studies com.). Home and Office: Biofeedback and Stress 19 Cloverfield Rd Valley Stream NY 11581-2421

SAPAN, CHRISTINE VOGEL, protein biochemist, researcher; b. New Eagle, Pa., Aug. 15, 1947; d. Norman Kay and Palma (Infante) Vogel; m. Robert Louis Sapan, Mar 10, 1979; 1 child, Sloane Whitney. BA, Beaver Coll., Glenside, Pa., 1969; MS, U. N.C., 1971, PhD, 1978. Cert. clin. lab. dir., Fla. Rsch. assoc. U. N.C., Chapel Hill, 1971-75, NIH fellow, 1976-78; postdoctoral fellow Duke U., Durham, N.C., 1975-76; mgr. hematology Coulter Corp., Miami, Fla., 1978-81; hematology cons. in pvt. practice Miami, 1981-87; dir. sci. affairs N.Am. Biologicals, Inc., Miami, 1987—; mem. adv. bd. Fla. Internat. U., Miami, 1990—; adj. prof. Union Inst., 1987—; chair Instl. Rev. Bd., Miami, 1990—; del. Nat. Coun. on Clin. Lab. Standards, 1987—. Contbr. articles to Circulation, Biochem. Biophys. Rsch. Comms., Biochemistry, Jour. Clin. Lab. Medicine, Internat. Conf. AIDS, others. Vol. Miami City Ballet Co., 1992—. Mem. AAAS, N.Y. Acad. Sci., Internat. Soc. for Interferon Rsch., Am. Heart Assn., Am. Assn. Blood Banks, Sigma Xi. Achievements include devel. of hyperimmune IV gamma globulin products for therapeutic market; multiple whole blood controls using stabilized human cellular components; line of tissue culture products for use in HLA and MLC; line of infectious diagnostic controls for HIV, hepatitis. Home: 385 Campana Ave Coral Gables FL 33156 Office: NAm Biologicals Inc 16500 NW 15th Ave Miami FL 33169

SAPHIRE, GARY STEVEN, podiatrist; b. Bklyn., July 10, 1952; s. Leonard and Dorothy (Henry) S.; m. Helene Frances Koolik, Sept. 12, 1982; children: Emily Laura, Erika Robyn. BS in Health & Sci., Bklyn. Coll., 1974; BS in Biol. Sci., Ill. Coll. Podiatric Medicine, 1978, D.P.M., 1978. Diplomate Am. Bd. Podiatric Surgery. Chief resident Coney Island Hosp., Bklyn., 1978-79; attending Coney Island Hosp., 1979-82, Maimonides Med. Ctr., Bklyn., 1979-86, Community Hosp., Bklyn., 1984—; dir. podiatric surgery Caledonian Hosp., Bklyn., 1986—; del. N.Y. State Podiatric Med. Assn., N.Y.C., 1988—; exec. bd. Kings County Podiatry Soc., Bklyn., 1987-88; chief podiatric surgery The Bklyn. Hosp. Ctr., 1989—. Author: Vascular Disease, 1982, Arthroscopy, 1989, Ankle Arthroscopy, 1989; editor, The Bull., 1988-89. Judge Sci. Fair, N.Y.C., 1985-89. Durlacher scholar Nat. Podiatric Honor Soc., 1978. Fellow Am. Coll. Foot Surgeons; mem. Am. Podiatric Med. Assn., Am. Diabetes Assn. Coun. Foot Health, N.Y. Acad. Sci., Westerliegh Tennis Club, Kappa Tau Epsilon (v.p. local chpt. 1977-78). Avocations: skiing, tennis, birding. Office: Parkway Podiatry Group 7516 Bay Pky Brooklyn NY 11214-1598

SAPIENZA, ANTHONY ROSARIO, physician, educator, dean ambulatory facilities; b. Chgo., Apr. 3, 1917; s. Carmelo and Catherine (Verdi) S.; m. Rose Ziegler, 1940 (div. 1957); children: John, Michael. BS, MD, U. Ill., 1943. Assoc. prof. Coll. of Medicine U. Ill., Chgo., 1972-79; assoc. prof. U. Health Sci. Chgo. Med. Sch., North Chicago, Ill., 1979—, assoc. dean med. edn. U. Health Sci., 1979-86, assoc. dean ambulatory facilities, 1987—. Mem. AMA, ACP, Chgo. Med. Soc. (univ. rep. 1982-84, continuing med. edn.c om. 1984-86), Ill. Med. Soc., Sigma Xi. Office: Chgo Med Sch 3333 Green Bay Rd North Chicago IL 60064-3095

SAPOFF, MEYER, electronics component manufacturer; b. N.Y.C., June 2, 1927; s. Benjamin and Mary (Charney) S. Student, Mohawk Coll., 1946-48, Poly. Inst. Bklyn., 1948-50, 52-53; BS in Elec. Engring. magna cum laude, Poly. Inst. Bklyn., 1950, postgrad., 1952-53; postgrad., MIT, 1951, U. Pa., 1951-52; MS in Elec. Engring., Drexel Inst. Tech., 1952. Rsch. engr. Franklin Inst. Labs., Phila., 1950-52; rsch. fellow sr. grade Poly. Inst. Bklyn., 1952-53; dir. rsch. Victory Engring. Corp., Springfield, N.J., 1953-57; dir. engring. Victory Engring. Corp., Springfield, 1957-63, v.p., 1963-69; cons., sr. staff scientist Keystone Carbon Co., St. Mary's, Pa., 1969-70; pres. Thermometrics, Inc., Edison, N.J., 1970-86, chmn. bd., CEO, 1986—; bd. dir. Thermometrics, Inc.; cons. in field; program com., chmn. E20.08 Med. Thermometry subcom., chmn. session on thermistors 6th Symposium on Temperature, Measurement and Control in Sci. and Industry. Contbr. articles to profl. jours.; patentee in field. Active Citizens League West Orange, 1962-75, West Orange PTA, 1960-76. With USN, 1945-46. Recipient Indsl. Rsch. IR-100 award, 1974; State of NYU scholar, 1948-50; Poly. Inst. Bklyn. fellow, 1953. Mem. IEEE, ASTM, AAAS, Mfrs. Assn. of Union, N.J. Mfrs. Assn., Poly. Inst. Bklyn. Alumni Assn., Am. Ceramic Soc., Internat. Orgn. for Legal Metrology, Am. Nat. Standards Inst., Am. Vacuum Soc., Tau Beta Pi, Eta Kappa Nu. Home: 1137 Stuart Rd Princeton NJ 08540-1216 Office: 808 US Hwy 1 Edison NJ 08817-4695

SAPONARO, JOSEPH A., company executive; b. Boston, Sept. 24, 1939; s. Joseph A. and Helen (Carradonna) S.; m. Susan L. Saponaro, Apr. 1, 1960. BS in Nav., Mass. Maritime Acad., 1959; postgrad., MIT, 1967-69; MS in Math., Northeastern U., 1968. 3d officer Grace Lines, MMP, N.Y., 1959-60; analyst Harvard U., Cambridge, Mass., 1960-62; programmer Sylvania (now GTE), Needham, Mass., 1962-64; project mgr. Systems Devel. Corp., Lexington, Mass., 1964-69; mgr. Intermetrics Inc., Cambridge, 1969-79, v.p., gen. mgr., 1979-86, CEO, pres., 1986—. Comdr. USNR, 1960-78. Recipient appreciation award for Apollo work MIT, 1969, for space shuttle work NASA, 1973. Mem. Am. Electric Assn. (bd. dirs. 1989—), Nat. Security Industry Assn. (pres., bd. dirs. 1988—), Mass. High Tech. Coun. (bds. dirs. 1990—), Mass. Computer Software Coun. Roman Catholic. Avocations: tennis, horseback riding. Office: Intermetrics Inc 733 Concord Ave Cambridge MA 02138

SAPPEY, ANDREW DAVID, chemical physicist; b. Warren, Ohio, Feb. 14, 1960; s. Robert Henri and Sally Ann (Safford) S. AB, Kenyon Coll., 1982; PhD, U. Wis., 1988; postgrad., Stanford Rsch. Inst., 1988-89. Staff scientist Los Alamos (N.Mex.) Nat. Lab., 1989—. Vol. Big Bros./Big Sisters Am., Columbus, Ohio, 1979-82. Mem. AIAA, Am. Physical Soc., U.S. Masters Swimming. Achievements include recent implementation of novel diagnostics for laser ablation plume imaging and density determination. Office: Los Alamos Nat Lab Box 1663 Los Alamos NM 87545

SAPSE, ANNE-MARIE, chemistry educator; b. Bucharest, Rumania, Feb. 19, 1939; came to U.S., 1962; d. Aurel Ion and Rose Michaela (Acker) Luca; m. Marcel Sapse, June 24, 1957; 1 child, Danielle Elena Sarah. BS in Chemistry, U. Bucharest, 1959; MA in Chemistry, CCNY, 1966; PhD in Chemistry, CUNY, 1969. Rsch. assoc. Yeshiva U., N.Y.C., 1968-69; prof. John Jay Coll. and Grad. Ctr. CUNY, N.Y.C., 1969—; adj. faculty Rockefeller U., N.Y.C., 1986-91, 92—. Author: Computers in the Security Business, 1980; contbr. numerous articles to profl. jours. Mem. Republican Inner Circle, Washington, 1990-92. City Univ. grantee CUNY, 1986—, IBM grantee, 1990. Mem. Am. Chem. Soc., Capitol Hill Club. Achievements include research in the lithium bond, in the application of quantum chemistry calculations to cancer drugs. Office: CUNY John Jay Coll 445 W 59th St New York NY 10019

SAPSFORD, RALPH NEVILLE, cardiothoracic surgeon; b. Bloemfontein, South Africa, Oct. 30, 1938; s. Roland Geoffrey and Doreen Inel (Cooper) S.; m. Simone Andree Evard, Jan. 13, 1962; children: Wayne, Lance, Andrea. MB, ChB, U. Capetown, Cape Province, South Africa, 1962, ChM, 1976. Resident Mpilo Hosp., Bulawayo, Rhodesia, 1963-65, various tng.

hosps., Eng., 1966-67; sr. resident various cardiothoracic units, Manchester, Leeds, London, England, 1968-77; cons., sr. lectr. cardiothoracic surgery Hammersmith Hosp., London, 1977-90; hon. cons. St. Mary's Hosp., London, 1981-90; cons. cardiothracic surgery, Wellington Hosp., London, Harley St. Clinic, Princes Grace Hosp., London, 1977—. Co-author 5 med. books; contbr. articles to profl. jours. Freeman of the City of London, 1989—; mem. Worshipful Socs. of Apothecaries, 1989—, Worshipful Soc. of Loriners, 1992—. Recipient Merit award C, Nat. Health Svc. Eng., 1983, Merit award B, 1987. Fellow Royal Coll. Surgeons, Edinburgh, London; mem. Soc. of Thoracic and Cardiovascular Surgeons, Brit. Cardiac Soc., European Soc. of Cardiothoracic Surgeons, European Cardiovascular Soc., Harlech Club, Sancta Maria Lodge. Mem. Conservative Party. Avocations: antique pistols, classic Jaguars, golf, yachting and horseback riding. Office: 66 Harley St, W1N 1AE London England

SARACCO, GUILLERMO JORGE, optical and medical products executive; b. Buenos Aires, Nov. 15, 1964; s. Roberto Jorge and Maria Adelia (Crotta) S.; m. Roxana Noemi Valsangiacomo Andurandeguy, Aug. 19, 1988; children: Maria Roxana, Guillermo Jorge. Degree, Inst. Marianista, Buenos Aires, 1977, Inst. Marianista, Buenos Aires, 1982; diploma optician-contact lense adapter, U. Buenos Aires, 1985, Nat. Acct., 1988. Tech. work Optica Saracco S.A., Buenos Aires, 1982-84, dept. head., 1984-86; with acctg. Estudio Contable SRL, Buenos Aires, 1985-87; gen. dir. R. Jorge Saracco, Buenos Aires, 1986-92; mem. optical and contact lenses fitting seminar and conf., Buenos Aires, 1986; dir. gerente R. Jorge Saracco e hijos S.A.; optical and optometry courses; pres. Distar SRL, Buenos Aires; dir. Saracco Hnos.SRL, 1992; ad-honorem cons. for Edn. and Culture Nat. Ministry. Mem. Contact Leses Argentine Soc., Profl. Assn. Economy Scis. Capital Fed. Dist., Ocular Protection Campaign (Spain), Belgrano Athletic Club, Sheraton Club Internat., Olivos Golf Club. Roman Catholic. Office: Optica Saracco, Juncal 821, 1062 Buenos Aires Argentina

SARAF, DILIP GOVIND, electronics executive; b. Belgaum, India, Nov. 10, 1942; s. Govind Vithal and Indira Laxman (Divekar) S.; m. Mary Lou Arnold, July 25, 1970; 1 son, Rajesh Dilip. B. Tech with honors, Indian Inst. Tech., Bombay, 1965; M.S.E.E., Stanford U., 1969. Sr. mgmt. trainee Delhi Cloth and Gen. Mills Co. (India), 1965-68; sr. research engr. SRI Internat., Menlo Park, Calif., 1969-78; project dir. Kaiser Electronics, San Jose, Calif., 1978-87; sr. engring. mgr. Varian Assocs., Santa Clara, Calif., 1987-90, pres. TOTAL QUALITY, 1990— ; cons. teaching U. Santa Clara, 1972, 73. Bd. dirs. Peninsula Childrens' Ctr., Palo Alto, Calif. Mem. IEEE, Soc. Am. Inventors, Am. Soc. Quality Ctrl., Speakers' Bur. Contbr. articles to profl jours. Patentee in field. Club: Toastmasters. Home: 28050 Horseshoe Ln Los Altos CA 94022-1924 Office: 101 1st St # 203 Los Altos CA 94022-2706

SARAFOGLOU, NIKIAS, economist, educator; b. Piraeus, Attica, Greece, May 17, 1954; s. Lazarus and Maria (Jamugianni) S. BA, Athens Sch. Econs., Greece, 1976; PhD, Umea (Sweden) U., 1988. Researcher Umea (Sweden) U., 1987-88; asst. prof. Sundsvall (Sweden) Unive Coll., 1988—; vis. scholar U. Pa., Phila., 1987, Boston U., 1989, U. Mich., Ann Arbor, 1990, Oslo (Norway) U., 1993; vis. sr. fellow George Mason U., Fairfax, 1991. Contbr. chpt. in book and articles to profl. jours. Mem. Regional Sci. Assn., Am. Econ. Assn., Ops. Rsch. Soc. Am., Swedish Econ. Assn., Ops. Rsch. Soc. Sweden. Office: Sundsvall U Coll, PO Box 860, 85124 Sundsvall Sweden

SARANTOPOULOS, THEODORE, physician, cardiologist, pathologist; b. Hermoupolis, Syros, Greece, Nov. 5, 1961; s. Konstantin and Maria (Lignou) S.; m. Monica Livia Pinticanu, May 11, 1989; 1 child, Maria-Emilia. MSc in Gen. Medicine, Inst. Pharmacy and Medicine, Cluj-Napoca, Romania, 1987; degree in cardiology, Athens Sch. Medicine, 1987. 3d asst. Spili Health Ctr., Spili-Rethmnon, 1989-90; asst. dept. pathology Syros Gen. Hosp., Hermoupolis-Syros, Greece, 1990-92; 2nd asst. Cardiology Clinic Red Cross Hosp., Athens, 1992—; health inspector Pub. Health Dept. of Pub. Sanitary, Hermoupolis-Syros, 1990-92. Mem. Athens Med. Assn., Marlboro Club. Orthodox. Avocations: basketball, sea diving, music, jet skiing, movies. Home: Niovis 25, 10446 Athens Attiki, Greece Office: Red Cross Hosp, G Papndreou 2, Str 1 Ampelokipi, 84100 Athens Cyclade, Greece

SARAVANJA-FABRIS, NEDA, mechanical engineering educator; b. Sarajevo, Yugoslavia, Aug. 2, 1942; came to U.S., 1970; d. Zarko and Olga Maria (Majstorovic) Saravanja; m. Gracio Fabris, Nov. 4, 1967; children: Drazen Fabris, Nicole. Diploma in mech. engring., U. Sarajevo, 1965; MSME, Ill. Inst. Tech., 1972, PhD in Mech. Engring., 1976. Lectr. in mech. engring. U. Sarajevo, 1965-70; teaching asst. Ill. Inst. Tech., Chgo., 1970-76; lectr. Ill. Inst. Tech., Chgo., 1974-75; mem. tech. staff Bell Telephone Lab., Naperville, Ill., 1976-79; prof. mech. engring. Calif. State U., L.A., 1979—, chair mech. engring. dept., 1989-93; assoc. researcher Lab. for Machine Tools, Aachen, Fed. Republic Germany, 1966-67; cons. Northrop Corp., L.A., 1984; mem. adv. bd. East L.A. Coll., 1991—. Contbr. articles to profl. publs. Grantee NSF, 1986, Grown & Sharpe Co., 1989; German Acad. Exch. fellow DAAD, 1966-67, Amelia Earhart fellow Zonta Internat., 1973-74, 75-76; recipient Engring. Merit award San Fernanco Valley Engring. Coun., 1990. Mem. AAUW, Soc. for Engring. Edn., Soc. Women Engrs. (sr.), Soc. Mfg. Engrs. (sr., chpt. v.p. 1984-88). Home: 2039 Dublin Dr Glendale CA 91206 Office: Calif State U LA 5151 State University Dr Los Angeles CA 90032

SARAVOLATZ, LOUIS DONALD, epidemiologist, educator; b. Detroit, Feb. 15, 1950; s. Samuel and Saya Betty (Chonich) S.; m. Yvette Susanne Braymer, Oct. 6, 1990; children: Samuel Francis, Louis Donald II. BS, U. Mich., 1972, MD, 1974. Fellow Am. Coll. Epidemiology. Internship Henry Ford Hosp., 1974-75, residency, 1975-77, infectious disease fellowship, 1977-79; intern Henry Ford Hosp., Detroit, 1974-75, 1975-77, fellow, 1977-79; dir. hosp. epidemiology Henry Ford Hosp., 1979-82, divsn. head infectious diseases, 1982—, dir. infectious diseases rsch. lab., 1982—; clin. prof. medicine U. Mich. Med. Sch., Ann Arbor, 1986—; mem. AIDS clin. drug devel. com. NIH, 1990—. Contbr. over 100 articles to profl. publs. Active Blue Ribbon Com. on AIDS State of Mich., Detroit, 1990; chmn. physician com. on AIDS Greater Detroit Health Coun., 1989. Fellow ACP, Infectious Diseases Soc. Am. Office: Henry Ford Hosp 2799 W Grand Blvd Detroit MI 48202

SARBACH, DONALD VICTOR, retired chemist; b. Lincoln, Nebr., Nov. 3, 1911; s. Meyer Leon and Margaret Bell (Cummings) S.; m. Ruth Lucille Wimberly, June 23, 1934; children: Kathleen Sarbach Brown, Douglas Lowry, Vicki Sarbach Karlin. BSc, U. Nebr., 1934. Various positions The B.F. Goodrich Co., Akron, Ohio, 1937-56; dir. rsch. Hewitt-Robins Inc., Stamford, Conn., 1956-58; dir. rubber tech. Goodrich-Gulf Chems., Inc., Cleve., 1958-60; internat. mktng. & polymer rsch. Goodrich Chem. Co., Cleve., 1960-76. Cons. editor Rubber & Plastics News, 1976-86; contbr. articles to profl. jours. Fellow Am. Inst. Chemists; mem. AAAS, Am. Chem. Soc. (50 yr.), Phi Lambda Upsilon, Alpha Chi Sigma. Achievements include 31 U.S. and 8 foreign patents in field. Home: 242 River Rd Hinckley OH 44233-9628

SARCHET, BERNARD REGINALD, retired chemical engineering educator; b. Byesville, Ohio, June 13, 1917; s. Elmer C. and Nellie Myrtle (Huff) S.; m. Lena Virginia Fisher, Dec. 13, 1941; children: Renee Erickson, Dawne, Melanie Sarchet Koewing. BS in Chem. Engring., Ohio State U., 1939; MS in Chem. Engring., U. Del., 1941. From engr. to dir. comml. devel. Koppers Co., Inc., Pitts., 1941-67; prof. and founding chmn. dept. engring. mgmt. U. Mo., Rolla, 1967-88; mgmt. cons. Sarchet Assocs., Rolla, 1975—. Co-author: Supervisory Management (Essentials), 2nd edit. 1976, Management for Engineers, 1981; contbr. articles to profl. jours. Mem. Planning Commn. Beaver, Pa., 1955-58; dir. Billy Graham Film Crusades, Rolla, 1969-75; area dir. Here's Life America, Rolla, 1977. Recipient Profl. Achievement award U. Del., 1952, Freedom Found. awards, 1974-75, Fellow Mem. awd., Am. Soc. for Engineering Educ., 1992. Fellow Am. Soc. Engring. Mgmt. (founding pres., bd. dirs. 1979—), Am. Soc. Engring. Edn. (chmn. 1976). Achievements include patent on composition for producing detergent polyglycol condensation products; led in developing engineering management for engineers. Home: PO Box 68 Rolla MO 65401

SARDINA, RAFAEL HERMINIO, nuclear engineer; b. Baire, Cuba, Apr. 28, 1946; came to U.S., 1960; s. Rafael M. and Amina (Cardona) S.; m. Olga Alicia Rodriguez, Dec. 25, 1969; children: Rafael Francis, Rafael Charles. BSEE, U. P.R., 1969, MS in Nuclear Engring., 1971. Cert. profl. engr. Grad. asst. U. P.R. Sch. Engring., Mayaguez, 1969-71; nuclear licensing engr. P.R. Electric Power Authority, San Juan, 1971-74; consulting engr. United Engrs. & Constructors, Inc., Phila., 1974-76, Ctr. for Energy & Environment Rsch., San Juan, 1977-80; pres. Energy & Environment Dynamics, Inc., San Juan, 1980—; cons. U. P.R., 1987—, Bayamon Ctrl. U., 1983—, Govt. of U.S. Virgin Islands, 1984-85. Contbr. articles in profl. jours. Mem. Assn. Vecinos de Club Manor, Rio Piedras, 1987—. Recipient AEC fellowship, 1970. Mem. Assn. Energy Engrs. (sr.), Assn. Engrs. and Surveyors of P.R., Nat. Soc. Profl. Engrs., P.R. Soc. Electrical Engrs. (past pres.). Achievements include energy engring. cons. to most higher edn. insts. and hosps. in P.R., responsible for achieving millions of dollars in savings by improving efficiency of energy systems. Office: Energy & Environ Dynamics 305 Badajoz Valencia Rio Piedras PR 00923

SARGENT, DOUGLAS ROBERT, air force officer, engineer; b. Manchester, N.H., Jan. 15, 1953; s. Robert Charles and Hazel Marie (Dearborn) S.; m. Pauline Elizabeth Conn, June 7, 1975; 1 child, Amber Marie. BS, Worcester Polytech. Inst., 1975; postgrad. Squadron Officers Sch., Air U., Maxwell AFB, Ala., 1982, postgrad Air Command and Staff Coll., 1992. Logistics officer lst BN 3d Field Artillery, Ft. Hood, Tex., 1975-78; programs dir. Armed Forces Examining & Entrance Sta., Portland, Maine, 1978-81; chief contract mgmt. Shaw (S.C.) AFB, 1981-83; programs & engring. dir. Thule AFB, Greenland, 1983-84; chief contracting br. USAF, Ramstein AFB, Germany, 1984-87; chief engr. & environ. planning Aviano AFB, Italy, 1988-90; asst. dir. plans & environ. engr. USAF Electronic Systems Div., Hanscom AFB, Mass., 1991—; moderator Environ. Protection Com., Aviano, 1988-90, Space Planning Com., Aviano, 1988-90; advisor Mgmt. Assistance Team, Europe, 1985-87; lectr. 1986-88. Chmn. Protestant Parish Fund Coun., Aviano, 1989-90, mem. Ramstein, 1985-87, co-chair, Thule, 1984. Mem. ASCE (tech. com. N.H. sect. 1993—), Soc. Am. Mil. Engrs., Constrn. Specifications Inst. (sec. N.H. chpt., 1993—), Mt. Washington Obs. Office: Electronic Systems Ctr Acquisition Civil Engring Hanscom AFB MA 01731

SARGENT, WALLACE LESLIE WILLIAM, astronomer, educator; b. Elsham, Eng., Feb. 15, 1935; s. Leslie William and Eleanor (Dennis) S.; m. Anneila Isabel Cassells, Aug. 5, 1964; children: Lindsay Eleanor, Alison Clare. B.Sc., Manchester U., 1956, M.Sc., 1957, Ph.D., 1959. Research fellow Calif. Inst. Tech., 1959-62; sr. research fellow Royal Greenwich Obs., 1962-64; asst. prof. physics U. Calif., San Diego, 1964-66; mem. faculty dept. astronomy Calif. Inst. Tech., Pasadena, 1966—; prof. Calif. Inst. Tech., 1971-81, Ira S. Bowen prof. astronomy, 1981—. Contbr. articles to profl. jours. Alfred P. Sloan fellow, 1968-70. Fellow Am. Acad. Arts and Scis., Royal Soc. (London); mem. Am. Astron. Soc. (Helen B. Warner prize 1969, Dannie Heineman prize 1991), Royal Astron. Soc. (George Darwin lectr., 1987), Internat. Astron. Union. Club: Athenaeum (Pasadena). Home: 400 S Berkeley Ave Pasadena CA 91107-5062 Office: Calif Inst Tech Astronomy Dept 105-24 Pasadena CA 91125

SARIDIS, GEORGE NICHOLAS, electrical, computers and system engineering educator, robotics and automation researcher; b. Athens, Greece, Nov. 17, 1931; came to U.S., 1961, naturalized, 1971; s. Nicholas and Anna (Tsofa) S.; m. Panayota Dimarogona, Apr. 10, 1985. Diploma in Mech. and Elec. Engring., Nat. Tech. U., Athens, 1955; MS in Elec. Engring., Purdue U., 1962, PhD, 1965. Instr. Nat. Tech. U., 1955-63, Purdue U., West Lafayette, Ind., 1963-65, asst. prof., 1965-70, assoc. prof., 1970-75, prof., 1975-81; prof. elec., computer and system engring. Rensselaer Poly. Inst., Troy, N.Y., 1981—, dir. Robotics and Automation Lab., 1982—; dir. NASA Ctr. for Intelligent Robotic Systems for Space Exploration, 1988-92; engring. program dir. NSF, Washington, 1973. Author: Self-Organizing Control of Stochastic Systems, 1977; co-author: Intelligent Robotic Systems: Theory and Applications, 1992; also numerous articles, reports. Co-author: Intelligent Robotic Sys.; co-editor, contbg. author: Fuzzy Automata, 1977; editor, contbg. author: Advances in Automation and Robotics, Vol. 1, 1985, Vol. 2, 1990. Fellow IEEE (founding pres. robotics and automation council 1981-84, Centennial medal 1984, Disting. Mem. award 1989); mem. ASME, Soc. Mfg. Engrs./Robotics Internat.-Machine Vision Assn. (sr.), Am. Soc. Engring. Edn., N.Y. Acad. Scis. Home: 38 Londonwood Ave Loudonville NY 12211 Office: Rensselaer Poly Inst NASA Ctr Intelligent Robotic School of Engineering Troy NY 12180-3590

SARJEANT, WALTER JAMES, electrical and computer engineering educator; b. Strathroy, Can., Apr. 7, 1944; s. Walter Burns and Margaret (Laurie) S.; m. Ann Richards, June 30, 1972; children: Eric, Cheryl. BSc in Math, Physics, U. Western Ont., Can., 1966, MSc in Physics, 1967, PhD in Physics, 1971. Asst. dir. R&D Gen-Tec Inc., Quebec City, Que., Can., 1971-73; program mgr. Lumonics Rsch. Ltd., Ottawa, Ont., Can., 1973-75; staff scientist Nat. Rsch. Coun., Ottawa, Ont., Can., 1975-78; project leader Los Alamos (N.Mex.) Nat. Lab., 1978-81; James Clerk Maxwell prof. elec. and computer engring. SUNY, Buffalo, 1981—. Author: High Power Electronics, 1989. Fellow IEEE; mem. Electromagnetics Acad., Electrostatics Soc., N.Y. Acad. Scis., Rotary, Eta Kappa Nu. Office: SUNY Elec Engring Dept 312 Bonner Hall Buffalo NY 14260

SARMA, GABBITA SUNDARA RAMA, aerospace scientist; b. Coimbatore, Tamilnadu, India, Aug. 1, 1937; s. Somasundara Rao and Varalakshmi (Kunisi) G.; m. Lalitha Sista, Aug. 24, 1970; children: Ushasri, Sumitra, Anita. BSc (hons.), Andhra U., Waltair, India, 1958, MSc, 1959; MS, U. Ill., 1965; PhD, Case Inst. Tech., Cleve., 1969. Demonstrator Andhra U., Waltair, 1958-59, lectr., 1959-64; Smith-Mundt fellow U. Ill., Urbana, 1961-65; grad. asst. Case Inst. Tech., Cleve., 1965-68, rsch. assoc. Case Western Res. U., Cleve., 1968-70; rsch. scientist German Aerospace Rsch. Est., Freiburg, 1970-77, Göttingen, 1977—; mem. COSPAR ISC and Working Group Material Sci. Space, SPACELAB Adv. Com., Freiburg/Göttingen, 1975-82; adv. coun. DRL, Göttingen and Cologne, 1978-93; vis. prof. von Kármán Inst. for Fluid Dynamics, Brussels, 1993—. Contbr. articles to profl. jours. Recipient Sir R. Venkatarathnam Medal, Andhra U., 1955, Sripathi medal, 1958, Metcalf medal, 1959; NATO Office of Sci. Rsch. rsch. grantee, 1984-87. Mem. AIAA, Ges. Angew. Math. and Mech. Achievements include demonstration of the decisive role of interfacial waves in Bénard-Marangoni instability extending and modifying a well-known result of Nield on the coupling of destabilizing mechanisms; use of magnetic fields and rotation to stabilize liquid layers; identification of species enrichment zones in models of gas-centrifuge separation process; results on existence and uniqueness of solutions for species distribution problem. Office: von Kármán Inst for, Fluid Dynamics, 72 Chaussée de Waterloo, B-1640 Rhode-Saint-Genese Belgium

SARPHIE, DAVID FRANCIS, biomedical engineer; b. South Bend, Ind., Jan. 23, 1962; s. Claude Samuel and Ethel Carroll (Rhodes) S. BSME, U. Notre Dame, 1984; MSME, Stanford U., 1985; DPhil in Engring. Sci., Oxford U., 1992. Mech. engr. KFA, Jülich, Fed. Republic Germany, 1984; engr., scientist McDonnell-Douglas Astro. Co., Huntington Beach, Calif., 1985-88. Grad. fellow NSF, 1984. Mem. AIAA, Tau Beta Pi, Pi Tau Sigma. Democrat. Roman Catholic. Achievements include patent for Gas Propulsion Device for Genetic Transformation of Plant Cells. Home: 112 Oakland Trace Madison AL 35758

SARPKAYA, TURGUT, mechanical engineering educator; b. Aydin, Turkey, May 7, 1928; came to U.S., 1951, naturalized, 1962; s. Hasip and Huriye (Fetil) S.; m. Gunel Ataisik, Aug. 26, 1963. B.S. in Mech. Engring, Tech. U. Istanbul, 1950, M.S., 1951; Ph.D. in Engring, U. Iowa, 1954. Research engr. MIT, Cambridge, 1954-55; asst. prof. U. Nebr., 1957-59, assoc. prof., 1959-62, prof. mech. engring., 1962, distinguished prof., 1962-66; research prof. U. Manchester, Eng., 1966-67, U. Gottingen, Fed. Republic of Germany, 1971-72; prof. mech. engring., chmn. dept. mech. engring. U.S. Naval Postgrad. Sch., Monterey, Calif., 1967-71, 72—, Disting. prof. mech. engring., 1975—; Cons. aerospace industry, 1967—, petroleum industry, 1976—. Author: Mechanics of Wave Forces on Offshore Structures, 1981; mem. editorial bd.: Zentralblatt fur Mathematik; editor: Procs. Heat Transfer and Fluid Mechanics Inst, 1970. Served with C.E. AUS, 1955-57. Fellow Royal Instn. Naval Architects, AIAA, ASME (Lewis F. Moody award 1967, exec. bd. fluids engring. div., chmn. review com., Freeman Scholar award 1988, Engring. award 1991, Fluids Engring. award 1990); mem. ASCE (Collingwood prize 1957), Heat Transfer and Fluid Mechanics Inst. (chmn.), Am. Inst. Aeros. and Astronautics, Internat. Assn. Hydraulics Research, Am. Soc. Engring. Edn. Achievements include patent for fluidic elements. Home: 25330 Vista Del Pinos Carmel CA 93923-8804 Office: Naval Postgrad Sch Mech Engring Code ME-SL Naval Postgraduate School Monterey CA 93943

SARRIS, JOHN, mechanical engineering educator; b. Hania, Greece, July 22, 1948; came to U.S., 1967; s. John and Nikie (Voulgari) S.; m. Theodora Kanaris, July 30, 1978; children: Nikie, Alexander. BA in Physics, Hamilton Coll., 1970; MSME, Tufts U., 1972, PhDME, 1976. Teaching asst. Tufts U., Medford, Mass., 1970-72; rsch. asst. NASA/Tufts U., 1972-76; asst. prof. dept. mech. engring. U. New Haven, West Haven, Conn., 1977-80, assoc. prof., 1980-85, prof., 1985—, chair dept., 1983—, assoc. dean engring., 1989—; cons in field. Textbook reviewer Macmillan, Addison Wesley, 1987. Scholar Hamilton Coll., Clinton, N.Y., 1967-70. Mem. ASME (sec. treas. New Haven sect. 1978-80), Acoustical Soc. Am., Am. Soc. Engring. Edn. Democrat. Office: U New Haven 300 Orange Ave West Haven CT 06516-1999

SARTORELLI, ALAN CLAYTON, pharmacology educator; b. Chelsea, Mass., Dec. 18, 1931; m. Alice C. Anderson, July 7, 1969. B.S., New Eng. Coll. Pharmacy Northeastern U., 1953; M.S., Middlebury (Vt.) Coll., 1955; Ph.D., U. Wis., 1958; M.A. (hon.), Yale U., 1967. Rsch. chemist Samuel Roberts Noble Found., Ardmore, Okla., 1958-60; sr. rsch. chemist Samuel Roberts Noble Found., 1960-61; mem. faculty dept. pharmacology Yale Sch. Medicine, New Haven, Conn., 1961—, prof., 1967—, head devel. therapeutics program Comprehensive Cancer Center, 1974-90, chmn. dept. pharmacology, 1977-84, dep. dir. Comprehensive Cancer Ctr., 1982-84, dir. Comprehensive Cancer Ctr., 1984-93, Alfred Gilman prof. pharmacology, 1987—, prof. epidemiology, 1991—; Charles B. Smith vis. rsch. prof. Meml. Sloan-Kettering Ctr., 1979; William N. Creasy vis. prof. clin. pharmacology Wayne State U., 1983; Mayo Found. vis. prof. oncology Mayo Clinic, 1983; Walter Hubert lectr. Brit. Assn. Cancer Rsch., 1985; Pfizer lectr. in clin. pharmacology U. Conn. Health Ctr., 1985; William N. Creasy vis. prof. clin. pharmacology Bowman Gray Sch. Medicine, 1987; Wellcome vis. prof. basic sci. U. Pitts. Sch. Medicine, 1990; sci. adv. bd. ImmunoGen, Inc., 1981—, U. Ind. Cancer Ctr., 1992; chmn. cancer sci. adv. bd. The Liposome Co., 1986—, sci. adv. bd. ViraChem. Inc., 1986—, vis. sci. adv. com. Columbia U. Comprehensive Cancer Ctr., 1986—, cancer adv. bd. Fox Chase Cancer Ctr., 1992—; mem. cancer clin. investigation rev. com. Nat. Cancer Inst., 1968-72, mgmt. cons. to dir. div. cancer treatment, 1975-77, bd. sci. counselors, div. cancer treatment, 1978-81, chmn. com. to establish nat. coop. drug discovery groups, 1982-83, chmn. special review com. Outstanding Investigator grant applications, 1992; mem. instl. rsch. grants com. Am. Cancer Soc., 1971-76, coun. analysis and projection, 1978-79; cons. in biochemistry U. Tex. M.D. Anderson Hosp. and Tumor Clinic, Houston, 1970-76; cons. Sandoz Forschungs-Institut, Vienna, Austria, 1977-80; mem. exptl. therapeutics study sect. NIH, 1973-77, working cadre nat. large bowel cancer project, 1973-76; mem. adv. com. Cancer Rsch. Ctr., Washington U. Sch. Medicine, 1971-75, sci. adv. com. U. Iowa Cancer Ctr., 1979-83, adv. com. SLSB Partners, L.P., 1992—; mem. external adv. com. Wis. Clin. Cancer Ctr., 1978-79, Duke Comprehensive Cancer Ctr., 1983—; mem. external adv. bd. U. Ariz. Cancer Ctr., 1982—, U. So. Calif. Cancer Ctr., 1983—, Clin. Cancer Rsch. Ctr., Brown U., 1980-86; mem. nat. program com. 13th Internat. Cancer Congress, 1979-81; cons. Bristol-Myers Co., 1982—, mem. selection com. prize in cancer rsch., 1977-85, chmn., 1979-81, chmn. selection com. award for disting. achievement in cancer rsch., 1989-92; bd. advisors Drug and Vaccine Devel. Corp. (Ctr. for Pub. Resources), 1980-81, Specialized Cancer Ctr., Mt. Sinai Med. Ctr., 1981-90, Grace Cancer Drug Ctr., Roswell Park Meml. Inst., 1986-89; mem. med. and sci. adv. com. grants rev. subcom. Leukemia Soc. Am., 1984-88; bd. dirs. Metastasis Rsch. Soc., 1984-90; mem. program planning com. Mary Lasker-Am. Cancer Soc. Conf., 1986; mem. external sci. rev. com. Massey Cancer Ctr., 1989—; bd. visitors Moffit Cancer Ctr. U.S. Fla., 1989—; mem. ad hoc cons. group for cancer ctrs. program Nat. Cancer Adv. Bd., 1989—; dep. dir. Cancer Prevention Rsch. Unit for Conn., 1989—, acting dir., 1991-93; mem. nat. bd. Cosmetic Toiletry and Fragnance Assn.'s Look Good...Feel Better Program, 1989—. Regional editor Am. Continent Biochem. Pharmacology, 1968—; mem. editorial adv. bd. Cancer Research, 1970-71; assoc. editor, 1971-78; editor-in-chief Communications, 1989—; editorial bd. Internat. Ency. Pharmacology and Therapeutics, 1972—, Cancer Biotherapy, 1992, Cancer Research Therapy and Control, 1992; editor Handbuch der experimentellen Pharmakologie vols. on antineoplastic and immunosuppressive agts; exec. editor Pharmacology and Therapeutics, 1975—; mem. editorial bd. Seminars in Oncology, 1973-83, Chemico-Bions, 1975-78, Jour. Medicinal Chemistry, 1977-82, Cancer Drug Delivery, 1982—, Jour. Enzyme Inhibition, 1984—, Anti-Cancer Drug Design, 1984—, Jour. Liposome Research, 1986—, In Vivo, 1990—; adv. bd. Advances in Chemistry Series, ACS Symposium Series, 1977-80; editorial adv. bd. Current Awareness in Biological Sciences, Current Advances in Pharmacology and Toxicology, 1983—, Cancer Cells, 1989—; editor series on cancer chemotherapy Am. Chem. Soc. Symposium, 1976; editorial cons. Biol. Abstracts, 1984—; mem. exec. adv. bd. Ency. of Human Biology, 1987—, Dictionary of Sci. and Tech., 1989—; contbr. articles to profl. jours. Bd. dirs. Shubert Performing Arts Ctr., 1992—, Shubert Opera Bd., 1991—. Recipient Outstanding Alumni award Northeastern U., 1987, Mike Hogg award M.D. Anderson Cancer Ctr., U. Tex., 1989, Alumni Achievement award Middlebury Coll., 1990. Fellow AAAS, N.Y. Acad. Scis.; mem. Am. Assn. Cancer Rsch. (dir. 1975-78, 84-87, chmn. publs. com. 1981-88, v.p. 1985-86, fin. com. 1985-88, exec. com. 1985-89, chmn. exec. com. 1987, pres. 1986-87, chmn. awards com. 1987), Am. Chem. Soc., Am. Soc. Microbiology, Am. Soc. Biochemistry and Molecular Biology, Am. Soc. Cell Biology, Am. Soc. Pharmacology and Exptl. Therapeutics (award in exptl. therapeutics 1986, award com. 1988, chmn. 1992), Assn. Am. Cancer Insts. (v.p. 1986, bd. dirs. 1986-89, liaison rep. to Nat. Cancer Inst. 1986, pres. 87-88, chmn. bd. dirs. 1989), Inst. of Medicine of Nat. Acad. Scis. (com. on govt. industry collaboration in biomed. rsch. and edn. 1989, mem. Forum on Drug Devel. and Regulation 1989—), Conn. Acad. Sci. and Engring., Coun. Biology Editors. Home: 4 Perkins Rd Woodbridge CT 06525-1616 Office: Yale U Yale Comprehensive Cancer Ctr 333 Cedar St New Haven CT 06510-3289

SARTORI, DAVID EZIO, statistician, consultant; b. Pitts., Aug. 22, 1962; s. Ezio Antonio and Elizabeth Ann (McCall) S.; m. Christine Ann Westin, Oct. 4, 1986; children: Anna, Maria. BA, Thiel Coll., 1984; MA, U. Pitts., 1992. Analytical chemist Calgon Corp., Pitts., 1984; devel. chemist PPG Industries, Allison Park, Pa., 1984-90, statistician, 1990—; cons. Ctr. Stats., U. Pitts., 1991-92. Contbr. articles to profl. publs. Mem. Am. Soc. Quality Control, Am. Statis. Assn. Republican. Presbyterian. Home: 291 W Ingomar Rd Pittsburgh PA 15237 Office: PPG Industries 4325 Rosanna Dr Allison Park PA 15101

SARWER-FONER, GERALD JACOB, physician, educator; b. Volkovsk, Grodno, Poland, Dec. 6, 1924; arrived in Can., 1932, naturalized, 1935; s. Michael and Ronia (Caplan) Sarwer-F.; m. Ethel Sheinfeld, May 28, 1950; children: Michael, Gladys, Janice, Henry, Brian. B.A., Loyola Coll. U., Montreal, 1945, M.D. magna cum laude, 1951; D.Psychiatry, McGill U., 1955. Diplomate: Am. Bd. Psychiatry and Neurology. Intern. Univ. Hosps. U. Montreal Sch. Medicine, 1950-51; resident Butler Hosp., Providence, 1951-52, Hosps. Western Res. U., Cleve., 1952-53, Queen Mary Vets. Hosp., Montreal, 1953-55; lectr. psychiatry U. Montreal, 1953-55; lectr., assoc. prof. McGill U., 1955-70; prof. psychiatry U. Ottawa, Ont., 1971—, prof., chmn. psychiatry, 1974-86; dir. dept. psychiatry Ottawa Gen. Hosp., 1971-87; dir. Lafayette Clinic, Detroit, 1989-92; prof. psychiatry Wayne State U., Detroit, 1989—; cons. in psychiatry Ottawa Gen. Hosp., Royal Ottawa Hosp., Nat. Def. Med. Ctr., Children's Hosp. of Eastern Ont., Ottawa, Pierre Janet Hosp., Hull, Que., Windsor (Ont.) Western Hosp. Ctr., Ottawa Sch. Bd.; Z. Lebensohn lectr. Silbey Meml. Hosp. Cosmos Club, Washington, 1991. Editor: Dynamics of Psychiatric Drug Therapy, 1960, Research Conference on the Depressive Group of Illnesses, 1966, Psychiatric Crossroads-the Seventies, Research Aspects, 1972; editor in chief Psychiat. Jour. U. Ottawa, 1976-90, emeritus editor in chief, 1990—; mem. editorial bds. of numerous internat. and nat. profl. jours.; editor numerous audio-video tapes; contbr. numerous articles to profl. jours. Bd. govs. Queen Elizabeth Hosp., Montreal, 1966-71; life gov. Queen Elizabeth Hosp. Found.; cons. Protestant Sch. Bd., Westmount, Que., 1966-71; advisor Com. on Health, City of Westmount, 1969-71. Served to lt. col. Royal Can. Med. Corps, 1949-62.

Recipient Sigmund Freud award Am. Assn. Psychoanalytic Physicians, 1982, William V. Silverberg Meml. award Am. Acad. Psychoanalysis, 1990, Poca award Assn. Psychiat.c Out Point Ctrs. Am., 1990; Simon Bolivar lectr. Am. Psychiat. Assn., New Orleans, 1981; Can. Decoration Knight of Malta; Found. fellow Royal Coll. Psychiatry, U.K. Fellow AAAS, Royal Coll. Physicians and Surgeons (Can.), Am. Coll. Neuropsychopharmacology (charter), Internat. Coll. Psychosomatic Medicine (sec.-gen. 1979-83), Am. Psychiat. Assn. (life), Am. Orthopsychiat. Assn. (life), Am. Coll. Psychiatrists (bd. regents 1978-80), Am. Psychopathol. Assn., Am. Coll. Psychoanalysts (pres. elect 1983, pres. 1984-85, Henry Laughlin award 1986), Am. Coll. Mental Health Adminstrn., World Psychiat. Assn. (chair Sci. program VI World Congress 1974, v.p. sect. on edn. 1989—, mem. internat. adv. com. 9th World Congress Rio de Janiero 1993), Collegium Internat. Neuropsychopharmacological; mem. Am. Acad. Psychiatry and the Law (pres. 1977 recipient Silver Apple award), Soc. Biol. Psychiatry (H. Azina Meml. lectr. 1963, pres. 1983-84), Can. Psychoanalytic Soc. (pres. 1977-81), Can. Assn. Profs. of Psychiatry (pres. 1976-77, 82-86), Am. Assn. for Social Psychiatry (v.p. 1987-89, pres. elect 1990, pres. 1992-94), Mich. Psychoanalytic Soc. Clubs: Cosmos (Washington); Royal Can. Mil. Inst. (Toronto). Home and Office: 3220 Bloomfield Shores Dr West Bloomfield MI 48323-3513

SASAKI, MASAFUMI, electrical engineering educator; b. Akita, Japan, Apr. 16, 1929; s. Mahei and Sugi Sasaki; m. Shoko Kasai, Mar. 26, 1960; children: Masato, Saeko. BEd, Akita U., 1953; MSc, Hokkaido U., Sapporo, Japan, 1955, DSc, 1959. Asst. prof. Nat. Def. Acad., Yokosuka, Japan, 1960-61, assoc. prof., 1961-70, prof., 1970—, chmn. dept. elec. engring., 1987-90; rsch. fellow dept. of math. U. Ill., Urbana, 1964-65; vis. rsch. scientist dept. OR/IE NYU, N.Y.C., 1965-66; lectr. faculty liberal arts and sci. Nippon U., Tokyo, 1971—; mem. orgn. com. JUSE, Tokyo, 1977—, chmn. program com. symposium on reliability and maintainability, 1990—; dir., head secretariat Inst. for Def. Systems, Def. Agy., 1972-91; invited prof. reliability mgmt. U. Electronic Sci. and Tech., China, 1984. Author: Information Systems Engineering, 1990, also 6 other books; contbr. more than 300 articles to profl. publs. Mem. IEEE (sr. mem., chmn. Tokyo chpt. reliability soc. 1987-89), Reliability Engring. Assns. Japan (bd. dirs. 1987-90, chmn. program com. symposium 1988—, chmn. referee com. Transactions 1990—, v.p., dir. 1992—), Ops. Rsch. Soc. Japan, Japanese Soc. Quality Control, Inst. Electronics, Info. and Communication of Japan. Avocations: gardening, travel. Home: 3-1-9 Imaizumi-dai, Kamakura 247, Japan Office: Nat Def Acad, 1-10-20 Hishirimizu, Yokosuka 239, Japan

SASAKI, SHIN-ICHI, university president, microbiologist, educator; b. Tokyo, Japan, Nov. 25, 1925; s. Ryu-ichiro and Kimiyo (Onokubo) S.; m. Kotoji Saito, Oct. 15, 1951; children: Fumiko, Hiroshi. BS, Tohoku U., 1949, DSc, 1959. Rsch. assoc. Tohoku U., Sendai, Japan, 1949-62, assoc. prof., 1963-70; rsch. assoc. MIT, Cambridge, 1959-61; prof. Miyagi U. Edn., Sendai, 1970-79; prof. Toyohashi (Japan) U. Tech., 1979-84, v.p., 1984-90, pres., 1990—; pres. Internat. Union Microbiological Socs., Newcastle Upon Tyne, Eng., 1990—; trustee Nat. Inst. Molecular Sci., Okazaki, 1991—. Pres. Assn. Japan-Germany Toyohashi, 1991—; trustee Internat. Friendship Asns. Toyohashi, 1990—. Recipient Acad. award Japan Info. Ctr. Sci. and Tech., Tokyo, 1980; Prize, Internat. Conf. on Computers in Chem. Rsch. and Edn., Rivadel Garda, Italy, 1989. Mem. Japan Acad. Engring., Chem. Soc. Japan, Japan Soc. Analytical Chemistry (acad. award 1974), Am. Chem. Soc. Achievements include being a pioneer of computer chemistry; inventor of chemical structure elucidation system. Office: Toyohashi U Tech, Hibarigaoka Tempaku, Toyohashi Aichi 441, Japan

SASAKI, SHUSUKE, electromechanical engineer; b. Tokyo, Apr. 1, 1956; came to U.S. 1982; s. Toshihiro and Atsuko (Kaiho) S. BS in Physics, U. Tsukuba, Japan, 1980; MS in Engring., UCLA, 1984. Design engr. Tokico Ltd., Kawasaki, Japan, 1980-81; R&D engr. Anja Engring. Corp., Monrovia, Calif., 1985-89; sr. mech. design engr. Spectrol Electronics Corp., City of Industry, Calif., 1989-90; mech. engr. III Alps Elec. USA, Inc., Garden Grove, Calif., 1990—; pres., cons. MABEC Engring., LaVerne, Calif., 1989—. Office: MABEC Engring Inc 4154 Bradford St La Verne CA 91750

SASAKI, TAIZO, physicist; b. Kobe, Japan, Feb. 14, 1925; s. Yasufumi and Tazu (Hashimoto) S.; m. Fusako Sato, Oct. 17, 1953; 1 child, Masako. B. in Physics, Osaka (Japan) U., 1951, postgrad., 1951-56; PhD, U. Tokyo, 1973. Asst. U. Tokyo, 1956-63, prof., 1963-82; guest prof. U. Hamburg (Fed. Republic Germany), 1966-68; prof., div. dir. photon factory Nat. Lab. for High Energy Physics, Tsukuba, Japan, 1980-85, dir. gen. photon factory, 1984-85; hon. prof. U. Tokyo and Nat. Lab. for High Energy Physics, 1985—; prof. Osaka U., 1985-88; rsch. cons. Rsch. Inst. for Physics and Chemistry, Tokyo, 1988—; hon. mem. Internat. Adv. Bd. of VUV Radiation Physics, 1989—. Series editor: (handbook series) Synchrotron Radiation, 1982—; bd. editors Optics Communication, 1975-80; co-author: Synchrotron Radiation, 1986, Synchrotron Radiation in Biosciences, 1993. Mem. Japanese Phys. Soc., Japan Soc. Applied Physics (Award for Optical Publ. 1963), Japanese Soc. Synchrotron Radiation Rsch. (pres. 1989-90), Am. Phys. Soc., Optical Soc. Am. Avocations: music, sports. Office: JAERI/RIKEN Project Team Spring 8, 2-28-8 Honkomagome, Bunkyo 113, Japan

SASAKI, WATARU, physics educator; b. Tokyo, Jan. 24, 1923; s. Rikichiro and Umeko (Nakamura) S.; m. Hideko Kuribayashi, May 21, 1959; children: Ayako, Tai, Yo. B. in Engring., U. Tokyo, 1947, DSc, 1960. Rsch. officer Electrotech. Lab., Tokyo, 1947-68, chief solid state physics sect., 1957-60; head low temp. physics lab., 1960-68; prof. faculty sci. U. Tokyo, 1968-83, chmn. grad. sch. physics, 1971-73, chmn. dept. physics, 1974-77, prof. emeritus, 1983—; prof. faculty sci. Toho U., 1983-91, dean faculty sci., 1988-91. Editor Jour. Phys. Soc. Japan, 1965-74, Japanese Jour. Applied Physics, Solid State Communications, 1980-90. Recipient Nishina Meml. prize Nishina Meml. Fund, 1962. Avocations: oil painting, mountain hiking, skiing. Home: Hakusan 4-2-8, Tokyo 112, Japan Office: Toho U Dept Physics, Miyama 2-2-1, Funabashi 274, Japan

SASAO, TOSHIAKI, psychologist, researcher; b. Kamikawa, Hokkaido, Japan, Sept. 12, 1955; came to U.S., 1975; s. Toshio and Ryoko (Shiwa) S.; m. Masami Yamaki, Mar. 26, 1989. BS in Psychology, U. Wash., 1979, MEd in Ednl. Psychology, 1981; PhD in Social Psychology, U. So. Calif., 1988. Rsch. psychologist U. So. Calif. Med. Sch., L.A., 1985-87; asst. prof. Marymount Coll., Rancho Palos Verdes, Calif., 1987-89; rsch. psychologist U. Calif., L.A., 1989—; cons. Asian Am. Drug Abuse Program, L.A., 1989—, Union of Pan Asian Communities, San Diego, 1991—, Ctr. for Substance Abuse Prevention, Rockville, Md., 1990—. Contbr. articles to profl. jours. Recipient Postdoctoral fellowship U. Calif., L.A., 1988. Mem. APHA, APA (charter), Am. Psychol. Soc., Japanese Psychol. Assn. Republican. Presbyterian. Achievements include first to conduct large scale drug needs assessment for Asian Pacific Islanders in Calif. Office: Univ Calif Dept Psychology 405 Hilgard Ave Los Angeles CA 90024-1563

SASSEEN, GEORGE THIERY, aerospace engineering executive; b. New Rochelle, N.Y., July 21, 1928; s. George Thiery and Rosemary (Head) S.; m. Gertrude Bradford, Nov. 11, 1951 (div. Sept. 1980); children: Sharon Sasseen Hudek (dec.), George (dec.), John; m. Margret E. Katorsky, Apr. 1, 1982. BME, Yale U., 1949. Registered profl. engr., Tex. Br. chief-elect spacecraft ops. Kennedy Space Ctr. NASA, Fla., 1961, chief engring. div., 1967-75; Apollo ops. mgr. Kennedy Space Ctr. NASA, 1962-64, chief ground system div., 1964-66; dir. engring. shuttle ops. Kennedy Space Ctr. NASA, Fla., 1975-84, asst. mgr. space sta., 1984-86, engring. dir. shuttle ops. return to flight, 1986—; with space sta. skunk works Johnson Space Ctr., NASA, Houston, summer 1984. 1st lt. USAF, 1951-53. Named Engr. of Yr. Space Congress, 1971; recipient NASA awards for exceptional svc., outstanding leadership (2), disting. svc., Presdl. Meritorious Exec. award. Mem. Morys Assn. (New Haven), Patrick Officers Club, Cocoa Beach (Fla.) Pow Squadron (comdr. 1975), Moose, Tau Beta Pi. Democrat. Presbyterian. Avocations: boating, fishing. Home: 215 Mizzen Ct Merritt Island FL 32953-3059 Office: NASA Kennedy Space Ctr Orlando FL 32899

SASSEN, GEORGIA, psychologist; b. N.Y.C., July 27, 1949; d. Bernard Nicholas Sassen and Rose Ellen Joseph Benjamin; m. L.S. Laing, Aug. 27, 1983; 1 child, Tai. BA, Wesleyan U., 1971; EdM, Harvard U., 1977; MS, U. Mass., 1981, PhD, 1985. Lic. psychologist, Mass. Program assoc. Am. Friends Svc. Com., Cambridge, Mass., 1971-76; field faculty Goddard-Cambridge Grad. Program, 1974-76; dir. field study Hampshire Coll., Amherst, Mass., 1977-80; asst. prof. U. Mass. Med. Sch., Worcester, 1985-88, asst. clin. prof., 1988-91, assoc. in psychiatry, 1991—; affiliate asst. prof. psychology Clark U., Worcester, 1989—; pvt. practice Shrewsbury, Mass., 1988—; summer faculty Smith Coll. Sch. Social Wk., Northampton, Mass., 1985-90; cons. Mass. Sch. Profl. Psychology, Dedham, 1991, Syndicat Nat. des Psychologues, Paris, 1985, Mass. Dept. Elder Affairs and Other Mental Health Agencies, 1981—; dir. Women's Relational Devel. Group, Boston, 1990-92. Co-author: Corporations and Child Care, 1974, The Abortion Business, 1975; contbr. articles to profl. jours. Vice pres. Women's Rsch. Action Project, Boston/Cambridge, 1975—. U. Mass. fellow in gerontology, 1986-87; Joseph P. Healey Found. grantee, 1986, AARP/Andrus Found. grantee, 1985, Sigma Xi grantee, 1985; NIMH and Harvard U. traineeship. Mem. APA, Assn. for Women in Psychology. Office: 12 Tatum Rd Shrewsbury MA 01545

SASTRY, PADMA KRISHNAMURTHY, electrical engineer, consultant; b. Bangalore, Karnataka, India, Mar. 18, 1960; came to U.S., 1984; d. S. and Saraswathi Krishnamurthy; m. Sudhir K. Sastry, July 18, 1983; children: Amit Vinay, Nevin Vishnu. B of Engring., Bangalore U., 1983; MBA, Pa. State U., 1987; MSEE, Ohio State U., 1991. Grad. assoc. Pa. State U., State College, 1986-87, Ohio State U., Columbus, 1989-90; tech. analyst Litel Telecom. Inc., Dublin, Ohio, 1990-91; cons. Aldiscon Inc., Dublin, Ohio, 1991—; adj. faculty Columbus State Community Coll., 1992-93. Mem. IEEE, Optical Soc. Am., Sigma Xi.

SATHER, J. HENRY, biologist; b. Presho, S.D., July 21, 1921; s. Anton and Anna (Imster) S.; m. Shirley M. Johnson, Aug. 21, 1948; children: Kristi, Signe, Ingrid. BS, U. Nebr., 1943; MS, U. Mo., 1948; PhD, U. Nebr., 1952. Rsch. biologist Nebr. Game Forestation and Pks. Comms., Lincoln, 1948-55; prof. biology, grad. dean Western Ill. U., Macomb, 1955-79, prof. emeritus, cons., 1979—; wetland cons. U.S. Fish and Wildlife Svc, EPA, Corps Engr., Washington, 1979—; chmn. Rsch. Adv. Bd. Des Plaines River Project, Chgo., 1985—; mem. envron. adv. bd. chief engrs. U.S. Corps Engrs., Washington, 1978-83. Co-author: National Values of Wetlands, 1988, Restoration of River Wetlands, 1989. Sgt. USAF, 1942-45. Recipient Patriatic Civilian Svc. award U.S. Corps. Engrs., 1983, Spl. Recognition Svc. award The Wildlife Soc., 1987; Edward K. Love fellow, 1946-48. Mem. Ecological Soc. Am., Nat. Wetland TEch. Coun., Rsch. Adv. Bd. Des Plaines River Wetlands Demonstration Project (chmn. 1985—), Bombay Natural History Soc., Sigma Xi. Lutheran. Home: 103 Oakland Ln Macomb IL 61455

SATO, GORDON HISASHI, retired biologist, researcher; b. Los Angeles, Dec. 17, 1927; married; 6 children. BA, U. So. Calif., 1950; PhD in Biology, Calif. Inst. Tech., 1955; DSc (hon.), SUNY, Plattsburgh, 1987. Tchg. asst. in microbiology Calif. Inst. Tech., Pasadena, 1953-55; fellow in virology and jr. asst. virologist U. Calif., Berkeley, 1955-56; fellow, instr. biophysics U. Colo. Med. Sch., 1956-58; from asst. prof. to prof. biochemistry Brandeis U., 1958-69; prof. biology U. Calif., San Diego, 1969-83; dir., chief exec. officer W. Alton Jones Cell Sci. Ctr., Lake Placid, N.Y., 1983—; Hon. prof. biology Tsinghua U., Beijing, Xiamen U., Peoples Rep. China; Disting. Rsch. prof. Clarkson U., Potsdam, N.Y., 1987, dir. lab. molecular biology; mem. panel molecular biology study sect. NIH; mem. Coun. for Tobacco Rsch.; bd. sci. counsellors Nat. Inst. Diabetes, Digestive and Kidney Diseases. Editor in chief: In Vitro Cellular Developmental Biology. Mem. NAS, Am. Acad. Arts and Scis., Am. Assn. Immunologists, Endocrine Soc., Assn. Biol. Chemistry. Achievements include research of animal cell culture, endocrinology, bacteriophage. Office: W Alton Jones Cell Sci Ctr 10 Old Barn Rd Lake Placid NY 12946-1099

SATO, HIROSHI, materials science educator; b. Matsuzaka, Mie, Japan, Aug. 31, 1918; came to U.S., 1954; s. Masayoshi and Fusae (Ohhara) S.; m. Kyoko Amemiya, Jan. 10, 1947; children: Norie M., Nobuyuki Albert, Erika Michiko. BS, Hokkaido Imperial U., Sapparo, Japan, 1938, MS, 1941; DSc, Tokyo U., 1951. Rsch. assoc. faculty sci. Hokkaido Imperial U., Sapparo, 1941-42, asst. prof. Inst. Low Temperature Sci., 1942-43; rsch. physicist Inst. Phys. and Chem. Rsch., Tokyo, 1943-45; prof. Tohoku Imperial U., Sendai, Japan, 1945-57; rsch. physicist Westinghouse Rsch. Labs., Pitts., 1954-56; prin. scientist Sci. Lab., Ford Motor Co., Dearborn, Mich., 1956-74; prof. materials engring. Purdue U., West Lafayette, Ind., 1974—, Ross Disting. prof. engring., 1984-89; Ross Disting. prof. engring. emeritus Purdue U., West Lafayette, 1989—; affiliate prof. dept. materials sci. U. Washington, Seattle, 1986-89; collaborator Los Alamos (N.Mex.) Nat. Lab., 1989—; vis. prof. U. Grenoble, France, 1967, Tokyo Inst. Tech., 1979, Tech. U. Hannover, Fed. Republic Germany, 1980-81; cons. Oak Ridge (Tenn.) Nat. Lab., 1978, 80. Contbr. over 260 articles to profl. jours., chpts. to books. Recipient U.S. Sr. Scientist award Alexander von Humboldt Found., 1980; fellow John Simon Guggenheim Meml. Found., 1966, Japan Soc. for Promotion Sci., 1979. Fellow Am. Phys. Soc.; mem. Japan Phys. Soc., Am. Ceramic Soc., Metall. Soc.-AIME, Japan Inst. Metals (hon. 1985—, Prize of Merit 1951). Democrat. Congregationalist. Office: Purdue U Sch of Materials Engring 1289 MSEE Bldg West Lafayette IN 47907-1289

SATO, KAZUYOSHI, pathologist; b. Shibata, Niigata, Japan, Apr. 3, 1930; came to U.S., 1968; s. Katsueita and Kyo (Sakagawa) S.; m. Ann Marie Farrenkopf, July 5, 1964 (dec. Aug. 1983); children: P.T. Sachiko, P. Miyoko, Michael T., Phillip K. Student, Niigata U., Japan, 1954, MD, 1958. Diplomate Am. Bd. Pathology, Anatomic and Clin. Pathology. Intern USAF Hosp., Tachikawa, Japan, 1958-59; intern Ellis Hosp., Schenectady, N.Y., 1959-60, asst. resident in pathology, 1960-61; resident in pathology Free Hosp. for Women, Brookline, Mass., 1961-62; resident in pathology The Children's Hosp. Med. Ctr., Boston, 1962-63, resident in neuropathology, 1963-64; resident fellow in pathology Mayo Grad. Sch. Medicine, Rochester, Minn., 1968-70; dir. labs. Falmouth (Mass.) Hosp., 1972—; dir. Falmouth Hosp. Service Lab., Sandwich, Mass., 1986—; pathologist and rsch. assoc. Atomic Bomb Casualty Commn., Nagasaki, Japan, 1964-68; pathologist, chief of pathology USPHS Hosp., Norfolk Va., 1970-72, Falmouth (Mass.) Hosp., 1972—. Recipient Fulbright scholarship, 1959. Fellow Coll. Am. Pathologists, Am. Soc. Clin. Pathologists; mem. Assn. Mil. Surgeons U.S. Home: 88 Two Ponds Rd Falmouth MA 02540-2225 Office: Falmouth Hosp 100 Ter Heun Dr Falmouth MA 02540-2599

SATO, MOTOAKI, geologist, researcher; b. Tokyo, Japan, Oct. 11, 1929; came to U.S., 1955, 63.; s. Iwazo and Kyoko (Ito) S.; m. Ellen B. Levinson, Feb. 11, 1961 (div. Sept. 1978); children: Emily Coates, Alice Isomé, Thomas Bartlett. BS in Geology, U. Tokyo, Japan, 1953, MS in Geology, 1955; PhD in Geology, U. Minn., 1959. Research asst. dept. geophysics Univ. Minn., Mpls., 1956-58; rsch. fellow in geophysics dept. geol. scis. Harvard Univ., Cambridge, Mass., 1958-61; assoc. prof. geology Inst. Thermal Springs Research, Misasa, Tottori, Japan, 1961-63; research geologist U.S. Geological Survey, Washington, 1963-65; geologist, project chief U.S. Geological Survey, Washington/Reston, Va., 1965—; prin. investigator Lunar Sample & Sci. Program, NASA, 1971-80. Contbg. author books and articles in profl. jours. Fulbright/Smith-Mundt fellow Inst. Internat. Edn., 1955-57, Gilbert fellow U.S. Geol. Survey, Reston, Va., 1982-83. Mem. Am. Geophysical Union, Geochemical Soc., Geological Soc. Washington (2d v.p. 1982-83), Geochemistry Div. Am. Chem. Soc. Home: 11173 Lake Chapel Ln Reston VA 22091-4308 Office: US Geol Survey 959 National Ctr Reston VA 22092

SATO, NORIAKI, engineering educator; b. Tokyo, July 29, 1928; s. Isamo and Fukuju S.; m. Kyoko Kojima, May 28, 1957; children: Norihito, Tomoko Sato Masubuchi, Yasuhiko. B in Engring., Tokyo Inst. Tech., 1953, PhD in Engring., 1965. First Class Chief Elect. Engr. Engr. Toshiba Corp., Tokyo, 1953-61, sub-chief engr., 1961-62, chief engr., 1962-66; assoc. prof. Nagoya U., 1967-72, prof., 1972-76; prof. Tokyo Inst. Tech., 1976-89, Tokyo Met. Inst. Tech., 1989-90, Saitama (Japan) U., 1991—. Recipient Authorship Prize Teshima Indsl. Edn. Soc., Tokyo, 1981. Fellow IEEE (vice-chmn. IA chpt. Tokyo sect., 1981-83, chmn. 1984-86, chmn. student activities 1985-86, bd. dirs. 1985-87, paper award 1972). Avocation: hiking. Home: 1-45-2 Sakuradai, Nerimaku, Tokyo 176, Japan

SATO, TAKAMI, pediatrician, medical oncologist; b. Oita-shi, Japan, July 27, 1955; s. Takao and Miyoko (Tashima) S.; m. Chiyo Motoyoshi, Dec. 4, 1984; children: Takahiro, Shingo, Rino. MD, Jichi Med. Sch., 1980. Med. dr. Kiyokawa-mura Clin., Onogun, Japan, 1985-88; assoc. dir. div. pediatrics and internal medicine Oita Prefectural Hosp., Onogun, 1988-90; asst. prof. Jichi Med. Sch., Kawachi-gun, Japan, 1990-93; asst. prof. dept. medicine Thomas Jefferson U., Phila., 1993--. Contbr. articles to profl. jours. Mem. AMA, AAAS, AACR. Office: Thomas Jefferson U 1015 Walnut St Ste 1005 Philadelphia PA 19026

SATSUMABAYASHI, KOKO, chemistry educator; b. Niigata, Japan, Feb. 27, 1940; d. Yoshitoshi and Nori (Inoue) Tanabe; m. Sadayoshi Satsumabayashi, Mar. 30, 1969. BS, Sci. U. Tokyo, 1962, MS, 1964, PhD in Chemistry, 1968. Asst. Sci. U. Tokyo, 1964-71; lectr. Nippon Dental U., Tokyo, 1971-72; assoc. prof. Nippon Dental U., Niigata, 1972-82, prof. chemistry, 1982—; part-time lectr. Tokyo Music U., 1968—, Sci. U. Tokyo, 1971-72. Co-author: Handbook of Organic Chemistry, 1968, History of Chemistry, 1970, Epoca, 1975, Experiments in Organic Chemistry, 1976. Capable Woman Govt. Svc., Kawasaki, Kanagawa, Japan, 1988, Niigata, 1989—. Recipient Life Saving award Met. Police Bd., Tokyo, 1969. Mem. Am. Chem. Soc., Chem. Soc. Japan, Japan Soc. for Biosci., Biotech. and Agrochemistry. Buddhist. Avocations: calligraphy, tea ceremony, rakugo, travel, thinking. Office: Nippon Dental U, Hamaura-cho 1-8, Niigata 951, Japan

SATSUMABAYASHI, SADAYOSHI, dental educator; b. Nagano-Ken, Japan, Oct. 18, 1935; s. Kenjuu and Waguri (Iwata) S.; m. Koko Tanabe, Mar. 30, 1969. BS, Sci. U. Tokyo, 1962, MS, 1964, DSc, 1969. Asst. Sci. U. Tokyo, 1964-71, lectr., 1971-74; assoc. prof. Nippon Dental U., Tokyo, 1974-76, prof., 1976—, dean students, 1979-83, vice chmn. instructive office work, 1983-86, chmn. admission office, 1986—. Mem. Chem. Soc. Japan, Am. Chem. Soc. Buddhist. Avocations: appreciation of music, golfing, chess, traveling, reading. Office: Nippon Dental U, Fujimi 1-9-20, Chiyoda-ku 102, Japan

SATTLER, NANCY JOAN, math and physical science educator; b. Toledo, July 14, 1950; d. Thomas Joseph and Margaret Mary (Linenkugel) Ainsworth; m. Rudolph Henry Sattler, June 17, 1972; children: Cortlund, Clinton, Corinne. BS, U. Toledo, 1972, MEd, 1988. Office worker/bookkeeper Gilbert Mail Svc., 1967-71; computer typesetter Quality Composition, Toledo, 1971-89; instr. Terra Tech. Coll., Fremont, Ohio, 1988-89, dept. head, 1989—; co-chmn. kids coll., 1993; co-chmn. Kids Coll., Terra Tech. Coll., 1993; adj. instr. Terra Tech. Coll., Fremont, 1982-88, U. Toledo, 1988, Lucas County Bd. Edn. Gifted Program, Toledo, 1988—; computer coord. St. Joseph Elem. Sch., Fremont, 1987—, quiz bowl, 1993; externship quality control Atlas Crankshaft, Fostoria, Ohio, 1990; instr. devel. math. A.O. Smith, Bellvue, Ohio, 1991; adult edn. computer instr. St. Joseph Cen. Cath., Fremont, 1990—; instr. developmental math and sci. Whirlpool Corp., Findlay, Ohio, 1992; presenter Am. Math Assn. of Two-Yr. Colls., 1991, 92, Nat. Coun. Tchrs. Math Conf., 1993; co-presenter Continuous Improvements through Faculty Externship Program Innovation Leadership 2000 Conf., League for Innovation, 1992; co-chmn. Ohio Great Tchrs. Seminar, 1993; facilitator Mo. Great Tchrs. Seminar, 1993. Author: The Implication of Math Placement Testing in the Two Year College, 1988, Applied Math for Industrial Tech., 1989; co-author: Math and Science Made Easy, 1992, The Metric System, Preparing for the Future, 1992. Sec. St. Joseph Cen. Cath. Sch. Bd., 1989—, pres. 1991-93; Sun. sch. dir. St. Joseph Ch., Fremont, 1977-87; pres. Plant 'N Bloom Garden Club, Fremont, 1977-79; clk. Sandusky County Cath. Music Boosters, 1991-93, v.p., 1990-91; rep. for deanery Early Childhood Devel., 1982-84; parliamentarian Welcome Wagon, 1980; Eucharistic minister, 1991—. Mem. Ohio Assn. Garden Clubs, Ohio Math. Assn. of Two-Yr. Colls. (pres. 1992—, NSF grant com. mem. 1992), Am. Math. Assn. of Two-Yr. Colls. (assessment com. 1990—, program com. 1993—), Nat. Coun. Tchrs. Math., Ohio Coun. Tchrs. Math. Democrat. Roman Catholic. Avocations: quilting, gardening, canning, sewing. Home: 712 Hayes Ave Fremont OH 43420 Office: Terra Tech Coll 2830 Napoleon Rd Fremont OH 43420

SAUER, BARRY W., medical research center administrator, bioengineering educator; b. Washington, Mar. 27, 1939. DVM, U. Ga., 1966. Rsch. asst. Spring Grove State Hosp., Balt., 1962; postdoctoral fellow Nat. Heart and Lung Inst. Colo. State U., Ft. Collins, 1966-70; asst. prof. bioengineering Clemson (S.C.) U., 1970-74, assoc. prof. bioengineering, 1974-77; assoc. prof. surgery U. Miss. Med. Ctr., Jackson, 1977-80, assoc. prof. surgery and restorative dentistry, 1980-88, dir. orthopaedic rsch., 1977-88; pres., CEO Harrington Arthritis Rsch. Ctr., Phoenix, Ariz., 1988—; adj. prof. bioengineering Ariz. State U., Tempe, 1991—; co-preparer initial proposal mgr. biomedical engring. sect., acting dir. materials sci. and engring. divsn. Inst. Tech. Devel., Jackson, 1983-88; standing mem. instl. animal care and use com. U. Miss. Med. Ctr., 1984; chmn. Fourth So. Biomedical Conf., 1985; cons. Porex Med., Fairburn, Ga., 1971—, Johnson & Johnson, Braintree, Mass., 1978-83, Howmedica, Inc., Rutherford, N.J., 1983-84. Editor Biomedical Engring. IV--Recent Devels., 1985; mem. editorial bd. Jour. Investigative Surgery; contbr. over 85 rsch. articles and papers to sci. jours. Fellow Acad. Surg. Rsch. (founding, bd. dirs. 1980—, pres.-elect 1989-90, pres. 1990-91); mem. AAAS, Arthritis Found. Am. (bd. dirs. ctrl. Ariz. chpts. 1988—), Am. Coll. Rheumatology, Am. Soc. for Artificial Internal Organs, Rehabilitation Engring. Soc. N.Am., Am. Voice Input/Output Soc., Am. Soc. for Gravitational and Space Biology, Nat. Svc. Robot Assn. (charter), Soc. for Biomaterials (charter), Orthopaedic Rsch. Soc., Biomedical Engring. Soc., Miss. Acad. Scis. (life), Miss. Veterinary Med. Assn., Ariz. Consortium for Children with Chronic Illness (children's rehab. svcs. com. 1989—), Ariz. Rheumatology Assn., Ariz. Innovation Network, Gov.'s Coun. for Arthritis and Musculoskeletal Diseases (vice chmn. 1992), Phoenix Rheumatology Club, Sigma Xi. Achievements include patents for Canine Ear Implant, Porous Coated Hip Prosthesis, Mouthstick. Office: Harrington Arthritis Rsch Ctr 1800 E Van Buren St Phoenix AZ 85006

SAUER, GORDON CHENOWETH, physician, educator; b. Rutland, Ill., Aug. 14, 1921; s. Fred William and Gweneth (Chenoweth) S.; m. Mary Louise Steinhilber, Dec. 28, 1944; children: Elisabeth Ruth, Gordon Chenoweth, Margaret Louise, Amy Kieffer.; m. Marion Green, Oct. 23, 1982. Student, Northwestern U., 1939-42; B.S., U. Ill., 1943, M.D., 1945. Diplomate: Am. Bd. Dermatology and Syphilology. Intern Cook County Hosp., Chgo., 1945-46; resident dermatology and syphilology N.Y. U.-Bellevue Med. Center, 1948-51; dermatologist Thompson-Brumm-Knepper Clinic, St. Joseph, Mo., 1951-54; pvt. practice Kansas City, Mo., 1954—; mem. staff St. Luke's, Research, Kansas City Gen. hosps.; asso. instr. U. Kans., 1951-56, vice chmn. sect. dermatology, 1956-58, asso. clin. prof., 1960-64, clin. prof., 1964—, head sect. dermatology, 1958-70; clin. asso., acting head dermatology sect. U. Mo., 1955-59, cons. dermatology, 1959-67, clin. prof., 1967—; cons. Munson Army Hosp., Ft. Leavenworth, Kans., 1959-68; Mem. dermatology panel, drug efficacy panel Nat. Acad. Sci.-FDA, 1967-69. Author: Manual of Skin Diseases, 6th edit., 1991, Teen Skin, 1965, John Gould Bird Print Reproductions, 1977, John Gould's Prospectuses and Lists of Subscribers to his work on Natural History: with an 1866 Facsimile, 1980, John Gould The Bird Man, 1982; editor: Kansas City Med. Bull., 1967-69; contbr. articles to profl. jours. Bd. dirs. Kansas City Area coun. Camp Fire Girls Am., 1956-59, Kansas City Lyric Theatre, 1969-74, Kansas City Chamber Choir, 1969-74, Chouteau Soc., 1985—, U. Mo.-Kansas City Friends of Libr., 1988-92; bd. dirs. Mo. br. The Nature Conservancy, 1984-91. Sr. asst. surgeon USPHS, 1946-48. Named Dermatology Found. Practitioner of Yr., 1992. Fellow Am. Acad. Dermatology and Syphilology (dir. 1975-79, v.p. 1980); mem. Mo., Jackson County med. socs., Mo. Dermatol. Soc. (pres. 1974-75), Dermatology Found. (trustee 1978-83), Am. Ornithol. Union, Wilson Ornithol. Soc., Royal Australasian Ornithologists Union, Soc. Bibliography Natural History, Am. Dermatol. Assn., Alpha Delta Phi, Nu Sigma Nu. Presbyterian. Office: 6400 Prospect Ave Kansas City MO 64132-1181

SAUER, HAROLD JOHN, physician, educator; b. Detroit, Dec. 1, 1953; s. Peter and Hildegard (Muehlmann) S.; m. Kathleen Ann Iorio, Sept. 4, 1982; children: Angela Karin Ferrante, Peter Rolf Jan Muehlmann, Josef Andrew John Iorio. BS, U. Mich., 1975; MD, Wayne State U., 1979. Diplomate Am. Bd. Ob-Gyn. Resident in ob-gyn William Beaumont Hosp., Royal Oak, Mich., 1979-83; fellow in reproductive endocrinology and infertility William Beaumont Hosp., Royal Oak, 1983-85; asst. prof. dept. ob-gyn and reproductive biology Mich. State U., East Lansing, 1985-91, assoc. prof. ob.-gyn, 1991—; mem. staff St. Lawrence Hosp., Lansing, Mich., 1985—, E.W. Sparrow Hosp., Lansing, 1985—; cons. Mich. Dept. Social Svcs., Lansing,

1985—; mem. Mich. Bd. Medicine, 1992—; researcher in field. Fellow Am. Coll. Ob-Gyn. (sec. Mich. sect. 1990—); mem. AMA, Ingham County Med. Soc., Lansing Ob-Gyn. Soc., Am. Fertility Soc., Am. Assn. Gynecol. Laparoscopists, Wayne State U. Med. Alumni Assn., Mich. Soc. Reproductive Endocrinology (sec.-treas. 1991—). Roman Catholic. Avocations: classical piano, microcomputers, skiing. Home: 2601 Creekstone Trail Okemos MI 48864-2455 Office: Mich State U Dept Ob-Gyn Reproductive Biology B-316 Clinic Ctr East Lansing MI 48824-1315

SAUER, HARRY JOHN, JR., mechanical engineering educator, university administrator; b. St. Joseph, Mo., Jan. 27, 1935; s. Harry John and Marie Margaret (Witt) S.; m. Patricia Ann Zbierski, June 9, 1956; children: Harry John, Elizabeth Ann, Carl Andrew, Robert Mark, Katherine Anne, Deborah Elaine, Victoria Lynn, Valerie Joan, Joseph Gerard. B.S., U. Mo.-Rolla, 1956, M.S., 1958; Ph.D., Kans. State U., 1963. Instr. mech. engring. Kans. State U., Manhattan, 1960-62; sr. engr., cons. Midwest Rsch. Inst., Kansas City, Mo., 1963-70; mem. faculty dept. mech. and aerospace engring. U. Mo., Rolla, 1957—, prof., 1966—, assoc. chmn., 1980-84, dean grad. study, 1984-92; cons. in field; mem. Gov.'s Commn. on Energy Conservation, 1977; mem. Mo. Solar Energy Resource Panel, 1979-83. Co-author: Environmental Control Principles, 1975, 78, 82, 85, Thermodynamics, 1981, Heat Pump Systems, 1983, Engineering Thermodynamics, 1985, Principles of Heating, Ventilating and Air Conditioning, 1991; contbr. articles to profl. jours. Pres. St. Patrick's Sch. Bd., 1970-72, St. Patrick's Parish Council, 1975-76. Recipient Ralph R. Teetor award Soc. Automotive Engrs., 1968; Hermann F. Spoehrer Meml. award St. Louis chpt. ASHRAE, 1979; also disting. service award, 1981, E. K. Campbell award of merit, 1983. Mem. ASME, ASHRAE, NSPE, Soc. Automotive Engrs., Am. Soc. Engring. Edn., Mo. Soc. Profl. Engrs., Mo. Acad. Sci., Sigma Xi. Roman Catholic. Home: College Hills Rt 4 Rolla MO 65401 Office: U Mo Rolla MO 65401

SAUERS, ISIDOR, physicist, researcher; b. Linz, Austria, Aug. 3, 1948; came to U.S., 1950; s. David and Nancy (Schoenbach) S.; m. Megan Denise Weekley, June 19, 1977; children: Aaron, Elisha. BS in Physics, Ga. Tech., 1969, PhD, 1974. Assoc. rsch. scientist dept. chemistry U. Md., College Park, Md., 1974-76; rsch. assoc. dept. chemistry Johns Hopkins U., Balt., 1976-77; rsch. assoc. dept. physics U. Tenn., Knoxville, 1977-78; rsch. staff Oak Ridge (Tenn.) Nat. Lab., 1978—; vis. scientist Nat. Bur. Standards, Gaithersburg, Md., 1986. Editor: Gaseous Dielectrics VI, 1991; contbr. articles to Jour. of Physics D, Applied Physics, Plasma Chemistry and Plasma Processing, Jour Chem. Physics, Phys. Review, IEEE Transactions on Elec. Insulation; (over 75 pubs.). Mem. bd. Temple Beth El, Knoxville, 1990-92. Recipient Martin Marietta award, Oak Ridge, Tenn., 1986. Mem. IEEE (com. on gas dielectrics), AAAS, Am. Phys. Soc., Sigma Xi. Achievements include patent on measuring degradation of sulfur hexafluorine in high voltage systems; rsch. on physics and chemistry of gas discharges and gaseous dielectrics; on sulfur hexafluoride decomposition and chemistry; mass spectomery rsch. *. Office: Oak Ridge Nat Labs PO Box 2008 MS 6123 Oak Ridge TN 37831-6123

SAUL, WILLIAM EDWARD, academic administrator, civil engineering educator; b. N.Y.C., May 15, 1934; s. George James and Fanny Ruth (Murokh) S.; m. J. Muriel Held Eagleburger, May 11, 1976. BSCE, Mich. Tech. U., 1955, MSCE, 1961; PhD in Civil Engring., Northwestern U., 1964. Registerd profl. engr., Wis., Idaho, Mich., profl. structural engr., Idaho. Mech. engr. Shell Oil Co., New Orleans, 1955-59; instr. engring. mechanics Mich. Tech. U., Houghton, 1960-62; asst. prof. civil engring. U. Wis., Madison, 1964-67, assoc. prof., 1967-72, prof., 1972-84; dean Coll. Engring., prof. civil engring. U. Idaho, Moscow, 1984-90; prof. civil engring. Mich. State U., East Lansing, 1990—, chair dept. civil and environ. engring., 1990—; cons. engr., 1961—; bd. dirs. Idaho Rsch. Found., 1984-90; vis. prof. U. Stuttgart, Fed. Republic Germany, 1970-71. Co-editor Conf. of Methods of Structural Analysis, 1976. Fulbright fellow 1970-71; von Humboldt scholar, 1970-71. Fellow ASCE (pres. Wis. sect. 1983-84); mem. NSPE, Mich. Soc. Profl. Engrs., Internat. Assn. Bridge and Structural Engrs., Am. Concrete Inst., Am. Soc. Engring. Edn., Sigma Xi, Phi Kappa Phi, Tau Beta Pi, Chi Epsilon. Avocations: hiking, reading, travel, gadgets. Home: 1971 Cimarron Dr Okemos MI 48864-3905 Office: Mich State U A349 Engring Bldg East Lansing MI 48824

SAUNDERS, BARRY COLLINS, civil engineer; b. St. Louis, Dec. 17, 1931; s. William Flewellyn and Naomi Harriet (Kober) S.; m. Marjorie Ruth Nordholm, June 11, 1960; children: Kristin Ruth, Jennifer Ann Saunders Gerlach. BS, St. Louis U., 1953; MBA, U. Denver, 1961; MS, Utah State U., 1967. Registered prof. engr., Colo. Test engr. McDonnell Aircraft Corp., St. Louis, 1956-58; sr. engr. Stanley Aviation Corp., Denver, 1958-62, Thiokol Chem. Corp., Brigham City, Utah, 1962-66; assoc. dir. Utah State Divsn. Water Resources, Salt Lake City; chmn. engring. com. Upper Colo. River Commn.; Utah rep. Colo. River Endangered Fishes Mgmt. Group, Colo. River Basin Salinity Control Forum; prin. investigator Fed. State Coop. Atmospheric Mod. Rsch. Program; mgr. Utah water Edn./Conservation Program. With U.S. Army, 1954-56. Mem. ASCE, Am. Water Resources Assn. Lutheran. Office: Utah State Divsn Water Res 1636 West North Temple Salt Lake City UT 84116

SAUNDERS, GRADY F., biochemistry and biology educator, researcher; b. Bakersfield, Calif., July 11, 1938; m. Priscilla Saunders; 1 child, Nicole. BS, Oreg. State U., 1960, MS, 1962; PhD, U. Ill., 1965. USPHS postdoctoral fellow Inst. de Biology, Physics-Chemistry, Paris, 1965-66; asst. prof. biochemistry dept. devel. therapeutics U. Tex. Anderson Cancer Ctr., Houston, 1966-72, assoc. prof. biochemistry, 1972-77; assoc. prof. biochemistry dept. biochemistry U. Tex. Anderson Cancer Ctr., Houston, 1977-78, prof., 1978—, Ashbel Smith prof. biochemistry and molecular biology, 1986-90, Anise J. Sorrell prof., 1990—; exchange scientist US-USSR Inst. Molecular Biology Acad. Scis. USSR, Moscow, 1972, '82; acting head dept. biochemistry U. Tex. Med. Dept. Anderscon Cancer Ctr., Houston, 1982-83; vis. prof. pediatrics U. Tex. Med. Sch. at Houston, 1990—; mem. numerous nat. and local med. coms. Editor Anticancer Rsch., 1990—; contbr. over 130 articles to profl. jours., chpts. to 17 books; editor: Cell Differentiation and Neoplasia, 1978, Perspectives on Genes and the Molecular Biology of Cancer, 1983, Yearbook of Cancer 1985, '86, '87, '88.. Recipient of many grants for cancer rsch., 1986—. Mem. AAAS, Am. Soc. Biol. Chemists, Am. Soc. Cell Biology, Am. Soc. Human Geneticists, Biophys. Soc., Human Genome Orgn., Phi Sigma, Sigma Xi. Achievements include two U.S. patents: probe for detection of specific human leukemia, 1989, defensins, 1991. Home: 5134 Loch Lomond Houston TX 77096 Office: U Tex Anderson Cancer Ctr 1515 Holcombe Blvd Houston TX 77030

SAUNDERS, JIMMY DALE, aerospace engineer, physicist, naval officer; b. Bronte, Tex., Dec. 16, 1948; s. James Howard and Wanda Lee (Lackey) S.; m. Judy Karon Falconer, Aug. 2, 1969; children: Jennifer Rebecca, Rachel Lee, Jason Allan. BS in Physics, U. Miss., 1976; MS in Physics, Naval Postgrad. Sch., Monterey, Calif., 1986. Enlisted USN, 1970, advanced through grades to comdr., 1991; intelligence officer Comdr. Submarines Mediteranean, Naples, Italy, 1979-82; weapons officer USS George Bancroft, 1982-84; asst. for strategic weapons systems Strategic Systems Programs, Washington, 1987-88, asst. head missile ops., 1988, asst. head missile engring., 1988-90, asst. for advanced systems, 1990-91, head missile engring., 1991—; staff assoc. Chief Naval Ops. for Spl. Studies, 1989-90, Def. Sci. Bd., 1990; tech. advisor Strategic Arms Reductions Treaty, Geneva, 1990-91. Contbr. article to Phys. Rev. Mem. County Sch. Adv. Coun., Goose Creek, S.C., 1982-84; commr. Springfield (Va.) Youth Football Program, 1990—; bd. dirs. Springfield (Va.) Youth Club, 1990—. Mem. AAAS, Am. Phys. Soc., Am. Soc. Naval Engrs., Sigma Xi. Methodist. Home: 8118 Lake Pleasant Dr Springfield VA 22153-3009 Office: Strategic Systems Programs (SP-273) Washington DC 20376-5000

SAUNDERS, JOHN WARREN, JR., biology educator, consultant; b. Muskogee, Okla., Nov. 12, 1919; s. John Warren and Amanda Mary (Schlattweiler) S.; m. Lilyan Clayton, Feb. 27, 1942; children: Sarah Elizabeth Reeder, John Warren, Margaret Ann Geist, Mary Katherine Brown. BS, U. Okla., 1940, MS, 1941; PhD, Johns Hopkins U., 1948. Jr. instr. in biology Johns Hopkins U., Balt., 1941-43, 46-48; instr. zoology U. Chgo., 1948-49; from assoc. prof. to prof. Marquette U., Milw., 1949-66; prof. anatomy U. Pa., Phila., 1966-67; prof. biology SUNY, Albany, 1967-85; author and cons. in pvt. practice Waquoit, Mass., 1985—; mem. adv. panel devel. biology NSF, Washington, 1961-66; trustee Marine Biol. Lab., Woods Hole, Mass., 1969-72. Author: Animal Morphogenesis, 1968, Principles of Animal Development, 1970, Developmental Biology, 1982; contbr. numerous articles to profl. jours. Bd. dirs. Milw. div. Am. Cancer Soc. 1960-65; elected mem. Town Meeting, Falmouth, Mass., 1988—. Lt. (s.g.) USN, 1943-46, PTO. Recipient Joseph Rigge Disting. Svc. award Marquette U., 1988. Fellow AAAS; mem. Assn. Am. Anatomists, Soc. for Devel. Biology (pres. 1968-69), Am. Soc. Zoologists (sec. 1964-66). Democrat. Roman Catholic. Home and Office: PO Box 3381 Waquoit MA 02536

SAUNDERS, RICHARD WAYNE, pipeline engineer; b. Independence, Kans., Oct. 12, 1953; s. Richard John and Juanita Faye (Langston) S.; m. Janice Lynn Robertson, Mar. 19, 1977; children: Brandi Lynn, Richard Bradley. AA, Independence C.C., 1973; BS, Kans. State U., 1975. Engr., estimator Parmac, Inc.-Maloney-Crawford Tank, Tulsa, 1976-77; structural engr MEC Co., Neodesha, Kans., 1977-80; engr. Arco Pipe Line Co., 1980-87; ops. engr. Arco Pipe Line Co., Midland, Tex., 1987-89; sr. project engr. Arco Pipe Line Co., Houston, 1989-91; lead pipeline engr. John Brown E&C, Inc., Houston, 1991-92; project engr., pipeline engr. John Brown Saudi Arabia, Ltd., Al Khobar, 1992—. Mem. ASCE. Republican. Office: John Brown E&C Inc PO Box 720421 Houston TX 77272

SAUNDERS, WILLIAM HUNDLEY, JR., chemist, educator; b. Pulaski, Va., Jan. 12, 1926; s. William Hundley and Vivian (Watts) S.; m. Nina Velta Plesums, June 25, 1960 (dec. June 1982); children: Anne Michele, Claude William. B.S., Coll. William and Mary, 1948; Ph.D., Northwestern U., 1952. Research assoc. Mass. Inst. Tech., 1951-53; mem. faculty U. Rochester, 1953—, prof. chemistry, 1964-91, faculty sr. assoc., 1991—, chmn. dept., 1966-70. Author: Ionic Aliphatic Reactions, 1965, (with A.F. Cockerill) Mechanisms of Elimination Reactions, 1973, (with L. Melander) Reaction Rates of Isotopic Molecules, 1980; contbr. articles to profl. jours. Guggenheim fellow, 1960-61; Sloan Found. fellow, 1961-64; NSF sr. postdoctoral fellow, 1970-71. Mem. Am. Chem. Soc., Chem. Soc. (London), Phi Beta Kappa, Sigma Xi, Phi Lambda Upsilon. Home: 15 Parkwood Ave Rochester NY 14620-3401

SAURA-CALIXTO, FULGENCIO, food chemist, educator; b. Villanueva de la Fuente, Spain, Dec. 12, 1946; s. Saura Rios Fulgencio and Calixto Arenas Juana; m. Maria Jose Martinez de Toda, June 30, 1973; 1 child, Santiago. Degree in chemistry, U. Murcia, 1968; PhD, U. Oviedo, 1975. Tchr. chrmistry U. Oviedo, Spain, 1971-75, U. Baleares, Palma de Mallorca, Spain, 1976-84, U. Alcala de Henares, Madrid, Spain, 1985-86; food sci. researcher Nat. Rsch. Coun., Madrid, 1987-90, prof., 1991—; vis. scientist U.K., Poland, Sweden, Switzerland, 1985-90; participant various rsch. programs of European Community. Contbr. articles to profl. jours. Achievements include research in nutrition, food science and technology, carbohydrates. Office: Inst Nutricion Bromatologia, Facultad de Farmacia, 28040 Madrid Spain

SAURO, JOSEPH PIO, physics educator; b. New Rochelle, N.Y., Apr. 4, 1927; s. Francesco Giovanni and Lucia (Arrivebene) S.; m. Elizabeth Joann Schellman, May 2, 1948; children: Brian, Michael, Joseph. BS, Poly Inst. Bkyn., 1955, MS, 1958, PhD in Physics, 1966. Dir. coll. sci. improvement program U. Mass., North Dartmouth, 1969-71, dean grad. sch., 1969-71, interim dean Coll. of Engring., 1978-80, dean Coll. Arts and Scis., 1969-80, prof. physics, 1965—. With USN, 1944-46. Sci. Faculty fellow NSF, 1964; State War Svc. scholar State of N.Y., 1953. Mem. Am Assn. Physics Tchr., Sigma Xi, Sigma Pi Sigma. Avocations: photography, travel, music. Home: 8 Captain Wing Rd East Sandwich MA 02537-1122 Office: U Mass North Dartmouth MA 02747

SAUSMAN, KAREN, zoological park administrator; b. Chgo., Nov. 26, 1945; d. William and Annabell (Lofaso) S. BS, Loyola U., 1966; student, Redlands U., 1968. Keeper Lincoln Park Zoo, Chgo., 1964-66; tchr. Palm Springs (Calif.) Unified Sch., 1968-70; ranger Nat. Park Svc., Joshua Tree, Calif., 1968-70; zoo dir. The Living Desert, Palm Desert, Calif., 1970—; natural history study tour leader internat., 1974—; part-time instr. Coll. Desert Natural History Calif. Desert, 1975-78; field reviewer conservation grants Inst. Mus. Svcs., 1987—, MAP cons., 1987—, panelist, 1992—; internat. studbook keeper for Sand Cats, 1988—, for Cuvier's Gazelle, Mhorr Gazelle, 1990—; co-chair Arabian Oryx species survival plan propogation group, 1986—; spkr. in field. Author Survival Captive Bighorn Sheep, 1982, Small Facilities- Opportunities and Obligations, 1983; wildlife illustrator books, mags, 1970—; editor Fox Paws newsletter Living Desert, 1970—, ann. reports, 1976—; natural sci. editor Desert Mag., 1979-82; compiler Conservation and Management Plan for Antelope, 1992; contbr. articles to profl. jours. Past bd. dirs., sec. Desert Protective Coun.; adv. coun. Desert Bighorn Rsch. Inst., 1981-85; bd. dirs. Palm Springs Desert Resorts Convention and Visitors Bur., 1988—; bd. dirs., treas. Coachella Valley Mountain Trust, 1989—. Named Woman Making a Difference Soroptomist Internat., 1989, 93. Fellow Am. Assn. Zool. Parks and Aquariums (bd. dirs., accredation field reviewer, desert antelope taxon adv. group, caprid taxon adv. group, felid taxon adv. group, small population mgmt. adv. group, wildlife conservation and mgmt. com., chmn. ethics com. 1987, mem. com., internat. rels. com., ethics task force, pres'. award 1972-77, outstanding svc. award 1983, 88, editor newsletter, Zool. Parks and Aquarium Fundamentals 1982); mem. Internat. Species Inventory System (mgmt. com., policy adv. group 1980-88), Calif. Assn. Mus., Calif. Assn. Zoos and Aquariums, Internat. Union Dirs. Zool. Gardens, Western Interpretive Assn. (so. Calif. chpt.), Am. Assn. Mus., Arboreta and Botanical Gardens So. Calif. (coun. dirs.), Soc. Conservation Biology, Nat. Audubon. Soc., Jersey Wildlife Preservation Trust Internat., Nature Conservancy, East African Wildlife Soc., African Wildlife Found., Kennel Club Palm Springs (past bd. dirs., treas. 1978-80), Scottish Deerhound Club Am. (editor SCottish Deerhounds in N.A., 1983, life mem. U.K. chpt.). Avocations: pure bred dogs, dressage, painting, photography. Office: Living Desert 47-900 Portola Ave Palm Desert CA 92260

SAUSVILLE, EDWARD ANTHONY, medical oncologist; b. Albany, N.Y., Apr. 3, 1952; s. Edward Adolphus and Pauline (Zamenick) S.; m. Carol Ann Cassidy, Feb. 1, 1975; children: Justin, Brendan, Elizabeth, Rebecca, Paul. BS, Manhattan Coll., 1973; MD, PhD, Albert Einstein Coll. Medicine, 1979. Med. house staff Brigham & Women's Hosp., Boston, 1979-82; med. staff fellow Nat. Cancer Inst., Bethesda, Md., 1982-85; sr. investigator Nat. Cancer Inst., Bethesda, 1985-88, 90—; assoc. prof. medicine Georgetown U. Sch. Medicine, Washington, 1988-90. Author: (book chpt.) "Lung Cancer" in Kelley Textbook Internal Medicine, 1989; contbr. articles to New Eng. Jour. Medicine, Jour. Biol. Chemistry, Cancer Rsch. Mem. Am. Assn. Cancer Rsch., Am. Soc. Clin. Oncology, Am. Soc. Biochem. Molecular Biology, Am. Fedn. Clin. Rsch., Phi Beta Kappa, Alpha Omega Alpha. Achievements include research on mechanisms of bleomycin action; bombesin-related peptic gene expression and response in lung cancer; optimal treatment and staging of cutaneous T-cell lymphoma. Home: 1114 Nora Dr Silver Spring MD 20904-2136 Office: Nat Cancer Inst Lab Biol Chem Bldg 37 Rm 5002 Bethesda MD 20892

SAUTER, FRANZ FABIAN, structural engineer; b. San Jose, Costa Rica, Feb. 7, 1933; s. Federico and Hilda (Fabian) S.; Engring. degree U. Costa Rica, 1956; postgrad. Internat. Inst. Seismology and Earthquake Engring., Tokyo, 1963-64; m. Maria Angeles Ortiz, June 30, 1957; children: Arnold, Hans Peter, Krista Maria, Manfred, Helmuth. Structural engr. Leonhardt & Andrä, Stuttgart, W. Ger., 1957-58; chief engr., then mgr. in charge engring. and sales Productos de Concreto S.A., San Jose, 1958-63; pres., prin. partner Franz Sauter & Asociados S.A., Cons. Engrs., San Jose, 1964—; prof. structural engring. U. Costa Rica, 1958-70; dir. Atlas Eléctrica S.A., Productos de Concreto S.A., Ricalit S.A., San Jose; mem. seismic code com. Colegio Federado de Ingenieros y Arquitectos de Costa Rica; pres. Asociacion Centroamericana del Cemento y Concreto, 1967-72, Institucion Cultural Germano-Costarricense, San Jose, 1969-73. UNESCO grantee, 1963-64; decorated Bundesverdienstkreuz 1st class (W. Ger.). Hon. mem. Asociacion Centroamericana Cemento y Concreto; mem. Colegio Federado de Ingenieros y Arquitectos de Costa Rica, ASCE, Earthquake Engring. Research Inst., Am. Concrete Inst., Prestressed Concrete Inst. Roman Catholic. Club: Rotary (San Jose). Author: Fundamentals of Earthquake Engineering Part I: Introduction to Seismology, 1989; contbr. articles to profl. jours.; research on earthquake ins. Address: Apartado Postal 6260, San Jose Costa Rica

SAUVAGE, LESTER ROSAIRE, health facility administrator, cardiovascular surgeon; b. Wapato, Wash., Nov. 15, 1926; s. Lester Richard Sauvage and Laura Marie Brouillard; m. Mary Ann Marti, June 9, 1956; children: Lester Jr., John, Paul, Helen, Joe, Laura, William, Mary Ann. Student, Gonzaga U., 1942-43, DSc (hon.), 1982; MD, St. Louis U., 1948; Honoris Causa (hon.), Seattle U., 1976. Diplomate Nat. Bd. Med. Examiners, Am. Bd. Surgery, Am. Bd. Thoracic Surgery. Intern King County Hosp., Seattle, 1948-49, surg. resident, 1949-50, sr. resident, 1955-56; sr. resident Children's Med. Ctr., Boston, 1956-58; rsch. assoc. dept. surgery U. Wash., Seattle, 1950-52; sr. resident in thoracic surgery Boston City Hosp., 1958; pvt. practice Pediatric and Cardiovascular Surgeons, Inc., Seattle, 1959-91; founder, med. dir. Hope Heart Inst. (formerly Reconstructive Cardiovascular Rsch. Ctr.), Seattle, 1959—; clin. prof. surgery sch. medicine U. Wash.; chmn. dept. surgery, dir. surg. edn. Providence Med. Ctr., Seattle; dir. cardiac surgery Children's Orthopedic Hosp. and Med. Ctr.; presenter in field, 1974—. Author: Prosthetic Replacement of the Aortic Valve, 1972; mem. editorial bd. Annals Vascular Surgery; contbr. over 200 rsch. papers to profl. jours. Capt. M.C., U.S. Army, 1952-54. Recipient Vocat. Svc. award Seattle Rotary Club, 1977, Humanitarian award Human Life Found., 1977, Brotherhood award Nat. Conf. Christians and Jews, 1979, Clemson award Soc. Biomaterials, 1982, Jefferson award Am. Inst. Pub. Svc., Seattle Post-Intelligencer, 1983, Gov.'s Disting. Vol. award, 1983, Spotlight award Am. Soc. Women Accts., 1985, Wash. State Medal of Merit, 1987, Seattle 1st Citizen award, 1992. Mem. AMA, Am. Acad. Pediatrics (surg. sect.), Am. Assn. Thoracic Surgery, Am. Coll. Cardiology, Am. Coll. Chest Physicians, Am. Coll. Surgeons, Am. Heart Assn., Am. Pediatric Surg. Assn., Neurovascular Soc. N.Am. (founding mem.), Wash. State Heart Assn., Wash. State Med. Assn., North Pacific Pediatric Soc., North Pacific Surg. Soc., N.W. Soc. Clin. Rsch., Pacific Assn. Pediatric Surgeons, Pacific Coast Surg. Assn., New Eng. Soc. Vascular Surgery (hon.), Seattle Surg. Soc., King County Med. Soc., Internat. Cardiovascular Soc., Soc. Artificial Internal Organs, Soc. Clin. Vascular Surgery (hon.), Soc. Vascular Surgery, Acad. Surg. Rsch., Alpha Omega Alpha, Alpha Sigma Nu. Roman Catholic. Achievements include research in synthetic blood vessel grafts, vascular surgical techniques, prediction and prevention of thrombotic complications of atherosclerosis, endothelial cell function and vascular autografts.

SAUVÉ, GEORGES, surgeon; b. Paris, Sept. 10, 1925; s. Louis de Gonzague andMarie (Bourdon) S.; m. Monique Lemaigre, June 11, 1955; children: Frédérique, Jacques-Phillipe, Diane, Claire, Marie-Amelie, Bérengère. MD, U. Paris, 1956. Intern Hosp. de Paris, 1952-57, chief of surgery, 1975-62; practice surgery, Laval, France, 1962—. Author: Les fils de Saint Come, 1987, De Louis XV à Poincaré, 1989. Mem. Internat. Coll. Surgeons, Lauréat Acad. Médecine, Acad. Maine, Acad. Généalogie. Roman Catholic. Avocations: music, art, literature. Home: 22 Place Du Gast, 53000 Laval France Office: Polyclinique du Maine, Ave Francais, Laval France

SAVAGE, DEBORAH ELLEN, chemical engineer, environmental policy researcher; b. Chapel Hill, N.C., Dec. 4, 1961; d. Thomas Corbet and Joan Ruth (Milton) S. BS, Clemson (S.C.) U., 1984; PhD, Mass. Inst. Tech., 1992. Summer researcher Clemson U., 1983; teaching asst., rsch. asst. MIT, Cambridge, 1986-92; rsch. assoc. Tellus Inst., Boston, 1992—; environ. cons. Sci. Rsch. Lab., Inc., Somerville, Mass., 1990; environ. intern ARCO Alaska, Inc., Anchorage, 1991. Fulbright fellowship, Clausthal, W. Germany, 1984-85. Mem. AIChE, Sigma Xi, Tau Beta Pi. Episcopalian. Home: 1137 Osage West Columbia SC 29169 Office: Tellus Inst 11 Arlington St Boston MA 02116-3411

SAVAGE, JULIAN MICHELE, civil engineer; b. Alamagordo, N.Mex., Aug. 2, 1956; s. Nathan and Barbara Jo (McEwen) S.; m. Tommie Glynis Garrison, Mar. 16, 1991; 1 child, Jamila Rushen. BS in Civil Engring., Memphis State U., 1985; MS in Civil Engring., Miss. State U., 1989. Rsch. civil engr. U.S. Army Waterways Experiment Sta., Vicksburg, Miss., 1986-90; civil engr. U.S. Army Corps Engrs., Memphis, 1990-92, Huntsville, Ala., 1991--. Mem. Am. Soc. Civil Engrs., Kappa Alpha Psi (treas. 1986-90). Home: 509 Royal Gardens Dr Madison AL 35758 Office: US Army Corps Engrs 106 Wynn Dr Huntsville AL 35807

SAVARD, G. S., earth scientist. Recipient Falconbridge Innovation award Can. Inst. Mining and Metallurgy, 1992. Office: care Xerox Tower Ste 1210, 3400 de Maisonneuve Blvd W, Montreal, PQ Canada H3Z 3B8*

SAVATSKY, BRUCE JON, chemical engineer; b. Amityville, N.Y., Aug. 20, 1954; s. Morris and Annette (Fried) S.; m. Janis Sandra Bloom, Aug. 17, 1980; children: Matthew, Jason. BSChE, U. Mass., 1976; PhD, U. Calif., Berkeley, 1981. Rsch. engr. DuPont, Wilmington, Del., 1981-83, 86-88, Orange, Tex., 1983-86; project engr. Amoco Performance Products, Bound Brook, N.J., 1988—; sr. rsch. engr. Amoco Performance Products, Alpharetta, Ga., 1989—. Mem. AIChE, Am. Chemistry Soc., Soc. Plastics Engrs. Achievements include development of process technology that was successfully implemented on commercial scale equipment for production of polymers. Home: 235 Halverson Way Duluth GA 30136

SAVEKER, DAVID RICHARD, naval and marine architectural engineering executive; b. San Jose, Calif., Jan. 10, 1920; s. William Thomas and Bernice (Lloyd) S.; m. Jessie Mae Walters, June 19, 1941; children: William, Linda, Richard (dec.). AA, San Jose State Coll., 1939; AB, Stanford U., 1941; cert. in naval architecture, U.S. Naval Acad., 1942; SM, MIT, 1946. Registered profl. engr., Calif. Commd. ensign USN, 1941, advanced through grades to capt., 1960, ret., 1968; assoc. prof. Calif. Poly. State U., San Luis Obispo, 1969-80; pres. D.R. Saveker Naval Architecture, Pismo Beach, Calif., 1980—; assoc. W.F. Searle Jr. Consortium, Alexandria, Va., 1983—. Inventor sinusoidal structure and applications. Research grantee NASA/Stanford Lab., 1973. Mem. AAUP, Marine Tech. Soc., Soc. Am. Mil. Engrs., Soc. Naval Architects and Marine Engrs. Democrat. Methodist. Club: Cosmos (Washington). Avocations: history, music, water colors. Home: 711 Hanford St Pismo Beach CA 93449-2347

SAVIC, MICHAEL I., engineering educator, signal and speech researcher; b. Belgrade, Serbia, Yugoslavia, Aug. 4, 1967; came to U.S., 1967; s. Miodrag and Jelena (Milisic) S.; 1 child, Alice. Diploma in Engring., U. Belgrade, 1955, DEng, 1965. R&D engr. Kretztechnick, Zipf, Austria, 1956-57, Tungsram, Vienna, Austria, 1957-58; asst. prof. U. Belgrade, 1958-67; rsch. assoc. Yale U., New Haven, Conn., 1967-68; prof. Western New Eng. Coll., Springfield, Mass., 1968-81, Rensselaer Poly. Inst., Troy, N.Y., 1981—; mem. adv. bd. Am. Coll. Cryosurgery, 1977-78; cons. Milton-Bradley Co., 1970-80; chmn. adv. panel NSF, 1977. Contbr. articles to profl. jours. Recipient scholarship U.S. Govt., 1962-64. Mem. IEEE (sr., Cert. of Appreciation 1972), N.Y. Acad. Scis. Achievements include patent in method and device for impedance controlled cryosurgery of malignant tumors; development of new methods and devices for controlled destruction of malignant tumors, monitoring leaks in underground pipelines, detection of cholesterol deposits in blood vessels and others. Home: 39 Sweetbrier Dr Ballston Lake NY 12019 Office: Rensselaer Poly Inst ECSE Dept 110 8th St Troy NY 12180-3590

SAVIC, STANLEY DIMITRIUS, physicist; b. Belgrade, Yugoslavia, Dec. 30, 1938; came to the U.S., 1958; s. Dimitrius and Zorka (Vuckovic) S. BS, Roosevelt U., 1962; MS, U. Ill., 1969. Staff scientist Argonne Cancer Rsch. Hosp., Chgo., 1962-63; with radiology staff U. Chgo., 1963-64; v.p. Zenith Electronics Corp., Glenview, Ill., 1964—; chief U.S. del. Internat. Electrotech. Commn., Geneva, 1986-90; lectr. in field, 1984-91; mem. com. FDA, Washington, 1978-81; mem. faculty N.Y. Acad. Fire Scis., Albany, 1991. Author: X-Ray Conference Proceedings, 1968, co-author, 1968; contbr. chpt.: Standards Management, 1990. Div. rep. United Way & Jobs Programs, Ill., 1984, 86. Mem. IEEE (sr.), N.Y. Acad. Scis., Nat. Fire Protection Assn., ASTM, Electronic Industries Assn. (chmn. safety com. 1983-87, mem. engring. policy coun. 1992—, Disting. Svc. award 1987). Republican. Serbian-Orthodox. Achievements include patent for a safety-related electronic circuit. Office: Zenith Electronics Corp 1000 Milwaukee Ave Glenview IL 60025

SAVILLE, MICHAEL WAYNE, physician, scientist; b. Evanston, Ill., Apr. 16, 1962; s. John J. and Judith (Landsberg) S.; m. Julia Barrett, Aug. 15, 1987. BS in Medicine, Northwestern U., 1984, MD, 1986. Diplomate Am.

Bd. Internal Medicine. Intern and resident in internal medicine U. Minn., Mpls., 1986-89, lectr., 1989-90; fellow in oncology Nat. Cancer Inst., Bethesda, Md., 1990-93. Lt. comdr. USPHS, 1990—. Mem. AAAS, ACP. Achievements include research in immunology and treatment of AIDS and AIDS-related malignancies. Office: Nat Cancer Inst/NIH 9000 Rockville Pike Bethesda MD 20892-0010

SAVIN, RONALD RICHARD, chemical company executive, inventor; b. Cleve., Oct. 16, 1926; s. Samuel and Ada (Silver) S.; m. Gloria Ann Hopkins, Apr. 21, 1962; children: Danielle Elizabeth, Andrea Lianne. Student, U. Cin., 1944-46; BA in Chemistry and Literature, U. Mich., 1948; postgrad., Columbia U., 1948-49, Sorbonne, Paris, 1949-50; grad., Air War Coll., 1975, Indsl. Coll. Armed Forces, 1976. Pres., owner Premium Finishes, Inc., Cin., 1957-91; cons. aerospace and anti-corrosive coatings; inventor Hunting Indsl. Coatings. Contbr. articles on aerospace, marine industry and transp. to profl. jours.; adv. coun. Chem. Week mag.; patentee in field of aerospace and anti-corrosion coatings; 8 patents. With USAF, 1948-55, World War II and Korea, col. Res. 1979, ret. 1986. Mem. Nat. Assn. Corrosion Engrs., Air Force Assn., Am. Internat. Club (Geneva), Res. Officers Assn. Avocations: scientific development, photography, tennis. Home: L'Eden, Thonon-les-Bains France Office: PO Box 1169 Rancho Mirage CA 92270-1169

SAVINELL, ROBERT F., engineering educator; b. Cleve., May 26, 1950; s. Robert D. and Lotte R. (Papierkowski) S.; m. Coletta A. Savinell, Aug. 23, 1974; children: Teresa, Robert, Mark. B in Chem. Engring., Cleve. State U., 1973; MS, U. Pitts., 1974, PhD. Registered profl. engr., Ohio. Rsch. engr. Diamond Shamrock Corp., Painesville, Ohio, 1977-79; assoc. prof. U. Akron, Ohio, 1979-86; prof. Case Western Reserve U., Cleve., 1986—, dir. Case Ctr. Electrochem. Sci., 1991—. Divsn. editor Jour. Electrochem. Soc., 1988-91; N.Am. editor Jour. Applied Electrochemistry, 1991—; contbr. articles to profl. jours. Named Presd.l. Young Investigator, NSF, Washington, 1984-89, Outstanding Engring. Alumnus, Cleve. State U., 1984. Mem. AICE (program chmn. 1986-92), Electrochem. Soc. (divsn. officer 1992—). Avocations: sailing, skiing. Office: Case Western Reserve U Dept Chem Engring AW Smith Bldg Cleveland OH 44106

SAVIR, ETAN, mathematics educator; b. N.Y.C., Oct. 10, 1961; s. David and Elizabeth Tamar (Miller) S.; m. Cathleen Mary O'Shea, May 1, 1986; children: Nathan, Stephanie Raphaela. AB in Classics, Princeton U., 1983; MA in Classics, U. N.C., 1986. Latin tchr. U. N.C., Chapel Hill, 1986; classics tchr. St. Swithun's Sch., Winchester, U.K., 1987-88; Latin tchr. Irvington (N.Y.) High Sch., 1988-89; dir. computer ctr. Foxcroft Sch., Middleburg, Va., 1989-91; computer coord. The John Cooper Sch., The Woodlands, Tex., 1991-92; math. tchr. The John Cooper Sch., The Woodlands, 1991—. Recipient John M. Morehead fellowship Morehead Found., Chapel Hill, 1983-86. Mem. Internat. Soc. for Philos. Enquiry (assoc.), Nat. Coun. Tchrs. Math., Am. Philol. Assn., Am. Classical League, Tex. Classical Assn., Am. Mensa, Cum Laude Soc., Phi Beta Kappa. Jewish. Office: The John Cooper Sch 3333 Cochrans Crossing Dr The Woodlands TX 77381

SAVITZ, MARTIN HAROLD, neurosurgeon; b. Boston, Mass., Jan. 20, 1942; s. Nathan and Bernice Beatrice (Segal) S.; m. Susan Rayna Gordon, June 23, 1968 (div. Sept. 1977); 1 child, Sean Isaac; m. Harmony Gwynne Keys, Oct. 28, 1979; 1 child, Ariel Austryn. AB, Harvard U., 1963; MD, Hahnemann, 1969. Diplomate Am. Bd. Neurol. Surgery, Am. Bd. Clin. Neurosurgery, Nat. Bd. Med. Examiners. Intern Boston City Hosp., 1969-70; resident Mount Sinai Hosp., N.Y.C., 1970-74; clin. instr. dept. neurosurgery Dept. Neurosurgery, Mount Sinai Sch. Medicine, 1974-82; asst. clin. prof. Dept. Neurosurgery, Mount Sinai Sch. Medicine, Nanuet, N.Y., 1982-86, assoc. clin. prof., 1986-93; attending neurosurgeon Nyack (N.Y.) Hosp., Good Samaritan Hosp., Rockland County, N.Y., 1974—; mem. pres.'s coun. Harvard Coll., 1991—; alumni bd. trustees Hahnemann U., 1991—. Contbg. editor Mt. Sinai Jour. Medicine, 1976-90, asst. editor, 1990—; mem. editorial bd. Jour. Orthopaedic Neurol. Medicine and Surgery, 1991—; contbr. 2 chpts. to textbooks; contbr. over 60 articles to profl. jours. Fellow ACS, Internat. Coll. Surgeons (chmn.-elect U.S. sect. neurosurgery 1992—), N.Y. Acad. Medicine, Phila. Coll. Physicians, Am. Acad. Neurol. Orthopaedic Surgery; mem. AMA, Am. Assn. Neurol. Surgeons, N.Y. Soc. Neurosurgery, Congress Neurol. Surgeons, N.Y. State Neurosurg. Soc., Am. Assn. Physicians, Internat. Soc. Minimal Intervention in Spinal Surgery, Hastings Ctr., Alpha Omega Alpha. Jewish. Home: Hobbit Hollow New City NY 10956 Office: 55 Old Turnpike Rd Ste 101 Nanuet NY 10954-2449

SAVONA, MICHAEL RICHARD, physician; b. N.Y.C., Oct. 21, 1947; s. Salvatore Joseph and Diana Grace (Menditto) S.; m. Dorothy O'Neill, Oct. 18, 1975. BS summa cum laude, Siena Coll., 1969; MD, SUNY, Buffalo, 1973. Diplomate Am. Bd. Internal Medicine. Intern in internal medicine Presbyn. Hosp. Columbia U., N.Y.C., 1973-74, resident in internal medicine, 1974-76; vis. fellow internal medicine Delafield Hosp./Columbia U. Coll. Physicians and Surgeons, N.Y.C., 1974-76; practice medicine specializing in internal medicine Maui Med. Group, Wailuku, Hawaii, 1976-87, gen. practice medicine, 1987—; dir. ICU, Maui Meml. Hosp., also dir. respiratory therapy, CCU., chmn. dept. medicine, 1980—; clin. faculty John A. Burns Sch. Medicine, U. Hawaii, asst. prof. medicine, 1985—, asst. rsch. prof., 1989—. Bd. dirs. Maui Heart Assn.; dir. profl. edn. Maui chpt. Am. Cancer Soc., mem. Maui County Hosp. Adv. Commn.; mem. coun. Community Cancer Program of Hawaii. Recipient James A. Gibson Wayne J. Atwell award, 1970, physiology award, 1970, Ernest Whitebsky award, 1971, Roche Lab. award, 1972, Pfiser Lab. award, 1973, Phillip Sang award, 1973, Hans Lowenstein M.D. Meml. award, 1973. Mem. AMA, Am. Thoracic Soc., Hawaii Thoracic Soc., Maui County Med. Assn. (past pres.), Hawaii Med. Assn., Hawaii Oncology Group, ACP, SW Oncology Coop. Group, Alpha Omega Alpha, Delta Epsilon Sigma. Office: 1830 Wells St Wailuku HI 96793-2334

SAWADA, HIDEO, polymer chemistry consultant, chemist; b. Kyoto, Japan, Jan. 29, 1934; s. Masao and Hiroko (Ohno) S.; m. Yoshiko Kasai, May 5, 1961; children—Yukari, Jun. B.S., Osaka U., 1956, D.S., 1965; student Cornell U., 1961-62. Cert. tchr. Research mgr. Daicel Ltd. Filter Lab., Osaka, Japan, 1973-80, gen. mgr., 1980-85, sr. exec. scientist rsch. ctr., 1986-89, gen, mgr. planning and devel. dept., 1989— ; chmn. technology com. Biodegradable Plastics Soc., 1990—. Author: Thermodynamics of Polymerization, 1976; Encyclopedia of Polymer Science and Engineering, 1985. Mem. AAAS, Am. Soc. Testing and Materials, Am. Chem. Soc., Soc. Polymer Sci. Japan, Chem. Soc. Japan, Soc. Fiber Sci. and Tech., N.Y. Acad. Scis., Internat. Union Pure and Applied Chemistry. Zen. Home: 2534 3-chome Sayama, Sayama Osaka 589, Japan Office: Biodegradable Plastics Soc, Dowa Bldg 10-5 Shimbashi 5-chome, Minato-ku Tokyo 150, Japan

SAWEIKIS, MATTHEW A., information scientist. BS in Acctg., Loyola U., Chgo., 1973; cert. telecom., DePaul U., 1990, postgrad. Consulting analyst Sara Lee Corp., 1974-79; dir. MIS Pettibone Corp., Chgo., 1979-82, Joanna We. Mills Co., Chgo., 1982-89; v.p. MIS Evans Inc., Chgo., 1989—. Office: Evans Inc 36 S State St Chicago IL 60603-2691

SAWH, LALL RAMNATH, urologist; b. Couva, Trinidad and Tobago, June 1, 1951; s. Ramnath Rooplal and Ramkumaria (Sinanan) S.; m. Sylvia Sheila Ragobar, Dec. 22, 1973; children: Sean Lall, Shane Stefan. MBBS, U. W.I., Mona, Jamaica, 1975. Intern Gen. Hosp. San Fernando, Trinidad, 1975-76, sr. housr officer, 1976-77, sr. registrar in urology, 1980-86, acting cons. urologist, 1987—; clin. asst. Royal Infirmary Edinburgh, Scotland, 1978-79; clin. attachment Inst. Urology, London, 1979; examiner Nursing Coun., Trinidad, 1984-86, examiner surgery U. of the West Indies, 1991-93, assoc. lectr. surgery, 1990-93; cons. urologist Gen. Hosp., Port of Spain, Trinidad, 1988; head local kidney transplant team, 1990; chmn. sic. com. Trinidad and Tobago Kidney Found., 1989-91; chmn. scientific com. Trinidad and Tobago Kidney Found., 1989-93; lectr. in urology to various Caribbean countries; surgeon Gen. Hosp., San Fernando, Port of Spain, Eric Williams Med. Sci. Comples, Mt. Hope, Trinidad. Author: Renal Hypothermic Surgery, 1982, Button-Hole Kidney Surgery, 1986; contbr. articles to profl. jours. Recipient award for Outstanding and Meritorious Svc. to Trinidad and Tobago, Caroni County Coun., 1989, Chaconia Gold medal Trinidad and Tobago, 1993; honored by city coun. for contbn. to medicine in San Fernando, 1989. Fellow Royal Coll. Surgeons; mem. Am. Urol. Assn. (corr. mem.), Med. Assn. Trinidad and Tobago (chmn. 1984-85), Endo-urol. Soc. U.S., Med. Bd. Trinidad and Tobago (specialist, med. officer), Caribbean Assn. of Nephrologists & Urologists, Caribbean Prostatic Health Coun. (Trinidad rep.), Surg. Edn. Com. (founder), Soc. Surgeons (treas. 1986-93). Club: Lawn Tennis (Point a Pierre, Trinidad). Achievements include performing first successful kidney transplant in Trinidad and Tobago, 1988, performed the first renal hypothermic surgery in Trinidad in 1981, performed the first button hole renal lithotrypsy in the West Indies in 1986. Avocations: lawn tennis, swimming. Office: Med Day Ctr, 15 Prince of Wales St, San Fernando Trinidad and Tobago

SAWLIVICH, WAYNE BRADSTREET, biochemist; b. Sidney, Maine, Sept. 8, 1957; s. Mitchell and Lucille (Bradstreet) S.; m. Mary Howard, Sept. 26, 1981; 1 child, Mitchell. BS, U. Maine, Orono, 1980; MS, Boston U., 1987. Rsch. asst. Mallory Inst. Pathology, Boston, 1981-83; mem. tech. staff Cambridge (Mass.) Rsch. Lab., 1983-84; sr. scientist Cambridge Med. Diagnostics, Billerica, Mass., 1987-89; rsch. scientist 2 Millipore, Milford, Mass., 1989-90; rsch. scientist 3 Millipore, Bedford, Mass., 1990—. Contbr. articles to profl. jours. Mem. N.Y. Acad. Scis. Achievements include patent for energy transfer in immunofluorescent assay.

SAWYER, DAVID NEAL, petroleum industry executive; b. Paducah, Ky., Sept. 27, 1940; s. David A. and Lois (Neal) S.; m. Mary Kirk Kelly, June 24, 1964; children: Laura Kathleen, Allen Neal. BS, La. State U., 1964; PhD, U. Wis., 1969. Registered profl. engr., Tex. Process engr. Shell Chem. Co., Norco, La., 1964-65; instr. chem. engring. U. Wis., Madison, 1969-70; sr. rsch. engr. Atlantic Richfield Co., Plano, Tex., 1970-73; staff reservoir engr. Arco Oil and Gas Co., Dallas, 1973-84; pres. Plantation Petroleum Co., Houston, 1984—; dir. Miss. CO2 Enhanced Oil Recovery Lab., 1990—; cons. in field; presenter, Beijing; presenter seminar internat. mktg. petroleum products, Lagos, Nigeria, 1986, 91; presenter seminar enhanced oil recovery, Lagos, 1991. Author of seminars. Panel mem. Leadership Devel. Seminar, 1986. Recipient Warrior Basin study, U.S. Dept. Energy, 1990-92; NSF fellow, 1965-69. Mem. Geosci. Inst. Oil and Gas Rsch. (state rep., bd. dirs. 1989—), Am. Soc. Engring. Edn. (presenter 1989), Soc. Petroleum Engrs. (bd. dirs.), AAAS, Tau Beta Pi, Sigma Xi. Avocations: coaching children's baseball, fishing. Home: 315 River Knoll Dr Atlanta GA 30328 Office: Miss State Univ PO Drawer PE Mississippi State MS 39762

SAWYER, JAMES LAWRENCE, architect; b. Bangor, Maine, Nov. 24, 1947; s. Maynard Wallace and Adelma Irene (Pascal) S.; m. Debralee Standley, June 6, 1980; children: Jason, Beverly, Joshua, John, Joseph. Ind. study, Instituto di Urbanistica, Florence, Italy, 1971; BArch, Pratt Inst., 1973. Ordained to ministry Glad Tidings Ch., 1987. Designer Sawhawk Inc., Bklyn., 1973-75; state architect Farmers Home Adminstrn., Orono, Maine, 1975—. Producer MPBN-TV Energy Independence on the Rise, 1988. Pres. One Nation Under God, Inc., Lighthouse Prison Ministry; bd. dirs. Gospel Tent Ministry; mem. adv. bd. Ea. Maine Tech. Coll., 1993; co-chmn. Guilding and Remodeling into the 90's conf., 1988, 93; mem. Nat. Trust Hist. Preservation, Maine Right To Life Com. Recipient Energy Innovation award U.S. Dept. Energy, 1984, Bangor Daily News Photography award, 1985. Mem. Maine Solar Energy Assn. (pres. 1988), Christian Civic League. Avocations: touring, photography. Office: Farmers Home Adminstrn USDA Bldg 444 Stillwater Ave Bangor ME 04401

SAWYER, LINDA CLAIRE, materials scientist; b. Toronto, Ont., Can., Jan. 2, 1944; came to U.S., 1945; d. Edward A. and Estelle D. Frija; m. David R. Sawyer; children: Michael D., Andrew L. MA, SUNY, 1971; BA, SUNY, New Paltz, 1965. Chemist Texaco, Beacon, N.Y., 1965-69; rsch. assoc. Geology dept., UCLA, 1970-71; sr. chemist Owens Corning Fiberglas, Granville, Ohio, 1972-74; rsch. supr. Hoechst Celanese Corp., Summit, N.J., 1974—. Author: Polymer Microscopy, 1987; contbr. chpts. to books and 30 articles to profl. jours. Editor ch. bull. Long Hill Chapel, Chatham, N.J., 1988-90. Regents scholarship SUNY, 1961. Fellow Royal Microscopy Soc. (Oxford, U.K., adv. bd. Jour. Microscopy, 1993—); mem. Electron Microscopy Soc. of Am. (bull. editor 1981-85, nat. dir. 1985-87), Am. Chem. Soc., Microanalysis Soc. Achievements include 5 patents on comfort fiber, in ceramics, in polymers. Office: Hoechst Celanese Corp 86 Morris Ave Summit NJ 07901

SAX, MARTIN, crystallographer; b. Wheeling, W.Va.. BS, U. Pitts., 1941, PhD in Phys. Chemistry, 1961. Rsch. chemist Trojan Powder Co., 1941-44, Glyco Products Co., 1951-59; asst. biochemist W. Pa. Hosp., 1946-57; postdoctoral crystallographer U. Pitts., 1961-63, asst. rsch. prof., 1963-66; rsch. chemist and dir. biocrystallography lab. Vets. Affairs Med. Ctr., Pitts., 1966—, assoc. chief staff R&D, 1972—; adj. assoc. prof. crystallographer U. Pitts., 1966-71, adj. prof., 1971—. Mem. AAAS, Am. Crystallography Assn., Am. Chem. Soc. Achievements include research in three dimensional structures and the functions of biological macromolecules as ascertained by x-ray diffractions from single crystals. Office: U Drive C Vets Adminstn Med Ctr Rsch Svc Pittsburgh PA 15240*

SAX, MARY RANDOLPH, speech pathologist; b. Pontiac, Mich., July 13, 1925; d. Bernard Angus and Ada Lucile (Thurman) TePoorten; m. William Martin Sax, Feb. 7, 1948. BA magna cum laude, Mich. State U., 1947; MA, U. Mich., 1949; Cert. clin. competence in speech and language pathology. Supr. speech correction dept. Waterford Twp. Schs., Pontiac, 1949-69; lectr. Marygrove Coll., Detroit, 1971-72; pvt. practice speech and lang. rehab., Wayne, Oakland Counties, Mich., 1973—; adj. speech pathologist Southfield, Mich.; lectr. on stroke Mich. Speakers Bur., Am. Heart Assn., 1990—; pub. speaking coach, 1989—; adj. faculty SS. Cyril and Methodius Sem., Orchard Lake, Mich., 1989-90, St. Mary's Preparatory Sch., Orchard Lake, Mich., 1990—; founder, mem. Stroke Project Task Force for Detroit, 1993—; com. mem. Charrette, study Architecture and Design for physical restructuring Franklin, Mich., 1993. Mem. sci. coun. stroke Am. Heart Assn. Grantee Inst. Articulation and Learning, 1969, others, project choices and funding Meadow Lake Community Coun., Birmingham, Mich., 1989—; christian svc. commn. St. Owen, Birmingham co-chmn. blood drive Red Cross, Franklin, Mich., 1991—; mem. natural resources adv. coun. Franklin (Mich.) Found., 1991—. Mem. AAUW, Am. Speech-Lang.-Hearing Assn. (clin. competence cert.), Mich. Speech-Lang.-Hearing Assn. (com. community and hosp. svcs.), Am. Heart Assn. of Mich. (mem. stroke awareness seminar, planning and operation ednl.), Stroke Com. of Am., Internat. Assn. Logopedics and Phoniatrics (Switzerland), Founders Soc. of Detroit Inst. Arts, Mich. Humane Soc., Theta Alpha Phi, Phi Kappa Phi, Kappa Delta Pi. Contbr. articles to profl. jours. including Language and Language Behavior Abstracts, Language Speech & Hearing Services, Speech Language Hearing Jour. Achievements include research in language and speech acquisition in children in reference to the development of and prediction of biological speech change; research interests in adult acquisition of language and speech relative to central and autonomic nervous systems. Home and Office: 31320 Woodside Dr Franklin MI 48025-2027

SAXENA, AMOL, podiatrist, consultant; b. Palo Alto, Calif., June 5, 1962; s. Arjun Nath and Veera Saxena; m. Karen Ann Palermo, Aug. 11, 1985; children: Vijay, Tara Ann. Student, U. Calif., Davis, 1980-82; BA, Washington U., St. Louis, 1984; D in Podiatric Medicine, William Scholl Coll. Podiatric Medicine, 1988. Diplomate Am. Bd. Pediatric Surgery; lic. podiatrist, Calif., Ill. Resident in podiatric surgery VA Westside Br., Chgo., 1988-89; cons. Puma U.S.A., Inc., Framingham, Mass., 1986—; pvt. practice Mountain View, Calif., 1989—; dir. Puma Sports Medicine, Framingham; mem. podiatry team St. Frances/Gunn Los Altos (Calif.) High Sch., Palo Alto, 1989—, Stanford (Calif.) U., 1989—; mem. med. staff El Camino Hosp., 1989—, team podiatrist Stanford U., 1989—. Contbr. articles to profl. jours. Vol. coach Gunn High Sch. Track and Cross County, Palo Alto, 1989—; podiatrist U.S. Olympic Track and Field Trials, New Orleans, 1992, 1993. Fellow Am. Acad. Podiatric Sports Medicine; mem. Am. Coll. Foot Surgeons (assoc.), Am. Podiatric Med. Assn., Calif. Podiatric Med. Assn., Am. Med. Soccer Assn., Aggie Running Club. Republican. Avocation: running. Office: 2204 Grant Rd # 104 Mountain View CA 94040-3877

SAXENA, ARJUN NATH, physicist; b. Lucknow, India, Apr. 1, 1932; s. Sheo and Mohan (Piyari) Shanker; came to U.S., 1956, naturalized, 1976; BSc, Lucknow U., 1950, MSc, 1952, profl. cert. in German, 1954; Post MS diploma, Inst. Nuclear Physics, Calcutta, India, 1955; PhD, Stanford U., 1963; m. Veera Saxena, Feb. 9, 1956; children: Rashmi, Amol, Varsha, Ashvin. Rsch. asst. Stanford U., 1956-60; mem. tech. staff Fairchild Semicondr. Co., Palo Alto, Calif., 1960-65; dept. head Sprague Electric Co., North Adams, Mass., 1965-69; mem. tech. staff RCA Labs., Princeton, N.J., 1969-71; pres., chmn. bd. Astro-Optics, Phila., 1972; pres. Internat. Sci. Co., Princeton Junction, N.J., 1973—; vis. scientist Centre de Récherches Nucléaires, Strasbourg, France, 1973, 77; sr. staff scientist, mgr. engring. Data Gen. Corp., Sunnyvale, Calif., 1975-80; mgr. process tech. Signetics Corp., Sunnyvale, 1980-81; Gould AMI scientist, dir. advanced process devel. Gould AMI Semicondrs., Santa Clara, Calif., 1981-87; dir. Ctr. for Integrated Electronics, prof. dept. elec. and computer system engring. Rensselaer Poly. Inst., Troy, N.Y., 1987—. Treas. Pack 66, Boy Scouts Am., W. Windsor, N.J., 1970-74. Recipient Disting. Citizen award State of N.J., 1975. Mem. Am. Phys. Soc., IEEE, Electrochem. Soc., Stanford Alumni Assn. (life). Contbr. articles on semicondr. tech., optics, nuclear and high-energy physics to sci. jours., 1953—; patentee in field. Home: 2 Birch Hill Rd Ballston Lake NY 12019-9370 Office: Rensselaer Poly Inst Ctr Integrated Electronics Dept Elec/Computer Sci Engring Troy NY 12180-3590

SAXENA, RENU, medical physicist; b. Delhi, India, Jan. 27, 1958; came to U.S., 1984; d. Balkrisman and Sarla S.; m. Amit Ghosh, Nov. 4, 1983. BS in Physics, Delhi U., 1976; MS in Biophysics, Moscow State U., 1982; PhD in Physics, U. N.H., 1990. Rsch. asst. Moscow State U., 1980-83; teaching asst. U. N.H., Durham, 1984, grad. rsch. assoc. Space Sci. Ctr., 1985-90; rsch. assoc. dept. of radiation oncology Wayne State U., Detroit, 1990-92, asst. prof. dept. radiation oncology, 1992—; chief med. physicist St. Francis Hosp., Evanston, Ill., 1992. Mem. Am. Assn. Med. Physics, Internat. Orgn. Neutron Capture Therapy. Office: St Francis Hosp Radiation Therapy 355 Ridge Ave Evanston IL 60202

SAXER, RICHARD KARL, metallurgical engineer, retired air force officer; b. Toledo, Aug. 31, 1928; s. Alexander Albert and Gertrude Minnie (Kuebeler) S.; m. Marilyn Doris Mersereau, July 19, 1952; children—Jane Lynette, Robert Karl, Kris Renee, Ann Luette. Student, Bowling Green State U., 1946-48; BS, U. S. Naval Acad., 1952; MS in Aero. Mechanics Engring., Air Force Inst. Tech., 1957; PhD in Metall. Engring., Ohio State U., 1962; grad., Armed Forces Staff Coll., 1966, Indsl. Coll. Armed Forces, 1971. Commd. 2d lt. U.S. Air Force, 1952, advanced through grades to lt. gen., 1976; electronics officer, mech. officer (4th Tactical Support Sqadron, Tactical Air Command), Sandia Base, N.Mex., 1953-54; electronics and mech. officer, spl. weapons assembly sect. supr. (SAC 6th Aviation Depot Squadron), French Morocco, 1954-55; project engr. mech. equipment br. Air Force Spl. Weapon's Center, Kirtland AFB, N.Mex., 1957-59; project officer Nuclear Safety div., 1959-60; assoc. prof. dept. engring. mechanics Air Force Inst. Tech., 1962-66; asso. prof., dep. dept. head USAF Acad., 1966-70; comdr., dir. Air Force Materials Lab., Wright-Patterson AFB, Ohio, 1971-74; dep. for Reentry System Space and Missile Systems Orgn., 1974-77; dep. for aero equipment Aero. Systems Div., 1977-80, dep. for tactical systems, 1980, vice comdr., 1981-83; aero. systems div. dir. Def. Nuclear Agy., 1983-85, ret., 1985; pres. R.K. Saxer & Assocs., 1985-91; CEO Universal Tech. Corp., Dayton, Ohio, 1991—; research and tech. com. materials and structures NASA, 1973-74; chmn. planning group aerospace materials Interagy. Council Materials, 1973-74; mem. Nat. Mil. Adv. Bd., 1971-74, NATO adv. group for research and devel., 1973-74. Contbr. articles to profl. jours. Decorated Def. Disting. Svc. medal, Legion of Merit, Meritorious Service medal USAF, D.S.M., Joint Svc. Commendation medal, Air Force Commendation medal with 3 oak leaf clusters, Army Commendation medal U.S., Def. Superior Service medal, Cross of Gallantry with palm Vietnam, Def. Meritorious Service medal; recipient Disting. award for systems mgmt. Air Force Assn., 1979; Disting. Alumnus award Ohio State U., 1986. Mem. Air Force Assn., Am. Def. Preparedness Assn. (pres. Dayton 1977-78), Sigma Xi, Phi Lambda Epsilon, Alpha Sigma Mu, Masons, Shriners. Home: 5916 Yarmouth Dr Dayton OH 45459-1450

SAYED, SAYED M., engineering executive; b. Nasser, Benisuef, Egypt, Oct. 2, 1951; s. Mourad Sayed Sayed and Wasila Abdelati Hussein; m. Hyatt A. Ibrahim, Aug. 12, 1977; children: Wael, Nahla, Halah, Aalaa, Maha. BSCE with honors, Cairo U., 1974, MSCE, 1977; PhD, Duke U., 1982. Registered profl. engr., La., Fla. Asst. prof. U. New Orleans, 1982-84, 85-88, assoc. prof., 1988; sr. project mgr., sr. cons., Jammal divsn. Profl. Svc. Industries, Winter Park, Fla., 1988-93; prin. Geotech Cons. Internat. Inc., Winter Park, Fla., 1993—; geotechnical cons./lectr. Internat. Consortium, Cairo, 1984-85, Tokyo Eng. Cons., Ltd., Cairo, 1984-85, SANYU Cons., Cairo, 1984-85, ATCO, Cairo, 1984-85, CDM/Cairo U., 1984-85; presenter Internat. Congress Theoretical and Applied Mechanics, Lyngby, Denmark, 1984, Midwestern Mechanics Conf., Columbus, Ohio, 1985, S.E. Conf. Theoretical and Applied Mechanics, Columbia, S.C., 1986, others; referee La. Engring. Advancement Program Minorities, New Orleans; reviewer ASCE, ASME. Author/editor: Geotechnical Modeling and Applications, 1987; contbr. articles to profl. jours. Recipient award Am. Pub. Works Assn., 1981, Cert. Appreciation, Kiwanis Club Honorable Elders, 1981, La. Engring. Advancement Program for Minorities, Inc., 1987; Postgrad. scholar Duke U., 1978-82; NORAD fellow Norwegian Inst. Tech., 1977-78. Mem. NSPE, ASCE (reviewer, co-chair geotech. group ea. ctrl. br. Fla. sect. 1989-90, chmn. 1990-91, chmn. nat. task com. on groundwater modeling in the area of solid and hazardous waste 1990-92, groundwater quality tech. com. 1992—, named Outstanding Geotech. Group chmn. 1990), Internat. Soc. Soil Mechanics and Found. Engring., Fla. Engring. Soc., La. Engring. Soc., Egyptian Assn. Profl. Engrs., Chi Epsilon, Sigma Xi. Achievements include research in soil-structure interaction modeling, constitutive models, soil stabilization, numerical models, in-situ testing. Office: Geotech Cons Internat Inc 2265 Lee Rd Ste 123 Winter Park FL 32789

SAYER, JOHN SAMUEL, information systems consultant; b. St. Paul, July 27, 1917; s. Arthur Samuel and Genevieve (Ollis) S.; m. Elizabeth Hughes, June 9, 1940; children: Stephen, Susan, Kathryn, Nancy. BSME, U. Minn., 1940. Registered profl. engr., Del. Sect. mgr. E.I. Du Pont de Nemours & Co, Wilmington, Del., 1940-60; exec. v.p. Documentation, Inc., Washington, 1960-63; v.p. corp. devel. Aurbach Corp., Phila., 1963-65; exec. v.p. Leasco Systems & Rsch., Bethesda, Md., 1966-70, Leasco Info. Products, Silver Spring, Md., 1971-74; pres. Remac Info. Corp., Gaithersburg, Md., 1975-82; cons. John Sayer Assocs., Gaithersburg, 1983—; numerous presentations in field. Contbr. numerous articles to profl. jours. Recipient Info. Product of Yr. award Info. Industries Assn., 1973. Mem. ASME, Assn. Info. and Image Mgmt., Am. Inst. Info. Sci. Achievements include direction of work leading to critical path method of planning and scheduling, technical word thesarus, microfiche, data base publishing. Office: 13209 Colton Ln Gaithersburg MD 20878-2103

SAYIGH, LAELA SUAD, biologist; b. New Haven, Feb. 11, 1963; d. Adnan Abdul-Rida and Anne Louise (Burchsted) S. BA, U. Pa., 1985; PhD, MIT/WHO1, 1992. Mem. faculty Mass. Maritime Acad., Buzzards Bay, 1992—; guest investigator Woods Hole (Mass.) Oceanographic Inst., 1992—. Contbr. articles to profl. jours. Ocean Ventures Fund grantee Woods Hole Oceanographic Inst., 1988-89; NSF grantee, 1990; Am. Cetacean Soc. grantee, 1991, Lerner Gray Fund for Marine Rsch. grantee 1993. Mem. Animal Behavior Soc., Soc. for Marine Mammology, Acoustical Soc. Am., Phi Beta Kappa, Sigma Xi. Democrat.

SAYKALLY, RICHARD JAMES, chemistry educator; b. Rhinelander, Wis., Sept. 10, 1947; s. Edwin L. and Helen M. S. BS, U. Wis., Eau Claire, 1970; PhD, U. Wis., Madison, 1977. Postdoctoral Nat. Bur. Standards, Boulder, Colo., 1977-79; prof. U. Calif., Berkeley, 1979—, vice chmn. dept. chemistry, 1988-91; Bergman Lectureship Yale U., 1987; Merck-Frost lectr. U. B.C., 1988; Bourke Lectureship Royal Soc. Chemistry, 1992; prin. investigator Lawrence Berkeley Lab., 1983-91; prin. investigator program Sci. for Sci NSF. Contbr. articles to profl. jours. Dreyfuss Found. fellow, 1979; presdl. investigator NSF, 1984-88; recipient Bomem Michelson prize for spectroscopy, 1989, E.K. Plyler prize for molecular spectroscopy, 1989, Disting. Alumnus award U. Wis., Eau Claire, 1987, E.R. Lippincott medal OSA, SAS, 1992, Disting. Teaching award U. Calif., Berkeley, 1992. Fellow Am. Phys. Soc., Royal Soc. Chem.; mem. AAAS, AAUP, Optical Soc. Am., Am. Chem. Soc. (Harrison Howe award 1992). Office: U Calif Dept Chemistry Berkeley CA 94720

SCAIRPON, SHARON CECILIA, information scientist; b. New Brunswick, N.J., May 7, 1946; d. Eric Christian and Erica Cecile (Smolar) Schreiber. Student, Trenton Jr. Coll., 1965-67; BSBA, Rider Coll., 1991.

Various clerical postions E.R. Squibb & Sons, Princeton, N.J., 1967-87, sr. interlibr. loan and reference technician, 1987-88; lit. resource assoc., lan adminstr. Bristol-Myers Squibb, Princeton, N.J., 1988-91, info. scientist, 1991—. Mem. Spl. Librs. Assn., Zonta (bd. dirs. Trenton 1990-91, chmn. Amelia Earhart svc. com. 1991-93), Sigma Iota Epsilon. Presbyterian. Avocations: volunteer work, bicycling, exploring historical sites. Home: PO Box 5041 Trenton NJ 08638-0041 Office: Bristol-Myers Squibb Rt 206 & Provinceline Rd Princeton NJ 08543-4000

SCALISE, OSVALDO HECTOR, physics researcher; b. Rojas, Argentina, Dec. 15, 1940; s. Carlos Alberto and Maria Rosario (Zambuto) S.; m. Evangelina Ana Cascardo, May 24, 1965; children: Augusto, Federico. BA, Nicolas Avellaneda, Rojas, Argentina, 1958; BS in Physics, Nat. La Plata U., Argentina, 1971, PhD in Physics, 1976. Teaching asst. physics dept. La Plata U., Argentina, 1971-78; rsch. assoc. math. & chemistry dept. Dalhousie U., Halifax, Nova Scotia, 1976-77, 79, IBM Rsch., San Jose, Calif., 1977; vis. asst. prof. Guelph U., Ontario, Can., 1984, vis. assoc. prof., 1987; researcher Buenos Aires Rsch. Coun., La Plata, 1978—; thesis supr. physics dept. La Plata U., 1984-88, subdir. inst. Inst. de Fisica de Liquidos y Sistemas Biológicos, 1989—. Contbr. articles to profl. jours. Aeronautic marine Navy, 1961-62. Recipient Nat. U. Student fellowship, 1970, Nat. Rsch. coun. fellowship CONICET, 1974-75. Mem. AAAS, Am. Phys. Soc.,. Avocations: outdoors activities, drawing, computing. Home: 39 No 483, 1900 La Plata Argentina Office: IFLYSIB, 59 No 789, 1900 La Plata Argentina

SCANCELLA, ROBERT J., civil engineer; b. Bristol, Pa., July 10, 1955; s. John Robert and MaryLou (Perkins) S.; m. Carol Ann Scannella, May 22, 1977; children: Jennifer, Dante, Joseph, John. BSCE, The Citadel, 1977; Cert. Pub. Mgr., Rutgers U., 1989. Transp. engr. Wash. Dept. Transp., Longview, 1977-80, N.J. Dept. Transp., Trenton, 1989—; task force mem. Constrn. Engring. Manpower Mgmt., Trenton, 1977-80; program chmn. Nat. Structural Materials Tech. Conf. Co-author tng. video: The Resident Engineer as a Witness, 1989; co-author manuals for N.J. Dept. Transp., 1982, 83, 86. Fellow ASCE, Am. Soc. Hwy. Engrs.; mem. Bayville Bus. Assn., KC (Grand Knight 1989-90). Republican. Roman Catholic. Home: 252 Point Pleasant Ave Bayville NJ 08721-1355

SCANDALIOS, JOHN GEORGE, geneticist, educator; b. Nisyros Isle, Greece, Nov. 1, 1934; s. George John and Calliope (Broujos) S.; m. Penelope Anne Lawrence, Jan. 18, 1961; children: Artemis Christina, Melissa Joan, Nikki Eleni. B.A., U. Va., 1957; M.S., Adelphi U., 1960; Ph.D., U. Hawaii, 1965; D.Sc. (hon.), Aristotelian U. Thessaloniki, Greece, 1986. Assoc. in bacterial genetics Cold Spring Harbor Labs., 1960-62; NIH postdoctoral fellow U. Hawaii Med. Sch., 1965; asst. prof. Mich. State U., East Lansing, 1965-70; assoc. prof. Mich. State U., 1970-72; prof., head dept. biology U. S.C., Columbia, 1973-75; prof., head dept. genetics N.C. State U., Raleigh, 1975-85; disting. univ. research prof. N.C. State U., 1985—; mem. Inst. Molecular Biology and Biotechnology, Research. Ctr. Crete, Greece; vis. prof. genetics U. Calif., Davis, 1969; vis. prof. OAS, Argentina, Chile and Brazil, 1972; mem. recombinant DNA adv. com. NIH. Author: Physiological Genetics, 1979; editor: Developmental Genetics, Advances in Genetics, Current Topics in Medical and Biological Research; co-editor: Isozymes, 4 vols., 1975, Monographs in Developmental Biology, 1968-86; molecular biology editor Physiol. Plant, 1988—. Served with USAF, 1957. Alexander von Humboldt travel fellow, 1976; mem. exchange program NAS, U.S.-USSR. Fellow AAAS; mem. Genetics Soc. Am., Am. Soc. Biochemistry and Molecular Biology, Am. Genetic Assn. (pres.), Soc. Devel. Biology (dir.), Am. Inst. Biol. Scis., Am. Soc. Plant Physiologists, N.Y. Acad. Scis., Sigma Xi. Office: NC State U Dept Genetics PO Box 7614 Raleigh NC 27695

SCANLON, ANDREW, structural engineering educator. Prof. structural engring. Pa. State U., University Pk. Recipient Le Prix P.L. Pratley award Can. Soc. Civil Engring., 1990. Office: Pa State U Structures Lab 212 Sackett Hall University Park PA 16802*

SCANNELL, WILLIAM EDWARD, aerospace company executive, consultant, educator; b. Muscatine, Iowa, Nov. 11, 1934; s. Mark Edward and Catharine Pearson (Fowler) S.; m. Barbara Ann Hoemann, Nov. 23, 1957; children: Cynthia Kay, Mark Edward, David Jerome, Terri Lynn, Stephen Patrick. BA in Gen. Edn., U. Nebr., 1961; BS in Engring., Ariz. State U., 1966; MS in Systems Engring., So. Meth. U., 1969; PhD, U.S. Internat. U., 1991. Commd. 2d lt. USAF, 1956, advanced through grades to lt. col., 1972; B-47 navigator-bombardier 98th Bomb Wing, Lincoln Air Force Base, Nebr., 1956-63; with Air Force Inst. of Tech., 1963-65, 68-69; chief mgmt. engring. team RAF Bentwaters, England, 1965-68; forward air contr. 20th Tactical Air Support Squadron USAF, Danang, Vietnam, 1970-71; program mgr. Hdqrs. USAF, Washington, 1971-74, staff asst. Office of Sec. Def., 1974-75, ret., 1975; account exec. Merrill Lynch, San Diego, 1975-77; program engring. chief Gen. Dynamics, San Diego, 1977-79, engring. chief, 1979-80, program mgr., 1980-83; mgr. integrated logistics support Northrop Corp., Hawthorne, Calif., 1984-88; mgr. B-2 program planning and scheduling Northrop Corp., Pico Rivera, Calif., 1988-91; pres. Scannell and Assocs., Borrego Springs, Calif., 1991—; mem. adj. faculty U.S. Internat. U., San Diego. Decorated DFC with three oak leaf clusters, Air medal with 11 oak leaf clusters. Mem. APA, Calif. Psychol. Assn., Soc. Indsl. and Orgnl. Psychology, Inst. Indsl. Engrs., Coronado Cays Yacht Club, Psi Chi. Republican. Roman Catholic. Home: PO Box 2392 717 Anza Park Trail Borrego Springs CA 92004 Office: Scannell and Assocs PO Box 2392 Borrego Springs CA 92004

SCARATT, DAVID J., marine biologist; b. Liverpool, Eng., Dec. 21, 1935; s. Josiah H. Scarratt and Eleanor M. Parr; m. Irene Scott Grant, Oct. 8, 1962; children: Michael Grant, Alison Margaret. BS, U. Wales, Aberystwyth, 1958, PhD, 1961. Rsch. scientist Govt. Can. Fisheries, St. Andrews, N.B., Can., 1961-85, Halifax, N.S., Can., 1985-92; cons. David Scarratt & Assocs., Halifax, 1992—. Office: David Scarratt & Assocs, PO Box 1564, Halifax, NS Canada B3J 2Y7

SCARBOROUGH, GEORGE EDWARD, aerospace engineer, researcher; b. Kansas City, Mo., June 16, 1945; s. George Edward and Elma Ann (Richardson) S.; m. Jackie Lynn McCracken, July 24, 1982; children: John Thomas Stephens, Amy Lynn Stephens. BA, Park Coll., Kansas City, 1980; MBA, So. Ill. U., 1983. Enlisted USAF, 1963, advanced through grades to sr. master sgt., 1980, aircraft environ. systems technician, 1963-73, mgmt. engr., 1973-83, ret., 1983; logistics rsch. analyst McDonnell Douglas Aircraft Co., St. Louis, 1983-86; mgr. product support McDonnell Douglas Electronics Co., St. Charles, Mo., 1985-87; mgr. logistics engring. McDonnell Douglas Astronautics Co., St. Charles, 1987-89; prin. staff specialist supportability tech. McDonnell Douglas Missile Systems Co., St. Louis, 1989-92; prin. staff specialist CALS/CITIS Data Integration, 1992—. Mem. sch. bd. Ft. Zumwalt Sch. Dist., O'Fallon, Mo., 1988-91; res. patrolman O'Fallon Police Dept., 1987-92; deacon Dardenne (Mo.) Presbyn. Ch., 1990-93. Assoc. fellow AIAA (sr., chmn. 1992-94); mem. Nat. Security Indsl. Assn. (chmn. 1990—), St. Louis Computer-Aided Acquisition and Logistics Support Interest Group, VFW (life). Republican. Achievements include devel. and implementation of use of large scale computer simulation models for qualification and quantification of logistics and ops. resources for USAF; contbns. to aeronautics and astronautics maintenance discipline. Office: McDonnell Douglas Aerospace Missilie Systems Co PO Box 516 Saint Louis MO 63166

SCARDELLATO, ADRIANO, software production company executive; b. Oderzo, Treviso, Italy, June 26, 1955; s. Egidio and Amelia (Dall'Ongaro) S.; m. Gianna Pasa, July 24, 1982; children: Giovanni, Carlo. Laurea in ingegneria civile, U. Padova, 1979; MCE, U. Calif., Berkeley, 1980. Registered profl. engr., Italy. Engring. profl. Treviso, 1981—; cons. El. da. Ingegneria S.p.A., Treviso, 1981-87, ptrn., mng. dir., 1987—; bd. dirs. S. Te.A. s.r.l. Feltre (BL). With Italian Army, 1980-81. Mem. Ordine Profl. degli Ingegneri di Treviso, ASCE (assoc.), IEEE Computer Soc. Avocations: skiing and teaching skiing, tennis, sailing. Office: El da Ingegneria Spa, Via Damiano Chiesa 5, 31100 Treviso Italy

SCARDERA, MICHAEL PAUL, air force officer; b. New Haven, Mar. 18, 1963; s. Michael and Georgette (Boilard) S.; m. Maria Ann Robinson, June 3, 1989. BS, MIT, 1985; MS, U. Md., 1986. Rsch. asst. MIT Space Systems Lab., Cambridge, Mass., 1981-85; rsch. engr. aerospace engring. dept. U. Md., College Park, 1985-86; commd. 2d lt. USAF, 1985, advanced through grades to capt., 1989; navigation analyst Navstar GPS USAF, Colorado Springs, Colo., 1986-91; chief navigation analyst 2d Space Ops. Squadron AF Space Command, Colorado Springs, Colo., 1991-92; astronaut. engr. U.S. Space Command/Space Control div. USAF, Colorado Springs, Colo., 1992—. Author papers in field. Mem. AIAA, Brit. Interplanetary Soc. Achievements include design netural buoyancy simulation of manned orbital work pod, proposed engine/systems design for Mars transportation, increased Navstar global positioning system navigation availability and accuracy.

SCARL, ETHAN ADAM, computer scientist; b. N.Y.C., Mar. 9, 1940; s. Aaron and Fannie (Meltzer) S.; m. Dona Jane Lethbridge, May 21, 1970; children: Austen Charles, Benjamin Ethan, Stefan Alan (dec.). BS, Reed Coll., 1961; MS, Washington U., 1964; PhD, U. British Columbia, 1973. Asst. prof. Boston U., 1977-79; with dept. staff MITRE Corp., Bedford, Mass., 1979-86; sr. computer scientist Boeing Corp., Seattle, 1986-92, Huntsville, Ala., 1992—. Contbr. articles to profl. jours. Mem. Am. Assn. Artificial Intelligence (chair 1989, chair 1st and 2d diagnosis workshops 1990, 3d internat. workshop on diagnosis 1992). Democrat. Unitarian-Universalist. Home: 1701 Fagan Circle Huntsville AL 35801 Office: Boeing Corp Huntsville AL 35824

SCARLATA, SUZANNE FRANCES, biophysical chemist; b. Phila., May 31, 1958; d. Angelo Louis and Theresa (Rocatto) S.; m. Walter Peter Zurawsky, Aug. 9, 1981; 1 child, Cassandra Louise. BA in Chemistry, Temple U., Phila., 1979; PhD in Chemistry, U. Ill., 1984. Mem. tech. staff AT&T Bell Labs., Princeton, N.J., 1984-86; adj. asst. prof. Mt. Sinai Sch. Medicine, N.Y.C., 1986—; asst. prof. biophys. chemistry Cornell U. Med. Coll., N.Y.C., 1986-91; asst. prof. biophysics, chemistry SUNY, Stony Brook, 1991—. Contbr. articles to profl. jours. Budget advisor United Way, Princeton, N.J., 1985. NIH grantee, 1988; Am. Chem. Soc.-Petroleum Rsch. Fund fellow, 1986; NIH grantee, 1986, 88, 91. Mem. AAAS, N.Y. Acad. Scis., Biophys. Soc. Achievements include a patent on fluorescence methods of monitor polymer cure. Office: SUNY Dept Physiology & Biophysic Stony Brook NY 11794

SCARPA, ANTONIO, medicine educator, biomedical scientist; b. Padua, Italy, July 3, 1942; s. Angelo and Elena (DeRossi) S. MD cum laude, U. Padua, 1966, PhD in Pathology, 1970; MA (hon.), U. Pa., 1978. Asst. prof. biochemistry, biophysics U. Pa., Phila., 1973-76, assoc. prof., 1976-80, prof., 1980-86, dir. biomed. instrumentation group, 1983-86; prof. dept. pathology Jefferson U., Phila., 1986—; prof., chmn. dept. physiology Case Western Res. U., Cleve., 1986—, dir. tng. ctr., program project, 1983—, prof. medicine, 1989—; cons. study sect. NIH, Bethesda, 1984—, Am. Heart Assn., Dallas, 1986-91. Editor (books): Frontiers of Biological Energetics, Calcium Transport and Cell Function, Transport ATPases, Membrane Pathology, Membrane and Cancer Cells; editor (jours.): Archives Biochemistry and Biophysics, Cell Calcium, Biochemistry Internat.; mem. editorial bd. Circulation Rsch., 1978-81, Biophys. Jour., 1979-82, Jour. Muscle Rsch., 1979—, Magnesium, 1982—, Physiol. Revs., 1982—, FASEB Jour., 1987—, Molecular Cellular Biochemistry, 1988—; contbr. numerous articles to profl. jours. Mem. Am. Soc. Physiologists, Am. Soc. Biolog. Chemistry, Biophys. Soc. (exec. council 1980-83), U.S. Bioenergetics Group (program chmn. 1974-75, 82, 83, exec. officer, 1985—), Physiology Assn. (pres., chmn.). Avocations: farming, sailing, painting. Office: Case Western Reserve Univ Dept of Physiology Cleveland OH 44106

SCARPA, ROBERT LOUIS, civil engineer; b. Revere, Mass., Feb. 21, 1950; s. Louis S. Scarpa and Caroline J. (Repucci) Samas; m. Denise Marie Tuminelli, Apr. 30, 1977; children: Jill Marie, Michael James. BS in Civil Engring., Rensselaer Poly. Inst., 1972. Registered profl. engr. Maine, Mass, Conn. Engr., project mgr. Metcalf and Eddy, Inc., Boston, 1975-85; project mgr. Metcalf and Eddy, Inc., Wakefield, 1985—; mem. A21 ductile iron pipe com. Am. Water Works Assn., Denver, 1989--. With USN, 1972-75, Vietnam. Mem. N.Eng. Water Works Assn., Construction Specifications Inst. Home: 40 Paige Farm Rd Amesbury MA 01913 Office: Metcalf and Eddy Inc 30 Harvard Mill Sq Wakefield MA 01880

SCARPONCINI, PAUL, computer scientist; b. Orange, N.J., Dec. 9, 1950; s. Anthony James and Elizabeth (Pizzano) S.; m. Julie Doranne Hebert, June 3, 1978; children: Timothy James, Amy Corinne. BS in Civil Engring., Rutgers U., 1973; MS in Archtl. Engring., Pa. State U., 1981; MS in Computer Sci., U. Mo., Rolla, 1988, PhD in Computer Sci., 1992. Registered profl. engr., Colo. Civil engr. City of Aurora (Colo.) Pub. Works, 1976-81; product mgr. Computer Sharing Svcs., Denver, 1981-82; cons. McDonnell Douglas Corp., St. Louis, 1982-91; project mgr. Electronic Data Systems, St. Louis, 1992—. Author conf. procs. NSF rsch. grantee, 1987. Mem. ASCE (mem. database com.), Am. Assn. for Artificial Intelligence. Republican. Roman Catholic. Achievements include planning and development of integrated database prototype for NAS-NRC Building Research Board; development of method of integrating computer graphic system with computer aided engineering, knowledge based, software, and database management systems. Home: 3530 Swaying Oaks Ln Wentzville MO 63385 Office: Electronic Data Systems 13736 Riverport Dr Maryland Heights MO 63043

SCARR, HARRY ALAN, federal agency administrator; b. Massillon, Ohio, May 4, 1934; s. Charles Edward and Helen May (Simonson) S.; m. Sandra Jane Wood, Dec. 26, 1961 (div. 1970); children: Phillip Ruxton, Karen Pelton, Rebecca Blackwell; m. Cecilia von Schantz, Nov. 19, 1971; children: Miriam Rachel, Sarah Eleanor. BA, U. Mich., 1956; PhD, Harvard U., 1963. Asst. rsch. sociologist Harvard U., Cambridge, Mass., 1961-63; staff fellow Dept. HEW, Washington, 1963-66; asst. prof. U. Pa., Phila., 1966-70; rsch. sci. HSR, Inc., McLean, Va., 1970-72; various positions Dept. Justice, Washington, 1972-82; dir. Bur. Justice Statistics, Washington, 1980-81; exec. asst. for statis. affairs Dept. Commerce, Washington, 1982-90, dep. asst. sec., 1990-92; dep. dir. Bur. Census, Washington, 1992—. Co-author: Variations in Value Orientations, 1961. Recipient Gold medal Dept. Justice, 1988, Presdl. Rank award Office of President U.S., 1992. Home: 1003 Orr Cir Leesburg VA 22075-4332 Office: Bureau of the Census Federal Center Washington DC 20233

SCARR, SANDRA WOOD, psychology educator, researcher; b. Washington, Aug. 8, 1936; d. John Ruxton and Jane (Powell) Wood; m. Harry Alan Scarr, Dec. 26, 1961 (div. 1970); children: Phillip, Karen, Rebbecca, Stephanie; m. James Callan Walker, Aug. 9, 1982. AB, Vassar Coll., 1958; AM, Harvard U., 1963, PhD, 1965. Lic. psychologist, Va. Asst. prof. psychology U. Md., College Park, 1964-67; assoc. prof. U. Pa., Phila., 1967-71; prof. U. Minn., Mpls., 1971-77, Yale U., New Haven, 1977-83; Commonwealth prof. U. Va., Charlottesville, 1983—, chmn. dept. psychology, 1984-90; mem. nat. adv. bd. Robert Wood Johnson Found., Princeton, N.J., 1985-91; mem. coordinating council psychology SUNY Bd. Regents, N.Y.C., 1984—. Author: Race, Social Class and Individual Diffrences in IQ, 1981, Mother Care/Other Care, 1984 (Nat. Book award Am. Psychol. Assn. 1985), Caring for Children, 1989; editor Jour. Devel. Psychology, 1980-86. Fellow Ctr. for Advanced Studies, Stanford U., Calif., 1976-77; grantee NIH, NSF, others, 1967—. Fellow AAAS, Am. Psychol. Assn. (chmn. com. on human research 1980-83, council of reps. 1984—, bd. dirs. 1988—, Award for Disting. Contbn. to Research on Pub. Policy 1988), Am. Psychol. Soc. (bd. dirs. 1991—, James McKeen Cattell award 1993); mem. Am. Acad. Arts and Scis., Behavior Genetics Assn. (pres. 1985-86, mem. exec. council 1976-79, 84-87), Soc. for Research in Child Devel. (governing council 1974-76, 87-93, chmn. fin. com. 1987-89, pres. 1989-91), Internat. Soc. for Study of Behavioral Devel. (exec. bd. 1987—). Democrat. Avocations: dogs, gardening. Home: 1243 Maple View Dr Charlottesville VA 22902 Office: U Va Dept Psychology Charlottesville VA 22903

SCASTA, DAVID LYNN, psychiatrist; b. Austin, Tex., Dec. 13, 1949; s. Albert Ray and Helen Pearl (Hennessy) S. BA, Baylor U., 1972; MD, Baylor Coll. of Medicine, 1977. Diplomate Am. Bd. Psychiatry and Neurology. Staff physician U. Houston, 1977-78; adminstr. Temple U. Med. Sch., Phila., 1982-83; residency in psychiatry Temple U. Hosp., Phila., 1982; dir. consultation svcs. Grad. Hosp., Phila., 1983-84; dir. outpatient programs Phila. Psychiat. Ctr., 1983-84; pvt. practice Grad. Hosp. Phila. Psychiat. Ctr., 1984-89; med. dir. Phila. Consultation Ctr., 1987-89; attending psychiatrist Hunterdon Med. Ctr., Flemington, N.J., 1989—; vice chmn. dept. mental health Hunterdon Med. Ctr., Flemington, 1989—; pvt. practice New Hope, Pa., 1989—; clin. assoc. prof. dept. psychiatry Temple U. Med. Sch., Phila., 1983—; researcher Assn. Gay and Lesbian Psychiatrists, Phila., 1989—; vice-chmn. dept. mental health Hunterdon Med. Ctr., Flemington, N.J. Editor Jour. of Gay & Lesbian Psychotherapy, 1987—, Newsletter of the Assn. of Gay & Lesbian Psychiatrists, 1984—. Dist. rep. Rep. Party of Tex., Houston, 1977, precinct sec., 1975-77. Named Ginsberg Fellow Group for Advancement of Psychiatry, 1980-82. Mem. Assn. Gay and Lesbian Psychiatrists, Am. Psychiat. Assn., Assn. Lesbian and Gay Psychologists, Parents and Friends of Lesbians and Gays. Republican. Baptist. Avocations: skiing, sailing, desktop publishing. Office: Hunterdon Med Ctr Dept Mental Health 2100 Wescott Dr Flemington NJ 08822-4606

SCEARCE, P. JENNINGS, JR., engineering executive. BS in Elec. Engring., N.C. State U. With ManTech Internat., Vitro Corp.; with Sequa Corp., Rockville, Md., 1976—, v.p. ops., sr. v.p., gen. mgr. def. systems divsn., ARC profl. svcs. group. Mem. Am. Soc. Naval Engrs., Am. Soc. Quality Control, Armed Forces Comm. and Electronics Assn., Quality and Productivity Mgmt. Assn. (v.p. Chesapeake coun.), Navy League U.S. (life). Office: Sequa Corp Def Systems Divsn 1375 Piccard Dr Rockville MD 20850*

SCEDROV, ANDRE, mathematics and computer science researcher, educator; b. Zagreb, Croatia, Aug. 1, 1955; came to U.S., 1977, naturalized, 1987; s. Oleg and Mira (Petric) S.; m. Bonnie Carol Hoke, July 23, 1983. BA, U. Zagreb, 1977; MA, SUNY, Buffalo, 1979, PhD in Math., 1981. T.H. Hildebrandt asst. prof. rsch. U. Mich., Ann Arbor, 1981-82; asst. prof. U. Pa., Phila., 1982-88, assoc. prof., 1988-92, prof., 1992—; vis. scholar U. Milan, 1982, McGill U., Montreal, 1985, U. Sydney, Australia, 1986, U. Catholique de Louvain, Louvain-La-Neuve, Belgium, 1988, U. Paris 7, 1992, Rijksuniv Utrecht, The Netherlands, 1993; vis. scientist Math. Scis. Inst. Cornell U., Ithaca, N.Y., 1987; vis. assoc. prof. Stanford (Calif.) U., 1988-90; cons. Odyssey Rsch. Assocs., Ithaca, 1987, HP Labs, Palo Alto, 1990; program chair 7th ann. IEEE Symposium on Logic in Computer Sci., Santa Cruz, Calif., 1992; mem. program com. Logical Found. Computer Sci., Tver, Russia, 1992, St. Petersburg, Russia, 1994; invited speaker Math. Founds. Programming Semantics, Oxford U., Eng., 1992, Computer Sci. Logic, San Miniato, Italy, 1992, Internat. Summer Sch. Logic Computer Sci., Chambery, France, 1993, Proof and Computation, Marktoberdorf, Germany, 1993. Author: (with P. Freyd) Categories, Allegories; editor Math. Structures in Computer Sci., 1989—, Annals Pure Applied Logic, 1993—; contbr. articles and rsch. papers to profl. publs. Recipient Young Faculty award Nat. Scis. Assn. U. Pa., 1987; rsch. grantee NSF, 1985—, Office Naval Rsch., 1988—. Mem. AAMS, Am. Math. Soc. (Centennial rsch. fellow 1993-94), Assn.for Symbolic Logic (editor jours. 1988—, chair nominating com. 1993, program com. 1988-90, coun. 1990—), Assn. for Computing Machinery, Math. Assn. Am. Office: U Pa Dept Math 209 S 33rd St Philadelphia PA 19104-6395

SCHAAD, NORMAN WERTH, plant pathologist; b. Myrtle Point, Oreg., Nov. 9, 1940; s. Harold S. and Mable (Parry) S.; m. Terry Barber, June 1966 (div. May 1985); children: Sandra Renee, Kristina Diane. BS in Plant Pathology, U. Calif., Davis, 1964, MS, 1966, PhD, 1969. Asst. prof. U. Ga., Griffin, 1971-77; assoc. prof. U. Ga., 1977-82, prof., 1982; assoc. prof. U. Idaho, Moscow, 1982-84; prof. U. Idaho, 1984-88; dir. biotechnology Harris Moran Seed Co., San Juan Bautista, Calif., 1988-92; rsch. leader foreign disease weed sci. rsch. unit USDA/ARS, Frederick, Md., 1992—; instr. workshops USAID/OICD, Kuala Lumpar, Malaysia, 1984, 86, UN/FAO, Bangkok, Thailand, 1987; NSF vis. prof. U. Brasilia, 1987; adj. prof. Hassen II U., Morocco, 1985—. Editor: Identification of Plant Pathogenic Bacteria, 1980, 2d edit., 1988; co-editor: Detection of Bacteria in Seed, 1989; assoc. editor: Phytopathology, 1980-83, Seed Sci.& Tech., 1992—; contbr. over 60 articles to profl. jours. Bd. dirs. Campus Christian Ctr., U. Idaho, 1983-88; bd. trustees United Meth. Ch., Moscow, 1985-88. Mem. Am. Phytopathology Soc. (chmn. bacteriology com. 1976-77, chmn. seed pathology com. 1988-89), Internat. Soc. Plant Pathology, Internat. Seed Testing Assn. (vice chmn. plant disease com. 1978-92, leader bacterial working group 1976—). Democrat. Methodist. Office: USDA/ARS Foreign Disease Weed Sci Bldg 1301 Fort Detrick MD 21702

SCHAAF, JOHN URBAN, communication management specialist; b. Boston, June 22, 1955; s. Robert Andrew and Margaritte Anne (Ostrowski) S.; m. Gail Ann Olsavsky, July 6, 1985. BSBA, George Mason U., 1979; MA in Computer Resource Mgmt., Webster U., 1990. Investigator MCI Telecommunications Corp., Washington, 1979-80; mgr. records mgmt. and distbn. Calvert Group, Bethesda, Md., 1980-84; info. systems specialist Dept. Def./USAF, Scott AFB, Ill., 1985-91; communication mgmt. specialist Nat. Park Svc., Dept. of Interior, Boston, 1992—. Mem. Beta Epsilon Phi. Democrat. Roman Catholic. Avocations: travel, photography, camping, hiking. Office: Nat Park Svc N Atlantic Regional Office 15 State St Boston MA 02109

SCHABERG, BURL ROWLAND, JR., engineering company executive; b. Ft. Smith, Ark., Sept. 18, 1944; s. Burl R. and Grace M. (Davis) S.; m. Sheila Bournival, Sept. 2, 1989; children: Mark Stanton Schaberg, Bradley J. Bournival, Laura Emily Bournival. BS in Archtl. Engring., U. Kans., 1967, Thermal Dynamics Engr., 1973. Owner Jayhawk Corp., Chanute, Kans., 1967-79; pres. Solar Power Sytems, Inc., Sarasota, Fla., 1979-83; tech. R&D head Helionetics Corp., Santa Ana, Calif., 1983-85; mktg. dir., tech. head Sealed Air Corp., Hayward, Calif., 1985-91; pres. Echo Innovations, Paso Robles, Calif., Sarasota, Fla., 1991—. Advisor Gov.'s Energy Office, Tallahassee, 1985-90. Mem. Fla. Solar Energy Industry Assn. (bd. dirs. 1986—). Republican. Achievements include development of the laser slicing process of the silicone ingot for the production of the photovoltaic solar chip; development of the DC/AC inverter for application with the photovoltaic solar electric system. Home: 3935 Higel Sarasota FL 34242 Office: Echo Innovations Inc 1405 Niderer Rd Paso Robles CA 93446

SCHACHTEL, BARBARA HARRIET LEVIN, epidemiologist, educator; b. Rochester, N.Y., May 27, 1921; d. Lester and Ethel (Neiman) Levin; m. Hyman Judah Schachtel, Oct. 15, 1941; children: Bernard, Ann.Mollie. Student Wellesley Coll., 1939-41; BS, U. Houston, 1951, MA in Psychology, 1967; PhD, U. Tex.-Houston, 1979. Psychol. examiner Meyer Ctr. for Devel. Pediatrics, Tex. Children's Hosp., Houston, 1967-81; instr. dept. pediatrics Baylor Coll. Medicine, Houston, 1967-81, asst. prof. dept. medicine, 1982—; asst. dir. biometry and epidemiology Sid W. Richardson Inst. for Preventive Medicine, Houston, 1981-88, dir. quality assurance, 1988—; mem. instl. rev. bd. for human rsch. Baylor Coll. Medicine, Houston, 1981-87; mem. devel. bd. U. Tex. Health Sci. Ctr., Houston, 1987—; mem. dean's adv. bd. Sch. Architecture U. Houston, 1987; bd. dirs. Tex. Medical Ctr. Contbr. articles to profl. jours. Vice pres., bd. dirs. Houston-Harris County Mental Health Assn., 1966-67; vice-chmn. bd. mgrs. Harris County Hosp. Dist., Houston, 1974-90, chmn. 1990-92, bd. dirs., 1970-93; trustee Inst. Religion in Tex. Med. Ctr., 1990—. Named Great Texan of Yr., Nat. Found. for Ilietis and Colitis, Houston, 1982, Outstanding Citizen, Houston-Harris County Mental Health Assn., 1985; recipient Good Heart award B'nai Brith Women, 1984, Women of Prominence award Am. Jewish Com., 1991. Mem. Am. Psychol. Assn., Am. Pub. Health Assn., Tex. Psychol. Assn., Houston Psychol. Assn. (psychol. assoc. rep. 1974), Wellesley Club of Houston (pres. 1968-70). Avocations: golf, tennis, books. Home: 2527 Glenhaven Blvd Houston TX 77030 Office: The Methodist Hosp Sid W Richardson Inst Preventive Med 6565 Fannin St Houston TX 77030-2707

SCHAD, THEODORE MACNEEVE, science research administrator, consultant; b. Balt., Aug. 25, 1918; s. William Henry and Emma Margaret (Scheldt) S.; m. Kathleen White, Nov. 5, 1944(dec. Aug., 1989); children: Mary Jane, Rebecca Christina. BSCE, Johns Hopkins U., 1939. Various positions water resources engring. U.S. Army C.E., U.S. Bur. Reclamation, Md., Colo., Oreg. Wash., 1939-54; prin. budget examiner water resources programs U.S. Bur. Budget, Exec. Office of Pres., 1954-58; sr. specialist engring. and pub. works, dep. dir. Congl. Research Service, Library of Congress, 1958-68; staff dir. U.S. Senate Com. Nat. Water Resources, 1959-61; exec. dir. Nat. Water Commn., 1968-73; exec. sec. Environ. Studies Bd., 1973-77; dep. dir. Commn. Natural Resources, Nat. Acad. Scis., Washington, 1977-83; exec. dir. Nat. Ground Water Policy Forum, 1984-86; sr. fellow Conservation Found., Washington, 1986-; U.S. commr. Permanent

Internat. Assn. Nav. Congresses, Brussels, 1963-70, commr. emeritus, 1987—; cons. U.S. Senate Com. Interior and Insular Affairs, 1963, U.S. Ho. of Reps. Com. Sci. and Tech., 1964-65, Fed. Council Sci. and Tech., 1962-65, U.S. Office Saline Water, 1965-67, A.T. Kearney, Inc., Alexandria, Va., 1979-80, Chesapeake Research Consortium, 1984, Ronco Cons. Corp., 1986—, Gambia River Basin Devel. Commn., Dakar, 1986-87, Apogee Rsch. Corp., 1987—, Office of Tech. Assessment U.S. Congress, 1992—. Contbr. articles to Ency. Brit. and profl. jours. Treas. Nat. Speleol. Found., 1961-65, trustee, 1965—; bd. dirs. Vets. Coop. Housing Assn., Washington, 1958-81, v.p., 1960-72. Recipient Meritorious Svc. award U.S. Dept. Interior, 1950, Icko Iben award Am. Water Resources Assn., 1978, Henry P. Caulfield medal, 1990. Fellow ASCE (treas. Nat. Capital chpt. 1952-55, v.p. 1967, pres. 1968, Julian Hinds prize 1991); mem. AAAS, Nat. Speleol. Soc., Am. Water Works Assn. (hon.), Am. Geophys. Union, Am. Acad. Environ. Engrs., Nat. Acad. Pub. Adminstrn., Permanent Internat. Assn. Nav. Congresses, Internat. Commn. Irrigation and Drainage. Clubs: Potomac Appalachian Trail; Cosmos (Washington); Colo. Mountain (Denver); Seattle Mountaineers. Home: 4540 25th Rd N Arlington VA 22207-4102 Office: The Conservation Found 1260-24th St NW Washington DC 20037

SCHADE, MARK LYNN, psychologist; b. Houston, Nov. 30, 1959; s. Alphonse Michael and Lillian Sue (Smith) S.; m. Shiree Elaine Plummmer, Mar. 10, 1984. BA, Rice U., 1982; MA, U. Houston, 1987, PhD, 1991. Lic. psychologist Tex. Dir. social learning program Austin (Tex.) State Hosp., 1989—; mem. admissions com. U. Houston, Clin. Psychology Program, 1985-87, admissions com. Camarillo (Calif.) State Hosp. Clin. Psychology Internship, 1989, budget adv. com. U. Houston, 1986-88, intern tng. com. Austin State Hosp. Dept. Psychology, 1991—, rsch. com., 1991—. guest editor Jour. Psychosocial Rehab. (spl. section), 1990; co-author: chpt. in Handbook Psychiatric Rehabilitation, 1992; contbr. articles to Internat. Rev. Psychiatry, New Directions for Mental Health Scvs.,. Recipient W. Jordan and Cora Jordan scholar grant, Rice U., 1980, Svc. award U. Houston Psychol. Rsch. and Svcs. Ctr., 1987. Mem. Am. Psychol. Assn., Am. Psychol. Soc., Tex. Psychol. Assn., Assn. for Advancement of Behavior Therapy. Achievements include direction of implementation of state-of-the-art psychosocial rehabilitation program; computerized observational data system in three units adult psychiatric patients in state hospital. Office: Austin State Hosp 4110 Guadalupe Austin TX 78751

SCHADEWALDT, HANS, medical educator; b. Kottbus, May 7, 1923; s. Johannes and Hedwig S.; m. Lotte, 1943. Educator Univs. Tûbingen, Wûrzburg and Kûnigsberg; lectr. U. Freiburg, 1961-63; prof. history of medicine U. Düsseldorf, 1963—, dean faculty of medicine, 1976-77. Author: Michelangelo und die Medizin seiner Zeit, 1965, Die berühmten Arzte, 1966, Kunst und Medizin, 1967, Der Medizinmann bei den Naturvölkern, 1968, Geschichte der Allergie, 1979-83, Die Chirurgie in der Kunst, 1983, Das Herz, ein Rätsel für die antike und mittelalterliche Welt, 1989, Betrachtungen zur Medizin in der bildenden Kunst, 1990, Pharmakologie bei Bayer 1890-1990, 1990. Mem. Rhine Westfalian Acad. Arts and Scis. (pres.1990), Officier Ordre des Palmes Academiques. Office: Rheinland-Westphalia Acad of Sci, Palmenstrasse 16, 4000 Düsseldorf Germany*

SCHADRACK, WILLIAM CHARLES, III, design engineer; b. Memphis, Feb. 22, 1950; s. William Charles Jr. and Marianne (Hardy) S.; m. Anedra Jean Lemons, Feb. 24, 1979; children: Amanda Jean, William Charles IV. BS in Biology, U. Tenn., Martin, 1972; BS in Chemistry, Memphis State U., 1974, BSME, 1975, MSME, 1977. Registered profl. engr., Ind., Tenn., Ala., La., Mo. Grad. asst. Memphis State U., 1977; devel. engr. Zimmer Co., Warsaw, Ind., 1977-82, Richards Med. Co., Memphis, 1982-85; mgr. power transp. dept. Lewis Supply Co., Memphis, 1985-87; spl. application engr. Dover Elevator Systems, Horn Lake, Miss., 1987—; part-time instr. Memphis State U., 1987. Mem. ASME, Nat. Soc. Profl. Engrs., Tenn. Soc. Profl. Engrs., Order of the Ring. Democrat. Roman Catholic. Achievements include patents for medical devices. Home: 639 Harwood Cove Memphis TN 38120-3003 Office: Dover Elevator Systems 6266 Hurt Rd Horn Lake MS 38637

SCHAEFER, CARL GEORGE, JR., aerospace engineer; b. Atlantic City, Mar. 29, 1962; s. Carl G. and Helen D. (Stearns) S.; m. Deborah Ann Manelski, Aug. 27, 1983; 1 child, Geoffrey Michael. BS in Aerospace/Ocean Engring., Va. Tech., 1985, MS in Systems Engring., 1990. Engring. intern David Taylor Rsch. Ctr., Carderock, Md., 1982-85; flight test engr. Naval Air Test Ctr., Patuxent River, Md., 1985-87; project engr. Naval Air Systems Command, Washington, 1987—; mem. Sr. Exec. Mgmt. Devel. Program, Washington, 1991—; tech. advisor F/A-18 System Safety Working Group, Washington, 1990-91; adj. lectr. No. Va. Community Coll., Arlington, 1990-91; del. to Internat. Coun. Aero. Scis., London, 1986. Contbr. articles to profl. jours. Recipient Sen. J. Warner fellowship Naval Aviation Exec. Inst., 1990. Mem. AIAA (head tech. paper judge 1990, Outstanding Paper Yr. 1986, Best Undergrad. Paper award 1986), Am. Helicopter Soc. (test and evaluation com. 1988-91), Am. Assn. for Artificial Intelligence, U.S. Naval Inst. Republican. Roman Catholic. Home: 12132 Beaverwood Pl Woodbridge VA 22192

SCHAEFER, HANS-ECKART, pathologist; b. Koblenz, Germany, Sept. 8, 1936; s. Hans and Mathilde (Sellerbeck) S.; m. Birgit Peters, Apr. 19, 1966. Degree in medicine, U. Mainz, Fed. Republic Germany, 1958; postgrad., U. Bonn, Fed. Republic Germany, 1962; MD, U. Bonn, 1963; PhD, U. Köln, Fed. Republic Germany, 1970. Med. diplomate. Resident Kantonales Hosp., Walenstadt, Switzerland, 1961-63; asst. physician Inst. Pathology U. Bonn, 1963-67; asst. physician Inst. Pathology U. Köln, 1967-73, head dept. ultrastructural pathology, 1973-83; mng. dir. Inst. Pathology U. Freiburg, Fed. Republic Germany, 1983—; also chair gen. and spl. pathology U. Freiburg. Author: Leukopoese and Myeloproliferative Erkrankungen, 1984; co-editor textbook Allgmeine und Spezielle Pathologie, 1989, 93; mem. editorial bd. Pathology Rsch. and Practice, Virchows Archiv B. Fellow Heidelberg Acad. Wissenschaften; mem. Gesellschaft Histochemie (pres. 1983-84), Internat. Acad. Pathology, Internat. Soc. Hematology, Deutsche Gesellschaft Pathologie, Soc. Europaea Pneumologica, Deutsche Gesellschaft Säugetierforschung, Deutsche Gesellschaft Arterioskleroseforschung (pres. 1989-92), Deutsche Gesellschaft Hämatologie Onkologie. Avocation: harpsichord playing. Home: Weinbergstrasse 15, D 79249 Merzhausen Federal Republic of Germany Office: Inst Pathology, Alberstrasse 19 PO Box 214, D 79002 Freiburg Germany

SCHAEFER, JOSEPH ALBERT, physics and engineering educator, consultant; b. Bellevue, Iowa, Dec. 24, 1940; s. Albert Francis and Eileen Clara (Schilling) S.; m. Carol Ruth Deppe, Nov. 20, 1965; children: Sarah Ellen, Amy Marie. BS, Loras Coll., 1962; MS, U. Toledo, 1964; PhD, Northwestern U., 1972. Profl. engr., Iowa. Physicist Inst. Gas Tech., Chgo., 1963, U.S. Naval Ordnance Lab., Silver Spring, Md., 1964; prof. physics & engring. sci. Loras Coll., Dubuque, Iowa, 1964—; team mem. U.S. AID Team in India/Tchrs. Coll. Columbia U., Agra, India, 1966; cons. applied mechs. group John Deere, Dubuque, 1980; assoc. rsch. scientist Iowa Inst. Hydraulic Rsch., Iowa City, 1985-91; cons., evaluator North Cen. Assn., Chgo., 1990—; reviewer NSF, Washington, 1981—; mem. organizing com. 8th Internat. Ice Conf., Iowa City, 1986. Author: Study Guide for Physics of Everyday Phenomena, 1992; contbr. articles to profl. jours.; author, reviewer Wm. C. Brown Pubs., Dubuque, 1988—. Mem. multiculturalism-nonsexism com. Dubuque Sch. Bd., 1991—; bd. dirs. Area Residential Care, Dubuque, 1977—; v.p. Dubuque Tchrs. Credit Union, 1987—. Named Iowa Prof. of Yr. Coun. for Advancement & Support of Edn., 1989; recipient Teaching Excellence & Campus Leadership award Sears-Roebuk Found., 1990; Danforth assoc. Danforth Found.; sci. faculty fellow NSF, 1968, 70; grantee NSF, 1990. Mem. Internat. Assn. for Hydraulic Rsch., Am. Assn. Physics Tchrs., Am. Soc. for Engring. Edn. Democrat. Roman Catholic. Achievements include patent for Bulk Polycrystalline Switching Materials for Threshold and/or Memory Switching; anomalies in induced torque in potassium. Home: 2675 Mineral Dubuque IA 52001-5645 Office: Loras Coll 1450 Alta Vista Dubuque IA 52004-0178

SCHAEFER, ROLAND MICHAEL, nephrologist, consultant; b. Wuerzburg, Germany, May 18, 1954; s. Gregor and Mathilda (Schaefer) S.; m. Liliana Barbara Denek Schaefer, Mar. 28, 1991; 1 child, Piotr. DMed, Sch. Med., Wuerzburg, Germany, 1973-80. Fellow internal med. and nephrology U. Hosp. Wuerzburg, Germany, 1982-90, asst. prof., 1991; cons. freelance in fields of biomaterials and genetechnology. Roman Catholic. Avocations: travel, literature, classical music. Home: Anne-Frank Str 29, Wuerzburg Germany Office: U Wuerzburg, Josef Schneider Strasse 2, Wuerzburg Germany

SCHAEFER, STEVEN DAVID, head and neck surgeon, physiologist; b. L.A., Mar. 25, 1945; s. Glen Arthur and Alice (Malerstein) S.; m. Phyllis Lois Clark, July 1, 1977; 1 child, Jessica Leigh. BA, U. Calif., Berkeley, 1967; MD, U. Calif., Irvine, 1972. Diplomate Am. Bd. Otolarnyology. Asst. prof. U. Tex. Southwestern and U. Tex. Dallas, 1972-82, assoc. prof., 1982-86, prof., 1986-92; prof, dept. chmn. N.Y. Med. Coll., N.Y.C., 1992—, N.Y. Eye and Ear Infirmary, N.Y.C., 1992—. Author: 2 books, 96 articles, 7 monographs, 15 abstracts. Dir. pub. edn. Tex. div., Am. Cancer Soc., Dallas, 1978-80. Named prin. investigator NIH, 1980-92. Fellow Am. Laryngol. Assn., Am. Acad. Otolarynology (Honor award 1990), Am. Laryngol, Rhinol., and Otol. Soc.; mem. Soc. Univ. Otolarynologists (pres. 1992-93). Office: NY Eye & Ear Infirmary 310 E 14th St New York NY 10003

SCHAFER, EDWARD ALBERT, JR., data processing executive; b. Johnstown, Pa., Sept. 10, 1939; s. Edward A. and Dorothy (Cook) S.; m. Christine A. Ferrazzi, Aug. 27, 1976; children: David E., Kimberly L., Jaime L. BSEE, U. Pitts., 1963. Sales engr. Westinghouse Electric Corp., Pitts., 1963-69; nat. sales mgr. Time Share Peripherals Corp., Bethel, Conn., 1969-75; dist. sales mgr. Versatec, Inc., Boston, 1975-80, Gerber Systems Tech., Boston, 1980-82; regional dir. Intergraph Corp., Reston, Va., 1982-88; v.p., gen. mgr. N.Am. ops. Structural Dynamics Rsch. Corp., Cin., 1988-92; pres., CEO Schafer Holdings, Inc., Palm Harbor, Fla., 1992—. Avocations: power boating, tennis. Office: Schafer Holdings Inc 33223 US Hwy 19 N Ste 601 Palm Harbor FL 34684-3153

SCHAFER, ROBERT LOUIS, agricultural engineer, researcher; b. Burlington, Iowa, Aug. 1, 1937; s. Marion Louis and Pansy (Head) S.; m. Carolyn Louise Henn, Aug. 1, 1959; 1 child, Elizabeth Diane. BS, Iowa State U., 1959, MS, 1961, PhD, 1965. Agrl. engr. Agrl. Rsch. Svc., USDA, Ames, Iowa, 1959-64, Auburn, Ala., 1964—. Contbr. articles to profl. jours. Fellow Am. Soc. Agrl. Engrs.; mem. Ruritan Nat. (various offices). Home: PO Box 57 Loachapoka AL 36865-0057 Office: USDA Agrl Rsch Svc PO Box 3439 Auburn AL 36831

SCHAFER, ROLLIE RANDOLPH, JR., neuroscientist; b. Denver, Feb. 17, 1942; s. Rollie R. and Marjorie June (Mason) S.; m. D. Sue Pestotnik, June 11, 1964; children: Anthony Boyd, Benjamin Charles. AB, U. Colo., 1964, MA, 1967, PhD, 1969. Asst. prof. Met. State Coll., Denver, 1968-69, N.Mex. Inst. Mining and Tech., Socorro, 1969-73, U. Mich., Ann Arbor, 1973-76; from asst. prof. to prof. U. North Tex., Denton, 1976—, asst. dean for rsch., 1978-83, assoc. dean sci. and tech., 1986-87, assoc. v.p. for rsch., grad. dean, 1987—; exec. dir. North Tex. Rsch. Inst., Denton, 1992—. Author: Experimental Neurobiology, 1978; contbr. articles to Jour. Exptl. Zoology, Sci., Chem. Senses, Comparative Biochem. Physiology. NSF grantee, 1972-73, 74-76, 81-85, 85-88. Mem. Soc. for Neurosci. (chpt. pres. 1983-86), Am. Soc. for Neurochemistry, AAAS, Sigma Xi (chpt. pres. 1970-73). Home: 3030 Hartlee Field Rd Denton TX 76208

SCHAFER, RONALD WILLIAM, electrical engineering educator; b. Tecumseh, Nebr., Feb. 17, 1938; s. William Henry and Esther Sophia (Rinne) S.; m. Dorothy Margaret Hall, June 2, 1960; children: William R., John C. (dec.), Katherine L., Barbara Anne. Student, Doane Coll., Crete, Nebr., 1956-59; BEE, U. Nebr., 1961, MEE, 1962; PhD in Elec. Engring., MIT, 1968. Mem. tech. staff Bell Labs., Murray Hill, N.J., 1968-74; John O. McCarty prof. elec. engring. Ga. Inst. Tech., Atlanta, 1974—; chmn. bd. Atlanta Signal Processors Inc., 1983—. Co-author: Digital Signal Processing, 1974, Digital Processing of Speech Signals, 1979, Speech Analysis, 1979, Discrete-Time Signal Processing, 1989. Recipient Class of 34 Disting. Prof. award Ga. Inst. Tech., 1985. Fellow IEEE (Emanuel R. Piore award 1980, Edn. medal 1992), Acoustical Soc. Am.; mem. IEEE Acoustics Speech and Signal Processing Soc. (soc. award 1982), AAAS. Democrat. Methodist. Lodge: Kiwanis. Office: Ga Inst of Tech Dept of Elec Engring Atlanta GA 30332-0250

SCHAFER, WALTER WARREN, dentist; b. Idalia, Colo., Nov. 22, 1919; s. Raymond Harris and Jennie Fern (Wooley) S.; m. Jennie A. Nixon, May 27, 1967. AA, Mesa Coll., 1948; DMD, Oreg. Health Scis. U., 1952. Pvt. practice dentistry Milwaukie, Oreg., 1952—; cons. forensic odontology; tchr., rschr. U. Oreg. Dental Sch., 1953-54, 55-57. Ret. engr. happy Valley Fire Dept.; mem. Clackamas County Fire Cause Investigating Team, Clackamas County Sheriff (Spl.); area coord. Oreg. Hunter Edn. Program, Portland, 1989-92; mem. fire bd. Happy Valley, Oreg. With USN, 1941-47, USNR, 1947-51. Mem. Pacific NW Forensic Study Group, Am. Acad. Forensic Sci. (mem. del. to Soviet Union 1988), Am. Soc. Forensic Odontology, Masons. Home: 12150 SE 117th St Clackamas OR 97015 Office: 3336 SE Harrison St Milwaukie OR 97222

SCHAFFER, BRUCE ALAN, plant physiologist, educator; b. Newark, Sept. 3, 1956; s. Ephraim and Judith (Wasserman) S.; m. Pamela April Moon, Nov. 1, 1981. BS, Colo. State U., 1978, MS, 1981; PhD, Va. Tech., 1985. Asst. prof. Tropical Rsch. and Edn. Ctr. U. Fla., Homestead, 1985-89, assoc. prof., 1990—. Mem. editorial rev. bd. Tree Physiology Jour., 1988-90; assoc. editor Am. Soc. Hort. Sci., 1991—; contbr. over 40 articles to profl. jours. Mem. Internat. Soc. Hort. Sci., Am. Soc. Hort. Sci., Sigma Xi, Gamma Sigma Delta, Phi Sigma. Office: U Fla Tropical Rsch & Edn Ctr 18905 SW 280th St Homestead FL 33031

SCHAFFER, PRISCILLA ANN, virologist; b. Dec. 28, 1941. BS, Hobart and William Smith Coll., 1964; PhD, Cornell U. Med. Coll., 1969; MS, Harvard U., 1981. Postdoctoral fellow Baylor Coll. Medicine, Houston, 1969-71, asst. prof. dept. virology and epidemiology, 1971-76; assoc. prof. dept microbiology and molecular genetics Harvard Med. Sch., Boston, 1976-81, prof., 1981—; sci. adv. bd. Vira Chem. Inc., N.Y.C., 1987—; bd. trustees Hobart and William Smith Coll., Geneva, N.Y., 1985—. Mem. editorial bd. Jour. of Virology, Virology; contbr. over 100 articles to profl. jours. and numerous book chpts. on molecular virology. Bd. sci. councellors Nat. Inst. Allergy and Infectious Diseases, 1983-86. Recipient Merit award Nat. Cancer Inst., 1990-2000; grantee NIH, Am. Cancer Soc., NSF, 1977—; named Found. lectr. Am. Soc. for Mircrobiology, 1981, 91. Mem. AAAS, Am. Soc. for Microbiology, Am. Soc. for Virology, Soc. for Gen. Microbiology. Achievements include determination of the functions of herpevirus genes associated with productive infection, latency, and transformation; contribution to understanding of mechanisms of actions of clin. useful antiherpes drugs. Office: Dana-Farber Cancer Inst 44 Binney St Boston MA 02115-6084

SCHAFFLER, MITCHELL BARRY, research scientist, anatomist, educator; b. Bronx, N.Y., Apr. 10, 1957; s. Walter and Shirley (Balter) S. BS, SUNY, Stony Brook, 1978; PhD, W.Va. U., 1985. Rsch. fellow in radiobiology U. Utah, Salt Lake City, 1985-87; asst. prof. U. Calif., San Diego, 1987-90; assoc. prof. Case Western U., Detroit, 1990—; sect. head anatomy, Bone and Joint Ctr. Henry Ford Health Scis. Ctr., Detroit, 1990—; adj. assoc. prof. Anatomy U. Mich., Ann Arbor, 1990—. Mem. editorial bd. Jour. Orthopaedic Rsch.; contbr. articles to profl. jours. Grantee Whitaker Found., 1988. Mem. Am. Assn. Anatomists, Am. Assn. Phys. Anthropology, Orthopaedic Rsch. Soc., Sigma Xi, Phi Kappa Phi. Achievements include research in skeletal biology and biomechanics. Office: Case Western U/Henry Ford Health Scis Ctr Bone and Joint Ctr 2799 W Grand Blvd Detroit MI 48202

SCHAFFNER, LINDA CAROL, biological oceanography educator; b. Freeport, N.Y., Dec. 8, 1954; d. John Charles Schaffner and Shirley Garnet Voges Sanders; m. Stephen Marshall Bennett, Apr. 7, 1979; 1 child, William Schaffner. BA, Drew U., 1976; MA, Coll. of William and Mary, 1981, PhD, 1987. Asst. prof. Va. Inst. Marine Sci., Coll. of William and Mary, Gloucester Point, 1988—; vis scientist Swedish Environ. Protection Bd., 1988. Contbr. articles to profl. jours. Scholar Drew U., 1975-76, Houston Underwater Club, 1981, U. Wash., 1983; grantee NOAA, 1987-91, U.S. Fish and Wildlife Svc., 1991, NOAA-EPA, 1991-93, 93—, Office of Naval Rsch., 1993—. Mem. Assn. of Women in Sci., Atlantic Estuarine Rsch. Soc. (treas. 1988-90), Estuarine Rsch. Fedn., Am. Soc. Limnology and Oceanography. Office: Sch Marine Sci RR 8 Gloucester Point VA 23062

SCHAFRIK, ROBERT EDWARD, materials engineer, information technologist; b. Cleve., Feb. 6, 1946; s. Edward E. and Sylvia E. (Farina) S.; m. Mary L. Schuhmann, Sept. 21, 1968; children: Robert E., Catherine M. Spage, Frances S., Steven J. Aerospace Engr., Air Force Inst. Tech., Dayton, Ohio, 1974; Metall. Engr., Ohio State U., 1979. Registered profl. engr., Ohio. Commd. 2d lt. U.S. Air Force, 1968, advanced through grades to lt. col., 1984; div. chief air superiority Hdqrs. Air Force Systems Command, Andrews AFB, Md., 1984-87; div. chief Strategic Def. Initiative Office, Washington, 1987-88; ret. U.S. Air Force, 1988; v.p. R&D Technology Assessment and Transfer, Inc., Annapolis, Md., 1988-91; dir. Nat. Materials Adv. Bd., Washington, 1991—. Contbr. articles to profl. jours. Chair Bicentennial Commn., Huber Heights, Ohio, 1975-77. Mem. IEEE, Am. Soc. for Metals Internat., Soc. for Advancement of Materials and Process Engring. Achievements include exploratory development research on titanium aluminides; program management for Air Force industrial modernization programs; program management F-16 engine programs; program management for Air Force integrated computer-aided manufacturing program. Office: NMAB/NRC 2101 Constitution Ave NW Washington DC 20418

SCHAIBLE, ROBERT HILTON, biologist; b. Horton, Kans., Apr. 30, 1931; s. Harold M. and Helen Grace (McCauley) S.; m. Catherine Allison Hirn, Mar. 3, 1973; children: Laura Ann, David Kent. BS in Fish Mgmt., Colo. State U., 1953; MS, Iowa State U., 1959, PhD in genetics, 1963. Diplomate Am. Bd. Med. Genetics. Rsch. assoc. U. Kans., Lawrence, 1962-63; postdoctoral fellow Yale U., New Haven, 1963-64; asst. prof. N.C. State U., Raleigh, 1964-68; spl. rsch. fellow pub. health svc. Ind. U., Bloomington, 1968-70; asst. prof. Ind. U.-Purdue U., Indpls., 1970-73, Ind. U. Sch. Medicine, Indpls., 1973-89; environ. scientist Ind. Dept. Environ. Mgmt., Indpls., 1991—. 1st lt. U.S. Army, 1953-55. Recipient Grant Nat. Inst. Gen. Med. Scis., Bethesda, Md., 1972-75, George Krejci Meml. award The Neurofibromatosis Inst., La Crescenta, Calif., 1989. Mem. AAAS, Am. Genetic Assn., Am. Soc. Human Genetics, Sigma Xi. Methodist. Achievements include research on the clonal devel. of the pigmentary system of the integument of mammals. Home: 3640 Willsee Ln Plainfield IN 46168 Office: Ind Dept Environ Mgmt OER/Superfund PO Box 6015 Indianapolis IN 46206-6015

SCHALLER, GEORGE BEALS, zoologist; b. Berlin, May 26, 1933; s. Georg Ludwig S. and Bettina (Byrd) Iwersen; m. Kay Suzanne Morgan, Aug. 26, 1957; children: Eric, Mar.. BS. in Zoology, BA in Anthropology, U. Alaska, 1955; Ph.D. in Zoology, U. Wis., 1962. Research assoc. Johns Hopkins U., Balt., 1963-66; research zoologist N.Y. Zool. Soc., Bronx, 1966—, dir. internat. conservation program, 1979-88; adj. assoc. prof. Rockefeller U., N.Y.C., 1966—; research assoc. Am. Mus. Natural History. Author: The Mountain Gorilla, 1963 (Wildlife Soc. 1965), The Year of the Gorilla, 1964, The Deer and the Tiger, 1967, The Serengeti Lion, 1972 (Nat. Bookaward 1973), Golden Shadows, Flying Hooves, 1973, Mountain Monarchs, 1977, Stones of Silence, 1980, The Giant Pandas of Wolong, 1985, The Last Panda, 1993. Ctr. Advanced Study in Behavioral Scis. fellow, Stanford U., 1962; fellow Guggenheim Found., 1971; decorated Order of Golden Ark (The Netherlands); recipient Gold medal World Wildlife Fund, 1980. Mem. Explorers Club (hon. dir. 1991). Office: The Wild Conservation Soc Bronx Park Bronx NY 10462-2272

SCHALLER, JANE GREEN, pediatrician; b. Cleve., June 26, 1934; d. George and May Alice (Wing) Green; children: Robert Thomas, George Charles, Margaret May. A.B., Hiram (Ohio) Coll., 1956; M.D. cum laude, Harvard U., 1960. Diplomate Am. Bd. Pediatrics, Am. Bd. Med. Examiners. Resident in pediatrics Children's Hosp.-U. Wash., Seattle, 1960-63; fellow immunology and arthritis Children's Hosp.-U. Wash., 1963-65; mem. faculty U. Wash. Med. Sch., 1965-83, prof. pediatrics, 1975-83; head div. rheumatic diseases Children's Hosp., Seattle, 1968-83; prof., chmn. dept. pediatrics Tufts U. Sch. Medicine/New Eng. Med. Ctr., 1983—; pediatrician-in-chief Boston Floating Hosp., 1983—; vis. physician Med. Research Council, Taplow, Eng., 1971-72; bd. visitors Sch. of Medicine U. Pitts., 1989—. Author articles in field.; Editorial bds. profl. jours. Bd. dirs. Seattle Chamber Music Festival, 1982-85; trustee Boston Chamber Music Soc., 1985—; mem. Boston adv. coun. UNICEF. Mem. Inst. Medicine of NAS, AAAS (sci. and human rights program)), Soc. Pediatric Research, Am. Pediatric Soc., Am. Acad. Pediatrics (chmn. subcom. on children and human rights 1989—, com. on internat. child health 1990—), Am. Coll. Rheumatology, New Eng. Pediatric Soc. (pres. 1991-93), Assn. Med. Sch. Pediatric Chmn. (exec. com. 1986-89, rep. to council on govt. affairs and council of acad. socs.), Com. Health in So. Africa (exec. com. 1986—), Physicians for Human Rights (exec. com. 1986—, pres. 1986-89) Aesculapian Club (pres. 1988-89), Harvard U. Med. Sch. Alumni Council (v.p. 1977-80, pres. 1982-83), Internat. Rescue Com. (med. adv. com., women's commn. for refugee women and children), Mass. Women's Forum, Internat. Women's Forum, Tavern Club, Saturday Club. Office: Tufts U Sch Medicine New Eng Med Ctr Floating Hosp 136 Harrison Ave Boston MA 02111

SCHALLER, JOHN WALTER, electrical engineer; b. Stamford, N.Y., June 3, 1963; s. John Blazius and Elizabeth Joan (Ritchfield) S. BS in Microelectronics, Rochester Inst. Tech., 1986; MEE, Poly. Inst., Farmingville, N.Y., 1989. Engr. Standard Microsystems Corp., Happauge, N.Y., 1986-89; sr. engr. Western Digital Corp., Irvine, Calif., 1989-92; assoc. prin. engr. Western Digital Corp., Irvine, 1992—. Mem. Soc. Photo-Optical Instrumentation Engineers. Achievements include devel. of projection aligner optimization for I-line resist system, 1988, customized exposure bandwidth for I-line resist systems, 1989.

SCHALLES, JOHN FREDERICK, biology educator; b. Pitts., Mar. 14, 1949; s. James Walter and Helen Louise (Lowry) S.; m. Nancy Ruth Edwards, Aug. 26,1972; children: Matthew David, James Michael. BS in Biology, Grove City (Pa.) Coll., 1971; MS in Zoology, Miami U., Oxford, Ohio, 1973; PhD in Biology, Emory U., 1979. Instr. biology Emory U., Atlanta, 1978-79; assoc. prof. biology Creighton U., Omaha, Nebr., 1979—; biologist USDA Agrl. Rsch. Svc., Durant, Okla., 1992; vis. researher Savannah River Ecology Lab., Aiken, S.C., 1979—; cons. EPA, 1985—; rsch. assoc. CALMIT, U. Nebr. Lincoln, 1987—; dir. Creighton U. Environ. Sci. Program. Author, editor: Bilogical Diversity: Problems and Challenges, 1993; contbr. chapts. to Fresh Water Wetlands and Wildlife, 1989, Wetland Ecology and Conservation, 1989. Mem. Sierra Club, Omaha, 1980—, past officer; Bd. mem. MEPA Gifford Ednl. Farm, Omaha, 1981-89; advisor Omaha City Planning Dept., 1990-91. Recipient Grad. Rsch. fellowship, Miami U., Oxford, 1972-73. Summer Rsch. fellowship Creighton U., Omaha, 1983, '89; Faculty Rsch. award, U.S. Dept. Energy, 1987—; grantee W.S. EPA, 1991—. Mem. Am. Soc. Photgrammetry and Remote Sensing, Nebr. Acad. Sci. (exec. bd. 1990—), Am. Soc. Limnology and Oceanography, Ecol. Soc. Am., Sigma Xi (pres. Omaha, 1990-91), Phi Sigma. Achievements include rsch. in the biol. productivity, community structure, geochemistry and hydrology of freshwater wetlands; rsch. achievments in ecol. application of remote sensing. Office: Creighton U Biology Dept Omaha NE 68178-0103

SCHALLY, ANDREW VICTOR, biochemist, researcher; b. Poland, Nov. 30, 1926; came to U.S., 1957; s. Casimir Peter and Maria (Lacka) S.; m. Maria Comaru, Aug. 1976; children: Karen, Gordon. B.Sc., McGill U., Can., 1955, Ph.D. in Biochemistry, 1957; 16 hon. doctorates. Research asst. biochemistry Nat. Inst. Med. Research, London, 1949-52; dept. psychiatry McGill U., Montreal, Que., 1952-57; research assoc., asst. prof. physiology and biochemistry Coll. Medicine, Baylor U., Houston, 1957-62; assoc. prof. Tulane U. Sch. Medicine, New Orleans, 1962-67, prof., 1967—; chief Endocrine Polypeptide and Cancer Inst. VA Med. Ctr., New Orleans; sr. med. investigator VA, 1973—. Author several books; contbr. articles to profl. jours. Recipient ; Van Meter prize Am. Thyroid Assn., 1969; Ayerst-Squibb award Endocrine Soc., 1970; William S. Middletown award VA, 1970; Ch. Mickle award U. Toronto, 1974; Gairdner Internat. award, 1974; Borden award Assn. Am. Med. Colls. and Borden Co. Found., 1975; Lasker Basic Research award, 1975; co-recipient Nobel prize for medicine, 1977; USPHS sr. research fellow, 1961-62. Mem. NAS, AAAS, Endocrine Soc., Am. Physiol. Soc., Soc. Biol. Chemists, Soc. Exptl. Biol. Medicine, Internat.

Soc. Rsch. Biology Reprodn., Soc. Study Reprodn., Soc. Internat. Brain Rsch. Orgn., Mex. Acad. Medicine, Am. Soc. Animal Sci., Nat. Medicine Brazil, Acad. Medicine Venezuela, Acad. Sci. Hungary, Acad. Sci. Russia. Home: 5025 Kawanee Ave Metairie LA 70006-2547 Office: VA Hosp 1601 Perdido St New Orleans LA 70146-3301

SCHALON, CHARLES LAWRENCE, psychologist; b. Greenville, Ohio, June 2, 1941; s. Arthur A. and Louise (Lutovsky) S.; m. Kristin L. Koontz, Aug. 1, 1964; children: Jill, Diane. Student, Ripon (Wis.) Coll., 1959-62; BS, U. Iowa, 1963, MA, 1965, PhD, 1966. Diplomate Internat. Acad. Profl. Counseling and Psychotherapy; lic. psychologist, Kans. Sr. psychologist Scott County Mental Health Ctr., Davenport, Iowa, 1970-75; clin. psychologist Wichita Psychiat. Ctr., 1976-84; clin. psychologist, ptnr. Wichita Clinic, PA, 1984—. Capt. U.S. Army, 1966-69. Mem. Am. Psychol. Assn., Kans. Psychol. Assn., Kans. Assn. Profl. Psychologists (exec. v.p. 1984, Pres. award 1984), Nat. Register Health Svc. Providers in Psychology. Office: Wichita Clinic PA 3311 E Murdock St Wichita KS 67208-3054

SCHAPIRO, JEROME BENTLEY, chemical company executive; b. N.Y.C., Feb. 7, 1930; s. Sol and Claire (Rose) S.; B.Chem. Engring., Syracuse U., 1951; postgrad. Columbia U., 1951-52; m. Edith Irene Kravet, Dec. 27, 1953; children: Lois, Robert, Kenneth. Project engr. propellents br. U.S. Naval Air Rocket Test Sta., Lake Denmark, N.J., 1951-52; with Dixo Co., Inc., Rochelle Park, N.J., 1954—, pres., 1966—; lectr. detergent standards, drycleaning, care labeling, consumers standards, orgns., U.S., 1968—; U.S. del. spokesman on drycleaning Internat. Standards Orgn., Newton, Mass., 1971, Brussels, 1972, U.S. del. spokesman on dimensional stability of textiles, Paris, 1974, Ottawa, 1977, Copenhagen, 1981; chmn. U.S. del. com. on consumer affairs, Geneva, 1974, 75, 76, spokesman U.S. del. on textiles, Paris, 1974, mem. U.S. del. on care labeling of textiles, The Hague, Holland, 1974, U.S. del., chmn. del. council com. on consumer policy, Geneva, 1978, 79, 82, Israel, 1980, Paris, 1981; leader U.S. del. com. on dimensional stability of textiles, Manchester, Eng., 1984; fed. govtl. appointee to Industry Functional Adv. Com. on Standards, 1980-81. Mem. Montclair (N.J.) Sch. Study Com., 1968-69; co-founder Jewish Focus, Inc., 1991, pub. Sullivan/Ulster Jewish Star. 2d lt. USAF, 1952-53. Mem. Am. Inst. Chem. Engrs., Am. Nat. Standards Inst. (vice chmn. bd. dirs., 1983-85 , exec. com. 1979-81, 83-85 , bd. dir. 1979-85, fin. com. 1982-85 , chmn. consumer council 1976, 79, 80, 81, mem. steering com. to advise Dept. Commerce on implementation GATT agreements 1976-77, mem. exec. standards coun., 1977-79), internat. standards coun., chmn. internat. consumer policy adv. com. 1978-86), Am. Assn. Textile Chemists and Colorists (mem. exec. com. on rsch. 1974-77, chmn. com. on dry cleaning 1976-88, vice chmn. internat. test methods com.) Am. Chem. Soc., Standards Engring. Soc. (cert.), ASTM (Fellow 1970, chmn. com. D-12 Soaps and Detergents, 1974-79, mem. standing com. on internat. standards 1980-84, hon. mem. award com. D-13, textiles), Internat. Standards Orgn. (mem. internat. standards steering com. for consumer affairs 1978-81), Nat. Small Bus. Assn. (assoc. trustee 1983-85). Jewish (v.p., treas. temple). Lodge: Masons. Home: PO Box 771 Wurtsboro NY 12790-0771 Office: 158 Central Ave PO Box 7038 Rochelle Park NJ 07662-4003

SCHARFF, MATTHEW DANIEL, immunologist, cell biologist, educator; b. N.Y.C., Aug. 28, 1932; s. Harry and Constance S.; m. Carol Held, Dec. 19, 1954; children—Karen, Thomas, David. A.B., Brown U., 1954; M.D., N.Y. U., 1959. House officer II and IV med. service Boston City Hosp., 1959-61; research asso. NIH, 1961-63; asst. prof. Albert Einstein Coll. Medicine, Yeshiva U., Bronx, N.Y., 1963-67; asso. prof. Albert Einstein Coll. Medicine, Yeshiva U., 1967-71, prof. dept. cell biology, 1971—, chmn. dept., 1972-83, dir. div. biol. scis., 1975-81; asso. dir. Cancer Center, 1975-86, dir., 1986—. Served with USPHS, 1961-63. Recipient Alumni Achievement award N.Y. U. Sch. Medicine, 1980. Mem. Am. Assn. Immunologists, Am. Soc. Clin. Investigation, Nat. Acad. Scis., Am. Acad. Arts and Sci., Phi Beta Kappa, Sigma Xi, Alpha Omega Alpha. Office: Cancer Rsch Ctr Albert Einstein Coll Med 1300 Morris Park Ave Bronx NY 10461-1924

SCHARPING, BRIAN WAYNE, mechanical engineer; b. Wichita Falls, Tex., Oct. 16, 1966; s. Stanley Howard and Ruby Marie (Stuchlik) S.; m. Jan Marlane Berggren, July 21, 1990. BSME, Kans. State U., 1990. Thermography coord. Commonwealth Edison Co., Zion, Ill., 1990—, vibration coord., 1991, safety related snubber coord., 1992—; tour guide, presenter Commonwealth Edison Co., Zion, 1991—. Home: 6337 59 Ave Kenosha WI 53142 Office: Commonwealth Edison Co 101 Shiloh Blvd Zion IL 60099

SCHARRER, BERTA VOGEL, anatomy and neuroscience educator; b. Munich, Dec. 1, 1906; d. Karl and Johanna V.; widowed. PhD in Zoology, U. Munich, 1930; MD (hon.), U. Giessen, Germany, 1976, U. Frankfurt, Germany, 1992; DSc (hon.), Northwestern U., 1977, U. N.C., 1978, Smith Coll., 1980, Harvard U., 1982, Yeshiva U., 1983, Mt. Holyoke Coll., 1984, SUNY, 1985, U. Salzburg, Austria, 1988; LLD, U. Calgary, Alta., Can., 1982. Research assoc. Research Inst. for Psychiatry, Munich, 1931-34, Neurol. Inst., Frankfurt-am-Main, 1934-37, U. Chgo. Dept. Anatomy, 1937-38, Rockefeller Inst., N.Y.C., 1938-40; instr., fellow Western Res. U. Dept. Anatomy, Cleve., 1940-46; John Guggenheim fellow U. Colo. Dept. Anatomy, Denver, 1947-48, spl. USPHS research fellow, 1948-50; asst. prof. (research) dept. anatomy U. Colo. Sch. Medicine, Denver, 1950-55; prof. anatomy Albert Einstein Coll. Medicine, 1955-77, acting chmn., 1965-67, 76-77, disting. prof. emeritus anatomy and neurosci., 1978—. Recipient Kraepelin Gold medal, 1978, F.C. Koch award Endocrine Soc., 1980, Nat. Medal Sci., 1983. Mem. NAS, Am. Acad. Arts and Scis., Deutsche Acad. Naturforscher Leopoldina (Schleiden medal 1983), Am. Assn. Anatomists (pres. 1978-79, Henry Gray award 1982), Am. Soc. Zoologists (hon. mem.), Soc. Neurosci., Endocrine Soc. (F.C. Koch award 1980). Achievements include rsch. in comparative neuroendocrinology, neurosecretion, neuroimmunology, neuropeptides. Home: 1240 Neill Ave Bronx NY 10461-1736 Office: Albert Einstein Coll Med Dept Anatomy 1300 Morris Park Ave Bronx NY 10461-1924

SCHATTEN, GERALD PHILLIP, cell biologist, educator; b. N.Y.C., Nov. 1, 1949; s. Frank and Sylvia Schatten. BS, U. Calif., Berkeley, 1971, PhD, 1975. Instr. U. Calif., Berkeley, 1975; postdoctoral fellow Rockefeller Found., 1976-77; successively asst. prof., assoc. prof., prof. Fla. State U., Tallahassee, 1979-86; prof. molecular biology and zoology U. Wis., Madison, 1986—, dir. integrated microscopy resource for biomedical rsch., 1986—; dir. gamete and embryo biol. tng. program, U. Wis., Madison, 1998—. Recipient Rsch. Career Devel. award NIH, 1981-86. Office: Univ Wis 1117 W Johnson St Madison WI 53706-1797

SCHATTEN, KENNETH HOWARD, physicist; b. N.Y.C., Mar. 1, 1944; s. Frank and Sylvia (Cohen) S.; m. Sharon Rappé Long, May 29, 1988; children: Lynnette, Miranda, Stephanie. BS, MIT, 1964; PhD, U. Calif., Berkeley, 1968. Rsch. physicist NASA, Greenbelt, Md., 1969-72, 79—; program dir. NSF, Washington, 1991—; sr. lectr. Victoria U., Wellington, New Zealand, 1972-76; sr. rsch. assoc. Stanford U., Palo Alto, Calif., 1977-78, Boston U., 1978-79; workshop leader; conf. organizer; U.S. del. solar activity studies, NAS, 1991. Contbr. articles to sci. jours. NSF grad. fellow, 1968. Office: NASA/GSFC Code 910 Greenbelt MD 20771

SCHATZEL, ROBERT MATHEW, logistics engineer; b. Bklyn., July 29, 1961; s. Richard William and Joan Matilda (Gehrken) S.; m. Tracy Christine Raines, Apr. 29, 1989. BS in Ocean Engring., Fla. Inst. Tech., 1984. Cert. profl. logistician. Logistic engr. I Newport News (Va.) Shipbuilding, 1984-87, logistic engr. II, 1987-89, logistic engr. III, 1989-90, logistic engring. supr., 1990—; mem. DOD/Industry Cals Concurrent Engring. Task Group, 1988—. Mem. Soc. Logistics Engrs. (vice chmn. logistics edn. found. 1991-92). Methodist. Achievements include research in reliability and maintainability engineering analysis integrations into the concurrent engineering design process. Home: 6241 Adams Hunt Dr Williamsburg VA 23188 Office: Newport News Shipbuilding Bldg 800 4101 Washington Ave Newport News VA 23607

SCHAUB, FRED S., mechanical engineer. BSME, Antioch Coll., 1952. Lic. profl. engr., Ohio. With Cooper-Bessemer divsn. Cooper Energy Svcs., Mt. Vernon, Ohio, 1952—, mgr. rsch. and devel., 1968—; Cooper reciprocating rep. Diesel Engine Mfrs. Assn. Contbr. tech. papers to profl. jours. Recipient Internal Combustion Engine award ASME, 1991, 2 Outstanding Paper awards. Mem. Combustion Inst. Achievements include patents for NOx Reduction for Spark-ignited Engines, NOx Reduction for Dual Fuel Engines; research in cylinder-scavenging and combustion-diagnostic techniques. Office: Cooper Energy Services Cooper Bessemer Division North Sandusky St Mount Vernon OH 43050*

SCHAUMBURG-LEVER, GUNDULA MARIA, dermatologist; b. Berlin, Feb. 10, 1942; d. Harald Ernst and Freia (Hellhoff) Schaumburg; m. Walter Frederick Lever, May 10, 1971; children: Insa Bettina, Mark Alexander. MD, Kiel U., 1969. Intern Karlsruhe (Germany) City Hosp., 1969-70; clin. and rsch. fellow Tufts-New Eng. Med. Ctr., Boston, 1971-74; asst. prof. Sch. Medicine Tufts U., Boston, 1971-83; rsch. asst. dept. dermatology Free U., Berlin, 1983-84; head of dermatopathology lab. dept. dermatology Eberhard Karl U., Tübingen, Germany, 1985—. Author: (with W.F. Lever) Histopathology of the Skin, 7th edit., 1990, Color Atlas of Histopathology of the Skin, 1988. Fellow Am. Acad. Dermatology; mem. Am. Soc. Dermatopathology, European Soc. Dermatol. Rsch., Japanese Soc. Investigative Dermatology, Soc. Investigative Dermatology, Soc. Cutaneous Ultrastructural Rsch., Internat. Soc. Dermatopathology, Dermatology Found., Women's Dermatology Soc., German Dermatol. Soc., Am. Orchid Soc., German Orchid Soc. Avocation: orchidist. Home: Im Kleeacker 29, D-772072 Tübingen Germany Office: Universitäts-Hautklinik, Calwerstr 7, D-72076 Tübingen Germany

SCHAUSS, ALEXANDER GEORGE, psychologist, researcher; b. Hamburg, Fed. Republic of Germany, July 20, 1948; came to U.S., 1953; s. Frank and Alla (Demjanov) S.; m. Laura Babin; children: Nova, Evan. BA, U. N.Mex., 1970, MA, 1972; PhD, Calif. Coast U., 1992. State probation/parole officer 2nd Judicial Dist. Ct., Albuquerque, 1969-73; criminal justice planner Albuquerque/Bernalillo County Criminal Justice Planning Com., 1973-75; state asst. adminstr. dept. corrections State of S.D., Pierre, 1975-77; dir. Pierce County Probation Dept., Tacoma, Wash., 1977-78; tng. officer IV Wash. State Criminal Justice Tng. Commn., Olympia, 1978-79; dir. Inst. Biosocial Rsch. City Univ. Grad. Sch., Seattle, 1979-80; exec. dir. Am. Inst. Biosocial Rsch. Inc., Tacoma, 1980-93, Am. Preventive Med. Assn., 1992—; mem., WHO Study Group on Health Promotion, Copenhagen, 1985; vis. lectr. pediatrics The John Radcliffe Hosp., Oxford U., England, summer 1985; sec. coun. on food policy Nat. Assn. Pub. Health Policy, 1990—; vis. scholar Kans. Community Coll. Consortium, 1982; vis. lectr. McCarrison Soc. Conf. at Oxford U., 1983. Author: Diet, Crime and Delinquency, 1980, rev., 1992, Nutrition and Behavior, 1986, Nutrition and Criminal Behavior, 1990; co-author: Zinc and Eating Disorders, 1989, Eating For A's, 1991; contbr. articles to profl. jours.; editor-in-chief Internat. Jour. Biosocial and Med. Rsch., 1979—; mem. editorial bd. 4 jours., 1979—. Master arbitrator Tacoma/Pierce County Better Bus. Bur., Tacoma, 1986—; mem. Pierce County N. Area Transp. Adv. Coun., Tacoma, 1991—; trustee Faith Homes Puget Sound Area Charity, 1991—; mem. Pierce County Pub. Safety Task Team, 1993. Recipient Rsch. award Wacker Found., 1983-85, 88; fellow Am. Coll. Nutrition, 1986-87. Fellow Am. Orthopsychiat. Assn.; mem. Brit. Soc. Nutritional Medicine (hon.), Am. Nutritionists Assn., Assn. Chemoreception Scis., Internat. Assn. Eating Disorders Profls., Am. Assn Correctional Psychologists, Am. Found. Preventative Medicine (treas. 1992), Acad. Criminal Justice Scis., Am. Soc. Criminology, Rotary (chmn. community scs. com. Tacoma chpt. 1989-90, chmn. civic affairs com. 1989-90, mem. Vladivostok com. 1991-93), N.Y. Acad. Scis. (emeritus.). Achievements include discovery of "Baker-Miller Pink" effect on muscular function and aggressivity; research in the relationship between zinc status and anorexia nervosa and bulimia nervosa, the relationship between diet and trace elements and criminal behavior. Office: Am Inst Biosocial Rsch Inc Divsn Life Scis PO Box 1174 Tacoma WA 98401-1174

SCHAWLOW, ARTHUR LEONARD, physicist, educator; b. Mt. Vernon, N.Y., May 5, 1921; s. Arthur and Helen (Mason) S.; m. Aurelia Keith Townes, May 19, 1951; children: Arthur Keith, Helen Aurelia, Edith Ellen. BA, U. Toronto, Ont., Can., 1941, MA, 1942, PhD, 1949, LLD (hon.), 1970; DSc (hon.), U. Ghent, Belgium, 1968, U. Bradford, Eng., 1970, U. Ala., 1984, Trinity Coll., Dublin, Ireland, 1986; DTech (hon.), U. Lund, Sweden, 1987; DSL (hon.), Victoria U., Toronto, 1993. Postdoctoral fellow, research asso. Columbia, 1949-51; vis. assoc. prof. Columbia U., 1960; research physicist Bell Telephone Labs., 1951-61, cons., 1961-62; prof. physics Stanford U., 1961-91, also J.G. Jackson-C.J. Wood prof. physics, 1978, prof. emeritus, 1991—, exec. head dept. physics, 1966-70, acting chmn. dept., 1973-74. Author: (with C.H. Townes) Microwave Spectroscopy, 1955; Co-inventor (with C.H. Townes), optical maser or laser, 1958. Recipient Ballantine medal Franklin Inst., 1962, Thomas Young medal and prize Inst. Physics and Phys. Soc., London, 1963, Schawlow medal Laser Inst. Am., 1982, Nobel prize in physics, 1981, U.S. Nat. Medal of Sci. NSF, 1991; named Calif. Scientist of Yr., 1973, Marconi Internat. fellow, 1977. Fellow Am. Acad. Arts and Scis., Am. Phys. Soc. (coun. 1966-70, chmn. div. electron and atomic physics 1974, pres. 1981), Optical Soc. Am. (hon. mem. 1983, dir.-at-large 1966-68, pres. 1975, Frederick Ives medal 1976); mem. NAS, IEEE (Liebmann prize 1964), AAAS (chmn. physics sect. 1979), Am. Philos. Soc., Royal Irish Acad. (hon.). Office: Stanford U Dept Physics Stanford CA 94305-4060

SCHECHTER, PAUL J., pharmacologist; b. Bklyn., Aug. 9, 1939; s. Harry and Frances (Maybloom) S.; m. Barbara F. Roth, Dec. 9, 1967; children: Rebecca, Christopher. BS, Columbia U., 1960; PhD, U. Chgo., 1966, MD, 1968. Rsch. asst. U. Chgo., 1966-68; intern Presbyn.-St. Luke's Hosp., Chgo., 1968-69; rsch. assoc. NIH, Bethesda, Md., 1969-71, clin. investigator, 1971-72; jr. asst. resident Georgetown U. Hosp., Washington, 1972-73; chief exptl. therapist Merrell Internat. Rsch. Ctr., Strasbourg, France, 1973-78, dir. clin. rsch., 1978-84; v.p. internat. clin. pharm. Merrell Dow Rsch. Inst., Strasbourg, 1984-88; v.p. clin. rsch. Merrell Dow Rsch. Inst., Cin., 1988-90; v.p. rsch. and devel. Fujisawa Pharm. Co., Deerfield, Ill., 1990—. Contbr. articles, abstracts to profl. publs.; patentee in field. Lt. comdr. USPHS, 1969-72. Recipient Clin. Pharmacol. award Internat. League Against Epilepsy. Mem. numerous profl. orgns. Avocations: tennis, skiing, music. Office: Fujisawa Pharm Co 3 Parkway N Deerfield IL 60015-2537

SCHECHTER, ROBERT SAMUEL, chemical engineer, educator; b. Houston, Feb. 26, 1929; s. Morris Samuel and Helen Ruth S.; m. Mary Ethel Rosenberg, Feb. 15, 1953; children: Richard Martin, Alan Lawrence, Geoffrey Louis. B.S. in Chem. Engring, Tex. A&M U., 1950; Ph.D. in Chem. Engring, U. Minn., 1956. Registered profl. engr., Tex. Asst. prof. chem. engring. U. Tex. at Austin, 1956-60, assoc. prof., 1960-63, prof., 1963—; adminstrv. dir. Ctr. Statis. Mechs. and Thermodynamics, 1968-72, chmn. dept. chem. engring., 1970-73, chmn. petroleum engring., 1975-78, E.J. Cockrell, Jr. prof. chem. and petroleum engring., 1975-81, Dula and Ernie Cockrell prof. engring., 1981-83, Getty prof. engring., 1984-85, Getty Oil Centennial chair in Petroleum Engring., 1985-89, W.A. (Monty) Moncrief Centennial Endowed chair in Petroleum Engring., 1989—; vis. prof. U. Edinburgh, Scotland, 1965-66; Disting. vis. prof. U. Kans., spring 1968; vis. prof. U. Brussels, 1969; Disting. Lindsay lectr. Tex. A&M U., 1993; cons. in field. Author: Variational Method in Engineering, 1967, (with G.S.G. Beveridge) Optimization—Theory and Practice, 1970, Adventures in Fortran Programming, 1975, (with B.B. Williams and J.L. Gidley) Acidizing Monograph, 1979, (with D.D. Shah) Enhanced Oil Recovery by Surfactants and Polymers, 1979; (with Maurice Bourrel) Microemulsions and Related Systems, 1988, Oil Well Stimulation, 1991; contbr. (with D.D. Shah) numerous articles to profl. jours. Served to 1st lt., Chem. Corps AUS, 1951-53. Decorated Chevalier Order Palmes Academique; recipient Outstanding Teaching award U. Tex., 1969, Outstanding Paper award, 1973, Gen. Dynamics award for Excellence in Engring. Teaching, Gen. Dynamics Corp., 1987, Sr. Rsch. award Engring. Rsch. Coun. of Am. Soc. Engring. Educators, 1991; Katz lectr., 1979; Disting. Lindsay lectr., 1993. Mem. AIME, Am. Inst. Chem. Engrs., Am. Chem. Soc., Soc. Petroleum Engrs., Nat. Acad. Engrs., Sigma Xi, Tau Beta Pi. Developer methods of measuring surface viscosity and ultra low inter-facial tensions; discoverer instability of thermal diffusion. Home: 4700 Ridge Oak Dr Austin TX 78731-4724

SCHEELEN, ANDRÉ JOANNES, chemical researcher; b. Bree, Belgium, July 2, 1964; s. Joannes Joseph and Marie (Schepers) S. M in Chemistry, Cath. U., Leuven, Belgium, 1986; B in Econs., Limburg U. Ctr., Diepenbeek, Belgium, 1989; PhD in Phys. Chemistry, Cath. U., Leuven, 1990. Researcher Cath. U. Leuven and Belgian Nat. Sci. Found., Leuven, 1986-90, Solvay & Cie SA, Brussels, 1991—. Contbr. articles to profl. jours. Mem. Jr. Chambers Internat., Brussels, Belgium, 1990. Recipient award Royal Acad. of Scis., Letters and Fine Arts of Belgium, 1990, award in chemistry and tech. Dutch State Mines, The Netherlands, 1990. Roman Catholic. Avocation: photography. Office: Solvay & Cie, Ransbeekstraat 310, 1120 Brussels Belgium

SCHEIBEL, ARNOLD BERNARD, psychiatrist, educator, researcher; b. N.Y.C., Jan. 18, 1923; s. William and Ethel (Greenberg) S.; m. Madge Mila Ragland, Mar. 3, 1950 (dec. Jan. 1977); m. Marian Diamond, Sept. 1982. B.A., Columbia U., 1944, M.D., 1946; M.S., U. Ill., 1952. Intern Mt. Sinai Hosp., N.Y.C., 1946-47; resident psychiatry Barnes and McMillan Hosp., St. Louis, 1947-48, Ill. Neuropsychiat. Inst., Chgo., 1950-52; asst. prof. psychiatry and anatomy U. Tenn. Med. Sch., 1952-53, assoc. prof., 1953-55; assoc. prof. UCLA Med. Center, 1955-67, prof. psychiatry and anatomy, 1967—, mem. Brain Rsch. Inst., 1960—, acting dir. Brain Rsch. Inst., 1987-90, dir., 1990—; cons. VA hosps., Los Angeles, 1956—. Contbr. numerous articles to tech. jours, chpts. to books.; editorial bd.: Brain Research, 1967-77, Developmental Psychobiology, 1968—, Internat. Jour. Neurosci., 1969—, Jour. Biol. Psychiatry, 1968—, Jour. Theoretical Biology, 1980—. Mem. Pres.'s Commn. on Aging, Nat. Inst. Aging, 1980—. Served with AUS, 1943-46; from lt. to capt. M.C. AUS, 1948-50. Guggenheim fellow (with wife), 1953-54, 59. Fellow Am. Acad. Arts and Scis., Norwegian Acad. Scis.; mem. AAAS, Am. Psychiat. Assn. (lifetime fellow), Am. Neurol. Assn., Soc. Neurosci., Psychiat. Research Assn., Am. EEG Assn., Am. Assn. Anatomists, Soc. Biol. Psychiatry, Am. Acad. Neurology, So. Calif. Psychiat. Assn. Home: 16231 Morrison St Encino CA 91436-1331 Office: UCLA Ctr for Health Sciences Rm 73364 Los Angeles CA 90024

SCHEINBERG, LABE CHARLES, physician, educator; b. Memphis, Dec. 11, 1925; s. Jacob and Ardie (Cohen) S.; m. Louise Goldman, Jan. 6, 1952; children—Susan, David, Ellen, Amy. A.B., U. N.C., 1945; M.D., U. Tenn., 1948. Intern Wesley Meml. Hosp., Chgo., 1949; resident psychiatry Elgin (Ill.) State Hosp., 1950; resident, asst. neurology Neurol. Inst., N.Y., 1952-56; mem. faculty Albert Einstein Coll. Medicine, 1956-93, prof. neurology, asst. dean, 1968-69, assoc. dean, 1969-70, prof. rehab. medicine and psychiatry, dean, 1970-72; dir. neurology Hosp., 1966-73; dir. dept. neurology and psychiatry St. Barnabas Hosp., Bronx, N.Y., 1974-79; prof. neurology Mt. Sinai Med. ctr., N.Y.C., 1993—. Cons. editor: N.Y. Acad. Scis, 1964, 84, editor-in-chief J. Neurologic Rehab., Rehab. Reports, Multiple Sclerosis Research Reports. Served as capt. M.C. USAF, 1951-52. Fellow Am. Acad. Neurology; mem. Am. Neurol. Assn., Am. Assn. Neuro-pathology, Am. Soc. Exptl. Pathology, Phi Beta Kappa, Alpha Omega Alpha. Home: 9 Oak Ln Scarsdale NY 10583-1621 Office: Mt Sinai Med Ctr 5 E 98th St New York NY 10029-6574

SCHEIRMAN, WILLIAM LYNN, chemical engineer; b. Kingfisher, Okla., Nov. 17, 1921; s. William R. and Stella B. (Lindsey) S.; m. Marian Joyce Thomson, Oct. 6, 1951; children: David, John, Margaret, Kathleen. BSME, Okla. State U., 1943, postgrad., 1954-56; postgrad., Kans. U., 1976-78, Avila Coll., 1987. Registered profl. engr., Okla. Liaison engr. Douglas Aircraft, Tulsa, 1943-44; chief product engr. Black, Sivalls & Bryson, Inc., Oklahoma City, Okla., 1947-68; sr. project engr. Rhodes Tech. Corp., Oklahoma City, 1968-69; sr. process engr. The Prichard Corp., Overland Park, Kans., 1969-92; cons. Overland Park, Kans., 1992-93; lectr. in field. Contbr. articles to profl. mags. and jours. including Oil and Gas Jour., Can. Oil and Gas Industries Mag., World Oil Mag., Erdol and Kohle (Germany), Oilweek, Hydrocarbon Processing; editor: (genealogical newsletter) Usu Leut. Asst. scoutmaster Boy Scout Troop 15, Oklahoma City, 1965-68. Jessie Ballard scholarship Okla. State U., 1941; recipient Product Engring Master Design award Product Engring. Mag., 1960. Mem. ASME (50 Yr. mem.), Am. Inst. Chem. Engrs., Kansas City Engrs. Club. Republican. Presbyterian. Achievements include 4 U.S. patents, 3 Canandian patents, 1 England patent i crude oil and natural gas processing equipment.

SCHELAR, VIRGINIA MAE, chemistry educator, consultant; b. Kenosha, Wis., Nov. 26, 1924; d. William and Blanche M. (Williams) S. BS, U. Wis., 1947, MS, 1953; MEd, Harvard U., 1962; PhD, U. Wis., 1969. Instr. U. Wis., Milw., 1947-51; info. specialist Abbott Labs., North Chgo., Ill., 1953-56; instr. Wright Jr. Coll., Chgo., 1957-58; asst. prof. No. Ill. U., DeKalb, 1958-63; prof. St. Petersburg (Fla.) Jr. Coll., 1965-67; asst. prof. Chgo. State Coll., 1967-68; prof. Grossmont Coll., El Cajon, Calif., 1968-80; cons. Calif., 1981—. Author: Kekule Centennial, 1965; contbr. articles to profl. jours. Active citizens adv. coun. DeKalb Consol. Sch. Bd.; voters svc. chair League Women Voters, del. to state and nat. convs., judicial chair, election laws chair. Standard Oil fellow, NSF grantee; recipient Lewis prize U. Wis. Fellow Am. Inst. Chemists; mem. Am. Chem. Soc. (membership affairs com., chmn. western councilor's caucus, exec. com., councilor, legis. counselor, chmn. edn. com., editor state and local bulletins). Avocations: swimming, folk dancing. Office: 5702 Baltimore Dr Apt 282 La Mesa CA 91942-1665

SCHELD, WILLIAM MICHAEL, internist, educator; b. Middletown, Conn., Aug. 15, 1947; s. William Herman and Lucille Laverne (Houchens) S.; m. Susan Ella Vaughan, June 14, 1969; 1 child, Sarah Walker. BS, Cornell U., 1969, MD, 1973. Diplomate Am. Bd. Internal Medicine. Intern, then resident U. Va. Sch. Medicine, Charlottesville, 1973-76, fellow in infectious diseases, 1976-79, asst. prof., 1979-82, assoc. prof., 1982-88, prof., assoc. chair dept. infectious diseases, 1988—; vice-chair Inter-sci. Conf. on Antimicrobial Agents and Chemotherapy. Editor: Infections of the Central Nervous System, 1991; contbr. sci. articles to profl. publs., chpts. to books. Fellow ACP, Infectious Diseases Soc. Am.; mem. Am. Soc. Clin. Investigation, Alpha Omega Alpha. Achievements include research on meningitis and other central nervous system infections, bacterial endocarditis, others. Home: 2075 Earlysville Rd Earlysville VA 22936 Office: Univ Va Health Sci Ctr Box 385 Charlottesville VA 22908

SCHELL, ALLAN CARTER, electrical engineer; b. New Bedford, Mass., Apr. 14, 1934; s. Charles Carter and Elizabeth (Moore) S.; m. Shirley T. Sardineer; children: Alice Rosalind, Cynthia Anne. B.S., MIT, 1956, M.S.E.E., 1956, Sc.D., 1961; student, Tech. U. Delft, Netherlands, 1956-57. Research physicist Air Force Cambridge Research Labs., Bedford, Mass., 1956-76; Guenter Loeser Meml. lectr. Air Force Cambridge Research Labs., 1965; dir. electromagnetics directorate Rome Air Devel. Ctr., Bedford, 1976-87; chief scientist Hdqrs. USAF Systems Command, 1987-92; chief scientist, dep. dir. sci. and tech. Hdqrs. USAF Materiel Command, 1992—; dir. Electro; vis. assoc. prof. MIT, 1974; chair dept. of elec. engring. adv. coun. U. Pa., 1992—. Contbr. articles to profl. jours.; patentee in field (9). Served as lt. USAF, 1958-60. Recipient Fulbright award, 1956-57, Meritorious Exec. award U.S. Govt., 1989; NSF fellow, 1955-56, 60-61. Fellow IEEE (bd. dirs. 1981-82, editor IEEE Press 1976-79, Proc. of IEEE 1990-92); mem. AIAA, Am. Def. Preparedness Assn., Antennas and Propagation Soc. of IEEE (pres. 1978, editor trans. 1969-71, John T. Bolljahn award 1966), Internat. Sci. Radio Union (U.S. nat. com.), Sigma Xi, Tau Beta Pi. Office: USAF Material Command Wright-Patterson AFB OH 45433

SCHELL, FARREL LOY, transportation engineer; b. Amarillo, Tex., Dec. 14, 1931; s. Thomas Phillip and Lillian Agnes (McKee) S.; m. Shirley Anne Samuelson, Feb. 6, 1955; children: James Christopher, Maria Leslyn Schell Peter. BS, U. Kans., 1954; postgrad., Carnegie-Mellon U., 1974. Registered profl. engr., Calif., Colo. Resident engr. Sverdrup & Parcel, Denver, 1957-61; project engr. Bechtel Corp., San Francisco, 1961-62, Parsons, Brinckerhoff-Tudor-Bechtel, San Francisco, 1962-67; mgr. urban transp. dept. Kaiser Engrs., Oakland, Calif., 1967-78; program dir. San Francisco Mcpl. Rwy I.C., 1978-80; project mgr. Houston Transit Cons., 1980-83, Kaiser Transit Group, Miami, 1983-85; mgr. program devel. Kaiser Engrs., Oakland, 1985-87; project mgr. O'Brien-Kreitzberg & Assocs., San Francisco, 1987-89; sr. project mgr. Bay Area Rapid Transit Dist., Oakland, Calif., 1989—; dir. Schelter Devel. Corp., Piedmont, Calif., 1982—. Contbr. articles to profl. jours. Lt. (j.g.) USN, 1954-57, PTO. Mem. ASCE, ASME, Nat. Soc. Profl. Engrs., Nat. Coun. Engring. Examiners, Am. Planners Assn., Am. Pub. Transit Assn., Lakeview Club, Scarab Club, Pachacamac Club, Sigma Tau, Tau Beta Pi. Avocations: fly fishing, camping. Home: 24 York Dr Piedmont CA 94611-4123 Office: Bay Area Rapid Transit Dist 800 Madison St Oakland CA 94607-4730

SCHELLENBERG, GERARD DAVID, geneticist, researcher; b. Reedly, Calif., Sept. 29, 1951; s. Ernest David and Clara (Hagen) S.; m. Mary Teresa Ensek, Aug. 23, 1986; children: Zachary David, Sierra Marie. BS in Biochemistry, U. Calif., Riverside, 1973, PhD in Biochemistry, 1978. Rsch. asst. biochemistry U. Calif., Riverside, 1973-78; sr. rsch. fellow med. genetics U. Wash., Seattle, 1978-79, sr. rsch. fellow neurology, 1979-82, sr. rsch. fellow genetics, 1982-83, rsch. asst. prof. neurology, 1983-90, rsch. assoc. prof. neurology, 1990—; mem. adv. bd. Jour. Neural Transmission, 1989—; mem. med./sci. adv. bd. Alzheimer's Assn., Chgo., 1990—; mem. Alzheimer's disease med./sci. adv. bd. Am. Health Assistance Found., 1992—. Achievements include identification of chromosome 14 Alzheimer's disease locus. Home: 7031 19th NW Seattle WA 98117 Office: U of Washington School of Medicine Neurology Div Seattle WA 98195

SCHELLHAAS, LINDA JEAN, scientist, consultant; b. South Haven, Mich., Apr. 27, 1956; d. Richard Louis and Virgene Frieda (Lietzke) Plankenhorn; m. Robert Wesley Schellhaas, May 27, 1990. BA in Biology, Albion Coll., 1978. Pathology rsch. asst. Internat. R&D Corp., Mattawan, Mich., 1978-80; toxicology rsch. coord. Borriston Labs., Inc., Temple Hills, Md., 1980-84; quality assurance coord. Tegeris Labs., Inc., Temple Hills, Md., 1984-85; good lab. practice compliance specialist, staff scientist Dynamac Corp., Rockville, Md., 1985-90; pres., regulatory compliance specialist Quality Reviews, Inc., Falling Waters, W.Va., 1990—; dir. quality assurance Pathology Assocs., Inc., Frederick, Md., 1992—; instr. regulatory compliance tng. seminars, 1990—. Contbr. articles to profl. jours. Mem. Soc. Quality Assurance, Albion Coll. Fellows, Pi Beta Phi, Phi Beta Kappa. Avocations: sheep and goat husbandry, raising sheep-herding dogs, needlework. Home: 1204 Berkeley Dr Falling Waters WV 25419 Office: Quality Reviews Inc PO Box 755 Falling Waters WV 25419

SCHELLHAAS, ROBERT WESLEY, data processing executive, composer, musician; b. Pitts., Feb. 27, 1952; s. Albert Wesley and Florence Elizabeth (Smiley) S.; m. Deborah Kathryn Ashcom, Mar. 25, 1972 (div. May 1990); children: Matthew L., Abigail K., David A.; m. Linda Jean Plankenhorn, May 27, 1990. BA, Thiel Coll., 1974; MDiv, Gordon-Conwell Theol. Sem., 1977; PhD, Calif. Grad. Sch. Theology, Glendale, 1986. Ordained minister Congl. Ch., 1978. Pastor Congl. chs., Everett, Peabody, Mass., 1976-80; chaplain U.S. Army, 1980-88; owner/founder Schellware, Falling Waters, W.Va., 1977-89; v.p., founder Quality Revs., Inc., Falling Waters, 1989—; also bd. dirs. Quality Reviews, Inc., Falling Waters; staff scientist Dynamac Corp., 1989-90; dir. Creative Alternatives Recording, Pub., 1992—; cons. in personal devel. Author: Personality Preference Inventory (P.P.I.), 1988; Ad editor/author: AIDS Ministry in Perspective, 1988, Toyosaurus Wrex, 1986, Intimacy: Theological and Behavioral Implications, 1986; Army monthly newsletters, 1980-85, various pub. articles; composer more than 150 songs; musician with 4 recordings on cassette. V.p. Ft. Stewart (Ga.) Sch. Bd., 1980-82; profl. counselor AIDS Ministry, leader Cub Scouts, 1981-85, dir. Childrens Surf and Sand Missions, 1976-79. Maj. USAR, 1988—. Recipient Mayor's award in Photography, Everett, 1976. Mem. DAV, SAR, Am. Philatelic Soc., Smithsonian Instn., Nat. Audubon Soc., Internat. Platform Assn. Avocations: music, songwriting, photography, philately, sheep- and goat-raising. Home: 1204 Berkeley Dr Falling Waters WV 25419-9657 Office: Quality Reviews Inc PO Box 755 Falling Waters WV 25419-0755

SCHELLMAN, JOHN A., chemistry educator; b. Phila., Oct. 24, 1924. AB, Temple U., 1948; MS, Princeton U., 1949, PhD, 1951; PhD (hon.), Chalmers U., Sweden, 1983, U. Padua, Italy, 1990. USPHS postdoctoral fellow U. Utah, 1951-52, Carlsberg Lab., Copenhagen, 1953-55; DuPont fellow U. Minn., Mpls., 1955-56; asst. prof. chemistry U. Minn., 1956-58; assoc. prof. chemistry, rsch. assoc. Inst. Molecular Biology, U. Oreg., Eugene, 1958-63; prof. chemistry, rsch. assoc. Inst. Molecular Biology, U. Oreg., 1963—; vis. Lab. Chem. Physics, Nat. Inst. Arthritis and Metabolic Diseases, NIH, Bethesda, Md., 1980; vis. prof. Chalmers U., Sweden, 1986, U. Padua, Italy, 1987. Contbr. articles to profl. jours. Served with U.S. Army, 1943-46. Fellow Rask-Oersted Found., 1954, Sloan Found., 1959-63, Guggenheim Found., 1969-70. Fellow Am. Phys. Soc.; mem. NAS, Am. Chem. Soc., Am. Soc. Biol. Chemists, Am. Acad. Arts and Scis., Biophys. Soc., Phi Beta Kappa, Sigma Xi. Democrat. Home: 780 Lorane Hwy Eugene OR 97405-2340 Office: Univ Oreg Inst Molecular Biology Eugene OR 97403

SCHELSKE, CLAIRE L., limnologist; b. Fayetteville, Ark., Apr. 1, 1932; s. Theodore J. and Ida S. S.; m. Betty Breukelman, June 2, 1957; children—Cynthia, John, Steven. A.B., Kans. State Tchrs. Coll., Emporia, 1955, M.S., 1956; Ph.D., U. Mich., 1961. Teaching and research asst. dept. biology Kans. State Tchrs. Coll., 1952-55, vis. instr., summer 1960; teaching fellow dept. zoology U. Mich., 1955-57; asst. prof. radiol. health dept. environ. health U. Mich. (Sch. Public Health); asst. research limnologist Gt. Lakes Research Div., Inst. Sci. and Tech., 1967-68, asso. research limnologist, 1969-71, research limnologist, 1971-87; asst. dir. Gt. Lakes Research Div., Inst. Sci. and Tech. (Gt. Lakes Research Div.), 1970-72, acting dir., 1973-76, assoc. prof. limnology, dept. atmospheric and oceanic sci., 1976-87; assoc. prof. natural resources Sch. Natural Resources, 1976-86, prof., 1986-87; Carl S. Swisher prof. water resources U. Fla., Gainesville, 1987—; research fellow Inst. Fisheries Research, Mich. Dept. Conservation, 1957-60; research assoc. U. Ga. Marine Inst., 1960-62; fishery biologist, supervisory fishery biologist, chief Estuarine Ecology Program, Bur. Comml. Fisheries, Radiobiol. Lab., Beaufort, N.C., 1962-66; adj. asst. prof. dept. zoology N.C. State U., Raleigh, 1964-66; tech. asst. Office Sci. and Tech., Exec. Office of Pres., Washington, 1966-67; cons. Ill. Atty. Gen., 1977-79. Author: (with J.C. Roth) Limnological Survey of Lakes Michigan, Superior, Huron and Erie, 1973. Recipient Disting. Alumnus award Emporia State U. (formerly Kans. State Tchrs. Coll.), 1989. Fellow AAAS, Am. Inst. Fishery Rsch. Biologists (regional and dist. dir. South-Cen. Gt. Lakes chpt. 1977-80); mem. Am. Soc. Limnology and Oceanography (sec. 1976-85, v.p. 1987-88, pres. 1988-90), Ecol. Soc. Am. (assoc. editor 1972-75), Internat. Assn. Gt. Lakes Rsch. (editorial bd. 1970-73, chmn. 20th Conf. 1977, assoc. editor 1984—), Phycological Soc. Am., Societas Internationalis Limnologiae. Home: 2738 SW 9th Dr Gainesville FL 32601-9003 Office: U Fla Dept Fisheries and Aquatic Scis 7922 NW 71st St Gainesville FL 32606-3071

SCHELTEMA, ROBERT WILLIAM, military officer; b. Grand Rapids, Mich., July 1, 1961; s. William Arthur and Beatrice Ann (Laansma) S.; m. Janet Lynn Laitola, Apr. 30, 1983; children: Brandon Robert, Daniel Jerik, Samantha Noelle. BS in Aerospace, U. Mich., 1983; postgrad., Embry-Riddle Coll., 1992—. Commd. 2d lt. USAF, 1984, advanced through grades to capt., 1988; laser optical systems engr. Air Force Weapons Lab., Albuquerque, 1984-87, weapons compatibility engr., 1987-88; chief engring. br. 384 Bombardment Wing, Wichita, Kans., 1988-92; weapons compatibility engr. Engring. Liaison Office, Ramstein, Germany, 1992—. Decorated Aircraft Maintenance badge, 1990, Space Comptroller badge, 1986; recipient Collier Trophy SAC, 1991. Mem. AIAA. Home: 1580 Airport Rd Muskegon MI 49444 Office: OL-EL/Engring Liaison Office PSC 2 Box 9494 APO AE 09012

SCHEMMEL, RACHEL ANNE, food science and human nutrition educator, researcher; b. Farley, Iowa, Nov. 23, 1929; d. Frederic August and Emma Margaret (Melchert) Schemmel. BA, Clarke Coll., 1951; MS, U. Iowa, 1952; PhD, Mich. State U., 1967. Dietitian, Children's Hosp. Soc., L.A., 1952-54; instr. Mich. State U., East Lansing, 1955-63, from asst. prof. to prof. food sci., human nutrition, 1967—. Author: Nutrition Physiology and Obesity, 1980. Contbr. articles on obesity, clin. nutrition to profl. jours. Recipient Disting. Alumni award Mt. Mercy Coll., 1971, Borden award for rsch. in applied nutrition, 1986, Disting. Faculty award Mich. State U., 1991. Mem. Am. Inst. Nutrition, Inst. Food Technologists, Am. Diet Assn. (pres. Mich. and Lansing 1975-76), Brit. Nutrition Soc., Soc. for Nutrition Edn., Sigma Xi (sr. rsch. award 1986, pres. Mich. State U. chpt. 1983-84), Phi Kappa Phi. Roman Catholic. Home: 1341 Red Leaf Ln East Lansing MI 48823-1339 Office: Mich State U Dept Food Sci and Human Nutrition East Lansing MI 48824

SCHEMMEL, TERENCE DEAN, physicist; b. Newport News, Va., Feb. 2, 1963; s. Ronald Joseph and Elizabeth Irene (McCord) S.; m. Laurel Ann Mitchell, Jan. 5, 1985. BS in Physics, Colo. State U., 1985. Mem. tech. staff Santa Barbara Rsch. Ctr., Goleta, Calif., 1986-88; thin film physicist Vec-Tec Systems, Inc., Boulder, Colo., 1988-89; thin film engr. Storage Tech. Corp., Louisville, Colo., 1989-92; sr. devel. engr. Rocky Mountain Magnetics, Louisville, 1992—. Mem. Am. Vacuum Soc. (chpt. chmn. 1991-92), Optical Soc. Am. Achievements include patent for cerium oxyfluoride antireflection coating. Office: Rocky Mountain Magnetics Corp 2270 S 88th St Louisville CO 80028-8188

SCHENA, FRANCESCO PAOLO, nephrology educator; b. Foggia, Puglia, Italy, Mar. 24, 1940; s. Oreste and Filomena (De Benedetto) S.; m. Adriana Santucci, Aug. 2, 1969; children: Stefano, Valentina. MD, U. Bari, Italy, 1964. Fellow Ministry of Edn., Rome, 1968-70, Louvain, Belgium, 1968-69; resident in internal medicine U. Bari, 1970, med. asst., 1970-71, asst. prof., 1972-82, resident in nephrology, 1974, assoc. prof., 1983-85, full prof., 1986—, chmn. nephrology, 1989—; fellow dept. nephrology U. Louvain, 1968-70; vis. prof. Inst. Pathology, Case Western Rev. U., Cleve., 1985; vis. prof. renal unit Guy's Hosp., London, 1986; chmn. REgional Com. Organ Transplants, Puglia. Author: Methods in Nephrology, 1983, Textbook of Nephrology, 1991. Recipient award Lepitit, 1964; grantee European Econ. Community, 1986, CNR, 1985—. Mem. European Renal Assn., Am. Kidney Found., Internat. Soc. Nephrology, Am. Soc. Nephrology, N.Y. Acad. Scis., Am. Assn. Immunologists, Rotary (mem. local com. 1991). Avocations: history, soccer. Home: Via Delle Murge 59/A, 70124 Bari, Puglia Italy Office: U Bari Polyclinic, Piazza G Cesare 11, 70124 Bari Italy

SCHENCK, JACK LEE, electric utility executive; b. Morgantown, W.Va., Aug. 2, 1938; s. Ernest Jacob and Virginia Belle (Kelley) S.; m. Rita Elizabeth Pietschmann, June 7, 1979; 1 son, Erik. B.S.E.E., B.A. in Social Sci., Mich. State U., 1961; M.B.A., NYU, 1975. Engr. AID, Tunis, Tunisia, 1961, Detroit Edison Co., 1962-63; engr., economist OECD, Paris, 1963-70; v.p. econ. policy analysis Edison Electric Inst., N.Y.C. and Washington, 1970-81; v.p., treas. Gulf States Utilities Co., Beaumont, Tex., 1981-92, sr. v.p., CFO, 1992—. Mem. Internat. Assn. Energy Econs., Triangle Club, Eta Kappa Nu. Republican. Office: Gulf States Utilities Co 350 Pine St Beaumont TX 77701-2437

SCHENERMAN, MARK ALLEN, research chemist, educator; b. Plainfield, N.J., Feb. 20, 1959; s. David Seymour and Toby (Gallanter) S.; m. Amy Bober, Oct. 20, 1991. BS, U. Md., Balt., 1980; PhD, U. Fla., 1986. Cert. med. technologist Am. Soc. Clin. Pathologists. Grad. assoc. U. Fla., Gainesville, 1981-86; postdoctoral assoc. Cornell U., Ithaca, N.Y., 1986-88; rsch. scientist Bristol-Myers Squibb Co., Syracuse, N.Y., 1988-91, project supr., 1991-92, sr. rsch. scientist, 1992—; adj. asst. prof. SUNY Health Sci. Ctr., Syracuse, 1989—. Contbr. articles to Biochromatography, Jour. Cellular Physiology, Procs. NAS, Biochimica Biophys. Acta. Active Big Bros. and Big Sisters, Syracuse, 1988—; asst. scoutmaster, merit badge counselor Boy Scouts Am., Syracuse, 1991—. Biotech. postdoctoral fellow Cornell U., 1986. Mem. AAAS, Am. Chem. Soc., N.Y. Acad. Scis., Sigma Xi. Democrat. Achievements include research on development of novel approaches to analysis of biotechnology-derived pharmaceuticals, utilizing techniques such as HPLC with pre-column derivitization, capillary electrophoresis and high voltage isoelectric focusing. Bioactivity is also analyzed using immunological techniques. Home: 125 Washington Blvd Fayetteville NY 13066 Office: Bristol-Myers Squibb Co PO Box 4755 Syracuse NY 13221

SCHENKEL, SUSAN, psychologist, educator, author; b. Wroclaw, Poland, Apr. 21, 1946; came to U.S., 1949; d. Leon and Siddi (Fiedleholz) S.; m. Alvin Helfeld, Apr. 8, 1984. BA, U. Wis., 1967; MA in Clin. Psychology, SUNY, Buffalo, 1970, PhD in Clin. Psychology, 1973. Lic. psychologist, Mass. Psychologist Fitchburg (Mass.) State Coll., 1972-75, instr. in psychology, 1973-74; staff psychologist div. of alcoholism Boston City Hosp., 1975-76; chief psychologist Cambridge (Mass.) Ct. Clinic, 1976-80; instr. in psychology dept. psychiatry Med. Sch. Harvard U., 1976-80; pvt. practice psychology Cambridge, 1976—; instr. in psychology U. Mass., Boston, 1978; speaker in field. Author: Giving Away Success, 1984, German edit., 1986, Brazilian edit., 1988, rev. edit. 1991, Chinese edit., 1991; contbr. articles to profl. jours. USPHS fellow, 1967-70; N.Y. State Regents scholar, 1968-70; SUNY Rsch. Found. grantee, 1971-72. Mem. Am. Psychol. Assn., Mass. Psychol. Assn., Am. Soc. Tng. and Devel., Assn. for Advancement of Behavior Therapy.

SCHENKER, MARC BENET, medical educator; b. L.A., Aug. 25, 1947; s. Steve and Dosella Schenker; m. Heath Massey, Oct. 8; children: Yael, Phoebe, Hilary. BA, U. Calif., Berkeley, 1969; MD, U. Calif., San Francisco, 1973; MPH, Harvard U., Boston, 1980. Instr. medicine Harvard U., Boston, 1980-82; asst. prof. medicine U. Calif., Davis, 1982-86, assoc. prof., 1986-92, prof., 1992—. Fellow ACP; mem. Am. Coll. Occupational Medicine, Am. Thoracic Soc., Am. Pub. Health Assn., Soc. Epidemiologic Rsch., Am. Coll. Epidemiology, Phi Beta Kappa, Alpha Omega Alpha. Office: U Calif Divsn Occupational & Environ Medicine ITEH Davis CA 95616

SCHEP, RAYMOND ALBERT, chemist; b. Pretoria, Republic of South Africa, Apr. 11, 1946; came to the U.S., 1976; s. Gerhardus Hermanus Schep and Magdalena Aletta (Alberts) Inman; m. Nadja Maria Sahliger, Aug. 17, 1981. MS, U. Pretoria, 1970, DSc, 1974. Sr. rsch. chemist Occidental Rsch. Corp., Irvine, Calif., 1974-82; asst. prof. U. Hawaii, Hilo, 1983-85; sr. rsch. assoc. UCLA, 1986-89; chemist, toxicologist Diagnostic Products Corp., L.A., 1989—. Contbr. articles to Jour. Analytical Toxicology, Jour. Am. Water Works Assn. Mem. Am. Soc. Mass Spectrometry, Audubon Soc. Achievements include 6 patents and research in analytical toxicology, gas chromatography, and mass spectrometry. Home: 10078 Westwanda Dr Beverly Hills CA 90210 Office: Diagnostic Products Corp 5700 W 96th St Los Angeles CA 90210

SCHER, ROBERT SANDER, instrument design company executive; b. Cin., May 24, 1934; s. Stanford Samuel and Eva (Ordan) S.; m. Audrey Erna Gordon, Oct. 21, 1961; children: Sarahh, Alexander, Aaron. SB, MIT, 1956, SM, 1958, Diploma in Mech. Engring., 1960, ScD, 1963. Rsch. and teaching asst. MIT, Cambridge, Mass., 1957-62; control system engr. RCA, Hightstown, N.J., 1963-65; engring. mgr. Sequential Info. System, Elmsford, N.Y., 1965-71; tech. dir. Teledyne Gurley, Troy, N.Y., 1971-78, v.p. engring., 1978-86, pres., 1986-92; pres. Encoder Design Assocs., Clifton Park, N.Y., 1993—. Co-author patent Linear Digital Readout, 1975. Mem. ASME, Optical Soc. Am. Jewish. Avocation: chamber music. Home: 2 Laurel Oak Ln Clifton Park NY 12065-4712

SCHERAGA, HAROLD ABRAHAM, physical chemistry educator; b. Bklyn., Oct. 18, 1921; s. Samuel and Etta (Goldberg) S.; m. Miriam Kurnow, June 20, 1943; children: Judith Anne, Deborah Ruth, Daniel Michael. B.S., CCNY, 1941; A.M., Duke U., 1942, Ph.D., 1946, Sc.D. (hon.), 1961; Sc.D. (hon.), U. Rochester, 1988, U. San Luis, 1992, Technion, 1993. Teaching, research asst. Duke U., 1941-46; fellow Harvard Med. Sch., 1946-47; instr. chemistry Cornell U., 1947-50, asst. prof., 1950-53, assoc. prof., 1953-58, prof., 1958-65, Todd prof. chemistry, 1965-92, Todd prof. chemistry emeritus, 1992—, chmn. dept., 1960-67; vis. assoc. biochemist Brookhaven Nat. Lab., summers 1950, 51, cons. biology dept., 1950-56; vis. lectr. div. protein chemistry Wool Research Labs., Melbourne, Australia, 1959; vis. prof. Soc. for Promotion Sci., Japan, Aug. 1977; mem. tech. adv. panel Xerox Corp., 1969-71, 74-79; Mem. biochemistry tng. com. NIH, 1963-65; mem. research career award com. NIGMS, 1967-71; commn. molecular biophysics Internat. Union for Pure and Applied Biophysics, 1965-69, mem. commn. macromolecular biophysics, 1969-75, pres., 1972-75, mem. commn. subcellular and macromolecular biophysics, 1975-81; adv. panel molecular biology NSF, 1960-62; Welch Found. lectr., 1962, Harvey lectr., 1968, Gallagher lectr., 1968, Lemieux lectr., 1973, Hill lectr., 1976, Venable lectr., 1981; co-chmn. Gordon Conf. on Proteins, 1963; mem. council Gordon Research Confs., 1969-71. Author: Protein Structure; Theory of Helix-Coil Transitions in Biopolymers; Co-editor: Molecular Biology, 1961-86; mem. editorial bd.: Physiol. Chemistry and Physics, 1969-75, Mechanochemistry and Motility, 1970-71, Thrombosis Research, 1972-76, Biophys. Jour, 1973-75, Macromolecules, 1973-84, Computers and Chemistry, 1974-84, Internat. Jour. Peptide and Protein Research, 1978—, Jour. Computational Chemistry, 1980—, Jour. Protein Chemistry, 1982—; corr. PAABS Revista, 1971-73; mem. editorial adv. bd. Biopolymers, 1963—, Biochemistry, 1969-74, 85—. Mem. Ithaca Bd. Edn., 1958-59; Bd. govs. Weizmann Inst., Israel, 1970—; mem. staff Naval Research Lab. Project, Air Force OSRD Project, World War II. Fulbright, Guggenheim fellow Carlsberg Lab., Copenhagen, 1956-57, Weizmann Inst., Israel, 1963; NIH Spl. fellow Weizmann Inst., 1970; Fogarty scholar NIH, 1984, 86, 88-91; recipient Townsend Harris medal CCNY, 1970, Chemistry Alumni Sci. Achievement award, 1977, Kowalski medal Internat. Soc. Thrombosis and Haemostasis, 1983, Linderstrøm-Lang medal Carlsberg Lab., 1983, Internat. Soc. of Quantum Chemistry and Quantum Pharmacology award in Theoretical Biology, 1993. Fellow AAAS; mem. NAS, Am. Chem. Soc. (chmn. Cornell sect. 1955-56, mem. exec. com. div. biol. chemistry 1966-69, vice chmn. div. biol. chemistry 1970, chmn. div. biol. chemistry 1971, Eli Lilly award 1957, Nichols medal 1974, Kendall award 1978, Pauling award 1985, Mobil award 1990, Repligen award 1990), Am. Soc. Biol. Chemists, Biophys. Soc. (coun. 1967-70), Am. Acad. Arts and Scis., N.Y. Acad. Scis. (hon. life), Hungarian Biophys. Soc. (hon.), Phi Beta Kappa, Sigma Xi, Phi Lambda Upsilon. Home: 212 Homestead Ter Ithaca NY 14850-6220

SCHERER, A. EDWARD, nuclear engineering executive; b. Bklyn., May 23, 1942; s. Samuel M. and Margie Scherer. BS in Mech. Engring., Worcester Poly. Inst., 1963; MS in Nuclear Engring., Pa. State U., 1965; MBA, Rensselaer Poly. Inst., Hartford, Conn., 1978. Registered profl. engr., Mass. Asst. to project mgr. Combustion Engring., Inc., Windsor, Conn., 1968-70, reactor project engr., test evaluation engr., 1970-73, asst. project mgr., 1973-75, mgr. nuclear licensing, 1975-80, dir. nuclear licensing, 1980-90, v.p. nuclear quality ABB Combustion Engring. Nuclear Power, 1990-92, v.p. regulatory affairs ABB Combustion Engring. Nuclear Fuel, 1992—; mem. tech. adv. com., mem. issues mgmt. com., mem. standardization oversight com. NUMARC, Washington. Contbr. articles to profl. jours. Capt. U.S. Army, 1965-67, South Vietnam. Fellow ASME; mem. Am. Nuclear Soc., Sigma Xi, Alpha Epsilon Pi (internat. pres. 1982-84, fiscal control bd. 1986—). Office: ABB Combustion Engring 1000 Prospect Hill Rd Windsor CT 06095

SCHERICH, ERWIN THOMAS, civil engineer, consultant; b. Inland, Nebr., Dec. 6, 1918; s. Harry Erwin and Ella (Peterson) S.; student Hastings Coll., 1937-39, N.C. State Coll., 1943-44; B.S., U. Nebr., 1946-48; M.S., U. Colo., 1948-51; m. Jessie Mae Funk, Jan. 1, 1947; children—Janna Rae Scherich Thornton, Jerilyn Mae Scherich Dobson, Mark Thomas. Civil and design engr. U.S. Bur. Reclamation, Denver, 1948-84, chief spillways and outlets sect., 1974-75, chief dams br., div. design, 1975-78, chief tech. rev. staff, 1978-79, chief div. tech. rev. Office of Asst. Commr. Engring. and Rsch. Ctr., 1980-84; cons. civil engr., 1984—. Mem. U.S. Com. Internat. Commn. on Large Dams. Served with AUS, 1941-45. Registered profl. engr., Colo. Fellow ASCE; mem. NSPE (nat. dir. 1981-87, v.p. southwestern region 1991-93), Profl. Engrs. Colo. (pres. 1977-78), Wheat Ridge C. of C. Republican. Methodist. Home and Office: 3915 Balsam St Wheat Ridge CO 80033-4449

SCHERR, LAWRENCE, physician, educator; b. N.Y.C., Nov. 6, 1928; s. Harry and Sophia (Schwartz) S.; m. Peggy L. Binenkorb, June 13, 1954; children: Cynthia E., Robert W. AB, Cornell U., 1950, MD, 1957. Diplomate Am. Bd. Internal Medicine (bd. dirs., sec.-treas. 1979-86). Intern Cornell Med. div. Bellevue Hosp. and Meml. Center, 1957-58, asst. resident, 1958-59, research fellow cardiorenal lab., 1959-60, chief resident, 1960-61, co-dir. cardiorenal lab., 1961-62, asst. vis. physician, 1961-63, asso. vis. physician, 1963-65, dir. cardiology and renal unit, 1963-67, asso. dir., 1964-67, vis. physician, 1966-68; physician to out-patients N.Y. Hosp., 1961-63, asst. attending physician, 1963-66, asso. attending physician, 1966-71, attending physician, 1971—; asst. attending physician Meml. Sloan-Kettering Cancer Center, 1962-71, cons., 1971—; chmn. dept. medicine North Shore Univ. Hosp., 1967—, dir. acad. affairs, 1969—; asst. in medicine Cornell U. Med. Coll., 1958-59; research fellow N.Y. Heart Assn., 1959-60, instr. medicine, 1960-63, asst. prof., 1963-66, assoc. prof., 1966-71, David J. Greene Disting. prof., 1971—, assoc. dean, 1969—; career scientist Health Research Coun., N.Y.C., 1962-66; teaching scholar Am. Heart Assn., 1966-67; pres. N.Y. State Bd. Medicine, 1974-75; bd. dirs. Nat. Bd. Med. Examiners, 1976-80; chmn. Accreditation Coun. for Grad. Med. Edn., 1988, N.Y. State Coun. on Grad. Edn., 1990—; chmn. N.Y. State Coun. on Grad. Med. Edn., 1990. Contbr. articles to profl. jours. Lt. USNR, 1950-53. Master ACP (chmn. and gov. Downstate N.Y. region II 1975-80, regent 1980—,chmn. bd. regents 1985-86; nat. pres.-elect 1986-87, pres. 1987-88, pres. emeritus); fellow Am. Heart Assn. (fellow council on clin. cardiology); mem. AMA, Am. Fedn. Clin. Research, Harvey Soc., N.Y. Med. Soc., Nassau County Med. Soc., Assn. Am. Med. Colls., Am. Clin. and Climatologic Assn. Home: 93 Hendrickson St Haworth NJ 07641-1801 Office: North Shore Univ Hosp Manhasset NY 11030

SCHERZER, NORMAN ALAN, medical educator; b. Bklyn., Nov. 29, 1939; s. Jack M. and Anna (Scherzer) S. BA, Hunter Coll., 1962; PhD, CUNY, 1973. Jr. bacteriologist King's County Hosp., Bklyn., 1960-61; instr. Hunter Coll., N.Y.C., 1973-77; prof. Essex County Coll., Newark, 1977—; clin. assoc. prof. U. Medicine and Dentistry of N.J., Newark, 1992—; vis. prof. Rutgers U., Newark, 1987—, NYU, 1991—; cons. State Panel of Sci. Advisors, N.J., 1981-83, NSF, Washington, 1989; AIDS educator/lectr. Co-editor: A Descriptive Dictionary and Atlas of Sexology, 1991; contbr. articles to profl. jours. Del. visitation to Cuba, Nat. League of Cities, 1979; mem. N.J.-Philippine Cultural Exchange Program, Manila, 1978. Sloan Found. fellow, 1963, Fisht for Sisht fellow, 1965. Mem. Am. Assn. Sex Educators, Counselors and Therapists, Soc. for Sci. Study of Sex. Home: 156 Glenside Trail Sparta NJ 07871 Office: Essex County Coll 303 University Ave Newark NJ 07102

SCHERZIGER, KEITH JOSEPH, civil engineer; b. Queens, N.Y., Jan. 28, 1968; s. Joseph Francis and Carolyn Lois (Cassidy) S. BSCE, U. Hartford, 1990. Registered engr.-in-tng., Conn. Assoc. staff engr. Yankee Gas Svcs. Co., Meriden, Conn., 1991—. Mem. ASCE. Republican. Roman Catholic. Home: 8 H Queen Terr Southington CT 06489

SCHEUER, PAUL JOSEF, chemistry educator; b. Heilbronn, Germany, May 25, 1915; came to U.S., 1938, naturalized, 1944; s. Albert and Emma (Neu) S.; m. Alice Elizabeth Dash, Sept. 5, 1950; children: Elizabeth E., Deborah A., David A.L., Jonathan L.L. BS, Northeastern U., 1943; MA, Harvard U., 1947, PhD with high honors, 1950. Asst. prof. U. Hawaii, Honolulu, 1950-55; assoc. prof. U. Hawaii, 1955-61, prof. chemistry, 1961-85, prof. emeritus, 1985—, chmn. dept., 1959-62; vis. prof. U. Copenhagen, 1977, 89; prof. Toyo Suisan U., Tokyo, 1992; Barton lectr. U. Okla., 1967; J.F. Toole lectr. U. N.B., 1977; Lilly lectr. Kansas U., 1993. Author: Chemistry of Marine Natural Products, 1973; editor: Marine Natural Products: Chemical and Biological Perspectives, vols. 1-2, 1978, vol. 3, 1980, vol. 4, 1981, vol. 5, 1983, Bioorganic Marine Chemistry, vol. 1, 1987, vol. 2, 1988, vol. 3, 1989, vol. 4, 1991, vols. 5, 6, 1992; mem. editorial bd. Toxin Revs., Jour. Natural Products. Served with AUS, 1944-46. Recipient Regents' award for excellence in rsch. U. Hawaii, 1972, Outstanding Alumni award Northeastern U., 1984, inaugural Paul J. Scheuer award in marine natural projects, 1992. Mem. AAAS, Am. Chem. Soc. (Ernest Guenther award 1994), Swiss Chem. Soc., Royal Soc. Chemistry, Sigma Xi, Phi Kappa Phi. Achievements include research in molecular structure of bioactive natural products from marine organisms, terrestrial plants. Home: 3271 Melemele Pl Honolulu HI 96822-1431 Office: U Hawaii 2545 The Mall Honolulu HI 96822

SCHEUING, RICHARD ALBERT, aerospace corporate executive; b. Lynbrook, N.Y., Aug. 19, 1927; s. Emil Conrad and Elise Marie (Blum) S.; m. Doris Elaine Heginger, May 27, 1950; children—Richard S., Alison I., Christopher J. B.S., M.S., M.I.T., Cambridge, 1948; Ph.D. in Aero-Astronautics, N.Y.U., 1971; Advanced Mgmt. Program, Harvard U., Cambridge, 1982. Research aerodynamicist Grumman Aero. Corp., Bethpage, N.Y., 1948-56, head fluid mechanics research, 1956-71, dep. dir. research, 1961-77, dir. research, 1977-83, v.p. research, 1983-85; v.p. corp. research ctr. Grumman Corp., Bethpage, N.Y., 1985—. Bd. trustees N.E. Solar Energy Ctr., Boston, 1980-83; mem. Action for Preservation of Long Island, 1970—, Cold Spring Harbor Whaling Mus., N.Y., 1981—. Grumman Engring. scholar, 1944-47; M.I.T. Grad. fellow, 1948. Assoc. Fellow AIAA; mem. Am. Phys. Soc., Am. Inst. Physics. AAAS, Sigma Gamma Tau, Sigma Xi. Episcopalian. Clubs: Lloyd Harbor Yacht (Huntington, N.Y.) (commodore 1982-84); Centerport Yacht (N.Y.). Avocations: Sailing; skiing. Office: Grumman Corp Research Ctr 1111 Stuart Ave Bethpage NY 11714*

SCHEURLE, JURGEN KARL, mathematician, educator; b. Schw. Gmund, Germany, Sept. 26, 1951; married; 2 children. Diploma in math., U. Stuttgart, Germany, 1974, D, 1975, habilitation, 1981. From wissmitarbeiter to privatdozent U. Stuttgart, Germany, 1975-85; assoc. prof., prof. Colo. State U., Fort Collins, 1985-87; prof. U. Hamburg, Germany, 1987—. Contbr. articles to profl. jours. Mem. Internat. Soc. for Interaction of Mechanics and Math., German Math. Assn., Assn. Angew. Math. and Mechanics, Math. Assn. Hamburg, Am. Math. Soc. Roman Catholic. Office: U Hamburg Angew Math Inst, Bundesstrasse 55, D-20146 Hamburg Germany

SCHEUZGER, THOMAS PETER, audio engineer; b. Evanston, Ill., Nov. 19, 1960; s. Peter and Ruth Erica (Hadorn) S. MusB in Music Prodn. and Engring., Berklee Coll. of Music, 1985. Audio engr. Eiger Engring., Watertown, Mass., 1983—; dir. audio New Eng. Conservatory, Boston, 1985—; producer, dir. Continental Cablevision, Watertown, 1989—. Producer, dir. Music from the Source, 1991. Recipient USA Video Festival award Nat. Fedn. Local Cable Programmers, 1992. Mem. Audio Engring. Soc. Avocations: photography, carpentry, hiking, camping. Home: 222 Palfrey St Watertown MA 02172-1836 Office: New Eng Conservatory 290 Huntington Ave Boston MA 02115-5000

SCHEYER, RICHARD DAVID, neurologist, researcher; b. Hempstead, N.Y., Nov. 28, 1958. BS, Stanford U., 1980; MD, SUNY, Syracuse, 1984. Intern Waterbury (Conn.) Hosp. Health Ctr., 1984-85; resident dept. neurology Yale U. Sch. Medicine, New Haven, 1985-88, fellow, 1988-89, instr., 1989-90, postdoctoral assoc., 1990-91, asst. prof. dept. neurology, 1991—. Mem. AMA, Am. Acad. Neurology, Am. Epilepsy Assn., Am. Med. Informatics Assn. Office: West Haven Med Ctr Neurology Svc 127 West Haven CT 06516

SCHIAVI, RAUL CONSTANTE, psychiatrist, educator, researcher; b. Buenos Aires, Argentina, Jan. 7, 1930; came to U.S. 1956; s. Constantino and Maria (Acquier) S.; m. Michelle deMiniac, Aug. 7, 1960; children: Isabelle, Nadine, Viviane. MD, U. Buenos Aires, 1953. Diplomate Am. Bd. Psychiatry and Neurology. Fgn. asst. psychiatry U. Paris, 1955-56; resident in psychiatry U. Pa., Phila., 1956-59; instr. psychiatry U. Pa., 1959-61; assoc. College de France, Paris, 1961-63; asst. prof. psychiatry Cornell U., N.Y.C., 1963-66; assoc. prof. psychiatry SUNY, Downstate Med. Ctr., Bklyn., 1966-71, Mt. Sinai Sch. Medicine, N.Y.C., 1971-78; prof. psychiatry Mt. Sinai Sch. Medicine, 1978—; fellow Found. Fund for Rsch. in Psychiatry, 1958-63; cons. NIMH, 1966-70, 77-81; dir. human sexuality program Mt. Sinai Sch. Medicine, 1973—; advisor World Health Orgn., 1989. Contbr. articles to profl. jours., chpts. to books; co-editor Jour. Sex and Marital Therapy; editorial bd. Archives of Sexual Behavior and Hormones and Behavior; mem. editorial bds. Revista Latinoamericana de Sexologia, Quaderni de Sessuologia Clinica, Revista Argentina de Sexualidad Humana, Annual Review Sex Research. Recipient Rsch. Sci. Devel. award, 1966, Masters and Johnson award Soc. for Sex Therapy and Rsch., 1991; grantee NIH, 1977-89, 87—, NIMH, 1976—, others. Fellow Am. Psychopathol. Assn., Psychiat. Rsch. Soc.; mem. AAAS, Am. Psychiat. Assn. (life, cons. 1989, Excellence in Edn. award 1992), Am. Psychosomatic Soc. (coun. 1985-88), Internat. Acad. Sex Rsch., Soc. Sex Therapy and Rsch. (pres. 1984-86), Sex Info. and Edn. Coun. of U.S. (bd. dirs. 1979-83), Internat. Soc. Psychoneuroendocrinology, Sigma Xi. Home: 25 E 86th St New York NY 10028-0553 Office: Mt Sinai Sch Medicine 1 Gustave L Levy Pl New York NY 10029-6504

SCHIDLOW, DANIEL, pediatrician, medical association administrator; b. Santiago, Chile, Oct. 23, 1947; m. Sally Rosen; children: David, Michael, Jessica. Grad., U. Chile, 1972. Diplomate Am. Bd. Pediatrics, Am. Bd. Pediatric Pulmonology; lic. in Washington, Pa., N.J. Rotating intern U. Chile Hosp., U. Chile Sch. Medicine, 1971-72, resident in internal medicine, instr. phys. diagnosis, 1972-73; resident, emergency rm. physician in pediatrics E.G. Cortes Hosp. Children, U. Chile, 1973-74; resident in pediatrics Albert Einstein Coll. Medicine Bronx (N.Y.)-Lebanon Hosp. Ctr., 1974-76; fellow pediatric pulmonary medicine St. Christopher's Hosp. Children, Temple U. Sch. Medicine, Phila., 1976-78; chief sect. pediatric pulmonology dept. pediatrics St. Christopher's Hosp. Children, Temple U. Sch. Medicine, 1983—; from asst. to assoc. prof. pediatrics sch. medicine Temple U., Phila., 1978-90, prof., 1990—, dep. chmn. dept. pediatrics, 1991—; attending physician St. Christopher's Hosp. Children, 1978—, dir. fellowship tng. and edn. program sect. pediatric pulmonology, 1979-91, assoc. dir. pediatric pulmonary and cystic fibrosis ctr., 1981-83, med. dir. dept. respiratory therapy, 1982-88, project dir. Phila. pediatric pulmonary ctr., 1983-86, dir. cystic fibrosis ctr., 1983—, chair capital campaign com. dept. pediatrics, 1987, mem. exec. com. med. staff, 1988—, mem. various coms.; courtesy staff Lancaster (Pa.) Gen. Hosp., 1980-82; cons. divsn. rehab. Pa. Dept. Health, 1983—; mem. promotions com. dept. pediatrics sch. medicine Temple U., 1986—, chmn. com. appointments clin.-educator track 1991—; attending staff no. divsn. Albert Einstein Med. Ctr., 1987—; cons. Nat. Ctr. Youth Disabilities, 1987—; mem. med. adv. coun. Cystic Fibrosis Found., Bethesda, Md., 1987—, trustee, 1990—, med. dir. home care svcs., 1991—, various other positons; consulting staff Temple U. Hosp., 1988—; mem. organizing com. N.Am. Cystic Fibrosis Conf., 1990-93, co-chmn., 1992—; co-chmn. Nat. Concensus Conf. Pulmonary Complications Cystic Fibrosis, McLean, Va., 1992; mem. adv. bd. Phila. Parenting Assocs., 1992—. Reviewer Jour. Pediatrics, Am. Jour. Diseases Children, others. Named Illustrious Guest, City of LaPlata, Argentina, 1992. Fellow Am. Acad. Pediatrics (Pa. chpt., sect. diseases chest), Am. Coll. Chest Physicians (sect. cardiopulmonary diseases children); mem. AAAS, Am. Thoracic Soc. (mem. nominating com. 1993—), Am. Fedn. Clin. Rsch., Chilean Pediatric Soc. (hon.), Pa. Thoracic Soc., Ea. Soc. Pediatric Rsch., Phila. Pediatric Soc. Home: 315 N Bowman Ave Merion Station PA 19066 Office: St Christopher's Hosp Children Pediatric Pulmonary & Cystic Fibrosis Ct Erie Ave at Front St Philadelphia PA 19134

SCHIEHLEN, WERNER OTTO, mechanical engineer; b. Heidenheim, Ger., May 19, 1938; s. Otto and Erika (Latzel) S.; m. Christine E. Heinzmann, Apr. 19, 1963; children: Joachim, Matthias, Michael, Annette. MSc, U. Stuttgart, 1963, PhD, 1966; Habilitation, Tech. U. Munich, 1971. Rsch. engr. Daimler-Benz AG, Stuttgart, 1963-64; asst. prof. U. Stuttgart, 1964-66; chief engr. Tech. U. Munich, 1966-72; rsch. assoc. NASA Marshall Space Flight Ctr., Huntsville, Ala., 1972-73; assoc. prof. Tech. U. Munich, 1973-77; prof. U. Stuttgart, 1977—; dean Faculty Prodn. Engring., U. Stuttgart, 1984-86; sec.-gen. Internat. Union of Theoretical and Applied Mechanics, Stuttgart, 1984-92. Author: Dynamics of Satellites, 1971, Lineare Schwingungen, 1976, Technische Dynamik, 1986, Fahzeugdynamik, 1993; editor: Multibody Systems Handbook, 1990. Mem. ASME, AIAA (sr. mem.), Verein Deutscher Ingenieure, Internat. Assn. vehicle System Dynamics (bd. dirs.). Achievements include software package NEWEUL for multibody system dynamics. Office: Univ of Stuttgart, Pfaffenwaldring 9, D-70550 Stuttgart 80, Germany

SCHIESS, KLAUS JOACHIM, mechanical engineer; b. Bern, Switzerland, Dec. 11, 1938; came to U.S., 1978; s. Werner Ernst and Elly Gretchen (Schmidt) S.; m. Margarethe von Mihalik, June 1, 1968; children: Nicoline, Janine. BSME, U. Witwatersrand, 1961. Registered profl. engr., South Africa, Md., Calif. Mechanical engr. Sulzer-Escher Wyss, Zurich, 1962-66; project engr. G.H. Marais & Ptnrs., Pretoria, South Africa, 1967-68; dir., ptnr. Everitt Germishuizen & Ptnrs., Pretoria, South Africa, 1969-78; project engr. RTKL Assocs. Inc., Balt., 1978-79, Gipe Assocs., Timonium, Md., 1979-80; assoc. RTKL Assocs., Inc., 1980-81; project engr. EBL Engrs., Inc., Towson, Md., 1981-83; v.p. Popov Engrs., Inc., Newport Beach, Calif., 1983-85; mech. engr. Verle A. Williams & Assocs., Inc., San Diego; pres. KSEngrs. Mech. Engring. Cons. and Energy Engrs., La Jolla, Calif., 1987—. Pres. Pretoria Athletic Club, 1967-68; mem. Swiss Nat. Track Team, 1962-66 (rec. holder for high hurdles 1962-68), South African Track Team, 1959.

SCHIFF, ERIC ALLAN, physics educator; b. L.A., Aug. 29, 1950; s. Gunther Hans and Katharine Sheperd (MacMillan) S.; m. Nancy Ruth Mudrick, Aug. 12, 1973; children: Nathan, Evan . BS, Calif. Inst. Tech., 1971; PhD, Cornell U., 1979. Rsch. assoc. U. Chgo., 1978-81; asst. prof. Syracuse (N.Y.) U., 1981-87, assoc. prof., 1987—; vis. Brown U., Providence, R.I., 1988-90. Contbr. articles to profl. jours. NSF Rsch. grantee, 1983-86. Mem. Am. Phys. Soc. (exec. com. N.Y. state chpt. 1991-94), Materials Rsch. Soc. (symposium organizing com. 1992—). Office: Syracuse U Dept Physics Syracuse NY 13244-1130

SCHIFF, GILBERT MARTIN, virologist, microbiologist, medical educator; b. Cin., Oct. 21, 1931; married, 1955; 2 children. BS, U. Cin., 1953, MD, 1957. Intern U. Hosp., Iowa City, 1957-58, resident internal medicine, 1958-59; med. officer lab br. Communicable Diseases Ctr., Ga., 1959-61; head tissue culture investigation unit, perinatal rsch. br. Nat. Inst. Neurol. Diseases and Blindness, 1961-64; dir. clin. virology lab. U. Cin., 1964-78, asst. prof. medicine and microbiology, 1964-67, asst. prof. microbiology, 1967-71, prof. medicine Coll. Medicine, 1971—; pres. James N. Gamble Inst. Medical Rsch., 1984—; attending physician dept. medicine Emory U., Atlanta, 1959-61; cons. com. maternal health Ohio State Med. Assn., 1964-70, Hamilton County Neuromuscular Diagnostic Clinic, 1966, 75, Contract Immunization Status in U.S., 1975-77; mem. com. viral hepatitis among dental pers. VA; mem. immunization practice adv. com. Surgeon Gen., 1971-75; dir. Christ Hosp Inst. Med. Rsch., Cin., 1974-83, chairperson libr. com., 1974—, mem. com. cancer programs, 1979—, mem. com. human rsch., 1980—, chairperson search com., dir. radiotherapy, 1980-82; mem. com. infection control, 1981—, mem. com. univ. liaisons, 1982—; mem. subcom. antimicrobial agents U.S. Pharmacopeia, 1977-80; mem. study sect., adv. com., review com. NIH; mem. com. Rubella immunization Ohio Dept. Health; com. Rubella control Cin. Dept. Health. Trustee Children's Hosp. Med. Ctr., rsch. com., 1985—; community adv. com. Hoxworth Blood Ctr., 1991—. Recipient career rsch. devel. award Nat. Inst. Child Health and Human Devel., 1970-74; grantee USPHS, 1964-67, Nat. found., 1965-67. Fellow ACP; mem. AAAS, Am. Soc. Microbiology, Am. Fedn. Clin. Rsch. (sec.-treas. 1967-70), Am. Pub. Health Assn., Sci. Rsch. Soc. Am., Ctrl. Soc. Clin. Rsch. (sec.-treas. 1977-81, v.p. 1983, pres. 1984), Infectious Disease Soc. Am. Am. Soc. Clin. Investigation, Sigma Xi. Office: James N Gamble Instit of Med Rsch 2141 Auburn Ave Cincinnati OH 45219*

SCHIFFMAN, LOUIS F., management consultant; b. Poland, July 15, 1927; s. Harry and Bertha (Fleder) S.;m. Mina R. Hankin, Dec. 28, 1963; children: Howard Laurence, Laura Lea. BChemE, NYU, 1948, MS, 1952, PhD, 1955. Rsch. engr. Pa. Grade Crude Oil Assn., Bradford, 1948-50; teaching fellow in chemistry NYU, 1950-54; rsch. chemist E.I. duPont de Nemours & Co., Wilmington, Del., 1954-56, Atlantic Refining Co., Phila., 1956-59; project leader, group leader, head corrosion sect. Amchem Products Inc., Ambler, Pa., 1959-70; pres. Techni Rsch. Assocs. Inc., Willow Grove, Pa., 1970—, real estate developer: ptnr. Bay Properties Co., Bay Club Marina, Margate, N.J., Willow Grove (Pa.) Assocs.; pub., editor Patent Licensing Gazette, 1968—, World Tech., 1975—; panelist on forum patents and inventions Delaware Valley Industry, 1973; mem. adv. oversight com. NSF, 1975, moderator energy conf. ERDA, Washington, 1976, Las Vegas, 1977; mem. adv. group in small bus. R&D programs Dept. Def., 1980. Editor: (with others) Guide to Available Technologies, 1985; contbr. to Encyclopedia of Chemical Technology, 1967; contbr. articles to profl. jours. Patentee in field. Recipient Founders Day award NYU, 1956. Fellow Am. Inst. Chemists; mem. Am. Chem. Soc., N.Y. Acad. Scis., Lic. Execs. Soc., Tech. Transfer Soc., Assn. Univ. Tech. Mgrs., Assn. Small Rsch. Cos. (editorial contbr. newsletter), Sigma Xi, Phi Lambda Upsilon. Home: 1837 Merritt Rd Abington PA 19001-4606 Office: Techni Rsch Assocs Inc Willow Grove Plz 102 N York Rd Willow Grove PA 19090

SCHIFFNER, CHARLES ROBERT, architect; b. Reno, Sept. 2, 1948; s. Robert Charles and Evelyn (Keck) S.; m. Iovanna Lloyd Wright, Nov. 1971 (div. Sept. 1981); m. Adrienne Anita McAndrews, Jan. 20, 1983. Student, Sacramento Jr. Coll., 1967-68, Frank Lloyd Wright Sch. Architecture, 1968-77. Registered architect, Ariz., Nev., Wis. Architect Taliesin Associated Architects, Scottsdale, Ariz., 1977-83; pvt. practice architecture Phoenix, 1983—; instr. Ariz. State U., Tempe, 1983-87. Prin. works include Ahwatukee House of the Future (cert. distinction Am. Architecture 1985), addition to Richard Black Residence, Encanto Park, Ariz. (1st place J. Brock 1986), The Pottery House, Paradise Valley, Ariz., Seventh Day Adventist Exec. Hdqrs., Scottsdale, Condominium Project, Phoenix, SRPMIC Replacement Housing Program, Outer Loop Hwy., Scottsdale; author (poem) Yellowstone Stream (2d prize Winter Wheat contest 1986). Named one of 35 Most Promising Young Americans Under 35, US mag., 1979; recipient Restoration award Sunset Mag. Western Home, 1989, 91. Democrat. Roman Catholic. Avocations: art, football. Home: 5202 E Osborn Rd Phoenix AZ 85018-6137 Office: Camelhead Office Ctr 2600 N 44th St # 208 Phoenix AZ 85008-1521

SCHIFFRIN, MILTON JULIUS, physiologist; b. Rochester, N.Y., Mar. 23, 1914; s. William and Lillian (Harris) S.; m. Dorothy Euphemia Wharry, Oct. 10, 1942; children: David Wharry, Hilary Ann. AB, U. Rochester, 1937, MS, 1939; PhD cum laude, McGill U., 1941. Instr. physiology Northwestern U. Med. Sch., Evanston, Ill., 1941-45; lectr. pharmacology U. Ill. Med. Sch., 1947-57, clin. asst. prof. anesthesiology, 1957-61; with Hoffmann-La Roche, Inc., Nutley, N.J., 1946-79, dir. drug regulatory affairs, 1964-71, asst. v.p., 1971-79; pres. Wharry Rsch. Assn., Seattle, Wash., 1979—; chmn. Everglades Health Edn. Ctr., 1986-87. Author: (with E.G. Gross) Clinical Analgetics, 1955; editor: Management of Pain in Cancer, 1957. Capt. USAAF, 1942-46. Mem. Am. Med. Writers Assn. (bd. dirs. 1967—, pres. N.Y. chpt. 1967-68, nat. pres. 72-73), Am. Physiol. Soc., Internat. Coll. Surgeons, Am. Therapeutic Soc., Coll. Clin. Pharmacology and Therapeutics, Am. Chem. Soc. Home and Office: Unit 401 1001 2d Ave W Seattle WA 98119

SCHILLING, FREDERICK AUGUSTUS, JR., geologist, consultant; b. Phila., Apr. 12, 1931; s. Frederick Augustus and Emma Hope (Christoffer) S.; m. Ardis Ione Dovre, June 12, 1957 (div. 1987); children: Frederick Christopher, Jennifer Dovre. BS in Geology, Wash. State U., 1953; PhD in Geology, Stanford U., 1962. Computer geophysicist United Geophys. Corp., Pasadena, Calif., 1955-56; geologist various orgns., 1956-61, U.S. Geol. Survey, 1961-64; underground engr. Climax (Colo.) Molybdenum Co., 1966-68; geologist Keradamex Inc., Anaconda Co., M.P. Grace, Ranchers Exploration & Devel. Corp., Albuquerque and Grants, N.Mex., 1968-84, Hecla Mining Co., Coeur d'Alene, Idaho, 1984-86, various engring. and environ. firms, Calif., 1986-91; prin. F. Schilling Cons., Canyon Lake, Calif., 1991—. Author: Bibliography of Uranium, 1976. Del. citizen amb. program People to People Internat., USSR, 1990-91. With U.S. Army, 1953-55. Fellow Explorers Club; mem. Geol. Soc. Am., Am. Assn. Petroleum Geologists, Soc. Mining Engrs., Internat. Platform Assn., Masons, Kiwanis, Sigma Xi, Sigma Gamma Epsilon. Republican. Presbyterian. Avocation: track and field. Office: F Schilling Cons 30037 Steel Head Dr Canyon Lake CA 92587

SCHILLING, HARTMUT, engineer; b. Dohna, Saxony, Germany, Aug. 10, 1951; s. Hans and Erika Schilling; m. Paula Parnes, Mar. 25, 1975. Diploma in engring., Tech. U. Darmstadt, Germany, 1977, D Natural Scis., 1979, Habilitation, 1981. Rschr. Tech. U. Darmstadt, 1981-84; rschr. Rheinmetall GmbH, Düsseldorf, Germany, 1984-88, head aeromechanics group, 1988-92, head systems simulation dept.; mgr. R&D RWE AG, Essen, Germany, 1993—. Contbr. articles to profl. jours. Mem. AIAA, Deutsche Gesellschaft für Luft-und Raumfahrt. Achievements include numerous patents in defense engineering. Home: Lichtenvoorder Strasse 22, D-41564 Kaarst 2, Germany Office: RWE AG, Kruppstr 5, D-45128 Essen 1, Germany

SCHILLING, WILLIAM RICHARD, aerospace engineer, research and development company executive; b. Manheim, Pa., Jan. 12, 1933; s. William Thomas and Ora Lee (Worley) S.; m. Patricia Elise Brigman, June 8, 1957; 1 child, Duane Thomas. BCE, Va. Poly. Inst., 1956; MS in Structural Engring., Pa. State U., 1959; MS in Aero. Engring., U. So. Calif., 1961, Engrs. Degree in Aerospace Engring., 1966. Aerodynamist Douglas Aircraft Co., Santa Monica, Calif., 1956-64; study chmn. Research Analysis Corp., McLean, Va., 1964-72; div. mgr. Sci. Applications, Inc., McLean, 1972-78; pres., chief exec. officer McLean Rsch. Ctr., Inc., 1978-89; exec. v.p. Wackenhut Applied Techs. Ctr., Fairfax, Va., 1989-91, Internat. Devel. and Resources, Inc., Falls Church, Va., 1991—; pres. Systems Rsch. Corp., Falls Church, 1991—; pres., bd. dirs. LaMancha Co., Santa Fe, 1985-89. Contbr. numerous articles to profl. jours. Chmn. com. Boy Scouts Am., McLean, 1968-78; vol. Am. Heart Assn., McLean, 1984-89; bd. dirs., chief exec. officer Internat. Housing Devel., McLean, 1986—. Mem. Assn. U.S. Army, Am. Def. Preparedness, Va. C. of C. Baptist. Avocations: classical lit., art, music, travel. Home: 6523 Old Dominion Dr Mc Lean VA 22101-4613 Office: Internat Devel & Resources 5107 Leesburg Pike # 2603 Falls Church VA 22041-3234

SCHILTZ, JOHN RAYMOND, biochemist, researcher; b. Saint Joseph, Mo., Nov. 5, 1941; s. George Francis Schiltz and Georgia Mae Slaughter; m. Jean McNeill, June 26, 1965; children: John Francis, Catherine Elizabeth. BS in Secondary Edn., N.W. Mo. State U., 1963; PhD, Kans. U., 1970. NIH postdoctoral fellow U. Pa., Phila., 1970-72; high sch. sci. tchr. L.A. Sch. Dist., 1963-65; assoc. prof. schs. medicine and dentistry Case Western Res. U., Cleve., 1974-82; sr. group leader Shulton rsch. dvsn. Am. Cyanamid, Clifton, N.J., 1982-88; sr. rsch. scientist Unilever Rsch. U.S., Edgewater, N.J., 1988—; reviewer various profl. jours., 1970—; ad hoc reviewer NIH rsch. grants and workshops for skin diseases, 1976-82; adj. faculty mem. U. Cinn., 1986-88. Contbr. articles to sci. jours., books. Sci. fair judge Shaker Heights, Ohio, 1978, Jersey City, 1985; mem. various pub. sch. parent orgs., Ramsey, N.J., 1982-91. Grantee NIH (U.S.), 1974-82. Mem. AAAS, Soc. Investigative Dermatology, Phi Lambda Upsilon, Sigma Xi. Republican. Achievements include patents in skin moisturization; research on biochemistry of cartilage differentiation, mechanisms of human blistering diseases of the skin, nucleic acid precursor transport in germinating fungi, regulation of collagen and polysaccharide metabolism in normal and diseased tissues, biochemical mechanisms of skin irritation, penetration of substances through skin. Home: 40 E Crescent Ramsey NJ 07446 Office: Unilever Rsch U S 45 River Rd Edgewater NJ 07020

SCHIMMEL, PAUL REINHARD, biochemist, biophysicist, educator; b. Hartford, Conn., Aug. 4, 1940; s. Alfred E. and Doris (Hudson) S.; m. Judith F. Ritz, Dec. 30, 1961; children: Kirsten, Katherine. A.B., Ohio Wesleyan U., 1962; postgrad., Tufts U. Sch. Medicine, 1962-63, Mass. Inst. Tech., 1963-65, Cornell U., 1965-66, Stanford U., 1966-67, U. Calif., Santa Barbara, 1975-76; Ph.D., Mass. Inst. Tech., 1966. Asst. prof. biology and chemistry Mass. Inst. Tech., 1967-71, assoc. prof., 1971-76, prof. biochemistry and biophysics, 1976-92, John D. and Catherine T. MacArthur prof. biochemistry and biophysics, 1992—; mem. NIH Study Sect. Physiol. Chemistry, 1975-79; John D. and Catherine T. MacArthur prof. biochemistry and biophysics, 1992—; indsl. cons. on enzymes and recombinant DNA. Author: (with C. Cantor) Biophysical Chemistry, 3 vols., 1980; mem. editorial bd. Archives Biochemistry, Biophysics, 1976-80, Nucleic Acids Rsch., 1976-80, Jour. Biol. Chemistry, 1977-82, Biopolymers, 1979-88, Internat. Jour. Biol. Macromolecules, 1983-89, Trends in Biochem. Scis., 1984—, Biochemistry, 1989—, Accounts of Chem. Rsch., 1989—, European Jour. Biochemistry, 1991, Protein Sci., 1991—, Proc. Nat. Acad. of Scis., 1993—. Alfred P. Sloan fellow, 1970-72. Fellow AAAS; mem. NAS, Am. Chem. Soc. (Pfizer award 1978, chmn. div. biol. chemistry 1984-85), Am. Soc. for Biochemistry and Molecular Biology, Am. Acad. Arts and Scis., Protein Soc. Office: MIT Dept Biology Cambridge MA 02139

SCHIPPA, JOSEPH THOMAS, JR., school psychologist, educational consultant, hypnotherapist; b. North Tarrytown, N.Y., Mar. 29, 1957; s. Joseph Thomas Sr. and Viola Elizabeth (De Marco) S. MusB, Manhattanville Coll., 1978, MA in Teaching, 1981; PD, Fordham U., 1989, postgrad., 1989—; D.C.H., Am. Inst. Hypnotherapy, 1991. Cert. sch. psychologist, clin. mental health counselor, addictions counselor, hypnotherapist, Nat. Cert. Counselor. Tchr. Sch. of St. Gregory the Gt., Harrison, N.Y., 1978-81; learning specialist Blind Brook High Sch., Rye Brook, N.Y., 1981-83; tchr. spl. edn. Ossining (N.Y.) High Sch., 1983-88; clin. intern in psychology Westchester County Med. Ctr. Psychiat. Inst., Valhalla, N.Y., 1988-89; sch. psychologist Putnam Valley (N.Y.) Elem. Sch., 1989—; dir. Learning Alternatives, North Tarrytown, N.Y. Theodore Presser Found. scholar, 1976, 77, 78. Mem. Am. Counseling Assn., Am. Psychol. Assn., Am. Bd. Hypnotherapy, Nat. Assn. Sch. Psychologists, N.Y. Fedn. Alcoholism and Chem. Dependency Counselors, Assn. for Ednl. and Psychol. Cons., Coun. for Exceptional Children, Orton Dyslexia Soc. Avocation: music. Office: Learning Alternatives 239 N Broadway Ste 6 North Tarrytown NY 10591

SCHIPPER, LEON, mechanical engineer; b. Cologne, Germany, Oct. 20, 1928; came to U.S. 1948; s. Osias and Mania (Mond) S.; m. Elise Diamant, Sept. 2, 1962; children: Kenneth-Henry, Gerard Charles. BSME, CCNY, 1958. Engr. USAF, Wright-Patterson AFB, Ohio, 1958-60; sr. engr. Electro Optical Systems, L.A., 1960-62, Rocketdyne, N.A., L.A., 1962-67; sr. program specialist Airsearch, L.A., 1967—. Contbr. articles to profl. jours. With U.S. Army, 1951-52. Home: 14644 Margate St Sherman Oaks CA 91411 Office: Allied Signal Aerospace Co 2525 W 190th St Mail Stop T-41 Torrance CA 90509-2960

SCHIRMACHER, PETER, molecular pathologist, educator; b. Saarbücken, Fed. Republic Germany, Nov. 4, 1961; s. Wolfgang H.E. and Sigrid A.H.E. (Brokate) S.; m. Roya Kamiar-Gilani, Aug. 21, 1991; 1 child, Daniel. MD, Gutenberg U., Mainz, Fed. Republic Germany, 1987; postgrad., Albert Einstein Coll., 1989-90. Postdoctoral rsch. fellow Inst. Pathology U. Mainz, 1987-89, asst. prof., 1991—; Belfer rsch. fellow Albert Einstein Coll. Medicine, Bronx, N.Y., 1989-90, cons., 1991—. Mem. Fed. Republic Germany Army, 1980-81. Mem. AAAS, Internat. Acad. Pathology. Lutheran. Avocations: tennis, long-distance running, history, philosophy. Office: Inst Pathology, Langenbeckstr 1, 55101 Mainz Germany

SCHIRMER, HOWARD AUGUST, JR., civil engineer; b. Oakland, Calif., Apr. 21, 1942; s. Howard August and Amy (Freuler) S.; m. Leslie May Mecum, Jan. 29, 1965; children: Christine Nani, Amy Kiana, Patricia Leolani. B.S., U. Calif., Berkeley, 1964, M.S., 1965. Registered profl. engr., Hawaii, Guam. Engr. in tng. materials and research dept. Calif. Div. Hwys., Sacramento, 1960-64; engring. analyst Dames & Moore, San Francisco, 1964-67; asst. staff engr. Dames & Moore, 1967-68, chief engr., 1969-72; asso. Dames & Moore, Honolulu, 1972-75; partner, mng. prin. in charge Dames & Moore, 1975-78; regional mgr. Dames & Moore, Pacific Far East and Australia, 1978-81; chief operating officer Dames & Moore, Los Angeles, 1981-83; mng. dir. Dames & Moore Internat., Los Angeles, 1983-89; mng. dir. CH2M Hill Internat. Ltd., Denver, 1989-90, pres., 1991; chmn. geotech. engring. com. Am. Cons. Engrs. Council, 1976-78, profl. liability com., 1984—; past chmn. adv. com. for engring. tech. Honolulu Community Coll. Important works include AFDM Berthing Wharf, Pearl Harbor, Aloha Stadium, Hawaii, Century Ctr., Honolulu, Manila Internat. Airport, Philippines. Past mem. intable UCLA Grad. Sch. Mgmt.; chmn. engring. sect., mem. budget com. Aloha United Way, 1974; founder Mauna Kea Ski Patrol, 1969. Fellow ASCE (past chmn. engring. mgmt. exec. com. 1986-87, Edmund Friedman Young Engr. award for profl. achievement 1974, pres. Hawaii sect., internat. activities com., internat. dir. 1989—); mem. Fedn. Internationale des Ingenieurs-Conseils (chmn. standing com. on profl. liability 1986-89, U.S. rep. to com., mem. task com. on constrn. ins. and law), Cons. Engrs. Coun. Hawaii (pres. 1972), Engring. Assn. Hawaii (past dir., 2d v.p.), Internat. Soc. Soil Mechanics and Found. Engring., Am. Public Works Assn. (sec. 1977-78, dir. 1979-80), Soc. Am. Mil. Engrs., Am. Soc. Profl. Engrs. (internat. contact dir. 1989—), Outrigger Canoe Club, Jonathan Club (L.A.), Met. Club (Denver), Chi Epsilon (hon. mem. U. Hawaii), Sigma Phi Epsilon. Republican. Episcopalian. Home: 4061 S Birch St Englewood CO 80110-5031 Office: CH2M Hill Internat Ltd 6060 S Willow Dr Englewood CO 80111-5142

SCHIRRA, WALTER MARTY, JR., business consultant, former astronaut; b. Hackensack, N.J., Mar. 12, 1923; s. Walter Marty and Florence (Leach) S.; m. Josephine Cook Fraser, Feb. 23, 1946; children: Walter Marty III, Suzanne Karen. Student, Newark Coll. Engring., 1940-42; B.S., U.S. Naval Acad., 1945; D. Astronautics (hon.), Lafayette Coll., U. So. Calif., N.J. Inst. Tech. Commd. ensign U.S. Navy, 1945, advanced through grades to capt., 1965; designated naval aviator, 1948; service aboard battle cruiser Alaska, 1945-46; service with 7th Fleet, 1946; assigned Fighter Squadron 71, 1948-51; exchange pilot 154th USAF Fighter Bomber Squadron, 1951; engaged in devel. Sidewinder missile China Lake, Calif., 1952-54; project pilot F7U-3 Cutlass; also instr. pilot F7U-3 Cutlass and FJ3 Fury, 1954-56; ops. officer Fighter Squadron 124, U.S.S. Lexington, 1956-57; assigned Naval Air Safety Officer Sch., 1957, Naval Air Test Ctr., 1958-59; engaged in suitability devel. work F4H, 1958-59; joined Project Mercury, man-in-space, NASA, 1959; pilot spacecraft Sigma 7 in 6 orbital flights, Oct. 1962; in charge operations and tng. Astronaut Office, 1964-69; command pilot Gemini 6 which made rendezvous with target, Gemini 7, Dec. 1965; comdr. 11 day flight Apollo 7, 1968; ret., 1969; pres. Regency Investors, Inc., Denver, 1969-70; chmn., chief exec. officer ECCO Corp., Englewood, Colo., 1970-73; chmn. Sernco, Inc., 1973-74; with Johns-Manville Corp., Denver, 1974-77; v.p. devel. Goodwin

Cos., Inc., Littleton, Colo., 1978-79; ind. cons., 1979-80; dir. Kimberly Clark, 1983-91. Decorated D.F.C.(3), Air medal (2), Navy D.S.M.; recipient Distinguished Service medal (2), also; Exceptional Service medal NASA. Fellow Am. Astronautical Soc., Soc. Exptl. Test Pilots. Home and Office: PO Box 73 Rancho Santa Fe CA 92067-0073

SCHLABACH, LELAND A., electrical engineer; b. Arthur, Ill., Sept. 12, 1931; s. Albert Edward and Pauline (Hershberger) S.; m. Lucille Anna Reagin, June 30, 1954 (div. Sept. 1976); children: Leon Arthur, Lou Ann, Larry Andrew; m. Helen Marie Lakly, May 23, 1981. BSEE, U. Ill., 1958; MSEE, U. Pitts., 1961, PhD, 1967. Design engr. Switchgear div. Westinghouse Electric Corp., Pitts., 1958-59, devel. engr. Rsch. Ctr., 1959-66, mgr. solid state inverters, 1967-69; engring. mgr. DC drives and systems Robicon Corp., Pitts., 1969—. Mem. staff Boy Scouts Am., Pitts., 1967-74, CAP, Pitts., 1974-76. Staff sgt. USAF, 1950-54. Lamme scholar Westinghouse Electric Corp., 1966. Mem. IEEE (chmn. local Industry Applications Soc. chpt. 1975-76, contbr. conf. and transaction papers), AIME, Assn. Iron and Steel Engrs., Tau Beta Pi, Phi Kappa Phi, Eta Kappa Nu. Republican. Presbyterian. Achievements include 5 patents, 8 patents pending and 30 patent disclosures on alternating current motor drive control circuits. Home: 4198 Gun Club Rd Murrysville PA 15668-9103 Office: Robicon Corp 100 Sagamore Hill Rd Pittsburgh PA 15239-2982

SCHLAG, EDWARD WILLIAM, chemistry educator; b. L.A., Jan. 12, 1932; s. Hermann and Milda (Noble) S.; m. Angela Grafin Zu Castell, june 15, 1955; children: Katherine, Karl, Elisabeth. BS, Occidental Coll., L.A., 1953; PhD, U. Wash., 1958; D (hon.) Causa, Hebrew U. Jerusalem, Israel, 1988. Rsch. scientist Films Dept., Buffalo, 1959; asst. prof. Northwestern U., Evanston, Ill., 1960-63, prof., 1969-70; dir. Inst. Phys. Chemistrty Tech. U., Munich, Germany, 1971—; Woodward lectr., 1987; Fritz-Haber lectr., 1988; Ames lectr., 1990. Contbr. over 250 articles to profl. jours. Alfred P. Sloan fellow, 1965. Mem. Deutsche Physikalische GeseUschaft, Arbeitsgemeinschaft Massenspektrometrie. Achievements include patents in field. Office: Technische U München, Lichtenbergstrasse 4, 8046 Garching Germany

SCHLEE, WALTER, mathematician, educator; b. Schwandorf, Bavaria, Germany, Sept. 12, 1942; s. Walter Konrad and Babette Maria (Birner) S.; Student Technische U. Munich; Diplom-Mathematiker, 1967, Dr.rer.nat., 1970; m. Walburga Koessler, Mar. 15, 1974. Researcher, mem. faculty dept. math. Technische U. Munich, 1967—. Fellow Royal Statis. Soc.; mem. Am. Statis. Assn., Deutsche Statistische Gesellschaft, Association pour la Statistique et ses Utilisations, Bernoulli Soc. for Math. Stats. and Probability, Soc. de Statistique de Paris et de France. Author works on math. optimization, graph theory and nonparametric stats. in English, French German; contbr. articles and revs. to profl. jours. Office: Arcisstrasse 21, D-80333 Munich 2, Germany

SCHLEGEL, JUSTIN J., psychological consultant; b. Mulhouse, Alsace, France, Apr. 6, 1922; s. Joseph and Justine (Meyer) S.; m. Brigitte Esther Biguet, Dec. 28, 1972; children: Claire, Beatrice. Licence de Psychologie, U. Strasbourg, 1957, Cert. d'Etudes pratiques de Psychologie, 1957, Cert. de Psychologie/Pathologique, 1958. Indsl. psychologist Societe Clemessy, Mulhouse, 1947-56; head psychol. ctr. Accident Prevention Svc., Social Security Agy., Strasbourg, 1957-84; psychol. cons. Strasbourg, 1984—; lectr. indsl. psychology U. Strasbourg, 1958-69; prof. Centre National des Arts et Metiers, Metz, 1963-67. Editor Bull. Internat. Test Commn., 1976-92, Bull. French Soc. Projective Techniques, 1972-90; contbr. articles to profl. jours. Mem. French Psychol. Soc. (Div. of Ea. France pres. 1960-67, Div. Occupational Psychology 1972-73), French Soc. for Projective techniques (coun. mem. 1958—), Internat. Test Commn. (coun. mem. 1976—). Avocations: literature, Greek numismatics. Home and Office: 55 Allee de la Robertsau, 67000 Strasbourg France

SCHLEGELMILCH, REUBEN ORVILLE, electrical engineer, consultant; b. Green Bay, Wis., Mar. 8, 1916; s. Raymond Adolf and Emma J. (Schley) S.; m. Margaret Elizabeth Roberts, Aug. 22, 1943; children: Janet R., Raymond J., Joan C., Margaret Ann. BS in Elec. and Agrl. Engring., U. Wis., 1938; MS in Elec. and Agrl. Engring., Rutgers U., 1940; postgrad. in elec. engring., Cornell U., 1940-41, Poly. Inst. Bklyn., 1947-51, U. Ill., 1941-42; SM in Indsl. Mgmt., MIT, 1955; postgrad. in elec. engring., Syracuse U., 1956-59. Registered profl. engr., N.J. Dir. research and devel. Rome Air (Elec.) Devel. Ctr., N.Y., 1955-59; tech. dir. def./space, Westinghouse Elec., Corp. Hdqrs., Washington, 1959-63; mgr. adv. tech. and missiles Fed Systems IBM, Owego, N.Y., 1963-68; gen. mgr., pres. Schilling Industries, Galesville, Wis., 1968-71; mgr. systems design U.S. Army Adv. Systems Concepts Agy., Alexandria, Va., 1971-74; mgr. gun fire control systems, Naval Sea Systems Command, Washington, 1974-80; tech. dir. office research and devel. U.S. Coast Guard Hdqrs., Washington 1980-86; cons. in field, 1986—; govt. cons. electronics, Dept. Def. Research and Devel. Bd., 1949-54; indsl. cons. missile/space, Aerospace Industries Assn., 1959-63; chmn. profl. sci. com. Rome Air Devel. Ctr., 1956-59; mem. nat. com. Engring. Mgmt. Inst. Elec. Engring., N.Y.C., 1956-59. Patentee target position indicator; author tech. reports and articles. Vol. Annandale Christian Community for Action (Va.), 1973-84; mem. Winterset Civic Assn., Annandale, 1971-84. MIT Alfred P. Sloan fellow, 1954-55. Mem. IEEE (sr. life, sec., vice chmn., chmn. 1956-59, Recognition award 1959), Am. Def. Preparedness Assn. (chmn. So. Tier Empire Post 1967-68, recognition award 1968), NSPE, N.Y. Acad. Scis., Soc. Sloan Fellows MIT. Lodges: Masons, Rotary, Shriners. Home: 8415 Frost Way Annandale VA 22003-2222

SCHLESS, JAMES MURRAY, internist; b. N.Y.C., May 27, 1918; s. Maurice J. and Adele (Harwood) S.; m. Jane Gratz, June 2, 1943 (div. 1959); m. Patsie Henery, Sept. 24, 1960; children: Barbara Jo Cassano, Michael James Schless. BA, Yale U., 1940; MD, SUNY, Bklyn., 1943. Diplomate Am. Bd. Internal Medicine. Intern Michael Reese Hosp., Chgo., 1944; resident in internal medicine N.Y. Med. Coll./Met. Hosp. of N.Y., N.Y.C., 1946-47; resident in pulmonary disease Treadau Sanitarium, Saronac Lake, N.Y., 1947-48; pvt. practice Denver, 1948-60; dir. tuberculosis hosps. State of Tex., Austin, 1960-66; dir. med. edn. MacNeal Meml. Hosp., Berwyn, Ill., 1966-71; dir. postgrad. med. edn. U. Minn. Med. Sch., Mpls., 1971-72; chief of staff, trainee VA Med. Ctr., Mpls., 1972-73; chief of staff VA Med. Ctr., Topeka, 1973-74, Allen Park, Mich., 1974-81, Shreveport, La., 1981-88; ind. cons. Austin, 1988—; instr., asst. prof. medicine, U. Colo., Denver, 1948-60; clin. asst. prof. medicine, Baylor U., Houston, 1960-66; asst., then assoc. prof. medicine, U. Ill., Chgo., 1966-71; assoc. prof. medicine U. Minn., Mpls., 1971-73; prof. medicine, asst. dean Wayne State U., Detroit, 1974-81; asst. dean., La. State U., Shreveport, 1981-88. Contbr. numerous articles on pulmonary medicine to profl. jours. Lt. USN, 1944-46, PTO. Fellow ACP, Am. Coll. Chest Physicians; mem. Am. Thoracic Soc., Assn. Am. Med. Colls., Soc. of Med. Cons. to Armed Forces of U.S., Assn. Mil. Surgeons U.S., Met. Club Austin. Home and Office: 6110 Mountain Villa Cv Austin TX 78731-3518

SCHLESSINGER, JOSEPH, pharmacology educator. BSc in Chemistry/Physics magna cum laude, The Hebrew U., Jerusalem, 1968, MSC in Chemistry magna cum laude, 1969; PhD, The Weizmann Inst. Sci., Rehovot, Israel, 1974. Postdoctoral assoc. dept. chemistry Sch. Applied Physics, Cornell U., 1974-76; vis. scientist immunology br. Nat. Cancer Inst., NIH, Bethesda, Md., 1977-78; sr. scientist dept. chem. immunology The Weizmann Inst. Sci., Rehovot, 1978-80, assoc. prof. dept. chem. immunology, 1980-84, prof. dept. chem. immunology, Ruth & Leonard Simon prof., 1984-91; dir. div. molecular biology Biotech. Rsch. Ctr. Meloy Labs., Inc., Rockville, Md., 1985-86, dir. Biotech. Rsch. Ctr., 1986-88; rsch. dir. Rorer Biotech., Inc., King of Prussia, Pa., 1988-90; prof., chmn. dept. pharmcology NYU Med. Ctr., N.Y.C., 1990—. Mem. editorial bds. European Molecular Biology Orgn. Jour., Jour. Cell Biology, Cell Regulation, Cancer Rsch., Receptors, Growth Factors, Cell Crowth & Differentiation, Protein Engineering, Oncogenes and Growth Factor Abstracts; contbr. articles to profl. jours. Recipient Sara Leedy Prize, Weizmann Inst. Sci., 1980, Levinson Prize, 1984; Hestrin Prize, Biochem. Soc. Israel, 1983. Mem. European Molecular Biology Orgn. Office: NYU Med Ctr Dept Pharmacology 550 1st Ave New York NY 10016-6402

SCHLINK, FREDERICK JOHN, retired mechanical engineer; b. Peoria, Ill.; s. Valentine Louis and Margaret (Brutcher) S.; m. Mary Catherine Phillips, May 28, 1932 (dec. 1981). BS, U. Ill., 1912, ME, 1917. Registered profl. engr., N.Y. Engr. Matthieson & Hegeler Zinc Co., LaSalle, Ill., 1912-13, Goldschmidt Detinning Co., East Chicago, Ind., 1912-13; assoc. physicist, tech. asst. to dir. Nat. Bur. of Standards, Washington, 1913-19; physicist Firestone Tire and Rubber Co., Akron, Ohio, 1919-20; mech. engr.-physicist Western Elec. Co., N.Y.C., 1920-22; asst. sec. Am. Nat. Standards Inst., N.Y.C., 1922-31, bd. dirs., 1977-81, mem. screening and rev. com., 1980-90; pres., tech. dir., founder Consumers' Rsch., Washington, N.J., 1931-82; cons. N.Y. Coun. of Jewish Philanthropies, N.Y.C., 1928, Macy Dept Store, N.Y.C., 1928; mem. consumer adv. coun. Underwriters Labs., Northbrook, Ill.; mem. Conf. Tech. Users of Consumer Products., Northbrook; lectr. grad. sch. U. Tenn., 1950's. Author: Eat, Drink and Be Wary, 1935; co-author: Your Money's Worth, 1927, 100,000,000 Guinea Pigs, 1933; editor reports Consumers Rsch. Mag., 1928-82; contbr. articles to profl. jours. Mem. rationing com. Nat. Recovery Adminstrn., Washington, 1943. Recipient Disting. Alumnus award U. Ill. Alumni Assn., 1969, Coll. Engring. Alumni Honor award, 1971. Fellow AAAS, ASME (life), Franklin Inst. (Edward Longstreth medal 1919), Am. Phys. Soc.; mem. IEEE (life), Am. Econ. Assn., Sigma Xi. Achievements include development of certain specialized weighing and measuring instruments; design of novel instruments and devices used in product testing activities of Consumers' Research. Home: RR 4 Box 209 Washington NJ 07882-9013

SCHLOBOHM, JOHN CARL, electrical engineer; b. San Francisco, Aug. 7, 1931; s. John Tietje and Magda (Scheidtmann) S.; m. Margaret Anne Schlobohm, Jan. 25, 1958 (div. 1981); children: Bonnie Jean, Julie Anne; m. Karin Scholbohm, Aug. 13, 1983. BS, Stanford U., 1953, MS, 1955. Assoc. lab. dir. SRI Internat., Menlo Park, Calif., 1953-85; sr. staff engr. ESL/TRW, Sunnyvale, Calif., 1985-87, 91—; program mgr. Maxim Techs., Santa Clara, Calif., 1987-90, Mirage Systems, Santa Clara, 1990-91. Home: 1234 McIntosh Ct Sunnyvale CA 94087

SCHLOEMANN, ERNST FRITZ (RUDOLF AUGUST), physicist, engineer; b. Borgholzhausen, Germany, Dec. 13, 1926; came to U.S., 1954, naturalized, 1965; s. Hermann Wilhelm and Auguste Wilhelmine (Koch) S.; m. Gisela Mattiat, June 19, 1955; children—Susan C., Sonia, Barbara. BS, U. Göttingen, Fed. Republic of Germany, 1951, MS, 1953, PhD, 1954. With rsch. div. Raytheon Co., Lexington, Mass., 1955—; cons. scientist, 1964—; vis. assoc. prof. Stanford U., 1961-62; vis. prof. U. Hamburg, Fed. Republic Germany, 1966. Assoc. editor: Jour. Applied Physics, 1974-76; contbr. numerous articles to profl. jours. Fulbright fellow, 1954-55. Fellow IEEE, Am. Phys. Soc., Sigma Xi. Democrat. Unitarian. Achievements include patents in field. Office: Raytheon Co Rsch Div 131 Spring St Lexington MA 02173-7803

SCHLOEMER, PAUL GEORGE, diversified manufacturing company executive; b. Cin., July 29, 1928; s. Leo Bernard and Mary Loretta (Butler) S.; m. Virginia Katherine Grona, Aug. 28, 1954; children: Michael, Elizabeth, Stephen, Jane, Daniel, Thomas. BSME, U. Cin., 1951; MBA, Ohio State U., 1955. R & D engr. Wright Patterson AFB, Dayton, Ohio, 1951-52; R & D officer Wright Patterson AFB, Dayton, 1952; resident engr. Parker Hannifin Corp., Dayton, 1957; also Ea. area mgr. Parker Hannifin Corp., Huntsville, Ala., 1957-65; v.p. aerospace group Parker Hannifin Corp., Irvine, Calif., 1965-77; pres. aerospace group Parker Hannifin Corp., Irvine, 1977, corp. v.p., 1978-81, exec. v.p., 1981; pres. Parker Hannifin Corp., Cleve., 1982-84, CEO, 1984-93, ret., 1993; bd. dirs. Soc. Corp., Cleve., Rubbermaid Inc., AMP Inc., Parker Hannifin Corp., N.A.M. Capt. USAF, 1952-53. Mem. Machinery and Allied Products Inst. (mem. exec. com.), Aerospace Industry Assn., Conf. Bd., Inc., The Country Club, Big Canyon Country Club, The Pepper Pike Club. Republican. Roman Catholic. Office: Parker Hannifin Corp 17325 Euclid Ave Cleveland OH 44112-1290

SCHLOM, JEFFREY BERT, research scientist; b. N.Y.C., June 22, 1942; s. David and Anna (Klein) S.; children: Amy Melissa, Steven Michael. BS (Pres.'s scholar), Ohio State U., 1964; MS, Adelphi U., 1966; Ph.D., Rutgers U., 1969. Instr. Columbia Coll. Phys. and Surg., 1969-71, asst. prof., 1971-73; chmn. breast cancer virus segment Nat Cancer Inst., NIH, Bethesda, Md., 1973-76; chief lab. tumor immunology and biology Nat Cancer Inst., NIH, 1983—, head molecular oncology sect., 1976-83; prof. George Washington U., Washington, 1975—; disting. lectr. Can. Cancer Soc., 1985. Mem. numerous editorial bds.; contbr. numerous articles to profl. jours. Recipient Dir.'s award NIH, 1977, 89. Mem. Am. Assn. Cancer Rsch. (Rosenthal award 1985), Am. Soc. Cytology (Basic Rsch. award 1987). Office: Insts of Health Bldg 10 Rm 8B07 Bethesda MD 20892

SCHLOSE, WILLIAM TIMOTHY, health care executive; b. West Lafayette, Ind., May 16, 1948; s. William Fredrick and Dora Irene (Chitwood) S.; m. Linda Lee Fletcher, June 29, 1968 (div. 1978); children: Vanessa Janine Schlose Hubert, Stephanie Lynn; m. Kelly Marie Martin, June 6, 1987; 1 child, Taylor Jean Martin-Schlose. Student, Bowling Green State U., 1966-68, Long Beach City Coll., 1972-75; teaching credential, UCLA, 1975. Staff respiratory therapist St. Vincent's Med. Ctr., L.A., 1972-75; cardio-pulmonary chief Temple Community Hosp., L.A., 1975-76; adminstrv. dir. spl. svcs. Santa Fe Meml. Hosp., L.A., 1976-79; mem. mktg. and pub. rels. staff Nat. Med. Homecare Corp., Orange, Calif., 1979-81, Medtech of Calif., Inc., Burbank, Calif., 1981-84; regional mgr. Mediq Health Care Group Svcs., Inc., Chatsworth, Calif., 1984-88; pres. Baby Watch Homecare, Whittier, Calif., 1988-90, Tim Schlose and Assocs., Anaheim, Calif., 1990—; staff instr., Montebello (Calif.) Adult Schs. Author: Fundamental Respiratory Therapy Equipment, 1977. With USN, 1968-72. Mem. Am. Assn. Respiratory Care, Calif. Soc. Respiratory Care (past officer), Nat. Bd. Respiratory Care, Nat. Assn. Apnea Profls., Am. Assn. Physicians Assts., L.A. Pediatric Soc., Calif. Perinatal Assn., Porsche Owners Club L.A., Porsche Club Am. Republican. Methodist. Avocations: boating, automobile racing, automobile restoration, basketball, travel. Office: Tim Schlose and Assocs 1290 E Katella Ave Anaheim CA 92805-6627

SCHLOSSER, HERBERT, theoretical physicist; b. Bklyn., Nov. 18, 1929; s. Abraham and Yetta (Lichtenberg) S.; m. Martha Chiterer, Aug. 7, 1960; children: Rachelle, Arthur. BS in Physics, Bklyn. Coll., 1950; MS in Physics, Poly. U., 1952; PhD in Theoretical Physics, Carnegie-Mellon U., 1960. Sr. physicist Gen. Tel. & Electronics, Bayside, N.Y., 1960-62; specialist, physicist Rep. Aviation Corp., Farmingdale, N.Y., 1962-63; asst. prof. Poly. U., Bklyn., 1963-68; assoc. prof. physics Cleve. State U., 1968-72, prof. physics, 1972—; Fulbright Hayes lectr. Dept. of State, Sao Carlos, Brazil, 1966-67; cons. Rep. Aviation Corp., Farmingdale, 1963-64. 2d Lt. USAF, 1953. IBM fellow Carnegie-Mellon U., 1958-59, Sr. Weizmann fellow Weizmann Inst. Sci., Rehovot, Israel, 1973-74; grantee Army Rsch. Office, 1964-66, NASA, 1986, 92, 93. Mem. Am. Phys. Soc. Achievements include research in the applicability of universality to a wide range of problems in condensed matter theory. Office: Cleveland State U 1983 E 24th St Cleveland OH 44115

SCHLOSSMAN, MITCHELL LLOYD, cosmetics and chemical specialties executive; b. N.Y.C., Dec. 30, 1935; s. Jack Lewis and Rae (Wernick) S.; m. Barbara Nadell, Dec. 24, 1956; childrn: David Scott, Edye Gail, Julie Ilene. BS, NYU, 1956; postgrad., Bklyn. Coll., 1956-59. Group leader Revlon, Inc., Bronx, N.Y., 1957-64; mgr. research and devel. Pfizer, Inc., Parsippany, N.J., 1964-69; dir. tech. ops. Paris Cosmetics, Inc., Jersey City, 1969-70; exec. v.p. Prince Industries Ltd., Linden, N.J., 1970-74; v.p. Emery Personal Products, Linden, 1974-78; pres. Tevco, Inc., South Plainfield, N.J., 1978—; pres. Presperse, Inc., South Plainfield, 1985—, also bd. dirs.; cons. Hibernia (N.J.) Labs., 1985—, Kobo Products, Inc., South Plainfield, N.J., 1986—. Co-author: Chemical Manufacture of Cosmetics, 1974; contbr. articles to profl. and trade jours; patentee in field. Fellow Soc. Cosmetic Chemists (merit award 1971, chmn. N.Y. chpt. 1982—, bd. dirs. Eastern sect.), Am. Inst. Chemists; mem. Cosmetics, Toiletries, Fragrance Assn. (chmn. sci. program planning com. 1982—). Republican. Jewish. Lodge: KP. Home: 454 Prospect Ave # 164 West Orange NJ 07052-4192 Office: Tevco Inc 110 Pomponio Ave South Plainfield NJ 07080-1925

SCHLOSSMAN, STUART FRANKLIN, physician, educator, researcher; b. N.Y.C., Apr. 18, 1935; s. Abe and Pearl (Susser) S.; m. Judith Seryl Rubin, May 25, 1958; children: Robert, Peter. BA magna cum laude, NYU, 1955, MD, 1958; MA, Harvard U., 1975. Intern in medicine med. div. III Bellevue Hosp., N.Y.C., 1958-59, asst. resident in medicine med. div. III, 1959-60; Nat. Found. fellow dept. microbiology Coll. Physicians Columbia U., N.Y.C., 1960-62; asst. physician med. svc. Vanderbilt Clinic, Coll. Physician USPHS, Washington, 1960-62; Ward hematology fellow dept. internal medicine Sch. Washington U., St. Louis, 1962-63; rsch. assoc. lab. biochemistry Nat. Cancer Inst. USPHS, Washington, 1963-65; clin. instr. in medicine Sch. of Medicine George Washington U., 1964-65; assoc. in medicine, dir. blood bank Beth Israel Hosp., Boston, 1965-66; instr. Med. Sch. Harvard U., Boston, 1966-68, asst. physician, 1967-68, chief clin. immunology, 1971-73; physician Beth Israel Hosp., Boston, 1968—; from asst. to assoc. prof. medicine Harvard Med. Sch., Boston, 1968-77, prof., 1977—, Baruj Benacerraf prof. medicine, 1990—, chief div. tumor immunology and immunotherapy Dana-Farber Cancer Inst., 1973—; sr. physician Brigham and Women's Hosp., Boston, 1976—. Mem. editorial bd. Jour. of Immunology, 1969-74, Cellular Immunology, 1970—, Human Immunology, 1979-84, Clin. Immunology and Immunopathology, 1979—, Hybridoma, 1980—, Cancer Investigation, 1981, Stem Cells, 1981, Cancer Revs., 1984—, Internat. Jour. of Cell Cloning, 1983-86; mem. adv. bd. Cancer Treatment Reports, 1976-80; assoc. editor Human Lymphocyte Differentation, 1980-82; contbr. numerous articles to profl. jours. Recipient Solomon Berson Achievement award, 1984, Robert Koch prize and medal, 1984. Fellow AAAS; mem. NAS, Am. Soc. Hematology, Am. Soc. Immunologists, Am. Soc. Clin. Investigation, Assn. Am. Physicians, Alpha Omega Alpha, IOM. Office: Dana-Farber Cancer Inst 44 Binney St Boston MA 02115-6084

SCHLUETER, GERALD FRANCIS, mechanical engineering educator; b. Glendale, Ohio, Oct. 26, 1961; s. Joseph Clarence and Mary Elizebeth (Weisbrod) S.; m. Kimberlie Sue Mauck, Aug. 12, 1983; children: James David, Jason Andrew. BSME, U. Mo., Rolla, 1984. Engring. coop. Conoco, Inc., Houston, 1983; maintenance mgr. A.E. Staley Mfg. Co., Decatur, Ill., 1984—; key rep. A.E. Staley Plant Engring. and Maintenance Mgr. Conf., Indpls., 1989—. V.p. South Shores Basebal Players, Decatur, 1992; advisor, coach Decatur YMCA Soccer League, Decatur, 1990—; com. chair Holy Family PTG, Decatur, 1991-92. Recipient Scholastic scholarship Cin. Plumbers, 1979. Mem. ASME, Am. soc. Profl. Engrs. (Ctrl. Ill. chpt. sec./treas. 1992—). Republican. Roman Catholic. Achievements include modification of a "Shirco" furnace to regenerate powdered carbon increasing reliability from 40% to 86% in three months. Home: 2797 Deerpath Pk Dr Decatur IL 62521 Office: A E Staley Mfg Co 2200 E Eldorado St Decatur IL 62525

SCHMALBROCK, PETRA, medical physicist, researcher, educator; b. Münster, Fed. Republic of Germany, Sept. 14, 1954; came to U.S., 1983; d. Franz-Josef and Irmgard (Schmitz) S.; m. Evan R. Sugarbaker, Feb. 15, 1985; 1 child, Alex. F. MS, Westfälische-Wilhelms U., Münster, 1978, PhD, 1982. Postdoctoral fellow Ohio State U., Columbus, 1983-85, rsch. assoc., 1985-88; cons. GE Med. Systems, Milw., 1989-90; rsch. scientist Ohio State U., Columbus, 1990—; lectr. GE Med. Systems, Milw., 1987. Co-author: Fundamentals of MRI, 1991. Mem. Am. assn. for Physicists in Medicine, Soc. for Magnetic Resonance in Medicine, Soc. for Magnetic Resonance Imaging. Achievements include development of methods for magnetic resonance angiography; of methods of ultra-high resolution MRI of the inner ear and other small analomic structures. Office: Ohio State U MRI Facility 1630 Upham Dr Columbus OH 43210

SCHMAUS, SIEGFRIED H. A., consulting engineer; b. Muelheim/Ruhr, W. Ger., Dec. 23, 1915; s. Wilhelm Friedrich and Hedwig (Flader) S.; student Staatliche Ingineur Schule, Duisburg, W. Ger., 1940-41, Esslingen, W. Ger., 1945-46; m. A. Babette Schmid, Aug. 17, 1946. Apprentice-designer Demag A.G., Duisburg, 1930-36; designer/supr. Meissner, Cologne, W. Ger., 1936-38; designer aircraft engines Daimler-Benz A.G., Stuttgart, W. Ger., 1943-45; designer Fischer & Porter, Warminster, Pa., 1948-53, Ametek Inc., Sellersville, Pa., 1954-65; staff research engr. Fischer & Porter, Warminster, 1966-80; pres. Sensor Devel. Inc., Broomall, Pa., 1977-90, Sensor Research Inc., Phila., 1980-90. Patentee in field. V.p. Friends Hist. Rittenhouse Town. Served with German Luftwaffe, 1938-42. Recipient Hess Ingenuity award, 1962. Mem. Franklin Inst. (sr., silver mem.), Instrument Soc. Am. (sr.), Am. Soc. Mfg. Engrs., German Soc. Pa. (v.p. 1984, Founders medal 1987, Officer's Cross of the Gov. of Germany 1988), Masons. Republican. Lutheran. Home and Office: Penfield Downs 806 Powder Mill Ln Wynnewood PA 19096

SCHMEIDLER, NEAL FRANCIS, engineering executive; b. Hays, Kans., Feb. 29, 1948; s. Cyril John and Mildred Mary (Karlin) S.; m. Lorrinda Mary Brungardt, Jan. 31, 1950; children: Lori Ann, LaNette Renee, Lance Edward, LeAnna Karleen. BS in Math., Fort Hays State U., 1970; MS in Indsl. Engring., Kans. State U., 1973. Master engr. Trans World Airlines, Inc., Kansas City, Mo., 1973-78; chief indsl. engr. U.S. Dept. of the Army, Fort Carson, Colo., 1978-80; staff indsl. engr. U.S. Dept. of Agriculture, Washington, 1980-83; tech. dir. Tech. Applications, Inc., Alexandria, Va., 1983-86; v.p. engring. and tech. svcs. div. Standard Tech., Inc., Bethesda, Md., 1986-88; dir. indsl. engring. and ops rsch. svcs. Operational Technologies Svcs., Inc., Vienna, Va., 1989-91; pres. OMNI Engring. and Tech., Inc., McLean, Va., 1989—; dir. No. Va. Tech. Coun., 1993—; bd. mgrs. Washington Acad. Scis., 1993—. Guest (radio talk show) Basically Business, 1991; contbr. articles to profl. jours. Named Sr. Engr. of Yr., D.C. Coun. of Engring. and Archtl. Socs., 1991; recipient Spl. Act or Svc. award U.S. Dept. of Army, 1980. Mem. ABA (small bus. com. 1991-92), Inst. Indsl. Engrs. (sr., nat. capital chpt. bd. dirs. 1986-93, Award of Excellence 1982), Assn. Small Rsch., Engring. and Tech. Svcs. Cos., Air Traffic Control Assn., Fairfax County C. of C., No. Va. Tech. Coun., Va. Asset Fin. Corp., Washington Acad. Scis., Kappa Mu Epsilon, Sigma Pi Sigma. Office: OMNI Engring & Tech Inc 7921 Jones Branch Dr # 530 Mc Lean VA 22102-3306

SCHMELL, ELI DAVID, biotechnologist; b. Balt., Jan. 5, 1950; s. Abraham Judah and Shirley (Rosenfeld) S.; m. Judith Ann Davidowitz, June 20, 1971; children: Rivka Adina, Ronnit Miriam. BS in Physics and Math., CCNY, 1971; PhD in Cellular and Molecular Biology, Johns Hopkins U., 1976. Postdoctoral fellow Johns Hopkins U., 1977-79; sr. staff fellow NIH, Bethesda, Md., 1979-81; program mgr. molecular biology Office Naval Rsch., Arlington, Va., 1981-87; R&D project mgr. Interpharm Labs Ltd., Nes-Ziona, Israel, 1987-90; regulatory affairs mgr. Interpharm Labs Ltd., Nes-Ziona, 1990—; vis. scientist neurobiology Weizmann Inst. Sci., Rechovot, Israel, summer 1985; vis. scientist biology Haifa: Israel Inst. Tech., Israel, summer 1983. Contbr. chpts. to books, scientific papers, review articles to profl. jours. Recipient Assoc. Mems. Rsch. prize Am. Fertility Soc., 1981, staff and sr. staff fellowships NIH, 1979-81, postdoctoral fellowship Leukemia Soc. Am., 1976-78, predoctoral fellowship NIH, 1971-75. Mem. AAAS, Am. Soc. for Cell Biologists, Am. Soc. for Biochemistry and Molecular Biology, Regulatory Affairs Profls. Soc. Jewish. Achievements include identification of specific receptor for sperm on egg cell surface, development of nat. rsch. program (Office Naval Rsch.) for non-med. applications of biotechnology. Home: 8A Herzog St, Rechovot Israel 76482 Office: Interpharm Labs Ltd, Science Based Indsl Park, Nes Ziona Israel 76110

SCHMELTZ, EDWARD JAMES, engineering executive; b. Newark, June 22, 1949; s. Edward Leo and Loretta (Pittman) S.; m. Donna Hoppi Schmeltz, Sept. 28, 1974; children: Leigh Erin Wildes, Erik Edward. BS, N.J. Inst. Tech., 1971; M in Engring., Tex. A&M U., 1972. Registered profl. engr., N.Y., N.J., Conn. Rsch. asst. Tex. A&M U., College Station, 1971-72; coastal/ocean engr. F.R. Harris, Lake Success, N.Y., 1972-74; sr. coastal engr. PRC Harris, Lake Success, 1974-76; dept. mgr. PRC Engring., Lake Success, 1976-79; project mgr. PRC Engring., N.Y.C., 1980-87; v.p., dep. dir. N.Y. ops. Frederic R. Harris, Inc., N.Y.C., 1987-93, sr. v.p., dir. N.Y. ops.; lectr. George Washington U., Lehigh U.; mem. Coastal Structures steering com., 1979, 83. Contbr. articles to technical and profl. jours. Mem. Flood and Erosion Control Bd., Rep. Town Meeting, Greenwich, Conn., 1988—. Recipient Adm. Harris award Frederic R. Harris, Inc., N.Y.C., 1989. Fellow Soc. Am. Mil. Engring.; mem. ASCE, NSPE, Permanent Internat. Assn. Navigation Congresses. Roman Catholic. Avocations: golf, flying. Office: Frederic R Harris Inc 300 E 42nd St New York NY 10017-5947

SCHMETZER, WILLIAM MONTGOMERY, aeronautical engineer; b. Bronx, N.Y., Aug. 15, 1924; s. Richard Edward and Margaret Roy (Montgomery) S.; m. Emma Frances Myers, June 21, 1952 (div. 1972);

children: Margaret, Karen, Charles, Ann. BS in Aero. Engring., Purdue U., 1952; postgrad., Va. Poly., 1960-61. Project engr. Douglas Aircraft Co., Santa Monica, Long Beach, Calif., 1952-66; v.p., asst. chief engr. Strato Engring. Cons., Burbank, Calif., 1966-69; S. Robbins assoc. Mech. & Engring. Cons., Van Nuys, Calif., 1969-73; unit project engr. Hughes Aircraft, Culver City, Calif., 1973-76; project engr. Marquardt Co., Van Nuys, Calif., 1976-78; sr. design specialist Lockheed Skunk Works, Burbank, 1978-82; sr. preliminary designer Northrop B-2 Div., Pico Rivera, Calif., 1982—; mem. material rev. bd. Hughes & Marquardt Co., Culver City and Van Nuys, Calif., 1973-78. 1st sgt. U.S. Army, 1943-46, ETO. Mem. AIAA, Soc. Automotive Engrs. Republican. Presbyterian. Achievements include patent for thrust and lift airfoil currently being evaluated; development of flying prototype of my design of first short take-off commuter jet of 10 ton payload capacity. Home and Office: Schmetz Aircraft Corp 15740 Sherman Way Van Nuys CA 91406

SCHMID, ANTHONY PETER, materials scientist, engineer; b. Springfield, Mass., Mar. 27, 1931; s. Anthony Peter and Emma Louise (Meyer) S.; m. Amanda Lee Wall, June 11, 1960; children: Anthony Peter (dec.), Alden Lee. BS in Math., U. Md., 1957; MS in Physics, Northeastern U., 1961. Mgr. dielectric lab. Owens Ill. Tech. Ctr., Toledo, 1963-74; mgr. process engring. EMCON div. ITW, San Diego, 1974-81; chief engr. Siltec Packaging Div., San Diego, 1979-81; prin. scientist Sci. Applications Internat. Corp., San Diego, 1981-86; mgr. ceramic prototype Unisys-Semiconductor Memory and Packaging Assn., San Diego, 1986-91; sr. tech. analyst Alcoa Electronic Packaging, San Diego, 1991-92; dir. ceramic tech. Silicon Video Corp., Cupertino, Calif., 1992—; instr. U. Toledo, 1969-74. Contbr. articles to profl. jours. Cpl. USMC, 1952-54, Korea. Mem. IEEE (sr., sec. San Diego chpt. 1990-92), Internat. Soc. Hybrid Microelectronics, Am. Phys. Soc., Electronic Packaging Soc. Achievements include 9 patents in underground power lines, semiconductor devices, semiconductor materials, furnaces; research on polaron theory, junction potential theory. Office: Silicon Video Corp 10460 Bubb Rd Cupertino CA 92075

SCHMID, CHARLES ERNEST, acoustician, administrator; b. Jamaica, N.Y., Oct. 30, 1940; s. Edson Scofield Schmid and Agatha Sofia Zimmermann; m. Linda Dexter, June 18, 1966; children: Andrew, Jenny. BSEE, Cornell U., 1963; MSEE, U. Conn., 1968; PhD, U. Wash., 1977. Systems engr. Gen. Dynamics/Electric Boat, Groton, Conn., 1963-66; fellow Honeywell, Seattle, 1966-90; exec. dir. Acoustical Soc. Am., Woodbury, N.Y., 1990—. Bd. dirs. Wash. State Diversification Com., Olympia, 1993. Congl. Sci. Engring. fellow AAAS, Washington, 1985-86. Fellow Acoustical Soc. Am.; mem. Am. Inst. Physics (governing bd. 1991-93, exec. com. 1993—). Achievements include research in simulation and analysis of underwater sound. Office: Acoustical Soc of Am 500 Sunnyside Blvd Woodbury NY 11797-2924

SCHMID, HANS DIETER, civil engineering consultant; b. Znaim, Czechoslovakia, Apr. 25, 1939; s. Hans and Emma (Ledl) S.; m. Isolde Meissner, Dec. 7, 1966; children: Manuela, Christian. Diploma, Vienna (Austria) Tech. U., 1963, doctorate, 1966. Chief engr. Suiselectra, Basel, Switzerland, 1967-69; site mgr. Golf Coast Aluminum, Lake Charles, U.S., 1969-71; project engr. Nabalco, Sidney, Australia, 1971; owner, cons. AJS S.A., Neuchatel, Switzerland, 1971—. Mem. Vereinigung Schweizerischer Strassenfachleute, Association Suisse des Ingénieurs-Conseils, Société Suisse des Ingénieurs et Architectes (pres. 1987-89). Office: AJS SA, Musee 4, 2001 Neuchâtel Switzerland

SCHMID, HARALD HEINRICH OTTO, biochemistry educator, academic director; b. Graz, Styria, Austria, Dec. 10, 1935; Came to U.S., 1962; s. Engelbert and Annemarie (Kletetschka) S.; m. Patricia Caroline Igou, May 21, 1977. MS, U. Graz, 1957, LLD, 1962, PhD, 1964. Rsch. fellow Hormel inst. U. Minn., Austin, 1962-65, rsch. assoc., 1965-66, asst. prof., 1966-70, assoc. prof., 1970-74, prof., 1974—; cons. NIH, Bethesda, Md., 1977—; acting dir. Hormel inst. U. Minn., 1985-87, exec. dir., 1987—; lectr. Mayo Med. Sch., Rochester, Minn., 1990—. Mng. editor Chemistry and Physics of Lipids, Elsevier Sci. Publs., Amsterdam, The Netherlands, 1984—; contbr. numerous articles to profl. jours. Rsch. grantee NIH, 1967-93. Mem. AAAS, Am. Soc. Biochemistry and Molecular Biology, Am. Chem. Soc., The Oxygen Soc. Avocations: yacht racing, downhill skiing, tennis, classical music. Home: RR 3 Box 246 Austin MN 55912 Office: U Minn Hormel Inst 801 16th St NE Austin MN 55912

SCHMIDLY, DAVID J., dean; b. Levelland, Tex., Dec. 20, 1943; m. Janet Elaine Knox, June 2, 1966; children: Katherine Elaine, Brian James. BS in Biology, Tex. Tech U., 1966, MS in Zoology, 1968; PhD in Zoology, U. Ill., 1971. From asst. prof. to prof. dept. wildlife fisheries scis. Tex. A&M U., College Station, 1971-82, prof., 1982—, head dept. wildlife, 1986-92; CEO, campus dean Tex. A&M U., Galveston, 1992—; chief curator Tex. Coop. Wildlife Coll., College Station, 1983-86; cons. Nat. Park Svc., Wildlife Assocs., Walton and Assocs., Environomics, Continental Shelf Assn., LGL.; press adv. com. Tex. A&M U., 1983—; charter mem. Tex. A&M U. Faculty Senate, 1983-85; chmn. Scholarship Com., 1978-82; bd. advisors Tex. Inst. Oceanography, 1992; lectr. various workshops and seminars. Author: The Mammals of Trans-Pecos Texas including Big Bend National Park and Guadalupe Mountains National Park, 1977, Texas Mammals East of the Balcones Fault Zone, 1983, The Bats of Texas, 1991; contbr. articles to profl. jours. Chmn. mammal subcom., Tex. Orgn. Endangerd Species, 1973-74; trustee Brazos Valley Mus. Nat. Sci., 1978-80, Tex. Nature Conservancy, 1991; tech adv. com., Tex. Parks and Wildlife Dept., 1991, blue ribbon elk panel, 1988; environ. adv. com. Tex. Gen. Land Office, Tex. Dept. Agr., 1990. Recipient Dist. Prof. award Assn. Grad. Wildlife and Fisheries Scis., 1985, Donald W. Tinkle Rsch. Excellence award Southwestern Assn. Naturalists, 1988, Diploma Recognition La Universidad Autonoma de Guadalajara, 1989, La Universidad Autonoma de Tamaulipas, 1990. Fellow Tex. Soc. Sci. (bd. dirs. 1979-81); mem. AAAS, Am. Soc. Mammalogists (life, editor Jour. Mammalogy 1975-78), Am. Inst. Biol. Scis. (bd. dirs. 1993—, coun. affiliate socs. 1989—), Am. Naturalist, Soc. Marine Mammalogy (charter mem.), Soc. Systematic Zoology, The Wildlife Soc. Soc. Conservation Biology, Nat. Geog. Sci. Soc., S.W. Assn. Naturalists (life mem., bd. govs. 1980-86, 91—, pres. 1981, trustee 1986—), Tex. Mammal Soc. (pres. 1985-86), Chihuahuan Desert Rsch. Inst. (v.p. bd. scientists 1982—, bd. dirs. 1991), Mexican Soc. Mammalogists, Sigma Xi (v.p. 1986-87, pres. 1987-88), Disting. Scientist award 1991), Beta Beta Beta. Phi Sigma, Phi Kapa Phi. Home: 4 Cadena Ct Galveston TX 77551 Office: Tex A&M U at Galveston Office of Dean PO Box 1675 Galveston TX 77553-1675

SCHMID-SCHOENBEIN, GEERT WILFRIED, biomedical engineer, educator; b. Albstadt, Baden-Wurttemberg, Germany, Jan. 1, 1948; came to U.S., 1971; s. Ernst and Ursula Schmid; m. Renate Schoenbein, July 3, 1976; children: Philip, Mark, Peter. Vordiplom, Liebig U., Giessen, Germany, 1971; MS & PhD in Bioengring., U. Calif., San Diego, 1973. Staff assoc. dept. physiology Columbia U., N.Y.C., 1976-77, sr. assoc., 1977-79; asst. prof. dept. AMES U. Calif., San Diego, 1979-84, assoc. prof., 1984-89, prof., 1989—. Editor: Frontiers in Biomechanics, 1986. Recipient Nelville medal ASME, 1990. Fellow Am. Inst. for Med. & Biol. Engring., Am. Heart Assn.; mem. ASME (Melville medal 1990), BMES (pres. 1991-92), Am. Microcirculatory Soc., European Microcirculatory Soc., Am. Physiol. Soc. Achievements include bioengineering research of cardiovascular disease. Office: Univ of California Bioengineering Division R-012 La Jolla CA 92093

SCHMIDT, CONSTANCE ROJKO, psychology educator, researcher; b. Washington, Nov. 19, 1954; d. Anthony S. and Genevieve L. (Brown) Rojko; m. Stephen R. Schmidt, May 24, 1975; children: Theresa L. and Kathleen E. BS in psychology, Univ. Va., 1976; MS in psychology, Purdue Univ., 1977, PhD in psychology, 1980. Asst. prof., assoc. prof. Va. Tech. Univ., Blacksburg, 1980-88; rsch. assoc. Vanderbilt Univ., Nashville, Tenn., 1988-90; from assoc. prof. to full prof. Middle Tenn. State U., Murfreesboro, 1989—. Contbr. articles to profl. jours. Recipient Rsch. Init. grant NSF, 1981-83. Mem. Soc. Rsch. in Child Devel., Am. Psychological Soc., Tenn. Acad. Sci. Democrat. Methodist. Office: Middle Tenn State U Box 508 Murfreesboro TN 37132

SCHMIDT, GREGORY MARTIN, osteopathic physician, independent oil producer; b. Oklahoma City, Dec. 30, 1952; s. William Charles and Anna Maude (Fore) S.; m. Linda Louise Seay, Dec. 23, 1978; 1 child, Austin Daniel. DO, Okla. State U., 1979. Pres., owner Woodland Hills Med. Clinic, Tulsa, 1980—; ind. oil producer. Mem. Am. Osteo. Assn., Okla. Osteo. Assn. Office: Woodland Hills Med Clinic 8212 73d St S Tulsa OK 74133

SCHMIDT, JAMES ROBERT, facilities engineer; b. Rome, N.Y., Sept. 22, 1932; s. Floyd Vincent and Sophia Louise (Halupka) S.; m. Suzanne Mae Thrasher, Mar. 27, 1965; children: Mark Adrian, Tara Lee. AAS, Mohawk Valley Community Coll., 1951; BS, Utica Coll., 1962; MBA, Syracuse U., 1967. Registered profl. engr., N.Y., Mass. Draftsman Rome (N.Y.) Cable Corp., 1951-55, designer, 1956-67, project coord., 1968-69; sr. project engr. Xerox Corp., Webster, N.Y., 1970-81; mgr. plant engring. Savin corp., Binghamton, N.Y., 1982-83; constrn. surveillance Horizons Tech./USAF, Thule, Greenland, 1983-84; sr. facilities engr. Horizons Tech., Billerica, Mass., 1984—. Chmn. Community Action Steering Com., Lee Center, N.Y., 1968; mem. ch. ofcl. bd. Meth. Ch., Shortsville, N.Y., 1979. With U.S. Army, 1951-54. Mem. Am. Inst. Plant Engrs., Am. Mil. Engrs., Cogeneration Inst. (charter), Assn. Energy Engrs., Nat. Soc. Profl. Engrs. Achievements include research on add-on system to control electrical surges in an electrical supply cable. Office: Horizons Tech Inc 700 Tech Park Dr Billerica MA 01821

SCHMIDT, JOHN THOMAS, neurobiologist; b. Louisville, Sept. 25, 1949; s. Adolph William and Olivia Ann (Hohl) S.; m. Marilyn Joan Gough, Jan. 6, 1979; children: Sarah, Benjamin. BS in Physics, U. Detroit, 1971; PhD in Biophysics, U. Mich., 1976. Postdoctoral assoc. Nat. Inst. for Med. Rsch., London, 1976-77, Vanderbilt U. Med. Sch., Nashville, 1977-80; asst. prof. biol. scis. SUNY, Albany, 1980-85, assoc. prof. biol. scis., 1985—, dir. Neurobiology Rsch. Ctr., 1988—. Editor: Activity-Driven CNS Changes, 1991. Mem. Soc. for Neuroscis. (treas. Hudson Berkshire chpt. 1981-83, pres. 1987-89), N.Y. Acad. Scis. Office: SUNY Dept Biol Sci 1400 Washington Ave Albany NY 12222

SCHMIDT, JOSEPH DAVID, urologist; b. Chgo., July 29, 1937; s. Louis and Marian (Fleigel) S.; m. Andrea Maxine Herman, Oct. 28, 1962. BS in Medicine, U. Ill., 1959, MD, 1961. Diplomate Am. Bd. Urology. Rotating intern Presbyn. St. Luke's Hosp., Chgo., 1961-62, resident in surgery, 1962-63; resident in urology The Johns Hopkins Hosp., Balt., 1963-67; faculty U. Iowa Coll. Medicine, Iowa City, 1969-76; faculty U. Calif., San Diego, 1976—, prof., head div. urology, 1976—; cons. U.S. Dept. Navy, San Diego, 1976—; attending urologist Vets. Affairs Dept., San Diego, 1976—. Author, editor: Gynecological and Obstetric Urology, 1978, 82, 93. Capt. USAF, 1967-69. Recipient Francis Senear award. U. Ill., 1961. Fellow Am. Coll. of Surgeons; mem. AMA, Am. Urol. Assn. Inc., Alpha Omega Alpha. Avocations: collecting antique medical books, manuscripts. Office: U Calif San Diego Med Ctr Div Urology 200 W Arbor Dr San Diego CA 92103-8897

SCHMIDT, MAARTEN, astronomy educator; b. Groningen, Netherlands, Dec. 28, 1929; came to U.S., 1959; s. Wilhelm and Antje (Haringhuizen S.; m. Cornelia Johanna Tom, Sept. 16, 1955; children—Elizabeth Tjimkje, Maryke Antje, Anne Wilhelmina. B.Sc., U. Groningen, 1949; Ph.D., Leiden U., Netherlands, 1956; Sc.D., Yale U., 1966. Sci. officer Leiden Obs., The Netherlands, 1953-59; postdoctoral fellow Mt. Wilson Obs., Pasadena, Calif., 1956-58; mem. faculty Calif. Inst. Tech., 1959—, prof. astronomy, 1964—, exec. officer for astronomy, 1972-75, chmn. div. physics, math. and astronomy, 1975-78; mem. staff Hale Obs., 1959-80, dir., 1978-80. Co-winner Calif. Scientist of Yr. award, 1964. Fellow Am. Acad. Arts and Scis. (Rumford award 1968); mem. Am. Astron. Soc. (Helen B. Warner prize 1964, Russell lecture award 1978), NAS (fgn. assoc., recip. James Craig Watson Medal, 1991), Internat. Astron. Union, Royal Astron. Soc. (assoc., Gold medal 1980). Office: Calif Inst Tech 105 24 Robinson Lab 1201 E California Blvd Pasadena CA 91125

SCHMIDT, ROBERT, mechanical and civil engineering educator; b. Ukraine, May 18, 1927; came to U.S., 1949, naturalized, 1956; s. Alfred and Aquilina (Konotop) S.; m. Irene Hubertine Bongartz, June 10, 1978; children: Ingbert Robert. B.S., U. Colo., 1951, M.S., 1953; Ph.D., U. Ill., 1956. Rsch. asst. U. Ill., 1953-56; asst. prof. mechanics U. Ill., Urbana, 1956-59; assoc. prof. U. Ariz., Tucson, 1959-63; prof. mechanics and civil engring. U. Detroit, 1963—, chmn. civil engring. dept., 1978-80; researcher in linear and nonlinear theory of elasticity and approximate methods of analysis. Editor: Indsl. Math., 1969—; book reviewer Applied Mechanics Rev., Indsl. Math. jour.; contbr. over 130 articles to profl. jours. With C.E., U.S. Army, 1951-52. Grantee NSF 1960-78. Mem. AAUP, ASCE, ASME (cert. recognition 1972), Am. Acad. Mechanics (a founder), Indsl. Math. Soc. (pres. 1966-67, 81-84, 1st Gold award 1986), Sigma Xi. Avocations: biosophy, walking, bicycling, swimming. Office: U Detroit Coll Engring Detroit MI 48221

SCHMIDT, ROBERT MILTON, physician, scientist, educator; b. Milw., May 7, 1944; s. Milton W. and Edith J. (Martinek) S.; children Eric Whitney, Edward Huntington. AB, Northwestern U., 1966; MD, Columbia U., 1970; MPH, Harvard U., 1975; PhD in Law, Medicine and Pub. Policy, Emory U., 1982. Diplomate Am. Bd. Preventive Medicine. Resident in internal medicine Univ. Hosp. U. Calif.-San Diego, 1970-71; resident in preventive medicine Ctrs. Disease Control, Atlanta, 1971-74; commd. med. officer USPHS, 1971; advanced through grades to comdr., 1973; dir. hematology div. Nat. Ctr. for Disease Control, Atlanta, 1971-78, spl. asst. to dir., 1978-79, inactive res., 1979—; clin. asst. prof. pediatrics Tufts U. Med. Sch., 1976-86; clin. asst. prof. medicine Emory U. Med. Sch., 1976-81, clin. asst. prof. community health, 1976-86; clin. assoc. prof. humanities in medicine Morehouse Med. Sch., 1977-79; attending physician dept. medicine Wilcox Meml. Hosp., Lihue, Hawaii, 1979-82; cons. physician dept. medicine California Pacific Presbyn. Med. Ctr., San Francisco, 1983—; dir. Ctr. Preventive Medicine and Health Research, 1983—, dir. Health Watch, 1983—; sr. scientist Inst. Epidemiology and Behavioral Medicine, Inst. Cancer Research, Calif. Pacific Med. Ctr., San Francisco, 1983-88; prof. hematology and gerontology, dir. Ctr. Preventive Medicine and Health Rsch., chair health professions program San Francisco State U., 1983—; cons. WHO, FDA, Washington, NIH, Bethesda, Md., Govt. of China, Mayo Clinic, Rochester, Minn., Northwestern U., Evanston, Ill., U. R.I., Kingston, Pan Am. Health Orgn., Inst. Pub. Health, Italy, Nat. Inst. Aging Rsch. Ctr., Balt., other univs., govts., profl. socs.; vis. rsch. prof. gerontology Ariz. State U., 1989-90; mem. numerous sci. and profl. adv. bds., panels, coms. Mem. editorial bd. Am. Jour. Clin. Pathology, 1976-82, The Advisor, 1988—, Generations, 1989—; book and film reviewer Sci. Books and Films, 1988—; author: 17 books and manuals including Hematology Laboratory Series, 4 vols., 1979-86, CRC Handbook Series in Clinical Laboratory Science, 1976—; assoc. editor: Contemporary Gerontology, 1993—; contbr. over 240 articles to sci. jours. Alumni regent Columbia U. Coll. Physicians and Surgeons, 1980—. Northwestern U. scholar, 1964-66; NSF fellow, 1964-66; Health Professions scholar, 1966-70; USPHS fellow, 1967-70; Microbiology, Urology, Upjohn Achievement, Borden Rsch. and Virginia Kneeland Frantz scholar awards Columbia U., 1970; recipient Am. Soc. Pharmacol. and Exptl. Therapy award in pharmacology, 1970, Commendation medal USPHS, 1973, Leadership Recognition awards San Francisco State U., 1984-89, 91-93, Meritorious Performance and Profl. Promise award, 1989, Meritorious Svc. award, San Francisco State U., 1992, Student Disting. Teaching and Svc. award Pre-Health Professions Student Alliance, 1992. Fellow ACP, AAAS, Royal Soc. Medicine (London), Gerontol. Soc. Am., Am. Geriatrics Soc., Am. Coll. Preventive Medicine (sci. com.), Am. Soc. Clin. Pathology, Internat. Soc. Hematology, Royal Soc. Medicine (London); mem. AMA, Am. Med. Informatics, Am. Pub. Health Assn., Internat. Commn. for Standardization in Hematology, Am. Soc. Hematology, Internat. Soc. Thrombosis and Hemostasis, Acad. Clin. Lab Physicians and Scientists, Am. Assn. Blood Banks, Nat. Assn. Advisors for Health Professions (bd. dirs.), Am. Assn. Med. Systems and Informatics, Assn. San Francisco Med. Soc., Calif. Med. Soc., Calif. Med. Assn., San Francisco Med. Soc., Calif. Coun. Gerontology and Geriatrics, Am. Coll. Occupational and Environ. Medicine, Assn. Tchrs. Preventive Medicine (edn. com., rsch. com.), Am. Soc. Microbiology, Am. Soc. Aging (book reviewer 1990—, Dychtwald Pub. Speaking award 1991), N.Y. Acad. Sci., Am. Soc. Microbiology, Internat. Health Evaluation Assn. (vice pres. for Ams. 1992—, bd. dirs.), Golden Key (hon. faculty mem.), Sigma Xi, others. Club: Army and Navy (Washington). Home: 25 Hinckley Walk San Francisco CA 94111-2303 Officelf Pacific Med Ctr PO Box 7999 San Francisco CA 94120-7999 also: San Francisco State U Sch Sci 1600 Holloway Ave San Francisco CA 94132-1789

SCHMIDT, STEFAN, mechanical engineer, economist; b. Darmstadt, Hessen, Germany, 1950; s. Josef and Elisabeth (Urnauer) S.; m. Elizabeth Anne Godwin-Schmidt, Nov. 11, 1983; children: Rebecca Elizabeth, Benjamin Stefan Godwin. Diploma mech. engring., Fachhochschule Frankfurt, Germany, 1972; diploma indsl. engring. and mgmt., Technische U. Berlin, Germany, 1980. Profl. mech. engr. Engr. Mech. Engring. Lab., Tokyo, 1973, C & P Telephone Co. of Md., Balt., 1978-79; planning engr. Systemtechnik, Darmstadt, Germany, 1980; rsch. assoc. Univ. Dortmund, Germany, 1981-84; logistics, quality assurance engr. BMW AG, Munich, 1984—; lectr. in field; mgmt. trainer Centre of Technol. Cooperation, Berlin, 1985—; seminar presenter C. of C., Passau, Germany, 1986—. Contbr. articles to profl. jours. Econ. commentator Internat. Newspapers and Jours., Frankfurt, Munich, Berlin, 1990—. Fellow Verein Deutscher Wirt. ing. Home: Fritzstrasse 41, D-82140 Olching Bayern, Germany Office: BMW AG, D-80788 Munich Germany

SCHMIDT, STEPHEN CHRISTOPHER, agricultural economist, educator; b. Isztimer, Hungary, Dec. 20, 1920; came to U.S., 1949, naturalized, 1965; s. Francis Michael and Anne Marie (Angeli) S.; m. Susan M. Varszegi, Dec. 20, 1945; children—Stephen Peter, David William. Dr.Sc., U. Budapest, Hungary, 1945; Ph.D., McGill U., Montreal, Que., Can., 1958. Asst. head dept. Hungary Ministry Commerce, Budapest, 1947-48; asst. prof. U. Ky., Lexington, 1955-57, Mont. State U., Bozeman, 1957-59; asst. prof. U. Ill., Urbana-Champaign, 1959-63, assoc. prof., 1963-70, prof. agrl. mktg. and policy, 1970—; cons. U.S. Tariff Commn., 1969-71; contract researcher U.S. Dept. Agr., 1979-81, 86-87; mem. faculty dept. agr. Lincoln Acad. Ill., 1966-75; cons. agriculture directorate of the Orgn. for Econ. Co-operation and Devel., 1992. Fulbright grantee Bulgaria, 1992-93; Ford Found. fellow, 1959; Agrl. Devel. Coun. grantee, 1966, U. Man. Rsch. fellow, 1968-69, Ford Found. rsch. grantee, 1973, 74, Whitehall found. grantee, 1979, Internat. Inst. Applied Systems Analyses (Laxenburg, Austria) rsch. scholar, 1976-77, USDA Intergovtl. Personnel Act grantee, 1983-84. Mem. Am. Agrl. Econs. Assn. (award 1979), Internat. Assn. Agrl. Economists, Am. Assn. Advancement Slavic Studies, Ea. Econ. Assn., Sigma Xi, Gamma Sigma Delta. Office: 1301 W Gregory Dr Urbana IL 61801-3608

SCHMIDT, WALTER FRIEDRICH, research chemist; b. Visbeck, Germany, June 6, 1948; came to U.S., 1951; s. Friedrich Wilhelm and Elizabeth (Barndt) S.; m. Donna Ann Emeigh, Feb. 14, 1978; 1 child, David. BS in Chemistry, U. Del., 1970; MS in Pharmacy, U. Wis., 1985; PhD, U. Ga., 1989. Chemist Princeton (Pa.) Applied Rsch., 1973-74; chemist FDA, Phila., 1974-77, Atlanta, 1977-83; rsch. chemist Wyeth Lab., Radnor, Pa., 1985-86; rsch. scientist Agrl. Rsch. Svc. USDA, Beltsville, Md., 1989—. Contbr. articles to profl. jours. Mem. Am. Assn. Pharm. Scis., Am. Pharm. Assn., Sigma Xi. Office: USDA Agr Rsch Svc ECL 1300 Baltimore Ave Beltsville MD 20705

SCHMIDT, WALTER J., exploration and mineral economist; b. Urbau, Austria, Aug. 11, 1923; s. Emil J. Schmidt and Gisela Hammerschmidt; m. Emma Rogner, Jan. 31, 1976. PhD, U. Vienna, 1949; D Habil., Tech. U., Vienna, 1952. Various positions Austrian Civil Svc., Internat. Mining & Petroleum Co.'s; prof. emeritus exploration and mineral econs. Univ. for Mining and Metallurgy, Leoben, Austria, 1990—; cons. various co.'s and internat. orgns. Contbr. articles to profl. jours. Trustee Austrian Nat. Sci. Found., 1982-90. Mem. Austrian Geol. Soc. (chmn. 1990—). Avocation: modern art. Home: Lustkandlgasse 44, A1090 Vienna Austria Office: U Mining and Metal, A8700 Leoben Austria

SCHMIDT, WILLIAM JOSEPH, engineering executive; b. Cin., July 27, 1946; s. Peter Joseph and Viola Barbara (Beckman) S.; m. Judy Ann Lokai; 1 child, Jo Ann; m. Susan Marie Osterday, Jan. 22, 1979; children: Jennifer Lynn, Kira Lee, Laura Su. BS in Mech. Engring., U. Dayton, 1969, MS in Mech. Engring., 1974; postgrad., Earlham Coll., 1982-83, U. Wis., 1987-88. Process engr. Dow Chem. Co., Cleve., 1968; engring. mgr. Inland div. GM, Dayton, Ohio, 1969-81; chief engr. Vernay Labs. Inc., Yellow Springs, Ohio, 1981-85, v.p. engring., 1985—. Patentee in field for Fluid Flow Controller, Metering Valve for Dispensing Aerosols, Variable Rate Flow Controller; over 32 U.S. and fgn. patents pending; developer di-electric curing system for wet-offset printing, efficient and reliable resistant welding systems for low carbon to stainless steel, mfg. systems for passive restraints (air bags) for automotive applications, indsl. utilization of impact machining of steel in closed die format; pioneer in rubber injection molding techniques for dynamic response automotive engine mounts. Foster parent Green County Children's Svcs., Xenia, Ohio, 1987—; advisor 4-H. Mem. ASME (all exec. 1971-79, appreciation award 1978), SAE, Am. Inst. Indsl. Engrs. (bd. dirs. Dayton sect. 1990—), Soc. Mfg. Engrs. Avocations: breeding, training Arabian horses, guitar, skiing. Office: Vernay Labs Inc 120 E South College St Yellow Springs OH 45387-1623

SCHMIDTMANN, LUCIE ANN, engineer; b. Jamaica, N.Y., Oct. 22, 1963; d. Otto Stanislaus and Nancy Dorothy (Koonmen) S. BS in Computer Sci., Siena Coll., Loudonville, N.Y., 1985; MS in Computer Sci., Stevens Inst. Tech., Hoboken, N.J., 1989; student, U.S. Coast Guard Acad., New London, Conn., 1981-82, St. John's U., 1980-81, 83; postgrad., Polytech U. Bklyn., Bklyn., 1990—. Computer cons. dept. computer sci. Siena Coll., Loudonville, N.Y., 1984-85; figure clk. King Kullen Grocery Co., Westbury, N.Y., 1983-85; project mgr. AT&T Bell Labs., Whippany, N.J., 1985-88; asst. to rsch. and devel. mgr. AT&T Bell Labs., 1988-89, system/software engr., 1990-93, systems test engr., 1993—; source selection cons. Highpoint Condominium Assn., Stanhope, N.J., 1989-92, bd. dirs. 1990-92; computer cons. Champcare Inc., Davenport, Iowa, 1989-91; head math judge North Jersey Regional Sci. Fair, 1990, head math and computer sci. judge, 1992-93, math. and computer sci. judge, 1992-93. Vol. N.J. Spl. Olympics, Area 3, Flanders, N.J., 1985—, vol. coord., 1985-93, design/graphic artist, 1989. With USCG, 1981-82. Recipient Vol. award, N.J. Spl. Olympics Area 3, 1989. Mem. ACM (vice chmn. 1984-85, capt. programming team 1984-85), IEEE, IEEE Computer Soc., Math. Assn. Am., Performance Mgmt. Assn. (mem. North Jersey chpt. planning com. 1990, sec. 1990-91), Upsilon Pi Epsilon. Republican. Roman Catholic. Home: 10-186 Dell Pl Stanhope NJ 07874 Office: AT&T Bell Labs 67 Whippany Rd PO Box 903 Whippany NJ 07981-0903

SCHMIDTMANN, VICTOR HENRY, instrumentation engineer; b. Benton HArbor, Mich., Nov. 8, 1917; s. Arthur Henry and Alma Amolia (Giese) S.; m. Agnes Patricia MacFarlane, June 20, 1948; children: Victoria, Warren. BA, U. Redlands, 1940. Analytical chemist Richfield Oil Corp. Rsch. & Devel., Anaheim, Calif., 1955-60; pres. Rsch. Mfg. Corp., San Diego, 1960-67; owner Hillcrest Auto Supply, 1968-73; svc. mgr. Monitor Labs. Inc., 1974-84; applications mgr. Telos Labs. Inc., Fremont, Calif., 1984-90; sr. engr. Data Systems Svcs., San Diego, 1990—; co. rep. Coord. Rsch. Coun., N.Y.C., 1948, 60. Moderator Bethel Bapt. Ch., Anaheim, 1955-56, 1st Bapt. Ch., La Mesa, Calif., 1978-83, Windsor Hills Bapt. Ch., La Mesa, 1991-92. Co.y rep. Coord. Rsch. Coun., N.Y.C., 1948, 60; moderator Bethel Bapt. Ch., Anaheim, 1955-56, 1st Bapt. Ch., La Mesa, Calif., 1978-83, Windsor Hills Bapt. Ch., La Mesa, 1991-92. Republican. Achievements include co-inventor underwater release device. Home: 7928 Pasadena Ave La Mesa CA 91941-7850 Office: Data Systems Svcs 9020 Danube Ln San Diego CA 92126

SCHMIDT-NIELSON, KNUT, physiologist, educator; b. Trondheim, Norway, Sept. 24, 1915. Mag. Sci., Copenhagen U., 1941, PhD in Zoophysiology, 1946; DM (hon.), Lund (Sweden) U., 1985. Rsch. assoc. Swarthmore Coll., 1946-48, Stanford U., 1948-49; asst. prof. coll. medicine U. Cin., 1949-52, prof., 1952-63; James B. Duke prof. physiology Duke U., Durham, N.C., 1963—; mem. biomedical engring. adv. com. Duke U., 1968-85; docent U. Oslo, 1947-49; cons. NSF, 1957-61; Brody Meml lectr. U. Mo., 1962; mem. sci. adv. com. New England Regional Primate Rsch. Ctr., Harvard Med. Sch., 1962-66; mem. nat. adv. bd. physiol. rsch. lab. Scripps Inst., U. Calif., 1963-69, chmn., 1968-69; mem. subcom. environ. physiology U.S. Nat. Com. Internat. Biol. Prog., 1965-67; mem. com. rsch. utilization uncommon animals divsn. biology and agriculture NAS, 1966-68; mem. U.S. Nat. Com. Internat. Union Physiol. Sci., 1966-78, vice-chmn., 1969-78; mem. animal resources adv. com. NIH, 1968; regent's lectr. U. Calif., 1963; Wellcome vis. prof. U. S.D., 1988—. Editor Jour. Exptl. Biology, 1975-79, 83-86; sect. editor Am. Jour. Physiology, 1961-64, Am. Jour. Applied Physiology, 1961-64. Trustee Mt. Desert Island Biol. Lab., 1958-61. Guggenheim fellow U. Algeria, 1953-54; recipient Rsch. Career award USPHS, 1964-85, Japan prize Sci. and Tech. Found. Japan, 1992. Fellow AAAS, Am. Acad. Arts and Scis., Am. Physiol. Soc., N.Y. Acad. Sci., Norwegian Acad. Sci.; mem. NAS, Internat. Union Physiological Soc. (pres. 1980-86), Royal Soc. London (fgn.), French Acad. Sci. (fgn.), Norwegian Acad. Sci. (fgn.), Royal Danish Acad. (fgn.). Achievements include research in comparative physiology, respiration and oxygen supply, water metabolism and excretion, tempurature regulation, physiology of desert animals. Office: Duke U Dept. Physiology Durham NC 27706*

SCHMITT, FREDERICK ADRIAN, gerontologist, neuropsychologist; b. Cin., July 22, 1953; s. Werner and L. Gerlinde (Adrian) S.; m. Melinda Greenlese, Oct. 16, 1984. B.S., Rensselaer Poly., 1975; Ph.D., U. Akron, 1982. Lic. psychologist, N.C., Ky. Postdoctoral fellow Duke Aging Ctr., Durham, N.C., 1981-83, fellow in geriatrics, 1983-84; vis. asst. prof. psychology, U. N.C., Chapel Hill, 1984-85; rsch. assoc. Duke Med. Ctr., Durham, 1984-85; dir. neuropsychology svc. U. Ky., Lexington, 1985—, assoc. prof. neurology, 1990—, adj. prof. psychiatry, adj. prof. dept. psychology; assoc. Sanders-Brown Ctr. on Aging; cons. Shaw U., Raleigh, N.C., 1983-84, NIMH Office AIDS Programs, Am. Found. for AIDS Rsch. Contbr. chpts. to books, articles to profl. jours. Mem. AAAS, APA, Am. Acad. Neurology, Nat. Acad. Neuropsychology, N.Y. Acad. Sci, Gerontol. Soc. Am., Internat. Neuropsychol. Soc., Soc. for Research Child Devel., Soc. for Neurosci. Mem. of Baha'i Faith. Office: U Ky Med Ctr Dept Neurology Lexington KY 40536

SCHMITT, HARRISON HAGAN, former senator, geologist, astronaut, consultant; b. Santa Rita, N.Mex., July 3, 1935; s. Harrison A. and Ethel (Hagan) S. BA, Calif. Inst. Tech., 1957; postgrad. (Fulbright fellow), U. Oslo, 1957-58; PhD (NSF fellow), Harvard U., 1964; hon. degree, Franklin and Marshall Coll., 1977, Colo Sch. Mines, 1971, Rensselaer Poly Inst., 1973. Geologist U.S Geol. Survey, 1964-65; astronaut NASA, 1965; lunar module pilot Apollo 17, 1972, spl. asst. to adminstr., 1974; asst. adminstr. Offcie Energy Programs, 1974; mem. U.S. Senate, N.Mex., 1977-83; cons., 1983—; mem. Pres.'s Fgn. Intelligence Adv. Bd., 1984-85, Army Sci. Bd., 1985—, Pres.'s Ethics Commn., 1989; bd. dir. Nord Resources, Sunwest Fin. Svcs., Orbital Scis. Corp. Trustee Lovelace Med. Found. Recipient MSC Superior Achievement award, 1970, Disting. Svc. medal NASA, 1973, Lovelace award NASA, 1989, Gilbert award GSA, 1989. Mem. AIAA, AAAS, Geol. Soc. Am., Am. Geophys. Union, Am. Assn. Petroleum Geologists. Home: PO Box 14338 Albuquerque NM 87191-4338

SCHMITT, HEINZ-JOSEF, physician; b. Wiesbaden, Germany, May 19, 1954; s. Alois and Elisabeth (Weil) S.; m. Barbara Josephine Zell, Aug. 13, 1982. MD, Gutenberg U., 1980. Cert. Bd. Pediatrics, Bd. Microbiology, Germany. Asst. physician Rodenwaldt Microbiology Lab., Koblenz, 1980-82; asst. psysician Childrens Hosp Gutenberg U., 1983-85; fellow in infectious diseases Meml. Sloan Kettering Cancer Ctr., N.Y.C., 1986-88, asst. physician, 1989-90, attending physician, 1990—; habilitation H. J. Gutenberg U, Mainz., Germany, 1993. Mem. Am. Soc. Microbiology, Deutsche Gesellschaft für Kinderheilkunde, Deutsche Gesellschaft für Pädiatrische Infektiologie, Am. Soc. Infectious Diseases. Roman Catholic. Avocations: piano, organ. Office: Childrens Hosp, Langenbeckstrasse 1, 55101 Mainz Germany

SCHMITT, LOUIS ALFRED, engineer; b. Pendleton, Oreg., June 24, 1941; s. Howard and Alice (Gove) S.; m. Sally Rohrback, July 27, 1963; children: Michael, Steven. BS, Oreg. State U., 1963; MBA, Portland State U., 1970. Registered profl. engr., Ariz. Engr. US Steel Corp., Pittsburg, Calif., 1963-65; div. engr. Reynolds Metals Co., Richmond, Va., 1965-73; mgr. engring. Motorola, Inc., Phoenix, 1973-75; dep. dir. Ariz. Dept. Transp., Phoenix, 1975-91; asst. county mgr. Maricopa County, Phoenix, 1991—; chmn. rail rsch. com. Transp. Rsch. Bd., Washington, 1977-80, chmn. weigh in motion com., 1983-85, mem. advanced vehicle systems tech., 1989—. Author numerous papers in field. Chmn. Phoenix Fgn. Trade Zone Com., 1990; chmn. faculty com. Theodore Roosevelt coun. Boy Scouts Am., Phoenix, 1989; mem. Phoenix Overall Econ. Devel. Com., 1982. Recipient Past Pres. award Ariz. Motor Transport Assn., 1990, Tech. Excellence award Ariz. Adminstrn. Assn. Mem. Inst. Transp. Engring., Nat. Assn. Profl. Engring., Ariz. Pub. Works Assn., Internat. City/County Mgrs. Assn. Achievements include design, development and implementation of heavy vehicle electronic license plate program, advanced vehicle command and control program. Home: 8525 E Mulberry Scottsdale AZ 85251 Office: Maricopa County Transp and Devel Agy 2901 W Durango Phoenix AZ 85009

SCHMITT, MICHAEL A., agronomist, educator. Prof. soil sci. dept. U. Minn., St. Paul. Recipient CIBA-GEIGY Agronomy award Am. Soc. Agronomy, 1992. Office: U Minnesota Soil Science Dept 439 Borlaug Hall Saint Paul MN 55108*

SCHMITT, NEIL MARTIN, biomedical engineer, electrical engineering educator; b. Pekin, Ill., Oct. 25, 1940; married; 2 children. BSEE, U. Ark., 1963, MSEE, 1964; PhD in Electrical Engring., So. Meth. U., 1969. Systems engr. IBM Corp., 1966-67; engr. Tex. Instruments, Inc., 1967-70; assoc. prof. U. Ark., Fayetteville, 1970-80, prof. electrical engring., 1980—. Grantee NSF, 1971-72. Mem. IEEE, Biomed. Engring. Soc. Achievements include research in health care delivery systems, early detection of heart disease. Office: U Ark Engring Rsch Ctr Fayetteville AR 72701*

SCHMITT, ROLAND WALTER, retired academic administrator; b. Seguin, Tex., July 24, 1923; s. Walter L. and Myrtle F. (Caldwell) S.; m. Claire Freeman Kunz, Sept. 19, 1957; children: Lorenz Allen, Brian Walter, Alice Elizabeth, Henry Caldwell. BA in Math, U. Tex., 1947, BS in Physics, 1947, MA in Physics, 1948; PhD, Rice U., 1951; DSc (hon.), Worcester Poly. Inst., 1985, U. Pa., 1985; DCL (hon.), Union Coll., 1985; DL (hon.), Lehigh U., 1986; DSc (hon.), U. S.C., 1988, Universite De Technologie De Compiegne, 1991; DL (hon.), Coll. St. Rose, 1992, Russell Sage, 1993. With GE, 1951-88; R & D mgr. phys. sci. and engring. GE, Schenectady, 1967-74; mgr. energy sci. and engring. R & D GE, 1974-78, v.p. corp. R & D, 1978-82, sr. v.p. corp. R & D, 1982-86, sr. v.p. sci. and tech., 1986-88, ret., 1988; pres. Rensselaer Poly. Inst., Troy, N.Y., 1988-93; bd. dirs. Gen. Signal Corp., 1987—; mem. tech. adv. bd. Chrysler Corp., 1990-93; chmn. Am. Inst. Physics, 1993—; past pres. Indsl. Rsch. Inst.; mem. energy rsch. adv. bd. U.S. Dept. Energy, 1977-83; chmn. CORETECH, 1988-93; mem. Com. on Japan, NRC, 1988-90, Comml. Devel. Ind. Adv. Group, NASA, 1988-90; exec. com. Coun. on Competitiveness, 1988-93; chmn. NRC Panel on Export Controls, 1989-91, Scientists' Inst. Pub. Info., 1993—; mem. Dept. Commerce Adv. Commn. on Patent Law Reform, 1990-92. Trustee N.E. Savs. Bank, 1978-84; bd. advisors Union Coll., trustee Schenectady, 1981-84, Argonne Univs. Assn., 1979-82, RPI, 1982-88; bd. govs. Albany Med. Ctr. Hosp., 1979-82, 88-90; bd. dirs. Sunnyview Hosp. and Rehab. Ctr., 1978-86, Coun. on Superconductivity for Am. Competitiveness, 1987-89; mem. exec. com. N.Y. State Ctr. for Hazardous Waste Mgmt., 1988-89; chmn. Office of Tech. Assessment adv. panel on industry and environment. With USAAF, 1943-46. Recipient RPI Community Svc. award, 1982, award for disting. contbns. Stony Brook Found., 1985, Rice U. Disting. Alumni award, 1985, IRI Medalist award, 1989, Royal Swedish Acad. of Engring. Sci., 1990; named Fgn. Assn. of Engring. Acad. of Japan. Fellow AAAS, IEEE (Centenial medal 1984, Engring. Leadership award 1989, Founder's medal 1992, Hoover medal 1993), Am. Phys. Soc. (Pake award 1993), Am. Acad. Arts and Scis.; mem. NAE (coun., com. tech., mgmt. and capital in smaller cos.), Am. Inst. Physics (chmn. com. on corp. assocs., governing bd. 1979-83, chmn. 1993—), Nat. Sci. Bd. (past chmn. 1982), Dirs. Indsl. Rsch., Rensselaer Alumni Assn. (Dist. Alumni award 1993), Cosmos. Office: Ste 459 400 Clifton Corporate Pkwy Clifton Park NY 12065

SCHMITT, WILLIAM HOWARD, cosmetics company executive; b. Sterling, Ill., Oct. 27, 1936; s. Alfred William and Katherine Henrietta (Skow) S.; m. Antionette Marie Payne, Mar. 22, 1960; children: Hilary Ann, Andrea Kay, Joseph Michael. BS in Pharmacy, Drake U., 1958. Rsch. assoc. G.D. Searle, Skokie, Ill., 1963-66; assoc. dir. rsch. Alberto Culver, Melrose Park, Ill., 1966-71; dir. product devel. Chesebrough-Pond's Inc., Trumbull, Conn., 1971-74, dir. internat., 1974-83, group dir. R&D, 1983-85, v.p. R&D, 1985-89, sr. v.p. R&D, 1989—; sci. adv. bd. Cosmetics, Toiletry and Fragrance Assn., Washington, 1986-89. Author: (with others) An Overview of World-Wide Regulatory Programs, 1984; patentee in toiletry and cosmetics field. Pres. Suburban Sportsman's Club. Lt. USAF, 1959-62. Mem. Soc. Cosmetic Chemists, Am. Assn. for Dental Rsch. Avocations: boating, hunting, fishing. Office: Chesebrough-Ponds Inc 40 Merritt Blvd Trumbull CT 06611-5494

SCHMUCKER, BRUCE OWEN, geotechnical engineer; b. Canton, Ohio, May 10, 1961; s. Owen Christian and Constance Fay (Hanel) S.; m. Margaret Mary Stitz, Nov. 23, 1991. BS in Civil Engring., U. Akron, 1984, MS in Geotech. Engring., 1987. Registered profl. engr., Ohio. Constrn. monitor Herron Cons., Cleve., 1981-84; staff engr. Malcolm Pirnie, White Plains, N.Y., 1985; co. engr. Snellman Constrn., Norwalk, Conn., 1985; teaching asst. U. Akron, 1985-87; asst. project engr. Woodward-Clyde Cons., Cleve., 1987-91; dist. engr. Browning-Ferris Industries Ohio, Cleve., 1991—; tng. instr. for Ohio EPA Landfill Operators Cert. program Belmont Tech. Coll., St. Clairsville, Ohio, 1992—. Mem. ASCE. Republican. Mem. United Ch. of Christ. Home: 1759 Orchard Dr Akron OH 44333 Office: Browning Ferris Industries 43502 Rte 20 E Oberlin OH 44074

SCHMUFF, NORMAN ROBERT, organic chemist; b. Balti., Mar. 9, 1952; s. Clayton Robert and Kathleen Louise (McNeill) S.; m. Elizabeth Ann Lane, Oct. 1, 1987; 1 child, Kirsten. MA, John Hopkins U., 1976; PhD, U. Wis., 1982. Mgr. chem. inf. svcs. Questel, Inc., Washington, 1983-85; sr. scientist Am. Cyanamid Co., Pearl River, N.Y., 1985-89; chemist Food and Drug Adminstrn., Rockville, Md., 1989--. Pres. South Rolling Rd. Community Assn., Catonsville, Md., 1991--. mem. Am. Chem. Soc., Am. Assn. Pharm. Scientists, Chem. Structure Assn. U.K. Office: FDA 5600 Fishers Ln HFD-530 Rockville MD 20852

SCHNAAR, RONALD LEE, biomedical researcher, educator; b. Detroit, Nov. 1, 1950; s. Herbert Norman and Faye (London) S.; m. Cynthia Roseman, June 24, 1972; children: Melissa Ann, Stephen Jeremy, Gregory Adam. BS summa cum laude, U. Mich., 1972; PhD, Johns Hopkins U., 1976. Postdoctoral fellow Johns Hopkins U., Balt., 1977-78, NIH, Bethesda, Md., 1978-79; asst. prof. Johns Hopkins Sch. Medicine, Balt., 1979-84, assoc. prof., 1984-90, full prof., 1990—; mem., chair adv. com. NIH, Bethesda, 1988—; mem. sci. adv. bd. Glycomed, Inc., Alameda, Calif., 1988—. Contbr. rsch. articles to profl. jours.; author more than 60 original rsch. reports, revs. and book chpts. Postdoctoral fellow Muscular Dystrophy Assn., 1978-79; recipient Jr. Faculty Rsch. award Am. Cancer Soc., 1981-84, Faculty Rsch. award Am. Cancer Soc., 1984-89. Mem. Am. Soc. for Biochemistry, Am. Soc. for Cell Biology, Soc. for Neurosci., Soc. for Complex Carbohydrates. Achievements include research on cellular recognition of and adhesion to complex carbohydrates, on ganglioside receptors in vertebrate brain; paten issued for Technique for Identification of Bioactive Carbohydrates. Home: 9094 Goldamber Garth Columbia MD 21045 Office: Johns Hopkins Sch Medicine 725 N Wolfe St Baltimore MD 21205

SCHNABLE, GEORGE LUTHER, chemist; b. Reading, Pa., Nov. 26, 1927; s. L. Irvin and Laura C. (Albright) S.; m. Peggy Jane Butera, May 4, 1957; children: Lee Ann, Joseph G. BS, Albright Coll., 1950; MS, U. Pa., 1951, PhD, 1953. Project engr. Lansdale (Pa.) Tube Co., 1953-58; engring. specialist Philco Corp., Lansdale, Pa., 1958-61; mgr. materials and processes Philco-Ford Corp., Blue Bell, Pa., 1961-71; head process rsch. RCA Labs., Princeton, N.J., 1971-80, head device physics and reliability, 1980-87; head device physics and reliability David Sarnoff Rsch. Ctr., Princeton, 1987-91; ind. tech. cons. Schnable Assocs., Landsdale, 1991—. Author: (with others) Advances in Electronics and Electron Physics, 1971, The Chemistry of the Semiconductor Industry, 1987, Microelectronics Reliability, 1989, Microelectronics Manufacturing Diagnostics Handbook, 1993; editor spl. issue RCA Rev., 1984; div. editor Jour. of the Electrochem. Soc., 1978-90; contbr. over 75 articles to profl. pubs. With U.S. Army, 1946-47. Fellow AAAS, Am. Inst. Chemists, Electrochem. Soc. (chm. Phila. sect. 1969-71); mem. IEEE (sr.) (assoc. guest editor proceedings 1974), Alpha Chi Sigma, Phi Lambda Upsilon, Sigma Xi. Achievements include 39 patents (several with others); contributions to semiconductor device fabrication technology and reliability. Home and Office: Schnable Assocs 619 Knoll Dr Lansdale PA 19446-2925

SCHNALL, EDITH LEA (MRS. HERBERT SCHNALL), microbiologist, educator; b. N.Y.C., Apr. 11, 1922; d. Irving and Sadie (Raab) Spitzer; AB, Hunter Coll., 1942; AM, Columbia U., 1947, PhD, 1967; m. Herbert Schnall, Aug. 21, 1949; children: Neil David, Carolyn Beth. Clin. pathologist Roosevelt Hosp., N.Y.C., 1942-44; instr. Adelphi Coll., Garden City, N.Y., 1944-46; asst. med. mycologist Columbia Coll. Physicians and Surgeons, N.Y.C., 1946-47, 49-50; instr. Bklyn. Coll., 1947; mem. faculty Sarah Lawrence Coll., Bronxville, N.Y., 1947-48; lectr. Hunter Coll., N.Y.C., 1947-67; adj. assoc. prof. Lehman Coll., City U. N.Y., 1968; asst. prof. Queensborough Community Coll., City U. N.Y., 1967, assoc. prof. microbiology, 1968-75, prof., 1975—, adminstr. Med. Lab. Tech. Program, 1985—; vis. prof. Coll. Physicians and Surgeons, Columbia U., N.Y.C., 1974; advanced biology examiner U. London, 1970—. Mem. Alley Restoration Com., N.Y.C., 1971—; mem. legis. adv. com. Assembly of the State of N.Y., 1972. Mem. Community Bd. 11, Queens, N.Y., 1974—, 3d vice-chmn., 1987-92, 2nd vice chmn., 1992—; public dir. of bd. dirs. Inst. Continuing Dental Edn. Queens County, Dental Soc. N.Y. State and ADA, 1973—. Rsch. fellow NIH, 1948-49; faculty rsch. fellow, grantee-in-aid Rsch. Found. of SUNY, 1968-70; faculty rsch. grant Rsch. Found. City U. N.Y., 1971-74. Mem. Internat. Soc. Human and Animal Mycology, AAAS, Am. Soc. Microbiology (coun., N.Y.C. br. 1981—, co-chairperson ann. meeting com. 1981-82, chair program com. 1982-83, v.p. 1984-86, pres. 1986-88), Med. Mycology Soc. N.Y. (sec.-treas. 1967-68, v.p. 1968-69, 78-79, archivist 1974—, fin. advisor 1983—, pres. 1969-70, 79-80, 81-82), Bot. Soc. Am., Med. Mycology Soc. Americas, Mycology Soc. Am., N.Y. Acad. Scis., Sigma Xi, Phi Sigma. Clubs: Torrey Botanical (N.Y. State); Queensborough Community Coll. Women's (pres. 1971-73) (N.Y.C.). Editor: Newsletter of Med. Mycology Soc. N.Y., 1969-85; founder, editor Female Perspective newsletter of Queensborough Community Coll. Women's Club, 1971-73. Home: 21406 29th Ave Flushing NY 11360-2622

SCHNAPF, ABRAHAM, aerospace engineer, consultant; b. N.Y.C., Aug. 1, 1921; s. Meyer and Gussie (Schaeffler) S.; m. Edna Wilensky, Oct. 24, 1943; children: Donald J., Bruce M. BSME, CCNY, 1948; MSME, Drexel Inst. Tech., 1953. Registered profl. engr., N.J. Devel. engr. on lighter-than-air aircraft Goodyear Aircraft Corp., Akron, Ohio, 1948-50; mgr. fire control system def. electronics RCA, Camden, N.Y., 1950-55, mgr. airbourne navigation system, aerospace weapon system, 1955-58; program mgr. TIROS/TOS weather satellite systems RCA Astro-Electronics, Princeton, N.J., 1958-70, mgr. satellite programs, 1970-79, prin. scientist, 1979-82; cons. Aerospace Systems Engring., Willingboro, N.J., 1982—; lectr., presenter on meteor. satellites, space tech., communication satellites. Sgt. USAF, 1943-46. Recipient award Nat. Press Club Washington, 1975, award Am. Soc. Quality Control-NASA, 1968, Pub. Svc. award NASA, 1969, cert. of appreciation U.S. Dept. Commerce, 1984; inducted into Space Tech. Hall of Fame, 1992. Fellow AIAA; mem. Am. Astro. Soc., Am. Meterol. Soc., Space Pioneers, N.Y. Acad. Scis. (mem. think tank week sessions 1980's), N.J. Arbitration Soc. Home and Office: 41 Pond Ln # 160 Willingboro NJ 08046-2756

SCHNARE, PAUL STEWART, computer scientist, mathematician; b. Berlin, N.H., Oct. 16, 1936; s. Herbert Stewart and Roma (Dahl) S.; m. Dorothy Hopkins, Jan. 11, 1960; children: Sigmund, Col, Kurt (dec.). BA, U. N.H., 1960, MS, 1961; PhD, Tulane U., 1967. Instr. U. New Orleans, 1961-66; asst. prof. U. Fla., Gainesville, 1967-74, Colby Coll., Waterville, Maine, 1974-75, Fordham U., Bronx, N.Y., 1975-76, Univ. Petroleum and Minerals, Dhahran, Saudi Arabia, 1976-80; assoc. prof. computer sci., math. Ea. Ky. U., Richmond, 1980—; cons. Ea. Cons. Assocs., Durham, N.H., 1961; NSF sci. faculty fellow Tulane U., New Orleans, 1966-67. Contbr. articles to Am. Math. Monthly, Fundamenta Mathematicae, Gen. Topology, other publs. Mem. IEEE Computer Soc., Assn. Computing Machinery, Am. Math. Soc., Math. Assn. Am., Nat. Computer Graphics Assn. (treas. Bluegrass chpt. 1990—). Office: Ea Ky U Dept Math-Stat-Computer Sci Wallace 402 Richmond KY 40475

SCHNECK, JEROME M., psychiatrist, medical historian, educator; b. N.Y.C., Jan. 2, 1920; s. Maurice and Rose (Weiss) S.; m. Shirley R. Kaufman, July 24, 1943. AB, Cornell U., 1939; MD, SUNY-Bklyn., 1943. Diplomate Am. Bd. Psychiatry and Neurology. Intern Interfaith Med. Ctr., 1943; mem. psychiat. staff Menninger Clinic, Topeka, 1944-45, L.I. Coll. Hosp., 1947-48, Kings County Hosp., 1948-70, SUNY Hosp., Bklyn., 1955-70; assoc. vis. psychiatrist Kings County Hosp., 1949-70; mem. psychiat. staff State U. Hosp., Bklyn., 1955-70; pvt. practice N.Y.C., 1947—; attending psychiatrist St. Vincent's Hosp. and Med. Ctr. N.Y., 1970—, hon. sr. psychiatrist, 1990—; psychiat. cons. VA Regional Office, 1947-48, N.Y. State Dept. Social Svcs., 1977-83, N.Y. State Dept. Civil Svc., 1978-84, N.Y. State Office Ct. Adminstrn., 1978-85, N.Y. State Dept. Edn., 1981-83; dir. Mt. Vernon Mental Hygiene Clinic, 1947-52; assoc. chief psychiatrist Westchester County Dept. Health, 1949-50, cons., 1951-52; clin. instr. L.I. Coll. Medicine, 1947-50; clin. assoc. SUNY Coll. Medicine, Bklyn., 1950-53, asst. prof., 1955-58, assoc. prof., 1958-70; supervising psychiatrist Community Guidance Svcs., 1955-70; cons. coun. on mental health AMA, 1956-58; cons. NBC, 1962, Ctr. Rsch. in Hypnotherapy, 1964-70; vis. lectr. N.Y. Med. Coll.-Met. Hosp., 1965; faculty Am. Inst. Psychotherapy and Psychoanalysis, 1970-85. Author: Hypnosis in Modern Medicine, 1953, 2d edit., 1959, Spanish lang. edit., 1962, 3rd edit., 1963, Studies in Scientific Hypnosis, 1954, A History of Psychiatry, 1960, The Principles and Practice of Hypnoanalysis, 1965 (Best Book award Soc. For Clin. and Exptl. Hypnosis 1965); editor: Hypnotherapy, Hypnosis and Personality, 1951; author over 400 med. and sci. publs., book chpts., articles; mem. bd. editors: Personality: Symposia on Topical Issues, 1960-61, Jour. Integrative and Eclectic Psychotherapy, 1986-89; contbg. editor Psychosomatics, 1961-75; mem. editorial bd. Voices—The Art and Science of Psychotherapy, 1965. Capt. AUS, 1945-47. Recipient Clarence B. Farrar award Clarke Inst. of Psychiatry, U. Toronto, 1976. Fellow AAAS, APA, Am. Med. Authors, Acad. Psychosomatic Medicine, Am. Psychiat. Assn. (life), Soc. for Clin. and Exptl. Hypnosis (life, founder, founding pres. 1949-56, exec. coun. 1949—, assoc. editor jour. 1953—, award of Merit 1955, Gold medal 1958, Bernard B. Raginsky award 1966, Shirley Schneck award 1970, Roy M. Dorcus award 1980 Spl. Presdl. award 1986), Am. Acad. Psychotherapists (co-founder, v.p. 1956-58), Am. Med. Writers Assn., Am. Soc. Psychoanalytic Physicians (founding fellow, bd. dirs. 1958-62), Am. Soc. Clin. Hypnosis, Internat. Soc. Clin. and Exptl. Hypnosis (co-founder, bd. dirs. 1958-68, founding fellow), Internat. Acad. Eclectic Psychotherapists (charter fellow); mem. AMA, Soc. Acad. Achievement (charter), Soc. Apothecaries London, Inst. Practicing Psychotherapists, Pan Am. Med. Assn. (v.p. sect. clin. hypnosis 1960-65, N.Am. v.p. 1966), N.Y. Soc. Med. History (exec. com. 1956-62), Am. Bd. Med. Hypnosis (founder, pres. 1958-60, life bd. dir.), Inst. Rsch. in Hypnosis Inc. (bd. dirs. and bd. editors 1957-70), Am. Assn. History Medicine, History of Sci. Soc., Assn. Advancement Psychotherapy (charter), Can. Med. History Assn., N.Y. Soc. for Clin. Psychiatry (chmn. com. on history of psychiatry), Sigma Xi. Address: 26 W 9th St New York NY 10011-8971

SCHNEEBERGER, EVELINE ELSA, pathologist, cell biologist, educator; b. The Hague, Holland, Oct. 2, 1934; came to U.S., 1952; d. Werner Friederich and Elsa (Graf) S. BA, U. Colo., 1956; MD, U. Colo., Denver, 1959; MA (hon.), Harvard U., 1990. Rsch. fellow in pathology Sir. William Dunn Sch., Oxford (Eng.) U.; rsch. fellow in pathology Harvard Med. Sch., Boston, 1966-67, instr. pathology, 1967-68; assoc. in pathology Harvard Med. Sch., 1968-70, asst. prof., 1970-74, assoc. prof., 1974-88, prof. pathology, 1988—; mem. pulmonary diseases adv. com. NIH, Bethesda, Md., 1975-78, mem. pathology A study sect., 1980-83, 87-90; mem. rev. com. Nat. Inst. Environ. Health Scis., Research Triangle, N.C., 1992—. Mem. editorial bd. Circulation Rsch., 1977-83, Tissue and Cell, 1983—, Am. Jour. Physiology, 1990—; contbr. chpts. to books, more than 100 articles to profl. jours. Fellow AAAS; mem. Am. Soc. for Cell Biology, Am. Assn. Pathologists, Am. Thoracic Soc., Microcirculatory Soc., Phi Beta Kappa, Sigma Xi, Alpha Omega Alpha. Office: Mass Gen Hosp Dept Pathology Fruit St Boston MA 02114

SCHNEEGURT, MARK ALLEN, biochemist, researcher; b. Bklyn., Jan. 18, 1962; s. Errol and Elaine Francis (Sacks) S. BS, Rensselaer Poly. Inst., 1984, MS, 1985; PhD, Brown U., 1989. Postdoctoral rsch. scientist Dow Elanco, Greenfield, Ind., 1989-91; postdoctoral fellow Purdue U., West Lafayette, Ind., 1992—. Contbr. articles to profl. jours. Recipient Nat. Rschr. Svc. award NIH, 1987; Nat. Merit scholar, 1980, N.Y. State Regents scholar, 1980. Mem. Am. Soc. Plant Physiologists, Sigma Xi (Rsch. grant 1988). Jewish. Achievements include research on protoheme and heme a made from glutamate in maize; the tRNA required for 5- aminolevulinic acid biosynthesis in a cyanobacterium is active in protein biosynthesis and the tRNA involved in 5- aminolevulinic acid biosynthesis from many species contains a UUC anticodon; definition of the site of action of piperalin, a sterol biosynthesis inhibitor in fungi; the aldehyde oxygen atom of chlorophyll b is derived from atmospheric oxygen. Office: Purdue Univ Hansen 135 West Lafayette IN 47907

SCHNEIDER, ARTHUR, computer graphics specialist; b. Chgo., June 17, 1947; s. George Joseph and Doris Hilda (Hirsberg) S.; m. Marita Rose Scherer (div. 1976); 1 child, Heather Patricia; m. Jayne Bangs, 1986. BA in Radio and TV, Ind. U., Bloomington, 1972. Corp. CAD system specialist Entre Computer Ctr.; computer graphic specialist, system builder ZML/McCann Erickson Worldwide; chief executive officer Applied Computer Graphics, Inc., Louisville; nat. sales leader, Entre Computer Ctr.; founder Electronic Imaging Soc. Inc. Mem. World Future Soc., Nat. Trust for Historic Preservation. Avocations: boxing, reading. Office: Applied Computer Graphics Inc 1347 S 3d St Louisville KY 40208-2344

SCHNEIDER, CALVIN, physician; b. N.Y.C., Oct. 23, 1924; s. Harry and Bertha (Green) S.; A.B., U. So. Calif., 1951, M.D., 1955; J.D., LaVerne (Calif.) Coll., 1973; m. Elizabeth Gayle Thomas, Dec. 27, 1967. Intern Los Angeles County Gen. Hosp., 1955-56, staff physician, 1956-57; practice medicine West Covina, Calif., 1957—; staff Inter-Community Med. Ctr., Covina, Calif. Cons. physician Charter Oak Found., Covina, 1960—. With USNR, 1943-47. Mem. AMA, Calif., L.A. County med. assns. Republican. Lutheran. Office: 224 W College St Covina CA 91723-1902

SCHNEIDER, EDWARD LEWIS, academic administrator, research administrator; b. N.Y.C., June 22, 1940; s. Samuel and Ann (Soskin) S. BS, Rensselaer Poly. Inst., 1961; MD, Boston U., 1966. Intern and resident N.Y. Hosp.-Cornell U., N.Y.C., 1966-68; staff fellow Nat. Inst. Allergy and Infectious Diseases, Bethesda, Md., 1968-70; research fellow U. Calif., San Francisco, 1970-73; chief, sect. on cell aging Nat. Inst. Aging, Balt., 1973-79, assoc. dir., 1980-84, dep. dir., 1984-87; prof. medicine, dir. Davis Inst. on Aging U. Colo., Denver, 1979-80; dean Leonard Davis Sch. Gerontology U. So. Calif., L.A., 1986—, exec. dir. Ethel Percy Andrus Gerontology Ctr., 1986—, prof. medicine, 1987—, William and Sylvia Kugel prof. gerontology, 1989—; sci. dir. Buck Ctr. for Rsch. in Aging, 1989—; cons. MacArthur Found, Chgo., 1985—, R.W. Johnson Found., Princeton, N.J., 1982-87, Brookdale Found., N.Y.C., 1985-89. Editor: The Genetics of Aging, 1978, The Aging Reproductive System, 1978, Biological Markers of Aging, 1982, Handbook of the Biology of Aging, 1985, Interrelationship Among Aging Cancer & Differentiation, 1985, Teaching Nursing Home, 1985, Modern Biological Theories of Aging, 1987, The Black American Elderly, 1988, Elder Care and the Work Force, 1990. Med. dir. USPHS, 1968—. Recipient Roche award, 1964. Fellow Gerontology Soc., Am. Soc. Clin. Investigation; mem. Am. Assn. Retired Persons, U.S. Naval Acad. Sailing Squadron (coach 1980-86). Office: U So Calif Andrus Gerontology Center Los Angeles CA 90089-0191

SCHNEIDER, ELEONORA FREY, physician; b. Basel, Switzerland, Jan. 17, 1921; came to U.S., 1952; d. Friedrich Ernst and Clara Melanie (Heiz) Frey; m. Jurg Adolf Schneider, Aug. 22, 1946; children: Andreas George, Daphne Eleanor, Diana Veronica, Claudia Elizabeth. MD, U. Basel, 1945. Lic. MD. Pharmacologist Sandoz Pharms., Basel, 1946-47; resident in anesthesiology U. Hosp., Basel, 1950-51; resident Pediatric Dept. Del. Div., Wilmington, 1971-73; physician Wilmington Pub Schs., 1973-79, Pub. Health, Wilmington, 1975-80; staff physician student health svc. U. Del., Newark, 1979—; v.p. Pharmacon, Inc., Wilmington, 1985—. Contbr. articles to profl. jours. V.p. del. Citizens for Clean Air, 1969-71; mem. com. Gov.'s Adv. Coun. for Exceptional Children, Dover, Del.; mem. adv. panel YWCA; vol. Girl Scouts U.S.A., ARC. Mem. AAUW (study group leader 1966-67, area chmn. community problems 1967-69, edn. com. 1968-69, new mems. advisor 1969-70, bd. dirs. 1967-70), AMA, Am. Acad. Pediatrics, Med. Soc. Del., New Castle County Med. Soc. Republican. Avocations: sailing, gardening, music.

SCHNEIDER, GEORGE WILLIAM, retired aircraft design engineer; b. Riley, Kans., Aug. 17, 1923; s. George William and Helen Juanita (Carey) S.; m. Marguerite Ann Bare, May 7, 1945 (div. Oct. 1977); children: Peggy Diane Schneider Tsolakopolous, Donald Lynn; m. L. Elaine Phillips, Oct. 22, 1977. Student, Wichita State U., 1952-58; BSME in Design, Kans. State U., 1962. Designer Ling Temco Voght, Dallas, 1962-65; lead designer 727 Boeing Airplane Co., Renton, Wash., 1965-66; lead designer 747 Boeing Airplane Co., Everett, Wash., 1966-72; designer 707, 727, 737, AWACS Boeing Airplane Co., Renton, 1972-75; designer DeHavilland Dash 7 Boening Airplane Co., Toronto, Ont., Can., 1975-77; design engr. Boeing Airplane Co., Morgantown, W.Va., 1977-79; lead designer 757 Boeing Airplane Co., Renton, 1980-81; sr. design engr. Boeing Airplane Co., Oak Ridge, Tenn., 1981-83; ret., 1983. Author books, articles, reports in field. Chmn. com. Explorer scouts Boy Scouts Am., Seattle, 1966-68. Mem. ASME (regional chmn. history and heritage 1991, regional nat. egenda bd. 1992-94, editor Dixie News regional news bull., chmn. govtl. rels. Greenville sect. 1990—, chmn. Greenville sect. 1989-93, chmn. awards and honors), S.C. Coun. of Engring. Soc. (sec., treas. 1992-93, v.p. 1993-94). Avocations: science, photography, travel, woodworking, fishing. Home: 32 W Hillcrest Dr Greenville SC 29609-4658

SCHNEIDER, ROBERT JAY, oncologist; b. Miami, Fla., May 31, 1949; s. Irving and Ethel (Pack) S.; m. Barbara Cunningham, June 1, 1974; children: Matthew, Kirsten. Student, Washington U., 1967-69; BA cum laude, Boston U., 1971; MD, Albert Einstein Coll. Medicine, N.Y.C., 1975. Diplomate Am. Bd. Internal Medicine, Am. Bd. Oncology; lic. physician, N.Y. Intern, jr. and sr. resident internal medicine Bronx Mcpl. Hosp., N.Y.C., 1975-78; fellow med. oncology Meml. Sloan-Kettering Cancer Ctr., N.Y.C., 1978-80, adj. attending physician/cons. dept. medicine, 1981—; asst. prof. medicine N.Y. Med. Coll., Valhalla, 1980-81; clin. instr. medicine Cornell U. Med. Coll., 1978-80; jr. clin. faculty fellow Am. Cancer Soc., 1980-81; mem. N.Y. Met. Breast Cancer Group, 1990—; cons. cancer program No. Westchester Hosp. Ctr., Mt. Kisco, N.Y., 1981-82; mem. staff Westchester County Med. Ctr., Valhalla, N.Y., No. Westchester Hosp. Ctr., Mt. Kisco, Meml. Sloan-Kettering Cancer Ctr., N.Y.C. Contbr. articles to profl. jours. Recipient Clin. Fellowship award Am. Cancer Soc., 1978-79. Mem. Am. Soc. Clin. Oncology, Westchester County Med. Soc., N.Y. State Med. Soc., Woodway Country Club. Republican. Presbyterian. Achievements include research in detection and treatment of early breast cancer, the human spirit in the fight against cancer, salvage chemotherapy with etoposide, ifosfamide and cisplatin in refractory germ cell tumors. Office: 439 E Main St Mount Kisco NY 10549

SCHNEIDER, STEPHEN HENRY, climatologist, researcher; b. N.Y.C., Feb. 11, 1945; s. Samuel and Doris C. (Swarte) S. BS, Columbia U., 1966, MS, 1967, PhD, 1971; DSc (hon.), N.J. Inst. Tech., 1990, Monmouth Coll., 1991. NAS, NRC postdoctoral research assoc. Goddard Inst. Space Studies NASA, N.Y.C., 1971-72; fellow advanced study program Nat. Ctr. Atmospheric Research, Boulder, Colo., 1972-73, scientist, dep. head climate project, 1973-78, acting leader climate sensitivity group, 1978-80, head visitors program and dep. dir. advanced study program, 1980-87; sr. scientist Nat. Ctr. Atmospheric Research, Boulder, 1980; head interdisciplinary climate systems sect. Nat. Ctr. Atmospheric Research, Boulder, Colo., 1987-92; prof. Stanford (Calif.) U., 1992—; affiliate prof. U. Corp. Atmospheric Research Lamont-Doherty Geol. Obs., Columbia U.; mem. Carter-Mondale Sci. Policy Task Force, 1976; sci. adviser, interviewee Nova, Sta. WGBH-TV, Planet Earth, Sta. WQED-TV; mem. internat. sci. coms. climatic change, energy, food and pub. policy; expert witness congl. coms.; mem. Def. Sci. Bd. Task Force on Atmospheric Obscuration. Author: (with Lynne E. Mesirow) The Genesis Strategy: Climate and Global Survival, 1976, (with Lynne Morton) The Primordial Bond: Exploring Connections between Man and Nature through Humanities and Science, 1981, (with Randi S. Londer) The Coevolution of Climate and Life, 1984, Global Warming: Are We Entering the Greenhouse Century?, 1989, (with W. Bach) Interactions of Food and Climate, 1981, (with R.S. Chen and E. Boulding) Social Science Research and Climate Change: An Interdisciplinary Appraisal, 1983, (with K.C. Land) Forecasting in the Social and Natural Sciences, 1987; editor: Climatic Change, 1976—; contbr. sci. and popular articles on theory of climate, influence of climate on soc., relation of climatic change to world food, population, energy and environ. policy issues, environ. aftereffects of nuclear war, carbon dioxide greenhouse effect, pub. understanding sci. Recipient Louis J. Battan Author's award Am. Meteorol. Soc., 1990; named one of 100 Outstanding Young Scientists in Am. by Sci. Digest, 1984; fellow MacArthur Found. 1993. Fellow AAAS (Westinghouse award 1991), Scientists Inst. for Pub. Info. Office: Nat Ctr Atmospheric Research PO Box 3000 Boulder CO 80307-3000 also: Stanford U Dept Biol Scis and Inst Internat Studies Stanford CA 94305

SCHNEIDER, THOMAS R(ICHARD), physicist; b. Newark, N.J., Nov. 14, 1945; s. Valentine William and Mary Bernadette (Scanlon) S.; m. Paula Doris Tulecko, June 8, 1968 (div. June 1984); 1 child, Laurie Ann. BS with high honors, Stevens Inst. Tech., Hoboken, N.J., 1967; PhD, U. Pa., 1971. Postdoctoral fellow U. Pa. Nat. Ctr. for Engergy Mgmt. and Power, Phila., 1971-72; prin. rsch. physicist Pub. Svc. Electric & Gas Co., Newark, 1972-77; various positions, dir. dept., exec. scientist Electric Power Rsch. Inst., Palo Alto, Calif., 1977—; pres. bd. dirs. Lighting Rsch. Inst., N.Y.C., 1990—; bd. dirs., chmn. energy task force Coun. on Superconductivity for Am. Competitiveness, Washington, 1991—; coord. for Ad hoc Working Group Report on Power Applications of Superconductivity for Dept. Commerce, 1990-91. Mem. IEEE, AAAS, Am. Phys. Soc., Illuminating Engring. Soc., Am. Soc. Assn. Execs. Achievements include advanced research in energy storage, conservation, and applications of superconductivity. Office: Electric Power Rsch Inst 3412 Hillview Ave PO Box 10412 Palo Alto CA 94303

SCHNEIDER, WOLF-DIETER, physicist, educator; b. Trondheim, Norway, Apr. 25, 1944; s. Willi Oswald and Margarethe (Pongraz) S.; m. Elke Reusch, May 22, 1970; children: Gesa, Britta. Diploma in physics, U. Bonn, Germany, 1971, PhD, 1975. Asst. prof. U. Campinas, Brazil, 1976-77; asst. prof. U. Berlin, 1977-83, docent, 1983-89; docent U. Neuchatel, Switzerland, 1983-89, sr. rsch. physicist, 1983-89; prof. physics U. Lausanne, Switzerland, 1989—; dir. Inst. Experimental Physics, U. Lausanne, 1991—. Author: (with Yves Baer) Handbook on the Physics and Chemistry of Rare Earths, 1987; contbr. articles to profl. jours. Lt. German Armed Forces, 1963-65. Fellow Deutscher Akademischer Austauschdienst, 1972. Mem. German Phys. Soc., Swiss Phys. Soc., European Phys. Soc. Avocations: tennis, music, guitar, wind-surfing, skiing. Office: U Lausanne, Inst Exptl Physics, CH-1015 Lausanne Switzerland

SCHNEIDER-CRIEZIS, SUSAN MARIE, architect; b. St. Louis, Aug. 1, 1953; d. William Alfred and Rosemary Elizabeth (Fischer) Schneider; m. Demetrios Anthony Criezis, Nov. 24, 1978; children: Anthony, John and Andrew. BArch, U. Notre Dame, 1976; MArch, MIT, 1978. Registered architect, Wis. Project designer Eichstaedt Architects, Roselle, Ill., 1978-80, Solomon, Cordwell, Buenz & Assocs., Chgo., 1980-82; project architect Gelick, Foran Assocs., Chgo., 1982-83; asst. prof. Sch. Architecture U. Ill., Chgo., 1980-86; exec. v.p. Criezis Architects, Inc., Evanston, Ill., 1986—. Graham Found. grantee MIT, 1977, MIT scholar, 1976-78; Prestressed Concrete Inst. rsch. grantee, 1981. Mem. AIA, Chgo. Archtl. Club, Chgo. Women in Architecture, Am. Solar Energy Soc., NAFE, Jr. League Evanston, Evanston C. of C. Roman Catholic. Avocations: tennis, swimming. Office: 1007 Church St Ste 101 Evanston IL 60201-3624

SCHNELL, GEORGE ADAM, geographer, educator; b. Phila., July 13, 1931; s. Earl Blackwood and Emily (Bernheimer) S.; m. Mary Lou Williams, June 21, 1958; children: David Adam, Douglas Powell, Thomas Earl. BS, West Chester U., 1958; MS, Pa. State U., 1960, PhD, 1965; postdoctoral study, Ohio State U., 1965. Asst. prof. Coll. SUNY, New Paltz, 1962-65, assoc. prof., 1965-68, prof. geography, 1968—, founding chmn. dept., 1968—; vis. assoc. prof. U. Hawaii, summer 1966; cons. community action programming, 1965; manuscript reader, cons. to several pubs., 1967—; founder, founding bd. mem. Inst. for Devel., Planning and Land Use Studies, 1986—; cons. Mid-Hudson Pattern for Progress, 1986, Open Space Inst., 1987, Mid-Hudson Regional Econ. Devel. Coun., 1989, Urban Devel. Corp., 1989-90, 93, Tech. Devel. Ctr., 1991, Catskill Ctr., 1991, Ednl. Testing Svc., 1993. Author: (with others) The Local Community: A Handbook for Teachers, 1971, The World's Population, Problems of Growth, 1972, Pennsylvania Coal: Resources, Technology, Utilization, 1983, West Virginia and Appalachia: Selected Readings, 1977, Hazardous and Toxic Wastes: Technology, Management and Health Effects, 1984, Environmental Radon: Occurrence, Control and Health Hazards, 1990, Natural and Technological Disasters: Causes, Effects and Preventive Measures, 1992, Conservation and Resource Management, 1993; co-author: (with M.S. Monmonier) The Study of Population: Elements, Patterns, Processes, 1983, (with Monmonier) Map Appreciation, 1988; editor: (with G.J. Demko and H.M. Rose) Population Geography: A Reader, 1970; contbr. articles to profl. jours.; presenter papers to more than 60 ann. meetings of scholarly and profl. socs. Appt. mem. local bds. and coms. Town and Village of New Paltz, and New Paltz Ctrl. Sch. Dist., 1965—; elder Reformed Ch. of New Paltz. With AUS, 1952-54. Recipient Excellence award N.Y. State/United Univ. Professions, 1990. Mem. Assn. Am. Geographers, Nat. Coun. Geog. Edn., Pa. Geog. Soc. (mem. editorial bd. Pa. Geographer), Pa. Acad. Sci. (assoc. editor jour. 1988—). Home: 29 River Park Dr New Paltz NY 12561-2636 Office: SUNY Coll at New Paltz Dept Geography Hamner House Rm 1 New Paltz NY 12561

SCHNELLER, EUGENE S., sociology educator; b. Cornwall, N.Y., Apr. 9, 1943; s. Michael Nicholas and Anne Ruth (Gruner) S.; m. Ellen Stauber, Mar. 24, 1968; children: Andrew Jon, Lee Stauber. BA, L.I. U., 1967; AA, SUNY, Buffalo, 1965; PhD, NYU, 1973. Rsch. asst. dept. sociology NYU, N.Y.C., 1968-70; project dir. Montefiore Hosp. and Med. Ctr., Bronx, N.Y., 1970-72; asst. prof. Med. Ctr. and sociology Duke U., Durham, N.C., 1973-75; assoc. prof., chmn. dept. Union Coll., Schenectady, 1975-79, assoc. prof., dir. Health Studies Ctr., 1979-85; prof., dir. Sch. Health Adminstrn. and Policy, Ariz. State U., Tempe, 1985-91, assoc. dean rsch. and adminstrn. Coll. Bus., 1992—; dir. L. William Seidman Rsch. Ctr., Tempe, 1992—; vis. rsch. scholar Columbia U., N.Y.C., 1983-84; chmn. Western Network for Edn. in Health Adminstrn., Berkeley, Calif., 1987-92; mem. Ariz. Medicaid Adv. Bd., 1990-92, Ariz. Data Adv. Bd., 1989-91, Ariz. Health Care Group Adv. Bd., 1989; mem. health rsch. coun. N.Y. State Dept. Health, 1977-85; fellow Accrediting Commn. on Edn. for Health Svcs. Adminstrn., 1983-84. Author: The Physician's Assistant, 1980; mem. editorial bd. Work and Occupations, 1975-93, Hosps. and Health Svcs. Adminstrn., 1989-92, Health Adminstrn. Press, 1991—; contbr. articles to profl. jours., chpt. to book. Trustee Barrow Neurol. Inst., Phoenix, 1989—. Mem. APHA, Am. Sociol. Assn., Assn. Univ. Programs Health Adminstrn. (bd. dirs. 1990—). Home: 9906 E Cinnabar Ave Scottsdale AZ 85258-4738 Office: Ariz State Univ Office of Dean Coll Bus Tempe AZ 85287

SCHNITZLER, PAUL, electronics engineering manager; b. N.Y.C., July 20, 1936; s. Jack and Mildred (Spero) S.; m. Carol Fay Neadle, Jan. 18, 1959 (div. May 1985); childen: Raymond Alan, Robin Kay, David Aaron. BSEE, NYU, 1957, MSEE, 1959; PhD, Poly. Inst. Bklyn., 1969. Mem. tech. staff RCA Lab., Princeton, N.J., 1958-69; engr. leader RCA Solid State div., Somerville, N.J., 1969-72; supr. Bell Labs., Holmdel, N.J., 1972-78; group head RCA Labs., Princeton, 1979-85; cons. Paul Schnitzler and Assocs., Inc., North Brunswick, N.J., 1985—; dir. tech. Time Inc. Mag. Co., N.Y.C., 1989-91. Contbr. articles to profl. jours. Hugo Gernsbach award NYU, 1956. Mem. IEEE (bylaws chmn. 1979-83, Svc. award 1982), Sigma Xi. Democrat. Jewish. Achievements include six patents; also first to use thin films to launch surface acoutic waves. Home and Office: 220 Hamlin Rd North Brunswick NJ 08902

SCHNITZLER, ROBERT NEIL, cardiologist; b. Bklyn., Dec. 12, 1940. BA, N.Y. Heights Coll. Arts & Sci., N.Y.C., 1961; MD, SUNY, Buffalo, 1965. Lic. physician, N.Y., Tex.; diplomate Am. Bd. Cardiovascular Diseases; cert. specialty Bd. Internal Medicine, sub-specialty Bd. Cardiovascular Medicine, Bd. Clin. Cardiac Electrophysiology. Intern Kings-County Downstate Med. Ctr., Bklyn., 1965-66, resident, 1966-67; resident N.Y. Hosp.-Cornell Med. Ctr., N.Y.C., 1967-68; asst. physician N.Y. Hosp., N.Y.C., 1967-68; trainee, fellow U. Rochester (N.Y.) Med. Ctr., 1968-70; asst. physician Strong Meml. Hosp., Rochester, 1968-69; dir. med. intensive care units Bexar County Hosp. and Audie L. Murphy VA Hosp., San Antonio, 1973-76, dir. cardiac intensive care units, 1976-77; pres. Cardiovascular Inst. for Continuing Med. Edn. and Rsch., San Antonio, 1983—; pvt. practice cardiology San Antonio, 1977—; asst. instr. U. Rochester Med. Sch., Strong Meml. Hosp., 1968-69, instr., 1969-70; rsch. assoc. USPHS Hosp., S.I., 1970-72; assoc. attending physician St. Vincent's Hosp., S.I., 1971-72; asst. prof. U. Tex. Health Sci. Ctr., San Antonio, 1972-76, assoc. prof., 1977, clin. assoc. prof., 1977-83, clin. prof., 1983—; dir. Graphics Method Lab. Bexar County Hosp., San Antonio, 1972-76; cons. electrophysiology USAF Wilford Hall Med. Ctr., San Antonio, 1979—; chief staff S.W. Tex. Meth. Hosp., 1992. Contbr. articles to profl. jours. Mem. adv. bd. health City of San Antonio, 1987—; med. coor. City of San Antonio Internat. Drug Summit, 1992. Surgeon USPHS, 1968-70. Fellow ACP, Am. Coll. Cardiology, Am. Coll. Chest Physicians, Coun. of Clin. Cardiology Am. Heart Assn.; mem. N.Y. Acad. Scis., N.Y. Heart Assn., Assn. for Advancement Med. Instrumentation, Am. Fedn. Clin. Rsch., Emergency Med. Svc. Coun., North Tex. Soc. Pacing and Electrophysiology, S.W. Tex. Angioplasty Soc., Am. Coll. Angiology, Internat. Soc. Heart and Lung Transplantation, Soc. Cardiac Angiography and Interventions, Am. Soc. Internal Medicine, Adv. Bd. of Schnedier, Alpha Omega Alpha. Achievements include research in internal medicine, cardiology and vascular diseases. Home and Office: 4330 Medical Dr Ste 400 San Antonio TX 78229-3324

SCHOBER, ROBERT CHARLES, electrical engineer; b. Phila., Sept. 20, 1940; s. Rudolph Ernst and Kathryn Elizabeth (Ehrisman) S.; m. Mary Eve Kanuika, Jan. 14, 1961; children: Robert Charles, Stephen Scott, Susan Marya. BS in Engring. (Scott Award scholar), Widner U., 1965; postgrad., Bklyn. Poly. Extension at Gen. Electric Co., Valley Forge, Pa., 1965-67, U. Colo., 1968-69, Calif. State U.-Long Beach, 1969-75, U. So. Calif., 1983-84. Engr. Gen. Electric Co., Valley Forge, 1965-68, Martin Marietta Corp., Denver, 1968-69; sr. engr. Jet Propulsion Lab., Pasadena, Calif., 1969-73, sr. staff, 1986—; mem. tech. staff Hughes Semiconductor Co., Newport Beach, Calif., 1973-75; prin. engr. Am. Hosp. Supply Corp., Irvine, Calif., 1975-83; sr. staff engr. TRW Systems, Redondo Beach, Calif., 1983-84; cons. Biomed. LSI, Huntington Beach, Calif. Mem. IEEE (student br. pres. 1963-65), Soc. for Indsl. and Applied Math., Assn. for computing Machinery, Tau Beta Pi. Republican. Patentee cardiac pacemakers. Current Work: Develop large scale integrated circuits for computer, spacecraft, and military, as well as commercial applications; design high speed signal processing integrated circuits. Subspecialties: application specific microprocessor archtl. design; ultra low power analog and digital systems, integrated circuits; focal plane array, electronics, cardiology and other implantable medical devices. Office: Jet Propulsion Lab 4800 Oak Grove Dr Pasadena CA 91109-8099

SCHOCK, WILLIAM WALLACE, pediatrician; b. Huntingdon, Pa., Aug. 15, 1923; s. Clarence and Mabel (Decker) S.; m. Doris Ann Wilson, Ju.y 1, 1944; 1 child, William Wallace. Student, Juniata Coll., 1941-43; MD, Temple U., 1947. Intern Conemaugh Valley Meml. Hosp., Johnstown, Pa., 1946-48; resident Women AFB, Cheyene, Wyo., 1951-52; pvt. practive medicine Huntingdon, 1948-50; pediatrician Warren AFB Hosp., 1951-52; chief outpatient svc. USAF, Cheyenne, Wyo., 951-52; pvt. practice medicine specializing in pediatrics Huntingdon, 1952—; pediatrician J. C. Blair Meml. Hosp.; local pub. health pediatrician. Pres. Huntingdon chpt. Am. Cancer Soc., 1955-57; bd. dirs. local chpt. Am. Heart Assn., 1955-62; mem. Am. Security Coun., Rep. Nat. Com., 2d Amendment Found. With AUS, 1942-45, USAF, 1950-52. Recipient Wisdom award Leon Gutterman, Wisdom Hall of Fame, 1970. Fellow Royal Soc. Health; mem. AMA, Pa. Med. Soc. Huntingdon County Med. Soc. (past pres.), Med. Alumni Assn. Temple U., Pa. Pediatric Soc., Huntingdon Pediatric Soc., Am. Assn. Mil. Surgeons U.S., Am. Acad. Gen. Practice (past pres. Huntingdon), Am. Acad. Pediatrics (assoc.), Internat. Platform Assn., Phi Rho Sigma, Huntingdon Country Club, Hiedelburg Country Club (Altoona, Pa.), U.S. Senatorial Club, Rotary. Republican. Presbyterian. Home and Office: RR 2 Box 69 Huntingdon PA 16652-9115

SCHOEFF, LARRY, educator; b. Bluffton, Ind., May 28, 1946; s. Keith E. and Martha N. (Williams) S.; m. Anita L. Reaser, Sept. 8, 1968; children: Andrew J., Kelly M. BS, Ind. U., 1968, MS, 1970. Med. technologist Marion County Gen. Hosp., Indpls., 1970-72; edn. coord. St. Joseph's Hosp., Ft. Wayne, Ind., 1972-75; asst. prof. U. Ill., Peoria, 1975-81, Wichita (Kans.) State U., 1981-82; assoc. prof. U. Ill., Chgo., 1982-90; dir., assoc. prof. U. Utah, Salt Lake City, 1990—; edn. dir. ARUP Inc., Salt Lake City, 1990—; mem. MLT program adv. bd. 3 area community colls., Chgo., 1984-90; MT program cons. Chiang Mai (Thailand) U., 1989. Author/editor: Principles of Laboratory Instruments, 1992; contbr. articles to profl. jours. Recipient Travel award to Thailand Midwest Univs. Consortium for Internat. Activities, 1989. Mem. Am. Soc. Clin. Pathology (Utah advisor 1990—, S.W. regional advisor 1993—, legis. liaison 1992—, tech. sample chemistry editor 1992—), Am. Assn. Clin. Chemistry, Am. Soc. Med. Tech. (Omicron Sigma award 1980, 81), Clin. Lab. Mgmt. Assn., ACLU, Nature Conservancy, Sierra Club (chair exec. com. 1992—). Achievements include first external articulation MLT to MT career mobility program in nation; criterion referenced equivalency examinations for med. lab. technician, med. technologist career mobility curriculum; comprehensive text lab. instruments. Office: U Utah 500 Chipeta Way Salt Lake City UT 84108

SCHOELER, GEORGE BERNARD, medical entomologist; b. Seattle, Aug. 5, 1961; s. Robert Gene and Inez (Stegman) S.; m. Jaime A. Rabdau, Aug. 31, 1985; children: Robert James, Jessika Ann. BS in Biology, Boise State Univ., 1989; MS in Entomology, Univ. Calif., 1992. Rsch. tech. USDA, ARS, ADRU, Caldwell, Idaho, 1986-89; rsch. asst. Univ. Calif., Berkeley, 1989-92; entomologist Navy Disease Vector Ecology and Control Ctr., Alameda, Calif., 1992—. Contbr. articles to profl. jours. Lt. USN, 1992—. Recipient Navy Ach. medal, 1993, grant Sigma Xi, 1991. Mem. Entomological Soc. Am., Sigma Xi (assoc.). Republican. Roman Catholic. Office: Disease Vector Ecology Control Ctr Naval Air Station Alameda CA 94501-5039

SCHOEN, ALLEN HARRY, aerospace engineering executive; b. N.Y.C., Mar. 10, 1936; s. Harry Alfred and Dorothy Julia (Browne) S.; m. Patricia Alice O'Madigan, June 1, 1958 (div. 1989); children: Theresa Mary, James Allen, Karen Linda. SB in Aero. Engring., MIT, 1958, postgrad., 1989. Aerodynamicist Douglas Aircraft Co., Santa Monica, Calif., 1958-61, United Aircraft Co., Farmington, Conn., 1961-66; with Boeing Helicopters, Phila., 1966—; technology mgr. Boeing Helicopters, 1980-84, dir. technology, 1984-86, dep. tech. dir. V-22 Osprey joint program, 1986-88, dir. preliminary design, 1988—; mem. aero. adv. com., NASA, Washington, 1985-90. Patentee propulsion system; author tech. papers. Fellow AIAA (assoc.), Am. Helicopter Soc. (hon.; pres. Phila. chpt. 1983-84, v.p. Mideast region 1986-88, dir.-at-large 1988-90). Republican. Episcopalian. Avocations: photography, gardening, woodworking, woodcarving. Home: 17 Mullray Ct Deptford NJ 08096-6713 Office: Boeing Helicopters PO Box 16858 Philadelphia PA 19142-0858

SCHOEN, ALVIN E., JR., environmental engineer; b. Milford, Conn., Jan. 3, 1945; s. Alvin E. and Thelma (Terrace) S.; m. Mary Ann Kosik; 1 child, Matthew S. BA in Math., U. Conn., 1968, BS in Engring., 1971; MSCE, Polytechnic Inst. of N.Y., 1977; PhD in Environ. Engring., U. Okla., 1994. Registered profl. engr., Ohio, Maine, N.Y., Conn., Mass., N.H., Vt., Okla., N.J. Field engr. Mobil Oil Corp., Scarsdale, N.Y., 1973-76, engring. supr., 1976-80; project engr. Mobil Oil Corp., Fairfax, Va., 1980-82; group leader, process engr. Mobil Oil Corp., Oklahoma City, 1985-89; environ. engr. rsch. and devel. Mobil Oil Corp., Princeton, N.J., 1989—; project engr. Arabian Am. Oil, Dhahran, Saudi Arabia, 1982-85; commr. Inland Wetlands Commn., Brookfield, Conn., 1978-80, Environ. Commn., Montgomery Twsp, N.J., 1992-93. 1st lt. C.E., U.S. Army, 1968-70, Vietnam. Fellow ASCE; mem. Am. Petroleum Inst. (mktg. terminal effluent task force, liner work group), Nat. Soc. Profl. Engrs., Water Pollution Control Fedn. Office: Mobil Rsch and Devel Corp PO Box 1026 Princeton NJ 08543

SCHOEN, HOWARD FRANKLIN, computer programmer, analyst; b. N.Y.C., Jan. 4, 1946; s. Sohl and Celia (Permut) S.; m. Althea Shepherd, June 15, 1986. BS, U. Pitts., 1965; PhD, CUNY, 1975. Fellow Mt. Sinai Sch. Medicine, N.Y.C., 1975-79; asst. prof. Downstate Med. Ctr., N.Y.C., 1979-89; sr. programmer, analyst Micro Healthsystems, Inc., W. Orange, N.J., 1989-92, project team leader, 1992—. Contbr. articles to profl. jours. Chmn. Somerset/Middlesex Area Libertarians, Edison, N.J., 1988-92; steering com. Libertarian Party of N.J., 1988-91, cand. for U.S. Congress, 1988, 90. NIH postdoctoral fellow, 1975-79, co-investigator, 1979-89. Mem. N.Y. Acad. Sci., Nat. Assn. Parliamentarians, Forth Interest Group. Achievements include development of instrumentation used in studying ion (salt) transport across epithelial cell membranes; demonstration of simple physical mechanism for ion transport in frog skin epithelium; demonstration of mechanisms for hormone and drug action. Office: Micro Healthsystems 414 Eagle Rock Ave West Orange NJ 07052-4211

SCHOEN, ROBERT DENNIS, civil engineer; b. Oak Park, Ill., Nov. 9, 1960; s. Jerry Egan and Patricia Ann (Ragen) S.; m. Marieke Lynn Dekker, July 6, 1985; 1 child, Alexander. BS in Civil Environ. Engring., U. Wis., Madison, 1983. Profl. Engr. Design engr. Donohue & Assoc., Schaumburg, Ill., 1983-90; dir. land devel. Marriott Corp., Schaumburg, 1990-91, Globe-Ill., Gurnee, 1991-92, Kennedy Homes, Arlington Heights, Ill., 1992—. Mem. Am. Soc. Civil Engrs. Home: 900 Forest Ave Oak Park IL 60302

SCHOENFELD, ELINOR RANDI, epidemiologist; b. Manhattan, N.Y., Apr. 9, 1956; d. Samuel and Helen (Goldstein) S. BS, SUNY, Stony Brook, 1977; MS, SUNY, Buffalo, 1980, PhD, 1988. Clin. assoc. Columbia U. Sch. Pub. Health, N.Y.C., 1980-82; data mgr. community oncology program Hackensack (N.J.) Med. Ctr., 1982-83; rsch. affiliate Roswell Park Cancer Inst., Buffalo, 1984-85, cancer rsch. scientist, 1985-88; epidemiology cons. Joel Bernstein, MD Otolaryngology, Buffalo, 1984-89; rsch. scientist SUNY Sch. Medicine, Stony Brook, 1988-93, rsch. instr., 1989-90, asst. prof., 1990—, dir. ops., 1992—, sr. rsch. scientist, 1993—; epidemiology cons. Univ. Hosp., Stony Brook, 1990—, Hosp. Joint Diseases, N.Y.C., 1990—; mem. admissions com. SUNY, Buffalo, 1985-87; presenter, invited speaker in field. Author: Applications of Diffusion Theory to Cancer Care in the United States: 1972-1981, 1990; author: (with others) On Breast Cancer, Cataracts, Otitis Media, 1988, 89, 91; contbr. articles to profl. jours. Predoctoral fellow NYU, 1977-78, Epidemiology Program fellow U. Minn., 1980, fellow in cancer epidemiology Columbia U., 1980-82; Nat. Cancer Inst. grantee NIH, 1987-88; Nat. Eye Inst. grantee, NIH, 1992—. Mem. APHA, Am. Assn. Diabetes Edn., Am. Diabetes Assn., Soc. Clin. Trials, Assn. for Rsch. in Vision & Ophthalmology, N.Y. Acad. Sci., N.Y. State Pub. Health Assn., Soc. for Epidemiologic Rsch., Assn. for Health Svcs. Rsch. Jewish. Office: SUNY Stony Brook Sch of Medicine Dept Preventive Medicine Stony Brook NY 11794-8036

SCHOENFELD, ROBERT LOUIS, biomedical engineer; b. N.Y.C., Apr. 1, 1920; s. Bernard and Mae (Kizelstein) S.; m. Helene Martens, Jan. 22, 1944 (div. 1965); children: David, Joseph, Paul; m. Florence Moskowitz, Dec. 11, 1965 (dec. 1989); children: Nedda, Bethany; m. Shulamith Stechel, July 8, 1990. BA, Washington Square Coll., 1942; BSEE, Columbia U., 1944; MEE, Poly. Inst. Bklyn., 1949, DEE, 1956. Registered profl. engr., N.Y. Rsch. assoc. Columbia U. Med. Sch., N.Y.C., 1947-51; rsch. fellow Sloan Kettering Cancer Rsch. Inst., N.Y.C., 1951-56; assoc. prof. Poly. Inst. Bklyn., 1947-54; assoc. prof. Rockefeller U., N.Y.C., 1957-90, prof. emeritus, 1990—. Contbr. articles to profl. jours. Lt. Signal Corps, U.S. Army, 1944-46, ETO. Fellow IEEE (edit. bd. 1965-75, Centennial medal 1985), Am. Inst. for Med. and Biological Engring. Democrat. Jewish. Achievements include being one of the first to apply computer automation to biological lab. experiments. Office: Rockefeller U 1230 York Ave New York NY 10002

SCHOENHARD, WILLIAM CHARLES, JR., health care executive; b. Kansas City, Mo., Sept. 26, 1949; s. William Charles S. and Joyce Evans (Thornsberry) Bell; m. Kathleen Ann Klosterman, June 3, 1972; children: Sarah Elizabeth, Thomas William. BS in Pub. Adminstrn., U. Mo., 1971; M of Health Adminstrn., Washington U., St. Louis, 1975. V.p., dir. gen. svcs. Deaconess Hosp., St. Louis, 1975-78; assoc. exec. dir. St. Mary's Health Ctr., St. Louis, 1978-81; exec. dir. Arcadia Valley Hosp., Pilot Knob, Mo., 1981-82, St. Joseph Health Ctr., St. Charles, Mo., 1982-86; exec. v.p. SSM Health Care System, St. Louis, 1986—; bd. dirs. Mark Twain Bank, 1986—.

Contbr. articles to profl. jours. pres. Shaw Neighborhood Improvement Assn., St. Louis, 1979-80; mem. adv. bd. St. Louis Area chpt. Lifeseekers, 1985—, bd. mgrs. Kirkwood-Webster (Mo.) YMCA, 1990—, Nat. Affairs Round Table Sen. Christopher Bond, St. Louis, 1990—, nat. adv. coun. Healthcare Forum, 1992—, health care adv. bd. Sanford Brown Colls., 1992—, excellence com. Catholic Health Assn. U.S. Ctr. Leadership, 1993—; alt. del. Am. Hosp. Assn. Ho. of Delegates, 1993—. With USN, 1971-72. Fellow Am. Coll. Health Care Execs.; mem. Am. Hosp. Assn. (ho. dels. 1993—), Univ. Club St. Louis, Phi Eta Sigma, Pi Omicron Sigma, Delta Upsilon, Delta Sigma Pi. Roman Catholic. Avocations: reading, jogging. Home: 420 Fairwood Ln Saint Louis MO 63122-4429 Office: SSM Health Care System 477 N Lindberg Blvd Saint Louis MO 63141

SCHOFIELD, KEITH, research chemist; b. Derby, England, July 27, 1938; came to U.S., 1963; s. Kenneth Schofield and Peggy (Hewitt) Furniss; divorced; children: Jeremy N., Clare, Susan A. BA, Cambridge U., 1960, MA and PhD, 1964. Rsch. scientist Nat. Ctr. Atmospheric Rsch., Boulder, Colo., 1965-67; rsch. physicist Cornell Aeronautical Rsch. Lab., Buffalo, 1967-68; sr. staff scientist GM Rsch. Lab., Santa Barbara, Calif., 1968-74; rsch. assoc. Quantum Inst. U. Calif., Santa Barbara, 1974-90; rsch. dir. Chemdata Rsch., Santa Barbara, 1974—; rsch. assoc. chemistry dept. U. Calif., Santa Barbara, 1991—; cons. GM Corp., Santa Barbara, 1975-88, Arnold Engring. Devel. Ctr., Arnold AFB, Tenn, 1988-92. Contbr. articles to profl. jours. Chmn. conservation Los Padres Chpt. Sierra Club, Santa Barbara, 1990-92; bd. dirs. Las Positas Friendship Park, Santa Barbara, 1993—. Postdoctoral Rsch. fellow U. Calif., 1964-65; U.K. State and Major scholar, Cambridge, 1957-60. Mem. Am. Chem. Soc., Am. Physical Soc., Combustion Inst. Home: PO Box 40481 Santa Barbara CA 93140

SCHOFIELD, WILLIAM HUNTER, research aerodynamicist; b. Perth, Australia, Oct. 22, 1940; s. Charles Campbell and Winifred Elvena (Hunter) S.; m. Hilary Lawrence Adams, May 22, 1965; children: Penelope Ellen, William. B Mech. Engring., U. Melbourne, Victoria, Australia, 1961, M Engring. Sci., 1962, PhD, 1969. Rsch. scientist Aero. Rsch. Lab., Melbourne, 1969-76, sr. rsch. scientist, 1976-81, prin. rsch. scientist, 1981-86, sr. prin. rsch. scientist, 1986-88, supt. aero propulsion div., 1988, chief flight mechanics and propulsion, 1988-92, chief airframes and engines, 1992—; invitational prof. Ariz. State U., Tempe, 1979; dir. Coop. Rsch. Ctr., Melbourne, 1992—, Defence Techs. of Australia, New South Wales, 1991—; external mem. faculty engring. U. Melbourne, 1991—. Contbr. articles to profl. jours. Mem. AIAA, ASME, Australian Inst. Co. Dirs. Achievements include research in new similarity law describing turbulent boundary layers in adverse pressure gradients, a unified theory for turbulent flow over a rough wall, a new unified theory for equilibrium flows in adverse pressure gradients, a theory for separation of two dimensional turbulent flow. Home: 18 Fawkner St, South Yarra Victoria, Australia 3141 Office: Aero Rsch Lab, 506 Lorimer St, Port Melbourne Victoria, Australia

SCHOLES, ROBERT THORNTON, physician, research administrator; b. Bushnell, Ill., June 24, 1919; s. Harlan Lawrence and Lura Zolene (Camp) S.; m. Kathryn Ada Tew, Sept. 3, 1948; 1 child, Delia. Student Knox Coll., 1937-38; BS, Mich. State U., 1941; MD, U. Rochester, 1950; postgrad. U. London, 1951-52, U. Chgo., 1953. Intern, Gorgas Hosp., Ancon, C.Z., 1950-51; lab. asst. dept. entomology Mich. State U., 1940-41; rsch. asst. Roselake Wildlife Exptl. Sta., 1941; rsch. assoc. Harvard U., 1953-57; served to med. dir. USPHS, 1954-71, med. officer, dep. chief health and sanitation div. U.S. Ops. Mission, Bolivia, 1954-57, chief health and sanitation div., Paraguay, 1957-60, internat. health rep. Office of Surgeon Gen., 1960-62; br. chief, research grants officer, acting assoc. dir. Nat. Inst. Allergy and Infectious Diseases, NIH, Bethesda, Md., 1962-71; co-founder, pres. The Bioresearch Ranch, Inc., Rodeo, N.Mex., 1977—; cons. Peace Corps, 1961, Hidalgo County Med. Services, Inc., 1979—, N.Mex. Health Systems Agy., 1980-86, N.Mex. Health Resources, Inc., 1981-93; Hidalgo County Health Coun., 1993—, Luna County Charitable Found., 1993—, Hidalgo County Health Coun., 1993, Luna County Charitable Found., 1993. Served to capt. USAAF, 1942-45. Commonwealth Fund fellow, 1953. Mem. AAAS, Am. Soc. Tropical Medicine and Hygiene, N.Y. Acad. Sci., Am. Pub. Health Assn., Am. Ornithologists Union, Sembot Hon. Soc. Contbr. papers to profl. publs. Achievements include research, writing and field test of first health survey indices detailing anthopological parameters; institution of first country wide malaria control project in Paraguay. Home and Office: PO Box 117 Rodeo NM 88056-0117

SCHOMAKER, VERNER, chemist, educator; b. Nehawka, Nebr., June 22, 1914; s. Edwin Henry and Anna (Heesch) S.; m. Judith Rooke, Sept. 9, 1944; children: David Rooke, Eric Alan, Peter Edwin. B.S., U. Nebr., 1934, M.S., 1935; Ph.D., Calif. Inst. Tech., 1938. With Union Carbide Research Inst., 1958-65, asst. dir., 1959-63, assoc. dir., 1963-65; prof. chemistry U. Wash., Seattle, 1965-84; prof. emeritus U. Wash., 1984—, chmn. dept., 1965-70; vis. assoc. Calif. Inst. Tech., 1984-92, faculty assoc., 1992—. Contbr. articles on molecular structure to chem. jours. John Simon Guggenheim Meml. Found. fellow, 1947-48; Recipient Am. Chem. Soc. award in pure chemistry, 1950. Fellow AAAS, N.Y. Acad. Scis.; mem. Am. Chem. Soc., Am. Crystallographic Assn. (pres. 1961-62), Sigma Xi. Home: 12506 26th Ave NE # 103 Seattle WA 98125

SCHOMBURG, DIETMAR, chemist, researcher; b. Braunschweig, Fed. Republic Germany, Apr. 21, 1950; s. Hermann A. and Ingeborg (Mahn) S.; m. Ida Maria Berlinghoff; children: Annika Carolyn, Karen Thyra. Diploma in chemistry, U. Braunschweig, 1974, PhD, 1976. Rsch. fellow U. Braunschweig, 1976-78, asst. prof., 1979-83, prof., 1989; rsch. fellow Harvard U., Cambridge, Mass., 1978-79; researcher GBF, Braunschweig, 1983-86, dept. head, 1987. Author, editor: Enzyme Handbook Volume 1-6, 1990-93. Corr. mem. task group biol. macromolecules CODATA. Achievements include development of methods in protein design and protein x-ray structure determinations. Office: GBF, Mascheroder Weg 1, 3300 Braunschweig Germany

SCHOMMER, GERARD EDWARD, mechanical engineer; b. West Bend, Wis., June 2, 1951; s. Marvin Edward and Marilyn Dorthy (Perkins) S.; m. Debra Lorraine Nielsen, Sept. 3, 1977; children: Ann, Sarah, Katie. BSME, U. Wis., 1974. Registered profl. engr., Wis. Jr. tire devel. engr. Firestone Tire & Rubber Co., Akron, Ohio, 1974; machine design engr. West Bend Co., 1974-75; design engr. Milsco Mfg., Milw., 1976-78; project engr. Autotrol Corp., Milw., 1978-83, supr. product engring., 1983-86, exec. engr., 1986-89, mgr. devel. engr., 1989—. Achievements include mechanism patent for agricultural/industrial vehicle seating. Home: W141 N6672 Memory Rd Menomonee Falls WI 53051

SCHONBACH, BERNARD HARVEY, engineering executive; b. Phila., Oct. 1, 1948; s. Frederik and Jeannette (Goldberg) S.; m. Sara Shnaper, Aug. 22, 1970; children: Joel Michael, Addie Lynn. BSME, Pa. State U., 1971; MBA, Lehigh U., 1982. Registered profl. engr., Pa. Design engr. Link-Belt Co., Colmar, Pa., 1971-72; mgr. product devel. Fuller Co., Bethlehem, Pa., 1972-85; dir. engring. Allen-Sherman-Hoff Co., Malvern, Pa., 1985-87; gen. mgr. product devel. Fuller Co., Bethlehem, 1987-90; v.p. engring. Pa. Crusher Corp., Broomall, Pa., 1990—. Patentee in field; contbr. articles to profl. jours. Avocations: collecting fountain pens, biking, calligraphy. Home: 1732 Brandywine Rd Allentown PA 18104-1704 Office: Pa Crusher Corp 600 Abbott Dr PO Box 100 Broomall PA 19008-0100

SCHÖNBERGER, WINFRIED JOSEF, pediatrics educator; b. Wiesbaden, Hessen, Germany, Apr. 2, 1940; s. Josef Adam and Thea (Brechmann) S.; m. Gisela Schönig, Apr. 8, 1971. MD, Gutenberg U., Mainz, Germany, 1966. Asst. prof. U. Mainz, 1972-75, pvt. docent, 1975-77, prof. pediatrics, 1977—. Author 2 books, 100 papers. Recipient Ann. award German Soc. Dentistry, 1977, Prize, Boehringer Preis, Ingelheim, 1980. Mem. German Soc. Pediatrics, South German Soc. Pediatrics, German Soc. Endocrinology. Home: Pfahlerstr 43, D-65193 Wiesbaden-Sonnenberg Hessen, Germany Office: Johannes Gutenberg U, Langenbeckstr 1, D-65 Mainz Germany

SCHONFELD, EUGENE PAUL, investment company executive, software designer; b. N.Y.C., July 9, 1943; s. Morris and Alice (Boyle) S.; m. Faith Johnson, Dec. 5, 1970. BA, U. Notre Dame, 1967; MS in Journalism, Northwestern U., 1968, PhD in Mgmt., 1975. Registered investment advisor SEC. Account exec. Paul A. Fergus Co., South Bend, Ind., 1964-68; project dir. N.W. Ayer, Phila., 1968-70; asst. prof. U. Ill., Chgo., 1973-75, Northwestern U., Evanston, Ill., 1975-76; pres., CEO Schonfeld & Assocs., Inc., Lincolnshire, Ill., 1977—; Author (book and software): Selling Marketing Risk Training System, 1971; contbr. articles to profl. jours., chpts. to books; author/designer (software): INGOT Modeling System, 1985. Pres., CEO Nat. Kidney Cancer Assn., Evanston, 1990—. Home: 67 Lakewood Pl Highland Park IL 60035 Office: 1234 Sherman Ave Evanston IL 60202

SCHONFELD, GUSTAV, medical educator; b. Mukacevo, Ukraine, May 8, 1934; came to U.S., 1946; s. Alexander Schonfeld and Helena Cottesmann; m. Miriam Steinberg, May 28, 1961; children: Joshua Lawrence, Julia Elizabeth, Jeremy David. BA, Washington U., St. Louis, 1956, MD, 1960. Diplomate Am. Bd. Internal Medicine. Intern. Bellevue Med. Ctr. NYU, 1960-61, resident in internal medicine, 1961-63; chief resident in internal medicine Jewish Hosp., St. Louis, 1963-64; NIH trainee in endocrinology and metabolism sch. medicine Washington U., St. Louis, 1964-66, instr. medicine, 1965-66, asst. prof. medicine sch. medicine, 1968-70, assoc. prof. preventive medicine and medicine, 1972-77, prof. preventive medicine dept. preventive medicine, 1977-86, prof. internal medicine, 1977—, William B. Kountz prof. medicine, 1987—, dir. atherosclerosis and lipid rsch. ctr,, 1972—, acting head dept. preventive medicine, 1983-86; rsch. assoc. Cochran VA Hosp., St. Louis, 1965-66, clin. investigator, 1968-70, cons. in internal medicine, 1972—; rsch. flight med. officer USAF Sch. Aerospace Medicine, Brooks AFB, Tex., 1966-68; asst. physician Barnes Hosp., Brooks AFB, 1969-70, 72-86, assoc. physician, 1986—; clin. instr. medicine med. sch. Harvard U., Boston, 1970-72; assoc. prof. metabolism and human nutrition, asst. dir. Clin. Rsch. Ctr. MIT, Boston, 1970-72; mem. rsch. com. Mo. Heart Assn., 1978-80; expert witness Working Group on Atherosclerosis NHLBI, 1979, Nat. Diabetes Adv. Bd., 1979; mem. endocrinologic and metabolic drugs adv. com. Pub. Health Svc., FDA, 1982-86; mem. nutritional study sect. NIH, 1984-88, spl. reviewer metabolism study sect.; mem. adult treatment guidelines panel Nat. Cholesterol Edn. Program, 1986—; mem. Consesus Devel. Conf. on Triglyceride, High Density Lipoprotein, and Coronary Heart Disease, 1992—; cons. Am. Egg Bd., Am. Dairy Bd., Inst. Shortening and Edible Oils, Ciba-Geigy, Sandoz, Fournier, Parke-Davis, Bristol-Meyers Squibb, Monsanto. Editor, mem. ad hoc review com. Atherosclerosis; past mem. editorial bd. Jour. Clin. Endocrinology and Metabolism, Jour. Clin. Investigation; mem. editorial bd. Jour. Lipid Rsch. Recipient Berg Prize in Microbiology, 1957, 58; grantee NIH. Fellow ACP; mem. Assn. Am. Physicians, Am. Soc. for Clin. Investigation, Am. Physiol. Soc., Am. Soc. Biol. Chemists, Am. Inst. Nutrition, Am. Diabetes Assn., Am. Heart Assn. (program com. coun. on atherosclerosis 1977-80, 86-88, nat. com. 1980-84, pathology rsch. com. 1980-83), Endocrine Soc., Internat. Union Immunological Socs. (standardization com. 1982—). Office: Washington University Sch Medicine 4566 Scott Ave Box 8046 Saint Louis MO 63110

SCHONHARDT, CARL MARIO, analytical chemist; b. Pitts., Sept. 26, 1947; s. Carl W. and Jill M. (Martinelli) S.; m. Barbara M. Huber, July 18, 1970; children: Melissa Anne, Aimee Jill. BS in Chemistry, Pa. State U., 1969. Nutrional chemist H.J. Heinz Rsch., Pitts., 1969-70; analytical lab. mgr. Arco Chem. R & D, Monaca, Pa., 1970-77; instrument sales Perkin-Elmer Corp., Pitts., 1977-81, Beckman Instruments, Pitts., 1981-84, Varian-Instrument Group, Pitts., 1984-88, Perkin-Elmer Corp., Pitts., 1988—. Mem. Spectroscopy Soc. Pitts. (scholarships and grants com. 1973—), Soc. Analytical Chemists Pitts (scholarships and grants com. 1969—), Am. Chem. Soc. Roman Catholic. Home: 8339 Van Buren Dr Pittsburgh PA 15237-4465 Office: Perkin-Elmer Corp 300 Alpha Dr Pittsburgh PA 15238

SCHONHOLTZ, GEORGE JEROME, orthopaedic surgeon; b. Bklyn., June 9, 1930; s. Morris and Rose (Stofsky) S.; m. Joan S. Hirsh, Aug. 21, 1951; children: Margot, Steven, Barbara. BA, NYU, 1950; MD, N.Y. State U., 1954. Diplomate Am. Bd. Orthopaedic Surgery, Nat. Bd. Med. Examination; lic. physician, Md. Intern, resident gen. surgery and orthopaedic surgery Walter Reed Gen. Hosp., Washington, 1956-59; asst. chief orthopaedic surgery Martin Army Hosp., Ft. Benning, Ga., 1960-63, asst. dir., dir. med. edn., 1962, 63; instr. human biology Am. U. Undergrad. Sch., Ft. Benning, 1962, 63; asst. clin. prof. orthopaedic surgery Howard U., Washington, 1964-66, Georgetown U., Washington, 1966-67, George Washington U., Washington, 1968—; pvt. practice Silver Spring, Md., 1964—; orthopaedic cons. VA Hosp., Martinsburg, W.Va., 1964-68; civilian cons. orthopaedic surgery Walter Reed Army Hosp., Washington, 1968—; chief orthopaedic surgery Holy Cross Hosp., Silver Spring, 1971-74, chmn. infection control com. 1975-76; v.p. med. and dental staff Washington Adventist Hosp., 1988-89, chmn. credential com., 1988-89; rep. Coun. of Musculoskeletal Soc., 1987-90. Author: Arthroscopy of the Shoulder, Elbow and Ankle, 1986, An Atlas of Arthroscopic Surgery of the Knee, 1988. Maj. U.S. Army, 1960-64. Mem. AMA, ACS, Am. Acad. Orthopaedic Surgery (mem. resolutions com. 1989—), Assn. Hosp. Dirs. Med. Edn., Soc. Mil. Orthopaedic Surgeons, Internat. Arthroscopy Assn., Ea. Orthopaedic Assn. (bd. incorporators, bd. dirs. 1970-79), Arthroscopy Assn. N.Am. (pres. 1988-89, bd. dirs. 1983-90), Montgomery County Med. Soc., Med. and Chirurgical Faculty Med., Washington Orthopaedic Soc., Internat. Soc. Knee. Republican. Avocations: golf, sailing. Office: Schonholtz & Magee 8830 Cameron St Silver Spring MD 20910-4114

SCHONHORN, HAROLD, chemist, researcher; b. N.Y.C., Apr. 2, 1928; s. Benjamin and Dorothy (Gitlin) S.; m. Esther Matesky, Jan. 17, 1954; children: Deborah, Jeremy. BS, Bklyn. Coll., 1950; PhD, N.Y. Polytech. U., 1959. Mem. tech. staff Bell Labs., Murray Hill, N.J., 1961-84; v.p. R & D Polyken Tech. div. Kendall Co., Lexington, Mass., 1984—. Contbr. over 100 articles to profl. jours. Pres. B'nai B'rith Lodge, Summit, N.J., 1970. With U.S. Army, 1953-55, Korea. Mem. Am. Chem. Soc. Achievements include 15 patents. Home: 12 Heathwood Ln Chestnut Hill MA 02167-2685 Office: Kendall Co Polyken Techs Div 17 Hartwell Ave Lexington MA 02173-3195

SCHOOLEY, ARTHUR THOMAS, chemical engineer; b. Plymouth, Pa., July 4, 1932; s. Arthur C. and Mary Ann (Thomas) S.; m. Dorothy Jean Ward, Sept. 18, 1955; children: Jay, David, Linda. BS, Carnegie-Mellon U., 1954; MS, Akron U., 1959. R&D group leader B.F. Goodrich, Brecksville, Ohio, 1960-66, sr. rsch. engr., 1966-68, rsch. assoc., 1968-79, sr. R&D assoc., 1979-88; engring. cons. Akron, Ohio, 1988—. Mem. Am. Inst. Chem. Engrs. (sect. chmn. Akron sect. 1969, Akron area Chem. Engr. of Yr. 1982). Presbyterian. Achievements include patent in field. Home and Office: 2015 Burlington Rd Akron OH 44313

SCHOOLMAN, ARNOLD, neurological surgeon; b. Worcester, Mass., Oct. 31, 1927; s. Samuel and Sarah (Koffman) Schulman; m. Gloria June Feder, Nov. 10, 1964; children: Hugh Sinclair, Jill. Student, U. Mass., 1945-46; BA, Emory U., 1950; PhD, Yale U., 1954, MD, 1957. Diplomate Am. Bd Neurol. Surgery, Nat. Bd. Med. Examiners. Intern U. Calif. Hosp., San Francisco, 1957-58; resident in neurol. surgery Columbia-Presbyn. Med. Ctr., Neurol. Inst. N.Y., N.Y.C., 1958-62; instr. neurol. surgery U. Kans. Sch. Medicine, Kansas City, 1962, asst. prof. surgery, 1964; assoc. prof. U. Mo. Sch. Medicine, Kansas City, 1976; chief sect. neurosurgery Research Med. Ctr., Kansas City, 1982; dir. Midwest Neurol. Inst., 1982-83. Patentee in field. Served with USN, 1946-48. Fellow ACS (mem. Mo. chpt.); mem. AMA, Mo. State Med. Assn., Kansas City Med. Soc., Kansas City Neurosurg. Soc. (pres. 1984-85), Kansas City Neurol. Soc., Rocky Mountain Neurosurg. Soc., Am. Assn. Neurol. Surgeons, AAAS, Mo. Neurol. Soc., Internat. Coll. Surgeons, Congress Neurol. Surgeons, Brit. Royal Soc. Medicine, Phi Beta Kappa, Sigma Xi. Avocation: pilot. Home: 8705 Catalina St Shawnee Mission KS 66207-2351 Office: 1000 E 50th St Ste 310 Kansas City MO 64110-2215

SCHOON, DAVID JACOB, electronic engineer; b. Luverne, Minn., May 6, 1943; s. Ted and Mildred (Thorson) S.; m. Maureen Perry, Mar. 15, 1969 (div. 1985); m. Patricia Mary Lentsch, Mar. 22, 1986; children: Christopher, Beth Schoon Washick. B.Physics, U. Minn., 1965, PhD, 1969. Sr. specialist 3M Co., St. Paul, 1969-86; staff cons. Printware, Inc., Mendota Heights, Minn., 1986-91; pres. Schoonscan Inc., Mendota Heights, Minn., 1991—. Mem. Imaging Sci. & Tech. Assn. (counselor 1991-93). Achievements include 16 patents. Home and Office: 871 Mendakota Ct Mendota Heights MN 55120

SCHOONEN, MARTIN ADRIANUS ARNOLDUS, geology educator; b. Hoogerheide, The Netherlands; came to U.S., 1985; s. Martinus Jacobus and Petronella (Dingemans) S.; m. Josephine Connolly, Dec. 22, 1990. BS in Geochemistry, U. Utrecht, The Netherlands, 1981, MS in Geochemistry, 1984; PhD in Geochemistry, Pa. State U., 1989. Asst. prof. SUNY, Stony Brook, 1989—. Contbr. articles to profl. jours. Grantee Am. Cancer Soc.-Petroleum Rsch. Fund, 1990, NSF, 1991. Mem. Am. Mineralogical Soc., Geochem. Soc. Achievements include research in sulfur geochemistry. Office: SUNY ESS 200 Stony Brook NY 11794

SCHOR, JOSEPH MARTIN, pharmaceutical executive, biochemist; b. Bklyn., Jan. 10, 1929; s. Aaron Jacob and Rhea Iress (Kay) S.; children: Esther Helen, Joshua David, Gideon Alexander; m. Laura Sharon Strumingher, June 14, 1992. B.S. magna cum laude, CCNY, 1951; Ph.D., Fla. State U., 1957. Sr. research chemist Armour Pharm. Co., Kankakee, Ill., 1957-59, Lederle Labs., Pearl River, N.Y., 1959-64; dir. biochemistry Endo Labs., Garden City, N.Y., 1964-76; v.p. sci. affairs Forest Labs., N.Y.C., 1977—. Editor, contbr.: Chemical Control of Fibrinolysis-Thrombolysis, 1970. Contbr. articles to profl. jours. Patentee in field. USPHS fellow, 1955-57. Fellow Am. Inst. Chemists (cert. profl. chemist), Internat. Soc. Hematology; mem. Am. Chem. Soc. (chmn. Nassau County subsect. 1971-72), Internat. Soc. on Thrombosis and Hemostasis, N.Y. Acad. Scis., AAAS, Phi Beta Kappa, Sigma Xi. Home: 28 Meleny Rd Locust Valley NY 11560-1221

SCHORR, MARTIN MARK, psychologist, educator, writer; b. Sept. 16, 1923; m. Dolores Gene Tyson, June 14, 1952; 1 child, Jeanne Ann. Student Balliol Coll., Oxford (Eng.) U., 1945-46; AB cum laude, Adelphi U., 1949; postgrad., U. Tex., 1949-50; MS, Purdue U., 1953; PhD, U. Denver, 1960; postgrad., U. Tex. Diplomate in psychology; lic. clin. pscyhologist. Chief clin. psychol. svcs. San Diego County Mental Hosp., 1963-67; clin. dir. human services San Diego County, 1963-76; pvt. practice, forensic specialist San Diego, 1962—; forensic examiner superior, fed. and mil. cts., San Diego, 1962—; prof. abnormal psychology San Diego State U., 1965-68; chief dept. psychology Center City (Calif.) Hosp., 1976-79; cons. Dept. Corrections State of Calif., Minnewawa, 1970-73, Disability Evaluation Dept. Health, 1972-75, Calif. State Indsl. Accident Commn., 1972-78, Calif. Criminal Justice Adminstrn., 1975-77, Vista Hill Found., Mercy Hosp. Mental Health, Foodmaker Corp., Convent Sacred Heart, El Cajon, FAA Examiner. Author: Death by Prescription, 1988; co-dir. Timberline Films, Inc. Recipient award for aid in developing Whistle Blower Law Calif. Assembly, 1986. Fellow Internat. Assn. Soc. Psychiatry; mem. AAAS, PEN, Am. Psychology Assn., Am. Acad. Forensic Scis., Qualified Med. Evaluator, Calif., 1993, Internat. Platform Assn., WOrld Mental Health Assn., Mystery Writers Am., Nat. Writers' Club., Mensa. Home and Office: 2970 Arnoldson Ave University City San Diego CA 92122-2114

SCHOTTENFELD, DAVID, epidemiologist, educator; b. N.Y.C., Mar. 25, 1931; m. Rosalie C. Schaeffer; children: Jacqueline, Stephen. AB, Hamilton Coll., 1952; MD, Cornell U., 1956; MS in Pub. Health, Harvard U., 1963. Diplomate Am. Bd. Internal Medicine, Am. Bd. Preventive Medicine. Intern in internal medicine Duke U., Durham, N.C., 1956-57; resident in internal medicine Meml. Sloan-Kettering Cancer Ctr., Cornell U. Med. Coll., N.Y.C., 1957-59; Craver fellow med. oncology Meml. Sloan-Kettering Cancer Ctr., 1961-62; clin. instr. dept. pub. health Cornell U., N.Y.C., 1963-67, asst. prof. dept. pub. health, 1965-70, assoc. prof. dept. pub. health, 1970-73, prof. dept. pub. health, 1973-86; John G. Searle prof., chmn. epidemiology sch. pub. health U. Mich., Ann Arbor, 1986—, prof. internal medicine, 1986—; vis. prof. epidemiology U. Minn., Mpls., 1968, 71, 74, 82, 86; W.G. Cosbie lectr. Can. Oncology Soc., 1987. Editor: Cancer Epidemiology and Prevention, 1982; author 7 books; contbr. over 150 articles to profl. jours. Served with USPHS, 1959-61. Recipient Acad. Career award in Preventive Oncology, Nat. Cancer Inst., 1980-85. Fellow AAAS, ACP, Am. Coll. Preventive Medicine, Am. Coll. Epidemiology, Armed Forces Epidemiology Bd.; mem. Phi Beta Kappa. Office: U of Mich Sch Pub Health Dept Epidemiology 109 Observatory St Ann Arbor MI 48109-2029

SCHRADER, ERNEST KARL, engineer; b. New Britain, Conn., Sept. 16, 1947; s. Ernest K. Schrader and Alma (Schultz) Longaker; m. Candice Carroll, May 7, 1988; 1 child, Ernest K. BSCE, Clarkson U., 1969, MSCE, 1971. Registered profl. engr. Oreg., Wash. Teaching asst. Clarkson U., Potsdam, N.Y., 1969-70; asst. supt. Turner Constrn., N.Y.C., 1970-71; materials engr. U.S. Army C.E., Walla Walla, Wash., 1974-83; cons. Schrader Cons., Walla Walla, 1989—. Co-author: Advanced Dam Engineering, 1991, Roller Compacted Concrete Dams, 1991; contbr. articles to profl. publs. Capt. C.E. U.S. Army, 1971-74. Recipient Tudor medal Soc. Am. Mil. Engrs., 1982. Mem. ASCE (Greenefelder Constrn. prize 1983), Am. Concrete Inst. (Wason medal 1984, Constrn. Practice award 1988), Internat. Assn. Concrete Repair Specialists, U.S. Commn. Large Dams. Achievements include patents on impact test method and device for concrete, pavement dowel and tie bar method, development of roller compacted concrete. Home and Office: Schrader Cons Blue Creek Rd Rte 4 Box 264 B Walla Walla WA 99362

SCHRAGER, GLORIA OGUR, pediatrician; b. N.Y.C., July 11, 1924; d. Ellis M. and Edith G. (Levine) Ogur; m. Alvin J. Schrager, May 25, 1952; children: Lewis K., Ralph M. BA, Bklyn. Coll., 1944; MD, Med. Coll. of Pa., 1948. Dir. pediatrics Overlook Hosp., Summit, N.J., 1972-89; clin. prof. pediatrics Columbia U. Coll. Physicians and Surgeons, N.Y.C., 1984—; pvt. practice Westfield, N.J., 1953-89. Contbr. articles to profl. jours. including Jour. Pediatrics, Lancet, Emergency Medicine and others. Bd. dirs. YM/YWCA, Westfield, 1992, bd. health, 1980-84. Home and Office: 1020 Summit Ave Westfield NJ 07090-2712

SCHRAMM, DAVID NORMAN, astrophysicist, educator; b. St. Louis, Oct. 25, 1945; s. Marvin M. and Betty Virginia (Math) S.; m. Judith J. Gibson, 1986; children from previous marriage: Cary, Brett. SB in Physics, MIT, 1967; PhD in Physics, Calif. Inst. Tech., 1971. Rsch. fellow in physics Calif. Inst. Tech., Pasadena, 1971-72; asst. prof. astronomy and physics U. Tex., Austin, 1972-74; assoc. prof. astronomy, astrophysics and physics Enrico Fermi Inst. and the Coll., U. Chgo., 1974-77, prof., 1977—, Louis Block prof. phys. scis., 1982—, prof. conceptual founds. of sci., 1983—, acting chmn. dept. astronomy and astrophysics, 1977, chmn., 1978-84; resident cosmologist Fermilab, 1982-84; cons., lectr. Adler Planetarium, Lawrence Livermore Lab., Los Alamos Nat. Lab.; organizer sci. confs.; frequent lectr. in field; chmn. bd. trustees Aspen Ctr. for Physics; bd. on physics and astronomy, exec. com. NRC, 1990—, vice chair, 1993—; bd. dirs. Astron. Rsch. Consortium, 1990—; pres. Big Bang Aviation, Inc.; bd. overseers Fermi Nat. Accelerator Lab., 1990—. Co-author: The Advanced Stages of Stellar Evolution, 1977, From Quarks to the Cosmos: Tools of Discovery, 1989, The Shadows of Creation: Dark Matter and the Structure of the Universe, 1991; co-editor: Supernovae, 1977, Fundamental Problems of Stellar Evolution, 1980, Essays in Nucleosynthesis, 1981, Gauge Theory and the Early Universe, 1988, Dark Matter in the Universe, 1990; editor profl. jours.; columnist Outside mag.; contbr. over 350 articles to profl. jours. Recipient Gravity Rsch. Found. prize, 1980, Humboldt award Fed. Republic Germany, 1987-88, Einstein medal Eotvos U., Budapest, Hungary, 1989. Fellow Am. Phys. Soc. (Lilienfeld prize 1993), Meteoritical Soc.; mem. Nat. Acad. Sci., Am. Astron. Soc. (Helen B. Warner prize 1978, exec. com. planetary sci. div. 1977-79, sec., treas. high energy astrophysics div. 1979-81), Am. Assn. Physics Tchrs. (Richtmeyer prize 1984), Astron. Soc. Pacific (Robert J. Trumpler award 1974), Internat. Astron. Union (commns. on cosmology, stellar evolution, high energy astrophysics), Aircraft Owners and Pilots Assn., British-N Am. Com., Alpine Club, Quadrangle Club, Sigma Xi. Achievements include development of the cosmological interface with particle physics and the use of cosmological arguments to constrain fundamental physics; use of big bang to form the principle argument regarding the cosmological density of normal matter. Home: 155 N Harbor Dr Apt 5203 Chicago IL 60601-7326 Office: Univ of Chicago AAC 140 5640 S Ellis Ave Chicago IL 60637-1467 also: 1163 Cemetery Ct Aspen CO 81611

SCHRECK, LISA TANZMAN, sensory analyst; b. Long Branch, N.J., Dec. 3, 1955; m. Ronald P. Schreck, Sept. 6, 1981; children: Daniel, Rose. BS, Rutgers U., 1977. Project scientist, mgr. sensory testing Internat. Flavors and Fragrances, Union Beach, N.J., 1980—. Mem. ASTM (task group chmn. 1987—), Inst. Food Technologists (Sensory Evaluation Div. exec.

com. mem.-at-large 1989-91). Achievements include patent for Schiff base reaction products of mixtures of aldehydes including helional and alkyl anthranilates; derivatives thereof; and organoleptic uses thereof. Office: Internat Flavors & Frgrnces 1515 Hwy 36 Union Beach NJ 07735

SCHRECK, RICHARD MICHAEL, biomedical engineer, consultant; b. Chgo., Mar. 31, 1943; s. Emil Michael and Emma Catherine (Kober) S.; m. Karen Elaine Tulin, Aug. 19, 1967; children: Peter Michael, Andrew Gregory. BSME, Northwestern U., 1966, PhD in Theoretical and Applied Mechanics, 1971. Engr. rsch. staff Ford Motor Co., Dearborn, Mich., 1963-66; sr. rsch. engr. vehicle rsch. dept. GM, Warren, Mich., 1971-75, staff rsch. engr. biomed. sci. dept., 1975-81, sr. staff rsch. engr., 1981-84, asst. dept. head, 1984-87, prin. rsch. scientist, 1987—; cons. U.S. Environ. Protection Agy., Research Triangle Park, N.C., 1990-93; reviewer U.S. Dept. Energy, Washington, 1992; lectr. Sch. Pub. Health U. Mich., Ann Arbor, 1985-93. Author: (with others) Inhalation Toxicology and Technology, 1981, Air Pollution: Physiological Effects, 1982; contbr. articles to Jour. Aerosol Sci. Pres. Birmingham Unitarian Ch., Bloomfield Hills, Mich., 1982-83; lectr. Focus: HOPE, Detroit, 1992, mentor, 1993. Mem. ASME (biomechanics divsn. 1973—) Am. Thoracic Soc. (air pollution subcom. 1983-85, environ. occupational health program com. 1987-1990), Am. Assn. Aerosol Rsch. Achievements include patents in field; rsch. in biomechanics of safety restraints, toxicity of gases and aerosols, biotech. for mfg. Office: GM Rsch Lab Biomed Sci Dept 30500 Mound Rd Warren MI 48090-9055

SCHREFLER, BERNHARD ARIBO, civil engineering educator; b. Merano, Italy, Oct. 4, 1942; s. Ludwig and Eleonora (Tirler) S.; m. Chantal Marie Madeleine Saint-Blancat, Sept. 30, 1971; 1 child, Lorna. Degree in civil engring., U. Padua (Italy), 1967; PhD, U. Wales, Swansea, 1984, DSc, 1992. Asst. lectr. in constrn. sci. U. Padua, 1969-70, asst. prof. structural mechanics, 1970-80, lectr. in computational mechanics, 1973-80, prof. structural mechanics, 1980-89, prof. constrn. sci., 1989—; dir. Inst. Structural Mechanics, U. Padua, 1988—. Author, editor 12 books; contbr. articles to profl. jours.; assoc. editor Internat. Jour. Environment and Pollution; mem. editorial bd. Internat. Jour. Communications in Applied Numerical Methods, Meccanica, Internat. Jour. Computer Applications in Tech., Internat. Jour. Numerical Methods in Engring., Jour. Marine Systems. Mem. N.Y. Acad. Scis., Internat. Assn. Computational Mechanics, Réunion Internat. des Laboratoires d'Essais et de Recherches sur les Matériaux et les Constructions, Rotary Padova Centro. Avocations: tennis, skiing, mountaineering. Home: Via Cappelli 7, 35123 Padua Italy Office: Ist. di Scienza e Tecnica delle Construzioni, Via Marzolo 9, 35131 Padua Italy

SCHREIBER, ANDREW, psychotherapist; b. Budapest, Hungary, Aug. 1, 1918; s. Alexander and Bella (Gruen) S.; m. Mona Schreiber, Aug. 6, 1950; children: Julie, Brad, Robin. BA, CCNY, 1941, MEd, 1943; MSW, Columbia U., 1949; PhD, Heed U., 1972. Diplomate Am. Bd. Sexology; lic. psychotherapist, Calif. Pvt. practice Belmont, Calif., 1970—; sales mgr. vibro ceramics dir. Gulton Industries, Metuchen, N.J., 1949-57; mktg. mgr. Weldotron Corp., Newark, 1957-63; head dept. spl. edn. San Mateo (Calif.) High Sch. Dist., 1964-70; mem. faculty Heed U., 1970-71, advisor to doctoral candidates on West Coast, 1971; lectr. spl. edn. U. Calif.-Berkeley, 1973. Art Students League of N.Y. scholar, 1933-35, San Francisco State U. grantee. Fellow Am. Acad. Clin. Sexology; mem. NEA, AACD, Learning Disabilities Assn., Am. Assn. Sex Educators, Counselors and Therapists, Calif. Assn. Marriage and Family Therapists, Calif. Tchrs. Assn. Home: 2817 San Ardo Way Belmont CA 94002-1341

SCHREIBER, EDWARD, computer scientist; b. Zagreb, Croatia, Mar. 17, 1943; came to U.S., 1956, naturalized, 1960; s. Hinko and Helen (Iskra) S.; m. Barbara Nelson, 1967 (div. 1969); m. Lea Lusia Hausler, Nov. 7, 1983. BSEE, U. Colo., Denver, 1970. Registered profl. engr. Colo.; cert. data processor. Sr. software scientist Autotrol, Denver, 1972-78; software engr. Sigma Design, Englewood, Colo, 1979-82; founder, v.p. Graphics, Info., Denver, 1982-86; chmn. Schreiber Instruments, 1987—; instr. computer sci. U. Colo., Denver, 1971-72, Colo. Women's Coll., Denver, 1972-73, U. Denver, 1983. Contbr. articles on computer graphics to profl. jours. Trustee 1st Universalist Ch., Denver, 1972-78; Dem. candidate for U.S. Ho. of Reps., 1980. Served with U.S. Army, 1960-66. Mem. IEEE, Assn. for Computing Machinery, Nat. Computer Graphics Assn., Mensa. Office: Schreiber Instruments Inc Ste # 250 4800 Happy Canyon Rd Denver CO 80237

SCHREIBER, EVERETT CHARLES, JR., chemist, educator; b. Amityville, N.Y., Nov. 13, 1953; s. Everett Charles Sr. and Mary Elizabeth (Johnston) S.; m. Jane Karen Sklenar, July 19, 1980. BS, Pace U., 1975; PhD, U. Nebr., 1980. Rsch. assoc. SUNY, Stony Brook, 1980-82; asst. dir. rsch. Muscular Dystrophy Assn., N.Y.C., 1983-84; rsch. assoc. SUNY, 1984-86; spectroscopist G.E. NMR Instruments, Fremont, Calif., 1986-87; quality assurance engr. Varian NMR Instruments, Palo Alto, Calif., 1987-89, tech. tng. specialist, 1989—. Author of tng. texts in engring. and computers. Vice pres. Old Bailey Place Home Owners Assn., Fremont, 1989, pres., 1990-93. Mem. Am. Chem. Soc., Biophys. Soc., N.Y. Acad. Sci., Soc. Magnetic Resonance in Medicine. Republican. Roman Catholic. Avocations: photography, computers, model trains, music. Office: Varian NMR Instruments 3120 Hansen Way Palo Alto CA 94304-1015

SCHREIBER, ROBERT JOHN, environmental engineer; b. Granite, Ill., Nov. 10, 1950; s. Robert John and Muriel Hady (Carver) S.; 1 child, Robert John III. BSChemE, U. Mo., 1972. Registered profl. engr., Mo., Kans., Calif., Ark., Ill., Ga., Md., Pa. Dir. Air Pollution Corp., Jefferson City, Mo., 1978-80; dir. divsn. environ. quality State of Mo., Jefferson City, 1980-85; v.p. Resource Recovery Inc., Hanizac, Mo., 1988-91, Solvent Recovery, Kansas City, Mo., 1988-90; pres. Schreiber Grant Yoncey, St. Louis, 1986—. Pres. Mo. Waste Control Coalition, 1985; appointee Gov.'s Task Force Coal, Jefferson City, 1980-81. Mem. Air and Waste Mgmt. Assn., Cement Kiln Recycling (bd. dirs.), Hazardous Waste Treatment. Achievements include patent for Use of Waste in Cement Kilms, for Kilns; development of research facility for dioxin. Home: 531 Goethe Kirkwood MO 63122 Office: Schreiber Grana Yonley 271 Wolfner Dr Saint Louis MO 63026

SCHREIBER, STUART L., chemist, educator; b. Feb. 6, 1956; m. Mimi Suzanne Packman, Aug. 9, 1981. BA in Chemistry, Harvard U., 1977; PhD in Organic Chemistry, U. Va., 1981. Asst. prof. dept. chemistry Yale U., New Haven, 1981-84, assoc. prof., 1984-86, prof. chemistry, 1986-88; prof. dept. chemistry Harvard U., Cambridge, Mass., 1988—; cons. Pfitzer, Inc., 1983-91; founder, mem. sci. adv. bd. Vertex Pharmaceuticals, 1989-90; mem. sci. adv. bd. Glytec, 1990, Cytel, 1991; founder, chief sci., mem. exec. com., mem. sci. and med. advisors bd. ARIAD Pharmaceuticals, 1991—; mem. vis. com. chemistry and structural biology Rockefeller U., 1992—; mem. adv. bd. Tables Rondes Roussel Uclaf. Co-editor: Chemistry and Biology; mem. editorial bd.: Comprehensive Organic Syntesis;; mem. bd. consulting editors: Tetrahedron Pubs.; mem. advisory editor: SynLett, Jour. Organic Chemsitry, Jour. Med. Chemistry, Biomed. Chemistry Lett, Biomed. Chemistry; contbr. articles to profl. jours. Mem. sci. cons. bd. Meml. Sloan-Kettering Cancer Ctr., 1993—. Recipient NSF Presdl. Young Investigator award, 1985, Excellence in Chemistry award ICI Pharmaceuticals, 1986, Arthur C. Cope Scholar award Am. Chem. Soc., 1986, Pure Chemistry award Am. Chem. Soc., 1989, Arun Guthikonda Meml. award, 1990, Biomed. Rsch. award Ciba-Geigy Drew, 1992, Thieme-IUPAC Synthetic Organic Chemistry award, 1992, NIH Merit award, 1992, Rhone-Poulenc Silver medal Royal Soc. Chemistry, 1992, Eli Lilly Biol. Chemistry award Am. Chem. Soc., 1993, Leo Hendrik Baekeland award, 1993, Creative Work in Synthetic Chemistry award Am. Chem. Soc., 1994. Office: Harvard U Dept Chemistry 12 Oxford St Cambridge MA 02138*

SCHREINER, CEINWEN ANN, mammalian and genetic toxicologist; b. Phila., May 27, 1943; s. Norman George and Ceinwen Mary (Heycock) S. BS in Biology, Muhlenberg Coll., 1965; MS in Zoology, U. N.H., 1969, PhD in Genetics, 1972. Rsch. scientist teratology E.R. Squibb and Sons, New Brunswick, N.J., 1967-69; group leader genetic toxicology McNeil Pharms., Fort Washington, Pa., 1972-79; supr. genetic toxicology Mobil Oil Corp., Princeton, N.J., 1979-86, mgr. pathology/immunology, 1986-88, mgr. biochem. toxicology, 1988-92; mgr. mammalian/genetic toxicology Mobil Oil Corp.-Environ. and Health Scis. Lab., Princeton, N.J., 1992—; chmn. Gordon Rsch. Conf. on Genetic Toxicology, New London, N.H., 1983; mem. rev. com. N.J. Commn. on Cancer Rsch., Trenton, 1986-90; chmn. neurotoxicology com. Am. Petroleum Inst., Washington, 1990—; mem. sci. com. Am. Indsl. Health Coun., Washington, 1990—; bd. dirs. Acad. of Toxicologic Scis. Contbg. author: Handbook of Experimental Pharmacology, 1983, Annals of N.Y. Academy of Sciences, 1983; contbr. numerous articles to profl. jours. Named fellow Acad. of Toxicologic Scis., 1987; recipient Womn in Industry Achievement award YWCA, 1992. Mem. Am. Coll. Toxicology (sec. 1992-95, councillor 1987-89), Environ. Mutagen Soc. (councilor 1987-89), Genetic Toxicology Assn. (co-founder 1978, treas. 1978-81), Teratology Soc. Achievements include co-devel. of modified Ames test to screen petroleum compounds for carcinogenesis; 4 patents on this and similar genetic toxicology methodology; devel. of testing and rsch. program tailored to phys. characteristics of petroleum compounds. Office: Mobil Oil Corp Environ & Health Scis Lab PO Box 1029 Princeton NJ 08543-1029

SCHREINER, CHRISTINA MARIA, emergency physician; b. Vienna, Austria; came to U.S., 1962; BS in Biology, George Wash. U., 1969, MS in Biology, 1972, PhD in Organic Chemistry, 1977, MD, 1980. Diplomate Am. Bd. Emergency Medicine. Resident in internal medicine SUNY, Buffalo, 1980-83; emergency physician N.W. Ark. Emergency Med. Group St. Mary-Rogers Meml. Hosp., 1983-85; emergency physician Coastal Emergency Svcs. Inc., Danville, Va., 1985-86; emergency physician Alexandria Physicians Group Alexandria (Va.) Hosp., 1987—, past mem. mass casualty com., past. mem. intensive care com. Past mem. City of Alexandria Med. Svcs. Coun., No. Va. Emergency Med. Svcs. Coun. Fellow Am. Coll. Emergency Physicians; mem. ACP, N.Y. Acad. Scis.

SCHREUDER, HEIN, chemical company executive, business administration educator; b. Djakarta, Indonesia, Dec. 24, 1951; s. Hendrikus and Cornelia G. (Kiesewetter) S.; 1 child, Hans Christiaan. MBA, Erasmus U., Rotterdam, Netherlands, 1976; PhD, Free U. Amsterdam (Netherlands), 1981. Bus. researcher Netherlands Econ. Inst., Rotterdam, 1975-76; head bus. rsch. Econ. and Social Inst., Amsterdam, 1976-81, dir., 1981-84; prof. bus. adminstrn. Maastricht, 1984—; fellow European Inst. Advanced Studies in Mgmt., Brussels, 1981-91; vis. scholar U. Wash., Seattle, 1982-83; disting. internat. lectr. Am. Acctg. Assn., 1985; pres. Ecozoek, The Dutch Found. for Pure Econ. Rsch., 1987-88; fellow Netherlands Inst. for Advanced Studies in Social Scis. and Humanities, 1989-90; pres. European Acctg. Assn., 1991-92; dir. planning and devel. DSM Polymers, 1991—. Author: (with Jan Klaassen) Corporate Reports, 1980; Social Responsibility, 1981; (with Jan Klaassen) Forecasting, 1982, (with Sytse Douma) Economic Approaches to Organizations, 1992; co-editor: European Accounting Research, 1984; General Economics and Business Economics, 1985; Interdisciplinary Perspectives on Organization Studies, 1993. Netherlands Orgn. Advancement of Pure Research grantee, 1982; Fulbright grantee, 1982. Mem. Strategic Mgmt. Soc., Acad. of Mgmt., European Acctg. Assn., Am. Acctg. Assn., European Group of Orgn. Studies. Home: Gr v Waldeckstraat 35, 6212AN Maastricht The Netherlands Office: DSM Polymers, PO Box 43, 6130 AA Sittard The Netherlands

SCHRIEFFER, JOHN ROBERT, physics educator, science administrator; b. Oak Park, Ill., May 31, 1931; s. John Henry and Louise (Anderson) S.; m. Anne Grete Thomsen, Dec. 30, 1960; children: Anne Bolette, Paul Karsten, Anne Regina. BS, MIT, 1953; MS, U. Ill., 1954, PhD, 1957, ScD, 1974; ScD (hon.), Tech. U., Munich, Germany, 1968, U. Geneva, 1968, U. Pa., 1973, U. Cin., 1977, U. Tel Aviv, 1987, U. Ala., 1990. NSF postdoctoral fellow U. Birmingham, Eng., also; Niels Bohr Inst., Copenhagen, 1957-58; asst. prof. U. Chgo., 1958-59; asst. prof., then assoc. prof. U. Ill., 1959-62; prof. U. Pa., Phila., 1962-79; Mary Amanda Wood prof. physics U. Pa., 1964-79; Andrew D. White prof. at large Cornell U., 1969-75; prof. U. Calif., Santa Barbara, 1980-91, Chancellor's prof., 1984-91, dir. Inst. for Theoretical Physics, 1984-89; Univ. prof. Fla. State U., Tallahassee, 1992—, chief scientist Nat. High Magnetic Field Lab., 1992—; chief scientist Nat. High Magnetic Field Lab., 1992—. Author: Theory of Superconductivity, 1964. Guggenheim fellow Copenhagen, 1967; Los Alamos Nat. Lab. fellow; Recipient Comstock prize Nat. Acad. Sci.; Nobel Prize for Physics, 1972; John Ericsson medal Am. Soc. Swedish Engrs., 1976; Alumni Achievement award U. Ill., 1979; recipient Nat. Medal of Sci., 1984; Exxon faculty fellow, 1979-89. Fellow Am. Phys. Soc. (Oliver E. Buckley solid state physics prize 1968); mem. NAS (coun. 1990—), Am. Acad. Arts and Scis., Coun. Nat. Acad. Sci., Royal Danish Acad. Scis. and Letters, Acad. Sci. USSR. Office: Fla State U NHMFL 1800 E Paul Dirac Dr Tallahassee FL 32306-4005

SCHRIESHEIM, ALAN, research administrator; b. N.Y.C., Mar. 8, 1930; s. Morton and Frances (Greenberg) S.; m. Beatrice D. Brand, June 28, 1953; children: Laura Lynn, Robert Alan. BS in Chemistry, Poly. Inst. Bklyn., 1951; PhD in Phys. Organic Chemistry, Pa. State U., 1954; DSc (hon.), No. Ill. U., 1991; PhD (hon.), Ill. Inst. Tech., Chgo., 1992. Chemist Nat. Bur. Standards, 1954-56; with Exxon Research & Engring. Co., 1956-83, dir. corp. research, 1975-79; gen. mgr. Exxon Engring., 1979-83; sr. dep. lab. dir., chief operating officer Argonne Nat. Lab., 1983-84; prof. chemistry dept. U. Chgo., 1984—; Karcher lectr. U. Okla., 1977; Hurd lectr. Northwestern U., 1980; Rosensteil lectr. Brandeis U., 1982; Welch Found. lectr., 1987; co-chmn. bd. on chem. scis. and tech. NRC, 1980-85; mem. com. to define future role of chemistry, 1983-84; vis. com. chemistry dept. MIT, 1977-82; mem. vis. com. mech. engring. and aerospace dept. Princeton (N.J.) U., 1983-87, mem. vis. com. chemistry dept., 1983-87; mem. Pure and Applied Chemistry Com.; del. to People's Republic of China, 1978; mem. Presdl. Nat. Commn. on Superconductivity, 1989-91, U.S.-USSR Joint Commn. on Basic Sci. Rsch, 1990—; mem. U.S. nat com. Internat. Union Pure and Applied Chemistry, 1982-85; mem. magnetic fusion adv. com. Div. Phys. Scis. U. Chgo. Magnetic Fusion adv. com. to U.S. DOE, 1983-86; mem. Dept Energy Rsch. Adv. Bd., 1983-85, Congl. Adv. Com. on Sci. and Tech., 1985 ; mem. com. on advanced fossil energy techs. NRC, 1983-85, com. on scientists and engrs. in the federal government, 1989—, commn. on physical scis., maths., applications, 1990—; mem. vis. coms. Stanford (Calif.) U., U. Utah, Tex. A&M U., Lehigh U.; bd. govs. Argonne Nat. Lab.; mem. adv. com. on space systems and tech. NASA, 1987—; mem. nuclear engring. and engring. physics vis. com. U. Wis., Madison; mem. Coun. Gt. Lakes Govs. Regional Econ. Devel. Commn. 1987—, rev. bd. Compact Ignition Tomamak Princeton U., 1988-91; advisor Sears Investment Mgmt. Co., 1988-89; bd. dirs. Petroleum Rsch. Fund, ARCH Devel. Corp., HEICO, Valley Indsl. Assn., Coun. on Superconductivity for Am. Competitiveness; mem. State of Ill. Commn. on the Future of Pub. Svc., 1990-92; co-chair Indsl. Rsch. Inst. Nat. Labs./Industry Panel, 1984-87; mem. Nat Acad. Engring. Adv. Commn. on Tech. and Soc., 1991-92, Nat. Rsch. Coun. Commn. on Environ. Rsch., 1991—, Sun Electric Corp. Bd., 1991-92, U.S. House of Reps. subcom. on Sci.-Adv. Group on Renewing U.S. Sci. Policy, 1992—. Bd. editors Chem. Tech.; mem. editorial bd. Rsch. and Devel., 1988, Superconductor Industry mag., 1988-92; patentee in field. Mem. spl. vis. com. Field Mus. of Natural History, Chgo., 1987-88; bd. dirs. LaRabida Children's Hosp. and Rsch. Ctr., Children's Meml. Hosp., Children's Meml. Inst. for Edn. and Rsch.; trustee The Latin Sch. of Chgo., 1990-92; adv. bd. WBEZ Chicagland Pub. Radio Community; mem. Conservation Found. DuPage County, 1983—, Econ. Devel. Adv. Commn. of DuPage County, 1984-88, State of Ill. Gov.'s Commn. on Sci. and Tech., 1986-90, Inst. for Ill. Council Advisors, 1988—, Ill. Coalition Bd. Dirs., 1989—, Inst. for Ill. Adv. Rev. Panel, 1986-88, NASA Sci. Tech. Adv. Com. Manpower Requirements Ad Hoc Rev. Team, 1988-91, State of Ill. Gov. Sci. and Exec. com., 1989—, U. Ill. Engring. Vis. com., Urbana-Champaign, 1986—; trustee Tchrs. Acad. for Math. and Sci. Tchrs. in Chgo., 1990—. Recipient Outstanding Alumni Fellow award Pa. State U., 1984; Disting. fellow Poly. U., 1989. Fellow AAAS (coun. del. chem. sect. 1986-92, bd. dirs. 1992—, sci. engring. & pub. policy com. 1992, standing com. audit 1992), N.Y. Acad. Scis.; mem. NAE (program adv. com. 1992—, chari study foreign participation in U.S. rsch. and devel. 1993—), Am. Chem. Soc. (Petroleum Chemistry award 1969, chmn. petroleum div., councilor, com. on chemistry and pub. affairs 1983-91, joint bd. coun. com. on sci. 1983-87, sci. adv. bd. coun. on chem. rsch. 1987-91), Am. Mgmt. Assn. (rsch. and d, Nat. Conf. Advancement Rsch. (conf. com. 1985—), Am. Petroleum Inst. (com. on refinery equipment), Am. Inst. Chem. Engrs. (AiChE awards com. 1992—), Gas Rsch. Inst. (rsch. coordination coun.), Am. Nuclear Soc., Rohm and Haas (bd. dirs.), Indsl. Rsch. Inst. (fed. adv. com. to Fed. Sci. and Tech. Com. 1992—, co-chmn. Nat. Labs. Indsl. Panel 1984-87), NRC (chair ad hoc panel on DOE Chem. and Rsch. 1980-82, mem. adv. com. to associateship programs of office of sci. and engring. pers. 1986—, com. on environ. rsch. 1991—, com. on scholarly comm. with China 1984-85, mem. govt.-univ.-indsl. roundtable 1985-87), Carlton Club (bd. govs. 1992—), Sigma Xi, Phi Lambda Upsilon. Club: Cosmos (Washington); Chicago, Economic, Commercial (Chgo.). Home: 1440 N Lake Shore Dr Apt 31ac Chicago IL 60610-1686 Office: Argonne Nat Lab 9700 S Cass Ave Argonne IL 60439-4832

SCHROCK, JOHN RICHARD, biology educator; b. Goshen, Ind., Oct. 23, 1946; s. Cletus Paul and Vera Idelle (Green) S.; m. Lois Sue West, Feb. 2, 1968; children: John Richard II, Donna Sue. BS in Biology, Ind. State U., 1971, MS in Sci. Edn., 1973; PhD in Entomology, U. Kans., 1983. Tchr. Campbell County Schs., Alexandria, Ky., 1968-73; instr. Lab. Sch. Ind. State U., Terre Haute, 1973-75; tchr. Hong Kong Internat. Sch., 1975-78; from asst. to assoc. prof. Emporia (Kans.) State U., 1986—. Co-author: Controlled Wildlife: State Wildlife Regulations, 1985; co-editor: A Guide to Museum Pest Control, 1988; contbr. over 20 articles to profl. jours. Fellow Ind. Acad. Sci.; mem. Kans. Assn. Biology Tchrs. (editor). Achievements include known authority in defense of dissection and animal use in the classroom. Office: Divsn Biol Scis Box 4050 Emporia KS 66801

SCHRODER, JOHN L., JR., retired mining engineer; b. Martinsburg, W.Va.. BS in Mining Engring., W.Va. Sch. Mines, MS, 1941. Registered profl. engr., W.Va., Ky. Jr. engr. H.C. Frick Coke Co. U.S. Steel, Uniontown, Pa., 1941-44; asst. engr. mine planning U.S. Steel, Gary, W.Va., 1949-51, asst. chief engr., 1951-1953; chief engr. U.S. Steel, Lynch, Gary, Ky., 1953-1958; gen. supt. U.S. Steel, Lynch, 1958-70; gen. mgr. coal ops. U.S. Steel, Pitts., 1970-79, v.p. coal ops. resource devel., 1979-81, pres. subsidiary U.S. Steel Mining Co., Inc., 1981-83, ret., 1983; prodn., safty engr. Gay Coal and Coke, Gay Mining Cos., 1948-49; spl. asst. to the pres. Am. Mining Congress, 1983-84; dean Coll. Mineral and Engry Resources W.Va. U., 1984-91, ret., 1991; chmn. Mine Insps. Exam. Bd.; mem Govs. Moore's Energy Task Force. Lt. j.g. USN, 1944-48. Recipient Howard N. Evanson award Soc. Mining, Metallurgy and Exploration, 1991. Mem. AIME, Nat. Mine Rescue Assn., W.Va. Coal Mining Inst., Old Timers Club, King Coal Club. office: 228 Maple Ave Morgantown WV 26505*

SCHROECK, FRANKLIN EMMETT, JR., mathematical physicist; b. San Antonio, Jan. 15, 1942; s. Franklin Emmett and Dorothy Helen (Bauers) S.; m. Janet Monica Allen, Aug. 10, 1968; children: Theodore Richard, Geoffrey Alan. BA in Physics, Rice U., 1964; PhD in Physics, U. Rochester, 1971. Grad. asst. U. Rochester, 1964-70; interim asst. prof. Fla. Atlantic U., Boca Raton, 1970-71, asst. prof. math., 1971-75, assoc. prof. math., 1975-82, prof. math., 1982—; vis. prof. math. U. Denver, 1979-80, Inst. Theoretical Physics, U. Cologne, Germany, 1987; co-organizer Second Internat. Wigner Symposium Goslar, Germany, 1989-91; sci. adv. bd. Symposium on the Found. of Modern Physics, Cologne, 1992-93; adv. bd. Third Internat. Wigner Symposium, Oxford, Eng., 1992-93. Referee Jour. Found. of Physics, Jour. Math. Physics, SIAM Jour. Applied Math., Hadronic Jour.; contbr. articles to profl. jours. Grantee Deutsche Forschungsgemeinschaft/U. Cologne, 1987, Internat. Wigner Symposium, Office of Naval Rsch., 1991, Internat. Assn. Math. Physics, 1990, Internat. Union Pure and Applied Physics, 1990. Mem. Am. Math. Soc., Am. Phys. Soc., Internat. Assn. Math. Physics, Internat. Quantum Structures Assn., Sigma Xi. Achievements include co-founder of the new mathematical formalism "Stochastic Quantum Mechanics" which allows for the unification of quantum theory and relativity. Office: Fla Atlantic Univ Dept Math 500 NW 20th St Boca Raton FL 33431

SCHROEDER, ALFRED CHRISTIAN, electronics research engineer; b. West New Brighton, N.Y., Feb. 28, 1915; s. Alfred and Chryssa (Weishaar) S.; m. Janet Ellis, Sept. 26, 1936 (dec.); 1 dau., Carol Ann Schroeder Castle.; m. Dorothy Holloway, Nov. 21, 1981. B.S., Mass. Inst. Tech., 1937, M.S., 1937. Mem. tech. staff David Sarnoff Rsch. Ctr. RCA, Princeton, N.J., 1937—. Contbr. articles to profl. jours. Recipient RCA Lab. awards, 1947, 50, 51, 52, 57, 70. Fellow IEEE (Vladimir Zworykin award 1971); mem. AAAS, Optical Soc. Am., Soc. Motion Picture and TV Engrs. (David Sarnoff Gold medal 1965), Soc. Info. Display (Karl Ferdinand Braun prize 1989), Sigma Xi. Quaker. Achievements include 75 patents for color TV products including shadow mask tube. Home: Apt I-114 Pennswood Village Newtown PA 18940 Office: David Sarnoff Rsch Ctr RCA Princeton NJ 08540

SCHROEDER, DAVID HAROLD, health care facility executive; b. Chgo., Oct. 22, 1940; s. Harry T. and Clara D. (Dexter) S.; m. Clara Doorn, Dec. 27, 1964; children: Gregory D., Elizabeth M. BBA, Kans. State Coll., 1965; MBA, Wichita State U., 1968; postgrad., U. Ill., 1968-69. CPA, Ill. Supt. cost acctg. Boeing Co., Wichita, Kans., 1965-68; sr. v.p., treas. Riverside Med. Ctr., Kankakee, Ill., 1971—; treas. Riverside Health System, 1982—, Kankakee Valley Health Inc., 1985—, Health Info. Systems Coop., 1991—; v.p., treas. Oakside Corp., Kankakee, 1982—; bd. dirs. Harmony Home Health Svc., Inc., Naperville, Ill.; mem. faculty various profl. orgns.; adj. prof. econs. div. health adminstrvn. Gov.'s State U., University Park, Ill., 1990—; trustee Riverside Found. Trust, 1989—, RMC Found., 1989—, Sr. Living Ctr., 1989—. Contbg. author: Cost Containment in Hospitals, 1980; contbr. articles to profl. jours. Pres. Riverside Employees Credit Union, 1976-79; founder Kankakee Trinity Acad., 1980, Riverview Hist. Dist., Kankakee, 1982; pres. Kankakee County Mental Health Ctr., 1982-84, United Way Kankakee County, 1984-85; chmn. Ill. Provider Trust, Naperville, 1983-85; mem. adv. bd. Students in Free Enterprise, Olivet Nazarene U., Kankakee, 1989—; pres. adv. coun. div. health adminstrn. Gov.'s State U., University Park, preceptor, 1987—; trustee, treas. Am. Luth. Ch. Capt. U.S. Army, 1969-71. Fellow Am. Coll. Healthcare Execs., Healthcare Fin. Mgmt. Assn. (pres. 1975-76, cert. mgr. patient accounts 1981), Fin. Analysts Fedn.; mem. AICPA, Ill. Hosp. Assn. (chmn. coun. health fin. 1982-85), Inst. Chartered Fin. Analysts, Nat. Assn. Accts., Fin. Exec. Inst., Ill. CPA Soc., Healthcare Fin. Mgmt. Assn. (William G. Follimer award 1977, Robert H. Reeves award 1981, Muncie Gold award 1987, Founders medal of honor 1990), Investment Analysts Soc. Chgo., Inc., Classic Car Club Am., Packard Club, Kiwanis (pres.), Masons, Alpha Kappa Psi, Sigma Chi. Avocations: classic automobile restoration, archl. preservation, computers. Home: 901 S Chicago Ave Kankakee IL 60901-5236 Office: Riverside Med Ctr 350 N Wall St Kankakee IL 60901-2901

SCHROEDER, HERMAN ELBERT, scientific consultant; b. Bklyn., July 6, 1915; s. Henry W. and Caroline (Schmidt) S.; m. Elizabeth Barnes, June 13, 1938; children: Nancy Schroeder Tarczy, Edward L., Peter H., Martha L. Schroeder Lewis. A.B. summa cum laude, Harvard, 1936, A.M., 1937, Ph.D., 1939. With E.I. du Pont de Nemours & Co., Wilmington, Del., 1938-80; asst. dir. R&D E.I. du Pont de Nemours & Co., 1957-63, dir. R&D, 1963-80; pres. Schroeder Sci. Svcs., Inc., 1980—; sci. cons. Met. Mus. Art, N.Y.C., Smithsonian Instn. Mem. Chester County Sch. Bd., Unionville, Pa., 1950-56; pres. Assn. Harvard Chemists, 1955-56; mem. vis. com. Harvard Chemistry Dept., 1960-72; mem. sci. adv. com. Winterthur Mus.; trustee, chmn. research com. U. Del. Research Found., 1976-84, former v.p. Recipient award Internat. Inst. Synthetic Rubber Producers, 1979, Lavoisier medal DuPont, 1992. Mem. AAAS, Am. Chem. Soc. (Charles Goodyear medal 1984), Faraday Soc., Soc. Chem. Industry, Phi Beta Kappa, Alpha Chi Sigma. Home and Office: 74 Stonegates 4031 Kennett Pike Greenville DE 19807

SCHROEDER, RICHARD PHILIP, quality control executive, consultant; b. Scranton, Pa., Mar. 10, 1951; s. Philip Richard and Pearl Marion (Maier) S.; m. Donna Lee Elliott; children: Derek, Kyle. BSBA and Engring., Pa. State U., 1973, BA, 1974. Bus. mgr. H.R. Imbt, Inc., Lake Ariel, Pa., 1973-75; quality engr. Fisher Body div. GM, Trenton, N.J., 1975-77; asst. quality control mgr. Pitts. Forgings, Phila., 1977-78; divisional quality assurance mgr. Bindery Systems div. Harris Corp., Melbourne, Fla., 1978-83; corp. dir. productivity and quality Harris Graphics Corp., Melbourne, 1978-85; dir. operational and total quality improvement practice Coopers and Lybrand, Boston, 1985-86; v.p. corp. quality assurance and customer svc. Fed. Govt. Compliance, Codex Corp. subs. Motorola, Inc., Canton, Mass., 1986-91; v.p. quality, time-based mgmt. ABB, Stamford, Conn., 1991—; sr. examiner Malcolm Baldridge Award, Washington, 1987-93. Contbr. articles to profl. publs.; inventor 3D inspection for CMM. Mem. ASTM, Am. Soc. for Quality Control (cert. quality engr.), Machine Vision Assn. (charter), Soc. Mfg. Engrs., Am. Gear Mfrs. Assn., Pa. State U. Alumni Assn. (alumni advisor 1980—), Pi Kappa Phi. Republican. Avocations: all sports, fishing, electronics, woodworking, automobiles. Home: 34 Shadow Ln Wilton CT 06897-3529 Office: ABB 900 Long Ridge Rd Stamford CT 06902-1194

SCHROEDER, STANLEY BRIAN, chemist, coating application engineer; b. Shelby, Mich., Feb. 6, 1941; s. Loyal Johnson and Elnora (Porter) S.; m. Pauline Geneva DeVos, June 14, 1963; children: Lynette Joy, Sandra Brooke. BS in Chemistry, Mich. Tech. U., 1963. R & D chemist E.I. DuPont deNemours, Inc., Montague, Mich., 1963-69; tech. svc. chemist E.I. DuPont deNemours, Inc., Wilmington, Del., 1969-70; coatings devel. chemist Boise Cascade, Inc., International Falls, Minn., 1970-78, coatings devel. mgr., 1978-85; group leader Glidden Co., Strongsville, Ohio, 1985—. Mem. ASTM (chmn. subcom., mem. exec. subcom., Henry Gardner award 1990), Am. Hardboard Assn. (field finishing task group 1991—), Cleve. Soc. Paint Tech., Am. Chem. Soc., Forest Products Rsch. Soc. (speaker, procs. author), Assn. Finishing Processes, Soc. Mfg. Engrs. Office: Glidden Co 16651 Sprague Rd Strongsville OH 44136

SCHROEPFER, GEORGE JOHN, JR., biochemistry educator; b. St. Paul, June 15, 1932; s. George John and Catherine Rita (Callaghan) S.; children: Lisa Marie Schroepfer Schwartz, Cynthia Marie Schroepfer Winsenried, Stephanie Marie, Jeanine Marie Schroepfer Smith, Dana Marie Schroepfer Rethwisch. BS, U. Minn., 1955, MD, 1957, Phd, 1961. Intern U. Minn. Hospitals, 1957-58; rsch. fellow depts. biochemistry and internal medicine U. Minn., 1958-61, asst. prof. biochemistry, 1963-64; rsch. fellow chemistry dept. Harvard U., Cambridge, Mass., 1962-63; asst. prof. biochemistry U. Ill., 1964-67, assoc. prof., 1967-70; dir. Sch. Basic Med. Scis. U. Ill., Urbana, 1968-1970; prof. U. Ill., 1970-72; prof. biochemistry and chemistry Rice U., Houston, 1972-83, Ralph and Dorothy Looney prof. biochemistry, prof. chemistry, 1983—, dir. lab. basic med. sci., 1987—, chmn. biochemistry dept., 1972-84, sci. dir. Inst. Biosci. and Bioengring., 1987—; vis. scientist Med. Rsch. Coun. Unit Hammersmith Hosp., London, 1961-62; mem. subcom. on biochem. nomenclature NAS, 1965-68; mem. biochemistry. tng. com. Nat. Inst. Gen. Med. Scis. of NIH, 1970-73, ad-hoc. com. on tng. in biochemistry, 1974; cons. Am. Cyanamid Co., 1984-90, undergrad. biol. scis. edn. panel Howard Hughes Med. Inst., 1988. Assoc. editor Lipids, 1969-76; mem. editorial bd. Jour. Biol. Chemistry, 1974-79, Jour. Lipid Rsch., 1983-88. Fellow AAAS, Arteriosclerosis Coun. of Am. Heart Assn.; mem. Am. Chem. Soc., Am. Soc. Biochemistry and Molecular Biology, Am. Soc. Mass Spectrom, Sigma Xi, Alpha Omega Alpha. Office: Rice U Dept Biochemistry PO Box 1892 Houston TX 77251-1892

SCHROER, BERNARD JON, industrial engineering educator; b. Seymour, Ind., Oct. 11, 1941; s. Alvin J. and Selma A. (Mellencamp) S.; m. Kathleen Dittman, July 5, 1963; children: Shannon, Bradley. BSE, Western Mich. U., 1964; MSE, U. Ala., 1967; PhD, Okla. State U., 1972. Registered profl. engr., Ala. Mech. designer Sandia Labs., Albuquerque, 1962-63; engr. Teledyne Co., Huntsville, Ala., 1964-67, Boeing Co., Huntsville, 1967-70, Computer Sci. Corp., Huntsville, 1970-72; dir. Johnson Ctr. U. Ala., 1972-91, prof., 1991—; adv. coun. Energy Dept., Montgomery, Ala., 1980—; bd. dirs. So. Solar Energy Ctr., Atlanta, 1980-83; mem. gov.'s cabinet State Ala., Montgomery, 1982. Author: Modern Apparel Manufacturing Systems and Simulation, 1991; contbr. articles to profl. jours. Named Outstanding Engr., Robotics Internat., 1986, Outstanding Engr., Ala. Soc. Profl. Engrs., 1987; recipient summer traineeship NSF, 1971. Fellow Inst. Indsl. Engr. (pres. 1972-86, Outstanding Engr. 1973, 77); mem. NSPE, Soc. Computer Simulation, Tech. Transfer Soc., Huntsville Rotary. Lutheran. Home: 716 Owens Dr Huntsville AL 35801 Office: Coll Engring U Ala in Huntsville Huntsville AL 35899

SCHROETER, DIRK JOACHIM, mechanical engineer; b. Solingen, Germany, Mar. 2, 1949; came to U.S., 1957; s. Joachim Willi and Doris Irmgard (Kroeber) S.; m. Melissa Dickerson, Apr. 20, 1974; 1 child, Keira Melissa. BSME, SUNY, Buffalo, 1971, MSME, 1973. Commd. 2d lt. USAF, 1971, advanced through grades to lt. col., 1992; acquistion project officer Guided Bomb Unit-System Program Office, Eglin AFB, Fla., 1972-77; requirements officer Launch Vehicles Deputate, L.A. Air Force Sta., Calif., 1977-79; mem. tech. staff radar systems Hughes Aircraft Co., El Segundo, Calif., 1979-81; mem. profl. staff Martin Marietta Missile Systems, Orlando, Fla., 1981—; acquistion officer space div. Air Force Res., L.A. Air Force Sta., 1979-81; electronic engr. aero. systems ctr. Air Force Materiel Command, Wright-Patterson AFB, 1981—. Editor: Computer Aided Engring. Bulletin, 1982-86; author: (guide) Industrial Modernization Improvement Program-Quick Reference to Modernization, 1986. Judge Orange County Sci. and Engring. Fair, Orlando, Fla., 1985-88. Decorated Commendation medal USAF, L.A., 1979. Mem. AIAA (interactive computer graphics tech. com. 1983-86), Soc. Automotive Engrs. Republican. Methodist. Achievements include developed neutral (IGES and text-based) database: Martin Integrated Neutral Graphics and Engring. Libr. used for computer graphics and file exchange for computer aided acquistion and logistic support and with vendors. Home: 4309 Winderlakes Dr Orlando FL 32835 Office: Martin Marietta PO Box 5837 Mail Point 135 Orlando FL 32855-5837

SCHRYVER, BRUCE JOHN, safety engineer; b. Newark, Aug. 14, 1944; s. Francis Henry and Ann Laura (Hart) S.; m. Lorraine Patricia Simodis, Oct. 8, 1966; children: Holly Lynn, Wendy Marie. BA in Occupational Safety and Health, Western States U., 1984, MS in Safety Mgmt., 1989, PhD in Safety Mgmt., 1989. Cert. safety profl.; cert. products safety mgr.; cert. hazard control mgr.; cert. hazardous materials mgr.; cert. healthcare safety profl. Inspector Lansing B. Warner Inc., Chgo., 1968-69; engring. rep. Glens Falls Ins. Co., Newark, 1969; safety dir. Hillside Metal Products, Newark, 1969-70; loss prevention specialist Warner Ins. Group, Chgo., 1970-79, regional loss control mgr., 1979-82, nat. loss control coordinator, 1982-85; mgr., asst. v.p. loss control svcs. Ins. Co. of the West, San Diego, 1985-90; v.p. loss control svcs. Ins. Co. of the West, 1990—; v.p. mcpl. law enforcement svcs. Ins. Co. of the West, San Diego, 1992—. Inventor Emergency Light Mount, 1971. Mem. Town of Clay (N.Y.) Pub. Safety Com., 1976-78, Beacon Woods East Homeowners Assn., Hudson, Fla., 1979-85, Meadowridge Homeowners Assn., La Costa, Calif., 1986—; cons. Town of Clay Police Dept., 1975-78. With USCG, 1964-68. Recipient lettter of appreciation Town of Clay, 1977, cert. of appreciation DAV, 1968, Golden State award, 1990. Mem. Am. Soc. Safety Engrs., Soc. Fire Protection Engrs., Nat. Safety Mgmt. Soc., Vets. Safety, Nat. Fire Protection Assn., San Diego Safety Coun., Calif. Conf. Arson Investigators. Republican. Roman Catholic. Avocations: auto racing, boating, bowling, photography, electronics. Home: 3047 Camino Limero Carlsbad CA 92009-4525 Office: Ins Co of the West 10140 Campus Point Dr San Diego CA 92121-1592

SCHUBEL, JERRY ROBERT, marine science educator, scientist, university dean and official; b. Bad Axe, Mich., Jan. 26, 1936; s. Theodore Howard and Laura Alberta (Gobel) S.; m. Margaret Ann Hostetler, June 14, 1958; children: Susan Elizabeth, Kathryn Ann. BS, Alma Coll., 1957; MA in Teaching, Harvard U., 1959; PhD, Johns Hopkins U., 1968. Research assoc. Chesapeake Bay Inst., Johns Hopkins U., Balt., 1968-69, research scientist, 1969-74, adj. research prof., assoc. dir., 1973-74; dir. Marine Scis. Research Ctr., SUNY, Stony Brook, 1974-83, dean, leading prof., 1983—, Disting. Svc. prof., acting vice provost for research and grad. studies, 1985-86, dir. COAST Inst., 1989, provost, 1986-89, acting dir. Waste Mgmt. Inst., 1985-87; hon. prof. East China Normal U., Shanghai, 1985—. Author: The Living Chesapeake, 1981, The Life and Death of the Chesapeake Bay, 1986; (with H.A. Neal) Solid Waste Management and the Environment, 1987, Garbage and Trash: Can We Convert Mountains Into Molehills?, 1992; editor: (with B. C. Marcy Jr.) Power Plant Entrainment, 1978; (with others) The Great South Bay, 1991; sr. editor Coastal Ocean Pollution Assement News, 1981-86; co-editor in chief Estuaries, 1986-88; editorial bd. CRC Revs. in Aquatic Scis.; contbr. articles to profl. jours. Mem. Environ. Sci. Com. Outer Continental Shelf Adv. Bd., Minerals Mgmt. Svc., 1984-86, chmn., 1986; bd. dirs. N.E. Area Remote Sensing System, 1983-85, L.I. Incubator Corp.; v.p. L.I. Forum for Tech., 1989-92. Recipient L.I. Sound Am. Environment Edn. award, 1987, Stony Brook U. medal, 1989, Matthew Fontaine Maury award, 1990; Alfred P. Sloan fellow, 1959. Mem. NAS (marine bd. 1989—, exec. com. 1990, vice chair 1991—, chair 1992— com. on Coastal Ocean 1989—), Nat. Assn. State Univ. and Land Grant Colls. (bd. dirs. marine div., chmn. 1986-88), L.I. Environ. Council, L.I. Marine Resources Adv. Coun. (chair 1990—), L.I. Rsch. Inst. (bd. dirs. 1992—), L.I. Environ.-Econ. Roundtable (co-chair 1991-92), Suffolk County Recycling Commn., (chmn. 1987-88), Estuarine Rsch. Fedn. (v.p. 1982-83, pres. 1985-87), N.Y. Sea Grant Inst. (chmn. governing bd. 1988-90, gov.'s task force on coastal resources 1990-91), The Nature Conservancy (trustee L.I. chpt. 1991—), Franklin Electronic Pubs. (bd. dirs. 1991—), Taproot (bd. dirs. 1988-93, vice chair 1990-93), Sigma Xi, Phi Sigma Pi. Avocation: photography. Home: 4 Hiawatha Ln East Setauket NY 11733-2216 Office: SUNY at Stony Brook Marine Sci Rsch Ctr Stony Brook NY 11794-5000

SCHUBERT, GERALD, planetary and geophysics educator; b. N.Y.C., Mar. 2, 1939; s. Morris and Helen (Nelson) S.; m. Joyce Elaine Slotnick, Jan. 16, 1960; children: Todd, Michael, Tamara. BS in Engring. Physics, MS in Aero. Engring., Cornell U., 1961; PhD in Aero. Sci. Engring., U. Calif., Berkeley, 1964. Head heat transfer dept. U.S. Naval Nuclear Power Sch., Mare Island, Calif., 1961-65; mem. tech. staff Bell Telephone Research Labs., Whippany, N.J., 1965; asst. prof. planetary and geophysics UCLA, 1966-70, assoc. prof., 1970-74, prof., 1974—. Co-author: Geodynamics, 1982; contbr. articles to profl. jours. Served to lt. USN, 1961-65. Alexander von Humboldt fellow, 1969, Guggenheim fellow, 1972, Berman fellow Hebrew U. Jerusalem, 1982-83, Nat. Acad. Scis. Nat. Research Council fellow, Cambridge, Eng., 1965-66. Fellow Am. Geophys. Union (James B. Macelwane fellow 1975); mem. AAAS, Div. Planetary Sci. Am. Astron. Assn. Avocations: skiing, handball. Office: UCLA Dept Earth and Space Sci Los Angeles CA 90024

SCHUBERT, GUENTHER ERICH, pathologist; b. Mosul, Iraq, Aug. 17, 1930; s. Erich Waldemar and Martha Camilla (Zchitzschmann) S.; m. Gisela Schultz, June 13, 1959; children: Frank, Marion, Dirk. MD, University, Heidelberg, Germany, 1957; pvt. docent in pathology, University, Tuebingen, Germany, 1966. Asst. med. dir. University Tuebingen, Fed. Republic of Germany, 1966-76; head Inst. Pathology, Wuppertal, Fed. Republic of Germany, 1976—; chair of pathology U. Witten-Herdecke, Fed. Republic of Germany, 1985—. Co-author: Coloratlas of Cytodiagnosis of the Prostate, 1975, Endoscopy of the Urinary Bladder, 1989; author: Textbook of Pathology, 1981, 87. Mem. Wissenschaftlicher Beirat, Bundesarztekammer, Bonn, Germany, 1976-85; pres. Medizinisch Naturwissenschaftliche Gesellschaft, Wuppertal, 1984-85, Onkologischer Schwerpunkt, Wuppertal, 1985-93, OSP Bergisch-Land, 1992—, Bergische Arbeitsgemeinschaft fur Gastroenterologie, Wuppertal, 1987-88, 1990-91. Mem. Deutsche Gesellschaft fur Pathologie, Deutsche Gesellschaft fur Nephrologie, Deutsche Gesellschaft fur Urologie, Internat. Acad. of Pathology, Lions. Avocations: music, diving, photography. Home: Am Anschlag 71, D-42ii3 Wuppertal 1, Germany Office: Inst of Pathology, Heusner Strasse 40, 42 283 Wuppertal 2, Germany

SCHUBERT, WILLIAM KUENNETH, hospital medical center executive; b. Cin., July 12, 1926; s. Wilfred Schubert; m. Mary Jane Pamperin, June 5, 1948; children: Carol, Joanne, Barbara, Nancy. BS, U. Cin., 1949, MD, 1952. Diplomate Am. Bd. Pediatrics. Pvt. practice specializing in pediatrics Cin., 1956-63; dir. clin. research ctr. Children's Hosp. Med. Ctr., Cin., 1963-76; dir. div. gastroenterology Children's Hosp., Cin., 1968-79; prof. pediatrics U. Cin., 1969—; chief of staff Children's Hosp. Med. Ctr., Cin., 1972-88; chmn. dept. pediatrics U. Cin., 1979-93; dir. Children's Hosp. Rsch. Found., Cin., 1979-93; pres., CEO Children's Hosp. Med. Ctr., Cin., 1983—; v.p. Ohio Solid Organ Transplant Consortium, Columbus, 1986-87, pres., 1987-88, alt. trustee, 1988—; regional bd. dirs. Ameritrust Corp., 1988-91; trustee med. rsch. James N. Gamble Inst., Cin., 1989—. Contbr. over 100 articles to profl. jours. Trustee Greater Cin. Hosp. Council, 1986—, Assn. of Ohio Children's Hosp., Columbus, 1986—; chmn. Greater Cin. Hosp. Coun., 1989; co-chmn. Citizen's Com. for Med. Ctr., Cin., 1980-81; chmn. Hosp. Div. 1988 Fine Arts Fund, Cin., 1987. With USN, 1944-46. Fellow Am. Acad. Pediatrics (exec. bd. 1984—); mem. Am. Pediatric Soc. (councillor 1986-93), Soc. Pediatric Research, Assn. Med. Sch. Pediatric Dept. Chmn., Cin. Acad. Medicine, AMA, Midwestern Soc. for Pediatric Research, Am. Assn. for Study of Liver Diseases, Central Soc. Clin. Research, Am. Gastroenterological Assn., N.Am. Soc. Pediatric Gastroenterology, Nat. Reye's Syndrome Found. (med. dir. 1976-87), Internat. Assn. Study Liver Diseases. Club: Queen City (Cin.). Office: Childrens Hosp Med Ctr Elland and Bethesda Aves Cincinnati OH 45229-2899

SCHUCK, TERRY KARL, chemist, environmental consultant; b. Washington Twp., Pa., Mar. 15, 1961; s. Ronald and Ann (Schrecter) S.; m. Zoann Parker, Sept. 26, 1986. BS, Pa. State U., 1984. Lab. technician Pa. Fish Commn., Bellefonte, 1984; chemist Lancaster (Pa.) Labs., 1984—; cons. waste water treatment Sewer Authorities, Pa., Lancaster, 1988—. Adviser Lancaster area 4-H, 1986—. Achievements include design and initiation of wastewater monitoring programs for municipal sewer authorities. Home: 264 Stony Hill Rd Quarryville PA 17566

SCHUERMANN, MARK HARRY, chemist, educator; b. St. Louis, Nov. 12, 1946; s. Francis Joseph and Helen Amelia (Pohlmann) S. BA, St. Mary's U., 1969; MS, Notre Dame U., 1974. Cert. chemistry and physics tchr. Tchr. Chaminade Coll. Preparatory Sch., Creve Coeur, Mo., 1969-71, Daniel J. Gross High Sch., Omaha, 1971-74, Parkway Cen. High Sch., Chesterfield, Mo., 1974-92; lectr. U. Mo., St. Louis, 1974—. Recipient Tapestry award Toyota and Nat. Sci. Tchrs. Assn., 1991-93; NSF grantee, 1971-74. Mem. NEA (bldg. rep. 1975-76), Sierra Club (exec. com. Ea. Mo. group 1989—, chmn. 1990-91, vice chair 1991-92, exec. com. Ozark chpt. 1989-90, Outstanding Achievement for New Mem. 1990). Home: 1227 Weatherton Pl Ballwin MO 63021 Office: Parkway Cen High Sch 369 N Woods Mill Rd Chesterfield MO 63017

SCHUETTE, OSWALD FRANCIS, physics educator; b. Washington, Aug. 20, 1921; s. Oswald Frances and Mary (Moran) S.; m. Kathryn E. Cronin, June 7, 1947; children: Patrick Thomas, Mary Kathryn, Elizabeth Anne. B.S., Georgetown U., 1943; Ph.D., Yale U., 1949. Assoc. prof. physics Coll. William and Mary, 1948-53; Fulbright research prof. physics Max Planck Inst. for Chemistry, 1953-54; sci. liaison officer U.S. Naval Forces, Germany, 1954-58; mem. staff Nat. Acad. Scis., Washington, 1958- 60; dep. spl. asst. for space Office Sec. Def., Washington, 1961-63; prof. physics dept. U. S.C., 1963-92, dit. prof. emeritus, 1992—; vis. prof. Inst. Exptl. Physics, U. Vienna, 1980. Lt. USNR, 1944-46. Mem. AAAS, Am. Phys. Soc., Am. Assn. Physics Tchrs., S.C. Acad. Sci., Sigma Xi. Home: 4979 Quail Ln Columbia SC 29206-4624

SCHUH, MERLYN DUANE, chemist, educator; b. Avon, S.D., Feb. 21, 1945; s. Edward Arthur and Amelia (Rueb) S.; m. Judy Anne Swigart, June 1, 1969. BA, U. S.D., 1967; PhD, Ind. U., 1971. Asst. prof. Middlebury (Vt.) Coll., 1971-75; prof. chemistry Davidson (N.C.) Coll., 1975—, James G. Martin prof. chemistry, 1992; adj. prof. chemistry Syracuse (N.Y.) U., 1982-83. Contbr. articles on gas and solution phase photophysics, laser spectroscopy and protein dynamics to tech. jours. Chmn., bd. dirs. Metrolina Assn. for Blind, Charlotte, N.C., 1980-81. Recipient profl. devel. award NSF, 1981; Dreyfus Found. scholar, 1992—. Mem. Davidson Lions (all offices). Democrat. Presbyterian. Achievements include indirect measurement of upper time limit for appearance of transient molecular openings in heme protein surfaces; developed the use of phosphorescent molecular probes of heme accessibility in heme proteins. Home: PO Box 704 Davidson NC 28036 Office: Davidson Coll PO Box 1749 Davidson NC 28036

SCHUH, SANDRA ANDERSON, ethics educator; b. Hartford, Conn., Aug. 22, 1947; d. Axel Magna and Dorothy Catherine (Spring) Anderson: m. Edward Walter Schuh, Apr. 10, 1978; Children: Matthew, Bradley. AM, U. Miami, Coral Gables, 1982, PhD, 1986. Philosophy instr. U. Miami, 1986-88, Fla. Atlantic U., Boca Raton, 1988-90; asst. prof. philosophy U. Tampa, Fla., 1990—; ethics cons. U.S. Govt. Dept. of Energy, Washington, 1987, Med. Ethics, Miami. Investigator Dept. of Human Resources, 1988; evaluator Fla. Endowment for Humanities, 1990, panelist, Tampa, 1990. Dana Grantee U. Tampa, 1991. Mem. AAAS (pres. 1993), Fla. Philosophic Assn., Am. Philosophic Assn., Am. Soc. for Engring. Edn., N.Y. Acad. Sci., Soc. Women in Sci. Achievements include research in environmental ethics. Office: U of Tampa Kennedy Blvd Tampa FL 33606-1411

SCHULER, MARTIN LUKE, geotechnical engineer; b. Kansas City, June 15, 1958; s. Martin Nugent and Patricia (Morrison) S. BS in Civil Engring., U. Kans., 1982, MS in Civil Engring., 1989. Project mgr. J.E. Dunn Constrn. Co., Kansas City, 1983-85; project engr. Hayes Drilling, Inc., Kansas City, 1985—; edn. commn. chmn. Am. Soc. Civil Engrs., Kansas City, 1992—; adv. bd. mem. U. Kans. Civil Engrs. Dept., Lawrence, 1988—. Mem. Sci. Pioneers, Kansas City., 1992. Mem. Am. Soc. Civil Engrs., U. Kansas Alumni Assn., Internat. Assn. Drilled Shaft Contractors. Office: Hayes Drilling Inc 8845 Prospect Kansas City MO 64132

SCHULER, ROBERT HUGO, chemist, educator; b. Buffalo, Jan. 4, 1926; s. Robert H. and Mary J. (Mayer) S.; m. Florence J. Forrest, June 18, 1952; children: Mary A., Margaret A., Carol A., Robert E., Thomas C. BS, Canisius Coll., Buffalo, 1946; PhD, U. Notre Dame, 1949. Asst. prof. chemistry Canisius Coll., 1949-53; asso. chemist, then chemist Brookhaven Nat. Lab., 1953-56; staff fellow, dir. radiation research lab. Mellon Inst., 1956-76, mem. adv. bd., 1962-76; prof. chemistry, dir. radiation research lab. Carnegie-Mellon U., 1967-76; prof. chemistry, dir. radiation lab. U. Notre Dame, Ind., 1976—; John A. Zahm prof. radiation chemistry U. Notre Dame, 1986—; Raman prof. U. Madras, India, 1985-86; vis. prof. Hebrew U., Israel, 1980. Author articles in field. Recipient Curie medal Poland, 1992. Fellow AAAS; mem. Am. Chem. Soc., Am. Phys. Soc., Chem. Soc., Radiation Research Soc. (pres. 1975-76), Sigma Xi. Club: Cosmos. Office: U Notre Dame Radiation Lab Notre Dame IN 46556

SCHULER, RONALD THEODORE, agricultural engineering educator; b. Valders, Wis., Dec. 26, 1940; s. Alphonse Henry and Margaret Mary (Basler) S.; m. Barbara Lee Howell, Jan. 21, 1967; children: Robert Howell, Sarah Howell. BS in Agr., BSME, U. Wis., 1958-63, MS, PhD in Agrl. Engring., 1965-70. Registered profl. engr., N.D. Rsch. asst. U. Wis., Madison, 1965-69, instr., 1969-70; chair, prof. U. Wis., Platteville, 1981-84, Madison, 1984—; asst. prof. N.D. State U., Fargo, 1970-75, assoc. prof., 1975-76; assoc. prof. U. Minn., St. Paul, 1976-81; pres. Farm Health and Safety Coun. Wis., 1992—. Author: (with others) Displacement Velocity, Acceleration and Sound, 1991. Bd. dirs. Kiwanis Club West Madison, 1989-92, Resource Ctr. for Farmers with Disabilities, Madison, 1990—. Capt. U.S. Army, 1963-65. Recipient Meritorious Svc. award Wis. 4-H Agts., Madison, 1990, Hon State degre Wis. Future Farmers Am., Madison, 1991. Mem. SAE (chair farm machine com. 1991-92, Recognition award), Am. Soc. Agrl. Engrs. (chair 12 coms., 9 Blue Ribbon awards 1985-91, Pres.'s Club 1989—). Office: U Wis Agr Engring Dept 460 Henry Mall Madison WI 53706

SCHULER, THEODORE ANTHONY, civil engineer, city official; b. Louisville, July 1, 1934; s. Henry R. and Virginia (Meisner) S.; B.C.E., U. Louisville, 1957, M.Engring., 1973; m. Jane A. Bandy, July 29, 1979; children: Marc, Elizabeth, Eric, Ellen. Design, constrn. engr. Brighton Engring. Co., Frankfort, Ky., 1960-65; design engr. Hensley-Schmidt Inc., Chattanooga, 1965-68, assoc. mem., 1969-73, sr. assoc. mem., 1973-75, prin., asst. v.p., head Knoxville office, 1975-81; chief planning engr. engring. dept. City of Knoxville, 1981—. Served to lt. (j.g.) USNR, 1957-60. Registered profl. engr., Ky., Tenn.; registered land surveyor, Ky., Tenn. Fellow ASCE. Home: 5907 Adelia Dr Knoxville TN 37920-5801 Office: Dept Engring City County-Bldg Rm 483 Knoxville TN 37901

SCHULL, WILLIAM J., genetecist, educator; b. Louisiana, Mo., Mar. 17, 1922; married, 1946. BS, Marquette U., 1946; PhD in Genetics, Ohio State U., 1949. Head dept. genetics Atomic Bomb Casualty Comsn., Japan, 1949-51; jr. genetecist Inst. Human Biology, U. Mich., 1951-53, asst. genetecist, 1953-56, from asst. prof. to prof. human genetics, med. sch., 1956-72, prof. anthropology, 1969-72; prof. human genetics U. Tex. Grad. Sch. Biomedical Sci., Houston, 1972—; vis. fellow Australian Nat. U., 1969; cons. Atomic Bomb Casualty Comsn., 1954, 56, NIH, 1956—, chmn. genetics study sect., 1969-72; dir. Child Health Survey, Japan, 1959-60; vis. prof. U. Chgo., 1963, U. Chile, 1975; German Rsch. Assn. guest prof. U. Heidelberg, 1970; mem. com. atomic casualties Nat. Rsch. Coun., 1951, subcom. biology Com. Dentistry, 1951-55, com. on collaborative project Nat. Inst. Neurol. Disease and Stroke, 1957—, panel in genetic effects of radiation, WHO, 1958—, panel experts human heredity, 1961—, nat. adv. com. radiation USPHS, 1960-64, bd. sci. counselors Nat. Inst. Dentistry Rsch., 1966-69; dir. Radiation Effects Rsch. Found. and head dept. epidemiology and Japan, 1978-80; adv. Nat. Heart and Lung Inst.; mem. subcom. biology and medicine AEC, human biology coun. Soc. Study Human Biology. Recipient Centennial award Ohio State U., 1970. Mem. AAAS, U.S.Mex. Border Health Assn., Japanese Soc. Human Genetics (hon.), Peruvian Soc. Human Genetics (hon.), Genetic Soc. Chile (hon.), Sigma Xi. Achievements include research in biometry. Office: U Tex Hlth Sci Ctr Med Genetics Ctr PO Box 20334 Houston TX 77225*

SCHULLER, DIANE ETHEL, allergist, immunologist, educator; b. Bklyn., Nov. 27, 1943; d. Charles William and Dorothy Schuller. AB cum laude with honors in Biology, Bryn Mawr Coll., 1965. Diplomate Am. Bd. Allergy and Immunology, Am. Bd. Pediatrics, Nat. Bd. Med. Examiniers. M.D., SUNY Downstate Med. Sch., Bklyn., 1970. Intern, then resident in pediatrics Roosevelt Hosp., N.Y.C., 1970-72, resident in allergy Cooke Inst. Allergy, 1972-74; asso. in pediatrics Geisinger Med. Center, Danville, Pa., 1974—, dir. dept. pediatric allergy, immunology and pulmonary diseases 1978—; asst. clin. prof. pediatrics Hershey Med. Coll., Pa. State U., 1974-79, assoc. clin. prof., 1979-88; clin. prof. Jefferson Med. Coll., Phila., 1989—; mem. Columbia-Montour Home Health Services Adv. Group of Profl. Personnel, 1975—. Bd. dirs. Central Pa. Lung and Health Assn.; bd. dirs., exec. com. Am. Lung Assn. of Pa., sec., 1992— ; chmn. Susquehanna Valley Lung Assn., 1983—; mem. scholarship com. Bryn Mawr Club, N.Y., 1970-75. Recipient Physicians Recognition award AMA, 1973-76, 74-76, 75-78, 79-82, 83-86, 1987-90, 91—. Fellow Am. Acad. Pediatrics, Am. Coll. Allergy and Immunology (2d v.p., bd. regents 1989-92, exec. com. 1990-932, treas. joint coun. of allergy and immunology 1991—, editorial bd. annals of allergy, v.p.1992-93), Am. Assn. Clin. Immunology and Allergy (regional dir., exec. com.), Joint Coun. Allergy and Immunology (bd. dirs. 1986—), Am. Acad. Allergy and Immunology; mem. Pa., N.Y. State allergy socs., N.Y. State, N.Y. County med. socs. Office: Geisinger Med Ctr Danville PA 17821

SCHULMAN, MARVIN, chemical engineer; b. N.Y.C., Oct. 28, 1934; m. Renee Schulman, Feb. 12, 1955; children: Cathy, Debby, Donna, KItt. BS in Engring., NYU, 1954. Engring. specialist Gen. Foods Corp., Cranbury, N.J., 1956-87; project leader chem. engring. natural flavor ingredients Internat. Flavor & Fragrance, Union Beach, N.J., 1987—. Mem. I.F.T., Am. Inst. Chem. Engrs. Achievements include 17 patents for food and flavor industries. Office: Internat Flavor & Fragrance 1515 Hwy 36 Union Beach NJ 07735

SCHULTES, RICHARD EVANS, ethnobotanist, museum executive, educator, conservationist; b. Boston, Jan. 12, 1915; s. Otto Richard and Maude Beatrice (Bagley) S.; m. Dorothy Crawford McNeil, Mar. 26, 1959; children: Richard Evans II, Neil Parker and Alexandra Ames (twins). AB cum laude, Harvard U., 1937, AM, 1938, PhD, 1941; MH (hon.), Universidad Nacional de Colombia, Bogotá, 1953; D Sci (hon.), Mass. Coll. Pharmacy, 1987. Plant explorer, NRC fellow Harvard Bot. Mus., Cambridge, Mass., 1941-42; research asso. Harvard Bot. Mus., 1942-53; curator Orchid Herbarium of Oakes Ames, 1953-58, curator econ. botany, 1958-85, exec. dir., 1967-70, dir., 1970-85; Guggenheim Found. fellow, collaborator U.S. Dept. Agr., Amazon of Colombia, 1942-43; plant explorer in South Am. Bur. Plant Industry, 1944-54; prof. biology Harvard U., 1970-72, Paul C. Mangelsdorf prof. natural scis., 1973-81, Edward C. Jeffrey prof. biology, 1981-85, emeritus prof., 1985—; adj. prof. pharmacognosy U. Ill., Chgo., 1975—; Hubert Humphrey vis. prof. Macalaster Coll., 1979; field agt. Rubber Devel. Corp. of U.S. Govt., in S.Am., 1943-44; collaborator Instituto Argronômico Norte, Belem, Brazil, 1948-50; hon. prof. Universidad Nacional de Colombia, 1953—, prof. econ. botany, 1963; bot. cons. Smith, Kline & French Co., Phila., 1957-67; mem. NIH Adv. Panel, 1964; mem. selection com. for Latin Am. Guggenheim Found., 1964-85; mem. sci. adv. bd. Palm Oil Rsch. Inst., Malaysia, 1980-89; chmn. Ethnobotany Specialist Group, Internat. Union Nature, Switzerland; chmn. on-site visit U. Hawaii Natural Products Grant NIH, 1966, Deutche Gesellschaft für Arzneiphanzenforchunge, Berlin, 1967, III Internat. Pharm. Congress, Sao Paulo, I Amazonian Biol. Symposium, Belem, Brazil, Symposium Ethnpharmacologic Search for Psychoactive drugs, San Francisco, others; Laura L. Barnes Annual lectr. Morris Arboretum, Phila., 1969; Koch lectr. Rho Chi Soc., Pitts., 33d Internat. Congress Pharm. Sci., Stockholm, 1971, Chgo., 1974; vis. prof. econ. botany, plants in relation to man's progress Jardín Botánico, Medellín, Colombia, 1973; Cecil and Ida H. Green vis. lectr. U. B.C., Vancouver, Can., 1974; co-dir. phase VII Alpha Helix Rsch. plant medicines Witoto, Bora Indianas, Peru, 2d Philip Morris Symposium, Richmond, Va., 1975, I Solanacese Conf., U. Birmingham, Eng., 1977, Internat. Symposium Erythdroxylon, Equador, 1979, Soc. of Americanists, Manchester, Eng., 1982, Salah Workshop for Conservation of Worldlands, Sultanate of Omen, 1983, Symposium of Environ. Protection, INDERENA Bogota, 1988, others; vis. scholar Rockefeller Study Conf. Centre, Bellagio,

Italy, 1980, 88; cons. Rubber Rsch. Inst., Malaysia, 1988—. Author: (with P. A. Vestal) Economic Botany of the Kiowa Indians, 1941, Native Orchids of Trinidad and Tobago, 1960, (with A. F. Hill) Plants and Human Affairs, 1960, rev. edit., 1968, (with A. S. Pease) Generic Names of Orchids—their Origin and Meaning, 1963, (with A. Hofmann) The Botany and Chemistry of Hallucinogens, 1973, rev. edit., 1980, Plants of the Gods, 1979, Plant Hallucinogens, 1976, (with W.A. Davis) The Glass Flowers at Harvard, 1982, Where the Gods Reign, 1987, El Reino de los Dioses, 1988, The Healing Forest, 1990, Vine of the Soul: Medicine Men of the Colombian Amazon—Their Plants and Rituals, 1992; contbg. author: Ency. Biol. Scis, 1961, Ency. Brit, 1966, 83, Ency. Biochemistry, 1967, McGraw-Hill Yearbook Sci., Tech, 1937-85, New Royal Horticulture Dictionary (ethnobotany chpt., 1992); author numerous Harvard Bot. Mus. Leaflets.; Asst. editor: Chronica Botanica, 1947-52; editor: Bot. Mus. Leaflets, 1957-85, Econ. Botany, 1962-79; mem. editorial bd. Lloydia, 1965-76, Altered States of Consciousness, 1973—, Jour. Psychedelic Drugs, 1974—; co-editor: series Psychoactive Plants of the World, Yale U. Press, 1987—; mem. adv. bd.: Horticulture, 1976-78, Jour. Ethnopharmacology, 1978—, Soc. Pharmacology, 1976—, Elaeis, 1988—, Flora of Ecuador, 1976—, Jour. Latin Am. Folklore, Ethnobotony (India) (also founding mem.), 1989—, Environ. Awareness (India) 1988—, Bol. Mus. Goeldi (Brazil), 1987—, Environ. Conservation (Geneva), 1987—; contbr. numerous articles to profl. jours. Mem. governing bd. Amazonas 2000, Bogotá; assoc. in ethnobotany Museo del Oro, Bogotá, 1974—; chmn. NRC panels, 1974, 75; mem. NRC Workshops on Natural Products, Sri Lanka, 1975, participant numerous sems., congresses, meetings; mem. adv. bd. Fitzhugh Ludlow Libr., Native Land Found., 1980, Morgan Meml. Archives Chadron State Coll., 1987—, Albert Hofmann Found.; v.p. Margeret Mee Amazon Trust, Royal Bot. Gardens. Decorated Orden de la Victoria Regia in recognition of work in Amazon by Colombian Govt., 1969, Cruz de Boyacá Govt. of Colombia, 1983, Gold medal for conservation presented by Duke of Edinburgh, 1984; recipient Tyler prize for Environ. Achievement, 1984, Gov. of Mass. recognition award Nat. Sci. Week, 1985, Linnean medal, 1992, cert. of merit Botanical Soc. Am., 1988, Lindbergh award, 1991, Harvard U. medal, 1992, Martin de la Cruz medallion. 1992, Janaki Ammal medal, India, 1992. Fellow Am. Acad. Arts and Scis., Am. Coll. Neuropsychopharmacology (sci. speaker San Juan, P.R. 1984), Third World Acad. Scis., Acad. Scis. India, Internat. Soc. Naturalists; mem. NAS, Linnean Soc., Academia Colombiana de Ciencias Exactas, Fisico-Quimicas y Naturales, Instituto Ecuatoriano de Ciencias Naturales, Sociedad Cientifica Antonio Alzate (Mexico), Argentine Acad. Scis., Am. Orchid Soc. (life hon.), Pan Am. Soc. New Eng. (gov.), Asociación de Amigos de Jardines Botánicos (life), Soc. Econ. Botany (organizer annual meeting 1961, Disting. Botanist of Yr. award 1979), New Eng. Bot. Club (pres. 1954-60), Internat. Assn. Plant Taxonomy, Am. Acad. Achievement, Am. Soc. Pharmacognosy, Phytochem. Soc. N.Am., Socieded Colombiana de Orquideologia, Soc. Ethnobot. India, Assn. Tropical Biology, Internat. Soc. Ethnobiology, Sociedad Cubana de Botánica, Sigma Xi (pres. Harvard chpt. 1971-72, medal 1985), Phi Beta Kappa (Harvard chpt.), Beta Nu chpt. Phi Sigma (first hon.). Unitarian (vestryman Kings Chapel, Boston 1974-76, 82-85). Home: 78 Larchmont Rd Melrose MA 02176-2906 Office: Bot Museum Harvard U Cambridge MA 02138

SCHULTZ, ALBERT BARRY, engineering educator; b. Phila., Oct. 10, 1933; s. George D. and Belle (Seidman) S.; m. Susan Resnikov, Aug. 25, 1955; children—Carl, Adam, Robin. B.S., U. Rochester, 1955; M.Engring., Yale U., 1959, Ph.D., 1962. Asst. prof. U. Del., Newark, 1962-65; asst. prof. U. Ill., Chgo., 1965-66, assoc. prof., 1966-71, prof., 1971-83; Vennema prof. U. Mich., Ann Arbor, 1983—. Contbr. numerous articles to profl. jours. Served to lt. USN, 1955-58. Research Career awardee NIH, 1975-80; Javits neurosci. investigator award NIH, 1985-92. Mem. NAE, Internat. Soc. for Study of Lumbar Spine (pres. 1981-82), ASME (chmn. bioengring. div. 1981-82, H.R. Lissner award 1990), Am. Soc. of Biomechanics (pres. 1982-83), U.S. Nat. Com. on Biomechanics (chmn. 1982-85), Phi Beta Kappa. Office: U Mich 3112 GG Brown Lab Ann Arbor MI 48109-2125

SCHULTZ, CHARLES WILLIAM, chemistry educator; b. Detroit, Sept. 15, 1942; s. William F. and Helen (Laitis) S.; m. Marybelle Alice Seavitte, Aug. 22, 1964; children: Cara, Christopher. MS, Ohio State U., 1966; PhD, U Mich., 1969. Asst. prof. Glassboro (N.J.) State Coll., 1970-77, assoc. prof., 1977-83, prof., 1983—; cons. Norell, Inc., Mays Landing, N.J., 1974—; analytical devel. DuPont, Jackson Labs, N.J., 1990; faculty fellow Princeton (N.J.) U., 1992-93. Contbr. articles to profl. jours. Faculty advisor Student Affiliate of Am. Chem. Soc., Glassboro State Coll., 1974-89; asst. scoutmaster Troop 289, Boy Scouts Am., Glassboro, 1990—; vol. fireman Glassboro Co. # 1, 1971-83. Mem. Am. Chem. Soc. (chmn. South Jersey sect. 1984, Outstanding Student Affiliate award 1978-83), Soc. Applied Spectroscopy. Office: Bosshart Hall Rowan Coll of NJ Glassboro NJ 08028

SCHULTZ, DALE HERBERT, chemical process control engineer; b. Ft. Wayne, Ind., May 1, 1948; s. Calvin Frederick and Doris Elizabeth (Siebold) S.; m. Cynthia Kay Kaetzel, June 6, 1970 (div. Jan. 1983); children: David, Christina; m. Jennifer Jane Charboneau, Dec. 28, 1985; 1 child, Andrew. BSChemE, MS in Indsl. Adminstrn., Purdue U., 1970. Registered U.S. Patent Agt. Spl. assignments engr. Dow Chem. Co., Midland, Mich., 1970-72, rsch. engr., 1972-76, sr. supr. mfg., 1976-82, sr. process control systems specialist, 1982-93; intellectual assets mgr. Dow Chem. Co., 1993—. Office: Dow Chem Co GPC Bldg Midland MI 48667

SCHULTZ, GERALD ALFRED, chemical company executive; b. Lockport, N.Y., Jan. 22, 1941; s. Alfred Henry and Lucy Vivian (Proctor) S.; m. Barbara Joan Beals, July 13, 1962; children: Amy Lynn Schultz Poole. AAS, Erie County Tech. Inst., Buffalo, 1961; BA in Chemistry, SUNY, Buffalo, 1969; student, Harvard U., 1979. Rsch. technician Occidential Chem. Corp., Niagara Falls, N.Y., 1961-63; rsch. engr. Nat. Gypsum Co. Inc., Buffalo, 1963-66; chemist, devel. mgr., gen. mgr. to v.p. Akzo Chems., Burt, N.Y. and Chgo., 1966-86; exec. v.p. VandeMark Chem. Co. Inc., Lockport, N.Y., 1986—. Contbr. articles to profl. jours; patentee in field. Fund raiser United Way, Newfane, N.Y., 1982-84; treas., bd. dirs. Newtane Intercommunity Meml. Hosp., 1980-84; bd. dirs. ARC, Lockport, N.Y., 1980-84. Mem. Soc. Plastic Engrs., Soc. Plastics INdustry (bd. dirs. 1974-76), Organic Peroxide Prodn. Safety Div. (chmn. 1974-76), Tuscawar Club, Olcott Yacht Club (past comdr. 1975), Synthetic Organic Chem. Mfrs. Assn. (bd. dirs. 1991—), N.Y. State Bus. Coun., N.Y. State Chem. Alliance, Lockport Indsl. Coun. (treas. 1991—), Ea. Niagara C. of C. (bd. dirs. 1992—), Rotary, Lockport, Town and Country Club. Republican. Episcopalian. Avocations: golf, gardening, computers, boating. Home: 4025 Hartland Rd Gasport NY 14067-9318 Office: VanDeMark Chem Co Inc One North Transit Rd Lockport NY 14094

SCHULTZ, HYMAN, analytical chemist; b. Bklyn., July 11, 1931; s. Joseph Israel and Mary (Reifen) S.; m. Shirley Miriam Befferman, Aug. 10, 1957; children: Richard Hirsch, Daniel Brian. BS, Bklyn. Coll., 1956; PhD, Pa. State U., 1962. Sr. rsch. engr. N.Am. Aviation, L.A., 1962-67; head gas analysis Teledyne, Isotopes, Inc., Westwood, N.J., 1967-70; sr. tech. advisor U.S. Dept. Energy, Pitts., 1971—; speaker in field. Contbg. author: Trace Elements in Coal, 1975, Chemistry of Coal Utilization, 1981; contbr. articles to tech. publs. Mem. State College (Pa.) Theatre Group, 1958-63; chmn. Friends of Libr., Rivervale, N.J., 1969-70; violinist South Hills Symphony Orch., Mt. Lebanon, N.J., 1975-77. With U.S. Army, 1952-53, Korea. Mem. Soc. Analytical Chemists Pitts. (fin. affairs com. 1993-94), Am. Chem. Soc. (head libr. com. Pitts. sect. 1979—), Pitts. Conf. (registration chmn. 1994—). Achievements include patent for method of determining low levels of oxygen in carbonaceous materials. Office: US Dept Energy PETC MS 84-202 PO Box 10940 Pittsburgh PA 15236

SCHULTZ, JEROLD MARVIN, materials scientist, educator; b. San Francisco, June 21, 1935; s. Ernst and Florence (Rubin) S.; m. Peggy June Ostrom, July 30, 1960; children: Carrie, Timothy, Peter, Anna. BS, U. Calif., Berkeley, 1957, MS, 1958; PhD, Carnegie-Mellon U., 1965. Intermediate engr. Westinghouse Rsch. Labs., Pitts., 1959-61; mem. faculty U. Del., Newark, 1964—. Author: Polymer Materials Science, 1974, Diffraction for Materials Scientists, 1982; editor: Properties of Solid Polymeric Materials, 1977, Solid State Behavior of Linear Polyesters and Polyamides, 1990; contbr. over 150 articles to profl. jours. Flutist Newark Symphony Orch., 1966—. Recipient Sr. Am. Scientist award Alexander von Humboldt Stiftung, Germany, 1977, 82, Kliment Ohridzki medal Peoples Republic of Bulgaria, 1986. Mem. Am. Phys. Soc., Polymer Processing Soc. Democrat. Episcopalian. Achievements include 8 patents on polymer processing and materials. Office: U Del Materials Sci Program Newark DE 19716

SCHULTZ, KIRK R., pediatric hematology-oncology educator; b. Seattle, Feb. 13, 1956; arrived in Can., 1992; BS, Nebr. Wesleyan, 1978; MD, U. Nebr., 1982. Intern/resident in pediatrics Coll. of Medicine, Houston, 1982-85; fellow in pediat. hematology/oncology U. Wash., Seattle, 1985-89; asst. prof. pediatrics and medicine Wayne State U. and Children's Hosp. of Mich., Detroit, 1989-92; asst. prof. pediatrics U. B.C. and B.C.'s Children's Hosp., Vancouver, 1992—. Office: BC Childrens Hosp/Oncology, Dept Pediatrics UBC 4480 Oak St, Vancouver, BC Canada V6H 3V4

SCHULTZ, LINDA JANE, epidemiologist; b. Newton, Mass., Oct. 8, 1962; d. James Howard and Suzanne Jane (Walter) S. BS, U. Calif., Davis, 1986, DVM, 1988, M in Preventive Vet. Medicine, 1990. Vet. asst. Nepal Social Svc. Fund, 1986-87; pvt. practice vet. Calif., 1988-89; large animal vet. Vols. in Overseas Coop. Assistance, Tunisia, 1989, Niger, 1989; resident in preventive medicine Calif. Dept. Health Svcs., Sacramento, 1990-91; epidemiologist malaria br. Ctrs. for Disease Control, Atlanta, 1991—. Mem. APHA, Am. Vet. Med. Assn., Am. Soc. Tropical Medicine and Hygiene. Avocations: gardening, music, photography, bicycling, backpacking. Office: Ctrs for Disease Control Atlanta GA 30333

SCHULTZ, PER-OLOV, occupational health physician, consultant; b. Sävsjö, Sweden, Mar. 1, 1932; s. Per Birger and Disa Karlsson; m. Ulla-Britt öster, Jan. 29, 1955; children: Per Johan, Ulla Kristina. MD, Upsala (Sweden) U., 1961. Resident Serafimer Hosp., Stockholm, 1959-60; physician Nässjö County Hosp., Nässjö, Sweden, 1961-67; physician, chief med. officer City of Nässjö, 1967-72, Nässjö Occupational Health Ctr., 1973—; cons. Nässjö Bd. Health, 1973—, Nat. Health Ins., Nässjö, 1977—. Lt. M.C., Swedish Army, 1959-67. Fellow Swedish Soc. Medicine; mem. AAAS, Am. Coll. Occupational Medicine, N.Y. Acad. Scis. Round Table Club (pres. 1969-70), Rotary (pres. Nässjö 1986-87). Avocations: fishing, reading, mountain climbing, collecting books. Home: Vattugatan 11, S-57140 Nässjö Sweden Office: Nässjöhälsan, S-57141 Nässjö Sweden

SCHULTZ, PETER G., chemistry educator. PhD, Calif. Inst. Tech., 1983. Mem. faculty chemistry dept. U. Calif., Berkeley, 1985—, now assoc. prof. NIH Postdoctoral fellow, MIT; recipient Nobel Laureate Signature award for Grad. Edn. in Chemistry, Pure Chemistry award Am. Chem. Soc., Alan T. Waterman award NSF, 1988, Denkewalter lectr. award Loyola U., 1990, Ernest Orlando Lawrence Meml. award U.S. Dept. Energy, 1991. Office: U Calif Dept Chemistry Dept Chemistry Berkeley CA 94720

SCHULTZ, PHILIP STEPHEN, engineering executive; b. Chgo., July 12, 1947; arrived in France, 1989; s. George Philip and Margaret Mary (Balog) S.; m. Sandra Arlene Levy, Sept. 4, 1976; children: Natalie, Ashley, Kelly. BS in Physics, DePaul U., 1970, MS in Physics with honors, 1971; PhD in Geophysics, Stanford U., 1976. Rsch. geophysicist Digicon Geophys. Corp., Houston, 1976-84; engring. mgr. Schlumberger, Ltd., Tokyo, 1984-87, London, 1987-89, Paris, 1989-93; tech. cons. Schlumberger-Geoquest, Houston, 1993—. Contbr. articles to profl. jours.; inventor data correction techniques. Mem. Soc. Exploration Geophysicists (assoc. editor Geophysics Jour. Houston and Tokyo 1984-85, Outstanding Presentation award 1980 ann. conv.), Soc. Petroleum Engrs., European Assn. Exploration Geophysicists, Lambda Tau Lambda (pres. 1967-68). Avocations: tennis, guitar. Office: Schlumberger-Geoquest 5858 Westheimer Ste 800 Houston TX 77057

SCHULTZ, RICHARD MICHAEL, biochemistry educator; b. Phila., Oct. 28, 1942; s. William and Beatrice (Levine) S.; m. Rima M. Lunin, Mar. 7, 1965; children: Carl M., Eli J. BA, SUNY, Binghamton, 1964; PhD, Brandeis U., 1969. Rsch. fellow Harvard U. Med. Sch., Boston, 1969-71; asst. prof. Loyola U. Stritch Sch. of Medicine, Maywood, Ill., 1971-78, assoc. prof., 1978-84, prof., 1984—, chmn. dept. molecular and cellular biochemistry, 1984—; mem. adv. med. bd. Leukemia Rsch. Found., Chgo., 1987-91. Contbr. articles to profl. jours. and chpts. to books. Recipient Rsch. grants NIH. Achievements include in vivo evidence for the role of profease enzymes and their inhibitors in regulating tumor cell metastasis; obtaining evidence on the nature of the transition-state in enzyme catalysis. Office: Loyola U Sch of Medicine Dept of Molecular & Cellular Biochem Maywood IL 60153

SCHULTZ, STANLEY GEORGE, physiologist, educator; b. Bayonne, N.J., Oct. 26, 1931; s. Aaron and Sylvia (Kaplan) S.; m. Harriet Taran, Dec. 25, 1960; children: Jeffrey, Kenneth. A.B. summa cum laude, Columbia U., 1952; M.D., N.Y. U., 1956. Intern Bellevue Hosp., N.Y.C., 1956-57; resident Bellevue Hosp., 1957-59; research assoc. in biophysics Harvard U., 1959-62, instr. biophysics, 1964-67; assoc. prof. physiology U. Pitts., 1967-70, prof. physiology, 1970-79; prof., chmn. dept. physiology U. Tex. Med. Sch., Houston, 1979—, prof. dept. internal medicine, 1979—; cons. USPHS, NIH, 1970—; mem. physiology test com. Nat. Bd. Med. Examiners, 1974-79, chmn., 1976-79. Editor: Am. Jour. Physiology, Jour. Applied Physiology, 1971-75, Physiol. Revs., 1979-85; mem. editorial bd.: Jour. Gen. Physiology, 1969-88; assoc. editor: Ann. Revs. Physiology, 1977-81, News in Physiol. Scis., 1989—; editorial bd. Ann. Revs. Physiology, 1974-81, Current Topics in Membranes and Transport, 1975-81, Jour. Membrane Biology, 1977—, Biochim. Biophys Acta, 1987-89; editor: Handbook of Physiology: The Gastrointestinal Tract, 1989-91; contbr. articles to profl. jours. Served to capt. M.C. USAF, 1962-64. Recipient Research Career award NIH, 1969-74; overseas fellow Churchill Coll., Cambridge U., 1975-76. Mem. AAAS, Am. Heart Assn. (established investigator 1964-69), Am. Physiol. Soc. (councillor 1989-91, pres.-elect 1991-92, pres. 1992-93), Fed. Am. Soc. Exptl. Biology (exec. bd. 1992—), Biophys. Soc., Soc. Gen. Physiologists, Internat. Cell Rsch. Orgn., Internat. Union Physiol. Scis. (chmn. internat. com. gastrointestinal physiology 1977-80, chmn. U.S. nat. com. 1992—), Assn. Am. Physicians, Am. Assn. Ob-Gyn. (hon. fellow), Assn. Chmn. Depts. Physiology (pres. 1985-86), Sigma Xi, Phi Beta Kappa. Home: 4955 Heatherglen Dr Houston TX 77096-4213

SCHULTZ, HELMUT WILHELM, chemical engineer, environmental executive; b. Berlin, July 10, 1912; came to U.S., 1924; s. Herman Ludwig Wilhelm and Emilie (Specka) S.; m. Colette Marie Francoise Prieur, Mar. 6, 1954; children: Raymond A., Caroline P., Roland W., Robert B., Thomas F. BS, Columbia U., 1933, ChE, 1934, PhD, 1942. Rsch. engr. to mng. dir. Union Carbide Corp., Charleston, W.Va., 1934-69; spl. asst. to dir. def. rsch. and engring. U.S. Dept. of Def., Washington, 1964-67; spl. asst. to U.S. commr. of edn. U.S. Dept. of Edn., Washington, 1971; sr. rsch. scientist, adj. prof. Columbia U., N.Y.C., 1972-85; pres., chief exec. officer Dynecology, Inc., Harrison, N.Y., 1974—; chmn. Brandenburg Energy Corp., Harrison, 1979—; chmn. emeritus Global Energy Inc., Cin., 1989—. Contbr. articles to profl. jours. Mem. N.Y.C. Mayor's Sci. and Tech. Adv. Coun., 1973-74; bd. dirs. Charleston Symphony Orch., 1956-62, Am. Cancer Soc., W.Va., 1954-58; chmn. W.Va. AEC, Charleston, 1962-64. Grantee in field. Fellow Am. Inst. Chem. Engrs.; mem. N.Y. Acad. Sci., Am. Chem. Soc. (emeritus), N.Y. Yacht Club, Cosmos Club. Achievements include patents for laser catalysis; centrifugation cascade for enrichment of fissionable uranium isotope; high acceleration rocket motor; tar-free, slagging coal/waste gasifier; enhanced oil recovery process; synthesis of ethanol from ethylene and steam; waste-to-energy conversion processes, and 60 others. Home: 611 Harrison Ave Harrison NY 10528-1406

SCHULZ, MICHAEL, physicist; b. Bremerhaven, Germany, July 17, 1959; came to U.S. 1988; s. Gerhard Richard and Anne Marie (Emisch) S. Diplom, U. Heidelberg, Germany, 1984, PhD, 1987. Rsch. assoc. Oak Ridge (Tenn.) Nat. Labs., 1988-89, Kans. State U., manhattan, 1989-90; asst. prof. U. Mo., Rolla, 1990--; refree Nat. Sci. Found., Washington, 1991--, Nuclear Instruments and Methods, Holland, 1990--. Contbr. articles to profl. jours. Nat. Sci. Found. grantee 1991, 92, 93, NATO grantee 1991, 93. Mem. Am. Phys. Soc. Home: RR4 Box 314-C Rolla MO 65401 Office: U Mo Physics Dept Rolla MO 65401

SCHULZ, PAUL, physicist; b. Rostock, Germany, Jan. 31, 1911; s. Heinrich and Marie (Bahrdt) S.; student U. Munich, U. Rostock, 1929-34; Ph.D., U. Rostock, 1934; Dr. Rer. Nat. Habil, U. Bonn, 1943; m. Irene Kerp, 1938; children—Ingrid, Beate, Gabriele. Asst. phys. inst. U. Bonn (Germany), 1935-37; head Osram Elec. Lighting Research Lab., Berlin, 1937-46; dir. Gas Discharge Research Inst., German Acad. Sci., Greifswald, 1946-49; prof. physics U. Greifswald, 1948-49; faculty U. Karlsruhe (W. Ger.), 1949—, prof., 1949—, rector (pres.), 1963-65. Recipient Elenbaas prize, 1969. Mem. Deutsche Physikalische Gesellschaft (chmn. fachausschuss fur plasma-physik 1950-60), Deutsche Lichttechnische Gesellschaft (pres. 1954-57, hon. mem. 1976—), Am. Phys. Soc., Commn. Internat. Photobiologie, Commn. Internat. de l'Eclairage, Heinrich-Hertz Soc. Mem. Evangelical Ch. Author: Elektronische Vorgange in Gasen und Festkorpern, 1968, rev. edit., 1974. Contbr. articles to profl. publs. Inventor xenon and metallhalide discharge lamps 1943, 49, patentee in field. Home: 5 Aschenbrodelweg, 7500 Karlsruhe 51, Federal Republic of Germany Office: 12 Kaiserstrasse, 7500 Karlsruhe 1, Germany

SCHULZ, RAYMOND CHARLES, electrical engineer; b. St. Louis, Nov. 2, 1942; s. Charles Edward and Evelyn Augusta (Moeller) S.; m. Ana G. Schulz, May 10, 1980; children: Kara Sun and Michelle Yung (both adopted from South Korea). BSEE, Washington U., 1965, MSEE, 1968. Registered profl. engr., Mo. Sr. engr. Electronics and Space div. Emerson Electric, St. Louis, 1968-70; assoc. dir. stategic bus. devel. Southwestern Bell Corp., St. Louis, 1970-91; v.p. Brooks Telecomm. Corp., St. Louis, 1993—. Mem. IEEE (sr.), Comm. Soc. IEEE, Eta Kappa Nu, Tau Beta Pi. Home: 3616 White Bark Ct Saint Louis MO 63129-2249

SCHULZE, MATTHIAS MICHAEL, chemist; b. Bremen, Germany, Apr. 4, 1964; s. Paul Werner and Ursula Ingeborg (Abicht) S. Vordiplom, Tech. U., Berlin, 1985, Diplom, 1988, PhD in Chemistry, 1991. Fellow Gradviertenkolleg Berlin, 1988-91; assoc. researcher U. Okla., Norman, 1990-91; assoc. researcher in chemistry Ohio State U., Columbus, 1991—. Mem. Am. Chem. Soc., German Soc. Chemists. Avocations: stamp collecting, cooking, skiing, furniture design. Home: Innstr 8, 12045 Berlin Germany Office: Tech U, Strasse des 17 Juni 122, 1000-12 Berlin Germany

SCHULZE, NORMAN RONNIE, aerospace engineer; b. Roanoke, Va., May 12, 1936; s. Heinrich Ernst Karl and Euqual Uvula (Baker) S.; m. Joan Frances Lipps, June 18, 1960; children: Erich Karl, Ronald Clinton, Byron Kirk. BS in Physics, U. Chgo., 1958. Propulsion rsch. engr. NASA/Langley Rsch. Ctr., Hampton, Va., 1958-62; mgr. Gemini propulsion NASA/Johnson Space Ctr., Houston, 1962-68, mgr. system safety, 1968-75; mgr. safety shuttle NASA, Washington, 1975-80, chief SRQ office shuttle, 1980-83, dep. dir. engring. div., 1985-87, program mgr. applied tech., 1987—. Pres. Homeowners Assn., Clifton, Va., 1977-78; asst. scoutmaster Boy Scouts Am., Clifton, 1976-86; coach Chantilly Soccer Club, 1979-80. Recipient Exceptional Svc. medal NASA, 1981. Mem. AIAA. Achievements include initiating applied technology programs related to space; organizing NASA-DOD Tech. Steering Coms.; findings related to attention of importance of fusion energy for space. Home: 12805 Knollbrook Dr Clifton VA 22024 Office: NASA Washington DC 20546

SCHULZE, RICHARD HANS, engineering executive, environmental engineer; b. Buffalo, May 28, 1933; s. Hans Joachim and Lucy (Kawczynska) S.; m. Jacqueline Van Luppen, Nov. 2, 1957 (div. Aug., 1979); children: Richard H Jr., Linda Schulze Keefer, John; m. Ineke Grooters, Aug. 29, 1987. BSME, Princeton U., 1954; MBA, Northwestern U., 1958. Registered profl. engr., Tex. Rsch. analyst U.S. Steel Corp., Pitts., 1958-60; chief engr. G&H Rsch. and Devel., McKeesport, Pa., 1960-62; cons. Mgmt. and Mktg. Inst., N.Y.C., 1962-63, IITRI mgmt. consulting div., N.Y.C., 1963-64; mkt devel. mgr. Mobil Chem. Co. plastics div., N.Y.C., Jacksonville, Ill., 1964-71; pres. Ecology Audits, Inc. (Core Labs.), Dallas, 1971-74, Trinity Cons,, Inc., Dallas, 1974—. Contbr. articles to Jour. of Air and Waste Mgt. Assn., Atmospheric Environment, Environmental Careers, others; presented papers at sci. symposiums, seminars, confs. Mem. Dallas Symphony Assn., Mus. of Art; bd. dirs. Dallas Opera, 1993—. Mem. ASME, TAPPI, Am. Acad. Environ. Engrs. (diplomate), Am. Chem. Soc., Am. Meteorologocal Soc., Air and Waste Mgmt. Assn. (bd. dirs. 1986-89. '90-93, v.p. 1988-89, 1 st v.p. 1990-91, pres. 1991-92, past pres. 1992-93), Nat. Soc. for Clean Air (UK), Soc Petroleum Engrs., Soc. for Risk Analysis, Semi-Conductor Safety Assn., Verein Deutscher Ingenieure, Assn. Francaise des Ingénieurs et Techniciens de L'environment. Presbyterian. Home: 7619 Marquette Dallas TX 75225 Office: Trinity Cons Inc Ste 1200 12801 N Central Expy Dallas TX 75243

SCHUMACHER, H(ARRY) RALPH, internist, researcher, medical educator; b. Montreal, Que., Can., Feb. 14, 1933; s. H. Ralph and Dorothy (Shreiner) S.; m. Elizabeth Jean Swisher, July 13, 1963; children: Heidi Ruth, Kaethe Beth. B.S., Ursinus Coll., 1955; M.D., U. Pa., 1959. Intern Denver Gen. Hosp., 1959-60; resident in medicine Wadsworth VA Hosp., L.A., 1960-62, fellow in rheumatology, 1962-63; fellow in rheumatology Robert B. Brigham Hosp. and Harvard U. Med. Sch., Boston, 1965-67; chief arthritis-immunology ctr. VA Med. Ctr., Phila., 1967—; faculty mem. U. Pa. Sch. Medicine, Phila., 1967—, prof. medicine, 1979—, acting arthritis div. chief, 1978-80, 91—. Author: Gout and Pseudogout, 1978, Essentials of a Differential Diagnosis of Rhematoid Arthritis, 1981, Rheumatoid Arthritis, 1988, Case Studies in Rheumatology for the House Officer, 1989, Atlas of Synovial Fluid and Crystal Identification, 1991, A Practical Guide to Synovial Fluid Analysis, 1991; editor: Primer on Rheumatic Diseases, 1981—; mem. editorial bd. Jour. Rheumatology, 1973—, Arthritis and Rheumatism, 1981-00, Revue du Rhumatisme, 1992—; contbr. articles to profl. jours.; lectr., author gardening. Pres. Eastern Pa. chpt. Arthritis Found., 1980-82; chmn., founder Phila. Garden Tours, 1987—; bd. dirs. Hemochromatosis Research Found., 1984—, Am. Bd. Med. Advancement China, 1983—. Served with M.C. USAF, 1963-65. Recipient VanBreeman award The Netherland Rheumatism Soc., 1988; Deposition VA grantee, 1967—, NIH grantee, 1981. Fellow ACP; mem. Am. Coll. Rheumatology (pres. Southeastern region 1981-82), Phila. Rheumatism Soc. (pres. 1980), Phila. Electron Microscopy Soc. (chmn. 1975-76), Rheumatism Soc. Mex., Rheumatism Soc. Australia, Rheumatism Soc. Colombia, Rheumatism Soc. Chile, Rheumatism Soc. Republic of China, Assn. Mil. Surgones (Philip Hench award 1986), Fedn. Clin. Rsch., AAAS. Office: Hosp U Pa Ravdin Bldg 3400 Spruce St Philadelphia PA 19104-4220

SCHUMAN, STANLEY, epidemiologist, educator; b. St. Louis, Dec. 29, 1925; married, 1952; 8 children. MD, Wash. U., 1948; MPH, U. Mich., 1960, PhD, 1962. Diplomate Am. Bd. Pediatrics. Clin. instr. pediatrics, sch. medicine Wash. U., 1954-59; from asst. prof. to prof. epidemiology, sch. pub. health U. Mich., Ann Arbor, 1962-73; prof. epidemiology in family practice, coll. medicine Med. U. S.C., Charleston, 1974—, prof. pediatrics, 1976—; med. dir. agromedical program Clemson Med. U., S.C., 1984—; project dir. S.C. Pesticide Study Ctr., EPA, 1981-84. Author: (textbook) Epidemiology, 1986. Mem. APHA, Am. Acad. Family Practice, Am. Epidemiol. Soc., Soc. Epidemiol. Rsch., Coun. Agrl. Sci. Tech., Sigma Xi. Achievements include research on epidemiology of heat waves in U.S. cities; human sweat studies in population survey; screening for cystic fibrosis; population surveys of injuries due to accidents; accident prevention; field trials with young drivers; epidemiology in family practice; computers in medicine; research in toxicology, cancer of the esophagus agricultural medicine. Office: Med U SC Agromedicine Program 171 Ashley Ave Charleston SC 29425*

SCHUMAN, WILLIAM JOHN, JR., mechanical engineer; b. Balt., Jan. 23, 1930; s. William John and Susie Iona (Kendrick) S.; m. Kathleen Elizabeth Sisson, Aug. 22, 1952; children: Todd Loring, Craig Lee, Grant William. BS, U. Md., 1952; MS, Pa. State U., 1954, PhD, 1965. Registered profl. engr., Md. Asst. prof. mechanics USAF Inst. Tech., Dayton, Ohio, 1954-57; sr. engr. Martin Marietta Corp., Balt., 1957; aero research engr./research physicist U.S. Army Ballistic Research Lab., Aberdeen Proving Ground, Md., 1957-83; rsch. physicist U.S. Army Harry Diamond Labs., Adelphi, Md., 1983-85; v.p., chief scientist SI/Div. of Spectrum 39, Balt., 1986—; cons. in field. Contbr. articles to profl. jours. Capt. USAF, 1952-57. Mem. ASME (chmn. com. on shock & vibration 1966), Am. Acad. Mechanics, Soc. for Advancement of Matl. and Process Engring., Sigma Xi. Democrat. Lutheran. Avocations: numismatics, squash, fishing. Office: SI Div of Spectrum 39 PO Box 10970 Baltimore MD 21234-0970

SCHUMANN, STANLEY PAUL, mechanical engineer; b. Columbus, Ohio, Sept. 25, 1954; s. Paul Otto and Joan (McCambridge) S.; m. Debra Lynn Kollmar, Mar. 22, 1980; children: David, Alexander. BSME, Ohio State U., 1976. Sr. engr. Owens Corning Fiberglas, Granville, Ohio, 1977-86; project engr. Thwing-Albert Instrument Co., Phila., 1987; sr. project engr. Copeland Corp., Sidney, Ohio, 1987—; instr. adult edn. Licking County Jr. Vocat. Sch., Newark, Ohio, 1982-86, Upper Valley Jr. Vocat. Sch., Piqua, Ohio, 1988—. Contbr. tech. papers to XVIII Internat. Congress of Refrigeration, ASHRAE Transactions, 2d Internat. Symposium on Fluid Flow, ASTM. Mem. ASME, Am. Soc. Heating, Refrigerating and Air Conditioning Engrs. Office: Copeland Corp 1675 W Campbell Rd Sidney OH 45365

SCHUMANN, THOMAS GERALD, physics educator; b. L.A., Mar. 15, 1937; s. Bernard Stricker and Anne (Mandel) S.; m. Syril Ann Feltquate, June 1967 (div. 1973). BS, Calif. Inst. Tech., 1958; PhD, U. Calif., Berkeley, 1965. Rsch. assoc. Brookhaven (N.Y.) Nat. Lab., 1965-67; asst. prof. CCNY, 1967-70; lectr. Calif. State U., Hayward, 1970; from asst. prof. to prof. physics Calif. Poly. State U., San Luis Obispo, 1971—. Contbr. articles to profl. jours. Mem. Am. Phys. Soc. Office: Calif Poly State U Dept Physics San Luis Obispo CA 93407

SCHUMER, DOUGLAS BRIAN, physicist; b. Passaic, N.J., Mar. 22, 1951; s. William and Janet (Levine) S.; m. Barbara Lee Witte, May 23, 1976; children: Ariel Diana, Suzanne Meryl. BS, Carnegie-Mellon U., 1973, MS, 1974; PhD, Rensselaer Poly. Inst., 1977. Mem. tech. staff AT&T Bell Labs., Holmdel, N.J., 1977-80; v.p. engring. and rsch. Ohaus Corp., Florham Park, N.J., 1980-89; v.p. rsch. and devel. Cardiac Pacemakers Inc., St. Paul, 1990—. Contbr. numerous articles to sci. and profl. jours. Mem. Am. Phys. Soc., Optical Soc. Am., IEEE. Office: Cardiac Pacemakers Inc 4100 Hamline Ave N Saint Paul MN 55112

SCHUR, WALTER ROBERT, physician; b. Webster, Mass., June 17, 1914; s. Robert O. and Alma L. (Gatzke) S.; student Valparaiso U., 1931-34; M.D., Middlesex U., Waltham, Mass., 1940; m. Delta Jean Newman, June 17, 1944; children—Paul, David, Jonathan, Ruth, Timothy, Peter, Stephen, Mary, Joel, Daniel, Rhoda. Resident, Milford (Del.) Meml. Hosp., 1940-41, Grace Hosp. Cleve., 1942-43; intern Lutheran Hosp., Cleve., 1941-42; pvt. practice, Oxford, Mass., 1944—; bd. dirs., pres. Doctors Hosp., Worcester, Mass., chmn. bd., 1978-87; bd. dirs. AdCare Hosp., 1987—, chmn. bd. dirs., 1987-91, Atlantic dist. Luth Ch.-Mo. Synod, 1978-87, mem., sec. edn. com., missions com., 1960-77, mem. stewardship com., youth com., edn. com., 1951-57, chmn. edn. com. Atlantic dist., 1954-57, mem. commn. on mission and ministry in ch., named Dist. Layman of Year, 1966, chmn. com. on ministry Atlantic dist., 1970; , bd. dirs. Luth. Assn. Works of Mercy, Assn. Evang. Luth. Chs.; bd. dirs. Valparaiso U., 1969—, sec., 1984—; pres., scholarship chmn. N.E. dist. Luth. Laymen's League, 1946-57; nat. bd. govs. Nat. Luth. Laymen's League, 1957; vice chmn. Luth. Hour Oper. Com., 1958, chmn. 1959-61; New Eng. bd. dirs. Assn. Evang. Luth. Chs., 1977-87, trustee East Coast Synod, 1977-87, mem. nat. bd. dirs., 1979-88; mem. council New Eng. Synod Evangelical Lutheran Ch. Am. 1988—; bd. dirs., vice chmn. French River Edn. Ctr., 1985—; mem. Oxford Sch. Com., 1961-86, Mass. Commn. on Christian Unity; assoc. charter mem. Park Ridge Ctr., 1986. Recipient award of Merit, Internat. Luth. Laymen's League, 1963. Fellow Am. Acad. Gen. Practice, Am. Acad. Family Physicians (charter); mem. AMA, Mass., Worcester Dist. med. socs., Am. Geriatrics Assn., New Eng. Obstet. and Gynecol. Soc., Valparaiso U. Alumni Assn. (past pres.), Luth. Acad. for Scholarship (bd. dirs. 1977-86), Concordia Hist. Inst., New Eng. Luth. Hist. Soc. (charter), Internat. Platform Assn., Rotary (past pres.). Home: Charlton Rd Oxford MA 01540 Office: 367 Main St Oxford MA 01540

SCHUSTER, FRANK FEIST, neurologist, social services administrator; b. Fulda, Germany, Sept. 26, 1928; came to U.S., 1928; s. Isaac and Rosel (Katzenstein) S.; m. Susan Bernstein, May 20, 1954; children: Roslyn Schuster Tayne, Lisa Schuster Klein, Paul R. AB, Johns Hopkins U., 1950; MD, N.Y. Med. Coll., 1955. Diplomate Am. Bd. of Pediatrics, 1964. Intern, then asst. resident in pediatrics Sinai Hosp. Balt., 1955-58; sr. asst. resident in pediatrics Johns Hopkins Hosp., Balt., 1958-59, acting chief pediatric neurology, head birth defects clinic, 1966-69, Nat. Inst. Neuro. Diseases & Blindness fellow, resident in neurology, 1959-62; pvt. practice neurology Balt., 1962-90; chief childhood br. Office Med. Evaluation, Office Disability, Social Security Adminstrn., Balt., 1990—. Contbr. articles to profl. jours. Mem. Am. Acad. Neurology (sr.), Child Neurol. Assn. (sr.). Jewish. Home: 2503 Shelleydale Dr Baltimore MD 21209 Office: Social Security Adminstrn Office Disability Rm 2090 1500 Woodlawn Dr Baltimore MD 21241-4999

SCHUSTER, LAWRENCE JOSEPH, mechanical engineer; b. San Francisco, July 29, 1946; s. Samuel and Esther (Kraus) S.; m. Judith Lynn Brown, Aug. 18, 1972; children: Rachel Anne, David Andrew. B of Univ. Studies, U. N.Mex., 1972, B of Mech. Engring., 1978, M of Mech. Engring., 1983. Registered profl. engr., N.Mex.; cert. energy mgr., testing and balancing engr. Sales engr. Johnson Controls, Inc., Albuquerque, N.Mex., 1969-75; project engr. Coupland-Moran & Assocs., Albuquerque, 1975-76; mgr. bldg. ops. U. N.Mex., Albuquerque, 1976-78, energy conservation engr., 1978-83, plant engr., 1985—; v.p. Universal Dynamics, Albuquerque, 1983-85; pres. Systems Cons., Albuquerque, 1979—; mem. N.Mex. State Energy Adv. Group, Santa Fe, 1991-92. Activities chmn. Cub Scouts Am., Albuquerque, 1989-92; soccer coach Duke City Soccer League, Albuquerque, 1991-92; religious ednl. com. chmn. 1st Unitarian Ch., Albuquerque, 1992. Recipient Nat. Energy award U.S. Dept. Energy, 1984. Mem. ASHRAE (nat. tech. com. 1981-83), Assn. Energy Engrs. (sr., Regional Energy Engr. of Yr. 1991), N.Mex. Assn. Energy Engrs. (pres. 1991), Assn. Phys. Plant Adminstrs. (assoc.). Home: 6904 Rosewood NE Albuquerque NM 87111 Office: U New Mexico Ford Utility Ctr Albuquerque NM 87131-3520

SCHUTT, JEFFRY ALLEN, physicist; b. Sheboygan, Wis., Aug. 18, 1959; s. Robert A. and Kathleen H. (Lutz) S.; m. Anne M. Meurer, Apr. 19, 1986; children: Kendra, Jonathan, Luke. BS in Physics, Marquette U., 1981; MBA, U. Wis., Milw., 1984. Quality assurance test engr. CAI Div. of RECON/Optical, Barrington, Ill., 1984-88; gen. mgr. Trace Labs., Chgo., 1988—; pres., nat. dir., v.p. Inst. Environ. Scis., Chgo., 1984— (John Martin Outstanding Younger Mem. award 1990). Contbr. articles to profl. jours. Achievements include implementation of multi-channel nanosecond event detection in test lab. Office: Trace Labs 4611 N Olcott Ave Chicago IL 60656

SCHUTTE, PAUL CAMERON, research scientist; b. Richmond, Va., Nov. 15, 1958; s. George John Schutte and Nell Virginia (Wood) Kinney; m. Kathleen Margaret Sikorski, July 8, 1989; 1 child, John Paul Schutte. BS, Mary Washington Coll., 1981; MS, Coll. William and Mary, 1988. Aerospace technologist instrument rsch. div. NASA Langley Rsch. Ctr., Hampton, Va., 1981-84, rsch. scientist flight mgmt. div., 1984-90, rsch. team leader flight mgmt. div., 1990—; mem. program com. Indsl. and Engring. Applications AI&ES, Tullahoma, Tenn., 1988-90. Recipient R&D 100 award R&D Mag., 1991. Mem. Am. Assn. for Artificial Intelligence. Achievements include development of an intelligent fault monitoring concept for aircraft applications called Monitaur. Office: NASA Langley Rsc Ctr Mail Stop 152 Hampton VA 23681

SCHUYLER, PETER R., biomedical electrical engineer, educator, researcher; b. Orange, N.J., Aug. 31, 1965; s. Raymond J. and Helen M. Schuyler; m. Joy E. Landcastle, Aug. 8, 1992. BS in Bioengineering, Syracuse U., 1987, MS in Electrical Engring., 1990, CAS, 1992. Rsch. engr. Inst. for Sensory Rsch., Syracuse, N.Y., 1987-90; instr. Syracuse (N.Y.) U., 1988-92; asst. prof. New England Inst. Tech., Warwick, R.I., 1992—; tutor Empire State Coll., Syracuse, 1991-92. Scholar Syracuse U., 1987-92. Mem. IEEE (jour. rschr.), Alpha Chi Rho (pres. 1985-86), Sigma Xi. Home: 2 Abby Ln Foster RI 02825

SCHWAB, GLENN ORVILLE, retired agricultural engineering educator, consultant; b. Gridley, Kans., Dec. 30, 1919; s. Edward and Lizzie (Sauder) S.; married; children: Richard, Lawrence,Mary Kay. BS, Kans. State. U., 1942; MS, Iowa State U., 1947, PhD, 1951; postdoctoral, Utah State U., 1966. Registered profl. engr., Ohio. Instr. to prof. agrl. engring. Iowa State U., Ames, 1947-56; prof. agrl. engring. Ohio State U., Columbus, 1956-85, prof. emeritus, 1985—; cons. Powell, Ohio, 1985—; bd. dirs. Internat. Water Mgmt. Program, Columbus. Co-author: Soil and Water Conservation Engineering, 4th edit., 1993, Elementary Soil and Water Engineering, 3d edit., 1985, Agricultural and Forest Hydrology, 1986; contbr. articles to profl jours. Served to capt. U.S. Army, 1942-46. Fellow Am. Soc. Agrl. Engrs. (bd. dirs. soil and water div. 1976-78, Hancock Brick and Tile Drainage Engr. 1968, John Deere medal 1987), Am. Soc. Engrs. Edn., Am. Soc. Testing Materials, Soil and Water Conservation Soc. Am., Am. Geophysical Union, Internat. Commn. Irrigation and Drainage. Avocations: rock polishing, wood working, photography, traveling. Home: 2637 Summit View Rd Powell OH 43065-8879 Office: Ohio State U 590 Woody Hayes Dr Columbus OH 43210-1057

SCHWALM, FRITZ EKKEHARDT, biology educator; b. Arolsen, Hesse, Germany, Feb. 17, 1936; came to U.S., 1968; s. Fritz Heinrich and Elisabeth Agnes (Wirth) S.; m. Renate Gertrud Streichhahn, Feb. 10, 1962; children—Anneliese, Fritz-Uwe, Karen. Ph.D., Philipps U., Marburg, Fed. Republic Germany, 1964; Staatsexamen, Philipps U., 1965. Educator boarding sch., Kiel, Fed. Republic Germany, 1956-57; lectr. Folk Universitetet, Stockholm, Sweden, 1959-60; research assoc. U. Witwatersrand, Johannesburg, South Africa, 1966-67, U. Notre Dame, South Bend, Ind., 1968-70; asst. prof., then assoc. prof. Ill. State U., Normal, Ill., 1970-82; assoc. prof. biology, then prof., chair dept. Tex. Woman's U., Denton, 1982—, dir. Animal Care Facility, 1990—, chmn. pro tem grad. coun., 1991-92; coordinator, chmn. S.W. Conf. for Devel. Biology, Denton, 1985, 90. Author: (monograph) Insect Morphogenesis, 1988; contbr. articles to profl. jours. Vice pres. PTA, Normal, 1975. Fellow Anglo-Am. Corp. South Africa, 1966, 67; NATO advanced research fellow, Freiburg, Fed. Republic Germany 1977. Mem. AAAS, Am. Soc. Zoologists, Deutsche Zoologische Gesellschaft, Soc. for Devel. Biology. Club: Univ. Denton (pres. 1983-85). Lodge: Kiwanis (v.p. 1987-88, pres. 1988-89). Home: 1116 Linden Dr Denton TX 76201-2721 Office: Tex Woman's U Biology Sci Rsch Lab Denton TX 76204

SCHWAN, HERMAN PAUL, electrical engineering and physical science educator, research scientist; b. Aachen, Germany, Aug. 7, 1915; came to U.S., 1947, naturalized, 1952; s. Wilhelm and Meta (Pattberg) S.; m. Anne Marie DelBorello, June 15, 1949; children: Barbara, Margaret, Steven, Carol, Cathryn. Student, U. Goettingen, 1934-37; Ph.D., U. Frankfurt, 1940; Doctor habil. in physics and biophysics, 1946; D.Sc. (hon.), U. Pa., 1986. Research scientist, prof. Kaiser Wilhelm Inst. Biophysics, 1937-47, asst. dir., 1945-47; research sci. USN, 1947-50; prof. elec. engring., prof. elec. engring. in phys. medicine, assoc. prof. phys. medicine U. Pa., Phila., 1950—, Alfed F. Moore prof. emeritus, 1983—; dir. electromed. div. U. Pa., 1952-73, chmn. biomed. engring., 1961-73, program dir. biomed. engr. tng. program, 1960-77; vis. prof. U. Calif.-Berkeley, 1956, U. Frankfurt Fed. Republic Germany, 1962, U. Würzburg, Fed. Republic Germany, 1986-87; lectr. Johns Hopkins U., 1962-67, Drexel U., Phila., 1983—; W.W. Clyde vis. prof. U. Utah, Salt Lake City, 1980; 10th Lauristan Taylor lectr. Nat. Council Radiation Protection and Measurements, 1986; Fgn. sci. mem. Max Planck Inst. Biophysics, Germany, 1962—; cons. NIH, 1962-90; chmn. nat. and internat. meetings biomed. engring. and biophysics, 1959, 61, 65; mem. nat. adv. council environ. health HEW, 1969-71; mem. Nat. Acad. Scis.-NRC coms., 1968-87, Nat. Acad. Engring., 1975—. Co-author: Advances in Medical and Biological Physics, 1957, Therapeutic Heat, 1958, Physical Techniques in Medicine and Biology, 1963; editor: Biol. Engring, 1969; co-editor: Interactions Between Electromagnetic Fields and Cells, 1985; mem. editorial bd. Environ. Biophysics, IEEE Transactions Med. Biol. Engring., Jour. Phys. Med. Biol., Nonionizing Radiation, Bioelectromagnetics; contbr. articles to profl. jours. Recipient Citizenship award Phila., 1952, 1st prize AIEE, 1953, Achievement award Phila. Inst. Radio Engring., 1963, Rajewsky prize for biophysics, 1974, U.S. sr. scientist award Alexander von Humboldt Found., 1980-81, Biomed. Engring. Edn. award Am. Soc. Engring. Edn., 1983, d'Arsonval award Bioelectromagnetics Soc., 1985. Fellow IEEE (Morlock award 1967, Edison medal 1983, Centennial award 1984, Phila. Sect. award 1991, chmn. and vice chmn. nat. profl. group biomed. engring. 1955, 62-68), AAAS; mem. Am. Standards Assn. (chmn. 1961-65), Biophys. Soc. (publicity com., council, constn. com.), German Biophys. Soc. (hon.), Soc. for Cryobiology, Internat. Fedn. Med., and Biol. Engring., Bioelectromagnetics Soc., Biomed. Engring. Soc. (founder, dir. 1968-71), Sigma Xi, Eta Kappa Nu. Achievements include discovery of counterion relaxation; dielectric spectroscopy of cells and tissues; nonlinearity law of electrode polarization; research on nonionizing radiation biophysics; fundamentals electromagnetic bioengineering; first standard for safe exposure to electrical fields. Home: 99 Kynlyn Rd Wayne PA 19087-2849 also: 162 59th St Avalon NJ 08202 Office: U Pa Dept Bioengring D2 Philadelphia PA 19104

SCHWANDT, LESLIE MARION, psychologist, researcher; b. Cleve.; d. Donald Theodore and June Rose (Vogt) S. BA, Baldwin-Wallace Coll., 1985; PhD, U. Minn., 1991. Rsch. asst. Psychol. Corp., Berea, Ohio, 1985; rsch. asst. Dept. Psychology U. N.C., Chapel Hill, 1986-87; teaching asst., instr. Dept. Psychology U. Minn., Mpls., 1989-91; postdoctoral rsch. assoc. Psychoneuroimmunology program U. Minn., St. Paul, 1991-93; rsch. fellow Addiction Rsch. Ctr., Balt., 1993—. Recipient NSF fellowship, 1988-91, U. Minn. Grad. Sch. fellowship, 1987-88, Ohio Bd. Regents scholarship, 1982-85. Mem. AAAS, APA, Am. Psychol. Soc., Psychonomic Soc. (assoc.). Office: Addiction Rsch Ctr NIDA-NIH Baltimore MD 21224

SCHWANK, JOHANNES WALTER, chemical engineering educator; b. Zams, Tyrol, Austria, July 6, 1950; came to U.S., 1978; s. Friedrich Karl and Johanna (Ruepp) S.; m. Lynne Violet Duguay; children: Alexander Johann, Leonard Friedrich. Diploma in chemistry, U. Innsbruck, Austria, 1975, PhD, 1978. Mem. faculty U. Mich., Ann Arbor, 1978—, assoc. prof. chem. engring., 1984-90, acting dir. Ctr. for Catalysis and Surface Sci., 1985-90, prof., interim chmn. dept. chem. engring., 1990-91, assoc. dir. Electron Microbeam Analysis Lab., 1990-91; chmn. dept. chem. engring., 1991—; vis. prof. U. Innsbruck, 1987-88, Tech. U. Vienna, 1988; cons. in field. Patentee bimetallic cluster catalysts, hydrodesulfurization catalysts and microelectronic gas sensors; contbr. over 70 articles to sci. jours. Fulbright-Hays scholar, 1978. Mem. AAAS, Am. Chem. Soc., Am. Inst. Chem. Engrs., Mich. Catalysis Soc. (sec.-treas. 1982-83, v.p. 1983-84, pres. 1984-85), Am. Soc. Engring. Edn. Home: 2335 Placid Way Ann Arbor MI 48105-1205 Office: U Mich Dept Chem Engring 2300 Hayward St Ann Arbor MI 48109-2136

SCHWARTZ, DAVID ALAN, infectious disease and placental pathologist; b. Phila., May 20, 1953; s. Harold Martin and Thelma (Bell) S; m. Stephanie Baker, May 16, 1993. BA, U. Pitts., 1974, MS in Hygiene, 1977; D in Medicine, Far Eastern U., Manila, The Philippines, 1984. Chief resident Hahnemann U. Sch. Medicine, Phila., 1987-88; instr. Harvard U. Sch. Medicine, Boston, 1988-89; asst. prof. Emory U. Sch. Medicine, Atlanta, 1989—; rsch. scientist Ctrs. for Disease Control, Atlanta, 1992—; vis. prof. Univ. Mayor San Simon, Bolivia, 1993—; cons. in AIDS Ctrs. for Disease Control, Atlanta, 1992—; cons. in chagas disease U.S. Agy. for Internat. Devel., Washington, Bolivian Min. Health, La Paz, 1992—; chmn. path subcom. WITS, NIH; pathology cons. Bangkok-CDC HIV study, 1993—. Editor: Diagnostic Pathology of Infectious Diseases; contbr. articles to profl. jours. Del. AMA, 1985—. Recipient Pathology Resident Rsch. award Am. Soc. Clin. Pathologists, 1985; AmFAR Pediatric AIDS Found. scholar, 1993—; Syphilis Rsch. grantee Ctrs. for Disease Control, 1991, Placental Infections Rsch. grantee Emory Med. Care Found., 1991—, Placental HIV Rsch. grantee NIH, 1991—. Fellow Coll. Am. Pathologists, Assn. Clin. Scientist, Sigma Xi; mem. Am. Soc. Tropical Medicine. Jewish. Achievements include characterization of pathologic features of human microsporidiosis; development of new diagnostic methods for pathologic identification of infectious agts.; description of molecular pathologic techniques for identification of viruses and parasites in the placenta; evaluation of mechanisms of transplacental infection utilizing placental perfusion, chorionic villus culture and placental cell cultures; identification of new opportunistic infections in AIDS patients. Office: Grady Meml Hosp 80 Butler St SE Atlanta GA 30335

SCHWARTZ, DREW, geneticist, educator; b. Phila., Nov. 15, 1919; s. Isaac and Miriam (Bonn) S.; m. Pearl Freeman, July 26, 1942 (dec. May 1985); 1 child, Rena Ann. BA, Pa. State Coll., 1942; MA, Columbia U., 1948, PhD, 1951. Sr. scientist Oak Ridge (Tenn.) Nat. Lab., 1951-62; prof. Western Res. U., Cleve., 1962-64; prof. Ind. U., Bloomington, 1964-91, prof. Emeritus, 1991—. With U.S. Army, 1943-46. Mem. AAAS, Genetics Soc. Am. Home: 4001 Saratoga Dr Bloomington IN 47408 Office: Ind U Bloomington IN 47405

SCHWARTZ, ELIAS, pediatrician; b. N.Y.C., Aug. 30, 1935; s. Rubin and Dusha (Premysler) S.; m. Esta Rosenberg, June 12, 1960; children: Samuel, Robert. AB, Columbia Coll., 1956; MD, Columbia U., 1960; MA (hon.), U. Pa., 1972. From asst. prof. to assoc. prof. pediatrics Jefferson Med. Coll., Phila., 1967-72; prof. pediatrics Sch. Medicine U. Pa., Phila., 1972—, prof. pediatrics in human genetics Sch. Medicine, 1979—, chmn. dept. pediatrics Sch. Medicine, 1990—; physician-in-chief Children's Hosp. Phila., 1990—; med. adv. bd. Cooley's Anemia Found., N.Y.C., 1976—; policy bd. sickle cell program NIH, Bethesda, Md., 1987—, reviewer's res., 1991—; sci. adv. bd. Enzon, Inc., South Plainfield, N.J., 1992—. Editor Hemoglobinopathies in Children, 1980; contbr. 190 articles to med. and sci. jours. Active Phila. Healthcare Congress, 1991—. Capt. USAF, 1963-65. Grantee NIH, 1968—. Mem. Am. Soc. Pediatric Hematology/Oncology (pres. 1989-91), Am. Soc. Clin. Investigation, Assn. Am. Physicians, Am. Pediatric Soc., Soc. for Pediatric Rsch. Achievements include rsch. in transfusion treatment of strokes in sickle cell disease, comprehensive program for removal of excess iron in chronically transfused patients, delineation of abnormalities of globin synthesis in several inherited hemoglobin disorders, discovery of genetic defects in several types of thalassemia, cloning and analysis of several human megakaryocyte genes. Office: Children's Hosp Phila 34th St & Civic Center Blvd Philadelphia PA 19104

SCHWARTZ, GEORGE R., physician; b. Caribou, Maine; m. Kathleen Schwartz; children: Ruth, Rebekah, Rachel, Moses, Abigail, John Gabriel. BS in Chemistry with honors, Hobart Coll., 1963; MD magna cum laude, SUNY, Bklyn., 1967. Diplomate Am. Bd. Family Practice, Am. Bd. Emergency Medicine; cert. CPR instr. Intern King County Hosp., Seattle, 1967-68; instr. dept. medicine U. Wash., Seattle, 1967-68; resident in psychiatry Hillside Hosp., Glen Oaks, N.Y., 1968-69; resident in surgery Ind. U. Med. Ctr., Indpls., 1971-72; instr. emergency medicine Med. Coll. Pa., Phila., 1972-76, dir. emergency svcs., asst. dir. emergency medicine program, 1972-74; dir. emergency medicine West Jersey Hosp., 1974-76; pvt. practice, 1977; assoc. prof., dir. divsn. emergency medicine U. N.Mex., Albuquerque, 1978-83; staff mem. emergency medicine Heights Gen. Hosp., Albuquerque, 1983-85; with Los Alamos (N.Mex.) Med. Ctr., 1985-90; vis. assoc. prof. Med. Coll. Pa., 1991. Author: Geriatric Emergencies, 1984; Co-author: (with Tandberg) Emergency Medicine Continuing Edn. Rev., 1981, 2d edit. 1984, (with Bosker) Geriatric Emergency Medicine, 1990; editor: Principles and Practice Emergency Medicine, 1978, 3d edit., 1992; co-editor Trauma Rounds, 1973-75; editorial bd. Annals Emergency Medicine, 1972-81, Resident and Staff Physician, 1978—, Emergency Med. Abstractrs, 1978-85, Med. Exam. Publ. Co., 1981-87; contbr. articles to profl. jours., chpts. to textbooks. Med. dir. The Bridge Counseling Ctrs., Los Alamos, N.Mex., 1988—; dir. planning com. disaster exercise Phila. Internat. Airport, 1974, Camden County Poison Ctr., 1974-76. Recipient Gallup award, 1973, Giraffe award, 1990. Mem. AAAS, AMA, Am. Coll. Emergency Physicians (pres. N.Mex. chpt. 1980-81), N.Mex. Med. Soc., Univ. Assn. Emergency Physicians (chmn. socio-econ. com. 1976-77), Internat. Emergency Care Assn., Am. Trauma Soc. (founding mem.), Am. Acad. Clin. Toxicology, Internat. Assn. for Study of MSG and Food Additives (pres. 1988). Achievements include patent for use of a pharmaceutical agent in male impotence; research on computer applications in medicine, new medical diagnostic instruments. Address: 257 Hyde Park Estates Santa Fe NM 87501

SCHWARTZ, GORDON FRANCIS, surgeon, educator; b. Plainfield, N.J., Apr. 29, 1935; s. Samuel H. and Mary (Adelman) S.; m. Rochelle DeG. Krantz, Sept. 5, 1959; children—Amory Blair, Susan Leslie. A.B., Princeton U., 1956; M.D., Harvard U., 1960; MBA, U. Pa., 1990. Intern N.Y. Hosp.-Cornell Med. Ctr., N.Y.C., 1960-61; resident in surgery Columbia-Presbyterian Med. Ctr., N.Y.C., 1963-68; instr. surgery Columbia U., N.Y.C., 1966-68; assoc. in surgery U. Pa., Phila., 1968-70; dir. clin. services Breast Diagnostic Ctr. Jefferson Med. Coll., 1973-78, asst. prof. surgery, 1970-71, assoc. prof., 1971-78, prof., 1978—; practice medicine specializing in surgery and diseases of breast, Phila., 1968—; founder, chmn. acad. com. Breast Health Inst., 1990—. Editor: Cancer Therapy Abstracts, 1976-80; author: (with R.H. Guthrie Jr.) Reconstructive and Aesthetic Mammoplasty, 1989; co-editor: Breast Disease, an Internat. Jour., 1987—. Mem. Pa. Gov.'s Task Force on Cancer, 1976-82; mem. breast cancer task force Phila. chpt. Am. Cancer Soc.; mem. clin. investigation rev. com. Nat. Cancer Inst., 1992—. Served to capt. AUS, 1961-63. NIH Cancer Control fellow, 1968-69. Mem. ACS, Assn. for Acad. Surgery, Allen O. Whipple Surg. Assn., Soc. Surg. Oncology, Internat. Cardiovascular Soc., Soc. for Surgery Alimentary Tract, Am. Soc. Clin. Oncology, AMA, Soc. for Study of Breast Diseases (pres. 1981-83), Soc. Internat. Senologie (treas. 1982-90, v.p. 1990-92, sci. comm. 1992—), Philadelphia County Med. Soc., Italian Soc. Senology (hon.), Pa. Med. Soc., Am. Soc. Transplant Surgeons, N.Y. Acad. Scis., AAUP, Am. Soc. Artificial Internal Organs, Am. Radium Soc., N.Y. Met. Breast Cancer Group, Union League, Locust Club (Phila.), Princeton Club (pres. Phila. 1989-91), Princeton Club (N.Y.C.), Princeton Terrace Club, Nassau Club, Phi Beta Kappa, Sigma Xi, Alpha Omega Alpha, Nu Sigma Nu. Republican. Jewish. Home: 1805 Delancey Pl Philadelphia PA 19103-6606 Office: Ste 510 1015 Chestnut St Philadelphia PA 19107

SCHWARTZ, JOEL, epidemiologist; b. N.Y.C., Dec. 12, 1947; s. Theodore and Gertrude (Greenbaum) S.; m. Ronnie Levin, Feb. 17, 1985; 1 child, Yuri Levin-Schwartz. BA, Brandeis U., 1969, PhD, 1980. Economist U.S. EPA, Washington, 1979-84; vis. scientist Harvard Sch. Pub. Health, Boston, 1987-88, U. Basel, Switzerland, 1989, U. Wuppertal, Germany, 1990; lectr. Harvard Sch. Pub. Health, Boston, 1991—; epidemiologist U.S. EPA, Washington, 19856; mem. NAS Com. on Lead, 1989—; mem. NAS Com. on Environ. Epidemiology, 1990—. Author: Costs and Benefits of Reducing Lead in Gasoline, 1985. MacArthur Found. fellow, Chgo., 1991; recipient Silver medal U.S. EPA, Washington, 1984, 86, Sci. Achievement award, 1989, 90, 92. Mem. Am. Statis. Assn., Am. Thoracic Soc., Soc. Epidemiologic Rsch. Democrat. Jewish. Achievements include research in getting lead out of gasoline, showing lead increase blood pressure, associating air pollution with daily mortality rate. Home: 1207 4th St SW Washington DC 20024 Office: US EPA PM221 401 M St SW Washington DC 20460

SCHWARTZ, JOHN HOWARD, poultry science and extension educator; b. Gettysburg, Pa., Nov. 26, 1948; s. John W. and Stella B. (Brown) S.; m. Kathryn A. Bentz, Sept. 21, 1986; children: Kathryn A., John L. BS, Pa. State U., 1970, MEd, 1975; PhD, Colo. State U., 1985. 4-H extn. agt. Pa. State U., West Chester, 1970-75; agrl. extn. agt. Pa. State U., Gettysburg, 1975-83; rsch. asst. Colo. State U., Ft. Collins, 1983-84; exec. v.p. Colo. Quality Rsch., Ft. Collins, 1984-85; asst. prof. Clemson (S.C.) U., 1986-88; poultry extn. agt. Pa. State U., Lancaster, 1988-91, county extension dir., 1991—; assoc. Thurmond Inst., Clemson, 1987. Mem. Nat. Assn. County Agrl. Agts. (Achievement award 1979), Poultry Sci. Assn., Coun. Agrl. Sci. and Tech., Pa. Poultry Fedn., MAsons, Phi Kappa Phi, Gamma Sigma Delta, Alpha Tau Alpha. Home: 2560 Lori Dr York PA 17404-1261 Office: Pa State U 1383 Arcadia Rd Rm 1 Lancaster PA 17601-3149

SCHWARTZ, LAWRENCE, aeronautical engineer; b. N.Y.C., Nov. 30, 1935; s. Harry and Fanny (Steiner) S.; m. Cherie Anne Karo, Aug. 12, 1979; children: Ronda, Daran. SB in Aero. Engring., MIT, 1958, SM in Aero. Engring., 1958; postgrad. Ohio State U., 1960, U. Dayton, 1962-63; PhD in Engring., UCLA, 1966. Electronics design engr. MIT. Instrumentation Lab., Cambridge, 1959; aerospace engr., Wright-Patterson AFB, Ohio, 1962-63; mem. tech. staff Hughes Aircraft Co., Culver City, Calif., 1963-65, staff engr., 1965-67, sr. staff engr., 1967-72, sr. scientist, 1972-79, chief scientist lab., 1979-93, tech. mgr. 1985-87; chmn., tech. adv. bd., 1987-88, prin. scientist/engr. 1993—; cons., tchr. in field. With USAF, 1959-62. Registered profl. engr., Colo., Calif. Mem. IEEE, AAAS, Sigma Xi, Sigma Gamma Tau, Tau Beta Pi. Contbr. articles to profl. jours. Home: 996 S Florence St Denver CO 80231-1952 Office: 16800 E Centretech Pky Aurora CO 80011-9046

SCHWARTZ, LYLE H., materials scientist, government official; b. Chgo., Aug. 2, 1936; s. Joseph K. Schwartz and Helen (Shefsky) Bernards; divorced; children—Ara, Justin; m. Celesta Sue Jurkovich, Sept. 1, 1973. B.S. in Sci.

Engring., Northwestern U., 1959, Ph.D. in Materials Sci., 1964. Prof. materials sci. Northwestern U., Evanston, Ill., 1964-84, dir. Materials Research Ctr., 1979-84; dir. materials sci. and engr. lab. Nat. Inst. Standards and Tech., Dept. Commerce, Gaithersburg, Md., 1984—; cons. Argonne Nat. Labs., Ill., 1965-79; vis. scientist Bell Telephone Labs., Murray Hill, N.J., 1971-73. Author: (with J.B. Cohen) Diffraction From Materials, 1977, 2d edit., 1987; also numerous articles and papers. NSF fellow, 1962-63; recipient Presdl. Rank Award of Meritorious Exec. for outstanding govt. svc., 1990. Fellow Am. Soc. for Metals; mem. AAAS, AIME, Am. Phys. Soc., Am. Crystallography Assn., Materials Rsch. Soc., Sigma Xi. Office: Nat Inst Standards and Tech Materials Bldg Rte 270 Gaithersburg MD 20899

SCHWARTZ, MARK WILLIAM, ecologist; b. Tulsa, Okla., July 20, 1958; s. Gaylord Paul Schwartz and Judy Granahan; m. Sharon Yehudit Strauss, Sept. 23, 1986; children: Eli Daniel Strauss, Ari Calé Schwartz. MS, U. Minn., 1986; PhD, Fla. State U., 1990. Regional land steward The Nature Conservancy, Tallahassee, Fla., 1988-90; conservation biologist Ill. Natural History Survey, Champaign, 1990—. Mem. AAAS, Soc. Conservation Biology, Ecol. Soc. Am. Achievements include research in population biology of rare plants, conservation of biodiversity through natural lands management, global climate change and biodiversity. Office: Ill Natural History Survey 607 E Peabody Dr Champaign IL 61820-6970

SCHWARTZ, MELVIN, physics educator, laboratory administrator; b. N.Y.C., Nov. 2, 1932; s. Harry and Hannah (Shulman) S.; m. Marilyn Fenster, Nov. 25, 1953; children: David N., Diane R., Betty Lynn. A.B., Columbia U., 1953, Ph.D., 1958, DSc honoris causa, 1991. Assoc. physicist Brookhaven Nat. Lab., 1956-58; mem. faculty Columbia U., N.Y.C., 1958-66; prof. physics Columbia U., 1963-66, Stanford U., Calif., 1966-83; cons. prof. Stanford U., 1983-91; chmn. Digital Pathways, Inc., Mountain View, Calif., 1970-91; assoc. dir. high energy and nuclear physics Brookhaven Nat. Lab., Upton, N.Y., 1991—; prof. physics Columbia U., N.Y.C., 1991—. Co-discoverer muon neutrino, 1962. Bd. govs. Weizmann Inst. Sci. Recipient Nobel prize in physics, 1988, John Jay award Columbia Coll., 1989; Guggenheim fellow, 1968. Fellow Am. Phys. Soc. (Hughes award 1964); mem. NAS. Home: 61 S Howells Point Rd Bellport NY 11713-2621 Office: Brookhaven Nat Lab Bldg 510F Upton NY 11973

SCHWARTZ, RICHARD JOHN, electrical engineering educator, researcher; b. Waukesha, Wis., Aug. 12, 1935; s. Sylvester John and LaVerne Mary (Lepien) S.; m. Mary Jo Collins, June 29, 1957; children: Richard, Stephan, Susan, Elizabeth, Barbara, Peter, Christopher, Margaret. BSEE, U. Wis., 1957; SM, MIT, 1959, ScD, 1962. Mem. tech. staff Sarnoff Rsch. Labs. RCA, Princeton, N.J., 1957-58; instr. MIT, Cambridge, 1961-62; v.p. Energy Conversions, Inc., Cambridge, 1962-64; assoc. prof. Purdue U., West Lafayette, Ind., 1964-71, prof., 1972—, head dept., 1985—, dir. Optoelectronic Ctr., 1986-89; bd. dir. Nat. Elec. Engring. Dept. Heads Assn.; solar cells cons., 1965—. Contbr. chpts. to books, articles to profl. jours. Served to 2nd lt. U.S. Army, 1957-58. Recipient Disting. Svc. medal U. Wis., 1989, Centennial medal, 1991. Fellow IEEE. Achievements include development of high intensity solar cells, of surface charge transfer device, and of numerical models for solar cells. Office: Purdue U 1285 Elec Engring Elec Engring Rsch Lab West Lafayette IN 47907

SCHWARTZ, ROBERT DAVID, fermentation microbiologist, bioengineer; b. Bklyn., Apr. 3, 1941; s. Solomon and Esther Hanna (Sachs) S.; m. Maxine Finkelstein, June 21, 1964 (div. 1987); children: Jeff, Scott; m. Ellen Marie Johnson, Feb. 7, 1992; children: Adam, Shawn, Peyton. BS, Bklyn. Coll., 1964; MS, Long Island U., 1967; PhD, Rutgers U., 1969. Rsch. scientist Exxon Rsch. & Engring. Co., Linden, N.J., 1970-76; project scientist Union Carbide Corp., South Charleston, W.Va., 1976-79; rsch. assoc. Stauffer Chem. Co., Richmond, Calif., 1979-87; sr. devel. scientist Abbott Laboratories, North Chicago, Ill., 1987—; adv. bd. Enzyme and Microbiology Tech., 1989-93, assoc. editor, 1993—. Mem. editorial bd. Applied Environ. Microbiology, 1980-85, Jour. Indsl. Microbiology, 1986—; contbr. articles to Applied Environ. Microbiology, Applied Microbiology Biotech. Mem. AAAS, Soc. for Indsl Microbiology (pres. 1991—, bd. dirs. 1983-86, Charles Porter award 1989), Am. Soc. for Microbiology, Am. Chem. Soc., Sigma Xi, Phi Sigma. Democrat. Jewish. Achievements include patents in field; commercialized inventions; development of fermentation processes based on dairy solids for products used in food industry, for specialty and commodity chemicals, for biopesticide fermentation devel. and process support. Office: Abbott Labs Fermentation Devel D451 North Chicago IL 60064

SCHWARTZ, RODNEY JAY, civil, mechanical and electrical engineer; b. Lincoln, Nebr., Sept. 9, 1944; s. Rodney John and Shirley Helen (Kennedy) S.; m. Donna Jean Ideus, July 4, 1969; children: Tara Noel, Theron Jay. BSME, U. Nebr., 1969; postgrad., Air Inst. Tech., 1978, 87, 93; MPA, U. Nebr., Omaha, 1988. Registered profl. engr., Ohio, Nebr.; cert. bldg. official. Dep. city engr. City of Lincoln, 1976-81, supt. codes, 1981-88, chief bldg. inspector, 1988-89, environ. engr., 1989; program mgr. U.S. Army Corps Engrs., Omaha, 1989-90, special asst. to chief engring. div., 1990-93; program mgr. EPA Superfund, 1993—; mem. Nebr. Gov.'s Task Force for State Bldg. Code, 1986-87. Mem. ednl. adv. com. Southeast C.C., Lincoln, 1985-88; auction co-chair 17th Ann. Cerebral Palsy Auction, Lincoln, 1990; chmn. 18th Ann. Cerebral Palsy Auction, Lincoln, 1991; team leader Combined Fed. Campaign, 1992. Col. USAFR, 1969—. Recipient Air Force Commendation medal, 1988, Air Force Individual Achievement medal, 1986, Air Force Meritorious Svc. medal, 1991, Engring. Excellence award Nat. Coun. Cons. Engrs., 1978; named to Citizen Amb. Program, People's Republic of China, 1987, 90, Australia and New Zealand, 1991. Mem. ASCE, Nebraskaland Conf. Bldg. Officials (past bd. dirs., sec.-treas. 1987-88, v.p. 1988-89, pres. 1989-90), Nat. Soc. Profl. Engrs., Lincoln Engrs. Club, Am. Pub. Works Assn. (past chmn. Nebr. chpt. coms., past bd. dirs.), Am. Soc. Heating, Refridgeration and Air Conditioning (mem. grade), Am. Soc. Mil. Engrs. (program chmn. 1993, mem. exec. bd.), Res. Officers Assn. (life), Air Force Assn. (life). Achievements include co-development of fastening method to allow precast concrete boxes to transmit rail car loading at joints, laser-activated construction equipment remotely operated by rotating laser and HUD loan program for low income housing; design of hardened aircraft shelter, for alert aircraft. Home: 1010 Anthony Ln Lincoln NE 68520 Office: US Army Corps of Engrs 215 N 17th St Omaha NE 68102-4910

SCHWARTZ, TERI J(EAN), clinical psychologist; b. N.Y.C., Dec. 30, 1949; d. Jerome and Shirley Ruth (Dushkind) Kraus; m. Raymond C. Schwartz; children: Rachel, Michael, Daniel. BA, Queens Coll., 1971; MS, C.W. Post Ctr., 1974; MA, New Sch. for Social Rsch., 1977, PhD, 1980. Staff psychologist to chief psychologist New Hope Guild, Howard Beach, N.Y., 1982-85, 85—; staff psychologist Queens Child Guidance Ctr., Flushing, N.Y., 1985-88; staff psychotherapist Adelphi Univ.-Postdoctoral Psychotherapy Ctr., Garden City, N.Y., 1984—; pvt. practice, clin. psychologist Briarwood and Floral Park, N.Y., 1984—; adj. clin. supr. Yeshiva Univ., Bronx, 1987—; clin. supr. Adelphi Univ., Garden City, 1991—, assoc. clin. prof., 1992—. Mem. exec. bd. Briarwood (N.Y.) Civic Assn., 1984—; adv. bd. Queens Community Mental Health Ctr. and Area D Subcom.-Queens Hosp. Ctr., Jamaica, N.Y., 1984—. Mem. Am. Psychol. Assn., N.Y. State Psychol. Assn., Adelphi Soc. for Psychoanalysis and Psychotherapy, Nassau County Psychol. Assn., Queens County Psychol. Assn. Avocations: reading, gourmet cooking. Office: 1 Holland Ave Floral Park NY 11001

SCHWARTZBERG, JOANNE GILBERT, physician; b. Boston, Nov. 30, 1933; d. Richard Vincent and Emma (Cohen) Gilbert; m. Hugh Joel Schwartzberg, July 7, 1956; children: Steven Jonathan, Susan Jennifer. BA magna cum laude, Radcliffe Coll., 1955; MD, Northwestern U., 1960. Diplomate Am. Bd. Quality Assurance and Utilization Rev. Physicians. Founder, med. dir. Chgo. Home Health Svc., 1972—; founder, bd. dirs., v.p., med. dir. Suburban Home Health Service, Chgo. area, 1975-87; clin. asst. prof. preventive medicine and community health U. Ill. Coll. Medicine, 1985—; dir. Dept. of Geriatrics Healh AMA, 1990—. Mem. Health Planning Commn. Chgo., 1961-63; mem. Community Adv. Bd. Joint Youth Devel. Commn. Chgo., 1963-67; pres. Near North Montessori Sch., Chgo., 1972-75, bd. dirs., 1970—. Recipient Mayor's Citation City of Chgo., 1963. Fellow Inst. Medicine of Chgo. (bd. dirs. 1990—, pres.-elect 1993), Am. Acad. of Home Care Physicians (pres., Physician of Yr. award 1992), Am. Coll. Utilization Rev. Physicians; Ill. Geriatrics Soc. (pres. 1990-92), Chgo. Geriatrics Soc. (pres. 1990-92), Ill. Med. Soc., Chgo. Med. Soc., Am. Acad. Med. Dirs., Am. Geriatrics Soc., Ill. Geriatrics Soc. (founding dir. 1984, pres. 1990-92), Am. Med. Women's Assn., Am. Pub. Health Assn., Ill. Pub. Health Assn., Alexander Graham Bell Assn. for Deaf (bd. dirs. 1984-90; chmn. internat. parents orgn. 1988-90, gen. chmn. internat. conv. 1986). Editor. Contbr. articles to profl. jours. Home: 853 W Fullerton Ave Chicago IL 60614-2412 Office: 515 N State St Chicago IL 60610

SCHWARTZBERG, LEO MARK, civil engineer, consultant; b. Miami, Fla., May 18, 1953; s. David and Mildred (Schauben) S.; m. Rhea Lynn Zirkes, Sept. 5, 1976; children: Hanan Yakov, Dina Yael, Shira Tamar. AS in Civil Engring. Tech., Miami Dade C.C., 1975; BS in Civil Engring. Tech., Fla. Internat. U., 1976. Registered profl. engr., Fla. Civil engr. Fla. Dept. Transp., Ft. Lauderdale, Fla., 1976-79; project engr. Robert H. Miller & Assocs. Inc., Davie, Fla., 1979-81; project mgr. Robert H. Miller & Assocs. Inc., Davie, 1981-83, chief of design, 1983-85, dir. ops., 1985-87; v.p. ops. Robert H. Miller & Assocs. Inc., Pembroke Pines, Fla., 1987-89; v.p. bus. devel. Robert H. Miller & Assocs. Inc., Pembroke Pines, 1989—; lectr. Fla. Bar Assn., 1990, 92, South Fla. sect. ASCE, 1985, 92. Mem. bd. govs. Samuel Scheck Hillel Community Day Sch., North Miami Beach, Fla., 1986—. Mem. ASCE (Young Engr. of Yr. Broward, Fla. br. 1988), Fla. Engring. Soc. (Young Engr. of Yr. Broward chpt. 1987), Fla. Assn. Environ. Profls., Builders Assn. South Fla. (bd. dirs., Pres. award 1988). Jewish. Office: Robert H Miller & Assocs Inc 1800 N Douglas Rd Ste 200 Pembroke Pines FL 33024

SCHWARTZBERG, ROGER KERRY, osteopath, internist; b. Bklyn., Mar. 30, 1948; s. Erwin and Edna (Kuchlik) S.; m. Linda Lurie, July 1, 1972 (div. Nov. 1974); m. Vicki Ann Davis, Nov. 28, 1976; children: Jeremy Dylan, Joshua Ryan. BA in Psychology, Syracuse U., 1970; DO, Mich. State U., 1973. Diplomate Am. Acad. Osteopathic Internists. Intern, sr. asst. surgeon USPHS Hosp., S.I., N.Y., 1973-74; med. resident Southeastern Med. Ctr., North Miami Beach, Fla., 1974-77, chief resident, 1975-77; pvt. practice Seminole, Fla., 1977—; active staff Univ. Gen. Hosp., Seminole 1978—, chmn. dept. internal medicine, 1981-82, governing bd. 1981-86, 88—, vice chmn. 1986, 88—, Met. Gen. Hosp., Pinellas Park, Fla., 1980—, chief of staff, 1985-86, Women's Med. Ctr., Seminole, 1989—; adj. clin. faculty U. New Eng. Coll. Osteo. Medicine, Biddeford, Maine, 1985—; clin. asst. prof. internal medicine Southeastern U. of Health Scis., 1987—; clin. asst. prof. internal medicine Kirksville (Mo.) Coll. Osteo. Medicine, 1989—; mem. cons. staff Horizon Hosp., Clearwater, 1980—; mem. affiliate staff Suncoast Osteo. Hosp., Largo, Fla., 1982—; mem. active staff Largo Med. Ctr. Hosp., 1990—, St. Petersburg Gen. Hosp., 1991—; clin. assoc. prof. coll. osteo. medicine U. Health Scis., 1992—. Named Educator of Yr. Met. Gen. Hosp., Pinellas Park, Fla., 1985. Fellow Am. Acad. Osteo. Internists (certification bd. 1990—); mem. Am. Osteo. Assn., Am. Assn. Osteo. Specialists, Am. Coll. Osteo. Internists, Fla. Osteo. Med. Assn. (trustee 1985-89), Pinellas County Osteo. Med. Soc. (trustee 1985-89, gov. 1989-90, v.p. 1991—). Jewish. Avocations: photographer, singing, keyboards. Office: Oakhurst Med Clinic 13020 Park Blvd Seminole FL 34646-9999

SCHWARZ, FERDINAND (FRED SCHWARZ), ophthalmologist, ophthalmic plastic surgeon; b. Trenton, N.J., Dec. 13, 1939; s. Ferdinand and Laura Francis Schwarz; m. Carol Ann Snyder, Feb. 26, 1966; children: Lesley Ann, Jeffrey Ryan, Jason Bradley, Allyson Larner. BSEE, Lafayette Coll., 1961; MD, N.J. Coll. Medicine and Dentistry, 1970; MBA, Winthrop Coll., 1991. Diplomate Nat. Bd. Med. Examiners. Project engr. Astro div. RCA, Princeton, N.J., 1961-66; intern Robert Packer Hosp., Sayre, Pa., 1970-71; resident Guthrie Clinic Ophthalmology/Robert Packer Hosp., Sayre, Pa., 1971-74; Heed fellowship in ophthalmic plastic and reconstructive surgery Hahnemann Med. Coll. and Temple Med. Sch. U. Pa., Phila., 1974-75; pvt. practice Columbia, S.C., 1975—; vice chief med. staff Providence Hosp., Columbia, 1984, chief med. staff, 1985; cons. Dorn VA Hosp., Columbia, 1975—, Moncrief Hosp., Columbia, 1975-89. Contbr. numerous articles to sci. engring. and med. publs., 1962—. Heed (Found.) Ophthalmic fellow, Chgo., 1975. Mem. AMA, S.C. Med. Soc., S.C. Ophthalmology Soc., Cen. S.C. Ophthalmology Soc. (pres. 1983), Palmetto Club, Spring Valley Country Club. Roman Catholic. Avocations: tennis, skiing, model trains, computers. Home: PO Box 23038 Columbia SC 29224-3038 Office: Drs Schwarz and Milne PA 1655 Brabham Ave Ste 100 Columbia SC 29204-2023

SCHWARZ, JOHN HENRY, theoretical physicist, educator; b. North Adams, Mass., Nov. 22, 1941; s. George and Madeleine (Haberfeld) S.; m. Patricia Margaret Moyle, July 11, 1986. AB, Harvard U., 1962; PhD, U. Calif., Berkeley, 1966. Instr. physics Princeton (N.J.) U., 1966-69, asst. prof., 1969-72; research assoc. Calif. Inst. Tech., Pasadena, 1972-85, prof. theoretical physics, 1985—, Harold Brown prof. theoretical physics, 1989—. Co-author: Superstring Theory, 1987. Trustee Aspen (Colo.) Ctr. for Physics, 1982—. Guggenheim fellow, 1978-79, MacArthur Found. fellow, 1987, Guggenheim fellow; recipient Dirac medal Internat. Centre for Theoretical Physics, 1989. Fellow Am. Phys. Soc., Phi Beta Kappa (vis. scholar 1990—). Office: Calif Inst Tech 452-58 Pasadena CA 91125

SCHWARZ, RICHARD HOWARD, obstetrician, gynecologist, educator; b. Easton, Pa., Jan. 10, 1931; s. Howard Eugene and Blanche Elizabeth (Smith) S.; m. Patricia Marie Lewis, Mar. 11, 1978; children by previous marriage: Martha L., Nancy Schwarz Tedesco, Paul H., Mary Katherine Schwarz Murray. MD, Jefferson Med. Coll., 1955; MA (hon.), U. Pa., Phila., 1971. Diplomate Am. Bd. Ob-Gyn. (examiner 1977—). Intern, then resident Phila. Gen. Hosp., 1955-59; prof. U. Pa., Phila., 1963-78; prof., chmn. Downstate Med. Ctr., Bklyn., 1978-90, dean, v.p. acad. affairs, 1983-89, provost, v.p. clin. affairs, 1988—, interim pres., 1993—. Author: Septic Abortion, 1968. Editor: Handbook of Obstetric Emergencies, 1984, mem. editorial bd. jour. Ob-Gyn., Milw., 1983-87; contbr. articles to profl. jours. Bd. dirs. March of Dimes, N.Y.C., 1985—. Capt. USAF, 1959-63. Mem. Am. Coll. Obstetricians and Gynecologists (chmn. dist. 2 1984-87, v.p. 1989-90, pres. elect 1990-91, pres. 1991-92). Republican. Presbyterian. Office: SUNY Health Sci Ctr 450 Clarkson Ave Brooklyn NY 11203-2098

SCHWARZ, ROBERT CHARLES, aerospace engineer, consultant; b. N.Y.C., Feb. 5, 1940; s. Charles Nicholas and Charlotte Agnes (Gunther) S.; m. Regina Neves, May 26, 1966 (div. 1987); children: Maria, Richard; m. Christine Bernhard, Nov. 23, 1990. B of Aerospace Engring., Poly. Inst., Bklyn., 1961; MS in Applied Mechanics, Poly. Inst., N.Y.C., 1970, MS in Ops. Rsch., 1978. Registered profl. engr., N.Y. Engr. Curtiss-Wright Corp., Woodridge, N.J., 1963-67, Grumman Aircraft Corp., Bethpage, N.Y., 1967—. Head coach Cath. Youth Orgn., Kings Park, N.Y., 1984. 1st lt. U.S. Army, 1961-63. Mem. Soc. for Exptl. Mechanics (chmn. applications com. 1989-91, tech. editor Exptl. Techniques 1991—). Achievements include patents in field of experimental stress analysis; patent pending for coldworked holes by caustics. Home: 5 Roslyn St Huntington NY 11743-4951

SCHWARZ, SAUL SAMUEL, neurosurgeon; b. Newark, Jan. 6, 1954; s. Martin Allen and Phyllis (Burdeau) S.; m. Julie Brief, Nov. 23, 1977; children; Tara Frances, Jesse Ira. BA, Princeton U., 1976; MD, Albert Einstein Coll., 1981. Diplomate Am. Bd. Neurol. Surgery. Intern Nat. Naval Med. Ctr, Bethesda, Md., 1981-82, resident in neurosurgery, 1983-88; head dept. neurosurgery Naval Hosp., Oakland, Calif., 1988-90; chief neurosurgery svc. Army Med. Ctr., Hawaii, 1990-92; neurosurgeon Johnson Neurol. Clinic, High Point, N.C., 1992—. Contbr. articles to profl. jours. Fellow Am. Coll. Surgeons; mem. Am. Assn. Neurol. Surgeons, Congress of Neurol. Surgeons. Office: Johnson Neurol Clinic 606 N Elm St High Point NC 27262

SCHWARZ, STEVEN ALLAN, electrical engineer; b. Chgo., Sept. 15, 1953; s. David and Beverly (Rothman) S.; m. Janet Bauer, Nov. 1, 1981; children: Daniel, Michael, Jonathan. BSEE, Mich. State U., 1974, MSEE, 1975; PhD in Elec. Engring., Stanford U., 1980. Mem. tech. staff AT&T Bell Labs., Murray Hill, N.J., 1980-83, Bellcore, Red Bank, N.J., 1984—; adj. prof. Queens Coll. Flushing, N.Y., 1991-92, SUNY, Stony Brook, 1993—. Contbr. articles to profl. jours. Mem. IEEE (sr.), Am. Vacuum Soc., Materials Rsch. Soc. Achievements include secondary ion mass spectrometry applied to optoelectronic materials; models of electron mobility; sputtering and secondary electron emission. Office: Bellcore Rm 3Z281 Red Bank NJ 07701-7040

SCHWARZBEK, STEPHEN MARK, physicist; b. Defiance, Ohio, Aug. 12, 1961; s. Robert Elmer and Donna Belle (Shinners) S. BS in Physics, Mich. State U., 1983; PhD in Physics, U. Notre Dame, 1990. Tech. staff TRW Superconductivity Area, Redondo Beach, Calif., 1989—. Monitor, Recording for the blind, Inc., South Bay, 1992—; poll clk. Election Bd., County of L.A., 1992—. Recipient Chmn.'s award/Innovation, TRW. Mem. AAAS, Am. Phys. Soc., Am. Statis. Assn. Achievements include demonstration of working logic gates based on High Temperature Superconducting materials. Office: TRW Space & Electronics R1/2136 One Space Pk Redondo Beach CA 90278

SCHWARZE, ROBERT FRANCIS, osteopath, dermatologist; b. St. Louis, Aug. 13, 1949; s. William Casper and Mary Constance (Glaser) S.; m. Donna Lea Jakubiak, Nov. 3, 1990. BS, Maryville Coll., 1971; DO, U. Health Sciences, Kansas City, 1980; Cert. in Dermatology, 1990. Intern Normandy Osteo. Hosp., St. Louis, 1980-81; resident in surgery Deaconess West Hosp., St. Louis County, 1982-83, resident in dermatology, 1986-89; emergency physician St. Louis Regional, Dexter Meml., Met. and other hosps., 1981-89; dir. emergency dept. Lincoln County Meml. Hosp., 1985-87; preceptor in dermatology Met. Hosp., St. Louis, 1986-89; dermatologist St. Louis, 1989—; asst. trainer dermatology residence program Deaconess West Hosp., St. Louis. Contbr. articles to profl. jours. Vol. Variety Club, St. Louis, 1989—; lectr. Jr. League, St. Louis, 1989; bd. dirs. Maryville Coll., St. Louis, 1971-74; fundraiser Incarnate Word Hosp., 1973-74, Chaminade Coll. Prep. Mem. Am. Osteo. Assn., St. Louis Met. Med. Soc., Am. Acad. Dermatology, Mo. Med. Soc., Mo. Osteo. Soc., St. Louis Osteo. Assn. Roman Catholic. Avocations: breeding and raising horses, orchids, restoring old cars, model trains. Home: 17 Godwin Ln Saint Louis MO 63124-1524 Office: North County Dermatology 1120 Graham Rd Florissant MO 63031-8013

SCHWARZSCHILD, MARTIN, astronomer, educator; b. Potsdam, Germany, May 31, 1912; came to U.S., 1937, naturalized, 1942; s. Karl and Else (Rosenbach) S.; m. Barbara Cherry, Aug. 24, 1945. Ph.D., U. Goettingen, 1935; D.Sc. (hon.), Swarthmore Coll., 1960, Columbia U., 1973; DSc, Princeton U., 1992. Research fellow Inst. Astrophysics, Oslo (Norway) U., 1936-37, Harvard U. Obs., 1937-40; lectr., later asst. prof. Rutherford Obs., Columbia U., 1940-47; prof. Princeton U., 1947-50, Higgins prof. astronomy, 1950-79. Author: Structure and Evolution of the Stars. Served to 1st lt. AUS, 1942-45. Recipient Dannie Heineman prize Akademie der Wissenschaften zu Goettingen, Germany, 1967, Albert A. Michelson award Case Western Res. U., 1967, Newcomb Cleveland Prize AAAS, 1987, Rittenhouse Silver medal, 1966, Prix Janssen Société astronomique de France, 1970, Medal from l'Assn. Pour le Developpement Internat. de l'Observatoire de Nice, 1986, Gerlach-Adolph von Muenchausen Medaille Goettingen U., 1987, Dirk Brouwer award Am. Astron. Soc., 1991. Fellow Am. Acad. Arts and Scis.; mem. Internat. Astron. Union (v.p. 1964-70), Akademie der Naturforscher Leopoldina, Royal Astron. Soc. (asso., Gold medal 1969, Eddington medal 1963), Royal Astron. Soc. Can. (hon.), Am. Astron. Soc. (pres. 1970-72), Nat. Acad. Scis. (Henry Draper medal 1961), Soc. Royale des Sciences de Liege (corr.), Royal Netherlands Acad. Sci. and Letters (fgn.), Royal Danish Acad. Sci. and Letters (fgn.), Norwegian Acad. Sci. and Letters, Astron. Soc. Pacific (Bruce medal 1965), Am. Philos. Soc., Sigma Xi.

SCHWASS, GARY L., utilities executive; b. Ludington, Mich., Sept. 30, 1945; s. Philip V. and Greta (Beebe) S.; m. Peggy Ann McElroy, Nov. 29, 1968; children: John P., Jeree A. BS, Western Mich. U., 1968; MBA, Eastern Mich. U., 1971. Tchr. maths. Annapolis High Sch., Dearborn Heights, Mich., 1968-71; corp. systems analyst Consumers Power Co., Jackson, Mich., 1971-73; prin. planning analyst Consumers Power Co., Jackson, 1974-78, dir. fin. planning, 1979-80, dir. fin. planning and projects, 1981-83, exec. dir. fin. planning and projects, 1984-85; treas. Duquesne Light Co., Pitts., 1985-88, v.p., 1987—, v.p. finance group, CFO, 1989—; v.p. and treas. DQE, Pitts., 1989—; pres., bd. dirs. Montauk, 1990—, Western Pa. Devel. Credit Corp., Pitts., 1993; chmn., bd. dirs. Custom Coals Internat., 1991-92; bd. dirs. Duquesne Enterprises, Pitts.; mem. steering com. Edison Electric Inst. Fin., 1991—. V.p., treas., bd. dirs. Cath. Social Svcs., Jackson, 1945-85; bd. dirs. Holy Family Inst., Pitts., 1988-92; v.p. and treas., bd. dirs. Holy Family Found., Pitts., 1992—, Mercy Hosp. Svcs., Altoona, 1992—. V.p., treas. bd. dirs. Cath. Social Services, Jackson, 1984-85; bd. dirs. Mercy Health Svcs., Altoona, 1992—. Methodist. Avocations: skiing, reading, woodworking. Office: Duquesne Light Co One Oxford Centre 301 Grant St Pittsburgh PA 15279-0001

SCHWEDOCK, JULIE, molecular biologist; b. N.Y.C., Oct. 10, 1964; d. Herbert Alvin and Estelle (Lermseider) S.; m. Brian T. White, Apr. 1, 1990. BS in Life Scis., MIT, 1985; PhD in Biol. Scis., Stanford U., 1992. Postdoctoral fellow Harvard U., Cambridge, Mass., 1992—. Contbr. articles to Molecular Plant-Microbe Interactions, Nature, Genetics, Genes and Development. Scholar Nabisco, 1981-85; grantee NIH, 1985-87; fellow ARC, 1989-90. Mem. AAAS, Am. Soc. for Microbiology. Achievements include discovery of biochem. function of two Rhizobium meliloti nodulation genes and functional and genetic redundancy of those genes. Home: 21 Piedmont St Arlington MA 02174 Office: Harvard Biol Labs 16 Divinity Ave Cambridge MA 02138

SCHWEICKART, RUSSELL L., communications executive, astronaut; b. Neptune, N.J., Oct. 25, 1935; s. George L. Schweichart; children: Vicki Louise, Russell and Randolph (twins), Elin Ashley, Diana Croom; m. Nancy Kudriavetz Ramsey; step-children: Matthew Forbes Ramsey, David Scot Ramsey. B.S. in Aero. Engring, Mass. Inst. Tech., 1956, M.S. in Aeros. and Astronautics, 1963. Former research scientist Mass. Inst. Tech. Exptl. Astronomy Lab.; astronaut Johnson Manned Spacecraft Center, Houston, lunar module pilot (Apollo 9, 1969); dir. user affairs Office of Applications, NASA, sci. adv. to Gov. Edmund G. Brown, Jr. State of Calif., 1977-79; chmn. Calif. Energy Commn., 1979-83, commr., 1979-85; pres., founder Assn. Space Explorers, 1985—; pres. NRS Communications, San Francisco, 1991—; cons. and lectr. in field. Served as pilot USAF, 1956-60, 61; Capt. Mass. Air N.G. Recipient Distinguished Service medal NASA, 1970, Exceptional Service medal NASA, 1974, De La Vaulx medal FAI, 1970, Spl. Trustees award Nat. Acad. TV Arts and Scis., 1969. Fellow Am. Astronautical Soc.; mem. Soc. Exptl. Test Pilots, AIAA, Sigma Xi. Club: Explorers. Office: NRS Communications 423 Washington St # 4 San Francisco CA 94111-2339

SCHWEICKERT, RICHARD JUSTUS, psychologist, educator; b. Madison, Wis., July 19, 1946; s. Carl E. and Marie E. (Dilzer) S.; m. Carolyn M. Jagacinski, Dec. 27, 1980; children: Patrick, Kenneth. BS in Math., U. Santa Clara, 1968; MA in Math., Ind. U., 1972; PhD in Psychology, U. Mich., 1979. Statistician Bellevue Psychiatric Hosp., N.Y.C., 1969-71; asst. prof. Purdue U., West Lafayette, Ind., 1978-83; assoc. prof. Purdue U., West Lafayette, 1984-91, prof., 1992—. Author (with others) Handbook of Human Factors, 1987; assoc. editor Psychological Bulletin and Review, 1993—; mem. edit. bd. Jour. Exptl. Psychology: Learning, Memory & Cognition, 1985-89, 91—, Jour. Math. Psychology, 1986—; contbr. articles to profl. jours. Grantee NSF, 1981-84, 92—, NIMH, 1983-86, 87-89. Fellow AAAS, Am. Psychol. Soc.; mem. Soc. for Math. Psychology (pres. 1990-91, bd. dirs.), Psychonomic Soc., Ops. Rsch. Soc. Am. Office: Purdue U Dept Psychol Scis West Lafayette IN 47907

SCHWEIGHARDT, FRANK KENNETH, chemist; b. Passaic, N.J., May 12, 1944; s. Frank and Anne (Mester) S.; m. Yvonne Marie, Aug. 10, 1968; children: Brian, Jennine. BS in Chemistry, Seton Hall U., 1966; PhD in Phys. Chemistry, Duquesne U., 1970. Asst. dean rsch. Pharmacy Sch. Duquesne U., Pitts., 1970; chemist Allegheny County Morgue, Pitts., 1970; NSF postdoctoral fellow U.S. Bur. Mines, Pitts., 1970-71, chemist, 1971-79; sr. rsch. chemist Air Products and Chems., Inc., Allentown, Pa., 1979—, mgr. R & D, 1979—, lectr. speaker bur., 1985—; cons. in field, Pitts., 1975-79. Contbr. 57 articles to profl. publs. Bd. dirs. Lehigh Valley March of Dimes, Allentown, 1987—, chmn. health profl. com., 1990—. U.S. Dept. Energy grantee, 1979-82. Mem. AAAS, Am. Che. Soc. (bd. dirs. fuel div., dir. advt. 1973-84), Spectroscopy Soc., Analytical Chemistry Soc. (Pitts. con com. 1970—), Pa. Acad. Sci. (editor jour. 1987—). Roman Catholic. Achievements include patents for Process for Solvent Refining of Coal Using a Denitrogenated and Dephenolated Solvent, Automated Apparatus for Solvent Separation of Coal Liquefaction Product Stream, Corrosion Inhibition for a Commerical Distillation Apparatus, 22 others. Office: Air Products and Chems Inc 7201 Hamilton Blvd R & D 1 Allentown PA 18195-1501

SCHWEIKERT, DANIEL GEORGE, electrical engineer, administrator; b. Bemidji, Minn., June 15, 1937; s. George and Viola (Brecht) S.; m. Judith Butler Johnson, Aug. 12, 1961; children: Eric, Karl, Kristen. B of Engring., Yale U., 1959; PhD, Brown U., 1966. Group head Gen. Dynamics/Electric Boat, Groton, Conn., 1961-64; supr. Bell Telephone Labs., Murray Hill, N.J., 1966-80; dir. United Techs. Microelectronics Ctr., Colorado Springs, Colo., 1980-88, Cadence Design Systems, San Jose, Calif., 1988—; gen. chair Design Automation Conf., Anaheim, Calif., 1992, mem. exec. com., 1986—; speaker in field. Contbr. articles to profl. jours. Mem. Berkeley Heights (N.J.) Planning Bd., 1974-80. Fellow IEEE; mem. Assn. for Computing Machinery. Home: 1578 Eddington Pl San Jose CA 95129 Office: Cadence Design Systems 555 River Oaks Pkwy 4A2 San Jose CA 95134

SCHWEIKERT, EDGAR OSKAR, dentist; b. Heidelberg, Fed. Republic Germany, Aug. 30, 1938; came to U.S., 1972; s. Oskar and Priska (Zehr) S.; m. Mary Lou Como, Apr. 7, 1969; 1 child, Marisa. Degree, Hamburg Dental Sch., 1966; Dr. Med. Dentistry, U. Munich, 1969. Lic. dentist, Calif., N.Y. Dentist, U.S. Army, Frankfurt, Fed. Republic Germany, 1969-72; gen. practice dentistry, L.A., 1972-73, Bklyn., 1973—; lectr. in field. Author Multiple Cantilevers in Fixed Prosthesis, 1988, Spanish edit., 1990; contbr. articles to profl. jours. Served as capt. German Air Force, 1967-69. Mem. ADA, German Dental Assn., Second Dist. Dental Assn., Bay Ridge Dental Soc., Guild Dental Craftsmen. Home and Office: 429 77th St Brooklyn NY 11209

SCHWEITZER, GEORGE KEENE, chemistry educator; b. Poplar Bluff, Mo., Dec. 5, 1924; s. Francis John and Ruth Elizabeth (Keene) S.; m. Verna Lee Pratt, June 4, 1948; children: Ruth Anne, Deborah Keene, Eric George. BA, Central Coll., 1945, ScD in History, 1964; MS, U. Ill., 1946, PhD in Chemistry, 1948; MA, Columbia U., 1959; PhD in Philosophy, NYU, 1964. Asst. Central Coll., 1943-45; fellow U. Ill., 1946-48; asst. prof. chemistry U. Tenn., 1948-52, assoc. prof., 1952-58, prof., 1960-69, Alumni Distinguished prof., 1970—; cons. to Monsanto Co., Proctor & Gamble, Internat. Tech., Am. Cyanamid Co., AEC, U.S. Army, Massengill; lectr. colls. and univs. Adv. bd. Va. Intermont Coll. Author: Radioactive Tracer Techniques, 1950, The Doctorate, 1966, Genealogical Source Handbook, 1979, Civil War Genealogy, 1980, Tennessee Genealogical Research, 1981, Kentucky Genealogical Research, 1981, Revolutionary War Genealogy, 1982, Virginia Genealogical Research, 1982, War of 1812 Genealogy, 1983, North Carolina Genealogical Research, 1983, South Carolina Genealogical Research, 1984, Pennsylvania Genealogical Research, 1985, Georgia Genealogical Research, 1987, New York Genealogical Research, 1988, Massachusetts Genealogical Research, 1989, Maryland Genealogical Research, 1991, German Genealogical Research, 1992; also articles. Faculty fellow Columbia U., 1958-60. Mem. Am. Chem. Soc., Am. Philos. Assn., History Sci. Soc., Soc. Profl. Genealogists, Phi Beta Kappa, Sigma Xi. Home: 407 Ascot Ct Knoxville TN 37923-5807

SCHWEITZER, MARK ANDREW, civil engineer; b. Benton Harbor, Mich., Apr. 11, 1968; s. John Anton and Elaine Jeanette (Bomke) S. BS, Worcester Poly., 1990, MS, 1993. Field engr. Stone & Webster Engring. Corp., Boston, 1990—. Mem. ASCE. Lutheran. Home: 70 King St Danbury CT 06811-2729

SCHWEIZER, KENNETH STEVEN, physics educator; b. Phila., Jan. 20, 1953; s. Kenneth Paul and Grace Norma (Fischer) S.; m. Janis Eve Pelletier, Oct. 18, 1986; 1 child, Gregory Michael. BS, Drexel U., 1975; MS, U. Ill., 1976, PhD, 1981. Postdoctoral rsch. assoc. AT&T Bell Labs., Murray Hill, N.J., 1981-83; sr. mem. tech. staff Sandia Nat. Labs., Albuquerque, 1983-91, cons., 1991—; prof. materials sci. engring. and chemistry U. Ill., Urbana, 1991—. Contbr. articles to profl. jours. Mem. Am. Phys. Soc. (John H. Dillon medal 1991), Am. Chem. Soc., Sigma Xi. Office: U Ill Dept Materials Sci Engring 1304 W Green St Urbana IL 61801-2920

SCHWEMMER, DAVID EUGENE, structural engineer; b. Pitts., Apr. 9, 1956; s. Eugene Joseph and Janet Edith (Drodge) S.; m. Dorothy Jane Rivi, Apr. 16, 1983; children: David Angelo, Stephanie Marie, Kristen Ann-Beatrice. BSCE, U. Pitts., 1978, MSCE, 1983. Registered profl. engr., Pa. Design engr. Leo J. Noker Engring., Pitts., 1978-85; rsch. engr. Boeing Co., Pitts., 1985-88; lead engr. Indsl. Design Corp., Pitts., 1988-89; job engr. Eichleay Engrs., Inc., Pitts., 1989—. Contbr. articles to profl. jours.; author profl. reports. Mem. NSPE, ASCE, Am. Inst. Steel Constrn., Am. Concrete Inst., Nat. Acad. Bldg. Inspection Engrs. Democrat. Roman Catholic. Achievements include documentation of behavior and anomalies in longwall mining equipment and other roof support systems, development of guidelines in use of transition elements in modelling discrete cracks in linear elastic fracture mechanics finite element system representations. Home: 364 Goldsmith Rd Pittsburgh PA 15237 Office: Eichleay Engrs Inc 6585 Penn Ave Pittsburgh PA 15206

SCHWENDEMAN, LOUIS PAUL, civil engineer, retired; b. Lowell, Ohio, Dec. 22, 1918; s. John Andrew and Alice Mary (Long) S.; m. Anne Frances Jarowey, May 18, 1945; children: David, Judith, Erik, James. BSCE, Ohio U., 1941. Registered profl. engr., Pa. Draftsman Am. Bridge Co., Ambridge, Pa., 1941-43; structural engr. Am. Bridge Co., Ambridge, 1945-48; engr. designer Green Engring. Co., Sewickley, Pa., 1948-51; bridge designer Richardson Gordon & Assocs., Pitts., 1951-83, ptnr., 1979-83; cons. HDR Engring. Inc., Pitts., 1983—. Lt. (j.g.) USNR, 1943-45, PTO. Decorated Purple Heart medal. Mam. ASCE, Pa. Soc. Profl. Engrs., Soc. Am. Mil. Engrs. Republican. Roman Catholic. Achievements include devel. of sealed welded steel box truss members for bridges. Home: 2308 Clearvue Rd Pittsburgh PA 15237

SCHWERDTFEGER, PETER ADOLF, research chemist; b. Stuttgart, Germany, Sept. 1, 1955; arrived in New Zealand, 1987; s. Adolf Otto and Gertrud Maria (Kreuzer) S.; m. Ulrike Maria Bucher, May 13, 1983; children: Laura Maria, Roman Arthur. Chem. engr. degree, Fachhhochschule Aalen, Germany, 1976; MSc in Chemistry, Stuttgart (Germany) U., 1980, BSc in Math., 1983, PhD in Chemistry, 1986. Sch. tchr. Intermediate Sch. Stuttgart, 1980-81; computer software analyst Stuttgart U., 1981-87; Feodor-Lynen fellow Alexander von Humboldt Found., Bonn, Germany, 1987-89; rsch. fellow Australian Nat. U., Canberra, Australia, 1989-91; sr. lectr. U. Auckland (New Zealand), 1991—. Contbr. articles to profl. jours. Mem. Am. Chem. Soc., Am. Phys. Soc., New Zealand Inst. Chem., Gesellschaft Deutscher Chemiker. Achievements include research on relativistic effects in molecules. Office: U Auckland Dept Chemistry, Private Bag 92019, Auckland New Zealand

SCHWERDTFEGER, WALTER KURT, public health official, researcher; b. Karlsruhe, Fed. Republic of Germany, Apr. 17, 1949; s. Walter Georg Hermann and Anna (Jooss) S.; m. Renate Schenk, May 4, 1984; children: Wolfgang, Cristina, Ines. Diploma in biology, U. Frankfurt, Fed. Republic of Germany, 1973, Dr.Sc., 1977, Privatdozent, 1983. Biol. diplomate. Sci. employee biol. faculty U. Frankfurt, 1974-77; scientist with tenure Max Planck Inst. for Brain Rsch., Frankfurt, 1977-88; head morphology and pathology div. Paul Ehrlich Inst., Langen, Fed. Republic of Germany, 1988-92; head rsch. coordination unit Fed. Ministry Health, Bonn, 1992—; lectr. med. faculty U. Frankfurt, 1978—; vis. prof. U. Valencia (Spain), 1985-87; sci. symposia organizer; referee for sci. jours. and founds. Author: Structure and Fiber Connections of the Hippocampus, 1984; co-author: The Brain of the Common Marmoset, 1980; editor: The Forebrain in Reptiles, 1988, The Forebrain in Nonmammals, 1990; co-editor: Current Topics in Biomedical Research, 1992; translator/editor: Histologie, 1990; translator: Farbatlas der Histologie, 1987; contbr. numerous articles to internat. sci. jours. and handbooks. Mem. Working Party on Waste Mgmt. and Environ., Frankfurt, 1991-92. Recipient Dr. Paul and Cilli Weill prize Dr. Paul and Cilli Weill Found., Frankfurt, 1985; grantee Deutsche Forschungsgemeinschaft, Bonn, 1984-88, Comisión Asesora de Investigación Cientifica y Técnica, Madrid, 1985-86, Ministry of Sci. and Arts, Wiesbaden, Fed. Republic of Germany, 1987. Mem. Deutsche Zoologische Ges., Anatomische Ges., European Neurosci. Assn., Internat. Brain Rsch. Orgn., Ges. der Freunde und Förderer der Johann Wolfgang Goethe U. Frankfurt a.M. Avocations: music, billiards, table tennis, fishing. Office: Fed Ministry Health, Belderberg 6, D-53111 Bonn 1, Germany

SCHWERTLY, HARVEY KENNETH, JR., computer electronics educator; b. Camden, N.J., Dec. 26, 1941; s. Harvey Kenneth Sr. and Marjorie Anna (Younghanns) S.; m. Barbara Ann Sills, Nov. 18, 1961; children: Barbara Anne, Catherine Anna, Mary Theresa. AS, SUNY, Albany, 1982, BS, 1987; MS, Nat. U., 1988. Cert. instr., Calif.; cert. electronics technician. Enlisted USN, 1960, advanced througth grades to chief petty officer, 1980, ret., 1980; instr. Telemedia, Inc., Alkhobar, Saudia Arabia, 1980-83, San Diego OIC, 1984, ITT Ednl. Svcs., La Mesa, Calif., 1985-86, San Diego Community Coll. Dist., 1986—; computer cons. 6PT Micro Maintenance, Lemon Grove, Calif., 1990—. Mem. Internat. Soc. Cert. Electronics Technicians (computer option com.), Calif. Coun. Electronic Instrs. Inc., Nat. U. Alumni Assn., Calif. State Electronics Assn., VFW, Fleet Res. Assn., Mil. Order of the Cootie, All Star Seam Squirrel, Am. Legion. Home: 3226 Harris St Lemon Grove CA 91945-2221 Office: San Diego CC 1400 Park Blvd San Diego CA 92101-4793

SCHWIMMER, DAVID, physician, educator; b. Gödényháza, Hungary, Dec. 8, 1913; came to U.S., 1921; s. George and Laura (Green) S.; m. Gertrude Alpha Dounn, Nov. 12, 1939; children: Betty Laura, Georgia, Mark Ian. B.S. cum laude, Lafayette Coll., 1935; M.D., N.Y. U., 1939; M.Med. Sci., N.Y. Med. Coll., 1944. Diplomate: Am. Bd. Internal Medicine. Intern Met. Hosp., N.Y.C., 1939-41; resident Met. Hosp., 1942-44; practice medicine specializing in internal medicine N.Y.C., 1944—; attending physician Flower Fifth Av. Hosp., pres. med. bd., 1970-71; attending physician Met., Bird S. Coler, Doctors, Manhattan Eye, Ear and Throat, Lenox Hill hosps.; mem. faculty N.Y. Med. Coll., 1944—, clin. prof. medicine, 1966—; dir. U.S. Quatermaster Survival Rations Study, 1945-48, 50-51; rsch. staff war rsch. div. Met. Hosp., Columbia U., 1944-46; dir. Pvt. Teaching Svc., mem., dir. multiple med. sch. exec. coms.; cons. internist Monmouth Med. Center, Long Branch, N.J., 1950—; cons. St. Luke's-Roosevelt Hosp. Med. Ctr., 1985—; spl. lectr. in medicine Columbia U. Coll. Physicians and Surgeons, 1980—; participant rsch. confs. including Gordon Rsch. Conf. 1948, NATO adv. study Inst.1962, Internat. Symposium 1967. Author: (with Morton Schwimmer) Role of Algae and Plankton in Medicine, 1955; Contbr. to profl. jours.; patentee in field. Research fellow N.Y. Med. Coll., 1944-51; fellow internal medicine, 1941-44. Fellow ACP, Royal Soc. Medicine, N.Y. Acad. Medicine, N.Y. Acad. Scis., Am. Coll. Angiology, N.Y. Cardiol. Soc.; mem. Endocrine Soc., Harvey Soc., AAAS, AAUP, Am. Soc. Internal Medicine, Alpha Omega Alpha. Home: 764 Carroll Pl Teaneck NJ 07666-3302 Office: 239 E 79th St New York NY 10021-0810

SCHWINGER, JULIAN, physicist, educator; b. N.Y.C., Feb. 12, 1918; s. Benjamin and Belle (Rosenfeld) S.; m. Clarice Carrol, 1947. A.B., Columbia U., 1936, Ph.D., 1939, D.Sc., 1966; D.Sc. (hon.), Purdue U., 1961, Harvard U., 1962, Brandeis U., 1973, Gustavus Adolphus Coll., 1975; LL.D., CCNY, 1972; D Honoris Causa, U. Paris, 1990. NRC fellow, 1939-40; research assoc. U. Calif.-Berkeley, 1940-41; instr., then asst. prof. Purdue U., 1941-43; staff mem. Radiation Lab., MIT, 1943-46; staff Metall. Lab., U. Chgo., 1943; asso. prof. Harvard U., 1945-47, prof., 1947-72, Higgins prof. physics, 1966-72; prof. physics UCLA, 1972-80, Univ. prof., 1980—; mem. bd. sponsors Bull. Atomic Sci.; sponsor Fedn. Am. Scientists; J.W. Gibbs hon. lectr. Am. Math. Soc., 1960. Author: Particles and Sources, 1969, (with D. Saxon) Discontinuities in Wave Guides, 1968, Particles, Sources and Fields, 1970, Vol. II, 1973, Vol. III, 1989, Quantum Kinematics and Dynamics, 1970, Einstein's Legacy, 1985; editor: Quantum Electrodynamics, 1958. Recipient C. L. Mayer nature of light award, 1949, univ. medal Columbia U., 1951, 1st Einstein prize award, 1951; Nat. Medal of Sci. award for physics, 1964; co-recipient Nobel prize in Physics, 1965; recipient Humboldt award, 1981, Monie A. Fest Sigma Xi award, 1986, Castiglione di Sicilia award, 1986, Am. Acad. of Achievement award, 1987; Guggenheim fellow, 1970. Mem. AAAS, ACLU, Nat. Acad. Scis., Am. Acad. Arts and Scis., N.Y. Acad. Scis. Office: U Calif Dept Physics 405 Higard Ave Los Angeles CA 90024

SCHWINN, DONALD EDWIN, environmental engineer; b. N.Y.C., July 18, 1935; s. Edwin William and Mary Louise (Conforti) S.; m. Mary Lou De Maria, June 25, 1960; children: Daniel, Michael, Brian, Christine. BS in Civil Engring., The Cooper Union, 1957; MS in Sanitary Engring., MIT, 1959. Diplomate Am. Acad. Environ. Engrs. Sanitary engr. Dorr-Oliver Inc., Stamford, Conn., 1957-62; sanitary engr. Metcalf & Eddy, Boston, 1962-66, exec. engr., 1966-70, v.p., 1970-74; ptnr. Stearns & Wheler, Cazenovia, N.Y., 1974-88, sr. ptnr., 1988—; mem. study com. water quality policy NAS, 1975-76; advisor to environ. engring. program Cornell U., Ithaca, N.Y., 1976, Syracuse (N.Y.) U., 1987—; mem. US/USSR joint tech. commn. external utility systems HUD, Washington, 1979-85. Sr. author: U.S. EPA Process Design Manual for Upgrading Wastewater Treatment Plants, 1974; contbg. author: Design Manual for Municipal Wastewater Treatment Plants, 1991, U.S. EPA Process Design Manual for Nitrogen Control, 1992; contbr. articles to profl. jours. Co-chmn. Cazenovia Youth Soccer Program 1980-85; mem. Cazenovia Conservation Commn., 1984—, St. James Choir, Cazenovia, 1990—. Recipient Arthur Sidney Bedell award Water Pollution Control Fedn., 1974, 1st award Consulting Engrs. Coun. N.Y., 1977, Chmn.'s award Water Pollution Control Fedn., 1988. Mem. ASCE, NSPE, Am. Water Works Assn., APWA, Water Environ. Fedn., N.Y. Water Pollution Control Assn. (Kenneth Allen award 1985). Roman Catholic. Achievements include invention and devel. of cyclical nitrogen removal process for removal of nitrogen from sewage.

SCHWITTERS, ROY FREDERICK, physicist, educator; b. Seattle, Wash., June 20, 1944; s. Walter Frederick and Margaret Lois (Boyer) S.; m. Karen Elizabeth Chrystal, June 18, 1965; children: Marc Frederick, Anne Elizabeth, Adam Thomas. S.B., MIT, 1966, Ph.D., 1971. Research asso. Stanford U. Linear Accelerator Center, 1971-74, asst. prof., then asso. prof., 1974-79; prof. physics Harvard U., 1979—; scientist Fermi Nat. Accelerator Lab., 1980-88, dir. Superconducting Super Collider, 1989—. Author papers on high energy physics; asso. editor: Ann. Rev. Nuclear and Particle Sci; div. asso. editor: Phys. Rev. Letters. Recipient Alan T. Waterman award NSF, 1980. Fellow AAAS, Am. Phys. Soc., Am. Acad. Arts and Scis. Home: 1121 Frost Hollow Dr De Soto TX 75115-7416 Office: Superconducting Super Collider Lab 2550 Beckleymeade Ave Dallas TX 75237-3974

SCIABICA, VINCENT SAMUEL, chemist, researcher; b. Greensburg, Pa., July 4, 1959; s. Samuel Vincent and Sallie (Scichilone) S. BS in Chemistry, St. Vincent Coll., 1981; MBA, W.Va. U., 1984. Registered rep. SEC. Project chemist Kennametal Inc., Latrobe, Pa., 1980-81; rsch. chemist U. Pitts., 1981; environ. chemist Wheeling (W.Va.)-Pitts. Steel Corp., 1982-85; environ. gas chromatography/mass spectrometry chemist Internat. Tech., Pitts., 1985-89; lab. mgr. Radian Corp. Research Trinagle Park, Raleigh, N.C., 1989-92; pres., lab. dir. CTM Analytical Labs., Albany, N.Y., 1992—; lighting cons. Perfect Image Lighting, 1986—; audio cons. Ears Ahead Audio/Ear Force Sound, Greensburg, 1979—; v.p. S.S.V., Inc., Greensburg, 1981—; pres. Ear Force Sound, Inc., Greensburg, 1988—. Mem. Soc. Audio Cons. (cert.), Nat. Assn. MBA's, Am. Chem. Soc. Republican. Roman Catholic. Office: CTM Analytical Labs Latham NY 12110

SCIACCA, KATHLEEN, psychologist; b. N.Y.C., Jan. 19, 1943; d. Rosario and Angela (Pucciarelli) S.; children: Kenneth Mortellaro, Cheryl Ann Mortellaro. BS in Psychology, SUNY, Empire State, 1976; MA in Psychology, New Sch. for Social Rsch., 1980, postgrad., 1980. Pvt. practice, 1976—; primary therapist, vocat. coord., program developer Bronx-Lebanon Hosp., South Bronx, N.Y., 1977-84; mental health program specialist N.Y. State Office of Mental Health, N.Y. statewide, 1984-92; founding exec. dir. Sciacca Comprehensive Svc. Devel. Mental Illness, Drug Addiction and Alcoholism, N.Y.C., 1990—, nat. lectr., program developer for multiple disorders; cons. program devel. Am. Assn. for Partial Hospitalization, Washington, 1987-91, Columbia U., N.Y.C., 1991, numerous mental health depts. across the country; nat. presenter and cons. in field. Author/developer: (manual-book) Mental Illness, Drug Addiction and Alcoholism (MIDAA) Service Manual, 1990; contbr. articles to profl. jours. Mem. APA (assoc.), N.Y. State Psychol. Assn. (assoc., com. on Alcoholism, Drug Abuse and other Addictions), Soc. Psychologists in Addictive Behaviors. Home and Office: 299 Riverside Dr #3E New York NY 10025-5278

SCIANCE, CARROLL THOMAS, chemical engineer; b. Okemah, Okla., Feb. 16, 1939; s. Carroll Elmer and Winifred (Black) S.; BS in Chem. Engring., U. Okla., 1960, M in Chem. Engring., 1964, PhD, 1966; m. Anita Ruth Fischer, Jan. 30, 1960; children: Steven, Frederick, Thomas, Erica. With E.I. duPont de Nemours & Co., Inc., 1966—, planning mgr. nylon intermediates div., petrochem. dept., Wilmington, Del., 1978-80, tech. mgr., 1980-83, dir. engring. research, engring. dept., 1983-87, prin. cons. corp. research and devel. planning div., 1987-89; mgr. petroleum products R&D div., Conoco, Inc., 1989-93; dir. Environ. Tech. Partnerships, ctrl. R & D dept. DuPont, 1993—. Mem. math. scis. and edn. bd. Nat. Rsch. Coun., 1987-89, adv. bd. NIST for Chem. Sci. & Tech., 1988—. Served as officer USAR, 1961-63. Fellow Am. Inst. Chem. Engrs. (bd. dirs. materials engring. and scis. div. 1986-92, chmn. new tech. com. 1990-92, govt. policy steering com. 1993—), mem. Fedn. Materials Socs. (v.p. 1988-92, pres. 1993—), Am. Chem. Soc., N.Y. Acad. Scis., Sigma Xi. Home: 9 Aston Circle Hockessin DE 19707 Office: Exp Sta 304/A306 Wilmington DE 19880

SCIFRES, DONALD RAY, physics and engineering administrator; b. Lafayette, Ind., Sept. 10, 1946; s. Ray E. and Ruth L. Scifres; m. Carol D. Scifres. BSEE, Purdue U., 1968; MS, U. Ill., 1970, PhD, 1972. Rsch. fellow, mgr. optoelectronics Xerox Palo Alto (Calif.) Rsch. Ctr., 1972-83; pres., CEO, chmn. SDL, Inc., San Jose, Calif., 1983—. Contbr. over 280 tech. papers to profl. publs., chpt. to book. Fellow IEEE (pres. 1992, bd. govs. lasers and electrooptics soc. 1988, Jack Morton medal 1985, Optical Soc. Am.; mem. Am. Phys. Soc., Am. Inst. Physics, (lasers and electrooptics soc., 1988—, Jack Morton medal 1985), Optical Soc. Am.; mem. Am. Phys. Soc., Am. Inst. Physics, Lasers and Electrooptics Mfrs. Assn. (bd. govs. 1992—). Achievements include 96 patents in field; first demonstration of the distributed feedback injection laser, of CW laser diode arrays, of electronic beam steering in semiconductor laser; founding of Spectra Diode Labs., Inc. Office: Spectra Diode Labs Inc 80 Rose Orchard Way San Jose CA 95134

SCIOLY, ANTHONY JOSEPH, chemistry and physics educator; b. Spokane, Wash., Oct. 10, 1950; s. Joseph Arthur and Bettie Jane (Cole) S.; m. Claudia Jeanne Eminger, Dec. 11, 1982; 1 child, Meredith. BS in Chemistry, U. Wash., 1973; PhD in Chemistry, U. Mich., 1985. Asst. prof. chemistry Siena Heights Coll., Adrian, Mich., 1985-92, assoc. prof. chemistry, 1992—. Contbr. publs. to Chem. Physics Letters, 1987, Polyhedron, 1987. Recipient Teaching Excellence award Mich. Sect. Am. Chem. Soc., 1981. Mem. Am. Chem. Soc., Math. Assn. Am., Coun. on Undergrad. Rsch., Mich. Acad. Sci., Arts, and Letters. Office: Siena Heights Coll Adrian MI 49221

SCIPIO, L(OUIS) ALBERT, II, aerospace science engineering educator, architect, military historian; b. Juarez, Mex., Aug. 22, 1922; s. Louis Albert and Marie Leona (Richardson) S.; m. Katherine Ruth Jones, Aug. 15, 1942; children: Louis Albert, Karen R. B.S., Tuskegee Inst., 1943; B.Civil Engring., U. Minn., 1948, M.S., 1950, Ph.D., 1958. Archtl. draftsman McKissack & McKissack, Tuskegee, Ala., 1943; instr. Tuskegee Inst., 1946; designer Long & Thorshov, Mpls., 1948-50; lectr. U. Minn., Mpls., 1950-59; research physicist Hughes Aircraft Co., Culver City, Calif., 1954; Fulbright prof. Cairo U., Giza, Egypt, 1955-56; assoc. prof. mechanics Howard U., Washington, 1959-61; Fulbright prof. Cairo U., Giza, Egypt, 1955-56; dir. grad. studies for engring. and architecture, prof. aerospace engring. Howard U., Washington, 1967-70, Univ. prof. space scis., 1970-87, Disting. Univ. prof. emeritus, 1987—; prof. phys. scis. U. P.R., Mayaguez, 1961-63; prof. aerospace engring. U. Pitts., 1963-67; pub. Roman Publs., Silver Springs, Md., 1981—; cons. in field; author Compendium of Aircraft Stress Analysis and Design, 1956. Author: Principles on Continua with Applications, 1966, Structural Design Concepts, 1967, E.M. Collar Insignia, 1907-1926, 1981, Last of the Black Regulars, 1983, With the Red Hand Division, 1985, The 24th Infantry at Fort Benning, 1986, Pre-War Days at Tuskegee, 1987. Bd. visitors Air Force Inst. Tech., 1979-83. Served with AUS, 1943-46. Mem. N.Y. Acad. Scis., Internat. Assn. Bridge and Structural Engrs., Soc. Natural Philosophy, AIAA, AAAS; mem. Am. Phys. Soc.; mem. NSPE, Co. of Mil. Historians, Coun. on Am. Mil. Past, Phi Beta Kappa, Sigma Xi, Alpha Kappa Mu, Pi Mu Epsilon, Sigma Pi Sigma, Sigma Gamma Tau, Pi Tau Sigma. Home: 12511 Montclair Dr Silver Spring MD 20904-2053

SCLATER, JOHN GEORGE, geophysics educator; b. Edinburgh, Scotland, June 17, 1940; s. John George and Margaret Bennett (Glen) S.; children: Iain Andrew, Stuart Michael. B.Sc., Edinburgh U., 1962; Ph.D., Cambridge (Eng.) U., 1966. Research geophysicist Scripps Inst. Oceanography, La Jolla, Calif., 1965-72, prof., 1991—; asso. prof. MIT, 1972-77, prof., 1977-83; dir. Joint Program Oceanography Woods Hole Oceanographic Inst., 1981-83; Shell Cos. chair in geophysics U. Tex., Austin, 1983-91; prof. Scripps Instn. Oceanography, U. Calif., San Diego, 1991—; Sweeney lectr. Edinburgh U., 1976. Contbr. articles to profl. jours. Recipient Rosenstiel award oceanography, 1979, numerous award for publs. Fellow Geol. Soc. Am., Royal Soc. London, Am. Geophys. Union (Bucher medal 1985); mem. NAS (mem. ocean studies bd., 1985-92, chair 1988-91). Home: 10869 Portobelo Dr San Diego CA 92124 Office: Scripps Instn of Oceanography La Jolla CA 92093-0215

SCLOVE, RICHARD EVAN, technology educator, writer, consultant; b. N.Y.C., Nov. 3, 1953; s. Abraham Bernard and Louise (Herzog) S.; m. Marcie Abramson, July 16, 1988; 1 child, Lena. BA, Hampshire Coll., 1975; SM in Nuclear Engring., MIT, 1978, PhD in Polit. Sci., 1986. Postdoctoral fellow in resource econs. Univ. Calif., Berkeley, 1986-87; vis. prof. Clark Univ., Worcester, Mass., 1989; rsch. fellow Inst. for Resource and Security Studies, Cambridge, Mass., 1989—; vis. prof. Rensselaer Poly. Inst., Troy, N.Y., 1991-92; exec. dir. The Loka Inst., Amherst, Mass., 1987—; rsch. fellow Inst. for Policy Studies, Washington, 1993—; cons. Energy Rsch. Adv. Bd., Washington, 1980, John D. and Catherine T. MacArthur Found., 1993, Sci. Tech. and Soc. Program, U. Mass., 1993—; advisor Telecomm. and Democracy project Rutgers U., 1989-92; Copeland fellow Amherst Coll., 1990; steering com. 21st Century Project, Cambridge, 1991—. Co-author: Uncertain Power, 1983, Critical Perspectives on Non-academic Science and Engineering, 1991, Democracy in a Technological Society, 1992, Technology for the Common Good, 1993, Preparing for Nuclear Power Plant Accidents, 1993; contbr. articles to profl. jours. Prodr. and moderator Amherst Forum, 1989; mem. Town Meeting, Amherst, 1973-74, 91-92. Mem. AAAS, Fedn. Am. Scientists, Soc. Philosophy and Tech., Soc. Social Studies of Sci., Soc. History of Tech., Computer Profls. for Social Responsibility, Coun. for Responsible Genetics, Graylyn Group. Achievements include research in democratizing technology, science, architecture and design; social and political impacts of science and technology. Office: Loka Inst PO Box 355 Amherst MA 01004-0355

SCOFIELD, LARRY ALLAN, civil engineer; b. Niskyuna, N.Y., Mar. 23, 1952; s. Jack Dewayne and Edythe Mae (Van Wie) S.; m. Apr. 3, 1981 (div. Aug. 1982). BSE, Ariz. State U., 1975, MSE. Registered profl. engr., Ariz. Engr.-in-tng. Ariz. Dept. of Transp., Phoenix, 1976-78, constrn. project engr., 1978-80, pavement design engr., 1980-81, resident engr., 1981-82, geologic and found. invest. engr., 1982-84, transp. engr. supr., 1984-90, mgr. transp. rsch., 1990—; mem. panel Nat. Coop. Hwy. Rsch. Program, Washington, 1986—; mem. Transp. Rsch. Bd. Com., Washington, 1990—; mem. expert task group Strategic Hwy. Rsch. Program, Washington, 1989—; mem. adv. panel Fed. Hwy. Adminstrn., Washington, 1986—. Contbr. articles to profl. jours. Mem. ASCE, ASTM, Asphalt Paving Technologists. Republican. Baptist. Achievements include research in highway planning, safety issues, crash testing, pavement analysis, traffic engineering, and structural design activities. Home: 807 W Keating Ave Mesa AZ 85210-7611

SCOGGIN, JAMES FRANKLIN, JR., electrical engineering educator; b. Laurel, Miss., Aug. 3, 1921; s. James Franklin and Berenice Evans (Phares) S.; m. Madeline Eve Lannelle, Mar. 6, 1948; children: Tracy Catherine, Beryl Evans, James Franklin. BS in Math., Miss. State U., 1941; BS in Mil. Engring., U.S. Mil. Acad., 1944; MA in Physics, Johns Hopkins U., 1951; PhD in Physics, U. Va., 1957. Registered profl. engr., S.C. Commd. 2nd lt. U.S. Army, 1944, advanced through grades to col., 1966, editor German WWII mil. history, R&D mgr., ret., 1968; prof. elec. engring. The Citadel, Charleston, S.C., 1968-91, prof. emeritus, 1991—. Fellow Radio Club of Am.; mem. IEEE (Centennial medal), Am. Nuclear Soc., Precision Measurements Assn., Tau Beta Pi (chief faculty adviser Citadel's chpt. 1979-

92). Presbyterian. Office: The Citadel Elec Engring Dept Charleston SC 29409

SCOLES, GIACINTO, chemistry educator; b. Torino, Italy, Apr. 2, 1935; came to Can., 1971; to U.S., 1987; s. Mario and Maria (Fiorio) S.; m. Giok-Lan Lim, Oct. 20, 1964; 1 child, Gigi. Degree in chemistry, U. Genoa, Italy, 1959, Libero Docente, 1968. Asst. prof. U. Genoa, 1960-61, 64-68, assoc. prof., 1968-71; research assoc. U. Leiden, The Netherlands, 1961-64; prof. chemistry and physics U. Waterloo, Ont., Can., 1971-86; Donner prof. sci. Princeton (N.J.) U., 1987—. Editor: Atomic and Molecular Beam Methods, 1987; contbr. numerous articles to profl. jours. Killam fellow Sci. Council Can., 1986. Fellow The Chem. Inst. Can.; mem. Can. Assn. Physicists, Am. Phys. Soc., Am. Chem. Soc., Optical Soc. Am. Office: Princeton U Dept Chemistry Princeton NJ 08544

SCOLNICK, EDWARD MARK, science administrator; b. Boston, Aug. 9, 1940; s. Barbara (Chasen) Scolnick; m. Barbara Bachrach; children: Laura, Jason, Daniel. AB, Harvard U., 1961; MD, Harvard U. Med. Sch., 1965. Intern Mass. Gen. Hosp., 1965-66, asst. resident internal medicine, 1966-67; research assoc. USPHS, 1967-69; sr. staff fellow lab. biochem. genetics NIH, 1969-70; instr. NIH Sem., 1970; sr. staff fellow viral leukemia and lymphoma br. Nat. Cancer Inst., 1970-71, spl. advisor to spl. virus cancer program, 1973-78, mem. coordinating com. for virus cancer program, 1975-78, chief. lab. tumor virus genetics, head. molecular virology sect., 1975-82; exec. dir. basic research virus and cell biology research Merck Sharp & Dohme Rsch. Labs., West Point, Pa., 1982-83, v.p. virus and cell biology research, 1983-84, sr. v.p., 1984, pres., 1985-93; sr. v.p., pres. rsch. Merck & Co., Inc., 1991—, v.p. rsch., 1993—; adj. prof. microbiology Sch. Medicine U. Pa., 1983-86. Editor-in-chief Jour. Virology; mem. editorial bd. Virology; contbr. numerous articles to profl. jours.;. Served with USPHS, 1965-67. Recipient Arthur S. Fleming award, 1976, PHS Superior Svc. award, 1978, Eli Lilly award, 1980, Indsl. Rsch. Inst. medal, 1990. Mem. NAS, Am. Soc. Biol. Chemists, Am. Soc. Microbiologists. Home: 811 Wickfield Rd Wynnewood PA 19096-1610 Office: Merck Rsch Labs PO Box 2000 Rahway NJ 07065-0900

SCOPATZ, STEPHEN DAVID, engineering executive, educator; b. Redondo Beach, Calif., Jan. 19, 1957; s. John Anthony and Mary (Porpiglia) S.; m. Jan A. Michael, Feb. 14, 1981; children: Anthony Michael, Shane Elliot, Julian Alexander. BS, U. So. Calif., 1979; MS in Engring., Purdue U., 1993. Mem. tech. staff Hughes Aircraft Co., L.A., 1978-81; rsch. supr. Pennwalt Corp., Three Rivers, Calif., 1981-85; optical engr. View Engring., Simi Valley, Calif., 1985-87; project engr., engring. supr. Boehringer Mannheim Corp., Indpls., 1987—; lab. instr. Moorpark (Calif.) Coll., 1986-87; physics instr. Ind. U./Purdue U., Indpls., 1989-91. Patentee in field. Mem. Soc. Photo-Optical Engrs., Soc. Mfg. Engrs., Machine Vision Assn. Avocations: photography, camping. Home: 734 E 73rd St Indianapolis IN 46240-3010 Office: Boehringer Mannheim Corp 9115 Hague Rd Indianapolis IN 46256-1045

SCOPP, IRWIN WALTER, periodontist, educator; b. N.Y.C., Dec. 8, 1909; s. Leon and Anne S.; B.S., CCNY, 1930; D.D.S., Columbia U., 1934; m. Edith Halprin, Dec. 25, 1941; 1 son, Alfred. Pvt. practice periodontics, N.Y.C., 1934-42; chief dental service VA Med. Center, N.Y.C., 1945-80; prof. periodontics N.Y.U. Coll. Dentistry, 1945-80, dir. continuing dental edn., 1980-85, dir. course in medicine Coll. Dentistry, 1982-90, clin. prof. periodontics, 1980—. Bd. govs. div. oral hygiene N.Y.C. Tech. Coll., 1960-79, chmn., 1979. Served to capt. Dental Corps, AUS, 1942-45. Recipient Disting. Profs. award N.Y.U., 1981. Mem. Northeastern Soc. Periodontics (sec.-treas. 1955-90), ADA (chmn. research 1978), Am. Acad. Periodontology, Am. Acad. Oral Medicine, Am. Coll. Dentists, Am. Acad. Sci., Internat. Assn. Dental Research, Am. Pub. Health Assn., Assn. Mil. Surgeons, Research Soc. Am., Hosp. Dental Chiefs Assn., Am. Acad. Sci., Internat. Assn. Dental Research. Author: Oral Medicine - A Clinical Approach with Basic Science Correlation, 1969, 2d edit., 1973. Contbr. chpts. to books. Home: 110 Bleecker St Apt # 29-B New York NY 10012 Office: 345 E 24th St New York NY 10010

SCOTT, ALAN, electrical engineer, consultant; b. June 15, 1964. BSEE, Widener U., 1986. Coop. engr. E. I. DuPont de Nemours and Co., Newark, Del., 1983-85; electromagnetic engr. David Taylor Naval Ship R&D, Annapolis, Md., 1986-87; product design engr. Chessell Corp., Newtown, Pa., 1989—; self-protection program Chester County Court House, West Chester, Pa., 1989—; prospective pres. Firehawk Rsch. Found., Exton, 1989—. Mem. IEEE, NSPE, Toastmasters, Tau Beta Pi. Home and Office: 290 King Road Box 223 Exton PA 19341

SCOTT, ALEXANDER ROBINSON, engineering association executive; b. Elizabeth, N.J., June 15, 1941; s. Marvin Chester and Jane (Robinson) S.; m. Angela Jean Kendall, July 17, 1971; children: Alexander Robinson, Jennifer Angela, Ashley Kendall. B.A. in History, Va. Mil. Inst., 1963; M.A. in Personnel and Counseling Psychology, Rutgers U., 1965. Sales mgr. Hilton Hotels, 1967-70; meetings mgr. Am. Inst. Mining Engrs., N.Y.C., 1971-73; exec. dir. Minerals, Metals and Materials Soc., 1973—. Served with U.S. Army, 1965-67. Decorated Bronze Star. Mem. Am. Soc. Assn. Execs. Republican. Baptist. Home: 107 Staghorn Dr Sewickley PA 15143-9506 Office: TMS 420 Commonwealth Dr Warrendale PA 15086-7511*

SCOTT, AMY ANNETTE HOLLOWAY, nursing educator; b. St. Albans, W.Va., Apr. 10, 1916; d. Oliver and Mary (Lee) Holloway; m. William M. Jefferson, June 22, 1932, (div. Oct. 1933); 1 child, William M. Jefferson, m. Vann Hyland Scott, Mar. 15, 1952, (dec. Dec. 1972). BS in Nursing Edn., Cath. U., Washington, 1948; cert. in psychiat. nursing, U. Paris, Paris, 1959. Indsl. nurse Curtiss Wright Air Plane Co., Lambert Field, St. Louis, 1941-44; faculty St. Thomas U., Manila, Philippines Island, 1948-50; pub. health nurse St. Louis Health Dept., 1951-56; capt. USAF Nursing Corps, Paris, 1956-60; resigned as maj. USAF (Nurse Corps), 1960, 1960; faculty St. Louis State Hosp., 1960-67; dept. head St. Vincents Hosp., St. Louis, 1967-68; faculty RN, creator psychiat. program Sch. of Nursing Jewish Hosp., 1968-72; adminstrv. nurse St. Louis State Hosp., 1972-84; initiated first psychiatric program sch. nursing, Jewish Hosp. Author: (short story) Two Letters (hon. mention) 1962, Storms, 1987. Past bd. dirs. county bd. Mo. U., 1984-88; hon. citizen Colonial Williamsburg Va.; chmn. Rep. Presdl. Task Force. Recipient Key to Colonial Williamsburg, Va., Medal of Merit, Rep. Presdl. Task Force, 1992; named to Rep. Presdl. Task Force Honor Roll, 1993. Mem. AAUW, NAFE, Internat. Fedn. Univ. Women, Internat. Soc. Quality Assurance in Health Care, Am. Biog. Inst. (life, governing bd.), Cambridge Centre Eng., Internat. Platform Assn. Roman Catholic. Avocations: music, boating, horseback riding, dog sled riding, travel.

SCOTT, ANGELA FREEMAN, civil engineer; b. Vicksburg, Miss., Sept. 26, 1967; d. George and Demmer Lee (Rollins) F.; m. Medgar Demetrius Scott, Aug. 15, 1992. BSME, Howard U., 1990. Civil engr. U.S. Army C.E., Vicksburg, Miss., 1991—. Mem. ASCE, Soc. Am. Mil. Engrs. Office: US Army CE 3909 Halls Ferry Rd Vicksburg MS 39180

SCOTT, BRUCE DOUGLAS, physicist; b. Kingsville, Tex., Sept. 3, 1958; s. Douglas L. and Patricia G. (MacGregor) S. BS in Physics, Oreg. State U., 1979; MS in Physics, U. Md., 1982, PhD, 1985. Postdoctoral fellow U. Tex., Inst. for Fusion Studies, Austin, 1985-87; Fulbright scholar Inst. Superior Tecnico, Lisboa, Portugal, 1987-88; vis. scientist Max Planck Inst. für Plasmaphysik, Garching bei Munich, Germany, 1988-89, postdoctoral fellow, 1989-90, rsch. scientist, 1990—. Contbr. articles to profl. jours. Mem. Com. for Internat. Freedom of Scientists, N.Y.C., 1991. Fulbright scholar, 1987-88; recipient fellowship NASA, 1980-82. Mem. Am. Phys. Soc., Am. Geophys. Union, Amnesty Internat., The Planetary Soc., The Cousteau Soc. Avocations: history, science fiction, chess, hiking, travel. Office: Max Planck Int Plasmaphysik, Boltzmannstrasse 2, W-8046 Garching Germany

SCOTT, CARL DOUGLAS, aerospace engineer; b. San Antonio, Nov. 14, 1937; s. Sparkman and Mildred (Snell) S.; m. Marcia Margaret Campbell, June 10, 1960; children: Stephen, Jacquelyn, James, Erin. BA, Rice Inst., 1960; PhD, U. Tex., 1969. Aerospace engr., physicist NASA Johnson Space Ctr., Houston, 1963-85, aerothermodynamics group leader, 1985—; vis. prof. U. Paris Nord, Villetaneuse, France, 1991-92; adj. prof. U. Houston, 1990—. Assoc. editor Jour. Thermophysics and Heat Transfer, 1987-90; co-editor: Thermophysical Aspects of Reentry Flows, 1986; contbr. articles to profl. jours. including Jour. Thermophysics and Heat Transfer. Lt. (j.g.) USN, 1960-63. Named Tech. Person of Yr., Coun. Tech. Socs., 1987. Fellow AIAA (assoc., thermophysics tech. com. 1984-86). Methodist. Achievements include patent for an aeroassisted orbital transfer vehicle; first to determine temperature dependent catalytic recombination coefficients for oxygen and nitrogen on space shuttle tiles. Home: 1014 Lynn Cir Friendswood TX 77546 Office: EG3 NASA Johnson Space Ctr Houston TX 77058

SCOTT, DAVID KNIGHT, physicist, university administrator; b. North Ronaldsay, Scotland, Mar. 2, 1940; married, 1966; 3 children. BSc, Edinburgh U., 1962; DPhil in Nuclear Physics, Oxford U., 1967. Rsch. officer nuclear physics lab. Oxford U., 1970-73; rsch. fellow nuclear physics Balliol Coll., 1967-70, sr. rsch. fellow, 1970-73; physicist Lawrence Berkeley Lab. U. Calif., 1973-75, sr. scientist nuclear sci., 1975-79; prof. physics, astronomy and chemistry Nat. Superconducting Cyclotron Lab. Mich. State U., East Lansing, 1979-93; Hannah disting. prof. physics, astronomy and chemistry Mich. State U., East Lansing, 1979-86, assoc. provost, 1983-86, provost, v.p. acad. affairs, 1986-92; Hannah Disting. prof. learning, sci. and soc. Nat. Superconducting Cyclotron Lab. Mich. State U., East Lansing, 1992-93; chancellor U. Mass., Amherst, 1993—. Fellow Am. Phys. Soc. Office: U Mass Office of Chancellor 374 Whitmore Adminstrn Blvd Amherst MA 01003

SCOTT, DAVID LAWRENCE, mechanical engineer; b. Omaha, Mar. 29, 1956; s. Lawrence Eugene and Donna Grace (Webb) S.; m. Julie Ann Viola, July 3, 1982; children: Anthony Michael, Zachary Mathew. BSME, Mich. State U., 1986. Staff engr. Raytheon Co., Sudbury, Mass., 1986; plant engr. Curtis Metal Finishing Co., Sterling Heights, Mich., 1987—. Mem. Soc. Mfg. Engrs., Assn. Energy Engrs. Achievements include development of torque/tension test equipment. Office: Curtis Metal Finishing 6645 Sims Dr Sterling Heights MI 48313

SCOTT, EUGENE RAY, electrical engineer, consultant; b. Belgrade, Nebr., Feb. 9, 1934; s. Leonard Allen and Lorene Grace (Schoening) S.; m. Carole Mae Trussell, Aug. 30, 1953; children: Blake, Shawn (dec.), Derek, Gena. BSEE, U. Nebr., 1962. Registered profl. engr., Colo., Ga., Tex., Nebr. Project engr. GE, San Jose, Calif., 1962-71; dir. projects Nebr. Pub. Power Dist., Columbus, 1971-74; mgr. projects Stone and Webster Engring. Corp., Boston, 1974-82; v.p. Commonwealth Elec., Lincoln, Nebr., 1982-85; dir. projects Cleve. Elec., Atlanta, 1985-87; cons. Scott and Assocs., Vail, Colo., 1989-90; prin. Kellogg, Littleton, Colo., 1990-91; cons. Scott and Assocs., Littleton, Colo., 1991-92; pres. Advanced Analytical Solutions, Inc., Englewood, Colo., 1992—. With USN, 1954-58. Mem. NSPE, Am. Nuclear Soc., Project Mgmt. Inst., C. of C. Achievements include records in installations of electrical materials at nuclear power plant. Home: 931 Mt Rose Way Golden CO 80401 Office: Advanced Analytical Solutions Inc 7400 E Orchard Rd Ste 160 Englewood CO 80111

SCOTT, H(ERBERT) ANDREW, retired chemical engineer; b. Marion, Va., Mar. 29, 1924; s. Charles Wassum and Carolyn Enyde (Snider) S.; widowed; children: Mark Andrew, Paul Ethan; m. Helen R. LaFollette, July 21, 1984. BSChemE, Va. Tech. Inst. and State U., 1944, MSChemE, 1947. Registered profl. engr., Tenn. Chem. engr. Tenn. Eastman Co., Kingsport, 1947-55, asst. to works mgr., 1955-60, supt. glycol dept., 1960-64, supt. polymers div., 1964-67, dir. engring. div., 1970-87; plant mgr. Holston Def. Corp., Kingsport, 1967-70, dir. systems devel., 1971-73, dir. engring. dvsn., 1974-87; vis. prof. U. Alaska, Fairbanks, 1988-89; mem. standards com. Eastman Kodak Co., Rochester, N.Y., 1982-87. Mem. mayor's adv. com. City of Kingsport, 1971-76; chmn. Kingsport Park Commn., 1976-80. Sgt. AUS, 1944-46. Named Engr. of yr., Tenn. Soc. Profl. Engrs., 1984. Fellow AICE (mem. Inst. Chem. Process Safety 1984-87, vocat. guidance com. 1960-64, chmn. local sect. 1956); mem. Am. Soc. Engring. Edn., Am. Soc. Engring. Mgmt., Kiwanis. Republican. Presbyterian. Achievements include patent in process for manufacture of acetic anhydride, implement pioneering quality management for knowledge workers, design for manufacturing chemicals from coal. Home: 4512 Chickasaw Rd Kingsport TN 37664-2110

SCOTT, IAN LAURENCE, biomedical engineer; b. Driffield, Yorkshire, Eng., Apr. 22, 1958; came to U.S., 1989; s. Laurence and Barbara Mary S.; m. Susan Schofield, July 16, 1983; children: Thomas Andrew Scott, Richard David Scott. BS in Physiology, U. Bristol, Eng., 1979; PhD, U. Oxford, 1983. Product devel. engr. Viggo-BOC Health Care, Swindon, Eng., 1983-85; product devel. mgr. Viggo-BOC Health Care, Swindon, 1985-87; lead engr. BOC Group Tech. Ctr., Murray Hill, N.J., 1987-89; sect. mgr. BOC Group Tech. Ctr., Murray Hill, 1989—. Contbr. articles to profl. jours. Post doctoral fellow (rsch. award) Wellcome Trust, Oxford, 1983. Mem. Biol. Engring Soc. London, Biomed. Engring. Soc. Calif. Achievements include British patent for improvement in sensors (co-inventor), devel. of fiber optic sensors for PO2, pH and pressure measurement for intavascular use.

SCOTT, J(AMES) MICHAEL, research biologist; b. San Diego, Sept. 20, 1941; s. James Melvin Scott and Eileen May (Rose) Scott Busby; m. Sharon Louise Middleton, Dec. 18, 1966; children: Kevin Charles, Heather Ann. BS in Biology, San Diego State U., 1966, MA in Biology, 1970; PhD in Zoology, Oreg. State U., 1973. Project dir. U.S. Fish and Wildlife Service, Mauna Loa Station, Hawaii Nat. Park, 1974-84, Condor Research Ctr., Ventura, Calif., 1984-86; leader Coop. Fish and Wildlife Research Unit, Moscow, Idaho, 1986—. Author: (monograph) Forest Bird Communities of the Hawaiian Islands, 1986 (Wildlife Publs. award Wildlife Soc. 1986); editor: Estimating Numbers of Terrestrial Birds, 1981, Hawaii's Terrestrial Ecosystems: Protection and Management, 1985; also articles. Vol. Peace Corps, Colombia, 1963-65; mem. Centennial Commn., State of Idaho, 1989—. Recipient R.E. Dimick award, Oreg. chpt. Wildlife Soc., 1973, Spl. Achievement award U.S. Fish and Wildlife Service, 1970, 81, 87. Mem. Ecol. Soc. Am., Am. Ornithologists Union (elective), Pacific Seabird Group (pres. 1973-75), The Nature Conservancy (bd. dirs. Idaho chpt.). Avocations: bird watching, body surfing, skin diving, stamp collecting. Home: 1130 Kamiaken St Moscow ID 83843-3855 Office: U Idaho Coop Fish and Wildlife Rsch Inst Coll of Forestry Moscow ID 83843

SCOTT, JAMES NOEL, quality assurance professional; b. Pottsville, Pa., June 14, 1939; s. James F. and Ethel (Noel) S.; m. Marie D. Laelas, Oct. 5, 1963 (div. 1988); children: James N., Christopher N.; m. Connie Harris Johnson, Aug. 8, 1989; children: Cody Johnson, Chris Slone, Greg Slone. BS in Metallurgy, Pa. State U., 1963; MS in Quality, Ea. Mich. U., 1992. Registered profl. engr., N.Y. Purchasing mgr. SCM Corp., Cortland, N.Y., 1969-80; gen. mgr. Hinderliter Heat Treating, Tulsa, 1980-83; plant mgr. Airco Temescal, Clairmore, Okla., 1983-85; quality mgr. Wolverine Tube Co., Decatur, Ala., 1985-87; gen. mgr. Conn. Metall. Processes, Bridgeport, Conn., 1987-89; rsch. metallurgist Fed. Mogul, Ann Arbor, Mich., 1989-91; quality mgr. Associated Spring, Saline, Mich., 1991-92; mem. adv. bd. indsl. tech. quality Ea. Mich. U., Ypsilanti, 1991—. Producer, actor Chelsea (Mich.) Area Players, 1990—; pres., campaign chmn. United Way Cortland County, 1975-78. Mem. ASTM (mem. various coms.), NSPE, ASM, Am. Soc. Quality Control (sr., editor newsletter 1991—, chairperson edn. 1992—). Home: 618 Grant St Chelsea MI 48118

SCOTT, LARRY MARCUS, aerospace engineer, mathematician; b. Bingham Canyon, Utah, June 14, 1945; s. Wright Marcus Scott and Margaret Ruth (Jackson) Sturzenegger; m. Paula Inger Elisabeth Kjellman, Aug. 21, 1972; children: Paul Marcus, Laura Elizabeth. BS, Brigham Young U., 1971; MS, Boston U., 1983. Engr. lifting body re-entry mechs. Douglas Missile and Space Systems, Sant Monica, Calif., 1965-66; missionary and dist. leader Ch. of Jesus Christ of Latterday Saints, Finland, 1966-69; researcher Math. Dept. Brigham Young U. for Air Force, Provo, Utah, 1969-71; researcher Russian Translation Linguistics Brigham Young U. for Nat. Security Agy., Provo, 1971-72; internship in Plasma Physics, Physics Dept. Brigham Young U., Provo, 1972-73; sci. programmer Lockheed Elect. Co., Edwards AFB, Calif., 1973-75; mathematician 6521 Range Squadron, 6250 Test Wing, Edwards AFB, 1976-80 1982-85; exchange sci. Inst. for Flight Dynamics, DFVLR, Federal Republic Germany, 1980-82; mathematician A.F. Logistics Command, Hill AFB, Utah, 1985-87; software engineer Ball Systems Engring. Div., Edwards AFB, 1987-89; nav. engr. Short Range Attack Missile, 1989-91; avionics engr. F-15E, Edwards AFB, Calif., 1992—. Co-author: Invention disclosure on laser nozzle, 1976, Evaluation of F-15E Operational Flight Program, AF Tech. Report TR 92-021; author: report for DFVLR in Germany, 1982, USAF Disclosure, 1984. Quorum Pres. Ch. of Latter Day Saints, Palmdale, Calif. 1973-74, Lancaster, Calif. 1984-85, Quorum Counselor, Lancaster 1988—; asst. varsity scout, Bountiful, Utah 1985-86. Recipient Bank of Am. Language Award, Hawthorne, Calif. 1963, Scholarship to Brigham Young U. 1963, Hon. Cert. Defense Language Inst., Monterey, Calif. 1980. Republican. Avocations: computers, classical music, playing baroque music, musical improvisation, mathematics, genealogy.

SCOTT, LEONARD LEWY, JR., mathematician, educator; b. Little Rock, Oct. 17, 1942; s. Leonard Lewy and Mary Ella (Simcoe) S.; m. Mary Ellena Broach; children: Mary Lisette, Walter Lewy. BA, Vanderbilt U., 1964; PhD, Yale U., 1968. Instr. U. Chgo., 1968-70; asst. prof. Yale U., New Haven, 1970-71; from assoc. prof. to prof. U. Va., Charlottesville, 1971-87, McConell/Bernard prof., 1987—; vis. fellow All Souls Coll., Oxford, 1987. Mem. editorial bd. Jour. Algebra, 1990—; contbr. articles to profl. jours. including Annals Math., Inventionés, Publ. Math IHES. Rsch. grantee NSF, 1972—, Sci. and Engring. Rsch. Coun. (Eng.), 1983, 92. Mem. Am. Math. Soc. (chmn. S.E. speaker sect. 1985-86, nominating com. 1988-89, editorial bd. univ. lectr. series 1987—). Achievements include origination of maximal subgroups program in finite group theory; development of isomorphism problem for group rings; invention of quasi hereditary algebras; proof of Carlson's conjecture on group cohomology; demonstration of generic cohomology for finite and algebraic groups. Home: 3250 Tearose Ln Charlottesville VA 22903 Office: U Va Dept Math New Cabell Dr Charlottesville VA 22903

SCOTT, MARY ELLEN ANN, chemist; b. Cleve., Aug. 17, 1949; d. Clarence Eugene and Ruth Marie (Sweeney) S.; m. Gene Alan Nelson, Sept. 28, 1985; children: Jennifer Ann, Alice Marie. B in Chemistry, U. Dayton, 1971; Masters, U. Ariz., 1977; PhD, U. Akron, 1993. With Technicon Instrument Inc., Tarrytown, N.Y., 1982-87, Biophysics Co., Cleve., 1990-. tcht. Pima C.C., 1974-78, Cuyahoga C.C., 1992—. Mem. Young Scientist Network. mem. Am. Chem. Soc., Am. Assn. Clin. Chemistry, Ohio Acad. Scis., Protein Soc. Roman Catholic. Achievements include patents for Phospholipid Conjugates and Thier Preparation, and New and Sensitive Substrates for Beta-Galactosidase. Home: 7374 Brookside Pkwy Middleburg Heights OH 44130-5468 Office: U Akron Akron OH 44325-3601

SCOTT, MATTHEW PETER, biology educator; b. Boston, Jan. 30, 1953; s. Peter Robert and Duscha (Schmid) S.; m. Margaret Tatnall Fuller, May 13, 1990; 1 child, Lincoln Fuller. BS, MIT, 1975, PhD, 1980. Postdoctoral tng. Ind. U., Bloomington, 1980-83; from asst. prof. to assoc. prof. U. Colo., Boulder, 1983-90; prof. Stanford (Calif.) U., 1990—. Recipient Passano Young Investigator award Passano Found., 1990. Achievements include research in developmental genetics, in particular, homeotic genes. Office: Stanford U Sch Med Dept Developmental Biology Stanford CA 94305-5427

SCOTT, MURRAY LESLIE, aerospace engineer; b. Melbourne, Victoria, Australia, Feb. 18, 1958; s. Reginald Henry and Rene Price (Miles) S.; m. Francisca Margaretha deRuwe, Dec. 18, 1982; children: Karen Victoria, Richard Murray. BEng (Aero.), Royal Melbourne Inst. Tech., 1979; MSc (Aircraft Design), Cranfield Inst. Tech., Bedford, Eng., 1982. Chartered profl. engr., Australia. Tutor Royal Melbourne Inst. Tech., 1979; engr. Commonwealth Aircraft Corp., Melbourne, 1981-85; structural analyst Northrop Corp., L.A., 1986; sr. lectr. aerospace engring. Royal Melbourne Inst. Tech., 1987—. Contbr. articles to profl. jours. Mem. AIAA (sr.), Royal Aero. Soc. (hon. sec. Melbourne br. 1984-85, councillor 1987—), Instn. of Engrs. Australia (sr.), Soc. for Advancement of Material and Process Engring., Soc. for Exptl. Mechanics, Internat. Com. Composite Materials, Australian Composite Structures Soc. (chmn. Melbourne br.). Office: Coopertive Rsch Ctr Aerospace Structures, 506 Lorimer St, Fishermen's Bend Victoria 3207, Australia

SCOTT, NORMAN L., engineering consultant; b. Meadow Grove, Nebr., Oct. 17, 1931; s. Laurence Ray Scott and Ruth Louise Braun; m. Joan Culbertson, Jan. 21, 1956; 1 child, Douglas Jay. BS in Civil Engring., U. Nebr., 1954. Registered profl. engr., Ill., Fla., Md., Minn., Va., Tex.; registered structural engr., Ill. Sales engr. R.H. Wright & Son, Ft. Lauderdale, Fla., 1956-58; mgr. Wright of Palm Beach, West Palm Beach, Fla., 1958-59; exec. sec. Prestressed Concrete Inst., Chgo., 1959-63; gen. mgr. Wiss, Janney, Elstner & Assoc., Northbrook, Ill., 1963-66; pres., chmn. The Consulting Engrs. Group Inc., Mt. Prospect, Ill., 1966—. 1st lt. USAF, 1954-56. Mem. Am. Concrete Inst. (pres. 1983-84, Henry C. Turner medal 1993), Ill. Soc. Profl. Engrs. (pres. North Shore chpt. 1962). Republican. Home: Consulting Engineers Gp 701 Chatham Rd Glenview IL 60025 Office: The Consulting Engrs Group 55 E Euclid Ave Mount Prospect IL 60056

SCOTT, NORMAN ROSS, electrical engineering educator; b. N.Y.C., May 15, 1918; s. George Norman and Lillias B.H. (Ogg) S.; m. Marjorie M. Fear, Apr. 6, 1950; children: Mari, George, Ian, Charles. BS, MS, MIT, 1941; PhD, U. Ill., 1950. Asst. prof. elec. engring. U. Ill., Urbana, 1946-50; asst. prof. to prof. elec. engring. U. Mich., Ann Arbor, 1951-87, assoc dean Coll. Engring., 1965-68; dean Dearborn Campus U. Mich., 1968-71, prof. emeritus of elec. engring. and computer sci., 1987—; cons. Nat. Cash Register Co., Dayton, 1956-65; mem. math. and computer sci. rsch. adv. com. AEC, Washington, 1961-63. Editor-in-chief IEEE Trans. on Computers, N.Y.C., 1961-65; author: Analog and Digital Computer Technology, 1959, Electronic Computer Technology, 1970, Computer Number Systems and Arithmetic, 1985. Maj. U.S. Army, 1941-46. Fellow IEEE. Home: 2260 Gale Rd Ann Arbor MI 48105-9512 Office: U Mich EECS Dept Ann Arbor MI 48109

SCOTT, OWEN MYERS, JR., nuclear engineer; b. Birmingham, Ala., Oct. 15, 1952; s. Owen Myers and Sarah (Watson) S.; m. Eleanor Eason, July 15, 1978, 1 child, Owen Myers III. BCE, Auburn U., 1977, MBA, U. Ala., Birmingham, 1981; MS Nuclear Engring., Ga. Inst. Tech., 1986. Registered profl. engr., Ala., Fla., Ga., Miss. Physics lab. instr. Auburn (Ala.) U., 1977; civil/structural design engr. So. Co. Svcs., Inc., Birmingham, 1977-84, nuclear analysis engr., 1984-90; sr. nuclear engr. So. Nuclear Co., Birmingham, 1991—; treas. So. Investors, Birmingham, 1988—. Co-author (computer program) radiol. shielding analysis, 1986. Instr., advisor Jr. Achievement/Project Bus., Birmingham, 1982. Mem. Am. Nuclear Soc., ASCE, Nat. Mgmt. Assn., Omicron Delta Epsilon. Presbyterian. Avocations: golf, music, computer programming, clockmaking, woodworking. Home: 3876 Timberline Way Birmingham AL 35243-2452 Office: So Nuclear Co 40 Inverness Center Pkwy PO Box 1295 Birmingham AL 35201

SCOTT, PAMELA MOYERS, physician assistant; b. Clarksburg, W.Va., Jan. 5, 1961; d. James Edward and Norma Lee (Holbert) Moyers; m. Troy Allen Scott, July 19, 1986. BS summa cum laude, Alderson-Broaddus Coll., 1983. Cert. physician asst. Physician asst. Weston (W.Va.) State Hosp., 1983-84, Rainelle (W.Va.) Med. Ctr., 1984—; speaker W. Va. Rural Health Conf., Morgantown, 1991; keynote speaker Alderson-Broadous Coll. Ann. Physician Assn. Banquet, 1992; presenter W. Va. Task Force on Adolescent Pregnancy and Parenting State Meeting; guest on Lifetime TV med. program Physician Jour. Update. Mem. W. VA. State Task Force on Adolescent Pregnancy and Parenting, 1992—. Named Young Career Woman of Yr. Rainelle chpt. and Dist. V. of W.Va., Citation of Honor at State Level of Competition, Bus. and Profl. Woman Clubs, 1986, Outstanding Physician Asst. of Yr. Am. Acad. Physician Assts., 1991; nominated for Rural Physician Asst. of Yr., 1992. Fellow Am. Acad. Physicians Asst. (del. mem. to People's Rep. China, 1992, W.Va. chief del. Ho. of Dels. at Nat. Conv. 1992, W. Va. del. 1993, mem. rural health caucus 1991—, pub. edn. com. 1992—). W. Va. Assn. Physician Assts. (chair membership com. 1989-91, nominations and elections com. 1990-91, pres. 1991—, presenter Continued Med. Edn. Conf. 1993). Republican. Baptist. Avocations: reading, handicrafts, shopping. Home: PO Box 43 Williamsburg WV 24991 Office: Rainelle Med Ctr 645 Kanawha Ave Rainelle WV 25962

SCOTT, RICHARD LYNN, data processing executive; b. Ora, Ind., Mar. 1, 1941; s. Harold Hophius and Maxine Louise (Strevey) S.; m. Karen Louise Kamp, Aug. 9, 1963; 1 child, Jonathon William. Student, Purdue U., 1959-61. Design engr. Kaydon, Muskegon, Mich., 1961-63; mgr. data processing Bastian Blessing, Grand Haven, Mich., 1963-79; dir. data programming

Oliver Machine Co., Grand Rapids, Mich., 1979-82; mgr. CAD/CAM-CIM Steelcase, Inc., Grand Rapids, 1983—. Cons. United Way Kent County, Grand Rapids, 1986. Mem. Data Processing Mgmt. Assn. (pres. 1984-85), Am. Prodn. and Inventory Control Soc. (regional dir. 1984-85), Soc. Mfg. Engrs. (sr.), Computer and Automated Systems Assn. (sr.). Avocations: hiking, reading, Civil War re-enactment. Home: 18063 Lake Hills Dr Spring Lake MI 49456-9412 Office: 901 44th St SE Grand Rapids MI 49508-7594

SCOTT, ROLAND BOYD, pediatrician; b. Houston, Apr. 18, 1909; s. Ernest John and Cordie (Clark) S.; m. Sarah Rosetta Weaver, June 24, 1935 (dec.); children—Roland Boyd, Venice Rosetta, Estelle Irene. BS, Howard U., 1931, MD, 1934, DSc (hon.), 1987. Diplomate Nat. Bd. Med. Examiners, Am. Bd. Pediatrics. Gen. Edn. Bd. fellow U. Chgo., 1936-39; faculty Howard U., Washington, 1939—, prof. pediatrics, 1952-77, disting. prof. pediatrics and child health, 1977—, chmn. dept. pediatrics, 1945-73; dir. Ctr. for Sickle Cell Anemia, 1973—; chief pediatrician Freedmen's Hosp., 1947-73; professorial lectr. in child health and devel. George Washington U. Sch. Medicine, 1971-75; staff Children's Hosp., Providence Hosp., Columbia Hosp., D.C. Gen. Hosp., Washington Hosp. Center; cons. in pediatrics to NIH, hosps.; Mem. com. Pub. Health Adv. Council, 1964—, U.S. Children's Bur., 1964-70; mem. Nat. Com. for Children and Youth.; mem. sickle cell adv. com. NIH, 1983-88. Author: (with Althea D. Kessler) Sickle Cell Anemia and Your Child, (with C.G. Uy) Guidelines For Care of Patients With Sickle Cell Disease; Editor: Procs. 1st Internat. Conf. on Sickle Cell Disease: A World Health Problem, 1979; Mem. editorial bd.: Advances in the Pathophysiology, Diagnosis and Treatment of Sickle Cell Disease, 1982, Clin. Pediatrics, 1962-80, Jour. Nat. Med. Assn, 1978; cons. editor: Medical Aspects of Human Sexuality; editorial bd.: Annals of Allergy, 1977-82, Pediatrics, 1988. Recipient Sci. and Community award Medico-Chirurgical Med. Soc. D.C., 1971, Community Service award Med. Soc. D.C., 1972, award for contbns. to sickle cell research Delta Sigma Theta, 1973, 34 years Dedicated and Disting. Service award in pediatrics dept. pediatrics Howard U., 1973, Faculty award for excellence in research, 1974, Alumni award for Service and Dedication, 1984, spl. recognition plaque for contbns. to research and edn., 1985, Ronald McDonald Children Charities award, 1987; certificate of appreciation Sickle Cell Anemia Research and Edn., 1977; Mead Johnson award D.C. chpt. Am. Acad. Pediatrics, 1978; Percy L. Julian award We Do Care, Chgo., 1979; Abraham Jacobi Meml. award AMA and Acad. Pediatrics, 1985; also plaques for work in sickle cell disease various orgns. including Columbia U. Sickle Cell Ctr., 1984, Elks, Nat. Assn. Med. Minority Educators, NIH, 1980-82. Fellow Am. Coll. Allergists (Disting. Service award 1977); mem. AMA, Am. Hematology Soc., Am. Pediatric Soc. (John Howland award 1991), Soc. Pediatric Research, Am. Acad. Allergy (v.p. 1966-67), Am. Acad. Pediatrics (mem. com. on children with handicaps, cons. head start program), Am. Fedn. Clin. Research, Nat. Med. Assn. (Disting. Service medal 1966), AAAS, Internat. Corr. Soc. Allergists, Assn. Ambulatory Pediatric Services, AAUP, Internat. Congress Pediatrics, Can. Sickle Cell Soc. (hon. life), Phi Beta Kappa, Sigma Xi (Percy L. Julian award Howard U. chpt. 1977), Kappa Pi, Beta Kappa Chi, Alpha Omega Alpha, others. Achievements include research and publications on sickle cell anemia, growth and devel. of infants and children, allergy in children. Home: 1316 Fenwick Ln Apt 601 Silver Spring MD 20910-3502 Office: Howard U 2121 Georgia Ave NW Washington DC 20059-0001

SCOTT, STEVEN DONALD, geology educator, researcher; b. Fort Frances, Ont., Can., June 4, 1941; s. Donald West and Shirley Margaret (Casselman) S.; m. Barbara Joan Armstrong, Dec. 20., 1962; children: Susan Joan, Donald Montgomery. BSc, U. Western Ont., 1963, MSc, 1964; PhD, Pa. State U., State College, 1968. Rsch. assoc. Pa. State U., State Coll., 1968-69; asst. prof. geology U. Toronto, Ont., 1969-72, assoc. prof., 1972-79, prof., 1979—, chmn. geol. engring., 1988—; pres. 507999 Ont. Ltd., Toronto, 1982—; bd. dirs. Econ. Geology Pub. Co.; lectr. in field. Contbr. numerous articles to profl. jours. Fellow Royal Soc. Can. (convenor for earth scis. 1989-90, Bancroft award 1990), Soc. Econ. Geologists (mem. coun. 1986-88, Lindgren award 1978), Mineral. Soc. Am., Geol. Assn Can. (tech. program chmn. ann. meeting 1991), Mineral. Assn. Can. (Past Pres.' award 1988), others. Avocations: travel, fishing, swimming; researcher in marine geology and ore deposits. Office: U of Toronto, Dept of Geology, Toronto, ON Canada M5S 3B1

SCOTT, STEVEN MIKE, mechanical engineer; b. Albuquerque, Jan. 11, 1955; s. Ivy Louis and Anna Lou (Robertson) S.; m. Sheila Ann Hodge, Dec. 17, 1973; children: Shannon Elizabeth, Tamara Nicole, Nathan Robert, Taylor Steven. BSME, So. Meth. U., 1984, M Engring. Mgmt., 1986. Mech. engr. Tex. Instruments, Inc., Dallas, 1975-91; sr. mech. engr. NewCo Partnership, Amelia, Ohio, 1991-92, Steart Filmscreen Corp., Amelia, 1992—. Contbr. articles to tech. publs. Mem. Am. Soc. Precision Engrs., Internat. Soc. Optical Engrs. Presbyterian. Achievements include development of processes for machining with single point diamond tools, machining of fresnel for projection television screens, laser rangefinder into thermal sight for military tank system. Office: Stewart Filmscreen Corp 3919 Bach Buxton St Amelia OH 45102

SCOTT, WILLIAM LEONARD, research chemist, educator; b. Phila., Nov. 7, 1945; s. Michael and Catherine (Fadde) S.; children: Stephen, Judy. BA, Williams Coll., 1967; PhD, UCLA, 1972. Postdoctoral Rockefeller U., N.Y.C., 1972-73, Calif. Tech. Inst., Pasadena, 1973-74; sr. chemist Eli Lilly Rsch. Labs., Indpls., 1974-80, rsch. scientist, 1980—; adj. prof. chemistry Butler U., Indpls., 1978—, Ind./Purdue U., Indpls., 1993—. Contbr. articles to profl. jours. Mem. AAAS, Am. Chem. Soc., Sigma Chi. Roman Catholic. Achievements include 7 patents in fields of chemistry and biochemistry. Office: Eli Lilly Rsch Labs Cancer Div DC 0444 Indianapolis IN 46285

SCOUTEN, WILLIAM HENRY, chemistry educator, academic administrator; b. Corning, N.Y., Feb. 12, 1942; s. Henry and M. Anna (Kimble) S.; m. Nancy Jane Coombs, July 16, 1965; children: Lisa, Linda, Michael, William Jr., Thomas, David. BA, Houghton Coll., 1964; PhD, U. Pitts., 1969. NIH postdoctoral fellow SUNY, Stony Brook, 1969-71; asst. prof. Bucknell U., Lewisburg, Pa., 1971-77; assoc. prof. Bucknell U., Lewisburg, 1977-83, prof., 1983-84; prof., chmn. dept. chemistry Baylor U., Waco, Tex., 1984-93; dir. biotech. ctr. Utah State U., Logan, 1993—; cons. DuPont de Nemours; vis. scientist for minority inst. Fed. Am. Soc. Exptl. Biology, Washington. Author: Affinity Chromatography, 1981; editor: Solid Phase Biochem, 1983; mem. editorial bd. Bioseparation. Fulbright fellow, 1976; Dreyfus Tchr. scholar, Dreyfus Found., 1976; NSF Sci. Devel. NSF, 1978; Lindbach Disting. Tchr. Bucknell U., 1975. Mem. Am. Soc. Biol. Chemists, Am. Chem. Soc., Internat. Soc. for Biorecognition Tech., Internat. Soc. for Molecular Recognition (pres. 1990-93), Assn. for Internat. Practical Tng. (bd. dirs. 1991—). Republican. Baptist. Office: Biotechnology Ctr Utah State U Logan UT 84322-4700

SCOZZIE, JAMES ANTHONY, chemist; b. Erie, Pa., Nov. 3, 1943. AB, Gannon Coll., 1965; MS, Case Western Res. U., 1968, PhD in Chemistry, 1970. Jr. rsch. chemist ctrl. rsch. dept. Lord Corp., 1965; rsch. chemist Diamond Shamrock Corp., 1970-72, sr. rsch. chemist, 1972-76, rsch. supr. pharmaceutics, 1976-78, group leader agrl. chemistry, 1978-81, assoc. dir. agrl. chemistry rsch., 1981-83; dir. agrl. chemistry rsch. SDS Biotech Corp., 1983-85, dir. corp. rsch., 1985—; pres. Ricerca, Inc., Painesville, Ohio, 1986—. Mem. Am. Chem. Soc., Am. Mgmt. Assn. Achievements include research in structure and chemistry of peptide antibiotics, synthesis of biologically active compounds, pesticides, process studies of organic compounds, commercial evaluation, nutrition and animal health, herbicides, plant growth regulants, cardiovascular agents and antiinflammatory agents. Office: Ricerca Inc 7528 Auburn Rd Painesville OH 44077*

SCRANTON, MARY ISABELLE, oceanographer; b. Atlanta, Feb. 28, 1950; d. Robert Lorentz and Louise (Capps) S.; m. Roger Donald Flood, Jan. 3, 1981; 1 child, Stephen David Flood. BA, Mt. Holyoke Coll., 1972; PhD, MIT and Woods Hole, Oceanographic Inst., 1977. Postdoctoral assoc. Naval Rsch. Lab. Nat. Acad. Sci., Washington, 1977-79; from asst. to assoc. prof. SUNY, Stony Brook, 1979-93, prof., 1993—; marine bd. com. on undersea vehicles and nat. needs NRC, 1992—; editorial bd. Limnology & Oceanography, 1986-89, Marine Chemistry, 1993—. Contbr. articles to profl. jours. Grantee NSF, Office Naval Rsch., EPA, Hudson River Found. Mem. Am. Geophys. Union, Geochem. Soc., Am. Soc. Limnology & Oceanography, Oceanography Soc. Office: SUNY Marine Scis Rsch Ctr Stony Brook NY 11794-5000

SCREPETIS, DENNIS, consulting engineer; b. Hoboken, N.J., Feb. 12, 1930; s. George and Athanasia (Stasinos) S.; m. Betty Pravasilis, Sept. 17, 1960. Student, Stevens Inst. Tech., Bklyn. Poly. Inst., Cooper Union, Rutgers U. Registered profl. engr., N.J., N.Y. Nuclear engr. Vitro Corp. Am., N.Y.C., 1957-60; project engr. Gen. Cable Corp., Bayonne, N.J., 1960-63; project mgr. AMF Atomics, York, Pa., 1963-65; sr. staff engr. nuclear div. Combustion Engring. Corp., Windsor, Conn., 1965-66; corp. engr. Standard Packaging Corp., N.Y.C., 1966-68; v.p. engring. Eastern Schokbeton, Bound Brook, N.J., 1968-74; cons. engr., Ft. Lee, N.J., 1974—. Patentee in nuclear sci. Mem. Am. Concrete Inst., Pre-Stressed Concrete Inst., Concrete Reinforcing Inst., Nat. Safety Coun., Am. Nat. Standards Inst., Nat. Fire Protection Assn., Internat. Platform Assn., Am. Water Works Assn., Am. Inst. Steel Constrn., ASTM, Am. Welding Soc., Concrete Industry Bd., Bldg. Ofcls. and Code Adminstrs. Soc. of Am. Mil. Engrs., Nat. Forensic Ctr., Am. Biog. Inst. Rsch. Assn. (bd. dirs.), Internat. Biog. Ctr. (bd. dir.), Masonry Soc. Greek Orthodox. Home and Office: 2200 N Central Rd Fort Lee NJ 07024-7523

SCRETTAS, CONSTANTINOS GEORGE, chemistry educator; b. Thessaloniki, Macedonia, Greece, Jan. 29, 1933; s. George B. and Vassiliki C. (Samarides) S.; m. Maria A. Micha, Dec. 26, 1972; children: Vassiliki, George. Chemistry diploma, U. Thessaloniki, 1957; MS in Chemistry, Tex. A&M U., 1961; PhD in Chemistry, U. Tenn., 1966. Rsch. chemist Chemetron Corp., Newport, Tenn., 1961-63; acting dir. rsch. Lithium Corp. Am., Bessemer City, N.C., 1966-68; rsch. prof. Nat. Hellenic Rsch. Found., Athens, Greece, 1968—. Contbr. sci. articles to profl. jours. including Jour. Organic Chemistry, Jour. Am. Chem. Soc., Jour. Phys. Chemistry. Chmn. Xth Fechem Conf. on Organometallic Chemistry, Crete, 1993. 2d lt. Greek Army, 1957-59. Mem. Am. Chem. Soc., Internat. Adv. Bd. Main Group Metal Chemistry, European Fedn. Chem. Soc. (working party on organometallic chemistry). Achievements include patents in field; pioneered single electron transfer mechanisms in organic chemistry. Office: Nat Helenic Rsch Found, 48 V Constantinou Av, 116 35 Athens Greece

SCRIMGER, JOSEPH ARNOLD, research company executive; b. South Shields, Durham, Eng., Aug. 5, 1924; s. John William and Winifred (Ryles) S.; m. Stella Arkley, June 17, 1950; children: Paul, Joan Margaret, Elizabeth Anne, Malcolm George. BSc, U. Durham, 1949; MSc, U. Sask., Can., 1954, PhD, 1956. Scientist Def. Rsch. Establishment Pacific, Victoria, B.C., Can., 1956-61, 63-67, 1968-73; scientist Hudson Lab., Dobbs Ferry, N.Y., 1961-63, Def. Rsch. Establishment Valcartier, Que., Can., 1967-68, Saclant ASW Ctr., Laspezia, Italy, 1973-76, Def. Hdqrs., Ottawa, Ont., 1976-80; pres. Jasco Rsch. Ltd., Victoria, B.C., 1982—. Contbr. articles to profl. jours. With Royal Navy, 1943-46. Can. Govt. grantee, 1982-92, U.S. Office Naval Rsch., 1982-92. Mem. Acoustical Soc. Am. Office: Jasco Rsch Ltd, 205 9865 W Saanich Rd, Sidney, BC Canada V8L 3S1

SCRIVNER, JAMES DANIEL, electrical engineer, consultant; b. Neptune, N.J., June 23, 1951; s. James Jr. and Janice Hull (Carpenter) S.; m. Melinda Carol Lawson, June 29, 1982. BSEE, Tex. Tech. U., 1974. Registered profl. engr., Tex. Mgr. Burstein-Applebee, Dallas, 1975-76; owner Audioworks, Arlington, Tex., 1976-78; engr. Barry Sales, Richardson, Tex., 1978-81; systems engr. CECO, Dallas, 1982-83, Compucon, Richardson, 1983-84; owner J. Daniel Scrivner, P.E., Arlington, Tex., 1984-86; engring. mgr. City of Dallas, Tex., 1986—; com. mem. Nat. Pub. Safety Planning Adv. Com., Washington, 1987-88; sub-com. chmn. Region 40 Plan Com., Arlington, 1988-90; com. mem. Region 40 Plan Review Com., Arlington, 1990—. Mem. Associated Pub. Safety Communications Officers (com. vice-chair 1990—), Internat. Brotherhood of Magicians (sec. 1992—), Mensa. Home: 502 Eldoro Dr Arlington TX 76006 Office: City of Dallas 3131 Dawson Dallas TX 75226

SCRUDDER, EUGENE OWEN, chemist, environmental specialist; b. Chattanooga, Nov. 25, 1918; s. Owen Oscar and Ruth (Webb) S.; m. Billie Marie Card, Apr. 26, 1947; children: Delia Marie, Roy Owen. BS with honors, U. Chattanooga, 1940; MS, U. Tenn., 1943. Mem. faculty, then acting head chemistry dept. Centre Coll. Ky., Danville, 1941-43; shift supr. Hercules Powder Co., Parlin, N.J., 1943; rsch. chemist Tenn. Products and Chem. Corp., Chattanooga, 1946-49, Ind. Rsch. Inst., U. Chattanooga, 1950-52; analytical chemist TVA, Chattanooga, 1952-57; fuels lab. supr. Combustion Engring., Inc., Chattanooga, 1957-61; plant chemist Olin Corp., Charleston, Tenn., 1962-70; chemist., environ. specialist State of Tenn., Chattanooga, 1971—; adv. bd. Chattanooga State Tech. Community Coll., 1972-73. Tech. sgt. U.S. Army, 1943-46, ETO. Fellow Am. Inst. Chemists; mem. Am. Chem. Soc. Episcopalian. Achievements include design and construction of equipment for testing corrosion resistance of materials in liquid and vapor; development of time-saving method for calculating composition of complex chlor-alkali mixtures, of unique system of handling blank corrections in analyses of low level parameters by molecular absorption spectrophotometry. Home: 225 Greenleaf St Chattanooga TN 37415-5020

SCRUGGS, DAVID WAYNE, aerospace engineer; b. Denver, Sept. 11, 1940; s. R.D. and Eldean Jula (Turner) S.; m. Sonja Fay Carroll, Dec. 21, 1963; 1 child, Christopher Wayne. BA, Hardin-Simmons U., 1962. Mgr. Johnson Space Ctr. NASA, Houston, 1964-80; asst. dir. hdqrs. NASA, Washington, 1980-82; dir. Orbital Scis. Corp., Denver, 1982-85; mgr. avionics systems Martin Marietta Aerospace, Denver, 1985-88; dir. govt. programs Comml. Titan Inc., Denver, 1988-90; mgr. upper stage program Martin Marietta Civil Space and Communications, Denver, 1990-91, mgr. advance programs, 1991—. Contbr. articles to ORSA-TIM, Am. Astronautical Soc., AIAA. Deacon Bapt. ch., Houston, Washington, 1975. 1st lt. U.S. Army, 1962-64, Korea. Mem. AIAA, Am. Astronautical Soc. Home: 8026 S Ammons St Littleton CO 80123

SCUDERI, LOUIS ANTHONY, climatology educator; b. Bklyn., Sept. 13, 1954; s. Louis Samuel Scuderi and Edith Grace Ambrosio; m. Joan E. Drake, Aug. 4, 1979; children: Louis, Benjamin. BA in Geography and Econs., UCLA, 1976, MA in Geography, 1978, PhD in Geography, 1984. Rsch. scientist Calif. Rsch. & Tech., Northridge, 1984-85; rsch. geographer, database mgr. Earth Techs. Corp., Long Beach, Calif., 1985; rsch. scientist Itujhes Aircraft Co., Long Beach, 1985-88; rsch. assoc. Isotope Lab. UCLA, 1986-88; asst. prof. Boston U., 1988—. Contbr. articles to Nature, Quaternavy Rsch., Science, Arctic and Alpine Rsch. Mem. Internat. Soc. for Optical Engring., Assn. of Am. Geographers, Nat. Geographic Soc., Sigma Xi. Achievements include research in solar-climate relationships, effects of volcanic eruptions on climate; creation of three-dimensional computer database for flight simulation and real time infrared simulation. Office: Boston U Dept Geography 675 Commonwealth Ave Boston MA 02215

SCULFORT, JEAN-LOU, chemistry and physics educator; b. Liesse, Picardie, France, June 20, 1944; s. Jean and Bernadette (Desson) S.; m. Françoise Hardy, Nov. 12, 1966; 1 child, Philippe. BS, U. Champagne, France, 1966; PhD in Physics, U. Paris, 1972, DSc, 1976. Asst. prof. U. Amiens, France, 1968-76, lectr., 1976-87; lectr. U. Paris, 1983-87; prof. U. Reims, 1987—; cons. C.N.R.S. Lab., Meudon, 1970—. Inventor photoelectrochm. cell; contbr. 64 articles to profl. jours. Served with French Marines, 1970-71. Mem. Am. Electrochem. Soc., France Chem. Soc., Am. Vacuum Soc., France-Swedish Assn., Am. Inst. Physics, I.U.P.A.C., Internat. Photochem. Soc., France Phys. Soc., European Phys. Soc. Mem. Polit. Liberal Party. Avocations: golf, bridge. Home: Versailles, 73 Ave de St Cloud, F-78000 Ile De France France Office: CNRS, 1 Pl A Briand, F-92190 Meudon France also: Inst U Tech, 9 Rue de Quebec, F-10026 Troyes France

SEABORG, GLENN THEODORE, chemistry educator; b. Ishpeming, Mich., Apr. 19, 1912; s. H. Theodore and Selma (Erickson) S.; m. Helen Griggs, June 6, 1942; children: Peter, Lynne Seaborg Cobb, David, Stephen, John Eric, Dianne. AB, UCLA, 1934; PhD, U. Calif.-Berkeley, 1937; numerous hon. degrees; LLD, U. Mich., 1958, Rutgers U., 1970; DSc, Northwestern U., 1954, U. Notre Dame, 1961, John Carroll U., Duquesne U., 1968, Ind. State U., 1969, U. Utah, 1970, Rockford Coll., 1975, Kent State U., 1975; LHD, No. Mich. Coll., 1962; DPS, George Washington U., 1962; DPA, U. Puget Sound, 1963; LittD, Lafayette Coll., 1966; DEng, Mich. Technol. U., 1970; ScD, U. Bucharest, 1971, Manhattan Coll., 1976; PhD, U. Pa., 1983. Rsch. chemist U. Calif., Berkeley, 1937-39, instr. dept. chemistry, 1939-41, asst. prof., 1941-45, prof., 1945-71, univ. prof., 1971—, leave of absence, 1942-46, 61-71, dir. nuclear chem. research, 1946-58, 72-75, asso. dir. Lawrence Berkeley Lab., 1954-61, 71—; chancellor Univ. (U. Calif.-Berkeley), 1958-61; dir. Lawrence Hall of Sci. U. Calif., Berkeley, 1982-84, chmn. Lawrence Hall of Sci., 1984—; sect. chief metall. lab. U. Chgo., 1942-46; chmn. AEC, 1961-71, gen. adv. com., 1946-50; research nuclear chemistry and physics, transuranium elements.; chmn. bd. Kevex Corp., Burlingame, Calif., 1972-87, Advance Physics Corp., Santa Barbara, Calif., 1988—; mem. Pres.'s Sci. Adv. Com., 1959-61; mem. nat. sci. bd. NSF, 1960-61; mem. Pres.'s Com. on Equal Employment Opportunity, 1961-65, Fed. Radiation Council, 1961-69, Nat. Aeros. and Space Council, 1961-71, Fed. Council Sci. and Tech., 1961-71, Nat. Com. Am.'s Goals and Resources, 1962-64, Pres.'s Com. Manpower, 1964-69, Nat. Council Marine Resources and Engring. Devel., 1966-71; chmn. Chem. Edn. Material Study, 1959-74, Nat. Programming Council for Pub. TV, 1970-72; dir. Ednl. TV and Radio Center, Ann Arbor, Mich., 1958-64, 67-70; pres. 4th UN Internat. Conf. Peaceful Uses Atomic Energy, Geneva, 1971, also chmn. U.S. del., 1964, 71; U.S. rep. 5th-15th gen. confs. IAEA, chmn., 1961-71; chmn. U.S. del. to USSR for signing Memorandum Cooperation Field Utilization Atomic Energy Peaceful Purposes, 1963; mem. U.S. del. for signing Limited Test Ban Treaty, 1963; mem. commn. on humanities Am. Council Learned Socs., 1962-65; mem. sci. adv. bd. Robert A. Welch Found., 1957—; mem. Internat. Orgn. for Chem. Scis. in Devel., UNESCO, 1981-92, pres., 1981-92, pres. emeritus, 1992—; mem. Nat. Commn. on Excellence in Edn., Dept. Edn., 1981-83; co-discoverer elements 94-102 and 106: plutonium, 1940, americium, 1944-45, curium, 1944, berkelium, 1949, californium, 1950, einsteinium, 1952, fermium, 1953, mendelevium, 1955, nobelium, 1958, element 106, 1974; co-discoverer nuclear energy isotopes Pu-239, U-233, Np-237, other isotopes including I-131, Fe-59, Te-99m, Co-60; originator actinide concept for placing heaviest elements in periodic system. Author: (with Joseph J. Katz) The Actinide Elements, 1954, The Chemistry of the Actinide Elements, 1957, (with Joseph J. Katz and Lester R. Morse) 2d ed. Vols. I & II, 1986, The Transuranium Elements, 1958, (with E.G. Valens) Elements of the Universe, 1958 (winner Thomas Alva Edison Found. award), Man-Made Transuranium Elements, 1963, (with D.M. Wilkes) Education and the Atom, 1964, (with E.K. Hyde, I. Perlman) Nuclear Properties of the Heavy Elements, 1964, (with others) Oppenheimer, 1969, (with Ben Loeb) Stemming the Tide, 1987, (with W.R. Corliss) Man and Atom, 1971, Nuclear Milestones, 1972, (with Ben Loeb) Kennedy, Khruschev and the Test Ban, 1981, (with Walt Loveland) Elements beyond Uranium, 1990; editor: (with Ben Loeb) The Atomic Energy Commission Under Nixon, 1992, Transuranium Elements: Products of Modern Alchemy, 1978, (with W. Loveland) Nuclear Chemistry, 1982; assoc. editor Jour. Chem. Physics, 1948-50; mem. editorial adv. bd. Jour. Inorganic and Nuclear Chemistry, 1954-82, Indsl. Rsch., Inc, 1967-75; mem. adv. bd. Chem. and Engring. News, 1957-59; mem. editoril bd. Jour. Am. Chem. Soc, 1950-59, Ency. Chem. Tech., 1975—, Revs. in Inorganic Chemistry, 1977—; mem. hon. editorial adv. bd. Internat. Ency. Phys. Chemistry and Chem. Physics, 1957—, Nuclear Sci. and Techniques, Chinese Nuclear Soc.; mem. panel Golden Picture Ency. for Children, 1957-61; mem. cons. and adv. bd. Funk and Wagnalls Universal Standard Ency, 1957-61; mem.: Am. Heritage Dictionary Panel Usage Cons., 1964—; contbr. articles to profl. jours. Trustee Pacific Sci. Center Found., 1962-77; trustee Sci. Service, 1965—, pres., 1966-88, chmn. 1988—; trustee Am.-Scandinavian Found., 1968—, Ednl. Broadcasting Corp., 1970-72; bd. dirs. Swedish Council Am., 1976—, chmn. bd. dirs., 1978-82; bd. dirs. World Future Soc., 1969—, Calif. Council for Environ. and Econ. Balance, 1974—; bd. govs. Am. Swedish Hist. Found., 1972—. Recipient John Ericsson Gold medal Am. Soc. Swedish Engrs., 1948; Nobel prize for Chemistry (with E.M. McMillan), 1951; John Scott award and medal City of Phila., 1953; Perkin medal Am. sect. Soc. Chem. Industry, 1957; U.S. AEC Enrico Fermi award, 1959; Joseph Priestley Meml. award Dickinson Coll., 1960; Sci. and Engring. award Fedn. Engring. Socs., Drexel Inst. Tech., Phila., 1962; named Swedish Am. of Year, Vasa Order of Am., 1962; Franklin medal Franklin Inst., 1963; 1st Spirit of St. Louis award, 1964; Leif Erikson Found. award, 1964; Washington award Western Soc. Engrs., 1965; Arches of Sci. award Pacific Sci. Center, 1968; Internat. Platform Assn. award, 1969; Prometheus award Nat. Elec. Mfrs. Assn., 1969; Nuclear Pioneer award Soc. Nuclear Medicine, 1971; Oliver Townsend award Atomic Indsl. Forum, 1971; Disting. Honor award U.S. Dept. State, 1971; Golden Plate award Am. Acad. Achievement, 1972; John R. Kuebler award Alpha Chi Sigma, 1978; Founders medal Hebrew U. Jerusalem, 1981; Henry DeWolf-Smyth award Am. Nuclear Soc., 1982, Great Swedish Heritage award, 1984, Ellis Island Medal of Honor, 1986, Vannevar Bush award NSF, 1988, Nat. Medal of Sci. NSF, 1991, Royal Order of the Polar Star Sweden, 1992; decorated officier Legion of Honor France; Daniel Webster medal, 1976. Fellow Am. Phys. Soc., Am. Inst. Chemists (Pioneer award 1968, Gold medal award 1973), Chem. Soc. London (hon.), Royal Soc. Edinburgh (hon.), Am. Nuclear Soc., Calif., N.Y., Washington acads. scis., AAAS (pres. 1972, chmn. bd. 1973), Royal Soc. Arts (Eng.); mem. Am. Chem. Soc. (award in pure chemistry 1947, William H. Nichols medal N.Y. sect. 1948, Charles L. Parsons award 1964, Gibbs medal Chgo. sect. 1966, Madison Marshall award No. Ala. sect. 1972, Priestley medal 1979, pres. 1976), Am. Philos. Soc., Royal Swedish Acad. Engring. Scis. (adv. council 1980), Am. Nat., Argentine Nat., Bavarian, Polish, Royal Swedish, USSR acads. scis., Royal Acad. Exact, Phys. and Natural Scis. Spain (acad. fgn. corr.), Soc. Nuclear Medicine (hon.), World Assn. World Federalists (v.p. 1980), Fedn. Am. Scientists (bd. sponsors 1980), Deutsche Akademie der Naturforscher Leopoldina (East Germany), Nat. Acad. Pub. Adminstrn., Internat. Platform Assn. (pres. 1981-86), Am. Hiking Soc. (bd. dirs. 1979-84, v.p. 1980, adv. com. 1984—), Phi Beta Kappa, Sigma Xi, Pi Mu Epsilon, Alpha Chi Sigma (John R. Kuebler award 1978), Phi Lambda Upsilon (hon.); fgn. mem. Royal Soc. London, Chem. Soc. Japan, Serbian Acad. Sci. and Arts. Clubs: Bohemian (San Francisco); Chemists (N.Y.C.); Cosmos (Washington), University (Washington); Faculty (Berkeley). Office: U Calif Lawrence Berkeley Lab Bldg 70A RM 3307 Berkeley CA 94720

SEABORN, CAROL DEAN, nutrition researcher; b. Mountainview, Okla., July 6, 1947; d. Clara (Allen) Jarnagin; m. Jodie L. Seaborn, June 5, 1965. MS, Tex. Tech. U., Lubbock, 1984; PhD, Okla. State U., 1990. Home econs. tchr. Olton (Tex.) High Sch., 1971-78; tchr., researcher Tex. Tech. U., Lubbock, 1981-87; instr. South Plains Coll., Levelland, Tex., 1987; rsch. assoc. Okla. State U., Stillwater, 1987-90; rsch. nutritionist U.S. Dept. Agr. Agrl. Rsch. Svc. Human Nutrition Rsch. Ctr., Grand Forks, N.D., 1990-93; asst. prof. U. Wis., Stout, 1993—. With USN, 1978-79. Gen. Foods fellow Okla. State U., Stillwater, 1989. Mem. Am. Inst. Nutrition, Am. Dietetic Assn., Am. Home Econs. Assn., Nutrition Soc. Achievements include finding of simple sugar feeding depleted chromium stores of mice, antacids decreased chromium absorption. Office: U Wis-Stout Coll Home Econs Dept Nutrition Menomonie WI 54751

SEABROOK, BARRY STEVEN, environmental scientist, consultant; b. Wyandotte, Mich., Jan. 18, 1966; s. Charles Newton and Hazel Ilene (Foote) S.; m. Rita Lou Adams, June 18, 1988. BA, Olivet Coll., 1988. Environ. scientist Holcomb Environ. Svcs., Olivet, Mich., 1988—. Mem. ASHRAE (com. 1989—), ASTM (com. 1989—), Air and Waste Mgmt. Assn. Methodist. Home: 23951 15 Mile St Apt 63 Bellevue MI 49021 Office: Holcomb Environ Svcs 17375 Garfield Rd Olivet MI 49021

SEADEN, GEORGE, civil engineer; b. Cracow, May 26, 1936; s. Simon and Mary (Guttman) S.; m. Linda Helen Mutch, Mar. 18, 1978; children: Amy Elisabeth, Maia Claire. BE, McGill U., Montreal, Que., Can., 1958; MS, Harvard U., 1968; postgrad., Northwestern U., 1992. Engr. Gatineau Power, Hull, Que., 1958-59, Ent. Fougerolle, Paris, 1960-62; mgr. Warnock Hersey Ltd., Montreal, 1959-60; assoc. Cartier, Coté, Piette, Montreal, 1962-67; sr. advisor Ministry Urban Affairs, Ottawa, Ont., Can., 1969-71; pres. Archer, Seaden & Assoc., Inc., Montreal, 1971-84; dir. gen. Inst. Rsch. in Constrn. Nat. Rsch. Coun., Ottawa, 1985—; mem. Constrn. Industry Devel., Can., 1988—, Can. Constrn. Rsch. Bd., 1985-91; pres. Conseil Internat. du Bâtiment, Rotterdam, The Netherlands, 1989-92; vis. prof. U. Ottawa, 1969-73; mem. jury to select best Can. constrn. projects and engring. design; lectr. numerous univs., rsch. ctrs. Co-editor: Trends in Building Construction Worldwide, 1989; mem. editorial bd. Bldg. Rsch. and Practice, 1990—, Constrn. Bldg. Rev., 1991—; contbr. numerous articles to profl. publs. Chmn. bd. dirs. St. Andrew's Sch., Westmount, Que., 1975-82. Home: 80

Lyttleton Gardens, Rockcliffe Park, ON Canada K1L 5A6 Office: Nat Rsch Coun, Montreal Rd Bldg M-20, Ottawa, ON Canada K1A 0R6

SEADLER, STEPHEN EDWARD, business and computer consultant, social scientist; b. N.Y.C., 1926; s. Silas Frank and Deborah (Gelbin) S.; AB in Physics, Columbia U., 1947, postgrad. in atomic and nuclear physics, 1947; postgrad. with George Gamow in relativity, cosmology and quantum mechanics, George Washington U., 1948-50; m. Ingrid Linnea Adolfsson, Aug. 7, 1954; children: Einar Austin, Anna Carin. Legal rsch. asst., editor AEC, Washington, 1947-51; electronic engr. Cushing & Nevell, Warner, Inc., N.Y.C., 1951-54; seminar leader, leader trainer Am. Found. for Continuing Edn., N.Y.C. 1955-57; exec. dir. Medimetric Inst., 1957-59; mem. long range planning com., chmn. corporate forecasting com., mktg. rsch. mgr. W. A. Sheaffer Pen Co., Ft. Madison, Iowa, 1959-65; founder Internat. Dynamics Corp., Ft. Madison and N.Y.C., 1965, pres., 1965-70; orginator DELTA program for prevention and treatment of violence, 1970; founder ID Center, Ft. Madison, now N.Y.C., 1968, pres., 1968—; mgmt. cons. in human resources devel. and conflict reduction, N.Y.C., 1970-73; pres. UNICON-SULT computer-based mgmt. and computer scis., N.Y.C., 1973—; speaker on decision support systems, internat. affairs and ideological arms control; author/speaker (presentation) Holocaust, History and Arms Control; originator social scis. of ideologics and computer based knowledge systems sci. of ideotopology; spl. works collection accessible via On-line Computer Libr. Ctr. Instr. polit. sci. Ia. State Penitentiary, 1959-62. Served with AUS, 1944-46. Mem. Am. Phys. Soc., Am. Statis. Assn., Acad. Polit. Sci., Am. Sociol. Assn., IEEE, N.Y. Acad. Sci., Am. Mgmt. Assn. (lectr. 1963-68), Internat. Platform Assn. Unitarian. Lodges: Masons (32 deg.), Shriners. Author: Savagery in Modern Marriage and Its Relation to Social and International Violence, 1990, Holocaust, History and Arms Control II, 1990; contbr. Ideologics and ideotopology sects. to Administrative Decision Making, 1977, Societal Systems, 1978; Management Handbook for Public Administrators, 1978; also articles profl. jours. Statement on ideological arms control in Part 4 of Senate Fgn. Relations Com. Hearings on Salt II Treaty, 1979, ideologics extended to treat ethnic, racial, religious conflict, 1992, with first call for Western Ecumenical Reformation at Morristown, N.J., Unitarian Ch., 1993. Office: 521 5th Ave Ste 1700 New York NY 10175-0003

SEAGLE, EDGAR FRANKLIN, environmental engineer, consultant; b. Lincolnton, N.C., June 27, 1924; s. Franklin Craig and Lillie Mae (James) S.; m. Doris Elaine Long, Mar. 23, 1958; children: Rebecca Jane, Mary Elaine, James Craig, William Franklin. AB in Chemistry, U. N.C., 1949, MS in Pub. Health, 1954; BCE, U. Fla., 1961; DPH, U. Tex., 1974. Registered profl. engr., Ala. Sr. sanitarian Health Dept., City of Charlotte, N.C., 1950-52, chief indsl. hygiene sect., 1956-59; sanitation cons. N.C. State Bd. Health, Raleigh, 1954-56; engr. dir. USPHS, Rockville, Md., 1961-78; asst. dir. Fellowship Office Nat. Acad. Scis., Washington, 1978-83; pub. health engr. Dept. of Environ., State of Md., Balt., 1985-88; ind. engring. cons. Rockville, 1984-85, 88—. Contbr. articles to profl. publs. With USN, 1943-46, PTO. Mem. ASCE, APHA, Am. Acad. Environ. Engrs. (diplomate). Methodist. Home and Office: 14108 Heathfield Ct Rockville MD 20853

SEAGO, JAMES LYNN, biologist, educator; b. Alton, Ill., June 2, 1941; s. James Lynn and Dorothy Florence (Watkins) S.; m. Jill Penton Dobbs, Dec. 24, 1969 (div. 1977); children: Kirstjan Erika, Robert Maclean; m. Marilyn Ann Meiss, Nov 25, 1982. BA, Knox Coll., 1963; MA, Miami U., Oxford, Ohio, 1966; PhD, U. Ill., 1969. From asst. prof. to assoc. prof. SUNY, Oswego, 1968-91, prof. biology, 1991—; cons. many book pubs., 1978—, Monstanto Agrochem., St. Louis, 1987-88. Mem. editorial bd.: Environ. and Exptl. Botany, 1992—; reviewer: Am. Jour. Botany, Canadian Jour. Botany, Internat. Jour. Plant Sci., Bull. Torrey Botan. Club; contbr. articles to Am. Biology Tchr., Am. Jour. Botany. NSF grantee, 1981, 92. Mem. Botan. Soc. Am., Torrey Botan. Club, Wilderness Soc., Nature Conservancy, Nat. Wildlife Soc., Sigma Xi. Achievements include research in ecological and developmental importance of roots to success of certain cattails in marshes, environmental and population biology education, use of research as teaching tool in undergraduate biology education. Home: PO Box 316 Minetto NY 13115 Office: SUNY Dept Biology Oswego NY 13126

SEAL, MICHAEL, physicist, industrial diamond consultant; b. Weston Super Mare, Eng., Apr. 15, 1930; s. Carl Cyril and Ina May (Hurford) S.; m. Cynthia Ida Austin Leach, Aug. 7, 1954; children: David, Anne, Rosemary, Susan, Christopher. BA, Cambridge U., 1952, MA, 1956, PhD, 1957. Head diamond rsch. Engelhard Industries Inc., Newark, 1959-67, tech. coord. rsch. and devel. div., 1965-67; dir. rsch. Amsterdam Diamond Test and Devel. Ctr., D. Drukker & Zn. N.V., Amsterdam, 1967-90, adj. dir., 1970-90; dir. Dubbeldee Diamond Corp., N.Y.C., 1969-88; pres. Sigillum B.V., Amsterdam, 1990—. Contbr. articles to profl. jours. Served with RAF, 1948-49. Postdoctoral rsch. fellow Cavendish Lab., Cambridge, 1957-59. Fellow Explorers Club, Inst. Physics of London, Mineral Soc. Am.; mem. IEEE, N.Y. Acad. Scis. (life), Am. Chem. Soc., Dutch Phys. Soc., Royal Instn. of Great Britain, Dutch Abrasives Soc. (chmn. 1977—), European Fedn. Abrasives Mfrs. (premier del. for Netherlands, chmn. diamond grains subcom. 1976-91), United Oxford and Cambridge Univs. Club. Home: 5 Guido Gezelle St, 1077 WN Amsterdam The Netherlands Address: PO Box 7129, 1007 JC Amsterdam The Netherlands

SEALE, JAMES LAWRENCE, JR., agricultural economics educator, international trade researcher; b. Memphis, Mar. 12, 1949; s. James Lawrence and Mary Helen (Keefe) S.; m. Colleen Welberry, Sept. 18, 1951. BA, U. Miss., 1972; postgrad., U. Chgo., 1978-79; PhD, Mich. State U., 1985. Agrl. vol. Peace Corps, Tondo, Zaire, 1973-75; agrl. advisor Harvard Inst. for Internat. Devel., Abyei, Sudan, 1978; specialist Mich. State U., Fayoum, Arab Republic of Egypt, 1980-83; asst. prof. agrl. econs. U. Fla., Gainesville, 1985-90, assoc. prof. agrl. econs., 1990—; vis. prof. U. of Leicester, Eng., 1992. Author: (with H. Theil and C.F. Chung) International Evidence on Consumption Patterns, 1989; contbr. articles to profl. jours. NIMH scholar U. Chgo., 1978-79; rsch. fellow Cairo U., 1980-83; McKethan-Matherly rsch. fellow, 1986-88, McKethan-Matherly sr. rsch. fellow, 1991—. Mem. Am. Econs. Assn., Am. Agrl. Econs. Assn., Internat. Assn. Agrl. Economists, Econometrics Soc., Caribbean Agro-Econ. Soc., Internat. Agrl. Trade Rsch. Consortium, Gamma Sigma Delta. Episcopalian. Avocation: scuba diving. Home: 1621 NW 65th St Gainesville FL 32605-4127 Office: U Fla Dept Food and Resource 1095 McCarty PO Box 110240 Gainesville FL 32611

SEAMAN, DUNCAN CAMPBELL, civil engineer; b. Norfolk, Va., Sept. 17, 1957; s. Harold Duncan and Catherine Claire (Goff) S.; m. Tammy Jean Quattlebaum, Dec. 19, 1981; children: Duncan Russell, Katharine Jeanne. BSCE, The Citadel, 1980; MSCE, Clemson U., 1986. Registered profl. engr., S.C. Commd. 2d lt. U.S. Army Corps Engrs., 1980, advanced through grades to capt.; co. comdr. U.S. Army Corps Engrs., Ft. Leonard Wood, Mo., 1983-84; civil engr. U.S. Army Corps Engrs., St. Louis, 1986-87, dep. resident engr., 1987-89; resigned U.S. Army Corps Engrs., 1989; project mgr. Harbert Constrn. Co., Birmingham, Ala., 1989—; presenter, mem. continuing engring. edn. program U. Mo.-Colombia, St. Louis, 1987-88. Campaign vol. United Way, Birmingham, 1990; fund raiser Boy Scouts Am., Birmingham, 1990. Mem. ASCE, Am. Assn. Cost Engrs., Project Mgmt. Inst., Soc. Am. Mil. Engrs., Chi Epsilon. Republican. Home: 329 Parkside Dr Simpsonville SC 29681 Office: Harbert Constrn Co 100 Concourse Pky Birmingham AL 35244-1868

SEAMAN, EDWIN DWIGHT, physician, laboratory director; b. Guymon, Okla., Aug. 7, 1934; s. Marvin C. and Florence (Washburn) S.; m. Yolanda Rodriquez, June 5, 1979; children: Stacey Elizabeth, Monique Louis, David Edwin, Adolfo Robert. BS in Chemistry, West Tex. State U., 1956; MD, U. Tex., Dallas, 1960. Intern St. Louis City Hosp., 1961-62; resident Stanford U. Hosp., Palo Alto, Calif., 1962-65, Meml. Hosp. for Cancer and Allied Diseases, 1965-66; asst. prof. U. Tex., Dallas, 1966-69; med. dir. labs. Vista Hills Med. Ctr., El Paso, Tex., 1984—. Capt. Med. Corp U.S. Army, 1966-68. Home: 520 Wild Willow El Paso TX 79922-7211

SEAMAN, WILLIAM DANIEL, biology educator, clergyman; b. Norlina, N.C., Dec. 15, 1943; s. Albert Carl and Katherine Lena Seaman; m. Susan Elizabeth McCreary, Dec. 27, 1966; children: Mark McCreary, Krista Marie, Karin Elizabeth. MA in Botany, U. N.C., 1976; D Ministry, Concordia Sem., Ft. Wayne, Ind., 1982. Ordained to ministry Luth. Ch., 1971. Psychiat. technician Renard Hosp., St. Louis, 1969-71; asst. pastor St. Paul's Luth. Ch., Rockford, Ill., 1972-78; instr. biology Rockford Luth. High Sch., 1971-78; pastor Our Savior Luth. Ch., Hickory, N.C., 1978—; instr. biology Catawba Valley Luth. High Sch., Hickory, 1984-92; instr. computers St. Stephens Luth. Sch., Hickory, 1979—; instr. biology Catawba Valley C.C., Hickory, 1992—; ind. computer cons., Hickory, 1978—; chmn. curriculum com. Catawba Valley Luth. High Sch., Hickory, 1982-92. Editor: God's Master Plan for the Ages, 1989. Chmn. Winnebago County Right to Life, Rockford, 1971-78; v.p. Catawba Valley Luth. Ministerial Assn., Hickory, 1988-89; bd. dirs. Catawba Valley Luth. High Sch., 1982-86. Mem. AAAS, N.C. Sci. Tchrs. Assn., Sigma Xi, St. Stephens Kiwanis (chmn. 1981-82). Home: 3420 21st St Dr NE Hickory NC 28601-9257 Office: Our Savior Luth Ch 2160 35th Ave Dr NE Hickory NC 28601-9257

SEAQUIST, ERNEST RAYMOND, astronomy educator; b. Vancouver, B.C., Can., Nov. 19, 1938; s. Egron Emanuel and Sigrid Alice (Back) S.; m. Gloria Stewart Jenkins, June 11, 1966; children: Jonathan William, Carolyn Suzanne. BASc, U. B.C., Vancouver, 1961; MA, U. Toronto, Ont., Can., 1962, PhD, 1966. Lectr. astronomy U. Toronto, 1965-66, asst. prof., 1966-72, assoc. prof., 1972-78, prof., 1978—, assoc. chmn. dept., 1974-88, chmn., 1988—; dir. David Dunlap Obs. U. Toronto, Richmond Hill, Ont., 1988—. Contbr. author: Classical Novae, 1989; also over 100 articles. Rsch. grantee Natural Scis. and Engring. Rsch. Coun. Can., 1967—. Mem. Internat. Astron. Union, Am. Astron. Soc., Can. Astron. Soc. (pres. 1986-88). Avocations: painting and sketching, collecting antiques. Office: U Toronto Dept Astronomy, 60 St George St, Toronto, ON Canada M5S 1A7

SEARL, JOHN ROY ROBERT, engineering executive; b. Wantage, Berkshire, England, May 2, 1932; s. Robert Henry and Violet Gertrude Maud (Pearce) S. BA (hon.), Russell-Cotes Nautical Sch., Dorset, England, 1943-46; student, Shenley Hosp., Hertfordshire, England, 1948-50; grad., Open U., London, 1980, Open U., London, 1987. Trainee pharmacist St. Peter's Cannings Pharmacy, London, 1946-48; elec./electronic fitter Midlands Electricity Bd., Tipton, England, 1953-55; prototype wireman Handley Page Aircraft Co., Reading, England, 1955-57; machine setter GKN Vandervell Ltd., Maidenhead, England, 1957-83; chmn. Searl Internat. Space Consortium, Mortimer, England, 1968-83; chmn., head R & D Direct Internat. Sci. Consortium, London, 1983—; lectr. Author: The Law of the Squares, Book 1, 1990, Book 1A, 1990, Book 1B, 1990, Book 2, 1990, Book 3, 1993. With Royal Air Force, 1950-53. Named prof. math Strukturen der Schöpfung und Energie, Munich, 1989. Mem. Blackbushe Aero Club (pilot 1964-83), Patentee & Inventors, Aerospace Club (chmn. 1968—). Avocations: pvt. flying, writing tech. books, electronic control systems designs, research devel. of magnetic motors, teaching maths. Home: 373 Rock Beach Rd Rochester NY 14617 Office: Direct Internat Sci Consort, 13 Blackburn Lower Strand, London NW9 5NG, England

SEARLE, LEONARD, astronomer, researcher. Dir. Carnegie Inst. Washington, Pasadena, Calif. Office: Carnegie Inst of Washington 813 Santa Barbara St Pasadena CA 91101-1232

SEARLE, ROGER BLAINE, manufacturing computer system consultant; b. Scranton, Pa., Mar. 22, 1953; s. John Barry and Barbara Jean (Bates) S.; m. Anne C. McLaughlin, Nov. 29, 1986. BS in Chemistry, Ind. U. of Pa., 1975; MS in Computer Sci., Villanova (Pa.) U., 1984, MBA, 1991. Sr. rsch. chemist Pennwalt Corp., King of Prussia, Pa., 1977-83, computer engr., 1983-88, mgr. scientific systems, 1988-90; cons. Cimtek Am., New Caste, Del., 1990—. Contbr. articles to profl. jours. including Analytical Chemistry. Mem. IEEE, Inst. Ind. Engring. Achievements include a patent for a method for measuring low levels of ammonia i process streams. Home: 918 Bay Lowell Dr Westchester PA 19380 Office: Cimtek Am 55 Reads Way New Castle DE 19720

SEARLS, DONALD TURNER, statistician, educator; b. Brookings, S.D., Nov. 22, 1930; s. Daniel Burdette and Florence (Glen) S.; m. Sondra Lynn Hohwieler, Mar. 17, 1956; children: Steven, Scott, Lori, Trace, Jeffrey. BS, S.D. State U., 1952, MS, 1957; PhD, N.C. State U., 1963. Statistician Rsch. Triangle Inst., Durham, N.C., 1959-63; v.p. Westat, Inc., Denver, 1963-71; dir. rsch. Edn. Commn. of States, Denver, 1971-83; prof. U. No. Colo., Greeley, 1983—; cons. Nat. Acad. Scis., Washington, 1986; adv. mem. Com. for Energy Statistics, Washington, 1987—. Contbr. articles to Jour. Am. Statis. Assn., Am. Statistician, Jour. Experimental Edn., Jour. Reading, Am. Ednl. Rsch. Jour., Jour. Pediatrics. Cpl. U.S. Army, 1952-54. Mem. Am. Statis. Assn. (Cert. of Appreciation 1987), Internat. Assn. Survey Statisticians. Office: U No Colo Dept Math Scis Greeley CO 80639

SEARS, CATHERINE MARIE, osteopath, radiologist; b. Chgo., Oct. 23, 1955; d. James Francis and Patricia Ann (Theiss) S. BA, St. Xavier Coll., Chgo., 1977; DO, Chgo. Coll. Osteo. Medicine, 1985. Commd. 2d lt. U.S. Army, 1981, advanced through grades to maj., 1991—; intern William Beaumont Army Med. Ctr.; gen. med. officer U.S. Army, South Korea, 1986-87; resident in radiology Brooke Army Med. Ctr. U.S. Army, San Antonio, 1987-91; staff radiologist Womack Army Med. Ctr. U.S. Army, Ft. Bragg, N.C., 1991; chief diagnostic radiology svc. Womack Army Med. Ctr. U.S. Army, Ft. Bragg, 1992—. Recipient U.S. Army Commendation medal. Mem. AMA, Am. Osteo. Assn., Mil. Osteo. Physicians and Surgeons, Am. Coll. Radiology, Am. Osteo. Coll. Radiology, Radiologic Soc. N.Am. Roman Catholic. Avocations: marathon running, golf, guitar. Office: USA MEDDAC Womac AMC Dept Radiology Fort Bragg NC 28307

SEARS, DAVID O'KEEFE, psychology educator; b. Urbana, Ill., June 24, 1935; s. Robert R. and Pauline (Snedden) S.; divorced; children: Juliet, Olivia, Meredith. BA in History, Stanford U., 1957; PhD in Psychology, Yale U., 1962. Asst. prof. to prof. psychology and polit. sci. UCLA, 1961—, dean social scis., 1983-92. Author: Public Opinion, 1964, Politics of Violence, 1973, Tax Revolt, 1985, Political Cognition, 1986, Social Psychology, 1993. Fellow Am. Acad. Arts and Scis.; mem. Soc. for Advancement Socio-Econs. (pres. 1991-92), Internat. Soc. Polit. Psychology (pres. 1994-95). Office: UCLA Psychology Dept Los Angeles CA 90024

SEARS, FREDERICK MARK, research manager, mechanical engineer; b. Houston, Jan. 18, 1952; s. B. T. and Dorothy M. (Bradley) S.; m. Karen Louise Callahan, Nov. 3, 1984; children: Marcus F., Brandon D. BS, MS, MIT, 1975; PhD, U. Calif., Berkeley, 1980. Mem. tech. staff AT&T Bell Labs., Murray Hill, N.J., 1980-84, Norcross, Ga., 1984-90; project mgr. Siecor Corp., Hickory, N.C., 1990—; adj. prof. Clark Atlanta U., 1987-90; mem. Catawba County 2000 Task Force, Hickory, 1992-93; mem. tech. com. Conf. on Optical Fiber Comms., Washington, 1993-96. Contbr. articles to Jour. Lightwave Tech., presenter papers at confs. and sci. proceedings. Bd. dirs. Plainfield Sci. Ctr., Plainfield, N.C., 1982-84. Mem. Optical Soc. Am., Pi Tau Sigma. Achievements include patents for mechanical connection for polarization - maintaining optical fiber and methods of making. Office: Siecor Corp Dept R & D 489 Siecor Pk Hickory NC 28603

SEARS, SANDRA LEE, computer consultant; b. Rochester, N.Y., Apr. 25, 1952. AB with distinction, Cornell U., 1974; MA, U. Conn., 1976, postgrad., 1976-81. Cert. in data processing, 1983. Tng. cons. Ins. Crime Prevention Inst., Westport, Conn., 1977-78; systems analyst Data Directions, Bloomfield, Conn., 1978-79; prin. S. S. Prindle Consulting, Manchester, Conn., 1979-81; dir. info. svcs. Conn. Attys. Title Ins., Rocky Hill, Conn., 1981-85; mgr., systems, programming Community Health Care Plan, Inc., Wallingford, Conn., 1985-87; assoc. dir. Mass. Mutual Life Ins., Springfield, Mass., 1987-91; cons. Computer Assistance div. Coopers & Lybrand, West Hartford, Conn., 1991—; adj. faculty U. New Haven, West Haven, Conn., 1976-77, Eastern Conn. State U., Willimantic, 1986—, Manchester Community Coll., 1989—; participant Tex. Instruments' Case Satellite Seminar, 1989. Mentor Career Beginnings, Hartford, 1991—. Presdl. scholar Nat. Merit Program, 1970, William Stout scholar Cornell U., 1973, AAUW fellow U. Conn., 1981. Mem. Cornell Club of Greater Hartford (mem. admissions vol. programs alumni adv. com., exec. bd., book award chair 1987—), Cornell Alumni Admissions Amb. Network (chair 1983-86), Mortar Board, Phi Kappa Phi, Pi Mu Epsilon. Office: Coopers & Lybrand Solutions Through Tech Practice 333 East River Dr East Hartford CT 06108

SEARS, WILLIAM REES, engineering educator; b. Mpls., Mar. 1, 1913; s. William Everett and Gertrude (Rees) S.; m. Mabel Jeannette Rhodes, Mar. 20, 1936; children—David William, Susan Carol. BS in Aero. Engring, U. Minn., 1934; Ph.D., aeronautics, Calif. Inst. Tech., 1938; DSc (hon.), U. Ariz., 1987. Asst. prof. Calif. Inst. Tech., 1939-41; chief aerodynamics Northrop Aircraft, Inc., 1941-46; dir. Grad. Sch. Aero. Engring., Cornell U., Ithaca, N.Y., 1946-63; dir. Center Applied Math., 1963-67, J.L. Given prof. engring., 1962-74; prof. aerospace and mech. engring. U. Ariz., Tucson, 1974-88, prof. emeritus, 1988—; F. W. Lanchester lectr. Royal Aero. Soc., 1973, Gardner lectr. MIT, 1987, Guggenheim lectr. Internat. Congress Aero. Scis., 1988; cons. aerodynamics. Author: The Airplane and its Components, 1941; Editor: Jet Propulsion and High-Speed Aerodynamics, vol. VI, 1954, Jour. Aerospace Scis, 1956-63, Ann. Revs. of Fluid Mechanics, Vol. I. Recipient Vincent Bendix award Am. Soc. Engring. Edn., 1965, Prandtl Ring Deutsche Gesellschaft für Luft-und Raumfahrt, 1974, Von Karman medal ARGARD (NATO), 1977, ASME medal, 1989. Fellow Internat. Acad. Astronautics, Am. Acad. Arts and Scis., Am. Inst. Aeros. and Astronautics (hon., G. Edward Pendray award 1975, S.A. Reed aeros. award 1981, Von Karman lectr. 1968); mem. Nat. Acad. Engring. Mexico (fgn.), Am. Phys. Soc. (Fluid Dynamics prize 1992), Nat. Acad. Scis., Sigma Xi. Home: 6560 N Skyway Rd Tucson AZ 85718-1933 Office: U Ariz Aerospace & Mech Engring Dept Tucson AZ 85721

SEASHORE, STANLEY E(MANUEL), social and organizational psychology researcher; b. Swedeburg, Nebr., Sept. 4, 1915; s. August Theodore and Jennie Caroline (Rose) S.; m. Eva Virginia Danielson, Aug. 29, 1940; children: Karen Rose, Christine Sigrid. MA in Anthropology, U. Minn., 1939; PhD in Social Psychology, U. Mich., 1954. Diplomate Am. Bd. Profl. Psychologists. Rsch. psychologist U.S. Steel Corp., Pitts., 1939-45; cons. A.T. Kearney & Co., Chgo., 1945-50; instr. U. Mich., Ann Arbor, 19952-55, asst. prof., 1955-58, assoc. prof., 1958-60, prof. psychology, 1960-83; program dir. Inst. for Social Rsch., Ann Arbor, 1960-83. Author: Cohesiveness in the Work Group, 1954; co-author: Changing the Structure and Functioning of an Organization, 1963, Management by Participation, 1967, Assessing Organizational Change, 1983; co-editor: Management of the Urban Crisis: Government and the Behavioral Sciences. Named Fulbright Fellow U.S. Edn. Commn., Norway, 1957, Guggenheim Fellow, Guggenheim Found., Eng., 1965, Disting. Scientist, Acad. Mgmt., 1984. Fellow Am. Pschol. Assn. (div. pres. 1968, chmn. policy com. 1974, Am. Sociol. Assn., Soc. for Psychol. Study of Social Issues, Netherlands Inst. for Advanced Study. Achievements include basic contributions to study of organizational structures, dynamics, change and effectiveness; rsch. assessing quality of work life. Home: 8102 Highwood Dr B132 Bloomington MN 55438

SEATON, MICHAEL JOHN, physicist; b. Bristol, Eng., Jan. 16, 1923; s. Arthur William Robert and Helen Amelia (Stone) S.; m. Olive May Singleton, June 10, 1943 (widowed); children: Jane, Richard John; m. Joy Clarice Balchin, Oct. 20, 1960; 1 child, Anthony Michael. BSc, U. London, 1948, PhD, 1951; Dr. hc, Observatoire de Paris, 1976; DSc, Queen's U., Belfast, Ireland, 1982. Lectr. U. Coll., London, 1950-54, reader, 1954-63, prof., 1963-89, emeritus prof., 1989—. Contbr. over 250 articles to profl. jours. Flight lt. RAF, 1942-46. Recipient Guthrie medal Inst. of Physics, 1984. Fellow Royal Soc. London (Hughes medal 1992); mem. Royal Astron. Soc. (pres. 1979-81, Gold medal 1983), Nat. Acad. Sci. (assoc.), Am. Astron. Soc. (hon.). Home: 51 Hall Dr, London SE26 6XL, England Office: U Coll London, Dept Physics & Astronomy, Gower St, London WC1E 6BT, England

SEATON, VAUGHN ALLEN, veterinary pathology educator; b. Abilene, Kans., Oct. 11, 1928; m. Clara I. Bertelrud; children: Gregory S., Jeffrey T. BS, Kans. State U., 1954, DVM, 1954; MS, Iowa State U., 1957. Pvt. practice Janesville, Wis., 1954; instr. pathology Vet. Diagnostic Lab. Iowa State U., Ames, 1954-57, from asst. to assoc. prof. pathology Vet. Diagnostic Lab., 1957-64, prof., head Vet. Diagnostic Lab., 1964—; lab. coord. regional emergency animal disease eradication orgn. Animal and Plant Health Inspection Svc. USDA, 1974—; mem. rsch. com. Iowa Beef Industry Coun., 1972-85; mem. adv. bd. Iowa State Water Resources Rsch. Inst., 1973-80; cons. several orgns. Co-author: (monographs) Feasibility Study of College of Veterinary Medicine, 1972, Veterinary Diagnostic Laboratory Facilities-State of New York, 1970; bd. dirs. Iowa State U. Press, 1985-88, mem. manuscript com., 1982-85; contbr. articles to profl. jours. Trustee Ames Pub. Libr., 1979-85; mem. Iowa State Bd. Health, 1971-77, v.p., 1976-77. Mem. AVMA, Am. Assn. Vet. Lab. Diagnosticians (bd. govs. 1973-88, pres. 1968, E. P. Pope award 1980), Am. Coll. Vet. Toxicologists, U.S. Animal Health Assn., Iowa Vet. Med. Assn. (pres. 1971), North Cen. Assn. Vet. Lab. Diagnosticians, Western Vet. Conf. (exec. bd. 1986-90), World Assn. Vet. Lab. Diagnosticians (pres. 1980-86), Masons (bd. dirs. 1985-88), Ames C. of C. (bd. dirs. 1970-73), Phi Kappa Phi, Phi Zeta (pres. 1964), Alpha Zeta, Gamma Sigma Delta. Office: Iowa State U Coll Vet Medicine Vet Diagnostic Lab Ames IA 50011

SEBASTIAN, JAMES ALBERT, obstetrician/gynecologist, educator; b. Milw., Feb. 20, 1945; s. Milton Arthur and Bernice Marian (Friske) S.; m. Jacqulin Victoria Johnson, June 14, 1969; children: Mila, Joel, Jon, Marnie. BS, U. Wis., 1966, MD, 1969. Diplomate Am. Bd. Ob-Gyn. Commd. officer USN, 1965, advanced through grades to lt. comdr., 1972; intern U.S. Naval Hosp., St. Albans, N.Y., 1969-70; resident in ob-gyn Naval Regional Med. Ctr., Portsmouth, Va., 1970-72, mem. staff, 1976-77; mem. staff Naval Hosp., Taipei, Republic of Taiwan, 1972-76; resigned, 1977; pvt. practice Duluth, Minn., 1977—; assoc. clin. prof. ob-gyn U. Minn. Med. Sch., Duluth, 1977—. Fellow Am. Coll. Ob-Gyn. (best rsch. paper award Armed Forces dist. 1976); mem. Am. Fertility Soc., Minn. Perinatal Assn. (bd. dirs. 1978-87, pres. 1985-86), Kiwanis (pres. 1989-90). Office: Duluth Ob-Gyn Assocs 1000 E 1st St Duluth MN 55805

SEBASTIAN, KUNNAT JOSEPH, physics educator; b. Ampara, Kerala, India, May 18, 1942; came to U.S., 1964; s. Joseph Mathen and Elizabeth (Devasia) Eleckattu; children: Minu, Anita. BS, U. Kerala, Trivandrum, 1961, MS, 1963; PhD, U. Md., 1969. Asst. prof. physics Lowell (Mass.) Tech. Inst., 1969-74; assoc. prof. physics U. Mass., Lowell, 1974-80, prof., 1980—; vis. assoc. prof. U. Rochester, N.Y., 1978-79' vis. prof. Tbilisi State U., Georgia, USSR, 1983, Pa. State U., University Park, 1987. Contbr. 32 articles to sci. jours. Grantee NSF, 1984-86. Mem. Am. Phys. Soc. Achievements include discovery of new way of calculating relativistic corrections to various decay processes and energy spectra; research on new approach to the interaction of a composite system with the radiation field. Office: U Mass Lowell 1 University Ave Lowell MA 01854

SEBASTIAN, MARCUS, software developer; b. Singapore, Nov. 16, 1963; s. Sebastian Antony and Gnanasundri (Sebastian) Udayar. B in Civil and Structural Engring., Nat. U. Singapore, 1988. CAD/CAM software developer, sr. engr. programmer ATS Software Ctr., Singapore, 1988-92; system engr. Canon Singapore (PTE) Ltd., 1992—; cons. coord. of projects with various univs., insts., including Nanyang Tech. U., GINTIC, Singapore, 1988—. Translator devel. DXF, IGES & ISIF HPGL, API devel. native CAD database access. Cpl. Singapore Med. Corps, 1981-84. Mem. Inst. Engrs. (Singapore), Nat. Computer Graphics Assn. Roman Catholic. Avocations: hiking, football, music, flying. Home: Apt Blk 23A, Queenclose # 19-187, Singapore 0314, Singapore

SEBASTIAN, SCOTT JOSEPH, civil engineer; b. Chgo., Dec. 13, 1954; s. Edmund Joseph and Sally Celine (Shukis) S.; m. Kimberly Kay Ignoffo, Nov. 15, 1985; children: Danielle, Ashley. BSCE, Ill. Inst. Tech., 1978; MSCE, U. Tex., 1979. Registered profl. engr., Ill. Project mgr. Walsh Constrn. Co. of Ill., Chgo., 1979—. Mem. ASCE, Ill. Soc. Profl. Engrs. Republican. Roman Catholic. Home: 11220 West 73rd Place Burr Ridge IL 60525 Office: Walsh Constrn Co of Ill 929 W Adams St Chicago IL 60607

SEBEK, KENNETH DAVID, aerospace engineer, air force officer; b. Fairmont, Minn., Aug. 27, 1960; s. David E. and Erma R. (Horner) S.; m. Millie T. Townsend, Nov. 4, 1989; 1 child, Timothy R. A of Aviation Maintenance Tech., Western Nebr. Tech. Coll., 1980; BS in Aerospace Engring., U. Kans., 1987. Commd. 2d lt. USAF, 1987, advanced through grades to capt., 1991; aircraft design engr. Wright Lab., Wright-Patterson AFB, Ohio, 1988-91; structural design engr. 4950 Test Wing, Wright-Patterson AFB, Ohio, 1991—. Author tech. manual; contbr. papers to AIAA. Mem.

AIAA (1st Pl. team aircraft design competition 1987), Exptl. Aircraft Assn. Baptist.

SEBESTYÉN, ISTVÁN, computer scientist, educator, consultant; b. Budapest, Hungary, Aug. 14, 1947; arrived in Austria, 1977; s. János and Hedvig (Vadnai) S.; m. Eszter Molnár, Oct. 25, 1975; children: Adam, David Richard, Sylvia Anna. MS in Elec. Engring., Tech. U. Budapest, 1970, PhD in Elec. Engring., 1974. Diplomate in elec. engring. Research asst. dept. high voltage techniques Tech. U. Budapest, 1968-70; sci. collaborator Inst. for Coordination Computer Techniques, Budapest, 1970-77; mgmt. cons. UN Indsl. Devel. Orgn., Vienna, Austria, 1977-78; scientist, research scholar Internat. Inst. Applied Systems Analysis, Laxenburg, Austria, 1978-84; systems planner, telecommunications adviser, standardization expert Siemens AG, Munich, 1985—; vis. prof. Tech. U. Graz, Austria, 1983-85; lectr., tchr. U. Klagenfurt, Austria, 1985—; cons. to internat. orgns. Author: Experimental and Operational East-West Computer Networks, 1983, Transborder Data Flows and Austria, 1986; mem. adv. bd. Jour. Transnat. Data Report, Washington, 1983—; contbr. over 80 articles to sci. jours. Mem. Austrian Computer Soc. Avocations: travel, sports. Home: Hunkelestrasse 35, 8000 Munich 71, Germany Office: Siemens AG, Hofmannstrasse 51, 8000 Munich 70, Germany

SECCO, ANTHONY SILVIO, chemistry educator; b. Antigonish, Can., Aug. 10, 1956; s. Etalo A. and Margaret F. (Brehaut) S.; m. M. Loretta Cameron, Jan. 2, 1979; 1 child, Melanie F. BS, St. F.X. U., Antigonish, Nova Scotia, 1978; PhD, U. British Columbia, Vancouver, 1982. Postdoctoral fellow U. Pa., Phila., 1982-84; assoc. prof. U. Manitoba, Winnipeg, Can., 1984—. Mem. AAAS, Can. Soc. Chemistry, Am. Crystallographic Assn. (sec. Can. divsn. 1992—), Chem. Inst. Can. (vice-chmn. Man. sect. 1992-93). Achievements include research in DNA crystal structures; modified nucleosides; protein structure and function. Office: U Manitoba, Dept Chemistry, Winnipeg, MB Canada R3T 2N2

SECOR, STEPHEN MOLYNEUX, physiological ecologist; b. Syracuse, N.Y., June 14, 1958; s. Robert Wesley and Elaine Zeda (Molyneux) S.; m. Diana Elizabeth Lenard, Aug. 1, 1981; 1 child, Everett Molyneux Secor. BS, SUNY, Syracuse, 1980; MS, U. Okla., 1985; PhD, U. Calif., L.A., 1992. Teaching fellow Dept. Zoology, U. Okla., Norman, 1981-84; rsch. asst. Okla. Biol. Survey, Norman, 1985; teaching fellow Dept. Biology, UCLA, 1985-91, lectr., 1992; postdoctoral rschr. Dept. Physiology, UCLA, 1992—. Author: (with others) Biology of the Pit Vipers, 1992; contbr. articles to profl. jours. Recipient F. Durham Meml. award So. Calif. Acad. Scis. Mem. Am. Soc. of Ichthyologists and Herpelologist (com. on grad. student participation 1989-91, Myvanwy Dick Rsch. award 1991), Ecol. Soc. of Am., Herpetologists League, Sigma Xi. Office: UCLA Dept Physiology Los Angeles CA 90024

SEDER, RICHARD HENRY, physician; b. Worcester, Mass., Sept. 17, 1938; s. Leonard and Sarah C. (Shapiro) S.; m. Margery Ann Lewis, June 22, 1963; children: David Bennett, Jonathan Lewis; m. Susan Mary Cotter, Aug. 28, 1983. AB, Harvard U., 1960, MD, 1965, MPH, 1966. Diplomate Am. Bd. Clin. Pathology, Am. Bd. Preventive Medicine. Epidemiologist USPHS, Bethesda, Md., 1966-68; health care planner Hahnemann Med. Sch., Phila., 1969-70; dir. planning Mass. Dept. Pub. Health, Boston, 1970-72; resident Brigham and Women's Hosp., 1972-74; resident VA Med. Ctr., Boston, 1974-76, clin. pathologist, 1974—; dir. clin. rsch. Cellcor Therapeutics, Boston, 1988-89; acting chief lab medicine Univ. Hosp., Boston, 1990-92; med. dir. Imugen, Norwood, Mass., 1989—. Author: (series) Apheresis in Cancer, 1981-88. Col. USAR, 1980—. Mem. Mass. Med. Soc. (com. blood banking 1991—), Boston Cancer Rsch. Assn. (pres. 1988-89), Am. Soc. for Apheresis (com. on awards 1990—), Am. Assn. Blood Banks, Phi Beta Kappa, Delta Omega. Jewish. Achievements include patent for treatment of HIV infections. Home: 11 Valley Rd Southborough MA 01772 Office: Imugen 315 Norwood Park S Norwood MA 02062

SEDEROFF, RONALD ROSS, geneticist; b. Montreal, Quebec, Can., Dec. 22, 1939; s. Hyman and Florence (Venor) S.; m. Margaret Kantor, Dec. 18, 1966; children: Kim, Sara. PhD, UCLA, 1967. Acting asst. prof. dept. zoology U. Calif., L.A., 1967; asst.. prof. dept. biol. scis. Columbia U., N.Y.C., 1967-75; asst. prof., assoc. prof. dept. biology U. Oreg., Eugene, 1975-78; assoc. prof. dept. genetics N.C. State U., Raleigh, 1978-84; st. scientist, plant molecular geneticist U.S. Forest Svc., Berkeley, Calif., 1984-87; prof. dept. forestry N.C. State U., 1987—. Contbr. 60 articles or book chpts. to jours. of molecular genetics. Postdoctoral fellow U. Geneva, Switzerland, 1967-69. Achievements include 2 patents in molecular genetics of forest trees. Office: N C State U Forestry Box 8008 Raleigh NC 27695

SEDGWICK, JULIE BETH, immunologist; b. Milw., Sept. 13, 1952; d. Donald Duane and Carol Ellen (Walters) S. BA in Chemistry, Lawrence U., Appleton, Wis., 1974; MS in Chemistry, Case Western Res. U., 1976; PhD in Immunology, U. Tex., 1981. Asst. scientist U. Wis., Madison, 1988-90; assoc. scientist U. Wis., 1990—. Achievements include research in histamine and H2 antagonist, eosinophils: biol. and clin. aspects. Office: U Wis CSC H6/355 Allergy 600 Highland Ave Madison WI 53792

SEDLAK, BONNIE JOY, university official; b. Oak Park, Ill., Jan. 30, 1943; d. Raymond Erwin and Eleanore Mildred (Rada) S. BA, Northwestern U., 1965, PhD, 1974; MA, Case Western Res. U., 1968. Rsch. asst. Case Western Res. U., Cleve., 1965-68; rsch. assoc. Northwestern U., Evanston, Ill., 1965-74; postdoctoral researcher Rush Med. Coll., Chgo., 1974-75; asst. prof. Smith Coll., Northampton, Mass., 1975-77; asst. prof., then assoc. prof. SUNY, Purchase, 1977-81; assoc. rsch. scientist U. Calif., Irvine, 1981-85; sales rep. N.Am. Sci. Assn., Irvine, 1986-87; program mgr. Microbics Corp., Carlsbad, Calif., 1987-88; sr. analyst Fritzsche, Pambianchi Assocs. Inc., Del Mar, Calif., 1988-90; bus. devel. and licensing mgr. Becton Dickinson Advanced Cellular Biol., San Jose, 1991-92; licensing officer office of tech. transfer U. Calif., Alameda, 1992—; biotechnology cons., Del Mar, 1990-91. Contbr. articles, tech. revs. to profl. publs. Mem. Licensing Exec. Soc., Assn. Univ. Tech. Mgrs. Home: PO Box 3021 Half Moon Bay CA 94019

SEE, PAUL DEWITT, geology educator; b. Seaside, Oreg., June 30, 1926; s. Harold Dewhit and Violet (West) S.; m. Shirley Evelyn Ball, Sept. 6, 1949; children: Dana Suzanne, Warren Scott. BS, Oreg. State U., 1949; MS, Portland State U., 1967. Registered profl. geologist, Oreg. Paleontologist Shell Oil Co., Long Beach, Calif., 1950-58; stratigrapher Shell Oil Co., Durango, Colo., 1958-63; prof. Clatsop Community Coll., Astoria, Oreg., 1967-87, dean instrn., 1970-87; geologic cons. Paul D. See & Assoc., Inc., Seaside, Oreg., 1985—; mem. Coastal Hazards Policy Working Group, State of Oreg., 1992—. Contbg. author: Guide to Geology of LA. and Ventura Regions, 1956, Shelf Carbonates of Paradox, San Juan Basins, 1963. Mem., chmn. Sch. Dist. #10, Seaside, 1970-79, Lewis & Clark Hist. Pageant, Seaside, 1987-92. Petty officer USN, 1943-46. Mem. AAAS, Am. Assn. Petroleum Geols. Republican. Home and Office: Paul D See Assocs Inc 300 Surf Pines Rd Seaside OR 97138

SEEFELDT, WALDEMAR BERNHARD, chemical engineer; b. Milw., Apr. 4, 1925; s. August Bernhard and Frieda Anne (Kuerschner) S.; m. Marie Hildegarde Henriksen, July 8, 1950; children: Laurel Marie, Lynn Noel, Lisa Kay. BS, Purdue U., 1947, MS, 1948. Chem. engr. Argonne (Ill.) Nat. Lab., 1948—; cons. Atomic Energy Commn. Author: (with others) Progress in Nuclear Energy, 1963, Plutonium: Facts and Inferences, 1976, Nuclear Engineering Handbook, 1958; editor: Scientifc Basis for Nuclear Waste Management, 1987. Bd. dirs. Lutheran Ch. Missouri Synod, Hillside, Ill., 1982-88. With USN, 1944-46. Mem. Am. Inst. Chem. Engrs., Am. Nuclear Soc., Inst. Nuclear Materials Mgmt. Republican. Home: 417 S Kensinton LaGrange IL 60525 Office: Argonne Nat Lab CMT205 9700 S Cass Ave Argonne IL 60439

SEELENBERGER, SERGIO HERNAN, chemical company executive; b. Santiago, Chile, Oct. 3, 1942; s. Gustav Julius and Sonia (Posternack) S.; m. Annabella Farba, Apr. 6, 1968 (separated 1993); children: Martha, Alberto, Alexander. Diploma in Architecture, Cath. U., Santiago, 1967, M in Regional Planning, 1968; M in City Planning, U. Calif., Berkeley, 1970. Registered profl. architect. Researcher U. Calif., Berkeley, 1969-70; prof. Cath. Univ., Santiago, 1970-73; chief planning advisor Sec. of State of Transp., Rio De Janeiro, Brazil, 1973-79; rsch. dir. Ibam-Brasil, Rio De Janeiro, 1975-81; prof. Federal Univ.-Coppe, Rio De Janeiro, 1975-77, Getulio Vargas Found., Rio De Janeiro, 1977-81; ptnr., dir. Planpur Ltd., Rio De Janeiro, 1977-81, Icham Ltd., Santiago, 1983-90, Clinitest Ltd./Labatria S.A., Santiago, 1985-93; ptnr., dir., pres. Biomerieux Chile S.A., Santiago, 1993—; ptnr., dir. S. Seelenberger & Sons Holding Co., Santiago, 1992—; cons. InterAm. Devel. Bank, Washington, 1988-90, World Bank, Washington, 1988-89, UN/Cepal, Rio de Janeiro, 1980-82. Author: Low Income Housing National Contest-Brasil, 1975 (Negrao de Lima prize 1976); contbr. articles to profl. jours. Internat. v.p. B'nai B'rith Internat., Washington/Santiago, 1990-92, pres. dist. 27, Chile-Bolivia, 1988-90, mission officer for help to Russian Refuseniks, Leningrad-Moscow, 1989; policy advisor Ministry of Housing, Santiago, 1970. Recipient grad. scholarship Cath. U., 1967, grad. studies grant Ford Found., Berkeley, 1968-70, rsch. grant Govt. Brasil, 1976. Mem. AAAS, InterAm. Planning Soc., Am. Inst. Planners, Am. Planning Assn., Am. Assn. for Clin. Chemistry. Jewish. Avocations: social work, chess, computer graphics, soccer, tennis. Home: Cabildo 6150 Apt 111 Las Condes, Santiago Chile Office: Clinitest Ltd, Ricardo Lyon 1899, Santiago Chile

SEELEY, JOHN GEORGE, horticulture educator; b. North Bergen, N.J., Dec. 21, 1915; s. Howard Wilson and Lillian (Fiedler) S.; m. Catherine L. Cook, May 28, 1938; children: Catherine Ann, David John, Daniel Henry, George Bingham, Thomas Dyer. B.S., Rutgers U., 1937, M.S., 1940; Ph.D., Cornell U., 1948. Research asst. N.J. Agrl. Exptl. Sta., 1937-40, foreman ornamental gardens, 1940-41; instr. floriculture Cornell U., Ithaca, N.Y., 1941-43, 45-48, asst. prof., 1948-49, prof. floriculture, 1956-83, prof. emeritus, 1983—, head dept. floriculture, 1956-70; prof. floriculture Pa. State U., 1949-56; D.C. Kiplinger chair floriculture, prof. horticulture Ohio State U., 1984-85; asst. agronomist Bur. Plant Industry Dept. Agr., 1943-44; chemist Wright Aero. Corp., Paterson, N.J., 1944-45. Trustee Kenneth Post Found., 1956-84, Fred C. Gloeckner Found., 1970—. Recipient S.A.F. Found. for Floriculture Rsch. & Edn. award, 1965, Cornell Edgerton Career Teaching award, 1983. Fellow AAAS, Am. Soc. Hort. Sci. (pres. 1982-83, chmn. bd. 1983-84, Leonard H. Vaughan rsch. award 1950, Bittner Extension award 1982); mem. Am. Acad. Floriculture (hon.), Soc. Am. Florists (Hall of Fame 1979), Mass. Hort. Soc. (Silver medal 1980), Am. Hort. Soc., Internat. Soc. Hort. Sci. (hon, Appreciation award 1986), Am. Carnation Soc., N.Y. Acad. Scis., N.Y. Flower Growers Assn., Pa. Flower Growers Assn., Sigma Xi, Phi Kappa Phi, Alpha Zeta, Pi Alpha Xi (pres. 1951-53), Epsilon Sigma Phi, Phi Epsilon Phi. Presbyterian. Lodge: Rotary Internat. (dist. gov. 1973-74). Home: 1344 Ellis Hollow Rd RD2 Ithaca NY 14850-9601

SEELEY, MARK, agronomist; b. Gary, Ind., May 3, 1942; s. Clayton Barron and Margaret Louise (Cook) S.; B.S., Purdue U., 1967; M.A. in Edn., Austin Peay State U., 1971. Staff asst. Purdue U., 1962; sci. tchr. Lake Central Sch. Corp., St. John, Ind., 1967-68; sci. tchr., Gary, Ind., 1972-73; mgr. agronomic crops R.L. Schultz Farms, Hobart, Ind., 1973—, dir. Lupin introduction and devel., 1980—; bd. dirs. On Line Electric Inc., mem. exec. steering com. corp. svcs. Mem. Lake Area United Way Vol. Service, Lake County Health Fair, 1974; sci. and engring. judge 26th and 27th Calumet Regional Sci. Fairs. Mem. AAAS, Am. Inst. Biol. Scis., Am. Soc. Hort. Sci. (Food Quality and Nutrition Working Group), Am. Soc. Agrl. Engrs. (pres.'s club 1980—), Nat. Soc. Profl. Engrs., Ind. Soc. Profl. Engrs. (scholarship com. Calumet chpt. 1982-83, co-chmn. 1984-85), Am. Soc. Agronomy, Am. Soc. Plant Physiologists, Council Agrl. Sci. and Tech. (mem. Century Club 1983), Crop Sci. Soc. Am., Fedn. Am. Scientists, Internat. Soc. Hort. Sci., Soil Sci. Soc. Am., Lake Michigan Flyers Assn., U.S. Hang Gliding Assn. Address: 6126 Sykes Rd Route 1 Hobart IN 46342

SEELYE, EDWARD EGGLESTON, psychiatrist; b. White Plains, N.Y., Feb. 8, 1924; s. Elwyn Eggleston and Eleanore Hartshorn (Janeway) S. BA, Columbia Coll., 1948; MD, Albany Med. Coll., 1955. Diplomate Am. Bd. Psychiatry and Neurology. Attending psychiatrist N.Y. Hosp.-Westchester, White Plains, 1959-93, dir. alcohol svc., 1970-72, asst. dir. admission svcs., 1991—. Contbr. articles to profl. jours. Fellow Am. Psychiatric Assn. (life); mem. AMA, Med. Soc. State of N.Y., Westchester County Med. Soc. Home: 21 Bloomingdale Rd White Plains NY 10605 Office: NY Hosp Westchester 21 Bloomingdale Rd White Plains NY 10605

SEEM, ROBERT CHARLES, plant pathologist; b. Allentown, Pa., Oct. 22, 1948; s. Joseph Trout and Kathryn Seibert (Shelly) S.; m. Carolyn Jean Poplawski, Sept. 18, 1971; children: Jessica Elizabeth, Peter Robert. BS, Pa. State U., 1970, MS, 1972, PhD, 1976. Asst. prof. Cornell U., Geneva, N.Y., 1975-82, assoc. prof., 1982—; assoc. dir. Cornell U., N.Y. State Agrl. Expt. Sta., Geneva, 1990—; assoc. editor Can. Jour. Plant Pathology, 1984-90, Phytopathology, 1985-87, sr. editor, 1990—; sr. vis. fellow Norwegian Agrl. Rsch. Coun., Aas, Norway, 1989. Co-author: Annual Review of Phytopathology, 1984, Techniques in Plant Disease Epidemiology, 1987. Mem. AAAS, Internat. Soc. Plant Pathology, Am. Phytopath. Soc. Democrat. Presbyterian. Achievements include research in mgmt. systems for diseases of apples and grapes, incidence and severity relationships in plant diseases. Home: 2445 Traver Rd Seneca Falls NY 13148 Office: NY State Agrl Expt Sta Jordan Hall Geneva NY 14456

SEEMAN, PHILIP, pharmacology educator, neurochemistry researcher; b. Winnipeg, Man., Can., Feb. 8, 1934; s. Jacob and Fanny (Wigdor) S.; m. Mary V. Szwarc; children: Marc, Bob, Neil. BS, McGill U., Montreal, Can., 1955, MS, 1956, MD, 1960; PhD, Rockefeller U., 1966. Prof. pharmacology U. Toronto, Can., 1967—, chmn. dept. pharmacology, 1977-87. Author over 300 publs. on dopamine receptors, their discovery, and their changes in brain diseases. Recipient Upjohn award Can. Pharmacology Soc., 1980, John Dewan award Ont. Mental Health Found., 1981, Heinz Lehmann award Can. Coll. Neuropsychopharmacology, 1985, Lieber award Nat. Alliance for Rsch. in Schizophrenia and Depression, 1990, Dean award Am. Coll. Psychiatrists, 1990, Tanenbaum Schizophrenia Rsch. award Can. Psychiat. Rsch. Fedn., 1991. Fellow Royal Soc. Can.; mem. Am. Soc. Pharmacology, Soc. Neurosci., Am. Coll. Neuropsychopharmacology. Home: 32 Parkwood Ave, Toronto, ON Canada M4V 2X1 Office: U of Toronto Med Scis Bldg, Dept of Pharmacology, Toronto, ON Canada M5S 1A8

SEGAL, IRVING EZRA, mathematics educator; b. N.Y.C., Sept. 13, 1918; s. Aaron and Fannie Segal. A.B., Princeton U., 1937; Ph.D., Yale U., 1940. Instr. Harvard U., 1941; research asst. Princeton U., 1941-42, assoc., 1942-43; asst. to O. Veblen, Inst. for Advanced Study, 1945-46; asst. prof. to prof. U. Chgo., 1948-60; prof. MIT, Cambridge, 1960—; vis. assoc. prof. Columbia U., 1953-54; vis. fellow Insts. Math. and Theoretical Physics, Copenhagen, 1958-59; vis. prof. Sorbonne, Paris, France, 1965, U. Lund, Sweden, 1971, Coll. de France, 1977. Author: Mathematical Problems of Relativistic Physics, 1963, (with R.A. Kunze) Integrals and Operators, 1968, Mathematical Cosmology and Extragalactic Astronomy, 1976, (with J.C. Baez and Z. Zhou) Introduction to Algebraic and Constructive Quantum Field Theory, 1992; editor: (with W.T. Martin) Analysis in Function Space, 1964, (with Roe Goodman) Mathematical Theory of Elementary Particles, 1988; contbr. articles to profl. jours. Served with AUS, 1943-45. Guggenheim fellow, 1947, 51-52, 67-68. Mem. Am. Math. Soc., Am. Phys. Soc., Am. Acad. Arts and Sci., Royal Danish Acad. Scis., Am. Astron. Soc., Nat. Acad. Sci. Home: 25 Moon Hill Rd Lexington MA 02173-6139 Office: MIT Rm 2-244 Cambridge MA 02139

SEGGEV, JORAM SIMON, allergist, clinical immumologist; b. Tel-Aviv, Israel, Apr. 7, 1944; came to U.S., 1986; s. Rudolf and Elizabeth (Staub) Landsberg; m. Varda Alexandri, Oct. 3, 1968; children: Guy, Michal, Itai. MD, Hebrew U., Jerusalem, 1970. Internal Medicine, Allergy and Clin. Immunology. Asst. prof. U. Mo., Columbia Sch. MEdicine, 1986-90, Med. Coll. Wis., Milw., 1990-91; assoc. prof. U. Nev. Sch. Medicine, Las Vegas, 1991—; sec., treas. Mo. State Allergy Soc., 1989-90; mem. commn. of rsch. Am. Acad. Allergy and Immunology, 1989-91; mem. commn. on Asthma Am. Coll. Allergy and Immunology, 1989—; mem. sub. committee on Rehab. in Asthma, Acad. Allergy and Immunology, 1992; mem. com. quality care in asthma Am. Acad. Allergy, 1993. Author 2 chpts. in Pulmonary Diseases, 1988. Lectr. Better Breathers Club, Columbia, Mo., 1988; participant Children's Miracle Telethon, Las Vegas, 1991. Recipient Rsch. award Winthrop Pharm., U. Nev. Sch. Medicine, 1991-93, Merit review award Dept. Veterans Affairs, Columbia, Mo., Miwl., 1988-91. Fellow Am. Acad. Allergy and Immunology, Am. Coll. of Allergy and Immunology; mem. ATS, AAI, Am. Thoracic Soc. Achievements include discovery of assay for diagnosis of Mycoplasma Pneumoniae Infection. Office: U Nevada Sch Medicine 2040 W Charleston Blvd #503 Las Vegas NV 89102

SEGRE, DIEGO, veterinary pathology educator, retired; b. Milan, Italy, Feb. 3, 1922; came to U.S. 1951; s. Ulderico and Corinna (Corinaldi) S.; m. Mariangela Bertani, July 22, 1952; children: Carlo, Alberto. DVM, U. Milan, 1947; MS, U. Nebr., 1954; PhD, U. Wis., 1957. Asst. prof. U. Milan, 1947-52; instr. U. Nebr., Lincoln, 1952-55; rsch. asst. U. Wis., Madison, 1955-57, asst. prof., 1957-60; prof. vet. pathology U. Ill., Urbana, Ill., 1960-92; retired, 1992; mem. Bd. Sci. Counselors,, Nat. Inst. on Aging, Bethesda, 1982-86, chair, 1984-86; cons. NIH. Contbr. approx. 140 sci. papers and books/book chpts. in field of immunology. Democrat.

SEHGAL, JAWAHARLAL, environmental company executive; b. Ludhiana, Punjab, India, July 9, 1937; s. Hem Raj and Devki (Devi) S.; m. Uma Devi; children: Sanjay, Aditya, Vivek. BsC, Govt. AG. Coll. PAU, 1957, MsC, 1960; MS, State Univ., Ghent, Belgium, 1967, DsC, 1970. Researcher Katholiek University, Leuven, Belgium, 1970-71; assoc. prof. Punjab Agril Univ., Ludhiana, Punjab, 1971-85; soils expert States Soil Orgn., Baghdad, 1976-80; soil-site evaluation Forestry Inst., Baghdad, 1980-81; sr. pedologist Punjab Agril. Univ., Ludhiana, 1985-86; dir. NBSS & LUP, Nagpur, 1986—; soil correlation and classification expert FAO-UN, Vietnam, 1989; v.p. Soil Micromorphology sub commn., 1982-86. Author: Introductory Pedology, 1986; contbr. articles to profl. jours; editor Pedology An Internat. Jour., 1982-86. Recipient award Govt. Iraq, 1978. Mem. Pedology An Internal. Journ., Indian Soc. of Soil Sci.(pres. survey and land use planning), Agropedology, J. Remote Sensing. Achievements include resource maps of all states of India., agro-ecological zoning of India., Land degradation cstatus and challenges. Home: 316 Model Town, Ludhiana India Office: NBSS LUP, Amravati Rd, Nagpur India

SEIBERLING, LUCY ELIZABETH, physicist; b. Fayetteville, Ark., Feb. 1, 1951; d. Harry Albert and Mary Lucianne (Malcolm) S. BS in Physics/Math., Calif. State U., L.A., 1975; PhD in Physics, Calif. Inst. Tech., 1980. Asst. prof. physics U. Pa., Phila., 1981-88; assoc. prof. physics U. Fla., Gainesville, 1988—; rev. panel for young investigator awards NSF, Washington, 1992. Contbr. articles to profl. jours. including Nuclear Instruments and Methods, Jour. of Crystal Growth, Surface Sci. Letters. Rsch. grantee NSF, 1987-90, 91—. Mem. Am. Phys. Soc., Materials Rsch. Soc., Böhmische Phys. Soc. Achievements include discovery of bonding pattern for low temperature deposition of films on silicon; rsch. on atomic structure of thin metal and semiconductor layers on silicon. Office: U Fla Physics Dept 215 Williamson Hall Gainesville FL 32611

SEICHTER, DANIEL JOHN, electrical engineer; b. New Holstein, Wis., Mar. 18, 1934; s. Martin John and Helen Kathryne (Schneider) S.; m. Joan Elizabeth Turgeon Zander, Mar. 16, 1982; children: Debra Zander, Daniel Zander. BSEE, U. Wis., 1961. Registered profl. engr., Wis. Field svc. engr. Giddings and Lewis Machine Tool Co., Fond du Lac, Wis., 1961-64, quality engr., 1964-69, R&D engr., 1969-71, mfg. engr., 1971-72, mgr. mech. engr., 1972-85, mgr. engring. Davis Tool div., 1985—. Sgt. USMC, 1952-55. Home: 614 8th St Fond Du Lac WI 54935-5375 Office: Davis Tool Div Giddings and Lewis 475 S Seymour St Fond Du Lac WI 54935-4754

SEIDEL, FRANK ARTHUR, chemical engineer; b. Raton, N.Mex., May 29, 1958; s. Frank Marion and Sarah Frances (Hazen) S.; m. Jennifer Lynn Hass, June 5, 1982; children: Hailey Lauren, Caroline Frances. BS in Chem. Engring., N.Mex. State U., 1981. Drilling engr. Am. Prodn. Co., Hobbs, N.Mex., 1981-87; sr. drilling engr. Am. Prodn. Co., Houston, 1987-91; staff drilling engr. Am. Oil Co., Denver, 1991—. Mem. Soc. Petroleum Engrs. (program chmn. Denver Drilling Study Group 1991). Office: Amoco Production Co 1670 Broadway Denver CO 80201

SEIDMAN, ROBERT HOWARD, computer information specialist, educator; b. Bklyn., Sept. 24, 1946. BS, Rutgers U., 1968; MS, Syracuse U., 1973, PhD, 1980. Asst. prof. SUNY, Oswego, 1973-75, Syracuse (N.Y.) U., Stony Brook, 1978-81; sr. rsch. assoc. Syracuse (N.Y.) U., 1975-78; prof. computer info. systems Grad. Sch. N.H. Coll., Manchester, 1981—. Exec. editor Jour. Ednl. Computing Rsch., N.Y.C., 1983—; co-author: Predicting the Behavior of the Educational System, 1980; contbr. chpts. to books. Office: NH Coll Grad Sch 2500 N River Rd Manchester NH 03104-1045

SEIDMAN, STEPHANIE LENORE, lawyer; b. Bklyn., July 13, 1949; 010BS. BBS, William Smith Coll., 1971; MS, Ind. U., 1976, PhD, 1979; JD, Cath. U., 1985. Bar: Washington 1986, U.S. Patent Office 1989. Postdoctoral staff fellow NIH, Washington, 1979-82; patent examiner U.S. Patent and Trademark Office, Washington, 1985-89; patent atty. Fleit, Jacobson et al., Washington, 1989-91; ptnr. Fitch, Even, Tabin & Flannery, San Diego, 1991—. Mem. ABA, D.C. Bar Assn., Am. Chem. Soc., Patent Bar Assn., AAAS. Home: 8538 Villa La Jolla Dr #180 La Jolla CA 92037 Office: Fitch Even Tabin & Flannery 4250 Executive Sq Ste 510 La Jolla CA 92037

SEIDMAN, STEPHEN BENJAMIN, computer scientist; b. N.Y.C., Apr. 13, 1944; s. Sylvan and Anne (Levine) S.; m. Barbara Heidemarie Koppe, Aug. 24, 1969; children: Miriam, Naomi. BS, CCNY, 1964; AM, U. Mich., 1965, PhD, 1969. Asst. prof. math. NYU, 1969-72; asst. prof. math. George Mason U., Fairfax, Va., 1972-76, assoc. prof. math., 1976-84, prof. computer sci., 1984-90; prof., dept. head computer sci., engring. Auburn (Ala.) U., 1990--. Author: Assembly Language programming in Compass, 1987. Mem. IEEE Computer Soc., Assn. for Computing Machinery. Home: 2675 Windy Hill Circle Auburn AL 36830 Office: Auburn U 107 Dunstan Hall Auburn AL 36849

SEIDMON, E. JAMES, urologist; b. N.Y.C., Apr. 26, 1947; s. Edward Edgar and Dorothy (Solomon) S.; m. Nancy F. Friedman, Nov. 11, 1979; children: Eric Matthew, Emily Ann. BS, Hobart Coll., 1969; MA, SUNY, Buffalo, 1971; MD, U. Guadalajara, Mex., 1975. Diplomate Am. Bd. Urology. Intern Rochester (N.Y.) Gen. Hosp., 1976-77; resident in surgery Beth Israel Med. Ctr., N.Y.C., 1977-78; resident in urology Albany (N.Y.) Med. Ctr., 1978-80; fellow in urology Montifiore Med. Ctr., Bronx, N.Y., 1980-82; fellow in urologic oncology Roswell Park Meml. Inst., Buffalo, 1982-83; assoc. prof. urology Temple U. Hosp., Phila., 1983—. Contbr. articles to profl. jours. including Jour. Urology. Fellow ACS; mem. Am. Urol. Assn., Mid Atlantic Urol. Assn., Soc. Univ. Urologists, Radiation Therapy Oncoloyg Group, So. Coop. Oncology Group, Soc. Urologic Oncology. Achievements include development of antegrade stent, ureteral safety wire guide, corporal balloon dilator. Office: Temple U Hosp Dept Urology Broad and Ontario St Philadelphia PA 19140

SEIERSEN, NICHOLAS STEEN, data processing executive; b. Geneva, Switzerland, June 23, 1955; s. Ove Steen and Kamini Shoshiela (Bhandari) S.; m. Sylvie Jacqueline Fenouillet, Nov. 7, 1981. BSc with honors, U. Sussex, 1976; MBA, Pacific State U., L.A., 1987. Project chief Metra Proudfoot Internat. Mgmt. Cons., Brussels, Belgium, 1977-81; unit mgr. Auchan Hypermarkets, Paris, 1981-83; internat. controller Pain Jacquet Group, Paris, 1983-85; sr. cons. A.T. Kearney Mgmt. Cons., Paris, 1985-88; European mktg. mgr. Digital Equip. Corp., Paris, 1988—. Contbr. articles to profl. jours. Mem. ASLOG, French Bus. Logistics Assn. (mgmt. com. 1993—), U.S. Bus. Logistics Assn. (coun. logistics mgmt.). Office: Digital Equip Corp, 5 Rue de la Renaissance, Antony France 92187

SEIFERT, ALVIN RONALD, psychologist; b. Wheatland, Iowa, Oct. 22, 1941; s. Alvin A. and Gladys M. (Brinkman) S.; m. Julie Ann Hodgkins, Mar. 29, 1988; 1 child, Matthew Charles. BA, Iowa Wesleyan Coll., 1967; postgrad., Drake U., 1968-70; PhD, U. Tenn., 1975. Diplomate Am. Acad. Pain Mgmt.; cert. biofeedback profl. Asst. prof. Coll. Medicine U. South Fla., Tampa, 1974-88; psychologist VA Med. Ctr., Tampa, 1974-88; pvt. practice Tampa, 1981-88; psychologist Behavioral Inst. Atlanta, 1988—; cons. VA Rsch., Topeka, 1977, East Tenn. State U. Med. Sch., 1981, So. Saw Co., Atlanta, 1989-92. Editor: International Perspectives on Self-Regulation and Health, 1991. With USN, 1960-64. Grantee Knoxville Children's Found., 1970-74; fellow Japan Soc. for Promotion of Sci., 1991. Mem. APA, AAAS, Assn. for Applied Psychophysiology and Biofeedback (Presdl. award

1992), Am. Psychol. Soc., Soc. for Behavioral Medicine, Soc. for Pscyhophysiol. Rsch. Office: Behavioral Inst Atlanta Ste 106 Med Quarters 5555 Peachtree Dunwoody Rd Atlanta GA 30342

SEIFERT, AVI, aerodynamic engineering educator and researcher; b. Rechovot, Israel, Aug. 8, 1959; s. Yitshak and Betty Seifert; m. Tamy Serfaty, Sept. 4, 1983; children: Meirav, Adi. MSc in Engring., Tel-Aviv U., 1986, PhD in Engring., 1992. Lectr, rschr engring dept. fluid mechanics and heat transfer Tel-Aviv U. Faculty Engring., 1990—; cons. engr. "Ramot" Tel-Aviv U. R&D, 1987—. Contbr. articles to Jour. Fluid Mechanics, Procs. of the RAS B.L. Stability Conf., AIAA Jour., Jour. Engring. Math. Mem. AIAA. Achievements include first to construct a fully automated wind tunnel boundary layer stability experiment; research in the insight to the prospects of point source disturbances non linear interaction and control in boundary layers, and proving the superior relative efficiency of separation control by oscillatory blowing. Home: 8/16 Achimeir Aba St, Tel Aviv 69126, Israel Office: Dept Fluid Mech Engring, Klosner St Ramat Aviv, Tel Aviv 69978, Israel

SEIFERT, JOSEF, chemist, educator; b. Prague, Czech Republic, Sept. 21, 1942; came to U.S., 1977; s. Josef and Marta (Nova) S.; m. Yukari Takeuchi, Dec. 26, 1988; 1 child, Daniela. MSc, Prague Inst. Chem. Tech., 1964, PhD, 1973; postgrad., U. Calif., Berkeley, 1977-78. Rsch. specialist Prague Inst. Chem. Tech., 1973-77; from asst. to assoc. rsch. specialist pesticide chemistry and toxicology lab. U. Calif., Berkeley, 1978-85; prof., pesticide chemist dept. environ. biochemistry U. Hawaii, Honolulu, 1986—; adj. prof. Coll. Agr., Prague, 1976; vis. assoc. prof., chemist U. Calif., Davis, 1992-93; speaker in field, 1971—. Contbr. 52 articles to profl. jours. Grantee USDA, 1987, 1988-92. Mem. Am. Chem. Soc. (div. agrochems, program com. 1987—, exec. com. 1992—), U.S. Tennis Assn., Soc. Toxicology, Soc. Neurosci., Sigma Xi. Achievements include research in analytical methods (biochemistry, pesticide toxicology, analytical chemistry) in novel mechanisms of organophosphorous acid triesters and methylcarbamate action (toxicology). Home: 1545 Nehoa # 401 Honolulu HI 96822 Office: U Hawaii Dept Environ Biochemistry 1800 E-W Honolulu HI 96822

SEIFERT, ROBERT P., agricultural products company executive; b. 1927; married. BS, Kans. State U., 1950, MS, 1952. With Pioneer Hi-Bred Internat., Inc., Des Moines, 1951—, dir. dept. corn breeding, 1977-84, dir. plant breeding div., 1984-87, dir. rsch. region I plant breeding divsn., 1987-89, v.p. rsch., 1989-90, sr. v.p. rsch., 1990—; div. coord. cereal rsch. and data mgmt. depts., Pioneer Hi-Bred Internat., Inc. 1984. Office: Pioneer Hi-Bred Internat 700 Capital Sq 400 Locust Des Moines IA 50309-2340*

SEIFERT, TIMOTHY MICHAEL, infosystems specialist; b. Marengo, Iowa, Jan. 17, 1951; s. Henry George and Margy Elaine (Gerard) S. BS, U. Iowa, 1973. Sr. clerical asst. Sears, Roebuck & Co., Chgo., 1973-74, asst. div. head, 1973-76, div. head, 1976-77; operator word processing Arnstein & Lehr, Chgo., 1978-80, supr. word processing dept., 1980-85, mgr. systems and procedures, 1985-89, mgr. tech. svcs., 1989-91; mem. adminstrv. com., 1988—, dir. of ops., 1991—. Mem. Assn. Legal Adminstrs., C/T 3000 User's Group, Internat. HP User's Group. Democrat. Methodist. Avocations: computers, exercise, traveling. Home: 2150 N Lincoln Park W Apt 1312 Chicago IL 60614-4647 Office: Arnstein & Lehr 120 S Riverside Plz Ste 1200 Chicago IL 60606-3913

SEIFF, ALVIN, planetary scientist, atmosphere physics and aerodynamics consultant; b. Kansas City, Mo., Feb. 26, 1922; s. Harry Louis and Sara Dorothy (Silverstone) S.; m. Robbye Walker, Mar. 27, 1948 (div. Oct. 1959); children: David Wilson, Deborah Ellen Seiff Hedgecock; m. Julia Gwynne Hill, June 23, 1968; children: Michael Harry, Geoffrey Bernard. BS ChemE, U. Mo., 1942; postgrad., U. Tenn., 1946-48, Stanford U., 1959-60. Chem. engr. TVA, Florence, Ala., 1942-43; tech. supr. uranium isotope separ. Tenn. Eastman Corp., Oak Ridge, 1944-45; instr. physics U. Tenn., Knoxville, 1945-48; aero. rsch. scientist NACA Ames Aero. Lab., Moffett Field, Calif., 1948-57; chief supersonic free flight rsch. br. NACA, Moffett Field, 1952-63; chief vehicle environment div. NASA Ames Rsch. Ctr., Moffett Field, 1963-72, sr. staff scientist dir.'s office, 1972-77, sr. staff scientist space sci. div., 1977-86; sr. rsch. assoc. San Jose (Calif.) State U. Found., 1987—; mem. entry sci. team Viking Mars Mission Langley Rsch. Ctr., NASA, Hampton, Va., 1972-77, mem. sci. steering group Pioneer Venus Project Ames Rsch. Ctr., Moffett Field, 1972-82, Galileo Project, sci. group Jet Propulsion Lab., Pasadena, 1979—, mem. sci. team Soviet-French Vega Venus Balloon Mission, 1984-87; mem. basic Rsch. Coun., NASA, Washington, 1973-76; Von Karman lectr., 1990. Author and editor: Ballistic Range Technology, 1972; (with others) Venus, 1983; contbr. articles to profl. jours. Recipient Exceptional Scientific Achievement medals NASA, 1978, 81, H. Julian Allen award Ames Rsch. Ctr., 1982. Fellow AIAA (assoc.); mem. Am. Astron. Soc. (div. planetary sci.), Am. Geophys. Union. Avocations: music, piano, gardening, home design and construction. Office: Ames Rsch Ctr Mail Stop 245-2 Moffett Field CA 94035

SEIGEL, ARTHUR MICHAEL, neurologist, educator; b. Rochester, N.Y., Oct. 9, 1944; s. Hyman and Judith (Hyman) S.; B.A. in Biology with distinction, SUNY, Buffalo, 1966, M.D, 1970; m. Ellen May Streitfeld, June 1, 1969; children—Daniel Aaron, Mark Louis. Intern, SUNY Affiliated Hosps., Buffalo, 1970-71; resident Yale-New Haven Hosp., 1973-76; asst. prof. pediatrics and neurology Yale U. Sch. Medicine, New Haven, 1976-77, clin. instr., 1977-81, clin. asst. prof., 1981—; cons. in neurology Gaylord Rehab. Hosp., Wallingford, Conn., 1976-86; practice medicine specializing in neurology, New Haven, 1977—; attending physician Hosp. St. Raphael, New Haven, Yale-New Haven Hosp. Served with USPHS, 1971-73. Diplomate Am. Bd. Psychiatry and Neurology. Fellow Royal Soc. Medicine (affiliate); mem. Conn. Neurol. Soc. (v.p. 1987-92), Am. Acad. Neurology, Conn. Med. Soc., New Haven County Med. Soc. Home: 38 Vineyard Ave Guilford CT 06437-3235 Office: 60 Temple St New Haven CT 06510

SEILACHER, ADOLF, biologist, educator. Prof., dept. biology Yale U., New Haven, Conn. Recipient Crafoord prize Royal Acad. Scis., Stockholm, Sweden, 1992. Office: Yale University Dept of Biology POB 6666 New Haven CT 06511*

SEILER, DAVID GEORGE, physicist; b. Green Bay, Wis., Dec. 17, 1940; s. George August and Esther Victoria Elizabeth (Gustafson) S.; m. Nancy Sarah Cowdrick, June 15, 1963; children: Laurel Elaine Seiler Brunvoll, Rebecca Jeanne. BS in Physics, Case Western Res., 1963; MS in Physics, Purdue U., 1965, PhD in Physics, 1969. Asst. prof. physics U. North Tex., Denton, 1969-72, assoc. prof. physics, 1973-80, prof. physics, 1980-88, regent's prof. physics, 1988; physicist Nat. Bur. Standards, Boulder, Colo., 1972-73; group leader Nat. Inst. of Standards and Tech., Gaithersburg, Md., 1988—; cons. Battelle Columbus (Ohio) Labs., 1981-82, Nat. Oceanic and Atmospheric Adminstrn., Silver Spring, Md., 1990-92. Author and editor: Semiconductors and Semimetals, vol. 36, 1992; editorial adv. bd. Jour. Semiconductor Sci. & Tech., Bristol, Eng., 1989—, Am. sub-bd. chmn., 1992—; contbr. articles to jours. Physics & Chemistry. Fellow Am. Phys. Soc.; mem. AAAS, Optical Soc. Am., Materials Rsch. Soc. Presbyterian. Achievements include rsch. in narrow gap semiconductors, mercury cadmium telluride characterization, optical and electrical characterization of semiconductors. Office: NIST Bldg 225 Rm A 305 Gaithersburg MD 20899

SEINFELD, JOHN HERSH, chemical engineering educator; b. Elmira, N.Y., Aug. 3, 1942; s. Ben B. and Minna (Johnson) S. BS, U. Rochester, 1964; PhD, Princeton U., 1967. Asst. prof. chem. engring. Calif. Inst. Tech., Pasadena, 1967-70, assoc. prof., 1970-74, prof., 1974—, Louis E. Nohl prof., 1980—, exec. officer for chem. engring., 1973-90, chmn. engring. and applied sci. div., 1990—; Allan P. Colburn meml. lectr. U. Del., 1976; Camille and Henry Dreyfus Found. lectr. MIT, 1979; Donald L. Katz lectr. U. Mich., 1981; Reilly lectr. U. Notre Dame, 1983; Dean's Disting. lectr. U. Rochester, 1985; Katz lectr. CUNY, 1985; McCabe lectr. N.C. State U., 1986; Lewis lectr. MIT, 1986; Union Carbide lectr. SUNY, Buffalo; Van Winkle lectr. U. Tex., 1988; Bicentennial lectr., La. State U., 1988; Ida Beam lectr. U. Iowa, 1989, David Mason lectr. Stanford U., 1989; Julian Smith lectr. Cornell U., 1990; Merck lectr. Rutgers U., 1991; Henske Disting. lectr. Yale U., 1991; mem. sci. adv. bd. EPA; lectr. Am. Inst. Chem. Engrs., 1980; mem. council Gordon Research Confs., 1980-83. Author: Numerical Solution of Ordinary Differential Equations, 1971, Mathematical Methods in Chemical Engineering, Vol. III, Process Modeling, Estimation and Identification, 1974, Air Pollution: Physical and Chemical Fundamentals, 1975, Lectures in Atmospheric Chemistry, 1980, Atmospheric Chemistry and Physics of Air Pollution, 1986, Fundamentals of Air Pollution Engineering, 1988, Distributed Parameter Systems - Theory and Applications, 1989; assoc. editor Environ. Sci., Tech., 1981—; mem. editorial bd. Computers, Chem. Engring, 1974—, Jour. Colloid and Interface Sci, 1978—, Advances in Chem. Engring, 1980—, Revs. in Chem. Engring, 1980—, Aerosol Sci. and Tech., 1981—, Large Scale Systems, 1982—; assoc. editor: Atmospheric Environment, 1976—. Recipient Donald P. Eckman award Am. Automatic Control Coun., 1970, Pub. Svc. medal NASA, 1980, Disting. Alumnus award U. Rochester, 1989; Camille and Henry Dreyfus Found. Tchr. Scholar grantee, 1972. Fellow Japan Soc. for Promotion Sci.; mem. AICE (bd. dirs. 1988-91, editorial bd. jour. 1985—, Allan P. Colburn award 1976, William H. Walker award 1986), NAE, Am. Soc. Engring. Edn. (Curtis W. McGraw Rsch. award 1976, George Westinghouse award 1987), Assn. Aerosol Rsch. (bd. dirs. 1983—, v.p. 1988-90, pres. 1990-92), Am. Acad. Arts and Scis. Air Waste Mgmt. Assn., Am. Chem. Soc. (Svc. through Chemistry award 1988, Creative Advanced in Environ. Sci. and Tech. award 1993), Sigma Xi, Tau Beta Pi. Home: 363 Patrician Way Pasadena CA 91105-1027 Office: Calif Inst Tech Div Engring and Applied Sci Pasadena CA 91125

SEITZ, FREDERICK, university president emeritus; b. San Francisco, July 4, 1911; s. Frederick and Emily Charlotte (Hofman) S.; m. Elizabeth K. Marshall, May 18, 1935. AB, Leland Stanford Jr. U., 1932; PhD, Princeton U., 1934; Doctorate Hon. Causa, U. Ghent, 1957; DSc (hon.), U. Reading, 1960, Rensselaer Poly. Inst., 1961, Marquette U., 1963, Carnegie Inst. Tech., 1963, Case Inst. Tech., 1964, Princeton U., 1964, Northwestern U., 1965, U. Del., 1966, Poly. Inst. Bklyn., 1967, U. Mich., 1967, U. Utah, 1968, Brown U., 1968, Duquesne U., 1968, St. Louis U., 1969, Nebr. Wesleyan U., 1970, U. Ill., 1972, Rockefeller U., 1981; LLD (hon.), Lehigh U., 1966, U. Notre Dame, 1962, Mich. State U., 1965, Ill. Inst. Tech., 1968, N.Y. U., 1969; LHD (hon.), Davis and Elkins Coll., 1970, Rockefeller U., 1981, U. Pa., 1985, U. Miami, 1989. Instr. physics U. Rochester, 1935-36, asst. prof., 1936-37; physicist research labs. Gen. Electric Co., 1937-39; asst. prof. Randal Morgan Lab. Physics, U. Pa., 1939-41, assoc. prof., 1941-42; prof. physics, head dept. Carnegie Inst. Tech., Pitts., 1942-49; prof. physics U. Ill., 1949-57, head dept., 1957-64, dir. control systems lab., 1951-52, dean Grad. Coll., v.p. research, 1964-65; exec. pres. Nat. Acad. Scis., 1962-69; pres. Rockefeller U., N.Y.C., 1968-78; U. Miami (Fla.), 1989; trustee Ogden Corp., 1977—; dir. tng. program Clinton Labs., Oak Ridge, 1946-47; chmn. Naval Rsch. Adv. Com., 1960-62; vice chmn. Def. Sci. Bd., 1961-62, chmn., 1964-68; sci. adviser NATO, 1959-60; mem. nat. adv. com. Marine Biomed. Inst. U. Tex., Galveston, 1975-77; mem. adv. group White House Conf. Anticipated Advances in Sci. and Tech., 1975-76; mem. adv. bd. Desert Rsch. Inst., 1975-79, Ctr. Strategic and Internat. Studies, 1975-81; mem. Nat. Cancer Adv. Bd., 1976-82; dir. Akzona Inc. Author: Modern Theory of Solids, 1940, The Physics of Metals, 1943, Solid State Physics, 1955, The Science Matrix, 1992. Trustee Rockefeller Found., 1964-77, Princeton U., 1968-72, Lehigh U., 1970-81, Rsch. Corp., 1966-82, Inst. Internat. Edn., 1971-78, Woodrow Wilson Nat. Fellowship Found., 1972-82, Univ. Corp. Atmospheric Rsch., Am. Mus. Natural History, 1975—; trustee John Simon Guggenheim Meml. Found., 1973-83, chmn. bd., 1976-83; mem. Belgian Am. Edn. Found.; bd. dirs. Richard Lounsberry Found., 1980—. Recipient Franklin medal Franklin Inst. Phila., 1965, Hoover medal Stanford U., 1968, Nat. Medal of Sci., 1973, James Madison award Princeton U., 1978, Edward R. Loveland Meml. award ACP, 1983, Vannevar Bush award Nat. Sci. Bd., 1983, J. Herbert Holloman award Acta Metallurgica, 1993. Fellow Am. Phys. Soc. (pres. 1961); mem. NAS, Am. Acad. Arts and Scis., AIME, Am. Philos. Soc., Am. Inst. Physics (chmn. governing bd. 1954-59), Inst. for Def. Analysis, Finnish Acad. Sci. and Letters (fgn. mem.), Phi Beta Kappa Assos. Address: Rockefeller U 1230 York Ave New York NY 10021

SEITZ, WILLIAM RUDOLF, chemistry educator; b. Orange, N.J., May 5, 1943; s. Rudolf Oskar and Evelyn Lucretia (Miller) S.; m. Anna Florence Woo, Feb. 8, 1969; children: Tasha, William. BA in Chemistry, Princeton U., 1965; PhD in Chemistry, MIT, 1970. Rsch. scientist EPA, Athens, Ga., 1970-73; instr. chemistry U. Ga., Athens, 1973-75, asst. prof., 1975-76; asst. prof. chemistry U. N.H., Durham, 1976-80, assoc. prof., 1980-84, prof., 1984—. Mem. editorial bd. Biosensors and Bioelectronics, 1988—; editor Talanta: Chem. Sensors, 1991—; co.-editor: Chemical Sensors and Microinstrumentation, 1990; contbr. articles to profl. jours. Mem. Am. Chem. Soc., Soc. Applied Spectroscopy. Achievements include research showing how fluorescence and chemiluminescence may be applied to analytical chemistry; research to develop fiber optic chem. sensors. Office: U NH Dept Chemistry Durham NH 03824

SEIZINGER, BERND ROBERT, molecular geneticist, physician, researcher; b. Munich, Germany, Dec. 27, 1956; came to U.S., 1984; s. August and Mathilde (Haselbeck) S. MD, Ludwig-Maximilians U. Med. Sch, Munich, Germany, 1983; PhD summa cum laude, Max-Planck Inst., Munich, Germany, 1984. Postdoctoral rsch. assoc. Max Planck Inst. for Psychiatry, Munich, Germany, 1984; postdoctoral rsch. assoc. Harvard Med. Sch., Boston, 1984-86, instr. neurology, 1986-88, asst. prof., 1988-90, assoc. prof. neuroscience, 1990—; assoc. geneticist Mass. Gen. Hosp., Boston, 1990—; v.p. oncology drug discovery Bristol-Myers Squibb Pharm. Rsch. Inst., Princeton, N.J., 1992—, dir. dept. molecular genetics and cell biology, 1992—; co-chmn. Internat. Consortium on Neurofibromatosis, N.Y.C., 1988—; chmn. Comm. on Gene Loss in Human Cancers Intern Human Gene Mapping Conf., Oxford, 1991, Rsch. Adv. Bd. Nat. Neurofibromatosis Found., N.Y.C., 1991—; vis. rsch. scientist, assoc. mem. faculty Princeton (N.J.) U., 1993—. Recipient Otto Hahn Medal Max Planck Soc., Munich, Germany, 1983, Jr. Investigators award Nat. Neurofibromatosis Found., N.Y.C., 1985, Wilson s. Stone Meml. award Cancer Rsch., 1987, Rsch. Faculty Scholar award Am. Cancer Soc., Atlanta, 1989. Mem. Nat. Neurofibromatosis Found., Am. Soc. Human Genetics, Am. Soc. Neuroscience, Am. Assn. Cancer Rsch. Achievements include discovery of fundamental genetic mechanisms of tumor formation in the human nervous system, chromosomal location of the tumor suppressor genes causing neurofibromatosis type I and II, and von Hippel-Lindau disease; co-discovery of gene for neurofibromatosis type II. Office: Bristol-Myers Squibb Pharm Rsch Inst PO Box 4000 Princeton NJ 08543-4000

SEJNOWSKI, TERRENCE JOSEPH, science educator; b. Cleve., Aug. 13, 1947; s. Joseph Francis and Theresa (Cudnik) S.; m. Beatrice Alexandra Golomb, Mar. 24, 1990. BS, Case Western Res. U., 1968; PhD, Princeton U., 1978. Rsch. fellow Harvard Med. Sch., Boston, 1979-82; prof. biophysics Johns Hopkins U., Balt., 1982-90; prof. U. Calif. San Diego, Salk Inst., La Jolla, 1988—; investigator Howard Hughes Med. Inst., 1991—; bd. dirs. San Diego McDonnell-Pew Ctr. for Cognitive Neurosci., 1990—, Inst. for Neural Computation, U. Calif. San Diego. Editor-in-chief Neural Computation; co-inventor: (with others) the Boltzmann machine and NET talk. Recipient Presdl. Young Investigator award NSF, 1984; Sherman Fairchild Disting. scholar Calif. Inst. Tech., 1993. Mem. Soc. for Neurosci., Am. Phys. Soc., Internat. Neural Network Soc. (governing bd. 1988—), IEEE, Am. Math. Soc., Assn. Rsch. in Vision and Ophthalmology, Am. Assn. Artificial Intelligence. Achievements include co-invention of the Boltzmann machine, of NETtalk, a neural network for text-to-speech. Office: Salk Inst PO Box 85800 San Diego CA 92186-5800

SEKAR, M. CHANDRA, pharmacology researcher, educator; b. Pudukkotai, Tamil Nadu, India, Nov. 30, 1954; came to U.S., 1984; s. T.S.M. and Saraswathy Sundaram; m. Padmini Sekar, Oct. 24, 1982; children: Nitin, Samantha. Bs in Pharmacy, Birla Inst. Tech. and Sci., Pilani Rajasthan, India, 1974, M of Pharmacy, 1976; MS, U. British Columbia, Can., 1980, PhD, 1984. Post doctoral fellow U. Wis., Madison, 1984-87; rsch. asst. prof. U. Ala., Birmingham, 1988—; reviewer Endocrinology, 1989—, Jour. Physiology, 1990—. Contbr. articles and reviews to Biochem. Jour. and Jour. Biol. Chemistry. Vol. Big Brothers, Big Sisters, Madison, Birmingham, 1984—. Recipient Rsch. fellowship Univ. British Columbia, Vancouver, 1980-83, Rsch. and Devel. award Am. Diabetes Assn., Birmingham, 1988-91, FIRST award Nat. Cancer Inst., Birmingham, 1991-96. Mem. Am. Diabetes Assn., Endocrine Soc., Am.Soc. Pharmacology and Therapeutics. Achievements include significant contbr. in area of inositol lipid derived second messengers. Office: U Ala Dept Pathology Birmingham AL 35294

SEKERIS, CONSTANTINE EVANGELOS, biochemistry educator; b. Nauplia, Argolis, Greece, June 12, 1933; s. Evangelos G. and Polyxeni E. (Eulambiou) S.; m. Kalliope N. Platsouka, Jan. 2, 1965 (div. 1986); 1 child, Evangelos. Diploma, U. Athens (Greece), 1962; D in Medicine, U. Munich, 1962. Rsch. asst. Med. Sch. U. Munich, 1962-64; asst. Med. Sch. U. Marburg (Fed. Republic Germany), 1964-66, sr. asst., 1966-69, assoc. prof., 1969-73; head. sect. molecular biology of cell German Cancer Rsch. Ctr., Heidelberg, 1974-78; dir. Inst. Biol. Rsch. Nat. Hellenic Rsch. Found., Athens, 1979—, v.p., 1986—; prof. biochemistry U. Athens, 1979—; hon. prof. U. Heidelberg, 1980. Contbr. articles to profl. jours. 2d lt. Greek med. unit, 1957-59. Mem. Hellenic Biochemistry Biophys. Soc. (past pres.), Hellenic Soc. Biology Rsch. Ges. Biologische Chemie, Biochem. Soc., European Molecular Biology Orgn. Gas Zellbiologie, Am. Soc. Cell Biology, Academia Europaea (sect. biochemistry 1992). Greek Orthodox. Home: Ithakis 9, 11369 Athens Greece Office: Nat Hellenic Rsch Found, 48 Vas Constantinou Ave, 11635 Athens Greece

SEKI, HIROHARU, political science educator, researcher; b. Tokyo, Mar. 31, 1927; s. Haruo and Tekeko Seki; m. Keiko Seki; 1 child, Kuara. LLD, U. Tokyo, 1953. Assoc. prof. Kokugakuin U., Tokyo, 1964-67; rsch. assoc. Northwestern U., Evanston, Ill., 1966; assoc. prof. U. Tokyo, 1967-71, prof., 1971-87; prof. Hiroshima (Japan) U., 1975-83; prof. internat. relations Ritsumeikon U., Kyoto, Japan, dean faculty internat. relations, 1987-88; prof. emeritus U. Tokyo, 1987—; dir. inst. internat. relations and area studies Ritsumeikon U., Kyoto, Japan, 1989—; vis. prof. U. Malaysia, 1972-73, Johns Hopkins U., Washington, 1982-83, U. Calif., Berkeley, 1983, U. Hawaii, 1983; exec. mem. Sci. Coun. Japan, 1985-91, chmn. peace studies com., 1988-91; prof. joint program U. B.C.-Ritsumeikan U., 1991-92. Author: Foundations of International System Theory, 1970, Understanding International Politics, 1981, The Asia Pacific in the Global Transformation, 1987, Idea of Global Politics, 1977. Organizer Conf. for Unification of Korea in Asian Peace, 1986. Recipient award Inst. for Eastern Philosophy, 1987. Mem. Peace Studies Assn. Japan (exec. mem., founder, pres. 1973-75), Inst. for Peace Sci. (bd. dirs. 1975-79), Japan Assn. Simulation Gaming (v.p. 1989-93), Internat. Simulation and Gaming Assn. (pres. 1991-92), Internat. Inst. for Korean Juche (exec.), others. Home: 521-84 Sonnocho, Chiba-shi, Inageku 263, Japan Office: Ritsumeikon U, Faculty Internat Rels, Tojuin-kita-machi kita-ku, Kyoto Japan

SEKIMOTO, TADAHIRO, electronics company executive; b. Kobe, Japan, Nov. 14, 1926; s. Taichiro and Tomi (Katayama) S.; B.S. in Physics, U. Tokyo, 1948, D.Engring., 1962; m. Mayako Mori, Apr. 16, 1956; children: Masakazu, Sumito, Misako. With NEC Corp., Ltd., 1948-65, 67—, dir., 1974—, assoc. sr. v.p., then sr. v.p., 1974-78, exec. v.p., 1978-80, pres., 1980—, also acting dir. subsidiaries; mgr. communications process lab. COMSAT, 1965-67; vice chmn. Japan Com. Econ. Devel.; chmn. New Bus. Conf. Recipient Japanese Govt. prize, 1976, Purple Ribbon Medal, 1982, Blue Ribbon Medal, 1989, Aerospace Communications award AIAA, 1992. Fellow IEEE (Edwin Haward Armstrong Achievement award, 1982, Satellite Hall of Fame award); mem. NAE (fgn. assoc.), Japan Electronics Devel. Assn. (chmn.), Keidanren (vice chmn.). Buddhist. Author, patentee in field. Home: 29-6 Higashi Yukigaya 1-chome, Ohta-ku Tokyo 145, Japan Office: NEC Corp, 7-1 Shiba, 5-Chome, Minato-ku Tokyo 108-01, Japan

SEKIMURA, TOSHIO, theoretical biophysics educator; b. Akasaki, Tottori, Japan, Jan. 5, 1947; m. Akemi Sekimura; children: Mitsuyo, Yasuyo, Hiroyo. MS, Hiroshima (Japan) U., 1973, DSc, 1980. Postdoctoral fellow Kyoto (Japan) U., 1980-83; lectr. Osaka (Japan) Inst. Tech., 1983-86; assoc. prof. theoretical biophysics Chubu U., Kasugai, Aichi, Japan, 1986-89, prof., 1989—; vis. scholar U. Wash., 1993. Editor Forma, 1989—; contbr. articles to profl. jours. Mem. AAAS, Soc. for Math. Biology, Japanese Assn. Math. Biology, Biophys. Soc. Japan, Japanese Soc. Devel. Biologists, Phys. Soc. Japan. Office: Chubu U Coll Engring, 1200 Matsumoto-cho, Kasugai Aichi 487, Japan

SEKIOKA, MITSURU, geoscience educator; b. Tokyo, Jan. 5, 1930; s. Hajime and Teiko (Nishimura) S.; m. Fujiko Komori, Oct. 4, 1958; children: Atsuko, Setsuko. MS, Kyoto (Japan) U., 1952, DSc, 1961. Rsch. asst. Kyoto U., 1952-58, Osaka (Japan) Kyoiku U., 1958-63; asst. prof. Nat. Def. Acad., Yokosuka, Japan, 1963-65; assoc. prof. Nat. Def. Acad., Yokosuka, 1965-73, prof., 1973—; chmn. com. Iwate Geothemal Project, Morioka, Japan, 1980—, Kazuno (Japan) Geothermal Project, 1980—; mem. com. NEDO, 1981—, organizing com. IAHS, 1992—. Author: Meteorology, 1981; co-author: Earth Sciences, 1980; chief editor Geothermal Energy, 1992—; inventor Helicopter-Borne Radio Sonde. Mem. Japan Geothermal Energy Assn. (bd. dirs. Tokyo chpt. 1984—), Geothermal Rsch. Soc. Japan (councilor Tokyo chpt. 1978— ,pres. 1990—). Avocations: reading, travel. Home: 15-14 Kotsubo 7 Zushi, Kanagawa 249, Japan Office: Nat Def Acad, 10-20 Hashirimitzu 1, Yokosuka Kanagawa 239, Japan

SEKITANI, TORU, otolaryngologist, educator; b. Kochi, Japan, May 5, 1932; s. Fusaharu and Miyoko (Tokushige) S.; m. Miyoko Uejo, Dec. 26, 1960; children: Miwako, Yoshiko, Tetsuko. MD, Yamaguchi Med. Sch., 1957. Intern Yamaguchi Med. Sch. Hosp., Ube, Japan, 1957-58, instr., 1962, assoc. prof., 1971, prof., chmn. dept. otolaryngology, 1976—; rsch. assoc. Univ. Iowa, Iowa City, 1969-71; head ENT clinic Yamaguchi Cen. Hosp., Hofu, Japan, 1972-74. Author: (with others) Vertigo: Basic and Clinic, 1976, Vestibular Mechanism in Health and Disease, 1978; editor: Vestibular Ganglia and Vestibular Neuronitis, 1988. Mem. Barany Soc., Prosper Meniere Soc., Otorhinoloaryngological Soc. Japan (exec. com.), Japan Soc. Equilibrium and Rsch. (exec. com. 1975—). Office: Yamaguchi U Sch Medicine, Kogushi, Ube 755, Japan

SEKYRA, HUGO MICHAEL, industrial executive; b. Mannersdorf, Austria, June 10, 1941; s. Hugo and Gertrude (Ertl) S.; m. Elfriede Steiner; children: Markus, Anna. Dr.Iuris, U. Vienna, 1964. Managerial asst. VMW, Ranshofen, Austria, 1965-67; product mgr. Chemie Linz, Austria, 1967-71; mng. dir. Rigips Austria, Bad Aussee, 1971-85; mem. bd., chief exec. officer Papierfabrik Laakirchen AG, Austria, 1985; chief exec. officer Austrian Industries ÖIAG, Vienna, 1986—; pres. supervisory bd. VA Stahl AG/AIT AG/AMAG, Linz, Austria, 1986—, Forschungszentrum Seibersdorf, 1987—; mem. supervisory bd. ÖMV AG, Vienna, 1986—, Bank Austria/Siemens Öterreich AG, Vienna, 1987—; bd. dir. Fed. Mogul Corp., Detroit. Recipient hon. medal Repulic Austria. Avocations: golf, hunting, skiing. Office: Austrian Industries AG, Kantgasse 1, 1015 Vienna Austria*

SELBERHERR, SIEGFRIED, university dean, educator, researcher, consultant; b. Klosterneuburg, Austria, Sept. 3, 1955; s. Johannes and Josefine (Henninger) S.; m. Margit Leonhard, Oct. 12, 1979; children: Andreas, Julia. Dipl., Ing. Tech. U., Vienna, Austria, 1978, Dr. techn., 1981, venia docendi, 1984. Research assoc. Tech. U., Vienna, 1978-79, asst. prof. microelectronics, 1979-84, prof. computer-aided design, 1984-88, dean microelectronics, 1988—; cons. to bus. and industry. Author: Analysis and Simulation of Semiconductor Devices, 1984; editor Jour. Transactions of the Soc. for Computer Simulation, 1983—, Jour. Electrosoft, 1988-92, Jour. Mikroelektonik, 1986—, book series Computational Microelectronics, 1985—; contbr. articles to profl. jours. Recipient Dr. Ernst Fehrer award Tech. U. Vienna, 1983, Heinz Zemanek award, 1988, Dr. Herta Firnberg Fed. award. Fellow IEEE; mem. Assn. Computing Machinery, Soc. Indsl. and Applied Math., Nachrichtentechnische Gesellschaft (award 1985). Home: Fasanstrasse 1, A 3430 Tulln Austria Office: Tech U Vienna, Gusshausstrasse 27-29, A 1040 Vienna Austria

SELIG, PHYLLIS SIMS, architect; b. Topeka, Nov. 16, 1931; d. Willis Nolan and Victoria Clarinda (Oakley) Sims; m. James Richard Selig, Mar. 31, 1957; children: Lin Ann, Susan Nan, Sarah Jo. BS in Architecture, U. Kans., 1956. Realtor Assoc. Realty, Lawrence, Kans., 1965-70; v.p. finance and housing Alpha Phi Internat. Fraternity, Inc., Evanston, Ill., 1968-74, chief exec. officer, internat. pres., 1974-78, trustee, 1978-80; sr. engr. tech. Nebr. Pub. Power, Columbus, 1980-86, staff architect, 1986-89, archtl. supr., 1989—. Republican. Lutheran. Avocations: wood working, painting. Office: Nebr Pub Power 1414 15th St Columbus NE 68601-5226

SELINGER, ROSEMARY CELESTE LEE, medical psychotherapist; b. Phila., Dec. 9, 1945; d. Dah Yen and Ming G. (Lum Young) Lee; m. Daniel Steven Selinger, Dec. 31, 1969; children: David Lee, Nicole Lee. SB in

Chemistry with high distinction, Simmons Coll., 1967; MSH in Radiobiology, Harvard U., 1969; MD, U. N.Mex., 1977. Diplomate Am. Bd. Med. Psychotherapists. Surg. rsch. fellow Harvard U.-Boston City Hosp., 1969; project fellow Chinatown Boston Student Health Project, 1969; chemist Nat. Heart, Lung, Blood Inst., NIH, Bethesda, Md., 1970, VA Hosp., Long Beach, Calif., 1971; dir. field svcs. Pneumococcal Surveillance Project-U. N.Mex., Gallup, 1973; rsch. fellow VA Hosp., Albuquerque, 1978, mem. house staff, 1979; pvt. practice Grants Pass, Oreg., 1979—; mem. liaison com. Oreg. Med. Assn./Bd. Med. Examiners, Portland, 1991—. Vol. tng. bd. Lovejoy Hospice, Grants Pass, 1980-90. Mem. AAAS (life), AAUW (life), Josephine County Med. Soc. (del. 1993, pres. 1990), Oreg. Med. Assn., Oreg. Bd. Med. Examiners, Josephine County Med. Alliance (chair epionean endowment 1985—). Jewish. Office: 125 Manzanita Grants Pass OR 97526-8846

SELIVANSKY, DROR, chemistry educator, resarcher; b. Tel Aviv, Israel, Apr. 18, 1950; s. Yehosua and Penina (Wachiel) S.; m. Smadar Ben-Dor, July 20, 1977; children: Gilad, Omer. BSc in Chemistry, Hebrew U., Jerusalem, Israel, 1974, MSc in Polymer Chemistry, 1976; PhD in Fiber and Polymer Sci., N.C. State U., 1983. Chief chemist The Rogosin Enterprises, Ashdon, Israel, 1977; rsch. assoc. Israeli Fibers Inst., Jerusaalem, 1978; sr. rsch. chemist Monsanto Chem. Co., Pensacola, Fla., 1983-86; sr. lectr Hebrew U., Jerusalem, 1987-88; mgr. R&D Nilit Ltd., Migdal Haemek, Israel, 1989—; sr. lectr. Shenkar Coll. Textiles, Ramatgan, Israel, 1987-91; cons. Tami Rsch. Inst. Haifa, Israel, 1991-92. Contbr. articles to profl. jours. Tchr. Beth Meyer Synagogue, Raleigh, N.C., 1979-82. Recipient scholarship Monsanto Fibers Co., Pensacola, 1979-82. Mem. Am. Chem. Soc., Israeli Chem. Soc., Israeli Textiles Soc., Israeli Plastics Assn. Achievements include two U.S. patents on superior nylon yarns; invention, product devel. and commercialization of Acrilan II, 2nd generation acrylic fibers. Home: 68 Jabotinsky, Givatym 32504, Israel Office: Nilit Ltd, PO Box 276, Migdal Haemek 10505, Israel

SELKER, HARRY PAUL, medical educator. BA, Reed Coll., Portland, Oreg., 1974; MD, Brown U., 1978; MSPH, UCLA, 1984. Diplomate Am. Bd. Internal Medicine. Intern in internal medicine UCLA/Cedars-Sinai Med. Ctr., L.A., 1978-79; jr. resident in internal medicine Boston City Hosp., 1979-80, sr. resident in internal medicine, 1980-81; chief med. resident Boston U. Med. Ctr./Univ. Hosp., 1981-82; Robert Wood Johnson clin. scholar dept. medicine UCLA Sch. Medicine, 1982-84; instr. medicine divsn. cardiology Sch. Medicine Boston U., 1982-84; asst. vis. physician cardiology dept. Boston City Hosp., 1981-84; spl. affiliate attending physician divsn. gen. internal medicine/med. intensive care, dept. medicine Cedars-Sinai Med. Ctr., L.A., 1984-85; asst. prof. medicine divsn. gen. internal medicine and health svcs. rsch. Sch. Medicine UCLA, 1984-85; asst. prof. medicine divsns. gen. medicine and clin. decision making Sch. Medicine Tufts U., Boston, 1985-91, assoc. prof. medicine, 1991—; asst. attending physician divsns. gen. medicine and clin. decision making, dept. medicine New Eng. Med. Ctr., Boston, 1985-91, attending physician, 1991—, dir. Ctr. for Cardiovascular Health Svcs. Rsch. (formerly Multictr. Cardiology and Health Svcs. Rsch. Unit), 1985—, dir. Ctr. for Health Svcs. Rsch. and Study Design, 1987—; staff privileges dept. emergency medicine Newton-Wellesley Hosp., Newton, Mass., 1992—; mem. coordinating com. and sci. base subcom., Soc. Gen. Internal Medicine rep. NIH Nat. Heart Attack Alert Program, 1992—; mem. criteria, objectives com. Mass. Peer Review Orgn., 1989—, quality of health care com. Tufts Associated Health Plan, 1988—; mem. spl. project study sect. Nat. Heart, Lung and Blood Inst. NIH, 1987-90, Ctrs. of Excellence VA HSR&D Field Program, 1992—; cons. Agy. for Health Care Policy and Rsch., Health Care Fin. Adminstrn., Inst. Medicine of Nat. Acad. Sci., RAND; reviewer Am. Jour. Cardiology, Am. Jour. Medicine, Annals Internal Medicine, Chest, Jour. Electrocardiography, Jour. Gen. Internal Medicine, Med. Care, New Eng. Jour. Medicine. Contbr. articles to profl. jours.; creator (with Griffith, Beshansky and MacLeod) Cardiac Severity System Software, 1989. Fellow Am. Coll. Physicians (teaching and rsch. scholar 1986-89); mem. AAAS (life), AMA (jour. reviewer), APHA, Am. Fedn. Clin. Rsch. (sec./treas. ea. region 1991—, nat. coun., pub. policy com. 1991—), Am. Heart Assn. (cardiopulmonary and critical care coun., clin. cardiology coun.), Internat. Soc. Computerized Electrocardiology, Assn. Health Svcs. Rsch., Soc. Gen. Internal Medicine (assoc. editor SGIM News 1990-93, editor 1993—), Soc. Med. Decision Making (jour. reviewer). Achievements include patents and patents pending for computerized electrocardiography for diagnosis and treatment of acute cardiac ischemia. Office: New Eng Med Ctr Dept Medicine Div Gen Medicine & Clin Decision Making 750 Washington St NEMCH # 1031 Boston MA 02111*

SELL, FRIEDRICH LEOPOLD, economics educator, researcher; b. Munich, May 26, 1954; s. Hans Joachim and Gertrud (Von Werthern) S. Diploma in polit. econs., Albert-Ludwigs U., Freiburg, Germany, 1979, D Polit. Econs., 1981, Habilitation, 1987. Rsch. asst. dept. math. econs. Inst. Econ. Rsch., Freiburg, 1979-81; rsch. fellow Inst. Devel. Policy, Freiburg, 1981-85; scholar German Rsch. Soc., Freiburg, 1985-87; div. chief Kiel (Germany) Inst. World Econs., 1987-89; prof. gen. and devel. econs. Justus-Liebig U., Giessen, Germany, 1989—; prof. internat. econs. U. Dresden, Germany, 1992—. Author: Geld-und Währungspolitik in Schwellenländern, 1991 (Franz Vogt prize U. Giessen); contbr. articles to profl. publs. Mem. Union for Social Politics, European Econ. Assn., Ausschuss Entwicklungsländer. Evangelical. Avocations: tennis, jogging, swimming, sailing. Home: Goethestrasse 52, D-6300 Giessen Germany also: U Dresden Faculty Econs, Mommsen str 13, D-8027 Dresden Germany

SELL, JEFFREY ALAN, physicist; b. Anderson, Ind., Sept. 18, 1952; s. John M. and Jean (Shake) S.; m. Lynne C. Johannessen, Sept. 18, 1982; children: Stephanie, Jason. BS with highest honors, Purdue U., 1974; PhD, Calif. Inst. Tech., Pasadena, 1979. Rsch. scientist and sect. mgr. physics dept. GM Rsch. Labs., Warren, Mich., 1978—. Editor: Photothermal Investigations of Solids and Fluids, 1989; contbr. over 50 articles to profl. jours. Mem. Am. Phys. Soc., Am. Chem. Soc., Materials Rsch. Soc., Soc. Automotive Engrs., Phi Beta Kappa, Sigma Xi. Achievements include 10 patents on laser spectroscopy, laser processing of materials. Office: GM R & D Ctr Physics Dept 30500 Mound Rd Warren MI 48090-9055

SELL, ROBERT EMERSON, electrical engineer; b. Freeport, Ill., Apr. 23, 1929; s. Cecil Leroy and Ona Arletta (Stevens) S.; m. Ora Lucile Colton, Nov. 7, 1970. B.S., U. Nebr., 1962. Registered profl. engr., Nebr., Mo., Ill., Ind., Ohio, W.Va., Ky., Ark., Tex., Oreg., Wash., Calif. Chief draftsman Dempster Mill Mfg. Co., Beatrice, Nebr., 1949-53; designer-engr. U. Neb., Lincoln, 1955-65; elec. design engr. Kirkham, Michael & Assos., Omaha, 1965-67; elec. design engr. Leo A. Daly Co., Omaha, St. Louis, 1967-69; mech. design engr. Hellmuth, Obata, Kassabaum, St. Louis, 1969-70; chief elec. engr. Biagi-Hannan & Assos., Inc., Evansville, Ind., 1971-74; elec. project engr. H.L. Yoh Co., under contract to Monsanto Co., Creve Coeur, Mo., 1974-77; elec. project engr. Dhillon Engrs., Inc., Portland, Oreg., 1978-85; project coordinator Brown-Zammit-Enyeart Engring., Inc., San Diego, 1985-88; elec. engr. Morgen Design, Inc., San Diego, 1988; lead elec. engr. Popov Engrs., Inc., San Diego, 1988-89; mech. and elect. specialist Am. Engring. Labs., Inc. div. Prof. Svc. Industries, Inc., San Diego, 1990—; instr. Basic Inst. Tech., St. Louis, 1971. Mem. ASHRAE, IEEE. Home: PO Box 261578 San Diego CA 92196-1578 Office: AEL/PSI 7940 Arjons Dr Ste A San Diego CA 92126-6303

SELL, STEWART, pathologist, immunologist, educator; b. Pitts., Jan. 20, 1935; s. Oliver Martin and Mary Myra (Stewart) S.; m. Patricia Damon King, June 20, 1958 (div. 1985); children: Sherri Lynn Phillips, Stacy L. Klinke, Sean Stewart, Stephaine King; m. Ilze Mara Klavins, Feb. 16, 1991; 1 child, Philip Janus. BS, Coll. of William and Mary, 1956; MD, U. Pitts., 1960. Diplomate Am. Bd. Pathology, Am. Bd. Med. Lab. Immunologists. Intern, asst. resident in pathology Mass. Gen. Hosp., Boston, 1960-62; rsch. assoc. germfree animal rsch. lab. NIAID, NIH, Bethesda, Md., 1962-64; spl. fellow dept. exptl. pathology U. Birmingham (Eng.) Med. Sch., 1964-65; instr., asst. prof., then assoc. prof. pathology U. Pitts. Sch. Medicine, 1965-69; assoc. prof., then prof. pathology U. Calif., San Diego, 1970-82; prof. U. Tex. Med. Sch., Houston, 1982—, chmn. dept. pathology, 1982-87; adj. prof. lab. medicine U. Tex.-M.D. Anderson Cancer Ctr., Houston, 1983; mem. pathology B study sect. NIH, 1972-77; mem. immunology adv. com. Am. Cancer Soc., 1983-88; mem. bd. sci. counsel divsn. cancer biol. diagnosis Nat. Cancer Inst., 1982-86. Author: Immunology, Immunopathology & Immunity, 1972, 4th edit., 1987, Basic Immunology, 1987; editor: Serological Cancer Markers, 1992, Monoclonal Antibodies in Cancer, 1985. NIH grantee, 1964—. Mem. Am. Assn. Immunologists, Am. Soc. Exptl. Pathology, Am. Assn. Cancer Rsch., Internat. Soc. Oncodevelopmental Biology and Medicine (bd. dirs. 1972—), Internat. Acad. Tumor Marker Oncology (bd. dirs. 1984—). Achievements include discovery of immunoglobulin on surface of lymphocytes, activation of B-cells by antibody to surface immunoglobulin; identification of promoter-enhancer region of alphafetoprotein in gene expression, of liver stem cell, of role of delayed hypersensitivity in immunity to syphilis. Office: U Tex Health Sci Ctr Box 20708 6431 Fannin Houston TX 77225

SELLA, GEORGE JOHN, JR., chemical company executive; b. West New York, N.J., Sept. 29, 1928; s. George John and Angelina (Dominoni) S.; m. Janet May Auf-der-Heide, May 14, 1955; children: George Caldwell, Jaime Ann, Lorie Jean, Michael Joseph, Carlie. B.S., Princeton U., 1950; M.B.A., Harvard U., 1952. With Am. Cyanamid Co., Wayne, N.J., 1954—; corp. v.p. Am. Cyanamid Co., 1977, sr. v.p., 1977-78, vice chmn., 1978-79, pres., 1979-90, chief exec. officer, 1983—, chmn. bd., 1984—; bd. dirs. Union Camp Corp., Equitable Life Assurance Soc. U.S. Bd. dirs. Multiple Sclerosis Soc. With USAF, 1952-54. Mem. NAM (bd. dirs.), Pharm. Mfrs. Assn. (bd. dirs.). *

SELLE, BURKHARDT HERBERT RICHARD, physicist; b. Pössneck, Thuringen, Germany, May 9, 1938; s. Willi Karl and Anna Helene (Beckert) S.; m. Heide Hannemann, Sept. 1, 1965; 1 child, Martin. Diploma Physics, Humboldt U., Berlin, 1962, D Natural Scis., 1971; DSc, Acad. Scis., Berlin, 1984. Staff scientist Inst. Luminescence Rsch., Liebenwalde, German Democratic Republic, 1962-71, Cen. Inst. Electron Physics, Berlin, German Democratic Republic, 1971-91, Hahn-Meitner-Inst., Berlin, 1991—; guest tutor Humboldt U., Berlin, 1971-90. Contbr. articles to sci. jours. Recipient Lessing Gold medal Ministry Edn. German Democratic Republic, 1956, Gustav Hertz Prize, Phys. Soc. German Democratic Republic, 1983. Mem. German Phys. Soc. Avocations: touring, cycling, classical music. Home: Walter-Friedrich-Str 17, D-13125 Berlin Germany Office: Hahn-Meitner-Inst Berlin, Rudower Chaussee 5, D-12489 Berlin Germany

SELLERS, LUCIA SUNHEE, systems engineer; b. Taegu, Korea, Mar. 26, 1949; came to U.S., 1965; d. Chongin and Elizabeth (Min) Kim; m. Gregory J. Sellers, Nov. 26, 1983; 1 child, Kristin. BS, Rutgers U., 1973. With MIS dept. N.J. Bell, Madison, 1972-84; with computer systems AT&T, Lisle, Ill., 1985-90; with govt. markets group AT&T Bell Labs., Naperville, 1991—. Home: 7S 515 Oak Trails Dr Naperville IL 60540

SELLERS, MACKLYN RHETT, JR., architect; b. Cheraw, S.C., July 9, 1962; s. Macklyn Rhett and Jackie Rae (Dickens) S.; 1 child, Macklyn Rhett III. BS in Design, Clemson U., 1984, MArch, 1987. Registered architect, S.C.; cert. Nat. Coun. Archtl. Registration Bds. Intern architect Overstreet Archtl. Assn., Anderson, S.C., 1984-85, Freeman-White Architects, Charlotte, N.C., 1985, 86; health care specialist The Edge Group, P.A., West Palm Beach, Fla., 1987; health care specialist Louis P. Batson III Architects, Greenville, S.C., 1987-89, architect, health care specialist, 1989-92; health facilities planner Freeman White Architects, Inc., Charlotte, 1992—; teaching asst. Coll. Architecture, Clemson (S.C.) U., 1986-87, health care components grad. teaching asst., 1987. Editor Pike to Pike, 1982-83. Mem. memls. bldg. com. Christ Ch. Episcopal, Greenville, 1991-92. Mem. AIA, Pi Kappa Alpha (newsletter editor 1982-83, 1st pl. award 1982-83). Episcopalian. Avocations: basketball, water and snow sports. Home: 1908 Paces Landing Ave Apt 1824 Rock Hill SC 29732 Office: Freeman White Architects Inc 8001 Arrowridge Blvd Charlotte NC 28273

SELLIN, M. DEREK, aerospace engineer; b. Oslo, Apr. 14, 1968; s. Theodore and Taru Anna (Jarvi) S. BS in Aerospace Engring. magna cum laude, U. Mich., 1990, MS in Aerospace Engring., 1991. Aerospace engr. Wright Lab. USAF, Dayton, Ohio, 1991—. Mem. AIAA, Air Force Assn. Office: Wright Lab USAF WL/POTX Wright Patterson AFB OH 45433-6563

SELLMYER, DAVID JULIAN, physicist, educator; b. Joliet, Ill., Sept. 28, 1938; s. Marcus Leo and Della Louise (Plumhoff) S.; m. Catherine Joyce Zakas, July 16, 1962; children: Rebecca Ann, Julia Maryn, Mark Anthony. BS, U. Ill., 1960; PhD, Mich. State U., 1965. Asst. prof. MIT, Cambridge, 1965-72, assoc. prof., 1972; assoc. prof. U. Nebr., Lincoln, 1972-75, prof., 1975—, chmn. dept. physics, 1978-84, George Holmes disting. prof., 1987, dir. Ctr. Materials Rsch., 1988—; cons. Dale Electronics, Norfolk, Nebr., 1980—. Contbr. articles, book revs. to refereed jours. Recipient tech. award NASA, 1972; disting. vis. prof. S.D. Sch. Mines and Tech., Rapid City, 1981. Fellow Am. Phys. Soc. Office: U Nebr Ctr Materials Rsch Lincoln NE 68588-0113

SELM, ROBERT PRICKETT, consulting engineer; b. Cin., Aug. 9, 1923; s. Frederick Oscar and Margery Marie (Prickett) S.; m. Rowena Imogene Brown, Nov. 25, 1945 (div. Jan. 1975); children: Rosalie C. Selm Pace, Linda R. Selm Partridge, Robert F., Michael E.; m. Janis Claire Broman, June 24, 1977. BSChemE, U. Cin., 1949. Registered profl. engr. Enlisted U.S. Army, 1943; advanced through grades to sgt. U.S. Army, CBI Marianas, 1943-46; commd. capt. U.S. Army, 1949, resigned, 1954; design engr. Wilson & Co., Salina, Kans., 1954-67, gen. ptnr., 1967-81, sr. ptnr., 1981-89; ptnr. in charge Wilson Labs., Salina, Kans., 1956-88, chmn. bd. dirs.; ind. investor Salina, Kans., 1989—. Contbr. articles to profl. jours.; patentee in field. Mem. Gov.'s Adv. Commn. on Health and Environ. Named Engr. of Yr. Kans. Engring. Soc., Topeka, 1986. Fellow AIChE; mem. NSPE (state chmn. environ. resource com., nat. legis. and govt. affairs com. 1988-91), Am. Chem. Soc., Am. Water Works Assn., Water Pollution Control Fedn., Am. Acad. Environ. Engrs. (diplomate), Petroleum Club, Salina Country Club (pres. 1986), Elks, Shriners. Republican. Episcopalian. Avocations: golf, lapidary arts. Home: 135 Mt Barbara Dr Salina KS 67401-3414 Office: Wilson & Co 631 E Crawford St Salina KS 67401-5116

SELTSER, RAYMOND, epidemiologist, educator; b. Boston, Dec. 17, 1923; s. Israel and Hannah (Littman) S.; m. Charlotte Frances Gale, Nov. 16, 1946; children: Barry Jay, Andrew David. MD, Boston U., 1947; MPH, Johns Hopkins U., 1957. Diplomate Am. Bd. Preventive Medicine (trustee, sec.-treas. 1974-77), Am. Bd. Med. Specialties (mem. exec. com. 1976-77). Asst. chief med. info. and intelligence br. U.S. Dept. Army, 1953-56; epidemiologist div. internal health USPHS, 1956-57; from asst. prof. to prof. epidemiology Johns Hopkins U. Sch. Hygiene and Pub. Health, 1957-81, assoc. dean, 1967-77, dep. dir. Oncology Ctr., 1977-81; dean U. Pitts. Grad. Sch. Pub. Health, 1981-87, prof. epidemiology, 1981-88, emeritus dean, prof. epidemiology, 1988—; assoc. dir. USPHS Ctrs. for Disease Control, Rockville, Md., 1988-90; assoc. dir. Ctr. for Gen. Health Svcs. Extramural Rsch. Agy. for Health Care Policy and Rsch., Rockville, 1990—; cons. NIMH, 1958-70, also various govtl. health agys., 1958-79; expert cons. Pres.'s Commn. on Three Mile Island, 1979-80; mem. Three Mile Island Adv. Panel Health, Nat. Cancer Inst. Cancer Control Grant Rev. Com., Pa. Dept. Health Preventive Health Service Block Grant Adv. Task Force, Gov.'s VietNam Herbicide Info. Commn. Pa.; chmn. Toxic/Health Effects Adv. Com., 1985-87. Trustee, mem. exec. com., chmn. profl. adv. com. Harmarville Rehab. Ctr., Pitts., 1982-87; bd. dirs. Health Edn. Ctr., Media Info. Service. Served to capt. AUS, 1951-53, Korea. Decorated Bronze Star; recipient Centennial Alumni citation Boston U. Sch. Medicine, 1973; elected to Johns Hopkins Soc. of Scholars, 1986. Fellow AAAS, Am. Pub. Health Assn. (mem. governing council 1975-77, chmn. EPI sect. council 1979-80), Pa. Pub. Health Assn. (bd. dirs. 1985-88, pres.-elect 1986-88), Am. Coll. Peventive Medicine, Am. Heart Assn.; mem. Am. Epidemiol. Assn., Internat. Epidemiol. Assn., Am. Soc. Preventive Oncology, Am. Cancer Soc. (bd. dirs. Pa. div. 1985-87, mem. exec. com. 1986-87), Assn. Schs. Pub. Health (sec. 1969-71, exec. com., chmn. edn. com. 1983-87), Soc. Med. Cons. Armed Forces, Soc. Epidemiologic Research, Nat. Council Radiation Protection and Measurements (consociate), Johns Hopkins Alumni Coun. (bd. dirs. 1991—), Delta Omega, Sigma Xi. Office: Agy for Health Care Policy and Rsch Ctr for Gen. Health Svcs Extramural Rsch 2101 Jefferson St Rockville MD 20852

SELVADURAI, ANTONY PATRICK SINNAPPA, civil engineering educator, applied mathematician, consultant; b. Matara, Sri-Lanka, Sept. 23, 1942; arrived in Can., 1975; s. Kanapathiyar Sinnappa and W. Mary Adeline (Fernando) S.; m. Sally Joyce; children: Emily, Paul, Mark, Elizabeth. Diploma in Engring., Brighton Poly., U.K., 1964; Diploma, Imperial Coll./London U., 1965; MS, Stanford U., 1967; PhD in Theoretical Mechanics, U. Nottingham, 1971; DSc, U. Nottingham, Eng., 1986. Registered profl engr., Can. Staff rsch. engr. Woodward Clyde Assocs., Oakland, Calif., 1966-67; rsch. assoc. dept. theoretical mechanics U. Nottingham, 1969-70; lectr. dept. civil engring. U. Aston, Birmingham, Eng., 1971-75; asst. prof. civil engring. Carleton U., Ottawa, Ont., Can., 1975-76, assoc. prof., 1976-81, prof., 1982-93, chmn. dept., 1982-90, Davidson Dunton Rsch. lectr., 1987; prof., chmn. dept. civil engring./applied mechanics McGill U., Montreal, 1993—; vis. rsch. scientist Bechtel Group, Inc., San Francisco, 1981-82; vis. prof. U. Nottingham, 1986, Inst. de Mécanique de Grenoble, France, 1990; cons. Atomic Energy of Can. Ltd., Pinawa, Man., 1983—, Ministry of Transp. Ont., Toronto, 1984—, Fleet Tech., Ottawa, 1988—, Atomic Energy Control Bd., 1987—. Author: Elastic Analysis of Soil Foundation Interaction, 1979; editor: Mechanics of Structured Media, 1981, Mechanics of Material Interfaces, 1986, Developments of Mechanics, 1987. King George VI Meml. fellow English Speaking Union of Commonwealth, 1965, rsch. fellow SRC, U.K., 1969, Erskine fellow U. Canterbury, New Zealand, 1992. Fellow Am. Acad. Mechanics, Can. Soc. Civil Engring. (Leipholz medal 1991), Engring. Inst. Can., Inst. Math. and Its Applications. Roman Catholic. Office: McGill U, Dept Civil Engring, Montreal, PQ Canada H3A 2K6

SELVAM, RATHINAM PANNEER, civil engineering educator; b. Tiruvannamalai, India, Nov. 17, 1955; came to U.S., 1981; s. Rathina and Viruthambal (Rathinam) Mudaliar; m. Chitra Vasudevan, June 14, 1989; 1 child, Sanjay. BSCE, U. Madras, 1978, M in Engring., 1980; MSCE, S.D. Sch. Mines and Tech., 1982; PhD, Tex. Tech. U., 1985. Registered profl. engr., Ark. Asst. engr. Tamil Nadu Water Supply and Drainage Bd., Madras, 1980; assoc. lectr. dept. civil engring. Coll. Engring., Guindy, India, 1980-81; grad. rsch. asst. S.D. Sch. Mines and Tech., Rapid City, 1981-82; rsch. and teaching asst. Tex. Tech. U., Lubbock, 1982-84, rsch. assoc., part time instr., 1984-85; asst. prof. civil engring. U. Ark., Fayetteville, 1986-91, assoc. prof. civil engring., 1991—; mem. Govs.' Earthquake Adv. Com., Ark., 1988—, Ark. Seismic Network Task Force, Ark., 1991—; vis. rsch. scientist div. bldg., constrn. and engring. CSIRO, Australia, 1990. Contbr. articles to Jour. Wind Engring. and Indsl. Aerodynamics, Jour. Solar Energy Engring., Instrument Soc. Am. Mem. ASCE (assoc.), Wind Engring. Rsch. Coun., Order of the Engr., Sigma Xi, Chi Epsilon, Tau Beta Pi, Lions Club, Toastmasters. Achievements include rsch. in numerical modeling of linear, nonlinear and dynamic behavior in structural mechanics, fluid dynamics and acoustics using boundary element, finite element and finite difference methods. Office: U Ark Dept Civil Engring Bell 4190 Fayetteville AR 72701

SELWYN, DONALD, engineering administrator, researcher, inventor, educator; b. N.Y.C., Jan. 31, 1936; s. Gerald Selwyn and Ethel (Waxman) Selwyn) Moss; m. Delia Nemec, Mar. 11, 1956 (div. Mar. 1983); children—Laurie, Gerald, Marcia; m. Myra Rowman Markoff, Mar. 17, 1986. B.A., Thomas A. Edison Coll. N.J., 1977. Service engr. Bendix Aviation, Teterboro, N.J., 1956-59; service mgr. Bogue Electric Mfg. Co., Paterson, N.J., 1959; proposal engr. advanced design group Curtiss-Wright Corp., East Paterson, N.J., 1960-64; ind. bioengr., rehab. engring. cons. N.Y.C., 1964-67; pres. bd. trustees, exec. tech. and tng. dir. Nat. Inst. for Rehab. Engring., Hewitt, N.J., 1967—; cons. N.Y. State Office Vocat. Rehab., 1964—, Pres.'s Com. on Employment of Handicapped, 1966—, bus. and industry and for Am. with Disabilities Act compliance, also numerous state rehab. agys., health depts., vol. groups, agys. for handicapped in fgn. countries; cons., trainer computer applications. Contbr. articles on amateur radio, rehab. of severely and totally disabled to profl., gen. mags. Trustee Nat. Inst. for Rehab. Engring., Rehab. Research Center Trust. Recipient Humanitarian award U.S. Ho. of Reps., 1972, Bicentennial Pub. Service award, 1975; named knight Malta, 1973. Mem. Am. Acad. Consultants, I.E.E.E. (sr.), Soc. Tech. Writers and Pubs. (sr.), Nat. Rehab. Assn., N.Y. Acad. Scis., Mensa. Achievements include being the developer or co-developer field-expander glasses for hemianopsia, tunnel and monocular vision, electronic speech clarifiers, electronically guided wheelchairs, off-road vehicles and cars for quadriplegics, others; patentee indsl., mil. and handicapped rehab. inventions; expert, cons. on handicapped product safety including design, manufacture, labelling and user instrnl. material, 1990—. Office: Nat Inst Rehab Engring PO Box T Hewitt NJ 07421-1020

SELZER, MICHAEL EDGAR, neurologist; b. Buenos Aires, Argentina, Feb. 14, 1943; s. Hans and Claire S.; m. Ruth (div.); children: Molly Beth, Carl Jacob; m. Mary Frances Morrison, Aug. 4, 1991. MD, N.Y.U., 1962-68, PhD, 1964-68. Am. Bd. Psychiatry and Neurology, Pa., 1976. Asst. prof. Dept. Neurology U. Pa. Sch. Medicine, 1974-80, assoc. prof. Dept. Neurology, 1980-86; vis. prof. U. Pa., Jerusalem, Israel, 1982-83; prof. Dept. Neurology U. Pa. Sch. Medicine, 1986—; dir. Ctr. for Rehab. Ctr. for Neuro. Rehab., 1991—; prof. Rehab. Medicine U. Pa. Sch. Medicine, 1992—; mem. NIH Study Section, Neurology B2, Bethesda, Md., 1987-91; assoc. editor Annals of Neurology, 1989-92. Recipient Founders Day award N.Y.U., 1969, NIH Acad. Career Devel. award, 1975; fellow Fogarty Sr, Internat. at Hebrew U., Jerusalem, Israel, 1982-83; named Am. Paralysis Assn. Honoree as Doctor of Yr., 1985. Mem. Am. Acad. Neurology, Am. Epilepsy Soc., Am. Neurol. Assn., Assn. for Rsch. in Nervous and Mental Diseases, Soc. for Neurosci., John Morgan Soc., David Mahoney Inst. Neurol. Scis., Phila. Neurol. Soc., Coll. Physicians Phila. Home: 720 Vernon Rd Philadelphia PA 19119 Office: Hosp U of Pa Dept of Neurology 3400 Spruce St 3W Gates Philadelphia PA 19104

SEMADENI, ZBIGNIEW WLADYSLAW, mathematician, educator; b. Warsaw, Poland, Mar. 1, 1934; s. Tadeusz and Irena (Konopacka) S.; m. Ewa Wierzchleyska, Apr. 25, 1958; children: Beata, Dorota, Monika. MA in Physics, U. Poznan, Poland, 1955, MA in Math., 1956, PhD in Math., 1959; habilitation Inst. Math., Polish Acad. Scis., Warsaw, 1963. Asst., adj. dept. math. U. Poznan, Poland, 1954-61; vis. asst. prof. dept. math. U. Wash., Seattle, 1961-62; docent, prof. Polish Acad. Scis., Warsaw, 1962-85; prof., dir. Inst. Math. U. Warsaw, 1986—; vis. prof. York U., Toronto, Can., 1982-83, U. Sydney, 1984, U. Calif., Davis, 1989-90. Author: Banach Spaces of Continuous Functions, vol. 1, 1971, Primary Math. Education (tv lectures), 1975-78; editor-in-chief Wiadomosci Matematyczne Jour., 1967—. V.p. Internat. Commn. Math. Instrn., 1983-86. Mem. Polish Math. Soc. (pres. Warsaw br. 1969-71, Sierpinski prize 1972), Am. Math. Soc., Warsaw Sci. Soc. Avocations: classical music, gardening. Home: Falata St 6 Apt 24, 02-534 Warsaw Poland Office: U Warsaw Inst Math, Banach St 2, 02-097 Warsaw Poland

SEMERARO, MICHAEL ARCHANGEL, JR., civil engineer; b. Paterson, N.J., Dec. 15, 1956; s. Michael Archangel and Ann Ruth (Windish) S.; m. Diane Cathleen Hartley, Oct. 12, 1986; children: Michael Archangel III, Laura Nicole. BCE, Lehigh U., 1979; MCE, MIT, 1982; MBA, Rutgers U., 1989. Registered profl. engr., N.J., N.Y., Pa., Conn., Va.; registered profl. planner, N.J. Engr. DeGrace and Assocs., Wayne, N.J., 1978; sr. assoc., v.p. Langan Engring. and Environ. Svcs., Elmwood Park, N.J., 1979—; Presenter in field. Pres. Passaic County (N.J.) 4-H Assn., 1992—; chmn. exploring com. Passaic County Boy Scouts Am., 1986-89; leader Preakness Aggies 4-H Club, Wayne, 1977—; founding chmn. Passaic County Fair, 1988. Mem. North Am. MOSS Users Group (pres. 1989-91), GDS Nat. Users Group (chmn. civil engring. spl. interest group 1990—), ASCE, AAAS, MIT Club No. N.J. (pres. 1992—), Chi Epsilon. Roman Catholic. Home: 2 Warner Way Wayne NJ 07470 Office: Langan Engring Environ Svcs River Drive Center II Elmwood Park NJ 07407

SEMERJIAN, HRATCH GREGORY, research and development executive; b. Istanbul, Turkey, Oct. 22, 1943; came to U.S., 1966; s. Krikor and Diruhi (Semerciyan) S.; m. Sona Kohar Kurkciyan, July 12, 1969 (dec. 1983); children: Tamar, Ara; m. Ayda Karabal, Feb. 8, 1986. BSME, Robert Coll., Istanbul, 1966; MSc in Engring., Brown U., 1968, PhD in Engring., 1972. Rsch. asst. div. engring. Brown U., Providence, 1966-70; lectr. chemistry U. Toronto, Ont., Can., 1971-73; rsch. engr. Pratt & Whitney Aircraft United Technologies Corp., East Hartford, Conn., 1973-77; group leader Ctr. for Chem. Tech. Nat. Bur. Standards (now Nat. Inst. Standards and Tech.), Gaithersburg, Md., 1977-87, divsn. chief Chem. Sci. and Tech. Lab., 1987-92, dir. Chem. Sci. and Tech. Lab., 1992—; organizer tech. sessions, confs.

and symposiums for various profl. orgns., 1978—. Contbr. rsch. articles to profl. publs.; editor numerous conf. procs. Mem. parish coun. St. George Armenian Apostolic Ch., Hartford, Conn., 1975-77; chmn. parish coun., dir. choir St. Mary Armenian Apostolic Ch., Washington, 1977—; coach youth soccer Montgomery Soccer, Inc., Rockville, Md., 1978-81; mem., treas. Ani Armenian Choral Group, Washington, 1988—. Hagopian scholar Robert Coll., 1961-64, A.M.&F. corp. fellow, 1965, C.B. Keen fellow Brown U., 1969; recipient Silver medal Dept. Commerce, Washington, 1984; named Fed. Engr. of Yr., NSPE, Washington, 1991. Mem. AAAS, AIAA, ASME, Am. Inst. Chem. Engrs., Am. Chem. Soc., Combustion Inst. Avocations: soccer, singing, boating. Office: Nat Inst Standards and Tech Bldg 222 Route 270 Gaithersburg MD 20899

SEMMLOW, JOHN LEONARD, biomedical engineer, research scientist; b. Chgo., Mar. 12, 1942; s. John Leonard and Marylin (Fischer) S.; m. Claire McKnight, June 20, 1967 (dec. Sept. 1977). BSEE, U. Ill., 1964; PhD, U. Ill., Chgo., 1970. Instr. physiol. optics U. Calif., Berkeley, 1970-71; asst. prof. bioengring. U. Ill., Chgo., 1971-77, Rutgers U., Piscataway, 1977-80; assoc. prof. bioengring. Rutgers U./U. Medicine and Dentistry of N.J., New Brunswick, 1980-91, prof., 1991—. Editor Annals of Biomed. Engring.; contbr. chpts. to books, articles to profl. jours. NSF/CNRS Exch. fellow, Marseille, France, 1985. Mem. IEEE Engring. Medicine and Biology (sr.), Biomed. Engring. Soc., Sigma Xi. Achievements include patents in field. Home: 81 Louis St New Brunswick NJ 08901 Office: Rutgers U Biomed Engring Piscataway NJ 08855

SEMON, MARK DAVID, physicist, educator; b. Milw., Mar. 27, 1950; s. Milton K. and Joyce Gloria (Kupper) S. Student, Imperial Coll., London, 1973-74; AB magna cum laude, Colgate U., 1971; PhD, U. Colo., 1976. Rsch. asst. Kitt Peak Nat. Obs., Tucson, Ariz., 1970, Los Alamos (N.Mex.) Sci. Lab., 1974; asst. prof. physics Bates Coll., Lewiston, Maine, 1976-83, assoc. prof., 1983-88, prof. physics, 1990—; vis. prof. physics Amherst (Mass.) Coll., 1988-90; accident reconstructionist Med. and Tech. Cons., Portland, Maine, 1986—; referee Am. Jour. Physics, 1988—, Founds. of Physics, 1989—. Asst. editor Am. Jour. Physics, 1988-90; contbr. articles to Phys. Rev., Il Nuovo Cimento, other profl. jours. Woodrow Wilson fellow, 1971; grantee NSF, 1980, Nat. Rsch. Corp., 1978. Mem. Am. Phys. Soc., Coun. Undergrad. Rsch., Soc. Woodrow Wilson Fellows. Achievements include evaluation of expectation values in Aharonov-Bohm Effect; co-authoring new equation of state for liquid/gas systems near critical point, alternative formulation of quantum electrodynamics. Office: Bates Coll Dept Physics 334 Carnegie Sci Bldg Lewiston ME 04240

SEMONIN, RICHARD GERARD, state official; b. Akron, Ohio, June 25, 1930; s. Charles Julius and Catherine Cecelia (Schooley) S.; m. Lennie Stuker, Feb. 3, 1951; children: Cecelia C., Richard G. Jr., James R., Patricia R. BS, U. Wash., 1955. With Ill. State Water Survey, Champaign, 1955-91, chief, 1986-91, chief emeritus, 1991—; adj. prof. U. Ill., 1975-91. Contbr. chpts. to books and articles to profl. jours.; co-editr: Atmospheric Deposition, 1983. Staff sgt. USAF, 1948-52. Grantee NSF, 1957-76, U.S. Dept. Energy, 1965-90. Fellow AAAS, Am. Meteorol. Soc. (councilor 1983-86); mem. Nat. Weather Assn. (councilor 1978-81), Weather Modification Assn., Ill. Acad. Scis., Sigma Xi. Roman Catholic. Avocations: Civil war, golf, fishing, geneology. Home: 1902 Crescent Dr Champaign IL 61821-5826 Office: Ill State Water Survey 2204 Griffith Dr Champaign IL 61820-7495

SEMPLE-ROWLAND, SUSAN LYNN, neuroscientist; b. Ft. Riley, Kans., Mar. 30, 1955; d. George Thomas Semple and Virginia Lee (Whitlock) Johnston; m. Neil Edward Rowland; children: Matthew Eric, Nilsson Stanley, Jena Marie. BA, Gustavus Adolphus Coll., 1977; MSc, U. Pitts., 1979; PhD, U. Fla., 1986. Teaching fellow U. Pitts., 1979-81; postdoctoral rsch. fellow U. Fla., Gainesville, 1986-89, asst. rsch. scientist, 1989—; referee sci jours. Exptl. Eye Rsch., Jour. Neurochemistry;. Contbr. articles to Exptl. Eye Rsch., Brain Rsch., Glia, Jour. Comparative Neurology, Biochem. and Biophys. Rsch. Communication, Electophoresis, Current Eye Rsch.. Grantee Fight-for-Sight Prevent Blindness, 1989-90, Nat. Eye Inst., NIH, 1990-93. Mem. Assn. for Rsch. in Vision and Ophthalmology, Internat. Soc. for Eye Rsch., Soc. for Neurosci., Sigma Xi. Office: U Fla Coll Medicine Dept Neurosci PO Box 100244 Gainesville FL 32610-0244

SEN, JYOTIRMOY (JOE SEN), control engineer; b. Calcutta, India, Jan. 9, 1933; s. Satyendra Mohan and Suniti Rani (Dasgupta) S.; m. Sreelekha Sen, July 12, 1963; children: Jayashree S. Doyle, Monish. BS in Physics, Wadia Coll., Poona, India, 1956; BSEE, U. Coll. Tech., Calcutta, 1957, MSEE, 1962. Registered prof. engr. Calif. Instr. engr. Mahindra and Mahindra Ltd., Calcutta, 1963-69; sr. engr. Bechtel, Inc., San Francisco, 1969-75, The Ralph M. Parsons Co., Pasadena, Calif., 1976-77; project mgr. Oasis Oil Co., Tripoli, Libya, 1977-80; engring. supr. Bechtel Inc., San Francisco, 1980-85; prin. engr. Kaiser Engring. Hanford Co., Richland, Wash., 1985-90, mgr. process control engring., 1990--; co. rep. Doe Metrication Com. Richland, 1991--; mem. electronics adv. bd. Pasco Community Coll., Wash., 1991--. Mem. steering com. Kaiser Engring. United Way Campaign, Richland, 1991-92; co. rep. Doe Contractors Asian-Pacific Month, Richland, 1991, 92; founder, mem. Prabasi, San Francisco, 1973; mem. allocation panel United Way, 1991-92, 92-93. Mem. Instrument Soc. Am. (Richlands sect. pres. 1992-93, bd. dirs. 1991-92). Democrat. Hindu. Achievements include patents in India. Home: 2500 George Washington Way # 222 Richland WA 99352 Office: Kaiser Engrs Hanford Co PO Box 888 Richland WA 99352

SEN, PABITRA N., physicist, researcher; b. Calcutta, India, Sept. 5, 1944; came to U.S., 1968; s. Bibudh N. and Uma (Sen) S.; m. Susan Shu, Feb. 18, 1964; children: India, Maya. MS, Calcutta U., 1966, PhD, U. Chgo., 1972. Mem. profl. staff Xerox, Palo Alto, Calif., 1973-76; sr. scientist Xonics, Santa Monica, Calif., 1976-78; sci. adv. Schlumberger, Ridgefield, Conn., 1978—; visiting prof. Univ. de Provence, Marseille, France, 1985; guest rsch. fellow Royal Soc., Eng., 1988-89. Fellow Am. Phys. Soc. Achievements include explanation of laws of conduction in porous media. Home: 52 Woodlawn Dr Ridgefield CT 06877-5120

SENA, KANAGA NITCHINGA, neurologist; b. Jaffna, Sri Lanka, Aug. 5, 1944; came to U.S., 1969; s. Kanagarajanayagam Mathavar and Parvathapathiny (Vaithilingam) Kanagarajah; m. Thilaga Velauthar, Nov. 10, 1969. MD, U. Sri Lanka, 1969. Diplomate Am. Bd. Psychiatry and Neurology. Intern Gen. Hosp., Kandy, Sri Lanka, 1969-70, Somerset Hosp., Somerville, N.J., 1971; resident in medicine Bridgeport (Conn.) Hosp., 1972-73, chief neurology svc., 1981—; clin. fellow neurology Yale U., New Haven, 1973-76; resident in neurology Yale New Haven Hosp., 1973-76; pvt. practice specializing in neurology Stratford, Conn.; assoc. clin. prof. neurology Yale U. Fellow Am. Assn. Disability Evaluation Physicians, Stroke Coun. Am. Heart Assn.; mem. Am. Soc. Neurorehabilitation (cert.), Am. Acad. Neurology, Am. Electroencepahlographic Soc., N.Y. Acad. Scis. Office: Neurol Specialists 2590 Main St Stratford CT 06497

SENDAX, VICTOR IRVEN, dentist, educator, dental implant researcher; b. N.Y.C., Sept. 14, 1930; s. Maurice and Molly R. S.; m. Deborah deLand Cobb, Dec. 17, 1969 (div. June 1976); 1 child, Jennifer Reiland; m. Marcia Ayer Pearson, Dec. 13, 1986; children: Anneliese Chase, Cordelia Ayer. G-rad., Tanglewood Music Ctr., 1953; BA, NYU, 1951, DDS, 1955; postgrad., Harvard U. Sch. Dental Medicine, 1969-72. Diplomate Am. Bd. Oral Implantology/Implant Dentistry. Commr. N.Y. State Dental Service Corp., 1969-73; pres. BioDental Research Found., Inc., N.Y.C., 1975—, Victor I. Sendax, D.D.S., P.C., N.Y.C., 1972—, Sendax Dental Implant Magnetics Inc., N.Y.C., 1985—; attending implantologist St. Lukes-Roosevelt Hosp. Dental Implant Ctr., N.Y.C., 1979—, Beth Israel Hosp., N.Y.C., 1991—, Doctors Hosp., N.Y.C., 1991—; adj. assoc. prof. implant prosthodontics Columbia U. Sch. Dental and Oral Surgery, N.Y.C., 1974-92; vis. lectr. HA-coated implant systems dept. implant dentistry NYU Coll. Dentistry; faculty 1st dist. Dental Soc. Sch. for Continuing Dental Edn.; mem. dental implant rsch. programs adv. com. Nat. Inst. Dental Rsch., HHS; cons. Julliard Sch. Voice and Drama, N.Y.C., 1972—, Vocal Dynamics Lab. Dept. Otolaryngology, Lenox Hill Hosp., N.Y.C., 1970—; founder Sendax Seminars; 1st dir. implant prosthodontics resident program Columbia U. Sch. Dental and Oral Surgery and Columbia Presbyn. Hosp. Editor: Dental Clinics of North America: HA-Coated Dental Implants, 1992; Mem. editorial bd. Oral Implantology, 1979—; patentee in oral implant magnetics, mini-implants and abutments. Bd. dirs. City Ctr. Music and Drama, Inc. div. Lincoln Ctr. Performing Arts, N.Y.C., 1966-75; mem. adv. bd. Amagansett (N.Y.) Hist. Assn., 1969—; trustee Leukemia Soc. Am., N.Y.C., 1967; bd. dirs. Schola Cantorum, 1980-90, Soc. Asian Music, 1965-76. Served to capt. Dental Corps USAF, 1955-57. Recipient Spl. Recognition, Am. Fund Dental Health, 1981, Cert. of Honor, Brit. Dental Implant Assn., 1988. Fellow Am. Coll. Dentists, Internat. Coll. Dentists, Am. Acad. Implant Dentistry (nat. pres. 1981), Royal Soc. Medicine Gt. Britain; mem. ADA (ho. of dels. 1969), Am. Assn. Dental Schs. (chmn. interdisciplinary group on dental implant edn.), Acad. of Osseointegration (active), Am. Prosthodontic Soc., Am. Equilibration Soc., Am. Analgesia Soc., Fedn. Dentaire Internat., Am. Assn. Dental Rsch. (implant group), Internat. Assn. Dental Rsch., N.Y. Acad. Scis., Japan Soc., Century Assn., Players Club (N.Y.C.), Sigma Epsilon Delta. Home: 70 E 77th St Apt 6A New York NY 10021-1811 Office: Victor I Sendax DDS PC 30 Central Park S Ste 14B Parkview Dental Implt Condo New York NY 10019

SENESAC, ANDREW FREDERICK, weed scientist; b. New Bedford, Mass., Oct. 23, 1952; s. Archibald Edward and Louise (Salles) S.; m. Anne Marie Hedges, May 28, 1979; children: Nina, Hanna, Lucy. BS, U. Mass., 1977; MS, Cornell U., 1979, PhD, 1985. Weed sci. specialist Cornell Coop. Extension, L.I. Hort. Rsch. Lab., Riverhead, N.Y., 1985—. Co-editor, author: Weed Facts Series, 1991-92; contbr. articles on weed tech. to profl. jours. Recipient Tools for Teaching Superior Performance award Epsilon Sigma Phi, N.Y.C., 1992. Mem. Am. Soc. Hort. Sci., Weed Sci. Soc. Am., Northeastern Weed Sci. Soc. Democrat. Roman Catholic. Achievements include weed control research of chemical and alternative methods for horticultural crops. Office: LI Hort Rsch Lab 39 Sound Ave Riverhead NY 11901

SENGERS, JAN VINCENT, physicist; b. Heiloo, Netherlands, May 27, 1931; came to U.S., 1963; s. Adriaan and Cornelia Alida (Van Schie) S.; m. Johanna M.H. Levelt, Jan. 21, 1963; children: Rachel Teresa, Adriaan Jan, Maarten Willem, Phoebe Josephine. PhD cum laude, U. Amsterdam, Netherlands, 1962; D honoris causa, Tech. U. Delft, 1992. Teaching asst. U. Amsterdam, 1952-53, rsch. asst., 1953-55, rsch. assoc., 1955-63; physicist Nat. Inst. Stds. and Tech., Gaithersburg, Md., 1963—; assoc. prof. physics U. Md., College Park, 1968-74, prof., 1974—, dir. chem. physics, 1978-85; affiliate prof. chem. engring., 1991—; vis. prof. Tech. U. Delft, Netherlands, 1974-75. Mem. editorial adv. bd. Physica A, Amsterdam, 1975—; assoc. editor Internat. Jour. Thermophysics, 1989—; contbr. over 200 articles to profl. jours. Recipient Touloukian medal ASME, 1991. Fellow AAAS, Am. Phys. Soc.; mem. Discaleed Carmelite Secular Order, Cath. Com. on Intellectual and Cultural Affairs, Royal Netherlands Acad. Sci. (corr.). Roman Catholic. Achievements include research in transport properties of gases, exptl. studies of static and dynamic critical phenomena of fluids, nonequilibrium fluctuations in fluids; development of nonclassical equations for the thermodynamic and transport properties of fluids in the critical region. Home: 110 N Van Buren St Rockville MD 20850-1861 Office: U Md Inst Phys Sci and Tech College Park MD 20742

SENGUPTA, MRITUNJOY, mining engineer, educator; b. Cuttack, Orissa, India, Oct. 24, 1941; came to U.S., 1968; s. Chandi P. and Bani S.; m. Nupur Bagchi, Jan. 15, 1981; children: Shyam S. ME, Columbia U., 1971, MS, 1972; PhD, Colo. Sch. of Mines, 1983. Mining engr. Continental Oil Co., Denver, 1977-78, United Nuclear Corp., Albuquerque, 1978-80, Morrison-Knudson Co., Boise, Idaho, 1975-77, 80-82; assoc. prof. U. Alaska, Fairbanks, 1983-88, prof., 1989—; cons. UN Devel. Program, 1987. Author: Mine Environmental Engineering, vols. I and II, 1989, Environmental Impacts of Mining, 1992; contbr. articles to profl. pubs. Recipient Gold medal Mining Metall. Inst. of India, 1976, Nat. Merit scholarship Govt. of India, 1959-63. Mem. NSPE, So. Mining Engrs. Achievements include development of new concepts for mine design in oilshale in Colo. Home: 421 Cindy Dr Fairbanks AK 99701 Office: U Alaska Fairbanks AK 99775

SENGUPTA, SUBRATA, dean, engineering educator; b. India, June 29, 1948; came to U.S., 1969; s. Pabitra Kumar and Shanti (Gupta) S.; m. Mala Dasgupta, Jan. 3, 1979; children: Vikram, Vivek. MS, Case Western Res. U., 1972, PhD, 1974. Rsch. asst. prof. U. Miami, Coral Gables, Fla., 1974-77, asst. prof., 1977-78, assoc. prof., 1978-81, prof., 1981-90, chmn. mech. engring., 1986-90; dean Sch. Engring. U. Mich., Dearborn, 1990—; mem. Mich. Ctr. for High Tech., Detroit, 1991—. Editor 9 books; contbr. over 130 articles to pubs. Fellow ASME; mem AIAA, Am. Geophys. Union, Soc. Automotive Engrs. Achievements include patent for microencapsulated phase change material heat sinks. Office: U Mich 4901 Evergreen Dearborn MI 48128

SENITZKY, ISRAEL RALPH, physicist; b. Vilna, Poland, Feb. 28, 1920; came to U.S., 1932; s. Sender and Henna (Bloch) S.; m. Lillian Levit, June 18, 1944; children: Judith Reichman, Naomi Galili. BS, CUNY, 1941; PhD, Columbia U., 1950. Physicist U.S. Army Elec. Command, Ft. Monmouth, N.J., 1942-72; prof. physics Israel Inst. Tech., Haifa, 1972-83; sr. rsch. scientist U. Southern Calif., L.A., 1984-85, rsch. prof. physics, 1985-91; vis. prof. physics U. Southern Calif., L.A., 1983-84. Contbr. 51 articles to scientific jours. Fellow Am. Physical Soc. Democrat. Jewish. Office: U Southern Calif Ctr for Laser Studies University Park Los Angeles CA 90089

SENKAN, SELIM M., chemical engineering educator; b. Sept. 11, 1950. BS, METU, Ankara, Turkey, 1973; MS, MIT, 1973, PhD, 1977. Dir. MIT Practice Sch., Cambridge, 1977-79; asst. prof. MIT, Cambridge, 1979-82; assoc. prof. Ill. Inst. Tech., Chgo., 1982-89, prof., 1989-90; prof. U. Calif., L.A., 1990--; dir. High Temp. Kinetics Combustion Rsch. Lab. U. Calif., L.A., 1990--; cons. Chem. and Petroleum Cos., Calif., 1990--. Contbr. articles to profl. jours. NATO Science fellow 1973-77. Mem. AAAS, Am. Inst. Chem. Engrs. (dir. environ. div. 1991--, Lawrence K.Cecil award 1991), Am. Chem. Soc., The Combustion Inst., Sigma Xi. Achievements include over 100 papers and 5 patents on reaction kinetics. Office: UCLA Chem Engring Dept 403 Hilgard Ave Los Angeles CA 90024-1592

SENKAYI, ABU LWANGA, environmental soil scientist; b. Mpigi, Uganda, Oct. 16, 1943; came to U.S., 1973; s. Alamanzane Buza and Manjeri (Nalwoga) Abalyawo; m. Sunajeh Nansamba, Dec. 27, 1969; children: Ali K., Sala N. BS, Makerere U., Kampala, Uganda, 1971, MS, 1973; PhD, U. Calif., Davis, 1977. Rsch. scientist Tex. A&M U., College Station, 1977-87; sr. soil chemist Ebasco Environ. Svcs., Dallas, 1987-90; soil chemist PRC Environ. Mgmt., Inc., Dallas, 1990—; tech. cons. U.S. EPA Region 6. Author 7 book chpts.; contbr. articles to Soil Sci. Soc. Am. Jour., Soil Sci. Jour., Clays and Clay Minerals Jour. Recipient PRC-EMI Exceptional Performance award, 1991. Mem. Soil Sci. Soc. Am., Clay Mineral Soc., Mineral. Soc. Great Britain, Sigma Xi. Achievements include investigation of mineralogical weathering processes in soils and problems associated with reclamation of surface-mined lands. Home: 1122 De Havilland Ave Duncanville TX 75137-4742 Office: PRC Environ Mgmt Inc 350 N St Paul St Ste 2600 Dallas TX 75201-4218

SENS, MARY ANN, pathology educator; b. Berea, Ohio, July 26, 1949; d. William Richard and Mary Margaret (Griffin) Miller; s. Donald A. Sens, July 15, 1972. PhD, U. So. Calif., Columbia, 1976; MD, Med. U. S.C., 1981. Diplomate Am. Bd. Pathology. Chief resident Med. U. S.C., Charleston, 1983-84, dep. chief med. examiner, 1984-90, asst. prof., 1984-90, assoc. prof. pathology, 1990—; grants reviewer Johns Hopkins Alternatives to Animal Testing, Balt., 1990—. Reviewer Jour. Am. Acad. Dermatology, Toxicology and Applied Pharmacology; contbr. articles to profl. jours. Advisor Gov.'s Com. Suicide Prevention, Columbia, S.C., 1985-86; advisor, mem. Gov.'s Com. Prevention Sexual Assault, Columbia, 1974-76; med. advisor Ptnrs. of Ams., Cali, Colombia, 1985-89. Grantee NIH, 1980-83, 85-89, 90—, Am. Heart Assn., 1989-90, Johns Hopkins U., 1985-89. Fellow Am. Assn. Pathologists (women's com. 1990—), Internat. Acad. Pathologists, Assn. Clin. Scientists, Coll. Am. Pathologist (lab inspection com. 1991). Achievements include development of microscopic technique for trace crime evidence evaluation and documentation, characterization and use of human kidney cells in serum-free media. Office: Med U SC Pathology 171 Ashley Ave Charleston SC 29425

SENSENIG, DAVID MARTIN, surgeon; b. Gladwyne, Pa., May 4, 1921; s. Wayne and Elizabeth Long (Crawford) S.; B.S., Haverford Coll., 1942; postgrad. Sch. Medicine, U. Pa., 1942-43; M.D., Harvard U., 1945; children—Philip Campbell, David Martin, Andrew Wilson, Thomas O'Brien; m. 2d, Bernice Evans, Dec. 20, 1975. Rotating intern Allentown (Pa.) Hosp., 1945-46; surg. house officer, jr. asst. resident Peter Bent Brigham Hosp., Boston, 1948-50; sr. asst. resident, resident surgeon New Eng. Center Hosp., Boston, 1950-52; surg. resident Westfield (Mass.) State Sanatorium, 1952-53; asst. chief surg. service, dir. surg. research lab. VA Med. Teaching Group Hosp., Memphis, 1953-55; asst. chief surg. service VA Hosp., Albany, N.Y., 1955-57; resident in thoracic and cardiac surgery Univ Hosp., State U. Iowa, Iowa City, 1957-59, instr. in surgery, 1957-58, asso. in surgery, 1958-59, asst. prof., asso. prof. surgery, 1960-62; chief thoracic surgery sect. VA Hosp., Phila., 1959-60, asst. chief surg. service, 1963-66; cardiothoracic surgeon Pa. Hosp., Phila., 1962-63; asst. prof. surgery U. Pa., Phila., 1962-66, supr. Animal Research Lab., 1963-66; pvt. practice medicine specializing in surgery, Bangor, Maine, 1966-88; attending surgeon Eastern Maine Med. Center, Bangor, 1966-88; attending surgeon St. Joseph Hosp., Bangor, 1966-88, chief surg. service, 1974-79, chief thoracic surgery sect., 1979-88; chief surg. service VA Hosp., Togus, Maine, 04330, 1988—. Served to capt. M.C., U.S. Army, 1943-48. Diplomate Am. Bd. Surgery, Am. Bd. Thoracic Surgery. Mem. ACS (gov. at large 1985-91), Pa. Assn. Thoracic Surgery, Penobscot County Med. Soc. (pres. 1974), Maine, Am. thoracic socs., AAAS, Internat. Cardiovascular Soc., Am. Geriatric Soc., Iowa Acad. Surgery, Phila. Acad. Surgery, Am. Coll. Chest Physicians, Bangor Med. Club (pres. 1970), Maine Vascular Soc. (pres. 1978), New Eng. Surg. Soc., N.Y. Acad. Scis., New Eng. Soc. Vascular Surgery. Republican. Episcopalian. Contbr. articles to sci. publs. Home: 55 Ridgewood Dr Augusta ME 04330-4337 Office: VA Hosp Togus ME 04330

SENTANDREU, RAFAEL, microbiologist; b. Castello de la Ribera, Valencia, Spain, July 26, 1937; s. Enrique and Asuncion (Ramon) S.; m. Maria Victoria Elorza. 1 child, Maria. BA in Pharmacy, U. Barcelona, 1961; PhD in Microbiology, U. Madrid, 1965; postgrad. cert., Cambridge U., U.K., 1966, PhD in Biochemistry, 1968. Pharmaceutical Diplolmate. Tech. officer Park & Davis, Madrid, 1962-63; PhD student Spanish Rsch. Coun., Madrid, 1963-65; postdoctoral fellow Cambridge U., Eng., 1965-68; rsch. assoc. Rutgers U., 1969-70; Fulbright scholar, rsch. assoc. Spanish Rsch. Coun., Spain, 1970-75; assoc. prof. U. Salamanca, Spain, 1975-78; full prof. U. Valencia, Spain, 1979—, dean faculty pharmacy, 1979-81, head microbiology sect., 1979—; mem. Coll. Rsch. Program NATO, Brussels, 1988-92, Risk Evaluation in Biotechnology, Spain, 1988—; expert Biot echnol. Vet Pharmacy E.C., Brussels, 1989—. Author over 100 papers in profl. jours. Active Fundacion V. Estudios Avanzados, Valencia, 1984—, Inst. V. Estudis i Investigacio, Valencia, 1987—; mem. com. sci. tech. policy Generalitat Valenciana, 1989—; exec. pres. Fundacion Valenciana de Investigacion en Biomedicina, 1991-93, dir. 1993—. Recipient: Severo Ochoa Prize, Fundacion Ferrer Spain, 1988, several grants from scientific bodies. 1972—. Mem. Biochemical Soc. U.K., Am. Soc. Microbiology, Internat. Yeast Commn., Sociedad Espanola de Bioquimica, Micology Group (pres. 1982-91), Nat. Acad. Pharmacy (corr.), Tennis Valencia, Fishing Valencia. Avocations: tennis, music, fishing. Home: Avgda Perez Galdos 92, Valencia Spain 46008 Office: Avgda Vicente Andres Estelles, s/n 46100 Burjassot, Valencia Spain

SENTELL, KAREN BELINDA, chemist; b. Charleston, S.C., Jan. 28, 1957; d. Bobby Gene and Ruth Evelyn (Weinberg) S.; m. Daniel A. Coffman, Apr. 29, 1988. BS in Chemistry magna cum laude, U. S.C., 1982; PhD, U. Fla., 1987. Undergrad. rsch. asst. U. S.C., 1982; grad. rsch., teaching asst. U. Fla., Gainesville, 1983-87, project coord. pesticide rsch. lab., 1984-85, Leopold Schepp Found. postdoctoral fellow, 1987-88; asst. prof. chemistry U. Vt., Burlington, 1989—; organizer symposia, confs. in field.; mem. anticancer drug devel. program Vt. Cancer Ctr. Mem. editorial bd. Jour. Chromatography; contbr. articles to profl. publs. Grantee U. Vt., Vt. Lung Assn., Vt. Initiative for Summer Intensive Tng., 1992, Am. Cancer Soc., 1992-93, U.S. EPA, 1992—, ICI Pharms., 1992, Alcoa Found., 1990-92, Hughes Endeavor for Life Sci. Excellence, 1990-92, others. Mem. Am. Chem. Soc., Assn. Women in Sci., Phi Beta Kappa, Iota Sigma Pi, Pi Mu Epsilon. Achievements include patent for ultrasound driven synthesis of reversed and normal phase stationary phases for liquid chromatography. Office: U Vt A-218 Cook Bldg Burlington VT 05405-0125

SEPHEL, GREGORY CHARLES, biochemist, educator; b. Ft. Wayne, Ind., May 24, 1950; s. Robert Charles and Lois Jean (Simpson) S.; m. Stephanie Louise Roth, July 31, 1976; children: Alysse Nicole, Kara Michelle and Erin Blake (twins). BS, U. Calif., Irvine, 1973; PhD, U. Utah, 1986. Lic. med. technologist. Med. technologist Green Hosp., Scripps Clinic and Rsch. Found., La Jolla, Calif., 1976-79, Hosp. Coop. Utah, Salt Lake City, 1979-81; teaching/rsch. asst. U. Utah, Salt Lake City, 1981-86, biologist, 1986; biologist NIH, Bethesda, Md., 1986-88; asst. prof. Vanderbilt U. Sch. Medicine, Nashville, 1988—; clin. chemist Vanderbilt U. Vet. Affairs Med. Ctr., Nashville, 1988—; mem. ancillary testing task force Dept. Vets. Affairs, Washington, 1989, quality improvement checklist cons., 1991-92. Vol. Sci. By Mail, Nashville, 1990; bd. dirs. Park Manor Retirement Apts., Nashville, 1991—. Grantee NIH, 1989, Dept. Vets. Affairs, 1990, 91. Mem. AAAS, Am. Assn. Clin. Chemistry, Am. Soc. Clin. Pathologists, Am. Soc. Cell Biology. Presbyterian. Achievements include research in genetically inherited abnormality of elastic tissue in cutis laxa, Hutchinson's Gilford Progeria involves a connective tissue defect, synthetic peptide of basement membrane protein supports cell attachment and neurite growth, basement membrane thickening of diabetic blood vessels occurs rapidly during blood vessel growth in diabetic wounds. Office: Vanderbilt U Med Ctr 1161 21st Ave S Nashville TN 37232-2561

SEPMEYER, LUDWIG WILLIAM, systems engineer, consultant; b. East St. Louis, Ill., Nov. 8, 1910; s. William Henry and Anne Antoinette (Brandt) S.; m. Inez Hopkins, July 3, 1936; 1 child, Adrienne. BS, U. Calif., Berkeley, 1933; postgrad., UCLA, 1934-42. Registered profl. engr., Calif. Rsch. asst. UCLA, 1934-37; ind.cons. engr. L.A., 1934-41, 63—; engr. Lansing Mfg. Co., L.A., 1939-40; systems engr. elec. rsch. products Western Electric Co., L.A., 1941-42; engr. war rsch. program U. Calif., San Diego, 1942-45; engr. Calif. Inst. Tech., Pasadena, 1945, elec. engr., then cons. U.S. Naval Ordnance Test Sta., Pasadena, Calif., 1945-57; engr. Rand Corp., Santa Monica, Calif., 1951-56, Systems Devel. Corp., Santa Monica, 1956-63; mem. sci. adv. coun. on noise control Calif. State Environ. Quality Study Coun. Fellow Acoustical Soc. Am., Audio Engring. Soc.; mem. ASTM, IEEE (sr. mem., life mem.), Inst. Noise Control Engring., Sigma Xi, Eta Kappa Nu. Achievements include patent on means for measuring root mean square value of complex electrical waves, for diversity system for noise masking. Home and Office: 1862 Comstock Ave Los Angeles CA 90025

SEPUCHA, ROBERT C., chemical physicist, optics scientist; b. Salem, Mass., June 12, 1943. BS, MIT, 1965, MS, 1967; PhD in Engring. Physics, U. Calif., San Diego, 1971. Sr. rsch. sci. Aerodyne Rsch. Inst., 1971-76; physicist, high energy laser system project office U.S. Army, 1976-78; prof. mgr. Space Defense Tech. Divsn. Directed Energy Office Defense Adv. Rsch. Project Agy., 1979-80, dep. dir., 1980-84; v.p. space tech. W.J. Schafer Assoc. Inc., 1984—. Mem. Am. Phys. Soc., Optical Soc. Am., Sigma Xi. Achievements include research in advanced high energy laser technology programs which may lead to eventual space-based weapon systems, high power chemical lasers, aquisition tracking and precision pointing, large spec-ture beam control. Office: W J Schfer & Assocs 321 Billerica Rd Chelmsford MA 01824-4100*

SEPULVEDA, EDUARDO SOLIDEO, chemical engineer; b. Loay, Philippines, Jan. 5, 1945; came to U.S., 1981; m. Consuelo S. Araneta, May 18, 1977; children: Edward, Josephus, Blaise. BSChE, De La Salle U., 1966; M of Chem. Engring., U. Philippines, 1972. Lic. profl. engr., Calif. Instr. U. Philippines, Quezon City, 1968-70; sr. instr. De La Salle U., Manila, Philippines, 1970-71, asst. prof., 1971-73; process engr. Philippines Petroleum Corp., Makati, 1973-74, sr. process engr., 1974-75, mgr. tech. svc., 1975-81; sr. engr. C F Braun & Co., Alhambra, Calif., 1981; prin. engr. Brown & Root Braun, Alhambra, Calif., 1987—; reviewer chem. bd. De La Salle U., 1970-79. Mem. St. John Corregidor Lodge, Manila, 1979. Mem. AICE, Pasadena Consolidated Lodge #272 Project Mgmt. Inst., Phi Kappa Phi. Achievements include designed and engineered a fluid catalyst transport system to hydrocracker reactor. Home: 3700 Oaklawn Ln Pico Rivera CA 90660 Office: Brown & Root Braun 1000 South Fremont Ave Alhambra CA 91803

SERAFIN, ROBERT JOSEPH, science center administrator, electrical engineer; b. Chgo., Apr. 22, 1936; s. Joseph Albert and Antoinette (Gazda) S.; m. Betsy Furgerson, Mar. 4, 1961; children: Katherine, Jenifer, Robert Joseph Jr., Elizabeth. BSEE, U. Notre Dame, 1958; MSEE, Northwestern U., 1961; PhDEE, Ill. Inst. Tech., 1972. Engr. Hazeltine Rsch. Corp. Ill. Inst. Tech. Rsch. Inst., 1960-62; assoc. engr., rsch. engr., sr. rsch. engr. Nat. Ctr. for Atmospheric Rsch., Bouulder, Colo., 1962-73, mgr. field observing facility, 1973-80, dir. atmospheric tech. div., 1981-89, dir. ctr., 1989—. Author: Revised Radar Handbook, 1989; also numerous articles; editorial founder Jour. Atmospheric and Oceanic Tech.; patentee in field. Chmn. Citizens Adv. Com., Boulder; speaker various civic groups in U.S. and internationally. Mem. NAS (coms.), IEEE (sr.), Am. Meteorol. Soc. (exec. com.), Boulder C. of C. Avocations: cycling, golf, fishing, skiing. Office: Nat Ctr Atmospheric Rsch PO Box 3000 1850 Table Mesa Dr Boulder CO 80303*

SERENO, PAUL C., paleontologist, educator; b. 1958. Ph.D. in paleontology, Columbia U, NY. Asst. prof. U Chicago, 1987—. Recipient Player of the Yr., Tempo All-Professor Team, Chicago Tribune, 1993. Discoverer of Eoraptor, the earliest known dinosaur, in the Argentine foothills of the Andes mountains. Office: U Chgo Dept Organismal Biology/ Anatomy 5801 S Ellis Ave Chicago IL 60637

SERGI, ANTHONY ROBERT, physician, surgeon; b. Elizabeth, N.J., May 1, 1963; s. Martin and Agnes (Kraus) S. BS in Biology, Fairleigh Dickinson U., 1985; DPM, Pa. Coll. Medicine, 1989. Surg. resident Kennedy Meml. Hosp., Saddle Brook, N.J., 1989-91; chief surg. resident Kennedy Meml. Hosp., Saddle Brook, 1990-91; physician, surgeon Foot and Ankle Treatment Ctr. of N.J., Paramus, N.J., 1991—. Contbr. articles to profl. jours. Named one of Outstanding Young Men of Am., 1989. Roman Catholic. Avocations: tennis, golf, windsurfing. Office: Foot and Ankle Treatment Ctr 241 Oradell Ave Paramus NJ 07652-4808

SERLING, JOEL MARTIN, educational psychologist; b. Seneca Falls; N.Y., Feb. 8, 1936; s. Philip and Cecil Serling; A.A., U. Buffalo, 1957; B.S. in Edn., Ohio No. U., 1959; M.A., Columbia U., 1960; children—Meredith Anne, Rebecca Lynne, Heather Lee. Instr. psychology West Liberty (W.Va.) State Coll., 1961-63; vocat. psychologist, div. child welfare, Cleve., 1963-64; sch. psychologist Steuben County Bd. Coop. Ednl. Services, Bath, N.Y., 1964-65, Chenango County Bd. Coop. Ednl. Services, Norwich, N.Y., 1965-67, Delaware County Bd. Coop. Ednl. Services, Walton, N.Y., 1967-68; sch. psychologist Vestal (N.Y.) Central Sch., 1968-70, Whitesboro (N.Y.) Central Sch., 1970—; adj. prof. psychology Utica Coll., Syracuse U., 1971-75, 86—, SUNY Coll. Tech., Utica-Rome, 1975—, Mohawk Valley Community Coll., 1971—; cons., mentor Empire Coll., SUNY, 1975—; instr. psychology Am. Inst. Banking, 1971—. Bd. edn., bd. dirs. Hillel Day Sch., Utica-Rome, 1971-75; mem. bd. profl. advs. Mohawk Valley Learning Disabilities Assn., 1972-76. Recipient cert. of recognition Mohawk Valley Learning Disabilities Assn., 1973; cert. sch. psychologist, N.Y. State, N.C. Mem. Am. Psychol. Assn., Nat. Assn. Sch. Psychologists (charter), N.Y. Assn. Sch. Psychologists (cert. of recognition, 1977), Sch. Psychologists of Upper N.Y., Central N.Y. Psychol. Assn., United Univ. Professions, N.Y. State United Tchrs. Assn., Whitesboro Tchrs. Assn., Phi Delta Kappa. Jewish. Clubs: Odd Fellows, Zeta Beta Tau. Co-author, co-developer; Early Identification Screening Index, 1971; contbr. articles to profl. publs., presentations to profl. confs. Home: 432 Upper Valley Rd Rochester NY 14624 Office: Whitesboro Ctrl Sch Whitesboro NY 13492

SERRAGLIO, MARIO, architect; b. Bassano, Veneto, Italy, Apr. 13, 1965; came to U.S., 1972; s. Luciano G. and Maria P. (Bellon) S. BS in Architecture, Ohio State U., 1988. Real estate agent Four Star Realty, Columbus, Ohio, 1984—; treas. Columbus Masonry, Inc., 1985-86; v.p. Serraglio Masonry, Inc., Columbus, 1986—; pres. Serraglio Builders, Inc., Columbus, 1987—; residential designer Gary A. Bruck, SGR, Inc., Columbus, 1988-89, Sullivan Gray Ptnrs., Columbus, 1989-92; project mgr. John Regan Architects, Columbus, 1992—. Recipient 1st pl. House Design competion, Columbus Builders Industry Assn., 1989. Mem. Columbus Bd. Realtors. Office: John Regan Architects 177 E Beck St 2d Fl Columbus OH 43206

SERRANO, LUIS FELIPE, biochemist, researcher; b. Madrid, July 2, 1959; s. Luis and Elvira (Pubul) S.; m. Isabel Vernos, June 29, 1985; children: Miguel, Laura. BS in Biochemistry, U. Complutense, Madrid, 1981; PhD, U. Autonoma, Madrid, 1985. Postdoctoral staff C.B.M. (CSIC-UAM), Madrid, 1985-87; postdoctoral staff M.R.C., Cambridge, Eng., 1987-91; group leader EMBL, Heidelberg, Germany, 1992—. Author: Methods in Enzimology, 1986. 2d lt. Spanish Army (infantry), 1982-83. Mem. N.Y. Acad. Scis. Achievements include determination of the tubulin domain responsible for the regulation of its polymerization into microtubules and the use of protein engineering to determine proteim folding pathways. Office: EMBL, Mayerhofstrasse 1, Heidelberg D-6900, Germany

SERRANO, MYRNA, materials scientist, chemical engineer; b. San Sebastian, P.R., Aug. 21, 1954; d. Francisco Serrano and Obdulia Méndez. B in Chem. Engring. magna cum laude, U. P.R., 1977; M in Chem. Engring., U. Del., 1980; PhD in Chem. Engring., U. Mass., 1986. Chem. engr. Hercules Inc., Wilmington, Del., 1980-82; sr. rsch. engr. The Dow Chem. Co., Midland, Mich., 1986-87; sr. rsch. engr. western div. Applied Plastic Materials, The Dow Chem. Co., Walnut Creek, Calif., 1987-89; project leader cen. rsch. The Dow Chem. Co., Midland, 1990-92; project leader styrenics, ETP, barrier resins TS&D Dow Chemical Co., Midland, 1992—. Contbr. articles to profl. jours. Named Outstanding Young Woman Am., 1988. Mem. Am. Inst. Chem. Engrs., Soc. for Advanced Material Plastics Engrs. Avocations: travel, classical music, camping. Office: The Dow Chem Co Bldg 433 Dow Plastics TS&D Midland MI 48640

SERRIE, HENDRICK, anthropology and international business educator; b. Jersey City, July 2, 1937; s. Hendrick and Elois (Edge) S.; m. Gretchen Tipler Ihde, Sept. 3, 1959; children: Karim Jonathan, Keir Ethan. BA with honors, U. Wis., 1960; MA, Cornell U., 1964; PhD with distinction, Northwestern U., 1976. Dir. Solar Energy Field Project, Oaxaca, Mex., 1961-62; instr. U. Aleppo, Syria, 1963-64; asst. prof. Beloit (Wis.) Coll., 1964-69, Calif. State U., Northridge, 1969-70, Purdue U., West Lafayette, Ind., 1970-72, New Coll./U. South Fla., Sarasota, 1972-77; tchr. Pine View Sch., Sarasota, 1978; prof. anthropology, internat. bus. Eckerd Coll., St. Petersburg, Fla., 1978—; dir. internat. bus. overseas programs Eckerd Coll., 1981—; sr. rsch. assoc., Human Resources Inst., St. Petersburg, 1988—. Author, editor: Family, Kinship, and Ethnic Identity Among the Overseas Chinese, 1985, Anthropology and International Business, 1986; writer, dir. films: Technological Innovation, 1962, Something New Under the Sun, 1963; contbr. articles to Wall Street Jour. and Wall Street Jour. Europe. Tchr. Sunday sch., North United Methodist Ch., Sarasota, 1977—. Exxon scholar, So. Ctr. for Internat. Issues, Atlanta, 1980-81; recipient Leavy award, Freedoms Found., Valley Forge, Pa., 1989. Fellow Am. Anthropol. Assn., Soc. Applied Anthropology; mem. So. Ctr. Internat. Issues, Acad. Internat. Bus., Tampa Bay Internat. Trade Coun., Internat. Soc. Intercultural Edn., Tng. and Rsch. Democrat. Avocations: singing, drawing, photography, cycling, sailing. Home: 636 Mecca Dr Sarasota FL 34234-2713 Office: Eckerd Coll Dept Internat Bus Saint Petersburg FL 33733

SERSTOCK, DORIS SHAY, retired microbiologist, educator, civic worker; b. Mitchell, S.D., June 13, 1926; d. Elmer Howard and Hattie (Christopher) Shay; B.A., Augustana Coll., 1947; postgrad. U. Minn., 1966-67, Duke U., summer 1969, Communicable Disease Center, Atlanta, 1972; m. Ellsworth I. Serstock, Aug. 30, 1952; children—Barbara Anne, Robert Ellsworth, Mark Douglas. Bacteriologist, Civil Service, S.D., Colo., Mo., 1947-52; research bacteriologist U. Minn., 1952-53; clin. bacteriologist Dr. Lufkin's Lab., 1954-55; chief technologist St. Paul Blood Bank of ARC, 1959-65; microbiologist in charge mycology lab. VA Hosp., Mpls., 1968-93; instr. Coll. Med. Scis., U. Minn., 1970-79, asst. prof. Coll. Lab. Medicine and Pathology, 1979-93. Mem. Richfield Planning Commn., 1965-71, sec., 1968-71. Extended ministries commn. Wood Lake Luth. Ch., Richfield, Minn., 1993—; rep. religious coun. Mall Am., Bloomington, Minn., 1993. Fellow Augusta Coll.; named to Exec. and Profl. Hall of Fame; recipient Alumni Achievement award Augustana Coll., 1977; Superior Performance award VA Hosp., 1978, 82, Cert. of Recognition, 1988; Golden Spore awards Mycology Observer, 1985, 87. Mem. Am. Soc. Microbiology, N.Y. Acad. Scis., Minn. Planning Assn. Republican. Clubs: Richfield Women's Garden (pres. 1959), Wild Flower Garden (chmn. 1961). Author articles in field. Home: 7201 Portland Ave Minneapolis MN 55423-3218

SESHADRI, RANGASWAMY, engineering dean; b. Madras, India, Aug. 27, 1945; s. Rajendram Seshadri and Shantha (Krishnamachari) Rangaswamy; m. Sherry Middleton, June 19, 1975; children: Jagan Nathan, Jana Kaye. B in Engring with honors, U. Jabalpur, India, 1967; ThM, Indian Inst. Tech., Madras, 1969; MSc, PhD, U. Calgary, Alta., 1974. Pres., dir. Dynatek Engring. Corp., Regina, Sask., Can.; project engr. EBA Engring. Cons., Edmonton, Alta., Can., 1974; intermediate engr. Assoc. Engring. Scv. Ltd., Edmonton, Alta., Can., 1977; sr. engr. Syncrude Can. Ltd., Edmonton, Ft. McMurray, Alta., Can.; engr. assoc. Syncrude Can. Ltd., Ft. McMurray, Alta., Can., 1978; assoc. prof. engring. U. Regina, 1987-88, prof., 1988—, dean, 1989—. Co-author: Group Invariance in Engineering Boundary Value Problems', 1985. Mem. ASME, Can. Soc. Mech. Engrs., Assn. Profl. Engrs. Sask. Home: 43 Wilkie Rd, Regina, SK Canada S4S 5Y3 Office: University of Regina, Dean of School of Engineering, Regina, SK Canada S4S 0A2*

SESHAN, KULATHU IYER, chemistry educator, researcher; b. Cochin, Kerala, India, May 4, 1951; s. Kulathu Iyer and Lakshmi S.; m. Jayanthi Seshan, Sept. 19, 1980. BSc in Edn., U. Mysore, India, 1972; MSc, U. Baroda, India, 1974; PhD, Indian Inst. Tech., Bombay, 1980. Rsch. officer Indian Petro Chem. Corp. Ltd., Baroda, India, 1980-86; sr. rsch. officer Indian Petro Chem. Corp. Ltd., Baroda, 1984-87; asst. prof. U. Twente, Enschede, The Netherlands, 1987—; correspondent News Brief Applied Catalysts, 1991—. Mem. Am. Chem. Soc., Dutch Catalysts Soc. Office: U Twente Chemistry Dept, Faculty Chemistry Tech, 7500AF Enschede The Netherlands

SESSOMS, ALLEN LEE, academic administrator, former diplomat, physicist; b. N.Y.C., Nov. 17, 1946; s. Albert Earl and Lottie Beatrice (Leff) S.; m. Csilla Manette von Csiky, Apr. 18, 1990; children: Manon Elizabeth, Stephanie Csilla. BS, Union Coll., Schenectady, N.Y., 1968; PhD, Yale U., 1972. Sci. assoc. CERN, Geneva, Switzerland, 1973-78; asst. prof. physics Harvard U., Cambridge, Mass., 1974-81; sr. tech. advisor OES, State Dept., Washington, 1980-82; dir. Office Nuclear Tech. & Safeguards, State Dept., Washington, 1982-87; counselor for sci. and tech. U.S. Embassy, Paris, 1987-89; polit. minister, counselor U.S. Embassy, Mexico City, 1989-91, dep. chief of mission, 1991-93; exec. v.p. U. Mass., Boston, 1993—. Contbr. articles to profl. jours. Ford Found. travel/study grantee, 1973-74; Alfred P. Sloan Found. fellow, 1977-81. Mem. AAAS, Am. Phys. Soc., N.Y. Acad. Sci., Cosmos Club. Office: U Mass Pres Office 18 Tremont St Ste 800 Boston MA 02108

ŠESTÁK, JIŘÍ VLADIMÍR, mechanical engineering educator; b. Brno, Czechoslovakia, Oct. 4, 1930; s. Antonín and Vlasta (Bumbalová) S.; m. Alena Komárková, Nov. 4, 1959; 1 child, Petra. MSc, Czechoslovakia Tech. U., Prague, 1954, PhD in Mech. Engring., 1965. Asst. prof. mech. engring. Czechoslovakia Tech. U., Prague, 1955-68, assoc. prof., 1969-88, dean faculty of mech. engring., 1990-91; mem. Czech Com. for Sci. Degrees, Govt. of Czech. Republic, Prague, 1990—; vis. prof. U. Toronto, Ontario, Can., 1982-83. Author, co-author, editor, books, textbooks, jour. articles in field. Sworn interpreter of English lang. Mcpl. Ct., Prague, 1967—. Named Sr. Rsch. fellow, Ford Found. U. Toronto, Can., 1966-67; recipient Felber Gold medal Czech Tech. Univ. in Prague, 1990. Mem. Czecoslovak Group of Rheology (chmn. 1982—), Czech. Soc. for Adv. Tech. Lit. (vice chmn. 1991—), Soc. Rheology U.S. Roman Catholic. Office: Fakulta strojni Czech Tech U, Technická 4, 166 07 Prague Czech Republic

SESTINI, VIRGIL ANDREW, biology educator; b. Las Vegas, Nov. 24, 1936; s. Santi and Merceda Francesca (Borla) S. BS in Edn., U. Nev., 1959; postgrad., Oreg. State U., 1963-64; MNS, U. Idaho, 1965; postgrad., Ariz. State U., 1967, No. Ariz. U., 1969; cert. tchr., Nev. Tchr. biology Rancho High Sch., 1960-76; sci. chmn., tchr. biology Bonanza High Sch., Las Vegas, 1976-90; ret., 1990; part time tchr. Meadows Sch., 1987—. Served with USAR, 1959-65. Recipient Rotary Internat. Honor Tchr. award, 1965, Region VIII Outstanding Biology Tchr. award, 1970, Nev. Outstanding Biology Tchr. award Nat. Assn. Biology Tchrs., 1970, Nat. Assn. Sci. Tchrs., Am. Gas Assn. Sci. Teaching Achievement Recognition award, 1976, 1980, Gustov Ohaus award, 1980, Presdl. Honor Sci. Tchr. award, 1983; Excellence in Edn. award Nev. Dept. Edn., 1983; Presdl. award excellence in math. and sci. teaching, 1984, Celebration of Excellence award Nev. Com. on Excellence in Edn., 1986, Hall of Fame award Clark County Sch. Dist., 1988, Excellence in Edn. award, Clark County Sch. Dist., 1987, 88, Spl. Edn. award Clark County Sch. Dist., 1988, NSEA Mini-grants, 1988, 89, 92, World Decoration of Excellence medallion World Inst. Achievement, 1989, Cert. Spl. Congl. Recognition, 1989, Senatorial Recognition , 1989, minigrant Jr. League Las Vegas., 1989, Excellence in Edn. award, Clark Country Sch. Dist., 1989; named Nev. Educator of Yr., Milken Family Found./Nev. State Dept. Edn., 1989; grantee Nev. State Bd. Edn., 1988, 89, Nev. State Edn. Assn., 1988-89. Author: Lab Investigations For High School Honors Biology, 1989, Microbiology: A Manual for High School Biology, 1992, Laboratory Investigations in Microbiology, 1992; co-author: A Biology Lab Manual For Cooperative Learning, 1989, Metrics and Science Methods: A Manual of Lab Experiments for Home Schoolers, 1990; contbr. articles to profl. jours. Mem. NEA, Nat. Sci. Tchrs. Assn. (Nev. State chpt. 1968-70), Nat. Assn. Biology Tchrs. (OBTA dir. Nev. State 1991—), Am. Soc. Microbiology, Coun. for Exceptional Children, Am. Biographic Inst. (rsch. bd. advisors 1988), Nat. Audubon Assn., Nat. Sci. Suprs. Assn., Am. Inst. Biol. Scis.

SETCHELL, JOHN STANFORD, JR., color systems engineer; b. Bklyn., Dec. 4, 1942; s. John Stanford and Elisa (Muenzfeld) S.; m. Cynthia Florance Andreasen, Feb. 27, 1965; 1 child, Anitra Lesa. BS in Physics, Rensselaer Polytech Inst., 1963; MS in Physics, U. Ill., 1969. Rsch. physicist Mfg. Tech. Kodak Apparatus Div., Rochester, N.Y., 1969-82; supr. project engr. Copy Products Eastman Kodak Co., Rochester, N.Y., 1982-84; supr. engr. Electronic Photography Div. Eastman Kodak Co., Rochester, N.Y., 1985-88; product planning dir. Printer Products Div. Eastman Kodak Co., Rochester, N.Y., 1989-90; color systems engr. Systems Products Div. Eastman Kodak Co., Rochester, N.Y., 1991—; lectr. in field. Contbr. articles to profl. jours. Advisor Sea Exploring Ship 303, Webster, N.Y., 1980—; leader Pioneer Club, Rochester, 1990—; worship leader Covenant Presbyn. Ch., Rochester, 1987—; singer Cathedral Choir Sch., Rochester, 1991—. Lt. USN, 1963. Mem. Am. Soc. Testing and Materials (chmn. com. B-4 1981-82), Am. Scientific Affiliation, Assn. Old Crows, Chi Gamma Iota. Republican. Achievements include research includes color image processing in desktop computers, digital color printing, environmental effects on electronic materials. Home: 376 English Rd Rochester NY 14616-2425 Office: Eastman Kodak 1447 St Paul St Rochester NY 14653-3700

SETHURAMAN, SALEM VENKATARAMAN, economist; b. Madras, India, May 23, 1935; m. Rajam T. Arunachalam, Apr. 29, 1963; children: Kavita, Ganesh. BSc, Mysore (India) U., 1955, BSc with honors, 1957; MA, Delhi (India) U., 1962; PhD in Econs., U. Chgo., 1969. Jr. economist Nat. Coun. Applied Econ. Rsch., New Delhi, 1958-64; rsch. economist U.S. AID, New Delhi, 1969-72; sr. fellow East West Ctr., U. Hawaii, Honolulu, 1972-73; sr. economist ILO, Geneva, 1973—; cons. World Bank, Washington, Internat. Fund for Agrl. Devel., also other internat. orgns. Author: Urban Development and Employment in Jakarta, 1976; editor: The Urban Informal Sector in Developing Countries, 1981, (with E. Chuta) Rural Small Industries and Employment in Africa and Asia, 1984, (with C. Maldonada) Technological Capability in the Informal Sectors of Developing Countries, 1991; contbr. articles to profl. jours., referee profl. jours. Pres. Indian Assn. Geneva, 1987. Mem. profl. assns. Avocations: gardening, bridge. Office: ILO, Rte du Morrillons 4, CH-1211 Geneva Switzerland

SETLOW, RICHARD BURTON, biophysicist; b. N.Y.C., Jan. 19, 1921; s. Charles Meyer and Elsie (Hurwitz) S.; children—Peter, Michael, Katherine, Charles; m. Neva Delihas, Mar. 3, 1989. A.B., Swarthmore Coll., 1941; Ph.D., Yale U., 1947; DSc, U. Toronto, 1985; D Medicine, U. Essen, 1993. Asso. prof. Yale U., 1956-61; biophysicist Oak Ridge Nat. Lab., 1961-74, sci. dir. biophysics and cell physiology, 1969-74; dir. U. Tenn.-Oak Ridge Grad. Sch. Biomed. Scis., 1972-74; sr. biophysicist Brookhaven Nat. Lab., Upton, N.Y., 1974—; chmn. biology dept. Brookhaven Nat. Lab., 1979-87, assoc. dir. life scis., 1985—; prof. biomed. scis. U. Tenn., 1967-74; adj. prof. biochemistry SUNY-Stony Brook, 1975—. Author: (with E.C. Pollard) Molecular Biophysics, 1962; editor: (with P.C. Hanawalt) Molecular Mechanisms for Repair of DNA, 1975. Recipient Finsen medal Internat. Assn. Photobiology, 1980, Enrico Fermi award U.S. Dept. Energy, 1988. Mem. NAS, Am. Acad. Arts and Scis., Boiphys. Soc. (pres. 1969-70), comité Internat. de Photobiologie (pres. 1972-76), Radiation Rsch. Soc., Am. Soc. Photobiology, Am. Soc. Biochemistry and Molecular Biology, Am. Soc. Cancer Rsch., Environ. Mutagen Soc., 11th Internat. Cong. Photobiology (hon. pres. 1992), Phi Beta Kappa. Home: 4 Beachland Ave East Quogue NY 11942-4941 Office: Biology Dept Brookhaven Nat Lab Upton NY 11973

SETTLES, GARY STUART, fluid dynamics educator; b. Maryville, Tenn., Oct. 9, 1949; s. Charles Henderson and Stella Mae (Schultz) S. BS in Aerospace Engring., U. Tenn., 1971; PhD in Aerospace and Mech. Scis., Princeton (N.J.) U., 1976. Rsch. scientist Princeton Combustion Labs., 1975-77; rsch. engr., lectr. M.A.E. Dept., Princeton (N.J.) U., 1977-83; assoc. prof. Mech. Engring. Dept., Pa. State U., University Park, 1983-86, prof., dir. gas dynamics lab., 1986—. Author: Handbook of Flow Visualization, 1989. Mem. AIAA (nat. undergrad. student award 1970), ASME, Optical Soc. of Am., Am. Phys. Soc. (Fluid Dynamics br.), Soc. of Photo-Optical Instrumentation Engrs. Achievements include rsch. on gas dynamics, flow visualization, shock wave/turbulent-flow interactions, optical flow instruments, compressible turbulent mixing and the application of gas dynamics to materials synthesis, materials processing and industrial problems. Home: 2053 Valley View Rd Bellefonte PA 16823 Office: Pa State U Dept Mech Engring 303 Reber Bldg University Park PA 16802

SEWELL, ADRIAN CLIVE, clinical biochemist; b. Salisbury, Wiltshire, Eng., Aug. 5, 1950; s. Reginald Alfred Leonard and Winifred Joan (Siffleet) S.; m. Janice Elizabeth Dodd, Feb. 12, 1977; children: Rebecca, Timothy. BS in Biochemistry, U. Bath, Eng., 1968, PhD, 1976. Rsch. asst. U. Childrens Hosp., Mainz, Fed. Republic of Germany, 1977-80; clin. biochemist Royal Gwent (Eng.) Hosp., 1980-84; dir. metabolic lab. U. Childrens Hosp., Mainz, 1984-89, Frankfurt, Fed. Republic of Germany, 1989—. Contbr. articles to profl. jours. Mem. Soc. for Study of Inborn Errors of Metabolism, Soc. for Inherited Metabolic Disorders, N.Y. Acad. Sci. Office: U Childrens Hosp, Theodor Stern Kai 7, Frankfurt 6000, Germany

SEXTON, AMY MANERBINO, computer analyst; b. Denver, Sept. 21, 1957; d. George Anthony and Victoria Violet (Marolt) Manerbino; m. Lindel Scott Sexton, May 29, 1982; 1 child, Monica Marie. Student, Colo. Sch. Mines, 1975-79; BS, Ft. Lewis Coll., 1980; M in Computer Info. Systems, U. Denver, 1989. Phys. sci. aid U.S. Geol. Survey, Golden, Colo., 1977-79; geologist Am. Stratigraphic Co., Denver, 1981-89; cons. geologist Denver, 1989-90; programmer Covia Partnership, Denver, 1990-91, computer analyst, trainee mentor, 1991-92; computer tech. analyst Covia Partnership, 1992—. Republican. Roman Catholic. Avocations: sewing, crafts, programming, skiing, home remodeling. Office: Covia Partnership 5351 S Valentia Way Englewood CO 80012

SEXTON, KEN, environmental health scientist; b. Moscow, Idaho, Nov. 6, 1949. BS, USAF Acad., 1972; MS, Wash. State U., 1977; MA, Tex. Tech. U., 1979; PhD, Harvard U., 1983; MBA, U. Pa., 1994. Rsch. asst. Wash. State U., Pullman, 1975-77, environ. engr., 1977-79; staff engr. Acurex Corp., Mt. View, Calif., 1979-80; rsch. asst. Harvard U., Cambridge, Mass., 1980-83; dir. indoor air Calif. Dept. Health, Berkeley, 1983-85; dir. sci. rev. Health Effects Inst., Boston, 1985-87; dir. health rsch. EPA, Washington, 1987—; chmn. Fed. Interagency task force on environ. cancer and heart and lung disease, Washington, 1987—, task force on air pollution rsch., 1991—. Contbr. articles to environ. jours. 2d lt. USAF, 1972-74. DuPont fellow Harvard U., 1980-81; Clin. Epidemiology Tng. grantee Nat. Heart, Lung and Blood Inst., Harvard U., 1981-82. Office: EPA Health Rsch Office 401 M St SW Washington DC 20460-0002

SEYDEN-PENNE, JACQUELINE, research chemist; b. Aubervilliers, France, July 4, 1930; d. Abram and Germaine Sara (Levy) Penne: m. Robert Seyden, Dec. 18, 1953; children: Brigitte, Eric. BS in Pharmacy, Paris U., 1952, DSc, 1956. Rsch. fellow Nat. Ctr. Sci. Rsch., Paris, 1952-64; group leader CRNS, Paris, 1964-68, Nat. Ctr. Sci. Rsch., Thiais, France, 1968-81; dept. leader Nat. Ctr. Sci. Rsch./Paris Sud U., Orsay, France, 1982-91, rsch. fellow, 1991—; lectr. Paris Sud U., Orsay, 1978-91; pres. organic chemistry sect. Nat. Ctr. Sci. Rsch. Nat. Com., Paris, 1980-83. Author: Reduction by Alumino and Borohydrides, 1988, (English version) 1991; editor Jour. of Chem. Rsch., 1981—. Mem. Soc. Chem. de France (editor bull. 1981-91, Lebel prize 1988), Am. Chem. Soc. Home: Le Vallat de Vermenoux, 84220 Goult France Office: U Paris Sud, Bat 420, 91405 Orsay France

SEYMOUR, BRIAN RICHARD, mathematics educator, researcher; b. Chesterfield, Derby, Eng., Sept. 25, 1944; came to U.S., 1968, Can., 1973; s. Douglas and Hilda (Ball) S.; m. Rosemary Jane Pembleton, Sept. 23, 1943; children—Mark, Jane, Richard. B.Sc. with honors, U. Manchester, 1965; Ph.D., U. Nottingham, 1968. Asst. prof. Lehigh U., 1969-70, N.Y. U., 1970-73, U. B.C., 1973-76; assoc. prof. U. B.C., 1976-81, Vancouver, Can., prof. math., 1981—, dir. inst. applied math, 1986—. Contbr. research papers to profl. jours. Sci. Research Council sr. research fellow Oxford U., 1978; Killam sr. fellow Killam Trust, Monash U., 1984. Mem. Can. Applied Math. Soc., Can. Meterological Ocean Soc. Avocation: field hockey. Office: U BC, 222-1984 Mathematics Rd, Vancouver, BC Canada V6T 1Z2

SEYMOUR, CHARLES WILFRED, electronics engineer; b. South Kingstown, R.I., July 26, 1965; s. Frederick Earnest Seymour and Sandra (Roberts) Hutto. BSEE, George Mason U., 1990. Engring. asst. Advanced Tech., Inc., Arlington, Va., 1986-88; systems analyst Emhart, Inc., Arlington, Va., 1988-89; project mgr. systems engring. PRC, Inc., Arlington, Va., 1990—. Optomist Club scholar, 1983. Mem. Nat. Parks Conservation. Democrat. Office: PRC Inc 2121 Crystal Dr Arlington VA 22202

SEYMOUR, FREDERICK PRESCOTT, JR., industrial engineer, consultant; b. Oak Park, Ill., June 19, 1924; s. Frederick Prescott and Ivy Louise (Horder) S.; m. Janet Mary Stocking, Oct. 15, 1960; children: Robert Prescott, Bruce Stocking, Mary Janet. BS, Cornell U., 1948; MS in Commerce, U. Ill., 1951; MBA, U. Chgo., 1957. Indsl. engr., dir. planning, exec. salesman R.R. Donnelley and Sons Co./Lakeside Press, Chgo., 1951-72; regional dir. U.S. Postal Svc., Chgo., 1972-76; dir. advt. Spiegel, Inc., Chgo., 1976-80; pres. Frederick P. Seymour and Assocs., Inc., Winnetka, Ill., 1980—; pres. Cornell Univ. Club. Chgo., 1960-61, Exec. Program Club, Chgo., 1972-73; mem. Postmaster Gen.'s adv. com., Washington, 1973—. Contbg. editor: Gravure mag., 1982—; contbr. articles on pub. and printing to trade mags. Precinct capt. New Trier Rep. Orgn., Winnetka, 1970-72. With USN, 1944-46, PTO. Mem. ASME (life), Cornell Soc. Engrs., Graphic Communications Assn. (Innovator award 1988), Gravure Assn. Am., Graphic Arts Industry Rsch. and Engring. Coun. (exec. com.). Office: Frederick Seymour & Assocs 303 Sheridan Rd Winnetka IL 60093-4227

SEYMOUR, HARLAN FRANCIS, computer services company executive; b. East St. Louis, Jan. 25, 1950; s. Harlan Edward and Agnes Wilhelmina (Noakes) S.; m. Ellen Katheleen Schmitt, Aug. 17, 1973; children: Melissa Ann, Harlan Francis Jr. BA in Math., U. Mo., 1973; MBA, Keller Grad. Sch. Mgmt., 1980. Corp. v.p. Statis. Tabulating Corp., Chgo., 1973-80; dist. mgr. Datacorp, Chgo., 1980-83; exec. v.p. First Fin. Mgmt. Corp., Atlanta, 1983—; pres., chief exec. officer First Health Svcs. Corp., Richmond, Va., 1989—; mem. mgmt. info. systems adv. bd. U. Ga., Athens, 1986, 87, Va. Commonwealth Univ., 1991—; bd. dirs. J. Sargeant Reynolds Community Coll., 1991—. V.p. St. Joseph's Home and Sch., Marietta, Ga., 1987—. Mem. Nat. Assn. Bank Servicers, Bank Mktg. Assn., Richmond Metro C. of C., Fin. Mgrs. Soc. Roman Catholic. Avocations: sailing, tennis, jogging. Home: 12106 Country Hills Ct Glen Allen VA 23060-5347 Office: First Health Svcs Corp 4300 Cox Rd Glen Allen VA 23060

SEYMOUR, JANET MARTHA, psychologist; b. Mineola, N.Y., June 13, 1957; d. John Andrews and Eileen (Brudie) S.; m. Mark Charles Adams, Aug. 27, 1983; children: Heidi Lynn, Hartley Ann. BA in Psychology and Music, Wheaton Coll., 1979; MA in Clin. Psychology, Rosemead Sch. Psychology, La Mirada, Calif., 1981, PsyD in Clin. Psychology, 1988. Lic.

psychologist, Calif. Psychology intern Colmery Oneil VA Med. Ctr., Topeka, 1985-86; psychotherapist Concord (N.H.) Psychol. Assocs., 1987-88; psychologist Jolliffe & Assocs., Long Beach, Calif., 1989—. Sunday sch. tchr. 1st Evang. Free Ch. of Fullerton, 1990-91, flutist orch., 1980—; baseball coach Pony Baseball, Whittier, Calif., 1989. Mem. APA. Republican. Avocations: stamp collecting, sewing, baseball. Office: Jolliffe & Assocs 3740 Atlantic Ave Ste 200 Long Beach CA 90807-3440

SEYMOUR, RONALD CLEMENT, entomologist, researcher; b. Smithville, Mo., Jan. 31, 1959; s. Charles K. and Rita A. (Hrenchir) S.; m. Tana L. Warren, May 26, 1990; 1 child, Melanie M. Kuehl. BS, Colo. State U., 1981; MS, U. Nebr., 1986. Agrl. cons. Crop Svc. Co., Yuma, CO, 1982-83; extension asst.-IPM U. Nebr., North Platte, 1987—. Mem. Entomol. Soc. Am. (Outstanding Student Presentation 1991), Sigma Xi, Gamma Sigma Delta. Achievements include discovery of new host parasitoid relationshp between the Staphylinid beetle Aleochara lacertina and stable and house flies. Office: U Nebr Rt 4 Box 46A North Platte NE 69101

SEZGIN, MESUT, environmental engineer; b. Babakale, Turkey, Oct. 13, 1948; came to U.S., 1971; s. Selahattin Eyup and Serife (Can) S.; m. Pamela J. Dorn, Sept. 16, 1990. BS, Tech. U. Istanbul, Turkey, 1970, M Engring., 1971; MS, U. Calif., Berkeley, 1974, PhD, 1977. Registered profl. engr., Ga. Asst. prof. Tech. U. Istanbul, 1978-79; rsch. assoc. Ga. Inst. Tech., Atlanta, 1979-82; civil engr. dept. pub. works City of Atlanta, 1982—; referee NSF, Washington, 1981, Internat. Assn. Water Quality, London, 1987-90, Water Environ. Fedn., Alexandria, Va., 1989—. Editor Ayyildiz newsletter, 1988-89; reviewer publs.; contbr. articles, reports to tech. publs. Bd. dirs. Turkish Am. Cultural Assn., Atlanta, 1986-87, treas., 1987-88, v.p., 1988-89, 90-91, organizer Turkish Festival, 1989. Recipient Engr. of the Yr. in Govt. award Engring. Soc. Ga., 1993. Mem. NSPE, Water Environment Fedn. (organizing com. 1990-92), Ga. Soc. Profl. Engrs., Ga. Water and Pollution Control Assn. (organizing com. indsl. waste conf. 1990-92). Home: 2290 Mitchell Rd Marietta GA 30062 Office: City of Atlanta Dept Pub Works 2440 Bolton Rd NW Atlanta GA 30318

SHACKS, SAMUEL JAMES, physician; b. Pine Bluff, Ark., May 20, 1939; s. Samuel and Bessie (Blackwell) S.; m. Janice Priscilla, Mary 20, 1982; 1 child, Dwight Jourdain, Jr. BS in Biology/Chemistry, U. Ark., Pine Bluff, 1960; PhD in Biology, U. Calif., Irvine, 1972, MD, 1977. Internship Martin Luther King Jr. Gen. Hosp., L.A., 1977-78; residency pediatrics Martin Luther King, Jr. Gen. Hosp., L.A., 1978-80; med. dir. The Do It Now Found., Hollywood, Calif., 1980-82, The Sunrise Community Counseling Ctr., L.A., 1981-82; dir. Minority Biomed. Rsch. Support Program, L.A., 1984—, Minority Access to Rsch. Careers Program, L.A., 1984—, Minority High Sch. Student Rsch. Apprentice Program, L.A., 1985—, Rsch. Tng. Inst., L.A., 1986—; assoc. dean rsch. Drew U., L.A., 1987-92; cons. Comprehensive Sickle Cell Disease Ctrs., Parent Rev. Com. Nat. Heart, Lung and Blood Inst., Bethesda, Md., 1986—; mem. Sickle Cell Disease Task Force, Nat. Heart, Lung and Blood Inst., NIH, Bethesda, 1990-91; liaison/coord. AMHPS Nat. Cancer Inst., NIH, Bethesda, 1987—; mem. Task Force for Grad. and Career Placement NSF Access Grant; mem. exec. adv. com. Nat. AIDS Minority Info. and Edn. Program, Howard U., Washington. Mentor Ptnrs. in Sci. and Industry, L.A., 1991—; mem. Rsch. Ctrs. in Minority Instns., Nat. Adv. Com. Fla. A&M U., Tallahassee, 1989—; commr. L.A. County Commn. on AIDS, L.A., 1989—; bd. mem. Nat. Cancer Control Rsch. Network, Inc., Nat. Cancer Inst./NIH, Sickle Cell Disease Rsch. Found., Drew-Meharry-Morehouse Cancer Consortium-Meharry Med. Coll., Nashville. With M.C., U.S. Army, 1961-64. Recipient fellowship Robert B. Brigham Hosp., 1972-73; named Rsch. fellow in Immunology and Microbiology, UCLA, 1973-74, MARC Faculty Fellow Clin. scholar in Pediatric Immunology, Harbor/UCLA Med. Ctr., 1980-83. Mem. AAAS, Assn. Minority Health Professions Schs., Nat. Med. Assn., Am. Acad. Allergy and Immunology, Assn. of Minority Health Professions Schs. Baptist. Achievements include research in allergy, immunology, pediatric infectious diseases, hematology and blood. Home: 5374 W Amberwood Dr Inglewood CA 90302 Office: Charles R Drew U Med and Sci 1621 E 120th St MP 27 Los Angeles CA 90059

SHACTER, JOHN, manager, technology/strategic planning consultant; b. Vienna, Austria, Sept. 26, 1921; came to U.S., 1938; s. Jacob and Regina (Bursten) S.; m. Kathleen Williams, Mar. 6, 1947; children: Suzanne, Linda. BSChemE, U. Pa., 1943. Rsch. devel. engr. Manhattan atomic project Union Carbide Corp., N.Y.C., 1943-44; ops. supr. Union Carbide Corp., Oak Ridge, Tenn., 1944-46, process design, analysis mgr., 1946-55; corp. projects and new ventures Union Carbide Corp., N.Y.C., 1956-58, mgr. corp. planning and mgmt. systems, 1958-66; founder, dir. multi-co. combined ops. planning Atomic En. Commn. Union Carbide Corp., Oak Ridge, 1966-72, asst. to pres., cons. Man. systems, 1972-76, sr. cons., 1976-83; founder, pres. JS Assocs. Mgmt., Tech. and Strategic Planning Cons., Kingston, Tenn., 1983—; adj. prof. grad. engring. design, econs. U. Tenn., U.S. Acad. Scis.; apptd. mem. of mgmt. and technology panel advising IIASA, Vienna. News columnist and host TV and radio discussions; contbr. numerous articles to profl. jours.; contbr. basic designs for new multi-billion dollar U.S. uranium enrichment plants, new methods of corp. planning investment analyses and performance evaluation. Named Outstanding Boss of Yr. Soc. of Profl. Secs. Fellow AIChE (past pres.), AAAS, Am. mgmt. Assn. (seminars chmn., lectr.); mem. Tenn. Soc. Profl. Engrs. (State's Outstanding Engr. 1992), Torch Club (past pres.), Rotary Club (past pres.), Profl. Soc. (past pres.), Knoxville Tech. Soc., Soc. Profl. Journalists, Tenn. Soc. Ret. Tchrs. Avocations: vol. tutor, watersports, soccer, music. Office: JS Assocs Rte 4 Box 332 Kingston TN 37763

SHADARAM, MEHDI, electrical engineering educator; b. Tehran, Iran, Apr. 19, 1954; came to U.S., 1976; s. Ali and Masoumeh (Bayram) S.; m. Luz Elena Inungaray, Mar. 24, 1990. BSEE, U. Sci. and Tech., Tehran, 1976; MSEE, U. Okla., 1981, PhD in Elec. Engring., 1984. Registered profl. engr., Tex. Lab. asst. U. Okla. Elec. Engring. Dept., Norman, 1979-81, lab. instr., 1982-84; project engr. Ra Nav Lab., Oklahoma City, 1981-82; asst. prof. U. Tex. Elec. Engring. Dept., El Paso, 1984-90, assoc. prof., 1990—. Contbr. articles to profl. jours. Recipient Faculty Fellowship award Assoc. Western U., 1990, 1991, Advising the Best Thesis award U. Tex. El Paso, 1990. Mem. IEEE (chmn. 1988-90, treas. 1987), Optical Soc. Am., Soc. Photo Optical and Instrumentation Engrs., Eta Kappa Nu. Home: 7113 Cerro Negro Dr El Paso TX 79912-2714 Office: U Tex at El Paso Elec Engring Dept El Paso TX 79968

SHAEIWITZ, JOSEPH ALAN, chemical engineering educator; b. Bklyn., Oct. 12, 1952; s. Herman and Rose Sarah (Richardson) S. BSChemE, U. Del., 1974; MSChemE, Carnegie Mellon U., 1976, PhD, 1978. Asst. prof. chem. engring. U. Ill., Urbana, 1978-84; assoc. prof. chem. engring. W.Va. U., Morgantown, 1984—. Contbr. articles to profl. publs. Mem. AICE, Am. Chem. Soc., Am. Soc. Engring. Edn. Home: 756 Northwest Dr Morgantown WV 26505-2675 Office: W Va U Dept Chem Engring PO Box 6101 Morgantown WV 26506-6101

SHAFER, BERMAN JOSEPH, oil company executive; b. Wooster, Ohio, July 15, 1927; s. David and Helen (Jacobs) S.; m. Miriam Ruth Stern, Apr. 8, 1951; 1 child, Leslie Faye. BS in Petroleum Engring., Ohio State U., 1950. Pvt. practice cons. Wooster, Ohio, 1953-57; owner Shafer Drilling, Wooster, Ohio, 1957-65; v.p. David Shafer Oil Producers Inc., Wooster, Ohio, 1965-85, pres., 1985—; mem. Ohio Oil & Gas Regulatory Rev. Commn., Columbus, 1986; exec. com. Ind. Petroleum Assn. Am., Washington, 1986-88; pres. Hillcrest Mgmt. Inc., 1990—. Contbr. articles to profl. jours. Co-chmn. United Way-Oil Div., 1980-82. Sgt. U.S. Army, 1950-52. Mem. Interstate Oil Compact Commn., Soc. Petroleum Engrs., Ohio Oil & Gas Assn. (pres. 1980-82, hon. trustee 1983—). Republican. Avocations: reading, golf. Office: David Shafer Oil Producers 459 S Hillcrest Dr PO Box 582 Wooster OH 44691

SHAFER, JOHN MILTON, hydrologist, consultant, software developer; b. Findlay, Ohio, Mar. 18, 1951; s. Paul Eugene and Mary Ethel (Schwyn) S.; m. Elise Ann Dunne, Apr. 11, 1980; children: Paul Emery, Jessica Elise, Elise Ann. BS in Earth Sci., Pa. State U., 1973; MS in Resource Devel., Mich. State U., 1975; PhD in Civil Engring., Colo. State U., 1979. Cert. hydrologist #218. Asst. rsch. prof. Colo. State U., Fort Collins, 1979-80; rsch. engr. Battelle Meml. Inst., Richland, Wash., 1980-83, sr. rsch. engr., 1983-84; hydrologist Ill. State Water Survey, Champaign, 1984-85, asst. head ground water sect., 1985-90, prin. hydrologist, 1988-91, head hydrology div., 1990-92; assoc. dir., rsch. prof. Earth Scis. and Resources Inst., U. S.C., Columbia, 1992—; cons. pvt. cos., 1984—; owner GWPATH, Columbia, S.C., 1990—; v.p. Environ. and Archtl. Signage, Findlay, Ohio. Developer software, 1987; contbr. articles to profl. jours. Recipient John C. Frye Meml. award in geology, 1991, Ill. Groundwater Sci. Achievement award, 1993. Mem. Intergovt. Coord. Com. Groundwater, Am. Geophys. Union, Am. Inst. Hydrology (pres. Ill. sect. 1985-92), Nat. Ground Water Assn., Ill. Groundwater Assn., Sigma Xi. Republican. Presbyterian. Avocations: tennis, handball, woodworking, model building. Home: 321 Lake Front Dr Columbia SC 29212-9999 Office: Earth Scis Resouces Inst U SC 901 Sumter St Columbia SC 29208-9999

SHAFER, JULES ALAN, pharmaceutical company executive; b. N.Y.C., Nov. 21, 1937; s. Samuel Z. and Ada (Gams) S.; m. Marcia Anolik, June 21, 1959; children: Toby, Howard, Neil. BChemE, CCNY, 1959; PhD in Chemistry, Poly. Inst. N.Y., 1963. Postdoctoral fellow Harvard U., Cambridge, Mass., 1963-64; asst. prof. U. Mich., Ann Arbor, 1964-69, assoc. prof., 1969-77, prof., 1977-90; exec. dir. Merck Rsch. Labs., West Point, Pa., 1990—. Mem. editorial bd. Jour. Biol. Chemistry, Archives of Biochemistry and Biophysics. Mem. AAAS, Am. Chem. Soc., N.Y. Acad. Scis., Am. Soc. for Biochemistry and Molecular Biology. Achievements include research in enzyme technology, biophysical chemistry, mechanisms for the action of enzymes, protein-protein interactions. Office: Merck Rsch Labs Sumneytown Pike 16-101 West Point PA 19486

SHAFER, KEVIN LEE, water resources engineer; b. Champaign, Ill., Jan. 4, 1961; s. Berton Earl and Norma Lynn (Funkhouser) S.; m. Carole Marie Nurczyk, Aug. 20, 1982; 1 child, Zachary Arvel. BSCE, U. Ill., 1982; MSCE, U. Tex., Arlington, 1988. Registered profl. engr., Tex., Ill, Ind. Hydrologic engr. U.S. Army Corps of Engrs., Fort Worth, 1982-87; water resources engr. Parsons Brinckerhoff, Chgo., 1989—. Mem. ASCE, SAME, AWWA. Office: Parsons Brinckerhoff 230 W Monroe Chicago IL 60606

SHAFFER, JILL, clinical psychologist; b. Columbus, Ohio, May 18, 1958; d. Melvin Warren and Emily (White) S.; m. Robert K. Yost, Jan. 9, 1991. BS in Psychology with honors, Wright State U., 1984, PsyD, 1988. Lic. psychologist, Ohio. Psychology talk show producer/participant Sta. WHIO-AM, Dayton, Ohio, 1981-83; psychology asst. and organizer Terrap S.W. Ohio, Dayton, 1981-83; psychology trainee Oakwood Forensic Ctr., Lima, Ohio, 1984-85, Wright State U., Dayton, 1984-87; predoctoral resident South Community Mental Health Ctr., Dayton, 1987-88; postdoctoral trainee Fulero and Assoc., Dayton, 1988-89; pvt. practice Dayton, 1989—; supervising psychologist GERI-Tech of Dayton, 1990-92; cons. psychologist disability evaluations for worker's compensation and social security disability, 1989—; state examiner Indsl. Commn. of Ohio, 1989—; owner, mgr. rental properties, 1988—. Author: (article) Strategic Intervention with Transvestism, 1989. Recipient scholarship Sch. of Profl. Psychology, 1985. Mem. APA, NOW, Ohio Psychol. Assn., Dayton Area Psychol. Assn. Avocations: home improvement, gardening, camping, sailing, biking. Office: 2705 Far Hills Ave Ste 4 Dayton OH 45419-1606

SHAFFER, STEPHEN M., computer system analyst; b. Lancaster, Pa., Feb. 14, 1954; s. Charles Henry and Martha Frances (Shorb) S. BSEE, Pa. State U., 1976; MBA, Frostburg (Md.) State U., 1983. Electronic technician Grove Mfg., Shady Grove, Pa., 1978-83; system engr. Solarex, Rockville, Md., 1983-85; analyst Western Union Govt. Networks, Upper Saddle River, N.Y., 1985-86, Am. Satellite Co., Rockville, 1986-88, Contel Fed. Systems, Chantilly, Va., 1988-92, GTE Fed. Systems Div., Fort Detrick, Md., 1991—; cons. Applied Computer Products, Waynesboro, Pa., 1984—. Mem. Franklin County Art Alliance (treas. 1989-92), Franklin Centre For the Arts. Democrat. Home: 253 Garfield St Waynesboro PA 17268-1152 Office: GTE Fed Systems Bldg 1671 PO Box 1069 Fort Detrick MD 21702

SHAFIR, ELDAR, psychology educator; b. Tel-Aviv, Israel, Mar. 9, 1959; came to U.S., 1980; s. Uri J. and Miri S.; m. Amy E. Pierce, June 9, 1990. BA, Brown U., 1984; PhD, MIT, 1988. Asst. prof. Princeton (N.J.) U., 1989—. Contbr. articles to profl. jours. Postdoctoral scholar Stanford U., 1988-89; recipient Best Paper award Internat. Assn. Rsch. in Econ. Psychology, 1991. Mem. Am. Psychol. Soc., Soc. Judgement & Decision Making (Hillel Einhorn New Investigator award 1992), Psychonomic Soc., Phi Beta Kappa, Sigma Xi. Office: Princeton U Psychology Dept Princeton NJ 08544

SHAFRITZ, DAVID ANDREW, physician, research scientist; b. Phila., Oct. 5, 1940; s. Saul and Ethel (Kohn) S.; m. Sharon C. Klemow, Aug. 16, 1964; children: Gregory S., Adam B., Keith M. AB in Chemistry with honors, U. Pa., 1962, MD, 1966. Diplomate Nat. Bd. Med. Examiners, Am. Bd. Internal Medicine. Intern, asst. resident U. Md. Hosp., Balt., 1966-68; research assoc. NIH, Bethesda, Md., 1968-71; clin. and research fellow Mass. Gen. Hosp., Boston, 1971-73; instr. Harvard Med. Sch., Boston, 1971-73, asst. prof. medicine, 1973; asst. prof. medicine and cell biology Albert Einstein Coll. Medicine, Bronx, N.Y., 1973-76, assoc. prof., 1976-81, prof. medicine and cell biology, 1981—, dir. Marion Bessin Liver Research Ctr., 1985—; cons. Integrated Genetics Inc., Framingham, Mass., 1981-86, Immuno, Vienna, Austria, 1986-91, Innovir, Inc., N.Y.C., 1991—, Eugenetech Internat., Inc., Ramsey, N.J., 1991—; temporary advisor WHO, Geneva, 1983; mem. Nat. Com. for Clin. Lab. Standards, Villanova, Pa., 1983—; mem. sci. adv. com. liver cancer program Inst. for Cancer Research, Fox Chase and Phila., 1987—, rev. panel C, study sect. Nat. Inst. Diabetes and Digestive and Kidney Diseases, 1988-92; mem. cen. coordination com. Liver Tissue Procurement and Distbn. System, 1986—. Co-author: The Liver: Biology and Pathobiology, 1982, 2d edit., 1988, Hepatobiliary Diseases, 1991; assoc. editor: Hepatology, 1981-86; mem. editorial bd. Jour. Med. Virology, 1982—, Hepatology, 1990—, Jour. Virology, 1992—; contbr. numerous research articles, revs., and book chpts.; patentee in field. Trustee Westchester Jewish Ctr., Mamaroneck, N.Y., 1980-86. Served to lt. comdr. USPHS, 1968-71. Recipient Merck award U. Pa., 1962, Morton McCutcheon Meml. Research prize U. Pa. Sch. Medicine, 1966, spl. research fellowship NIH, 1971-73, Career Scientist award Irma T. Hirschl Trust, N.Y.C., 1974-79, Research Career Devel. award NIH, 1975-80, fellowship European Molecular Biology Orgn., 1978; research grantee NIH, 1974—. Mem. Am. Assn. for Study of Liver Diseases, Internat. Assn. for Study of Liver, Am. Gastroenterol. Assn., Am. Soc. Biochemistry and Molecular Biology, Am. Soc. Clin. Investigation, Assn. Am. Physicians, N.Y. Acad. Scis., Harvey Soc., Interurban Clin. Club. Democrat. Jewish. Avocations: jogging, tennis. Home: 4 Pheasant Run Larchmont NY 10538-3423 Office: Yeshiva U Albert Einstein Coll Med Marion Bessin Liver Rsch Ctr 1300 Morris Park Ave Bronx NY 10461-1924

SHAH, BANKIM, software consultant; b. 1961. MS in Mech. Engr., Indian Inst. Tech. Kanpur, 1985; MS in Computer Sci., IIT, 1991. Program analyst automobile co., India; sr. program analyst health ins. co., Chgo., 1987; system analyst Abbott Labs., Chgo., 1988—. Author: Implemented Knowledge Acquisition System for Medas Expert System, 1991; co-author: Automation of Spindle Design, 1985. Home: 3241 Drew St Downers Grove IL 60515

SHAH, HARISH HIRALAL, mechanical engineer; b. Ahmadabad, India, Sept. 6, 1942; came to the U.S., 1969; s. Hiralal J. and Ramangauri H. Shah; m. Hemal H. Shah, Feb. 11, 1969; children; Tanvi, Nirvi. BS in Mech. Engring., Tuskegee Inst., 1971; MS, U. South Fla., 1978. Registered profl. engr., Fla., Tenn. Jr. mech. engr. FMC Corp., Lakeland, Fla., 1971-72; mech. engr. Tampa Bay Engring. Co., St. Petersburg, Fla., 1972-74; sr. mech. engr. Davy McKee Corp., Lakeland, 1974-82, Watson & Co., Tampa, Fla., 1982-84, Reynolds, Smith and Hills, Tampa, 1984—. Mem. ASHRAE, Fla. Engring. Soc. (pres. Ridge chpt. 1983-84, v.p. 1988-89), NSPE, Nat. Fire Protection Assn., India Assn. Lakeland (pres. 1979-80). Home: 3816 Eric Ct Lakeland FL 33813 Office: Reynolds Smith and Hills 1715 N Westshore Blvd Tampa FL 33607

SHAH, HASH N., plastics technologist, researcher; b. Radhanpur, Gujarat, India, Mar. 25, 1934; came to U.S., 1973; s. Navinchandra M. and Taraben Shah; m. Hansa Shah, Dec. 25, 1959; children: Saumil, Viral. BS with honors, U. Bombay, 1955, BS in Tech., 1957, MS in Tech., 1960; degree, S. German Plastic Centre, 1970. Rsch. engr. Info-Chem, Fairfield, N.J., 1975-79, Becton Dickinson, Fairfield, 1980-83; with rsch. and tech. divsn. Shah Plasti-coats and Prints, Bombay, 1983-85; cons., engr. Johnson & Johnson, Warren, N.J., 1989-90; rsch. scientist, engr. Convatec, Bristol Meyers & Squibb, Skillman, N.J., 1990—. Sr. editor Bombay Technologist Centennial, 1960. Rep. freeholder, New Brunswick, N.J., 1986. Indian Govt. scholarm 1958-60; German Govt. scholar, 1968-70; recipient award Materials of Engring. mag., 1976. Mem. AAAS, Soc. Plastics Engrs. (sr., bd. dirs. 1979-80), Chem. Engring. Democrat. Mem. Jainistic Temple. Achievements include patent for disposable cassette for monitoring infusion pump; patents pending for surgical sharp disposable kit, heat sinker pannels. Home: 10 Landing Ln # 7G New Brunswick NJ 08901 Office: Convatec Bristol Meyers & Squibb Co 200 Headquarters Park Dr Skillman NJ 08558

SHAH, JAYPRAKASH BALVANTRAI, civil engineer; b. Jamshedpur, India, Nov. 2, 1946; came to U.S., 1969; s. Balvantrai Talakchand and Mangala (Kamani) S.; m. Bharti Shah, Nov. 29, 1972; children: Bejal, Rupal. MS, Wayne State U., 1970; MBA, U. Detroit, 1981. Profl. engrs., Mich. Construction engr. George Jerome and Co., Detroit, 1970-71; ops. engr. Oakland County DPW, Pontiac, Mich., 1971-79; vice chmn., asst. CEO Spalding, Dedecker and Assoc., Inc., Madison Heights, Mich., 1979--. Contbr. articles to profl. jours. Planning commr. Waterford Twp., Mich., 1987—; dir. Oakland County Econ. Devel. Corp., Pontiac, 1982—, vice chmn. 1993; mem. Water Twp. Zoning Bd., 1989—. Recipient Best Design award Walchand Coll. Bd., 1968. Fellow ASCE (chmn. Mich.-Ohio Dist. Coun. 1984-85, chmn. engring. mgmt. divsn. 1990-91, pres. Mich. sect. 1980-81, Outstanding Engr. of Yr. 1993), Am. Soc. Engrs. from India (life, pres. 1982). Office: Spalding Dedecker and Assocs 655 W 13 Mile Rd Madison Heights MI 48071

SHAH, KIRTI JAYANTILAL, electrical engineer; b. Patan, North Gujarat, India, Dec. 10; came to the U.S., 1969; s. Jayantilal Khemchand Shah and Bhagwatiben Jayantilal; m. Sudha K. Shah, Dec. 9, 1980; children: Binal K., Kruti K. BEE, U. Bombay, 1968; MEE, U. Pitts., 1971. Field engr. and suport engr. Westinghouse Electric Corp., Pitts., Calif., 1973-86; field engr. Siemens Corp., Hayward, Calif., 1986-87, Reliance Electric Co., Anaheim, Calif., 1987-88; applications engr. Teledyne Inet Inc., Torrance, Calif., 1988-92; sales engr. Wahl Instruments Co., Culver City, Calif., 1992; cons. engr. Magnetek Corp., Anaheim, Calif., 1992—. Contbr. articles to profl. jours. Mem. IEEE, Instrument Soc. Am., Eta Kappa Nu, Sigma Xi. Jain Religion. Home: 23015 Madison St Torrance CA 90505 Office: Magnetek Corp 901 E Ball Rd Anaheim CA 92805

SHAH, MUBARIK AHMAD, surgeon; b. Lahore, Punjab, Pakistan, Dec. 19, 1949; came to U.S., 1976; s. Tufail Mohammad and Amna (Bibi)) S.; m. Mansoora Nazli, Nov. 10, 1974; children: Rizwan, Maliha, Numaan. FSc, T.I. Coll., Rabwah, Pakistan, 1966; MBBS, Punjab U., Multan, Pakistan, 1971. Diplomate Am. Bd. Surgery. Instr. King Edward Med. Coll., Pakistan, 1975-76; house surgeon N.Y. Polyclinic Hosp., N.Y.C., 1976-77; resident surgery St. Mary Hosp., Waterbury, Conn., 1977-78, New Rochelle (N.Y.) Hosp., 1978-79; resident surgery Presbyn.-U. Pa. Med. Ctr., Phila., 1979-82, fellow vascular surgery, 1982-83; clin. instr. U. Pa. Sch. Medicine, Phila., 1979-83; attending physician, surgeon St. Mary Hosp., Langhorne, Pa., 1990—; dir. trauma svcs. Med. Coll. Hosps., Bucks County Campus, Warminster, Pa., 1988-90. Med. advisor Trevose (Pa.) Rescue, 1988-90; pres. Ahmadiyya Youth Orgn., Phila., 1989-90. Capt. Pakistan Army, 1971-75. Fellow ACS, Internat. Coll. Surgeons, Am. Soc. Abdominal Surgeons; mem. AMA, Am. Trauma Soc., Pa. Med. Soc., Bucks County Med. Soc. Avocations: photography, fishing, camping, rowing. Office: St Mary Med Office Bldg Ste 302 1205 Langhorne Newtown Rd Langhorne PA 19047

SHAH, NATVERLAL JAGJIVANDAS, cardiologist; b. Godhra, Gujarat, India, Aug. 3, 1926; s. Jagjivandas Purshottamdas and Mahalaxmi (Jagjivandas) S.; m. Sundri Choitram Malkaney, Mar. 15, 1956; children: Sailesh N., Sarina N. M.D., U. Bombay, 1954. Med. registrar, sr. med. tutor Cardiology King Edward Meml. Hosp., Bombay, 1954-56; hon. assoc. physician Gokuldas Tejpal Hosp., also hon prof. medicine Grant Med. Coll., Bombay, 1959-74; hon. physician and cardiologist, head ICC unit, mobile coronary CCU, exercise stress test Bombay Hosp., 1960—; patron 2d Internat. Conf. on Hypertension, 1985. Author: A Handbook of Endocrine Disorders, 1960, Prevent a Heart Attack, 2d edit., 1977, An Approach to Electrocardiography, 3d edit., 1978, Heart-Before, During, After, 1984, Advanced Electrocardiography, 1986, Clinical Electrocardiography, 1988; contbr. articles to profl. jours. Hon. spl. exec. magistrate Govt. of Maharashtra, 1980—. Mem. Internat. Congress Hypertension (pres. 1977), Cardiol. Soc. Indian (chmn. Bombay br. 1977-78), Internat. Conf. Advances Internat. Medicine (v.p.), Internat. Conf. Prevention Heart Disease and Cardiac Rehab. (v.p. 1978), Nat. Coun. Hypertension (pres., founder 1978-81), Nat. Soc. Prevention Heart Disease and Rehab. (pres. 1988—), Internat. Congress Directions in Cardiovascular Advances (chmn. 1989, Bombay), Wellington Sports Club, Nat. Sports Club (Bombay). Home: 4/D Ananta, Rajaballi Patel Rd, Bombay 400026, India Office: Med Rsch Centre, New Marine Lines, Bombay 400020, India

SHAH, NIPULKUMAR, principal design engineer; b. Nairobi, Kenya, Feb. 8, 1953; came to U.S., 1987; s. Gulabchand and Kanchan S.; m. Renuka Shah, Nov. 8, 1976; 1 child, Amar. BSc, Imperial Coll., London, 1975, PhD, 1979. Rsch. asst. Imperial Coll., London, 1980-83; head of methods combustion Rolls Royce PLC, Bristol, U.K., 1983-86; heat transfer engr. Sundstrand Power Systems, San Diego, 1986—. Author: (symposium proceedings) A New Radiation Solution Method for Incorporation in General Combustion Prediction Procedures, 1981. Mem. ASME, AIAA (sr. mem.). Achievements include patents for Spectral Infrared Traversable Absorption Radiometer, Spiral Cooling and Attachment of Vanes, Compact Stored Energy Combustor, Gas Turbine Sealing providing enhanced pattern factor, Inlet Particle Separator and Particle Inlet Protection System; research on discrete transfer method to calculate radiative heat transfer. Office: Sundstrand Power Systems 4400 Ruffin Rd San Diego CA 92186

SHAH, NIRODH, nuclear engineer; b. Sydney, Ohio, Aug. 19, 1967; s. Harish Chanderlal and Sulochana Shah. BS in Nuclear Engring., U. Ill., 1989. Health physics tech. collaborator Brookhaven Nat. Lab., Upton, N.Y., 1989; engring. cons. Dames & Moore, Pearl River, N.Y., 1989; health physics inspector Nuclear Regulatory Commn., Glen Ellyn, Ill., 1990—. Mem. NSPE, Am. Nuclear Soc. Office: US Nuclear Regulatory Commn 799 Roosevelt Rd Glen Ellyn IL 60137

SHAH, SHIRISH KALYANBHAI, computer science educator, chemistry and environmental science educator; b. Ahmedabad, India, May 24, 1942; came to U.S., 1962, naturalized, 1974; s. Kalyanbhai T. and Sushilaben K. S.; B.S. in Chemistry and Physics, St. Xavier's Coll., Gujarat U., 1962; PhD in Phys. Chemistry, U. Del., 1968; cert. in bus. mgmt. U. Va., 1986; PhD in Cultural Edn. (hon.) World U. West, 1986; m. Kathleen Long, June 28, 1973; 1 son, Lawrence. Asst. prof. Washington Coll., Chestertown, Md., 1967-68; dir. quality control Vita Foods, Chestertown, 1968-72; asst. prof., assoc. prof. sci., adminstr. food, marine sci. and vocat. programs Chesapeake Coll., Wye Mills, Md., 1968-76; assoc. prof., prof. sci., chmn. dept. tech. studies Community Coll. of Balt., 1976-91; assoc. prof. chemistry Coll. Notre Dame of Md., 1991—, advisor to Young Republicans, 1992—; chmn. computer systems and engring. techs., 1982-89, coord. tech. studies, 1989-91; mem. Balt. City Adult Edn. Adv. Com., 1982-89; chmn. Coll. wide computer user com., 1985-91; permanent mem. Rep. Senatorial Com.; charter mem. Rep. Presdl. Task Force. Mem. com. Am. Lung Assn., 1971-80; mem. Congl. Adv. Com., 1983—. Fellow Am. Inst. Chemists; mem. IEEE, Am. Chem. Soc., Data Processing Mgmt. Assn., Nat. Environ. Tng. Assn., Nat. Sci. Tchrs. Assn., Nat. Assn. Indsl. Tech. (dir. local region, bd. accreditors), Am. Vocat. Assn., Am. Tech. Edn. Assn., Am. Fedn. Tchrs., Md. State Tchrs. Assn., Md. Assn. Community and Jr. Colls. (v.p. 1977-78, pres. 1978—), Sigma Xi, Epsilon Pi Tau, Iota Lambda Sigma Nu. Jain, Roman Catholic. Contbr. articles on sci. and tech. to profl. jours. Home: 5605 Purlington Way Baltimore MD 21212-2950 Office: Coll Notre Dame Dept Chem 4701 N Charles St Baltimore MD 21210

SHAH, SURENDRA POONAMCHAND, engineering educator, researcher; b. Bombay, Aug. 30, 1936; s. Poonamchand C. and Maniben (Modi) S.; m. Dorothie Crispell, June 9, 1962; children: Daniel S., Byron C. BE, B.V.M.

Coll. Engring., India, 1959; MS, Lehigh U., 1960; PhD, Cornell U., 1965. Asst. prof. U. Ill., Chgo., 1966-69, assoc. prof., 1969-73, prof., 1973-81; prof. civil engring Northwestern U., Evanston, Ill., 1981—, dir. Ctr. for Concrete and Geomaterials, 1987—; dir. NSF Sci. and Tech. Ctr. for Advanced Cement-Based Materials Northwestern U., 1989—, Walter P. Murphy prof. of engring., 1992—; cons. govt. agys. and industry, U.S.A., UN, France, Switzerland, People's Republic China, Denmark, The Netherlands; vis. prof. MIT, 1969, Delft U., The Netherlands, 1976, Denmark Tech. U., 1984, LCPC, Paris, 1986, U. Sidney, Australia, 1987; NATO vis. sci. Turkey, 1992. Contbr. over 300 articles to profl. jours.; editor 12 books; mem. editorial bd. 4 internat. jours.; editor-in-cheif Journal Of Advanced Cement Based Materials; co-author Fiber Reinforced Cement Composites, 1992. Recipient Thompson award ASTM, Phila., 1983, Disting. U.S. Vis. Scientist award Alexander von Humboldt Found., 1989. Fellow Am. Concrete Inst. (chmn. tech. com., Anderson award 1989); mem. ASCE (past chmn. tech. com.), Internat. Union Testing & Rsch. Labs. Materials and Structures (chmn. tech. com. 1989—, Gold medal 1980). Home: 921 Isabella St Evanston IL 60201-1773 Office: Northwestern U Tech Inst Rm A130 2145 Sheridan Rd Evanston IL 60208-0002

SHAHAB, SALMAN, systems engineer; b. Karachi, Pakistan, Oct. 18, 1962; came to U.S., 1985; s. Mohammed F. Siddiqui and Zarina Zarrin. B Engring., N.E.D. Univ., Karachi, Pakistan, 1984; MS in Computer Sci., Northrop U., 1988. Registered profl. engr., Pakistan. Exec. engr. Pasmic, Karachi, 1983-84; sr. test engr. Veritest, Inc., Inglewood, Calif., 1988-89; product engr. C. Itoh Electronics, Irvine, Calif., 1989; asst. prof. Nat. Univ., L.A., 1990—; dir. mgmt. info. systems UCLA, 1989—; cons. Northrop U., L.A., 1988—. Advisor Pakistani Student Assn., L.A., 1992. Fellow Inst. Engrs.; mem. Pakistani Engring. Coun. Office: UCLA JCCC 10833 Le Conte Los Angeles CA 90024-1781

SHAHEEN, ELI ESBER, environmental engineer; b. Cushing, Okla., Dec. 11, 1968; s. Esber Ibrahim and Shirley Ann (King) S. BS in Chem. Engring., U. Mo., 1990. Rsch. lab. technician U. Mo., Columbia, 1988-89, rsch. asst. dept. chem. engring., 1989; rsch. assoc. Internat. Inst. Tech., Inc., 1990; engr. Star Enterprise, Joplin, Mo., 1991—. Home: 1440 Lila Ave Baton Rouge LA 70820

SHAHEEN, WILLIAM A., civil engineering educator, consultant; b. Newburyport, Mass., Jan. 12, 1957; s. Richard Arthur and Millicent (Turko) S.; m. Eileen Regina Rausch, June 23, 1990 (div. 1992); 1 child, Devin. BSCE, U. Mass., 1983, MSCE, 1985; PhD, U. Conn., 1991. Registered profl. engr., Mass., Conn. Broadcast engr. Sta. WGGB-TV, Springfield, Mass., 1978-83; geotech. engr. Goldberg-Zoino & Assocs., Vernon, Conn., 1984-85; lectr. U. Conn., Storrs, 1988-89, 92—; pres. Analytical Engring., Inc., Granby, Mass., 1989—; asst. prof. civil engring. U. Hartford, West Hartford, 1989-93; faculty senate U. Hartford, West Hartford, 1992-93; tech. advisor Am. Radio Relay League, Newington, Conn., 1985—; tech. cons. Westover AFB, Mass., 1992—. Contbr. articles to profl. jours. Chmn. Granby Mass. Bd. Health, 1989—; mem. Landfill Oversite Com., 1990—, mem. Emergency Planning Com., 1993—. Mem. IEEE, ASCE, U. Conn. Engring. Alumni Assn. (bd. dirs. 1991—), Am. Radio Relay League, Sigma Xi, Pi Tau Sigma, Chi Epsilon, Phi Kappa Phi, Tau Beta Pi. Achievements include research in earth anchors, septic system performance, transmission tower design, groundwater flow modeling, foundation engineering. Office: Analytical Engring Inc Granby MA 01033

SHAHIN, M. Y., engineering. Recipient Fed. Engr. of the Year award Nat. Soc. Profl. Engrs., 1991. Home: 1207 Wilshire Ct Champaign IL 61821*

SHAHIN, MAJDI MUSA, biologist; b. Jerusalem, Palestine, Dec. 5, 1936; arrived in France, 1978; s. Musa and Mariam (Abu-Zayyad) S.; mm. Edith Heun Anneliese, Mar. 8, 1962; 1 child, Mariam. Pre-diploma in Biology, Free U. Berlin, Berlin, Fed. Republic Germany, 1962, diploma in Biology, 1965, PhD, 1969. Postdoctoral fellow Atomic Energy of Canada, Ltd., Chalk River, Ont., Canada, 1970-73; visiting fellow Nat. Inst. Eviron. Health and Scis., Rsch. Triangle Pk., N.C., U.S.A., 1973-74; rsch. assoc. dept. genetics U. Alta., Edmondton, Can., 1974-78; dir. dept. mutagenesis L'Oreal Rsch. Labs., Aulnay-sous-Bois, France, 1978-91; dir. dept. chem. protection and photobiol. rsch. in vitro L'Oréal Advanced Rsch. Ctr., 1991—. Contbr. articles to profl. jours., chpts. to books; editor spl. issue Mutation Rsch. on Structured Activity Relationships in Chem. Mutagenesis. Mem. AAAS, Am. Soc. Microbiology, European Environ. Mutagen Soc., The Genetics Soc. Canada, Am. Soc. Radiation Rsch., Environ. Mutagen Soc. (U.S.), N.Y. Acad. Scis., Am. Soc. Photobiology. Office: L'Oreal Rsch Labs, 1 Ave Eugéne Schueller, 93601 Aulnay-sous-Bois France

SHAHNAVAZ, HOUSHANG, ergonomics educator; b. Tehran, Iran, Mar. 12, 1936; arrived in Sweden, 1980; s. Mohamad Nasser Shahnavaz; m. Azita Shahnavaz; children: Rasmy-Nive, Shadi. Diploma in Mech. Engring., Darmstadt, 1966; MSc in Ergonomics, Birmingham (Eng.) U., 1973, PhD in Ergonomics, 1976. Tech. dir. Tehran Opera House, 1966-68, asst. prof., 1977-80; lectr. Tehran Tech. U., 1968-73; research fellow Birmingham U., 1976-77; acting prof. indsl. ergonomics Luleå (Sweden) U., 1980-88, prof., 1989—, dir. Ctr. for Ergonomics of Developing Countries, 1983—; head coordinating dept. Royal Commn., Tehran, 1978; cons. ergonomics WHO, Internat. Labour Office and World Bank. Spl. editor Internat. Jour. Indsl. Ergonomics, Internat. Jour. Applied Ergonomics; contbr. numerous articles to profl. jours. Fellow Ergonomics Soc.; mem. Human Factors Soc., Internat. System Safety Soc., S.E. Asian Ergonomics Soc., Nordiska Ergonomisällskapet. Office: Lulea U, Dept Ergonomics, 951-87 Lulea Sweden

SHAIK, SASON SABAKH, chemistry educator; b. Bagdad, Iraq, Dec. 17, 1948; s. Salim and Nazima (Zadik) S. m. Sara Yadid, June 29, 1980; 1 child, Yfaat. BSc and MSc, Ramat Gan, Israel, 1974; PhD, U. Wash., Seattle, 1978. Lectr. Ben Gurion U., Beer-Sheva, Israel, 1979-80; sr. lectr. Ben Gurion U., Beer-Sheva, 1980-84, assoc. prof., 1985-88; prof. Ben Gurion U., Beer Sheva, Israel, 1988-92, The Hebrew U., Jerusalem, Israel, 1992—; editorial bd. Israel Jour. Chemistry, 1988—; adviser Unit for Gifted Children, Beer-Sheva, Israel, 1988-91. Author: Structural Theory of Organic Chemistry, 1978, Teoretical Aspects of Physical Organic Chemistry, 1992. Sgt. Israeli Army, 1965-68. Named Fulbright Fellow, Fulbright Found., 1974-78. Mem. Am. Chem. Soc., Israel Chem. Soc. (Outstanding Young Chemist 1987), Swiss Chem. Soc., Soc. for Advancement of Arts and Sci. for Iraqi Jews (steering com. 1991—). Jewish. Achievements include creation of a gen. model for understanding the reactivity of molecules, the origins of barriers and the mechanism of transition state formation. Office: The Hebrew Univ, Dept Chemistry, Jerusalem 91904, Israel

SHAIKH, MUZAFFAR ABID, management science educator; b. Bombay, Jan. 5, 1946; came to U.S., 1966; s. Shaikh A. Razzaque and Khudaija R. Shaikh; m. Farhat Anjum, Dec. 29, 1968; children: Mahjabeen, Shahbaaz, Shoaib. BS, U. Bombay, 1966; MS, Kans. State U., 1968; PhD, U. Ill., 1983. Mgmt. sci. rsch. analyst Caterpillar Inc., Peoria, Ill., 1968-85; sr. staff engr. Harris Corp., Melbourne, Fla., 1985-87; assoc. prof. Fla. Inst. Tech., Melbourne, 1987-92, prof. mgmt. sci., 1992—; assoc. dean grad. programs Sch. of Bus. Fla. Inst. Tech., Melbourne, 1992—; cons. Harris Corp., Melbourne, 1987-92, Grumman Corp., Melbourne, 1987-92. Assoc. editor Trans. of Simulation, 1989; reviewer Internat. Jour. Computer and Indsl. Engring., 1990—; contbr. over 57 articles to profl. jours. Fellow Sigma Xi; sr. mem. Ops. Rsch. Soc. Am., Inst. Mgmt. Sci., Inst. Indsl. Engrs. Islam. Achievements include research in consumer behavior. Home: 409 Crystal Lake Rd Melbourne FL 32940 Office: Fla Inst Tech 150 W University Blvd Melbourne FL 32901

SHAIKH, NAIMUDDIN, medical physicist; b. Khurda, Orissa, India, Feb. 10, 1951; s. Nasibuddin Shaikh and Mahemuda Bibi; m. Najmul Akhtar, Oct. 29, 1978; children: Arif, Alim, Amin. MS, Utkal U., 1974; MA, CCNY, 1981; MS, U. Wis., 1983, PhD, 1986. Diplomate Am. Bd. Radiology, Am. Bd. Med. Physics. Lectr. in physics Govt. Orissa, Bhubaneswar, Orissa, 1974-79; rsch. asst. U. Wis., Madison, 1986-87; chief diagnostic physicist Lahey Clinic Med. Ctr., Burlington, Mass., 1987—; mem. Lahey Clinic Radiation Safety Com., Burlington, 1987—; nat. examiner Am. Bd. Med. Physics, Reston, Va., 1990—. Recipient Nat. scholarship state govt., India, 1968-74, fellowship CCNY, 1980-81. Mem. Am. Assn. Physicists in Medicine, Soc. magnetic Resonance in Medicine, Am. Coll. Med. Physicists, Am. Coll. Radiology. Home: 67 Tyngsboro Rd Westford MA 01886 Office: Lahey Clinic Med Ctr 41 Mall Rd Burlington MA 01805

SHAKES, DIANE CAROL, developmental biologist, educator; b. Oberlin, Ohio, Dec. 23, 1961; d. Joseph Stanley Shakes and Julia Almanrode Hansen. BA in Biology, Pomona Coll., 1983; PhD in Biology, Johns Hopkins U., 1988. Postdoctoral assoc. genetics & devel. Cornell U., Ithaca, N.Y., 1988-91; asst. prof. dept. biology U. Houston, 1991—; faculty mentor U. Houston, 1992—, Community in Schs. Program, Houston, 1992. Contbr. articles to profl. jours. Vol. Loaves and Fishes, Ithaca, 1990. Grnatee NSF, 1992—, U. Houston, 1992; Carnegie Corp. fellow Carnegie Inst. Washington, Balt., 1986-88. Mem. Nat. Sci. Tchrs. Assn., Soc. for Developmental Biology, Sigma Xi. Achievements include genetic dissection of spermatogenesis in C elegans as a model system for how cells are able to change their shape and fuction; identification of the few known paternal effect genes which indicates a role for sperm in development beyond that of membrane stimulation and nuclear contributions. Office: U Houston Dept Biology Houston TX 77204-5513

SHAKHMUNDES, LEV, mathematician; b. Leningrad, USSR, Dec. 29, 1933; came to Can., 1975; s. Yudel and Alexandra (Voitsekhovskaya) S.; children: Nadia, Daniel. MS., Leningrad U., 1957; PhD, Leningrad Poly. Inst., 1965. Engr., rsch. assoc., cons., various instns., Leningrad, 1957-74; rsch. asst. U. Toronto, 1976-78; sr. cons. analyst Union Gas Ltd., Chatham, Ont., 1978—. Co-author: Economic Efficiency of Capital Expenditures (in Russian), 1969; contbr. articles to Soviet and U.S. periodicals; patentee Ministry Sci. and Tech. USSR. Mem. Assn. Profl. Engrs. Ont., Canadian Econs. Assn., Am. Econs. Assn. Avocations: Sports. Home: PO Box 383, Chatham, ON Canada N7M 5K5

SHALABI, MAZEN AHMAD, chemical engineer, educator; b. Taif, Saudi Arabia, Sept. 10, 1946; s. Ahmad Khalid Shalabi and Sobhia Mohammad Doukhi; m. Hanan Nadim, July 21, 1975; children: Thuraya, Essam, Rafah, Yasmin. BS with honors, Cornell U., 1970, MS, 1972; PhD, Colo. Sch. Mines, 1977. Chmn. dept. chem. engring. King Fahd U. Petroleum and Minerals, Dhahran, Saudi Arabia, 1989—; vis. prof. Stanford U., 1984-85, U. Cambridge, Eng., summers 1991, 92, Colo. Sch. Mines, 1993—. Mem. Am. Chem. Soc., Am. Inst. Chem. Engrs., Tau Beta Pi.

SHALACK, JOAN HELEN, psychiatrist; b. Jersery City, Mar. 6, 1932; d. Edward William and Adele Helen (Karski) S.; m. Jerome Abraham Sheill. Student, Farleigh Dickinson U., 1950-51; BA cum laude, NYU, 1954; MD, Women's Med. Coll. Pa., 1958. Intern Akron (Ohio) Gen. Hosp., 1958-59; resident in psychiatry Camarillo (Calif.) State Hosp., 1959-62; resident in psychiatry UCLA Neuropsychiat. Inst., 1962, U. So. Calif., L.A., 1963; pvt. practice Beverly Hills, Calif., 1963-83, Century City L.A., Calif., 1983-86, Pasadena, Calif., 1986—; pres., chair bd. dirs. Totizo Inc., Beverly Hills, 1969-71; mem. staff Westwood Hosp., 1970-75. Mem. AMA, Calif. Med. Assn., L.A. County Med. Assn., L.A. County Med. Women's Assn., Physicians for Social Responsibility, Am. Med. Women's Assn., Phi Beta Kappa, Mu Chi Sigma. Avocations: art, archaeology, gardening, tennis, bicycling. Home and Office: 1405 Afton St Pasadena CA 91103-2703

SHALHOUB, VICTORIA AMAN, molecular biochemist, researcher; b. Melbourne, Australia, Aug. 10, 1951; came to U.S., 1980; d. Emil Nassim and Laure Naomi (Moujaes) S. BSc, Am. U. Beirut, 1974, diploma in sci. edn., 1977, MSc, 1979; PhD, Boston U., 1987. Postdoctoral assoc. U. Mass. Med. Ctr., Worcester, 1987-90, instr., 1990—. Contbr. articles to sci. jours. Mem. AAAS, Am. Soc. Bone and Mineral Rsch. Greek Orthodox. Office: U Mass Med Ctr Dept Cell Biology 55 Lake Ave N Worcester MA 01655

SHALIT, BERNARD LAWRENCE, dentist; b. Quincy, Mass., Feb. 17, 1920; s. L. Melville and Mildred (Kolb) S.; m. Helen L. Shoener, Oct. 11, 1951; children: Barbara L., William L. DMD, Tufts U., 1944; grad. East Coast Aero Tech. Sch., 1981. Pvt. practice dentistry, Quincy. Dir. maintenance Manning Aviation. Served from lt. (j.g.) to lt. Dental Corps. USNR, 1944-46. Author: Doc's Stock Investment Letter. Life fellow seminars on Hypnosis Found.; mem. Mass. Dental Soc. (life), Am. Dental Assn. (life), Am. Soc. Clin. Hypnosis (life), New Eng. Soc. Clin. Hypnosis, NRA (life), Norfolk County Beekeepers Assn., Masons. Home and Office: 14 Walker St Quincy MA 02171-1924

SHALITA, ALAN REMI, dermatologist; b. Bklyn., Mar. 22, 1936; s. Harry and Celia; m. Simone Lea Baum, Sept. 4, 1960; children: Judith and Deborah (twins). AB, Brown U., 1957; BS, U. Brussels, 1960; MD, Bowman Gray Sch. Medicine, 1964; DSc (hon.), L.I. U., 1990. Intern Beth Israel Hosp., N.Y.C., 1964-65; resident dept. dermatology NYU Med. Ctr., 1967-68; NIH tng. grant fellow dept. dermatology Med. Ctr. NYU, 1968-70, instr. dermatology Med. Ctr., 1970-71, asst. prof. Med. Ctr., 1971-73; asst. prof. dermatology Columbia U., N.Y.C., 1973-75; assoc. prof. medicine, head div. dermatology SUNY Downstate Med. Ctr., Bklyn., 1975-79; prof. medicine, head div. dermatology SUNY Downstate Med. Ctr., 1979-80; prof. and chmn. dept. dermatology Med. Ctr. SUNY Downstate Med. Ctr., Bklyn., 1980—; asst. dean Downstate Med. Ctr. SUNY, 1977-83, acting dean Queen's campus Downstate Med. Ctr., 1983-84; assoc. dean clin. affairs SUNY Health Sci. Ctr., Bklyn., 1989-92; assoc. provost for clin. affairs, 1992—; asst. attending in dermatology Univ. Hosp., N.Y.C., 1970-73, Bellevue Hosp. Ctr., 1970-73, Manhattan VA Hosp., 1971-73, Presbyn. Hosp., 1973-75; mem. med. bd. Kings County Hosp. Ctr.; cons. dermatology Bklyn. VA Hosp., 1975—; chief dermatology Brookdale Med. Ctr., 1977-90; chief dermatology Univ. Hosp. of Bklyn., 1975—; chief dermatology Kings County Hosp. Ctr., Bklyn., 1975—, acting med. dir., 1989-92; med. dir. Univ. Hosp. of Bklyn., 1992—. Pres. Temple Shaaray Tefila, N.Y.C., 1982-86, chmn. bd. trustees, 1987—. Lt. M.C. USNR, 1965-67. Recipient Torch of Liberty award Anti-Defamation League, 1987, Surg. and Pediatric awards Beth Israel Hosp., N.Y.C., 1965; spl. fellow NIH, 1970-73. Mem. AMA, ACP, AAAS, Am. Acad. Dermatology (bd. dirs. 1983-87), Soc. Investigative Dermatology, Dermatology Found. (past trustee), Am. Dermatol. Assn., N.Y. Acad. Scis., Am. Soc. Dermatol. Surgery (past bd. dirs.), Internat. Soc. Tropical Dermatology, N.Y. State Med. Soc., Soc. Cosmetic Chemists, N.Y. Acad. Medicine, Dermatol. Soc. Greater N.Y. (pres. 1980-81), N.Y. State Dermatol. Soc., Assn. Profs. Dermatology (sec-treas. 1988—), N.Y. Dermatol. Soc. (pres. 1989-90). Republican. Home: 70 E 77th St New York NY 10021-1811 Office: 450 Clarkson Ave Brooklyn NY 11203-2098

SHAM, LU JEU, physics educator; b. Hong Kong, Apr. 28, 1938; s. T.S. and Cecilia Maria (Siu) Shen; m. Georgina Bien, Apr. 25, 1965; children: Kevin Shen, Alisa Shen. GCE, Portsmouth Coll., Eng., 1957; BS, Imperial Coll., London U., Eng., 1960; PhD in Physics, Cambridge U., Eng., 1963. Asst. rsch. physicist U. Calif. at San Diego, La Jolla, 1963-66, assoc. prof., 1968-76, prof., 1975—, dean div. natural scis., 1985-89; asst. prof. physics U. Calif. at Irvine, 1966-67; rsch. physicist IBM Corp., Yorktown Heights, N.Y., 1974-75; reader Queen Mary Coll. U. of London, 1967-68; summer visitor neutron physics Chalk River Nuclear Lab. AEC Inst., Can., 1969; cons. Bell Tel. Labs., Murray Hill, N.J., 1968, Bellcore, Redbank, N.J., 1988-90. Contbr. sci. papers. to profl. jours. Recipient Churchill Coll. studentship, Eng., 1960-63, Sr. U.S. Scientist award Humboldt Found., Stuttgart, Germany, 1978; fellow Guggenheim Found., 1984. Fellow Am. Phys. Soc.; mem. AAAS. Democrat. Avocation: tennis, folk dancing. Office: U Calif San Diego Dept Physics 0319 La Jolla CA 92093-0319

SHAMAIENGAR, MUTHU, chemist; b. Bangalore, Mysore, India, Jan. 17, 1925; came to U.S., 1953; s. Srinivas and Vedamma S.; m. M. Shantha, Mar. 29, 1953; children: Baiju, Rita, Lata, Gita, Ravi. BS in Chemistry, U. Mysore, Bangalore, India, 1944; MS in Chem. Technology, Indian Inst. Sugar Tech., Kanpur, India, 1948; MS in Chemistry, Indian Inst. Sci., Bangalore, India, 1952; PhD in Sci., U. Bombay, 1959. Technologist USX Corp., Pitts., 1960-62; sr. rsch. chemist Quaker Chem. Corp., Conshohocken, Pa., 1962-65; tech. advisor Chrysler Corp., Detroit, 1965-68; sr. environ. chemist U.S. EPA, Washington, 1971-81; sr. engring. assoc. Engelhard, Newark, 1981-82; rsch. dir. Solson Chem. Labs., Washington, 1982-87; sr. chemist U.S. EPA, Washington, 1987-89; chemist Dept. Def., NAVSEA, Materials Engring. Dept., Washington, 1987—; advisor alcohol fuels Dept. of Energy, Washington, 1979-80; rsch. chemist API Project 6/Carnegie Tech., Pitts., 1955-59; high pressure technologist Ministry of Edn., Khargpur, India, 1952-53. Author: Production of Ethanol from Corn, 1979, Advances in Poetic Expressions on America, 1987, Neo-Vedic Prayer Book; contbr. 20 sci. articles to profl. jours. and 180 tech. reports. Recipient Safety at Work award U.S. EPA, 1972, Scholarship U. Mysore, 1940-44, sr. rsch. fellowship Ministry of edn., 1951, grad. assistantship G.S.I.A. Carnegie Mellon U., 1954, fellowship Northwestern U., 1953. Mem. ASTM (sec. 1990—), Am. Soc. Lubricating Engrs., Am. Chem. Soc., Am. Inst. Steel Engrs. Achievements include patent and patents pending on lubricants/scientific test instruments; design and development of recreation activity and center for aged; conducted worldwide consulting services and U.S. production of lubricants, fuels and metal working compounds. Home: 2301 S Jefferson Davis Hwy # 1026 Arlington VA 22202 Office: Dept of Def USN NAVSEA Nat Ctr # 2 Washington DC 20362

SHAMLAYE, CONRAD FRANCOIS, health facility administrator, epidemiologist; b. Victoria, Seychelles, Nov. 10, 1952; s. Francois and Marthe (Ponwaye) S.; m. Heather Elizabeth Thompson, Sept. 19, 1981; children: Catriona Michelle, Julie Anne. B Medicine and B Surgery, U. Glasgow, Scotland, 1978; MSc in Epidemiology, U. London, 1984. Medical officer Ministry Health, Victoria, Seychelles, 1979-84, epidemiologist, 1984-86, prin. sec., 1986—; exec. bd. dirs. WHO, 1990—; adv. com. Commonwealth Regional Health Community, 1990—. Founder, nat. com. mem. Red Cross Soc. Seychelles, 1990. Home: Danzil, Bel Ombre Mahe, Seychelles Office: Ministry of Health, PO Box 52, Victoria Mahe, Seychelles

SHAMMAS, NAZIH KHEIRALLAH, environmental engineering educator; b. Homs, Syria, Feb. 18, 1939; came to U.S., 1991; s. Kheirallah Hana and Nazha Murad (Hamwi) S.; m. Norma Massouh, July 28, 1968; children: Sarmed Erick, Samer Sam. Engring. degree with distinction, Am. U., Beirut, Lebanon, 1962; MS in Sanitary Engring., U. N.C., 1965; PhD in Civil Engring., U. Mich., 1971. Instr. Civil Engring. Am. U., Beirut, Lebanon, 1965-68; asst. prof. Civil Engring. Am. U., Beirut, 1972-76; teaching fellow U. Mich., Ann Arbor, 1968-71; asst. prof. Civil Engring. King Saud U., Riyadh, Saudi Arabia, 1976-78; assoc. prof. King Saud U., Riyadh, 1978-91; prof. Environ. Engring. Lenox (Mass.) Inst. Water Tech., 1991-92, dean edn., 1992—; cons., ptnr. Cons. and Rsch. Engrs., Beirut, 1973-76; advisor, cons. Riyadh Water and Sanitary Drainage Authority, 1979-83, Ar-Riyadh Devel. Authority, 1977—. Author: Environmental Sanitation, 1988, Wastewater Engineering, 1988; contbr. articles to profl. jours. Recipient block grant U. Mich., 1968-70. Mem. ASCE, Water Environ. Fedn., Am. Water Works Assn., European Water Pollution Control Assn., Internat. Assn. Water Quality, Assn. Environ. Engring. Profs. Achievements include math. modeling of nitrification process, water and wastewater mgmt. in developing countries, water conservation, wastewater reuse, appropriate tech. for developing countries. Home: 63 Leona Dr Pittsfield MA 01201 Office: Lenox Inst Water Tech 101 Yokun Ave Lenox MA 01240

SHANAHAN, THOMAS CORNELIUS, immunologist; b. Buffalo, Dec. 21, 1952; s. Thomas F. and Eileen T. (Guerin) S.; m. Kathleen McDonald, Oct. 20, 1978; 1 child, Niall Thomas. MS, SUNY, Buffalo, 1979; PhD, Temple U., 1985. Diplomate Am. Bd. Med. Lab. Immunologists. Rsch. assoc. Temple U., Phila., 1983-85; postdoctoral fellow in clin. immunology Royal Victoria Hosp./McGill U., Montreal, Quebec, Canada, 1985-86; dir. histocompatibility and immunogenetics Erie County Med. Ctr., Buffalo, 1986—. Author: (with others) Complement Deficiencies and Diseases, 1990, Histocompatibility Antigens and Hearing Disorders, 1988; contbr. articles to profl. jours. Bd. dirs. Upstate N.Y. Transplantation Svcs., Buffalo, 1987—. Mem. Am. Soc. for Histocompatibility and Immunogenetics (regional commr. 1991), Assn. Med. Lab. Immunologists. Roman Catholic. Achievements include development of new procedures in transplantation immunology; improvements in transplantation related immunological testing; elucidation of immunological responses to exptl. tumors. Office: Erie County Med Ctr 462 Grider St Buffalo NY 14215-3021

SHANBAKY, IVNA OLIVEIRA, physicist; b. Cruz das Almas, Brazil, Sept. 10, 1942; came to U.S., 1968; d. Antonio Candido and Maria Lais (Medeiros) Oliveira; m. Mohamed Mahmoud Shanbaky, Sept. 10, 1971; children: Isis M. M., Ramsey M. M. MS, Purdue U., 1971, PhD, 1975. Researcher Rsch. Inst., Bahia, 1965-68; grad. asst. Purdue U., West Lafayette, Ind., 1973-75; radiation health physicist Bur. Radiation Protection/Dept. Environ. Resources, Reading, Pa., 1982-86; radiation protection mgr. Bur. Radiation Protection/DER, Conshohocken, Pa., 1986—; coms. on regulations Bur. Radiation Protection/DER, Conshohocken, 1992. Mem. Am. Assn. Med. Physicists, Conf. of Radiation Control Program Dirs., Inc., Delaware Valley Soc. Radiation Safety, Purdue Alumni Assn. (life), Iota Sigma Pi, Sigma Xi. Moslem. Home: 413 Park Rd Downingtown PA 19335 Office: Bur Radiation Protection DER 555 North Ln Ste 6010 Conshohocken PA 19428

SHANE, WILLIAM WHITNEY, astronomer; b. Berkeley, Calif., June 3, 1928; s. Charles Donald and Mary Lea (Heger) S.; B.A., U. Calif., Berkeley, 1951, postgrad., 1953-58; Sc.D., Leiden U. (Netherlands), 1971; m. Clasina van der Molen, Apr. 22, 1964; children—Johan Jacob, Charles Donald. Research asso. Leiden (Netherlands) U., 1961-71, sr. scientist, 1971-79; prof. astronomy, dir. Astron. Inst., Cath. U. Nijmegen, Netherlands, 1979-88; guest prof. astronomy Leiden U., The Netherlands, 1988—. Served with USN, 1951-53. Mem. Internat. Astron. Union (commns. 33, 34), Am. Astron. Soc., Astron. Soc. Netherlands, Astron. Soc. of the Pacific, AAAS, Phi Beta Kappa. Research on structure and dynamics of galaxies, radio astronomy. Home: Postbus 43, 6580 AA Malden The Netherlands Office: Sterrewacht Leiden, Postbus 9513, 2300 RA Leiden The Netherlands

SHANK, CHARLES VERNON, science administrator, educator; b. Mt. Holly, N.J., July 12, 1943; s. Augustus Jacob and Lillian (Peterson) S.; m. Brenda Buckhold, June 16, 1969. BS, U. Calif., Berkeley, 1965, MS, PhD, 1969. Mem. tech. staff AT&T Bell Labs., Holmdel, N.J., 1969-76, head quantum physics and electronics dept., 1976-83, dir. Electronics Rsch. Lab., 1983-89; dir. Lawrence Berkeley Lab., faculty mem. chemistry, physics, elec. engring. and computer scis. U. Calif., Berkeley, 1989—. Numerous patents in field. Recipient E. Longstreth medal Franklin Inst., Phila., 1982, Morris E. Leeds award IEEE, 1982, David Sarnoff award IEEE, 1989. R.W. Wood prize. Fellow AAAS, Am. Phys. Soc., Optical Soc. Am. (R. W. Wood prize 1981); mem. NAS, NAE, Am. Acad. Arts and Scis. Home: 9 Ajax Pl Berkeley CA 94708-2119 Office: U Calif Lawrence Berkeley Lab Doris Pl Ofc 1 Berkeley CA 94720-0001*

SHANK, FRED ROSS, federal agency administrator; b. Harrisonburg, Va., Oct. 11, 1940; m. Peggy Anne Westbrook, June 1967; children: Virginia Anne, Fred Ross III. BS in Agriculture, U. Ky., 1962, MS in Nutrition, 1964; PhD, U. Md., 1969. Dep. dir. Office Nutrition and Food Sci. FDA, Washington, 1979-86, dir. Office Phys. Sci., 1986-87, dep. dir. Ctr. for Food SAfety and Applied Nutrition, 1987-89, dir., 1989—. Fellow Inst. Food Technologists; mem. Am. Assn. Cereal Chemists, Am. Inst. Nutrition, Am. Soc. for Clin. Nutrition, Assn. Food and Drug Ofcls. Home: 2621 Steeplechase Dr Reston VA 22091-2130 Office: FDA Ctr Food Safety and Applied Nutrition 200 C St SW Washington DC 20204

SHANKAR, VIJAYA V., aeronautical engineer. Dir. Rockwell Internat. Corp. Sci. Ctr., Thousand Oaks, Calif. Recipient Hugh L. Dryden Rsch. lectureship Am. Inst. Aeronautics and Astronautics, 1991. Office: Rockwell Intl Corp-Science Ctr POB 1085 1049 Camino Dos Rios Thousand Oaks CA 91358*

SHANKLIN, DOUGLAS RADFORD, physician; b. Camden, N.J., Nov. 25, 1930; s. John Ferguson and Muriel (Morgan) S.; student Wilson Tchrs. Coll., 1949; A.B. in Chemistry, Syracuse U., 1952; M.D., SUNY, Syracuse, 1955; m. Virginia McClure, Apr. 7, 1956; children—Elizabeth, Leigh, Lois Virginia, John Carter, Eleanor. Intern in pathology Duke U., 1955-56, resident, 1958; resident in pathology SUNY, Syracuse, 1958-60; practice medicine specializing in pathology, Gainesville, Fla., 1960-67, 78-83; mem. faculty U. Fla., 1960-67; prof. pathology, ob-gyn U. Chgo., 1967-78; pathologist-in-chief Chgo. Lying-In Hosp., 1967-78; prof., vice chmn. dept. pathology U. Tenn.-Memphis, 1983-90, prof. obstetrics, 1986—; vis. prof. U. Okla., 1967, Duke U., Mich. State U., 1969, Leeds U., Dundee U., Karolinska, 1974, Leeds U., 1978, 85, Emory U., 1980, London U., Edinburgh U., 1981, 85, U. Brit. Coll., 1987; jr. investigator Marine Biol. Lab., Woods Hole, Mass., 1951-54, sr. investigator, 1966—, mem. corp.,

1970—; parliamentarian, 1990—; mem. Marine Resources Adv. Com., 1988-90; chmn. nat. adv. com. W-I-C evaluation U.S. Dept. Agr., 1979-86; lectr. Coll. Law U. Fla., 1963-67, 77-83; cons. Pan Am. Health Orgn., 1973-89; sr. cons. Santa Fe Found., 1976-79, exec. dir., 1979-83; course dir. Center Continuing Edn., U. Chgo., 1980-82. Trustee Coll. Light Opera Co., Falmouth, Mass., 1970—, Hippodrome Theatre, Gainesville, 1975-83, Opera Memphis, 1989-92. With M.C., USNR, 1956-58. Recipient Best Basic Sci. Teaching award U. Fla., 1967; named freeman citizen of Glasgow, 1981. Fellow Royal Soc. Medicine (London); mem. AMA, AAAS, Am. Assn. Pathologists, Soc. Pediatric Research, Soc. Exptl. Pathology, Internat. Acad. Pathologists, So. Soc. Pediatric Research, So. Med. Assn., N.Y. Acad. Scis., Am. Coll. Ob-Gyn, Physicians Social Responsibility, Internat. Physicians for Prevention Nuclear War, Coll. Physicians and Surgeons Costa Rica, Tenn. Med. Assn., Pediatric Pathology Club (sec.-treas. 1970-75, pres. 1981-82), Phi Beta Kappa. Sigma Xi. Author: Syllabus for Study of Gynecologic-Obstetric-Pediatric Disease, 1961; Diseases of Woman, Pregnancy, Child, 1964; Maternal Nutrition and Child Health, 1979; Tumors of Placenta and Umbilical Cord, 1990; editor: Interscience Devel. Disorders, 1971-80; assoc. editor Jour. Reproductive Medicine, 1986-70, 79-85, editor in chief, 1970-75; contbr. articles to profl. jours. Home: 1238 NW 18th Ter Gainesville FL 32605-5370 Office: 134 Grove Park Cir Memphis TN 38117-3115

SHANKS, STEPHEN RAY, engineer, consultant; b. San Antonio, Nov. 1, 1956; s. Leroy and Jane Adams (Coats) S.; m. Vickie Lynn Morrow, Aug. 6, 1977; 1 child, Erin Monette. Student pub. schs., Corpus Christi, Tex. Engring. technician Gulf Coast Testing Lab., Inc., Corpus Christi, part-time, 1971-75, full-time, 1975-78; projects mgr., quality control adminstr. Shilstone Engring. Testing Lab. div. Profl. Service Industries, Tex. and La., 1978-86; sr. cons., quality assurance mgr., sr. project mgr.; corp. radiation safety officer Bhate Engring. Corp., 1987—. Author: Procedures and Techniques for Construction Materials Testing, 1978, Inspection and Testing of Asphaltic Concrete, 1979, Concrete Barges: Construction and Repair Techniques, 1984, Management and Marketing Strategies for Branch Offices with Rural Influences, 1985, How to Improve Profitability and Increase the Quality of Services, 1986, Quality and Process Control Systems for Federal Highway Administration Projects, 1990, Causes and Prevention of Efflorescence in Masonry Construction, 1992, It's Coming Unglued: Case Studies in the Failure of White Marble Veneers, 1992, Ultrasonic Examination and Evaluation of Tubular T-, Y-, and K-Connections, 1993; editor: (lit. mag.) Viva!, 1975; contbr. articles to profl. jours. Lay Eucharistic min. St. Francis of Assisi Episcopal Ch., Pelham, Ala. Mem. ASTM, Am. Soc. of Nondestructive Testing (dir. Birmingham sect.), Am. Concrete Inst., Am. Mgmt. Assn., Am. Welding Soc., Constrn. Specifications Inst., Constrn. Mgmt. Assn. Am. Democrat. Episcopalian. Lodge: Rotary. Home: 19 King Valley Rd Pelham AL 35124-1915 Office: 5217 5th Ave S Birmingham AL 35212-3515

SHANN, FRANK ATHOL, paediatrician; b. Melbourne, Victoria, Australia, Oct. 25, 1944; s. Frank and Enid Isabell (Wilson) S.; m. Angela Helen Mackenzie, June 30, 1980; children: Helen Mackenzie, Alice Mackenzie, Frank Mackenzie. MB, BS, U. Melbourne, 1968, MD, 1985. Jr. resident Wangaratta Hosp., Victoria, 1969; sr. resident Royal Melbourne Hosp., Victoria, 1970, med. registrar, 1971-72; house officer Nairobi (Kenya) Hosp., 1973; med. registrar Royal Children's Hosp., Victoria, 1974-76, staff specialist, 1982-84, dir. intensive care, 1986—; specialist paediatrician Goroka Hosp., Papua New Guinea, 1977-81; sr. lectr. U. Melbourne, 1985-86; mem. internat. adv. bd. The Lancet; mem. expert adv. panel on acute respiratory infections WHO. Assoc. editor Jour. Paediatrics and Child Health. Fellow Royal Australasian Coll. Physicians, Royal Soc. Tropical Medicine and Hygiene; mem. Australian Coll. Paediatrics, Australian and New Zealand Intensive Care Soc., Australian Med. Assn., British Med. Assn., Australasian Soc. Infectious Diseases. Avocations: reading, computers, tennis. Office: Royal Children's Hosp, Intensive Care Unit, Parkville Victoria 3052, Australia

SHANNON, JAMES EDWARD, water chemist, consultant; b. Fort Collins, Colo.; s. Dale E. and Arlene G. (Sims) S.; m. Anne Hagemeyer, June 2, 1962; children: Mary Shannon Shay, Nancy Anne. BS in Chemistry, Colo. State U., 1962; MS in Water Chemistry, U. Wis., 1965. From project leader to tech. dir. NALCO Chem. Co., Naperville, 1970-84, from internat. mgr. to mkt. devel. mgr., 1984-92; pres. Global SOS, Naperville, 1992—; water chemistry presenter on cooling water systems worldwide; toured seven continents. Capt. U.S. Army, 1965-67. Mem. Am. Chem. Soc., Am. Water Works Assn., Cooling Tower Inst., Sigma Xi. Achievements include U.S. patent for filtration device. Home and Office: Global SOS Inc 709 Buttonwood Cir Naperville IL 60540-6308

SHANNON, LARRY JAMES, civil engineer; b. Mansfield, Ohio, Mar. 27, 1948; s. James Kenneth and Elizabeth (Fisher) S.; m. Cathy Lee Bullert, June 6, 1970; children: Penny Kay, Charissa Lynn, Heather Renee. Student, Ohio State U., 1966-68, 71. Registered profl. engr., Ohio. Engr. technician Richland Engring. Ltd., Mansfield, 1972-80, staff engr., 1980-85; project engr. K.E. McCartney and Assocs., Mansfield, 1985-89; geometrics engr. Dept. Transp., State of Ohio, Columbus, 1989—. Contbr. (reference manual) Location and Design Manual for State of Ohio, 1991. Chmn. MathCounts Com. of North Cen. Chpt., Ohio Soc. Profl. Engrs., 1985-89. With U.S. Army, 1969-71. Mem. NSPE (chpt. pres. 1987-88), ASCE. Republican. Baptist. Office: Ohio Dept Transp 25 S Front St Columbus OH 43215

SHANNON, ROBERT RENNIE, optical sciences center administrator, educator; b. Mt. Vernon, N.Y., Oct. 3, 1932; s. Howard A. and Harriebell (Rennie) S.; m. Helen Lang, Feb. 13, 1954; children: Elizabeth, Barbara, Jennifer, Amy, John, Robert. B.S., U. Rochester, 1951, M.A., 1957. Dir. Optics Lab., ITEK Corp., Lexington, Mass., 1959-69; prof. Optical Scis. Ctr., U. Ariz., 1969—, dir., 1983-92, prof. emeritus, 1992—; cons. Lawrence Livermore Lab., 1980-90; mem. commn. next generation currency, 1992—; trustee Aerospace Corp., 1985—; mem. Air Force Sci. Adv. Bd., 1986-90; mem. NRC Commn. on Next Generation Currency, 1992—, mem. com. on def. space tech. Air Force Studies Bd., 1989—, Hubble Telescope recovery panel, 1990; bd. dirs. Precision Optics Corp., Schott Glass Techs. Editor: Applied Optics and Optical Engineering, Vol. 7, 1980, Vol. 8, 1981, Vol. 9, 1983, Vol. 10, 1987, Vol. 11, 1992. Fellow Optical Soc. Am. (pres. 1985, mem. engring. coun. 1989-91), Soc. Photo-Optical Instrumentation Engrs. (pres. 1979-80, recipient Goddard award 1982); mem. NAE, Tucson Soaring Club (past pres.), Sigma Xi. Home: 7040 E Taos Pl Tucson AZ 85715-3344 Office: U Ariz Optical Scis Ctr Tucson AZ 85721

SHANTEAU, JAMES, psychology educator, researcher; b. Glendale, Calif., May 11, 1943; s. Clairborne O. and Helen (Wolf) s.; m. Doreen Spotts, Sept. 3, 1966; children: Karen Louise, David James, Jill Kathleen. BA, San Jose State U., 1966; PhD, U. Calif., San Diego, 1970. Postdoctoral fellow U. Mich., Ann Arbor, 1970-71; asst. prof. Kans. State U., Manhattan, 1971-75, assoc. prof., 1975-80, prof., 1980—; vis. rsch. assoc. U. Colo., Boulder, 1979-80; vis. prof. Cornell U., Ithaca, N.Y., 1976; program dir. NSF, Washington, 1989-90; fed. grant coord. Kans. State U., Manhattan, 1992—, interim assoc. vice provost for rsch., 1993—. Author: Concepts in Judgement and Decision Research, 1981; editor: Organ Donation and Transplantation, 1990; contbr. articles to profl. jours. Leader Woodwinds Anonymous Music Group, Manhattan, Kans., 1976—. Recipient Contract award Army Rsch. Inst., 1980, Conf. award NIH, 1988; fellow NIH, 1967-80; grantee NSF, 1975, EPA, 1993, NIH, 1972, 74, 78, 88. Fellow APA, Am. Psychol. Soc. (rep. 1989—); mem. Judgment Decision Making Soc. (founder, pres. 1986-87), Assn. Consumer Rsch., Psychonomic Soc., Math. Psychol. Soc., Soc. for Consumer Psychology. Achievements include development of procedure identifying expert decision makers, computer program to conduct functional measurement analyses; identification of factors influencing decision to donate organs; research on influence of farmer's crop insurance, training procedures to improve nursing decision making, systems of organizational behavior and human decision making. Office: Kans State U Bluemont Hall Manhattan KS 66506

SHANTZ, CAROLYN UHLINGER, psychology educator; b. Kalamazoo, Mich., May 19, 1935; d. James Roland and Gladys Irene (Jerrett) Uhlinger; m. David Ward Shantz, Aug. 17, 1963; children: Catherine Ann, Cynthia Anne. BA, DePauw U., 1957; MA, Purdue U., 1959, PhD, 1966. Rsch. assoc. Merrill-Palmer Inst., Detroit, 1965-71; prof. Wayne State U., Detroit, 1971—; com. mem. grant rev. panel NIMH, NIH, Washington, 1979-81, 84-86; reviewer grant proposals NSF, Washington, 1978—; cons. Random House, Knopf, Guilford, others. Editor Merrill-Palmer Quar., 1981—; contbr. articles to profl. jours. Rsch. grantee NSF, NICHHD, OEO, Edn., Spencer Found., 1966-89. Fellow Am. Psychol. Assn. (pres. div. on devel. psychology 1983-84), Am. Psychol. Soc.; mem. Soc. for Rsch. in Child Devel., Sigma Xi, Phi Beta Kappa. Office: Wayne State U Dept Psychology Detroit MI 48202

SHAO, WENJIE, librarian; b. Beijing, May 4, 1931; s. Shangming and Junhua (Jiang) S.; m. Wenxian Teng, Aug. 14, 1957; 1 child, Lei. Grad., Nat. Tsinghua U., 1952; postgrad., Inst. Obrobki Skrawaniem, Poland, 1965-67. Engr. No. 1 Motor Vehicle Plant, Changchun, Peoples Republic of China, 1952-56; lectr. Machinery Industry Sch., Shenyang, Peoples Republic of China, 1956-58, Inst. Mech. Engring., Beijing, 1959-78; head acquisition com. Nat. Libr. China, Beijing, 1979-87, dep. dir., rsch. libr., 1988—; head editorial bd. Chinese Libr. Book Classification, Beijing, 1988-92; assoc. dir. China Nat. Tech. Com. Standardization for Info. and Documentation, Beijing, 1988-92. Contbr. articles to Bull. Nat. Libr. China, China Soc. Libr. Sci. Bull., UAP in China, UBC of Chinese Documentation, PAC of Chinese Librs. Office: Nat Libr China, 39 Baishiqiao Rd, Beijing 100081, China

SHAPERO, HARRIS JOEL, pediatrician; b. Winona, Minn., Nov. 22, 1930; s. Charles and Minnie Sara (Ehrlichman) S.; m. Byong Soon Yu, Nov. 6, 1983; children by previous marriage: Laura, Bradley, James, Charles. A.A., UCLA, 1953; B.S., Northwestern U., 1954, M.D., 1957. Diplomate and cert. specialist occupational medicine Am. Bd. Preventive Medicine; cert. aviation medicine FAA. Intern, Los Angeles County Harbor Gen. Hosp., 1957-58, resident in pediatrics, 1958-60, staff physician, 1960-64; attending physician Perceptually Handicapped Children's Clinic, 1960-63; disease control officer for tuberculosis, L.A. County Health Dept., 1962-64; pvt. practice medicine specializing in pediatrics and occupational medicine, Cypress, Calif., 1965-85; pediatric cons. L.A. Health Dept., 1963-85, disease control officer sexually transmitted diseases, 1984-85; emergency room dir. AMI, Anaheim, Calif., 1968-78; mem. med. staff Anaheim Gen. Hosp., Beach Community Hosp., Norwalk Community Hosp.; courtesy staff Palm Harbor Gen. Hosp., Bellflower City Hosp.; pediatric staff Hosp. de General, Ensenada, Mex., 1978—; primary care clinician Sacramento County Health, 1987-88; pvt. practice medico-legal evaluation, 1986-92; founder Calif. Legal Evaluation Med. Group; apptd. med. examiner in preventive and occupational medicine State of Calif. Dept. of Indsl. Rels., 1989; health care provider, advisor City of Anaheim, City of Buena Park, City of Cypress, City of Garden Grove, Cypress Sch. Dist., Magnolia Sch. Dist., Savanna Sch. Dist., Anaheim Unified Sch. Dist., Orange County Dept. Edn.; pediatric and tuberculosis cons. numerous other orgns.; FAA med. examiner, founder Pan Am. Childrens Mission. Author: The Silent Epidemic, 1979. Named Headliner in Medicine Orange County Press Club, 1978. Fellow Am. Coll. Preventive Medicine; mem. L.A. County Med. Assn., L.A. County Indsl. Med. Assn., Am. Pub. Health Assn., Mex.-Am. Border Health Assn. Republican. Jewish. Avocations: antique books and manuscripts, photography, graphics, beekeeper. Home: PO Box 228 Wilton CA 95693-0228

SHAPERO, SANFORD MARVIN, hospital executive, rabbi; b. Cin., Mar. 4, 1929; s. David Theodore and Leah Freda (Adler) S.; m. Evelyn Leavitt, Jan. 24, 1982. BA, U. Dayton, 1950; BHL, Hebrew Union Coll., 1952, MHL, 1955, DHL, 1959, DD (hon.), 1981. Rabbi Elmira, N.Y., 1957-59, Bridgeport, Conn., 1959-64, Beverly Hills, Calif., 1964-68; exec. v.p. Alliance Med. Industries, Straford, Conn., 1968-69; v.p. N.Am. Biologicals, Miami, 1969-72; regional dir., nat. dir. gerontology Union Am. Hebrew Coll., Miami, 1972-79; pres., chief exec. officer City of Hope, L.A., 1979—. Contbr. articles to profl. jours. Lt. (j.g.) USNR, 1955-57. Mem. CCAR, Nat. Adv. Coun. on Aging, Rotary (bd. dirs. 1987—). Avocations: private flying, anthropology, travel, classical music, biking. Office: Hope City Beckman Rsch Inst Div Neuroscis 1450 E Duarte Rd Duarte CA 91010-3011*

SHAPIRO, BENNETT MICHAELS, biochemist, educator; b. Phila., July 14, 1939; s. Simon and Sara (Michaels) S.; m. Fredericka Foster, Mar. 13, 1982; children: Lisa, Lise, Jonathan. BS, Dickinson Coll., 1960; MD, Jefferson Med. Coll., 1964. Research assoc. NHLI, NIH, 1965-68, med. officer, 1970-71; vis. scientist Inst. Pasteur, Paris, 1968-70; from assoc. prof. to full prof. biochemistry U. Wash., 1971-90, chmn. biochemistry dept., 1985-90; exec. v.p. for worldwide basic rsch. Merck Rsch. Labs., Rahway, N.J., 1990—. Contbr. articles to profl. jour. Served as surgeon USPHS, 1968-70. John S. Guggenheim fellow, 1982; Japan Soc. for Promotion Sci., 1984. Mem. Am. Soc. Biol. Chemists, Am. Soc. Cell Biology, Am. Soc. Devel. Biology, Phi Beta Kappa, Alpha Omega Alpha. Office: Merck Rsch Labs PO Box 2000 Rahway NJ 07065-0900

SHAPIRO, IRWIN IRA, physicist, educator; b. N.Y.C., N.Y., Oct. 10, 1929; s. Samuel and Esther (Feinberg) S.; m. Marian Helen Kaplun, Dec. 20, 1959; children: Steven, Nancy. A.B., Cornell U., 1950; A.M., Harvard U., 1951, Ph.D., 1955. Mem. staff Lincoln Lab. MIT, Lexington, 1954-70; Sherman Fairchild Distinguished scholar Calif. Inst. Tech., 1974; Morris Loeb lectr. physics Harvard, 1975; prof. geophysics and physics MIT, 1967-80, Schlumberger prof., 1980-84; Paine prof. practical astronomy, prof. physics Harvard U., 1982—; sr. scientist Smithsonian Astrophys. Obs., 1982—; dir. Harvard-Smithsonian Ctr. for Astrophysics, 1983—; cons. NSF, NASA. Contbr. articles to profl. jours. Recipient Albert A. Michelson medal Franklin Inst., 1975, award in phys. and math. scis. N.Y. Acad. Scis., 1982; Guggenheim fellow, 1982. Fellow AAAS, Am. Geophys. Union (Charles A. Whitten medal 1991, William Bowie medal 1993), Am. Phys. Soc.; mem. Am. Acad. Arts and Scis., Nat. Acad. Scis. (Benjamin Aprhorp Gould prize 1979), Am. Astron. Soc. (Dannie Heineman award 1983, Dirk Brouwer award 1987), Internat. Astron. Union, Phi Beta Kappa, Sigma Xi, Phi Kappa Phi. Home: 17 Lantern Ln Lexington MA 02173-6029 Office: Harvard U 60 Garden St Cambridge MA 02138-1596*

SHAPIRO, ISADORE, materials scientist, consultant; b. Mpls., Apr. 25, 1916; s. Jacob and Bessie (Goldman) S.; m. Mae Hirsch, Sept. 4, 1938; children: Stanley Harris, Jerald Steven. BChemE. summa cum laude, U. Minn., 1938, PhD, 1944. Asst. instr. chemistry U. Minn., 1938-41, rsch. fellow, 1944-45; rsch. chemist E. I. duPont de Nemours and Co., Phila., 1946; head chem. lab. U.S. Naval Ordnance Test Sta., Pasadena, Calif., 1947-52; dir. rsch. lab. Olin-Mathieson Chem. Corp., 1952-59; head chemistry Hughes Tool Co., Aircraft div., Culver City, Calif., 1959-62; pres. Universal Chem. Systems Inc. 1962—, Aerospace Chem. Systems, Inc., 1964-66; dir. contract rsch. HITCO, Gardena, Calif., 1966-67; prin. scientist Douglas Aircraft Co. of McDonnell Douglas Corp., Santa Monica, Calif., 1967; prin. scientist McDonnell Douglas Astronautics Co., 1967-70; head materials and processes AiResearch Mfg. Co., Torrance, Calif., 1971-82, cons., 1982—; inaugurated dep. gov. Am. Biog. Inst. Rsch. Assn., 1988; dep. dir. gen. Internat. Biog. Ctr., 1989, Eng. Rater U.S. Civil Svc. Bd. Exam., 1948-52. Served 1st lt. AUS, 1941-44. Registered profl. engr., Calif. Fellow Am. Inst. Chemists, Am. Inst. Aeros and Astronautics (assoc.); mem. AAAS, Am. Ordnance Assn., Am. Chem. Soc., Soc. Rheology, Soc. Advancement Materials and Process Engring., Am. Inst. Physics, AIM, Am. Phys. Soc., N.Y. Acad. Sci., Am. Assn. Contamination Control, Am. Ceramic Soc., Nat. Inst. Ceramic Engrs., Am. Powder Metallurgy Inst., Internat. Plansee Soc. for Powder Metallurgy, Sigma Xi, Tau Beta Pi, Phi Lambda Upsilon. Author articles in tech. publs. Patentee, discoverer series of carborane compounds; created term carborane; formulator of universal compaction equation for powders (metals, ceramics, polymers, chemicals). Home: 5624 W 62d St Los Angeles CA 90056

SHAPIRO, JAMES A., bacterial geneticist, educator; b. Chgo., May 18, 1943; m. Joan E. Shapiro, June 14, 1964; children: Jacob N., Danielle E. BA in English magna cum laude, Harvard U., 1964; PhD in Genetics, Cambridge U., 1968. Postdoctoral fellow Svc. de Genetique Cellulaire, Inst. Pasteur, Paris, 1967-68; rsch. fellow dept. bacteriology and immunology Harvard Med. Sch., 1968-70; invited prof. dept. genetics Sch. Biol. Scis., U. Havana, Cuba, 1970-72; rsch. assoc. Rosenstiel Basic Med. Scis. Rsch. Ctr., Brandeis U., Waltham, Mass., 1972-73; asst. prof. dept. microbiology U. Chgo., 1973-78, assoc. prof. dept. microbiology, 1978-82, prof. dept. microbiology, 1982-84, prof. microbiology dept. molecular genetics and cell biology, 1984-85, prof. microbiology dept. biochemistry and molecular biology, 1985—; vis. prof. dept. microbiology Tel Aviv U., 1980; chmn. midwest regional selection com. Marshall Scholarship, 1991—. Author: Mobile Genetic Elements, 1983, (with others) DNA Insertion Elements, Plasmids and Episomes, 1977; mem. editorial bd. Jour. Bacteriology, 1976-83, 86-88, Enzyme and Microbial Tech., 1981-88, Biotechnology (series), 1981-88, FEMS Microbiological Reviews, 1985-91; contbr. articles to profl. jours. Bd. dirs. capital com. KAM-Isaiah Israel Congregation, 1990—. Recipient Marshall scholarship, 1966, Wellcome Rsch. Tng. scholarship, 1966-67, Jane Coffin Childs fellowship, 1967-70. Mem. Am. Soc. Microbiology, Soc. Gen. Microbiology, Genetics Soc. Am., Genetical Soc., Am. Soc. Biochemistry and Molecular Biology, Phi Beta Kappa. Achievements include discovery of bacterial insertion sequences; first purification of defined genetic segment; research in molecular mechanism of transposition; in regulated DNA rearrangements in bacterial populations; in pattern formation in bacterial colonies. Office: U Chgo 920 E 58th St Chicago IL 60637

SHAPIRO, LYNDA P., biology educator, director; b. Bklyn., June 11, 1938; d. Nathan and Evelyn (Zetler) S. BA, U. Ark., 1960, MS, 1963; postgrad., Columbia U., 1966-67; PhD, Duke U., 1974. Instr. biology La. State U., 1963-66; rsch. teaching asst. Columbia U., N.Y.C., 1967-68; tech. assoc. med. ctr. Duke U., 1968-69; postdoctoral investigator Woods Hole Oceanographic Inst., 1975-75, asst. scientist, 1975-79; rsch. scientist Bigelow Lab. for Ocean Scis., 1979-90; prof. biology, dir. Inst. Marine Biology U. Oregon, Charleston, 1990—; commr. South Slough Nat. Estuarine Rsch. Reserve, 1990—; invited speaker various workshops, symposiums. Contbr. articles to profl. jours. Woodrow Wilson fellow. Mem. AAAS, PSA, TOS, Am. Soc. Limnology and Oceanography, Phi Sigma. Achievements include research in ecology, biogeography and physiology of marine phytoplankton; distributions of ultraphytoplankton species, effects of environmental factors on those distributions, and effects of those distributions on the various marine environments; predation on ultraplankton by pelagic protists. Office: Univ Oreg Oreg Inst Marine Biology Charleston OR 97420

SHAPIRO, MAURICE MANDEL, astrophysicist; b. Jerusalem, Israel, Nov. 13, 1915; came to U.S., 1921; s. Asher and Miriam R. (Grunbaum) S.; m. Inez Weinfield, Feb. 8, 1942 (dec. Oct. 1964); children: Joel Nevin, Elana Shapiro Ashley Naktin, Raquel Tamar Shapiro Kislinger. B.S., U. Chgo., 1936, M.S., 1940, Ph.D., 1942. Instr. physics and math. Chgo. City Colls., 1937-41; chmn. dept. phys. and biol. scis. Austin Coll., 1938-41; instr. math. Gary Coll., 1942; physicist Dept. Navy, 1942-44; lectr. physics and math. George Washington U., 1943-44; group leader, mem. coordinating council of lab. Los Alamos Sci. Lab., U. Calif., 1944-46; sr. physicist, lectr. Oak Ridge Nat. Lab., Union Carbon and Carbide Corp., 1946-49; cons. div. nuclear energy for propulsion aircraft Fairchild Engine & Aircraft Corp., 1948-49; head cosmic ray br. nucleonics div. U.S. Naval Research Lab., Washington, 1949-65, supt. nucleonics div., 1953-65, chief scientist Lab. for Cosmic Ray Physics, 1965-82, apptd. to chair of cosmic ray physics, 1966-82, chief scientist emeritus, 1982—; lectr. U. Md., 1949-50, 52—, assoc. prof., 1950-51, vis. prof. physics and astronomy, 1986—; vis. prof. physics and astronomy U. Iowa, 1981-84; vis. prof. astrophysics U. Bonn, 1982-84; vis. scientist Max Planck Inst. für Astrophysik, W. Ger., 1984-85; cons. Argonne Nat. Lab., 1949; cons. panel on cosmic rays U.S. nat. com. IGY; lectr. physics and engring. Nuclear Products-Erco div. ACF Industries, Inc., 1956-58; lectr. E. Fermi Internat. Sch. Physics, Varenna, Italy, 1962; vis. prof. Weizmann Inst. Sci., Rehovoth, Israel, 1962-63, Inst. Math. Scis., Madras, India, 1971; Inst. Astronomy and Geophysics Nat. U. Mex., 1976; vis. prof. physics and astronomy Northwestern U., Evanston, Ill., 1978; cons. space research in astronomy Space Sci. Bd., Nat. Acad. Scis., 1965; cons. Office Space Scis., NASA, 1965-66, 89; prin. investigator Gemini S-9 Cosmic Ray Expts., NASA, 1964-69, Skylab, 1967-76, Long Duration Exposure Facility, 1977—; mem. Groupe de Travail de Biologie Spatiale, Council of Europe, 1970—; mem. steering com. DUMAND Consortium, 1976—, mem. exec. com., 1979-82, mem. sci. adv. com., 1982—; lectr. Summer Space Inst., Deutsche Physikalische Gesellschaft, 1972; dir. Internat. Sch. Cosmic-Ray Astrophysics, Ettore Majorana Centre Sci. Culture, Erice, Italy, 1977—, also sr. corr., 1977—; chmn. U.S. IGY com. on interdisciplinary research, mem. nuclear emulsion panel space sci. bd.; Nat. Acad. Scis., 1959—; chief U.S. rep., steering com. Internat. Coop. Emulsion Flights for Cosmic Ray Research; cons. CREI Atomics, 1959—; vis. com. Bartol Research Found., Franklin Inst., 1967-74; mem. U.S. organizing com. 13th and 19th Internat. Confs. on Cosmic Rays; mem. sci. adv. com. Internat. Confs. on Nuclear Photography and Solid State Detectors, 1966—; mem. Com. of Honor for Einstein Centennial, Acad. Naz. Lincei, 1977; mem. Internat. Organizing com. Tex. Symposia on Relativistic Astrophysics, 1976—; Regents lectr. U. Calif. Riverside, 1985; Edison lectr. Naval Rsch. Lab award, 1990. Mem. editorial bd. Astrophysics and Space Sci., 1968-75; assoc. editor: Phys. Rev. Letters, 1977-84; editor (NATO) ASI Series on Cosmic-Ray Astrophysics; contbr. to Am. Inst. Handbook of Physics, various encys. Mem. exec. bd. Cong. Beth Chai, Washington, 1987—; trustee Nat. Capital Astronomers, Washington, 1989—; mem. internat. panel Chernobyl World Lab., 1988. Recipient Disting. Civilian Svc. award Dept. Navy, 1967, medal of honor Soc. for Encouragement au Progrès, 1978, publs. award Naval Rsch. Lab., 1970, 74, 76, Dir.'s Spl. award, 1974, sr. U.S. Scientist award Alexander von Humboldt Found., 1982, Profl. Achievement Citation U. Chgo., 1992, Disting. Career in Sci. award Washington Acad. Scis., 1993. Fellow Am. Phys. Soc. (chmn. organizing com. div. cosmic physics, chmn. 1971-72, com. on publs. 1977-79), AAAS, Washington Acad. Scis. (past com. chmn., Disting. Career in Scis. award, 1993); mem. Am. Astron. Soc. (exec. com. div. high-energy astrophysics 1978—, chmn. 1982), Philos. Soc. Washington (past pres.), Am. Technion Soc. (Washington bd.), Assn. Los Alamos Scientists (past chmn.), Assn. Oak Ridge Engrs. and Scientists (past chmn.), Fedn. Am. Scientists (past mem. exec. com., nat. council), Internat. Astron. Union (organizing com. commn. on high energy astrophysics), Phi Beta Kappa, Sigma Xi (Edison lectr. 1990). Club: Cosmos (Washington). Achievements include patents in field; discovery of first definitive evidence for production of cosmic ray secondaries in the interstellar medium; research in cosmic radiation, composition, origin, propagation, and nuclear transformations; in high-energy astrophysics, in particles and fields, in nuclear physics, neutron physics and fission reactors; in hydrodynamics and gamma-ray and neutrino astronomy. Office: 205 S Yoakum Pky Ste 1514 Alexandria VA 22304-3838

SHAPIRO, MICHAEL HAROLD, health care executive, consultant, publisher; b. Moscow, May 29, 1949; came to U.S., 1974; s. Oscar Shapiro and Betty Karolik; m. Mitra Roshodesh, Dec. 29, 1984; children: Daniel, Emilia. MS in Phys. Chemistry, SUNY, Buffalo, 1978; MBA in Econs., NYU, 1985. Chemist Hooker Chems., Niagara Falls, N.Y., 1974-78; tech. forecaster Brown & Williamson, Louisville, 1978-80; bus. mgr. Humana, Louisville, 1980-81, Technicom/Revlon, Tarrytown, N.Y., 1981-84; esxec. Robert S. First, Inc., White Plains, N.Y., 1984-85; pres. Venture Planning Group Inc., N.Y.C., 1985—. Editor, pub. (industry report series) The Top 10 World's Leading Companies, The Infectious Disease Testing Market: U.S., Europe and Japan, Emerging Opportunities in Cancer Testing, DNA Probes, Biosensors, and others. Office: Venture Planning Group Inc 350 5th Ave Ste 3308 New York NY 10118-0069

SHAPIRO, MURRAY, structural engineer; b. N.Y.C., July 5, 1925; s. Samuel and Fannie (Korman) S.; m. Florence Morrison, June 16, 1951; children: Fred Richard, Alan Neil. BCE, CCNY, 1947. Registered profl. engr., N.Y., N.J., Pa., Md., Ga., N.C., Mass., Conn. Steel detailer Knopf & Amron, N.Y.C., 1947-48; asst. engr. N.Y.C. Bd. Transp., 1948-50; designer James Ruderman cons. engrs., N.Y.C., 1950-53, sr. engr., 1953-58, assoc., 1958-65; jr. ptnr. Office of James Ruderman, N.Y.C., 1965-66, sr. ptnr., 1966—. Structural designer many highrise office buildings, including GM Bldg., N.Y.C., Pan Am Bldg., N.Y.C., also schs., apartment houses, theaters. With U.S. Army, 1943-45, ETO. Decorated Purple Heart, Bronze Star. Mem. N.Y. Cons. Engrs. Assn. (trustee 1972-77, sec. 1974-76), Am. Concrete Inst., N.Y. Acad. Scis., Glen Head Country Club. Republican. Jewish. Home: 60 Fern Dr East Hills NY 11576 Office: Office of James Ruderman 15 W 36th St New York NY 10018

SHAPIRO, NATHAN, acoustical engineer, retired; b. Worcester, Mass., Mar. 25, 1915; s. Menachem Mendel and Emma (Rudashevski) S.; m. Rose Edythe Turbow, Jan. 10, 1943; children: Matthew, Joel, David, Lainie. AB, Clark U., 1939; MS, Cath. U. Am., 1947. Physicist Nat. Bur. Stds., Washington, 1941-46, David Taylor Model Basin, Carderock, Md., 1946-49, Naval Ordnance Lab., Silver Spring, Md., 1949-52, Armour Rsch. Found., Chgo.,

1952-55; rsch. engr., group leader Lockheed-Calif. Co., Exterior Noise Group, Burbank, 1955-85. Fellow AIAA (assoc.), Acoustical Soc. Am.; mem. Aerospace Industries Assn. (airplane noise control com. 1972-85, chmn. 1980), Soc. Automotive Engrs. (aircraft noise com. 1964-85), Inst. Noise Control Engring., Sigma Xi. Achievements include development of acoustical design of Lockheed L-1011 Tristar. Home: 1407 Chautauqua Blvd Pacific Palisades CA 90272

SHAPIRO, ROBERT M., electronics company executive; b. San Diego, June 13, 1945; s. Oscar J. and Mary (Schneider) S.; m. Nancy J. Sattinger, July 1, 1966 (div.); children: Scott H., Todd M.; m. Judith Ann Gable, Aug. 22, 1975. BA magna cum laude, U. San Diego, 1967. Mktg. rep. Proctor and Gamble, Riverside, Calif., 1967-68; adminstrv. ops. mgr. IBM, Riverside, 1968-71; adminstrn. mgr. IBM Co., Oakland, Calif., 1971-73; regional mgr. IBM Co., Detroit, 1973-75; fin. mgr. IBM Co., Franklin Lakes, N.J., 1975-81; mgr. resource IBM Co., Rye Brook, N.Y., 1981-84; dir. human resources Prodigy Services Co. (partnership IBM and Sears), White Plains, N.Y., 1984-87; v.p. market rels. and mgmt. svcs., officer Prodigy Svcs. Co. (partnership IBM and Sears), White Plains, N.Y., 1987-91; sr. v.p. comml. mktg. Prodigy Svcs. Co., White Plains, 1991—; mem. The Conf. Bd., N.Y.C.; officer Prodigy Svc. Co.; lectr. sch. bus. Stanford U., grad. sch. Northwestern U.; spl. project worker Northwestern U., Harvard U. Patron Met. Opera. Mem. Am. Mgmt. Assn. Avocations: music, photography. Office: Prodigy Svcs Co IBM/Sears Ptnrs 445 Hamilton Ave White Plains NY 10601-1814

SHAPIRO, SANDOR SOLOMON, hematologist; b. Bklyn., July 26, 1933. BA, Harvard U., 1954, MD, 1957. Intern Harvard med. svc. Boston City Hosp., 1957-58, asst. resident, 1960-61; asst. surgeon divsn. biol. std. NIH, USPHS, 1958-60; NIH spl. fellow MIT, 1961-64; from instr. to assoc. prof. Cardeza found. Jefferson Med. Coll., Phila., 1964-72, prof. medicine, 1972—, assoc. dir., 1978-85, dir., 1985—; mem. hematology study sect. NIH, 1972-76, 78-79; mem. med. adv. coun. Nat. Hemophilia Found., 1973-75; chmn. Pa. State Hemophilia Adv. Com., 1974-76. Mem. Am. Soc. Clin. Investigation, Am. Soc. Hematology, Am. Assn. Immunologists, Am. Assn. Physicians, Internat. Soc. Thrombosis and Hemostasis. Achievements include research in hemostasis and thrombosis, prothrombin metabolism, hemophilia, lupus anticoagulants, endothelial cells. Office: Thomas Jefferson U Cardeza Found Hematologic Rsch 1015 Walnut St Philadelphia PA 19107*

SHAPIRO, THEODORE, chemical engineer; b. Boston, Feb. 17, 1923; s. Philip and Sadie (Diamond) S.; m. Selma Gertrude Kravit, Oct. 3, 1945; children: Sandra, Susan, Rhonda, Philip. BA, Boston U., 1944. Registered profl. engr., Tenn. Rsch. engr. Carbide & Carbon Chem. Co., N.Y.C., 1944-45; ops. supr. Carbide & Carbon Chem. Co., Oak Ridge, Tenn., 1945-48, process engr., 1948-52; process/project engr. nuclear div. Union Carbide, Oak Ridge, 1952-87; prin. engr. Martin Marietta Energy Systems, Oak Ridge, 1987-91; assoc. task leader, project mgr. PAI Co., Oak Ridge, 1991—. Contbr. articles on decontamination of process equipment and recovery of enriched uranium to profl. publs. Pres. Highland Rim coun. Girl Scouts U.S.A., 1960, Oak Ridge chpt. Tenn. Soc. Profl. Engrs., 1975-76, United Fund of Anderson County, Tenn., 1976. Recipient Letter of Commendation, Westinghouse Savannah River Plant for design and fabrication uranium solidification facility, 1991, Martin Marietta Energy Systems, 1991. Mem. Am. Soc. Chem. Engrs., Nat. Soc. Profl. Engrs. (bd. govs. 1989-91, del.), Profl. Engrs. in Industry (state chmn. 1976-77). Achievements include patent for design of double-wall heat exchanger for hazardous gas service; development of technology for enriched uranium processing technology. Home: 435 East Dr Oak Ridge TN 37830

SHAPPIRIO, DAVID GORDON, biologist, educator; b. Washington, June 18, 1930; s. Sol and Rebecca (Porton) S.; m. Elvera M. Bamber, July 8, 1953; children: Susan, Mark. B.S. with distinction in Chemistry, U. Mich., 1951; A.M., Harvard U., 1953, Ph.D. in Biology, 1955. NSF postdoctoral fellow in biochemistry Cambridge U., Eng., 1955-56; research fellow in physiology Am. Cancer Soc.-NRC, U. Louvain, Belgium, 1956-57; mem. faculty U. Mich., Ann Arbor, 1957—, prof. zool. and biology, 1967—, Arthur F. Thurnau prof., 1989—; assoc. chmn. div. biol. scis. U. Mich., 1976-83, acting chmn., 1978, 79, 80, 82, 83, coordinator NSF undergrad. sci. edn. program, 1962-67, dir. honors program Coll. Lit. Sci. and Arts, 1983-91; vis. lectr. Am. Inst. Biol. Scis., 1966-68; cons. on textbook devel.; reviewer grant proposals and papers for publ. in profl. jours. Author, editor, research on biochemistry and physiology growth, devel., dormancy; invited speaker, rsch. symposia of nat. and internat. orgns. in field. Recipient Disting. Teaching award U. Mich., 1967, Excellence in Edn. award, 1991, Bausch & Lomb Sci. award, 1974; Lalor Found fellow, 1953-55; Danforth Found. Fellow AAAS; mem. Am. Inst. Biol. Scis. (vis. lectr. 1966-68), Am. Soc. Cell Biology, Biochem. Soc., Entomol. Soc. Am., Am. Soc. Zoologists, Royal Entomol. Soc., Lepidopterists Soc., Soc. Exptl. Biology, Soc. Gen. Physiologists, Assn. Biol. Lab. Edn., Xerces Soc., Phi Beta Kappa. Office: U Mich Dept Biology 3065 Natural Sci Bldg Ann Arbor MI 48109-1048

SHARBAUGH, W(ILLIAM) JAMES, plastics engineer, consultant; b. Pitts., Apr. 13, 1914; s. Oliver Michael and Sarah Marie (Wingenroth) S.; m. Eileen Carey, May 14, 1938; children: William James Jr., Eileen Sharbaugh Pinkerton, Susan Sharbaugh Coté. BS in Engring., Carnegie Inst. Tech., 1935. Project engr. MSA Corp., Pitts., 1935-46; founder, gen. mgr. ENPRO, Inc., St. Louis, 1947-62; mgr. plastics div. Vulcan Rubber and Plastic, Morrisville, Pa., 1962-67; v.p. engring. and mfg. FESCO div. Celanese, Pitts., 1967-72; exec. v.p. plastics div. Lenox, Inc., St. Louis, 1970-72; div. mgr. Crown Zellerbach, Inc., San Francisco, 1972-77; pres. Plastics Assocs., Cons., Newport Beach, Calif., 1977—; founder ISOBET USA, Inc., Newport Beach; dir. devel. and tech. Crown Zellerbach, Inc.; pres. Western Plastics Pioneers; cons. nat. and internat. plastics cos. Author tech. papers, reports. Mem. Soc. of Plastics Engrs. (first pres. Pitts. sect.), Soc. Plastics Industry (profl.). Republican. Roman Catholic. Achievements include patents for military products, consumer items and the development of ISOBET construction materials. Home: 1516 Seacrest Dr Corona del Mar CA 92625 Office: 4400 MacArthur Blvd # 500 Newport Beach CA 92660

SHAREEF, IQBAL, mechanical engineer, educator; b. Hyd, India, Jan. 8, 1954; came to the U.S., 1978; s. Omar Shareef and Jeelani Begum; m. Maliha Jabeen, June 15, 1986; children: Farah, Sarah. PhD, Ill. Inst. Tech., 1983. Registered profl. engr., Ill. NASA/ODU fellow Old Dominion U., Norfolk, Va., 1978-79; rsch. asst., instr. Ill. Inst. Tech., Chgo., 1979-83, postdoctoral rsch. assoc., 1983-84; asst. prof. Bradley U., Peoria, Ill., 1984-89, assoc. prof., 1989—; manuscript reviewer ASME, SME, ASM, Addison Wesley, 1984—. Contbr. articles to profl. jours. Fundraiser numerous orgns., 1986—. Mem. AAES, ASME, ASEE, SME, ASM, NCEES, SEM, ISA, TMS, Sigma Xi, Tau Beta Pi. Home: 1420 W Glen # 410 Peoria IL 61614 Office: Bradley U Dept Mfg 1308 Bradley Ave Peoria IL 61625

SHARIFF, ASGHAR J., geologist; b. Haft Kel, Iran, July 28, 1941; came to U.S., 1964, naturalized, 1978; s. Abdulwahab and Sakineh (Kamiab) S.; m. Kay L. Schoenwald, Aug. 9, 1969; 1 child, Shaun. B.Sc., Calif. State U., Northridge, 1971, M.Sc., 1983. Cert. profl. geologist, Va., Wyo. Petroleum geologist Iranian Oil Exploration and Producing Co., Ahwaz, 1971-74; geol. cons. D.R.L., Inc., Bakersfield, Calif., 1974-76, Stratā-log, Inc., 1976-79, Energy Log, Inc., Sacramento, 1979-80; geologist U.S. Dept. Energy, Washington, 1980-81, Bur. Land Mgmt. Dept. Interior, Washington, 1981-89, asst. dist. mgr., Rawlins, Wyo., 1989—. Contbr. articles to profl. jours. Mem. Am. Assn. Petroleum Geologists, Soc. Profl. Well Log Analysts, Soc. Petroleum Engrs.

SHARIFI, IRAJ ALAGHA, organic chemistry educator; b. Gorgan, Mazandaran, Iran, Feb. 11, 1938; s. Abolfath Sharifi and Shokat (Heravi) S.; m. Zahra Dardashty Salehian, Nov. 26, 1965; children: Azalea, Neda. BS, MS in Engring., Tehran (Iran) Poly., 1963; MS, Hays Kans. State U., 1969, Colo. State U., 1972; PhD, Colo. State U., 1976. Asst. prof. U. Isfahan (Iran), 1971-81; prin. Preuniversity Sch., Isfahan, 1977-79; dep. dir. gen. Ministry of Plan and Budget, Tehran, 1981-83; project dir. Ministry of Industry, Tehran, 1983-85; tchr., rschr. Inst. Nutrition and Food Scis., Tehran, 1985—, Free Islamic U. Iran, Tehran, 1990—; owner, bd. dirs. Natural Food Product Rsch. Ctr., Tehran, 1991—. Contbr. articles to profl. jours. including Phytochemistry; author: Food Preservation 2000, 1993, 8th World Congress of Food Sciences & Technology, 1991, Six International Congress of Heterocyclic Chemistry, 1977, First Seminar on the Oceanographic Issues. Iranian National Commission for UNESCOo, 1990, First Joint Symposium Phytochemical Society of European and Phytochemical Society of North America, Belgium, 1977, others. Rsch. grantee Ministry Higher Edn., Isfahan,1 973, NIH, 1970-72, U. Tchr. Tng., Tehran, 1989, 90, project grantee Ministry of Industry, Tehran, 1983. Mem. Iranian Food Sci. and Tech. (bd. dirs. 1987—), Am. Chem. Soc., Iranian Chem. Soc. Muslim. Achievements include research on separation and identification of 5 new alkaloids from zanthoxylum, synthesis of new sorbophenone derivatives, making gelatin from fishery wastes of Iran, preparation of fruit sugar for diabetic patients, standard date syrup. Home: Ostad Hasan Banna St, Sharifi St No 10, Tehran 16677, Iran

SHARMA, ARJUN DUTTA, cardiologist; b. Bombay, June 2, 1953; came to U.S., 1981; s. Hari D. and Gudrun (Axelsson) S.; m. Carolyn D. Burleigh, May 9, 1981; chldren: Allira, Eric, Harison. BSc, U. Waterloo, Ont., Can., 1972; MD, U. Toronto, Ont., 1976. Intern Toronto Gen. Hosp., 1976-77, resident in medicine, 1978-80; resident in medicine St. Michael's Hosp., Toronto, 1980-81; residency medicine Toronto Gen. Hosp., 1977-78; Rsch. assoc. Washington U., St. Louis, 1981-83; asst. prof. pharmacy and toxicology U. Western Ont., London, 1985-89, asst. prof. medicine, 1983-89, assoc. prof. medicine, 1989-90; dir. interventional electrophysiology Sutter Meml. Hosp., Sacramento, 1990—; abstract reviewer, faculty of ann. scientific sessions North Am. Soc. for Pacing and Electrophysiology, 1993; assoc. clin. prof. U. Calif., Davis, 1990—; cons. Medtronic Inc., Mpls., 1985-89, Telectronics Pacing Systems, Inc., 1990—; mem. rsch. com. Sutter Inst. Med. Rsch.; mem. exec. com. Sutter Heart Inst. Reviewer profl. jours., including Circulation, Am. Jour. Cardiology; contbr. articles to profl. publs. Mem. coun. for basic sci. Am. Heart Assn., chmn. ann. sci. session, 1989; active Crocker Art Mus. Recipient John Melady award, 1972, Dr. C.S. Wainwright award, 1973-75, Rsch. prize Toronto Gen. Hosp., 1979, 80, Ont. Career Scientist award Ont. Ministry of Health, 1983-89; Med. Rsch. Coun. Can. fellow, 1981-83. Avocations: skiing, tennis, philately. Office: 3941 J St Ste 260 Sacramento CA 95819

SHARMA, BRAHAMA DATTA, chemistry educator; b. Sampla, Punjab, India, June 5, 1931; s. Des Raj and Kesara Devi (Pathak) S.; m. Millicent M. Hewitt, Dec. 22, 1956; children: Nalanda V. Sharma Bowman, Renuka D.; BS with honors, U. Delhi, India, 1949, MS, 1951; PhD, U. So. Calif., 1961. Chemist Govt. Opium Factory, Ghazipur, India, 1951-52; lab. assoc., sci. asst. Nat. Chem. Lab., Poona, India, 1952-55; lab. assoc. U. So. Calif., Los Angeles, 1955-61; research fellow Calif. Inst. Tech., Pasadena, 1961-65; asst. prof. chemistry U. Nev., Reno, 1963-64, Oreg. State U., Corvallis, 1965-70; asst. prof. chemistry Calif. State U., Northridge, 1973-75, assoc. prof., 1975-76; prof. Los Angeles Pierce Coll., Woodland Hills, Calif., 1976—; part-time assoc. prof. chemistry Calif. State U., L.A., 1973-85, prof., 1985—; vis. assoc. Calif. Inst. Tech., 1979, 82; pres. L.A. Pierce Coll. Senate, 1981-82, chmn. profl. and acad. stds., 1989-92. Contbr. articles to profl. jours. Grantee E.I. duPont de Nemours, L.A., 1961, Am. Chem. Soc. Petroleum Rsch. Fund, Washington, 1965-69, NSF, Washington, 1967-69. Mem. AAAS, Am. Chem. Soc. (chmn. edn. com. So. Calif. chpt. 1981-82), Royal Soc. Chemistry, Am. Crystallog Assn., Sigma Xi. Avocations: playing bridge, reading, history, classical music, crystal models. Office: Los Angeles Pierce Coll Woodland Hills CA 91371

SHARMA, KULDEEPAK BHARDWAJ, pharmaceutical scientist; b. New Delhi, India, June 21, 1956; came to U.S., 1981; s. Shyam Sunder and Bhagqati (Devi) S.; m. Mala Palaria, Mar. 12, 1986; children: Chriag, Divya. Diploma in Pharmacy, Delhi Bd. Tech. Edn., New Delhi, 1973; BS in Pharmacy, Delhi U., 1977; PhD, U. Utah, 1987. Head dept. drug delivery Synorex, Inc., Newport Beach, Calif., 1987-90; staff scientist, project leader Cygnus Therapeutic Systems, Redwood City, Calif., 1990—. Author: Topics in Pharmaceutical Science, 1983; contbr. articles to profl. jours. Mem. Am. Assn. Pharm. Scientist, Am. Pharm. Assn., Controlled Release Soc. Achievements include patents in field; development of new transdermal system for various drugs, e.g. Buprenorphine and Alprazolam. Office: Cygnus Therapeutic Systems 400 Penobscot Dr Redwood City CA 94063

SHARMA, MAHENDRA KUMAR, chemist; b. Etah, India, Oct. 15, 1948; came to U.S., 1981; s. Makhan Lal and Shree (Devi) S.; m. Rama Sharma, Dec. 6, 1978; children: Amol, Anuj. BS in Chemistry, Agra U., India, 1968, MS in Phys. Chemistry, 1970, PhD in Colloid and Surface Sci., 1975. Grad. rsch. asst. Agra U., 1971-75, rsch. scientist, assoc., 1975-79; vis. rsch. assoc. Basel (Switzerland) U., 1979-81; rsch. assoc. U. Fla., Gainesville, 1981-83, asst. rsch. prof., scientist, 1983-85; sr. rsch. chemist, scientist Eastman Kodak Co., Kingsport, Tenn., 1986-91, prin. rsch. chemist, scientist, 1991—. Editor: Surface Phenomena and Fine Particles in Water-Based Coatings and Printing Technology, 1991, Surface Phenomena and Additives in Water-Based Coatings and Printing Technology, 1991, Particle Technology and Surface Phenomena in Minerals and Petroleum, 1991; contbr. over 50 articles to profl. jours. Mem. Am. Chem. Soc., Fine Particle Soc. (mem. exec. coun., bd. dirs. 1989—), Hindi Literary Assn. Achievements include 10 patents in water-dispersible polymeric compositions, liquid-dispersible, polymeric colorant compositions and aqueous dispersions and process for preparation, solid-form additive systems, numerous others. Home: 2600 Brighton Ct Kingsport TN 37660-4762 Office: Eastman Kodak Co PO Box 1972 Kingsport TN 37662-1972

SHARMA, MINOTI, chemist, researcher; b. Shillong, Assam, India, May 19, 1940; came to U.S., 1971; d. Indreswar and Kamini Sharma; m. Moheswar Sharma, Feb. 8, 1963; children: Mirand, Moira. MS, Tufts U., 1965; PhD, Southampton (Eng.) U., 1970. Asst. rsch. prof. SUNY, Buffalo, 1977-80; cancer rsch. scientist Roswell Park Cancer Inst., Buffalo, 1981—; rsch. asst. prof. grad. div. SUNY-Buffalo, 1984—; vis. scholar Merrifield's Lab., Rockefeller U., N.Y.C., 1988. Contbr. articles to profl. jours. NIH grantee, 1989—. Mem. AAAS, Am. Chem. Soc., Analytical Chemistry Soc., Fedn. Am. Socs. for Exptl, Biology. Achievements include development of nonisotopic detection technique to assay DNA damage by combining HPLC and HPCE with laser induced fluroscence detection. Office: Roswell Park Cancer Inst Dept Biophysics Elm and Carlton Sts Buffalo NY 14263

SHARON, MICHAEL, endocrinologist; b. N.Y.C., Apr. 7, 1955; s. Rebecca (Menaged) S.; m. Linda E. Kupfer; 1 child, David G. BS, Cornell U., 1976; MD, N.Y. Med. Coll., 1980. Dipolomate Am. Bd. Internal Medicine. Intern Mt. Sinai Hosp., N.Y.C., 1980-81, resident, 1981-83, chief med. resident, 1983-84; med. staff fellow, Inter-Inst. Endocrinology Tng. Program NIH, Bethesda, Md., 1984-87; med. staff fellow, Cell Biology and Metabolism Br. NICHD NIH, Bethesda, 1985-87, sr. staff fellow, Cell Biology and Metabolism Br. NICHD, 1987-89; assoc. med. dir. Diabetes Treatment Ctr. Georgetown U. Hosp., Washington, 1989-90, clin. asst. prof. medicine, 1990—; assoc. med. dir. Diabetes Treatment Ctr. Washington Hosp. Ctr., 1990—; jr. attending Washington Hosp. Ctr., 1989—, acting co-dir. divsn. endocrinology, 1989-90; assoc. Tanenberg and Vigersky, P.C., pvt. practice endocrinology, 1989-93; diabetes and endocrine cons., Levy & Sharon, M.D., P.C., pvt. practice; guest scientist Cell Biology and Metabolism Br., NICHD, 1989-91, clin. con. Developmental Endocrinology Br., 1989—. Contbr. numerous articles to profl. jours. Named Outstanding Jr. Student Interested in Acad. Medicine N.Y. Med. Coll. Dept. Medicine, 1979; recipient Dr. and Mrs. David Harrison scholarship N.Y. Med. Coll., 1980, Cor et Manus citation for svc. to N.Y. Med. Coll., 1980, Am. Fedn. for Clin. Rsch. award to trainees in clin. rsch., 1987. Fellow ACP; mem. Endocrine Soc., Am. Diabetes Assn., Alpha Omega Alpha. Achievements include first description of the Beta sub unit of the IL-2 receptor. Office: 344 University Blvd Ste 328 Silver Spring MD 20901

SHARP, CHARLES PAUL, electronics technician; b. Tucson, Apr. 1, 1942; s. Charles Edwin and Anna Louise (Nash) S.; m. Lucille Ann Chach, July 14, 1962; 1 child, Charles P. Jr. AA in Elec. Engring., MTI, 1964. Cert. NEC. Field engr. Western Union, Kansas City, Mo., 1964-75, Ft. Dodge, Iowa, 1975-88; electronics technician U. Iowa, Iowa City, 1988—. Mem. U.S. Pilot Assn. (bd. dirs. 1988—). Achievements include design of computer room air flow, design of ion transmitters. Home: 1800 12th Ave Coralville IA 52241 Office: U Iowa Telecom Dept 230 Madison Iowa City IA 52242

SHARP, DAVID PAUL, computer technologist, researcher; b. Mt. Carmel, Ill., Feb. 16, 1950; s. Raymond C. and Frances K. (Reburn) S.; m. Rita Gail Lucas, Nov. 5, 1971; children: Mindy Suzanne, David Paul II. Student, LaSalle U., Chgo. Designer Audio Video Designs, Inc., 1971-73, Fisk Telephone Systems, 1973-75; technologist Harris County Data Processing, 1975-80; designer NL Industries, Houston, 1980-81, Comspec Inc., Houston, 1981-83; various Compaq Computer Corp., Houston, 1983—; corp. LAN mgr., 1990—. Author: Map of the System, 1986, Map Version 2.0, 1990; tech. cons. (TV movie) The Hunting Ground, 1989; cons. adv. com. Electronics Internat., 1976; reviewer Tex. Coll. Physics book, 1984, 85. With U.S. Army, 1968-71. Mem. IEEE, Audio Engring. Soc. Am., Soc. Profl. Well Logging Engrs. Republican. Methodist. Achievements include patents and patents pending on fiber optics, computing tech. and related areas; major research on 20 Gigabyte data bases on PC's for business and law. Office: Compaq Computer Corp 20555 SH 249 Houston TX 77070

SHARP, DEXTER BRIAN, organic chemist, consultant; b. Chgo., June 14, 1919; s. Mahlon Earl and Olive Marie (Smith) S.; m. Peggy Elizabeth Person, July 8, 1945; children: Peggy Lynn, Judith Ann, Janice Kaye. BA, Carleton Coll., 1941; MA, U. Nebr., 1943, PhD, 1945. Chemist DuPont, Wilmington, Del., 1945-46; postdoctoral fellow U. Minn., Mpls., 1946-47; assoc. prof. Kans. State U., Manhattan, 1947-51; chemist Monsanto, Dayton, Ohio, 1951-53; group leader Monsanto, Dayton, St. Louis, 1953-68; mgr. rsch. Monsanto, St. Louis, 1968-75; dir. rsch. Monsanto Agrl. Co., St. Louis, 1975-85, ret., 1985; cons. Stewart Pesticide Registration Assn., Washington, 1986—; mem. Delaney com. Nat. Acad. Sci., 1982-84, agrochem commn. Internat. Union Pure and Applied Chemistry, 1984-88. Author: (book chpt.) Alachlor/Herbicide Metabolism, 1985; contbr. articles to profl. jours. Mem. Am. Chem. Soc. (emeritus), Nat. Agrl. Chemicals Assn., Weed Sci. Soc. Am. Achievements include 21 patents. Home: 13042 Weatherfield Dr Saint Louis MO 63146-3646 Office: Dexter B Sharp Inc 13042 Weatherfield Dr Saint Louis MO 63146-3646

SHARP, JAMES J., civil engineer. Recipient Les Prix Camiile A. Dagenais award Can. Soc. Civil Engring., 1990. Home: 2127 Styles Cr E, Regina, SK Canada S4V 0P8*

SHARP, JOHN MALCOLM, JR., geology educator; b. St. Paul, Mar. 11, 1944. BGeoE, U. Minn., 1967; MS, PhD., U. Ill., 1974. Civil engring. officer USAF, 1967-71; with geology dept. U. Mo., Columbia, 1974-82; Chevron Centennial prof. geology U. Tex., Austin, 1982—. Alexander von Humbolt fellow 1981-83. Fellow Geol. Soc. Am. (O.E. Meinzer award 1979); mem. Phi Kappa Phi, Sigma Xi, Tau Beta Pi. Office: U Tex Coll Natural Scis Dept Geol Scis Dept Geol Scis Austin TX 78712

SHARP, KEVAN DENTON, civil engineer; b. Ft. Collins, Colo., May 18, 1957; s. Bobby Ray and Yvonne June (Brashier) S.; m. Margaret Sue Birch, Oct. 16, 1987. BS magna cum laude, Utah State U., 1980, MS, 1981, PhD, 1982. Grad. rsch. asst. Utah State U., Logan, 1979-82, rsch. asst. prof., 1982-84; geotech. engr. CH2M Hill, Bellevue, Wash., 1984—, mgr. geotech. engring., 1992—. Editor: Soil Improvement Using Fly Ash, 1993; contbr. articles to profl. jours. Com. mem. Boy Scouts Am., 1988—; vol. LDS Ch., 1987—. Mem. ASCE (mem. com. soil improvement and geosynthetics), ASFE (chair practice environment com. 1993—), Assn. Engring. Firms Practicing in Geoscis. Office: CH2M Hill 777 108th Ave NE Bellevue WA 98004

SHARP, LOUIS, scientist. Recipient Excellence in Adhesion Sci. award Adhesion Soc., 1993. Office: AT & T Bell Laboratories 600 Mountain Ave New Providence NJ 07974*

SHARP, PHILLIP ALLEN, academic administrator, biologist, educator; b. Ky., June 6, 1944; s. Joseph Walter and Katherin (Colvin) S.; m. Ann Christine Holcombe, Aug. 29, 1964; children: Christine Alynn, Sarah Katherin, Helena Holcombe. BA, Union Coll., Barbourville, Ky., 1966, LHD (hon.), 1991; PhD, U. Ill., 1969. NIH postdoctoral fellow Calif. Inst. Tech., 1969-71; sr. research investigator Cold Spring Harbor (N.Y.) Lab., 1972-74; assoc. prof. MIT, Cambridge, 1974-79, prof. biology, 1979—, head dept. biology, 1991—, dir. Ctr. Cancer Rsch., 1985-91; Co-founder, mem. sci. bd., dir. BIOGEN, 1978—; chmn. sci. bd., 1987—. Mem. editorial bd. Cell, 1974—, Jour. Virology, 1974-86, Molecular and Cellular Biology, 1974-85. Recipient awards Am. Cancer Soc., 1974-79, awards Eli Lilly, 1980, awards Nat Acad. Sci./U.S. Steel Found., 1980, Howard Ricketts award U. Chgo., 1985, Alfred P. Sloan Jr. prize Gen. Motors Research Found., 1986, award Gairdner Found. Internat., 1986, award N.Y. Acad. Scis., 1986, Louisa Horwitz prize, 1988, Albert Lasker Basic Med. Rsch. award, 1988, Dickson prize U. Pitts., 1990, Nobel Prize in Medicine, Nobel Foundation, 1993; awarded Class of '41 chair, 1986-87, John D. MacArthur chair, 1987-92, Salvador E. Luria chair, 1992—. Fellow AAAS; mem. Am. Chem. Soc., Am. Soc. Microbiology, NAS (councilor 1986), Am. Acad. Arts and Scis, European Molecular Biology Orgn. (assoc.), Am. Soc. Biochemistry and Molecular Biology (elected mem. coun.), Am. Philos. Soc. (elected mem.), Inst. of Medicine of NAS (elected mem.). Home: 119 Grasmere St Newton MA 02158-2212 Office: MIT Rm E17 529B 40 Ames St Cambridge MA 02139-4307

SHARP, ROBERT PHILLIP, geology educator, researcher; b. Oxnard, Calif., June 24, 1911; s. Julian Hebner Sharp and Alice Sharp Darling; m. Jean Prescott Todd, Sept. 7, 1938; adopted children—Kristin Todd, Bruce Todd. B.S., Calif. Inst. Tech., Pasadena, 1934, M.S., 1935; M.A., Harvard U., Cambridge, Mass., 1936, Ph.D., 1938. Asst. prof. U. Ill., Urbana, 1938-43; prof. U. Minn., Mpls., 1946-47; prof. Calif. Inst. Tech., Pasadena, 1947-79, chmn., 1952-67, prof. emeritus, 1979—. Author: Glaciers, 1960, Field Guide-Southern California, 1972, Field Guide-Coastal Southern California, 1978, Living Ice-Understanding Glaciers and Glaciation, 1988, (with A.F. Glazner) Geology Under Foot in Southern California, 1993. Served to capt. USAF, 1943-46. Recipient Exceptional Sci. Achievement medal NASA, 1971, Nat. Medal Sci., 1989, Charles P. Daly medal Am. Geog. Soc., 1991; Robert P. Sharp professorship Calif. Inst. Tech., 1978. Fellow Geol. Soc. Am. (councillor, Kirk Bryan award 1964, Penrose medal 1977), Am. Geophys. Union; hon. fellow Internat. Glaciological Soc.; mem. NAS. Republican. Avocations: flyfishing, snorkeling, camping. Home: 1901 Gibraltar Rd Santa Barbara CA 93105-2326 Office: Calif Inst Tech 1200 E California Blvd Pasadena CA 91125-0001

SHARP, VICTORIA LEE, medical director; b. L.A., Feb. 19, 1947; d. William Carmen and Laura (Gile) S. BA, U. Calif., 1973; MD, Free U. Brussels, 1983. Cert. Am. Bd. Internal Medicine. Intern Albany (N.Y.) Med. Ctr. Hosp., 1983-84, sr. resident, 1985-86, chief resident, 1986-87; jr. resident St. Luke's Roosevelt Hosp., N.Y.C., 1984-85; med. dir. AIDS Treatment Ctr. Albany Med. Ctr., 1987-89; med. dir. Spellman Ctr. for HIV Related Disease St. Clare's Hosp. and Health Ctr., N.Y.C., 1989—; trustee Terence Cardinal Cooke Health Care Ctr., N.Y.C., 1990—; coun. advisors Ministry of Health, Prague, Czechoslovakia, 1990—; mem. task force on tuberculosis in the criminal justice system N.Y.C. Dept. Health, 1992—, mem. med. care criteria com. AIDS intervention mgmt. systems AIDS Inst, 1988—, mem. AIDS Ctr. Standards Com., 1990—; mem. HIV/TB working group United Hosp. Fund, 1992—; mem. Health Care Exec. Forum, 1992—; mem. HIV task force Hosp. Assn. N.Y. State, 1992—; mem. med. dirs. designated AIDS Ctrs. forum N.Y. Acad. Medicine, 1991—; mem. N.Y. Med. Soc. AIDS Task Force, 1990—; bd. dirs. Hell's Kitchen AIDS Project, Inc., N.Y.C., 1990-93. Editor: Clinician Revs., 1990—; mem. edit. bd. Audio Jour. of AIDS Mgmt., 1990—; contbr. articles to profl. jours. Ryan White CARE grantee, 1991, Robert Wood Johnson grantee, 1988. Mem. Am. Coll. Physicians (assoc.), Am. Pub. Health Assn. Office: St Clare's Hosp and Health 415 W 51st St New York NY 10019

SHARP, WILLIAM CHARLES, systems engineer; b. Cambridge, N.Y., Dec. 2, 1953; s. William Leland and Phyllis Evelyn (Burns) S.; children: William Welsey Leland, Natasha Nicole Nativa. BS in System Engring., Rensselaer Poly. Inst., 1976. Engr. Applicon, Burlington, Mass., 1976-78, Xylogics, Burlington, 1978, McDon, Long Beach, Calif., 1978-81, Hughes Aircraft Co., Fullerton, Calif., 1981-82; engring. mgr. Able Computer, Irvine, Calif., 1982-84; engr. Sierra Cybernetics, Brea, Calif., 1984-86, Midcom Corp., Anaheim Hills, Calif., 1986-91, Jet Propulsion Lab., Pasadena, Calif., 1986-91; pres. Glacier Blue, Rancho Santa Margarita, Calif., 1986—. Mem.

Armed Forces Comm. and Electronics Assn., Digital Equipment Corp. User's Soc., Order of DeMolay (adult advisor, chevalier). Libertarian. Avocations: hang gliding, skiing, juggling. Home and Office: 1 San Pablo Rancho Santa Margarita CA 92688-2518

SHARP, WILLIAM WHEELER, geologist; b. Shreveport, La., Oct. 9, 1923; s. William Wheeler and Jennie V. (Benson) S.; m. Rubylin Slaughter, 1958; children: Staci Lynn, Kimberly Cecile; 1 child from previous marriage, John E. BS in Geology, U. Tex., Austin, 1950, MA, 1951. Lic. pvt. pilot. Geol. Socony-Vacuum, Caracas, Venezuela, 1951-53; surface geol. chief Creole, 1953-57; dist. devel. geologist, supr. exploration, devel., and unitization of 132 multi-pay oil and gas fields, expert geol. witness, coll. recruiter, research assoc. ARCO, 1957-85; discovered oil and gas at Bayou Boullion, Bayou Sale, Jeanerette, La.; petroleum exploration in Alaska, Aus., Can., U.S. and S.A. Contbr. articles to profl. jours. Past dir. and chmn. U.S. Tennis Assn. Tournaments, 12th Nat. Boys Tournament; pres. Lafayette Tennis Adv. Com., 1972; past dir. Jr. Achievememt and United Fund Programs. Served as sgt. USAF, 1943-46, PTO. Winner and finalist more than 75 amateur tennis tournaments including Confederate Oil Invitational, Gulf Coast Oilmen's Tournament, So. Oilmen's Tournament, Tex.-Ark.-La. Oilmen's Tournament; named Hon. Citizen of New Orleans, 1971. Mem. Dallas Geol. Soc., Lafayette Geol. Soc. (bd. dirs. 1973-74), Am. Assn. Petroleum Geologists (co-author Best of SEG conv. 1982), VFW, Am. Legion, Appaloosa Horse Club. Republican. Methodist. Avocations: sports, music, horses.

SHARPE, MITCHELL RAYMOND, science writer; b. Knoxville, Tenn., Dec. 22, 1924; s. Mitchell Raymond and Katie Grace (Hill) S.; m. Virginia Ruth Lowry, 1952 (div.); children: Rebecca, Rachel, David. BS, Auburn U., 1948, MA, 1954; postgrad., Emory U., 1955, U.S. Army Command and Gen. Staff Coll., 1967. Supervisory tech. writer U.S. Army Missile Command, Redstone Arsenal, Ala., 1955-60; tech. writer, historian Marshall Space Flight Ctr., Huntsville, Ala., 1960-74; cons. U.S. Space and Rocket Ctr., Huntsville, 1970—, Nat. Air and Space Mus., Washington, 1965-80, Coupole d'Helfaut-Wizernes, Arques, France, 1989—, Gemeindererwaltung Peenemuende His-tech., Informationzentrum, Peenemuende, Germany, 1991—. Author: It Is I, Seagull, Valentina Tereshkova, First Woman in Space, 1975, Living in Space, The Astronaut and His Environment, 1969, Yuri Gagarin, First Man in Space, 1969, Satellites and Probes, The Development of Unmanned Space Flight, 1970; co-author: Applied Astronautics, An Introduction to Space Flight, 1963, Basic Astronautics, An Introduction to Space Science, Engineering, and Medicine, 1962, Dividends from Space, 1974, The Rocket Team, 1979, 82, 92; also articles. With U.S. Army, World War II, Korean War; col. USAR ret. Recipient Goddard Essay award Nat. Space Club, 1969, 75, Gold medal Tsiolkovsky Hist. Mus., Kaluga, USSR, 1973. Fellow Brit. Interplanetary Soc.; mem. AIAA (sr., tech. com. on history), Internat. Acad. Astronautics (corr., history com. 1985—), Soc. for History of Tech., Nat. Assn. Sci. Writers, Nat. Geog. Soc. Avocations: travel, photography. Home and Office: 7302 Chadwell Rd SW Huntsville AL 35802-1718

SHARPE, WILLIAM NORMAN, JR., mechanical engineer, educator; b. Chatham County, N.C., Apr. 15, 1938; s. William Norman and Margaret Horne (Womble) S.; m. Margaret Ellen Strowd, Aug. 21, 1959; children: William N., J. Ashley. BS, N.C. State U., 1960, MS, 1961; PhD, Johns Hopkins U., 1966. Registered profl. engr., Mich., La., Md. Assoc. prof. Mich. State U., East Lansing, 1970-75, prof., 1975-78; prof., chmn. dept. mech. engring. La. State U., Baton Rouge, 1978-83; prof., dept. mech. engring. Johns Hopkins U., Balt., 1983—, Decker prof. mech. engring., 1985—. Recipient Alexander von Humboldt award, Fed. Republic Germany, 1989. Fellow ASME, Soc. Exptl. Mechanics (Tatnall award, exec. bd. 1979-81, pres. 1984-85); mem. Am. Soc. Engring. Edn., ASTM. Home: 220 Ridgewood Rd Baltimore MD 21210-2539 Office: Johns Hopkins U Dept Mech Engring Latrobe Hall Rm 127 Baltimore MD 21218

SHARPLESS, K. BARRY, chemist; b. Phila., Apr. 28, 1941; m. Jan Dueser, Apr. 28, 1965; children: Hannah, William, Isaac. BA, Dartmouth Coll., 1963; PhD, Stanford U., 1968. Postdoctoral assoc. Harvard U.; postdoctoral assoc. Stanford U., to 1970, faculty dept. chemistry, 1977-80; faculty MIT, Cambridge, 1970—, prof. chemistry, 1975-77, 80-90; prof. chemistry Scripps Rsch. Inst., La Jolla, Calif., 1990—. Recipient Paul Janssen Prize for Creativity in Organic Sunthesis, Rolf Sammet prize vis. lectureship Johann W. Goethe Univ., Frankfurt, Fed. Republic Germany, 1988, Chem. Pioneer award Am. Inst. Chemists, 1988, Prelog Medal Eidenössische Technische Hochschuee, Zurich, 1988, Scheele medal and prize Swedish Acad. Pharm. Scis.; A.P. Sloan fellow, Guggenheim fellow, 1987-88; Camille and Henry Dreyfus Tchr. scholar. Fellow AAAS, Am. Acad. Arts and Scis.; mem. NAS, Am. Chem. Soc. (Creative Work in Synthetic Organic Chemistry award 1983, Arthur C. Cope scholar's award 1986, Harrison Howe award Rochester chpt. 1987, Remsen award Md. sect. 1989). Office: Scripps Rsch Inst CVN-2 10666 N Torrey Pines Rd La Jolla CA 92037-1027

SHARTLE, STANLEY MUSGRAVE, consulting engineer, land surveyor; b. Brazil, Ind., Sept. 27, 1922; s. Arthur Tinder and Mildred C. (Musgrave) S.; m. Anna Lee Mantle, Apr. 7, 1948 (div. 1980); 1 child, Randy. Student Purdue U., 1947-50. Registered profl. engr., land surveyor, Ind. Chief dep. surveyor Hendricks County, Ind., 1941-42; asst. to hydrographer Fourteenth Naval Dist., Pearl Harbor, Hawaii, 1942-44; dep. county surveyor Hendricks County (Ind.), Danville, 1944-50, county engr., surveyor, 1950-54, county hwy. engr., 1975-77; staff engr. Ind. Toll Rd. Commn., Indpls., 1954-61; chief right of way engring. Ind. State Hwy. Commn., Indpls., 1961-75; owner, civil engr. Shartle Engring., Indpls., 1977-89; prin. Parsons & Shartle Engrs., Inc., Indpls., 1990—, right of way engring. cons. Gannett Fleming Transp. Engrs., Inc., Indpls., 1983-88; part-time lectr. Purdue U. for Ind. State Hwy. Commn., 1965-67. Prin. works include residential subdiv., 1989 (named Indiana's Most Sucessful Ind. Bus. Jour., 1989). Author: Right of Way Engineering Manual, 1975, Musgrave Family History, 1961, Shartle Genealogy, 1955; contbr. tech. articles in sci. jours. Ex-officio mem., charter mem. exec. sec. Hendricks County (Ind.) Plan Commn., 1951-54; mem. citizen adv. com. Hendricks County Subdivision Control Ordinance, 1988—. Recipient Outstanding Contbn. award Hendricks County Soil and Water Conservation Dist., 1976. Mem. Am. Congress Surveying and Mapping (life), Nat. Soc. Profl. Surveyors, Ind. Soc. Profl. Land Surveyors (charter, life, bd. dirs. 1979), Nat. Geneal. Soc. (Quarter Century club), Ind. Toll Road Employees Assn. (pres. 1959-60), The Pa. German Soc. Republican. Avocations: astronomy, genealogy, geodesy. Home and Office: 19 Kings Ct Danville IN 46122-9729

SHASTRI, RANGANATH KRISHNA, materials scientist; b. Gokarn, Mysore, India, Feb. 24, 1951; came to U.S., 1974; s. Krishna Ganesh and Radha K. (Kurse) S. BE in Chem. Engring., U. Mysore, Surathkal, India, 1973; MS in Chem. Engring., U. Cin., 1976, PhD in Materials Sci., 1979. Rsch. assoc. dept. orthopaedic surgery/materials sci. U. Cin., 1979-81; polymer scientist Revere Rsch., Inc., Edison, N.J., 1981-82; v.p. J.R. Enterprises, Greenwich, Conn., 1983; sr. rsch. engr. Dow Chem. Co., Midland, Mich., 1983-86, project leader, 1986-89, project leader database devel. Materials Engring. Ctr., 1989-91, devel. leader, 1991—. Contbr. articles to profl. jours.; contbr. articles to Jour. Bone & Joint Surgery, Jour. Biomed. Material Rsch., Jour. Coll. Interface Sci.; author chpt. in book; holds 3 patents. Bd. dirs. Voluntary Action Ctr. Midland County, 1986-88. NIH fellow, 1976-79. Mem. ASM Internat. (chmn. polymer composites session Indpls. 1989, polymer composites com. of composites tech. div. 1987—, World Material Congress chmn. advanced composites tutorial/panel discussion session, engring. materials achievement award selection com., 1990-92), ASTM (chmn. task groups 1990—), Am. Chem. Soc. (chmn. materials sci. info. session Atlanta 1990), Soc. Plastics Engrs. Home: 2913 Jeffrey Ln Midland MI 48640-2472 Office: Dow Chem Co MEC/433 Bldg Midland MI 48667

SHASTRY, SHAMBHU KADHAMBINY, scientist, engineering executive; b. Kerala, India, Aug. 29, 1954; came to U.S., 1978; s. Mahalinga K. and Parameshwari (Laxmi) S.; m. Suma S. Shastry, Dec. 30, 1984; children: Disha Laxmi, Divya Gowri. MS, Rensselaer Poly. Inst., 1980, PhD, 1982. Leader Microwave Semiconductor Corp., Somerset, N.J., 1982-85; sr. mem. tech. staff GTE Labs., Waltham, Mass., 1985-88; mgr. advanced tech. Kopin Corp., Taunton, Mass., 1988-91, dir. III-V Products, 1991—. Reviewer: Prepublication Review of S.K. Ghandhi's Book on VLSI Fabrication Principles; contbr. articles to profl. jours.; referee Applied Physics Letters, Jour. Applied Physics, 1985—. Vol. New Eng. Hindu Temple, Inc., Ashland, Mass., 1984—. Mem. IEEE, Am. Phys. Soc., Electrochem. Soc., Sigma Xi. Achievements include patents on heteroepitaxy of dissimilar materials, such as gallium arsenide on silicon. Office: Kopin Corp 695 Myles Standish Blvd Taunton MA 02780-1042

SHATKIN, AARON JEFFREY, biochemistry educator; b. Providence, July 18, 1934; s. Morris and Doris S.; m. Joan A. Lynch, Nov. 30, 1957; 1 son, Gregory Martin. A.B., Bowdoin Coll., 1956, D.Sc. (hon.), 1979; Ph.D., Rockefeller Inst., 1961. Sr. asst. scientist NIH, Bethesda, Md., 1961-63; research chemist NIH, 1963-68; vis. scientist Salk Inst., La Jolla, Calif., 1968-69; assoc. mem. dept. cell biology Roche Inst. Molecular Biology, Nutley, N.J., 1968-73; full mem. Roche Inst. Molecular Biology, 1973-77, head molecular virology lab., 1977-86, head dept. cell biology, 1983-86; dir. N.J. Ctr. Advanced Biotech. Medicine, 1986—; prof. molecular genetics UMDNJ, 1986—; univ. prof. molecular biology Rutgers U., New Brunswick, N.J., 1986—; adj. prof. cell biology Rockefeller U.; vis. prof. molecular biology Princeton U. Mem. editorial bd.: Jour. Virology 1969-82, Archives of Biochemistry & Biophysics, 1972-82, Virology, 1973-76, Comprehensive Virology, 1974-82, Jour. Biol. Chemistry, 1977-83; editor: Advances Virus Research, 1983—, Jour. Virology, 1973-77; editor-in-chief: Molecular and Cellular Biology, 1980-90. Served with USPHS, 1961-63. Recipient U.S. Steel Found. prize in molecular biology, 1977, N.J. Sci. and Tech. Pride award, 1989, Thomas Edison Sci. award State of N.J., 1991; Rockefeller fellow, 1956-61. Fellow Am. Acad. Microbiology, N.Y. Acad. Scis.; mem. NAS, AAAS, Am. Soc. Microbiology, Am. Soc. Biol. Chemists, Am. Soc. Virology, Am. Chem. Soc., Am. Soc. Cell Biology, Harvey Soc. Home: 1381 Rahway Rd Scotch Plains NJ 07076 Office: CABM 679 Hoes Ln Piscataway NJ 08854-5638

SHATNEY, CLAYTON HENRY, surgeon; b. Bangor, Maine, Nov. 4, 1943; s. Clayton Lewis and Regina (Cossette) S.; m. Deborah Gaye Hansen, Apr. 5, 1977; children: Tony, Andy. BA, Bowdoin Coll., 1965; MD, Tufts U., 1969. Asst. prof. surgery U. Md. Hosp., Balt., 1979-82; assoc. prof. U. Fla. Sch. Medicine, 1982-87; clin. assoc. prof. Stanford (Calif.) U. Sch. Medicine, 1987—; dir. traumatology Md. Inst. Emergency Med. Svcs., Balt., 1979-82; dir. trauma U. Hosp., Jacksonville, 1982-85; assoc. dir. trauma Santa Clara Valley Med. Ctr., 1992—; cons. VA Coop. Studies Program, Washington, 1980—. Editorial bd. Circulatory Shock, 1989—, Shock Research, 1993—; writer, actor med. movie. Maj. U.S. Army, 1977-79. State of Maine scholar Bowdoin Coll., 1961-65. Fellow ACS, Southeastern Surg. Congress, Southwestern Surg. Congress, Soc. Surg. Alimentary Tract, Am. Assn. Surg. Trauma, Soc. Critical Care Med., Societe Internat. de Chirurgie, Phi Kappa Phi. Home: 900 Larsen Rd Aptos CA 95003-2605 Office: Valley Med Ctr Dept Surgery 751 S Bascom Ave San Jose CA 95128-2604

SHATTO, ELLEN LATHEROW, mathematics educator; b. Harrisburg, Pa., Aug. 2, 1946; d. Andrew Edward and Frances Sara (Muckler) Latherow; m. Clair Shatto, Jr., Oct. 7, 1967; children: John David, Deborah Susan, Andrew Eugene, Ann Marie. BA, Lebanon Valley Coll., 1967; MEd, Penn State U., 1979. Cert. math. educator. Tchr. Cen. Dauphin Sch. Dist., Harrisburg, Pa., 1967-68; instr. Pa. Jr. Coll. Med. Arts, Dept. Nuclear Medicine, Harrisburg, 1969-70; instr. math. Penn State U.-Harrisburg, 1979-92; asst. prof. math. Harrisburg Area C.C., 1992—. Contbr. articles to profl. jours. Recipient Max Lehamn Meml. Math. Prize Lebanon Valley Coll., 1965. Mem. Math. Assn. Am., Nat. Coun. Tchrs. Math., Am. Assn. Adult and Continuing Edn., Pa. State Math. Assn. Two-Year Colls. (newsletter editor 1985-89, pres.-elect 1991-93, pres. 1993—). Avocation: collector "Precious Moments". Home: 3910 Mark Ave Harrisburg PA 17110-3636 Office: Harrisburg Area CC 223-H Whitaker Hall 1 HACC Dr Harrisburg PA 17110

SHATTUCK, DOUGLAS E., mechanical engineer, consultant; b. Paterson, N.J., Aug. 11, 1948; s. Norman G. and Caroline A. (Zurcher) S.; m. Jo-Ann Wolff, Feb. 6, 1977; children: Lindsay, Allison. BSME, N.J. Inst. Tech., 1970. Registered profl. engr., N.J. Heating, ventilation, air conditioning engr. Prudential, Newark, 1970-79; engring. mgr. A.J. Celiano, Inc., Cranford, N.J., 1979-87; cons. engr., prin. Douglas E. Shattuck, P.E., Manalapan, N.J., 1987-88; sr. mech. engr. Clayton Environ. Cons., Inc., Edison, N.J., 1988—. Contbr. to profl. publs. Mem. ASHRAE, ASTM. Achievements include development of protocols for evaluating commercial buildings for causes of sick building syndrome and building-related illness. Home: 6 Hillside Rd Manalapan NJ 07726 Office: Clayton Environ Cons 160 Fieldcrest Ave Edison NJ 08837

SHAVER, MARC STEVEN, aerospace engineer, air force executive officer; b. Ft. Leonardwood, Mo., Sept. 30, 1963; s. Robert Donald and Marilyn Joyce (Goodwin) S.; m. Tracey Wood, June 6, 1987; 1 child, Emily Ann. BS in Astronautical Engring., USAF Acad., 1987. Commd. 2d lt. USAF, 1987, advanced through grades to capt., 1991; F-16 performance engr. 6516 Test Squadron, Edwards AFB, Calif., 1987-89; YF-22A ops. engr. 6511 Test Squadron, Edwards AFB, Calif., 1989-91; spacecraft systems project engr. Space Test and Transp. Office, L.A., 1991-93; exec. officer Space Experimentation and Test Program Office, L.A., 1993—. Mem. AIAA. Baptist. Office: SMC/CUE 160 Skynet Ste 1536A Los Angeles AFB CA 90245-9999

SHAW, BREWSTER HOPKINSON, JR., astronaut; b. Cass City, Mich., May 16, 1945; m. Kathleen Mueller; children: Brewster H. III, Jessica Hollis, Brandon Robert. BS in Engring. Mechanics, U. Wis., 1968, MS in Enginrg. Mechanics, 1969. Commd. USAF, 1969, advanced through grades to col.; student pilot Craig AFB, Ala., 1970; assigned to F-100 replacement tng. unit Luke AFB, Ariz., 1970-71; served as F-100 combat fighter pilot, 352d Tactical Fighter Squadron Phan Rang Air Base, Republic of Vietnam, 1971; assigned to F-4 Replacement Tng. Unit George AFB, Calif., 1971; served with 25th Tactical Fighter Squadron Ubon Republic Thailand AFB; F-4 flight instr. 20th Tactical Fighter Tng. Squadron George AFB, Calif., 1973; student USAF Test Pilot Sch., Edwards AFB, Calif., 1975-76, instr., operational test pilot 6512th Test Squadron, 1977-78; astronaut NASA, 1978—, mem. support crew and Entry CAPCOM for STS-3 and STS-4, head Orbiter fleet, astronaut office liaison with Dept. Def., staff mem. Roger's Presdl. Commn. investigating STS 51-1, Challenger accident, pilot STS-9/Spacelab-1, 1983; spacecraft comdr. STS-61B, 1985, STS-28, 1989, completing total of 534 hours in space; dep. dir. Space Shuttle Ops., repsonsible all operational aspects of Space Shuttle Program, chmn. Mission Mgmt. Team Kennedy Space Ctr., Fla., 1989—. Decorated DFC with 7 oak leaf clusters, Def. Superior Svc. medal, Def. Meritorious Svc. medal, Air medal with 20 oak leaf clusters, Cross of Gallantry Republic Vietnam, NASA Space Flight medals, 1983, 85, 89, numerous other mil. and NASA awards; recipient Flight Achievement award Am. Astronautical Soc., 1983, Nat. Space award VFW, 1984. Avocations: running, skiing, flying, hunting. Office: care JF Kennedy Space Ctr NASA Kennedy Space Center FL 32899

SHAW, BRYCE ROBERT, author; b. Mansfield, Pa., Feb. 22, 1930; s. Wilford Walter and Genevieve (Cox) S.; m. Sally Ruth Prutsman, June 29, 1952; children: David Bryce, Jody Lynn McMillin, Erin Suzanne Hunsinger. AB, Muhlenberg Coll., Allentown, Pa., 1952; MA, U. Mich., 1953, postgrad., 1959-64. Cert. secondary edn. tchr. Teaching fellow U. Mich., Ann Arbor, 1953-55; math. analyst Willow Run Rsch. Ctr., Ypsilanti, Mich., 1954-59; tchr. math. Mt. Morris (Mich.) Bd. Edn., 1959-60, Flint (Mich.) Bd. Edn., 1960-64; lectr. math. U. Mich, Flint, 1961-68; coord. math. Flint Bd. Edn., 1964-75, dir. math. and computer systems, 1975-77; author in residence Houghton Mifflin Co., Boston, 1977-81, sr. author, 1970—; rsch. mathematician sch. dists., univs., Houghton Mifflin, others, Flint, Boston, L.A., 1961—; math. cons. sch. bds., state edn. depts., univs., U.S., Can., Mex., 1965—, Am. Sch. Found., Mexico City, 1965-68; lectr. edn. Ea. Mich. U., Ypsilanti, 1964-67; lectr. math. Nat. and Regional Math. Couns., Mich. State U., East Lansing, 1965-67; chmn. math. dept. Flint Bd. Edn., 1960-64; lectr. math. curriculum Flint Bd. Edn., 1960-80; lectr. math. curriculum devel. bds. edn., sch. math. staffs. Author: (textbooks) General Math I, 2d edit., 1979, Mathematics Plus, 1980, Fundamentals of Mathematics, 2d edit., 1986, Personalized Computational Skills Program, Vol. I, 1980, vol. II, 1981, Personalized Computational Skills Program--Skills and Applications, 1982, Personalized Computational Skills Program: Module A, B, C, D, E, and F, 3d edit., 1982, Mathematics Plus!, 1982, Computer Math Program, Computational Skills Program, 1988; contbr. articles to profl. jours. Recipient Internat. Man of Yr. award, 1991-92, Key of Success: Profl. Performance Achievement Rsch. and Notable Author award, Biographical Hon. award in Math.; named Disting. Lectr. Coll. Mathematicians. Mem. Math. Assn. Am., Nat. Coun. Tchrs. Math., Textbook Authors Assn., Phi Delta Kappa, Omicron Delta Kappa, Alpha Kappa Alpha. Avocations: music, chess. Home and Office: PO Box 531 Venice FL 34284-0531

SHAW, DEAN ALVIN, architect; b. El Paso, Tex., Jan. 22, 1954; s. Harold Alvin and Ann May (Glass) S.; m. Wendy June Hudgens, May 11, 1985; 1 child, Deanna Marie. BArch, Tex. Tech. U., 1979. Registered architect, Tex., Calif. Draftsman J.V. Scoggins Engring., Lubbock, Tex., 1978-79; intern R.S. Colley Architects, Corpus Christi, Tex., 1979-82; intern Campbell Taggart, Inc., Dallas, 1982-83, archtl. sec. leader, 1983, mgr. facilities design, 1983—. Chmn. youth edn. com. Wilshire Bapt. Ch., Dallas, 1990—. Named Best Citizen, Ozona Women's League, 1972, Outstanding Young Am., 1985. Mem. Constrn. Specifications Inst., AIA, Tex. Soc. of Architects, Nat. Coun. Archtl. Registration Bd.. Republican. Baptist. Avocations: photography, gardening, racquetball, softball, tennis. Office: Campbell Taggart Inc 6211 Lemmon Ave Dallas TX 75209-5788

SHAW, HENRY, chemical engineering educator; b. Paris, Oct. 25, 1934; came to U.S, 1947; s. Joseph B. and Sadie (Milstein) S.; m. Evelyn Goodman, Aug. 11, 1963; children: Laura Rachel, David Michael, Jessica Anne. PhD, Rutgers U., 1967; MBA, Rutgers U., Newark, 1976. Nuclear chem. engr. Bacbcock & Wilcox Co., Lynchburgh, Va., 1957-60, Mobil Oil Co., Princeton, N.J., 1961-65; instr. Rutgers U., New Brunswick, 1965-67; environ. mgr. Exxon Rsch. and Engring., Florham Park, N.J., 1967-86; prof. N.J. Inst. Tech., Newark, 1986—; invited mem. Nat. Commn. on Air Quality, Washington, 1980, com. mem. Nat. Rsch. Coun., Washington, 1980-93; chmn. of bd. Engring. Found., N.Y.C., 1988-90; chmn. adv. com. Oak Ridge (Tenn.) Nat. Lab., 1984-87. Developer university courses on design of non-polluting chemical processes; presenter sci. papers at profl. confs., seminars, symposia. Capt. U.S. Army, 1958. Recipient Rsch. grants Hazardous Substance Rsch. Ctr., Newark, 1989-92, EPA, DOE, DOD, Linden, N.J., 1973-80, fellowship, Allied Corp., Florham Park, N.J., 1966. Fellow AICE; mem. AAAS, Am. Chem. Soc., Engring. Found. (bd. mem., chmn. 1988-90). Achievements include development of air pollution control methods treating nitrogen oxides from power plants and gas turbines; 3 patents; plasma production of fine powders, control of particulates, catalytic NOx control. Home: 2 Gary Ct Scotch Plains NJ 07076 Office: NJ Inst Tech 138 Warren St Newark NJ 07102

SHAW, JOHN ANDREW, information systems executive; b. London, Ont., Can., July 15, 1957; arrived in Germany, 1986; m. Ute Behrendt, Aug. 11, 1986; 1 child, Julia. AA, Cen. Fla. Coll.; BS, Internat. Coll., Cayman Islands. Teaching asst. Control Data Inst., Toronto, Can., 1981; cons. C.T.C. Systems, Milan, 1982-83; gen. mgr. Cayman Corp. Computer Svcs., Cayman Islands, 1983-84; mktg. mgr. C.T.C. Group, Milan, 1985; sales mgr. Projekte-Software-Orgn., Cologne, Germany, 1986; cons. Interprogram GmbH, Langenfeld, Germany, 1987; mgr. Europe Mallinckrodt Med. GmbH, Hennef, Germany, 1987-92; pres. Shaware, Bonn, Germany, 1992—. Developer software. Sponsor Greenpeace. Mem. IEEE. Avocations: chess, karate. Home and Office: Rheinallee 51a, 53173 Bonn Germany

SHAW, JOHNNY HARVEY, civil engineer; b. Cullman, Ala., Apr. 18, 1938; s. John Nelson and Mittie Jane (Gardner) S.; m. Carolyn Sue Hays, Oct. 31, 1976 (dec. 1981); 1 child, Timothy; m. Catherine Elaine Barber, June 1, 1990; 1 child, Richard. BSCE, U. Ala., Birmingham. Registered profl. engr., Ala. With Jefferson County Commn., Birmingham, Ala., 1963—, chief draftsman, 1968-72, civil engr., 1972—; mem. Sewer Moratorium Com., Birmingham, 1972—. Mem. Elks, 1976, Eagles, 1980. Mem. NSPE, Am. Congress of Surveying and Mapping (Cert.), Soc. Mil. Engrs., Ala. Soc. Profl. Land Surveyors. Mem. Full Gospel Ch. Home: 211 Forest Dr Leeds AL 35094 Office: Jefferson County Commn Courthouse 716 N 21st St Birmingham AL 35236

SHAW, KEITH MOFFATT, engineer; b. Sydney, N.S.W., Australia, Sept. 16, 1944; s. Harold and Edna Mildred (Swain) S. B.E. in Mech. Engring., Sydney U., 1965, Diploma in Bldg. Sci., 1972. Engr., Elec. Engring. div. Email Ltd., Sydney, N.S.W., 1966-67, Air Conditioning and Refrigeration div., 1967-68, Rankine & Hill, Sydney, 1968; design engr. dept. arts and adminstrv. svcs. Australian constrn. svcs., Sydney, 1969—. Mem. Australian Inst. Refrigeration, Air-Conditioning and Heating, Bldg. Sci. Forum of Australia. Club: Sydney Rowing. Address: Australia Sq, PO Box H-189, 2000 Sydney Australia

SHAW, LAWRANCE NEIL, agricultural engineer, educator; b. Hallock, Minn., Mar. 15, 1934; s. Alexander and Marguerite (Ash) S.; m. Elizabeth A. Todd, July 9, 1966. BS, N.D. State U., 1957; MS, Purdue U., 1959; PhD, Ohio State U., 1969. Registered profl. engr., Fla., Maine. Extension agrl. engr. U. Maine, Presque Isle, 1959-67; grad. rsch. assoc. Ohio State U., Columbus, 1967-69; asst. prof. agrl. engring. U. Fla., Gainesville, 1969-74, assoc. prof. agrl. engring., 1974-79, prof. agrl. engring., 1979—; vis. prof. Nat. Inst. Agrl. Engring., Silsoe, Eng., 1976, Friedrich-Wilhelms U., Bonn, Germany, 1986. Contbr. papers to profl. publs. Named Tchr. of Yr., U. Fla. chpt. Alpha Zeta, 1989, Eminent Engr., U. Fla. chpt. Tau Beta Pi, 1989; USDA-SBIR grantee, 1990, 91. Mem. Am. Soc. Agrl. Engrs., Am. Soc. Hort. Sci., Fluid Power Soc., Kiwanis. Achievements include 4 patents in field. Home: 8715 NW 4th Pl Gainesville FL 32607 Office: U Fla IFAS Rogers Hall Gainesville FL 32611

SHAW, MARTIN ANDREW, clinical and research psychologist; b. N.Y.C., Jan. 27, 1944; s. Aaron S. and Betty Shaw; m. Dorothy Korot, Nov. 7, 1971; 1 child, Anatole Bernard. BS, NYU, 1966; MA, Dalhousie U., 1972; PhD, U. Wis., 1977; postgrad. Advanced Inst. Analytic Psychotherapy, 1977-82. Art tchr. N.Y.C. Schs., 1966-69; art therapist Kingsbridge VA Hosp., Bronx, N.Y., 1969-71; grad. teaching asst. Dalhousie U., 1971-72, Killam Children's Hosp., Halifax Sch. for Blind, N.S., Can., 1971-72; psychometrician, cons. N.Y.C. Bd. Edn., 1977-80; staff therapist, staff psychologist Advanced Ctr. for Psychotherapy, Jamaica, N.Y., 1977-82; clin./child psychologist Health Ins. Plan, Mental Health Service, N.Y.C., 1980-84; pvt. practice clin. and clin. child psychology, N.Y.C., 1981-83, Gt. Neck, N.Y., 1981—; cons. psychologist Hearing and Speech Ctr. of L.I. Jewish-Hillside Med. Ctr., New Hyde Park, N.Y., 1986-87, Trinity-Pawling (N.Y.) Sch., 1986—; founder, pres. ORT Inst., Inc.; trustee Signal Hill Edn. Ctr., Inc., 1983-87. Recipient Founders Day award, N.Y.U., 1966, VA commendation, 1970; lic. psychologist, N.Y. State. Fellow Soc. for Personality Assessment, Brit. Soc. for Projective Psychology (hon.); mem. Am. Psychol. Assn., N.Y. State Psychol. Assn., N.Y. Soc. Clin. Psychologists, Nassau County Psychol. Assn., AAAS, N.Y. Acad. Scis., Internat. Rorschach Soc., Soc. Psychoanalytic Rsch., Pi Lambda Theta. Author: Object Relations Technique: Objectified Assessment/Basic Rationale, 1993; contbr. papers to profl. jours. and confs. Office: 29 Barstow Rd Ste 102 Great Neck NY 11021-2209

SHAW, MARY ANN, psychologist; b. Dallas, July 5, 1937; d. Leon V. and Mabel (Bartlett) S.; B.S., U. Tex., 1959; M.Ed., U. Houston, 1966, Ed.D., 1973. Tchr. educable mentally retarded Spring Branch, Tex., 1959-64; vocat. counselor, Houston, 1964-66; psychometrist pvt. psychol. clinic, Houston, 1966-70; coordinator research Tex. Edn. Agency grant project, 1970-72; dir. psychol. services Tex. Scottish Rite Hosp. for Crippled Children, Dallas, 1972-82; dir. Dean Evaluation Ctr., Dallas, 1982-84; pvt. practice, 1982—; mem. clin. staff U. Tex. Health Sci. Center; cons. pvt. and public schs. Mem. Am. Psychol. Assn., Dallas Psychol. Assn., Assn. Pediatric Psychologists. Author: What Do I Do When, Because I Said So; contbr. article to profl. jour.; research in field.

SHAW, MARY ELIZABETH, plant pathologist; b. Champaigne, Ill., Mar. 28, 1950; d. Raymond Wallace and Elizabeth Jean (Pretzer) Rall; 1 child, Jesse Andrew Sorrell. MS, U. Calif., 1981, PhD, 1991. Technician Calif. Dept. Food and Agr., Sacramento, 1981-89; rsch. asst. U. Calif., Davis, 1989-91; rsch. assoc. Okla. State U., Stillwater, 1991-93; asst. prof. biology N.Mex. Highlands U., Las Vegas, 1993—. Recipient T.H. Shalla award U. CAlif. Davis Dept. Plant Patholog, 1991; USDA grantee 1992; Jasto Shields Rsch. fellow, 1990, U. CAlif. Davis fellow 1989. Mem. Am. Phytopathological Soc., Sierra Club (conservation chair 1992), Sigma Xi. Unitarian. Office: NMex Highlands U Life Sci Dept Las Vegas NM 87701

SHAW, MILTON CLAYTON, mechanical engineering educator; b. Phila., May 27, 1915; s. Milton Fredic and Nellie Edith (Clayton) S.; m. Mary Jane Greeninger, Sept. 6, 1939; children—Barbara Jane, Milton Stanley. B.S. in Mech. Engring, Drexel Inst. Tech., 1938; M.Eng. Sci., U. Cin., 1940, Sc.D., 1942; Dr. h.c., U. Louvain, Belgium, 1970. Research engr. Cin. Milling Machine Co., 1938-42; chief materials br. NACA, 1942-46; with Mass. Inst. Tech., 1946-61, prof. mech. engring., 1953-61, head materials processing div., 1952-61; prof., head dept. mech. engring. Carnegie Inst. Tech., Pitts., 1961-75; univ. prof. Carnegie Inst. Tech., 1974-77; prof. engring. Ariz. State U., Tempe, 1977—; Cons. indsl. cos.; lectr. in, Europe, 1952; pres. Shaw Smith & Assos., Inc., Mass., 1951-61; Lucas prof. Birmingham (Eng.) U., 1961; Springer prof. U. Calif. at Berkeley, 1972; Distinguished guest prof. Ariz. State U., 1977; mem. Nat. Materials Adv. Bd., 1971-74; bd. dirs. Engring. Found., 1976, v.p. conf. com., 1976-78. Recipient Outstanding Research award Ariz. State U., 1981; Am. Machinist award, 1972; P. McKenna award, 1975; Guggenheim fellow, 1956; Fulbright lectr. Aachen T.H., Germany, 1957; OECD fellow to Europe, 1964—. Fellow Am. Acad. Arts and Scis., ASME (Hersey award 1967, Thurston lectr. 1971, Outstanding Engring. award 1975, ann. meeting theme organizer 1977, Gold medal 1985, hon. 1980), Am. Soc. Lubrication Engrs. (hon., nat. award 1964), Am. Soc. Metals (Wilson award 1971, fellow 1981); mem. Internat. Soc. Prodn. Engring. Research (pres. 1960-61, hon. mem. 1975), Am. Soc. for Engring. Edn. (G. Westinghouse award 1956), Soc. Mfg. Engrs. (hon. mem. 1970, Gold medal 1958, internat. edn. award 1980), Nat. Acad. Engring., Polish Acad. Sci., Am. Soc. Precision Engrs. (hon.). Home: 2625 E Southern Ave Tempe AZ 85282-7601 Address: Arizona State Univ Tempe AZ 85287-6106

SHAW, MONTGOMERY THROOP, chemical engineering educator; b. Ithaca, N.Y., Sept. 11, 1943; s. Robert William and Charlotte (Throop) S.; m. Stephanie Habel, Sept. 5, 1966 (dec. 1989); 1 child, Steven Robert. B-ChemE, Cornell U., 1966, MS, 1966; MS, Princeton (N.J.) U., 1968, PhD, 1970. Engr., project scientist Union Carbide Corp., Bound Brook, N.J., 1970-76; assoc. prof. Dept. Chem. Engring., U. Conn., Storrs, 1977-83, prof., 1983—; sabbatical prof. Sandia Nat. Labs., Albuquerque, 1983-84; adv. bd. Jour. of Applied Polymer Sci., 1984-89. Co-author: Polymer-Polymer Miscibility, 1977. Grantee Alcoa Found., 1985, Exxon Edn. Found., 1986. Mem. IEEE (assoc. editor transactions on elec. insulation), Soc. Rheology (sec. 1977-81), Am. Chem. Soc., Am. Phys. Soc. Achievements include patents on rheological measurement method and apparatus and low density microcellular foams. Office: U Conn IMS 97 S Eagleville Rd Mansfield CT 06269-3136

SHAWHAN, PETER SVEN, physicist; b. Stockholm, Sweden, July 22, 1968; s. Stanley Dean and Susan (Jenkins) S.; m. Julia Ann Giardina, Aug. 22, 1992. AB in Physics, Washington U., 1990; MS in Physics, U. Chgo., 1992. Rsch. asst. Enrico Fermi Inst., Chgo., 1991--. Cellist U. Chgo. Symphony Orch., 1990--. U.S. Presdl. scholar, 1986, A.H. Compton scholar, 1986-90; Grad. Rsch. fellow NSF, 1990-, R.R. McCormick fellow U. Chgo., 1990-. Mem. Am. Phys. Soc. Home: 5455 S Blackstone Ave # 5A Chicago IL 60615 Office: Enrico Fermi Inst 5640 S Ellis Ave Chicago IL 60637

SHAY, EDWARD GRIFFIN, chemical engineer; b. Mt. Kisco, N.Y., Nov. 18, 1928; s. Edward Barnabus and Marjorie (Griffin) S.; m. Nan Donnelly, May 8, 1954 (dec. 1992); children: Christopher, Griffin, Guenever, Thaddeus, Hillary. BS in Chemistry, Fordham U., 1951; MS in Chem. Engring., Columbia U., 1953. Sect. leader R&D Atlantic Richfield Co., Phila., 1953-62; tech. dir. Onyx Chem. Co., Jersey City, 1962-69; mgr. R&D Avon products, Suffern, N.Y., 1969-78; sr. program officer Nat. Acad. Scis., Washington, 1978—; staff study dir. various books. With USMC, 1946-47. Achievements include 34 patents on chemicals and chemical processes. Home: 12769 Potomac Overlook Ln Leesburg VA 22075 Office: Nat Acad Scis 2101 Constitution Ave Washington DC 20418

SHEA, MARY ELIZABETH CRAIG, psychologist, educator; b. Gainesville, Fla., May 16, 1962; d. Charles Poe and Dolores Jean (Osborn) Craig; m. Steven John Shea, Sept. 1, 1991. BA, Ohio Wesleyan U., 1980; MA, Columbia U., 1986; PhD, U. S.C., 1990. Lic. clin. psychologist. Med./geriatric psychologist VA Med. Ctr., Columbia, S.C., 1990—; asst. prof. U. S.C. Sch. Medicine, Columbia, 1991—; pvt. practice, Chapin, S.C., 1991—. Contbr. articles to Archives Sexual Behavior, Clin. Psychology Rev., Jour. Sex Rsch., Jour. Social and Clin. Psychology, others. Mem. Am. Psychol. Assn., Am. Psychol. Soc., Southeastern Psychol. Assn., NOW, Greenpeace, Natural Resources Def. Coun., Save the Manatee Club, Psi Chi. Methodist. Office: VA Med Ctr Psychiatry Svc 116 Columbia SC 29201

SHEA, WILLIAM RENE, historian, philosopher of science, educator; b. Gracefield, Que., Can., May 16, 1937; s. Herbert Clement and Jeanne (Lafreniere) S.; m. Evelyn Fischer, May 2, 1970; children—Herbert, Joan-Emma, Louisa, Cecilia, Michael. B.A., U. Ottawa, 1958; L.Ph., Gregorian U., Rome, 1959; L.Th., Gregorian U., 1963; Ph.D., Cambridge U., Eng., 1968. Assoc. prof. U. Ottawa, Ont., Can., 1968-73; fellow Harvard U., Cambridge, Mass., 1973-74; prof. history and philosophy of sci. McGill U., Montreal, 1974—; dir. d'etudes Ecole des Hautes Etudes, Paris, 1981-82; sec.-gen. Internat. Union of history and Philosophy of Sci., 1983-89, pres., 1990-93; mem. gen. com. Internat. Coun. of Sci. Union, Paris, 1983-89; cons. Killam Found., Ottawa, Ont., 1983-85; mem. McGill Centre for Medicine, Ethics and Law, 1990—, Hydro-Quebec prof. Environ. Ethics, 1992—; vis. prof. Univ. Rome, 1992. Author: Galileo Intellectual Revolution, 1972, The Magic of Numbers and Motion, 1991; co-author: Galileo Florentine Residences, 1979; editor: Nature Mathematized, 1983, Otto Hahn and the Rise of Nuclear Physics, 1983, Revolutions in Science, 1988, Creativity in the Arts and Science, 1990, Persuading Science: The Art of Scientific Rhetoric, 1991, Interpreting the World, Science and Society, 1992. Can. Coun. fellow, 1965-68, Can. Cultural Inst. fellow, Rome, 1973, Social Scis. and Humanities Rsch. Coun. Can., 1980-81, Inst. of Advanced Studies in Berlin fellow, 1988-89; recipient The Alexandre Koyre medal Internat. Acad. of History of Sci., 1993, Knight of the Order of Malta, 1993. Fellow Royal Soc. Can.; mem. History of Sci. Soc. (coun. 1973-76), European Sci. Found. (standing com. for humanities 1989—), Can. Nat. Com. of History and Philosophy Assn., Internat. Acad. History of Sci. (ordinary), Order of Malta (knight 1993). Club: McGill Faculty. Home: 217 Berkely, Saint Lambert, PQ Canada J4P 3C9 Office: McGill Univ-McGill Centre, 3690 Peel St, Montreal, PQ Canada H3A 1W9

SHEALY, Y. FULMER, medicinal and organic chemist; b. Chapin, S.C., Feb. 26, 1923; s. L. Yoder and L. Essie (Fulmer) S.; m. Elaine Curtis, Oct. 5, 1950; children: Robin T., Nancy G., Priscilla B. BS, U. S.C., 1943; PhD, U. Ill., 1949. Chemist Office of Sci. R&D, 1943-45; postdoctoral fellow U. Minn., Mpls., 1949-50; chemist Upjohn Co., Kalamazoo, 1950-56; asst. prof. U. S.C., 1956-57; sr. chemist So. Rsch. Inst., Birmingham, 1957-59, sect. head, 1959-66, head medicinal chem. div., 1966-90, disting. scientist, 1990—; speaker 12th Internat. Cancer Congress, 3d Internat. Conf. on Prevention of Cancer. U.S. and fgn. Patentee for medicinal agents; pioneer in carbocyclic analogs of purine and pyrimidine nucleosides including carbodine, 2'-CDG and aristeromycin; synthesized anti-cancer drugs dacarbazine, clomesone and BCTIC; synthesized antifungal and antibacterial agents including the first monocyclic 1,2,5-Selenadiazoles and triazenylimidazole esters; synthesized antiviral, anticancer and cancer preventive agents including MCTIC (which laid the foundations for mitozolomide and temozolomide) and chloroethylnitrosocarbamates; synthesized cancer chemopreventative retinoids such as retinoylamino acids, retinyl ethers and 4-oxoretinoic acid derivatives. Contbr. more than 120 articles to profl. jours.; awarded 18 patents. Mem. AAAS, Am. Chem. Soc., Am. Assn. Cancer Rsch., Internat. Soc. Nutrition and Cancer, N.Y. Acad. Scis., Internat. AIDS Soc., Am. Pharm. Assn., Internat. Soc. Heterocyclic Chemistry, Internat. Soc. Antiviral Rsch., European Retinoid Rsch. Group, Pharm. Soc. Japan, Sigma Xi, Phi Beta Kappa, Phi Lambda Upsilon, Pi Mu Epsilon. Office: So Rsch Inst 2000 9th Ave S Birmingham AL 35205-2708

SHEAR, WILLIAM ALBERT, biology educator; b. Coudersport, Pa., July 5, 1942. BA, Coll. Wooster, 1963; MS, U. N.Mex., 1965; PhD in Evolutionary Biology, Harvard U., 1971. Asst. prof. Concord Coll., 1970-74; assoc. prof. Hampden-Sydney (Va.) Coll., 1974-80, prof. Biology, 1981—; rsch. assoc. Am. Mus. Natural History, 1978—. Recipient John Peter Mettauer award, 1980, Cabell award 1985. Mem. Am. Arachnol. Soc., Sigma Xi. Achievements include research in behavior, taxonomy and biogeogrpahy of arachnids and myriapods, early evolution of land animals, revisions of families adn genera of Opiliones and Diplopoda, especially North American forms, web-building behavior os spiders, Devonian fossils of arachnids and myriapods. Office: Hampden Sydney Coll Dept Biology Hampden Sydney VA 23943*

SHEARD, CHARLES, III, dermatologist; b. Toronto, Ontario, Can., Nov. 22, 1914; came to U.S., 1945; s. Charles Jr. and Alice Elizabeth (Ramsay) S.; m. Katherine Patricia Murphy, Nov. 19, 1937; children: Joan Virginia Sheard Cumming, Pamela Carol Sheard McGuiness, Wendy Alice Sheard Geyer. Sr. matriculation, Upper Can. Coll., Toronto, 1933; MD, U. Toronto, 1939. Diplomate Am. Bd. Dermatology. Intern Toronto Gen. Hosp., 1939-40; instr. physiology, anatomy U. Toronto Med. Faculty, 1940-41; surgical asst. resident Hosp. for Sick Children, Toronto, 1945; chief resident dermatologist Columbia Presbyn. Hosp., N.Y.C., 1945-49; assoc. prof. medicine Cornell U. Med. Coll., N.Y.C., 1950—. Author: (textbook) Treatment in Dermatology, 1948; contbr. articles to profl. jours. Flight lt. RCAF, 1941-45. Fellow ACP, Royal Coll. Physicians of Can.; mem. Metro-Manhattan Dermatol. Soc.N.Y.C. (sec. 1970-80), Can. Club of N.Y., Royal Can. Yacht Club, St. George's Club, Wee Burn Golf & Country Club Darien. Republican. Episcopalian. Avocations: sailing, fishing, golf. Office: Charles Sheard MD PC 440 Bedford St Stamford CT 06901-1599

SHEARER, WILLIAM T., pediatrician, educator; b. Detroit, Aug. 23, 1937. BS, U. Detroit, 1960; PhD, Wayne State U., 1966; MD, Washington U., St. Louis, 1970. Diplomate Am. Bd. Pediatrics, Am. Bd. Allergy & Immunology, Nat. Bd. Med. Examiners; cert. in diagnostic lab. immunology. Post-doctoral fellow in biochemistry dept. chem. Indiana U., Bloomington, 1966-67; intern in pediatrics St. Louis Children's Hosp., 1970-71, resident in pediatrics, 1971-72, resident in immunology in pediatrics, 1972-74; dir. divsn. allergy and immunology St. Loius Children's Hosp., 1974-78; spl. USPHA sci. rsch. fellow in medicine dept. medicine Washington U., 1972-74, assoc. prof., 1978, prof., 1978; prof. pediatrics and microimmunology Baylor Coll. Medicine, Houston, 1978—, dir. AIDS rsch. ctr., 1991—; head sect. allergy & immunology Tex. Children's Hosp., Houston, 1978—; mem. ACTU community adv. bd. Tex. Children's Hosp., 1991; ad hoc reviewer NIAID-NIH, Bethesda, Md., 1991; mem. AIDS ad hoc working group NHLBI-NIH, Bethesda, 1991; chair Am. Bd. Allergy & Immunology, 1990—; dir. pediatric HIV/AIDS Clin. Rsch. Ctr., Houston, 1988—. Guest editor Seminar Pediatric Infectious Disease, 1990; contbr. introduction: Allergy: Principles and Practice, 1992; contbr. articles to profl. jours. including New Eng. Jour. Medicine. AIDS cons. Houston Ind. Sch. Dist., 1986—; med. adv. Spring Branch Ind. Sch. Dist., Houston, 1987—; chmn. community HIV/AIDS adv. group Tex. Med. Ctr., 1991—. NIH grantee, 1988—; recipient Rsch. Scholar award Cystic Fibrosis Found., 1974-77, Faculty Rsch. award Am. Cancer Soc., 1977-79, Myrtle Wreath award Hadassah, 1985. Mem. Am. Soc. Clin. Investigation, Am. Acad. Pediatrics (mem. exec. com. sect. allergy & immunology) 1991—). Achievements include research in half-matched T-cell-depleted bone marrow transplants, in membrane signal pathway of human B lymphcytes. Office: Baylor Coll Medicine Dept Pediatrics 1 Baylor Plz Houston TX 77030

SHEARWIN, KEITH EDWARD, biochemist; b. Brisbane, Australia, Mar. 9, 1965; came to U.S., 1990; s. Edward John and Evelyn Joyce (Norup) S. BSc with honors, Griffith U., 1986; PhD, U. Queensland, Brisbane, 1990. Rsch. asst. biochemistry dept. U. Queensland, 1987-90; postdoctoral rsch. assoc. biochemistry dept. Brandeis U., Waltham, Mass., 1990-93; with dept. biochemistry Adelaide U., South Australia, 1993—. Contbr. articles to profl. jours. Mem. Am. Soc. for Advancement of Sci. Achievements include study of protein-protein interactions using techniques of physical biochemistry. Office: Adelaide U, Dept Biochemistry North Terr, Adelaide 5001, Australia

SHEATH, ROBERT GORDON, botanist; b. Toronto, Ont., Can., Dec. 26, 1950; came to U.S., 1978; s. Harry Gordon and Shirley Irene (Rose) S.; B.Sc., U. Toronto, 1973, Ph.D., 1977. NRC Can. postdoctoral fellow U. B.C., 1977-78; asst. prof. aquatic biology U. R.I., Kingston, 1978-82, assoc. prof., 1982-86, chmn. dept. botany, 1986-91; head dept. botany Meml. U., St. Johns, Nfld., Can., 1991—. NSF grantee, 1980—. Recipient G.A. Cox Gold medal, U. Toronto, 1973. Mem. Phycological Soc. Am. (Bold award 1976), Am. Inst. Biol. Scis., Am. Soc. Limnology and Oceanography, Arctic Inst. N.Am., Brit. Phycological Soc., Internat. Phycological Soc., N.Y. Acad. Sci., Nature Conservancy, Sigma Xi, Phi Kappa Phi. Author papers in field. Office: Meml U, Dept Biology, Saint John's, NF Canada A1B 3X9*

SHEBUSKI, RONALD JOHN, pharmaceutical company executive; b. Green Bay, Wis., Feb. 26, 1953; m. Cynthia C. Wilson, Aug. 11, 1979; children: Jessica, Ronald Jr., Neil. BS in Bacteriology, U. Wis., 1976; PhD in Pharmacology, U. Minn., 1985. Cardiovascular rsch. asst. Dept. Cardiology, Univ. Hosps., U. Wis., Madison, 1976-77; biology/biochemistry asst. II exptl. biology rsch. Upjohn Co., Kalamazoo, 1977-81; assoc. sr. investigator cardiovascular pharmacology Smith Kline & French Labs., King of Prussia, Pa., 1985-86, sr. investigator cardiovascular pharmacology, 1986-88; sr. rsch. fellow dept. pharmacology Merck Sharp & Dohme Rsch. Labs., West Point, Pa., 1988-90; assoc. dir. cardiovascular diseases rsch. Upjohn Co., Kalamazoo, 1990-92, dir. cardiovascular diseases rsch., 1992—. Contbr. articles to profl. jours.; reviewer Jour. Biochem. Pharmacology, Jour. Cardiovascular Pharmacology, Jour. of Pharmacology and Exptl. Therapeutics, Circulation, Blood, Protaglandins. Recipient Travel award Am. Soc. Pharmacology and Exptl. Therapeutics, 1984, Bacaner Basic Sci. Rsch. award for outstanding grad. rsch. in pharmacology, 1985. Mem. AAAS, Am. Heart Assn., Am. Soc. Pharmacology and Exptl. Therapeutics, Hematology Soc. Phila., N.Y. Acad. Scis., Physiol. Soc. Phila., Thombosis Coun. of Am. Heart Assn. Home: 4334 Sunnybrook Dr Kalamazoo MI 49008 Office: Upjohn Labs Cardiovascular Disease Rsch 301 Henrietta St Kalamazoo MI 49001

SHEDD, DONALD POMROY, surgeon; b. New Haven, Aug. 4, 1922; s. Gale and Marion (Young) S.; m. Charlotte Newson, Mar. 17, 1946; children—Carolyn, David, Ann, Laura. B.S., Yale U., 1944, M.D., 1946. Diplomate Am. Bd. Surgery. Intern Yale New Haven Hosp., 1946-47, asst. resident, resident, 1949-53; instr. surgery Yale U. Med Sch., New Haven, 1953-54, asst. prof., 1954-56, assoc. prof., 1956-67; chief dept. head and neck surgery Roswell Park Meml. Inst., Buffalo, 1967—. Co-editor: Surgical and Prosthetic Speech Rehabilitation, 1980, Head and Neck Cancer, 1985; contbr. numerous articles to profl. jours. Founding bd. dirs. Hospice Buffalo, Inc., 1973-83. Served to capt. U.S. Army, 1947-49. Mem. Soc. Univ. Surgeons, New Eng. Surg. Soc., Soc. Head and Neck Surgeons (pres. 1976-77). Avocations: sailing; windsurfing; tennis. Home: 671 Lafayette Ave Buffalo NY 14222-1435 Office: Roswell Park Meml Inst 666 Elm St Buffalo NY 14263-0001

SHEDDEN, JAMES REID, avionics engineer, consultant; b. Englewood, N.J., Apr. 2, 1956; s. Charles Hamilton and Helena Catherine (Kluk) S. BSEE, N.J. Inst. Tech., Newark, 1978. Project engr. Underwriter's Labs., Melville, N.Y., 1978-80; product engr. Merganthaler Linotype, Melville, 1980-81; project engr. Veeco Instruments, Plainview, N.Y., 1981-82; staff engr. GEC Marconi Electronic Systems, Totowa, N.J., 1982—; owner, operator Saturday Night Software Design. Mem. Turbo Users Group. Achievements include design engr. on F14D/E2-C interface unit for Joint Tactical Info. Distbn. System; design of data acquisition system for mass spectrometer leak detector; design of software database system to accurately index, formulate and price automotive paints; design of normalized relational model and data translator for access to chemotherapy survey information. Home: 88 Lakeside Trl Kinnelon NJ 07405-2872 Office: GEC-Marconi Electronic Systems MS11B51 PO Box 975 164 Totowa Rd Wayne NJ 07474-0975

SHEDLOWSKY, JAMES PAUL, engineer; b. Pontiac, Mich., Feb. 10, 1937; s. Joseph Harold and Tommie Lee (Vickory) S.; m. Nancy Jean Cece, Sept. 12, 1964 (div. Apr. 1971); children: Jamie Lynn, Stanley Thomas; m. Janice Emily Brooks, Nov. 28, 1975; stepchildren: Brian Allen Tate, Jeffrey Scott Tate, Christopher Edward Tate. BS in Engring. Physics, U. Mich., 1960. Noise/vibration devel. engr. GM Corp., Milford, Mich., 1963-68; test & devel. engr. GM Corp., Pontiac, Mich., 1968-75, design reslease & devel. engr., 1975-85, engring. systems mgr., 1985-89; staff devel. engr., tech. mgr. GM Corp., Milford, Mich., 1989—. Contbr. articles to profl. jours. Lt. U.S. Army, 1961-63. Mem. Am. Inst. Physics, Acoustical Soc. Am., Soc. Automotive Engrs., Inst. Noise COntrol Engring. Roman Catholic. Achievements include developed the original concept and initial implementation of the "Hush Panel"; contbr. tothe innovative revisions of automobile accoustical packaging. Home: 3033 Shawnee Ln Waterford MI 48329 Office: GM Corp MCD/NAO 9040 Bldg 16 GM Proving Ground Milford MI 48380-3726

SHEEHAN, JERRARD ROBERT, technology policy analyst, electrical engineer; b. Manhasset, N.Y., Sept. 30, 1964; s. Jerry F. and Patricia E. (Langton) S.; m. Elizabeth Anne Murphy, Aug. 12, 1990. SB in Electrical Engring., MIT, 1986, SM in Technology and Policy, 1991. Mem. rsch. staff System Planning Corp., Arlington, Va., 1986-89; rsch. asst. MIT, Cambridge, Mass., 1989-91; staff assoc. Nat. Rsch. Coun., Washington, 1991; cons. U.S. Congress, Office of Tech. Assessment, Washington, 1991—. MacArthur Found. grad. rsch. fellow, 1989-91. Mem. Tau Beta Pi, Eta Kappa Nu, Sigma Xi (assoc.). Achievements include published research on U.S. industrial competitiveness, defense conversion and technology policy in the electronics industry. Office: US Congress Office of Tech Assessment Washington DC 20510-8025

SHEEHAN, RICHARD LAURENCE, JR., packaging engineer; b. Detroit, Aug. 26, 1946; s. Richard and Ethel S.; m. Melinda Elledge, Oct. 28, 1982; children: Timothy, Christopher. BS, Mich. State U., 1968. Cert. quality engr.; cert. packaging profl. Packaging engr. 3M, St. Paul, 1972—. Lt. (j.g.) USN, 1968-72. Mem. Am. Soc. Testing and Materials (edit. bd. 1979—, cive chmn. 1984-90), Am. Soc. Quality Control, Inst. Packaging Profls. Office: 3M Co 3M Center 230-IE-04 Saint Paul MN 55144

SHEEHAN, WILLIAM FRANCIS, chemist educator; b. Chgo., Oct. 19, 1926; s. William Francis and Catherine Henriette S. BS with Honors, Loyola U., 1948; PhD, Calif. Inst. Tech., 1952. Chemist Shell Devel. Co., Emeryville, Calif., 1952-55; prof. Santa Clara (Calif.) U., 1955-91, prof. emeritus, 1991—. Contbr. articles to profl. publs. Roman Catholic. Home: 2614 Via Berrenda Santa Fe NM 87505

SHEELER, JOHN BRIGGS, chemical engineer; b. Anita, Iowa, Oct. 25, 1921; s. Ivan Howard and Mildred Lucile (Briggs) S.; m. Charlsee A. Pitt (div. 1960); m. Mary Irene Squire, July 18, 1970; children: John R., Daniel, Anne, RObert, William, Diane, James, John E. BS, Iowa State U., 1950, PhD, 1956. Rsch. asst. Iowa Engring. Experiment Sta., Ames, 1950-56; asst. prof. Iowa State U., Ames, 1956-59, assoc. prof., 1959-88, prof. emeritus, 1988—; cons. Iowa Dept. Transp., Ames, 1950-88. Contbr. scientific articles to profl. jours. Coach, vol. NROTC (Rifle and Pistol Team) Iowa State U., 1947-88. Sgt. USMC, 1941-47. Fellow Iowa Acad. Sci. (emeritus); mem. Masons, Moose Lodge, SIgma Xi (emeritus). Presbyterian. Home: 505 Bel Aire Dr Marshalltown IA 50158

SHEEM, SANG KEUN, optical engineering researcher; b. Seoul, Korea, Mar. 20, 1944; s. Eung-Taek and Ki-Jik (Oh) S.; m. Susan Kim, Mar. 22, 1970; children: Edward J., Shana J. MS in Engring., U. Calif., 1973, PhD in Engring., 1975. Rsch. physicist U.S. Naval Rsch. Lab., Washington, 1976-81; mgr. Rockwell Internat., Dallas, 1981-86; project engr. Lawrence Livermore (Calif.) Nat. Lab., 1986--; mem. corp. optical panel Rockwell Internat., Dallas, 1982-86; cons. Katron Fiber Optic Co., Palo Alto, Calif., 1987-88, Amaco Rsch. Ctr., Naperville, Ill., 1986-87. Contbr. articles, referee to profl. jours. Mem. IEEE, Optical Soc. Am., Korean Scientist and Engr. Assn. (pres. No. Calif. chpt. 1990-91). Achievements include 15 patents in fiber optics and integrated optics. Office: Lawrence Livermore Nat Lab 7000 East Ave L-407 Livermore CA 94550

SHEEN, PORTIA YUNN-LING, retired physician; b. Republic of China, Jan. 13, 1919; came to U.S., 1988; d. Y. C. and A. Y. (Chow) Sheen; m. Kuo, 1944 (dec. 1970); children: William, Ida, Alexander, David, Mimi. MD, Nat. Med. Coll. Shanghai, 1943. Intern, then resident Cen. Hosp., Chungking, Szechuan, China, 1943; with Hong Kong Govt. Med. and Health Dept., 1948-76; med. supt. Kowloon Hosp., Kowloon, Hong Kong, 1948-63, Queen Elizabeth Hosp., Kowloon, Hong Kong, 1963-73, Med. and Health Hdqrs. and Health Ctr., Kowloon, Hong Kong, 1973-76, Yan Chai Hosp., New Territories, Hong Kong, 1976-87. Fellow Hong Kong Coll. Gen. Practitioners; mem. AAAS, British Med. Assn., Hong Kong Med. Assn., Hong Kong Pediatric Soc., N.Y. Acad. Sci. Methodist. Avocations: reading, music. Home: 1315 Walnut St Berkeley CA 94709-1408

SHEESLEY, JOHN HENRY, statistician; b. Harrisburg, Pa., May 26, 1944; s. Norman Austin and Helen Kirk (Brown) S.; m. Lynn Thomas, June 1, 1968; children: Samantha, Emily Corinn, Amanda Lorin. BS in Math., Rutgers U., 1968, Lafayette Coll., 1966. Registered profl. engr., Calif. Quality planning engr. Western Electric Co., Allentown, Pa., 1968-73; sr. statistician GE, Cleve., 1973-85; lead statistician Air Products and Chemicals, Inc., Allentown, 1985—; chmn. SQC Roundtable, Allentown, 1990; presenter Pitts. Conf., 1991. Contbg. author: Experiments in Industry; editor ASSIST, 1986; contbr. articles to Jour. Quality Tech. Math. coun. Cedar Crest Coll., Allentown, 1988-92. Mem. ANSI (mem. subcom. Z11 1988-89), Am. Soc. for Quality Control (sr. mem., mem. statis com. 1974-76, Brumbaugh award), Am. Statis. Assn. Republican. Presbyterian. Home: 1713 Laurel Ln Orefield PA 18069 Office: Air Products and Chems 7201 Hamilton Blvd Allentown PA 18195

SHEETS, ROBERT CHESTER, meteorologist; b. Marion, Ind., June 7, 1937; 3 children. BS, Ball State Tchrs. Coll., 1961; MS, U. Okla., 1965, PhD, 1972. Chief forcaster USAF Air Weather Svc., Ft. Knox, Ky., 1961-64; asst. meteorologist U. Okla, 1964-65; rsch. meteorologist Nat. Oceanic & Atmospheric Adminstrn., U.S. Dept. Commerce, 1965-74; supervisory meteorologist, acting chief hurricane group Environ. Rsch. Lab., Nat. Hurricane & Experimental Lab., 1975-88; dir. Nat. Hurricane Ctr., 1988—. Mem. Am. Meteorol. Soc., Weather Modification Assn. Achievements include rsch. in experimental and theoretical rsch. on the formation, motion, intensity, scale, interactions and structure of hurricanes and other tropical storms; emphasis on hurricane modification schemes and evaluation techniques. Office: Nat Weather Svc Nat Hurricane Ctr 1320 S Dixie Hwy-IRE Bldg Miami FL 33146*

SHEFCHICK, THOMAS PETER, forensic electrical engineer; b. Trenton, N.J., Apr. 17, 1947; s. Peter and Ann Shefchick. BSEE, Rutgers U., 1969. Registered profl. engr., N.J., Pa., Del., N.Y., Calif., Hawaii. Elec. field engr. Westinghouse Electric Corp., Phila., 1969-79; elec. project mgr. Am. Biltrite Corp., Trenton, 1979-80; sr. elec. design engr. Air Products and Chems. Inc., Trexlertown, Pa., 1980-81; forensic elec. engr. Shefchick Assocs., West Trenton, N.J., 1983-92, Honolulu, 1989—. Fellow Am. Acad. Forensic Scientists (chair mem. com. 1991—); mem. IEEE, Soc. Am. Mil. Engrs., Nat. Acad. Forensic Engrs. (sr., diplomat), Internat. Assn. Arson Investigators (chmn. engring. com. 1987-93, bd. dirs. N.J. chpt. 1983-89), Internat. Assn. Forensic Scis. (chmn. engring. com. 1993—), Mercer County Profl. Engrs., Honolulu Profl. Engrs. Office: 733 Bishop St Ste 2057 Honolulu HI 96813

SHEFFIELD, JOEL BENSEN, biology educator; b. Bklyn., Dec. 30, 1942; m. Lucy Paige, June 20, 1965; 1 child, Jennifer. AB, Brandeis U., 1963; PhD, U. Chgo., 1969. Rsch. scientist Inst. Med. Rsch., Camden, N.J., 1971-77; prof. biology Temple U., Phila., 1977—. Editor book series: Ocular and Visual Development; contbr. articles to profl. jours. Mem. AAAS, Am. Soc. Cell Biology, Soc. for Neurosci., Assn. Rsch. in Vision and Ophthalmology. Office: Temple U Dept Biology Philadelphia PA 19122

SHEFFIELD, LEWIS GLOSSON, physiologist; b. Adel, Ga., Oct. 30, 1957; s. Eugene Davis and Martha Sue (Sinclair) S.; m. Mary Frances Tanner, July 18, 1980. MS, Clemson U., 1980; PhD, U. Mo., 1983. Rsch. asst. Clemson (S.C.) U., 1978-80, U. Mo., Columbia, 1980-83; postdoctoral assoc. Mich. State U., East Lansing, 1983-86; asst. prof. dairy sci. dept. U. Wis., Madison, 1986-91, assoc. prof., 1991—, dir. endocrinology-reproductive physiology program, 1990—. Contbr. articles to profl. jours. Recipient

First award NIH, 1988. Mem. Am. Dairy Sci. Assn. (milk synthesis chair 1991-92), Com. on Mammary Gland Biology, Endocrine Soc., Sigma Xi. Achievements include demonstration that epidermal growth factor interacts with estrogen and progesterone to regulate mammary devel. and are working to understand the cellular and molecular basis of that interaction; that prolactin causes a decrease in epidermal growth factor-induced growth responses, which appears to be related to mammary gland differentiation. The molecular regulation of this response is also under investigation. Office: U Wis 1675 Observatory Dr Madison WI 53706

SHEIKH, SUNEEL ISMAIL, aerospace engineer, researcher; b. Bristol, Gloucester, Eng., Jan. 21, 1966; came to U.S., 1975, U.S. Citizen 1987; s. Hyder Ismail and Joan Mary (Duncan) S. BS in Aerospace Engring, Maths., U. Minn., 1988; MS in Aeronautics and Astronautics, Stanford U., 1990. Student intern Honeywell, Inc., Mpls., 1989-90; assoc. engr. Martin Marietta Corp., Denver, 1990-91; rsch. scientist Honeywell, Inc., Mpls., 1991—; cons., engr. Honeywell, Inc., Mpls., 1991. Recipient Honorable Mention award NSF, 1988. Mem. AIAA, Planetary Soc. Home: 1012 Thomas Ave S Minneapolis MN 55405 Office: Honeywell Inc 3660 Technology Dr Minneapolis MN 55418

SHEIN, SAMUEL T., clinical psychologist; b. Salem, Mass., Mar. 19, 1932; s. William and Bessie (Resnick) S.; m. Susan Lamer; children: Pamela, Karen, Meredith, Danielle. AB, Boston U., 1954; MA, PhD, NYU, 1961; postgrad., Psychoanalytical Tng. Inst. Nat. Psychol. Assn. for Psychoanalysis, N.Y.C., 1962-70. Lic. psychologist, N.J., Pa., N.Y. Instr. psychology NYU, N.Y.C., 1957-58; intern Bergen Pines County Hosp., Paramus, N.J., 1958-59, staff psychologist, 1959-61, supervising psychologist, 1961-63; chief psychologist, adminstrv. dir. Community Mental Health Ctr., Dumont, N.J., 1963-66; psychologist Learning Lab., Spl. Edn., Wood-Ridge, N.J., 1966-69; pvt. practice Teaneck, N.J., 1962—; founding bd. dirs., officer Delta Inst. for Prevention of Violence, Abuse and Victimization, Essex and Bergen Counties, N.J., 1990—; guest clin. psychologist TV show Nine Broadcast Plaza, Cable and Sta. WWOR, N.Y.; guest lectr. Bergen County Conf. on Learning Disabilities, Paramus, 1991, JCC-on-Palisades, Tenafly, N.J., 1988—, on Communication and Intimate Relationships. Cited in book Star Treatment (Dick Stelger), 1977. Sec., v.p. profl. adv. com. Bergen County Mental Health Bd., 1978-81; bd. dirs., co-founder Cliffwood Community Mental Health Ctr., Englewood, N.J., 1980-82. Mem. APA, N.J. Psychol. Assn., N.J. Acad. Psychology, Assn. Lic. Psychologists, Bergen County Psychol. Assn. (past pres.), Wilton Psychotherapy Ct. (founder, dir. 1982—). Office: Wilton Med Bldg North 757 Teaneck Rd Teaneck NJ 07666

SHELANSKI, MICHAEL L., cell biologist, educator; b. Phila., Oct. 5, 1941; s. Herman Alder and Bessie B.; m. Vivien Brodkin, June 9, 1963; children: Howard, Samuel, Noah. Student, Oberlin Coll., 1959-61; M.D. (Life Ins. Med. Research Fund fellow), U. Chgo., 1966, Ph.D., 1967. Intern in pathology Albert Einstein Coll. Medicine, N.Y.C., 1967-68; fellow in neuropathology Albert Einstein Coll. Medicine, 1968-70, asst. prof. pathology, 1969-74; staff scientist NIH, Bethesda, Md., 1971-73; vis. scientist Inst. Pasteur, Paris, 1973-74; assoc. prof. neuropathology Harvard U., Cambridge, Mass., 1974-78; sr. research assoc., asst. neuropathologist Children's Hosp. Med. Center, Boston, 1974-78; prof., chmn. dept. pharmacology N.Y. U. Med. Center, N.Y.C., 1978-86; Delafield Prof., chmn. dept. pathology Coll. Physicians and Surgeons, Columbia U., N.Y.C., 1987—; dir. pathology services Presbyn. Hosp., N.Y.C., 1987—; mem. Neurology A study sect. NIH, 1974-78, Pharmacological Scis. study sect., 1986-90; mem. sci. and med. adv. bd. Alzheimer's Disease and Related Disorders Assn., 1985-92, sec., 1987-92, bd. dirs., 1991-92; chmn. overhead powerline adv. panel State of N.Y., 1981-87; dir. Alzheimer's disease rsch. ctr. Columbia U., 1989—; mem. Am. Cancer Soc. IRG Panel, 1989-93, sci. adv. bd. Dystonia Assn., Amyotrophic Lateral Sclerosis Assn. Mem. editorial bd. Jour.Neurochemistry, 1982-90, Jour. Neuropathology and Exptl. Neurology, 1983-85, Neuroscis., 1985—, Neurobiology of Aging, 1988—, Lab. Investigation, 1989—, Brain Pathology, 1990—. Served as sr. asst. surgeon USPHS, 1971-73. Guggenheim fellow, 1973-74. Mem. Am. Soc. Cell Biology, Am. Assn. Neuropathologists, Assn. Med. Coll. Pharmacologists, Am. Soc. Neurochemistry. Achievements include research on fibrous proteins of brain, aging of human brain, devel. neurobiology. Office: Columbia U Coll Physicians and Surgeons Dept Pathology 630 W 168th St New York NY 10032-3702*

SHELBY, BEVERLY JEAN, quality assurance professional; b. Ft. Wayne, Ind., Oct. 29, 1965; d. Thomas Alfred Shelby and Juanita Jean (Goodnight) Evans. BS in Indsl. Engring., Purdue U., 1988. Indsl. engr. Ft. Wayne Foundry, Columbia City, Ind., 1989-90; quality assurance engr. Albion Wire, Inc., Kendallville, Ind., 1990-91, quality assurance mgr., 1991—. Mem. Am. Soc. Quality Control, Inst. Indsl. Engrs. Office: Albion Wire Inc 811 Commerce Dr Kendallville IN 46755

SHELBY, JAMES ELBERT, materials scientist, educator; b. Memphis, Mar. 11, 1943; s. James Elbert and Jessie Mae (Morris) S.; divorced; 1 child, Stephanie. BS, U. Mo., 1965, MS, 1967, PhD, 1968. Mem. tech. staff Sandia Nat. Lab., Livermore, Calif., 1968-82; prof. glass sci. N.Y. State Coll. Ceramics, Alfred, 1982—; cons. Corning (N.Y.) Inc., 1987-90, Delco, Milw., 1991-92, Prexair, N.Y., 1991-92, Molten Metal Tech., Mass., 1992-93; chair Gordon conf. on glass Gordon Rsch. Found., 1985. Contbr. over 190 articles to profl. jours. Fellow Am. Ceramic Soc. (George W. Morey award 1975); mem. Soc. Glass Tech. Republican. Achievements include leading authority on interaction of gases with glasses and melts; recognized authority on preparation and properties of glasses; discoverer of many new glass-forming systems. Office: NY State Coll Ceramics 2 Pine St Alfred NY 14802

SHELBY, JAMES STANFORD, cardiovascular surgeon; b. Ringgold, La., June 15, 1934; s. Jesse Audrey and Mable (Martin) S.; BS in Liberal Arts La. Tech. U., 1956; MD, La. State U., 1958; m. Susan Rainey, July 15, 1967; children: Bryan Christian, Christopher Linden. Intern, Charity Hosp. La., New Orleans, 1958-59, resident surgery and thoracic surgery, 1959-65; fellow cardiovascular surgery Baylor U. Coll. Medicine, Houston, 1965-66; practice medicine specializing in cardiovascular surgery, Shreveport, La., 1967—; mem. staff Schumpert Med. Ctr., Highland Hosp., Willis-Knighton Med. Ctr.; assoc. prof. surgery La. State U. Sch. Medicine, Shreveport, 1967—. With M.C., AUS, 1961-62. Diplomate Am. Bd. Surgery, Am. Bd. Thoracic Surgery. Recipient Tower of Medallion award La. Tech. U., 1982. Mem. Am. Coll. Cardiology, AMA, Soc. Thoracic Surgeons, Am. Heart Assn., Southeastern Surg. Congress, So. Thoracic Surg. Assn. Home: 6003 E Ridge Dr Shreveport LA 71106-2425 Office: 2751 Virginia Ave Ste 3B Shreveport LA 71103-3987

SHELBY, WILLIAM RAY MURRAY, logistics engineer; b. Mt. Pleasant, Pa., Nov. 30, 1946; s. Joseph J. and Louella (Murray) S. BA, U. Pa., 1968; BSME, U. Mich., 1981. Liaison engr. Chevrolet div. GM, Flint, Mich., 1975-81; logistics engr. GDE Systems, Inc., San Diego, 1981—. Mem. Soc. Logistics Engrs. Achievements include developing computer models to analyze maintenance and supply support requirements for state of art defense electronics. Home: 13532 Maryearl Ln Poway CA 92064

SHELDON, MARK SCOTT, research engineer; b. Orange, Calif., May 19, 1959; s. Howard Lezurn and Vida Louise (Winegar) S.; m. Marti Reisman, Aug. 8, 1986. BS in Engring. and Applied Sci., Calif. Inst. Tech., 1981; MSME, Cornell U., 1985. Rsch. engr. Energy and Environ. Rsch. Corp., Irvine, Calif., 1985-91, sr. rsch. engr., 1991—. Mem. ASME (assoc.). Mem. Reorganized LDS Ch. Office: Energy and Environ Rsch Corp 18 Mason Irvine CA 92718-2798

SHELDRICK, GEORGE MICHAEL, chemistry educator, crystallographer; b. Huddersfield, Great Britain, Nov. 17, 1942; s. George and Elizabeth S.; m. Katherine E. Herford, 1968; 4 children. Student, Huddersfield New Coll., Jesus Coll., Cambridge. Lectr. Cambridge U., Eng., 1966-78; prof. inorganic chemistry U. Göttingen, Germany, 1978—. Contbr. articles to profl. jours. Recipient Meldola and Corday-Morgan medals Royal Soc. Chemistry, Leibniz prize Deutsche Forschungsgemeinschaft, A.L. Patterson award Am. Crystallographic Assn., 1993. Achievements include design of widely used computer program for crystal structure determination. Office: U Göttingen-Facultat Chemie, Postfach 3744, D-3400 Göttingen Germany*

SHELKE, KANTHA, cereal chemist; b. Bangalore, Mysore, India, Mar. 9, 1957; d. Hari Hara and Saroja (Ramaswamy) Subramanian; children: Tara, Nikhil. MSc in Chemistry, Bangalore U., India, 1978; PhD in Cereal Sci., N.D. State U., 1986. Instr. N.D. State U., Fargo, 1986-87; instr., rsch. assoc. Kans. State U., Manhattan, 1987-92; scientist Grand Met. Food Sector, Mpls., 1992-93; rsch. assoc. Continental Baking Co., St. Louis, 1993—; cons. in field, 1987-92. Columnist: (monthly column) Sunflower Mag., 1986-87; contbr. articles to profl. jours. Mem. AAUW, Manhattan, Kans., 1989-92. Recipient Nat. Sci. Talent award Commonwealth Sci. and Indsl. Rsch. Orgn./Nat. Coun. on Edn., Rsch. and Tng., New Delhi, 1973-78, Nat. Bean Coun. award Bean Coun. Am., Mich., 1981. Mem. Am. Assn. Cereal Chemists (com. chair 1991—), Am. Chem. Soc., Inst. Food Technologists, Sigma Xi. Achievements include several patents in field. Office: Continental Baking Co 1 Checkerboard Sq Saint Louis MO 63164

SHELTON, JOEL EDWARD, clinical psychologist; b. Havre, Mont., Feb. 7, 1928; s. John Granvil and Roselma Fahy (Ervin) S.; m. Maybelle Platzek, Dec. 17, 1949; 1 child, Sophia. AB, Chico (Calif.) State Coll., 1951; MA, Ohio State U., 1958, PhD, 1960. Psychologist Sutter County Schs., Yuba City, Calif., 1952-53; tchr., vice prin. Lassen View Sch., Los Molinos, Calif., 1953-55; tchr. S.W. Licking Schs., Pataskala, Ohio, 1955-56; child psychologist Franklin Village, Grove City, Ohio, 1957; clin. psychologist Marion (Ohio) Health Clinic, 1958; intern Children's Mental Health Ctr., Columbus, Ohio, 1958-59; acting chief research psychologist Children's Psychiat. Hosp., Columbus, 1959-60; cons. to supt. schs. Sacramento County, Calif., 1960-63; mem. faculty Sacramento State Coll., 1961-69; clin. psychologist DeWitt State Hosp., Auburn, Calif., 1965; exec. dir. Children's Ctr. Sacramento, Citrus Heights, Calif., 1963-64, Gold Bar Ranch, Garden Valley, Calif., 1964-72; clin. psychologist El Dorado County Mental Health Ctr., Placerville, Calif., 1968-70, Butte County Mental Health Dept., Oroville, Calif., 1970—; dir. dept. consultation, edn. and community services Butte County Mental Health Ctr., Chico, 1974-85, outpatient supr., 1985-86; MIS cons., 1986—; mgmt. cons., 1972—; advisor to pres. Protaca Industries, Chico, 1974-80; exec. sec. Protaca Agrl. Rsch., 1974-80; small bus. cons., 1983—; cons. on coll. scholarships and funding, 1991-92. Mem. APA, Western Psychol. Assn. Home: 1845 Veatch St Oroville CA 95965-4742 Office: Butte County Mental Health 18C County Center Dr Oroville CA 95965-3317

SHELTON, KEVIN L., geology educator. Prof. geology U. Mo., Columbia. Recipient Lindgren award Soc. Economic Geologists, 1991. Office: Univ of Missouri Columbia Dept of Geology 101 Geology Building Columbia MO 65211*

SHELTON, THOMAS MCKINLEY, electronics engineer; b. Ft. Worth, Aug. 19, 1939; s. Thomas McKinley and Margaret Leslie (Pringle) S.; m. Jeanne Louise DaCosta, July 18, 1966. BS in Physics magna cum laude, St. Mary's U., 1961; MS in Physics, Rutgers U., 1969. Teaching asst. Rutgers U., New Brunswick, N.J., 1966-69; engr. astroelectronics divsn. RCA, Hightstown, N.J., 1961-65; electronics engr. missile and surface radar divsn. RCA, Moorestown, N.J., 1965-66; hybrid packaging engr. missile and surface radar divsn. RCA, Moorestown, 1969-87; hybrid tech. engr. govt. electronics systems dept. GE, Moorestown, 1987-93, Martin Marietta, Moorestown, 1993—. Capt. U.S. Army, 1962-64. Mem. Internat. Soc. for Hybrid Microelectronics. Republican. Scientologist. Home: 255 Northampton Dr Willingboro NJ 08046-1351

SHELTON, WILLIAM CHASTAIN, retired government statistician, investor; b. Athens, Ga., May 5, 1916; s. William Arthur and Effie Clyde (Landrum) S.; m. Helen Higgins, Dec. 17, 1938; children: Stuart H., Terry Ann Shelton Coble, Jean R. Shelton Jaffray, Alvin C. AB, Princeton U., 1936; postgrad. U. Chgo., 1937-38. Economist, statistician Fed. Govt., Washington, 1936-48; chief stats. sect. USRO-Marshall Plan, Paris, France, 1948-55; mgr. bus. research Fla. Devel. Com., Tallahassee, 1956-60; asst. com. foreign labor Bur. Labor Stats., Washington, 1960-75; spl. asst. statis. policy div. Office Mgmt. and Budget, 1975-77. Author: (with Joseph W. Duncan) Revolution in U.S. Government Statistics, 1926-76, 1978; contbr. articles to profl. jours. Mem. Am. Statis. Assn., Washington Soc. Investment Analysts, Nat. Economists Club, Sigma Xi, Phi Beta Kappa. Republican. Presbyterian. Home: 8401 Piney Branch Rd Silver Spring MD 20901-4353

SHEN, BENJAMIN SHIH-PING, scientist, engineer, educator; b. Hangzhou, China, Sept. 14, 1931; s. Nai-cheng and Chen-chiu (Sun) S.; m. Lucia Elisabeth Simpson, July 31, 1971; children: William, Juliet. AB, Assumption Coll., Mass., 1954, ScD (hon.), 1972; AM in Physics, Clark U., 1956; DSc d'Etat in Physics, U. Paris, 1964; MA (hon.), U. Pa., 1971. Asst. prof. physics SUNY, Albany, 1956-59; assoc. prof. space sci., dept. aeros. and astronautics Engring. Sch., NYU, 1964-66; assoc. prof. U. Pa., Phila., 1966-68, prof., 1968-72, Reese W. Flower prof. astronomy and astrophysics, 1972—; mem. Ctr. for Energy and Environment U. Pa., 1976-93, chmn. Roundtable on Sci., Industry and Policy, 1976—, assoc. provost, 1979-80, chmn. council grad. deans, 1979-81, provost, 1980-81, chmn. dept. astronomy and astrophysics, 1973-79, dir. Flower and Cook Obs., 1973-79, prof. Sch. Engring and Applied Sci., 1980-85; mem. U.S. Nat. Sci. Bd., 1990— (chmn. U.S. Sci. and Engring. Indicators 1990-92), chmn. Nat. Sci. Bd. Task Force Sci. Literacy, 1992—; cons. Gen. Electric Co., 1961-68, Office of Tech. Assessment, U.S. Congress, 1977-78; sci. and tech. advisor U.S. Senate Budget Com., 1976-77; guest staff mem. Brookhaven Nat. Lab., 1963-64, 65-70; chmn. Com. on Pub. Understanding of Sci., N.Y. Acad. Scis., 1972-75. Author: Nuclear Problems in Radiation Shielding in Space, 1963, Passage des Protons dans des Milieux Condensés, 1964; editor, co-author: High-Energy Nuclear Reactions in Astrophysics, 1967, co-editor, co-author: Spallation Nuclear Reactions and Their Applications, 1976; mem. editorial bd. Earth and Extraterrestrial Scis., 1974-78, assoc. editor, 1978-79; assoc. editor: Comments on Astrophysics, 1979-85; contbr. articles to profl. jours. Mem. Hayden Planetarium com. of bd. trustees Am. Mus. Natural History, 1978—; mem. sci. adv. bd. Children's TV Workshop, N.Y.C., 1977, 79—; trustee Sci. Inst. Pub. Info., 1993 ; mem. adv. com. Mt. John Obs., New Zealand, 1978-84; mem. ABA-AAAS Nat. Conf. Bd. Lawyers and Scientists, 1986-92; former mem. governing bd. N.Y. Acad. Scis., University City Sci. Ctr. Research Park (Phila.), U. Pa. Research Found., Morris Arboretum, Phila., Univ. Mus., Phila., The Pa. Ballet Co. Decorated Ordre des Palmes Académiques (France); recipient Vermeil medal for sci. Soc. d'Encouragement au Progrès, France, 1978. Fellow Am. Phys. Soc., AAAS (com. on sci. engring. and pub. policy 1978-84, chmn. subcom on fed. research and devel. budget 1978-81), Royal Astron. Soc. (U.K.); mem. Internat. Astron. Union. Office: U Pa David Rittenhouse Lab Philadelphia PA 19104-6394

SHEN, GENE GIIN-YUAN, organic chemist; b. Taipei, Taiwan, Republic of China, Apr. 12, 1957; came to U.S., 1981; s. Chi and Su-Chin (Huang) S.; m. Grace Hsiao-Fen Chien, July 31, 1982; 1 child, Jennifer Iting. BS in Chemistry, Nat. Taiwan U., 1979; PhD in Organic Chemistry, U. Calif., Riverside, 1986. Postdoctoral fellow U. Calif., Riverside, 1986-87; rsch. chemist Nucleic Acid Rsch. Inst. ICN, Costa Mesa, Calif., 1987-88; prin. investigator Pharm-Eco Labs., Inc., Simi Valley, Calif., 1988-91; staff scientist Beckman Instruments, Inc., Brea, Calif., 1991—. Contbr. articles to Jour. Am. Chem. Soc., Jour. Steroir Biochemistry and Molecular Biology, Nucleosides and Nucleotides, others. Mem. Am. Chem. Soc., Phi Beta Kappa. Achievements include research in antisense oligonucleotides, near infrared fluorescent dyes and their application to fluoroimmuno assay, deoxynucleosides as anti-AIDS drugs, avidin-biotin chemistry, turbidimetric and nephelometric immunoinhibition assay. Office: Beckman Instruments Inc 200 S Kraemer Blvd Brea CA 92621

SHEN, HSIEH WEN, civil engineer, consultant, educator; b. Peking, China, July 13, 1931; s. Tsung Lien and Bick Men (Jeme) S.; m. Clare Tseng, Oct. 20, 1956; children: Eveline, Anthony. BS, U. Mich., 1953, MS, 1954; PhD, U. Calif.-Berkeley, 1961. Hydraulic engr. Harza Engring. Co., Chgo., 1961-63; mem. faculty Colo. State U., Ft. Collins, 1964-86; prof. civil engring. U. Calif., Berkeley, 1986—; cons. World Bank, UN, Harza, Stone & Webster, U.S. Army C.E. Author, editor: River Mechanics 1971; Sedimentation, 1973; Modeling of Rivers, 1979. Recipient Horton award Am. Geophys. Union, 1976, Joan Hodges Queneau award Am. Assn. Engring. Socs. and Nat. Audubon Soc., 1992; Guggenheim Found. fellow, 1974. Mem. U.S. Nat. Acad. Engring., Internat. Assn. Hydraulic Research (pres. fluvial hydraulics 1984-86), ASCE (Freeman scholar 1966, chmn. probability approach 1983-84, Einstein award 1990). Office: U Calif 412 O'Brien Hall Berkeley CA 94720

SHEN, KANGKANG, research physicist; b. Hangzhou, Zhejiang, China, Oct. 28, 1958; came to U.S., 1985; s. Z.J. Jiang and K.N. Shen; m. Xiao-Qin Lu, Jan. 7, 1987; 1 child, Lucy. BS, Zhejiang U., Hangzhou, 1982; MS, San Diego State U., 1987; PhD, Ga. Inst. Tech., 1993. Engr. Zhejiang U., 1982-85; teaching asst. San Diego State U., 1985-87; rsch. asst. Ga. Inst. Tech., Atlanta, 1987—. Contbr. articles to sci. jours. Recipient award for outstanding tech. paper Sci. Applications Internat. Corp., 1991. Mem. Optical Soc. Am., Soc. for Photo-Optical Instrumentation Engrs. Achievements include development of chemically pumped sodium laser; research on quantum well excitons, visible chemical laser. Home: 251 10th St NW Apt 33 Atlanta GA 30318 Office: Ga Inst Tech Sch Physics Atlanta GA 30332

SHEN, LIANG CHI, electrical engineer, educator, researcher; b. China, Mar. 17, 1939; came to U.S., 1962; s. Kuang Huai and ting Chin (Yu) S.; m. Grace Liu, June 26, 1965; children: Michael, Eugene. BSEE, Nat. Taiwan U., Taipei, 1961; PhD, Harvard U., 1967. Registered profl. engr., Tex. Prof., chmn. electrical engring. dept. U. Houston, 1977-81, prof., dir. well logging lab., 1978—. Author: Applied Electromagnetism, 1987. Fellow IEEE (assoc. editor geosci. and remote sensing 1986—). Office: U Houston Dept Elec Engring Houston TX 77204-4793

SHEN, MASON MING-SUN, pain and stress management center administrator; b. Shanghai, Jiang Su, China, Mar. 30, 1945; came to U.S., 1969; s. John Kaung-Hao and Mai-Chu (Sun) S.; m. Nancy Hsia-Hsian Shieh, Aug. 7, 1976; children: Teresa Tao-Yee, Darren Tao-Ru. BS in Chemistry, Taiwan Normal U., 1963-67; MS in Chemistry, S.D. State U., 1971; PhD in Biochemistry, Cornell U., 1977; MS in Chinese Medicine, China Acad., Taipei, Taiwan, 1982; OMD, San Francisco Coll Acupuncture, 1984. Diplomate Nat. Commn. for Cert. of Accupuncturists; lic. acupuncturist. Rsch. assoc. Lawrence Livermore (Calif.) Lab., 1979-80; assoc. prof. Nat. Def. Med. Coll., Taipei, 1980-82; prof. Inst. of Chinese Medicine China Acad., Taipei, 1981-82, San Francisco Coll. Acupuncture, 1983-85; chief acupuncturist Acupuncture Ctr. of Livermore, Calif., 1982-93; prof. Acad. Chinese Culture & Health Scis., Oakland, Calif., 1985-86; chief acupuncturist Acupuncture Ctr. of Danville, Calif., 1985-89; dir. Pain & Stress Mgmt. Ctr., Danville, Calif., 1989-90; chmn. adminstrn. subcom., 1991-92, acupuncture com. State of Calif., 1988-92. Contbr. articles to profl. jours. Rep. Rep. Party, Danville, 1988—; bd. dirs. Asian Rep. Assembly, 1989—. 2d lt. Rep. of China Army, 1966-69. Recipient Nat. Rsch. Svc. award NIH, 1977. Mem. AAAS, N.Y. Acad. Scis., Calif. Cert. Acupuncturists Assn. (bd. dirs. 1984-88, pres. 1984-85), Acupuncture Assn. Am. (bd. dirs. 1986-90, v.p. 1987-89), Am. Assn. Acupuncture and Oriental Medicine (bd. dirs. 1987-92, pres. 1989-90), Nat. Acupuncture Detoxification Assn. (cons. 1987—), Presdl. Round Table (presdl. adv. com.), Hong Kong and Kowloon Chinese Med. Assn. (hon. life pres. 1985). Republican. Avocations: travel, horse back riding, rifles. Home: 3240 Touriga Dr Pleasanton CA 94566-6966 Office: Pain and Stress Mgmt Ctr 185 Front Ste 207 Danville CA 94526-3323

SHEN, TSUNG YING, medicinal chemistry educator; b. Beijing, China, Sept. 28, 1924; came to U.S., 1950; s. Tsu-Wei and Sien-Wha (Nieu) S.; m. Amy T.C. Lin, June 20, 1953; children: Bernard, Hubert, Theodore, Leonard, Evelyn, Andrea. B.Sc., Nat. Central U., Chung King, China, 1946; diploma, Imperial Coll. Sci. and Tech., London, 1948; Ph.D., U. Manchester, Eng., 1950, D.Sc., 1978. Research assoc. Ohio State U., Columbus, 1950-52, MIT, Cambridge, 1952-56; sr. research chemist Merck, Sharp & Dohme Research Labs., Rahway, N.J., 1956-65, dir. synthetic chem. research, 1966-76, v.p. membrane chem. research, 1976-77, v.p. membrane and arthritis research, 1977-86; A. Burger Prof. Medicinal Chemistry U. Va., Charlottesville, 1986—; vis. prof. U. Calif.-Riverside, 1973, U. Calif.-San Francisco, 1985, Harvard Med. Sch., 1986; adj. prof. Stevens Inst. Tech., Hoboken, N.J., 1982-85. Mem. editorial bd. Clinica Europa Jour., 1977, Prostaglandins and Medicine, 1978, Medicinal Rsch. Revs., 1979, Jour. Medicinal Chemistry, 1980-83, Medicinal Chem. Rsch., 1991; patentee in field. Recipient Outstanding Patent award N.J. Research and Devel. Council, 1975, Rene Descartes medal U. Paris, 1977, medal of Merit Giornate Mediche Internazionali del Collegium Biologicum Europea, 1977, cert. of merit Spanish Soc. Therapeutic Chemistry, 1983, achievement award Chinese Inst. Engrs.-U.S.A., 1984. Mem. Am. Chem. Soc. (1st Alfred Burger award in medicinal chemistry 1980), N.Y. Acad. Scis., AAAS, Internat. Soc. Immunopharmacology, Acad. Pharm. Scis. (hon.), Am. Assn. Pharm. Sci. Home: 303 Ednam Dr Charlottesville VA 22903-4715 Office: Chem Dept U Va Charlottesville VA 22901

SHEN, YUAN-YUAN, mathematics educator; b. Yenchao, Taiwan, Republic of China, June 30, 1954; came to U.S., 1979; s. Wu-Hsing and Jin-Fon (Kung) S.; m. Joan Chyong-Jue Chen, June 25, 1988; 1 child, Angela Rachel. BA, Tunghai U., Taiwan, 1976; PhD, U. Md., 1988. Asst. prof. Cath. U. of Am., Washington, 1988—. 2d lt. Republic of China mil., 1976-78. Mem. Am. Math. Soc., Math. Assn. Am., Phi Tau Phi. Home: 3149 Beethoven Way Silver Spring MD 20904-6860 Office: Cath U of Am 620 Michigan Ave NE Washington DC 20064-0001

SHEN, YUEN-RON, physics educator; b. Shanghai, China, Mar. 25, 1935; came to U.S.; BS, Nat. Taiwan U., 1956; MS, Stanford U., 1959; PhD, Harvard U., 1963. Rsch. asst. Hewlett-Packard Co., Palo Alto, Calif., 1959; rsch. fellow Harvard U., Cambridge, Mass., 1963-64; asst. prof. U. Calif., Berkeley, 1964-67, assoc. prof., 1967-70, full prof., 1970—; prin. investigator Lawrence Berkeley Lab., 1964—. Author: The Principles of Nonlinear Optics, 1984. Sloan fellow, 1966-68; recipient Guggenheim Found. fellowship, 1972-73, Charles Hard Townes award, 1986, Arthur L. Schawlow prize Am. Phys. Soc., 1992, Alexander von Humboldt award, 1984. Fellow Am. Phys. Soc., Optical Soc. Am.; mem. Am. Acad. Arts and Scis., Academic Sinica. Achievements include outstanding rsch. in solid state physics, DOE-MRS rsch., 1983, sustained outstanding rsch. in solid state physics, DOE-MRS rsch., 1987. Office: U Calif Berkeley Dept Physics Berkeley CA 94720

SHENEFELT, PHILIP DAVID, dermatologist; b. Colfax, Wash., July 31, 1943; s. Roy David and Florence Vanita (Cagle) S.; m. Debrah Ann Levenson; children: Elizabeth, Sara, Shaina. BS with honors, U. Wis.-Madison, 1966, MD, 1970, MS in Adminstrv. Medicine, 1984. Intern U.S. Naval Hosp., Bethesda, Md., 1970-71; general practice Oregon (Wis.) Clinic, 1975; resident in dermatology U. Wis. Hosp., Madison, 1975-78, staff, 1978-87; asst. prof. dermatology sect. Dept. Internal Medicine U. South Fla., 1987—; chief dermatology sect. VA Hosp., Bay Pines, Fla., 1987-89, asst. chief, Tampa, 1988—; dermatologist Univ. Health Svc., U. Wis.-Madison, 1978-87, VA Hosp., Madison, 1982-85. Served to lt. comdr. USN, 1969-74; capt. USNR (ret.). Kellogg fellow, 1980-82. Mem. AMA, Am. Acad. Dermatology, Fla. Dermatol. Soc., Fla. West Coast Dermatol. Soc., Am. Acad. Med. Dirs. Episcopalian. Home: 3508 W Tacon St Tampa FL 33629-7929 Office: U South Fla Internal Medicine / Dermatology 12901 Bruce B Downs Blvd Box 19 Tampa FL 33612-4799

SHENHAR, JORAM, mechanical engineer; b. Tel Aviv, Israel, Feb. 23, 1940; s. Jacob and Regina (Fischlevitch) Schitenberg; m. Rose-Mary Romy Rohrlich, Oct. 30, 1967; children: Dana, Amit. BSc, Technion-Israel Inst. Tech., 1966, MSc, 1971; PhD, Va. Poly. Inst. and State U., 1983. Registered mech. engr., Israel. Mech. engr. Israeli Ministry of Def., Tel Aviv, 1966-69; R&D mech. engr. Israeli Aircraft Industries, Yahud, 1973-78; instr. Va. Poly. Inst. and State U., Blacksburg, 1979-82, asst. prof., 1984-86; engring. specialist PRC-Kentron, Hampton, Va., 1986-89; staff engr., supr., mgr. controls lab. Lockheed Engring. and Sci., Hampton, 1989—. Author solution manual McGraw Hill Book Co., 1986; contbr. articles to Jour. of Optimization Theory and Application, others. 1st sgt. Israeli Air Force, 1959-61. Fellow AIAA (assoc.); mem. Sigma Xi. Jewish. Home: 120 Kohler Crescent Newport News VA 23606 Office: Lockheed Engring and Scis 144 Research Dr Hampton VA 23666

SHENK, THOMAS EUGENE, molecular biology educator; b. Bklyn., Jan. 1, 1947; s. Eugene Richard and Helen Marie (Deffenbaugh) S.; m. Susan Mary Hillman, July 4, 1979; children—Christopher Thomas, Gregory Thomas. BS, U. Detroit, 1969; PhD, Rutgers U., 1973; postgrad., Stanford

U., 1973-75. Asst. prof. molecular biology U. Conn.-Farmington, 1975-80; prof. molecular biology SUNY-Stony Brook, 1980-84; prof. molecular biology Princeton U., 1984—, prof. Am. Cancer Soc., 1986—. Mem. Am. Soc. Microbiology (Eli Lilly award 1982). Office: Princeton U Dept Molecular Biology Princeton NJ 08544

SHENOUDA, GEORGE SAMAAN, engineering executive, consultant; b. Alexandria, Egypt, June 7, 1943; came to U.S., 1972; s. Samaan and Sania S. Shenouda. BSc, Alexandria U., 1963, MSc, 1969. Rsch. assoc. Dept. Rsch., Cairo, 1963-72; engr. Unitrode Corp., Watertown, Mass., 1973-78; mgr. Raytheon Corp., Waltham, Mass., 1978-80, Analog Devices Inc., Wilmington, Mass., 1980-81, Kollsman Instrument Co., Merrimack, N.H., 1981-85; br. mgr. precision products div. Northrop Corp, Norwood, Mass., 1985-89; sect. mgr. Northrop Corp., PPD, Norwood, Mass., 1989-90; sr. mgr. electronic systems div. Northrop Corp., ESD-N, Norwood, Mass., 1990-91; sr. tech. staff Baxter Corp., Miami, Fla., 1992—. Mem. IEEE, Am. Soc. Safety Engrs., Inst. Environ. Scis. (sr.), Internat. Soc. Pharm. Engrs. Avocations: traveling, tennis, photography, music. Office: Baxter Corp PO Box 520672 Miami FL 33152-0672

SHEPARD, ALAN BARTLETT, JR., astronaut, real estate developer; b. East Derry, N.H., Nov. 18, 1923; s. Alan Bartlett and Renza (Emerson) S.; m. Louise Brewer, Mar. 3, 1945; children: Laura, Juliana. Student, Admiral Farragut Acad., 1940; B.S., U.S. Naval Acad., 1944; grad., Naval War Coll., 1958; M.S. (hon.), Dartmouth Coll.; D.Sc. (hon.), Miami U. Commd. ensign USN, 1944, advanced through grades to rear adm., 1971, designated naval aviator, 1947; assigned destroyer U.S.S. Cogswell, Navy Test Pilot Sch., Pacific, World War II, Fighter Squadron 42, aircraft carriers in Mediterranean, 1947-49; with USN Test Pilot Sch., 1950-53, 55-57, took part in high altitude tests, expts. in test and devel. in-flight refueling system, carrier suitability trials of F2H3 Banshee; also trials angled carrier deck ops. officer Fighter Squadron 193, Moffett Field (Calif.), carrier U.S.S. Oriskany, Western Pacific, 1953-55; test pilot for F4D Skyray, 1955, F3H Demon, F8U Crusader, F11F Tigercat, 1956; project test pilot F5D Skylancer, 1956; instr. Naval Test Pilot Sch., 1957; aircraft readiness officer staff Comdr.-in-Chief, Atlantic Fleet, 1958-59;; joined Project Mercury man in space program NASA, 1959; first Am. in space, May 5, 1961, chief of astronaut office, 1965-74, selected to command Apollo 14 Lunar Landing Mission, 1971, became 5th man to walk on moon, hit 1st lunar golf shot; pres. Seven Fourteen Enterprises, 1986, Windward Coors Co., Deer Park, Tex., 1974; presdl. appointee, del. 26th Gen. Assembly UN, 1971. Decorated D.S.M., D.F.C., Presdl. unit citation, NASA Disting. Service medal, Congressional Medal of Honor, 1978; recipient Langley medal Smithsonian Instn., 1964. Fellow Soc. Exptl. Test Pilots; mem. Order Daedlians, Soc. Colonial Wars, Lions, Kiwanis, Rotary.

SHEPARD, MARK LOUIS, animal scientist; b. Cedar Rapids, Iowa, June 22, 1949; s. Charles Frederick and Dorothy Maxine (White) S.; m. Suzanne Fisher, Jan. 24, 1970; 1 child, Justin Noel. BS, Angelo State U., San Angelo, Tex., 1977, MS, 1981. Wastewater treatment specialist City of San Angelo, 1975-77, quality control chemist, 1977-78; grad. fellow Angelo State U., 1978-79; tech. asst. Tex. A&M Rsch. Ctr., San Angelo, 1979-81; cons. Shotwell & Carr, Inc., Dallas, 1981—; lectr. Inst. for Applied Pharm. Sci., Kansas City, Mo., 1983, East Brunswick, N.J., 1984, others. Assoc. editor: Veterinary Practitioner's Guide to Approved New Animal Drugs, 1987, 2d edit. 1988; editor: Complete Handbook of Approved New Animal Drug Applications in the United States, 1981—. With USAF, 1968-72. Decorated Air medal with 3 oak leaf clusters; Angelo State U. grad. fellow, 1978-79. Mem. Coun. for Agrl. Sci. and Tech. Home: 1747 Arledge Carrollton TX 75007 Office: Shotwell & Carr Inc 3003 LBJ Ste 100 Dallas TX 75234

SHEPARDSON, JED PHILLIP, computer consultant; b. Pitts., Nov. 10, 1932; s. Vene Phillip and Ione Marguerite (von Drokowsky) S.; m. Mildred Humphrey, Mar. 29, 1958 (dec. Sept. 1984); 1 child, Eric William; m. Junille Glenda Wieting, Dec. 27, 1987. BSChEng, Ga. Inst. Tech., 1955. Profl. engr. Ky., Maine. With plant ops. Dow Corning Corp., Midland, Mich., 1955-65; project mgr. Dow Corning Corp., Carrollton, Ky., 1965-84; free lance computer cons. Glen Cove, Maine, 1984—. Pres. Optimist Club, Carrollton, 1982, Band Booster Club, Carrollton, 1983. Lt. U.S. Army, 1955. Methodist. Achievements include patent for unique diffusion pulp fluids. Home and Office: 155 Warrenton St Glen Cove ME 04846

SHEPHARD, BRUCE DENNIS, obstetrician, medical writer; b. San Francisco, Apr. 21, 1944; s. Richard G. and Madelyn (Rogers) S.; m. Carroll Anne Swanson; children: Christopher, Carleton, Elizabeth. BA in History, U. Calif., Berkeley, 1966; MD, U. Calif., San Francisco, 1970. Diplomate Am. Bd. Obstetrics and Gynecology. Intern Jackson Meml. Hosp.-U. Miami (Fla.), 1970-71, resident in ob-gyn., 1971-74; obstetrician Tampa (Fla.) Ob-Gyn Assocs., 1976—; clin. assoc. prof. obstetrics U. So. Fla. Sch. Medicine, Tampa, 1976—; bd. dirs. Ctr. of Excellence, 1983-90, Humana Women's Hosp., Tampa, Fla., 1983-90, Gulf Coast Health Systems Agy., 1980-83; mem. midwifery adv. com. Fla. Dept. Health and Human Resources, Tallahassee, 1982-86. Author: (with Carroll Shephard) The Complete Guide to Women's Health, 1982, rev., 1990; prin., writer, spokesperson (series of TV comms.) The Healthy Woman (Gold Link award 1987); contbr. articles to profl. jours. and women's mags. Served as maj. USAF, 1974-76. Mem. AMA, Am. Coll. Obstetricians and Gynecologists (patient edn. com. 1984-86, John McCain fellow 1981), Am. Med. Writers Assn., Acad. Radio and TV Health Communicators, Phi Beta Kappa. Democrat. Lutheran. Avocations: tennis, photography, golf, antique collector. Home: 4201 Carrollwood Village Dr Tampa FL 33624-4609 Office: Tampa Ob-Gyn Assocs 2901 W St Isabel St Tampa FL 33607-6345

SHEPHERD, FREEMAN DANIEL, physicist; b. Boston, June 7, 1936; s. Freeman D. and Cora I. (Donnelly) S.; m. Carol A. Smith, Jan. 11, 1959; children: Freeman III, Suzanne, Mark. BSEE, MSEE, MIT, 1959; PhD in Elec. Engring., Northeastern U., 1965. Student trainee AF Cambridge Rsch. Ctr., Hanscom AFB, Mass., 1956-59; electronic scientist AF Cambridge Rsch. Ctr., Hanscom AFB, 1959-75; div. chief, supervisory physicist Rome Air Devel. Ctr., Hanscom AFB, 1975-91; sr. scientist Infrared Arrays & Sensors, Rome Lab., Hanscom AFB, 1991—; chmn. adv. elec. tech. panel MIT Lincoln Lab., Lexington, Mass., 1982—; mem. passive sensing steering group Dept. Def., 1986—. Contbr. articles to profl. jours. Recipient Harold Brown award USAF, Washington, 1989. Fellow IEEE (Harry Diamond award 1992), SPIE, Sigma Xi. Achievements include 6 patents in field. Office: Rome Lab Dept of Air Force RL/ER Hanscom AFB MA 01731

SHEPP, LAWRENCE ALAN, mathematician, educator; b. Bklyn., Sept. 9, 1936; s. Benjamin and Francis (Stein) S.; m. Britt Louise Gunnarsdotter; children: Steven Lawrence, Linda Louise, Eric Brian. BS in Applied MAth., Polytech. Inst. Bklyn., 1958; MA in Math., Princeton U., 1959, PhD in Math, 1961. Instr. U. Calif., Berkeley, 1962; mem. tech. staff AT&T Bell Labs., Murray Hill, N.J., 1962—; prof. Stanford (Calif.) U., 1980—. Recipient W.L. Putnam prize, 1958, Paul Levy prize, 1967, Disting. Scientist award IEEE Soc. Nuclear Sci. Soc., 1982. Fellow Inst. Math. Statistics; mem. NAS, Inst. Medicine, Am. Math. Soc., B'nai Brith. Avocations: chess, ping-pong, Russian language. Office: AT&T Bell Labs 600 Mountain Ave Rm 2374C New Providence NJ 07974-2010

SHEPPARD, JOHN WILBUR, computer research scientist; b. Pitts., Aug. 21, 1961; s. Harry Reid and Mary Jane (Amon) S.; m. Justine Anne Pape, Oct. 29, 1988. BS, So. Meth. U., 1983; MS, Johns Hopkins U., 1989, postgrad., 1990—. Systems analyst Sheppard Internat., Inc., Hermitage, Pa., 1979-86; rsch. analyst ARLINC Rsch. Corp., Annapolis, Md., 1986—. Contbr. articles to profl. jours. Mem. YMCA, Hermitage, 1979-85, Md. Hall for the Creative Arts, Annapolis, 1988—; pres. Univ. Chapel Campus Ministry, Dallas, 1982-83. Mem. IEEE, Am. Assn. for Artificial Intelligence, Mensa, Internat. Neural Network Soc., Internat. Assn. Knowledged Engrs., Soc. Logistics Engrs., Kappa Mu Epsilon. Republican. Lutheran. Achievements include U.S. and foreign patent pending for methods and apparatus for diagnostic testing; development of explanation-based learning approach for fault diagnosis. Home: Unit 2D 104 Mountain Rd Glen Burnie MD 21060-7929 Office: ARINC Rsch Corp 2551 Riva Rd Annapolis MD 21401-7435

SHEPPARD, LOUIS CLARKE, biomedical engineering educator; b. Pine Bluff, Ark., May 28, 1933; s. Ellis Allen and Louise (Clarke) S.; m. Nancy Louise Mayer, Feb. 8, 1958; children—David, Susan, Lisa. B.S. in Chem. Engring., U. Ark., 1957; Ph.D. in Elec. Engring., U. London, 1976. Registered profl. engr., Ala., Tex. Devel. staff supr. Diamond Alkali Co., Deer Park, Tex., 1957-63; staff engr. IBM, Rochester, Minn., 1963-66; assoc. prof. surgery dept. U. Ala.-Birmingham, 1966-88, sr. scientist Cystic Fibrosis Research Ctr., 1981-87, prof., chmn. biomed. engring. dept., 1979-88; prof. phsiology and biophysics, asst. v.p. rsch. U. Tex., Galveston, 1988-90, assoc. v.p. bioengring. and biotech., 1990—, prof. biomed. engring., Austin; adj. prof. elec. engring. U. Houston; mem. med. adv. bd. Hewlett Packard, 1980-84; cons. IMED Corp., 1982-83, Oximetrix, 1982, 86-88, MiniMed, 1986-88; dir. FBK Internat.; pres. S.E.A. Corp.; mem. editorial bd. Med. Progress Through Tech., Springer-Verlag, Berlin; cons. Nat. Heart, Lung and Blood Inst. Bd. dirs. Birmingham Met. Devel. Bd. Served with AUS, 1958-66. Recipient Ayerton Premium award IEE (U.K.), 1984. Recipient Disting. Alumnus citation, U. Ark., 1987, Lifetime Achievement award M.D. Buyline, 1987. Fellow IEEE, Am. Inst. for Med. and Biol. Engring., Am. Soc. Med. Info., Am. Coll. Med. Informatics; mem. Brit. Computer Soc., Cardiovascular System Dynamics, Biomed. Engring. Soc. (dir.), IEEE, Am. Inst. Chem. Engrs., Am. Med. Informatics Assn., Acad. Med. Arts and Scis., Blue Key, Sigma Xi, Tau Beta Pi, Alpha Pi Mu, Eta Kappa Nu, Theta Tau. Club St. Andrews Soc. of Middle South, The Houstonian. Contbr. abstracts, chpts. to books, editorials; patentee method and system for estimation of arterial pressure. Home: 5 Broad Oaks Ln Houston TX 77056-1201 Office: U Tex Med Br 5 108 Adminstrn Bldg Galveston TX 77555-0133

SHEPPARD, WILLIAM VERNON, transportation engineer; b. Harlan, Ky., Apr. 18, 1941; s. Vernon L. and Margaret M. (Montgomery) S.; m. Charlotte A. McGehee, Nov. 6, 1981; children: W. Kevin, Candice Gaye. BCE, The Citadel, 1964. Registered profl. engr., Pa., Calif. and 10 other states. Hwy. engr. Howard Needles, Tammen & Bergendoff, Kansas City, Mo., 1964-65; with Wilbur Smith & Assos., Columbia, S.C., 1967-80, various positions to western regional v.p., so. regional v.p.; v.p., dir. transp. Post Buckley, Schuh & Jernigan, Inc., Columbia, 1980-85; sr. v.p., Sverdrup Civil, 1985—, also bd. dirs.; guest lectr. U. So. Calif. Sch. Architecture and Urban Planning. Mem. engring. adv. bd. Clemson U., 1977-80. Served to capt. U.S. Army, 1965-67. Decorated AEM medal. Fellow ASCE, Inst. Transp. Engrs. (pres. S.C. div. 1979); mem. Nat. Soc. Profl. Engrs., S.C. Coun. Engring. Socs. (pres. 1978), Tau Beta Pi. Republican. Roman Catholic.

SHERIDAN, ANDREW JAMES, III, ophthalmologist; b. Washington, Aug. 19, 1944; s. Andrew J. and Mildred (Stohlman) S.; m. Carol Dinkelacker, Oct. 23, 1971; children—Elizabeth, Margaret. A.B., Villanova U., 1966; M.D., Georgetown U., 1970. Diplomate Am. Bd. Ophthalmology. Intern Nassau County Med. Ctr., N.Y., 1970-71, resident in ophthalmology, 1973-76; practice medicine specializing in ophthalmology Eye Assocs., Arlington, Va., 1976—; chief of ophthalmology Arlington Hosp., 1980-84, 90—; clin. instr. ophthalmology Georgetown U., 1982—. Bd. dirs. Va. Med. Polit. Action Com., 1985-89, Polit. Action Com. for Ophthalmology in Va., 1985—, chmn, 1991—. Lt. comdr. USPHS, 1971-73; pres. Georgetown Clin. Soc., 1987-88. Fellow ACS, Am. Acad. Ophthalmology; mem. AMA, Med. Soc. Va., Arlington County Med. Soc. (exec. bd.), No. Va. Acad. Ophthalmology (v.p. 1988-90, pres. 1990-92, bd. dirs.), Brent Soc., Am. Soc. Cataract and Refractive Surgery. Republican. Roman Catholic. Lodge: Lions. Office: Eye Assocs 1715 N George Mason Dr Ste 206 Arlington VA 22205-3649

SHERIDAN, PHILP HENRY, pediatrician neurologist; b. Washington, June 29, 1950; s. Andrew James and Mildred Adele (Stohlman) S.; m. Margaret Mary Williams, Oct. 3, 1987; children: Gerard Andrew, Philip Henry, Kathleen Mary, Patrick Gerard. BS magna cum laude, Yale U., 1972; MD cum laude, Georgetown U., 1976. Diplomate Am. Bd. Pediatrics, Am. Bd. Psychiatry and Neurology, Am. Bd. Qualification in Electroencephalography. Resident in pediatrics Children's Hosp. Phila., 1976-79; fellow in pediatric neurology Hosp. of U. Pa., Phila., 1979-82; med. staff fellow NIH, Bethesda, Md., 1982-84, neurologist, epilepsy br. Nat. Inst. Neurol. Disorders and Stroke, 1984—, health scientist adminstr., guest worker researcher, 1984-89, chief Devel. Neurology Br., NIH, 1989—; cons., lectr. Nat. Naval Med. Ctr., Bethesda, 1984—; med. dir. U.S. Pub. Health Svc. Contbr. articles on clin. and rsch. neurology to med. jours. Neurologist Div. Children's Splty. Svcs., Fairfax, Va., 1984—. Mem. Am. Acad. Neurology, Child Neurology Soc. (invited reviewer), Soc. for Neurosci., Am. Epilepsy Soc. (invited reviewer), Alpha Omega Alpha. Roman Catholic. Current work: Planning and administering a comprehensive research program concerning pediatric neurology, developmental neurobiology, and neuromuscular disorders. Subspecialties: Neurology, Pediatrics. Office: NIH Fed Bldg Rm 8C10 Bethesda MD 20892

SHERIF, S. A., mechanical engineering educator; b. Alexandria, Egypt, June 25, 1952; came to U.S., 1978; s. Ahmed and Ietedal H. (Monib) S.; m. Azza A. Shamseldin, Feb. 6, 1977; children: Ahmed S., Mohammad S. BSME, Alexandria U., 1975, MSME, 1978; PhD in Mech. Engring., Iowa State U., 1985. Teaching asst. mech. engring. Alexandria U., 1975-78; teaching assoc. mech. engring. U. Calif., Santa Barbara, 1978-79; rsch. asst. mech. engring. Iowa State U., Ames, 1979-84; asst. prof. No. Ill. U., Dekalb, 1984-87; asst. prof. civil and archtl. engring. and mech. engring. U. Miami, Coral Gables, Fla., 1987-91; assoc. prof. mech. engring. U. Fla., Gainesville, 1991—; cons. Dade Power Corp., Miami, Fla., 1988—, Ind. Energy Systems, Miami, 1988—, Carey Dwyer Eckhart Mason Spring & Beckham, P.A. Law Offices, Miami, 1988—, Michael G. Widoff, P.A., Attys. at Law, Ft. Lauderdale, Fla., 1989—, Law Offices of Pomeroy and Betts, Ft. Lauderdale, 1991-92; resident assoc. Argonne (Ill.) Nat. Lab., Tech. Trasfer Ctr., summer 1992; faculty fellow NASA Kennedy Space Ctr., Cape Canaveral, Fla., summer 1993. Co-editor: Industrial and Agricultural Applications of Fluid Mechanics, 1989, The Heuristics of Thermal Anemometry, 1990, Heat and Mass Transfer in Frost and Ice, Packed Beds, and Environmental Discharges, 1990, Industrial Applications of Fluid Mechanics, 1990, rev. edit. 1991, Mixed Convection and Environmental Flows, 1990, Measurement and Modeling of Environmental Flows, 1992, Industrial and Environmental Applications of Fluid Mechanics, 1992, Thermal Anemometry-1993, 1993, Devices for Flow Measurement and Control-1993, 1993, Heat Transfer in Turbulent Flows, 1993; contbr. articles to profl. jours. Mem. environ. awareness adv. com., Dade County Sch. Bd., 1989-91, lab. dir. community lab. rsch. program, 1989-91. Mem. ASME (mem. coord. group fluid measurements ASME fluids engring. divsn. 1987—, vice chmn. 1990-92, chmn. 1992—, mem. fluid mech. tech. com., 1990—, mem. fluid mech. com. 1987—, mem. environ. heat transfer com. ASME heat transfer divsn. 1987—, mem. fluid applications and systems tech. com. 1990—, mem. systems analysis tech. com. 1989—, mem. fundamentals and theory com., 1990—, chmn. CGFM nominating com 1992—, chmn. profl. devel. com. ASME Rock River Valley sect. 1987), ASHRAE (mem. heat transfer fluid flow com. 1988-92, corr. mem. 1992—, mem. thermodynamics and psychrometrics com. 1988-92, corr. mem. 1992—, vice chmn. 1990-92, mem. liquid-to-refrigerant heat exchangers com. 1989-93, sec. 1990-92, chmn. standards project com. on measurement of moist air properties, 1989—), AIAA, Am. Inst. Chem. Engrs., Internat. Assn. Hydrogen Energy, Internat. Solar Energy Soc., Am. Solar Energy Soc., Internat. Energy Soc. (mem. scientific coun.), Assn. Energy Engrs. (sr.), European Assn. Laser Anemometry (ASME/FED rep., mem. steering com.), Internat. Inst. Refrigeration (assoc.), Sigma Xi. Moslem. Avocations: reading, soccer, basketball, history. Home: Apt A-3 5400 NW 39th Ave Gainesville FL 32606 Office: U Fla Dept Mech Engring Gainesville FL 32611-2050

SHERMAN, ALAN THEODORE, computer science educator; b. Cambridge, Mass., Feb. 26, 1957; s. Richard Beatty and Hanni Fey (Fechenbach) S.; m. Tomoko Shimakawa, Aug. 2, 1986. ScB in Math. magna cum laude, Brown U., 1978; SM in Elec. Engring and Computer Sci., MIT, 1981, PhD in Computer Sci., 1987. Instr. Tufts U., Medford, Mass., 1985-86, asst. prof., 1986-89; asst. prof. U. Md. Balt. County, Catonsville, 1989—; mem. Inst. for Advanced Computer Studies U. Md., College Park, 1989-92; rsch. affiliate MIT Lab. for Computer Sci., Cambridge, 1985-88. Author: VSLI Placement and Routing: The PI Project, 1989; co-editor: Advances in Cryptology: Proceedings of Crypto 82, 1983; contbr. articles to profl. jours. Mem. Assn. for Computing Machinery, IEEE, Internat. Assn. for Cryptologic Rsch., AAUP, Soc. for Indsl. and Applied Maths., Phi Beta Kappa, Sigma Xi. Avocations: tennis, Aikido, piano, chess. Home: 4706 Hallowed Stream Ellicott City MD 21042-5965 Office: U Md Computer Sci Dept Baltimore County Baltimore MD 21228

SHERMAN, DAVID MICHAEL, geologist; b. Redwood City, Calif., Aug. 2, 1956. BA in Chemistry, BS in Earth Sci., U. Calif., Santa Cruz, 1980; PhD in Geochemistry, MIT, 1984. Geologist mineral and geochemistry U.S. Geol. Survey, Denver, 1984—. Recipient F.W. Clarke medal Geochemical Soc., 1991. Mem. Mineral Soc. Am., Am. Geophys. Union. Achievements include research on applications of quantum chemistry and spectroscopy, optical, infrared and Mossbauer, problems in mineralogy and geochemistry, electronic structures of transition metal oxides and silicates. Office: US Geological Survey Mail Stop 964 Box 25046 DFC Denver W Bld Denver CO 80225*

SHERMAN, JAMES OWEN, psychologist; b. Iron Mountain, Mich., June 29, 1942; s. James William and Gwendolyn (Eslinger) S.; m. Marcia Anne Butler, June 3, 1963; children: Katherine Layne, Kelly Anne. BS in Psychology, Trinity U., 1965, MS in Psychology, 1966; PhD in Psychology, U.S. Internat. U., San Diego, 1975. Lic. psychologist, Tex.; cert. psychologist, correctional healthcare profl. Lectr. dept. psychology U. Guelph, Ont., Can., 1967-68; psychometrist, therapist Ont. Tng. Sch. for Girls, Galt, 1968-69; instr. psychology dept. Our Lady of the Lake U., San Antonio, 1969-73; staff tng. specialist San Antonio State Hosp., 1974-75; psychologist Psychiat./Psychol. Office, San Antonio, 1975-83; dir. psychol. svcs. Bexar County Med./Psychiat. Dept., San Antonio, 1983-85; dir. med. psychiat. svcs. Bexar County Juvenile Ctr., San Antonio, 1985—; dir., behavioral intervention program, juvenile probation and detention ctr., San Antonio, 1977-79; adj. asst. prof. Trinity U., San Antonio, 1977-83; clin. asst. prof. dept. pediatrics U. Tex. Health Sci. Ctr., San Antonio, 1990—. Mem. Am. Psychol. Assn., Southwestern Psychol. Assn., Tex. Psychol. Assn., San Antonio Kennel Club (pres. 1983-90). Office: Bexar County Juvenile Ctr 660 Mission Rd San Antonio TX 78210

SHERMAN, LOUIS ALLEN, biologist, researcher; b. Chgo., Dec. 16, 1943; s. Stanley E. and Sarah R. Sherman; m. Debra Meddoff, June 15, 1969; children: Daniel, Jeff. BS in Physics, U. Chgo., 1965, PhD in Biophysics, 1970. Postdoctoral fellow Cornell U., Ithaca, N.Y., 1970-72; asst. prof. U. Mo., Columbia, 1972-78, assoc. prof., 1978-83, prof., 1983-88, dir. biol. scis., 1985-88; prof., head dept. biol. scis. Purdue U., West Lafayette, Ind., 1989—. Contbr. articles to profl. jours. NIH fellow, 1965-72; Fulbright Hays scholar, The Netherlands, 1979-80; NSF travel grantee, Fed. Republic Germany, Japan; grantee NIH, USDA, Dept. Edn. Fellow AAAS, Am. Acad. Microbiology; mem. AAUP, Am. Soc. Microbiology, Biophys. Soc., Plant Molecular Biology Soc. Office: Purdue U Dept Biol Scis Lilly Hall West Lafayette IN 47907

SHERMAN, SIGNE LIDFELDT, securities analyst, former research chemist; b. Rochester, N.Y., Nov. 11, 1913; d. Carl Leonard Broström and Herta Elvira Maria (Thern) Lidfeldt; m. Joseph V. Sherman, Nov. 18, 1944 (dec. Oct. 1984). BA, U. Rochester, 1935, MS, 1937. Chief chemist Lab. Indsl. Medicine and Toxicology Eastman Kodak Co., Rochester, 1937-43; chief rsch. chemist Chesebrough-Pond's Inc., Clinton, Conn., 1943-44; ptnr. Joseph V. Sherman Cons., N.Y.C., 1944-84; advisor Signe L. Sherman Cons., Troy, Mont., 1984—. Author: The New Fibers, 1946. Fellow Am. Inst. Chemists; mem. AAAS, AAUW (life), Am. Chem. Soc., Am. Econ. Assn., Am. Assn. Ind. Investors (life), Fedn. Am. Scientists (life), Union Concerned Scientists (life), Western Econ. Assn. Internat., Earthquake Engring. Rsch. Inst., Nat. Ctr. for Earthquake Engring. Rsch., N.Y. Acad. Scis. (life), Cabinet View Country Club. Avocations: photography, water and land sports, writing, landscaping. Office: Signe L Sherman Cons Angel Island 648 Halo Dr Troy MT 59935-9415

SHERMAN, WILLIAM MICHAEL, physiology educator, researcher; b. Columbus, Ohio, Feb. 12, 1955; s. George W. and Ruth D. (Thiergartner) S.; m. Betty L. Rider, Aug. 30, 1986; children: Lindsay Kaitlin. BS, Ohio U., 1977; MS, Ball State U., 1981; PhD, U. Tex., 1984. Asst. prof. Tex. A&M U., College Station, 1984-85; from asst. prof. to assoc. prof. Ohio State U., Columbus, 1985-92, prof., 1993—; adv. bd. nutrition com. U.S. Olympic Com., Colorado Springs, 1991; vis. rsch. scholar Flinders U., South Australia, 1988. Conbtr. chpts. to books. Mem. Am. Diabetes Assn., Am. Coll. Sports Medicine (vis. scholar award 1991), Ctrl. Ohio Diabetes Assn. (rsch. com. 1989—). Achievements include rsch. on glycogen supercompensation, preexercise carbohydrate feedings and diets for athletes, exercise intervention in diabetes. Office: Ohio State U Sch HPER 337 W 17th Ave Columbus OH 43210

SHERR, RUBBY, nuclear physicist, educator; b. Long Branch, N.J., Sept. 14, 1913; s. Max and Anna (Jacobson) S.; m. Rita P. Ornitz, Sept. 11, 1936; children: Elizabeth Sherr Sklar, Frances Sherr. BA, NYU, 1934; PhD, Princeton U., 1938. Instr. Harvard U., Cambridge, Mass., 1938-42; mem. staff Radiation Lab. MIT, Cambridge, 1942-44; mem. staff Los Alamos Lab., Santa Fe, N.Mex., 1944-46; asst. prof. physics Princeton (N.J.) U., 1946-49, assoc. prof. physics, 1949-58, prof. physics, 1958-82, prof. emeritus physics, 1982—. Contbr. articles to profl. jours. Home: 73 McCosh Cir Princeton NJ 08540 Office: Princeton U Physics Dept Princeton NJ 08540

SHERREN, ANNE TERRY, chemistry educator; b. Atlanta, July 1, 1936; d. Edward Allison and Annie Ayres (Lewis) Terry; m. William Samuel Sherren, Aug. 13, 1966. BA, Agnes Scott Coll., 1957; PhD, U. Fla.-Gainesville, 1961. Grad. teaching asst. U. Fla., Gainesville, 1957-61; instr. Tex. Woman's U., Denton, 1961-63, asst. prof., 1963-66; rsch. participant Argonne Nat. Lab., 1973-80; assoc. prof. chemistry N. Cen. Coll., Naperville, Ill., 1966-76, prof., 1976—. Clk. of session Knox Presbyn. Ch., 1976—, ruling elder, 1971—. Mem. Am. Chem. Soc., Am. Inst. Chemists, AAAS, AAUP, Ill. Acad. Sci., Sigma Xi, Delta Kappa Gamma, Iota Sigma Pi (nat. pres. 1978-81, nat. dir. 1972-78, nat. historian 1989—). Presbyterian. Contbr. articles in field to profl. jours. Office: North Ctrl Coll Dept Chemistry Naperville IL 60566

SHERRILL, RONALD NOLAN, pharmacist, consultant; b. Sheffield, Ala., Jan. 25, 1949; s. Guy Nolan and Anna Mazie (Moore) S.; m. Eleanor Ray Wilson, Mar. 23, 1974; children: Kimberly Raye, Wilson Nolan. BS in Econs., U. Tenn., 1971; BS in Pharmacy, Auburn U., 1974; MPH in Planning, U. Tenn., 1977. Lic. nursing home adminstr. Pharmacy intern Vanderbilt Hosp. and VA Med. Ctr., Nashville, 1974-75; staff pharmacist Vanderbilt Hosp., Nashville, 1975; pharmacy supr. VA Outpatient Clinic, Knoxville, Tenn., 1975—; preceptor Coll. Pharmacy, U. Tenn., Knoxville, 1976—; dir. pharmacy N.W. Gen. Hosp., Knoxville, 1980-90; v.p., owner First Pharmacy Mgmt., Inc., Knoxville, 1985—; chmn. bd. dirs. First Pharmacy Mgmt., Inc., Knoxville, cons., 1985—. Named one of Outstanding Young Men of Am., 1983, Tenn. Preceptor of Yr., 1992. Fellow Am. Coll. Apothecaries (state pres. 1984), Am. Soc. Cons. Pharmacists; mem. Am. Soc. Hosp. Pharmacists, Am. Pharm. Assn., Tenn. Pharm. Assn. (LTC com. 1988-92, infusion therapy com. 1992—), Knoxville Pharmacy Assn. (pres. 1985), Knoxville Soc. Hosp. Pharmacists (pres. 1986), Sigma Nu Alumni Assn. (sec. bd. 1986—), Rotary Breakfast Club (bd. dirs. 1990-91), Omicron Delta Kappa, Phi Lambda Sigma, Kappa Psi. Presbyterian. Home: 502 Briar Creek Dr Knoxville TN 37922 Office: VA Outpatient Clinic 9031 Cross Park Dr Knoxville TN 37923

SHERRY, CAMERON WILLIAM, occupational hygienist, consultant; b. Asbestos, Que., Can., June 9, 1939; s. Homer Kent and Elizabeth Tait (Elliot) S.; m. Winnifred Rose Dixon, May 4, 1963; children: Wendy, Kim, Craig, Kathleen. BSc, Queen's Coll., Kingston, Ont., Can., 1962. Registered profl. engr., cert. indsl. hygienist, Que. Pulp and paper researcher Howared Smith, Cornwall, Ont., 1961-62; constrn. material researcher Domtar, Senneville, Que., 1963-69, group leader, 1970-78; product mgr. Domtar Constrn. Materials, Montreal, Que., 1978-82, mgr. indsl. hygiene, 1982-90; pres. Enviro-Risque, Pointe-Claire, Que., 1991—. Contbr. articles to profl. publs. Sch. commr. Lakeshore Sch. Bd., Beaconsfield, Que., 1989—. Mem. ASTM (chmn. subcoms.), Can. Acoustical Assn. (pres. 1983-86), Inst. Noise Control Engring., Am. Acad. Indsl. Hygienists, Can. Legion (curling pres. 1990). Achievements include patent for magnesium oxychloride cement, demountable partition system. Home: 445 Bellevue Ave, Dorion, PQ Canada J7V 2B1 Office: Enviro-Risque, 78 Lucerne Ave, Pointe Claire, PQ Canada H9R 2V2

SHERWIN, ROGER WILLIAM, medical educator; b. Macclesfield, Chesire, Eng., Aug. 23, 1931; came to U.S., 1961; s. William Kenneth and Marjorie (Chrimes) S.; divorced. BA, Cambridge U., Eng., 1953; Bachelor of Medicine B. of Chirurgy, Cambridge U., 1958. Intern and resident in internal medicine Whittington and Fulham Hosps., London, 1957-61; fellow in internal medicine Johns Hopkins U., Balt., 1961-66, asst. prof., 1968-70; asst. prof. Sch. Medicine, U. Md., Balt., 1970-73, assoc. prof., 1973-80, prof., 1980—. Co-editor: Manual of Clinical Nutrition, 1982; contbr. 80 articles to profl. jours. Fellow Am. Coll. Epidemiology; mem. Am. Epidemiology Soc. Achievements include design of first randomized clinical trial of maternal nutritional supplementation during pregnancy, clinical trials in the areas of heart disease and diabetes, batch-randomized clinical trials. Home: 12 W Mount Vernon Pl Baltimore MD 21201 Office: U Md Sch Medicine Dept Epidemiology & Preventive Medicine Baltimore MD 21201

SHESTAKOV, SERGEY VASILIYEVICH, geneticist, biotechnologist; b. Leningrad, Russia, Nov. 23, 1934; s. Vasiliy Ivanovich and Ludmila Alexandrovna (Krumholtz) S.; m. Galina Andreevna Grigorieva, Sept. 9, 1964; 1 child, Alexander Sergeevich. MS in Biochemistry, Moscow State U., 1957, PhD, 1964, DSc (hon.), 1974. Rschr. Moscow State U., 1961-77, prof. dept. genetics, 1977-80, dep. dean faculty biology, 1978-80, chmn. dept. genetics, 1980—, dir. Internat. Biotechnology Ctr., 1991—; postdoctoral scholar Princeton (N.J.) U., 1966-67; vis. prof. Pa. State U., University Park, 1975; vis. rschr. Pasteur Inst., Paris, 1981; dir. N. Vavilov Inst. Gen. Genetics, Russian Acad. Scis., Moscow, 1988-91;. Author: (textbook) Experimental Techniques in Microbial Genetics, 1984; editor: Molecular Basis of Genetic Processes, 1981. Advanced scholar Fullbright-Hays, 1975; recipient award USSR Min. Higher Edn., 1981, USSR State prize, 1988; UNESCO fellow, 1985. Mem. Russian Acad. Scis. (corr., chmn. sci. coun. genetics 1988—), Biotechnological Acad. Russia (coun. 1990), Russian Soc. Genetics and Breeding (coun. 1971—), Internat. Coun. for Sci. Devel., Internat. Soc. Plant Molecular Biology. Achievements include patents for plant controlled cultivation of microorganisms, mutants of cyanobacteria superproducing molecular hydrogen and ammonia, mutants of photosyntheic bacteria-producers of biologically active compunds; discovery of genetic transformation in blue-green algae; discovery and molecular analysis of novel genes in cyanobacteria and those involved in photosynthesis; development of gene transfer systems in purple bacteria; formulation of key role of repair of DNA-membrane complex and compact chromosome in postirradiation cell restoration, the requirement of two homologous DNA duplexes for repair of DNA double-strand breaks leading to the radiation death of cells; decoding of recombinational repair systems in cyanobacteria and fungi; isolation and investigation of mutants altered in cell division; construction of integrative and birepliconic vectors and cloning of different genes in cyanobacteria; isolation of mutants depressed for nitrogenase and excreting ammonia; research in genetics and genetic engineering of cyanobacteria, recombination and repair systems, hydrogen and nitrogen metabolism, photoautotrophy, genetic control and molecular mechanisms of nitrogen fixation in phototrophic bacteria, cloning of genes involved in the regulation of nitrogen fixation. Office: Vavilov Inst Gen Genetics, Ulitsa Gubkina 3, 117809 Moscow Russia

SHETH, ATUL CHANDRAVADAN, chemical engineering educator, researcher; b. Bombay, India, Dec. 2, 1941; came to U.S., 1967; s. Chandravadan N. and Tanman C. (Randeri) S.; m. Sheila A. Amin, Feb. 16, 1965; children: Roma, Archana. BSChemE, U. Bombay, 1964; MSChemE, Northwestern U., 1969, PhD in Chem. Engring., 1973. Prodn. supr. Esso Standard Ea., Inc., Bombay, 1964-67; chem. operator Riker Labs., Northridge, Calif., 1967; process engr. Armour Indsl. Chems. Co., McCook, Ill., 1969; asst. plant mgr. Charlotte Charles, Inc., Evanston, Ill., 1971-72; chem. engr. Argonne Nat. Lab., Chgo., 1972-80; staff engr. Exxon Rsch. & Engring. Co., Baytown, Tex., 1980-84; assoc. prof. U. Tenn. Space Inst., Tullahoma, 1984—; cons. Dept. Energy, Chgo. & Washington, 1978-80; team leader Argonne Nat. Lab., 1978-80; sec. mgr. U. Tenn. Space Inst., Tullahoma, 1985—. Contbr. articles to profl. jours. Judge Internat. Sci. Fair, Knoxville & Nashville, 1988, 92; advisor Tullahoma High Sch., 1990—; penpal scientist Cumberland Sci. Mus., Nashville, 1989-90. Recipient Anil P. Desai prize Elphinstone Coll., Bombay, 1960; grantee U.S. Dept. Energy, 1990-93. Mem. AIChE, Air and Waste Mgmt. Assn., Sigma Xi (pres. U. Tenn. Space Inst. club 1992-93, pres.-elect 1991-92, v.p. membership 1988-91). Democrat. Hindu. Achievements include 4 patents on desulfurization/dechlorination of alkali metal compounds using anion-exch. resins in U.S.; C12/HC1 emmissions control by MHD seed and preventing chloride build-up in MHD, others. Home: 204 Regwood Dr Tullahoma TN 37388 Office: U Tenn Space Inst B H Goethert Pkwy Tullahoma TN 37388-8897

SHETLER, STANWYN GERALD, botanist, museum official; b. Johnstown, Pa., Oct. 11, 1933; s. Sanford Grant and Florence Hazel (Young) S.; m. Elaine Marie Retberg, Feb. 2, 1963; children: Stephen Garth, Lara Suzanne. BS with distinction, Cornell U., 1955, MS, 1958; PhD, U. Mich., 1979. Asst. curator dept. botany Nat. Mus. Natural History, Smithsonian Instn., Washington, 1962-63, assoc. curator, 1963-81, curator, 1981—, asst. dir., 1984-91, dep. dir., 1991—; program sec., then program dir. Flora N.Am. program, Smithsonian Instn., 1966-78; bd. dirs. Audubon Naturalist Soc. of Cen. Atlantic States, Washington, 1971-74, 79-82, pres., 1974-77; exec. com. U.S. Com. Man and Biosphere program, Washington, 1985—; mem. Open Space Adv. Com., Loudoun County, Va., 1980-90, chmn., 1982; bd. dirs. Piedmont Environ. Coun., Va., 1985-88. Author: Komarov Botanical Institute: 250 Years of Russian Research, 1967, Variation and Evolution of the Nearctic Harebells (Campanula subsect: Heterophylla), 1982, Portraits of Nature: Paintings by Robert Bateman, 1986; contbr. articles to scholarly and popular publs. Recipient citation for innovative leadership Audubon Naturalist Soc., Washington, 1977, Individual Conservation award Piedmont Environ. Coun., 1981. Mem. AAAS, Am. Inst. Biol. Scis., Bot. Soc. Am., Washington Biologists' Field Club, numerous other profl. orgns. Democrat. Mennonite. Achievements include botanical field work in Australia, Siberia, Europe, South America, North America, organization of major exhibition of nature paintings of Robert Bateman at the Smithsonian Institution. Home: 142 Meadowland Ln E Sterling VA 20164 Office: Nat Mus Natural History Smithsonian Instn Rm 421 MRC 106 Washington DC 20560

SHETTY, MULKI RADHAKRISHNA, oncologist, consultant; b. Hiriadka, Karnataka, India, July 10, 1940; came to U.S., 1974; s. Mulki Sunderram and Kusumavati Shetty. MBBS, Stanley Med. Coll., Madras, 1964; DTM, U. Liverpool, Eng., 1968; LMCC, Med. Coun., Can., 1975. House surgeon and physician Bombay Hosp., 1965-66; sr. house officer Manor Pk. Hosp., Bristol, Eng., 1966-67, Torbay Hosp., 1967-68, St. Lukes Hosp., Huddersfield, 1969-70; sr. resident Gen. Hosp. Meml. U., New Foundland, 1971-72; intern Ottawa Gen. Hosp., 1972-73; fellow in chemotherapy Ont. Cancer Found., Ottawa, Can., 1973-74; fellow in clin. oncology U. Fla., Gainesville, 1974-75; attending oncologist N.W. Community Hosp., Arlington Heights, Ill., 1975—; cons. N.W. Community Hosp., 1975—; ex-chmn. Vijaya Bank Ltd. Author: Lung Cancer, 1980, Recent Advances in Chemotherapy, 1985; contbr. numerous articles to profl. jours.; coined new word, calcifectomy. Recipient Cert. for Outstanding Svc., Am Cancer Soc., 1982. Fellow Royal Soc. Medicine; mem. Internat. Assn. for Study of Lung Cancer, Chgo. Med. Soc., Kusumavati Soc. Hindu. Office: NW Community Hosp 800 W Central Rd Arlington Heights IL 60005-2392

SHEU, LIEN-LUNG, chemical engineer; b. Kaohsuing, Taiwan, China, Apr. 15, 1959; came to U.S., 1985; s. Yao-Kai and Yu-Lan S.; m. Pey-Hwa Lee, July 11, 1987; 1 child, Ashley Shaomin. BS, Nat. Taiwan U., 1981; PhD, Northwestern U., 1989. Rsch. asst. Process Synthesis Lab., Taipei, Taiwan, 1983-85; rsch. assoc. Northwestern U., Evanston, Ill., 1985-89; sr. engr. Tech. Ctr./The BOC Group, Inc., Murray Hill, N.J., 1989—; referee Applied Catalysis, Amsterdam, 1991—, Gas Seperation & Purification, Oxford, 1991—. Contbr. articles to profl. jours. Mem. Am. Soc. Testing and Material (com. mem. 1991—), Am. Inst. Chem. Engrs., N.Am. Catalysis Soc. Achievements include synthesize cubooctahedral Pd carbonyl inside zeolite cages at room tempatrue; find a universal relationship between metal particle size and its absorbed CO's IR spectra; develope an oxygen based process to produce anhydride. Home: 2424 Hill Rd Scotch Plains NJ 07076 Office: The BOC Group Inc 100 Mountain Ave Murray Hill NJ 07974

SHEVCHENKO, SERGEY MARKOVICH, organic chemist; b. Leningrad, USSR, Apr. 19, 1952; s. Tatiana S. (Shitova) Shevchenko; m. Tatiana S. Rotnova, June 22, 1978; children: Anna, Sergey. BS in Chemistry, Leningrad U., 1974, PhD in Organic Chemistry, 1980; DSc in Wood Chemistry, Acad. Forestry, St. Petersburg, Russia, 1992. Engr. Leningrad U., 1974-80; jr. rsch. assoc. Acad. Forestry, Leningrad, 1981-82, sr. rsch. assoc., 1982-89; prin. rsch. assoc. Acad. Forestry, St. Petersburg, 1990—; vis. scientist Rugjer Boskovic Inst., Zagreb, Yugoslavia, 1986-87, Comenius U., Bratislava, Czechoslovakia, 1989, N.C. State U., Raleigh, 1990-91; referee Kymiya Drevesiny, Riga, Latvia, 1986—; mem. program com. Biannual Conf. on Redox Reactions of Wood, Arkhangelsk, Russia, 1991—; cons. East European Tour TAPPI, Atlanta, 1990-91; exec. sec. UNESCO Internat. Expert Coun. on Chemistry of Vegetal Resources. Author: A Molecule in Space, 1986, The Image of a Molecule, 1989; contbr. over 120 articles to profl. publs.; author satiric column to Novosti YaMR v Pismakh, Moscow, 1990—. Chmn. Coun. Employees, Acad. Forestry, St. Petersburg, 1988-90. Mem. Mendeleev Chem. Soc., Assn. NMR Spectroscopists, Am. Chem. Soc. (affiliate). Achievements include patents in heterocyclic and wood chemistry; first synthesis of simplest hydrazinothiol, reaction of thiiranes with hydrazones, minimal conformational graph, new conformational effects, chemistry of guinone methides, chemical reactivity of lignin. Home: Volodarskogo 38 Kv 2 Sestroretsk, 189640 Saint Petersburg 189640, Russia Office: Acad Forestry, Dept Organic Chemistry, Institutsky per 5, 194018 Saint Petersburg Russia

SHI, ZHENG, radiological physicist; b. Shang Hai, China, Dec. 11, 1947; came to U.S., 1982; s. Bin Shi and Yiying Zhou; m. Susan S. Chen, Feb. 24, 1978; 1 child, Veronica J. Shi. MS, U. Mo., 1984, PhD, 1987. Diplomate Am. Bd. Radiology, Am. Bd. Sci. in Nuclear Medicine. Chief radiol. physicist Truman Med. Ctr., Kansas City, Mo., 1987—, radiation safety officer, 1988—; asst. prof. Sch. Medicine U. Mo., Kansas City, 1987—; adj. prof. Nuclear Engring. U. Mo., Columbia, 1988—. Mem. Am. Assn. of Physicists in Medicine, Health Physics Soc., Am. Coll. Radiology, Sigma Xi. Office: Truman Med Ctr 2301 Holmes St Kansas City MO 64108

SHIBATA, AKIKAZU, semiconductor scientist; b. Sendai, Japan, Feb. 25, 1935; s. Kitaro and Toshi (Tsushima) S.; m. Saeko Margaret Kurakake; children: Yasushi, Izumi. BA, Internat. Christian U., Mitaka, Japan, 1957; MA, Cornell U., Ithaca, N.Y., 1959; D in Engring., Nagoya (Japan) U., 1967. Rsch. asst. Cornell U., Ithaca, 1957-59; rsch. sci. Sony Corp., Tokyo, 1959-61, Sony Rsch. Ctr., Yokohama, 1976-84; gen. mgr. engring. Sony Trading Corp., Tokyo, 1980-83; staff sci. Sony Rsch. Ctr., Tokyo, 1984-86; mgr., internat. standardization Sony Corp., Tokyo, 1987—, gen. mgr., internat. standardization, 1989—. Contbr. articles to profl. jours.; patentee in field. Mem. IEEE, Soc. Applied Physics of Japan, Internat. Electotechnical Commn. Home: 914 Serizawa, Chigasaki 253, Japan Office: Sony Corp Tech Rels Divsn, 6-7-35 Kitashinagawa, Shinagawa Tokyo 141, Japan

SHIBATA, ERWIN FUMIO, cardiovascular physiologist; b. Honolulu, Apr. 22, 1950; s. James Shigeo and Harriet (Ide) s.; m. Aimee Emiko Sodetani, Aug. 18, 1978; children: Jonathan Fumio Alika, Jason Senkichi Kalani, Jeremy James Douglas, Jenna Emi Kehaulani. BSEE, U. Wash., 1974, MSEE, 1980; PhD, U. Tex., Galveston, 1984. Rsch. asst. U. Wash., Seattle, 1974-79; rsch. assoc. U. Calgary, Alb., Can., 1984-86; asst. prof. U. Iowa, Iowa City, 1986—. Editorial bd. Circulation Rsch., 1993—; referee Am. Jour. of Physiology; contbr. articles to profl. jours. including Am. Jour. of Physiology, Circulation Rsch., and Jour. Clin. Investigation. Den leader Cub Scouts Am., Iowa City, 1988—. Recipient Established Investigator award Am. Heart Assn., 1992—. Mem. AAAS, Biophys. Soc., Soc. for Neuroscience. Achievements include research on receptor regulation of cardiac sodium current, characterization of ion channels in single coronary artery smooth muscle cells. Office: U Iowa Dept Psychology 5-472 Bowen Science Bldg Iowa City IA 52242

SHIBAYAMA, MITSUHIRO, materials science educator; b. Shikatsu, Aichi, Japan, June 27, 1954; s. Tatsuaki and Suzuko (Tanaka) S.; m. Sumiko Takinaka, May 5, 1980; children: Motoki, Yuki, Naoyuki. B in Engring., Kyoto U., 1977, Kyoto U., 1979; D in Engring., Kyoto U., 1983. Fellow Japan Soc. Promotion of Sci., Tokyo, 1982-83; rsch. assoc. U. Mass., Amherst, 1983-84; asst. prof. Kyoto (Japan) Inst. Tech., 1984-88, assoc. prof. polymer materials sci., 1988—; vis. prof. MIT, Cambridge, Mass., 1991-92. Grantee Tokuyama Sci. Found. Japan, 1989. Mem. Am. Chem. Soc., Soc. Polymer Sci. (grantee 1990), Soc. Fiber Sci. and Tech. (Sakurada Takeshi award 1991). Avocations: tennis, social dancing. Home: 15-3 Seiwa-cho, Otsu 520-02, Japan Office: Kyoto Inst Tech, Matsugasaki, Sakyoku, Kyoto 606, Japan

SHIBER, MARY CLAIRE, biochemist; b. Paterson, N.J., July 8, 1960; d. Fred and Claire Catherine (Jenkinson) S. BS in Biology, Fairfield U., 1982; MSc in Microbiology, Rutgers U., 1985, PhD in Microbiology, 1988. Postdoctoral fellow dept. biochemistry Johns Hopkins U., Balt., 1988-90; rsch. fellow dept.molecular biology Princeton (N.J.) U., 1990—. Contbr. to profl. publs. Charles and Johanna Busch fellow Rutgers U., 1985-88; Sterling rsch. fellow Princeton U., 1990-92. Mem. AAAS, Am. Soc. Microbiology, Am. Littoral Soc., Nat. Wildlife Fedn., Sigma Xi. Home: 28 Nathan Way Wayne NJ 07470 Office: Princeton U Lewis Thomas Lab Princeton NJ 08544-1014

SHIELDS, LAWRENCE THORNTON, orthopedic surgeon; b. Boston, Oct. 2, 1935; s. George Leo and Catherine Elizabeth (Thornton) S.; AB, Harvard U., 1957; MD, Johns Hopkins U., 1961; m. Karen S. Kraus, Sept. 21, 1968; children: Elizabeth Coulter, Laura Thornton, Sarah Daly, Michael Lawrence. Intern, Barnes Hosp., Washington U., St. Louis, 1961-62, resident, 1962-63; resident orthopedic surgeon Children's Hosp. Med. Ctr., Boston, 1966-67, Mass. Gen. Hosp., Boston, 1967-68, Peter Bent Brigham, Robert Breck Brigham hosps., Boston, 1968-69; resident orthopedic surgeon Harvard Med. Sch., Boston, 1965-69, instr., 1969—; orthopedic surgeon Peter Bent Brigham & Women's Hosp., Children's hosps., 1969—; orthopedic surgeon Waltham (Mass.)-Weston Hosp. and Med. Ctr., 1969—, also chief orthopedic surgery, pres. med. staff; mem. Waltham-Weston Orthopedic Assos.; proprietor Boston Athenaeum; mem. staffs Hahnemann Hosp., Boston, Newton-Wellesley (Mass.) hosps.; cons. orthopedic surgeon VA Hosp., Boston; mem. faculty Harvard Med. Sch.; bd. dirs. Wal-West Health Systems, 1986—; pres. Massachusetts Bay Investment Trust; dir. Waltham Investment Group. Bd. dirs. Mass. Acad. Emergency Med. Technicians, Waltham Boys' Club; bd. of overseers Boston Lyric Opera, 1993—; trustee, exec. com. Waltham-Weston Hosp. and Med. Ctr. Lt. M.C., USNR, 1963-65. Diplomate Am. Bd. Orthopedic Surgery. Fellow ACS, Am. Acad. Orthopedic Surgeons, Mass. Hist. Soc. Libr., Mass. Hist. Soc.; mem. N.Y. Acad. Scis., Royal Soc. Medicine, Mass. Orthopaedic Assn. (bd. dirs., sec. 1986—), New Eng., Boston orthopedic clubs, Charles River Dist. (treas., exec. com., pres. 1982-83) Mass. (councillor; v.p. 1982-83) med. socs., R. Austen Freeman Soc. (v.p.), Thomas B. Quigley Sports Medicine Soc., Titanic Hist. Soc., Boston Opera Assns. (bd. dirs.), Harvard Mus. Assn., Thoreau Soc., Emerson Soc., Trollope Soc. (founding mem., London), Handel and Hayden Soc. (bd. overseers), Waltham Hist. Soc., St. Botolph Club (Boston), Les Amis d'Escoffier Soc., Harvard Club, Algonquin Club of Boston (bd. dirs., pres. 1990—), St. Crispin's Soc. Boston (founding mem., pres. 1991—), English Speaking Union (bd. dirs.), Union Club of Boston, St. Botolph Club of Boston, Boston Lyric Opera (bd. overseers 1993), Rotary, Pi Eta (Harvard). Contbr. articles to med. jours. Home: 9 Beverly Rd Newton MA 02161-1112 Office: 721 Huntington Ave Boston MA 02115 also: 20 Hope Ave Ste 314 Waltham MA 02154

SHIGESADA, NANAKO, mathematical biology educator, researcher; b. Kurashiki, Okayama, Japan, July 7, 1941; d. Yutaro and Toshiko Inoue; m. Katsuya Shigesada, May 15, 1966; 1 child, Yukihiko. BS, Kyoto (Japan) U., 1964, MS, 1966, DSc (hon.), 1971. Instr. Kyoto U., 1971-92; vis. prof. SUNY, Stony Brook, 1979-80; vis. scientist Stanford (Calif.) U., 1980-81; prof. info. and computer sics. Nara (Japan) Women's U., 1992—. Author: Mathematical Perspective of Life Sciences, 1975, Mathematical Models of Biological Invasions, 1992; contbr. articles to profl. jours. Mem. Japanese Assn. Math. Biology (sec. gen. 1990-92). Home: 503 Haitsu-Empera-Shugakuin, Yamabana-Ichodacho, Kyoto 606, Japan Office: Nara Women's U, Dept Info and Computer Scis, Kita-Uoya Nishimachi, Nara 630, Japan

SHIH, JING-LUEN ALLEN, medical physicist; b. May 31, 1961. MS, U. Mo., 1987, PhD, 1991. Rsch. asst. Mo. U. Rsch. Reactor, Columbia, 1987-90; tech. collaborator Brookhaven Nat. Lab., Upton, N.Y., 1990-92; med. physicist Henry Ford Hosp., Detroit, 1991—. Contbg. author: Progress in Neutron Capture Therapy for Cancer, 1992; contbr. articles to profl. jours. Mem. Am. Assn. Physicists in Medicine. Achievements include research in neutron induced brachytherapy with Gd seeds and needles; in neutron capture therapy with Gd-157 compounds; in neutron autoradiography of trace amounts of gadolinium. Office: Henry Ford Hosp Radiation Oncology Dept 2799 W Grand Blvd Detroit MI 48202

SHIKARKHANE, NAREN SHRIRAM, laser scientist; b. Pune, India, Nov. 13, 1954; s. Shriram Keshav and Indumati Shriram (Thombare) S.; m. Aruna Naren Mandke; children: Gauri, Durga. BS, Pune U., 1975. Counter salesman Prakash Gen. Stores, Pune, 1973-74; hotel billing clk. Shreyas Hotel, Pune, 1975-76; scientific asst. Bhabha Atomic Rsch. Ctr., Bombay, 1976-88, scientific officer, 1988-92; with Scantech, 1992—. Contbr. articles to profl. jours. Mem. Optical Soc. Am., Indian Laser Assn. (life). Hindu. Achievements include research into double power extraction from pulse TEA CO2 laser by use of doubled ended resonater, first indigenous Nd-YAG laser, CW, 100W made for industrial purposes in India. Home: 2/40 Sahakar Nagar No 3, Chembur Bombay 400 071, India

SHIKATA, JUN-ICHI, surgery educator; b. Tokyo, Japan, Jan. 28, 1926; s. Keichi and Masago (Suminokura) S.; m. Sakiko Mita, May 12, 1960. MD, U. Tokyo, 1950, PhD, 1960. Intern and surg. resident U. Tokyo Hosp., 1950-56; rsch. fellow faculty medicine U. Tokyo, 1956-58; rsch. fellow Ind. U. Med. Ctr., Indpls., 1958-59; asst. surgeon U. Tokyo Hosp., 1959-62; chief surgeon Tokyo Met. Bokuto Hosp., 1962-71; clin. assoc. prof. U. Tokyo, 1962-71; prof., chmn. First Dept. Surgery Teikyo U. Sch. Medicine, Tokyo, 1971-92, prof. emeritus, 1992—; cons. Aisei Hosp., Tokyo, 1980—, Icho (Gastrointestinal) Hosp., Tokyo, 1986—. Author: Diagnosis of the Acute Abdomen, 1968, Abdominal Trauma, 1971, Intestinal Obstruction, 1990. Sponsor World Ecol. Rsch. Found., Tokyo, 1989—. Mem. ACS, Japan Surg. Soc., Japan Soc. Gastrointestinal Surgery, Soc. for Surgery of the Alimentary Tract, Soc. Internat. de Chirurgie. Avocations: golfing, skiing, fishing, swimming. Home: 18-1-302 Ichibancho, Chiyoda-ku, Tokyo 102, Japan Office: Teikyo U Sch Medicine, 2-11-1 Kaga, Itabashi-ku, Tokyo 173, Japan

SHIM, JACK V., civil engineer, consultant; b. Kingston, Jamaica, Feb. 14, 1955; came to U.S., 1977; s. Dudley Adolph and Verna Angela (Lee) S. B in Engring., McGill U., Montreal, Can., 1977; MBA, U. So. Fla., 1983. Registered profl. engr.; cert. gen. contractor. Design engr. Peabody & Childs, Inc., Pompano Beach, Fla., 1977-79, Briel, Rhame, Poynter & Houser, Inc., Melbourne, Fla., 1979-80, Davy McKee Corp., Lakeland, Fla., 1980-83, Badger Engrs., Tampa, Fla., 1983-86; sr. engr. South Fla. Water Mgmt., West Palm Beach, Fla., 1986-90; project mgr. Broward County Pub. Works, Ft. Lauderdale, Fla., 1990—; mgmt. cons. Video Awards, Inc., Oakland Park, Fla., 1989—, Golden Gate Restaurant, Margate, Fla., 1989-91; pres. Shim & Assocs., Inc., Tampa, 1986-87. Author computer programs in beam design and analysis. Fund raiser Big Bros./Sisters of Broward, Ft. Lauderdale, 1991. Mem. ASCE, Fla. Engring. Soc., Beta Gamma Sigma. Achievements include research and development of prefabricated concrete residential structures for low income families. Home: 1880 NW Ninety Four Ave Plantation FL 33322

SHIMA, MIKIKO, polymer scientist; b. Kyoto, Japan, Feb. 26, 1929; d. Goro and Ayako (Takayasu) S. Diploma in teaching math. and scis., Tokyo Higher Normal Sch. Women, 1950; BSc, Tokyo U. Edn., 1953, DSc, 1958. Rsch. assoc. Tokyo U. Edn., 1958-65; assoc. prof. Tokyo Woman's Christian U., 1965-71, prof. of chemistry, 1971—; postdoctoral fellow Polymer Rsch. Ctr., SUNY, Syracuse, 1961-63; vis. acad. staff mem. Imperial Coll. Sci. and Tech., U. London, 1980-81. Co-author: Text Book of Polymer Science, 1978; contbr. articles to profl. jours. Recipient award AAUW, 1961; grantee Fulbright Found., 1961, NSF, 1962, 63, Japanese Ministry of Edn., Sci. and Culture, 1974, 77, 82, 83, Japanese Assn. for Promotion of Scis., 1978. Mem. Chem. Soc. Japan (mem. coun. 1967-69, editor Chemistry & Industry 1973-75), Soc. Polymer Sci. (editor High Polymer 1968-70, 82-84, sec. 1986—), Am. Chem. Soc., Japanese Assn. Univ. Women (bd. dirs. 1966-69, 76-80), Soc. Japanese Women Scientists (sec. 1963-76). Home: 3-19-13 Nishioginminami, Suginami-ku, Tokyo 167, Japan Office: Tokyo Woman's Christian U, 2-6-1 Zempukuji, Suginami-ku, Tokyo 167, Japan

SHIMADA, SHINJI, dermatology educator, researcher; b. Kyoto, Japan, Apr. 8, 1952; s. Koichi and Teruko (Hara) S.; m. Eri Minato, Feb. 26, 1978. MD, U. Tokyo, 1977, PhD, 1982. Assoc. U. Tokyo, 1977-86; assoc. prof. Yamanashi (Japan) Med. Coll., 1986-91, U. Tokyo, 1991—; vis. fellow NIH, Bethesda, Md., 1983-87. Recipient Minami prize for Med. Rsch., 1981. Avocations: golf, baseball, tennis, travelling. Home: 1-1 Midoricho # 1-302, Tanashi-shi Tokyo 188, Japan Office: Tokyo U Br Hosp Dept Dermat, 3-28-6 Mejirodai, Bunkyo-ku Tokyo 112, Japan

SHIMAHARA, KENZO, applied microbiology educator; b. Tokyo, Mar. 22, 1928; s. Itsuzo and Mikie (Miyagawa) S.; m. Yoko Tsujii; 1 child, Taku. B. Engring., Keio U., Tokyo, 1950, PhD, 1974. Chem. engr. Nitto-Kagaku Co. Ltd., Yokohama, Japan, 1950-54; tchr. Musashi-Kogyo U. High Sch., Tokyo, 1955-62; lectr. Seikei U., Tokyo, 1962-64, assoc. prof., 1964-74, prof., 1974-93, prof. emeritus, 1993—; guest prof. Liaoning Normal U., Dalian, 1990—. Editor: Lecture Demonstrations in Chemistry, 1982; author: Biochemistry, 1991, (with others) Chitin, Chitosan and Related Enzymes, 1984; editor, author (with others): Erroneous Descriptions in the History of Chemistry, 1988. Mem. Japanese Soc. for the History of Chemistry (bd. dirs. 1982—), Japanese Soc. for Chitin and Chitosan (v.p. 1989-92). Home: 6-27-43 Higashi-Koigakubo, Kokubunji 185, Japan

SHIMANEK, RONALD WENZEL, engineering executive; b. Algoma, Wis., June 29, 1945; s. Wenzel J. and Viola M. (Massart) S.; m. Jeanne F. Kotchi; Apr. 20, 1968; 1 child, Stephen A. BSEE with hons., U. Wis., 1968; MSEE, Purdue U., 1974. Registered profl. engr. Ind. Engr. design Delco Electronics Corp., Kokomo, Ind., 1968-76, sr. engr., 1976-80, supr. quality control, 1984-87, staff engr., 1987-90, mgr. customer satisfaction engr., 1990--; mgr. quality control and engring. GM Singapore Pte Ltd., 1980-84; lectr. Ind. U., Kokomo, 1978-80; instr. Ivy Tech. Sch., Kokomo, 1978-80. speaker in field. Treas. Montessori Children's House, Kokomo, 1974; pres. Kokomo Photo Guild, 1984-92. Mem. Tau Beta Pi, Eta Kappa Nu. Achievements include development of first digital frequency synthesized automobile receiver and patent for Electronic Indicator. Home: 123 S Conradt Ave Kokomo IN 46904 Office: Delco Electronics One Corporate Ctr Kokomo IN 46904-9005

SHIMAZAKI, YASUHISA, cardiac surgeon; b. Kagoshima, Japan, Feb. 19, 1946; s. Hisashi and Sumiko (Tsukada) S.; m. Yuko Inomata, Oct. 10, 1982; children: Taisuke John, Kyosuke, Yasuro. MD, Osaka U., 1970, DMS, 1982. Resident in surgery Osaka U. Hosp., 1970-77; asst. Med. Sch. Osaka U., 1977-82, sr. assoc., 1986—; Graham Traveling Fellow Am. Assn. for Thoracic Surgery, Birmingham, Ala., 1982-83; cardiovascular surgery fellow U. Ala., Birmingham, 1983-84; staff surgeon Osaka Boshi Ctr. Hosp., Sakai, 1984-86; cardiovascular surgery fellow U. Ala., 1988-89; asst. prof. surgery Osaka U., 1991. Mem. Japan Assn. Thoracic Surgery, Japan Assn. Pediatric Surgery. Home: 1-9-21 Inaba, Higashiosaka 578, Japan Office: Osaka U Med Sch, Yamadaoka 2-2, Suita Osaka 565, Japan

SHIMBO, MASAKI, technology company executive; b. Utsunomiya, Japan, July 19, 1914; s. Toraji and Nigi (Sonoyama) Shimbo; m. Michiyo Fujino, Feb. 11, 1944; 1 child, Hirohiko. B., Osaka (Japan) Imperial U., 1940, PhD, 1960. Asst. prof. Osaka Imperial U., 1940-47; lab. mgr. Mizuno Co. Ltd., Osaka, 1947-59; lectr. Kansai U., Osaka, 1959-61, prof., 1961-67, dean faculty engring., 1967-74, dean grad. sch., 1974-85, prof. emeritus, 1985—; pres. Inst. of Poly. Technology, Osaka, 1985—. Mem. Japan Chem. Soc. (bd. dirs.), Am. Chem. Soc. Home: 3-2-106 Naruko Kita-ku, Kobe 651-11, Japan Office: Inst Polymer Technology, 1-3-11 Minami Shinmachi, Osaka 540, Japan

SHIMOJI, SADAO, applied mathematics educator, engineer; b. Tokyo, Apr. 22, 1932; s. Keijo and Kiyo (Miyakuni) S.; m. Tamiko Matsumoto, Dec. 15, 1961; children: Yuichi, Atsushi. MS in Physics, Tokyo U., 1957; PhD in Physics, Osaka (Japan) U., 1966. Cert. engring. Engr. Mitsubishi

Electric Corp., Amagasaki, Japan, 1957-65, group mgr., 1965-79, mgr., 1970-75; concurrent lectr. Inst. of Space Tech. and Scis., Tokyo U., 1972-74; sr. scientist Mitsubishi Electric Corp., Amagasaki, Japan, 1975-86; prof. Tokyo Engring. U., Hachioji, Japan, 1986—; mem. program sub-com. 9th, 10th, 13d, 14th ISTS, Tokyo, 1970-84. Co-author: AIAA Progress, Vol. 65, 1978; contbr. articles to Jour. Phys. Soc. Jaoan, Jour. of AIAA, Japanese Jour. Applied Physics, Jour. Applied Physics. Mem. a coun. of city admninstrn. Hachioji city, 1987—. Liberal Democrat. Buddhist. Achievements include discovery of optimum F-value found for the statistical regression analysis of thermal-vacuum test data of artificial satellites; development of theory that the behavior of the Carreau viscosity model for the long spinning time in the spin coating process was clarified by the numerical method and an analytical model of the solvent evaporation of polymer solutions. Home: Kitanodai 3-28-8, Hachioji 192, Japan Office: Tokyo Engring U, Katakura 1404, Hachioji 192, Japan

SHINA, SAMMY GOURGY, engineering educator, consultant; b. Baghdad, Iraq, Sept. 28, 1944; s. George and K. (Ezra) S.; m. Jacqueline Marks, Nov. 28, 1968; children: Michael, Gail, Nancy, Jonathan. BSEE, BS in Indsl. Mgmt., MIT, 1966; MS in Computer Sci., Worcester Poly. Inst., 1972. Registered profl. engr., Mass. Mfg. engr. Union Carbide, Florence, S.C., 1967-68; mfg. engring. mgr. RCA, Marlboro, Mass., 1968-71; productivity mgr. Hewlett Packard, Waltham, Mass., 1971-88; assoc. prof. U. Mass, Lowell, 1988—; bd. dirs. Mass. Quality Award, Lowell, Inventors Assn. New Eng., Lexington; design judge Milton Kiver Awards Competition, Anaheim, Calif., 1992; tech. session chmn. on quality function deployment NEPCON Confs., Boston, L.A., 1990-93; cons. numerous sci. cos. and corps. Author: Concurrent Engineering, 1991; contbr. over 40 articles to profl. jours. Chief Parent Child Program YMCA, Framingham, Mass., 1980; coach Youth Soccer Program, Framingham, 1976-82. Mem. IEEE (sr.), Am. Soc. Quality Control (sr.), Soc. Mfg. Engrs. (chmn. 1990-91, Leadership award 1992). Home: 19 Swanson Rd Framingham MA 01701 Office: U Mass 1 University Ave Lowell MA 01857

SHINE, KENNETH I., cardiologist, educator; b. Worcester, Mass., 1935. Grad., Harvard Coll., 1957; MD, Harvard U., 1961. Diplomate Am. Bd. Internal Medicine. Intern Mass. Gen. Hosp., 1961-62, resident, 1962-63, 65-66, fellow in cardiology, 1966-68; surgeon USPHS, 1963-65; assoc. in medicine Beth Israel Hosp., Resont, Va., from 1969; instr. Harvard Med. Sch., from 1968; asst. prof. medicine UCLA Sch. Medicine, 1971-73, assoc. prof., 1973-77, prof., 1977-92; dir. CCU UCLA Sch. Medicine, Washington, 1971-75; chief div. cardiology UCLA Sch. Medicine, 1975-79, vice chmn. dept. medicine, 1979-81, exec. chmn., 1981-86, dean, 1986-92, provost for med. scis., 1991-92; pres. Inst. Medicine, Washington, 1992—. Mem. Am. Heart Assn. (pres. 1986, 87), Assn. Am. Med. Colls. (adminstrv. bd. coun. deans 1989-92, exec. bd. 1990-92, chmn. coun. deans 1991-92). Office: Inst of Medicine 2101 Constitution Ave NW Washington DC 20418*

SHINER, JOHN STEWART, biophysicist educator; b. Front Royal, Va., Apr. 9, 1949; arrived in Switzerland, 1980; s. Ernest Thompson and Amy Lee (Ritter) S.; m. Anita Dina Jorns, Dec. 27, 1987. AB, Duke U., 1971; postgrad., U. Ala., 1971-72; PhD, Med. Coll. Va., 1978. Rsch. assoc. U. Stuttgart, Fed. Republic Germany, 1978-79; rsch. fellow U. Calif., San Francisco, 1979-80; instr. U. Basel, Switzerland, 1980-84; fellow U. Wuerzburg, Fed. Republic Germany, 1984-85; asst. prof. U. Bern, Switzerland, 1985—; mem. editorial bd. Info. in Phys. and Life Sci. Jour., Warsaw, Poland, 1990—. Contbr. articles to profl. jours. NIH fellow, 1979-80, Alexander von Humboldt Found. fellow, 1984-85; grantee Swiss Nat. Sci. Foun., 1988-91, Digital Equipment Corp., European Contbns. Program, 1990, Sandoz Found. for the Advancement of the Med.-Biol. Scis., 1990-91. Mem. AAAS, Am. Phys. Soc., Biophys. Soc., Swiss Physiol. Soc., European Soc. Muscle Rsch., Internat. Ctr. Thermodynamics (founding mem.). Avocations: riding, literature, music. Office: U Bern, Dept Physiology, Buehlplatz 5, CH-3012 Bern Switzerland

SHINNAR, REUEL, chemical engineering educator, industrial consultant; b. Vienna, Austria, Sept. 15, 1923; came to U.S., 1962; s. Abraham Emil and Rosa (Storch) Bardfeld; m. Miryam Halpern, June 22, 1948; children—Shlomo, Meir. Diploma in Chem. Engring., Technion, Haifa, Israel, 1945, M.Sc. in Chem. Engring., 1954; Dr. Engring. Sci., Columbia U., 1957. Various position in chem. engring. Israel, 1945-58; adj. assoc. prof. Technion, Haifa, Israel, 1958-62; visiting research fellow Guggenheim Labs., Princeton (N.J.) U., 1962-64; prof. chem. engring. CCNY, 1964—, disting. prof., 1979—; Pinhas Naor lectr. Technion U., 1974; Wilhelm Meml. lectr. Princeton U., 1985, Kelly lectr. Purdue U., 1991; cons. to various oil and chem. cos. Assoc. editor: Chem. Engring. Revs., Jour. Am. Inst. Chem. Engrs. Contbr. numerous articles to profl. jours. Patentee in field. Fellow AICE (Founders award 1992); mem. AIAA, Am. Chem. Soc., Nat. Acad. Engring., Alpha Chi Sigma. Office: City Coll NY Dept Chem Engring 140th St and Convent Ave New York NY 10031

SHINODA, KŌZŌ, chemistry educator, researcher; b. Tokushima, Japan, Aug. 29, 1926; s. Junzo and Shizuko (Maki) S.; m. Mieko Hirose, Mar. 8, 1954; children: Mitsuko, Mary, Haruko. BS, U. Tokyo, 1951, DSc, 1956. Asst. Yokohama (Japan) Nat. U., 1951-54, instr., 1955-56, asst. prof., then assoc. prof., 1957-63, prof. of chemistry, 1964-92; prof. Lund (Sweden) U., 1992—; vis. prof. Stanford (Calif.) U., 1961, Ytkemiska Inst., Stockholm, Sweden, 1972, Lund (Sweden) U., 1984, 86-91; mem. adv. bd. Surface Sci., N.Y., 1969—, Jour. Dispersion Sci. and Tech., N.Y., 1980—. Author: Colloidal Surfactants, 1963, Principles of Solution and Solubility, 1978 (with Friberg) Emulsions and Solubilization, 1986; author, editor: Solvent Properties of Surfact, 1967. Mem. Chem. Soc. Japan (Porgressive prize 1960, Highest Soc. award 1982), Am. Chem. Soc. (sr. mem., adv. bd. jour. 1985-89), Japan Oil Chemist's Soc. (Society award 1975). Avocations: swimming, surfing, gardening, carpentry, hiking. Office: Lund U Phys Chem 1, POB 124, S-22100 Lund Sweden

SHIOIRI, TAKAYUKI, pharmaceutical science educator; b. Yokohama, Kanagawa, Japan, Aug. 15, 1935; s. Hideji and Hisako (Iyoda) S.; m. Haruko Terashima, Mar. 13, 1966; children: Azusa, Akane. BSc, U. Tokyo, 1959, MSc, 1961, PhD, 1967; Diploma, Imperial Coll., London, 1970. Lic. pharmacist. Instr. U. Tokyo, 1962-64, rsch. assoc., 1964-77, assoc. prof., 1977; vis. academics Imperial Coll., London, 1968-70; prof. Sch. Pharm. Scis. Nagoya (Japan) City U., 1977—. Regional editor Tetrahedron, 1990—. Recipient Suzuki award Pharm. Soc. Japan, 1974, Abbott award, 1978, Aichi Pharm. award Aichi Pharmacist Soc., 1981, acad. award Pharm. Soc. Japan, 1993. Avocations: growing vegetables, travel, walking, reading. Home: 1-18-12 Minamigaoka, Nisshin-Cho, Aichi-Gun, Aichi-Ken 470-01, Japan Office: Nagoya City U Sch Pharm Sci, 3-1 Tanabe-Dori, Mizuho-Ku, Nagoya 467, Japan

SHIOTSU, MASAHIRO, engineering educator; b. Osaka, Japan, June 15, 1942; m. Yoshi Oyama, July 9, 1972. B in Engring., Kyoto U., 1966, D of Engring., 1978. Rsch. assoc. Kyoto U., Uji, 1968-79, assoc. prof., 1979—. Mem. ASME (Best Paper award 1990, Melville medal 1991), Atomic Energy Soc. of Japan, Cryogenic Energy Soc. Japan, Heat Transfer Soc. Japan. Office: Inst Atomic Energy Kyoto U, Gokashe, Uji Kyoto, Japan 611

SHIOYA, SUTEAKI, biochemical engineering educator; b. Fukui, Japan, Feb. 16, 1945; s. Tatsuo and Masae Shioya; m. Noriko Katayama, Oct. 24, 1971; children: Akiko, Makoto. MS, Kyoto (Japan) U., 1969, PhD, 1975. Cert. engring. Asst. prof. Kyoto U., 1971-86; asst. prof. Osaka U., Suita, Japan, 1986-87, assoc. prof., 1987—; Assoc. editor Jour. Fermentation and Bioengring., 1986—; editor Process Control and Quality, 1990—. Recipient Terui prize Soc. Fermentation Technology, Osaka, Japan, 1991. Home: Charano Nishikyo, 1-6-7 Nishisakaidani, Oharano Nishikyo, Kyoto 610 11 610 11, Japan Office: Osaka U Dept Biotechnology, 2-1 Yamadaoka Ste 565, Suita 565, Japan

SHIOZAKI, MASAO, synthetic and organic chemist; b. Saginomiya, Tokyo, Japan, Oct. 25, 1941; s. Masaichi and Chiyo (Suzuki) S.; m. Tomoko Oubuchi, Mar. 21, 1971; children: Ko, Yu. BS, U. Tokyo, 1966, MSc, 1968, PhD, 1971. Researcher Sankyo Co., Ltd., Tokyo, 1971-80, asst. chief researcher, 1980-86, chief researcher, 1986—. Vis. rsch. scholar Duke U., Durham, N.C., 1984-85. Mem. Am. Chem. Soc., Chem. Soc. Japan, Agrl. Chem. Soc. Japan. Home: Saginomiya 3-31-2, Nakano-Ku Tokyo 165, Japan Office: Sankyo Co Ltd, Hiromachi 1-2-58, Shinagawa-ku Tokyo 140, Japan

SHIOZAWA, DENNIS KENJI, zoology educator; b. Logan, Utah, Dec. 16, 1949; s. Kenji and Helen Teruko (Shiratori) S.; m. Janet Yamashita, Aug. 16, 1972; children: Natalie Hana, Becky Akiko. BA, Weber State Coll., 1972; MS, Brigham Young U., 1975; PhD, U. Minn., 1978. Asst. prof. zoology Brigham Young U., Provo, Utah, 1978-85, assoc. prof. zoology, 1985—. Contbr. articles to profl. jours. NSF traineeship, 1972-73; Bush Found. fellow U. Minn., 1975-76; Nat. Geographic Soc. grantee, 1987—. Mem. AAAS, Desert Fishes Coun., Am. Fisheries Soc., Am. Soc. Naturalists, Am. Soc. Limnology and Oceanography, N.Am. Benthological Soc., Ecol. Soc. Am., Internat. Soc. Theoretical and Applied Limnology. Democrat. LDS. Achievements include research in population dynamics of benthic invertebrates, stream microcrustaceans, trophic cascades and benthic linkages, molecular phylogenetics of trout-starch-gel electrophoresis, mitochondrial DNA, ribosomal DNA, slainity. Home: 995 W 5000 S Palmyra UT 84660 Office: Brigham Young U Dept Zoology 574 WIDB Provo UT 84602

SHIPBAUGH, CALVIN LEROY, physicist; b. Huntington, Ind., Aug. 28, 1958; s. Paul and Marguerite (Pinkerton) S. BA, Rice U., 1980; PhD, U. Ill., 1988. Rsch. asst. U. Ill., Champaign-Urbana, 1981-88; analyst RAND Corp., Santa Monica, Calif., 1988—; space and surface power panel mem. Rand Support to NASA Project Outreach, Santa Monica, 1990; vis. scientist Fermilab, Batavia, Ill., 1982-85; workshop leader biotechnology Group, RAND; team mem. POET, Arlington, Va., 1989-92. Contbr. articles to Phys. Rev. Letters, Physics Letters and RAND Pub. Series. Mem. AIAA, Am. Phys. Soc. Achievements include research to measure charm particles' decay and hadronic production properties; evaluation of proposals from the public to the Space Exploration Initiative; led biotechnology effort for workshop at RAND. Office: The RAND Corp 1700 Main St Santa Monica CA 90401-3297

SHIPE, GARY THOMAS, ceramic engineer; b. San Antonio, July 24, 1960; s. Larry Lee and Jane Louise (Laudenslager) S.; m. Donna Marie Schooner, Apr. 9, 1988; children: Christine Marie, Sarah Nicole. BS in Ceramic Engring., Pa. State U., 1982. Plant supr. Schaefer Industries, Inc., Allentown, Pa., 1984-86; process engr. Frenchtown (N.J.) Ceramics Co., 1986-89; design engr., lead engr. Philips Techs./Airpax, Cambridge, Md., 1989—. Mem. Pa. State Alumni Assn., University Park. Recipient Pres.'s award for tech. innovation Stackpole Corp., 1987. Mem. Am. Ceramic Soc., ASM Internat. Office: Airpax Philips Techs PO Box 520 Woods Rd Cambridge MD 21613

SHIPKIN, PAUL M., neurologist; b. Long Branch, N.J., Feb. 2, 1945; m. Barbara S. Gray, 1973; children: Rachael Gray Shipkin, Zachary Gray Shipkin. AB in Physics magna cum laude, Rutgers U., 1967; MD, Albert Einstein Coll. Medicine, 1971. Diplomate Nat. Bd. Med. Examiners. Intern Bronx (N.Y.) Mcpl. Hosp. Ctr., 1971-72, asst. resident medicine, 1972-73; resident in neurology Hosp. of the Univ. of Pa., Phila., 1973-76; neuro-ophthalmology fellow Bascom Palmer Eye Inst., U. of Miami (Fla.) Sch. Medicine, 1978-79; attending neurologist Jeanes Hosp., Phila., 1980—, div. chief dept. neurology, med. dir. EEG, 1985—; asst. instr. dept. medicine Albert Einstein Coll. Medicine, Bronx, 1972-73; asst. instr. dept. neurology U. Pa. Sch. Medicine, Phila., 1973-76; asst. prof. neurology and neuro-ophthalmology dept. neurology Med. Coll. Va., Richmond, 1979-80; clin. asst. prof. dept. neurology Hosp. of the U. of Pa., Phila., 1981—; attending neurologist Am. Oncologic Hosp., Rolling Hill Hosp., Nazareth Hosp., all Phila., 1980—; cons. physician Comprehensive Epilepsy Ctr., Grad. Hosp., Phila., 1981-82. Contbr. articles to Neurology, Biomedical Founds. of Ophthalmology, Annals of Neurology, Lancet, Neuro-Ophthalmology, Blood, Am. Jour. Hematology. Lt. comdr. USN, 1976-78. Recipient Heed Ophthalmic fellowship, 1978-79. Mem. AAAS, Am. Heart Assn., Am. Acad. Neurology, Phila. Neurol. Soc., The Soc. of Heed Fellows, Nat. Stroke Assn., Epilepsy Found. Am., Peripheral Neuropathy Inst., Queen Square Alumni Assn., Phi Beta Kappa, Phi Beta Kappa Assn. of the Delaware Valley. Home: 1527 Washington Ln Rydal PA 19046 Office: 7602 Central Ave Ste 203 Philadelphia PA 19111

SHIPKOWITZ, NATHAN L., microbiologist; b. Chgo., Mar. 29, 1925; children: Reuben, Abigail Tanya, Eirene, Gita. BS, U. Ill., 1949, MS, 1950; PhD, Mich. State U., 1952. Asst. prof. vet. sci. U. Mass., Amherst, 1952-54; asst. rsch. bacteriologist Hooper Found. U. Calif. Med. Ctr., San Francisco, 1954-58; clin. bacteriologist and virologist Good Samaritan Med. Ctr., Portland, Oreg., 1959-62; assoc. rsch. fellow Abbott Labs., Abbott Park, Ill., 1963—. Cpl. U.S. Army, 1943-46, ETO. Mem. Am. Soc. Microbiology, AAAS, Sigma Xi. Achievements include development of anti-infective drugs. Office: Abbott Labs Dept 47T Bldg AP3 Abbott Park IL 60064

SHIPLEY-PHILLIPS, JEANETTE KAY, aquatic biologist; b. Bentonville, Ark., Oct. 7, 1954; d. Walter Sherman and Ila Christine (Mason) Shipley; m. David Steven Phillips, Dec. 3, 1983. PhD in Botany, U. Ark., 1991. Tchr. biology Springdale (Ark.) High Sch., 1979-83; teaching asst. U. Ark., Fayetteville, 1983-91, instr. zoology, 1986-88; microbiologist Ark. Water Resources-Water Quality Lab., Fayetteville, 1991; asst. prof. biology Valparaiso (Ind.) U., 1991—. Contbr. articles to profl. jours. Recipient Trael award U. Ark., 1991; rsch. grantee Valparaiso U., 1993; Huron Mt. Wildlife Found. grantee, 1993. Mem. Phycological Soc. Am. (rsch. grantee 1990), Botan. Soc. Am., Internat. Phycological Soc., Ind. Acad. Sci., Sigma Xi (rsch. grantee 1990). Office: Valparaiso U Biology Dept Valparaiso IN 46383

SHIPMAN, JAMES MELTON, propulsion engineer; b. Madison, Tenn., July 7, 1939; s. J. T. and Mary Nadine (Jennings) S.; m. Jean Branham, Nov. 5, 1966. BS in Aero. Engring., Auburn U., 1961. Maj. Operational Fighter Units and Hqrs.-USAF, 1962-83; project engr. United Technologies Pratt & Whitney, West Palm Beach, Fla., 1983—; aircraft accident investigator USAF, 1964-72. Recipient Commendation medal USAF, 1964, 74, 78, Meritorious Svc. medal USAF, 1972, 78, 83. Mem. AIAA, Air Force Assn. (v.p. 1984—). Achievements include devel. of most comprehensive vulnerability assessment for multiple threats ever produced for man rated turbine engines. Home: 12930 N Normandy Way Palm Beach Garden FL 33410 Office: Pratt & Whitney PO Box 109600 West Palm Beach FL 33410-9600

SHIRAI, TAKESHI, physician; b. Tensin, Republic of China, May 14, 1928; s. Yasushi and Miyo S.; m. Katsuko Shirai, Aug. 6, 1962; children: Takamitsu Shirai, Ryoko Shirai. MD, Nippon Med. Coll., 1954; Dr.Med.Sci., Kyushu U., 1958; PhD in Biochemistry, Bantride Forest Sch., London, 1980. Med. diploma of Japan. Intern Kyushu U. Hosp., Fukuoka, 1954-55; resident Kyushu U Hosp., Fukuoka, 1955-58, clin. and rsch. asst., 1958-62, assoc., instr. of medicine, 1966-74; trainee in metabolism Dept. of Medicine, Coll. Med. U. of Vt., 1962-66; prof. of medicine Fukuoka Dental Coll., 1974—; physician in chief Fukuoka Dental Coll. Hosp., 1974—; dir. Libr. of Fukuoka Dental Coll., 1989—; instr. medicine Faculty of Dentistry Kyushu U., Fukuoka, 1971—; vis. prof. biochemistry Bantride Forest Sch., London, 1980—. Author: Medicine for Dentist, 1990; translator: The Kidney, 1987; contbr. articles to jours. including Minerva Nefrologica, Annals of Internal Medicine, Diabetologia. Mem. Japanese Soc. of Internal Medicine, Internat. Soc. of Nephrology, Japanese Soc. of Nephrology (coun.), Japanese Soc. of Clin. Physiology (coun.), Sigma Xi. Home: Chuouku, 4-1-807 Josui Dori, Fukuoka 810, Japan Office: Fukuoka Dental Coll Dept Mdcn, 2-15-1 Tamuna Sawara-ku, Fukuoka 814 01, Japan

SHIRAI, YASUTO, information science educator, researcher; b. Kooriyama, Fukushima, Japan, Aug. 9, 1961; s. Keiji and Yoko (Ishii) S.; m. Miki Ito, May 15, 1993. BS, U. Toronto, Toronto, Can., 1984, MS, 1987. Rsch. assoc. U. Tokyo, Tokyo, Japan, 1989-90; assoc. prof. Shizuoka (Japan) U., 1990—. Author: Progress in Fuzzy Sets and Systems, 1990. Mem. IEEE, Computer Graphics Soc., Assn. for Computing Machinery, Info. Processing Soc. Japan, Japan Soc. Fuzzy Theory and Systems. Avocations: reading, gardening. Office: Shizuoka U, Faculty Edn, 836 Ohya, Shizuoka Shizuoka 422, Japan

SHIRAIWA, KENICHI, mathematician, educator; b. Tottori, Japan, Dec. 29, 1928; s. Shoichi and Hisayo (Tanaka) S.; m. Ikuko Tanaka, Mar. 30, 1953; children: Junko Yamada, Kuniko Yamada. B.S., Nagoya U., 1953, Sc.D., 1960. Asst., Nagoya U., Japan, 1953-59; research assoc. Columbia U., N.Y.C., 1958-59; lectr. Nagoya U., 1960-63, assoc. prof., 1963-67, prof. math., 1967-91; prof. Sci. U. Tokyo, 1991—; vis. prof. U. Bahia, Salvador, Brazil, 1963. Author: Dynamical Systems, 1974; Differential Equations, 1975; Linear Algebra, 1976; Caculus, 1981, Dynamical Systems and Entropy, 1985. Fulbright grantee, 1956-57. Mem. Math. Soc. Japan, Am. Math. Soc. Avocations: music; go-play. Home: Sunadabashi 2-chome 1 A-1011, Higashi-ku Nagoya 461, Japan Office: Sci U Tokyo, Fac Ind Sci and Tech, Yamazaki Noda 278, Japan

SHIRES, GEORGE THOMAS, surgeon, physician, educator; b. Waco, Tex., Nov. 22, 1925; s. George Thomas and Donna Mae (Smith) S.; m. Robbie Jo Martin, Nov. 27, 1948; children: Donna Blain, George Thomas III, Jo Ellen. MD, U. Tex., Dallas, 1948. Diplomate Am. Bd. Surgery (dir. 1968-74, chmn. 1972-74). Intern Mass. Meml. Hosp., Boston, 1948-49; resident Parkland Meml. Hosp., Dallas, 1950-53; mem. faculty U. Tex. Southwestern Med. Sch. at Dallas, 1953-74, assoc. prof. surgery, acting chmn. dept., 1960-61, prof., chmn. dept., 1961-74; surgeon in chief surg. services Parkland Meml. Hosp., 1960-74; prof., chmn. dept. surgery U. Wash. Sch. Medicine, Seattle, 1974-75; chief of service Harborview Med. Center, Seattle, Univ. Hosp., Seattle, 1974-75; chmn. dept. surgery N.Y. Hosp.-Cornell Univ. Med. Coll., 1975-91; dean and provost for med. affairs Cornell U. Med. Coll., 1987-91; prof., chmn. surgery Tex. Tech U., Lubbock, 1991—; cons. to Surgeon Gen., Nat. Inst. Gen. Med. Sci., 1965—, Surgeon Gen. Army, 1965-75, Jamaica Hosp., 1978-91; mem. com. metabolism and trauma Nat. Acad. Sci./NRC, 1964-71, com. trauma, 1964-71; mem. rsch. program evaluation com., reviewer clin. investigation applications career devel. program VA, 1972-76; mem. gen. med. rsch. program projects com. NIH, 1965-69; mem. Surgery A study sect., 1970-74, chmn., 1976-78; mem. Nat. Adv. Gen. Med. Scis. Coun., 1980-84; cons. editorial bd. Jour. Trauma, 1968—. Mem. editorial bd. Year Book Med. Publs., 1970-92, Annals of Surgery, 1972—, Surg. Techniques Illustrated: An International Comparative Text, 1974-76, Am. Jour. Surgery, 1968—, Contemporary Surgery, 1973-87; assoc. editor-in-chief Infections in Surgery, 1981; mem. editorial coun. Jour. Clin. Surgery, 1980-82; editor Surgery, Gynecology and Obstetrics, 1982—. Lt. M.C. USNR, 1949-50, 53-55. Life Ins. Med. Rsch. fellow, 1947. Mem. ACS (bd. regents 1971—, chmn. bd. regents 1978-80, pres. 1981-82), AMA, Dallas Soc. Gen. Surgeons (pres.-elect, pres. 1972-74), Am. Assn. Surgery Trauma, Am. Surg. Assn. (sec. 1969-74, pres. 1980), Digestive Disease Found. (founding mem.), Halsted Soc., Internat. Soc. Burn Injuries, Internat. Surg. Soc. (sec. 1978-81, pres. U.S. chpt. 1982-84), Pan-Am. Med. Assn. (surgery council 1971—), Pan Pacific Surg. Assn., Soc. Clin. Surgery, Soc. Surgery Alimentary Tract, Soc. Surg. Chairmen (pres. 1972-74), Soc. Univ. Surgeons (chmn. publs. com. 1969-71), So. Surg. Assn., Surg. Biology Club (sec. 1968-70), Western Surg. Assn., Allen O. Whipple Surg. Soc., James IV Assn. Surgeons (bd. dirs. 1980-81, sec. 1981-87, pres. 1987-90), Alpha Omega Alpha, Alpha Pi Alpha, Phi Beta Pi. Office: Tex Tech U Med Coll Lubbock TX 79430

SHIRILAU, JEFFERY MICHEAL, engineering executive; b. Honolulu, Aug. 30, 1953; s. Cornelius Afai Lauliiuokalani and Dolores Bennett (Bezanson) Lau; life ptnr. Mark Steven Shirey. Cert. data processing, ITT Peterson, Seattle, 1982; AS, Rancho Santiago, Santa Ana, Calif., 1984. Ordained deacon The Ecumenical Cath. Ch., Sept. 23, 1984. Owner Lau & Assocs., Seattle, 1979-82; div. mgr. M.S.E., Santa Ana, 1982-86; supr. U.S. Postal Svc., Alhambra, Calif., 1986-87; pres., CEO Aloha Systems, Inc., Villa Grande, Calif., 1987—, also bd. dirs. With airborne rangers and Green Beret U.S. Army, 1971-74. Charter mem. Assn. of Demand-Side Mgmt. Profls.; mem. IEEE, Ainahau Hawaiian Orange County, Calif. Civic Club, Pacific Bears Leather Motorcycle Club. Republican. Home: 20200 River Blvd Monte Rio CA 95462 Office: Aloha Systems Inc PO Box 32 Villa Grande CA 95486-0032

SHIRILAU, MARK STEVEN, utilities executive; b. Long Beach, Calif., Dec. 13, 1955; s. Kenneth Eugene and Marjorie Irene (Thorvick) Shirey; m. Jeffery Michael Lau, Nov. 25, 1984. BSEE, U. Calif., Irvine, 1977, MS Bus. Adminstrn., 1980; M in Engring., Calif. Poly. State U., 1978; Diploma in Theology, Episc. Theol. Sch., Claremont, Calif., 1984; MA in Religion, Sch. Theology at Claremont, 1985; PhD, U. Calif., Irvine, 1988. Ordained priest Ecumenical Cath. Ch., 1987; consecrated bishop, 1991. Grad. asst. Electric Power Inst., 1977-78; pres., chief exec. officer M.S.E., Santa Ana, Calif., 1977—; adminstrv. mgr. EECO Inc., Santa Ana, Calif., 1979-83; fin. engr. So. Calif. Edison Co., Rosemead, 1983-84, conservation engr., 1984-85, conservation supr., 1985-89; exec. v.p. Aloha Systems, Inc., 1989—, bd. dirs.; part-time instr. Santa Ana Coll., 1982-84; lectr. engring. West Coast U., Orange, Calif., 1984—; bd. dirs. Am. Electronics Assn. Credit Union. Sweetwater Springs Water Dist., Heat Pump Coun. So. Calif. Pastor St. Michael's Ecumenical Cath. Ch., Villa Grande, Calif., now bishop. Mem. IEEE (sr. mem.), Assn. Energy Engrs. (sr. mem.), Assn. Demand-Side Mgmt. Profls. (bd. dirs., charter mem., exec. v.p.), Pacific Bears Club (v.p.), Dignity Integrity (life mem.), Eta Kappa Nu. Democrat. Author: Triune Love: An Insight into God, Creation, and Humanity, 1983, Salvation, Scripture and Sexuality, 1992. Home: 20200 River Blvd Monte Rio CA 95462 Office: PO Box 32 Villa Grande CA 95486-0032

SHIRK, KEVIN WILLIAM, fiber optic product manager; b. Anderson, Ind., Mar. 13, 1957; s. Gilbert Edward and Jean Laverne (Carroll) S. BA, Ind. U., 1980; postgrad., Ind. U., Indpls., 1980-82. Rsch. assoc. RCA, Indpls., 1980-83; engr. Litton Industries, L.A., 1984-88; product mgr. Phys. Optics Corp., Torrance, Calif., 1988—. Contbr. articles to Jour. IEEE, WDM Applications Mag. Achievements include patent in video transmission through wavelength division multiplexing; research in optical interconnects in volume hold. Office: Phys Optics Corp 2545 W 237th St Ste B Torrance CA 90505-5228

SHIUE, GONG-HUEY, chemist; b. Taiwan, Republic of China, Nov. 16, 1947; s. Chun-Tung Shiue and Han-Hsiu Hsieh; m. Rita Ming-Yung Tang, Feb. 10, 1976; children: Andrew, Eric. MSc, Taiwan U., 1974; PhD, Ohio State U., 1983. Rsch. assoc. Princeton (N.J.) U., 1983-84, Am. Health Found., Valhalla, N.Y., 1984-87; sr. scientist United Biomed. Inc., Lake Success, N.Y., 1987-88, Bachem Biosci. Inc., Phila., 1988-89; dir. peptide chemistry Synthecell Corp., Rockville, Md., 1989—; cons. Yes-Biotech, Inc., Toronto, Can., Shanghai Slic Kehua Biochem., Shanghai, China, Large Scale Biologies, Inc., Rockville, Md. Editor: Selected Organic Syntheses, 1984; contbr. articles to profl. jours. Mem. Am. Protein Soc., Am. Peptide Soc., Am. Chem. Soc., Sigma Psi. Achievements include DNA binding of Nitro-polynuclear aromatic hydrocarbons in vitro and in vivo; a novel strategy for solid phase peptide synthesis; modification of oligonucleotides on phosphate backbonds. Office: Synthecell Corp 7101 Riverwood Dr Columbia MD 21046

SHIVE, RICHARD BYRON, architect; b. Cleve., Jan. 16, 1933; s. Roy Allen and Mary Elizabeth (Thompson) S.; m. Patricia Butler, Aug. 28, 1954; children: Lisa Ann, Laura Mary, John Thompson, Nancy Butler. BS, Rensselaer Poly. Inst., Troy, N.Y., 1954; postgrad., Newark (N.J.) Coll. Engring., 1957, Rutgers U., 1960-63. Registered architect, N.J., N.Y., Pa., Vt.; lic. profl. planner, N.J. Field engr. Wigton-Abbott Corp., Plainfield, N.J., 1954-55, The Glenwal Co., Rochelle Park, N.J., 1955; asst. supt. Wigton-Abbott Corp., Plainfield, 1955-57; archtl. draftsman Raymond B. Flatt, Architect, Bloomfield, N.J., 1957-58; chief draftsman Raymond B. Flatt, Architect, 1958-60; project architect Scrimenti/Swackhamer/Perantoni Architects, Somerville, N.J., 1960-66; assoc. Scrimenti/Swackhamer/Perantoni Architects, 1966-69; ptnr. Scrimenti, Shive, Spinelli, Perantoni Architects, Somerville, 1969-86, Shive/Spinelli/Perantoni & Assocs., Architects & Planners, Somerville, 1986—; adv. com. First Fidelity Bank, Bound Brook, N.J., 1989-91; chmn. bd. Somerset Health Care Corp., 1987-91. Contbr. articles to profl. jours. Bd. dirs., exec. com. N.J. Hosp. Assn., Princeton, 1986-92; chmn. bd. trustees Somerset Med. Ctr., Somerville, 1973—; mem. Nat. Trust for Hist. Preservation; bd. dirs. Ctr. for Health Affairs, Inc., 1992—. Recipient award James F. Lincoln Arc Welding Found., 1973, President's award for outstanding svc. Rolling Hills coun. Girl Scouts U.S.A., 1988, Trustee of Yr. award N.J. Hosp. Assn., 1993, Outstanding Citizen of Yr. award Somerset County C. of C., 1993; Paul Harris fellow Bound Brook-Middlesex Rotary Club, 1993. Mem. AIA, ASTM, ASHRAE, ACI (chpt. bd. dirs. 1978-83), N.J. Soc. Architects, Illuminating

Engring. Soc., Congress of Hosp. Trustees of Am. Hosp. Assn., Nat. Fire Protection Assn., Greater Somerset County C. of C. (v.p. 1985-86, 92—, Outstanding Citizen of Yr. award 1993), Rotary (pres. 1969-70, Paul Harris fellow 1993), Wash. Campground Assn. (pres. 1975-76, v.p. 1977-78, sec. 1978—), Chi Phi (sec. 1973). Republican. Congregationalist. Avocations: fishing, photography, skiing, canoeing, backpacking. Home: 1001 N Mountain Ave Bound Brook NJ 08805-1451 Office: Shive Spinelli Perantoni & Assocs PO Box 758 148 W End Ave Somerville NJ 08876-0758

SHLIAN, DEBORAH MATCHAR, physician; b. Balt., Jan. 2, 1948; d. Joseph Charles and Evelyn (Wegman) Matchar; m. Joel Nathan Shlian, Nov. 8, 1970. BA, U. Mich., 1968; MD, U. Md., Balt., 1972; MBA, UCLA, 1988. Diplomate Am. Bd. Family Practice. Intern Sinai Hosp., Balt., 1972-73; resident Kaiser Found. Hosp., L.A., 1973-75; ptnr. So. Calif. Permanente Med. Group, L.A., 1973-83; med. dir. primary care UCLA Student Health Svc., 1983-90; pres. Deborah Shlian & Assoc., L.A., 1990—; clin. faculty div. family practice UCLA, 1976—, adv. bd. biotechnology assessment, 1992—. Author: Karen Evans, MD #3 Space Medicine, 1982, Nursery, 1984, Wednesday's Child, 1986, (screenplay) Double Illusion, 1983; editor: College Health Guide, 1989; contbr. articles to profl. jours. including Jour. AMA, New Eng. Hour., Jour. Infection and Disease, Coll. Health, F.R. Rsch. Jour., Western Jour. Bd. dirs. Children's Pl., Santa Monica, Calif., 1990-91. Grantee Pfizer Pharm., 1985, Cyanamid Co., 1986, Norwich Eaton Pharm., 1988, Bristol Meyers, 1988, Abbott, 1989. Mem. Am. Coll. Physician Execs. Achievements include rsch. in patterns of healthcare utilization, screening & immunology of Rubella, toxicity of BHT, TB screening, cost effectiveness of measles immunizations. Office: Deborah Shlian & Assoc 1390 Miller Dr Los Angeles CA 90069

SHOCKLEY, CAROL FRANCES, psychologist, psychotherapist; b. Atlanta, Nov. 24, 1948; d. Robert Thomas and Frances Lavada (Scrivner) S. BA, Ga. State U., 1974, MEd, 1976; PhD, U. Ga., 1990. Cert. in gerontology. Counselor Rape Crisis Ctr., Atlanta, 1979-80; emergency mental health clinician Gwinnett Med. Ctr., Lawrenceville, Ga., 1980-86; psychotherapist Fla. Mental Health Inst., Tampa, 1987-89, Tampa Bay Acad., Riverview, Fla., 1990-91; sr. psychologist State of Fla. Dept. of Corrections, Bushnell, 1991-92; pvt. practice psychology St. Marys, Brunswick, Ga., 1992—. Author: (with others) Relapse Prevention with Sex Offenders, 1989. Vol. Ga. Mental Health Inst., Atlanta, 1972; leader Alzheimer's Disease Support Group, Athens, Ga., 1984; vol. therapist Reminiscence Group for Elderly, Athens, 1984-85. Recipient Meritorious Svc. award Beta Gamma Sigma, 1975. Mem. Am. Counseling Assn., Am. Psychol. Assn., Ga. Psychol. Assn., Sigma Phi Omega, Psi Chi. Avocations: astronomy, archeology, music, travel. Office: 2475 Village Dr Ste 107 Kingsland GA 31548

SHOCKMAN, GERALD DAVID, microbiologist, educator; b. Mt. Clemens, Mich., Dec. 22, 1925; s. Solomon and Jennie (Madorsky) S.; m. Arlyne Taub, June 2, 1949; children—Joel, Deborah. B.S., Cornell U., 1947; Ph.D., Rutgers U., 1950. Predoctoral fellow Rutgers U., 1947-50; research asso. U. Pa., 1950-51; research fellow, research asso. Inst. Cancer Research, Phila., 1951-60; asso. prof. Temple U. Sch. Medicine, Phila., 1960-66; prof. dept. microbiology and immunology Temple U. Sch. Medicine, 1966—, chmn. dept., 1974-90. Contbr. articles in field to profl. jours. Served with U.S. Army 1942-44. Recipient Research Career Devel. award NIH, 1965-70, Titular de la Chaire d'Actualité Scientifique U. Liège, Belgium, 1971-72; NRC fellow, 1954-55. Mem. Am. Soc. Biol. Chemists, Am. Acad. Microbiology, Am. Soc. Microbiology, AAAS, Sigma Xi. Home: 901 Rodman St Philadelphia PA 19147-1247 Office: Temple U Sch Medicine 3400 N Broad St Philadelphia PA 19140-5196

SHOEMAKER, CLARA BRINK, retired chemistry educator; b. Rolde, Drenthe, The Netherlands, June 20, 1921; came to U.S., 1953; d. Hendrik Gerard and Hendrikje (Smilde) Brink; m. David Powell Shoemaker, Aug. 5. 1955; 1 child, Robert Brink. PhD, Leiden U., The Netherlands, 1950. Instr. in inorganic chemistry Leiden U., 1946-50, 51-53; postdoctoral fellow Oxford (Eng.) U., 1950-51; rsch. assoc. dept. chemistry MIT, Cambridge, 1953-55, 58-70; rsch. assoc. biochemistry Harvard Med. Sch., Boston, 1955-56; project supr. Boston U., 1963-64; rsch. assoc. dept. chemistry Oreg. State U, Corvallis, 1970-75, rsch. assoc. prof. dept. chemistry, 1975-82, sr. rsch. prof. dept. chemistry, 1982-84, prof. emerita, 1984—. Sect. editor: Structure Reports of International Union of Crystallography, 1967, 68, 69; co-author chpts. in books; author numerous sci. papers. Bd. dirs. LWV, Corvallis, 1980-82, bd. dirs., sec., Oreg., 1985-87. Recipient fellowship Internat. Fedn. Univ. Women, Oxford U., 1950-51. Mem. Metall. Soc. (com. on alloy phases 1969-79), Internat. Union of Crystallography (commn. on structure reports 1970-90), Am. Crystallographic Assn. (crystallographic data com. 1975-78, Fankuchen award com. 1976), Sigma Xi, Iota Sigma Pi (faculty adv. Oreg. State U. chpt. 1975-84), Phi Lambda Upsilon. Avocation: outdoor activities. Office: Oreg State U Dept Chemistry Corvallis OR 97331

SHOEMAKER, EUGENE MERLE, geologist; b. Los Angeles, Apr. 28, 1928; s. George Estel and Muriel May (Scott) S.; m. Carolyn Jean Spellmann, Aug. 18, 1951; children: Christine Carol, Patrick Gene, Linda Susan. B.S., Calif. Inst. Tech., 1947, M.S., 1948; M.A., Princeton U., 1954, Ph.D., 1960; Sc.D., Ariz. State Coll., 1965, Temple U., 1967, U. Ariz., 1984. Geologist U.S. Geol. Survey, 1948, exploration uranium deposits and investigation salt structures Colo. and Utah, 1948-50, regional investigations geochemistry, vulcanology and structure Colorado Plateau, 1951-56, research on structure and mechanics of meteorite impact and nuclear explosion craters, 1957-60, with E.C.T. Chao, discovered coesite, Meteor Crater, Ariz., 1960, investigation structure and history of moon, 1960-73, established lunar geol. time scale, methods of geol. mapping of moon, 1960, application TV systems to investigation extra-terrestrial geology, 1961—, geology and paleomagnetism, Colo. Plateau, 1969—, systematic search for planet-crossing asteroids and comets, 1973—, Trojan asteroids, 1985—, geology of satellites of Jupiter, Saturn, Uranus and Neptune, 1978—, investigating role of large body impacts in evolution of life, 1981—; impact craters of Australia, 1983—; organized br. of astrogeology U.S. Geol. Survey, 1961; co-investigator TV expt. Project Ranger, 1961-65; chief scientist, center of astrogeology U.S. Geol. Survey, 1966-68, research geologist, 1976—; prin. investigator geol. field investigations in Apollo lunar landing, 1965-70, also television expt. Project Surveyor, 1963-68; prof. geology Calif. Inst. Tech., 1969-85, chmn. div. geol. and planetary scis., 1969-72. Recipient (with E.C.T. Chao) Wetherill medal Franklin Inst., 1965; Arthur S. Flemming award, 1966; NASA medal for exceptional sci. achievement, 1967; honor award for meritorious service U.S. Dept. Interior, 1973; Disting. Service award, 1980; Disting. Alumni award Calif. Inst. Tech., 1986, co-recipient Rittenhouse medal, 1988, Nat. Medal of Sci. NSF, 1992. Mem. NAS, Am. Acad. Arts and Scis., Geol. Soc. Am. (Day medal 1982, Gilbert award 1983), Mineral Soc. Am., Soc. Econ. Geologists, Geochem. Soc., Am. Assn. Petroleum Geologists, Am. Geophys. Union (Whipple award 1993), Am. Astron. Soc. (Kuiper prize 1984), Internat. Astron. Union, Meteoritical Soc. (Barringer award 1984, Leonard medal 1985). Achievements include co-discovery of Comet Shoemaker-Levy 9. Home: PO Box 984 Flagstaff AZ 86002-0984 Office: US Geol Survey 2255 N Gemini Dr Flagstaff AZ 86001-1698

SHOEMAKER, HAROLD DEE, mechanical engineer; b. Argos, Ind., Mar. 4, 1942; s. Roy H. and Hazel I. (Snyder) S.; m. Sandra L. Cale, June 16, 1962; children: Susan R., Jonathan E. BS in Agrl. Engring., W.Va. U., 1969, MS in Agrl. Engring., 1971, PhD in Mech. Engring., 1976. Project mgr. U.S. Dept. Energy, Morgantown, W.Va., 1974—; instr. Pa. State U.-Fayette Campus, Uniontown, 1973-74, W.Va. U., Morgantown, 1971-75. Co-autohr: Strenght of Materials, 1975; contbr. to profl. publs. Bd. dirs. Grace brethren Ch., Uniontown, 1973-82, Calvary Bapt. Ch., Uniontown, 1984—, Chestnut Ridge Christian Acad., Uniontown, 1990—. Mem. ASME (Arctic com. 1984-85, organizing com. internat. conf. on port and ocean engring. under Arctic conditions 1986-87), Soc. Petroleum Engrs., Tau Beta Pi. Republican. Achievements include patent for belt fertilizer spreader. Home: 15 Clinton Dr Uniontown PA 15401 Office: US Dept Energy PO Box 880 Morgantown WV 26507

SHOEMAKER, HAROLD LLOYD, infosystem specialist; b. Danville, Ky., Jan. 3, 1923; s. Eugene Clay and Amy (Wilson) S.; A.B., Berea Coll., 1944; postgrad. State U. Ia., 1943-44, George Washington U., 1949-50, N.Y. U., 1950-52; m. Dorothy M. Maddox, May 11, 1947 (dec. Feb. 1991). Research physicist State U., Ia., 1944-45, Frankford Arsenal, Pa., 1945-47; research engr. N.Am. Aviation, Los Angeles, 1947-49, Jacobs Instrument Co., Bethesda, 1949-50; asso. head systems devel. group The Teleregister Corp., N.Y.C., 1950-53; mgr. electronic equipment devel. sect., head planning for indsl. systems div. Hughes Aircraft Co., Los Angeles, 1953-58; dir. command and control systems lab. Bunker-Ramo Corp., Los Angeles, 1958-68, v.p. Data Systems, 1968-69, corp. dir. data processing, 1969-75; tech. staff R & D Assocs., Marina Del Rey, Calif., 1975-85; info. systems cons., 1985—. Served with AUS, 1945-46. Mem. IEEE. Patentee elec. digital computer. Home: PO Box 3385 Granada Hills CA 91394-0385

SHOEMAKER, PATRICK ALLEN, engineer; b. Montebello, Calif., Nov. 6, 1955; s. Frederick Thomas and Mary Elizabeth (Patterson) S.; m. Gae L. Stimmel, May 4, 1985 (div. 1992); children: Lauren A., Alanna C. BA, U. Calif., San Diego, 1977, PhD, 1984. Clin. bioengr. Children's Hosp. and Health Ctr., San Diego, 1977-78; engr. Naval Command, Control and Ocean Surveillance, San Diego, 1984—. Contbr. articles to profl. jours. Mem. IEEE, Internat. Neural Network Soc. Achievements include patents for non-volatile analog memory, CMOS analog four-quadrant multiplier. Office: NCCOSC RDT & E Divsn 551 53475 Strothe Rd San Diego CA 92152-6365

SHOHET, JUDA LEON, electrical and computer engineering educator, researcher, high technology company executive; b. Chgo., June 26, 1937; s. Allan Sollman and Frannye Ina (Turner) S.; m. Amy Lenore Scherz, Sept. 5, 1969; children: Aaron, Lena, William. BS, Purdue U., 1958; MS, Carnegie Mellon U., 1960; PhD, Carnegie-Mellon U., 1961. Registered profl. engr., Wis. Asst. prof. Johns Hopkins U., Balt., 1961-66; assoc. prof. U. Wis., Madison, 1966-71, prof., 1971—, chmn. dept. elec. and computer engring., 1986-90, dir. Torsatron/Stellarator Lab., 1974—, dir. Engring. Rsch. Ctr. for Plasma-Aided Mfg., 1986—; pres. Omicron Tech., Inc., Madison, 1985—; cons. Hewlett-Packard Co., Abbott Labs, Trane Co., Oak Ridge Nat. Lab., McGraw-Edison Co., Princeton U., Los Alamos Sci. Lab., Nicolet Instruments, Lawrence Livermore Nat. Lab., Westinghouse Elec. Corp. Author: The Plasma State, 1971, Flux Coordinates and Magnetic Field Structure. 1991; contbr. over 100 articles to profl. jours; author over 350 conf. papers; patentee in field. Recipient Frederick Emmons Terman award Am. Soc. for Engring. Edn., 1978, John Yarborough Meml. medal British Vacuum Coun., 1993. Fellow IEEE (Centennial medal 1984, Richard F. Shea Disting. Mem. award 1992), Am. Phys. Soc.; mem. IEEE Nuclear and Plasma Scis. Soc. (pres. 1980-82, Merit award 1978, Plasma Sci. and Applications prize 1990). Avocations: skiing, sailing, hiking. Home: 1937 Arlington Pl Madison WI 53705-4001 Office: U Wis Dept Elec & Computer Engring 2414 Engring Bldg Madison WI 53706

SHOHOJI, TAKAO, statistician; b. Kure, Japan, Jan. 1, 1938; d. Mitsuto and Takeko (Ishida) S.; m. Yoshie Nagasawa, May 25, 1967; children: Takashige, Mio. BA, Hiroshima U., 1962, MA, 1965; PhD, NYU, 1972. Analytic statistician Atomic Bomb Casualty Commn., Hiroshima, 1965-72; assoc. rsch. scientist NYU Med. Ctr., 1968-72; asst. prof. Hiroshima U., 1972-74, acting head computing ctr., 1972-78, head lab computing ctr., 1978-80, assoc. prof., 1974-85, prof., 1985—; Senogawa Hosp., Hiroshima, 1966-90. Contbr. articles to profl. jours. Recipient Founders Day award NYU, 1973. Fellow Human Biology Coun.; mem. Japan Math. Soc., Japan Statis. Assn. (coun. 1988—), Chugoku New Media Assn. (chmn. 1987-90), Japanese Soc. Applied Statis. (coun. 1986—), Biometric Soc. (coun. 1988—), Am. Statis. Assn., Internat. Statis. Inst., Japanese Soc. Co. Statis. Assn. (coun. 1987—, assoc. editor 1987-92, v.p. 1993—), Biometric Soc. Japan (sec. 1986-90, treas., 1991-93, coun. 1982—), Japanese Classification Soc. (coun. 1991—), Info. Processing Soc. Japan, N.Y. Acad. Scis. Club: Gehnan Tennis. Home: 10-46, 2 chome Shimizu, Kure 737, Japan Office: Hiroshima U Integrated Arts Scis, 1-7-1 Kagamiyama, Higashi Hiroshima 724, Japan

SHOJI, EGUCHI, organic chemistry educator; b. Aichi-Ken, Japan, Oct. 27, 1934; married, 1963. BS, Nagoya U., 1957, PhD, 1962. Rsch. assoc. Johns Hopkins U., Balt., 1964-66; rsch. assoc. Nagoya U., 1962-76, assoc. prof., 1977-84, prof., 1985—. Co-author: Modern Synthetic Reaction, 1969, New Problems in Organic Chemistry, 1982; co-translator: Modern Synthetic Reaction, 1974, Physical Methods in Organic Chemistry, 1972; assoc. editor Bull. Chem. Soc. Japan, Tokyo, 1986-88; com. bd. Chem. Soc. Japan, Tokai Br., 1985-86, 93—, pres. 1993. Suzuken grantee Suzuken Meml. Found., 1989. Fellow Royal Soc. Chemistry; mem. ACS, ISHC, Chem. Soc. Japan (bd. com. Tokai br. 1985-86), Synthetic Organic Chemistry, Japan (v.p. Toaki br. 1992-93, pres. 1993—). Office: Nagoya U, Furo-cho Chikusa-ku, Nagoya 464-01, Japan

SHOJI, SADAO, soil scientist; b. Sendai, Miyagi, Japan, Nov. 17, 1931; s. Teiji and Tamichi Shoji; m. Yasuko Koide, May 30, 1960; children: Chikako, Chiyo, Kinu. BAgr, Tohoku U., Sendai, 1956; PhD, Tohoku U., 1966; MSD, Mich. State U., 1958. Researcher Hokkaido Nat. Agr. Exptl. Sta., Sapporo, 1954-62; from asst. prof. to prof. faculty of agr. Tohoku U., Sendai, 1962—. Author: Controlled Release Fertilizers, 1992, Volcanic Ash Soils, 1993; contbr. articles to profl. jours.; pres. Soil Sci. Plant Nutrition, 1992—; editorial bd. Geoderma, 1992—. Mem. Japanse Soc. Soil Sci. and Plant Nutrition (pres. 1992—, award 1972, Agrl. Sci. award 1993, Yomiuri Shimbun award 1993), Am. Soc. Agronomy. Avocation: fishing. Home: 5-13-27 Nishitaga, Taihakuku, Sendai 982, Japan Office: Tohoku U Faculty Agr, Amamiya-machi Tautsumidori, Aobaku Sendai 981, Japan

SHOLDER, JASON ALLEN, medical products company executive; b. Boston, Dec. 13, 1944; s. Samuel and Muriel (Weiss) S.; m. Nancy Kate Oyer, Dec. 20, 1992; 1 child by previous marriage: Adam Craig. BSEE, Northeastern U., Boston, 1967; ME in Physics, MIT, 1969. Sr. v.p. tech. Siemens Pacesetter, Inc., Sylmar, Calif., 1977—. Contbr. articles to profl. jours. Mem. N.Am. Soc. Pacing and Electrophysiology. Achievements include 51 patents in field of cardiac pacemakers. Office: Siemens Pacesetter Inc 15900 Valley View Ct Sylmar CA 91342-3577

SHOLTIS, JOSEPH ARNOLD, JR., nuclear engineer, retired military officer; b. Monongahela, Pa., Nov. 28, 1948; s. Joseph and Gladys (Frye) S.; m. Cheryl Anita Senchur, Dec. 19, 1970; children: Christian Joseph, Carole Lynne. BS in Nuclear Engring. (Disting. Mil. Grad.), Pa. State U., 1970; diplomas Air Univ., 1975, 78; MS in Nuclear Engring., U. N.Mex., 1977, postgrad., 1978-80. Lic. sr. reactor operator NRC, 1980-84. Mathematician, statistician, mine safety analyst U.S. Bur. Mines, Pitts., 1968-70; commd. 2d lt. USAF, 1970, advanced through grades to lt. col. 1988, ret., 1993; nuclear rsch. officer Fgn. Tech. Div., USAF, Wright-Patterson AFB, Ohio, 1971-74; chief space nuclear system safety sect. Air Force Weapons Lab., Kirtland AFB, N.Mex., 1974-78; mil. mem. tech. staff, project officer Sandia Nat. Labs., Albuquerque, 1978-80; chief radiation sources div., reactor facility dir. Armed Forces Radiobiology Rsch. Inst., Bethesda, Md., 1980-84; program mgr. SP-100 space reactor power system tech. devel. program Air Force Element U.S. Dept. Energy, Germantown, Md., 1984-87; chief analysis and evaluation br., 1988-91, chief nuclear power and sources div., 1991-92, chief nuclear energy systems Air Force Safety Agency, Kirtland AFB, N.Mex., 1987-93; sr. v.p., gen. mgr. we. ops. Oakton Internat. Corp., Va., 1993—; cons. in field; space shuttle nuclear payload safety assessment officer Air Force Weapons Lab., Kirtland AFB, 1976-78; instr. med. effects nuclear weapons Armed Forces Radiobiology Rsch. Inst., Bethesda, 1980-85, mem. reactor and radiation facility safety com., 1980-85; faculty, lectr. Uniformed Svcs. Univ. Health Scis., Bethesda, 1982-87; chmn. Power System Subpanel Interagency Nuclear Safety Rev. Panel risk assessments of Galileo, Ulysses, Cassini, and TOPAZ-II nuclear-powered space missions, 1987-92; Dept. of Def. chmn. Interagency Nuclear Safety Rev. Panel evaluation of Ulysses, Cassini and Topaz-II nuclear-powered space missions for the office of the pres.; instr. Inst. for Space Nuclear Power Studies U. N.Mex., 1987-91; U.S. del., tech. advisor UN Sci. and Tech. Subcom. and Legal Subcom. Working Group on Nuclear Power Sources in Outer Space, 1984-88; mem. U.S. contingent U.S. and U.S.S.R. discussions on nuclear space power system safety, 1989-90; mem. adv. com., tech. program com. Symposia on Space Nuclear Power and Propulsion, U. N.Mex., 1989—; mem. Multimegawatt Space Reactor Power Project safety working group, 1988-91; mem. SP-100 Space Reactor Project safety adv. com., 1990-93; mem. space exploration initiative Nuclear Safety Policy Working Group, 1990-91; mem. Air Force Thermionic Space Power Program Safety com., 1990—; mem. safety com., 1990-93; mem. Strategic Def. Initiative Orgn. Ind. Evaluation Group, 1991-93; mem., ind. advisor U.S. Dept. Energy Ind. Safety Assessment of TOPAZ-II space reactor power system, 1993; mem. program com. Reactor Safety Divsn. Am. Nuclear Soc., 1992—; lectr. N.Mex. Acad. of Sci. Vis. Scientist Program, 1991—. Author: (with others) LMFBR Accident Delineation, 1980, Military Radiobiology, 1987, Power System Subpanel Report for Galileo Space Mission, 1989, Power System Subpanel Report for Ulysses Space Mission, 1990, Safety Evaluation Report for Ulysses Space Mission, 1990; contbr. articles, chpts. in books. Pres. Fort Detrick Cath. Parish Community, Md., 1984; charter mem. N.Mex. Edn. Outreach Com., 1989—; Decorated Def. Meritorious Service medal (2), Air Force Meritorious Svc. medal (2) Air Force Commendation medal (3), Nat. Def. Svc. medal (2), U.S. Army Reactor Comdr. Badge, U.S. Air Force Missileman Badge, Air Force Master Space Systems Badge, Nat. Aeronautics and Space Administration Achievement awards (3). Mem. Am. Nuclear Soc. (Best Paper 1977), ASME, AIAA, AAAS, N.Mex. Acad. Scis., Sigma Xi. Republican. Avocations: hunting, fishing, camping, golfing, motorcycle touring. Office: PO Box 910 Tijeras NM 87059-0910

SHOOP, BARRY LEROY, electrical engineer; b. Fairbanks, Alaska, Nov. 25, 1957; s. Roy Alvin and Ruth Isabelle (Eister) S.; m. Linda Dorothy Wolfgang, Dec. 15, 1979; children: Brandon Roy, Aubrey Lynn. BSEE, Pa. State U., 1980; MSEE, U.S. Naval Postgrad. Sch., Monterey, Calif., 1986; PhD in EE, Stanford U., 1992. Commd. 2d lt. U. S. Army, 1980, advanced through grades to maj., 1992; comm. engr. Tobyhanna (Pa.) Army Depot, 1980-84; electronics engr. U.S. Army Fgn. Sci. and Tech. Ctr., Charlottesville, Va., 1986-89; sr. rsch. scientist, asst. prof. U.S. Mil. Acad., West Point, N.Y., 1992—; sci. advisor Tech. Bd. Dept. of Def., Washington, 1986-89, mem. steering com. Sensor Fusion Conf., Orlando, Fla., 1987-89. Contbr. articles to profl. jours. Mem. IEEE, Optical Soc. Am. (Stanford chpt. pres. 1991-92), Sigma Xi, Eta Kappa Nu. Achievements include development of a new method of optical analog-to-digital conversion based on oversampling and error diffusion coding techniques; a novel method of laser stabilization using multiple quantum well modulators. Office: US Mil Acad Dept Elec Engring West Point NY 10096

SHOPE, ROBERT ELLIS, epidemiology educator; b. Princeton, N.J., Feb. 21, 1929; s. Richard Edwin Shope and Helen Madden (Ellis) Flemer; m. Virginia Elizabeth Barbour, Dec. 27, 1958; children—Peter, Steven, Deborah, Bonnie. B.A., Cornell U., 1951, M.D., 1954. Intern then resident Grace-New Haven Hosp., 1954-58; mem. staff Rockefeller Found., Belem, Brazil, 1959-65; dir. Belem Virus Lab., Brazil, 1963-65; from asst. to assoc. prof. epidemiology Yale Sch. Medicine, New Haven, 1965-75, prof., 1975—; mem. adv. bd. Gorgas Inst., Panama City, 1972—; mem. WHO Expert Panel Arboviruses, Geneva, Switzerland, 1974—; mem. U.S. del. U.S.-Japan Coop. Med. Scis. Program, Washington, 1977—; mem. Pan Am. Health Orgn. Commn. for Dengue, Washington, 1980—. Served to capt. U.S. Army, 1955-57, Southeast Asia. Fellow Am. Acad. Microbiology; mem. Am. Soc. Tropical Medicine and Hygiene (pres. 1980, Bailey K. Ashford award 1974), Am. Soc. Virology, Am. Soc. Epidemiology, Infectious Diseases Soc. Am. Democrat. Home: 249 Pine Orchard Rd Branford CT 06405-5537 Office: Yale Arbovirus Rsch Unit Yale Sch Medicine PO Box 3333 New Haven CT 06510-0333*

SHORT, DENNIS RAY, engineering educator; b. Lafayette, Ind., Feb. 3, 1954; s. Jesse Harold and Oma Dell (Cook) S. BA, Purdue U., 1980, MS in Edn., 1983. Cert. mfg. engr., Ind. From instr. to assoc. prof. Sch. Tech. Purdue U., West Lafayette, 1983—; vis. scholar dept. civil engring. Fukuoka U., Kyushu, Japan, 1991; referee profl. publs. Contbr. articles to profl. jours. Grantee IBM, 1990-92, 91-92, AutoDesk, Inc., 1991—, CadKey, Inc., 1991—. Mem. Am. Soc. Engring. Edn., Soc. Mfg. Engrs. (sr.). Office: Purdue U 1419 Knoy Hall Rm 363 West Lafayette IN 47907-1419

SHORT, ELIZABETH M., physician, educator, federal agency administrator; b. Boston, June 2, 1942; d. James Edward and Arlene Elizabeth (Mitchell) Meehan; m. Herbert M. Short, Sept. 2, 1963 (div. 1969); 1 child, Timothy Owen; m. Michael Allen Friedman, June 21, 1976; children: Lia Gabrielle, Hannah Ariel, Eleanor Elana. BA Philosophy magna cum laude, Mt. Holyoke Coll., 1963; MD cum laude, Yale U., 1968. Diplomate Am. Bd. Internal Medicine, Am. Bd. Med. Genetics. Intern, jr. resident internal medicine Yale New Haven Hosp., 1968-70; postdoctoral fellow in human genetics Yale Med. Sch., 1970-72; postdoctoral fellow in renal metabolism U. Calif., San Francisco, 1972-73; sr. resident in internal medicine Stanford (Calif.) Med. Sch., 1973-74, chief resident in internal medicine, 1974-75; staff physician Palo Alto Veterans Med. Ctr., Stanford, Calif., 1975-80; asst. prof. of medicine Stanford Med. Sch., 1975-83, asst. dean Student Affairs, 1978-80, assoc. dean Students Affairs/Medical Education, 1980-83; dir. biomed. rsch. and faculty devel. Assn. Am. Med. Colls., Washington, 1983-87, dep. dir. dept. acad. affairs, 1983-87, dep. dir. biomedical rsch., 1987-88; dep. assoc. chief med. dir. for acad. affairs VA, Washington, 1988-92, assoc. chief medical dir. for acad. affairs, 1992—; vis. prof. Human Biology, Stanford U., 1983-86; resource allocation com. Veteran's Health Adminstrn., 1989-91; budget planning and policy review coun. 1991—; planning review com. Veterans Health Adminstrn., 1991—; chair resident work limit task force 1991—; managed care task force, 1993—; co-chair com. status women Am. Fedn. Clin. Rsch., 1975-77; mem. numerous adminstrv. coms., Yale Med. Sch., Stanford U., Va.; accredation coun. grad. med. edn., 1988—; mem. public policy com. Am. Soc. Human Genetics, 1984—, chair, 1986—; mem. White House Task Force on Health Care Reform, 1993—. assoc. editor Clin. Rsch. Jour., 1976-79, editor elect, 1979-80, editor 1980-84; contbr. articles to profl. jours. Chief med. dir's. AIDS steering com., 1988-89; pres. Bethesda Jewish Congregation, 1993—; nat. child health adv. coun. NIH, 1991—; com. edn. and human resources Office Sci. and Tech. policy, 1991—. Recipient Maclean Zoololgy award Mt. Holyoke Coll.; Munger scholar, Markle Scholar, Sara Williston scholar Mt. Holyoke Coll., 1959-63, Yale Men in Medicine scholar, 1964-68; Bardwell Meml. Med. fellow, 1963. Mem. AAAS, Am. Soc. Human Genetics (chair pub. policy 1986—), Am. Fedn. Clin. Rsch. (bd. dirs. 1973-83, editor 1978-83, nat. coun., exec. com., pub. policy com. 1977-87), Am. Assn. Women in Scis., Western Soc. Clin. Investigation, Calif. Acad. Medicine, Phi Beta Kappa, Alpha Omega Alpha. Home: 6807 Bradley Blvd Bethesda MD 20817 Office: Dept Veterans Affairs 810 Vermont Ave Washington DC 20420

SHORT, STEVE EUGENE, engineer; b. Crockett, Calif., Oct. 17, 1938; s. Roger Milton and Ida Mae (Mills) S.; B.S. in Gen. Engring. with honors, U. Hawaii, 1972, M.B.A., 1973; M.S. in Meteorology, U. Md., 1980; m. Yumie Sedaka, Feb. 2, 1962; children—Anne Yumie, Justine Yumie, Katherine Yumie. With Nat. Weather Service, NOAA, 1964—, govt. exec. Silver Spring, Md., 1974-81, program mgr. ASOS, 1981—, transition dir. 1991—; cons. engring. and mgmt.; cons. SBA. Contbr. articles to sci. jours. Served with USMC, 1956-60. Registered profl. engr., Hawaii. Recipient Gold Medal award U.S. Dept. Commerce, 1992. Mem. VFW, Am. Meteorol. Soc., Japan-Am. Soc., Am. Soc. Public Adminstrn. Home: 3307 Rolling Rd Chevy Chase MD 20815 Office: Nat Weather Svc 1325 E West Hwy Silver Spring MD 20910

SHORTELLE, KEVIN JAMES, systems engineer, consultant; b. Wallingford, Conn., May 7, 1955; s. James William and Louise Marie (DeFilipo) S.; m. Ann Marie Bergquist, July 5, 1985; children: Janet, Jennifer. BSEE, U. Notre Dame, 1977, MSEE, 1984. Registered profl. engr., Calif., Mass. Mem. staff harpoon program McDonnell Douglas Co., St. Louis and Concord, Calif., 1977-82; sect. mgr. strategic systems div. TASC, Reading, Mass., 1984-88; mgr. integrated systems group System Dynamics Internat., Inc., Gainesville, Fla., 1988—. Author tech. reports, procs. Recipient letter of commendation USN, 1982. Mem. IEEE, Tau Beta Pi, Eta Kappa Nu. Office: System Dynamics Internat Ste E 4140 NW 27th Ln Gainesville FL 32606

SHORTER, JOHN, chemistry lecturer; b. Redhill, Surrey, England, June 14, 1926; s. Frank Side and Edith Mabel (Doney) S.; m. Mary Patricia Steer, July 28, 1951; children: Christopher John, Stephen James, Caroline Mary. BA, Oxford (England) U., 1947, BSc, 1949, DPhil, 1950. Asst. lectr. in chemistry U. Coll. Hull, England, 1950-52, lectr. in chemistry, 1952-54; lectr. in chemistry U. Hull, 1954-63, sr. lectr. in chemistry, 1963-72, reader in phys. organic chemistry, 1972-82, emeritus reader in chemistry, 1982—; R.T. French vis. prof. U. Rochester, N.Y., 1966-67; sec. Internat. Group for Correlation Analysis in Organic Chemistry, Hull, 1982—. Author: Correlation Analysis in Organic Chemistry, 1973, Correlation Analysis of Organic

Reactivity 1982; co-editor: Advances in Linear Free Energy Relationships, 1972, Correlation Analysis in Chemistry, 1978, Similarity Models in Organic Chemistry, Biochemistry, and Related Fields, 1991; contbr. articles to profl. jours. Fellow Royal Soc. Chemistry (sec. then chmn. history of chemistry group 1982—); mem. Assn. Univ. Tchrs., Internat. Union of Pure and Applied Chemistry. Mem. Ch. of England. Avocations: mountaineering, family history researching. Home: 10 Cross St, Beverley HU17 9AX, England Office: U Hull, Sch of Chemistry, Hull HU6 7RX, England

SHOTT, EDWARD EARL, engineer, researcher; b. Harrisburg, Pa., Dec. 26, 1946; s. Russell Edward and Evangelene Cora (Topper) S. BS, Rochester Inst. Tech., 1968; postgrad., York Coll., 1982, Pa. State U., York, 1983. Owner Teen Photo, Dillsburg, Pa., 1963-65; med. photographer Strong Meml. Hosp., Rochester, N.Y., 1968-69; pvt. practice oil painting N.Y.C., 1969-71; owner Shott Mfg., Rsch. and Constrn., Wellsville, Pa., 1971—; v.p. engring. dept. Iris Imaging Systems, York, 1986—; technician AMTECH, Camp Hill, Pa., 1987-89; mem. engring. group Indsl. Controls, Mechanicsburg, Pa., 1989—. Author: (book) Rain Dance, 1968, Neo-Rocking Horse Art, 1970, Prosthetic Vision-Review and Prospectus, 1991, Type R, 1991, Prosthetic Visions: Art and Science, 1993; editor-contbr.: (book) District 45 Archives, York, Pa., 1990. East coast dir. Nat. Socialist White Workers Party, San Francisco, 1978-82. Mem. AAAS, Soc. Photographic Scientists and Engrs, 12 & 12/ Anchora Club (pres. 1988-90). Achievements include research leading to practical prosthetic visual device for the blind utilizing wireless transmission from receptor to cortex. Home: PO Box 31 Wellsville PA 17365-0031

SHOTTS, WAYNE J., nuclear scientist, federal agency administrator; b. Des Plaines, Ill., Mar. 20, 1945; s. Norman Russell Shotts and Winnifred Mae (Averill) Shotts Goeppinger; m. Melinda Maureen Antilla, June 24, 1967 (div. Feb. 1975); children: Kenneth Wayne Shotts, Jeffery Alan Shotts; m. Jacquelyn Francyle Willis, Aug. 11, 1979;. BA in Physics, U. Calif., Santa Barbara, 1967; PhD, Cornell U., 1973. Rsch. physicist E.I. duPont deNemours & Co., Wilmington, Del., 1973-74; physicist U. Calif., Livermore, Calif., 1974—; physicist Lawrence Livermore Nat. Lab., Livermore, Calif., 1974-79, group leader, thermonuclear design divsn., 1979-85, divsn leader, nuclear chemistry, 1985-86, divsn. leader, prompt diagnostics, 1986-88, prin. dep. assoc. dir., military applications, 1988-92, prin. dep. assoc. dir., defense systems/nuclear design, 1992—. Recipient Ernest Orlando Lawrence Meml. award U.S. Dept. Energy, Washington, 1990. Mem. Am. Phys. Soc., Am. Assn. Advancement Sci. Office: Lawrence Livermore Nat Lab PO Box 808 Livermore CA 94550

SHOUB, EARLE PHELPS, chemical engineer; b. Washington, July 19, 1915; m. Elda Robinson; children: Casey Louis, Heather Margaret Shoub Dills. BS in Chemistry, Poly. U., 1938, postgrad. 1938-39. Chemist, Hygrade Food Products Corp., N.Y.C., 1940-41, Nat. Bur. Standards, 1941-43; regional dir. U.S. Bur. Mines, 1943-62; chief div. Accid Prev & Health, 1962-70; dep. dir. Appalachian Lab. Occupational Respiratory Diseases, Nat. Inst. Occupational Safety and Health, Morgantown, W.Va., 1970-77, dep. dir. div. safety research, 1977-79; mgr. occupational safety, indsl. environ. cons., safety products div. Am. Optical Corp., Southbridge, Mass., 1979, cons., 1979—; assoc. prof. dept. anesthesiology W.Va. U. Med. Center, Morgantown, 1977-82, prof. Coll. Mineral and Energy Resources, 1970-79. Recipient Disting. Service award Dept. Interior and Gold medal, 1959. Registered profl. engr.; cert. safety profl. Fellow Am. Inst. Chemists; mem. AIME, ASTM, NSPE, Am. Indsl. Hygiene Assn., Vets. of Safety, Am. Soc. Safety Engrs., Nat. Fire Protection Assn., Am. Conf. Govtl. Indsl. Hygienists, Internat. Soc. Respiratory Protection (past pres.), Am. Nat. Standards Inst., Sigma Xi. Methodist. Contbr. articles to profl. jours. and texts. Home: 5850 Meridian Rd Apt 202C Gibsonia PA 15044-9605

SHOUMAN, AHMAD RAAFAT, mechanical engineering educator; b. Mashtul, Sharkiah, Egypt, Aug. 8, 1929; came to U.S., 1953; s. Mohammad Said and Om El-Saad Shouman; m. Marjorie Louise Bevan, June 18, 1960; children: Ahmad Radey, Kamal Farid, Suzanne, Mariam, Ramsey Omar. BSME, Cairo U., 1950; MS, U. Iowa, 1954, PhD, 1956. Instr. mech. engring. Cairo U., 1950-53; asst. prof. U. Wash., Seattle, 1956-60; assoc. prof., prof. N.Mex. State U., Las Cruces, 1960-86, prof. emeritus, 1986—; prof. U. Petroleum & Minerals, Dhahran, Saudi Arabia, 1981-83, U. Qatar, Doha, 1989-91; cons. EER Systems Corp., Seebrook, Md., 1992; sr. postdoctoral rsch. fellow NAS-NRC, NASA, Huntsville, Ala., 1966-67; vis. scientist Argonne (Ill.) Nat. Lab., 1969; mem. coll. faculty Sandia Nat. Lab., Albuquerque, 1973-74. Mem. ASME (sect. chief, dir. 1963-65), Am. Soc. Engring. Edn., Sigma Xi, Phi Tau Sigma. Republican. Muslim. Achievements include patents for methods and means for reducing power requirements of supersonic wind tunnels, very high efficiency hybrid steam/gas turbine power plant with bottoming vapor rankine cycle. Home: 1006 Bloomdale Dr Las Cruces NM 88005 Office: N Mex State U Mech Engring Dept Las Cruces NM 88001

SHOUP, CARL SUMNER, retired economist; b. San Jose, Calif., Oct. 26, 1902; s. Paul and Rose (Wilson) S.; m. Ruth Snedden, Sept. 27, 1924; children: Dale, Paul Snedden, Donald Sumner (dec.). AB, Stanford U., 1924; PhD, Columbia U., 1930; PhD (hon.), U. Strasbourg, 1967. Mem. faculty Columbia U., 1928-71; dir. Internat. Econ. Integration Program and Capital Tax Project, 1962-64; Editor Bull. Nat. Tax Assn., 1931-35; staff mem. N.Y. State Spl. Tax Commns., 1930-35; tax study U.S. Dept. Treasury, June-Sept. 1934, Aug.-Sept. 1937, asst. to sec. Treasury, Dec. 1937-Aug. 1938, research cons., 1938-46, 62-68; interregional adviser, tax reform planning UN, 1972-74; sr. Killam fellow Dalhousie U., 1974-75; staff Council of Econ. Advisers, 1946-49; dir. Twentieth Century Fund Survey of Taxation in U.S., 1935-37, Fiscal Survey of Venezuela, 1958, Shoup Tax Mission to Japan, 1949-50, Tax Mission to Liberia, 1969; co-dir. N.Y.C. finance study, 1950-52; pres. Internat. Inst. Pub. Finance, 1950-53; cons. Carnegie Ctr. for Transnat. Studies, 1976, Harvard Inst. for Internat. Devel., 1978-83, Venezuelan Fiscal Commn., 1980-83, Jamaica Tax Project, 1985, World Bank Value-Added Tax Study, 1986-87, Duke U. Tax Missions Study, 1987-88; vis. prof. Monash U., 1984. Author: The Sales Tax in France, 1930, (with E.R.A. Seligman) A Report on the Revenue System of Cuba, 1932, (with Robert M. Haig and others) The Sales Tax in the American States, 1934, (with Roy Blough and Mabel Newcomer) Facing the Tax Problem, 1937, (with Roswell Magill) The Fiscal System of Cuba, 1939, Federal Finances in the Coming Decade, 1941, Taxing to Prevent Inflation, 1943, Principles of National Income Analysis, 1947, (with others) Report on Japanese Taxation, 1949, (with others) The Fiscal System of Venezuela, 1959, Ricardo on Taxation, 1960,reprinted, 1992, The Tax System of Brazil, 1965, Federal Estate and Gift Taxes, 1966, Public Finance, 1969 (transl. into Japanese 1974, Spanish 1979), (with others) The Tax System of Liberia, 1970; Editor: Fiscal Harmonization in Common Markets, 1966. Decorated Order Sacred Treasure (Japan), Grand Cordon. Disting. fellow Am. Econ. Assn.; mem. Nat. Tax Assn. (pres. 1949-50, hon. mem.), Phi Beta Kappa. Address: 48 Heard Rd Center Sandwich NH 03227

SHOUP, JAMES RAYMOND, computer systems consultant; b. McKees Rocks, Pa., Apr. 9, 1932; s. Jacob Daniel and Violet May Shoup; student U. Md., 1953-54, U. Miami, 1957-58, Palm Beach Jr. Coll., 1964-68; AA, Fla. Jr. Coll., 1978, AS, 1980; m. Caren Michelle Gagner, Nov. 20, 1988; children—Emily Ruth, Rhonda Lou, Richard Eugene, Sean Jason, Amy Marisa, Rodney Warren. With Fla. Power and Light Co., Delray Beach 1954-68; pres. JSE Corp., 1954-68; fin. cons. area bus., 1954-68; with FAA, 1968-72; with sales and mgmt. depts. Montgomery Ward Co., 1972-75; project mgr. JR Shoup & Assocs., Jacksonville, Fla., 1975—; v.p. rsch. and devel. JP Computing Co., Jacksonville Beach, 1981-86, Alken Computer Systems Co., Flower Mound, Tex., 1979-82; with U.S. Postal Svc., Jacksonville, 1975—; systems instr. microcomputer sci. Duval County Community Schs., Jacksonville, 1980-84; cons. on EDP acctg. applications analysis and EDP systems engring., 1976—; fin. cons., 1967—. Author manuals on computer applications in indsl. and transp. mgmt., 1975-84. Asst. chief, pres., dir. Tri-Community Fire Dept., 1955-57; Sunday sch. tchr., deacon, treas., elder, local Presbyn. chs., 1955—, chmn. pulpit com., 1990; pub. rels. officer N.B. Forrest High Sch. Band Parents Assn., 1973-79; mem. Rep. Presdl. Task Force. With USAF, 1950-54. Mem. EDP Auditors Assn., Jacksonville C. of C. (com. of 100-1982), Mensa. Designer Alken computers, disk patch for tiny Pascal, system 8000 computers, IMAS and IMASNET acctg. systems for microcomputers, MIC series computers; oil painter represented in pvt. collections, Fla. Home and Office: 1832 Lane Ave N Jacksonville FL 32254-1526

SHOUP, TERRY EMERSON, university dean, engineering educator; b. Troy, Ohio, July 20, 1944; s. Dale Emerson and Betty Jean (Spoon) S.; m. Betsy Dinsomore, Dec. 18, 1966; children: Jennifer Jean, Matthew David. BME, Ohio State U., 1966, MS, 1967, PhD, 1969. Asst. prof. to assoc. prof. Rutgers U., New Brunswick, N.J., 1969-75; assoc. prof. to prof. U. Houston, 1975-80; asst. dean, prof. Tex. A&M U., College Sta., 1980-83; dean, prof. Fla. Atlantic U., Boca Raton, 1983-89; dean, Sobrato prof. Santa Clara (Calif.) U., 1989—; cons., software specialist Numerical Methods in Engring. Author: (books) A Practical Guide to Computer Methods for Engineers, 1979, Resheniye Ingenyernikh Zadach NA EVM Prakticheskoye rukovodstvo, 1982, Narichnik Po Izchislitelni Methodi Za Ingeneri, 1983, Numerical Methods for the Personal Computer, 1983, Applied Numerical Methods for the Microcomputer, 1984, (with L.S. Fletcher) Introduction to Engineering with FORTRAN Programming, 1978, Solutions Manual for Introduction to Engineering Including FORTRAN Programming, 1978, Introduccion a la ingenieria Incluyendo programacion FORTRAN, 1980, (with L.S. Fletcher and E.V. Mochel) Introduction to Design with Graphics and Design Projects, 1981, (with S.P. Goldstein and J. Waddell) Information Sources, 1984, (with Carl Hanser Verlag) Numerische Verfahren fur Arbeitsplatzrechner, 1985, (with F. Mistree) Optimization Methods with Applications for Personal Computers, 1987; (software) Numerical Methods for the Personal Computer-Software User's Guide, Version 2, 1983, Optimization Software for the Personal Computer, 1986; editor in chief Mechanism and Machine Theory, 1977—; contbr. more than 100 articles to profl. jours. Fellow ASME (chmn. Design Engring. div. 1987-88, Mech. Engring. div. 1980-81, Centennial medal 1980, Gustus Larson award 1981); mem. Am. Soc. for Engring. Edn. (Dow Outstanding Faculty award 1974, Western Electric award 1984), Fla. Engring. Soc. Home: 1310 Quali Creek Circle San Jose CA 95120 Office: Santa Clara Univ Coll of Engring Office of the Dean Santa Clara CA 95053

SHPILRAIN, VLADIMIR EVALD, mathematician, educator; b. Moscow, Sept. 11, 1960; arrived in Israel, 1991; s. Evald and Valentina (Titushina) S.; m. Elen Burov, Jan. 22, 1983. MA in Math., Moscow State U., 1982, PhD in Math., 1989. Asst. prof. Moscow Inst. for Computer Sci., 1982-86, Moscow State U., 1989-90; prof. Haifa, Israel, 1991—. Contbr. articles to profl. jours. Recipient scholarship Internat. Math. Union, Kyoto, Japan, 1990, Rsch. grants Israeli Acad. Sci., Haifa, 1991, '92, German-Israeli Found., Haifa, 1992, fellowship, Minerva Found. fellow, Bochum, Germany, 1993. Mem. Moscow Math. Soc., Moscow Club of Scientists, Am. Math. Soc. Avocation: Alpine skiing. Office: Technion Dept Math, Haifa 32000, Israel

SHRAGE, SYDNEY, aeronautical engineering consultant; b. Bklyn., Nov. 6, 1927; m. Joyce Aronovitz, June 24, 1956; children: Stephen, Nevin, Allison. BS in Aero. Engring., U. Mich., 1951, BS Engring. in Math., 1951. Sr. propulsion engr. Chance Vought Aircraft, Dallas, 1951-55; group propulsion engr. Martin Co., Balt., 1955-58, asst. to prin. scientist, 1958-60; sect. head TRW Inc., Cleve., 1960-65; sr. project engr. McDonnell Douglas Astronautics, St. Louis, 1965-74, 78-87; engring. mgr. RCA Globcom, Piscataway, N.J., 1974-77; cons. Shrage & Assoc., Ltd., St. Louis, 1987—; adj. lectr. Case Inst. Tech., Cleve., 1960-61, John Carroll U., Cleve., 1963-65. Contbr. articles to profl. publs. With USN, 1945-46. Recipient Skylab Achievement award NASA, 1974. Fellow AIAA (assoc.); mem. Inventors Assn. St. Louis (chmn. bd. dirs. 1992—, pres. 1991-92). Home: 534 Bonhomme Woods Saint Louis MO 63132-3403

SHRESTHA, BIJAYA, nuclear scientist; b. Kathmandu, Nepal, July 8, 1955; came to U.S., 1985; s. Kalidas and Kamala M. (Joshi) S.; m. Puja, May 7, 1975; children: Anjana, Anjaya, Srijana, Samjhana. MS in Plasma Physics, Tribhuvan U., Kathmandu, Nepal, 1978; MS in Nuclear Physics, La. State U., 1988. Asst. lectr. Tribhuvan U., Kathmandu, Nepal, 1979-82, lectr., 1982-85; teaching asst. La. State U., Baton Rouge, 1985-88; rsch. teaching asst. U. Mo., Rolla, 1988—. Author: Campus Physics, 1982. Treas. Univ. Tchrs. Assn., Kathmandu, 1982. Recipient Fulbright scholarship, 1985. Mem. Am. Phys. Soc., Am. Nuclear Soc., Nepal Phys. Soc., Sigma Pi Sigma, Alpha Nu Sigma. Home: 204 E 11th Rolla MO 65401 Office: U Mo Dept Nuclear Engring Rolla MO 65401

SHRINER, ROBERT DALE, economist, management consultant; b. Hobart, Okla., Nov. 28, 1937; s. William Dale and Mildred Ellen (Goodson) S.; m. Nancy Lee Thompson, June 6, 1961; 1 child, Leslie Annette. BA, U. Okla., 1965, MA, 1967; PhD, Ind. U., 1974. Asst. to chief ops. Gen. Dynamics Astronautics, Altus, Okla., 1961-63; dir. Wyo. tech. asst. program U. Wyoming, 1966-69; research assoc. Ind. U. Bur. Bus. Research, 1969-71; asst. prof. Ind. U. Sch. Pub. and Environ. Affairs, 1972-77; assoc. dir. resource devel. internship program Council of State Govt., 1970-72; dir. aerospace research application ctr. Ind. U., 1972-76; mng. assoc., sr. economist Booz Allen & Hamilton, Washington, 1977-79; dir. Washington ops. Chase Econometrics, Washington, 1979-82; mng. ptnr. Shriner-Midland Co., Washington, 1982—; cons. Aerospace Industries Assocs., Nat. Endowment for Arts., Nat. Restaurant Assn., Presl. Commn. on Social Security, U.S. Cath. Conf., YMCA of U.S.; also cons. to various major corps. and nat. assns. Editor, pub. Managing Technology and Change, 1972-75, 86-89; creator computer programs, 1982, 91; contbr. articles to profl. jours. Pres. grad. students assn. U. Okla., Norman, 1965-66; chmn. Rocky Mountain Tech. Svcs. Coun., Wyo., 1967-69; sci. advisor Wyo. Gov., 1968-69; vice chmn. YMCA Fairfax County, Va., 1978-82; bd. dirs. YMCA of Metro Washington, 1982-89, fin. com., 1983-89, treas., 1986-89; exec. com. Gettysburg Coll. Parents Coun., 1985-89. Served with USAF, 1957-61. Recipient Disting. Service award YMCA Metro Washington, 1985. Mem. AAAS, Nat. Assn. Bus. Economists, Am. Mgmt. Assn., Am. Econs. Assn., Bus. Planning Forum, Northern Va. Advanced Tech. Assn. (v.p. programs, 1986-89). Club: Nat. Economists. Lodge: Rotary (pres. 1976-77). Home and Office: Shriner-Midland Co 6432 Quincy Pl Falls Church VA 22042-3117

SHROFF, RAMESH NAGINLAL, research scientist, physicist; b. Jambusar, Gujarat, India, Apr. 27, 1937; came to U.S., 1959; s. Naginial P. Shroff and Motanben N. Shah; m. Manjari M. Majmudar, Oct. 1965 (dec. Oct. 1972); children: Monica R., Bella R.; m. Nirmala G. Vora, Jan. 20, 1974. BS, Bombay U., 1959; MS, Lehigh U., 1961, PhD, 1966. Sr. rsch. physicist Goodyear Tire and Rubber Co., Akron, Ohio, 1965-68; sr. rsch. chemist Chemplex Co., Rolling Meadows, Ill., 1968-78; asst. mgr. Chemplex Co., Rolling Meadows, 1978-80, rsch. assoc., 1980-86; assoc. scientist Enron Chem./Quantum Chem., Rolling Meadows, 1986-90; rsch. scientist Quantum Chem., Cin., 1990—. Contbr. articles to Modern Plastics, Jour. of Rheology, Jour. Applied Polymer Sci., Polymer Engring. and Sci., Plastics Tech. Achievements include rsch. in expertise in structure property relationships in polymer, polymer rheology, effect of molecular structure on melt rheology and end-use properties, coextrusion, screw and die design, computer programming polyolefin product devel. Office: Quantum Chem 11530 Northlake Dr MSN 44 Cincinnati OH 45249

SHROPSHIRE, DONALD GRAY, hospital executive; b. Winston-Salem, N.C., Aug. 6, 1927; s. John Lee and Bess L. (Shouse) S.; m. Mary Ruth Bodenheimer, Aug. 19, 1950; children: Melanie Shropshire David, John Devin. B.S., U. N.C., 1950; Erickson fellow hosp. adminstrn., U. Chgo., 1958-59; LLD (hon.), U. Ariz., 1992. Personnel asst. Nat. Biscuit Co., Atlanta, 1950-52; asst. personnel mgr. Nat. Biscuit Co., Chgo., 1952-54; adminstr. Eastern State Hosp., Lexington, Ky., 1954-62; assoc. dir. U. Md. Hosp., Balt., 1962-67; adminstr. Tucson Med. Ctr., 1967-82, pres., 1982-92, pres. emeritus, 1992—; pres. Tucson Hosps. Med. Edn. Program, 1970-71, sec., 1971-86; pres. So. Ariz. Hosp. Council, 1968-69; bd. dirs. Ariz. Blue Cross, 1967-76, chmn. provider standards com., 1972-76; chmn. Healthways Inc., 1985-92; chmn. Healthways Inc., 1985-92; bd. dirs. First Interstate Bank of Ariz., Tucson Electric Power Co., Tucson-Rincon & Sonoran Insts. Bd. dirs., exec. com. Health Planning Council Tucson, 1969-74; chmn. profl. div. United Way Tucson, 1969-70, vice chmn. campaign, 1988, Ariz. Health Facilities Authority, 1992—, bd. dirs., 1992—; chmn. dietary services com., vice chmn., 1988. Md. Hosp. Council, 1966-67; bd. dirs. Ky. Hosp. Assn., 1961-62, chmn. council profl. practice, 1960-61; past pres. Blue Grass Hosp. Council; trustee Assn. Western Hosps., 1974-81, pres., 1979-80; mem. accreditation Council for Continuing Med. Edn., 1982-87, chmn., 1986; bd. govs. Pima Community Coll., 1970-76, sec., 1973-74, chmn., 1975-76, bd. dirs. Found., 1978-82, Ariz. Bd. Regents, 1982-90, sec., 1983-86, pres. 1987-88; mem. Tucson Airport Authority, 1987—, bd. dirs., 1990— ; v.p. Tucson Econ. Devel. Corp., 1977-82; bd. dirs. Vol. Hosps. Am., 1977-88, treas., 1979-82; mem. Ariz. Adv. Health Coun. Dirs., 1976-78, Ariz. Health Facilities Authority Bd., 1992—; bd. dirs. Tucson Tomorrow, 1983-87, Tucson Downtown Devel. Corp., 1988—, Rincon Inst., 1992—; dir. Mus. No. Ariz., 1988—; nat. bd. advisors Coll. Bus. U. Ariz, 1992—, chmn. Dean's Bd. Fine Arts, 1992—. Named to Hon. Order Ky. Cols.; named Tucson Man of Yr. 1987; recipient Disting. Svc. award Anti-Defamation League B'nai B'rith, 1989, Salisbury award Ariz. Hosp. Assn. Mem. Am. Hosp. Assn. (nominating com. 1983-86, trustee 1975-78, ho. dels. 1972-78, chmn. coun. profl. svc. 1973-74, regional adv. bd. 1969-78, chmn. joint com. with Nat. Assn. Social Workers 1963-64, Disting. Svc. award 1989), Ariz. Hosp. Assn. (Salisbury award, bd. dirs. 1967-72, pres. 1970-71), Ariz. C. of C. (bd. dirs. 1988-93), Assn. Am. Med. Colls. (mem. assembly 1974-77), Tucson C. of C. (bd. dirs. 1968-69), United Comml. Travelers, Nat. League Nursing, Ariz. Acad. (bd. dirs. 1982, treas. 1985), Pima County Acad. Decathlon Assn. (dir. 1983-85), Tucson Community Coun. Baptist (ch. moderator, chmn. finance com., deacon, ch. sch. supt., trustee, bd. dirs. ch. found.). Home: 6734 N Chapultapec Circle Tucson AZ 85715 Office: TM Care 2195 River Rd Ste 202 Tucson AZ 85718

SHROUT, PATRICK ELLIOT, psychometrician, educator; b. Appleton, Wis., Dec. 9, 1949; s. Herbert Kirby and Marie Ann (Fitzgerald) S.; m. Jane M. Hanson, June 16, 1973; children: Anelise, Hanson Timothy. AB, St. Louis U., 1972; PhD, U. Chgo., 1976. Instr. Dept. Psychology, Roosevelt U., Chgo., 1975-76; instr. pub. health Div. Biostatistics, Columbia U. Sch. Pub. Health, N.Y.C., 1976-77, asst. prof. pub. health, 1977-83, assoc. prof. clin. pub. health, 1983-89, assoc. prof. pub. health and psychiatry, 1989—; rsch. scientist V N.Y. State Psychiat. Inst., N.Y.C., 1980-92; prof. psychology NYU, N.Y.C., 1992—; cons. Drug Abuse Resource Devel. Rev. Com., Nat. Inst. Drug Abuse, 1981; spl. review com. WHO Study of Schizophrenia, NIMH, 1982; rsch. methods review group Task Force on DSM-IV, Am. Psychiatric Assn., 1988—; mem. epidemiologic and svcs. rsch. review com. NIMH, 1982-86. Contbr. numerous articles to profl. jours., chpts. to books; reviewer Am. Jour. Psychiatry, Biometrics, Evaluation and Program Planning, Jour. Cons. and Clin. Psychology, Jour. Personality, Jour. Psychosomatic Medicine, Multivariate Behavioral Rsch., Psychology and Aging, Psychometrika, Psychol. Bull., Pub. health Reports; editorial bd. Jour. of Personality, 1987—. Grantee NIMH, 1980—, NIH, 1981-83. Mem. AAAS, APA, Am. Assn. Applied and Preventive Psychology, Am. Psychol. Soc., Am. Psychopathologic Assn., Am. Pub. Health Assn., Am. Statis. Assn., Biometric Soc., N.Y. Acad. Sci., Psychometric Soc., Soc. for Multivariate Behavioral Rsch., Phi Beta Kappa, Sigma Xi. Home: 39 Marion Rd Upper Montclair NJ 07043 Office: NYU Dept Psychology 6 Washington Pl Mail Rm 401 New York NY 10003

SHTEINFELD, JOSEPH, electrical engineer; b. Riga, Latviya, Russia, Oct. 17, 1941; s. Isidor and Sarah (Itigina) S.; m. Velina Kraizel, Oct. 5, 1984; 1 child, Tanya Kraizel. BSEE, Marine Engring. Coll., Russia, 1969. Cert. engring. operations exec., facilities mgr. Asst. chief engr. Hotel Roosevelt, N.Y.C.; bldg. mgr. Art Leather Mfg., N.Y.C., 1987-88; asst. dir. engring. Hotel Pierre-Four Seasons Hotel, N.Y.C., 1988—. Mem. Am. Soc. Energy Engrs., Am. Soc. Heat, Ventilation & Air Conditioning, AIPE. Home: 104-40 Quees Blvd Apt 11F Flushing NY 11375

SHU, SUN, geologist. Dir. Inst. Geology Acad. Scis., China, 1985—. Office: Academia Sinica, 52 San Li He Rd, Beijing 100864, China*

SHUART, MARK JAMES, aerospace engineer; b. Niagara Falls, N.Y., Sept. 23, 1954; s. James Campbell and Kalleope (Philosophos) S.; m. Jane Elizabeth Nelson, June 19, 1976; children: Amy, Emily. BS, Va. Poly. Inst. and State U., 1976, MS, 1978; PhD, U. Del., 1986. Rsch. engr. NASA Langley Rsch. Ctr., Hampton, Va., 1977-89, asst. head aircraft structures br., 1989-92, asst. chief structural mechanics div., 1992—. Contbr. articles to profl. jours. Chmn. adminstrv. bd. First United Meth. Ch.-Fox Hill, Hampton, 1987-90; pres. Asbury Sch. PTA, Hampton, 1992—. Mem. AIAA (sr.). Achievements include original research conducted for the response and failure of composite structures for aerospace applications.

SHUCH, H. PAUL, engineering educator; b. Chgo., May 23, 1946; s. Ben Aaron Wakes and Phyllis Anita (Greenwald) S.; m. M. Suk Chong, Mar. 24, 1969 (div. 1990); children: Erika Chong Shuch, Andrew Pace Shuch; m. Janet A. Sherman, July 14, 1991. AS, West Valley Coll., Saratoga, Calif., 1972; BS, San Jose State U., 1975, MA, 1986; PhD, U. Calif., Berkeley, 1990. Lic. FAA comml. pilot and instrument flt. instr. Engring. technician TransAction Systems, Inc., Palo Alto, Calif., 1969-71; rsch. and devel. engr. Applied Tech., Sunnyvale, Calif., 1972-75; electronics instr. West Valley Coll., Saratoga, Calif., 1973-77; sr. engring. instr. Lockheed Missiles and Space Co., Sunnyvale, 1975-77; avionics lectr. San Jose (Calif.) State U., 1984-87; head microwave instr. San Jose City Coll., 1977-90; prof. electronics Pa. Coll. Tech., Williamsport, 1990—; chief engr. Microcomm, San Jose, 1975-90; accident prevention counselor FAA, San Jose and Harrisburg, Pa., 1983—. Co-author: ARRL UHF/Microwave Experimenter's Manual, 1990; contbr. Radio Handbook, 1975; contbr. articles to profl. jours. Chmn. Santa Clara County Airport, San Jose, 1983-87; faculty sen. San Jose City Coll., 1979-81; chmn. ARRL/CSVHFC Joint FCC Briefing Com., Washington, 1991—; guest lectr. Air Safety Found., Frederick, Md., 1992. Sgt. USAF, 1965-69, SE Asia. Named Flt. Instr. of Yr., FAA, Harrisburg, 1992; recipient Doctoral Thesis prize Fannie & John Hertz Found., Livermore, Calif., 1990, Prof. Robert Horonjeff Meml. grant U. Calif. Berkeley Inst. Transp. Studies, 1990, Robert Goddard Meml. scholarship Nat. Space Club, Washington, 1988, fellowship in applied phys. sci. Fannie & John Hertz Found., Livermore, Calif., 1989. Mem. Am. Coun. Edn. (mil. program evaluator 1991—), Am. Radio Relay League (adv. com. chmn. 1983-87), WESCON (tech. session organizer 1979). Democrat. Jewish. Achievements include design and prodn. of world's first comml. home satellite TV receiver, 1978; patent for aircraft binaural doppler collision alert system, 1987. Office: Pa Coll Tech One College Ave Williamsport PA 17701

SHUGART, HERMAN HENRY, environmental sciences educator, researcher; b. El Dorado, Ark., Jan. 19, 1944; s. Herman Henry and Katherine Luvois (Rich) S.; m. Ramona Jeanne Kozel, Aug. 27, 1966; children: Erika Christine, Stephanie Laurel. BS, U. Ark., 1966, MS, 1968; PhD, U. Ga., 1971. Research scientist Oak Ridge (Tenn.) Nat. Lab., 1971-84; asst. prof. U. Tenn., Knoxville, 1971-75, assoc. prof., 1975-82, prof., 1982-84; W.W. Corcoran prof. U. Va., Charlottesville, 1984—, chmn. dept. environ. sci., 1990-93. Author: Time Series Analysis, 1978, Systems Ecology, 1979, Forest Succession, 1981, A Theory of Forest Dynamics, 1984, A Systems Analysis of the Global Boreal Forest, 1992, Vegetation Dynamics and Global Change, 1993. Fellow AAAS; mem. Ecol. Soc. Am. (mem. editorial bd. 1981-83). Democrat. Unitarian/Universalist. Avocation: gardening. Home: 107 Cannon Pl Charlottesville VA 22901-2109 Office: Univ Va Dept Environ Sci Charlottesville VA 22903

SHUI, XIAOPING, materials scientist; b. Shanghai, Peoples Republic of China, Dec. 28, 1953; came to U.S., 1989; s. Qianhong and Mingihu (Hu) S.; m. Penny Jin, Aug. 23, 1982; 1 child, Steven. BS, Jiaotong U., Shanghai, 1982, MS, 1985; postgrad, SUNY, Buffalo. Rsch. scientist Nat. Lab. for Matel Matrix Composite, Shanghai, 1985-89; lectr. Jiaotong U., Shanghai, 1987-89. Contbr. articles to profl. jours. Mem. Minerals, Metals, Materials Soc., Soc. for Advancement of Material and Process Engring. Achievements include patents pending in the area of processing and application of submicron carbon filaments. Office: SUNY Buffalo Buffalo NY 14260

SHUKLA, KAPIL P., medical physicist; b. Morbi, Gujarat, India, Mar. 15, 1943; came to U.S., 1974; s. Purshottam D. and Chandrakala P. (Dave) S.; m. Annapurna Trivedi, Dec. 12, 1967; 1 child, Urvashi. BSc, Gujarat U., 1966, MSc, 1968; MS in Physics, U. Scranton, 1976; postgrad., U. Pitts., 1978. Prof. physics Saurashtra U., India, 1967-74; asst. prof. clin. Wheeling (W.Va.) Coll., 1979-81; clin. coord. U. Mass. Med. Ctr., Worcester, 1981-82; med. physicist Morton Plant Hosp., Clearwater, Fla., 1983-85, West Penn Hosp., Pitts., 1985-88, VA Med. Ctr., Tucson, 1988-89, Wilkes-Barre (Pa.) Gen. Hosp., 1989—. Co-author: (textbooks) Physics Lab Guide for

Freshman, 1971, Physics Problems for Sophomores, 1972, Physics Textbook for Juniors, 1973, (rsch. publ.) Emmision Computed Tomography, 1983. Bhabha Atomic Energy scholar for postgrads., Bombay, India, 1966-68; grad. fellow U. Scranton, 1976; spl. vice-chancellor grantee for med. physics studies U. Pitt., 1978. Mem. Am. Assn. Physicist in Medicine (nuclear medicine com. 1980, radiation protection com. 1987-90, diagnostic x-ray imaging task group 1990), Health Physics Soc. (plenary mem.). Democrat. Hindu. Home: 202 Indian Creek Dr Wilkes Barre PA 18702 Office: Wilkes Barre Gen Hosp N River and Auburn Sts Wilkes Barre PA 18764

SHUKLA, MAHESH, structural engineer; b. Morvi, India, Jan. 21, 1927; came to U.S., 1949; s. Girijashanker and Umaben (Dave) S.; m. Minaxi Rawal, May 28, 1967; children: Kapil, Jagruti. BS in Civil Engring., U. Calif., Berkeley, 1951; MS in Civil Engring., U. So. Calif., 1957. Various positions, 1951-64; structural engring. assoc. II Dept. Bldg. & Safety, L.A., 1964—. Mem. ASCE. Hindu.

SHULL, HARRISON, chemist, educator; b. Princeton, N.J., Aug. 17, 1923; s. George Harrison and Mary (Nicholl) S.; m. Jeanne Louise Johnson, 1948 (div. 1962); children: James Robert, Kathy, George Harrison, Holly; m. Wil Joyce Bentley Long, 1962; children: Warren Michael Long, Jeffery Mark Long, Stanley Martin, Sarah Ellen. A.B., Princeton U., 1943; Ph.D., U. Calif. at Berkeley, 1948. Assoc. chemist U.S. Naval Research Lab., 1943-45; asst. prof. Iowa State U., 1949-54; mem. faculty Ind. U., 1955-79, research prof., 1961-79, dean Grad. Sch., 1965-72, vice chancellor for research and devel., 1972-76, dir. Research Computing Center, 1959-63, acting chmn. chemistry dept., 1965-66, acting dean arts and scis., 1969-70, acting dean faculties, 1974; mem. faculty, provost, v.p. acad. affairs Rensselaer Poly. Inst., 1979-82; chancellor U. Colo., Boulder, 1982-85; prof. dept. chemistry U. Colo., 1982-88; provost Naval Postgrad. Sch., 1988—; asst. dir. research, quantum chemistry group Uppsala (Sweden) U., 1958-59; vis. prof. Washington U., St. Louis, 1960, U. Colo., 1963; founder, supr. Quantum Chemistry Program Exchange, 1962-79; chmn. subcom. molecular structure and spectroscopy NRC, 1958-63; chmn. Fulbright selection com. chemistry, 1963-67; mem. adv. com. Office Sci. Personnel, 1957-60; chmn. First Gordon Research Conf. Theoretical Chemistry, 1962; mem. com. survey chemistry Nat. Acad. Sci., 1964-65; mem. adv. panel chemistry NSF, 1964-67; mem. adv. panel Office Computer Activities, 1967-70, cons. chem. information program, 1965-71, mem. adv. com. for research, 1974-76; mem. vis. com. chemistry Brookhaven Nat. Lab., 1967-70; mem. adv. com. Chem. Abstracts Service, 1971-74; provost, now chief academic dean Naval Postgraduate Sch, Monterey, Calif.; dir. Storage Tech. Corp.; chief of Naval Ops. Exec. Panel, 1984-88. Assoc. editor: Jour. Chem. Physics, 1952-54; editorial adv. bd.: Spectrochimica Acta, 1957-63, Internat. Jour. Quantum Chemistry, 1967—, Proc. NAS, 1976-81; contbr. articles to profl. jours. Trustee Argonne U. Assn., 1970-75, Asso. Univs., Inc., 1973-76, U. Rsch. Assn., 1984-89, Inst. Defense Analysis, 1984—. Served as ensign USNR, 1945. NRC postdoctoral fellow phys. scis. U. Chgo., 1948-49; Guggenheim fellow U. Uppsala, 1954-55; NSF sr. postdoctoral fellow, 1968-69; Sloan research fellow, 1956-58. Fellow Am. Acad. Arts and Scis. (v.p. 1976-83, chmn. Midwest Center 1976-79), Am. Phys. Soc.; mem. Nat. Acad. Scis. (com. on sci. and public policy 1969-72, council, exec. com. 1971-74, chmn. U.S.-USSR sci. policy subgroup for fundamental research 1973-81, naval studies bd. 1974-79, chmn. Commn. on Human Resources 1977-81, nominating com. 1978), Am. Chem. Soc., AAAS, Assn. Computing Machinery, Royal Swedish Acad. Scis. (fgn. mem.), Royal Acad. Arts and Scis. Uppsala (corr. mem.), Phi Beta Kappa, Sigma Xi, Phi Lambda Upsilon. Club: Cosmos (Washington). Office: Naval Postgrad Sch Chief Academic Dean Monterey CA 93943-5000

SHULL, JULIAN KENNETH, JR., biology educator; b. Anniston, Ala., Oct. 15, 1941; s. Julian Kenneth and Ruth Dale (Cook) S.; m. Carolyn June Saxon, Aug. 12, 1967; children: Julian Kenneth III, Gwyndolen Dale. BS in Chemistry, U. Ala., 1963, MS in Biology, 1967; PhD in Genetics, Fla. State U., 1973. From asst. to assoc. prof. biology Loyola U., New Orleans, 1973-84; prof biology Appalachin State U., Boone, N.C., 1984—; organizer Internat. Boone Chromosone Conf., 1986, '89, '92. Deacon First Bapt. Ch., Boone, N.C.,. Mem. AAAS, Assn. Southeastern Biologists (v.p. 1981-82), Genetics Soc. of Can. Office: Appalachin State U Dept Biology Boone NC 28608

SHULMAN, LAWRENCE EDWARD, biomedical research administrator, rheumatologist; b. Boston, July 25, 1919; s. David Herman and Belle (Tishler) S.; m. Pauline K. Flint, July 19, 1946; 1 son, Lawrence E.; m. Reni Trudinger, Mar. 20, 1959; children: Kathryn Verena, Barbara Corina. A.B., Harvard U., 1941, postgrad., 1941-42; Ph.D., Yale U., 1945, M.D., 1949. Nat. Bd. Med. Examiners. Intern Johns Hopkins Hosp., 1949-50, resident and fellow in internal medicine, 1950-53; dir. connective tissue div. Johns Hopkins U., 1955-75, assoc. prof. medicine, 1964—; assoc. dir. div. arthritis, musculoskeletal and skin diseases NIH, Bethesda, Md., 1976-82, dir., 1982—; dir. Nat. Inst. Arthritis and Musculoskeletal and Skin Diseases NIH, 1986—; chmn. med. adminstrn. com. Arthritis Found., Atlanta, 1974-75, exec. com., 1972-77; dir. Lupus Found. Am.; med. adv. bd. United. Scleroderma Found., Watsonville, Calif., 1977-88. Discoverer: Eosinophilic Fasciitis, 1974, new med. sign friction rubs in scleroderma, 1961. Recipient Sr. Investigator award Arthritis Found., 1957-62, Disting. Svc. award 1979, Heberden medal for rsch., London, 1975, Superior Svc. award USPHS, 1985; Spl. Recognition award Nat. Osteoporosis Found., 1991, Spl. award Am. Acad. Orthopaedic Surgeons, 1992; W.R. Graham Meml. Lectr., 1973. Fellow ACP; mem. Am. Rheumatism Assn. (pres. 1974-75, master 1987—), Soc. Clin. Trials, Pan-Am. League Against Rheumatism (pres. 1982-86), Am. Soc. for Bone and Mineral Rsch., WHO (chmn. sci. group on rheumatism diseases, Geneva, 1989). Home: 6302 Swords Way Bethesda MD 20817-3350 Office: NIH Rm 4C32 9000 Rockville Pike Bethesda MD 20892-0001

SHULMAN, STANFORD TAYLOR, pediatrics educator, infectious disease researcher; b. Kalamazoo, May 13, 1942; s. Edward I. and Clara (Portnoff) S.; m. Claire Elaine Zaner, June 21, 1964; children: Deborah, Elizabeth, Edward. BS, U. Cin., 1963; MD, U. Chgo., 1967. Diplomate Am. Bd. Pediatrics. Resident, chief resident in pediatrics U. Chgo., 1967-70; fellow in pediatric infectious disease U. Fla., Gainesville, 1970-73, asst. prof., 1973-76, assoc. prof., 1976-79; prof. Northwestern U. Med. Sch., Chgo., 1979—, assoc. dean, 1989—; chief pediatric infectious disease Children's Meml. Hosp., Chgo., 1979—; chmn. com. on rheumatic fever and endocarditis Am. Heart Assn., Dallas, 1982-88. Editor: Pharyngitis, 1984, Kawasaki Disease, 1986, Biologic and Clinical Basis of Infectious Diseases, 1991, Handbook of Pediatric Infectious Disease and Antimicrobial Therapy, 1993; mem. edit. bd. Pediatric Infectious Disease Jour., 1988—; contbr. over 150 articles to profl. jours. Recipient numerous grants to study new antibiotics, streptococci and forms of treatment of Kawasaki disease from pharm. cos., 1979—. Mem. Am. Pediatric Soc., Soc. for Pediatric Rsch., Am. Assn. Immunologists, Infectious Disease Soc. Am. Jewish. Achievements include findings related to diagnosis and treatment of streptococcal pharyngitis or sore throat (strep throat), rheumatic fever and Kawasaki disease. Office: Children's Meml Hosp 2300 Children's Plz Chicago IL 60614

SHUMATE, CHARLES ALBERT, retired dermatologist; b. San Francisco, Aug. 11, 1904; s. Thomas E. and Freda (Ortmann) S.; B.S., U. San Francisco, 1927, H.H.D., 1976; M.D., Creighton U., 1931. Pvt. practice dermatology, San Francisco, 1933-73, ret., 1973; asst. clin. prof. dermatology Stanford U., 1956-62; pres. E Clampus Vitus, Inc., 1963-64; hon. mem. staff St. Mary's Hosp. Mem. San Francisco Art Commn., 1964-67, Calif. Heritage Preservation Commn., 1963-67; regent Notre Dame Coll. at Belmont, 1965-78, trustee, 1977-93; pres. Conf. Calif. Hist. Socs., 1967; mem. San Francisco Landmarks Preservation Bd., 1967-78, pres., 1967-69; trustee St. Patrick's Coll. and Sem., 1970-86. Served as maj. USPHS, 1942-46. Decorated knight comdr. Order of Isabella (Spain); knight Order of the Holy Sepulchre, knight of St. Gregory, knight of Malta. Fellow Am. Acad. Dermatology; mem. U. San Francisco Alumni Assn. (pres. 1955), Calif. Book Club (pres. 1969-71), Calif. Hist. Soc. (trustee 1958-67, 68-78, pres. 1962-64), Soc. Calif. Pioneers (dir. 1979—). Clubs: Bohemian, Olympic, Roxburghe (pres. 1958-59) (San Francisco); Zamorano (Los Angeles). Author: Life of George Henry Goddard; The California of George Gordon, 1976, Jas. F. Curtis, Vigilante, 1988, Francisco Pacheco of Pacheco Pass, 1977; Life of Mariano Malarin, 1980; Boyhood Days: Y. Villegas Reminiscences of California 1850s, 1983, The Notorious I.C. Woods of the Adams Express, 1986, Rincon Hill and South Park, 1988, Captain A.A. Ritchie, Pioneer, 1991. Home: 1901 Scott St San Francisco CA 94115-2613 Office: 490 Post St San Francisco CA 94102-1401

SHUMICK, DIANA LYNN, computer executive; b. Canton, Ohio, Feb. 10, 1951; d. Frank A. and Mary J. (Mari) S.; 1 child, Tina Elyse. Student, Walsh Coll., 1969-70, Ohio U., 1970-71, Kent State U., 1971-77. Data entry clk. Ohio Power Co., Canton, 1969-70; clk. City of Canton Police Dept., 1971-73; system engr. IBM, Canton, 1973-81; adv. market support rep. IBM, Dallas, 1981-89; system engr. mgr. IBM, Madison, Wis., 1989-93; mktg. customer satisfaction mgr. IBM, Research Triangle Park, N.C., 1993—. Author: Technical Coordinator Guidelines, 1984. Pres., bd. dirs. Big Bros. and Sisters of Denton (Tex.) County, 1989, v.p., 1988, sec., 1987; mem. St. Philip Parish Coun., Lewisville, Tex., 1988-89, Western Stark County Red Cross, Canton, 1980; v.p. Parents without Ptnrs., Madison, 1991; founding bd. mem. Single Parents Network, 1991; vol. ARC, 1985—; mem., bd. dirs. Rape Crisis Ctr. Dane County, sec. 1990-91; vol. Paint-A-Thon, Dane County, 1990, Badger State Games Challenge, 1992, Cystic Fibrosis Found. Great Strides, 1992, 93.

SHUMILA, MICHAEL JOHN, electrical engineer; b. Newark, Aug. 30, 1947; s. Michael John and Anne (Zavocki) S. AS, Essex County Coll., Newark, 1971; BSEE, N.J. Inst. Tech., 1974, MSEE, 1977. Systems engr. Lockheed Electronics, Plainfield, N.J., 1977-78; engr. group VI Reeves Teletape, N.Y.C., 1978-79; sr. mem. engring. staff G.E. Am. Communications, Princeton, N.J., 1979-81; mem. tech. staff David Sarnoff Rsch. Ctr., Princeton, N.J., 1981—. Author: NAB Engineering Handbook, 1985; contbr. articles to profl. jours. Recipient RCA Outstanding Tech. Achievement award David Sarnoff Rsch. Ctr., 1985, First Cut Silicon award VLSI Tech., Inc., 1991. Mem. Aircraft Owners and Pilots Assn., Exptl. Aircraft Assn., Soc. Motion Picture and TV Engrs. Lutheran. Achievements include contribution to the architecture and design of the RCA Digital Audio Transmission system; patents on real time median filter design, adaptive control of median filters. Home: 16 Pintinalli Dr Trenton NJ 08619-1538

SHUMWAY, NORMAN EDWARD, surgeon, educator; b. Kalamazoo, Mich., 1923. M.D., Vanderbilt U., 1949; Ph.D. in Surgery, U. Minn., 1956. Diplomate: Am. Bd. Surgery, Am. Bd. Thoracic Surgery. Intern U. Minn. Hosps., 1949-50, med. fellow surgery, 1950-51, 53-54, Nat. Heart Inst. research fellow, 1954-56, Nat. Heart Inst. spl. trainee, 1956-57; mem. surg. staff Stanford U. Hosps., 1958—, asst. prof. surgery, 1959-61, assoc. prof., 1961-65, prof., 1965—, head div. cardiovascular surgery Sch. Medicine, 1974—; Frances and Charles D. Field prof. Stanford U., 1976—. Served to capt. USAF, 1951-53. Mem. AMA, Soc. Univ. Surgeons, Am. Assn. Thoracic Surgery, Am. Coll. Cardiology, Transplantation Soc., Samson Thoracic Surg. Soc., Soc. for Vascular Surgery, Alpha Omega Alpha. Office: Stanford U Med Ctr Dept Cardiovascular Surgery 300 Pasteur Dr Palo Alto CA 94304-2203

SHUMWAY, SANDRA ELISABETH, shellfish biologist; b. Taunton, Mass., Mar. 29, 1952; d. Alonzo Harrison and Lois Elisabeth (Tyndal) S. BS summa cum laude, Southampton Coll., 1974; PhD, Univ. Coll. North Wales, 1976; DSc, 1993. Rschr. Marine Sci. Labs., Menai Bridge, Wales, 1974-77, Portobello (New Zealand) Marine Lab., 1978-79; rsch. asst. dept. ecology and evolution SUNY, Stony Brook, 1979-82; scientist Dept. Marine Resources, Boothbay Harbor, Maine, 1983-93, Bigelow Lab., Boothbay Harbor, 1984—; prof. of marine scis. Southampton Coll., 1994—. Editor: Jour. Shellfish Rsch., 1986—; mem. editorial bd. Revs. in Aquatic Scis., Jour. Med. and Applied Malacology; editor 2 books on scallop biology; contbr. articles to sci. publs. Grantee, Internat. Pectinid Workshop, 1989, Marine Aquaculture Innovation Ctr., 1990, New England Fisheries Devel. Assn., 1990-92, NOAA/Sea grantee Nat. Marine Fisheries Svc. Mem. Marine Biol. Assn. U.K., Nat. Shellfisheries Assn. (pres. 1991-92), Can. Aquaculture Assn., Am. Malacological Union, Am. Soc. Zoologists, Assn. Women in Sci., Nature Conservancy, Coun. Biology Editors, European Assn. Sci. Editors. Home: 3 Union St Boothbay Harbor ME 04538 Office: Southampton Coll Southampton NY 11968

SHUMWAY, SARA J., cardiothoracic surgeon; b. Lake Charles, La., Dec. 21, 1952. BS in Biol. Scis., Stanford U., 1975; MD, Vanderbilt U., 1979. Diplomate Am. Bd. Surgery, Am. Bd. Thoracic Surgery; lic., Minn., Md. Resident in gen. surgery Vanderbilt U. Affiliated Hosps., Nashville, 1979-82, 83-85; resident in cardiothoracic surgery, chief resident The Johns Hopkins Hosp., Balt., 1985-88; assoc. prof. surgery, assoc. mem. grad. faculty U. Minn. Hosp. and Clinic, Mpls., 1988-93; with Mpls. VA Hosp., St. Paul Ramsey Med. Ctr.; vis. scientific worker Transplantation Biology sect. Sir Peter Medawar Clin. Rsch. Ctr., Harrow, U.K., 1982-83; hon. sr. registrar Brompton Hosp., London, 1986; spl. primary reviewer Am. Heart Assn. Mich. Rsch. Grant Com., 1991. Author numerous book chpts., abstracts, presentations in field; guest reviewer The Annals of Thoracic Surgery, 1992, Clin. Transplantation, 1989-92, mem. editorial bd., 1992; contbr. articles to profl. jours. Named one of Two Thousand Notable Am. Women, 1993. Fellow Am. Coll. Surgeons, Am. Coll. Cardiology; mem. AMA, AAAS, Am. Soc. Transplant Surgeons, Am. Coll. Chest Physicians, Am. Heart Assn., Brit. Soc. Immunology, Brit. Transplantation Soc., Internat. Soc. for Heart and Lung Transplantation, N.Y. Acad. Scis., Mpls. Surg. Soc., Twin City Thoracic and Cardiovascular Surg. Soc., Ctrl. Surg. Soc., Transplantation Soc., Johns Hopkins Med. and Surg. Assn., H. William Scott Jr. Soc., Alpha Omega Alpha (sec.-treas. Mpls. chpt. 1990-92). Home: 1925 James Ave S Minneapolis MN 55403 Office: U Minn Hosp and Clinic Harvard St at East River Rd Box 207 UMHC Minneapolis MN 55455

SHUPE, LLOYD MERLE, chemist, consultant; b. Amanda, Ohio, July 6, 1918; s. Lloyd M. and L. Marie (Balthaser) S.; m. Mary Noecker, June 7, 1941; children: David, Richard, Jane, Douglas. BS, Capital U., 1940. Chemist chem. warfare svc. U.S. Army, Columbus, Ohio, 1941-45; crime lab supr Columbus Police Dept., 1945-70; police planner State of Ohio, Columbus, 1970-82; forensic chemist Columbus, 1982—. Editor: What's New in Forensic Criminalistics, 1966; contbr. articles to 30 scientific and tech. publs. Pres. Redeemer Luth. Ch., Columbus, 1958-63. Fellow Am. Acad. Forensic Sci.; mem. Masons (master 1957-58), Rose Croix. Home: 2600 Floribunda Dr Columbus OH 43209

SHUPLER, RONALD STEVEN, environmental engineer; b. Bklyn., Dec. 15, 1954; s. M. and Marilyn S.; m. Karen Ann Shupler, June 9, 1979; 1 child, Matthew Stephen. BSCE, Ind. Inst. Tech., Ft. Wayne, Ind., 1975; MBA, Fairleigh Dickinson U., Teaneck, N.J., 1980. Registered profl. engr., Fla. Sales engr. Faber Mech. Products, Clifton, N.J., 1976-80; sales mgr. William Steinen Mfg., Parsippany, N.J., 1980-81; product mgr. Parkson Corp., Ft. Lauderdale, Fla., 1981—. Contbr. articles to profl. jours. Mem. ASCE, Mfrs.' Alliance Productivity and Innovation, Fla. Profl. Engring. Soc. Alpha Sigma Phi.

SHURINA, ROBERT DAVID, educator; b. Curtisville, Pa., July 8, 1957; s. John and Jennie Elizabeth (Soster) S.; m. Frances Ruth Zoretich, Oct. 1, 1983; children: Adrianne Renée, Katharine Rose, Matt Zoretich. BS in Biochemistry, Pa. State U., 1981; PhD in Biochemistry, Thomas Jefferson U., 1991. Technician Pa. State U., State College, 1980-81; rsch. specialist U. Pa., Phila., 1981-83; technician The Wistar Inst., Phila., 1983-84; rsch. assoc. Lankenau Med. Rsch. Ctr., Wynnewood, Pa., 1990-93; asst. prof. LaSalle U., Phila., 1993—; lector DuPont-Merck Rsch. Sta., Glenolden, Pa., 1992—, Wilmington, 1992—; instr. Thomas Jefferson U., Phila., 1992—. Author: (with others) Biological Reactive Intermediates IV, 1919; contbr. articles to profl. jours. Recipient Award Mid-Atlantic Soc. of Toxicology 1990. Mem. AAAS, Phila. Cancer Rsch. Assn., Am. Assn. for Cancer Rsch. (assoc.), Sigma Xi (Award of Recognition 1989). Office: Lankenau Med Rsch Ctr 100 Lancaster Ave Wynnewood PA 19023

SHURTLEFF, MALCOLM C., plant pathologist, consultant, educator, extension specialist; b. Fall River, Mass., June 24, 1922; s. Malcolm C. and Florence L. (Jewell) S.; m. Margaret E. Johnson, June 14, 1950; children: Robert Glen, Janet Lee, Mark Steven. BS in Biology, U. R.I., 1943; MS in Plant Pathology, U. Minn., 1950, PhD in Plant Pathology, 1953. Asst. plant pathologist Conn. Agrl. Expt. Sta., New Haven, 1942, R.I. Agrl. Expt. Sta., Kingston, 1943; asst. extension prof. U. R.I., Kingston, 1950-54; assoc. extension prof. Iowa State U., Ames, 1954-61; prof. plant pathology U. Ill., Champaign-Urbana, 1961-92; cons., writer Urbana, 1992—. Author: How to Control Plant Diseases, 1962, 66 (Am. Garden Guild award 1962, 66), How to Control Lawn Diseases and Pests, 1973, How to Control Tree Diseases and Pests, 1975, Controlling Turfgrass Pests, 1987; editor-in-chief Phytopathology News, 1966-69, Plant Disease, 1969-72; contbr. articles to encys.; contbr. over 1400 rsch., extension, mag. articles. Lt. (j.g.) USN, 1943-46, PTO. Recipient Disting. Svc. award U.S. Dept. Agriculture, Washington, 1986. Fellow Am. Phytopathological Soc. (councilor at large 1970-71, Excellence in Extension Plant Pathology award 1991); mem. Internat. Soc. Plant Pathology (chmn. extension com. 1975-80), Am. Phytopathological Soc. (mem. various coms.). Avocation: photography. Home: 2707 Holcomb Dr Urbana IL 61801 Office: U Ill Dept Plant Pathology N-427 Turner Hall Urbana IL 61801

SHURYAK, EDWARD VLADIMIROVICH, physicist; b. Odessa, USSR, Sept. 6, 1948; came to U.S., 1990; s. Volf D. and Valentina I. (Kocherzhinskaya) S.; m. Marina Sheikot, Dec. 27, 1975; 1 child, Igor. Degree, Budker Inst. Nuclear Physics, Novosibirsk, USSR, 1974; PhD, Inst. Nuclear Physics, Novosibirsk, USSR. Jr., sr. then chief researcher Budker Inst. Nuclear Physics, Novosibirsk, 1970-89; scientist Brookhaven Nat. Lab., Upton, N.Y., 1990; prof. physics SUNY, Stony Brook, 1990—. Author: The QCD Vacuum, Hadrons and the Superdense Matter, 1986; editorial bd. Phys. Revs., Nuclear Physics. Dept. Energy grantee. Home: 45 William Penn Dr Stony Brook NY 11790 Office: SUNY Physics Dept Stony Brook NY 11794

SHUSHKEWICH, KENNETH WAYNE, structural engineer; b. Winnipeg, Man., Can., Sept. 22, 1952; m. Valdine Cuffe, Sept. 28, 1980. BSCE, U. Man., Winnipeg, 1974; MS in Structural Engring., U. Calif., Berkeley, 1975; PhD in Structural Engring., U. Alta., Edmonton, Can., 1985. Engr. Wardrop and Assocs., Winnipeg, 1974-78, Preconsult Can., Montreal, Que., Can., 1978-80; prof. U. Alta., 1981-85, U. Man., 1985-87; engr. T.Y. Lin Internat., San Francisco, 1988-90, H.J. Degenkolb Assocs., San Francisco, 1990-92, Ben C. Gerwick, Inc., San Francisco, 1993—; assoc. mem. bridge design com., prestressed concrete com. ASCE-Am. Concrete Inst., 1988—. Contbr. to tech. publs. Recipient award for design of Vierendeel truss bridge, Man. Design Inst., 1977. Mem. ASCE, Am. Concrete Inst., Prestressed Concrete Inst., Internat. Assn. Bridge and Structural Engrs. Achievements include design of prestressed concrete segmental bridges, seismic strengthening of San Francisco Ferry Building damaged in Loma Prieta earthquake. Office: Ste 5024 4 Embarcadero Ctr San Francisco CA 94111

SHUTTLEWORTH, ANNE MARGARET, psychiatrist; b. Detroit, Jan. 17, 1931; d. Cornelius Joseph and Alice Catherine (Rice) S.; A.B., Cornell U., 1953, M.D., 1956; m. Joel R. Siegel, Apr. 19, 1959; children: Erika, Peter. Intern, Lenox Hill Hosp., N.Y.C., 1956-57; resident Payne Whitney Clinic-N.Y. Hosp., 1957-60; practice medicine, specializing in psychiatry, Maplewood, N.J., 1960—; cons. Maplewood Sch. System, 1960-62; instr. psychiatry Cornell U. Med. Sch., 1960; mem. Com. to Organize New Sch. Psychology, 1970. Mem. AMA (Physicians Recognition award 1975, 78, 81, 84, 87, 90, 93), Am. Psychiat. Assn., Am. Med. Women's Assn., N.Y. Acad. Scis., Acad. Medicine N.J., Phi Beta Kappa, Phi Kappa Phi. Home: 46 Farbrook Dr Short Hills NJ 07078-3007 Office: 2066 Millburn Ave Maplewood NJ 07040-3715

SHYY, WEI, aerospace, mechanical engineering educator; b. Tainan, Taiwan, China, July 19, 1955; came to U.S. 1979; s. Chiang-Chen and June-Hua (Chao) S.; m. Yuchen Shih; children: Albert, Alice. BS, Tsin-Hua U., Taiwan, 1977; MSE, U. Mich., 1981, PhD, 1982. Postdoctoral rsch. scholar U. Mich., Ann Arbor, 1982-83; rsch. scientist GE Corp. Rsch. and Devel. Ctr., Schenectady, N.Y., 1983-88; faculty mem. of aeronautics and astronautics Nat. Cheng-Kung U., Taiwan, 1987; assoc. prof. aerospace engring., mechanics and engring. sci. U. Fla., Gainesville, 1988-92, prof. aerospace engring., mechanics and engring. sci., 1992—; cons. Ford Motor Co., 1982-83, GE, 1982-83, 88—; lectr. in field. Contbr. numerous articles to profl. jours.; reviewer for NSF, Nat. Sci. Coun. Taiwan, AIAA Jour., ASME Jour. Fluids Engring., Jour. Fluid Mechanics, Numerical Heat Transfer, Jour. Computational Physics, and numerous others; editor: Recent Advances in Computational Fluid Dynamics, 1989. Bd. advisors Am. Biog. Inst., 1991—. Recipient GE Rsch. and Devel. Ctr. 1986 Pubs. award, Chinese Soc. of Mech. Engrs. 1987 Rsch. Paper award, NASA/ASEE 1991 Cert. of Recognition. Mem. ASME (Compustion and Fuel Com. 1984 Hon. Paper award), AIAA, Combustion Inst., NASA Consortium of Computational Fluid Dynamics for Propulsion. Achievements include research in computational fluid dynamics, combustion and propulsion, gravity-induced thermofluid transport processes, materials processing and solidification, microgravity sciences. Office: U Fla Dept Aerospace Engring 231 Aero Bldg Gainesville FL 32611

SIATKOWSKI, RONALD E., chemistry educator; b. Newark, Oct. 30, 1950; s. Edward H. and Eleanora (Rozanski) S. BS in Biology, Pa. State U., 1972; MA in Molecular Biology, Temple U., 1978, PhD in Biophys. Chemistry, 1985. Teaching asst. biology dept. Temple U., Phila., 1976-78, teaching assoc. chemistry dept., 1978-85; coop. rsch. assoc. NRC-Naval Rsch. Lab., Washington, 1985-87; cons., rsch. scientist Geo-Ctrs., Inc., Ft. Washington, Md., 1987-88; asst. prof. chemistry U.S. Naval Acad., Annapolis, Md., 1988—; assoc. supr. grad. record exams. Ednl. Testing Svc., Princton, N.J., 1988—; mem. organizing com. 3rd Internat. Symposium on Molecular Electronic Devices, Arlington, Va., 1988; rsch. scientist Imclone Systems, Inc., N.Y.C., 1989; McNeil Consumer Products Co., Ft. Washington, Pa., 1990; rsch. scientist Purdue Frederick Res. Ctr., Yonkers, N.Y., 1991, 92. Co-editor: Molecular Electronic Devices Proc., 1988; contbr. articles to profl. jours. Capt. USMCR, 1973-75. Recipient fellowships Temple U. (summers) 1980, '81. '82; grantee 1983, '84. Fellow Am. Inst. Chemists; mem. AAAS, Am. Chem. Soc., Am. Phys. Soc., Internat. Union Pure and Applied Chemistry, N.Y. Acad. Scis., Royal Soc. Chemistry, Soc. for Applied Spectroscopy, Sigma Xi. Republican. Home: 1033 Martha Ct Apt 3B Annapolis MD 21403-1764 Office: US Naval Acad Chemistry Dept Annapolis MD 21402

SIBECK, DAVID G., geophysicist. Ph.D. atmospheric sciences, U California, Los Angeles, 1984. Postdoctoral research assoc. Applied Physics Lab., Johns Hopkins U., Baltimore, Md., 1985-87; senior physicist, 1987—. Recipient James B. Macelwane medal, Am. Geophysical Union, 1992. Office: Johns Hopkins U Applied Physics Lab 3400 N Charles St Baltimore MD 21218*

SIBILSKI, PETER JOHN, chemical engineer; b. Passaic, N.J., Mar. 10, 1959; s. Casimir John and Helga (Waller) S.; m. Kandy Leogrande, Feb. 16, 1980 (div. July 1990); children: Kristin, Laura; m. Catherine Pargeans, Apr. 20, 1991. BS in Chem. Engring., N.J. Inst. Tech., 1981. Process engr. Diamond Shamrock, Harrison, N.J., 1981-87; plant mgr. Scher Chems., Clifton, N.J., 1987-89; sr. process engr. Olin Hunt Specialty Products, West Paterson, N.J., 1989-93; sr. process specialist EI Assocs., East Orange, N.J., 1993—. Mem. Am. Inst. Chem. Engrs., Nat. Assn. Corrosion Engrs. Republican. Roman Catholic. Office: EI Assocs 115 Evergreen Pl East Orange NJ 07018

SIBLEY, CHARLES GALD, biologist, educator; b. Fresno, Calif., Aug. 7, 1917; s. Charles Corydon and Ida (Gald) S.; m. Frances Louise Kelly, Feb. 7, 1942; children: Barbara Susanne, Dorothy Ellen, Carol Nadine. A.B., U. Calif.—Berkeley, 1940, Ph.D., 1948; M.A. (hon.), Yale U., 1965. Biologist, USPHS, 1941-42; instr. zoology U. Kans., 1948-49; asst. prof. San Jose (Calif.) State Coll., 1949-53; assoc. prof. ornithology Cornell U., 1953-59, prof. zoology, 1959-65; prof. biology Yale U., 1965-86, prof. emeritus, 1986—, William Robertson Coe prof. ornithology, 1967-86, emeritus, 1986—; dir. div. vertebrate zoology, curator birds Peabody Mus., 1965-86, dir. mus., 1970-76; Dean's prof. sci., prof. biology San Francisco State U., 1986-92; cons. systematic biology, 1963-65; mem. adv. com. biol. medicine NSF, 1968—; exec. com. biol. agr. NRC, 1966-70. Co-author: (with Jon Ahlquist) Phylogeny and Classification of Birds, 1990, (with Burt Monroe) Distribution and Taxonomy of Birds of the World, 1990; mem. editorial bd. Jour. Molecular Evolution, 1983—, Molecular Biology and Evolution, 1986—. Served to lt. USNR, 1943-45. Guggenheim fellow, 1959-60. Fellow AAAS; mem. NAS (Daniel Giraud Elliot medal 1988), Soc. Study Evolution, Am. Soc. Naturalists, Soc. Systematic Biology, Am. Ornithologists' Union (pres. 1986-88, Brewster Meml. medal 1971), Royal Australian Ornithol. Union, Deutsche Ornithol. Gesellschaft, Internat. Ornithol. Congress (sec.-gen. 1962, pres. 20th congress 1986-90). Home: 433 Woodley Pl Santa Rosa CA 95409

SIBLEY, DEBORAH ELLEN THURSTON, immunochemist; b. Clearwater, Fla., Nov. 9, 1957; d. James Paul and Betty Carmen (Hyman) Thurston; m. Timothy Avard Sibley, Aug. 6, 1988; children: Dacey Renee, Nancilyn Katrina. BS in Biology, Newcomb Coll. Tulane U., 1979. Dance instr. dept. phys. edn. Newcomb Coll., New Orleans, 1977-81; rsch. asst. U.S. VA Med. Ctr. Tulane Med. Sch., New Orleans, 1981-82; rsch. assoc. U.S. VA Med. Ctr., New Orleans, 1987-89; data processing geophys. exploration technologist Amoco Oil R&D, New Orleans, 1982-85; rsch. assoc. La. State U. Eye Ctr., Med. Ctr., New Orleans, 1985-87; rsch. biochemist, immunochemist Universal Sensors, Inc., Metairie, La., 1990—. Choreographer, dancer Komenka Ethnic Dance Ensemble, New Orleans, 1982-86; active Spina Bifida Assn., New Orleans, 1989-92. Democrat. Baptist. Achievements include patent pending for natural marine adhesive-based immunosensor; development of immunoassay for Escherichia coli with detection of 100 cells/mL. Home: 244 Jeffer Dr Waggaman LA 70094-2190 Office: Universal Sensors Inc 5258 Veterans Blvd Ste D Metairie LA 70006

SICHERL, PAVLE, economics educator, consultant; b. Ljubljana, Slovenia, Yugoslavia, Feb. 16, 1935; s. Janko and Ana (Debeljak) S.; m. Jana Milava Primc, Aug. 11, 1962; children: Igor, Borut. Diploma in Econs., U. Ljubljana, 1960; MA, Williams Coll., Williamstown, Mass., 1962; PhD, U. Ljubljana, 1967. Economist Fed. Planning Bur., Belgrade, Yugoslavia, 1960-62, Yugoslav Inst. Econ. Rsch., Belgrade, 1962-66; Hallsworth rsch. fellow U. Manchester, Eng., 1966-67; dep. dir. Yugoslav Inst. Econ. Rsch., Belgrade, 1967-68, acting dir., 1968-69; macroecon. advisor Harvard Inst. Internat. Devel., Addis Ababa, Ethiopia, 1970-74; prof. econs. U. Ljubljana Law Sch., 1975—; vis. prof. Williams Coll., Williamstown, 1980; vis. fellow Yale U., New Haven, 1986-87; vis. scholar Centre for the Study of Public Policy, Strathclyde U., Glasgow and Centre for Econ. Performance, London Sch. Econs., 1992-93, London sch. econs. London Inst. World Econs., Kiel, 1992-93; mem. Fed. Coun. Econ. Advisors, Belgrade, 1980-84; cons. UN, World Bank, Washington, 1987, 92, OECD, 1991; chmn. Subcom. Commn. Econ. Reform, 1988. Author: Capital as a Factor of Economic Growth, 1971, Personality of Public Enterprise, 1981, Methods of Measuring Disparity Between Men and Women, 1989, Slovenia Now, 1990, Integrating Comparisons Across Time and Space, 1993. Recipient Sr. Fulbright rsch. award Fulbright Commn., 1986. Mem. Internat. Assn. Rsch. Income and Wealth, Am. Econ. Assn., Slovenian Statis. Soc., Slovenian Econ. Assn. Avocation: chess. Home: Brajnikova 19, 61000 Ljubljana Slovenia Office: U Ljubljana Faculty Law, Kongresni trg 12, 61000 Ljubljana Slovenia

SICHEWSKI, VERNON ROGER, physician; b. Winnipeg, Man., Can., Dec. 10, 1942; came to U.S., 1980; s. Nicholas and Helen (Sabanski) S. BS, U. Man., 1963; MD, Cairo U., 1979. Diploamte Am. Bd. Emergency Medicine. Resident Charity Hosp. La., New Orleans, 1980-83, Bellevue Hosp., N.Y.C., 1980-83; pvt. practice Broward Gen. Med. Ctr., Ft. Lauderdale, Fla., 1983-86, Trauma Care Assocs., North Miami, Fla., 1986—; flight physician Nat. Jets, Ft. Lauderdale, 1986—; attending physician trauma unit Jackson Meml. Hosp. U. Miami, 1989—. Flight lt. RCAF, 1963-74. Fellow Am. Coll. Emergency Physicians; mem. AMA, So. Med. Assn. Republican. Roman Catholic. Avocations: stamp collecting, hunting, fishing, antiques. Home: 1108-2841 N Ocean Blvd Fort Lauderdale FL 33308-2323 Office: Trauma Care Assocs 1175 NE 125th St Ste 612 Miami FL 33161-5013

SIDDAYAO, CORAZON MORALES, economist, educator; b. Manila, July 26, 1932; came to U.S., 1968; d. Crispulo S. and Catalina T. (Morales) S. Cert. in elem. teaching, Philippine Normal Coll., 1951; BBA, U. East, Manila, 1962; MA in Econs., George Washington U., 1971, MPhil, PhD, 1975. Tchr. pub. schs. Manila, 1951-53; asst. pensions officer IMF, Washington, 1968-71; cons. economist Washington, 1971-75; rsch. assoc. Policy Studies in Sci. and Tech. George Washington U., Washington, 1971-72, teaching fellow dept. econs., 1972-75; natural gas specialist U.S. Fed. Energy Adminstrn., Washington, 1974-75; sr. rsch. economist, assoc. prof. Inst. S.E.A. Studies, Singapore, 1975-78; sr. rsch. fellow energy/economist East-West Ctr., 1978-81, acad. staff coord. energy and industrialization, 1981-86; vis. fellow London Sch. Econ., 1984-85; sr. energy economist in charge energy program Econ. Devel. Inst., World Bank, Washington, 1986—; vis. prof. U. Montpellier, France, 1992; affiliate prof. econs. U. Hawaii, 1979—; cons. internat. orgns.; vis. prof. econs. U. Philippine, intermittent 1989—. Author: Increasing the Supply of Medical Personnel, 1973, The Offshore Petroleum Resources of Southeast Asia: Some Potential Conflicts and Related Economic Factors, 1978, Round Table Discussion on Asian and Multinational Corporations, 1978, The Supply of Petroleum Reserves in Southeast Asia: Economic Implications of Evolving Property Rights Arrangements, 1980, Critical Energy Issues in Asia and the Pacific: the Next Twenty Years, 1982, Criteria for Energy Pricing Policy, 1985, Energy Demand and Economic Growth, 1986; editor: Energy Policy and Planning series, 1990-92, Energy Investments and the Environment, 1993; co-editor: Investissements Energetiques et Environnement; contbr. chpts. to books, articles to profl. jours. Grantee in field. Mem. Am. Econ. Assn., Internat. Assn. Energy Economists, Pan Am. Clipper Club, Alliance Francaise, Omicron Delta Epsilon. Roman Catholic.

SIDDIQUI, MAQBOOL AHMAD, engineering consultant and executive; b. Sangla Hill, Punjab, Pakistan, Sept. 18, 1941; s. Pir Bakhash and Sahib Nisa Siddiqui; m. Ceyla, Oct. 1964 (div.); m. Robina Anjum, Aug. 22, 1985; children: Sahib Nisa, Ertan Maqbool, Khalid Maqbool. BSc, Punjab U., Lahore, Pakistan, 1961; postgrad. diploma, Bettersea Coll., London, 1964; MPhil, London U., 1976. Registered profl. engr., Pakisan, U.K. Civil svc. Army Tech. Liaison Office Pakistan Embassy, London, 1962-63; project engr. Film Cooling Towers, Ltd., Richmond-London, 1964-66; chem. engr. Atomic Power Constrn., Ltd., Sutton-London, 1966-68; process engr. Badger Ltd./Badger N.V., London and The Hague, 1968-73; cons. process engring. Pritchard Rhodes, Ltd., London, 1973; lead process engr. Lummus Crest Co./Monsanto Co., London and St. Louis, 1973-75; mgr. engring. Alarko A.S., Istanbul, Turkey, 1975-76, dep. mng. dir., 1976-83; chmn. Pirsons Chem. Engring. (Pvt) Ltd., Multan, Pakistan, 1984—. Contbr. articles to profl. jours. Chmn. planning and policies Pakistan Saraiki Party, Multan, 1990. Mem. Inst. Chem. Engrs. (Pakistan, London, U.S., Turkey), Engring. Coun. London (chartered engr.), Pakistan Engring. Coun. (corp. mem.), SEC, Pakistan-Turkish Soc., Svcs. Club. Mem. Pakistan Saraiki Party. Muslim. Avocations: photography, research and development, stamp collection, travel. Home: 321A Sher Shah Rd, Multan Punjab, Pakistan Office: Pirsons Chem Engring Ltd, Siddiqui Lodge Sher Shah Rd, Multan Punjab, Pakistan

SIDLE, ROY C., research hydrologist; b. Quakertown, Pa., Oct. 31, 1948; s. Carl G. and Isabel (Ziegler) S.; children: Shelley, Bradley. BS, U. Ariz., 1970, MS, 1972; PhD, Pa. State U., 1976. Hydrologist Wright Water Engrs., Denver, 1972; rsch. asst. Pa. State U., University Park, 1972-76; rsch. soil scientist USDA-ARS, Morgantown, W.Va., 1976-78; watershed extension specialist Oreg. State U., Corvallis, 1978-80; rsch. hydrologist Forest Svc., PNW Rsch. Sta., USDA, Juneau, 1980-86; project leader (hydrologist) Forest Svc., INT Rsch. Sta., USDA, Logan, Utah, 1986—; adj. prof. Utah State U., Logan, 1986—; symposium chmn. Am. Soc. Agronomy, Anaheim, Calif. and Las Vegas, Nev., 1982; vis. fellow Japanese Forestry and Forest Products Rsch. Inst., 1991. Sr. author: Hillslope Stability and Land Use, 1985; assoc. editor Jour. of Environ. Quality, 1990—; contbr. numerous articles to profl. jours. Grantee, Water Resources Rsch. Inst., Oreg. State U., 1979, USDA-Forest Svc., Portland, Oreg., 1979, Am. Geophys. Union, 1982, Am. Philos. Soc., 1982; recipient Cert. of Merit, Chief, USDA-Forest Svc., 1989. Mem. Am. Soc. Agronomy, Soil Sci. Soc. Am., Am. Geophys. Union, Am. Water Resources Assn., Internat. Soc. Soil Sci., Sigma Xi, Gamma Sigma Delta. Avocations: skiing, weight lifting, baseball card collecting, antique collecting, biking. Office: USDA Forest Svc INT Sta 860 N 1200 E Logan UT 84321-3699

SIDMAN, KENNETH ROBERT, chemical engineer; b. Cambridge, Mass., Apr. 22, 1945; s. Marshall Benjamin and Evelyn (Persky) S.; m. Margaret Schnaper, June 9, 1968; children: Deborah, Pamela. BSCE, MIT, 1967, MSCE, 1968. Staff engr. Dynatech R & D Co., Cambridge, Mass., 1968-71; mgr. applied sci. unit Arthur D. Little Inc., Cambridge, Mass., 1971-83; mgr. advanced market planning Norton Co. Performance Plastics Div., Wayne, N.J., 1984-86, dir. mktg. & bus. devel., 1986-91, v.p. mktg., 1992—. Contbr. articles to profl. jours. Mem. AICE, Am. Chem. Soc., Soc. Advancement Material and Process Engring, Sigma Xi. Achievements include 4 patents and 3 cert. of recognition from NASA. Office: Norton Performance Plastics 150 Dey Rd Wayne NJ 07470

SIEBERT, KARL JOSEPH, food science educator, consultant; b. Harrisburg, Pa., Oct. 29, 1945; s. Christian Ludwig and Katharine (Springer) S.; m. Sui Ti Atienza, Mar. 14, 1970; children: Trina, Sabrina. BS in Biochemistry, Pa. State U., 1967, MS in Biochemistry, 1968, PhD in Biochemistry, 1970. Chemist Applied Sci. Labs., State College, Pa., 1968-70; rsch. assoc. Stroh Brewery Co., Detroit, 1971, head R & D sect., 1971-73, mgr. R & D lab., 1973-82, dir. rsch., 1982-90; v.p. Strohtech, Detroit, 1986-90; prof., chmn. Dept. Food Sci. and Tech. Cornell U., 1990—. Contbr. articles to profl. jours. Bd. visitors Oakland U. Biology Dept., Rochester, Mich., 1985-89; bd. dirs. Cornell Rsch. Found., 1990—, Geneva Concerts Inc., 1991—. Capt. USAR, 1967-75. Recipient Presdl. award Master Brewers Assn., 1986, 90. Fellow NSF; mem. Am. Chem. Soc. (divsn. agrl. and food chemistry), Master Brewer Assn. Ams., Am. Soc. Brewing Chemists (chmn. tech. com. 1986-88, mem. edit. bd. 1983-91), Inst. Food Technologists, Internat. Chemometrics Soc. (N.Am. chpt.). Avocations: computers, electronics. Home: 9 Parkway St Geneva NY 14456-9765 Office: NY State Agrl Expt Sta Cornell U Dept Food Sci Geneva NY 14456

SIEBRAND, WILLEM, theoretical chemist, science editor; b. Ijsselmuiden, Overijsel, Netherlands, Aug. 12, 1932; came to Can., 1963; s. Gerrit and Pietje (Brienne) S.; m. Elisabeth Kroon, Dec. 20, 1962; children—Barbara, Saskia. B.Sc., U. Amsterdam, 1956, M.Sc., 1960, Ph.D., 1963. Asst. research officer Nat. Research Council Can., Ottawa, Ont., 1963-66, assoc. research officer, 1966-72, sr research officer, 1972-83, prin., 1983—. Editor Can. Jour. Chemistry, 1985—; assoc. editor Jour. Chem. Physics, 1977-79; mem. editorial bd. Chem. Physics Letters, Amsterdam, 1971—, Jour. Raman Spectroscopy, London, 1974—; author numerous articles. Trustee E.W.R. Steacie meml. fund. Fellow Royal Soc. Can. (Henry Marshall Tory medal 1991); mem. Am. Phys. Soc., (corr. mem.) Royal Dutch Soc. Arts and Scis., Can. Inst. Chemistry. Home: 752 Lonsdale Rd, Ottawa, ON Canada K1K 0K2 Office: NRC Canada, Steacie Inst for Molecular Scis, Ottawa, ON Canada K1A 0R6*

SIEGAL, BURTON LEE, product designer, consultant; b. Chgo., Sept. 27, 1931; s. Norman A. and Sylvia (Vitz) S.; m. Rita Goran, Apr. 11, 1954; children: Norman, Laurence Scott. BS in Mech. Engring., U. Ill., 1953. Torpedo designer U.S. Naval Ordnance, Forest Park, Ill., 1953-54; chief engr. Gen. Aluminum Corp., Chgo., 1954-55; product designer Chgo. Aerial Industries, Melrose Park, Ill., 1955-58; chief designer Emil J. Paidar Co., Chgo., 1958-59; founder, pres. Budd Engring. Corp., Chgo., 1959—; dir. Dur-A-Case Corp., Chgo.; design cons. to numerous corps. Holder more than 90 patents in over 20 fields including multimemory for power seats and electrified office panel systems; contbr. articles to tech. publs. Mem. math., sci. and English adv. bds. Niles Twp. High Schs., Skokie, Ill., 1975-79; electronic cons. Chgo. Police Dept., 1964. Winner, Internat. Extrusion Design Competition, 1975; nominated Presdl. Medal Technology Sen. Paul Simon and Rep. Dan Rostenkowski, 1986; named Inventor of Yr. Patent Law Assn. Chgo., 1986. Mem. No. Ill. Indsl. Assn., Inventor's Coun., Soc. Automotive Engrs., Pres.'s Assn. Ill., Ill. Mfg. Assn. Office: 8707 Skokie Blvd Skokie IL 60077-2269

SIEGBAHN, KAI MANNE BÖRJE, physicist, educator; b. Lund, Sweden, Apr. 20, 1918; s. Manne and Karin (Hogbom) S.; m. Anna-Brita Rhedin, May 23, 1944; children: Per, Hans, Nils. B.S., 1939, Licentiate of Philosophy, 1942; Ph.D., U. Uppsala, 1944; D.Sc. honoris causa, U. Durham, 1972, U. Basel, 1980, U. Liege, 1980, Upsala Coll., 1982, U. Sussex, 1983. Research assoc. Nobel Inst. Physics, 1942-51; prof. physics Royal Inst. Tech., Stockholm, 1951-54; prof., head physics dept. U. Uppsala, Sweden, 1954-84. Author: Beta and Gamma-Ray Spectroscopy, 1955; Alpha, Beta and Gamma-Ray Spectroscopy, 1965; ESCA-Atomic, Molecular and Solid State Structure Studies by Means of Electron Spectroscopy, 1967; ESCA Applied to Free Molecules, 1969. Recipient Lindblom Prize, 1945, Bjorken Prize, 1955, 77, Celsius medal, 1962, Sixten Heyman award, 1971, Harrison Howe award, 1973, Maurice F. Hasler award, 1975, Charles Frederick Chandler medal, 1976, Torbern Bergman medal, 1979, Nobel Prize, 1981, Pitts. award spectroscopy, 1982, Röntgen medal, 1985, Fiuggi award, 1986, Humboldt award, 1986, Premio Castiglione Di Sicilia, 1990. Mem. Royal Swedish Acad. Sci., Royal Swedish Acad. Engring. Scis., Royal Soc. Sci., Royal Acad. Arts and Sci. Uppsala, European Acad. Arts, Scis. and Humanities, Academia Europaea, Royal Physiographical Soc. Lund, Societas Scientairum Fennica, Norwegian Acad. Sci., Royal Norwegian Soc. Scis. and Letters, Am. Acad. Arts and Scis. (hon.), Comite des Poids et Mesures, Internat. Union Pure and Applied Physics (pres. 1981-84), Pontifical Acad. Scis., NAS (fgn. assoc.).

SIEGEL, JACK S., federal official; b. Long Branch, N.J., June 2, 1946; s. Joseph and Idene (Boss) S.; m. Shari Diane Goodman, June 12, 1971; children: Erica Robyn, Adam Michael. BSChemE, Worcester Poly. Inst., 1968; postgrad., George Washington U. Chem. engr. Dept. of the Navy, Indian Head, Md., 1968-71; supervisory chem. engr. EPA, Washington, 1971-76; asst. dir. for environ. planning and assessment U.S. Energy R & D Adminstrn., Washington, 1976-77; dir. office of environ. policy Dept. of Energy, Washington, 1977-83, dep. dir. for coal utilization, advanced combustion and gasification, 1983-84, dep. asst. sec. for coal tech., 1984—; fed. rep. to Nat. Coal Assn., 1986—; dir. tech. panel Innovative Control Tech. Adv. Panel, Washington, 1987-89; speaker in field. Mem. Am. Inst. Chem. Engrs., Washington Coal Club. Jewish. Avocations: exercise, racquetball. Office: Dept Energy Coal Tech 1000 Independence Ave SE Washington DC 20585-0001

SIEGEL, MICHAEL ELLIOT, nuclear medicine physician, educator; b. N.Y.C., May 13, 1942; s. Benjamin and Rose (Gilbert) S.; m. Marsha Rose Snower, Mar. 20, 1966; children: Herrick Jove, Meridith Ann. A.B., Cornell U., 1964; M.D., Chgo. Med. Sch., 1968. Diplomate Nat. Bd. Med. Examiners. Intern Cedars-Sinai Med. Ctr., L.A., 1968-69; resident in radiology, 1969-70; NIH fellow in radiology Temple U. Med. Ctr., Phila., 1970-71; NIH fellow in nuclear medicine Johns Hopkins U. Sch. Medicine, Balt., 1971-73, asst. prof. radiology, 1972-76; assoc. prof. radiology, medicine U. So. Calif., L.A., 1976—; prof. radiology, 1989—; dir. div. nuclear medicine, 1982—; dir. Sch. Nuclear Medicine, L.A. County-U. So. Calif. Med. Ctr., 1976—; dir. div. nuclear medicine Kenneth Norris Cancer Hosp. and Rsch. Ctr., L.A., 1983—; dir. nuclear medicine Orthopaedic Hosp., L.A., 1981—, dir. dept. nuclear medicine Intercommunity Hosp., Covina, Calif., 1981—; cons. dept. nuclear medicine Rancho Los Amigos Hosp., Downey, Calif., 1976—. Author: Textbook of Nuclear Medicine, 1978, Vascular Surgery, 1983, 88, and numerous others textbooks; editor: Nuclear Cardiology, 1981, Vascular Disease: Nuclear Medicine, 1983. Mem. Maple Ctr., Beverly Hills. Served as maj. USAF, 1974-76. Recipient Outstanding Alumnus award Chgo. Med. Sch., 1991. Fellow Am. Coll. Nuclear Medicine (sci. investigator 1974, 76, nominations com. 1980, program com. 1983, bd. trustees 1993, disting. fellow, 1993, bd. reps., 1993—); mem. Soc. Nuclear Medicine (sci. exhbn. com. 1978-79, program com. 1979-80, Silver medal 1975), Calif. Med. Assn. (sci. adv. bd. 1987—), Radiol. Soc. N.Am., Soc. Nuclear Magnetic Resonance Imaging, Alpha Omega Alpha. Lodge: Friars So. Calif. Research on devel. of nuclear medicine techniques to: evaluate cardiovascular disease and diagnose and treat cancer, clinical utilization of video digital displays in nuclear medicine development; inventor pneumatic radiologic pressure system. Office: U So Calif Med Ctr PO Box 693 1200 N State St Los Angeles CA 90033

SIEGEL, RICHARD ALLEN, economist; b. Chgo., Mar. 11, 1927; s. Mandel Irving and Mary Marsha (Shulman) S.; m. Shirley Platin, Dec. 17, 1950 (dec. 1980); children: Joel, Robert, Peter; m. Rosalyn Sandra Miller, June 28, 1981. AB, UCLA, 1953, MBA, 1959, PhD, 1961. Asst. prof. econs. SUNY, Buffalo, 1962-64; economist Bank of Am., San Francisco, 1964-66, Calif. Dept. Fin., Sacramento, 1967-68, Arthur D. Little, Inc., Cambridge, Mass., 1968-70; pres., economist Richard Siegel Assocs., Boston, 1970-79; prin., economist Econ. Research Assocs., Boston, 1979-83; pres., economist Applied Econs., Inc., Boston, 1983—. Contbr. articles to profl. jours. Served with USN, 1945-46, PTO. Jewish. Club: Appalachian Mountain (chmn. Boston chpt. 1983-84). Avocations: sailing, skiing, European history. Office: Applied Econs Inc 2 Liberty Sq Boston MA 02109-4867

SIEGEL, RICHARD STEVEN, water resource specialist; b. N.Y.C., Nov. 23, 1955; s. Chester and Helen (Lefkowitz) S.; m. Janet Maria Wallace, Apr. 7, 1984; children: Jennifer, Paige. BS, Ariz. State U., 1980, MS, 1985. Lab. technician USDA, Phoenix, 1979-81; lab. asst. U. Ariz., Mesa, 1981-83; rsch. asst. Ariz. State U., Tempe, 1983-85; water resource specialist Ariz. Dept. Water Resources, Phoenix, 1985-88; sr. analyst Salt River Project, Tempe, 1988—; team mem. Modified Roosevelt Dam Operating Agreement, 1993, Granite Reef Underground Storage Project, 1993. Mem. Salt River Project Polit. Involvement Com., Tempe, 1988—, Scottsdale Leadership, 1993. Mem. ASCE, Ariz. Hydrological Soc. Office: Salt River Project PO Box 52025 Phoenix AZ 85072-2025

SIEGER, EDWARD REGIS, software engineer; b. Pitts., Apr. 20, 1957; s. Edward Regis Sr. and Eleanor (McConville) S.; m. Helen Marquerite Zang, Apr. 26, 1986; children: Caroline Marguerite, David Edward, Emily Catherine, Daniel Thomas. BS in Math., Carnegie Mellon U., 1978. Assoc. engr. WABCO, Pitts., 1978-79; engr. rshc. and devel. Westinghouse, Pitts., 1979-80, sr. engr. productivity and quality, 1980-87; prin. engr. AEG Automation Systems Corp., Pitts., 1987-91, mgr., 1991--. Vice chmn. Jackson Twp. Water Auth., 1991—. Republican. Roman Catholic. Achievements include development of artificial intelligence based customer order engineering systems and development of AEG's Unicell XL FMS software product. Office: AEG Automation Systems Corp 1010 McKee Rd Oakdale PA 15071

SIEGLAFF, CHARLES LEWIS, chemist; b. Waterloo, Iowa, Sept. 30, 1927; s. Walter Frank and Dorthy (Mergy) S.; m. Donna J. Rippen, Oct. 20, 1985 (div. June 1989); children: Diane, Susan. MS, U. Cin., 1953; PhD, U. Iowa, 1956. Sr. researcher Dow Chem., Midland, Mich., 1956-60; rsch. fellow Diamond Shamrock, Painesville, Ohio, 1960-84; chmn. Gordon Conf., Santa Barbara, Calif., 1979. Cpl. U.S. Army Air Corps, 1945-46. Office: DMC 4399 Hamann Pkwy Willoughby OH 44094

SIEGLER, MARK, internist, educator; b. N.Y.C., June 20, 1941; s. Abraham J. and Florence (Sternlieb) S.; m. Anna Elizabeth Hollinger, June 4, 1967; children:Dillan, Alison, Richard, Jessica. AB with honors, Princeton U., 1963; MD, U. Chgo., 1967. Diplomate Am. Bd. Internal Medicine. Resident, chief resident internal medicine U. Chgo., 1967-71; hon. sr. registrar in medicine Royal Postgrad. Med. Sch., London, 1971-72; asst. prof. medicine U. Chgo., 1972-78, assoc. prof. medicine, 1979-85, acting dir. div. gen. internal medicine, 1983-85, dir. Ctr. Clin. Med. Ethics, 1984—, prof. medicine, 1985—, dir. nat. leadership tng. program in clin. med. ethics, 1986—; vis. asst. prof. medicine U. Wis., Madison, 1977; vis. assoc. prof. medicine U. Va., Charlottesville, 1981-82. Co-author: Clinical Ethics, 1981, 2d edit., 1986, 3d edit., 1992, An Annotated Bibliography of Medical Ethics, 1988, Institutional Protocols for Decisions about Life-Sustaining Treatment, 1988; co-editor: Changing Values in Medicine, 1985, Medical Innovations and Bad Outcomes, 1987; editorial bd. Am. Jour. Medicine, Archives Internal Medicine, 1979-90, Bibliography of Bioethics, Jour. Med. Philosophy, 1978-89, Humane Medicine, Jour. Clin. Ethics; contbr. articles to profl. publs. Bd. govs. Josephson Inst. for Advancement of Ethics, L.A., 1986-92. Grantee Andrew W. Mellon Found., Henry J. Kaiser Family Found., Pew Charitable Trusts, Field Found. Ill., Ira De Camp Found.; Phi Beta Kappa vis. scholar, 1991-92, others. Fellow ACP (human rights com., ethics com. 1985-90), Hastings Ctr.; mem. Assn. Am. Physicians, Am. Geriatrics Soc. (ethics com. 1988—), Chgo. Clin. Ethics Program (pres. 1989-90). Office: U Chgo Ctr Clin Med Ethics MC 6098 5841 S Maryland Ave Chicago IL 60637-1470

SIEGLER, MELODY VICTORIA STEPHANIE, neuroscientist; b. L.A., Dec. 19, 1948; d. Ralph Henry and Margaret Don (Bray) S. BA in Zoology, Pomona Coll., 1969; PhD in Biology, U. Calif., Santa Cruz, 1975; MA, Cambridge U., 1975. Rsch. scientist Cambridge (England) U., 1975-85; asst. prof. Emory U., Atlanta, 1986-92, assoc. prof., 1992—. Contbr. articles to Sci., Nature, Jour. Neurosci., Jour. Comparative Neurology, Jour. Neurophysiology. Fellow Grass Found., 1975, Tucker-Price fellow Girton Coll., 1975; grantee Med. Rsch. Coun., NIH, NSF, Whitehall Found., 1982—. Mem. Internat. Soc. for Neuroethology (councillor 1992—), Soc. for Neurosci. (councillor Atlanta chpt. 1987-91), AAAS, Sigma Xi. Achievements include research in structure of nonspiking neurons and their function in sensation and movement, local neuronal circuits underlying tactile and proprioceptive reflexes, mechanisms underlying phenotypic diversity in neuronal lineages. Office: Emory U Dept Biology 1510 Clifton Rd Atlanta GA 30322

SIEGMAN, ANTHONY EDWARD, electrical engineer, educator; b. Detroit, Nov. 23, 1931; s. Orra Leslie and Helen Salome (Winnie) S.; (married). AB summa cum laude, Harvard U., 1952; MS, UCLA, 1954; PhD, Stanford U., 1957. Mem. faculty Stanford (Calif.) U., 1957—, assoc. prof. elec. engring., 1960-65, prof., 1965—; dir. Edward L. Ginzton Lab., 1978-83; cons. Lawrence Livermore Labs., Coherent Inc., GTE; mem. Air Force Sci. Adv. Bd.; vis. prof. Harvard U., 1965. Author: Microwave Solid State Masers, 1964, An Introduction to Lasers and Masers, 1970, Lasers, 1986; contbr. over 200 articles to profl. jours. Recipient Schawlow award Laser Inst. Am., 1991; Guggenheim fellow IBM Rsch. Lab., Zurich, 1969-70; Alexander von Humboldt Found. sr. scientist Max Planck Inst. Quantum Optics, Garching, Fed. Republic Germany, 1984-85. Fellow AAAS, IEEE (W.R.G. Baker award 1971, J.J. Ebers award 1977), APS, LIA, Optical Soc. Am. (R.W. Wood prize 1980), IEEE Laser Electro-Optics Soc. (Quantum Electronics award 1989), Am. Acad. Arts and Scis.; mem. NAS, NAE, AAUP, Phi Beta Kappa, Sigma Xi. Patentee microwave and optical devices and lasers, including the unstable optical resonator. Office: Stanford U Ginzton Lab Stanford CA 94305-4085

SIEKMANN, JÖRG HANS, computer science educator; b. Heidelberg, Germany, Aug. 5, 1941; s. Hans and Helene (Hemme) S.; married, 1983; 1 child, Helen Maureen. Abitur, Braunschweig (Germany) Coll., 1969; Vordiplom in Math./Physics, Göttingen U., 1972; MSc, U. Essex, Eng., 1973, PhD, 1976. Rsch. asst. U. Karlsruhe, Germany, 1976-83, asst. prof., 1983-91; prof. computer sci. and artificial intelligence U. Kaiserslautern, Germany, 1986-91, U. des Saarlandes, Saarbrucken, Germany, 1991—; prof. computer sci. German Rsch. Inst. for Artificial Intelligence, Saarbrucken, 1991—; active in founding artificial intelligence in Germany. Contbr. articles to artificial intelligence, deduction systems and unification theory to profl. jours. Office: U Saarlandes DFKI, Stuhlsatzenhausweg 3, 6600 Saarbrücken Germany

SIELOFF, CHRISTINA LYNE, nurse; b. Detroit, Dec. 6; d. Louis F. and Doris C. (Bakewell) S. BSN, Wayne State U., 1970, MS in Nursing, 1977. Clin. specialist Henry Ford Hosp., Detroit, 1979-81; extension program coord. Wayne State U., Detroit, 1981-82, faculty, 1982-83; asst. dir. nursing Kingswood Hosp., Ferndale, Mich., 1983-87, specialist nursing adminstrn., 1987-90; clin. dir. Oakland Gen. Hosp., Madison Heights, Mich., 1990-92; supr. nurse Wayne Ctr., Detroit, 1992-93; instr. Oakland U. Sch. Nursing, Rochester, Mich., 1993—; reviewer Aspen Systems, Rockville, Md., 1985—, Jour. of Child Adolscent Nursing, 1988—. Author: Imogene King, 1992, Memory Bank for Medications, 1986, (1 chpt.) Future Practice Environment and the Nurse Administrator, 1991; co-editor: Psychiatric and Mental Health Nursing with Children and Adolscents, 1992 (Am Jour. Nursing Book of Yr., 1993). Pres. Kingstowne Manor Condo Assn., Commerce Twp., Mich., 1990; sec. Holy Spirit Luth. Ch., West Bloomfield, Mich., 1992—. Mem. ANA (councilor psychiat. and mental health nursing), Mich. Nurses Assn. (Outstanding Psychiat. and Mental Health Nursing, chair elect 1992—). Independent.

SIEMIATYCKI, JACK, epidemiologist, biostatistician, educator; b. Innsbruck, Austria, Dec. 31, 1946; emigrated to Canada, 1948; s. Marek and Tauba S.; m. Lesley Richardson-Young, Nov. 6, 1982; children: Emma, Kate. B.S., McGill U., 1967, M.S., 1971, Ph.D., 1976. Research fellow

McGill U., Montreal, Que., Can., 1967-71; research dir. Pointe St. Charles Clinic, Montreal, 1971-72; cons. Internat. Agy. Research Cancer, Lyon, France, 1978, 83; asst. prof. dept. epidemiology and health Inst. Armand-Frappier, Laval, Que., 1978-79, assoc. prof., 1979-83, prof., 1983—; mem. sci. coun. health com. Internat. Joint Commn., Windsor, Ont., 1982-89; chair rev. panel Nat. Health Rsch. and Devel. Programme of Can., 1989—; mem. sci. coun. task force on reproductive health Govt. N.B., 1983; mem. panel on priority substances Can. Environ. Protection Act, 1988; mem. panel on human health effects of electromagnetic fields Health Effects Inst., 1990. Author of book on occupational causes of cancer; assoc. editor Am. Jour. Epidemiology, 1989—; contbr. articles to sci. publs. Research grantee in cancer and environment, health survey methods, juvenile onset diabetes. Mem. Am. Pub. Health Assn., Soc. Epidemiologic Research, Internat. Epidemiology Assn., Que. Pub. Health Assn. Home: 106 Columbia St, Westmount, PQ Canada H3Z 2C3 Office: Institut Armand-Frappier, 531 des Prairies Blvd, Ville de Laval, PQ Canada H7V 1B7

SIENER, JOSEPH FRANK, utilities supervisor; b. North Vernon, Ind., Nov. 25, 1938; s. Arthur Edward and Helen Louise (Federer) S.; m. Rita Sue Rhinehart, May 9, 1959; children: Theresa Thompson, Andrew J., Marjorie Secrest. Grad., N. Vernon High Sch., 1957. Supr. N. Vernon Wastewater Plant, Ind., 1960-73; utilities supt. City N. Vernon, 1973--; mem. natural resources com. State Ind., Indpls., 1987--. Pres. N. Vernon Fire Dept., 1990-91. Recipient Sagamore of Wabash award Gov. Ind., 1986, 88; named Fire Man of Yr. N.Vernon Fire Dept., 1987. Mem. Ind. Water Pollution Control Assn., Am. Water Works Assn. (pres south eastern dist. 1978), Ind. Vol. Fireman's Assn., KC, Jaycees (pres., sen. 1967, 75). Roman Catholic. Home: RR #2 Box 41 Bl 1010 W Private Rd 25 S North Vernon IN 47265 Office: City North Vernon 275 Main St North Vernon IN 47265

SIENKIEWICZ, FRANK FREDERICK, observatory curator; b. Westfield, Mass., May 1, 1965; s. Willard Ernest and Sally Ann (Christianson) S.; m. Sandie Christine Kirby, Oct. 10, 1992. BA in Astronomy, Boston U., 1988. Curator observatory Boston U., 1988—. Grantee NSF 1991. Mem. Am. Astron. Soc. Democrat. Lutheran. Home: 25 Appleton St Malden MA 02148 Office: Boston U Astronomy Dept 725 Commonwealth Ave Boston MA 02215

SIERACKI, MICHAEL EDWARD, biological oceanographer; b. Columbia, S.C., Apr. 28, 1955; s. Edward Robert and Isa Irene (Jensen) S.; m. Pamela Ann Staveley, June 6, 1981; children: Rita Elizabeth, Hannah Margaret. BA in Biol. Sci., U. Del., 1977; MS in Microbiology, U. R.I., 1980, PhD in Biol. Oceanography, 1985. Asst. prof. Va. Inst. Marine Sci./Coll. of William and Mary, Gloucester Point, Va., 1985-91, assoc. prof., 1991; rsch. scientist Bigelow Lab. for Ocean Scis., West Boothbay Harbor, Maine, 1991—; lectr. in field various confs., workshops; dir. J.J. MacIsaac Individual Particle Analysis Facility, Bigelow Lab. for Ocean Scis., 1991—. Contbr. articles to profl. jours. NSF grantee, 1988-90, 1989-91, Va. Sea grantee, 1990-92, NSF-Sml. Bus. Innovation Rsch. grantee, 1991, NSF-Biol. Oceanography grantee, 1991-93, NSF-U.S. JGOFS grantee, 1991—. Mem. AAAS, Internat. Soc. Analytical Cytology, Am. Soc. Limnology and Oceanography, Oceanography Soc., Sigma Xi. Achievements include six major oceanographic cruises in North Atlantic and Gulf of Maine; developed advanced image analysis techniques for automated measurement of microbial biomass in natural samples. Office: Bigelow Lab for Ocean Scis McKown Point Rd West Boothbay Harbor ME 04575

SIERY, RAYMOND ALEXANDER, laboratory administrator; b. Passaic, N.J., Nov. 26, 1951; s. Alexander and Violet (Hejda) S.; m. Pamela Landsberg, Sept. 27, 1975; children: Alexis, Iain, Nathan. BA, Newark State U., 1974. Lab. mgr. Barringer Labs., Inc., Golden, Colo., 1977-85, v.p., 1985-86; lab. mgr. Battelle Columbus (Ohio) Div., 1986-88, Twin City Testing, St. Paul, 1988-90; sr. sect. mgr. Roy F. Weston, Inc., Lionville, Pa., 1990—. Mem. Am. Chem. Soc., Soc. Applied Spectropists. Republican. Office: Roy F Weston Inc 208 Welsh Pool Rd West Chester PA 19341

SIESS, ALFRED ALBERT, JR., engineering executive, management consultant; b. Bklyn., Aug. 16, 1935; s. Alfred Albert and Matilda Helen (Suttmeier) S.; m. Gale Murray Scholes, Dec. 17, 1966; children: Matthew Alan, Daniel Adam. BCE, Ga. Inst. Tech., 1956; postgrad. in bus. Boston Coll., 1968; MBA, Lehigh U., 1972. With fabricated steel constrn. div. Bethlehem Steel Corp. (Pa.), 1958-76, project mgr., 1969-76, engr., projects and mining div., 1976-86; sr. cons. T.J. Trauner Assocs., Phila., 1986-87; assoc. S.T. Hudson Internat., Phila., 1987-90; dir. mktg. SWIN Resource Systems, Inc., Bloomsburg, Pa., 1989-90; mem. adj. faculty Drexel U., 1976—. Weekly columnist Economic and Environmental Issues, East Pa. edit. The Free Press, 1981-86; co-patentee suspension bridge erection equipment. Founder S.A.V.E. Inc., Coopersburg, Pa., 1969, pres., 1970, 75, 81, bd. dirs., 1970—. Served with C.E., USN, 1956-58. Recipient Environ. Action award S.A.V.E. Inc., 1975. Mem. ASCE (chmn. environ. tech. com. Lehigh Valley sect. 1971-83), Chi Epsilon. Republican. Mem. United Church of Christ. Lodge: Lions. Home: 6460 Blue Church Rd Coopersburg PA 18036-9357 Office: C E Resource Group PO Box 39 Coopersburg PA 18036

SIETSEMA, WILLIAM KENDALL, biochemist; b. Evanston, Ill., Sept. 18, 1955; s. Jacob William and Marilyn Joyce (Lafferty) S.; m. Brenda Lou Hodge, Mar. 17, 1979; children: Jessica Stephanie, Laura Diane. BA in Chemsitry magna cum laude, U. Colo., 1977; PhD in Biochemistry, U. Wis., 1982. Metabolism chemist Mobay Chem., Kansas City, Mo., 1982-84; staff scientist Procter & Gamble, Cin., 1984-87; pharm. project mgr. Procter & Gamble Pharms., Norwich, N.Y., 1987—; lectr. in field. Author: (with others) Bisphosphonates: Current Status and Future Prospects, 1992; contbr. 8 articles to profl. jours. Mem. Am. Chem. Soc., Am. Soc. Bone and Mineral Rsch. Achievements include patent for Novel Dose Form; research in pharmacology of bisphosphonates, pharmaceutical development processes and strategies. Home: RR 3 Box 466A Norwich NY 13815 Office: Procter & Gamble Pharms Woods Corners Labs Norwich NY 13815

SIEVER, RAYMOND, geology educator; b. Chgo., Sept. 14, 1923; s. Leo and Lillie (Katz) S.; m. Doris Fisher, Mar. 31, 1945; children—Larry Joseph, Michael David. B.S., U. Chgo., 1943, M.S., 1947, Ph.D., 1950; M.A. (hon.), Harvard U., 1960. With Ill. Geol. Survey, 1943-44, 47-56, geologist, 1953-56; research asso., NSF sr. postdoctoral fellow Harvard U., 1956-57, mem. faculty, 1957—, prof. geology, 1965—, chmn. dept. geol. scis., 1968-71, 76-81; assoc. geology Woods Hole (Mass.) Oceanographic Instn., 1957-65; cons. to industry and govt., 1957—. Author: (with others) Geology of Sandstones, 1965, Sand and Sandstone, 1972, 2d edit., 1987, Earth, 4th edit, 1986, Planet Earth, 1974, Energy and Environment, 1978, Sand, 1988; also numerous articles, papers. Served with USAAF, 1944-46. Recipient Pres.'s award Am. Assn. Petroleum Geologists, 1952. Fellow AAAS, Am. Acad. Arts and Scis., Geol. Soc. Am.; mem. Geochem. Soc. (pres. organic geochemistry group 1965), Am. Geophys. Union, Soc. Sedimentary Geology (Best Paper awards 1957, 91, hon. 1990). Home: 38 Avon St Cambridge MA 02138-1525 Office: Hoffman Lab Harvard Univ Cambridge MA 02138

SIEVERT, LYNNETTE CARLSON, biologist, educator; b. Albia, Iowa, May 31, 1957; d. Delmar Duane and Zona Marie (Goodrich) Carlson; m. Gregory Arthur Sievert, July 30, 1983. BS, Buena Vista Coll., 1979; MS, Ea. Ky. U., 1983; PhD, U. Okla., 1988. Postdoctorate U. Okla., Norman, 1989-90, Auburn (Ala.) U., 1990-91; asst. prof. Maryville (Tenn.) Coll., 1991—. Author: Reptiles of Oklahoma, 1988; contbr. articles to Comparative Biochemistry and Physiology, Copeia, Jour. Thermal Biology, Herpetologica and Biochemical Jour. Recipient Student Rsch. award Sigma Xi, 1987. Mem. Am. Soc. Ichthyologists and Herpetologists, Herpetologists' League. Office: Maryville Coll Dept Biology Maryville TN 37801

SIGALÉS, BARTOMEU, chemical engineering educator; b. Barcelona, Catalunya, Spain, Oct. 26, 1934; s. José Maria and Carmen (Pueyo) Sigalés; m. Maria Jesùs Romero, Aug. 12, 1962; children: Nuria, Montserrat, Mercé, Meritxell. Ingeniero indsl., Escuela Tecnica Superior de Ingenieros Industriales, Spain, 1961; D Ingeniero, Escuela Tecnica Superior de Ingenieros Industriales, 1971. Researcher Instituto Petrolquimica Aplicada, UPC, Barcelona, 1972-79; prof. Escuela Tecnica Superior de Ingenieros Industriales, Barcelona, 1979—; dir. high tech. div. Ingest, S.A., Barcelona, 1973-77; corp. planning sec. and head; ESSO Petroleos Españoles, S.A., Madrid, 1965-73; process engring. dir. TECPLANT-Ingest, Barcelona, 1977-83. Author: Transmisión de calor, vol. 1, 1983, rev. edit. 1992, Transmissió de calor vol. 3, 1990, Transmisión vol. 4, 1986; contbr. articles to profl. jours. Ensign, 1962, Spain. Mem. AIChE, ASTM, AIAA, ASHRAE. Office: Laboratori de Termotecnia ETSEIB, Av Diagonal 647, 08028 Barcelona Spain

SIGFRIED, STEFAN BERTIL, software development company executive, consultant; b. Stockholm, Apr. 14, 1955; s. Bertil and Dagmar Viola Sigfried. MS, Royal Inst. Tech., Stockholm, 1982. System programmer Ericsson, Stockholm, 1982-84; test leader Saab Instruments, Jönköping, Sweden, 1984-86; methods cons. Peab, Uppsala, Sweden, 1986-89; owner Objective Ideas AB, Knivsta, Sweden, 1989—; speaker Uniforum, Stockholm, 1991, 92, Nordic Forum for Info., 1992; cons. Assn. Swedish Engring. Industries, Stockholm, 1990—. Author: Understanding Object-Oriented Engineering, 1993; contbr. articles to profl. jours. Sgt. Swedish Air Force, 1974-75. Mem. IEEE (sr.), Fedn. European Nat. Assn. Engrs. (group 1). Home and Office: Högåsvägen 29, 74141 Knivsta Sweden

SIGINER, DENNIS A., mechanical engineering educator, researcher; b. Ankara, Turkey, July 10, 1943; came to U.S., 1976; s. Kazim Siginer and Emine Turkoz. ScD, Tech. U. Istanbul, 1971; PhD, U. Minn., 1982. Rsch. assoc. U. Minn., Mpls., 1976-80; asst. prof. U. Ala., Tuscaloosa, 1981-83; assoc. prof. Auburn (Ala.) U., 1984—; organizer several internat. and nat. confs.; invited speaker to several internat. mtgs.; reviewer Jour. Non-Newtonian Fluid Mechanics, Jour. Engring. Sci., Rheological Acta, Jour. Fluids and Structures, Jour. Fluids Engring., Jour. Heat Transfer, Jour. Dynamic Systems Measurement and Ctrl., Jour. Applied Mechanics, book revs. for pubs. Author 4 books; contbr. over 90 articles to profl. jours. Recipient three univ. teaching awards. Mem. ASME, Am. Soc. for Engring. Edn. (rsch. award 1992), Soc. Rheology, Am. Acad. Mechs., Am. Inst. Physics, Soc. of Engring. Sci., N.Y. Acad. Sci., Sigma Xi. Home: 3809 Flintwood Ln Opelika AL 36801 Office: Auburn U Dept Mech Engring Auburn AL 36849-5341

SIGLER, PAUL BENJAMIN, molecular biology educator, protein crystallographer; b. Richmond, Va., Feb. 19, 1934; s. George and Florence (Kaminsky) S.; m. Althea Jo Martin, Oct. 2, 1958; children—Jennifer, Michele, Jonathan, Deborah, Rebecca. A.B. in Chemistry summa cum laude, Princeton U., 1955; M.D., Columbia U., 1959; Ph.D. in Biochemistry, Cambridge U., 1967. Intern and resident dept. medicine Columbia-Presbyn. Med. Ctr., N.Y.C., 1959-61; research assoc. NIAMD, 1961-63, staff Lab. Molecular Biology, 1963-64; vis. fellow MRC Lab. Molecular Biology, Cambridge, Eng., 1964-67; assoc. prof. biophysics U. Chgo., 1967-73, prof. biophysics and theoretical biology, 1973-84, prof. biochemistry and molecular biology, 1984-88; prof. molecular biophysics and biochemistry Yale U., New Haven, 1989—; investigator Howard Hughes Med. Inst. Served with USPHS, 1961-64. Recipient Research Career Devel. award USPHS, 1971-75; Guggenheim fellow, 1974; Katzir fellow, 1975. Fellow Am. Acad. Arts and Scis.; mem. NAS, Am. Crystallographic Assn. Jewish. Avocations: painting; bicycling. Office: Yale U Dept Biophysics New Haven CT 06510

SIGLER, WILLIAM FRANKLIN, environmental consultant; b. LeRoy, Ill., Feb. 17, 1909; s. John A. and Bettie (Homan) S.; m. Margaret Eleanor Brotherton, July 3, 1936; children: Elinor Jo, John William. B.S., Iowa State U., 1940, M.S., 1941, Ph.D., 1947; postdoctoral studies, UCLA, 1963. Conservationist Soil Conservation Service, Ill., 1935-37; cons. Central Engring. Co., Davenport, Iowa, 1940-41; research assoc. Iowa State U., 1941-42; 1945-47; asst. prof. wildlife sci. Utah State U., 1947-50, prof., head dept., 1950-74; pres. W.F. Sigler & Assocs. Inc., 1974-86; cons. U.S. Surgeon Gen., 1963-67, FAO, Argentina, 1968. Author: Theory and Method of Fish Life History Investigations, 1952, Wildlife Law Enforcement, 1956, 3d edit., 1980, Fishes of Utah, 1963, Fisheries of the Great Basin, 1987, (with J.W. Sigler) Fishes of the Great Basin, Fishery Management: Theory and Application, 1990; contbr. numerous articles to profl. jours. Mem. Utah Water Pollution Control Bd., 1957-65, chmn., 1963-65. Served as lt. (j.g.) USNR, World War II. Named Wildlife Conservationist of Yr. Nat. Wildlife Fedn., 1970, Outstanding Educator of Year, 1971; recipient Disting. Service cert. recognition Iowa Coop. Wildlife Research Unit, 1982, Outstanding Service award Utah State U., 1986, Alumni Achievement award Coll. Natural Resources Utah State U., 1987, award of Merit Bonneville chpt. Am. Fisheries Soc., 1990. Fellow Internat. Acad. Fishery Scientists, AAAS; mem. Ecol. Soc. Am., Wildlife Soc. (hon.), Am. Fisheries Soc., AAUP, Outdoor Writers Am., Sigma Xi, Phi Kappa Phi. Home: 309 E 2d S Logan UT 84321

SIGNORELLI, JOSEPH, control systems engineer; b. Kansas City, Mo., Apr. 18, 1961; s. John Joseph and Carmen Marie (Fiorella) S.; m. Denise Levin, Nov. 15, 1986. BS in Electronics Engring. Tech., DeVry Inst. Tech., 1985. Field engr. Microtel, Overland Park, Kans., 1984-87; control systems engr. Black & Veatch, Kansas City, Mo., 1987—; mem. ISA, Research Triangle Park, N.C., 1987—. Lutheran. Home: 12116 W 63rd Ter Shawnee KS 66216-2757 Office: Black & Veatch 8400 Ward Pkwy Kansas City MO 64114

SIGNORETTI, RUDOLPH GEORGE, propulsion engineer, consultant; b. N.Y.C., Dec. 8, 1930; s. Arnold John and Helen Catherine (Myers) S.; m. Shirley Jean Doble, May 5, 1950 (dec. 1988); children: Sherry, Frank, Arnold, Brian. BSME, Fairleigh Dickinson U., 1959. Flight ops. engr. Curtiss-Wright Corp., Woodridge, N.J., 1952-66; product support engr. Pratt and Whitney Aircraft, East Hartford, Conn., 1966-70; mgr. power plant engring. Flying Tiger Line, L.A., 1970-85, dir. engine mgmt., 1985-89; program mgr. Fed. Express Corp., Memphis, 1989—, power plant cons. USN Accident Team, Washington, 1960-61; com. mem. FAA, Washington, 1980-82. Contbr. articles to profl. pubs. With USN, 1948-52. Republican. Roman Catholic. Office: Fed Express Corp 7401 World Way W Los Angeles CA 90045-5836

SIGURDSSON, HARALDUR, oceanography educator, researcher; b. Stykkisholmur, Snaefellsnes, Iceland, May 31, 1939; came to U.S., 1974; s. Sigurdur and Anna (Oddsdottir) Steinthorsson; m. Jean Marie Bloom; children: Bergljot, Ashildur. BS in Geology, Queen's U., Belfast, Northern Ireland, 1965; PhD in Geology, Durham (Eng.) U., 1970. Geologist Univ. Rsch. Inst., Reykjavik, Iceland, 1965-67; volcanologist U. West Indies, St. Augustine, Trinidad and Tobago, 1970-74; prof. oceanography U. R.I., Kingston, 1974—; cons. U.S. State Dept., 1979, 84, 86. Editor Bull. Volcanology; assoc. editor Jour. Geophys. Rsch.; contbr. over 100 articles to sci. jours., popular sci. jours. including Nat. History mag. Recipient 20 rsch. grants NSF, 5 rsch. grants Nat. Geog. Soc., also Iceland Rsch. Coun. grantee. Fellow Icelandic Acad. Sci., Explorers Club; mem. Am. Geophys. Union, Internat. Assn. Volcanology and Chemistry of Earth's Interior, Glaciological Soc. Iceland. Avocations: sailing, mountaineering, art. Office: Univ RI Grad Sch Oceanography Kingston RI 02881

SIGURDSSON, THORDUR BALDUR, data processing executive; b. Reykjavik, Iceland, July 9, 1929; s. Sigurdur and Olafia (Hjaltested) Thordarson; m. Anna Hjaltested, Nov. 30, 1951; children—Magnus, Bjoern, Sigurdur, Anna, Ingveldur, Olafur, Katrin. Grad., Comml. Coll. Iceland, 1949; postgrad. U. Iceland, 1949-52. Chief acct. Icelandic State Land Reclamation, Reykjavik, 1947-72; mng. dir. Raftaekjaverzlunin Ltd., 1959-65; EDP mgr. Agrl. Bank of Iceland, Reykjavik, 1972-77, br. mgr., Stykkisholmur, 1974-75; tchr. math. Vogaskoli, Reykjavik, 1966-71; mng. dir. Icelandic Banks Data Ctr., Reykjavik, 1977—, mem. adv. bd., 1973-77. Editor Verzlunarskolabladid, 1947, Studentabladid, 1949, Ithrottabladid, 1967-68; maj. acting role Nord-deutsce Rundfunk's TV series Paradise Regained, 1979-80. Vestryman, Langholt Parish, Reykjavik, 1969-76. Recipient Gold Emblem, Athletic Union Iceland, 1968, Iceland Sports Fedn., 1972; Ace-Emblem, Athletic Union Iceland, 1967. Club: Reykjavik Football (Emblem Gold/Laures 1974). Home: Langholtsvegur 179, 104 Reyjavik Iceland Office: Icelandic Banks Data Ctr, Kalkofnsvegur 1, 150 Reykjavik Iceland

SIH, ANDREW, biologist, educator; b. N.Y.C., Mar. 10, 1954; s. Peter and Helen (Chiu) S.; m. Marie-Sylvie Baltus, Oct. 14, 1983; children: Loric. BS in Biology, SUNY, Stony Brook, 1974; PhD in Biology, U. Calif., Santa Barbara, 1980. Postdoctoral fellow Ohio State U., Columbus, 1980-81, Mich. State U., Hickory Corners, 1981-82, U. Calif., Berkeley, 1982; asst. prof. U. Ky., Lexington, 1982-87, assoc. prof., 1987-91, prof. biology, 1991—; vis. scientist Oxford (England) U., 1990; panel mem. NSF, Washington, 1991, 92. Editor: Predation: Direct and Indirect Impacts on Aquatic Communities, 1987; contbr. articles to profl. jours. Rsch. grantee NSF, 1985-88, 88-91, 91-93, 93—, Training grantee NSF, 1988-91. Mem. Am. Soc. Naturalists, Ecol. Soc. Am. (edn. com. 1980-83, eminent ecologist com. 1991-92, Buell award 1980), Internat. Soc. Behavioral Ecology, Soc. for Study of Evolution. Achievements include research on effects of conflicting demands on behavior, population and community dynamics of fresh water organisms, the roles of natural selection, evolutionary history and behavioral genetics. Office: U Ky Sch Biol Scis Lexington KY 40506

SIH, CHARLES JOHN, pharmaceutical chemistry educator; b. Shanghai, China, Sept. 11, 1933; s. Paul Kwang-Tsien and Teresa (Dong) S.; m. Catherine Elizabeth Hsu, July 11, 1959; children—Shirley, Gilbert, Ronald. A.B. in Biology, Caroll Coll., 1953; M.S. in Bacteriology, Mont. State Coll., 1955; Ph.D. in Bacteriology, U. Wis., 1958. Sr. research microbial biochemist Squibb Inst. for Med. Research, New Brunswick, N.J., 1958-60; mem. faculty U. Wis.-Madison, 1960—, Frederick B. Power prof. pharm. chemistry, 1978, Hilldare prof., 1987—. Recipient 1st Ernest Volwiler award, 1977; Roussel prize, 1980, Am. Pharm. Assoc. award 1987. Mem. Am. Chem. Soc., Soc. Am. Biol. Chemists, Acad. Pharm. Scis., Soc. Am. Microbiologists. Home: 6322 Landfall Dr Madison WI 53705-4309

SIIROLA, JEFFREY JOHN, chemical engineer; b. Patuxent River, Md., July 17, 1945; s. Arthur Raymond and Nancy Ellen (Harris) S.; m. Sharon Ann Atwood, Apr. 24, 1971; childen: John Daniel, Jennifer Ann. BS in Chem. Engring., U. Utah, 1967; PhD, U. Wis., 1970. Sr. rsch. assoc. Eastman Chem. Co., Kingsport, Tenn., 1972--; trustee CACHE Corp., Austin, Tex., 1983--. Co-author: Process Synthesis, 1973. Vol. Kingsport C. of C. Recycling, 1988--; chmn. Appalachian tr. maintenance Eastman Hiking Club, Kingsport, 1983--. With U.S. Army, 1970-72. Mem. AIChE (A.E. Marshall award 1967, Computing Practice award 1991, CAST div. programming chair 1988—), Am. Chem. Soc., Assn. for Computing Machinery, Am. Assn. for Artificial Intelligence. Achievements include development of the AIDES chem. process flowsheet invention procedure. Home: 2517 Wildwood Dr Kingsport TN 37660 Office: Eastman Chem Co PO Box 1972 Kingsport TN 37662-5150

SIKKEMA, DOETZE JAKOB, chemist; b. Rotterdam, Netherlands, Jan. 31, 1944; s. Romko Koert and Catharina Elisabeth (Vander) S.; m. Carla Rolina Smit, Dec. 30, 1970; children: Arjan Pieter, Friso Doetze, Wouter Jakob. Degree, Leiden U., 1966, Leiden U., 1969. Rsch. chemist Akzo Rsch. Labs., Arnhem, The Netherlands, 1970-82, sr. scientist, 1982—. Contbr. articles to profl. publs., including Macromolecules, Polymer, Chemtech, Jour. Applied Polymer Sci., Makromolekulare Chemie, Synthesis. Achievements include invention of sympatex breathing watertight membrane for rainwear; research in thermoplastic aramids, all-aliphatic thermotropic liquid crystal polyether carboxymethylcellulose with xanthan gumlike rheology in solution, mechanisms in cationic polymerization. Office: Akzo Rsch Labs, Velperweg 76, 6800 SB Arnhem The Netherlands

SILAGE, DENNIS ALEX, electrical engineering educator; b. Trenton, N.J., June 16, 1946; s. Alex K. and Mary A. (Marames) S.; m. Kathleen Rooney, May 10, 1975; children: Marisa, Matthew. MSE, U. Pa., 1972, PhD, 1975. Assoc. prof. medicine U. Pa., Phila., 1974-84; prof. elec. engr. Temple U., Phila., 1984—; rsch. scientist Mt. Sinai Med. Ctr., N.Y.C., 1984-89; biomed. specialist VA Med. Ctr., Phila., 1988—. Co-author: Experimental Methods in Lung Research, 1989; contbg. author: Pulmonary Diseases and Disorders, 1988. Mem. IEEE (sr.), Am. Radio Relay League (tech. coord. 1992), Eta Kappa Nu, Tau Beta Pi, Phi Kappa Phi, Sigma Xi. Achievements include devel. of computerized pulmonary function clin. instrumentation, licensed to Sensor Medics Corp. Office: Temple U Dept Elec Engring Philadelphia PA 19122

SILANI, VINCENZO, neurology and neuroscience educator; b. Brescia, Lombardy, Italy, Apr. 24, 1952; s. Cesare Silla and Onorina (Squillante) S.; m. Sabine M. Pabisch, Mar. 23, 1991; 1 child, Francesco M. MD with honors, U. Milan, 1977. Intern Inst. Pathology, U. Milan, 1974-77, intern Inst. Neurology, 1975-77, med. intern, 1977-78, tng. in neurology, 198l, tng. in neurosurgery, 1989, asst. prof. neurology, 198l-88, assoc. prof., 1989—; rsch. dir. neurobiology Dino Ferrari Ctr., 198l—; rsch. dir. in molecular genetics Inst. Sieroterapico Milanese, 1987-90; postdoctoral fellow in neurology Baylor Coll. Medicine, Houston, 1979-80, vis. prof. Baylor Coll. Neurology, 1989; examiner Italian Med. Lic., Milan, 1983—. Author more than 60 sci. papers, several book chpts., more than 100 internat. communications in basic and clin. neurobiology; first neurotransplant of adrenal after pre-cavity in a Parkinsonian patient, 1988; inventor human neuron freezing, human adrenal freezing, adult neuron cultured, nerve growth factor gene sequence and expression in human CNS. Italian Rsch. Coun. grantee, 1984-88, Maggiore Hosp. neurotransplant grantee, 1985—, Am. Parkinson's Disease Assn. grantee, 1986-90. Mem. AAAS, European Neurosci. Assn., Soc. for Neurosci., Italian Neurosci. Assn., Italian Neuropathology Assn. (founder, counselor 1985-87), Network of European CNS Transplantation and Restoration (sec. 1991, pres. 1992), N.Y. Acad. Scis. Roman Catholic. Avocations: tennis, ice hockey, skiing, riding. Home: Via G. Galeazzo 16, I-20136 Milan Italy Office: U Milan Inst Neurology, Via F Sforza 35, I-20122 Milan Italy

SILBERBERG, DONALD H., neurologist; b. Washington, Mar. 2, 1934; s. William Aaron and Leslie Frances (Stone) S.; m. Marilyn Alice Damsky, June 7, 1959; children—Mark, Alan. M.D., U. Mich., 1958; M.A. (hon.), U. Pa., 1971. Intern Mt. Sinai Hosp., N.Y.C., 1958-59; clin. assoc. in neurology NIH, Bethesda, Md., 1959-61; Fulbright scholar Nat. Hosp., London, 1961-62; NINDB spl. fellow in neuro-ophthalmology Washington U., St. Louis, 1962-63; assoc. neurology U. Pa., 1963-65, asst. prof., 1965-67, assoc. prof., 1967-71, prof., 1971-73, acting chmn. dept., 1973-74, prof., vice chmn. neurology, 1974-82, chmn., 1982—; inpatient staff Hosp. U. Pa.; cons. Children's Hosp., both Phila. Contbr. articles to profl. jours., abstracts, chpts. in books. Recipient grants in study of multiple sclerosis. Mem. Am. Acad. Neurology, Am. Assn. Neuropathologists, Am. Neurol. Assn., Am. Soc. Neurochemistry, Assn. Rsch. in Nervous and Mental Disease, Coll. Physicians Phila., Internat. Brain Rsch. Orgn., Internat. Soc. Devel. Neuroscis., Internat. Soc. Neurochemistry, John Morgan Soc. U. Pa. (pres. 1974-75), N.Y. Acad. Scis., Nat. Multiple Sclerosis Soc. (chmn. med. adv. bd.), Assn. Univ. Profs. Neurology (pres.-elect 1993), Phila. Neurol. Soc. (pres. 1978-79), Soc. Neurosci., Alpha Omega Alpha. Office: Hosp U Pa Dept Neurology 3400 Spruce St Philadelphia PA 19104-4220

SILBERBERG, REIN, nuclear astrophysicist, researcher; b. Tallinn, Estonia, Jan. 15, 1932; came to U.S., 1950; s. Jüri and Elisabeth (Linkvest) S.; m. Ene Liis Rammul, Aug. 28, 1965; children: Hugo Valter, Ingrid Kaja. MA, U. Calif., Berkeley, 1956, PhD, 1960. Postdoctoral rsch. Naval Rsch. Lab., Washington, 1960-62, rsch. physicist, 1962-81, head cosmic ray sect., 1981-85, dep. br. head cosmic and gamma ray, 1985-90; co-dir., editor, lectr. Internat. Sch. Cosmic Ray Astrophysics, Erice, Italy, 1978-93; cons. Univs. Space Rsch. Assn., Washington, 1990—. Author: (with others) Albert Einstein 100-Year Memorial Volume, 1979; co-editor: Currents in High Energy Astrophysics, 1993, Cosmic Rays and the Interstellar Medium, 1991, Particle Astrophysics and Cosmology, 1993; contbr. chpts. to Ann. Revs. of Nuclear Sci., articles to Astrophys. Jour. Recipient Meritorious Civil Svc. award U.S. Govt., 1980, Handicapped Employee of Yr. award U.S. Govt., 1985. Fellow Am. Phys. Soc., Am. Astron. Soc., Am. Geophys. Union, Radiation Rsch. Soc., Internat. Astron. Union. Achievements include development of Silberberg-Tsao cross section equations; derivation and explanation of cosmic ray source composition; pioneering development of theoretical high-energy neutrino astronomy, gamma-ray astrophysics; formulation of radiation protection requirements for lunar base and for manned Mars mission; calculation of single event upsets on shielded spacecraft. Home: 7507 Hamilton Spring Rd Bethesda MD 20817-4541

SILBERBERG, STEVEN RICHARD, meteorology educator; b. N.Y.C., Dec. 15, 1956; s. Sol and Millie (Neugeboren) S.; m. Michelle Suzanne Vandall, July 27, 1991. BS in Atmospheric Sci, Math., Geography, SUNY, Albany, 1978, MS in Atmospheric Sci., 1980; PhD in Meterology, U. Wis., 1991. Researcher Naval Environ. Prediction Rsch. Facility, Monterey,

Calif., 1981-82; rsch. assoc. Cires/U. Colo., Boulder, 1991-92; asst. prof. Creighton U., Omaha, 1992—. Contbr. articles to profl. jours. Recipient Teaching award U. Wis. Dept. Meteorology, 1990. Mem. Royal Meteorol. Soc., Am. Meteorol. Soc., Sigma Xi. Achievements include identifying underprediction of oceanic cyclones, overprediction of lee cyclones, subsynoptic structure, associated weather in Middle East storms, role of sensible and latent heat transfer from Great Lakes in cyclone development; diagnosing how atmospheric heat sources/sinks maintain atmosphereic kinetic energy, how ocean sensible and latent heat transfer to atmosphere is the energy source that maintains the northern hemisphere winter extratropical circulation; showing that atmospheric heating over the Brazil rainforest maintains the Atlantic Ocean and North African subtropical jetstream, that atmospheric heating over the tropical east Pacific Ocean maintains the North American subtropical jet stream. Home: 107 Gregg Cir Bellevue NE 68005-4956 Office: Creighton U Dept Atmospheric Scis 2500 California Pla Omaha NE 68178-0110

SILBERGELD, ELLEN KOVNER, environmental epidemiologist and toxicologist; b. Washington, July 29, 1945; d. Joseph and Mary (Gion) Kovner; m. Alan Mark Silbergeld, 1969; children: Sophia, Nicholas. AB, Vassar Coll., 1967; PhD, Johns Hopkins U., 1972. Kennedy fellow Johns Hopkins Med. Sch., Balt., 1974-75; scientist NIH, Bethesda, Md., 1975-81; chief toxics scientist Environ. Def. Fund, Washington, 1982-90; prof. epidemiology, toxicology and affll. prof. environ. law U. Md., Balt., 1990—; adj. prof. Johns Hopkins Med. Inst., 1990—; guest scientist NIH, 1982-84; mem. sci. adv. bd. EPA, 1983-89; bd. on environ. sci. and toxicology Nat. Acad. Sci.- NRC, 1983-89; bd. sci. counselors Nat. Inst. Environ. Health Scis., 1987-93; cons. Oil & Chem. Atomic Workers, 1970, NSF, 1974-75, OECD, 1987—. Contbr. articles to profl. jours.; mem. editorial bd. Neurotoxicology, 1981-86 , Neurobehavioral Toxicology, 1979-87, Environ. Rsch., 1983—, Am. Jour. Indsl. Medicine, 1980—, Hazardous Waste, 1985—, Archives Environ. Health, 1986—. Mem. Homewood Friends Meeting. Fulbright fellow London, 1967, Nat. Acad. Sci. exch. fellow, Yugoslavia, 1976, MacArthur Found. fellow, 1993; Baldwin scholar Coll. Notre Dame; recipient Wolman award Md. Pub. Health Assn., 1991, Barsky award APHA, 1992, Md. Gov. Excellence citation, 1990, 93. Mem. AAAS, Am. Soc. Pharmacology and Exptl. Therapeutics, Soc. for Occupational and Environ. Health (sec.-treas. 1983-85, pres. 1987-89), Soc. Toxicology, Soc. for Neurosci., Am. Pub. Health Assn., Collegium Ramazzini, Phi Beta Kappa. Office: U Md Med Sch Dept Epid Prev Medicine Howard Hall 104 Baltimore MD 21201

SILECCHIA, JEROME A., mechanical engineer; b. Bklyn., Oct. 25, 1941; s. Charles and Concetta (Yorio) S.; m. Collette Ambrico, June 27, 1964; children: Suzanne, Charles, Jerome. BME, Villanova U., 1963; MBA, Baruch Sch. Bus., 1973. Registered profl. engr., N.Y., N.J.; cert. plant engr., real property adminstr. Design engr. The Austin Co., N.Y.C., 1963-64, Worthington Corp., Harrison, N.J., 1964-67, York Div. of Borg Warner Corp, N.Y.C., 1967-69; mgr. H.K. Porter/Marlo Coil Div., N.Y.C., 1969-74; exec. v.p. engr. Nat. Engring. Maintenance Co., Inc., N.Y.C., 1974—; adj. lectr. NYU, 1990—. Contbr. articles to profl. jours. Mem. NSPE, Am. Inst. Plant Engrs., Am. Soc. Heating, Refrigeration and Air Conditioning Engrs., Soc. Real Property Adminstrs. Roman Catholic. Home: 92 Riverside Dr Rockville Centre NY 11570

SILINS, ANDREJS ROBERTS, physics educator; b. Riga, Latvia, Oct. 12, 1940; s. Roberts Peteris and Marta Karlis (Vitols) S.; m. Elga Janis Balta, Aug. 14, 1965; children: Lelde, Antra, Andrejs. MS in Physics, Moscow State U., 1966; PhD, U. Latvia, 1972, D of Physics and Math., 1984, Dr. hab. Phyusics, 1991. Asst. prof. physics and math. U. Latvia, Riga, 1966, jr. scientist semiconductor physics lab., 1966-67, aspirant, 1967-70, div. head, 1971-78, vice dir. sci. Inst. Solid State Physics, 1978-84, dir., 1984-92; prof. Inst. Solid State Physics, 1991—; sec. gen. Latvian Acad. Scis., 1992—; exch. student McMaster U., Hamilton, Can., 1973-74; exch. scientist Brown U., Providence, 1980-81; adviser to prime minister Coun. of Ministers of Republic of Latvia, 1990—. Author: Point Def. and Elem. Ex. in Cryst. and Glas. Sio2; editor: Spectrometry of Glas. Syst., 1988; contbr. over 122 articles to sci. jours. Mem. Coun. Experts in Physics, Riga, 1990—, Sci. Coun. of Latvia, Riga, 1990—; mem. Coun. Univ. Latvia, Riga, 1984—. Recipient award of Honor for Achievements in Tng. Young Scientists, 1989, Medal for Achievements in the People Education, 1990. Mem. Latvian Phys. Soc., Sci. Union of Latvia, Am. Phys. Soc., Latvian Acad. Sci. Avocations: volleyball, gardening, child edn., apiculture. Home: 3/1-101 Dienvidu St, Salaspils LV-2121, Latvia Office: U Latvia Inst Sol State Phy, Inst Solid State Physics, Kengaraga iela 8, Riga LV-1063, Latvia

SILK, MARSHALL BRUCE, emergency physician; b. Providence, Apr. 3, 1955; s. Marvin and Ruth Helen (Kenner) S. BA, Drake U., 1977; DO, Coll. Osteo. Medicine, 1981. Diplomate Am. Bd. Emergency Medicine. Pvt. practice, owner Silk Emergency Care, 1984—. Mem. Am. Coll. Emergency Physicians, Am. Osteo. Assn., N.Mex. Osteo. Med. Assn., Am. Assn. Osteo. Specialists. Avocations: private aviation, international travel, skiing, music. Home: Villa Muro Di Grani Star Rte 610 Placitas NM 87043

SILLER, CURTIS ALBERT, electrical engineer. BSEE, U. Tenn., 1966, MS, 1967, PhD, 1969. Mem. tech. staff Bell Tel. Labs., Whippany, N.J., 1969-71, North Andover, Mass., 1971-84; disting. mem. tech. staff AT&T Bell Labs., North Andover, 1984--. Contbr. articles to profl. jours. AT&T Bell Labs. fellow, 1989. Mem. IEEE (chmn. signal processing and communication electronics tech. com. 1988-92). Achievements include 4 patents in field. Office: AT&T Bell Labs 1600 Osgood St North Andover MA 01845

SILLS, RICHARD REYNOLDS, scientist, educator; b. N.Y.C., Sept. 19, 1946; s. Leonard Harold and Carol (Rudin) S. BA, Boston U., 1968. Tchr. N.Y.C. Pub. Schs., 1968-70, 79-81; v.p. Plutronics, Inc., N.Y.C., 1981-85; pvt. practice N.Y.C., 1985—. Author: (children's book) Jonny the Jester, 1977; contbr. articles to profl. jours.; patentee method and apparatus for encoding and decoding signals. Mem. Rep. Nat. Com., Washington, 1981—; rep. Presdl. Task Force, Washington, 1982—. Named Educator of Decade, Found. for Universal Brotherhood Inc., 1978. Avocations: running, weight lifting.

SILTANEN, PENTTI KUSTAA PIETARI, cardiologist; b. Tampere, Finland, June 6, 1926; s. Lauri A. and Saimi E. (Silventoinen) S.; m. 1950 (div. 1975); children: Marjukka, Riitta, Juha P.; m. Pirkko H. Parviainen, Oct. 4, 1975; children: Helena, Juha K. and Timo Saarelainen. MB, Helsinki U., 1949, lic. of Medicine, 1954, MD, 1968, prof. (hon.), 1983. Diplomate in Medicine, 1954, Internal Medicine, 1961, Cardiology, 1965. Med. officer of health various rural communities, Finland, 1950-57; asst. physician Salus Hosp. of the Wihuri Rsch. Inst., Helsinki, 1959; asst. physician, 3d dept. medicine Helsinki U., 1960-62, sr. cardiologist, dept. thoracic surgery, 1963-72, sr. lectr. in cardiology, 1970—, dir. cardiovascular lab., 1st dept. medicine, 1972-89, dir., lectr. postgrad. course in cardiology for gen. practitioners, 1979-90; cons. cardiologist The Finnish Nat. Bd. Health, Helsinki, 1981-91, Ministry Health, 1991—, The Finnish Heart Assn., Helsinki, 1961-68, The Dist. Hosp. Kiljava, 1960-76, The North Carelia Cen. Hosp., Joensuu, 1974-86; mem. med. expert coms. Red Cross of Finland, The Finnish Heart Assn., Valio Co. of Dairy Products, 1960—; cons. physician Life Ins. Co. Suomi-Salama, Helsinki, 1962-72; dir. rsch. dept. Finnish Heart Assn., 1969-78; dir. Helsinki Coronary Register, 1969-78; med. advisor World Health Orgn., 1970-76; lectr. med. tech. Helsinki U. Tech., 1985. Editor various books; contbr. articles to profl. jours., textbooks on clin. cardiology, pathology, biochemistry, epidemiology, psychosomatics, electrocardiography. Chmn. exec. com. Coun. Med. Tech. on the Helsiki Dist., 1982-89; mem. expert coms. Finnish Red Cross Orgn., Helsinki, 1970—; mem. planning and sci. coms. The Norh Carelia Project, 1972-76; vice chmn. bd. Finnish Heart Assn., 1991—. Grantee Acad. Finland, various pvt. founds. Fellow European Soc. Cardiology; mem. Finnish Soc. Clin. Physiology (founding mem., hon.), Finnish Soc. Ins. Medicine, Finnish Soc. Angiology (bd. dirs. 1985-90), Finnish Cardiac Soc. (founding mem., treas. 1968-70, pres. 1988-90, hon. mem. 1991), Lions. Evangelic-Lutheran. Avocations: music, drawing (cartoonist). Home: Leppakertuntie 4C, 02120 Espoo Finland Office: Helsinki U Cen Hosp, Meilahti Med Ctr, 00290 Helsinki 29, Finland

SILVA, BENEDICTO ALVES DE CASTRO, surgeon, educator; b. Salvador, Bahia, Brazil, June 26, 1927; s. Octacílio Alves de Castro and Nathercia Crusoé Silva; m. Maria Guanaes, Dec. 20, 1958; children: Catia Maria, Marta Maria, Gloria Maria. Degree, Bahia U., Salvador, 1952. Asst. prof. faculty odontology Bahia U., 1962-72, adj. prof. faculty odontology, 1972—, maxillar buco surgeon, 1962—; maxillar buco surgeon Santa Izabel Hosp., Salvador, 1953-72, Hosp. Martagão Gesteira, Salvador, 1958-60; coord. Bahia Oral Cancer Ctr., Salvador, 1988—; coord. Oncology Ctr.-Mouth-Bahia, Salvador, 1988—. Author: Patients of High Risk, 1988; contbr. chpt. to book: Phamacology, 1980. Pres. Bahia Dental Coun., Salvador, 1981-85. Officer Brazilian Army, 1944-45. Mem. Brit. Assn. Oral Maxillar Surgery (assoc.), Bahia Dental Acad., Pierre Fuchard Acad. (medal 1990), Minas Gerais Dental Acad. (medal 1990), Brazilian Soc. Cancer, European Sch. Oncology, Bahia Acad. Odontology (pres. 1985—). Home: Padre Daniel Lisboa # 5-A, 40 285-560 Salvador Brazil Office: Med Ctr Graça, Humberton de Campos St # 11, 40150 Salvador Brazil

SILVA, FERNANDO ARTURO, civil engineer; b. Guatemala City, Guatemala, Dec. 11, 1942; s. Fernando Jose and Alicia (Galvez) Silva-Peña; m. Rosa Aida Gavarrete, Aug. 2, 1969; children: Maria Gabriela, Jose Fernando, Ana Cecilia, Rodrigo. CE, San Carlos U., Guatemala City, 1968; MSCE, Purdue U., 1969; Magister Artium, Francisco Marroquin U., Guatemala City, 1983. Engr., supr. Inst. Fomento Mcpl., Guatemala City, 1967-68; prof. U. San Carlos, 1970-74; gen. contractor Oficina Ingeniero Silva, Guatemala City, 1975-78; prodn. mgr. Constructora de Occidente, Guatemala City, 1979; project engr. Hidroelectrica Chulac, Panzos, Guatemala, 1980-82; fin. mgr. Municipality of Guatemala City, 1983-85; pvt. practice civil engring., Guatemala City, 1985—; pres. Inversiones Drema, Guatemala City, 1986—, Ensambles y Servicios, Guatemala City, 1991—; treas. Subcommn. Informatics and Electronics, 1990—. Author: Introduccion al Simplex, 1968. Sec. bd. dirs. Liceo Javier, Guatemala City, 1987-89. Mem. ASCE (assoc.), Guatemalan Coll. Engrs., Gremial Exportadores de Productos no Tradicionales. Avocations: fishing, camping, boating, hunting, target shooting. Home: 3 Avenida 4-77 Zona 8, 01057 Mixco Guatemala

SILVA, NORBERTO DEJESUS, biophysicist; b. Pueblo, Colo., Feb. 18, 1963; s. Norberto Rojano and Ofelia Beatriz (Alcala) S.; m. Anne N.L. Okuku, July 15, 1989. BA in Physics and Math. magna cum laude, Cornell U., 1985; MS in Physics, U. Ill., 1986, PhD in Physics, 1992. Teaching asst. U. Ill., Urbana, 1985-86, rsch. asst., 1986-92; rsch. fellow Mayo Clinic, Rochester, Minn., 1992—. Contbr. articles to profl. jours. Fellow NSF, 1985. Mem. Phi Kappa Phi. Home: 1221 1st St SW # 2B Rochester MN 55902 Office: Mayo Clinic Guggenheim 14 200 1st St SW Rochester MN 55905

SILVA, PAUL CLAUDE, botanist; b. San Diego, Oct. 31, 1922; s. Roy Arthur and May (Henson) S. BA, U. So. Calif., L.A., 1946; MA, Stanford (Calif.) U., 1948; PhD, U. Calif., Berkeley, 1951. Faculty mem. Dept. Botany, U. Ill., Champaign, 1952-60; vis. prof. dept. botany U. Calif., Berkeley, 1960-61, sr. Herbarium botanist, 1961-67, rsch. botanist, 1967—; mem. adv. com. Cordell Expeditions, Walnut Creek, Calif., 1982—. Editor Phycologia, I.P.S., 1961-68. Lt. (j.g.) USNR, 1942-46. Recipient Darbaker Prize Botanical Soc. Am., 1959; John Simon Guggenheim Meml. fellowship, 1958-59, fellowship Calif. Acad. Scis., 1972; hon. professorship Universidad Nacional Federico Villarreal, 1983. Mem. Internat. Assn. for Plant Taxonomy (editorial com. 1981—, chmn. com. for algae, 1954—). Home: 1516 Westview Dr Berkeley CA 94705 Office: U Calif Herbarium Berkeley CA 94720

SILVA, PAUL DOUGLAS, reproductive endocrinologist; b. Durban, Natal, Republic South Africa, Oct. 29, 1956; came to U.S., 1968; s. George Douglas and Georgette Marie (Schedivetz) S.; m. Diane Elisabeth Deterville, June 28, 1980; children: Julie Renee, Jennifer Marie, Dawn Elisabeth. BA in Biology, UCLA, 1976; MD, U. Calif., Davis, 1981. Diplomate Am. Bd. Ob-Gyn, Am. Bd. Reproductive Endocrinology. Resident in ob-gyn U. Calif., Irvine, 1981-85; fellow in reproductive endocrinology U. So. Calif., L.A., 1985-87; reproductive endocrinologist Gundersen/Luth. Med. Ctr., La Crosse, Wis., 1987—; med. researcher Gundersen Med. Found., La Crosse, 1987—; cons. St. Francis Med. Ctr., La Crosse, 1988—. Coontbr. articles to Jour. Am. Acad. Dermatology, Am. Jour. Ob-Gyn, Jour. Clin. Endocrinology and Metabolism, Acta Endocrinology, also others. Lectr. to community orgns. Recipient Geog. Acad. award U. Calif., Irvine, 1984, rsch. award Soc. for Gynecologic Investigation, 1987, svc. award Pacific Coast Fertility Soc., 1987; Gundersen Med. Found. grantee, 1989-93. Fellow Am. Coll. Obstetricians and Gynecologists, Am. Fertiltiy Soc.; mem. Am. Assn. Gunecologic Laparoscopists, Soc. Reproductive Endocrinologists. Roman Catholic. Achievements include development of outpatient methods for surgical treatment of reproductive diseases which were previously treated by inpatient methods; demonstration that androstenedione may be a more important androgen in women than testosterone. Office: Gundersen Clinic 1836 South Ave La Crosse WI 54601-5494

SILVA-RUÍZ, SERGIO ANDRÉS, biochemist; b. San Juan, P.R., Jan. 12, 1944; s. Sergio A. and América (Ruíz) Silva-Izquierdo; m. Iris M. Piñero, Dec. 28, 1973; children: Maite Ira, Javier Juan, Siris Anya. BS in Chemistry, U. P.R., Rio Piedras, 1965, MS in Biochemistry, 1977, PhD, 1978. Scientist Schering-Plough, Manati, P.R., 1978-79, sr. scientist, 1979-82; mgr. R & D lab. Schering Corp., Manati, P.R., 1982-85; sr. scientist Schering Manati, Inc., Manati, P.R., 1986-91; mgr. fermentation plant Schering-Plough Products, Manati, P.R., 1991-93, plant mgr. antibiotics, 1993—; mem. adv. com. for biotech. Bachelors Degree, U. P.R., Rio Piedras, 1986-87; adv. bd. U. P.R. Minority Rsch. Ctr. for Excellence, Mayaquez, P.R., 1991-93; mem. program com. Am. Soc. Microbiology Biotech. Conf., 1991; mem. planning com. Am. Soc. Microbiology Conf. on Water Quality in Western Hemisphere, P.R., 1992-93. Pres. Sci. Rev. Com. Arecibo's Edn. Region, Dept. Edn., 1992-93. Recipient Dept. Def. fellowship, 1973-74, Merck Manual, U. P.R. Med Scis. Campus, 1978. Mem. AAAS (Caribbean Div. pres. 1992-93, coun, mem. 1993—), Parenteral Drug Assn. (v.p. P.R. chpt. 1991-93), Sociedad Microbiólogos de P.R. (pres. 1982-83), Am. Soc. for Microbiology (mem. coun. 1983-85), Am. Soc. Quality Control, N.Y. Acad. Sci., Sigma Xi. Home: RR6 B9228 Rio Piedras PR 00928 Office: Schering-Plough Products PO Box 486 Manati PR 00674-0486

SILVA-TULLA, FRANCISCO, civil engineer; b. San Juan, P.R., Oct. 7, 1950; s. Francisco Silva and Haydeé Tulla.; m. Arlyn Sanchez-Carbó, July 16, 1971; children: Arlene, Lisette Marie, Jacqueline Frances, Carolyn Anne. BS of Civil Engring. with high honors, U. Ill., 1971; MS, MIT, 1975, DSc, 1977. Registered profl. engr., Mass., Fla., P.R. Activity civil engr. USN Sta., Key West, Fla., 1971-72; asst. dir. civic action and rural devel. Inter-Am. Naval Tng. Ctr., Key West, 1972-73; rsch. asst. MIT, Cambridge, Mass., 1973-75; consulting geotech. engr. T. William Lambe Group, Lexington, Mass., 1975—; participant confs. The George Washington U., Found. Deformation Symposium; lectr. Seminario Venezolano de Geotēcnia, Caracas, symposium on art and sci. of geotech. engring. at the dawn of the 21st century U. Ill., U.S. Corps Engrs., conf. on natural disasters NSF. Contbr. articles and papers to conf. proceedings, article to ency. Panelist ednl. issues for Hispanics in Boston The Boston Found., 1985; advisor beach erosion Amuay (Venezuela) Fishermen's Orgn., 1987—; Hispanic role model greater Boston Schs. The Network, Andover, Mass., 1983—. With LCDR, Civil Engineering Corps., USNR. Mem. ASCE (embankment dams and slopes com., presenter confs.), Internat. Soc. Soil Mechs. and Found. Engring., Am. Soc. Testing and Materials, U.S. Com. on Large Dams. Sociedad Venezolana Mecánica de Suelos (corr.), S.E. Asian Geotech. Soc. (presenter confs.), Boston Soc. Engrs. Achievements include research in and development of strength of natural and compacted soils, soil mechanics, landslides, earth dams, earth structures, hazardous waste storage facilities, geoenvironmental engineering. Office: Geotechnics 12 Baskin Rd Lexington MA 02173

SILVER, BARNARD STEWART, mechanical engineer, consultant; b. Salt Lake City, Mar. 9, 1933; s. Harold Farnes and Madelyn Cannon (Stewart) S.; m. Cherry Bushman, Aug. 12, 1963; children: Madelyn Stewart Palmer, Cannon Farnes, Brenda Picketts Call. BS in Mech. Engring., MIT, 1957; MS in Engring. Mechanics, Stanford U., 1958; grad. Advanced Mgmt. Program, Harvard U., 1977. Registered profl. engr., Colo. Engr. aircraft nuclear propulsion div. Gen. Electric Co., Evandale, Ohio, 1957; engr. Silver Engring. Works, Denver, 1959-66, mgr. sales, 1966-71; chief engr. Union Sugar div. Consol. Foods Co., Santa Maria, Calif., 1971-74; directeur du complexe SODESUCRE, Abidjan, Côte d'Ivoire, 1974-76; supt. engring. and maintenance U and I, Inc., Moses Lake, Wash., 1976-79; pres. Silver Enterprises, Moses Lake, 1971-88, Salt Lake, 1990—, Silver Energy Systems Corp., Moses Lake, 1980—, Salt Lake, 1990— ; pres., gen. mgr. Silver Chief Corp., 1983—; pres. Silver Corp., 1984-86, 93—; chmn. bd. Silver Pubs., Inc., 1986-87, 89—; v.p. Barnard J. Stewart Cousins Land Co., 1987-88, 92—; dir. Isle Piquant Sugar Found., 1993—; mem. steering com. World Botanical Inst., 1993—; instr. engring. Big Bend C.C., 1980-81. Explorer adviser Boy Scouts Am., 1965-66, 89-90, chmn. cub pack com., 1968-74, chmn. scout troop com., 1968-74, vice chmn. Columbia Basin Dist., 1986-87; pres. Silver Found., 1971-84, v.p., 1984—; ednl. counselor MIT, 1971-89; pres. Chief Moses Jr. High Sch. Parent Tchr. Student Assn., 1978-79; missionary Ch. of Jesus Christ of Latter-day Saints, Can., 1953-55, West Africa, 1988, Côte d'Ivoire, 1988-89, Zaire, 1989, Holladay Stake, 1991; 2d counselor Moses Lake Stake Presidency, 1980-88; bd. dirs. Columbia Basin Allied Arts, 1986-88; mem. Health Sci. Coun. U. Utah, 1991—; mem. Sunday sch. gen. bd. Ch. of Jesus Christ of Latter-Day Saints, 1991—, com. for mems. with disabilities, 1992—. Served with Ordnance Corps, U.S. Army, 1958-59. Decorated chevalier Ordre National (Republic of Côte d'Ivoire). Mem. ASME, Assn. Energy Engrs., AAAS, Am. Soc. Sugar Beet Technologists, Internat. Soc. Sugar Cane Technologists, Am. Soc. Sugar Cane Technologists, Sugar Industry Technicians, Nat. Fedn. Ind. Bus.; Utah State Hist. Soc. (life), Mormon Hist. Assn., G.P. Chowder and Marching Soc., Western Hist. Assn., Univ. Archeol. Soc. (life), Kiwanis, Sigma Xi (life), Pi Tau Sigma, Sigma Chi, Alpha Phi Omega. Republican. Mormon. Home: 4391 Carol Jane Dr Salt Lake City UT 84124-3601 Office: Silver Energy Systems Corp 4390 S 2300 E Salt Lake City UT 84117 also: Silver Enterprises 4391 South 2275 East Salt Lake City UT 81424-3601 also: Silver Pubs Inc PO Box 17755 Salt Lake City UT 84117-0755

SILVER, GORDON HOFFMAN, metallurgical engineering consultant; b. Cin., Aug. 14, 1921; s. Charles and Nettie (Hoffman) S.; m. Lottie Sobel, Aug. 26, 1948 (dec. Feb. 1987); children: Carol Theise, Resa Waldman; m. Barbara R. Sacks, Jan. 11, 1990. Metall. Engr., U. Cin., 1948. Registered profl. engr., Ohio, Mass., Fla. Chief metallurgist Monarch Machine Tool, Sidney, Ohio, 1948-51; dir. materials and tech. processes Titeflex, Inc., Newark and Springfield, Mass., 1951-56; asst. chief engr. Nortronics div. of Northrop, Norwood, Mass., 1956-62; pres. G.H. Silver & Assocs., Inc., Newton, Mass., 1962-87; cons. engr. Gordon H. Silver, P.E., Palm Beach Gardens, Fla., 1987—; arbitrator Am. Arbitration Assn., 1980—. 1st lt. AUS, 1941-46, PTO. Recipient scholar Am. Foundrymen's Assn., U. Cin., Sigma Xi, 1948. Mem. ASTM, Am. Soc. Materials, Am. Welding Soc. Jewish. Achievements include devel. of precision thin metal welding for instrumentation and procedure for plating on berylium. Home and Office: 12870 Briarlake Dr D 103 Palm Beach Gardens FL 33418

SILVER, HULBERT K.B., physician, educator; b. Montreal, Que., Can., July 15, 1941; s. Arthur Disraeli and Nora Joanna (Belford) S.; m. Susan Daphne Andrew, Oct. 8, 1967; children: Hulbert Jr., Signe, William. BS, Bishops U., 1962; MD, McGill U., 1966, PhD, 1975. Rsch. oncologist UCLA, 1973-74, asst. prof. surgery, 1974-76; from asst. prof. medicine to assoc. prof. medicine U. B.C., Vancouver, 1976-84, prof. medicine, 1984—. Fellow ACP, Royal Coll. Physicians and Surgeons Can. (Medal of Medicine 1975). Office: BC Cancer Agy, 600 W 10th Ave, Vancouver, BC Canada V52 4F6

SILVER, LEE MERRILL, science educator; b. Phila., Apr. 27, 1952; s. Joseph and Ethel (Goodman) S.; m. Susan Remis, Aug. 25, 1985; children: Rebecca, Ari, Maxwell. BA, MS, U. Pa., 1973; PhD, Harvard U., 1978. Postdoctoral fellow Sloan-Kettering Inst., N.Y.C., 1977-80; asst. prof. Cornell Med. Divsn., N.Y.C., 1979-80; sr. scientist Cold Spring Harbor (N.Y.) Lab., 1980-84; asst. prof. SUNY, Stony Brook, N.Y., 1981-84; prof. Princeton (N.J.) U., 1984—. Editor Mammalian Genome, 1990—. Office: Princeton U Princeton NJ 08544

SILVERBERG, JAMES MARK, anthropology educator, researcher; b. N.Y.C., Dec. 16, 1921; s. Mark Silverberg and Pauline Kathryn (Nauheim) Obermeyer; m. Donna Marie Crothers, Aug. 27, 1949; children: Matthew Crothers, Conrad Mark, Erica Jean Silverberg Sandberg. BA in Hispanic Studies with high honors, U. Wis., 1947, MS in Anthropology, 1950, PhD in Anthropology, 1962. Prof. U. Cen. Venezuela, Caracas, 1953-56; instr. U. Wis., Milw., 1956-61, asst. prof., 1961-64, assoc. prof., 1964-67, prof., 1967-90, emeritus prof., 1990—; asst. to dir. Columbia U. Rsch. India, Gujarat and Uttar Pradesh, 1950-51; rsch. assoc. Bur. Social Sci. Rsch., Lucknow, India, 1952; field dir. Social Use of Solar Energy Field Rsch., Mex., 1962; instr., field dir. Changing Colombia Rsch., Tolu, 1964; acting dir. Whitemarl Site Archaeology Field Project, Jamaica, 1965. Author (with others), editor: Social Mobility in the Caste System in India, 1968; author: (with others) Main Currents in Indian Sociology, vol. 3, 1978, Peace and War in Cross-Cultural Perspectives, 1986; author, co-editor: Discourse and Inference in Cognitive Anthropology, 1978, Sociobiology: Beyond Nature/Nurture?, 1980; author (with others), co-editor: Agression/Peacefulness in Humans/Other Primates, 1992. Mem. Union of Concerned Scientists, Com. of Concerned South Asian Scholars, ACLU, Planned Parenthood. 2d lt. U.S. Army, 1942-45. Recipient Ford Found. fellowship, 1952-53, Gorjanovich-Kramberger plaque, Croation Anthropology Soc., 1988. Fellow AAAS (chmn. sect. H 1984-86, del. sect. H to governing coun. 1973-78, 86), Am. Anthrop. Assn. (rep. to AAAS 1972-73), Soc. for Applied Anthropology (exec. bd. 1969-72); mem. Internat. Union of Anthrop./Ethnol. Scis. (U.S. del., permanent coun. 1978-87), Cen. States Anthrop. Soc. (exec. bd. 1969-72). Home: 2515 N Terrace Ave Milwaukee WI 53211 Office: U Wis Dept Anthropology PO Box 413 Milwaukee WI 53201

SILVERMAN, BENJAMIN K., pediatrician, educator; b. Balt., July 21, 1924; s. Charles and Flora (Krulewich) S.; m. Beverly Miller, Dec. 28, 1948; children: Richard, Steven, Robert, Jonathan. BA, Johns Hopkins U., 1947; MD, U. Md., 1948. Lic. Pa., Calif., Am. Bd. Pediatrics, 1955. Intern rotating Sinai Hosp., Balt., 1948-49; intern pediatrics with Dr. Harold Harrison Balt. City Hosps., 1949-50; resident pediatrics Children's Hosp., Boston, 1950-51; fellow pediatric cardiology Nat. Heart Inst., Children's Hosp., Boston, 1953-54; rsch. fellow pediatrics Harvard Med. Sch., 1954; pvt. practice Princeton, N.J., 1954-85; coord. heart disease control program N.J. Dept. Health, 1955-59; chmn. dept. pediatrics Med. Ctr. Princeton, 1960-62, 72-74, dir. pediatric edn., 1974-80; pediatrician Children's Hosp. Orange County; cons. student health Princeton U., 1965-85; consulting pediatrician Premature Infant Health and Devel. Program, CHOP, 1986-90; coord. life support courses Am. Acad. Pediatrics, 1990-91; adv. office of rsch. reporting NICHD, NIH, 1978-80; mem. panel antenatal diagnosis consensus devel. conf. on fetal distress, NICHD, NIH, 1978-79, patient info. pamphlet evaluation adv. FDA, Inst. Medicine, Nat. Acad. Scis.; co-editor Clin. Pediatrics, 1979-88, editor, 1988-91; attending physician, preceptor, lectr., course dir., instr. Advanced Pediatric Life Support, instr. Pediatric Advanced Life Support, course dir. Pediatric Emergency Medicine, Children's Hosp. Phila., Emergency Care of Very Ill Child, 1983-91; clin. assoc. prof. Pediatrics, U. Pa. Sch. Medicine, 1982-92, prof., 1992; vis. prof. Clin. Pediatrics U. Southern Calif. Sch. Medicine, 1992—. Author (with others), consulting editor: Textbook of Pediatric Emergency Medicine, 1983, 87, 93; author: Pediatric Physical Diagnosis, 1985, Dialogues in Pediatric Management series, 1985, Primary Care of the Preterm Infant, 1991; author, editor: (course manual) Advanced Pediatric Life Support, 1989, 93; author 5 original papers, 4-part review of Antenatal Consensus Devel. Conf. Cosulting pediatrician Trenton Neighborhood Health Ctr., 1968-70; physician Head Start, Ctrl. N.J., 1970-76; vol. sch. physician Stuart Sch. of Sacred Heart, Princeton, 1974-85; mem. planning com. surgeon gen.'s confs. Handicapped Children and their Families, Children's Hosp. Phila., 1982, HIV Infection in Children, 1987; cons. on site Program for delivery health care to Polish Ministry Health, Project Hope, 1983-91, Program for delivery Health Care, Island Grenada, 1984-85, 85-86, 90; cons. on site Evaluation of Emergency Svcs., Seguro-Social System, Project Hope, Costa Rica, 1989. Capt. USAF 1951-53. Fellow Am. Acad. Pediatrics (mem. sect. of emergency pediatrics, exec. bd. dirs. N.J. chpt. 1964-72, chmn. pediatric practice com. N.J. chpt. 1968-70); mem. Ambulatory Pediatric Assn., Soc. Behavioral Pediatrics. Democrat. Jewish. Achievements include research in comparison of efficacy of bid vs. qid ten-day dosing of oral penicillin in Group A Beta hemolytic Streptococcal infection, comparative serological changes following treated

Group A streptococcal pharyngitis (first clinical evaluation of streptozyme now in standard usage), evaluation of erythromycin succinate (400 mgm) in treatment of Group A Beta hemolytic Streptococcal Pharyngitis, evaluation and comparison of analgesics in preverbal children (developed a model and tested in the 17-27 month age group), suprofen concentrations in human breast milk (developed a model and tested). Office: Children's Hosp Orange County 455 S Main St Orange CA 92668

SILVERMAN, DAVID CHARLES, materials engineer; b. Newark, July 24, 1947; s. Arthur and Lillian Silverman; m. Joyce Beverly Goldberg, Jan. 8, 1977; children: Ari, Roshelle. BS, MIT, 1970, MS, 1970; PhD, Stanford U., 1976. From sr. engr. to assoc. fellow Monsanto Co., St. Louis, 1975-87, fellow, 1987—. Co-editor: Flow Induced Corrosion: Fundamental Studies and Industrial Experiments, 1991, Electrochemical Impedance Analysis and Interpretation, 1993; mem. editorial bd. Corrosion Jour.; contbr. articles to profl. jours. Active Parents Adv. for Gifted Edn. Parkway Sch. Dist., 1991—. Mem. ASTM (chmn. numerous coms.), AIChE, Am. Chem. Soc., Nat. Assn. Corrosion Engrs. (Tech. Achievement award 1991, chmn. numerous coms.), Sigma Xi. Achievements include 2 U.S. patents. Office: Monsanto Co 800 N Lindbergh Blvd Saint Louis MO 63167

SILVERMAN, HARVEY FOX, engineering educator, dean. BS with Honors in Engring., Trinity Coll., 1963, BSE, 1966; ScM, Brown U., 1968, PhD, 1971. Rsch. assoc. Gerber Sci. Instrument Co., Hartford, Conn., 1964-66; various rsch. and mgmt. positions T.J. Watson Rsch. Ctr. IBM, Yorktown Heights, N.Y., 1970-80; prof. engring. Brown U., Providence, R.I., 1980-91, dir. undergrad. engring. program, 1988-90, dean engring., 1991—; cons. submarine signal divsn. Raytheon Co., Portsmouth, R.I., 1968. Contbr. articles to profl. jours. Grantee IBM 1981-85, Analog Devices, 1982-87, GLAK 1982, AMP, 1984-85, Tektronix, 1985, US West, 1988; IBM fellow 1984-87, Metrabyte, 1992. Achievements include research in home designed networks of processing nodes, each of which has a high-speed RISC host coupled to a large, reconfigurable system, general C functions, time-varying speech analysis, talker independent connected-speech recognition algorithms and systems, and non-linear optimization. Office: Brown U Divsn Engring 182 Hope St PO Box D Providence RI 02912

SILVERMAN, LESTER PAUL, economist, energy industry consultant; b. N.Y.C., Feb. 28, 1947; s. Eli and Irene B. (Karp) S.; m. Janit Roslyn Smith, June 14, 1969 (dec.); 1 child, Leigh. BS in Adminstrn. and Mgmt. Sci., Carnegie-Mellon U., 1969, MS in Indsl. Adminstrn., 1969, PhD in Econs., 1973. Economist Ctr. for Naval Analyses, Arlington, Va., 1969-74; assoc. exec. dir. NAS, Washington, 1974-78; dir. policy analysis Dept. Interior, Washington, 1978-80; prin. dep. asst. sec. Dept. Energy, Washington, 1980-81; exec. v.p. Dist. Heat & Power, Inc., Washington, 1981-82; dir. McKinsey & Co., Inc., Washington, 1982—; cons. in field, 1966-78. Author (with others) govt. report: Reducing U.S. Oil Vulnerability, 1981; editor: Population Redistribution and Public Policy, 1978; contbr. articles to profl. publs. Mem. exec. coun. Am. Jewish Com., Washington, 1983-84. Recipient Spl. Achievement award Dept. Interior, 1979, Outstanding Svc. award Dept. Energy, 1981. Mem. NAS (panel on natural gas stats., 1983-84, exploratory com. on future of nuclear power, 1984, alternative energy R&D com., 1989), Am. Econ. Assn., Internat. Assn. Energy Economists, Omicron Delta Epsilon, Omicron Delta Kappa. Home: 3728 Military Rd NW Washington DC 20015-1766 Office: McKinsey & Co Inc 1101 Pennsylvania Ave NW Washington DC 20004-2504

SILVERMAN, PAUL HYMAN, parasitologist, former university official; b. Mpls., Oct. 8, 1924; s. Adolph and Libbie (Idlekope) S.; m. Nancy Josephs, May 20, 1945; children: Daniel Joseph, Claire. Student, U. Minn., 1942-43, 46-47; B.S., Roosevelt U., 1949; M.S. in Biology, Northwestern U., 1951; Ph.D. in Parasitology, U. Liverpool, Eng., 1955, D.Sc., 1968. Research fellow Malaria Research Sta., Hebrew U., Israel, 1951-53; research fellow dept. entomology and parasitology Sch. Tropical Medicine, U. Liverpool, 1953-56; sr. sci. officer dept. parasitology Moredun Inst., Edinburgh, Scotland, 1956-59; head dept. immunoparasitology Allen & Hanbury, Ltd., Ware, Eng., 1960-62; prof. zoology and veterinary pathology and hygiene U. Ill., Urbana, 1963-72; chmn. and head dept. zoology U. Ill., 1963-68; sr. staff mem. Center for Zoonoses Research, 1964; prof. biology, head div. natural scis. Colo. Women's Coll., Denver, 1970-71; prof., chmn. dept. biology, v.p., assoc. provost for rsch. U. N.Mex., 1972-77; provost for research and grad. studies SUNY, Central Adminstrn., Albany, 1977-79; pres. Research Found., SUNY, Albany, 1979-80, U. Maine, Orono, 1980-84; fellow bio. and med. div. Lawrence Berkeley Lab., U. Calif. Berkeley, 1984-86, acting div. head, 1986-87; adj. prof. med. parasitology Sch. Pub. Health U. Calif.-Berkeley, 1986, assoc. lab. dir. for life scis., dir. Donner Lab., 1987-90, dir. Systemwide Biotech. Rsch. and Edn. Program, 1989-90; dir. Beckman's Scientific Affairs, Fullerton, Calif., 1990—; cons., Commn. Colls. and Univs., North Central Assn. Colls. and Secondary Schs., 1964—; chmn. Commn. on Instns. Higher Edn., 1974-76; Fulbright prof. zoology Australian Nat. U., Canberra, 1969; adjoint prof. biology U. Colo., Boulder, 1970-72; examiner for Western Assn. Schs. and Colls., Accrediting Commn. for Sr. Colls. and Univs., Calif., 1972—; mem. bd. Nat. Council on Postsecondary Accreditation, Washington, 1975-77; faculty apointee Sandia Corp., Dept. Energy, Albuquerque, 1974-81; project dir. research in malaria immunology and vaccination AID, 1965-76; project dir. research in Helminth immunity USPHS, NIH, 1964-72; sr. cons. to Ministry Edn. and Culture, Brasilia, Brazil, 1975—; cons. to U.S. Senator George Mitchell, Maine; adv. on malaria immunology WHO, Geneva, 1967; bd. dirs. Inhalation Toxicology Research Inst., Lovelace Biomed. and Environ. Research Inst., Albuquerque, 1977-84; mem. N.Y. State Gov.'s High Tech. Opportunities Task Force; chmn. research and rev. com. N.Y. State Sci. and Tech. Found.; mem. pres.'s council New Eng. Land Grant Univs.; mem. policies and issues com. Nat. Assn. State Univs. and Land Grant Colls.; bd. advs. Lovelace-Bataan Med. Center, Albuquerque, 1974-77; adv. com. U.S. Army Command and Gen. Staff Coll., Ft. Leavenworth, Kans., 1983-84. Contbr. articles to profl. jours. Chmn. Maine Gov.'s Econ. Devel. Conf.; chmn. research rev. com. N.Y. State Sci. and Tech. Found. Fellow Royal Soc. Tropical Medicine, Hygiene, N. Mex. Acad. Sci.; mem. Am. Soc. Parasitologists, Am. Soc. Tropical Medicine and Hygiene, Am. Soc. Immunologists, Brit. Soc. Parasitology (council), Brit. Soc. Immunologists, Soc. Gen. Microbiology, Soc. Protozoologists, Am. Soc. Zoologists, Human Genome Orgn., Am. Inst. Biol. Scis., AAAS, N.Y. Acad. Scis., N.Y. Soc. Tropical Medicine, Sigma Xi, Phi Kappa Phi. Club: B'nai B'rith. Office: Beckman Instruments Inc 2500 N Harbor Blvd Fullerton CA 92634-3100

SILVERN, LEONARD CHARLES, engineering executive; b. N.Y.C., May 20, 1919; s. Ralph and Augusta (Thaler) S.; m. Gloria Marantz, June 1948 (div. Jan. 1968); 1 child, Ronald; m. Elisabeth Beeny, Aug. 1969 (div. Oct. 1972); m. Gwen Taylor, Nov. 1985. BS in Physics, L.I. U., 1946; MA, Columbia U., 1948, EdD, 1952. Registered profl. consulting engr., Calif. Tng. supr. U.S. Dept. Navy, N.Y.C., 1939-49; tng. dir. exec. dept. N.Y. Div. Safety, Albany, 1949-55; resident engring. psychologist Lincoln Lab. MIT for Rand Corp., Lexington, 1955-56; engr., dir. edn., tng., rsch. labs. Hughes Aircraft Co., Culver City, Calif., 1956-62; dir. human performance engring. lab., cons. engring. psychologist to v.p. tech. Northrop Norair, Hawthorne, Calif., 1962-64; cons. engr., 1969—; prin. scientist, v.p., pres. Edn. and Tng. Cons. Co., L.A., 1964-80, Sedona, Ariz., 1980, pres. Systems Engring. Labs. div., 1980—; cons. hdqrs. Air Tng. Command USAF, Randolph AFB, Tex., 1964-68, Electronic Industries Assn., Washington, 1963-69, Edn. R and D Ctr., U. Hawaii, 1970-74, Ctr. Vocat. and Tech. Edn., Ohio State U., 1972-73, Coun. for Exceptional Children, 1973-74, Canadore Coll. Applied Arts and Tech., Ont., Can., 1974-76, Centro Nacional de Productividad, Mexico City, 1973-75, N.S. Dept. Edn., Halifax, 1975-79, Aeronutronic Ford-Ford Motor Co., 1975-76, Nat. Tng. Systems Inc., 1976-81, Nfld. Pub. Svc. Commn., 1978, Legis. Affairs Office USDA, 1980, Rocky Point Techs., 1986; adj. prof. edn., pub. adminstrn. U. So. Calif. Grad. Sch., 1957-65; vis. prof. computer scis. U. Calif. Extension Div., L.A., 1963-72. Dist. ops. officer, disaster communications svc. L.A. County Sheriff's Dept., 1973-75, dist. communications officer, 1975-76; bd. dirs. SEARCH, 1976—; mem. adv. com. West Sedona Community Plan of Yavapai County, 1986-88; councilman City of Sedona, 1988-92; rep. COCOPAI, 1988-89; vol. earth team Soil Conservation Svc., U.S. Dept Agr., 1989-92; Verde Resource Assn., 1988-90, Group on Water Logistics, 1989-90; chair publs. com. Ariz. Rural Recycling Conf., 1990. With USNR, 1944-46. Mem. IEEE (sr.), Am. Psychol. Assn., Am. Radio Relay League (life), Nat. Solid Waste Mgmt. Symposium (chmn. publs. com. 1988-89), Ariz. Rural Recycling Conf. (chair publs. com. 1990), Friendship Vets. Fire Engine Co. (hon.), Soc. Wireless Pioneers (life), Quarter Century Wireless Assn. (life), Sierra Club (treas. Sedona-Verde Valley Group 1991-93), Sedona Westerners., Assn. Bldg. Coms., Vox Pop (chmn. bd. dirs. Sedona, 1986-93, dir. 1993—), Nat. Parks and Conservation Assn., Wilderness Soc., Ariz. Ctr. Law in Pub. Interest. Contbg. editor Ednl. Tech., 1968-73, 81-85; reviewer Computing Revs., 1962-92. Contbr. numerous articles to profl. jours. Office: PO Box 2085 Sedona AZ 86339-2085

SILVERSTEIN, ALAN JAY, physician; b. Bklyn., June 20, 1946; s. Louis Leonard and Mildred (Abrams) S.; m. Robin Jane Ringler, Feb. 1, 1969; children: Todd Micah, Brad Matthew, Daniel Charles, Jamie Erin. BA, Cornell U., 1968; MD cum laude, Union U., Albany, N.Y., 1972. Intern Duke Hosp., Durham, N.C., 1972-73; resident ob-gyn. Magee-Women's Hosp., Pitts., 1975-78; staff ob-gyn., clin. instr. Mercy Hosp., Pitts., 1978; staff ob-gyn., clin. asst. Magee-Womens Hosp., Pitts., 1978-91; staff ob-gyn. St. Clair Meml. Hosp., Pitts., 1978-91; staff ob-gyn., co-dir. ons. clinic Ohio Valley Hosp., McKees Rocks, Pa., 1989-91; staff ob-gyn. Wetzel County Hosp., New Martinsville, W.Va., 1992; med. cons. Regional Alcoholism Program, Mercy Hosp., Pitts.; examining physician Pitts. Summer Youth Employment, 1974, Pre-employment Examination Svc., 1974-75; cons. physician Career Assessment and Devel. Svc. of Pitts. Pastoral Inst., 1974-78; gynecologist, gynecologist-in-chief U. Pitts. Student Health Svc., 1976-79; bd. advisors Pitts. Orgn. for Childbirth Edn., 1979, 83-85, Am. Soc. Psychoprophylaxis in Obstet., 1979-82; speaker numerous workshops and symposiums. Co-author: Poultry Science, 1969, Fertility and Sterility, 1978, Obstetrics and Gynecology, 1979, 81. Fellow Am. Coll. of Ob-Gyn., Pitts. Ob-Gyn. Soc., Am. Assn. of Gynocol. Laparoscopists; mem. Pa. Med. Soc., Allegheny County Med. Soc., Alpha Omega Alpha. Jewish. Home: 6 Jaycee Dr Pittsburgh PA 15243

SILVERSTEIN, SETH, physician; b. N.Y.C., May 20, 1939; s. Reuben B. and Mollie (Silver) S. BS, L.I. U., 1963; MS, Hofstra U., 1967; MD, U. Cen. Del Este, Dominican Republic, 1980. Lic. physician N.Y., Pa., La. Diagnostic radiology resident Queens Hosp. Ctr., Jamaica, N.Y., 1983-84, Newark (N.J.) Beth Israel Med. Ctr., 1984-85; physician nuclear medicine VA Med. Ctr., Northport, N.Y., 1985-86, SUNY U. Hosp., Stony Brook, 1986-87; diagnostic radiologist La. State U., New Orleans, 1987-89; pvt. practice in diagnostic radiology nuclear medicine Roslyn, N.Y., 1990—; parole officer N.Y.C.; tchr. N.Y. City Bd. Edn. Mem. AMA, Radiol. Soc. N.Am., N.Y. Med. Soc., N.Y. Roentgen Soc., King's County Med. Soc., Soc. Nuclear Medicine. Jewish. Avocations: amateur radio, photography, skiing, sailing. Home: PO Box 136 Roslyn NY 11576

SILVERSTONE, DAVID EDWARD, ophthalmologist; b. N.Y.C., Feb. 16, 1948; s. Sidney Milton and Estelle (Cohen) S.; m. Linda Carol Thalberg, June 19, 1969; 1 child, Scott. AB, Columbia Coll., 1969; MD, NY Med. Coll., 1973. Cert. Ophthalmology, Am. Bd. Ophthalmology, 1977. Acad. internat. eye fellow Albert Schweitzer Hosp., Deschapples, Haiti, 1976; instr. Dept. Ophthalmology and Visual Scis. Yale Sch. Medicine, Newhaven, Conn., 1976-77, asst. clin. prof. Dept. Ophthalmology and Visual Scis., 1977-86, assoc. clin. prof. Dept. Ophthalmology and Visual Scis., 1986-91, clin. prof. Dept. Ophthalmology and Visual Scis., 1991—; chief ophthalmology VA Hosp., West Haven, Conn., 1977-85; attending physician Yale-New Haven Hosp., New Haven, Conn., 1976—, asst. chief ophthalmology, 1988—; dir. continuing edn. Am. Soc. Cataract and Refractive Surgery, Washington, 1991—; mem. Bd. Permanent Officers Yale Sch. Medicine, New Haven, 1991—. Author: Automated Visual Field Testing, 1986; contbr. articles to profl. jours. Recipient Med. Student Essay award Am. Sc. Pharmacology and Experimental therapeutics, 1971, Moshy Book award N.Y. Med. Coll., N.Y.C., 1973, Physician's regognition award AMA, Chgo., 1976, 79, 82, 85, Honor award Am. Acad. Ophthalmology, San Francisco, 1990. Fellow Am. Acad. Ophthalmology; mem. New England Ophthalmological Soc., AMA, Conn. State Med. Soc., Conn. Soc. Eye Physicians, New Haven County Med. Assn., Yale Alumni Ophthalmology, Assn. for Rsch. in Vision and Ophthalmology. Avocation: computers. Office: Temple Eye Physicians 60 Temple St New Haven CT 06510

SILVESTRI, ANTONIO MICHAEL (TONY), electrical engineer; b. Shanghai, China, Aug. 1, 1940; came to U.S., 1957; s. Amelio and Raffaela T. (Sangalan) S.; m. Therese Ann Hanzelka, June 21, 1969; children: Rick, Dave, Andy. BSEE, MIT, 1961; MSEE, Northeastern U., 1964; PhD in Elec. Engring., Stanford U., 1971. Rsch. engr. Melpar, Inc., Watertown, Mass., 1961-62, LFE Electronics, Boston, 1962-63; sr. scientist Tech Ops, Burlington, Mass., 1963-66; sr. staff scientist Elkonix Corp., Burlington, 1969-75; sr. staff engr. ESL (subs. of TRW), Sunnyvale, Calif., 1975-79, engring. dept. mgr., 1979-88, program mgr., 1989—; Author, editor, reviewer tech. reports; contbr. articles on psychophysics of vision, image processing to tech. publs. Mem. IEEE, Sigma Xi. Democrat. Roman Catholic. Achievements include in-situ determination of modulation transfer function, resectioning algorithm (image perspective transformation), modelling of numerous processes, softcopy stereo mensuration. Home: 5155 Glentree Dr San Jose CA 95129 Office: ESL Inc 495 Java Dr Sunnyvale CA 94089

SIM, AH TEE, engineer; b. Singapore, July 6, 1944; s. Kim Pow Lim; m. Elizabeth Alice Baptist, June 26, 1965; children: Seng Chye, Siew Cheng, Syew Chye. Supr. Gen. Insulation Pte Ltd., Singapore, 1970-72; mgr. Sun Yew Engring. Works Pte Ltd., Singapore, 1973-82; mng. dir. Sun Indsl. Coatings Pte Ltd., Singapore, 1982—. Inventor, patentee Drag Plating System, Drag Soldering System, DIP Carrier, Quad Pack Carrier, SOIC Carrier. mem. Singapore Mfrs.' Assn. (mem. com.), Assn. Electronic Industries in Singapore, Semiconductor Equipment & Material Internat. Home: 00-02 Windy Heights, 80 Jalan Daud 1441, Singapore Office: SUN Indsl Coatings Pte Ltd, No 30 Pioneer Rd N Jurong, Singapore 2262, Singapore

SIMAAN, MARWAN A., electrical engineering educator; b. July 23, 1946; m. Rita Simaan. MSEE, U. Pitts., 1970; PhD in Elec. Engring., U. Ill., 1972. Registered profl. engr., Pa. Rsch. engr. Shell Devel. Co., Houston, 1974-76; assoc. prof. elec. engring. U. Pitts., 1976-85, prof., 1985-89, Bell of Pa./Bell Atlantic prof., 1989—, chmn. dept. elec. engring., 1991—; cons. Gulf Rsch. and Tech., Pitts., 1979-85, ALCOA, Pitts., 1986-89. Editor: Vertical Seismic Profiles, 1984, Two-dimensional Transforms, 1985, Artificial Intelligence in Petroleum Exploration, 1989, Expert Systems in Exploration, 1991, (series) Advances in Geophysical Signal Processing; co-editor jour. Multidimensional Systems and Signal Processing; contbr. over 200 articles on signal processing and control to profl. jours. Grantee NSF, ONR, Ben Franklin, Gulf, ALCOA. Fellow IEEE (Best Paper award 1985); mem. Soc. Exploration Geophysics, Am. Assn. Artificial Intelligence, Eta Kappa Nu, Sigma Xi (Best Paper award ALCOA chpt. 1988). Achievements include patent in application of signal processing technology in aluminum manufacturing. Office: Univ Pitts Dept Elec Engring Pittsburgh PA 15261

ŠIMÁNEK, VILÍM, chemist, educator; b. Olomouc, Czechoslovakia, Oct. 12, 1942; s. Vilím and Judit (Gruberova) Š; m. Eva Šimáčková, Apr. 30, 1971; children: Vilím, Veronika. Degree in chemistry, Palacky U., 1966, MD, 1982; PhD, Charles U., 1972; DSc, Czechoslovakian Acad. Sci., 1986. Asst. Palacky U., Olomouc, 1968-76, asst. prof., 1976-88, prof., 1988—; vice dean med. faculty Palacky U., 1985-90. Author: Alkaloids, The Alkaloids, Vol. 26, 1985; contbr. articles to Planta Medica, Heterocycles, Phytochemistry. Recipient Honour medal Hungarian Pharm. Soc., 1990. Mem. Czech Chem. Soc. (v.p. 1990—). Roman Catholic. Achievements include research in anti-inflammatory activity of quaternary benzophenanthridine alkaloids. Home: Na Struze 11, 772 00 Olomouc Czech Republic

SIMATOS, NICHOLAS JERRY, aerospace sciences educator, consultant; b. Argostoli, Kefalonia, Greece, Aug. 21, 1948; came to U.S., 1955; s. Jerry Nicholas and Jenny (Kostantakis) S.; m. Maria Pantazies, Oct. 13, 1979; children: Alexander Nicholas, Diana Lindsay. AD, Hudson Valley Community Coll., Troy, N.Y., 1972; BS, Embry Riddle Aero. U., Daytona, Fla., 1974; MS, Embry Riddo Aero. U., Miami, Fla., 1978, M in Mgmt., 1978. Charter pilot Piedmont Aviation, Miami, 1977-78; instr. pilot Flight Safety Acad., Vero Beach, Fla., 1978-79; commd. 2d lt. U.S. Air Force, 1979, advanced through grades to capt.; space systems dir. 19th Surveillance SON, Diyarbakir, Turkey, 1983-84; astronaut, instr. NASA Johnson Space Ctr., Houston, 1984-87; resigned U.S. Air Force, 1987; aggressor pilot Flight Internat., Jacksonville Naval Air Sta., Fla., 1987-88; asst. prof. Embry-Riddle Aero. U., Daytona, 1989—; mem. Life Support Systems for Space Exploration/Ops., 1993; cosmonaut candidate Aerospace Ambs., Huntsville, Ala., 1990—; aerospace cons. Dept. Edn., State of Fla., 1989—; recert. dir. NASA, Houston, 1986-87. Author: Futuristic Spacecraft Systems, 1989, Space Transportation Systems, 1990, Space Flight Technologies, 1991. State advisor U.S. Congressional Adv. Bd., Washington, 1982; chmn. Hellenic Profl. Soc. Scholarship Program, Houston, 1986. Decorated Recognition award Turkish Air Force, award of excellence Dutch Air Force, others. Mem. AIAA (sr.). Republican. Eastern Orthodox. Avocations: snow and water skiing, flying, swimming, jogging. Home: 1948 Spruce Creek Landing Daytona Beach FL 32124-6869 Office: Embry Riddle Aero U Space Sci Dept Daytona Beach FL 32114

SIME, JAMES THOMSON, consultant psychologist; b. Elgin, Moray, Scotland, Aug. 9, 1927; s. James Alexander and Jessie Ann (Scott) S.; divorced; children: James A., Julie-Ann; m. Jaya Rani Sinha, June 3, 1992. MA, U. Edinburgh, Scotland, 1954, diploma in edn., 1955; cert. in teaching, Moray House, Edinburgh, 1955; MEd, U. Glasgow, Scotland, 1957. Chartered psychologist. Psychologist in charge various psychol. svcs., 1960-64, 67-80, pvt. practice as psychotherapist, 1960-84; cons. psychologist on brain damage and post-traumatic stress U.K., 1984—; lectr., 1964-67. Served with Royal Air Force, 1945-48. Fellow Brit. Psychol. Soc. (assoc.); mem. Internat. Auster Pilot Club (founder 1973, hon. v.p.). Avocations: flying, hill walking, travel, human behavior, languages. Home and Office: 19 Cedric Rise, Livingston EH54 6JR, England Office: 8 Lindley St, Rotherham S65 1RT, England

SIMEÓN NEGRÍN, ROSA ELENA, veterinary educator; b. Havana, Cuba, June 17, 1943; married; 1 child. Student, U. Havana. Chief dept. virology Nat. Ctr. Scientific Rsch. (CENIC), 1968-73, chief microbiol. divsn., 1974-76, prof. sch. vet. medicine, 1969-73; prof. sch. vet. medicine Nat. Hosp. and Nat. Ctr. Scientific Investigations, 1975, Nat. Inst. Vet. Medicine, 1977-78, 81; dir. Nat. Ctr. Agrl. Health (CENSA), 1985; pres. Acad. Scis. Cuba, 1985—; mem. numerous profl. coms. Contbr. articles to profl. jours. •

SIMINOVITCH, LOUIS, biophysics educator, scientist; b. Montreal, Que., Can., May 1, 1920; s. Nathan and Goldie (Watchman) S.; m. Elinore Esther Faierman, July 2, 1944; children: Harriet Jean, Katherine Ann, Margaret Ruth. B.Sc., McGill U., 1941, Ph.D., 1944; D honoris causa, U. Montreal, 1990, McGill U., 1990, U. Western Ont., 1990. Mem. staff Nat. Research Council Can., 1944-47; Canadian Royal Soc. fellow Pasteur Inst., Paris, 1947-49; mem. staff Centre Nationale de la Recherche Scientifique, 1949-53; Nat. Cancer Inst. Can. fellow U. Toronto, Ont., Can., 1953-56; asst. prof. dept. med. biophysics U. Toronto, 1956-58, assoc. prof. med. biophysics, 1958-60, prof. med. biophysics, 1960—, Univ. prof., 1976—, assoc. prof. pediatrics, 1972-78, chmn. dept. med. cell biology, 1969-72, med. genetics, 1972-79; dir. rsch. Samuel Lunenfeld Rsch. Inst., Mt. Sinai Hosp., Toronto, 1983—; head microbiology sect. biol. research div. Ont. Cancer Inst., Toronto, 1957-63, head div. biol. research, 1963-69; geneticist-in-chief Hosp. Sick Children, Toronto, 1970-85; mem. virology and rickettsiology sect. NIH, 1966-68; mem. health research com. Ont. Council Health, 1966-82; mem. research adv. group Nat. Cancer Inst. Can., 1969-74, chmn., 1970-72, bd. dirs., 1975-85, pres., 1982-84; bd. dirs. Canadian Weizmann Inst. Sci., 1972—; mem. adv. bd. Ont. Mental Health Found., 1974-78 ; chmn. Ont. Health Research and Devel. Com., 1974-82; task force on genetic services Ont. Ministry Health, 1974-76; mem. Ont. Task Force On Health Research Requirements, 1974-76; chmn. com. on guidelines for Recombinant DNA, Med. Research Council Can., 1975-77; bd. dirs. Mount Sinai Hosp., Toronto, 1975-82; mem. United Ch. Can. Gen. Council Commission on Genetic Engring., 1974-78; bd. advisors Clin. Research Inst. Montreal Center Bioethics, 1976-80; chmn. adv. com. on genetic services Ont. Ministry Health, 1976-82; G. Malcolm Brown Meml. lectr. Royal Coll. Physicians and Surgeons, 1978; bd. dirs. Ont. Cancer Treatment and Rsch. Found., 1979-93, chmn. rsch. adv. panel, 1986—; mem. Alfred P. Sloan, Jr. selection com. Gen. Motors Cancer Rsch. Found., 1980-81, 83-84; nat. bd. dirs. Canadian Cancer Soc., 1981-84; mem. sci. adv. bd. Huntingtons Soc. Can., 1984-89; adv. com. Coll. Biol. Scis., Guelph, Ont., 1986-90; mem. rsch. coun. Can. Inst. Advanced Rsch., Toronto, 1982-91, chmn. adv. com. evolutionary biology 1986—; mem. bd. govs. Baycrest Centre Geriatric Care, 1987—; chmn. external adv. com., Loeb Inst. Med. Rsch., 1987—; chmn. adv. bd. Allelix Inc., 1987—; mem. sci. tech. svcs. sub-com. Sci. Coun. Can., 1988-89; chmn. steering com. for evaluation of MRC grants program, MRC, Ottawa, 1989-91; chmn. sci. adv. com. Rotman Rsch. Inst., Baycrest Centre, 1990—, mem. Can. Inst. of Acad. Medicine, 1992—; mem. Montreal Neurol. Inst. adv. bd. Montreal, 1992—. Editor Virology, 1960-80, Bacteriological Revs, 1969-72, Jour. Molecular and Cellular Biology, 1980-90; founding mem., pres. editorial bd. Science Forum, 1966-79; mem. editorial bd. Cell, 1973-81, Somatic Cell Genetics, 1974-84, Jour. Cytogenetics and Cell Genetics, 1974-80, Mutation Research, 1976-82, Jour. de Microscopie et de Biologie Cellulaire, 1976-86, Cancer Genetics and Cytogenetics, 1979-84, Jour. Cancer Surveys, 1980-89; corr. editor Proc. Royal Soc. B, 1989—; contbr. numerous articles to sci. jours. Decorated officer Order of Can.; recipient Queen Elizabeth II Silver-Jubilee award 1977, Izaak Walton Killam meml. price, 1981, Wightman award Gairdner Found., 1981, Medal of Achievement, Inst. de Recherches Cliniques de Montreal, 1985, Environ. Mutagen Soc. award, 1986, R.P. Taylor award Can. Cancer Soc., Nat. Cancer Inst., 1986, Disting. Rsch. award Can. Soc. Clin. Investigation, 1990, Toronto Biotech. Initiative Community Svc. award, 1991; named Companion of the Order of Can., 1989, The Gov. Gen. Commemortive medal for the 125th Anniversary of Can. Confederation, 1992. Fellow Royal Soc. Can. (Centennial medal 1967, Flavelle l 1978, mem. adv. com. on evaluation of rsch. 1989—), Royal Soc. (Can.); mem. AAAS. Home: 106 Wembley Rd, Toronto, ON Canada M6C 2G6 Office: Samuel Lunenfeld Rsch Inst, Mt Sinai Hosp 600 University Ave, Toronto, ON Canada M5G 1X5

SIMMINS, JOHN JAMES, ceramic engineer, consultant; b. Hornell, N.Y., May 10, 1961; s. Charles Henry and Marianna Rita (DeLaney) S. BS in Ceramic Sci., Alfred U., 1984, PhD in Ceramic Sci., 1990. Staff Nat. Bur. Standards, Gaithersburg, Md., 1982-83; devel. engr. Trans Tech, Inc., Adamstown, Md., 1984; sr. ferrite scientist Titan Advanced Materials, Valparaiso, Ind., 1988-89; sr. ceramic engr. Trans Tech, Inc., Adamstown, Md., 1990—; guest scientist Oak Ridge (Tenn.) Nat. Lab., 1989, Brookhaven Nat. Lab., Upton, N.Y., 1990. Contbr. articles to profl. jours. Recipient grant NASA, Huntsville, 1990. Mem. Am. Ceramic Soc. (program chair 1991—, Balt.-Washington sect. chmn. 1993, magnetic symposium chmn. Ann. Meeting 1994), Material Rsch. Soc., Am. Crystalographic Assn., NRA. Republican. Roman Catholic. Achievements include development of search routines for the Crystal Data Base to locate epitaxial substates; determination of several crystal structures and phase relationships in the Y-Ba-Cu-O-Co2 system, activation energies for CO2 decomposition. Home: 12 Sunny Way Thurmont MD 21788 Office: Trans Tech Inc 5520 Adamstown Rd Adamstown MD 21710-9697

SIMMONS, HARRY DADY, pathologist; b. Chgo., June 10, 1938; s. Harry Dady and Ruth (Finkelberg) S. BS, U. Ill., 1950; PhD in Chemistry, MIT, 1966; MD, SUNY, Bklyn., 1977. Resident Mt. Sinai Med. Ctr., N.Y.C., 1985-87; staff pathologist, clin. asst. prof. Huntington (W.Va.) VA Med. Ctr., Marshall U. Med. Ctr., 1991—. Rsch. fellow Rensselaer Polytech Inst., Troy, N.Y., 1981-85, Montefiore Med. Ctr., Bronx, 1987-91. Mem. AMA, Am. Soc. Clin. Pathologists, Coll. Am. Pathologists. Office: Marshall U Med Sch 1542 Spring Valley Dr Huntington WV 25704

SIMMONS, HOWARD ENSIGN, JR., chemist, research administrator; b. Norfolk, Va., June 17, 1929; s. Howard Ensign and Marie Magdalene (Weidenhammer) S.; m. Elizabeth Anne Warren, Sept. 1, 1951; children: Howard Ensign III, John W. BS in Chemistry, MIT, 1951, PhD in Organic Chemistry, 1954; DSc (hon.), Rensselaer Poly. Inst., 1987. Mem. rsch. staff cen. rsch. and devel. dept. E.I. du Pont de Nemours & Co., Wilmington, Del., 1954-59, rsch. supr., 1959-70, assoc. dir., 1970-74, dir. rsch., 1974-79, dir., 1979-83, v.p., 1983-90, v.p., sr. sci. advisor, 1990—; adj. prof. chemistry U. Del., 1974—; Sloan vis. prof. Harvard U., 1968; Kharasch vis. prof. U. Chgo., 1978; mem. Nat. Sci. Bd., 1990. Author: (with R.E. Merrifield) Topological Methods in Chemistry, 1989. Trustee Gordon Rsch. Confs., 1974-77. Recipient Chandler medal Columbia U., 1991, Nat. medal of sci. NSF, 1992. Fellow N.Y. Acad. Scis., Am. Acad. Arts and Scis.; mem. NAS, AAAS, Soc. Chem. Industry, Am. Chem. Soc. (Priestley medal 1994), Indsl.

Rsch. Inst., Nat. Sci. Bd., Delta Kappa Epsilon. Home: PO Box 3874 Wilmington DE 19807-0874 Office: E I du Pont de Nemours & Co 1007 N Market St Wilmington DE 19898-0001

SIMMONS, JAMES, geography educator. Prof. geography U. Toronto, Ont., Can. Recipient CAG award for Scholarly Distinction in Geography, Can. Assn. Geographers, 1992. Office: U of Toronto/Dept of Geography, 100 St George St, Toronto, ON Canada M5S 1A1*

SIMMONS, LEE GUYTON, JR., zoological park director; b. Tucson, Feb. 20, 1938; s. Lee Guyton and Dorothy Esther (Taylor) S.; m. Marie Annette Geim, Sept. 6, 1959; children: Lee Guyton, Heather, Heidi. Student, Cen. State Coll.; DVM, Okla. State U. Resident veterinarian Columbus Zoo, Powell, Ohio, 1963-66, Henry Doorly Zoo, Omaha; research cons. VA Hosp.; assoc. instr. U. Nebr. Med. Ctr., Omaha; assoc. clin. prof. Creighton U. Sch. Dentistry. Contbr. articles to profl. jours. Bd. dirs. Nebr. State Mus., Lincoln. Served with USAR. Recipient Nat. Idealism award City of Hope, 1979; named Man of Yr., Lions Club, 1978. Fellow AVMA, Am. Assn. Zool. Veterinarians (pres.), Am. Assn. Zool. Parks, Nebr. Vet. Med. Assn. (Veterinarian of Yr. 1979). Lodge: Rotary. Office: Office of the Director Henry Doorly Zoo 3701 S 10th St Omaha NE 68107-2200

SIMMONS, ROBERT ARTHUR, engineer, consultant. With Simmons Consulting Ltd., Lethbridge, Alta., Can. AIC fellowship award Agrl. Inst. Can., 1992. Office: Simmons Consulting Ltd, 1010-29 St A South, Lethbridge, AB Canada T1K 2X7*

SIMMONS, ROBERT MARVIN, environmental scientist, consultant; b. Dallas, Dec. 27, 1959; m. Sharla M. Miller; children: Zandra, Rhiannon. AAS in Occupational Safety and Health, Lamar U., Beaumont, Tex., 1983, BS in Indsl. Tech., 1983. Ordained and lic. to ministry Internat. Ministerial Fellowship; ordained to ministry New Beginnings Ministries, Inc. Health and safety officer Region VI field investigation team EPA, Dallas, 1984-85; health and safety officer Internat. Tech., Baton Rouge, 1985-86; mgr. environ. health and safety Layne-Western Co. Inc., Shawnee Mission, Kans., 1987-88; safety and indsl. hygiene engr. Armco Midwestern Steel, Kansas City, Mo., 1988-89; sr. environ. and safety specialist Environ. & Safety Svcs. Inc., Kansas City, 1989—. Home: 11011 W 48th St Shawnee KS 66203 Office: Environ & Safety Svcs Inc Dept 141 PO Box 7305 Kansas City MO 64116

SIMMONS, SAMUEL WILLIAM, retired public health official; b. Benton County, Miss., June 5, 1907; s. Britt L. and Ida E. (Pegram) S.; BSc with honors, Miss. State U., 1931; A.M., George Washington U., 1934; PhD, Iowa State U., 1938; m. Lois Grantham, Aug. 5, 1928; children: Samuel William, Grant P. With U.S. Dept. Agr., Bur. Entomology, 1931-44; with USPHS, 1944-71, scientist to scientist dir., dir. Carter Meml. Lab., 1944-47, chief tech. devel. br., 1947-53, chief tech. br. communicable disease center 1953-66; dir. rsch. labs. and operational programs on communicable diseases and their control, chief pesticides program Nat. Ctr. for Disease Control, Atlanta, 1966-68; dir. div. pesticide community studies FDA, 1968-71; dir. div. pesticide community studies EPA, 1971-72, ret.; vis. lectr. tropical pub. health Harvard U., 1952-67; assoc. preventive medicine and community health Emory U., 1957-72; USPHS rep. Fed. Com. on Pest Control. Recipient Alumni Achievement award George Washington U., 1946, Alumni Centennial Citation award Iowa State U., 1958, Disting. Svc. medal USPHS, 1965, William Crawford Gorgas medal Assn. Mil. Surgeons U.S., 1968, Disting. Career award EPA, 1972. Hon. cons. Army Med. Library, 1940-53; adv. bd. Inst. Agrl. Medicine, U. Iowa Sch. Medicine, U.S.-Japan Com. on Sci. Cooperation. Diplomate Am. Bd. Microbiology. Fellow Am. Soc. Tropical Medicine and Hygiene (councilor 1953), Chem. Spltys. Mfrs. Assns. (interdepartmental com. pest control, chmn. inter-agy. com. water resources 1964-66), U.S.-Mex. Border Health Assn., WHO (chmn. com. on pesticides Geneva 1951, 56, 57), AMA (com. on insecticides 1950-59, com. on toxicology 1960), Rsch. Soc. Am., Entomol. Soc. Am., Nat. Malaria Soc. (sec.-treas. 1951), Nat. Environ. Health Assn., Agrl. Rsch. Inst., Am. Mosquito Control Assn., Armed Forces Pest Control Bd., Nat. Rsch. Coun., Nat. Assn. Watch and Clock Collectors (fellow, nat. dir. 1979-83, mem. awards com., patron), Sigma Xi, Phi Kappa Phi, Gamma Sigma Delta, Los Hidalgos. Contbr. over 80 articles to profl. jours.; editor and co-author: The Insecticide DDT and Its Significance, vol. II; contbr. to Human and Veterinary Medicine, 1959; author: (genealogies) Descendants of John Simmons of North Carolina, 1760, 1979, Samuel W. Simmons: An Autobiography, 1979; The Pegrams of Virginia and Descendants, 1688-1984, 1985. Avocations: early Am. furniture, horology, photography. Home: 2050 Black Fox Dr NE Atlanta GA 30345-4138

SIMMONS, SHALON GIRLEE, electrical engineer; b. Oklahoma City, Apr. 9, 1966; d. Joe Douglass and Lillie May (Hudspeth) S. BSEE, Okla. State U., 1989; MBA, Oklahoma City U., 1991. Registered engr. intern. Rsch. tech. JG AMOCO Rsch., Tulsa, 1985-86; coop. elec. engr. LTV Aerospace and Def. (Missiles), Grand Prairie, Tex., 1987-88; mgr. computer ops. Southwestern Bell Telephone, Oklahoma City, 1989-90, mgr. current planning, 1990—, mgr. transmission engring. Southwestern Bell Telephone corp. rep. Soc. Women Engrs. (Okla. State), Stillwater, 1991, 92.; bd. dirs. Nat. Inst. Abuse Okla. State Epid. Work Group, Oklahoma City, 1991-92; steering com. mem. Oklahoma City Found. Future Fund, Oklahoma City, 1991—; mem. adv. bd. Millwood Sch. Dist. Improvement Plan Com., Oklahoma City, 1991-92. Named one of Outstanding Young Women of Am., 1989. Mem. Nat. Soc. Black Engrs. (nat. coll. init. 1991—, pres. Oklahoma City chpt. 1991-92, Southwestern Bell Telephone corp. rep. L.A. conf. 1991, Outstanding Exec. Bd. Mem. 1992, alumni extension), Okla. Soc. Profl. Engrs. (v.p. pub. rels. cen. chpt. 1991-92, v.p. intersoc. cen. chpt. 1992—, v.p. programs cen. chpt.).

SIMMS, JOHN CARSON, logic, mathematics and computer science educator; b. Columbus, Ind., Oct. 24, 1952; s. Roberta Ann (Cooke) S.; m. Florence Chizue Miyamoto, June 22, 1974; 1 child, Carson Chizumi. BA with highest distinction, Ind. U., 1972, MA, 1974; PhD, Rockefeller U., 1979. Assoc. instr. Ind. U., Bloomington, 1973-74; grad. fellow Rockefeller U., N.Y.C., 1974-78; vis. lectr. Tex. Tech U., Lubbock, 1978-80; v.p. Custom Computation, Inc., Lubbock, 1980-81; computer programmer and analyst Furr's Inc., Lubbock, 1981-82; contract computer programmer Lubbock, 1982-83; asst. prof. math. and computer sci. Marquette U., Milw., 1983-92, assoc. prof. math. and computer sci., 1992—. Assoc. editor Modern Logic, Modern Logic Books, 1990—; contbr. articles to profl. jours. Mem. AAAS, Assn. for Symbolic Logic, Am. Math. Soc., Kurt-Gödel-Gesellschaft (Collegium Logicum lectr. 1990), Phi Beta Kappa. Achievements include research on new second-order semantics, on a natural argument against the continuum hypothesis, on a natural argument against the axiom of choice, on applications of natural probabilistic notions to the foundations of mathematics. Home: 8969 N Pelham Pky Bayside WI 53217-1954 Office: Marquette U Dept Math Stats and Computer Sci Milwaukee WI 53233-2189

SIMON, ALBERT, physicist, engineer, educator; b. N.Y.C., Dec. 27, 1924; s. Emanuel D. and Sarah (Leitner) S.; m. Harriet E. Rubinstein, Aug. 17, 1947 (dec. June 1970); children: Richard, Janet, David; m. Rita Shiffman, June 11, 1972. B.S., CCNY, 1947; Ph.D., U. Rochester, 1950. Registered profl. engr., N.Y. State. Physicist Oak Ridge Nat. Lab., 1950-54, asso. dir. neutron physics div., 1954-61; head plasma physics div. Gen. Atomic Co., San Diego, 1961-66; prof. dept. mech. engring. U. Rochester, N.Y., 1966—; prof. physics U. Rochester, 1968—, chmn. dept. mech. engring., 1977-84; mem. Inst. for Advanced Study, Princeton, 1974-75; sr. vis. fellow U.K. Sci. Research Council, Oxford U., 1975. Author: An Introduction to Thermonuclear Research, 1959; contbr. to: Ency. Americana, 1964, 74; Editor: Advances in Plasma Physics, 1967—. Recipient Univ. Mentor award, 1988-89; John Simon Guggenheim fellow, 1964-65. Fellow Am. Phys. Soc. (chmn. plasma physics div. 1963-64); mem. ASME, ASEE (chmn. nuclear engring. div. 1985-86). Home: 263 Ashley Dr Rochester NY 14620-3327

SIMON, FREDERICK EDWARD, chemist; b. Port Hueneme, Calif., Oct. 3, 1953; s. Frederick A. and Joan H. (Harwin) S.; m. Mary Randzo (div. 1988); children: Katherine, Philip Daniel. BA, Rutgers U., 1976; PhD, SUNY, 1981. Chemist Naval Rsch. Lab., Washington, 1982-85; sr. scientist Campbell Soup Co., Camden, N.J., 1985—; mem. Campbell Microwave Inst., Camden, 1985—. Contbr. articles to profl. jours. N.J. State scholar, 1972. Mem. Am. Chem. Soc., Internat. Microwave Power Inst., Inst. Food Technologists. Achievements include patents on conformable wrap susceptor, microwave releasable seal package, microwave oven preparation of waffle. Office: Campbell Soup Co Campbell Pl Camden NJ 08103

SIMON, GARY LEONARD, internist, educator; b. Bklyn., Dec. 18, 1946; s. Bernard and Dorothy (Ligeti) S.; m. Vicki Thiessen, Aug. 29, 1970; children: Jason, Jessica. BS, U. Md., 1968; PhD, U. Wis., 1972; MD, U. Md., 1975. Diplomate Am. Bd. Internal Medicine. Asst. prof. dept. medicine George Washington U., Washington, 1980-84, assoc. prof., 1984-89, prof., assoc. chmn., 1989—; cons. on AIDS Assn. Am. Med. Coll., Washington, 1990—. Contbr. articles to profl. jours. on AIDS and infectious diseases. Fellow Am. Coll. Physicians, Infectious Disease Soc. Am.; mem. Am. Soc. Microbiology, Assn. Program Dirs. in Internal Medicine. Office: George Washington U 2150 Pennsylvania Ave NW Washington DC 20037

SIMON, MELVIN I., molecular biologist, educator; b. N.Y.C., Feb. 8, 1937; s. Hyman and Sarah (Liebman) S.; m. Linda, Jan. 7, 1959; children—Joshua, David, Rachel. B.S., CCNY, 1959; Ph.D., Brandeis U., 1963. Postdoctoral fellow Princeton U., N.J., 1963-65; prof. biology U. Calif.-San Diego, La Jolla, 1965-82, Calif. Inst. Tech., Pasadena, 1982—; pres., dir. Agouron Inst., La Jolla, 1980—. Contbr. articles to profl. jours. Mem. Nat. Acad. Scis. (Selman A. Waksman microbiology award 1991), Am. Soc. Microbiology.

SIMON, MICHELE JOHANNA, computer systems specialist; b. Reading, Pa., Nov. 28, 1957; d. Joseph Aloysius and Mildred Adella (White) S. BS, BA, Albright Coll., 1983. Adminstr. Dept. Def., Washington, 1979-81, Dept. Labor, Washington, 1981-83; programmer Berkshire Health Systems, Reading, 1983-85; computer systems analyst Reading Hosp. and Med. Ctr., 1985—; toy inventor, owner Designs by Natural Selection, 1992—; owner, mgr. Intervilla Apts., West Lawn, Pa., 1979—. Co-author (with Richard Murphy, screenplays) Fat Chance, 1987, Base Instinct, 1991. Vol. Meals on Wheels, Reading, 1992. Nat. Merit scholarship finalist, 1974. Mem. Supporters of Emo Philips (pres. 1986—). Democrat. Avocation: wildlife observation.

SIMON, SOLOMON HENRY, artificial intelligence manager; b. Charleston, S.C., Jan. 2, 1955; s. Gus and Naomi Simon. MS in Nuclear Physics, Tex. A&M U., 1983, PhD in Artificial Intelligence, 1985. Rsch. scientist Chem. Abstracts Svc., Columbus, Ohio, 1985-88; artificial intelligence R&D dept. Loral Vought Systems Corp., Dallas, 1988—. Welsh Found. fellow, 1981; recipient Pubs. award NASPA, 1988, Dirs. award, 1990. Mem. AIAA (sr.), Am. Assn. Artificial Intelligence, SPIE. Achievements include first smart structure hardware prototype and neural network sensor processing. Home: 3712 Pimlico Dr Arlington TX 76017 Office: Loral Vought Systems Corp PO Box 650003/PT-88 Dallas TX 75265-0003

SIMON, THEODORE RONALD, physician, medical educator; b. Hartford, Conn., Feb. 2, 1949; s. Theologos Lingos and Lillian (Faix) S.; m. Marcia Anyzeski, Apr. 5, 1974; children: Jacob T., Theodore H., Mark G. BA cum laude, Trinity Coll., Hartford, 1970; MD, Yale U., 1975. Diplomate Am. Bd. Nuclear Medicine, Diplomate Nat. Bd. Med. Examiners; lic. Calif., Tex. Intern in surgery Strong Meml. Hosp., Rochester, N.Y., 1975-76; resident in diagnostic radiology U. Calif., San Francisco, 1976-78; resident in nuclear medicine Yale-New Haven Hosp., Conn., 1978-80, chief resident, 1979-80; asst. prof. nuclear medicine U. Tex. Southwestern Med. Ctr., Dallas, 1980-88, assoc. prof., 1990—; cons. nuclear medicine St. Paul's Hosp., Dallas, 1981-88; cons. internal medicine Presbyn. Hosp., Dallas, 1981-88, 90, Humana Hosp., Dallas, 1989—; cons. nuclear medicine VA Med. Ctr., Dallas, 1981-82, chief nuclear medicine svc., 1982-88; nat. dep. dir. nuclear medicine VA, 1985-88; dep. chief nuclear medicine NIH, Bethesda, Md., 1988-90; mem. del. Taiwan Atomic Energy, U.S. State Dept., 1990. Mem. editorial bd. Jour. History of Med. and Allied Scis., 1974-75; contbr. articles to Internat. Jour. Radiol. Applications, Jour. Nuclear Medicine, Am. Jour. Cardiology, Clin. Nuclear Medicine, Circulation, Yale Jour. Biol. Medicine, Radiology, Surg. Radiology, and others. Pres. Christ Lutheran Ch., University Park, Tex. Mem. Soc. Nuclear Medicine (treas. correlative imaging coun. 1988-90, mem. exec. com 1988—). Achievements include patent for Complex Motion Device to Enhance Single Photon Emission Computed Tomography Uniformity; research in single photo emission computed tomography as it related to substance abuse, schizophrenia, depression, neurotoxicity and chronic fatigue syndrome. Home and Office: 4429 Southern Ave Dallas TX 75205-2622

SIMON, WAYNE EUGENE, engineer, mathematician; b. Dupree, S.D., Dec. 7, 1928; s. Max Alfred Ernst and Ruth Leona (Pittam) S.; m. Miriam June Honenberger, Sept. 23, 1956; children: Max Boehm, Jeb Heilig, Tavey Ann. BSME, Wash. State U., 1949; MS in Aero. Engring., Ohio State U., 1959; PhD in Aero. Engring. Sci., U. Colo., 1967. Registered profl. engr., Minn., Colo. Engr. Am. Gas Machine Co., Albertlea, Minn., 1950-51, Conlon Moore Corp., Joliet, Ill., 1951-52; engring. specialist North Am. Aviation, Columbus, Ohio, 1952-54, 1955-60; prin. scientist Martin Marietta Corp., Denver, 1960-93; pres. Tactical Tech. Solutions, Broomfield, Colo., 1993—. Contbr. articles to profl. jours. With U.S. Army, 1954-55. Mem. ASME, AIAA (Engr. of Yr. Rocky Mountain sect. and Region V 1993). Lutheran. Achievements include patents for explosive welding, fluid dynamic ejectors, mechanisms. Home: PO Box 125 Evergreen CO 80439 Office: Tactical Tech Solutions PO Box 656 Broomfield CO 80038

SIMONAITIS, RICHARD AMBROSE, chemist; b. Chgo., Dec. 7, 1930; s. George Peter and Sofija Constance (Woijkiewicz) S.; m. Vera Sandra Hall, Sept. 17, 1960; children: Steven, Rachel, Laura. Student Loyola U., Chgo., 1948-50; BS, U. Ill., 1952; postgrad. Ohio State U., 1952-55, MS, 1957, PhD, 1962. Chemist, Aerojet-Gen. Corp., Nimbus, Calif., 1962-64; rsch. chemist, Gulf Oil Corp., Merriam, Kans., 1964-66; analytical chemist, Gen. Electric Co., Liverpool, N.Y., 1966-69; rsch. chemist, Agrl. Rsch. Svc., U.S. Dept. Agr., Savannah, Ga., 1970—; abstractor, Chem. Abstracts, 1965—. Bd. dirs. Savannah coun. Girl Scouts U.S.A., 1978-84, exec. com., 1980-84, neighborhood chmn., Oleander Neighborhood, 1980-89; booth chmn., Night in Old Savannah Ethnic Festival, 1977—; usher, Nativity of Our Lord Ch., 1974—, capt. ushers, 1977—, sec., Men's Club, 1976, Sunday sch. tchr., 1977-81; bd. dirs., Savannah Young People's Theater, 1980-85, treas., 1983-85; bd. dirs., Savannah Theatre Co., 1990—, treas., 1991—. With U.S. Army, 1955-56. Mem. Am. Chem. Soc. (exec. com. 1979-83, disting. contbn. plaque, 1978, cert. recognition Chem. Abstract Service 1975, sec., treas., 1979, chmn. elect., 1980, chmn., 1981, counselor, 1981), Entomol. Soc. Am., Rsch. Soc. Am., Ga. Entomol. Soc., Assn. Ofcl. Analytical Chemists, ASTM, Chem. Analysts Central N.Y., Wilmington Island Pleasure and Improvement Assn. (treas. 1975—), Tybee Light Power Squadron. Sigma Xi, Phi Lambda Upsilon. Roman Catholic. Lodge: K.C. Contbr. numerous articles to sci. jours. Office: USDA Agrl Rsch Svc PO Box 22909 3401 Edwin St Savannah GA 31403

SIMONE, JAMES NICHOLAS, physicist; b. Mineola, N.Y., Aug. 26, 1956; s. James J. and Genevieve M. Simone; m. Georgia P. Schwender, Dec. 28, 1991. MS, Calif. State U., L.A., 1980; PhD, UCLA, 1991. Lectr. physics Calif. State U., L.A., 1980-85; software engr. Oasys, Cambridge, Mass., 1986-87; rsch. asst. physics dept. UCLA, 1987-89; rsch. assoc. Brookhaven Nat. Lab., Upton, N.Y., 1989-91; postdoctoral researcher Edinburgh (Scotland) U., 1991—; cons. Greenhills Software, Glendale, Calif., 1987-89. Achievements include lattice study of nonleptonic charm decays. Home: 131 Comiston Rd, Edinburgh EH10 6AQ, Scotland Office: Edinburgh U Dept Physics, Kings Bldgs, Edinburgh EH9 3JZ, Scotland

SIMONETTI, IGNAZIO, cardiologist; b. Taranto, Italy, May 5, 1949; s. Leonida Simonetti and Rita (Dionisio) Catalano; m. Amalia Magnavacca, Oct. 10, 1976; 1 child, Stefano. BA in Sic., Sci. Sch., 1969; MD, U. Pisa, Italy, 1976, Subspecialty in Cardiology, 1979. Med. Diplomate Cardiology. Resident internal medicine U. Pisa, Italy, 1976-77, fellow cardiology, 1977-79, univ. investigator, 1981—; vis. asst. prof. U. Iowa, Iowa City, 1986-88; asst. dir. cardiac catheterization lab. C.N.R. Inst. Clin. Physiology, Pisa, 1989—; staff coronary care unit C.N.R. Inst. Clin. Physiology, Pisa, 1979-85, staff cardiac catheterization lab., 1980—, Univ. of Iowa Hosps., 1986-88; cons. in field. Contbr. 35 articles to sci. papers, 28 chpts. to sci. books, 70 abstracts presented to sci. meetings. Recipient EURATOM fellowship, 1976-78, C.N.R. Fellowship award, 1978-81, NATO-CNR Sr. Fellowship award, 1986, European Community grant, 1989. Fellow Am. Fedn. for Clin. Rsch.; internat. fellow Am. heart Assn.; assoc. fellow Am. Coll. Cardiology; mem. Italian Soc. Cardiology, European Soc. Cardiology. Avocations: tennis, fishing. Home: Via Lavagna # 9, 56100 Pisa Italy Office: CNR Inst Clin Physiology, Via Savi # 8, 56100 Pisa Italy

SIMONIAN, SIMON JOHN, surgeon, scientist, educator; b. Antioch, French Ter., Apr. 20, 1932; came to U.S., 1965, naturalized, 1976.; s. John Simon and Marie Cecile (Tomboulian) S.; m. Arpi Ani Yeghiayan, July 11, 1965; children: Leonard Armen, Charles Haig, Andrew Hovig. MD, U. London, 1957; BA in Animal Physiology, St. Edmund Hall, U. Oxford, Eng., 1964; MA in Physiology, U. Oxford, Eng., 1969; MSc, Harvard U., 1967, Sc.D., 1969. Diplomate Am. Bd. Surgery. Rsch. asst. immunology unit Lister Inst. Preventive Medicine, Elstree, Essex, U.K., 1952; intern Univ. Coll. Hosp., London, 1957; intern Edinburgh (Scotland) Royal Infirmary, 1957-58, resident, 1961-62; clin. clk. Nat. Hosp. London, 1958; resident Edinburgh Western Gen. Hosp., 1958-59, City Hosp., Edinburgh, Birmingham Accident and Burns Hosp., U. Birmingham, Eng., 1959-60; demonstrator dept. anatomy Edinburgh U., 1960-61; rsch. fellow in pathology Harvard U., Boston, 1965-68; trainee NIH Harvard U., 1967; instr. immunology Harvard Med. Sch., Boston, 1966-70; instr., assoc. in surgery Harvard Med. Sch., 1968-70, surg. dir. course on transplantation, biology and medicine, 1968-70; vis. prof. Harvard Med. Sch., Mass. Gen. Hosp., Brigham and Womens Hosp., New Eng. Deaconess Hosp., 1982; dir. transplantation immunology unit, asst. in surgery Brigham and Womens Hosp., Boston, 1968-70; resident in surgery Boston City Hosp., 1970-74; attending surgeon in transplantation and gen. surgery services U. Chgo. Med. Ctr., 1974-77; asst. prof. surgery, mem. com. immunology U. Chgo., 1974-77; head div. renal transplantation Hahnemann U. Sch. Medicine and Hosp., 1978-87, prof. surgery, 1978-88, chmn. Transplantation Com., 1983-88, chmn. quality assurance of surgery com., 1986-88; dept. surgery coord. with joint commn. for accreditation of hosps. Hahnemann U. Sch. Medicine, 1986; chief and chmn. dept. surgery St. John Hosp. and Med. Ctr., Detroit, 1988-89, chmn. credentials com. of surgery and oper. rm. com., 1988-89, assoc. v.p. for med. affairs, 1989-90; pres., CEO Vein Inst. of Met. Washington, Inc., 1990—; assoc. Fairfax Hosp., Falls Church, Va., 1990-92, active faculty, 1992—; clin. assoc. prof. surgery Georgetown U. Sch. Medicine, Washington, 1992—; lectr. in field; vis. prof. dept. biochemistry Vanderbilt U., 1968, Cedars-Sinai Med. Ctr., UCLA, 1977, Addenbroke's Hosp., Cambridge U., 1977, Karolinska Inst., 1977, Huddinge Hosp., U. Stockholm, 1977, Med. Coll. Pa. and Hosp., 1980, 81, 85, Grad. Hosp., U. Pa., 1981, 85, U. Athens, 1981, Univ. Coll. Hosp., U. London, 1981, VA Hosp., Tufts U., 1982, John Radcliffe Hosp., Oxford U., 1982, Western Gen. Hosp., U. Edinburgh, 1982; cons. in gen. surgery City of Phila., 1986-88; chief med. team support for U.S. presdl. visits to Detroit, 1988, 89; vis. surgeon Inst. Vein Disease, Mich., 1989-90; vis. scientist Argonne (Ill.) Nat. Lab., 1969, 74-77; guest lectr. Eight Internat. Congress of Nephrology, Athens, Greece, 1981, Internat. Soc. Edn. and Rsch. in Vascular Disease, San Diego, bd. dirs., 1992; session chmn. 5th Armenian Medical World Congress, Paris France, 1992, session chmn. 11th World Congress Internat. Union of Phlebology, Montreal, 1992; session chmn. 6th Annual Meeting N.Am. Soc. of Phlebology, Lake Buona Vista, Fla., 1993, sec., bd. dirs. Woodrock Inc., 1992. Co-author: Manual of Vascular Access Procedures, 1987; cons. to editorial bd. dateline: Issues in Transplantation, 1985-87; mem. editorial bd. Phila. Medicine, 1988, Transplantation Proc., 1987—, Jour. Transplantation Abstracts, 1968-70; mem. rev. bd. New England Jour. Medicine, 1993—, Jour. Am. Med. Assn., 1993; contbr. articles to profl. jours. and books; appeared in med. movie Giving. Co-founder Armenian Youth Soc., Eng., 1953, pres., 1953-54; Armenian Studies program U. Chgo., 1975; bd. govs. Friends Sch., London, 1964-65; Mass. del., co-founder Armenian Assembly, Washington, 1970-74; trustee, fellow, co-founder Entry into Manhood of Armenian Youth at Age 13, 1981; co-founder ArmenianHealth Assn. of Greater Washington, 1992, mem. pharmaceuticals com., 1992—; mem. Am. Friends of St. Edmund Hall, U. Oxford, 1992—; mem. Rep. Presdl. Task Force; mem. St. Mary's Armenian Apostolic Ch., Washington; bd. dirs. Arlington (Va.) Symphony Orch., 1992. Nairn scholar, 1949-52; Middlesex scholar, 1952-57; recipient Suckling prize, 1956, Brit. Med. Research Council award, 1962-64, Alt prize, 1973, Thompson award, 1974-77, Johnson award, 1975-77, Presdl. Medal of Merit, 1982, Kabakjian award Armenian Student Assn. Am., 1986; named outstanding new citizen of Citizenship Coun. of Met. Chgo. and Dept. Justice, Washington, 1976-77, Jonathan E. Rhoads ann. orator, 1984; co-endowed The John and Marie J. Simonian Award, St. Nerces Sem., 1981, John R. Pfeifer, MD, Rsch. Award, Providence Hosp., Southfield, Mich., 1992; endowed the Dennis Knight prize Royal Acad. Music, London, 1991; endowed The Marie J. Simonian Prize, Georgetown U. Med. Sch., 1991 (prize com. 1991—); established The John N.D. Kelly Prize in Med. Studies St. Edmund Hall, U. Oxford, 1992, The Simon J. and Arpi A. Simonian Prize for scholastic excellence for doctoral candidates, Harvard U., 1992, The Carys M. Bannister Prize in Med. Rsch. U. Manchester Med. Sch., U.K., 1993; recognized for philanthropy to Hahnemann U. by placques in med. sch. and hosp. lobbies.; grantee U.S. Govt., industry cos., founds. Fellow Royal Coll. Surgeons Edinburgh, ACS (Phila., Mich. and Washington chpts.), Phila. Acad. Surgery (Jonathan E. Rhoads ann. orator 1984—, Samuel D. Gross prize com. 1988, councillor 1988); mem. AAAS, AMA (mem. jour. rev. bd. 1993), AAUP, Royal Coll. Surgeons of Eng., Royal Coll. Physicians of London Licentiates, Nat. Assn. Armenian Studies and Rsch., Armenian Gen. Benevolent Union (pres.' club 1990—), Knights of Vartan, Am. Armenian Med. Assn. (co-founder 1972, treas. 1972-74), Brit. Med. Assn., Immunology Club Boston, Cancer Rsch. Assn. Boston, Physicians for Social Responsibility, Am. Pub. Health Assn., Assn. for Study of Med. Edn., Armenian Med. and Dental Assn. Greater Phila. (co-founder 1983, pres. 1983-85, Outreach award 1986), Assn. Acad. Surgery, Transplantation Soc. (mem. membership com. 1980-82), Am. Fedn. Clin. Research, N.Y. Acad. Scis., Am. Soc. Transplant Surgeons (co-founding mem. 1974, chmn. immunosuppression study com. 1974-77, membership com. 1985-87), Am. Venous Forum, Assn. of Ill. Transplant Surgeons, Chgo. Assn. Immunologists, Chgo. Soc. Gastroenterology, Phila. Acad. Scis. (co-chmn. membership com. 1980-88), Greater Delaware Valley Soc. Transplant Surgeons (councillor 1978-80, 85-88, pres. elect 1980-82, pres. 1982-85), Phila. County Med. Soc. (rep. City Ctr. br. 1981-83, pres. 1984, bd. dirs. 1985-87, chmn. long range planning com. 1986-88), Pa. Med. Soc., Samuel Hahnemann Surg. Soc., Am. Technion Soc., Am. Soc. Artificial Internal Organs, European Soc. Organ Transplant, Oxford and Cambridge Soc. of Phila. and Washington, Internat. Cardiovascular Soc. (N.Am. chpt.), North Am. Soc. Phlebology (curriculum devel. projects com. 1992—, faculty 1993), End Stage Renal Disease Network 24 (mem. med. rev. bd. 1980-82, 1986-87), Am. Coll. Physician Execs., Detroit Acad. Surgery, Detroit Surgical Assn., Transplantation Soc. Mich., Organ Procurement Agy. Mich. (adv. bd. 1988-89), Wayne County Med. Soc., Mich. State Med. Soc., Fairfax County Med. Soc., Med. Soc. Va., Met. Vascular Conf., Soc. Brigham Surg. Alumni, Greater Washington Telecomm. Assn., Chesapeake Vascular Soc., Washington Acad. Surgery, Harvard Club (Phila. and Washington), Med. Club (Phila.), U. Chgo. Club (Washington), Sigma Xi. Mem. Soc. of Friends. Achievements include demonstration that after bilateral lung reimplantation and vagus section in dogs lung function recovers without vagus nerve regeneration, that staged-en-masse cardiopulmonary reimplantation in dogs results in survival with recovery of heart and lung function, that zinc deficiency in rats depressed zinc dependeyme activities, that solubilized specific donor antigen pretreatment of recipient dogs and rats resulted in prolonged survival of renal allografts due to measured quantities of enhancing antibody, that IgG fractionation by column chromatography of antilymphocyte globulin enhances potency for prolonging renal allograft survival in dogs; demonstration of the reversal of renal transplant rejection with IgG fraction of antithymocyte globulin in a randomized prospective clinical trial, of a method of quantitatively measuring and controlling gastric pH in prevention and treatment of massive gastroduodenal hemorrhage from hemorrhagic gastritis in the human, of the successful conversion of an arteriovenous shunt, for initiation of acute hemodialysis, to an arteriovenous fistula for immediate use in maintenance chronic hemodialysis; demonstration of the non-labile synthesis of a protein carrier with a cytotoxic 211 Astatine Conjugate as a model with antibody carrier for eventual use in the specific control of transplant rejection, cancer and other diseases, of the ureter as a new component of renal allograft rejection; co-discovery of genetic control of antibody formation in the rat, co-discovery that essential aminoacids, phenylalanine and tryptophan, are absolute requirements for normal antibody formation in the rat; assistance in the lyophilization of the smallpox vaccine, which, in large part, helped WHO to eliminate smallpox as an epidemic by 1978; collaboration and co-authorship with Joseph E. Murray pioneer of human kidney transplantation; research on the advantages, disadvantages and prevention of splenectomy in renal transplant recipients; confirmation of the diagnostic usefulness of quantitative air plethysmography in venous insufficiency, confirmation that appropriate prophlaxis leads to significant reduction in deep vein thrombosis. Office: 3301 Woodburn Rd Annandale VA 22003-1229

SIMONS, DAVID STUART, physicist; b. Columbus, Ohio, July 29, 1945; s. Eugene Morris and Ruth (Lazar) S.; m. Rachel Cropsey, June 13, 1976; children: Daniel, Julia. BS in Physics, Carnegie-Mellon U., 1967; PhD in Physics, U. Ill., 1973. Rsch. assoc. Materials Rsch. Lab. U. Ill., Urbana, 1973-76; chemist Knolls Atomic Power Lab., Schenectady, N.Y., 1976-79; rsch. physicist Nat. Inst. Stds. and Tech., Gaithersburg, Md., 1979—. Editor: Secondary Ion Mass Spectrometry SIMS V, 1986; contbr. articles to profl. jours. Recipient Bronze medal U.S. Dept. Commerce 1984. Mem. Am. Soc. Mass Spectrometry, Am. Vacuum Soc., Microbeam Analysis Soc. (sec. 1992—), Sigma Xi. Achievements include developing quantitative data reduction procedures for laser microprobe mass spectrometry and secondary ion mass spectrometry. Office: Nat Inst Stds and Tech Bldg 222 Rm A 113 Gaithersburg MD 20899

SIMONS, ELWYN LAVERNE, physical anthropologist, primatologist, paleontologist, educator; b. Lawrence, Kans., July 14, 1930; s. Verne Franklin and Verna Irene (Cuddeback) S.; m. Mary Hoyt Fitch, June, 1964(div. 1972); 1 child, David Brenton Simons; m. Friderum Annursel Ankel, Dec. 2, 1972; children: Cornelia Verna Mathilde Simons, Verne Franklin Herbert Simons. BS, Rice U., 1953; MA, Princeton U., 1955, PhD, 1956; D Philosophy, Oxford U., England, 1959; MA (hon.), Yale U., 1967. Demonstrator, exhibitor Oxford U., England, 1956-58; asst. prof. zoology U. Pa., Phila., 1959-61; vis. assoc. prof., curator Yale U., New Haven, Conn., 1960-61, head, divsn. vert. paleontology, 1960-77, prof. paleontology, 1967; prof. anthropology, anatomy Duke U., Durham, N.C., 1977-82; dir, Duke Primate Ctr., Duke U., Durham, N.C., 1977-91, James B. Duke prof., 1982—, sci. dir., 1991—; bd. dirs., Ctr. Tropical Conservation, N.C., Malagasy Fauna Group; lectr. Dept. Geology Princeton U., N.J., 1959. Author: Primate Evolution, 1972, Simons Family History, 1975; contbr. over 200 articles to profl jours. Gen. George Marshall scholar Oxford Univ., 1956-59; Recipient Anadale Mem. medal Asiatic Soc. of Calcutta, 1973; Alexander van Humboldt Sci. award, Germany, 1975-76; Named Hon. citizen by Fayum Province, Egypt, 1983. Democrat. Achievements include discoveries primarily concerned with fossil and living primates; discovered Gigantopithecus on India, 1968; Named earliest higher primates from Egypt: Oligopithecus 1962, Aegyptopithecus, 1965, Patrania, 1983, Catopithecus and Proteopithecus, 1989, Serapia, Plesiopithecus and Arsinoea, 1992. Home: 2518 Lanier Pl Durham NC 27705 Office: Duke Univ Primate Ctr 3705 Erwin Rd Box 90385 Durham NC 27705

SIMONS, GALE GENE, nuclear engineering educator, university administrator; b. Kingman, Kans., Sept. 25, 1939; s. Robert Earl and Laura V. (Swartz) S.; m. Barbara Irene Rinkel, July 2, 1966; 1 child, Curtis Dean. BS, Kans. State U., 1962, MS, 1964, PhD, 1968. Engr. Argonne Nat. Lab., Idaho Falls, Idaho, 1968-77, mgr. fast source reactor, head exptl. support group, 1972-77; prof. nuclear engring. Kans. State U., Manhattan, 1977—, assoc. dean for rsch., dir. rsch. coun. Coll. Engring., 1988—, bd. dirs. Rsch. Found., 1988—, Presdl. lectr., 1983—, career counselor, 1984—; cons. to pvt. and fed. agys., 1983—; bd. dirs. Kans. Tech. Enterprise Corp., Topeka; com. mem. Kans. Gov.'s Energy Policy Com., Topeka, 1992—; numerous presentations in field; reviewer proposals fed. agys. Contbr. over 100 articles to sci. jours.; patentee radiation dosimeter. Expert witness State of Kans., Topeka, 1986. Fellow AEC, 1964-67; numerous grants from fed. agys., 1979—. Mem. AAAS, IEEE, Am. Nuclear Soc., Health Physics Soc., Am. Soc. for Engring. Edn., Masons, Rotary, Phi Kappa Phi, Tau Beta Pi, Pi Mu Epsilon. Home: 2395 Grandview Terr Manhattan KS 66502 Office: Kans State U Durland Hall Rm 148 Manhattan KS 66506-5103

SIMONSEN, JOHN CHARLES, exercise physiologist; b. Rochester, N.Y., Mar. 21, 1955; s. David Raymond and Lillie Mae (Fuller) S.; m. Mary Virginia Alexander, Oct. 10, 1992. BA, Furman U., 1977, MBA, 1980; MS, Tex. Tech. U., 1985; PhD, Ohio State U., 1989. Lectr. Ohio State U., Columbus, 1989-90; work physiologist Ohio Indsl. Rehab. Commn., Columbus, 1990-91; exercise physiologist Martinat Outpatient Rehab. Ctr., Winston-Salem, N.C., 1991—. Contbr. articles to profl. jours. With U.S. Army, 1978-83. Mem. Am. Coll. Sports Medicine, U.S. Orienteering Fedn. (bd. dirs. 1990-91, v.p. competition 1991—), Phi Kappa Phi, Sigma Xi. Home: 235 Oakwood Ct Winston-Salem NC 27103 Office: Martinat Outpatient Rehab Ctr 1903 S Hawthorne Rd Winston Salem NC 27103

SIMPERS, GLEN RICHARD, aerospace engineer; b. Balt., Oct. 20, 1952; s. Robert Berwagner and Annie Catherine (Patterson) S.; m. Bonnie Fern Clarke, May 4, 1974; children: Scot Matthew, Lori Kristin. B of Engring., Va. Poly. Inst., 1973; MA, MS in Engring., Princeton U., 1976. Rsch. asst. Princeton (N.J.) U., 1973-76; assoc. engr. Grumman Aerospace Corp., Bethpage, N.Y., 1976-80; sr. engr. Naval Ordnance Sta., Indian Head, Md., 1980-86, U.S. Dept. Def., Washington, 1986—. Cubmaster Boy Scouts Am., Bryan Rd., Md., 1980-92; coun. mem. Peace Luth. Ch., Waldorf, Md., 1992. Mem. AIAA (sr. mem., author papers). Home: 367 Fawn Ln White Plains MD 20695 Office: US Dept Def Bolling AFB Washington DC 20340

SIMPKINS, PETER G., applied physicist, researcher; b. London; came to U.S., 1959; s. William Fredrick and Catherine Elizabeth (Bush) S.; m. Marion R. Hazell. MS, Calif. Inst. Tech., 1960; PhD, Imperial Coll., London, 1965. Apprentice engr. Handley Page Aircraft Co., Radlett, Eng., 1953-58; rsch. asst. Imperial Coll. Sci., London, 1960-65; cons. staff scientist AVCO Corp., Wilmington, Mass., 1965-68; mem. tech. staff Bell Telephone Labs., Whippany, N.J., 1968-73; sr. fellow U. Southampton, Eng., 1973-75; mem. tech. staff AT&T Bell Labs., Murray Hill, N.J., 1975—. Author 60 tech. papers. AGARD scholar Von Karman Inst., Brussels, 1958-59; NATO scholar Calif. Inst. Tech., 1959-60. Mem. ASME, Am. Phys. Soc. Achievements include 1 patent in field. Office: AT&T Bell Labs 600 Mountain Ave Murray Hill NJ 07974-0636

SIMPSON, BERYL B., botany educator; b. Dallas, Apr. 28, 1942; d. Edward Everett and Barbara Frances (Brintnall) S.; children: Jonathan, Meghan. AB, Radcliffe Coll., 1964; MA, Harvard U., 1968, PhD, 1968. Rsch. fellow Arnold Arboretum/Gray Herbarium, Cambridge, Mass., 1969-71; curator Smithsonian Instn., Washington, 1971-78; prof. U. Tex., Austin, 1978—; chmn. U.S. Com. to IUBS, 1985-88; co-pres. Internat. Congress Systematic and Evolutionary Biology, 1980-85. Author: Economic Botany, 1986; editor: Mesquite, 1977; contbr. over 100 articles and notes to profl. jours. Fellow, AAAS, 1983; recipient Greenman award Mo. Bot. Garden, 1970, Cooley award Am. Soc. Plant Taxonomists. Mem. Soc. for Study Evolution (pres. 1985-86), Bot. Soc. Am. (pres. 1990-91, BSA Merit award 1992), Bot. Soc. Washington (v.p. 1975), Soc. for Study Evolution (coun. 1975-80). Office: U Tex Dept Botany BIO 308 Austin TX 78713

SIMPSON, DENNIS DWAYNE, psychologist, educator; b. Lubbock, Tex., Nov. 9, 1943; s. Homer Arnold and Georgie Lee (Barrett) S.; m. Sherry Ann Johnson, Aug. 20, 1965; children: Jason Renn, Jeffrey Todd, Jennifer Lynn. BA, U. Tex., 1966; PhD, Tex. Christian U., 1970. Asst. prof. psychology Tex. Christian U., Ft. Worth, 1970-74, assoc. prof., 1974-79, prof., 1979-82, dir., prof., 1989—, S.B. Sells prof. psychology, 1992—; dir., prof. Tex. A&M U., College Station, 1982-89; sci. adv. bd. NIDA Rsch. Ctrs., Washington, 1992; adv. bd. Nat. Drug Treatment Evaluation Studies, Washington, 1992; expert advisor U.S. Acctg. Office, Health and Human Svcs., others; cons. WHO, fgn. govts. regarding drug rsch. Editorial bd. Am. Jour. of Drug & Alcohol Abuse, 1992; contbr. over 125 articles to profl. jours.; author 5 books. Recipient Disting. Rsch. Achievement award Tex. Commn. on Alcohol and Drug Abuse, 1987; recipient numerous grants. Mem. APA, Am. Psychol. Soc., Am. Evaluation Assn., Soc. of Psychologists in Addictive Behaviors, Southwestern Psychol. Assn., Sigma Xi. Achievements include research emphasis on the process of treatment service delivery in relation client attributes and how they related to retention rates, relapse, posttreatment outcomes; research on drug use in the workplace, other areas. Office: Tex Christian U Inst Behavioral Rsch Box 32880 Fort Worth TX 76129

SIMPSON, GENE MILTON, III, mechanical engineer, manufacturing engineer; b. Oak Park, Ill., Oct. 8, 1967; m. Debbie J. Tomaino, Apr. 4, 1992. BSME, No. Ill. U., 1991. Mfg. engr. A.B. Dick Co., Chgo., 1989-91; mech. engr., mfg. engr. Semblex Corp., Elmhurst, Ill., 1991—. Evans scholar No. Ill. U., DeKalb, 1991. Office: Semblex Corp 199 W Diversey Elmhurst IL 60126

SIMPSON, HENRY KERTAN, JR., psychologist, educator; b. Chgo., Mar. 15, 1941; s. Henry Kertan and Miriam Rebecca (Marvill) S.; m. Roberta Lee Jackson, Jan. 28, 1965; children: Henry III, Elizabeth. BSEE, San Diego State U., 1964; MA in English, U. Calif., Santa Barbara, 1974, PhD in Ednl. Psychology, 1987. Rsch. assoc. Human Factors Rsch., Santa Barbara, Calif., 1964-77; west coast editor Digital Design mag., Costa Mesa, Calif., 1977; sr. scientist Anacapa Scis., Santa Barbara, 1978-84; freelance author, cons. Santa Barbara, 1984-87; prin. investigator Navy Personnel R&D Ctr., San Diego, 1987—. Author numerous computer textbooks, 1981—; contbr. articles to profl. jours. Cpl. USMC, 1960-65. Mem. IEEE, Assn. Computer Machinery, Human Factors Soc. (cons. 1984—). Democrat. Office: Navy Personnel R&D Ctr San Diego CA 92152-6800

SIMPSON, JOANNE MALKUS, meteorologist; b. Boston, Mar. 23, 1923; d. Russell and Virginia (Vaughan) Gerould; m. Robert H. Simpson, Jan. 6, 1965; children by previous marriage—David Starr Malkus, Steven Willem Malkus, Karen Elizabeth Malkus. B.S., U. Chgo., 1943, M.S., 1945, Ph.D., 1949; D.Sc. (hon.), SUNY, Albany, 1991. Instr. physics and meteorology Ill. Inst. Tech., 1946-49, asst. prof., 1949-51; meteorologist Woods Hole Oceanographic Instn., 1951-61; prof. meteorology UCLA, 1961-65; dir. exptl. meteorology lab. NOAA, Dept. Commerce, Washington, 1965-74; prof. environ. scis. U. Va., Charlottesville, 1974-76; W.W. Corcoran prof. environ. scis. U. Va., 1976-81; head Severe Storms br. Goddard Lab. Atmospheres, NASA, Greenbelt, Md., 1981-88, chief scientist for meteorology, 1988—; Goddard sr. fellow, earth scis. dir. Goddard Space Flight Ctr., NASA, 1988-93; project scientist tropical rainfall measuring mission, 1986—; bd. dirs. Atmospheric Scis. and Climate, NRC/NAS, 1990—. Author: (with Herbert Riehl) Cloud Structure and Distributions Over the Tropical Pacific Ocean; assoc. editor: Revs. Geophysics and Space Physics, 1964-72, 75-77; contbr. articles to profl. jours. Mem. Fla. Gov.'s Environ. Coordinating Coun., 1971-74. Recipient Disting. Authorship award NOAA, 1969; Silver medal, 1967, Gold medal, 1972 both from Dept. Commerce; Vincent J. Schaefer award Weather Modification Assn., 1979; Community Headliner award Women in Comm., 1973; Profl. Achievement award U. Chgo. Alumni Assn., 1975, 1992, Lifetime Achievement award Women in Sci. Engring, 1990; Exceptional Sci. Achievement award NASA, 1982; named Woman of Yr. L.A. Times, 1963; Guggenheim fellow, 1954-55, Goddard Sr. fellow, 1988—. Fellow Am. Meteorol. Soc. (Meisinger award 1962, Rossby Rsch. medal 1983, coun. 1975-77, 79-81, exec. com. 1977, 79-81, commr. sci. and tech. activities 1982-88, pres.-elect 1988, pres. 1989, Charles Franklin Brooks award 1992, publs. commr. 1992—; mem. NAE, Am. Geophys. Union, Oceanography Soc., Cosmos Club, Phi Beta Kappa, Sigma Xi. Home: 540 N St SW Washington DC 20024-4557 Office: NASA Goddard Space Flight Ctr Earth Scis Dir Greenbelt MD 20771

SIMPSON, JOHN ALEXANDER, physicist; b. Portland, Oreg., Nov. 3, 1916; s. John A. and Janet (Br) S.; m. Elizabeth Alice Hilts, Nov. 30, 1946 (div. Sept. 1977); children: Mary Ann, John Alexander; m. Elizabeth Scott Johnson, Aug. 23, 1980. A.B., Reed Coll., 1940, D.Sc. (hon.), 1981; M.S., NYU, 1942, Ph.D., 1943. Research assoc. OSRD, 1941-42; sci. group leader Manhattan Project, 1943-46; instr. U. Chgo., 1945-47, asst. prof., 1947-49, assoc. prof., 1949-54, chmn. com. on biophysics, 1951; prof. physics, dept. physics and Fermi Inst. Nuclear Studies, 1954- 68, Edward L. Ryerson Disting. Service prof. physics, 1968-74, Arthur H. Compton Disting. Service prof. physics, 1974—, Arthur H. Compton Disting. prof. emeritus, 1987—; also dir. Enrico Fermi Inst., 1973-78; Mem. Internat. Com. IGY; chmn. bd. Ednl. Found. Nuclear Sci.; mem. tech. panel cosmic rays NRC; mem. Internat. Commn. Cosmic Radiation, 1962—; mem. astronomy missions bd. NASA, 1968; vis. assoc. physics Calif. Inst. Tech.; vis. scholar U. Calif., Berkeley.; founder Lab. Astrophysics and Space Research in Enrico Fermi Inst., 1962, Space Sci. Working Group, Washington, 1982. Bd. overseers, vis. com. astronomy Harvard; mem. Pres. Ford's Sci. Adv. Group on Sci. Problems, 1975-76; life trustee Adler Planetarium, 1977—. Recipient medal for exceptional sci. achievement NASA, Quantrell award for excellence in teaching, Gagarin medal Nat. Soviet Socialists Rep. Acad. of Scis., Cospar award UN Com. on Space Rsch., 1990; fellow Ctr. Policy Study, U. Chgo., Guggenheim fellow, 1972, 84-85; Nora and Edward Ryerson lectr., 1986; A. H. Compton Centennial lectr., 1992. Fellow Am. Acad. Arts and Scis., Am. Geophys. Union (Parker lectr. 1992), Am. Phys. Soc. (chmn. cosmic physics div. 1970-71); mem. NAS (mem. space sci. bd., Henryk Arctowski medal and premium 1993), Internat. Union Pure and Applied Physics (pres. cosmic ray commn. 1963-67), Atomic Scientists Chgo. (chmn. 1945-46, bd. bull. 1945—, pres. bull. bd. sponsors 1993—), Am. Astron. Soc. (Bruno Rossi prize 1991), Internat. Acad. Astronautics, Smithsonian Inst. (Martin Marietta chair in history of space sci. 1987-88, Glennan, Webb, Seamans Group 1986—), Phi Beta Kappa, Sigma Xi. Achievements include research in nuclear radiation and instrumentation, also origin of cosmic radiation, solar physics, magnetospheric physics, high energy astrophys. problems, and acceleration and isotopic and elemental composition of charged particles in space; prin. investigator for 31 expts. in earth satellites and deep space probes, also 1st probes to Mercury, Mars, Jupiter and Saturn, fly by at Venus, comet dust expts. on the 2 Vega spacecraft to Halley's comet, 1986; Ulysses space craft experiments over poles of the sun, 1990—, Pioneer 10 space craft outside solar system. Office: Fermi Inst 5630 S Ellis Ave Chicago IL 60637-1433

SIMPSON, JOHN AROL, retired government executive, physicist; b. Toronto, Ont., Can., Mar. 30, 1923; came to U.S., 1926; naturalized, 1938; s. Henry George and Verna Lavinia (Green) S.; m. Arlene Badel, Feb. 11, 1948; 1 child, George Badel. BS, Lehigh U., 1946, MS, 1948, PhD, 1951. Rsch. physicist Nat. Bur. Standards, Washington, 1948-62, supervisory physicist, 1962-69, dep. chief optical physics div., 1969-75, chief mechanics div., 1975-78; dir. Ctr. for Mfg. Engring. Nat. Bur. Standards, Gaithersburg, Md., 1978-91; dir. Mfg. Engring. Lab., Nat. Inst. Standards and Tech., Gaithersburg, 1991—; ret. Contbr. articles on electron optics to profl. jours. With U.S. Army, 1943-46. Recipient Silver medal Dept. Commerce, 1964, Gold medal, 1975; Allen V. Austin Measurement Sci. award, 1984; Disting. Exec. award Sr. Exec. Svc., 1985, Am. Machinist award, 1986. Fellow Am. Phys. Soc.; mem. NAE, Sigma Xi. Home: 312 Riley St Falls Church VA 22046-3310

SIMPSON, MYLES ALAN, acoustics researcher; b. N.Y.C., Mar. 15, 1947; s. Jacob and Sarah (Rubinstein) S.; m. Gail Helen Mezey, Aug. 16, 1970; children: Ian, Howard, Eric. BS in Physics, UCLA, 1967; MS in Physics, Brown U., 1970. Supervisory cons. Bolt Beranek & Newman, L.A., 1968-86; sr. prin. scientist and engr. McDonnell Douglas Corp., Long Beach, Calif., 1986—; cons. Agoura Hills, Calif., 1986—. Contbr. articles to profl. jours. Recipient Group Achievement award NASA 1991. Mem. AIAA, Acoustical Soc. Am., Inst. Noise Control Engring., Sigma Pi Sigma. Jewish. Achievements include specializing in community and transportation noise assessment and control and aircraft interior noise. Investigated active noise cancellation techniques and performed first demonstration of global cabin noise reduction on jet transport aircraft. Home: 28758 Aries St Agoura Hills CA 91301 Office: McDonnell Douglas Aerospace 3855 Lakewood Blvd MC 36-60 Long Beach CA 90846

SIMPSON, RAYMOND WILLIAM, electronics engineer; b. Merrick, N.Y., Aug. 4, 1944; s. Raymond G. and Lucy (Scheier) S.; m. Carole Irene Hyams, May 23, 1971; children: Raymond M., David W.M. BSEE, Poly. Inst. Bklyn., 1966, MSEE, 1966; PhD, U. Pa., 1976. Detection engr. Grumman Aerospace, Bethpage, N.Y., 1968-70; sr. engr. EG&G Princeton (N.J.) Applied Rsch., Inc., 1974-79; mem. tech. staff RCA Astro-Space, Princeton, 1979-80; sr. engr. Bactomatic, Inc., Princeton, 1980-81; cons. PA Cons. Group, Princeton, 1981-88; v.p. engring. O'Neill Comms. Inc., Princeton, 1988-91; engring. mgr. Princeton Instruments, Inc., Trenton, N.J., 1991—. Contbr. articles to profl. publs. 1st lt. U.S. Army, 1966-68. Mem. IEEE (panelist), Am. Radio Relay League (life). Achievements include 3 patents in spread spectrum radio and 1 in cellular radio. Home: 62 Wesleyan Dr Hamilton Square NJ 08690 Office: Princeton Instruments Inc 3660 Quakerbridge Rd Trenton NJ 08619

SIMPSON, STEPHEN LEE, consulting civil engineer; b. Roanoke, Va., Sept. 26, 1964; s. Irvin Warren and Ella Mae (Roman) S.; m. Helen Louise Steyer, May 23, 1987. BS in Civil Engring., Va. Poly. Inst. and State U., 1986. Registered profl. engr., Va., Md.; cert. bldg. plans examiner, plumbing plans examiner. Staff engr. Black & Veatch, Gaithersburg, Md., 1986-89, project engr., 1989—. Mem. Greenbriar Civic Assn., Fairfax, Va., 1988—. Mem. ASCE, Am. Water Works Assn., Am. Concrete Inst., Phi Kappa Phi, Golden Key, Chi Epsilon. Republican. Presbyterian. Achievements include research in ozone system design for water and wastewater treatment. Home: 4210 Penner Ln Fairfax VA 22033 Office: Black & Veatch # 500 18310 Montgomery Village Gaithersburg MD 20879

SIMRALL, DOROTHY VAN WINKLE, psychologist; b. Morris, Ill., Dec. 20, 1917; d. Lapsley Ewing and Madge (Van Winkle) Simrall. BA, Grinnell (Iowa) Coll., 1940; MA, U. N.C., 1942; PhD, U. Ill., 1945. Instr. Mt. Holyoke Coll., S. Hadley, Mass., 1945-48; asst. prof. Tulane U., New Orleans, 1948-51, Albion (Mich.) Coll., 1951-57, Drake U., Des Moines, 1957-62; assoc. prof. psychology St. Bonaventure U., Olean, N.Y., 1965-70; dir. and psychologist Cresson Ctr., Cresson, Pa., 1972-84, Polk (Pa.) Ctr., 1972-84; cons. in field; lectr. in field. Contbr. articles to profl. jours. Mem. Am. Psychol. Assn., S.C. Psychol. Assn., Sigma Xi. Avocations: writing, reading, statistics. Home: 702 Edwards Rd Apt 62 Greenville SC 29615-1203

SIMRIN, HARRY S., aerospace operations specialist; b. Oakland, Calif., July 19, 1945; s. David and Jean (Frey) S.; m. Vilma Grimany Manteiga, Sept. 15, 1977; 1 child, Felipe Manteiga. BS in Math, Harvey Mudd Coll., 1967; MS in Math., U. Fla., 1970; PhD in Math., U. North Tex., 1980. Vol. U.S. Peace Corps, Zamboanga, The Philippines, 1967-69; ops. analyst Gen. Dynamics, Ft. Worth, 1975-77, ops. analyst sr., 1977-80, engring. specialist, 1980-83, engring. chief, 1983-88, engring. specialist sr., 1988-90, engring. staff specialist, 1990-93; engring. staff specialist Lockheed Fort Worth Co., 1993—; speaker in field. Co-author text: Geometry for Teachers, 1969; author documents in field. Achievements include development for numerous operations analysis computer models. Home: 9901 Ivy Leaf Ln Fort Worth TX 76108 Office: Lockheed Fort Worth Co PO Box 748 Fort Worth TX 76101

SINAI, ALLEN LEO, economist, educator; b. Detroit, Apr. 4, 1939; s. Joseph and Betty Paula (Feinberg) S.; m. Lee Davis Etsten, June 23, 1963; children—Lauren Beth, Todd Michael. A.B., U. Mich., 1961; M.A., Northwestern U., 1966, Ph.D., 1969. Asst. prof. to assoc. prof. econs. U. Ill.-Chgo., 1966-75; chmn. fin. info. group, chief fin. economist Data Resources, Lexington, Mass., 1971-83; chief economist, mng. dir. Lehman Bros. and Shearson Lehman Bros. Inc., N.Y.C., 1983-87; chief economist, exec. v.p. The Boston Co. Inc., 1988-93; pres., CEO The Boston Co. Econ. Advisors Inc., Boston and N.Y.C., 1988-93; mng. dir. Lehman Bros., N.Y.C., 1993—; pres., CEO, chief economist Econ. Advisors, Inc., 1993—; cons. Laural Cons., Lexington and Evanston, Ill., 1966—; vis. assoc. prof. econs. and fin. MIT, Cambridge, 1975-77; adj. prof. econs. Boston U., 1977-78, 81-83, NYU, 1984-88; adj. prof. econs. and fin. Lemberg Sch., Brandeis U., 1988—, vis. faculty Sloan Sch., MIT, 1989-91. Contbr. articles to profl. jours. and books. Mem. reducing the fed. budget deficit task force Roosevelt Ctr., Washington, 1984; bd. govs. Com. on Developing Am. Capitalism, 1984—, chmn., 1990—. Recipient Alumnus Merit award Northwestern U., 1985. Mem. Am. Econ. Assn., Econometric Soc., Ea. Econs. Assn. (v.p. 1988-89, pres. 1990-91, Otto Eckstein prize 1988), Western Econ. Assn. Jewish. Avocations: tennis; skiing. Home: 16 Holmes Rd Lexington MA 02173-1917 Office: Econ Advisors Inc 15th Fl 260 Franklin Boston MA 02208

SINAI, YAKOV G., theoretical mathematician, educator; b. Moscow, Sept. 21, 1935. BS, Moscow State U., 1957, Ph.D. in Math., 1960, Doctor Degree, 1963; Dr. Honoris Causa (hon.), Warsaw U., 1993. Sci. rschr. lab. probabilistic and statis. methods Moscow State U., 1960-71; sr. rschr. Landau Inst. Theoretical Physics Acad. Scis., Moscow, USSR, 1971—; prof. math. Moscow State U., 1971-93; prof. math. dept. Princeton (N.J.) U., 1993—; Loeb lectr. Harvard U., 1978; plenary speaker Internat. Congresses Math. Physics, Berlin, 1981, Marseille, 1986, Internat. Congress Math., Kyoto, 1990; disting. lectr., Israel, 1989; S. Lefshetz lectr., Mex., 1990. Recipient Boltzman Gold medal, 1986, Heineman prize, 1989, Markov prize, 1990, Paul Adrian Maurice Dirac medal Internat. Centre for Theoretical Physics, 1992. Mem. Am. Acad. Arts and Sci. (fgn. hon.), Russian Acad. Scis., Hungarian Acad. Scis. (fgn.), London Math. Soc. (hon.). Office: Princeton University Dept of Mathematics Princeton NJ 08544*

SINATRA, FRANK RAYMOND, pediatric gastroenterologist; b. Glendale, Calif., June 7, 1945; s. Vincent and Opal (Rice) S.; m. Roberta Lee Blasquez-Hill, Aug. 17, 1968; children: Gina Maria, Vincent Robert. BA in Chemistry, Whittier Coll., 1967; MD, U. So. Calif., 1971. Diplomate Nat. Bd. Med. Examiners, Am. Bd. Pediatrics, Sub-bd. Pediatric Gastroenterology. Fellow Stanford (Calif.) U., 1974-76; div. head Children's Hosp., L.A., 1976—; prof. pediatrics U. So. Calif., L.A., 1976—; div. head L.A. County-U. So. Calif. Med. Ctr., 1988-92; adv. bd. Children's Liver Found., 1980-86, Am. Liver Found., 1987-88, United Liver Found., 1988-92. Contbr. over 100 sci. articles to profl. jours. and chpts. to books. Coach Glendale (Calif.) Little League, 1982-87, Diamond Busters Softball, Glendale, 1988-89. Recipient Resident award Children's Hosp. of L.A., 1972, Press Humanism award, 1982, Children's Liver Found. award Antelope Valley Chpt., 1985. Fellow Am. Acad. Pediatrics, Am. Coll. of Nutrition; mem. Am. Gastroenterol. Assn., North Am. Soc. for Pediatric Gastroenterology and Nutrition. Roman Catholic. Home: 264 Sleepy Hollow Ter Glendale CA 91206 Office: Children's Hosp of LA 4650 Sunset Blvd Los Angeles CA 90027

SINCLAIR, A(LBERT) RICHARD, petroleum engineer; b. Oklahoma City, Feb. 5, 1940; s. Albert Leonce and Elizabeth Serene (Turner) S.; m. Carol Sue Hodam, Aug. 20, 1962 (div. 1971); children: Steven Richard, David Carlisle; m. Carolyn Pittard, Mar. 22, 1975. BS in Mech. Engring., U. Okla., 1963, MS in Mech. Engring., 1964. Registered profl. engr., Tex. Rsch. scientist Ames Rsch. Ctr., NASA, Moffit Field, Calif., 1964-67; sr. rsch. engr. Exxon Prodn. Rsch., Houston, 1967-76; rsch. assoc. Maurer Engring., Houston, 1976-81; pres. Santrol, Inc., Houston, 1978-89, exec. v.p., 1989—; pres. Well Stimulation, Inc., Houston, 1981—; mem. rheology com. Am. Petroleum Inst., Houston, 1976-80. Author: Petrocalc 6 "Wellbore Stimulation", 1985; contbr. articles to profl. jours. 1st lt. USAF, 1964-67. Recipient Raymond Best Tech. Paper award AIME, N.Y.C., 1983. Mem. Soc. Petroleum Engrs. (Ferguson Best Tech. Paper award 1981). Achievements include 13 patents in oilfield improvements in technology, methods of stimulating oil and gas wells by hydraulic fracturing with new materials, high strength particles for oilfield operations. Office: Santrol 11757 Katy Fwy # 1260 Houston TX 77079

SINCLAIR, JOHN DAVID, psychologist; b. Bluefield, W.Va., Mar. 28, 1943; s. John Thornton and Carolyn June (Biddle) S.; m. Kirsti Kaarina Laine, May 18, 1973; children: Stephanie, Joanna, Pamela, Annette. BA, U. Cin., 1965, MA, 1967; PhD, U. Oreg., 1972. Teaching asst. U. Cin., 1963-67, rsch. asst., 1964-67; NDEA fellow U. Oreg., Eugene, 1967-70; NSF trainee U. Oreg., 1970-71; coord. psychol. rsch. Alko Ltd., Helsinki, Finland, 1972—; lectr. U. Helsinki, 1978; vis. scientist Ctr. for Advanced Study in Theoretical Psychology, Edmonton, 1979, U. N.C., Chapel Hill, 1980, Ind. Univ. Sch. Medicine, Indpls., 1988, 90. Co-author: Analyzing Data, 1970; editor: Animal Models in Alcohol Research, 1980; author: The Rest Principle, 1981; inventor in field of alcohol rsch.; permanent exhibit designer Finnish Sci. Ctr., Vantaa, Finland, 1984-92. Mem. AAAS, N.Y. Acad. Advancement of Sci. Avocations: art, popular writing. Home: Nokkalanniemi 7, FIN-02230 Espoo Finland Office: Alko Ltd Biomed Rsch Ctr, PO Box 350, SF-00101 Helsinki Finland

SINCLAIR, WARREN KEITH, radiation biophysicist, organization executive, consultant; b. Dunedin, New Zealand, Mar. 9, 1924; came to U.S., 1954; naturalized, 1959; s. Ernest W. and Jessie E. (Craig) S.; m. Elizabeth J.

Edwards, Mar. 19, 1948; children: Bruce W., Roslyn E. Munn. B.Sc., U. Otago, N.Z., 1944, M.Sc., 1945; Ph.D., U. London, 1950. Cert. Am. Bd. Health Physics. Radiol. physicist U. Otago, 1945-47; radiol. physicist U. London Royal Marsden Hosp., 1947-54; chmn. dept. physics, prof. U. Tex. M.D. Anderson Hosp., 1954-60; sr. biophysicist Argonne (Ill.) Nat. Lab., 1960—, div. dir., 1970-74, assoc. lab. dir., 1974-81; prof. radiation biology U. Chgo., 1964-85, prof. emeritus, 1985—; mem. Internat. Commn. on Radiation Units and Measurements, 1969-85, Internat. Commn. on Radiol. Protection, 1977—; alt. del. UN Sci. Com. on Effects of Atomic Radiation, 1979—; mem. Nat. Council on Radiation Protection and Measurements, 1967-91, hon. mem., 1991—, pres., 1977-91, pres. emeritus, 1991—, L.S. Taylor lectr., 1993; mem. expert panel WHO; sec. gen. 5th Internat. Congress Radiation Research, 1974; cons. in field. Author: Radiation Research: Biomedical, Chemical and Physical Perspectives, 1975; Contbr. numerous articles to profl. jours., also chpts. to books. Served with N.Z. Army, 1942-43. Nat. New Zealand scholar, 1942-45. Fellow Inst. Physics; mem. Am. Assn. Physicists in Medicine (pres. 1961-62, Coolidge award 1986), Radiation Research Soc. (council 1964-67, pres. 1978-79, Failla award 1987), Brit. Inst. Radiology (council 1953-54), Internat. Assn. Radiation Research (council 1966-70, 76-83), Radiol. Soc. N.Am., Biophys. Soc., Soc. Nuclear Medicine., Bioelectromagnetics Soc., Health Physics Soc., Soc. Risk Analysis, Hosp. Physicists Assn. Clubs: Innominates (U. Chgo.); Cosmos (Washington). Home: 2900 Ascott Ln Olney MD 20832-2626 Office: 7910 Woodmont Ave Ste 800 Bethesda MD 20814-3095

SINDELAR, ROBERT ALBERT, civil engineer; b. Friendship, Wis., Apr. 6, 1943; s. Harold Anthony and Mary Caroline (Pavlicek) S.; m. Dorothy Ann Allen, June 5, 1965; children: Jillene Jo, Cheryl Ann, Harold Mark. BA and Sci., U. Wis., Platteville, 1965; BS, U. Wis., 1968. Bldg. constrn. engr., planner Fruin Colon Constrn., St. Louis, Mo., 1968-70; hwy. constrn. engr., planner Dept. Transp. State of Wis., Green Bay, 1970-75; project coord. Barrientor & Assocs., Madison, Wis., 1975-78; hwy. engr. Dodge County Hwy. Commn., Juneau, Wis., 1978—. Recipient Award of Excellence Wis. Asphalt Pavers Assn., 1991. Mem. ASCE. Achievements include patents on method and apparatus for asphalt paving, asphalt plant with segmented drum and zonal heating. Home: 606 E South St Beaver Dam WI 53916 Office: Dodge County Hwy Commn 211 E Center St Juneau WI 53039

SINDONI, ELIO, physics educator; b. Merate, Como, Italy, Nov. 1, 1937; s. Adolfo and Amalia (Tocco) S. Laurea in Physics, U. Milan, 1961, PhD in Nuclear Physics, 1966. Asst. U. Milan, 1961-66, assoc. prof., 1967-90; prof. U. Udine, Italy, 1990-91, prof. gen. physics, 1991—; postdoctoral fellow Princeton (N.J.) U., 1969-70; prof. environ. sci. U. Milan, 1991—; sci. sec. Internat. Sch. Plasma Physics, Piero Caldirola, Milan-Varenna, Italy, 1971-84, pres., dir., 1984—. Author: Electromagnetismo I and II, 1976, Il Fuoco della Fusione, 1984; editor various courses and workshops and confs.; contbr. over 50 rsch. papers on plasma physics, thermonuclear fusion. Mem. Italian Phys. Soc., Am. Phys. Soc., European Phys. Soc. Roman Catholic. Avocation: classical music. Home: Via Della Sila 15, Milan 20131, Italy Office: Dept Physics U Milan, Via Celoria 16, Milan 20133, Italy

SINDORIS, ARTHUR RICHARD, electronics engineer, government official; b. Boston, Sept. 24, 1943; s. Joseph Charles and Frances Mildred (Ruka) S.; m. Lynne Meryl Pauker, June 16, 1968; children: Rebecca Wendy, Samantha Tasha. SB in Elec. Engring., MIT, 1965; MS in Elec. Engring., NYU, 1967, PhD in Elec. Engring., 1971. Rsch. engr. Harry Diamond Labs., Adelphi, Md., 1971-74; team leader Harry Diamond Labs., Adelphi, 1975-77, program mgr., 1980-82, br. chief, 1983-86; vis. assoc. prof. N.C. State U., Raleigh, 1978-79; spl. asst. Office Asst. Sec. Army, Dept. Army, Washington, 1987; vis. lectr. in electronics Capitol Coll., Laurel, Md., 1982-84; dep. dir. S3 tech. orgn. U.S. Army Lab. Command, Adelphi, 1988-92; assoc. chief signal processing div. Army Rsch. Lab., Adelphi, 1993—; exec. sec. InterSvc. Antenna Group, Adelphi, 1976-81; chair Automatic Target Recognition Subpanel of Joint Dirs. of Labs., 1990-93. Contbr. articles on antenna design to profl. jours.; patentee unique microwave antennas. Judge Sci. Fair Com., Montgomery County, Md., 1980—; sci. advisor Joint Bd. on Sci. Edn., Washington, 1981—; chmn. traffic safety Colesville (Md.) Civic Assn., 1989-92. Mem. IEEE (sr. exec. com. Washington sect. 1980-83, def. R&D com. 1983—, tech. policy conf. com. 1987—, chmn. MTT Soc. chpt. 1975), AAAS, Am. Phys. Soc., Am. Soc. Engring. Edn., N.Y. Acad. Scis., Robin Hood Tennis Club. Achievements include conformal antenna research and patriot missile electronics research. Office: SS-I Army Rsch Lab 2800 Powder Mill Rd Hyattsville MD 20783-1197

SINFELT, JOHN HENRY, chemist; b. Munson, Pa., Feb. 18, 1931; s. Henry Gustave and June Lillian (McDonald) S.; m. Muriel Jean Vadersen, July 14, 1956; 1 son, Klaus Herbert. B.S., Pa. State U., 1951; Ph.D., U. Ill., 1954, D.Sc. (hon.), 1981. Research engr. Exxon Research Engring. Co., Linden, N.J., 1954-57; sr. research engr. Exxon Research Engring. Co., 1957-62, research assoc., 1962-68, sr. research assoc., 1968-72, sci. advisor, 1972-79, sr. sci. advisor, 1979—; vis. prof. chem. engring. U. Minn., 1969; Lacey lectr. Calif. Inst. Tech., 1973; Reilly lectr. U. Notre Dame, 1974; Francois Gault lectr. catalysis Council Europe Research Group Catalysis, 1980; Mobay lectr. in chemistry U. Pitts., 1980; disting. vis. lectr. in chemistry U. Tex., 1981; Robert A. Welch Found. lectr. Confs. on Chem. Research, 1981; Camille and Henry Dreyfus lectr. UCLA, 1982; Edward Clark Lee Meml. lectr. U. Chgo., 1983; Dow Disting. lectr. in chemistry Mich. State U., 1984; Arthur D. Little lectr. Northeastern U., 1985; Vollmer W. Fries lectr. Rensselaer Poly. Inst., 1986; Disting. lectr. Ctr. Chem. Physics U. Fla., 1988. Contbr. articles to sci. jours. Recipient Dickson prize Carnegie-Mellon U., 1977, Internat. prize for new materials Am. Phys. Soc., 1978, Nat. Medal of Sci., 1979, Chem. Pioneer award Am. Inst. Chemists, 1981, Gold medal Am. Inst. Chemists, 1984, Perkin medal in chemistry Soc. Chem. Industry, 1984. Fellow Am. Acad. Arts and Scis.; mem. Am. Inst. Chem. Engrs. (Alpha Chi Sigma award 1971, Profl. Progress award 1975), Am. Chem. Soc. (Carothers lectr. Del. sect. 1982, Petroleum Chemistry award 1976, Murphree award 1986), Catalysis Soc. (Emmett award 1973), Nat. Acad. Scis., Nat. Acad. Engring. Methodist. Achievements include introduction and development of concept of bimetallic clusters as catalysts; invention of polymetallic cluster catalysts used commercially in petroleum reforming. Office: Exxon Research Engineering Co Clinton Township Route 22 East Annandale NJ 08801

SINGER, BURTON HERBERT, statistics educator; b. Chgo., June 12, 1938; married; 3 children. B.S., Case Inst. Tech., 1959, M.S., 1961; Ph.D. in Stats., Stanford U., 1967. From asst. to assoc. prof. stats. Columbia U., N.Y.C., 1967-77, prof. math. stats., 1977-85, chmn. dept. math. stats., 1985-89; then chmn. biostats., then chmn. and assoc. dean pub. health dept. epidemiology and pub. health Yale U., New Haven, 1989-91, chmn. epidemiology and pub. health, 1991-93, Ira Vaughan Hiscock prof. epidemiology and pub. health, 1991—, prof. econs. and stats. dept. epidemiology and pub. health, 1991—; research assoc. statistician Princeton U., 1972-73. Mem. AAAS, Am. Statis. Assn., Psychometric Soc. Office: Yale U Sch Medicine Dept Epidemiology/Health 60 College St # 3333 New Haven CT 06510-3210

SINGER, EDWARD NATHAN, engineer, consultant; b. Phila., Jan. 20, 1917; s. David and Esther (Levy) S.; (widowed Apr. 1965); 1 child, Gary L.; m. Hilda Gofstein, Sept. 7, 1966. BS, CCNY, 1938; MEE, Polytech. Inst. Bklyn., 1959. Registered profl. engr., N.Y. Electronic scientist Watson Labs., Eatontown, N.J., 1946-48; field engr. FCC, N.Y.C., 1948-54; radio engr. Naval Applied Sci. Lab., N.Y.C., 1954-70, N.Y. Fire Dept., N.Y.C., 1970-85; pvt. practice cons. radio engr. N.Y.C., 1985—. Author: Land Mobile Radio Systems, 1989; contbr. articles to profl. jours. Pres. Home Owners Assn., S.I., 1987-88. Capt. USAF, 1941-46, CBI. Fellow Radio Club Am.; mem. IEEE, N.Y. Acad. Scis., Sigma Xi. Jewish. Achievements include patents for pulse statistical distribution analyzer, pulse percent indicator, time controlled switching system, adjustable cam, and automatic peak level indicator system; development of broad band antenna for field intensity meters. Home and Office: 68 Claradon Ln Staten Island NY 10305-2809

SINGER, ISADORE MANUEL, mathematician, educator; b. Detroit, May 3, 1924; married; 5 children. BS, U. Mich., 1944; MS, U. Chgo., 1948, PhD in Math., 1950; ScD (hon.), Tulane U., 1981; LLD (hon.), U. Mich., 1989, U. Ill., Chgo. Moore instr. math. MIT, Cambridge, 1950-52, prof. math., 1956-70, Norbert Wiener prof., 1970-79, John D. MacArthur prof. math. (1st holder), 1983—, Inst. prof., 1987—; asst. prof. UCLA, 1952-54; vis. prof. math U. Calif., Berkeley, 1977-79, prof., 1979-83, Miller prof. math., 1982-83, prof. math., 1977-83; vis. asst. prof. math. Columbia U., N.Y.C., 1954-55; mem. Inst. Advanced Study, 1955-56; past steering com. Ctr. for Nonlinear Scis., Los Alamos Nat. Labs; adv. bd. Inst. Theoretical Physics, U. Calif., Santa Barbara; bd. dirs. Santa Fe Inst.; mem. various organizing coms. and editor proc. for confs. in field. Former editor profl. jours. Alfred P. Sloan fellow, 1959-62; Guggenheim fellow, 1968-69, 75-76; recipient Nat. Medal of Sci., 1983. Mem. NAS (past councillor, former mem. com. math. and phys. scis., other coms.), Am. Philos. Soc., Am. Acad. Arts and Scis., Am. Math. Soc. (v.p. 1970-72, past exec. com., Bocher Meml. prize 1969), Am. Phys. Soc., Internat. Congress Mathematicians (program com. 1986, Wigner prize 1989). Office: MIT Dept Math RM 2=174 Cambridge MA 02139-4307

SINGER, JEFFREY MICHAEL, organic analytical chemist; b. N.Y.C., Feb. 2, 1949; s. Samuel and Theresa (Pohl) S.; m. Linda Arlene Prizer, Oct. 13, 1972; 1 child, Sarah. BA, CUNY, 1971; MS, Rensselaer Poly. Inst., 1976; MA, CUNY, 1979; PhD, Poly. U., Bklyn., 1987. Analytical chemist Equitable Environ. Health Inc., Woodbury, N.Y., 1979-80; group leader Chemtech Cons. Group, N.Y.C., 1980-81; sr. chemist/lab. supr. Revlon Health Care, Tuckahoe, N.Y., 1981-86; analytical devel. chemist Lederle Labs., Pearl River, N.Y., 1986-87; sr. chemist PepsiCo Inc., Valhalla, N.Y., 1987-89; lab. mgr. Pall Corp., Glen Cove, N.Y., 1989-90; mgr. analytical tech. support Du Pont Pharms., Garden City, N.Y., 1990—. Author: Analytical Profiles of Drug Substances, 1985. Charter mem. N.Y. Hall of Sci., Flushing, 1985; judge borough competition N.Y.C. Annual Sci. Fair, Flushing, 1987, 88, 90. Mem. Assn. Official Analytical Chemists (program chair 1988-90, pres. N.Y.-N.J. sect. 1990-92), ACS, AAAS, ASTM, Am. Assn. Pharm. Scientists. Achievements include research in Chromatographic Analytical Methods Development of biologically and pharmacologically active molecules; in Pharmacognosy of Novel Natural Products; in Laboratory Information Management Systems. Office: DuPont Merck Pharm Co 1000 Stewart Ave Garden City NY 11530-4888

SINGER, JOSEF, aerospace engineer, educator; b. Vienna, Austria, Aug. 24, 1923; arrived in Israel, 1933; s. Zvi and Etel (Isler) S.; m. Shoshana, June 29, 1954; children: Gideon, Tamar, Uri. BSc Mech. Engring. (1st class hon.), Univ. London, 1948, DIC Aero. Engring., 1949; MAero. Engring., Poly Inst. Bklyn., 1953, DAero. Engring., 1957; DSc (hon.), Poly. Inst., N.Y., 1983; D honoris causa, Univ. d'Aix, Marseilles, 1986; D Engring. honoris causa, U. Glasgow, 1993. Engring. officer Israel Air Force, 1949-55, head test and devel. sect., 1953-55; from sr. lectr. to prof. Technion IIT Aeronautical Engring., Haifa, Israel, 1955-91, prof. emeritus, 1991—, head dept., 1958-60, 65-67; pres. Technion IIT, Haifa, Israel, 1982-86; vis. assoc. prof. aeronautics, Stanford (Calif.) U., 1963-64; vis. prof. aeronautics Calif. Inst. Tech., Pasadena, Calif., 1968-69, U. London, 1988, RWTH Aachen, Germany, U. Calif., 1990-91; sr. v.p. engring. Israel Aircraft Industries, 1971-73, chmn. rsch. and devel. com., 1979-82, 1986-87, chmn. bd. dirs., 1986-87; cons. in field; rschr. in field. Editor Congress ICAS Proceedings, 1974, 78, 80, Babcock Meml. Volume, 1989; edit. adv. bd., reviewer various sci. jours.; contbr. articles to profl. jours. Mem. Israel Coun. Higher Edn., 1975-81. Major RAF Israel Air Force, 1949-55. Vinton Hayes sr. fellow Harvard U., 1976-77; Sherman Fairchild Disting. scholar Calif. Inst. Tech., 1987-88, 89. Fellow AIAA, RAeS, Inst. Mech. Engring., City and Guilds London Inst.; mem. Internat. Coun. Aero. Scis. (pres. 1982-86, internat. program com., chmn. 1978-82, exec. com. 1974—, Maurice Roy medal 1990), U.S. Nat. Acad. Engring. (fgn. assoc.), Internat. Acad. Astronautics, Académie Nat. de l'Air et d'Espace (fgn. assoc.), Deutsche Gesellschaft Luft & Raumfahrt (hon. mem.), Israel Soc. Aeronautics and Astronautics (founding mem., pres.), Soc. Exptl. Stress Analysis. Home: 9 Malal St, Haifa Neve Shaanan 32714, Israel Office: Technion IIT, Aerospace Engr, Haifa 32000, Israel

SINGER, KENNETH DAVID, physicist, educator; b. Cleve., Aug. 1, 1952. BS in Physics, Ohio State U., 1975; PhD in Physics, U. Pa., 1981. Postdoctoral fellow U. Pa., Phila., 1981-82; mem. tech. staff AT&T Bell Lab., Princeton, N.J., 1982-90; assoc. prof. physics Case Western Res. U., Cleve., 1990—, Warren E. Rupp assoc. prof., 1990. Editorial bd. Nonlinear Optics, 1990—. Achievements include patent on polymeric nonlinear optical materials. Office: Case Western Res U Dept Physics Cleveland OH 44106-7079

SINGER, MAXINE FRANK, biochemist; b. N.Y.C., Feb. 15, 1931; d. Hyman S. and Henrietta (Perlowitz) Frank; m. Daniel Morris Singer, June 15, 1952; children: Amy Elizabeth, Ellen Ruth, David Byrd, Stephanie Frank. AB, Swarthmore Coll., 1952, DSc (hon.), 1978; PhD, Yale U., 1957; DSc (hon.), Wesleyan U., 1977, Swarthmore Coll., 1978, U.Md.-Baltimore County, 1985, Cedar Crest Coll., 1986, CUNY, 1988, Brandeis U., 1988, Radcliffe Coll., 1990, Williams Coll., 1990, Franklin and Marshall Coll., 1991, George Washington U., 1991, NYU, 1992, Lehigh U., 1992, Dartmouth Coll., 1993. USPHS postdoctoral fellow NIH, Bethesda, Md., 1956-58; rsch. chemist biochemistry NIH, 1958-74; head sect. on nucleic acid enzymology Nat. Cancer Inst., 1974-79; chief Lab. of Biochemistry, Nat. Cancer Inst., 1979-87, rsch. chemist, 1987-88; pres. Carnegie Inst. Washington, 1988—; Regents vis. lectr. U. Calif., Berkeley, 1981; bd. dirs. Johnson & Johnson; mem. sci. coun. Internat. Inst. Genetics and Biophysics, Naples, Italy, 1982-86; Chulabhorn Rsch. Inst. (adv. bd. 1990—). Mem. editorial bd. Jour. Biol. Chemistry, 1968-74, Sci. mag., 1972-82; chmn. editorial bd. Procs. of NAS, 1985-88; author (with Paul Berg) 2 books on molecular biology; contbr. articles to scholarly jours. Trustee Wesleyan U., Middletown, Conn., 1972-75, Yale Corp., New Haven, 1975-90; bd. govs. Weizmann Inst. Sci., Rehovot, Israel, 1978—; bd. dirs. Whitehead Inst., 1985—; chmn. Smithsonian Coun., 1992—. Recipient award for achievement in biol. scis. Washington Acad. Scis., 1969, award for research in biol. scis. Yale Sci. and Engring. Assn., 1974, Superior Service Honor award HEW, 1975, Dirs. award NIH, 1977, Disting. Service medal HHS, 1983, Presdl. Disting. Exec. Rank award, 1987, U.S. Disting. Exec. Rank award, 1987, Mory's Cup Bd. Govs. Mory's Assn., 1991, Wilbur Lucius Cross Medal of Honor Yale Grad. Sch. Assn., 1991, Nat. Medal Sci. NSF, 1992. Fellow Am. Acad. Arts and Scis.; mem. NAS (coun. 1982-85, com. Sci., Engring and Pub. Policy 1989-91), AAAS (Sci. Freedom and Responsibility award 1982), Am. Soc. Biol. Chemists, Am. Soc. Microbiologists, Am. Chem. Soc., Am. Philos. Soc., Inst. Medicine of NAS, Pontifical Acad. of Scis, Human Genome Org, Smithsonian coun., N.Y. Acad. Scis. Home: 5410 39th St NW Washington DC 20015-2902 Office: Carnegie Inst Washington 1530 P St NW Washington DC 20005

SINGER, PETER, physician, researcher, consultant; b. Danzig, Germany, Mar. 15, 1937; s. Fritz and Gertrud (Everhan) S.; m. Brigitta Kleff, Apr. 20, 1963 (div. June 1974); children: Franziska, Konrad. MD, Med. Sch. Charité, Berlin, 1962; DSc, Acad. Scis. German Dem. Republic, Berlin, 1979. Intern County Hosp., Berlin, 1961-62; asst. Clinic of Diabetes and Metabolic Diseases, Berlin-Kaulsdorf, German Dem. Republic, 1962-65; asst. lectr. med. dept. Charité, Berlin, 1965-69; sr. physician Clinic of Diabetes and Metabolic Diseases, Berlin, 1969-75; sr. cons. Ctr. Inst. Cardiovascular Rsch., Acad. Scis. German Dem. Republic, Berlin-Buch, 1975-88, head Clin. Lipid Rsch. Group, 1984-88; pvt. practice Heppenheim, Fed. Republic of Germany, 1991—; cons. Sci. Coun. Diabetes Rsch., Karlsburg, German Dem. Republic, 1981-88, Sci. Coun. Nutrition Rsch., Rehbrücke, German Dem. Republic, 1985-88; guest prof. diabetes, hypertension, lipids, omega-3 fatty acids, Can. and U.S.A., 1989. Contbr. over 300 articles to profl. jours.; contbr. chpts. to books; patentee lipids and polyunsaturated fatty acids fields, ISFE-Price, 1991. Lutheran. Avocations: sports, classical music, theatre. Home: Melibokusstr 14, 64625 Bensheim Germany Office: Gemeinschaftsprax, Lehrstr 26-28, 64646 Heppenheim Germany

SINGER, SHERWIN JEFFREY, theoretical chemist, chemistry educator; b. Chgo., Mar. 20, 1954; s. Alex and Jennie (Coretsky) S.; m. Madeline Rivera, June 9, 1984; children: Ana, Alex. BA, U. Chgo., 1976, PhD, 1984. Postdoctoral rschr. U. Pa., Phila., 1983-85, AT&T Bell Labs., Murray Hill, N.J., 1985-86; asst. prof. Ohio State U., Columbus, 1987-92, assoc. prof., 1992—. Contbr. articles to profl. jours. Recipient grant NSF, 1991; recipient Disting. New Faculty award Dreyfus Found., 1987. Mem. Am. Chem. Soc. (grant 1987), Am. Phys. Soc. Achievements include research in the theory of molecular dynamics; of statistical mechanics of pattern-forming fluids; of electronic processes in clusters and condensed phases. Office: Ohio State U Dept Chemistry 120 W 18th Ave Columbus OH 43210

SINGER, WILLIAM HARRY, computer/software engineer, expert systems designer, consultant, entrepreneur; b. Lancaster, Pa., Jan. 25, 1947; s. Wilbur Weitzel and Mildred (Myers) S.; m. Nanette Platt Willis, July 28, 1989. BS, U. Pitts., 1973; MS, Drexel U., 1982. Rsch. asst. Dept. Molecular Biology, E.P.P.I., Phila., 1976-81; rsch. fellow Dept. Biophys. Chemistry, Biozentrum, Basel, Switzerland, 1981-83; cons. to industry in applied artificial intelligence Palmyra, Pa., 1983—; ltd. ptnr. Tech. Svcs. Software, 1983-86; co-founder Singer, Stewart and New, Inc., 1987; gen. ptnr. Singer and Singer Assocs., 1993—; software cons. REORG, Darmstadt, 1984, Kroeplin GmbH, 1983-86; co-founder, prin. ptnr. Singer Cons., Inc., 1993—. Contbr. articles to profl. jours.; co-author: Diabetic's Daily Diary, 1987. Swiss Nat. Sci. Found. fellow, 1981-83. Mem. ACM, IEEE Computer Soc., Math. Assn. Am., Am. Assn. for Artificial Intelligence, Okinawa Isshinryu Karate Assn. (life), Wasserfahrverein Horburg of Basel (Switzerland) (life mem.). Achievements include research in drug and chemically induced membrane fusion, and statistical analysis and control of serum glucose levels in Type II diabetics. Home and Office: 1533 Cambridge Ct Palmyra PA 17078-9375

SINGH, ALLAN, psychology educator; b. New Amsterdam, Berbice, Guyana, Dec. 29, 1956. BA, U. Tex., 1984; PhD, Princeton U., 1990. Tchr. English Loscher-Ebbinghaus Lang. Inst., Caracas, Venezuela, 1977-80; lectr. Princeton (N.J.) U., 1989; instr. Frostburg (Md.) State U., 1989-90; asst. prof. psychology Trenton (N.J.) State Coll., 1990—. Recipient Tex. Good Neighbor award U. Tex., Austin, 1982-84. Mem. Am. Psychol. Soc., Ea. Psychol. Assn., Capital Area Social Psychol. Assn. (Dissertation award 1992), Phi Beta Kappa. Achievements include research to assess people's understanding of the contributions of situational and personality variables on behavior. Office: Trenton State Coll Hillwood Lakes Trenton NJ 08650-4700

SINGH, DEVENDRA PAL, polymer scientist; b. Shahjahanpur, India, Sept. 2, 1954; s. Darshan Singh and Shivinder Kaur; m. Khushwant Kaur, Mar. 30, 1986; children: Simreen Kaur, Mohineesh Singh. MSc, Loughborough U. Tech., 1977; PhD, U. West London, 1982. Sr. materials engr. Specnat Ltd., Leominster, Eng., 1982-86; scientist Ctr. Indsl. Rsch. (name now SINTEF-SI), Oslo, 1986—. Author: 1st Scandinavian SAMPE Symposium, 7th Internat. Conf. SAMPE Europe, France, 1987, 9th Scandinavian Symposium, NTH, 1987, others, Polymer Testing 1985. Mem. Soc. Advancement of Materials and Process Engring. (bd. dirs. 1986—). Sikh. Achievements include classified British patent for Radar Absorbing Materials, for Method and Apparatus for Forming, in Particular Hydroforming, a Composite Sheet Material; patent pending for Manufacturing Techniques for 3 Dimensional Formed Sandwich Structures; British patents for Method and Apparatus for Making Articles from Foamed Thermoplastic Materials, for Production of Cross-Linked Thermoplastic Foam. Home: Ullernchausseen 80A, 0381 Oslo Norway Office: SINTEF-SI, Forsknings Veien 1, N0314 Oslo Norway

SINGH, GAJENDRA, agricultural engineering educator; b. Bhogpur, India, Aug. 3, 1944; s. Sardar and Bhuro (Devi) S.; m. Vimlesh, July 30, 1971; 1 child, Arti Singh. BS in Agrl. Engring., Pantnagar U., India, 1966; MS, Rutgers U., 1968; PhD, U. Calif., Davis, 1973. Assoc. prof. Pantnagar U., U.P., India, 1973-75; asst. prof. Asian Inst. Tech., Bangkok, Thailand, 1975-77; assoc. prof. Asian Inst. Technol., Bangkok, Thailand, 1978-84, Rutgers U., New Brunswick, N.J., 1984-86; v.p. acad. affairs Asian Inst. Technol., Bangkok, 1986-88; vis. prof. Iowa State U., Ames, 1991; prof. Asian Inst. Technol., Bangkok, 1984—; cons. Asian Devel. Bank, Manila, Philippines, 1981-83, Food & Agriculture Orgn. of UN, Rome, 1990, Govt. Bhutan, Thimpu, 1990-91. Editor: Conf. Proceedings Rural Devel., 1977, Agrl. Engring. and Agrl. Industries, 1981, and Agrl. Engring., 1992; contbr. over 70 articles to profl. jours. Recipient John Gilmore award U. Calif., 1970, Emil E. Mark Internat. award, 1991; Commendation medal Indian Soc. Agrl. Engrs., 1985. Fellow India Soc. Agrl. Engring., Instn. Engrs., Asian Assn. Agrl. Engring. (pres. 1990—); mem. Am. Soc. Agrl. Engrs. (Kishida Internat. award 1990). Hindu. Avocations: reading, hiking. Office: Asian Inst Tech, GPO Box 2754, Bangkok 10501, Thailand

SINGH, HARKISHAN, chemist, educator; b. Lyallpur, India, Nov. 25, 1928; s. Subeg and Kirpal (Kaur) S.; m. Gian Kaur, June 15, 1958; children: Tript Pal, Manjeet Kaur. B in Pharmacy, Panjab U., 1950; PhD, Banaras Hindu U., 1956. Lectr. Banaras Hindu U., Uttar Pradesh, India, 1952-56; asst. prof. Saugar U., Madhya Pradesh, India, 1956-64; postdoctoral fellow U. Md., Balt., 1958-61; reader Panjab U., Chandigarh, India, 1964-72, prof., 1972-88, emeritus fellow, 1989-92; vis. prof. U. Miss., 1967-68; acad. staff fellow London U., 1971-72; nat. fellow Univ. Grants Commn., New Delhi, 1985-87; gen. pres. Indian Pharm. Congress, Calcutta, 1981-82. Co-author: Organic Pharmaceutical Chemistry, 1982, 87, 91, Medicinal Chemistry Research in India, 1985; contbr. chpts. to Progress in Medicinal Chemistry on Heterosteroids and Drug Research, 1979, 91; contbr. articles to profl. jours. Recipient Rsch. award Mody Rsch. Found., 1975, Ranbaxy Rsch. Found., 1987, G.P. Srivastava award Assn. Pharm. Tchrs. India, 1983. Mem. Am. Chem. Soc., Internat. Soc. Heterocyclic Chemistry, Indian Pharm. Assn. (P.C. Ray Gold medal 1982), Am. Inst. History Pharmacy. Sikh. Achievements include 14 patents in synthetic heterosteroids; research in chandonium iodide. Home: 1135 Sector 43, Chandigarh 160022, India

SINGH, JASWANT, environmental company executive; b. Gurne Kalan, Punjab, India, Aug. 12, 1937; came to U.S., 1963; s. Chanan and Ripudaman (Khara) S.; m. Mary Kathryn Todd, Feb. 5, 1939; children: David Paul, Monica Maria. PhD, Punjab U., 1963. Diplomate Am. Acad. Indsl. Hygiene. Rsch. assoc. U. So. Calif., L.A., 1963-65; rsch. fellow NRC of Can., Ottawa, 1965-68; dir. rsch. Pollution Dynamics Corp., Rochester, N.Y., 1968-70; tech. dir. Galson Tech. Svcs., Syracuse, N.Y., 1970-77; tech. dir., sr. v.p. Clayton Environ. Cons., Novi, Mich., 1977-89, Cypress, Calif., 1989—; bd. dirs. Am. Bd. Indsl. Hygiene, 1987—; leader Citizen Ambassador Program del. to China, U.S. Indsl. Hygiene, 1989. Author: Standard Handbook of Plant Engineering, 1983; contbr. articles to profl. jours. Mem. Am. Chem. Soc., Am. Indsl. Hygiene Assn., Brit. Occupational Hygiene Soc., Air. Pollution Control Assn., Applied Indsl. Hygiene Assn. (editor jour. 1985—). Avocations: sailing, traveling. Office: Clayton Environ Cons 22345 Roethel Dr Novi MI 48375

SINGH, KRISHNA DEO, toxicologist; b. Basti, India, Nov. 30, 1934; came to the U.S., 1958; s. Ram Kumar and Sampat (Devi) S.; children: Kalpana, Vandana, Satyendra. MS, Agra U., 1956; PhD, U. Mo., 1964. Diplomate Am. Bd. Forensic Toxicology. Sr. staff chemist S.B. Penick & Co., Lyndhurst, N.J., 1964-68; asst. mem. Inst. for Muscle Disease, N.Y.C., 1968-74; sr. toxicologist Office Chief Med. Examiner, N.Y.C., 1974—. Contbr. 20 articles to profl. jours. Rsch. fellow Govt. of India, 1956-58. Mem. Soc. Forensic Toxicologists, Acad. Forensic Scis., N.J. Assn. Forensic Scis. Home: 6 Callaway Terr Somerset NJ 08873

SINGH, MANMOHAN, orthopedic surgeon, educator; b. Patiala, Punjab, India, Oct. 5, 1940; came to U.S., 1969; s. Ajmer and Kartar (Kaur) S.; m. Manjit Anand, Jan. 1, 1974; children: Kirpal, Gurmeet. MB, BS, Govt. Med. Coll., Patiala, 1964; MSurgery, Panjab U., Chandigarh, India, 1964. Diplomate Am. Bd. Orthopaedic Surgery. Rsch. fellow Inst. Internat. Edn., Chgo., 1969-74; resident in orthopedic surgery Michael Reese Hosp. and Med. Ctr., Chgo., 1974-78, mem. attending staff, dir. orthopedic rsch., 1979—; fellow in orthopedic oncology Mayo Clinic and Mayo Found., Rochester, Minn., 1978-79; asst. prof. U. Ill., Chgo., 1983—; mem. vis. faculty Mayo Grad. Sch., Rochester, 1969. Developer x-ray method (Singh Index) and bone density method (Radius Index) for diagnosis of osteoporosis. Fulbright travel grantee, 1968. Fellow Am. Acad. Orthopaedic Surgeons, Am. Orthopedic Foot and Ankle Soc.; mem. Orthopaedic Rsch. Soc., Am. Soc. for Bone and Mineral Rsch. Democrat. Sikh. Avocations: stamp collecting, photography, tennis, hiking. Office: Michael Reese Hosp and Med Ctr 2929 S Ellis Ave Chicago IL 60616-3302

SINGH, NIRBHAY NAND, psychology educator, researcher; b. Suva, Fiji, Jan. 27, 1952; arrived in New Zealand, 1970; s. Shiri Ram and Janki Kumari (Singh) S.; m. Judy Daya, May 17, 1973; children: Ashvind Nand, Subhashi Devi. Ph.D., U. Auckland, New Zealand, 1979. Sr. clin. psychologist, head psychology dept. Mangere Hosp. and Tng. Sch., Auckland, 1976-81; assoc. in clin. psychology U. Auckland, 1977-80; lectr. psychology U. Canterbury, Christchurch, New Zealand, 1981-82, sr. lectr. psychology, 1983-87; sr. rsch. scientist Ednl. Research and Services Ctr., De Kalb, Ill., 1987-89; prof. psychiatry Med. Coll. of Va., Richmond, 1989—; dir. rsch. Commonwealth Inst. for Child and Family Studies, Richmond, 1989—; cons. Project MESH, U. Otago, Dunedin, New Zealand, 1982-87, external examiner, diploma in edn., 1982-87; cons. Kimberley Hosp. and Tng. Sch., Levin, N.Z., 1984-87; cons. adv. com. tng. officers Dept. Health, Wellington, 1984-87; cons. curriculum adv. com. Vol. Welfare Agy. Tng. Bd., Wellington, 1984-87; cons. spl. edn. adv. com. Christchurch Tchrs.' Coll., 1986-87; expert cons. Dept. Justice, Washington, 1988—. Co-author: I Can Cook. Editor: Mental Retardation in New Zealand: Research and Policy Issues, 1983; Mental Retardation in New Zealand: Provisions, Services and Research, 1985; Exceptional Children in New Zealand, 1987, Psychopharmacology of the Developmental Disabilities, 1988, Perspective on the Use of Non-aversive and Aversive Interventions for Persons With Developmental Disabilities, 1990, Learning Disabilities: Nature Theory and Treatment, 1992, Self-Injury: Analysis, Assessment and Treatment, 1992; editor-in-chief Journal of Behavioral Education, Jour. of Child and Family Studies; contbr. chpts. to books; editorial bd. numerous jours. Winifred Gimblett scholar 1974; Med. Rsch. Coun. postgrad. scholar, 1975-76; Erskine fellow U. Canterbury, 1984. Fellow APA, Behavior Therapy and Research Soc.; mem. Am. Assn. Mental Deficiency, Assn. Severely Handicapped, Assn. Child Psychology and Psychiatry, Assn. Advancement Behavior Therapy, Soc. Advancement Behavior Analysis, Psychonomic Soc., N.Y. Acad. Scis. Avocations: squash, racquetball. Office: Med Coll Va Dept Psychiatry PO Box 489 Richmond VA 23298-0489

SINGH, RAJ KUMAR, biochemist, researcher; b. Hardoi, U.P., India, Jan. 31, 1955; came to U.S., 1982; s. Hari and Ram Vati S.; m. Kusum, May 28, 1981; children: Sonia, Sumeet. BS, Lucknow U., Lucknow, U.P., India, 1974, MS, 1976; PhD, Kanpur U., Kanpur, U.P., India, 1981. Cert. recombinant DNA tng. Rsch. assoc. Ctrl. Drug Rsch. Inst., Div. of Endocrinology, Lucknow, U.P., India, 1981-82; postdoctoral fellow St. Louis U. Med. Ctr., Dept. Physiology, 1982-84, rsch. assoc., 1984-86; rsch. biochemist So. Rsch. Inst., Biochemistry Dept., Birmingham, Ala., 1986-91; sr. rsch. assoc. U. Ala., Dept. Pathology, Birmingham, Ala., 1991-92, rsch. instr., 1992—. Contbr. articles to profl. jours. and chpts. to books. Recipient Jr. Rsch. fellowship Indian Coun. Med. Rsch., 1977, Sr. fellowship, Indian Coun. Med. Rsch., 1980; grantee So. Rsch. Inst., 1988, NIH, 1992. Mem. N.Y. Acad. Scis., Biol. Chemists of India, Sigma Xi. Achievements include characterization of Estrogen/Antiestrogen receptor-chromatic sites; first to isolate native Retinoic Acid receptors from chick embryos; devel. of novel extracellular matrix from human tissues. Office: U of Alabama Dept Pathology LHRB 701 S 19th StRm 533 Birmingham AL 35294-0007

SINGH, RAJENDRA, mechanical engineering educator; b. Dhampur, India, Feb. 13, 1950; came to U.S., 1973; s. Raghubir and Ishwar (Kali) S.; m. Veena Ghungesh, June 24, 1979; children: Rohit, Arun. BS with honors, Birla Inst., 1971; MS, U. Roorkee, India, 1973; PhD, Purdue U., 1975. Grad. instr. Purdue U., West Layfayette, Ind., 1973-75; sr. engr. Carrier Corp., Syracuse, N.Y., 1975-79; asst. prof. Ohio State U., Columbus, 1979-83, assoc. prof., 1983-87, prof., 1987—; adj. lectr. Syracuse (N.Y.) U., 1977-79; bd. dirs. Nat. Conf. Fluid Power, Milw.; gen. chmn. Nat. Noise Conf., Columbus, 1985; vis. prof. U. Calif., Berkeley, 1987-88; cons., lectr. in field. Contbr. articles to profl. jours. Recipient Gold medal U. Roorkee, 1973, R. H. Kohr Rsch. award Purdue U., 1975, Excellence in Teaching award Inst. Noise Control Engring., 1989, George Westinghouse award Am. Soc. Engring. Edn., 1993. Fellow ASME, Acoustical Soc. Am.; mem. Soc. for Exptl. Mechanics, Inst. Noise Control Engring., Am. Soc. Engring. Edn. (George Westinghouse award 1993). Achievements include patent for rolling door; development of new analytical and experimental techniques in machine dynamics, acoustics, vibration and fluid control. Home: 4772 Belfield Ct Dublin OH 43017 Office: Ohio State U 206 W 18th Ave Columbus OH 43210-1107

SINGH, RAKESH KUMAR, process engineer, educator; b. Chauja, India, July 1, 1952; came to U.S., 1977; s. Ram Chandra and Mewati (Devi) S.; m. Sunita Kumari, May 2, 1985; children: Rahul Kumar, Supriya Kumari. B-Tech., U. Agr. and Tech., Pant Nagar, India, 1975; MS, U. Man., Winnipeg, Can., 1977; PhD, U. Wis., 1983. Rsch. asst., then rsch. assoc. U. Wis., Madison, 1977-85; asst. prof. Tech. U. N.S., Halifax, Can., 1983-85; asst. prof. Purdue U., West Lafayette, Ind., 1985-91, assoc. prof. dept. food sci., 1991—; environ. engr. DKI Group Engrs., Inc., PMO, Milw., 1982-83. Contbr. articles to profl. publs.; co-editor: Advances in Aseptic Processing Technologies, 1992. Grantee Dairy Promotion and Rsch. Bd., Arlington, Va., 1988-92, State of Ind., 1989-92, USDA,1992-95. Mem., ASTM, Am. Soc. Agrl. Engrs., Inst. Food Technologists. Achievements include development of models to predict and optimize microbial lethality in continuous heat exchangers and new method for lactose crystallization in one step. Office: Purdue U 1160 Smith Hall West Lafayette IN 47907-1160

SINGH, REEPU DAMAN, civil engineer; b. Diamond, Guyana, Feb. 16, 1952; came to U.S., 1984; s. Jagdeo and Rajdai (Avatar) S.; m. Savitrie Singh, Oct. 21, 1978; children: Vinoo, Ganesh, Arun. BSCE, Concordia U., Montreal, Que., Can., 1978; MS in Ocean Engring., U. Conn., 1990. Registered profl. engr., Fla., Conn. Owner, cons. Reepu D. Singh, Profl. Engr., Brooklyn, Conn., 1988—; civil engr. Town of Plainfield, Conn. Mem. ASCE, Brooklyn Bus. Assn. Republican. Home: 34 Tripp Hollow Rd Brooklyn CT 06234 Office: Town of Plainfield Town Hall 8 Community Ave Plainfield CT 06374

SINGH, SAHJENDRA NARAIN, electrical engineering educator, researcher; b. Patna, India, Jan. 7, 1943; came to U.S., 1969; s. Shyam N. and Yashoda Singh; m. Sobha Sinha, June 25, 1973; children: Himanshu Kumar, Manish Kumar. ME, Indian Inst. Sci., Bangalore, 1968; PhD, Johns Hopkins U., 1972. Asst. lectr. Regional Inst. Tech., Jamshedpur, India, 1965-66; rsch. scientist Indian Space Rsch. Orgn., Trivandrum, 1973-77; rsch. assoc. NASA Langley Rsch. Ctr., Hampton, Va., 1977-78; vis. prof. Fed. U. Santa Maria, Brazil, 1978-79; prof. Fed. U. Santa Catarina, Florianopolis, Brazil, 1980-83; sr. scientist Vigyan Rshc. Assocs., Hampton, 1983-86; prof. U. Nev., Las Vegas, 1986—; rsch. assoc. AFOSR Summer Faculty, Edwards AFB, Calif., 1991. Contbr. articles to profl. jours. NASA-NRC resident assoc. NRC, Hampton, 1977. Mem. AIAA (sr.), IEEE (sr.), IEEE Aerospace & Electronic Systems Soc. (control systems panel). Achievements include research on nonlinear systems and control theory, stability and control of aerospace vehicles & robotics. Office: U Nev Las Vegas 4505 Maryland Pkwy Las Vegas NV 89154

SINGH, SARDUL, physicist; b. Village-Raur Khera, Punjab, Jan. 15, 1950; came to Nigeria, 1978; s. Darshan Singh and Harbans Kaur; m. Bhupinder Kaur, Aug. 26, 1979 (dec. Jan. 1990); children: Jaspreet, Inderjeet, Mstr Gurdev; m. Hardeep kaur, Feb. 5, 1990. MSc in Physics, Meerut (India) U., 1971; PhD in Physics, U. Ilorin, Nigeria, 1989. Asst. State Bank India, Dehradun, 1971-76; lectr. Coll. Tchr. Edn., Addis Ababa, Ethiopia, 1976-78; edn. officer Kwara State Edn. Mgmt. Bd., Ilorin, 1976-86; lectr. U. Ilorin, 1987—; vis. scientist Ctr. Space Scis., U. Tex., Dallas, 1992-93; mem. adv. inspection team Ministry Edn., Ilorin, 1985. Contbr.: Science Quiz, 1978. In charge Red Cross Soc., Ilorin, Grammar Sch., Ilorin, 1979. Mem. Nigerian Inst. Physics, Sci. Tchrs. Assn. Nigeria (lecture com. 1978). Sikh. Achievements include discovery of pronounced asymmetries in the worldwide occurrence distbn. of inonosphoric plasma bubbles and a clear linkage between their occurrence levels and global thuderstorm activity, assn. of plasma bubbles of greater depths with range spread F and of smaller depths with frequency spread F, indication of an eastward drift of bubbles and of the presence of a windshear in the F region ionosphere, the disappearance of plasma bubbles by diffusive healing, a strong role for gravity waves in the generation of bubbles. Office: U Ilorin Dept Physics, PMB 1515, Ilorin Kwara, Nigeria

SINGH, TARA See TARA

SINGHAL, AVINASH CHANDRA, engineering educator; b. Aligarh, India, Nov. 4, 1941; came to U.S., 1960, naturalized, 1979; s. Shiam Sunder and Pushpa Lata (Jindal) S.; . Uma Rani Sharma, Sept. 5, 1967; children: Ritu Chanchal, Anita, Neil Raj Dave. BSc, Agra U., Kanpur, India, 1957; BSc in Engring. with honors, St. Andrews U., Dundee, Scotland, 1960; MS, MIT, 1961, CE, 1962, ScD, 1964. Registered profl. engr., N.Y., Que., Ariz. Rsch. engr. Kaman Aircraft, Burlington, Mass., 1964-65; prof. Laval U., Quebec, Can., 1965-69; asst. program mgr. TRW, Redondo Beach, Calif., 1969-71; mgr. GE, Phila., 1971-72; mgr. tech. svcs. Engrs. India Ltd., New Delhi, 1972-74; project engr. Weidlinger Assocs., N.Y.C., 1974-77; prof. Ariz. State U., Tempe, 1977—; dir. Cen. Bldg. Rsch. Inst., Govt. of India, 1992-93; dir. Earthquake Rsch. Lab., Tempe, 1978-89; grad. coord. structural engring. Ariz. State U., Tempe, 1991-92; cons. McDonnell Aircraft Corp., St. Louis, 1977-78, 90, Kaiser Engrs., Tudor, 1991-92; reviewer of proposals NSF, Washington, 1980-91; U.S. del. U.S./China Workshop on Arch Dams, Beijing, 1987, Can. del. Shell Structures, USSR, 1964; session chmn. Fifth Internat. Conf. on Soil Dynamics and Earthquake Engring., Karlsruhe, Fed. Republic of Germany, 1991; rsch. prof. Nat. Cen. U. Taiwan, Republic of China., 1990; vis. prof. U. Melbourne, Australia, 1983-84, U. Auckland, New Zealand, 1983-84; Nodal dir. wood substitute rsch. program, India, 1992-93. Mem. editorial bd. Soil Dynamics and Earthquake Engring., 1991—; contbr. Nuclear Waste Storage, 1986, (proceeding publ.) Earthquake Behavior of Buried Pipelines, 1989, Wood Substitute: A New Priority, 1992, System Flexibility and Reflected Pressures, 1993, Simulation of Blast Pressures on Flexible Panels, 1993; editor: Seismic Performance of Pipelines & Storage Tanks, 1985, Recent Advances in Lifeline Earthquake Engineering, 1987; contbr. articles to Jour. Performance of Constructed Facilities, Am. Soc. of Civil Engrs., Jour. Computers and Structures, Jour. ASME, Jour. Aerospace Engring. ASCE; reviewer, bd. editors Jour. Earthquake Engring. and Structural Dynamics, Structural Engring. Papers Jour. ASCE. Chm. bd. dirs. India Assn. Greater Phoenix, 1985-86; pres. India Assn. Greater Boston, 1964-65; v.p., treas. Dobson Ranch Homeowners Assn., Mesa, Ariz., 1988-91; founding mem. Asian Am. Assn. Ariz., Phoenix, 1987-89; founding mem., pres. Asian Am. Faculty Assn., Ariz. State U., Tempe, 1986-88; mem. sci. adv. com. 5th Internat. Conf. on Soil Dynamics, Germany, 1991; con. UN Devel. Program/Govt. of India, 1991-92. McLintock fellow MIT, 1960, Carnegie fellow MIT, 1960-63, fellow Royal Astron. Soc., London, 1961-64, rsch. fellow Kobe U., Japan, 1990; Denninson scholar Instn. Civil Engrs., London, 1959; Henry Adams Rsch. medal Structural Engrs., London, 1972; grantee Can. Def. Rsch. Bd., 1966-69, NSF, 1978-82, Engring. Found., 1978-79, U.S. Army Corps Engrs., 1984-86, U.S. Dept. Interior, 1986-88; recipient Henry Adams medal, 1st prize bridge bldg. Instn. Structural Engrs., Merit award Inst. Engrs., India. Fellow ASCE (chmn. subcom. materials); mem. Am. Soc. Mech. Engrs. (editor), Sigma Xi, Tau Beta Pi, Chi Epsilon. Republican. Achievements include research in computer solutions, in structural engineering, in lifeline earthquake engineering, strengthening of deteriorated arch dams, buildings, bridges, building materials, and wood substitution. Home: 2631 S El Marino Mesa AZ 85202-7302 Office: Ariz State U Dept Civil Engring 5306 Tempe AZ 85287-5306

SINGHELLAKIS, PANAGIOTIS NICOLAOS, endocrinologist, educator; b. Athens, Attiki, Greece, June 24, 1941; s. Nicholaos John Singhellakis and Eugenia Panagiotis Kastanis. Degree in medicine, U. Athens, 1965, MD, 1971. Fellow in internal medicine Evanghelismos Hosp., Athens, 1966-69, sr. fellow, 1969-71, fellow in endocrinology, 1972-78, sr. fellow, 1978-81; lectr. in endocrinology U. Athens, 1981-82, assoc. prof. medicine, 1987; head dept. endocrinology Praeus (Greece) Gen. Hosp., 1982-86, Agios Savas Hosp. Greek Anticancer Inst., Athens, 1987—. Contbr. articles to profl. jours. Maj. Greek Spl. Forces, 1965-66. Recipient 1st award 4th Greek Congress in Nephrology, Nicosia, 1986, 1st award 15th Greek Congress in Endocrinology, 1988; named Hon. Prof. U. Bucharest (Romania) Inst. Parhon, 1987. Mem. N.Y. Acad. Scis., European Soc. Clin. Investigation. Mem. New Democracy Party. Christian Orthodox. Home: 2 Grammou, 172 34 Athens Greece Office: Agios Savas Hosp, Greek Anticancer Inst, Alexandras Ave, 115 22 Athens Greece

SINGLETON, JOY ANN, quality systems professional; b. Pitts., Sept. 23, 1953; d. Joseph Francis and Steffie (Dabroski) Solomey; m. Gerald Hovey Singleton, Mar. 29, 1985; children: David Scott, Amanda Joy. AS, Pa. State U., 1973; BSME, Point Park Coll., 1978. Project engr. PPG Industries, Inc., Creighton, Pa., 1973-78, Fiber Industries, Inc. (Celanese), Greenville, S.C., 1978-82; project engr. Union Carbide Corp., Greenville, 1982-83, sr. buyer, purchasing engr., 1983-85; process engr. Union Carbide Corp. (Amoco Performance Products, Inc.), Greenville, 1985-87; sr. quality specifications engr. Amoco Performance Products, Inc., Greenville, 1987-92; quality systems mgr. Asten Monotech, Inc., Summerville, S.C., 1992—. Fin. adviser Jr. Achievement, New Kensington, Pa., 1975-76, Greenville, 1980, exec. adviser, Greenville, 1983, 84. Mem. NAFE, Am. Soc. Quality Control, Am. Soc. Test Methods. Office: Asten Monotech Inc 230 Deming Way Summerville SC 29483

SINGLEY, MARK ELDRIDGE, agricultural engineering educator; b. Delano, Pa., Jan. 25, 1921; s. Maurice and Clara (Rhodes) S.; m. Janet Twichell, Oct. 3, 1942; children: Donald Heath, Frances Marvin, Jeremy Mark, Paul Victor. BS, Pa. State U., 1942; MS, Rutgers U., 1949. Adminstrv. asst. UNRRA, 1946; prof. II biol. and agrl. engring. Rutgers U., New Brunswick, N.J., 1947-87; chmn. dept. Rutgers U., 1961-71; v.p. rsch. and devel. Bedminster Bioconversion Corp., Haddonfield, N.J., 1987—; bd. dirs. Agriplane. Chmn. Hillsborough Twp., Somerset County (N.J.) Planning Bd., 1956-73; pres. Hillsborough Twp. (N.J.) Democratic Club, 1979-80, Agrl. Mus. State of N.J., 1983—, pres. bd. trustee, 1984-89. With USNR, 1942-46. Named Prof. of Yr. Cook Coll., 1985. Fellow AAAS (sect. com. O), Am. Soc. Agrl. Engrs. (chmn. North Atlantic region 1966, bd. dir. 1973-75, Massey-Ferguson medal 1987); mem. Am. Forage and Grassland Council (bd. dir. 1966-69). Home and Office: 335 Amwell Rd Belle Mead NJ 08502-9802

SINHA, AGAM NATH, engineering management executive; b. Srinagar, Bihar, India, Aug. 14, 1947; came to U.S., 1968; s. Ayodhya Nath and Sushila (Shrivastava) S.; m. Diane Yvonne Gardell, Jan. 10, 1974; children: Savita Sinha, Akash K. Sinha. BS, Indian Inst. of Tech., 1968; MS in Mgmt. of Tech., Am. U., 1985; MS, U. Minn., 1970, PhD, 1974. Rsch. analyst U. Minn., Mpls., 1970-72; mem. tech. staff The Mitre Corp., McLean, Va., 1972-75, task leader, 1975-77, group leader, 1977-82, assoc. dept. head, 1982-85, dept. head, 1985—; chmn. Nat. Rsch. Coun./Transp. Rsch. Bd. Com., Washington, 1990—; mem. Aviation Week Rsch. Adv. Panel, N.Y.C., 1990-91, Nat. Rsch. Coun. Grad. Rsch. Program Selection Panel, Washington, 1989—. Referee ASCE, ION, IEEE; contbr. articles to Jour. of Aircraft, Jour. of the Inst. of Engrs. of India. Com. mem. PTA, Chantilly, Va., 1986-90. Alumni grad. fellowship U. Minn., 1971, grad. sch. rsch. ctr. fellowship, 1970-72. Mem. AIAA, Ops. Rsch. Soc. of Am. (referee), Transp. Rsch. Bd. (com. chmn.). Home: 13718 Southernwood Ct Chantilly VA 22021 Office: The Mitre Corp 7525 Colshire Dr Mc Lean VA 22102

SINHA, MAHADEVA PRASAD, chemist, consultant; b. Darbhanga, Bihar, India, Jan. 1, 1944; came to U.S., 1968; s. Raj Kishore Prasad and Ram Jyoti Devi; m. Chandra Prabha, June 11, 1966; children: Lakshmi Rinu, Sangeeta, Ravi. BS in Chemistry with honors, Bihar U., Muzaffardur, India, 1961; MS, Bihar U., 1964, NYU, 1971; PhD, Columbia U., 1974. Asst. prof. of Chemistry Bihar U., Muzaffardur, India, 1964-67; grad. rsch. asst. NYU, 1968-70, Columbia U., N.Y.C., 1970-73; postdoctoral rsch. assoc. & lectr. Yale U., New Haven, Conn., 1974-77; mem. tech. staff Jet Propulsion Lab., Pasadena, Calif., 1978—; vis. assoc. resident engr. UCLA Dept. Chemistry, 1984-90; lectr. Calif. State Poly. U., Pomona, 1983—; cons. Perkin-Elmer Corp., Pomona, 1991, Cetac Technology,Inc., Omaha, 1991—. Author: (book chapters) Rapid Microbiological Analysis, 1985, Particles in Gasesand Liquids, 1990; contbr. articles to profl. jours. Mem. PTA Adv. Coun. Temple City (Calif.) Unified Sch. Dist., 1991. Recipient Fulbright-Hayes grant U.S. Dept. State, 1968, rsch. grants NSF, U.S. Army, EPA, IBM, 1980—, NYU Predoctoral Merit Scholarship, 1970, seven NASA Tech. Briefs awards, 1980—, NASA Exceptional Svc. medal, 1988, NASA award for U.S. patents, 1989. Achievements include two inventions, Method to Analyze Particles in Real-Time (patented), A Miniaturized Lightweight Magnetic Sector for a Field Portable Mass Spectrograph (patent pending). Office: Jet Propulsion Lab 4800 Oak Grove Dr Pasadena CA 91109

SINHAROY, SAMAR, physicist, researcher; b. Bengal, India, Dec. 16, 1940; came to U.S., 1966; s. Bhaktibrata and Madhabi (Choudhuri) S.; m. Semahat Dengi, May 14, 1976; children: Sheela, Sinan. MSc in Physics, Calcutta (India) U., 1963; PhD in Physics, Poly. U., Bklyn., 1972. Rsch. assoc. Tech. U. Clausthal, Fed. Republic Germany, 1973-75; rsch. asst. prof. U. Mo., Rolla, 1975-78; sr. scientist Westinghouse Sci. and Tech. Ctr., Pitts., 1978-92, fellow scientist, 1992—; Contbr. articles to profl. jours. including Jour. Vacuum Sci. and Tech., Jour. Applied Physics. Mem. Am. Vacuum Soc. Democrat. Hindu. Achievements include patent for thin films of ferroelectric barium magnesium fluoride (BAMg F4) on silicon for nonvolatile memory device application. Office: Westinghouse Sci and Tech Ctr 1310 Beulah Rd Pittsburgh PA 15235-5098

SINKO, PATRICK J., pharmacist, educator; b. Passaic, N.J., Jan. 7, 1959; s. Patrick and Patricia (Anderson) S.; m. Noreen Marie Seccia, Apr. 29, 1989; 1 child, Patrick. BS in Pharmacy, Rutgers U., 1982; PhD in Pharmaceutics, U. Mich., 1988. Rsch. scientist U. Mich./TSRL, Inc., Ann Arbor, 1988-91; asst. prof. pharmaceutics Rutgers U., Piscataway, N.J., 1991—. Contbr. articles to profl. publs. Grantee NIH, 1992, Unigene Labs., Inc., 1992, TSRL, Inc., 1992; recipient Young Investigator award Eli Lilly and Co., 1992. Mem. Am. Assn. Pharm. Scis. (chair sect. publs. com., pharmaceutics and drug delivery sect.), N.Y. Acad. Sci., Nat. Eagle Scout Assn., Sigma Xi. Achievements include patents for enhancement of bioavailability of proteolytically labile therapeutic agents; research on improved delivery of anti-AIDS drugs. Office: Rutgers U Coll Pharmacy PO Box 789 Piscataway NJ 08855

SINNETT, PETER FRANK, physician, geriatrics educator; b. Sydney, Australia, Dec. 16, 1934; s. Sydney Thomas and Minty (Pottinger) S. MB, BChir, U. Sydney, 1960, D of Medicine, 1973. Resident med. officer Sydney Hosp., 1960-62, cardiology fellow, 1964, clin. rsch. registrar, 1965; tng. fellow U. Sydney, 1962-64; Nat. Heart Found. rsch. fellow Australian Nat. U./U. Papua New Guinea, 1965-71; coun. mem., prof. rep. U. Papua New Guinea, 1971-73, found. prof. human biology, 1971-76, mem. working party on future of univ., 1973, mem. acad. planning and rsch. coms.; found. prof. geriatrics U. NSW, Woden, Australia, 1979—; clin. dir. rehab. and aged care svc. Woden Valley Hosp., Canberra; cons. to auditor-gen. on med. adminstrv. aspects of nursing home programs, 1980; chmn. bd. censors, mem. coun. Australian Coll. Rehab., 1980-82; mem. Consultative Com. on Social Welfare, 1980-81; dir. rehab. and aged care svc. Australian Capital Ter. Authority, 1984—; vis. fellow Australian Nat. U., 1987—; coun. mem. Papua New Guinea Inst. for Med. Rsch., 1973-75; chair of sci. and rsch. com. Continence Found. of Australia, 1993. Author: The People of Murapin, 1978; contbr. articles to profl. publs. Mem. med. rsch. adv. com., mem. food and nutrition adv. com. Govt. of Papua New Guinea, 1974-75. Fellow Royal Australian Coll. Physicians, Royal Australian Coll. Rehab. Medicine; mem. Am. Acad. Phys. Medicine and Rehab., Australian Epidemiol. Assn. (mem. found. 1987). Office: Univ NSW Dept Rehab/ Geriatr, PO Box 11, Woden Australia 2606

SINNETTE, JOHN TOWNSEND, JR., research scientist, consultant; b. Rome, Ga., Nov. 4, 1909; s. John T. Sinnette and Katherine Alice Lyon. BS, Calif. Inst. Tech., 1931, MS, 1933. Chemist Met. Water Dist., Banning, Calif., 1937-39, Boulder City, Nev., 1939-40; physicist U.S. Bur. Reclamation, Boulder City, 1940-41; rsch. scientist Nat. Adv. Com. for Aeronautics, Langley Field, Va., 1941-43, Cleve., 1943-51; cons. physicist U.S. Naval Ordnance Test Sta., Pasadena, Calif., 1951-58, Cleve. Pneumatic Industries, El Segundo, Calif., 1958-60; tech. dir. Hydrosystems Co., El Segundo, 1960-62; physicist Thrust Systems Corp., Costa Mesa, Calif., 1963-64; lectr. compressor design Case Inst. Tech., 1946-48; cons. many firms in aeronautical and related industries, 1950-79. Contbr. papers to sci. meetings and confs. Vol. Am. Cancer Soc., Costa Mesa, Calif., 1976-80, Cancer Control Soc., 1979-82; contbr. Action on Smoking and Health, Washington, 1985-93. Mem. AAAS, Am. Statistical Assn., Am. Math Soc., Nat. Health Fed. Democrat. Achievements include patent dealing with a novel jet engine design (U.S., Britain); made first detailed measurement of silt flow (density currents) in Lake Mead and the deposition of the silt behind Boulder Dam, 1940; detailed rsch. on first government multi-stage axial-flow compressor suitable for jet engines; orginated and proved possibilities of extending useful operating range of axial-flow compressors by use of adjustable stator blades; used aerodynamic theory to design high-performance centrifugal compressors; development of statistical theory and demonstrated its usefulness in predicting the distribution of primes and factors of Fermat and related numbers. Home: 135 N B St Tustin CA 92680-3110

SINNING, ALLAN RAY, anatomy educator, researcher; b. Miller, S.D., Jan. 19, 1957; s. Elvin Lee and Fern Harriet (Drake) S.; m. Christina Rita Litzlbeck, Aug. 13, 1983; children: Gregory Charles, Geoffrey Allan. BS, U. Wis., Platteville, 1979; MS, U. N.D., 1983, PhD, 1985. Postdoctoral fellow Med. Coll. Wis., Milw., 1985-90; asst. prof. U. Miss. Med. Ctr., Jackson, 1990—. Author: (chpt.) Molecular Biology of the Cardiovascular System, 1990; contbr. articles to profl. jours. Recipient Hamre fellowship Dept. Anatomy, 1984, Postdoctoral fellowship Am. Heart Assn. Wis., 1986, Grant-in-Aid, Am. Heart Assn. Miss., 1991. Roman Catholic. Achievements include research in the role of extracellular matrix in development. Office: U Miss Med Ctr 2500 N State St Jackson MS 39216

SINON, JOHN ADELBERT, JR., electrical engineer; b. Burien, Wash., May 1, 1956; s. John Adelbert and Eileen Margaret (Anderson) S.; m. Christine Ellen Morgan, July 20, 1985; children: Michael James, Timothy John, Alexander Evan. BS in Physics, U. Wash., 1985, MS in Physics, 1993. Mfg. engr. Boeing Comml. Airplane, Seattle, 1979-83; ordinance engr Boeing Aerospace, Seattle, 1985-86; electromagnetic analyst Boeing Comml. Airplane, Seattle, 1986—. Author, pub. Investigators Information Sources, 1992. With U.S. Army, 1973-75. Mem. Soc. Physics Students. Democrat. Lutheran. Achievements include four patents pending for electromagnetic measurement and analysis tools. Home: 21807 12th Ave S Des Moines WA 98198

SINTON, CHRISTOPHER MICHAEL, neurophysiologist; b. Beckenham, Kent, Eng., Sept. 10, 1946; came to U.S., 1983; s. Leslie George and Evelyn Mabel (Burn) S. BA, Cambridge U., Eng., 1968, MA, 1977; BSc, London U., 1978; PhD, U. Lyon, France, 1981. Rsch. fellow U. Lyon, 1980-83; rsch. assoc. Princeton (N.J.) U., 1983-84; sr. scientist Ciba-Geigy Corp., Summit, N.J., 1984-88; dir. electrophysiology Neurogen Corp., Branford, Conn., 1988—; rsch. asst. prof. med. NYU, N.Y.C., 1986—. Contbr. 38 articles to profl. jours. Med. Rsch. Coun. vis. scholar, France, 1983. Mem. N.Y. Acad. Scis., Soc. Neurosci., European Neurosci. Assn. Achievements include research on fetal effects of in-utero caffeine exposure, possible functional role of REM sleep and neuropeptide modulation of synaptic input. Office: Neurogen Corp 35 NE Industrial Rd Branford CT 06405-2844

SINTZOFF, MICHEL, computer scientist, educator; b. Brussels, Aug. 12, 1938; s. Serge and Anna (Emeljanoff) S.; m. Jeanne de Strycker, Dec. 26, 1964; children: André, Marie, Ivan, Catherine, Paul. BS in Math., U. Louvain, Belgium, 1962; Dr.h.c., U. J.F. Fourier, Grenoble, 1993. Teaching asst. U. Lubumbashi, Zaire, 1962-64; researcher Philips Rsch. Lab., Brussels, 1964-82; prof. U. Louvain, Louvain-la-Neuve, Belgium, 1982—; vis. lectr. U. Brussels, 1971-74, U. Louvain, 1972-82, U. P. and M. Curie, Paris, 1973-75; assoc. prof. U. Edmonton, Can., 1972-73; hon. prof. U. Brussels, 1974—; vis. sr. researcher Ctr. Nat. Rsch. Sci., Nancy, France, 1977-78; mem. numerous editorial bds., sci. bds. Co-author: Manuel du Langage Algorithmique ALGOL68, 1975, Report on the Algorithmic Language ALGOL68, 1976, Mes Premières Constructions de Programmes, 1977, Raisonner pour Programmer, 1986; founding editor-in-chief Sci. Computer Programming, 1981—; contbr. articles to profl. jours. Recipient Soc. for Worldwide Interbank Fin. Telecommunications award Fonds Nat. de Rsch. Sci., Belgium, 1989; named Officer of the Order of Léopold, 1989. Avocations: reading, movies, music, sports. Home: 7 av du Château, B-1330 Rixensart Belgium Office: U Cath Louvain, Informatique 2 pl Ste Barbe, B-1348 Louvain-la-Neuve Belgium

SIPINEN, SEPPO ANTERO, obstetrician/gynecologist; b. Helsinki, Finland, Aug. 11, 1946; s. Uno Emil Rafael and Martta Liisa (Knuuttila) S.; m. Taru Katriina, Oct. 19, 1968; children: Samuel, Susanne. MD, U. Freiburg, Fed. Republic Germany, 1971, U. Helsinki, 1972; cert. specialist ob.-gyn., U.

Helsinki, 1979, D in Med. Sci., 1981. Resident in Ob-Gyn. and Surgery U. Helsinki, State Maternity Hosp., Helsinki, 1973-80; sr. physician Ob-Gyn. State Maternity Hosp., Helsinki, 1981-83; commd. comdr. Finnish Navy Med. Corps., 1989; surgeon gen. Finnish Navy Med. Corps., Helsinki, 1983—; head. naval dept. Rsch. Inst. of Mil. Medicine, Helsinki, 1983—; cons. ob-gyn. Finnish Def. Forces, 1983, Subway in Helsinki, 1976-77; rsch. group Dept. Med. Chemistry, U. Helsinki, 1977-86, lectr. diving and hyperbaric physiology and medicine, 1977—; cons. devel. group State Dept. Finland, 1984-86; head diving and hyperbaric med. treatment of State Salv. Edn. Inst., Finland, 1985-86; cons. devel. group of Profl. Diving Nat. Bd. Labor Protection, Finland, 1989-90, Compressed Air Work of Subway, 1976-77; mem., rep. for Finland European Diving Tech. Com., 1986—; mem. sci. bd. Diving Alert Network Europe, 1991; mem. European Commn. Hyperbaric Medicine, 1991—. Contbr. 80 articles in endocrinology, bacteriology, serology, diving, hyperbaric medicine to profl. jours. Recipient medal for Mil. Merits, 1988, Silver medal Finnish Sportdivers Fedn., 1988, named Diver of Yr., 1989. Mem. AAAS, European Underwater and Baromed. Soc. (at-large, exec. com.), Finnish Med. Assn., Finnish Soc. of Ob-Gyn. (pres.), The Finnish Soc. Perinatal Medicine, The Finnish Soc. of Diving and Hyperbaric Medicine, Undersea and Hyperbaric Med. Soc., European Undersea Biomed. Soc., Finnish Sport Divers Fedn. (safety com. 1976-79, pres. 1977-79, exec. bd. 1977-79, med. com., pres. 1980, Silver medal 1988, Diver of Yr. 1989), Espoo Gymnastics Team (v.p.), The Planetary Soc. Achievements include construction of diving support vessel; development of decompression tables for air diving, of oxygen-nitrogen mixed gas diving. Office: Finnish Naval Hdqrs, Pohjoiskaari 36, 00200 Helsinki Finland

SIPOS, CHARLES ANDREW, manufacturing executive; b. Bridgeport, Conn., July 2, 1946; s. Charles Frank and Mary Elizabeth (Kilmer); m. Juanita Marie Coffey, Aug. 30, 1969 (div. Mar. 1980); children: William Todd, Christopher Charles; m. Helen Elaine Thompson, June 26, 1982; 1 stepchild, Jason Christopher Pappas. BA in Math., Fla. Atlantic U., 1972; MBA in Aviation, Embry-Riddle Aero. U., 1989; MS in Mgmt. of Tech., U. Miami, 1992. Sr. program analyst Pratt and Whitney Aircraft, West Palm Beach, Fla., 1972-74, Bethesda Meml. Hosp., Boynton Beach, Fla., 1974-75; systems programmer, analyst Nat. Enquirer, Lantana, Fla., 1975-82; supr. McDonnell Douglas Automation Co., St. Louis, 1982-84; sr. mgr. McDonnell Douglas Missile Systems Co., Titusville, Fla., 1984—. Sgt. USMC, 1966-72. Sr. mem. AIAA (mem. computer systems tech. com.); mem. Nat. Mgmt. Assn., Internat. Assn. for Mgmt. of Tech. Office: McDonnell Douglas Missile Systems Co 701 Columbia Blvd PO Box 600 Titusville FL 32780

SIRAT, GABRIEL YESHOUA, physics educator; b. Toulouse, Haute Garr, France, Feb. 13, 1955; s. Rene-Samuel and Colette (Salamon) S.; m. Dinah Esther Moatti, Aug. 29, 1991; 1 child, Emmanuel. BS, Jerusalem U., 1973, PhD, 1983. Asst. U. Jerusalem, Israel, 1973-78; post doctoral fellow Calif. Inst. Tech., Pasadena, 1983-85; assoc. prof. Telecom, Paris, 1986-87; prof. Telecom, 1988-92; co-organizer Neuro-88 1st European meeting on neural networks, Paris, 1988; founder Le Conoscope S.A., 1988, pres., 1992—. Inventor conoscopic holography, 1984, conoscope, 1988; patentee in field. Mem. French-Israeli Assn. for Rsch. and Tech. (dep. sec.-gen.), Optical Soc. Am., Photoelectric Soc., Soc. Mfg. Engrs., Soc. Française d'Optique. Jewish. Office: Le Conoscope SA, 59 Bld VACIN, F-75015 Paris France

SIRAUT, PHILIPPE C., watch and electronics company executive; b. Etterbeek, Brabant, Belgium, June 22, 1961; s. André Prosper and Monique Clara (Clérin) S. BS in Philosophy, Cath. U. Louvain-la-Neuve, 1982; MSEE, Cath. U. Louvain-La-Neuve, 1984. Rsch. engr. Cath. U. Louvain-la-Neuve, 1985-88; tech. officer tech. staff Belgian Army, 1987-88; devel. engr. ETA SA Fabriques d'Ebauches, Grenchen, Switzerland, 1988-91, Swatch-pager prodn. mgr., 1991—. Mem. IEEE, Belgian Soc. Telecommunications and Electronics Engrs. Avocation: trekking. Home: Rue du Poujet 2, CH-2800 Delémont Switzerland Office: ETA SA Fabriques d'Ebauches, Schild-Ruststrasse 17, CH-2540 Grenchen Switzerland

SIRCAR, RATNA, neurobiology educator, researcher; b. Benaras, India, Aug. 15, 1952; came to U.S., 1981; d. Hem Chandra and Anjali (Sen) Roy Chowdhury; m. Krisha Prasad Sircar, Mar. 8, 1980; children: Monica, Debashish. BSc, U. Delhi, New Delhi, 1972; MSc in Physiology, All India Inst. Med. Scis., New Delhi, 1975, PhD in Physiology, 1981. Rsch. assoc. dept. pharmacology Columbia U., N.Y.C., 1982; postdoctoral rsch. fellow depts. psychiatry and neurosci. Albert Einstein Coll. Medicine, Bronx, N.Y., 1982-85, rsch. assoc. dept. psychiatry, 1985-88, asst. prof. psychiatry and neurology, 1988—, dir. Lab. Devel. Neurosci., 1992—; presenter, speaker in field, 1985—; reviewer Brain Rsch., Neurosci. Letters, Jour. Neurochemistry, Neurotoxicology and Teratology, 1987—; ad hoc reviewer spl. rev. com. Nat. Inst. on Drug Abuse, 1991. Contbr. articles and abstracts to Indian Jour. Physiology and Pharmacology, Life Scis., Brain Rsch., Jour. Pharmacology and Experimental Therapeutics, Neurosci. Letters, European Jour. Pharmacology; also chpts. in books. Recipient Milton Rosenbaum rsch. award Albert Einstein Coll. Medicine, 1986, 87, 88, Young Investigator award Internat. Congress on Schizophrenia Rsch., 1987, Nat. Alliance for Rsch. on Schizophrenia and Depression, 1990, Young Scientist award Upjohn Co., 1988, Young Neurosci. Achievement award Assn. Scientists Indian Origin, 1989, Stanley Found. Rsch. award Nat. Alliance for Mentally Ill, 1991; grantee Nat. Inst. on Drug Abuse, 1992-95; also others. Mem. AAAS, Am. Soc. for Pharmacology and Exptl. Therapeutics, Internat. Soc. for Devel. Neurosci., Internat. Soc. for Devel. Psychobiology, Internat. Brain Rsch. Orgn., Soc. for Neurosci., Internat. Narcotic Rsch. Soc., N.Y. Acad. Sci. (sci. fair judge 1991—). Achievements include research on developmental neurobiology, cellular and molecular mechanisms underlying mental and neurological diseases, drug-receptor interactions, mechanisms of action of psychotomimetic drugs including phencyclidine (PCP) and PCP-like behaviorally active drugs; specific PCP receptors in the brain (identification in rat and human brain biochemical characterization, anatomical location, ontogeny, interaction with the N-methyl-D-asparate type of excitatory amino acid receptor); significance of the PCP-NMDA receptor-ion channel complex in physiological and pathological conditions; opiate and sigma receptor subtypes; receptor binding with computerized data analysis, quantitative autoradiography, image analysis, animal surgery, stereotaxy, stimulation of specific regions of the brain, recording EEG, Kindling procedure, neurohistology. Office: Albert Einstein Coll Med Dept Psychiatry 1300 Morris Park Ave F-109 Bronx NY 10461

SIRÉN, ANNA-LEENA KAARINA, neuroscientist; b. Oulu, Finland, Sept. 12, 1955; came to U.S., 1984; d. Ingmar Eric Torsten and Sirkka Kaarina (Pako) S. MD, U. Oulu, Finland, 1979, PhD, 1982. Pharmacology instr. U. Finland, Oulu, 1979-82; asst. tchr. pharmacology U. Helsinki, 1983-84; rsch. asst. prof. Uniformed Svcs. Health Sci. U., Bethesda, Md., 1984-90, rsch. assoc. prof. dept. neurology, 1990—, dir. lab. support div., 1989—. Contbr. articles to Circ. Rsch., Jour. Pharmacology and Exptl. Therapeutics, Am. Jour. Physiology, Hypertension. Recipient Neuropeptide award Annual Neuropeptide Meeting, Colo., 1986. Mem. AAAS, N.Y. Acad. Scis., Scandinavian Pharm. Soc., Soc. for Neurosci. Achievements include research in role of neuropeptides in central autonomic control, stress, hypertension, and stroke. Office: USUHS Dept Neurology 4301 Jones Bridge Rd Bethesda MD 20814-4799

SIRI, WILLIAM E., physicist; b. Phila., Jan. 2, 1919; s. Emil Mark and Caroline (Schaedel) S.; m. Margaret Jean Brandenburg, Dec. 3, 1949; children: Margaret Lynn, Ann Kathryn. B.Sc., U. Chgo., 1942; postgrad. in physics, U. Calif.-Berkeley, 1947-50. Licensed profl. engr., Calif. Research engr. Baldwin-Lima-Hamilton Corp., 1943; physicist Manhattan Project Lawrence-Berkeley Lab., U. Calif., Berkeley, 1943-45, prin. investigator biophysics and research, 1945-74, mgr. energy analysis program, 1974-81, sr. scientist emeritus, 1981—; cons. energy and environment, 1982—; lectr. U. Calif. Summer Inst., 1962-72; vis. scientist Nat. Cancer Inst., 1970; exec. v.p. Am. Mt. Everest Expdn., Inc.; field leader U. Calif. Peruvian Expdns., 1950-52; leader Calif. Himalayan Expdn., 1954; field leader Internat. Physiol. Expdn. to Antarctica, 1957; dep. leader Am. Mt. Everest Expdn., 1963. Author: Nuclear Radiations and Isotopic Tracers, 1949, papers on energy systems analyses, biophys. research, conservation and mountaineering. Pres. Save San Francisco Bay Assn., 1968-88; bd. dirs. Sierra Club Found., 1964-78; gov. gen. Mountain Medicine Inst., 1988—; vice chmn. The Bay Inst., 1985—; bd. dirs. San Francisco Bay-Delta Preservation Assn., 1987—, treas., 1987—. Lt. (j.g.) USNR, 1950-59. Co-recipient Hubbard medal Nat. Geog. Soc., 1963, Elsa Kent Kane medal Phila. Geog. Soc., 1963, Sol Feinstone Environ. award, 1977, Environ. award East Bay Regional Park Dist., 1984. Mem. Am. Phys. Soc., Biophys. Soc., Am. Assn. Physicists in Medicine, Sigma Xi. Democrat. Lutheran. Clubs: Sierra (dir. 1955-74, pres. 1964-66, William Colby award 1975), American Alpine (v.p.), Explorers (certificate of merit 1964). Home: 1015 Leneve Pl El Cerrito CA 94530-2751 Office: U Calif Lawrence Berkeley Lab 1 Cyclotron Rd Berkeley CA 94720

SIRICA, ALPHONSE EUGENE, pathology educator; b. Waterbury, Conn., Jan. 16, 1944; s. Alphonse Eugene and Elena Virginia (Mascolo) S.; m. Annette Marie Murray, June 9, 1984; children: Gabrielle Theresa, Nicholas Steven. MS, Fordham U., 1968; PhD in Biomed. Sci., U. Conn., 1976. Asst. prof. U. Wis., Madison, 1979-84; assoc. prof. Med. Coll. Va., Va. Commonwealth U., Richmond, 1984-90, prof. of pathology, 1990—, divsn. chair grad. pathology rsch. edn., 1992—; regular mem. sci. adv. com. on carcinogenesis and nutrition Am. Cancer Soc., Atlanta, 1989-92, metabolic pathology study sect. NIH, Bethesda, 1991—. Editor, author: The Pathobiology of Neoplasia, 1989, The Role of Cell Types in Hepatocarcinogenesis, 1992; mem. editorial bd. Pathobiology and Hepatology; rev. bd. In Vitro Cellular and Devel. Biology; contbr. rsch. papers to Am. Jour. Pathology, others. Nat. Cancer Inst./NIH grantee, 1981—. Mem. AAAS, Am. Soc. Cell Biology, Am. Assn. Cancer Rsch. (chmn. Va. State Legis. Com.), Tissue Culture Assn., Assn. clin. Scientist, Am. Soc. Investigative Pathology (chair-elect program com. 1993—), Am. Assn. Study Liver Diseases, N.Y. Acad. Scis.,Soc. Expt. Biology and Med., Hans Popper Hepatopathology Soc. Democrat. Roman Catholic. Achievements include discovery of collagen gel-nylon mesh system for culturing hepatocytes; first to establishment and characterization of hyperplastic bile ductular epithelial cells in culture; research in hepato and biliary carcinogenesis, pathobiology of hepatocyte and biliary epithelial cells. Office: Med Coll Va Va Commonwealth U PO Box 662 Richmond VA 23205-0662

SIRIGNANO, WILLIAM ALFONSO, aerospace and mechanical engineer, educator; b. Bronx, N.Y., Apr. 14, 1938; s. Anthony P. and Lucy (Caruso) S.; m. Lynn Haisfield, Nov. 26, 1977; children: Monica Ann, Jacqueline Hope, Justin Anthony. B.Aero.Engring., Rensselaer Poly. Inst., 1959; Ph.D., Princeton U., 1964. Mem. research staff Guggenheim Labs., aerospace, mech. scis. dept. Princeton U., 1964-67, asst. prof. aerospace and mech. scis., 1967-69, assoc. prof., 1969-73, prof., 1973-79, dept. dir. grad. studies, 1974-78; George Tallman Ladd prof., head dept. mech. engring. Carnegie-Mellon U., 1979-85; dean Sch. Engring., U. Calif.-Irvine, 1985—; cons. industry and govt., 1966—; lectr. and cons. NATO adv. group on aero. research and devel., 1967, 75, 80; chmn. nat. and internat. tech. confs.; chmn. acad. adv. council Indsl. Research Inst., 1985-88; mem. space sci. applications adv. com. NASA, 1985-90, chmn. combustion sci. microgravity disciplinary working group, 1987-90; chmn. com. on microgravity rsch. space studies bd. NRC. Assoc. editor: Combustion Sci. and Tech, 1969-70; assoc. tech. editor Jour. Heat Transfer, 1985-92; contbr. articles to nat. and internat. profl. jours., also research monographs. United Aircraft research fellow, 1973-74; Disting. Alumni Rsch. award U. Calif. Irvine, 1992. Fellow AIAA (Pendray Aerospace Lit. award 1991, Propellants and Combustion award 1992), ASME (Freeman Scholar award 1992, IDERS Oppenheim award 1993), AAAS; mem. Combustion Inst. (treas. internat. orgn., chmn. Eastern sect.), Soc. Indsl. Applied Math. Office: U Calif Sch Engring 305 Rockwell Engrg Ct Irvine CA 92717

SISSOM, JOHN DOUGLAS, systems engineer; b. Manchester, Tenn., Apr. 7, 1937; s. Jesse Spain and Gladys Helen (White) S.; m. Janice Darlene Denham, May 31, 1959; children: Jay Douglas, Joy Darlene, Jolie Deann. AS, Le Tourneau Coll., 1963; BS in Elec. Tech., Le Tourneau U., 1964. Jr. machine designer Delco Electronics, Kokomo, Ind., 1964-66, machine designer, 1966-68, supr., 1968-69, project engr., 1970-80, sr. project engr., 1981—. Author: The Sissom Family History, 1989. Bd. dirs. Evang. Bapt. Mission, Kokomo, 1973—. With USNG, 1955-63. Home: 1312 Corvair Ct Kokomo IN 46902-2531 Office: Delco Electronics 700 E Firmin St Kokomo IN 46902-2340

SITU, MING, aerospace engineering researcher; b. Shanghai, People's Republic China, May 4, 1937; s. Jin and Fong Mang (Wang) S.; m. Zhi Fang Jiang, Mar. 15, 1965; children: Haiyan, Hailing, Hailou. BS, Tsinghua U., Beijing, 1961; PhD, Va. Poly. Inst. and State U., 1989. Rsch. fellow China Aerospace Industry Corp., Beijing, 1961—; vis. scholar aerospace dept. Va. Poly. Inst. and State U., Blacksburg, 1981-83. Contbr. articles to AIAA Jour., Jour. Jet Propulsion China, ASME Jour. Fluids Engring. Mem. AIAA. Achievements include research on Ramjet, combustion, multiphase fluid dynamics, turbulent model, Scramjet, hypersonic flows. Office: Aerospace Corp, PO Box 7208-9, Beijing China

SIVAM, THANGAVEL PARAMA, aerospace engineer; b. Madukkur, India, June 12, 1944; s. Thangavel Arumugam Chettiar and Sivapackiam Thangavel (Vadivel) Achi; m. Maheswari Uthirapathy, Sept. 14, 1972; children: Uma Devi, Senthil Kumar, Sunthosh Kumar. MS, Indian Inst. Tech., Madras, 1971; PhD, SUNY, Buffalo, 1978. Project engr. Indian Space Rsch. Orgn., Trivandrum, 1971-74; rsch. asst. SUNY, Buffalo, 1974-78; asst. prof. W.Va. Tech., Montgomery, 1978; sr. structures engr. Gates Learjet Corp., Wichita, Kans., 1978-82; asst. prof. Wichita State U., 1982-87; prin. engr. Boeing Mil. Airplane, Wichita, 1987-89; sr. staff specialist Chrysler Techs. Airborne Systems, Inc., Waco, Tex., 1989—; cons. Precision Composites, Wichita, 1984-86; rsch. assoc., adj. faculty Inst. for Aviation Rsch. at Wichita State U., 1986-89. Contbr. articles to profl. jours. Merit scholar Govt. of India, 1965; grantee NASA, 1982-84, U.S. Army, 1987-90, USAF/Raytheon, 1989—. Mem. AIAA, Soc. Automotive Engrs. (chmn. structures com. 1985—). Achievements include invention of natural fiber/epoxy composites; originated steel wire-reinforced fiberglass/epoxy cylindrical shells. Office: Chrysler Techs Airborne Systems 7500 Maehr Rd MS 1135 Waco TX 76715

SIVASUBRAMANIAN, KOLINJAVADI NAGARAJAN, neonatologist, educator; b. Coimbatore, Madras, India, May 9, 1945; came to U.S., 1971; s. Kolinjavadi Ramaswamy and Sukanthi (Subramanian) Nagarajan; m. Kalyani Hariharier, Feb. 5, 1975; children: Ramya, Rajeev, Ranjan. BSc, Madras U., 1964, MD, 1969. Diplomate Am. Bd. Pediatrics and Neonatal-Perinatal Medicine. Intern in pediatrics Jewish Hosp. and Med. Ctr., Bklyn., 1971-72; resident in pediatrics U. Md. Hosp., Balt., 1972-74; fellow in neonatology Georgetown U. Hosp., Washington, 1974-76; attending neonatologist Goergetown U. Hosp., Washington, 1976—, dir. nurseries, cheif neonatology, 1981—, vice chair pediatrics, 1988—; prof. pediatrics Georgetown U. Hosp., Washington. Editor: Trace Elements/Mineral Metabolism During Development, 1993; editor pub. SIDS Series, 1985; editor jour. Current Concepts in Neonatology, India, 1990—; internat. editor Indian Jour. Pediatrics, India, 1988—. Chmn. Siva Vishnu Temple, Lanham, Md., 1981-91; mem. Fetus and New Born Com., Washington, 1988. Recipient "Preemies" cover article Newsweek, 1988. Fellow Am. Acad. Pediatrics; mem. AAAS, N.Y. Acad. Sci. Hindu. Achievements include research in neonatology, trace elements kinetics, reduction in infant mortality and bioethics. Office: Georgetown U Hosp 3 South Hospital 3800 Reservoir Rd NW Washington DC 20007-2196

SIVATHANU, YUDAYA RAJU, aerospace engineer; b. Trivandrum, Kerala, India, Mar. 7, 1962; came to U.S., 1986; s. Sivathanu Arunachalam and Agasthiar Annamalai. BS in Mech. Engring., Kerala U., Trivandum, 1984; MS in Aero. Engring., U. Mich., 1988, PhD in Aero. Engring., 1990. Scientist Indian Defence R&D Orgn., Kerala, Cochin, 1984-86; rsch. engr. U. Md., Balt., 1990-91; rsch. assoc. Purdue U., Lafayette, Ind., 1991—. Referee Jour. Heat Transfer, Combustion & Flame, Internat. Symposium in Combustion. Mem. AIAA, ASME, Combustion Inst., Amnesty Internat., Am. Contract Bridge League (life master 1992, McKinney award 1990, 91), Sigma Xi. Hindu. Achievements include research in generalized state relationships for hydrocarbon flames, discrete probability function methods for turbulent flows, intrusive emission absorption spectroscopy. Home: 2421-E Kestral Blvd West Lafayette IN 47906-6542 Office: Purdue U Mech Engring Dept TSPC West Lafayette IN 47907-1003

SIWEK, DONALD FANCHER, neuroanatomist; b. Portchester, N.Y., Apr. 15, 1954; s. Manuel and Pauline (Fancher) S.; m. Melissa Katwa Sophia Vlahos, Dec. 31, 1986; children: Kelsey Fancher, Maxwell Riesner. BA, Hampshire Coll., 1977; PhD, Boston U., 1988. Postdoctoral fellow Boston U., 1988-91; rsch. health scientist Edith Nourse Rogers Meml. VA Hosp., Bedford, Mass., 1991—. Contbr. articles to Jour. Comparative Neurology, Cerebral Cortex, Visual Neurosci. Mem. Am. Assn. Anatomists, Soc. Neurosurgery, Brain Wave Info. Exch. Assn. (chmn. 1989—). Office: Edith Nourse Rogers Meml VA Hosp Rsch 151 200 Springs Rd Bedford MA 01730

SIYAN, KARANJIT SAINT GERMAIN SINGH, software engineer; b. Mauranipur, India, Oct. 16, 1954; came to U.S., 1978; s. Ahal Singh and Tejinder Kaur (Virdi) S.; m. Dei Gayle Cooper, Apr. 8, 1987. B in Tech. Electronics, Indian Inst. Tech., 1976, M in Tech. Computer Sci., 1978; MS in Engring., U. Calif., Berkeley, 1980, postgrad in computer sci. Cert. enterprise netware engr.; cert. microsoft profl. Sr. mem. tech. staff Rolm Corp., San Jose, Calif., 1980-84; cons. Siyan Cons. Svcs., L.A., 1985-86, Emigrant, Mont., 1987—. Author, sr. instr.: Learning Tree Internat., L.A., 1985—; author: Network - The Professional Reference, 1992; co-author: Downsizing Netware, LAN Connectivity; author seminars on Novell Networking, TCP/IP networks, Windows NT, Solaris-PC Network Integration. Mem. IEEE, ACM, CNEPA, ECNE, Windos NT MCP, Kappa Omicron Phi.

SIZEMORE, CAROLYN LEE, nuclear medicine technologist; b. Indpls., July 22, 1945; d. Alonzo Chester and Elsie Louise Marie (Osterman) Armstrong; m. Jessie S. Sizemore Sr., June 9, 1966; 1 child, Jessie S. Jr. AA in Nuclear Medicine, Prince George's Community Coll, Largo, Md., 1981; BA in Bus. Adminstrn., Trinity Coll., 1988. Registered technologist (nuclear medicine); cert. nuclear medicine technologist, Md.; lic. nuclear med. technologist. Nuclear med. technologist Washington Hosp. Ctr., 1981-88; chief technologist, mem. com. Capitol Hill Hosp., Washington, 1988-91; chief technologist, asst. radiation safety officer Nat. Hosp. Orthopaedics and Rehab., Arlington, Va., 1991—; mem. Am. Registry of Radiologic Technologists Nuclear Medicine Exam. Com., 1990-93. Contbr. articles to profl. jours. Mem. com. Medlantic Rsch. Found., Washington, 1989-93. Mem. Md. Soc. Radiologic Technologists, Md. Soc. Nuclear Medicine Technologists, Soc. Nuclear Medicine (chmn. membership 1983-85, sec. 1985-87, 88-89, co-editor Isotopics 1991, editor Isotopics 1992—), Nuclear Medicine Adv. Bd., Am. Legion Aux. (exec. com. 1975-76). Republican. Lutheran. Avocations: various crafts, reading, jazzercise. Home: 6700 Danford Dr Clinton MD 20735-4019

SIZEMORE, ROBERT CARLEN, immunologist; b. Lexington, Ky., Sept. 30, 1951; s. Dewey and Juanita (Peel) S.; m. Katherine Killelea, Sept. 29, 1990; children: Katherine Peel, Robert Carlen Jr. BS, U. Ky., 1973, MS, 1975; PhD, U. Louisville, 1982. Postdoctoral rsch. assoc. U. Miss. Med. Ctr., Jackson, 1982-84; dir. cellular immunology IMREG, Inc., New Orleans, 1984—; adj. asst. prof. Tulane U. Sch. Medicine, New Orleans, 1985—. Patentee in field; contbr. articles to profl. jours. Recipient Project award U. Louisville, 1978, Grad. Dean's Citation U. Louisville, 1983; named Outstanding Young Men of Am., 1984. Mem. Am. Assn. Immunologists, Internat. AIDS Soc., Am. Soc. Tropical Medicine & Hygiene, Internat. Soc. Devel. & Comparative Immunology, Fedn. Am. Scientists, AAAS. Avocations: music, composing, photography, travel. Home: 4401 Copernicus St New Orleans LA 70131-3615

SJOBERG, BERNDT OLOF HARALD, chemist, research administrator; b. Jonkoping, Sweden, May 24, 1931; s. Gunnar and Ruth (Magnusson) S.; m. Elisabet Carlsen, June 25, 1955; children: Jan Mikael, Ann Helen. MA, U. Lund, Sweden, 1954; PhD, U. Uppsala, Sweden, 1960; postgrad., Wayne State U., 1958-59. Rsch. asst. dept. chemistry U. Uppsala, 1956-58, 59-60, asst. prof. organic chemistry, 1960-74; rsch. chemist AB Astre, Sodertalje, Sweden, 1960-65; dir. R & D Astra Lakemedel, Sodertalje, 1965-79; dir. Vitrum Inst. Human Nutrition, Stockholm, 1979-81; exec. v.p. Vitrum AB, Stockholm, 1979-81; v.p. R&D Kabi Vitrum AB, Stockholm, 1981-90; dir. R&D, corp. sr. scientist KabiPharmacia AB, 1990—; adj. prof. U. Uppsala, 1974-83; cons. Swedish Bd. Tech. Devel., Stockholm, 1975-79. Mem. edit. bd. Jour. Antibiotics, 1974-80, Acta Pharmaceutica Suecia, 1970—; contbr. articles to profl. jours.; patentee medical field. Mem. Swedish Acad. Engring. Scis. (bd. dirs. 1972-75, 82, pres. biotech. div. 1987-90), N.Y. Acad. Scis., Swedish Acad. Pharm. Scis., Swedish Soc. Medicine, Royal Soc. Medicine (London), Am. Soc. Chemistry, Assn. Swedish Pharm. Industry (pres. R & D com. 1985-90), Royal Yachting Assn., Sodertalje Yachting Club (bd. dirs. 1980-90). Home: Torekallgatan 13, 15173 Sodertalje Sweden Office: KabiPharmacia AB, Lindhagensgatan 133, 11287 Stockholm Sweden

SJOSTRAND, FRITIOF STIG, biologist, educator; b. Stockholm, Sweden, Nov. 5, 1912; s. Nils Johan and Dagmar (Hansen) S.; m. Marta Bruhn-Fahraeus, Mar. 24, 1941 (dec. June 1954); 1 child, Rutger; m. Ebba Gyllenkrok, Mar. 28, 1955; 1 child, Johan; m. Birgitta Petterson, Jan. 23, 1969; 1 child, Peter. M.D., Karolinska Institutet, Stockholm, 1941, Ph.D., 1945; Ph.D. (hon.), U. Siena, 1974, North-East Hill U., Shillon, India, 1989. Asst. prof. anatomy Karolinska Institutet, 1945-48, assoc. prof., 1949-59, prof. histology, 1960-61; research assoc. MIT, 1947-48; vis. prof. UCLA, 1959, prof. zoology, 1960-82, prof. emeritus molecular biology, 1982—. Author: Über die Eigenfluoreszenz Tierischer Gewebe Mit Besonderer Berücksichtigung der Sägertierniere, 1944, Electron Microscopy of Cells and Tissues, Vol. I, 1967, Deducing Function from Structure, Vols. I and II, 1990; also numerous articles. Decorated North Star Orden Sweden; recipient Jubilee award Swedish Med. Soc., 1959, Anders Retzius gold medal, 1967; Paul Ehrlich-Ludwig Darmstaedter prize, 1971. Fellow Royal Micros. Soc. (hon., London), Am. Acad. Arts and Scis.; mem. Electron Microscopy Soc. Am. (hon., Disting. Scientist award 1992), Japan Electron Microscopy Soc. (hon.), Scandinavian Electron Microscopy Soc. (hon.). Achievements include development technique for high resolution electron microscopy of cells, fluorescence microspectrography; inventor ultramicrotome.

SKAAR, STEVEN BAARD, engineering educator; b. Syracuse, N.Y., June 15, 1953; s. Christen and Dorothy (Arpert) S.; m. Emily Holliday, June 16, 1979; children: Emily Kristen, Stephanie Cade. AB, Cornell U., 1975; MS, Va. Poly. Inst. and State U., 1978, PhD, 1982. Assoc. engr. Babcock and Wilcox, Lynchburg, Va., 1978-79; instr. Va. Poly. Inst. and State U., Blacksburg, 1979-82; assoc. prof. Iowa State U., 1982-89, U. Notre Dame, Ind., 1989—; vis. lectr. Inst. Technologico de Estudies Superiores de Monterry, Mex., 1987, 89. Contbr. articles to profl. jours. including Internat. Jour. of Robotics Rsch., Transactions on Automatic Control, Jour. of Applied Mechanics. Grantee U.S. Office of Naval Rsch. grant, 1990—, 87-89, NSF, 1984-86. Mem. AIAA. Republican. Presbyterian. Achievements include patent for camera space manipulation; developed unique and uniquely-capable method for vision-based control of manipulators, a new approach to the guidance of autonomous vehicles. Home: 17358 Woodhurst Rd Granger IN 46530 Office: U Notre Dame Fitzpatrick Hall Notre Dame IN 46556

SKAGGS, RICHARD WAYNE, agricultural engineering educator; b. Grayson, Ky., Aug. 20, 1942; s. Daniel M. and Gertrude (Adkins) S.; m. Judy Ann Kuhn, Aug. 25, 1962; children: Rebecca Diane Skaggs Ramsey, Steven Glen. BS in Agr. Engring., U. Ky., 1964, MS in Agr. Engring., 1966; PhD, Purdue U., 1970. Registered profl. engr., N.C. Grad. asst. U. Ky., Lexington, 1964-66; grad. instr. in rsch. Purdue U., West Lafayte, Ind., 1966-70; asst. prof. agrl. engring. N.C. State U., Raleigh, 1970-74, assoc. prof., 1974-79, prof., 1979-84, William Neal Reynolds prof., 1984—, disting. univ. prof., 1991—; cons. on drainage U.S. AID, Egypt, 1989—, cons., lectr. on water mgmt., India, 1992. Contbr. over 200 articles on water mgmt. and hydrology to profl. jours. Recipient Outstanding Young Scientist award N.C. State U. chpt. Sigma Xi, 1978; alumni rsch. award N.C. State U. Alumni Assn., 1983, Alumni Disting. Prof. award for grad. teaching, 1991; Superior Svc. award USDA, 1986, 90; named to Drainage Hall of Fame, Ohio State U., 1984, Outstanding Alumnus Agrl. Engr., U. Ky., 1985. Fellow Am. Soc. Agrl. Engrs. (chmn. nat. drainage symposium com. 1976, mem. nominating com. 1979, Hancor Soil and Water Engring. award 1986, bd. dirs. 1992-94, John Deere Gold medal 1993); mem. NAE. Avocations: basketball, golf, reading. Home: 2824 Sandia Dr Raleigh NC 27607 Office: NC State U Dept Biol-Agrl Eng PO Box 7625 Raleigh NC 27695-7625

SKALA, GARY DENNIS, electric and gas utilities executive management consultant; b. Bay Shore, N.Y, Oct. 15, 1946; s. Harry A. and Emily Skala. BS in Mgmt. Engring., Rensselaer Polytech. Inst., 1969; MA in

Psychology, Hofstra U., 1972. Engr. L.I. Lighting Co., Hicksville, N.Y., 1969-71; labor rels. coord. L.I. Lighting Co., 1971-73; mgmt. cons. Gilbert/Commonwealth, N.Y.C., 1973-74; sr. mgmt. cons. Booz, Allen & Hamilton, San Francisco, 1974-78; mgr. utility cons. A.T. Kearney, Chgo., 1978-81; mng. cons. Cresap, div. Towers Perrin, Chgo., 1981-85; pres. Gary D. Skala & Assocs. Mgmt. Cons., Chgo., 1985—; lectr. on utility bus. issues; subcontracting cons. Arthur D. Little Inc., Liberty Cons. Group, Ernst & Young, Cresap, A.T. Kearney, Michael Paris Assocs. Ltd., Planmetrics, Am. Inst. Indsl. Engrs. Contbr. articles to profl. jours. Mem. Inst. Indsl. Engrs. (sr. mem. utility div. 1978—, charter), Am. Inst. Indsl. Engrs. (chmn. Midwest chpt. utility div. 1980-81). Avocations: managing the Jerry Lee Lewis Archives, Jason D. Williams Archives.

SKALAK, RICHARD, engineering mechanics educator, researcher; b. N.Y.C., Feb. 5, 1923; s. Rudolph and Anna (Tuma) S.; m. Anna Lesta Allison, Jan. 24, 1953; children: Steven Leslie, Thomas Cooper, Martha Jean, Barbara Anne. BS, Columbia U., 1943, CE, 1946, PhD, 1954; MD (hon.), Gothenburg U., Sweden, 1990. Instr. civil engring. Columbia U., N.Y.C., 1948-54, asst. prof., 1954-60, assoc. prof., 1960-64, prof., 1964-77, James Kip Finch prof. engring. mechanics, 1977-88, emeritus, 1988—, dir. Bioengring. Inst., 1978-88; prof. bioengring. U. Calif., San Diego, 1988—, dir. Inst. for Mechs. and Materials, 1992—; panel mem. Gov.'s Conf. on Sci. and Engring. Edn., R & D, 1989-90. Contbr. articles to sci. jours. Bd. dirs. Biotech. Inst., Gothenburg, Sweden, 1978—. With USN, 1944-46. Recipient Great Tchr. award Columbia Coll. Soc. of Older Grads., 1972, Merit medal Czechoslovakian Acad. Scis., 1990. Fellow ASME (Centennial medal 1980, Melville medal 1990, editor jour. 1984), Am. Acad. Mechanics, Soc. Engring. Sci., Am. Inst. Med. and Biol. Engring. (founding); mem. NAE, AAAS, Soc. Rheology, Am. Heart Assn., Microcirculatory Soc., Internat. Soc. Biorheology (Poiseuille medal 1989), Biomed. Engring. Soc. (Alza medal 1983), Cardiovascular System Dynamics Soc., Am. Soc. for Engring. Edn., Tau Beta Pi, Sigma Xi. Democrat. Presbyterian. Office: U Calif San Diego Ames Dept Bioengring La Jolla CA 92093-0412

SKALECKI, LISA MARIE, aerospace engineer; b. Cleve., June 27, 1962; d. Edward Joseph and Colleen Shirley (Lipps) S. BS in Aerospace Engring., U. Cin., 1985; MS in Aeronautical Engring., Ohio State U., 1988. Aerospace engring. co-op NASA Johnson Space Ctr., Houston, 1981-84; controls engr. Hughes Aircraft Co., L.A., 1985-86; teaching asst. Ohio State U., Columbus, 1986-88; software cons. Columbus, 1986-88; guidance specialist engr. Boeing Co., Seattle, 1988-91; software cons. Seattle, 1991—. Contbr. articles to profl. jours. Princeton U. fellow, 1992. Mem. AIAA. Achievements include development of "FAST" spacecraft guidance algorithm, a nonlinear programming approach that has been internationally recognized in the guidance, navagation and control community. Office: Princeton U Mech and Aerospace Engring Princeton NJ 08544

SKALKA, ANNA MARIE, molecular biologist, virologist; b. N.Y.C., July 2, 1938. AB, Adelphi U., 1959; PhD in Microbiology, NYU, 1964. Am. Cancer Soc. fellow molecular biology genetics rsch. unit Carnegie Inst., 1964-66, fellow, 1966-69; asst. mem. dept. cell biology lab. molecular and biochemical genetics Roche Inst. Molecular Biology, 1969-71, assoc. mem., 1971-76, mem., 1976-80, head, 1980—; now dir. Inst. Cancer Rsch., Phila.; vis. prof. dept. molecular biology Albert Einstein Coll. Medicine, 1973—, Rockefeller U., 1975. Mem. AAAS, Am. Soc. Microbiology, Am. Soc. Biol. Chem., Assn. Women Sci., Sigma Xi. Achievements include research in the structure and function of DNA, host and viral functions in the synthesis of viral DNA and RNA, phage DNA as a vehicle for the amplification and study of eukaryotic genes, molecular biology of avian retroviruses. Office: Inst for Cancer Rsch Fox Chase Cancer Ctr 7701 Burholme Ave Philadelphia PA 19111-2497*

SKALKA, HAROLD WALTER, ophthalmologist, educator; b. N.Y.C., Aug. 22, 1941; s. Jack and Sylvia Skalka; m. Barbara Jean Herbert, Oct. 2, 1965; children: Jennifer, Gretchen, Kirsten. AB with distinction, Cornell U., 1962; MD, NYU, 1966. Intern Greenwich (Conn.) Hosp., 1966-67; resident in ophthalmology Bellevue Hosp., Univ. Hosp., Manhattan VA Hosp., 1967-70; fellow in retinal physiology and ultrasonography, 1970-71; cons. in ophthalmology St. Jude's Hosp., Montgomery, Ala., 1971-73; asst. prof. ophthalmology U. Ala., Birmingham, 1973-75, assoc. prof., 1975-80, prof., 1980-81, assoc. prof. dept. medicine, 1980—, prof., chmn. combined program in ophthalmology, 1981—; acting chmn. combined program ophthalmology U. Ala., 1974-76; ophthalmologist Lowndes County Bd. Health Community Health Project, 1972. Contbr. articles to Am. Jour. Ophthalmology, Eye, Ear, Nose and Throat Monthly, Annals of Ophthalmology, Ophthalmic Surgery, Jour. Clin. Ultrasound, Jour. Pediatric Ophthalmology and Strabismus, The Lancet, AMA Archives of Ophthalmology, Jour. So. Med. Assn., Acta Ophthalmologica, Metabolic and Pediatric Ophthalmology, Applied Radiology, British Jour. Ophthalmology, Blood, Neuro-Ophthalmology; editorial bd.: Ala. Jour. Med. Sci. Major USAFMC, 1971-73. Mem. AAAS, AMA, ACS, SIDUO, Ala. Sight Conservation Assn., Ala. Conservancy, Ala. Wildlife Fedn., Eye Bank Bd., Am. Acad. Ophthalmology, Am. Inst. Ultrasound in Medicine, Internat. Soc. for Clin. Electrophysiology of Vision, Internat. Soc. on Metabolic Eye Disease, Assn. for Rsch. in Vision and Ophthalmology, AAUP, Am. Intraocular Implant Soc., Am. Assn. Ophthalmology, Pan Am. Assn. Ophthalmology, So. Med. Assn., Rsch. to Prevent Blindness, Ala. Acad. Ophthalmology, Ala. Med. Assn., Jefferson County Med. Soc., Contact Lens Assn. Ophthalmologists, Ala. Ultrasound Soc., Royal Soc. Medicine, N.Y. Acad. Scis., Am. Soc. Standardized Ophthalmic Echography (charter exec. bd. mem.), Am. Coll. Nutrition. Office: Eye Found Hosp U Ala 700 18th St S Ste 300 Birmingham AL 35233-1859

SKATRUD, DAVID DALE, physicist; b. Conrad, Mont., May 9, 1956; s. Dale W. and Elaine R. (Larson) S. BA, St. Olaf Coll., 1979; PhD, Duke U., 1984. Program mgr. Rsch. Office U.S. Army, Research Triangle Park, N.C., 1985-90, assoc. dir. Physics Div., 1991—; adj. asst. prof. Physics Dept., Duke U., Durham, N.C., 1986-90, adj. assoc. prof., 1990—. Office: US Army Rsch Office PO Box 12211 Chapel Hill NC 27514

SKEATH, ANN REGINA, mathematics educator; b. Frackville, Pa., Jan. 31, 1935; d. Charles Henry and Helen Regina (Tokarczyk) Wertz; m. James Edward Skeath, June 23, 1962 (dec. Apr., 1990); children: Susan Emily, James Benjamin. BS, Ursinus Coll., 1956; MA, U. Ill., 1961. Tchr. Lansdowne (Pa.) Aldan Sch. Dist., 1956-60, Lower Marion Sch. Dist., Ardmore, Pa., 1961-62, Champaign (Ill.) Sch. Dist., 1962-63; instr. Widener U., Chester, Pa., 1976-79, West Chester (Pa.) U., 1969-75, '79—; mem. Sch. Dist. Curriculum Devel. Com., Wallingford-Swarthmore (Pa.) Sch. Dist., 1988-89. Mem. Refugee Com. Swarthmore (Pa.) Prebyn. Ch. 1979—, chmn. 1985-87; mem. Bd. Deacons Swarthmore Presbyn. Ch., 1980-83, Swarthmore Refugee House Com., 1983-84. Recipient NSF fellowship U. Ill., 1960-61. Mem. Nat. Coun. Tchrs. Math., Pa. Tchrs. Math., Tchrs. of Math Phila. and Vicinity, Math. Assn. Am. Avocations: reading, foreign langs. Office: West Chester U 13-15 University Ave West Chester PA 19383

SKEEN, DAVID RAY, computer systems administrator; b. Bucklin, Kans., July 12, 1942; s. Claude E. and Velma A. (Birney) S.; BA in Math., Emporia State U., 1964; MS, Am. U., 1972, cert. in Computer Systems, 1973; grad. Fed. Exec. Inst., 1983, Naval War Coll., 1984; postgrad. George Washington U., 1989—; m. Carol J. Stimpert, Aug. 23, 1964; children: Jeffrey Kent, Timothy Sean, Kimberly Dawn. Cert. office automation profl. Computer systems analyst to comdr.-in-chief U.S. Naval Forces-Europe, London, 1967-70; computer systems analyst Naval Command Systems Support Activity, Washington, 1970-73; dir. data processing Office Naval Rsch., U.S. Navy Dept., Arlington, Va., 1973-78, dir. mgmt. info. systems Naval Civilian Pers. Command, Washington, 1978-80; dep. dir. manpower, pers. tng. automated systems Dept. Naval Mil. Pers. Command, Washington, 1980-85; dir. manpower, personnel & tng. info. resource mgmt. div. Chief Naval Ops., 1985-91; assoc. dir. Office of IRM, USDA, Washington, 1992—; lectr. Inst. Sci. and Pub. Affairs, 1973-76; cons. Electronic Data Processing Career Devel. Programs, 1975—; detailed to Pres.'s Reorgn. Project for Automated Data Processing, 1978, Spl. Navy IRM Studies, SECNAV, 1991 and USDA/Office of Mgmt. and Budget IRM, 1993, Pres.'s Fed. Automated Data Processing Users Group, Washington, 1978-80; assoc. prof. Sch. Engring. and Applied Sci., George Washington U. Served with USN, 1964-67, capt. Res. ret. Recipient Outstanding Performance award Interagy. Com. Data Processing, 1976. Mem. Sr. Exec. Assn., Am. Mgmt. Assn., Assn. Computing Machinery, Data Processing Mgmt. Assn., Naval Res. Assn., Navy League. Contbr. articles to profl. jours. Home: 707 Forest Park Rd Great Falls VA 22066-2908 Office: Dept Agriculture Washington DC 20250

SKELTON, GORDON WILLIAM, data processing executive, educator; b. Vicksburg, Miss., Oct. 31, 1949; s. Alan Gordon and Martha Hope (Butcher) S.; m. Sandra Lea Champion, May 1974 (div. 1981); m. Janet Elaine Johnson, Feb. 14, 1986; 1 stepchild, Brian Quarles. BA, McMurry Coll., 1974; MA, U. So. Miss., 1975, postgrad., 1975-77, MS, 1987. Cert. in data processing. Systems analyst Criminal Justice Planning Commn., Jackson, Miss., 1978-80; cord. Miss. Statis. Analysis Ctr., Jackson, 1980-83; data processing mgr. Dept. Adminstrn. Fed.-State Programs, Jackson, 1983-84; mgr. pub. tech. So. Ctr. Rsch. and Innovation, Hattiesburg, Miss., 1985-87; internal cons. Sec. of State, State of Miss., Jackson, 1987; system support mgr. CENTEC, Jackson, 1987-88; instr. dept. computer sci. Belhaven Coll., Jackson, 1988—; v.p. info. svcs. Miss. Valley Title Ins. Co., Jackson, 1988—. Author: (with others) Trends in Ergonomics/Human Factors, 1986. Treas. Singles and Doubles Sunday Sch. Class, Jackson, 1989, 91. With U.S. Army, 1970-73, Vietnam. Recipient Cert. of Appreciation, U.S. Dept. Justice/Bur. Justice Stats., 1982. Mem. IEEE Computer Soc., Data Processing Mgmt. Assn. (chpt. pres. 1991, 92, program chair 1990), Assn. Computing Machinery, Am. Soc. Quality Control. Presbyterian. Avocations: gardening, baseball card collecting. Office: Miss Valley Title Ins Co 315 Tombigbee St Jackson MS 39201-4605

SKELTON, JOHN EDWARD, computer science educator, consultant; b. Amarillo, Tex., May 10, 1934; s. Floyd Wayne and Lucille Annabelle (Padduck) S.; m. Katherine Dow, Mar. 22, 1959; children: Laura Ann, Jeanette Kay, Jeffrey Edward. BA, U. Denver, 1956, MA, 1962, PhD, 1971. Mathematician U.S. Naval Ordnance Lab., Corona, Calif., 1956-59; various sales support and mktg. positions Burroughs Corp., Denver, Detroit, Pasadena, Calif., 1959-67; asst. prof. U. Denver, 1967-74; dir. Computer Ctr., U. Minn., Duluth, 1974-85; prof., dir. computing svcs. Oreg. State U., Corvallis, 1985—; cons. World Bank, China, 1988, Educom Cons. Group, 1985. Author: Introduction to the Basic Language, 1971; co-author: Who Runs the Computer, 1975; also articles. Mem. Assn. for Computing Machinery (pres. Rocky Mountain chpt. 1971, faculty advisor U. Minn. 1980-82, peer rev. team 3 regions 1981-90), Assn. for Spl. Interest Group on Univ. Computing (bd. dirs. 1987-91), Rotary (dist. youth exch. com. 1991—), Sigma Xi (chpt. pres. 1983-84), Phi Kappa Phi (chpt. pres. 1989-90). Episcopalian. Avocations: travel, photography, hiking. Office: Oreg State U Computer Ctr Corvallis OR 97331

SKELTON, JOHN GOSS, JR., psychologist; b. Columbus, Ga.; s. John Goss and Willie Mae (Langford) S.; B.A., Emory U., 1950; M.Ed., Our Lady of Lake U., 1964; Ph.D., Tex. Tech. U., 1967. Positions with advt. agy., newspaper, trade assn., 1950-63; staff psychologist San Antonio State Hosp., 1963-65; resident in clin. psychology U. Tex. Med. Br., Galveston, 1966-67; dir., clin. psychologist Psychol. Services Clinic, Harlingen, Tex., 1967-68; clin. psychologist Santa Rosa Med. Center, San Antonio, 1968-71; pvt. practice clin. psychology San Antonio, 1971—; chmn. dept. psychology Park North Gen. Hosp., San Antonio, 1971-83; clin. asst. prof. U. Tex. Health Sci. Center, San Antonio, 1971-78; cons. S.W. Ind. Sch. Dist. San Antonio, 1974—, Parkside Lodge Ctr., San Antonio; instr. Our Lady of Lake U., San Antonio, summers 1968-71, Tex. Tech U., summer 1967. Pres., Vis. Nurses Assn. Bexar County, 1976-77; bd. dirs. Halfway House San Antonio, 1964-65, Mental Health Assn. Bexar County, 1969-75; dir. steering com. San Antonio Area Crisis Center, 1971-74. Served with USN, 1945-46. Vocat. Rehab. fellow, 1965-67. Mem. Am. Psychol. Assn., Bexar County Psychol. Assn. (pres. 1970-71), Biofeedback Soc. Tex., Tex. Psychol. Assn., Soc. Behavioral Medicine, San Antonio Mus. Assn. Episcopalian. Office: 4402 Vance Jackson Rd Ste 242 San Antonio TX 78230-5334

SKETCH, MICHAEL HUGH, cardiologist, educator; b. Paris, June 25, 1931; came to U.S, 1953; s. Wilfred George and Myriam (Seymour) S.; m. Nancy Ann Wilcox, Sept. 1, 1956; children: Michael H., Peter G., Sarah A., James C., Martin H. B.S. cum laude, Creighton U., 1959, M.D., 1963. Am. Bd. Internal Medicine Nat. Bd. Med. Examiners. Intern Creighton Meml.-St. Joseph Hosp., Omaha, 1963-64, resident fellow in cardiology, 1964-68; asst. prof. medicine Creighton U., Omaha, 1968-71, assoc. prof., 1971-77, prof., 1977—, dir. cardiac labs., 1969—, assoc. dir. div. cardiology, 1971-72, co-dir., 1972-78, dir., 1978—. Contbr. articles to profl. jours. Served with U.S. Army, 1954-56. Fellow Am. Coll. Cardiology (gov. 1981-84), ACP, Am. Heart Assn. (council clin. cardiology), Am. Coll. Chest Physicians; mem. Alpha Sigma Nu, Alpha Omega Alpha. Republican. Roman Catholic. Home: 9744 Fieldcrest Dr Omaha NE 68114-4933 Office: Creighton U Divsn Cardiology 601 N 30th St Omaha NE 68131-2100*

SKIBA, AURELIA ELLEN, parochial school educator; b. Chgo., May 17, 1943; d. Anthony J. and Josephine (Cyza) Trojnar; m. Edward S. Skiba, June 11, 1966; children: Robert E. Randall A., Jeffrey W. BA, DePaul U., 1965; M.Math. Edn., U. Ill., Champaign, 1990. Cert. tchr., Ill. Tchr. Resurrection High Sch., Chgo., 1965-72, head math. dept., 1967-72, tchr., 1979—. Contbr. articles to profl. jours. Mem. parish liturgy bd. St. Isaac Jogues, Niles, Ill., 1990—. Geometry fellow U. Ill., 1987, stats. fellow U. Ill., 1988, calculus fellow U. Ill., 1989; NSF grantee, 1988, 91, 92, 93. Mem. Nat. Coun. Tchrs. Math., Ill. Coun. Tchrs. Math. (conf. speaker 1987), Met. Math. Club Chgo., Math. Tchrs. Assn. Chgo. Roman Catholic. Avocations: choir, cake decorating, knitting, church lector, reading. Office: Resurrection High Sch 7500 W Talcott Ave Chicago IL 60631-3742

SKIBITZKE, HERBERT ERNST, JR., hydrologist; b. Benton Harbor, Mich., Mar. 30, 1921; s. Herbert Ernst and Jennie (Richie) S.; m. Eva Hegel, Mar. 22, 1943; 1 child, Herbert William (dec.). Student, Ariz. State U., 1947; Hon. ScD, U. Ariz., 1988. Registered profl. engr. 11 states; registered ground water hydrologist. Mathematician U.S. Geol. Survey, Phoenix, 1948-54, rsch. hydrologist, 1954-76; pres., sr. hydrologist Hydro Data, Inc., Tempe, Ariz., 1976-85; cons. Skibitzke Engrs. & Assocs., Phoenix, 1986-90; pres., sr. hydrologist Hydro Analysis, Inc., Tempe, 1990—; co-founder, instr. Sch. Hydrology, U. Ariz., Tucson, 1960-64; adv. bd. Sch. Hydrology, Tarleton State U., Stephensville, Tex., 1986-90, Am. Inst. Hydrology, Mpls., 1989—. Contbr. articles to profl. jours., chpts. to books. Lt. USN, 1943-46. Fellow ASCE; mem. NSPE, Am. Inst. Hydrology (C.V. Theis award 1991), Assn. Ground Water Scientists and Engrs., Geol. Soc. Washington. Achievements include developed first computer (electric analog) applied to GW studies; patent on electronic flowmeter for U.S. Geological Survey; pioneered development/application of remote sensing techniques to collect information on water resources; methods to analyze contaminant movement in ground water systems. Office: Hydro-Analysis Inc PO Box 27334 Tempe AZ 85285-7334

SKIBSTED, LEIF HORSFELT, food chemistry researcher; b. Silkeborg, Denmark, Aug. 28, 1947; s. Arne and Grethe (Sørensen) S.; m. Benedicte Kruse, Dec. 20, 1983; children: Marie, Anders, Peter. PharmM, Royal Sch. of Pharmacy, Copenhagen, 1972; PhD, Royal Sch. of Pharmacy, 1976. From asst. prof. to docent, prof. Royal Vet. and Agrl. U., Frederiksberg, Denmark, 1974-90; rsch. dir. ctr. food rsch. Royal Vet. and Agrl. U., Frederiksberg, 1991—; rsch. assoc. U. Calif., Santa Barbara, 1978-79. Contbr. articles to profl. jours. Recipient Ole Rømer award Ministery of Edn., Copenhagen, 1981, Bjerrum Gold medal Ellen and Niels Bjerrum Found., Copenhagen, 1986, Ulrik Brinks award Brinks Found., 1993. Mem. Am. Chem. Soc., Danish Chem. Soc. Avocations: railroad history, botany, horse breeding. Home: Ørstedvej 54 B, DK-4130 Copenhagen Denmark Office: Royal Vet/Agrl U-Food Rsch, 40 Thorvaldsensvej, DK-1871 Frederiksberg Denmark

SKIDMORE, DUANE RICHARD, chemical engineering educator, researcher; b. Seattle, Mar. 5, 1927; s. Everett Sylvester and Ruth (Butler) S.; m. Joann Rebecca Mataleno, Feb. 24, 1962; children: David Louis, Carla Marie, Lara Marie, Richard Duane. BSChemE, U. N.D., 1949; MSChemE, U. Ill., 1951; PhD in Phys. Chemistry, Fordham U., 1960. Rsch. chemist E. I. DuPont, Wilmington, Del., 1961-64; from asst. to assoc. prof. U. N.D., Grand Forks, 1964-72; prof. mineral processing engring. W.Va. U., Morgantown, 1972-78; prof. chem. engring. Ohio State U., Columbus, 1978-90, prof. emeritus, 1990—; cons. DuPont, Occidental Petroleum, 1964-90. Contbr. chpts. to books. Dist. co-chmn. Rep. Com., Wilmington, 1962-64. With U.S. Army, 1945-46. Recipient Victory medal, 1945. Mem. Am. Chem. Soc., AICE (nat. program com. 1973-77). Republican. Roman Catholic. Achievements include development of coal liquefaction, coal gasification, coal desulfurization processes. Home: 960 Lynbrook Rd Columbus OH 43235 Office: Ohio State U 140 W 19th Ave Columbus OH 43210

SKIDMORE, ERIC ARTHUR, industrial engineer; b. Colville, Wash., Apr. 4, 1952; s. Donald Lee and P. Jean (Kifer) S.; m. Brenda Carol White, Mar. 22, 1975 (div. June 1981); children: Lynn Danae, Walter Neil; m. Susan Kae Dillon, Apr. 28, 1984. AS, Spokane Community Coll., Wash., 1973. Journeyman electrician Tower Electric, Spokane, 1973-75; panel wireman Electro-Power Corp., Spokane, 1975-76; roller mill operator Cominco American, Inc., Spokane, 1976-78; assembly supr. R.A. Pearson Co., Spokane, 1978-80; elec. designer Electro Engring., Spokane, 1981-82, R.A. Pearson Co., 1983-86; indsl. engr. ASC Machine Tools, Inc., Spokane, 1986-90; chief exec. officer, computer systems analyst Consol. Bus. Resources, Inc., Harrison, Idaho, 1990—; indsl. engr. Wagstaff Inc., Spokane, 1991—. Avocations: bowling, shooting, computers. Office: Consol Bus Resources Inc HC 2 Box 6 Harrison ID 83833-9601

SKIEST, EUGENE NORMAN, food company executive; b. Worcester, Mass., Feb. 2, 1935; s. Hyman Arthur and Dorothy Ida (Brickman) S.; m. Toby Aisenberg, Aug. 14, 1957 (div. 1973); children: Jody, Daniel, Nancy; m. Carol Tata, Nov. 26, 1974. BS, Mass. Coll. Pharmacy, 1956; MS, U. Mich., 1958; PhD, 1961. Rsch. chemist Foster Grant Co., Leominster, Mass., 1961-62; sr. rsch. chemist Thompson Chem. Co., Atteboro, Mass., 1962-64; pres. C&S Polymers, Westminster, Mass., 1965-66; group leader Borden Chem. Co., Leominster, 1966-69, devel. mgr., 1969-77, dir. devel. and applications, 1976-78; assoc. dir. quality assurance and compliance Borden, Inc., Columbus, Ohio, 1978-79, dir. quality assurance and compliance, 1979-81, corp. tech. dir. chems., 1981-84, corp. tech. dir., 1984-88, corp. dir. rsch., devel. and technology, 1989-91, corp. v.p. sci. and tech., 1992—, co. rep. to Ind. Rsch. Inst., 1983—. Contbr. articles, papers to profl. publs. Patentee in field. Bd. dirs. Pickawillany Assn., Westerville, Ohio, 1981; rep. to Ind. Res. Inst., 1985—; mem. GMA tech. regulatory com., 1992—; coun. mem. Nat. Food Process, 1993—; trustee Food Update, 1993, Opera Columbus; rsch. dirs. Roundtable, 1992—. Mem. Am. Chem. Soc., Soc. Plastics Engrs., Formaldehyde Inst. (chmn. tech. com. 1979-84, bd. dirs. 1981, vice chmn. exec. com. 1986-88), Chem. Mfrs. Assn. (chmn. task force 1983-88), Soc. Plastics Industries (vinyl toxicology subcom. 1978-82, vinyl acetate task force 1980-82, chmn. ad hoc packaging risk assessment com. 1978-81), Continental Tennis Club, Lakes Country Club, York Golf Culb, Capitol Club. Avocations: tennis, golf, reading, sports. Home: 134 Green Springs Dr Columbus OH 43235-4643 Office: Borden Inc 1105 Schrock Rd Ste 401 Columbus OH 43229-0410*

SKIMINA, TIMOTHY ANTHONY, electrical engineer, computer consultant; b. Elmhurst, Ill., June 13, 1960; s. Vitus Anthony and Rindalee M. (Ralick) S.; m. Margaret Ann Hein, Jan. 9, 1988. BA in Physics, Wabash Coll., 1982; BSEE, Purdue U., 1985, MS in Biology, 1986. Lab. mgr. Purdue Univ., Hammond, Ind., 1982-87; sr. programmer analyst Northwestern U., Chgo., 1987-90, tech. svcs. specialist, 1990—; computer cons. Ill., Ind., Fla., Mo., 1985—; active apple support coords. adv. coun. Apple Computer, 1990-91, co-chmn., 1991. Co-author: (lab. manual) Transmission Electron Microscopy Procedures Manual, 1983; author: (lab. manual) 2-Dimensional Electrophoresis Procedures Manual, 1986. Recipient All Inland scholarship Inland-Ryerson Found., Chgo., 1978-82, Pres.'s scholarship Wabash Coll., 1978, Am. Legion scholarship, 1978, Robert S. Harvey Journalism award Wabash Coll., 1980. Mem. IEEE, IEEE Computer Soc., Passiflora Soc. Internat. (charter mem., assoc. sec. 1993—), Kennicott Natural Sci. Club (v.p. 1989-90). Home: 8941 Kennedy Ave Highland IN 46322-1911 Office: Northwestern U 303 E Chicago Ave Chicago IL 60611-3008

SKINNER, G(EORGE) WILLIAM, anthropologist, educator; b. Oakland, Calif., Feb. 14, 1925; s. John James and Eunice (Engle) S.; m. Carol Bagger, Mar. 25, 1951 (div. Jan. 1970); children: Geoffrey Crane, James Lauriston, Mark Williamson, Jeremy Burr; m. Susan Mann, Apr. 26, 1980; 1 dau., Alison Jane. Student, Deep Springs (Calif.) Coll., 1942-43; B.A. with distinction in Far Eastern Studies, Cornell U., Ithaca, N.Y., 1947, Ph.D. in Cultural Anthropology, 1954. Field dir. Cornell U. S.E. Asia program, also Cornell Research Center, Bangkok, Thailand, 1951-55; rsch. assoc. in Indonesia, 1956-58; asso. prof., then prof. anthropology Cornell U., Ithaca, N.Y., 1960-65; asst. prof. sociology Columbia, 1958-60; sr. specialist in residence East-West Ctr. Honolulu, 1965-66; prof. anthropology Stanford, 1966-89; Barbara Kimball Browning prof. humanities and scis., 1987-89; prof. anthropology U. Calif., Davis, 1990—; vis. prof. U. Pa., 1977, Duke U., spring, 1978, Keio U., Tokyo, spring 1985, fall 1988, U. Calif.-San Diego, fall 1986; field research China, 1949-50, 77, S.E. Asia, 1950-51, Thailand, 1951-53, 54-55, Java and Borneo, 1956-58, Japan, 1985, 88; mem. joint com. on contemporary China Social Sci. Research Council-Am. Acad. Learned Socs., 1961-65, 80-81, internat. com. on Chinese studies, 1963-64, mem. joint com. on Chinese studies, 1981-83; mem. subcom. research Chinese Soc. Social Sci. Research Council, 1961-70, chmn., 1963-70; dir. program on East Asian Local Systems, 1969-71; dir. Chinese Soc. Bibliography Project, 1964-73; asso. dir. Cornell China Program, 1961-63; dir. London-Cornell Project Social Research, 1962-65; mem. com. on scholarly communication with People's Republic of China, Nat. Acad. Scis., 1966-70, mem. social scis. and humanities panel, 1982-83; mem. adv. com. Center for Chinese Research Materials, Assn. Research Libraries, 1967-70. Author: Chinese Society in Thailand, 1957, Leadership and Power in the Chinese Community of Thailand, 1958; also articles; Editor: The Social Sciences and Thailand, 1956, Local, Ethnic and National Loyalties in Village Indonesia, 1959, Modern Chinese Society: An Analytical Bibliography, 3 vols, 1973, (with Mark Elvin) The Chinese City Between Two Worlds, 1974, (with A. Thomas Kirsch) Change and Persistence in Thai Society, 1975, The City in Late Imperial China, 1977, The Study of Chinese Society, 1979. Served to ensign USNR, 1943-46. Fellow Center for Advanced Study in Behavioral Scis., 1969-70, Guggenheim fellow, 1969; NIMH spl. fellow, 1970. Mem. NAS, AAAS, Am. Anthrop. Assn., Am. Sociol. Assn., Assn. Asian Studies (bd. dirs. 1962-65, chmn. nominating com. 1967-68, pres. 1983-84), Soc. for Cultural Anthropology, Internat. Union for Sci. Study of Population, Social Sci. History Assn., Am. Ethnol. Soc., Population Assn. Am., Siam Soc., Soc. Qing Studies, Soc. Econ. Anthropology, Phi Beta Kappa, Sigma Xi. Office: Dept Anthropology U Calif Davis CA 95616

SKINNER, HELEN CATHERINE WILD, biomineralogist; b. Bklyn., Jan. 25, 1931; d. Edward Herman and Minnie (Bertsch) Wild; m. Brian John Skinner, Oct. 9, 1954; children: Adrienne W.S. Scott, Stephanie Skinner, Thalassa Skinner. BA, Mt. Holyoke Coll., 1952; MA, Radcliffe/Harvard, 1954; PhD, Adelaide (Australia) U., 1959. Mineralogist sect. molecular structure Nat. Inst. Arthritis and Metabolic Diseases, NIH, 1961-65; mem. sect. crystal chemistry Lab. Histology and Pathology Nat. Inst. Dental Rsch., NIH, 1965-66; lectr. dept. geology and geophysics Yale U., 1967-69, rsch. assoc. dept. surgery, 1967-72, sr. rsch. assoc. dept. surgery, 1972-75; Alexander Agassiz vis. lectr. dept. biology Harvard U., 1976-77; lectr. dept. biology Yale U., 1977-83; assoc. prof. biochemistry in surgery Yale U., New Haven, 1978-84, lectr. dept. orthopaedic surgery, 1977—, rsch. assoc. in geology and geophysics, 1985—; pres. Conn. Acad. Arts and Scis., 1986—; mineralogist AEC, summer 1953; master Jonathan Edwards Coll., Yale U., 1977-82; vis. prof. sect. ecology and systematics dept. biology Cornell U., 1980-83; dental adv. com. Yale-New Haven Hosp., 1973-80; mem. faculty adv. com. Yale-New Haven Tchrs. Inst., 1983—; chmn. site visit team Nat. Inst. Dental Rsch., 1974-75. Author: (with others) Asbestos and other Fibrous Materials: Mineralogy, Crystal Chemistry and Health Effects, 1988; co-editor: Biomineralization Process of Iron and Manganese: Modern and Ancient Environments, 1992; tech. abstractor Geol. Soc. Am., 1961-65; sect. editor Am. Mineralogist, 1978-82; contbr. over 40 articles to profl. jours. Mem. bd. edn. com. Conn. Fund for Environ., 1983-89, mem. sci. adv. com., 1989-92; founder, pres. Investor's Strategy Inst., New Haven, 1983-85; trustee Miss Porter's Sch., Farmington, Conn., mem. edn. com., 1986-88, mem. salaries and benefits com., 1988-91; treas. YWCA, New Haven, 1983-84. Fellow Geol. Soc. Am., Mineral. Soc. Am. (fin. counsel 1881-86, devel. com. 1986-88, Pub. Svc. award 1991); mem. Am. Soc. Bone and Mineral Rsch., Am. Assn. Crystal Growth, Am. Assn. Dental Rsch., Internat. Assn. Dental Rsch., Mineral. Soc. Can. Home: PO Box 894 Woodbury CT 06798-

0894 Office: Yale U Dept Geology Geophysics PO Box 6666 New Haven CT 06511-8101

SKINNER, JAMES LAURISTON, chemist, educator; b. Ithaca, N.Y., Aug. 17, 1953; s. G. William and Carol (Bagger) S.; m. Wendy Moore, May 31, 1986; children: Colin Andrew, Duncan Geoffrey. AB, U. Calif., Santa Cruz, 1975; PhD, Harvard U., 1979. Rsch. assoc. Stanford (Calif.) U., 1980-81; from asst. prof. to prof. chemistry Columbia U., N.Y.C., 1981-90; Hirschfelder prof. chemistry U. Wis., Madison, 1990—; vis. scientist Inst. Theol. Physics U. Calif., Santa Barbara, 1987; vis. prof. physics U. Jos. Fourier, Grenoble, France, 1987. Contbr. articles to profl. jours. Recipient Fresenius award Phi Lambda Upsilon, 1989, Camille and Henry Dreyfus Tchr.-Scholar award, 1984, NSF Presdl. Young Investigator award, 1984, Humboldt Sr. Scientist award, 1993; NSF grad fellow, 1975, NSF postdoctoral fellow, 1980, Alfred P. Sloan Found. fellow, 1984, Guggenheim fellow, 1993. Mem. AAAS, Am. Chem. Soc., Am. Phys. Soc. Achievements include fundamental research in condensed phase theoretical chemistry. Office: U Wis Dept Chemistry Theoretical Chem Inst 1101 Univ Ave Madison WI 53706

SKINNER, JAMES STANFORD, physiologist, educator; b. Lucedale, Miss., Sept. 22, 1936; married, 1963; 2 children. BS, U. Ill., 1958, MS, 1960, PhD in Phys. Edn. and Physiology, 1963. Assoc. physiologist sch. medicine George Washington U., 1964, asst. prof. lectr., 1964-65; rsch. assoc. cardiologist sch. medicine U. Wash., Seattle, 1965-66; asst. prof. lab. human performance rsch. Pa. State U., University Park, 1966-70; rsch. assoc. med. clin. U. Freiburg, Germany, 1970-71; assoc. prof. phys. edn., rsch. assoc. inst. cardiology U. Montreal, 1971-77; prof. phys. edn. U. Western Ont., 1977-82, Ariz. State U., Tempe, 1982—; dir. exercise and sport rsch. inst. Ariz. State U., 1983—. Fellow Am. Coll. Sports Medicine (pres. 1979-80, Citation award 1986), Am. Heart Assn.; mem. Can. Assn. Sports Sci. (sec. 1976-78), Am. Assn. Health, Phys. Edn. and Recreation, Am. Acad. Phys. Edn. (sec.-treas. 1990-92). Achievements include research in physiology of exercise, especially pertaining to cardiovascular disorders, in exercise, training and genetics. *

SKINNER, MARK ANDREW, astrophysicist; b. Amityville, N.Y., Mar. 17, 1960; s. Richard Alvin and Doris Elizabeth (Edwards) S.; m. Alison J. Ressler. SB, MIT, 1984; PhD, U. Wis., 1991. Rsch. asst. physics dept. U. Wis., Madison, 1984-91, postdoctoral researcher, 1991; rsch. assoc. dept. astronomy and astrophysics Penn State U., University Park, 1991—; tech. cons. Space Sci. and Engring., U. Wis., Madison, 1991-92. Contbr. articles to profl. jours. Grantee U. Wis., 1989. Mem. Am. Phys. Soc., Internat. Soc. for Optical Engrs., AIAA. Office: Pa State U Dept Astronomy 525 Davey Labs University Park PA 16802

SKINNER, MAURICE EDWARD, IV, information security system specialist; b. Balt., Aug. 31, 1962; s. Maurice Edward III and Patricia U. (O'Connor) S. BSc in Computer Sci., BBA, Lynchburg (Va.) Coll., 1986. Cert. info. systems auditor. Systems analyst Lynchburg Coll., 1985-86; ptnr. Bates Marine Basin, Oxford, Md., 1985-87; mgr. info. tech. audit svcs. Coopers & Lybrand, Washington, 1986-92; systems mgr. faculty practice info. systems Pasquerilla Healthcare Ctr., Georgetown U., Washington, 1992—. Mem. Assn. for Computing Machinery, Electronic Data Processing Auditors Assn., Info. Systems Security Assn., Digital Equipment Corp. Users Soc. (chmn. mid-Atlantic area security local users group). Republican. Methodist. Avocations: sailboat racing, computer programming, target shooting. Home: 5818 Inman Park Cir # 200 Rockville MD 20852-5474 Office: Pasquerilla Healthcare Ctr Faculty Practice Info 3800 Reservior Rd NW Washington DC 20007

SKINNER, RAY, JR., earth science educator, consultant; b. Junction City, Ohio, Apr. 21, 1926; s. Ray and Naomi Tresa (Edgerly) S.; m. Patricia Frame, Aug. 24, 1952 (div. Feb. 1981); children: Sarah Jane Skinner-Morris, Suzanne Kay; m. Marilyn Lamont, July 23, 1983. BS, Ohio State U., 1949, MA, 1954; PhD, Kent State U., 1966. Cert. secondary sci. and math. tchr., Ohio. Tchr. biology, math. Margaretta High Sch., Castalia, Ohio, 1950-53, Upper Arlington High Sch., 1953-54; tchr. math. Bellevue (Ohio) High Sch., 1954-56; tchr. biology, math. Milan (Ohio) High Sch., 1956-60; tchr. earth sci. Norwalk (Ohio) High Sch., 1960-63; rsch. asst., part-time instr. Kent (Ohio) State U., 1963-66; prof. Ohio U., Athens, 1966-80, prof. emeritus, 1980—; dir. Upward Bound, Ohio U., Athens, 1967-68, NSF Inst., 1968-73, Jr. Sci. and Humanities Symposium, 1973-83, Regional Sci. Olympiad, 1986-91, Gov.'s Scholars Talented and Gifted Program, 1985-92, Energy-Environment Workshop, 1992. Contbr. author: Readings on the Environment, 1974; contbr. articles to profl. jours. Coach Castalia Margaretta's Res. and Varisty Basketball Teams, Erie County, 1950's; pres. Athens County Hist. Soc. and Mus., 1988-92. With U.S. Army Air Corps, 1944-45. Named Man of Yr. in field of sci. S.E. Regional Coun. C. of C., 1975; recipient Spirit of '87 award Ohio's Bicentennial Commn., 1987, Faculty Svc. award Nat. Univ. Continuing Edn. Assn., 1989, Svc. to Edn. award Soc. Alumni and Friends Coll. Edn. Ohio U., 1991, Disting. Svc. award Ohio Alliance for the Environment, 1992, svc. medal Sons of the Am. Revolution. Mem. Ohio Assn. for Gifted Children (Disting. Svc. award 1989), S.E. Ohio Consortium Coords. for Gifted (Meritorious Svc. award 1988), Ohio State U. Scholarship Dormitory Alumni Assn. (Oustanding Alumni award 1988), Ohio U. Emeriti Assn. (pres.-sec. 1986-92), Athens Kiwanis Club (pres. 1977-78, Disting. award 1978). Home: 2 Coventry Ln Athens OH 45701 Office: Ohio U McCracken Hall Athens OH 45701

SKINNER, SAMUEL BALLOU, III, physics educator, researcher; b. Russellville, S.C., Sept. 24, 1936; s. Samuel Ballou Jr. and Mary (Timmons) S.; m. Beverly Corinne Jones, Dec. 21, 1958; children: Teresa Lynn, Curtis Ballou, Mary Angela. BS, Clemson U., 1958; MA in Teaching, U. N.C., 1963, PhD, 1970. Sci. tchr. Franklin (Tenn.) High Sch., 1959-60, Irmo (S.C.) High Sch., 1960-62; asst. prof. physics St. Andrews Presbyn. Coll., Laurinburg, N.C., 1964-67; assoc. prof. physics Columbia (S.C.) Coll., 1970-72; prof. physics U. S.C. Coastal Carolina Coll., Conway, S.C., 1972—; dir. acad. affairs U. S.C. Coastal Carolina Coll., Conway, 1972-74; advisor Gov.'s Nuclear Adv. Coun., Columbia, 1977-80, S.C. Joint Legis. Com. on Energy, Columbia, 1980-85; U.S.A. rep. Internat. Symposium on Nuclear Waste, Vienna, Austria, 1980; researcher in current aerospace problems, nuclear radiation physics and critical thinking devel.; project counselor Space Life Sci. Tng. Program, Kennedy Space Ctr., 1990. Author: Education and Psychology, 1973, Energy and Society, 1981; contbr. articles to profl. jours. Pres. Horry County Am. Cancer Soc., Conway, 1974-75, Horry County Literacy Coun., 1989-93; mem. Horry County Assessment Appeals Bd., 1981-91. Recipient fellowships NASA (6), 1987-93, Dept. Energy (4), 1972-86, USAF, 1985; grantee Dept. Energy, 1984; physics edn. del. to Vietnam, Citizen Amb. Program, 1993; participant Institut Teknologi MARA Mucia Program, Malaysia, 1993—. Mem. Am. Assn. Physics Tchrs., Am. Assn. Higher Edn., S.C. Acad. Sci., U. S.C. Coastal Carolina Athletic Club (bd. dirs.), Lions (pres. Conway Club 1977-78). Presbyterian. Avocations: tennis, camping, sports spectator, travel. Home: 126 Citadel Dr Conway SC 29526 Office: Coastal Carolina U Physics Dept Conway SC 29526

SKIPPER, ADRIAN, chemical engineer; b. Picayune, Miss., June 2, 1951; s. H.A. and Gladys Ivell (Smith) S.; m. Sharron Rose Griffith, Aug. 4, 1973; children: Jason Troy, Jennifer Kay, Justin Ian. BSCE, Miss. State U., 1974. R&D lab. technician Ciba-Geigy Chems., McIntosh, Ala., 1971-73; process engr. Conoco Chems., Aberdeen, Miss., 1974; process engr. Chevron USA, Pascagoula, Miss., 1974-75, refinery planning analyst, 1975-77, oper. asst., 1979-82, chief chemist, 1982-89, coord. environ., safety and health, 1989—. Mem. adv. com. Miss. Legis. Environ. Protection Coun., Jackson, 1989—, Miss. Gov.'s Environ. Task Force, Jackson, 1991. Mem. Mid-Continent Oil and Gas Assn. (chmn. hazardous waste com. 1990—), Miss. Mfrs. Assn. (environ. com. 1989—), Miss. Water Resources Assn. Republican. Baptist. Office: Chevron USA Products Co PO Box 1300 Pascagoula MS 39568-1300

SKITKA, LINDA JEAN, psychology educator; b. Gaylord, Mich., Jan. 31, 1961; d. Paul Roger Skitka and Elaine Marie (Hoyt) Deneen; m. Mark Eliot Nussbaum, May 14, 1983; children: Joshua Paul, Samantha Glenn. BA, U. Mich., 1983; MA, U. Calif., Berkeley, 1986, PhD, 1989. Asst. prof. So. Ill. U., Edwardsville, 1989—; cons. Family Svcs. and Vis. Nurses Assn., Alton, Ill., 1992—. Contbr. articles to profl. jours. Recipient Outstanding Faculty award Psi Chi, 1992; Fourth Quarter fellow So. Ill. U., 1991, 92, 93. Mem. Am. Psychol. Assn., Internat. Soc. Polit. Psychology, Soc. for Study Social Issues, Soc. for Personality and Social Psychology. Office: So Ill U Dept Psychology Dept Psychology Bldg III 0118 Edwardsville IL 62026-1121

SKJAERSTAD, RAGNAR, electronics company executive; b. Oslo, Norway, Mar. 15, 1944; s. Willy Ragnar and Esther (Schulstad) S.; m. Toril Elisabeth Nyquist, June 24, 1967; children: Richard, Nina Elisabeth. Diploma in engring., Nordic Corr. Inst., Oslo, 1967, European Fedn. Nat. Engring., Paris, 1987; student, London Sch. Econs., 1976, Norwegian U. Tech. Studies, 1976. Test engr. STK/ITT, Oslo, 1969-70, 84-85; project mgr. STK/ITT, 1970-72, prodn. mgr., 1972-78, mfg. mgr., 1978-85, quality mgr. Alcatel STK, 1985-87; mng. dir. Lintron A/S, Lindas, Norway, 1987-90; also bd. dirs. Lintron A/S; ops. mgr. Alcatel Telecom Norway A/S, Oslo, 1990—; bd. dirs. Tech. Rev. Weekly, Oslo, 1980-92, chmn. editors coun., 1992—. Bd. dirs. Norwegian Conservative Polit. Party, 1992—. With Norwegian armed forces, 1967. Mem. Norwegian Engr.'s Orgn. (v.p.), Norwegian Poly. TEch. Coun., Norwegian Quality Coun. Avocations: sailing, photography, old cars. Home: Bjerkealleen 43, 0487 Oslo 4, Norway Office: Alcatel Telecom Norway A/S, PO Box 310 Økern, 0511 Oslo Norway

SKLAR, ALEXANDER, electric company executive, educator; b. N.Y.C., May 18, 1915; s. David and Bessie (Wolf) S.; student Cooper Union, N.Y.C., 1932-35; M.B.A., Fla. Atlantic U., 1976; m. Hilda Rae Gevarter, Oct. 27, 1940; 1 dau., Carolyn Mae (Mrs. Louis M. Taff). Chief engr. Aerovox Corp., New Bedford, Mass., 1933-39; mgr. mfg., engring. Indsl. Condenser Corp., Chgo., 1939-44; owner Capacitron, Inc., 1944-48; exec. v.p. Jefferson Electric Co., Bellwood, Ill., 1948-65; v.p., gen. mgr. electro-mech. div. Essex International, Detroit, 1965-67; adviser, dir. various corp., 1968—; vis. prof. mgmt. Fla. Atlantic U., Boca Raton, 1971-92; ret. 1993; lectr. profl. mgmt. U. Calif. at Los Angeles, Harvard Grad. Sch. Bus. Adminstrn., U. Ill. Mem. Acad. Internat. Bus., Soc. Automotive Engrs. Address: 4100 Galt Ocean Dr Fort Lauderdale FL 33308

SKLAREW, ROBERT JAY, biomedical research educator, consultant; b. N.Y.C., Nov. 25, 1941; s. Arthur and Jeanette (Laven) S.; m. Toby Willner, July 15, 1970; children: David Michael, Gary Richard. BA in Zoology, Cornell U., 1963; MS, NYU, 1965, PhD in Biology, 1970. Assoc. rsch. scientist Sch. of Medicine NYU, N.Y.C., 1965-70, rsch. scientist Sch. of Medicine, 1971-73, sr. rsch. scientist Sch. of Medicine, 1973-79; rsch. asst. prof. pathology Sch. of Medicine Goldwater Meml. Hosp., N.Y.C., 1979-87, rsch. assoc. prof. pathology Sch. of Medicine, 1987-88, dir. cytokinetics and imaging lab. NYU Rsch. Svc., 1980-88; prof. cell biology, anatomy and medicine N.Y. Med. Coll., Valhalla, 1988—; rsch. assoc. dept. pathology Lenox Hill Hosp., N.Y.C., 1981-88, pres., CEO R. J. Sklarew Imaging Assoc., Inc., Larchmont, N.Y., 1990—. Author: Microscopic Imaging of Steroid Receptors, 1990; sr. author: Cytometry, Jour. Histochem. Cytochem., Cancer, Exptl. Cell Rsch. Mem. Beth Emeth Synagogue, Larchmont, 1974—; group leader Boy Scouts Am., Larchmont, 1978-80. Grantee Am. Cancer Soc., Nat. Cancer Inst./NIH Conc. for Tobacco Rsch., R.J. Reynolds Industries Found., NYU. Mem. AAAS, Cell Kinetics Soc. (sec. 1983-85, 85-87, v.p. 1987-88, pres. 1988-89, chmn. nominations 1991, 93), N.Y. Acad. Sci., Soc. for Analytic Cytology, Soc. for Cell Biology, Tissue Culture Assn., Union Concerned Scientists. Democrat. Achievements include development of methodology, algorithms and Receptogram analytic software for application of microscopic imaging in medical research and in pathodiagnosis of cancer, imaging methods for simultaneous densitometry and autoradiographic analysis; research in diagnostic imaging of steroid receptors, oncogenes and DNA ploidy in cancer, proliferative patterns and cell cycle kinetics of human solid tumors. Home: 8 Vine Rd Larchmont NY 10538-1247 Office: NY Med Coll Cancer Rsch Inst 100 Grasslands Rd Elmsford NY 10523-1110

SKLOVSKY, ROBERT JOEL, pharmacology educator; b. Bronx, N.Y., Nov. 19, 1952; s. Nathan and Esther (Steinberg) S.; m. Michelle Sklovsky-Welch, Dec. 21, 1985. BS, Bklyn Coll., 1975; MA in Sci. Edn., Columbia U., 1976; PharmD, U. of Pacific, 1977; D in Naturopathic Medicine, Nat. Coll. Naturopathic Medicine, 1983. Intern Tripler Army Med. Ctr., Honolulu, 1977; prof. pharmacology Nat. Coll. Naturopathic Medicine, Portland, Oreg., 1982-85; pvt. practice specializing in naturopathic medicine Milwaukie, Oreg., 1983—; cons. State Bd. Naturopathic Examiners, Oreg., Hawaii, Clackamas County Sherriff's Dept.; cons. Internat. Drug Info. Ctr., N.Y.C., 1983—; cons. Albert Roy Davis Scientific Research Lab, Orange Park, Fla. 1986. Recipient Bristol Labs. award, 1983. Fellow Am. Coll. Apothecaries; mem. N.Y. Acad. Sci., Soc. for Study of Biochem. Intolerance, Internat. Bio-oxidative Med. Found. Avocations: classical and jazz music, tap dance, art, botany, acting. Office: 6910 SE Lake Rd Portland OR 97267-2196

SKOGEN, HAVEN SHERMAN, oil company executive; b. Rochester, Minn., May 8, 1927; s. Joseph Harold and Elpha (Hemphill) S.; m. Beverly R. Baker, Feb. 19, 1949; 1 child, Scott H. BS, Iowa State U., 1950; MS, Rutgers U., 1954, PhD, 1955; MBA, U. Chgo., 1970. Registered profl. engr., Wis. Devel. engr. E.I. duPont, Wilmington, Del., 1955-57; prof. Elmhurst (Ill.) Coll., 1957-58; chief engr. Stackpole, St. Marys, Pa., 1958-62; plant mgr. Magnatronics, Elizabethtown, Ky., 1962-65; mgr. Allen-Bradley, Milw., 1965-70; v.p. Dill-Clithrow, Chgo., 1970-74; chief chemist Occidental Oil Co., Grand Junction, Colo., 1974—. Author: Synthetic Fuel Combustion, 1984; inventor radioactive retort doping, locus retorting zone. Naval Rsch. fellow, 1951-55. Fellow Am. Inst. Chemists; mem. Masons, Elks, Sigma Xi, Phi Beta Kappa, Phi Lambda Upsilon. Republican. Avocations: fly fishing, travel, reading, teaching. Home: 3152 Primrose Ct Grand Junction CO 81506-4147 Office: PO Box 2399 Grand Junction CO 81502-2399

SKOLNICK, ANDREW ABRAHAM, science and medical journalist, photographer; b. N.Y.C., Oct. 21, 1947; s. Solomon and Blanche (Blidner) S. Cert. profl. photography, Paier Art Sch., Hamden, Conn., 1974; BA in Liberal Arts, Charter Oaks State Coll., Hartford, Conn., 1978; MS in Journalism, Columbia U., 1981. Photographer biology dept. Yale U., New Haven, 1975-76; sci. writer March of Dimes Birth Defects Found., White Plains, N.Y., 1981-85; life scis. editor-news bur. U. Ill., Urbana-Champaign, 1985-87; assoc. sci. news editor AMA, Chgo., 1987-89; assoc. news editor Jour. of AMA, Chgo., 1989—; vis. instr. Trumbull Coll. Yale U., New Haven, Conn., 1976-77. Contgb. author Collier's Ency., Merit Students' Ency., Macmillan, Inc., N.Y.C., 1984—; contbr. articles, op-eds and photographs to mags., newspapers and books. Recipient 1st Ann. Community Health award for journalistic excellence Nat. Assn. Community Health Ctrs., 1992; co-recipient Responsibility in Journalism award Com. for Sci. Investigation of Claims of the Paranormal, Buffalo, 1992; Nate Haseltine Meml. fellow in sci. writing Coun. for Advancement of Sci. Writing, 1980-91. Mem. Nat. Assn. Sci. Writers, Soc. Am. Magicians. Avocations: skiing, fishing, archery, tennis, magic. Home: 328 Des Plaines Forest Park IL 60130 Office: Jour of AMA 515 N State St Chicago IL 60610

SKOLNIK, LEONARD, chemical engineer; b. Cleve., June 13, 1921; s. Louis and Fannie (Koschenik) S.; m. Evelyn Wold, Feb. 27, 1944; children: Eileen, Joel Ian, Daniel Z. BChE, Ohio State U., 1943; MSChE, U. Mich., 1947; PhD, Case Inst. Tech., 1952. Rsch. engr. Dow Chem. Co., Midland, Mich., 1943-46, C.F. Prutton & Assocs., Cleve., 1947-48; rsch. asst. Case Inst. Tech., Cleve., 1948-51; rsch. engr. Harshaw Chem. Co., Cleve., 1951-52, Indsl. Rayon, Cleve., 1952-59; sr. rsch. assoc. B.F. Goodrich, Brecksville, Ohio, 1959-81; cons. in field Cleve., 1981—. Contbr. articles to profl. jours. Bd. dirs. Cleve. Hebrew Schs., 1956—. Mem. Am. Chem. Soc. (emeritus, rubber div.), Fiber Soc., AAAS (emeritus), Michelson Club (bd. dirs., pres. 1956-59), Cleve. Labor Zionist Am. (pres. 1950-56). Jewish. Achievements include patents for defouling of double effect evaporators, stretch cord, radial tire manufacturing; developed tests for industrial textiles and adhesion to rubber, flatspot testing of tire cords. Home: 3826 Meadowbrook Blvd University Heights OH 44118

SKOOG, FOLKE KARL, botany educator; b. Fjärås, Sweden, July 15, 1908; came to U.S., 1925, naturalized, 1935; s. Karl Gustav and Sigrid (Person) S.; m. Birgit Anna Lisa Bergner, Jan. 31, 1947; 1 dau., Karin. B.S., Calif. Inst. Tech., 1932, Ph.D., 1936; Ph.D. (hon.), U. Lund, Sweden, 1956; D.Sc. (hon.), U. Ill., 1980; D. Agr. Sci. (hon.), U. Pisa, Italy, 1991; D.Sc. (hon.), Swedish U. Agrl. Scis., Uppsala, 1991. Teaching asst., research fellow biology Calif. Inst. Tech., 1934-36; NRC fellow U. Calif., Berkeley, 1936-37, summer 1938; instr., tutor biology Harvard U., 1937-41, research assoc., 1941; assoc., assoc. prof. biology Johns Hopkins U., 1941-44; chemist Q.M.C.; also tech. rep. U.S. Army ETO, 1944-46; asso. prof. botany U. Wis.-Madison, 1947-49, prof., from 1949, C. Leonard Huskins prof. botany, now emeritus.; vis. physiologist Pineapple Research Inst., U. Hawaii, 1938-39; assoc. physiologist NIH, USPHS, 1943; vis. lectr. Washington U., 1946, Swedish U. Agrl. Scis., Ultuna, 1952; v.p. physiol. sect. Internat. Bot. Congress, Paris, 1954, Edinburgh, 1964, Leningrad, 1975. Editor: Plant Growth Substances, 1951, 80; contbr. articles to profl. jours. Track and field mem. Swedish Olympic Team, 1932. Recipient certificate of merit Bot. Soc. Am., 1956, Nat. Medal of Sci. NSF, 1991, Cosimo Ridolfi medal, 1991, John Ericsson medal, 1992. Mem. NAS, Bot. Soc. Am. (chmn. physiol. sect. 1954-55), Am. Soc. Plant Physiologists (v.p. 1952-53, pres. 1957-58, Stephen Hales award 1954, Reid Barnes life membership award 1970), Soc. Developmental Biology (pres. 1971), Am. Soc. Gen. Physiologists (v.p. 1956-57, pres. 1957-58), Internat. Plant Growth Substances Assn. (hon. life mem., v.p. 1976-79, pres. 1979-82), Am. Soc. Biol. Chemists, Am. Acad. Arts and Scis., Deutsche Akademie der Naturforscher Leopoldina, Swedish Royal Acad. Scis., Tissue Culture Assn. (hon. life, Life Achievement award 1992). Achievements include patents in field. Home: 2820 Marshall Ct Madison WI 53705-2270 Office: U Wis Dept Botany Madison WI 53706

SKOOG, WILLIAM ARTHUR, oncologist; b. Culver City, Calif., Apr. 10, 1925; s. John Lundeen and Allis Rose (Gatz) S.; A.A., UCLA, 1944; B.A. with gt. distinction, Stanford U., 1946, M.D., 1949; m. Ann Douglas, Sept. 17, 1949; children—Karen, William Arthur, James Douglas, Allison. Intern medicine Stanford Hosp., San Francisco, 1948-49, asst. resident medicine, 1949-50; asst. resident medicine N.Y. Hosp., N.Y.C., 1950-51; sr. resident medicine Wadsworth VA Hosp., Los Angeles, 1951, attending specialist internal medicine, 1962-68; practice medicine specializing in internal medicine, Los Altos, Calif., 1959-61; pvt. practice hematology and oncology Calif. Oncologic and Surg. Med. Group, Inc., Santa Monica, Calif., 1971-72; pvt. practice med. oncology, San Bernardino, Calif., 1972—; assoc. staff Palo Alto-Stanford (Calif.) Hosp. Center, 1959-61, U. Calif. Med. Center, San Francisco, 1959-61; asso. attending physician U. Calif. at Los Angeles Hosp. and Clinics, 1961-78; vis. physician internal medicine Harbor Gen. Hosp., Torrance, Calif., 1962-65, attending physician, 1965-71; cons. chemistry Clin. Lab., UCLA Hosp., 1963-68; affiliate cons. staff St. John's Hosp., Santa Monica, Calif., 1967-71, courtesy staff, 1971-72; courtesy attending med. staff Santa Monica Hosp., 1967-72; staff physician St. Bernardine (Calif.) Hosp., 1972—, San Bernardino Community Hosp., 1972-90, courtesy staff, 1990—; chief sect. oncology San Bernardino County Hosp., 1972-76; cons. staff Redlands (Calif.) Community Hosp., 1972-83, courtesy staff, 1983—; asst. in medicine Cornell Med. Coll., N.Y.C., 1950-51; jr. research physician UCLA Atomic Energy Project, 1954-55; instr. medicine, asst. research physician dept. medicine UCLA Med. Center, 1955-56, asst. prof. medicine, asst. research physician, 1956-59; clin. asso. hematology VA Center, Los Angeles, 1956-59; co-dir. metabolic research unit UCLA Center for Health Scis., 1955-59, 61-65; co-dir. Health Scis. Clin. Research Center, 1965-68, dir., 1968-72; clin. instr. medicine Stanford, 1959-61; asst. clin. prof. medicine, assoc. research physician U. Calif. Med. Center, San Francisco, 1959-61; lectr. medicine UCLA Sch. Medicine, 1961-62, assoc. prof. medicine, 1962-73, assoc. clin. prof. medicine, 1973—. Served with USNR, 1943-46, to lt. M.C., 1951-53. Fellow ACP; mem. Am., Calif. med. assns., So. Calif. Acad. Clin. Oncology, Western Soc. Clin. Research, Am. Fedn. Clin. Research, Los Angeles Acad. Medicine, San Bernardino County Med. Soc., Am. Soc. Clin. Oncology, Am. Soc. Internal Medicine, Calif. Soc. Internal Medicine, Inland Soc. Internal Medicine, Phi Beta Kappa, Alpha Omega Alpha, Sigma Xi, Alpha Kappa Kappa. Episcopalian (vestryman 1965-70). Club: Redlands Country. Contbr. articles to profl. jours. Home: 30831 Miradero Dr Redlands CA 92373-7429 Office: 401 E Highland Ave Ste 552 San Bernardino CA 92404-3801

SKOVIRA, ROBERT JOSEPH, information scientist, educator; b. Mt. Pleasant, Pa., May 4, 1943; s. Robert Joseph and Genevieve (Budney) S.; m. Mary Elizabeth Machuga, Aug. 21, 1971; 1 child, Suzanne Marie. BA, St. Vincent Coll., 1966; MA, U. Pitts., 1972, MS in Info. Scis., 1986, PhD, 1977. Cert. tchr., Pa.; cert. in computer programming and ops. Tchr. Greensburg (Pa.) Cen. Cath. High Sch., 1967-75; asst. visiting prof. U. Va., 1977-78; archives fieldworker U. Pitts., 1979; asst. visiting prof. U. Houston, Victoria, Tex., 1980-81; instr. St. Vincent Coll., Latrobe, Pa., 1982-84; assoc. prof. Robert Morris Coll., Coraopolis, Pa., 1983—. Fellow Philosophy of Edn. Soc., John Dewey Soc.; mem. am. soc. for Info. Sci. (chair. SIG FIS 1989-90, mem. SIG cabinet steering com. 1989-92, chair Pitts. chpt. 1991), Assn. Computer Educators, Assn. for Computing Machinery, Internat. Assn. Knowledge Engrs., Ohio Valley Philosophy Edn. Soc., Slovak Studies Assn. Democrat. Byzantine Catholic. Avocations: fishing, hiking, reading. Office: Robert Morris Coll Narrows Run Rd Coraopolis PA 15108-1189

SKOVRONEK, HERBERT SAMUEL, environmental scientist; b. Bklyn., Apr. 19, 1936; s. Alex and Gertrude (Drillings) S.; children: Eric Stephen, Alex Harold. BS in Chemistry, Bklyn. Coll., 1956; PhD in Chemistry, Pa. State U., 1961. Rsch. chemist Texaco Inc., Beacon, N.Y., 1961-62, Rayonier Inc., Whippany, N.J., 1962-67; group leader J.P. Stevens, Garfield, N.J., 1967-71; tech. adv. to dir. U.S. EPA, Edison, N.J., 1971-78; mgr. environ. control Allied, Morristown, N.J., 1978-82; cons. Environ. Svcs., Morris Plains, N.J., 1982-92; sr. environ. scientist Sci. Applications Internat. Corp., Hackensack, N.J., 1988—; rsch. assoc., adj. prof. N.J. Inst. Tech., Newark, 1984-88. Contbr. articles to profl. jours., chpts. to books. Recipient Gold medal Am. Electroplaters Soc., 1977. Fellow Hazardous Materials Rsch. Inst.; mem. Am. Chem. Soc. Jewish. Achievements include 3 patents on lubricant additives; innovative use of animals for hazardous chem. discovery. Home: 88 Moraine Rd Morris Plains NJ 07950 Office: Sci Applications Internat Corp 411 Hackensack Ave Hackensack NJ 07601

SKOWRON, TADEUSZ ADAM, physician; b. Czestochowa, Poland, Dec. 17, 1950; came to U.S., 1976; s. Stanislaw and Genowefa (Widera) S.; m. Elizabeth Sliwowska, Feb. 17, 1990; 1 child, Sebastian Adam. MD, Med. Acad., Lodz, Poland, 1975. House physician Bklyn.-Cumberland Med. Ctr., 1979-80, fellow in neurology, 1981-83; resident in medicine Marshall U. Sch. Medicine, Huntington, W.Va., 1983-86, instr., 1986-87; pvt. practice, Bridgeport, Conn., 1987—; clin. specialist II, State Sch., Newark, 1981; advisor Congress Med. Polonia, Czestochowa, 1990—. Mem. Polish cultural events com. Sacred Heart U., Fairfield, Conn., 1990—. Mem. ACP, AMA, AAAS, N.Y. Acad. Scis. Home: 47 Mcquillan Ave Stratford CT 06497-4626 Office: 50 Ridgefield Ave Ste 317 Bridgeport CT 06610-3103

SKRAMSTAD, PHILLIP JAMES, chemist; b. Milw., Dec. 8, 1941; s. Charles Hans and Gladys (Babcock) S.; m. Sandra Sue Careraos, Aug. 25, 1962; children: Peter, Paul, Jennifer. BS in Chemistry, U. Wis., Milw., 1965; MA in Chemistry, U. No. Iowa, 1970. Sci. tchr. Milw. Pub. Schs., 1965-67; tchr. chemistry Green Bay (Wis.) Pub. Schs., 1967-78; tech. asst. Wis. Electric Power-Point Beach Nuclear Plant, Two Creeks, Wis., 1978-80; radiochemist Point Beach Nuclear Plant, Two Creeks, 1980-81; supt. chemistry and radiation protection Bench Nuclear Plant, Two Creeks, 1981-84; supt. chemistry and radiation protection Crystal River Nuclear Plant - Fla. Power Corp., 1984-90, adminstr. master schedule, 1990—. Republican. Baptist. Office: Fla Power Corp SA2C PO Box 219 Crystal River FL 34423

SKRAMSTAD, ROBERT ALLEN, oceanographer; b. Montevideo, Minn., Apr. 3, 1937; s. Vernon Donald and Ann May (Tollefsen) S. Student, St. Olaf Coll., 1956-60; BS in Geol. Engring., S.D. Sch. Mines and Tech., 1965. Geologist Naval Oceanographic Office, Washington, 1965-70, oceanographer, 1970-75; oceanographer Naval Oceanographic Office, Bay St. Louis, Miss., 1975-82, phys. scientist, 1982—. With U.S. Army, 1957-60. Mem. Geol. Soc. Am., Am. Soc. Photogrammetry and Remote Sensing, Nat. Geographic Soc. Republican. Avocations: photography, jogging, travel, mineral collecting. Home: 2012 W 2d St Apt 8F Long Beach MS 39560-5552 Office: Naval Oceanographic Office Stennis Space Ctr Bldg 1002 Bay Saint Louis MS 39522-5001

SKRITEK, PAUL, electrical engineering educator, consultant; b. Vienna, Austria, Sept. 9, 1952; s. Otto and Grete (Eisenmagen) S.; m. Gertrud Stoeckinger, Oct. 8, 1982; children: Sebastian, Bernhard. Grad., Acad. Music, Vienna, 1976; diploma in engring., Tech. U., Vienna, 1977. Rsch.

asst. Tech. U., 1977-79, asst. prof., 1979-88; prof. Hoehere Technische Bundeslehranstalt, Vienna, 1988—; lectr. Tech. U., 1982—; acoustic and audio cons., Vienna, 1980—; curriculum commn. Tech. U., 1984-88, faculty elec. engring., 1985-86; mem. Austrian Schoolbook Commn., 1990—. Author: Handbook of Audio Circuit Design, 1988, Shielding of Audio Circuits, 1989; contbr. numerous articles to profl. jours. Mem. Audio Engring. Soc., IEEE, Oesterreichischer Verein fuer Elektrotechnik, Union Press Radio et Electronique, Internat. Gesellschaft fuer Ingenieurpädagogik, Oesterreichische Computer Gesellschaft. Mem. Social Democratic Party. Avocations: skiing, fine arts. Office: HTLBA-22, Donaustadtstrasse 45, A-1220 Vienna Austria

SKROBACZ, EDWIN STANLEY, structural engineer, educator; b. Piscataway, N.J., Jan. 2, 1961; s. Edwin Stanley and Agnes (Dolgos) S.; m. Susan Elizabeth Shaw, Oct. 18, 1986; 1 child, Megan Kate. AAS, Middlesex County Coll., 1981; BSCE, N.J. Inst. Tech., 1984, MSCE, 1988. Registered profl. engr., N.J., Pa. Drafter Elizabethtown Gas Co., Elizabeth, N.J., 1982; constrn. estimator Turner Constrn. Co., N.Y.C., 1983; field engr. Soils Engring. Svcs., Whippany, N.J., 1983; engr. aide Hardesty & Hanover Inc., N.Y.C., 1984; staff engr. T&M Assocs. Inc, Red Bank, N.J., 1985-91; sr. structural engr. Parsons Brinckerhoff, West Trenton, N.J., 1991—; instr. Middlesex Coll., Edison, N.J., 1989—. Mem. NSPE, ASCE. Roman Catholic. Home: 19 Emory St Howell NJ 07731 Office: Parsons Brinckerhoff 830 Bear Tavern Rd West Trenton NJ 08628

SKRZYPCZAK-JANKUN, EWA, crystallographer; b. Poznan, Poland, Sept. 14, 1948; came to U.S., 1982; d. Jozef and Maria (Byczynska) S.; m. Jerzy Jankun, Mar. 8, 1972; children: Monika, Hanna. MS, A. Mickiewicz U., 1972, PhD, 1976. Asst. A.Mickiewicz U., Poznan, Poland, 1972-73; sr. asst. A.Mickiewicz U., Poanan, Poland, 1973-76; rsch. assoc. Mich. State U., East Lansing, 1982-90; chem. instr. supr. U. Toledo, 1990—; adj. A. Mickiewicz, 1976-82. Bd. trustees Toledo-Poznan Alliance, 1992; mem. Solidarity, 1980-82. Mem. Am. Crystallographic Assn. Office: U Toledo Chemistry Dept Toledo OH 43606-3390

SKUJA, ANDRIS, physics educator; b. Riga, Latvia, Mar. 1, 1943; came to U.S., 1976; s. Edvins Martins and Rita (Ozolnieks) S. BSc, U. Toronto, Can., 1966; PhD, U. Calif., Berkeley, 1972. Rsch. officer U. Oxford, Eng., 1972-76; asst. prof. U. Md., College Park, 1976-81, assoc. prof., 1981-89, prof., 1989—; vis. prof. McGill U., Montreal, Que., Can., 1981; vis. scientist DESY, Hambury, Fed. Republic Germany, 1983. Contbr. articles to profl. jours. Mem. Am. Phys. Soc. Achievements include rsch. on the study of structure of nuclei by deep inelastic lepton scattering, study of the decay of the Zo; pioneer of the first electronic measurement of neutrino-electron scattering; study of gamma gamma interactions. Home: 7711 Lake Glen Dr PO Box 702 Glenn Dale MD 20769 Office: U Md Dept Physics College Park MD 20742

SKUP, DANIEL, molecular biologist, educator, researcher; b. Chgo., Apr. 1, 1951; came to Can., 1951; s. Paul and Bella (Sheer) S.; m. Magaly Sala, June 9, 1972; children: Eric, Alionka. BS, Moscow State U., 1973, MS in Molecular Biology with first degree honors, 1974; PhD in Biochemistry, McGill U., Montreal, Que., Can., 1980. Postdoctoral researcher biology sect. Inst. Curie, Orsay, France, 1980-82; asst. prof. rsch. Inst. du Cancer de Montréal U. Montreal, 1982-88, assoc. prof. rsch., 1987-88, assoc. prof., 1988—, sci. dir. Inst. du Cancer de Montréal, 1987-88, dir., 1988-92; cons. New England Nuclear, Boston, 1979; mem. adv. bd. Biccapital, Inc., Montreal, 1989—, Investissements R&D Martlet, Inc., Montreal, 1991—; Med. Rsch. Coun. vis. prof. faculty of medicine U. Ottawa, Ont., Can., 1983, Meml. U., 1991; mem. com. vis. USSR Med. Rsch. Coun. Can., 1990, mem. molecular biology grants panel, 1987-89, mem. cancer panel, 1989-92; mem. Que. Govt. Com. on biotechs. vis. Israel, 1991; mem. grants panel Nat. Cancer Inst. Can., 1984-88, mem. fellowship panel, 1988-92. Contbr. articles to profl. jours. including Jour. Molecular Biology, Procs. NAS, EMBO Jour., Nucleic Acids Rsch., Cancer Rsch. Grantee Med. Rsch. Coun., 1982—, Nat. Cancer Inst. Can., 1983—. Mem. Am. Soc. Microbiology, Internat. Soc. Interferon Rsch., Réseau Interhospitalier de Cancérologie de l'Université de Montréal, Clin. Rsch. Club Que. Achievements include determination of mechanism of viral take-over of protein synthesis by reovirus; cloning and characterization of murine interferon genes and tissue inhibitor of metalloproteinases gene. Office: Montreal Cancer Inst, 1560 E Sherbrooke St, Montreal, PQ Canada H2L 4M1

SKURNICK, JOAN HARDY, biostatistician, educator; b. Mt. Vernon, N.Y., Dec. 8, 1942; d. Glendon Day and Ethel Marie (Pritchett) Hardy; m. David Skurnick, Dec. 27, 1964; children: Jennifer Frances, Sarah Marie. BA in Math., Wellesley Coll., 1964; MS in Biometry, Temple U., 1975; PhD in Biostats., U. Calif., Berkeley, 1983. Statistician Stanford Rsch. Inst., Menlo Park, Calif., 1978-83; biostatistician VA Med. Ctr., Palo Alto, Calif., 1983-86; instr. N.J. Med. Sch., Newark, 1986-89, asst. prof., 1989—; sci. adviser Community Rsch. Initiative, Newark, 1989—; mem. pub. health edn. com. N.J. chpt. Am. Cancer Soc., New Brunswick, 1991—. Mem. Am. Statis. Assn., Am. Pub. Health Assn., Soc. Controlled Clin. Trials, Biometrics Soc., N.J. Wellesley Club. Democrat. Home: 4 Grimes Terr Montville NJ 07045 Office: U Medicine Dentistry NJ NJ Med Sch 185 S Orange Ave Newark NJ 07103

SKYLV, GRETHE KROGH, rheumatologist, anthropologist; b. Copenhagen, Denmark, May 31, 1938; d. Aage Krogh and Herdis Fischer (Lindeskov) Christoffersen; m. Axel Skylv, Jan. 12, 1962; children: Lise, Kirsten, Mikael. MD, U. Copenhagen, 1967, MA in Anthropology, 1990. Resident various hosps., Copenhagen, 1967-79; pvt. practice Hillerod, Denmark, 1979-84; dept. head Rehab. Ctr. for Torture Victims, Copenhagen, 1985-92; cons. Danish Red Cross, 1993—; cons. Orgn. for Manual Therapy, Denmark, 1985—, Internat. Rehab. Medicine Assn., 1991—; bd. dirs. Network for Interdisciplinary Qualitative Rsch., Denmark, 1989—. Guest editor: Danish Soc. for Anthropology Soc. Jour., 1988-89; contbr. articles to profl. jours. Recipient Honorary award Cranio-Facial Pain Ctr. 1990. Fellow European Assn. Social Anthropologists, Danish Med. Assn., Danish Assn. for Manual Medicine, Danish Manual Therapy Orgn., Danish Assn. for Rheumatology, Danish Assn. Internal Medicine, Danish Soc. for Social Anthropology. Avocation: black belt in jujitsu. Home: Borrebyvej 62, 2700 Bronshoj Denmark

SLACK, MARION KIMBALL, pharmacy educator; b. Casper, Wyo., Oct. 10, 1944; d. Emory James and Nona Lee (Graves) Kimball; m. Donald Carl Slack, Dec. 19, 1964; children: Jonel Marie, Jennifer Michelle. BS in Pharmacy, U. Ky., 1969; MA in Edn., U. Minn., 1984; PhD in Pharmacy Adminstrn., U. Ariz., 1989. Registered pharmacist, Ky. Staff pharmacist Good Samaritan Hosp., Lexington, Ky., 1974-75, Met. Med. Ctr., Mpls., 1976-81; rsch. assoc. Coll. of Pharmacy, Tucson, Ariz., 1985-89; adj. lectr. Coll. of Pharmacy, Tucson, 1989-92, asst. rsch. scientist, 1992—; cons. Maranatha Clinic, Khon Kaen, Thailand, 1970-73. Author: Inadvertent Dispensing Errors, 1988, New Look at Medication Errors, 1988, Recognition Errors by Healthcare Professionals: Information Processing Failures or Moral Failure, 1991. Mem. ASTM, Am. Pharm. Assn., Am. Assn. Colls. of Pharmacy, Am. Inst. History of Pharmacy, Am. Soc. Hosp. Pharmacists, Phi Beta Kappa, Rho Chi. Home: 9230 E Visco Pl Tucson AZ 85710-3167

SLÁMA, KAREL, biologist, zoologist; b. Tichá, Czechoslovakia, Czech Republic, Dec. 17, 1934; s. Vladimír Sláma and Marie (Michlová) Slámová; m. Věra Ležatková, June 25, 1960; children: Pavla, Martina, Tereza. MSc, Masaryk U., Brno, Czechoslovakia, 1957; PhD, Czechoslovakian Acad. Scis., Prague, 1961. Rsch. asst. Czechoslovakian Acad Scis., Prague, 1961-64, rschr. Entomology Inst., 1965-85, rschr. Inst. Organic Chemistry, 1985-90; rsch. fellow Harvard U., Cambridge, Mass., 1964-65; dir. rsch. Lab. Ecol. Pharm. Intereco, Prague, 1990-92; dir Entomology Inst. Czech Acad Scis., Ceske Budějovie, Czech Republic, 1993—. Co-author: Insect Hormones and Bioanalogues, 1974; mng. dir. European Jour. Entomology; contbr. over 150 sci. papers on insect hormones, chpt. to book. Achievements include discovers of paper factor with insect hormone activity in American paper products. Home: Evropska 674, 160 00 Prague Czech Republic Office: Czech Acad Scis, Entomol Inst Branisovska 31, 370 05 Ceske Budejovice Czech Republic

SLATER, JAMES MUNRO, radiation oncologist; b. Salt Lake City, Jan. 7, 1929; s. Donald Munro and Leone Forestine (Fehr) S.; m. JoAnn Strout, Dec. 28, 1948; children: James, Julie, Jan, Jerry, Jon. B.S. in Physics, U. Utah, Utah State U., 1954; M.D., Loma Linda U., 1963. Diplomate: Am. Bd. Radiology. Intern Latter Day Saints Hosp., Salt Lake City, 1963-64; resident in radiology Latter Day Saints Hosp., 1964-65; resident in radiotherapy Loma Linda U. Med. Ctr., White Meml. Med. Center, Los Angeles; fellow in radiotherapy Loma Linda U. Med. Ctr., White Meml. Med. Center, 1967-68, U. Tex.-M.D. Anderson Hosp. and Tumor Inst., Houston, 1968-69; mem. faculty Loma Linda (Calif.) U., 1975—, prof. radiology, 1979—, chmn. dept., 1979—, dir. nuclear medicine, 1970—, dir. radiation oncology, 1975-79; co-dir. community radiation oncology program Los Angeles County/U. So. Calif. Comprehensive Cancer Ctr. 1978-83; mem. cancer adv. coun. State Calif., 1980-85; clin. prof. U. So. Calif., 1982—; founding mem. Proton Therapy Coop. Group, 1985—, chmn., 1987—; cons. Charged Particle Therapy Program, Lawrence Berkeley Lab., 19856—; cons. R & D monoclonal antibodies Hybritech Inc., 1985—, bd. dirs., 1985—; cons. Sci. Applications Internat. Corp., 1970—. Bd. dirs. Am. Cancer Soc., San Bernardino/Riverside, 1976—, exec. com., 1976—; pres. Inland Empire chpt., 1981-83. NIH fellow, 1968-69; recipient exhbn. awards Radiol. Soc. N.Am., 1973, exhbn. awards European Assn. Radiology, 1975, exhbn. awards Am. Soc. Therapeutic Radiologists, 1978. Fellow Am. Coll. Radiology; mem. AMA, ACS (liaison mem. to commn. on cancer 1976—), Am. Radium Soc., Am. Soc. Clin. Oncology, Am. Soc. Therapeutic Radiologists, Assn. Univ. Radiologists, Calif. Med. Assn., Calif. Radiol. Soc., Gilbert H. Fletcher Soc. (pres. 1981-82), Loma Linda U. Med. Sch. Alumni Assn., Radiol. Soc. N.Am., San Bernardino County Med. Soc., Soc. Chmn. Acad. Radiology Depts., So. Calif. Radiation Therapy Soc., Western Assn. Gynecologic Oncologists, Alpha Omega Alpha. Home: 1210 W Highland Ave Redlands CA 92373-6659 Office: Loma Linda U Radiation Medicine Loma Linda CA 92350

SLATOPOLSKY, EDUARDO, nephrologist, educator; b. Buenos Aires, Argentina, Dec. 12, 1934; (parents Am. citizens); married, 1959; 3 children. BS, Nat. Coll. Nicolas Avellaneda, 1952; MD, U. Buenos Aires, 1959. Postdoctoral renal USPHS, renal divsn., Dept. Internal Medicine Washington U. Sch. Medicine, 1963-65, instr. med. nephrology, 1965-67, from asst. prof. to assoc. prof. medicine dept. nephrology, 1967-75; dir. Chromalloy Am. Kidney Ctr., Washington U. Sch. medicine, St. Louis, 1967—, co-dir. renal divsn., 1972—, prof. medicine, nephrology dept., 1975—, Joseph Friedman Prof. renal disease medicine, 1991—; adv. mem. regional med. program, renal program sch. medicine Washington U., 1970-75; chmn. transplantation com. Barne Hosp., 1975—; fellow com. Kidney Found. Ea. Mo. and Metro.-E., 1978; mem. adv. com. artificial kidney-chronic uremia program NIH, 1978-90, rep. Latin-Am. nephrology, 1983-88; mem. study sect. Gen. Med., NIH, 1984-88. Recipient Frederick C. Bartter award 1991. Mem. AAAS, Am. Fedn. Clin. Rsch., Internat. Soc. Nephrology, Am. Soc. Nephrology, Endocrine Soc., Sigma Xi. Achievements include pathogenesis and treatment of secondary hyparathyroidism and bone disease in renal failure; studies conducted at both levels: clinical, on patients maintained on chronic dialysis and on animals with experimentally induced renal feilure; detailed studies of the effects of calcitriol on PTH MRNA and the extra-renal production of calcitriol by macrophages; vitro studies in primary culture of bovine parathyroid cells used to understand the mechanisms that control the secretion of PTH. Office: Washington U Chromalloy Am Kidney Ctr 1 Barnes Hosp Plz Box 8129 Saint Louis MO 63110*

SLAUGHTER, FREEMAN CLUFF, dentist; b. Estes, Miss., Dec. 30, 1926; s. William Cluff and Vay (Fox) S.; student Wake Forest Coll., 1944; student Emory U., 1946-47, DDS, 1951; m. Genevieve Anne Parks, July 30, 1948; children: Mary Anne, Thomas Freeman, James Hugh. Practice gen. dentistry, Kannapolis, N.C., 1951-89; ret.; mem. N.C. Bd. Dental Examiners, 1966-75, pres., 1968-69, sec.-treas., 1971-74; chief dental staff Cabarrus Meml. Hosp., Concord, N.C., 1965-66, 75; mem. N.C. Adv. Com. for Edn. Dental Aux. Personnel-N.C. State Bd. Edn., 1967-70; adviser dental asst. program Rowan Tech. Inst., 1974-76; Duke Med. Ctr. Davison Century Club. Trustee N.C. Symphony Soc., 1962-68, pres. Kannapolis chpt., 1961; mem. Cabarrus County Bd. Health, 1977-83, chmn., 1981-83, acting health dir., 1981; vice chmn. Kannapolis Charter Commn., 1983-84; mem. City Council Kannapolis, 1984-85; Mayor protem, Kannapolis, 1984-85; active Boy Scouts Am., Eagle scout with silver palm. Served with USN, 1944-46, WW II. ETO, MTO. Recipient Kannapolis Citizen of Yr. award, 1982; lic. real estate broker. Fellow Am. Coll. Dentists; mem. Am. Legion, Kannapolis Jr. C. of C. (v.p. 1952), Toastmasters Internat. (pres. Kannapolis 1963-64), ADA (life), Am. Assn. Dental Examiners (Dentist Citizen of Year 1975; v.p. 1977-79), So. Conf. Dental Deans and Examiners (v.p. 1969), N.C. Dental Soc. (resolution of commendation 1975), N.C. Dental Soc. Anesthesiology (pres. 1964), Southeastern Acad. Prosthodontics, So. Acad. Oral Surgery, Am. Soc. Dentistry for Children (pres. N.C. unit 1957), Internat. Assn. Dental Research, Cabarrus County Dental Soc. (pres. 1953-54, 63-64, 69), N.C. Assn. Professions (dir. 1976-80), Omicron Kappa Upsilon, Alpha Epsilon Upsilon. Clubs: Masons, Shriners, Kannapolis Music (pres. 1962-63), Rotary (dir. 1977-80).

SLAUGHTER, JOHN BROOKS, university president; b. Topeka, Mar. 16, 1934; s. Reuben Brooks and Dora (Reeves) S.; m. Ida Bernice Johnson, Aug. 31, 1956; children: John Brooks, Jacqueline Michelle. Student, Washburn U., 1951-53; BSEE, Kans. State U., 1956, DSc (hon.), 1988; MS in Engring., UCLA, 1961; PhD in Engring. Scis, U. Calif., San Diego, 1971; D Engring. (hon.), Rensselaer Poly. Inst., 1981; DSc (hon.), U. So. Calif., 1981, Tuskegee Inst., 1981, U. Md., 1982, U. Notre Dame, 1982, U. Miami, 1983, U. Mass., 1983, Tex. So. U., 1984, U. Toledo, 1985, U. Ill., 1986, SUNY, 1986; LHD (hon.), Bowie State Coll., 1987; DSc (hon.), Morehouse Coll., 1988, Kans. State U., 1988; LLD (hon.), U. Pacific, 1989; DSc (hon.), Pomona Coll., 1989; LHD (hon.), Alfred U., 1991, Calif. Luth. U., 1991, Washburn U., 1992. Registered profl. engr., Wash. Electronics engr. Gen. Dynamics Convair, San Diego, 1956-60; with Naval Electronics Lab. Center, San Diego, 1960-75, div. head, 1965-71, dept. head, 1971-75; dir. applied physics lab. U. Wash., 1975-77; asst. dir. NSF, Washington, 1977-79; dir. NSF, 1980-82; acad. v.p., provost Wash. State U., 1979-80; chancellor U. Md., College Park, 1982-88; pres. Occidental Coll., Los Angeles, 1988—; bd. dirs., vice chmn. San Diego Transit Corp., 1968-75; mem. com. on minorities in engring. Nat. Rsch. Coun., 1976-79; mem. Commn. on Pre-Coll. Edn. in Math., Sci. and Tech. Nat. Sci. Bd., 1982-83; bd. dirs. Monsanto Co., ARCO, Avery Dennison Corp., IBM, Northrop Corp., Music Ctr. L.A. Country. Editor: Jour. Computers and Elec. Engring, 1972—. Bd. dirs. San Diego Urban League, 1962-66, pres., 1964-66; mem. Pres.'s Com. on Nat. Medal Sci., 1979-80; trustee Rensselaer Poly. Inst., 1982; chmn. Pres.'s Com. Nat. Collegiate Athletic Assn., 1986-88; bd. govs. Town Hall of Calif., 1990; bd. dirs. L.A. World Affairs Coun., 1990. Recipient Engring. Disting. Alumnus of Yr. award UCLA, 1978, UCLA medal, 1989, Roger Revelle award U. Calif. at San Diego, 1991; Disting. Svc. award NSF, 1979,. Svc. in Engring. award Kans. State U., 1981, Disting. Alumnus of Yr. award U. Calif., San Diego, 1982; Naval Electronics Lab. Ctr. fellow, 1969-70. Fellow IEEE (chmn. com. on minority affairs 1976-80), Am. Acad. Arts and Scis.; mem. NAE, Nat. Collegiate Athletic Assn. (chmn. pres. commn.), Tau Beta Pi, Eta Kappa Nu, Phi Beta Kappa (hon.). Office: Occidental Coll 1600 Campus Rd Los Angeles CA 90041-3314

SLAVKIN, HAROLD C., biologist; b. Chgo., Mar. 20, 1938; m. Lois S. Slavkin; children: Mark D., Todd P. BA (hon.), U. So. Calif., 1961, DDS (hon.), 1965; Doctorate (hon.), Georgetown U., 1990. Mem. faculty grad. program in cellular and molecular biology U. So. Calif., L.A., 1968—, mem. faculty gerontology inst., 1969, prof. sch. dentistry, 1974—, chmn. grad. program in craniofacial molecular biology, 1975-85; dir. Ctr. for Craniofacial Molecular Biology, L.A., 1989—; vis. prof. Insrael Inst. Tech., Haifa, 1987-88; cons. U.S. News and World REport, 1985—, L.A. Edn. Partnership, 1983—, Torstar Books, Inc., 1985—. Contbr. articles to profl. jours. Mem. scientific adv. bd. Calif. Mus. Sci. and Tech., 1985—. Rsch. scholar U. Coll. London, 1980. Mem. Am. Assn. for the Advacement of Sci., Am. Assn. Anatornisk, Am. Insts. Biological Scis., Am. Soc. for Cell Biology, Internat. Assn. for Dental Rsch., N.Y. Acad. Scis., L.A. County Art Mus. Assocs. Office: U So Calif Sch Dentistry University Park Los Angeles CA 90007-3284*

SLAWIATYNSKY, MARION MICHAEL, biomedical electronics engineer, software consultant; b. Phila., Nov. 21, 1958; s. Walter Wasyl and Maria Margaret (Sauer) S. BA in Biology, LaSalle U., 1980; MS in Biomed. Engring., Drexel U., 1984, BS in Electronics Engring. (hon.), 1982. Quality control technician Colocraft, Bensalem, Pa., 1978-80; quality control engr. Sentinel Electronics, Bristol, Pa., 1980-82; quality control engr. Innovative Med. Systems, Ivyland, Pa., 1982-84, electronics engr., 1984-90, sr. electronics engr., 1990—. Soloist Male Chorus Prometheus, 1976—; mem. Steuben Soc. Am., Phila., 1990—. Mem. IEEE, Assn. for the Advancement of Med. Instrumentation, Vereinigung Erzgebirge (mem. exec. bd., dir. 1990-91). Republican. Roman Catholic. Achievements include patent for electro-optical lock-in amplifier detector for coagulation instrument; development of medical quality control for FDA certification of electronic medical instruments, of high precision peristalitic pump with electronic autocalibration, of software quality control guidelines for FDA certification; design of electro-optical circuitry for medical coagulation analyzer and flourescence polarization instrument. Home: 503 S Warminster Rd # 5 Hatboro PA 19040-4101 Office: Innovative Med Systems 55 Steam Whistle Dr Warminster PA 18974-1477

SLAYTON, RANSOM DUNN, consulting engineer; b. Salem, Nebr., Mar. 10, 1917; s. Laurel Wayland and Martha Ellen (Fisher) S.; B.S. with distinction, U. Nebr., 1938; postgrad. Ill. Inst. Tech., 1942, DePaul U., 1945-46; m. Margaret Marie Ang, Sept. 25, 1938; children—R. Duane, David L., Sharon J. Slayton Manz, Karla M. Slayton Fogel, Paul L. With Western Union Telegraph Co., Lincoln, Nebr., 1937-38, St. Paul, 1938-40, Omaha, 1940, Chgo., 1940-45; asst. prof. elec. engring. Chgo. Tech. Coll., 1945-46; with Teletype Corp., Chgo. and Skokie, Ill., 1946-82, lectr., China and Japan, 1978, 79, 80. Active vol. civic orgns., numerous ch. offices. Mem. IEEE (sr., life, numerous coms.), IEEE Communications Soc. (parliamentarian 1972-80, 82—, vice chmn. terminals com. 1980-82, chmn. 1983-84). Patentee in field. Home: 1530 Hawthorne Ln Glenview IL 60025-2261

SLECHTA, JIRI, theoretical physicist; b. Havlickuv Brod, Bohemia, Czechoslovakia, Apr. 26, 1939; came to England, 1984; s. Josef and Marie (Posikova) S.; m. Miriam Vydrarova, July 17, 1971; children: Vera, Martin. Dr. rer. nat., Charles U., Prague, Czechoslovakia, 1962. Sr. lectr. dept. theoretical physics Charles U., Prague, 1964-69; rsch. fellow dept. physics U. Warwick, Conventry, Eng., 1969-71; sr. rsch. assoc. Sch. Math. and Physics U. East Anglia, Norwich, Eng., 1971-74; rsch. fellow dept. physics U. Leeds (Eng.), 1976-77; chair 3 Symp. 13th Internat. Congress on Cybernetics, Namur, Belgium, 1992; researcher in field. Author 62 papers and 42 contbns. at nonpub. confs.; editor Informatica; patentee in field. 2d lt. Czechoslovakia mil., 1962-64. Benevolent Fund IOP ann. grantee, London, 1979—. Assoc. fellow Inst. Math. and Applications; mem. Am. Phys. Soc., Internat. Acad. Scis. San Marino, Inst. Physics, European Phys. Soc., Brit. Cybernetic Soc., Internat. Assn. Cybernetics, N.Y. Acad. Scis., Czechoslovak Math. Phys. Union. Mem. Conservative party. Achievements include theory of disordered materials and self-organizing systems (brain, economy, society) and cybernetics. Home: 18 Lidgett Hill, Leeds LS8 1PE, England

SLEDGE, CLEMENT BLOUNT, orthopedic surgeon, educator; b. Ada, Okla., Nov. 1, 1930; s. John B. and Mollie D. (Blount) S.; m. Georgia Kurrus, Apr. 13, 1957; children—Margaret, John, Matthew, Claire. M.D., Yale U., 1955; M.A. (hon.), Harvard U., 1970; ScD (hon.), U. The South, 1987. Diplomate: Am. Bd. Orthopedic Surgery. Intern Barnes Hosp., St. Louis, 1955-56; resident in orthopedic surgery Harvard U., 1960-63; fellow in orthopedic pathology Armed Forces Inst. Pathology, 1963; vis. scientist Strangeways Research Lab., Cambridge (Eng.) U., 1963-66; asst. prof. orthopedic surgery Harvard U., 1963-67, asso. prof., 1967-70, prof., 1970—, chmn. dept., 1970—; chmn. dept. orthopedic surgery Brigham and Women's Hosp. Editor: Textbook of Rheumatology, 1981, 85, 89; contbr. more than 100 articles to sci. jours. Active Arthritis Found.; chmn. Nat. Arthritis Adv. Bd., 1978-80. Served with M.C. USNR, 1956-58. Fellow Med. Found. Boston, 1963-66; Gebbie research fellow, 1968; NIH grantee, 1967—. Mem. Am. Acad. Orthopedic Surgeons (pres. 1985-86), Orthopedic Rsch. Soc. (pres. 1978-80), Am. Rheumatism Assn., Inst. of Medicine, Nat. Acad. Sci., Interurban Orthopedic Club, The Hip Soc. (pres. 1985). Episcopalian. Office: Brigham & Women's Hosp 75 Francis St Boston MA 02115-6195 also: Harvard Med Sch 25 Shattuck St Boston MA 02115

SLEEZER, PAUL DAVID, organic chemist; b. Chgo., Jan. 26, 1936; s. Paul Edward and Esther Ethyl S.; m. Nancy Elaine Sleezer, Oct. 12, 1963; children: David Edward, Lucinda Elaine, Karen Amelia. BS in Chemistry, U. Rochester, 1958; PhD in Chemistry, UCLA, 1963. Exploratory researcher Allied Chem. Co., Syracuse, N.Y., 1963-66; process devel. chemist Bristol-Myers Squibb, Syracuse, 1966—. Contbr. articles to profl. publs. Achievements include 12 patents including composition of matter for intermediates as potential new pharmaceuticals as well as process patents. Office: Bristol Myers Squibb PO Box 4755 Syracuse NY 13221-4755

SLEMMONS, ROBERT SHELDON, architect; b. Mitchell, Nebr., Mar. 12, 1922; s. M. Garvin and K. Fern (Borland) S.; AB, U. Nebr., 1947, BArch, 1948; m. Dorothy Virginia Herrick, Dec. 16, 1945; children: David (dec.), Claire, Jennifer, Robert, Timothy. Draftsman, Davis & Wilson, architects, Lincoln, Nebr., 1947-48; chief designer, project architect Office of Kans. State Architect, Topeka, 1948-54; asso. John A. Brown, architect, Topeka, 1954-56; partner Brown & Slemmons, architect, Topeka, 1956-69; v.p. Brown-Slemmons-Kreuger, architects, Topeka, 1969-73; owner Robert S. Slemmons, A.I.A. & Assos., architects, Topeka, 1973—. Cons. Kans. State Office Bldg. Commn., 1956-57; lectr. in design U. Kans., 1961; bd. dirs. Kaw Valley State Bank & Trust Co., Topeka, 1978-92. Bd. dirs. Topeka Civic Symphony Soc., 1950-60, Midstates Retirement Communities, Inc., 1986-92, Topeka Festival Singers; cons. Ministries for Aging, Inc., Topeka, 1984—; mem. Topeka Bd. Bldg. and Fire Appeals, Kans. Com. for Employer Support of the Guard and Res., With USNR, 1942-48. Mem. AIA (Topeka pres. 1955-56, Kans. dir. 1957-58, mem. com. on architecture for justice, com. for hist. resources), Internat. Conf. Bldg. Ofcls., Topeka Art Guild (pres. 1950), Am. Corrections Assn., Kans. Council Chs. (dir. 1961-62), Shawnee County Hist. Soc. Topeka, Greater Topeka C. of C. (mil. affairs task force), Downtown Topeka, Inc. (v.p. 1992—), St. Andrews Soc. (pres.), SAR (recording sec., pres. chpt.), U. Nebr. Alumni Assn. (life), Band Alumni Assn., Kiwanis (pres. 1966-67), Topeka Knife and Fork Club (dir.). Presbyterian (elder, chmn. trustees). Prin. archtl. works include: Kans. State Office Bldg., 1954, Topeka Presbyn. Manor, 1960-74, Meadowlark Hills Retirement Community, 1979, Shawnee County Adult Detention Facility, 1985. Office: Slemmons Assocs Architects 534 S Kansas Ave Topeka KS 66603-3432

SLEPECKY, NORMA B., cell biologist; b. Montgomery, Ala., Jan. 2, 1944. BS, Syracuse U., 1965, MS, 1968; PhD, SUNY, Syracuse, 1985. Rsch. scientist dept. physiology Karolinska Inst., Stockholm, 1985-86, NIH, Nat. Inst. Deafness and Other Communicative Disorders, Bethesda, Md., 1991—; assoc. prof. dept. bioengring. and neurosci. Syracuse U, 1986—. Contbr. over 40 articles to profl. jours. Mem. Am. Soc. Cell Biology, Electron Microscope Am., Assn. Rsch. in Otolaryngology. Achievements include research in cell biology of the sensory cells of the inner ear. Office: Syracuse U Inst Sensory Rsch Syracuse NY 13244-5290

SLICHTER, CHARLES PENCE, physicist, educator; b. Ithaca, N.Y., Jan. 21, 1924; s. Sumner Huber and Ada (Pence) S.; m. Gertrude Thayer Almy, Aug. 23, 1952 (div. Sept. 1977); children: Sumner Pence, William Almy, Jacob Huber, Ann Thayer; m. Anne FitzGerald, June 7, 1980; children—Daniel Huber, David Pence. AB, Harvard U., 1946, MA, 1947, PhD, 1949. Research asst. Underwater Explosives Research Lab., Woods Hole, Mass., 1943-46; mem. faculty U. Ill., Urbana, 1949—, prof. physics, 1955—, prof. Ctr. for Advanced Study, 1968—, prof. chemistry, 1986—; Morris Loeb lectr. Harvard U., 1961; dir. Polaroid Co.; mem. Pres.'s Sci. Adv. Com., 1964-69, Com. on Nat. Medal Sci., 1969-74, Nat. Sci. Bd., 1975-84, Pres.'s Com. Sci. and Tech., 1976. Author: Principles of Magnetic Resonance, 1963, 3d edit., 1989; Contbr. articles to profl. jours. Former trustee, mem. corp. Woods Hole Oceanographic Instn. Recipient Langmuir award Am. Phys. Soc., 1969; Alfred P. Sloan fellow, 1955-61. Mem. Nat. Acad. Scis. (Comstock prize 1993), Am. Acad. Arts and Scis., Am. Philos. Soc., Internat. Soc. Magnetic Resonance (pres. 1987-90, Trienniel prize 1986), Harvard Corp. Home: 61 Chestnut Ct Champaign IL 61821-7121

SLIGER, REBECCA NORTH, nuclear engineer; b. Coventry, England, Sept. 23, 1967; came to U.S., 1967; d. Paul and Margaret Ann (Horner) North; m. David Matthew Sliger, Mar. 23, 1991. BS, U. Utah, 1991. Analyst Tenera, Idaho Falls, Idaho, 1991—. Diamond scholar, U. Utah, 1990-91. Mem. ASME, Am. Nuclear Soc., Tau Beta Pi, Pi Tau Sigma.

SLIPMAN, (SAMUEL) RONALD, hospital administrator; b. New Orleans, Aug. 24, 1939; s. Jake and Esther (Steinman) S.; m. Carole Marie Green, July 1, 1961 (div. Feb. 1982); children: Susan Rachel, Lawrence Jay; m. Marilyn Morais, Feb. 5, 1983 (dec. June 1985); m. Lelia Ruth Foster, Jan. 12, 1986; children: Ronald Andrew, Brian Edward. BS, Tulane U., 1961; cert. in supervision techniques, La. State U., 1984; postgrad., NE La. U., 1978-79, 80-81. Design progress estimator Boeing Co., New Orleans, 1964-66; interviewer Tex. Employment Commn., Tyler and Lufkin, 1977-78; pers. technician State of La., Baton Rouge, 1961-62, 63-64, 73-75, 81; rsch. statistician La. Ins. Commn., Baton Rouge, 1967-68; labor market analyst La. Dept. Labor, Baton Rouge, 1969-70, 77; pers. dir. Royal Orleans Hotel, New Orleans, 1966-67; mgmt. analyst for quality assurance Earl K. Long Hosp., Baton Rouge, 1981-84, dir. ancillary svcs., 1984-86; mgmt. analyst, spl. asst. to dir. for total quality mgmt. Dept. Vets. Affairs Med. Ctr., Alexandria, La., 1987-88, 89-90; mgmt. cons., 1990—. Mem. adminstrv. bd. 1st United Meth. Ch. Mem. La. Soc. Hosp. Pharmacists, S.W. La. Bridge Assn. (pres., bd. dirs.), Nat. Mgmt. Assn. Republican. Methodist. Avocations: duplicate bridge, tennis, horseback riding. Home and Office: Rt 2 105 Fox Fire Ln Alexandria LA 71302-8638

SLOAN, EARLE DENDY, JR., chemical engineering educator; b. Seneca, S.C., Apr. 23, 1944; s. Earle Dendy and Sarah (Bellotte) S.; m. Marjorie Nilson, Sept. 7, 1968; children: Earle Dendy III, John Mark. BSChemE, Clemson U., 1965, MSChemE, 1972, PhD in Chem. Engring., 1974. Engr. Du Pont, Chattanooga, 1965-66, Seaford, Del., 1966-67; cons. Du Pont, Parkersburg, W.Va., 1967-68; sr. engr. Du Pont, Camden, S.C., 1968-70; postdoctoral fellow Rice U., 1975; prof. chem. engring. Colo. Sch. Mines, Golden, 1976—, Gaylord and Phyllis Weaver dist. prof. chem. engring., 1992—; pres. faculty senate, Colo. Sch. Mines, 1989-90. Author: Clathrate Hydrates of Natural Gases, 1990; chmn. pub. bd. Chem. Engring. Edn., 1990—. Scoutmaster local Cub Scouts; elder Presbyn. Ch., Golden, Colo., 1977-79, 92—. Mem. Am. Soc. for Engring. Edn. (chmn. ednl. rsch. methods div. 1983-85, chem. engring. div. 1985), Am. Inst. Chem. Engrs. (chmn. area Ia thermodynamics and transport 1990-93), Am. Chem. Soc. Avocations: long distance running, piano, philosophy. Home: 2121 Washington Ave Golden CO 80401-2377

SLOAN, HAROLD DAVID, chemical engineering consultant; b. Olney, Tex., Jan. 4, 1949; s. James Robert Jr. and Laura Faye (Riddle) S.; m. Barbara Ellen Wilson, Dec. 17, 1970 (div. 1982); m. Maureen Ann Moriarity, Mar. 17, 1983; 1 child, Christa Lauren. BSChemE, Tex. Tech U., 1972. Registered profl. engr., Tex. Field engr. Halliburton Svcs., Corpus Christi, Tex., 1972-73; mgr. tech. svc. Engelhard Corp., Houston, 1987-90; systems engr., process engr., then process mgr. M.W. Kellogg Co., Houston, 1973-87, sr. product tech. cons., 1990—. Contbr. articles to tech. jours. and mags. Pres. Sagemeadow Civic Club, Houston, 1978; v.p. West Harris County Mcpl. Utility Dist. 10, Houston, 1985; Sunday sch. tchr. Met. Bapt. Ch., Houston, 1992—. Mem. AICE, Tex. Soc. Profl. Engrs. (pres. Sam Houston chpt. 1980, Outstanding Young Engr. award 1978), NRA (life), Sigma Xi. Achievements include research on role of delayed coking in a clean fuels environment, economic options for heavy crude upgrading, processing heavier crude blends. Home: 16631 Avenfield Rd Tomball TX 77375 Office: MW Kellogg Co 601 Jefferson Ave Houston TX 77210

SLOAN, MARY JEAN, media specialist; b. Lakeland, Fla., Nov. 29, 1927; d. Marion Wilder and Elba (Jinks) Sloan. BS, Peabody Coll., Nashville, 1949; MLS, Atlanta U., 1978, S.L.S., 1980. Cert. libr. media specialist. Music dir. Pinecrest Sch., Tampa, Fla., 1949-50, Polk County Schs., Bartow, Fla., 1950-54; pvt. music tchr. Lakeland, 1954-58; tchr. Clayton County Schs., Jonesboro Ga., 1958-59; media specialist Eastualley Sch., Marietta, Ga., 1959—; coord. conf. Ga. Library Media Dept., Jekyll Island, 1982-83, sec., Atlanta, 1982-83, com. chmn. ethnic conf., Atlanta, 1978, pres., 1984-85, state pres., 1985-86; program chmn. Ga. Media Orgns. Conf, Jekyll Island, 1988. Contbr. to bibliographies. Recipient Walter Bell award Ga. Assn. Instructional Tech., 1988, Disting. Svc. award, 1991. Mem. ALA (del. 1984, 85, 90), NEA, Southeastern Library Assn., Am. Assn. Sch. Librarians, Soc. for Sch. Librarians, Internat., Ga. Assn. Educators (polit. action com. 1983), Beta Phi Mu, Phi Delta Kappa. Republican. Methodist. Home: 797 Yorkshire Rd NE Atlanta GA 30306-3264 Office: Eastvalley Elem Sch 2570 Lower Roswell Rd Marietta GA 30067

SLOAN, MICHAEL LEE, physics and computer science educator; b. Chgo., Jan. 24, 1944; s. Robert Earl Sloan and Cyril (Lewis) Glass; m. Claudia Ann Schultz, Sept. 27, 1969. BS in Physics, Roosevelt U., 1966, MS, 1971. Tchr. physics Glenbard West High Sch., Glen Ellyn, Ill., 1966-79; computer cons. Midwest Visual, Chgo., 1979-82; sr. engr. Apple Computer, Rolling Meadows, Ill., 1982-85; asst. prof. Roosevelt U., 1971-73; tchr. computer sci./physics Ill. Math. and Sci. Acad., Aurora, 1987—; instr. Harper Coll., Palatine, Ill., 1984. Author: AppleWorks: The Program For the Rest of Us, 1985, 2nd edit., 1988, Working with Works, 1987, Word Power, 1989, Working with PC Works, 1989, Working With Works 2.0., 1990. Bd. dirs. Youth Symphony Orch., Chgo., 1977-78; trustee Body Politic Theatre, Chgo., 1981-82. Home: ON008 Evans Ave Wheaton IL 60187

SLOAN, PAULA RACKOFF, mathematics educator; b. Manchester, N.H., July 3, 1945; d. Herbert I. and Wilma (Weill) Rackoff; m. Frank Allen Sloan, June 22, 1969; children: Elyse, Richard. BA, CUNY, 1966; MA in Math., Harvard U., 1967, UCLA, 1971. High sch. math. instr. Mass. and Fla., 1967-78; jr. colls. math. instr. Fla. and Tenn.; lectr. Vanderbilt U., Nashville, 1979-87, asst. prof. math. mgmt., 1988-92, assoc. prof. math. mgmt., 1992—. Textbook reviewer; contbr. articles to profl. jours. Meals coord. Room in the Inn, Nashville, 1986—; trustee The Temple, Nashville, 1990—. Mem. Assn. for Women Math., Assn. for Women Sci., Nat. Coun. Tchrs. Math., Math. Assn. Am. Office: Vanderbilt U Owen Grad Sch Mgmt Nashville TN 37203

SLOAN, TOD BURNS, anesthesiologist, researcher; b. Washington, Dec. 21, 1948; s. Emerald Foster and Gwendolin (Burns) S.; m. Celia Kaye, June 26, 1973; children: Gwendolyn, Heather. BSEE, BSChem, Calif. Poly. State U., 1972; PhD, Northwestern U., 1978, MD, 1979. Diplomate Am. Bd. Anesthesiology. Medicine intern Northwestern U., 1979-80; anesthia resident, 1980-82, asst. prof., 1982-88, assoc. prof., 1988—, dir. neurophysiologic monitoring, 1986-89; dir. neurophysiology U. Tex., San Antonio, 1987—; exec. coun. Am. Soc. Neurophysiologic Monitoring, 1991—. Mem. IEEE, Internat. Anesthesiology Rsch. Soc., Am. Soc. Anesthesiologists, N.Y. Anesthesiologists Soc., Am. Chem. Soc. Office: U Tex Health Sci Ctr 7703 Floyd Curl San Antonio TX 78284

SLOAN, WAYNE FRANCIS, mechanical engineer; b. Kansas City, Mo., Dec. 17, 1950; s. Hershel Francis and Elizabeth Fay (Wade) S.; m. Doris Elain Weach, Jan. 8, 1978; children: Charles Edward, Casey Elizabeth. BSME, Kans. State U., 1977. Registered profl. engr., Tex., La. Heating ventilating and air conditioning design engr. Wilson & Co., Engrs. & Architects, Salina, Kans., 1978-79; prodn. engr. Rockwell Internat., Richardson, Tex., 1979-89; design engr. Rockwell Internat., Shreveport, La., 1989—. With USN, 1969-73. Mem. ASME (section vice chmn. publicity 1991-93, section chmn. 1993-94), AIAA, The Planetary Soc., La. Engring. Soc., Kans. State Alumni Assn. Home: 9524 Pitch Pine Shreveport LA 71118 Office: Rockwell Internat AMC 6990 Challenger Dr Shreveport LA 71109-7754

SLOANE, BEVERLY LEBOV, writer, consultant, academic administrator; b. N.Y.C., May 26, 1936; d. Benjamin S. and Anne (Weinberg) LeBov; m. Robert Malcolm Sloane, Sept. 27, 1959; 1 child, Alison Lori. AB, Vassar Coll., 1958; MA, Claremont Grad. Sch., 1975, doctoral study, 1975-76; cert. in exec. mgmt., UCLA, 1982, grad. exec. program., UCLA, 1982; grad. profl. pub. course, Stanford U., 1982; grad. intensive bioethics course Kennedy Inst. Ethics, Georgetown U., 1987, advanced bioethics course, 1988; grad. sem. in Health Care Ethics, U. Wash. Sch. Medicine, Seattle, summer 1988, 89, 90; grad. Summer Bioethics Inst. Loyola Marymount U., summer, 1990; grad. Annual Summer Inst. on Advanced Teaching of Writing, Columbia U. Tchrs. Coll., summer, 1993; cert. in ethics corps tng. program, Josephson Inst. of Ethics, 1991; ethics fellow Loma Linda U. Med. Ctr., 1989; cert. clin. intensive biomedical ethics, Loma Linda U., 1989. Circulation librarian Harvard Med. Library, Boston, 1958-59; social worker Conn. State Welfare, New Haven, 1960-61; tchr. English, Hebrew Day Sch., New Haven, 1961-64; instr. creative writing and English lit. Monmouth Coll., West Long Branch, N.J., 1967-69; freelance writer, Arcadia, Calif., 1970—; v.p. council grad. students, Claremont Grad. sch., 1971-72; mem. adv. council tech. and profl. writing Dept. English, Calif. State U., Long Beach, 1980-82; mem. adv. bd. Calif. Health Rev., 1982-83; mem. Foothill Health Dist. Adv. Council L.A. County Dept. Health Svcs., 1987—, pres., 1989-91, immediate past pres., 1991-92. Ann. Key Mem. award, 1990. Author: From Vassar to Kitchen, 1967, A Guide to Health Facilities: Personnel and Management, 1971, 3d edit., 1992. Mem. pub. relations bd. Monmouth County Mental Health Assn., 1968-69; mem. task force edn. and cultural activities, City of Duarte, 1987-88, strategic planning task force com., campaign com. for pre-eminence, Claremont Grad. Sch., 1986-87; Vassar Coll. Class rep. to Alumnae Assn. Fall Coun. Meeting, 1989,, class corr. Vassar Coll. Quarterly Alumnae Mag., 1993—; grad. AMA Ann. Health Reporting Conf., 1992, 93; mem. exec. program network UCLA Grad. Sch. Mgmt., 1987—; trustee Ctr. for Improvement of Child Caring, 1981-83; mem. League Crippled Children, 1982—, bd. dirs., 1988-91, treas. for gen. meetings, 1990-91, chair hostesses com., 1988-89, pub. rels. com., 1990-91; bd. dirs. L.A. Commn. on Assaults Against Women, 1983-84; v.p. Temple Beth David, 1983-86; mem. community relations com. Jewish Fedn. Council Greater Los Angeles, 1985-87; del. Task Force on Minorities in Newspaper Bus., 1987-89; community rep. County Health Ctrs. Network Tobacco Control Program, 1991. Recipient cert. of appreciation City of Duarte, 1988, County of L.A., 1988; Coro Found. fellow, 1979; named Calif. Communicator of Achievement, Woman of Yr. Calif. Press Women, 1992. Fellow Am. Med. Writers Assn. (dir. 1980—, Pacific S.W. del. to nat. bd. 1980-87, 89-91, chmn. various conv. coms., chmn. nat. book awards trade category 1982-83, chmn. Nat. Conv. Networking Luncheon 1983, 84, chmn. freelance and pub. relations coms. Nat. Midyr. Conf. 1983-84, workshop leader ann. conf. 1984-87, 90, 91, 92, nat. chmn. freelance sect. 1984-85, gen. chmn. 1985, Asilomar Western Regional Conf., gen. chmn. 1985, workshop leader 1985, program co-chmn. 1987, speaker 1985, 88-89, program co-chmn. 1989 nat. exec. bd. dirs. 1985-86, nat. adminstr. sects. 1985-86, pres.-elect Pacific S.W. chpt. 1985-87, pres. 1987-89, immediate past pres. 1989-91, bd. dirs., 1991—, moderator gen. session nat. conf. 1987, chair gen. session nat. conf., 1986-87, chair Walter C. Alvarez Meml. Found. award 1986-87, Appreciation award for outstanding leadership 1989, named to Workshop Leaders Honor Roll 1991); mem. Women in Comm. (dir. 1980-82, 89-90, v.p. community affairs 1981-82, N.E. area rep. 1980-81, chmn. awards banquet 1982, sem. leader, speaker ann. nat. profl. conf., 1985, program adv. com. L.A. chpt. 1987, v.p. activities 1989-90, chmn. eles chpt. 1st ann. Agnes Underwood Freedom of Info. Awards Banquet 1982, recognition award 1983, nominating com. 1982, 83, com. Women of the Press Awards luncheon 1988, Women in Comm. awards luncheon 1988), Am. Assn. for Higher Edn., AAUW (legis. chmn. Arcadia br. 1976-77, books and plays chmn. Arcadia br. 1973-74, creative writing chmn. 1969-70, 1st v.p. 1975-76, networking chmn. 1981-82, chmn. task force promoting individual liberties 1987-88, Woman of Yr., Woman of Achievement award 1986, cert. of appreciation 1987), Coll. English Assn., Am. Pub. Health Assn., Am. Soc. Law and Medicine, Calif. Press Women (v.p. programs L.A. chpt. 1982-85, pres. 1985-87, state pres. 1987-89, past immediate past state pres. 1989-91, chmn. state speakers bur. 1989—, del nat. bd. 1989—, moderator ann. spring conv., 1990, 92, chmn. nominating com. 1990-91, Calif. lit. dir. 1990-92, dir. state lit. com. 1990-92, dir. family literary day Calif., 1990, Cert. of Appreciation, 1991, named Calif. Communicator of Achievement 1992), AAUP, Internat. Comm. Assn., N.Y. Acad. Scis., Ind. Writers So. Calif. (bd. dirs. 1989-90, dir. Specialized Groups 1989-90, dir. at large 1989-90, bd. dirs. corp. 1988-89, dir. Speech Writing Group, 1991-92), Hastings Inst., AAAS, Am. Med. Writers Assn. (pres. 1987-89, nat. adminstr. sects. 1985-86, nat. exec. bd. dirs. 1985-86, chmn. nominating com. Pacific S.W. chpt., 1987-89, workshop leader ann. conf. 1984-87, 90, 91, steering com. seminar on med. writing 1988, program planning com. for daylong program on med. writing in collaboration with Ind. Writers So. Calif., 1988-89, topic leader Nat. Conf. Networking Breakfast 1988, del. nat. bd. 1989-91, Appreciation award Outstanding Leadership, 1989, Presdl. award Pacific S.W. chpt. 1990, Am. Med. Writers Workshop Leaders Honor Roll 1991), Nat. Fedn. Press Women, (bd. dirs. 1987-89, nat. co-chmn. task force recruitment of minorities 1987-89, del. 1987-89, nat. dir. of speaker bur. 1989-93, editor of speakers bur. directory 1991, cert. of appreciation, 1991, 93, Plenary of Past Pres. state 1989—, workshop leader-speaker ann. nat. conf. 1990, chair state women of achievement com. 1986-87, editor Speakers Bur. Addendum Directory, 1992, editor Speakers Bur. Directory 1991, 92, named 1st runner up Nat. Communicator of Achievement 1992), AAUW (chpt. Woman of Achievement award 1986, chmn. task force promoting individual liberties 1987-88, speaker 1987, Cert. of Appreciation 1987, Woman of Achievement-Woman of Yr. 1986), Internat. Assn. Bus. Communicators, Soc. for Tech. Comm. (workshop leader, 1985, 86), Kennedy Inst. Ethics, Soc. Health and Human Values, Assoc. Writing Programs. Clubs: Women's City (Pasadena), Vassar of So. Calif., Claremont Colls. Faculty House, Petroleum (L.A.), Pasadena Athletic, Town Hall of Calif. Faculty (vice chair community affairs sect. 1982-87, speaker 1986, faculty-instr. Exec. Breakfast Inst. 1985-86, mem. study sect. council 1986-88). Lodge: Rotary (chair Duarte Rotary mag. 1988-89, mem. dist. friendship exch. com. 1988-89, mem. internat. svc. com. 1989-90, info. svc. com. 1989-90). Home and Office: 1301 N Santa Anita Ave Arcadia CA 91006-2419

SLOANE, BONNIE FIEDOREK, pharmacology and cancer biology educator, researcher; b. Pitts., Aug. 12, 1944; d. Leo Anthony and Bettie Thorburn (Findlay) Fiedorek; m. David E. E. Sloane, June 18, 1966 (div. 1976); m. Douglas Roy Yingst, Aug. 21, 1987. BS, Duke U., 1966, MA, 1968; PhD, Rutgers U., 1976. NIH fellow U. Pa., Phila., 1976-78, asst. prof. rsch., 1979; asst. prof. Mich. State U., East Lansing, 1979-80; asst. prof. sch. medicine Wayne State U., Detroit, 1980-84, assoc. prof., 1984-89, prof., 1989—; mem. NIH Pathology B Study Sect. 1987-91, 92—; exec. bd. Internat. Com. on Proteolysis, 1990—. Contbr. articles to profl. jours. Busch fellow Rutgers U., 1974-75, Nat. Rsch. Svc. award postdoct. fellow NIH-Child Health and Human Devel., Phila., 1977-79, Rsch. Career Devel. award fellow NIH-Nat. Cancer Inst., Detroit, 1984-89, Gershenson Dist. Faculty fellow Wayne State U., Detroit, 1991-93; recipient Academic Achievement award Probus club, 1984. Mem. AAAS, Am. Soc. Cell Biology, Am. Physiol. Soc., Am. Assn. Cancer Rsch. (chair membership com. 1991-92). Achievements include research on role of cysteine proteases, cysteine protease inhibitors in malignant progression, invasion, metastasis. Office: Wayne State U Sch Med 540 E Canfield Detroit MI 48201

SLOCUM, LESTER EDWIN, environmental engineer; b. Cadillac, Mich., Jan. 4, 1950; s. Lester George and Anneliese Elizabeth (Fey) S.; m. Sheryl Margaret Van Dam, June 13, 1969; children: Bradley S., Brent J., Katherine A. AS, Oakland Community Coll., Auburn Hills, Mich., 1977; BS, U. Mich., Flint, 1981. Mfg. engr. GMC Fisher Body Divsn., Pontiac, Mich., 1977-82; sr. mech. engr. GMC Pontiac Motors Divsn., 1982-86; ind. svcs. coord. Antor Mech. Contractors, Howard City, Mich., 1987-89; project engr. ABB Paint Finishing, Troy, Mich., 1989—. Home: 10868 Forest Rd Marion MI 49665 Office: ABB Paint Finishing 1400 Stephenson Hwy Troy MI 48099

SLOGGY, JOHN EDWARD, engineering executive; b. Mpls., Aug. 2, 1952; s. William Edwin and Dorthea Ann (Darling) S.; m. Vivian Arlene Gilles, Aug. 9, 1975; children: Cheryl Ann, Jo Anna Ellen, Laura Ann, William Harrison, Sarah Ellen, Andrew Harrison. Assoc. in Mech. Design, Superior Inst. of Tech., 1974; BS in Indsl. Tech. summa cum laude, U. Wis., Stout, 1976; MBA, U. Wis., 1982. Mech. design engr. Gilman Engring., Janesville, Wis., 1977-79, sr. mech. design engr., 1979-80; systems project mgr. Giddings & Lewis, Janesville, 1982-84; sr. project engr. Black & Decker, Fayetteville, N.C., 1984-85, mfg. engring. supr., 1985-86, mfg. engring. mgr., 1986-89, adv. mfg. engring. mgr., 1989-90, productivity engring. mgr., 1990-92; modernization engring. mgr., 1992-93; engring. mgr. Gen. Corp., Marion, Ind., 1993—. V.p. country club North Community Assn., Fayetteville, 1985—, Community Watch Assn., Fayetteville, 1985—; advisor PTA, Fayetteville, 1986—. Mem. Soc. Mfg. Engrs., Am. Soc. Quality Control, Soc. Plastics Engrs., Am. Soc. Value Engrs., Am. Soc. Materials, Inst. Indsl. Engrs., N.Am. Diecasting Assn. Avocations: public speaking, community association involvement. Home: 522 Hilliard Dr Fayetteville NC 28311-2677

SLONIM, ARNOLD ROBERT, biochemist, physiologist; b. Springfield, Mass., Feb. 15, 1926; s. Sam and Esther (Kantor) S.; married, 1951; 3 children; m. 1984. BS, Tufts Coll., 1947; AM, Boston U., 1948; PhD, Johns Hopkins U., 1953. Rsch. asst. nutrition Sterling-Winthrop Rsch. Inst., Rensselaer, N.Y., 1948-49; rsch. asst. pharmacology George Washington U. Med. Sch., Washington, 1949-50; rsch. asst., jr. instr. biology Johns Hopkins U., Balt., 1950-53; rsch. assoc. chemotherapy Children's Cancer Rsch. Found. Harvard U., Boston, 1953-54; head chem. lab. Lynn (Mass.) Hosp., 1955-56; various positions including chief applied ecology, supervisory rsch. biologist, physiologist & biochemist, phys. sci. adminstr., biotech. mgr. Aerospace Med. Rsch. Lab., Wright-Patterson AFB, Ohio, 1956-86; cons., pres. ARSLO Assocs., Columbus, Ohio, 1987—; lectr. Mass. Sch. Physiotherapy, Boston, 1955-56, Antioch U., 1984-85; mem. internat. bioastronautics com. Internat. Astronautical Fedn., 1966—; mem. environ. carcinogens program Internat. Agy. for Rsch. on Cancer/WHO, Paris, 1981—. Mem. com. on biol. handbooks Fedn. Am. Socs. for Exptl. Biology, 1966-71; mem. editorial bd. Aerospace Medicine, 1967-71; contbr. articles to profl. jours. Served with USN, 1944-46. Mem. Aerospace Med. Assn., Am. Soc. Biochemistry and Molecular Biology, Am. Physiol. Soc., N.Y. Acad. Sci., Internat. Acad. Aviation and Space Medicine, Sigma Xi, Masons, Scottish Rite, Shriners. Office: 630 Cranfield Pl Columbus OH 43213-3407

SLONIM, RALPH JOSEPH, JR., cardiologist; b. Plainfield, N.J., Nov. 10, 1925; s. Ralph J. and Gertrude (Posner) S.; m. Roberta Raymond, Aug. 2, 1958; children: Lloyd, Suzanne. BA, Yale U., 1948; MS, Rutgers U., 1950; MD, Hahnemann U., 1954. Diplomate Am. Bd. Internal Medicine. Intern U. Hosp., Balt., 1954-55, resident internal medicine, 1955-56; NIH postdoctoral trainee in cardiology Hahnemann Med. Ctr., Miami, Fla., 1956-57; resident internal medicine Jackson Meml. Hosp., Miami, Fla., 1957-59; Physician pvt. practice, Miami, Fla., 1961-90; faculty mem. U. Miami Sch. Medicine, 1961—. Contbr. articles to profl. jours. With U.S. Army, 1944-46. U. Miami Sch. Medicine postdoctoral rsch. fellow NIH, 1959-61. Mem. AAAS, Fla. Med. Assn., Dade County Med. Assn. Home: 1581 Brickell Ave # 1801 Miami FL 33129

SLOT, LARRY LEE, molecular biologist; b. Grand Rapids, Mich., Nov. 8, 1947; s. Russell Lee and Vivian June (Wolfert) S.; m. Pamela Chronis, Nov. 25, 1948; children: Franchot, Jason, Lara. AS, Grand Rapids Jr. Coll., 1971; BS, Mich. State U., 1982. Pub. Grand Rapids Interpreter, Mich., 1975-78; bush pilot Palacios, Honduras, 1978-80; pres. Genemsco Corp., Kingston, Mass., 1984—. Sculpture include The Pontibus, 1986; creator, copywriter (cell cloning and gene splicing module) The Dr. Cloner's Genetic Engineering Home Cloning Kit, 1984. With USMC, 1966-69. NSF fellow, 1982. Mem. AAAS, Am. Soc. Microbiology, Airplane Owners and Pilots Assn., Pontibus Soc. (founder 1986, pres. 1986-87), Golden Key Soc., Phi Kappa Phi. Achievements include design of first Tandem Cortege horizontal gel electrophoresis system, over 90 different mutant clones of the Moloney Leukemia Virus, builder molecular-level structural models of sea water carbonate and silicate crystals impregnated in fibrous proteins of conchiolin; research in biological structures, processes and functions associated with the production of carbonate-silicate-impregnated fibrous proteins; discovery of first intercontinental tetrahedralized matrix of silicate and carbonate impregnated fibrous protein. Home: 10 Braintree Ave Kingston MA 02364-1714 Office: Genemsco Corp Genemsco Beach Kingston MA 02364-1714

SLOVACEK, RUDOLF EDWARD, biochemist; b. Bloomington, Ind., Jan. 4, 1948; s. Rudolf Edward and Dolores (Jean) S.; m. Ronnie Diane Miller, Aug. 15, 1970 (div. 1982; children: Joel Evan, Gregory Heath, Caitlyn Alexis; m. Patricia D. Fici, June 8, 1993. BS, U. Rochester, 1970, PhD, 1975. Rsch. assoc. Nat. Rsch. Coun., Washington, 1975-76; rsch. assoc. Brookhaven (N.Y.) Lab., L.I., 1976-78, asst. biophysicist, 1979-80; tech. group leader Corning (N.Y.) Glass Work, 1980-83; sr. scientist Ciba Corning Diagnostics Corp., Cambridge, Mass., 1983-86; sr. scientist Ciba Corning Diagnostics Corp., Medfield, Mass., 1987-89, sr. staff scientist, 1990—; Author: (book chpts.) Biosensor with Fiber Optics, 1991, Non-isotopic Immunoassay, 1988. Recipient scholarship U. Rochester, 1966-70, Resident associateship Nat. Rsch. Coun., 1976, Rsch. grant USDA, 1980. Mem. N.Y. Acad. Scis., Electrochemistry Soc., Internat. Soc. for Optical Engring. Achievements include patents in field of Evanescent Wave Senors; Evanescent Wave Sensor Shell and Apparatus. Office: Ciba Corning Diagnostics 63 North St Medfield MA 02052

SLOVIC, STEWART PAUL, psychologist; b. Chgo., Jan. 26, 1938; s. Jacob S. and Blanche (Cohen) S.; m. Roslyn Judith Resnick, Aug. 30, 1959; children: Scott, Steven, Lauren, Daniel. BA, Stanford U., 1959; MA, U. Mich., 1962, PhD, 1964. Rsch. assoc. Oreg. Rsch. Inst., Eugene, 1964-76; rsch. assoc. Decision Rsch., Eugene, 1976-86, pres., 1986—; prof. dept. psychology U. Oreg., Eugene, 1986—; bd. sci. dirs. Risk Sci. Inst., Washington, 1987-91; cons. EPA, Washington, 1987-90; adviser WHO, Geneva, 1991; bd. dirs. Nat. Coun. Radiation Protection and Measurement. Author: Acceptable Risk, 1981; editor: Judgment Under Uncertainty, 1982; contbr. articles to profl. publs. J.S. Guggenheim fellow, 1986-87. Fellow AAAS, APA (Disting. Sci. Contbn. award 1993), Am. Psychol. Soc. (charter), Soc. Risk Analysis (pres. 1983-84, Disting. Contbn. award 1991). Achievements include development of methods for describing and understanding public perceptions of risk and incorporating those perceptions into public policy and risk-management decisions. Office: Decision Rsch 1201 Oak St Eugene OR 97401

SLY, WILLIAM S., biochemist, educator; b. East St. Louis, Ill., Oct. 19, 1932. MD, St. Louis U., 1957. Intern, asst. resident Ward Med Barnes Hosp., St. Louis, 1957-59; clin. assoc. nat. heart inst. NIH, Bethesda, Md., 1959-63, rsch. biochemist, 1959-63; dir. divsn. med. genetics, dept. medicine and pediatrics, sch. medicine Washington U., 1964-84, from asst. prof. to prof. medicine, 1964-78, from asst. prof. to prof. pediatrics, 1967-78, prof. pediatrics, medicine and genetics, 1978-84; prof. biochemistry, chmn. E. A. Doisy dept. biochemistry, prof. pediatrics sch. med. St. Louis U., 1984—; vis. physician Nat. Heart Inst., 1961-63, pediatric genetics clinic U. Wis., Madison, 1963-64; Am. Cancer Soc. fellow lab. enzymol Nat. Ctr. Sci. Rsch., Gif-sur-Yvette, France, 1963, dept. biochemistry and genetics U. Wis., 1963-64; attending physician St. Louis County Hosp., Mo., 1964-84; asst. physician Barnes Hosp., St. Louis, 1964-84, St. Louis Children's Hosp., 1967-84; genetics cons. Homer G. Philips Hosp., St. Louis, 1969-81; mem. genetics study sect. divsn. rsch. grants NIH, 1971-75; mem. active staff Cardinal Glennon Children's Hosp., St. Louis, 1984—; mem. med. adv. bd. Howard Hughes Med. Inst., 1989—. Recipient Merit award NIH, 1988; named Passano Found. laureate, 1991; Travelling fellow Royal Soc. Medicine, 1973. Mem. NAS, AMA, AAAS, Am. Soc. Human Genetics (mem. steering com. human cell biology program 1971-73, com. genetic counseling 1972-76), Am. Soc. Clin. Investigation, Am. Chem. Soc., Genetics Soc. Am., Am. Soc. Microbiology, Soc. Pediatric Rsch., Sigma Xi. Achievements include research in biochemical regulation, enveloped viruses as membrane probes in human diseases, lysosomal enzyme replacement in storage diseases, somatic cell genetics. Office: St Louis U Med Sch Dept Biochemistry 1402 S Grand Blvd Saint Louis MO 63104-1004*

SMAGORINSKY, JOSEPH, meteorologist; b. N.Y.C., Jan. 29, 1924; s. Nathan and Dinah (Azaroff) S.; m. Margaret Knoepfel, May 29, 1948; children: Anne, Peter, Teresa, Julia, Frederick. BS, NYU, 1947, MS, 1948, PhD, 1953; ScD (hon.), U. Munich, 1972. Research asst., instr. meteorology N.Y. U., 1946-48; with U.S. Weather Bur., 1948-50, 53-65, chief gen. circulation research sect., 1955-63; meteorologist Inst. Advanced Study, Princeton, N.J., 1950-53; acting dir. Inst. Atmospheric Scis. Environ. Scis. Services Adminstrn., Washington, 1965-66; dir. Geophys. Fluid Dynamics Lab. Environ. Scis. Services Adminstrn.-NOAA, Washington and Princeton, 1964-83; cons., 1983—; vice chmn. U.S. Com. Global Atmospheric Research Program, Nat. Acad. Sci., 1967-73, 80-87, officer, 1974-77, mem. climate bd., 1978-87, chmn. com. on internat. climate programs, 1979, bd. internat. orgns. and programs, 1979-83, chmn. climate research com., 1981-87; chmn. joint organizing com. Global Atmospheric Research Program, Internat. Council Sci. Unions/World Meteorol. Orgn., 1976-80, officer, 1967-80; chmn. Joint Sci. Com. World Climate Research Program, 1980-81; chmn.

climate coordinating forum Internat. Council Sci. Unions, 1980-84; vis. lectr. with rank of prof. Princeton U., 1968-83, vis. sr. fellow, 1983—; Sigmx Xi nat. lectr., 1983-85; Brittingham vis. prof. U. Wis., 1986. Contbr. to profl. publns. 1st lt. USAAF, 1943-46. Decorated Air medal; recipient Gold medal Dept. Commerce, 1966, award for sci. research and achievement Environ. Sci. Services Adminstrn., 1970, U.S. Presdl. award, 1980, Buys Ballot Gold medal Royal Netherlands Acad. Arts and Scis., 1973, IMO prize and Gold medal World Meteorol. Orgn., 1974. Fellow AAAS, Am. Meteorol. Soc. (hon. mem., councilor 1974-77, assoc. editor jour. 1965-74, Meisinger award 1967, Wexler Meml. lectr. 1969, Carl-Gustaf Rossby Research Gold medal 1972, Cleveland Abbe award for disting. service to atmospheric sci. 1980, pres. 1986, Charles Franklin Brooks award 1991); mem. Royal Meteorol. Soc. (hon., Symons Meml. lectr. 1963, Symons Meml. gold medal 1981). Home: 21 Duffield Pl Princeton NJ 08540-2605

SMAGULA, CYNTHIA SCOTT, molecular biologist; b. Cleve., May 12, 1943; d. James Henry Strauch and Florence Elizabeth (Downs) Turner; m. Howard John Smagula, Oct. 7, 1966 (div. 1977); 1 child, Stefan John. BA, U. Tex., 1965; BS, U. Tex., San Antonio, 1980; PhD, U. Tex., Dallas, 1988. Assoc. Howard Hughes Med. Inst., Dallas, 1988-92; molecular cardiology fellow Southwestern Med. Ctr. U. Tex., Dallas, 1992—. Co-author: Intracellular Trafficking of Proteins, 1991. Welch Found. scholar, 1980, predoctoral fellow, 1981. Mem. AAAS, Am. Crystallographic Assn., N.Y. Acad. Scis. Achievements include identification of an import determinant of the mitochondrial ADP/ATP carrier protein; crystallization of the conserved domain of the neuroprotein synapsin; ultrastructural localization of the MRP RNA. Home: 5353 Keller Springs Apt 713 Dallas TX 75248

SMALDONE, GERALD CHRISTOPHER, physiologist; b. N.Y.C., Sept. 1, 1947; s. Gerald J. and Theresa (Petrolino) S.; m. Arlene Merne, July 29, 1972; children: Marc, Lauren. BSChE, NYU, 1969, MD, PhD, 1975. Intern, resident Strong Meml. Hosp. U. Rochester, N.Y., 1975-77; fellow in medicine Johns Hopkins U., Balt., 1977-80, fellow in environ. health sci., 1977-80; asst. prof. medicine, physiology and biophysics SUNY, Stony Brook, 1980-86, assoc. prof., 1986—; cons. to pvt. industry and govt., 1988—. Contbr. articles to profl. jours.: editor: Jour. Aerosol Medicine, 1987—. Mem. Internat. Soc. for Aerosols in Medicine (bd. dirs. 1991—). Avocations: bicycling, sports cars, tropical fish. Office: SUNY Pulmonary Critical Care Div HSC T-17040 Stony Brook NY 11794-8172

SMALL, ALBERT HARRISON, engineering company executive; b. Washington, Oct. 15, 1925; s. Albert and Lillian S.; BChemE., U. Va., 1946; student George Washington U. Law Sch., 1947-48, Am. U. Grad. Sch. Bus. Adminstrn., 1949-51; m. Shirley Schwalb, Sept. 14, 1952; children: Susan Carol, Albert H., James H. Founder, So. Engring. Corp., Washington, 1952—, pres., chief exec. officer, 1968—. Bd. dirs. Nat. Symphony Orch., Va. Engring. Found., U. Va.; bd. visitors U. Va. Served with USNR, 1943-46; bd. trustees Meridian House Internat., Folger Shakespeare Library Com. Mem. Nat. Assn. Real Estate Bds., Nat. Assn. Home Builders, Urban Land Inst. Republican. Clubs: Georgetown, Pisces, Harmonie, Army-Navy, Internat. Home: 7116 Glenbrook Rd Bethesda MD 20814-1225 Office: 1050 Connecticut Ave NW Ste 444 Washington DC 20036-5309

SMALL, PETER MCMICHAEL, physician, researcher; b. Bethesda, Md., Sept. 18, 1959; s. Parker Adams and Natalie Edmea (Settimelli) S. BA, Princeton U., 1981; MD, U. Fla., 1985. Intern, resident U. Calif., San Francisco, 1985-88, chief med. resident, 1988-89; fellow Div. Infectious Diseases, Stanford, Calif., 1990-92; assoc. Howard Hughes Med. Inst., Stanford, 1990—; cons. Muhimbilli Med. Ctr., Dar es Salaam, Tanzana, 1990. Author: Endocarditis, 1990, Tuberculosis, 1992, Molecular Epidemiology, 1992. Mem. ACP, Am. Thoracic Soc., Physicians for Social Responsibility, Sigma Xi (assoc.), Alpha Omega Alpha. Democrat. Achievements include research in treatment of Tb in patients with AIDS. Office: Howard Hughes Med Inst Beckman Ctr Rm 251 Stanford CA 94305

SMALL, WILFRED THOMAS, surgeon, educator; b. Boston, June 13, 1920; s. Fred Wentworth and Isabelle (Scott) S.; BS, Bowdoin Coll., 1943; MD, Tufts U., 1946; m. Muriel Yoe Gratton, Sept. 25, 1948; children: Wilfred Thomas, Richard Gratton, James Stewart, John Wentworth. Intern surg. svc. The Boston Children's Hosp., 1946-47, then research fellow; assoc. in surgery Peter Bent Brigham Hosp., Harvard U., 1949-50; resident, chief resident in surgery New Eng. Med. Ctr., Tufts U., 1950-53; practice medicine specializing in surgery, Worcester, Mass., 1953-88; assoc. prof. surgery U. Mass., from 1973, now prof. surgery; mem. staff Meml. Hosp., 1953-88, chief div. surgery, 1973-81; instr. Harvard U., 1949-50, Tufts U., 1952-60. Bd. dirs. Worcester Boys Club, Vis Nurse Assn., Vero Beach, Fla.; mem. Worcester Art Museum, Worcester County Music Assn. Served to lt. (j.g.) USN, 1947-49. Diplomate Am. Bd. Surgery. Fellow ACS (pres. Mass. chpt. 1979); mem. New Eng. Surg. Soc., New Eng. Cancer Soc., Soc. Surgery Alimentary Tract, Mass., Pan Am. med. socs., AMA, Am. Trauma Soc., Worcester Econs. Club (past pres.), Worcester Council on Fgn. Rels., Tatnuck Country Club, Sakonnet Golf Club, Riomar Golf Club (Vero Beach, Fla.). Episcopalian. Contbr. articles to profl. jours. Home: 2733 Ocean Dr Vero Beach FL 32963-2059

SMALLEY, ARTHUR LOUIS, JR., engineering and construction company executive; b. Houston, Jan. 25, 1921; s. Arthur L. and Ebby (Curry) S.; m. Ruth Evelyn Britton, Mar. 18, 1946; children: Arthur Louis III, Tom Edward. BSChemE, U. Tex., Austin, 1942. Registerd profl. engr., Tex. Dir. engring. Celanese Chem. Co., Houston, 1964-72; mktg. exec. Fish Engring. Co., Houston, 1972-74; pres. Matthew Hall & Co., Inc., Houston, 1974-87; cons. Davy McKee Corp., Houston, 1987—; dir. Walter Internat., 1991. Life mem. Houston Livestock and Rodeo; mem. Engring. Found. Adv. Coun. U. Tex. Recipient Silver Beaver award Boy Scouts Am., 1963; named Disting. Engring. Grad., U. Tex., 1987. Mem. Am. Inst. Chem. Engrs., Am. Petroleum Inst., Pres. Assn., Petroleum Club (Houston), Chemists Club of N.Y., Oriental Club (London), Houston Club, Traveler's Century Club, Rotary. Republican. Episcopalian. Mem. internat. adv. bd. Ency. Chem. Processing and Design. Home: 438 Hunterwood Dr Houston TX 77024-6936 Office: 2925 Briarpark Dr Houston TX 77042-3715

SMALLEY, RICHARD ERRETT, chemistry and physics educator, researcher; b. Akron, Ohio, June 6, 1943; s. Frank Dudley and Virginia (Rhodes) S.; m. Judith Grace Sampieri, May 4, 1968 (div. July 1979); 1 child, Chad; m. Mary Lynn Chapieski, July 10, 1980. BS in Chemistry, U. Mich., 1965; MA in Chemistry, Princeton U., 1971, PhD in Chemistry, 1973; Doctor Honoris causa, Univ. Liege, Belgium, 1991. Assoc. The James Franck Inst., Chgo., 1973-76; asst. prof. William Marsh Rice U., Houston, 1976-80, prof., 1981-82, Gene & Norman Hackerman prof. chemistry, 1982—; prof. Dept. Physics Rice U., Houston, 1990—; chmn. Rice Quantum Inst., Houston, 1986—. Contbr. numerous articles to profl. jours. Harold W. Dodds fellow Princeton U., 1973, Alfred P. Sloan fellow Sloan Found., 1973-80; recipient E.O. Lawrence award U.S. Dept. Energy, 1991, Welch Found. award, 1992. Fellow Am. Phys. Soc. (Internat. New Materials prize 1992, Robert A Welch award in chemistry 1992, J. S. Kilby award 1992, Nichols medal 1993); mem. AAAS, NAS, Am. Chem. Soc. (Irving Langmuir Chem. Physics prize 1991, William H. Nichols medal, 1993), Materials Rsch. Soc., Am. Acad. Arts and Scis., Sigma Xi. Office: Rice U Dept Chemistry Mail stop 100 PO Box 1892 Houston TX 77251-1892

SMARANDACHE, FLORENTIN, mathematics researcher, writer; b. Balcesti-Vilcea, Romania, Dec. 10, 1954; came to U.S., 1990; s. Gheorghe and Maria (Mitroiescu) S.; m. Eleonora Niculescu; children: Mihai-Liviu, Silviu-Gabriel. MS, U. Craiova, 1979; postgrad., Ariz. State U., 1991. Mathematician I.U.G., Craiova, Romania, 1979-81; math. prof. Romanian Coll., 1981-82, 1984-86, 1988; math. tchr. Coop. Ministry, Morocco, 1982-84; French tutor pvt. practice, Turkey, 1988-90; software engr. Honeywell, Phoenix, 1990-93. Author: Nonpoems, 1990, Only Problems, Not Solutions, 1991, numerous other books; contbr. articles to profl. jours. Mem. U.S. Math. Assn., Romania Math. Assn., Zentralblatt fur Math. (reviewer). Achievements include a function in the Number Theory called Smarandache Function: f(n) is the smallest integer m such that m! is divisible by n. Home: PO Box 42561 Phoenix AZ 85080-2561

SMARDON, RICHARD CLAY, landscape architecture/environmental studies educator; b. Burlington, Vt., May 13, 1948; s. Philip Albert and Louise Gertrude (Peters) S.; m. Anne Marie Graveline, Aug. 19, 1973; children: Regina Elizabeth, Andrea May. BS cum laude, U. Mass., 1970, MLA, 1973; PhD in Environ. Planning, U. Calif., Berkeley, 1982. Environ. planner, landscape architect Wallace, Floyd, Ellenzweig, Inc., Cambridge, Mass., 1972-73; assoc. planner Exec. Office Environ. Affairs, State of Mass., Boston, 1973-75; environ. impact assessment specialist USDA extension svc. Oreg. State U., Corvallis, 1975-76; landscape architect USDA Pacific S.W. Forest and Range Expt. Sta., Berkeley, 1977; rsch. landscape architect U. Calif., Berkeley, 1977-79; prof. landscape architecture, sr. rsch. assoc. SUNY Coll. Environ. Sci. and Forestry, Syracuse, 1979-86, prof. environ. studies, 1987—, dir. Inst. for Environ. Policy and Planning, 1987—; co-dir. Gt. Lakes Rsch. Consortium, Syracuse, 1986—; guest lectr. numerous univs.; adj. asst. prof. U. Mass., Amherst, 1974-75; Sea Grant trainee Inst. for Urban and Regional Devel., Berkeley, 1976; condr., presenter numerous seminars and workshops; cons. to numerous orgns.; mem. com. on environ. design and landscape Transp. Rsch. Bd.-NAS, 1985; mem. tech. adv. bd. Wetlands Rsch., Inc., Chgo., 1985; mem. adv. bd. Wetlands Fund, N.Y., 1985; v.p. Integrated Site Inc., Syracuse. Co-editor: Out National Landscape, 1979, spl. issue Coastal Zone Mgmt. Jour., 1982, The Future of Wetlands, 1983, Foundations for Visual Project Analysis, 1986, The Legal Landscape, 1993; mem. edit. bd. Northeastern Environ. Sci. Jour., 1981, Landscape and Urban Planning, 1991; contbr. over 100 articles to profl. jours. Bd. dirs. Sackets Harbor Area Hist. Preservation Found., Watertown, N.Y., 1984—; pres. Save the County, Inc., Fayetteville, N.Y., 1986-88; appointed to Great Lakes (N.Y.) Adv. Commn.. N.Y. Recipient Beatrice Farrand award U. Calif., 1979, Am. Soc. Landscape Architects award, 1972, Pub. Svc. award in edn., 1990, Progressive Architecture mag. award, 1992. Mem. AAAS, Am. Land Resource Assn. (charter), Internat. Assn. for Impact Assessment, Coastal Soc., Alpha Zeta (life), Sigma Lambda Alpha. Avocations: folk guitar, hiking, skiing, travel. Office: SUNY Inst for Environ Policy and Planning Syracuse NY 13210 Office: Integrated Site Inc 886 E Brighton Ave Syracuse NY 13205

SMARR, LARRY LEE, science administrator, educator, astrophysicist; b. Columbia, Mo., Oct. 16, 1948; s. Robert L. Jr. and Jane (Crampton) S.; m. Janet Levarie, June 3, 1973; children: Joseph Robert, Benjamin Lee. BA, MS, U. Mo., 1970; MS, Stanford U., 1972; PhD, U. Tex., 1975. Rsch. asst. in physics U. Tex., Austin, 1972-74; lectr. dept. astrophys. sci. Princeton U., 1974-75; rsch. assoc. Princeton U. Obs., 1975-76; rsch. affiliate dept. physics Yale U., New Haven, Conn., 1978-79; asst. prof. astronomy dept. U. Ill., Urbana, 1979-81, asst. prof. physics dept., 1980-81, assoc. prof. astronomy and physics dept., 1981-85, prof. astronomy and physics dept., 1985—; dir. Nat. Ctr. for Supercomputing Applications, Champaign, Ill., 1985—; cons. Lawrence Livermore Nat. Lab., Calif., 1976—, Los Alamos (New Nex.) Nat. Lab., 1983—; mem. Commn. on Phys. Sci., Math. and Resources, NRC, Washington, 1987-90, commn. on Geoscience, Environ. and Resources, 1990—, adv. panel on Basic Rsch. in the 90's Office Tech. Assesment, 1990—. Editor: Sources of Gravitational Radiation, 1979; mem. editoral bd. Science mag., 1986-90; contbr. over 50 sci. articles to jours. in field. Cofounder, co-dir. Ill. Alliance to Prevent Nuclear War, Champaign, 1981-84. Recipient Fahrney medal Franklin Inst., Phila., 1990; NSF fellow Stanford U., 1970-73, Woodrow Wilson fellow, 1970-71, Lane Scholar U. Tex., Austin, 1972-73, jr. fellow Harvard U., 1976-79, Alfred P. Sloan fellow, 1980-84. Fellow Am. Phys. Soc.; mem. AAAS, Am. Astron. Soc., Govt. Rsch. Roundtable U. Ind. Avocations: marine aquarium, gardening. Office: NCSA at UIUC 152 Computing Applications Bldg 605 E Springfield Ave Champaign IL 61820-5577

SMART, MELISSA BEDOR, environmental consulting company executive; b. St. Johnsbury, Vt., Mar. 5, 1953; d. Leslie Oscar and Helen Catherine (Kenney) Bedor; m. Glenn Robin Smart, Oct. 1, 1983; children: Catherine Jean, Jenny Laura. BS in Ecology and Environ. Conservation, U. N.H., 1975; MS in Water and Land Use Planning, SUNY, Syracuse, 1981. Environ. instr. NSF, Hooksett, N.H., 1975; environ. scientist, planner Parsons Brinckerhoff Quade & Douglas, Inc., Boston, 1976-78, sr. environ. scientist, planner, 1981-82; research asst. SUNY Coll. Environ. Sci. and Forestry, Syracuse, 1978-79; environ. planner St. Lawrence Eastern Ontario Commn., Watertown, N.Y., 1979; sr. environ. scientist, mktg. dir. VTN Consolidated, Inc., Boston, 1982-83; pres. The Smart Assocs., Inc., Contoocook, N.H., 1984-90, Concord, N.H., 1990—; water resource cons. Soc. Protection N.H. Forests, Concord, 1984—; water resource lectr. Harris Ctr. Conservation Edn., Hancock, N.H., 1985; mem. steering com. N.H. Rivers Campaign, Concord, 1986. Author: Directory of Water Testing Expertise in New Hampshire, 1985. Pastor's aide So. Congl. Ch., Concord, 1988-90; mem. N.H. Gov.'s Task Force on Wetlands; bd. trustees Centennial Home. Am. Field Service scholar, Australia, 1970. Mem. Am. Water Resources Assn., N.H. Water Works, N.H. Assn. Wetland Scientists (past pres.), Assn. State Wetlands Mgrs., Women's Transp. Seminar. Democrat. Congregationalist. Avocations: reading, skiing, golf, bird-watching, quilting. Home: 81 Cedar St Contoocook NH 03229-9212 Office: The Smart Assocs Inc 72 N Main St Concord NH 03301-4915

SMELSER, NEIL JOSEPH, sociologist; b. Kahoka, Mo., July 22, 1930; s. Joseph Nelson and Susie Marie (Hess) S.; m. Helen Thelma Margolis, June 10, 1954 (div. 1965); children: Eric Jonathan, Tina Rachel; m. Sharin Fateley, Dec. 20, 1967; children: Joseph Neil, Sarah Joanne. B.A., Harvard U., 1952, Ph.D., 1958; B.A., Magdalen Coll., Oxford U., Eng., 1954; M.A., Magdalen Coll., Oxford U., 1959; grad., San Francisco Psychoanalytic Inst., 1971. Mem. faculty U. Calif.-Berkeley, 1958—, prof. sociology, 1962—, asst. chancellor ednl. devel., 1966-68; assoc. dir. Inst. Internat. Rels., 1969-73, 80-89, Univ. prof. sociology, 1972; dir. edn. abroad program for U. Calif., 1977-79; bd. dirs. Found. Fund for Rsch. in Psychiatry, 1967-70, Social Sci. Rsch. Coun., 1968-71, chmn., 1971-73; trustee Ctr. for Advanced Study in Behavioral Scis., 1980-86, 87-93, chmn., 1984-86, Russell Sage Found., 1990—; mem. subcom. humanism Am. Bd. Internal Medicine, 1981-85; mem. com. econ. growth Social Sci. Rsch. Coun., 1961-65; chmn. sociology panel Behavioral and Social Scis. survey NAS and Social Sci. Rsch. Coun., 1967-69; mem. com. on basic rsch. in behavioral and social scis. NRC, 1980-89, chmn., 1984-86, co-chmn., 1986-89. Author: (with T. Parsons) Economy and Society, 1956, Social Change in the Industrial Revolution, 1959, Theory of Collective Behavior, 1962, The Sociology of Economic Life, 1963, 2d edit., 1975, Essays in Sociological Explanation, 1968, Sociological Theory: A Contemporary View, 1971, Comparative Methods in the Social Sciences, 1976, (with Robin Content) The Changing Academic Market, 1980, Sociology, 1981, 2d edit., 1984, 3d edit. 1987, 4th edit. 1991, Social Paralysis and Social Change, 1991; editor: (with W.T. Smelser) Personality and Social Systems, 1963, 2d edit., 1971, (with S.M. Lipset) Social Structure and Mobility in Economic Development, 1966, Sociology, 1967, 2d edit., 1973, (with James Davis) Sociology: A Survey Report, 1969, Karl Marx on Society and Social Change, 1973, (with Gabriel Almond) Public Higher Education in California, 1974, (with Erik Erikson) Themes of Work and Love in Adulthood, 1980, (with Jeffrey Alexander et al) The Micro-Macro Link, 1987, Handbook of Sociology, 1988, (with Hans Haferkamp) Social Change and Modernity, 1992, (with Richard Munch) Theory of Culture, 1992; editor Am. Sociol. Rev., 1962-65, 89-90; adv. editor Am. Jour. Sociology, 1960-62. Rhodes scholar, 1952-54; jr. fellow Soc. Fellows, Harvard U., 1955-58; fellow Russell Sage Found., 1989-90. Mem. NAS, Am. Sociol. Assn. (coun. 1962-65, 67-70, exec. com. 1963-65), Pacific Sociol. Assn., Internat. Sociol. Assn. (exec. com. 1986—, v.p. 1990—), Am. Acad. Arts and Scis., Am. Philos. Soc. Home: 109 Hillcrest Rd Berkeley CA 94705-2808

SMELSER, RONALD EUGENE, mechanical engineer; b. Celina, Ohio, Nov. 17, 1947; s. Raymond Harold and Gertrude Celina (Conner) S.; m. Barbara Ann Schmalz, Sept. 2, 1972; children: Elizabeth Ann, Peter Christopher. BSME, U. Cin., 1971; MSME, MIT, 1972; PhD, Carnegie-Melon, 1978. Asst. prof. U. Pitts., 1978-81; rsch. scientist U.S. Steel Rsch. Labs., Pitts., 1981-83; scientific assoc. Alcoa Tech. Ctr., Alcoa Center, Pa., 1983-93. Co-editor: Macro/Macro Scale Phenomena in Solid Fraction Processes, 1992. Mem. ASME, Am. Soc. Engring. Edn., Tau Beta Pi, Pi Tau Sigma, Sigma Xi.

SMERDON, ERNEST THOMAS, academic administrator; b. Ritchey, Mo., Jan. 19, 1930; s. John Erle and Ada (Davidson) S.; m. Joanne Duck, June 9, 1951; children: Thomas, Katherine, Gary. BS in Engring., U. Mo., 1951, MS in Engring., 1956, PhD in Engring., 1959. Registered profl. engr., Ariz. Chmn. dept. agrl. engring. U. Fla., Gainesville, 1968-74, asst. dean for rsch., 1974-76; vice chancellor for acad. affairs U. Tex. System, Austin, 1976-82; dir. Ctr. for Rsch. in Water Resources U. Tex., 1982-88; dean Coll. Engring. and Mines U. Ariz., Tucson, 1988-92; vice provost, dean Engring U. Ariz., 1992—; mem. bd. sci. and tech. for internat. devel. NRC, Washington, 1990—. Editor: Managing Water Related Conflicts: The Engineer's Role, 1989. Mem. Ariz. Gov.'s Sci. and Tech. Coun., 1989—; bd. dirs. Greater Tucson Econ. Coun., Tucson, 1990. Recipient Disting. Svc. in Engring. award U. Mo., 1982. Fellow AAAS, Am. Soc. Agrl. Engrs.; mem. ASCE (Outstanding Svc. award irrigation and drainage div. 1988, Royce Tipton award 1989), NAE (peer com. 1986—, acad. adv. bd. 1989-92, tech. policy options co. 1990-91), Am. Water Resources Assn. (Icko Iben award 1989), Am. Geophys. Union, Univ. Coun. on Water Resources, Ariz. Soc. Profl. Engrs. (Engr. of Yr. award 1990), Sigma Xi, Phi Kappa Phi, Tau Beta Pi, Pi Mu Epsilon. Avocations: jogging, hiking, golf, scuba diving, painting. Office: University of Arizona Engineering Experiment Station 100 Civil Engineering Bldg Tucson AZ 85721

SMETHERAM, HERBERT EDWIN, company executive; b. Seattle, Sept. 9, 1934; s. Francis Edwin and Grace Elizabeth (Warner) S.; m. Beverly Joan Heckert, Sept. 7, 1963; children: Alice, Helen, Charles. BA, U. Wash., 1956; diploma, Naval Intelligence Sch., 1962; MA, U. Md., 1971; MBA, Rollins Coll., 1991. Ensign USN, 1956, advanced through grades to capt., 1976; comdr. USS Lind (DD-703), 1971-73; attache to Sweden USN, Stockholm, 1978-81; comdr. Naval Adminstrn. Command, Orlando, Fla., 1981-84; strategic planner electronics, info. and missiles group Martin Marietta Corp., Orlando, 1985-93; exec. dir. Naval Tng. Ctr. Re-Use Com., Orlando; Mem. Senator Hawkins Naval Acad. Nominating Com., Orlando, 1991, Orlando Naval Tng. Ctr. Retention Com., 1991-93, Fla. Gov.'s Def. Reinvestment Task Force, 1992-93; mem. USO Coun. Ctrl. Fla., Orlando, 1981-93, pres., 1991-93. Mem. Senator Hawkins Naval Acad. Nominating Com., Orlando, 1991, Orlando Naval Tng. Ctr. Retention Com., 1993—, Fla. Gov.'s Def. Reinvestment Task Force, 1992-93; mem. USO Coun. Ctrl. Fla., Orlando, 1981—, pres., 1991-93. Mem. AIAA, SAR, Electronics Industry Assn. (requirements com. 1985-93), Navy League, Delta Kappa Epsilon. Republican. Episcopalian. Avocation: tennis. Home: 3985 Lake Mira Dr Orlando FL 32817-1643

SMILEY, JOSEPH ELBERT, JR., evaluation engineer, librarian; b. Cin., Dec. 21, 1922; s. Joseph Elbert and Esther Marie (Lentz) S.; m. Leona Caroline Besenfelder, Aug. 23, 1953 (dec. Aug. 1986); 1 child, Mary Susan Smiley Freud; m. Betty Concklin, May, 9, 1987. A.A., Edison Community Coll., 1978; B.A., U. S. Fla., 1981, M.A., 1983. Expediter VA, Miami, Fla., 1948-51, analyzer, 1953; evaluation engr. photographic equipment, CIA, Washington, 1953-75. Second v.p. pub. relations Country Club Estates Assn. Lehigh Acres, Inc., 1981-84, pres., 1984-86 ; coordinator, acting zone capt. Lehigh Acres Emergency Preparedness Com., 1983-84, chmn., 1984-86 . Served with U.S. Army, 1942-45, with USAF, 1951-52. Recipient commendation ribbon USAF, 1951; cert. of merit CIA, 1974, certs. of appreciation, 1975, 1981; letter of congratulations Gerald R. Ford, CIA, 1975. Mem. ALA, Fla. Library Assn., Internat. Platform Assn., Phi Theta Kappa, Kappa Delta Pi, Phi Kappa Phi, Beta Phi Mu. Republican. Roman Catholic. Club: KC. Home: 107 Riviera St Lehigh Acres FL 33936-5349

SMILEY, RICHARD WAYNE, research center administrator, researcher; b. Paso Robles, Calif., Aug. 17, 1943; s. Cecil Wallace and Elenore Louise (Hamm) S.; m. Marilyn Lois Wenning, June 24, 1967; 1 child, Shawn Elizabeth. BSc in Soil Sci., Calif. State Poly. U., San Luis Obispo, 1965; MSc in Soils, Wash. State U., 1969, PhD in Plant Pathology, 1972. Asst. soil scientist Agrl. Rsch. Svc., USDA, Pullman, Wash., 1966-69; rsch. asst. dept. plant pathology Wash. State U., Pullman, 1969-72; soil microbiologist Commonwealth Sci. and Indsl. Rsch. Orgn., Adelaide, Australia, 1972-73; rsch. assoc. dept. plant pathology Cornell U., Ithaca, N.Y., 1973-74, asst. prof., 1975-80, assoc. prof., 1980-85; supt. Columbia Basin Agr. Rsch. Ctr., prof. Oreg. State U., Pendleton, 1985—; vis. scientist Plant Rsch. Inst., Victoria Dept. Agr., Melbourne, Australia, 1982-83. Author: Compendium of Turfgrass Diseases, 1983, 2d edit., 1992; contbr. more than 200 articles to profl. jours.; author slide set illustrating diseases of turfgrasses. Postdoctoral fellow NATO, 1972. Mem. Am. Phytopath. Soc. (sr. editor APS Press 1984-87, editor-in-chief 1987-91), Am. Soc. Agronomy, Internat. Turfgrass Soc., Am. Sod Producers Assn. (hon. life), Coun. Agrl. Sci. and Tech., Rotary (Pendleton, pres. 1991-92). Achievements include discovery of the etiology of a serious disease of turfgrasses, which led to a redefinition of studies and disease processes in turfgrasses. Office: Oreg State U Columbia Basin Agr Rsch Ctr PO Box 370 Pendleton OR 97801-0370

SMINCHAK, DAVID WILLIAM, civil engineer, consultant; b. Cleve., July 5, 1949; s. William Charles and Gertrude Helen (Lanese) S.; m. Pamela Marie Latessa, July 17, 1971; children: Jon David, Michael Joseph. BSCE, Purdue U., 1971. Registered profl. engr., Ohio. Civil engr. Babcock & Wilcox Co., Barberton, Ohio, 1971-72, Paul A. Frank & Assocs., Parma Heights, Ohio, 1972-76; mgr. market svcs. Price Bros. Co., Dayton, Ohio, 1976-82; mgr. bus. devel. R.E. Warner & Assocs., Westlake, Ohio, 1982-86, v.p., 1986—. Pres. Forest Hills Homeowners Assn., Brunswick, Ohio, 1982. Mem. ASTM (com. mem. 1987—), ASCE (com. chair 1986-87), Ohio Assn. Consulting Engrs. (com. chair 1986—), Am. Concrete Pipe Assn. (Richard C. Longfellow award 1978), Chi Epsilon, Tau beta Pi. Republican. Roman Catholic. Home: 203 Melbourne Dr Brunswick OH 44212 Office: RE Warner & Assocs 2001 Crocker Rd Westlake OH 44145

SMITH, AARON, health researcher, clinical psychologist; b. Boston, Nov. 3, 1930; s. Harry and Anne (Gilgoff) S.; m. Sept. 7, 1952 (div.); children—Naomi E., Jeffrey O., David G., Andrew H.; m. D. Sharon Casey, Jan. 7, 1972. A.B., Brown U., 1952; Ph.D., U. Ill., 1958. Cert. clin. psychologist, Pa. Co-dir., Northeast Psychol. Clinic, Phila., 1959-75; dir. research Haverford State Hosp., Pa., 1962-73, asst. hosp. dir., 1973-75; assoc. rsch. prof. U. Nev., Reno, 1975—; dir. rsch VA Med. Ctr., Reno, 1975—; chmn. Nev. Legislature Mental Health Task Force, Carson City, 1978; sci. adviser Gov.'s Com. on Radiation Effects, Carson City, 1979-82. Co-author: Anti-depressant Drug Studies 1956-66, 1969; Medications and Emotional Illness, 1976; co-editor: Goal Attainment Sealing: Application, Theory, and Measurement, 1993; contbr. chpts. to books and articles to profl. jours. Grantee Squibb Inst. Med. Research, 1965-69, NIMH, 1965-69, Smith Kline & French Labs., 1968-69, VA Health Services Research, 1976—. Mem. Am. Psychol. Assn., Western Psychol. Assn., Gerontol. Soc. Am. Home: 12790 Roseview Ln Reno NV 89511-8652 Office: VA Med Ctr 1000 Locust St Reno NV 89520

SMITH, AL JACKSON, JR., environmental engineer, lawyer; b. Meridian, Miss., Aug. 26, 1935; s. Al Jackson and Katherine (Felker) S.; m. Patricia Scruggs, Dec. 20, 1957; children: Johnny, Vicki, Katherine. BSCE, Miss. State U., 1958; MS in Environ. Engring., Vanderbilt U., 1969; JD, Atlanta Coll. Law, 1977; LLM, Woodrow Wilson Coll., 1980. Bar: Ga. 1979, U.S. Dist. Ct. (no. dist.) Ga. 1979, U.S. Ct. Appeals (11th cir.). Engr. City of Vicksburg, Miss., 1964-66; dir. br. emergency Region IV EPA, Atlanta, 1966-86, dep. dir. div. water, 1986-90; counsel Hurt, Richardson, Todd, Garner and Caddenhead, Atlanta, 1990-93, McRae Secrest & Fox, Atlanta, 1993—; solicitor City of Stockbridge, Ga., 1984-87; lectr. Nat. Emergency Tng. Ctr., Emmitsburg, Md. 1980—; city judge Locust Grove, Ga., 1988—. Author: Managing Hazardous Substance Accidents, 1981; Oil Pollution Control, 1973; contbg. author: Hazardous Materials Handbook, 1982; contbr. articles to profl. jours. Served to capt. USAR, 1958-70. Mem. Internat. Assn. Chiefs Police, Ga. Bar Assn., N.C. Assn. Fire Chiefs. Baptist. Home: 1550 S Ola Rd Locust Grove GA 30248-9472

SMITH, ALAN PAUL, plant ecologist and physiologist; b. Madison, N.J., Mar. 31, 1945; s. Glenn Wilson and Ruth Geraldine (Hadley) S. BA, Earlham Coll., 1967; MA, Duke U., 1970, PhD, 1974. Asst. prof. U. Pa., Phila., 1974-81; assoc. prof. U. Miami, Coral Gables, Fla., 1982-87; staff scientist Smithsonian Tropical Rsch. Inst., Balboa, Panama, 1974-93, asst. dir., 1989-93. Contbr. articles to Jour. Ecology, Occologia Biotropica, Nature, Jour. Tropical Ecology. Mem. Am. Soc. Naturalists, Ecol. Soc. Am. Achievements include first to initiate plant physiology research program in the tropical forest canopy and laboratory. Home: Balboa Panama *Died Aug. 26, 1993.*

SMITH, ALBERT CHARLES, biologist, educator; b. Springfield, Mass., Apr. 5, 1906; s. Henry Joseph and Jeanette Rose (Machol) S.; m. Nina Grönstrand, June 15, 1935; children: Katherine (Mrs. L. J. Campbell),

Michael Alexis; m. Emma van Ginneken, Aug. 1, 1966. AB, Columbia U., 1926, PhD, 1933. Asst. curator N.Y. Bot. Garden, 1928-31, asso. curator, 1931-40; curator herbarium Arnold Arboretum of Harvard U., 1940-48; curator div. phanerogams U.S. Nat. Mus., Smithsonian Instn., 1948-56; program dir. systematic biology NSF, 1956-58; dir. Mus. of Natural History, Smithsonian Instn., 1958-62, asst. sec., 1962-63; prof. botany, dir. research U. Hawaii, Honolulu, 1963-65; Gerrit Parmile Wilder prof. botany U. Hawaii, 1965-70, prof. emeritus, 1970—; Ray Ethan Torrey prof. botany U. Mass., Amherst, 1970-76; prof. emeritus U. Mass., 1976—; editorial cons. Nat. Tropical Bot. Garden, Hawaii, 1977-91; bot. expdns., Colombia, Peru, Brazil, Brit. Guiana, Fiji, West Indies, 1926-69; del. Internat. Bot. Congresses, Amsterdam, 1935, Stockholm, 1950; v.p. systematic sect., Montreal, 1959, Internat. Zool. Congress, London, 1958; pres. Am. Soc. Plant Taxonomists, 1955, Bot. Soc. Washington, 1962, Biol. Soc. Washington, 1962-64, Hawaiian Bot. Soc., 1967. Author: Flora Vitiensis Nova: a New Flora of Fiji, Vol. I, 1979, Vol. II, 1981, Vol. III, 1985, Vol. IV, 1988, Vol. V, 1991; also tech. articles; Editor: Brittonia, 1935-40, Jour. Arnold Arboretum, 1941-48, Sargentia, 1942-48, Allertonia, 1977-88; editorial com.: International Code Botanical Nomenclature, 1954-64. Recipient Robert Allerton award for excellence in tropical botany, 1979, Asa Gray award Am. Soc. Plant Taxonomists, 1992; Bishop Mus. fellow Yale U., 1933-34, Guggenheim fellow, 1946-47. Fellow Am. Acad. Arts and Scis., Linnean Soc. London; mem. NAS, Bot. Soc. Am. (Merit award 1970), Assn. Tropical Biology (pres. 1967-68), Internat. Assn. Plant Taxonomy (v.p. 1959-64), Fiji Soc. (hon.). Club: Washington Biologists' Field (pres. 1962-64). Home: 2232 Halekoa Dr Honolulu HI 96821

SMITH, ALEXANDER GOUDY, physics and astronomy educator; b. Clarksburg, W.Va., Aug. 12, 1919; s. Edgell Ohr and Helen (Reitz) S.; m. Mary Elizabeth Ellsworth, Apr. 19, 1942; children: Alexander G. III, Sally Jean. B.S., Mass. Inst. Tech., 1943; Ph.D., Duke U., 1949. Physicist Mass. Inst. Tech., Radiation Lab., Cambridge, 1943-46; research asst. Duke U., Durham, 1946-48; asst. prof. to prof. physics U. Fla., Gainesville, 1948-61; asst. dean grad. sch. U. Fla., 1961-69, acting dean grad. sch., 1971-73, chmn. dept. astronomy, 1962-71, prof. physics and astronomy, 1956—, Disting. prof., 1981—; dir. U. Fla. Radio Obs., 1956-85; cons. USN, USAF. Author: (with others) Microwave Magnetrons, 1958, (with T.D. Carr) Radio Exploration of the Planetary System, 1964 (also Swedish, Spanish and Polish edits), Radio Exploration of the Sun, 1966; also numerous articles in field. Fellow AAAS, Optical Soc. Am., Am. Phys. Soc., Royal Micros. Soc.; mem. Am. Astron. Soc. (editor Photo-Bull. 1975-87), Astron. Soc. Pacific, Internat. Astron. Union, Internat. Sci. Radio Union, Fla. Acad. Scis. (treas. 1957-62, pres. 1963-64, medal 1965), Assn. Univs. for Rsch. in Astronomy (dir., cons.), S.E. Univs. Rsch. Assn. (trustee 1981-91), Soc. Photog. Scientists and Engrs., Sigma Xi (nat. lectr. 1968, past pres. Fla. chpt.), Phi Kappa Phi, Sigma Pi Sigma. Republican. Christian Scientist. Clubs: Athenaeum (past pres.), Woodside Racquet, Gainesville Golf and Country. Home: 1417 NW 17th St Gainesville FL 32605-4014 Office: U Fla Dept Astronomy 211 Space Scis Bldg Gainesville FL 32611

SMITH, AMOS BRITTAIN, III, chemist, educator; b. Lewisburg, Pa., Aug. 26, 1944; s. Amos Brittain and Mildred (Cornelius) S.; m. Janet L. Duyckinck; children: Amos Matthew MacMillan, Kathryn Schuyler. BS, MS, Bucknell U., 1966; PhD, Rockefeller U., 1972; MA (hon.), U. Pa., 1978. Rsch. assoc. Rockefeller U., N.Y.C., 1972-73; asst. prof. U. Pa., Phila., 1973-78, assoc. prof., 1978-81, prof., 1981—, chmn. dept. chemistry, 1988—, Rhodes-Thompson prof. chemistry, 1990—; vis. dir. Kitasato Inst., Tokyo, 1990—; ad hoc mem. medicinal chemistry A and bioorganic and natural products chemistry rev. panel, NIH, 1980-83, mem. medicinal chemistry study sect., 1983-87, chmn. study sect. workshop on medicinal chemistry, 1986, mem. study sect. for AIDS and related rsch., 1989; vis. prof. Columbia U., N.Y.C., 1980, Cambridge (Eng.) U., 1982; Arthur D. Little lectr. Northeastern U., 1986, H. Martin Friedman lectr. Rutgers U., 1988, Merck-Frosst lectr., Ottawa, Can., 1988, U. Sherbrooke, Québec, Can., 1993; Phillips lectr. Haverford Coll, 1992; Soc. Francaise de Chimie lectr., Paris, 1988, lectr. at French-speaking univs., Switzerland, 1989. Mem. editorial bd. Jour. Organic Chemistry, 1982-86, Jour. Am. Chem. Soc., 1988—, Organic Reactions, 1987—, Organic Synthese, 1990—, Jour. Chem. Soc., Perkin I, 1992—, Fullerene Sci. and Tech., 1993—; contbr. numerous articles to profl. jours., chpts. to books., revs., etc. Recipient Camille and Henry Dreyfus Tchr.-Scholar award, 1978-83, Career Devel. award NIH, 1980-85, Arthur C. Scholar award, 1991, Alexander von Humboldt Rsch. award for sr. U.S. scientists, 1992, Ernest Guenther award in the chemistry of essential oils, 1993; J.S. Guggenheim Found. fellow, 1985-86, Japan Soc. for Promotion Sci. fellow, 1986-87. Mem. Am. Chem. Soc. (chmn. nominating com. organic chemistry div. 1985, exec. com. 1987-90, chmn. fellowship com. 1988-89, symposium exec. officer Nat. Organic Symposium 1993, canvasing com. 1989—, Phila. sect. award 1986, Ernest Guenther award 1993), Internat. Soc. Chem. Ecology (counselor), Gordon Rsch. Conf. (vice chmn. organic reaction and processes 1979, chmn. 1980), Franklin Inst. (com. on scis. and arts 1985—), Phila. Organic Chemists Club (chmn. 1978, award 1991). Avocations: hunting, fishing. Office: U Pa Dept Chemistry Philadelphia PA 19104

SMITH, ARTHUR JOHN STEWART, physicist, educator; b. Victoria, B.C., Can., June 28, 1938; s. James Stewart and Lillian May (Geernaert) S.; m. Norma Ruth Askeland, May 20, 1966; children: Peter James, Ian Alexander. B.A., U. B.C., 1959, M.Sc., 1961; Ph.D., Princeton U., 1966. Postdoctoral fellow Deutsches Electronen-Synchrotron, Hamburg, W. Germany, 1966-67; mem. faculty dept. physics Princeton U., 1967—, prof., 1978—, Class of 1909 prof., 1992—, assoc. chmn. dept., 1979-83, chmn. dept. physics, 1990—; vis. scientist Brookhaven Nat. Lab., 1967—, Fermilab, 1974—, Superconducting Supercollider Lab., 1990—. Assoc. editor Phys. Rev. Letters, 1986-89; contbr. articles to profl. jours. Fellow Am. Phys. Soc. (vice-chmn. div. of particles and fields 1990, chmn. div. of particles and fields 1991). Achievements include research on experimental high-energy particle physics; kaon decays and quark structure of hadrons. Home: 4 Ober Rd Princeton NJ 08540-4918 Office: PO Box 708 Princeton NJ 08544-0708

SMITH, AUGUSTINE JOSEPH, physicist, educator; b. Serabu, Sierra Leone, Dec. 3, 1949; came to U.S., 1991; s. Karimu Sundufu and Hawa Manu (Yagbaji) Smith-Bonai; m. Saffiatu Saadatu, Sept. 27, 1979 (div. 1989); children: Boakai James, Taiyou Augustine; m. Miatta Nyambe Labor, May 7, 1990; children: Salu Christoph, Kehma John. MS, Oreg. State U., 1973, PhD, 1979. Lectr. U. Sierra Leone, Njala, 1979-83, sr. lectr., head dept. physics, 1983-91; asst. prof. physics Lock Haven (Pa.) U., 1991—; assoc. Internat. Ctr. Theoretical Physics, Trieste, Italy, 1989. Contbr. articles to profl. jours. UN bursary Internat. Atomic Energy Agy., Vienna, 1972; UN univ. fellow, U. Malaysia, 1985; Alexander von Humboldt Found. fellow U. Duesseldorf, Fed. Republic Germany, 1989. Mem. Am. Phys. Soc., Am. Assn. Physics Tchrs., Nat. Assn. Black Physicists. Achievements include design of experiment to measure current, resistance and inductance of a nitrogen laser channel, measurement of the K-beta spectrum of helium-like iron Fe XXV from TFTR plasmas; detemination of cross sections for collisions of High Rydberg atoms with neutral atoms. Office: Lock Haven U Dept Chem/Physics/Geoscis Lock Haven PA 17745

SMITH, BODRELL JOER'DAN, architect, city planner; b. Little Rock, Ark., Sept. 21, 1931; s. Robert Stanslaw and Neva (Long) S.; m. Ingrid Elin Kehlet, Oct. 20, 1979; children: Cameron Corbin, Astrid Johannah, Walker Darel. BArch, U. So. Calif., 1956; M in City Planning, U. Paris, France, 1957. Registered architect, Calif. and 17 other states. Designer LeCorbusier, Paris, 1956-57; chief designer W.L. Pereira Assocs., Los Angeles, 1960-61; prin. Bodrell Joer'dan Smith Architects, Los Angeles, 1961-78; pres., chief exec. officer USTEC Architects, Redwood City, Calif., 1978-87; pres. Bodrell Joer'dan Smith Partnership, Mountain View, San Mateo and L.A., Calif., 1987—; bd. dirs. Native Am. Design Collaborative. Bldgs. featured in Time, Fortune, House and Home, Progressive Architecture and other mags. Trustee Buckley Sch., Encino, Calif., 1965-70. Served to maj. C.E., U.S. Army, 1958-60. Recipient Fulbright Scholarship Dept. of State, Paris, 1956. Mem. AIA (nat., regional and urban design com.), Am. Soc. Mil. Engrs., Am. Soc. Planning Ofcls., Nat. Orgn. Minority Architects, Native Am. Design Collaborative(bd. dirs.). Republican. Congregationalist. Avocations: skiing, sky diving, backpacking.

SMITH, BRIAN RICHARD, hematologist, oncologist, immunologist; b. Glen Cove, N.Y., May 7, 1952; s. Frank C. and Gloria R. S.; A.B. in Chemistry summa cum laude, Princeton U., 1972; M.D., Harvard U., 1976. Clin. fellow in medicine Harvard U., also med. house officer Peter Bent Brigham Hosp., Boston, 1976-77, asst. resident physician, 1977-78, clin. fellow in hematology-oncology, 1978-81; instr. medicine Harvard Med. Sch. and asso. Brigham and Women's Hosp., Children's Hosp., Dana-Farber Cancer Inst., Boston, 1981-88; asst. prof. medicine Harvard Med. Sch., 1985-88; assoc. prof. medicine, lab. medicine and pediatrics Sch. of Medicine Yale U., New Haven, 1988—; dir. rsch. Yale Bone Marrow Transplant Unit, dir. clin. immunohematology; DeCamp lectr. biomed. ethics, 1992—. Recipient George A. Howe prize Princeton U., 1976; Am. Cancer Soc. fellow, 1981-84; Leukemia Soc. fellow, 1982-88, scholar Leukemia Soc. Am., 1989—. Diplomate Am. Bd. Internal Medicine. Fellow ACP; mem. AAAS, NIH (recombinant DNA adv. com., 1992—), N.Y. Acad. Sci., Phi Beta Kappa, Sigma Xi, Alpha Omega Alpha. Roman Catholic. Office: Yale U Sch Medicine 333 Cedar St New Haven CT 06510-3289

SMITH, C. D., civil engineering educator. Design engr. PFRA; cons. hydraulic engr., prof. civil engring. U. Sask., Saskatoon, Can. Author: two books; contbr. to over 70 profl. jours. Recipient T.C. Keefer medal Can. Soc. Civil Engring., 1991. Mem. Can. Soc. Civil Engring. (pres.). Office: Univ of Saskatchewan, Dept of Civil Engiring, Saskatoon, SK Canada S7N 0W0*

SMITH, CHARLES HAYDEN, utilities executive; b. Danville, Ill., Oct. 25, 1933; s. W. Hayden and Minnie M. (Smith) S.; m. Alice Joan Rhodes, Sept. 11, 1955 (div. June 1974); children: Catherine Lynn, Timothy Hayden, Letitia Joan; m. Thelma Lynette East, June 7, 1975. BSCE, U. Ill., 1955, MS in Sanitary Engring., 1963. Profl. engr. Ill., Mo.; water treatment plant operator Ill., Mo.; waste plant operator, Ill.; diplomate Am. Acad. Environ. Engrs. Resident engr. Ill. Div. Hwy., Paris, 1955-56; project engr. Clark, Daily, Dietz Engr., Urbana, Ill., 1958-62; prodn. mgr. No. Ill. Water Corp., Champaign, Ill., 1962-74; v.p. prdon., distbn. St. Louis County Water Co., St. Louis, 1974-79; v.p. ops. Kankakee (Ill.) Water Co., 1979-84; pres. Consumers Ill. Water Co., Kankakee, 1984—, Inter-State Water Co., Danville, Ill., 1986—; bd. dirs., chmn. Riverside Med. Ctr., Kankakee; bd. dirs. 1st of Am. Bank, Kankakee. Bd. dirs., chmn. Kankakee Indsl. Devel. Assn., 1989—; mem. Kankakee River Basin Commn., 1987—. Lt. col. USAF Res., 1956-58. Mem. ASCE (pres. east br. Ill. sect. 1983-84), NSPE, Am. Acad. Environ. Engrs., Am. Water Works Assn. (chmn. 1988-89 Ill. sect. 1963—, bd. dirs. 1990—), Nat. Assn Water Cos. (bd. dirs. 1984-90, Kiwanis (v.p. 1993-94), Kankakee Country Club, U. Ill. Alumni Assn., Ill. C. of C. (bd. dirs. 1987—, chmn. polit. action com. 1989—). Republican. Mormon. Avocations: hunting, golf, flying, fishing. Office: Consumers Ill Water Co 1000 S Schuyler PO Box 152 Kankakee IL 60901-0152

SMITH, CHARLES ROBERT, conservationist, naturalist, ornithologist, educator; b. Johnson City, Tenn., June 30, 1948; s. Charles Oliver and Ruby (Maston) S. BS, East Tenn. State U., 1970; PhD, Cornell U., 1977. Lectr. dept. natural resources Cornell U., Ithaca, N.Y., 1976-77; dir. edn. and info. svcs. Cornell Lab. Ornithology, Ithaca, 1977-87, acting exec. dir., 1979-81, sr. extension assoc., dir. spl. projects, 1987-92; adj. asst. prof. wildlife sci. dept. natural resources Cornell U., 1979-88, adj. assoc. prof. wildlife sci., 1988-92, sr. rsch. assoc. dept. natural resources, 1992—; adviser nongame and endangered species mgmt. Commr. of N.Y. Dept. Environ. Conservation, 1984—, commr. N.Y. Office of Parks, Recreation and Historic Preservation, 1987—; mem. Freshwater Wetlands Adv. Com., 1988—, Adv. Com. for No. Montezuma Wetlands Complex, N.Am. Waterfowl Mgmt. Plan, 1989—, Cornell Plantations Adv. Bd., 1985—; statewide coord. N.Y. State Breeding Bird Surveys, 1986—; assoc. chmn. N.Y. State Breeding Bird Atlas Steering Com., 1979-88; del. Internat. Coun. Bird Preservation. Tech. editor The Living Bird Quar., 1981-92; mem. editorial adv. com. Atlas of Breeding Birds in New York State, 1985-88; contbr. articles to profl. and refereed publs. Grantee N.Y. State Dept. Environ. Conservation, 1983-84, 84-85, 85-86, 86-87, 87-88, 88-89, USDA, 1988-91, 91-92, Nat. Fish and Wildlife Found., 1988-89, 92-93, U.S. Forest Svc., 1989-90, 90-91, U.S. Fish and Wildlife Svc., 1991-94. Mem. Am. Ornithologists' Union, Cooper Ornithol. Soc., Ecol. Soc. Am., Fedn. N.Y. State Bird Clubs, Internat. Assn. Landscape Ecology, Soc. Conservation Biology (charter), Tenn. Ornithol. Soc. (life), Wildlife Soc., Wilson Ornithol. Soc. (life). Office: Cornell U Dept Natural Resources Fernow Hall Ithaca NY 14853-3001

SMITH, CHARLES WILLIAM, mechanical engineer, educator; b. Christiansburg, Va., Jan. 1, 1926; m. Doris Burton, Sept. 9, 1950; children: Terry Smith Kelley, David. BS in Civil Engring., Va. Polytech Inst., 1947, MS in Applied Mechanics, 1949. Registered profl. engr., Va. Teaching fellow Va. Poly. Inst., Blacksburg, 1947, instr., 1948, asst. prof., 1949-52; assoc. prof. Va. Poly. Inst. and State U., Blacksburg, 1953-57, prof., 1958-81, alumni disting. prof., 1982—; cons. and lectr. in field. Author: (with others) Experimental Techniques in Fracture Mechanics, vol. 2, 1973, Inelastic Behavior of Composite Materials, vol. 3, 1975, Mechanics Fracture, vol. 6, 1981, Handbook of Experimental Mechanics, 1979; guest editor Jour. for Optics and Lasers in Engring., 1989-90; regional editor Internat. Jour. Fracture Mechanics Tech., 1982—. Recipient NASA Exceptional Sci. Achievement award, 1986, Engring. Educator of the Yr. award State of Va., 1991. Fellow Am. Acad. Mechanics, Soc. Exptl. Mechanics (chmn. fracture com. 1978-80, exec. bd. 1988-90, chmn. all tech. divs. 1988, M. M. Frocht award 1983, Murray medal 1993); mem. ASME, ASTM (organizing com. 1978-84), Nat. Soc. Profl. Engrs. (local and state com. 1950-70), Soc. Exptl. Mechanics (chmn. com. of fellows 1991—), Internat. Assn. Structural Mechanics in Reactor Tech., Soc. Engring. Sci. (gen. chair ann. meeting 1984), Sigma Xi, Phi Kappa Phi, Tau Beta Pi, Omicron Delta Kappa, Chi Epsilon. Achievements include development of optical methods for evaluating stress intensity distributions in three dimensional cracked body problems, applied to nuclear reactors, rocket motors, and aircraft structures. Home: 107 College St Christiansburg VA 24073 Office: Va Poly Inst and State U Blacksburg VA 24061

SMITH, CHARLES WILLIAM, physicist; b. Balt., Nov. 18, 1955; s. Charles William and Kathleen Ruth (Mayer) S.; m. Sandra Louise MacWilliams, May 3, 1986. BS, U. Md., 1977; PhD, Coll. William and Mary, 1981. Rsch. scientist U. N.H., Durham, 1981-87; rsch. scientist Bartol Rsch. Inst., Newark, Del., 1987-91, sr. rsch. scientist, 1991—. Contbr. articles on space plasma physics, cosmic ray propagation, interplanetary medium and planetary shocks to sci. publs. Grantee NASA, NSF. Mem. Am. Phys. Soc., Am. Geophys. Union, Sigma Xi. Office: U Del Bartol Rsch Inst Newark DE 19716

SMITH, CHRISTIE PARKER, operations researcher; b. Ft. Campbell, Ky., Nov. 8, 1960; d. Herbert Gerald and Florida (Fisher) P.; m. Gerald Smith Jr., Aug. 31, 1991. BS, U. Fla., 1982; M in Health Sci., Johns Hopkins U., 1984. Ops. rsch. analyst USAITAC, Washington, 1986—. Mem. APHA, Ops. Rsch. Soc. Am., Inst. Mgmt. Sci., Mil. Ops. Rsch. Soc., N.Y. Acad. Scis. Democrat. Baptist.

SMITH, CLINTON W., civil engineer, consultant; b. Paducah, Tex., Dec. 9, 1952; s. Arnold T. and V. Pauline (Ramsom) S.; m. Mona B. Robinson, Apr. 10, 1971 (div. Jan. 1982); children: Matthew W., Christina N.; m. J. Renee Howell, Sept. 3, 1982; children: Michael Barrett, Latrece Barrett, Danny Barrett. BSCE, N.Mex. State U., 1974. Registered profl. engr., Tex., N.Mex., Okla., Kans., Colo. Structural engr. Southwestern Pub. Svc., Amarillo, Tex., 1974-79; supervisory structural engr. Southwestern Pub. Svc., Amarillo, 1979-82, sr. structural engr., 1982-87; project mgr. Utility Engring., Amarillo, 1987-92, project dir., 1992—; adv. bd. mem. West Tex. State U. Engring. Program, Canyon, 1992—. Mem. Tex. Soc. Profl. Engrs. (Young Engr. of Yr. 1987). Republican. Baptist. Home: 4511 Wild Dunes Ct Austin TX 78747 Office: Utility Engring 5601 I-40 W Amarillo TX 79106

SMITH, DANIEL MONTAGUE, engineer; b. Gainesville, Tex., Oct. 17, 1932; s. Alex Morton and Mary Louise (Shriver) S.; married, Oct. 15, 1953; children: Gregory M., Timothy D., Christopher E. BA, North Tex. State Coll., 1952, MS, 1953; PhD, U. Tex., 1959. Mem. tech. staff Oak Ridge (Tenn.) Nat. Lab., 1958-61, Tex. Instruments Inc., Dallas, 1961-75; sr. engr. Nitron div. McDonnell Douglas, Cupertino, Calif., 1975-77; mgr. product engring. Nat. Semiconductor Corp., Santa Clara, Calif., 1977-79, Motorola Semiconductor Sector, Austin, Tex., 1979—. Cpl. U.S. Army, 1953-55. Mem. Am. Phys. Soc. Home: 10301 Parkfield Dr Austin TX 78758

SMITH, DAVID, chemistry educator; b. Fall Rivers, Mass., Nov. 7, 1939; s. Jacob M. and Bertha (Horvitz) S.; m. Renee Gutfreund, Nov. 23, 1968; children: Aliza, Miriam, Shimon, Shmuel, Yanky, Leeba, Aharon, Bryna, Esther. BS, Providence Coll., 1961; PhD, MIT, 1965. Instr. Bklyn. Coll., 1965-68; prof. Pa. State U., Hazleton, 1969—. Contbr. articles to profl. jours. including Jour. of Chem. Physics. Mem. Am. Chem. Soc., Am. Phys. Soc. Jewish. Home: 68 2d Ave Kingston PA 18704 Office: Pa State U Hazleton PA 18201

SMITH, DAVID CARR, organic chemist; b. Ft. Wayne, Ind., Aug. 9, 1944; s. James Nolan and Kathryn Ellen (Mefford) S.; m. Dolores Joan Kurz, July 9, 1966; children: David James, Daniel Paul. BS in Chemistry, Clarkson Coll., 1969, PhD in Chemistry, 1975. Postdoctoral fellow Utah State U., Logan, 1974-75, U. S.C., Columbia, 1975-77; sr. rsch. chemist Ash Stevens Inc., Detroit, 1977-80; group leader Sterling Organics, Rensselaer, N.Y., 1980-84; mgr. bus. devel. Sterling Organics, N.Y.C., 1984-85; mgr. pharm. technology John Brown Inc., Stamford, Conn., 1985-93; sr. project mgr. Lockwood Greene Engring., Atlanta, 1993—. Contbr. articles to profl. jours. Mem. AAAS, Am. Chem. Soc., Licensing Exec. Soc., Internat. Soc. Pharm. Engrs., Regulatory Affairs Profls. Soc. (cert.), Sigma Xi. Office: Lockwood Greene Engring INFORUM Ste 4000 250 Williams St Atlanta GA 30303-1036

SMITH, DAVID DOYLE, engineer, consultant; b. Newport, Tenn., Aug. 17, 1956; s. Doyle E. and Lena Maude (Clemmons) S.; m. Debra Turnbull, Dec. 11, 1982 (div. May 1990); 1 child, Adam Gabriel; m. Judith Ann Craig, Nov. 1, 1991; stepchildren: Christine, James. BSEE, U. Tenn., 1981. Registered profl. engr., Tenn., Ga. Engring. apprentice E.I. DuPont, Brevard, N.C., 1976; field engr. IBM Corp., Knoxville, Tenn., 1977-79; rsch. asst. Office of Naval Rsch. U. Tenn., Knoxville, 1980-81; systems test engr. Tex. Instruments, Inc., Johnson City, Tenn., 1981-82, product engr., 1982-83, product mgr., 1983-87; missile design engr., supr. Tex. Instruments, Inc., Lewisville, Tex., 1987-89; control systems engr. U.S. Data Corp., Richardson, Tex., 1989-90; cons. AGS Info. Systems, Inc., Atlanta, 1991-93; sr. engr. Harris DTS, Atlanta, 1993—; presenter, mem. Computer Integrated Mfg. Spl. Interest Group, Atlanta, 1991. Co-author profl. papers. Mem. IEEE (pres. student chpt. 1980-81, asst. sec. 1991—, vol. Atlanta olympic games project 1996), Plasma Sci. Soc. of IEEE, Tenn. Gas. Assn., Etta Kappa Nu. Avocation: archaeology. Home: 1080 Allenbrook Ln Roswell GA 30075-2983 Office: Harris Corp DTS Divsn 3772 Pleasantdale Rd Ste 165 Atlanta GA 30340

SMITH, DAVID FLOYD, chemical engineer; b. Memphis, Tenn., May 18, 1956; s. James F. and Evelyn (Pelham) S.; m. Ginger Marie Chambliss, May 27, 1977; 1 child, Brandon N. BSChE, U. South Ala., 1983. With R&D dept. Union Carbide Corp., Chickasaw, Ala., 1977-84; prodn. engr. Kay Fries Chems., Theodore, Ala., 1984-87; process engr., prodn. engr. Dynamit Nobel Chems., Theodore, 1987-90; process controls mgr. Huls Am., Theodore, 1990-93, plant engring. mgr., 1993—. Mem. Instrumentation Soc. Am. Office: Huls Am PO Box 889 Theodore AL 36590

SMITH, DAVID MITCHELL, systems and software researcher, consultant; b. Malden, Mass., May 9, 1960; s. Lester and Claire Smith; m. Debra A. Love. BA, Clark Univ., 1982; MS, Boston Univ., 1985. Engr. Digital Equipment, Littleton, Mass., 1984-87; cons. Digital Equipment, Nashua, N.H., 1987-91; dir. Internat. Date Corp., Framingham, Mass., 1991—. Mem. Phi Beta Kappa.

SMITH, DAVID WELTON, mechanical engineer, director research; b. Detroit, Mar. 27, 1942; s. Welton John and Miriam Elizabeth (Anthony) S.; m. Mary Ellen Alexander, Aug. 29, 1964; children: David Livingston, Drew Alexander. BSME, Gen. Motors Inst., 1965; MSME, Stanford U., 1969. Registered profl. engr. Iowa, Mich. Mech. engr. Saginaw (Mich.) Steering Gear, 1960-70; project mgr. KLC Enterprises, Frankenmuth, Mich., 1970-72; dir. engring. rsch. and devel. Townsend Engring., Des Moines, 1972—. Author: Electrochemical Machining, 1965. Dir. ptnr. with youth YMCA, West Des Moines, 1973-76, sec. bd. dirs., 1975; active Christmas Spree for Underprivilaged Children, West Des Moines. Republican. Lutheran. Achievements include patents on method for stuffing sausage products, method & means for encasing sausage or the like, a machine to seperate sausage links, a hand held meat skinning device. Home: 4307 Aspen Dr West Des Moines IA 50265 Office: Townsend Engring Co 2425 Hubbell Ave Des Moines IA 50305

SMITH, DEANE KINGSLEY, JR., mineralogy educator; b. Berkeley, Calif., Nov. 8, 1930; s. Deane Kingsley and Anna Virginia (Long) S.; m. Patricia Ann Lawrence, July 24, 1956; children: Paula Lynn, Jeanette Diane, Kingsley Lawrence, Dana Elroy, Sharon Rene. BS, Calif. Inst. Tech., 1952; PhD, U. Minn., 1956. Rsch. asst. U. Minn., Mpls., 1952-56; chemist Portland Cement Assn. Fellowship, Washington, 1956-60, Lawrence Livermore (Calif.) Lab., 1960-68; prof. mineralogy Pa. State U., State College, 1968—; mem., chmn. bd. JCPDS Internat. Ctr. Diffraction Data, Newtown Square, Pa., 1965—. Editor Powder Diffraction, 1985—; contbr. over 100 articles to profl. jours. Recipient C.S. Barrett award Denver Analytical X-Ray Conf., 1991. Fellow Am. Mineralogical Soc. (sec. 1975-78), Geol. Soc. Am.; mem. Am. Crystallographic Soc. (sec. 1975-78). Home: 1652 Princeton Dr State College PA 16803 Office: Pa State U 239 Deike Bldg University Park PA 16802

SMITH, DONALD ARTHUR, mechanical engineer, researcher; b. Hartford, Conn., Apr. 9, 1945; s. Winfred Arthur and Marguerite Elisabeth (Johnson) S.; m. Marianne Carol Taverna, June 17, 1967; 1 child, Adam James. BS in Mech. Engring., U. Hartford, 1968. Rsch. engr. Combustion Engring. Inc., Windsor, Conn., 1968-71; supr. fluid rsch. Combustion Engring. Inc., Windsor, 1971-77, mgr. combustion rsch., 1977-84; dir. R&D Hartford Steam Boiler Inspection & Ins. Co., 1984—; co. rep. Indsl. Rsch. Inst., Washington, 1989—; treas. Am. Flame Rsch. Com., 1983—. Tech. editor: HSB Locomotive, 1990-91. Haddam (Conn.) Planning and Zoning Commn., 1991-95; pres. Sherwood Camp Assn., Haddam, 1971-81. Named Engr. Yr. ASME (Hartford sect.), 1989. Mem. U. Hartford Alumni Bd., Lions. Republican. Roman Catholic. Achievements include patents in spray atomizers, burners, ignitors and flame scanning systems for indsl. application. Established Combustion Engring.'s fluid mechanics and combustion rsch. facilities, Hartford Steam Boiler's corp. R&D program. Home: PO Box 95-42 Smith Hill Rd Haddam CT 06438 Office: Hartford Steam Boiler One State St Hartford CT 06102

SMITH, DONALD DEAN, biologist, educator; b. Independence, Mo., Aug. 31, 1946; s. William Richard Jr. and NeLeta Madge (Thompson) S.; m. Phyllis Ann Bredehoft, Feb. 9, 1991. BS, U. Mo., 1968, MS, 1974. Quality control technician Trainin Egg Products, Kansas City, Mo., 1966; rsch. asst. dept. biology U. Mo., Kansas City, 1966; lab. technician Midwest Rsch. Inst., Kansas City, 1966-68; rsch. assoc. lab. nuclear medicine UCLA, 1970-73; rsch. assoc. dept. pathology U. Kans. Med. Ctr., Kansas City, 1974-88, rsch. assoc. dept. medicine, 1988—; adj. faculty Avila Coll., 1978-88, instr. dept. natural sci., 1988—; cons. Mid-Am. Poison Control Ctr., Kansas City, Kans., 1982—. Contbr. articles, abstracts, reports to sci. publs. Judge Kansas City (Mo.) Sci. and Engring. Fair, 1986—. Trainee NSF, 1968-69. Mem. Soc. for Study of Amphibians and Reptiles, Mo. Herpetological Assn., Lake Lotawana Sportsmen's Club (sec. 1993), Sigma Xi. Achievements include patents for processes to prevent shedding of Toxoplasma gondii oocysts by cats and to immunize cats against such shedding. Home: 501 N 39th St Blue Springs MO 64015 Office: U Kans Med Ctr Divsn Allergy/Rheumatology 3901 Rainbow Blvd Kansas City KS 66160

SMITH, DONALD NORBERT, engineering executive; b. Ft. Wayne, Ind., June 12, 1931. BS in Indsl. Mgmt., Ind. U., 1953; Diploma Grad. Sch. Bus., U. N.C., 1960. Asst. mgr. Ann Arbor (Mich.) Rsch. Labs/Burroughs Corp., 1961-64; dir. Indsl. Devel. Divsn. U. Mich., Ann Arbor, 1964-93; assoc. dir.-mfg. systems rsch. Office for Study Auto Transp. U. Mich. Transp. Rsch. Inst., Ann Arbor, 1993—. Co-author: Management Standards for Computers and Numerical Controls, 1977; contbr. articles to profl. jours. Recipient Peace award, Israel, 1967, Mfg. Tech. award ASTME, 1968, Engring. Merit award, San Fernando Valley Engrs. Coun., 1971, Man of Yr. award Great Lakes Chpt. Numerical Control Soc., 1975, Archimedes Engring. award Calif. Soc. Profl. Engrs., Disting. CAD/CAM Achievements award L.A. Coun. Engrs., 1982, Achievement award for promoting Swedish/Am. trade, Kingdom of Sweden, 1983, Tech. Transfer award NASA/Rockwell Internat., 1984. Fellow Soc. Mfg. Engrs. (Indsl. Tech. Mgmt. award 1978, internat. awards com. 1986, 88, 89, Joseph A. Siegel Internat. Svc. award 1988, Pres.'s award Robotics Internat. 1985). Roman Catholic. Avocation: boating. Office: Office Study Automotive Transp 2901 Baxter Rd Ann Arbor MI 48109-2150

SMITH, EDWARD JOHN, geophysicist, physicist; b. Dravosburg, Pa., Sept. 21, 1927; married, 1953; 4 children. BA, UCLA, 1951, MS, 1952, PhD in Physics, 1960. Rsch. geophysicist Inst. Geophysics UCLA, 1955-59; mem. tech. staff Space Tech. Labs., 1959-61, Jet Propulsion Lab, 1961—. Recipient medal Exceptional Sci. Achievement NASA. Mem. AAAS, Internat. Sci. Radio Union, Am. Geophys. Union, Am. Astron. Soc., Sigma Xi. Achievements include research in planetary magnetism, space physics, interplanetary physics, wave-particle interactions in plasmas, propagation of electromagnetic waves, solar-terrestrial relations. Office: California Institute Technology Jet Propulsion Lab Pasadena CA 91109*

SMITH, EDWIN DAVID, electrical engineer; b. Cape Town, South Africa, Feb. 17, 1938; s. Hermanus Wilhelm and Dorothea Louise (Johl) S.; m. Stella van Nierop, Nov. 25, 1961; children: Hermine Mariana, Paul Albert. BSc, U. Cape Town, 1959, PhD, 1961. Registered profl. engr., South Africa. Rschr., cons. Coun. Sci. and Industry Rsch., Pretoria, South Africa, 1989—; cons. U. Pretoria, 1989—. Fellow South Africa Inst. Elec. Engrs. (premium award 1975), South Africa Acoustics Inst. (pres. 1986-91). Achievements include research leading to construction of the first economically viable electric shark barrier for the protection of bathers at beaches. Home: 473 Kings Hwy, Lynnwood Transvaal 0081, South Africa Office: Shark Technologies Ltd, PO Box 11325, Brooklyn Transvaal 0011, South Africa

SMITH, ERNEST KETCHAM, electrical engineer; b. Peking, China, May 31, 1922; (parents Am. citizens); s. Ernest Ketcham and Grace (Goodrich) S.; m. Mary Louise Standish, June 23, 1950; children: Priscilla Varland, Nancy Smith, Cynthia Jackson. BA in Physics, Swarthmore Coll., 1944; MSEE, Cornell U., 1951, Ph.D., 1956. Chief plans and allocations engr. Mut. Broadcasting System, 1946-49; with Nat. Bur. Standards, 1951-65; chief ionosphere research sect. Nat. Bur. Standards, Boulder, Colo., 1957-60; div. chief Nat. Bur. Standards, 1960-65; dir. aeronomy lab. Environ. Sci. Services Adminstrn., Boulder, 1965-67; dir. Inst. Telecommunication Scis., 1968, dir. univ. relations, 1968-70; assoc. dir. Inst. Telecommunications Scis. Office of Telecommunications, Boulder, 1970-72, cons., 1972-76; mem. tech. staff Jet Propulsion Lab. Calif. Inst. Tech., Pasadena, 1976-87; adj. prof. dept. Elec. and Computer Engring. U. Colo., Boulder, 1987—; vis. fellow Coop. Inst. Research on Environ. Scis., 1968; assoc. Harvard U. Coll. Obs., 1965-75; adj. prof. U. Colo., 1969-78, 87—,; internat. vice-chmn. study group 6, Internat. Radio Consultative Com., 1958-70, chmn. U.S. study group, 1970-76; mem. U.S. commn. Internat. Sci. Radio Union, mem.-at-large U.S. nat. com., 1985-88. Author: Worldwide Occurrence of Sporadic E, 1957; (with S. Matsushita) Ionospheric Sporadic E, 1962. Contbr. numerous articles to profl. jours. Editor: Electromagnetic Probing of the Upper Atmosphere, 1969; assoc. editor for propagation IEEE Antennas and Propagation Mag., 1989—. Served with U.S. Army, 1944-45. Recipient Diploma d'honneur, Internat. Radio Consultative Com., Internat. Telecommunications Union, 1978. Fellow IEEE (fellow com. 1993), AAAS; mem. Am. Geophys. Union, Electromagnetics Acad., Svc. Club, Kiwanis. Mem. First Congregational Ch. Clubs: Harvard Faculty; University (Boulder); Athenaeum (Pasadena); Boulder Country. Home: 5159 Idylwild Trl Boulder CO 80301-3618 Office: U Colo Dept Elec and Computer Engring Campus Box 425 Boulder CO 80309

SMITH, ESTHER THOMAS, editor; b. Jesup, Ga., Mar. 13, 1939; d. Joseph H. and Leslie (McCarthy) Thomas; m. James D. Smith, June 2, 1962; children: Leslie, Amy, James Thomas. BA, Agnes Scott Coll., 1962. Staff writer, Sunday women's editor Atlanta Jour.-Constn., 1961-62; mng. editor Bull. of U. Miami Sch. Medicine, 1965-66; corr. Atlanta Jour.-Constn. and Fla. Times-Union, 1964, 67-68; founding editor Bus. Rev. of Washington, 1978-81; founding editor, gen. mgr. Washington Bus. Jour., 1982; pres. Tech News, Inc., 1986—; editor Washington Tech., 1986-93; editor Tech. Transfer Bus. Mag., 1992—; bd. dirs. MIT Enterprise Forum of Washington/Balt., 1981-82; mem. Women's Forum, Washington, 1981—, adv. bd. Va. Math Coalition, 1991—; bd. dirs. Ctr. for Excellence in Edn., 1993—. Mem. Assn. Tech. Bus. Couns. (chmn. bd. advisors 1989—), Pres.'s Forum, Mid-Atlantic Venture Assn., No. Va. Tech. Coun., Suburban Maryland High Tech. Coun. Office: 1953 Gallows Rd Ste 130 Vienna VA 22182-3932

SMITH, EUAL RANDALL, mechanical engineer; b. Chattanooga, Jan. 12, 1947; s. Eual Clyde and Velma Drucilla (Brown) S.; m. Cynthia Elaine Cooper, May 31, 1969; children: Christopher Randall, Traci Elaine. BSME, U. Tenn., 1973. Cert. engr.-in-tng., Tenn. Design technician Lockheed Ga. Co., Marietta, 1968-70; engring. assoc. TVA Engring. Design, Knoxville, Tenn., 1970-73, mech. engr., 1973-77; mech. engr. TVA Power Prodn., Chattanooga, 1977-81, TVA Ops. Support, Chattanooga, 1981-87, TVA Nuclear Engring., Soddy Daisy, Tenn., 1987—. Mem. recreation com. Dallas Bay Elem. PTO, Hixson, Tenn., 1982-83. Named Disting. Citizen, City of Chattanooga, 1983. Mem. Soc. Fire Protection Engrs., TVA Engrs. Assn. Republican. Baptist. Home: 5235 Cassandra Smith Rd Hixson TN 37343 Office: TVA Sequoyah Nuclear Plant Sequoyah Access Rd PO Box 2000 DSDIA-SQN Soddy Daisy TN 37379

SMITH, GARY CHESTER, meat scientist, researcher; b. Ft. Cobb, Okla., Oct. 25, 1938; s. William Chester and Aneta Laura (Lisk) S.; m. Carol Ann Jackson (div. 1965); children: Todd, Toni; m. Kay Joy Camp, Feb. 12, 1965; children: Leaneta, Stephanie, Kristl, Leland. BS, Calif. State U., Fresno, 1960; PhD, Tex. A&M U., 1968. Asst. prof. dept. animal sci. Wash. State U., Pullman, 1960-69, from assoc. prof. to prof. dept. animal sci. Tex. A&M U., College Station, 1969-82, head dept. animal sci., 1982-90; prof. Nat. Meat Inspection Tng. Ctr., College Station, 1987-90; Monfort Endowed prof. dept. animal sci. Colo. State U., Ft. Collins, 1990—, univ. disting. prof., 1993—; chmn. irradiation com. NAS, 1977-79, mem. packaging com. Office Tech. Assessment, 1973-74. Author: Laboratory Exercises in Meat Science; contbr. over 280 articles to Jour. Animal Sci., Jour. Food Sci., Meat Sci. Bd. dirs. Internat. Stockmen's Edn. Found., Houston, 1983-91. Recipient Disting. Svc. award Nat. Livestock Grading & Mktg. Assn. Mem. Am. Meat Sci. Assn. (pres. 1976-77, Disting. Rsch. award 1982, Disting. Teaching award 1984), Am. Soc. Animal Sci. (Meat Rsch. award 1974, Disting. Teaching award 1980), Inst. Food Technologists, Coun. Agrl. Sci. & Tech. Republican. Baptist. Achievements include development of USDA grading standards for beef, pork, lamb carcasses; process for electrical-stimulation tenderization of beef carcasses; discovery of relationship between fat-cover insulation and tenderness of beef and lamb; research on first shipments of chilled beef to Europe and Japan. Home: 1102 Seton St Fort Collins CO 80525-9498 Office: Colo State U Dept Animal Scis Fort Collins CO 80523

SMITH, GARY LEE, chemical engineer; b. Roanoke, Va., Nov. 9, 1959; s. John Thomas Jr. and Anna Mae (Sult) S. AS, Va. Western Community Coll., 1980, AS in Edn., 1980; BS in Chemistry & BA in Philosophy, Roanoke Coll., 1982; AS in Bus. Adminstrn., Va. Western Community Coll., 1985. Environ. lab. technician Centec Corp. Analytical Svcs. Div., Salem, Va., 1979-81; asst. med. lab. tech. Community Hosp., Roanoke, Va., 1982-83; environ. chemist Centec Corp. Analytical Svcs. Div., 1985; process ops. engr. IT&T Electro Optical Products Div., Roanoke, 1985-87; environ. chemist Environ. Options Inc., Rocky Mount, Va., 1987-88; sr. rsch. & devel. engr. Ni-Tec Optic Elec. Corp., Garland, Tex., 1988-91; sr. process engr. Silicon Materials Svc./Air Products, Garland, Tex., 1991—; rsch. & devel. engr. Nitec-Optic Elecs., 1991-92. Trauma med. tech. Roanoke Life Saving Crew, Inc., 1973-79; woodland search & rescue Civil Air Patrol, Roanoke, 1971-73. With U.S. Army Med. Corps., 1982-84. Recipient Freshman Chemistry & Physics award Chem. Rubber Co., 1980. Mem. IEEE, AM. Geophys. Union (life mem.), Am. Chem. Soc., NRA, Am. Vacuum Soc. Home and office: 5818 Brahma Rd SW Roanoke VA 24018-9999

SMITH, HAMILTON OTHANEL, molecular biologist, educator; b. N.Y.C., N.Y., Aug. 23, 1931; s. Bunnie Othanel and Tommie Harkey S.; m. Elizabeth Anne Bolton, May 25, 1957; children: Joel, Barry, Dirk, Bryan, Kirsten. Student, U. Ill., 1948-50; A.B. in Math, U. Calif., Berkeley, 1952; M.D., Johns Hopkins U., 1956. Intern Barnes Hosp., St. Louis, 1956-57; resident in medicine Henry Ford Hosp., Detroit, 1959-62; USPHS fellow dept. human genetics U. Mich., Ann Arbor, 1962-64; rsch. assoc. U. Mich., 1964-67; asst. prof. molecular biology and genetics Sch. Medicine Johns Hopkins U., Balt., 1967-69; assoc. prof. Johns Hopkins U., 1969-73, prof., 1973—; asso. Inst. für Molekularbologie der U. Zurich, Switzerland, 1975-76; assoc. Rsch. Inst. Molecular Pathology, Vienna, 1990-91. Contbr. articles to profl. jours. Served to lt. M.C. USNR, 1957-59. Recipient Nobel Prize in medicine, 1978; Guggenheim fellow, 1975-76. Mem. Am. Soc. Microbiology, AAAS, Am. Soc. Biol. Chemists, Nat. Acad. Sci. Office: Johns Hopkins U Sch Med Dept Molecular Biology/Gen 725 N Wolfe St Baltimore MD 21205-2105

SMITH, HARRY JAMES, mathematician; b. Shelbyville, Ind., Jan. 27, 1932; s. Joseph J. and Dessie A. (Fastlaben) S.; m. E. Joan Dale, June 9, 1952; children: Michael (dec.), Todd, Dana, Jeannine. BS, U. Dayton, 1954; MS, Purdue U., 1956. Rsch. asst. Statis. Lab. Purdue U., West Lafayette, Ind., 1954-56; sr. scientist Lockheed Missiles and Space Co., Sunnyvale, Calif., 1956-61; div. prin. engr. software Litton Computer Svcs., San Jose, Calif., 1961—. Roman Catholic. Achievements include computation of largest known juggler number. Home: 19628 Via Monte Dr Saratoga CA 95070-4522 Office: Litton Computer Svcs 4747 Hellyer Ave PO Box 210059 San Jose CA 95151-0059

SMITH, HOWARD DUANE, zoology educator; b. Fillmore, Utah, June 25, 1941; married; 2 children. BS, Brigham Young U., 1963; MS, U. Ill., 1966, PhD, 1969. Instr. biology U. Ill., Urbana, 1968-69; from asst. to assoc. prof. Brigham Young U., Salt Lake City, 1969-80, prof. zoology, 1981—; collaborator Intermountain Forest & Range Experiment Sta., U.S. Forest Svc., 1965—; cons. Wilderness Assocs., 1974—, terrestrial wildlife. NSF fellow, 1969—. Mem. AAAS, Am. Soc. Mammalogy, Wildlife Soc., Ecol. Soc. Am., Am. Mus. Natural History. Achievements include rsch. in small mammal populations; demography; bioenergetics; environ. impact studies on man; environ. impact of coal generating power plants or pesticide applications on the biota; wildlife biology. Office: Am Soc of Mammalogists Brigham Young University Dept of Zoology 501 Widtsoe Bldg Provo UT 84602*

SMITH, IAN CORMACK PALMER, biophysicist; b. Winnipeg, Man., Can., Sept. 23, 1939; s. Cormack and Grace Mary S.; m. Eva Gunilla Landvik, Mar. 27, 1965; children—Brittmarie, Cormack, Duncan, Roderick. BS, U. Man., 1961, MS, 1962; PhD, Cambridge U., England, 1965; Filosophie Doktor (hon.), U. Stockholm, 1986; DSc (hon.), U. Winnipeg, 1990. Fellow Stanford U., 1965-66; mem. rsch. staff Bell Tel. Labs., Murray Hill, N.J., 1966-67; rsch. officer divsn. biol. scis. NRC, Ottawa, 1967-87, dir. gen., 1987-91; dir.-gen. Inst. Biodiagnostics, Winnipeg, 1992—; adj. prof. chemistry and biochemistry Carleton U., 1973-90, U. Ottawa, 1976-92; adj. prof. chemistry, physics and anatomy U. Man., 1992—; adj. prof. biophysics U. Ill., Chgo., 1974-80; adj. prof. chemistry, physics, and anatomy U. Man., 1992—; allied scientist Ottawa Civic Hosp., 1985-92, Ottawa Gen. Hosp., 1989-92, Ont. Cancer Found., 1989-92, St. Boniface Hosp., 1992—. Contbr. chps. to books, articles in field to profl. jours. Recipient Barringer award Can. Spectroscopy Soc., 1979, Herzberg award, 1986, Organon Teknika award Can. Soc. Clin. Chemists, 1987. Fellow Chem. Inst. Can. (Merck award 1978, Labatt award 1984), Royal Soc. Can.; mem. Am. Chem. Soc., Biophys. Soc., Canadian Biochem. Soc. (Ayerst award 1978), Biophys. Soc. Can., Soc. Magnetic Resonance Medicine (exec. com. 1989-94). Office: Inst Biodiagnostics, Winnipeg, MB Canada R3B 1Y6

SMITH, JACQUELINE HAGAN, toxicologist; b. Cheverly, Md., Dec. 3, 1954; d. Walter Hagan and Norma JEan (Timmons) S.; m. Jerry Bruce Hook, May 18, 1985. BS in Biology, Dickinson Coll., 1975; MS in Biomed. Scis., Hood Coll., 1978; PhD in Pharmacol. & Environ. Toxicology, Mich. State U., 1983. Diplomate Am. Bd. Toxicology. Lab. tech. Frederick (Md.) Cancer Rsch. Ctr., 1975-79; NIH predoctoral trainee Mich. State U. Dept. Pharm. & Toxicology, East Lansing, 1979-83; pharm. rsch. assoc. trainee Nat. Cancer Inst., NIH, Bethesda, Md., 1983-85; cons. toxicologist Exxon Biomed. Scis. Inc., East Millstone, N.J., 1985-87, performance products toxicology group head, 1987-88, petroleum & synfuels cons., toxicology sect. head, 1988-92, chems. cons. toxicology sect. head, 1992—. Contbr. articles to profl. jours. Mem. Mid-Atlantic Chpt. Soc. Toxicology, Soc. Toxicology. Democrat. Achievements include interest in risk assessment and communication, non-genotoxic mechanisms of carcinogenicity, food-grade mineral hydrocarbons toxicity issues, mechanisms of nephrotoxicity and in vitro models of nephrotoxicity. Office: Exxon Biomed Scis Inc Mettlers Rd CN 2350 East Millstone NJ 08875-2350

SMITH, JAMES BIGELOW, SR., electrical engineer; b. Hamilton, Ont., Can., Dec. 18, 1908; s. Percy Merrihew and Ethel Burgess (Torrey) S.; m. Elizabeth Perry Dane, Mar. 1936 (div. 1954); children: James Bigelow Jr., Sylvia Perry, David Prentiss; m. Catharine Dean Temple, Jan. 17, 1959. BSEE, MIT, 1932, MSEE, 1933. Registered profl. engr., Mass. Quality control engr. Sylvania Electric Products Co., Salem, Mass., 1933; claims adjuster Liberty Mutual Ins. Soc., N.Y.C. and Lynn, Mass., 1933-35; lab. engr. engr.-in-charge sect. Factory Mutual Rsch. Corp., Norwood, Mass., 1935-74, asst. chief engr. to chief engr., v.p., mgr. applied rsch.; assoc. cons. engr. Lement & Assocs., Waltham, MA, 1974—; chair black liquor recovery boiler adv. com. for pulp and paper industry, 1960's. Patentee fire protection methods and systems apparatus. Fellow Soc. Fire Protection Engrs. (life, charter mem., pres. New Eng. chpt. 1962-63, historian 1980—, bd. dirs. 1980—, Richard E. Stevens award 1985), Nat. Fire Protection Assn. (life, past chair oven and furnace com.), Fire Detection Inst. (past chair). Home and Office: 25 Mitchell Grant Way Bedford MA 01730-1258

SMITH, JAMES EDWARD, petroleum engineer, consultant; b. Phoenix, Aug. 11, 1935; s. Edward Tracy and Willie (Neel) S.; widowed, June, 1992; children: Paula, James David. Student, Arlington State Coll., 1955-57; BS in Petroleum Engring., Tex. A&M U., 1958. Registered profl. engr., Tex. Field engr. RRC, Abilene, Tex., 1958-60, asst. dist. dir., 1960-61, dist. dir., 1962-67; dist. dir. RRC, Kilgore, Tex., 1967-74; field ops. dir. RRC, Austin, Tex., 1974-76; v.p. Petro-Mgmt., Inc., Tyler, Tex., 1976-78; pres. Smith and Harman Engring. Co., Tyler, 1978-81, James E. Smith and Assocs., Inc., Tyler, 1981—; past chmn. East Tex. Chpt. Joint Engrs. Banquet, East Tex. Symposium on Hydrogen Sulfide Opers.; instr. Petro-Mgmt., Inc.; advisor, cons. in field, guest speaker in field; developer, conductor safety tng. seminars Tex. Pub. Safety Personnel. Contbr. articles to profl. jours.; pub. papers in field. Congregation Chmn. Our Saviour Luth. Ch., Kilgore, 1972; former scout master Boy Scouts Am.; active adv. com. Tex. Railroad Commn., 1984. Recipient Cert. Appreciation award West Ctrl. Tex. Mple. Water Dist., Desk and Derrick Club, Nat. Profl. Engrs. Hall of Fame, Lubbock Area Firemen Conf.; named Hon. Citizen of Kilgore. Mem. Nat. Soc. Profl. Engrs., Am. Assn. Drilling Engrs., Soc. Profl. Engrs., N.Y. Acad. Scis. Achievements include development of Texas Railroad Commission Rule 36 for Hydrogen Sulfide Public Safety, H2S Safety Plan, H2S System on H2S operation for drilling, completion and production; research in development of the Cotton Valley Geopressure Zone in Panola County, Texas using air/N2 drilling and openhole completion techniques, texas regulations for sour gas operations, regulations of offshore and onshore disposal, pollution control practices in East Texas, application of engineering concepts to the solution of pollution problems, application of steam for crude oil recovery. Office: James E Smith and Assocs Inc 1324 S Beckham # 147 Tyler TX 75701

SMITH, JAMES LANNING, engineering executive; b. Bellefonte, Pa., Feb. 2, 1947; s. Raymond Lanning and Nancy Marlene (Fassero) S.; m. Christine Inez Beaulieu, Apr. 22, 1972; 1 child, Jennifer Lanning. AS, Contra Costa Coll., 1976; BS, U. Md., 1981. Quality assurance supr. Bechtel Corp., San Francisco, 1973-77, Gaithersburg, Md., 1977-79; sr. tech. advisor NUSAC Inc., McLean, Va., 1979-80; mgr. quality assurance Gasser Assocs., Olney, Md., 1980-90; prin. staff advisor BDM Internat., Germantown, Md., 1990-91; divsn. mgr. Performance Devel. Corp., Germantown, 1991—; working group chmn. nuclear quality assurance com. ASME, N.Y.C., 1981—. With USN, 1967-73. Mem. Am. Soc. Quality Control, Am. Nuclear Soc. Roman Catholic.

SMITH, JANET SUE, systems specialist; b. Chgo., Jan. 15, 1945; d. Curtis Edwin and Margaret Louise (Yost) Smith; B.A., Ind. U., 1967. Sales mgr. Marshall Field & Co., Chgo., 1968-70, programmer, 1970-72; sr. programmer, analyst Trailer Train Co., Chgo., 1972-75; mgr. data base and systems devel. RAILINC-Assn. Am. R.R., Washington, 1975-85, asst. v.p., corp. sec., 1985—. Nat. student v.p. YWCA, 1966-67; bd. dirs., v.p. planning and fin. Guide Internat., Friends of the Nat. Zoo; advisor Jr. Achievement. Mem. Am. Council R.R. Women, Ind. U. Alumni Assn. (life), Women's Transp. Seminar. Home: 2000 N St NW Washington DC 20036-2336 Office: 50 F St NW Washington DC 20001

SMITH, JEAN, interior design firm executive; b. Oklahoma City; d. A. H. and Goldy K. (Engle) Hearn; m. W. D. Smith; children: Kaye Smith Hunt, Sidney P. Student Chgo. Sch. Interior Design, 1970. v.p. Billco-Aladdin Wholesale, Albuquerque, 1950-92, v.p. Billco Carpet One of Am, 1970. Pres. Albuquerque Opera, 1979-83, advisor to bd. dirs.; active Civic Chorus, Cen. Meth. Ch.; pres. Inez PTA, 1954-55, life mem.; hon. life mem. Albuquerque Little Theater, bd. dirs. Republican. Clubs: Albuquerque Country, Four Hills Country, Daus. of the Nile (soloist Yucca Temple). Home: 1009 Santa Ana Ave SE Albuquerque NM 87123-4232 Office: Billco-Aladdin Wholesale 7617 Menaul Blvd NE Albuquerque NM 87110-4647

SMITH, JEFFREY ALAN, occupational medicine physician, toxicologist; b. Plainfield, N.J., Dec. 13, 1953; s. John Oliver and Regina Delores (Rudnicki) S. BSChemE with high honors, N.C. State U., 1974; MD with honors, U. N.C., 1979; MS in Toxicology, W.Va. U., 1990. Environ. engr. U.S. EPA, Durham, N.C., 1974-75; mem. staff, cons. PEDCo Environ., Inc., Durham, 1975-80; dir. environ. health PEDCo Environ., Inc., Cin., 1981-82; intern New Hanover Meml. Hosp., Wilmington, N.C., 1980-81; ind. contractor various urgent care ctrs. Cin., 1982-84; dir. Marion (Ohio) Correctional Inst., F.C. Smith Clinic, Inc., 1984-88; resident W.Va. U., Morgantown, 1988-91; med. officer Nat. Inst. Occupational Safety and Health, Morgantown, 1991; dir. occupational medicine Concord (N.C.) Family Medicine/Monroe Urgent Care, 1991—. Co-author: Environmental Assessment of the Domestic Industries, 1976, Dioxins, 1980. Mem. Am. Coll. Occupation and Environ. Medicine, Am. Conf. Govt. Indsl. Hygienists, Am. Coll. Physician Execs., N.C. Med. Soc., So. Med. Assn., Phi Kappa Phi, Alpha Omega Alpha. Avocations: book collecting, philately, mountain climbing, white water canoeing. Home: 8215 Golf Ridge Dr Charlotte NC 28277 Office: Monroe Urgent Care 613 E Roosevelt Blvd Monroe NC 28112

SMITH, JEFFRY ALAN, public health administrator, physician, consultant; b. L.A., Dec. 8, 1943; s. Stanley W. and Marjorie E. S.; m. Jo Anne Hague. BA in Philosophy, UCLA, 1967, MPH, 1972; BA in Biology, Calif. State U., Northridge, 1971; MD, UACJ, 1977. Diplomate Am. Bd. Family Practice. Resident in family practice WAH, Takoma Park, Md., NIH, Bethesda, Md., Walter Reed Army Hosp., Washington, Children's Hosp. Nat. Med. Ctr., Washington, 1977-80; occupational physician Nev. Test Site, U.S. Dept. Energy, Las Vegas, 1981-82; dir. occupational medicine and environ. health Pacific Missile Test Ctr., Point Mugu, Calif., 1982-84; dist. health officer State Hawaii Dept. Health, Kauai, 1984-86; asst. dir. health County of Riverside (Calif.) Dept. Health, 1986-87; regional med. dir. Calif. Forensic Med. Group, Salines, 1987—; med. dir. Community Human Svcs. Project, Monterey, Calif., 1987—. Fellow Am. Acad. Family Physicians; mem. AMA, Am. Occupational Medicine Assn., Flying Physicians, Am. Pub. Health Assn. Avocations: pvt. pilot. Home: 27575 Via Sereno Carmel CA 93923

SMITH, JEREMY OWEN, chemist; b. Springfield, Mass., May 3, 1960. BS in Chemistry, U. Tex., San Antonio, 1983. Analytical chemist Raba-Kistner Cons., San Antonio, 1984-88; chemist Dept. of Army, Corpus Christi, Tex., 1988—. Robert A Welch Found. scholar, 1983. Mem. ASM, Am. Chem. Soc., Am. Electroplaters and Surface Finishers Soc. (cert.). Office: Corpus Christi Army Depot SDS-QLC Stop 27 Corpus Christi TX 78419

SMITH, JOE MAUK, chemical engineer, educator; b. Sterling, Colo., Feb. 14, 1916; s. Harold Rockwell and Mary Calista (Mauk) S.; m. Essie Johnstone McCutcheon, Dec. 23, 1943; children—Rebecca K., Marsha Mauk. B.S., Calif. Inst. Tech., 1937; Ph.D., Mass. Inst. Tech., 1943. Chem. engr. Texas Co., Standard Oil Co. of Calif., 1937-41; instr. chem. engring. Mass. Inst. Tech., 1943; asst. prof. chem. engring. U. Md., 1945; prof. chem. engring. Purdue U., 1945-56; dean Coll. Tech., U. N.H., 1956-57; prof. chem. engring. Northwestern U., 1957-59, Walter P. Murphy prof. chem. engring., 1959-61; prof. engring. U. Calif., 1961—, chmn. dept. chem. engring., 1964-72; hon. prof. chem. engring. U. Buenos Aires, Argentina, 1964—; Fulbright lectr., Eng., Italy, Spain, 1965, Argentina, 1963, 65, Ecuador, 1970, Brazil 1990, Mudaliar Meml. lectr. U. Madras, India, 1967; UNESCO cons., Venezuela, 1972-82; spl. vis. professorship Yokohama Nat. U., Japan, 1991. Author: Introduction to Chemical Engineering Thermodynamics, 1949, 4th edit, 1986, Chemical Engineering Kinetics, 1956, 3d edit., 1981. Guggenheim research award for study in Holland; also Fulbright award, 1953-54. Mem. Am. Chem. Soc., Am. Inst. Chem. Engrs. (Walker award 1960, Wilhelm award 1977, Lewis award 1984), Nat. Acad. Engring., Sigma Xi, Tau Beta Pi.

SMITH, JOHN JAMES, JR., environmental engineering laboratory director; b. Franklin, N.J., Dec. 6, 1936; s. John James Smith and Estelle Mary (Gurka) Cook; m. Sondra L. Smith, Dec. 19, 1956; 1 child, James H. BS in Chemistry, U. Fla., 1967. Mgr. div. water and wastewater CH2M Hill SE Inc., Gainesville, Fla., regional mgr. Gainesville, mgr lab dist. Author conf. proc.; contbr. articles to profl. jours. Bd. dirs. Fla. Arts Celebration, Coun. on Econ. Outreach. With USAF. Mem. Am. Chem. Soc., Am. Water Works Assn. (nat. chmn. water quality monitoring com., trustee Fla. sect. 1979-82, chmn. Fla. sect. 1985-86, George Warren Fuller award 1988), Fla. Inst. Cons. Engrs., Fla. Chem. Industry Coun., N.E. Fla. Regional Fla. Engring. Ednl. Delivery System Coun., U. Fla. Engring. Adv. Coun., Internat. Assn. Water Pollution Rsch., Water Pollution Control Fedn., Rotary, Sigma Xi. Achievements include development, design and research in numerous water and wastewater systems and facilities. Office: CH2M Hill SE Inc Ste C 2772 NW 43rd St S Gainesville FL 32606

SMITH, JOHN JULIAN, agronomist; b. Reading, Berkshire, U.K., Oct. 18, 1957; came to U.S. 1987; s. Michael and Dorothy (Partridge) S.; m. Maureen McDonnell, May 26, 1979; children: Victoria Ann, Steven James. BS, U. Leeds, U.K., 1978; PhD, U. Stirling, U.K., 1981. Chartered biologist, Inst. Biology of U.K. Agronomic svcs. mgr. J.W. Chafer Ltd., U.K., 1981-87; exec. v.p., rsch. dir. Fluid Fertilizer Found., St. Louis, 1987-89; v.p. rsch., edn. and mkt. devel. Nat. Fertilizer Solutions Assn., St. Louis, 1989-90; mgr. agronomy and product devel. J.R. Simplot Co., Mineral & Chem. Group, Pocatello, Idaho, 1990-92, v.p. environment & agronomy, 1992—; bd. dirs. Fluid Fertilizer Found., St. Louis, 1991—; mem. Calif. Dept. Food and Agr. Fertilizer Adv. & Inspection Bd., 1993—. Contbr. articles to profl. jours.; editorial adv. bd. Solutions Mag., 1988-90. Coach West County Soccer, St. Louis, 1988-90, Gate City Soccer League, Pocatello, 1990—. Sci. Rsch. Coun. grantee, 1978-81; Ministry of Agr. study grantee, 1985-88. Mem. Inst. Biology U.K., Fertilizer Soc. U.K., Am. Soc. Agronomy (agronomic ind. com. 1990), The Fertilizer Inst. (agronomy task force 1990—), Soil Sci. Soc. Am., Crop Sci. Soc. Am. Achievements include development of novel fluid fertilizer application techniques and agronomic product development; research on management and information dissemination relating to agronomic, economic and environmentally sound use of crop fertilizers, on mechanism of cereal leaf resistance to bacterial infection. Office: JR Simplot Co PO Box 912 Pocatello ID 83204-0912

SMITH, JOHN MARVIN, III, surgeon, educator; b. San Antonio, July 31, 1947; s. John M. and Jane (Jordan) S.; m. Jill Jones, Aug. 1, 1981. M.D., Tulane U., 1972. Diplomate Am. Bd. Surgery, Am. Bd. Thoracic Surgery. Intern, U. Tex. Southwest Med. Sch., Dallas, 1972-73; resident in surgery U. Tex., San Antonio, 1973-77, resident in thoracic and cardiovascular surgery Tex. Heart Inst., Houston, 1977-79; practice medicine specializing in cardiovascular and thoracic surgery, San Antonio, 1979—; mem. Staff Bapt. Med. Ctr., S.W. Tex. Meth. Hosp., Santa Rosa Med. Center, Met. Hosp., Nix Meml. Hosp.; clin. asst. prof. surgery U. Tex. Health Sci. Ctr., San Antonio, 1979—. Contbr. articles to profl. jours. Bd. mgrs. Bexar County Hosp. Dist.; bd. dirs. Tex. Ranger Assn., San Antonio Med. Found., Meth. Hosp. Found.

Internat. Affairs Coun. Served to maj. USAF, 1979-81. Fellow Am. Coll. Cardiology, ACP, ACS; mem. AMA, Tex. Med. Assn., Bexar County Med. Soc., Denton A. Cooley Cardiovascular Surg. Soc., Cooley Hands, J. Bradley Aust Surg. Soc., Soc. Air Force Clin. Surgeons, San Antonio Surg. Soc., San Antonio Cardiology Soc., Soc. of Thoracic Surgeons, Tex. Surgical Soc., Tex. Hist. Soc., Sigma Alpha Epsilon, Nu Sigma Nu. Episcopalian. Clubs: Tex. Cavaliers, San Antonio Country, The Argyle, Giraud, Order Alamo, German, Christmas Cotillion, Sons Republic Tex., San Antonio Gun. Home: 204 Zambrano Rd San Antonio TX 78209-5459 Office: 4330 Medical Dr # 300 San Antonio TX 78229

SMITH, JOHN ROBERT, materials scientist; b. Salt Lake City, Oct. 1, 1940; married; 2 children. BS, Toledo U., 1962; PhD in Physics, Ohio State U., 1968. Aerospace engr. surface physics Lewis Rsch. Ctr. NASA, 1965-68; sr. rsch. physicist, head surface and interface physics group Gen. Motors, Warren, Mich., 1972-80, sr. staff scientist, head solid state physics group, 1980-86, prin. rsch. scientist rsch. lab., 1986—; adj. prof. dept. physics U. Mich., 1983—. Air Force Office Sci. Rsch., Nat. Rsch. Coun. fellow U. Calif., 1970-72. Fellow Am. Phys. Soc. (David Adler Lectureship award in field of materials sci. 1991); mem. Am. Vacuum Soc., Sigma Xi. Achievements include research in the theory of solid surfaces, electronic properties, magnetic properties and chemisorption, adhesion, metal contact electronic structure, defects and universal features of bonding in solids. Office: General Motors Research Lab Box 9055 30500 Mound Rd Warren MI 48090*

SMITH, JOHN STEPHEN, educational administrator; b. Wheeling, W.Va., Aug. 16, 1938; s. Carl Edward and Martha Ellen (Van Meter) S.; m. Bernice E. Eichenlaub, Sept. 7, 1963; children: Karl Bartholomew, Ann Kathryn, Suzanne Lee. BS, Wheeling Jesuit Coll., 1960; MS, Duquesne U., 1965; PhD, U. Pitts., 1971. Instr. biochemistry U. Pitts., 1962-69, mgr. budget systems, 1969-72; asst. dean, asst. to chancellor La. State U. Med. Ctr., New Orleans, 1972-74; asst. to v.p. health affairs U. Ala. Birmingham Med. Ctr., 1974-78, asst. v.p. health affairs, 1978-88, assoc. v.p. health affairs, 1988—; dep. dir. internat. program U. Ala. Birmingham, 1980-90. Contbr. articles to acad. jours. Active Boy Scouts Am., 1978—; chmn. spl. events Vulcan dist., Birmingham, 1986-90, chmn. comms. com. Birmingham Area coun., 1991—; bd. dirs. Red Mountain Mus. Soc., Birmingham, 1982. Mem. Assn. Am. Med. Colls. (chmn. so. region planning 1979-80, nat. chmn. instl. plannig 1980-81), Assn. Instl. Rsch., Soc. Coll. and Univ. Planning, Com. Fgn. Rels., Phi Beta Delta. Avocations: hiking, golf, reading. Office: U Ala Birmingham 102-H Mortimer Jordan Hall Birmingham AL 35294-2010

SMITH, JOSEPH FRANK, spacecraft electronics and systems engineer; b. Hinsdale, Ill., July 16, 1953; s. Joseph Matthew and Genevieve Theresa (Mikrut) S.; m. Keri Lizbeth Oda, June 28, 1986; 1 child, Kelsey Akiko. BSEE, Purdue U., 1976, MSEE, 1978. Design engr. Jet Propulsion Lab., Pasadena, Calif., 1978-84, systems engr., 1984-87, tech. group supr., 1987-88; systems engr. McDonnell Douglas Aerospace, Huntington Beach, Calif., 1988—. Mem. IEEE, AIAA (digital avionics tech. com.). Office: McDonnell Douglas Aerospace Systems Co 5301 Bolsa Ave MIC17-B Huntington Beach CA 92647

SMITH, JULIAN CLEVELAND, JR., chemical engineering educator; b. Westmount, Que., Can., Mar. 10, 1919; s. Julian Cleveland and Bertha (Alexander) S.; m. Joan Elsen, June 1, 1946; children: Robert Elsen, Diane Louise Smith Brook, Brian Richard. B.Chemistry, Cornell U., 1941, Chem. Engr., 1942. Chem. engr. E. I. duPont de Nemours and Co., Inc., 1942-46; mem. faculty Cornell U., 1946—, prof. chem. engring., 1953-86, prof. emeritus, 1986—, dir. continuing engring. edn., 1965-71; assoc. dir. Cornell U. (Sch. Chem. Engring.), 1973-75, dir., 1975-83; vis. lectr. U. Edinburgh, 1971-72; cons. to govt. and industry, 1947—; UNESCO cons. Universidad de Oriente, Venezuela, 1975. Author: (with W. L. McCabe and P. Harriott) Unit Operations of Chemical Engineering, 1956, 5th edit., 1993, also articles; sect. editor: Perry's Chemical Engineers Handbook, 1963. Fellow Am. Inst. Chem. Engrs.; mem. Am. Chem. Soc., Sigma Xi, Tau Beta Pi, Phi Kappa Phi, Alpha Delta Phi. Clubs: Ithaca Country, Statler (Ithaca). Home: 711 The Parkway Ithaca NY 14850-1546

SMITH, KEITH BRUNTON, civil engineer; b. Drexel Hill, Pa., June 15, 1960; s. Kenneth Brunton and Elizabeth (Grassie) S.; m. Jean Arlotta, May 10, 1991; 1 child, Taylor Brunton. BCE, U. Dayton, 1983. Registered profl engr., N.J.; registered profl. planner; cert. mcpl. engr. Civil engr. trainee N.J. Dept. Transp., Trenton, 1983-84; jr. engr. Hardesty & Hanover Cons. Engrs., N.Y.C., 1984-85; sr. project mgr., assoc. Schoor & Canger Group Inc., Manalapan, N.J., 1985—. Mem. ASCE (assoc.), Inst. Transp. Engrs. (assoc.), Soc. Mcpl. Engrs. Home: 71 Essex Ave West Keansburg NJ 07734

SMITH, KEITH SCOTT, electrical engineer; b. Kansas City, Mo., May 17, 1958; s. Robert Houston and Norma Lee (Scott) S.; m. Elizabeth Hull, Aug. 27, 1983; 1 child, Margaret Ann. BSEE, U. Mo., Rolla, 1980; MBA, Fairleigh Dickinson U., 1986. Registered profl. engr., Colo., N.J. Dist. engr. Mo. Pub. Svc. Co., Grandview, 1980-83; mgr. engring. and ops. Sussex (N.J.) Rural Elec. Coop., 1983-87; mgr. engring. Intermountain Rural Elec. Assn., Sedalia, Colo., 1987—. Mem. NSPE (Denver sect.), Denver IEEE Power Engrs. Republican. Home: 9605 E Coronado Ct Parker CO 80134

SMITH, KENNETH CARLESS, electrical engineering educator; b. Toronto, Ont., Can., May 8, 1932; s. Reginald Thomas and Viola Evelyn (Carless) S.; children—K. David, Kevin A. B.A.Sc., U. Toronto, 1954, M.A.Sc. in Elec. Engring., 1956, Ph.D. in Physics, 1960. Transmission engr. Can. Nat. Telegraphs, Toronto, 1954-55; rsch. engr. U. Toronto at U. Ill., Urbana, 1956-58; asst. prof. elec. engring. U. Ill., Urbana, 1961-64, assoc. prof., 1964-65; asst. prof. elec. engring. U. Toronto, 1960-61, assoc. prof. elec. engring. and computer sci., 1965-70, prof., 1970-81, prof. depts. elec. engring., computer sci., library and info. sci., mech. engring., 1981—, chmn. dept. elec. engring., 1976-81; adv. prof. Shanghai Inst. Rwy. Tech., People's Republic China, 1989—; pres. Elec. Engring. Consociates Ltd., Toronto, 1974-76; dir. several small firms, U.S. and Can.; mem. Ont. Task Force on Microelectronic Tech., 1980-81. Author: (with A.S. Sedra) Microelectronic Circuits, 1982, 2d edit., 1987, Spanish edit., 1985, 3d edit., 1991, Hebrew edit., 1990, Korean edit, 1989, Lab. Explorations, 1991, Trial and Success, 1992; contbr. articles to profl. jours., chpts. to books; patentee in field. Fellow IEEE, Circuits and Systems Soc. (chmn. confs. procedures and planning, publs. coun.), Can. Soc. for Profl. Engrs. (bd. dirs. 1984—, pres. 1988-91), Internat. Solid State Circuit Conf. (exec. com., awards chmn. 1975—). Liberal. Mem. Anglican Ch. Home: 56 Torbrick Rd, Toronto, ON Canada M4J 4Z5 Office: U Toronto Dept Elec Engring, 10 King's College Rd, Toronto, ON Canada M5S 1A4

SMITH, LAWRENCE ABNER, aeronautical engineer; b. Rivera, N.Mex., Jan. 16, 1921; s. Abner Dee and Laura (Cox) S.; m. Judith evelyn Roberson, Dec. 21, 1942; children: Lawrence Jr., Ronald, Samuel, Kenneth, Rena, David, Daniel, Laura, Randall. BSME, N.Mex. State U., 1942; postgrad., Washington U., 1948-68. Design engr. Curtis Wright Corp., Louisville, 1943-44; design engr. McDonnell Douglas, St. Louis, 1946-56, project engr., 1958-68, v.p. engring., 1982-84, v.p. advanced design, 1984-86, program mgr. advanced aircraft, 1983-86; mem. mechanical engring. educator N.Mex State U., 1988—; owner pvt. bus., 1986—; aircraft design cons. AIAA, 1986—. Author: Harrier on the Guam, 1970. Sponsor Am. Youth Fair, New Haven, Mo., 1986—; elder Presbyn. Ch., Chesterfield, Mo., 1950—. With USN, 1944-46. Named Disting. Alumnus N.Mex. State U., 1979; recipient Paul E. Haueter award Am. Helicopter Soc., 1984. Fellow AIAA (assoc., chmn. internat. aeronautics com. 1984-86, vice chair aircraft design com. 1980-84), McDonnell Aircraft Ret. Engring. Coun., N.Mex. State U. Mech. Engring. Acad. Alumni Assn. Republican. Presbyterian. Achievements include patents in landing gear design, aircraft design. Home and Office: 379 Madewood Ln Chesterfield MO 63017

SMITH, LLOYD HILTON, independent oil and gas producer; b. Pitts., July 9, 1905; s. Roland Hilton and Jane (Lloyd) S.; m. Jane Clay Zevely, Sept. 7, 1931; children—Camilla; m. Elizabeth Keith Wiess, May 25, 1940; children: Sandra Keith, Sharon Lloyd, Sydney Carothers. Ph.B., Yale U., 1929. Statistician Biggs, Mohrman & Co., N.Y.C., 1932; mgr. New York office Laird & Co., 1933-34; v.p. Argus Research Corp., 1934-35; chmn., dir. Paraffine Oil Corp., 1949-81; past dir. 1st City Nat. Bank Houston, Nat. Rev., Curtiss-Wright Corp., Info. Storage Systems, Falcon Seaboard, Inc. Mem. Houston Mus. Sci., Com. on Present Danger; trustee Pine Manor Coll., 1963-74; past dir. DeBakey Med. Found. Lt. comdr. USNR, 1942-45. Republican. Clubs: Bayou, Ramada (Houston); Everglades, Bath and Tennis (Palm Beach); Racquet and Tennis (N.Y.C.), Brook (N.Y.C.), River (N.Y.C.); Nat. Golf Links of America, Southampton, Meadow. Office: 2210 First City Tower 1001 Fannin St Houston TX 77002-6707

SMITH, LOUIS, maintenance engineer; b. Shreveport, La., Nov. 2, 1934; s. Louis and Savannah (Durham) S.; m. Velma Smith, Jan. 1, 1961; 1 child, Gerald W. Student, Rancho Los Amigos, Downey, Calif., 1976. Maintenance engr. L.A. Dept. Water and Power, 1968—; chauffeur Cowboy Limosine Svc., Pasadena, Calif., 1988—; show horseman Com. for Altadena (Calif.) Old Fashioned-Day Parades; leading rider monty police Palm Spring Parade, 1991; 1st pl. Western singleman rider, Lancaster, 1991; featured rider Palm Springs, 1993. Mem. Tournament of Roses Com., Pasadena Coun. Parade of Roses. With U.S. Army, 1957-58. Recipient Trophies, Altadena Town Coun., 1982, Desert Circus, 1985, 86, Palm Springs C. of C., 1981, 87, City of Barstowe (Calif.), 1988, 89, Golden West Parader Assn., 1989, San Bernardino City Coun., 1990. Mem. Internat. Platform Assn., Golden West Parader Assn., Friends of the Friendless, First Travel Club. Democrat. Baptist. Avocations: tournament of dominoes, deer hunting, fresh water fishing, country-western guitarist. Home: 1980 Santa Rosa Ave Pasadena CA 91104-1127

SMITH, LUTHER A., statistician; b. Little Rock, Sept. 5, 1953; s. Huie Haskell and Adeline (Schattle) S.; m. Marjolein Estella van der Vaart, May 26, 1979; 1 child, Daniel Pieter. BA cum laude, Vanderbilt U., 1975; MS, N.C. State U., 1977, PhD, 1985. Statistician Atmospheric Impacts Rsch. Program N.C. State U., Raleigh, 1985-87, ManTech Environ. Tech., Research Triangle Park, N.C., 1988—; Contbr. articles to profl. jours.; author reports. Mem. Ridge Rd. Bapt. Ch., Raleigh. Mem. Ecol. Soc. Am., Soc. Math Biol. Democrat. Home: 2501 Anne Carol Ct Raleigh NC 27603 Office: ManTech Environ Tech 2 Triangle Dr Research Triangle Park NC 27709

SMITH, MARC KEVIN, mechanical engineering educator; b. Pontiac, Mich., Sept. 3, 1954; s. Donald Charles and Evelyn Agnes (Evans) S.; m. Suzanne Marie Maitland, Aug. 10, 1985; 1 child, Logan Charles Maitland Smith. BS, Mich. State U., 1976; MS, Stanford U., 1977; PhD, Northwestern U., 1982. Structural engr. GM, Detroit, 1977-79; instr. MIT, Cambridge, 1982-83; post-doctoral fellow U. Cambridge, Eng., 1983-84; asst. prof. Johns Hopkins U., Balt., 1984-91; assoc. prof. Ga. Inst. Tech., Atlanta, 1991—. Contbr. articles to Jour. Fluid Mechanics & Physics of Fluids. Named Presdl. Young Investigator NSF, 1985. Mem. ASME, Am. Phys. Soc., Soc. for Indsl. & Applied Math. Office: Ga Inst Tech Sch Mech Engring Atlanta GA 30332-0405

SMITH, MARTIN JAY, physician, biomedical research scientist; b. Bklyn., May 21, 1934; s. I. Richard and Marilyn (Bernard) S.; m. Joyce Ellen Gleason, June 26, 1960 (div. Nov. 1968); children: Danielle, Robert, Alexander; m. Ruby Helen Rhodes, Apr. 7, 1972. BA, Hofstra Coll., 1955; MD, Columbia U., 1959. Diplomate Am. Bd. Internal Medicine, Am. Bd. Internal Medicine in Hematology, Am. Bd. Pathology in Clin. Pathology, Am. Bd. Pathology in Immunopathology. Intern in medicine Meth. Hosp., N.Y.C., 1959-60, resident in medicine, 1960-61; resident in medicine Montefiore Hosp., N.Y.C., 1963-64; rsch. fellow in medicine Harvard Coll., Cambridge, Mass., 1964-66; clin. and rsch. fellow in medicine Mass. Gen. Hosp., Boston, 1964-66; physician Gundersen Clinic and Luth. Hosp., La Crosse, Wis., 1966—, chmn. dept. internal medicine, 1971-73; dir. spl. hematology lab. Gundersen Clinic, 1967—, chmn. dept. lab. medicine, 1973—; dir. lab. medicine Luth. Hosp., La Crosse, 1973—; dir. rsch. Gundersen Med. Found., 1975-88; med. dir. Med. Lab. Tech. Program Western Wis. Tech. Inst., 1978—. Contbr. articles to New Eng. Jour Medicine, Jour. Lab. Clin. Medicine, Blood, Ann. Internal Medicine, Am. Jour. Hematology, Clin. Chemistry, others. Capt. U.S. Navy, 1961-63. Fellow ACP, Coll. Am. Pathologists (inspector labs. 1983—); mem. Am. Assn. for Cancer Rsch., Am. Soc. Hematology, Internat. Soc. Hematology, Assn. Med. Lab. Immunologists, Phi Beta Kappa. Home: 1428 Main St La Crosse WI 54601-4225 Office: Gundersen Clinic Ltd 1836 South Ave La Crosse WI 54601-5494

SMITH, MARVIN SCHADE, civil engineer, land surveyor; b. Perth Amboy, N.J., July 12, 1905; s. Forrest Leigh and Jeannette Banks (Schade) S.; m. Harriet Lillian Lease, Nov. 30 (dec. 1969); children: Roger Alan, David Leigh, Margaret Jean, Deborah Ann; m. Elizabeth Robbins, Oct. 9, 1971. BS in Civil Engring., MIT, 1926. Registered profl. engr., Del. Cons. engr. Port Raritan Dist. Commn., Perth Amboy, N.J., 1926-30; estimator R & H Chem. Co., Perth Amboy, N.J., 1930-31; profl. engr. and land surveyor Perth Amboy, N.J., 1931-33; project engr. C.W.A. & E.R.A., Perth Amboy, N.J., 1933-35; engring. draftsman Port of N.Y. Authority, N.Y.C., 1935-39; field engr. Panama R.R. Co., Balboa, Panama, 1939-41; design engr. E.I.DuPont de Nemours & Co., Inc., Wilmington, Del., 1941-49; cons. engr. Wilmington, Del., 1949-77; ret. Pres. Heritage Farm Civic Assn., Wilmington, 1983-85, New Castle chpt. AARP, 1988-90. Mem. ASCE, NSPE, Del. Assn. Cons. Engrs. (pres. 1970-71), Masons (Meritorious Svc. award 1972). Republican. Presbyterian. Home: 2606 Grendon Dr Wilmington DE 19808

SMITH, MARY KAY WILHELM, safety engineer; b. Random Lake, Wis., Jan. 27, 1962; d. Eugene Paul and Ruth Esther (Wetor) Wilhelm; m. Reginald O. Smith, Nov. 28, 1987. BS in Aerospace Engring., U. Notre Dame, 1984; MS in Aerospace Engring., U. Dayton, 1987. Cert. safety profl. System safety engr. USAF, Dayton, Ohio, 1984-86, system safety mgr., 1986-88; test mgr. USAF, Dayton, 1988-89; chief integration divsn. USAF, Bedford, Mass., 1989-90; cost engr. Walk, Haydel and Assocs., Inc., New Orleans, 1990-91; safety engr. Sverdrup Tech., Inc., Stennis Space Center, Miss., 1991—. Big sister Big Bros./Big Sisters, Dayton, 1986-87; com. chair Altrusa, Dayton, 1987-89; youth counselor Aldersgate United Meth. Ch., Slidell, La., 1991—. Capt. USAF, 1984-90. Mem. AIAA (sr., systems effectiveness and safety tech. com. 1993—), Nat. Mgmt. Assn. (mgmt. devel. com. 1992—). United Methodist. Office: Sverdrup Tech Inc Bldg 1100 Rm 1002 Stennis Space Center MS 39529

SMITH, MARY-ANN HRIVNAK, meteorologist, geophysicist; b. Johnstown, Pa., Dec. 12, 1951; d. John and Ann (Demchak) Hrivnak; m. Charles Frederick Smith, June 9, 1973; children: Catherine Ann, Michael John. BS in Meteorology, Pa. State U., 1973; PhD in Geophys. Scis., U. Chgo., 1978. Rsch. asst. prof. Physics Dept. Coll. William and Mary, Williamsburg, Va., 1978-80; sr. rsch. scientist NASA Langley Rsch. Ctr., Hampton, Va., 1980—; Mem. Editorial Bd. Jour. Molecular Spectroscopy, 1990—, Adv. Bd. Ohio State Molecular Spectroscopy Symposium, Columbus, 1992—. Co-author: (book) Atlas of Ozone Parameters from Microwave to Medium Infrared, 1990; author: (book chpts.) Molecular Spectroscopy, Modern Research, vol. III, 1985, Spectroscopy of the Earth's Atmosphere and Interstellar Medium, 1992; contbr. 67 articles to profl. jours. Recipient Nat. Merit Scholarship, 1969, 2nd Place Macelwane award, Am. Meteorol.Soc., 1972, Spl. Achievement awards, NASA, 1982, 86. Mem. AAAS, Am. Meteorol. Soc., Am. Geophys. Union, Assn. for Women in Sci., Hampton Roads Sect. of Women in Aerospace, Sigma Xi. Achievements include one-year assignment as staff scientist in the Upper Atmosphere Rsch. Program Office at NASA Hdqs., Washington. Organized NASA workshops onSpectroscopic Parameters for Upper Atmospheric Measurements. Office: NASA Langley Rsch Ctr Mail Stop 401A Hampton VA 23681-0001

SMITH, MATTHEW JAY, chemist; b. N.Y.C., Nov. 11, 1961. BA, Colby Coll., 1983; PhD, Duke U., 1988. Applications scientist Nicolet Instrument Corp., Madison, Wis., 1987-91; sales engr. Nicolet Instrument Corp., Cary, N.C., 1991—. Contbr. articles to profl. jours. Mem. Am. Chem. Soc., Soc. Applied Spectroscopy, Coblentz Soc., Sigma Xi. Office: Nicolet Instrument Corp 5225 Verona Rd Madison WI 53711

SMITH, MICHAEL, biochemistry educator; b. Blackpool, Eng., Apr. 26, 1932. BSc, U. Manchester, Eng., 1953, PhD, 1956. Fellow B.C. Rsch. Coun., 1956-60; rsch. assoc. Inst. Enzyme Rsch., U. Wis., 1960-61; head chem. sect. Vancouver Lab. Fisheries Rsch. Bd. Can., 1961-66; med. rsch. assoc. Med. Rsch. Coun. Can., 1966—; assoc. prof. biochem. U. B.C., Vancouver, 1966-70, prof., 1970—. Recipient Gairdner Found. Internat. award, 1986, Nobel Prize in Chemistry, Nobel Foundation, 1993. Fellow Chem. Inst. Can., Royal Soc. Chemistry, Sigma Xi. Achievements include rsch. in nucleic acid and nucleotide chemistry and biochemistry using in-vitro mutagenesis gene expression. Office: 300-2466 W 3rd Ave, Vancouver, BC Canada V6K 1L8 Office: U BC Biotech Lab, Wesbrook Bldg Rm 237, Vancouver, BC Canada V6T 1W5*

SMITH, MICHAEL HOWARD, ecologist; b. San Pedro, Calif., Aug. 30, 1938; s. William Smith and Frances Harriet (Ryder) Roe; m. Irma Beatrice Summers, Apr. 9, 1958; children: Michael William, Karen Marie. BA, San Diego State U., 1960, MA, 1962; PhD, U. Fla., 1966. Asst. prof. U. Ga., Athens, 1966-71, assoc. prof., 1971-77, prof. ecology, 1977—; dir. Savannah River Ecology Lab., Aiken, S.C., 1973—; vis. assoc. prof. U. Tex., Austin, 1970-71; hon. lectr. U. London, 1977-78; rsch. assoc. U. Calif.-Berkeley, 1987; rev. panel NSF, 1974-77; adv. group Clemson (S.C.) U., 1983—. Editor: Mineral Cycling in Southeastern Ecosystems, 1975, Mammalian Population Genetics, 1981; contbr. chpt. to Population Genetics of the White-Tailed Deer, 1984, Genetic Variability and Antler Development, 1990, Population Genetics of the Slider Turtle, 1990. Mem. Am. Soc. Mammalogists (life, bd. dirs. 1978-81, 86-89, Merriam award 1985), Evolution Soc., Ecol. Soc. Am., Sigma Xi (life). Home: 1129 Parsons Ln Aiken SC 29803-5323 Office: Savannah River Ecology Lab Drawer E Aiken SC 29801*

SMITH, MICHAEL LAWRENCE, knowledge engineer, researcher; b. Sheboygan, Wis., Aug. 29, 1958; s. Lawrence Eugene Patrick and Velma Mary (Baltus) S. BS in Computer Sci., U. Wis., 1980; MS in Computer Sci., U. So. Calif., 1988. Lab. asst. McArdle Lab. for Cancer Rsch., Madison, Wis., 1978-80; systems programmer/analyst Controls and Data Systems, Belvidere, Ill., 1980-82; sr. systems software engr. Ex-Cell-O Mfg. Systems Co., Rockford, Ill., 1982-85; artificial intelligence cons. McDonnell-Douglas Artificial Intelligence Ctr., Cypress, Calif., 1985-88; sr. artificial intelligence engr. Eaton Corp., Milw., 1988-92; sr. knowledge engr. Inference Corp., L.A., 1992—. Author: (chpt.) Cooperating Artificial Neural and Knowledge-Based Systems in a Truck Fleet Brake-Balance Application, 1991. Founder, chmn. Computer Profls. for Social Responsibility, Milw., 1989-90. Mem. AAAS, Am. Assn. for Artificial Intelligence, N.Am. Fuzzy Info. Processing Soc., Internat. Neural Network Soc. Achievements include patents for fuzzy logic controlled washing machine; for fuzzy logic controlled automobile engine coolant valve; for knowledge-based ion-beam deposition semi conductor mfg. system; for blackboard-based intelligent agts.; research in ability to train practical, real world artificial neural systems using scarce data and back propagation algorithm and cooperating knowledge bases. Home and Office: Inference Corp 9098 N 75th St Milwaukee WI 53223

SMITH, MICHAEL STEVEN, data processing executive; b. San Antonio, May 7, 1956; s. Columbus and Mary Patricia (Leahy) S.; m. Lynda M. Gillen, July 30, 1992. Student, San Bernardino Valley (Calif.) Coll., 1974-76, AS in Computer Scis., 1983; student, L.A. Community Coll., 1978-79, U. Md., 1980-81, City Colls. Chgo., 1980-81. Communications cons. Telephone Products Corp., San Bernardino, 1974-76; student svcs. advisor computer scis. lab. San Bernardino Valley Coll., 1982-83; assoc. programmer Aerojet ElectroSystems Corp., Azusa, Calif., 1983-85; mgr. data processing. Bonita Unified Sch. Dist., San Dimas, Calif., 1985—, dir. computer info. svcs., 1989—; analyst computer mktg. Pentamation Enterprises, Bethelehem, Pa., 1987—; cons. computer systems San Dimas, 1985—. With USN, 1976-82. Mem. Assn. for Computing Machinery, Digital Equipment Computer Users Soc., Calif. Assn. Sch. Bus. Ofcls., Calif. Ednl. Data Processing Assn. Avocations: computer systems design, scuba diving, tennis, motorcycles, classic automobiles. Office: 115 W Allen Ave San Dimas CA 91773-1437

SMITH, MILDRED CASSANDRA, systems engineer; b. Rocky Mount, N.C.; d. Naaman and Mildred (Laws) Foster; m. Edward B. Smith III, July 22, 1967 (div. 1976); children: Camille Eileen, Regina Dar. BA, Howard U., 1966, MS, Georgetown U., 1973, PHD, 1979. Cert. computer programmer. Programmer IBM Corp., Gaithersburg, Md., 1966-76; asst. prof. Howard U., Washington, 1976-80; engr./analyst VITRO Corp., Silver Spring, Md., 1980-82; analyst U.S. Dept. Agriculture, Washington, 1983-86; systems engr. MITRE Corp., McLean, Va., 1986—. Contbr. article to Software Reengring. Vol. D.C. Pub. Schs., 1978-80; judge Alice Deal Jr. High Sci. Fair, Washington, 1987; mentor corp. engring. enrichment program T.C. Williams High Sch., 1991. Named one of Outstanding Young Women Am., 1977, 78. Mem. D.C. Assn. for Computing Machinery (chmn. local interest group on mgmt. of data 1989—), Assn. for Computing Machinery, Assn. for Computational Linguistics. Democrat. Presbyterian. Avocation: duplicate bridge.

SMITH, NEVILLE VINCENT, physicist; b. Leeds, Eng., Apr. 21, 1942; came to U.S., 1966; s. Horace J.H. and Ethel (Stevens) S.; m. Elizabeth Jane Poulson, Mar. 21, 1970; children: Katherine, Elizabeth. BA, Cambridge (Eng.) U., 1963, MA, 1967, PhD, 1967. Rsch. assoc. Stanford (Calif.) U., 1966-68; mem. staff AT&T Bell Labs., Murray Hill, N.J., 1969—, head condensed state physics rsch. dept., 1978-81. Contbr. articles to jours. in field. Fellow Am. Phys. Soc. (Davisson-Germer prize 1991); mem. AAAS. Home: 510 Church St Bound Brook NJ 08805-1729 Office: AT&T Bell Labs 600 Mountain Ave New Providence NJ 07974-2010

SMITH, NOEL WILSON, psychology educator; b. Marion, Ind., Nov. 2, 1933; s. Anthony and Mary Louise (Wilson) S.; m. Marilyn C. Coleman, June 17, 1954; children: Thor and Lance (twins). AB, Ind. U., 1955, PhD, 1962; MA, U. Colo., 1958. Asst. prof. psychology Wis. State U., Platteville, 1962-63; asst. prof. psychology SUNY, Plattsburgh, 1962-66, assoc. prof., 1966-77, prof., 1971—. Co-author: The Science of Psychology: Interbehavioral Survey, 1985; sr. editor: Reassessment in Psychology, 1983; author: Greek and Interbehavioral Psychology, 1990, rev. edit., 1993, An Analysis of Ice Age Art: Its Psychology and Belief, 1992; editor Interbehavioral Psychology newsletter, 1970-77; contbr. articles to profl. jours. Fellow APA; mem. AAUP (pres. SUNY coun. 1980-82), Am. Psychol. Soc., Cheiron Internat. Soc. History of Behavior Sci., Sigma Xi. Home: 7 W Court St Plattsburgh NY 12901 Office: SUNY Dept Psychology Beaumont Hall Plattsburgh NY 12901

SMITH, NOVAL ALBERT, JR., systems engineer; b. Huntington, W.Va., Apr. 24, 1949; s. Noval Albert and Palmaneda (Baker) S.; m. Sheryl Lynn Mueller, Aug. 22, 1970; children: James, Joy. BS in Nuclear Engring., U. Fla., 1971, M in Engring., 1972. Registered profl. engr., Va. Thermal design engr. Knolls Atomic Power Lab., Niskayuna, N.Y., 1972-77; engr. staff Va. Power Co., Glen Allen, Va., 1977—; mem. system analysis working group Elec. Power Rsch. Inst., Palo Alto, Calif., 1977-82; lectr. in field. Contbr. articles to profl. jours. Bd. dirs. Christian Counselors Tng. Ctr., Richmond, Va., 1988-91; pastor Grace Christian Fellowship, Richmond, 1982—. Mem. ASME, Am. Nuclear Soc. Achievements include development of Va. Power Reactor System transient analysis capability. Office: Va Power Co 500 Dominion Blvd Glen Allen VA 23060

SMITH, ORIN ROBERT, chemical company executive; b. Newark, Aug. 13, 1935; s. Sydney R. and Gladys Emmett (DeGroff) S.; m. Stephanie M. Bennett-Smith; children: Lindsay, Robin; 1 stepchild, Brendan. B.A. in Econometrics, Brown U., 1957; M.B.A. in Mgmt., Seton Hall U., 1964; PhD (hon.) in Econs., Centenary U., 1991. Various sales and mktg. mgmt. positions Allied Chem. Corp., Morristown, N.J., 1959-69; dir. sales and mktg. Richardson-Merrell Co., Phillipsburg, N.J., 1969-72; with M&T Chems., Greenwich, Conn., 1972-77, pres., 1975-77; with Engelhard Minerals & Chems. Corp., Menlo Park, Edison, N.J., 1977-81, corp. sr. v.p., 1978-81, pres. div. minerals and chems., 1978-81, also bd. dirs., 1979-81, pres., dir. various U.S. subs., 1979-81; exec. v.p., pres. div. minerals and chems. Engelhard Corp., Menlo Park, Edison, 1981-84; pres., chief exec. officer Engelhard Corp., Iselin, N.J., 1984—, also bd. dirs.; bd. dirs. Summit Trust Co., The Summit Bancorp, Vulcan Materials Co., La. Land. and Exploration Co., N.J. Mfrs. Ins. Co., N.J. Reins. Co. Trustee Mfrs. Alliance for Productivity and Innovation; bd. overseers N.J. Inst. Tech., Plimoth Plantation and Exec. Coun. Fgn. Diplomacy; vice chmn. bd. trustees Centenary Coll.; mem. adv. bd. Watchung Area coun. Boy Scouts Am.; trustee Welkind Rehab. Hosp.; past chmn. Ind. Coll. Fund N.J.; past dir.-at-large U. Maine Pulp and Paper Found. Lt. (j.g.) USN, 1957-59. Mem. Chem. Mfrs. Assn. (past bd. dirs.), Am. Mgmt. Assn. (gen. mgmt. coun.), N.J. State C. of C. (bd. dirs.),

N.J. Bus. and Industry Assn. (trustee), Econ. Club N.Y.C., Union League Club N.Y.C., Roxiticus Golf Club (Menham, N.J.), N.Y. Yacht Club. Office: Engelhard Corp 101 Wood Ave Iselin NJ 08830-0770

SMITH, PAUL CHRISTIAN, mechanical engineer; b. Pitts., Jan. 25, 1950; s. John Henry and Catherine Christine (Magel) S.; m. Monica Louise Hawrot, Apr. 4, 1987; children: Harrison R., Sara K. AS in Basic Engring., C.C. of Allegheny County, Pitts., 1971; BSME, Gannon U., 1973; MSME, Carnegie Mellon U., 1981. Registered profl. engr., Pa., N.Y., Ind., W.Va., Ohio, Ala., Wash. Design engr. Koppers Co., Inc., Pitts., 1974-77, project engr., 1977-78, engring. supr., 1978-82, sr. design engr., 1982-84; sr. design engr. Raymond Kaiser Engr., Pitts., 1984-85; assoc. engr. Kaiser Engrs., Pitts., 1985-86, design supervising engr., 1986-87; prin. engr. ICF Kaiser Engrs., Pitts., 1987-92; lead piping engr. Continental Design and Mgmt. Group, Pitts., 1993—. Mem. NSPE, ASHRAE, Ea. States Blast Furnace and Coke Oven Assn. Republican. Roman Catholic. Avocation: running (10K races). Office: Continental Design and Mgmt Group 1 Gateway Ctr Pittsburgh PA 15222

SMITH, PAUL JOHN, plastic and reconstructive surgeon, consultant; b. Bangor, Ireland, July 25, 1945; s. John Joseph and Mary Patricia (Maher) S.; m. Anne Westmoreland Snowdon, July 14, 1972; children: Mark, Jaime, Victoria, Francesca. MB BS, Newcastle (Eng.) U., 1968. Lectr. in anatomy Glasgow (Scotland) U., 1969-71; surgical trainee Western Infirmary, Glasgow, 1971-76; rsch. asst. Dept. of Microsurgery, U. Louisville, 1978; Christine Kleinert Fellow in hand surgery U. Louisville, 1978-79; resident and clin. instr. in plastic surgery Duke U., Durham, N.C., 1979-80; sr. registrar in plastic surgery Mt. Vernon Hosp., London, 1980-82, cons. plastic surgeon, 1982—; cons. plastic surgeon The Hosp. for Sick Children, Great Ormond St., London, 1988—; sec. Royal Soc. of Medicine, London, 1988; editorial com. Jour. of Hand Surgery (British volume), London, 1987-91. Author: Principles of Hand Surgery, 1989; contbr. articles to profl. jours. Mem. rsch. com. Restoration of Appearance and Function Trust Found. Recipient 1st prize resident competition, Am. Assn. of Hand Surgery, Toronto, 1979. Fellow Royal Coll. Surgeons; mem. British Assn. Plastic Surgeons (organizing com. advanced courses plastic surgery, Hayward found. scholarship 1978), British Soc. for Surgery of the Hand (Pulvertaft prize, 1986), British Assn. of Aesthetic Plastic Surgeons. Office: Mt Vernon Hosp, Rickmansworth Rd, Northwood HA6 2RN, England

SMITH, PAUL LETTON, JR., research scientist; b. Columbia, Mo., Dec. 16, 1932; s. Paul Letton and Helen Marie (Doersam) S.; m. Mary Barbara Noel; children: Patrick, Melody, Timothy, Christopher, Anne. BS in Physics, Carnegie Inst. Tech., 1955, MSEE, 1957, PhD in Elec. Engring., 1960. From instr. to asst. prof. Carnegie Inst. Tech., Pitts., 1955-63; sr. engr. Midwest Rsch. Inst., Kansas City, Mo., 1963-66; from rsch. engr. to sr. scientist and group head Inst. Atmospheric Scis., S.D. Sch. Mines and Tech., Rapid City, 1966-81; vis. prof. McGill U., Montreal, Que., Can., 1969-70; chief scientist Air Weather Svc. USAF, Scott AFB, Ill., 1974-75; dir. Inst. Atmospheric Scis., S.D. Sch. Mines and Tech., Rapid City, 1981—; lectr. Tech. Svc. Corp., Silver Spring, Md., 1972—; dir. S.D. Space Grant Consortium, Rapid City, 1991—. Contbr. over 40 articles to profl. jours. Fellow Am. Meteorol. Soc. (Editor's award 1992); mem. IEEE (sr.), Am. Soc. Engring. Edn., Weather Modification Assn., Sigma Xi. Home: 2107 9th St Rapid City SD 57701 Office: SD Sch Mines & Tech 501 E Saint Joseph St Rapid City SD 57701

SMITH, PEGGY O'DONIEL, physicist, educator; b. Lakeland, Fla., Nov. 27, 1920; d. John Arthur and Carrie Mattie (Jackson) O'Doniel; m. Fenton Frederick Smith, Oct. 11, 1943; children: James Scott, Stephen Arthur, Melody Ann, Candy Lou. Aviation Pilot Lic., Stetson U., Deland, Fla., 1941; BS in Sci. and Math., Fla. So. Coll., 1942; MA in Edn., U.S. Internat. U., San Diego, 1968. Physicist degausser U.S. Navy, Key West, Fla., 1942; physicist compass compensator U.S. Navy, Charleston, S.C., 1943; physicist magnetic signature analyst U.S. Navy, Washington, 1944; tchr. Chula Vista (Calif.) Sch. Dist., 1963-73, math specialist, 1974-77; owner Mineral Store, Chula Vista, 1977-82; ret.; leader math. workshops for girls, 1992-93. contbr. articles to profl. jours. Del. White House Conf. on Edn., 1956; chmn. Orphans of Italy, 1957-58. Recipient Kazanjian award, Joint Coun. Econ. Edn., Chula Vista, 1972; Chula Vista Sch. Dist. math grantee, 1975. Mem. AAUW (v.p. 1989), Inner Circle, Calif. Ret. Tchrs. Assn., San Diego Gem and Mineral Soc. Republican. Avocations: golf, mineral collecting, coin collecting, bridge, travel. Home: 87 K St Chula Vista CA 91911-1409

SMITH, PETER GUY, neuroscience educator, researcher; b. Boston, July 22, 1950; s. Harvey James and Susan Alta (Muto) S.; m. Ellen Penny Averett, Feb. 22, 1986; 1 child, Harrison Jesse. BA, U. N.H., 1973; PhD, Duke U., 1978. Rsch. assoc. dept. pharmacology Duke U., Durham, N.C., 1978-82, asst. med. rsch. prof., 1982-87; assoc. prof. dept. physiology U. Kans. Med. Ctr., Kansas City, 1987-93, prof. dept. physiology, 1993—; mem. neurology B2 study sect. NIH, 1991—, chmn., 1993—; mem. rsch. adv. com. Kans. affiliate Am. Heart Assn., Topeka, 1991-94. Contbr. articles to Brain Rsch., Neurosci., Hypertension, Jour. Comparative Neurology, Exptl. Neurology. Recipient Excellence in Edn. award Student Voice, U. Kans. Sch. Medicine, 1991, 92; fellow Pharm. Mfrs. Assn. Found., 1978-80; grantee NIH, 1986—, Marrion Merrel Dow Found., 1991-93. Mem. Soc. for Neurosci., World Fedn. Neurology (rsch. group on autonomic nervous system), Am. Heart Assn. (high blood pressure rsch. coun., established investigator 1983-88). Achievements include contributions to fields of development plasticity of the autonomic nervous system. Office: U Kans Med Ctr Physiology Dept 3901 Rainbow Rd Kansas City KS 66160-7401

SMITH, PETER LLOYD, physicist; b. Victoria, B.C., Can., Apr. 28, 1944; came to U.S., 1966; s. Lloyd Wood and Joan Mary Smith; m. Lois Elaine Hodgson, July 1968 (div. June 1990); children: Alexandra C.H., Colin W.H.; m. Donna Jeanne Coletti, Apr. 27, 1991. BSc, U. B.C., Vancouver, 1965; PhD, Calif. Inst. Tech., 1972. Postdoctoral fellow Calif. Inst. Tech., Pasadena, 1972-73; asst. prof. Harvey Mudd Coll., Claremont, Calif., 1972-73; postdoctoral fellow Obs. Harvard U., Cambridge, Mass., 1973-76, rsch. assoc. Obs., 1976—, lectr. in astronomy, 1979-83. Editor: Atomic and Molecular Data for Space Astronomy, 1992; contbr. articles to profl. jours. Mem. Am. Astron. Soc., Internat. Astron. Union (sec. commn. 14 1991—), Am. Phys. Soc., Optical Soc. Am. Office: Harvard U Smithsonian Ctr Astrophys 60 Garden St MS-50 Cambridge MA 02138

SMITH, PHILIP LUTHER, research biochemist; b. Milan, Ind., Dec. 23, 1956; s. Donald Walter and Evelyn Emma (Vornheder) S.; m. Mary Ann Radike, Feb. 9, 1985; children: Martha Jesse, Philip Benjamin. BS, Purdue U., 1980. Rsch. asst. III U. Cin. Coll. of Medicine, Cin., 1981-84; sr. phys. biochemist Med. Coll. of Ohio, Toledo, 1985-89; rsch. biochemist Marion Merrell Dow Rsch. Inst., Cin., 1990—; rsch. cons. Med. Coll. of Ohio Dept. of Biochemistry, Toledo, 1990. Contbr. articles to profl. jours. Mem. AAAS, Am. Chem. Soc., The Prot. Soc. Roman Catholic. Achievements include rsch. in synthesis, purification and characterization of DNA/RNA oligonucleotides, peptides, proteins and synthetic compounds with a pharmaceutical significance; rsch. in protein chemistry with an emphasis on protein structure. Office: Marion Merrell Dow 2110 E Galbraith Rd Cincinnati OH 45215

SMITH, PHILIP MEEK, research organization executive; b. Springfield, Ohio, May 18, 1932; s. Clarenc Mitchell S. and Lois Ellen (Meek) Dudley. B.S., Ohio State U., 1954, M.A., 1955; DSc (hon.), N.C. State U., 1986. Mem. staff U.S. Nat. Com. for Internat. Geophys. Yr., Nat. Acad. Scis., 1957-58; program dir. NSF, 1958-63, dir. ops. U.S. Antarctic Research program, 1964-69, dep. head div. polar programs, 1970-73; chief gen. sci. br. Office Mgmt. and Budget Exec. Office of Pres., 1973-74; exec. asst. to dir. and sci. advisor to pres. NSF, 1974-76; exec. sec. Pres.'s Com. Nat. Medal of Sci., 1974-76; assoc. dir. Office Sci. and Tech. Policy, Exec. Office of Pres., 1976-81; spl. asst. to chmn. Nat. Sci. Bd., 1981; exec. officer NRC-Nat. Acad. Scis., Washington, 1981—; corp. mem. Woods Hole Oceanographic Instn., 1983-89; pres. Cave Research Found., Yellow Springs, Ohio, 1957-63; chmn. tech. panels Fed. Coordinating Council Sci. and Engring. and Tech., 1976-80. Author: (with others) Defrosting Antarctic Secrets, 1962; The Frozen Future, a Prophetic Report from Antarctica, 1973; contbr. numerous articles to profl. jours. Bd. dirs. Washington Project for Arts, 1983-84, Washington Sculptors Group, 1983-84. 1st lt. U.S. Army, 1955-57. Decorated Commendation medal and Antarctic Service medal U.S. Navy, 1957; recipient Meritorious Service medal NSF, 1972. Mem. AAAS, Antarctican Soc., Cosmos Club (Washington), Am. Alpine Club (N.Y.C.), Coun. Excellence Govt. (prin.). Office: NRC-NAS 2101 Constitution Ave NW Washington DC 20418-0001

SMITH, RALPH EARL, virologist; b. Yuma, Colo., May 10, 1940; s. Robert C. and Esther C. (Schwarz) S.; m. Sheila L. Kondy, Aug. 29, 1961 (div. 1986); 1 child, Andrea Denise; m. Janet M. Keller, 1988. BS, Colo. State U., 1961; PhD, U. Colo., 1968. Registered microbiologist Am. Soc. Clin. Pathologists. Postdoctoral fellow Duke U. Med. Ctr., Durham, N.C., 1968-70, asst. prof., 1970-74, assoc. prof., 1974-80, prof. virology, 1980-82; prof., head dept. microbiology Colo. State U., Ft. Collins, 1983-88, prof. microbiology, assoc. v.p. rsch., 1989—, interim v.p. rsch., 1990-91, prof. microbiology, assoc. v.p. rsch., 1991—; cons. Bellco Glass Co., Vineland, N.J., 1976-80, Proctor & Gamble Co., Cin., 1985-86, Schering Plough Corp., Bloomfield, N.J., 1987-89. Contbr. articles to profl. jours.; patentee in field. Bd. dirs. Colo. Ctr. for Environ. Mgmt., v.p. for rsch.; mem. pollution prevention adv. bd. Colo. Dept. Health; mem. Rocky Mountain U. Consortium on Environ. Restoration; asst. scoutmaster Boy Scouts Am., Durham, 1972-82, com. mem., Ft. Collins, 1986—; mem. adminstrv. bd. 1st United Meth. Ch., Ft. Collins. Eleanor Roosevelt fellow Internat. Union Against Cancer 1978-79. Mem. AAAS, Am. Soc. Microbiology, N.Y. Acad. Scis., Am. Soc. Virology, Am. Assn. Immunologists, Am. Assn. Avian Pathologists, Am. Assn. Cancer Rsch., Gamma Sigma Delta. Democrat. Methodist. Avocations: photography, hiking. Home: 2406 Creekwood Dr Fort Collins CO 80525-2034 Office: Colo State U VP Rsch Fort Collins CO 80523

SMITH, RANDOLPH RELIHAN, plastic surgeon; b. Augusta, Ga., Apr. 13, 1944; s. Lester Vernon and Maxine (Relihan) S.; m. Becky Jo Hardy; children: Katherine, Randolph, Rececca, Michael. B.S., Clemson U., 1966; M.D., Coll. Ga., 1970. Intern Bowman Gray Sch. Medicine, Winston-Salem, N.C., 1970-71; resident in surgery and otolaryngology Duke U., Durham, N.C., 1971-75; resident in plastic and reconstructive surgery Med. Coll. Ga., 1975-77; Christine Kleinert fellow in hand surgery U. Louisville, 1977; attending physician Univ. Hosp., Augusta, Ga., 1977—; asst. clin. prof. plastic surgery Med. Coll. Ga., 1978—; pres. med. staff Univ. Hosp., Augusta. Bd. dirs. Ga. Bank and Trust Co. of Augusta, Richmond County Hosp. Authority; vestryman St. Paul's Ch. Served to maj. M.C., U.S. Army, 1971-77. Diplomate Am. Bd. Otolaryngology, Am. Bd. Plastic Surgery. Fellow ACS, Am. Acad. Otolaryngology; mem. Am. Soc. Plastic and Reconstructive Surgeons, Am. Soc. Aesthetic Plastic Surgery, Ga. Soc. Plastic and Reconstructive Surgeons, Southeastern Soc. Plastic and Reconstructive Surgeons. Episcopalian. Clubs: Exchange, Augusta Symphony League. Contbr. articles in field to profl. jours. Office: Univ Hosp 820 St Sebastian Way Ste 2F Augusta GA 30910-2399

SMITH, RAYMOND EDWARD, health care administrator; b. Freeport, N.Y., June 17, 1932; s. Jerry Edward and Madelyn Holman (Jones) S.; B.S. in Edn., Temple U., 1953; M.H.A., Baylor U., 1966; m. Lena Kathryn Jernigan Hughes, Oct. 28, 1983; children: Douglas, Ronald, Kevin, Doris Jean, Raymond. Commd. 2d lt. U.S. Army, 1953, advanced through grades to lt. col., 1973; helicopter ambulance pilot, 1953-63; comdr. helicopter ambulance units, Korea, 1955, Fed. Republic of Germany, 1961; various hosp. adminstrv. assignments, 1963-73; personnel dir. Valley Forge (Pa.) Gen. Hosp., 1966; adminstr. evacuation hosp., Vietnam, 1967; dep. insp. Walter Reed Gen. Hosp., Washington, 1970; dir. personnel div. Office of Army Surgeon Gen., Washington, 1971-73, ret., 1973; adminstr. Health Care Centers, Phila. Coll. Osteo. Medicine, 1974-76; dir. bur. hosps. Pa. Dept. Health, Harrisburg, 1976-79; contract mgr. Blue Cross of Calif., San Diego, 1979-88, Community Care Network, San Diego, 1989—. Decorated Bronze Star, Legion of Merit. Mem. Am. Hosp. Assn., Am. Legion, Ret. Officers Assn., Kappa Alpha Psi. Episcopalian. Club: Masons. Home: 7630 Lake Adlon Dr San Diego CA 92119-2518 Office: Community Care Network 8911 Balboa Ave San Diego CA 92123-1584

SMITH, REGINALD BRIAN FURNESS, anesthesiologist, educator; b. Warrington, Eng., Feb. 7, 1931; s. Reginald and Betty (Bell) S.; m. Margarete Groppe, July 18, 1963; children—Corinne, Malcolm. M.B., B.S., U. London, 1955; D.T.M. and H., Liverpool Sch. Tropical Medicine, 1959. Intern Poole Gen. Hosp., Dorset, Eng., 1955-56, Wilson Meml. Hosp., Johnson City, N.Y., 1962-63; resident in anesthesiology Med. Coll. Va., Richmond, 1963-64; resident in anesthesiology U. Pitts., 1964-65, clin. instr., 1965-66; asst. prof., 1969-71, asso. clin. prof., 1971-74, prof., 1974-78, acting chmn. dept. anesthesiology, 1977-78; prof., chmn. dept. U. Tex. Health Sci. Center, San Antonio, 1978—; anesthesiologist in chief hosps. U. Tex. Health Sci. Ctr., 1978—; dir. anesthesiology Eye and Ear Hosp., Pitts., 1971-76; anesthesiologist in chief Presbyn. Univ. Hosp., Pitts., 1976-78. Contbg. editor: Internat. Ophthalmology Clinics, 1973, Internat. Anesthesiology Clinics, 1983; contbr. articles to profl. jours. Served to capt. Brit. Army, 1957-59. Fellow ACP, Am. Coll. Anesthesiologists, Am. Coll. Chest Physicians; mem. AMA, Internat. Anesthesia Rsch. Soc., Am. Soc. Anesthesiologists (pres. Western Pa. 1974-75), Tex. Soc. Anesthesiologists, San Antonio Soc. Anesthesiologists (pres. 1990), Tex. Med. Assn., Bexar County Med. Soc. Home: 213 Canada Verde St San Antonio TX 78232-1104 Office: 7703 Floyd Curl Dr San Antonio TX 78284-7700

SMITH, RICHARD ALAN, neurologist, medical association administrator. Grad., Brandeis U., 1961, U. Miami, 1965. Intern in medicine Jackson Meml. Hosp., Miami, Fla., 1965-66; resident in neurology Stanford U. Hosp., Palo Alto, Calif., 1966-69; head neurology br. Navy Neuropsychiatric Rsch. Unit, San Diego, 1969-71; mem. assoc. staff microbiology Scripps Clinic and Rsch. Found., La Jolla, Calif., 1972-79, mem. assoc. staff neurology, 1972-82; dir. Ctr. Neurologic Study, San Diego, 1979—; mem. sr. staff Scripps Meml. Hosp., La Jolla, 1982—; mem. CTNF protocol com. Regeneron Pharm. Corp., med. adv. bd. Multiple Sclerosis Soc., San Diego; animal welfare com. Whittier Inst., fitness clinic mentally disabled San Diego State U. Editor: Interferon Treatment for Neurologic Disorders, 1988, Handbook of Amyotrophic Lateral Sclerosis, 1992; contbr. articles to profl. jours. Recipient Henry Newman award San Francisco Neurologic Soc., 1968. Mem. AAAS, Am. Acad. Neurology (assoc.). Achievements include patents in methodologies of enhancing the systemic delivery of Dextromethorphan for the treatment of neurological disorders, of reducing emotional lability in neurologically impaired patients, and in the use of Cytochrome Oxidase Inhibitor to increase the cough-suppressing activity of Dextromethophan. Office: Ctr for Neurologic Study 11211 Sorrento Valley Rd Ste H San Diego CA 92121

SMITH, RICHARD DALE, chemist, researcher; b. Lawrence, Mass., July 1, 1949; s. Clarence Richard and Marion Ruth (Peters) S.; m. Elaine Grace Chapman, Jan. 5, 1985; 1 child, Jeffrey Scott. BS, Lowell Tech. Inst., 1971; PhD, U. Utah, 1975. Post-doctoral Naval Rsch. Lab., Washington, 1975-76; rsch. scientist Pacific N.W. Lab., Richland, Wash., 1976-78, sr. rsch. scientist, 1978-84, staff scientist, 1984-88, sr. staff scientist, 1988—. Contbr. articles to profl. jours. Recipient R & D 100 award R & D Mag., 1983, 88. Achievements include patents for supercritical fluid film deposition/powder formation process; capillary electrophoresis-mass spectrometry; reverse micelle and microemulsion formation in supercritical fluids; devel. new methods for characterization of large molecules using electrospray ionization-mass spectrometry. Home: 402 Scot St Richland WA 99352 Office: Pacific NW Lab 326 Bldg MS P8-19 Richland WA 99352

SMITH, RICHARD MELVYN, government official; b. Lebanon, Tenn., May 2, 1940; s. Roy D. and V. Ruth (Draper) S.; m. Patti Hawkins, Feb. 29, 1964; 1 child, Douglas. B.S.E.E., Tenn. Technol. U., 1963. Asst. Engr.-in-charge FCC, Phila., 1972, Balt., 1971-72; chief investigations br. FCC, Washington, 1974-77, asst. chief enforcement div., 1977-80, dep. chief Field Ops. Bur., 1980-81, chief Field Ops. Bur., 1981—. Recipient sr. exec. service award FCC, 1983, 84, 85, 86, 87, 88. Avocations: instrument-rated pvt. pilot. Office: FCC Rm 734 1919 M St NW Washington DC 20554

SMITH, RICK A., mechanical engineer, consultant; b. Shelby, Ohio, Sept. 10, 1948; s. Reginald A. and Ella Mae (Bolin) S.; m. Rhea Dawn Wilcox, Dec. 15, 1973. BSME, Purdue U., 1976; M of Engring., Ohio State U., 1988. Registered profl. engr., Ohio. Project engr. Armour-Dial, Inc., Montgomery, Ill., 1976-77, Purdue U., West Lafayette, Ind., 1977-79; plant energy engr. ALCOA, Lafayette, Ind., 1979-81; facility project mgr. Cummins Engine Co., Columbus, Ind., 1981-83; project mgr., sr. engr. Ohio State U., Columbus, 1983-88; pres., cons. mech. engr. Applied Thermal Engring., Ostrander, Ohio, 1988—. Mem. Mayor's Dist. Heating Task Force, Columbus, Ohio, 1984-86. 1st lt. USMC, 1968-72, Vietnam. Mem. ASME, NSPE, Am. Cons. Engrs. Coun. (mem. profl. affairs com. Ohio assn. 1990—), Am. Inst. Plant Engrs., Pi Tau Sigma. Republican. Avocations: flying, shooting, Dalmatians, physical fitness. Office: Applied Thermal Engring Inc PO Box 212 Ostrander OH 43061-9313

SMITH, ROBERT F., JR., civil engineer; b. Oneida, N.Y., Apr. 17, 1949; s. Robert F. and Lucy (Rice) S.; m. Lane K. McDonald, Nov. 21, 1984 (div. 1989); children: Sean Michael, Kevin Robert. BCE, Clarkson U., 1971. Registered profl. engr., N.Y., Ky. Asst. city engr. City of Oneida, 1971-78; chief stormwater mgmt. engr. Met. Sewer Dist., Louisville, 1978—. V.p. United Way, Oneida, 1976-77. Named Ky. Col., 1982. Mem. Nat. Soc. Profl. Engrs. (chmn. profl. engrs. in govt., 1993-94), Ky. Soc. Profl. Engrs. (v.p. 1989-91, D.V. Terrell award 1990, Disting. Engr. 1983, 88), ASCE (chpt. pres. 1984, Zone II Govt. Civil Engr. of Yr. 1989), Am. Pub. Works Assn. Democrat. Roman Catholic. Achievements include development of stormwater utility for city of Louisville and Jefferson County, Kentucky. Office: Met Sewer Dist 400 S 6th St Louisville KY 40202-2397

SMITH, ROBERTS ANGUS, biochemist, educator; b. Vancouver, B.C., Can., Dec. 22, 1928; came to U.S., 1953, naturalized, 1963; s. Alvin Roberts and Hazel (Mather) S.; m. Mary Adela Marriott, Aug. 22, 1953; children—Roberts H.A., James A.D., Eric J.M., Richard I.F. B.S.A., U. B.C., 1952, M.Sc., 1953; Ph.D., U. Ill., 1957. Instr. U. Ill., 1957-58; mem. faculty UCLA, 1958—, prof. biochemistry, 1968-87; pres. Viratek Inc.; dir. ICN Pharms., Inc. Guggenheim fellow, 1963-64. Mem. Am. Soc. Biol. Chemists, Am. Chem. Soc. Home: 221 17th St Santa Monica CA 90402-2221

SMITH, ROY EDWARD, mechanical engineer, rail vehicle consultant; b. Enfield, Eng., Jan. 7, 1940; arrived in Can., 1965; s. John Edward and Ethel (Thompson) S.; m. Sally Rosemary Willis, Sept. 8, 1962; children: Barnaby, Tiffany, Emma. BSc, U. Birmingham, 1963; M in Engring., U. Toronto, Ont., Can., 1970. Grad. apprentice Rolls-Royce and Assocs. Ltd., Derby, Eng., 1963-65; design engr. Atomic Energy of Can. Ltd., Mississauga, Ont., 1965-71; chief engr. hydraulics Husky Injection Molding Systems Inc., Toronto, 1971-75; pres., cons. Dynamic Drives Ltd., Burlington, Ont., 1975-76; truck specialist Urban Transportation Devel. Corp. Ltd., Kingston, Ont., 1976-89; pres., cons. Resco Engring., Kingston, 1989—; adj. assoc. prof. Queen's U., Kingston, 1990—; curriculum adv. bd. St. Lawrence C.C., Kingston, 1982-85; presenter papers at profl. confs. Contbr. to profl. publs. Mem. bd. govs. Ongwanda Hosp., Kingston, 1983—, pres., 1985-89; troop leader Kingston area Boy Scouts Can., 1978-82. Mem. ASME, Assn. Profl. Engrs. and Geoscientists of B.C., Assn. Profl. Engrs. Ont. Achievements include creation of first freely steered railway trucks, floating frame truck, swing frame truck, 23 patents. Office: Resco Engring, 823 Overlea Ct, Kingston, ON Canada K7M 6Z8

SMITH, RUSSELL LAMAR, biochemistry educator; b. Westminster, S.C., Jan. 25, 1959; s. John Sam and Audrey Jean (Stansell) S. BS, Clemson U., 1980; MA, U. Tex., 1984, PhD, 1986. Postdoctoral assoc. Centre d'Etudes Nucleaires de Grenoble, France, 1986-87, Pa. State U., State College, 1987-89, Memphis State U., 1989-90; asst. prof. biochemistry U. Tex., Arlington, 1990—; reviewer Idaho Specific. Rsch. Grant Program, Boise, 1990—. E. J. Lund Meml. fellow U. Tex., 1984-86. Mem. AAAS, Am. Soc. for Microbiology, Am. Chem. Soc., Am. Soc. for Plant Physiology., Sigma Xi. Office: U Tex Arlington Dept Chemistry and Biochemistry 502 Yates St Arlington TX 76019

SMITH, STAN LEE, biology educator; b. Wolf Lake, Ind., June 8, 1947; s. Donald Ray and Margaret Ellen (Hutsell) S.; m. Beth Ellen Grimme, Aug. 12, 1972; children: Daniel Nathan, Michael David. BS in Biology, Purdue U., 1970, MS in Biology, 1973; PhD in Biology, Northwestern U., 1978. Bacteriologist City-County Bd. Health, Ft. Wayne, Ind., 1972-74; lectr. Northwestern U., Evanston, Ill., 1978-79, NIH postdoctoral fellow, 1979-80; from asst. prof. to assoc. prof. Bowling Green (Ohio) State U., 1980-92, prof., 1992—; reviewer Scott Foresman and Co., Willard Grant Press, Houghton Mifflin Co., Times Mirror Mosby Coll. Publ., West Ednl. Publ., D.C. Heath Co., NSF, McGraw Hill, W.H. Freeman, Prentice Hall. Contbr. articles to Nature, Insect Biochemistry, Jour. Insect Physiology, Molecular and Cellular Endocrinology, Biochem. and Biophys. Rsch. Comms., Experientia. Recipient Nat. Rsch. Svc. award NIH, 1979-81; rsch. grantee NIH, 1984-86, Ohio Bd. Regents Bowling Green State U., 1986-89, Faculty Rsch. Com., 1980-83, 92-93. Mem. AAAS, Am. Soc. Zoologists, Entomol. Soc. Am., Sigma Xi (Outstanding Young Scientist Bowling Green State U. chpt. 1989). Democrat. Achievements include rsch. and discoveries in areas of insect biochemistry and endocrinology, biochemistry and physiology of ecdysteroids and cytochrome P-450 enzyme systems, effects of plant allelochemicals on ecdysteroidogenesis. Office: Bowling Green State U Dept Biol Scis Bowling Green OH 43403

SMITH, STEVEN COLE, computer scientist; b. Idaho Falls, Idaho, Oct. 3, 1952; s. Merrell Cordon and Myrtle Jean (McArthur) S.; m. Gay Lynn Pendleton, May 2, 1975; children: Jennifer, Melinda, Gregory, Aimilee. BS, Brigham Young U., 1977; MS, West Coast U., 1992. Engr. Gen. Dynamics, San Diego, 1978-81, Hughes Aircraft, L.A., 1981-82; cons. CAD/CAM Splty., L.A., 1982-83; system mgr. Solar Turbines, San Diego, 1983-87; mktg. support Evans and Sutherland, Costa Mesa, Calif., 1987-88; mktg. mgr. Computervision, San Diego, 1988—; instr. Southwestern C.C., San Diego, 1986-87. Mem. AIAA. Mormon. Office: Computervision Ste 160 9805 Scranton Rd San Diego CA 92121

SMITH, STEVEN SIDNEY, molecular biologist; b. Idaho Falls, Idaho, Feb. 11, 1946; s. Sidney Ervin and Hermie Phyllis (Robertson) S.; m. Nancy Louise Turner, Dec. 20, 1974. BS, U. Idaho, 1968; PhD, UCLA, 1974. Asst. research scientist Beckman Research Inst. City of Hope Nat. Med. Ctr., Duarte, Calif., 1982-84, staff Cancer Ctr., 1983—, asst. research scientist depts. Thoracic Surgery and Molecular Biology, 1985-87, assoc. research scientist, 1987—, dir. dept. cell and tumor biology, 1990—; cons. Molecular Biosystems Inc., San Diego, 1981-84. Contbr. articles to profl. jours. Grantee NIH, 1983—, Coun. for Tobacco Rsch., 1983-92, March of Dimes, 1988-91, Smokeless Tobacco Rsch. Coun., 1992—; Swiss Nat. Sci. Found. fellow U. Bern, 1974-77, Scripps Clinic and Rsch. Found., La Jolla, Calif., 1978-82, NIH fellow Scripps Clinic, 1979-81. Mem. Am. Soc. Cell Biology, Am. Assn. Cancer Rsch., Am. Crystallographic Assn., Am. Chem. Soc., Am. Weightlifting Assn., Phi Beta Kappa. Avocation: backpacking. Office: City of Hope Nat Med Ctr 1500 Duarte Rd Duarte CA 91010-3012

SMITH, STEWART EDWARD, physical chemist; b. Balt., Oct. 5, 1937; s. Ambrose Jefferson and Gladys Ruth (Stewart) S.; children: Nicole Catherine, Stewart Bradford. BS, Howard U., 1960; PhD, Ohio State U., 1969. Chemist Du Pont, Gibbstown, N.J., 1963-64; rsch. chemist Du Pont, Wilmington, Del., 1972-74; tech. svc. rep. Du Pont, Wilmington, 1974-78; rsch. chemist Sun Oil, Marcus Hook, Pa., 1969-71; coal chemist Exxon Rsch. and Engring., Baytown, Tex., 1976-81, group head, 1981-82; relocation coord. Exxon Rsch. and Engring., Clinton, N.J., 1982-84; sr. staff chemist Exxon Rsch. and Engring., Baytown, 1984-86; advisory engr. Westinghouse-Bettis, West Mifflin, Pa., 1986—. Contbr. articles to profl. jours. Mem. jr. high sch. adv. bd., Wilmington, 1975; coach Little League Baseball, Clear Lake City, Tex., 1979. Lt. U.S. Army, 1961-63. Recipient Pres.'s award Howard U. Alumni, Wilmington, 1976. Mem. AAAS, Am. Chem. Soc., Sigma Xi, Kappa Alpha Psi. Avocations: physical fitness, jogging, gourmet cooking. Home: 125 Amberwood Ct Bethel Park PA 15102-2252 Office: Westinghouse Elec Corp PO Box 79 West Mifflin PA 15122-0079

SMITH, SUSAN FINNEGAN, computer management coordinator; b. Charlottesville, Va., Feb. 18, 1954; d. Marcus Bartlett and Betsy Neil (Hammer) Finnegan; m. William Clark Smith, June 21, 1975; 1 child, Morgan Elizabeth. Student, Cornell U., 1972-73; BSc magna cum laude, Brown U., 1976, MSc, 1982. From assoc. through sr. software engr. Raytheon Corp., Portsmouth, R.I., 1976-84; mem. sr. tech. staff Oracle Corp., Braintree, Mass., 1984; mgr. DBMS project team Brown U., Providence, 1984-86, mgr. user svcs., 1986-87, mgr. info. resources, 1987-89; mgr. Learning Ctr. Sybase Inc., Emeryville, Calif., 1989-90; coord. data and sytems integration Brown U., Providence, 1990—.

SMITH, THOMAS BRADFORD, engineering administrator; b. Teaneck, N.J., Sept. 4, 1955; s. Bernard Lawrence and Virginia Ann (Wohlfarth) S.; m. Maria Firlej, Aug. 27, 1983; children: Lisa Ann, Ian Bradford. BSME, Rutgers U., 1977. Registered profl. engr., N.J. Sales engr. Johnson Controls, Long Island City, N.Y., 1977-78; v.p. Mayflower Vapor Seal Corp., Little Ferry, N.J., 1978-81; pres. Mayflower Electronics, Little Ferry, 1982-89; facility mgr. NASA, Houston, 1989—. Home: 1102 Larkspur Ln Seabrook TX 77586 Office: NASA Mail Code SP52 Houston TX 77058

SMITH, THOMAS GRAVES, JR., image processing scientist, neurophysiologist; b. Winnsboro, S.C., Mar. 22, 1931; s. Thomas G. and Mary Eula (Mungo) S.; m. Jo Ann Horn, June 16, 1956. BA, Emory U., 1953; BA, MA, Oxford (Eng.) U., 1956; MD, Columbia U., 1960; postgrad., MIT, 1964-66. Intern Bronx (N.Y.) Mcpl. Hosp. Ctr., 1960-61; sect. chief NIH, Bethesda, Md., 1966—; referee various sci. jours., including Nature, Sci., 1961—; referee, reviewer NSF, Washington, 1961—, NIH, 1970—. Contbr. articles to sci. jours. Capt. USPHS, 1961-63. Rhodes scholar, 1953. Mem. Am. Physiol. Soc., Soc. for Neurosci., Phi Beta Kappa, Alpha Omega Alpha. Democrat. Achievements include introducing concepts of fractal geometry to cytology; analyzing and designing voltage-clamp amplifiers; studying and analyzing neuronal conuctances; confocal microscopy. Home: 5512 Oakmont Ave Bethesda MD 20817 Office: NIH Bldg 36 Rm 2C02 Bethesda MD 20892

SMITH, THOMAS HUNTER, ophthalmologist, ophthalmic plastic and orbital surgeon; b. Silver Creek, Miss., Aug. 10, 1939; s. Hunter and Wincil (Barr) S.; m. Michele Ann Campbell, Feb. 27, 1982; 1 child, Thomas Hunter IV. BA, U. So. Miss., 1961; MD, Tulane U., 1967; BA in Latin Am. Studies, Tex. Christian U., 1987. Diplomate Am. Bd. Ophthalmology. Intern Charity Hosp., New Orleans, 1967-68; resident in ophthalmology Tulane U., New Orleans, 1968-71; dir., sec. bd. dirs. Ophthalmology Assocs., Ft. Worth, 1971—; clin. prof. Tex. Tech. U. Med. Sch., Lubbock, 1979—; bd. examiners Am. Bd. Ophthalmology, 1983-90; guest lectr., invited speaker numerous schs., confs., symposia throughout N.Am., Cen.Am., South Am. and India. Contbr. articles to profl. jours. Bd. dirs. Planned Parenthood Assn. Tarrant County; cons. ophthalmologist Hellen Keller Internat.; deacon South Hills Christian Ch. Recipient Tex. Chpt. award Am. Assn. Workers for the Blind, 1978, Recognition award Lions Club Sight & Tissue Found., Cen. Am., 1977-79; named to Alumni Hall of Fame U. So. Miss., 1989. Fellow Am. Acad. Ophthalmology, Am. Acad. Facial Plastic and Reconstructive Surgery, ACS; mem. AMA (com. socio-econs.), Pan-Am. Assn. Ophthalmology (adminstr. 1988—), Am. Soc. Contemporary Ophthalmology, Internat. Glaucoma Congress, Internat. Cos. Cryosurgery, Royal Soc. Medicine (affiliate), Tex. Soc. Ophthalmology and Otolaryngology, Peruvian Ophthalmol. Soc. (hon. mem.), Tex. Ophthalmol. Assn. (past mem. exec. coun., treas.), Tex. Med. Assn., Tarrant County Med. Soc., Byron Smith Ex Fellows Assn., Tarrant County Ophthalmology Soc., (past pres.), Tarrant County Multiple Sclerosis Soc. (past pres.), Tarrant County Assn. for Blind, Colonial Country Club, Petroleum Club Ft. Worth., Omicron Delta Kappa, Sigma Xi. Mem. Disciples of Christ. Avocations: hunting, fishing, world travel. Office: Ophthalmology Assocs 308 S Henderson St Fort Worth TX 76104-1015

SMITH, TROY ALVIN, aerospace research engineer; b. Sylvatus, Va., July 4, 1922; s. Wade Hampton and Augusta Mabel (Lindsey) S.; m. Grace Marie Peacock, Nov. 24, 1990. BCE, U. Va., 1948; MS in Engring., U. Mich., 1952, PhD, 1970. Registered profl. engr., Va., Ala. Structural engr. U.S. Army C.E., Norfolk, Va., Wilmington, N.C., Washington, 1948-59; chief structural engr. Brown Engring. Co., Inc., Huntsville, Ala., 1959-60; structural rsch. engr. U.S. Army Missile Command, Redstone Arsenal, Ala., 1960-63, aerospace engr., 1963-80, aerospace rsch. engr., 1980—. Contbr. articles to AIAA Jour. With USNR, 1942-46, PTO. Fellow Dept. Army, 1969. Mem. N.Y. Acad. Scis., Assn. U.S. Army, Elks, Sigma Xi. Achievements include research on procedures for analysis of structures. Home: 2406 Bonita Dr SW Huntsville AL 35801-3907 Office: US Army Missile Command Redstone Arsenal AL 35898

SMITH, VERNON SORUIX, neonatologist, pediatrician, educator; b. Dacoma, Okla., June 10, 1938; s. Guy Edward and Helen Marie (Rexroat) S.; children: Michelle Marie, Brian Patrick, Carol Wannette, Juanita Ann, Russell Wayne. BSBA, Phillips U., 1964; MS in Physiology, U. Okla., 1970; DO, Kansas City Coll. Osteo. Medicine, 1974. Dir. subhuman primate rsch. U. Okla. Coll. Medicine, Oklahoma City, 1963-70; staff pediatrician Indian Health Svc., Pawnee, Okla., 1977-79; prof. pediatrics, dir. sr. program Okla. State U. Coll. Osteo. Medicine, Tulsa, 1978-82; staff physician Dept. Corrections, Tulsa, 1980-92; asst. clin. prof. pediatrics U. Okla. Tulsa Med. Coll., 1984—; asst. dir. Ea. Okla. Perinatal Ctr., St. Francis Hosp., Tulsa, 1984-87; dir. spl. care nusery Hillcrest Med. Ctr., Tulsa, 1987—, sect. chief pediatrics, 1993—; dir. newborn ICU Tulsa Regional Med. Ctr., 1991—; owner Maverick Ranch, Hulbert, Okla.; speaker med. continuing edn. various orgns. Contbr. articles to profl. jours. Bd. mem. Faith United Meth. Ch., Tulsa, 1986-90; physician, missionary Nicaragua, 1990. Lt. commdr. U.S. Coast Guard, 1977-79. Fellow Am. Coll. Osteo. Pediatricians (dir. resident writing 1991); mem. Okla. Osteo. Assn., AM. Osteo. Assn., Cimarron Valley Osteo. Assn. Avocations: ranching, gardening. Office: Hillcrest Neonatal Assocs 1120 S Utica Ave Tulsa OK 74104-4090

SMITH, WALTER J., engineering consultant; b. Climax, Kans., Feb. 8, 1921; s. Jacob Walter and Thelma Christina (Stark) S.; m. Wanda Jean Sandys, Apr. 20, 1944 (div. 1965); children: Walter Brooke, Judith Jean; m. Evadean Louise Smith, Sept. 21, 1965; stepchildren: Stephen Henslee, Kimberly Ann; 1 adopted child, Nancy Louise. BEE, Cleve. State U., 1948; postgrad., UCLA, 1955-58, Western State U. Law, Anaheim, Calif., 1970-71. Lic. profl. engr., Ohio, Calif. Field tech. rep. to Air Force Jack & Heintz, Inc., Maple Hts., Ohio, 1942-44; rsch. engr. Jack & Heintz, Inc., 1948-50, N. Am. Aviation Inc., Downey, Calif., 1950-54; asst. chief engr. Ala. Engring. & Tool Co., Huntsville, Ala., 1954-55; rsch. specialist to dir. prodn. ops. N. Am. Aviation Inc./Rockwell Internat., Anaheim, 1955-86; engring. mgmt. cons. Anaheim, 1986-93; engring. mgmt. cons., Bermuda Dunes, Calif., 1993—. Contbr. articles to profl. jours. Mem. Anaheim Indsl. Devel. Bd., 1982-86, Anaheim Pub. Utilities Bd., 1987-92; bd. dirs. Rep. Cen. Com. of Orange County, 1976-78; pres., bd. dirs. Galerie Homeowners Assn., 1987-93; bd. dirs. Coun. on Environ. Edn. and Econ. Through Devel., Inc., 1974-86, Action Com. to Inform Orage Now, Inc. Mem. Anaheim C. of C. (bd. dirs. 1983-90, pres. 1983-84), Gladhanders Acad. Hospitality Internat. (bd. dirs. 1989—, Man of Yr. 1989). Republican. Religious Science. Avocations: golf, jewelry design, gardening, dancing. Home and Office: 78615 Purple Sagebrush Ave Bermuda Dunes CA 92201-9051

SMITH, WENDY ANNE, biologist, educator; b. Pitts., July 12, 1954; d. Martin B. and Marjorie A. (Rothenburg) Smith; m. Stephen P. Soltoff, Dec. 5, 1987; 1 child, Benjamin D. BA, New Coll., Sarasota, Fla., 1975; PhD, Duke U., 1981. Rsch. assoc. Duke U., Durham, N.C., 1981-83, U. N.C., Chapel Hill, 1983-85; asst. prof. biology Northeastern U., Boston, 1985-91, assoc. prof., 1991—. Contbr. articles to profl. jours. NIH grantee, 1986—. Mem. Am. Soc. Zoologists, Soc. for Cell Biology, Soc. for Neurosci., Sigma Xi. Office: Northeastern U Dept Biology 360 Huntington Ave Boston MA 02115

SMITH, WILLIAM LEO, meteorologist, researcher, educator; b. Detroit, May 13, 1942; s. Rolland Francis and Ruth Mary (Gerhardstein) S.; m. Marcia Jean Simmerer, Aug. 11, 1962; children: William Jr., Jeanne, Steven, Julie, Joanna, Jonathon, Sarah, Kiara. BS, St. Louis U., 1963; MS, U. Wis., 1964, PhD, 1966. Chief br. radiation NOAA/NESDIS U.S. Dept. Commerce, Washington, 1966-73, dir. devel. lab. NOAA/NESDIS, 1973-84; dir. CIMSS U. Wis., Madison, 1984—; Sec. Internat. Radiation Commn., 1989—. Author: (with others) Handbook of Applied Meteorology, 1983; contbr. over 100 articles to profl. jours. Fellow Am. Meteorol. Soc. Roman Catholic. Achievements include development of first algorithms for operational processing of satellite sounding data, of high special resolution interferometer sounder, of infrared temperature profile radiometer, of earth radiation budget radiometer, and of high resolution infrared radiation sounder. Office: U Wis 1225 W Dayton St Madison WI 53706-1695

SMITH, WILLIAM RAY, biophysicist, engineer; b. Lyman, Okla., June 26, 1925; s. Harry Wait and Daisy Belle (Hull) S. BA, Bethany Nazarene Coll., 1948; MA, Wichita State U., 1950; postgrad. U. Kans., 1950-51; PhD, UCLA, 1967. Engr., Beech Aircraft Corp., Wichita, Kans., 1951-53; sr. group engr. McDonnell Aircraft Corp., St. Louis, 1953-60; sr. engr. Lockheed Aircraft Corp., Burbank, Calif., 1961-63; sr. engr. scientist McDonnell Douglas Corp., Long Beach, Calif., 1966-71; mem. tech. staff Rockwell Internat., L.A., 1973-86, CDI Corp.-West, Costa Mesa, Calif., 1986-88, McDonnell Douglas Aircraft Corp., Long Beach, 1988—; tchr. math. Pasadena Coll. (now Point Loma Coll., San Diego), 1960-62, Glendale Coll., Calif., 1972; asst. prof. math. Mt. St. Mary's Coll., L.A., 1972-73. Contbr. articles to sci. jours. L.A. World Affairs Coun.; docent Nature Mus. Will Rogers State Pk. Recipient citation McDonnell Douglas Corp., 1968; Tech. Utilization award Rockwell Internat., 1981; cert. of recognition NASA, 1982. Mem. UCLA Chancellor's Assocs., Internat. Visitors Coun. L.A., Town Hall Calif., Yosemite Assocs., Sigma Xi, Pi Mu Epsilon. Republican. Presbyterian (deacon). Office: McDonnell Douglas Corp 3855 N Lakewood Blvd # 4159 Long Beach CA 90846-0001

SMITHIES, OLIVER, pathologist, educator; b. Halifax, Eng., July 23, 1925; naturalized citizen; PhD in Biochemistry, Oxford U., Eng., 1951. Postdoctoral fellow phys. chemistry U. Wis., Madison, 1951-53, from asst. prof. to prof. genetics and med. genetics, 1960-63, Leon J. Cole prof., 1971-80, Hilldale prof., 1980-88; rsch. asst., assoc. Connaught Med. Rsch. Lab., Toronto, Can., 1953-60; Excellence prof. pathology U. N.C., Chapel Hill, 1988—; mem. nat. adv. med. sci. coun. NIH, 1985. Contbr. articles to profl. jours. Recipient William Allen Meml. award Am. Soc. Human Genetics, 1964, Karl Landsteiner Meml. award Am. Assn. Blood Banks, 1984, Internat. award Gairdner Found., 1990, 93; Merkel scholar, 1961. Fellow AAAS; mem. NAS, Am. Acad. Arts & Sci., Genetics Soc. Am. (v.p. 1974, pres. 1975). Achievements include research on targetted modification of specific genes in living animals. Office: Univ of N C Dept of Pathology Chapel Hill NC 27514*

SMITH-KAYODE, TIMI, food technologist; b. Ogbomoso, Oyo, Nigeria, Jan. 17, 1956; came to U.S., 1992; s. Daniel Smith-Kayode and Comfort Folorunso; m. Helen Dickson; children: Tobi Aniekan, Dami Aniettie. BS in Biochemistry, U. Lagos, Nigeria, 1980; PhD in Food Tech., U. Ibadan, Nigeria, 1993. cert. post harvest technologist. Rsch. asst. U. Ibadan, Nigeria, 1987-88; rsch. officer Nihort, Nigeria, 1988-91; vis. scientist U. Hawaii, Honolulu, 1991-92; rsch. fellow East West Ctr., Inc., Honolulu, 1992; cons. RMRDC, Lagos, Nigeria, 1991-92; tech. advisor Naseni, Lagos, Nigeria, 1990-92. Grad. scholar U. Ibadan, Nigeria, 1987; fellow African-Am. Inst., 1992, Agy. Internat. Devel., Washington/Laos, Nigeria, 1992, IEDP East West Ctr., Hololulu, 1992; recipient Hawaii Children award Fuji Film, Honolulu, 1992. Mem. AAAS, Inst. Food Technologists (prof.), Soc. Pub. Analysts, Smithsonian Inst.. Episcopalian. Achievements include development of clarified tropical fruit juice concentrate, the use of plantains and banaba for table wine processing, upgrading of packaging techniques of tropical spicy food. Office: U Hawaii CTAHR 3050 Maile Way Honolulu HI 96822

SMOAK, KARL RANDAL, civil engineer; b. Shreveport, La., Oct. 21, 1959; s. Albert Madison Jr. and Irene (Rhoads) S.; m. Sandie Gay Majors, Aug. 2, 1985; children: Victoria Joy, Julianne Hope. Student, Northwestern State U., Natchitoches, La., 1977-78; BS in Civil Engring., La. Tech. U., 1981, postgrad., 1981-84. Registered civil engr., La., Ark. Staff engr. McDermott Inc., New Orleans, 1981-82; field engr. McDermott Internat. Inc., Dubai, United Arab Emirates, 1982; project engr. Coyle Engring. Co., Inc., Bossier City, La., 1983-92; owner, operator Smoak Engring., Dixie, La., 1992—. Editor: (newsletter) Civil Talk, 1991-92. Bossier C. of C. diplomat, 1987-88. Mem. ASCE (officer 1991-92), NSPE, La. Engring. Soc. (officer 1984-85), Rotary. Mem. United Pentecostal Ch. Home and Office: 8255 Dixie Shreveport Rd Dixie LA 71107

SMOKOVITIS, ATHANASSIOS A., physiologist, educator; b. Thessaloniki, Greece, June 2, 1935; s. Aristotle D. and Vassiliki Ch. (Karapali) S.; m. Despina K. Andreadis, Aug. 1, 1965. Diplomate in Vet. Medicine, Aristotelian U., 1957, diplomate in biology, 1966, PhD, 1968. Lectr. in physiology Faculty Veterinary Medicine Aristotelian U., Thessaloniki, 1970-73, prof., head dept. physiology, 1981—; rsch. assoc. Sch. Medicine Ind. U., Gary, 1975-76; vis. investigator Inst. Med. Rsch., Mitchell Found., Washington, 1973-75, Gaubius Inst. Health Rsch. Orgn., Leiden, Holland, 1976-77, Med. Sch. U. Vienna, 1977-81, hon. prof. physiology Med. Sch. U. Vienna, 1979. Author: Physiology, 1985, Topics in Physiopathology, 1992, and others; contbr. articles to profl. jours. Mem. Am. Physiol. Soc., Am. Heart Assn., others. Avocations: lit., philosophy, history. Home: Kon Melenikou 27, Thessaloniki 54635, Greece Office: Aristotelian U Dept Physiology, Faculty Vet Medicine, Thessaloniki 54006, Greece

SMOL, JOHN PAUL, limnologist, educator; b. Montreal, Que., Can., Oct. 10, 1955. BSc, McGill U., Can., 1977; MSc, Brock U., 1979; PhD in Paleolimnology, Queen's U., Kingston, Ont., 1982. Fellow Nat. Sci. and Engring. Rsch. Coun., 1982-83; asst. prof. biology Queen's U., 1984—; vis. scientist in paleolimnology Nat. Sci. and Engring. Rsch. Coun., Geol. Survey Can., 1983-84. Recipient Darbaker prize Botanical Soc. Am., 1991. Mem. Am. Soc. Limnology and Oceanography, Am. Phycological Assn., Soc. Can. Limnologists, Internat. Assn. for Theoetical and Applied Limnology, Internat. Phycological Assn., Freshwater Biol. Assn. Achievements include research in limnology and paleoecology of lakes, lake acidification, entrophication, high Artic and alpine lakes. Office: Queen's University, Kingston, ON Canada K7L 3N6*

SMOLEK, MICHAEL KEVIN, optics scientist; b. Knox, Ind., July 21, 1955; s. Frank David and Anna Marie (Wappel) S. BA, Ind. U., 1977, PhD, 1986. Rsch. assoc. visual sci. dept. Ind. U., Bloomington, 1981-82, assoc. instr. Optometry Sch., 1982-86; NIH rsch. fellow Emory U. Eye Ctr., Atlanta, 1987-89, assoc. ophthalmology, 1989-90; owner, cons. MKS Sci. Svcs., The Woodlands, Tex., 1990-91; NIH rsch. fellow La. State U. Eye Ctr., New Orleans, 1991-92, rsch. scientist, 1992—; manuscript referee major ophthalmol. jours., 1988—; NASA tech. briefs reader Opionion Panel Program Participant, 1992-93. Mus. vol. Confederate Air Force, Atlanta, 1989-90; artist, cons. Tchrs. Energy Conservation Program McDonald's Corp. and Ball State U., Muncie, Ind., 1979-80. Ezell fellow Am. Optometric Found., 1983-86; grantee in aid of rsch. Ind. U. Found., Bloomington, 1986, Sigma Xi, Nat. Acad. Scis., 1986. Mem. AAAS, Assn. Rsch. in Vision and Ophthalmology (cornea sect. moderator 1989), Sigma Xi, Sigma Pi Sigma. Roman Catholic. Achievements include development of real-time holographic interferometry for testing biomech. properties of the eye and applied holography to the surg. eye, derived elasticity modulus based on surface area; development of models to measure corneal tissue strength, and demonstrated anisotropic conditions in the human eye; discovered weak structural binding in the human cornea; derived new coefficients for describing corneal topography; contributed to the development of automatic keratoconus detection computer programs. Office: La State U Eye Ctr 2020 Gravier St Ste B New Orleans LA 70112

SMOOKE, MITCHELL DAVID, mechanical engineering educator, consultant; b. Hartford, Conn., Aug. 10, 1951; s. Irving Smooke and Jeannette (Rappaport) Kaufman; m. Deanna Lee Richards, Dec. 8, 1977; 1 child, Eric. BS, Renssalear Poly. Inst., 1973; PhD, Harvard U., 1978; MBA, U. Calif., Berkeley, 1983. Scientist Sandia Nat. Lab., Livermore, Calif., 1978-84; prof. mech. engring. Yale U., New Haven, 1984—; mem. microgravity working group NASA, Cleve., 1991—. Author, editor: Reduced Mechanisms and Asymptolic Approximations for Methane-Air Flames, 1991. Rsch. grantee NASA, 1991, Office Naval Rsch., 1991, Dept. Energy, 1992. Mem. AIAA, Combustion Inst. (mem. governing bd. ea. states sect. 1991—), Soc. Indsl. and Applied Math. (organizing bd. meeting numerical combustion 1987—), AFOSR. Achievements include research in combustion sci. and tech. and supercomputer applications. Office: Yale U 9 Hillhouse Ave New Haven CT 06520

SMOORENBURG, GUIDO FRANCISCUS, biophysicist, educator; b. Haarlem, The Netherlands, Aug. 2, 1943; s. Henk F. and Felicia (Dronkers) S.; m. Deana Olyslager, Apr. 3, 1968; children: Jeroen V., Claire C. MS, Utrecht U., 1967, PhD, 1971. Postdoctoral fellow U. Wis., Madison, 1971-72; cons., head dept. perception TNO Inst. for Human Factors, Soesterberg, The Netherlands, 1971—; prof. exptl. audiology Utrecht U., The Netherlands, 1980—; rep. The Netherlands to European Community Program Tech. for Communication in the Hearing Impaired, 1987-90. Editor: Frequency Analysis and Periodicity Detection in Hearing, 1970, Hearing Impairment and Signal Processing Hearing Aids, 1989, Signal Processing Hearing Aids, 1993; mem. editorial bd. Hearing Rsch., 1989—; assoc. editor: Acta Acustica; contbr. articles to profl. jours. Recipient postdoctoral stipend Dutch Sci. Found., The Hague, 1971-72. Mem. Acoustical Soc. Am., Assn. Rsch. Otolaryngology, Dutch Biophys. Soc. (chmn. auditory chpt. 1992—), Netherlands Biophysics Soc. (bd. dirs. 1987-91). Achievements include cofounder Utrecht Biophysics Inst., 1990; early description of nonlinear mechanics in the cochlea; internationally recognized damage risk criteria for impulse noise; methods for assessing speech perception by hearing impaired. Home: Vosseveldlaan 13a, Soest The Netherlands 3768 gk Office: TNO Inst for Human Factors, PO Box 23, 3769 ZG Soesterberg The Netherlands

SMOOT, GEORGE FITZGERALD, III, astrophysicist; b. Yukon, Fla., Feb. 20, 1945. BS in math., BS in physics, MIT, 1966, Ph.D. in physics, 1970. Rsch. physicist MIT, 1970; rsch. physicist Univ. Calif., Berkeley, Calif., 1971—, Lawrence Berkeley Lab., 1974—; team leader, differential microwave radiometer experiment, COBE (Cosmic Background Explorer) satellite. Recipient Space/Missiles Laurels award Aviation Week & Space Technology, 1993. Mem. Internat. Astron. Union, Am. Phys. Soc., Am. Astron. Soc., Sigma Xi.

SMOOT, JOHN ELDON, mechanical engineer; b. Wellington, Kans., Dec. 14, 1960; s. Delbert Joe and Deanna Kay (Dague) S.; m. Catherine Ann Schulze, Sept. 12, 1987; children: Christian Vann, Connor Alexander. AA, Independence (Kans.) C.C., 1981; BSME, Kans. State U., 1983; postgrad., U. Kans., 1984. Cert. engr.-in-tng., Mo. Staff engr. Allied-Signal Aerospace, Kansas City, Mo., 1984—. Mem. Internat. Soc. Hybrid Microelectronics. Republican. Lutheran. Achievements include patent applications. Home: 415 E 72d St Kansas City MO 64131 Office: Allied-Signal Aerospace 2000 E 95th St Kansas City MO 64141

SMOOT, LEON DOUGLAS, university dean, chemical engineering educator; b. Provo, Utah, July 26, 1934; s. Douglas Parley and Jennie (Hallam) S.; m. Marian Bird, Sept. 7, 1953; children: Analee, LaCinda, Michelle, Melinda Lee. BS, Brigham Young U., 1956, B in Engring. Sci., 1957; MS, U. Wash., 1958, PhD, 1960. Registered profl. engr., Utah. Engr. Boeing Corp., Seattle, 1956; teaching and research asst. Brigham Young U., 1954-57; engr. Phillips Petroleum Corp., Arco, Idaho, 1957; engr., cons. Hercules Powder Co., Bacchus, Utah, 1961-63; asst. prof. Brigham Young U., 1960-63; engr. Lockheed Propulsion, Redlands, Calif., 1963-67; vis. asst. prof. Calif. Inst. Tech., 1966-67; asso. prof. to prof. Brigham Young U., 1967—, chmn. dept. chem. engring., 1970-77, dean Coll. Engring. and Tech., 1977—; dir. Advanced Combustion Engring. Rsch. Ctr., 1986—; expert witness on combustion and explosions; dir. Advanced Combustion Engring. Research Ctr. (NSF), 1986—; cons. Hercules, Thiokol, Lockheed, Teledyne, Atlantic Research Corp., Raytheon, Redd and Redd, Billings Energy, Ford, Bacon & Davis, Jaycor, Intel Com Radiation Tech., Phys. Dynamics, Nat. Soc. Propellants and Explosives, France, DFVLR, West Germany, Martin Marietta, Honeywell, Phillips Petroleum Co., Exxon, Nat. Bur. Standards, Eyring Research Inst., Systems, Sci. and Software., Los Alamos Nat. Lab., others. Contbr. over 200 articles to tech. jours.; Author 5 books on coal combustion. Mem. Am. Inst. Chem. Engrs., Am. Inst. Aeros. and Astronautics, Am. Soc. Engring. Edn., Combustion Inst., Research Soc. Am., Tau Beta Pi, Phi Lambda Epsilon, Sigma Xi. Republican. Mem. LDS Ch. Home: 1811 N 1550 Provo UT 84604-5709 Office: Brigham Young U Advanced Combustion Engring Rsch Ctr 45 Crabtree Tech Bldg Provo UT 84602

SMULSKI, STEPHEN JOHN, wood scientist, consultant; b. Haverhill, Mass., Dec. 29, 1955; s. Stanley John and Constance Beatrice (Leathers) S.; m. Meryl Ann Mandell, Aug. 17, 1983. BS in Wood Sci. and Tech., U. Mass., 1977; MS in Environ. and Resource Engring., SUNY, Syracuse, 1980; PhD in Forestry and Forest Products, Va. Tech, 1985. Rsch. intern Scott Paper Co., Phila., 1977; vis. lectr. U. Mass., Amherst, 1985-87, asst. prof., 1987-92; pres. Wood Sci. Specialists Inc., Shutesbury, Mass., 1992—; adj. asst. prof. U. Mass, Amherst, 1992—. Contbr. numerous articles to profl. jours. Mem. Internat. Assn. Wood Anatomists, Forest Products Soc. (1st Place Wood award 1986, exec. bd. mem. 1991—), Soc. Wood Sci. and Tech., Sigma Xi. Office: Wood Sci Specialists Inc 3 Old Wendell Rd Shutesbury MA 01072

SMYRL, WILLIAM H., chemistry educator; b. Brownfield, Tex., Dec. 12, 1938; s. Garvin H. and Opal Faye (Coor) S.; m. Donna Kay Clayton, Nov. 29, 1964; children: Elliot K., Clifford G. BS in Chemistry, Tex. Tech U., 1961; PhD in Chemistry, U. Calif., Berkeley, 1966. Asst. prof. U. Calif., San Francisco, 1966-68; mem. tech. staff Boeing Sci. Rsch. Lab., Seattle, 1968-72, Sandia Nat. Lab., Albuquerque, 1972-84; prof., dir. corrosion rsch. ctr. U. Minn, Mpls., 1984—. Contbr. over 100 articles to refereed jours. Mem. AAAS, Electrochemical Soc. (chair corrosion divsn. 1990-92), Sigma Xi. Democrat. Baptist. Office: U Minn Corrosion Rsch Ctr 221 Church St SE Minneapolis MN 55455-0157

SMYTHE, WILLIAM RODMAN, physicist, educator; b. Los Angeles, Jan. 6, 1930; s. William Ralph and Helen (Keith) S.; m. Carol Richardson, Nov. 27, 1954 (dec. Dec. 1987); children: Stephanie, Deborah, William Richardson, Reed Terry; m. Judith Brean Travers, Jan. 1, 1989. B.S., Calif. Inst. Tech., 1951, M.S., 1952, Ph.D., 1957. Engr. Gen. Electric Microwave Lab., Palo Alto, Calif., 1956-57; asst. prof. U. Colo., 1958-63, assoc. prof., 1963-67, prof., 1967—, chmn. nuclear physics lab., 1967-69, 81-83, 90-92. Group leader Rocky Mountain Rescue Group, 1967-68. Mem. Am. Phys. Soc. Club: Colorado Mountain (Boulder). Achievements include inventing negative ion cyclotron, fractional turn cyclotron. Home: 2106 Knollwood Dr Boulder CO 80302-4706

SNAPER, ALVIN ALLYN, engineer; b. Hudson City, N.J., Sept. 9, 1929; m. Kathleen M. Scovel, Apr. 17, 1964; children: Sheryl, Curtis. BS, McGill U., Montreal, Can., 1949. Registered profl. engr., Calif. Sr. chemist Bakelite/Union Carbide, Bound Brook, N.J., 1949-52; chief chemist McGraw Colorgraph Co., Burbank, Calif., 1952-56; chief engr. Houston Fearless Corp., L.A., 1958-62; sr. engr. Marquardt Corp., Van Nuys, Calif., 1962-65; v.p. Advanced Patent Tech., Las Vegas, Nev., 1968-73; pres. Neo-Dyne Rsch., Inc., Las Vegas, 1979--; cons. Sumitomo/JCC, Kanagawa, Japan, 1991—, Multi-Arc Vacuum Systems Inc., St. Paul, Minn., 1981-85, SGC, Internat., St. Petersburg, Russia, 1992—; dir. Advanced Patent Tech., Inc., Las Vegas, 1968-79. Recipient Patent of Yr. award Design News Mag., 1968, 70, 73. Achievements include over 600 U.S. and foreign patents. Home: 2800 Cameo Circle Las Vegas NV 89107 Office: Neo-Dyne Rsch Inc 1000 W Bonanza Rd Las Vegas NV 89106

SNAPPER, ERNST, mathematics educator; b. The Netherlands, Dec. 2, 1913; came to U.S., 1938, naturalized, 1942; s. Isidore and Henrietta (Van Buuren) S.; m. Ethel Lillian Klein, June 1941; children—John William, James Robert. MA, Princeton U., 1939, PhD, 1941; MA (hon.), Dartmouth Coll., 1964. Instr. Princeton, 1941-45, vis. asso. prof., 1949-50, vis. prof., 1954-55; asst. prof. U. So. Calif., 1945-48, asso. prof., 1948-53, prof., 1953-55; NSF post-doctoral fellow Harvard, 1953-54; Andrew Jackson Buckingham prof. math. Miami U., Oxford, Ohio, 1955-58; prof. math. Ind. U., 1958-63; prof. math. Dartmouth, 1963—, Benjamin Pierce Cheney prof. math., 1971—. Mem. Am. Math. Soc., Math. Assn. Am. (pres. Ind. sect. 1962-63, Carl B. Allendoerfer award 1980), Assn. Princeton Grad. Alumni (governing bd.), Soc. for Preservation Bridges of Konigsburg, Phi Beta Kappa (hon.), Pi Mu Epsilon (hon.). Home: PO Box 67 Norwich VT 05055-0067

SNAREY, JOHN ROBERT, psychologist, researcher, educator; b. New Brighton, Pa., Jan. 12, 1948; s. John Herbert and Esther Snarey; m. Carol Dunn Snarey, June 11, 1970; children: Johnny, Elizabeth. BS, Geneva Coll.,

1969; MA, Wheaton (Ill.) Coll., 1973; EdD, Harvard U., 1982. Postdoctoral rsch. fellow dept. psychiatry Harvard U., Cambridge, Mass., 1982-84; assoc. rsch. psychologist Wellesley (Mass.) Coll., 1984-85; assoc. prof. human devel. Northwestern U., Evanston, Ill., 1985-87, Emory U., Atlanta, 1987—. Mem. editorial bd. Harvard Ednl. Rev., 1979-81; mem. editorial adv. bd. Lawrence Erlbaum Assocs., 1988-90; contbr. numerous articles to profl. jours. Mem. APA, Am. Ednl. Rsch. Assn. (div. E exec. bd. 1990—), Assn. for Moral Edn. (exec. bd. 1986—), Soc. for Rsch. in Child Devel., Nat. Coun. on Family Rels. Avocation: bird watching. Home: 2165 Pine Forest Dr NE Atlanta GA 30345-4184

SNAVELY, WILLIAM PENNINGTON, economics educator; b. Charlottesville, Va., Jan. 25, 1920; s. Tipton Ray and Nell (Aldred) S.; m. Alice Watts Pritchett, June 4, 1942; children: Nell Lee, William Pennington, Elizabeth Tipton. Student, Hampden-Sydney Coll., 1936-37; BA with honors, U. Va., 1940, MA, 1941, PhD, 1950; postgrad. (Bennett Wood Green fellow), Harvard U., 1946-47. Mem. faculty U. Conn., 1947-73, prof. econs., 1961-73, chmn. dept., 1966-72, economist econ. edn. workshop, summers 1954, 55, 56; prof. econs. George Mason U., 1973-86, chmn. dept., 1973-81; acting dean CAS, summer 1981, 85-86, assoc. dean, 1982-85; prof. econs. Liberty U., 1986-93; cons. Ford Found., Jordan Devel. Bd., Amman, 1961-62, Ministry of Planning, Beirut, Lebanon, 1964-65, Saudi Arabian Cen. Planning Orgn., Riyadh, 1964-65, Am. U. Beirut, 1969-70, Bahrain Ministry Fin. and Nat. Economy, 1974, 75, 76, UN, Jordan Nat. Planning Coun., Amman, 1972; mem. Danforth Workshop, summer 1966; mem. adv. com. Willimantic Trust Co., Conn., 1968-73; v.p. Contemporary Econs. & Bus. Assn., 1988—. Author: (with W.H. Carter) Intermediate Economic Analysis, 1961, Theory of Economic Systems, 1969, (with M.T. Sadik) Bahrain, Qatar and the United Arab Emirates, 1972; contbr. articles to jours.; articles Ency. Americana. Bd. regents Liberty U. Capt. AUS, 1942-46. Fellow Fund Advancement Edn. Harvard, 1951-52; faculty-bus. exchange fellow Chase Nat. Bank, N.Y.C., summer 1952; fellow Merrill Center Econs., summer 1957; Fulbright research fellow Rome, 1958-59. Mem. Am. Econ. Assn., So. Econ. Assn., Assn. Christian Economists, Assn. Comparative Econs., Va. Assn. Economists (pres. 1979-80), Phi Beta Kappa, Phi Kappa Phi. Home: 1551 Dairy Rd Charlottesville VA 22903-1303

SNEARY, MAX EUGENE, physician; b. Zanesfield, Ohio, Oct. 13, 1930; s. Kenneth Douglas and Grace Agnes (Yeiser) S.; m. Joy Ann Preston, Apr. 4, 1950; children: Candice Barbulesco, Jennifer Laur. Student, Wabash Coll., 1949-52; MD, Ind. U., 1956. Pvt. practice Avilla, Ind., 1957—; coroner Noble County (Ind.) Coroner's Office, 1960-64, 72-76; pres. bd. dirs. McCray Meml. Hosp., Kendallville, Ind., 1961-64; mem. Noble County Bd. of Health, 1962-65, 76-80, health officer, 1965-67. Bd. dirs. Kendallville Bank & Trust Co., 1980-88, chmn., 1987-88. Named Citizen of the Yr., Town of Avilla, 1988, Indian Family Physician of the Yr., 1990. Fellow Am. Acad. Family Physicians; mem. AMA, Am. Bd. Family Physicians (charter), Ind. Med. Assn., Noble County Med. Soc., Alpha Omega Alpha. Avocations: reading. Home and Office: 205 Baum St # 140 Avilla IN 46710

SNEEGAS, STANLEY ALAN, air force officer, aerospace engineer; b. Lawrence, Kans., May 31, 1950; s. Byron Carl and Mary Jeanne (Johnson) S.; m. Barbara Kline, Aug. 16, 1980; children: Andrew Kenneth, Alan Byron. BS in Aero. Engring., U. Kans., 1973; MS in Systems Mgmt., Air Force Inst Tech., 1982. Commd. 2d lt. USAF, 1974, advanced through grades to lt. col., 1991; engine mgr. Joint Cruise Missile Project Office USAF, Washington, 1978-81; F-15 systems mgr. Air Force Systems Command USAF, Andrews AFB, Md., 1982-86; chief spacecraft div. Space Sytstems Div. USAF, Los Angeles AFB, Calif., 1986-91; chief space test div. Phillips Lab. USAF, Kirtland AFB, N.Mex., 1991—. Mem. AIAA (sr.), Air Force Assn. (life), Sigma Iota Epsilon, Alpha Phi Omega. Office: PL SXS 3550 Aberdeen Ave SE Kirtland AFB NM 87117-5776

SNELL, GEORGE DAVIS, geneticist; b. Bradford, MA, Dec. 19, 1903; s. Cullen Bryant and Katharine (Davis) S.; m. Rhoda Carson, July 28, 1937; children: Thomas Carleton, Roy Carson, Peter Garland. B.S., Dartmouth Coll., 1926; M.S., Harvard U., 1928, Sc.D., 1930; M.D. (hon.), Charles U., Prague, 1967; LL.D. (hon.), Colby Coll., 1982; Sc.D. (hon.), Dartmouth Coll., 1974, Gustavus Adolphus Coll., 1981, U. Maine, 1981, Bates Coll., 1982, Ohio State U., 1984. Instr. zoology Dartmouth Coll., 1929-30, Brown U., 1930-31; asst. prof. Washington U., St. Louis, 1933-34; rsch. assoc. Jackson Lab., 1935-73, sci. adminstr., 1949-50, sr. staff scientist, 1957-73, sr. staff scientist emeritus, 1973—. Author: Search for a Rational Ethic, 1988, (with others) Histocompatibility, 1976; also sci. papers in field; editor: The Biology of the Laboratory Mouse, 1941. Recipient Bertner Found. award in field cancer research, 1962; Griffin award Animal Care Panel, 1962; career award Nat. Cancer Inst., 1964-68; Gregor Mendel medal Czechoslovak Acad. Scis., 1967; Internat. award Gairdner Found., 1976; Wolf Found. prize in medicine, 1978; award Nat. Inst. Arthritis and Infectious Disease-Nat. Cancer Inst., 1978; Nobel prize in medicine (with Dausset and Benacerraf), 1980; NRC fellow U. Tex., 1931-33; NIH health research grantee for study genetics and immunology of tissue transplantation, 1950-73 (allergy and immunology study sect. 1958-62); Guggenheim fellow, 1953-54. Mem. Nat. Acad. Scis., Transplantation Soc., Am. Acad. Arts and Sci., French Acad. Scis. (fgn. asso.), Am. Philos. Soc., Brit. Transplantation Soc. (hon.), Phi Beta Kappa. Home: 21 Atlantic Ave Bar Harbor ME 04609-1703

SNELL, KAREN BLACK, audiologist, educator; b. Chgo., May 30, 1949; d. William Mitchell and Earla (Musselman) Black; m. Stephen Conrad Snell, Aug. 28, 1971; children: Emily, Albert. BA, U. Chgo., 1970; PhD, U. Iowa, 1984. Cert. Clin. Competence-Audiology. Cons. Rochester (N.Y.) Inst. Tech., 1983-86, asst. prof., 1986-93, assoc. prof., 1993—. Contbr. articles to sci. jours. Program Project grantee Nat. Inst. in Aging, 1992—. Mem. Acoustical Soc. Am., Am. Speech Lang. and Hearing Assn., Endl. Audiology Assn. (N.Y. State rep. 1992—), Sigma Xi. Achievements include psychoacoustics, speech perception and deafness. Home: 228 Oakdale Dr Rochester NY 14618 Office: Rochester Inst Tech 52 Lomb Mem Dr Rochester NY 14618

SNIDER, JAMES RHODES, radiologist; b. Pawnee, Okla., May 16, 1931; s. John Henry and Gladys Opal (Rhodes) S.; B.S., U. Okla., 1953, M.D., 1956; m. Lynadell Vivion, Dec. 27, 1954; children—Jon, Jan. Intern, Edward Meyer Meml. Hosp., Buffalo, 1956-57; resident radiology U. Okla. Med. Center, 1959-62; radiologist Holt-Krock Clinic and Sparks Regional Med. Center, Ft. Smith, Ark., 1962—. Dir. Fairfield Community Land Co., Little Rock, 1968-87, Fairfield Communities, Inc., 1968-87. Mem. Ark. Bd. Pub. Welfare, 1969-71. Bd. dirs. U. Okla. Assn., 1967-70, U. Okla. Alumni Devel. Fund, 1970-74; bd. visitors U. Okla. Served to lt. comdr. USNR, 1957-62. Mem. Am. Coll. Radiology, Radiol. Soc. N.Am., Am. Roentgen Ray Soc., AMA, Phi Beta Kappa, Beta Theta Pi (trustee corp.), Alpha Epsilon Delta. Asso. editor Computerized Tomography, 1976-88. Home: 5814 S Cliff Dr Fort Smith AR 72903-3845 Office: 1500 Dodson Ave Fort Smith AR 72901-5193

SNIDERMAN, MARVIN, dentist; b. Pitts., Oct. 23, 1923; s. Abraham and Rebecca (Hecht) S.; B.S. in Pharmacy, U. Pitts., 1943, D.M.D., 1947; m. Eleanore Jessie Cohen, Dec. 25, 1947; 1 child, Abby Milstein. Pvt. dental practice Pitts., 1947-50, 53—; chief oral surgery dept. Pitts. Skin and Cancer Found., 1947-67. Mem. dental adv. com. Allegheny County Dept. Health, div. dental health, 1958-70; mem. undergrad. and postgrad. faculty U. Pitts. Sch. Dental Medicine, 1962—, assoc. prof. oral medicine, 1972-89, clin. assoc. prof. diagnostic svcs., 1989—; mem. charter council, 1986—; chief dental service Rehab. Inst. Pitts., 1948—. Mem. health adv. com. Pitts. Bd. Pub. Edn., 1965-70, dental health com. Mayor's Com. on Human Resources, Operation Head Start, 1965, adv. com. on health Mayor's Com. on Human Resources, 1965-70, charter council U. Pitts., 1986—; dental cons. USPHS, 1965-78; bd. dirs. Delta Dental of Pa., 1971-78; health edn. com. Allegheny County Adv. Council, 1972-74, 78—; mem. dental adv. com. Office Med. Programs Pa. Dept. Pub. Welfare, 1976-89; bd. dirs., sec. editorial com. Jewish Chronicle of Pitts., 1987—; vis. lectr. Emory U., U. Pa., Polyclinic and French Med. Sch., N.J. Coll. Medicine Dentistry, Temple Dental Sch. Albert Einstein Coll. Medicine, NYU Dental Sch., N.Y.C.; bd. dirs. Health/Edn. Commn. of Allegheny County, 1978—. Editor Odontological Bull. Western Pa., 1960-70, Pa. Dental Assn., 1970-72; cons. editor Jour. Dental Practice Adminstrn., 1980—; abstractor Jour. Oral Research Abstracts; contbr. articles in field. Served with AUS, 1943-44, to capt., Dental Corps, 1950-53. Recipient Bicentennial Medallion of Distinction, 1987 U. Pitts. Fellow Am. Coll. Dentists (pres. Pitts. sect. 1972), Acad. Dentistry for Handicapped (charter), Acad. Gen. Dentistry (master), (charter), Soc. Oral Physiology and Occlusion, Internat. Coll. Dentists (dep. regent 1970-76, counselor 1981—), Am. Endodontic Soc. (charter), Internat. Coll. Applied Nutrition, Acad. Oral Medicine (academic, charter), Acad. Dentistry Internat. (charter), Am. Soc. Dentistry for Children, Am. Acad. Craniomandibular Orthopedics (charter), Acad. Stress and Chronic Disease (charter), Am. Equilibration Soc.; mem. AAUP, ADA (councils on journalism and dental rsch. and coun. on dental practice 1989, cons. editor jour.), Pa. Dental Assn. (vis. lectr., editor 1970-92, editor emeritus 1992—), Pierre Fauchard Acad. (Annual award 1992), Am. Soc. Acupuncture, Am. Assn. Functional Orthodontics, Am. Acad. Oral Medicine (charter), Am. Soc. Assn. Execs., N.Y. Acad. Scis., Am. Assn. Dental Editors, Internat. Assn. Study Pain, Am. Pain Soc. (Charter), Odontological Soc. Western Pa. (pres. 1965, Albert R. Pechan award of excellence, 1981), Am. Internat. Acad. Preventive Medicine, Am. Prosthodontic Soc., Internat. Assn. Dental Research, AAAS, Am. Assn. Hosp. Dentists, Am. Assn. Dental Sch., Am. Acad. Implant Dentistry, Fedn. Prosthodontic Assns., Am. Endodontic Soc., Am. Dental Soc. Anesthesiology, Am. Analgesia Soc., Am. Acad. Dental Practice Adminstrn., Internat. Coll. of Oral Implantologists, U. Pitts. Dental Alumni Assn. (pres. 1973-74, Alumnus Distinction 1986), Am. Med. Writers Assn. Jewish. Home: 5633 Callowhill St Pittsburgh PA 15206-1452 Office: 204 5th Ave Pittsburgh PA 15222

SNIFFEN, PAUL HARVEY, electronic engineer; b. Red Bank, N.J., Jan. 2, 1942. AA, Thomas Edison Coll., BA, 1976. Cert. tchr., N.J. Sr. assoc. AT&T Bell Labs., Holmdel, N.J., 1981-84; sr. engr. U.S. Army CECOM, Ft. Monmouth, N.J., 1985-90; electronic engr. Aero. Test Techs., Red Bank, 1990—. With USN, 1959-65. Recipient Energy Edn. award N.J. Edn. Assn., 1976. Achievements include design and testing of first audio-video codec for commercial television broadcast. Office: Aero Test Tech PO Box 124 Red Bank NJ 07701

SNIPES, JOSEPH ALLAN, research scientist, physicist; b. Atlanta, Sept. 2, 1959; s. Robert Paul and Alice Catherine (Dunn) S.; m. Francesca Bombarda, Dec. 28, 1991. AB in Physics, U. Chgo., 1981; PhD in Physics, U. Tex., 1985. Rsch. asst. Enrico Fermi Inst., Chgo., 1978-81; teaching asst. U. Tex. Dept. Physics, Austin, 1981-82; rsch. asst. U. Tex. Fusion Rsch. Ctr., Austin, 1982-85; rsch. assoc. Culham Lab., Abingdon, Oxon, U.K., 1985-86; rsch. scientist JET Joint Undertaking, Abingdon, Oxon, U.K., 1986-89, ENEA-Frascati, Frascati, Italy, 1989-90, MIT Plasma Fusion Ctr., Cambridge, Mass., 1990—. Contbr. articles to profl. jours. Mem. Am. Phys. Soc., Charles River Wheelmen. Office: MIT Plasma Fusion Ctr 175 Albany St Cambridge MA 02139

SNITCH, THOMAS HAROLD, science educator, consultant; b. Cleve., July 14, 1954; s. Harold and Betty (Siek) S.; m. Mary Leslie Lassiter, Oct. 13, 1990. BA in Asian Studies, Bowling Green State U., 1975; MA in Internat. Econs., The Am. U., 1977, PhD in Internat. Econs., 1981. Dir. fgn. policy programs The Am. U., Washington, 1977-82; sr. polit. advisor U.S. Arms Control and Disarmament Agy., Washington, 1982-87; dir. strategic studies The Applied Scis. Corp., Arlington, Va., 1987-89; dir. study NAS, Washington, 1989-91; dir. internat. programs Applied Rsch. Lab., Arlington, 1991; CEO Little Falls Assocs. Inc., Bethesda, Md., 1992—; guest scholar The Brookings Instn., Washington, 1976. Author: International Terrorism, 1982, (study) Finding Common Ground, 1991. Pres. Green Acres-Glen Cove Citizen's Assn., Bethesda, 1989—; coord. Nat. Cathedral Homeless Program, Washington, 1989—; dist. coord. Congresswoman Connie Morella, 1988—. Internat. scholar NASA, 1977; doctoral scholar The Am. U., 1979. Mem. AIAA (sr. mem.), Nat. Space Club (sr. mem.). Republican. Epsicopalian. Home: 5202 Little Falls Dr Bethesda MD 20816 Office: Little Falls Assocs Inc 5205 Little Falls Dr Bethesda MD 20816

SNITZER, ELIAS, physicist; b. Lynn, Mass., Feb. 27, 1925; s. Isaac and Jenny (Sussman) S.; m. Shirley Ann Wood, Nov. 22, 1950; children—Sandra, Barbara, Peter, Helen, Louis. B.S.E.E., Tufts U., 1946; M.S. in Physics, U. Chgo., 1950, Ph.D., 1953. Research physicist Honeywell Corp., Phila., 1954-56; assoc. prof. Lowell Technol. Inst., Mass., 1956-58; dir. research Am. Optical Co., Southbridge, Mass., 1959-76; mgr. applied physics United Technologies Research, East Hartford, Conn., 1977-84; mgr. fiber optics Polaroid, Cambridge, Mass., 1984-88; prof. Rutgers U., 1989—. Contbr. articles to profl. jours. Inventor glass laser. Served with USN, 1943-46. Fellow Optical Soc. Am., Ceramic Soc.; mem. NAE, IEEE (George Morey award 1971, Quantum Electronics award 1979, Charles Townes award 1991), Am. Phys. Soc. Democrat. Jewish. Home: 8 Smoke Tree Close Piscataway NJ 08854-5109 Office: Rutgers U Fiber Optics Materials Rsch Program PO Box 909 Piscataway NJ 08855-0909

SNODGRASS, SAMUEL ROBERT, neurologist; b. Sept. 13, 1937; s. Samuel R. and Margaret (Kinney) S.; m. Kay B. Kessler, Aug. 20, 1960; children: Bryan R., Wayne K., Angela. BA, Harvard Coll., 1959; MD, Harvard Med. Sch., Boston, 1963. Resident trainee dept. neurology Boston Children's Hosp., 1967-70; postdoctoral trainee MRC Neurochem. Pharm. Unit, Cambridge, Eng., 1971-73; asst. prof. neurology Harvard Med. Sch., Boston, 1973-76, assoc. prof., 1976-79; chief child neurology Children's Hosp. L.A. Sch. Medicine, 1979-92, U. Miss. Med. Ctr., Jackson, 1992—; physician mem. Venezuela Huntington's Disease Rsch. Group, Santa Monica, Calif., 1983—. Author: (with M.J. Bresnan & K.K. Nakano) Pediatric Neurology Review, 1976; contbr. articles to profl. jours. Treas. Parents After Sch. Day Care Program, Newton, Mass., 1973-79; organizer Asyo Soccer League, Pasadena, Calif., 1982-90. Lt. USNR, 1964-66. Recipient Tchr. Investigator award NIH, Bethesda, Md., 1970-75. Mem. Soc. Neurosci., Child Neurology Soc., Am. Acad. Neurology, Soc. Neurology and Philosophy. Achievements include research on epilepsy and neurotransmitter receptors. Office: U Miss Med Ctr 2500 N State St Jackson MS 39216

SNOW, BLAINE ARLIE, language educator; b. Seattle, Sept. 19, 1956; s. Donald Dugeon and Shirley (Auer) S.; m. Tuula Sorsa, June 26, 1982; children: Emilia, Saara (twins). AA, North Seattle C.C., 1976; BS, Evergree State Coll., 1988. Engring. asst. Tone Commdr. Systems, Redmond, Wash., 1976-82; lang. instr. Vantaa Mcpl. Sch., Helsinki, Finland, 1982-86, EF Internat. Sch., Olympia, Wash., 1990—; dir. Sys. Edn. Rsch. Project. Contbr. articles to profl. jours. Mem. The Nature Conservancy, Environ. Def. Fund. Mem. Am. Soc. for Cybernetics, The Elmwood Inst. (cons. 1989—), Internat. Soc. for the System Scis. (systems edn. com. 1991, cons. 1990—). Office: EF Lang Sch Evergreen State Coll Olympia WA 98505

SNOW, CLYDE COLLINS, anthropologist; b. Ft. Worth, Tex., Jan. 8, 1928; married; 5 children. BS, Ea. NMex. U., 1950; MS, Tex. Tech. Coll., 1955; PhD, U. Ariz., 1967. Diplomate Am. Bd. Forensic Anthropology. Rsch. asst. anatomy Med. Sch. S.C., 1960-61, rsch. anthropologist, 1961-65; chief application biology sect. Civil Aeromed. Inst. Fed. Aviation Agy., 1965-69, chief phys. anthropology rsch., 1969-79; forensic anthrop. cons., 1979—; from adj. instr. to adj. asst. prof. anthropology U. Okla., 1962-80, adj. prof., 1980—, rsch. assoc. Sch. Medicine, 1964—; trustee Forensic Sci. Found., 1973-79; forensic anthrop. cons. Okla. State Med. Examiner, 1978—; med. exaniner Cook County, Ill., 1979—; cons. select com. assassinations U.S. Ho. of Reps., 1978-79; pres. Forensic Sci. Edn., Inc., 1982-86. Mem. Am. Acad. Forensic Sci. (v.p. 1978-79), Am. Anthrop. Assn., Am. Assn. Phys. Anthropology, Am. Soc. Forensic Odontology, Soc. Study Human Biology, Sigma Xi. Achievements include research in forensic anthropology, study of human skeletal remains to establish personal identification and cause of death. Office: Oklahoma State Medical Examiner 901 N Stonewall Oklahoma City OK 73117*

SNOW, JAMES BYRON, JR., physician, research administrator; b. Oklahoma City, Mar. 12, 1932; s. James B. and Charlotte Louise (Andersen) S.; m. Sallie Lee Ricker, July 16, 1954; children: James B., John Andrew, Sallie Lee Louise. B.S., U. Okla., 1953; M.D. cum laude, Harvard U., 1956; M.A. (hon.), U. Pa., 1973. Diplomate: Am. Bd. Otolaryngology (dir. 1972-90). Intern Johns Hopkins Hosp., Balt., 1956-57; resident Mass. Eye and Ear Infirmary, Boston, 1957-60; prof., head dept. otorhinolaryngology Sch. Medicine U. Okla., Oklahoma City, 1962-72; prof., chmn. dept. otorhinolaryngology and human communication U. Pa. at Phila., 1972-90; dir. Nat. Inst. on Deafness and Other Communication Disorders, Bethesda, Md., 1990—; Mem. nat. adv. council neurol. and communicative disorders and stroke NIH, 1972-76, 82-86 ; chmn. Nat. Com. Research Neurol. and Communicative Disorders, 1979-80. Editor: Am. Jour. Otolaryngology, 1979-83; Contbr. articles to sci. and profl. jours. Served with M.C. AUS, 1960-62. Recipient Regents award for superior teaching U. Okla., 1970, Golden award Internat. Fedn. Otorhinolaryngological Socs., 1989; named to Soc. of Scholars Johns Hopkins U., 1991. Fellow Japan Broncho-Esophagological Soc. (hon.); mem. ACS (regent 1982-90), AMA (coun. on sci. affairs 1975-86), Soc. Univ. Otolaryngologists (pres. 1975), Am. Acad. Otolaryngology-Head and Neck Surgery, Assn. Acad. Depts. Otolaryngology (pres. 1981-82), Am. Laryngol., Rhinol., and Otol. Soc., Am. Otol. Soc., Am. Laryngol. Assn. (editor 1983-89, pres. 1990-91), Am. Broncho-esophagol. Assn. (editor trans. 1973-77, pres. 1979), Collegium Otorhinolaryngologicum, Phi Beta Kappa, Alpha Omega Alpha. Home: 119 Discoll Way Gaithersburg MD 20878 Office: Nat Inst on Deafness and Other Communication Disorders Bethesda MD 20892

SNOW, STEVEN ASHLEY, chemist; b. Albuquerque, N.Mex., Sept. 30, 1959; s. Benjamin Franklin and Geraldine (Davis) S.; m. Sarah Jean Severson, July 7, 1984; children: Andrew Roland, Jessica Ashley. BS, U. N.Mex., 1979; PhD in Chemistry, U. Utah, 1985. Chemist CORE Labs., Albuquerque, 1980; teaching and rsch. asst. U. Utah, Salt Lake City, 1980-85; sr. rsch. specialist Dow Corning Corp., Midland, Mich., 1985—. Contbr. articles to profl. jours. U. N.Mex. Presdl. Scholar, 1976-79. Mem. Am. Chem. Soc., N.Y. Acad. Sci., Sigma Xi. Presbyterian. Achievements include patents on the discovery and application of siloxane based surfactants; discovery of novel Zwitterionic and cationic siloxane surfactants, silicone vesicles (liposomes), novel metal complexes of neutral boron hydrides. Office: Dow Corning Corp Mail # C042A1 Midland MI 48686-0994

SNOW, THEODORE PECK, JR., astrophysicist, author; b. Seattle, Jan. 30, 1947; s. Theodore P. and Louise (Wertz) S.; s. Constance M. Snow, Aug. 23, 1969; children: McGregor A., Tyler M., Reilly A. BA, Yale U., 1969; MS, U. Wash., 1970, PhD, 1973. Mem. rsch. staff Princeton (N.J.) U., 1973-77; prof. U. Colo., Boulder, 1977—, dir. Ctr. for Astrophysics and Space Astronomy, 1986—. Author: (textbook) The Dynamic Universe, 1983, 4th edit., 1991, Physics, 1986; contbr. over 200 articles to profl. jours. Mem. Am. Astron. Soc., Astron. Soc. of the Pacific, Sigma Xi. Achievements include discovery, through observations in ultraviolet visible, and infrared monolengths, of several important processes involving interstellar gas and dust, and their roles in star formation and late stages of stellar evolution. Office: U Colo Ctr Astrophysics Space Astronomy Campus Box 389 Boulder CO 80309

SNOW, W. STERLING, biology and chemistry educator; b. Devils Lake, N.D., Feb. 14, 1947; s. Morgan Williams and Josephine Elizabeth Ann (Erickstad) S.; m. Barbara Kay Jolley, Aug. 29, 1976; 1 child, Michelle Rene. AB, U. Calif., Santa Cruz, 1970; tchr. credential, U. Calif., Santa Barbara, 1971; MA, Chapman Coll., 1976. Cert. secondary sch. tchr., Calif., Alaska; cert. adminstrn., Calif. Tchr., coach Monterey (Calif.) Peninsula Unified Sch. Dist., 1972-76; tchr., coach Anchorage (Alaska) Sch. Dist., 1976—, athletic dir., 1987-92, tchr., 1992—; conf. asst. U. Calif., Santa Cruz, 1971-78. Bd. dirs. Dimond Alumni Found., Anchorage, 1987-92. Mem. AAAS, ASCD, Nat. Assn. Biology Tchrs., Nat. Interscholastic Athletic Adminstrs. Assn. (life), Nat. Assn. Basketball Coaches, Alaska Sci. Tchrs. Assn., N.Y. Acad. Scis. Lutheran.

SNYDER, CHARLES THEODORE, geologist; b. Powell, Wyo., July 19, 1912; s. Lee G. and Eda Belle (Hansen) S.; m. Marion Ruth Harris, Dec. 22, 1945 (dec. 1973); children: Anita Maria, Kristin Eileen; m. Alberta Irene Dangel, Oct. 15, 1973. BS, U. Ariz., 1948. Registered profl. geologist, Calif. Hydrologist U.S. Geol. Survey, Menlo Park, Calif., 1948-75; dir. Scotts Valley (Calif.) Water Dist., 1980-84; vis. scientist Carter County Mus., Ekalaka, Mont., 1983-84; researcher Resurgent Lakes in Western U.S. Author: Effect of Off-Road Vehicles, 1976. Disaster chmn. ARC, 1982-83. Mem. AAAS, Arctic Inst. N.Am., Soc. Vertebrate Paleontologists. Republican. Presbyterian. Home: 552-17 Bean Rd Scotts Valley CA 95066

SNYDER, DONALD BENJAMIN, biology educator; b. N. Manchester, Ind., Oct. 6, 1935; s. Benjamin Franklin and Eva Katherine (Speicher) S.; m. Wilma Frankie Simpson, Aug. 8, 1965; children: Douglas, Jonn. BS, Manchester Coll., Ind., 1957; MS, Ohio State U., 1959, PhD, 1963; postgrad., U. Puerto Rico, 1966. Cert. wildlife biologist. From grad. asst. in zoology to rsch. fellow in wildlife Ohio State U., 1957-63; biology instr. Houghton (N.Y.) Coll., 1963; asst. prof. biology Central (S.C.) Wesleyan Coll., 1963-64, Geneva Coll., Beaver Falls, Pa., 1964-69; prof. biology Edinboro (Pa.) U., 1969—, Pymatuning Lab. of Ecology U. Pitts., Pitts., 1982, 88, 89; bird records com. Presque Isle Audubon Soc., Erie, Pa., 1975—; vol. for wildlife Pa. Game Commn., Harrisburg, 1990—; ornithol. tech. com. Pa. Biol. Survey, Harrisburg, 1991—. Contbr. numerous articles to profl. jours. Committeeman Boy Scouts Am., Laketon, Ind., 1965-70; elder Christian & Missionary Alliance Ch., Erie, 1987-90; trustee Purple Martin Conservation Assn., Edinboro, Pa., 1988—. Equipment grantee Atomic Energy Commn., Oak Ridge, Tenn., 1966; recipient Meritorious Svc. award Edinboro U. Pa., 1974. Mem. Wildlife Soc., Assn. Field Ornithologists, Assn. Pa. State Coll. and Univ. Faculty, Beta Beta Beta. Avocations: hiking, biking, canoeing. Home: 13190 Cambridge Rd Edinboro PA 16412-2837 Office: Edinboro U Dept Biology Edinboro PA 16444

SNYDER, DONALD CARL, JR., physics educator; b. Phila., Aug. 24, 1954; s. Donald Carl and Gloria (Nicklous) S. BS, Temple U., 1976, MS, 1980. Cert. tchr., Pa. Gen. sci. tchr. James Russell Lowell Sch., Phila., 1977-89; physics tchr. South Phila. High Sch., 1989—, biol. tchr., 1990-92; rep. Phila Sci. Tchrs. Assn., Nat. Assn. Biol. Tchrs., 1992; mem. project 2061 Sch. Dist. Phila., 1992-93. Author: Science Assessment in the Service of Reform, 1991. Active U.S. Olympic Soc., 1980—. Mem. AAAS, Pa. Acad. Sci., Nat. Sci. Tchrs. Assn., Nat. Sci. Suprs. Assn., Nat. Assn. Biology Tchrs. (state rep. 1992—), ASCD, Fulbright Alumni Assn. (no. Ireland 1983-84), Phi Delta Kappa. Democrat. Lutheran. Avocations: traveling, reading, photography, theater. Home: 5210 Westford Rd 1st Fl Philadelphia PA 19120 Office: South Phila High Sch Broad and Snyder Ave Philadelphia PA 19148

SNYDER, FRED LEONARD, health sciences administrator; b. New Ulm, Minn., Nov. 22, 1931. BS in Chemistry and Biology, St. Cloud State Coll., 1953; MS in Biochemistry, U. N.D., 1955, PhD in Biochemistry, 1958, DSc (hon.), 1983. Chief scientist med. and health scis. div. Oak Ridge (Tenn.) Associated Univs., 1958-75, asst. chmn. med. and health scis. div., 1975-79, assoc. chmn. med. and health scis. div., 1979-88, corp. disting. scientist, 1988—, vice chmn. med. scis. div., 1988-92; assoc. dir. med. scis. div. Oak Ridge Inst. for Sci. and Edn., Oak Ridge Associated U., 1992—; vis. scientist U. N.C., Chapel Hill, 1964; prof. biochemistry U. Tenn. Ctr. for Health Scis., Memphis, 1964-86; prof. medicinal chemistry U. N.C., Chapel Hill, 1966—; adj. prof. U. Tenn. Oak Ridge Graduate Sch. Biomedical Scis., 1972—; vis. lectr. Cardiovascular Rsch. Inst., U. Calif., San Francisco, 1979; vis. prof. The Grad. Sch. and Univ. Ctr. of CUNY, 1989; mem. internat. adv. com. 7th Internat. Conf. on Prostaglandins and Related Compounds, Florence, Italy, 1990; others. Exec. editor, mem. editorial bd.: Archives of Biochemistry and Biophysics; editor: Handbook of Lipid Research, 1987—; assoc. editor: Cancer Research, 1971-78; mem. editorial bd.: Jour. of Lipid Rsch., 1966-82, Archives of Biochemistry and Biophysics, 1972-78, Reviews on Cancer, 1973-80, Biochimica et Biophysica Acta, 1973-80, 89—, Lipids, 1986—, Jour. of Biol. Chemistry, 1986-91, Jour. of Lipid Mediators, 1987—; contbr. over 350 articles to profl. jours. Recipient Predoctoral fellowship NIH, 1955-58, U. N.D. Sioux award for disting. svc. and outstanding achievements, 1974. Mem. Am. Soc. for Biochemistry and Molecular Biology, Am. Assn. for Cancer Rsch., Soc. for Exptl. Biology and Medicine (sectional councilor southeastern sect. 1965-68, sec.-treas. 1969-70), Sigma Xi. Office: Oak Ridge Associated Univs Med Scis Div PO Box 117 Oak Ridge TN 37831-0117

SNYDER, GEORGE ROBERT, engineer; b. Pontiac, Mich., June 3, 1939; s. George Vincent and Elsa Gertrude (Niederschmidt) S.; m. Ellen Lorraine Rogler, Nov. 5, 1966 (div. 1982); children: Sharon, Lisa; m. Mary Anne Eagen, June 3, 1988; 1 child, Susan. Student, U. Mich., 1956-60. Indsl. engr. Electric Autolite Co., Inc., Port Huron, Mich., 1960-61; engr. Capac

(Mich.) Mfg. Corp., 1961-62; chief engr. Inflated Products Co., Inc., Beacon, N.Y., 1962-66; plant mgr. Marmac Industries, Inc., Marysville, Mich., 1966-78, X-Tyal Internat. Corp., North Adams, Mass., 1980-85; owner Ohio Rustproofing, Inc., Bowling Green, Ohio, 1978-80; gen. mgr. Stemaco Products, Inc., Port Huron, 1987—. Recipient Cert. of Appreciation, U.S. Def. Logistics Agy., 1991. Achievements include development of product and process designs for manufacture of rail transportable tank, combat helmet, chemical and biological agents protective field hospital shelter, protective headgear. Home: 4075 Wilson Dr Fort Gratiot MI 48059 Office: Stemaco Products Inc 5139 Lapeer Rd Smiths Creek MI 48074

SNYDER, JED COBB, foreign affairs specialist; b. Phila., Mar. 24, 1955; s. David and Lynn S. BA, Colby Coll., 1976; MA, U. Chgo., 1978, postgrad., 1978-79. Rsch. asst. U. Chgo., 1979; asst. researcher Pan Heuristics div. R&D Assocs., Marina del Rey, Calif., 1979-80, assoc. researcher, asst. div. mgr., 1980-81, cons., 1982-83; cons. Sci. Applications, Inc., 1979-81, Rand Corp., Santa Monica, Calif., 1979-81, Los Alamos Nat. Lab., 1984; sr. spl. asst. to dir. Bur. of Politico-Mil. Affairs, Dept. State, Washington, 1981-82; rsch. assoc. Internat. Security Studies Program, Woodrow Wilson Internat. Ctr. for Scholars, Smithsonian Instn., Washington, 1982-84; founder, chmn. Washington Strategy Seminar, 1984-90, pres., 1984-93, corp. dir., 1993—; dep. dir. nat. security studies Hudson Inst., 1984-87; sr. rsch. fellow Nat. Strategy Info. Ctr., 1988-90; appointee v.p. Bush's adv. task force on Mid. East; appointee sr. fellow Inst. for Nat. Strategic Studies, Nat. Def. U., 1992—; cons. Office of Sec. of Def., 1988-92, Rand Corp., 1983-88; fellow Brit.-Am. Project for Successor Generation, 1991—; apptd. sr. fellow Inst. for Nat. Strategic Studies Nat. Def. U., 1992—. Contbr. articles on U.S. fgn. policy and mil. def. to profl. publs. Trustee Kents Hill (Maine) Sch. Guest scholar Sch. Advanced Internat. Studies, Johns Hopkins U., 1982-83; fellow U. Chgo., 1979, Inter-Univ. Seminar on Armed Forces and Soc., 1980, MacArthur Sr., 1985-86, Herman Kahn, 1985-86, Smith Richardson, 1987-88, John M. Olin, 1987-88, British-Am. Successor Generation Project, 1991; selected as a Young Am. Leader, Am. Coun. on Fed. Republic of Germany, 1984. Mem. Internat. Inst. for Strategic Studies, Internat. Studies Assn., Coun. on European Studies, Mil. Ops. Rsch. Soc., U.S. Naval Inst., Fgn. Policy Rsch. Inst., AIAA, Am. Polit. Sci. Assn., Coun. on Fgn. Rels. Home: 2201 L St NW Apt 602 Washington DC 20037-1412 Office: Inst Nat Strategic Studies Nat Def Univ Ft Lesley McNair Washington DC 20319-6000

SNYDER, JOHN MENDENHALL, medical administrator, retired thoracic surgeon; b. Slatington, Pa., Aug. 1, 1909; s. James Wilson and Gertrude Winifred (Mendenhall) S.; m. Betty June Wiltrout, Feb. 14, 1942 (dec. May 1991); children: Sue Anne Snyder-Alexy, John Sanford. BS in Biology, Bucknell U., 1930; MD, U. Pa., 1934; MS in Surgery, U. Minn., 1941. Diplomate Am. Bd. Surgery, Am. Bd. Thoracic Surgery. Rotating intern Bryn Mawr (Pa.) Hosp., 1934-35; asst. resident in medicine Univ. Hosps. of Cleve.-Western Res. U., 1935-36; fellow in surgery Mayo Clinic, Rochester, Minn., 1936-41; practiced thoracic surgery Pa., 1945-77; emeritus asst. chief surg. svc., in charge thoracic surgery St. Luke's Hosp., Bethlehem, Pa., 1977—; med. dir. Bur. of Health, Bethlehem, 1981—; formerly on staff St. Luke's Hosp., Sacred Heart Mt. Trexler San.; formerly thoracic surg. cons. Sacred Heart Hosp., Allentown, Pa., Allentown State Hosp., Easton (Pa.) Hosp., Graden Huetten Hosp., Lehighton, Pa., Muhlenberg Med. Ctr., Bethlehem. Contbr. articles to profl. jours. Dir. med. sect. Bethlehem CD; mem. Bethlehem Air Pollution Coun.; chmn. Lehigh Valley Med. Adv. Com.; chmn. case finding com. Lehigh Valley Tb and Health Assn., pres., 1978-80. Lt. col. U.S. Army, 1942-45, ETO, Col. USAR, 1945-58. Decorated Silver Star, Legion of Merit, Bronze Star. Mem. Masons. Republican. Episcopalian. Avocations: singing, cooking. Home: 139 E Market St Bethlehem PA 18018-6225 Office: Bur of Health 10 E Church St Bethlehem PA 18018

SNYDER, MARTIN BRADFORD, mechanical engineering educator; b. Evergreen Park, Ill., Dec. 19, 1942; s. Bernard A. and Helena M. (Piro) S. BS in Physics, MIT, 1964; PhD in Nuclear Engring., Northwestern U., 1972; PhD in Bioengring., U. Mich., 1985. Presdl. intern Argonne Nat. Lab., Chgo., 1972-73; staff engr. Sargent and Lundy Co., Chgo., 1973-74; Parker B. Francis fellow U. Fla., Gainesville, 1979-81; vis. scholar U. Mich., Ann Arbor, 1981-82, asst. rsch. scientist Sch. Medicine, 1984-85; biomed. engr. VA Hosp., Ann Arbor, 1982-84; assoc. prof. dept. mech. engring. U. Nev., Reno, 1985—. Contbr. articles on nuclear engring., physiology and mech. engring. to profl. publs. Sci. tchr. U.S. Peace Corps, India, 1965-67. NSF fellow, 1971; recipient Mark Mills award Am. Nuclear Soc., 1973; NIH trainee, 1975; rsch. fellow Whitaker Found., 1985. Mem. Am. Phys. Soc. Office: U Nev Reno Dept Mech Engring Reno NV 89557

SNYDER, MELISSA ROSEMARY, biochemist; b. Windber, Pa., May 14, 1970; d. Donald C. and Mabel Carole (Hunt) S. BS, Juniata Coll., 1992; postgrad., Mayo Grad. Sch., 1992—. Lab. asst. Juniata Coll., Huntingdon, Pa., 1989-92; undergrad. researcher Penn State U., State Coll., Pa., 1990; organic chem. teaching aide Juniata Coll., Huntingdon, 1990-92; undergrad. researcher Mayo Clinic, Rochester, Minn., 1991; grad. student, researcher Mayo Grad. Sch., Rochester, Minn., 1992—; presenter in field. Recipient Organic Chemistry award Juniata Coll., 1990, Founder's award Juniata Coll., 1988, Merck award, 1992. Mem. Am. Chem. Soc., Sigma Xi Rsch. Soc.

SNYDER, MICHAEL, biology educator; b. Phoenixville, Pa., Oct. 3, 1955; s. Kermith C.G. and Phyllis Snyder. BA, U. Rochester, 1977; PhD, Calif. Inst. Tech., 1982. Postdoctoral fellow dept. biochemistry Stanford U., Calif., 1982-86; prof. dept. biology Yale U., Newhaven, Conn., 1986-90, assoc. prof. dept. biochemistry, 1990--; assoc. prof. dept. molecular biophysics and biochemistry JT Appointment, 1992--; review panel mem. NIH, 1989, 90. Contbr. articles to profl. jours. Helen Hay Whitney fellow, 1982-85, Yale Jr. Faculty fellow, 1989; recipient United Scleroderma award, 1986, Pew Scholar award, 1987-91. Mem. Genetics Soc. Am., Am. Assn. for the Advancement Sci., Am. Soc. Cell Biologists. Achievements include discovery of molecules involved in polarized growth, divsn. and proposed models by which polarized divsns. occur, eucaryotic nucleus in nonrandomly organized in diploid cells, carbon source as the critical nutrient for stimulating cell growth of nonproliferating cells. Office: Yale U Dept Biology 219 Prospect St New Haven CT 06511

SNYDER, PETER JAMES, nuclear engineer; b. McKeesport, Pa., Aug. 27, 1966; s. Peter John and Kathryn Ann (Snyder) S. BSME, Pa. State U., 1988. Nuclear engr. Norfolk Naval Shipyard, Portsmouth, Va., 1988—. Mem. Easter Seal Soc. Mem. ASME, NSPE (engr.-in-tng.), N.Y. Acad. Scis. (active). Democrat. Roman Catholic. Home: 1216 Waterfront Dr Apt 301 Virginia Beach VA 23451 Office: Norfolk Naval Shipyard Bldg 1500 Code 2310 Portsmouth VA 23709

SNYDER, PETER RUBIN, neuropsychologist; b. Ann Arbor, Mich., Nov. 18, 1964; s. Daniel Raphael Snyder and Susan (Jacobs) Etkind; m. Bonnie Lynn, June 3, 1990. AB with high hons., U. Mich., 1986; PhD, Mich. State U., 1992. Pre-doctoral fellowship Epilepsy Found. Am./Yale U., New Haven, Conn., 1987; lectr. in neuropsychology, neurosurgery Yale U., New Haven, 1988; neuropsychology intern Hillside Hosp./L.I. Jewish Med. Ctr., Glen Oaks, N.Y., 1991-92; clin. neurosci. fellow Hillside Hosp./Albert Einstein Coll. Medicine, Glen Oaks, 1992-93. Contbr. articles to profl. jours. Wilder Penfield fellow Epilepsy Found. Am., 1992, Tarkus Rsch. fellow, 1993; recipient Dissertation Rsch. award Am. Psychol. Assn., 1990. Mem. Internat. Neuropsychol. Soc., Soc. Neurosci., Am. Epilepsy Soc., N.Y. Acad. Scis., Sigma Xi, Mortar Bd. Democrat. Jewish. Office: Hillside Hosp/Rsch PO Box 38 Glen Oaks NY 11004

SNYDER, ROBERT LYMAN, ceramic scientist, educator; b. Plattsburgh, N.Y., June 5, 1941; s. George Michael and Dorothy (Lyman) M.; m. Sheila Nolan, Sept. 1, 1963; children: Robert N., Kristina N. BA, Marist Coll., 1963; PhD, Fordham U., 1968. Postdoctoral fellow NIH U. Pitts., 1968; NRC fellow NASA Elec. Rsch. Ctr., Cambridge, Mass., 1969; asst. prof. ceramic sci. Alfred (N.Y.) U., 1970-77, assoc. prof., 1977-83, prof., 1983—, dir. Inst. Ceramic Superconductivity, 1987—; vis. prof. Lawrence Livermore (Calif.) Lab., 1977, 78, U.S. Nat. Bur. Standards, Gaithersburg, Md., 1980, 81, Siemens AG (Cen. Rsch. Labs.), Munich, 1983, 91. Author: X-Ray Materials Science and Technology, 1992; contbr. over 175 articles to profl. jours. Deputy mayor Village of Alfred, 1973-77; pres. Alfred Vol. Fire Co., 1979-88. Recipient Chancellor's award SUNY, 1980, numerous research grants; named Faculty Exch. scholar SUNY, 1978—. Mem. NAS (U.S. Nat. Com. of Crystallography 1991—), Am. Ceramic Soc., Nat. Inst. Ceramic Engrs., Am. Crystallography Assn. (chmn. applied crystallography div. 1988—), Materials Rsch. Soc., Ceramic Ednl. Coun., Internat. Ctr. Diffraction Data (bd. dirs. 1986—), Alfred and Allegany County Fire Assn., Sigma Xi, Phi Kappa Phi. Democrat. Achievements include numerous patents for glass-ceramic superconductors. Home: 56 Pine Hill Dr Alfred NY 14802-1327 Office: Alfred U Coll Ceramic-Inst Ceramic Superconductivity Alfred NY 14802

SNYDER, SOLOMON HALBERT, psychiatrist, pharmacologist; b. Washington, Dec. 26, 1938; s. Samuel Simon and Patricia (Yakerson) S.; m. Elaine Borko, June 10, 1962; children: Judith Rhea, Deborah Lynn. M.D. cum laude, Georgetown U., 1962, D.Sc. (hon.), 1986; D.Sc. (hon.), Northwestern U., 1981; PhD (hon.), Ben Gurion U., 1990. Intern Kaiser Found. Hosp., San Francisco, 1962-63; research asso. NIMH, Bethesda, Md., 1963-65; resident psychiatry Johns Hopkins Hosp., Balt., 1965-68; asso. prof. psychiatry and pharmacology Johns Hopkins Med. Sch., 1968-70, prof., 1970-77, disting. Service prof. psychiatry and pharmacology, 1977-80, disting. Service prof. neurosci., psychiatry, and pharmacology, 1980—, dir. dept. neurosci., 1980—; NIH lectr., 1979. Author: Uses of Marijuana, 1971, Madness and the Brain, 1973, Opiate Receptor Mechanisms, 1975, The Troubled Mind, 1976, Biologic Aspects of Mental Disorder, 1980, Drugs and the Brain, 1986, Brainstorming, 1989; editor Perspectives in Neuropharmacology, 1971, Frontiers in Catecholamine Research, 1973, Handbook of Psychopharmacology, 1974; contbr. articles to profl. jours. Served with USPHS, 1963-65. Recipient Outstanding Scientist award Md. Acad. Scis., 1969; John Jacob Abel award Am. Pharmacology Soc., 1970; A.E. Bennett award Soc. Biol. Psychiatry, 1970; Gaddum award Brit. Pharm. Soc., 1974; F.O. Schmitt award in neuroscis. MIT, 1974; Nicholas Giarman lecture award Yale U., 1975; Rennebohm award U. Wis., 1976; Salmon award, 1977; Stanley Dean award Am. Coll. Psychiatrists, 1978; Harvey Lecture award, 1978; Lasker award, 1978; Wolf prize, 1983; Dickson prize, 1983; Sci. Achievement award AMA, 1985; Ciba-Geigy-Drew award, 1985; Strecker prize, 1986; Edward Sachar Meml. award Columbia U., 1986; Paul K. Smith Meml. lecture award George Washington U., 1986; Sense of Smell award Fragrance Research Found., 1987; Julius Axelrod lecture award CUNY, 1988; John Flynn Meml. lecture award Yale U., 1988; V. Erspamer lecture award Georgetown U., 1990; J. Allyn Taylor prize, 1990; Pasarow Found. award, 1991; Bower award Achievement Sci. Franklin Inst., 1991; Chauncey Leake Lecture award, 1992; William Veatch lectr. award Harvard Med. Sch., 1992; Joseph Priestley prize Dickinson Coll., 1992; Konrad Bloch lectr. award Harvard U., 1992; Basic Neurochem. lectr. award Am. Soc. Neurochem., 1993; Nanine Duke lectr. award Duke U., 1993; Salvador Luria lectr. award MIT, 1993. Fellow Am. Coll. Neuropsychopharmacology (Daniel Efron award 1974), Am. Psychiat. Assn. (Hofheimer award 1972, Disting. Svc. award 1989), Am. Acad. Arts and Scis., Am. Philosophical Soc.; mem. Psychiat. Research Soc., Nat. Acad. Scis., Soc. for Neuroscis. (pres. 1979-80), Am. Soc. Biol. Chemists, Am. Pharmacology Soc., Inst. Medicine. Home: 3801 Canterbury Rd Apt 1001 Baltimore MD 21218-2315 Office: Johns Hopkins U Med Sch Dept Neurosciences 725 N Wolfe St Baltimore MD 21205

SNYDERMAN, RALPH, medical educator, physician; b. Bklyn., Mar. 13, 1940; m. Judith Ann Krebs, Nov. 18, 1967; 1 child, Theodore Benjamin. B.S., Washington Coll., Chestertown, Md., 1961; M.D., SUNY-Bklyn., 1965. Diplomate Am. Bd. Internal Medicine, Am. Bd. Allergy and Immunology. Med. intern Duke U. Hosp., Durham, N.C., 1965-66, med. resident, 1966-67, asst. prof. medicine and immunology, 1972-74, assoc. prof., 1974-77, chief, div. rheumatology and immunology, 1975-87, prof. medicine and immunology, 1980-84, Frederic M. Hanes prof. medicine, prof. immunology, 1984-87, James B. Duke prof. medicine, dean sch. medicine, chancellor for health affairs, 1989—; surgeon USPHS, NIH, Bethesda, Md., 1967-69; sr. staff fellow Nat. Inst. Dental Research, NIH, Bethesda, Md., 1969-70, sr. investigator immunology sect. lab. microbiology and immunology, 1970-72; chief, div. rheumatology Durham VA Hosp., Bethesda, Md., 1972-75; v.p. med. rsch. and devel. Genentech, Inc., South San Francisco, Calif., 1987-88, sr. v.p. med. rsch. and devel., 1988-89; chancellor for health affairs, dean Sch. Medicine Duke U., Durham, 1989—, James E. Duke prof. medicine, 1989—; adj. asst. prof. oral biology U. N.C. Sch. Dental Medicine, Chapel Hill, 1974-75; dir. Lab Immune Effector Function, Howard Hughes Med. Inst., Durham, 1977—; adj. prof. medicine U. Calif., San Francisco, 1987—. Editor: Contemporary Topics in Immunobiology, 1984, Medical Clinics of North American, 1985, Inflammation: Basic Concepts and Clinical Correlates, 1988; contbr. articles to profl. jours. Recipient Alexander von Humboldt award Fed. Republic Germany, 1985. Mem. Assn. Am. Physicians, Am. Assn. Immunologists, Am. Soc. Clin. Investigation, Am. Acad. Allergy, Am. Assn. Cancer Research, Am. Soc. Exptl. Pathology, Am. Fedn. Clin. Research, Am. Assn. Pathologists, Reticuloendothelial Soc., Am. Rheumatism Assn., Sigma Xi. Office: Duke U Sch Medicine PO Box 3701 Durham NC 27710

SOARES, EUSEBIO LOPES, anesthesiologist; b. Lisbon, Portugal, Oct. 20, 1918; s. Jose Lopes and Rosaria (Sousa) S. M.D., U. Lisbon, 1942; m. Edviges Velasques Monteiro, Aug. 7, 1951; 1 child, Maria Helena. Tng. in anesthesia, U.K., 1947-48; asst. prof. U. Lisbon Faculty Medicine, 1949-50; dir. dept. anesthesia Hosp. do Ultramar, Lisbon, 1952-56; sr. anesthesiologist Lisbon Civil Hosp., 1956-69; dir. dept. anesthesia Hosp. St. Ant. Capuchos, Lisbon, 1969—, also clin. dir.; vis. prof. Ibero-Latin-Am. Ctr. Anesthesiology, Cen. U. Venezuela, 1969. Officer M.C., Portuguese Army, 1945-46. Decorated grand officer Order of Merit, Govt. of Portugal. Fellow Royal Coll. Surgeons Eng., Royal Soc. Medicine; mem. World Fedn. Socs. Anesthesiologists (v.p. 1964), European Acad. Anesthesiology, Portuguese Soc. Anesthesiology (founder 1955, pres. 1955-57, 59-60), Soc. Anesthesiology Brazil, Soc. Anesthesiology Argentina, Soc. Anesthesiology Spain (hon.), Soc. Anesthesiology Great Britain and Ireland, Soc. Anesthesiology Belgium, U.S. Contbr. articles to profl. jours. Home: 15-6 deg-D, D Estefania, 1 100 Lisbon Portugal Office: Hosp St Ant Capuchos, Dept Anesthesia, 1 100 Lisbon Portugal

SOBALLE, DAVID MICHAEL, limnologist; b. Annapolis, Md., Aug. 5, 1950; s. Verner Jensen and Vivian Cecilia (Erickson) S.; m. Constance Belinda Bingham, Aug. 5, 1978; children: Erik, Jens. BS, U. Notre Dame, 1972; MS, Mich. Tech. U., 1978; PhD, Iowa State U., 1981. Postdoctoral rsch. assoc. U. Okla. Biol. Sta., Kingston, Okla., 1981-82; vis. asst. prof. Bowling Green (Ohio) State U., 1982-83; faculty rsch. assoc. Oak Ridge (Tenn.) Nat. Lab., U. Tenn., 1983-87; sr. environ. scientist South Fla. Water Mgmt. Dist., West Palm Beach, Fla., 1987-91; limnologist U.S. Fish and Wildlife Svc., Onalaska, Wis., 1991—; assoc. editor North Am. Lake Mgmt. Soc., 1989—. Author: (book chpt.) Biodiversity of S.E. U.S. Aquatic Communities, 1992; contbr. articles to profl. jours. Lt. USN, 1972-76. Mem. Am. Soc. Limnology and Oceanography, Phycological Soc. of Am., Ecol. Soc. of Am., North Am. Lake Mgmt. Soc. Roman Catholic. Office: US Fish and Wildlife Svc 575 Lester Ave Onalaska WI 54650

SOBEL, KENNETH MARK, electrical engineer, educator; b. Bklyn., Oct. 3, 1954; s. Seymour Phillip and Marilyn (Nanus) S. BSEE, CCNY, 1976; MEngring, Rensselaer Poly. Inst., 1978, PhD, 1980. Sr. rsch. specialist Lockheed Calif. Co., Burbank, 1980-87; assoc. prof. dept. elec. engring. CCNY, 1987-93, prof., 1993—; adj. asst. prof. U. So. Calif., L.A., 1982-87; prin. investigator USAF, 1989, 91, mem. summer faculty fellowships, 1987, 88, 90; mem. exec. com. PhD program in engring. CUNY, 1989—. Assoc. editor Jour. Guidance, Control and Dynamics, 1993—; contbr. articles to profl. jours., chpts. to books. Program vice-chmn. 1986 Am. Control Conf., Seattle. Recipient Prof. of Yr. award Beta Pi chpt. Eta Kappa Nu, 1991-92; PSC-CUNY rsch. grantee, 1988-90. Fellow AIAA (assoc.); mem. IEEE (sr., exhibits chmn. 23d conf. on decision and control 1984, registration chmn. 27th conf. on decision and control 1988, tech. assoc. editor Control Systems mag., 1986—), Am. Radio Relay League (life), Control System Soc. (bd. govs. 1990), Sigma Xi, Alpha Phi Omega. Office: CCNY Dept Elec Engring New York NY 10031

SOBEN, ROBERT SIDNEY, systems scientist; b. Corpus Christi, Tex., Feb. 7, 1947; s. Sydney Robert and Rose Mary (Bailey) S.; 1 child, Dena Dianne. BS in Electrical Engring., La. Tech. U., 1973; MA in Communication, U. Okla., 1982; MS in Mgmt. Scis., Troy (Ala.) State U., 1988; PhD in Engring. Mgmt. Sci., LaSalle U., 1990. Digital computer sci. USAF Air Training Command, Keesler AFB, Miss., 1966-71; command pilot USAF, worldwide, 1971-82; NATO instr. pilot 80th Fighter Training Wing, Sheppard AFB, Tex., 1978-82; electro-optics br. chief Electronics Systems Test Div., Eglin AFB, Fla., 1982-84; mission ops. officer Deputate for Testing Engring., Eglin AFB, Fla., 1984-85, test support div. chief, 1985—; adj. prof. Troy State U., Ft. Walton Beach, Fla., 1987—, St. Leo's Coll., Eglin AFB, 1988—. Author: Digital Computer Basics, 1970, Application of Expert Systems to Scientific and Technical Information Command, Control and Communication Management, 1990; author USAF tech. report Video Augmentation, 1984, tng. manual and system test engring., 1988. Avocations: sailing, scuba diving, writing, racing cars. Home: 1301 Windward Cir Niceville FL 32578-4310

SOBERING, GEOFFREY SIMON, biomedical scientist; b. Chgo., June 30, 1960; s. Simon Edgar and Patricia (Moran) S.; m. Denise Suzanne Harmer, June 4, 1983. BA, Drew U., 1982; PhD, U. Wis., 1989. Researcher analytical div. corp. R&D Allied Signal Corp., Morristown, N.J., 1981-82; teaching asst. dept. chemistry Drew U., Madison, N.J., 1981-82; teaching asst. dept. chemistry U. Wis., Madison, 1982-83, tech. asst. Instrument Ctr., dept. chemistry, 1983-88; researcher Chem Solve, Inc., Morristown, 1982; sr. staff fellow In Vivo NMR Rsch. Ctr., NIH, Bethesda, Md., 1989—. Contbr. articles to profl. publs. Grantee Dow Chem. Corp., 1985. Office: NIH Bldg 10 Rm B1D 125 Bethesda MD 20892

SOBERS, DAVID GEORGE, environmentalist; b. Washington, Dec. 15, 1940; s. Peter Paul and Catherine Cecilia (Popp) S.; m. Karen Ann Nelson, Mar. 18, 1967; 1 child, Scott. BS in Agronomy, U. Md., 1962, MS in Econs., 1965. Environ. planner Md.-Nat. Capital Park and Planning Commn., Silver Spring, 1969-70, Office of Program Coordination, Montgomery County, Md., 1971; asst. dir. Office of Planning and Capital Programming, Montgomery County, 1972-73; dir. Office Environ. and Energy Planning, Montgomery County, 1974-79; chief Div. Environ. Planning and Monitoring, Montgomery County, 1980-90; acting dir. Dept. Environ. Protection, Montgomery County, 1991; chief Div. Environ. Policy and Compliance, Montgomery County, 1992; v.p. for Solid Waste Mgmt. Woodward-Clyde Cons., Gaithersburg, Md., 1992—. Home: 11513 Brandy Hill Ln Gaithersburg MD 20878 Office: Woodward Clyde Cons 904 Wind River Ln Gaithersburg MD 20878

SOBIE, WALTER RICHARD, semiconductor engineer; b. Newark, Mar. 30, 1943; s. Walter K. and Dorothy (Barr) S.; m. Janet Piazza, Nov. 23, 1963 (div. Sept. 1971); 1 child, Gregory Alan; m. Janet Lassek, July 28, 1973; 1 child, Adrian Richard. BSEE, Newark Coll. Engring., 1968; degree in mgmt., Northeastern U., 1978. Process engr. RCA Solid State, Somerville, N.J., 1965-72; process mgr. KMC/MACOM, Long Valley, N.J., 1972-73, MACOM, Burlington, Mass., 1973-79; v.p. microwave div. Millis (Mass.) Corp., 1980-82; process tech. mgr. Raytheon, Northboro, Mass., 1982-85; tech. cons. Tech. Assocs., Westford, Mass., 1985-88; tech. mktg. mgr. Loral Microwave FSI, Chelmsford, Mass., 1988—; ind. tech. cons., Westford, 1979—. Mem. Am. Vacuum Soc., Am. Inst. Physics. Achievements include patent for flip chip bonding technique. Home: 37 Buckboard Dr Westford MA 01886-2752

SOBOL, BRUCE J., internist, educator, researcher; b. N.Y.C., June 10, 1923; s. Ira J. and Ida S. (Gelula) S.; B.A., Swarthmore Coll., 1947; M.D., N.Y.U., 1950; m. Barbara Sue Gordon, Apr. 30, 1951; children: Peter Gordon, Scott David. Intern, Bellevue Hosp., N.Y.C., 1950-51, resident, 1951-52, N.Y. Heart Assn. fellow, 1953-55; resident VA Hosp., Boston, 1952-53; practice medicine specializing in internal medicine, White Plains, N.Y., 1955-59; dir. cardio-pulmonary lab. Westchester County (N.Y.) Med. Ctr., Valhalla, 1959-78; rsch. prof. medicine N.Y. Med. Coll., 1970-78; dir. med. rsch. Boehringer Ingelheim, Ltd., Ridgefield, Conn., 1978-83. Bd. dirs. Westchester Community Svcs. Coun., 1977-79; pres. Westchester Heart Assn., 1976-79. Served with inf. AUS, World War II; ETO. Diplomate Am. Bd. Internal Medicine. Fellow ACP, Am. Coll. Allergy, Am. Coll. Chest Physicians, N.Y. Acad. Scis.; mem. Am. Physiol. Soc., Am. Heart Assn., N.Y. Trudea Soc., Am. Thoracic Soc., Am. Fedn. Clin. Rsch. Contbr. numerous articles to profl. publs. Office: 275 Ridgebury Rd Ridgefield CT 06877-1410

SOBOL, WLAD THEODORE, physicist; b. Katowice, Poland, June 27, 1949; came to U.S., 1986; s. Teodor Feliks and Lucja (Rojek) S.; m. Barbara Krzanowska, Oct. 29, 1976; 1 child, Anna. MS cum laude, Silesian U., 1972; PhD magna cum laude, Jagiellonian U., 1978. Asst. prof. Silesian U., Katowice, 1972-82; postdoctoral fellow U. Waterloo, Can., 1983-85; asst. prof. Bowman Gray Sch. of Medicine, Winston-Salem, N.C., 1986-91; assoc. prof. U. Ala. Sch. Medicine, Birmingham, 1991—; grad. faculty Bowman Gray Sch. Medicine, Winston-Salem, 1988-91; biochemistry assoc., 1988-91. Editor Molecular Physics, 1982; contbr. articles to profl. jours. including Theory and Experiment in Nuclear Magnetic Resonance, Tunneling in Solids, Theory and Practice of Magnetic Resonance Imaging, Nuclear Magnetic Resonance Tissue Characterization, Image Processing in Magnetic Resonance Imaging, Diagnostic Radiol. Physics. Recipient Polish Min. of Sci. award Polish Ministry of Sci., 1979. Mem. Soc. of Magnetic Resonance Imaging, Am. Assn. of Physicists in Medicine, Internat. Soc. of Magnetic Resonance. Office: U Ala Hosp Radiology Dept JT1107 619 S 19th St Birmingham AL 35233-1924

SOCOLOW, ROBERT HARRY, mechanical and aerospace engineering educator, scientist; b. N.Y.C., Dec. 27, 1937; s. A. Walter and Edith (Gutman) S.; m. Elizabeth Anne Sussman, June 10, 1962 (div. Apr. 27, 1982); children: David, Seth; m. Jane Ries Pitt, May 25, 1986; stepchildren—Jennifer, Eric. B.A., Harvard U., 1959, M.A., 1961, Ph.D., 1964. Asst. prof. physics Yale U., New Haven, 1966-71; assoc. prof. mech. and aerospace engring. Princeton U. (N.J.), 1971-77, prof. mech. and aerospace engring., 1977—; mem. bd. global change Nat. Rsch. Coun., 1993—; mem. Inst. Advanced Study, Princeton, 1971; dir. Center for Energy and Environmental Studies, Princeton, 1978—. Author: (with John Harte) Patient Earth, 1971, (with K. Ford, G. Rochlin, M. Ross) Efficient Use of Energy, 1975, (with H.A. Feiveson, F.W. Sinden) Boundaries of Analysis: An Inquiry into the Tocks Island Dam Controversy, 1976, Saving Energy in the Home: Princeton's Experiments at Twin Rivers, 1979; editor Ann. Rev. of Energy and Environment, 1992—; dir. Global Change Inst. "Industrial Ecology and Global Change", 1992. Chmn. bd. Am. Coun. for Energy Efficient Econ., 1989-93. John Simon Guggenheim fellow, 1976-77; German Marshall Fund fellow, 1976-77; NSF Postdoctoral fellow, 1964-66; NSF Predoctoral fellow, 1960-64. Fellow AAAS, Am. Phys. Soc.; mem. Nat. Audubon Soc. (bd. dirs.). Jewish. Home: 34 Westcott Rd Princeton NJ 08540-3060 Office: Princeton U H102 Engineering Quad Princeton NJ 08544

SODARO, EDWARD RICHARD, psychiatrist; b. Glen Cove, N.Y., Oct. 3, 1947; s. Edward Richard and Mae Florence (Culp) S.; m. Denise Roberta Stetch; 2 children. BS, Siena Coll., Loudonville, N.Y., 1969; MD, Georgetown U., 1973; MA, Grad. Faculty of New Sch., N.Y.C., 1976. Diplomate Am. Bd. Psychiatry and Neurology, Am. Bd. Adolescent Psychiatry; cert. addiction specialist Am. Soc. Addiction Medicine; qualified in geriatric and addiction psychiatry. Resident in psychiatry L.I. Jewish Hosp., New Hyde Park, N.Y., 1973-76; staff psychiatrist N.Y. Hosp./Cornell Med. Ctr., White Plains, N.Y., 1979-81; faculty N.Y. Hosp./Cornell Med. Ctr., 1979-81; sr. psychiatrist South Oaks Hosp., Amityville, N.Y., 1981—; mem. exec. com. South Oaks Hosp., Amityville, 1991—. Mem. Am. Psychiat. Assn. (bd. pres. 1990), Am. Acad. Clin. Psychiatrists, N.Y. Acad. Scis., N.Y. Psychiat. Assn. (legis. rep. 1987—), Am. Coll. Physician Execs., Med. Soc. State of N.Y. (com. for physicians, health), Am. Soc. Addiction Medicine, Am. Coll. Physicians Execs. Democrat. Roman Catholic. Office: South Oaks Hosp 400 Sunrise Hwy Amityville NY 11701-2508

SODERBERG, DAVID LAWRENCE, chemist; b. Evergreen Park, Ill., Jan. 28, 1944; s. Arthur Lawrence and Jean Van Norden (Freeman) S. AB in Chemistry, Ripon (Wis.) Coll., 1969. Rsch. asst. Ripon Coll., 1968-69, Pomona Coll., Claremont, Calif., 1969-70; chemist animal and plant health inspection svc. USDA, N.Y.C., 1972-73; chemist food safety and quality svc. USDA, Athens, Ga., 1974-83; supervisory chemist food safety and inspection svc. USDA, St. Louis, 1983-87; chemist, program devel. and tech. mgmt. USDA, Washington, 1987—. Contbr. articles to profl. jours. With U.S. Army, 1965-67, Vietnam. Recipient awards USDA, 1976, 87, cert. of merit, 1980. Mem. AAAS, ASTM, Am. Chem. Soc., Assn. Official Analytical Chemists (gen. referee meat and poultry products). Office: USDA FSIS CD DTMB 300 12th St SW Rm 524 Washington DC 20250-0001

SÖDERSTRÖM, HANS TSON, economist; b. Stockholm, Feb. 25, 1945; s. Torkel A.R. and Elisabet (Zielfelt) S.; Ekon dr. (Ph.D.), Stockholm Sch. Econs., 1974; children: Christofer, Ebba, Marie. Sr. fellow Inst. Internat. Econ. Studies, U. Stockholm, 1971-84; assoc. dir., 1979-84; exec. dir. SNS Ctr. Bus. and Policy Studies, Stockholm, 1985—; adj. prof. macroecons. Stockholm Sch. Econs., 1992—; dir. Stockholm Stock Exch., Hagströmer Qviberg AB. Mem. Royal Swedish Acad. Engring. Scis., Swedish Econ. Assn. (dir. 1990—), Am. Econ. Assn., Liberal Econ. Club (chmn. 1984-85), Jan Wallander Found. for Social Rsch. (dir. 1981—, vice chmn. 1993—). Editor: Ekonomisk Debatt, 1977-78, mem. editorial bd. 1979—; author: Microdynamics of Production, 1974, Finland's Economic Crisis: Causes, Present Nature, and Policy Options, 1993; editor: Sweden-the Road to Stability, 1985, Getting Sweden Back to Work, 1986, One Global Market, 1989, The Swedish Economy at the Turning Point, 1991, Disinflation, adjustment and growth: The Swedish Economy 1992 and Beyond, 1992, Sweden's Economic Crisis: Diagnosis and Cure, 1993; contbr. articles to sci. jours. Home: Valhallavägen 77, S-114 27 Stockholm Sweden Office: SNS, Box 5629, S-11486 Stockholm Sweden

SOEKEN, KAREN LYNNE, research methods educator, researcher; b. Petoskey, Mich., Sept. 24, 1944; d. Walter L. and Helen L. (Fischer) Gienapp; m. Donald R. Soeken, Feb. 27, 1965; children: Jeffrey, Elizabeth. BA, Valparaiso U., 1965; MA, U. Md., 1970, PhD, 1979. Asst. prof. U. Md. Sch. Nursing, Balt., 1979-85, assoc. prof., 1985—. Mem. editorial bd. Evaluation and the Health Professions, 1989—; reviewer several jours., 1979—; contbr. articles to profl. jours. Mem., officer Luth. Women's Missionary League, 1965—. Named one of Outstanding Young Women of Am., 1973. Office: U Md 655 W Lombard Baltimore MD 21201

SOETANTO, KAWAN, biomedical-electrical engineering educator; b. Surabaya, Jawa, Indonesia, Mar. 10, 1951; came to U.S., 1987; s. Kerlim and Suin S.; m. Jennie Herman, Aug. 31, 1976; children: Nerrie, Jun, Ainie. M Engring., Tokyo U. of A&T, 1982; DEng, Tokyo Inst. Tech., 1984; D in Medicine, Tohoku U., Sendai, Japan, 1987. Mgr. fgn. div. R&D Pacific Electronic Co., Surabaya and Tokyo, 1972-80; rsch. asst. surgery dept. Nissei Hosp., Osaka, Japan, 1979-82; rsch. engr./cons. Tokin, Ricoh, Aloka, Tokyo, 1979-88; rsch. fellow diagnostic imaging dept. Kanto Cen. Hosp., Tokyo, 1982-87; sr. rsch. scientist Med. Engring. Lab. Toshiba Corp., Nasu, Japan, 1985-87; rsch. scholar Inst. Chest Diseases & Cancer, Tohoku U., Sendai, 1984-88; vis. prof. Univ. Sci. and Tech. of China, Beijing, 1987, Indonesia U., 1988, Tokyo Inst. Tech., 1989, 90, Calif. Inst. Tech., 1990, Duke U., 1991, U. Calif., 1992; assoc. prof. biomed. engring. Drexel U., Phila., 1987-93; prof. dept. control and system engring. Toin Univ. Yokohama, Japan, 1993—; adj. assoc. prof. radiology dept. Thomas Jefferson U., Phila., 1989—; vis. prof. T.I.T., Tokyo, 1989, 90; mem. US-Japan Collaborating on Sci. Project, Sendai, 1984-86; liaison, exec. mem. AF-SUMB, Tokyo, Bali, 1985-88; cons. advanced tech. lab. ATL, Inc., Seattle, 1988; exec. com. Human Sci. and Tech. Ctr., 1993—. Author: (with others) Medical Ultrasonic Measurement Technique, 1984, Invasive/Noninvasive Technique, 1988, Ultrasound Imaging & Signal Processing, 1986, Ultrasound Speckle Analysis; contbr. articles to EMCJ, Elec. Engring. Jpn., Inst. Electronic Com. Jpn. US, Bull. PME, Japan Jour. Med. Ultra., Japanese Jour. Apply Phys., Jour. Acoustics Japan, Jour. Acoustic Soc. Am., Jour. Ultrasound in Medicine and Biology, IEEE Transaction on Ultrasonics, Ferroelectrics and Frequency Control. Adv. mem. Pan Asian Assn. of Greater Phila., 1989-91, bd. dirs. 1991—; observer Internat. Electrotechnical Commn., 1988, Mayor's Asian Adv. Bd., Phila., 1990; v.p. Indonesian Communities of Delaware Valley, 1990—; sch. affair mem. Japanese Assn. Greater Phila., 1989—; mem. Pa. Heritage Affairs Commn., 1993—. Tokyu Found. fellow, 1985-88, Japan Ministry Edn. predoctoral fellow, 1980-82, doctoral fellow, 1982-85; Japan Ministry Edn. Undergrad. scholar, 1978-80; recipient Outstanding Achievement awards Pan Asian Assn., 1990, medals Asian Fedn. Socs. of Ultrasound in Medicine and Biology, 1987, 89, rsch. awards Toshiba Corp. Med. Engring. Lab., Japan, 1986-88, Outstanding Rsch. award Tokyu Found., 1985-88, Best Rsch. award Computer & Communication Found. NEC, 1986-87, Cert. of Honor, Japan and Indonesian Soc. Ultrasound in Medicine, 1985, 87; Nat. Cancer Inst. grantee NIH, 1990-93. Fellow Am. Inst. Ultrasound in Medicine (ethics and profl. standards com. 1993—); mem. IEEE/EMBS (co-chmn. Phila. chpt. 1990-92, chmn. 1992—, internat. program com. 1991—), IEEE (sr.), Acoustical Soc. Japan, Acoustical Soc. Am., Japan Soc. Ultrasound in Medicine. Achievements include development of pocket noise simulator, image signal processing technique, medical imaging, ultrasonic medical equipment. Office: Toin U Yokohama, 1614 Kurogane-Cho Midori-Ku, Yokohama 225, Japan

SOFIA, R. D., pharmacologist; b. Ellwood City, Pa., Oct. 8, 1942. BS, Geneva Coll., 1964; MS, Fairleigh-Dickenson U., 1969; PhD in Pharmacology, U. Pitts., 1971. Rsch. biologist Lederle Labs., N.Y., 1964-67; rsch. assoc. pharmacology Union Carbide Corp., 1967-69; sr. pharmacologist Pharmakon Labs., Pa., 1969; sr. rsch. pharmacologist Pharmakon Labs., 1971-73, dir. dept. Pharmacology and Toxicology, 1973-76, v.p. biology rsch., 1976-80, v.p. R&D, 1980-82; v.p. pre-clin. rsch. Wallace Labs., Cranbury, 1982—; cons. Pharmakon Labs., 1969-71. Mem. Am. Soc. Pharmacology and Experimental Therapeutics, Soc. Toxicology, SOc. Neuroscience, Internat. Soc. Study Pain, Am. Rheumatism Assn. Achievements include research in pharmacology and toxicology of various constituents of marijuana, development of new drugs for cardiovascular pulmonary and central nervous system diseases and pain relief. Office: Carter-Wallace Inc Half-acre Rd # 1001 Cranbury NJ 08512*

SOHAILI, ASPI ISFANDIAR, physiologist, educator; b. Nairobi, Kenya, June 5, 1966; came to U.S., 1986; s. Mehraban and Rezwan S. Cert., U. London, 1985; BS, Okla. State U., 1989; MS, Ark. State U., 1991; postgrad., U. Nebr., 1991—. Lab. asst. Reproductive Physiology Lab. Okla. State U., Stillwater, 1987-89; grad. asst. Coll. Agr. Ark. State U., State University, 1990-91; grad. fellow dept. animal sci. U. Nebr., Lincoln, 1992—. Vol. in rural area social and econ. devel. plan projects, Baha'i Faith Community, Kenya, 1985-86; v.p. African student orgn., Okla. State U., 1988-89, v.p. Baha'i Faith Club, 1987-89; grad. student adv. coun. rep. Coll. Agrl. Ark. State U., 1990-91; v.p. Baha'i Assn. U. Nebr., 1993—. Mem. Am. Soc. Animal Sci., Ark. Cattleman's Assn., Nat. Geographic Soc., African Wildlife Found., Soc. for Study of Reproduction, Assn. Baha'i Studies-Can., Baha'i Assn. U. Nebr. (pres. 1993—). Baha'i. Avocations: reading, hiking, computers, wildlife and environmental conservation. Home: 3300 Huntington Ave Apt 33 Lincoln NE 68504 Office: U Nebr Dept Animal Sci Lincoln NE 68583 also: PO Box 44262, Nairobi Kenya

SOHAL, IQBAL SINGH, structural engineer, educator; b. Ramidi Kapurthala, Punjab, India, Dec. 15, 1956; s. Gurdip Singh and Jaswant Kaur (Bal) S.; m. Kulvinder Kaur Grewal, Sept. 12, 1981; children: Jessica Kaur, Robinder Singh. BSCE, Punjab U., 1978; MS in Structural Engring., Purdue U., 1983, PhD in Structural Engring., 1986. Engr.-in-tng. Postdoctoral rsch. assoc. Purdue U., West Lafayette, Ind., 1987, vis. asst. prof. structural engring., 1987-88; asst. prof. Rutgers U., Piscataway, N.J., 1988—; presenter nat. and internat. confs., 1986—; invited lectr. Alcoa Labs., Alcoa Ctr., Pa., 1988. Author: (with W. F. Chen) Cylindrical Members in Offshore Structures, Thin-Walled Structures, 1988, symposium proceedings 1980—; contbr. articles to profl. jours. Recipient Rsch. Initiation award NSF, 1990, James F. Lincoln Arc Welding Found. award, 1984. Mem. ASCE (div. engring. mechanics, com. structural stability 1990—), Am. Soc. Engring. Edn., Am. Concrete Inst., Am. Acad. Engring. Mechanics, Tau Beta Pi. Sikh. Avocation: jogging, badminton. Home: 39 Hunt Dr Piscataway NJ 08854-6270 Office: Rutgers U Dept Civil Engring PO Box 909 Piscataway NJ 08855-0909

SOHIE, GUY ROSE LOUIS, electrical engineer, researcher; b. Antwerp, Belgium, Nov. 8, 1956; came to U.S., 1978; s. Andre and Lydia (Boussery) S.; m. Angela M. Sloman, Apr. 9, 1987; children: Oliver A., Harry N. Ind. Ingenieur, Industriele Hogeschool Antwerpen Mechelen, Antwerp, 1978; PhD, Pa. State U., 1983. Asst. prof. Ariz. State U., Tempe, 1985-87; applications engr. Motorola Inc., Austin, Tex., 1988-89; tech. staff Gen. Electric, Schenectady, 1989—. Co-author: Elements of Digital Systems, 1993; contbr. articles to profl. jours. Ensign Belgian Navy, 1983-84. Fulbright fellow, Brussels, 1978. Mem. IEEE, Sigma Xi, Phi Kappa Phi, Eta Kappa Nu. Achievements include 3 patents in emergency signal warning system; patent pending in image processing system for detection and tracking. Home: 5956 Curry Rd Extension Schenectady NY 12303 Office: Gen Electric CRD 1 River Rd Schenectady NY 12345

SOHN, HONG YONG, chemical and metallurgical engineering educator, consultant; b. Kaesung, Kyunggi-Do, Korea, Aug. 21, 1941; arrived U.S., 1966; s. Chong Ku and Soon Deuk (Woo) S.; m. Victoria Bee Tuan Ngo, Jan. 8, 1972; children: Berkeley Jihoon, Edward Jihyun. B.S. in Chem. Engring., Seoul (Korea) Nat. U., 1962; M.S. in Chem. Engring., U. N.B., Can., 1966; Ph.D. in Chem. Engring., U. Calif.-Berkeley, 1970. Engr., Cheil Sugar Co., Busan, Korea, 1962-64; research assoc. SUNY-Buffalo, 1971-73; research engr. DuPont Co., Wilmington, Del., 1973-74; prof. metall. engring., adj. prof. chem. engring. U. Utah, Salt Lake City, 1974—; cons. Lawrence Livermore Nat. Lab., 1976—, Kennecott Co., Salt Lake City, 1976—, Cabot Corp., 1984—, DuPont Co., 1987—, Utah Power and Light Co., 1987—. Co-author: Gas-Solid Reactions, 1976; co-editor: Rate Processes of Extractive Metallurgy, 1979, Extractive Metallurgy of Refractory Metals, 1980, Advances in Sulfide Smelting, 1983; Recycle and Secondary Recovery of Metals, 1985, Gas-solid Reactions in Pyrometallurgy, 1986, Flash Reaction Processes, 1988, Metallurgical Processes for the Year 2000 and Beyond, 1988; patentee process for treating sulfide-bearing ores; contbr. numerous articles to sci., tech. jours. Camille and Henry Dreyfus Found. Tchr. Scholar awardee, 1977; Fulbright Disting. lectr., 1983; Japan Soc. for the Promotion of Sci. fellow, 1990. Mem. The Minerals, Metals and Materials Soc. (past dir., Extractive Metallurgy Lectr. award, 1990, Champion H. Mathewson Gold Medal award, 1993, Extractive metallurgical sci. award 1990), Am. Inst. Chem. Engrs., Korean Inst. Chem. Engrs. Office: U Utah 412 Browning Bldg Salt Lake City UT 84112

SOHRABI, MORTEZA, chemical engineering educator; b. Tehran, Iran, Aug. 28, 1945; s. Hossein and Ghodrat Azam (Vali) S.; m. Tahereh Kaghazchi, Oct. 25, 1968; children: Maryam, Manijeh. BSChemE, U. Tehran, 1968; PhDChemE, Bradford (U.K.) U., 1972. Asst. prof. Amirkabir U., Tehran, 1973-77, assoc. prof., 1977-84, prof. chem. engring., 1984—, chmn. dept., 1975-77, dean of faculty engring. and tchr. tng., 1977-80; bd. trustees Tehran U. and Amirkabir U., Tehran, 1991—. Author: Mass Transfer, 1992; chief editor Amirkabir Jour. of Engring., Tehran, 1985—; contbr. articles to Chemtech, Jour. Chem. Tech. Biotechnol., Chimia, Chem. Engring. Tech. Recipient Medal for Ednl. Achievements Iranian Govt., 1968; named Disting. Prof. all Iranian U. Ministry of Culture and Higher Edn., 1992. Mem. AICE, Iranian Acad. Scis., Am. Chem. Soc., Soc. of Chem. Industry (U.K.). Achievements include estimation of the behavior of non ideal solutions by spectrospopic methods, modelling of multi-phase chem. reactions and synthesize of catalysts. Office: Amirkabir U Tech, Hafez, Tehran 15, Iran

SOILEAU, KERRY MICHAEL, aerospace technologist, researcher; b. New Orleans, June 8, 1956; s. Donald and Heloise Marie (LeBourgeois) S. BS, U. Cen. Fla., 1976; MS, La. State U., 1980. Aerospace technologist Johnson Space Ctr., NASA, Houston, 1980—; trajectory officer, flight dynamics officer Space Shuttle Mission Control NASA. Contbr. articles to profl. jours.; presenter sci. papers to confs.; developer (computer program) GradePlus. Newscaster Houston Taping for the Blind, 1984-90. Mem. Am. Astron. Soc. Office: NASA JSC/DM34 Houston TX 77058

SOKOLIĆ, MILENKO, biotechnologist, researcher; b. Novivinodolski, Croatia, Apr. 10, 1949; s. Tomislav and Rozalija Sokolić; m. Jan. 15, 1977; children: Saša, Nuša. BS in Biology, U. Zagreb, Croatia, 1974; MS in Biotechnology, U. Zabreb, Croatia, 1989. Microbiologist dept. control KRKA P.O., Novo Mesto, Slovenia, 1975-83, high performance liquid chromatography analyst dept. biotechnological devel., 1983—. Contbr. articles to profl. jours. Office: KRKA PO, Cesta Herojev 45, Novo Mesto Slovenia 68000

SOKOLIK, IGOR, physicist; b. Yessentuki, Russia, July 2, 1944; came to U.S., 1990; s. Anatoliy and Rozaliya (Schtokhammer) S.; m. Tatyana Kolosova, Feb. 12, 1967 (div. 1982); children: Katya, Anatoliy; m. Nina Vinogradskaya, May 31, 1989. MS in Physics, Moscow Phys. Engr. Inst., 1967; PhD in Chem. Physics, Inst. Chem. Physics, Moscow, 1972. From rsch. scientist to sr. staff scientist Inst. Chem. Physics, Moscow, 1967-88; sr. staff scientist Inst. Energy Problems Chem. Physics, Moscow, 1988-90; sr. scientist U. Mass., Amherst, 1991—; NYU vis. scientist Soros Found., 1988. Contbr. over 40 articles to profl. jours. Adminstrv. com. Russian Ednl. Soc., Moscow, 1980-90. Mem. Am. Chem. Soc., Am. Phys. Soc. Achievements include patents for Magnetoresistive sensors; discovery of magnetic field effects on the rate of photooxidation of molecular crystals and on the conductivity of doped heat-treated or irradiated organic solids, development of polymer light-emitting diodes. Office: U Mass Rm 701 GRC Amherst MA 01003

SOKOLOFF, LOUIS, physiologist, neurochemist; b. Phila., Oct. 14, 1921; married; 2 children. BA, U. Pa., 1943, MD, 1946. Intern Phila. Gen. Hosp., 1946-47; rsch. fellow in physiology U. Pa. Grad. Sch. Medicine, 1949-51, instr., then assoc., 1951-56; assoc. chief, then chief sect. cerebral metabolism NIMH, Bethesda, Md., 1953-68; chief lab. cerebral metabolism NIMH, 1968—. Chief editor Jour. Neurochemistry, 1974-78. Served to capt. M.C. U.S. Army, 1947-49. Recipient F.O. Schmitt medal in neurosci., 1980, Albert Lasker clin. med. research award, 1981, Karl Spencer Lashley award Am. Philos. Soc., 1987, Disting. Grad. award U. Pa., 1987, Nat. Acad. Scis. award in Neurosci., 1988, Georg Charles de Hevesy Nuclear Medicine Pioneer award Soc. Nuclear Medicine, 1988, Mihara Cerebrovascular Disorder Rsch. Promotion award, 1988. Mem. Am. Physiol. Soc., Assn. Rsch. Nervous and Mental Diseases, Am. Biophys. Soc., Am. Acad. Neurology, Am. Neurol. Assn., Am. Soc. Biol. Chemists, Am. Soc. Neurochemistry, U.S. Nat. Acad. Scis. Achievements include development of methods for measurement of cerebral blood flow and metabolism in animals and man. Office: NIMH Bldg 36 Rm 1A-05 Bethesda MD 20892

SOKOLSKY, ANDREJ GEORGIYEVICH, mathematician, academic administrator; b. Moscow, Feb. 18, 1950; s. Georgiy Alexandrovich and Liubov Vasilievna Sokolsky; m. Olga Mikhailovna Subbotina, July 11, 1972; children: Marina, Irina. Dr., Moscow Phys.-Tech. Inst., 1976; dr. higher degree, Moscow U., 1987. Registered profl. engr. in applied math. From asst. to asst. prof. to prof. Moscow Aviation Inst., 1976-88; dir. Inst. of Theoretical Astronomy of Russian Acad. Scis., St. Petersburg, 1989—; pres. Internat. Inst. of Problems of Asteroid Hazard, St. Petersburg, 1991—. Co-author: Methods and Algorithms of Normalization of Differential Equations, 1985; contbr. more than 150 articles to profl. jours. including Celestial Mechanics, Astronomical Jour., Applied Math. and Mechanics, Cosmic Investigations. Mem. NASA USA, Internat. Astron. Union, Assn. for Computing Machinery, Russian Acad. of Cosmonautics (academician). Achievements include theory of stability of Hamiltonian systems of differential equations and resonances; new theory of periodic motions of Hamiltonian systems and applications; proof of stability of Lagzangian solutions of 3 body problems; conception of global monitoring of near earth natural objects in the problem of asteroid hazard. Office: Inst Theoretical Astronomy, nab Kutuzova 10, Saint Petersburg 191187, Russia

SOLAJA, BOGDAN ALEKSANDAR, chemist, educator; b. Belgrade, Yugoslavia, July 8, 1951; s. Aleksandar Bogdan and Vera Branko (Kalember) S.; m. Anica Dragoliub Melikijan, Mar. 14, 1991; 1 child, Olga-Kosara. BS, U. Belgrade, 1975, PhD, 1984. Asst. faculty of chemistry U. Belgrade, 1978-85, scientific co-worker faculty of chemistry, 1987-89, asst. prof., 1989—; scientific co-worker Organic Chemistry Inst. U. Zurich, 1985-87. Editorial bd.: Jour. Serbian Chem. Soc., 1992—; contbr. articles to Tetrahedron, Helv. Chem. Acta, Jour. Serbian Chem. Soc., Jour. Chem. Soc. Perkin Trans 2. Fellow Legerlotz Found., 1986-87. Mem. Am. Chem. Soc., Swiss Chem. Soc., Serbian Chem. Soc. Achievements include research in partial and total synthesis of natural and related molecules, new reactions of aluminum alkyls. Office: U Belgrade, PO Box 550, Studentski trg 16, YU11001 Belgrade Yugoslavia

SOLBERG, JAMES JOSEPH, industrial engineering educator; b. Toledo, May 27, 1942; s. Archie Norman and Margaret Jean (Olsen) S.; m. Elizabeth Alice Snow, May 28, 1966; children: Kirsten Kari, Margaret Elizabeth. BA, Harvard U., 1964; MA, U. Mich., 1967, MS, 1967, PhD, 1969. Asst. prof. U. Toledo, 1969-71; assoc. prof. Purdue U., West Lafayette, Ind., 1971-81, prof., 1981—, dir. engring. rsch. ctr., 1986—. Author: Operations Research, 1976 (Book of the Yr. 1977); contbr. over 100 articles to profl. jours. Mem. NAE, Inst. Indsl. Engrs. (Disting. Rsch. award 1982), Soc. Mfg. Engrs., AAAS. Achievements include invention of CAN-Q which is a method for predicting performance of manufacturing systems used by hundreds of companies. Office: Purdue U AA Potter Engring Ctr West Lafayette IN 47907

SOLBERG, MYRON, food scientist, educator; b. Boston, June 11, 1931; s. Alexander and Ruth (Graff) S.; m. Rona Mae Bernstein, Aug. 26, 1956; children: Sara Lynn, Julie Sue, Laurence Michael. BS in Food Tech, U. Mass., 1952; PhD, MIT, 1960. Commd. 2d lt. USAF, 1952, advanced through grades to lt. col., 1973, ret., 1991; cons. to food industry, 1956-60, 64—; mem. rsch. staff food tech. MIT, 1954-60; quality control mgr. Colonial Provision Co., Inc., Boston, 1960-64; sci. editor Meat Processing mag., Chgo., 1968-69; mem. faculty Rutgers U., 1964—, prof. food sci., 1970—, dir. Ctr. for Advanced Food Tech., 1984—; UN expert on food product quality control, 1973-74; vis. prof. Technion, Israel Inst. Tech., Haifa, 1973-74. Co-editor Jour. Food Safety, 1977-88; contbr. articles to profl. jours. Pres. Highland Park (N.J.) Bd. Health, 1971-72. Recipient numerous research grants. Fellow AAAS, Am. Chem. Soc., Inst. Food Technologists (pres. N.Y. sect. 1971-72, Food Scientist of Yr. N.Y. sect. award 1981, Nicholas Appert award 1990); mem. Am. Soc. Microbiology, Am. Soc. Quality Control, Am. Meat Sci. Assn., N.Y. Acad. Scis., N.J. Acad. Sci. Home: 415 Grant Ave Highland Park NJ 08904-2705 Office: Rutgers U Nabisco Advanced Food Tech Inst New Brunswick NJ 08903

SOLDATOS, KOSTAS P., mathematician; b. Athens, Sept. 9, 1951; s. Panayiotis and Maria (Drakou) S.; m. Mariangela Gasparatou, Feb. 28, 1981. BSc in Math., U. Ioannina, Greece, 1974, PhD in Mechanics, 1980. Asst. lectr. U. Ioannina, 1977-80, lectr., 1980-85, asst. prof., 1985-90; rsch. fellow U. Nottingham, Eng., 1981-82, 89-90, lectr., 1990—; rsch. fellow Carleton U., Ottawa, Ont., Can., 1982-83; Contbr. articles to profl. jours. Mem. AIAA, Hellenic Soc. Theoretical and Applied Mechanics, Nottingham U. Composites Club. Achievements include research in mathematical modelling and analytical study of linear and non-linear mechanical behaviour of homogeneous and laminated anisotropic structures and structural elements (beams, plates and shells) of reinforced materials. Office: U Nottingham, Univ Campus, Nottingham England NG7 2RD

SOLÉ, PEDRO, chemical engineer; b. Guatemala, Guatemala, Sept. 4, 1936; came to U.S., 1974; s. Pedro Solé and Luisa Raquel (Castellanos) de Solé; m. Dorothy Tuteur, Mar. 19, 1961; children: Tania Dolores, Jeanne Marguerite, Pedro Ernesto. Degree chem. engring., San Carlos U., Guatemala, 1958; PhD in Chem. Engring., Polytech U., N.Y., 1965. Ops. mgr. Alimentos Kern de Guatemala, 1968-72; gen. dir. Riviana España S.A., Seville, Spain, 1972-74; plant mgr. Casera Foods Inc., San Juan, P.R., 1974-78; gen. mgr. processed bananas Tela RR Co., La Lima, Honduras, 1979-82; v.p. bus. devel. Numar Processed Foods Group, San José, Costa Rica, 1982-86; dir. tech. svcs. Chiquita Brands Inc., N.Y.C., 1986-89; v.p. quality assurance/control Chiquita Brands Internat., Inc., Cin., 1989—. Contbr. articles to profl. jours. Exch. scholar U.S. Dept. State, 1959-60, Food Tech. scholar, Karlsruhe (Fed. Republic Germany) U., 1966-67. Mem. Inst. Food Tech. (profl.), Am. Inst. Chem. Engrs., Am. Chem. Soc., Nat. Food Processors Assn. (aseptic processing and packaging com. 1989—). Roman Catholic. Achievements include patents for individual coffee extractor; for recovery of vegetable oil; for banana processing; for banana peel processing; several fgn. patents issued and pending. Home: PO Box 18134 Cincinnati OH 45218-0134 Office: Chiquita Brands Internat 250 E 5th St Cincinnati OH 45202-4103

SOLIDUM, EMILIO SOLIDUM, electronics and communications engineer; b. Manila, May 28, 1936; s. Agapito and Olympia Solidum; m. Adela Villanueva, Dec. 24, 1961; children: Ademil, Mildred, Mary Rose Adelle. BS in Communications Engring., Capitol Radio Engring. Inst., Manila, 1972; BSEE, M.L. Quezon U., Manila, 1976. Lic. electronics and communications engr., Manila. Radio and TV maintenance technician ABS-CBN Broadcasting Corp., Quezon City, The Philippines, 1962-65, chief maintenance dept., 1965-67, chief engr., 1967-69, engring. dir., 1969-72; engring. dir. GMA Radio-TV Arts, Quezon City, The Philippines, 1972-78, v.p. engring., 1978-88, sr. v.p., 1988—; electronics cons., Transradio Broadcasting Corp., Manila, 1973-76; broadcast engr., cons. Vicor Music Corp., Quezon City, 1976-78; tech. liaison officer Asia-Pacific Broadcasting Union, 1988-93; electronics instr. Capitol Tech. Inst., Manila. Author: KBP Distance Study Course, 1976 (plaque of appreciation 1979). Colombo Plan grantee KBP-NEDA-JICA, 1977. Mem. Inst. Electronics and Communications Engrs. Philippines (bd. dirs. 1988-89, outstanding engr. award 1990), Soc. Broadcast Engrs. (chmn. mem. com.), Kapisanan ng mga Brodkaster sa Pilipinas (sub-com. chmn. TV tech. stds. com.), Torres High Alumni Assn. (bd. dirs.), KC, Rotary (internat. svc. dir.). Avocations: electronics projects, bowling, swimming, chess, weightlifting. Home: 14 Penguin, St Francis Village, Meycauayan, Bulacan The Philippines Office: GMA Radio-TV Arts, Corner Edsa/Timog, Quezon City The Philippines

SOLIMANDO, DOMINIC ANTHONY, JR., pharmacist; b. Bklyn., Apr. 4, 1950; s. Dominic Anthony and Grace Evelyn (Phillips) S. BS, Phila. Coll. Parm. and Sci., 1976; MA, Cen. Mich. U., 1980; postgrad., Purdue U., 1986—. Pharmacist Walter Reed AMC, pharmacy svc., Washington, 1977; chief pharmacy svc. Andrew Rader USA Health Clinic, Ft. Myer, Va., 1977-79; oncology pharmacist Walter Reed Army Med. Ctr., Washington, 1979-82; clin. preceptor sch. pharmacy Med. Coll. Va., 1980; chief hem./oncology pharmacy Tripler Army Med. Ctr., Honolulu, 1983-86, Letterman Army Med. Ctr., San Francisco, 1989-90, 91-92; chief pharmacy svc. 28th Combat Support Hosp. Operation Desert Shield/Desert Storm, Saudi Arabia, 1990-91; chief hematology/oncology pharmacy treatment sect. Walter Reed Army Med. Ctr., 1992—; adj. prof. Coll. Pharmacy, U. Pacific, Stockton, Claif., 1983-86, 1989-91, editorial panel Drug Intelligence and clin. pharm., Cin., 1984-88; clin. asst. prof. Coll. Pharmacy, U. Md., 1992—; clin. preceptor Coll. Pharmacy, Howard U., 1992—. Maj. U.S. Army, 1976—. Recipient Upjohn rsch. grant Am. Coll. Clin. Pharmacy, 1988, Acad. of Pharmacy Practice and Mgmt. fellow Am. Pharm. Assn., 1987, Bristol awd. Phila. Coll. Pharmacy and Sci., 1976. Fellow Am. Soc. of Hosp. Pharmacists; mem. AAAS, Am. Coll. Clin. Pharmacy (constn., by-laws com, 1989-90, clin. practice affairs com. 1990-91, vice chmn. 1993), Am. Pharm. Assn. (various offices), Am. Inst. Hist. Pharmacy, Am. Soc. Hosp. Pharmacists, Assn. Mil. Surgs. U.S., Assn. of U.S. Army, N.Y. Acad. Sci., Pa. Pharm. Assn., Res. Officers Assn., Met. Wash. Soc. Hosp. Pharmacists, Va. Soc. Hosp. Pharmacists, Va. Pharm. Assn., No. Va. Soc. Hosp. Pharmacists, Rho Chi, Fedn. Internat. de Pharm.(hosp. and mil. sects.), Kappa Psi. Avocations: bicycling, chess, history, cooking, travel. Office: Walter Reed Army Med Ctr Pharmacy Svc Washington DC 20307-5001

SOLLITT, CHARLES KEVIN, ocean engineering educator, laboratory director; b. Mpls., Aug. 8, 1943; s. Charles Clinton and Wilma Shirley (Bong) S.; m. Melissa Ann Jones, June 17, 1967; children: Katherine, Thomas. BS, U. Wash., 1966, MS, 1968; PhD, MIT, 1972. Rsch. and teaching asst. U. Wash., Seattle, 1966-68, MIT, Cambridge, 1968-72; asst. prof. Oregon State U., Corvallis, 1972-78, assoc. prof., 1978—, chmn. ocean engring., 1981-90, dir. Wave Rsch. Lab., 1981—; mem. sci. and tech. ctrs. rev. team com. NSF, Washington, 1989; cons. CH2M-Hill, Corvallis, 1973-86. Contbr. articles to profl. jours. including Am. Soc. Civil Engrs. Jour., Ocean Engring. Jour. Grantee NOAA, 1978-91, Office Natural Resources, 1986-91. Mem. AAAS, ASCE (chmn. tech. com. ocean engring. 1986-87). Republican. Unitarian. Home: 113 NW 28th St Corvallis OR 97330-5304 Office: Oreg State U Civil Engring Dept Apperson Hall 202 Corvallis OR 97311

SOLNIT, ALBERT JAY, commissioner, physician, educator; b. Los Angeles, Aug. 26, 1919; s. Benjamin and Bertha (Pavin) S.; m. Martha

Benedict, 1949; children—David, Ruth, Benjamin, Aaron. B.A. in Med. Scis., U. Calif., 1940, M.A. in Anatomy, 1942, M.D., 1943; M.A. (hon.), Yale U., 1964. Rotating intern L.I. Coll. Hosp., 1944, asst. resident in pediatrics, 1944-45; resident in pediatrics and communicable diseases U. Calif. div. San Francisco Hosp., 1947-48; asst. resident dept. psychiatry and mental hygiene Yale U., 1948-49, sr. resident, 1949-50, fellow in child psychiatry, 1950-52, instr. pediatrics and psychiatry, 1952-53, asst. prof., 1953-60, assoc. prof., 1960-64, prof., 1964-70, Sterling prof., 1970—, dir. Child Study Ctr., 1966-83; commr. dept. of mental health State of Conn., Hartford, 1991—; tng. and supervising analyst Western New Eng. Inst. Psychoanalysis, 1962—, N.Y. Psychoanalytic Inst., 1962—; cons. Childrens Bur., HEW; mem. adv. coun. Erikson Inst. for early Childhood Edn., 1966—; nat. adviser Children, publ. of Children's Bur., 1965—; mem. com. on publs. Yale U. Press, 1971—; adv. bd. Action for Children's TV, Newtonville, Mass., 1973—; mem. div. med. scis. Assembly Life Scis., NRC, 1974—; cons. div. mental health svc. program NIMH, 1974—; Sigmund Freud Meml. prof. U. Coll. London, 1983-84; Sigmund Freud prof., dir. Freud Ctr. Psychoanalytic Studies Hebrew U., Jerusalem, 1985-87. Author: (with M.J.E. Senn) Problems in Child Behavior and Develpment, 1968), (with A. Freud, J. Goldstein) Beyond the Best Interests of the Child, 19732, (with Goldstein) Divorce and Your Child, 1983, (with R. Lord, B. Nordhaus) When Home Is No Haven, 1992; The Many Meanings of Play, 1993; editor: (with S. Provence) Modern Perspectives in Child Development, 1963; mng. editor Psychoanalytic Study of the Child, 1971—; mem. editorial bd. Israel Annals Psychiatry and Related Disciplines, 1969—. WHO prof. psychiatry and human devel. U. Negev, Beer-Sheva, Israel, 1973-74. With USAAF, 1945-47. Recipient Disting. Svc. award Am. Psychiatric Assn., 1992. Mem. AAAS, Inst. Medicine of NAS, Am. Orthopsychiatric. Assn. (editorial bd. jour. 1974-82), Am. Psychoanalytic Assn. (past pres., editorial bd. jour. 1972-74), Am. Acad. Child and Adolescent Psychiatry (past pres., editorial bd. jour. 1975), Internat. Pediatric Soc., Am. Assn. Child Psychoanalysis (past pres.), Am. Acad. Pediatrics (editorial bd. jour. 1968-76, task force pediatric edn.), Internat. Psychoanalytic Assn., Internat. Assn. Child and Adolescent Psychiatry (pres. 1974-76), N.Y. Psychoanalytic Soc., Soc. Profs. Child Psychiatry. Home: 107 Cottage St New Haven CT 06511-2465 Office: 333 Cedar St New Haven CT 06510-3289 also: 90 Washington St Hartford CT 06106

SOLOMON, ALLEN LOUIS, chemist; b. Pitts., Oct. 1, 1922; s. Harold Louis and Cecelia (Shrager) S.; married, Apr. 15, 1951; children: Ruth Ann, Susan Carol. BS, Yale U., 1942, MS, 1944, PhD, 1948. Rsch. chemist RCA Labs., Princeton, N.J., 1948-56; mgr. solid state devices GTE Labs., Bayside, N.Y., 1956-72; CCD program mgr. Fairchild Semiconductor R&D, Palo Alto, Calif., 1972-75; sr. scientist N.Y. Inst. Tech. Rsch. Ctr., Dania, Fla., 1975-77; prin. staff engr. McDonell Douglas Astronautics Co., Huntington Beach, Calif., 1977-82; dir. advanced systems HTL K West, Allegheny Internat., Santa Ana, Calif., 1982-84; project mgr. Hughes Aircraft Co., El Segundo, Calif., 1984-85; engring. specialist Grumman Space Systems Div., Irvine, Calif., 1986-91; pvt. practice tech. cons. Fullerton, Calif., 1991—. Contbr. articles to profl. publs. Mem. IEEE (sr.), Electrochem. Soc., Am. Chem. Soc., Sigma Xi. Achievements include patents for detector interface device, multilayer integrated curcuit module, method of making a trench gate complimentary metal oxide semiconductor transistor, trench JFET integrated circuit elements, and others. Home and Office: 1800 Fairford Dr Fullerton CA 92633-1511

SOLOMON, DANIEL, psychologist; b. Chgo., May 11, 1933; s. Isadore Albert and Esther (Aaron) S.; m. Jean Ann Soerens, Apr. 13, 1963; children: Nicholas J., Paula K. BA, Antioch Coll., 1956; MA, U. Mich., 1956, PhD, 1960. Rsch. assoc. Ctr. for Study of Literal Edn. for Adults, Chgo., 1959-63; sr. rsch. assoc. Inst. Juvenile Rsch., Chgo., 1963-71; social psychologist Montgomery County Pub. Schs., Rockville, Md., 1971-79; social scientist U.S. Bur. Census, Suitland, Md., 1979-80; dir. rsch. Devel. Studies Ctr., San Ramon, Calif., 1980—; ptnr. Social Rsch. Cons., Gaithersburg, Md., 1976-80. Author: Children in Classrooms, 1979; contbr. articles to profl. jours. Mem. Am. Psychol. Assn., Am. Ednl. Rsch. Assn., Soc. Pshcol. Study of Social Issues, Soc. Rsch. Child Devel. Home: 7210 Blake St El Cerrito CA 94530 Office: Devel Studies Ctr 2000 Embarcadero Ste # 305 Oakland CA 94606

SOLOMON, DAVID, sales representative; b. Springfield, Mass., July 17, 1952; s. Jack and Florence (Bengelsdorf) S.; m. Pamela Peckham, June 15, 1973 (div. Sept. 1978); m. Alix Kyle, Mar. 23, 1986; children: Sasha Parsons, Ayla Cheyenne, Zachary Nathan. AS in Agrl. Bus. Mgmt., Stockbridge Coll., 1972; BS in Biology, U. Mass., North Dartmouth, 1977. Analyzer application engr. The Foxboro (Mass.) Co., 1977-88; sales rep. Varian Assocs., Palo Alto, Calif., 1988—; pres. S.A.S., Inc., Marion, Mass., 1982-86, S.A.S., Inc., Richmond, Va., 1986—. Contbr. articles to Analysis Instrumentation, Med. Device and Diagnostic, Roundel. V.p. Stonequarter Neighborhood Assn., Richmond, 1990-92. Mem. Am. Chem. Soc. (assoc.), Instrument Soc. Am. (sr.). Democrat. Jewish. Achievements include application, mktg. and sales of world's only pneumatic gas chromatograph. Home: 2104 Stonequarter Ct Richmond VA 23233 Office: Varian Assn Ste 150 505 Julie Rivers Rd Sugar Land TX 77478

SOLOMON, DAVID HARRIS, physician, educator; b. Cambridge, Mass., Mar. 7, 1923; s. Frank and Rose (Roud) S.; m. Ronda L. Markson, June 23, 1946; children: Patti Jean (Mrs. Richard E. Sinaiko), Nancy Ellen (Mrs. Marvin Evans.). A.B., Brown U., 1944; M.D., Harvard U., 1946. Intern Peter Bent Brigham Hosp., Boston, 1946-47, resident, 1947-48, 50-51; fellow endocrinology New Eng. Center Hosp., Boston, 1951-52; faculty UCLA Sch. Medicine, 1952—, prof. medicine, 1966—, vice chmn. dept. medicine, 1968-71, chmn. dept., 1971-81, assoc. dir. geriatrics, 1982-89; dir. Ctr. on Aging UCLA, 1991—, prof. emeritus, 1993—; chief med. service Harbor Gen. Hosp., Torrance, Calif., 1966-71; cons. Wadsworth VA Hosp., Los Angeles, 1952—, Sepulveda VA Hosp., 1971—; cons. metabolism tng. com. USPHS, 1960-64, endocrinology study sect., 1970-73; mem. dean's com. Wadsworth, Sepulveda VA hosps. Editor: Jour. Am. Geriatric Soc., 1988-93; contbr. numerous articles to profl. jours. Recipient Mayo Soley award, 1986. Master ACP; mem. Assn. Am. Physicians, Am. Soc. Clin. Investigation, Am. Fedn. Clin. Research, Western Soc. Clin. Research (councillor 1963-65), Endocrine Soc. (Robert H. Williams award 1989), Am. Thyroid Assn. (pres. 1973-74, Disting. Service award 1986), Inst. Medicine Nat. Acad. Scis., AAAS, Assn. Profs. Medicine (pres. 1980-81), Western Assn. Physicians (councillor 1972-75, pres. 1983-84), Am. Fedn. Aging Rsch. (Irving S. Wright award), Am. Geriatrics Soc. (bd. dirs. 1985-93, Milo Leavitt award 1992), Phi Beta Kappa, Sigma Xi, Alpha Omega Alpha. Home: 863 Woodacres Rd Santa Monica CA 90402-2107 Office: UCLA Sch Medicine Dept Medicine Los Angeles CA 90024-1687

SOLOMON, ELINOR HARRIS, economics educator; b. Boston, Feb. 26, 1923; d. Ralph and Linna Harris; m. Richard A. Solomon, Mar. 30, 1957; children: Joan S. Griffin, Robert H., Thomas H. AB, Mt. Holyoke Coll., 1944; MA, Radcliffe U., 1945; PhD, Harvard U., 1948. Jr. economist Fed. Res. Bank Boston, 1945-48; economist Fed. Res. Bd. Govs., Washington, 1949-56; internat. economist U.S. State Dept., Washington, 1957-58; professorial lectr. Am. U., Washington, 1964-66; sr. economist antitrust div. U.S. Dept. Justice, Washington, 1966-82; prof. econs. George Washington U., Washington, 1982—; econ. cons., Washington, 1982—; expert witness antitrust, electronic funds transfer cases, Washington, 1988—. Author, editor: Electronic Funds Transfers and Payments, 1987, Electronic Money Flows, 1991; contbr. articles on econs., banking and law to profl. jours. Mem. Am. Econs. Assn., Nat. Economist Club. Home: 6805 Delaware St Bethesda MD 20815-4164 Office: George Washington U Dept Econs Washington DC 20052

SOLOMON, JULIUS OSCAR LEE, pharmacist, hypnotherapist; b. N.Y.C., Aug. 14, 1917; s. John and Jeannette (Krieger) S.; student Bklyn. Coll., 1935-36, CCNY, 1936-37; BS in Pharmacy, U. So. Calif., 1949; postgrad. Long Beach State U., 1971-72, Southwestern Colls., 1979, 81-82; PhD, Am. Inst. Hypnotherapy, 1988; m. Sylvia Smith, June 26, 1941 (div. Jan. 1975); children: Marc Irwin, Evan Scott, Jeri Lee. Cert. hypnotherapist; cert. hypnoanaesthesia therapist. Dye maker Fred Fear & Co., Bklyn., 1935; apprentice interior decorator Dorothy Draper, 1936; various jobs, N.Y. State Police, 1940-45; rsch. asst. Union Oil Co., 1945; lighting cons. Joe Rosenberg & Co., 1946-49; owner Banner Drug, Lomita, 1949-53, Redondo Beach, Calif., 1953-72, El Prado Pharmacy, Redondo Beach, 1961-65; pres. Banner Drug, Inc., Redondo Beach, 1953-72, Thrifty Drugs, 1972-74, also Guild Drug, Longs Drug, Drug King, 1976-83; pres. Socoma, Inc. doing bus. as Lee & Ana Pharmacy, 1983-86, now Two Hearts Help Clinic, 1986-91. Charter commr., founder Redondo Beach Youth Baseball Council; sponsor Little League Baseball, basketball, football, bowling; pres. Redondo Beach Boys Club; v.p. South Bay Children's Health Ctr., 1974, Redondo Beach Coordinating Coun., 1975; founder Redondo Beach Community Theater, 1975; active maj. gift drive YMCA, 1975; mem. SCAG Com. on Criminal Justice, 1974, League Calif. Environ. Quality Com., 1975; mem. Dem. State Cen. Com., Los Angeles County Dem. Cen. Com.; del. Dem. Nat. Conv., 1972; chmn. Redondo Beach Recreation and Parks Commn.; mem. San Diego County Parks Adv. Commn., 1982; mem. San Diego Juvenile Justice Commn., 1986—; mem. San Diego County Adv. Com. Adult Detention, 1987—; mem. human resource devel. com., pub. improvement com. Nat. League of Cities; v.p. Redondo Beach Coordinating Coun.; councilman, Redondo Beach, 1961-69, 73-77; treas. 46th Assembly Dist. Coun.; candidate 46 Assembly dist. 1966; nat. chmn. Pharmacists for Humphrey, 1968, 72; pres. bd. dirs. South Bay Exceptional Childrens Soc., Chapel Theatre; bd. dirs. so. div. League Calif. Cities, U.S.-Mex. Sister Cities Assn., Boy's Club Found. San Diego County, Autumn Hills Condominium Assn. (pres.), Calif. Employee Pharmacists Assn. (pres. 1985), Our House, Chula Vista, Calif., 1984—; mem. South Bay Inter-City Hwy. Com., Redondo Beach Round Table, 1973-77; mem. State Calif. Commn. of Californias (U.S.-Mexico), 1975-78; mem. Chula Vista Safety Commn., 1978, chmn., 1980-81; chmn. San Diego County Juvenile Camp Contract Com., 1982-83; mem. San Diego County Juvenile Delinquency Prevention Commn., 1983-85, 89—, San Diego County Juvenile Justice Commn., 1986—, San Diego County Adv. Com. for Adult Detention, 1987—; spl. participant Calif. Crime and Violence Workshop; mem. Montgomery Planning Commn., 1983-86; mem. Constnl. Observance Com., 1990—, Troubled Teenagers Hypnosis Treatment Program, 1989—. With USCGR, 1942-45. Recipient Pop Warner Youth award, 1960, 1962, award of merit Calif. Pharm. Assn., 1962, award Am. Assn. Blood Banks, 1982. Diplomate Am. Bd. Diplomates Pharmacy Internat., 1977-81; Fellow Am. Coll. Pharmacists (pres. 1949-57); mem. South Bay Pharm. Assn. (pres.), South Bay Councilman Assn. (founder, pres.), Palos Verdes Peninsula Navy League (charter), Am. Legion, U. So. Calif. Alumni Assn. (life), Assn. Former N.Y. State Troopers (life), AFTRA, Am. Pharm. Assn., Nat. Assn. Retail Druggists, Calif. Pharmacists Assn., Calif. Employee Pharmacist Assn. (bd. dirs. 1980-81), Hon. Dep. Sheriff's Assn., San Ysidro C. of C. (bd. dirs. 1985-87), Fraternal Order of Police, San Diego County Fish and Game Assn., Rho Pi Phi (pres. alumni). Club: Trojan (life). Lodges: Elks (life), Masons (32 deg.; life), Lions (charter mem. North Redondo). Established Lee and Ana Solomon award for varsity athlete with highest scholastic average at 10 L.A. South Bay High Schs. in Los Angeles County and 3 San Diego area South Bay High Schs.

SOLOMON, PATRICK MICHAEL, mechanical engineer; b. Chgo., Feb. 27, 1967; s. William Burton and Barbara Jean (Casey) S.; m. Dawn Marie Stralis, Sept. 3, 1989. BS in Mech. Engring., Northwestern U., 1989. Licensed profl. engr. in tng. Facilities engr. U.S. Naval Facilities Engring. Command, Alexandria, Va., 1989-91; tech. staff engr. Commonwealth Edison, Zion, Ill., 1991—. Am. Cancer Soc. Summer Scholar in Medicine, 1983, scholar Soc. Am. Mil. Engrs., 1987; fed. jr. fellow USN, 1985. Democrat. Roman Catholic. Home: 291 Penny Ln Grayslake IL 60030 Office: Commonwealth Edison 101 Shiloh Blvd Zion IL 60099

SOLOMON, RICHARD STREAN, psychologist, educator; b. Montreal, Que., Can., June 7, 1954; s. Frederick Lawrence Solomon and Joan Hope (Strean) Lazarus; m. Karen Ann Kaplan, June 18, 1977; children: Benjamin, Victoria. BS, Hobart Coll., 1975; PhD, U. R.I., 1982. Cert. clin. psychologist, sch. psychologist, Mass. Staff psychologist R.I. Youth Guidance Ctr., Pautucket, 1982-83; cons. psychologist R.I., Conn. and Mass. sch. depts., 1982—, R.I. and Mass. child protective depts., 1987—; clin. dir., chief psychologist Delta Cons., Providence and Attleboro, Mass., 1987—; part0time instr. dept. psychology U. R.I., Providence, 1982—. Co-author: Paternal Deprivation and Child Maltreatment, 1986, also article. Mem. rev. com. Child Advocate's Office R.I., Providence, 1990-91. Recipient Commr.'s award Dept. Health and Human Svcs., 1988. Mem. APA, Mass. Psychol. Assn., Am. Humane Assn., Nat. Com. for Prevention of Child Abuse (pres. R.I. chpt. 1986-88, co-pres. 1990-91). Jewish. Home: 2 Hollow Ct Warwick RI 02886

SOLOMON, ROBERT DOUGLAS, pathology educator; b. Delavan, Wis., Aug. 28, 1917; s. Lewis Jacob and Sara (Ludgin) S.; m. Helen Fisher, Apr. 4, 1943; children: Susan, Wendy, James, William. Student, MIT, 1934-36; BS in Biochemistry, U. Chgo., 1938; MD, Johns Hopkins U., 1942. Intern John's Hopkins Hosp., 1942-43; resident in pathology Michael Reese Hosp., 1947-49; lectr. U. Ill., Chgo., 1947-50; fellow NIH pathology U. Ill., 1949-50; asst. prof. U. Md., Balt., 1955-60; assoc. prof. U. So. Calif., L.A., 1960-70; chief of staff City of Hope Nat. Med. Ctr., 1966-67; prof. U. Mo., Kansas City, 1977-78, SUNY, Syracuse, 1978-88; chief of staff The Hosp., Sidney, N.Y., 1985-86; adj. prof. U. N.C., Wilmington, 1988—; cons. VA Hosp., Balt., 1955-60, Med. Svc. Lab., Wilmington, 1989—. Co-author: Progress in Gerontological Research, 1967; contbr. papers to profl. jours. V.p. Rotary, Duarte, Calif., 1967; v.p. and pres. Force for an Informed Electorate. Capt. Med. Corps, AUS, 1943-46, PTO. Grantee NIH, Fleischmann Found., Am. Heart Assn., Nat. Cancer Inst., 1958-70. Fellow Am. Coll. Physicians (pres. Md. chpt.); mem. Coll. Am. Pathologists (past pres. Md. chpt.), Am. Soc. Clin. Pathologists, Assn. Clin. Scientists, Am. Chem. Soc., Royal Soc. Medicine (London). Avocations: cruising, astronomy, mathematics, fishing, stamps. Home: 113 S Belvedere Dr Hampstead NC 28443-9212

SOLOMON, SAMUEL, biochemistry educator; b. Brest Litovsk, Poland, Dec. 25, 1925; s. Nathan and Rachel (Greenberg) S.; m. Sheila R. Horn, Aug. 11, 1953 (div. 1974); children—David Horn, Peter Horn, Jonathan Simon; m. Augusta M. Vineberg, July 12, 1974. B.S. with honors, McGill U., 1947, M.S., 1951, Ph.D. in Biochemistry, 1953. Research asst. Columbia, 1953-55, asso. in biochemistry, 1958-59, asst. prof., 1959-60; asso. prof. biochemistry and exptl. medicine McGill U., 1960-66, prof., 1967—, prof. obstetrics and gynecology, 1976—; dir. endocrine lab. Royal Victoria Hosp., Montreal, Que., 1965—, dir. research inst., 1982-85; mem. endocrinology and metabolism grants com. Med. Rsch. Coun. Can., 1967-71, regional dir. for Quebec, 1993—; vis. prof. endocrinology U. Vt., 1964; cons. in field; Joseph Price orator, 1982; mem. steering com. Pharm. Mfg. Assn. Am.-Med. Rsch. Coun. Can. Partnership, 1993—; Med. Rsch. Coun. Can. dir. for McGill U., 1993—. Co-editor: Chemical and Biological Aspects of Steroid Conugation, 1970; Editorial bd.: Endocrinology, 1962; asso. editor: Can. Jour. Biochemistry, 1967-71, Jour. Med. Primatology, 1971; Contbr. articles profl. jours. Mem. bd. govs. McGill U., 1975-78; mem. steering com. European Study Group on Steroid Hormones, 1974—, chmn. steering com., 1983—, chmn. programe com. 1990-91; mem. Dubin Commn. of Inquiry Drugs in Athletes, 1988-90. Recipient McLaughlin medal Royal Soc. Can., 1989. Fellow Chem. Inst. Can., Am. Ob-Gyn. Soc. (hon.), Perinatal Rsch. Soc. Am. (pres. 1976), Soc. Gynecol. Investigation (program chmn. 1980), Endocrine Soc. (publ. com. 1986-89). Home: 239 Kensington Ave 603, Montreal, PQ Canada H3Z 2H1 Office: Royal Victoria Hosp Dept Endocrinology, 687 Pine Ave W, Montreal, PQ Canada H3A 1A1

SOLOMON, SUSAN, chemist, scientist; b. Chicago, Ill., Jan. 19, 1956; d. Leonard Marvin and Alice (Rutman) Solomon; m. Barry Lane Sidwell, Sept. 20, 1988. BS in Chemistry, Ill. Inst. Tech., 1977; MS in Chemistry, U. Calif., Berkeley, 1979, PhD in Chemistry, 1981. Rsch. chemist aeronomy lab. NOAA, Boulder, Colo., 1981-88, program leader middle atmosphere group aeronomy lab., 1988—; head project sci. Nat. Ozone Expedition, McMurdo Sta., Antarctica, 1986, 1987; adj. faculty U. Colo., 1982—. Co-author: Aeronomy of the Middle Atmosphere, 1984; contbr. articles to sci. jours. Recipient Gold medal U.S. Dept. Commerce, 1989. Fellow Royal Meteorol. Soc., Am. Meteorol. Soc., Am. Geophys. Union (J.B. McElwane award 1985); mem. NAS, Am. Acad. Arts and Scis. Avocations: creative writing, crafts, scuba diving.

SOLTANOFF, JACK, nutritionist, chiropractor; b. Newark, Apr. 24, 1915; s. Louis and Rose (Yomteff) S.; m. Esther Katcher, Sept. 29, 1939; children: Howard, Ruth C. Soltanoff Jacobs, Hillory Soltanoff Seaton. N.M.D. Mecca Coll. Chiropractic Medicine, 1928, U.S. Sch. Naturopathy and Allied Scis., 1951; D.Chiropractic, Chiropractic Inst. N.Y., 1956; postgrad. Atlantic States Chiropractic Inst., 1962-63, Nat. Coll. Chiropractic, 1964-65; PhD, diplomate in nutrition Fla. Natural Health Coll., 1982. Gen. practice chiropractic medicine, cons. in nutrition, N.Y.C., 1956-75, West Hurley, N.Y. and Singer Island, Fla., 1975—; lectr., cons. in field. Author: Natural Healing; pub. Warner Books; contbr. articles to profl. jours. Syndicated newspaper columnist. Fellow Internat. Coll. Naturopathic Physicians; mem. Am. Chiropractic Assn., Internat. Chiropractic Assn., Brit. Chiropractic Assn., N.Y. Acad. Scis., Am. Council on Diagnosis and Internal Disorders, Council on Nutrition, Ethical Culture Soc. Unitarian. Instrumental in instituting chiropractic care in union contracts for mems. of Teamsters Union. Home: Rte 28A PO Box 447 West Hurley NY 12491 also: Martinique II 4100 N Ocean Dr Singer Island FL 33404 Office: Rte 28 and Van Dale Rd West Hurley NY 12491

SOLTAU, RONALD CHARLES, chemical engineer; b. Chgo., Apr. 29, 1944; s. Charles Otto Soltau and Julia Elizabeth (Dienes) Vogel; m. Sharan Oliver, Apr. 29, 1971; 1 child, Eric. BSChemE, Ill. Inst. Tech., 1968. Registered profl. engr., Tex. Ops. engr. Shell Chem. Co., Houston, 1968-70; systems engr. M.W. Kellogg Co., Houston, 1970-72; engring. mgr. Arabian Am. Oil Co., Dhahran, Saudi Arabia, 1972-76; sr. project mgr. Ford, Bacon & Davis, Dallas, 1978-86; cons. Solomon & Assocs., Dallas, 1986-88; project engr. Bechtel Corp., Houston, 1988-90; sr. project mgr. Jacobs Engring. Group Inc., Houston, 1990—. Mem. NSPE, Tex. Soc. Profl. Engrs., Jacobs Chatterers Toastmasters (pres. 1991-92). Achievements include development, installation and start-up of a methyl mercartan recovery process for Shell Chem. Co.; development of factors to relate the cost of building refineries and petrochem. plants in 60 U.S. and fgn. locations. Home: 506 Flaghoist Houston TX 77079 Office: Jacobs Engring Group Inc 4848 Loop Ctrl Dr Houston TX 77081

SOLURSH, MICHAEL, biology educator, researcher; b. L.A., Dec. 22, 1942; s. Louis and Helen (Schwartz) S.; m. Victoria R. Raskin, Mar. 21, 1964; 1 child, Elizabeth. BA, UCLA, 1964; PhD, U. Wash., 1969. Asst. prof. U. Iowa, Iowa City, 1969-73, assoc. prof., 1973-79, prof., 1979—. Mem. Am. Soc. Zoologists (program officer 1990-93), Soc. for Developmental Biologists, Am. Assn. Anatomists (program chairperson 1992-93), Tissue Culture Assn., Am. Soc. for Cell Biology. Achievements include research in morphogenetic mechanisms in mesenchyme. Office: U Iowa Dept Biol Scis Iowa City IA 52242

SOMADASA, HETTIWATTE, mathematician, educator; b. Matara, Sri Lanka, June 30, 1937; m. Chitra Indrani Philip, Sept. 8, 1967; children: Sujeeva, Amalka, Panduka. BA with honors, U. Ceylon, Colombo, Sri Lanka, 1960; PhD, U. Wales, U.K., 1966. Lectr. U. Kelaniya, Sri Lanka, 1960-79; sr. lectr. U. Calabar, Nigeria, 1979-82; sr. lectr. Open U., Sri Lanka, 1982-89, head dept. math., 1991—; prof. Inst. Tech. Studies, Sri Lanka, 1990-91. Home: 37/2 Kulatunga Rd, Panadura Sri Lanka Office: Open U Dept Math, Nawala, Nugegoda Sri Lanka

SOMANI, ARUN KUMAR, electrical engineer, educator; b. Beawar, India, July 16, 1951; came to the U.S., 1985; s. Kanwar Lal and Dulari Devi (Mundra) S.; m. Deepa-Toshniwal, Jan. 21, 1976 (dec. 1985); children: Ashutosh, Paritosh; m. Manju-Kankani, July 6, 1987; 1 child, Anju. BS with honors, B.I.T.S., Pilani, India, 1973; MTech, IIT, Delhi, 1979; MSEE, McGill U., 1983, PhD, 1985. Tech. officer Electronics Corp. India, Hyderabad, 1973-74; scientist Dept. Electronics, Delhi, 1974-82; asst. prof. dept. elec. engring. U. Wash., Seattle, 1985-90, assoc. prof. elec. engring. and computer sci. and engring., 1990—. Designer Proteus multi computer system for automated classification of objects; patentee in field; contbr. over 60 articles to profl. jours. and chpts. to books. Mem. IEEE (sr.), Assn. for Computing Machinery, Eta Kappa Nu. Hindu. Avocations: squash, tennis, Indian cooking, bridge. Home: 16609 126th Ave NE Woodinville WA 98072-7979 Office: U Wash Dept Elec Engring Ft-10 Seattle WA 98195

SOMANI, SATU MOTILAL, pharmacologist, toxicologist, educator; b. Hingoli, India, Mar. 14, 1937; came to U.S., 1961; s. Motilal B. and Tulsabal Somani; m. Shipra Somani, Nov. 5, 1966; children: Indira, Sheila. MSc in Biochemistry, Poona (India) U., 1959; MS in Pharmacy, Duquesne U., 1964; PhD in Pharmacology, Liverpool (Eng.) U., 1969. Rsch. fellow U. Pitts., 1969-70, instr., 1970-71; asst. prof. pharmacology Univ. Pitts., 1971-74; assoc. prof. So. Ill. U. Sch. Medicine, Springfield, 1974-82, prof. pharmacology and toxicology, 1982—; cons. St. John's Hosp. Poison Ctr., Springfield, 1974—. Editor, contbg. author: Environmental Toxicology: Principles and Policies, 1981, Chemical Warfare Agents, 1992; contbr. articles to profl. jours. Founder, treas. First Hindu Temple in U.S., Pitts., 1970-74; chair Asian Indian Polit. Action Com. Ctrl. Ill., 1989-92. Ellis T. Davies fellow, 1967-69; NIH fellow, 1969-70; grantee U.S. EPA, U.S. Army, Am. Heart Assn. Mem. AAAS, Am. Soc. Pharmacology and Exptl. Therapeutics, Soc. Toxicology, Am. Chem. Soc. (chair Decatur-Springfield chpt. 1981-83), Assn. Scientists of Indian Origin (chair 1987-88). Hindu. Achievements include first to show the metabolic pathway of quaternary ammonium compound (neosigmine) and glucuronide formation; discoveries include physostigmine as an antidote (pretreatment) for soman-nerve agent pharmacokinetics and pharmacodynamics studies; showed the efectiveness of caffeine and theophylline for treatment of apneic episodes in premature infants. Home: 81 Interlacken Springfield IL 62704 Office: So Ill U Sch Medicine 801 N Rutledge Springfield IL 62702

SOMASUNDARAN, PONISSERIL, mineral engineering and applied science educator, consultant, researcher; b. Pazhookara, Kerala, India, June 28, 1939; came to U.S., 1961; s. Kumara Moolayil and Lakshmikutty (Amma) Pillai; m. Usha N., May 25, 1966; 1 child, Tamara. BS, Kerala U., Trivandrum, India, 1958; BE, Indian Inst. Sci., Bangalore, 1961; MS, U. Calif., Berkeley, 1962, PhD, 1964. Rsch. engr. U. Calif., 1964; research engr. Internat. Minerals & Chem. Corp., Skokie, Ill., 1965-67; rsch. chemist R.J. Reynolds Industries, Inc., Winston-Salem, N.C., 1967-70; assoc. prof. Columbia U., N.Y.C., 1970-78, prof. mineral engring., 1978-83, La Von Duddleson Krumb prof., 1983—; chmn. Henry Krumb Sch. Chem. Engring., Materials Sci. and Mineral Resources Engring., Columbia U., 1988—, dir. Langmuir Ctr. for Colloids and Interfaces, 1987—; cons. numerous agys., cos., including NIH, 1974, B.F. Goodrich, 1974, NSF, 1974, Alcan, 1981, UNESCO, 1982, Sohio, 1984-85, IBM, 1984, Am. Cyanamd, 1988-89, Duracell, 1988-89, DuPont, 1989, Canmet, 1990—, Unilever, 1991—, Engelhard, 1991—, UoP, 1991—, Alcoa, 1991—; mem. panel NRC; chmn. numerous internat. symposia and NSF workshops; mem. adv. panel Bur. Mines Generic Ctr., 1983—; keynote and plenary lectr. internat. meetings.; hon. prof. Cen. South U. Tech., China; Brahm Prakash chair in metallurgy and material sci. Indian Inst. Sci., Bangalore, 1990; hon. rsch. advisor Beijing Gen. Rsch. Inst., 1991—. Editor books, including Fine Particles Processing, 1980 (Publ. Bd. award 1980); editor-in-chief Colloids and Surfaces, 1980—; Henry Krumb lectr. AIME, 1988; contbr. numerous articles to profl. publs., patentee in field. Pres. Keralasamajam of Greater N.Y., N.Y.C., 1974-75; bd. dirs. Fedn. Indian Assocs., N.Y.C., 1974—, Vols. in Service to Edn. in India, Hartford, Conn., 1974—. Recipient Disting. Achievement in Engring. award, AINA, 1980, Antoine M. Gaudin award Soc. Mining Engrs.-AIME, 1983, Achievements in Applied Sci. award 2d World Malayalam Conf., 1985, Robert H. Richards award, AIME, 1986, Arthur F. Taggart award Soc. Mining Engrs.-AIME, 1987, honor award Assn. Indian in Am., 1988, VHP award of Excellence, Ellis Island medal of Honor, 1990, Commendations citation State of N.J. Senate, 1991; named Mill Man of Distinction, Soc. Mining Engrs.-AIME, 1983, Disting. Alumnus award Indian Inst. Sci., Bangalore, 1989, Outstanding Contbns. and Achievement award Cultural Festival India, 1991, Recognition award SIAA, 1992. Fellow Instn. Mining and Metallurgy (U.K.); mem. AICE, NAE, Soc. Mining Engrs. (bd. dirs. 1982-85, Disting. Mem. award, also others), Engring. Found. (chmn. conf. com. 1985-88, bd. exec. com. 1985-88, bd. dirs. 1991—, Frank Aplan award 1992), Am. Chem. Soc., N.Y. Acad. Scis., Internat. Assn. Colloid and Surface Scientists (councillor 1989—), Indian Material Rsch. Soc. (hon.), Sigma Xi. Office: Columbia U 911 SW Mudd Bldg New York NY 10027

SOMERVILLE, MARGARET ANNE GANLEY, law educator; b. Adelaide, Australia, Apr. 13, 1942; d. George Patrick and Gertrude Honora (Rowe) Ganley; divorced. A.u.A. (pharm.), U. Adelaide, 1963; LLB (hon. I), U. Sydney, 1973; D.C.L., McGill U., 1978. Registered pharmacist; Bar: Supreme Ct. New South Whales 1975, Quebec 1982. Pharmacist So. Australia, New South Whales, 1963-69; atty. Mallesons, Sydney, Australia, 1974-75; cons. Law Reform Com. Can., 1976-85; asst. prof., faculty law Inst. of COmparative Law, 1978, assoc. prof., 1979, prof., 1984—; prof., faculty law, faculty medicine McGill U., 1984—; dir. McGill Ctr. Medicine Ethics & Law, 1986—; Gale prof. law McGill U., 1989—; vis. prof. Sydney U., 1984, 86, 90, Ctr. for Human Bioethics, Monash U., 1985-86; cons. to numerous orgns. Contbr. numerous articles, review, lectrs. and papers; reviewer Jour. Clin. Epidemiology, 1988—, Can. Jour. of Family Law, 1988—, Jour. of Pharmacy Practice, 1988—, Community Health Studies, 1987—. Clin. ethics com. Royal Victoria Hosp., 1980—; prin. investigator Nat. Health Res. & Devel. Program (aspects of AIDS in Can.), 1986-89; assoc. mem. McGill AIDS Ctr., 1990—, Nat. Adv. Com. of AIDS in Can., 1986—; chmn. Nat. Rsch. Coun. Can., Ethics Com., 1991. Australian Commonwealth scholar, McGill U., 1975; recipient U. Sydney medal, 1976, Joseph Dainow prize McGill U., 1976, Disting. Svc. award Am. Soc. Law & Medicine, 1985. Mem. Am. Soc. Pharm. Law, Inst. Soc., Ethics & Life Sic., Hastings Ctr., Am. Soc. Law & Medicine, Assn. des Prof. de Droit du Que, Can. Bar Assn., Can. Pharm. Assn., World Assn. Med. Law, Can. Law Tchrs. Assn., Soc. Health & Human Values, Internat. Acad. Com. Law. Office: McGill Ctr Medicine Ethics & Law, 3690 Peel St, Montreal, PQ Canada H3A 1W9

SOMERVILLE, RICHARD CHAPIN JAMES, science educator; b. Washington, May 30, 1941. BS in Meteorology, Pa. State U., 1961; PhD in Meteorology, NYU, 1966. Postdoctoral fellow Nat. Ctr. Atmospheric Rsch., Boulder, Colo., 1966-67; rsch. assoc. geophysical fluid dynamics lab. NOAA, Princeton, N.J., 1967-69; rsch. scientist Courant Inst. Math. Scis., N.Y.C., 1969-71; meteorologist Goddard inst. space studies NASA, N.Y.C., 1971-74; adj. prof. Columbia U., NYU, 1971-74; head numerical weather prediction sect. Nat. Ctr. Atmospheric Rsch., Boulder, 1974-79; prof. meteorology, dir. climate rsch. divsn. Scripps instn. oceanography U. Calif. San Diego, La Jolla, 1979—; mem. adv. bd. Aspen Global Change Inst.; mem. adv. com. atmospheric scis. NSF; mem. panel climate and global change Nat. Oceanic & Atmospheric Adminstrn.; chmn. bd. trustees Univ. Corp. Atmospheric Rsch., mem.'s rep. Scripps instn. oceanography U. Calif. San Diego, mem. office interdisciplinary earth studies steering com. Mem. editorial bd. Climate Dynamics. Fellow Am. Meteorol. Soc.; mem. AAAS, Am. Geophysical Union, Oceanography Soc. Office: U Calif San Diego Scripps Instn Oceanography A-0224 La Jolla CA 92093-0224

SOMES, GRANT WILLIAM, statistician, biomedical researcher; b. Bloomington, Ind., Jan. 30, 1947; s. William Henry and Margaret Juanita (Sparks) S.; m. Brenda Sue Weddle, Sept. 2, 1967; children: Anthony William, Joshua Michael, Meghan Elizabeth. AB, Ind. U., 1968; PhD, U. Ky., 1975. Asst. prof. dept. community medicine U. Ky., Lexington, 1975-79; assoc. prof., dir. Biostats./Epidemiology Rsch. Lab. East Carolina U., Greenville, N.C., 1979-84; prof., chmn. dept. biostats. & epidemiology U. Tenn., Memphis, 1984—; cons. Community Health Mgmt. Info. System, Memphis, 1992—, Mid-South Found. for Med. Care, Memphis, 1992—. Contbr. 72 articles to profl. jours.; author 70 presented papers/abstracts. Coach Little League baseball, Lexington, 1976-79, Aydon, N.C., 1979-84. Recipient Outstanding Alumni award U. Ky., 1993. Mem. Am. Statis. Assn. (v.p. West Tenn. chpt. 1985-86), Biometric Soc., Sigma Xi, Pi Mu Epsilon. Achievements include research in cardiovascular risk factors, epilepsy, psychosocial factors and illness, smoking and behavior, dentistry and nutrition, statistical theory mainly in categorical data analysis and nonparametric statistics.

SOMMER, ALFRED, public health professional, epidemiologist; b. N.Y.C., Oct. 2, 1942; s. Joseph and Natalie Sommer; m. Jill Abramson, Sept. 1, 1963; children: Charles Andrew, Marni Jane. BS summa cum laude, Union Coll., 1963; MD, Harvard U., 1967; MHS in Epidemiology, Johns Hopkins U., 1973. Diplomate Am. Bd. Ophthalmology, Nat. Bd. Med. Examiners. Teaching fellow in medicine Harvard U. Med. Sch., Boston, 1968-69; assoc. in medicine Emory U., Atlanta, 1969-70; dir. Nutritional Blindness Prevention Rsch. Program, Bandung, Indonesia, 1976-79; vis. fellow Inst. Ophthalmology U. London, Eng., 1979-80; founding dir., Dana Ctr. for Preventive Ophthalmology Johns Hopkins Med. Insts., Balt., 1980-90; assoc. prof. Johns Hopkins U., Balt., 1981-85, prof. ophthalmology, epidemiology and internat. health, 1985—, dean Johns Hopkins Sch. Hygiene and Pub. Health, 1990—; vis. prof. ophthalmology U. Padjadjaran, Indonesia, 1976-79; cons., advisor Helen Keller Internat., N.Y.C., 1973—; cons., chmn. com. NIH, Bethesda, Md., 1981—; bd. dirs. Internat. Agy. for the Prevention of Blindness, Geneva, Switzerland, 1978—; cons., com. mem. Nat. Acad. Scis., Washington, 1989; chmn. program adv. group on blindness prevention WHO, Geneva, 1989-90, com. mem., 1978-90, expert com., 1990—; steering com. Internat. Vitamin A Cons. Group, Washington, 1975—; pres. Internat. Fedn. of Tissue Banks. Author: Epidemiology and Statistics for the Ophthalmologist, 1980, Nutritional Blindness: Xerophthalmia and Keratomalacia, 1982; contbr. articles to Lancet, Brit. Med. Jour., other profl. jours. Charles A. Dana Found. award for Pioneering Achievements in Health, 1988, Disting. Svc. award for Contbrn. to Vision Care APHA, 1988, E.V. McCollum Internat. Lectureship in Nutrition Am. Inst. Nutrition, 1988, Second Annual Am. Coll. Advancememnt in Medicine Achievement award in Preventive Medicine, 1990, Disting. Contbn. to World Ophthalmology award Internat. Fedn. Opthalmol. Socs., 1990, Smaedel award Infection Disease Soc. Am., 1990. Mem. Inst. Medicine of NAS, Am. Acad. Ophthalmology (chmn. pub. health com. 1982-88, chmn. Quality of Care/Clin. Guidelines 1986-90, Hon. award 1986), Nat. Soc. to Prevent Blindness (bd. dirs. 1989), Internat. Assn. to Prevent Blindness (bd. dirs. 1978). Achievements include first to detail and publish epidemiologic approach to disaster assessment; demonstration that nutritional indices predict subsequent mortality in free-living children, surveillance and containment is effective intervention strategy for controlling and eradicating small pox, vitamin A deficiency increases childhood mortality, vitamin A supplementation decreases childhood mortality, six-month, large dose vitamin A supplementation protects from nutritional blindness, nerve fiber layers are valuable diagnostic and prognostic sign of early glaucoma; routine preventive services cost-effective in eye disease; clinical guideline development and importance of outcome assessment; research in epidemiologic and public health approaches to ophthalmology and blindness prevention. Office: Johns Hopkins Sch Hygiene and Pub Health 615 N Wolfe St Rm 1041 Baltimore MD 21205-2103

SOMMER, ALFRED HERMANN, retired physical chemist; b. Frankfurt, Germany, Nov. 19, 1909; came to U.S., 1953; s. Julius J. and Paula Henriette (Wormser) S.; m. Rosemary Hulm, July 16, 1938; children: Jane, Julia, Helen. DrPhil in Phys. Chemistry, Berlin U., 1934. Rschr. photoemission Patin & Co., Berlin, 1934-35, Baird TV Co. (name changed to Cinema TV Co.), London, 1936-45, EMI Rsch. Lab., Hayes, England, 1946-53, RCA Labs., Princeton, N.J., 1953-74; cons. Thermoelectron Co., Waltham, Mass., 1974-90; ret., 1990. Author: Photoemissive Materials, 1968, 2d edit., 1980; contbr. chpts. to books and articles to profl. jours. Recipient Gaede-Langmuir medal Am. Vacuum Soc., Balt., 1982, Gold medal Internat. Soc. Optical Engring., 1993. Fellow IEEE; mem. Am. Phys. Soc. Achievements include over 30 British and US patents; invented four new photocathodes; improved and explained the mechanism of the cathodes. Home: 37 Dogwood Ln Northampton MA 01060

SOMMER, HENRY JOSEPH, III, mechanical engineering educator; b. Springfield, Ill., Oct. 10, 1952; s. Henry Joseph and Mary Bell (Stephens) S.; m. Jane ALice McGrath, Aug. 12, 1978; children: Henry Joseph IV, Edward Jordan, Rachel Mary. BS with highest honors, U. Ill., 1974, PhD, 1980. Asst. prof. Pa. State U., University Park, 1980-84, assoc. prof., 1984-91, prof., 1991—; vis. scientist NIH Biomechanics Lab., 1992-93. Contbr. articles to profl. jours. Recipient Outstanding Advisor Pa. State Engring. Soc., 1985, Outstanding Tchr., 1991. Mem. ASME, Am. Soc. Biomechanics, Sigma Xi. Roman Catholic. Home: 649 Belmont Cir State College PA 16803 Office: Pa State U 336C Reber Bldg University Park PA 16802

SOMMERFIELD, THOMAS A., process engineer; b. Pitts., Sept. 22, 1958; s. Watson J. and Loretta D. (Mroz) S.; m. Laura L. Parks. BS in Chem. Engring., Ind. Inst. Tech., Ft. Wayne, 1980. Process engr. ER Carpenter Co., Russellville, Ky., 1980-84; lead process engr. GE Plastics, Mt. Vernon, Ind., 1984-91; sr. process engr. GE Mt. Vernon (Ind.) Phenol Plant, 1991—. Mem. Am. Inst. Chem. Engrs., Am. Chem. Soc., Sigma Pi. (treas. 1978-79). Roman Catholic. Achievements include key member of GE Plastics Mt. Vernon Sara emissions 75% reduction team; research in methylene chloride biological degradation. Home: RR # 3 Box 257C Mount Vernon IN 47620 Office: GE Plastics 1 Lexan Ln Mount Vernon IN 47620

SOMMERMAN, KATHRYN MARTHA, retired entomologist; b. New Haven, Jan. 11, 1915; d. George VanName and Anna Hilda (Sperling) S. BS, Conn. State Coll., 1937; MS, U. Ill., 1941, PhD, 1945. Instr. botany Wells Coll., Aurora, N.Y., 1945; entomologist Army Med. Ctr. 1948 Alaska Insect Project, Washington, 1946-51, U.S. Dept. Agr., Washington, 1951-53; rsch. entomologist Arctic Health Rsch. Ctr., USPHS, Anchorage, 1955-67; rsch. entomologist Arctic Health Rsch. Ctr., USPHS, Fairbanks, 1967-73, ind. rsch. entomologist, 1973-92; cons. in field. Author: Airborne Pollen: Father of Lichens and Fungi, 1992; contbr. articles to profl. jours. Recipient Exceptional Civilian Svc. award Dept. of Army, 1950, Sustained Superior Performance award U.S. Dept. HEW, 1960. Fellow Entomol. Soc. Am.; mem. N.Y. Entomol. Soc. Achievements include discovery that airborne pollen is the father of lichens and fungi. Home: SR 76 Box 384 Greenville ME 04441

SOMMERMANN, JEFFREY HERBERT, mechanical engineer; b. Ridley Park, Pa., Oct. 18, 1958; s. Herbert Ludvig and Joan Marie (Tempone) S.; m. Linda Mary Watkins, Oct. 18, 1986. BSME, Drexel U., 1982. Maintenance engr. GPU Nuclear Corp., Parsipanny, N.J., 1982-83, Forked River, N.J., 1983-86; maintenance and engring. mgr. Formosa Plastics Corp., Delaware City, Del., 1986-91; plant mgr. Formosa Plastics Corp., Delaware City, 1989-91; maintenance and engring. mgr. Allied Signal Corp., Claymont, Del., 1991—; chmn. Chem. Industry Coun. Del., Wilmington, 1989-92. Mem. ASME, NSPE, Am. Inst. Plant Engrs. Republican. Home: 113 Washington Ave Wilmington DE 19805 Office: Allied Signal Corp PO Box 607 6100 Philadelphia Pike Claymont DE 19703

SOMMERS, ADELE ANN, engineering specialist, technical trainer; b. L.A., Jan. 21, 1955; d. Morris Samuel Sommers and Elizabeth Noreen (Wilson) Bixler. BA in Social Psychology, Antioch U., 1975; MA in Bus. Mgmt., U. Redlands, 1988. Computer operator and programmer TRM Acctg. Svcs., Santa Barbara, Calif., 1979-81; info. systems analyst Raytheon ESD, Goleta, Calif., 1981-84; tech. support and mktg. specialist Archtl. Computer Software, Santa Barbara, 1984-85; tech. support and tng. specialist Softool Corp., Goleta, 1985-86; sr. configuration mgmt. specialist Santa Barbara Rsch. Ctr. (GM-Hughes) Goleta, 1986—; instr. U. Calif., Santa Barbara, 1988—, West Coast U., L.A., 1988—; pres. Open Door Enterprises, tng. publs., Goleta, 1991—. Mem. ASTD, Am. Soc. Quality Control, Electronic Industries Assn. Office: SBRC GM-Hughes 75 Coromar Dr Santa Barbara CA 93117-3088

SOMMERS, SHELDON CHARLES, pathologist; b. Indpls., July 7, 1916; s. Charles Birk and Leonore Sommers; m. Edith Briggs, Nov. 9, 1943 (dec.); Bernice Lang, Nov. 11, 1990. BS, Harvard Coll., 1937; MD, Harvard Med. Sch., 1941. Asst. resident, resident in pathology New England Deaconess Hosp., Boston, 1946-48, staff pathologist, 1950-53; resident Free Hosp. Women, Boston Lying-In Hosp., 1948-49, Henry Ford Hosp., Detroit, 1949-50; pathologist Mass. Meml. Hosps., Boston, 1953-61, Scripps Meml. Hosp., La Jolla, Calif., 1961-63; assoc. dir., dir. labs. Delafield Hosp., N.Y.C., 1963-68, Lenox Hill Hosp., N.Y.C., 1968-81; ret., 1989; mem. breast cancer task force Nat. Cancer Inst., Bethesda, Md., 1955; mem. scientific adv. bd. COun. for Tobacco Rsch., N.Y.C., 1966-89, Smokeless Tobacco Rsch. Coun., N.Y.C., 1990—. Editor, co-editor Pathology Annuals, 1966-86, Digestive Disease Pathology, 1989-90, Am. Jour. Diagnostic Gynecology and Obstetrics, 1979-82; contbr. 349 articles to profl. jours. Chmn. Bd. Health, Alpine, N.J., 1972-76, N.Y. State Mental Hygiene Med. Rev. Bd., N.Y.C., 1976-85. Capt. U.S. Army, 1942-46. Decorated Silver star, Bronze star, Croix de Guerre. Mem. New England Pathol. Soc. (pres. 1957-58), N.Y. Pathol. Soc. (pres. 1977-79), N.Y. County Med. Soc., N.Y. Acad. Medicine, Am. Soc. Clin. Pathologists, Arthur Purdy Stout Soc. Surg. Pathologists (pres. 1983-85). Achievements include research in endocrine pathology of multiple primary human cancer, endometrial carcinoma in situ as a pathologic entity, kidney pathology of essential hypertension. Home: Cambridge Way PO Box 1115 Alpine NJ 07620

SOMORJAI, GABOR ARPAD, chemist, educator; b. Budapest, Hungary, May 4, 1935; came to U.S., 1957, naturalized, 1962; s. Charles and Livia (Ormos) S.; m. Judith Kaldor, Sept. 2, 1957; children: Nicole, John. B-SChemE, U. Tech. Scis., Budapest, 1956; PhD, U. Calif., Berkeley, 1960; D (hon.), Tech. U. Budapest, 1989, U. Paris, 1990, Free Univ Brussels, Belgium, 1992. Mem. research staff IBM, Yorktown Heights, N.Y., 1960-64; dir. Surface Sci. and Catalysis Program Lawrence Berkeley Lab., Calif., 1964—; mem. faculty dept. chemistry U. Calif.-Berkeley, 1964—, assoc. prof., 1967-72, prof., 1972—, Miller prof., 1978; Unilever prof. dept. chemistry U. Bristol, Eng., 1972; vis. fellow Emmanuel Coll., Cambridge, Eng., 1989; Baker lectr., Cornell U., Ithaca, N.Y., 1977; mem. editorial bds. Progress in Solid State Chemistry, 1973—, Jour Solid State Chemistry, 1976—, Nouveau Jour de Chemie, 1977-80, Colloid and Interface Sci., 1979—, Catalysis Revs., 1981, Jour. Phys. Chm. 1981—, Langmuir, 1985, Jour. Applied Catalysis, Molecular Physics, 1992—. Author: Principles of Surface Chemistry, 1972, Chemistry in Two Dimensions, 1981, Introduction to Surface Chemistry and Catalysis, 1993; editor in chief Catalysis Letters, 1988—; contbr. articles to profl. jours. Recipient Emmett award Am. Catalysis Soc., 1977, Kokes award Johns Hopkins U., 1976, Albert award Precious Metal Inst., 1986, Sr. Disting. Scientist award Alexander von Humboldt Found., 1989, E.W. Mueller award U. Wis.; Guggenheim fellow, 1969. Fellow AAAS, Am. Phys. Soc.; mem. NAS, Am. Acad. Arts and Scis., Am. Chem. Soc. (chmn. colloid and surface chemistry 1981, Surface and Colloid Chemistry award 1981, Peter Debye award 1989), Am. Phys. Soc., Catalysis Soc. N.Am., Hungarian Acad. Scis. (hon. 1990). Office: U Calif Dept Chemistry D 58 Hildebrand Hall Berkeley CA 94720 Home: 665 San Luis Rd Berkeley CA 94707-1725

SON, KI SUB, health facility administrator, surgeon; b. Sacheon, Kyungnam, Korea, July 24, 1928; s. Byoung Dong and Deoksuaek (Lee) S.; m. Sun Dal Cho, Oct. 17, 1961; children: So Ung, So Hun, So Young, So Joung. MD, Seoul Nat. U., Korea, 1957, PhD, 1969. Fellow Rig's Hosp., Copenhagen, 1970-71; chmn., chief gen. surgery Chungnam Nat. U. Med. Coll., Daejeon, Korea, 1972—; hosp. dir., 1972-76, dean med. coll., 1980-84, dir. cancer rsch. inst., 1990—, Changham Nat. U., Daejeon; poet, essayist, med. critic. Author: Textbook of Modern Surgery, Intestinal Obstruction, 1987, also 6 collections of poems, 1978—; contbr. over 120 articles to profl. jours. Maj. Korean Air Force, 1957-65. Recipient Cultural prize acad. achievement City of Daejeon, 1992. Fellow Internat. Coll. Surgeons; mem. Korean Surg. Soc. (v.p. 1992—, exec. com. 1988—), Coll. Internat. Chirurgie Digestive, Japanese Soc. Gastroenterology, Asian Assn. Hepatobiliary Pancreatic Surgery, Internat. PEN (Korean Ctr. mem. 1982—), Korean Coloprotological Soc. (pres. 1986-87), Korean Assn. Med. Edn. (pres. 1982-83), Korea and India Cultural Assn. (v.p. 1984—), Korean Soc. Poets (cen. com. 1974—). Home: 92-11 Busadong ChungKu, Daejeon 301-030, Republic of Korea Office: Chungnam Nat U, 640 Daesadong ChungKu, Daejeon 301-040, Republic of Korea

SONDAK, STEVEN DAVID, electrical engineer; b. N.Y.C., June 28, 1957; s. Arthur and Sylvia (Mayran) S.; m. Denise Ann Gingras, June 15, 1986; children: Samuel, Chelsea. BS, Farileigh Dickinson, 1980; MBA, Bryant Coll., 1988. Nuclear test trainee Gen. Dynamics, Groton, Conn., 1980-82, nuclear test supr., 1982-84; engr. Stone & Webster Engring., Boston, 1984-86; sr. engr. Gen. Dynamics, 1986-87, engring. supr., 1987-89, chief engring., 1989—. Commn. mem. Planning & Zoning Commn., Brooklyn, Conn., 1991—. Home: 116 Fortin Dr Danielson CT 06239

SONE, MASAZUMI, electrical engineer; b. Tokyo, Aug. 31, 1949; came to U.S., 1988; s. Kanjiro and Ayako (Sekido) S.; m. Rey Akai, Oct. 1, 1977; 1 child, Momoko. BEE, U. Hokkaido, Japan, 1973, MEE, 1975. Rsch. engr. electronics rsch. lab. Nissan Motor Co., Yokosuka, Kanagawa, Japan, 1975-82; strategic planner technol. strategy Nissan Motor Co., Atsugi, Kanagawa, Japan, 1982-87; dir. rsch. Nissan Rsch. & Devel., Inc., Ann Arbor, Mich., 1988-93; dir. tech. United Techs. Automotive, Dearborn, Mich., 1993—; exec. com. internat. participation Internat. Congress on Transp. Electronics, 1992—; bd. dirs. Inst. Magnesium Tech., Quebec, Can.; lectr. in field. Contbr. articles to profl. jours. Mem. N.Y. Acad. Sci. Achievements include patents in field; research in global sci. and engring. work force, automotive diagnosis, electromagnetic compatibility. Office: United Techs Automotive 5200 Auto Club Dr Dearborn MI 48126

SONE, TOSHIO, acoustical engineering educator; b. Furukawa, Japan, May 14, 1935; s. Kikichi and Michio (Haga) S.; m. Noriko Tanaka, Sept. 5, 1964; children: Yasutomo, Atsushi, Susumu. B of Engring., Tohoku U., Sendai, Japan, 1958, M of Engring., 1960, DEng, 1963. Rsch. assoc. engring. faculty Tohoku U., 1963-64, assoc. prof. elec. engring., 1964-79, prof., 1979-81, prof. elec. communications engring., 1981—; sec.-gen. Western Pacific Commn. Acoustics, Sendai, 1988—. Author: Electroacoustic Engineering, 1963, Practice in Electromagnetics, 1973, Foundations of Acoustics, 1990, Life and Sound, 1991. Active Sendai Mcpl. Coun. Environ. Pollution Control, 1978—, Miyagi Prefectural Coun. Environ. Pollution Control, Sendai, Japan, 1982—, Iwate Prefectural Coun. Environ. Pollution Control, Morioka, Japan, 1982—. Rsch. grantee Kajima Sci. Promotion Found., Tokyo, 1984, Sound Tech. Promotion Found., Tokyo, 1987, Hoso-Bunka Found., Tokyo, 1991. Fellow Acoustical Soc. Am.; mem. Inst. Electronics, Info. and Communication Engrs. Japan, Acoustical Soc. Japan (bd. dirs. 1989—, v.p. 1991-93, pres. 1993—, Sato prize 1992), Inst. Noise Control Engring. Japan (bd. dirs. 1980-94, v.p. 1990-93), Inst. Noise Control Engring. U.S.A. (corr.), Japan Soc. Mech. Engrs., Japan Audiological Soc. Avocation: calligraphy. Home: 4-9-5 Midorigaoka, Taihaku-ku, Sendai Japan 982 Office: Tohoku U Rsch Inst Elec Com, 2-1-1 Katahira, Aoba-ku, Sendai Japan 980

SONEIRA, RAYMOND M., computer company executive, scientist; b. N.Y.C., July 10, 1949; s. Raymond Mario and Amelia (Rodriguez) S.; m. Julia Lobsitz, Feb. 11, 1988; 1 child, Lauren. BA, Columbia Coll., 1972; PhD, Princeton U., 1978. Long term mem. Inst. for Advanced Study, Princeton, N.J., 1978-83; prin. investigator, computer systems rsch. lab. AT&T Bell Labs, Holmdel, N.J., 1983-89; v.p. devel. Cactus Computers, Inc., Rumson, N.J., 1989—; cons. engring. and devel. CBS TV Network, N.Y.C., 1967-70. Author: DisplayMate Reference, 1990, DisplayMate Professional Reference, 1992; contbr. articles to profl. jours. Recipient Best Utility award COMDEX Computer Industry, Atlanta, 1991. Mem. IEEE (sr.), Am. Phys. Soc., Sigma Xi. Achievements include patent in color convergence of color television cameras and monitors; discovery of hierarchial clustering of galaxies, laser range finder using the Parallax Principle; designer of stellar model of the Milky Way galaxy. Office: Sonera Techs PO Box 565 Rumson NJ 07760

SONG, LIMIN, acoustical engineer; b. Hebei, China, Apr. 26, 1960; m. Duanli Yan, Feb. 2, 1986; 1 child, Victoria. PhD, Pa. State U. Sr. engr. Mobil R&D Corp., Princeton, N.J., 1990—. Contbr. articles to profl. jours. Mem. Acoustic Soc. Am. Achievements include devel. of a simulation model of acoustic agglomeration of fine particles; devel. of a numerical approach to compute acoustical radiation of a complex structure. Office: Mobil R&D Corp PO Box 1026 Princeton NJ 08543

SONG, OHSEOP, research engineer; b. South Korea, Nov. 27, 1954; came to U.S., 1984; s. Young S.S. and Yeol H. (Lee) S.; m. Bo K. Chun, Apr. 5, 1983; 1 child, Jae-Young. MSME, N.J. Inst. Tech., 1986; PhD, Va. Poly. Inst. & State U., 1990. Rsch. assoc. Va. Poly. Inst. and State U., Blacksburg, 1991—. Contbr. articles to AIAA Jour., Internat. Jour. Engring. Sci., Jour. of Sound and Vibration Composite Engring. Mem. AIAA, ASME, Tau Beta Phi. Achievements include research of smart material system and composites for aerospace structures. Office: Va Poly Inst & State U ESM Dept Blacksburg VA 24061-0219

SONG, XIAOTONG, physicist; b. Taizhou, Jiangsu, People's Republic of China, Oct. 18, 1934; came to U.S., 1989; s. Hoshu Song and Jingying Wang; m. Chuchu Zhu, 1966; 1 child, Jianyang. BS in Physics, Fudan U., Shanghai, China, 1955, PhD in Physics, 1963. Rsch. fellow Dept Def., China, 1955-58; rsch. assoc. Hangzhou U., China, 1958-63; lectr., 1963-66, 77-83, assoc. prof., 1983-86, prof., 1986—; rsch. cons. dept. physics Inst. Nuclear and Particle Physics, U. Va., Charlottesville, 1989-90, rsch. prof., cons., 1990—; mem. adv. com. for professorship exam., Zhejiang, China; dep. dir. theory div. dept. physics Hangzhou U., China, 1984-89; referee Phys. Rev., High Energy Physics and Nuclear Physics, Nat. Natural Sci. Found., China; vis. scientist Tech. U. Munich, 1986, 88-89, CERN, Switzerland, 1986-87, Instituto Nazionale di Fisica Nucleare Turin Sect., Italy, 1986-88, Internat. Ctr. Theoretical Physics, Italy, 1986-88, Los Alamos Nat. Lab., 1987, Utah State U., 1987, Kans. State U., 1989, Brookhaven Nat. Lab., 1987-89, others. Contbr. articles to profl. jours. Recipient Prize of Natural Sci., Com. Sci. and Tech., Zhejiang Province, China, 1983, 84; grantee Nat. Natural Sci. Found., 1984-87. Mem. AAAS, Internat. Ctr. Theoretical Physics (sr. assoc.), Am. Phys. Soc., Chinese High Energy Physics Soc., Chinese Phys. Soc., N.Y. Acad. Sci., Sigma Xi. Achievements include research in theoretical nuclear and particle physics. Office: U Va Physics Dept McCormick Rd Charlottesville VA 22901

SONNABEND, JOSEPH ADOLPH, microbiologist; b. Johannesburg, Republic of South Africa, Jan. 6, 1933; came to U.S., 1969; s. Henry and Fira (Sandler) S. MB BCh, U. Witwatersrand, Johannesburg, 1956. Med. diplomate. Intern St. Bartholomew's Hosp., Rochester, Kent, Eng., 1957, Royal Free Hosp., London, 1957-58; resident Cntl. Middlesex Hosp., London, 1959-61; rsch. fellow dept. bacteriology U. Edinburgh, 1961-63; mem. sci. staff Nat. Inst. Med. Rsch., London, 1963-69; assoc. prof. microbiology Mt. Sinai Med. Sch., N.Y.C., 1969-72; assoc. prof. medicine Downstate Med. Ctr., SUNY, Bklyn., 1973-77; dir. continuing med. edn. Bur. of Venereal Disease Control, N.Y. Dept. Health, 1977-79; clin. assoc. prof. Uniformed Svcs. of Health Scis., Bethesda, Md., 1982-85; assoc. rsch. scientist Columbia U., N.Y.C., 1983—; mem. adv. com. Hepatitis B Vaccine Trial, N.Y. Blood Ctr., 1978-80; bd. dirs. Community Rsch. Initiative, 1987-91; mem. instl. rev. bd. AIDS Med. Found., 1983-85; mem. Community Rsch. Initiative on AIDS, N.Y., 19916; founding mem. bd. dirs. People With AIDS Health Group, N.Y., 1987-92. Founder, editor AIDS Rsch. Jour., 1983-87; contbr. articles to profl. jours. Polio Rsch. Found. fellow Edinburg U., 1962; Med. Rsch. Coun. tng. fellow Albert Einstein Coll. Medicne, 1967; recipient Nellie Westerman Prize Am. Fedn. Clin. Rsch., 1983. Jewish. Achievements includes studies on interferon action showing requirement for synthesis protein; effects on interferon on uninfected cells, effect of interferon on translation of viral RNA; discovery of serum interferon in AIDS, role of EBV and allogenic. Office: Community Rsch Initiative AIDS 275 7th Ave New York NY 10001

SONNTAG, BERNARD H., agrologist, research executive; b. Goodsoil, Sask., Can., June 27, 1940; s. Henry R. and Annie (Heesing) S.; m. Mary L. Ortman, Aug. 10, 1963; children: Calvin, Galen, Courtney Anne. BSA, Sask. U., Saskatoon, 1962, MSc, 1965; PhD, Purdue U., 1971. Economist Agriculture Can., Saskatoon, 1962-66; cons. D.W. Carr & Assoc., Ottawa, Ont., Can., 1966-68; economist Agriculture Can., Lethbridge, Alta., 1968-79, Saskatoon, 1979-80; dir. rsch. sta. Agriculture Can., Brandon, Man., 1980-86, Swiftcurrent, Sask., 1986-89, Lethbridge, 1989—; pres. Man. Inst. Agrologists, Brandon, 1984. Recipient Leadership award Bell Can., 1993. Mem. Rotary. Roman Catholic. Home: Box 3000, Lethbridge, AB Canada T1J 4B1 Office: Agriculture Canada, Research Station Box 3000, Lethbridge, AB Canada T1J 4B1

SONOGASHIRA, KENKICHI, chemistry educator; b. Kagoshima, Japan, Oct. 25, 1931; s. Kiyoto and Soe Sonogashira; m. Seiko Oka, May 31, 1964; children: Akane, Madoka. BS in Chemistry, Osaka U., Japan, 1956, MS in Organic Chemistry, 1958, DSc, 1961. Rsch. assoc. Osaka U., 1961-68, assoc. prof. chemistry, 1968-81; prof. Osaka City U., 1981—, dir. rsch. inst. for atomic energy, 1987-89. Contbr. numerous articles to profl. jours. Alexander von Humboldt fellow, 1966-67. Mem. Chem. Soc. Japan, Soc. Synthetic Organic Chemistry Japan, Catalysis Soc. Japan, Soc. Polymer Sci. Japan, Am. Chem. Soc. Home: Ao-matani, Higashi 6-12-9, Minou, Osaka 562, Japan Office: Osaka City U Faculty Engring, Sumiyoshi-ku Sugimoto 3-3-138, Osaka 558, Japan

SONSTEBY, KRISTI LEE, healthcare consultant; b. Anoka, Minn., Nov. 16, 1958; d. Glenn and Rosella (Rebischke) S. Charge nurse Baylor U. Med. Ctr., Dallas, 1980-81; clin. nurse specialist ARA Living Ctrs., Houston, 1981-86; pres., owner KristiCare Inc., Dallas, 1986-89; healthcare cons. SDG Ent., Inc., Austin, Tex., 1989-90; pres., owner NursePlus Inc., Mpls., 1991—;

judge Provider Mag., Washington, 1988; cons. in field; lectr. in field; conductor workshops in field. Patentee in field; author: Handbooks for Nurses, Vols. I-X, 1991; contbr. articles to profl. jours. Vol. to elderly various civic orgns. Avocations: violinist, chamber music. Office: Nurse Plus Inc 716 Hwy 10 NE Ste 163 Minneapolis MN 55434-2331

SONTAG, GLENNON CHRISTY, electrical engineering consultant, travel industry executive; b. St. Louis, Nov. 17, 1949; s. Robert Matthew and Agnes Marie (Bueckendorf) S.; m. Victoria Ann Dill, May 17, 1986; children: Christy Victoria, Jonathon Tyler. AS in Engring. with honors, Florissant Valley Community Coll., St. Louis, 1975; BSEE, U. Mo., 1977. Elec. engr. Sverdrup Corp., St. Louis, 1974-81; project engr. Anheuser Busch Cos., Inc., St. Louis, 1981-87; co-founder, prin. Indusl. Control Concepts, Inc., St. Louis, 1986—; prin., v.p. West Travel Ltd., St. Louis, 1986—; agent Thomas Realty, St. Louis, 1975-87; co-owner Front Page restaurant, St. Louis, 1979-80. Sgt. USAF, 1969-73. Mem. Instrument Soc. Am., Chesterfield C. of C. Republican. Roman Catholic. Avocations: snow skiing, golf, tennis, jogging. Office: Indsl Control Concepts Inc 707 N 2nd St Ste 520 Saint Louis MO 63102

SONTHEIMER, HARALD WOLFGANG, scientist, cell biology researcher; b. Aalen, Federal Republic of Germany, Dec. 31, 1960; came to U.S., 1989; s. Alfred and Gudrun Elizabeth S.; m. Marion Christiane; 1 child, Melanie Simone. BS, U. Ulm, Federal Republic of Germany, 1982; PhD, U. Heidelberg, Federal Republic of Germany, 1989. Diplomate in cell biology. Assoc. rsch. scientist Heidelberg U., 1988-89; asst. prof. neurology Yale U. Sch. Medicine, New Haven, 1991—, asst. prof. neurobiology, 1992—. Contbr. articles to Sci., Nature, Neuron, Jour. Neurophysiology, Jour. Clin. Neuophysiology, others. Mem. Soc. for Neurosci., Neurosci. Soc. Germany, German Zool. Soc. Office: Yale Sch Medicine Dept Neurology 333 Cedar St LCI 704 New Haven CT 06510

SOOMRO, AKBAR HAIDER, ophthalmologist, educator; b. Larkana, Sindh, Pakistan, Sept. 18, 1947; s. Ghulam Haider Soomro and Arbab Khatoon; m. Fahmida Famy, July 13, 1977; 1 child, Noushervan Haider. MB, BS, Liaquat Med. Coll., Hyderabad, Pakistan, 1969, M of Optometry, 1973; D of Ophthalmology, Sind U., Hyderabad, 1973. Ophthal. house surgeon Liaquat Med. Coll. Eye Hosp., Hyderabad, 1970-71, resident in ophthalmology, 1971-73, ophthal. registrar, 1973-75; asst. prof. ophthalmology Nawabshah Med. Coll., Pakistan, 1975-76; asst. prof. Chandka Med. Coll., Larkana, 1976-82, assoc. prof., 1982-86, prof., 1986—, prin., 1985—; med. supr. Chanka Med. Coll. Hosp., 1984—; hon. dir., prin. med. research officer Pakistan Med. Research Council, Larkana, 1982-85, bd. dirs., 1985—. Contbr. articles to profl. jours. Chmn. Pakistan Eye Bank Soc. Chandka Br., Larkana, 1984, Patients Welfare Orgn., Larkana, 1985. Fellow Pakistan Acad. Ophthalmology, Internat. Coll. Surgeons, Internat. Coll. Angiology; mem. Pakistan Ophthal. Soc. Avocations: badminton, socializing. Home: Haider St, Larkana Sind, Pakistan Office: Chandka Med Coll, Office of the Principal, Larkana Sind, Pakistan

SOOMRO, ELLAHI BUKHSH, Pakistani federal minister; s. Haji Maula Bukhsh Soomro; ed. as engr. Mng. dir. Sind Indsl. Estate, chmn. Karachi Devel. Authority; sr. diplomat Pakistan High Commn. in U.K.; minister for industries Govt. of Pakistan, Islamabad. Office: Ministry of Science & Technology, Block S Pakistan Secretariat, Islamabad Pakistan

SOPER, THOMAS SHERWOOD, biochemist; b. Plattsburgh, N.Y., Mar. 13, 1947; s. Harold Sherwood and Martha Pauline (Vass) S. AB, SUNY, Cortland, 1969; PhD, Purdue U., 1974. Rsch. assoc. Purdue U., West Lafayette, Ind., 1974-75; rsch. assoc. Rockefeller U., N.Y.C., 1975-82, asst. prof., 1982-84; sr. staff scientist Oak Ridge (Tenn.) Nat. Lab., 1984-90, cons., 1990-91; scientist Amvax, Inc., Beltsville, Md., 1991—. Contbr. articles to profl. jours. including Jour. Biochemistry, 1990, 92, Pro Engring., 1990, Gene, 1993; pub. numerous scientific papers. Recipient Pub. award Martin Marietta Energy System, 1980; Predoctoral fellow NIH, 1969. Mem. Am. Soc. Biochemistry and Molecular Biology, Am. Chem. Soc. (biol. chemistry chpt.). Achievements include research in cloning expression and in vitro refolding of bacterial porins, structural-functional relationships at the active site of RuP2 carboxlose (a key plant enzyme), mechanics of action of novel suicide substrates of D-amino acid transaminase. Office: Amvax Inc 12040 Indian Creek Ct Beltsville MD 20705-1260

SOPKO, STEPHEN JOSEPH, structural engineer; b. Troy, N.Y., Dec. 22, 1950; s. Joseph Stanly and Helen Olga (Abrams) S. BCE, Manhattan Coll., Bronx, N.Y., 1973; MCE, RPI, Troy, N.Y., 1975. Profl. Engr., N.Y., Vt., N.H. Assoc. Ryan-Biggs Assocs., P.C., Troy, N.Y., 1974—; structural group, exec, com. ASCE, Mohawk-Hudson, 1977-81. Contbr. articles to profl. jours. Mem. Am. Soc. Civil Engrs., Am. Concrete Inst. Office: Ryan-Biggs Assocs PC 291 River St Troy NY 12180

SORACCO, REGINALD JOHN, microbiological biochemist; b. Liberty, N.Y., Mar. 27, 1946; s. Louis and Elma Elizabeth (Rutledge) S. AAS, Orange County Community Coll., Middletown, N.Y., 1965; BS, SUNY, Albany, 1967, MS, 1976; PhD, Rensselaer Poly Inst., 1981. Grad. rsch. asst. Rensselaer Polytechnic Inst., Troy, N.Y., 1980; rsch. assoc. Rensselaer Fresh Water Inst., Troy, 1981-84; program mgr. Bioindustrial Techs., Inc., Troy, 1986-87; sci. and adminstr. cons. Rensselaer Fresh Water Inst., Troy, 1987-92; pres. BCM Consulting, Inc., Troy, 1982—; adj. prof. computer sci. Rensselaer Polytech Inst., 1985; bd. dirs. Rensselaer-Taconic Land Conservancy, 1989—, Rensselaer Inst. Limnolgy, 1992—; mem. Hudson River Environ. Soc., Poughkeepsie, N.Y., 1990—. Author: (book chpt.) Autecological Studies in Microbial Limnology, 1986. Vol. State Assembly Race, Liberty, 1992. With USN, 1969-74. Recipient image processing equipment Hasselblad Found., 1989, AA spectrophotometer NSF, 1990. Mem. Nat. Assn. Underwater Instrs., Am. Soc. Microbiology, Am. Water Works Assn., Mason (jr. warden 1978-79, master mason 1975), Sigma Xi. Democrat. Roman Catholic. Achievements include determining effectiveness and mode of action of biocides on Legionnaire's disease bacteria; three-yr. study of Rensselaer County water quality data base. Home: 2431 21st St Apt #1 Troy NY 12180 Office: BCM Consulting Inc 2431 21st St Rom #1 Troy NY 12180

SORCI-THOMAS, MARY GAY, biomedical researcher, educator; b. New Orleans, Mar. 9, 1956; d. Leon Philip and Ethel Mae (Andrews) S.; m. Michael J. Thomas, June 9, 1979. BS, La. State U., 1979; PhD, Wake Forest U., 1984. Rsch. technician La. State U., Baton Rouge, 1976-79; instr. Bowman Gray Sch. Medicine/Wake Forest U., Winston-Salem, N.C., 1987-88, asst. prof., 1988—, assoc. in biochemistry, 1989—; ad hoc reviewer NIH, 1992. Contbr. articles to Jour. Lipid Rsch., Jour. Biol. Chemistry. Spl. fellow R.J. Reynolds Co., 1983-84; NIH NRSA postdoctoral fellow, 1984-86. Mem. Am. Heart Assn. (arteriosclerosis coun., coord. N.C. affiliate 1992, rsch. rev. subcom. 1989-92, bd. dirs. 1991-93, rsch. com. 1992-95, Louis N. Katz finalist in Basic Sci. Rsch., 1991, nutrition com. 1992-95 grantee 1983-84), Am.Soc. Biochemists and Molecular Biols., Sigma Xi (chpt. treas. 1991-92, chpt. 1992-93, chpt. pres. 1993-94), NIH awardee, 1989-93. Office: Bowman Gray Sch Medicine Dept Comparative Medicine Medical Center Blvd Winston Salem NC 27157

SOREGAROLI, A(RTHUR) E(ARL), mining company executive, geologist; b. Madrid, Iowa, Jan. 4, 1933; arrived in Can., 1962; s. Arthur Samuel and Margaret Alice (Teasdale) S.; m. Rosalie Ann Lawrick, Dec. 22, 1962; children: Carla Jean, Brian Arthur. B.Sc. in Geology, Iowa State U., 1959; M.Sc. in Geology, U. Idaho, 1961; Ph.D. in Geology, U. B.C., Vancouver, 1968. Geologist Idaho Bur. Mines and Geology, Moscow, 1961-62; geologist Noranda Exploration Co. Ltd., Vancouver, 1963-68, chief geologist western dist., 1968-72; asst. prof. geology U. B.C., Vancouver, 1972-74; research scientist Geol. Survey Can., Ottawa, Ont., 1974-76; v.p. exploration Westmin Resources Ltd., Vancouver, 1976—. Contbr. papers to sci. lit. Served with U.S. Army, 1952-54, Korea. Fellow Geol. Assn. Can. (chmn. Robinson Fund com. 1983—); mem. Soc. Econ. Geologists (pres. 1985), Assn. Exploration Geochemists (councillor 1983—, v.p. 1987-88), Can. Inst. Mining and Metallurgy (chmn. student essays com. 1978—, chmn. geology div. 1978, Dist. Proficiency Gold medal 1986, Julian Boldy Meml. award 1989), Geol. Assn. Can. (Duncan R. Derry Gold medal, 1987), Mineral. Assn. Can. Club: Engrs. (Vancouver). Avocations: sports, mineral collecting. Office: Westmin Resources Ltd, Box 49066 Bentall Ctr, Vancouver, BC Canada V7X 1C4

SOREN, DAVID, archaeology educator, administrator; b. Phila., Oct. 7, 1946; s. Harry Friedman and Erma Elizabeth (Salamon) Soren; m. Noelle Louise Schattyn, Dec. 22, 1967. B.A., Dartmouth Coll., 1968; M.A., Harvard U., 1972, Ph.D., 1973. Cert. Rome Classics Ctr. Curator of coins Fogg Art Mus., Cambridge, Mass., 1972; asst. prof. U. Mo., Columbia, 1972-76, assoc. prof., dept. head, 1976-81; prof. archaeology U. Ariz., Tucson, 1982-83, dept. head, 1984-89; guest curator Am. Mus. Natural History, N.Y.C., 1983-90, lectr., 1993—; creator dir. Kourion excavations, Cyprus, 1982-89, Portugal, 1983-84, pot cons., field dir. Tunisia Excavations Chgo. Oriental Inst./Smithsonian Instn., 1973-78; creator/dir. Am. Excavations at Lugnano, Italy, 1988—; dir. U. Ariz. humanities program, 1992. Author: (books) Unreal Reality, 1978, Rise and Fall of Fantasy Film, 1980, Carthage, 1990; co-author: Kourion: Search for a Lost Roman City, 1988, Corpus des Mosaiques de Tunisie, 1972, 3rd rev. edit., 1986, Carthage: A Mosaic of Ancient Tunisia, 1987; editor: Excavations at Kourion I, 1987; producer: (film) Carthage: A Mirage of Antiquity, 1987; creator and guest curator: (internat. traveling exhbn.) Carthage: A Mosaic of Ancient Tunisia, 1987-92; editor, founder Roscius, 1993—; contbr. articles to profl. jours. Subject of National Geographic spl. Archeological Detectives, 1985; work subject of feature articles in Newsweek, Conoisseur, National Geographic and others; recipient Cine Golden Eagle, 1980, Angenieux Film award Industrial Photography mag., 1980, Outstanding American Under 40 award C. Johns Hopkins-Britain's Royal Inst. Internat. Affairs, 1985; named Outstanding American Under 40 Esquire mag., 1985, hon. Italian citizen Lugnano, Italy, 1989; grantee NEH, 1979, 87, Fulbright, Lisbon, 1983. Mem. Nat. Geog. Soc. (project dir. 1983-84), Am. Sch. Oriental Rsch. (dept. rep. 1981-85), Archaeol. Inst. Tucson (pres. 1983-86), Luso-Am. Commn. (citation 1983-84), Explorer's Club. Office: U Ariz Dept Classics 371 MLB Tucson AZ 85721

SORENSEN, HENRIK VITTRUP, electrical engineering educator; b. Skanderborg, Denmark, Jan. 17, 1959; came to U.S. 1983; s. Evan Anton and Anne Marie (Vittrup) S.; m. Karen Ann Taylor, Mar. 5, 1988; 1 child, Amanda Elisabeth. MS, Aalborg U. Ctr., 1983; PhD, Rice U., 1988. Asst. prof. Dept. Electrical Engring. U. Pa., Phlia., 1988—; cons. AT&T Bell Labs., Murray Hill, N.J., 1990-. Author: Handbook for Digital Signal Processing, 1992; contbr. articles to profl. jours. Fellow Rotary; mem. IEEE (editor 1990-, vice chmn. Phila. sect. 1991-), Sigma Xi, Eta Kappa Nu. Lutheran. Achievements include development of fast algorithms for the split radix fast Fourier transform and for the fast Hartley transform. Home: 379 Yarnall Dr Springfield PA 19064 Office: U Pa 200 S 33d St Philadelphia PA 19104

SORENSEN, JOHN NOBLE, mechanical and nuclear engineer; b. Mpls., Jan. 2, 1934; s. Alfred Noble and Helen Viola (Baker) S.; m. Joan Elizabeth Reiche, Sept. 15, 1954; children: Laura Elizabeth, Nancy Helen, Karen Lynn. BSME, U. N.D., 1955; MSME, U. Pitts., 1958. Cert. engr. Sr. engr. Westinghouse Electric, Pitts., 1955-67; v.p., gen. mgr. NUS Corp., Rockville, Md., 1967-86; v.p., dir. Grove Engring., Inc., Rockville, 1986-. Mem. Am. Nuclear Soc., Am. Soc. Mechanical Engrs., Nat. Soc. Profl. Engrs. Home: 629 Crocus Dr Rockville MD 20850 Office: Grove Engring Inc 15215 Shady Grove Rd Rockville MD 20850

SORENSEN, RAYMOND ANDREW, physics educator; b. Pitts., Feb. 27, 1931; s. Andrew J. and Dora (Thuesen) S.; m. Audrey Nickols, Apr. 2, 1953; 1 dau., Lisa Kirsten. B.S., Carnegie Inst. Tech., 1953, M.S., 1955, Ph.D., 1958. Mem. faculty Columbia, 1959-61; asst. prof. Carnegie-Mellon U., Pitts., 1961-65, assoc. prof., 1965-68; prof. physics Carnegie-Mellon U., 1968—, chmn. dept., 1980-89. NSF sr. postdoctoral fellow, 1965-66. Mem. Am. Phys. Soc. Home: 1235 Murdoch Rd Pittsburgh PA 15217-1234

SORIA, MARCO RAFFAELLO, molecular and cellular biologist; b. Tunis, Tunisia, Dec. 29, 1945; came to Italy, 1946; s. Guido and Suzette (Perez) S.; m. Orietta Elena Sternfeld, July 27, 1975; children: Alex, Daniel. MD, U. Naples, Italy, 1969; PhD, Harvard U., 1975. Rsch. scientist NRC, Pavia, Italy, 1978-82; lab. chief R&D dept. Farmitalia Carlo Erba, Milan, 1982-86, dept. head dept. biotech., 1986-90; sr. mem. San Raffaele Rsch. Inst., Milan, 1990—; mem. exec. sci. com. Italfarmaco, Milan, 1991—; mem. working party on applied molecular genetics European Fedn. Biotech., 1988—. Mem. Harvard Bus. Sch. Club Italy. Jewish. Achievements include first cloning of type 1 ribosome-inactivating protein reported. Home: 13 Via Guerrini, 20133 Milan Italy Office: San Raffaele Rsch Inst, 60 Via Olgettina, 20132 Milan Italy

SORKIN, ROBERT DANIEL, psychologist, educator; b. N.Y.C., May 24, 1937; s. Harry and Cynthia (Erdreich) S.; m. Nancy Jayne Sloan, July 3, 1960; children: David, Susan. BEE, Carnegie Inst. Tech., 1958; PhD, U. Mich., 1965. Assoc. rsch. engr. Cooley Labs. U. Mich., Ann Arbor, 1960-65; asst. prof. psychology Purdue U., West Lafayette, Ind., 1965-68, assoc. prof., 1968-73, prof. dept. psychol. scis., 1973-88; prof., chair dept. psychology U. Fla., Gainesville, 1988—; asst. dean sch. humanities, social scis. and edn. Purdue U., 1973-75; dir. psychobiology program NSF, Washington, 1975-76; mem. exec. bd. Coun. Grad. Depts. Psychology, Blacksburg, Va., 1992—; mem. com. on hearing and bioacoustics Nat. Rsch. Coun., Washington, 1987-90. Co-author: Human Factors: Understanding People-System Relationships, 1983; contbr. articles to Jour. Acoustical Soc. Am., Perception and Psychophysics, Jour. Exptl. Psychology, Human Factors, others. With U.S. Army, 1960. Fellow Acoustical Soc. Am., Am. Psychol. Assn., Am. Psychol. Soc.; mem. Human Factors Soc. Office: U Fla Dept Psychology Gainesville FL 32611

SORNETTE, DIDIER PAUL CHARLES ROBERT, physicist; b. Paris, June 25, 1957; s. Christian and Nicole (Henry) S.; m. Anne Sauron, June 28, 1986; 1 child, Paul-Emmanuel. PhD, U. Nice, France, 1985; M Physics, U. Pierre et Marie Curie, Paris, 1978; degree in phys. scis., Ecole Normale Supérieure, 1981; PhD, U. Nice, France, 1981-85. Student prof. Ecole Normale Supérieure, 1977-81; rsch. fellow CNRS, Nice, France, 1981-82, 83-90; dir. rsch. CNRS, Nice, 1990—; rsch. scientist rsch. lab. Thomson-Sintra Co., 1982-83; cons. Thomson-Sintra Co., Nice-Sophia Antipolis, France, 1984—; dir. rsch. X-RS Co., Orsay, France, 1986—; speaker at confs. in field, worldwide; visitor dept. applied math. Australian Nat. U., Canberra, 1984-85; part-time staff mem. Ctr. for Theoretical Physics, Ecole Poly. Paris, 1986-90. Contbr. numerous articles to profl. jours. Grantee French Nat. Ctr. for Sci. Rsch., Direction de la Recherche des Etudes Techniques of French Army; Recipient Science et Defence National award 1985. Mem French Phys. Soc., European Phys. Soc., Am. Acoustical Soc., Geophys. Union. Achievements include research in statistical and condensed matter physics, waves in random media, earthquakes, quantum chaos and high frequency vibrations. Home: Levalrose B, 27 Ave Caravadossi, 06000 Nice France

SOROTA, STEVE, biomedical researcher; b. Middletown, Conn., Oct. 22, 1954; s. Stanley Peter and Zona Mae (Baumgartner) S. BS in Pharmacy, U. Conn., 1977; PhD in Pharmacology, U. Conn., Farmington, 1986. Postdoctoral trainee dept. pharmacology Columbia U., N.Y.C., 1986-88, assoc. rsch. scientist dept. pharmacology, 1988-90, asst. prof. dept. pharmacology, 1990—. Contbr. articles to profl. jours. Recipient First award Nat. Heart, Lung and Blood Inst., 1992; grantee Am. Heart Assn., N.Y.C. affiliate, 1992, investigatorship, 1992. Mem. Am. Heart Assn. Basic Sci., Biophysical Soc., Cardiac Electrophysiology Soc. Achievements include discovery of pertussis toxin sensitivity of atrial potassium channel regulation concurrently with other research groups and description of a swelling-induced chloride current in mammalian atrial cells. Office: Columbia U Dept Pharmacol 630 W 168th St New York NY 10032

SOROUSH, MASOUD, chemical engineer; b. Mahallat, Iran, Aug. 26, 1960; came to the U.S., 1986; s. Morteza and Zahra (Hendi) S.; m. Azam Soroush, May 27, 1988; 1 child, Ali. MS in Engring., U. Mich., 1988, PhD, 1992. Instr. Abadon Inst. Tech., Ahwaz, Iran, 1985-86; rsch. asst. U. Mich., Ann Arbor, 1988-92, instr., rsch. fellow, 1992-93; asst. prof. Drexel U., Phila., 1993—. Contbr. articles to AIChE, Indsl. Engring., Chemistry Rsch. Mem. AIChE, Sigma Xi.

SORREL, WILLIAM EDWIN, psychiatrist, educator, psychoanalyst; b. N.Y.C., May 27, 1913; s. Simon and Lee (Lesenger) S.; m. Rita Marcus, July 1, 1950; children: Ellyn Gail, Joy Shelley, Beth Mara. BS, NYU, 1932; MA, Columbia U., 1934, MD, 1939; PhD, NYU, 1963. Diplomate Am. Bd. Med. Psychotherapists (profl. adv. coun. 1992—); qualified psychiatrist, also cert.examiner N.Y. State Dept. Mental Hygiene. Intern Madison (Tenn.) Sanitarium and Hosp., 1939; resident physician Alexian Bros. Hosp., St. Louis, 1940; officer instrn. St. Louis U. Sch. Medicine, 1940-41; asst. psychiatrist Central State Hosp., Nashville, 1941; assoc. psychiatrist Eastern State Hosp., Knoxville, 1942-44; assoc. attending neuropsychiatrist, chief clin. psychiatry Jewish Meml. Hosp., N.Y.C., 1946-59; assoc. attending neuropsychiatrist, chief clin. child psychiatry Lebanon Hosp., Bronx, N.Y., 1947-65; psychiatrist-in-chief Psychiatry Clinic, Yeshiva U., 1950-66, asst. prof. psychiatry, 1952-54, assoc. prof., 1954-58, prof., 1959-62, psychiatrist-in-chief, assoc. dir. Psychol. Center,, 1957-67; prof. human behavior Touro Coll., 1974—; attending psychiatrist St. Clare's Hosp., N.Y.C., 1983—; asst. prof. clin. psychiatry Albert Einstein Coll. Medicine, 1986—; psychiat. cons. SSS, 1951, N.Y. State Workmens Compensation Bd., 1951—, Bronx-Lebanon Med. Ctr., 1985—; vis. psychiatrist Fordham Hosp., N.Y.C., 1951; attending neuropsychiatrist, chief mental hygiene svc. Beth-David Hosp., 1950-60; attending neuropsychiatrist Grand Central Hosp., 1958-66, Morrisania Hosp., 1959-72; psychiatrist-in-chief Beth Abraham Hosp., 1954-60; psychiat. cons. L.I. U. Guidance Ctr., 1955-60, Daytop Village, 1970-71; assoc. psychiatrist Seton City Hosp., 1955; guest lectr. U. London, 1947; vis. prof. Jerusalem, Israel Acad. Med., 1960, Hebrew U., 1960; mem. psychiat. staff Gracie Sq. Hosp., 1960—; chief psychiatry Trafalgar Hosp., 1962-72; vis. prof. psychiatry Tokyo U. Sch. Medicine, 1964; adj. prof. N.Y. Inst. Tech., 1968; vis. lectr. in psychiatry N.Y. U., 1971-73; Am. del. Internat. Conf. Mental Health, London, 1948; mem. Am. Psychiat. Commn. to USSR, Poland and Finland, 1963, Empire State Med., Sci. and Ednl. Found. Author: (booklets) Neurosis in a Child, 1949, A Psychiatric Viewpoint on Child Adoption, 1954, Shock Therapy in Psychiatric Practice, 1957, The Genesis of Neurosis, 1958, The Prejudiced Personality, 1962, The Schizophrenic Process, 1962, The Prognosis of Electroshock Therapy Success, 1963, Psychodynamic Effects of Abortion, 1967, Violence Towards Self, 1971, Basic Concepts of Transference in Psychoanalysis, 1973, A Study in Suicide, 1972, Masochism, 1973, Emotional Factors Involved in Skeletal Deformities, 1977, Cults & Cult Suicide, 1979; assoc. editor: Jour. Pan Am. Med. Assn., 1992—; contbr. articles on the psychoses. Vice pres. Golden Years Found.; N.Y.C. chmn. Com. Med. Standards in Psychiatry, 1952-54. Recipient Sir William Osler Internat. Honor Med. Soc. Gold Key; 3d prize oil paintings N.Y. State Med. Art Exhibit, 1954; NYU Founders Day award, 1963; Presdl. Achievement award, 1984; others. Fellow Am. Psychiat. Assn. (life, pres. Bronx dist. 1960-61, other offices, Gold medal 1974), Am. Assn. Psychoananlytic Physicians (pres. 1971-72, bd. govs. 1972—); mem. AMA, Ea. Psychiat. Assn., N.Y. State Soc. Med. Rsch., Am. Med. Writers Assn., N.Y. Med. Soc., N.Y. County Med. Soc., N.Y. Soc. for Clin. Psychiatry, Pan Am. Med. Assn. (various offices including pres. 1989—, assoc. editor jour. 1992—), Assn. for Advancement Psychotherapy, Bronx Soc. Neurology and Psychotherapy (pres. 1960-61, Silver medal 1970), Mensa. Home: 23 Meadow Rd Scarsdale NY 10583-7642 Office: 263 West End Ave New York NY 10023

SORRELL, WILFRED HENRY, astrophysics educator; b. Birmingham, Ala., July 2, 1944; s. John Henry and Inez (Vernal) S. Student, Stillman Coll., 1962-66; BS in Physics, U. Wis., 1966, MS in Physics, 1972, PhD in Astronomy, 1989. Jr. engring. asst. for Apollo space program IBM Corp., Huntsville, Ala., 1966-67; teaching asst. physics U. Wis., Madison, 1967-68, teaching asst. math., 1981-83, math. instr. for minority summer engring. program, 1981-90, teaching asst. astronomy, 1987; asst. prof. astrophysics U. Mo., St. Louis, 1990—. Contbr. articles to profl. publs. Recipient St. Louis Am. Salute Excellence Merit award, 1992; grantee Inst. fur Astrophysik, 1983, U. Wis., NSF, 1984, 86, 88, NASA, 1987-88; Vilas fellow, 1966-68. Mem. Am. Astron. Soc. Achievements include research on theoretical astrophysics with emphasis on nature of interstellar and circumstellar dust grains, central energy source in quasi-stellar objects and Seyfert nuclei, models for formation of low mass stars. Office: U Mo Physics Dept 8001 Natural Bridge Rd Saint Louis MO 63121-4499

SORRENTINO, DARIO ROSARIO, medical educator; b. Alghero, Sassari, Italy, July 25, 1957; came to U.S., 1983; s. Antonio and Rosa (Messina) S. BS, Liceo Sci., Alghero, 1976; MD magna cum laude, U. Sassari, 1982. Cert. ECFMG. Vis. scientist King's Coll. Hosp., London, 1978, Royal Free Hosp., London, 1979-80; rsch. fellow U. Calif., San Francisco, 1983-86; asst. prof. medicine Mt. Sinai Sch. Medicine, N.Y.C., 1986—, clin. fellow medicine, 1992; asst. prof. medicine U. Udine, Italy, 1992—; vis. prof. Cath. U., Louvain, Belgium, 1988, U. Queensland, Brisbane, Australia, 1990-91, McGill U., Montreal, 1992; reviewer Jour. Clin. Investigation, Am. Jour. Physiology, Jour. Hepatology, Gastroenterology, 1988—; cons. NIH, 1993. Mem. editorial bd. Hepatology, 1991—; contbr. articles to Biochem. Biophysics Rsch. Communications, Proceedings NAS, Jour. Clin. Investigation, Baillere Clin. Gastroenterology, Am. Jour Physiology, Progress in Liver Diseases, Jour. of Hepatology, Hepatology, Jour. Biol. Chemistry. Grantee U. Calif. Acad. Senate, 1985, Italian Ministry Edn., 1987-90, Mt. Sinai Sch. Medicine, 1990-91, European Soc. Clin. Investigation/Bayer, 1991-93. Mem. Internat. Assn. Study Liver, Am. Assn. Study Liver Disease, European Assn. Study Liver, N.Y. Acad. Scis. Office: Medicina Interna Policlinico Univ, Pza Santa Maria Misericordia # 1, 33100 Udine Italy

SORRENTINO, JOHN ANTHONY, environmental economics educator, consultant; b. Bklyn., Sept. 1, 1948; s. John Anthony and Monica Denise (Bissonnette) S.; m. Margaret Mary Johnson, Oct. 11, 1969 (div. Jan. 1981); children: Rachel Denise, Demian Andrew; m. Judith Ann King, July 14, 1986. BBA, Baruch Coll.-CUNY, 1969; MS, Purdue U., 1971, PhD, 1973. Instr. Purdue Univ., Lafayette, Ind., 1969-72; asst. prof. econs. Temple Univ., Phila., 1973-79, assoc. prof. econs., 1979—; consultant, Glenside, Pa., 1986—. Contbr. chpts. to books and articles to profl. procs. and jours. Mem. energy task force Consumer Coun. Phila., 1975; v.p. dir. Citizens Comm. on Environ. Control, Cheltenham, Pa., 1975-77; lectr. Am. Youth Hostels, Phila., 1978; mem. Energy Edn. Adv. Coun., Phila. Electric Co., 1987—. Recipient traineeship NSF, Purdue Univ., 1971-72, summer fellowships Nat. Aero. and Space Adminstrn., Huntsville, Ala., 1975, Energy R&D Adminstrn., Washington, 1976, Dept. Energy, Livermore, Calif., 1984. Mem. Am. Econ. Assn., Assn. Environ. and Resource Economists, Internat. Soc. for Ecol. Econs., Earthright. Avocations: phys. fitness, classical and popular music. Office: Temple Univ Ambler 580 Meetinghouse Rd Ambler PA 19002-3989

SORROWS, HOWARD EARLE, executive, physicist; b. Hewitt, Tex., Aug. 10, 1918; s. George Jefferson and Lillian Nora (Gregory) S.; m. Martha Jane Summerville, Dec. 10, 1943; children: Mary Margaret Hughes, Carolyn Clare Stump, Joyce Jean Jimerson, Lynne Louise Moon, Bryan Bruce. BA in Physics and Math., Baylor U., 1940; postgrad., George Washington U., 1944, MA in Physics, 1948; PhD in Physics, Math. and Civil Engring., Cath. U. Am., 1958. Tchr. math. and physics Sabinal (Tex.) Pub. High Sch., 1940-41; electronic scientist Nat. Bureau Standards, Washington, 1941-52; missile engr. U.S. Bur. Ordnance, Washington, 1952-54; sci. adminstr. U.S. Office Naval Rsch., Washington, 1954-59; dir. long range planning, tech. intelligence, new product devel. Tex. Instruments, Inc., Dallas, 1959-62, mgr. space and environ. sci., 1962-65; sr. exec. Nat Bur. Standards, Gaithersburg, Md., 1965-87; pvt. practice Potomac, Md., 1987-88; dir. bd. assessment Nat. Inst. Standards & Tech. Nat. Rsch. Coun., Washington, 1988—; mem. adv. com. elec. engring. and applied sci. U. Pa., Phila., 1977-82; mem. adv. coun. La. Sea Grant Coll., Baton Rouge, 1979, 91—, bd. regents, 1988—; founding mem. La. Partnership for Tech. and Innovation, 1989—; instr. Ministry Fed. Sci. and Tech., Lagos, Nigeria, 1982; adj. prof. Am. U., Washington, 1984-86; nat. tour speaker on technol. forecasting Dept. Commerce; mem. program com. Nat. Conf. for Advancement Sci., 1985-87. Pres. Standards Alumni Assn., Gaithersburg, Md., 1989; charter chmn. Fed. Profl. Assn. Nat, Bur. Standards chpt. Gaithersburg, 1963. Recipient Superior Performance award Dept. Commerce, Washington, 1966, 67, 74, 79, 84, 85. Fellow Washington Acad. Sci.; mem. IEEE (sr. mem., chmn. Dallas-Ft. Worth chpt. 1963-64, Outstanding Contbn. award 1958), AAAS, KCCH, Masons, Sigma Xi (pres. Nat. Bur. Standards chpt., Outstanding Contbn. award 1988), Sigma Pi Sigma (sr. mem.). Avocations: tennis, gardening. Home: 8820 Maxwell Dr Rockville MD 20854-3122 Office: Nat Rsch Coun HA Bldg Rm 550 2101 Constitution Ave NW Washington DC 20418-0001

SORSTOKKE, SUSAN EILEEN, systems engineer; b. Seattle, May 2, 1955; d. Harold William and Carrol Jean (Russ) S. BS in Systems Engring., U. Ariz., 1976; MBA, U. Wash., Richland, 1983. Warehouse team mgr. Procter and Gamble Paper Products, Modesto, Calif., 1976-78; quality assurance engr. Westinghouse Hanford Co., Richland, Wash., 1978-80; supr. engring. document ctr. Westinghouse Hanford Co., Richland, 1980-81; mgr. data control and adminstrn. Westinghouse Electric Corp., Madison, Pa., 1981-82, mgr. data control and records mgmt., 1982-84; prin. engr. Westinghouse Elevator Co., Morristown, N.J., 1984-87; region adminstrn. mgr. Westinghouse Elevator Co., Arleta, Calif., 1987-90; ops. rsch. analyst Am. Honda Motor Co. Inc., Torrance, Calif., 1990—; adj. prof. U. LaVerne, Calif., 1991-92. Advisor Jr. Achievement, 1982-83; literacy tutor Westmoreland Literacy Coun., 1983-84, host parent EF Found., Saugus, Calif., 1987-88, Am. Edn. Connection, Saugus, 1988-89, 91,; instr. Excell, L.A., 1991-92. Mem. Soc. Women Engrs., Am. Inst. Indsl. Engrs., Nat. Coun. on Systems Engring., Optimists Charities Inc. (bd. dirs. Acton, Calif. 1991—). Republican. Methodist. Home: 21647 Spice Ct Santa Clarita CA 91350-1656 Office: Am Honda Motor Co Inc Dept Parts Rsch and Planning 1919 Torrance Blvd Torrance CA 90501-2746

SOSA-RIERA, RAUL, air transportation executive; b. Montevideo, Uruguay, Oct. 29, 1944; arrived in Spain, 1969; s. Damaso and Agueda (Riera) S.; m. Celia Tejerina, Apr. 15, 1979; 1 child, Elena. Grad., London Bus. Sch., 1987. Check pilot ops. Iberia Airlines, Madrid, 1980-84, v.p. tech. ops., 1984-89, sr. v.p. flight ops., 1987-89; dir. flight ops. Binter Mediterraneo, Valencia, Spain, 1989—; mem. tech. com. IAT, Geneva, 1988, flight ops. adv. com., 1987-88, Assn. European Airlines Tech., Brussels, 1986-87. Contbr. articles to airline safety mags. Lt. Uruguyan Air Force, 1964-66. Mem. AIAA, Inst. Iberoamericano de Derecho Aeronautico. Home: Calle 136 No 10 La Canada, 46182 Paterna Spain Office: Binter Mediterraneo, Bloque Tecnico Aeropuerto, 46940 Manises Spain

SOSLOW, ARNOLD, quality consultant; b. Phila., Nov. 13, 1938; s. Samuel and Betty (Goldfine) S.; m. Frances Isen, May 15, 1960; children: Michael Allan, Beverly Ruth Soslow Warner. AS in Bus. Adminstrn., Temple U., 1974, BBA in Ind. Mgmt., 1976. Design draftman Unisys, Blue Bell, Pa., 1959-66; process control/quality specialist Gen. Electric Co., Phila., 1966-69; mgr. ops. B & F Instruments, Cornwells Heights, Pa., 1969-75; quality engr. Ronson Corp., Bridgewater, N.J., 1976-78; quality assurance mgr. Kooltronics Inc., Hopewell, N.J., 1978-83; quality mgr. Electro-Sci. Labs., King of Prussia, Pa., 1983-90; pres., quality advisor The Quality Mgmt. Co., Phila., 1988—. Mem. Am. Soc. for Quality Control, Internat. Electronic Package Soc., Internat. Soc. for Hybrid Microelectronics (program chmn. joint symposium 1987-89, pres. Keystone chpt. 1989, gen. chmn. joint symposium 1990-92, gen. chmn. quality workshop 1992-93), Surface Mt. Tech. Assn. Republican. Jewish. Achievements include research in field. Home and Office: The Quality Mgmt Co 8844 Manchester St Philadelphia PA 19152-1515

SOSLOWSKY, LOUIS JEFFREY, bioengineering educator, researcher; b. Bklyn., Apr. 4, 1964; s. Martin and Phyllis (Popowitz) S. BS, Columbia U., 1986, MS, 1987, PhD, 1991. Rsch. asst. Bioengring. Inst., Columbia U., N.Y.C., 1983-86, rsch. fellow, 1986-91; asst. prof. bioengring., mech. engring., orthopedic surgery U. Mich., Ann Arbor, 1991—; reviewer Jours. Biomech. Engring., Biomechanics, Orthopaedic Rsch., Surg. Rsch., 1991—; panelist Shoulder Workshop, Am. Acad. Orthopaedic Surgeons-ASES, NIH, Vail, Colo., 1992. Contbr. numerous articles to Biorheology Jour., Biomechanics Jour. Orthopaedic Rsch., Clinics in Sports Medicine, numerous chpts. in Biomechanics of Diarthrodial Joints, Basic Orthopedic Biomechanics, also others. Grantee Orthopaedic Rsch. and Edn. Found., 1991—, NSF, 1992—, Whitaker Found., 1992—. Mem. ASME, AAUP, Orthopaedic Rsch. Soc., Am. Soc. Biomechanics, Sigma Xi, Tau Beta Pi, Chi Epsilon. Office: U Mich Orthopaedic Rsch Labs Rm G-161 400 N Ingalls Ann Arbor MI 48109-0486

SOTIRAKOS, IANNIS, civil engineer; b. Athens, Attiki, Greece, Nov. 12, 1962; s. Triadis and Titika (Biliraki) S. BSc in Civil Engring., Nat. Tech. U., Athens, 1986; MSc in Urban Planning, Sorbonne Paris IV U., 1987; degree in bus. adminstrn., Hellenic Inst. Productivity, 1990. Asst. engr. Orgn. Athens, 1983-84, mgr. rsch. EEC project., 1988-90; mgr. rsch. project Sorbonne Paris IV U., 1986-87; mgr. s/w devel. Navy Hdqrs., Athens, 1987-88; dir. Tekton Engring., Kifissia, Greece, 1987-89; founder, dir. civil engring. Engring. Office, Athens, 1989—; cons. Computer mag., Athens, 1988-90, Mayor of Athens, Athens Light Railway Project; co-founder, dir. F-Zein Adventures Ltd., Athens, 1990—. Co-editor, cons. Estate Internat., 1991—; contbr. articles to profl. jours. Mem. IEEE, Tech. Chamber Greece, Ekali Club Athens. Avocations: sailing, enduro racing, reading, cooking, photography. Home: 50 Parnithos Str, 145 65 Ekali Athens, Greece

SOTOMORA-VON AHN, RICARDO FEDERICO, pediatrician, educator; b. Guatemala City, Guatemala, Oct. 22, 1947; s. Ricardo and Evelyn (Von Ahn) S.; m. Eileen Marie Holcomb, May 9, 1990. M.D., San Carlos U., 1972; M.S. in Physiology, U. Minn., 1978; m. Victoria Monzon, Nov. 26, 1971; children—Marisol, Clarisa, Ricardo, III, Charlotte Marie. Rotating intern Gen. Hosp. Guatemala, 1971-72; pediatric intern U. Ark., 1972-73, resident, 1973-75; fellow in pediatric cardiology U. Minn., 1975-78; research assoc. in cardiovascular pathology United Hosps., St. Paul, 1976; fellow in neonatal-perinatal medicine St. Paul's Children's Hosp., 1977-78, U. Ark., 1981-82; instr. pediatrics U. Minn., 1978-79; pediatric cardiologist, unit cardiovascular surgery Roosevelt Hosp., Guatemala City, 1979-81; asst. prof. pediatrics (cardiology and neonatology), U. Ark., Little Rock, 1981-83; practice medicine specializing in pediatric cardiology-neonatology, 1983— Diplomate Am. Bd. Pediatrics, Sub-Bd. Pediatric Cardiology, Neonatal-Perinatal Medicine. Fellow Am. Acad. Pediatrics, Am. Coll. Cardiology, Am. Coll. Chest Physicians, Am. Coll. Angiology; mem. AMA, AAAS, Ark. Med. Soc., N.Y. Acad. Scis., Am. Heart Assn., Soc. Pediatric Echocardiography, Guatemala Coll. Physicians and Surgeons, Central Ark. Pediatric Soc., So. Soc. Pediatric Research, Soc. Critical Care Medicine. Clubs: Pleasant Valley Country (Little Rock); American (Guatemala). Home: #1 Shaw Bridge Ln Little Rock AR 72212 Office: Med Towers II Ste 800 Little Rock AR 72205

SOUDER, EDITH IRENE, information scientist; b. DeKalb County, Ind., July 22, 1937; d. Lucias L. and Dorothy (Taylor) Love; m. William E. Souder, Mar. 2, 1957; children: Dianna, Denneta, Dorene. BS, U. Pitts., 1977, MS, 1979. Info. specialist Compaign for Gov. Dick Thornburgh, Pitts., 1977-78; tech. researcher Info Source, Pitts., 1978-79; pres., prin. investigator INFRA, Pitts., 1978-90; sr. program analyst East Coast Credit, Pitts., 1988-90; farm specialist Blooming Acres, Gurley, Ala., 1990—. Mem. Am. Soc. Info. Sci. Home and Office: INFRA 315 Blooming Acres Ln Gurley AL 35748

SOUKIASSIAN, PATRICK GILLES, physics educator, physicist; b. Châlons, France, June 1, 1944; s. Pascal Haroutioun and Berthe Berdjouhie (Djamdjian) S.; m. Monique Nicole Ménage, Sept. 10, 1972; children: Méliné, Laetitia, Tatiana. BS in Physics, U. Reims (France), 1969, MS in Electron Optics summa cum laude, 1971, PhD in Electron Optics summa cum laude, 1974; PhD in Physics summa cum laude, U. Paris, Orsay, France, 1985. Asst. prof. U. Reims, 1971-74, assoc. prof., 1974-88; assoc. prof. U. Wis., Madison, 1985-86; prof. No. Ill. U., Dekalb, 1986-88, U. Paris, Orsay, 1988—; lectr. in field; cons. No. Ill. U., Dekalb, 1985-86; rsch. group leader Synchrotron Radiation Ctr., Madison, Wis., 1986—, Commissariat Energie Atomique, Saclay, France, 1988—; assoc. dir. for indsl. and pub. rels. Master Applied Physics program U. Paris, Orsay, 1988—. Author: (with others) Metallization and Metal/Semiconductor Interfaces, 1989, Physics and Chemistry of Alkali Metal Adsorption, 1989, Fundamental Approach to New Materials Phases, 1991; mem. editorial bd. Surface Review and Letters; contbr. articles to profl. jours.; inventor, patentee in field. With French Army Reserve, 1970—, major, 1985—. Mem. Am. Phys. Soc., European Phys. Soc., Soc. Francaise de Physique. Roman Catholic. Office: Commissariat à l'Energie Atomique, CEA DRECAM SRSIM Bât 462, 91191 Saclay France

SOULE, THAYER, documentary film maker. With Associated Film Artists, Camp Connell, Calif. Recipient Centennial award Nat. Geographic Soc., Wash., 1988. Office: Assoc Film Artists PO Box 4437 Camp Connell CA 95223*

SOURES, JOHN M., physicist, researcher; b. Galatz, Romania, Jan. 2, 1943; came to U.S., 1954; s. Michael Ioannis and Dia (Petrakis) S.; m. Diana Carrousos, June 29, 1969; children: Mandy M., Nicholas J., Eleni C., Alexander J., Sophia A. BS in Physics, U. Rochester, 1965, PhD in Mech. and Aerospace Scis., 1970. Rsch. assoc. Lab. for Laser Energetics, Rochester, N.Y., 1970-72, sr. scientist, 1972—, group leader, 1976-81, div. dir., 1979—, deputy dir., 1983—; cons. Oak Ridge (Tenn.) Nat. Lab., 1970-71, Lawrence Livermore (Calif.) Nat. Lab., 1988-90. Contbr. articles to profl. jours. Treas., v.p. Greek Orthodox Ch. of Annunciation, Rochester, 1990-92. Recipient APS Divsn. of Plasma Physics Excellence award in Plasma Physics Rsch., 1993; NASA trainee, 1967-69. Fellow Am. Physical Soc. (conf. chmn. 1986, reward for excellence in plasma physics divsn. plasma physics 1993), Optical Soc. Am. (conf. organizer 1984-86, 92-93). Republican. Orthodox. Achievements include co-invention of the laser beam smoothing system and a Nd: glass amplifier system. Home: 146 E Brook Rd Pittsford NY 14534 Office: Lab for Laser Energetics 250 E River Rd Rochester NY 14623

SOURKES, THEODORE LIONEL, biochemistry educator; b. Montreal, Que., Can., Feb. 21, 1919; s. Irving and Fannie (Golt) S.; m. Shena Rosenblatt, Jan. 17, 1943; children: Barbara, Myra. B.Sc., McGill U., 1939, M.Sc. magna cum laude, 1946; Ph.D., Cornell U., 1948; D.U. honoris causa, U. Ottawa, Can., 1990. Asst. prof. pharmacology Georgetown U. Med. Sch., 1948-50; research asso. dept. enzyme chemistry Merck Inst. Therapeutic Research, Rahway, N.J., 1950-53; sr. research biochemist Allan Meml. Inst., Montreal, 1953-65; dir. lab. neurochemistry Allan Meml. Inst. Psychiatry, 1965—; mem. faculty McGill U., Montreal, 1954—; prof. biochemistry McGill U., 1965—, prof. psychiatry, assoc. dean of medicine for research Faculty Medicine, 1972-75; prof. pharmacology, 1990—, emeritus, 1991; Mem. Que. Med. Research Council, 1971-77; sr. fellow Parkinson's Disease Found., N.Y.C., 1963-66; assoc. mem. McGill Ctr. for Medicine, Ethics and Law, 1991. Author: Biochemistry of Mental Disease, 1962, Nobel Prize Winners in Medicine and Physiology, 1901-1965, 1967. Decorated Officer Order of Canada. Fellow Royal Soc. Can.; mem. Canadian Biochem. Soc., Pharmacol. Soc. Can., Canadian Coll. Neuropsychopharmacology (Heinz Lehmann award 1982, medal 1990), Am. Soc. Biol. Chemists, Am. Soc. Pharmacology and Exptl. Therapeutics, Am. Soc. Neurochemistry, Internat. Soc. Neurochemistry, Internat. Brain Research Orgn., Venezuelan Order Andrés Bello, Sigma Xi. Research and publs. on drugs for treatment high blood pressure; 1st basic research on methyldopa; elucidation of role of dopamine and other monamines in nervous system; first trials of L-dopa in Parkinson's disease, biochemistry of mental depression, pathways of stress in the nervous system, imaging serotonin in brain, history of biochemistry. Home: 3033 Sherbrooke St W #303, Montreal, PQ Canada H3Z 1A3

SOUSTELLE, MICHEL MARCEL PHILIPPE, chemistry educator, researcher; b. St. Etienne, Loire, France, July 15, 1937; s. Emile and Jeanne (Brunet) S.; m. Monique Guerin, Aug. 12, 1961; children: Philippe, Jean-Pierre. Lic., Ecole Nationale Superieure d'Electrochemie et d'Electrometallurgie, Grenoble, France, 1960, Engr., 1961; PhD, U. Grenoble, 1967. Asst. prof. U. Grenoble, 1960-66; asst. instr. Ecole Mines, St. Etienne, 1966-67, rsch. instr., 1967-73, prof. chemistry, 1973—, dir. dept. phys. chemistry, 1973-93, rsch. dir., 1993—; cons. F-ench Ministry Rsch., Paris, 1979-82. Author: Traite de Chimie Minerale, 1970, Modelisation Macroscopique des Transformations Physics-Chiminques, 1990; contbr. over 200 articles to sci. jours. Recipient several sci. prizes, France. Mem. Soc. Chemique France, N.Y. Acad. Scis. Roman Catholic. Home: 4 Rue de la Convention, 42100 Saint Etienne France Office: Ecole des Mines, 158 Cours Fauriel, 42023 Saint Etienne France

SOUTHARD, JAMES HEWITT, biochemist, researcher; b. Greenfield, Mass., Jan. 28, 1943; s. Herman Charles and Eloise (Bangs) S.; m. Alexandria Kozikowski, June 12, 1965; children: Jennifer, Elizabeth. BS, Springfield Coll., 1964; MS, Univ. Mass., 1968, PhD, 1970. Asst. rsch. prof. Inst. for Enzyme Rsch., Madison, 1972-76; assoc. rschr. dept. surgery Univ. Wis., Madison, 1976-78, asst. scientist dept. surgery, 1978-84, asst. prof. dept. surgery, 1984-88, assoc. prof. dept. surgery, 1988-92, prof. dept. surgery, 1992—. Contbr. articles to profl. jours.; author 12 book chapters. Mem. Soc. for Cryobiology (pres. 1992-94). Achievements include 3 patents on organ preservation solutions. Office: Univ Wis Dept Surgery 600 Highland Ave Madison WI 53792

SOUTHWICK, CHARLES HENRY, zoologist, educator; b. Wooster, Ohio, Aug. 28, 1928; s. Arthur F. and Faye (Motz) S.; m. Heather Milne Beck, July 12, 1952; children: Steven, Karen. B.A., Coll. Wooster, 1949; M.S., U. Wis., 1951, Ph.D., 1953. NIH fellow, 1951-53; asst. prof. biology Hamilton Coll., 1953-54; NSF fellow Oxford (Eng.) U., 1954-55; faculty Ohio U., 1955-61; assoc. prof. pathobiology Johns Hopkins Sch. Hygiene and Pub. Health, Balt., 1961-68; prof. Johns Hopkins Sch. Hygiene and Pub. Health, 1968-79; assoc. dir. Johns Hopkins Internat. Ctr. for Med. Rsch. and Tng., Calcutta, India, 1964-65; chmn. dept. environ., population and organismic biology U. Colo., Boulder, 1979-82, prof. biology, 1979—; researcher and author publs. on animal social behavior and population dynamics, influences animal social behavior on demographic characteristic mammal populations, primate ecology and behavior, estuarine ecology and environmental quality; mem. primate adv. com. Nat. Acad. Sci.-NRC, 1963-75, com. primate conservation, 1974-75; mem. Gov.'s Sci. Adv. Com. State of Md., 1975-78; mem. com. on rsch. and exploration Nat. Geog. Soc., 1979—; mem. adv. bd. Caribbean Primate Rsch. Ctr., 1987—, Wis. Primate Rsch. Ctr., 1990—; mem. Integrated Conservation Rsch., 1989—. Editor: Primate Social Behavior, 1963, Animal Aggression, 1970, Nonhuman Primates in Biomedical Research, 1975, Ecology and the Quality of Our Environment, 1976, Global Ecology, 1985; Ecology and Behavior of Food-Enhanced Primate Groups, 1988. Recipient Fulbright Rsch. award India, 1959-60. Fellow AAAS, Acad. Zoology, Animal Behavior Soc.; mem. Am. Soc. Zoologists, Ecol. Soc. Am., Am. Soc. Mammalogists, Am. Soc. Primatology, Internat. Primatology Soc., Am. Inst. Biol. Scis., Primatology Soc. Great Britain, Internat. Soc. Study Aggression.

SOUTO BACHILLER, FERNANDO ALBERTO, chemistry educator; b. Andújar, Jaēn, Spain, Mar. 27, 1951; s. Manuel and Marīa Encarnaciōn (Bachiller Mora) Souto Garcīa; m. Josefina Melgar Gōmez, July 20, 1974; children: Antonio Alberto, Fernando José, Natacha. Licenciate of Sci., U. Granada, Spain, 1973; PhD, U. Alta., Can., 1978; ScD, Ministerio de Educaciōn, Madrid, 1988. Asst. prof. chemistry U. P.R., Mayagüez, 1979-82; assoc. prof. U. P.R., Mayaguez, 1983-88, prof., 1988—, dir. CRIL, 1984—; vis. prof. U. Granada, 1988, EPFL, Lausanne, Switzerland, 1989, U. Málaga, Spain, 1990. Contbr. articles to profl. jours. Queen Elizabeth scholar, 1975; Fundaciōn Juan March rsch. fellow, 1979. Fellow Royal Soc. Chemistry (London), Am. Inst. Chemists; mem. Am. Chem. Soc. (councilor 1989-91, chmn. 1987), Sigma Xi. Roman Catholic. Achievements include development of two interrelated areas of research in organic molecular photophysics and photochemistry of aromatic natural products and production of natural products by in vitro biosynthesis, rotating optical disk ring electrode; elucidation of geometry and spin multiplicity of cyclobutadiene C4H4 and C4D4. Home: H-21 Calle Almirante Alturas de Mayaguez Mayaguez PR 00680 Office: U of PR Chemistry Dept Mayaguez PR 00681

SOUW, BERNARD ENG-KIE, physicist; b. Pekalongan, Java, Indonesia, Jan. 7, 1942; came to U.S., 1984, naturalized citizen, 1990; s. Tjwan-Ling and Pek-Liang (Kwee) S.; m. Martha Tjoei-Lioe Lim, July 17, 1967; children: Victor, Verena. Diplom Physiker, Tech. U. of Clausthal, Zellerfeld, Fed. Republic of Germany, 1972; Dr. Rer.-Nat., U. of Duesseldorf, Fed. Republic of Germany, 1981. Rsch. assoc. U. Duesseldorf, 1973-84; rsch. scientist Isotope Rsch. Inst., Haan, Fed. Republic of Germany, 1983; univ. asst. Free U. of Berlin, 1984; vis. scientist A. F. Wright Aero. Labs., Dayton, Ohio, 1984-85; rsch. scientist Brookhaven Nat. Lab., Upton, N.Y., 1985—; cons. plasma and laser applications. Contbr. articles to Jour. Applied Physics, Jour. Quantitative Spectroscopy, Jour. Plasma Physics, Physica. Mem. Am. Phys. Soc., L.I. Optical Soc. Office: Brookhaven Nat Lab Bldg 130 Upton NY 11973

SOUZA, MARCELO LOPES, aerospace engineer, researcher, consultant; b. Sao Luis, Brazil, Dec. 27, 1951; s. José Martins and Aracy Lopes (Ferreira) S.; m. Rose Mary Almeida, Jan. 3, 1981. EE in Electronics, Tech. Inst. Aeronautics, São José dos Campos, Brazil, 1976; MSc in Space Mechanics and Control, Nat. Inst. for Space Rsch., São José dos Campos, Brazil, 1980; PhD in Aeronautics and Astronautics, MIT, 1985. Asst. rschr. Nat. Inst. for Space Rsch., Sao Jose dos Campos, 1977-85, assoc. rschr., 1985-91, sr. rschr., 1991—, head dept. space mechanics and control, 1989-91; orbit and attitude control rschr. Nat. Ctr. Space Studies, CNES, Toulouse, France, 1979; digital control cons., lectr., Avibras-Industria Aeroespacial, Sao Jose dos Campos, 1987; digital control cons., lectr. Embraer-Empr. Brasileira de Aeronautica, Sao Jose dos Campos, 1991; co-organizer 1st Brazilian Symposium Aerospace Tech., 1990, VI Japan-Brazil Symposium on Sci. and Tech., 1988, 7th Congresso Brasileiro de Automatica, 1988; 4th Braziliam Coloquium Orbital Dynamics, 1988. Co-editor: Multilingual Dictionary of Astronomy and Space Sciences, 1988. Tchr., dir. Curso Casd, Sao Jose dos Campos, 1972-76; tchr., mgr. Curso ATP, Sao Luis, 1969-70; politics and strategy trainee Escola Superior de Guerra, Rio de Janeiro, 1992. Lt. Brazilian Air Force, 1972-73. Mem. IEEE, AIAA, Tech. Inst. Aeronautics Alumni Assn., ESG Alumni Assn., IEEE Control Systems Soc., MIT Alumni Assn. Achievements include first to exactly solve the weighted time plus fuel optimal control of an undamped harmonic oscillator and to use it to improve control of a flexible spacecraft; two quasilinerarization methods to quantify Vander Velde's limit cycles; estimate of gas consumption during derived rate feedback (DRF) limit cycles. Home: Apt 302, Rua Marechal Rondon 593, 12215070 Sao Jose dos Campos SP Brazil Office: Inst Nac de Pesquisas Espac, Ave dos Astronautas 1758, 12227010 Sao Jose dos Campos Brazil

SOUZA, MARCO ANTONIO, civil engineer, educator; b. Valença, Rio de Janeiro, Sept. 26, 1951; s. Oldemar and Elza Souza; m. Silvia Marina Pinto Souza, May 6, 1982. Degree in structural engring., Pontificia U. Cath. Rio de Janeiro, 1975, MS, 1978; PhD, Univ. Coll. London, 1982. Design engr. Montreal Engenharia S.A., Rio de Janeiro, 1975-76; assoc. prof. civil engring. dept. Pontificia U. Cath. Rio de Janeiro, 1982-89, assoc. prof., 1991—; vis. prof. Va. Poly. Inst. and State U., Blacksburg, 1989-91, U. Alberta, Edmonton, Can., 1987; cons. Ecopetrol, Bogota, Colombia, 1988; chmn. local arrangements com. 1st Pan Am. Congress Applied Mechanics, 1989; organizing and editorial com. 2d and 3d Pan Am. Congresses Applied Mechanics, 1991, 93; reviewer jours.; leader presentations and seminars in field. Assoc. editor Ocean Engring., 1990—; contbr. numerous articles to profl. jours. Coun. Nat. Desenvolvimento Tech. and Sci. Scholar, 1976-78, 79-82, grantee, 1983—. Mem. AIAA, ASCE, ASME (reviewer Jour. Applied Mechanics), Am. Soc. Engring. Edn., Am. Acad. Mechanics (steering com. Pan Am. Congresses), Brazilian Soc. Mech. Sci. (reviewer jour.), Soc. Engring. Sci., Structural Stability Rsch. Coun., Sigma Xi. Office: Pontificia U Católica do Rio De Janeiro, Rua Marquês de São Vicente 225, 22453-900 Rio de Janeiro Brazil

SOUZA MENDES, PAULO ROBERTO DE, mechanical engineering educator; b. Rio de Janeiro, Brazil, Aug. 15, 1954; came to U.S., 1991; s. Joao Jose de and Ernestina de (Von Have) S.; m. Denise Cloria Marchesini, Aug. 15, 1979; children: Paula, Luisa, Carla, Aline, Michelle. BSME, Rio de Janeiro Cath. U., 1976, MSME, 1979; PhD, U. Minn., 1982. Teaching asst. Rio de Janeiro Cath. U., Brazil, 1977-79; rsch. asst. U. Minn., Mpls., 1980-82; asst. prof. Rio de Janeiro Cath. U., 1983-86, assoc. prof., 1987—, dir. grad. studies, 1985-86, dept. head, 1987-88; vis. assoc. prof. U. Minn., 1991-93. Editor Internat. Jour. Experimental Heat Transfer, 1988-91; contbr. articles to profl. jours. Mem. ASME, Brazilian Soc. Mech. Engrs. (chmn. com. 1988-91), Soc. Rheology, Polymer Processing Soc.

SOVDE-PENNELL, BARBARA ANN, sonographer; b. McPherson, Kans., Sept. 27, 1955; d. Benton Ellis and Mary Ann (Ball) Sovde; m. Paul Edwin Pennell, June 5, 1982; 1 child, Eric Louis. AA in Radiologic Tech., Hutchinson Community Jr. Coll., 1977; BS in Radiologic Tech., U. Okla., 1993. Registered diagnostic med. sonographer, radiological technol. Radiographer Hertzler Clinic, Halstead, Kans., 1977-78; radiographer Mercy Health Ctr., Okla. City, 1978-81, sonographer, supr. ultrasound dept., 1981-83; mobile sonographer Sun Med. Systems, Okla. City, 1983-84, Diagnostic Radiology, Edmond, Okla., 1984-87; prin., owner, pres. of corp. Ultrasound Unltd., Inc., Edmond, 1987—; part-time clin. specialist ultrasound Circadian Can. Ultrasound Equipment Co., 1991—. Active neighborhood recycling, Edmond, 1990—; mem. Greenpeace. Named Outstanding Leader in S.W. Nat. Allied Health Assn., 1981; recipient Outstanding Alumnus award U. Okla. Coll. Allied Health, 1990. Mem. Soc. Diagnostic Med. Sonographers (state rep. 1981-87, regional dir., bd. dirs. 1987-90), Okla. Sonographers Soc. (pres. 1982-84, steering com. 1984—). Democrat. Avocations: reading, biking, camping, environmental issues.

SOWA, PAUL EDWARD, research engineer; b. Chgo., Dec. 10, 1952; s. Edward Joseph and Gladys Angela (Bogdas) S.; m. Nanette Elizabeth Raddatz, Dec. 17, 1977; children: Adam, Alexander. BSME, U. Ill., 1976; MBA, Keller Grad. Sch., 1990. Mech. engr. Commonwealth Edison, Chgo., 1976-79; product specialist Signode Corp., Glenview, Ill., 1979-81, packaging engr., 1981-85, product mgr., 1985-86; sr. packaging engr. ITW/Signode, Glenview, 1986—; del. Nat. Safe Transit Assn., Chgo., 1988—. Inst. Packaging Profls. (cert.), ASTM (mem. com. D-10 1990—). Democrat. Roman Catholic. Office: ITW/Signode 3640 W Lake Ave Glenview IL 60025

SOWDER, DONALD DILLARD, chemicals executive; b. Rocky Mt., Va., Mar. 28, 1937; s. Roman Dillard and Virginia (Dowdy) S.; m. Beverly Reid, Nov. 29, 1957; children: Reid Dillard, Susan Allison, Donald Stuart. BS, Va. Tech., 1959; cert. in sales mgmt., Columbia U., 1976, cert. in fin., 1984; diploma, U.S. Army Command & Gen. Staff Coll., 1978. Sales rep. Sealtest Foods, Norfolk, Va., 1962-64; med. sales rep. Lederle Labs. div. Am. Cyanamid Co., Norfolk, 1964-69; dist. sales mgr. Lederle Labs. div. Am. Cyanamid Co., Washington, 1969-74; nat. mgr. sales tng. Lederle Labs. div. Am. Cyanamid Co., Pearl River, N.Y., 1974-76; mgr. fed. govt. affairs Lederle Labs. div. Am. Cyanamid Co., Washington, 1976-81; nat. sales mgr. hosp. div. Lederle Labs. div. Am. Cyanamid Co., Wayne, N.J., 1981-85, nat. sales mgr. oncology div., 1985-88; dir. govt. sales Lederle Labs. div. Am. Cyanamid Co., Fairfax, Va., 1988—; instr. U.S. Army Command & Gen. Staff Coll., Washington, 1977-81; govt. sales advisor Nat. Wholesale Drug Assn., Alexandria, Va., 1991. Contbr. articles to profl. jours. Bd. dirs. Shadow Walk Devel. Assn., 1990—. Col. USAR. Instr. of Yr. USAR, 1979. Mem. Assn. Mil. Surgeons U.S. (chmn. sustaining mems. 1980-81, lectr. 1989), Am. Soc. Hosp. Pharmacists, Res. Officers Assn., Va. Tech. Soc. Alumni Assn. (bd. dirs.), Mil. Dist. of Washington Officers Club System. Republican. Methodist. Avocations: golf, tennis, water sports. Home: 10415 Dominion Valley Dr Fairfax VA 22039-2415 Office: Am Cyanamid Co Lederle Govt Sales 12701 Fair Lakes Cir # 380 Fairfax VA 22033-4910

SOWDER, ROBERT ROBERTSON, architect; b. Kansas City, Kans., Dec. 29, 1928; s. James Robert and Agnes (Robertson) S.; m. Joan Goddard, July 26, 1954; 1 dau., Lisa Robertson Lee. B.A., U. Wash., 1953; B.Arch., U. Va., 1958; grad. diploma in Architecture, Ecole Des Beaux Arts, Fontainebleau, France, 1952. Designer Architects Collaborative, Boston, 1958-59, Peirce & Pierce (architects), Boston, 1959-63; asso. Fred. Bassetti & Co. (architects), Seattle, 1963-67; partner Naramore, Bain, Brady & Johanson (architects), Seattle, 1967-81; pres. NBBJ Internat., 1976-81; architect TRA, Seattle, 1981-83; v.p. Daniel, Mann, Johnson & Mendenhall, San Francisco, 1983-93; with RRS Consulting, San Francisco, 1993—; archtl. design critic Boston Archtl. Ctr., 1961-62. Important works include Ridgeway III Dormitories, Bellingham, Wash. (Dept. Housing and Urban Devel. Honor award), Seattle Rapid Transit (HUD Excellence award), Safeco Ins. Co. Home Office Complex, Seattle, King County Stadium, Balt. Conv. Ctr., Oreg. Conv. Ctr., San Francisco (Moscone) Conv. Ctr. Expansion, Honolulu Conv. Ctr., Wilmington (Del.) Conv. Ctr. Served with CIC U.S. Army, 1954-56. Recipient Premier Prix D'Architecture Ecole Des Beaux Arts, Fontainebleau, 1951, 52, Prix D'Remondet Fontainebleau, 1952. Mem. AIA, Internat. Assn. Auditorium Mgrs., Nat. Assn. Expo. Mgmrs., Scarab, Sigma Chi. Episcopalian. Clubs: Seattle Tennis, Rainier. Home: 2390 Hyde St San Francisco CA 94109-1505 Office: RRS Consulting 2390 Hyde St San Francisco CA 94109

SOWELL, JAMES ADOLF, quality assurance professional; b. West Bay, Fla., Apr. 25, 1943; s. George Leslie and Henrietta (Herndon) S.; m. Linda Jane Dupre, Apr. 24, 1965; children: Charles V., Marcelene L. BS, SUNY,

Albany, 1979; MBA, U. Hartford, 1982. Account exec. A.G. Edwards & Sons, Panama City, Fla., 1969-71; mgr. ops. Hartford (Conn.) Steam Boiler Inspection and Ins. Co., 1971-85; pres. Inspection Assn. Am., Panama City, 1985—; instr. The Ctr. for Profl. Advancement, E. Brunswick, N.J., 1976—. Contbr. articles to profl. jours. Charter mem. Young Republicans, Panama City, 1970-73, North Bay Alert, Panama City, 1992—. With USN, 1962-68. Mem. Am Soc. Quality Control (sr.), Am. Soc. Nondestructive Testing. Office: Inspection Assn Am 2313 Mound Ave Panama City FL 32405

SOWERS, EDWARD EUGENE, lawyer; b. Crawfordsville, Ind., Nov. 26, 1942; s. Eugene B. and Gladys L. (Newnum) S.; m. Margaret R. Sayre, July 1, 1967; children: William E., Jodi L. AB, Wabash Coll., 1964; PhD, Tufts U., 1970; JD, Ind. U., 1990. Bar: Ind. 1991, U.S. Dist. Ct. (no. and so. dists.) Ind. 1991, U.S. Patent Office 1992. Rsch. chemist Reilly Industries Inc., Indpls., 1969-73, mgr. product devel., 1973-78, sr. sect. head, 1978-92, atty. intellectual properties, 1992—. Inventor and patentee in field, 1983—. Mem. ABA, Ind. Bar Assn., Am. Chem. Soc. Avocations: reading, gardening, woodworking, music, restoring old autos. Home: 3280 State Rd 39 Mooresville IN 46158 Office: Reilly Industries Inc 1500 S Tibbs Ave Indianapolis IN 46242

SOX, HAROLD CARLETON, JR., physician, educator; b. Palo Alto, Calif., Aug. 18, 1939; s. Harold Carleton and Mary (Griffiths) S.; m. Carol Helen Hill, Aug. 26, 1962; children: Colin Montgomery, Lara Katherine. B.S., Stanford U., 1961; M.D. cum laude, Harvard U., 1966. Diplomate: Am. Bd. Internal Medicine. Intern and resident Mass. Gen. Hosp., Boston, 1966-68; clin. assoc. Nat. Cancer Inst., Bethesda, Md., 1968-70; instr. Dartmouth Med. Sch., Hanover, N.H., 1970-73; assoc. chief staff for ambulatory care VA Med. Ctr., Palo Alto, Calif., 1976-88; asst. prof. medicine to prof. Stanford U. Sch. Medicine, Calif., 1973-88; Joseph Huber prof., chmn. dept. medicine Dartmouth Med. Sch., 1988—; panel mem. Nat. Bd. Med. Examiners, Physicians Assts. Nat. Certifying Examination, 1973-76, Physicians Computer-based Certifying Examination, 1984-86; chair com. on priority-setting for health tech. assessment Inst. Medicine, 1990-91, chair U.S. preventive svcs. task force, 1990—, pretest writing com. Am. Bd. Internal Medicine, 1992—. Author: Medical Decision Making, 1988; editor: Common Diagnostic Tests, 1st edit. 1987, 2d edit. 1990; contbr. numerous articles to med jours., chpts. to books; mem. editorial bd.: Med. Decision Making, 1980-87, Jour. Gen. Internal Medicine, 1985-87, New Eng. Jour. Medicine, 1990—; cons. assoc. editor Am. Jour. Medicine, 1988—. Fellow ACP (clin. efficacy assessment subcom. 1985-92, chair 1990-92, bd. regents 1991—), Soc. for Gen. Internal Medicine (mem. coun. 1980-83, chmn. com. on health tech. evaluation 1982-87), Soc. for Med. Decision Making (pres. 1983-84, trustee 1980-83), Am. Fedn. Clin. Rsch., Assn. Am. Physicians, Assn. Profs. Medicine, Inst. Medicine, NAS, Alpha Omega Alpha. Home: Faraway Ln Hanover NH 03755-2312 Office: Darthmouth-Hitchcock Med Ctr Dept Lebanon NH 03756

SOYDEMIR, CETIN, geotechnical engineer, earthquake engineer; b. Izmir, Turkey, July 15, 1935; s. Sevket and Mediha Soydemir; m. Shirley Louise Franklin, Nov. 23, 1963; 1 child, Kemal. BS, Robert Coll., Istanbul, Turkey, 1957; MS, Harvard U., 1958, ME, 1960; PhD, Princeton U., 1967. Registered profl. engr., Mass., N.H. Assoc. prof. civil engring. Mid. East Tech. U., Ankara, Turkey, 1971-75, prof., 1975-76; vis. prof. civil engring. Rensselaer Poly. Inst., Troy, N.Y., 1976-77; prof. civil engring. U. Lowell, Mass., 1977-80; sr. specialist Haley & Aldrich, Inc., Cambridge, Mass., 1980-89, v.p., assoc., 1989—; Advisor, mem. Mass. Seismic Adv. Com., Boston, 1989—. Contbr. articles to profl. jours. Asst. pres. Mid. East Tech. U., Ankara, 1969-71. Terzaghi Libr. fellow Norwegian Geot. Inst., 1971. Fellow ASCE (editorial bd. Jour. Geotech. Engring. 1990—); mem. Earthquake Engring. Rsch. Inst., Can. Geotech. Soc., Deep Founds. Inst., Assn. Drilled Shaft Contractors, Boston Soc. Civil Engrs. (Clemens Hershel award 1970). Achievements include research and practice in the area of geotech. engring. and geotech. earthquake engring. Home: 55 Cannongate Rd Tyngsboro MA 01879 Office: Haley & Aldrich Inc 58 Charles St Cambridge MA 02141

SOYFER, VALERY NIKOLAYEVICH, biology educator; b. Gorky, RSFSR, USSR, Oct. 16, 1936; came to U.S., 1988; s. Nikolay Ilya Soyfer and Anna A. Kuznetsova; m. Nina I. Yakovleva, Aug. 12, 1961; children: Marina, Vladimir. BA in Agronomy, Timiryazev Agrl. Acad., Moscow, 1957; MA in Biophysics, Lomonosov State U., Moscow, 1961; PhD in Molecular Genetics, Kurchatov Inst. Atomic Energy, Moscow, 1964. Head Group Inst. Gen. Genetics, Moscow, 1966-70; dir. Lab. Molecular Genetics, Moscow, 1970-79; sci. dir. USSR Inst. Applied Molecular Biology and Genetics, Moscow, 1974-76; pres. Moscow Ind. U., 1985-88; disting. prof. Ohio State U., Columbus, 1988-90; Robinson prof. George Mason U., Fairfax, Va., 1990-93, disting. prof. genetics, 1993—; sci. sec. Coun. on Molecular Biology and Genetics, Moscow, 1972-80; mem. USSR Govtl. Coun. on Molecular Biology and Molecular Genetics, 1974-80; invited lectr. Halle-Wittenburg U., German Democratic Republic, 1975; prin. investigator USSR State Com. on Sci., 1972, 74, 78, NIH, 1990. Author: Molecular Mechanisms of Mutagenesis, 1969, History of Molecular Genetics, 1970, Molekulare Mechanismen der Mutagenese und Reparatur, 1976, Power and Science, History of the Crushing of Soviet Genetics, 1989; contbr. more than 200 articles on molecular genetics, biophysics and history of sci. to Nature, Mutation Rsch., Nucleic Acids Rsch., others. Chmn. Bd. Friends of Leningrad Ind. U., N.Y., 1990—; pres. USSR Amnesty Internat. Group, Moscow, 1983-88. Mem. USSR Soc. Geneticists and Breeders (founding), Gt. Britain Genetical Soc., USSR Biochem. and Microbiol. Soc., Internat. Soc. for History, Philosophy and Socal Studies of Biology (charter), European Culture Club (charter), Internat. Sci. Fedn. (bd. dirs. 1992—). Achievements include discovery of DNA Repair in higher plants; establishment of correlation between structural damages in DNA and mitagenesis rate in higher plants; co-development of the method of photofootprinting of DNA triplexes. Office: George Mason U 200 East Bldg Fairfax VA 22030-4444

SPAAN, WILLY JOSEPHUS, molecular virologist; b. Geleen, Limburg, Netherlands, Aug. 18, 1954; s. Leo J. and Tony (Beekes) S.; m. Annelies E. Souren, May 31, 1977; children: Annemieke, Maryke, Gonneke, Willem. Doctorandus cum laude, U. Utrecht, 1980, doctorate cum laude, 1984; postdoctoral, Salk Inst., La Jolla, Calif., 1984-85. Rsch. assoc. vet. faculty U. Utrecht, 1985-87, assoc. prof. vet. faculty, 1987-90; prof. med. faculty U. Leiden, 1990—; bd. dirs. study sect. NWO, 1988—, mem. study groups ICTV, 1988—. Contbr. articles to profl. jours. Mem. Soc. Gen. Microbiology, Am. Soc. of Microbiology, Am. Soc. Virology. Avocations: classical music, squash, gardening, biking. Home: Jacoba Van Beierenweg 7, 2215 KS Voorhout The Netherlands Office: Med Faculty Dept Virology, Rijnsburgerweg PO Box 320, 3200 AH Leiden The Netherlands

SPACU, PETRU GEORGE, chemistry educator; b. Charlottenburg, Germany, June 6, 1906; s. George and Margareta Alexandrina (Genoi) S.; m. Aurelia Cecilia Botez, Oct. 19, 1938. Grad. chemist, U. Cluj, Romania, 1929, assistent, 1930, dr., 1932; postgrad., Sorbonne U., Paris, 1934-36; postgrad, Munich Poly., 1936-37. Cert. chemist. Reader Lab. Inorganic Chemistry, Cluj, 1929-32, asst. prof., 1932-34; prof. Poly. Inst., Bucharest, 1937-55, Oil and Gas Inst., Bucharest, 1948-51; prof. faculty of chemistry U. Bucharest, 1955-72; cons. prof. Faculty of Chem. U., Bucharest, 1973—, Poly. U., Bucharest, 1992—; dep. dean Faculty of Ind. Chem.-Poly. Inst., Bucharest, 1944-48, dean, 1951-55; pro-rector U. Bucharest, 1966-69. Author: Chemistry of Coord. Compounds, 1969, rev. edit., 1974, Chemistry of Metals, 1978; contbr. 350 rsch. papers to profl. jours. Decorated Order of Labour, 1964, Star of SRR, 1966, Order of Sci. Merit, 1967. Mem. Romanian Acad. (corr., pres. chem. sect. 1990—), Acad. Scis. in Gôttingen, Mediterranean Acad. Scis.-Catania, IUPAC (inorganic chemistry divsn. 1967-71). Achievements include 6 patents for production of $SiCl_4$ from native diatomite, syntheses and properties of semiconductors, obtention of melting and degassing additives from indigenous raw materials, re-use of platinum from spent catalysts, obtention of boric acid from native ores. Home: Academia Romana, Naum Rimniceanu 2, 71102 Bucharest 71229, Romania Office: Romanian Acad, Calea Victoriei 125, Bucharest Romania

SPADA, MARIANNE RINA, medicinal chemist; b. Boston, July 1, 1955; d. John Bartholomew and Anna Dorothy (DePaula) S. BA in Chemistry, U. Mass., 1978; PhD in Medicinal Chemistry, U. R.I., 1987. Rsch. assoc. Sloan-Kettering Inst. Cancer Rsch., Rye, N.Y., 1987-89; rsch. fellow Inst. Phys. and Chem. Rsch., Wako-Shi, Japan, 1989-90; sr. rsch. chemist Rsch. Biochems. Inc., Natick, Mass., 1990—; instr. Framingham (Mass.) State Coll., 1991—; NSF/Sci. and Tech. Agy. rsch. fellow, Wako-shi, Japan, 1989. Mem. Am. Chem. Soc. (div. medicinal chemistry 1990—). Roman Catholic. Achievements include research in total synthesis of nucleoside antibiotic, synthesis of anticancer drug analogs of cyclophosphamide, nucleoside, neurochems. Home: 89 Trenton St East Boston MA 02128 Office: Rsch Biochems Inc 1 Strathmore Rd Natick MA 01760-2418

SPADE, GEORGE LAWRENCE, scientist; b. Sioux City, Iowa, Dec. 14, 1945; s. Walter Charles and LaVancha May (Green) S.; m. Carol Margaret Deaton, Mar. 14, 1966 (div. June 1985); children: Aaron Michael, Margaret. Mem. earthquake study group for China, U.S. Citizen Amb. Programs, 1989. Contbr. articles to profl. jours. Mem. AAAS, Am. Math. Soc., Math. Assn. Am., N.Y. Acad. Scis., Mensa. Avocations: poetry, painting, music. Home and Office: PO Box 2260 Columbia Falls MT 59912-2260

SPAEPEN, FRANS AUGUST, applied physics researcher, educator; b. Mechelen, Belgium, Oct. 29, 1948; came to U.S., 1971; s. Jozef F. M. and Ursula (Roppe) S.; m. Moniek Steemans, Aug. 21, 1973; children—Geertrui M., Elizabet U., Hendrik J.L. Burgerlijk Metaalkundig Ingenieur, U. Leuven, Belgium, 1971; Ph.D., Harvard U., 1975. IBM postdoctoral fellow Harvard U., Cambridge, Mass., 1975-77; asst. prof. applied physics Harvard U., 1977-81, assoc. prof., 1981-83, Gordon McKay prof. applied physics, 1983—; vis. prof. U. Leuven, Belgium, 1984; chmn. Gordon Conf. on Phys. Metallurgy, 1988; dir. Harvard Materials Rsch. Lab., 1990—; mem. solid state scis. com. NRC, 1990-93; Sauveur Meml. lectr., 1992. Co-editor: Solid State Physics; contbr. numerous articles to profl. jours., chpts. to books. Fellow Am. Phys. Soc. (chmn. divsn. materials physics 1992); mem. TMS, AIME, ASM, Materials Rsch. Soc. (councillor 1986-88, 90-92, co-chmn. fall meeting Boston 1990), Koninklijke Vlaamse Ingenieurs Vereniging, Böhmische Physikalische Gesellschaft, Orde van den Prince. Office: Harvard U Div Applied Scis 29 Oxford St Cambridge MA 02138-2901

SPAKE, ROBERT WRIGHT, safety engineer, consultant; b. Chgo., Jan. 3, 1923; s. Lloyd C. and Mary Perren (Wright) S.; m. Elizabeth Hird Spake, Sept. 11, 1964. BSc, Purdue U., 1942. Compliance officer N.Mex. OSHA, Santa Fe, 1977-81; cons. Safety Advisors, Inc., Albuquerque, 1981-82; safety officer N.Mex. Engring. Rsch. Inst., Albuquerque, 1982-88, rsch. engr. IV, 1988-91; sr. safety engr. Scientech Inc., Albuquerque, 1991—; adv. bd. Com. on Oil and Gas Drilling Guidelines in N.Mex., Santa Fe, 1980-81. Contbr. articles to profl. jours. 1st lt. USA, 1943-46, ETO. Mem. Am. Soc. Safety Engrs. (Scrivener award, 1990), Internat. Assn. Elec. Inspectors. Achievements include patents for Packaged Thermoplastics and Pre-Failure Tension Warning Device. Office: Scientech Inc 5121 Indian School Rd NE Albuquerque NM 87110-3931

SPALDING, GEORGE ROBERT, acoustician, consultant; b. Lancaster, Pa., Dec. 1, 1927; s. George Ness and Nellie (Wallace) S.; m. Marian Elizabeth Ehler, Aug. 23, 1952; children: Robert, Georgia, Edith. BS, Pa. State U., 1953, MS, 1955. Rsch. scientist Willow Run Rsch. Ctr., Ann Arbor, Mich., 1955; rsch. scientist Armstrong World Industries, Inc., Lancaster, Pa., 1955-63, sr. rsch. scientist, 1963-83, prin. scientist, 1983—. Office: Armstrong World Ind Inc PO Box 3511 Lancaster PA 17604-3511

SPALTRO, SUREE METHMANUS, chemist, researcher; b. Bangkok, Thailand, Aug. 8, 1953; came to U.S. 1975; d. Tia-Seng Sue and Haruthai Methmanus. BS, Mansfield State Coll., 1979; PhD, Lehigh U., 1988. Rsch. asst. Lehigh U., Bethlehem, Pa., 1984-87, rsch. assoc., 1988; chemist Unilever Rsch. U.S., Edgewater, N.J., 1989-91, rsch. scientist, 1991—. Contbr. articles to profl. jours. Mem. Am. Chem. Soc. Achievements include patent pending. Office: Unilever Rsch US 45 River Rd Edgewater NJ 07020

SPAMER, EARLE EDWARD, museum executive; b. Phila., July 28, 1952; s. Edward Lawrence and Jeannette Leda (Blouin) S.; m. Donna Alvin, 1977 (div. 1984). Student, Rutgers U., 1973-78; BA, Thomas Edison Coll., 1982. Tech. N.J. State Mus., Trenton, 1985-86; collection mgr. Acad. Natural Scis., Phila., 1986—. Author: Geology of the Grand Canyon: An Annotated Bibliography, Vols. 1-5, 1983-92, bibliography of Grand Canyon and Lower Colorado River From 1540, 1991, The Grand Canyon Fossil Record, 1992. Trustee Wagner Free Inst. Sci., Phila., 1988—. Mem. Geol. Soc. Am., Grand Canyon Natural Hist. Assn. (life), Ariz.-Nev. Acad. Sci. (life), Delaware Valley Paleontol. Soc. (bd. mem. 1982—). Office: Acad Natural Scis 1900 Benjamin Franklin Pkwy Philadelphia PA 19103-1195

SPANGLER, MILLER BRANT, science and technology analyst, planner, consultant; b. Stoyestown, Pa., Sept. 1, 1923; s. Elbert Bruce and Raye Isabel (Brant) S.; m. Claire Labin Kussart, Sept. 20, 1947; children: Daryl Claire, Philip Miller, Coreen Sue. BS with honors, Carnegie-Mellon U., 1950; MA, U. Chgo., 1953, PhD, 1956. Chem. engr. Gulf Rsch. Corp., Harmarville, Pa., 1950-51; assoc. engr. rsch. corp. IBM, Yorktown Heights, N.Y., 1956-60; mgr., market rsch. fed. systems div. IBM, Rockville, Md., 1960-63; program economist U.S. Agy. for Internat. Devel., Turkey, India, 1963-66; dir. ctr. for techno-econ. studies Nat. Planning Assn., Washington, 1966-72; chief, cost benefit analysis br. U.S. Atomic Energy Commn., Washington, 1972-75; spl. asst. policy devel. U.S. Nuclear Regulatory Commn., Washington, 1975-89; pres. Techno-Planning, Inc., Bethesda, Md., 1989—; mem. adv. bd. NSF Sea Grant Program, Washington, 1969, Environ. Profl. Jour., L.A., 1981-88. Author: New Technology and the Supply of Petroleum, 1956, New Technology and Marine Resource Development, 1970, The Role of Research and Development in Water Resources Planning, 1972, U.S. Experience in Environmental Cost-Benefit Analysis, 1980; contbr. numerous articles and papers to profl. jours. Recipient Planning Rsch. award Program Edn. and Rsch. in Planning U. Chgo., 1953. Mem. N.Y. Acad. Scis., Am. Assn. for the Advancement Sci., Nat. Assn. Environ. Profls., Soc. for Risk Analysis, Internat. Assn. for Impact Assessment., Tau Beta Pi. Republican. Methodist. Avocations: oriental gardening, traveling, photography, fishing. Home: 9115 McDonald Dr Bethesda MD 20817-1941

SPANIER, GRAHAM BASIL, university administrator, family sociologist; b. Capetown, South Africa, July 18, 1948; s. Fred and Rosadele (Lurie) S.; m. Sandra Kay Whipple, Sept. 11, 1971; children: Brian Lockwood, Hadley Alison. BS, Iowa State U., 1969, MS, 1971; PhD, Northwestern U., 1973. Assoc. dean, prof. in charge Pa. State U., University Park, 1973-82; vice provost, prof. SUNY, Stony Brook, 1982-86; provost, v.p. for acad. affairs Oreg. State U., Corvallis, 1986-91; chancellor U. Nebr., Lincoln, 1991—; social sci. analyst U.S. Bur.of Census, Washington, 1978; clin. intern Family Counseling Service, Evanston, Ill., 1972-73. Author of books and numerous articles to profl. jours. Pres., chmn. bd. dirs. Christian Children's Fund, Richmond, Va., 1985—; del. White House Conf. on Families, Washington, 1980; host Pub. Broadcast TV programs, 1973-76. Recipient Moran award Am. Home Econ. Assn., 1987; named Outstanding Young Alumnus Iowa State U., 1982; Am. Assn. Marriage and Family Therapy fellow, 1983—, Woodrow Wilson fellow, 1972. Mem. Nat. Coun. on Family Rels. (pres. 1987-88, Outstanding Grad. Student award 1972), Internat. Acad. Sex Rsch., Population Assn. Am., Am. Sociol. Assn. (family sect. chmn. 1983-84), Internat. Sociol. Assn. Democrat. Avocations: aviation, magic, athletics. Office: University of Nebraska-Lincoln Chancellors Office Administration Bldg 201 Lincoln NE 68588-0419

SPANOS, POL DIMITRIOS, engineering educator; b. Messini, Peloponnisos, Greece, Feb. 27, 1950; came to U.S., 1973; s. Dimitrios Constandin Spanos and Aicaterine Polychronis Bonaros; m. Olympia Constandin Critikou, Mar. 22, 1976; children: Demetri, Eudokia. Diploma in mech. engring., Nat. Tech. U., Athens, 1973; MS in Civil Engring., Calif. Inst. Tech., 1974, PhD in Applied Mechanics, 1976. Registered profl. engr., Tex., Greece. Rsch. asst. Calif. Inst. Tech., Pasadena, 1973-76, rsch. fellow, 1976-77; from asst. prof. to assoc. prof. U. Tex.-Austin, 1981-84, P.D. Henderson assoc. prof. engring., 1983-84; prof. mech. engring. and civil engring. Rice U., Houston, 1984-88, L.B. Ryon endowed chair in engring., 1988—; cons. on analytical and numerical applications of theory of dynamics and vibrations, worldwide. Author or editor 10 books and jour. issues devoted to dynamics and vibrations; mem. editorial bd. 8 jours.; editor or co-editor 2 primary jours. on mechanics. Recipient European award of sci. N.V. Phillipps Co., Eindhoven, Netherlands, 1969, Presdl. Young Investigator award in earthquake engring. NSF, 1984-89, Cert. merit McDonnell Douglas Astronautics Co., Houston, 1987; scholar Greek Scholarships Instn., 1968-72. Fellow ASME (participant tech. confs. and coms., Pi Tau Sigma Gold medal 1982, G.L. Larson Meml. award 1991), ASCE (participant tech. confs. and coms., W.L. Huber Civil Engring. Rsch. prize 1989, Alfred M. Freudenthal medal 1992); mem. Am. Acad. Mechanics, Earthquake Engring. Rsch. Inst., Internat. Assn. for Structural Safety and Reliability, Hellenic Profl. Soc. (sponsor scholarship com.). Office: Rice U PO Box 1892 Houston TX 77251

SPARKS, LARRY LEON, physicist; b. Flagler, Colo., Jan. 11, 1940; s. Lundie Leon Sparks and Ruby Ethyl (Dorsey) Hollenbaugh; m. Patricia Ruth Heid, Aug. 16, 1959; children: Lundy Lane, Jacquelyn Heidi Sparks Fesenmeyer. BS in Engring. Physics, U. Colo., 1962. Physicist Nat. Bur. Standrads (now Nat. Inst. Standards and Tech.), Boulder, Colo., 1961-84, group leader, 1984-88, div. chief, 1988-91, group leader, 1991—; chmn. Internat. Thermal Expansion Symposium, Boulder, 1989; organizer Internat. Thermaphys. Properties Conf., Boulder, 1991; chmn. local arrangements Internat. Cryogenic Materials Conf., Colorado Springs, Colo., 1983. Contbg. author: Materials at Low Temperatures, 1983, ASTM Spl. Publ., 1993; contbr. articles to tech. pubIs.; editor Internat. Jour. Thermophysics, 1991. Chmn. Good Samaritan Found., Longmont, Colo., 1971; moderator United Congl. Ch., Longmont, 1975-76. Recipient cert. of recognition NASA, 1989. Mem. ASTM, ASME, Am. Soc. Metals, Internat. Thermal Expansion Symposium (governing bd. 1988—). Democrat. Achievements include establishment of cryogenic calibrations for thermocouples. Office: Nat Inst Standards/Tech 325 Broadway Boulder CO 80303

SPARKS, SHERMAN PAUL, osteopathic physician; b. Toledo, Jan. 23, 1909; s. Earnest Melvin and Nancy Jane (Keller) S.; m. Helen Mildred Barnes, Aug. 1, 1930 (div. July 1945); 1 child, James Earl; m. Billie June Wester, Feb. 20, 1946 (dec. Apr. 1959); children: Randal Paul, Robert Dale; m. Joyce Marie Sparks, Jan. 23, 1965 (dec.); 1 child, David Paul. BS, U. Ill., 1932, MS, 1938; DO, Kirksville Coll. Osteopathy and Surgery, 1945. Diplomate Am. Bd. Osteo. Medicine. Tchr. high schs., Kincaid, Pesotum, Mt. Olive, Ill., 1930-42; intern Sparks Hosp. and Clinic, Dallas, 1945-46; pvt. practice, Rockwall, Tex., 1946—; team physician Rockwall High Sch., 1946-78; med. examiner Am. Cancer Soc., Rockwall, 1946-76. Coord. CD, Rockwall, 1946-80; chmn. Rockwall Centennial Assn., 1954, Rockwall County Rep. Com., 1964-78; pres. Rockwall PTA, 1961. Recipient Spirit of Tex. award, TV sta., Dallas, 1985; named Hon. State Farmer, Future Farmer's Am., 1980. Mem. Am. Coll. Gen. Practitioners, Tex. Osteo. Med. Assn. (chmn. chmn. 1946-54, numerous offices), Rockwall C. of C. (pres. 1961), SAR (pres. Plano chpt. 1989), Masons, Psi Sigma Alpha, Alpha Phi Omega. Avocations: travel, inventing. Home: 710 W Rusk St Rockwall TX 75087-3625 Office: Rockwall Osteo Clinic 106 N 2D St Rockwall TX 75087

SPARKS, WILLIAM SIDNEY, biologist; b. L.A., Jan. 14, 1951; s. William Franklin and Genevieve Nancy (Small) S.; m. Gretchen Louise Bender, June 7, 1980; children: Sarah Louise, Kristen Michelle, Claire Genevieve. BS in Psychobiology, U. Calif., Riverside, 1972. Biologist Sesame Found/GENESA, Fallbrook, Calif., 1973-74; biochemist PKS Rsch., Fallbrook, 1974-75; rsch. biologist Biogenics Internat., Houston, 1982-84; v.p. R&D Biotics Rsch. Corp., Houston, 1975—; mem. sci. bd. Life Scis., Austin, TEx., 1991-92. Mem. Internat. Soc. Free Radical Rsch., The Oxygen Soc., Am. Assn. Pharm. Scientists, Am. Chem. Soc., N.Y. Acad. Scis. Presbyterian. Achievements include 2 patents pending on enhanced uptake of emulsified Co enzyme Q10; development of antioxidant enzyme fortication for food mixtures; specialist-plant sources superoxide dismutase; genetic technologist performing study for Love Canal (Buffalo, N.Y.) which lead to removal of population from that site; tracing mineral effects on legumes. Office: Biotics Rsch 4850 Wright Rd # 150 Stafford TX 77477

SPARLIN, DON MERLE, physicist; b. Joplin, Mo, Mar. 29, 1937; s. Sidney Merle and Valetta Elaine (Darr) S.; m. Linda Kay Plake, May 29, 1959; children: Amber Megann, Jennifer Elaine, Jessica Nell, Nancy Elizabeth. BS Engring. Physics, U. Kans., 1959; PhD Physics, Northwestern U., 1964. Asst. prof. Case Western Reserve U., Cleve., 1964-68; from asst. prof. to assoc. prof. U. Mo., Rolla, 1968-91, prof., 1990—. Co-author Physics Lab Manual, various articles in profl. jours. Pres. World's Finest Rolla Town Band, 1986-92, World's Finest Rolla German Band, 1986-92. Mem. Am. Inst. Physics, Am. Assn. Physics Tchrs., Mo. Acad. Sci., Sigma Xi. Office: U Mo Physics Dept Rolla MO 65401

SPARROWE, ROLLIN D., wildlife biologist. BS in Wildlife Biology and Mgmt., Humboldt State Coll., 11964; MS, S.D. State U., 1966; PhD, Mich. State U., 1969. Asst. leader Mo. coop. wildlife rsch. unit Wildlife Mgmt. Inst., 1969-75, supr., 1976-79, chief divsn. coop. fish and wildlife rsch. units, 1979-83, chief divsn. wildlife rsch., 1983-84, chief office migratory bird mgmt., 1984-89, dep. asst. dir. refuges and wildlife, 1989-91; pres. Wildlife Mgmt. Inst., Washington, 1991—; leader planning team Coto Donana Nat. Park, Spain; negotiating team U.S. Can. and Mex.; speaker in field. Contbr. articles to profl. jours. Office: Wildlife Manag Inst Ste 725 1101 14th St NW Washington DC 20005*

SPATARO, VINCENT JOHN, electrical engineer; b. Jersey City, Apr. 11, 1953; s. Anthony Frank and Julia Mary (Starovich) S.; m. Linda Suzanne Lazar, July 19, 1992; children: Jaime A., David A. BS in Physics, Fairleigh Dickinson U., 1980; MS in Engring. Physics, Stevens Inst. Tech., 1984. Rsch. assoc. Philips Labs., Briarcliff Manor, N.Y., 1979-84; project engr. GEC Marconi, Wayne, N.J., 1984—; pres. Technology Solutions, Oakland, N.J., 1988—. Author conf. procs. in field. Recipient Innovative Tech. award Discover Mag., 1990. Mem. IEEE, Mensa. Achievements include patents for current limited quasi-resonant voltage converting power supply, magnetic tranformer switch and combination thereof with a discharge lamp. Office: Technology Solutions PO Box 393 Oakland NJ 07436

SPATER-ZIMMERMAN, SUSAN, psychiatrist, educator; b. Brookline, Mass.; m. Seth Allan Zimmmerman, Apr. 29, 1984. BA, U. Rochester, 1977; premed. cert., Columbia U., 1979; MD, Albert Einstein Coll. Medicine, 1983. Diplomate Nat. Bd. Med. Examiners. Intern in internal medicine Greenwich (Conn.) Hosp.-Yale U., 1983-84; resident in psychiatry N.Y. Med. Coll.-Westchester County Med. Ctr., Valhalla, 1984-87; pvt. practice, Garden City, N.Y., 1990-91, Manhasset, N.Y., 1991—; acting med. dir. Madonna Heights Svcs. Residential Treatment Facility Ctr., Dix Hills, N.Y., 1992-93; med. dir. Woodward Mental Health Ctr., Freeport, N.Y., 1993—; psychiatrist Madonna Heights Svcs. Residential Treatment Facility, Dix. Hills, N.Y., 1990; cons., med. dir. outpatient svcs. Seafield Ctr., Mineola, N.Y., 1991-92; mem. provisional attending staff North Shore Univ. Hosp.-Cornell U. Med. Coll. Manhasset, 1991; clin. instr. psychiatry Cornell U. Med. Coll., N.Y.C., 1991—. Mem. Am. Psychiatr. Assn., Am. Med. Womens Assn., Acad. Orgnl. and Occupational Psychiatry. Office: Community Drive Med Ctr 444 Community Dr Ste 208 Manhasset NY 11030-3889

SPATZ, KENNETH CHRIS, statistics educator; b. Tyler, Tex., Mar. 25, 1940; s. Kenneth Christopher and Mary E. (Harton) S.; m. Thea Siria, May 31, 1961; children: Mark C., Kenneth S., Elizabeth A. BA, Hendrix Coll., 1962; PhD, Tulane U., 1966. Asst. prof. U. of the South, Sewanee, Tenn., 1966-69; assoc. prof. U. Ark., Monticello, 1971-73; prof., chair Hendrix Coll., Conway, Ark., 1973—. Author: Basic Statistics: Tales of Distributions, 1st edit., 1976, 5th edit., 1993. Fellow U. Calif., Berkeley, 1969-71. Office: Dept of Psychology Hendrix Coll Conway AR 72032

SPEAKER, EDWIN ELLIS, retired aerospace engineer; b. Washington, June 24, 1928; s. Charles Richard and Clare Jean (Auerbach) S.; m. Donna June Hendershot, June 4, 1950; children: Gail Elaine Speaker Carr, Gregory Charles. BS, U.S. Naval Acad., 1950; MS, MIT, 1956; PhD, U. Mich., 1968. Commd. 2d lt. USAF, 1950, advanced through grades to lt. col., 1970; aerospace engr. various locations, 1950-70; rsch. staff scientist Rand Corp., Washington, 1970-73; div. chief Def. Intelligence Agy., Washington, 1974-80; project mgr. Goddard Space Flight Ctr. NASA, Greenbelt, Md., 1980-84; chief plans integration NASA Hdqs., Office of Space Sta., Washington, 1984-88; dir. space rsch. Fla. Inst. Tech., Melbourne, 1988-91; cons. Sci. and Tech. Corp., Satellite Beach, Fla., 1991—. Mem. Brevard County R&D Bd., Melbourne, Fla., 1991. Recipient Personal Communication award

Dir. CIA, 1980. Fellow AIAA, Am. Astronautical Soc. (pres. 1981-83). Home: 417 Red Sail Way Satellite Beach FL 32937

SPEARMAN, TERENCE NEIL, biochemist; b. Winnipeg, Man., Can., Dec. 26, 1948; s. Robert Andrew and Isabel (Sobotkowich) S.; m. Janet Steffensen, Sept. 1, 1973; children: Scott Craig, Heather Marie. BS, U. Man., 1970, PhD, 1979. Post doctoral fellow dept. biochemsitry W. Va. U., Morgantown, 1979-81, rsch. instr. dept. biochemistry, 1981-84, rsch. asst. prof. dept. biochemistry, 1984-90; dir. monoclonal antibody and protein analysis facility Mary Babb Randolf Cancer Ctr., Morgantown, 1988-90; dir. hybridoma lab. CBA Internat., Lexington, Ky., 1990-91, dir. biotherapy, 1991--; reviewer Jour. Dental Rsch., Winnipeg, 1984-90. Recipient Cand. Dental Rsch. Found award, 1978, New Investigator award Cystic Fibrosis Found., 1981-83. Mem. Am. Assn. for the Advancement Sci. Home: 4668 Spring Creek Dr Lexington KY 40515 Office: CBA Internat 600 Perimeter Dr Lexington KY 40517

SPEARS, RANDALL LYNN, civil engineer; b. Birmingham, Ala., May 27, 1960; s. Cecil Thomas and Rose Ann (Slovensky) S.; m. Linda Kaye Parker, June 11, 1983; children: Margaret Ann, Sara Beth. BSCE, U. Ala., Birmingham, 1982, MBA, 1989. Registered profl. engr., Ala., Ark., Ky., La., Mass., Tenn. Coop. student Mosher Steel Co., Birmingham, 1980-82; designer, detailer Mosher Steel Co., Houston, 1983-84; design engr. Rust Internat., Birmingham, 1984, Butler Mfg. Co., Kansas City, Mo., 1984-85; project engr. Butler Mfg. Co., Birmingham, 1985-89, regional engring. mgr., 1989—. Deacon Dora (Ala.) First Bapt. Chr., 1988—, Sunday sch. tchr., 1988—. Mem. NSPE, So. Bldg. Code Congress Internat. Office: Butler Mfg Co 931 Ave W Birmingham AL 35214

SPEAS, ROBERT DIXON, aeronautical engineer, aviation company executive; b. Davis County, N.C., Apr. 14, 1916; s. William Paul and Nora Estelle (Dixon) S.; m. Manette Lansing Hollingsworth, Mar. 4, 1944; children: Robert Dixon, Jay Hollingsworth. BS, MIT, 1940; grad., Boeing Sch. Aero., 1938. Aviation reporter Winston Salem Jour., 1934; sales rep. Trans World Airlines, 1937-38; engr. Am. Airlines, 1940-44, asst. to v.p., 1944-46, dir. maintenance and engring., cargo div., 1946-47, spl. asst. to pres., 1947-50; U.S. rep. A.V. Roe Can., Ltd., 1950-51; pres., chmn. bd. R. Dixon Speas Assocs., Inc. (aviation cons.), 1951-76; chmn., chief exec. officer Speas-Harris Airport Devel., Inc., 1974-76; chmn. bd., pres. Aviation Consulting, Inc., 1976-84; pres. PRC Aviation, 1984—; Mem. aeros. and space engring. bd. Nat. Research Council, 1980-84. Author: Airplane Performance and Operations, 1945, Pilots' Technical Manual, 1946, Airline Operation, 1949, Technical Aspects of Air Transport Management, 1955, Financial Benefits and Intangible Advantages of Business Aircraft Operations, 1989. Recipient 1st award Ann. Nat. Boeing Thesis Competition, 1937; research award; Am. Air Transport Assn., 1942, William A. Downes Airport Operators Coun. Internat. award, 1992. Fellow AIAA (treas. 1963-64, council 1963-64, chmn. ethics com. 1989—), Royal Aero. Soc., Soc. Automotive Engrs. (v.p. 1955, mem. council 1964-66); mem. ASME, Flight Safety Found. (bd. govs. 1958-71, 79-90, exec. com. 1979-90), Inst. Aero. Scis. (past treas.; council 1959-62, exec. com. 1962), Coll. Aeronautics (trustee 1967—), Soc. Aircraft Investigators, 1964—. Manhasset C. of C. (pres. 1962), Wings Club (sight lectr. 1992), Wings (pres. 1968-69, coun. 1966-71, 73-90, chmn. devel. com. 1989—), Skyline Country. Home: 4771 E Country Villa Dr Tucson AZ 85718-2640 Office: 6262 N Swan Rd Tucson AZ 85718-3600

SPECHT, ELIOT DAVID, physicist; b. N.Y.C., Oct. 17, 1959; s. Harry and Riva (Genfan) S. BA, U. Calif., 1981; PhD, MIT, 1987. Rsch. staff Oak Ridge (Tenn.) Nat. Lab., 1987—; mem. bd. govs. U. Nat. Lab. Ind. Collaborative Access Team, Advanced Photon Source, Argonne, Ill., 1991-92. Contbr. articles to profl. jours. Mem. Am. Phys. Soc. Office: Oak Ridge Nat Lab PO Box 2008 MS 6118 Oak Ridge TN 37831-6118

SPECHT, GORDON DEAN, retired petroleum executive; b. Garner, Iowa, June 3, 1927; s. Reuben William and Gladys (Leonard) S.; m. Cora Alice Emmert, May 24, 1952; children: Mary Ellen, Grant. BS in Chem. Engring., Iowa State U., 1950, MS in Chem. Engring., 1951; SM in Chem. Engring., MIT, 1954. Engr. Exxon Corp. Bayway Refinery, Linden, N.J., 1951-59, systemn services div. mgr., 1960-61, engring. services div. mgr., 1962-63, chem. coordinating div. mgr., 1964; mgr. systems dept. Exxon Corp.-Exxon Chem. Co., N.Y.C., 1965-70; sr. advisor communications and computer scis. dept. Exxon Corp., Florham Park, N.J., 1971-76, assoc. cons., 1977-85; retired, 1986. Patentee in field. Asst. scoutmaster Boy Scouts Am., Westfield, N.J., 1986—; sr. qualified observer Sperry Obs., Cranford, N.J., 1986—; celestial navigation instr. U.S. Power Squadrons, 1990—. With U.S. Army, 1945-46, 1st lt. C.E., 1952-53, Korea. Decorated Bronze Star. Mem. Am. Inst. Chem. Engrs., Amateur Astronomers, Inc., No. N.J. Power Squadron, MIT Club of No. N.J., Nat. Eagle Scout Assn., Tau Beta Pi, Phi Lambda Upsilon, Phi Kappa Phi, Tau Kappa Epsilon. Republican. Methodist. Avocations: astronomy, sailing, canoeing, swimming, bicycling. Home: 15 Normandy Dr Westfield NJ 07090-3431

SPECK, KENNETH RICHARD, materials scientist; b. Balt., Apr. 28, 1961; s. Fay Edward and Eleanor Alfreda (Czaczka) S.; m. Dawn Marie Carter, Mar. 19, 1983; 1 child, Joseph Paul. BS, Loyola Coll., 1983; MSE, Johns Hopkins U., 1987, PhD, 1989. Rsch. scientist KMS Fusion, Ann Arbor, 1989-91; rsch. assoc. Norton Co., Salt Lake City, 1992--. Contbr. articles to profl jours. Mem. Am. Vacuum Soc. Republican Roman Catholic. Achievements include patents infield. Home: 419 E 10185 S Sandy UT 84070 Office: Norton Co 2532 S 3270 West Salt Lake City UT 84119

SPECTOR, HARVEY M., osteopathic physician; b. Phila., July 10, 1938; s. Philip and Sylvia (Rischall) S.; m. Rochelle Fleishman, June 16, 1963; children: Jill, Larry. DO, Phila. Coll. Osteo. Medicine, 1963. Osteopathic physician Phila., 1964—; preceptor Hershey (Pa.) Med. Sch., 1987—, Phila. Coll. Osteopathic Medicine, 1989—; assoc. prof. medicine Med. Coll. Pa., 1991—. Recipient Humanitarian award, Chapel of Four Chaplains, Phila., 1984. Mem. Am. Osteo. Assn. (del.), Pa. Osteo. Med. Assn. (del.), Am. Acad. Osteo. Gen. Practitioners, Phila. County Osteo. Med. Soc., Abington Dolphins Aquatic Club (pres. 1984-86), B'nai B'rith. Jewish. Avocations: golf, swimming. Office: 1220 Cottman Ave Philadelphia PA 19111-3694

SPECTOR, JONATHAN MICHAEL, research psychologist, cognitive scientist; b. Pensacola, Fla.; s. Joseph and Dorothy Margaret (Givens) S. (dec.); children: Julia May, Samuel Dylan, David Elijah, Miriam Pearl. BS with honors, USAF Acad., 1967; PhD, U. Tex., 1978. Commd. 2d lt. USAF, 1967, advanced through grades to 1st lt., 1968; intelligence officer USAF, Philippines, 1967-71; resigned USAF, 1971; systems analyst systems devel. divsn. lab. IBM, Boulder, Colo., 1971-72; instr. philosophy U. Tex.-El Paso C.C., 1972-81; systems analyst Air Combat Maneuvering Instrumentation System Cubic Corp., Holloman AFB, N.Mex., 1981-83; sr. programmer Tower Telescope, Sunspot, N.Mex., 1983-84; prof. computer sci. Jacksonville (Ala.) State U., 1984-91; sr. scientist Instrnl. Design Br., Brooks AFB, Tex., 1991—; cons. Spector and Assocs., San Antonio, 1988—; organizer, participant Advanced Rsch. Workshop, NATO, Spain, 1992, Advanced Study Inst., Norway, 1993. Editor: Automating Instructional Design, 1992, The Promise and Potential of Distance Learning, 1992; contbr. to profl. publs. Vol. Shertu La'am, Kiryat Shemona, Israel, 1971, Kitty Stone Elem. Sch., Jacksonville, 1984-87. Mem. Am. Assn. Artificial Intelligence, Am. Edn. Rsch. Assn., Data Processing Mgmt. Assn. (chpt. pres. 1985-87), Assn. Devel. CBI Systems, Assn. Computing Machinery. Achievements include design of Advanced Instructional Design Advisor. Home: 3435 Ridge Country St San Antonio TX 78247-3456 Office: Armstrong Lab HRTC 7909 Lindbergh Dr Brooks AFB TX 78235-5352

SPEER, JAMES RAMSEY, developmental psychologist; b. Easton, Md., Nov. 5, 1936; s. James Ramsey Jr. and Irene Louise (Trippe) S.; m. Patsy Ruth Moore, Aug. 30, 1958; children: Hillary Ann, John Brookes. BA, U. Houston, 1976; PhD, Stanford (Calif.) U., 1980. Gen. mgr. OTS, Ltd., London, 1970-75; prof. Stephen F. Austin State U., Nacogdoches, Tex., 1980—; assoc. dean, 1993—. Editorial cons. to various scientific jours., 1980—; contbr. articles to profl. jours. Bd. dirs. Tex. Faculty Assn., Austin, 1991-92. Fulbright award Coun. for Internat. Exch. of Scholars, U.S.S.R., 1986, 87. Mem. Soc. for Rsch. in Child Devel., Am. Psychol. Soc. Democrat. Episcopalian. Achievements include rsch. in development of metacognition and theory of mind. Home: Rte 5 Box 3395 Nacogdoches TX 75961 Office: Stephen F Austin State U PO Box 13046 Nacogdoches TX 75962

SPEIER, JOHN LEO, JR., chemist; b. Chgo., Sept. 29, 1918; s. John L. and Mary Jane (Dickman) S.; m. A. Louise Kimmel, Oct. 21, 1944; children—Susan, Genevieve, Dorothy, Margaret, John L. III, Thomas J. B.Sc., St. Benedict's Coll., 1941; M.Sc., U. Fla., 1943; Ph.D., U. Pitts., 1947. Naval Stores research fellow U. Fla., 1941-43; research fellow Mellon Inst., Pitts., 1943; sr. fellow Mellon Inst., 1947-56; mgr. organic research Dow Corning Corp., Midland, Mich., 1956-69; scientist in corp. research Dow Corning Corp., 1969-75, sr. scientist in corp. research, 1975—. Contbr. numerous articles to profl. jours., 1950—; holder 100 patents prodn. organosilicon compounds and allied products. Named Indsl. Research and Devel. Scientist of Yr. Indsl. Research/Devel. mag., 1978. Mem. AAAS, Am. Chem. Soc. (Frederick Stanley Kipping award 1990), Sigma Xi. Office: Dow Corning Corp Dept Research Midland MI 48640

SPEIER, PETER MICHAEL, mathematics educator; b. Bklyn., Nov. 4, 1946; s. Fred A. and Herta (Katz) S.; m. Patricia Carol Johnson, Nov. 27, 1976. BS, SUNY, Cortland, 1968; MEd, U. Ga., 1971; MS, Adelphi U., 1975. Dept. chair J. L. Mann High Sch., Greenville, S.C., 1971-72; tchr. Long Beach (N.Y.) Jr. High Sch., 1972-75; tchr., coord. Largo High Sch., Upper Marlboro, Md., 1975-89; tchr. Oxon Hill (Md.) High Sch., 1989—; adj. assoc. prof. Prince George's Community Coll., Largo, 1976—; tchr. Community Based Classroom, Lanham, Md., 1990—. With U.S. Army, 1968-70, Vietnam. N.Y. State Regents scholar SUNY, 1964-68. Mem. NEA, VFW, DAV, Nat. Coun. Tchrs. Math., Am. Math. Assn. Two Yr. Colls., Md. Tchrs. Assn., Md. Coun. Tchrs. Math., Prince George's County Educators Assn. Avocation: travel. Home: 6613 Pine Grove Dr Suitland MD 20746-3527

SPEIGHT, JAMES GLASSFORD, research company executive; b. Murton, Eng., June 24, 1940; came to U.S., 1980; s. George Madison and Elizabeth (Glassford) S.; m. Sheila Elizabeth Stout, Dec. 28, 1963; 1 child, James. BSc in Chemistry with honors, Manchester U., Eng., 1961, PhD in Organic Chemistry, 1965. Research fellow Manchester U., 1965-67; research officer Research Council, Edmonton, Alta., Can., 1967-80; research assoc. Exxon Corp., Linden, N.J., 1980-84; chief sci. officer Western Rsch. Inst., Laramie, Wyo., 1984-89, chief exec. officer, 1990—; adv. com. Grant McEwan Community Coll., Edmonton, 1975-80; chmn. petroleum-natural gas research task force, Alta. Research Council, 1978-79; search com. V.P. for Research and Grad. Studies, U. Wyo., 1985; external mem. promotions com. U. Mosul, Iraq, 1985; thesis examiner, Indian Inst. Techn., Bombay, 1974, U. Mosul, 1976, 77, 78; vis. lect. petroleum sci., U. Mosul, Iraq, 1978; lectr. petroleum sci., U. Alberta, Edmonton, Can., 1976-80, U. Calgary, Alta., 1979-80. Editor Fuel Sci. and Tech. Internat., 1983—; refereed numerous jours., manuscripts; contbr. over 150 sci. articles to profl. jours. Fellow Royal Soc. Chemistry (chartered chemist), Chem. Inst. Can. (treas. Edmonton sect. 1971-78, editor newsletter 1975-77); mem. Am. Chem. Soc. (program com. petroleum div. 1981-91, bus. mgr. petroleum div. 1982-85), Sigma Xi. Office: Western Rsch Inst PO Box 3395 University Sta Laramie WY 82071

SPEIR, JEFFREY ALAN, biophysicist; b. San Diego, Mar. 6, 1967; s. William Franklin and Audrey Catherine (Gilchrist) S. BA in Chemistry, U. Calif. San Diego, 1989; doctoral studies, Purdue U., 1989—. Jr. rsch. tech Scripps Rsch. Inst., La Jolla, Calif., 1988-89; undergrad. rsch. asst. U. Calif., La Jolla, 1988-89, undergrad. teaching asst., 1989; grad. teaching asst. Purdue U., West Lafayette, Ind., 1990, grad. rsch. asst., 1989—. Contbr. articles to profl. jours. Named Grad. Teaching Assn. of Yr. Purdue U., 1991; biophysics tng. grantee NIH, 1989, Grad. Student travel grantee Am. Soc. Virology, 1992. Mem. Am. Crystallographic Assn., Am. Soc. for Virology. Republican. Achievements include purification of cold agglutinin antibodies; purification of bacterial expressed firefly luciferase enzyme; identification of first crystals of plant bromoviruses that diffract x-rays to high resolution. Home: 1308 Rochelle Dr Lafayette IN 47905-3051

SPEIRS, ROBERT FRANK, logistics engineer; b. Middletown, N.Y., May 31, 1942; s. Robert S.Y. and Ruth A. (Decker) S.; m. Frances C. Crawford, Nov. 3, 1962; children: Robert S., Donna M., Laura L. A of Electronics, C. C. of Air Force, 1976. Cert. profl. logistician. Enlisted USAF, 1960, advanced through grades to sr. master sgt., ret., 1980; logistics mgr. USAF, Randolph AFB, Tex., 1975-80; logistics engr. GTE, Needham, Mass., 1980-82, ILS mgr., 1982-83; prin. logistics engr. Honeywell, Mpls., 1983—. Vol. Am. Cancer Soc., Mpls., 1985—. Mem. Soc. Logistics Engrs. (sr., vice-chmn. dist. 6, chpt. 6 1985-87, chmn. 1987-89). Home: 14513 96th Ave N Maple Grove MN 55369 Office: Honeywell Inc MN15-2019 1625 Zarthan Ave Saint Louis Park MN 55416

SPELLACY, WILLIAM NELSON, obstetrician-gynecologist, educator; b. St. Paul, May 10, 1934; s. Jack F. and Elmyra L. (Nelson) S.; m. Lynn Larsen; children: Kathleen Ann, Kimberly Joan, William Nelson. B.A., U. Minn., 1955, B.S., 1956, M.D., 1959. Diplomate: Am. Bd. Ob-Gyn, subsplty. cert. in maternal and fetal medicine. Intern Hennepin County Gen. Hosp., Mpls., 1959-60; resident U. Minn., Mpls., 1960-63; practice medicine specializing in ob-gyn. Mpls., 1963-67, Miami, Fla., 1967-73, Gainesville, Fla., 1973-79, Chgo., 1979-88; prof., head dept. U. Ill. Coll. Medicine, Chgo., 1979-88; prof., chmn. dept. U. So. Fla. Coll. Medicine, Tampa, 1988—; prof. dept. obstetrics and gynecology U. Miami, 1967-73; prof., chmn. dept. U. Fla., 1974-79. Contbr. articles to med. jours. Mem. AMA, Am. Gynecol. Soc., Am. Assn. Obstetricians and Gynecologists, Am. Gynecol. and Obstet. Soc., Soc. Gynecol. Investigation, Am. Coll. Obstetricians and Gynecologists, Endocrine Soc., Am. Fertility Soc., Assn. Profs. Gynecology and Obstetrics, Am. Diabetes Assn., Perinatal Research Soc., South Atlantic Soc. Obstetrics and Gynecology, Central Assn. Obstetrics and Gynecology, Soc. Perinatal Obstetricians, Ill. Med. Soc., Inst. of Medicine. Episcopalian. Club: Rotary. Home: 845 Seddon Cove Way Tampa FL 33602-5704 Office: U South Fla Coll Medicine Dept OBGYN 4 Columbia Dr Ste 514 Tampa FL 33606-3568

SPELLMAN, GEORGE GENESER, SR., internist; b. Woodward, Iowa, Sept. 11, 1920; s. Martin Edward and Corinne (Geneser) S.; m. Mary Carolyn Dwight, Aug. 26, 1942; children: Carolyn Anne Spellman Rambow, George G. Jr., Mary Alice, Elizabeth Spellman-Chrisinger, John Martin Pile-Spellman, Loretta Suzanne Spellman Hoffman. B.S., St. Ambrose Coll., 1940; M.D., State U. Iowa, 1943. Diplomate Am. Bd. Internal Medicine. Intern Providence Hosp., Detroit, 1944; resident in internal medicine State U. Iowa, Iowa City, 1944-46; practice medicine specializing in internal medicine Mitchell, S.D., 1948-50, Sioux City, Iowa, 1950—; instr. Coll. Medicine U. S.D., 1975-77, now clin. assoc. prof. medicine; mem. staff St. Joseph Mercy Hosp., 1950-91, chief of staff, 1963; mem. staff St. Vincent Hosp., 1950-91, chief of staff, 1954-77, also bd. dirs.; mem. staff St. Luke's Med. Ctr.; clin. assoc. prof. medicine State U. Iowa; ret., 1991; med. vol. Project U.S.A. of AMA, 1991—; mem. Iowa State Bd. Med. Examiners, 1989-92, 92—; instr. schs. nursing St. Vincent Hosp., Luth. Hosp.; co-founder, pres. Siouxland Mental Assn., 1968-75; vol. cons. Siouxland Community Health Clinic; bd. dirs. Mid-Step Mentally Handicapped, Hospice of Siouxland, Marian Health Ctr., 1974-80, 88-91, also co-founder, 1st pres. chmn. dependency unit, founder renal dialysis unit, 1964. Contbr. articles to med. jours. Ordained deacon Cath. Ch., 1988; vol. cons. Siouxland Community Health Ctr., 1993—. Capt. M.C., U.S. Army, 1946-48. Decorated Knight of St. Gregory (Vatican); named Internist of Yr., Iowa Soc. Internal Medicine, 1987; recipient Laureate award Iowa Chpt. ACP, 1991, Humanitarian award Siouxland Community, 1991. Fellow ACP; mem. AMA, Am. Acad. Scis., Iowa State Med. Soc., Woodbury Med. Soc., Am. Soc. Internal Medicine, Iowa Soc. Internal Medicine, Am. Thoracic Soc., Iowa Thoracic Soc., Am. Heart Assn., Iowa Heart Assn., Am. Geriatric Soc., Alpha Omega Alpha. Home: 3849 Jones St Sioux City IA 51104-1447

SPENCE, DONALD POND, psychologist, psychoanalyst; b. N.Y.C., Feb. 8, 1926; s. Ralph Beckett and Rita (Pond) S.; m. Mary Newbold Cross, June 2, 1951; children: Keith, Sarah, Laura, Katherine. AB, Harvard U., 1949; PhD, Columbia U., 1955. Lic. psychologist, N.Y., N.J. From rsch. asst. to prof. psychology NYU, 1954-74; prof. psychiatry Robert Wood Johnson Med. Sch., Piscataway, N.J., 1974—; vis. prof. psychology Stanford (Calif.) U., 1971-72, Princeton (N.J.) U., 1975—, Louvain-le-Neuve, Louvain, Belgium, 1980, William Alanson White Inst., N.Y.C., 1992; mem. personality and cognition rsch. rev. com. NIMH, 1969-73 (recipient rsch. scientist award, 1968-72). Author: Narrative Truth and Historical Truth, 1982, The Freudian Metaphor, 1987; mem. editorial bd. Psychoanalysis and Contemporary Thought, Psychol. Inquiry, Theory and Psychology; contbr. articles to profl. jours. With U.S. Army, 1944-46, ETO. Mem. APA (pres. theoretical and philos. divsn. 1992-93), Am. Psychoanalytic Assn., N.Y. Acad. Sci., Sigma Xi. Democrat. Home: 9 Haslet Ave Princeton NJ 08540-4913 Office: Robert Wood Johnson Med Sch Piscataway NJ 08854

SPENCER, DONALD CLAYTON, mathematician; b. Boulder, Colo., Apr. 25, 1912; s. Frank Robert and Edith (Clayton) S.; m. Mary Jo Halley (div.); children: Maredith (dec.), Marianne; m. Natalie Robertson (dec.); 1 child, Donald Clayton Jr. BA, U. Colo., 1934; BS, MIT, 1936; PhD, Cambridge (Eng.) U., 1939, ScD, 1963. Instr. MIT, Cambridge, 1939-42; assoc. prof. Stanford (Calif.) U., 1942-46, prof., 1946-50, 63-68; assoc. prof. Princeton (N.J.) U., 1950-53, prof., 1953-63, 68-78, Henry Burchard Fine prof. emeritus, 1978—. Co-author: (with A.C. Schaeffer, monograph) Coefficient Regions for Schlicht Functions, 1950, (with M. Schiffer) Functions of Finite Riemann Surfaces, 1954, (with A. Kumpera) Lie Equations, vol. I: General Theory, 1972, (with H.K. Nickerson and N.E. Steenrod, textbook) Advanced Calculus, 1959. Recipient Bocher prize Am. Math. Soc., 1948, Nat. medal of sci. Pres. of U.S., 1989, George Norlin award U. Colo., 1990. Fellow Am. Acad. Arts and Scis.; mem. NAS. Home: 943 County Rd 204 Durango CO 81301-8547

SPENCER, EDGAR WINSTON, geology educator; b. Monticello, Ark., May 27, 1931; s. Terrel Ford and Allie Belle (Shelton) S.; m. Elizabeth Penn Humphries, Nov. 26, 1958; children: Elizabeth Shawn, Kristen Shannon. Student, Vanderbilt U., 1949-50; B.S., Washington and Lee U., 1953; Ph.D., Columbia U., 1957. Lectr. Hunter Coll., 1954-57; mem. faculty Washington and Lee U., 1957—, prof. geology, head dept., 1962—, Ruth Parmly prof.; Pres. Rockbridge Area Conservation Council, 1978; NSF sci. faculty fellow, New Zealand and Australia; dir. grant for humanities and pub. policy on land use planning Va. Found., 1975; dir. grant Petroleum Research Fund, 1981-82; leader field trip Cen. Appalachian Mountains Internat. Geol. Congress, 1989. Author: Basic Concepts of Physical Geology, 1962, Basic Concepts of Historical Geology, 1962, Geology: A Survey of Earth Science, 1965, Introduction to the Structure of the Earth, 1969, 3d edit., 1988, The Dynamics of the Earth, 1972, Physical Geology, 1983, Geologic Maps, 1993. Recipient Va. Outstanding Faculty award Va. Coun. of Higher Edn., 1990. Fellow Geol. Soc. Am., AAAS; mem. Am. Assn. Petroleum Geologists (dir. field seminar on fold and thrust belts 1987, 88-91), Am. Inst. Profl. Geologists, Am. Geophys. Union, Nat. Assn. Geology Tchrs., Yellowstone-Bighorn Rsch. Assn., Phi Beta Kappa (hon.), Sigma Xi. Home: PO Box 1055 Lexington VA 24450-1055

SPENCER, FRANCIS MONTGOMERY JAMES, pharmacist; b. St. John's, Antigua, Mar. 11, 1943; came to U.S., 1974; s. Stanley M. and Sarah Jane Elizabeth (Spencer) James; m. Jean V. Cole, May 9, 1981; children: David, Frances, Weslie. BS in Pharmacy, Northeastern U., Boston, 1982. Registered pharmacist, Mass., N.H., Fla.; registered cons. pharmacist, pharmacy preceptor, Fla. Sr. dispensing druggist Holberton Hosp., Antigua, 1968-73, lectr. in pharmacy, 1970-73; pharmacist, intern Mount Auburn Hosp., Cambridge, Mass., 1978-82; staff pharmacist Centro-Asturiano Hosp., Tampa, Fla., 1986-90, Dr.'s Hosp., Tampa, Fla., 1991—; pharmacy mgr. Eckerd Drug Co., Tampa, Fla., 1983—; co-founder, pres., chief exec. officer Spenscott, Inc., Bronx, N.Y., 1989—; assoc. mem. Delta Search, Inc., Tampa, 1987—. Mem. profl. adv. panel Drug Topics mag., Oradell, N.J., 1987—. Fellow Am. Soc. Cons. Pharmacists (registered cons. Fla.), Internat. Biog Assn. (life, dept. dir. gen.); mem. AAAS, Fla. Pharmacy Assn., Am. Biog. Inst. Inc. (dep. gov., hon. mem. rsch. bd. advisors), N.Y. Acad. Scis., Am. Soc. Pharmacy Law, Mass. State Pharmacy Assn., N.H. Pharmacy Assn., Am. Coll. Heatlh Care Adminstrs. Methodist. Avocations: reading, travel, classical music. Home: PO Box 245 Mango FL 33550-0245

SPENCER, LEWIS VANCLIEF, retired physicist; b. Hillsdale, Mich., Nov. 29, 1924; s. William Gear and Dorothy (Burns) S.; m. Elizabeth Williams, Sept. 20, 1948; children: Dorothy Spencer Wagener, Betty Spencer Schiele, Carl, Mary Ellen Spencer Goree, Robert. BA, Franklin Coll., 1945; PhD, Northwestern U., 1948. Rschr. radiation transport Nat. Bur. of Standards, Washington, 1948-57; cons. Nat. Bur. of Standards, Gaithersburg, Md., 1957-68; sect. chief radiation physics Nat. Bur. of Standards, Washington, 1960-61; rsch. radiation shielding Nat. Bur. of Standards, Gaithersburg, 1969-84, ret., 1984; asst. prof. physics and math. Ottawa (Kans.) U., 1957-60, prof. physics and math., 1961-68; adj. prof. physics Hopkinsville C.C. of U. Ky., 1988—. Author: Structure Shielding Against Fallout Gamma Rays from Nuclear Detonations, 1980; contbr. numerous articles to profl. jours. Mem. Gov.'s Nuclear Energy Adv. Coun., Kans., 1964-69, chmn. civil def. subcom., 1967. Recipient Meritorious Svc. award U.S. Dept. Commerce, 1952, 62, 72, Disting. Svc. award U.S. Office Civil and Def. Mobilization, 1960, L.H. Gray medal for outstanding contbn. to radiation sci. Internat. Commn. on Radiation Units and Measurements, 1969, Disting. Jayhawker award Kans., 1964, Community Svc. award C. of C., 1964. Mem. AAAS. Baptist. Achievements include research in penetration and diffusion of x-rays, theory of electron penetration, theory of cavity ionization. Home: PO Box 87 Hopkinsville KY 42241-0087

SPENCER, PETER SIMNER, neurotoxicologist; b. London, Nov. 30, 1946; U.S. citizen; married; 2 children. BSc, U. London, 1968, PhD in Pathology, 1971. Rsch. asst. Nat. Hosp. Nervous Disorders U. London, 1968-70, rsch. fellow Royal Free Hosp. Sch. Medicine, 1970-71; fellow pathology Albert Einstein Coll. Medicine, 1971-73, asst. prof., 1973-81, assoc. prof. neurosci., 1977-83, prof. neurosci., 1983—, assoc. prof. pathology, dir. Inst. Neurotoxicology, 1979—; dir. Oreg. Health Sci. U.; cons. Nat. Inst. Occupational Safety & Health, 1976-77, EPA, 1977—; chmn. adv. bd. Jour. Neurotoxicology, 1978—; mem. adv. bd. Rutgers U. Toxicology Program, 1981, Howe & Assocs., 1985, Peripheral Nerve Repair & Regeneration, 1983, mem. bd. toxicol. and environ. health hazards NAS, 1984, Safe Drinking Water Com., 1985; sec. Third World Med. Rsch. Found., 1985—. Assoc. editor Jour. Neurocytology, 1977—. Fellow Joseph P. Kennedy Jr. Found., 1974-76. Mem. AAAS, Am. Assn. Neuropathologists (Weil award 1976), Am. Soc. Cell Biologists, Anatomic Soc. Gt. Britain & Ireland, Brit. Neuropathology Soc., World Fedn. Neurology, Royal Coll. Pathologists, Pan-Am. Neuroepidiology Found. (hon.). Achievements include research in cellular relationships in the nervous system and the effects of neurotoxic chemicals. Office: Oregon Hlth Sciences Univ 3181 SW Sam Jackson Park Rd L606 Portland OR 97201*

SPENCER, THOMAS C., mathmatician; b. Dec. 24, 1946. BA, U. Calif., Berkeley, 1968; PhD, NYU, 1972. Fellow Harvard U., Cambridge, Mass., 1974-75; assoc. prof. Rockefeller U., 1975-77; prof. Rutgers U., 1978-80; prof. Courant Inst. Math. NYU, 1980-86; prof. Inst. Advanced Study Princeton (N.J.) U., 1986—; vis. mem. Inst. for Advanced Study, Princeton U., 1972-74. Recipient Dannie H. Heineman Math. Physics prize Am. Phys. Soc., 1991; Sloan fellow. Office: Inst for Advanced Study Sch of Mathematics Olden Lane Princeton NJ 08540*

SPENCER, WILLIAM STEWART, radiologic technologist; b. Salem, Mass., July 29, 1949; s. Hugh Morrison and Virginia (Lord) S. AB, Bates Coll., 1971; postgrad., Salem State Coll., 1972-73; AS, North Shore Community Coll., Beverly, Mass., 1976. Cert. Am. Registry of Radiologic Technologist. Maintenance man Sterling Last Co., Lynn, Mass., 1972; nursing asst. Lynn Hosp., 1973-75, staff technologist, 1976-77; supr. CT scan Atlanticare Med. Ctr., Lynn, 1977—. Mem. Am. Soc. of Radiologic Technologists, Mass. Soc. of Radiologic Technologists. Republican. Avocations: reading, sword and sorcery, science fiction.

SPENCER-GREEN, GEORGE THOMAS, medical educator; b. Lima, Ohio, Sept. 8, 1946; s. Ormond George and Elizabeth Ann (Thomas) Spencer-Green; m. Linda Ann Jonas, Oct. 31, 1980; 1 child, Elizabeth Ann. BA, Oberlin (Ohio) Coll., 1969; MD, Columbia U., 1974. Diplomate Am. Bd. Internal Medicine, Am. Bd. Rheumatology, Am. Bd. Allergy and Immunology (Diagnostic Lab. Immunlology). Instr. in medicine Coll. of Medicine, U. Cin., 1980, asst. prof. medicine, pathology and lab. medicine, 1980-84; assoc. prof. medicine Dartmouth Med. Sch., Hanover, N.H., 1985—; assoc. med. dir. diagnostic immunology lab. U. Hosp., Cin., 1982-

84; chief of immunology Vets. Hosp., Cin., 1982-84, White River Junction, Vt., 1985—. Author: (chpt.) Textbooks on Rheumatic Diseases, 1985—; contbr. articles to profl. jours. Mem. adv. bd. N.H. chpt. Arthritis Found., Concord, 1988—. Fellow ACP, Am. Coll. Rheumatology; mem. Am. Fedn. Clin. Rsch., N.Y. Acad. Sci. Avocations: canoeing, photography, tennis. Home: RR 1 Box 522H Norwich VT 05055-9526 Office: Dartmouth-Hitchcock Med Ctr One Medical Center Dr Lebanon NH 03756

SPENGLER, KENNETH C., meteorologist, professional society administrator; b. Harrisburg, Pa.. BA, Dickinson Coll., 1936; MS, MIT; DSc (hon.), U. Nev., 1966. Exec. dir. Am. Meteorol. Soc., Boston, 1946-88, emeritus exec. dir., 1988—. Recipient Cleveland Abbe Disting. Svc. award Am. Meteorol. Soc., 1993. Mem. AAAS, AIAA, Am. Geophys. Union, Coun. Engring. & Sci. Soc. Office: American Meteorological Society 45 Beacon St Boston MA 02108-3693*

SPERBER, DANIEL, physicist; b. Vienna, Austria, May 8, 1930; came to U.S., 1955, naturalized, 1967; s. Emanuel and Nelly (Liberman) S.; m. Ora Yuval, Nov. 29, 1963; 1 son, Ron Emanuel. M.Sc., Hebrew U., 1954; Ph.D., Princeton U., 1960. Tng. and rsch. asst. Israel Inst. Tech., Haifa, 1954-55, Princeton U., 1955-60; sr. scientist, rsch. adviser Ill. Inst. Tech. Rsch. Inst., Chgo., 1960-67; assoc. prof. physics Ill. Inst. Tech., 1964-67, Rensselaer Poly. Inst., Troy, N.Y., 1967-72; prof. Rensselaer Poly. Inst., 1972—; Nordita prof. Niels Bohr Inst., Copenhagen, 1973-74, NATO research fellow, vis., prof., 1974-77; vis. prof. G.S.I., Darmstadt, Fed. Republic Germany, 1983; sr. Fulbright research scholar, Saha Inst. Nuclear Physics, Calcutta, India, 1987-88. Contbr. over 100 sci. papers to profl. jours. Served to capt. Israeli Army, 1948-51. Fellow Am. Phys. Soc.; mem. Israel Phys. Soc., N.Y. Acad. Scis., Sigma Xi. Jewish. Home: 1 Taylor Ln Troy NY 12180-7162 Office: Rensselaer Poly Inst Dept Physics Troy NY 12181

SPERBER, IRWIN, sociologist; b. N.Y.C.; s. Samuel and Emily (Brody) S.; children: Janette, Claudia. BA in Polit. Sci., Bklyn. Coll., 1959; MA in Sociology, CUNY, 1961; PhD in Sociology, U. Calif., Berkeley, 1975. Asst. prof. sociology Stanislaus State Coll., Turlock, Calif., 1968-70, Trent U., Peterborough, Ont., 1970-72; assoc. prof. sociology SUNY, New Paltz, 1972—; instr., advisor Prison Edn. program, SUNY, New Paltz, 1972-76; lectr. in field. Author: Fashions in Science, 1990; contbr. articles to profl. jours.; editor Berkeley Jour. of Sociology, 1970. Woodrow Wilson fellow, 1962. Achievements include research in psychoneuroimmunology and the sociology of medicine. Office: SUNY New Paltz Dept Sociology Faculty Tower 516 New Paltz NY 12561

SPERELAKIS, NICHOLAS, physiology and biophysics educator, researcher; b. Joliet, Ill., Mar. 3, 1930; s. James and Arestia (Kayadakis) S.; m. Dolores Martinis, Jan. 28, 1960; children: Nicholas Jr., Mark, Christine, Sophia, Thomas, Anthony. BS in Chemistry, U. Ill., 1951, MS in Physiology, 1955, PhD in Physiology, 1957. Teaching asst. U. Ill., Urbana, 1954-57; instr. Case Western Res. U., Cleve., 1957-59, asst. prof., 1959-66, assoc. prof., 1966; prof. U. Va., Charlottesville, 1966-83; Joseph Eichberg prof. physiology Coll. Medicine U. Cin., 1983—, chmn. dept., 1983-93; cons. Natural Products Scis., Salt Lake City, 1988—, Carter Wallace, Inc., Cranbury, N.J., 1988—; vis. prof. U. St. Andrews, Scotland, 1972-73, U. San Luis Potosi, Mex., 1986; Rosenblueth prof. Centro de Investigacion y Avanzades, Mex., 1972; mem. sci. adv. com. several internat. meetings, editorial bd. numerous sci. jours. Editor: Handbook of Physiology on Circulation: The Heart, 1979, Physiology and Pathophysiology of the Heart, 1984, 2d edit., 1988, 3rd edit., 1993, Calcium Antagonists: Mechanisms of Action on Cardiac Muscle and Vascular Smooth Muscle, 1984, Cell Interactions and Gap Junctions, vols. I and II, 1989, Frontiers in Smooth Muscle Research, 1990, Ion Channels in Vascular Smooth Muscle and Endothelial Cells, 1991, Physiology, 1992, Principles of Cell Physiology and Biophysics, 1993; assoc. editor Circ. Rsch., 1970-75, Molecular Cellular Cardiology; contbr. articles. Lectr. Project Hope, Peru, 1962. Sgt. USMC, 1951-53, Res., 1953-59. Recipient Disting. Alumnus award Rockdale (Ill.) Pub. Schs., 1958; U. Cin. Grad. fellow, 1989; NIH grantee, 1959—. Mem. Am. Psychol. Soc. (chair steering com., sect. 1981-82), Biophys. Soc. (coun. 1990-93), Am. Soc. Pharmacology and Exptl. Therapeutics, Internat. Soc. Heart Rsch. (coun. 1980-89, 92—), Am. Heart Assn. (established investigator 1961-66), Am. Hellenic Ednl. Progressive Assn. (pres. Charlottesville chpt. 1980-82), Ohio Physiol. Soc. (pres. 1990-91), Phi Kappa Phi. Democrat. Greek Orthodox. Avocations: ancient coins, stamp collecting. Office: U Cin Coll Medicine 231 Bethesda Ave Cincinnati OH 45267

SPERLING, GEORGE, cognitive scientist, educator; s. Otto and Melitta Sperling. B.S. in Math., U. Mich., 1955; M.A. in Psychology, Columbia U., 1956; Ph.D. in Psychology, Harvard U., 1959. Rsch. asst. in biophysics Brookhaven Nat. Labs., Upton, N.Y., summer 1955; rsch. asst. in psychology Harvard U., Cambridge, Mass., 1957-59; mem. tech. rsch. staff Acoustical and Behavioral Rsch. Ctr., AT&T Bell Labs., Murray Hill, N.J., 1958-86; prof. psychology and neural sci. NYU, N.Y.C., 1970-92; disting. prof. cognitive scis. U. Calif., Irvine, 1992—; instr. psychology Washington Sq. Coll., NYU, 1962-63; vis. assoc. prof. psychology Duke U., spring 1964; adj. assoc. prof. psychology Columbia U., 1964-65; acting assoc. prof. psychology UCLA, 1967-68; hon. rsch. assoc. Univ. Coll., U. London, 1969-70; vis. prof. psychology U. Western Australia, Perth, 1972, U. Wash., Seattle, 1977; vis. scholar Stanford (Calif.) U., 1984; mem. sci. adv. bd. USAF, 1988-92. Gomberg scholar U. Mich., 1953-54; Guggenheim fellow, 1969-70. Fellow AAAS, Am. Acad. Arts and Sci., Am. Psychol. Assn. (Disting. Sci. Contbn. award 1988), Optical Soc. Am.; mem. NAS, Assn. for Research in Vision and Opthalmology, Ann. Interdisciplinary Conf. (founder, organizer 1975—), Eastern Psychol. Assn. (bd. dirs. 1982-85), Soc. for Computers in Psychology (steering com. 1974-78), Psychonomic Soc., Soc. Exptl. Psychologists, Soc. for Math. Psychology (chmn. 1983-84, exec. bd. 1979-85), Phi Beta Kappa, Sigma Xi. Office: Dept Cognitive Sciences SS Tower Univ of California Irvine CA 92717

SPERRY, ROGER WOLCOTT, neurobiologist, educator; b. Hartford, Conn., Aug. 20, 1913; s. Francis B. and Florence (Kraemer) S.; m. Norma G. Deupree, Dec. 28, 1949; children: Glenn Tad, Jan Hope. AB, Oberlin Coll., 1935, M.A., 1937, D.Sc. (hon.), 1982; Ph.D., U. Chgo., 1941, D.Sc. (hon.), 1976; D.Sc. (hon.), Cambridge U., 1972, Kenyon Coll., 1979, Rockefeller U., 1980; Oberlin Coll., 1982. Rsch. fellow Harvard and Yerkes Labs., 1941-46; asst. prof. anatomy U. Chgo., 1946-52, sect. chief Nat. Inst. Neurol. Diseases of NIH, also asso. prof. psychology, 1952-53; Hixon prof. psychobiology Calif. Inst. Tech., 1954-84, Bd. Trustees prof. emeritus, 1984—; rsch. brain orgn., neurospecificity, split-brain rsch., hemispheric specialization, consciousness revolution. Author: Science and Moral Priority, 1983, Nobel Prize Conversations, 1985; contbr. articles to profl. jours. Recipient Oberlin Coll. Alumni citation, 1954, Howard Crosby Warren medal Soc. Exptl. Psychologists, 1969, Disting. Sci. Contbn. award Am. Psychol. Assn. 1971, Calif. Scientist of Year award Calif. Mus. Sci. and Industry, 1972, award Passano Found., 1973, Albert Lasker Basic Med. Rsch. award, 1979, co-recipient William Thomas Wakeman Rsch. award Nat. Paraplegia Found., 1972, Claude Bernard sci. journalism award, 1975, Disting. Rsch. award Internat. Visual Literacy Assn., 1979, Wolf Found. prize in medicine, 1979, Nobel prize in physiology or medicine, 1981, Realia award Inst. for Advanced Philos. Rsch., 1986, Mentor Soc. award, 1987, Nat. medal of sci., 1989; William James fellow Am. Psychol. Soc.,1990. Fellow NAS, AAAS, Am. Acad. Arts & Scis., Am. Philos. Soc.(Karl Lashley award), Am. Neurol. Assn., Royal Soc. (fgn. mem.), Pontifical Acad. Scis., USSR Acad. Scis. (fgn. mem.). Office: Calif Inst Tech Div Biology 156-29 1201 E. California St Pasadena CA 91125

SPESER, PHILIP LESTER, social scientist, consultant; b. Buffalo, N.Y., Mar. 17, 1951; s. David and Theodora (Cowen) S.; m. Nancy Jean Parafinczyk, Nov. 27, 1980; children: Arendt, Ariel. BA in Polit. Sci. and Journalism, Case Western Res. U., 1973; JD, SUNY, Buffalo, 1980, PhD in Polit. Sci., 1981. Spl. asst. for sci. and tech. Fedn. Am. Scientists, Washington, 1980-81; pres. Foresight Sci. and Tech., Port Townsend, Wash., 1981—; Wash. rep. Soc. Am. Archeology, 1982-89; exec. dir. Nat. Coalition for Sci. and Tech., Washington, 1985-89; session chair Nat. Biotech. Edn. Sharing Conf., Madison, Wis., 1991; cons. Office of Gov. State of N.Y., 1980; adj. prof. dept. anthropology Am. U., Washington, 1988; adv. panelist on univ. small bus. ctrs., NSF, Washington, 1985. Author: The Defense-Space Market, 1985, The Politics of Science, 1987, Technology Transfer Handbook, 1990, The Federal Laser and Optics Market, 1990, Small Business Guide to Federal Research and Development Funding, 1991, others; author, editor numerous reports, articles. Founding chair Glen Echo (Md.) Park Found., 1987-88; bd. dirs. Jefferson County Edn. Found., Port Townsend, 1991—, v.p., 1993—; bd. dirs., exec. com. Jefferson County Econ. Devel. Coun., Port Townsend, 1991—, v.p., 1992—; lead lobbyist Small Bus. Innovation Devel. Act of 1982; founding pres. Olympic Penninsula Found., 1993. Grantee USDA, Small Bus. Adminstrn., Dept. Energy, NSF, others. Mem. AAAS, Tech. Transfer Soc. (bd. dirs., chair task force on nat. tech. transfer policy 1988-91), Bar Assn. D.C. Democrat. Achievements include key role in many pieces of legislation; development of expert system for technology transfer services for small rural manufacturers; establishment of vocational-technical and science education programs in rural communities. Office: Foresight Sci and Tech 1200 W Simms Way Port Townsend WA 98368

SPETH, JAMES GUSTAVE, United Nations executive, lawyer; b. Orangeburg, S.C., Mar. 4, 1942; s. James Gustave and Amelia St. Clair (Albergotti) S.; m. Caroline Cameron Council, July 3, 1964; children: Catherine Council, James Gustave, Charles Council. BA summa cum laude, Yale U., 1964, LLB, 1969; MLitt, Oxford U., 1966. Bar: D.C. 1969. Law clk. to Justice Hugo L. Black U.S. Supreme Ct., 1969-70; sr. staff atty. Natural Resources Def. Council, Washington, 1970-77; mem. Council Environ. Quality, Washington, 1977-79, chmn., 1979-81; prof. law Georgetown U. Law Ctr., Washington, 1981-82; pres. World Resources Inst., Washington, 1982-93; adminstr. UN Devel. Program and Under Sec. Gen., UN, 1993—; founded World Rsch. Inst. Ctr. Internat. Devel. and Environ.; organized Western Hemisphere Dialogue environ. and devel., 1990; chaired U.S. Task Force internat. devel. and environ. security. Contbr. articles to profl. jours.; speaker in field. Bd. dirs. Com. for Environ. and Energy Study Inst., Environ. Law Inst., Keystone Ctr., Nat. Resources Def. Coun., S.C. Coastal Conservation League, Global Environ. Forum-Japan, Woods Hole Rsch. Ctr. Recipient Resources Def. award Nat. Wildlife Fedn., 1976, Barbara Swain award of honor Nat. Resources Coun. Am., 1992; named to Global 500 Honor Role United Nations Environ. Program, 1988; Rhodes scholar, 1964-66. Mem. Coun. on Fgn. Rels. (N.Y.C.), China Coun. for Internat. Coop. on Environment and Devel. Episcopalian. Home: 3237 Arcadia Pl NW Washington DC 20015-2329 Office: UNDP 1 United Nations Plz New York NY 10017

SPIBERG, PHILIPPE FREDERIC, research engineer; b. Neuilly sur Seine, France, Dec. 27, 1957; s. Charles and Arlette (Bramy) S. PhD in Physics, U. Paris VI, 1988. Rsch. assoc. U. Calif., Irvine, Calif., 1989-90; rsch. scientist Norton Diamond Filth, Northboro, Mass., 1991--. With French Army, 1982-84. mem. Optical Soc. Am. Office: Norton Diamond Film 14 Mason Irvine CA 92718

SPICER, JOHN AUSTIN, physicist; b. Rock Springs, W.Va., Sept. 25, 1930; s. Ernest Marvin and Ruth (Stevens) S.; m. Erika Gruendig, 1959; children: Cynthia, Michael, Marilynn. BS, U. Wyoming, 1956, MS, 1957; PhD, U. Freiburg, Germany, 1962. Mathematician Geotech. Corp., Laramie, Wyo., 1956-57; physicist Goodyear Aerospace Corp., Litchfield, Ariz., 1962-63; head engr. Aerojet Gen. Corp., Azusa, Calif., 1963-64; mathematical analyst North Am. Aviation Info. Systems Div., Downey, Calif., 1973-76; program mgr. Chrysler Space Systems Div., New Orleans, La., 1966-68; sr. research mathematician U. Dayton Research Inst., Ohio, 1968-70; ops. research analyst U. McCall Printing Corp. Systems Dept., Dayton, 1970-71; mathematician Systems Dyamics Br. AF Flight Dynamics La., 1971-72; physicist Radar and Microwave Tech. Br. Af Avionics lab., 1972-74, Analysis and Evaluation Br. AF Avionics Lab., 1974-89; physicist tech group, target recognition br. AF Avionics Lab., Dayton, Ohio, 1989—. Contbr. articles on neural networks, wavelets and fractal methodology to profl. jours.; inventor "exact stability," which gives a dead beat response, that is, always stable and controllable. Home: 4666 N St Rt 235 Conover OH 45317-9601 Office: US Avionics Lab WL AARA 2 Dayton OH 45433-7001

SPIEGEL, HERBERT, psychiatrist, educator; b. McKeesport, Pa., June 29, 1914; s. Samuel and Lena (Mendlowitz) S.; m. Natalie Shainess, Apr. 24, 1944 (div. Apr. 1965); children: David, Ann; m. Marcia Greenleaf, Jan. 29, 1989. B.S., U. Md., 1936, M.D., 1939. Diplomate: Am. Bd. Psychiatry. Intern St. Francis Hosp., Pitts., 1939-40; resident in psychiatry St. Elizabeth's Hosp., Washington, 1940-42; practice medicine specializing in psychiatry N.Y.C., 1946—; attending psychiatrist Columbia-Presbyn. Hosp., N.Y.C., 1960—; faculty psychiatry Columbia U. Coll. Physicians and Surgeons, 1960—; adj. prof. psychology John Jay Coll. Criminal Justice, CUNY, 1983—; mem. faculty Sch. Mil. Neuropsychiatry, Mason Gen. Hosp., Brentwood, N.Y., 1944-46. Author: (with A. Kardiner) War Stress and Neurotic Illness, 1947, (with D. Spiegel) Trance and Treatment: Clinical Uses of Hypnosis, 1978; subject of book: (by Donald S. Connery) The Inner Source: Exploring Hypnosis with Herbert Spiegel, M.D.; Mem. editorial bd.: Preventive Medicine, 1972; Contbr. articles to profl. jours. Mem. profl. advisory com. Am. Health Found.; mem. pub. edn. com., smoking and health com. N.Y.C. div. Am. Cancer Soc.; mem. adv. com. Nat. Aid to Visually Handicapped. Served with M.C. AUS, 1942-46. Decorated Purple Heart. Fellow Am. Psychiat. Assn., Am. Coll. Psychiatrists, Am. Soc. Clin. Hypnosis, Am. Acad. Psychoanalysis, Internat. Soc. Clin. and Exptl. Hypnosis, William A. White Psychoanalytic Soc., N.Y. Acad. Medicine, N.Y. Acad. Scis.; mem. Am. Orthopsychiat. Assn., Am. Psychosomatic Soc., AAAS, AMA, N.Y. County Med. Soc. Office: 19 E 88th St New York NY 10128-0500

SPIEGEL, RONALD JOHN, electrical engineer; b. Cleve., May 31, 1942; s. John Franz and Lila (Zion) S. BEE, Ga. Inst. Tech., 1964; PhD, U. Ariz., 1970. Antenna engr. Boeing Co., Seattle, 1972-74; rsch. engr. IIT Rsch. Inst., Chgo., 1974-76; sr. rsch. engr. S.W. Rsch. Inst., San Antonio, 1976-80; supervisory electronics engr. U.S. EPA, Research Triangle Park, N.C., 1980-87, sr. project engr., 1987—. Contbr. articles to profl. jours. Mem. IEEE, Bioelectromagnetics Soc., Sigma Xi. Achievements include patent on fuzzy logic integrated control method and apparatus to improve motor efficiency. Home: 19 Kyleway Dr Chapel Hill NC 27514 Office: US EPA Mail Drop 63 Research Triangle Park NC 27711

SPIELBERGER, LAWRENCE, physician, educator; b. Budapest, Hungary, Mar. 3, 1911; came to U.S., 1936; s. Frank and Frida (Spitzer) S.; m. Blanche Berman (div.); m. Greta Spanierman. MD, U. Rome, 1935. Chief dept. anesthesiology VA, Togus, Maine, 1949-51, Providence, 1951-53; acting chief dept. anesthesiology VA, Manhattan, N.Y., 1953-55; clin. assoc. prof. anesthesiology NYU, N.Y.C., 1972, ret., 1984. Capt. AUS, 1943-46, PTO. Jewish. Home: 201 E 77th St New York NY 10021

SPIELMAN, ANDREW IAN, biochemist; b. Tirgu Mures, Romania, June 23, 1950; arrived in Can., 1982; s. Joseph and Rachel (Sebestyén) S.; m. Kathy Szabó, Dec. 15, 1977; 1 child, Robert-Dan. DMD, U. Medicine and Pharmacy, Tirgu Mures, 1974; cert. specialist in oral surgery, Technion, Haifa, Israel, 1982; MSc, U. Toronto (Can.), 1985, PhD in Oral Biology and Biochemistry, 1988. Asst. mem. Monell Chem. Senses Ctr., Phila., 1988-89; assoc. clin. prof. U. Pa. Sch. Dental Medicine, Phila., 1989-92; affiliate mem. Monell Chem. Senses Ctr., Phila., 1989—; asst. prof. oral medicine and pathology NYU Coll. of Dentistry, N.Y.C., 1989; assoc. dir. rsch. NYC Coll. of Dentistry, N.Y.C., 1992—; faculty student rsch. advisor NYU Coll. of Dentistry, N.Y.C., 1990—. Author: (with others) Encyclopedia of Human Biology, 1991; contbr. articles to Brain Rsch., Chem. Senses, Jour. Dental Rsch., Archives of Oral Biology. Republican fellow Univ. of Medicine and Pharmacy, Tirgu-Mures, 1972, U. Toronto Open fellow, 1983, Med. Rsch. fellow Med. Rsch. Coun. can., 1983-88. Mem. Internat. Assn. for Dental Rsch. (faculty student advisor), AAAS, N.Y. Acad. Sci., Assn. for Chemoreception Scis., Sigma Xi. Jewish. Achievements include research on the molecular basis of bitter taste mechanisms; on the interaction of saliva and taste; on identification of sweat-odor binding proteins in human axillary secretion. Office: NYU Coll of Dentistry 345E 24th St New York NY 10010-4086

SPIER, RAYMOND ERIC, microbial engineer; b. Manchester, England, Dec. 3, 1938; s. Jack Leo and Fanny (Sher) S.; m. Merilyn Gail Wolfe, Apr. 4, 1971; children: Emmet Hyam, Avron Dyan, Dania Arielle. MA, Oxford U., 1961, PhD, 1964; diploma in biochem. engring., U. Coll., London, 1965. Chartered engr., biologist. Process engr. British Glues and Chems., London, 1965-68; sr. process engr. Westreco (Nestle), Marysville, Ohio, 1968-69, Merck Sharpe & Dohme, West Point, Pa., 1970-73; prin. scientific officer Animal Virus Rsch. Inst., Pirbright, United Kingdom, 1973-83; prof., dept. head U. Surrey, Guildford, United Kingdom, 1983-90. Editor: Vaccine, 1982—, Enzymes and Microbial Tech., 1986—, Animal Cell Biotechnology, 1985—. Recipient various awards SERC, Wolfson Found., others. Fellow Inst. Chem. Engrs., Inst. Biology, Royal Soc. Arts; mem. Inst. Dirs., European Soc. for Animal Cell Tech. (founder, first chmn. 1975—). Achievements include research in scale-up of anchorage dependent animal cell culture systems, control biology of animal cells in culture to enhance utility to industry, theoretical understanding of animal cell technology systems. Office: U Surrey, Guildford England

SPIES, JACOB JOHN, health care executive; b. Sheboygan, Wis., Jan. 27, 1931; s. Jacob Alfred and Julia Effie (Wescott) S.; m. Donna Dolores Jerale, June 17, 1954; children: Gary, Joni, Shari. BBA, U. Wis., 1955. Asst. v.p. health care systems Wausau (Wis.) Ins. Cos., 1972-77, v.p. mgmt. systems, 1977-79; dep. dir. Health Policy Inst. Boston U., 1979-85; pres., chief exec. officer Co-Med, Inc., Columbus, Ohio, 1984-85; sr. v.p. PARTNERS Nat. Health Plans, Irving, Tex., 1985-90; chmn. PARTNERS Health Plans of Colo., Denver, 1986-90; ptnr., pres. The Furst Group, Dallas, 1990—. Co-author: A Corporations Experience with IPA-HMO, 1981, Health Care Cost Containment, 1983. Sgt. U.S. Army, 1952-54, Korea. Decorated Bronze Star, 1953. Mem. Group Health Assn. Am. (com. mem. 1988, 91), Am. Med. Care and Rev. Assn. (bd. dirs.), Nat. Assn. Employers on Health Care Actions (chmn.), LaCima Club, Denton Country Club. Episcopalian. Home: 1492 Rockgate Rd Roanoke TX 76262-9804 Office: The Furst Group 5215 N O'Connor Blvd Ste 920 Irving TX 75039

SPIES, PHYLLIS BOVA, information services company executive; b. Syracuse, N.Y., Nov. 10, 1949; d. Ralph Anthony and Elizabeth Margaret (Caputo) Bova; m. John William Spies, June 28, 1980; children: Fletcher, Logan. BA in Art History, SUNY, Cortland, 1971; MLS in Libr. and Info. Sci., Syracuse U., 1972. Libr. systems analyst Ohio Coll. Library Ctr., Columbus, 1973-78; mgr. libr. systems analysis OCLC Online Computer Libr. Ctr., Dublin, Ohio, 1978-83, div. v.p., 1983-89, v.p. internat., 1989-92, v.p. mktg. and sales, 1992—; trustee Maps Micrographic Preservation Svc., Bethlehem, Pa., 1990—. Contbr. articles to profl. jours. Mem. Columbus Coun. World Affairs. Fellow The Gaylord Co., 1971. Mem. ALA, Internat. Fedn. Libr. Assns., Dublin Women in Bus. Avocations: gardening, cooking. Office: OCLC Online Computer Libr Ctr 6565 Frantz Rd Dublin OH 43017-5308

SPIESMAN, BENJAMIN LEWIS, mechanical design engineer; b. Geneva, Ohio, Sept. 16, 1961; s. David Leo and Mary Jane (Sandella) S.; m. Patrece Anne Siembor, Sept. 9, 1990. AAS in Mech. Engring., Kent State U., 1981; BSME, Ohio U., 1986. Engring. tech. Cleve. Electric Illuminating Co. Perry Nuclear Power Plant, Perry, Ohio, 1981-84, reliability engr., 1984-86, mech. design engr., 1986—. Mem. ASME (assoc.). Independent. Roman Catholic. Home: 7510 Lake Shore Blvd Madison OH 44057

SPIKES, JOHN JEFFERSON, SR., forensic toxicologist, pharmacologist; b. Grand Prairie, Tex., Jan. 30, 1929; s. George W. and Madge (Ballard) S.; m. Marilyn Ruth Tomlimson, Apr. 17, 1949; children: Juli Spikes Jensen, John J. Jr., James M., Jay S., Jerry D. BS, Union Coll., 1951; MS, George Washington U., 1959; PhD, U. Tex., Galveston, 1971. Diplomate Am. Bd. Forensic Toxicologists. Instr. U. Tex. Med. Br., Galveston, 1971-72; toxicologist Pathology Labs. Houston, 1972-73, Biochemical Procedures, North Hollywood, Calif., 1973-75; chief toxicologist Ill. Dept. Pub. Health, Chgo., 1975-85; forensic toxicologist Nat. Med. Svcs., Willow Grove, Pa., 1985—; mem. drug abuse adv. com. FDA, Washington, 1984-87; cons. U.S. Postal Svc., 1982-84, Chgo. Transit Authority, 1978-86; presenter in field. Co-author: (chpt.) Progress in Clinical Pathology, vol. III, 1981; contbr. articles to profl. jours. Cpl. U.S. Army Med. Corp 1952-54. Fellow Am. Acad. Forensic Scientists; mem. Soc. Forensic Toxicologists, Assn. Clin. Scientists, Sigma Xi. Office: Nat Med Svcs Inc 2300 Stratford Ave Willow Grove PA 19090

SPILHAUS, ATHELSTAN FREDERICK, JR., oceanographer, association executive; b. Boston, May 21, 1938; s. Athelstan F. and Mary (Atkins) S.; m. Sharon Brown, June 11, 1960; children—Athelstan F. III, Ruth Emily, Mary Christina. S.B. in Chem. Engring., MIT, 1959, S.M. in Geology and Geophysics, 1960, Ph.D. in Oceanography, 1965. Cert. meeting profl. Phys. scientist U.S. Govt., Washington, 1965-67; asst. exec. dir. Am. Geophys. Union, Washington, 1967-70, exec. dir., 1970—; bd. dirs. Renewable Natural Resources Found., Washington. editor newspaper EOS. Chmn. Conv. Liaison Council, Washington, 1981-82. Fellow AAAS, Washington Acad. Sci.; mem. Am. Soc. Limnology and Oceanography, Council Biology Editors, Am. Inst. Physics (mem. gov. bd. 1988—), Geol. Soc. Am., Philos. Soc. Washington (pres. 1982-83), Am. Geophys. Union, Geol. Soc. Washington (2nd v.p. 1975), Soc. Exploration Geophysicists, Am. Soc. Assn. Execs., Assn. Am. Pubs. (div. exec. com. 1980-82), Assn. Earth Sci. Editors (dir. 1972-78, pres. 1977), Council Engring. and Scientific Soc. Execs. (dir. 1976-82, pres. 1980-81), Internat. Un. Geodesy Geophysics (mem. fin. com. 1987—). Clubs: Cosmos (Washington, pres. 1992-93); Chesapeake Yacht (Md.). Home: 10900 Picasso Ln Rockville MD 20854-1710 Office: Am Geophys Union 2000 Fla Ave NW Washington DC 20009

SPILMAN, TIMOTHY FRANK, utilities engineer; b. Valley City, N.D., Mar. 31, 1961; s. James Frank and Juliann Frances (Zirbes) S. AA in Engring., Bismarck State Coll., 1980; BSEE, U. N.D., 1983, BSCE, 1985; M in Mgmt., U. Mary, Bismarck, N.D., 1990. Engr.'s asst. Basin Electric Power Coop., Bismarck, 1981-83; electric engr. Montana Dakota Utilities Co., Mobridge, S.D., 1984-89, mktg. supr., 1989-90; sr. gas engr. Montana Dakota Utilities Co., Bismarck, 1990—. Author: Mobridge Jaycee Individual Development, 1987 (1st place S.D. 1987). Actor, set worker Mobridge Showboat Theatre, 1986-88; project chmn. Mobridge Beef-an-Fun com. 1986-89; mem. Mobridge Retail Com., 1989. Mem. IEEE, NSPE (Young Engr. award for N.D. 1992), N.D. Soc. Profl. Land Surveyors, Mobridge Jaycees (v.p. 1986-88, bd. dirs., chpt. Jaycee of Yr 1987), KC (3d degree), Moose. Avocations: bowling, tennis, golf, house remodeling. Home: 812 W Avenue A Bismarck ND 58501-2445 Office: Montana-Dakota Utilities Co 909 Airport Rd Bismarck ND 58504-6114

SPINDEL, ROBERT CHARLES, electrical engineering educator; b. N.Y.C., Sept. 5, 1944; s. Morris Tayson and Isabel (Glazer) S.; m. Barbara June Sullivan, June 12, 1966; children—Jennifer Susan, Miranda Ellen. B.S.E.E., Cooper Union, 1965; M.S., Yale U., 1966, M.Phil., 1968, Ph.D., 1971. Postdoctoral fellow Woods Hole Oceanographic Instn, Mass., 1971-72, asst. scientist, 1972-76; assoc. scientist Woods Hole Oceanographic Instn., Mass., 1976-82; sr. scientist Woods Hole Oceanographic Instn, Mass., 1982-87, chmn. dept. ocean engring., 1982-87; dir. applied physics lab. U. Wash., 1987—. Contbr. articles to profl. jours.; patentee on underwater nav. Recipient A.B. Wood medal Brit. Inst. Acoustics, 1981, Gano Dunn medal The Cooper Union, 1989. Fellow IEEE (assoc. editor jour. 1982—), Acoustical Soc. Am. (exec. coun. 1985-86), Marine Tech. Soc. (pres. elect 1991-93, pres. 1993—). Democrat. Jewish. Avocations: automobile restoration, hiking. Home: 14859 SE 51st St Bellevue WA 98006-3515 Office: U Wash Applied Physics Lab 1013 NE 40th St Seattle WA 98105-6698

SPINDEL, WILLIAM, chemistry educator, scientist, educational administrator; b. N.Y.C., Sept. 9, 1922; s. Joseph and Esther (Goldstein) S.; m. Sara Lew, 1942 (div. 1966); children: Robert Andrew, Lawrence Marshall; m. Louise Phyllis Hoodenpyl, July 30, 1967. B.A., Bklyn. Coll., 1944; M.A., Columbia U., 1947, Ph.D., 1950. Jr. scientist Los Alamos Lab, Manhattan Dist., 1944-45; instr. Poly. Inst., Bklyn., 1949-50; assoc. prof. State U. N.Y., 1950-54; research assoc., vis. prof. Columbia, 1954-57, vis. prof., sr. lectr., 1962-74; assoc. prof., then prof. Rutgers U., 1957-64; prof., chmn. dept. chemistry Belfer Grad. Sch. Sci., Yeshiva U., 1964-74; exec. sec., office chemistry and chem. tech. NAS-NRC, 1974-81, also staff dir. bd. on chem. scis. and tech., prin. staff officer commn. phys. scis., math. and resources, 1982-90, sr. cons., 1990—; vis. Am. scientist, Yugoslavia, 1971-72. Contbr. articles to profl. jours. Served with AUS, 1943-46. Recipient prof. staff award NRC, 1985; Guggenheim fellow, 1961-62; Fulbright Research scholar,

1961-62. Fellow AAAS; mem. Am. Chem. Soc. Club: Cosmos. Achievements include patents in field. Home: 6503 Dearborn Dr Falls Church VA 22044-1116

SPINELLI, JULIO CESAR, biomedical engineer; b. La Plata, Argentina, Dec. 25, 1956; came to the U.S., 1989; s. Jorge Horacio and Elina Celia (Pinha) S.; m. Debora Fabiana Valentinuzzi, July 11, 1981; children: Natalia Elizabeth, Nicolas Rene, Horacio Eugenio. M in Telecommunications, Fed. U. La Plata, 1980; PhD in Bioengring., U. Tucuman, 1989. Jr. fellow Fed. Rsch. Coun., Tucuman, 1981-83, sr. fellow, 1983-85, staff mem., 1985-89; prin. scientist Cardiac Pacemakers Inc., St. Paul, 1989—; instr. U. Tucuman, 1981-83, chief instr., 1983-85, adj. prof., 1985-88, tenure prof., 1988-91; adv. bd. mem. Fed. U. Parana, 1986-88. Author: Defibrillation, 1986; contbr. articles to profl. jours. Recognized for Best Contbn. to Cardiology Catalina B. de Baron Cordic Found., 1985. Mem. Argentinian Soc. Bioengring. (bd. dirs., sec. 1986-89). Achievements include four patents pending. Office: Cardiac Pacemakers Inc 4100 Hamline Ave N Arden Hills MN 55112

SPINGARN, CLIFFORD LEROY, internist, educator; b. Bklyn., May 8, 1912; s. Alexander and Eleanor (Trinz) S.; m. Eleanor Harrison, June 9, 1937; children: John Harrison, Alexandra. AB, Columbia U., 1933, MD, 1937. Diplomate Am. Bd. Internal Medicine. Intern Mt. Sinai Hosp., N.Y.C., 1937-40, asst. attending physician, 1946-63, assoc. attending physician, 1963—, chief parasitology clinic, 1956-80; attending physician Doctors Hosp., N.Y.C., 1968—, chmn. com. on continuing med. edn., 1976—; pvt. practice internal medicine N.Y.C., 1946—; instr. pharmacology Columbia, 1940-42; asst. clin. prof. preventive medicine NYU, 1956-68; assoc. clin. prof. medicine Mt. Sinai Sch. Medicine, 1966-83, lectr. in medicine, 1983—. Author numerous papers. Trustee Milton Helpern Libr. Legal Medicine, 1982—; bd. dirs. N.Y. Faculty Continuing Med. Edn., 1982-86. Lt. (j.g.) to lt. comdr. M.C., USNR, 1942-46; lt. comdr. ret. res. Recipient Disting. Svc. award Doctors Hosp., 1987. Fellow ACP, N.Y. Acad. Medicine; mem. AAAS, N.Y. Soc. Tropical Medicine, Am. Soc. Tropical Medicine and Hygiene, Am. Soc. Parasitologists, Am. Soc. Internal Medicine, Med. Soc. County N.Y. (chmn. grievance com. 1969-72, chmn. bd. censors 1978-80, pres. 1981, trustee 1982-87, Disting. Svc. award 1986), Gerontol. Soc. Am., Soc. Internal Medicine County New York (pres. 1965-67), N.Y. State Soc. Internal Medicine, N.Y. Cardiological Soc. (bd. dir. 1971-73), Phi Beta Kappa, Sigma Xi, Alpha Omega Alpha. Home: 201 E 79th St New York NY 10021-0830 Office: 66 E 80th St New York NY 10021-0223

SPINILLO, PETER ARSENIO, energy engineer; b. Sant' Arsenio, Salerno, Italy, Jan. 18, 1952; naturalised in 1967; s. Umberto and Rosaria (DiMatteo) S.; m. Linda Ann Crudo, July 22, 1973 (div. Apr. 1984); children: Umberto Anthony, Dawn Marie. BS in Archtl. Tech. cum laude, N.Y. Inst. of Tech., 1991, MS in Energy Mgmt., 1992. Advanced cert. environ. mgmt. Owner SPA Home Improvement, Brentwood, N.Y., 1973-79; constrn. mgr. Calderone Constrn. Corp., Valley Stream, N.Y., 1979-86; energy engr. Environ.-Energy Contracting, Brentwood, 1986—; cons. Energy Assocs., Bklyn., 1991. Sec. Vol. Fire Dept., Brentwood, fireman. Recipient Bravery citation Brentwood Fire Dept., 1979, Suffolk Legislation, 1979. Mem. Assn. Energy Engrs., Cogeneration Inst., Environ. Inst. Achievements include rsch. in renewable energy, photovoltaics, cogeneration, biomass, waste to energy conversion. Home: 1021 Candlewood Rd Brentwood NY 11717 Office: Environ-Energy Contracting 1021 Candlewood Rd Brentwood NY 11717

SPINRAD, HYRON, astronomer; b. N.Y.C., Feb. 17, 1934; s. Emanuel B. and Ida (Silverman) S.; m. Bette L. Abrams, Aug. 17, 1958; children—Michael, Robert, Tracy. A.B., U. Calif. at Berkeley, 1955, M.A., 1959, Ph.D. (Lick Obs. fellow), 1961. Studied galaxies U. Calif. at Berkeley, 1960-61; planetary atmospheres work Jet Propulsion Lab., Pasadena, Calif., 1961-63; investigation atmospheres of coolest stars U. Calif. at Berkeley, 1964-70. Mem. Am. Astron. Soc., Astron. Soc. Pacific. Spl. research water vapor on Mars, molecular hydrogen on Jupiter, Saturn, Uranus and Neptune, temperature measurements on Venus atmosphere, spectra of galaxies and near-infrared observations, 71-72, location of faint radio galaxies, redshifts of galaxies, galaxy evolution and cosmology, 1973, spectroscopic observations of volatile gases in comets. Home: 7 Ketelsen Dr Moraga CA 94556 Office: Univ California Dept Astronomy Berkeley CA 94720

SPINRAD, ROBERT JOSEPH, computer scientist; b. N.Y.C., Mar. 20, 1932; s. Sidney and Isabel (Reiff) S.; m. Verna Winderman, June 27, 1954; children: Susan Irene, Paul Reiff. B.S., Columbia U., 1953, M.S. (Bridgham fellow), 1954; Ph.D. (Whitney fellow), MIT, 1963. Registered profl. engr., N.Y. Project engr. Bulova Research & Devel. Lab., N.Y.C., 1953-55; sr. scientist Brookhaven Nat. Lab., Upton, N.Y., 1955-68; v.p. Sci. Data Systems, Santa Monica, Calif., 1968-69; v.p. programming Xerox Corp., El Segundo, Calif., 1969-71; dir. info. scis. Xerox Corp., 1971-76, v.p. systems devel., 1976-78; v.p. research Xerox Corp., Palo Alto, 1978-83; dir. systems tech. Xerox Corp., 1983-87, dir. corp. tech., 1987-92, v.p. tech. analysis and devel., 1992—; cons. Contbr. articles to profl. jours. Mem. IEEE, SCM, Nat. Acad. Engring., Sigma Xi, Tau Beta Pi. Achievements include patents in field. Office: Xerox Corp 3333 Coyote Hill Rd Palo Alto CA 94304-1314

SPIRO, MELFORD ELLIOT, anthropology educator; b. Cleve., Apr. 26, 1920; s. Wilbert I. and Sophie (Goodman) S.; m. Audrey Goldman, May 27, 1950; children: Michael, Jonathan. B.A., U. Minn., 1941; Ph.D., Northwestern U., 1950. Mem. faculty Washington U., St. Louis, 1948-52, U. Conn., 1952-57, U. Wash., 1957-64; prof. anthropology U. Chgo., 1964-68; prof., chmn. dept. anthropology U. Calif., San Diego, 1968—; bd. dirs. Social Sci. Research Council, 1960-62. Author: (with E.G. Burrows) An Atoll Culture, 1953, Kibbutz: Venture in Utopia, 1955, Children of Kibbutz, 1958, Burmese Supernaturalism, 1967, Buddhism and Society: A Great Tradition and Its Burmese Vicissitudes, 1971, Kinship and Marriage in Burma, 1977, Gender and Culture: Kibbutz Women Revisited, 1979, Human Nature and Culture, 1993; editor: Context and Meaning in Culture Anthropology, 1965, Oedipus in the Trobriands, 1982, Burmese Brother or Anthropological Other?, 1992. Fellow Am. Acad. Arts and Scis., Nat. Acad. Scis.; mem. Am. Anthrop. Assn., Am. Ethnol. Soc. (pres. 1967-68), AAAS, Soc. for Psychol. Anthropology (pres. 1979-80). Home: 2500 Torrey Pines Rd La Jolla CA 92037-3431

SPIRO, THOMAS GEORGE, chemistry educator; b. Aruba, Netherlands Antilles, Nov. 7, 1935; s. Andor and Ilona S.; m. Helen Handin, Aug. 21, 1959; children—Peter, Michael. B.S., UCLA, 1956; Ph.D., M.I.T., 1960. Fulbright researcher U. Copenhagen, Denmark, 1960-61; NIH fellow Royal Inst. Tech., Stockholm, 1962-63; research chemist Calif. Research Corp., LaHabra, 1961-62; mem. faculty Princeton U., 1963—, prof. chemistry, 1974—, head dept., 1979-88, Eugene Higgins prof., 1981—. Author: (with William M. Stigliani) Environmental Issues in Chemical Perspective, 1980, Environmental Science in Perspective, 1980; contbr. articles to profl. jours. Recipient Bomem-Michelson award Bomem Corp., 1986; NATO sr. fellow, 1972, Guggenheim fellow, 1990. Fellow AAAS; mem. Am. Chem. Soc., Phi Beta Kappa, Sigma Xi. Office: Princeton U Dept Chemistry Princeton NJ 08544

SPITZ, ERICH, electronics industry executive; b. Brno, Czecholslovakia, Mar. 27, 1931; arrived in France, 1957; s. Emmerich and Anna (Bergler) S.; m. Sylvie Kayser, Mar. 23, 1963; children: Olivier, Claudia, Isabelle. Dipl. Engr., Tech. Univ., Prague, 1954, PhD, 1956; DSc honoris causa, U. Prague, 1990. Rsch. fellow Radioastronomic Observatory, Meudon, France, 1957-58; engr., mgr. Compagnie Générale de TSF, Orsay, France, 1958-68; dir. ctrl. rsch. lab. Thomson, Corbeville, France, 1975-83; tech. and rsch. dir. Thomson, Paris, France, 1983-86; sr. v.p. R & D Thomson S.A., Paris, France, 1986—; bd. dirs. Valeo, Thomson Ventures, Paris; pres. Thomson LCD, Grenoble, France, 1989—, European Indsl. Rsch. Mgmt. Assn., Paris, 1991; chmn. European Info. Tech. Round Table, Paris, 1992—; mem. High Coun. Rsch. and Tech., Paris, 1983-85. Author Broad Band Antennas, 1963, Optical Video Recording, 1977; contbr. articles to profl. jours. Fellow IEEE, Sté Frse des Electriciens et Electroniciens (pres. elctronics chpt. 1973—); mem. French Engring. Acad., French Acad. Sci. (corr.), chevalier de la Legion D'Honneur, officier de l'Ordre Nat. du Mérite. Achievements include 90 patents in field electronics and optics, pioneering work in three dimensional holographic memories, coherent light propagation in fibers, optical recording on discs. Home: 87 Bd de Port Royal, 75013 Paris France Office: Thomson, 175 Bd Haussmann, 75008 Paris France

SPIVACK, HERMAN M., fluid dynamicist, aerospace engineer; b. N.Y.C., Jan. 3, 1927; s. Jacob and Bessie (Rothman) S.; m. Helen Kohn, 1951; children: Bernhard, Richard, Robin. BA, George Washington U., 1944; MS, Cath. U., Washington, 1946. With mechanics div. Nat. Bur. Stds., Washington, 1946-47; electronic physicist Naval Ord. Test Sta., China Lake, Calif., 1947-51; instr. specialist N.Am. Aviation Aerophysics Lab., L.A., 1951-91; tech. dir. West Coast Rsch. Corp., L.A., 1992—; cons. in field. Contbr. articles to profl. jours. Chair organizing com. Union of Am. Scientists, Washington, 1946. Mem. IAS, AIAA, APS. Achievements include rsch. on vortex patterns, dynamometer development, weigh-in-motion system, ships propeller dynamometer, differential pressure transducer, universal wind tunnel design. Office: West Coast Rsch Corp 1527-26th St Santa Monica CA 90404

SPIVEY, HOWARD OLIN, biochemistry and physical chemistry educator; b. Gainesville, Fla., Dec. 10, 1931; s. Herman Everette and Havens Edna (Taylor) S.; m. Dorothy Eleanor Luke, June 19, 1959; children: Bruce Allen, Curt Olin, Diane Elizabeth. BS, U. Ky., 1954; PhD, Harvard U., 1962. Rsch. assoc. Rockefeller U., N.Y.C., 1962-64; NIH fellow MIT, Cambridge, Mass., 1964-65; asst. prof. U. Md., College Park, 1965-67; asst. prof. Okla. State U., Stillwater, 1967-69, assoc. prof., 1969-75, prof., 1975—. Contbr. articles to profl. jours. Grantee NIH, 1969-72, 75-78, 79-83, NSF, 1969-72, 86-89, Am. Heart Assn., 1988-91. Mem. AAAS, Am. Chem. Soc., Am. Soc. for Biochemistry & Molecular Biology, Phi Beta Kappa. Achievements include research on substrate channeling between several enzymes. Home: 2222 W 11th St Stillwater OK 74074 Office: Okla State U 246 NRC Dept Biochem Molecular Biology Stillwater OK 74078-0454

SPLINTER, WILLIAM ELDON, agricultural engineering educator; b. North Platte, Nebr., Nov. 24, 1925; s. William John and Minnie (Calhoun) S.; m. Eleanor Love Peterson, Jan. 10, 1953; children: Kathryn Love, William John, Karen Ann, Robert Marvin. BS in Agrl. Engring., U. Nebr., 1950; MS in Agrl. Engring., Mich. State U., 1951, PhD in Agrl. Engring., 1955. Instr. agrl. engring. Mich. State U., East Lansing, 1953-54; assoc. prof. biology and agrl. engring. N.C. State U., Raleigh, 1954-60, prof. biology and agrl. engring., 1960-68; prof., chmn. dept. agrl. engring. U. Nebr., Lincoln, 1968-84, George Holmes Disting. prof., 1984—, head dept. agrl. engring., 1984-88, assoc. vice chancellor for rsch., 1988-90, intermim vice chancellor for rsch., dean grad. studies, 1990-92, vice chancellor for rsch., 1992-93; George Holmes Disting. Prof. emeritus, 1993—; cons. engr. Mem. exec. bd. Am. Assn. Engring. Socs.; hon. prof. Shengyang (People's Republic of China) Agrl. U. Contbr. articles to tech. jours.; patentee in field. Served with USNR, 1946-51. Recipient Massey Ferguson gold medal; named to Nebr. Hall of Agrl. Achievement. Fellow AAAS, Am. Soc. Agrl. Engrs. (pres., adminstrv. council, found. pres.); mem. Nat. Acad. Engring., Soc. Automotive Engrs., Am. Soc. Engring. Edn., Nat. Soc. Profl. Engrs., Sigma Xi, Sigma Tau, Sigma Pi Sigma, Pi Mu Epsilon, Gamma Sigma Delta, Phi Kappa Phi, Beta Sigma Psi. Home: 4801 Bridle Ln Lincoln NE 68516-3436 Office: U Nebr 202 Biological Systems Engring Labs Lincoln NE 68583-0726

SPLITTSTOESSER, WALTER EMIL, plant physiologist; b. Claremont, Minn., Aug. 27, 1937; s. Waldemar Theodore and Opal Mae (Young) S.; m. Shirley Anne O'Connor, July 2, 1960; children: Pamela, Sheryl, Riley. B.S. with distinction (univ. fellow), U. Minn., 1958; M.S., S.D. State U., 1960; Ph.D., Purdue U., 1963. Plant breeder U. Minn., 1956-58; weed scientist S.D. State U., 1958-60; plant physiologist Purdue U., 1960-63, Shell Oil Co., Modesto, Calif., 1963-64; biochemist U. Calif., Davis, 1964-65; mem. faculty U. Ill., Urbana, 1965—; prof. plant physiology U. Ill., 1974—, head vegetable crops div., 1972-82; vis. prof. Univ. Coll., Dublin, Ireland, 1987, Univ. Coll., London, 1972; biologist Parkland Coll., Champaign, Ill., 1974; vis. research asso. Rothamsted Exptl. Sta., Harpenden, Eng., 1980; disting. vis. prof. Nagoya U. (Japan), 1982; biotechnologist U. Coll., Dublin, Ireland, 1987. Author: Vegetable Growing Handbook, 1979, 2d edit., 1984, 3d edit., 1990; contbr. over 200 articles to sci. jours.; rev. editor: Analytical Biochemistry, 1969-78, NSF, 1978-79; numerous others. Recipient J.H. Gourley award Am. Fruit Grower-Am. Soc. Hort. Sci., 1974, Outstanding Grad. Educator award, 1990; NIH fellow, 1964-65. Fellow Am. Soc. Hort. Sci. (rev. editor jour. 1969—), Japanese Soc. Promotion of Sci.; mem. Weed Sci. Soc. Am., Am. Soc. Plant Physiologists, Japanese Soc. Plant Physiologists, Sigma Xi (pres. 1990-91), Alpha Zeta, Gamma Sigma Delta, Delta Theta Sigma, Phi Kappa Phi. Home: 2006 Cureton Dr Urbana IL 61801-6226 Office: U Ill 1102 S Goodwin Ave Urbana IL 61801-4798

SPOCK, BENJAMIN MCLANE, physician, educator; b. New Haven, May 2, 1903; s. Benjamin Ives and Mildred Louise (Stoughton) S.; m. Jane Davenport Cheney, June 25, 1927 (div. 1976); children: Michael, John Cheney; m. Mary Morgan Councille, Oct. 24, 1976. B.A., Yale U., 1925, student Med. Sch., 1925-27; M.D., Columbia U., 1929. Intern in medicine Presbyn. Hosp., N.Y.C., 1929-31; in pediatrics N.Y. Nursery and Child's Hosp., 1931-32; in psychiatry N.Y. Hosp., 1932-33; practice pediatrics N.Y.C., 1933-44, 46-47; instr. pediatrics Cornell Med. Coll., 1933-47; asst. attending pediatrician N.Y. Hosp., 1933-47; cons. in pediatric psychiatry N.Y. City Health Dept., 1942-47; cons. psychiatry Mayo Clinic and Rochester Child Health Project, Rochester, Minn.; asso. prof. psychiatry Mayo Found., U. Minn., 1947-51; prof. child devel. U. Pitts., 1951-55, Western Res. U., 1955-67. Author: Baby and Child Care, 1946, (with J. Reinhart and W. Miller) A Baby's First Year, 1954, (with M. Lowenberg) Feeding Your Baby and Child, 1955, Dr. Spock Talks with Mothers, 1961, Problems of Parents, 1962, (with M. Lerrigo) Caring for Your Disabled Child, 1965, (with Mitchell Zimmerman) Dr. Spock on Vietnam, 1968, Decent and Indecent, 1970, A Teenagers Guide to Life and Love, 1970, Raising Children in a Difficult Time, 1974, Spock on Parenting, 1988, (with Mary Morgan) Spock on Spock: A Memoir of Growing Up With the Century, 1989. Presdl. candidate Peoples Party, 1972, advocator Nat. Com. for a Sane Nuclear Policy (SANE), co-chmn., 1962 . Served to lt. comdr. M.C., USNR, 1944-46. Home: POB 1890 Saint Thomas VI 00801 also: PO Box 1268 Camden ME 04843-1268

SPOHN, HERBERT EMIL, psychologist; b. Berlin, Germany, June 10, 1923; s. Herbert F. and Bertha S.; m. Billie M. Powell, July 28, 1973; children—Jessica, Madeleine. B.S.S., CCNY, 1949; Ph.D., Columbia U., 1955. Research psychologist VA Hosp., Montrose, N.Y., 1955-60; chief research sect. VA Hosp., 1960-64; sr. research psychologist Menninger Found., Topeka, 1965-80; dir. hosp. research Menninger Found., 1979—, dir. research dept., 1981—; mem. mental health small grant com. NIMH, 1972-76, mem. treatment assessment rev. com., 1983-86, chmn. 1986-87. Author: (with Gardner Murphy) Encounter with Reality, 1968; assoc. editor: Schizophrenia Bull, 1970-87, 91—; contbr. articles to profl. jours. Served with AUS, World War II. USPHS grantee, 1964—. Fellow Am. Psychopath. Assn.; mem. AAAS, N.Y. Acad. Sci., Soc. Psychopath. Research, Phi Beta Kappa, Sigma Xi. Office: Menninger Found PO Box 829 Topeka KS 66601-0829

SPOKANE, ROBERT BRUCE, biophysical chemist; b. Cleve., Aug. 5, 1952; s. Herbert Norman and Marjorie Ellen (Firsten) S.; m. Linda Carol Wright, June 20, 1976; children: Lea, Hannah, Tara. BS in Chemistry, Ohio U., 1975; MS in Biophys. Chemistry, U. Colo., 1978, PhD in Biophys. Chemistry, 1981. Cert. full cave diver. Teaching asst. Dept. Chemistry, U. Colo., Boulder, 1975-77, rsch. asst., 1977-81; staff scientist Procter & Gamble Co., Cin., 1981-84; rsch. scientist Dept. Neurophysiology, Children's Hosp., Cin., 1984-90, YSI Co., Rsch. Ctr., Yellow Springs, Ohio, 1990—; cons. Synthetic Blood Internat., Yellow Springs, 1992. Contbr. articles to profl. jours. Rescuer, treas. Boulder Emergency Squad, 1980; rescue diver Kitty Hawk Scuba, Dayton, Ohio, 1992. Recipient Merck Index award Ohio U., 1975. Mem. Am. Chem. Soc., N.Y. Acad. Sci., Am. Physiol. Soc., Nat. Speleological Soc., Sigma Xi. Achievements include research in implantable glucose sensors; oxygen tonometer for peritoneal oxygen measurements; interferant removal system for biosensors, water chemistry in submerged caves. Home: 1715 Garry Dr Bellbrook OH 45305 Office: YSI Co 1725 Brannum Ln Yellow Springs OH 45387

SPONABLE, JESS M., astronautical engineer, physicist; b. Madrid, Spain, Nov. 29, 1955; s. Edson J. and Helen D. (Williams) S.; m. Debra A. Hamlin, Apr. 16, 1983; 1 child, Christopher Jay. BS in Physics, USAF Acad., 1978; MS in System Mgmt., U. So. Calif., 1981; MS in Astronautical Engring., Air Force Inst. Tech., 1982. Commd. 2d lt. USAF, 1978; advanced through grades to maj., 1988; launch vehicle engr. 6595th Space Test Group, Vandenberg AFB, Calif., 1978-81; manned space flight engr. space Shuttle SPO, L.A. Air Force Sta., Calif., 1983-86; chief mission ops. div. Global Positioning System JPO, L.A. Air Force Sta., 1983-86; chief space application Nat. Aerospace Plan JPO, Wright-Patterson, Ohio, 1987-90; program mgr. single stage rocket tech. Strategic Def. Initiative, Washington, 1991—. Contbr. articles to profl. jours. Active Boy Scouts, Roseville, Calif., 1967-73. Recipient Am. Legion award, Roseville, Calif., 1970. Fellow AIAA; mem. Nat. Space Soc.

SPORES, JOHN MICHAEL, psychologist; b. Chgo., Mar. 15, 1957; s. Andrew John and Catherine (Mathos) S.; m. Billie Ladas, Feb. 29, 1992. BS in Psychology, Ill. State U., 1979, MS in Clin. Psychology, 1981; PhD in Clin. Psychology, Purdue U., 1990. Lic. psychologist, Ind. Youth and family counselor OMNI Youth and Family Counseling Svcs., Arlington Heights, Ill., 1981-85; predoctoral psychology intern Counseling and Psychol. Svcs. Ctr., Ball State U., Muncie, Ind., 1989-90; staff psychologist Southlake Ctr. for Mental Health, Merrillville and Hobart, Ind., 1992; asst. prof. psychology Purdue U. North Cen. Campus, Westville, Ind., 1991—; staff psychologist, dir. psychology internship tng. Vale Park Paychiat. Hosp./Porter-Starke Svcs., Valparaiso, Ind., 1992—; presenter in field. Contbr. articles to profl. jours. Mem. APA, New Eng. Psychol. Assn., Purdue U. Alumni Assn. (life), Mental Health Assn., Alpha Chi, Sigma Nu (scholarship award 1980, 81). Republican. Greek-Orthodox. Home: 1102 Monticello Park Dr Valparaiso IN 46383-4018 Office: Purdue U North Cen Campus 1401 South US 421 Westville IN 46391-9528 also: Porter/Starke Svcs Inc Vale Park Psychiat Hosp 701 Wall St Valparaiso IN 46383

SPORN, MICHAEL BENJAMIN, cancer etiologist; b. N.Y.C., Feb. 15, 1933; married; 2 children. MD, U. Rochester, 1959. Intern U. Rochester Sch. Medicine, 1959-60; mem. staff lab. neurochemistry Nat. Inst. Neurol. Diseases and Blindness, 1960-64; mem. staff Nat. Cancer Inst., Bethesda, Md., 1964-70, head lung cancer unit, 1970-73, chief lung cancer br., 1973-78, chief lab. chemoprevention, 1978—. Mem. Am. Assn. Cancer Rsch. (B.F. Cain Meml. award, 1991), Am. Assn. Biol. Chemistry, Am. Soc. Pharmacology and Experimental Therapeutics. Achievements include research in nucleic acids and cancer, vitamin A and related compounds, carcinogeneisis studies, retinoids and cancer prevention, peptide growth factors and transforming growth factor-beta. Office: Nat Cancer Inst Cancer Etiology 9000 Rockville Pike, Bldg 31 Bethesda MD 20892*

SPRADLIN, JOSEPH E., embryologist, psychologist, medical educator; b. Bloom, Kans., July 12, 1929; married; 3 children. BA, U. Kans., 1951; MS, Ft. Hays State Coll., 1954; PhD in Psychology, George Peabody Coll., 1959. Teaching asst. psychology Ft. Hays Coll., 1953-54; clin. psychologist Winfield State Sch., 1954-56; rsch. fellow psychology George Peabody Coll., 1956-58; rsch. assoc. Bur. Child Rsch. U. Kans., 1958-59, dir. Parsons Rsch. Ctr., 1959-69, 87—, dir. Kans. U. Affiliate Program, 1978-88, prof. human devel., 1969—; mem. mental retardation com. Nat. Inst. Child Health and Human Devel., 1974-78. Consulting editor Am. Jour. Mental Retardation, 1973-75, Jour. Speech and Hearing Disorders, 1975-78. Fellow Am. Psychol. Assn., Am. Psychol. Soc. Achievements include research in behavior analysis with a special emphasis on stimulus control with persons with retardation. Office: Parsons Research Ctr PO Box 738 Parsons KS 67357*

SPRAFKIN, ROBERT PETER, psychologist, educator; b. N.Y.C., Dec. 18, 1940; s. Benjamin R. and Dora M. (Berman) S.; m. Barbara Marcus, July 19, 1964; children: Jeffrey P., Neal R., Noah M. AB, Dartmouth Coll., 1962; MA, Columbia U., 1964; PhD, Ohio State U., 1968. Lic. psychologist, N.Y. Asst. prof. psychology Syracuse (N.Y.) U., 1968-71, adj. assoc. prof., 1973-88, adj. prof., 1989—; chief day treatment ctr. VA Med. Ctr., Syracuse, 1971—, dir. psychology tng. program, 1983—; clin. assoc. prof. dept. psychiatry SUNY Health Sci. Ctr., Syracuse, 1973—; cons. psychologist Assn. for Retarded Citizens, Syracuse, 1983—. Co-author: Skilltraining for Community Living, 1976, Skillstreaming the Adolescent, 1980, Social Skills for Mental Health, 1993. Mem. Onondaga County Legis. Coun. on Disabled, Syracuse, 1982—; mem. community svcs. bd. County Dept. Mental Health, 1987—. Mem. APA, Assn. Advancement of Behavior Therapy, Soc. Behavioral Medicine, Cen. N.Y. Psychol. Assn. (pres.), Dartmouth Club (pres.). Office: VA Med Ctr 800 Irving Ave Syracuse NY 13210-2796

SPRAGUE, NORMAN FREDERICK, JR., surgeon, educator; b. L.A., June 12, 1914; s. Norman F. and Frances E. (Ludeman) S.; m. Caryll E. Mudd, Dec. 27, 1941 (dec. Apr. 1978); children: Caryll (Mrs. Mingst), Norman Frederick III, Cynthia Sprague Connolly, Elizabeth (Mrs. Day); m. Erlenne Estes, Dec. 31, 1981. AB, U. Calif., 1933; MD, Harvard U., 1937. Intern Bellevue Hosp., N.Y.C., 1937, house surgeon, 1938-39; pvt. med. practice L. A., 1946—; mem. hon. staff Hosp. of Good Samaritan, L. A.; mem. staff St. Vincent Med. Ctr., L. A.; asst. clin. prof. surgery UCLA, 1951—; dir. emeritus Western Fed. Savs. & Loan Assn.; chmn. bd. dirs. Western Pioneer Co., 1961-63, Pioneer Savs. & Loan Assn., 1959-63; dir. Arden-Mayfair, Inc., 1966-69; also chmn. exec. com.; dir., mem. exec. com. Cyprus Mines Corp., 1959-79; trustee Mesabi Trust, 1964-76. Chmn. exec. com., v.p. Harvard Sch., 1954-65; mem. Community Redevel. Agy. City of L.A., 1966-69, vice chmn., 1967-69; mem. Calif. Regional Med. Programs Area IV Council, 1970-75; bd. dirs., v.p. Calif. Inst. Cancer Rsch., 1974-80, pres., 1980-82; bd. dirs. Cancer Assoc., 1975-80; trustee UCLA Found., Marlborough Sch., 1981-90, Mildred E. and Harvey S. Mudd Found., Hollywood Bowl Assn., 1962-66; hon. trustee Calif. Mus. Found.; mem. exec. com., trustee Youth Tennis Found., 1960-70; trustee, pres. mem. exec. com. S.W. Mus.; founding trustee, Harvey Mudd Coll.; chmn. bd. trustees Caryll and Norman Sprague Found., 1957—, Harvard Sch.; mem. bd. visitors UCLA Med. Sch.; mem. adv. com. Univs. Space Rsch. Assn., Div. Space Biomedicine, 1982—, nat. bd. dirs. Retinitis Pigmentosa Internat. Served to maj. M.C. AUS, 1941-46. Decorated Bronze Star; recipient Bishop's award of Merit Episc. Diocese L.A., 1966, Highest Merit award So. Calif. Pub. Health Assn., 1968. Mem. AMA, Calif. Med. Assn., Los Angeles County Med. Assn. (pres. jr. sect. 1953), SAR, Am. Cattlemen's Assn., Symposium Soc., Tennis Patrons Assn. (dir. 1960-70), Calif. Club, Harvard Club, L.A. Country Club, Regency Club (L.A.), Delta Kappa Epsilon. Home: 550 S Mapleton Dr Los Angeles CA 90024-1811 Office: Ste 2760 2049 Century Park E Los Angeles CA 90067-3202

SPRAGUE, VANCE GLOVER, JR., oceanography executive, naval reserve officer; b. Bellefonte, Pa., Oct. 28, 1941; s. Vance Glover and Elaine L. (Tottingham) S.; m. Claire Lewis, Nov. 26, 1982 (dec. Oct. 1987); m. Heather Munro Ferguson, June 3, 1989. BS, Pa. State U., 1963; MS, Salve Regina U., 1983, Naval War Coll., 1983. Commd. ensign USN, 1963; lt. j.g. USS Towhee, Norfolk, Va., 1963-65; oceanographer Naval Oceanographic Office, D.C., 1965-77, sect. head, 1977-80; br. head Naval Oceanographic Office, Stennis Space Ctr., Miss., 1980-85, dir. Phys. Oceanography Div., 1985—; exec. officer, naval oceanography res. unit Naval Air Sta., New Orleans, 1983-87, comdg. officer, 1987-89. Capt. USNR, 1987. Mem. U.S. Naval Inst., Am. Geophys. Union, Long Beach (Miss.) Yacht Club, Delta Chi. Avocations: woodworking, gardening, choral singing. Home: 114 Camelia Dr Pass Christian MS 39571-4706 Office: Naval Oceanographic Office Stennis Space Center MS 39522

SPRATT, BRIAN GEOFFREY, microbiologist, educator, researcher; b. Margate, Kent, Eng., Mar. 21, 1947; s. Clarence Albert and Marjory Alice (Jeffreys) S.; m. Jennifer Broome-Smith (div. 1993); 1 child, Timothy Peter. BSc, London U., 1968, PhD, 1972. Rsch. assoc. Princeton (N.J.) U., 1973-75; rsch. fellow Leicester (Eng.) U., 1975-80; lectr. Sussex U., Brighton, Eng., 1980-87, reader, 1987-89, prof., 1990—, wellcome trust prin. rsch. fellow, 1989—. Contbr. numerous articles to profl. jours. Recipient Fleming award Soc. Gen. Microbiology, 1982, Hoechst-Roussel award Am. Soc. Microbiology, 1993; Fellow Royal Soc. London, 1993. Achievements include research on mechanisms of action and mechanisms of resistance to antibiotics. Home: 10 4th Ave Flat 4, Hove BN3 2PH, England Office: Sussex U, Sch Biology, Brighton BN1 9QG, England

SPRAY, PAUL, surgeon; b. Wilkinsburg, Pa., Apr. 9, 1921; s. Lester E. and Phoebe Gertrude (Hull) S.; m. Mary Louise Conover, Nov. 28, 1943; children—David C., Thomas L., Mary Lynn (Mrs. Thomas Branham). B.S., U. Pitts., 1942; M.D., George Washington U., 1944; M.S., U. Minn., 1950. Diplomate Am. Bd. Orthopedic Surgery. Intern U.S. Marine Hosp., S.I., 1944-45; resident Mayo Found., Rochester, Minn., 1945-46, 48-50; practice medicine specializing in orthopedic surgery Oak Ridge, Tenn., 1950—; mem. staff Oak Ridge Hosp., Park West Hosp., Knoxville, Harriman Hosp., Tenn., Bapt. Hosp. of Roane County; vol. vis. cons., CARE Medico, Jordan, 1959, Nigeria, 1962, 65, Algeria, 1963, Afghanistan, 1970, Bangladesh, 1975, 77, 79, Peru, 1980, U. Ghana, 1982; AMA voluntary physician, Vietnam, 1967, 72; vis. assoc. prof. U. Nairobi, 1973; mem. teaching team of Internat. Coll. Surgeons to Khartoum, vis. prof. orthopedic surgery U. Khartoum, 1976; hon. prof. San Luis Gonzaga U., Ica, Peru; AmDoc vol. cons., U. Biafra Teaching Hosp., 1969; vis. prof. Mayo Clinic, 1988; sec. orthopedics overseas div. CARE Medico, 1971-76, sec. medico adv. bd., 1974-76, vice chmn., 1976, chmn., 1977-79, v.p. CARE, Inc., 1977-79, pub. mem. care bd. dirs., 1980-90, bd. overseers, 1991—; chmn. Orthopedics Overseas, Inc., 1982-86, treas., 1986-88; mem. U.S. organizing com. 1st Internat. Acad. Symposium on Orthopedics, Tianjin, China, 1983; mem. CUPP Internat. Adv. Council, 1986—; bd. dirs. East Tenn. Health Plan, 1987-90. Mem. editorial bd. Contemporary Orthopedics, 1984—. V.p. Anderson County Health Coun., 1975, pres., 1976-77, hon. bd. dirs., 1991; pres. health commn. Coun. Southern Mountains, 1958-65, sec. bd. dirs., 1965-66; Tenn. pres. UN Assn., 1966-67; vice chmn. bd. Camelot Care Ctr., Tenn., 1979-82, chmn., 1982-86; chmn. bd. dirs. Camelot Found., 1986-87; hon. mem. World Orthopedic Concern, 1990. Recipient Svc. to Mankind award Sertoma, 1967, Freedom Citation, 1978, Medico Disting. Svc. award, 1990, 1st Ann. Vocat. Svc. award Oak Ridge Rotary, 1979, Tech. Communication award East Tenn. chpt. Soc. for Tech. Communication, 1983, Individual Achievement award Meth. Med. Ctr. of Oak Ridge, 1991, Humanitarian award Orthopaedics Overseas, 1992. Fellow ACS, Internat. Coll. Surgeons (Tenn. regent 1976-80, bd. councillors 1980-84, hon. chmn. bd. trustees 1981-83, trustee 1983-84, v.p. U.S. sect. 1982-83, mem. surg. teams com. 1983-90, Humanitarian award 1992); mem. AMA (Humanitarian Svc. award 1967, 72), Société Internationale de Chirugie Orthopédique et de Traumatologie, So. Orthopedic Assn., Western Pacific Orthopedic Assn., Orthopedic Letters Club, Am. Fracture Assn., Am. Acad. Orthopedic Surgeons (mem. com. on injuries 1980-86), Tenn. Med. Assn. (com. on emergency med. svcs. 1978-88), Alumni and Friends of Medico (pres. 1975-77), Peru Acad. Surgery (corr.), Peruvian Soc. Orthopedic Surgery and Traumatology (corr.), Internat. Soc. for Fracture Repair (founding), Clin. Orthopaedic Soc., Mid-Am. Orthopaedic Soc., Lions (Humanitarian award 1968, Amb. Goodwill award 1979, Melvin Jones fellow 1993), Rotary Club (Oak Ridge chpt.). Home: 507 Delaware Ave Oak Ridge TN 37830-3902 Office: 145 E Vance Rd Oak Ridge TN 37830-6522

SPRAY, THOMAS L., surgeon, educator; b. Rochester, Minn., Aug. 28, 1948; married. BA in Molecular Biology, Haverford Coll., 1970; MD, Duke U., 1973. Diplomate Am. Bd. Surgery, Am. Bd. Thoracic Surgery. Intern in surgery Duke U. Med. Ctr., 1973, resident in surgery, 1974-75, staff assoc., 1975-77, sr. resident, 1977-80, rsch. fellow in surgery, 1980-81, chief resident, 1981-82; teaching scholar, 1982-83; from asst. prof. to prof. cardiothoracic surgery/pediatrics Washington U. Sch. Medicine, 1983-93; dir. pediatric cardiothoracic surgery St. Louis Children's Hosp., 1984—; prof. surgery and pediatrics Washington U. Sch. Medicine, 1992—; mem. staff St. Louis Children's Hosp., Jewish Hosp., St. Louis, Barnes Hosp., St. Louis Regional Med. Ctr.; cons. staff John Cochran VA Hosp. Contbr. chpts to books and articles to jours. Annals of Thoracic Surgry, Am. Heart Jour., Jour. Thoracic and Cardiovascular Surgery, Jour. of Cell Biology, Immunology, and others. With USPHS, 1975-77. ACS scholar, 1980-82; grantee Am. Heart Assn., 1984-85, NIH, 1984-85. Fellow ACS, Am. Coll. Chest Physicians, Am. Coll. Cardiology, Am. Assn. for Thoracic Surgry; mem. AMA, Assn. for Academic Surgery Barnes Hosp. Soc. David C. Sabiston Jr. Soc., Internat. Soc. Cardiovascular Surgery, Interant. Soc. for Heart Transplantation, So. Thoracic Surg. Soc., St. Louis Am. Heart Assn. (sci. peer rev. com. 1988-92), St. Louis Children's Med. Staff Soc., St. Louis Pediatric Soc., St. Louis Thoracic Surg. Soc., Soc. Thoracic Surgeons, St. Louis Cardiac Club. Achievements include rsch. in pediatric heart transplantation, transplantation for hypoplastic left heart syndrome, pediatric lung transplantation. Office: SW24 Saint Louis Childrens Hosp/1 Childrens Place Saint Louis MO 63110

SPRECHER, GUSTAV EWALD, pharmacy educator; b. Kupferzell, Germany, Nov. 17, 1922; s. Emil and Emma (Küstner) S.; m. Helga Mohr, Aug. 28, 1955; 1 child, Wolfram. BS in Pharmacy, Tech. U. Karlsruhe, 1953, Dr. rer. nat., 1955, Dr. Habil., 1960. Asst. prof. Tech. U. Karlsruhe, Fed. Republic Germany, 1955-64; assoc. prof. Tech. U. Karlsruhe, 1964-69; prof. U. Hamburg, Fed. Republic Germany, from 1969; now prof. emeritus U. Hamburg. Author: Arzneistoffproduktion, 1983, (with F. Deutschmann and B. Hohmann) Drogenaualyse I: Morph. Anatomie, 1992; contbr. articles to sci. publs. Mem. German Pharm. Soc. (Hermann-Thoms medal), Soc. Medicinal Plant Rsch. (pres. 1988-89). Home: Sandmoorweg 31, D 22559 Hamburg Germany

SPRENGNETHER, RONALD JOHN, civil engineer; b. St. Louis, July 28, 1944; s. William Francis Sprengnether and Roberta (Lucas) Harris. BSCE, St. Louis U., 1967; MSCE, U. Mo., 1984. Registered profl. engr., Mo. Civil engr. Sverdrup Corp., Maryland Heights, 1972—. Lt. (j.g.) USNR, 1968-71. Mem. NSPE, Am. Water Works Assn., Engrs. Club St. Louis. Roman Catholic.

SPRIGGS, RICHARD MOORE, ceramic engineer, research center administrator; b. Washington, Pa., May 8, 1931; s. Lucian Alexander and Kathryn (Aber) S.; m. Patricia Anne Blaney, Aug. 1, 1953; children—Carolyn Elizabeth Spriggs Muchna, Richard Moore, Alan David. B.S. in Ceramics, Pa. State U., 1952; M.S. in Ceramic Engring., U. Ill., 1956, Ph.D. in Ceramic Engring., 1958. Sr. research engr. Ferro Corp., Cleve., 1958-59; sr. staff scientist, group leader, ceramics rsch. AVCO Corp., Wilmington, Mass., 1959-64; assoc. prof. metall. engring. Lehigh U., Bethlehem, Pa., 1964-67, prof. metallurgy and materials sci. and engring., 1967-80, adminstrv. asst. to pres., 1970-71, asst. v.p. for adminstrn., 1971-72, v.p. for adminstrn., 1972-78, dir. phys. ceramics lab., 1964-70, assoc. dir. Materials Research Ctr., 1964-70; vis. sr. staff assoc. Nat. Materials Adv. Bd. NRC, Washington, 1979-80, sr. staff officer, staff scientist, 1980-87, staff dir. bd. on assessment of NBS programs, 1984-87; J.F. McMahon prof. ceramic engring., dir. NYS Ctr. Advanced Ceramic Tech. N.Y. State Coll. Ceramics, Alfred (N.Y.) U., 1987—, dir. office of sponsored programs, 1988—. Contbr. articles to profl. publs. Co-patentee in field. Pres., bd. dirs. YMCA, Bethlehem, Pa., 1978-79. Served to lt. USNR, 1952-56. Fellow Armco Steel Corp., 1956-58, Am. Council on Edn., 1970-71. Fellow Am. Ceramic Soc. (disting. life, pres. 1984-85, Ross Coffin Purdy award 1965, Hobart M. Kraner award Lehigh Valley sect. 1980, trustee pension trust fund 1979-84, Orton lectr. 1988, McMahon lectr. 1988, coord. programs and meetings 1991-92), Ceramic Soc. Japan (Centennial medal 1991), Brit. Inst. of Ceramics; mem. AAAS, N.Y. Acad. Scis., Internat. Inst. for Sci. of Sintering, Nat. Inst. Ceramic Engrs., Ceramic Ednl. Council, Brit. Ceramic Soc., Internat. Acad. Ceramics (trustee 1988—), Am. Soc. Engring. Edn., Materials Rsch. Soc., Fedn. of Materials Socs. (trustee 1978-84), Ceramic Assn. N.Y. (sec.-treas. 1988—), Serbian Acad. Scis. and Arts (fgn.). Lodge: Rotary (dir. 1982-87, pres. 1985-86). Office: Alfred U Ctr Advanced Ceramic Tech NY State College of Ceramics Alfred NY 14802

SPRIMONT, THOMAS EUGENE, computing executive; b. Bradley, Ill., Aug. 2, 1957; s. Eugene and Eugenia (Dades) S.; children: Preston Eugene Farnsworth. BS, Purdue U., 1978, So. Ill. U., 1979. CAD/CAM program analyst McAuto, Long Beach, Calif., 1980-82; software specialist Digital Equipment Corp., Culver City, Calif., 1982-84; sr. applications engr. Gould Sel, Torrance, Calif., 1984-85; cons. G.M. Advanced Concept Ctr., Newbury Park, Calif., 1985-86; CAE systems supr. Rogerson Kratos, Pasadena, Calif., 1986-88; v.p. corp. computing Rogerson Aircraft Corp., Irvine, Calif., 1988—; tchr. UCLA, Westwood, 1987—. Contbr. articles to profl. jours. Home: 19 Oak Spring Ln Laguna Hills CA 92656 Office: Rogerson Aircraft Corp 2201 Alton Ave Irvine CA 91105-2609

SPRINGER, ANDREA PAULETTE RYAN, physical therapist, biology educator; b. Uvalde, Tex., Nov. 20, 1946; d. William McKinley and Dora Edna (Garnett) Ryan; m. A. E. Springer IV; children: Alfred E. Springer V, Paul Ryan Springer. BS, Tex. Women's U., 1969, MS, 1973; PhD, Baylor U., 1977. Staff phys. therapist Dallas Soc. for Crippled Children, Tex., 1969-70; phys. therapist Convalescent Center, Dallas, 1971-72; chief phys. therapist Kerrville State Hosp., Tex., 1973-74; Coord. phys. therapy Kilgore Region VII Edn. Center, Tex., 1979-81; phys. therapist Longview Ind. School Dist., Tex., 1979-80, Texas Home Health, Longview, 1980-81; asst. prof. LeTourneau Coll., Longview, 1981-86; phys. therapist Smith County Coop., 1980-; instr./coord. Kilgore Coll., 1986-92; dir. phys. therapy East Tex. Treatment Ctr., Kilgore, 1993—. Contbr. articles to profl. jour. Mem. Evergreen Garden Club, East Tex. VHF-FM Soc., First Baptist Ch. Mem. Am. Phys. Therapy Assn., Tex. Public Employee Assn., Soc. for Neuroscience, Sigma Xi. Home: 411 Highland Dr Kilgore TX 75662-3915

SPRINGER, GEORGE STEPHEN, mechanical engineering educator; b. Budapest, Hungary, Dec. 12, 1933; came to U.S., 1959; s. Joseph and Susan (Grausz) S.; m. Susan Martha Flory, Sept. 15, 1963; children: Elizabeth Anne, Mary Katherine. B in Engring., U. Sydney, Australia, 1959; M in Engring., Yale U., 1960, MSc in Engring., 1961, PhD, 1962. Registered profl. engr., Mass. Asst. prof. mech. engring. MIT, Cambridge, Mass., 1962-67; prof. mech. engring. U. Mich., Ann Arbor, 1967-83; prof. mech. engring., chmn. dept. aeronautics/astronautics Stanford (Calif.) U., 1983—. Author: Erosion by Liquid Impact, 1975; co-author, co-editor 12 books; contbr. over 150 articles to scholarly and profl. jours. Recipient Pub. Svc. Group Achievement award NASA, 1988. Fellow AIAA, ASME; mem. Am. Physical Soc., Soc. Automotive Engrs. (Ralph Teeter award 1978), Soc. for the Advancement of Materials and Process Engring. (Del Monte award 1991). Achievements include patent in field. Office: Stanford U Dept Aeronautics and Astronautics Stanford CA 94305

SPRINGER-MILLER, JOHN HOLT, computer systems developer; b. Hanover, N.H., Mar. 20, 1955; s. Fred Frank and Glenn Abbott (Harden) S.-M. Student, U. N.H., 1973-74. Owner Major's Inn Dinner Theater, Gilbertsville, N.Y., 1976-79; cons. various dinner theaters, 1979-80; theater critic Gannett Westchester/Rockland publs., Spring Valley, N.Y., 1980-82; programmer various cons. projects, N.Y.C., 1982-83; owner, pres. Springer-Miller Systems, Inc., Stowe, Vt., 1983—; pres., head R&D Custom Verticals Software, Inc., Stowe, 1986—, John Holt Co., Inc., Stowe, L.A., Atlanta, Las Vegas, Dallas, 1988—. Author: A Gift, 1983; contbr. revs., critiques to profl. publs. Founder, pres. Stowe Playhouse, Inc., 1985-88; dir. Stowe High Sch. Theatre, 1985-89; mem. pres.'s forum U. Vt., Burlington, 1992. Recipient various awards from various civic orgns. Mem. Am. Hotel Motel Assn. (session leader 1983—, cert. of recognition 1988—), VLRA (cert. recognition 1983—), Vt. Software Developers Assn. Republican. Episcopalian. Achievements include development of computer software. Home and Office: 782 Mountain Rd PO Box 1547 Stowe VT 05672

SPRINGFIELD, J., civil engineer. Recipient Le Prix A. B. Anderson award Can. Soc. Civil Engring., 1991. Home: 194 Bayview Heights Dr, Toronto, ON Canada M4G 2Z2*

SPRINGSTEEN, ARTHUR WILLIAM, organic chemist; b. Milford, Conn., Oct. 30, 1948; s. Stanley Arthur and Catherine Magdalene (Schatz) S.; m. Kathryn Rose Mooney, June 12, 1971; 1 child, Anne Elizabeth. BS in Chemistry, St. Francis Coll., 1970; MS in Chemistry, Marshall U., 1972; PhD, W.Va. U., 1977. Asst. prof. chemistry Colby-Sawyer Coll., New London, N.H., 1979-86; dir. rsch. Labsphere, Inc., North Sutton, N.H., 1986—. Editor: Academic Press Handbook of Molecular Spectroscopy, 1993; author tech. manuals; contbr. articles to profl. publs. Recipient Top 100 award Photonics Spectra mag., 1990. Mem. ASTM, Am. Chem. Soc., Soc. Profl. Electro-optic Engrs., InterSoc. Color Coun., Coun. Optical Radiation Measurement (bd. dirs. 1991—). Democrat. Congregationalist. Achievements include development of Spectralon, InfraGold, PlasmaGold and related high reflectance materials used in optical instrumentation, Spectrafelct, Duraflect high reflectance optical coatings; patents in field. Office: Labsphere Inc PO Box 70 North Sutton NH 03260-0070

SPRINKLE, JAMES THOMAS, paleontologist, educator; b. Arlington, Mass., Sept. 2, 1943; s. Rex Thomas and Rose (Weiss) S.; m. G.K. Klizicki, Sept. 1, 1968; children: David, Diana. SB, MIT, 1965; MA, Harvard U., 1966, PhD, 1971. Postdoctoral assoc. NRC-U.S. Geol. Survey at Paleontolog. and Stratospheric Br., Denver, 1970-71; asst. prof. dept. geolog. scis. U. Tex., Austin, 1971-78, assoc. prof., 1978-83, prof., 1983-86, 1st Mr. and Mrs. Charles E. Yager prof., 1986—; mem. treatise adv. bd. U. Kans. Paleontolog. Inst., Lawrence, 1991—. Contbr. articles to sci. and profl. jours.; co-author monographs, lab. manual in paleobiology, textbook chpts. Grantee NSF, 1977, 89, 93. Fellow Geolog. Soc. Am. (treatise commn. 1982-86); mem. Paleontolog. Soc. (councilor 1981-83), Soc. for Sedimentary Geology, Soc. Systematic Zoologists. Achievements include proposal of new class of fossil echinoderms (Ctenocystoidea), 1969; new subphylum of fossil echinoderms (Blastozoa), 1973; co-author paper reporting discovery of fossilized eggs in late paleozoic blastoid, 1976. Home: 2801 Winston Ct Austin TX 78731 Office: U Tex Dept Geolog Scis Austin TX 78712

SPRINKLE, WILLIAM MELVIN, engineering administrator, audio-acoustical engineer; b. Washington, Sept. 2, 1945; s. Melvin Cline and Gladys Virginia (Miller) S.; div.; children: Timothy William, Allison Anne. BS in Chemistry, Randolph-Macon Coll., 1967; M in Engring. Adminstrn., Va. Poly. Inst. & State U., 1990. Registered profl. engr., Va. Sr. cons. Sprinkle & Assocs., Kensington, Md., 1973-76; audio systems engr. Robertshaw Controls Co., Richmond, Va., 1976-80; sr. engr. TDFB-Engrs. & Architects, Richmond, Va., 1980-85; property mgmt. officer Signet Bank, Richmond, Va., 1985-87; asst. dir. engring. Va. Dept. Corrections, Richmond, 1987—; mem. summer adj. faculty Eastman Sch. Music, Rochester, N.Y., 1974-83. Contbr. Time Saver Standards for Architectural Design Data, 1982. Newsletter editor Bicycle Orgn. of Southside, Richmond, 1992. Mem. Acoustical Soc. Am. Methodist. Office: Dept of Corrections 6900 Atmore Dr Richmond VA 23225

SPROULL, WAYNE TREBER, consultant; b. Racine, Wis., Aug. 3, 1906; s. John Coppess and Mabel Claire (Warner) S.; m. Ethel Lenore Miller, Aug. 18, 1934; children—Thomas Walter, Sally Ruth Sproull Hohn. B.S. in Physics, U. Akron, 1927; M.S. in Physics, Lehigh U., 1929; Ph.D. in Physics, U. Wis., 1933. Sr. scientist Gen. Motors Research Labs., Detroit, 1933-46, Lockheed Research Lab., Burbank, Calif., 1946; head liquid rocket sect. Jet Propulsion Lab., Calif. Inst. Tech., Pasadena, 1947; dir. research Western Precipitation Corp., Los Angeles, 1947-59; chief physicist W.P. Div. Joy Mfg. Co., Los Angeles, 1959-72; cons. indsl. gas cleaning, Glendale, Calif., 1972—; cons. Electric Power Research Inst., Palo Alto, Calif., Dresser Industries Advanced Tech. Ctr., Irvine, Calif., Carolina Power and Light Co., Raleigh, N.C. Author: X-rays in Practice, 1946; Air Pollution and Its Control, 1972, also tech. articles. Recipient Cert. Appreciation, Office Sci. Research and Devel., 1945. Mem. Am. Phys. Soc., AAAS (life), Air Pollution Control Assn. Republican. Methodist. Club: Verdugo Hills Chess (treas. Tujunga, Calif. 1960-84). Home and Office: RR # 3 Box 144A Coudersport PA 16915

SPROW, FRANK BARKER, oil company executive; b. Council Bluffs, Iowa, Nov. 20, 1939; s. Dwight Barker and Lucille (Tuttle) S.; m. Ann Bledsoe, Aug. 24, 1962; children: John, Diane. BS in Chem. Engring., MIT, 1962, MS in Chem. Engring., 1963; PhD in Chem. Engring., U. Calif., Berkeley, 1965. Jr. engr. Humble Oil and Refining Co., Baytown, Tex., 1962; sr. analyst long range supply Humble Oil and Refining Co., Houston, 1971-75; sr. rsch. chem. engr./Sp/Sh Esso Rsch. and Engr. Co., Baytown, 1965-71; tech. mgr. Bayway Ref. Exxon Co. U.S.A., Linden, N.J., 1975-77, ops. mgr. Bayway Ref., 1977-79; gr. mgr. petroleum programs Exxon Rsch. and Engring. Co., Florham Park, N.J., 1979-80, v.p. synthetic fuels rsch., 1980-82; v.p. tech. support Exxon Rsch. and Engring. Co., Annandale, N.J., 1982-86, v.p. corp. rsch., 1986—; mem. governing bd. Coun. for Chem. Rsch., Washington, 1984—; bd. dirs. Exxon Edn. Found., Dallas, 1986—. Contbr. articles to jours. in field. Bd. dirs. R & D Coun. N.J., Parsippany, 1985—, N.J. Safety Coun, Newark, 1983—; campaign officer Republican Party, various elections, Mercer County, N.J. Recipient award in computing Max Planck Soc., Fed. Republic Germany, 1988. Mem. Am. Inst. Chem. Engrs., Soc. Automotive Engrs., Nassau Club (Princeton, N.J.). Office: Exxon Research & Engring Co 180 Park Ave PO Box 101 Florham Park NJ 07932*

SPRUCH, LARRY, physicist, educator; b. Bklyn., Jan. 1, 1923. BA, Bklyn. Coll., 1943; PhD in Physics, U. Pa., 1948. From asst. instr. to instr. physics U. Pa., 1943-46; Atomic Energy Commn. fellow MIT, 1948-50; from asst. prof. to assoc. prof. NYU, N.Y.C., 1950-61, prof. physics, 1961—; cons. Lawrence Radiation Lab., 1959-66; vis. prof. inst. theoretical physics U. Colo., 1961, 68; mem. Inst. Advanced Study, 1981-82; del. China U.S. Physics Exam and Application, 1985, 86; mem. adv. bd. inst. theoretical atomic and molecular physics Harvard-Smithsonian Ctr. Astrophysics, 1989-91. Corr. Comments Atomic & Molecular Physics, 1972—. Recipient von Humboldt Sr. award, 1985, 88; Tyndale fellow U. Pa., 1946-48, NSF sr. fellow U. London, Oxford U., 1963-64. Fellow Am. Phys. Soc. (Davisson-Germer prize 1992). Achievements include research in Beta decay, nuclear moments, isomeric transitions, internal conversion, atomic and nuclear scattering, variational principles, astrophysics, charge transfer, Thomas-Fermi theory, radiative corrections, atoms in magnetic fields, Levinson's theorem, casimir interactions, semi-classical radiation theory. Office: NYU Dept of Physics Washington Sq New York NY 10003*

SPUR, GÜNTER, manufacturing engineering educator; b. Brunswick, Germany, Oct. 10, 1928; s. Wenzel and Martha (Held) S.; m. Maria Alberts. Student, Tech. U., Brunswick, Germany, 1948-54, DEng, 1960; hon. degree, Catholic U., Leuven, Belgium, 1983. Sci. asst. Tech. U., Brunwick, 1956, chief engr., 1959; prof., dir. Inst. Machine Tools and Prodn. Tech. Tech. U., Berlin, 1965—; dir. Gildemeister, Berlin; head Inst. Prodn. Tech. & Automation, FVG, Berlin; mem. supervisory bd. dirs. DIAG Deutsche Industrieanlagen GmbH, Berlin, Schorch AG, Mönchengladbach, Kraftwerk Union AG, Mühlheim. Author: Mehrspindeldrehautomaten, 1970, Optimierung Fertigungssystem Werkzeugmaschine, 1972, Spanende Werkzeugmaschinen, 1977; editor: Handbuch der Fertigungstechnik; editor Zeitschrift für wirtschaftliche Fertigung; contbr. articles to profl. jours. Recipient M. Eugene Merchant Mfg. medal ASME, 1992. Mem. SME (hon.), Assn. German Engrs. (chmn. Berlin dist., hon. ring, hon. medal), Phys-Tech. Fed. Inst. (bd. govs.), Nat. Acad. Engring. USA, Internat. Rsch. Assn. Mechanic Prodn. Tech., Rotary Club. Home: Richard Straub St 20, D-1000 Berlin 33, Germany Office: Fraunhofer Inst for Production, Tech Univ of Berlin, W-100 Berlin 10, Germany*

SPURGEON, EARL E., systems engineer; b. Jefferson City, Mo., Apr. 19, 1963; s. Doyle D. and Janice I. (Hesemann) S. BS in Elec. Engring., U. Mo., 1986, BS in Computer Engring., 1986. Registered profl. engr., Mo. Programmer U.S. Geol. Survey, Rolla, Mo., 1981-86; supr. Southwestern Bell Telephone, St. Louis, 1986-88, system adminstr., 1989-92; systems engr. Advanced Bus. Consultants, Kansas City, Mo., 1992—. Fellow NSPE, Tau Beta Pi. Home: Rt 4 Box 4596 Saint James MO 65559

SPURLIN, LISA TURNER, biologist; b. Gallatin, Tenn., Apr. 21, 1962; d. Henry Dodd and Jean (Brown) Turner; m. Max Steven Spurlin, June 27, 1992. BS in Biology, Memphis State U., 1985; MS in Biology, Tenn. Tech U., 1989. Sr. level trainee U.S. EPA Region IV, Athens, Ga., 1987; life scientist U.S. EPA Region IV, Atlanta, 1989-92, aquatic biologist, 1992—. Contbr. publs. to profl. jours. Recipient Outstanding Performance award U.S. EPA, 1990, 91, 92. Mem. Assn. Aoutheastern Biologists, Tenn. Acad. Sci., Soc. Environ. Toxicology and Chemistry, Sigma Xi, Omicron Delta Kappa. Achievements include serving as specialist in whole effluent toxicity for U.S. EPA Region IV. Home: 1066 Seaboard Ave NW Atlanta GA 30318 Office: US EPA Region IV 345 Courtland St NE Atlanta GA 30365

SPURLOCK, PAUL ANDREW, nuclear engineer; b. Atlanta, Aug. 4, 1960; s. Jack Marion and Phyllis Lowene (Ridgway) S. BS Nuclear Engring., Ga. Inst. Tech., 1985. Student rsch. engr. Ga. Inst. Tech., Nuclear Rsch. Facility, Atlanta, 1983-85; test engr. Dept. of the Navy, Portsmouth, Va., 1985-87; plant engr. Ga. Power Co., Vidalia, Ga., 1987-89; cons. Spurlock and Assocs., Boca Raton, Fla., 1989-90; sr. v.p. S & A Automated Systems, Inc., Boca Raton, 1990—; cons. Jet Propulsion Lab., Pasadena, Calif., 1990—; session organizer ICES Life Support Controls, Colorado Springs, Colo., 1992-93; session chmn. ICES Life Support Sensors, Seattle, 1992. Mem. AAAS, AIAA, Am. Nuclear Soc., N.Y. Acad. Scis. Achievements include research in process-control integration requirements, methodology for technology assessment. Office: S & A Automated Sytems Inc 123 NW 13th St Ste 222 Boca Raton FL 33432

SPURR, PAUL RAYMOND, organic chemist; b. Adelaide, Australia, Oct. 31, 1957; s. Thomas George and Jean Elizabeth (Sanders) S.; m. Lana El-Kahef, Feb. 9, 1990; 1 child, Adrian Wasim. LMusA, U. Adelaide, 1976, BSc, 1978, BSc with honors, 1979, PhD, 1983. Postdoctoral fellow U. Pa., Phila., 1983; rsch. officer U. Adelaide, 1984; Alexander von Humboldt fellow U. Freiburg, Germany, 1985-86; rsch. chemist Hoffmann-La Roche, Basel, Switzerland, 1987—. Contbr. articles to profl. jours. including Jour. of the Am. Chem. Soc., Chimia, Tetahedron Letters, Chem. Comm., Synthesis, Chem. Berichte, Angewandte Chemie. Univ. Adelaide scholar, 1978. Mem. N.Y. Acad. Sci., Am. Chem. Soc., Planetary Soc. Achievements include patent for new dehydration process in the field of carotenoid/vitamin A chemistry.

SQUIBB, SAMUEL DEXTER, chemistry educator; b. Limestone, Tenn., June 20, 1931; s. Benjamin Bowman and Lou Pearl S.; m. JoAnn Kyker, Dec. 15, 1951; children—Sandra Lavanne, Kevin Dexter. B.S., E. Tenn. State U., 1952; Ph.D., U. Fla., 1956. Asso. prof., dir. chemistry Western Carolina U., Cullowhee, N.C., 1956-60; asst. prof., dir. chemistry Eckerd Coll., St. Petersburg, Fla., 1960-63; asso. prof. Eckerd Coll., 1963-64; prof. chemistry U. N.C., Asheville, 1964—; chmn. dept. U. N.C., 1964—; vis. prof. U. N.C., Chapel Hill, 1976-81, 83-87, 92-93, Clemson U., S.C., 1982; cons. So. Assn. Colls. and Schs., State of W.Va. Author: Experimental Organic Chemistry, 1972, Understanding Chemistry One, 1979, Two, 1981, Three, 1981, Four, 1981, Five, 1981, Six, 1984, Chemistry One, 1976, Two, 1980, Experimental Chemistry One, 1976, Two, 1981; contbr. articles to profl. jours. Mem. Grose United Meth. Ch. Disting. Tchr. award U. N.C.-Asheville, 1983. Fellow Am. Inst. Chemists (life, nat. publs. bd. 1988—); mem. Am. Chem. Soc. (recipient Charles H. Stone award Carolina Piedmont sect. 1979, chmn. Tampa Bay subsect. 1963, Western Carolina sect. 1981, editor Periodic News Western Carolina sect. 1980—), N.C. Inst. Chemists (pres. 1977-79, sec. 1975-77, 1985-91, Disting. Chemist award 1986), Skyland Twirlers Square Dance Club, Silver Spurs Advanced Square Dance Club, Skylark Round Dance Club, Phi Beta Kappa, Gamma Sigma Epsilon, Sigma Xi, Alpha Chi Sigma. Office: U NC Dept Chemistry Asheville NC 28804

SQUINTO, STEPHEN PAUL, molecular biologist, biochemist; b. Cook County, Ill., Sept. 19, 1956; s. Leonard and Emily (Gorowski) S.; m. Adrienne L. Block, June 14, 1987; 1 child, Eric. BA, Loyola U., Chgo., 1978, PhD, 1984. Grad. fellow Loyola U., Chgo., 1978-84, lectr., 1984-86; postdoctoral fellow Northwestern U., Chgo., 1984-86; asst. prof. La. State U. Med. Ctr., New Orleans, 1986-89; sr. staff scientist Regeneron Pharm., Tarrytown, N.Y., 1989-92; co-founder program dir. Alexion Pharm., New Haven, Conn., 1992—; cons. Nat. Inst. for Drug Abuse, Rockville, Md., 1989-90, NIH, Rockville, 1991—. Author: (with others) Methods of Enzymology, 1988, Prospects for Antisense Nucleic Acid Therapy of Cancer and Viral Infections, 1990; contbr. articles to profl. jours. Recipient Cancer Rsch. award United Way and Cancer Assn., New Orleans, 1987-89, Leukemia Soc., New Orleans, 1988-89, Ind. Nat. Rsch. Svc. award NIH, 1984-86. Mem. AAAS, Am. Soc. for Microbiology, Am. Soc. for Biochemistry and Molecular Biology, Am. Soc. for Neuroscience. Achievements include co-discovery of neurotrophic factor (NT-3) related to nerve growth factor, which is currently being developed as a potential therapeutic for neurodegenerative disease; co-discovery that a neurotrophic factor related to nerve growth factor (BDNF) protects nerves from those neurotoxins that cause Parkinson's disease and is currently being developed as a potential therapeutic for Parkinson's disease. Office: Alexion Pharm 25 Science Pk New Haven CT 06511

SQUIRES, RICHARD FELT, research scientist; b. Sparta, Mich., Jan. 15, 1933; s. Monas Nathan and Dorothy Lois (Felt) S.; m. Else Saederup, 1 child, Iben. BS, Mich. State U., 1958; postgrad., Calif. Inst. Tech., 1961. Rsch. biochemist Pasadena Found. for Med. Rsch., 1961-62; chief biochemistry sect. rsch. dept. A/S Ferrosan Soeborg, Demark, 1963-78; neurochemistry group leader CNS Biology sect. Lederle Labs. div. Am. Cyanamid Co., Pearl River, N.Y., 1978-79; prin. rsch. scientist The Nathan

S. Kline Inst. for Psychiat. Rsch., Orangeburg, N.Y., 1979—. Contbr. articles to profl. jours.; patentee in field. Nat. Inst. Neurol. and Communication Disorders and Stroke grantee, 1981-84. Mem. AAAS, Collegium Internationale Neuro-Psychopharmacologicum, Internat. Soc. Psychoneuroendocrinology, Soc. Neurosci., Internat. Soc. Neurochemistry, European Neurosci. Assn., Am. Soc. Neurochemistry, Am. Chem. Soc., Am. Soc. Biochemistry and Molecular Biology, Am. Soc. Pharmacology and Exptl. Therapeutics. Home: 10 Termakay Dr New City NY 10956-6434 Office: Nathan S Kline Inst Psychiat Rsch Orangeburg NY 10962

SREENIVASAN, KATEPALLI RAJU, mechanical engineering educator; b. Kolar, India, Sept. 30, 1947; married 1980; 2 children. BE, Bangalore U., 1968; ME, Indian Inst. Sci., 1970, PhD in Aeronautical Engring., 1975. JRD Tata fellow Indian Inst. Sci., 1972-74, project asst., 1974-75; fellow U. Sydney, Australia, 1975, U. Newcastle, 1976-77; rsch. assoc. Johns Hopkins U., Balt., 1977-79; from asst. prof. to assoc. prof. Yale U., New Haven, 1982-85, prof. mech. engring., 1985—, Harold W. Cheel prof. mech. engring., 1988—, prof. physics, 1990—, prof. applied physics, 1993—; vis. scientist Indian Inst. Sci., 1979, vis. prof., 1982; vis. sci. DFVLR, Gottingen, Germany, 1983; vis. prof. Calif. Inst. Tech., Pasadena, 1986, Rockefeller U., 1989, Jawaharlal Nehru Ctr. Advancement Sci. Studies, 1992; chmn. Coun. Eng. Dept., 1987-92. Recipient Narayan Gold medal Indian Inst. Sci., 1975, Disting. Alumnus award Indian Inst. Sci., 1992; named Humboldt Found. fellow, 1983, Guggenheim fellow, 1989. Fellow Am. Phys. Soc., ASME; mem. AIAA, Am. math. Soc., Conn. Acad. Sci. and Engring., Sigma Xi. Achievements include research in origin and dynamics of turbulence; control of turbulent flows; chaotic dynamics; fractals. Office: Yale Univ Dept Mech Engring M6 ML New Haven CT 06520

SREENIVASAN, SREENIVASA RANGA, physicist, educator; b. Mysore, Karnataka, India, Oct. 20, 1933; came to U.S., 1959; s. Sreenivasachari and Alamelammal (Rangaswami) S.; m. Claire de Reineck, Nov. 16, 1963; children: Gopal, Govind, Gauri, Gayatri, Aravind. BS, U. Mysore, 1950, U. Mysore, 1952; PhD, Gujarat U., India, 1958. Lectr. St. Philomena's Coll., Mysore, 1952-54; rsch. fellow Harvard U., Cambridge, Mass., 1959-61; rsch. assoc. NASA Inst. for Space Studies, N.Y.C., 1961-64; vis. scientist Max Planck Inst. Physics and Astrophysics, Munich, Fed. Republic of Germany, 1964-66; prof. physics U. Calgary, Alta., Can., 1967—; vis. prof. Royal Inst. Tech., Stockholm, Sweden, 1974-75. Contbr. articles to profl. jours. Chmn. Coun. South Asians, Calgary, 1981-84; pres. Calgary Interfaith Community Action Assn., 1986; pri. Sch. East Indian Langs. and Performing Arts, Calgary, 1986. Recipient Govt. of India Sr. Rsch. scholar, Ahmedabad, 1955-58. Achievements include rsch. in force-free fields, electrostatic instabilities in plasmas, evolution of massive stars, size of convective cores in rotating stellar models. Home: 2110 30 Ave SW, Calgary, AB Canada T2T 1R4 Office: U Calgary, 2500 University Dr NW, Calgary, AB Canada T2N 1N4

SRERE, PAUL A., biochemist, educator; b. Davenport, Iowa, Sept. 1, 1925. BS in Chemistry, U. Calif., L.A., 1947; PhD in Comparative Biochemistry, U. Calif., Berkeley, 1951. Biochemist Mass. Gen. Hosp., Boston, Mass., 1951-53; postdoctoral fellow Jane Coffin Fund for Med. Rsch. Yale U., 1953-54; post doctoral fellow Pub. Health Rsch. Inst. City of N.Y., 1954-55; postdoctoral fellow U. Munich, Germany, 1955-56; asst. prof. dept. biol. chemistry U. Mich., 1956-61, assoc. prof., 1961-63; rsch. biochemist U. Calif., 1963-66; prof. dept. of biochemistry, Southwestern Med. Ctr. U. Tex., Dallas, 1966—; chief pre-clin. sci. unit Dept. V.A. Med. Ctr., Dallas, 1966—; prof. dept. internal medicine, Southwestern Med. Ctr. U. Tex., Dallas, 1973—; rsch. career scientist Dept. V.A. Med. Ctr., Dallas, 1975—; vis. scientist U. Lund, Sweden, 1972, Inst. Enzymology, Budapest, 1982, Sheffield U., England, U. Basel Biozentrum, Switzerland, 1985, U. Lund, 1992; mem. numerous editorial bds. Editor: (with others) Regulation of Cardiac Metabolism, 1976, Metabolism, Clinical, and Experimental, 1976, Microenvironments and Metabolic Compartmentation, 1978, Current Topics in Cellular Regulation, 1981, UCLA Symposia on Molecular and Cellular Biology, 1990. Recipient William S. Middleton award, 1974, Wilford T. Doherty awd., Am. Chemical Soc., 1993. Mem. AAAS, AAUP, Am. Chemical Soc. (treas. 1979-81), Am. Soc. for Cell Biology, Am. Soc. Biological Chemists, Am. Soc. Bichemistry and Molecular Biology, Protein Soc. Achievements include research in stucture and function, both chemical and biological of the three citrate lyase enzymes. Office: U Texas Southwestern Medical Ctr Dept Biochemistry & Internal Med 5323 Harry Hines Blvd Dallas TX 75235*

SRESTY, GUGGILAM CHALAMAIAH, environmental engineer; b. Perala, India, Nov. 11, 1954; came to U.S., 1976; s. Ramasubbarao and Suseela (Utakuri) Guggilam; m. Annapurnadevi Guggilam Madipalli, Nov. 7, 1979; children: Padma, Hema. B in Tech., Banaras Hindu U., Varanasi, India, 1975; MS, Columbia U., 1978. Assoc. engr. Ill. Inst. Tech. Rsch. Inst., Chgo., 1979-80, rsch. engr., 1980-85, sr. engr., 1985-91, mgr. environ. technologies sect., 1991-92, mgr. energy and environ. scis. dept., 1992—; mem., sr. tech. com. Nat. Inst. for Petroleum and Energy Rsch., Bartlesville, Okla., 1984-86. Author: Size Reduction, 1985; co-author: Particle Size Analysis, 1980. Pres. India Club of Columbia U., N.Y.C., 1978; student coord. Andhra Samithi, Varanasi, India, 1974. Recipient Best Presentation award Am. Inst. Mining, Metallurgy and Petroleum Engrs., 1978, Gold medal Banaras Hindu U., 1975, Hadfield medal, 1975. Mem. AICE, Am. Inst. of Mining, Metallurgy and Petroleum Engring. Hindu. Achievements include co-invention of a single well stimulation technology commercially used in several nations; co-invention of waste disinfection technology commercially used at several plants in the U.S. Home: 8241 S Mason Ave Burbank IL 60459 Office: Ill Inst Tech Rsch Inst 10 W 35th St Chicago IL 60616-3799

SRIDHARA, CHANNARAYAPATNA RAMAKRISHNA SETTY, health facility administrator, consultant; b. Channarayapatna, Karnataka, India, Aug. 15, 1948; came to U.S., 1975.; m. Vijaya Belagodu; children: Nethra, Rashmi. MD, Mysore (India) U., 1971. Intern K.R. Hosp., Mysore, 1972-73, sr. house officer medicine and dermatology, 1973-74; sr. house officer ENT Singleton Hosp., Swansea, Wales, 1974-76; chief resident and clin. instr., dept. phys. and med. rehab. Temple U. Hosp., Phila., 1978-79, clin. asst. prof., dept. phys. and med. rehab., 1979-82, asst. prof. dept. phys. and med. rehab., 1982-90, clin. assoc. prof. dept. phys. and med. rehab., 1990—; attending physiatrist/dir. pulmonary rehab. Moss Rehab. Hosp., Phila., 1979-82; acting unit dir., dept. phys. and med. rehab. Albert Einstein Med. Ctr., Phila., 1982-83, dir. clin. svcs., dept. phys. and med. rehab., 1983-87, 89-90, acting chmn., dept. phys. and med. rehab., 1987-89, assoc. chmn., 1990—; residency program coord. dept. phys. and med. rehab. Temple U. Hosp., 1985-87; cons. Belmont Ctr. for Comprehensive Treatment, Phila., 1985—; staff physiatrist Moss Rehab. Hosp., 1979—. Sec. Triveni Cultural Orgn., N.J., Pa., Del., 1985. Fellow Am. Acad. Phys. Medicine and Rehab. (grad. edn. com. 1987-89), Am. Assn. Electrodiagnostic Medicine (workshop com. 1990-93); mem. Phila. Soc. Phys. Medicine and Rehab. (pres. 1990-91), Phila. County Med. Soc. (inter splty. coun. 1990—). Achievements include standardization of nerve conduction studies of branches of superficial peroneal nerve and electromyography of entrapment neuropathy of superficial peroneal nerve; research in archives of phys. medicine and rehab. medicine/surgery. Office: Albert Einstein Med Ctr 5501 Old York Rd Paley Bldg 1st Flr Philadelphia PA 19141

SRINIVASACHARI, SAMAVEDAM, chemical engineer; b. Visakhapatnam, India, Oct. 5, 1926; came to U.S., 1958; s. Appalachari Srinivasa and Chudamani Samavedam; m. Vasanta S. Chari, Feb. 11, 1955; children: Sarita, Roger. M of Chem. Engring., NYU, 1959; PhD in Chem. Engring., Poly. Inst. Bklyn., 1967. Registered profl. engr., Pa. Teaching fellow Poly. Inst. Bklyn., 1960-63; sr. process devel. engr. Internat. Latex & Chemical, Dover, Del., 1966-73; prin. process engr. Catalytic, Inc., Phila., 1973-75; sr. process engr. Coalcon/Union Carbide, N.Y.C., 1976-77, Foster Wheeler Energy Corp., Livingston, N.J., 1977-82; chemical and environ. engr. Duro-Test Corp., Clifton, N.J., 1987-92, mgr. environ. engring., 1992—. Mem. AICE, Am. Chem. Soc. Democrat. Hindu. Achievements include design of chem. and petroleum plants, design of synthetic fuels plants; rsch. in environ. and regulatory problems of indsl. plants, environ. clean up of indsl. plant sites. Home: 12 The Terrace Rutherford NJ 07070 Office: Duro-Test Corp 185 Scoles Ave Clifton NJ 07012

SRINIVASAN, MANDAYAM PARAMEKANTHI, software services executive; b. Mysore City, India, July 1, 1940; s. Appalacharya Paramekanthi and Singamma Budugan; came to U.S., 1970, naturalized 1991; B.S., U. Mysore, 1959, B.E. in Mech. Engring., 1963; M.S. in Ops. Research, Poly. Inst. N.Y., 1974, M.S. in Computer Sci., 1983; m. Ranganayaki Srirangapatnam, June 18, 1967. Costing engr. Heavy Engring. Corp., Ranchi, Bihar, India, 1963-70; inventory analyst Ideal Corp., Bklyn., 1970-75; systems analyst Electronic Calculus, Inc., N.Y.C., 1975-76; cons. in software, project leader Computer Horizons Corp., N.Y.C., 1976-85; pres. Compmusic, Bellerose, N.Y., 1985—; tchr., cons. in-house tng. Founding mem. governing council Vishwa Hindu Parishad of U.S.A., 1973—, pres. N.Y. State chpt., 1977-86. Mem. Assn. for Computing Machinery, IEEE, Inst. Engrs. (India). Republican. Hindu. Office: Compmusic Inc 8229 251st St Jamaica NY 11426-2527

SRIVASTAVA, KAILASH CHANDRA, microbiologist; b. Varanasi, India, Sept. 4, 1947; came to U.S., 1969; s. Hari and Bimla (Varma) Kishore; m. Kumkum Chandra, Feb. 27, 1977; children: Mukta, Tarun K. BS, Punjab U., 1968; MS, Seton Hall U., 1971; PhD, Univ. Coll., London, 1975. Dir. lab. ops. Ivy Med. Lab., N.Y., 1975-78; assoc. prof. Dept. Food Sci. & Technology, Londrina, Brazil, 1978-80; prof. dept. food sci. and tech. Dept. Food Sci. & Technology, Londrina, 1980-84, Forest Products Lab., Madison, Wis., 1984-86, Mich. State U., East Lansing, 1986-87; vis. sr. scientist Mich. Biotech. Inst., Lansing, 1987-88, rsch. scientist, 1988—; adj. asst. prof. Mich. State U., East Lansing, 1990—. Contbr. articles to profl. jours.; co-author: (symposium volume) Internat. Yeast Symposium, Biochemical Pentose Fermenting Yeasts, 1986, Symposium on Fuels, Technology, Novel Sources For Cellulase Production, 1985. Recipient Cert. of Appreciation, USDA, 1985, Svc. award Mich. Biotechnology Inst., 1989, Edwina Mountbatten scholarship U. London, 1975, Yusuf Ali award, 1972, Brit. Coun. studentship, 1971-74. Mem. Inst. Food Technology, N.Y. Acad. Scis., Japan Soc. for Biosci., Biotechnology & Agrochemistry. Achievements include patent on alkaophilic, thermostable lipase; elucidation of styrene degradation pathway in thermophilic Bacillus; development of microbial consortia to degrade recalcitrant compounds; isolation of novel adicuric, thermostable glucose isomerase from new organism, of novel microorganisms and novel lipase, collagenase and thermostable pullulanase enzyme producing microorganisms, new collagenase from thermophilic bacillus; first to demonstrate mesosomal structures in yeasts; research on degradation of recalcitrant, xenobiotic compounds by microbes. Office: Arctech, Inc Ste 210 14100 Park Meadow Dr. Chantilly VA 22021*

SRIVASTAVA, OM PRAKASH, soil science and agricultural chemistry educator; b. Banda, India, Jan. 6, 1944; s. Brijmohan Lal and Shanti Devi Srivastava; m. Sharda Rani, June 12, 1966; children: Shikha, Prashant. MSc in Agrl. Chemistry, Agra U., India, 1962, PhD in Agrl. Chemistry, 1967. Asst. soil chemist Haryana Agrl. U., Hisar, Haryana, India, 1966-75; rice chemist, prof. Chnadra Shedhar Azad Agr. U., Kanpur, India, 1975-83; prof. soil sci. and agrl. chemistry Banaras Hindu U., Varanasi, India, 1984—; mem. rsch. Meerut, Agra & Purwanchal Univs., India, 1984-90; exptl. mem. selection com. various univs., India, 1984-92. Author bull. Efficient use of Fertilizers, 1972; editor: Jour. Ind. Soc. Agrl. Chemists, 1990-93, Agr. Rsch. Jour., 1992-93; contbr. articles to profl. jours. fellow Fertilizer Assn. India-Delhi Cloth Mills, 1963-66; grantee Univ. Grants commn., 1971-72, 91-92. Mem. Ind. Soc. Soil Scientist. Mem. Vedic Ch. Achievements include research in micro distillation apparatus for soil-N, neutral blue a new indicator, disciplines of N,P, organic matter, micro-nutrients, water-logged soils and salt affected soils. Home and Office: Banaras Hindu U, New D2 Tulsidas Colony, Varanasi UP 221005, India

STAAB, HEINZ A., chemist; b. Darmstadt, Germany, Mar. 26, 1926; m. Ruth Mueller, Aug. 22, 1953; children: Doris, Volker. BSc, U. Marburg, Fed. Republic Germany, 1949; Diploma in Chemistry, U. Tuebingen, Fed. Republic Germany, 1951; PhD, U. Frankfurt, Fed. Republic Germany, 1953; MD, U. Heidelberg, Fed. Republic Germany, 1960; PhD (hon.), Weizmann Inst., Rehovot, Israel, 1983. Research assoc. Max Planck Inst., Heidelberg, 1953-59; sucessively asst. prof, assoc. prof., prof. chemistry U. Heidelberg, 1959—, dir. Inst. Organic Chemistry, 1964-76, head dept. organic chemistry, 1976—; bd. dirs. Max Planck Inst., Bayer Ag Levekinsen, Defgussa Ag Frankfurt; pres. Max Planck Soc., 1984-90. Contbr. articles to profl. jours. Recipient Weizmann award in scis. and humanities, 1989, Grosses Bundesverdienstkreuz mit Stern, Govt. Fed. Republic Germany, 1990. Mem. German Chem. Soc. (pres. 1984, 85, Adolf v. Baeyer medal 1979), Gesellschaft Deutscher Naturforscher und Aerzte (pres. 1980, 81), Deutsche Forschungsgemeinschaft (senator 1976-82, 84-90), Internat. Union Pure and Applied Chemistry (v.p., pres.-elect 1991), Heidelberg Acad. Wissenschaften (nat. sci. sect. sec. 1970-77, pres.-elect 1993), Austrian Acad. Scis., Acad. Leopoldina (chmn. chem. com. 1974-90), Bavarian Acad. Scis., Indian Acad. Scis. (hon.), Academia Sinica (hon. prof.), Academia Europaea London (found. mem., exec. coun. 1988-89), Rotary. Avocations: music, travel. Home: Schloss-Wolfsbrunnenweg 43, 69118 Heidelberg Germany Office: Max Planck Inst Med Rsch, Jahnstr 29, 6900 Heidelberg Germany

STAATS, DEE ANN, toxicologist; b. Parkersburg, W.Va., Mar. 15, 1957; d. Edward Harlan and Ada Maxine (Kittle) S.; m. Daniel Patrick Leigh, May 9, 1992. BA in Chemistry, BS in Biology, W.Va. Wesleyan U., 1979; PhD in Pharmacology/Toxicology, W.Va. U., 1987. Postdoctoral fellow Baylor Coll. Medicine, Houston, 1987-88; sr. scientist NSI Tech. Svcs., Dayton, Ohio, 1988-90, Internat. Tech. Corp., Knoxville, Tenn., 1990-91; owner, operator Staats Creative Scis., Mt. Dora, Fla., 1991—. guest columnist Fla. Specified Mag., 1992—; contbr. articles to sci. jours. Recipient Travel award Procter & Gamble, Cin., 1984, Tng. Grant award NIH, Washington, 1983-86. Mem. Soc. Risk Analysis, Soc. Toxicology, Soc. Environ. Toxicology and Chemistry, N.Y. Acad. Scis. Achievements include first scientist to show localization of vitamin E in different physiol. zones of the adrenal cortex; found that depletion of vitamin E from the adrenal cortex due to vitamin deficiency affects steroidogenic enzymes and renders the organ more susceptible to toxic insult; developed physiologically based pharmacokinetic model for chloroform cytotoxicity/carcinogenicity. Home and Office: 1220 Oakland Dr Mount Dora FL 32757

STABENAU, M. CATHERINE, engineering executive; b. Indpls., May 12, 1948; d. Charles Harold and Marie Catherine (Scharfenberger) Bishop; m. Walter Frank Stabenau, Nov. 20, 1971; children: Elizabeth Ann, Derek Walter. BA in Math., Spaulding Coll., 1970; BEE, U. Dayton, 1979. System programmer Air Force Logistics, WPAFB, Ohio, 1970-74; sr. engr. GE Heavy Mil. Equipment Div., Syracuse, N.Y., 1980-84; mgr. Martin Marietta Astrospace Divsn., King of Prussia, Pa., 1984-91, Princeton, N.J., 1991—. Recipient Women in Engring. Fellowship, NSF, 1979. Home: 1543 Silo Rd Morrisville PA 19067-4240 Office: Martin Marietta Astrospace Divsn PO Box 800 Princeton NJ 08543-0800

STABENAU, WALTER FRANK, systems engineer; b. Cleve., Apr. 24, 1942; s. Walter Kurt and Helen (Koris) S.; m. Mary Catherine Bishop, Nov. 20, 1971; children: Elizabeth Ann, Derek Walter. BS in Physics, Case Inst. Tech., 1964; PhD in Nuclear Sci., Cornell U., 1969. Computer programmer Air Force Logistics Commd., Wright Patterson AFB, Ohio, 1970-74; navigation analyst Logicon, Inc., Dayton, Ohio, 1974-80; sonar engr. Gen. Electric, Syracuse, N.Y., 1980-84; prin. systems engr. RCA Corp., Moorestown, N.J., 1984-86, mgr. combat systems analysis, 1986-89, project mgr., 1989-91; cons. Sonalysts Inc., Willingboro, N.J., 1991—. Contbr. articles to profl. jours. Mem. Assn. Old Crows. Avocation: stamp collecting. Home: 1543 Silo Rd Morrisville PA 19067-4240 Office: Sonalysts Inc 624 Highland Dr Mount Holly NJ 08060

STACEY, THOMAS RICHARD, geotechnical engineer; b. Durban, Republic of South Africa, Oct. 17, 1943; s. Norman Burbeck and Gretta Elaine (Thomas) S.; m. Judy Ann Whitelaw, Jan. 6, 1968; children: Helen, Jean. BSME, U. Natal, Durban, 1965, MSME, 1968; DScEng, U. Pretoria, Republic South Africa, 1973; diploma, Imperial Coll., U. London, 1974. Registered profl. engr., Republic of South Africa, chartered engr., U.K. Rsch. engr. Coun. for Sci. and Indsl. Rsch., Pretoria, 1967-73; acad. vis. Imperial Coll. Sci. and Tech., London, 1973-74; geotech. engr. D.L. Webb & Assocs., Durban, 1974-75; prin. Steffen, Robertson & Kirsten, Johannesburg, 1976—; hon. prof. U. Natal, Durban; postgrad. external examiner U. Pretoria, 1982—, U. Witerwarsrand, Johannesburg, 1987—, U. Cape Town, Republic of South Africa, 1989. Author: Practical Handbook for Underground Rock Mechanics, 1986; contbr. articles, conf. procs. to profl. publs. Recipient Thomas Price award South Africa Instn. Mech. Engrs., 1969, Best Tech. Note award South Africa Instn. Civil Engrs., 1982, Shell Design award Design Inst., 1984. Mem. South African Inst. for Mining and Metallurgy (Silver medal 1989), Inst. Mining and Metallurgy, South African Nat. Group on Rock Mechanics, Internat. Soc. Rock Mechanics (v.p. Africa 1991—). Home: PO Box 699, Rivonia 2128, Republic of South Africa Office: Steffen Robertson & Kirsten, PO Box 55291, Northlands 2116, South Africa

STACHEL, JOHN JAY, physicist, educator; b. N.Y.C., Mar. 29, 1928; s. Jacob Abraham and Bertha Z. Stachel; m. Evelyn Lenore Wassermann, Feb. 8, 1953; children: Robert, Laura, Deborah. B.S., CCNY, 1956; M.S., Stevens Inst. Tech., 1959, Ph.D., 1962. Instr. physics Lehigh U., Bethlehem, Pa., 1959-61; instr. physics U. Pitts., 1961-62, research assoc., 1962-64; asst. prof. physics Boston U., 1964-69, assoc. prof., 1969-72, prof., 1972—; dir. Ctr. for Einstein Studies, Boston U., 1985—; vis. research assoc. Inst. Theoretical Physics, Warsaw, 1962; vis. prof. King's Coll., U. London, 1970-71, U. Paris, 1990-91; vis. sr. research fellow Dept. Physics, Princeton U., 1977-84. Editor: Selected Papers Leon Rosenfeld, 1979, Foundations of Space-Time Theories, 1977, Einstein Studies, 1989—, Collection Papers of Albert Einstein, Princeton U. Press, 1977-88. Office: Boston U Dept Physics Boston MA 02215

STACY, DENNIS WILLIAM, architect; b. Council Bluffs, Iowa, Sept. 22, 1945; s. William L. and Mildred Glee (Carlsen) S.; BArch., Iowa State U., 1969; postgrad. U. Nebr., 1972. Registered architect, Iowa, Tex., Colo., Mo.; m. Judy Annette Long, Dec. 28, 1968; 1 child, Stephanie. Designer Troy & Stalder Architects, Omaha, 1967, Architects Assocs., Des Moines, 1968-69, Logsdon & Voelter Architects, Temple, Tex., 1970; project architect Roger Schutte & Assos., Omaha, 1972-73; architect, assoc. Robert H. Burgin & Assocs., Coun. Bluffs, 1973-75, Neil Astle & Assocs., Omaha, 1975-78; owner, prin. Dennis W. Stacy, AIA, Architect, Glenwood, Iowa, 1978-81, Dallas, 1981—. Mem. City of Dallas Urban Design Adv. Com., 1992; chmn., Glenwood Zoning Bd. Adjustment, 1979-81; chmn. Mills County Plant Iowa Program, 1979-81; mem. S.W. Iowa Citizen's Adv. Com., Iowa State Dept. Transp., 1977-81; regional screening chmn. Am. Field Svc. Internat./Intercultural Programs, 1974-79, Iowa-Nebr. rep., 1978-80. With U.S. Army, 1969-71. Decorated Nat. Def. Svc. medal, Vietnam Svc. medal, Vietnam Campaign medal, Army Commendation medal. Mem. AIA (recipient Iowa Design Honor award 1981, commendation awards (2) 1990, citation of honor award 1991, 92, Dallas Design awards (2) 1991, Texas Design Honor award 1992, Dallas AIA Firm of Yr. award 1992; Dallas commr. design, 1991; chmn. Dallas Design Awards 1992;mem. Tex. architect publs.com. 1992), Nat. Coun. Archtl. Registration Bds., Mus. Modern Art., The 500 Inc. (outstanding mem. 1985), Glenwood Optimist (Disting. Svc. award 1982, pres. 1980-81), Masons. Archtl. works include Davies Amphitheater, 1980, Addison Nat. Bank Bldg., 1985, Fairview Recreation Complex, 1984, Computer Lang. Rsch. Corp. Learning Ctr., 1987., Villa Roma, 1988, Robbins Spa, 1989, Dallas Chpt. AIA Offices, 1990. Home: 4148 Cobblers Ln Dallas TX 75287-6725 Office: 2136 N Harwood St Ste 100 Dallas TX 75201-2246

STACY, PHERIBA, archaeologist; b. Jacksonville, Fla., Feb. 11, 1940; d. Robert Carson and Alberta (Carlton) S.; m. Merrill Ross Pritchett, Oct. 7, 1978. BA, Fla. State U., 1960; PhD, U. Ariz., 1974. Sci. curator Jacksonville Mus. Arts and Scis., 1960-64; asst. prof. Fla. State U., Tallahassee, 1968-69; archaeologist Nat. Park Svc., Tucson, 1970-74; survey archaeologist Ark. Archaeol. Survey, Fayetteville, 1974-78; asst. prof. U. Ark., Monticello, 1974-80; archaeol. permits Md. Hist. Trust, Annapolis, 1981-82; assoc. prof. U. Balt., 1986—; vis. prof. U. Nebr., 1979; participant INAH Conf. N.W. Mex., Hermosillo, Mex., 1974, Ariz. State U. Archaeology, 1980; cons. in field. Author Cerros de Trincheras in Arizona, 1981; contbr. articles to profl. jours. Founder So. Ark. chpt. Ark. Archaeol. Survey, Monticello, 1975. Faculty rsch. grantee U. Ark., 1979; NEH fellow UCLA, 1978. Mem. Sigma Xi. Achievements include perspectives on Brit. acculturation, stratigraphic interpretation of P. III ruin superimposed over excavated P. II pit house with burned log roof; directed Papago Indian archaeol. tng. program; excavated archaeol. evidence to confirm Caddoan interaction sphere in so. Ark. Home: 1624 N Calvert St Baltimore MD 21202 Office: U Balt 1420 N Charles St Baltimore MD 21202

STADDON, JOHN ERIC RAYNER, psychology, zoology, neurobiology educator; b. Grayshott, Hampshire, Eng.; came to U.S., 1960; s. Leonard John and Dulce Norine (Rayner) S.; m. Lucinda Paris. BSc, Univ. Coll., London, 1960; PhD, Harvard U., 1964. Asst. prof. psychology U. Toronto, Ont., Can., 1964-67; from asst. prof. to prof. Duke U., Durham, N.C., 1967-72, prof., 1972-83, J.B. Duke prof. psychology, prof. neurobiology and zoology, 1983—. Author: Adaptive Behavior and Learning, 1983; editor Behavioral Processes, 1979—; assoc. editor Jour. Exptl. Analysis of Behavior, 1979-82. Recipient von Humboldt prize, 1985. Fellow AAAS, N.Y. Acad. Scis., Soc. Exptl. Psychology; mem. Phi Beta Kappa (hon.), Sigma Xi. Avocations: history, philosophy of science, public policy. Office: Duke U Dept Exptl Psychology Durham NC 27706

STADLBAUER, HARALD STEFAN, engineer; b. Vöcklabruck, Austria, Feb. 9, 1963; s. Rupert and Brigitte (Ogerer) S. Diplom-Ingenieur, U. Linz, Austria, 1986; PhD, Tech. U. Vienna, Austria, 1991. Student asst. Risc-Linz, U. Linz, 1985-86; guest researcher Voest Linz, Automation & Info. Systems, 1986-87, Tech. U. INFA, Vienna, 1988; researcher in engring. Tech. U. INFA, 1988—; cons. in field. Avocations: sailing, philosophy. Home: Waldstrasse 2, Lenzing Austria A-4860 Office: Tech Univ, Gusshausstr 27, Vienna Austria A-1040

STADTMAN, EARL REECE, biochemist; b. Carrizozo, N.Mex., Nov. 15, 1919; s. Walter William and Minnie Ethyl (Reece) S.; m. Thressa Campbell, Oct. 19, 1943. B.S., U. Calif., Berkeley, 1942, Ph.D, 1949. With Alcan Hwy. survey Pub. Rds. Adminstrn., 1942-43; rsch. asst. U. Calif., Berkeley, 1938-49, sr. lab. technican, 1949; AEC fellow Mass. Gen. Hosp., Boston, 1949-50; chemist lab. cellular physiology Nat. Heart Inst., 1950-58, chief enzyme sect., 1958-62, chief lab. biochemistry, 1962—; biochemist Max Planck Inst., Munich, Germany, Pasteur Inst., Paris, 1959-60; faculty dept. microbiology U. Md.; prof. biochemistry grad. program dept. biology Johns Hopkins U.; adv. com. Life Scis. Rsch. Office, Am. Fedn. Biol. Sci., 1974-77; bd. dirs. Found. Advanced Edn. Scis., 1966-70, chmn. dept. biochemistry, 1966-68; biochem. study sect. rsch. grants NIH, 1959-63. Editor Jour. Biol. Chemistry, 1960-65, Current Topics in Cellular Regulation, 1968—, Circulation Rsch., 1968-70; exec. editor Archives Biochemistry and Biophysics, 1960—, Life Scis., 1973-75, Procs. NAS, 1975-81, Trends in Biochem. Rsch., 1975-78; mem. editorial adv. bd. Biochemistry, 1969-76, 81—. Recipient medallion Soc. de Chemie Biologique, 1955, medallion U. Pisa, 1966, Presdl. Rank award as Disting. Sr. Exec., 1981, Welch Found. award, 1991, Rsch. award Am. Aging Assn., 1992, Paul Glen award Am. Gerontology Soc., 1993. Mem. Am. Chem. Soc. (Paul Lewis Lab. award in enzyme chemistry 1952, exec. com. biol. div. 1959-64, chmn. div. 1963-64, Hillebrand award 1969), Am. Soc. Biol. Chemists (publs. com. 1966-70, coun. 1974-77, 82-84, pres. 1983—, Merckaward 1983), Nat. Acad. Scis. (award in microbiology 1970), Am. Acad. Arts and Scis., Am. Soc. Microbiology, Washington Acad. Scis. (award biol. chemistry 1957, Nat. medal sci. 1979, meritorious exec. award 1980, Robert A. Welch award in chemistry 1991). Office: Nat Heart and Lung Inst 9000 Rockville Pike Bethesda MD 20892-0001

STADTMAN, THRESSA CAMPBELL, biochemist; b. Sterling, N.Y., Feb. 12, 1920; d. Earl and Bessie (Waldron) Campbell; m. Earl Reece Stadtman, Oct. 19, 1943. BS, Cornell U., 1940, MS, 1942; PhD, U. Calif.-Berkeley, 1949. Rsch. assoc. U. Calif., Berkeley, 1942-47; Rsch. assoc. med. sch. Harvard U., Boston, 1949-50; biochemist Nat. Heart, Lung and Blood Inst. NIH, USPHS, HHS, Bethesda, Md., 1950—. Editor Jour. Biol. Chemistry, Archives Biochemistry and Biophysics, Molecular and Cellular Biochemistry; editor-in-chief Bio Factors; contbr. articles on amino acid metabolism, methane biosynthesis, vitamin B12 biochemistry, selenium biochemistry to profl. jours. Helen Haye Whitney fellow Oxford U., Eng., 1954-55; Rockefeller Found. grantee U. Munich, 1959-60; recipient Rose award, 1987, Klaus Schwarz medal, 1988. Mem. Am. Soc. Microbiology, Biochem. Soc., Soc. Am. Biochemists, Am. Chem. Soc., Nat. Acad. Scis., Am. Acad. Arts and Scis., Sigma Delta Epsilon (hon.). Home: 16907 Redland Rd Derwood MD 20855 Office: Nat Heart Lung & Blood Inst HHS Bethesda MD 20892

STAFF, ROBERT JAMES, JR., international economist; b. 1946; s. Robert J. and Harriet G. (Karber) S.; m. Martha Lee Coleman, 1976; children: Adrian , Marika. Student Wilhelms Universität, W.Ger., 1966; BA, Kalamazoo Coll., 1968; postgrad. Ateneo de Manila Univ., Philippines, 1969; MBA, Am. Mgmt. Assn., 1970; MPA, Harvard U., 1990. Sr. cons. George Odiorne Assocs., Ann Arbor, Mich., 1972-74; assoc. cons. Hutchings Orgn., Palo Alto, Calif., 1974-75; sr. mgmt. cons. E.H. White & Co., Inc., San Francisco, 1975-78; pres. The Wavelink Orgn., Honolulu, 1978—; internat. economist dept. fin. econ. Coll. Bus. Adminstrn U. Hawaii, 1991—; consulting analyst State of Hawaii, Honolulu, 1979—; research mgr. Inst. Philippine Culture, Manila, 1969; sector adviser U.S. Dept. State, Washington, 1978; fiscal project mgr. U. Hawaii, Honolulu, 1980; vis. lectr. Pacific Asian Mgmt. Inst., 1981; coordinator Pacific Devel. Program East-West Ctr., Honolulu, 1981; mem. staff Prime Minister's Disaster Relief Com., Fiji, 1981; mem. adv. staff Kahauale'a Geothermal Energy Project, 1982-83; mem. Gov.'s Adv. Com. on Criminal Justice Info. Systems, 1984; vis. lectr. Inst. Econ. Devel. and Policy, East-West Ctr., Honolulu, 1992; mem. state innovations task force Nat. Govs. Assn, 1992. Author: Political Aspects of Modernization: Buddhist Experiences in Southeast Asia, 1968, Assessment of International Health Manpower Planning, 3 vols., 1978-80; Consolidated Fiscal Procedures in Education, 1981, Trades and Tradeoffs: Strategic Policy Operants in Public-Private Partnerships, 1990, Bioregionalism as a Public Policy Operant, 1992; co-author: MBO Systems Manual, 1974; National Manpower Utilization Study, 1976. Dep. dir. Anchorage Econ. Opportunity Agy., Alaska, 1971-72; diplomatic liaison Coll. Fgn. Study Program, W.Ger., 1966. Served with USAFR, 1970-72. Hughes Meml. scholar Kalamazoo Coll., 1965-68; invited scholar Internat. Negotiations Workshop Harvard Law Sch., 1990. InterPacific fellow, 1989-90; recipient Rewick award, McMannis award Harvard U. Sch. Govt., 1989-90, Internat. Peace award Beyond War Found., 1987. Home: PO Box 1873 Honolulu HI 96805-1873 Office: State Capitol PO Box 150 Honolulu HI 96810-0150

STAFFA, JUDY ANNE, epidemiologist; b. Pittsfield, Mass., Aug. 27, 1960; d. Francis Albert and Doris Marilyn (Smith) Gingras; m. Edward Joseph Staffa, June 27, 1987; children: Anthony Joseph, Caroline Elizabeth. BS in Pharmacy, U. Conn., 1983; MS in Behavioral Sci., Harvard U., 1987; student, Johns Hopkins U., 1989—. Registered pharmacist, Conn., Mass., Md. Pharmacist People's Drug Stores, Inc., Alexandria, Va., 1983-89; epidemiologist The Degge Group, Ltd., Arlington, Va., 1988—; pharmacist Eaton Apothecaries, Waltham, Mass., 1986; rsch. asst. dept. behavioral scis. Harvard U., Boston, 1986, Med. Found., Inc., Boston, 1987; course leader Pharm. Mfrs. Assn. Edn. & Rsch. Isnt., Arlington, 1991-92. Author: (with others) Communication in Pharmaceutical Medicine: A Challenge for 1992, 1991; contbr. articles to Am. Jour. Hospital Pharmacy, Pharmacoepidemiology and Drug Safety, Fertility & Sterility. Vol. Am. Heart Assn., Washington, 1992. Scholar U. Conn. Alumni Assn., 1981, Rite Aid Corp., 1982. Mem. APHA, Internat. Soc. Pharmacoepidemiology, Soc. Epidemiologic Rsch., Drug Info. Assn., Rho Chi. Roman Catholic. Home: 8713 38th Ave College Park MD 20740 Office: The Degge Group Ltd 1616 N Ft Myer Dr Ste 1430 Arlington VA 22209

STAFFORD, FRED EZRA, science adminstrator; b. N.Y.C., Mar. 31, 1935; s. Frank H. and Eva Stafford; m. Barbara Maria Stafford, Aug. 24, 1963. AB, Cornell U., 1956; PhD, U. Calif., Berkeley, 1960. From asst. to assoc. prof. dept. chemistry Northwestern U., Evanston, Ill., 1961-75; spl. asst. NSF, Washington, 1974-75, program dir., 1975-87; dir. spl. projects U. Chgo., 1986—; mem. various panels NSF, Washington, 1987—. Contbr. articles to profl. jours. Del. Gov.'s Sci. Adv. Com. Edn. Task Force, Ill., 1990; bd. dirs. Chgo. unit Recording for the Blind, 1990—. NSF fellow, 1952-59, NSF postdoctoral fellow, 1959-61. Mem. Am. Chem. Soc., Am. Phys. Soc., Fed. Exec. Inst. Alumni Assn., Phi Beta Kappa, Palme Académique. Achievements include research in structure and reactive intermediates of gaseous boron hydrides. Office: U Chgo 970 E 58th St Chicago IL 60637

STAFFORD, PATRICK MORGAN, biophysicist; b. Roanoke, Va., June 3, 1950; s. Jess Woodrum and Georgine Elna (Morgan) S.; m. Kristina Lee Troyer, July 10, 1976; children: Kathryn Lee, Jess Walter. BS in Physics, Va. Poly. Inst., 1972; M Med. Sci. in Medical Physics, Emory U., 1979; PhD in Biophysics, U. Tex., Houston, 1987. Diplomate Am. Bd. Radiology. Assoc. engr. Duke Power Co., Charlotte, N.C., 1973-78; rsch. scientist U. N.Mex., Los Alamos, 1979-81; asst. physicist U. Tex., Houston, 1981-87; asst. prof. U. Pa., Phila., 1987-91; v.p. Radiation Care, Inc., Atlanta, 1991—; lectr. in field. Author: Dynamic Treatment with Pions at Lampf, 1980, Critical Angle Dependance of CR-39, 1986, Real-Time Portal Imaging; jour. reviewer Medical Physics, Internat. Jour. Radiation Oncology, Biology and Physics. Active Atlanta Emory Symphony, 1978; deacon Crabapple Bapt. Ch., Alpharetta, Ga., 1992. Rosalie B. Hite fellow U. Tex., 1983. Mem. Am. Assn. Physicists in Medicine (radiation therapy com. 1989-92), Am. Coll. Med. Physics (commn. on comm. 1991—), Am. Coll. Radiology, Am. Soc. for Therapeutic Radiology and Oncology. Home: 430 Kensington Farms Dr Alpharetta GA 30201 Office: Radiation Care Inc 1155 Hammond Dr Bldg A Atlanta GA 30328

STAFFORD, STEVEN WARD, civil engineer; b. Pearisburg, Va., Apr. 23, 1967. BSCE, Va. Mil. Inst., 1989. Design engr. Paciulli, Simmons and Assocs., Fairfax, Va., 1989-90; dir. pub. works Town of Pearisburg, 1990—; tech. adv. com. Va. Rural Water Assn., Pearisburg, 1991-92. Mem. NSPE, Va. Soc. Profl. Engrs., Bldg. Ofcls. and Code Adminstrs. Internat., Va. Bldg. and Code Ofcls. Assn. Office: Town of Pearisburg 112 Tazewell St Pearisburg VA 24134

STAFFORD, THOMAS PATTEN, retired military officer, former astronaut; b. Weatherford, Okla., Sept. 17, 1930; m. Linda A. Dishman; children: Dionne, Karin. BS, U.S. Naval Acad., 1952; student, USAF Exptl. Flight Test Sch., 1958-59; DSc (hon.), Oklahoma City U., 1967; LLD (hon.), Western State U. Coll. Law, 1969; D Communications (hon.), Emerson Coll., 1969; D Aero. Engring. (hon.), Embry-Riddle Aero. Inst., 1970. Commd. 2d lt. USAF, 1952; advanced through grades to lt. gen.; chief performance br. Aerospace Research Pilot Sch., Edwards AFB, Calif.; with NASA, Houston, 1962-75; assigned Project Gemini, pilot Gemini VI, 1965, command pilot Gemini IX, 1966, comdr. Apollo X, 1969, chief astronaut office, 1969-71; dep. dir. flight crew operations, comdr. Apollo-Soyuz flight, 1975; comdr. Air Force Flight Test Ctr., Edwards AFB, 1975; lt. gen., dep. chief staff Research, Devel. and Aquisition, 1979; ret., 1979; chair The White House/NASA Com. to Independently Advise NASA How to Return to Moon and Explore Mars, 1990-91; chmn. bd. Omega Watch Co. Am.; dir. F-117A Stealth Fighter program, 1978; co-founder tech. cons. firm Stafford, Burke, Hecker, Inc., Alexandria, Va.; adv. numerous govtl. agys. including Nat. Aeronautics and Space Adminstrn., Air Force Systems command; defense adv. to Ronald Reagan during presdl. campaign; bd. dirs. numerous cos. Co-author: Pilot's Handbook for Performance Flight Testing, Aerodynamics Handbook for Performance Flight Testing. Decorated DFC with oak leaf cluster, D.S.M. (3); recipient NASA Disting. Svc. medal (2), NASA Exceptional Svc. medal (2), Air Force Command Pilot Astronaut Wings; Chanute Flight award AIAA, 1976; VFW Nat. Space award, 1976; Gen. Thomas D. White USAF Space trophy Nat. Geog. Soc., 1976; Gold Space medal Fedn. Aeronautique Internationale, 1976, Congl. Space medal honor V.P. Dan Quayle, 1993, Rotary Nat. award Space Achievement, 1993, Goddard award, Astronaut Hall of Fame, 1992; co-recipient AIAA, 1966, Harmon Internat. Aviation trophy, 1966, Nat. Acad. Television Arts and Scis. spl. Trustees award, 1969. Fellow Am. Astronautical Soc., Soc. Exptl. Test Pilots; mem. AFTRA (hon. life). Holder all-time world speed record for space flight, 24,791.4 miles per hour. Address: 1006 Cameron St Alexandria VA 22314

STAGE, RICHARD LEE, utilities executive; b. Byesville, Ohio, Nov. 5, 1936; s. Clifford Earl Stage and Evelyn Virginia (Nunley) Rolston; m. Joan Eleanor Bednarz, Feb. 1, 1958; 1 child, Julie Marie. B in Mgmt., Malone Coll., 1987. Fleet office supr. Ohio Power Co., Canton, 1954-77; supr. automotive acctg. and leasing Am. Electric Power, Canton, 1977-83; dir. automotive equipment Am. Electric Power, Columbus, 1983—. Mem. Soc. Automotive Engrs. (utilities com. chmn. 1988-89, exec. com.), Edison Electric Inst. (fleet mgmt. com. 1983—), Masons. Republican. Avocations: golf, woodworking. Home: 1329 Davis St SW Canton OH 44706-4503 Office: Am Electric Power 1 Riverside Pla Columbus OH 43215

STAHEL, WALTER RUDOLF, industrial analyst, consultant; b. Zurich, Switzerland, June 5, 1946; s. Walter Max and Frida (Abrecht) S.; m. Christiane Andrée Collaud, Feb. 9, 1970; children: Dominique Halim, Thomas Baylon. Diploma, Swiss Fed. Inst. Tech., 1970. Architect, Bicknell & Hamilton, London, 1967-68, 71-72, Aebli & Sochalski, Zurich, 1970-71; design architect Obrist & Ptnr., St. Moritz, Switzerland, 1972-73; project mgr. Battelle Inst., Geneva, 1973-79; personal asst. to chief exec. officer Phymec Luxembourg S.A., Geneva, 1980-82; ind. researcher, cons., Geneva, 1983—; chmn. ECOM Corp. S.A., Geneva, 1981—; founder, dir. Product Life Inst., Geneva, 1982—; sec. gen. European chpt. Internat. Sci. Policy Found., Geneva, 1986—; treas. European Group Local Employment Initiatives, Brussels, 1985—; deputy sec. gen. Geneva Assn., Geneva, 1988—. Author: Langlebigkeit und Material-Recycling, 1991, The Limits to Certainty, 1989, 93, Jobs for Tomorrow, 1981; Unemployment-Occupation-Profession, 1980; contbr. articles to profl. jours.; advisor Change, internat. tech. newspaper, London, 1983—. Recipient 1st prize Deutsche Gesellschaft für Zukunftsfragen, West Berlin, 1978, 3d prize Mitchell Prize Competition, Houston, 1982. Mem. Assn. of Mitchell prize winners, Houston, Tex., Indsl. Designer Soc. Am., Internat. Sci. Policy Found. Ltd., Assn. Former Students of Swiss Fed. Inst. Tech. Home: 7 chemin des Vignettes, Conches, CH-1231 Geneva Switzerland Office: Inst de la Duree, PO Box 832, CH-1211 Geneva 3, Switzerland

STAHL, DESMOND EUGENE, systems engineer; b. Sacramento, Dec. 29, 1968; s. Fred and Nancy (Kavanagh) S. BS in Computer Engring., Santa Clara U., 1991. Test engr. Pendragon Labs., San Jose, Calif., 1991-92; systems engr. Hi-Tech Rsch. Co., San Jose, Calif., 1992—. Mem. Tau Beta Pi. Achievements include research in natural language processing, intelligent spread sheet and data base. Office: Hi-Tech Rsch Co 888 N 1st St Ste 311 San Jose CA 95112

STAHL, FRANK LUDWIG, civil engineer; b. Fuerth, Germany, 1920; came to U.S., 1946, naturalized, 1949; s. Leo E. and Anna (Regensburger) S.; m. Edith Cosmann, Aug. 31, 1947; children—David, Robert. BSCE, Tech. Inst. Zurich, Switzerland, 1945. With Ammann & Whitney, Cons. Engrs., N.Y.C., 1946—, project engr., 1955-67, assoc., 1968-76, sr. assoc., 1977-81, chief engr. Transp. div., 1982. Prin. works include: Verrazano-Narrows Bridge, Throgs Neck Bridge, Walt Whitman Bridge, Improvements to Golden Gate Bridge, rehab. of Williamsburg Bridge, N.Y.C. Royal Gorge Bridge, Colo., Interstate-10 Deck Tunnel, Phoenix, Ariz.; contbr. articles to profl. jours. Recipient Gold award The James F. Lincoln Arc Welding Found., 1986, John A. Roebling medal Internat. Bridge Conf., 1992. Fellow ASCE (Thomas Fitch Rowland prize 1967, Innovation in Civil Engring. award of merit 1983, Metro. Civil Engr. of Yr. award 1987, Roebling award 1990), ASTM (vice chmn. com. A-1 on steel, stainless steel and related alloys 1978-83, chmn. steel reinforce-subcom. 1971-82, award of merit 1982); mem. Am. Inst. Steel Constrn., Engring. Found. (rsch. coun. on structural connections), Internat. Assn. Bridge and Structural Engring., Internat. Bridge Tunnel and Turnpike Assn. Home: 20911 28th Rd Flushing NY 11360-2412 Office: 96 Morton St New York NY 10014-3326

STAHL, JOEL SOL, plastic-chemical engineer; b. Youngstown, Ohio, June 10, 1918; s. John Charles and Anna (Nadler) S.; B. in Chem. Engring., Ohio State U., 1939; postgrad. Alexander Hamilton Inst., 1946-48; m. Jane Elizabeth Anglin, June 23, 1950; 1 child, John Arthur. With Ashland (Ky.) Oil, Inc., 1939-50, mgr. spl. products, 1946-50; pres. Cool Ray Co., Youngstown, 1950-51, Stahl Industries, Inc., Youngstown, 1951—; CEO Stahl Cos., Orlando, Fla., 1992—. Pres.' cabinet bd. dirs. Ohio State U. Found.; chair dean's exec. adv. bd. Coll. Engring. U. Cen. Fla.; gov.'s coun. for High Tech. and Industry for State Fla. Named Ky. col., 1967. Mem. Regional Export Expansion Council, Soc. Plastics Engr., Soc. Plastics Industry, The Plastics Acad., Internat. Platform Assn., Ohio Soc. N.Y., Citrus Club, Masons, Shriners, Rotary, Toastmasters (pres. 1949), S.C. Yacht Club, Circumnavigators. N.Y. Acad. Sci., Sweetwater Country Club, Varsity O Club, Tau Kappa Epsilon, Phi Eta Sigma, Phi Lambda Upsilon. Republican. Chrisitan Scientist. Patentee insulated core walls, plastic plumbing wall, housing in continous process, encasement asbestos, lead. Contbr. articles to profl. jours. Home: 530 E Central Blvd Apt 1504 Orlando FL 32801-4306 Office: 20 Federal Pla W Ste 600 Youngstown OH 44503

STAHOVICH, THOMAS FRANK, mechanical engineer, researcher; b. Bellflower, Calif., June 8, 1966; s. Arthur E. and Marjorie L. (Kipka) S. BS, U. Calif., Berkeley, 1988; SM, MIT, 1990. Rsch. asst. MIT CADLab, Cambridge, Mass., 1988-92, MIT Artificial Intelligence Lab., Cambridge, 1992—. Recipient Rohsenow fellowship MIT, Cambridge, 1988. Mem. Sigma Xi, Golden Key Nat. Honor Soc., Tau Beta Pi, Pi Tau Sigma. Roman Catholic. Home: 154 Central St Apt 9 Somerville MA 02145

STAHR, HENRY MICHAEL, analytical toxicology; b. White, S.D., Dec. 10, 1931; s. George Conrad and Kathryn (Smith) S.; m. Irene Frances Sondey, July 27, 1952; children: Michael G., John C., Mary C., Patrick J., Matthew G. BS, S.D. State U., 1956; MS, Union Coll., 1961; PhD, Iowa State U., 1976. Analytical devel. chemist capacitor dept. GE, Hudson Falls, N.Y., 1956-61; analyt devel. chemist campacitor dept. GE, Irmo, S.C., 1961-65; sr. scientist Philip Morris Rsch., Richmond, Va., 1965-69; prof. analytical toxicology Iowa State U., Ames, 1969—, mem. exec. com. faculty senate, 1990—. Editor: Analytical Toxicology Methods Manual, 1975, 2d edit., 1982, Analytical Methods in Toxicology, 1991; contbr. articles to publs. Bd. dirs. Ogden (Iowa) Community Sch. Dist., 1977-89. With USMC, 1949-52, Korea. Fellow Am. Coll. Comparative and Vet. Toxicology, Assn. Officers and Analytical Chemists; mem. Am. Chem. Soc., Assn. Officers and Analytical Chemists Internat., KC, Lions. Roman Catholic. Achievements include research in method development using instrumental methods, chromatography, spectroscopy, genotoxicology, techniques in toxicology. Home: 208 NE 3d St Ogden IA 50212 Office: Iowa State U 1635 College Vet Med Ames IA 50011

STAINES, MICHAEL LAURENCE, oil and gas production executive; b. Guildford, Eng., May 30, 1949; came to U.S., 1958; s. John Richard and Myrra (Smith) S.; m. Laura Catherine Terdoslavich, May 11, 1974; children: Leslie Myrra, Claire Alexandra, Julia Wallis. BS, Cornell U., 1971; MBA, Drexel U., 1976. Asst. comptr. grants U. Pa., phila., 1976-78; sr. analyst Sun Co., Radnor, Pa., 1978-80, Penn Cen. Energy Group, Radnor, 1980-83; v.p., sec. Bryn Mawr Energy Co., Bala Cynwyd, Pa., 1983-88; sr. v.p., dir., sec. Resource Am., Inc., Phila. and Akron, Ohio, 1988—. Chmn. stewardship com. St. Mary's Episc. Ch., Radnor, 1990. Winner Silver medal in coxless pair rowing, 1976 Olympic Games, Montreal; named Oarsman of Yr. Schuylkill Navy, Phila., 1976; U.S. Nat. Rowing Champion, 1971, 72, 73. Mem. Soc. Corp. Secs., Ohio Oil and Gas Assn., Oil and Gas Assn. N.Y., Oil and Gas Assn. W.Va., Havre de Grace (Md.) Yacht Club, Bachelors Barge Club (Phila.), Vesper Boat Club (capt. 1973). Avocations: classic automobiles and motorcycles, sailing. Office: Resource Am Inc 2876 S Arlington Rd Akron OH 44312-4712

STALCUP, THOMAS EUGENE, SR., civil engineer; b. Jonesboro, Ark., July 24, 1948; s. Thomas Hardin and Lois Lucille (Cooper) S.; m. Barbara Renee Eubanks, Apr. 19, 1985; children: Stacey, Thomas Jr. BSAE, Ark. State U., 1976. Registered profl. engr., Ark. Draftsman Associated Engrs., Inc., Jonesboro, Ark., 1971-75; engr., prin. BCI Engrs./Surveyors, Inc., Jonesboro, 1975-84; engr. Brackett/Krennerich & Assoc. P.A., Jonesboro, 1984—; mem. Elec. Code Revision Com., Jonesboro, 1989—, Earthquake Damage Assessment Team, Jonesboro, 1990—, Craighead County On-Site Wastewater Disposal Adv. Com., Jonesboro, 1981-82. Co-author (papers): Alicia Wastewater System 1982 to 1986 A Case History, 1986, Proposed Design Parameters for Tri-City Wastewater Disposal System, 1986. Com. chmn. Boy Scouts Am. Troop 89, Brookland, Ark., 1988—. Recipient Engr. of Yr. award East Chpt. Ark. Soc. Profl. Engrs., 1989, 90, Pres.'s award Ark. Soc. Profl. Engrs., 1990, 81. Mem. Ark. Soc. Profl. Engrs. (pres. 1989-90), Nat. Soc. Profl. Engrs., ASCE, Order of Engr. Nazarene. Achievements include design and development of a small diameter septic tank effluent collection and treatment system, commonly referred to as the Alicia system. Office: Brackett Krennerick & Assoc 100 E Huntington Ave Ste D Jonesboro AR 72401-2900

STALHEIM-SMITH, ANN, biology educator; b. Garretson, S.D., Oct. 19, 1936; d. Oliver Theodore and O'dessa Beldina (Olson) Stalheim; m. Christopher Carlisle Smith, Aug. 24, 1960; children: Heather, Andrea, Jamie. BS, Augustana Coll., Sioux Falls, S.D., 1958; MS, U. Colo., 1960; PhD, No. Ariz. U., 1982. Teaching asst. Augustana Coll., 1955-58; teaching fellow U. Colo., Boulder, 1958-59, rsch. asst., 1959-60; instr. Pacific Luth. U., Tacoma, 1960-61, Fisk U., Nashville, 1967; instr. Kans. State U., Manahattan, 1970-86, asst. prof. biology, 1986—; mem. grant rev. panel NSF, Washington, 1979, 81, 91; textbook reviewer Harper & Row Pub., West Pub., Saunders, others, 1975-87. Author: (with Greg K. Fitch) Understanding Human Anatomy and Physiology, 1993; contbr. articles to profl. jours. Grantee NSF, 1979-81, 90—, Howard Hughes Found., 1992—. Mem. AAAS, Am. Soc. Zoologists, Sigma Xi. Democrat. Lutheran. Home: 1328 Fremont St Manhattan KS 66502 Office: Kans State U Div Biology Ackert Hall Manhattan KS 66506

STALL, TRISHA MARIE, mechanical engineer; b. Pittsfield, Mass., Feb. 14, 1965; d. Ralph Everett and Marjorie (Larabee) S. BSME, Western New Eng. Coll., 1987, postgrad., 1988—. Registered profl. engr., Mass. Heating, ventilation and air conditioning engr. Frosty Mech. Contractors, Inc., Pittsfield, 1987—. Mem. ASME, ASHRAE, Soc. Women Engrs. Home: PO Box 321 New Lebanon NY 12125-0321 Office: Frosty Mech Contractors Inc 55 S Merriam St Pittsfield MA 01201

STALLING, DAVID LAURENCE, research chemist; b. Kansas City, Mo., Oct. 24, 1941; s. C. Lawrence and Genevieve F. (Simmons) S.; m. Dorothy A. Borgman, May 26, 1962; children: Sheila K., Mark D., Michael. BS, Mo. Valley Coll., 1962; MS, U. Mo., 1964, PhD, 1967. Chief chemist Columbia Nat. Fisheries Contaminat Rsch. Ctr., Columbia, Mo., 1968-85, sr. scientist, 1985-89; sr. v.p. rsch. and devel. ABC Labs., Inc., Columbia, 1989--; co-founder, sec., bd. dirs. ABC Labs., Columbia, 1967-93; bd. dirs. Nat. Inst. Std. Tech. Caals Consortium. Contbr. articles to profl. jours. Mem. Mayor's Columbia 2000 com., Mo., 1992; mem. Cosmopolitan Internat. Svc. Cl ub, Columbia, 1967--. Recipient Alumni Faculty award U. Mo., 1974, Dept. Interior Medal, 1989; NASA Pre-doctoral fellow, 1965. Mem. Am. Chem. Soc., Assn. Official Analytical Chemists, Soc. Environ. Toxicology and Chemistry, Chemometrics Soc., Applied Spectroscopy Soc., Electrochem. Soc., Sigma Xi. Presbyterian. Achievements include patents for siylating reagent and GPC Cleanup for pesticides, surface linked fullevents, new materials for chromatography, a high sensitivity chromatography system. Office: ABC Labs PO Box 1097 Columbia MO 65205

STALLINGS, FRANK, JR., industrial engineer; b. Concord, N.C., Aug. 21, 1954; s. Frank and Theresa Ann (Iorlano) S. BS in Indsl. Engring., N.C. State U., 1976; MS in Adminstrn., George Washington U., 1979. Jr. indsl. engr. Naval Air Rework Facility, Norfolk, Va., 1974-75; indsl. engr. Babcock & Wilcox, Lynchburg, Va., 1977-79; sr. prin. engr. NCR Corp., Columbia, S.C., 1979-82; mgr. indsl. engring. Mars Electronics, Ltd., Reading, Eng., 1987-88; liaison between European/U.S. mfg. divs., sr. indsl. engr. M&M/MARS, Inc., Waco, Tex., 1982-84, mgr. quality assurance, 1984-87, mgr. indsl. engring., 1988-91, logistics mgr., 1991—. Coach Heart of Tex. Soccer League, Waco, 1985-87; v.p. Sugar Creek Homeowners Assn., Waco, 1986-87; Sunday sch. tchr. Columbus Ave. Bapt. Ch., Waco, 1988—, mem. Missions com., 1991—, Bapt. Youth leader, 1990—, mem. ch. singles coun., 1989-91; exec. mem. singles coun. Waco Bapt. Assn.; counselor Royal Ambs., 1990—. Mem. Inst. Indsl. Engrs. (sr.), Am. Soc. Quality Control, Am. Prodn. and Inventory Control Soc. Republican. Avocations: camping, boating, volleyball, running, cycling. Home: 1241 Woodland West Dr Waco TX 76712-3407 Office: M&M/MARS Inc 1001 Texas Central Pkwy Waco TX 76712

STALNAKER, JOHN HULBERT, physician; b. Portland, Oreg., Aug. 29, 1918; s. William Park II and Helen Caryl (Hulbert) S.; m. Louise Isabel Lucas, Sept. 8, 1946; children: Carol Ann, Janet Lee, Mary Louise, John Park, Laurie Jean, James Mark. Student, Reed Coll., Portland, 1936-38; AB, Willamette U., Salem, Oreg., 1941; MD, Oreg. Health Scis. U., 1945. Diplomate Am. Bd. Internal Medicine. Intern Emanuel Hosp., Portland, 1945-46; resident in internal medicine St. Vincent Hosp., Portland, 1948-51; clin. instr. U. Oreg. Med. Sch., 1951-54, 60-62; staff physician VA Hosp., Vancouver, Wash., 1970-79; cons. in internal medicine, 1951-79. Contbr. articles to profl. jours. Pianist various civic and club meetings, Portland; leader Johnny Stalnaker's Dance Orch., 1936-39. Lt. (j.g.) USNR, 1946-48. Fellow ACP; mem. AMA, Multnomah County Med. Soc., Oreg. State Med. Assn., N.Am. Lily Soc., Am. Rose Soc. Republican. Avocations: music, photography, horticulture. Home: 2204 SW Sunset Dr Portland OR 97201-2068

STALZER, JOHN FRANCIS, environmental biologist; b. N.Y.C., Apr. 25, 1959; s. Ernest Karl and Anna Dorothea (Belay) S. BS in Environ. Biology, L.I. U., Southampton, N.Y., 1983; BS in Computer Sci., L.I. U., Greenvale, N.Y., 1987. Tech. collaborator Brookhaven Nat. Lab., Upton, N.Y., 1981-82; computer graphics analyst Young & Young, Riverhead, N.Y., 1985; environ. analyst Ethan C. Eldon Assoc., Inc., Westbury, N.Y., 1987—. Mem. Am. Assoc. Biol. Scientists, N.Y. Acad. Scis. Home: 19 Albin St Glen Cove NY 11542 Office: Ethan C Eldon Assocs Inc 900 Ellison Ave Westbury NY 11590-5114

STAMBAUGH, JOHN EDGAR, oncologist, hematologist, pharmacologist, educator; b. Everett, Pa., Apr. 30, 1940; s. John Edgar and Rhoda Irene (Becker) S.; B.S. cum laude in Chemistry, Dickinson Coll., 1962; M.D., Jefferson Med. Coll., 1966, Ph.D., 1968; m. Shirley Louise Fultz, June 24, 1961; 4 children. Intern, Thomas Jefferson U. Hosp., Phila., 1968-69, resident, 1969-70, oncology fellow, 1970-72, instr. pharmacology, 1969-70, asst. prof., 1970-74, assoc. prof., 1974-82, prof., 1982—, asst. prof. medicine, 1976—; pvt. practice medical oncology Hematology and Chronic Pain, Woodbury, N.J.; staff physician Thomas Jefferson Hosp., Phila., 1972—, Cooper Med. Center, Camden, N.J., 1972—, Underwood Meml. Hosp., Woodbury, 1972—, Garden State Hosp., Marlton, N.J., 1973—, Cherry Hill (N.J.) Med. Center, 1978—, West Jersey Hosp., Camden. Fellow Am. Coll. Clin. Pharmacology; mem. AMA, N.J. Med. Soc., Camden County Med. Soc., Am. Soc. for Pharmacology and Exptl. Therapeutics, Am. Soc. Clin. Oncology, Am. Assn. for Cancer Research, Internat. Assn. for Study of Pain, Am. Pain Soc., Am. Assn. Clin. Research, Sigma Xi. Contbr. articles to profl. jours. Office: 17 Redbank Ave Ste 101 Woodbury NJ 08096 also: 1210 Brace Rd Cherry Hill NJ 08034

STAMLER, JEREMIAH, physician, educator; b. N.Y.C., Oct. 27, 1919; s. George and Rose (Baras) S.; m. Rose Steinberg, 1942; 1 son, Paul J. AB, Columbia U., 1940; MD, SUNY, Bklyn., 1943. Cert. specialist in clin. nutrition. Intern L.I. Coll. Medicine div. Kings County Hosp., Bklyn., 1944, fellow pathology, 1947; research fellow cardiovascular dept. Med. Research Inst., Michael Reese Hosp., Chgo., 1948, research assoc., 1949-55, asst. dir. dept., 1955-58; established investigator Am. Heart Assn., 1952-58; dir. heart disease control program Chgo. Bd. Health, 1958-74, dir. chronic disease control div., 1961-63, dir. div. adult health and aging, 1963-74; assoc. dept. medicine Med. Sch., Northwestern U., Evanston, Ill., 1958-59, asst. prof., 1959-65, assoc. prof., 1965-71, prof., chmn. dept. community health and preventive medicine, 1972-86, Harry W. Dingman prof. cardiology, 1973—; attending physician Northwestern Meml. Hosp., 1973-89; exec. dir. Chgo. Health Research Found., 1963-72, bd. dirs., 1972—; cons. medicine St. Joseph Hosp., Chgo., 1964—, Rush-Presbyn.-St. Luke's Hosp., Chgo., 1966—; professorial lectr. dept. medicine Pritzker Sch. Medicine, U. Chgo., 1970—; vis. prof. internal medicine Rush. Presbyn.-St. Luke's Med. Center, 1972—. Author: (with L. N. Katz) Experimental Atheroscleroses, 1953, (with others) Nutrition and Atherosclerosis, 1958, (with A. Blakeslee) Your Heart Has Nine Lives-Nine Steps to Heart Health, 1963, (with others) Epidemiology of Hypertension, 1967, Lectures on Preventive Cardiology, 1967, (with A. Blakeslee) Four Keys to a Healthy Heart, 1976; fgn. cons., editoral cons.: Heartbeat. Served to capt. AUS, 1944-46. Recipient award for outstanding efforts in heart research Am. Heart Assn., 1964, Howard W. Blakeslee award, 1964, award of merit, 1967; Albert and Mary Lasker Med. Journalism award, 1965, Conrad Elvehjem award Wis. Med. Soc., 1967, (with others) Albert Lasker Spl. Service award, 1980, Donald Reid medal, London, 1988, John Jay award Columbia College, N.Y., 1990, others. Fellow Am. Coll. Cardiology (Disting. Svc. award 1985), Am. Pub. Health Assn. (John M. Snow award 1986), AAAS; mem. Am. Fedn. Clin. Research, Am. Heart Assn. (bd. dirs., past vice-chmn. exec. com., fellow coun. arteriosclerosis, chmn. council on epidemiology 1979—, Rsch. Achievement award

1981, Disting. Achievement award 1988), Am. Physiol. Soc., Am. Soc. Clin. Investigation, Am. Soc. Clin. Nutrition, Am. Soc. Study Arteriosclerosis (past bd. dirs., past chmn. program com., past sec.-treas.), Am., Chgo. diabetes assns., Assn. Tchrs. Preventive Medicine, Am. Soc. Clin. Nutrition, Assn. Clin. Scientists, Middle States Pub. Health Assn., Central Soc. Clin. Research, Chgo. Heart Assn. (Coeur d'Or award 1979), Ill. Pub. Health Assn. (mem. exec. com.), Ill. Acad. Scis., Diabetes Assn. Greater Chgo. (dir.), Soc. Exptl. Biology and Medicine (sec. Ill. chpt.), Am. Inst. Nutrition, Chgo. Nutrition Assn., Chgo. Acad. Scis., Internat. Soc. and Fedn. Cardiology (chmn. sci. bd., council on epidemiology and prevention), Inst. Medicine Chgo. (Coleman award 1987), Phi Beta Kappa. Office: Ste 1102 680 N Lake Shore Dr Chicago IL 60611

STAMM, ROBERT FRANZ, research physicist; b. Mt. Vernon, Ohio, Mar. 28, 1915; s. John Frederick William and Alice Maude (Swartout) S.; m. Elizabeth Ona Ladd, June 1, 1947 (div. Oct. 1958); m. Isabel Golinski, Jan. 28, 1964. AB with 1st honors, Kenyon Coll., 1937; PhD, Iowa State U., 1942. Rsch. physicist Am. Cyanamid Co. Cen. Rsch. Lab., Stamford, Conn., 1942-72, Clairol Rsch. Lab., Stamford, 1973-82. Contbr. 35 tech. articles to profl. jours. Prin. flutist Norwalk (Conn.) Community Symphony Orch., 1948-79, Greenwich (Conn.) Philharm. Orch., 1954-65, Stamford Symphony Orch. Recipient cert. for work essential to prodn. of atomic bomb U.S. Army, War Dept., Armed Forces, Corps of Engrs., Manhattan Dist., 1945. Mem. Phi Beta Kappa, Sigma Xi. Republican. Episcopalian. Achievements include 25 U.S. and foreign patents for optical and spectrosopic instrumentation, for development of dividing engine (with interferometric control) for ruling diffraction gratings, and development of detailed specifications of optical and mechanical parameters required for ruling tiny trigonal pyramids to be replicated in order to yield an embossing tool for producing retroreflective sheets for highway signs. Home: 158 Rufous Ln Sedona AZ 86336-7116

STAMNES, JAKOB JOHAN, physicist educator; b. Roest, Norway, June 30, 1943; s. Alfred Johannes and Petra Amanda (Antonsen) S.; m. Sigrid Kristin Eriksen, Aug. 28, 1970; children: Sonja, Monica. MS, Norwegian Inst. Tech., Trondheim, Norway, 1969; PhD, U. Rochester, 1975. Researcher Norwegian Def. Rsch. Est., Kjeller, Norway, 1974-76, Ctr. for Indsl. Rsch., Oslo, Norway, 1976-85; rsch. mgr. Norwawe Tech. A.S., Oslo, Norway, 1985-88; mng. dir. Norwawe Devel. A.S., Oslo, Norway, 1988—; prof. U. Bergen, Norway, 1990—. Author: Waves in Focal Regions, 1986; contbr. over 50 articles to profl. jours. Lt. Infantry, 1963-64, Norway. Recipient Simrad's award for Achievements in Electro-Optics, Norwegian Phys. Soc., 1981. Fellow Optical Soc. Am.; mem. Norwegian Phys. Soc., European Optical Soc. (sec. gen. 1991-92, bd. dirs. 1991—). Home: Kvernkallv 12A, 0382 Oslo 3 Norway Office: U Bergen Dept of Physics, Allegaten 55, 5007 Bergen Norway

STAMPER, DAVID ANDREW, psychologist; b. Cleve., Dec. 22, 1941; s. Lee E. and Margaret Ann Stamper; m. Judith Clarke, Feb. 2, 1968 (div. Aug. 1971); m. Karen Ann Elder, May 22, 1972; children: Cori, John, Victoria. BS, Sioux Falls Coll., 1965; MA, U. Colo., 1975. Psychol. technician U.S. Army Med. Rsch. and Nutrition Lab., Denver, 1966-74; rsch. psychologist Letterman Army Inst. Rsch., San Francisco, 1974-92, Army Med. Rsch. Detachment, San Antonio, Tex., 1992—. Race dir. Half Moon Bay (Calif.) Pumpkin Festival Run, 1986-91. With U.S. Army, 1966-68. Mem. Am. Psychol. Soc., Toastmasters Internat. (dist. treas. 1990-91). Office: Army Med Rsch Detachment 7914 A Dr Brooks AFB TX 78235-5138

STANBACK, MARK THOMAS, behavioral ecologist, researcher; b. Salisbury, N.C., May 18, 1962; s. William Charles and Anne Elizabeth (Ragland) S.; m. Nancy Jane Popkin, July 22, 1990. BS, Davidson Coll., 1984; PhD, U. Calif., Berkeley, 1991. NIMH postdoctoral fellow U. Wash., Seattle, 1991-92, NSF postdoctoral fellow, 1992-94. Contbr. numerous articles to sci. jours., chpt. to book. NSF predoctoral fellow, 1985. Democrat. Achievements include research of hatching, asynchrony, cooperative breeding, behavior endocrinology, sperm storage in birds. Office: U Wash Dept Zoology NJ-15 Seattle WA 98195

STANBURY, JOHN BRUTON, physician, educator; b. Clinton, N.C., May 15, 1915; s. Walter A. and Zula (Bruton) S.; m. Jean F. Cook, Jan. 6, 1945; children: John Bruton, Martha Jean, Sarah Katherine, David McNeill, Pamela Cook. A.B., Duke U., 1935; M.D., Harvard U., 1939; M.D. (hon.), U. Leiden (Netherlands), 1975. House officer Mass. Gen. Hosp., 1940-41, asst. resident, 1946, chief med. resident, 1948, mem. med. staff, 1949—; research fellow pharmacology Harvard Med. Sch., 1947; vis. prof. medicine U. Leiden, 1955; prof. exptl. medicine MIT, Cambridge, 1966-80; emeritus MIT, 1980—; cons. Pan Am. Health Orgn., WHO, UNICEF, U.S. AEC. Author: Endemic Goiter: The adaptation of man to iodine deficiency, 1954, Metabolic Basis of Inherited Disease, 5th edit., 1984, The Thyroid and Its Diseases, 5th edit., 1984, Endemic Goiter, 1969, Human Development and the Thyroid, 1972, Endemic Goiter and Endemic Cretinism, 1980, Prevention and Control of Iodine Deficiency Disorders, 1987, A Constant Ferment, 1991. Served from lt. (j.g.) to comdr. USNR, 1941-45. Recipient Delmar S. Fahrney medal Franklin Inst., 1993. Mem. Am. Assn. Physicians, Soc. Clin. Investigation, Am. Thyroid Assn. (pres. 1969), Am. Acad. Arts and Scis., Endocrine Soc., endocrine socs. Finland, Colombia, Peru and Argentina, Internat. Coun. for Control of Iodine Deficiency Disorders (chmn.). Democrat. Episcopalian. Home: 43 Circuit Rd Chestnut Hill MA 02167-1802

STANCIL, DANIEL DEAN, electrical engineering educator; b. Raleigh, N.C., Jan. 11, 1954; s. David Hadley and Sue Ellen (Ray) S.; m. Katherine Elaine Campbell, Sept. 6, 1975; children: Brian Alan, Michael Adam. BSEE, Tenn. Tech. U., 1976; MS, MIT, 1978, diploma in elec. engring., 1979, PhD, 1981. Engring. trainee NASA Langley Rsch. Ctr., Hampton, Va., 1974-75; rsch. asst., teaching asst. MIT, Cambridge, Mass., 1977-80; part-time staff mem. Lincoln Lab. MIT, Lexington, Mass., 1980-81; asst. prof. N.C. State U., Raleigh, 1981-86; from assoc. prof. to prof., assoc. dept. head Carnegie Mellon U., Pitts., 1986—; cons. IBM, Yorktown Heights, N.Y., 1982, 85, Naval Rsch. Lab., Washington, 1988; local chmn. 5th joint MMM-INTERMAG Conf., Pitts., 1991. Author: Theory of Magnetostatic Waves, 1993; contbr. articles to Jour. Applied Physics, others. Scholar Tenn. Tech. U., 1976. Mem. IEEE (adminstrv. com., mag. soc. 1989—, contbr. articles to jour.), Am. Phys. Soc., Tau Beta Pi, Eta Kappa Nu. Achievements include patent for optical frequency shifter using magnetostatic waves. Office: Dept Elec Comp Engring 5000 Forbes Ave Pittsburgh PA 15213

STANDLER, RONALD B., physicist; b. Osaka, Japan, Dec. 11, 1949; s. Roy F. and Cornelia (Echols) S. BS in Physics, U. Denver, 1971; PhD in Physics, N.Mex. Tech, 1977. Vis. asst. prof. elec. engring. U. Fla., Gainesville, 1977-79; sr. engr. Xerox Corp, Rochester, N.Y., 1979-81; asst. prof. elec. engring. Rochester (N.Y.) Inst. Tech., 1981-83; rsch. scholar USAF Weapons Lab., Kirtland AFB, N.Mex., 1983-84; assoc. prof. elec. engring. The Penn State U., State College, Pa., 1984-90; cons. Lexington, Ky., 1990—. Author: Protection of Electronic Circuits, 1989; contbr. more than 30 articles to profl. publs. Recipient Langmuir award for Excellence in Rsch., N.Mex. Tech, 1990. Mem. IEEE, AAAS, Verband Deutscher Elektrotechniker, Sigma Xi. Achievements include patents in field; contributions to origin, propagation, and mitigation of transient overvoltages; explanation of relationship between electric field at surface of earth beneath thunderstorm and field aloft. Office: HVTRL Inc PO Box 24463 Lexington KY 40524

STANDLEY, PAUL MELVIN, chemist; b. East Liverpool, Ohio, Nov. 8, 1943; s. Paul N. and Ruth J. (Barnhart) S.; m. Rebecca S. Nentwick, Nov. 8, 1969 (div. Jan. 1988); children: Celeste N., Matthew J., Zachary P.; m. Monica Lee Sneller, July 28, 1990. BS in Chemistry, Kent State U., 1970. Compounder Gen. Tire Devel., Akron, Ohio, 1970-74; from rsch. chemist to sr. rsch. chemist Dayco Tech. Ctr., Springfield, Mo., 1974-81, mgr. tech. svc., 1981-83, mgr. adv. tech. group, 1983-86; bus. mgr. auto accessories Dayco Products Plant, Springdale, Ariz., 1986-89; v.p. R & D Dayco Products, Inc., Dayton, Ohio, 1989—. With USAF, 1962-66. Mem. Rubber Div. of Am. Chem. Soc. (chmn. symposium 1984, program planning com. 1985-88). Achievements include patents for elastomeric products, for process equipment and product applications, for pulleys, for power transmission products, for material compositions. Office: Dayco Products Inc Box 1004 1 Prestige Pl Dayton OH 45401-1004

STANEK, DONALD GEORGE, JR., computer analyst; b. Elgin, Ill., Apr. 27, 1959; s. Donald George Sr. and Darlene Charlotte (Dells) S. Asst. quality mgr. INTEC Inc., Palatine, Ill., 1983-87; quality mgr. Delta Tech Mold, Arlington Heights, Ill., 1987-90; ops. mgr. Crystal Die and Mold, Algonquin, Ill., 1990—. Home: 828 N River Rd McHenry IL 60050 Office: Crystal Die and Mold 905 W Algonquin Rd Algonquin IL 60102

STANEVICH, KENNETH WILLIAM, mechanical design engineer; b. Berwyn, Ill., June 12, 1958; s. William George and Susanna Regina (Logins) S.; m. Jill Elaine Dahlberg, June 2, 1990. AS, Morton Coll., 1981; student, U. Ill., Chgo., 1981-85. Designer Grayhill, Inc., LaGrange, Ill., 1983; product engr. Molex, Inc., Lisle, Ill., 1985-88; supr. product devel. engring. Auto Divsn. Augat Inc., Augat Automotive Div., Mt. Clemens, Mich., 1988-92; mgr. mech. design Robertshaw Controls Co.-Simicon Divsn., Holland, Mich., 1992—. Patentee in field. Co-chmn. study-a-thon Muscular Dystrophy Assn., 1981. Mem. Soc. Plastics Engrs., Soc. Auto. Engrs. (student chpt. pres. 1984-85). Avocations: running, auto-cross, skiing, golf.

STANFORD, DENNIS JOE, archaeologist, museum curator; b. Cherokee, Iowa, May 13, 1943; s. William Erle and Mary L. (Fredenburg) S.; m. Margaret Brierty, June 4, 1988; 1 dau., Brandy L. B.A., U. Wyo., 1965; M.A., U. N.Mex., 1967, Ph.D., 1972. Archeologist, curator Smithsonian Instn., Washington, 1972—, head div. archeology, 1990-92, chmn. dept anthropology, 1992—; v.p., dir. Taraxacum Press, 1981—; mem. adv. bd. Ctr. for the Study of Early Man, 1985—; rsch. assoc. Denver Mus. Natural History, 1989—. Author: The Walakpa Site, Alaska, 1975; editor: (with Robert L. Humphrey) Pre-Llano Cultures of the Americas, 1979, (with George C. Frison) The Agate Basin Site, 1982, (with Jane Day) Ice Age Hunters of the Rockies, 1992. Mem. Anthrop. Soc. Washington (gov. 1974-77), Soc. Am. Archeology, Am. Quaternary Assn. Achievements include research and publications on Paleo-Indian Studies, N. and S. Am., China, especially Western U.S., Arctic. Home: 1350 Massachusetts Ave Washington DC 20003-0001 Office: Smithsonian Instn Washington DC 20560

STANFORD, JACK ARTHUR, biological station administrator; b. Delta, Colo., Feb. 18, 1947; s. LeRoy and Wilma (Tucker) S.; children: Jake, Chriss. BS in Fisheries Sci., Colo. State U., 1969, MS in Limnology, 1971; PhD in Limnology, U. Utah, 1975. Fisheries biologist Alaska-Fish and Game, Dillingham, 1968-69; rsch. biologist and limnologist instr. U. Mont., Missoula, 1973-74; dir. Flathead Lake Biol. Sta. U. Mont., Polson, 1980—; research prof. zoology U. Mont., Missoula, 1983-86; prof. N. Tex. State U., Denton, 1974-81; panelist div. biotic system NSF, Washington, 1985-89. Editor: Ecology of Regulated Streams, 1979; mem. bd. editors: Regulated Rivers: Research and Management, 1985—; contbr. over 75 articles to profl. jours. Advisor Nature Conservancy, Boulder, Colo., 1982—. Named Bierman Prof. Ecology U. Mont., 1986—; grantee N. Tex. State U., EPA, U.S. Army, U.S. Bur. Reclamation, NSF, U.S. Nat. Park Svc. Mem. Mont. Acad. Sci., Am. Soc. Limnology and Oceanography, Ecol. Soc. Am., N.Am. Benthological Soc. (exec. com. 1979, 1988-89), AAAS. Avocation: fly fishing, skiing. Home and Office: U Mont Flathead Lake Biol Sta 311 Bio Station Ln Polson MT 59860-9659

STANFORD, MICHAEL FRANCIS, scientist, engineer; b. Newark, July 19, 1948; s. Richard Alexander and Edith (Dunn) S.; m. Patricia Faith Wonderley, Aug. 11, 1984. BA in Liberal Studies, Stockton State Coll., 1974, BS in Physics, 1975; PhD in Biophysics, SUNY, Bklyn., 1982. Radiation specialist Naval Aerospace Med. Labs., Pensacola, Fla., 1983-86; sr. staff mem. BDM Corp., Albuquerque, 1986-88; sr. rsch. scientist KRUG Internat., Houston, 1988-89; chief scientist, engr. McDonnell Douglas Space Systems Co., Houston, 1989-93; sr. rsch. scientist Houston Advanced Rsch. Ctr., Space Tech. and Rsch. Ctr., The Woodlands, Tex., 1993—; adj. prof. Pensacola (Fla.) Jr. Coll., 1983-86; presenter in field. Contbr. articles to Engring. Construction and Ops. in Space II, Electromagnetic Fields and Neuro-Behavioral Function, Annals of the N.Y. Acad. Sci., Sci., Physiol. Chem. Phys. Lt. comdr. USNR, 1986—. Mem. AIAA, AAAS, Am. Geophysical Union, N.Y. Acad. Scis., Naval Res. Officers Assn. Office: HARC/STAR 4800 Research Forest Dr The Woodlands TX 77381

STANG, PETER JOHN, organic chemist; b. Nûrnberg, Germany, Nov. 17, 1941; came to U.S., 1956; s. John Stang and Margaret Stang Pollman; m. Christine Schirmer, 1969; children: Antonia, Alexandra. BS, DePaul U., Chicago, 1963; Ph. D., U. California, Berkeley, 1966; hon. degr., Moscow State Lomonossov U., 1992, Russian Academy of Sciences, 1992. Instr. Princeton (N.J.) U., 1967-68; from asst. to assoc. prof. U. Utah, Salt Lake City, 1969-79, prof., 1979-92, Disting. prof. chemistry, 1992—. Co-author: Organic Spectroscopy, 1971; author: (with others) Vinyl Cations, 1979; contbr. 250 articles to sci. publs. Humboldt-Forschungspreis, 1977; JSPS Fellowship, 1985; Fulbright-Hays Sr. Scholarship, 1988. Fellow AAAS; mem. Am. Chem. Soc. (assoc. editor jour. 1982—). Office: Univ Utah Dept Chemistry Salt Lake City UT 84112

STANIFER, ROBERT DALE, system analyst, manager; b. Hamilton, Ohio, Mar. 25, 1961; s. Robert Milas and Alene (Slusher) S.; m. Jacqulene Ruth Settles, Sept. 20, 1986; 1 child, Amber Etta. AS, So. Ohio, 1982. Asst. vax systems mgr. Appalachian Computer Svcs., London, Ky., 1984-88; vax systems mgr. analyst Cincom Systems Inc., Cin., 1988—; mem. DECUS, Cin., 1986—. Sun. sch. dir. First Bapt. Ch. of Pine Grove, London, 1986-88; mem. PTO, Ross, Ohio, 1992—. Republican. Baptist. Home: 101 Jeff Scott Ln Hamilton OH 45013 Office: Cincom System Inc 2300 Montana Ave Cincinnati OH 45211

STANISLAO, JOSEPH, engineering educator, academic administrator, industrial consultant; b. Manchester, Conn., Nov. 21, 1928; s. Eduardo and Rose (Zaccaro) S.; m. Bettie Chloe Carter, Sept. 6, 1960. BS, Tex. Tech. U., 1957; MS, Pa. State U., 1959; DEng, Columbia U., 1970. Registered profl. engr., Mass. Asst. engr. Naval Ordnance Research, University Park, Pa., 1958-59; asst. prof. N.C. State U., Raleigh, 1959-61; dir. research Darlington Fabrics Corp., Pawtucket, R.I., 1961-62; from asst. prof. to prof. U. R.I., Kingston, 1962-71; prof., chmn. dept. Cleve. State U., 1971-75; prof., dean N.D. State U., Fargo, 1975-93, acting v.p. agrl. affairs, 1983-85, asst. to pres., 1983—; dir. Engring. Computer Ctr. N.D. State U., 1984—; pres. XOX Corp., 1984-90; chmn. bd., chief exec. officer ATSCO, 1989—, chief engr., 1993—. Contbr. chpts. to books, articles to profl. jours.; patentee pump apparatus, pump fluid housing. Served to sgt. USMC, 1948-51. Recipient Sigma Xi award, 1968; Order of the Iron Ring award N.D. State U., 1972, Econ. Devel. award, 1991; USAF recognition award, 1979, ROTC appreciation award, 1982. Mem. Am. Inst. Indsl. Engrs. (sr.; v.p. 1964-65), ASME, Am. Soc. Engring. Edn. (campus coord. 1979-81), Acad. Indsl. Engrs. Tex. Tech U., Lions, Elks, Am. Legion, Phi Kappa Phi, Tau Beta Pi (advisor 1978-79). Roman Catholic. Home: 3520 Longfellow Rd Fargo ND 58102-1227 Office: ND State U Engring Expt Sta U Sta PO Box 5285 Fargo ND 58105

STANISZEWSKI, ANDRZEJ MAREK, surgeon, educator; b. Wroclaw, Poland, July 15, 1952; s. Zdzislaw and Maria Eugenia (Zagorska) S.; 1 child, Piotr; m. Jadwiga Maria Trojanowska, Aug. 24, 1991. 1st med. degree, Med. Acad. Wroclaw, 1977, Grade I specialization, 1981, Grade II specialization, 1986, MD, 1986. Asst. Dept. Pathology, Med. Acad., Wroclaw, 1977-78; asst. physician State Teaching Hosp. No. 1, Wroclaw, 1979-82; rsch. asst. dept. thoracic surgery Med. Acad., Wroclaw, 1982-86; vis. physician Vienna-Lainz City Hosp., Vienna, Austria, 1983-84; vis. fellow U. Coll. Hosp., London, 1989; asst. Hosp. for Tuberculosis and Pulmonary Diseases, Wroclaw, 1984—; lectr. dept. thoracic surgery Med. Acad. Wroclaw, 1986—; vis. physician Chest Hosp. Heckeshorn, Berlin, 1993; trainee Toranomon Hosp., Tokyo, 1993; researcher nat. rsch. programmes Ministry of Health, Polish Acad. Scis., Wroclaw, 1984—; dir. rsch. programme Dept. Thoracic Surgery, Med. Acad. Wroclaw, 1990—. Contbr. articles to profl. jours.; co-author proceedings (book): Breast and Pulmonary Surgery, 1990. Mem. coun. of faculty of postgrad. edn. Med. Acad. Wroclaw, 1992—; bd. dirs. Lower Silesian Med. Diagnostic Ctr. "Dolmed", Wroclaw, Regional Children's Orthopedic Hosp., Trzebnica, 1993—. Austrian Fed. Ministry of Sci. and Rsch. scholar, 1983-84, Brit. Coun. fellow, 1989, European Sch. Oncol. alumnus, 1990, Deutscher Akademischer Austanschdienst scholar, 1993; trainee Japanese Coun. for Med. Tng., 1993; recipient Individual Rsch.prize Rector of Med. Acad. Wroclaw, 1986. Mem. Polish Cybernetic Soc. (sec. local com. 1978), Polish Med. Soc., Polish Oncol. Soc., Assn. of Polish Surgeons, Internat. Coll. Surgeons (mem. Polish sect. 1986—), European Soc. Mastology. Avocations: film, theatre, reading, art, travel. Home: Grabiszynska 101/7, Wroclaw Poland PL53439 Office: Med Acad Thoracic Surgery, Grabiszynska 105, Wroclaw Poland PL53439

STANKER, LARRY HENRY, biochemist; b. San Mateo, Calif., July 25, 1948; s. Henry Joseph and Loretta Emma (Heppler) S.;m. Joanne H. Heintzberger, Jan. 6, 1979; children: Michael, Christy. BA, San Francisco State U., 1971; MS, U. Ill., 1973, PhD, 1980. Postdoctoral fellow M.D. Anderson Hosp. & Tumor Inst., U. Tex., Houston, 1980-82; sr. biomed. scientist Lawrence Livermore Nat. Labs., U. Calif., 1982-91; project leader food and animal protection rsch. lab. U.S. Dept. Agr., Agrl. Rsch. Svc., College Station, Tex., 1991—. Editor: Human Exposure to Toxic Chemicals, 1991. Mem. AAAS, Assn. Ofcl. Analytical Chemists. Achievements include patents on devel. of antibodies and immunoassays; devel. of numerous immunoassays for environ. contaminants such as dioxin, heptachlon, others; devel. of rapid method to purify immunoglobulins. Office: US Dept Agr Agrl Rsch Svc FAPRL Rt 5 Box 810 College Station TX 77845

STANKIEWICZ, ANDRZEJ JERZY, physician, biochemistry educator; b. Lidzbark, Poland, Sept. 28, 1948; came to U.S., 1981; s. Wincenty and Zofia (Plawgo) S. MD, Med. Sch., Gdansk, Poland, 1972, PhD, 1976. Asst. prof. Med. Sch., Gdansk, 1972-77; adj. prof., lectr. Med. Sch., 1978-81; rsch. fellow Med. Sch. Harvard U., Boston, 1981-84; resident Sch. of Medicine Brown U., Providence, 1984-87, fellow in oncology Sch. of Medicine, 1987-90; pvt. practice Providence, 1990—. Contbr. over 49 articles to profl. jours. Mem. Physicians for Social Responsibility, Internat. Phys. Rev. Nuclear War, Union Concerned Scientists. Fellow Internat. Union Biochemistry; mem. AAAS, ACP, AMA, Societas Scientiarum Gedanensis. Roman Catholic. Achievements include evolution of adenine metabolizing systems, rare abnomalities of blood coagulation interactions between hemostasis and complement system. Office: St Josephs Hosp 200 High Service Ave North Providence RI 02904-5113

STANKO, RONALD THOMAS, physician, medical educator, clinical nutritionist; b. Tarentum, Pa., Mar. 24, 1947; s. Andrew Anthony and Mary Louise (Staricek) S.; m. Rita Ann Kaniecki, June 27, 1970. BS in Biol. Scis., Carnegie-Mellon U., 1969; MD, U. Pitts., 1973. Diplomate Am. Bd. Clin. Nutrition. Asst. prof. medicine U. Pitts., 1979-88, assoc. prof. medicine, 1988—; mem. med. adv. bd. Caremark, Inc., Chgo., 1987-91; sci. advisor Am. Coun. Sci. and Health, 1985—; staff Montefiore Univ. Hosp., Presbyn. Univ. Hosp. Author: Medicine for You; contbr. numerous articles to profl. jours. Fellow Am. Coll. Nutrition; mem. ACP, Tau Beta Pi, Phi Kappa Phi. Roman Catholic. Achievements include patents for preventing body fat deposition, improving glucose metabolism, preventing fatty liver, decreasing plasma cholesterol, and enhancing exercise endurance capacity. Home: 795 Scrubgrass Rd Pittsburgh PA 15243 Office: Montefiore Univ Hosp 3459 5th Ave Pittsburgh PA 15213

STANLEY, BRUCE ALAN, molecular biologist; b. Washington, Apr. 7, 1949; s. Richard Jameison and Gloria (Hesiler) S.; m. Anne Czermak, Sept. 27, 1986. MS, Cornell U., 1984, PhD, 1987. Asst. prof. dept. cellular and molecular physiology M.S. Hershey Med. Ctr./Pa. State Coll. Medicine, Hershey, 1989—. Contbr. articles to Jour. of Biol. Chemistry. Chateaubriand fellow Govt. of France, 1984-85. Mem. AAAS, Am. Soc. Biochemistry and Molecular Biology. Home: 1703 Fishburn Rd Hershey PA 17033-9742 Office: MS Hershey Med Ctr Dept Cellular & Molecular Physiology Hershey PA 17033-0850

STANLEY, H(ARRY) EUGENE, physicist, educator; b. Norman, Okla., Mar. 28, 1941; s. Harry Eugene and Ruth S.; m. Idahlia Dessauer, June 2, 1967; children: Jannah, Michael, Rachel. BA in Physics (Nat. Merit scholar), Wesleyan U., 1962; postgrad. (Fulbright scholar), U. Cologne, W. Ger., 1962-63; PhD in Physics, Harvard U., 1967. NSF predoctoral rsch. fellow Harvard U., 1963-67; mem. staff Lincoln Lab MIT, 1967-68, asst. prof. physics, 1969-71, assoc. prof., 1971-73; Miller rsch. fellow U. Calif., Berkeley, 1968-69; Hermann von Helmholtz assoc. prof. health scis. and tech. Harvard U.-MIT Program in Health Scis. and Tech., 1973-76; vis. prof. Osaka (Japan) U., 1975; univ. prof., prof. physics, prof. physiology Sch. Medicine, dir. Ctr. Polymer Studies Boston U., 1976—; Joliot-Curie vis. prof. Ecole Superieure de Physique et Chimie, Paris, 1979; vis. prof. Peking U., 1981, Seoul Nat. U., 1982, 30th Ann. Saha Meml. Lecture, 1992; dir. NATO Advanced Study Inst., Cargese, Corisca, 1985, 88, 90; dir. IUPAP Internat. Conf. on Thermodynamics and Statis. Mechanics, 1986; cons. Sandia Nat. Lab., 1983—, Dowell Schlumberger Co., 1982—, Elscint Co., 1983-85; nat. co-chmn. Com. of Concerned Scientists, 1974-76. Author: Introduction to Phase Transitions and Critical Phenomena, 1971, From Newton to Mandelbrot: A Primer in Theoretical Physics, 1990, Fractal Forms, 1991; editor: Biomedical Physics and Biomaterials Science, 1972, Cooperative Phenomena Near Phase Transitions, 1973, On Growth and Form: Fractal and Non-Fractal Patterns in Physics, 1985, Statistical Physics, 1986, Random Fluctuations and Pattern Growth, 1988, Correlations and Connectivity: Geometric Aspects of Physics, Chemistry and Biology, 1990; assoc. editor Physica A, 1988-91, editor, 1991—. Recipient Choice award Am. Assn. Book Pubs., 1972, Macdonald award, 1986, Venture Rsch. award British Petroleum, 1989, Mass. Prof. of Yr. award Coun. Advancement and Support of Edn., 1992; John Simon Guggenheim Meml. fellow, 1979-80. Fellow Am. Phys. Soc. (chmn. New Eng. sect. 1982-83); mem. NAS (non-linear sci. panel). Home: 30 Metacomet Rd Newton MA 02168-1463 Office: Boston U Ctr for Polymer Studies Boston MA 02215

STANLEY, RICHARD HOLT, consulting environmental engineer; b. Muscatine, Iowa, Oct. 20, 1932; s. Claude Maxwell and Elizabeth Mabel (Holthues) S.; m. Mary Jo Kennedy, Dec. 20, 1953; children: Lynne Elizabeth, Sarah Catherine, Joseph Holt. BSEE, Iowa State U., 1955, B.S. in Mech. Engring., 1955; M.S. in San. Engring., U. Iowa, 1963. Registered profl. engr., Iowa, other states. With Stanley Cons. Inc., Muscatine, Iowa, 1955—, pres., 1971-87, chmn., 1984—; also bd. dirs. Stanley Cons. Inc.; bd. dirs. HON Industries, Inc., 1964—, vice chmn., 1979—; chmn. Nat. Constrn. Industry Coun., 1978, Com. Fed. Procurement Archtl-Engring. Svcs., 1979; bd. dirs. N.E.-Midwest Inst., 1989—, treas., 1991—; pres. Ea. Iowa C.C., Bettendorf, 1966-68; mem. indsl. adv. coun. Iowa State U. Coll. Engring., Ames, 1969—, chmn., 1979-81. Contbr. articles to profl. jours. Bd. dirs. Stanley Found., 1956—, pres., 1984—; bd. dirs. Muscatine Health Support Found., pres. 1984—; bd. dirs. Muscatine United Way, 1969-75, Iowa State U. Meml. Union, 1968-83, U. Dubuque, Iowa, 1977-93, Inst. Social and Econ. Devel., 1992—; bd. govs. Iowa State U. Achievement Found., 1982—. With C.E., U.S. Army, 1955-57. Recipient Young Alumnus award Iowa State U. Alumni Assn., 1966, Disting. Svc. award Muscatine Jaycees, 1967, Profl. Achievement citation Coll. Engring., Iowa State U., 1977, Anson Marston medal Iowa State U., 1991; named Sr. Engr. of Yr., Joint Engring. Com. Quint Cities, 1973. Fellow ASCE, Am. Cons. Engrs. Coun. (pres. 1976-77), Iowa Acad. Sci.; mem. IEEE (sr.), ASME, Am. Soc. Engring. Edn., Nat. Soc. Profl. Engrs., Cons. Engrs. Coun. Iowa (pres. 1967), Iowa Engring. Soc. (pres. 1973-74, John Dunlap-Sherman Woodward award 1967, Disting. Svc. award 1980, Voice of Engr. award 1987, Herbert Hoover Centennial award 1989), Muscatine C. of C. (pres. 1972-73), C. of C. of U.S. (constrn. action coun. 1976-91), Tau Beta Pi, Phi Kappa Phi, Pi Tau Sigma, Eta Kappa Nu. Presbyterian (elder). Club: Rotary. Home: 601 W 3D St Muscatine IA 52761 Office: Stanley Cons Inc Stanley Bldg Muscatine IA 52761

STANLEY, S. J., civil engineering educator. Prof. civil engring. U. Alberta, Edmonton, Alta., Can. Recipient T.C. Keefer medal Can. Soc. Civil Engring., 1992. Office: Univ of Alberta, Dept of Civil Engering, Edmonton, AB Canada T6G 2G7*

STANLEY, THOMAS P., chief engineer. BSEE, Johns Hopkins U.; MA, Princeton U., PhD. With Bell Tel. Labs., U.S. Army Singal Corps, Inst. Def. Analysis; various pos. including chief scientist office engring and tech. FCC, Washington, 1981-86, chief engr., 1986—. Office: FCC Engring & Tech 2025 M St NW Washington DC 20554-0001*

STANNARD, WILLIAM GEORGE, civil engineer; b. Concordia, Kans., Jan. 4, 1952; s. William Albert and Geraldine Francis (Chapman) S.; m. Susan Marie Hansen, Aug. 23, 1974; 1 child, Andrew Christopher. BSCE, BSBA, Kans. State U., 1975. Registered profl. engr., Kans., Mich., Ind., Ohio. Staff engr. Black & Veatch, Kansas City, Mo., 1975-78, project engr., 1978-80, project mgr., 1980-87, project dir., 1988-90, div. mgr., 1990—, ptnr., 1992—; cons. in utility fin. and mgmt.; expert witness; presenter in field at state and nat. confs. Contbr. articles to profl. publs. Pres. Black and Veatch Credit Union, Overland Park, Kans., 1987-92, treas. 1993—. Mem. ASCE, Am. Water Works Assn., Water Environ. Fedn., Urban and Regional Info. Systems Assn., Govt. Fin. Officers Assn., AM/FM Internat. Lutheran. Office: Black and Veatch 8400 Ward Pky Kansas City MO 64114-2031

STANNERS, CLIFFORD PAUL, molecular biologist, cell biologist, biochemistry educator; b. Sutton, Surrey, Eng., Oct. 19, 1937; married; 3 children. BSc, McMaster U., 1958; MSc, U. Toronto, 1960, PhD, 1963. Fellow molecular biology MIT, Cambridge, Mass., 1962-64; from asst. prof. to prof. med. biophysics U. Toronto, Can., 1964-82; sr. sci. biol. rsch. Ont. (Can.) Cancer Inst., 1964-82; prof. biochemistry McGill U., Montreal, 1982—, dir. Cancer Ctr., 1988—; mem. grants Med. Rsch. Coun. & Nat. Cancer Inst. Can., 1965—, U.S. Nat. Cancer Inst., 1973-79, Multiple Sclerosis Soc. Can., 1979—; mem. sci. adv. bd. Amyotropic Lateral Sclerosis Soc. Can., 1977-79. Assoc. editor: Jour. Cell. Physiology, 1973—, Cell, 1975-84. Mem. Can. Biochem. Soc., Can. Soc. Cell. Biology. Achievements include rsch. in growth control of animal cells; protein synthesis somatic cell genetics; molecular genetics; cell virus interactions; persitent infection with vesicular stomatitis virus; human cancer; human carcinoembryonic snitgen. Office: McGill U, 845 Sherbrooke St W, Montreal, PQ Canada H3A 2M5*

STANNY, GARY, infosystems specialist, rocket scientist; b. Detroit, Aug. 2, 1953; s. Richard Telesfor and Gertrude Mildred (Eisenbach) S. AS, Washtenaw Community Coll., 1973; B in Computer Sci., Ea. Mich. U., 1975. Programmer Ann Arbor (Mich.) Terminals, 1976-78; sr. programmer, analyst Mfg. Data Systems Inc., Ann Arbor, 1978-83; sr. software specialist Digital Equipment Corp., Farmington Hills, Mich., 1983-85; sr. systems cons. Tierra del Fuego Ltd, Whitmore Lake, Mich., 1985-87, v.p. rsch. and devel., 1987—, also bd. dirs.; bd. dirs. RTS Enterprises, South Lyon, Mich. Author: Ingres RMS Benchmarks, 1986, The S&P Premium Matrix, 1991; inventor in field; developer software TDF-lib, 1993. Counselor Drug Help, Ann Arbor, 1971-77, Rep. candidate senate, Ann Arbor, 1976; mem. Students Dem. Soc. State of Mich. Competitive Sci. grantee, 1971. Mem. Am. Assn. Artificial Intelligence Rsch. (computing machinery), Digital Equipment Computer Users Soc., Decus Artificial Intelligence Spl. Interest Group, Soc. Machine Intelligence, Mensa. Avocations: geopolitics, backpacking, travel, computers, artificial intelligence research. Office: Tierra del Fuego Ltd 7725 Shady Beach St Whitmore Lake MI 48189-9514

STANOVSKY, JOSEPH JERRY, aerospace engineering educator; b. Galveston, Tex., Mar. 4, 1928; s. Joseph and Hedvicka (Kaderka) S.; m. Barbara Jo Behal, Aug. 31, 1951 (div. 1958); children: Joseph Jerry, Matthew Neil; m. Juanita Bridges, Sept. 8, 1961 (dec.); children: Clinton Sebastian, Derek Joseph. BS, So. Meth. U., 1948; MS, U. Tex., 1951; PhD, Pa. State U., 1966. Registered profl. engr., Tex. Asst. prof. Tex. Tech. U., Lubbock, 1958-59; sr. structural engr. Boeing Airplane Co., Seattle, 1959-60; instr. Pa. State U., University Park, 1961-66; assoc. prof. aerospace and engring. mechs. U. Tex., Arlington, 1966—; vis. prof. civil engring. U. Petroleum and Minerals, Dhahran, Saudi Arabia, 1974-76. Author: Fatal Flaws Compelling Concepts, 1993; contbr. articles to Machine Design Mag. With USNR, 1946-47. Fellow Ford Found., 1964-65, Teetor fellow Soc. Automotive Engrs., 1973. Mem. AAAS, Am. Assn. Astronomy and Aerospace, N.Y. Acad. Scis. Home: 1816 Michael Ct Arlington TX 76010 Office: U Tex Mech and Aerospace Engring Arlington TX 76019

STANTON, JOSEPH ROBERT, physician; b. Boston, Aug. 8, 1920; s. Joseph S. and Mary Elizabeth (Sullivan) S.; m. Mary Frances Gordon, May 10, 1950; children: Michael, Anne Marie, Joseph, John, Mark (dec.), Paul, William, Kathleen, Matthew, Luke, Thomas. AB, Boston Coll., 1942; MD, Yale U., 1945; LLD, St. Anselm Coll., 1973, Our Lady of Elms, 1974. Diplomate Am. Bd. Internal Medicine. Asst. surg. and medicine Boston U. Sch. Medicine, 1946-51; rsch. fellow Evans Meml. Hosp., 1946-51; instr. medicine Tufts Med. Sch., 1951-58, assoc. clin. prof. medicine, 1958-85; mem. attending staff Holy Ghost Hosp., Cambridge, Mass., 1948-55; attending physician Bethany Infirmary, Framingham, Mass., 1948-75; vis. physician St. Elizabeth's Hosp., Boston, 1952-85; del. White House Conf. on Aging, 1981; grant reviewer HHS, 1983; cons. Medico Moral Commn. Mass Cath. Conf., 1978-91. Author or co-author 30 papers on medicine. Founding mem., sec.-treas. Americans United for Life, Chgo., 1971-85; founding mem., dir. Value of Life Com., Boston, 1970-91; founding mem., dir., v.p. Mass. Citizens for Life, 1974-91. Recipient Alumni Sci. award Boston Coll., 1980, Poverello award U. Steubenville (Ohio), 1989, Pro Vita award Archdiocese Boston, 1991. Fellow AMA, ACP, Mass. Med. Soc.; mem. Boston Coll. Alumni Assn., Yale Med. Alumni Assn. Roman Catholic. Avocations: family, reading, travel, politics. Home: 760 Highland Ave Needham MA 02194-1635 Office: Value of Life Com 637 Cambridge St Brighton MA 02135-2899

STANTON, TIMOTHY KEVIN, physicist; b. Fayetteville, Ark., May 29, 1953; s. Harold Stanley and Joyce June (Blinn) S.; m. Cynthia Francene Simmons, Aug. 20, 1978; children: Liam Gregory, Emma Dalbey. BS in Physics, Oakland U., 1974; MS, Brown U., 1976, PhD, 1978. Rsch. asst. Brown U., Providence, 1975-78; sr. engr. Raytheon Co., Portsmouth, R.I., 1978-80; project assoc. U. Wis., Madison, 1980-82, asst. scientist, 1982-85, assoc. scientist, 1986-88; assoc. scientist Woods Hole (Mass.) Oceanographic Inst., 1988—. Guest editor: IEEE Jour. Oceanic Engring., 1989; assoc. editor: Jour. Acoustical Soc. Am., 1989-92; contbr. articles to profl. jours. Grantee Office Naval Rsch., 1984—, NSF, 1992—; recipient A.B. Wood medal and prize for Significant Contbn. to Underwater Acoustics, 1985. Mem. Acoustical Soc. Am. (chair various coms.). Achievements include research in underwater acoustics describing sound scattering by seafloor, sea ice, and zooplankton. Office: Woods Hole Oceanographic In Woods Hole MA 02543

STANZIONE, KAYDON AL, aerospace industry executive, advisor; b. Phoenixville, Pa., Dec. 16, 1956; s. Dominic John and Katherine (Mitropolous) S.; 1 child, Alyssa Marie. BSME, Rutgers U., 1978, MSME, 1979. Cert. comml. instrument pilot, multi-engine pilot. Chief engr. C.A. Tech., Oceanport, N.J., 1974-79; sr. engr. advanced design Boeing Vertol Co., Phila., 1979-84; chief engr. Korax Corp., Haddon Heights, N.J., 1984-86; chief exec. officer/chief engr. Praxis Techs. Corp., Woodbury, N.J., 1986—; expert advisor various domestic and internat. cos. including IBM, LTV Mil. Aircraft Div., Aerospatiale, Westland among others, 1986—, Inst. Def. Analyses, Alexandria, Va., 1987—; expert advisor/lectr. Coll. Engring. U. Md., College Park, 1986—, US Spl. Ops. Forces, 19990—. Author: Systems Approach to Helicopter Design and Technology Assessment, 1987, Aerospace Engineering Encyclopaedia Britannica, 1987, Helicopter Design, 1990, V/STOL Design, Systems and Technology Assessment, 1993; inventor and patentee in field; contbr. numerous articles to mags., newspapers and profl. jours. Mem. ASME, AIAA (chmn. students programs 1984-87, tech. V/STOL systems 1985—, tech. aircraft design 1987—, first place design award 1980, U.S. Space Shuttle Challenger award 1984), Am. Helicopter Soc. (design competition and edn. com. 1982—, bd. dirs., 1992—, Oustanding Mem. award. 1981), Aircraft Owners and Pilots Assn., U.S. Naval Inst, Washington C. of C., Rotary. Republican. Roman Catholic. Avocations: flying, skiing, weightlifting, piano. Home: PO Box 247 Woodbury NJ 08096-3902 Office: Praxis Technologies Corp Ste 200 1047 N Broad St Woodbury NJ 08096

STAPLETON, SUSAN REBECCA, biochemistry educator; b. Reading, Pa., May 14, 1957; d. John Russell and Stella Mary (Nowoczynski) S.; m. David Scott Reinhold, May 14, 1987; children: Collin John, Jacob Paul. BS, Juniata Coll., 1979; PhD, Miami U., Oxford, Ohio, 1983. Rsch. assoc. U. Wis., Madison, 1984; postdoctoral fellow Case Western Res. U., Cleve., 1985-87; rsch. assoc. U. Iowa, Iowa City, 1988-89; asst. prof. Western Mich. U., Kalamazoo, 1990—. Contbr. articles to profl. jours. Postdoctoral fellowship Am. Heart Assn., 1987, NIH, 1985-86; grantee Diabetes Rsch. and Edn. Found., 1992, NIH, NIAAA. Mem. AAAS, Am. Chem. Soc., Am. Soc. for Biochem. and Molecular Biology, Sigma Xi. Office: Western Mich U Dept Chemistry Kalamazoo MI 49008

STAPLEY, EDWARD OLLEY, retired microbiologist, research administrator; b. Bklyn., Sept. 25, 1927; s. Charles Olley and Helen Beulay (Mirrielees) S.; m. Helen Alberta Strang, July 2, 1949; children: Susan Jean, Robin Lynn, Janice Carol. BS, Rutgers U., 1950, MS, 1954, PhD, 1959. Microbiologist Merck & Co., Inc., Rahway, N.J., 1950-58; sr. rsch. microbiologist Merck Sharp & Dohme Rsch. Labs., Rahway, 1959-64, rsch. fellow in microbiology, 1965-68, asst. dir. microbiology, 1969-74, dir. microbiology, 1974-77, sr. dir. microbiology, 1978-83, exec. dir. microbiology, 1984-92; vis. biologist program speaker Am. Inst. Biol. Scis., 1969-72. Mem. editorial bd. Jour. of Antibiotics, 1974—. Mem. Spotswood (N.J.) Bd. of Edn., 1965, pres., 1967-68. Named to Selman A. Wakeman Lectureship, Theobald Smith Soc., 1990. Fellow Am. Acad. Microbiology, Sigma Xi; mem. Soc. for Indsl. Microbiology (speaker's bur. 1968-71). Republican. Episcopalian. Achievements include 28 patents; discovery of many antibiotics and microbial chemotherapeutics including Fosfomycin, Cephamycin, Thienamycin, Avermectin and Mevinolin. Home: 110 Highland Ave Metuchen NJ 08840-1913

STAPP, JOHN PAUL, surgeon, former air force officer; b. Bahia, Brazil, July 11, 1910; s. Charles Franklin and Mary Louise (Shannon) S.; m. Lillian Lanese, Dec. 23, 1957. B.A., Baylor U., 1931, M.A. cum laude, 1932, D.Sc., 1956; Ph.D., U. Tex., 1939; M.D., U. Minn., 1943; grad., Army Field Service Sch., 1944, Sch. Aviation Medicine, 1945, Indsl. Med. Course, 1946; D.Sc., N.Mex. State U., 1979. Diplomate Am. Bd. Preventive Medicine. Commd. 1st lt. U.S. Army, 1944; advanced through grades to col. M.C. USAF, 1957; research project officer (Aero Med. Lab.), Wright Field, Ohio, 1946; chief lab. (Aero Med. Lab.), 1958-60; chief scientist aerospace med. div. Brooks AFB, Tex., 1960-65; chief impact injury br. (Armed Forces Inst. Pathology), Washington, 1965-67; chief med. scientist Nat. Hwy. Safety Bur., 1967-70; ret., 1970; cons. Dept. Transp., Washington, 1970—; adj. prof. Safety and Systems Mgmt. Center, U. So. Calif., 1972—; Systems Mgmt. Center, Los Angeles, 1973—; cons. accident epidemiology and pathology Armed Forces, Bur. Standards, NIH, Nat. Acad. Scis., Gen. Services Adminstrn.; chief Aero Med. Field Lab., Holloman AFB, Alamogordo, N.Mex., 1953-58; cons. N.Mex. State U. Phys. Scis. Lab., Las Cruces, 1972—; mem. subcom. on flight safety NACA; permanent chmn. Ann. Stapp Car Crash Conf.; chmn. Gov.'s Commn. Internat. Space Ctr., 1986; pres. N.Mex. Rsch. Inst., 1987. Mem. N.Mex. Gov.'s Commn. Internat. Space Hall of Fame, 1974—; mem. N.Mex. Planning Bd., 1975—; Bd. dirs. Kettering Found.; v.p. Internat. Astronautical Fedn., 1959-60. Decorated D.S.M. with bronze oak leaf, Legion of Merit (for crash research) with bronze oak leaf; recipient award for outstanding research by Air Force officer Nat. Air Council, 1951; John Jeffries award for med. research Inst. Aero. Sci., 1953; Air Power award for sci. Air Force Assn., 1954; Flight Safety Found. award for contbns. Air Transp. Safety, 1954; Air Force Cheney award, 1955; Gorgas award Assn. Mil. Surgeons, 1956; Med. Tribune award for automotive safety, 1965; award for contbns. to automotive safety Am. Assn. for Automotive Medicine, 1972; Cresson medal Franklin Inst., 1973; Excalibur award safety research, 1975; cert. of achievement Nat. Space Club, 1976; Lovelace award NASA Assn. Flight Surgeons, 1982; Outstanding Service award Aviation/Space Writers Assn., 1984; Honda medal ASME, 1984; Disting. Alumnus award Baylor U., 1986, Nat. medal for tech. Pres. Bush, 1991; elected to Internat. Space Hall of Fame, 1979, Nat. Aviation Hall of Fame, 1985, Safety and Health Hall of Fame, Internat., 1991; recipient Nat. Medal Tech., 1991; annual John Paul Stapp medal biomechanics Aerospace Med. Assn., 1993—. Fellow Aero. Med. Assn. (Liliencrantz award for deceleration research 1957), Am. Astronautical Soc., Am. Rocket Soc. (pres. 1959, Wyld award 1955, Leo Stevens medal 1956), Soc. Automotive Engrs.; mem. U.S. Mil. Surgeons Assn., Internat. Acad. Astronautics, Internat. Acad. Aviation Medicine, Civil Aviation Medicine Assn. (pres. 1968), Am. Soc. Safety Engrs. (hon.), Order Daedalians (hon.), Sigma Xi. Achievements include research rocket sled experiments reproducing aircraft crash forces to determine human tolerance limits, 1947-51. Home: PO Box 553 Alamogordo NM 88311-0553 Office: NMex Rsch Inst PO Box 454 Alamogordo NM 88311-0454

STAREK, RODGER WILLIAM, chemist; b. Cin., Aug. 5, 1953; s. Robert Alfred and Mary Lou Starek. BA, U. Mass., 1975. Svc. engr. Spectra Metrics, Andover, Mass., 1978-81; chemist Metals Processing Industries, San Jose, Calif., 1981-83; sr. scientist Mineralab, Hayward, Calif., 1983-85; ind. cons. Pleasanton, Calif., 1985-87; applications chemist Fisons Instruments, Danvers, Mass., 1987—. Mem. Soc. for Applied Spectroscopy (treas. So. Calif. sect. 1990).

STARK, BENJAMIN CHAPMAN, biology educator, molecular biologist; b. Saginaw, Mich., Nov. 22, 1949; s. Robert Ellis and Gladys Marian (Beck) S.; m. Carol Ann Hunt, May 19, 1979; children: Rebecca Anna, Sarah Jane. BS, U. Mich., 1971; MPhil, Yale U., 1974, PhD, 1977. Postdoctoral fellow dept. botany Wash. State U., Pullman, 1977-79; postdoctoral fellow dept. biology Ind. U., Bloomington, 1979-82, vis. asst. prof. dept. biology, 1982-83; asst. prof. dept. biology Ill. Inst. Tech., Chgo., 1983-89, assoc. prof. dept. biology, 1989—; cons. Met. Water Reclamation Dist. Greater Chgo., 1989—. Mem. AAAS, Sigma Xi, Phi Beta Kappa. Jewish. Achievements include first discoverer of enzyme proved to have required RNA component; first proof that an enzyme can have a required RNA component; discovered and named 1st RNA that was eventually proved to be catalytic, others. Office: Ill Inst Tech Dept Biology IIT Ctr Chicago IL 60616

STARK, GEORGE EDWARD, software reliability engineer, statistician; b. Pueblo, Colo., July 21, 1961; s. Mark Martin and Beatrice Catherine (Morris) S.; m. Patricia Kathleen Coen, Aug. 19, 1984; children: Jeremy Martin, Kelley Marie. BS, Colo. State U., 1983; MS, U. Houston, Clear Lake, 1988. Statistician Ultrasystems, Inc., Sierra Vista, Ariz., 1983-84; mem. tech. staff Mitre Corp., Houston, 1984-87; lead scientist MITRE Corp., Houston, 1990—; mgr. S.W. reliability Raynet Corp., Menlo Park, Calif., 1987-89; cons. U. Houston-Clear Lake, 1992—. Contbr. articles to profl. jours.; referee various jours. Mem. maintenance com. Imperial Estates Homeowners Assn., Friendswood, Tex., 1992—. Recipient NASA Quality Partnership award NASA Johnson Space Ctr., Houston, 1992. Mem. AIAA (vice chmn. software reliability stds. com. 1989—). Achievements include development of NASA-Johnson Space Center ground systems software measurement guidelines and metric sets. Office: Mitre Corp 1120 NASA Rd One Ste 600 Houston TX 77058

STARK, JOHN DAVID, pesticide toxicologist; b. Sept. 4, 1956; s. John Charles and Sally Ann (Macdonald) S.; m. Laura Kealaonapua Keolanui, Dec. 15, 1979; children: John Keliipomaikai, Sarah Kealaonapua. BS, SUNY, Syracuse, 1978; MS, La. State U., 1981; PhD, U. Hawaii, 1987. Rsch. asst. La. State U., Baton Rouge, 1979-81; rsch. fellow U. Hawaii, Honolulu, 1984-86; rsch. entomologist/toxicologist USDA, Kapaa, Hawaii, 1987-90; asst. prof. environ. toxicology and entomology Wash. State U., Puyallup, 1990—; environ. advisor burrowing shrimp com. State of Wash., Olympia, 1991—; mem. adv. panel Ecol. Monitoring for Pesticides program EPA, Seattle, 1991—. Contbr. articles to profl. jours. Mem. Am. Chem. Soc., Soc. Environ. Toxicology and Chemistry, Entomol. Soc. Am., Sigma Xi, Phi Kappa Phi, Gamma Sigma Delta. Achievements include development of more environmentally sound methods of insect control. Office: Wash State U 7612 Pioneer Way E Puyallup WA 98371

STARKEY, RUSSELL BRUCE, JR., utilities executive; b. Lumberport, W.Va., July 20, 1942; s. Russell Bruce and Dorotha Mable (Field) S.; m. Joan McClellan, May 27, 1966; children: Christine, Pamela, Joanne. BS, Miami U., Oxford, Ohio, 1964; grad. student U. New Haven, 1972-73, N.C. State U., 1974-75, U.S. Navy Schs., 1964-66, 68. Sr. engr., nuclear generation sect. Carolina Power & Light Co., Raleigh, N.C., 1973-74, sr. engr. ops. quality assurance, 1974, prin. engr., 1974-75, quality assurance supr. Brunswick Steam Electric Plant, Southport, N.C., 1975-76, supt. tech. and adminstrn., 1976, supt. ops. and maintenance, 1976-77, plant mgr. H.B. Robinson Steam Electric Plant, Hartsville, S.C., 1977-83, mgr. environ. services, Raleigh, 1984-85, mgr. nuclear safety and environ. services dept., 1985-88; mgr. Brunswick Nuclear Project Dept., 1988-89, v.p., 1989-92; v.p. Nuclear Svcs. Dept., 1992—; exec. dir. nuclear prodn. Pub. Service Ind., Jeffersonville, 1983-84. Served with USN, 1964-73. Mem. Am. Nuclear Soc., Rotary. Home: PO Box 306 Wilsons Mills NC 27593-0306 Office: PO Box 1551 One Hanover Sq 7th Fl Raleigh NC 27602

STARKS, MICHAEL RICHARD, stereographer, consultant; b. Iron Mount, Mich., June 23, 1941; s. Verna E. (Haensgen) S. BA, U. Calif., Santa Barbara, 1964. V.p. Stereographics Corp., San Rafael, Calif., 1978-83; sr. rsch. scientist UME Corp., Larkspur, Calif., 1985-87; pres. 3 DTV Corp., San Rafael, Calif., 1989—. Contbr. articles to profl. jours. Achievements include patent, creation and market of indsl. and home stereoscopic TV and stereoscopic computer graphics systems. Office: 3 DTV Corp PO Box Q San Rafael CA 94913-4316

STARKWEATHER, GARY KEITH, optical engineer, computer company executive; b. Lansing, Mich., Jan. 9, 1938; married; 2 children. BS, Mich. State U., 1960; MS, U. Rochester, 1966. Engr. Bausch & Lomb, Inc., 1962-64; area mgr. optical systems Xerox Palo Alto (Calif.) Rsch. Ctr., 1964-80, sr. rsch. fellow, 1980-88; Apple fellow advanced tech. group Apple Computer, Inc., Saratoga, Calif., 1988—; dir. Apple Computer, Inc., Cupertino, Calif.; instr. optics Monroe Community Coll., 1968-69. Mem. Optical Soc. Am. (David Richardson medal 1991), Soc. Photog. Inst. Engrs. Achievements include research in optics and electronics and their specific system interaction, involving display and hard copy image systems. Office: Apple Computer Inc 20525 Mariani Ave Cupertino CA 95014*

STARMACK, JOHN ROBERT, mathematics educator; b. McKeesport, Pa., Mar. 10, 1942; s. Walter Paul and Amelia Fanny (Hager) S.; m. Kathryn M. Skweres, June 8, 1963; children: John Jr., Michelle, Thomas, Lana. BS in Edn., California (Pa.) State Coll., 1964; MEd in Math., Indiana U. of Pa., 1968; PhD in Math. Edn., U. Pitts., 1991. Tchr. math. West Mifflin (Pa.) Area Sch. Dist., 1964-69; prof. math. C.C. Allegheny County, West Mifflin, 1969—, coord. secondary honors program, 1979—, coord. ann. gifted and talented conf. and expn., 1983—; adj. instr. U. Pitts., 1989—. Recipient cert. of excellence Presdl. Scholars Program, 1988, NE Regional Faculty Mem. award Assn. Community Coll. Trustees, 1988. Mem. Pa. State Math. Assn. Two-Year Colls. (co-chair curriculum com. 1982—, chair student award com. 1990—), Pa. Coun. Tchrs. Math., Am. Math. Assn. Two-Year Colls. (Pa. del. 1986—), Pa. Assn. for Gifted Edn., Western Pa. Coun. Tchrs. Math., Nat. Coun. for Tchrs. Math. Democrat. Roman Catholic. Avocations: aerobics, golf, teaching. Home: 616 Shadyside Dr West Mifflin PA 15122-3229 Office: Community Coll Allegheny County South Campus 1750 Clairton Rd West Mifflin PA 15122-3097

STARNES, WILLIAM HERBERT, JR., chemist, educator; b. Knoxville, Tenn., Dec. 2, 1934; s. William Herbert and Edna Margaret (Osborne) S.; m. Maria Sofia Molina, Mar. 4, 1986. BS with honors, Va. Poly Inst., 1955; PhD, Ga. Inst. Tech., 1960. Rsch. chemist Esso Rsch. & Engring. Co., Baytown, Tex., 1960-62, sr. rsch. chemist, 1962-64, polymer additives sect. head, 1964-65, rsch. specialist, 1965-67, rsch. assoc., 1967-71; instr. and rsch. assoc. dept. chemistry U. Tex., Austin, 1971-73; mem. tech. staff AT&T Bell Labs., Murray Hill, N.J., 1973-85; prof. chemistry Poly. U., Bklyn., 1985-89, head dept. chemistry and life scis., 1985-88, assoc. dir. polymer durability ctr. 1987-89, Floyd Dewey Gottwald Sr. prof. chemistry Coll. William and Mary, Williamsburg, Va., 1989—; invited lectr. several fgn. countries and U.S.; vis. scientist Tex. Acad. Scis., 1964-67; mem. bd. doctoral thesis examiners Indian Inst. Tech., New Delhi, 1988, McGill U., Montreal, 1989, MacQuarie U., Sydney, 1991; panelist, reviewer NSF Acad. Rsch. Facilities Modernization Program, 1990; cons. numerous indsl. cos.; course dir. continuing edn. Mem. adv. bd. and bd. reviewers Jour. Vinyl Tech., 1981-83; mem. editorial bd. Jour. of Chemical and Biochemical Kinetics, 1992—; contbr. articles to profl. jours., chpts. to books; patentee in field. NSF fellow 1958-60; recipient Profl. Progress award Soc. Profl. Chemists and Engrs. 1968, Disting. Tech. Staff award AT&T Bell Labs. 1982, Polymer Sci. Pioneer award Polymer News, 1988, Honor Scroll award N.J. Inst. Chemists, 1989; NSF grantee, Nat. Bur. Standards Ctr. for Fire Rsch. grantee, Internat. Copper Rsch. Assn. grantee, Va. Ctr. Innovative Tech. grantee. Fellow AAAS (Project 2061 1985-86, chmn. chemistry subpanel 1985-86, mem. panel on phys. scis. and engring., 1985-86), Am. Inst. Chemists (life); mem. Am. Chem. Soc. (bd. dirs. southeastern Tex. sect. 1970, speakers bur. div. polymer chemistry 1976—), Soc. Plastics Engrs., N.Y. Acad. Scis. (life), Va. Acad. Sci., Sigma Xi (M.A. Ferst award Ga. Inst. Tech. chpt. 1960), Phi Kappa Phi, Phi Lambda Upsilon (pres. Va. Poly. Inst. chpt. 1954-55). Current work: Degradation, stabilization, flammability, microstructures, and polymerization mechanisms of synthetic polymers, especially poly (vinyl chloride); free radical chemistry; carbon-13 nuclear magnetic resonance and organic synthesis. Subspecialties: Organic chemistry; Polymer chemistry. Office: Coll William and Mary Dept Chemistry PO Box 8795 Williamsburg VA 23187-8795

STAROSOLSZKY, ÔDÔN, civil engineer; b. Veszprèm, Hungary, Dec. 26, 1931; s. Sándor and Irma (Benko) S.; m. Erzsébet Zilahi-Kiss, Apr. 1, 1961. Dipl.Ing., T.U., Budapest, 1954, Dr(Eng.), 1968. Engring. diplomate. Rsch. engr. Rsch. Inst. Water Rsch., Budapest, 1954-60, head sect., dept., 1960-71; head dept. Nat. Water Authority, Budapest, 1971-76; dir. Inst. ofr Hydraulics, Budapest, 1976-89; dep. gen. dir. Water Res. Rsch. Ctr., Budapest, 1989-91, gen. dir., 1991—; chmn. com. on water scis. Hungarian Acad. Scis., Budapest, 1990—; pres. Commn. for Hydrology, Geneva, 1984-92. Author: Civil Engineering Hydraulics, 1971; co-author/editor: Hydraulic Engineering, 1973, Applied Surface Hydrology, 1987; co-editor: Hydrology of Disaster, 1989. Mem. Hungarian Hydrol. Soc. (chmn. com. internat. affairs 1990—, mem. presidium, Vásárhelyi prize 1986, 92), Internat. Assn. Hydrol. Rsch. (v.p. 1988), Hungarian Soc. Environ. Roman Catholic. Achievements include rsch. results in the field of hydrology, hydraulics, hydraulic engineering and water managment. Office: Water Rsch Ctr VITUKI, Kvassay Jenout 1, H-1095 Budapest Hungary

STARR, JAMES LEROY, soil scientist; b. Almont, Mich., Aug. 14, 1939; m. Loretta M. Young, June 25, 1960; two children. BS, Mich. State U., 1961, MS, 1970; PhD, U. Calif., Davis, 1973. Soil scientist Agrl. Exptl. Sta. State of Conn., New Haven, 1974-78, USDA-ARS, Beltsville, Md., 1979—. Contbr. numerous articles to profl. jours. Mem. Soil Sci. Soc. Am., Am. Geophys. Union, Internat. Soc. Soil Sci., Sigma Xi. Office: USDA Agrl Rsch Svc NRI ECL Bldg 007 BARC-W NRI ECL Bldg 007 BARC-W Beltsville MD 20705

STARR, RICHARD CAWTHON, botany educator; b. Greensboro, Ga., Aug. 24, 1924; s. Richard Neal and Ida Wynn (Cawthon) S. BS in Secondary Edn., Ga. So. Coll., 1944; MA, George Peabody Coll., 1947; postgrad. (Fulbright scholar), Cambridge (Eng.) U., 1950-51; PhD, Vanderbilt U., 1952. Faculty Ind. U., 1952-75, prof. botany, 1960-76; founder, head culture collection algae U. Tex., Austin, prof. botany, 1976—; Head course marine botany Marine Biol. Lab., Woods Hole, Mass., 1959-63. Algae sect. editor: Biol. Abstracts, 1959—; editorial bd.: Jour. Phycology, 1965-68, 76-78; assoc. editor: Phycologia, 1963-69; Contbr. articles to profl. jours. Trustee Am. Type Culture Collection, 1962-68, 80-85. Guggenheim fellow, 1959; sr. fellow Alexander von Humboldt-Stiftung, 1972-73; recipient Disting. Tex. Scientist award Tex. Acad. Sci., 1987. Fellow AAAS, Ind. Acad. Sci.; mem. NAS (Gilbert Morgan Smith Award 1985), Am. Inst. Biol. Scis. (governing bd. 1976-77, exec. com. 1980), Bot. Soc. Am. (sec. 1965-69, v.p. 1970, pres. 1971, Darbaker prize 1955), Phycological Soc. Am. (past pres., v.p., treas.), Soc. Protozoologists, Internat. Phycological Soc. (sec. 1964-68), Brit. Phycological Soc., Akademie Wissenschaft zu Göttingen (corr.), Sigma Xi. Office: U Tex Dept Botany Austin TX 78713

STARR, STUART HOWARD, systems engineer, long range planner; b. N.Y.C., Jan. 29, 1942; s. Benjamin and Beatrice (Danson) S.; m. Gloria Fuenmayor, Dec. 20, 1970; children: Alexandra Lee, Adrienne Beatrice. BSEE, Columbia U., 1963; MSEE, U. Ill., 1965, PhD, 1969. Mem. tech. staff MITRE, McLean, Va., 1969-73; sr. project leader Inst. for Def. Analyses, Alexandria, Va., 1973-80; dir. long range planning and systems evaluation C3I Office Sec. Def., Dept. of Def., Washington, 1980-82; asst. v.p., C3I systems M/A-COM Govt. Systems, Vienna, Va., 1982-85; dir. plans MITRE, McLean, 1985—; cons. Office Sec. Def., Newport, R.I., 1992-93; mem. sr. adv. panel Office Tech. Assessment, Washington, 1985-86; cons., gen. officer tng. course Army War Coll., Carlisle, Pa., 1984-86. Coauthor: Science of Command and Control: Part II--Coping with Complexity, 1989; contbr. articles to profl. jours. Sec. Jeb Stuart High Sch. PTA,

Fairfax, Va., 1989-91; coll. recruitment Columbia U., N.Y.C., 1987-92. Recipient fellowship NSF, 1966-69. Mem. IEEE (sr.), AIAA (sr.), Mil. Ops. Rsch. Soc. (bd. dirs.), Armed Forces Communications Electronics Assn., Phi Beta Kappa, Pi Mu Epsilon. Achievements include development and design of Friend, Foe or Neutral Testbed; co-developer of mission oriented approach to long range planning; of applied approach to planning initiatives in NATO, Dept. of Def.; research in deriving C3I lessons learned from Desert Storm in support of Joint Chiefs of Staff. Home: 3422 Rusticway Ln Falls Church VA 22044 Office: MITRE 7525 Colshire Dr Mc Lean VA 22102

STARRS, JAMES EDWARD, law and forensics educator, consultant; b. Bklyn., July 30, 1930; s. George Thomas and Mildred Agatha (Dobbins) S.; m. Barbara Alice Smyth, Sept. 6, 1954; children: Mary Alice, Monica, James, Charles, Liam, Barbara, Siobhan, Gregory. BA and LLB, St. John's U., Bklyn., 1958; LLM, NYU, 1959. Bar: N.Y. 1958, D.C. 1966, U.S. Ct. Mil. Appeals 1959, U.S. Dist. Ct. (so. and ea. dists.) N.Y. 1960. Assoc. Lawless & Lynch, N.Y.C., 1958; teaching fellow Rutgers U., Newark, 1959-60; asst. prof. law DePaul U., Chgo., 1960-64; assoc. prof. law George Washington U., Washington, 1964-67, prof. law, 1967—, prof. forensic scis., 1975—; cons. Nat. Commn. Reform Fed. Criminal Laws, Washington, 1968, Cellmark Diagnostics, Germantown, Md., 1987—, Time-Life Books, 1993; participant re-evaluation sci. evidence and trial of Bruno Richard Hauptmann for Lindbergh murder, 1983; participant reporting sci. re-analysis of firearms evidence in Sacco and Vanzetti trial, 1986; project dir. Alfred G. Packer Victims Exhumation Project, 1989, A Blaze of Bullets: A Sci. Investigation into the Deaths of Senator Huey Long and Dr. Carl Austin Weiss, 1991, Meriwether Lewis Exhumation Project, 1992. Author: (with Moenssens and Inbau) Scientific Evidence in Criminal Cases, 1986; editor: The Noiseless Tenor, 1982; co-editor: (review) Scientific Sleuthing, 1976—; mem. editorial bd. Jour. Forensic Sci., 1980—; contbr. articles to profl. jours. Served to sgt. U.S. Army, 1950-53, Korea. Recipient Vidocq Soc. award, 1993; Ford Found. fellow, 1963; vis. scholar in residence USMC, 1984. Fellow Am. Acad. Forensic Sci. (chmn. jurisprudence sect. 1984, bd. dirs. 1986-89, Jurisprudence Sect. award 1988); mem. ABA, Mid-Atlantic Assn. Forensic Sci., Assn. Trial Lawyers Am., Internat. Soc. Forensic Sci. (chmn. jurisprudence sect. 1988—). Roman Catholic. Home: 8602 Clydesdale Rd Springfield VA 22151-1301 Office: George Washington U Nat Law Ctr 720 20th St NW Washington DC 20052-0001

STARZER, MICHAEL RAY, petroleum engineer; b. Tulsa, Okla., May 23, 1961; s. Richard Ray and Marilyn Lea (Hamm) S.; m. Patricia Suzanne Kilthau, Jan. 7, 1984; 1 child, Moriah Suzanne. BS in Petroleum Engring., Colo. Sch. Mines, 1983; MS in Engring. Mgmt., U. Alaska, 1990. Registered petroleum engr., Alaska, Calif. Petroleum engr. Unocal Union Oil and Gas Co., Anchorage, 1983-90; sr. engr., supr. Unocal Union Oil and Gas Co., Kenai, Alaska, 1990-91; engring. supr. State Lands Commn., Long Beach, Calif., 1991—; cons. Okla. Chem. Cons., Long Beach, 1991—, Pep Engring., Irvine, Calif., 1991—. Contbr. articles to Jour. Petroleum Tech. Transactions. Mem. Soc. Petroleum Engrs., Soc. Petroleum Evaluation Engrs., L.A. Basin Soc. Petroleum Engrs. (dir. awards 1992—), NSPE, Pi Epsilon Tau. Achievements include development of well stimulation/scale inhibition treatment design, techniques for improved reservoir management, maximization of recovery. Office: UNOCAL 1800 30th St Bakersfield CA 93301

STASIOR, WILLIAM F., engineering company executive; b. 1941. BSEE, Northwestern U., MSEE. With Booz Allen & Hamilton Inc., N.Y.C., 1967—, pres., COO, 1990—, CEO, chmn., 1991—. Office: Booz Allen & Hamilton Inc 101 Park Ave New York NY 10178-0002*

STASKUN, BENJAMIN, retired chemistry educator; b. Kovno, Lithuania, Aug. 29, 1925; s. Sender and Dora (Iserowitz) S.; m. Mina Gertrude Friedman, Apr. 12, 1959; 1 child, Jonathan Harold. MSc, U. Witwatersrand, Johannesburg, Republic South Africa, 1951, PhD, 1955, DSc, 1992. Lectr. U. Witwatersrand, 1962-65, sr. lectr., 1965-69, reader in organic chemistry, 1970-73, reader and assoc. prof. in chemistry, 1973-92; postdoctoral fellow U. Sask., Can., 1961; postdoctoral assoc. Stanford U., Palo Alto, Calif., 1968; vis. prof. U. Del., Newark, 1975, UCLA, 1981, Va. Poly. Inst. and State U., Blacksburg, 1989. Contbr. articles to profl. publs. Mem. Am Chem. Soc., South African Chem. Inst. Jewish. Home: 3 Talton Rd Forest Town, Johannesburg 2193, South Africa Office: U Witwatersrand, Dept Chemistry, Johannesburg South Africa

STATLER, IRVING CARL, aerospace engineer; b. Buffalo, N.Y., Nov. 23, 1923; s. Samuel William and Sarah (Strauss) S.; m. Renee Roll, Aug. 23, 1953; children—William Scott, Thomas Stuart. B.S. in Aero. Engring., U. Mich., 1945, B.S. in Engring. Math., 1945; Ph.D., Calif. Inst. Tech., 1956. Research engr. flight research dept. Cornell Aero. Lab., Inc., Buffalo, 1946-53; prin. engr. flight research dept. Cornell Aero. Lab., Inc., 1956-57, asst. head aero-mechanics dept., 1957-63, head applied mechanics dept., 1963-70, sr. staff scientist aeroscis. div., 1970-71; research scientist U.S. Army Air Mobility Research and Devel. Lab., Moffett Field, Calif., 1971-73; dir. Aeromechanics Lab. U.S. Army Air Mobility Research and Devel. Lab., 1973-85, dir. AGARD, 1985-88; sr. staff scientist NASA Ames Rsch. Ctr., 1988-92, chief Human Factors Rsch. Divsn., 1992—; research scientist research analysis group Jet Propulsion Lab., Pasadena, Calif., 1953-55; chmn. flight mechanics panel adv. group aerospace research and devel. NATO, 1974-76; lectr. U. Buffalo, Millard-Fillmore Coll., Buffalo, 1957-58. Served with USAAF, 1945-46. Fellow AIAA (Internat. Cooperation award 1992), AAAS, Royal Aero. Soc.; mem. Am. Helicopter Soc., Sigma Xi. Home: 1362 Cuernavaca Circulo Mountain View CA 94040-3571 Office: NASA Ames Rsch Ctr MS262-1 Moffett Field CA 94035

STAUFFER, LARRY ALLEN, mechanical engineer, educator, consultant; b. Portsmouth, Va., Nov. 16, 1954; m. Susan P. Paider, Nov. 25, 1991; children: Will Allen, Carson William. BS, MS, Va. Polytech Inst. & State U., 1979; PhD, Oreg. State U., 1987. Registered profl. engr., Idaho. Design engr. Westinghouse Elec. Corp., Idaho Falls, Idaho, 1979-83; rsch. asst., instr. Oreg. State U., Corvallis, 1983-87; prof. mech. engr. U. Idaho, Moscow, 1987—. Contbr. articles to profl. jours. Recipient rsch. award NASA, 1989-95, NSF, 1988-91, 92-95. Mem. ASME (conf. chmn. 1991—), Am. Soc. Engring. Edn. Achievements include research in model fundamental processes of mechanical design, design tool for concept evaluation. Office: Univ Idaho Dept Engring Moscow ID 83843

STAUFFER, RONALD JAY, project engineer, aerospace engineer; b. Decatur, Ind., June 15, 1961; s. Roger Jay Jr. and Carol Sue (Yoder) S.; m. Joni Sue Gerhardt, Aug. 19, 1989; 1 child, Robert Joshua. BS in Aero./Astron. Engring., Ohio State U., 1984; MS in Aero. Mgmt., Embry-Riddle Aero. U., 1991. Commd. 2d lt. USAF, 1984, advanced through grades to capt., 1988; asst. program mgr. Electronic Combat Systems Program Office, Wright-Patterson AFB, Ohio, 1984-85, program mgr., 1985-87, sr. program mgr., 1987-89; aerospace engring. program integrator Defense Plant Representative Office Gen. Electric Aircraft Engines, Cin., 1989-91, chief program support br., 1991-92; spl. projects officer Unmanned Aerial Vehicles Joint Project Office, Washington, 1992—. Area rep. young astronaut program Dayton (Ohio) Pub. Schs., 1985-89. Decorated Achievement medal with oak leaf cluster USAF, 1990; recipient Commendation medal USAF, 1989, Joint Svc. Achievement medal Def. Logistics Agy., 1992, Nat. Def. Svc. medal Def. Logistics Agy., 1991, Def. Meritorious Svc. Medal, 1992. Mem. AIAA, Assn. for Unmanned Vehicles, Air Force Assn. (life), Ohio State U. Alumni Assn. Mem. United Ch. of Christ. Office: PEO (CU) UPQ1 304 Crystal Gateway 4 1213 Jefferson Davis Hwy Washington DC 20361

STAVELY, HOMER EATON, JR., psychologist educator; b. New Brunswick, N.J., Dec. 27, 1939; s. Homer Eaton Sr. and Elizabeth (Williams) S.; m. Linda Lee Finch, June 10, 1967 (div. June 1983); children: Jotham Reid, Rachel Alena; m. Mary Walton Mayshark, July 31, 1988; stepchildren: May Walton Mantell, Ila Nell Harriet Mantell. BA, DePauw U., 1961; PhD, Princeton U., 1964. Asst. prof. psychology Hamilton Coll., Clinton, N.Y., 1964-68, Albion (Mich.) Coll., 1968-69, Windham Coll., Putney, Vt., 1969-71; from asst. prof. to prof. psychology Keene (N.H.) State Coll., 1972—. Contbr. articles to profl. jours. Sec. Northfield Hist. Commn., 1991—, Northfield Dem. Com., 1992—. Mem. Am. Psychol. Soc., AAAS, Am. Assn. Higher Edn., Keene State Coll. Edn. Assn. (pres. 1988-90).

STAVNEZER, JANET MARIE, immunologist; b. Mpls., Mar. 9, 1944; d. Robert and Vivian (Wrabek) Nordgren; m. Edward Stavnezer, Dec. 30, 1967 (div. July 15, 1991); 1 child, Jessica Rachel. BA, Swarthmore Coll., 1966; PhD, Johns Hopkins U., 1971. Postdoctoral fellow U. Calif. Med. Sch., San Francisco, 1972-77; asst. prof. Sloan Kettering Inst., N.Y.C., 1977-85; assoc. prof. U. Mass. Med. Sch., Worcester, 1985-92, prof., 1992—; dep. editor Jour. of Immunology, 1992—; rsch. adv. com. Am. Cancer Soc., Atlanta, 1990—. Contbr. articles to profl. jours. Grantee NIH, 1978—, Leukemia Soc., 1975-80, Am. Cancer Soc., 1992—. Office: U Mass Med Sch Dept Molecular Genetics 55 Lake Ave N Worcester MA 01655

STEADMAN, ROBERT KEMPTON, oral and maxillofacial surgeon; b. Mpls., July 8, 1943; s. Henry Kempton and Helen Vivian (Berg) S.; m. Susan E. Hoffman; children: Andrea Helene, Darcy Joanne, Richard Kempton, Michael Dean. BS, U. Wash., Seattle, 1969, DDS, 1974. Diplomate Am. Bd. Oral and Maxillofacial Surgery. Residency USAF, Elgin AFB, Fla., 1974-75; resident oral and maxillofacial surgery U. Okla., 1977-80, La. State U., Shreveport, 1980-81; pvt. practice Spokane, Wash., 1981—; cons. Group Health Coop., 1989—; mem. adv. bd. Osteoporosis Awareness Resource, 1988—. Select recruiting ptnr. U. Wash. Sch. Dentistry, 1990. Lt. col. USAFR, 1974—. Fellow Am. Coll. Oral & Maxillofacial Surgery, Am. Soc. Oral & Maxillofacial Surgery, Acad. Gen. Dentistry; mem. Internat. Soc. Plastic, Aesthetic and Reconstructive Surgery, Delta Sigma Delta (pres. 1987-88). Office: 801W E 5th Ave # 212 Spokane WA 99204-2823

STEARNS, FRED LEROY, controls engineer; b. Westerly, R.I., May 17, 1947; s. Robert L. and Margaret (Jones) S.; m. Melissa Whun-Ying Mok, Dec. 20, 1980; children: Phillip David, Alice Kimberly. BS in Mech. Engring., U. Tex., 1981. Registered profl. engr., Tex. Controls engr. Lower Colo. River Authority, Austin, Tex., 1982—. Mem. ASME (chmn. Austin chpt., treas. Austin chpt.), Instrument Soc. Am. (sr.), Indsl. Computing Soc. (founding). Home: 14811 Yellowleaf Trl Austin TX 78728 Office: Lower Colo River Authority 3701 Lake Austin Blvd Austin TX 78703

STEARNS, STEPHEN RUSSELL, civil engineer, forensic engineer, educator; b. Manchester, N.H., Feb. 28, 1915; s. Hiram Austin and Elisabeth Scribner (Brown) S.; m. Eulalie Moody Holmes, Jan. 1 1939; children: Marjorie Elisabeth, Stephen James, Jonathan David. A.B., Dartmouth Coll., 1937; C.E., Thayer Sch. Engring., 1938; M.S., Purdue U., 1949. Civil engr. Gannett, Eastman, Fleming, Harrisburg, Pa., 1938-40; marine egr. Bur. Ships, Phila. Navy Yard, 1940-41; engr. Dry Dock Assocs., Phila Navy Yard, 1941-43; instr. Thayer Sch. Engring., Dartmouth, 1943-45, asst. prof., 1945-53, prof. civil engring., 1953-80; UN cons., Poland, 1974, 78; engr. Ops. Research, Inc., Washington., 1962-64; phys. reconnaisance in Alaska Boston U. Phys. Research Labs., 1953; chief applied snow and ice research br. Snow, Ice, Permafrost Research Establishment, U.S. Army Engrs., 1954-55; Mem. Sch. Bd., Hanover, N.H., 1951-60, chmn., 1957-59; mem. N.H. Gov.'s Transp. Com., 1966-67, Task Force, 1969; chmn. Lebanon (N.H.) Regional Airport Authority, 1966-69; mem. N.H. Bd. Registration Profl. Engrs., 1975-83. Mem. Am. Soc. Engring. Edn., ASCE (dir. 1978-81, pres. 1983-84), Sigma Chi. Congregationalist. Home: 10 Barrymore Rd Hanover NH 03755-2402

STEBBINGS, WILLIAM LEE, chemist, researcher; b. Santa Ana, Calif., Mar. 1, 1945; s. Richard Lee and Virginia Grace (Penquite) S.; m. Deanna Lou Harlow, Jan. 23, 1968; 1 child, Kenneth Edward. BS, Iowa State U., 1966; PhD, U. Wis., 1972. Chemist 3M Co., St. Paul, Minn., 1972-82; rsch. mgr. 3M Co., St. Paul, 1982-90, sr. rsch. specialist, 1992—. Mem. Am. Chem. Soc., Soc. for Applied Spectroscopy, Am. Soc. for Mass Spectroscopy, North Am. Thermal Analysis Soc. Office: 3 M Co PO Box 33221 Saint Paul MN 55133-3221

STEBBINS, GEORGE LEDYARD, research botanist, retired educator; b. Lawrence, N.Y., Jan. 6, 1906; s. George Ledyard and Edith Alden (Candler) S.; m. Margaret Goldsborough Chamberlaine, June 14, 1931; children: Edith Candler Paxman, Robert Lloyd, George Ledyard (dec.); m. Barbara Jean Brumley, July 27, 1958 (dec. Feb. 1993); stepson, Marc C. Monaghan. AB, Harvard U., 1928, AM, 1929, PhD, 1931; DSc (hon.), U. Paris, 1962; ScD (hon.), Carleton Coll., 1983, Ohio State U., 1983, U. Mass., 1993. Instr. biology Colgate U., Hamilton, N.Y., 1931-35; jr. geneticist U. Calif., Berkeley, 1935-39, asst. prof., 1939-41, assoc. prof., 1941-47, prof. genetics, 1947-50; prof. genetics U. Calif., Davis, 1950-73, prof. emeritus, 1973—; vis. exchange prof. U. Chile, Santiago, 1973, vis. prof. Carleton Coll., Northfield, Minn., 1977, 86, San Francisco State U., 1977-78, Ohio State U., Columbus, 1978-79. Author: (with C.W. Young) The Human Organism and the World of Life, 1983; author: Variation and Evolution in Plants, 1950, Processes of Organic Evolution, 1966, 2 edit., 1971, 3 edit., 1977, The Basis of Progressive Evolution, 1969, Higher Plant Evolution Above the Species Level, 1974, Darwin to DNA: Molecules to Humanity, 1982, Evolucion Hacia Una Nueva Sintesis, 1989. Recipient Nat. medal of Sci., 1980; Guggenheim Found. research fellow, Algeria, Europe, 1954, 60-61, Australian Am. Exchange fellow, Canberra, Australia, 1974-75, Ctr. for Advanced Behavioral Studies fellow, Palo Alto, Calif., 1968-69; named Disting. Vis. Scientist, Smithsonian Instn., Washington, 1981-82. Mem. Nat. Acad. Scis., Am. Philos. Soc. (Lewis prize 1959), Am. Acad. Arts and Scis., Bot. Soc. Am. (pres. 1962), Soc. for Study of Evolution (pres. 1959), Am. Soc. Naturalists (pres. 1969). Democrat. Unitarian. Avocations: hiking, mountain climbing, music listening. Home: 216 F St # 165 Davis CA 95616 Office: Univ of Calif Dept of Genetics Davis CA 95616

STEBBINS, RICHARD HENDERSON, electronics engineer, peace officer, security consultant; b. Pittsburgh, Pa., Dec. 2, 1938; s. Earl Carlos and Esther Frances (Kusluch) S.; m. Rosemary Tanneberger, Aug. 12, 1984; children from previous marriage: Richard Earl, Susan Elizabeth. BSEE with high honors, U. Md., 1965; postgrad., Trinity U., 1973-74. Cert. peace officer, Tex. Engring. tech. Nat. Security Agy., Ft. Meade, Md., 1960-65; design engr. Page Communications Engr., Washington, 1965-66, Electromechanical Rsch., Inc., College Park, Md., 1966-67, Honeywell, Inc., Annapolis, Md., 1967-68; electronics engr., intelligence rsch. specialist Fed. Civil Svc., San Antonio, 1968-91; pvt. cons. San Antonio, 1991—; comdr.'s advisor Air Force Cryptologic Support Ctr., San Antonio, 1988-91; deputy dir. countermeasures ops., intelligence rsch. specialist USAF HQ Electronic Security Command, San Antonio, 1981-88; mem. blue ribbon com. on ops. security & comm. security roles & relationships for command and svc. Author, lectr. in field; contbr. articles to profl. jours. With USN, 1956-59. Mem. NRA, Tau Beta Pi, Eta Kappa Nu, Phi Kappa Phi. Republican. Episcopalian. Avocations: hunting, fishing, target shooting, family history rsch. Home: 9602 Clear Falls San Antonio TX 78250-5067

STECICH, JOHN PATRICK, structural engineer; b. Chgo., Nov. 1, 1949; s. William Frank and Margaret Mary (Hanrahan) S.; m. Rita Louise Fahey, July 1, 1972; children: Eric John, Thomas John. BSCE, Ill. Inst. Tech., 1971, MSCE, 1972. Registered profl. engr., Ill., Ind., Pa.; lic. structural engr., Ill. Design engr. Chgo. Bridge and Iron Co., Oak Brook, Ill., 1971-79; sr. cons. Wiss, Janney, Elstner Assocs., INc., Chgo., 1979—; speaker in field. Contbr. papers to prof. publs. Mem. ASCE (design of steel bldg. structures com.), ASTM (dimension stone com.), Am. Inst. Steel Constrn., Am. Concrete Inst., Am. Soc. for Metals, Chgo. Com. on High Rise Bldgs., Structural Engrs. Assn. Ill. (1st Prize award 1988, Meritorious Publ. award 1991, award of merit for Amoco Bldg. Facade Recladding 1993). Home: 11306 S Central Park Ave Chicago IL 60655-3416 Office: Wiss Janney Elstner Assocs 29 N Wacker Dr Chicago IL 60606-3203

STECKEL, RICHARD J., radiologist, academic administrator; b. Scranton, Pa., Apr. 17, 1936; s. Morris Leo and Lucille (Yellin) S.; m. Julie Raskin, June 16, 1960; children: Jan Marie, David Matthew. BS magna cum laude, Harvard U., 1957, MD cum laude, 1961. Diplomate: Am. Bd. Radiology. Intern UCLA Hosp., 1961-62; resident in radiology Mass. Gen. Hosp., Boston, 1962-65; clin./rsch. assoc. Nat. Cancer Inst., 1965-67; mem. faculty UCLA Med. Sch., 1967—, prof. radiol. scis. and radiation oncology, dir. Jonsson Comprehensive Cancer Ctr., 1974—; pres. Assn. Am. Cancer Insts., 1981. Author two books; contbr. over 130 articles on radiology and cancer diagnosis to profl. publs. Fellow Am. Coll. Radiology; mem. Radiol. Soc. N. Am., Am. Roentgen Ray Soc., Assn. Univ. Radiologists. Office: UCLA Med Ctr Jonsson Comp Cancer Ctr 10833 Le Corte Ave Los Angeles CA 90024-1602

STEDINGER, JERY RUSSELL, civil and environmental engineer, researcher; b. Oakland, Calif., June 22, 1951; s. Russell Phillip and Vivian Lavina (Nelson) S.; m. Robin Lee Gray, June 30, 1973; children: Matthew, Carolyn. BA, U. Calif., Berkeley, 1972; AM, Harvard U., 1974, PhD, 1977. Math. programmer Lawrence Livermore Lab., Livermore, Calif., 1973; rsch. asst., teaching fellow Engr. and Applied Physics, Harvard U., Cambridge, Mass., 1974-77; asst. prof. Civil and Environ. Engr., Cornell U., Ithaca, N.Y, 1977-83; hydrologist U.S. Geol. Survey, Resten, Va., 1983-84; assoc. prof. Civil and Environ. Engr., Cornell U., Ithaca, N.Y., 1989-93, prof., 1989—; cons. entomology dept. U. Maine, Orono, 1977, Pacific Electric and Gas Co., San Francisco, 1989-93; chair bd. cons. New Eng. Power Svc. Corp., Westborough, Mass., 1986-88. Author: Water Resources Systems Planning and Analysis, 1981; contbr. articles to profl. jours. Asst. scoutmaster Troop 2, Boy Scouts Am., Ithaca, N.Y., 1988-93. Recipient Editor's Citation for Excellence in Reviewing award Am. Geophys. Union, 1983-90; named Presdl. Young Investigator, NSF, 1984-90, CEE Prof. of Yr., Chi Epsilon, 1979-80. Mem. ASCE (Huber Civil Engring. Rsch. prize 1989(, Am. Geophys. Union, Inst. Mgmt. Scis., Soc. for Risk Analysis. Office: Cornell U Sch Civil Environ Engring 213 Hollister Hall Ithaca NY 14853-3501

STEELE, CLAUDE MASON, psychology educator; b. Chgo., Jan. 1, 1946; s. Shelby and Ruth (Hootman) S.; m. Aug. 27, 1967; children: Jory, Claude Benjamin. BA, Hiram Coll., 1967; PhD, Ohio State U., 1971. Asst. prof. U. Utah, Salt Lake City, 1971-73; from asst. to prof. U. Washington, Seattle, 1978-87; prof. U. Mich., Ann Arbor, 1987-91, Stanford U., Calif., 1991—. Recipient numerous rsch. grants. Mem. Soc. Exptl. Social Psychology (chmn. 1988-89), Am. Psychol. Aos. (bd. dirs. 1991—). Home: 562 Junipero Serra Stanford CA 94305 Office: Stanford U Psychol Dept Stanford CA 94305

STEELE, GLENN DANIEL, JR., surgical oncologist; b. Balt., June 23, 1944; m. Diana; 1 child, Joshua; m. Lisa; children: Kirsten, Lara. AB magna cum laude, Harvard Coll., 1966; MD, NYU, 1970; PhD, Lund U., Sweden, 1975. Intern, then resident Med. Ctr. U. Colo., Denver, 1970-76; fellow NIH in immunology Univ. Lund, Sweden, 1973-75; asst. surgeon Sidney Farber Cancer Inst., Boston, 1976-78; clin. assoc. surgical oncology Sidney Farber Cancer Inst., 1978-79; jr. assoc. in surgery Peter Bent Brigham Hosp., Boston, 1976-82; instr. surgery Med. Sch. Harvard, Boston, 1976-78; asst. prof. surgery Med. Sch. Harvard Coll., 1978-81; asst. physician surgical oncology Sidney Farber Cancer Inst., 1979-82; assoc. prof. surgery Med. Sch. Harvard Coll., 1981-84; surgeon Brigham & Women's Hosp., 1982-84; assoc. physician surgical oncology Dana-Farber Cancer Inst., 1982-84, physician surg. oncology, 1984—; chmn. dept. surgery, deaconess Harvard Surg. Svc. New England Deaconess Hosp., Boston, 1985—; William V. McDermott prof. surgery Med. Sch. Harvard Coll., 1985—; cons. surgeon Boston Hosp. for Women, 1977-80. assoc. editor Jour. of Clin. Oncology, 1986—, Jour. of Hepatobiliary-Pancreatic Surgery, 1993—; mem. editorial bd. Annals of Surgery, Annals of Surg. Oncology, British Jour. of Surgery, Surgery, Surgical Oncology; contbr. numerous articles to profl. jours. Recipient NIH fellow 1973-75, Am. Cancer Soc. fellow 1972-73, 76-79, various other rsch. grants. Fellow Am. Coll. Surgeons (chmn. patient care and rsch. com. commn. on cancer 1989-91, mem. bd. govs. 1991—, chmn. commn. on cancer 1991-93, exec. com. 1993—); mem. Am. Assn. Immunologists, Am. Bd. Surgery (dir. 1993—), Am. Bd. Med. Specialties, Am. Soc. Clin. Oncology, Am. Surg. Assn., Assn. Program Dirs. in Surgery, Assn. for Surgical Edn., Internat. Fedn. Surg. Colls., Internat. Surg. Group, New England Cancer Soc., and numerous other mems. Office: New England Deaconess Hospital 110 Francis St Ste 3A Boston MA 02215

STEELE, JOHN HYSLOP, marine scientist, oceanographic institute administrator; b. Edinburgh, U.K., Nov. 15, 1926; s. Adam and Annie H.; m. Margaret Evelyn Travis, Mar. 2, 1956; 1 son, Hugh. B.Sc., Univ. Coll., London U., 1946, D.Sc., 1964. Marine scientist Marine Lab., Aberdeen, Scotland, 1951-66; sr. prin. sci. officer Marine Lab., 1966-73, dep. dir., 1973-77; dir. Woods Hole Oceanographic Instn., Mass., 1977-89, pres., 1986-91; mem. NAS/NRC Ocean Sci. Bd., 1978-88, chmn., 1986-88; mem. rsch. and exploration com. Nat. Geog. Soc.; mem. Arctic Rsch. Commn., 1988-92; trustee Univ. Corp. Atmospheric Rsch., 1987-91, Bermuda Biol. Sta., R.W. Johnson Found.; del. Internat. Coun. Exploration Sea; bd. dirs. Exxon Corp. Author: The Structure of Marine Ecosystems, 1974; Contbr. articles to profl. jours. Served with Brit. Royal Air Force, 1947-49. Recipient Alexander Agassiz medal Nat. Acad. Sci., 1973. Fellow Royal Soc. London, AAAS, Royal Soc. Edinburgh, Am. Acad. Arts and Scis. Home: PO Box 25 Woods Hole MA 02543-0025 Office: Woods Hole Oceanographic Instn Woods Hole MA 02543

STEELE, JOHN WISEMAN, pharmacy educator; b. Motherwell, Scotland, May 27, 1934; emigrated to Can., 1958; s. James F. H. and Janet H. M. (Ogilvie) S.; m. Muriel Grace Gibbon, Dec. 27, 1958; children: Colin, Alison, Graham, Alistair. B.Sc. in Pharmacy (hon.), U. Glasgow, 1955, Ph.D., 1959. Lectr. in pharmacy U. Man. (Can.), Winnipeg, 1958-59; asst. prof. U. Man. (Can.), 1959-63, assoc. prof., 1963-68, prof., 1968—, dean Faculty of Pharmacy, 1981-92. Fellow Chem. Soc. London; mem. Man. Pharm. Assn. (Centennial award 1979-80), Can. Pharm. Assn., Royal Inst. Chemistry London, Fedn. Internat. Pharmaceutique. Club: Winnipeg Lawn Tennis. Home: 61 Agassiz Dr, Winnipeg, MB Canada R3T 2K9 Office: University of Manitoba, Faculty of Pharmacy, Winnipeg, MB Canada R3T 2N2

STEELE CLAPP, JONATHAN CHARLES, chemical engineer; b. St. Paul, Jan. 26, 1968; s. Charles E. and Betty J. (Huff) Clapp; m. Laura J. Steele, Aug. 3, 1991. BS, U. Minn., 1991. Tech. aide 3M Co., Mpls., 1988-91; prodn. control engr. ICI Americas, Inc., Bayonne, N.J., 1991—. Mem. NSPE, Tau Beta Pi. Office: ICI Americas Inc 229 E 22d St Bayonne NJ 07002

STEELMAN, SANFORD LEWIS, research scientist, biochemist; b. Hickory, N.C., Oct. 11, 1922; s. John Avery and Blanche O.; m. Margaret E. Abee, Jan. 10, 1945; children: Sanford, Jr., Brian L. BS in Chemistry, Lenoir-Rhyne Coll., 1943; PhD in Biochemistry, U. N.C., 1949. Head biochem. rsch. Armour Labs, Chgo., 1953-56; assoc. prof. Baylor Univ./Coll. of Medicine, Houston, 1956-58, U. Tex. Coll. of Medicine, Houston, 1956-58; dir. endocrinology Merck Inst., Rahway, N.J., 1958-72; sr. investigator Merck Inst., Rahway, 1978-86; dir. clin. pharm. Merck Sharp and Dohme Labs, Rahway, 1972-78; cons. Armour Labs., Kankakee, Ill., 1956-58. Contbr. 100 articles to profl. jours., publs. Mem. sch. bd. Watchung (N.J.) Bd. Edn., 1964-67; exec. com. Boy Scouts Am., 1964-70. Recipient Trustees award Lenoir-Rhyne Coll., Hickory, N.C., 1980. Fellow AAAS; mem. Endocrine Soc., Am. Soc. Pharm. Exptl. Therapeutics, Am. Chem. Soc., Kiwanis Club. Achievements include five U.S. patents relating to preparation and use of hormonal products. Home: PO Box 5358 Hickory NC 28603

STEENBERGEN, GARY LEWIS, computer aided design educator; b. Detroit, Jan. 23, 1959; s. Edwin P. and Pauline (Bright) S.; m. Barbara Earlene Snyder, June 2, 1979; children: Christopher Lewis, Stephanie Brooke. AAS in Pre-engring., Sue Bennett Coll., London, Ky., 1987; AS in Vocat., Indsl. and Tech. Edn. with high distinction, Eastern Ky. U., 1993. Archtl. draftsman Laminated Timbers, London, 1977; archtl. designer Ea. Ky. Homes, Corbin, 1979-80; detail draftsman East Ky. Steel, Barbourville, 1979-80; quality control technician Elicon divsn. Nat. Standard, Corbin, Ky., 1980-82, plant engring. technician, 1984-90; CAD instr. Ky. Tech., Pineville, 1990—; chmn. Craft Adv. Com. for Vo-Tech Edn., Corbin, 1980-90. Scoutmaster Boy Scouts Am., Corbin, 1982—; chess coach U.S. Chess Found., Corbin, 1988—. Recipient Scouters award Boy Scouts Am., 1985, Scoutmaster award, 1985. Mem. NEA, Ky. Edn. Assn., Sue Bennett Alumni, Ea. Ky. U. Alumni, Kappa Delta Pi, Golden Key Honor Soc. Republican. Pentecostal Ch. Home: Rt 1 Box 537E Corbin KY 40701-4605

STEFANOVITS, PÁL, agriculturalist; b. Nov. 24, 1920. Prof., dir. dept. pedology U. Agrl. Sci. Author: Magyarország talajai, vol. 1, 2, 1956-63, Talajlan, 1975-81. Mem. Hungarian Acad. Scis. Office: Magyar Tudományos Akadémia, Roosevelt-tér 9, 1051 Budapest Hungary*

STEFFAN, WALLACE ALLAN, entomologist, educator, museum director; b. St. Paul, Aug. 10, 1934; m. Sylvia Behler, July 16, 1966; 1 child, Sharon. B.S., U. Calif.-Berkeley, 1961, Ph.D., 1965. Entomologist dept. entomology Bishop Mus., Honolulu, 1964-85, head diptera sect., 1966-85, asst. chmn.,

1979-85; dir. Idaho Mus. Natural History, Idaho State U., Pocatello, 1985-89, U. Alaska Mus., 1989-92; prof. biology U. Alaska Fairbanks, 1989-92; exec. dir. Great Valley Mus. Natural History, 1992—; mem. grad. affiliate faculty dept. entomology U. Hawaii, 1969-85; liaison officer Bishop Mus., Mus. Computer Network, 1980-85; reviewer NSF, 1976—; mem. internat. editorial adv. com. World Diptera Catalog, Systematic Entomology Lab., U.S. Dept. Agr., 1983-85; mem. affiliate faculty biology, Idaho State U., 1986-89; bd. dirs. Idaho State U. Fed. Credit Union, 1986-89; mem. adv. coun. Modesto Conv. & Visitors Bureau, 1992—; mem. Ft. Hall Replica Commn., 1986-89. Acting editor Jour. Med. Entomology, 1966; assoc. editor Pacific Insects, 1980-85. Judge Hawaii State Sci. and Engring. Fair, 1966-85, chief judge sr. display div., 1982, 83, 84; advisor to bd. Fairbanks Conv. and Visitors Bur., 1989-91; mem. vestry St. Christophers Episcopal Ch., 1974-76, St. Matthew's Episcopal Ch., Fairbanks, 1990-91; pres. Alaska Visitors Assn., Fairbanks, 1991; advisor Fairbanks Conv. Visitors Bur. Bd., 1989-91; bd. dirs. Kamehameha Fed. Credit Union, 1975-77, chmn., mem. supervisory com., 1980-84. Served with USAF, 1954-57. Grantee NIH, 1962, 63, 67-74, 76-81, 83-85. U.S. Army Med. Research and Devel. Command, 1964-67, 73-74, NSF, 1968-76, 83-89, City and County of Honolulu, 1977, U.S. Dept. Interior, 1980, 81. Mem. Entomol. Soc. Am. (mem. standing com. on systematics resources 1983-87), Am. Mosquito Control Assn., Pacific Coast Entomol. Soc., Soc. Systematic Zoology, Hawaiian Entomol. Soc. (pres. 1974, chmn. coms. 1966-85, editor procs. 1966), Hawaiian Acad. (councillor 1976-78), Entomol. Soc. Wash., Fairbanks C. of C. (adv. bd. Conv. Visitors Bur. 1989), Alaska Visitors Assn. (pres. Fairbanks chpt. 1991), Modesto Convention and Visitors Bur. (adv. coun. 1992—) Sigma Xi (pres.-elect San Joaquin chpt. 1993). Office: Great Valley Museum of Natural Hist 1100 Stoddard Ave Modesto CA 95350

STEFFEN, ALAN LESLIE, entomologist; b. Ansonia, Ohio, Feb. 27, 1927; s. Henry William and Maude Moiselle (DuBois) S.; m. Genevieve Carlyle, Dec. 27, 1950 (dec. Jan. 1989); m. Doris Mae Rable, Jan. 20, 1990. AB, Miami U., 1948; MSc in Entomology, Ohio State U., 1949; diploma, Malaria Tng. Ctr., 1959; postgrad., WHO, Sri Lanka and The Philippines, 1967, 68. Registered profl. entomologist. Malaria specialist Agy. for Internat. Devel., Jakarta, Indonesia, 1959-65; chief malaria advisor Agy. for Internat. Devel., Kathmandu, Nepal, 1966-72, Addis Ababa, Ethiopia, 1972-76, Kathmandu, 1976-78, Islamabad, Pakistan, 1978-80; malaria specialist Ctr. Disease Control, Sonfkhla, Thailand, 1965-66; tropical disease cons. Belleville, Ill., 1981—; cons. U.S. AID, Port Au Prince, Haiti, 1981, WHO, Geneva, 1981—, Tifa, Ltd., Millington, N.J., 1982, John Snow, Inc., Boston, 1984, Vector Biology and Control Project, Arlington, Va., 1986. Mem. Nature Conservancy, Washington, 1986-88. With U.S. Army, 1945-46, ETO. Recipient Meritorious Honor award U.S. Dept. State, 1972. Fellow Royal Soc. Tropical Medicine; mem. Entomological Soc. Am., Am. Registry Profl. Entomologists, Nat. Assn. Retired Fed. Employees (life). Avocations: stamp collecting, study of Asian art. Home and Office: 303 Hickory Bend Belleville IL 62223

STEFFENS, FRANZ EUGEN ALOYS, computer science educator; b. Bonn, Fed. Republic Germany, Mar. 2, 1933; s. Aloys Josef and Maria (van Hoff) S.; m. Heike Eger, May 4, 1968; 1 child, Martina-Brigitte. Diploma, U. Cologne, Fed. Republic Germany, 1959, doctorate, 1965; degree in habilitation, U. Frankfurt, Fed. Republic Germany, 1972. Rsch. asst. Cologne U., 1960-66; mgr. dept. devel. IBM, Stuttgart, Fed. Republic Germany, 1966-72; lectr. Mannheim (Fed. Republic Germany) U., 1967-71, prof. indsl. mgmt. and data processing, 1972—; lectr. Frankfurt U., 1969-71. Mem. Rotary. Home: Hebelstrasse 3, 69257 Wiesenbach Germany Office: U Mannheim, PO Box 10 34 62, 68131 Mannheim Germany

STEFFY, JOHN RICHARD, nautical archaeologist, educator; b. Lancaster, Pa., May 1, 1924; s. Milton Grill and Zoe Minerva (Fry) S.; m. Esther Lucille Koch, Oct. 20, 1951; children: David Alan, Loren Craig. Student, Pa. Area Coll., Lancaster, 1946-47, Milw. Sch. Engring., 1947-49. Ptnr. M.G. Steffy & Sons, Denver, Pa., 1950-72; ship reconstructor Kyrenia Ship Project, Cyprus, 1972-73, Inst. Nautical Archaeology, College Station, Tex., 1973—; from lectr. to prof. anthropology Tex. A&M U., College Station, 1976-91, prof. emeritus, 1991—; lectr. on ship constrn. Contbr. articles to profl. publs. Sec. Denver Borough Authority, Pa., 1962-72. Served with USN, 1942-45. MacArthur Found. fellow, 1985. Mem. Archaeol. Inst. Am., Soc. Nautical Research, N.Am. Soc. Oceanic History. Republican. Methodist. Lodge: Lions. Office: Tex A&M U Inst Nautical Archaeology College Station TX 77843

STEGER, EDWARD HERMAN, chemist; b. New Orleans, Dec. 11, 1936; s. Herman Christoph and Katherine (Walther) S.; m. Amy Patricia Duvall, July 29, 1960; children: David B., Sandra E. BS, Tulane U., 1958. Analytical chemist Atlantic Rsch. Corp., Gainesville, Va., 1960-64, head control lab., 1964—; presenter at profl. confs. Contbr. articles to Fine Particle Soc. Jour. Lt. USNR, 1958-60. Mem. Am. Chem. Soc., Fine Particle Soc., Phi Beta Kappa, Phi Eta Sigma, Alpha Chi Sigma. Baptist. Home: 4311 Alta Vista Dr Fairfax VA 22030-5302 Office: Atlantic Rsch Corp 5945 Wellington Rd Gainesville VA 22065-1699

STEGER, RALPH JAMES, chemist; b. Meridian, Okla., Jan. 24, 1940; s. Daniel Bose and Opal Creola (Brothers) S. BS in Chemistry and Math., Langston U., 1962. Cartographer Aeronautical Chart and Info. Ctr. ACIC USAF, St. Louis, 1962-63; lab. technician Sigma Chem. Co., St. Louis, 1963; phys. scientist U.S. Army Chem. Corps, Edgewood Arsenal, Md., 1963-65; rsch. chemist Chem. Rsch., Devel. and Engring. Ctr. SMCCR Rsch. Lab., Analytical Div., Aberdeen Proving Ground, Md., 1965-86; chemist Chem. Rsch., Devel. and Engring. Ctr. SMCCR-Detection, Detection Technology, Aberdeen Proving Ground, 1986—; adv. com. Garrison Gents, Balt., 1980—; ACOR monitoring govt. contracts, Balt., 1987—. Contbr. articles to profl. publs. Mem. Okla. Hist. Soc. Mem. AAAS, N.Y. Acad. Sci., Okla. Hist. Soc. Office: CBDA-RTE Aberdeen Proving Ground MD 21010-5423

STEGINK, LEWIS DALE, biochemist, educator; b. Holland, Mich., Feb. 8, 1937; s. Benjamin and Reka (Brandsma) S.; m. Carol N. Stegink, June 30, 1962; children: David W., Daniel E. BA, Hope Coll., Holland, Mich., 1958; MS, U. Mich., 1963, PhD, 1965. Postdoctoral fellow U. Iowa, Iowa City, 1965-67, asst. prof., 1965-71, assoc. prof., 1971-76, prof. biochemistry, 1976-82; prof. pediatrics and biochemistry, 1982—; mem. adv. bd. Abbott Labs., North Chicago, Ill., 1972-80, G.D. Searle, Skokie, Ill., 1975-85. Editor, author: Aspartame: Physiology and Biochemistry, 1984; contbr. articles to profl. jours. Elder Trinity Christian Reformed Ch., Iowa City, 1967. Mem. Am. Chem. Soc., Am. Inst. Nutrition, Am. Soc. Biochemistry and Molecular Biology, Am. Pediatric Soc. Achievements include research in neurotoxicity of aspartate and MSG; development of new parenteral regimen for infants, metabolism of diketopiperazine. Home: 2 Bedford Ct Iowa City IA 52240 Office: U Iowa S385 Hospital Sch Iowa City IA 52242-1011

STEGMANN, THOMAS JOSEPH, physician; b. Hannover, Fed. Republic Germany, Nov. 20, 1946; s. Adalbert and Marianne (Grindel) S.; m. Dagmar Reddert, Sept. 7, 1984; 1 child, Christoph-Alexander. Student med. scis., U. Bonn, from 1968; MD, U. Heidelberg, Fed. Republic Germany, 1974. Fellow Med. High Sch., Hannover, 1976-82, docent, 1982, resident, 1982-84; dir., dept. thoracic and cardiovascular surgery Fulda Med. Ctr., 1985—, prof. surgery, 1989—. Author: Heart Hypertrophy, 1974, Coronary and Cerebral Air Embolism, 1983, Surgery for Aortic Dissection, 1987, Heart Transplantation, 1988, Pacemaker-Implant, 1989; editor Jour. for Heart Medicine, 1985. Mem. German Soc. for Thoracic, Cardiac and Vascular Surgery (Rudolf-Nissen Meml. award 1982), German Soc. for Surgery, Internat. Soc. for Heart Transplantation, European Soc. for Cardiovascular Surgery, European Assn. for Thoracic and Cardiovascular Surgery, Lions (Fulda). Avocation: sailing. Home: Spiegelstr 10, D-36100 Petersberg Germany Office: Fulda Med Ctr, Pacelli Allee 4, D-36043 Fulda Germany

STEHLIN, JOHN SEBASTIAN, JR., surgeon; b. Brownsville, Tenn., June 16, 1923; s. John Sebastian and Princess (King) S.; m. Mary Elizabeth Cleary, Sept. 19, 1950 (div. 1962); 1 child, Mary Cleary. Student, Vanderbilt U., 1941-42, Notre Dame U., 1943-44; M.D., Med. Coll. Wis., 1947. Diplomate: Am. Bd. Surgery. Intern Milw. Hosp., 1947-48; resident pathology Bapt. Hosp., Memphis, 1948-49; resident surgery Milw. Hosp., 1949-52; fellow surgery Lahey Clinic, Boston, 1952-53; sr. fellow surgery U. Tex., M.D. Anderson Hosp. and Tumor Inst., Houston, 1955-56; fellow surgery Lahey Clinic, Boston, 1956; mem. surg. staff U. Tex., M.D. Anderson Hosp. and Tumor Inst., Houston, 1957-67; asst. surgeon U. Tex., M.D. Anderson Hosp. and Tumor Inst., 1957-60, asso. surgeon, 1961-67; asst. prof. surgery U. Tex. Postgrad. Sch. Medicine, Houston, 1957-60; asso. prof. U. Tex. Postgrad. Sch. Medicine, 1961-63; asso. prof. surgery U. Tex. Postgrad. Sch. Medicine (Grad. Sch. Biomed. Scis.), 1963-67; clin. asso. prof. surgery Baylor Coll. Medicine, Houston, 1967—; mem. surg. staff St. Joseph Hosp., Houston, 1967—; hon. prof. faculty medicine U. Republic Uruguay, 1965; founder, sci. dir. Stehlin Found. Cancer Research, Houston, 1969—. Contbr. over 100 articles to sci. jours. Served to capt. USAF, 1953-55. Recipient humanitarian award B'nai B'rith, 1982; named to City of Houston Hall of Fame, 1985. Fellow ACS; mem. Am. Assn. Cancer Research, AAAS, AMA, Cancer Assn. Argentina (hon.), Cancer Soc. Chile (hon.), Internat. Platform Assn., Soc. Surg. Oncology, Inc., Pan Am. Med. Assn., Soc. Dermatology Uruguay (hon.), Surg. Soc. Chile (hon.), Royal Soc. Medicine, Western Surg. Assn., Southwestern Surg. Congress, So. Med. Assn., Tex. Med. Assn., Tex. Surg. Soc., N.Y. Acad. Scis., Salem Surg. Soc. (hon.), Phoenix Surg. Soc. (hon.), Harris County Med. Soc., Houston Surg. Soc., Am. Judicature Soc. Office: Stehlin Foundation for Cancer 1315 Calhoun St Suite 1800 Houston TX 77002

STEIGBIGEL, ROY THEODORE, infectious disease physician and scientist, educator; b. Bklyn., Nov. 23, 1941; s. Samuel and Lillian I. (Parker) S.; m. Julia Ann Enterline, June 10, 1967 (div. 1983), children: Keith D., Glenn N.; m. Sidonie Ann Morrison, Oct. 15, 1985; 1 child, Andrew M. BA, Carleton Coll., 1962; MD, U. Rochester, 1966. Diplomate Am. Bd. Internal Medicine, Am. Bd. Infectious Disease. Resident U. Rochester, N.Y., 1966-68; resident Stanford U., Palo Alto, Calif., 1970-71, fellow, 1971-73; from asst. to assoc. prof. U. Rochester, N.Y., 1973-83; prof. SUNY, Stony Brook, 1983—; mem. adv. bd. infectious disease U.S. Pharmacopea, Rockville, Md., 1980—; mem. adv. panels NIH, Bethesda, Md., 1985-87. Contbr. over 8 chpts. to books, over 55 articles to profl. jours. Served in USPHS, 1968-70. Fellow NIH, 1971-73, grantee, 1985—. Fellow ACP, Infectious Disease Soc. Am. Office: SUNY School of Medicine HSC-T-15-080 Stony Brook NY 11794-8153

STEIGER, FRED HAROLD, chemist; b. Cleve., May 11, 1929; s. Jacob and Helen (Gross) S.; m. Claire Geller, Sept. 7, 1952 (dec. Mar. 1985); children: Eden Linda Steiger Fisher, Susan Leigh Steiger Baron; m. Estelle Dubin, Jan. 6, 1991. BA, U. Pa., 1951; MA, Temple U., 1956. Rsch. chemist Rohm & Haas Co., Phila., 1951-60; group leader Personal Products Co. div. Johnson & Johnson, Milltown, N.J., 1960-62, sr. rsch. chemist, 1962-68, sr. rsch. scientist, 1968-74, sr. rsch. assoc., 1974-76, mgr. product devel., 1976-85, mgr. tech. assessment, 1985-88; pvt. practice cons. East Brunswick, N.J., 1988—. Editor Chem. Abstracts, Columbus, Ohio, 1968—; contbr. articles to profl. jours. Fellow Am. Inst. Chemists; mem. Am. Assn. Textile Chemists and Colorists (chmn. rsch. com. Lowell, Mass. chpt. 1958-60), N.J. Inst. Chemists (v.p. 1975-77), Am. Chem. Soc., Sigma Xi, Phi Lambda Upsilon. Achievements include patents in the fields of textile finishing and sanitary protection devices. Home and Office: 10 Tompkins Rd East Brunswick NJ 08816-1709

STEIGER, GRETCHEN HELENE, marine mammalogist, research biologist; b. Williamsport, Pa., May 7, 1960; d. Robert Folk and Helene (Moltz) S.; m. John Calambokidis, July 29, 1989; 1 child, Alexei Steiger Calambokidis. BS in Zoology, U. N.C., 1982. Rsch. biologist Cascadia Rsch. Collective, Olympia, Wash., 1982—, also bd. dirs.; pres. Cascadia Rsch., 1988—; whale census technician, Barrow, Alaska, 1988; rsch. assoc. U. Alaska, Fairbanks, 1991; presenter in field. Author (with others) numerous govt. reports and publs.; contbr. articles to profl. jours. Instr. Feminists Self-def. Tng., Olympia, 1984-92. Mem. Soc. Northwestern Vertebrate Biology, Soc. Marine Mammalogy (charter), Wildlife Soc. (Wash. chpt.). Office: Cascadia Research Collective 218 1/2 W 4th Ave Olympia WA 98501

STEIN, ARTHUR OSCAR, pediatrician; b. Bklyn., Apr. 3, 1932; s. Irving I. and Sadie (Brander) S.; A.B., Harvard U., 1953; M.D., Tufts U., 1957; postgrad. U. Chgo., 1963-66; m. Judith Lenore Hurwitz, Aug. 27, 1955; children: Susan, Jeffrey, Benjamin. Intern U. Chgo. Hosps., 1957-58, resident, 1958-59; resident N.Y. Hosp.-Cornell U. Med. Center, 1959-61; practice medicine specializing in pediatrics, 1963—; instr. pediatrics U. Chgo., 1963-66, asst. prof. pediatrics, 1966-70; mem. Healthguard Med. Group, San Jose, Calif., 1970-72; mem. Permanente Med. Group, San Jose, 1972—; asst. chief pediatrics Santa Teresa Med. Center, 1979-87; clin. instr. Santa Clara Valley Med. Center, Stanford U., 1970-72. Served to capt., M.C., AUS, 1961-63. USPHS Postdoctoral fellow, 1963-66. Fellow Am. Acad. Pediatrics. Jewish (v.p. congregation 1969-70, pres. 1972-73). Clubs: Light and Shadow Camera (pres. 1978-80) (San Jose); Central Coast Counties Camera (v.p. 1980-81, pres. 1981-82), Santa Clara Camera. (pres. 1991). Co-discoverer (with Glyn Dawson) genetic disease Lactosylceramidosis, 1969. Home: 956 Redmond Ave San Jose CA 95120-1831 Office: Kaiser/ Permanente Med Group 260 Internat Circle San Jose CA 95119

STEIN, GERALD, metallurgical engineer; b. Cochem, Fed. Republic Germany, May 24, 1943; s. Paul A. and Amanda E. (Hess) S.; m. Ingeborg E. Pika; children: Thomas, Philipp. Diploma in engring., GSH, Duisburg, 1969. Fellow engr. heat-treatment shop Fried, Krupp Schmiede & Giesserei, Essen, Fed. Republic Germany, 1969-72, fellow Engr. forging shop, 1972-74; head of NDT and testhouse Krupp Metall- und Schmiedewerke, Essen, Fed. Republic Germany, 1974-79; head quality engring. Krupp Stahl AG Schmiede und Bearbeitung, Essen, 1979-84; head quality dept. Schmiedewerke Krupp-Klöckner GmbH, Essen, 1984-88, Vereinigte Schmiedewerke GmbH, Essen, 1988-91; mem mgmt. bd. Deutsche, Titan GmbH Essen., 1991—; mem. div. bd. Vereinigte Schmiedewerke GmbH Essen. Editor: New Metallurgical Alloys, 1985, 90; contbr. 50 articles to profl. jours. Recipient Honorable Medal in Silver for Merits in Metall. Rsch., Swiss Fed. Inst. Tech., Zürich, Switzerland, 1990. Mem. ASTM, Verein Deutscher Eisenhüttenleute, Internat. Secretariat Cath. Engrs., Agronomists and Economists (pres. 1979-88, v.p. 1988—). Christian Democrat. Roman Catholic. Achievements include 10 patents on material and processes of a new generation of steels: Nitrogen Alloyed Austenitic and Ferritic-Martensitic Steels. Office: Vereinigte Schmiedewerke, Altendorfer Str 104, D 4300 Essen Germany

STEIN, RICHARD LOUIS, chemist, educator; b. Washington, July 7, 1944; s. Edward John and Margaret Ann (Hickey) S.; m. J. Teresa Taylor, Apr. 12, 1968. BS in Chemistry, George Washington U., 1966; PhD in Medicinal Chemistry, Med. Coll. Va., 1970. Rsch. asst. Med. Coll. Va., Richmond, 1966-70; asst. prof. chemistry Germanna C.C., Locust Grove, Va., 1970-72, assoc. prof., 1973-75, prof. chemistry, 1975—, program head dept. natural sci., 1970-84; vis. scientistVa. Acad. Sci., Richmond, 1980—; com. sci. curricula Va. C.C. System, Richmond, 1976; design cons. J. Sargent Reynolds C.C., Richmond, 1974. Author: Experiments in Chemistry, 1985; contbr. articles to chem. publs. including Jour. Medicinal Chemistry, Jour. Chem. Edn. Bd. dirs. Am. Cancer Soc., Culpeper, Va., 1972; drug abuse cons. Rappahannock Regional Coun. Drug Abuse, Culpeper, 1974; judge Rappahannock Sci. Fair, Orange, Va., 1978. Grantee Inst. Instrn. Excellence, 1985; summer faculty fellow Oak Ridge (Tenn.) Nat. Atomic Rsch. Lab., 1982; grantee Autodesk, Inc., Sausalito, Calif., 1992. Fellow AICE; mem. AAAS, Am. Chem. Soc., Sigma Xi. Office: Germanna C C Box 339 Locust Grove VA 22508

STEIN, RICHARD MARTIN, mechanical engineer; b. Buffalo, June 11, 1948; s. Thorvald Martin and Jean Grace (Pond) S.; m. Penelope Lynn Rowe, Sept. 18, 1971; children: Christopher M., Jill K., Matthew R. BSME, SUNY, Buffalo, 1976. Registered profl. engr., N.Y. Chief product engr. Strippit Houdaille, Akron, N.Y., 1980-83, chief engr. systems, 1983-86, engring. mgr., 1986-87; dir. engring. Eastman Machine, Buffalo, 1987-88; dir. engring. Wysong & Miles, Greensboro, N.C., 1988-89, v.p. engring., 1989-92; pres. RMS Engring., Greensboro, 1992—. Mem. ASME (tech. com. B5 stds. com. 1990—), Soc. Mfg. Engrs. Achievements include 2 patents in field; specializing in devel. of sheet metal fabrication machinery. Home: 4521 Tower Dr Greensboro NC 27410 Office: RMS Engring PO Box 19773 Greensboro NC 27419

STEIN, RUTH ELIZABETH KLEIN, physician; b. N.Y.C., Nov. 2, 1941; d. Theodore and Mimi (Foges) Klein; m. H. David Stein, June 9, 1963; children: Lynn Andrea Stein Melnick, Sharon Lisa, Deborah Michelle. AB, Barnard Coll., 1962; MD, Albert Einstein Coll. Medicine, 1966. Instr. dept. pediatrics George Washington U., Washington, 1969-70; with Albert Einstein Coll. of Medicine, Bronx, 1970—, assoc. prof. dept. pediatrics, 1977-83, prof. dept. pediatrics, 1983-92; vice chmn. dept. pediatrics Albert Einstein Coll., 1992—; vis. prof. pub. health dept. epidemiology Yale U. Sch. of Medicine, New Haven, 1986-87; dir., prin. investigator Preventive Intervention Rsch. Ctr. for Child Health, N.Y., Nat. Child Health Assessment Planning Project, N.Y., Behavioral Pediatric Tng. Program, N.Y.; dir. gen. pediatrics Pediatric Div., N.Y.; pediatrician in chief Albert Einstein Coll. Medicine; pediatric primary care center Bronx Municipal Hosp. Ctr. Editor: Caring for Children with Chronic Illness: Issues and Strategies, 1989; contbr. articles to profl. jours. Fellow Am. Acad. Pediatrics; mem. Ambulatory Pediatric Assn. (bd. dirs. 1980-88, pres. 1986-87), Am. Pediatric Soc., Soc. for Pediatric Rsch., Am. Pub. Health Assn., Alpha Omega Alpha, Soc. for Behavioral Pediatrics. Jewish. Home: 91 Larchmont Ave Larchmont NY 10538-3748 Office: Albert Einstein Coll Medicine 1300 Morris Park Ave Bronx NY 10461-1924

STEIN, THEODORE ANTHONY, biochemist, educator; b. St. Louis, Aug. 30, 1938; s. Leonard A. and Mathilda M. (Ellwangen) S.; widowed. BS, St. Louis U., 1960; MS, So. Ill. U., 1970, PhD, CUNY, 1987. Rsch. instr. surgery Washington U. Sch. Medicine, St. Louis, 1972-75; rsch. supr. surgery L.I. Jewish-Hillside Med. Ctr., New Hyde Pk., N.Y., 1975-76, rsch. coord. surgery, 1977—; asst. prof. surgery SUNY, Stony Brook, 1978-89, Albert Einstein Sch. of Medicine, Bronx, N.Y., 1989—; biostats. cons. NIH grantee, 1962; Am. Liver Found. grantee. Mem. AAAS, N.Y. Acad. Scis., Am. Fedn. Clin. Rsch., Am. Pub. Health Assn., Am. Gastroenterol. Assn., Sigma Xi. Republican. Roman Catholic. Contbr. articles to profl. jours. Achievements include development of chromatographic methods to determine prostaglandin and leukotriene content in tissues using fluorescent agents to increase sensitivity, elastase activity in the aorta with disease, and active anabolites of 5-fluorouracil in tumors; improvement of regulation of liver growth after surgery by diet; demonstration of diagnostic value of liver function tests, surgery on obese patients interferes with sugar metabolism and intestinal function; research in etiology of pancreatitis and pharmacological modification of pancreatic function; effect of stress on the stomach and colon; investigation of inflammatory bowel disease. Home: 10 Glamford Rd Port Washington NY 11050-2437 Office: LI Jewish Hillside Med Ctr New Hyde Park NY 11042-1433

STEINBERG, FRED LYLE, radiologist; b. Chgo., Jan. 12, 1957. BS, Northwestern U., Evanston, Ill., 1979; MS, U. Calif., San Francisco, 1981; MD, Northwestern U., Chgo., 1982. Radiologist Mass. Gen. Hosp., Harvard Med. Sch., Boston, 1988-90; dir. cardiac vascular and body Cedars-Sinai Med. Ctr. MRI, L.A., 1990-92; dir. Ctr. Vascular and Advanced Body Magmetoc Resonance Imaging, Beverly Hills, Calif., 1992—; pres. Steinberg Imaging Med. Group, 1992—. Mem. Radiological Soc. N.Am., Am. Roentgen Ray Soc., Soc. Magnetic Resonance Imaging, Soc. Magnetic Resonance in Medicine. Achievements include research in applications of body mangetic resonance angiography.

STEINBERG, MALCOLM SAUL, biologist, educator; b. New Brunswick, N.J., June 1, 1930; s. Morris and Esther (Lerner) S.; children—Jeffery, Julie, Eleanor, Catherine; m. Marjorie Campbell, 1983. B.A., Amherst Coll., 1952; M.A., U. Minn., 1954, Ph.D., 1956. Postdoctoral fellow dept. embryology Carnegie Instn., Washington, 1956-58; asst. prof. Johns Hopkins, Balt., 1958-64; assoc. prof. Johns Hopkins, 1964-66; prof. biology Princeton U., 1966-90, Henry Fairfield Osborn prof. biology, 1975—, prof. molecular biology, 1990—; instr.-in-charge embryology course Marine Biol. Lab., 1967-71, trustee, 1969-77; chmn. Gordon Research Conf. on Cell Contact and Adhesion, 1985; appointed to NAS/NRC Bd. on Biology, 1986-92. Mem. editorial bd. Bioscience, 1976-82; contbr. articles to profl. jours. Fellow AAAS; mem. AAUP, Am. Soc. Zoologists (program officer div. developmental biology 1966-69, chmn.-elect, then chmn. 1982-85), Am. Soc. Cell Biology, Internat. Soc. Developmental Biologists, Internat. Soc. Differentation, Soc. Developmental Biology (trustee, sec. 1970-73), Sigma Xi. Home: 86 Longview Dr Princeton NJ 08540-5642

STEINBERG, MARCIA IRENE, science foundation program director; b. Bklyn., Mar. 7, 1944; d. Solomon and Sylvia (Feldman) S.; 1 child, Eric Gordon. BS, Bklyn. Coll., 1964, MA, 1966; PhD, U. Mich., 1973. Rsch. scientist Meth. Hosp. Bklyn. Dept. Pathology, Bklyn., 1966-67, U. Mich. Dept. Surg. Rsch., Ann Arbor, Mich., 1967-68; post doctoral fellow Syracuse U. Dept. Biology, Syracuse, N.Y., 1973-76; from post doctoral fellow to assoc. prof. SUNY, Syracuse, N.Y., 1976-89; with NSF, Washington, 1990—; reviewer NATO fellowship NSF, San Francisco, 1988, Ad Hoc NSF, Syracuse, N.Y., manuscripts in field; vis. scientist Weizmann Inst. Renal Rsch. Fund, Weizmann Inst., Israel, 1987, 1988. Contbr. articles to profl. jours. Recipient Wellcome Rsch. Travel Grant, Wellcome Found. Cambridge U., U.K., 1985, Regents scholarship SUNY Bd. Regents, Bklyn., 1960-64. Mem. AAAS, Am. Soc. Biochemistry Molecular Biology, Assn. Women Sci., Sigma Xi. Office: 1800 G St NW Washington DC 20550-0002

STEINBERG, MEYER, chemical engineer; b. Phila., July 10, 1924; s. Jacob Louis and Freda Leah S.; m. Ruth Margot Elias, Dec. 24, 1950; children: David Martin, Jay Louis. B.Chem.Engr., Cooper Union, 1944; M.Chem.Eng., Bklyn. Poly. Inst., 1949. Registered profl. engr. N.Y. Jr. chem. engr. Manhattan dist., Kellex Corp., Oak Ridge, Los Alamos, 1944-46; asst. chem. engr. Deutsch & Loonam, 1947-50; chem. engr. Guggenheim Brothers, Mineola, N.Y., 1950-57; head process sci. div. Brookhaven Nat. Lab., Upton, N.Y., 1957—; expert in fossil and nuclear energy. Contbr. articles to profl. jours. Served with AUS, 1944-46. Recipient IR-100 award, 1970; Wasson award Am. Concrete Inst., 1972, Engr. of Year award, 1985, Ind. award Quest, 1985. Fellow Am. Nuclear Soc., Am. Inst. Chem. Engrs. (dir. L.I. sect.); mem. Am. Chem. Soc., AAAS, Am. Concrete Inst., Inst. Assos. Hydrogen Energy, Sigma Xi. Democrat. Jewish. Research on nuclear and fossil energy. Home: 15 Alderfield Ln Melville NY 11747-1724 Office: Brookhaven Nat Lab Upton NY 11973

STEINBERG, REUBEN BENJAMIN, utility management engineer; b. Ludz, Poland, Aug. 24, 1946; s. Wolf and Chaja Ita (Stecher) S.; m. Pnina Friend, Aug. 15, 1971; children: Steven, Ayal. BSME, NYU, 1970; M Engring. Mgmt., N.J. Inst. Tech., 1986. Registered profl. engr., N.J. Field engr. IBM, N.Y.C., 1967-68; sr. valuation engr. N.Y. State Pub. Svc. Commn., Albany, 1970-81; lead engr. Pub. Svc. Elec. & Gas, Newark, 1981-84, sr. staff engr., 1984—. Author: Feasibility of Burning Hazardous Waste in Utility Boiler, 1986. Mem. IEEE, Assn. of Energy Engrs. Home: 319 Stoughton Ave Cranford NJ 07016 Office: Pub Svc Elec & Gas Fl 14A 80 Park Plaza Newark NJ 07102

STEINBERGER, JACK, physicist, educator; b. Bad Kissingen, Fed. Republic Germany, May 25, 1921; came to U.S., 1935; s. Ludwig Lazarus and Berta (May) S.; m. Joan Beauregard, 1943 (div. 1962); children: Joseph, Richard Ned; m. Cynthia Eva Alff; children: Julia Karen, John Paul. BS in Chemistry, U. Chgo., 1942, PhD in Physics, 1948; hon. degree, Ill. Inst. Tech., 1989, U. Glasgow, 1990, Dortmund U., 1990, Columbia U., 1990, U. Autonoma de Barcelona, Spain, 1992. Mem. Inst. for Advanced Study, Princeton, N.J., 1948-49; asst. U. Calif., Berkeley, 1949-50; prof. Columbia U., N.Y.C., 1950-68, Higgins prof., 1968-72; staff mem. European Orgn. for Nuclear Research, Geneva, 1968-86, dir., 1969-72; Gallilean prof. physics Scuola Normale, Pisa, Italy, 1986—. Pfc. U.S. Army, 1943-46. Co-recipient Nobel prize in physics, 1988; recipient Nat. Medal of Sci., 1988, Mateuzzi medal Societa Italiane delle Scienze, 1991; fellow Guggenheim Found., Sloan Found. Mem. NAS, Am. Acad. Arts and Scis., Heidelberg Acad. Scis., Academia Europea. Home: 25 Chemin des Merles, CH 1213 Onex Switzerland Office: European Ctr for Nuclear Rsch, CH 1211 Geneva 23, Switzerland

STEINBERGER, YOSEF, ecologist, biologist; b. Seini, Rumania, Jan. 2, 1947; immigrated to Israel, 1964; m. Lea Rotman; children: Hila, Merav, Zvi. BS, Bar-Ilan Univ., Ramat-Gan, Israel, 1974, MS, 1976, PhD, 1980. Rsch. asst. Avdat Agrl. Rsch. Farm, 1968-70; coord. ecology courses Sede-Boker Field Sch., Israel, 1969-79; rsch. asst. N.Mex. State U., Las Cruces,

1981-82; lectr., researcher Bar-Illan U., 1982-86; rsch. assoc. Jacob Blaustein Inst. for Desert Rsch./Ben Gurion U. Negev, Israel, 1983-90; prof. Bar-Ilan Univ., Israel, 1990—; sr. lectr., researcher, Bar-Illan U., 1986; presenter seminars in field, confs. in field. Co-editor: Environmental Quality and Ecosystem Stability, 1986, 89, 92; contbr. articles to profl. jours. Rsch. grantee Nature Res. Authority, Israel, 1977, Binational Sci. Found., 1982, Israel Acad. Scis. and Humanities, 1984, German-Israel Res. Found., 1989; recipient Quinney Vis. Scholar award Coll. Natural Resources, Utah State U., Logan, 1989, DAAD Scholarship award, Germany, 1993. Mem. British Ecol. Soc., Israel Ecol. Soc., Ecol. Soc. Am., Acarol. Soc. Am., Soc. Nematologists, European Soc. Nematologists, Soil Ecology Soc., Sigma Xi. Achievements include rsch. in ecol. systems, mass and energy flow through trophic levels, physiol. ecology, mineral cycling and decomposition processes, soil ecology. Office: Bar-Ilan Univ, Biology Dept, Ramat Gan 52 100, Israel

STEINBORN, E(RNST) OTTO H., physicist, educator; b. Dresden, Germany, May 8, 1932; s. Heinrich and Gertrud (Thomas) S.; diploma Tech. U. Dresden, 1959; Dr. phil. nat., U. Frankfurt-Main, 1965; habilitation, Tech. U. Berlin, 1970; m. Gudrun Mnich, Sept. 14, 1968. Research asst. U. Frankfurt-Main, W.Ger., 1961-67; research asso. Iowa State U., Ames, 1967-69; instr. phys. chemistry Tech. U. Berlin, 1969-70, prof., 1970-71; prof. U. Regensburg (W.Ger.), 1971—. Lutheran. Author articles on theoretical and phys. chemistry, nuclear physics. Office: U Regensburg Chemistry Inst, Universitätsstrasse 31, D-93053 Regensburg Germany

STEINBRECHER, DONALD HARLEY, electrical engineer; b. Umatilla, Fla., July 1, 1936; s. Herman D. and Luada (Hales) S.; m. Brenda C. Rowland, 1967 (div. 1980); children: Brian Lee, Karl Walter, Sheri Lea, Andrew Donald; m. Rochelle L. Robbins, Mar. 27, 1983; children: Jeffrey Robbins Steinbrecher, Gregory Robbins Steinbrecher. BSEE, U. Fla., 1960; MSEE, MIT, 1963, PhD, 1966. Assoc. prof. elec. engring. MIT, Cambridge, Mass., 1967-72; pres., founder Steinbrecher Corp., Woburn, Mass., 1972-91, chmn., 1991—. Contbr. articles to profl. jours. Achievements include development and patenting of technology which led to introduction of Accuverter product line. The Accuverter is an air interface for computers, permitting access to electromagnetic signals in a radiated state. Office: Steinbrecher Corp 185 New Boston St Woburn MA 01801

STEINER, DIRK DOUGLAS, psychology educator; b. Dayton, Ohio, Sept. 7, 1959; s. Donald Eugene and Elizabeth (Ridgeway) S. BA, Ohio State U., 1980; MS, Pa. State U., 1982, PhD, 1985. Asst. prof. psychology La. State U., Baton Rouge, 1985-91, assoc. prof., 1991—. Contbr. chpts. to books, articles to profl. jours. Recipient Bourse Chateaubriand, French Govt., Paris, 1983-84. Mem. Am. Psychol. Soc., Soc. for Indsl./Orgnl. Psychology, Southeastern Psychol. Assn., Acad. of Mgmt. Office: La State U Audubon Hall Baton Rouge LA 70803-5501

STEINER, DONALD FREDERICK, biochemist, physician, educator; b. Lima, Ohio, July 15, 1930; s. Willis A. and Katherine (Hoegner) S. BS in Chemistry and Zoology, U. Cin., 1952; MS in Biochemistry, U. Chgo., 1956, MD, 1956; D Med. Sci. (hon.), U. Umea, 1973, U. Ill., 1984, Technische Hochschule, Aachen, 1993, U. Uppsala, 1993. Intern King County Hosp., Seattle, 1956-57; USPHS postdoctoral research fellow, asst. medicine U. Wash. Med. Sch., 1957-60; mem. faculty med. sch. U. Chgo., 1960—, chmn. dept. biochemistry, 1973-79, A.N. Pritzker prof. biochemistry, molecular biology and medicine, 1985—, sr. investigator Howard Hughes Med. Inst., 1986—; Jacobaeus lectr., Oslo, 1970; Luft lectr., Stockholm, 1984. Co-editor: The Endocrine Pancreas, 1972, discoverer proinsulin. Recipient Gairdner award Toronto, 1971, Hans Christian Hagedorn medal Steensen Meml. Hosp., Copenhagen, 1970, Lilly award, 1969, Ernst Oppenheimer award, 1970, Diaz-Cristobal award Internat. Diabetes Fedn., 1973, Banting medal Am. Diabetes Assn., 1976, Banting medal Brit. Diabetes Assn., 1981, Passano award, 1979, Wolf prize in medicine, 1985, Frederick Conrad Koch award Endocrine Soc., 1990. Mem. Nat. Acad. Scis., Am. Soc. Biochemists and Molecular Biologists, AAAS, Am. Diabetes Assn. (50th Anniversary medallion 1972), European Assn. Study Diabetes, Am. Acad. Arts and Scis., Sigma Xi, Alpha Omega Alpha. Home: 2626 N Lakeview Ave Apt 2508 Chicago IL 60614-1821

STEINER, GEORGE, information systems and management science educator, researcher; b. Budapest, Hungary, Sept. 11, 1947; arrived in Can., 1974; s. Aladar and Magda (Klein) S.; m. Judit Csizmazia, Sept. 21, 1974; children: Adam, David. Diploma in math. cum laude, Eötvös Lorand U., Budapest, 1971; PhD in Math., U. Waterloo, Ont., Can., 1982. Ops. rsch. analyst Infelor Systems Engring. Inst., Budapest, 1970-73; systems analyst Steel Co. Can., Hamilton, Ont., 1974-80; asst. prof. mgmt. sci. McMaster U., Hamilton, 1981-85, assoc. prof., 1985-92, prof., 1992—, chmn. mgmt. sci. and info. systems dept., 1989—; vis. prof. U. Bonn, Germany, 1987. Contbr. articles on math., ops. rsch. and combinatorial optimization to profl. jours. Grantee Natural Scis. Engring. Rsch. Coun. Can., 1981-82, 1983—; fellow Social Scis. and Humanities Rsch. Coun. Can., 1986-87, German Acad. Exch. Svc., 1987. Fellow Inst. of Combinatorics and Its Applications. Avocations: racquetball, theatre, books, music, chess. Office: McMaster U, 1280 Main St W, Hamilton, ON Canada L8S 4M4

STEINER, JEFFERY ALLEN, project engineer, executive; b. Longview, Wash., June 22, 1954; s. Glyn Elmer and Betty Jean (Shuster) S.; m. Cynthia Gene Schoppey, June 5, 1976; children: Peter, David, Scott. BSME, U.S. Naval Acad., 1976; MSCE, Oreg. State U., 1982. Registered profl. engr., Calif., Minn. Commd. ensign USNR, 1976; gunnery officer U.S.S. Bradley, San Diego, 1976-79; maintenance officer Naval Base, Guantanamo Bay, Cuba, 1979-81; co. comdr. Amphibious Constrn., San Diego, 1982-85; engr. City of Chula Vista (Calif.), 1985-86; project mgr. Mayo Clinic, Rochester, Minn., 1986—; tech. advisor Minn. Pollution Control Agy., St. Paul, 1988—. Bd. trustees Ronald McDonald House, Rochester, Minn., 1990—; bd. govs. Grace Evang. Free Ch., Stewartville, Minn., 1990—. Decorated Navy Achievement, Navy Battle Excellence. Mem. Minn. Soc. Profl. Engrs. (sec.-treas. 1992—), Toastmasters. Achievements include development of concepts, drawings and environmental review for the waste management facility for the Mayo Clinic, Rochester. Office: Mayo Clinic 200 1st St SW Rochester MN 55905

STEINER, MARK DAVID, engineering manager; b. Erie, Pa., Feb. 27, 1961; s. William George and Joyce Elaine (Alexis) S.; m. Melissa Ann Suain, Dec. 30, 1983; children: Jonathan, Michelle. BSEE, Penn State U., 1983. Electronics engr. Spl. Payloads Div., NASA-Goddard, Greenbelt, Md., 1983-86, 88-89; elec. systems mgr. Satellite Servicing Project, NASA-Goddard, Greenbelt, 1986-88; sect. head. Payload Devel. Sect., NASA-Goddard, Greenbelt, 1989-91, Payload Design Sect., NASA-Goddard, Greenbelt, 1991—; mgr. Spartan 204 Mission, NASA-Goddard, Greenbelt, 1992—. Recipient Spl. Achievement award NASA-Goddard Space Flight Ctr., 1985, Group Achievement award, 1985, Outstanding Achievement award, 1984; named to Dean's List. Mem. IEEE, Tau Beta Pi, Eta Kappa Nu. Lutheran. Office: NASA Code 743 Greenbelt MD 20771

STEINER, ROBERT FRANK, biochemist; b. Manila, Philippines, Sept. 29, 1926; came to U.S., 1933; s. Frank and Clara Nell (Weems) S.; m. Ethel Mae Fisher, Nov. 3, 1956; children: Victoria, Laura. A.B., Princeton U., 1947; Ph.D., Harvard U., 1950. Chemist Naval Med. Research Inst., Bethesda, Md., 1950-70; chief lab. phys. biochemistry Naval Med. Research Inst., 1965-70; prof. chemistry U. Md., Balt., 1970—; chmn. dept. chemistry U. Md., 1974—, dir. grad. program in biochemistry, 1985; mem. biophysics study sect. NIH, 1976. Author: Life Chemistry, 1968, Excited States of Proteins and Nucleic Acids, 1971, The Chemistry of Living Systems, 1981, Excited States of Biopolymers, 1983; editor Jour. Biophys. Chemistry, 1972—, Jour. of Fluorescence, 1991; contbr. more than 150 artices to profl. jours. Served with AUS, 1945-47. Recipient Superior Civilian Achievement award Dept. Def., 1966; NSF rsch. grantee, 1971-77, NIH, 1973-93. Fellow Washington Acad. Sci., Japan Soc. for Promotion Sci.; mem. Am. Soc. Biol. Chemists. Club: Princeton (Washington). Achievements include development of fluorescence techniques for studying proteins. Home: 2609 Turf Valley Rd Ellicott City MD 21042-2021 Office: 5401 Wilkens Ave Baltimore MD 21228-5329

STEINERT, LEON ALBERT, mathematical physicist; b. Shattuck, Okla., May 2, 1930; m. Emanuela Giovanna Montauti, May 21, 1988. PhD, U. Colo., 1962. Theoretical physicist Nat. Bur. Standards, Boulder, Colo., 1953-65; sr. rsch. engr. Lockheed Missiles & Space Co., Sunnyvale, Calif., 1972-79, 83-86; staff physicist IRT Corp., San Diego, 1979-81; prin. engr.-scientist McDonnell Douglas Corp., Huntington Beach, Calif., 1967-70, 88-92; cons. scientist Phys. Synergetics Inst., Sunnyvale, Calif., 1981—. Contbr. theoretical physics articles to profl. jours.

STEINFELD, JEFFREY IRWIN, chemistry educator, consultant, author; b. Bklyn., July 2, 1940; s. Paul and Ann (Ravin) S. B.Sc., MIT, 1962; Ph.D., Harvard U., 1965. Postdoctoral fellow U. Sheffield, Yorkshire, Eng., 1965-66; asst. prof. chemistry MIT, Cambridge, 1966-70; assoc. prof. MIT, 1970-79, prof., 1980—; mem. sci. adv. bd. Lasertechnics, Inc., Albuquerque, 1982—. Author: Molecules & Radiation, 1974; co-author: Chemical Kinetics and Dynamics, 1989; editor: Laser and Coherence Spectroscopy, 1977, Laser-Induced Chemical Processes, 1981; co-editor: Spectrochimica Acta, 1983—; contbr. articles to profl. jours. Treas. Ward 2 Democratic Com., Cambridge, 1972-73. NSF fellow Harvard U., Cambridge, 1962-65; NSF fellow Sheffield U., 1965-66; Alfred P. Sloan Found. research fellow MIT, 1969-71; Guggenheim fellow, 1972-73. Fellow Am. Phys. Soc.; mem. AAAS, Fedn. Am. Scientists, Sigma Xi, Phi Lambda Upsilon. Jewish. Office: MIT Room 2-221 Cambridge MA 02139

STEINFINK, HUGO, chemical engineering educator; b. Vienna, Austria, May 22, 1924; s. Mendel and Malwina (Fiderer) S.; m. Cele Intrator, Mar. 21, 1948; children: Dan E., Susan D. B.S., CCNY, 1947; M.S., Columbia U., 1948; Ph.D., Bklyn. Poly. Inst., 1954. Research chemist Shell Devel. Co., Houston, 1948-51, 53-60; T. Brockett Hudson prof. chem. engring. U. Tex., Austin, 1960—. Contbr. articles to profl. jours. Served with AUS, 1944-46. Fellow Am. Mineral Soc.; mem. Am. Chem. Soc., Am. Crystallographic Soc., Am. Inst. Chem. Engrs., Materials Research Soc., Phi Beta Kappa, Sigma Xi, Phi Lambda Epsilon. Home: 3811 Walnut Clay Dr Austin TX 78731-4011 Office: U Tex Coll Engring Austin TX 78712

STEINGLASS, PETER JOSEPH, psychiatrist, educator; b. N.Y.C., Mar. 1, 1939; s. Sam and Bella Sarah (Bernstein) S.; m. Abbe Stahl, July 1, 1962; children: Matthew Aaron, Joanna Eowyn. AB, Union Coll., 1960; MD, Harvard U., 1965. Diplomate, Am. Bd. Psychiatry and Neurology. Head clin. rsch. program Nat. Inst. Alcohol Abuse and Alcoholism, Washington, 1971-74; asst. prof. psychiatry George Washington U., Washington, 1974-77; assoc. prof. psychiatry George Washington U., 1977-81, prof. psychiatry and behavioral sci., 1981-90; dir. Ackerman Inst. for Family Therapy, N.Y.C., 1990—; vis. prof. psychiatry Hebrew U., Jerusalem, 1981-82; clin. prof. psychiatry Cornell U. Med. Coll., 1993—. Author: The Alcoholic Family, 1987; contbr. articles to sci. publs. Lt. comdr. USPHS, 1969-71. Fellow Am. Psychiat. Assn., Am. Assn. Marriage and Family Therapy, Assn. Clin. Psychosocial Rsch.; mem. Am. Family Therapy Acad. (charter, bd. dirs. 1987-89, v.p. 1989-91, Disting. Contbn. award 1987), Aesculapian Soc., Phi Beta Kappa. Democrat. Jewish. Avocations: photography, classical music. Office: Ackerman Inst for Family Therapy 149 E 78th St New York NY 10021-0405

STEINKAMP, KEITH KENDALL, electrical engineer; b. Beatrice, Nebr., May 14, 1953; s. Herman Henry Frederick and Inez Irene (Schoneweise) S.; m. Brenda Lee Cook, Oct. 30, 1982; 1 child, Eric Ryan. AA, Nebr. Tech. Coll., 1972. Elec. engr. Genex Corp., Beatrice, 1972-76; mgr. Ray's Inc., Crete, Nebr., 1976-84; plant mgr. NVS Inc., Lincoln, Nebr., 1984-86; svc. mgr. SMG Inc., Lincoln, 1986—; pres., prin. K & B Electronics Inc., Lincoln, 1980—; pres. Nebr. Waveguide Corp., Lincoln, 1990-92. Recipient Svc. award Lincoln Amateur Radio Club, 1990. Mem. Nat. Dart Assn. (coord. 1989), Amateur Radio Relay League (amateur radio emergency svc. 1990—), Nebr. Civil Def., 5.19 Repeater Assn. (v.p. 1991-92). Republican. Lutheran. Achievements include rsch. in high voltage radio wave transmission, relay logic control design. Home: 3700 Sweetbriar Ln Lincoln NE 68576-4510 Office: K & B Electronics Inc PO Box 22181 Lincoln NE 68502

STEINKE, RONALD JOSEPH, aerospace engineer, consultant; b. Chgo., June 13, 1939; s. Joseph John and Mae Ann (Janovsky) S.; m. Susanne Walker, June 26, 1968; 1 child, Jennifer Lynn. BS, Ill. Inst. Tech., 1961, MS, 1964. Engr. No. Ill. Gas Co., Chgo., 1960-61; rsch. engr. Ill. Inst. Tech., Chgo., 1961-64; engr. Revcor Inc., Chgo., 1964-65; aerospace engr. NASA Lewis Rsch. Ctr., Cleve., 1965—; cons. DOT/FAA, Atlantic City, 1980—. Contbr. articles to profl. jours. Named Outstanding Alumni Ill. Inst. Tech., 1986; NSF rsch. grantee, 1961; recipient DOD fellow, 1961-63, NASA fellow, 1963-64, Apollo Achievement award NASA, 1969. Mem. AIAA (Sustained Contbns. award 1987), Toastmasters Internat., Nat. Stuttering Project, Tau Beta Pi, Pi Tau Sigma. Achievements include design and computer simulation of flow fluid and losses in jet engine compressors; research on jet engine performance with rainwater ingestion. Office: NASA Lewis Rsch Ctr Mail Stop 5-11 21000 Brookpark Rd Cleveland OH 44135

STEINKRAUS, KEITH HARTLEY, microbiology educator; b. Bertha, Minn., Mar. 15, 1918; s. Henry Frank Steinkraus and Alice Elizabeth Hartley; m. Maxine Grace Curtiss, Aug. 26, 1941; children: Bonnie, Nancy, Donald, Anna, Karen. BA cum laude, U. Minn., 1939; PhD, Iowa State U., 1951. Chemist Am. Crystal Sugar Co., Chaska, Minn., 1939; microbiologist Joseph Seagram & Sons, Lawrenceburg, Ind., 1941-42; instr. electronics U.A. Army Air Corp, Sioux Falls, S.D., 1942-43; rsch. microbiologist Gen. Mills, Inc., Mpls., 1943-47; asst. prof. microbiology Iowa State U., Ames, 1951-52; prof. microbiology Cornell U., Geneva and Ithaca, N.Y., 1952-88, prof. emeritus, 1988—; prof. U. Philippines, Los Banos, 1967-69, Southbank Polytechnic, London, 1972-77; rsch. scientist Nestle Cen. Rsch., Switzerland, 1979-80, Singapore, 1986-87, cons., New Milford, Conn., 1988—; conductor nutrition surveys Ecuador, South Vietnam, Korea, Philippines, Thailand, Indonesia, others, 1959-65. Editor: Handbook of Indigenous Fermented Foods, 1983, Industrialization of Indigenous Fermented Foods, 1989; contbg. author several books in field; contbr. articles to scientific publications. Recipient Alumni Merit award Iowa State U., 1987, Internat. award Inst. Food Techs., 1985. Fellow AAAS, Am. Acad. Microbiology, Inst. Food Techs., Sigma Xi. Achievements include seven patents in field. Home: 15 Cornell St Ithaca NY 14850 Office: Cornell Univ Comstock / Dyce Ithaca NY 14853

STEINMANN, JOHN COLBURN, architect; b. Monroe, Wis., Oct. 24, 1941; s. John Wilbur and Irene Marie (Steil) S.; m. Susan Koslosky, Aug. 12, 1978 (div. July 1989). BArch., U. Ill., 1964; postgrad. Ill. Inst. Tech., 1970-71; Project designer C.F. Murphy Assocs., Chgo., 1968-71, Steinmann Architects, Monticello, Wis., 1971-73; design chief, chief project architect State of Alaska, Juneau, 1973-78; project designer Mithun Assos., architects, Bellevue, Wash., 1978-80; owner, prin. John C. Steinmann Assos., Architect, Kirkland, Wash., 1980—; bd. dirs. Storytell Internat.; lectr. Ill. Inst. Tech., 1971-72; prin. works include: Grant Park Music Bowl, Chgo., 1971, Menomonee Falls (Wis.) Med. Clinic, 1972, Hidden Valley Office Bldg., Bellevue, 1978, Kezner Office Bldg., Bellevue, 1979, The Pines at Sunriver, Oreg., 1980, also Phase II, 1984, Phase III, 1986, The Pines at Sunriver Lodge Bldg., 1986, 2d and Lenora highrise, Seattle, 1981, Bob Hope Cardiovascular Research Inst. lab. animal facility, Seattle, 1982, Wash. Ct., Bellevue, 1982, Anchorage Bus. Park, 1982, Garden Townhouses, Anchorage, 1983, Vacation Internationale, Ltd. corp. hdqrs., Bellevue, 1983, Vallarta Torres III, Puerto Vallarta, Mex., 1987, Torres Mazatlan (Mex.) II, 1988, Canterwood Townhouses, Gig Harbor Wash., 1988, Inn at Ceres (Calif.), 1989, Woodard Creek Inn, Olympia, Wash., 1989, Northgate Corp. Ctr., Seattle, 1990, Icicle Creek Hotel and Restaurant, Leavenworth, Wash., 1990, Bellingham (Wash.) Market Pl., 1990, Boeing Hot Gas Test Facility, Renton, Wash., 1991, Boeing Longacres Customer Svc. Tng. Ctr. Support Facilities, Renton, 1992, also pvt. residences. Served to 1st lt. C.E., USAR, 1964-66; Vietnam. Decorated Bronze Star. Registered architect, Wash., Oreg., Calif., N.Mex., Ariz., Utah, Alaska, Wis., Ill. Mem. AIA, Am. Mgmt. Assn., Nat. Council Archtl. Registration Bds., Alpha Rho Chi. Republican. Roman Catholic. Clubs: U. Wash. Yacht, Columbia Athletic. Address: 4316 106th Pl NE Kirkland WA 98033

STEINVALL, KURT OVE, physicist; b. Bureå Västerbotten, Vaslerbotten, Sweden, Sept. 20, 1944; s. Ernst Johannes and Valborg Elizabeth (Persson) S.; m. July 5, 1974 (div. 1986); children: Maria, David, Jacob. M in Physics, Uppsala U., Sweden, 1969; PhD, Chalmers Inst. Tech., Gothenburg, Sweden, 1974. Rsch. engr. Swedish Nat. Defense Rsch. Inst., Stockholm, 1970-72; head laser tech. group Swedish Nat. Defense Rsch. Inst., Linkoping, Sweden, 1978-90, rsch. dir., 1990—. Contbr. articles to profl. jours. Recipient Swedish Navy award, Stockholm, 1990, Swede-Ocean award, Stockholm, 1991. Mem. Optical Soc. Am., Soc. Photo-optical Engrs., Royal Acad. Defence Sci. (award 1989). Avocations: summer house, wild life, skiing, skating. Office: Swedish Defence Rsch Inst, PO Box 1165, Linköping Sweden

STEINWACHS, DONALD MICHAEL, public health educator; b. Boise, Idaho, Sept. 9, 1946; s. Don Peter and Emma Bertha (Weisshaupt) S.; m. Sharon Kay Carlson, Aug. 25, 1972. MS, U. Ariz., 1970; PhD, Johns Hopkins U., 1973. From asst. to assoc. prof. pub. health adminstrn. Johns Hopkins U., Balt., prof. health policy and mgnt., 1982—, dir. Health Svcs. Rsch. Ctr., 1982—; sec. adv. com. Dept. Vets. Affairs, Washington, 1991-92; mem. Inst. Medicine, NAS, Washington, 1993—; bd. dirs. Health Outcomes Inst., Inc., Mathematica Policy Rsch., Inc. Contbr. articles to profl. jours. Mem. Gov.'s Commn. on Health Policy Rsch. and Fin., Md., 1988-90. Capt. U.S. Army, 1973. Grantee NIMH, Agy. for Health Care Policy and Rsch., Robert Wood Johnson Found. Mem. Ops. Rsch. Soc. Am., Assn. Health Svc. Rsch. (bd. dirs., pres.), Found. for Health Svc. Rsch. (bd. dirs., pres.). Achievements include development of methods for using management information systems to examine patterns of medical care, costs, and indicators of the quality of care. Office: Johns Hopkins U 624 N Broadway Baltimore MD 21205-1901

STEITZ, JOAN ARGETSINGER, biochemistry educator; b. Mpls., Jan. 26, 1941; d. Glenn D. and Elaine (Magnusson) Argetsinger; m. Thomas A. Steitz, Aug. 20, 1966; 1 child, Jonathan Glenn. B.S., Antioch Coll., 1963; Ph.D., Harvard U., 1967; D.Sc. (hon.), Lawrence U., Appleton, Wis., 1982, Rochester U. Sch. Medicine, 1984, Mt. Sinai Sch. Medicine, 1989, Bates Coll., 1990; DSc (hon.), Trinity Coll., 1992, Harvard U., 1992. Postdoctoral fellow MRC Lab. Molecular Biology, Cambridge, Eng., 1967-70; asst. prof. molecular biophysics and biochemistry Yale U., New Haven, 1970-74; assoc. prof. Yale U., 1974-78, prof. molecular biophysics and biochemistry, 1978—. Recipient Young Scientist award Passano Found., 1975, Eli Lilly award in biol. chemistry, 1976, U.S. Steel Found. award in molecular biology, 1982, Lee Hawley, Sr. award for arthritis rsch., 1984, Nat. Medal of Sci., 1986, Dickson prize for Sci. Carnegie-Mellon U., 1988, Warren Triennial prize Mass. Gen. Hosp., 1989, Christopher Columbus Disc. award in biomed. rsch., 1992. Fellow AAAS; mem. Am. Acad. Arts and Sci., Nat. Acad. Arts and Sci., Am. Phil. Soc. Home: Stone Creek 45 Prospect Hill Rd Branford CT 06405 Office: Yale U Sch Medicine Dept Blochem & Biophsics 333 Cedar St New Haven CT 06510

STEKEL, FRANK DONALD, physics educator; b. Hillsboro, Wis., Aug. 26, 1941; s. Frank A. and Agnes E. (Heidenreich) S.; m. Shirley L. Dow, Aug. 24, 1967; children: Sharon I., Sandra M. MS, U. Wis., 1965; EdD, Ind. U., 1970. Instr. U. Wis., Whitewater, 1965-68, asst. prof., 1968-70, assoc. prof., 1970-76, prof. physics, 1976—. Author: Sourcebook for Chemistry and Physics, 1972. Fellow AAAS; mem. Am. Assn. Physics Tchrs., Nature Conservancy. Office: U Wis-Whitewater Physics Dept Whitewater WI 53190

STELL, WILLIAM KENYON, neuroscientist, educator; b. Syracuse, N.Y., Apr. 21, 1939; emigrated to Can. landed immigrant, 1980; s. Henry Kenyon and Edith Doris (Lawson) S.; m. Judith Longbotham, June 27, 1974; children: Jennifer Susan, Sarah Ruth. B.A. in Zoology with high honors, Swarthmore Coll., 1961; Ph.D. in Anatomy, U. Chgo., 1966, M.D. with honors (E. Gellhorn prize 1967), 1967. Staff fellow, then sr. staff fellow Nat. Inst. Neurol. Diseases and Stroke, NIH, 1967-72; asso. prof., then prof. ophthalmology and anatomy UCLA Med. Sch., 1972-80; assoc. dir. Jules Stein Eye Inst., UCLA, 1978-80; prof. anatomy U. Calgary (Can.) Faculty Medicine, 1980—, head dept., 1980-85; dir. Lions Sight Centre, Calgary, 1980—; chercheur invité Lab. Physiolique Nerveuse, CNRS, Grif-sur-Yvette, France, 1985-86. Served with USPHS, 1967-69. William and Mary Greve Internat. research scholar, 1979-80; grantee USPHS; grantee Med. Research Council Can.; grantee Alberta Heritage Found. Med. Research.; grantee Natural Scis. and Engring. Research Council of Can., NATO grantee. Mem. Assn. Research Vision and Ophthalmology, Soc. Neurosci., Am. Soc. Cell Biology, Am. Assn. Anatomists, Am. Physiol. Soc., Soc. Neurochemistry, Can. Assn. Anatomy, Can. Soc. Neurosci. Home: 5917 Bow Crescent NW, Calgary, AB Canada Office: 3330 Hospital Dr NW, Calgary, AB Canada T2N 4N1

STELLA, VALENTINO JOHN, pharmaceutical chemistry educator; b. Melbourne, Victoria, Australia, Oct. 27, 1946; came to U.S., 1968; s. Giobatta and Mary Katherine (Sartori) S.; m. Mary Elizabeth Roeder, Aug. 16, 1969; children: Catherine Marie, Anne Elizabeth, Elise Valentina. B of Pharmacy, Victorian Coll. Pharmacy, Melbourne, 1967; PhD, U. Kans., 1971. Lic. pharmacist, Victoria. Pharmacist Bendigo (Victoria) Base Hosp., 1967-68; asst. prof. Coll. Pharmacy U. Ill., Chgo., 1971-73; from asst. prof. to assoc. prof. to prof. Sch. Pharmacy U. Kans., Lawrence, 1973-90, Univ. disting. prof., 1990—; dir. Ctr. for Drug Delivery Rsch.; cons. to 15 pharm. cos., U.S, Japan, Europe. Co-author: Chemical Stability of Pharmaceuticals, 2d edit., 1986; co-editor: Prodrugs as Novel Drug Delivery Systems, 1976, Directed Drug Delivery, 1985, Lymphatic Transport of Drugs, 1992; author numerous papers, revs., abstracts. Fellow AAAS, Am. Assn. Pharm. Scientists, Am. Acad. Pharm. Scientists. Roman Catholic. Achievements include 11 U.S. patents; rsch. in application of phys./organic chemistry to the solution of pharm. problems. Home: 1324 Lawrence Ave Lawrence KS 66049 Office: U Kans Dept Pharm Chemistry 3306 Malott Hall Lawrence KS 66045

STELLE, KELLOGG SHEFFIELD, physicist; b. Washington, Mar. 11, 1948; s. Charles Clarkson and Jane Elizabeth (Kellogg) S. BA, Harvard Coll., 1970; PhD, Brandeis U., 1977. Field observer Bartol Research Found., South Pole, Antarctica, 1970-72; lectr. math. King's Coll., U. London, Eng., 1977-78; sci. assoc. Cern, Geneva, 1980-81, 87; rsch. fellow Imperial Coll., London, 1978-80, advanced fellow, 1982-87, lectr. physics, 1987-88, reader, 1988—; mem. Inst. for Advanced Study, Princeton, N.J., 1986; vis. fellow Ecole Normale Supérieuve, Paris, 1981-82. Editor: Classical and Quantum Gravity, 1984—; contbr. articles to profl. jours. Mem. AAAS, Am. Phys. Soc., Fedn. Am. Scientists. Office: Blackett Lab Imperial Coll, Prince Consort Rd, London SW7 2BZ, England

STELLING, JOHN HENRY EDWARD, chemical engineer; b. Charleston, S.C., Oct. 3, 1952; s. John Henry Edward Jr. and Louise Ferguson (Myers) S.; divorced. BSChemE, Clemson U., 1974; MSChemE, U. Tenn., 1979. Registered profl. engr., N.C., S.C., Fla., Ga., Mass., Mich., Conn., N.Y., Tenn., Ky., Ind., Ala., Maine; NCEES, REA, Calif. Summer tech. worker Tenn. Eastman Co., Kingsport, 1973; rsch. engr. Dow Chem. USA, Plaquemine, La., 1974-77; teaching asst. U. Tenn., Knoxville, 1978-79; staff engr. Radian Corp., Durham, N.C., 1979-83; dept. head Radian Corp., Research Triangle Park, N.C., 1985—; v.p. Radian Engring. of N.Y., Rochester, 1990—, The Herbert Co., Charleston, S.C., 1983-86. Mem. AIChE, Nat. Water Well Assn., Air and Waste Mgmt. Assn., Sigma Xi, Tau Beta Pi, Phi Kappa Phi. Home: 1319 Arnette Ave Durham NC 27707 Office: Radian Corp PO Box 13000 Research Triangle Park NC 27709

STELLMAN, STEVEN DALE, epidemiologist; b. Toronto, Ont., Can., May 7, 1945; s. Samuel David and Lillian (Mandlsohn) S.; m. Jeanne Esther Mager, Sept. 10, 1967; children: Andrew, Emma. BS in Chemistry, Ohio State U., 1966; PhD in Phys. Chemistry, NYU, 1971; MPH in Health Policy and Mgmt., Columbia U., 1992. Rsch. assoc. biochem. sci. Princeton (N.J.) U., 1971-73; lectr. in chemistry U. Colo., Denver, 1973-74; chief div. computing and biostats. Am. Health Found., N.Y.C., 1975-80, sr. rsch. scientist, 1991—; asst. v.p. epidemiology Am. Cancer Soc., N.Y.C., 1980-88; asst. commr. biostat. and epidemiol. rsch. N.Y.C. Dept. Health, 1988-91; adj. assoc. prof. dept. community medicine Mt. Sinai Sch. Medicine, N.Y.C., 1981—; sci. cons. agt. orange vet. payment program U.S. Dist. Ct., Bklyn., 1985—; mem. adv. bd. pub. health grad. program Robert Wood Johnson Sch. Medicine, Piscataway, N.J., 1986—; cons. in epidemiology and biostatistics Meml. Sloan-Kettering Cancer Ctr., N.Y.C., 1993—. Author Women and Cancer, 1986; editor Vital Stats. Summaries, N.Y.C., 1988-91; assoc. editor Women and Health, 1991—; contbr. articles to profl. publs.; co-author spl. issue Environ. Rsch., 1988. Condr. DeRossi singers Kane St.

Synagogue. Fogarty Sr. Internat. fellow NIH, 1992—. Mem. APHA, Am. Coll. Epidemiology, Soc. for Epidemiologic Rsch., Am. Chem. Soc. Democrat. Jewish. Achievements include study of health effects of agent orange, cancer prevention study of 1.2 million Ams. Home: 117 St Johns Pl Brooklyn NY 11217-3496 Office: Am Health Found 320 E 43d St New York NY 10017

STELLRECHT BURNS, KATHLEEN ANNE, virologist; b. Schenectady, N.Y., Nov. 5, 1959; d. Robert Howard Stellrecht and Margaret Lucille (Myers) Osterlitz; m. Christopher P. Burns. BS, SUNY, Buffalo, 1981; MS, Rensselaer Poly. Inst., 1986, PhD, 1990. Med. technologist Albany (N.Y.) Med. Ctr. Hosp., 1981-85; teaching asst. Rensselaer Poly. Inst., Troy, N.Y., 1984-88; postdoctoral fellow Mt. Sinai Med. Ctr., N.Y.C., 1989-91, Albany Med. Ctr. Hosp., 1991—. Contbr. articles to profl. jours. Recipient Travel award Sigma Xi, 1987, 88. Mem. AAAS, Am. Assn. Clin. Pathologists, Am. Soc. Virology (travel award 1988), Am. Soc. Microbiology (grad. student rsch. award 1988 Ea. br.). Home: 8 Cambridge Rd Albany NY 12203 Office: Albany Med Ctr Hosp Dept Clin Microbiology Mailstop A-22 Albany NY 12208

STELSON, ARTHUR WESLEY, research chemical engineer, atmospheric scientist; b. Pitts., May 12, 1955; s. Thomas Eugene and Constance Anne (Semon) S. B Chem. Engring., Ga. Inst. Tech., 1975, MSChemE, 1976; PhD, Calif. Inst. Tech., 1982. Registered engr.-in-tng., Ga. Grad. rsch. and teaching asst. Calif. Inst. Tech., Pasadena, 1976-81; project engr. Exxon Rsch. and Engring. Co., Florham Park, N.J., 1981-84; sr. rsch. scientist Atlanta Univ. Ctr., Inc., 1985-90; assoc. prof. Clark Atlanta U., 1990-91; sr. rsch. scientist Ga. Inst. Tech., Atlanta, 1991—; mem. Univ. Coal Rsch. peer rev. com. U.S. Dept. Energy, Pitts., 1986—. Contbr. articles to Environ. Sci. and Tech., Atmospheric Environ., Jour. Air Pollution Control Assn., Energy and Fuels, chpt. to book. Mem. Rep. Nat. Com., Washington, 1985—; mem. adminstrv. bd. Trinity United Meth. Ch., Austell, Ga., 1987—. Mem. AIChE, NSPE, Am. Chem. Soc., Am. Inst. Chemists, Sigma Xi, Phi Kappa Phi, Tau Beta Pi, Chi Epsilon Sigma. Achievements include being one of first engineers to apply classical thermodynamics to urban air pollution problems and modelling. Office: Sch Earth-Atmospheric Scis Ga Inst Tech Atlanta GA 30332-0340

STELZER, JOHN FRIEDRICH, nuclear engineer, researcher; b. Leipzig, Saxony, Germany, Feb. 26, 1928; s. Karl and Johanna (Richter) S.; m. Marianne Rost, May 12, 1951; children: Mechthild, Roderich, Hermann. Dipl.-ing., Technische Hochschule, Aachen, Fed. Republic of Germany, 1964, Dr.-ing., 1971. Registered profl. engr. Farmer Altmittweida, Saxony, 1951-60; project engr. Nuclear Rsch. Ctr., Juelich, Fed. Republic of Germany, 1964-66; head of group thermodynamics Nuclear Rsch. Ctr. Inst. for Reactor Experiments, Juelich, 1967-72; head of tech. analysis subdivision Rsch. Ctr. KFA, Juelich, 1973-92; lectr. Fachhochschule Juelich, 1964-80, House of Techniques, Essen, Germany, 1977—; mem. organizing com. Conf. on Numerical Methods in Thermal Problems, Stanford, Calif., 1991. Author: Heat Transfer and Fluid Flow, 1971, Physical Property Algorithms, 1984; contbg. author to numerous books; co-editor: Engineering Computations, Communications in Applied Numerical Methods, Numerical Methods for Heat and Fluid Flow; contbr. numerous papers to profl. jours.; patentee in field. Co-founder PROFEM GmbH, Aachen, 1984. Achievements include developments in the finite element method and computer graphics.

STEMLER, ALAN JAMES, plant biology educator; b. Chgo., July 29, 1943; s. Carl R. and Adeline (Jurecki) S.; m. Elisabeth Marie Creach, 1985; children: Kenneth A., Chloe M. BS, Mich. State U., 1965; PhD, U. Ill., 1974. Prof. U. Calif., Davis, 1978—. Office: Section of Plant Biology U Calif Davis CA 95616

STEMPLE, ALAN DOUGLAS, aerospace engineer; b. Elkins, W.Va., July 19, 1963; s. Stephen Warren and C. Phyllis (Cavalier) S. BS cum laude, Davis and Elkins Coll., 1984; BS in Aerospace Engring. cum laude, W.Va. U., 1985; MS, U. Md., 1986, PhD, 1989. Rotorcraft fellow Ctr. for Rotorcraft Edn. and Rsch., U. Md., College Park, 1985-89; structures rsch. engr. McDonnell Douglas Helicopter Co., Mesa, Ariz., 1989—; reviewer tech. papers, 1990—. Contbr. articles to profl. publs. Army Rotorcraft fellow U. Md., 1985-89. Mem. AIAA, Am. Helicopter Soc. (Vertical Flight Found. scholar 1988). Home: 1233 N Mesa Dr # 2084 Mesa AZ 85201-2763 Office: McDonnell Douglas Mail Stop 530/B337 5000 E McDowell Rd Mesa AZ 85205-9797

STEN, JOHANNES WALTER, controls system engineer, consultant; b. Balt., Aug. 1, 1934; s. Johannes Adolf and Aili Augusta (Bohm) S.; m. Elizabeth Eleanore Mackey, May 6, 1961; children: Nathan Allen, Curtis John, Theresa Jean, David Hal. B Engring. Sci., Johns Hopkins U., 1957; postgrad., u. Pa., 1962-65. Elec. engr. Johns Hopkins Physics Lab., Silver Spring, Md., 1957-58; R&D engr. E.I DuPont de Nemours Co., Wilmington, Del., 1960-64; instruments engr. E.I DuPont de Nemours Co., Newark, Del., 1964-74, sr. instruments engr., 1974, systems cons., 1974-79, project engr., 1979-90, sr. cons. polymers engring., 1990—. Contbr. to profl. publs. 1st lt. U.S. Army, 1958-60. Mem. Instrument Soc. Am. (charter mem. batch standards com., cited as pioneer in computer-based control 1992), Delaware Valley Finnish Am. Soc. Lutheran. Achievements include establishing method of defining chemical batch process characteristics and complexity for controls system configuration, hardware and software. Office: EI Du Pont de Nemours & Co PO Box 6090 Newark DE 19714

STENDAHL, STEVEN JAMES, mathematician, system engineer; b. Mpls., June 11, 1955; s. Everett S. and Florence M. Stendahl. BS in Physics, Purdue Univ., 1977; MS, Cornell Univ., 1980; DSc, Washington Univ., 1983. Staff mem. The BDM Corp., McLean, Va., 1983-87; system engr. W.J. Schafer Assocs., Arlington, Va., 1987-89, Sparta, Inc., McLean, 1989—. Mem. Am. Assn. Artificial Intelligence, Soc. Industrial & Applied Mathematics, Assn. Computing Machinery, Mathematical Assn. Am., Phi Beta Kappa, Phi Kappa Phi, Sigma Pi Sigma. Achievements include development of several computer simulations used to support analyses of ballistic missile defense including kinetic and directed energy weapons, optical sensors; application of functional integrals to stochastic control problems; simulation of the dynamics of a Kosterlitz-Thouless phase transition. Home: 7400 Colshire Dr Apt 6 McLean VA 22102 Office: Sparta Inc 7926 Jones Branch Dr Ste900 Mc Lean VA 22102

STENGEL, EBERHARD FRIEDRICH OTTO, botanist; b. Freiburg, Germany, Oct. 18, 1936; s. Otto Friedrich and Hanna Helene Marie (Hoefft) S.; m. Margarete Heuing, Nov. 8, 1974 (div. 1986); children: Daniel, Tobias. PhD in Limnology, Albert Ludwigs U., Freiburg, 1968. Scientific asst. Kohlenstoffbiologische Forschungsstation, Dortmund, Germany, 1967-74; group leader Gesellschaft fuer Strahlen und Umweltforschung, Dortmund, 1974-79, Forschungszentrum Juelich, Germany, 1979—; mem. Wetlands Rsch. Lab. and Inst. for Coastal and Estuarine Rsch., U. West Fla., Pensacola, 1993—. Editorial adv. bd.: Aquadocumenta Verlag, 1983—; contbr. articles to profl. jours. Named Best Soloist German Jazz Fedn., 1958. Mem. Internat. Assn. Theoretical and Applied Limnology, Deutsche Botanische Gesellschaft, Internat. Assn. on Water Quality, Aquatic Plant Mgmt. Soc. Avocations: jazz drumming, mountaineering, botanic historic research. Home: Agricola str 5, D-52445 Titz Roedingen Germany Office: Forschungszentrum Juelich, Institut Biotechnologie, D-52425 Jülich Germany

STENGEL, ROBERT FRANK, mechanical and aerospace engineering educator; b. Orange, N.J., Sept. 1, 1938; s. Frank John and Ruth Emma (Geidel) S.; m. Margaret Robertson Ewing, Apr. 8, 1961; children: Brooke Alexandra, Christopher Ewing. SB, MIT, 1960; MS in Engring., Princeton U., 1965, MA, 1966, PhD, 1968. Aerospace technologist NASA, Wallops Island, Va., 1960-63; tech. staff group leader C.S. Draper Lab., Cambridge, Mass., 1968-73, Analytic Scis. Corp., Reading, Mass., 1973-77; assoc. prof. Princeton (N.J.) U., 1977-82, prof. mech. and aerospace engring., 1982—; cons. GM, Warren, Mich., 1985—; mem. com. strategic tech. U.S. Army NRC, 1989-92; vice chmn. Congl. Aero. Adv. Com., Washington, 1986-89; mem. com. on trans-atmospheric vehicles USAF Sci. Adv. Bd., 1984-85; mem. com. on low altitude wind shear and its hazard to aviation Nat. Rsch. Coun., 1983. Author: Stochastic Optimal Control: Theory and Application, 1986; N.Am. editor Cambridge Aerospace Series, 1993—, Cambridge Univ. Press, 1993—; contbr. over 100 tech. papers to profl. publs.; patentee wind probing device. Lt. USAF, 1960-63. Recipient Apollo Achievement award NASA, 1969, Cert. of Commendation, MIT, 1969. Fellow IEEE; mem. AIAA (assoc. fellow), Soc. Automotive Engrs. (mem. aerospace guidance and control systems com.). Avocations: photography, music, bicycling. Home: 329 Prospect Ave Princeton NJ 08540-5330 Office: Princeton U D202 Engineering Quadrangle Princeton NJ 08544

STENWICK, MICHAEL WILLIAM, internist, geriatric medicine consultant; b. Red Wing, Minn., Nov. 12, 1941; s. Vincent Ferdinand and Geraldine Frances (Veith) S.; m. Judith Ann Nelson, June 10, 1961; children: Scott Michael, Gregg William. BS cum laude, Hamline U., 1963; MD, U. Minn., 1969. Diplomate Am. Bd. Internal Medicine. Fellow dept. pharmacology U. Minn., Mpls., 1966-68; intern in internal medicine Northwestern Hosp., Mpls., 1969-70, resident in internal medicine, 1970-73; sr. internist internal medicine sect. Bloomington Lake Clinic, Mpls., 1973—; bd. dirs. Bloomington Lake Clinic, Mpls., pres., 1977, v.p. 1989—, fin. com., 1987—, chmn. properties, 1984—, chmn. trustees profit sharing; med. adviser Kimberly Quality Care, St. Paul, 1990—; internal medicine cons. Fairview Multiple Sclerosis Ctr. and Rehab. Unit, Mpls., 1986—; informal adviser internal medicine sect. Minn. Relative Value Index, Mpls., 1971; mem. task force Riverside Med. Ctr., Mpls., 1988-91, chmn. critical care com., 1986-91, reviewer quality assurance subcom., 1989-90. Contbr. articles to profl. jours. Mem., co-organizer, 1st pres. Cyrus Barnum Soc., U. Minn. Med. Sch., Mpls.; bd. dirs. Signal Inn Beach and Racquetball Club, Sanibel Island, Fla., 1983-84, 89—, Signal Inn Condominium Assn., Sanibel Island, 1983-84, 89—; co-emcee Nursing Talent Show, Northwestern Hosp., Mpls., 1969; 1st med. dir. Beltrami Health Ctr., Mpls., 1970-72. Recipient scholarship Charles and Alora Allis Found., 1960-63, Walter Kenyan award, 1963, grant U. Minn., 1963. Fellow Am. Coll. Physicians; mem. AMA, Am. Soc. Internal Medicine, Minn. Med. Assn., Hennepin County Med. Assn., Mpls. Soc. Internal Medicine. Republican. Lutheran. Achievements include research in drug specificity that could be defined even in an alkylating agent. Office: Bloomington Lake Clinic 3017 Bloomington Ave Minneapolis MN 55407

STEPHAN, CHARLES ROBERT, retired ocean engineering educator, consultant; b. N.Y.C., Sept. 30, 1911; s. Charles Albert and Ella (Wallendorf) S.; m. Eleanor Grace Storck, Feb. 14, 1937 (dec. July 1992); children: Yvonne Stephan Brown, Joan Stephan Cathcart, Charles Royal, Robert W. BS in Engring., U.S. Naval Acad., 1934; D Engring. (hon.), Fla. Atlantic U., 1978. Commd. ensign U.S. Navy, 1934, advanced through grades to capt.; served various capacities including WWII, South Pacific and Korean war areas U.S. Navy, various locations, 1941-52; ret. U.S. Navy, 1963; prof. ocean engring. Fla. Atlantic U., Boca Raton, 1964-76; prof. emeritus Fla. Atlantic U., 1976—; assoc. prof. naval sci., Rensselaer Poly. Inst., Troy, N.Y., 1944-46. Contbr. articles to various publs. Bd. dirs. Legion of Valor of USA, 1985-92, mem. chmn., 1985—. Decorated Navy Cross, 2 Bronze Star medals. Fellow Marine Technology Soc.; mem. U.S. Naval Inst., U.S. Navy League (v.p. Delray Beach coun.), Pearl Harbor Survivors (pres. Fla. Gold Coast chpt.), Kiwanis. Republican. Lutheran. Avocations: photography, travel, swimming, bowling. Home and Office: 1136 York Ln Virginia Beach VA 23451

STEPHANEDES, YORGOS J., transportation educator; b. Athens, Greece; came to U.S., 1970; BA in Engring. Sci., Dartmouth Coll., 1973; MSEE, Carnegie-Mellon U., 1974; PhD, Dartmouth-Thayer Sch., 1978. Registered profl. engr., Minn. From asst. to assoc. prof. dept. civil engring. U. Minn., Mpls., 1978-89, prof. dept. civil engring., 1989—, dir. grad. studies dept. civil engring., 1991—; cons. Gamma Inst., Athens, Mpls., 1986—; editorial bd. Transp. Rsch. Jour., 1993—, Jour. Advanced Trans., 1990—; evaluator, auditor Drive, Shrp-Idea, 1990—. Editor: Application of Advanced Technologies in Transportation Engineering; contbr. over 100 articles to profl. jours. Mem. Minn. Intelligent Vehicle Hwy. Systems Planning Group, St. Paul, 1989—. Grantee NSF, U.S. Dept. Transp., Minn. Dept. Transp., Fed. Hwy. Adminstrn., DOE, Minn. Supercomputer Inst.; recipient D. Grant Mickle award NRC, 1993. Mem. IEEE, ASCE (chmn. com. advanced tech. in transp. 1990—), Intelligent Vehicle Hwy. Systems Am. (founding mem., coun. standards orgns., steering com. standard and specifications), TRB (applications emerging tech. com., com. econ. analysis to tranp. 1990—), Inst. Transp. Engrs,. Achievements include research in automatic incident detection, real-time optimal control in freeways and arterials, real-time traffic prediction.

STEPHANICK, CAROL ANN, dentist, consultant; b. South Amboy, N.J., Feb. 5, 1952; d. Edward Eugene and Gladys (Pionkowski) S. BS, Rutgers U., 1974; MS, Med. Coll. Pa., 1980; DMD, Temple U., 1984. Lic. dentist, Pa., N.J., Vt. Med. technologist Jersey Shore Med. Ctr., Neptune, N.J., 1975-76, South Amboy Meml. Hosp., 1976-78, Smith-Kline Clin. Labs., King of Prussia, Pa., 1981; instr. dept. biology St. Peter's Coll., Jersey City, 1976-78; instr., edn. coord. Coll. Allied Health, Hahnemann U., Phila., 1978-80; instr. dept. oral radiology Sch. Dentistry, Temple U., Phila., 1984-87; assoc. dentist Personal Choice Dental Assocs., South Amboy, 1985-86, Marcucci and Marcucci, P.C., Phila., 1986-90, Gwynedd Dental Assocs., Springhouse, Pa., 1990-92; spl. events coord. Liberty Dental Conf., Phila., 1990—. Neighbor patrol Sprague St. Neighbors Town Watch, Phila., 1986—. Named to Legion of Honor, Chapel of Four Chaplains, 1987. Mem. ADA, Pa. Dental Assn., Philadelphia County Dental Soc. (publicity coord. 1990—, pub. info. coord. 1991, semi-finalist judge sr. smile contest 1990—, com. on concerns of women dentists, select com. 1988—), Delaware Valley Assn. Women Dentists, Am. Assn. for Functional Orthodontics, Am. Soc. Clin. Pathologists (med. technologist), Delta Sigma Delta. Roman Catholic. Avocations: reading, weight training, walking, sailing, dog training. Home: 6939 Sprague St Philadelphia PA 19119-1308 Office: 777 S Whitehorse Pike Hammonton NJ 08037

STEPHANOPOULOS, GREGORY, chemical engineering educator, consultant, researcher; b. Kalamata, Greece, Mar. 10, 1950; came to U.S., 1973; s. Nicholas and Elizabeth (Bitsanis) S.; m. Maria Flytzani; children—Nicholas-Odysseas, Alexander, Rona-Elisa. B.S., Nat. Tech. U., Athens, Greece, 1973; M.S., U. Fla., Gainesville, 1975; Ph.D., U. Minn., Mpls., 1978. Registered profl. engr., Greece. Asst. prof. chem. engring. Calif. Inst. Tech., Pasadena, 1978-83, assoc. prof. chem. engring., 1983-85; prof. chem. engring. MIT, Cambridge, 1985—. Editor: Kinetics and Thermodynamics of Biological Systems, 1983. Mem. editorial bd. Mathematical Biosciences, 1984—, Biotech. Progress, 1984—. Contbr. articles to profl. jours. Dreyfus Tchr. scholar Camille and Henry Dreyfus Found., 1982; recipient Pres. Young Investigator award NSF, 1984; NSF grantee, 1980—. Mem. Am. Inst. Chem. Engrs. (programming coordinator 1983), Am. Chem. Soc. Greek Orthodox. Avocations: chess; music; travel. Office: Mass Inst Tech Dept Chem Engring 66-552 77 Massachusetts Ave Cambridge MA 02139-4307

STEPHENS, DAVID BASIL, civil engineer; b. Corpus Christi, Tex., Oct. 22, 1963; s. Ronald Theron and Dora Coleen (Isaac) S.; m. Sheri Lyn Kemp, Nov. 19, 1988; children: Whitney Carroll, Jessica Lyn. BS, Tex. A&M U., 1986. Registered profl. engr., Okla. Engr.-in-tng. Okla. Dept. Transp., Oklahoma City, 1987; engr. intern Okla. Dept. Transp., Ardmore, Okla., 1987-88; staff engr. Okla. Dept. Transp., Duncan, Okla., 1988-90, asst. resident engr., 1990-92; area maintenance engr. Okla. Dept. Transp., Ada, Okla., 1992—. Deacon First Bapt. Ch., Marlow, Okla., 1992, pres. adult choir, 1990; participant Leadership Duncan, 1989-90. Recipient Disting. Student award Tex. A&M U., 1986. Mem. Assn. Former Students Tex. A&M U. Baptist. Office: Okla Dept Transp PO Box 549 Ada OK 74820

STEPHENS, DEBORAH LYNN, health facility executive; b. Newton, Iowa, May 30, 1952; d. Clarence Harry and Nancy Elizabeth (Gass) Wright; m. David K. Brender, Dec. 18, 1971 (div.); m. Michael E. Stephens, May 21, 1988. BS, U. Iowa, 1974; postgrad., U. Wis., Milw., 1978-80, U. Calif., Berkeley, 1987. Asst. to dean of fin. U. Iowa Coll. Medicine, Iowa City, 1975-77; contract audit acct. Miller Brewing Co., Milw., 1977-79; asst. controller Unicare Health Facilities, Milw., 1979-81; v.p. fin. Sacred Heart Rehab. Hosp., Milw., 1981-84; exec. v.p., chief operating officer Sacred Heart Rehab. Hosp., Med. Rehab. Inst., Milw., 1984-88; prin. founding mem., pres., chief exec. officer Behavioral Health Systems, Birmingham, Ala., 1989—, also bd. dirs.; cons. on rehab., corp. planning and zero-base budgeting, Birmingham, 1988; bd. dirs. Rehab. Mgmt. Svc., Inc., Milw., 1986-88, Med. Rehab. Equipment, Inc., Milw., 1986-88; founding mem. Am. Rehab. Network, Inc., Washington, 1986-87; mem. oral exam. bd. City of Milw., 1984-86; mem. prospective payment adv. com. HHS, Washington, 1986; nat. presenter on zero-base budgeting, corp. reorgns., managed care, and planning. Contbr. articles to profl. jours. Mem. healthcare cost containment com. Bus. Coun. Ala., Birmingam. Mem. Hosp. Fin. Mgmt. Assn. (governing bd. 1981-88), Nat. Forensic League (life), Nat. Assn. Accts., Nat. Assn. Rehab. Facilities (prospective payment adv. bd. 1986-88, com. on med. oriented facilities 1983-88), Birmingham C. of C., Kappa Kappa Gamma Alumnae Assn, Venture Club. Avocations: dancing, skiing, jogging, travel, reading. Office: Behavioral Health Systems 2 Metroplex Dr Ste 503 Birmingham AL 35209

STEPHENS, DOUGLAS KIMBLE, chemical engineer; b. Monticello, Ark., June 22, 1939; s. Vardeman King and Lila Belle (McMurtery) S.; m. Mary Joan John, Dec. 4, 1957; children: Kenneth R., David B. BSChemE, U. Ark., 1962. Registered profl. engr., Tex.; cert. safety profl. Sr. engr., safety supt. Monsanto Co., Alvin, Tex., 1967-73, mfg. supt., 1973-78; ptnr. Robert T. Bell & Assocs., Houston, 1978-80; v.p. Tech. Inspection Svcs., Inc., Houston, 1980-84, Bell & Stephens Labs., Inc., Houston, 1980-84; pres. Stephens Engring. Labs., Inc., Webster, Tex., 1984—. Contbr. articles to profl. publs. Capt. U.S. Army, 1962-67, Vietnam. Decorated Bronze Star, Air medal. Mem. ASTM (mem. coms., cons.), Am. Inst. Chem. Engrs. (sect. chmn. 1976-77), Am. Soc. Safety Engrs. (cons.), Tex. Soc. Profl. Engrs. (chpt. chmn. 1991-92, pres. 1990-91, Engr. of Yr. 1992), Nat. Assn. Corrosion Engrs. (mem. com. 1978-93, cons.). Methodist. Achievements include development of ethylene cracking furnaces decoking process, waste oil recovery process, propylene storage process, depropanizer computor control process, differential thermal analysis methods for fusion bond epoxy coatings extent of cure. Home: 1116 Deats Rd Dickinson TX 77539-4426 Office: 100 E Nasa Blvd Ste 203 Webster TX 77598-5330

STEPHENS, JACK EDWARD, civil engineer, educator; b. Eaton, Ohio, Aug. 17, 1923; s. Harry M. and Mary Elizabeth (Galloway) S.; m. Virginia May Ives, June 19, 1948; children: Jay Edward, Jerry Edward, Jill Louise, Jana Lynn. BS in Engring., U. Conn., 1947; MS in Engring., Purdue U., 1955, PhD, 1959. Registered profl. engr., Conn. Jr. hwy. engr. Conn. Dept. Hwys., New Haven, 1949-50; instr. U. Conn., Storrs, 1947-48, asst. prof., then assoc. prof. civil engring., 1950-62, prof. civil engring., 1962-88, head civil engring. dept., 1965-72, prof. emeritus, 1989—; soils cons. A.J. Macchi Engrs., Hartford, Conn., 1958-65; pavement cons. Conn. Dept. Hwys., Hartford, 1962-63, Consumers Union Auto Test Facility, Colchester, Conn., 1991—; prin. Jack E. Stephens Soil and Materials Test Lab., Storrs, 1958—. Contbr. jour. articles to Procs. Assn. Asphalt Paving Tech., Trans. Rsch. Bd., others. Cpl. U.S. Army, 1943-46, ETO. Fellow Automobile Safety Found., Washington, 1958-59; recipient citation for teaching excellence Western Electric Fund, Washington, 1974. Mem. ASCE (life, B. Wright award Conn. sect. 1989), NSPE, AAUP, Assn. Asphalt Paving Tech. (life), Conn. Acad. Sci. and Engring. (chmn. transp. com. 1984—), Am. Rd. and Transp. Bldrs. Assn., Transp. Rsch. Bd., Am. Assn. Engring. Edn., Am. Soc. for Photogrammetry and Remote Sensing, Sigma Xi, Phi Kappa Phi, Chi Epsilon. Office: Univ Conn Dept Civil Engring Box U 37 Storrs CT 06269-3037

STEPHENS, JOHN WALTER, facility engineer; b. Castle AFB, Calif., Feb. 6, 1968; s. Joe Autrey and Emogene (Wallis) S. BS in Engring., Calif. State U., Northridge, 1992. Cert. engr.-in-tng., Calif. Civil engring. technician PL/TOE Civil Svc., Edwards, Calif., 1989-92; facility engr. Honda R & D, Cantil, Calif., 1992—. Mem. ASCE (treas. 1991-92), Toastmasters Internat. Mem. Christian Ch. Home: PO Box 1022 Rosamond CA 93560

STEPHENS, LARRY DEAN, engineer; b. Sterling, Colo., Sept. 1, 1937; s. John Robert and Shirley Berniece (Rudel) S.; m. Carol Ann Wertz, Sept. 1, 1957 (div. May 1975); children: Deborah Lynn, Janell Diane, Dana Larry, Hilary Elizabeth Melton. BS in Engring., Colo. State U., 1960; MBA, U. Colo., 1967. Registered profl. engr., Colo. Engr. Bur. Reclamation, Denver, 1960-90, cons., 1991—; exec. v.p. U.S. Com. on Irrigation and Drainage, Denver, 1971—; exec. dir. U.S. Com. on Large Dams, Denver, 1986—. V.p. Internat. Commn. on Irrigation and Drainage, 1989-92. With USNG, 1961-62. Mem. Am. Soc. Agrl. Engrs., Assn. State Dam Safety Officials, Colo. River Water Users Assn., Coun. on Engring. and Sci. Soc. Execs. Republican. Methodist. Home: 1625 Larimer St Apt 1505 Denver CO 80202-1532 Office: USCID 1616 17th St Ste 483 Denver CO 80202-1277

STEPHENS, LAURENCE DAVID, JR., linguist, consultant; b. Dallas, July 26, 1947; s. Laurence D. Sr. and Amy Belle (Schickram) S.; m. Susan Leigh Foutz, Apr. 16, 1988; 1 child, Laurence David III. MA, Stanford U., 1972, PhD, 1976. Vis. fellow Yale U., New Haven, Conn., summer 1979; rsch. fellow U. S.C., Columbia, 1980; asst. prof. U. N.C., Chapel Hill, 1982-88, assoc. prof., 1989—. Co-author: Two Studies in Latin Phonology, 1977, Language and Metre, 1984; editor (annual vol.) L'Année Philologique, 1987-92; contbr. over 60 articles to profl. jours. Mem. Nat. Trust for Hist. Preservation, Washington, 1989—, Dallas Opera Guild, 1992—, Metro. Opera Guild, N.Y.C., 1992—. Recipient L'Année Philologique, NEH, 1987-89, 89-91, 91-93. Mem. Greek and Latin Linguistic Assn. (chmn. 1987-92), Linguistic Soc. Am., N.Y. Acad. Scis., Indogermanische Gesellschaft, Société Internationale de Bibliographie Classique, Sigma Xi. Achievements include discovery of language universal regularities concerning labiovelar phonemes, laws of palatalization, the law of catathesis in Greek (pitch lowering); co-developer of the Justeson-Stephens probability distribution for chance cognates between unrelated languages, the Justeson-Stephens probability distribution of the numbers of vowels, consonants, and total phonological inventory size in the languages of the world; research in the law of the quantitative form of diachronic polysemy growth, semantic universals of aspect and modality. Home: 3319 Greenbrier Dr Dallas TX 75225 Office: Univ NC Chapel Hill Dept Classics CB # 3145 212 Murphey Hall Chapel Hill NC 27599

STEPHENS, OLIN JAMES, II, naval architect, yacht designer; b. N.Y.C., N.Y., Apr. 13, 1908; s. Roderick and Marguerite (Dulon) S.; m. Florence Reynolds, Oct. 21, 1930; children: Olin James III, Samuel R. Student, MIT, 1926-27; M.S. (hon.), Stevens Inst. Tech.; M.A. (hon.), Brown U.; D laurea ad honorem in Arch., Universario Architetura, Venice, Italy. Draftsman Henry J. Gielow, N.Y.C., 1927-28; draftsman P.L. Rhodes, N.Y.C., 1928; formed with Drake H. Sparkman firm of Sparkman & Stephens, 1928, Inc., 1929, chief designer, 1929-78; faculty mem. Royal Designers for Ind., London. Yachts designed include Dorade, 1930, Stormy Weather, 1934, Lulu, Ranger, (with W. Starling Burgess), 1937, Baruna, Blitzen, Goose, 1938, Vim, Gesture, 1939, Llanoria, 1948, Finisterre, 1954, Columbia, 1958, Constellation, 1964, Intrepid, 1967, Charisma, Morning Cloud, 1971, Courageous, 1974, Enterprise, 1977, Freedom, 1979, others; design agt. U.S. Navy, 1939—. Recipient David Taylor medal Soc. Naval Architects & Marine Engrs., 1959, Beppe Circle award Internat. Yacht Racing Union, 1992, Gibbs Brothers medal NAS, 1993. Fellow Soc. Naval Architects and Marine Engrs. (David W. Traylor medal 1959); mem. Am. Boat and Yacht Coun. (pres. 1959, 60), N.Am. Yacht Racing Union, Offshore Racing Coun. (chmn. internat. tech. com. 1967-73, 76-79), N.Y. Yacht (tech. com. 1989—), Manhasset Bay Club (hon.), Cruising Club of Am. (tech. com. 1989—), Royal Ocean Racing Club (Eng.), Royal Thames Yacht Club (London). Clubs: N.Y. Yacht (tech. com. 1989—), Manhasset Bay (hon.), Cruising Club of Am. (tech. com. 1980—); Royal Ocean Racing (Eng.); Royal Thames Yacht (London). Home: 80 Lyme Rd Apt 160 Kendal at Hanover Hanover NH 03755 Office: 79 Madison Ave New York NY 10016-7802

STEPHENS, PAUL ALFRED, dentist; b. Muskogee, Okla., Feb. 28, 1921; s. Lonny and Maudie Janie (Wynn) S.; m. Lola Helena Byrd, May 7, 1950; children: Marsha Stephens Wilson, Paul Alfred Jr., Derek M. BS cum laude, Howard U., 1942, DDS, 1945. Instr. dentistry Howard U., Washington, 1945-46; gen practice dentistry Gary, Ind., 1947—; chmn. bd. Assocs. Med. Ctr., Inc., Gary; Sec. Gary Ind. Sch. Bldg. Corp., 1967-85; pres. Bd. Health, 1973-81; Ind. State Bd. Dental Examiners, 1975-83. Mem. adv. bd. Ind. U.-Purdue U. Calumet Campus, 1973; bd. dirs. Urban League Northwest Ind.; pres. Gary Ednl. Devel. Found., 1990—. With AUS, 1942-44. Fellow Internat. Coll. Dentists, Acad. Dentistry Internat., Acad. Gen. Den-tistry (pres. chpt. 1973, nat. chmn. dental care com. 1977, Midwestern v.p., nat. bd. dirs. 1984-89, v.p. 1990-91, pres. 1992-93), Am. Coll. Dentists; mem. ADA, Nat. Dental Assn., N.W. Ind. Dental Assn. (bd. dirs., pres. 1976-77, Disting. Svc. award 1993), Am. Soc. Anesthesia in Dentistry, Am. Acad. Radiology, Gary C. of C., Alpha Phi Alpha (pres. Gary Ednl. Found. 1988, pres. Gary Ednl. Devel. Found. 1990—). Baptist. Home: 1901 Taft St Gary IN 46404-2759 Office: 2200 Grant St Gary IN 46404-3439

STEPHENS, SHERYL LYNNE, family practice physician; b. Huntington, W.Va., Dec. 11, 1949; d. William Clayton Stephens and Virginia Eleanor (Hatten) Stephens Terry; 1 child, William Earl Hicks III (dec.); m. Lannie Dale Rowe, Jan. 17, 1981; 1 child, Seton Christopher. BA, U. Ky., 1972; MA, Marshall U., 1982, MD, 1988. Tchr. Wayne County Bd. Edn., Ceredo, W.Va., 1973-83; real estate developer Huntington, 1981-88; resident in family practice Grant Med. Ctr., Columbus, Ohio, 1988-91; gen. practice physician Columbus (Ohio) Health Dept., 1991—; med. dir. Billie Brown Jones Family Health Ctr., 1992—; researcher, 1976-81. Counselor, instr. Contact of Huntington, 1975-88; polit. activist pro choice movement and ratification of equal rights amemdment, 1976-81. Recipient Leadership award Marshall U., 1985. Mem. Am. Assn. Family Practitioners (pres. 1984-85, Leadership award 1985), Am. Med. Women's Assn. (sec. 1985-86), NOW (pres. 1976-78, 79-81, v.p. Huntington 1978-79, sec. 1981-82), Nat. Abortion Rights Action League. Democrat. Avocations: horseback riding, reading, boating, skiing (snow and water), travel. Home: 703 French Dr Columbus OH 43228-2907 Office: Columbus Health Dept 181 S Washington Ave Columbus OH 43215-5327

STEPHENS, SIDNEY DEE, chemical manufacturing company executive; b. St. Joseph, Mo., Apr. 26, 1945; s. Lindsay Caldwell and Edith Mae (Thompson) S.; m. Ellen Marie Boeh, June 15, 1968 (div. 1973); m. Elizabeth Ann Harris, Sept. 22, 1973; 1 child, Laura Nicole. BS, Mo. Western State U., 1971; MA, U. Houston, 1980. Assoc. urban planner Met. Planning Commn., St. Joseph, Mo., 1967-71; prodn. acctg. assoc. Quaker Oats Co., St. Joseph, 1971-72, office mgr., pers. rep., Rosemont, Ill., 1972-73, employee and community rels. mgr., New Brunswick, N.J., 1973-75, Pasadena, Tex., 1975-80; site pers. mgr. ICI Americas, Inc., Pasadena, Tex., 1980-90, regional mgr. human resources agrl. products div., 1990-93; regional mgr. human resources Zeneca Inc., 1993—; pvt. practice mgmt. cons., Houston, 1981—. Contbr. articles to profl. jours. With USNR, 1963-65. Mem. Soc. for Human Resources Mgmt., Houston Pers. Assn. (community and govtl. affairs com. 1984-85, 85-86). Republican. Methodist. Home: 16446 Longvale Dr Houston TX 77059-5420 Office: ICI Americas Inc 5757 Underwood Rd Pasadena TX 77507-1031

STEPHENS, THOMAS WESLEY, biochemist; b. Shreveport, La., July 3, 1950; s. Gordon Wesley Stephens and Urma Juanita (Harmon) Mundell; m. Brenda Sue Bruton, July 7, 1972; children: Alissa Leigh, Adam Wesley. BS, Rose Hulman Inst. Tech., 1972; PhD, Ind. U., Indpls., 1985. Med. technologist St. Vincent Hosp., Indpls., 1972-74; mgr. tech. svc. Medico Electronic Inc., Indpls., 1974-78; supr. rsch. Am. Monitor Corp., Indpls., 1978-85; sr. scientist Rice U., Houston, 1985-87; rsch. scientist Eli Lilly & Co., Indpls., 1987—; adj. asst. prof. dept. biochemistry Sch. Medicine U. Ind., Indpls., 1989—. Contbr. articles to profl. jours. Mem. AAAS, Am. Soc. Clin. Pathologists, Am. Diabetes Assn., Biochem. Soc. Achievements include patents for method of detection of chloride in plasma, method of detection of creatinine in plasma, method or analytical numerical determination of carryover. Home: 7412 Riverbirch Ln Indianapolis IN 46236 Office: Lilly Rsch Labs Lilly Corp Ctr Indianapolis IN 46285

STEPHENS, WILLIAM EDWARD, electronics engineer; b. Bellevue, Nebr., Nov. 8, 1953; s. Lawrence Edward and Virginia Gladys (Lund) S.; m. Monica Louise Blunt, July 9, 1983; children: Thomas Edward, Karen Elizabeth. BSEE, U. So. Calif., L.A., 1977; MSEE, U. So. Calif., 1980, PhD in Elec. Engring., 1981. Elec. engr. Gen. Dynamics-Convair Div., San Diego, 1977-78; rsch./teaching asst. U. So. Calif., L.A., 1978-81; mem. tech. staff TRW Inc., Redondo Beach, Calif., 1981-85, staff engr., 1985-86; dir. Bell Comm. Rsch., Morristown, N.J., 1986—; adj. prof. U. Calif., Irvine, 1992, Berkeley, 1993. Contbr. articles to profl. jours. Biegler Meml. awardee U. So. Calif., 1977. Mem. IEEE (sr., mem. tech. program com. Globecom conf. 1992—, tech. program com. Electronic Components and Tech. Conf. 1988-90, tech. editor Lightwave Telecomm. 1988-92), Sigma Xi, Eta Kappa Nu, Tau Beta Pi. Republican. Presbyterian. Achievements include research in 1.3 Mm fiber optical link operating at 5 GHZ; demonstration of HDTV over 622 mbps optical communications link; demonstration of 1.3 Mm optical switch using acousto-optic principals; demonstration of 155 mbps data signals over 100m of datagrade unshielded twisted pair copper cable. Office: Bellcore 445 South St Morristown NJ 07960

STEPHENSON, EDWARD THOMAS, consulting metallurgist; b. Atlantic City, Nov. 7, 1929; s. Edward Thomas and Alberta Phillips (Adams) S.; m. Irma Patricia Geiger, June 27, 1953; children: William, Mark, Thomas, Debra. BS in Metall. Engring., Lehigh U., 1951, PhD in Metall. Engring., 1965; MS in Metallurgy, MIT, 1956. Registered profl. engr., Pa. Rsch. engr. Bethlehem (Pa.) Steel Corp., 1956-73, sr. rsch. engr., 1973-80, sr. scientist, 1980-88, rsch. fellow, 1988-92; metallurgy cons., Bethlehem, 1992—; instr. Metal Engring. Inst., Cleve., 1988-92; adj. prof. Lafayette Coll., 1992—. Contbr. over 30 articles to tech. jours. Lt. (j.g.) USN, 1951-54. Fellow Am. Soc. for Metals Internat. (various publ. coms. 1965-91, Marcus A. Grossman award 1965, Bradley Stoughton award 1988); mem. AIME. Achievements include determination of strengthening mechanism in V-bearing HSLA steel, effect of residual elements on steel; improvement of steel for motors. Home and Office: 1852 Levering Pl Bethlehem PA 18017

STERGIOU, KONSTANTINOS, biological oceanographer, researcher; b. Athens, Greece, Mar. 19, 1959; s. Iordanis and Eleni (Akrivou) S. BSc in Biology, U. Thessaloniki, Greece, 1981; MSc in Oceanography, McGill U., Montreal, Can., 1984; PhD in Icthyology (hon.), U. Thessaloniki, 1991. Researcher Nat. Ctr. for Marine Rsch., Athens, 1985-87, 1991—; invited lectr. Hellenic Ctr. Productivity, Athens, 1986-93. Contbr. articles to sci. jours. Served with Hellenic Army, 1988-89. Mem. Am. Soc. Limnology and Oceanography, Am. Soc. Naturalists, Ecol. Soc. Am., Brit. Ecol. Soc., Fisheries Soc. Brit. Isles, Greek Union Oceanographers, Network Tropical Fisheries Scientists. Greek Orthodox. Achievements include research in climatic changes and their effect on fisheries, in fisheries forecasting, in fish biology and population dynamics. Office: Nat Ctr for Marine Rsch, Agios Kosmas Hellinikon, 16604 Athens Greece

STERMER, RAYMOND ANDREW, agricultural engineer; b. Barclay, Tex., July 22, 1924; s. Henry Jacob and Roseline (Voltin) S.; m. Gladys Freida Hoelscher, June 15, 1948; children: Linda Gayle, Nancy Louise. BS, Tex. A&M U., 1950, MS, 1958, PhD, 1971. Registered profl. engr., Tex. Agrl. engr. Soil Conservation Svc. USDA, various cities, Tex., 1950-55; agrl. engr. Agrl. Mktg. Svc. USDA, College Station, Tex., 1955-62, agrl. engr. Agrl. Rsch. Svc., 1962-89; agrl. engr. Tex. Agrl. Exptl. Sta.-Tex. A&M U., College Station, 1989—. Contbr. to profl. publs. Mem. Am. Soc. Agrl. Engrs., Sigma Xi. Office: Agrl Engr Dept Tex A&M Univ College Station TX 77843-2117

STERN, E. GEORGE, architect, educator; b. Wuerzburg, Germany, Aug. 10, 1912; came to U.S., 1936; s. Bruno and Frida (Hellman) S.; m. Marianne Erlichman, 1940 (dec. 1942); 1 child, Jean Carol; m. Marianne Stamm, Apr. 7, 1943; children: John Peter, Robert Jay. Dipl.Ing., Tech. U., Munich, 1936; MS in Architectural Engring., Pa. State U., 1938, PhD in Architectural Engring., 1941. Registered architect, Va., Pa. Rsch. engr. Va. Poly. Inst. and State U., Blacksburg, 1941-46, assoc. prof., 1946-48, prof., 1948-53, Earle B. Norris rsch. prof. wood constrn., 1953-79, Earle B. Norris prof. emeritus, 1979—; cons. Nat. Wooden Pallet and Container Assn., Arlington, Va., 1950—, Nat. Ornamental and Miscellaneous Metals Assn., Forest Park, Ga., 1979-92. Author: All-Nailed Trussed Rafters, 1941; editor: Anchorage to Concrete; contbr. over 900 tech. papers to profl. publs., more than 160 rsch. reports in profl. bulls. Pres. Piedmont II Home Owners Assn., Blacksburg, 1988—. Goodwill amb. U.S. Dept. State, 1961, 62; recipient Disting. Svc. award Nat. Wooden Pallet and Container Assn., 1978, Cert. of Appreciation Va. Poly. Inst. and State U., 1989. Fellow ASTM (com. chmn., award of merit 1985); mem. ASCE (life), ASME (com. chmn.), Internat. Coun. Bldg. Rsch. Studies Documentation, Internat. Standard Orgn., Internat. Union Forestry Rsch. Orgn., Forest Products Rsch. Soc., Forest Products Soc. Achievements include development of more effective mechanical fasteners for wood construction, design of roof constructions, anchorage systems, railing systems and rails for buildings; improving designs for wooden pallets, testing procedures.*. Home: PO Box 361 Blacksburg VA 24063-0361 Office: Va Poly Inst and State U Brooks Ctr Ramble Rd Blacksburg VA 24061-0503

STERN, ERIC PETRU, chemist; b. Arad, Romania, Jan. 1, 1941; s. Oscar and Palma (Friedman) S.; 1 child, Erika Cindy. Student, Humboldt U., Berlin, German Dem. Republic; diploma, Bucharest (Romania) U., 1963; MBA, Concordia U., Montréal, Que., Can., 1984. Chemist ICECHIM Research Inst., Bucharest, 1962-69; plant supr. Industria Lechera/Nestlé, Cayambe, Ecuador, 1969-73; chemist BPCO Inc. (formerly Esso Bldg. Products), Lasalle, Que., 1974-90; cons. EXPERTECH CMSC BPCO Inc., 1989-90; cons., owner EXPERTECH CMSC, 1990—; mem. fire test com. Underwriter Can., Toronto, Ont., Can., 1975-77, ASTM Philadelphia, 1979—; abstractor Chem. Abstracts, Columbus, Ohio, 1973—. Contbr. articles to profl. jours.; patentee in field. Mem. Soc. Plastic Engrs., Am. Chem. Soc., Quebec Order of Profl. Chemists. Avocations: computers, networking, creativity, walking. Home and Office: 5386 W Broadway, Montreal, PQ Canada H4V 2A4

STERN, MARTIN O(SCAR), physicist, consultant; b. Essen, Germany, Mar. 25, 1924; came to U.S., 1940; s. Max and Beate (Herzberg) S.; m. Charlotte Gutmann, Aug. 19, 1951; children: Pia Renata, Marc Owen, Roger Michael. BA, U. Calif., Berkeley, 1947, PhD, 1951. Postdoctoral fellow Lawrence Lab., Berkeley, Calif., 1951-52; rsch. assoc. Carnegie-Mellon U., Pitts., 1952-56; sr. staff mem., asst. chmn. physics dept. Gen. Atomic, La Jolla, Calif., 1956-69; sr. staff mem. Internat. Rsch. and Tech., Washington, 1969-77; cons. San Diego, 1977—. Contbr. articles to profl. jours. Bd. dirs. Citizens Coordinate, San Diego. With U.S. Army, 1943-46. Home and Office: 3143 Bremerton Pl La Jolla CA 92037

STERN, MICHAEL, chemist; b. Moscow, Feb. 26, 1969; came to U.S., 1981; s. Walter Bella (Pritsker) S. BS in Biochemistry, Phila. Coll., 1991. Rsch. asst. Phila. Coll. of Pharmacy and Sci., 1989-91; chemist U.S. Dept. Agrl., Beltsville, Md., 1992—; safety officer U.S. Dept. Agrl., Beltsville, 1992—, environ. coord., 1992—. Contbr. articles to profl. jours. Vol. Phila. Engring. Soc. of New Ams., 1985, Golden Slipper Uptown Home for the Aged, Phila., 1984. Scholarship Phila. Coll. Pharmacy, 1987-91. Mem. AAAS, Am. Chem. Soc., Smithsonian Inst. Republican. Jewish. Office: USDA-ARS Bldg 161 Rm 104 BACR-East Beltsville MD 20705

STERN, ROBERT MORRIS, psychology educator and psychophysiology researcher; b. N.Y.C., June 18, 1937; s. Irving Dan and Nellie (Wachstetter) S.; m. Wilma Olch, June 19, 1960; children—Jessica Leigh, Alison Rachel. A.B., Franklin and Marshall Coll., 1958; M.S., Tufts U., 1960; Ph.D., Ind. U., 1963. Research assoc. dept. psychology Ind. U., 1963-65; asst. prof. psychology Pa. State U., 1965-68, assoc. prof., 1968-73, prof., 1973—, disting. prof., 1992—, head dept., 1978-87. Author: (with W.J. Ray) Biofeedback, 1977, (with W.J. Ray and C.M. Davis) Psychophysiological Recording, 1980, (with K.L. Koch) Electrogastrography, 1985; contbr. articles to profl. jours. Recipient Nat. Media award Am. Psychol. Found., 1978. Mem. Am. Psychol. Soc., Aerospace Med. Assn., Soc. Psychophysiol. Rsch., Am. Gastroent. Assn. Home: 1360 Greenwood Cir State College PA 16803-3232 Office: Pa State U 512 Moore Bldg University Park PA 16802

STERN, ROBIN LAURI, medical physicist; b. Urbana, Ill., Mar. 12, 1959; d. Morris Stern and Myrna (Tanzer) Stern Longenecker; m. Donald Neil Bittner, May 20, 1989. BA in Physics and German Studies, Rice U., 1981; MS in Physics, U. Mich., 1983, PhD in Physics, 1987. Rsch. assoc. Duke U., Durham, N.C., 1987-89; postdoctoral rsch. fellow U. Mich., Ann Arbor, 1989-91; asst. adj. prof. U. Calif., San Francisco, Davis, 1992—; cons. Scanditronix, Inc., Livonia, Mich., 1991. Contbr. articles to jours. Rev. of Sci. Instruments, Magnetic Resonance Imaging, Med. Physics. Argonne Nat. Lab. grantee, 1985-86. Mem. Am. Assn. Physicists in Medicine, Am. Soc. Therapeutic Radiology and Oncology, Sigma Pi Sigma, Phi Beta Kappa.

STERN, THOMAS LEE, physician, educator, medical association administrator; b. San Francisco, Jan. 14, 1920; s. Bernard Michael and Alice Sarah (Halberstadt) S.; m. Gladys Crawford, June 26, 1944; children: Donnel Bernard, Lee Crawford, Pamela Ann. BS, Willamette U., 1947; MD, U. Oreg., 1950; DSc (hon.), Med. Coll. Ohio, 1982. Rotating intern St. Vincents Hosp., 1950-51, resident in general surgery, 1951-52; tech. advisor Marcus Welby M.D.-TV, Universal City, Calif., 1970-75; residency dir. Santa Monica (Calif.) Hosp., 1969-74; lectr. preventive and social medicine UCLA, L.A., 1970-73; v.p. Am. Acad. Family Physicians, Kansas City, Mo., 1974-83; assoc. prof. family medicine Kans. U. Med. Ctr., Kansas City, 1975-83; pres. Internat. Ctr. for Family Medicine, Buenos Aires, 1982-83; prof. family medicine U. Fla. Med. Ctr., Gainesville, 1983-86; v.p. Am. Acad. Family Physicians Found., Kansas City, 1983-91, audio tape editor, 1991—; cons. U.S. Naval Med. Corps, U.S. Army Med. Corps, Washington, 1975-82. With USN, 1941-45, PTO. Recipient F. Marian Bishop award Soc. Tchrs. Family Medicine, 1991. Fellow Am. Acad. Family Physicians (Thomas Johnson award 1985, Award of Merit 1983, John Walsh award 1991); mem. AMA. Avocations: wine, computer. Home: 4441 W 124th Ter Shawnee Mission KS 66209-2280 Office: Am Acad Family Physicians Home Study 8800 Ward Pky Kansas City MO 64114-2707

STERNBERG, DAVID EDWARD, psychiatrist; b. Norfolk, Va., Jan. 18, 1946; s. Theodore and Bella (Rosenblatt) S.; m. Frances Toby Glazer; children: Jonathan Theodore, Daniel Alexander. BA in Biopsychology, U. Chgo., 1967; MD, Tufts U., 1971. Fellow in psychiatry Yale U., New Haven, 1972-75; staff psychiatrist, dir. alcohol rehab. Nat. Naval Med. Ctr., Bethesda, Md., 1975-77; rsch. coord., staff psychiatrist Biol. Psychiatry br. NIMH, Bethesda, 1977-79; asst. prof., chief clin. rsch. unit Yale U., New Haven, 1979-83; med. dir. Falkirk Hosp., Central Valley, N.Y., 1983-88, Kansas Inst., Olathe, 1988-90; dir. Assocs. for Psychiatry and Psychotherapy, Overland Pk., Kans., 1990—; lectr. Karl. Menninger Sch. Psychiatry, Topeka, 1988—, dept. psychiatry Yale U., New Haven, 1983—; assoc. clin. prof., U. Kans., Kansas City, 1988—. Author: Evaluation and Treatment of Drug Abuse, 1990, (with others) Dual Diagnosis: Addiction and Psychiatric Disorders, 1988; contbr. 87 articles to profl. jours. Lt. comdr. USN, 1975-77, comdr. USPHS, 1977-79. Mem. Am. Psychiat. Assn., Soc. for Biol. Psychiatry, Soc. Neurosci., Acad. Clin. Psychiatrists, Am. Acad. Psychiatrist in Alcoholism and Addictions. Avocations: running, swimming, tennis, classical music. Office: Assocs for Psychiatry Ste 850 6900 College Blvd Ste 850 Shawnee Mission KS 66211-1536

STERNBERG, MARK EDWARD, chemical engineer; b. Bklyn., Nov. 20, 1957; s. Irwin Martin and Hope Paula (Schimmel) S.; m. Diane Josephine Groszek, Aug. 4, 1984; 1 child, Jesse Michael. BS, Rutgers Coll., 1979. Chem. engr. Conoco, Ponca City, Okla., 1979-81; process engr. Conoco, Denver, 1981-82; staff engr. Conoco, Lake Charles, La., 1982-86; environ. coord. United Refining Co., Warren, Pa., 1986--; mem. com. for refinery environ. control Am. Petroleum Inst., Washington, 1987-92. Home: 511 Fourth Ave Warren PA 16365 Office: United Refining Co Bradley St Box 780 Warren PA 16365

STERNBERG, PAUL WARREN, biologist, educator; b. Queens, N.Y., June 14, 1956. BA in Biology and Math., Hampshire Coll., 1978; PhD in Biology, MIT, 1984. Rsch. asst. U. Pa., 1977-78; postdoctoral fellow U. Calif., San Francisco, 1984-87; asst. prof. biology Calif. Inst. Tech., Pasadena, 1987-92, assoc. prof. biology, 1992—; asst. investigator Howard Hughes Med. Inst., 1989-92, assoc. investigator, 1992—; adj. asst. prof. dept. anatomy and cell biology U. So. Calif. Sch. Medicine, 1989—. Mem. editorial bd. Molecular Biology of the Cell, 1989—, Mechanisms of Development, 1990—; contbr. articles, revs. to profl. publs., chpts. to books. Recipient Presl. Young Investigator award, NSF, 1988; named fellow Jane Coffin Childs Meml. Found. for Med. Rsch., 1984-85, Searle scholar, 1988-91. Fellow AAAS; mem. Am. Soc. Cell Biology, Genetics Soc. Am., Helminthological Soc. Washington, Internat. Soc. Neuroethology, Soc. for Devel. Biology, Soc. Nematologists. Office: Calif Inst Tech Mail Code 156-29 Pasadena CA 91125

STERNLIEB, CHERYL MARCIA, internist; b. N.Y.C.; d. Isidore and Ida (Stolpen) S.; m. Fred Kashan; children: Glenn, Melissa. BA, Harpur Coll., 1961; MD, N.Y. Med. Coll., 1965. Diplomate Am. Bd. Internal Medicine. Internship St. Joseph's Hosp., Syracuse, N.Y., 1965-66; resident in medicine Flower & Fifth Ave Hosps., Met. Hosp., N.Y.C., 1966-69; asst. attending physician Trafalgar Hosp., N.Y.C., 1969-74, assoc. attending physician, 1975-78; asst. adj. physician Bronx-Lebanon Hosp., 1970-72, adj. attending, 1972-73, assoc. attending, 1973-75; asst. attending physician St. Barnabus Hosp., 1972-75; attending physician Beth Israel Hosp.-North, 1974—; med. advisor Mandl Sch. of Med. Assts., N.Y.C., 1970-89; fellow N.Y. Acad. Scis., former chair biomed. sect. Contbr. articles to profl. jours. Recipient Am. Geriatrics Soc. fellowship, 1968-69. Fellow Am. Coll. Physicians, N.Y. Acad. Medicine. Office: 133 E 73rd St New York NY 10021

STEVENS, DALE MARLIN, civil engineer; b. Boyd, Minn., June 11, 1940; s. Leslie Dale Stevens and Hazel Margaret Anderson Neumann; m. Renee Joy Pflueger, Apr. 21, 1967 (div. 1977); children: Joan Marie, Jeffrey Michael; m. Marianne Jean Solting, Feb. 10, 1979. BSCE, S.D. State U., 1962; MSCE, S.D. Sch. Mines & Tech., 1973. Registered profl. engr., S.D., Ala. Commd. U.S. Army, 1962-83, advanced through grades to lt. col., 1979; staff engr. U.S. Army C.E., Huntsville, Ala., 1983-86; sr. engr. The BDM Corp., Huntsville, Ala., 1986-89; mgr. bus. devel. Wyle Labs., Inc., Huntsville, Ala., 1989; sr. systems engr. Teledyne Brown Engring., Huntsville, Ala., 1990—; mem. civil and environ. adv. coun. U. Ala., Huntsville, 1990—. Coun. mem. St. Mark's Luth. Ch., Huntsville, 1986-87, chmn., 1987. Named Civil Engr. of the Yr. Huntsville Assn. Tech. Socs., 1988-89. Mem. ASCE (br. pres. 1988), Am. Concrete Inst., Soc. Am. Mil. Engrs. (post v.p. 1991), Prestressed Concrete Inst., Ret. Officers Assn., Elks, Chi Epsilon. Republican. Lutheran. Home: 2527 Willena Dr Huntsville AL 35803 Office: Teledyne Brown Engring 300 Sparkman Dr Huntsville AL 35807-7007

STEVENS, (ERNEST) DONALD, zoology educator; b. Calgary, Alta., Can., July 5, 1941; s. Douglas Ernest and Lorna (Stuart) S.; m. Elinor Mae Hagborg, Feb. 21, 1964; children: Kenneth Harold, Wendy Leilani. BSc, U. Victoria, B.C., 1963; MSc, U. B.C., Vancouver, 1965, PhD, 1968. Rsch. assoc. Stanford (Calif.) U., 1968; from asst. to assoc. prof. U. Hawaii, Honolulu, 1968-75; assoc. prof. U. Guelph, Ont., 1975-88, prof., 1988—. Author over 80 sci. papers & book chpts. on comparative physiology. Mem. Soc. for Exptl. Biology, Can. Soc. Zoology. Office: U of Guelph, Guelph, ON Canada N1G 2W1

STEVENS, DONALD KING, aeronautical engineer, consultant; b. Danville, Ill., Oct. 27, 1920; s. Douglas Franklin and Ida Harriet (King) S.; B.S. with high honors in Ceramic Engring., U. Ill., 1942; M.S. in Aeros. and Guided Missiles, U. So. Calif., 1949; grad. U.S. Army Command and Gen. Staff Coll., 1957, U.S. Army War Coll., 1962; m. Adele Carman de Werff, July 11, 1942; children: Charles August, Anne Louise, Alice Jeanne Stevens Kay. Served with Ill. State Geol. Survey, 1938-40; ceramic engr. Harbison-Walker Refractories Co., Pitts., 1945-46; commd. 2d lt. U.S. Army, 1942, advanced through grades to col., 1963; with Arty. Sch., Fort Bliss, Tex., 1949-52; supr. unit tng. and Nike missile firings, N.Mex., 1953-56; mem. Weapons Systems Evaluation Group, Office Sec. of Def., Washington, 1957-61; comdr. Niagara-Buffalo (N.Y.) Def., 31st Arty. Brigade, Lockport, N.Y., 1963-65; study dir. U.S.A. ballistic missile def. studies DEPEX and X-66 for Sec. Def., 1965-66; chief Air Def. and Nuclear br. War Plans div. 1965-67, chief strategic forces div. Office Dep. Chief Staff for Mil. Ops., 1967-69; chief spl. weapons plans, J5, U.S. European Command, Fed. Republic Germany, 1969-72, ret., 1972; guest lectr. U.S. Mil. Acad. 1958-59; cons. U.S. Army Concepts Analysis Agy., Bethesda, Md., 1973—; cons. on strategy Lulejian & Assocs., Inc., 1974-75; cons. nuclear policy and plans to Office Asst. Sec. of Def., 1975-80, 84—; cons. Sci. Applications, Inc., 1976-78; Asst. camp dir. Piankeshaw Area coun. Boy Scouts Am., 1937; mem. chancel choir, elder First Christian Ch., Falls Church, Va., 1957-61, 65-69, 72—; elder, trustee Presbyn. Ch., 1963-65. Decorated D.S.M., Legion of Merit, Bronze Star. Mem. Am. Ceramic Soc., Assn. U.S. Army, U. Ill. Alumni Assn., U. So. Calif. Alumni Assn., Keramos, Sigma Xi, Sigma Tau, Tau Beta Pi, Phi Kappa Phi, Alpha Phi Omega. Clubs: Niagara Falls Country; Ill. (Washington); Terrapin. Lodge: Rotary. Contbr. articles to engring. jours.; pioneer in tactics and deployment plans for Army surface-to-air missiles. Address: 5916 5th St N Arlington VA 22203 also: Woodmont Bldg Rm # 300 Bethesda MD 20014

STEVENS, DONNA JO, nuclear power plant administrator; b. Chattanooga, Feb. 16, 1958; d. James Donald and JoAnne (Smartt) S. AS in Nuclear Technology, Chattanooga State Tech. C.C., 1980. Health physics technician TVA, Muscle Shoals, Ala., 1980; health physics technician Sequoyah Nuclear Plant TVA, Soddy Daisy, Tenn., 1981-86, sr. lead health physics technician, 1986-90, shift supr. field ops. Sequoyah Nuclear Plant, 1990—. Author, editor: Radiation Worker Handbook, 1992. Baptist. Home: 3418 Whittaker Ave Chattanooga TN 37415

STEVENS, ELIZABETH, psychotherapist; b. Evanston, Ill., Jan. 11, 1950; d. Kenneth M. and C. Jane (Reynolds) S.; m. David W. Handy, Oct. 3, 1986. BA in Psychology, U. Fla., 1973; MA in Clin. Psychology, Kent State U., 1976. Lic. prof. counselor, Tex., cert. chem. dependency specialist, lic. marriage and family therapist. Exec. dir. Genesis Women's Shelter, Dallas, 1986-87; dir. outpatient svcs. Green Oaks Hosp., Dallas, 1987-88; mgmt. cons. Houston, 1977—, pvt. practice, 1990—; cons. St. Joseph Hosp., 1977—; founder, N.E. Hospice, Med. Affiliates, Support N.E. Cancer Workers; co-founder Emergency Support System for Police, Fire Dept. and Ambulance Svc.; ambassador St. Josephs Hosp. Contbr. articles to profl. jours., mags., and newspapers. Vol. Mental Health Assn., Houston and Harris County, bd. dirs., 1988—, chair nominating com., membership com., sec. exec. com.; bd. advisors N.E. Hospice; co-founder Associated Mental Health Group. Named Exceptional Vol. of Yr. Mental Health Assn., Speakers Bur. award. Mem. Walden Country Club. Office: Stevens Counseling Ctrs 9810FM 1960 Ste # 205 Humble TX 77338

STEVENS, HERBERT HOWE, mechanical engineer, consultant; b. Gardiner, Maine, May 12, 1913; s. Herbert Howe and Marjorie (Allen) S.; m. Elaine Stevens, 1946; children: Jane, India, Charlotte. BSME, Ga. Tech. Inst., 1936; MA in Liberal Studies and Arts, New Sch., 1969; MS in Physics, Southeastern Mass. U., 1986; MA in Profl. Writing, U. Mass., North Dartmouth, 1993. Registered profl. engr., Mass. Housing heating researcher Pierce Found., N.Y.C., 1938-39; engr. NYU, Bronx, 1941-47, Walter Balfour & Co., Bklyn. and L.I. City, N.Y., 1947-53, 65-81, Schick Inc., Champion Inst., N.Y.C. and Stamford, Conn., 1947-58; nuclear fuels engr. M & C div. Tex. Instruments, Attleboro, Mass., 1958-65; cons. engr. H.H. Stevens R & D Engrs., Westport, Mass., 1941—. Contbr. articles to profl. publs. Chmn. Westport Sewer Study Com., 1978-92. Mem. ASME, Philosophy of Sci. Assn., Inst. of Religion in an Age of Sci. Mem. Soc. of Friends. Achievements include patents for sliding wall motor operator, air supported roof. Home: 218 Hix Bridge Rd Westport MA 02790

STEVENS, JOHN GEHRET, chemistry educator; b. Mt. Holly, N.J., Dec. 16, 1941; s. Robert Bucy and Helen Stevens; m. Virginia Entwistle; children: Shelly, Robby, John G. BS, N.C. State U., 1964, PhD, 1969. Rsch. prof. U. Nijimegan, The Netherlands, 1976-81; asst. prof. U. N.C., Asheville, 1963-68, assoc. prof., 1973-79, prof., 1979—; dir. Mossbauer Effect Data Ctr., Asheville, 1970—; nat. exec. officer Coun. on Undergrad. Rsch., Asheville, 1991—. Contbr. 100 articles to profl. jours. Deacon, elder Presbyn. Ch.; scoutmaster, organizer troop 24 Boy Scouts Am.; mem. parent coun. N.C. Sch. Sci. and Math.; coord. neighborhood playground devel. programs, Advocates for Nuclear Arms Freeze, Western N.C. Coalition of Social Concerns; chmn. Asheville Presbytery's Peacemaking Task Force and Campus Ministries Com.; past trustee Kings Coll. Mem. Internat. Commn. Applications of Mossbauer Effect, Sigma Pi Sigma, Sigma Xi. Office: U NC Dept Chemistry One University Heights Asheville NC 28804

STEVENS, JOHN LAWRENCE, quality assurance professional; b. Cleve., Apr. 28, 1948; s. John Paul and Harriett (Dunmire) S.; m. Diana Gayle Ferrero, Apr. 8, 1972; children: Danielle R., Kristen M. BA in Biol. Sci., Calif. State U., Fullerton, 1970; postgrad., Golden State U., 1982-83. Investigator, pub. affairs officer FDA, L.A., 1972-82; mgr. regulatory and clin. affairs Heyer Schulte div.Am. Hosp. Supply, Santa Barbara, Calif., 1982-84;

mgr. regulatory affairs Unitek div. Bristol Myers Co., Monrovia, Calif., 1984-86; v.p. regulatory affairs and quality assurance Retroperfusion Systems, Inc., Costa Mesa, Calif., 1986-90, Edwards LIS div. Baxter Healthcare, Irvine, Calif., 1990-93, Cardiac Pathways Corp., Sunnyvale, Calif., 1993—. Recipient commendable svc. award FDA, 1975. Mem. Regulatory Affairs Professional Soc. (pres. west sect. 1981-84), Am. Soc. for Quality Control, Am. Heart Assn., N.Y. Acad. Scis. Home: 551 Santa Rita Ave Palo Alto CA 94301 Office: Cardiac Pathways Corp 670 Almanor Ave Sunnyvale CA 94086

STEVENS, JOHN RICHARD, architectural historian; b. Toronto, Ont., Can., Mar. 19, 1929; came to U.S., 1954; s. Walter John and Florence Rosalie (Warr) S.; m. Marion Frances Moore, May 7, 1964. Student, Columbia U., 1966-67. Comml. artist, tech. illustrator Toronto, Ont., Can. and New Haven, Conn.; asst. to curator Mystic Seaport, 1957; curator Maritime Mus. Can., Halifax, 1960-63; with dept. no. affairs Hist. Sites Divsn., 1963-66; surveyor early bldgs. Halifax, Quebec, Fredericton, Woodstock, St. John River Valley, Ea. Twps. of Quebec; lighthouses of Great Lakes, Nova Scotia, New Brunswick; with Archtl. Heritage, Inc., 1967-70; prin. John R. Stevens Assocs., Greenlawn, N.Y., 1970—; cons. hist. restoration Old Bethpage Village Restoration, 1967—, Soc. for the Preservation L.I. Antiquities, Roslyn Preservation Corp., Smithtown Hist. Soc., Colonial Farmhouse Restoration Soc., numerous others; restorations include Van Nostrand-Starkins House, Revolutionary War "Arsenal", two c.1900 Am.-built streetcars for City of Detroit, 1976, 80, 1878 N.Y. elevated railroad car, 1983, first electric freight locomotive, 1988, c. 1880 horsecar for Rochester Mus., 1987; lectr. Dutch-Am. bldgs., street railway history. Author: Old Time Ships, 1949, H.M. Schooner Tecumseth, 1961, Ships of the North Shore, 1963, (guidebook) Ride Down Memory Lane, 1965, 2d rev. edit., 1984, Early History of Street Railways - The New Haven Area, 1982, The Derby Horse Railway and the World's First Electric Freight Locomotive, 1987; co-author/editor: Pioneers of Electric Railroading, 1991; contbr. articles on hist. bldg. tech., book revs. to profl. publs. Bd. trustees Roslyn Landmark Soc., 1980—. With U.S. Army, 1955-57. Mem. Branford Electric Rlwy. Assn. (bd. trustees 1957, 74, 80-81, 83-85, supt. equipment 1974-75, supt. bldgs. and grounds 1980-82, chmn. bd. trustees 1983, pres. 1984, 85, contbr. articles to jour.). Home and Office: 1 Sinclair Dr Greenlawn NY 11740

STEVENS, KENNETH NOBLE, electrical engineering educator; b. Toronto, Ont., Can., Mar. 23, 1924; came to U.S., 1948, naturalized, 1962; s. Cyril George and Catherine (Noble) S.; m. Phyllis Fletcher, Jan. 19, 1957; children: Rebecca, Andrea, Michael Hugh, John Noble. B.A.Sc., U. Toronto, 1945, M.A.Sc., 1948; Sc.D., MIT, 1952. Inst. U. Toronto, 1946-48; faculty MIT, Cambridge, 1948—; prof. elec. engring. MIT, 1963—, Clarence J. Lebel prof., 1977—; Vis. fellow Royal Inst. Tech., Stockholm, 1962-63; cons. to industry, 1952—; vis. prof. phonetics U. Coll., London, 1969-70; mem. Nat. Adv. Council on Neurol. and Communicative Disorders and Stroke NIH, 1982-86. Author: (with A.G. Bose) Introductory Network Theory; Contbr. articles to profl. jours. Trustee Buckingham Browne and Nichols Sch., 1974-80. Recipient Quintana award Voice Found., 1992; Guggenheim fellow, 1962. Fellow Acoustical Soc. Am. (exec. com. 1963-66, v.p. 1971-72, pres. elect 1975-76, pres. 1976-77), IEEE, Am. Acad. Arts and Scis.; mem. Nat. Acad. Engring. Home: 7 Hancock Pl Cambridge MA 02139-2208 Office: MIT 77 Massachusetts Ave Cambridge MA 02139

STEVENS, MARK GREGORY, immunologist; b. Rocky Ford, Colo., Dec. 7, 1955; s. Herbert Averal and Margie Burl (Harkness) S. BS, Kans. State U., 1978, DVM, 1982; PhD, Wash. State U., 1988. Diplomate Am. Coll. Veterinary Microbiologists. Postdoctoral fellow Wash., Oreg., Idaho Regional Program in Veterinary Medicine, Pullman, Wash., 1984-88; immunologist Nat. Animal Disease Ctr., USDA, Ames, Iowa, 1988—; bd. scientific reviewers Am. Jour. Veterinary Rsch., Schaumburg, Ill., 1992—. Contbr. articles to profl. jours. Mem. Am. Assn. Veterinary Immunologists, Am. Veterinary Med. Assn. (Harwal award 1982), Sigma Xi, Phi Zeta. Achievements include development of colorimetric assay for quantituting Bovine Neutrophil Bactericidal activity, identifying Interleukin-4 producing T-cell in cattle, testing and developing vaccines to eradicate Brucella Abortus infections in cattle. Office: Nat Animal Disease Ctr 2300 Dayton Ave Ames IA 50010

STEVENS, MARY ELIZABETH, biomedical engineer, consultant; b. Portland, Oreg., May 18, 1955; d. Robert Leroy and DeEtta Beatrice (Movius) Schulze; m. Bradley Miller Stevens, June 12, 1977; children: Lauren Elizabeth, Christopher James. MA, Oregon State U., 1979; PhD, U. Calif. Irvine, 1984. Postdoctoral fellow U. Calif., San Francisco, 1984-86, staff scientist Gladstone labs., 1986-88; mem. faculty U. Calif., Davis, 1988-90; rsch. specialist Stanford (Calif.) U., 1990—; scis. adv. com. DMSC, San Rafael, Calif., 1986—. Contbr. numerous articles to profl. jours., 1984—. V.p. PTA Bd. Dixie Sch., San Rafael, Calif., 1986—; asst. scoutmaster Boy Scouts Am., San Rafael, 1988—. Recipient Holcomb scholarship U. Calif. Irvine, 1984; grantee NATO, 1986, UICC, 1987. Mem. AAAS, Developmental Biology Soc. Achievements include new methodologies in creating and screening transgenic animals. Office: Howard Hughes Med Inst Stanford U B275 Beckman Ctr Stanford CA 94305

STEVENS, PRESCOTT ALLEN, environmental health engineer, consultant; b. Boston, July 19, 1922; s. Harold Wentworth and Helen Louise (Gustin) S.; m. Mary Golestaneh, Mar. 12, 1953; children: Katherine Maryam, Robert David, Rosalind Sara. BSCE, Worcester (Mass.) Poly. Inst., 1948; MS in Sanitary Engring., Harvard U., 1949. Registered profl. engr., Mass.; Diplomate Am. Acad. Environ. Engrs. Area hygiene officer Internat. Refugee Orgn., Italy, 1949-50; sanitary engr., Malaria advisor WHO, Iran, 1950-53, sanitary engr. insect-borne diseases control team, 1953-56; chief sanitary engr. Malaria eradication programme WHO, Geneva, 1957-63; sr. regional environ. health advisor WHO, Alexandria, Egypt, 1963-68; sanitary engr. wastes disposal programme WHO, Geneva, 1968-71, sanitary engr. pre-investment planning program, 1971-77, mgr. environ. health tech. and support programme, 1977-82; pvt. practice cons. La Rippe, Switzerland, 1982—. Contbr. articles to profl. jours. With U.S. Army, 1942-46, ETO. Fellow Am. Soc. Civil Engrs., APHA, Delta Omega. Avocation: piano. Home and Office: Chemin des Charbouilles, CH-1261 La Rippe Switzerland

STEVENS, ROBERT EDWARD, engineering company executive; b. Kansas City, Mo., Oct. 30, 1957; s. Kenneth E. and Nina (France) S. BS in Chem. Engring., U. Mo., Rolla, 1980; MS in Engring. Mgmt., U. Mo.-Rolla, 1985. Process design engr. The Pritchard Corp., Kansas City, Mo., 1981-83; process engr. Procter & Gamble, Cape Girardeau, Mo., 1986-87; tech. mgr. Procter & Gamble, 1987-90; project engring. mgr. Bechtel, 1990, mgr. engring., 1990—. Contbr. to Physical Properties of Gases and Liquids, 1987. Chairperson bd. dirs. Wesley Found., St. Louis, 1993. Recipient Stan Adams Reliability award P & G Paper Div., 1990; Nat. Merit scholar, 1976. Mem. AICE, Am. Soc. Engring. Mgmt., U. Mo. Rolla-Wesley Alumni Assn. (pres. 1988—), Alpha Chi Sigma (Cert. Appreciation 1991, pres. St. Louis Profl. chpt. 1993). Methodist. Home: 2908 Wind Flower Dr Florissant MO 63031-1042 Office: Hwy 111 And Madison St Wood River IL 62095

STEVENS, STEPHEN EDWARD, psychiatrist; b. Phila.; s. Edward and Antonia S.; BA cum laude, LaSalle Coll., 1950; MD, Temple U., Phila., 1954; LLB, Blackstone Sch. Law, 1973; m. Isabelle Helen Gallacher, Dec. 27, 1953. Intern, Frankford Hosp., Phila., 1954-55; resident in psychiatry Phila. State Hosp., 1955-58; practice medicine specializing in psychiatry Woodland Hills, Calif., 1958-63, Santa Barbara, Calif., 1970-77; asst. supt. Camarillo (Calif.) State Hosp., 1963-70; cons. ct. psychiatrist Santa Barbara County, 1974-77; clin. dir. Kailua Mental Health Ctr., Oahu, Hawaii, 1977—. Author: Treating Mental Illness, 1961. Served with M.C., USAAF. Diplomate Am. Bd. Psychiatry and Neurology. Decorated Purple Heart. Fellow Am. Geriatrics Soc. (founding); mem. Am. Acad. Psychiatry and Law, AMA, Am. Psychiat. Assn., Am. Legion, DAV (Oahu chpt. 1), Caledonia Soc., Am. Hypnosis Soc., Am. Soc. Adolescent Psychiatry, Hawaiian Canoe Club, Honolulu Club, Elks, Aloha String Band (founder and pres.). Home: PO Box 26413 Honolulu HI 96825-6413 Office: 2333 Kapiolani Blvd Honolulu HI 96826

STEVENS, WILLIAM FREDERICK, III, software engineer; b. Paducah, Ky., Aug. 13, 1954; s. William Frederick Jr. and Imogene (Outland) S. Student, Case Western Res. U., 1973-77. Software engr. Allen Bradley Co., Cleve., 1977-80; sr. analyst Mark Bus. Systems, Cleve., 1980-82, Datacomp Corp., Cleve., 1982-84; sr. cons. GE, Cleve., 1984-87; sr. engr. Micro Dimensions Inc., Cleve., 1987—; bd. dirs. N.E. Ohio Apple Corps. Mem. Assn. for Computing Machinery, Maths. Assn. of Am. Avocations: science fiction, computers, theater. Home: 9117 Chillicothe Rd Apt 101 Willoughby OH 44094-9228

STEVENSON, EARL, JR., civil engineer; b. Royston, Ga., May 8, 1921; s. Earl and Compton Helen (Randall) S.; B.S. in Civil Engring., Ga. Inst. Tech., 1953; m. Sue Roberts, Apr. 25, 1956; children—Catherine Helen, David Earl. Engr., GSA, Atlanta, 1959-60; engr., pres. Miller, Stevenson & Steinichen, Inc., Atlanta, 1960—; sr. v.p. Stevenson & Palmer, Inc., Camilla, 1984—; dir. Identification & Security Products, Inc., Atlanta. Served with USAAF, 1944-45. Registered profl. engr., Ga., Ala., S.C., Miss. Mem. Ga. Soc. Profl. Engrs., Water Pollution Control Fedn. Methodist. Home: 3163 Laramie Dr NW Atlanta GA 30339-4335 Office: 2430 Herodian Way Smyrna GA 30080-2906

STEVENSON, HAROLD WILLIAM, psychology educator; b. Dines, Wyo., Nov. 19, 1924; s. Merlin R. and Mildred M. (Stodick) S.; m. Nancy Guy, Aug. 23, 1950; children: Peggy, Janet, Andrew, Patricia. B.A., U. Colo., 1947; M.A., Stanford U., 1948, Ph.D., 1951. Asst. prof. psychology Pomona Coll., 1950-53; asst. to asso. prof. psychology U. Tex., Austin, 1953-59; prof. child devel. and psychology, dir. Inst. Child Devel., U. Minn., Mpls., 1959-71; prof. psychology, fellow Center for Human Growth and Devel., U. Mich., Ann Arbor, 1971—; dir. program in child devel. and social policy U. Mich., 1978-93; adj. prof. Tohoku Fukushi Coll., Japan, 1989—, Peking U., 1990—; mem. tng. com. Nat. Inst. Child Health and Human Devel., 1964-67; mem. personality and cognition study sect. NIMH, 1975-79; chmn. adv. com. on child devel. Nat. Acad. Scis.-NRC, 1971-73; exec. com. div. behavioral scis. NRC, 1969-72; mem. del. early childhood People's Republic of China, 1973, mem. del. psychologists, 1980; mem. vis. com. Grad. Sch. Edn., Harvard U., 1979-86; fellow Center Advanced Studies in Behavioral Scis., 1967-68, 82-83, 89-90. Fellow Am. Acad. Arts and Scis., Nat. Acad. Edn.; mem. APA (pres. divsn. devel. psychology 1964-65, G. Stanley Hall award 1988), Soc. Rsch. Child Devel. (governing coun. 1961-67, pres. 1969-71, chmn. long range planning com. 1971-74, social policy com. 1977-85, internat. affairs com. 1991—, Disting. Rsch. award 1993), Internat. Soc. Study Behavioral Devel. (exec. com. 1972-77, pres. 1987-91), Phi Beta Kappa, Sigma Xi. Home: 1030 Spruce Dr Ann Arbor MI 48104-2847

STEVENSON, J. ROSS, biological sciences educator, researcher; b. Canton, China, Sept. 4, 1931; came to U.S., 1932; s. Donald Day and Lois Elisabeth (Davis) S.; m. Nancy Ruth Hanson, June 19, 1954; children: Peter Day, Philip Ewing, Jay Ross. BA, Oberlin (Ohio) Coll., 1953; MS, Northwestern U., 1955, PhD, 1960. Instr. biology Chatham Coll., Pitts., 1956-59; rsch. assoc. U. Wash., Seattle, 1959-60; instr. biol. scis. Kent (Ohio) State U., 1960-62, asst. prof., 1962-66, assoc. prof. biology, 1966-71, prof. biol. scis., 1971—; assoc. marine biology Rockefeller U., N.Y.C. and Woods Hole, Mass., 1963-64; assoc. dean grad. coll. Kent State U., 1973-74; sabbatical leaves Tulane U., New Orleans, 1971, Cleve. Clinic, 1983; advisor 17 MS students Kent State U., 1962—, 10 PhD students, 1970—. Contbr. 36 articles to profl. jours. Rsch. grantee NIH, 1961-67, 79, 85-86, NSF, 1965-67, 69-71, Kent State U., 1963, 68, 70, 79, 86, State of Ohio, 1986-90. Fellow Ohio Acad. Sci. (membership chmn., 1971-72, v.p. zoology 1972-73), mem. AAAS, Am. Assn. Immunologists, Am. Soc. Cell Biology, Sigma Xi. Mem. Soc. of Friends. Achievements include developing criteria for molt staging in crayfish and finding how numerous biochemical changes in cuticle and epidermis of animal vary with molt stages and hormone changes; currently studing stress, hormones and immunity in rats. Home: 303 Valleyview Dr Kent OH 44240 Office: Kent State U Dept Biol Scis Kent OH 44242

STEVENSON, JAMES RALPH, school psychologist, author; b. Kemmerer, Wyo., June 29, 1949; s. Harold Ralph and Dora (Borino) S.; m. Alice M. Paolucci, June 17, 1972; children: Tiffany Jo, Brian Jeffrey. BA, U. No. Colo., 1971, MA, 1974, EdS, 1975. cert. elem. sch. counselor, Colo., nationally cert. sch. psychologist. Sch. psychologist Jefferson County Pub. Schs., Golden, Colo., 1975-87, 89-91, Weld County Sch. Dist. 6, Greeley, Colo., 1987-89, Weld Bd. Coop. Edn. Svcs., LaSalle, Colo., 1991—, Greeley, 1992—. Asst. coach Young Am. Baseball, Greeley, 1989, 90, head coach, 1992, 93; asst. basketball coach Recreation League for 6th-7th Grades, 1992, 93. U. No. Colo. scholar, 1974. Mem. NEA, NASP (alt. del. Colo. chpt. 1975-77, dir. Apple II users group Washington chpt. 1989—), Colo. Soc. Sch. Psychologists (chmn. task force on presch. assessment 1991—), Colo. Edn. Assn., Ft. Lupton Edn. Assn., Jefferson County Psychologists Assn. (sec. 1986-87). Democrat. Roman Catholic. Avocations: travel, reading, sports events, plays, music. Home: 1937 24th Ave Greeley CO 80631-5027 Office: Weld County BOCES PO Box 578 204 Main St LaSalle CO 80645

STEVENSON, JOHN O'FARRELL, JR., dean; b. Bklyn., Oct. 11, 1947; s. John O'Farrell Sr. and Vivian Eslie (Pemberton) S. BA in Math., Fordham U., 1968; PhD, MS in Math., Polytechnic U., 1976. Instr. Poly. Inst., Bklyn.; asst. prof. LaGuardia Community Coll., Queens, N.Y.; dean Empire State Coll., N.Y.C., 1977-80; pres. NSSFNS, N.Y.C., 1980-84; assoc. dean Bronx Community Coll., N.Y., 1984-92; Math. prof. LaGuardia C.C., 1992—, assoc. dean, 1992—. Danforth Found. scholar, 1968; Presdl. scholar, 1964. Mem. Soc. for Value in Higher Edn., Am. Assn. on Higher Edn., 100 Black Men, Phi Beta Kappa. Democrat. Roman Catholic. Office: LaGuardia Community Coll 31-10 Thompson Ave Long Island City NY 11101

STEVENSON, ROBERT EDWIN, microbiologist, culture collection executive; b. Columbus, Ohio, Dec. 2, 1926; s. Arthur Edwin and Mary Lucille (Beman) S. BS, Ohio State U., 1947, MS, 1950, PhD, 1954. Cert. Am. Bd. Microbiology. Virologist USPHS, Cin., 1954-58; head cell culture sect., Tissue Bank U.S. Naval Med. Sch., Bethesda, Md., 1958-60; head cell culture and tissue material sect. Nat. Cancer Inst., Bethesda, 1960-63, chief viral carcinogenesis br., 1963-67; mgr. biolog. scis., corp. devel. dept. Union Carbide Corp., Tarrytown, N.Y., 1967-72; v.p., gen. mgr., Frederick (Md.) div. Litton Bionetics, 1972-80; dir. Am. Type Culture Collection, Rockville, Md., 1980-93; dir. emeritus, 1993; dir. Large Scale Biology, Inc., Rockville, 1984—; chmn. biotech. adv. com. Dept. Commerce, Washington, 1985-93. With USN, 1944-45. Fellow Inst. for Soc., Ethics & Life Scis.; mem. Tissue Culture Assn. (pres. 1988-90), World Fedn. Culture Collections, U.S. Fedn. Culture Collections (pres. 1988-90), Am. Soc. Micrbiology, Cosmos Club (Washington). Episcopalian. Avocations: painting, cross country skiing. Home: 511 E Palace Ave Santa Fe NM 87501-2225

STEVENSON, WILLIAM JOHN, mechanical engineer; b. Florida, Transvaal, Republic of South Africa, Dec. 8, 1949; s. William Bernard and Aletta Rachel (Venter) S.; m. Isobel Anne Gilfillan, Dec. 20, 1974; children: William Alexander, Andrew John. BSc in Aero Engring., Stellenbosch U., Republic of South Africa, 1971, MEngring., 1983; PhD, U. Cin., 1993. Registered profl. engr., Republic of South Africa. Part-time lectr. Stellenbosch Engring., 1972-73; mem. staff Coun. for Sci. and Indsl. Rsch., Pretoria, Republic of South Africa, 1975-87, engr., 1991—. Author articles, conf. papers, reports in field. Mem., past sec. Bursary Fund, Lynnwood Bapt. Ch., Pretoria, 1983-87. Mem. AIAA, Internat. Neural Network Soc., South African Instn. Mech. Engrs. (Thomas Pringle award 1982). Home: 165 Swaardlelie Ave, The Willows Pretoria 0041, South Africa Office: Aerotek CSIR, PO Box 395, Pretoria 0001, South Africa

STEVER, HORTON GUYFORD, aerospace scientist and engineer, educator, consultant; b. Corning, N.Y., Oct. 24, 1916; s. Ralph Raymond and Alma (Matt) S.; m. Louise Risley Floyd, June 29, 1946; children: Horton Guyford, Sarah, Margarette, Roy. A.B., Colgate U., 1938, Sc.D. (hon.), 1958; Ph.D., Calif. Inst. Tech., 1941; LL.D., Lafayette Coll., U. Pitts., 1966, Lehigh U., 1967, Allegheny Coll., 1968, Ill. Inst. Tech., 1975; D.Sc., Northwestern U., 1966, Waynesburg Coll., 1967, U. Mo., 1975, Clark U., 1976, Bates Coll., 1977; D.H., Seton Hill Coll., 1968; D.Engring., Washington and Jefferson Coll., 1969, Widener Coll., Poly. Inst. N.Y., 1972, Villanova U., 1973, U. Notre Dame, 1974; D.P.S., George Washington U., 1981. Mem. staff radiation lab. MIT, Cambridge, 1941-42; asst. prof. MIT, 1946-51, asso. prof. aero. engring., 1951-56, prof. aero. and astro., 1956-65, head depts. mech. engring., naval architecture, marine engring., 1961-65, asso. dean engring., 1956-59, exec. officer guided missiles program, 1946-48; chief scientist USAF, 1955-56; pres. Carnegie-Mellon U., Pitts., 1965-72; dir. NSF, Washington, 1972-76; sci. adviser, chmn. Fed. Council Sci. and Tech., 1973-76; dir. Office Sci. and Tech. Policy, sci. and tech. adviser to Pres., 1976-77, sci. cons., corp. trustee, 1977—; mem. secretariat guided missiles com. Joint Chiefs of Staff, 1945; sci. liaison officer London Mission, OSRD, 1942-45; mem. guided missiles tech. evaluation group Research and Devel. Bd., 1946-48; mem. sci. adv. bd. to chief of staff USAF, 1947-69, chmn., 1962-69; mem. steering com. tech. adv panel on aeros. Dept. Def., 1956-62; chmn. spl. com. space tech. NASA, chmn. research adv. com. missile and spacecraft aerodynamics, 1959-65; mem. Nat. Sci. Bd., 1970-72, mem. ex-officio, chmn. exec. com., 1972-75; mem. Def. Sci. Bd., 1962-68; mem. adv. panel U.S. Ho. Reps. Com. Sci. and Astronautics, 1959-72; mem. Pres.'s Commn. on Patent System, 1965-67; chmn. U.S.-USSR Joint Commn. Sci. and Tech. Cooperation, 1973-77, Fed. Council Arts and Humanities, 1972-76; Pres. com. Nat. Sci. medal, 1973-77. Author: Flight, 1965; Contbr. articles to profl. publs. Past trustee Colgate U., Shady Side Acad., Sarah Mellon Scaife Found., Buckingham Sch.; truste Univ. Rsch Assn., 1977—, pres., 1982-85; trustee Woods Hole Oceanographic Inst., 1980—, Univ .Corp. for Atmospheric Rsch., 1980-83; bd. dirs. Saudi Arabia Nat. Ctr. for Sci. and Tech., 1978-81; bd. govs. U.S. Israel Binat. Sci. Found., 1972-76, chmn., 1972-73; mem. Carnegie Commn. on Sci., Tech. and Govt., 1988-93. Recipient Pres.'s Cert. of merit, 1948, Exceptional Civilian Svc. award USAF, 1956, Scott Gold medal Am. Ordinance Assn., 1960, Disting. Pub. Svc. medal Dept. Def., 1969, NASA, 1988, Nat. Medal of Sci. NSF, 1991; comdr. Order of Merit Poland. Fellow AIAA (hon., pres. 1960-62), AAAS, Royal Aero. Soc., Am. Acad. Arts and Scis., Royal Soc. Arts, Am. Phys. Soc.; mem. NAS (chmn. assembly engring. 1979-83), NAE (chmn. aero. and space engring. bd. 1967-69, fgn. sec. 1984-88), Acad. Engring. of Japan (fgn. mem.), Royal Acad. of Engring. of Great Britain (fgn. mem.), Cosmos Club, Bohemian, Phi Beta Kappa, Sigma Xi, Sigma Gamma Tau, Tau Beta Pi. Episcopalian. Home: 1528 33d St NW Washington DC 20007 Office: Carnegie Commn for Sci Tech & Govt 1616 P St NW Washington DC 20036-1434

STEWARD, A(LMA) RUTH, chemistry educator, researcher; b. Silverton, Oreg., Mar. 12, 1935; d. Willard Palmer and Bessie Leola (Lindsey) S. BA, Pomona Coll., 1956; MA, Yale U., 1957, MS, 1959; PhD, U. Calif., Davis, 1982. Asst. lectr. chemistry U. Nigeria, Nsukka, 1961-63; instr. phys. sci. El Camino Coll., Torrance, Calif., 1964-66; tchr. sci. L.A. City Schs., 1967-75; rsch. assoc. Med. Coll. Va., Richmond, 1982-84; rsch. scientist SUNY Coll., Buffalo, 1984—; adj. instr. environ. sci. SUNY, Buffalo, 1987. Reviewer Aquatic Toxicology, Seattle, 1992; contbr. articles to profl. jours. Monsanto Fund predoctoral fellow, 1980-81, NIH postdoctoral fellow, 1982-84; recipient Acad. Rsch. Enhancement award Nat. Inst. Environ. Health Sci., 1992. Mem. AAAS, Am. Chem. Soc., Soc. for Environ. Toxicology and Chemistry, Soc. Toxicology, Internat. Assoc. Great Lakes Rsch. Achievements include research in biochemical markers of environmental pollution, metabolism of polynuclear aromatics, arylamines and polyhalogenated aromatics by fish, genetic expression of cytochromes P-450. Office: SUNY Coll Divsn Environ Toxicology Chemistry 1300 Elmwood Ave Buffalo NY 14222

STEWARD, MOLLIE AILEEN, mathematics and computer science educator; b. Bridgeton, N.J., Dec. 10, 1952; d. E. Emerson and Eulalie (Loatman) S. BA in Liberal Arts, Villanova U., 1975; MA in Psychology, Glassboro (N.J.) State Coll., 1981, BA in Math., 1984, MA in Math. Edn., 1986. Title I summer tchr. Bridgeton Pub. Schs., 1976, substitute tchr., 1976-77, compensatory edn. tchr., 1977-78; substitute tchr. Upper Deerfield Schs., Seabrook, N.J., 1978-80; mem. adj. faculty Rowan Coll. of N.J., 1981-82, 83-84; grad. asst. in devel. edn. Glassboro State Coll., 1984-85, rsch. asst., 1985-86; coord. math. and sci. lab. Cumberland County Coll., Vineland, N.J., 1987-88, asst. prof. I math. and computer sci., 1988—. Mem. AAUW, Math. Assn. Am., Nat. Coun. Tchrs. Math., Assn. Math. Tchrs. N.J. Episcopalian. Avocations: photography, music. Office: Cumberland County Coll PO Box 517 Vineland NJ 08360-0517

STEWART, ALAN FREDERICK, electrical engineer; b. Epping, Eng., Aug. 24, 1948; came to U.S., 1950; s. John James Wilfred and Ida Elizabeth (Anderson) S.; m. Brenda Gail Tatum, Feb. 20, 1986. BSEE, Rutgers U., 1980. Registered profl. engr., Ga. Commd. 2d lt. USAF, 1967, advanced through grades to capt., 1984; energy engr. Hq. ATC USAF, Randolph AFB, Tex., 1984-88; sr. engr. Honeywell Inc., Atlanta, 1988—. Contbr. articles to profl. jours. Mem. Orphanage Support, Kunsan AFB, Korea, 1982, Helping the Homeless, Atlanta, 1991. Decorated 4 Air Force Commendations, Air Force Achievement. Mem. NSPE, ASHRAE, Assn. Energy Engrs. (coun. chair 1989-92), Soc. Am. Mil. Engrs. Achievements include identification of major hydronic flaw in the design of Air Force underground fuel storage tanks during explosion investigation; co-development of energy managment system plan. Home: 4732 Bentley Pl Duluth GA 30136

STEWART, ALEC THOMPSON, physicist; b. Windthorst, Sask., Can., June 18, 1925; s. Arthur and Nelly Blye (Thompson) S.; m. Alta Aileen Kennedy, Aug. 4, 1960; children—A. James Kennedy, Hugh D., Duncan R. B.Sc., Dalhousie U., Halifax, N.S., Can., 1946, M.Sc., 1949, LL.D. (hon.), 1986; Ph.D., Cambridge U., Eng., 1952. Research officer Atomic Energy Can., Chalk River, Ont., Can., 1952-57; assoc. prof. Dalhousie U., Halifax, 1957-60; assoc. prof. to prof. U. N.C., Chapel Hill, 1960-68; head physics Queen's U., Kingston, Ont., 1968-74, prof. physics, 1968—; vis. prof. various univs., Can., Europe, Japan, China, Hong Kong. Author 2 books and numerous articles in profl. jours. Guggenheim fellow, 1965-66. Fellow Am. Phys. Soc., Royal Soc. Can. (pres. Acad. Sci. 1984-87), Japan Soc. for Promotion Sci.; mem. Can. Assn. Physicists (pres., other offices 1970-74 CAP medal achievement in physics 1992). Achievements include research in solid state physics, behavior of phonons, electrons, positrons and postronium in crystals and liquids, public service, nuclear reactor safety, possible hazards of power frequency electric and magnetic fields. Avocations: sailboat racing and cruising. Home: 835 Wartman Ave, Kingston, ON Canada K7M 4M3 Office: Queens U, Dept Physics, Kingston, ON Canada K7L 3N6

STEWART, ARLENE JEAN GOLDEN, designer, stylist; b. Chgo., Nov. 26, 1943; d. Alexander Emerald and Nettie (Rosen) Golden; m. Randall Edward Stewart, Nov. 6, 1970; 1 child, Alexis Anne. BFA, Sch. of Art Inst. Chgo., 1966; postgrad., Ox Bow Summer Sch. Painting, Saugatuck, Mich., 1966. Designer, stylist Formica Corp., Cin., 1966-68; with Armstrong World Industries, Inc., Lancaster, Pa., 1968—, interior furnishings analyst, 1974-76, internat. staff project stylist, 1976-78, sr. stylist Corlon flooring, 1979-80, sr. exptl. project stylist, 1980-89, sr. project stylist residential DIY flooring floor div., 1989—, master stylist DIY residential tile, 1992—. Exhibited textiles Art Inst. Chgo., 1966, Ox-Bow Gallery, Saugatuck, Mich., 1966. Home: 114 E Vine St Lancaster PA 17602-3550 Office: Armstrong Innovation Ctr 2500 Columbia Ave Lancaster PA 17603-4117

STEWART, BRENT KEVIN, radiological science educator; b. Seattle, Nov. 2, 1957; s. Wendel Keith and Martha Erlene (Stevens) S.; m. Aliea Leeann Peterson, Sept. 8, 1979. BS, U. Wash., 1980; PhD, UCLA, 1988. Asst. prof. radiol. scis. U. Cin., 1988-90, UCLA, 1990—. Contbr. articles to profl. publs. NSF fellow, 1984-87. Mem. IEEE, IEEE Computer Soc., Am. Assn. Physicists in Medicine, Soc. Photo-Optical and Instrumentation Engrs. Office: UCLA Sch Medicine 10833 Le Conte Ave Los Angeles CA 90024-1721

STEWART, FRANK MAURICE, federal agency administrator; b. Okalona, Miss., Apr. 1, 1939; s. Frank Maurice Stewart and Henryne Annette (Walker) Goode; m. Regina Diane Mosley, Dec. 26, 1964; children: Lisa Ann, Dana Joy. BA, Wesleyan U., 1961, MA in Teaching, 1963, diploma further study, 1963; postgrad., Am. U., 1982-84. Dir. urban edn. corps N.J. State Dept. Edn., Trenton, 1969-70; dir. urban teaching intern program Sch. Edn. Rutgers U., New Brunswick, N.J., 1970-71; staff asst. White House Conf. on Aging, Washington, 1971-73; chief program devel. U.S. Office of Equal Edn. Opportunity, Washington, 1973-74; chief policy analysis U.S. Adminstrn. on Aging, Washington, 1974-75; asst. exec. sec. U.S. HEW, Washington, 1975-77; dir. govt. programs U.S. Dept. Energy, Washington, 1977-80, dir. instnl. conservation programs, 1980-84, dir. state and local assistance programs, 1984-90, dep. asst. sec. for tech. and fin. assistance, 1990—. Recipient Svc. Recognition award Assn. Phys. Plant Adminstrs., Washington, 1982, Svc. Appreciation award Nat. Assn. State Energy Officials, Washington, 1987; named Energy Exec. of Yr. Assn. Energy Engrs.,

Atlanta, 1988. Mem. Sr. Execs. Assn. Republican. Episcopalian. Home: 6221 Verne St Bethesda MD 20817-5930 Office: US Dept Energy 1000 Independence Ave SE Washington DC 20585-0001

STEWART, HARRIS BATES, JR., oceanographer; b. Auburn, N.Y., Sept. 19, 1922; s. Harris B. and Mildred (Woodruff) S.; m. Elise Bennett Cunningham, Feb. 21, 1959; children: Dorothy Cunningham, Harry Hasburgh; 2d m. Louise Conant Thompson, Dec. 22, 1988. Grad., Phillips Exeter Acad., 1941; AB, Princeton, 1948; MS, Scripps Instn. Oceanography, U. Calif., 1952, PhD, 1956. Hydrographic engr. U.S. Navy Hydrographic Office expdn. to, Persian Gulf, 1948-49; instr. Hotchkiss Sch., 1949-51; research asst. Scripps Instn. Oceanography, 1951-56; diving geologist, project mgr. Geol. Diving Cons., Inc., San Diego, 1953-57; chief oceanographer U.S. Coast & Geodetic Survey, 1957-65, dept. asst. dir., 1962-65; dir. Inst. Oceanography, Environmental Sci. Services Adminstrn., U.S. Dept. Commerce, 1965-69; dir. Atlantic Oceanographic and Meteorol. Labs., NOAA, 1969-78, cons., 1978-80; prof. marine sci., dir. Center for Marine Studies, Old Dominion U., Norfolk, Va., 1980-85; adj. prof. dept. oceanography Old Dominion U., 1986—; dir. S.E. Bank of Dadeland; chmn. Fla. Comm. Marine Sci. and Tech.; mem. exec. com., earth scis. div. Nat. Acad. Scis.; chmn. adv. bd. Nat. Oceanographic Data Center, 1965-66; chmn. survey panel interagy. com. oceanography Fed. Council Sci. and Tech., 1959-67; chmn. adv. com. underseas features U.S. Bd. Geog. Names, 1964-67; mem. sci. party No. Holiday Expdn., 1951; Capricorn Expdn., 1952-53; chief scientist Explorer Oceanographic Expdn., 1960, Pioneer Indian Ocean Expdn., 1964, Discoverer Expdn., 1968, NOAA-Carib Expdn., 1972, Researcher Expdn., 1975; mem. U.S. delegation Intergovtl. Oceanographic Commn., 1961-65; mem. Gov. Calif. Adv. Commn. Marine Resources; chmn. adv. council Dept. Geol. and Geophys. Scis. Princeton; v.p. Dade Marine Inst., 1976-77, pres., 1977-79; trustee, mem. exec. com. Assoc. Marine Insts.; mem. Fisheries Mgmt. Adv. Council Va. Marine Resources Commn., 1984-85; vice chmn. adv. council Univ. Nat. Oceanographic Lab. System, 1983-85; U.S. nat. assoc. to intergovtl. oceanographic commn. UNESCO program for Caribbean, 1964-89, vice chmn., 1974. Author: The Global Sea, 1963, Deep Challenge, 1966, The Id of the Squid, 1970, Challenger Sketchbook, 1972, No Dinosaurs on the Ark, 1988, Grungy George and Sloppy Sally, 1993. Bd. dirs. Vanguard Sch., Miami, 1974-76; trustee Metro Zoo, Miami, 1991—. Served as pilot USAAF, 1942-46, PTO. Decorated comendador Almirante Padilla (Colombia); recipient Meritorious award Dept. Commerce, 1960, Exceptional Service award, 1965. Fellow AAAS, Geol. Soc. Am., Nat. Tropical Botanical Gardens, Marine Tech. Soc. (v.p.); mem. Fla. Acad. Scis. (pres. 1978-79), Va. Acad. Sci., Am. Geophys. Union, Internat. Oceanographic Found. (v.p. 1974-80), Zool. Soc. Fla. (pres. 1970-73), Maine Hist. Soc., Marine Hist. Assn., Cape Ann Hist. Assn., Marine Coun. (Miami), Explorers Club (N.Y.), Prouts Neck Yacht Club (Maine), Cosmos Club. Presbyterian. Home (summer): 11 Atlantic Dr Scarborough ME 04074 Home (winter): 644 Alhambra Circle Coral Gables FL 33134

STEWART, JAMES IAN, agrometeorologist; b. San Diego, Jan. 9, 1928; s. Castle Elmore and Myrtle Catherine (Hasty) S.; m. Robbie Nell Oliver, Mar. 23, 1975; children: Virginia Lane Stewart Carton, Ian Castle Stewart, Kevin Scott Overby. BSc, U. Calif., Berkeley, 1950; PhD, U. Calif., Davis, 1972. Farm advisor Agrl. Extension Svc., U. Calif., Stockton and Merced, 1950-61; extension expert Irrigation, Food and Agrl. Orgn. UN, Nicosia, Cyprus, 1962-66; assoc. rsch. water scientist U. Calif., Davis, 1966-77; supervisory soil scientist USDA/Office for Internat. Cooperation and Devel., Nairobi, Kenya, 1977-83; team leader, agrometeorologist USAID/Kenya Mission, 1977-83; founder, pres. Found. for World Hunger Alleviation Through Response Farming (WHARF), Davis, 1984—; cons., agrometeorology AID, USDA, World Bank, FAO/UNDP, 35 countries of Ams., Europe, Asia, Africa, Australia, 1965—; sci. convocations, internat. 13 countries worldwide, 1969—. Author: Response Farming in Rainfed Agriculture, 1988; creator (computer programs) Wharf, Wharfdat, 1990; contbr. numerous articles to profl. jours. Mem. Am. Soc. Agronomy, Crop Sci. Soc. Am., Soil Sci. Soc. Am., Internat. Soil Sci. Soc., Internat. Com. for Irrigation and Drainage (life U.S. com.), Indian Soc. of Dryland Agriculture (life), Sigma Xi. Achievements include pioneering research on soil water extraction by crops; crop water requirements; relations between crop yield and water evapotranspired; impacts of water deficits in different crop growth stages; relations between season rainfall behavior and season dates of onset; findings of time/depth models of soil water extraction by major crops in drydown sequences; four-growth period linear model for estimating crop water requirements; linear and weighted growth stage models for estimating crop yields from actual evapotranspiration; response farming methodology for flexible dryland cropping management strategy, based on rainfall season date of onset as defined to meet crop establishment requirements. Home: 640 Portsmouth Ave Davis CA 95616-2738 Office: World Hunger Alleviation Through Response Farming PO Box 1158 Davis CA 95617-1158

STEWART, JANE, psychology educator; b. Ottawa, Ont., Can., Apr. 19, 1934; d. Daniel Wallace and Jessie Stewart; m. Dalbir Bindra, Aug. 5, 1959 (dec. 1981). BA with honours, Queen's U., Kingston, Ont., 1956; PhD, U. London, 1959; DSc (hon.), Queen's U., 1992. Sr. rsch. biologist Ayerst Labs., Montreal, Que., 1959-63; part-time instr. psychology Sir George, Montreal, 1962-63; assoc. prof. psychology Williams U., Montreal, 1963-69; prof., chmn. psychology SGW Univ. (now Concordia U.), Montreal, 1969-75; prof. psychology Concordia U., Montreal, 1975—; dir. Ctr. for Studies in Behavioral Neurobiology, Concordia U., Montreal, 1990—. Fellow AAAS, APA, Can. Psychol. Assn.; mem. Soc. for Neurosci., Corp. Psychologists Province of Que., N.Y. Acad. Sci., Sigma Xi. Office: Concordia University, 1455 de Maisonneuve Blvd W, Montreal, PQ Canada H3G 1M8

STEWART, KENNETH RAY, food scientist; b. Klamath Falls, Oreg., June 20, 1951; s. Robert Paul and Patrica Sarah (McFarland) S.; m. Beverly May Gates, June 15, 1974; children: Elizabeth, Thomas. BS, Oreg. State U., 1974. Mgr. mfg. Clermont, Inc., Hillsboro, Oreg., 1974-82; tech. svc. mgr. Seneca Foods Corp., Prosser, Wash., 1983-84; gen. mgr. North Marion Fruit Co., Woodburn, Oreg., 1983-84; project mgr. Lewis Packing Co., Gresham, Oreg., 1984-88; juice ops. mgr. N.W. Packing Co., Vancouver, Wash., 1988-91; cons. Currant Ideas, Hillsboro, Oreg., 1991--. Mem. Am. Chem. Soc., AUAC, Inst. Food Technologists. Home and Office: Currant Ideas 21949 NW West Union Rd Hillsboro OR 97124

STEWART, MARGARET MCBRIDE, biology educator, researcher; b. Guilford County, N.C., Feb. 6, 1927; d. David Henry and Mary Ellen (Morrow) S.; m. Paul C. Lemon, June 1962 (div. 1968); m. George Edward Martin, Dec. 19, 1969. AB, U. N.C.-Greensboro, 1948; MA, U. N.C.-Chapel Hill, 1951; PhD, Cornell U., 1956. Instr. biology Greensboro Evening Coll. U. N.C., Greensboro, 1950-51; instr. biology Catawba Coll., Salisbury, N.C., 1951-53; extension botanist Cornell U., Ithaca, N.Y., 1954-56; asst. prof. biology SUNY, Albany, 1956-59, assoc. prof., 1959-65, prof. vertebrate biology, 1965—, Disting. Teaching prof., 1977—; faculty rsch. participant Oak Ridge Assoc. Univs., 1983. Author: (with A.H. Benton) Keys to the Vertebrates of the Northeastern States, 1964, Amphibians of Malawi, 1967; contbr. numerous articles and revs. to profl. jours. Bd. dirs. E.N. Huyck Nature Preserve, Rensselaerville, N.Y., 1976-86; bd. dirs. Ea. N.Y. chpt. Nature Conservancy, 1983-88, 90—, N.Y. State chpt., 1987-90. Recipient Citizen Laureate award SUNY Found., 1987, Am. Philos. Soc. rsch. grantee, 1975, 81, NSF grantee, 1978-80, Oak Ridge Assocs. Univs. grantee, 1983—. Fellow Herpetologists League (bd. dirs. 1978-80); mem. Soc. for Study Amphibians and Reptiles (pres. 1979), Am. Soc. Ichthyologists and Herpetologists (bd. govs. 1975-80, 87-90, herpetology editor 1983-85), Ecol. Soc. Am., Assn. for Tropical Biologists, Soc. Study of Evolution, Sigma Xi, Sigma Delta Epsilon, Phi Kappa Phi. Democrat. Presbyterian. Avocations: photography, gardening, reading, travel. Office: SUNY Dept Biol Scis 1400 Washington Ave Albany NY 12222-0001

STEWART, MICHAEL KENNETH, quality assurance professional; b. Denver, Jan. 5, 1956; s. Harold Glen and Leola (Gass) S.; m. Gail Ann Juranek, Apr. 25, 1981; children: Matthew Brian, Adam Hilon. BS in Mfg., Colo. State U., 1978. Mfg. engr. Morton Thiokol, Brigham City, Utah, 1979-81; sr. product engr. Rockwell Internat., Golden, Colo., 1981-87; sr. process engr. UNC Naval Products, Uncasville, Conn., 1988-90; sr. quality engr. Westinghouse, Aiken, S.C., 1990-92; program mgr., sr. quality assurance engr. RUST Environment & Infrastructure, Aiken, 1992—. Mem. Am. Soc. Quality Control, Soc. Mfg. Engrs. Republican. Roman Catholic. Home: 3207 Roses Run Aiken SC 29803 Office: RUST Environment/Infrastruc 955 Millbrook Ave Aiken SC 29803

STEWART, RICHARD ALLAN, civil engineer; b. Pitts., Sept. 24, 1947; s. Walter James and Evelyn Dorothy (Neuner) S.; m. Susan DeVine, Nov. 29, 1975; children: William, Jocelyn. BSCE, West Va. U., 1970. Registered profl. engr., Pa., W.Va., Ohio. Design engr. Green Internat., Inc., Sewickley, Pa., 1970-75; design engr. NIRA Cons. Engrs., Inc., Corapolis, Pa., 1975-80, project engr., 1980-86, project mgr., 1986-89, partner, 1990—. Bd. dirs. Ohio Twp. Vol. Fire Co., Pitts.; chmn. Ohio Twp. Zoning Hearing Bd.; asst. scoutmaster Boy Scouts Am. Mem. ASCE, Am. Water Works Assn. United Presbyterian. Office: NIRA Cons Engrs Inc 950 5th Ave Coraopolis PA 15108-1887

STEWART, ROBERT JACKSON, software development engineer, researcher; b. Beaumont, Tex., Feb. 1, 1958; s. Hester Reid and Donna Dea (Saxe) S.; m. Theresa Marie Goluszek, May 12, 1984. BA summa cum laude, U. Tex., 1980; MS, U. So. Calif., 1981. Mem. tech. staff AT&T Bell Labs., Naperville, Ill., 1980-88; staff engr. Tellabs, Inc., Lisle, Ill., 1988-91; systems engr. Compaq Computers, Houston, 1991—. Mem. Jaycees, Phi Beta Kappa. Office: Compaq Computer Corp 20555 S H 249 Houston TX 77070

STEWART, RONALD, chemical engineer; b. Dobbs Ferry, N.Y., May 6, 1941; s. Alan Carlyle and Jacqueline (Shively) S.; m. Barbara Whitfield, June 18, 1967; children: Terrance Collins, Christopher David. BS, Washington & Lee Coll., 1964; BChemE, Rensselaer Poly. Inst., 1964. Registered profl. engr., N.C., Va. Prodn. engr. E.I. DuPont, Kinston, N.C., 1964-67; process design engr. BASF Corp., Williamsburg, Va., 1967-89; mgr. process engring. BASF Corp., Enka, N.C., 1990—. Mem. AICE, Mensa. Home: 6 Glen Cove Rd Arden NC 28704 Office: BASF Corp General Delivery Enka NC 28728

STEWART, SUE ELLEN, molecular biologist; b. Phillipsburg, N.J., June 25, 1955; d. Russell Harrison and Ruth (Hartman) S.; m. Joseph Lee Jones, Sept. 23, 1989; 1 child, Katherine. BS, MIT, 1977; PhD, Brandeis U., 1986. Postdoctoral fellow Yale U., New Haven, 1986-89; tech. asst. MIT, Cambridge, Mass., 1977-79, postdoctoral assoc., 1989-92; sr. scientist molecular biology T Cell Scis., Cambridge, 1992—. Democrat. Office: T Cell Scis 38 Sidney St Cambridge MA 02139

STEWART, WILLIAM KENNETH, JR., ocean engineer; b. Savannah, Ga., Aug. 27, 1950; s. William Kenneth and Jean (Fritts) S. AAS in Marine Tech., Cape Fear Tech. Inst., 1974; BS in Ocean Engring., Fla. Atlantic U., 1982; PhD in Oceanographic Engring., MIT, 1988. Marine constrn. Marine Constrn. & Engring. Co., Freeport, Nassau, Bahamas, 1965-68; pipeline foreman Freeport Constrn. Co., Bahamas, 1968-69; heavy equip. operator Nello H. Teer Co., Southport, N.C., 1972; asst. terminal mgr./computer electronics tech. Cape Fear Tech. Inst., Wilmington, N.C., 1973-75; asst. sta. mgr., equip. mgr. Marine Sci. Consortium, Wallops Island, Va., 1975-77; svc. mgr. Herald Office Systems, Wilmington, 1979; computer programmer/operator McQueen & Co., Wilmington, 1979; systems analyst Essential Software Concepts for Architects and Profl. Engrs., Boca Raton, Fla., 1979-82; systems analyst cons. S.E. Computer Concepts, Boca Raton, Fla., 1980-81; from grad. rsch. asst. to assoc. scientist Woods Hole (Mass.) Oceanographic Instn., 1982—; vis. asst. prof. elec. engring. U. Va., Charlottesville, 1991-92. Contbr. articles to profl. jours. With USN, 1969-72. Internat. Yachtsman Assn. scholar, 1981; Link fellow Harbor Br. Found., 1981, Office of Naval Rsch. grad. fellow, 1982-86. Mem. IEEE, Acoustical Soc. Am., Internat. Soc. Optical Engring., Marine Tech. Soc., Nat. Computer Graphics Assn., Oceanography Soc., Sigma Xi, Tau Beta Epsilon. Achievements include research in undersea robotics; real time acoustical/optical modeling and imaging; quantitative seafloor characterization; precision underwater mapping and surveying. Home: 486 Woods Hole Rd Woods Hole MA 02543 Office: Woods Hole Oceanographic In WHOI Blake 109 Woods Hole MA 02543

STEWART, WILLIAM TIMOTHY, environmental engineer; b. Nashville, June 18, 1965; s. William Blanchard and Helen (Puckett) S. BS in Civil Engring., Tenn. Tech. Coll., 1989. Environ. specialist Tenn. Army Nat. Guard, Nashville, 1989-90, Resource Cons., Inc., Brentwood, Tenn., 1990-93; Tenn. Div. of Superfund, 1993—. Vice-pres. Fellowship Christian Athletics, Cookeville, Tenn., 1986. With U.S. Army, 1987—. Mem. Am. Water Works Assn., Water Pollution Control Fedn., Assn. U.S Army, Am. Soc. Civil Engrs. Pentecostal. Home: 2014 Cedar Ln #4 Nashville TN 37212 Office: Dept Environ & Conservation Nashville Environ Field Off 537 Brick Church Pk Dr Nashville TN 37243-1550

STICK, ALYCE CUSHING, information systems consultant; b. N.J., July 13, 1944; d. George William and Adele Margaret (Wilderotter) Cushing; m. James McAlpin Easter, July, 1970 (div. Aug. 1986); m. T. Howard F. Stick, June, 1989. AA, Colby-Sawyer Coll., 1964; student, Boston U., 1964-65, Johns Hopkins U., 1972-74; cert., Control Data Inst. and Life Office Mgmt. Assn., 1976. Claims investigator Continental Casualty Co., Phila., 1967-69; data processing coord. Chesapeake Life Ins. Co., Balt., 1970-72; sr. systems analyst Comml. Credit Computer Corp., Balt., 1972-80; v.p. Shawmut Computer Systems, Inc., Owings Mills, Md., 1980-85; pres. Computer Relevance, Inc., Gladwyne, Pa., 1985—; cons. Sinai Hosp., Balt., 1982-85, AT&T, Reading, Pa., 1987-88, Dun and Bradstreet, Allentown, Pa., 1988, Arco Chem. Co., Newtown Square, Pa., 1990-91, Rohm and Haas Co., Phila., 1992-93. Designer/author: (computer software systems) Claim-Track, 1977, Property-Profiles, 1979, Stat-Model, 1989; co-designer/author: Patient-Profiles, 1983. Treas. Balt. Mus. Art, Sales and Rental Gallery, 1984; mem. exec. com. Springfield (Pa.) Twp. Concerned Citizens, 1989. Mem. Assn. for Systems Mgmt., Data Processing Mgmt. Assn., Ind. Computer Cons. Assn., Merion Cricket Club (Haverford, Pa.). Republican. Avocations: Am. antiques, Chinese export porcelain dealer. Office: Computer Relevance Inc 1501 Monticello Dr Gladwyne PA 19035

STICKELS, CHARLES ARTHUR, metallurgical engineer; b. Detroit, Apr. 6, 1933; s. Charles Henry and Alice Mable (Mayer) S.; m. Patricia Jane Forbes, June 16, 1957; 1 child, Charles Forbes. BSChemE, BSMetE, U. Mich., 1956, MSMetE, 1960, AM in Math., 1962, DMetE, 1963. Rsch. engr. Ford Motor Co., Dearborn, Mich., 1963-91; cons. ERIM, Ann Arbor, Mich., 1991—. Contbr. book chpts. and articles to profl. jours.; editor: Jour. Heat Treating, 1982-85; inventor, patentee in field. Lt. (j.g.) USN, 1956-59. Fellow Am. Soc. for Metals; mem. AIME, Minerals, Metals and Materials Soc., Sigma Xi. Congregationalist. Home: 2410 Newport Rd Ann Arbor MI 48103-2265

STIDD, BENTON MAURICE, biologist, educator; b. Bloomington, Ind., June 30, 1936; s. Benjamin David and Alma Mae (Selzer) S.; divorced; children: Beth, Laura, Kelly, Faye, Reva. BS, Purdue U., 1954-58; MS, Emporia (Kans.) U., 1963; PhD, U. Ill., 1968. Tchr. Wheatland (Ind.) High Sch., 1958-62, North Knox High Sch., Edwardsport, Ind., 1964-65; asst. prof. U. Minn., Mpls., 1968-70; prof. biology We. Ill. U., Macomb, 1970—. Contbr. articles to Am. Jour. Botany, Rev. of Paleobotany and Palynology, Sci., Paleontology, others. NSF rsch. grantee, 1974, 76; Equipment grantee NSF, 1986, U.S. Dept. Edn., 1987; Summer Inst. grantee NEH, 1982. Mem. Bot. Soc. Am., Soc. Systematic Biology, Internat. Soc. for the History, Philosophy and Social Studies of Biology, Ill. Acad. Scis. Home: 704 N Campbell Macomb IL 61455 Office: We Ill U Dept Biology Macomb IL 61455

STIEFEL, VERNON LEO, entomology educator; b. Ft. Carson, Colo., June 8, 1961; s. Werner Konrad and Maria (Frodl) S.; m. Elisa Ann Harris, Sept. 15, 1984; children: Jenna Mae, Clara Marie. BS in Botany, Colo. State U., 1983; MS in Entomology, Kans. State U., 1991, postgrad., 1992—. Lab., field asst. dept. weed sci. Colo. State U., Ft. Collins, 1982-84; agrl. intern Land Inst., Salina, Kans., 1985; grad. rsch. asst. Kans. State U., Manhattan, 1989-91; cons. botanist Autonomous U. Guadalajara, Jalisco, Mex., 1984. Contbr. articles to profl. publs. Mem. Entomol. Soc. Am. (student), Ctrl States Entomol. Soc. (student), Phi Beta Kappa, Sigma Xi, Phi Kappa Phi. Home: 813 Colorado St Manhattan KS 66502-6253

STIEGLER, KARL DRAGO, mathematician; b. Zagreb, Croatia, Oct. 24, 1919; s. Stephan and Catharina S.; m. Hildegard Sarco, Sept. 22, 1951; 1 child, Cornelia. MS in Math. and Theoretical Physics, U. Zagreb, 1946; PhD in Theoretical Physics, Tech. U., Munich, 1963; DSc (hon.), World U., 1990. Prof. math. Engring Coll., Zagreb, 1946-50; research mathematician, constructor optical instruments Ghetaldus Co., Zagreb, 1950-56, chief dept. ophthalmol. optics, 1957-59; fellow faculty of math. Tech. U., Munich, 1964-84; correspondence with Albert Einstein concerning the found. of Spl. Theory of Relativity, 1951; sci. cons. AFGA Camera Works, Munich, 1960—, Tech. U., Munich, 1964—; lectr. in field; active profl. internat. congs. Contbr. articles to profl. jours.; researcher in relativity, quantum physics, cosmology, theory of group representations, history and philosophy of math. scis.; discovery of the non-archimedean order of grandeurs incomparables in the calculus differentialis of G.W. Leibniz (1942), law of anomalous rotation of spherical cosmical bodies with application to the Sun, Jupiter and Saturn; discovered non-equivalency of right and left in the electrodynamics of moving bodies; comprehensive research on methaphysics of G.W. Leibniz and atomistic of Rogerius (Rudjer) Boscovich. Active participate Séminaire de Recherches Louis de Broglie and Séminaire Nicolas Bourbaki. UNESCO research fellow Inst. Henri Poincaré, Sorbonne, U. Paris, 1954-55; hon. fellow Research Inst. History of Sci. and Tech., Deutsches Mus., Munich, 1964; scholar sci. rsch. Joyce and Zlatko Balokovich Found. Harvard U., 1961-63. Fellow Royal Astron. Soc. (London), Nat. Acad. Scis. (India), N.Y. Acad. Scis.; mem. Soc. Astronomica Italiana, Soc. Math. de France, Internat. Math. Union, Deutsche Gesellschaft für Angewandte Optik. Office: Postfach 750 839, Munich 75, Germany

STIEMER, SIEGFRIED F., civil engineer. Recipient Le Prix E. Whitman Wright award Can. Soc. Civil Engring., 1991. Home: 5731 137A St, Surrey, BC Canada V3W 5E7*

STIENMIER, SAUNDRA KAY YOUNG, aviation educator; b. Abilene, Kans., Apr. 27, 1938; d. Bruce Waring and Helen E. (Rutz) Young; m. Richard H. Steinmier, Dec. 20, 1958; children: Richard, Susan, Julia, Laura. AA, Colo. Women's Coll., 1957; student, Temple Buell Coll., U. Colo., 1959, 69; ed., Embre Riddle Aviation U., Ramstein, Germany. Cert. FAA pilot. Dir. Beaumont Gallery, El Paso, Tex., 1972-77; mem. grad. studies faculty Embre Riddle Aviation U., 1979-80; mgr. Ramstein Aero Club, USAF, 1977-80, Peterson Aero Club, USAF, Peterson AFB, Colo., 1980—. Named Outstanding S.W. Artist. Mem. Internat. Platform Assn., Order of Eastern Star, Scottish Soc. Pikes Peak, Scots Heritage Soc., Internat. Women Pilots Assn., Beta Sigma Phi, Delta Psi Omega, Aircraft Owners & Pilots Assn., Nat. Pilots Assn., Colo. Pilots Assn., Soc. Arts and Letters, 99's Club. Office: PO Box 14123 Colorado Springs CO 80914-0123

STIFFEY, ARTHUR VAN BUREN, microbiologist; b. Burgettstown, Pa., Apr. 16, 1918; s. Homer F. and Ruby A. (Forsythe) S.; m. Helen F. Jancsik, June 14, 1941; children: Artis, Arthur, Thomas, Helen, Stuart, Anne. BS in Microbiology, U. Pitts., 1940; MS in Microbiology, Lehigh U., 1948; PhD in Environ. Sci., Fordham U., 1981. Scientist Lederle Labs., Pearl River, N.Y., 1952-76; asst. prof., chmn. biology dept. Ladycliff Coll., Highland Falls, N.Y., 1976-81; postdoctoral Naval Rsch. Lab., Washington, 1981-83; scientist Naval Oceanographic Rsch., Stennis Space Ctr., Miss., 1983-91; adj. assoc. prof. U. New Orleans, 1991-93; with Lumitox Gulf, New Orleans, 1993—. Contbr. articles to Jour. Agrl. and Food Chemistry, Jour. Bacteriology, Applied Microbiology, Jour. Immunology. Lt. col. U.S. Army M.C. Lehigh Inst. Rsch. fellow, 1947-48, Nat. Sci. fellow, 1981; grantee Fordham U., 1977, NSF grant Ladycliff Coll., 1977. Fellow Explorers Club; mem. Beta Beta Beta, Sigma Xi. Achievements include patents for production of lysine, solid-state photometer circuit; research in biosynthesis of lysine from adipic acid precursors by yeast, improvement of collagen sutures from regenerated collagen by incorporation of Guar Gum, bioluminescence assays, marine anti-fouling paints. Home: 811 Freedom Ln Slidell LA 70458

STIFFLER, KEVIN LEE, engineering technologist; b. Tulare, Calif., Nov. 11, 1968; s. Ronald Lee and Kathleen Ruth (Johnson) S.; m. Kristi Dawn Borum, Aug. 1, 1992. AS in Engring., Coll. of Sequoias, 1989; BS in Engring. Tech., Calif. Poly. State U., 1992. Cert. engr. in tng., Calif. Apprentice svc. technician McElmoyl, Inc., Tulare, 1983-90; prodn. asst. Dairyman's Coop. Creamery Assn., Tulare, 1990-91; apprentice journeyman BMI Mech., Inc., Tulare, 1991-92; air conditioning engr. a.a. Marthedal Co., Inc., Fresno, Calif., 1992—. Roy Poage Meml. scholar Calif. Poly. State U., San Luis Obispo, 1991, Assn. Energy Engrs. scholar, 1991. Mem. ASHRAE (scholar So. Calif. 1992). Home: 233 W Lexington # 208 Fresno CA 93711 Office: aa Marthedal Co Inc 1477 N Thesta Fresno CA 93703

STIGLER, STEPHEN MACK, statistician, educator; b. Mpls., Aug. 10, 1941; s. George Joseph and Margaret (Mack) S.; m. Virginia Lee, June 27, 1964; children: Andrew, Geoffrey, Margaret, Elizabeth. BA, Carleton Coll., 1963; PhD, U. Calif., Berkeley, 1967. Asst. prof. U. Wis., Madison, 1967-71, assoc. prof., 1971-75, prof., 1975-79; prof. U. Chgo., 1979—; chmn. dept., 1986-92; Ernest DeWitt Burton Disting. Svc. prof. U. Chgo., 1992—; trustee Ctr. for Advanced Study in the Behavioral Scis., Stanford, Calif., 1986-92, 93—. Author: The History of Statistics, 1986; contbr. articles to jours. in field. Guggenheim Found. fellow, 1976-77; Ctr. for Advanced Study in Behavioral Scis. fellow, 1978-79. Fellow AAAS, Am. Acad. Arts and Scis., Inst. Math. Stats. (Neyman lectr. 1988, pres. 1993), Am. Statis. Assn. (editor Jour. 1979-82, Outstanding Statistician award Chgo. chpt. 1993), Royal Statis. Soc. (Fisher lectr. 1986); mem. Internat. Statis. Inst., Bernoulli Soc., History Sci. Soc., Brit. Soc. for History Sci., Quadrangle Club, Sigma Xi. Office: U Chgo Dept Statistics 5734 S University Ave Chicago IL 60637-1546

STIGLITZ, MARTIN RICHARD, electrical engineer; b. Vienna, Austria, Mar. 24, 1920; came to U.S., 1939, naturalized, 1942; s. Georg Adolph and Maria (Brun) S.; BS, Northeastern U., 1957, MS in Electronics Engring., 1959, MBA in Mgmt., Western New Eng. Coll., 1977; m. Lenna Schoenberg, Dec. 10, 1950 (dec. Apr. 1991); m. Sachiko Sakimura, May 1, 1990. Mech. engr. S.A. Woods Machine Co., Boston, 1939-51; electronics engr., rsch. scientist Air Force Cambridge Rsch. Labs., Hanscom AFB, Bedford, Mass., 1945-75; rsch. electronics scientist Rome Air Devel. Command electromagnetic scis. div. U.S. Air Force, Bedford, Mass., 1985-88; tech. editor Horizon House-Microwave, Inc., Norwood, Mass., 1985—; dir. Solar Energy Tech. Inc., Bedford. With U.S. Army, 1942-45. Mem. IEEE, N.Y. Acad. Scis., Sigma Xi. Patentee solid state devices, med. instruments; contbr. over 50 articles to sci. and profl. jours. Home: 30 Woodpark Cir Lexington MA 02173-7208

STILES, JOHN CALLENDER, physicist; b. Paris, Feb. 21, 1927; came to U.S., 1927; s. William Callender Irvine and Ellen Douglass (Fillebrown) S; m. Virginia Taggert, June 17, 1949; children: Janet, Susan, Judith, William, Robert. BA in Physics, Princeton U., 1950. Jr. scientist Bendix Aviation, Teterboro, N.J., 1951-52; prin. scientist Kearfott Guidance & Navigation Co., Little Falls, N.J., 1952-68, head dept. applied physics, 1972-82, dir. rsch., 1982—; chief scientist Litton Industries, Woodland Hills, Calif., 1968-72. With U.S. Army, 1944-47. Mem. Aero Club Albatross. Achievements include 35 patents in field; development of tuned rotor gyros, of ring laser gyros, of digital accelerometers, of instrumentation for use in inertial navigation systems. Home: 10 Abingdon St Morris Plains NJ 07950-3006 Office: Kearfott Guidance & Navigation Co 1150 Mcbride Ave Little Falls NJ 07424-2564

STILL, EUGENE FONTAINE, II, plastic surgeon, educator; b. Rocky Mount, N.C., Sept. 2, 1937; s. Eugene Fontaine and Eva Ruth (Stevens) S.; m. Frances Davis, Aug. 14, 1965; 1 child, Eugene Fontaine III. BA, Vanderbilt U., 1959; MD, U. Ark., 1966. Diplomate Am. Bd. Cosmetic Surgery (examiner). Intern Univ. Hosp., Little Rock, 1966-67; resident in gen. surgery U. Tenn. Med. Ctr., Memphis, 1967-71; resident in plastic surgery U. Mo. Med. Ctr., Kansas City, 1971-73; instr. surgery U. Mo., Kansas City, 1972-73; sr. surgeon dept. plastic surgery Holt-Krock Clinic, Ft. Smith, Ark., 1973-87; asst. prof. surgery U. Ark., Little Rock, 1974—; chief surgery Sparks Regional Med. Cen., Ft. Smith, 1979; pres. Bd. Cert. Plastic Srugeons, Ft. Smith, 1979—; dir. Ark. Ctr. Plastic Surgery, Van Buren, 1984—; chief of surgery St. Edward Mercy Med. Ctr., Ft. Smith, Ark., 1989-93, Crawford Meml. Hosp., 1993; chief staff St. Edward Mercy Med. Ctr., 1990-91; examiner Am. Bd. Cosmetic Surgery, 1990-93. Served with U.S. Army, 1960-62. Merck Pharm. Co. scholar, 1966. Fellow ACS, Am. Acad. Cosmetic Surgery; mem. Am. Soc. Plastic Surgeons, Southeastern Soc. Plastic Surgeons, Ark. Soc. Plastic Surgeons (sec.-treas. 1983-85, pres.-elect 1986-88, co-founder, pres. 1988-93), Am. Soc. Lipo-suction Surgery, Am. Coll. Physician Execs., Am. Soc. Maxillofacial Surgeons, Ark. Med. Soc. (chmn. ins. com. 1984-90), Ft. Smith Town, Handscrabble County Club, Alpha Omega Alpha. Avocations: golf, clubmaking, woodworking. Home: 10101 Hwy 253 Fort Smith AR 72901-9107 Office: Plastic Surgery Specialists 2717 S 74th St Fort Smith AR 72903-5100

STILL, GERALD G., plant physiologist, research director; b. Seattle, Aug. 13, 1933; s. Edmond N. and Clara J. (Brown) S.; m. Carole J. Hall, June 27, 1954; children: Denise K., Kirk J., Carrie R. BS in Biochemistry, Wash. State U., 1959; MS in Biochemistry, Oreg. State U., 1963, PhD in Biochemistry, 1965. Research fellow Oreg. State U., Corvallis, 1959-65; research chemist USDA Agrl. Research Service CPRB-MRRL, Fargo, N.D., 1965-76; staff scientist USDA Agrl. Research Service, Nat. Park Service, Beltsville, Md., 1976-80, nat. program dir., 1982-84; chief scientist USDA SEA-OD, Washington, 1980-82; dir. USDA Agrl. Research Service, Pacific Gas & Electric Co., Albany, Calif., 1984—; asst. prof. biochemistry N.S. State U., Fargo, 1968-76. Mem. editorial bd. Jour. Agrl. and Food Chemistry, 1978-80; contbr. numerous articles to profl. jours. Served with U.S. Army, 1952-55. Mem. Am. Chem. Soc. (numerous offices div. pesticide chemisty, fellow 1977), Am. Soc. Plant Physiologists, Sigma Xi, Phi Lambda Upsilon. Office: USDA Agrl Research Service Plant Gene Expression Ctr 800 Buchanan St Berkeley CA 94710-1100

STILL, HAROLD HENRY, JR., engineering company executive; b. Beggs, Okla., Sept. 17, 1925; s. Harold Henry and Hannah Jane (Blackburn) S.; student U. Calif., Santa Barbara, 1946-49, USC, 1949-50, UCLA, 1975-76. Sr. specification writer Welton Becket and Assocs., L.A., 1968-71; dept. head Maxwell Starkman and Assocs., Beverly Hills Architect, 1971-72; project mgr. May Dept. Stores Co., L.A., 1972-74; sr. coordinator C.F. Braun and Co., Alhambra, Calif., 1975—, on leave, constrn. mgr. spl. project Runhau Evans Runhau Assocs., Riverside, Calif., 1978-79; architect, project adminstr. developing May Co. Dept. Store Complex for firm Leach Cleveland, Hyakawa, Barry & Assocs.; project mgr. constrn. Lyon Assocs., Inc.; pvt. cons. in constrn. practices, 1983—. With U.S. Army, 1943-45. Decorated Purple Heart. Mem. CSI, ICBO, ASTM, VFW (comdr. post 10965), Constrn. Inspection Assn. Republican. Congregationalist. Home: 503 Beverly Ave Paso Robles CA 93446-1227

STILL, MARY JANE (M. J. STILL), mathematics educator; b. Kingsport, Tenn., Apr. 14, 1940; d. James Charles and Allie Fair (Williams) S.; m. Thos L. Scruggs, 1972 (div. 1975); children: Amanda Fair, Jacob Charles. AB in English, Math., Edn-Psychology, Trevecca Nazarene Coll., 1962; MEd in Math., Statistics, Auburn U., 1969. File clk. FBI, Washington, 1958; tchr. Stratford Jr.-Sr. High Sch., Nashville, 1962-63; statistician Pub. Welfare Dept. State of Tenn., Nashville, 1963-65; math. and English tchr. Smiths Sta. High Sch., Smiths, Ala., 1965-66; math., English, psychology tchr. West Point High Sch., West Point, Ga., 1966-67; math. tchr. La Grange High Sch., La Grange, Ga., 1967-68; math. and English tchr. Townsend High Sch., Townsend, Tenn., 1968-72; math., English, physical edn. tchr. Northshore High Sch., West Palm Beach, Fla., 1974-75; prof. math. Palm Beach Community Coll., Lake Worth, Fla., 1975-78, Palm Beach Community Coll.-North Campus, Palm Beach Gardens, Fla., 1978—; cons. Fla. Power & Light Co., North Palm Beach, 1989; lectr. Palm Beach Community Coll. Speakers, Palm Beach Gardens, 1986-89. editor, advisor: College Mathematics, 1989; textbook editor, advisor Dellen Pub., Scott Foresman Pub., 1988—, Little Brown, McGraw Hill, 1989—; contbr. articles to profl. jours.; author math. booklets, children's stories, poetry; appeared in 2 TV commls., also producer. Scorekeeper baseball and softball leagues Palm Beach area, 1980-90; supporter Jackson polit. campaign, West Palm Beach, 1988, Children's Mus. and Turtle Soc., Juno Beach, Fla., 1986; coach, mgr., sponsor boys' baseball little league, girls' softball, ladies' softball, Lake Park, Palm Beach Gardens, Fla., 1980-92, active softball and basketball coll. and community leagues. Math. summer fellow NSF Northeastern U., Boston, 1988; grad. scholar NSF, Auburn, Ala., 1967-69, Shakespeare scholar Shakespearean Soc. Palm Beach, Stratford-on-Avon, Eng., 1975. Mem. NEA, Math. Assn. Am., Fla. Assn. Community Colls., Am. Statis. Assn., Dreher Sci. Mus., Bus. Women North Palm Beach, Animal Rescue League (West Palm Beach, life), Audubon Soc., NOW, Hist. Soc., Rwy. Club. Nazarene. Avocations: artist, music, drama, sports, church work. Office: Palm Beach Community Coll N 3160 P G A Blvd West Palm Beach FL 33410-2893

STILLINGER, FRANK HENRY, chemist, educator; b. Boston, Aug. 15, 1934; s. Frank Henry and Gertrude (Metcalf) S.; m. Dorothea Anne Keller, Aug. 18, 1956; children—Constance Anne, Andrew Metcalf. B.S., U. Rochester, 1955; Ph.D., Yale U., 1958. NSF postdoctoral fellow Yale U., 1958-59; with Bell Telephone Labs., Murray Hill, N.J., 1959—; head chem. physics dept. Bell Telephone Labs., 1976-79; mem. evaluation panel Nat. Bur. Standards, 1975-78; mem. adv. com. for chemistry NSF, 1980-83, mem. adv. com. for advanced scientific computing, 1984-86; disting. lectr. in chemistry U. Md., 1981; Karcher lectr. U. Okla., 1984; Trumbull lectr. Yale U., 1984; Washburn Meml. lectr. U. Nebr., 1986l Gucker lectr. U. Ind., 1987; W.A. Noyes lectr. U. Tex., 1988; Regents lectr. UCLA, 1990, Meek Indsl. lectr. Ohio State U., 1990, McElvane lectr. U. Wis., 1992. Assoc. editor Jour. Stat. Physics, Jour. Chem. Physics, Phys. Rev. Contbr. articles to profl. jours. Recipient Elliott Cresson medal Franklin Inst., 1978, Hildebrand award Am. Chem. Soc., 1986, Peter J. Debye award Am. Chem. Soc., 1992; Welch Found. fellow, 1974. Fellow Am. Phys. Soc. (Langmuir award 1989); mem. AAAS, Nat. Acad. Scis. Club: Early Am. Coppers Inc. Home: 216 Noe Ave Chatham NJ 07928-1548 Office: 600 Mountain Ave New Providence NJ 07974-2010

STILLINGS, DENNIS OTTO, research director; b. Valley City, N.D., Oct. 30, 1942; s. Harlow Cecil and Ruth Alice (Wolff) S. BA, U. Minn., 1965. Tchr. Henry (S.D.) Pub. Schs., 1965-66, Darby (Mont.) Pub. Schs., 1966-68; tech. rsch. libr., then mgr. hist. dept. Medtronic, Inc., Mpls., 1968-79; instr. humanities U. Minn., Mpls., 1970-72; founding dir., then curator Bakken Libr., Mpls., 1976-80; ind. antiquarian hist. cons. Mpls., 1979-81; project dir. Archaeus Project, Kamuela, Hawaii, 1981—, v.p., 1989—; cons. Ctr. for Sci. Anomalies Rsch., Ann Arbor, Mich., 1983—; bd. dirs. Dan Carlson Enterprises, Mpls. Columnist Med. Progress Through Technology, 1974—; columnist Med. Instrumentation, 1973-76, guest editor, 1975; editor: Cyberphysiology: The Science of Self-Regulation, 1988, Cyberbiological Studies of the Imaginal Component in the UFO Contact Experience, 1989, The Theology of Electricity: On the Encounter and Explanation of Theology and Science in the 17th and 18th Centuries, 1990. Fellow Am. Inst. of Stress; mem. Assn. Sci. Study Anomalous Phenomena, Bioelectromagnetics Soc., Internat. Soc. Biometeorology, Soc. Sci. Exploration. Avocations: Jungian psychology, golf, fishing, travel. Home and Office: Archaeus Project PO Box 7079 Kamuela HI 96743

STILLMAN, GERALD ISRAEL, electrical engineer, science administrator; b. Bklyn., Dec. 27, 1926; s. Morris and Rose (Schlyapin) S.; m. Mira Taube, May 20, 1951; children: Sandy, Joshua, Ezra. BEE cum laude, CCNY, 1946; MS in Elec. Engring., U. Pitts., 1950; postgrad., Poly. Inst. N.Y., 1963; cert. in thermodynamics, U. Wis., 1979. Registered profl. engr., N.Y. Cadet/engr. Westinghouse Electric Corp., 1946-47, d.c. motor design engr., 1947-48; elec. engr. Burns & Roe, Inc., 1948-52, JG White Engring. Corp., 1952-56; sr. planning engr. Pub. Svc. N.J., 1956-60; asst. prof. elec. engring. CCNY, N.Y.C., 1960-62; special projects sect. head Am. Elec. Power Svc. Corp., N.Y.C., 1964-70; supervisory elec. engr. FitzPatrick Nuclear Plant N.Y. Power Authority, N.Y.C., 1970-72, prin. engr. plant siting studies, 1972-77, acting prin. environ. engr., 1976-77, prin. R & D engr., 1977-82, v.p., dir. R & D, 1982-91; pres. Meysh Svcs., Inc., Maplewood, N.J., 1992—; vis. prof. MIT, Cambridge, 1967-68, Northeastern U., Boston, 1968-70; utility adv. bd. Brookhaven Nat. Lab., Upton, N.Y., 1988-90; mem. task force and coms. Elec. Power Rsch. Inst., Palo Alto, Calif., 1987—; mem. steering com. MIT Elec. Utility Program, 1985—; mem. Internat. Conf. Large Power Networks, Paris, 1983—, Empire State Elec. Energy Corp., N.Y.C., 1985—. Co-editor: Three Great Classic Modern Yiddish Writers, 1991; translator: The Parasite (Mendele Moykher Sforim), 1956, Fishke the Lame, 1960; editorial adv. bd. Jewish Currents (Janofsky award), 1979; contbr. articles to Geopolitics of Energy, Transmission and Distbn., others. Steering com. mem. SANE, Maplewood, N.J., 1985—; chmn. com. to compare nuclear and coal costs N.Y. Power Pool; election campaign organizer Democratic Party, Maplewood, 1972. With U.S. Army Signal Corps, 1953-55. Fellow NSF, 1963-64; recipient Translation from Yiddish Cash award Zhitlowsky Found., 1978. Mem. AAAS, IEEE (life, energy devel. subcom., Report Recognition award 1979), Empire State Electric Energy Rsch. Corp. (former chmn. adminstrv. and fossil fuel and advanced generation coms.), Conference Internationale des Grands Reseaux Electrques, N.Y. Acad. Sci., Tau Beta Pi, Eta Kappa Nu. Jewish. Achievements include patents for motor brushholder, for Confined Vortex Cooling Tower. Home and Office: Meysh Svcs Inc 31 Woodland Rd Maplewood NJ 07040-1239

STILLMAN, MICHAEL JAMES, neuroscientist; b. Boston, Sept. 7, 1962; s. Sidney George and Pearl Marion (Bates) S. AB, Bowdoin Coll., 1984; MSc, U. Lowell, 1986. Rsch. fellow Ctr. for Applied Social Sci., Boston, 1989; teaching fellow Psychology Dept., Boston U., 1987-90, rsch. fellow, 1989-90; instr. Mass. Bay C.C., Wellesley, 1991, Newbury Coll., Boston, 1989—; neuroscientist U.S. Army Rsch. Inst. of Environ. Medicine, Natick, Mass., 1989—; cons. edn., Malden, Mass., 1986—; mem. adj. faculty dept. psychology Boston Coll., 1992—. Contbr. articles to profl. jours. Recipient Cert. of Merit Dana-Farber Cancer Inst., 1981. Mem. AAAS, APA, Am. Psychol. Soc., Nat. Assn. of Biology Tchrs., Soc. for Neurosci., Sigma Xi. Home: 183 Bainbridge St Malden MA 02148-2939 Office: USARIEM-Mil Perf Neurosci div Kansas St Natick MA 01760

STIMSON, PAUL GARY, pathologist; b. Ogden, Utah, Jan. 11, 1932; s. Margaret Georgia (Payne) S.; m. Ardell Elizabeth Quiser, June 27, 1958; children: Gregory, Louise, Janiece. DDS, Loyola U., Chgo., 1961; MS, U. Chgo., 1966. Diplomat Am. Bd. Forensic Odontology (pres. 1990-91), Am. Bd. Oral Pathology. From assoc. prof. to prof. U. Tex. Dental Br., Houston, 1965—. Co-editor: Forensic Odontology, 1977. With USN, 1951-54, Korea. Fellow Am. Acad. Oral Pathology, Am. Acad. Forensic Scis.; mem. Masons (past master 1988-89, grand orator 1992), Omicron Kappa Upsilon. Presbyterian. Office: U Tex Dental Br PO Box 20068 Houston TX 77225

STINE, JEFFREY K., science historian, curator; b. San Diego, Calif., Feb. 25, 1953; s. Howard Henry and Dorothy (Graham) S.; m. Marcel Chotowski LaFollette, July 28, 1986. BA, U. Calif., Santa Barbara, 1975, PhD, 1984. Cons. House Com. on Sci. and Tech., Washington, 1984-85; cons. U.S. Army Corps of Engrs., WAshington, 1985-89; curator of engring. Smithsonian Inst., WAshington, 1989—; cons. Carnegie Commn. on Sci., Tech. and Govt., Washington, 1990; commr. U.S., Canadian, Mexican Trilateral Com. on Environ. Edn., 1992. Editor (book reviews) Tech. and Culture, 1987—; author: (Congl. report) A History of Science Policy in the United States, 1940-85, 1986; co-editor: Technology and Choice, 1991. Trustee Pub. Works Hist. Soc., 1990-93. Recipient Congl. fellow Am. Hist. Assn., 1984, Weyerhauser award Forest History Soc., 1984, James Madison prize Soc. for History in the Fed. Govt., 1992, Wesley Johnson prize for Nat. Coun. Pub. History. Mem. Soc. for History of Tech., Am. Soc. for Environ. History, History of Sci. Soc. Democrat. Office: Smithsonian Inst Nat Mus Am History 5014 Washington DC 21560

STINSON, MICHAEL ROY, physicist; b. Vancouver, B.C., Can., Aug. 9, 1949; s. Roy Albert and Florence Jane (Armstrong) S.; m. Susan Ellen Thompson, May 11, 1974; children: Kevin, Cheryl, Valerie, Christopher. BSc with honors, Simon Fraser U., Burnaby, B.C., 1971, MSc, 1973; PhD, Queens U., Kingston, Ont., 1979. Rsch. officer Nat. Rsch. Coun. of Can., Ottawa, Ont., 1979—; vis. assoc. rsch. scientist Columbia U., N.Y.C., 1986—. Contbr. articles to profl. jours. including Jour. Acoustical Soc. Am. and Jour. Phys. F: Metal Physics. Fellow Acoustical Soc. Am. (chmn. tech. program meeting 1993); mem. Inst. Noise Control Engring., Can. Acoustical Assn. Office: Nat Rsch Coun, Montreal Rd Bldg M36, Ottawa, ON Canada K1A 0R6

STIPANOVIC, ROBERT DOUGLAS, chemist, researcher; b. Houston, Oct. 28, 1939. BS, Loyola U., 1961; PhD, Rice U., 1966. Rsch. technician Stauffer Chem. Co., Houston, 1961; teaching asst. Rice U., Houston, 1961-62, rsch. asst., 1962-66; rsch. assoc. Stanford (Calif.) U., 1966-67; mem. grad. faculty Tex. A&M U., College Station, 1967—, asst. prof. chemistry, 1967-71; rsch. chemist Cotton Pathology Rsch. Unit USDA, College Station, 1971-87; rsch. leader USDA, College Station, 1987—; vis. rsch. scientist Agr. Can., Rsch. Ctr. London, Ont., 1985. Welch fellow Rice U., 1963-65, Grad. fellow, 1965-66. Mem. Sigma Xi. Home: 1103 Esther Blvd Bryan TX 77802-1924 Office: USDA Agrl Rsch Svc RR 5 Box 805 College Station TX 77845-9593

STIREWALT, EDWARD NEALE, chemist, scientific analyst; b. Hartsville, S.C., Nov. 29, 1918; s. Neale Summers and Evelyn (Fraser) S.; m. Marcia Marvin Winton, Nov. 21, 1947; children: James Neale, Evelyn Fraser, Marcia Winton. AB, High Point U., 1938; MA, U. N.C., 1942. Staff scientist U.S. AEC, Washington, 1948-53; physicist, supr. U.S. Naval Rsch. Lab., Washington, 1953-57; br. chief Analytic Svcs., Inc., Arlington, Va., 1957-63; ind. cons. def. and energy fields Washington, 1963-77; sr. assoc. Planning Rsch. Corp., McLean, Va., 1977-86; assoc. editor PV News, Casanova, Va., 1984—; mem. Fairfax (Va.) County Air Pollution Control Bd., 1980-85; bd. dirs. Fairfax Hosp., 1963-66. Co-author: Photovoltaics--Sunlight to Electricity in One Step, 1981, A Guide to the Photovoltaic Revolution, 1985. Chmn. Fairfax County Hosp. Commn., 1963-66; mem. Herndon (Va.) Planning Commn., 1989—; bd. dirs. Assn. to Unite Dems., Washington, 1992. Lt. (j.g.) USNR, Manhattan Project, PTO, 1944-46. Mem. AAAS, Washington Philos. Soc., Masons, Sigma Xi. Presbyterian. Home and Office: 762 Monroe St Herndon VA 22070

STIRRAT, WILLIAM ALBERT, electronics engineer; b. Syracuse, N.Y., Nov. 5, 1919; s. Robert William and Doris (White) S.; m. Bernice Amelia Wilson, July 13, 1958; children: Valerie Lynne, Dorothy Grace, William Ellsworth. Student, Triuna (Yaddo) Arts of the Theater Sch., 1936; BS in Physics, Rensselaer Poly. Inst., 1942, postgrad., 1949-50; postgrad., Rutgers U., 1951-58, Fairleigh Dickinson U., 1971. With GE, Schenectady, N.Y., 1941-44; instr. physics Clarkson Coll. Tech., 1947-49; electronic engr. rsch. and devel. U.S. Army, Fort Monmouth, N.J., 1950-87; prin. engr. Eagle Tech., Inc., Eatontown, N.J., 1987-92; pres. Stirrat Arts & Scis., Freehold, N.J., 1992—. Author: (with Alex North) Unchained Melody, 1936 (Top song of Yr., Acad. award nomination 1955), Why 3? (Army award 1985); assoc. editor IEEE Transactions on Electromagnetic Compatability, 1970-76; contbr. articles to profl. jours.; patentee in field. Chmn. pub. rels. Battleground dist. Monmouth coun. Boy Scouts Am., 1970-77; mem. Rep. Congl. Leadership Coun. Mem. IEEE (sr. editor N.J. Coast sect. Scanner 1974-75), Internat. Platform Assn., Assn. of Old Crows, Cen. Jersey Natural Food Club (pres. 1970). Episcopalian. Achievements include development of binomial pulse. Home and Office: 218 Overbrook Dr Freehold NJ 07728-1525

STITT, KATHLEEN ROBERTA, nutrition educator; b. Roanoke, Ala., Dec. 27, 1926; d. Mabrey and Bertha (Green) S. BS in Dietetics, U. Ala., 1946, MS in Food and Nutrition, 1955; PhD in Human Nutrition, Ohio State U., 1965. Adminstrv. dietitian Case Western Res. Hosp., Cleve., 1947-49; hosp. dietitian Selma, Ala., 1949-54; instr., asst. prof. U. Ala., Tuscaloosa, 1955-63, prof. human nutrition and hospitality mgmt. and chmn., 1965-80, coord. or rsch., 1980-82, prof., 1982—. Contbr. numerous articles to profl. jours. Hazel Lapp fellow, Mead Johnson fellow. Mem. Am. dietetic Assn. (Medallion award, mem. Plan V rev. com. 1989-92), Ala. Dietetic Assn. (pres., mem. exec. bd., named Outstanding Dietitian 1976), Am. Home Econs. Assn. (life), Am. Sch. Food Svc. Assn. (editorial adv. panel 1982-87, 89-93, advisor exec. bd. 1987-89), Ala. Home Econs. Assn. (pres. 1975-76), Am. Heart Assn. (ad hoc com. on nutrition edn. for young), AAAS, Am. Pub. Health Assn., N.Y. Acad. Scis., Soc. Nutrition Edn., Am. Soc. Parenteral and Enteral Nutrition, Sigma Xi, others. Office: Univ of Ala PO Box 870158 Tuscaloosa AL 35487-0158

STITTSWORTH, JAMES DALE, neuroscientist; b. Vallejo, Calif., Oct. 8, 1951; s. James Dale and Theresa (Marek) S.; m. Karen Koprowicz, Dec. 30, 1971 (div. 1983); children: Jessica, Antonia, Melissa; m. Kathleen Ann Thoma, July 6, 1985; children: Shannon, Shaun. BA in Biology, Sangamon State Coll., Springfield, Ill., 1975, MA in Psychobiology, 1979. Rsch. assoc. So. Ill. U., Springfield, 1974-81, U. Ill., Rockford, 1981-82, UHS/Chgo. Med. Sch., North Chicago, Ill., 1983-84; rsch. scientist Abbott Labs., Abbott Park, Ill., 1984-91, Searle, Skokie, Ill., 1991—; adj. prof. Oakton Community Coll., Des Plaines, Ill., 1987—; vis. asst. prof. Northwestern Ill. U., Chgo., 1990. Contbr. articles to profl. jours. Mem. Soc. for Neurosci., Sigma Xi. Home: 881 Oxford Pl Wheeling IL 60090 Office: Searl Neurol Disease 4901 Searle Pkwy Skokie IL 60077

STIVER, JAMES FREDERICK, pharmacist, health physicist, administrator, scientist; b. Elkhart, Ind., Jan. 27, 1943; s. Melvin Hugh and Pauline Anna (Schrock) S.; m. Joan Louise Trindle, Aug. 14, 1965; children: Gregory James, Richard Frederick, Kristin Louise, Elizabeth Ann. BS in Pharmacy and Pharm. Scis., Purdue U., 1966, MS, 1968, PhD, 1970. Lic. pharmacist, Ind., N.D. Asst. prof. N.D. State U., Fargo, 1969-73, assoc. prof., 1973-76, radiol. safety officer, 1969-76; radiation safety officer KMS Fusion Inc., Ann Arbor, Mich., 1976-80; mgr., pharmacist Kroger Sav-On Pharmacy Co., Elkhart, Ind., 1980-81; pharmacist Elkhart Gen. Hosp., 1981; environ. regulatory affairs adminstr. Upjohn Co., Kalamazoo, Mich., 1981-88; patent liaison scientist, 1988-92; sr. patent liaison scientist, 1992—; cons., lectr. Mem. Trinity Luth. Ch., Goshen, Ind. Named to Honorable Order Ky. Cols. Fellow Am. Inst. Chemists; mem. AAAS, Am. Pharm. Assn., Ind. Pharmacists Assn., N.D. Pharm. Assn., Am. Chem. Soc., Health Physics Soc., Internat. Radiation Protection Assn., Am. Biol. Safety Assn., N.Y. Acad. Scis., Kappa Psi, Rho Chi, Phi Lambda Upsilon, Sigma Xi. Contbr. articles, abstracts to publs. Home: 505 Skyview Dr Middlebury IN 46540-9427 Office: Upjohn Co Kalamazoo MI 49001

STIVERS, MARSHALL LEE, civil engineer; b. DeLand, Fla.; s. Leonard D. and Verna (Knott) S.; m. Donna Mowen, Sept. 5, 1959; children: Kathryn Ann, Kelly Sue. Registered profl. engr., Fla. With hwy. design Fla. Dept. Transp., DeLand, 1959-63, engr. trainee, 1963-65; final estimator Fla. Dept. Transp., Tallahassee, 1965-69, computer liaison engr., 1969-71, engr. maintenance systems, 1971-79, engr. roadway maintenance, 1979-87, state maintenance engr., 1987—; chmn. weight rev. bd. State of Fla., 1988—; mem. hwy. beautification coun. Fla. Dept. Transp., 1989—. Treas., sec., v.p., pres. Fla. Engring. Soc., Big Bend chpt., 1975-80. Mem. NSPE. Home: 3026 Hawks Glen Tallahassee FL 32312 Office: Fla Dept Transp 605 Suwannee St MS52 Tallahassee FL 32301

STIVERS, THEODORE EDWARD, food products executive, consultant; b. Cleveland, Tenn., Jan. 26, 1920; s. Theodore Edward and Eulalee (Rose) S.; m. Sara Jane Reid, Aug. 8, 1942 (div.); children—Samuel Reid, Karen Elaine Stivers Norman, Joanne Elizabeth Stivers Wheeler; m. Mary Jackson, May 24, 1973. B.S. in Milling Tech., Kans. State U., 1941; postgrad. in engring., Pa. State U., 1942. Registered profl. engr., 15 states. Flour miller Stivers Milling Co., Rome, Ga., 1938-40; trainee Quaker Oats Co., Sherman, Tex., 1941; research mgr. Quaker Oats Co., Akron, Ohio, 1945-53; pres. T.E. Stivers Orgn., Inc., Decatur, Ga., 1953-78, cons., 1978-85; chmn. bd. T.E. Stivers Assocs., Inc., Decatur, Ga., 1970-89; owner, operator Southeastern Mill Machinery, Decatur, Ga., 1955-63; owner Happyvale Mills, Griffin, Ga., 1985—; exec. v.p. K & S Foods, Inc., Decatur, Ga., 1981-89. Chpt. editor: Plant Layout and Design handbook, 1961, 70, 76. Co-chmn. fin. Ga. Citizens for Reagan, 1975-76; del. Republican Nat. Conv., Kansas City, Mo., 1976; chmn. Ga. Reagan for Pres., 1977-82, Reagan-Bush Campaign, 1980; del., chmn. del. Rep. Nat. Conv., Detroit, 1980; mem. nat. adv. com. Reagan-Bush Com., 1984; co-chmn. Nat. Com. Engrs. for Reagan-Bush, 1984. Served to lt. USNR, 1941-45. Decorated Navy Commendation medal; recipient Disting. Service award Nat. Inst. Plant Engrs., 1976. Mem. Am. Soc. Agrl. Engrs. (sr., Cyrus McCormick award 1982), Cons. Engrs. Council Ga. (pres. 1970-71, Engr. of Yr. award 1972, Life Mem. award 1985), Am. Soc. Agrl. Cons. (pres. 1972-73, Disting. Service award 1973), Nat. Council Engring. Examiners (pres. 1976-77), Nat. Council Profl. Services Firms (bd. dirs. 1973-81). Methodist. Clubs: Druid Hills (Atlanta); Capitol Hill (Washington). Avocations: golfing; fishing; travel. Home: PO Box 608 Decatur GA 30031-0608

STOCKARD, JOE LEE, public health service officer, consultant; b. Lees Summit, Mo., May 5, 1924; s. Joseph Frederick and Madge Lorraine (Jones) S.; m. Elsie Anne Chamberlain, Dec. 27, 1957. BS, Yale U., 1945; MD, U. Kans., 1948; MPH, Johns Hopkins U., 1961. Med. officer U.S. Army Med. Corps, Korea and Malaya, 1952-55; asst. prof. preventive medicine Sch. Medicine U. Md., Balt., 1955-58; dep. dir. Cholera Rsch. Lab., Dhaka, Bangladesh, 1960-63; advanced through grades to cap., epidemiologist USPHS, Washington, Md., 1960-76, 64-67; chief preventive medicine sect. USAID, Saigon, Vietnam, 1965-68; assoc. dir. Office Internat. Health, Office of Surgeon Gen. USPHS, Washington, 1967-69; epidemiologist, med. officer Agy. for Internat. Devel., Washington, 1969-87; mem. expert adv. com. WHO, Ouagadougou Burkina, 1987-92; mem. AID project officer Onchocerciasis Control Program, West Africa, 1975-87; med. officer AID Africa Bur., Washington, 1976-87; guest speaker profl. seminar on leptospirosis, 1957; organizer plague sect. meeting 8th Internat. Congress of Tropical Medicine, 1969. Author: (with others) Communicable and Infectious Diseases, 1964; contbr. articles to U.S. Armed Forces Med. Jour., N.Y. Acad. Scis. Jour. Fellow Royal Soc. Tropical Medicine and Hygiene; mem. APHA, Am. Soc. Tropical Medicine and Hygiene, Retired Officers Assn. Achievements include discovery that massive doses of benadryl will not prevent shock in Korean epidemic hemorragic fever, gangrene is not previously a recognized manifestation of bubonic plaque in S.E. Asia; discovery that leptospirosis is a significant problem in troops in Malayasia. Office: 17 Angelwing Dr Hilton Head Island SC 29926-1903

STOCKMAYER, WALTER H(UGO), chemistry educator; b. Rutherford, N.J., Apr. 7, 1914; s. Hugo Paul and Dagmar (Bostroem) S.; m. Sylvia Kleist Bergen, Aug. 12, 1938; children—Ralph, Hugh,. S.B., MIT, 1935, Ph.D., 1940; B.Sc. (Rhodes scholar), Oxford U., 1937; D.Sc., U. Louis-Pasteur, Strasbourg, France, 1972; L.H.D., Dartmouth Coll., 1983. Instr. M.I.T., 1939-41, asst. prof., 1943-46, assoc. prof., 1946-52, prof., 1952-61; prof. chemistry Dartmouth, 1961-79, prof. emeritus, 1979—; instr. Columbia, 1941-43; cons. E.I. duPont de Nemours & Co., Inc., 1945—; vis. com. Nat. Bur. Standards, 1979-84. Contbr. articles on phys. and macromolecular chemistry to sci. jours. Recipient Nat. Medal of Sci., 1987, MCA Coll. Chemistry Tchr. award 1960, ; Guggenheim fellow, 1954-55, hon. fellow Jesus Coll., Oxford, Eng., 1976, Alexander von Humboldt fellow, 1978-79. Fellow Am. Acad. Arts and Scis., Am. Phys. Soc. (Polymer Physics prize 1975); mem. NAS, Am. Chem. Soc. (assoc. editor Macromolecules 1968-74, 76—, chmn. polymer chem. div. 1968, Polymer Chemistry award 1965, Peter Debye award 1974, T. W. Richards medal 1988, polymer div. award 1988), Soc. Plastics Engrs. (Internat. award 1991), Soc. Polymer Sci. Japan (hon. 1991), Sigma Xi (William Procter prize 1993), Appalachian Mountain Club. Office: Dartmouth Coll Chemistry Dept Hanover NH 03755

STOCKTON, ANDERSON BERRIAN, electronics company executive, consultant, genealogist; b. Lithonia, Ga., Oct. 7, 1943; s. Berrian Henry and Mary Grace (Warbington) S.; m. Linda Arlene Milligan, June 9, 1963; 1 child, Christopher Lee. Cert. in cryptographic engring., USAF Acad., Wichita Falls, Tex., 1963. Supr. Western Union Telegraph Co., East Point, Ga., 1965-67; mgr. RCA Corp., Cherry Hill, N.J., 1967-72; v.p. Universal Tech., Inc., Verona, N.J., 1972-76, Siemens Am., Anaheim, Calif., 1976-84, Concorde, El Toro, Calif., 1984-85, Data Card Troy, Inc., Santa Ana, Calif., 1985-86; dir. laser engring. div. ITT, San Jose, Calif., 1986-87; v.p. S.T.A.R. Ricoh Corp., San Jose, 1988-93; exec. dir. mktg. QMS, Inc., Mobile, Ala., 1993—; cons. Hutchinson (Minn.) Tech. Corp., 1984-87. Author: Polled Network Communications, 1976, A Quest for the Past, 1991; patentee in field. With USAF, 1961-65. Mem. IEEE, Am. Electronics Assn. Avocations: classic car collecting, genealogical and historical research, sword, coin and stamp collecting. Home: 2086 Silence Dr San Jose CA 95148

STOCKWELL, ALBERT H., procurement professional; b. Niagara Falls, N.Y., Feb. 18, 1933; s. Frank and Amelia (Rochan) S.; children: Karyn L. Stockwell, Lori Wright. BS in Natural Scis., Niagara U., 1970. Cert. purchasing mgr. Buyer Hooker Chem. Co., Grand Island, N.Y., 1960-70; MRO buyer Dow Chem. Co., Midlands, Mich., 1970-74; purchasing mgr. Allied Chem. Co., Macon, Ga., 1974-78; sr. elec. buyer Polysius Corp., Atlanta, 1978-84; cons. Kennesaw Cons. and Constructors, Kennesaw, 1984-87; purchasing mgr. Tie Down Engring., Atlanta, 1987-90; procurement mgr.

Humboldt Wedag Inc., Norcross, Ga., 1990—. Office: Humboldt Wedag Inc 3200 Pointe Pkwy Norcross GA 33092

STODDARD, FORREST SHAFFER, aerospace engineer, educator; b. Eglin AFB, Fla., Nov. 4, 1944; s. Edward Forrest and Esther Grace (Shaffer) S.; SB, MIT, 1966, SM, 1968; PhD, U. Mass., 1979; m. Mary Anne Maher Matthews, June 16, 1979; children: Joshua Forrest, Nathan Edward. Partner, chief engr. U.S. Windpower Inc., Burlington, Mass., 1977-80 ; wind power engrng. cons., Amherst, Mass., 1980—; cons. Wind Systems Test Center, U.S. Dept. Energy Solar Energy Rsch. Inst., Commonwealth of Mass.; asst. prof. mech. engrng. U. Mass., 1982—; founder, pres. Pioneer Wind Power, Inc., 1982-87; cons., prin. investigator U.S. Dept. Energy, 1985-86; rsch. prof. West Tex. State U., 1986-92; design engr. Second Wind Inc., Somerville, Mass., 1992—. Served to capt. USAF, 1968-72. Co-author Wind Turbine Engineering Design, 1987. Mem. Am. Wind Energy Assn. (dir., sec.), Am. Helicopter Soc., AIAA, Am. Solar Energy Soc., Friends of Earth, Sigma Xi. Mem. United Ch. Christ. Acting editor Wind Tech. Jour., 1979-82. Home: 14 Whittemore St Arlington MA 02174 Office: Second Wind Inc 7 Davis Square Somerville MA 02144 Office: AEI/WTSU PO Box 248 Canyon TX 79016-0002

STOFFA, PAUL L., geophysicist, educator; b. Palmerton, Pa., July 9, 1948; married, 1968; 2 children:. BS, Rensselaer Poly. Inst., 1970; PhD in Geophysics, Columbia U., 1974. Research assoc. marine geophysics Lamont-Doherty Geol. Observatory, 1974-81; cons. Gulf Sci. Tech., 1981—; with Inst. for Geophysics U. Tex., Austin, now Wallace E. Pratt prof. geophysics, sr. research scientist Inst. for Geophysics, from 1978; adj. asst. prof. Columbia U., 1978—. Mem. IEEE, Am. Geophys. Union, Soc., Soc. Exploration Geophysicists, Sigma Xi. Office: U Tex at Austin Dept of Geol Scis Austin TX 78712

STOFFER, BARBARA JEAN, research laboratory technician; b. East Saint Louis, Mo., Mar. 19, 1946; d. William Eygene Calvert and Hazel Ray; m. Robert Lee Petty (div. Mar. 1969); 1 child, Natashia Mary Louise; m. Raymond Russel Stoffer, Sept. 11, 1970. Assoc., Ind. Tech. Coll., 1986. Dept. mgr. W.T. Grant Dept. Store, Terre Haute, Ind., 1965-68; teletype and flex writer Weston Paper, Terre Haute, 1968-70; chem. lab. technician Comml. Solvents, Terre Haute, 1974-76, micro biol. lab. technician, 1977-78; agronomics lab. technician IMC, Terre Haute, 1978-86; pharm. lab. technician Pittman Moore, Terre Haute, 1986—. Active Apt. Assn. Ind., Terre Haute, 1986—; foster parent Foster Parent Assn., Terre Haute, 1989—; bd. dirs. State Legal Aid Assn., Terre Haute, 1982-83. Mem. Am. Spectroscopy Assn. Home: 6623 N 36th St Terre Haute IN 47805 Office: Pittman Moore 1331 S 1st St Terre Haute IN 47808

STOHR, ERICH CHARLES, biomedical engineer; b. Rochester, N.Y., May 11, 1964; s. Herbert Charles and Eva-Maria (Graf) S. BS, Case Western Res. U., 1986; MS, Worcester Poly. Inst., 1992. Rsch. tech. Univ. Hosp., Cleve., 1985-86; engrng. cons. AMT, Cleve., 1986-90; clin. engr. MetroHealth Med. Ctr., Cleve., 1986-90; biomed. engr. Medtronic, Mpls., 1992—. Contbr. articles to profl. jours. Mem. WMUG, NCCETA. Home: 61 Cambridge Ct Fairport NY 14450

STOIA, DENNIS VASILE, industrial engineering educator; b. Aberdeen, S.D., Dec. 31, 1928; s. John and Seanna (Biliboca) S.; m. Margaret Ann Tyne, May 11, 1974; 1 child, Justin Michael. B of Indsl. Engrng., Ohio State U., 1954; MBA, U. Chgo., 1962. Indsl. engr. Sunbeam Corp., Chgo., 1953-64; v.p. mfg. Aerosol Rsch. Co., North Riverside, Ill., 1964-74; ops. mgr. Ethyl/VCA, Bridgeport, Conn., 1974-75; labor arbitrator Somonauk, Ill., 1975—; assoc. prof. No. Ill. U., DeKalb, 1978—, tech. dept. chmn., 1987—; arbitrator Fed. Mediation and Conciliation, Washington, 1981—, Coal Arbitration Svc., Washington, 1987—; hearing officer Ill. State Bd. Edn., Springfield, 1980—. Author Arbitrator award, Bur. Nat. Affairs, 1990. Bd. dirs. Somonauk Sch., sec., Sandwich Community Hosp., treas. Sgt. U.S. Army, 1946-48, Japan, 1950-51, Korea. Mem. Nat. Assn. for Indsl. Tech., Soc. Profls. in Dispute Resolution, Soc. Fed. Labor Rels. Profls., Am. Arbitration Assn., Am. Soc. for Engrng. Edn., Theta Tau, Epsilon Pi Tau. Home: 13767 Chicago Rd Somonauk IL 60552-9724 Office: No Ill U Dept Engrng Tech De Kalb IL 60115

STOJANOWSKI, WIKTOR J., mechanical engineer; b. Filipow, Poland, June 3, 1936; arrived in Canada, 1986; s. Stanislaw and Bronislawa (Mentel) S.; m. Danuta A. Wrona, Oct. 19, 1958; children: Dorota, Anna, Robert. MS, Acad. Mining and Metallurgy, 1958, PhD in Mech. Engrng., 1973. Chief engr. design Food Industry Equipment Plant, Cracow, Poland, 1958-63; sr. design engr. Chemistry Machine and Process Design, Cracow, 1963-65; tutor Acad. Mining and Metallurgy, Cracow, 1965-80, asst. prof., 1981-87; vis. lectr. U. Wis., Madison, 1980-81; project engr. Vibron Ltd., Mississauga, Ontario, Canada, 1988-90; ptnr. J.E. Coulter Assocs. Engrng., Willowdale, Ontario, 1991—; noise control expert Assn. Polish Mechanics Engrng., Cracow, 1973-87. Co-author: Lecture of AMM #454, 1974; contbr. articles to Jour. Mech. Engrng., Inter-Noise. Achievements include patent for noise eliminator of gas stream flow into atmosphere, valve for gas expansion. Home: 1 Markburn Ct, Etobicoke, ON Canada M9C 4Y6

STOKES, CHARLES ANDERSON, chemical engineer, consultant; b. Mohawk, Fla., Oct. 28, 1915; m. Constance Currier; children: Jeffrey Andrew, Harry Currier, Christopher Alden. BS, U. Fla., 1938; ScD, MIT, 1951. Registered profl. engr., Fla., Mass. Instr., asst. prof. MIT, Cambridge, 1940-45; cons. fuels and synthetic rubber materials War Prodn. Bd., Washington, 1944-55; dir. R&D Cabot Corp., Boston, 1945-55; v.p., tech. dir. Tex. Butadiene and Chem. Corp., Houston, 1955-59; v.p. tech. and planning Columbian Div. Cities Svc. Co., N.Y.C., 1960-69; chmn. Stokes Consulting Group, Naples, Fla., 1969—; chmn. adv. bd., conf. co-chmn. U. Pitts., 1977-83; mem. adv. bd. Fla. Solar Energy Ctr., Cape Canaveral, Fla., U.S. Rep. Porter Goss, Ft. Myers, Fla.; cons. in fields of fuel methanol, vapor recovery and resource recovery. Co-author: (with R. Williams) World Methanol Survey, 3d edit., 1981; contbr. articles to profl. jours. Named Engr. of Yr., Cen. Jersey Engrng. Coun., 1977, Outstanding Alumni of Yr., U. Fla., 1990. Fellow AICE; mem. Am. Chem. Soc., Fla. Engrng. Soc. (Outstanding Tech. Achievement award 1979, Engr. of Yr. 1984-85). Achievements include patents in field of carbon black; pioneering work carbon black process development.

STOKES, CHARLES JUNIUS, economist, educator; b. Washington, Aug. 17, 1922; s. Francis Warner and Vivienne E. (Cooke) S.; m. Anne Richardson Wood, June 13, 1946; children—Kevin Barrett, Keith Warner. A.B. with honor and distinction, Boston U., 1943, A.M., 1947, Ph.D., 1950. Mem. faculty Atlantic Union Coll., South Lancaster, Mass., 1946-60, dean coll., 1954-56; Charles A. Dana prof. econs. U. Bridgeport, Conn., 1960-89; Charles A. Dana univ. prof. U. Bridgeport, 1990-92, chmn. dept., 1960-72; prof. econs. Andrews U., Berrien Springs, Mich., 1990—; dir. econ. rsch., region I OPS, 1951-53; dir. Latin Am. case studies Brookings Instn., 1963-64; Fulbright prof., Ecuador, 1958-59, Argentina, 1960, Peru, 1964; lectr. Inter-Am. Def. Coll., 1977-78; Staley Disting. lectr. Andrews U., 1983, founder, dir. Chan Shun Ctr. for Bus. Rsch., 1991; E. A. Johnson Disting. lectr., 1989; vis. prof., lectr. U. Colo., U. Conn., Clark U., Andrews U., U.S. Naval Postgrad Sch., Yale U., Columbia U., So. Conn. State U., Atlantic Union Coll., and numerous countries; founder, dir. Conn. Small Bus. Devel. Ctr. U. Bridgeport, 1985-88; chmn., dir. Monroe Bank & Trust Co.; cons. to industry, founds. Author: Crecimiento Economico (Economic Growth), 1964, Transportation and Economic Development in Latin America, 1968, Managerial Economics: A Case Book, 1968, Managerial Economics: A Textbook, 1969, Historic Fairfield County Churches, 1969, Urban Housing Market Performance, 1975, Economics for Managers, 1978; editor: THRUST, 1978-89; columnist Christian Sci. Monitor; also articles to profl. jours., columns in regional newspapers. Chmn. Lancaster (Mass.) Housing Authority, 1957-61; chmn. econ. com. Greater Bridgeport Regional Planning Agy., 1961-72; asst. dir. U.S. GAO, 1972-73; mem. Instn. for Social and Policy Studies, Yale U., 1977-85; Trustee Pioneer Valley Acad., 1966-69, Andrews U., 1967-72, Atlantic Union Coll., 1968-73, Conn. Grand Opera Assn., 1980-86; mem. Conn. com. Regional Plan Assn., 1977—; bd. dirs. Greater Bridgeport Symphony Soc., 1978—, Adventist Living Ctrs., Inc., Adventist Health System/North, 1980-85, New England Trade Adjustment Assistance Ctr., Inc.; chmn. bd. dirs., chief exec. officer Geer Meml. Hosp., 1983-89.; Served with AUS, 1943-46; assoc. Kellogg Ctr., U. Notre Dame, 1990—. Decorated Medal of Honor Argentina; named to Collegium of Disting. Alumni Boston U. Coll. Liberal Arts, 1974; Sears Found. Fed. faculty fellow, 1972-73. Fellow New Eng. Bd. High Edn.; mem. Nat. Economists Club, Am. Econ. Assn. (pres. Conn. Valley 1966), Nat. Assn. Bus. Economists (pres. Fairfield County chpt. 1980-81), Phi Beta Kappa, Phi Kappa Phi, Phi Beta Kappa Assos., Delta Sigma Rho, Beta Gamma Sigma. Home: Pepper Crossing Stepney CT 06468 Office: Andrews U ChanShon Hall Berrien Springs MI 49104

STOKES, THOMAS LANE, JR., biologist, consultant; b. Norfolk, Va., May 29, 1957; s. Thomas Lane and Martha Ann (Kavanaugh) S.; m. Selina Leigh Basnight, Oct. 26, 1985; children: Thomas Lane III, Mary Lyall Stokes. BA in Biology, Hampden-Sydney Coll., 1979; MS, Old Dominion U., 1982; MBA in Statistics, NYU, 1988. Tech. water quality Hampton Rds. Sanitation Dist., Norfolk, 1982-84; scientist water quality N.Y.C. Dept. Environ. Protection, 1984-87, chief water quality, 1988-90; pres. Stokes Environ. Assocs., Ltd., Norfolk, 1990—; mem. N.Y. Harbor Restoration Program, 1988-90. Co-author: Wastewater Biology: Manual of Practice, 1990; contbr. articles to profl. jours. Mem. Hampton Rds. C. of C., AAAS, Assn. Groundwater Scientists and Engrs., Soc. Wetland Scientists, Water Pollution Control Fedn., Sigma Xi. Presbyterian. Achievements include development of statistical basis of industrial waste regulation. Office: Stokes Environ Assocs Ltd 550 E Main St Ste 408 Norfolk VA 23510

STOLBERG, ERNEST MILTON, environmental engineer, consultant; b. Balt., Aug. 7, 1913; s. Edward B. and Marie S. Stolberg; m. Doris Pearl Goodman, June 1, 1947; children: Barbara, Arnold. B of Engrng., Johns Hopkins U., 1937, M in Environ Engrng., 1969. Registered profl. engr., Pa., Ohio, Md. Engr., draftsman U.S. Army, Aberdeen, Md., 1939-42; chief mech. engr. Housing Authority Balt., 1947-56; chiel plant engr. long range planning Westinghouse Electric Corp., Pitts., 1957-59; chief plant engr. Comml. Shearing Co., Youngstown, Ohio, 1959-67; mech. engr. U.S. Army, Odenton, Md., 1967-72, Exxon, Odenton, 1972-76; chief mech. engr. Washington Govt., 1976-77; environ engr. value analysis U.S. VA, Washington, 1977-85; environ and energy cons. rsch. divsn. dept. legis. reference Md. State Legislature, Annapolis, 1987-90. Contbr. articles to profl. jours. Lt. comdr. USN, 1942-46, ETO, PTO, NATOUSA, USNR, ret. Named Distinguished Toastmaster, Toastmasters Internat., 1983. Mem. ASME (life, chmn. 1975-76, cert. of appreciation 1977), Am. Inst. Plant Engrs. (chmn. 1962), Nat. Wildlife Fedn., Nat. Geographic Soc., Md. Soc. Profl. Engrs. (chmn. 1972, Outstanding Svc. award 1983), The Cousteau Soc., People for the Am. Way, Nat. Assn. Ret. Fed. Employees (chmn. Randallstown, Md. chpt.), Common Cause, Sierra Club. Democrat. Jewish. Achievements include invention promoting boiler burner safety. Home: 902 Bare Branch Ct Baltimore MD 21208

STOLEN, ROGERS HALL, optics scientist; b. Madison, Wis., Sept. 18, 1937. BA, St. Olaf Coll., 1959; PhD in Physics, U. Calif., Berkeley, 1965. Fellow U. Toronto, 1964-66; mem. tech. staff solid state optics AT&T Bell Labs., Holmdel, N.J., 1966—. Mem. Am. Phys. Soc., Optical Soc. Am. (R. W. Wood prize 1990). Achievements include research in nonlinear properties of optical fibers, polarization preserving optical fibers, light scattering in glass. Office: AT & T Bell Labs 4B-421 Holmdel NJ 07733*

STOLLERY, ROBERT, construction company executive; b. Edmonton, Alta., Can., May 1, 1924; s. Willie Charles and Kate (Catlin) S.; m. Shirley Jean Hopper, June 11, 1947; children: Carol, Janet, Douglas. B.Sc. Eng., U. Alta., 1949, LL.D. (hon.), 1985; hon. LL.D., Concordia U., Montreal, Que., 1986. Field engr. Poole Constrn. Ltd., Edmonton, 1949-54, project mgr., 1954-64, v.p., 1964-69, pres., 1969-81; chmn. bd. PCL Constrn. Group Inc., Edmonton, 1979—; bd. dirs. TransCanada Pipelines, Fed. Industries Ltd., Winnipeg, Man., Toronto Dominion Bank. Chmn. bus. adv. coun. U. Alta., gov. of trustees; pres. Edmonton Community Found. Recipient Exec of Yr. award Inst. Cert. Mgmt. Cons. of Alta., 1988, Can. Businessman of Yr. award, 1993. Fellow Can. Acad. Engrng.; mem. Assn. Profl. Engrs. (Frank Spragins Meml. award 1981), Engrng. Inst. Can. (Julian C. Smith medal 1990), Conf. Bd. Can. (vice chmn. 1980-82), Constrn. Assn. Edmonton (pres. 1972, Claude Alston Meml. award), Can. Constrn. Assn. (v.p. 1970, Can. Businessman of the Yr. award 1993). Conservative. Mem. United Ch. of Canada. Club: Mayfair Golf and Country (Edmonton). Office: PCL Construction Group Inc, 5410 99 St, Edmonton, AB Canada T6E 3P4

STOLLNITZ, FRED, comparative psychologist, government program administrator; b. N.Y.C., Apr. 13, 1939; s. Henry Sande and Helen Cecile (Bessemer) S. m. Janet Louise Gabar, Aug. 6, 1961; children: Nancy Beth, Eric Joel. BA with high honors, Swarthmore Coll., 1959; MS, Brown U., 1961, PhD, 1963. Asst. prof. psychology Brown U., Providence, R.I., 1963-66, Cornell U., Ithaca, N.Y., 1966-71; asst. prog. dir. for psychobiology NSF, Washington, 1971-72, assoc. prog. dir., 1972-76, prog. dir. for psychobiology, 1976-88, prog. dir. for animal behavior, 1988—. Co-editor and contbg. author: (books) Behavior of Nonhuman Primates, 5 vols., 1965-74; contbr. articles to profl. jours. Recipient Crane prze Swarthmore (Pa.) Coll., 1958, predoctoral rsch. fellowship USPHS/Brown U., 1960-62, grad. fellowship NSF, 1962-63. Mem. Animal Behavior Soc., Psychonomic Soc., Am. Psychol. Assn., Am. Psychol. Soc., Internat. Primatol. Soc., Ea. Psychol. Assn., AAAS, Phi Beta Kappa, Sigma Xi. Office: NSF IBN Rm 679 4201 Wilson Blvd Arlington VA 22203

STOLTENBERG, CAL DALE, psychology educator; b. North Bend, Nebr., Apr. 1, 1953; s. Kenneth Rudolph and Betty Jean (Pribnow) S.; m. Peggy Ann Hofman, June 19, 1976; children: Braden, Ilea, Kara. MEd, U. Nebr., 1977; PhD, U. Iowa, 1981. Psychology intern Counseling & Cons. Svc. Ohio State U., Columbus, 1980-81; asst. prof. psychology, adj. sr. staff psychologist Tex. Tech. U., Lubbock, 1981-86; assoc. prof., program dir. counseling psychology program U. Okla., Norman, 1986-90, prof., chmn. edn. psychology, 1990—, adj. rsch. fellow Ctr. for Rsch. in Minority Edn., 1991—; adj. clin. faculty psychology residency program Okla. Health Dept., 1990—. Author and editor: Social Perception in Clinical and Counseling Psychology, 1984, Social Process in Clinical and Counseling Psychology, 1987; author: Supervising Counselors and Therapists, 1987; contbr. articles to Jour. of Counseling and Devel., Profl. Psychology, Rsch. and Practice, Jour. of Counseling Psychology. Mem. com. Univ. Luth. Ch., Norman, 1990, PTO, Norman, 1991. Fellow APA, Am. Psychol. Soc. (charter, div. 17, chmn. nominations com. 1992—); mem. ACA. Democrat. Lutheran. Achievements include developement of model of counselor/psychotherapist training; research in aplying social psychological research and practice to counseling psychology. Home: 320 Redwing Dr Norman OK 73071 Office: U Okla Dept Edn Psychology 820 Van Vleet Oval Norman OK 73019-0260

STONE, EDWARD CARROLL, physicist, educator; b. Knoxville, Iowa, Jan. 23, 1936; s. Edward Carroll and Ferne Elizabeth (Baber) S.; m. Alice Trabue Wickliffe, Aug. 4, 1962; children—Susan, Janet. AA, Burlington Jr. Coll., 1956; MS, U. Chgo., 1959, PhD, 1964; DSc (hon.), Washington U., Saint Louis, 1992, Harvard U., 1992, U. Chgo., 1992. Rsch. fellow in physics Calif. Inst. Tech., Pasadena, 1964-66, sr. rsch. fellow, 1967, mem. faculty, 1967—, prof. physics, 1976—, chmn. div. physics, math. and astronomy, 1983-88, v.p. for astron. facilities, 1988-90, v.p., dir. Jet Propulsion Lab., 1991—; Voyager project scientist, 1972—; cons. Office of Space Scis., NASA, 1969-85, mem. adv. com. outer planets, 1972-73; mem. NASA Solar System Exploration Com., 1983; mem. com. on space astronomy and astrophysics Space Sci. Bd., 1979-82; mem. NASA high energy astrophysics mgmt. operating working group, 1976-84, NASA Cosmic Ray Program Working Group, 1980-82, Outer Planets Working Group, NASA Solar System Exploration Com., 1981-82, Space Sci. Bd., NRC, 1982-85, NASA Univ. Relations Study Group, 1983, steering group Space Sci. Bd. Study on Major Directions for Space Sci., 1995-2015, 1984-85; mem. exec. com. Com. on Space Research Interdisciplinary Sci. Commn., 1982-86; mem. commn. on phys. scis., math. and resources NRC, 1986-89; mem. adv. com. NASA/Jet Propulsion Labs. vis. sr. scientist program, 1986-90; mem. com. on space policy NRC, 1988-89; chmn. adv. panel for The Astronomers, KCET, 1989—. Mem. editorial bd. Space Sci. Instrumentation, 1975-81, Space Sci. Rev., 1982-85, Astrophysics and Space Sci., 1982—. Recipient medal for exceptional sci. achievement NASA, 1980, Disting. Service medal, 1981; Am. Edn. award, 1981; Dryden award, 1983; Space Sci. award AIAA, 1984; Sloan Found. fellow, 1971-73; NASA Disting. Pub. Service medal, 1985, NASA Outstanding Leadership Medal, 1986; Aviation Week and Space Tech. Aerospace Laureate, 1989, Science award Nat. Space Club, 1990, Sci. Man of Yr. award ARCS Found., 1991, Pres.'s Nat. medal of Sci., 1991, Am. Philos. Soc. Magellanic award, 1992, Am. Acad. Achievement Golden Plate award, 1992, COSPAR award for outstanding contribution to space sci., 1992, LeRoy Randle Grumman medal, 1992, Disting. Pub. Svc. award Aviation/Space Writers Assn., 1993. Fellow AIAA (assoc.), Am. Phys. Soc. (chmn. cosmic physics divsn. 1979-80, exec. com. 1974-76), Am. Geophys. Union; mem. NAS, AAAS, Internat. Astron. Union, Internat. Acad. Astronautics, Am. Astron. Soc. (com. mem. divsn. planetary scis. 1981-84), Am. Assn. Physics Tchrs., Am. Philos. Soc., Calif. Assn. Rsch. in Astronomy (bd. dirs. 1985—, vice chmn. 1987-88, 90—), Astron. Soc. Pacific (hon.), Am. Philos. Soc. Office: Jet Propulsion Lab 4800 Oak Grove Dr M/S 180-904 Pasadena CA 91109

STONE, JACQUELINE MARIE, biotechnology patent examiner; b. Washington, Nov. 19, 1959; d. Clyde Orville and Alma Mae (Linn) Bradley; m. Michael Wayne Stone, June 6, 1981; children: Jessica, Shannon. BS, U. Md., 1981. Rsch. chemist USDA, Beltsville, Md., 1979-83; patent examiner in therapeutics U.S. Patent & Trademark Office, Washington, 1983—. Named to Dean's List, U. Md., 1978. Mem. AAAS, Nat. Honor Soc., Patent and Trademark Office Soc. Republican. Methodist. Achievements include development of patentability determinations in the field of gene therapy, lymphokine and enzyme therapy. Office: Patent & Trademark Office 10th Fl Rm A03 1911 Jefferson Davis Hwy Arlington VA 22202-3508

STONE, JAMES ROBERT, surgeon; b. Greeley, Colo., Jan. 8, 1948; s. Anthony Joseph and Dolores Concetta (Pietrafeso) S.; m. Kaye Janet Friedman, May 16, 1970; children: Jeffrey, Marisa. BA, U. Colo., 1970; MD, U. Guadalajara, Mex., 1976. Diplomate Am. Bd. Surgery, Am. Bd. Surg. Critical Care. Intern Md. Gen. Hosp., Balt., 1978-79; resident in surgery St. Joseph Hosp., Denver, 1979-83; practice medicine specializing in surgery Grand Junction, Colo., 1983-87; staff surgeon, dir. critical care Va. Med. Ctr., Grand Junction, 1987-88; dir. trauma surgery and critical care, chief surgery St. Francis Hosp., Colorado Springs, Colo., 1988-91; pvt. practice Kodiak, Alaska, 1991-92; with Surgical & Trauma Assocs., Englewood, Colo., 1992—; asst. clin. prof. surgery U. Colo. Health Sci. Ctr., Denver, 1984—; pres. Stone Aire Cons., Grand Junction, 1988—; owner, operator Jjnka Ranch, Flourissant, Colo.; spl. advisor CAP; med. advisor med. com. unit, 1990-92; recipient Bronze medal of Valor. Contbr. articles to profl. jours.; inventor in field. Bd. dirs. Mesa County Cancer Soc., 1988-89, Colo. Trauma Inst., 1988-91. Colo. Speaks out on Health grantee, 1988; recipient Bronze medal of Valor Civil Air Patrol. Fellow Denver Acad. Surgery, Southwestern Surg. Congress, Am. Coll. Chest Physicians, Am. Coll. Surgeons (trauma com. Colo. chpt.), Am. Coll. Critical Care; mem. Am. Coll. Physician Execs., Soc. Critical Care (task force 1988—). Roman Catholic. Avocations: horse breeding, hunting, fishing.

STONE, JENNINGS EDWARD, applications engineer; b. Charleston, W.Va., Oct. 30, 1942; s. Wirt Jennings and Thelma Jean Stone; m. Patricia Lynn Oldham; children: Laura Anne, Marlene Jeanette, Paris Edward, Aaron Michel Wirt. Grad., Stonewall Jackson High Sch., Charleston, W.Va., 1960; postgrad., DeVry Tech. Inst., Chgo., 1961-65. Cert. applications engr. Quality control Zenith Corp., Chgo., 1961-65; lineman C&P Telephone, Charleston, 1965; sales rep. RPS, Parkersburg, W.Va., 1965-78; store mgr. RPS, Charleston, 1978-79; sales rep. TMS, Beckley, W.Va., 1979-81; pres. Appalachian Tooling Inc., Charleston, 1981—. With USN, 1961. Mem. Charleston C. of C. Republican. Avocations: gardening, computers. Home and Office: Appalachian Tooling Inc 1203 Early St Charleston WV 25302-1020

STONE, MARVIN JULES, immunologist, educator; b. Columbus, Ohio, Aug. 3, 1937; s. Roy J. and Lillian (Bedwinek) S.; m. Jill Feinstein, June 29, 1958; children: Nancy Lillian, Robert Howard. Student, Ohio State U., 1955-58; SM in Pathology, U. Chgo., 1962, MD with honors, 1963. Diplomate Am. Bd. Internal Medicine, (Hematology, Med. Oncology). Intern ward med. svc. Barnes Hosp., St. Louis, 1963-64, asst. resident, 1964-65; clin. assoc. arthritis and rheumatism br. Nat. Inst. Arthritis and Metabolic Diseases, NIH, Bethesda, Md., 1965-68; resident in medicine, ACP scholar Parkland Meml. Hosp., Dallas, 1968-69; fellow in hematology-oncology, dept. internal medicine U. Tex. Southwestern Med. Sch., Dallas, 1969-70, instr. dept. internal medicine, 1970-71, asst. prof., 1971-73, assoc. prof., 1974-76, clin. prof., 1976—, chmn. bioethics com., 1979-81; mem. faculty and steering com. immunology grad. program, Grad. Sch. Biomed. Scis., U. Tex. Health Sci. Ctr., Dallas, 1975, adj. mem., 1976—; dir. Charles A. Sammons Cancer Ctr., chief oncology, dir. immunology, co-dir. div. hematology-oncology, attending physician Baylor U. Med. Ctr., Dallas, 1976—; adj. prof. biology So. Meth. U., Dallas, 1977—; v.p. med. staff Parkland Meml. Hosp., Dallas, 1982; cons. in internal medicine Dallas VA Hosp. Contbr. chpts. to books, articles to profl. jours. Chmn. com. patient-aid Greater Dallas/Ft. Worth chpt. Leukemia Soc. Am., 1971-76, chmn. med. adv. com., 1978-80, bd. dirs., 1971-80; mem. v.p. Dallas unit Am. Cancer Soc., 1977-78, pres., 1978-80; mem. adv. bd. Baylor U. Med. Ctr. Found.; mem. med. adv. bd. Dallas chpt. Lupus Found. Am., 1982-87. With USPHS, 1965-68. Named Outstanding Full Time Faculty Mem. Dept. Internal Medicine, Baylor U. Med. Ctr., 1978, 87. Fellow ACP (gov. No. Tex. 1993—); mem. AMA, Am. Coll. Rheumatology, Soc. Leukocyte Biology, Am. Assn. Immunologists, Am. Fedn. Clin. Rsch., Am. Soc. Hematology, Internat. Soc. Hematology, Internat. Soc. Preventive Oncology, N.Y. Acad. Scis., Am. Assn. Cancer Edn., Coun. on Thrombosis, Am. Heart Assn. (established investigator 1970-75), Am. Soc. Clin. Oncology, Am. Osler Soc., Am. Assn. for Cancer Rsch., So. Soc. Clin. Investigation, Tex. Med. Assn., Dallas County Med. Soc., Clin. Immunology Soc., Phi Beta Kappa, Sigma Xi, Alpha Omega Alpha. Office: Baylor U Med Ctr Charles A Sammons Cancer Ctr 3500 Gaston Ave Dallas TX 75246-2088

STONE, MICHAEL DAVID, landscape architect; b. Moscow, Idaho, Apr 11, 1953; s. Frank Seymour Stone and Barbara Lu (Wahl) Stone/Schonthaller; m. Luann Dobaran, Aug. 12, 1978; children: Stephanie Nicole, David Michael. B in Landscape Architecture, U. Idaho, 1976; postgrad., Oreg. State U., 1986, Harvard U., 1990; MA in Orgnl. Leadership, Gonzaga U., 1990. Registered landscape architect, Wash., cert. leisure profl. Landscape designer Robert L. Woerner, ASLA, Spokane, Wash., 1976-77; park planner Spokane County Pks. and Recreation, 1977-82; landscape architect City of Spokane Pks. and Recreation, 1982-84, asst. parks mgr., 1984-86; golf and community devel. mgr. City of Spokane Parks and Recreation, 1986—; cons. Lake Chelan (Wash.) Golf Course, 1988. Pres. Sacred Heart Parish Coun., Spokane, 1987-89; v.p. Cataldo Sch. Bd. Dirs., Spokane, 1987-89; pres. South Spokane Jaycees, 1977-86; active Leadership Spokane, 1989. Named Outstanding Young Man Am., 1980, 85, Outstanding Knight, Intercollegiate Knights, 1972-73, Jaycee of the Yr., South Spokane Jaycees, 1981, Vet. of the Yr., South Spokane Jaycees, 1984-85; recipient Holy Grail award Intercollegiate Knights, 1972-73. Mem. Nat. Recreation and Park Asns., Am. Soc. Landscape Architects, Wash. Recreation and Park Assn., N.W. Turfgrass Assn., Beta Chi, Delta Tau Delta. Roman Catholic. Avocations: golf, basketball, photography, travel. Home: S 5428 Arthur Spokane WA 99223 Office: City of Spokane 808 W Spokane Falls Blvd Spokane WA 99201-3333

STONE, MICHAEL PAUL, biophysical chemist, researcher; b. Berkeley, Calif., Aug. 23, 1955; s. William Thomas and Ruth Marie (Danielsen) S. BS, U. Calif., Davis, 1977; PhD, U. Calif., Irvine, 1981. Rsch. assoc. U. Rochester, N.Y., 1981-84; rsch. instr. Vanderbilt U., Nashville, 1984-86, rsch. asst. prof., 1986-91, rsch. assoc. prof., 1991—. Mem. AAAS, Am. Chem. Soc. (past chmn. Nashville chpt.), Biophysical Soc., Sigma Xi. Office: Vanderbilt U Dept Chemistry Nashville TN 37235

STONE, PHILIP M., physicist, nuclear engineer; b. Pitts., Nov. 23, 1933; s. Morris Denor and Marissa (Glickman) S.; m. Margaret Ann Johns, Sept. 10, 1955 (div. Jan. 1980); children: Dale, Deborah, David; m. Annie Lee, Oct. 25, 1980. MSE, U. Mich., 1956, PhD, 1962. Staff mem. Los Alamos (N.Mex.) Sci. Lab., 1956-63; rsch. mgr. Sperry Corp. Rsch. Ctr., Sudbury, Mass., 1963-75; br. chief divsn. Office of Fusion Energy U.S. Dept. Energy, Washington, 1975-87, dir. sci. and tech,. 1987—; vis. scientist CEA Nuclear Lab., Saclay, France, 1973-74; assoc. prof. SUNY-Buffalo, 1968-69; vis. assoc. prof. U. Pitts., 1967-68. Contbr. 15 articles to profl. jours. Rsch. Coun. fellow Univ. Coll., London, 1965-66, grad. study fellow Los Alamos Sci.

Lab., 1958-92. Mem. AAAS, IEEE (sr.), Am. Phys. Soc., Sigma Xi. Home: 7507 Whittier Blvd Bethesda MD 20817

STONE, RICHARD JOHN, physicist; b. Seattle, Nov. 3, 1955; s. Richard Emil Stone and Anita Sharon (Bell) Butler. BA in Physics, Johns Hopkins U., 1979; MS in Physics, U. Wash., 1982, PhD in Materials Sci., 1987. Scientist, electro-optical engr. BDM Internat., Huntsville, Ala., 1987—. Author, editor computer code Exoseek (a simulation of a staring focal array infrared seeker with signal and data processing algorithms), 1989. Leader Boy Scouts Am., Huntsville, 1987-92. Mem. Am. Phys. Soc., Am. Ceramic Soc., Soc. Photo-Optical Infrared Engrs., Soc. Automotive Engring., Sigma Xi. Home: 2027 Flagstone Dr # 2112 Madison AL 35758 Office: BDM Engring Svcs Co 950 Explorer Blvd Huntsville AL 35806

STONE, ROBERT JOHN, research psychologist; b. Plymouth, Devon, U.K., Apr. 2, 1958; s. William James and Mary Joan (Curtiss) S. BS in Psychology, Univ. Coll., London, 1979, MS in Ergonomics, 1981. Group leader ergonomics Brit. Aerospace, Filton, Bristol, U.K., 1980-89; tech. mgr., virtual reality and telepresence team leader U.K. Advanced Robotics Rsch. Ctr., Salford, U.K., 1989—; tech. evaluator Commn. for European Communities, Brussels, Belgium, 1989-90; vis. prof. virtual reality U. Salford, 1993—. Contbr. articles to profl. jours. Recipient Applications award U.K. Ergonomics Soc., 1985. Assoc. fellow Brit. Psychol. Soc.; mem. Ergonomics Soc. (Applications award 1985, Otto Edholm award, 1993). Achievements include rsch. in display of range image sensory data; virtual reality for remote driving, telepresence, ergonomics prototyping and sensor data visualization; head-slaved stereo audio visual systems; force/tactile feedback. Office: UK Advanced Robotics Rsch, University Rd, Salford M5 4PP, England

STONE, THOMAS ALAN, geologist; b. Portland, Maine, Apr. 26, 1947; s. George Ellis and Beatrice Ruth (Pennell) S.; m. Ann Elizabeth Nagel, Nov. 13, 1972; children: Debra Lee, Michael T. BA in History, Bates Coll., 1970; postgrad., Northeastern U., 1978-79; MA in Earth Scis., Dartmouth Coll., 1982. Social worker Dept. Human Svcs., State of Maine, Portland, 1972-76; field geologist, remote sensing interpreter Chiasma Cons., South Portland, 1979; lab. asst., geologist N.E. Geochem.-FM Beck, Inc., Yarmouth, Maine, 1980; rsch. asst. remote sensing, teaching asst. hydrology Dept. Earth Scis., Dartmouth Coll., Hanover, N.H., 1980-82; rsch. asst. remote sensing, adminstrv. officer pro tem The Ecosystems Ctr., Marine Biol. Lab., Woods Hole, Mass., 1982-84, rsch. assoc. remote sensing, 1984-87; rsch. asst. Woods Hole (Mass.) Rsch. Ctr., 1987-92, sr. rsch. assoc., 1992—. Contbr. articles to profl. jours. Dartmouth Coll. grad. fellow, 1980-82; recipient Best Social Inventions '89 award Communications Category, Inst. for Social Inventions; Presdl. Citation Am. Soc. Photogrammetry and Remote Sensing, 1991. Mem. IEEE Geosci. and Remote Sensing Soc. (affiliate), Internat. Soc. Tropical Foresters, Am. Soc. Photogrammetry and Remote Sensing, Am. Geophys. Union, Geol. Soc. Maine, Ecol. Soc. Am. Office: Woods Hole Rsch Ctr PO Box 296 Woods Hole MA 02543

STONE, WILLIAM C., mathematics educator. BA, Union Coll., 1942; MS, U. Chgo., 1949, PhD, 1952. With Union Coll., Schenectady, N.Y., 1942-44, 51—, now Mary Louise Bailey prof. Office: Union Coll Dept of Math Schenectady NY 12308

STONECIPHER, LARRY DALE, mathematics educator; b. Mt. Vernon, Ill., Nov. 18, 1949; s. Paul and Dixie E. (Mulch) S.; m. Brenda D. Vallowe, May 30, 1970; children: Andrea, Aaron. BS in Edn., Ea. Ill. U., 1971, MA, 1975; PhD, So. Ill. U., 1986. Math. tchr. Mt. Vernon City Schs., 1971-76, math. dept. chair, 1976-87, gifted edn. dir., 1978-87; prof. math., staff devel. dir. Sangamon State U., Springfield, Ill., 1987—; math. cons. Ill. State Bd. Edn., Springfield, 1980—, Edn. Svc. Ctr. 14, Springfield, 1987—. Contbr. articles. to profl. jours. Liaison to gov. on math. edn. State of Ill., 1986—. Mem. Ill. Coun. Tchrs. Math. (bd. dirs. 1986-89), Ill. Math. and Sci. Acad. (univ. affiliates bd. dirs. 1988—). Avocation: trick water skiing. Home: 14 Penacook Dr Rochester IL 62563-9401 Office: Sangamon State U Brookens 321 Springfield IL 62794

STONEKING, MARK ALLEN, anthropologist; b. Hawthorne, Calif., Aug. 1, 1956; s. James Allen and Phyllis Margot (Kepfer) S.; m. Geraldine Lee, Dec. 1, 1978 (div. Sept. 1989); 1 child, Cynthia Lee; m. Linda Anne Vigilant, Oct. 31, 1989; 1 child, Colin James. BA, U. Oreg., 1977; MS, Pa. State U., 1979; PhD, U. Calif., Berkeley, 1986. Postdoctoral fellow Biochemistry dept., U. Calif., Berkeley, 1986-88; staff scientist Lawrence Berekely Lab., 1989; assoc. rsch. scientist Cetus Corp., Emeryville, Calif., 1989-90; asst. prof. Anthropology dept. Pa. State U., University Park, 1990—. Contbr. articles to profl. jours. including Transactions of the Am. Fisheries Soc., Jour. of the Fisheries Rsch. Bd. Can., Biochem. Genetics, Genetics, Jour. of Heredity, Copeia, Biol. Jour. of the Linnaean Soc., Nature, Nucleic Acids Rsch., The Colonization of the Pacific: A Genetic Trail, Am. Jour. of Human Genetics, Jour. of Molecular Evolution, Science, Systematic Zoology, Am. Jour. of Human Biology, others. Named Outstanding Young Alumnus, U. Oreg., 1990. Mem. AAAS, Genetic Soc. of Am., Soc. for the Study of Evolution, Sigma Xi. Achievements include rsch. on departure of human mtDNA variation from neutral expectations: an alternative explanation, mitochondrial DNA and human evolution, genetic variation, inheritance and quaternary structure of malic enzyme in brook trout. Home: 1202 W Beaver State College PA 16802 Office: Pa State U Anthropology Dept University Park PA 16802

STONER, GARY DAVID, pathology educator; b. Bozeman, Mont., Oct. 25, 1942; married; 2 children. BS, Mont. State U., 1964; MS, U. Mich., 1968, PhD in Microbiol., 1970. Asst. rsch. scientist U. Calif., San Diego, 1970-72, assoc. rsch. scientist, 1972-75; cancer expert Nat. Cancer Inst., 1976-79; assoc. prof. pathology Med. Coll. Ohio, 1979—; cons. Nat. Heart Lung & Blood Inst., 1974—, EPA, 1979—, Cancer Inst., 1979—, Nat. Toxicol. Program, 1981—; lectr. W. Alton Jones Crell Sci. Ctr., 1978—; mem. study sect. NIH, 1980—, Am. Cancer Soc., Ohio, 1982—. Grantee Nat. Cancer Inst., EPA, U.S. Army R & D Command. Mem. AAAS, Am. Assn. Cancer Rsch., Am. Tissue Culture Assn., Am. Assn. Pathologists, Am. Soc. Cell Biology. Achievements include research in carcinogenesis studies in human and animal model respiratory and esophagael tissues, carcinogen metabolism, mutagenesis, in vitro transformation of epithelial cells. Office: Medical College of Toledo Dept of Pathology 3000 Arlington Ave Toledo OH 43699*

STONER, HENRY RAYMOND, retired chemical engineer; b. Ambridge, Pa., Jan. 23, 1928; s. Adam and Veronica (Ciapala) S.; married, July 19, 1950; children: Larry A., Karen Stoner Sweetland. BS in Chemistry, Geneva Coll., Beaver Falls, Pa., 1952. Sales engring. chemist Valvoline Oil Co., Freedom, Pa., 1952-56; lab. supr. Koppers Co. Inc, Kobuta, Pa., 1956-62; chief chemist coatings div. Koppers Co. Inc, Garwood, N.J., 1962-67, Newark, 1967-69; tech. dir. coatings devel. Koppers Co. Inc, Verona, Pa., 1969-72; tech. dir. coatings devel. Koppers Co. Inc, Newark, 1972-86, mgr. govt. compliance coatings div., 1986-88; ret., 1988; task group chmn. coal tar coatings and coal tar epoxies Steel Structures Painting Coun., Pitts., 1975-88; rep. on potable water coatings, protective barrier Nat. Sanitation Found., Ann Arbor, Mich., 1985-88. With U.S. Army, 1945-47. Mem. ASTM, Nat. Assn. Chem. Engrs., Am. Waterworks Assn. (chmn. task force steel water pipe com.), Steel Structures Painting Coun. Republican. Roman Catholic. Office: Henry R Stoner & Assocs 343 W End Ave North Plainfield NJ 07063-1734

STONNINGTON, HENRY HERBERT, physician, medical executive, educator; b. Vienna, Austria, Feb. 12, 1927; came to U.S., 1969; m. Constance Mary Leigh Hamersley, Sept. 19, 1953. MB, BS, Melbourne U., Victoria, Australia, 1950; MS, U. Minn., 1972. Diplomate Am. Bd. Phys. Medicine and Rehab. Pvt. practice Sydney, Australia, 1955-65; clin. tchr. U. New South Wales, Sydney, 1965-69; resident in Phys. Medicine and Rehab. Mayo Clinic, Rochester, Minn., 1969-72; staff Mayo Clinic, Rochester, 1972-83; assoc. prof. Mayo Med. Sch., Rochester, 1975-83; chmn. dept rehab. medicine Med. Coll. Va., Va. Commonwealth U., Richmond, 1983-88, prof. rehab. medicine, 1983—, dir. rsch. tng. ctr., 1988-89; v.p. med. svcs. Sheltering Arms Hosp., Richmond, 1985-92; prof. and chmn. dept. phys. medicine and rehab. U. Mo., Columbus, 1992—. Editor Brain Injury, 1987—; contbr. articles to profl. jours. Recipient award Rsch. Tng. Ctr. Model System Nat. Inst. Disability and Rehab. Rsch., Washington, 1987, 88. Fellow Australia Coll. Rehab. Medicine, Royal Coll. Physicians Edinburgh, Am. Assn. Phys. Medicine and Rehab., Am. Coun. Rehab. Medicine, Am. Assn. Acad. Physicians. Office: Rusk Rehab Ctr 2R01 1 Hospital Dr Columbia MO 65212

STOOKEY, GEORGE KENNETH, research institute administrator, dental educator; b. Waterloo, Ind., Nov. 6, 1935; s. Emra Gladison and Mary Catherine (Anglin) S.; m. Nola Jean Meek, Jan. 15, 1955; children—Lynda, Lisa, Laura, Kenneth. A.B. in Chemistry, Ind. U., 1957, M.S.D., 1962, Ph.D. in Preventive Dentistry, 1971. Asst. dir. Preventive Dentistry Research Inst., U. Ind., Indpls., 1968-70; assoc. dir. Oral Health Research Inst., U. Ind., 1974-81, dir., 1981—; assoc. prof. preventive dentistry Sch. of Dentistry, Ind. U., 1973-78, prof., 1978—, assoc. dean research, 1987—; cons. U.S. Air Force, San Antonio, 1973—, ADA, Chgo., 1972—, Nat. Inst. Dental Research, Bethesda, Md., 1978-82. Author: (with others) Introduction to Oral Biology and Preventive Dentistry, 1971, Preventive Dentistry for the Dental Assistant and Dental Hygienist, 1977, Preventive Dentistry in Action, 1972, 80 (Meritorious award 1973); contbr. articles to profl. jours. Mem. Internat. Assn. for Dental Research, European Orgn. Caries Research, Am. Assn. Lab. Animal Sci. Republican. Office: Oral Health Research Inst 415 Lansing St Indianapolis IN 46202-2876

STOOKEY, STANLEY DONALD, chemist; b. Hay Springs, Nebr., May 23, 1915; s. Stanley Clarke and Hermie Lucille (Knapp) S.; m. Ruth Margaret Watterson, Dec. 26, 1940; children—Robert Alan, Margaret Ann, Donald Bruce. B.A., Coe Coll., 1936, LL.D., 1959; M.S., Lafayette Coll., 1937; Ph.D. in Phys. Chemistry, MIT, 1940. With Corning Glass Works Research, N.Y., 1940-79, dir. fundamental chem. research, 1970-79. Contbr. articles to profl. jours.; patentee field of photosensitive glasses, glass ceramics, photochromatic and polychromatic glasses. Recipient Nat. Medal of Tech., 1986, World Materials Congress award, 1988. Fellow Am. Ceramic Soc.; mem. NAE, Am. Chem. Soc. (Inventor of Yr. award 1971), Sigma Xi. Republican. Methodist. Home: 12 Timber Ln Painted Post NY 14870-9340

STOOPLER, MARK BENJAMIN, physician; b. N.Y.C., Sept. 29, 1950; s. Alex and Blanche Sylvia (Kappel) S.; m. Lynn Sara Fruchter, Jan. 10, 1982; children: David Andrew, Emily Rachel, Jesse Bryan. BS, Tulane U., 1971; MD, Cornell U., 1975. Diplomate Am. Bd. Internal Medicine, Am. Bd. Oncology. Intern and resident in internal medicine North Shore U. Hosp., Manhasset, N.Y., 1975-78; intern and resident in internal medicine Meml. Sloan-Kettering Cancer Ctr., N.Y.C., 1975-78, asst. chief resident in medicine, 1978, fellow in med. oncology, 1978-80; asst. attending physician Presbyn. Hosp., N.Y.C., 1980—; asst. prof. of clin. medicine Columbia U. Coll. of Physicians and Surgeons, N.Y.C., 1980—. Contbr. articles to profl. jours. Recipient U. scholar Tulane U., 1970-71. Fellow ACP; mem. Am. Soc. of Clin. Oncology, Am. Fedn. for Clin. Research, Internat. Assn. for the Study of Lung Cancer, Phi Beta Kappa. Office: Columbia-Presbyn Med Ctr 161 Ft Washington Ave New York NY 10032-3713

STOPA, EDWARD GREGORY, neuropathologist; b. Newark, N.J., July 6, 1954; s. Peter Adelbert and Hedwig (Juchniewicz) S.; m. Karen Elisa Madras, May 22, 1980; children: Emily, Eva, Eliza, Arielle. BS, McGill U., 1976, MD, 1980. Diplomate Nat. Bd. Med. Examiners, Am. Bd. Pathology. Internship Framingham (Mass.) Union Hosp., 1980-81; residency anatomic pathology Brigham and Women's Hosp., Boston, 1981-83, residency neuropathology, 1983-85; asst. prof. pathology Tufts U. Sch. of Medicine, Boston, 1985-89; asst. prof. SUNY Health Sci. Ctr., Syracuse, N.Y., 1989-92, assoc. prof., 1992—; dir. neuropathology R.I. Hosp., Providence, 1993—; rsch. cons. McLean Hosp., Belmont, Mass., 1985—; neuropathology cons. in field. Contbr. articles to profl. jours. Recipient Physician Scientist award Nat. Inst. Aging, 1985-90; grantee Nat. Inst. Aging, 1989, 91. Mem. AAAS, Mass. Med. Soc., Am. Assn. Neuropathologists, Soc. for Neurosci., Soc. for Rsch. on Biol. Rhythms, Soc. for Light Treatment and Biol. Rhythms, Internat. Acad. Pathology. Office: RI Hospital Dept Pathology 593 Eddy St Providence RI 02903

STOPA, PETER JOSEPH, biochemist, microbiologist; b. Newark, Mar. 26, 1953; s. Peter Adalbert and Hedwig J. (Juchniewicz) S.; m. Carol Ann Liskowicz, June 30, 1979. BS, U. Scranton, 1975, MA, 1977. Rsch. scientist Johns Hopkins Med. Sch., Balt., 1979-80; immunologist Becton-Dickinson (BBL div.), Balt., 1980-83; chemist Med. Rsch. Inst. Infectious Disease U.S. Army, Ft. Detrick, Md., 1983-88; chemist Edgewood Rsch. Devel. Engring. Ctr. U.S. Army, Aberdeen Proving Ground, Md., 1988—; cons. SETA, Inc., 1975-76, Ctr. Indsl. and Instnl. Devel., Durham, N.H., 1977-78. Contbr. chpt. to Manual of Clinical Immunology, 1980, articles to profl. publs. Adult leader Texas (Md.) area Boy Scouts Am., 1980—; mem. architect com. Mays Chapel Homeowners, Timonium, Md., 1981-83. Recipient Spl. Act award U.S. Army,1 990, 91, Best Poster award U.S. EPA, 1991. Mem. Am. Chem. Soc., Am. Soc. Microbiology (sec. Md. br. 1985-86). Democrat. Roman Catholic. Achievements include patent on test for streptococcus A, description of application of fluorescent immunoassays in infectious disease diagnosis, development of fielded biowarfare detection system, military use of immunoassay technology. Office: US Army Edgewood RD&E Ctr SCBRD RTE Bldg E3549 Aberdeen Proving Ground MD 21010-5423

STOPHER, PETER ROBERT, transportation executive, consultant; b. Crowborough, Eng., Aug. 8, 1943; came to U.S., 1968; s. Harold Edward and Joan Constance (Salmon) S; m. Valerie Anne Alway, Apr. 11, 1964 (div. Feb. 1989); children: Helen Margaret Anne, Claire Elizabeth; m. Catherine Coville Jones July 7, 1990. BSCE, U. Coll., London, 1964, PhD, 1967. Research officer Greater London Council, London, 1967-68; asst. prof. transp. planning, applied statistics, math. modeling Northwestern U., Evanston, Ill., 1968-70, from assoc. prof. to prof., 1973-79, vis. prof., 1980-81; asst. prof. McMaster U., Hamilton, Ontario, 1970-71; assoc. prof. Cornell U., Ithaca, N.Y., 1971-73; tech. v.p. Schimpeler Corradino Assoc., Miami, Fla., and Los Angeles, 1980-84, v.p., 1984-87; dir., CFO Evaluation and Tng. Inst., 1987-90; prin., co-founder Applied Mgmt. and Planning Group, 1988-90; prof. civil engring., dir. La. Transp. Rsch. Ctr. La. State Univ., Baton Rouge, 1990—; spl. advisor Nat. Inst. Transp. and Rd. Research, Pretoria, S. Africa, 1976-77; vis. prof. U. Syracuse, N.Y., 1971-73, U. Louvain, Belgium, 1980. Co-author Urban Transportation Planning and Modeling, 1974, Transportation Systems Evaluation, 1976, Survey Sampling and Multivariate Analysis, 1978; contbr. articles to profl. jours. Recipient Fred Burgraaf prize Hwy. Research Bd., 1968, Jules Dupuit prize World Conf. on Transp. Rsch., 1992. Fellow Inst. Hwy. Engrs., Royal Stats. Soc.; mem. ASCE, Am. Stats. Assn., Transp. Research Bd. (com. chmn. 1970-77). Democrat. Methodist. Avocations: working out, gardening, photography, reading, classical music. Home: 3533 Granada Dr Baton Rouge LA 70810 Office: La Transp Rsch Cntr 4101 Gourrier Ave Baton Rouge LA 70808

STOPPANI, ANDRES OSCAR MANUEL, director research center, educator; b. Buenos Aires, Aug. 19, 1915; s. Oscar Carlos and Julia Severa (Bahia) S.; m. Antonia Emmy Delius, July 1, 1967. MD, U. Buenos Aires, Buenos Aires, 1941; PhD in Chemistry, Buenos Aires, 1945; PhD, U. Cambridge, Eng., 1953. Prof. biochemistry U. La Plata, Argentina, 1947-48; prof. biochemistry Sch. Medicine U. Buenos Aires, 1949-55, prof., dir. Inst. Biochemistry, Sch. Medicine, 1955-80, dir. dept. physiol. scis., 1970-80; bd. dirs. NRC, Argentina, 1963-66, 80-83; pres. Nat. Acad. Exact, Phys. and Natural Scis., Buenos Aires. Contbr. numerous articles to profl. jours. Recipient Weissman prize NRC, 1962, Campomar prize, 1970, Silver award Rotary, 1975, Bunge-Born prize, 1979, Am. States Orgn. Interam. Sci. prize, Bernardo A. Houssay, 1989, Biochemistry and Microbiology prize Konex Found., 1993; Rockefeller Found. grantee, 1957-60, WHO grantee, 1977-82. Fellow Third World Acad. Scis. Trieste, Chilean Acad. Scis., Latin Am. Acad. Scis., Am. Acad. Clin. Biochemistry; mem. Argentine Assn. for Advancement of Sci., Argentine Soc. for Biochem. Rsch. (pres. 1970-71), Argentine Soc. Biology (pres. 1980-83), Argentine Soc. Protozoologists (pres. 1980-81), Am. Chem. Soc., Biochem. Soc. U.K., Soc. Gen. Microbiology U.K., N.Y. Acad. Scis., Am. Soc. Biochemistry and Molecular Biology, Oxygen Soc., Soc. for Free Radical Rsch., Internat. Soc. for Study of Xenobiotics, Nat. Acad. Exact, Phys. and Natural Scis., Nat. Acad. Medicine, Buenos Aires Nat. Acad. Scis., Chilean Acad. Medicine (hon.), Chilean Acad. Scis. (hon.), Brazilian Acad. Scis. (corr.). Home: Viamonte 2295, 1056 Buenos Aires Argentina Office: Facultad de Medicina, Paraguay 2155, 1121 Buenos Aires Argentina

STORCH, JOEL ABRAHAM, mathematician; b. N.Y.C., Nov. 11, 1949; s. Sidney and Hermione (Kaufman) S.; m. Therese Weiss, Dec. 25, 1981. BS, CCNY, 1971; MS, Columbia U., 1972; PhD, MIT, 1974. Numerical analyst Goddard Inst. for Space Studies, N.Y.C., 1972-76, Brookhaven Nat. Lab., Upton, N.Y., 1976-78; dynamicist Draper Lab., Cambridge, Mass., 1978-88, Jet Propulsion Lab., Pasadena, Calif., 1988-91, Draper Lab., Cambridge, 1991—; lectr. CCNY, 1974-76; adj. prof. Northeastern U., Boston, 1980-88. Contbr. articles to profl. jours. Recipient Citation, NASA, 1986. Mem. AIAA, Soc. Indsl. and Applied Math. Jewish. Home: 1809 Beacon St Brookline MA 02146-4206 Office: Draper Lab 555 Technology Sq Cambridge MA 02139-3563

STOREY, GARY GARFIELD, agricultural studies educator; b. Davidson, Sask., Can., Mar. 9, 1939; s. Robert Lincoln and Burna Rose (Townsend) S.; m. Joelle Herbert, Aug. 7, 1965; children: Kristin Jennifer, Lauren Caroline. BSA, U. Sask., 1963, MSc, 1966; MA, U. Wis., 1968, PhD, 1970. Rsch. economist Govt. of Sask., 1964-66; asst. prof. U. Sask., Saskatoon, 1970-72, prof. agrl. econs., 1977—, asst. dean agr., 1992—; operator grain farm; dir. IPSCO, Agrl. Credit Corp. of Sask. Author, co-editor: The Political Economy of Agricultural Trade and Policy: Toward a New Order for Europe and North America, 1990; co-editor: International Agricultural Trade, Advanced Readings in Price Formation, Market Structure, and Price Instability, 1984. Fellow Agrl. Inst. Can., 1991. Mem. Can. Agrl. Econs. and Farm Mgmt. Soc. (pres. 1977-78), Agrl. Inst. Can. (nat. coun. 1980-82, AIC Fellowship award 1991), Assn. Faculties Agr. in Can. (pres. 1986-87). Office: Univ of Saskatchewan, Dept of Ag Economics, Saskatoon, SK Canada S7N 0W0*

STORGAARD, ANNA K., agriculturalist. Recipient AIC Fellowship award Agrl. Inst. Can., 1992. Home: 1054 Mulvey Ave, Winnipeg, MB Canada R3M 1J4*

STORK, GILBERT (JOSSE), chemistry educator, investigator; b. Brussels, Belgium, Dec. 31, 1921; s. Jacques and Simone (Weil) S.; m. Winifred Stewart, June 9, 1944 (dec. May 1992); children: Diana, Linda, Janet, Philip. B.S., U. Fla., 1942; Ph.D., U. Wis., 1945; D.Sc. (hon.), Lawrence Coll., 1961, U. Paris, 1979, U. Rochester, 1982, Emory U., 1988, Columbia U., 1993. Sr. research chemist Lakeside Labs., 1945-46; instr. chemistry Harvard U., 1946-48, asst. prof., 1948-53; assoc. prof. Columbia U., N.Y.C., 1953-55, prof., 1955-67, Eugene Higgins prof., 1967-92, prof. emeritus, 1992—, chmn. dept., 1973-76; plenary lectr. numerous internat. symposia, named Lectureships in U.S. and abroad; cons. Syntex, Internat. Flavors and Fragrances; chmn. Gordon Steroid Conf., 1958-59. Recipient Baekeland medal, 1961, Harrison Howe award, 1962, Edward Curtis Franklin Meml. award Stanford, 1966, Gold medal Synthetic Chems. Mfrs. Assn., 1971, Nebr. award, 1973, Roussel prize in steroid chemistry, 1978, Edgar Fahs Smith award, 1982, Willard Gibbs medal, Chgo. section Am. Chem. Soc., 1982, National Medal of Sci., 1982, Linus Pauling award, 1983, Tetrahedron prize, 1985, Remsen award, 1986, Cliff S. Hamilton award 1986, Mony Ferst award Sigma Xi, 1987, Roger Adams Award in Organic Chemistry, Am. Chem. Soc., 1991, George Kenner award, 1992, Chemical Pioneer award Am. Inst. of Chemistry, 1992, Welch Found. award, 1993; Guggenheim fellow, 1959. Fellow NAS (award in chem. sci. 1982), French Acad. Scis., Am. Acad. Arts and Scis.; mem. Am. Chem. Soc. (chmn. organic chemistry div. 1967, award in pure chemistry 1957, award for creative work in synthetic organic chemistry 1967, Nichols medal 1980, Arthur C. Cope award 1980, Roger Adams award in organic chemistry 1991), Royal Soc. Chemistry (hon., London), Pharm. Soc. Japan (hon.), Chemists Club (hon.). Home: 459 Next Day Hill Dr Englewood NJ 07631-1921 Office: Columbia U Dept of Chemistry Havemeyer Hall New York NY 10027

STORK, WILMER DEAN, II, physical chemist, researcher; b. Fairbury, Ill., Dec. 2, 1965; s. Wilmer Dean and Gwendolyn Lou (Hearne) S. BS in Chemistry, So. Ill. U., 1989, MS in Chemistry, 1992. Researcher Argonne (Ill.) Nat. Lab., summer 1990; chemist So. Ill. U., Carbondale, 1989-91, U. Ill., Urbana, 1991—. Contbr. articles to profl. jours. Recipient Ill. State Scholar award State of Ill., 1984, Presdl. scholarship So. Ill. U., 1984. Achievements include preparation of pure formyl cyanide in the gas phase and found it to be more stable than originally thought; used a novel method of pseudo upperstate level interweaving to discover transitions belonging to a dark state of DBF2. Office: U Ill 505 S Mathews Box 10 Urbana IL 61821

STORY, DAVID FREDERICK, pharmacologist, educator; b. Launceston, Australia, July 27, 1940; s. Frederick Joseph and Gladys Ray (Stone) S.; m. Margot Elizabeth Turner, May 25, 1966; children: Andrew David, Michael Barry. BS, U. Melbourne, 1966, PhD, 1972. Head pharmacology Riker Labs. Australia, Thornleigh, 1966-69; sr. rsch. officer dept. pharmacology U. Melbourne, Australia, 1972, lectr., 1973-75, dept. head, 1986-92, sr. lectr., 1976-86, reader, 1987-89, assoc. prof., 1989-93; prof. pharmacology, head dept. med. lab. sci. Royal Melbourne Inst. Tech., 1993—. Contbr. articles to profl. jours. Mem. Australian Physiol. and Pharmacological Soc., High Blood Pressure Rsch. Coun. Australia, Australasian Soc. Clin. and Experimental Pharmacologists (pres. 1992). Avocations: camping, swimming, gardening. Office: Royal Melbourne Inst Tech, Dept Med Lab Sci, Melbourne 3000, Australia

STORY, MICHAEL THOMAS, biomedical researcher, educator; b. Cedar Rapids, Iowa, May 6, 1940; s. Donald Leo and Mary Francis (Benda) S.; m. Carol Ann Blake, Aug. 3, 1963; children: Daniel Michael, Allison Leigh. BS, Loras Coll., 1962; MA, Drake U., 1964; PhD, Kansas U., 1972. Instr. biology Park Coll., Parkville, Md., 1964-68; instr. gyn.-ob. Med. Coll. Wis., Milw., 1978-79, asst. prof., 1980-87, assoc. prof. urology and biochemistry, 1987—; grad. sch. faculty Med. Coll. Wis., Milw., 1982—, mem. Cancer Ctr., 1986—; study sect. mem. NIH Reproductive Endocrinology, Washington, 1988-91. Contbr.: Cancer Surveys, vol. 11, 1991; contbr. articles to profl. jours. Recipient Rsch. grant NIH, 1982—. Mem. AAAS, Am. Urol. Assn. (nominating com. 1989—), Tissue and Culture Assn., Endocrine Soc. Roman Catholic. Achievements include discovery of basic fibroblast growth factor in human prostate; research in the role of growth factors in regulating normal and abnormal growth of the prostate. Office: Med Coll Wis Dept Urology 9200 W Wisconsin Ave Milwaukee WI 53226

STORY, RANDALL MARK, biochemist; b. Mountain View, Calif., Mar. 30, 1965; s. Roland M. and Alice T. (Warner) S. BA, Rice U., 1987; PhD, Yale U., 1991. Postdoctoral assoc. dept. molecular biophysics/biochemistry Yale U., New Haven, 1991-93; rsch. fellow divsn. biology Calif. Inst. Tech., Pasadena, 1993—. Contbr. articles to Nature. NSF predoctoral fellow, 1988-91. Office: Calif Inst Tech Divsn Biology 147-75 Pasadena CA 91125

STOTZKY, GUENTHER, microbiologist, educator; b. Leipzig, Germany, May 24, 1931; came to U.S., 1939; s. Moritz Stotzky and Erna (Angres) Kester; m. Kayla Baker, Mar. 17, 1958; children: Jay, Martha, Deborah. BS, Calif. Poly. State U., 1952; MS, Ohio State U., 1954, PhD, 1956. Spl. sci. employee Argonne Nat. Lab. USAEC, Lemont, Ill., 1955; rsch. assoc. Dept. Botany U. Mich., Ann Arbor, 1956-58; head soil microbiology Cen. Rsch. Labs. United Fruit Co., Norwood, Mass., 1958-63; chmn., microbiologist Kitchawan Rsch. Labs. Bklyn. Botanic Garden, Ossining, N.Y., 1963-68; assoc. prof. Dept. Biology NYU, 1967-70, prof., 1970—, chmn., 1970-77. Editor: Soil Biochemistry, 1990—; series editor Marcel Dekker, Inc., 1986-92; contbr. over 250 articles to profl. jours. and chpts. to books. With USCG, 1957. Recipient Selman A. Waksman Hon. Lecture award Theobald Smith Soc., 1989, Fisher Co. award for applied and environ. microbiology, 1990, Honored Alumnus of Yr. award Calif. Poly. State U., 1992; Disting. Vis. Scientist U.S. EPA, 1986-89. Fellow AAAS, Am. Acad. Microbiology, Am. Soc. Microbiology, Am. Soc. Agronomy, Soil Sci. Soc. Am. Jewish. Avocations: fishing, reading, music. Office: NYU Dept Biology 1009 Main New York NY 10003

STOUDT, THOMAS HENRY, research microbiologist; b. Temple, Pa., Apr. 6, 1922; s. Thomas Lester and Edith (Yocum) S.; m. Kathryn H. Hendel, Dec. 31, 1943; children: Thomas H., Frank E., Carol A. BS, Albright Coll., 1943; MS, Rutgers U., 1944; PhD, Purdue U., 1949. Sr. microbiologist Merck & Co., Rahway, N.J., 1949; sect. head Merch Sharp & Dohme Rsch. Lab., Rahway, N.J., 1958-68; dir. Merck Sharp & Dohme Rsch. Lab., Rahway, N.J., 1969-74; sr. dir. Merck Sharpe & Dohme Rsch. Lab., Rahway, N.J., 1975-76, exec. dir., 1977-81, exec. tech. advisor, 1981-

84; pres. Stoudt Assocs., Westfield, N.J., 1984—. Mem. adv. bd. Health and Environment Union County, Elizabeth, N.J., 1988—; mcpl. chmn. Westfield Dem. Com., 1964-66. With USN, 1944-46. Mem. Am. Chem. Soc., Am. Soc. for Microbiology (pres. Theobald Smith Soc. 1964-65), N.Y. Acad. Scis. Achievements include patents in the field of microbial transformations, microbial enzymes, and vaccines. Home and Office: Stoudt Assocs 857 Village Grn Westfield NJ 07090-3515

STOUT, GLENN EMANUEL, water resources center administrator; b. Fostoria, Ohio, Mar. 23, 1920. AB, Findlay U., 1942, DSc, 1973. Sci. coord. NSF, 1969-71; asst. to chief Ill. State Water Survey, Champaign, 1971-74; prof. Inst. Environ. Studies, Urbana, Ill., 1973—, dir. task force, 1975-79; dir. Water Resources Ctr. U. Ill., Urbana, 1973—; rsch. coord. Ill.-Ind. Sea Grant Program, 1987—; mem. Gov.'s Task Force on State Water Plan, 1980—; bd. dirs. Univ. Council Water Resources, 1983-86, chmn. internat. affairs, 1989—. Contbr. articles to profl. jours. Active Salvation Army. Mem. Am. Water Resources Assn., Internat. Water Resources Assn. (sec. gen. 1985-91, v.p., 1992—, exec. dir. 1984—), Am. Meteorol. Soc., Am. Geophys. Union, N.Am. Lake Mgmt. Soc., Ill. Lake Mgmt. Assn. (bd. dirs. 1985-88), Internat. Assn. Rsch. Hydrology, Am. Water Works Assn., Kiwanis (pres. local club 1979-80, lt. gov. 1982-83), Sigma Xi (pres. U. Ill. chpt. 1985-86). Home: 920 W John St Champaign IL 61821-3907 Office: Water Resources Ctr 205 N Mathews Ave Urbana IL 61801-2350

STOUT, LARRY JOHN, civil engineer, consultant; b. Tilden, Nebr., Aug. 19, 1950; s. John P. and Dora A. (West) S.; m. Elvira Avila, Oct. 7, 1972; children: Steven, David, Sandra, Laura. BS in Civil Engring., N.Mex. State U., 1978. Registered profl. engr., Fla., N.Mex., land surveyor, Fla. Civil engr. Internat. Boundary and Water Commn., El Paso, Tex., 1978, 90-93; water resource engr. South Fla. Water Mgmt. Dist., West Palm Beach, 1979-80, project mgr., 1989-90; civil engr. Hayes, Sea, Mattern & Mattern, West Palm Beach, 1980-81; project engr., dept. mgr. Gee & Jenson, Engrs., Architects & Planners, West Palm Beach, 1981-89; cons. South Fla. Regional Bldg. Com., West Palm Beach, 1980-90, N.Mex. Regional Bldg. Com. of Jehovah's Witnesses, Las Cruces, 1991-93. With U.S. Army, 1969-72. Mem. ASCE. Home: 3604 S Hwy # 28 Las Cruces NM 88005

STOUT, PHILIP JOHN, chemist; b. Chattanooga, Apr. 29, 1958; s. William Ward and Angeline Dorothy (Engeman) S.; m. Beverly Ann Cain, Apr. 14, 1988; 1 child, Zachary August. BS in Chemistry, Coll. Charleston, 1980; MS in Analytical Chemistry, U. Tenn., 1987. Applications chemist Digilab divsn. Bio-Rad, Cambridge, Mass., 1987—. Contbr. chpts. to books, articles to jours. Mem. Am. Chem. Soc., Soc. Applied Spectroscopy. Achievements include rsch. in chem. analysis of polycyclic ar. cmpls., in-situ epitaxial layer monitoring, superconductor characterization. Office: Bio-Rad Digilab divsn 237 Putnam Ave Cambridge MA 02139

STOUT, THOMAS MELVILLE, control system engineer; b. Ann Arbor, Mich., Nov. 26, 1925; s. Melville B. and Laura C. (Meisel) S.; m. Marilyn J. Koebnick, Dec. 27, 1947; children: Martha, Sharon, Carol, James, William, Kathryn. BSEE, Iowa State Coll., 1946; MSE, U. Mich., 1947, PhD, 1954. Registered profl. engr., Calif. Jr. engr. Emerson Electric Co., St. Louis, 1947-48; instr., then asst. prof. U. Wash., Seattle, 1948-54; rsch. engr. Schlumberger Instrument Co., Ridgefield, Conn., 1954-56; dept. mgr. TRW/Bunker-Ramo Corp., Canoga Park, Calif., 1956-65; pres. Profimatics, Inc., Thousand Oaks, Calif., 1965-83; pvt. practice cons. Northridge, Calif., 1984—; active profl. engring. registration and certification. Contbr. articles, revs., papers to profl. publs., chpts. to books. Ens. USN, 1943-46. Fellow, hon. mem. Instrument Soc. Am.; mem. IEEE (sr. mem.), NSPE, Am. Inst. Chem. Engrs., Am. Soc. for Engring. Edn., Calif. Soc. Profl. Engrs. Achievements include four patents in computer control of industrial processes; participant in early digital computer installations for industrial process control. Home and Office: 9927 Hallack Ave Northridge CA 91324-1120

STOVER, DONALD RAE, software engineering executive, retired; b. Ponca City, Okla., Oct. 6, 1934; s. Frederick Edward and Myrtle Inez (Grantham) S.; m. Velma June Kirkpatrick, June 10, 1961; 1 child, Roy James. BSEE with high distinction, U. Iowa, 1956; MA in Math., UCLA, 1959; PhD in EE, U. Iowa, 1968. Mem. tech. staff Hughes Aircraft Co., Culver City, Calif., 1956-59; logic designer Emerson Electric, St. Louis, 1959-62; project engr. Collins Radio Co., Cedar Rapids, Iowa, 1962-71; tech. staff mem. Rockwell Avionics Group, Cedar Rapids, 1971-79, software mgr., 1979-91; mem. Rockwell tech. panel on Artificial Intelligence, 1984-89, on Software, 1989-90; responsible for validated Ada Compiler devel., 1985-91. Contbr. articles to profl. jours. Tennis program coord. YMCA, Marion, Iowa, 1973, canvasser, Cedar Rapids, 1968. Mem. Pi Mu Epsilon, Eta Kappa Nu. Republican. Methodist. Avocations: tennis, skiing, computer music, bridge. Home: 2270 26th St Marion IA 52302-1639

STOVER, SAMUEL LANDIS, physiatrist; b. Bucks County, Pa., Nov. 19, 1930; 3 children. BA, Goshen Coll., 1952; MD, Jefferson Med. Coll., 1959. Diplomate Am. Bd. Pediatrics, Am. Bd. Phys. Medicine and Rehab. Intern St. Luke's Hosp., Bethlehem, Pa., 1959-60; pvt. practice Ark., 1960-61, Indonesia, 1961-64; resident pediatrics Children's Hosp., Phila., 1964-66; asst. med. dir. Children's Seashore House, Atlantic City, 1966-67; resident phys. medicine and rehab. U. Pa., 1967-69; assoc. prof. pediatrics, prof. phys. medicine and rehab. U. Ala., Birmingham, 1969-76, prof. rehab. medicine, chmn. dept. Univ. Hosp. and Clinics, 1976—. Mem. AMA, Am. Acad. Pediatrics, Am. Acad. Phys. Medicine and Rehab., Am. Congress Rehab. Medicine. Office: University of Alabama Spinal Cord Injury Care System UAb Sta Birmingham AL 35294*

STOWE, DAVID HENRY, JR., agricultural and industrial equipment company executive; b. Winston-Salem, N.C., May 11, 1936; s. David Henry and Mildred (Walker) S.; m. Lois Burrows, Nov. 28, 1959; children: Priscilla, David Henry. BA in Econs., Amherst Coll., 1958. V.p. First Nat. Bank Boston, 1961-68; mgr. Deere & Co., Moline, Ill., 1968-71; dir. Deere & Co., Moline, 1971-77, v.p., 1977-82, sr. v.p., 1982-87, exec. v.p., 1987-90, pres., chief oper. officer, 1990—. Home: 4510 5th Ave Moline IL 61265-1904 Office: Deere & Co John Deere Rd Moline IL 61265-8098

STOWELL, LARRY JOSEPH, agricultural consultant; b. San Pedro, Calif., June 12, 1952; s. James E. and Dorothy L. (Geiser) S.; m. Wendy D. Gelernter, Feb. 22, 1986. BS, U. Ariz., 1977, PhD, 1982. Rsch. assoc. U. Ariz., Tucson, 1976-82; postdoctoral rschr. U. Calif., Davis, 1982-84; group leader Mycogen Corp., San Diego, 1984-88; prin. Pace Cons., San Diego, 1988—. Author: (with others) Microbial Products For Medicine and Agriculture, Microbial Control of Weeds, Advanced Engineered Pesticides. U. Ariz. Alumni Assn. scholar, 1978; U. Ariz. grantee, 1981. Mem. Am. Phytopathol. Soc., Entomol. Soc. Am., Agronomy Soc. Am., Weed Sci. Soc. Am., Nat. Alliance Ind. Crop Cons. (bd. dirs. 1992—), Assn. Applied Insect Ecologists (bd. dirs. 1991—), Am. Registry Cert. Profls. in Agronomy Crops and Soils (bd. dirs. 1993—), Soc. Indsl. Microbiology. Achievements include co-patents for Bioherbicide for Florida Beggarweed, Synergistic Herbicidal Compositions. Home and Office: Pace Cons 1267 Diamond St San Diego CA 92109

STRAATSMA, BRADLEY RALPH, ophthalmologist, educator; b. Grand Rapids, Mich., Dec. 29, 1927; s. Clarence Ralph and Lucretia Marie (Nicholson) S.; m. Ruth Campbell, June 16, 1951; children: Cary Ewing, Derek, Greer Grusky. Student, U. Mich., 1947; MD cum laude, Yale U., 1951; DSc (hon.), Columbia U., 1984. Diplomate Am. Bd. Ophthalmology (vice chmn. 1979, chmn. 1980). Intern New Haven Hosp., Yale U., 1951-52; resident in ophthalmology Columbia U., N.Y.C., 1955-58; spl. clin. trainee Nat. Inst. Neurol. Diseases and Blindness, Bethesda, Md., 1958-59; assoc. prof. surgery/ophthalmology UCLA Sch. Medicine, 1959-63, chief div. ophthalmology, dept. surgery, 1959-68, prof. surgery/ophthalmology, 1963—, dir. Jules Stein Eye Inst., 1964—, chmn. dept. ophthalmology, 1968—; ophthalmologist-in-chief UCLA Med. Ctr., 1968—; lectr. numerous univs. and profl. socs. 1971—; cons. to surgeon gen. USPHS, mem. Vision Research Tng. Com., Nat. Inst. Neurol. Diseases and Blindness, NIH, 1959-63, mem. neurol. and sensory disease program project com., 1964-68; chmn. Vision Research Program Planning Com., Nat. Adv. Eye Council, Nat. Eye Inst., NIH, 1973-75, 75-77, 85-89; mem. med. adv. bd. Internat. Eye Found., 1970—, chmn. 1989—; mem. adv. com. on basic clin. research Nat. Soc. to Prevent Blindness, 1971-87; mem. med. adv. com. Fight for Sight, 1960-83; bd. dirs. So. Calif. Soc. to Prevent Blindness, 1967-77, Ophthalmic Pub. Co., 1975—, v.p. 1990—, Pan-Am. Ophthalmol. Found., 1985—; chmn. sci. adv. bd. Ctr. for Partially Sighted, 1984-87; mem. nat. adv. panel Found. for Eye Research, Inc., 1984—; mem. cons. com. Palestra Oftalmologica Panamericana, 1976-81; coord. com. Nat. Eye Health Edn. Program, 1989. Editor-in-chief Am. Jour. Ophthalmology, 1993—; mem. editorial bd. UCLA Forum in Med. Scis., 1974-82, Am. Jour. Ophthalmology, 1974-91, Am. Intra-Ocular Implant Soc. Jour., 1978-79, EYE-SAT Satellite-Relayed Profl. Edn. in Ophthalmology, 1982-86; mng. editor von Graefe's Archive for Clin. and Exptl. Ophthalmology, 1976-88; contbr. over 400 articles to med. jours. Trustee John Thomas Dye Sch., Los Angeles, 1967-72. Served to lt. USNR, 1952-54. Recipient William Warren Hoppin award N.Y. Acad. Medicine, 1956, Univ. Service award UCLA Alumni Assn., 1982, Miguel Aleman Found. medal, 1992, Benjamin Boyd Humanitarian award Pan Am. Assn. Ophthalmology, 1991, Lucian Howe medal, Am. Ophthalmological Soc., 1992. Fellow Royal Australian Coll. Ophthalmologists (hon.); mem. Academia Ophthalmologica Internationales, Am. Acad. Ophthalmology (bd. councillors 1981), Found. of Am. Acad. Ophthalmology (trustee 1989, chmn. bd. trustees 1989-92), Am. Soc. Cataract and Refractive Surgery, AMA (asst. sec. ophthalmology sect. 1962-63, sec. 1963-66, chmn. 1966-67, council 1970-74), Am. Ophthalmol. Soc. (coun. 1985-90, v.p. 1992, pres. 1993), Assn. Research in Vision and Ophthalmology (Mildred Weisenfeld award 1991), Assn. U. Profs. of Ophthalmology (trustee 1969-75, pres.-elect 1973-74, pres. 1974-75), Assn. VA Ophthalmologists, Calif. Med. Assn. (mem. ophthalmology adv. panel 1972—, chmn. 1974-79, sci. bd. 1973-79, ho. of dels. 1974, 77, 79), Chilean Soc. Ophthalmology (hon.), Columbian Soc. Ophthalmology (hon.), Contact Lens Assn. Ophthalmologists, Inc., Glaucoma Soc. Internat. Congress of Ophthalmology (hon.), Heed Ophthalmic Found. (chmn., bd. dirs. 1990—), Hellenic Ophthalmol. Soc. (hon.), Los Angeles County Med. Assn., Los Angeles Soc. Ophthalmology, The Macula Soc., Pacific Coast Oto-Ophthalmol. Soc., Pan-Am. Assn. Ophthalmology (council 1972—, pres. elect 1985-87, pres. 1987-89), Peruvian Soc. Ophthalmology (hon.), The Retina Soc. Republican. Presbyterian. Clubs: Internat. Intra-Ocular Implant, The Jules Gonin, West Coast Retina Study. Avocations: music, tennis, scuba diving. Home: 3031 Elvido Dr Los Angeles CA 90049-1107 Office: UCLA Jules Stein Eye Inst 100 Stein Pla Los Angeles CA 90024-7000

STRACK, HAROLD ARTHUR, retired electronics company executive, retired air force officer, planner, analyst, musician; b. San Francisco, Mar. 29, 1923; s. Harold Arthur and Catheryn Jenny (Johnsen) S.; m. Margaret Madeline Decker, July 31, 1945; children: Carolyn, Curtis, Tamara. Student, San Francisco Coll., 1941, Sacramento Coll., 1947, Sacramento State Coll., 1948, U. Md., 1962, Indsl. Coll. Armed Forces, 1963. Commd. 2d lt. USAAF, 1943; advanced through grades to brig. gen. USAF, 1970; comdr. 1st Radar Bomb Scoring Group Carswell AFB, Ft. Worth, 1956-59; vice comdr. 90th Strategic Missile Wing SAC Warren AFB, Cheyenne, Wyo., 1964; chief, strategic nuclear br., chmn. spl. studies group Joint Chiefs of Staff, 1965-67; dep. asst. to chmn. for strategic arms negotiations Joint Chiefs Staff, 1968; comdr. 90th Strategic Missile Wing SAC Warren AFB, Cheyenne, 1969-71; chief Studies, Analysis and Gaming Agy. Joint Chiefs Staff, Washington, 1972-74, ret., 1974; v.p., mgr. MX Program electronics div. Northrop, Hawthorne, Calif., 1974—. 1st clarinetist, Cheyenne Symphony Orch., 1969-71. Mem. Cheyenne Frontier Days Com., 1970-71. Decorated D.S.M., Legion of Merit, D.F.C., Air medal, Purple Heart, Order Pour le Merite. Mem. AIAA (sr. mem.), Inst. Nav., Am. Def. Preparedness Assn. (nat. v.p. far west region), Aerospace Edn. Found. (councillor), Am. Fedn. Musicians, Cheyenne Frontier Days "Heels". Home: 707 James Ln Incline Village NV 89451-9612

STRAEDE, CHRISTEN ANDERSEN, research center administrator; b. Kloster, Denmark, Dec. 19, 1952; s. Anders Andersen and Christine (Christensen) S.; m. Lili Bay Kristensen, Apr. 29, 1989. MSc in Physics and Math., U. Aarhus, Denmark, 1981; PhD in Physics, Cen. Bur. Nuclear Measurement, Geel, Belgium, 1985. Fellow CBNM European Atomic Energy Community, Geel, 1982-85; project mgr. Danish Technol. Inst., Aarhus, 1985—; mgr. Ctr. for Surface Tech., Aarhus, 1989—; sect. mgr. tribology ctr. Danish Tech. Inst., Aarhus, 1992—; EC expert, surface engr. Commn. European Communities-Basic Rsch. in Indsl. Techs. in Europe/European Rsch. on Advanced Materials, Brussels, 1989—; Danish nat. rep. Versailles Project on Advanced Materials and Standards Tech. Working Area-Wear Test Methods, 1991; conf. presenter in field. Contbr. articles to sci. jours. and conf. proc. Fellow CEC, 1982-85. Home: Lykkenshoej 30, DK-8220 Brabrand Denmark Office: Danish Technol Inst, Teknologiparken, DK-8000 Aarhus Denmark

STRAFUSS, DAVID LOUIS, mechanical engineer; b. Manhattan, Kans., Aug. 4, 1962; s. Herman Anton and Rita Marie (Tajchman) S.; m. Traci Deautaun Bartlett, Sept. 29, 1990. BSME Tech., Kans. State U., 1987. Mech. engr. Ea. Air Lines, Inc., Miami, Fla., 1987-88; sr. mech. engr. Am. Airlines, Inc., Tulsa, 1988—. Roman Catholic. Home: 7408 S 111th East Ave Tulsa OK 74133-3251 Office: Am Airlines Inc Maintenance and Engring Mail Drop 208-PO Box 582809 Tulsa OK 74158-2809

STRAGIER, CYNTHIA ANDREAS, pharmacist; b. Bellingham, Wash., Aug. 4, 1957; d. Carl B. Andreas Jr. and Katherine T. (Townsend) Prehm; m. James B. Chumbley, Oct. 12, 1980 (div. Sept. 1986); m. Pierre M. Stragier, Jan. 2, 1987; 1 child, Shane N. B in Pharmacy, Wash. State U., 1980; BS in Civil Engring., U. Alaska, 1987. Registered pharmacist, Alaska, Wash. Staff pharmacist Walla Walla (Wash.) Gen. Hosp., 1980-83, Fairbanks (Alaska) Mem. Hosp., 1983—. Bd. dirs. Play-N-Learn Corp., Fairbanks, 1990—. Mem. Am. Soc. Hosp. Pharmacists, Alaska State Pharmacists Assn., Wash. State Pharmacists Assn. Home: 1371 Chena Ridge Rd Fairbanks AK 99709

STRAHAN, JIMMIE ROSE, mathematics educator; b. Slate Springs, Miss., June 2, 1942; d. William Vance and Minnie Lee (Bridges) Earnest; m. Richard Denman Strahan, May 30, 1970. AA, Hinds Community Coll., 1961; BS, Miss. Coll., 1963; MCS, U. Miss., Oxford, 1968; EdD, U. Fla., 1980. Tchr. math. Gulfport (Miss.) High Sch., 1963-66, Murrah High Sch., Jackson, Miss., 1966-68; instr. Delta State U., Cleveland, Miss., 1968-75, asst. prof., 1975-79, asst. prof., acting chair, 1979-80, assoc. prof., chair dept., 1980-83, prof., chmn. dept., 1983—; author curriculum Dept. Edn. State of Miss., 1988; dir. Delta Math. Project; speaker Nat. Coun. Tchrs. Math., 1974, 76, 77, 79, 82, 86, 87, 91, conv. chair, 1993. Author (manual) Miss. Assn. Colls. Tchr. Edn., 1988. Democrat. Baptist. Avocations: golf, reading. Home: 1611 Bellavista Rd Cleveland MS 38732-2910 Office: Delta State U Box 3242 DSU Cleveland MS 38733

STRAHILEVITZ, MEIR, inventor, researcher, psychiatry educator; b. Beirut, July 13, 1935; s. Jacob and Chana Strahilevitz; m. Aharona Nattiv, 1958; children: Michal, Lior. MD, Hadassah Hebrew U. Med. Sch., 1963. Diplomate Am. Bd. Psychiatry and Neurology, Royal Coll. Physicians and Surgeons Can. Asst. prof. Washington U. Med. Sch., St. Louis, 1971-74; assoc. prof. So. Ill. U., Springfield, 1974-77, U. Chgo., 1977, U. Tex. Med. Br., Galveston, 1978-81; chmn. dept. psychiatry Kaplan Hosp., Rehovot, Israel, 1987-88; prof. U. Tex. Med. Sch., Houston, 1988-92. Contbr. articles to profl. jours. Fellow Am. Psychiat. Assn. Achievements include patents for immunological methods for removing species from the blood circulatory system, for treatment methods for psychoactive drug dependence, for immunological methods for treating mammals; invention of use of antibodies to receptors and their fragments as drugs, of immunoadsorption treatment of hyperlipidemia cancer, autoimmune disease and coronary artery disease. Office: PO Box 190 Hansville WA 98340

STRAHLE, WARREN CHARLES, aerospace engineer, educator; b. Whittier, Calif., Dec. 29, 1938; s. John Dunn and Josephine Irene (Hoffman) S.; m. Pamella Ann Liles, June 25, 1965 (div. 1969); 1 son, John Curtis; m. Jane Allen Couch, June 23, 1973 (div. 1988). B.S., Stanford U., 1959, M.S., 1960; M.A., Princeton U., 1964, Ph.D., 1964. Mem. tech. staff Aerospace Corp., San Bernardino, Calif., 1964-67; profl. staff mem. Inst. Def. Analysis, Washington, 1967-68; assoc. prof. Ga. Inst. Tech., Atlanta, 1968-71, prof., 1971-74, Regent's prof., 1974—; cons. in field. Reviewer for tech. jours. Recipient Pendray Aerospace Lit. award, 1985. Fellow AIAA (chmn. tech. com. 1969-81 cert. appreciation, jour. editor 1980-82); mem. Combustion Inst. (bd. dirs.), Am. Soc. Engring. Edn., Sigma Xi (chpt. pres. 1977-78 research award, sustained research award). Club: Stanford (Ga. chpt.) (pres. 1974-75). Home: 1515 Peachtree Battle Ave NW Atlanta GA 30327-1427 Office: Ga Inst Tech Sch Aerospace Engring Atlanta GA 30332

STRAIGHT, RICHARD COLEMAN, photobiologist; b. Rivesville, W.Va., Sept. 8, 1937. BA, U. Utah, 1961, PhD in Molecular Biology, 1967. Asst. dir. radiation biology summer inst. U. Utah, 1961-63; supervisory chemist med. svc. VA Hosp., 1965—; dir. VA Venom Rsch. Lab., 1975—; adminstrv. officer rsch. svc. VA Ctr., 1980—; dir. Dixon laser inst. U. Utah, Salt Lake City. Mem. AAAS, Am. Chem. Soc., Am. Soc. Photobiology, Biophysics Soc., Internat. Solar Energy Soc. Achievements include research in photodynamic action of biomonomers and biopolymers, tumor immunology, effect on antigens on mammary adenocarcinoma of C3H mice, ageing, biochemical changes in ageing, venom toxicology, mechanism of action of psychoactive drugs. Office: U Utah Dixon Laser Inst 391 Chipeta Way Ste G Salt Lake City UT 84108*

STRAIT, BRADLEY JUSTUS, electrical engineering educator; b. Canandaigua, N.Y., Mar. 17, 1932; s. Clifford Norman and Marjory (Pratt) S.; m. Nancy Elaine Brown, Sept. 14, 1957; children: Andrew W., Martha S. BS, Syracuse U., 1958, MS, 1960, PhD, 1965. Engr. Eastman Kodak Co., Rochester, N.Y., 1960-61; asst. prof. Syracuse (N.Y.) U., 1965-69, assoc. prof., 1969-74, prof. elec. engring., 1974—, chmn. dept. elec. engring., 1974-79, dean engring., 1981-84, 1989-92; pres. Northeast Consortium Engring. Edn., Syracuse, N.Y., 1975—; cons. Southeastern Ctr. Elec. Engring. Edn., Orlando, Fla., 1978—; dir. Syracuse Ctr. Advanced Tech., Syracuse U., 1983-91. Editor: Application of the Method of Moments to Electromagnetic Fields, 1980. Active Community Mus., Jamesville, N.Y., PTA, Jamesville-DeWitt, N.Y. Served with USN, 1951-55. Recipient Teaching Excellence award Gen. Electric Co., 1981, Data Processing Person of Yr. award CNY, 1986. Fellow IEEE (chmn. Syracuse sect. 1980, Centennial medal and cert.); mem. AAUP, Am. Soc. Engring. Edn. (Western Electric Fund award 1981). Republican. Methodist. Home: 4460 Carriage Cir Jamesville NY 13078-9511 Office: Syracuse U 2-212 Ctr for Science & Tech Syracuse NY 13244

STRAIT, JEFFERSON, physicist, educator; b. Washington, Mar. 27, 1953; s. Edward Bernard and Sally (Thompson) S.; m. Robin Dee Brickman, June 27, 1976; children: Jared Hillel, Caleb Edward. AB, Harvard U., 1975; PhD, Brown U., 1985. Postdoctoral fellow AT&T Bell Labs., Holmdel, N.J., 1984-85; asst. prof. dept. physics and astronomy Williams Coll., Williamstown, Mass., 1985-92, assoc. prof., 1992—. Instrumentation grantee NSF, Washington, 1989, PPG Industries Found. grantee Rsch. Corp., Tucson, Ariz., 1986. Mem. Am. Phys. Soc., Optical Soc. Am., Am. Assn. Physics Tchrs., Sigma Xi (pres. local club 1990-92). Achievements include observation of light-induced defects in hydrogenated amorphous silicon by picosecond photoinduced absorption; study of photorefractive effect in semiconductors. Office: Williams Coll Dept Physics Williamstown MA 01267

STRAITON, ARCHIE WAUGH, electrical engineering educator; b. Arlington, Tex., Aug. 27, 1907; s. John and Jeannie (Waugh) S.; m. Esther McDonald, Dec. 28, 1932; children: Janelle (Mrs. Thomas Henry Holman), Carolyn (Mrs. John Erlinger). BSEE, U. Tex., 1929, MA, 1931, PhD, 1939. Engr. Bell Telephone Labs., N.Y.C., 1929-30; from instr. to assoc. prof. Tex. Coll. Arts and Industries, 1931-41, prof., 1941-43, head dept. engring., 1941-43; faculty U. Tex., Austin, 1943—; prof. U. Tex., 1948-63, dir. elec. engring. research lab., 1947-72, Ashbel Smith prof. elec. engring., 1963-89, Ashbel Smith prof. emeritus, 1989—, chmn. dept., 1966-71, acting v.p., grad. dean, 1972-73. Contbr. articles to profl. jours. Fellow IEEE (Thomas A. Edison medal 1990); mem. NAE, Sigma Xi, Tau Beta Pi, Eta Kappa Nu. Home: 4212 Far West Blvd Austin TX 78731-2804

STRAKHOV, VLADIMIR NIKOLAYEVICH, geophysics educator; b. Moscow, May 3, 1932; s. Nikolai Michailovich and Alexandra (Vasilievna) S.; widowed; 1 child, Alexandr. Degree in engring. and geophysics, Moscow Inst. Geol. Prospecting, 1955. Registered profl. mining engr., geophysicist. Engr. Inst. of Physics of the Earth Russian Acad. Sci., Moscow, 1959-62, sr. sci. officer, 1962-74, head of lab., 1974-87, head of dept., 1987-89, dir., 1989—; Author: Methods of Interpretation of Gravity and Magnetic Anomalies, 1984, Handbook of Geophysicst, Gravity Prospecting, 1990; contbr. chpts. to books, over 450 sci. papers and monographs to profl. jours. including Physics of the Earth, Reports of the Academy of Science of the USSR. Achievements include reseach in theorems of uniqueness in the potential field theory, the theory of optimum regularization; development of new numerical methods of solution of linear and non-linear inverse problems; analysis of phenomenon of equivalence; the common methodolgy of interpretation of the gravity and magnetic fields. Office: Inst of Physics of Earth, Bolshaya Gruzinskaya 10, Moscow 123810, Russia

STRAND, KAJ AAGE, astronomer; b. Hellerup, Denmark, Feb. 27, 1907; came to U.S., 1938; s. Viggo Peter and Constance (Malmgren) S.; m. Ulla Nilson, Mar. 11, 1943 (div. Dec. 1948); 1 child, Kristina Ragna Strand; m. Emilie Rashevsky Strand, June 10, 1949; 1 child, Constance Vibeke Strand. AB, U. Copenhagen, 1931, MSc, 1931, PhD, 1938. Geodesist Royal Geodetic Inst., Copenhagen, 1931-33; asst. to dir. Univ. Obs., Leiden, Holland, 1933-38; rsch. assoc. astronomer Swarthmore (Pa.) Coll., 1938-42; assoc. prof. U. Chgo., 1946-47; prof. astronomy Northwestern U., Evanston, Ill., 1947-58; dir. astrometry and astrophysics U.S. Naval Obs., Washington DC, 1958-63, sci. dir., 1963-77; vis. prof. U. Copenhagen, 1954, The USSR Acad. Scis., 1959, Acad. Sinica, China, 1987; cons. NASA, 1977-79, Lincoln Lab., MIT, 1981-85; mem. adv. bd. NSF, 1953-56, Office of Naval Rsch., 1954-57, chmn. 1956-57, Nat. Rsch. Coun., 1954-57, 71-77, Nat. Bur. Standards, 1971-77, Sr. Fulbright-Hays Program Com., 1972-75. Contbr. over 100 articles in Am., European and USSR profl. jours. Served from pvt. to capt. USAAF, 1942-45. Danish Rask-Oersted Found. fellow, 1938-40, John Simon Guggenheim Found. fellow, 1946; decorated knight cross 1st class Royal Order Dannebrog (Denmark); honor cross 1st class Literis et Artibus (Austria); recipient Disting. Civilian Svc. award Dept. Navy, 1973. Mem. Am. Astron. Soc., Astron. Soc. Pacific, Internat. Astron. Union, Royal Danish Acad. Scis. and Letters, AAAS, Philosophical Soc. of Washington DC, Cosmos Club, Sigma Xi. Achievements include research on double stars and astrometric instrumentation; development of 1.51m Astrometric Reflector and the Strand Automatic Measuring Machine. Home: 3200 Rowland Pl NW Washington DC 20008-3223

STRAND, TENA JOY, civil engineer; b. Sedalia, Mo., Dec. 9, 1968; d. Garey Clarence and Betty Mae (Chaffins) S.; m. Troy Frank Campbell, June 16, 1992. AS, Weber State Coll., 1989; BS in Civil Engring., U. Utah, 1992. Engr. in tng. Civil engr. Def. Depot, Ogden, Utah, 1990—. Sterling scholar State of Utah, 1987, honors scholar Weber State Coll., 1988, scholar U. Utah, 1991-92. Mem. ASCE, Soc. Women Engrs. Office: Def Depot 500 W 12th St Ogden UT 84407

STRANGWAY, DAVID WILLIAM, university president; b. Can., June 7, 1934. BA in Physics and Geology, U. Toronto, 1956, MA in Physics, 1958, PhD, 1960; DLittS (hon.), Victoria U., U. Toronto, 1986; DSc (hon.), Meml. U. Nfld., 1986, McGill U., Montreal, Que., Can., 1989, Ritsumeikan U., Japan, 1990; D.Ag.Sc. (hon.), Tokyo U. Agr., 1991. Sr. geophysicist Dominion Gulf Co. Ltd., Toronto, 1956; chief geophysicist Ventures Ltd., 1956-57, sr. geophysicist, summer 1958; research geophysicist Kennecott Copper Corp., Denver, 1960-61; asst. prof. U. Colo., Boulder, 1961-64, M.I.T., 1965-68; mem. faculty U. Toronto, 1968-85, prof. physics, 1971-85, chmn. dept. geology, 1972-80, v.p., provost, 1980-83, pres., 1983-84; pres. U. B.C., 1985—; chief geophysics br. Johnson Space Ctr., NASA, Houston, 1970-72, chief physics br., 1972-73, acting chief planetary and earth sci. div., 1973; vis. prof. geology U. Houston, 1971-73; interim dir. Lunar Sci. Inst., Houston, 1973; vis. com. geol. scis. Brown U., 1974-76, Meml. U., St. John's Nfld., 1974-79, Princeton U.; v.p. Can. Geosci. Coun., 1977; chmn. proposal evaluating program Univs. Space Rsch. Assocs., 1977-78, Ont. Geosci. Rsch. Fund; Pahlavi lectr. Govt. Iran, 1978; cons. to govt. and industry, mem. numerous govt. and sci. adv. and investigative panels; hon. prof. Changchun Coll. Geology, People's Republic China, 1985, Guilin Coll. Geology, People's Republic China, 1987; bd. dirs. MacMillan Bloedel Ltd., Echo Bay Mines, Ltd., BC Gas, Corp.-Higher Edn. Forum, Internat. Inst. for Sustainable Devel. Author numerous papers, reports in field. Active Royal Trust

Adv. Coun. Recipient Exceptional Sci. Achievement medal NASA, 1972. Fellow Royal Astron. Soc., Royal Soc. Can.; mem. AAAS, Soc. Exploration Geophysicists (Virgil Kauffman Gold medal 1974), Geol. Assn. Can. (pres. 1978-79, Logan Gold medal), Can. Geophys. Union (chmn. 1977-79, J. Tuzo Wilson medal 1987), Am. Geophys. Union (sect. planetology sect. 1978-81), European Assn. Exploration Geophysicists, Soc. Geomagnetism and Geoelectricity Japan, Can. Geosci. Council (pres. 1980), Can. Exploration Geophysicists, Soc. Exptl. Geophysics (hon.), Canada-Japan Soc., Japan Soc. Can. (founding bd. dirs.), Internat. House Japan, Inc.

STRANSKY, ROBERT JOSEPH, JR., nuclear engineer; b. Bklyn., Nov. 7, 1964; s. Robert Joseph and Constance Regina (Eckenrode) S.; m. Jacqueline Marie Greer, Aug. 3, 1990. BS in Nuclear Engring., Pa. State U., 1986; postgrad., Johns Hopkins U., 1991—. Reactor systems specialist U.S. Nuclear Regulatory Commn., Washington, 1986-90, project engr., 1990-91, project mgr., 1991—. Co-author: Graf/X Mag., 1988-89. Mem. Am. Nuclear Soc., Alpha Nu Sigma. Republican. Office: US Nuclear Regulatory Commn 13 D 18 Washington DC 20555

STRATES, BASIL STAVROS, biomedical educator, researcher; b. Finiki-Filiates, Epirus, Greece, Aug. 16, 1927; came to U.S., 1947; s. Stavros Ch. and Alexandra (Sotiriou) S.; m. Margrith Muelheim, June 21, 1957; children: Chris, Helen, Felix. BS in Chemistry and Biology, Clark U., 1953; MS in Pharmacology and Biochemistry, U. Rochester, 1956; PhD in Biochemistry, U. Thessoloniki, Greece, 1963, MD, 1967. Medical diplomate, Greece. Rsch. fellow U. Rochester, 1955-57; from asst. to assoc. prof. U. Thessoloniki, 1958-67; postdoctoral fellow UCLA, 1967-68; from asst. to assoc. prof. UCLA Sch. Medicine, 1968-84; rsch. biochemist VA Med. Ctr., Omaha, 1987-90; rsch. biochemist, prin. investigator VA Med. Ctr., Gainesville, Fla., 1990—; assoc. prof. orthopaedic rsch. Creighton U./U. Nebr., Omaha, 1985-90; prof. orthopaedics rsch. U. Fla. Coll. Medicine, Gainesville, 1990—; rsch. coord. McGaw Labs., L.A., 1972-73; dir. med. dept. Dow Pharm.-Dow Chem., Indpls., 1974-81, Adria Labs., Columbus, Ohio, 1981-85. Author: (Greek) Clinical Chemistry, 1961, Chemical and Industrial Toxicology, 1967; co-author: Collagen, 1988, Bone, 1991, 92. Rsch. grantee NIH, 1956—, Dept. VA, 1989—; Clark U. scholar, U. Rochester scholar, 1950-55; U. Rochester fellow, 1955-57, UCLA fellow, 1967-68. Fellow Am. Coll. Clin. Pharmacology; mem. AAAS, Am. Soc. for Bone Mineral Rsch., N.Y. Acad. Scis., Orthopaedic Rsch. Soc. Achievements include rsch. on osteogenesis and bone repair stimulation by cell growth factors, microcrystalline hydroxyapatite-growth factor delivery systems, characterization of bone morphogenetic protein in bone matrix; enhanced osteogenesis and skeletal repair by demineralized bone matrix (DBM), bone morphogenetic protein (BMP), and TGFB superfamily. Office: Univ Florida PO Box 100246 JHMHC Gainesville FL 32610-0246 also: VA Med Ctr Gainesville FL 32608-1197

STRATTON, BRENTLEY CLARKE, research physicist; b. L.A., Jan. 23, 1956; s. Thomas Oliver and Carol Joyce (Wilson) S.; m. Elizabeth Gayle Mason, Sept. 14, 1985; 1 child, Elizabeth Kent. BA in Physics, Franklin and Marshall Coll., 1978; PhD in Physics, Johns Hopkins U., 1984. Postdoctoral fellow Johns Hopkins U., Balt., 1984-85; rsch. physicist Plasma Physics Lab., Princeton (N.J.) U., 1985—. Contbr. numerous articles to sci. jours. Mem. Am. Phys. Soc. Achievements include research on plasma spectroscopy. Office: Princeton U Physics Plasma Lab PO Box 451 Princeton NJ 08543

STRAUB, KARL DAVID, biochemist researcher; b. Louisville, Aug. 17, 1937; married; 3 children. BS in Math., Duke U., 1959, MD, 1965, PhD in Biochemistry, 1968. Lic. N.C., Ark. Intern Duke U., Durham, N.C., 1968-69; prof. physics Free Electron Laser Lab. Duke U., Durham, 1993—; resident in internal medicine U. Ark. Med. Ctr., Little Rock, 1972-76; from asst. prof. to prof. medicine U. Ark. for Med. Scis., Little Rock, 1972-93, from asst. prof. to prof. biochemistry, 1972-93; rsch. biophysicist and staff physician VA Hosp., Little Rock, 1972-74, assoc. chief staff rsch., 1974-93; sec. rsch. and devel. com. VA Med. Ctr. 1974—, clin. exec. bd., 1974—; mem. experimental program to stimulate competitive rsch., NSF, 1980—, acting chmn., 1983—; rsch. com. Am. Heart Assn., 1980—, bd. dirs., 1983—; mem. legis. NCTR task force, 1982—; cons. Fed. Applied Sci. Evaluation Ctr., Washington, 1982—; chmn. instl. review bd., UAMS, 1984—; chmn. basic sci. review bd., ASTA, 1986—. Contbr. articles to profl. jours. Lt. comdr. USN, 1970-72. Mem. Am. Soc. Biological Chemistry, Am. Chem. Soc., Am. Fedn. for Clin. Rsch., Biobhysical Soc., So. Soc. Clin. Investigation, N.Y. Acad. Sci., Internat. Soc. Heart Rsch., Royal Soc. Chemistry, Sigma Xi, Phi Eta Sigma, Phi Mu Epsilon, Sigma Pi Sigma, Phi Beta Kappa, Alpha Omega Alpha. Achievements include research in bioenergetics (active transport oxidative phosphorylation), solid state biophysics (semi-conduction in proteins, energy transfer in proteins), mechanisms of protein hormones, picosecond spectroscopy of porhyrins and hemoproteins. Office: Duke U Dept Physics Durham NC 27708-0319

STRAUS, DAVID JEREMY, hematologist, educator; b. Urbana, Ill., Apr. 22, 1944; s. Gerhard D. and Lois (Marin) S.; m. Karen Bassak, Aug. 20, 1966 (div. 1992); children: Jennifer, Emily. AB, U. Chgo., 1965; MS, Marquette (Med. Coll. Wis.), 1969. Intern Montefiore Hosp., N.Y.C., 1969-72; clin. fellow in hematology Beth Israel Hosp., Boston, 1972-73; clin. fellow in med. oncology Meml. Hosp., N.Y.C., 1975-77, asst. attending physician dept. medicine, 1977-82, assoc. attending physician, 1982—; asst. prof. dept. medicine Cornell U. Med. Coll., 1977-82, assoc. prof., 1982—; rsch. assoc. Sloan-Kettering Inst., 1982—, assoc. mem. Meml. Hosp., 1990, acting chief hematology svc. div. hematologic oncology dept. medicine, 1990. Contbr. articles to Am. Jour. Medicine, Med. Pediatric Oncology, Blood, Cancer Treatment Report, Cancer, Clin. Immunopathology, Medicine, Jour. Clin. Oncology, Cancer Chemotherapy Pharmacology, Jour. Clin. Exptl. Cancer Rsch., Annals Internal Medicine, others. Major U.S. Army, 1973-75. Grantee Nat. Cancer Inst. Fellow ACP; mem. Am. Soc. Hematologu, Am. Soc. Clin. Oncology, Am. Assn. for Cancer Rsch., Am. Fedn. for Clin. Rsch. Achievements include research in clinical hematology and hematologic malignancy particularly malignant lymphoma. Office: Meml Sloan Kettering Cancer 1275 York Ave New York NY 10021

STRAUS, LEON STEPHAN, physicist; b. Takoma Park, Md., May 29, 1943; s. Cheryl Sarran Straus, Apr. 4, 1970; children: Jonathan, Jennifer. BS in Physics, Antioch Coll., Yellow Springs, Ohio, 1965; M Physics, Georgetown U., 1970, PhD in Physics, 1971. Mem. rsch. staff Ctr. Naval Analyses, Alexandria, Va., 1973-75; field rep. CTF 69 Ctr. Naval Analyses, Naples, Italy, 1975-77; project mgr. Ctr. Naval Analyses, Alexandria, 1977-79; field rep. CTF 66/67 Ctr. Naval Analyses, Naples, Italy, 1979-82; assoc. dep. dir. Ctr. Naval Analyses, Alexandria, 1982-85; field rep. CTF 72 Ctr. Naval Analyses, Kamiseya, Japan, 1985-87; program mgr. Ctr. Naval Analyses, Alexandria, 1987-90; field rep. COMSIXTHFLT Ctr. Naval Analyses, Gaeta, Italy, 1990-92; project mgr. Ctr. Naval Analyses, Alexandria, 1992—; asst. AEC, Germantown, Md., 1968-71. Contbr. articles to profl. jours. Vol. Jewish lay leader USN, Naples, 1975-77, 79-82. Recipient Fellowship Georgetown U., Washington, 1965-68. Mem. Acoustical Soc. Am., Navy Submarine League. Jewish. Achievements include evaluting U.S. Navy and Joint strategy, tactics, comunications and technology. Office: Ctr Naval Analyses 4401 Ford Ave Alexandria VA 22302

STRAUSS, BRUCE PAUL, engineering physicist, consultant; b. Elizabeth, N.J., Aug. 19, 1942; s. Edward and Sylvia (Levine) S.; m. Judi Ann Schleimer, July 7, 1964 (div. Apr. 1979); children: Lori Ellen Strauss Feldman, Lisa Beth; m. Suzanne F.L. Geller, Oct. 30, 1983. SB, MIT, 1964, PhD, 1967; MBA, U. Chgo., 1972. Registered profl. engr., Ill., Mass. Prin. rsch. engr. Avco Everett (Mass.) Rsch. Lab., 1967-68; rsch. engr. Argonne (Ill.) Nat. Lab., 1968-69; assoc. div. dir. Fermi Nat. Lab., Batavia, Ill., 1969-79; dir. devel. Magnetic Corp. Am., Waltham, Mass., 1979-85; v.p. Powers Assocs., Inc., Salem, Mass., 1985-92, Cosine, Inc., Brookline, Mass., 1992—; mem. adv. com. Export Control Bd., Dept. of Def., Washington, 1987-91; founder, treas. Applied Superconductivity Conf., Inc., Batavia, 1985—. Editor proceedings Applied Superconductivity Conf., 1976; contbr. over 50 articles to profl. jours. Instr. Am. Heart Assn., Boston, 1979—; instr., trainer YMCA Underwater Activities Program, Atlanta, 1973—. Mem. Am. Soc. for Metals, Am. Phys. Soc., Materials Rsch. Soc. Achievements include 8 patents in field. Home: 232 Summit Ave 203 Brookline MA 02146 Office: Cosine Inc PO Box 1078 1037 Beacon St Brookline MA 02146

STRAUSS, SIMON WOLF, technical consultant; b. Bedzin, Keltz, Poland, Apr. 15, 1920; came to U.S., 1929; s. Israel Calvin and Anna (Hops) S.; m. Mary Jo Boehm, Dec. 27, 1957; children: Jack Calvin, Ruth Ann. BS in Chemistry, Polytech. Inst. of Bklyn., 1944, MS in Chemistry, 1947, PhD in Chemistry, 1950. Rsch. chemist Nat. Bur. Standards, Washington, 1951-55; from phys. chemist to head chem. metallurgy sect. Naval Rsch. Lab., Washington, 1955-63; sr. staff scientist Air Force Systems Command, Washington, 1963-80; pvt. practice tech. cons. Washington, 1980—; mem. bd. civil svc. examiners for sci. and tech. pers. U.S. Naval Dist. of Washington, 1959-63; chair rsch. steering com. Air Force Dir. of Sci. Tech., Washington, 1976-80. Prin. compiler 75 Years of Scientific Thought, 1987. Recipient Air Force Decoration for Exceptional Civilian Svc., 1980, Disting. Career in Sci. award Wash. Acad. Scis., 1988, Disting. Svc. award, 1990. Fellow AAAS, Wash. Acad. Scis. (pres. 1986-87), Am. Inst. Chemists, Cosmos Club, Sigma Pi Sigma, Phi Lambda Upsilon, Sigma Xi. Achievements include 3 patents for electrodeposition of Cadmium on high strength steel; research and development of advanced composites technology. Home: 4506 Cedell Pl Temple Hills MD 20748-3805

STRAUSS, ULRICH PAUL, educator, chemist; b. Frankfurt, Germany, Jan. 10, 1920; s. Richard and Marianne (Seligmann) S.; m. Esther Lipetz, June 20, 1943 (dec. Sept. 1949); children—Dorothy, David; m. Elaine Greenbaum, Nov. 23, 1950; children—Elizabeth, Evelyn. A.B., Columbia U., 1941; Ph.D., Cornell U., 1944. Sterling fellow Yale U., 1946-48; faculty Rutgers U., New Brunswick, N.J., 1948—, prof. phys. chemistry, 1960-90, prof. emeritus, 1990—; also dir. Sch. Chemistry, 1965-71, chmn. dept. chemistry, 1974-80; prof. emeritus Rutgers U., 1990—. Mem. editorial adv. bd. Macromolecules, 1990—; contbr. articles to profl. jours. Recipient Sci. achievement award Johnson Wax Co., 1986; NSF sr. fellow Nat. Center Sci. Research, Strasbourg, France, 1961-62; Guggenheim fellow U. Oxford, Eng., 1971-72. Fellow N.Y. Acad. Scis.; mem. Am. Chem. Soc. (chmn. phys. chemistry group N.J. sect. 1956, councillor 1961-72, honored by 1-day symposium at nat. meeting N.Y.C. 1986). Home: 227 Lawrence Ave Highland Park NJ 08904-1837 Office: Dept Chemistry Rutgers U New Brunswick NJ 08903

STRAW, WILLIAM RUSSELL, environmental scientist; b. Washington, June 19, 1957; s. Russell Don Straw and Joan Cornelia (White) Chester; m. Melbis Cinencia Rodriguez, May 30, 1981. BS in Earth Scis., U. Kans., 1981; BS in Geography, U. Md., 1991; postgrad. in ecology, U. Ga., 1992—. Cert. photogrammetrist. Earth scientist Greenhorne & O'Mara, Inc., Greenbelt, Md., 1986-89, environ. scientist, 1989-92; part-time environ. planner Fed. Emergency Mgmt. Agy., Atlanta, 1992—. Park vol. U.S. Nat. Park Svc., Greenbelt National Park, Md., 1987. Capt. U.S. Army, 1975-78, 81-85. Mem. Internat. Soc. Ecology, Ecol. Soc. Am., Sigma Gamma Epsilon (treas. 1991), Gamma Theta Upsilon, Beta Beta Beta. Roman Catholic. Achievements include environmental assessments, wetland studies and maps, land use/land cover maps, hydrogeologic studies and maps, soil studies and maps, meteorological and climatological studies in U.S., Europe, Africa and Asia. Office: Univ Ga Inst Ecology Athens GA 30602-2202

STRAWA, ANTHONY WALTER, researcher; b. Chgo., Apr. 22, 1950. BS, USAF Acad., Colorado Springs, Colo., 1973; PhD, MS, Stanford (Calif.) U., 1986. Rsch. asst. Stanford U., 1982-86; lead researcher ballistic range NASA-Ames Rsch. Ctr., Moffett Field, Calif., 1986-89, prin. investigator Aerosissit flight expt., 1990-91, rsch. scientist, 1991—; mem. NASA Aerodynamic Sensors Working Group, 1989—. Mem. AIAA (sec. aerodynamic measurement tech. com. 1989—), AAAS, Aeroballistic Range Assn. Office: NASA-Ames Rsch Ctr Mail Stop 245-4 Moffett Field CA 94035

STRAWDERMAN, WILLIAM E., statistics educator; b. Westerly, R.I., Apr. 25, 1941; s. Robert Lee and Alida Browning (Dow) S.; m. Susan Linda Grube; July 20, 1985; children: Robert Lee, William Edward, Heather Lynne. BS, U. R.I., 1963; MS, Cornell U., 1965, Rutgers U., 1967; PhD, Rutgers U., 1969. Mem. tech. staff Bell Telephone Labs., Holmdel, N.J., 1965-67; prof. Stanford (Calif.) U., 1969-70; instr. Rutgers U., New Brunswick, N.J., 1967-69, prof. stats., 1970—. Contbr. over 90 articles to profl. jours. Fellow Inst. Math. Stats., Am. Statis. Assn. Office: Rutgers U Statistics Dept Hill Ctr-Busch Campus New Brunswick NJ 08903

STREB, ALAN JOSEPH, government official, engineer; b. Balt., Mar. 12, 1932; s. H. Albert and Anna Marie (Minderlein) S.; m. Dorothy Anne Forestal, Apr. 14, 1956; children: John A., David A., Mark A., Marla A., Christopher A. BMechE, Johns Hopkins U., 1954; MS, Drexel U., 1961. Engr. Glenn L. Martin, Balt., 1951-60; tech. dir. Martin Marietta, Balt., 1960-63, mgr. adv. programs, 1963-67; v.p. mktg. Dynatherm Corp., Cockeysville, Md., 1967-76; dep. asst. sec. Dept. Energy, Washington, 1977—. Contbr. numerous articles on energy-related subjects to profl. publs.; patentee in field. Pres., treas. Homeowner's Assn., Glen Arm, Md., 1967-72; bd. dirs., mem. architecture com. Homeowner's Assn., Bethany Beach, Del., 1979—. Named Inventor of Yr., Martin Marietta Corp., 1964. Home: 26 Gunpowder Rd Glen Arm MD 21057-9460 Office: US Dept Energy Indsl Techs 1000 Independence Ave SE Washington DC 20585-0001

STRECKER, ROBERT EDWIN, neuroscientist, educator; b. Cin., Sept. 23, 1955; s. Robert Louis and Charlotte Josephine (Gilbert) S.; m. Ulla M. Larsen, Dec. 27, 1985; children: Laura, Marcus. BS with highest honors, U. Calif., Santa Barbara, 1977; PhD, Princeton U., 1985. Postdoctoral fellow Lund U., Sweden, 1985-87; staff scientist Hana Biologies, Alameda, Calif., 1987-89; asst. prof. SUNY, Stony Brook, 1989—; behavioral neuroscience com., 1991, behavioral neuroscience dir. search com., 1991-92. Contbr. articles to profl. jours. Bd. dirs. Stony Brook Child Care Svcs., 1991—. Recipient NIH grant, 1991-93, Health Found. grant, 1990, SUNY Stony Brook grant, 1991-92. Mem. AAAS, Soc. for Neuroscience, European Neuroscience Assn., N.Y. Acad. Sci. Democrat. Achievements include pending patent in methods of producing genetically modified astrocytes and uses. Home: 211 Michigan Ave Port Jefferson NY 11777 Office: SUNY Dept Psychiatry Putnam Hall Stony Brook NY 11794-8790

STREEPER, ROBERT WILLIAM, environmental chemist; b. N.Y.C., Dec. 17, 1951; s. Horace LeRoy and Jeannette Helene (Thompson) S.; m. Deborah Joan Brogan, June 15, 1979; children: Christopher, Benjamin, Michael. BS in Environ. Sci., Lamar U., 1977. Environ. chemist Mobil Oil Corp., Beaumont, Tex., 1990—. Scoutmaster Boy Scouts, Beaumont, 1985-86; chmn. com. Cub Scouts, Beaumont, 1987-89. Mem. Astronomical Soc., Am. Chem. Soc., Lamar U. Cardinal Club, Lamar U. Alumni, Lambda Chi Alpha (pres. 1972-73). Republican. Office: Mobil Oil Corp PO Box 3311 Beaumont TX 77704

STREETER, JOHN WILLIS, information systems manager; b. Topeka, Sept. 3, 1947; s. Jack and Edith Bernice (Vowels) S.; m. Nancy Ann Buck, June 15, 1968 (div. 1985); children: Sarah Beth, Timothy Paine; m. Linda Lea Wenrich Weisbender, Sept. 13, 1986; stepchildren: Michael Leon Weisbender II, Debra Ann Weisbender Johnson, Dawn Marie Weisbender. BS in Computer Sci., Kans. State U., 1973, MBA in Mgmt., 1974; postgrad., Harvard U., 1992. Computer programmer U.S.M.C., 1965-70, Kans. State U., Manhattan, 1970-74; cons., mgr., prin. Am. Mgmt. Systems, Inc., Arlington, Va., 1974-83; systems planning analyst Fed. Nat. Mortgage Assn., Washington, 1983-85; assoc. dir. computing and telecomm. Kans. State U., Manhattan, 1985—, dir. info. systems, 1991—. Author: Streeter Genealogy, 1985. Staff sgt. USMC, 1965-70. Recipient Navy Achievement medal in data processing Sec. Navy, 1971. Mem. ASM, IEEE Computer Soc. (affiliate), Am. Inst. Cert. Computer Profls., Am. Mgmt. Assn., Cause Inc. (Kans. State U. voting mem. rep. 1987—, mem. liaison com. 1987-89), Streeter Family Assn. (bd. dirs. 1988—, v.p. 1990-94), SOR, KC. Republican. Roman Catholic. Avocations: genealogy, history. Home: 6765 Salzer Rd Wamego KS 66547-9636 Office: Kans State U Info Systems 2323 Anderson Ave Ste 215 Manhattan KS 66502-2947

STREETMAN, BEN GARLAND, electrical engineering educator; b. Cooper, Tex., June 24, 1939; s. Richard E. and Bennie (Morrow) S.; m. Lenora Ann Music, Sept. 9, 1961; children: Paul, Scott. BS, U. Tex., 1961, MS, 1963, PhD, 1966. Fellow Oak Ridge Nat. Lab., 1964-66; asst. prof. elec. engring. U. Ill., 1966-70, asso. prof., 1970-74, prof., 1974-82; rsch. prof. Coordinated Sci. Lab., 1970-82; prof. elec. engring. U. Tex., Austin, 1982—, dir. Microelectronics Rsch. Ctr., 1984—, Dula D. Cockrell Centennial chair engring., 1989—; cons. in field. Author: Solid State Electronic Devices, 3d edit., 1990. Recipient Frederick Emmons Terman award Am. Soc. Engring. Edn., 1981, AT&T Found. award Am. Soc. Engring. Edn., 1987. Fellow IEEE (Edn. medal 1989), Electrochem. Soc.; mem. NAE, Tau Beta Pi, Eta Kappa Nu, Sigma Xi. Home: 3915 Glengarry Dr Austin TX 78731-3835 Office: U Tex Microelectronics Rsch Ctr BRC/MER Austin TX 78712

STREHBLOW, HANS-HENNING STEFFEN, chemistry educator; b. Berlin, Fed. Republic Germany, Sept. 21, 1939; s. Otto Alfred Wilhelm and Johanna Annemarie (Herold) S.; children: Kaja Ines, Alexander Dirk. Diplom in Chemistry, Freie U. Berlin, 1967, PhD, 1971, Habilitation in Phys. Che, 1977. Asst. Freie U. Berlin, 1969-73, asst. prof., 1973-79; cons. scientist Bell Labs., Murray Hill, N.J., 1976-77; akademischer rat Heinrich-Heine Univ., Düsseldorf, 1979-82; prof. phys. chemistry Heinrich-Heine Univ., 1982—; vis. prof. Johns Hopkins U., Balt., 1987-88, U. Pierre et Marie Curie Paris, 1993; guest scientist NIST, Gaithersburg, Md. Contbr. articles to profl. jours. Mem. Electrochem. Soc., Bunsen Gesellschaft, Gesellschaft Deutscher Chemiker, Fachgruppe Angewandte Elektrochemie. Avocation: gardening. Office: Heinrich Heine U Inst Phys/Elec Chem, Düsseldorf Geb 2642, 40225 Düsseldorf Germany

STREITWIESER, ANDREW, JR., chemistry educator; b. Buffalo, June 23, 1927; s. Andrew and Sophie (Morlock) S.; m. Mary Ann Good, Aug. 19, 1950 (dec. May 1965); children—David Roy, Susan Ann; m. Suzanne Cope Beier, July 29, 1967. A.B., Columbia U., 1949, M.A., 1950, Ph.D., 1952; postgrad. (AEC fellow), MIT, 1951-52. Faculty U. Calif., Berkeley, 1952-92, prof. chemistry, 1963-92, prof. emeritus, 1993—; researcher on organic reaction mechanisms, application molecular orbital theory to organic chemistry, effect chem. structure on carbon acidities, f-element organometallic chemistry; cons. to industry, 1957—. Author: Molecular Orbital Theory for Organic Chemists, 1961, Solvolytic Displacement Reactions, 1962, (with J.I. Brauman) Supplemental Tables of Molecular Orbital Calculations, 1965, (with C.A. Coulson) Dictionary of Pi Electron Calculations, 1965, (with P.H. Owens) Orbital and Electron Density Diagrams, 1973, (with C.H. Heathcock and E.M. Kosower) Introduction to Organic Chemistry, 4th edit., 1992; also numerous articles; co-editor: Progress in Physical Organic Chemistry, 11 vols., 1963-74. Recipient Humboldt Found. Sr. Scientist award, 1976, Humboldt medal, 1979. Fellow AAAS; mem. NAS, Am. Chem. Soc. (Calif. sect. award 1964, award in Petroleum Chemistry 1967, Norris award in phys. organic chemistry 1982, Cope scholar award 1989), Am. Acad. Arts and Scis., German Chem. Soc., Bavarian Acad. Scis. (corr.), Phi Beta Kappa, Sigma Xi. Office: U Calif Dept Chemistry Berkeley CA 94720

STRESEN-REUTER, FREDERICK ARTHUR, II, chemical company communications executive; b. Oak Park, Ill., July 31, 1942; s. Alfred Procter and Carol Frances (von Pohek) S.-R.; cert. in German, Salzburg (Austria) Summer Sch., 1963; BA, Lake Forest Coll., 1967. Mgr. advt. Stresen-Reuter Internat., Bensenville, Ill., 1965-70; mgr. animal products mktg. Internat. Minerals & Chem. Corp., Mundelein, Ill., 1971-79, dir. animal products mktg., 1979-87; dir. communications Pitman-Moore, Inc. subs. IMCERA, Inc., 1987-92; pres. Brit. Iron Ltd., Lake Forest, Ill., 1984-86 ; lectr. mktg. U. Ill., 1977, Am. Mgmt. Assn., 1978; cons. mktg. to numerous agrl. cos., 1973—. Trustee governing mem. Libr. Internat. Rels., Chgo., 1978; founding pres. Woodstock (Ill.) Mozart Festival, 1988-90; bd. dirs. Woodstock Opera House, Ill. Arts Alliance Found., Ill. Arts Alliance, 1993—. Recipient cert. of excellence Chgo. 77 Vision Show, 1977; Silver Aggy award, 1977; spl. jury gold medal V.I., N.Y. Internat. film festival awards, 1977; CINE Golden Eagle, 1980, 88; Bronze medal N.Y. Internat. Film Festival, 1981, Silver medal, 1982; Silver Screen award U.S. Indsl. Film Festival, 1981. Mem. Nat. Feed Ingredients Assn. (chmn. publicity and pubs. 1976), Nat. Agrl. Mktg. Assn. (numerous awards), Am. Feed Mfrs. Assn. (citation 1976, pub. rels. com., conv. com.), Mid-Am. Commodity Exch., USCG Aux., U.S. Naval Inst., Am. Film Inst., Bugatti Owners Club. Episcopalian. Club: Sloane (London), The Chgo. Farmer's Club. Contbr. articles to profl. jours. Home: Tryon Grove Farm 8914 Tryon Rd Ringwood IL 60072

STRICKLAND, CHRISTOPHER ALAN, statistician, systems analyst; b. Titusville, Fla., Oct. 2, 1957; s. G.B. and Martha Jane (Foster) S.; m. Kathleen Ann Winningham, Dec. 31, 1976; children: David Mathew, Brian Jeffrey. BS in Mathematics and Statistics, Fla. State U., 1980, MS in Statistics, 1981. Instr. math dept. Fla. State U., 1980-82; OSAD(c) budget systems analyst 1st Info. Systems Group, Arlington, Va., 1982-85; chief, def. systems analyst B-1B Test Team, Dyess AFB, Tex., 1985-89; systems analyst, statistician Space Station Support, Cocoa Beach, Fla., 1989—; pres. New Horizons Software, Titusville, 1991—. Contbr. articles to profl. jours. Com. chmn. Cub Scouts, Titusville, 1991-91. With USAF, 1982-89. Republican. Achievements include developing and copywriting statistical software. Office: Space Station Office 1355 N Atlantic Ave Cocoa Beach FL 32931-0220

STRICKLER, HOWARD MARTIN, physician; b. New Haven, Conn., Oct. 26, 1950; s. Thomas David and Mildred Laing (Martin) S.; m. Susan Hunter, May 2, 1982; children: Hunter Gregory, Howard Martin Jr. BA, Berea Coll., 1975; MD, Univ. Louisville, 1979. Diplomate Am. Bd. Family Practice. Resident Anniston (Ala.) Family Practice Residency, 1979-82; pvt. practice Monteagle, Tenn., 1982-85; fellow in addictive diseases Willingway Hosp., Statesboro, Ga., 1985-86; faculty devel. fellow Univ. N.C., Chapel Hill, 1985-86; pvt. practice Birmingham, Ala., 1986-90; pres. Employers Drug Program Mgmt., Inc., Birmingham, 1990—; med. dir. Bradford Facilities, Birmingham, 1987-90, New Life Clinic, Bessemer, Ala., Physicians Smoke Free Clinic, Birmingham, 1988-90; chmn. dept. family practice and emergency medicine Bessemer Carraway Med. Ctr., 1993—. With U.S. Army, 1969-72, Vietnam. Decorated Bronze Star, 1971, Vietnam Campaign medal, Vietnam Svc. medal 3 Stars, 1971. Fellow Am. Acad. Family Physicians; mem. Am. Soc. Addiction Medicine (cert.), Am. Coll. Occupational and Environ. Medicine, Am. Assn. Med. Rev. Officers (cert.), Med. Assn. of State of Ala., Jefferson County Med. Soc., Jefferson County Soc. Family Physicians, So. Med. Assn., Phi Kappa Phi. Methodist. Avocations: flying, tennis, golf. Home: 868 Tulip Poplar Dr Birmingham AL 35244-1633 Office: 616 S 9th St Birmingham AL 35233

STRICKLER, JOHN RUDI, biological oceanographer; b. Zurich, Nov. 5, 1938; came to the U.S., 1971; m. Eva C. Strickler, Aug. 16, 1963. Diploma, Swiss Fed. Inst. Tech., Zurich, 1965, D of Natural Scis., 1969. Prin. rsch. scientist Australian Inst. Marine Sci., Townsville, 1980-84; rsch. prof. U. So. Calif., L.A., 1984-86; dir., prof. marine program Boston U., 1986-90; Shaw disting. prof. biol. sci. U. Wis., Milw., 1990—. Author: Trophic Interactions Within Aquatic Ecosystems, 1984; contbr. over 50 articles to profl. jours. Grantee NSF. Office: Ctr for Great Lakes Studies 600 E Greenfield Ave Milwaukee WI 53204

STRICKLEY, ROBERT GORDON, pharmaceutical chemist; b. Sacramento, Jan. 15, 1961; s. Bruce R. and Jan E. (Bush) S. BS in Chemistry, U. Calif., Berkeley, 1985; postgrad., U. Utah, 1989—. Rsch. chemist Syntex Inc., Palo Alto, Calif., 1985-89. Contbr. articles to profl. jours. Elliot Adkinson scholar, 1981; predoctoral fellow, 1991. Mem. Am. Assn. Pharm. Scientists. Achievements include patent in concentrated aqueous solutions of carboxyl group containing non-steroidal anti-inflammatory drugs. Office: U Utah Dept Pharms 421 Wakara Way Ste 315 Salt Lake City UT 84108

STRIEGEL, ANDRE MICHAEL, chemist; b. Gary, Ind., Dec. 8, 1967; s. Daniel Jospeh and Maria Teresa (Elorza) S. BS in Chemistry, U. New Orleans, 1991. Gas chromotography chemist Analytical Assn. S. Inc., Kenner, La., 1990-91; environ. chemist Environ. Indsl. Rsch. Assn., St. Rosa, La., 1989-91; teaching asst. U. New Orleans, 1991-92, rsch. asst., 1992--. Mem. Am. Assn. for the Advancement Sci., Am. Chenm. Soc., Grad. Student Assn. Chemistry. Roman Catholic. Home: 4610 Knight Dr New Orleans LA 70127 Office: U New Orleans Lakefront New Orleans LA 70148

STRIER, MURRAY PAUL, chemist, consultant; b. N.Y.C., Oct. 19, 1923; s. Jack and Rose (Goldman) S.; m. Arlene Schimmel, Feb. 12, 1955; children: Sheri Jeanette, Karen Barbara, Robin Joy. BChemE, CCNY, 1944; MS, Emory U., 1947; PhD, U. Ky., 1952. Rsch. chemist Reaction Motors Inc. (named changed to Thiokol Co.), Denville, N.J., 1952-56; sect. head Air Reduction, Inc., Murray Hill, N.J., 1956-58; chief chemist Fulton-Irgon

Corp. (now Inc. with Lithium Corp.), Lake Denmark, N.J., 1958-59; supr. Rayonier, Inc., Whippany, N.J., 1959-61; rsch. chemist McGraw Edison Co., West Orange, N.J., 1961-64; sr. rsch. scientist McDonnell Douglas Corp., Newport Beach, Calif., 1964-69; rsch. assoc. Hooker Rsch. Ctr., Grand Island, N.Y., 1969-71; phys. scientist EPA, Washington, 1972-86; cons. Rockville, Md., 1986—; instr. analytical chem. Upsala Coll., East Orange, N.J., 1963-64; cons. electroplating NSF, Washington, 1973-75. Contbr. articles to Jour. Am. Chem. Soc., Jour. Electrochem. Soc., Jour. Environ. Sci. & Tech. Commr. sci. and tech. commn. City of Rockville, 1985-91; vol. office consumer affairs Montgomery County, Rockville, 1989-90, dept. environ. protection, 1989-90. With USNR, 1944-46. Recipient Gold medal EPA, Washington, 1979. Fellow Am. Inst. Chemists (cert.); mem. AAAS, ASTM, Am. Chem. Soc., Electrochem. Soc. Achievements include patents on advanced solid rocket propellants, new high energy density batteries and fuel cells, improved methods for removal of toxic chemicals from water by electrochemical methods; development of less toxic dielectrics for capacitors and more facile methods for evaluation, optimal treatment methods, for removal of toxic chemicals from waters by use of chemical structure-activity approach.

STRIKE, DONALD PETER, pharmaceutical research director, research chemist; b. Mt. Carmel, Pa., Oct. 24, 1936; s. Peter and Veronica (Dugan) S.; m. Sally Ann Cavanaugh, July 28, 1972; children: Brian, Samantha. BS in Chemistry, Phila. Coll. Pharmacy & Sci., 1958; MS, Iowa State U., 1960, PhD, 1963. Rsch. chemist Wyeth Labs., Radnor, Pa., 1965-69, group leader, 1969-77, mgr., 1977-87; assoc. dir. Wyeth-Ayerst Rsch., Princeton, N.J., 1987—. Contbr. articles to sci. jours.; author 53 patents in field. NIH postdoctoral rsch. fellow Southampton U., Eng., 1963-64. Mem. Am. Chem. Soc. (medicinal chemistry and organic chemistry sects.), Phila. Organic Chemist Club, Phi Delta Chi, Phi Lambda Upsilon. Democrat. Roman Catholic. Avocations: sailing, skiing, golf, hunting, camping. Home: 445 Iven Ave Wayne PA 19087-4828 Office: Wyeth-Ayerst Rsch CN-8000 Princeton NJ 08543

STRINGER, JOHN, materials scientist; b. Liverpool, Eng., July 14, 1934; came to U.S., 1977; s. Gerald Hitchen and Isobel (Taylor) S.; m. Audrey Lancaster, Feb. 4, 1957; children: Helen Caroline, Rebecca Elizabeth. BS in Engring., U. Liverpool, 1955, PhD, 1958, D in Engring., 1974. Chartered engr., U.K. Lectr. Univ. Liverpool, Eng., 1957-63; fellow Battelle Columbus (Ohio) Labs., 1963-66; prof. materials sci. Univ. Liverpool, 1966-77; sr. project mgr. Electric Power Rsch. Inst., Palo Alto, Calif., 1977-81; sr. program mgr. Electric Power Rsch. Inst., Palo Alto, 1981-87, dir. tech. support, 1987-91, dir. applied rsch., 1991—; chmn. Sci. and Tech. Edn., Merseyside, Liverpool, 1971-74; pres. Corrosion and Protection Assn., London, 1972. Editorial bd.: Oxidation of Metals Jour., 1971—; author: An Introduction to the Electron Theory of Solids, 1967; editor: (book) High Temperature Corrosion of Advanced Materials, 1989, Chlorine in Coal, 1991, Applied Chaos, 1992; contbr. over 300 articles to profl. jours. Fellow AAAS, NACE Internat., Inst. Energy, Royal Soc. Arts; mem. ASM Internat. Office: Electric Power Rsch Inst 3412 Hillview Ave Palo Alto CA 94304-0813

STRITTMATTER, PETER ALBERT, astronomer, educator; b. London, Eng., Sept. 12, 1939; came to U.S., 1970.; s. Albert and Rosa S.; m. Janet Hubbard Parkhurst, Mar. 18, 1967; children—Catherine D., Robert P. B.A., Cambridge U., Eng., 1961, M.A., 1963, Ph.D., 1967. Staff scientist Inst. for Astronomy, Cambridge, Eng., 1967-70; staff scientist dept. physics U. Calif.-San Diego, La Jolla, Calif., 1970-71; assoc. prof. dept. astronomy U. Ariz., Tucson, 1971-74, prof. dept. astronomy, 1974—; dir. Steward Observatory, Tucson, 1975—; mem. staff Max Planck Inst. Radio-astronomy, Bonn, W. Germany, 1981—. Contbr. articles to profl. jours. Recipient Sr. award Humboldt Found., 1979-80. Fellow Royal Astron. Soc.; mem. Am. Astron. Soc., Astronomische Gesellschaft. Office: U Ariz Steward Observatory Tucson AZ 85721

STROBEL, DAVID ALLEN, psychology educator; b. Madison, Wis., Jan. 17, 1942; s. Carl Herbert and Anita (Wells) S.; m. Harriet Hartshorne, June 12, 1964 (div.); 1 child, Laura Wells; m. Linda Zimmermann, Mar. 31, 1982. B.A., Lake Forest Coll., 1964; M.A., U. Wis., 1971; Ph.D., U. Mont., 1972. Asst. prof. Northwestern U., Evanston, Ill., 1972-73; asst. prof., assoc. prof. then prof. psychology, U. Mont., Missoula, 1973—, chmn. dept., 1981, 83-90, assoc. dean Grad. Sch., 1990—; dir. Ft. Missoula Primate Lab., 1973-92. Contbr. chpts. to books. Cons. Missoula Drug Treatment Program, 1978, Missoula Spl. Edn. Coop., 1983. Grantee NIH, 1973-77, Dept. Agr./Agrl. Research Service, 1977-84. Mem. Mont. Psychol. Assn., Rocky Mountain Psychol. Assn., Western Psychol. Assn., Am. Assn. Primatologists, Internat. Soc. Primatologists, Missoula Humane Soc., Mont. Wilderness Assn., N. Rocky Mountain Assn. Lab. Animal Sci., Psychologists for Ethical Treatment Animals, Phi Beta Kappa, Sigma Xi. Home: 11284 Grant Creek Rd Missoula MT 59802-9345 Office: U of Montana Dept Psychology Missoula MT 59812

STROH, OSCAR HENRY, agricultural, civil and industrial engineer, land surveyor; b. Harrisburg, Pa., Jan. 11, 1908; s. Simon Henry and Alice (Feaser) S.; BS, U. Fla., 1948; grad. Command and Gen. Staff Coll., 1944, Armed Forces Indsl. Coll., 1954; PhD in History, Internat. Inst. Advanced Studies, 1979; Registered profl. engr., Pa., Vt.; m. Geraldine Bradshaw, Dec. 18, 1936; children: Jon Robert, Dana Evelyn. Ofcl. photographer U. Fla., 1938-40; civil engr. U.S. Govt., 1952-67, Commonwealth of Pa., 1967-73; cons. engr., Harrisburg, 1973—. Pa. forest fire warden, 1931-71. Bd. dirs. Central Dauphin Sch. System, 1953-65; bd. govs. Daniel Boone Nat. Found., Birdsboro, Pa., 1969-83, Fishing Creek Valley Community Assn., 1949—; former v.p. Am. Coll. Heraldry; organizer, adminstr. U.S. Army Photographic Sch., 1940-41. Served to lt. col. AUS, 1940-47, 50-52, Korea. Named Grand Prior of the Ea. States of Am. in the Order of St. John of Jerusalem, Knights Hospitallers; Knight grand cross Order St. Eugene de Trebizonde; Knight Grand Cross Byzantine Order Holy Sepulchre. Mem. Nat. Soc. Profl. Engrs., Constrn. Specifications Inst., Co. Mil. Historians, Assn. Former Intelligence Officers, Palatines to Am. (past pres. Pa. chpt.), Mil. Order Fgn. Wars, SAR (past chpt. pres.), Mil. Order World Wars, U.S. Horse Cavalry Assn. (charter). Lodge: Masons (32 deg.). Club: Sojourners. Author: Thompson's Battalion, 1975; Paxton Rangers; Pennsylvania German Tombstone Inscriptions, Vols. I, II; Dauphin County Tombstone Inscriptions, Vols. I-III; Heraldry in the U.S. Army; Story of the Hospitaller Order of St. John of Jerusalem; The Last Combat Sword of the American Army; Bugle Calls and Pistol Shots. Contbr. biweekly hist. column Paxton Herald, 1973-78. Home and Office: 1531 Fishing Creek Valley Rd Harrisburg PA 17112

STROJNY, NORMAN, analytical chemist; b. Edwardsville, Pa., June 14, 1943; s. John M. and Brunislawa (Stawarz) S. BS in Chemistry, Wilkes Coll., 1966; MS in Chemistry, Montclair State Coll., 1974; PhD in Analyt. Chemistry, Rutgers U., 1980, MBA in Mgmt., 1985. From jr. chemist to sr. chemist Hoffmann LaRoche, Nutley, N.J., 1965-85; from sr. chemist to supr. Danbury Pharmacal, Carmel, N.Y., 1985—. Contbr. articles to profl. jours. Mem. APHA, Am. Chem. Soc., Am. Inst. Chemistry, Soc. Applied Spectroscopy (local chmn. 1983-84, bd. dirs. 1983-84). Achievements include research in analytical research and development. Office: Danbury Pharmacal PO Box 990 Stoneleigh Ave Carmel NY 10512

STROKE, GEORGE WILHELM, physicist, educator; b. Zagreb, Yugoslavia, July 29, 1924; came to U.S., 1952, naturalized, 1957, arrived in Fed. Republic of Germany, 1978, naturalized, 1988; s. Elias and Edith Mechner (Silvers) S.; m. Masako Haraguchi, Feb. 5, 1973. B.Sc., U. Montpellier, France, 1942; Ing.Dipl., Inst. Optics, U. Paris, 1949; Dr. ès Sci. in Physics, Sorbonne U., Paris, 1960. Mem. rsch. staff and def. rsch.staff MIT, 1952-63, lectr. elec. engring., 1960-63; asst. rsch. prof. physics Boston U., 1956-57; NATO rsch. fellow U. Paris, 1959-60; prof. elec. engring., head electro-optical sci. labs. U. Mich., 1963-67; prof. elec. scis. and med. biophysics SUNY-Stony Brook, 1967-79; mem. corp. mgmt. staff Messerschmitt-Bolkow-Blohm GmbH, Munich, W. Ger., 1980-84, chief scientist space div., 1984-86, chief scientist, Corp. Hdqrs.-Devel., 1986-89; sr. advisor corp. strategy and bus. devel. Deutsche Aerospace Corp., 1989-91; vis. prof. Harvard U. Med. Sch., 1970-73, Tech. U. Munich, 1978-79, Keio U. Med. Sch., Tokyo, 1992-93, faculty sci. and tech., 1993—; adviser laser task force USAAF Systems Command, 1964; govt. sci. cons. U.S. and abroad, 1964—; cons. NASA Electronics Rsch. Ctr., Cambridge, Mass., 1966—; mem. commn. I, Internat. Radio Sci. Union, Nat. Acad. Scis., 1965—; cons. Am. Cancer Soc., 1972—; mem. NSF blue ribbon task force on ultrasonic imaging, 1973-74; mem. U.S. Ho. of Reps. Select Com., photog. evidence panel on Pres. J. F. Kennedy's assasination, 1978-79. mem. Max-Planck Soc., 1982—; bd. advisors Max-Planck Soc. Inst. Quantum Optics, 1986—, Inst. Diagnostic and Interventional Radiology, U. Witten/Herdecke, 1993—, Bavarian Acad. Fgn. Trade, 1993—; dir. NATO-AGARD study group on lasers, 1989. invited speaker Japan Internat. Sci. and Tech. Exchange Ctr., Tokyo, 1992; Recipient Humboldt prize, 1978. Fellow Optical Soc. Am., Am. Phys. Soc., IEEE. Contbr. articles to profl. jours. Author: An Introduction to Coherent Optics and Holography, 1966, Diffraction Gratings, 1967, Optical Engineering, 1980; co-editor: Ultrasonic Imaging and Holography, 1974. Address: Ottostrasse 13, D-80333 Munich Germany

STROKE, HINKO HENRY, physicist, educator; b. Zagreb, Yugoslavia, June 16, 1927; came to U.S., 1943, naturalized, 1949; s. Elias and Edith (Mechner) S.; m. Norma Bilchick, Jan. 14, 1956; children: Ilana Lucy, Marija Tamar. BEE, N.J. Inst. Tech., 1949; MS, MIT, 1952, PhD, 1954. Rsch. asst. Princeton (N.J.) U., 1954-57, rsch. assoc., 1957; mem. research staff. Sponsored Research div., lectr. dept. physics MIT, 1957-63; asso. prof. physics NYU, N.Y.C., 1963-68, prof., 1968—, dept. chmn., 1988-91; assoc. prof. U. Paris, 1969-70, Ecole Normale Supérieure, 1976; vis. scientist Max Planck Inst. für Quantenoptik, Garching, U. Munich, 1977-78, 79, 81, 82, 92; cons. Atomic Instrument Co., MIT Sci. Translation Svc., Tech. Rsch. Group, Cambridge Air Force Rsch. Ctr., Am. Optical Corp., ITT Fed. Labs., NASA, also others; mem. com. on line spectra of elements NAS-NRC, 1976-82; sci. assoc. CERN, Geneva, 1983—. Contbg. author: Nuclear Physics, 1963, Atomic Physics, 1969, Hyperfine Interactions in Excited Nuclei, 1971, Francis Bitter: Selected papers, 1969, Atomic Physics 3, 1973, Nuclear Moments and Nuclear Structure, 1973, A Perspective of Physics, Vol. 1, 1977, Atomic Physics 8, 1983, Lasers in Atomic, Molecular, and Nuclear Physics, 1989; Editor: Comments on Atomic and Molecular Physics. Mem. Chorus Pro Musica, Boston, 1951-54, 57-63, Münchener Bach-Chor, Munich, 1977-82, 92; Choeur pro Arte, Lausanne, 1983-92; mem. Collegiate Chorale, N.Y., 1964—. With AUS, 1946-47. Recipient Sr. U.S. Scientist award Alexander von Humboldt Found., 1977; NATO sr. fellow in sci., 1975. Fellow Am. Phys. Soc. (publs. oversight com. 1991-93), Optical Soc. Am.; mem. AAAS, European Phys. Soc., Société Française de Physique, Tau Beta Pi, Sigma Xi, Omicron Delta Kappa. Home: 271 Old Army Rd Scarsdale NY 10583-2619 Office: NYU Dept Physics 4 Washington Pl New York NY 10003-6603

STROM, STEPHEN ERIC, astronomer; b. Bronx, N.Y., Aug. 12, 1942; s. Albert William and Beatrice (Reisinger) S.; m. Karen Marie Lewallen, Mar. 31, 1960; children: Robert Charles, Kathy Marie, David Michael, Julie Eileen. A.B., Harvard U., 1962, A.M., Ph.D., 1964. Lectr. Harvard U., 1964-68; astro-physicist Smithsonian Astrophysical Obs., Cambridge, Mass., 1964-68; assoc. prof. astronomy SUNY, Stony Brook, 1969-71; prof. SUNY, 1971-72, coordinator astronomy and astrophysics, 1969-72; astronomer Kitt Peak Nat. Obs., Tucson, 1972-83; chmn. galactic and extragalactic program Kitt Peak Nat. Obs., 1975-79; prof. physics and astronomy U. Mass., Amherst, 1983—; chmn. Five Coll. Astronomy Dept., 1984—; assoc. mng. editor Astrophys. Jour., 1979-85; vis. prof. U. Calif., Berkeley, 1979; mem. Space Sci. Bd., 1980-83; chmn. Com. on Space Astronomy and Astrophysics, 1980-83; chmn. large deployable reflector sci. coordination group NASA, 1983-86, mem. space and earth scis. adv. com., 1984-87, chmn. Infrared Processing and Analysis Ctr. Users Com., 1986-88; chmn. com. on future dirs. of nat. optical astronomy observatories Assn Univs. Research in Astronomy, chmn. optical panel, astronomy and astrophysics survey (decade rev.). Author: (with Joy Harjo) Secrets From The Center Of The World, 1989. Recipient Bok prize Harvard, 1970; Warner prize Am. Astronom. Soc., 1975; Woodrow Wilson fellow (hon.), 1962; Sloan Found. fellow, 1970-72. Mem. Am. Astronom. Soc. (councilor 1980-83, v.p. 1987-89), Astronom. Soc. Pacific, Internat. Astron. Union, AAAS (councilor 1983-85), Phi Beta Kappa, Sigma Xi. Home: 163 Heatherstone Rd Amherst MA 01002-1638 Office: U Mass Five Coll Astronomy Dept GRC Tower B Amherst MA 01003

STROMINGER, JACK LEONARD, biochemist; b. N.Y.C., Aug. 7, 1925. AB, Harvard U., 1944; MD, Yale U., 1948; DSc (hon.), Trinity Coll., Dublin, 1975, Washington U., 1988. From asst. prof. to prof. pharmacology sch. med. Washington U., St. Louis, 1955-61, prof. pharmacology and microbiology, 1961-64; prof. pharmacology and chem. microbiology med. sch. U. Wis., Madison, 1964-68; prof. biochemistry Harvard U., 1968-83, chmn. dept. biochemistry and molecular biology, 1970-73, Higgins prof. biochemistry, 1983—; head tumor virol. divsn. Dana-Farber Cancer Inst., Boston, 1977—. Recipient John J. Abel award, 1960, Paul-Lewis Lab award, 1962, Rose Payne award Am. Soc. Histocompat. & Immunogen., 1986, Hoechst-Roussel award, 1990, Pasteur medal, 1990; named Passano Found. laureate, 1993. Mem. NAS (mem. inst. medicine, Microbiology award 1968, Selman Waxman award 1968), AAAS, Am. Soc. Biol. Chemists, Am. Soc. Pharmacology & Exptl. Therapeutics, Am. Assn. Immunologists, Am. Soc. Microbiologists, Am. Chem. Soc., Am. Acad. Arts & Sci., European Molecular Biol. Orgn., Sigma Xi. Office: Dana Farber Cancer Inst Dept of Biochem 44 Binney St Boston MA 02115-6084*

STROM-PAIKIN, JOYCE ELIZABETH, nursing administrator; b. Syracuse, N.Y., Oct. 25, 1946; d. Paul H. and Elizabeth (Bartlett) Strom Black; m. Frank J. Iaconis, May 31, 1963 (div. Mar. 1974); children: Paul, Michael; m. Lester Paikin, June 26, 1982. AAS in Nursing, Cayoga Community Coll., Auburn, N.Y., 1976; BS, Nova U., Ft. Lauderdale, Fla., 1978, MS, 1980; postgrad., Saybrook Inst., San Francisco, 1988—. Diplomate Am. Bd. Med. Psychotherapy; cert. psychiat. nurse clinician. Surg. nurse Auburn Meml. Hosp., 1967-70; office nurse Auburn, 1971-74; charge nurse Mercy Rehab. Ctr., Auburn, 1974-75; nurse St. Joseph's Hosp., Syracuse, 1975-76, charge psychiat. nurse, 1976-77, staff psychotherapist, nursing adminstr. Pyscho-Awareness Inc., Tamarac, Fla., 1980—; adj. prof. Nova U., 1976-79, Broward Community Coll., Ft. Lauderdale, 1980—; cons. Impaired Profls., Tallahassee, 1983—. Author: Medical Treason, 1989; contbr. articles to profl. jours. Mem. Am. Psychol. Assn., Am. Assn. Counseling and Devel., Am. Nurses Acad., Nat. Psychiat. Assn., Broward County Mental health Assn. Democrat. Episcopalian. Lodge: Soroptimists. Avocations: knitting, boating, bodybuilding. Office: Psycho-Awareness Inc 5455 N State Rd 7 Fort Lauderdale FL 33319-2954

STRONG, RICHARD ALLEN, environmental safety engineer; b. Detroit, Apr. 11, 1930; s. Leo Marshall and Jean Liddle (Pettigrew) S.; m. Rosa Maria Amaya, Aug. 18, 1957; children: Harold, Edward, Randall, Marisa. BSc, Univ. Mich., 1964; MA, Central Mich. Univ., 1975. Registered profl. engr., Ohio. System safety engr. Air Force Systems Command I.G., Andrews AFB, Md., 1969-71; chief system safety engr. Deputy for Engring. HQASD, Wright-Patterson AFB, Ohio, 1972-74; system safety engr. Lockheed Missiles & Space Co., Sunnyvale, Calif., 1975; chief system safety 4950th Test WIng, HQ ASD, Wright-Patterson AFB, Ohio, 1975-83; chief safety div. Wright Lab. HQ ASD, Wright-Patterson AFB, Ohio, 1983-92; CEO Psychic Sci. Internat. S.I.G., Huber Heights, Ohio, 1977—, Safety Analysis Systems, 1992—; lectr., presenter on environ. safety, psychic science, 1970—. Author: (booklet) Star-Tracker: An Aid for Science Lerning, 1973, What's New in Psychic Science?, 1987, Cosmic Wisdom : Analysis and Planning with Personal Computers, Strongmobile Development Report; contbr. articles to profl. jours. With USAF, 1955-75. Decorated Silver Star USAF, 1966, Disting. Flying Cross with cluster USAF, 1966, Air medal with 13 clusters USAF, 1966. Fellow Am. Soc. Psychical Rsch.; mem. System Safety Soc. (sr.), Parapsychological Assn. (assoc.). Roman Catholic. Achievements include patents on autoplane designs and instrument displays. Home and Office: 7514 Belle Plaine Dr Huber Heights OH 45424-3229

STROUD, DEBRA SUE, medical technologist; b. Jacksonville, Fla., Feb. 15, 1954; d. Albert LeRoy and Jessie Nell (Igou) Brown; m. Stephen Ray Torok (div. 1975); m. Edward Lee Stroud, May 31, 1978. BA, U. North Fla., 1979. Lic. med. technologist, Fla. Office lab. technologist Women's Med. Group, P.A., Jacksonville, 1977-78; hematology and blood bank technologist Meml. Med. Ctr., Jacksonville, 1978-85; hematology supr. Humana Hosp., Orange park, Fla., 1985—; owner, ptnr. Stroud's Creative Designs, Callahan, Fla., 1990-92; continuing edn. provider, contact person Fla. State Dept. Health and Rehab. Svcs., Orange Park. Vol. Catfish One, Hook Kids in Fishing-Not Drugs, Hilliard-Callahan, Fla., 1991; vol. fundraiser Found for Cheryl Davis, Callahan, 1991. With U.S. Army, 1972-76. Recipient Most Admired Woman of the Decade award, Women of Yr. award, Silver Shield of Valor, 1992, 20th Century award of Achievement; named Internat. Woman of Yr., 1991, 92, Life fellow Am. Biog. Inst., fellow Internat. Biog. Assn. Mem. NAFE, Am. Soc. Clin. Pathologists, Fla. C. of C., West Nassau C. of C., Nat. Wildlife Fedn. (life), NWF Leaders Club. Baptist. Avocations: woodworking, tole painting, fishing, working with children. Home: RR 2 Box 398 Hilliard FL 32046-9408 Office: Humana Hosp Orange Park 2001 Kingsley Ave Orange Park FL 32073-5156

STROUD, JOHN FRANKLIN, engineering educator, scientist; b. Dallas, June 29, 1922; s. Edward Frank and Ethel A. Stroud; m. Dorcas Elizabeth Stroud, Feb. 4, 1944; children: Kevin, Karen, Richard. BSME, Stanford (Calif.) U., 1949, postgrad., 1949-53; cert. in fin. mgmt. for sr. execs., U. Pa., 1984. Aero. rsch. scientist NASA Ames Rsch. Lab., Moffett Field, Calif., 1949-53; thermodynamics engr. Lockheed, Burbank, Calif., 1953-55, group engr. propulsion, 1955-63, dept. engr. propulsion, 1963-70, from dept. engr. to divsn. engr. propulsion, 1970-83, chief engr. flight scis., 1983-85, divsn. engr., 1985-90, ret., 1990; cons. spl. studies in econs. and engring. sci., 1990—; mem. ad hoc adv. congrl. subcom. on high tech wind tunnels, 1985. Contbr. articles to profl. jours.; author and speaker in field. Charity fund raiser United Way, 1970-80, others. Lt., naval aviator USN, 1942-45, ETO. Decorated Battle of Atlantic. Fellow AIAA (assoc., airbreathing propulsion com. 1966-68, 80-83, chmn. many sessions, 1970—); mem. Soc. of Automotive Engrs. (aviation div. air transport com., propulsion com., chmn. many sessions nat confs. 1970—, co-chmn. AIAA/SAE nat. propulsion conf. 1978). Achievements include patent for low drag external compression supersonic inlet; designed and devel. integrated F-104A inlet and air induction system, integrated inlet/air inducation system into the total propulsion system and airframe; manged team that developed, designed, and integrated the aeropropulsion systems on the L1011 commercial transport; devised new theory for turbulent boundary layers in adverse pressure gradients; numerous other patents pending.

STROUD, SALLY DAWLEY, nursing educator, researcher; b. Ellensburg, Wash., Sept. 24, 1947; d. Lawrence Eugene and Theda Eva (Crowe) D.; m. Donald Lewis Stroud, June 26, 1970 (div. 1973). Diploma, U. Ala. Hosp. Sch. Nursing, 1968; BS in Nursing, Columbus Coll., 1977; MS in Nursing, Vanderbilt U., 1978; EdD, Auburn U., 1987. RN, Tex. Staff nurse U. Wash. Hosp., Seattle, 1968-70; staff nurse ICU Lee County Hosp., Opelika, Ala., 1970-77; critical care coord. E. Ala. Med. Ctr., Opelika, 1978-80; mem. faculty Sch. Nursing Auburn (Ala.) U., 1980-87, Ga. State U., Atlanta, 1987-91; part-time staff nurse Emory U. Hosp., Atlanta, 1987-91; dir. of rsch. Houston Immunological Inst., 1991-93, Med. Univ. S.C., Charleston, 1993—; expert witness, lectr., cons. in field. Contbr. articles to profl. jours. Mem. AACN, ANA, Tex. Nurses Assn., Southern Nursing Rsch. Soc., Sigma Theta (pres. local chpt. 1986-87, 89-91), Phi Kappa Phi, Omicron Delta Kappa. Roman Catholic. Avocations: white water rafting, cross-stitch. Home: 1133 Monaco Dr Mount Pleasant SC 29464 Office: Med Univ S C Coll Nursing 171 Ashley Ave Charleston SC 29425-2401

STROUP, DAVID RICHARD, architect; b. Asheville, N.C., Dec. 4, 1954; s. H.B. and Louise (Pinkerton) S. B Envirion Design in Architecture, N.C. State U., 1977. Lic. architect, N.C., S.C., Nat. Coun. Archtl. Registration Bd. Estimator Stroup Sheet Metal Works Inc., Asheville, N.C., 1973-74; student sales mgr. Southwestern Co., Nashville, 1974-77; draftsman Brown, Edwards & Miller, Raleigh, N.C., 1977-78; assoc. architect Flour/Daniel Internat. Corp., Greenville, S.C., 1978-82; project architect Hiller & Hiller, Architects, Greenville, 1982-83, Neal, Prince & Browning, Architects, Greenville, 1983-85, Craig, Gaulden & Davis Architects, Greenville, 1985-90; corp. architect Milliken & Co., Spartanburg, S.C., 1990—. Recipient Masonry Design award S.C. Masonry Assn., Spartanburg, 1986. Mem. AIA (pres. Greenville 1990, young architect forum rep. S.C. 1991), Constrn. Specifications Inst., Optimist Club Greenville (pres. 1985-86, bd. mem. 1985—, sec.-treas. 1991-92), S.C. Dist. Optimist Internat. (lt. gov. 1986-87, bd. mem. 1985-87, outstanding lt. gov. 1987). Republican. Baptist. Avocations: snow and water skiing, travel, coin collecting. Home: 106 Idonia Dr Taylors SC 29687-3879 Office: Milliken & Co 920 Milliken Rd Spartanburg SC 29303-9301

STRUBLE, DONALD EDWARD, mechanical engineer; b. Oakland, Calif., Oct. 10, 1942; s. Donald Edward and Marjorie E. (Griffin) S.; m. Allison Florence Dietrick, Dec. 20, 1964; children: Lisa Kathleen, Donald Lyman, John Dietrick. BS, Calif. Poly., 1964; MS, Stanford U., 1965; PhD, Ga. Inst. Tech., 1972. Asst. prof. Calif. Poly. State U., San Luis Obispo, Calif., 1970-74; sr. v.p. Minicars, Inc., Goleta, Calif., 1974-81; pres. Dynamic Sci., Inc., Phoenix, 1981-83; sr. engring. assoc. Cromack Engring. Assoc., Tempe, Ariz., 1983-85; cons. engr. Donald E. Struble, PhD, Phoenix, 1985-87; sr. engr. Collision Safety Engring., Phoenix, 1987—. Author: Fundamentals of Aerospace Structural Analysis, 1972; contbr. articles to profl. jours. Mem. Soc. Automotive Engrs., Assn. Advancement Automotive Medicine, Sigma Xi. Democrat. Episcopalian. Achievements include patent for Inflatable Restraint for Side Impactcs; research in experimental safety vehicles. Home: 564 W Moon Valley Dr Phoenix AZ 85023 Office: Collisions Safety Engring 2320 W Peoria Ave Ste B 147 Phoenix AZ 85023

STRUBLE, GORDON LEE, physicist, researcher; b. Cleve., Mar. 7, 1937; s. Fred Clarence and Elizabeth Ann (Francis) S.; m. Jean Louise Wells, June 14, 1961; children: Elizabeth Ann, Andrea Lynne, Stephanie Yvonne, Gretchen Suzanne. BS, Rollins Coll., Winter Park, Fla., 1960; PhD, Fla. State U., 1964. Asst. prof. chemistry U. Calif., Berkeley, 1966-71; group leader Lawrence Livermore Nat. Lab., Livermore, Calif., 1971-91, dep. dir. lab. directed R&D, 1991—; mem. nuclear data com. Dept. Energy, Washington, 1978-82. Contbr. articles to profl. jours. Bd. dirs., pres. Assn. for the Preservation of Danville Blvd., Danville, Calif., 1980-87; bd. dirs. Alamo (Calif.) Improvement Assn., 1986-87. Recipient 5 awards of distinction Soc. Tech. Commun., 1986-91; NATO rsch. grantee, 1969, 90; rsch. fellow German Fed. Republic, 1971. Mem. AAAS, Am. Phys. Soc., Am. Chem. Soc. Democrat. Episcopalian. Achievements include elucidation of the effects of particle and collective interaction in atomic nuclei; discovery of multiple shape coexistence in nuclei, and the gamma decay of fission isomers; incoporation of residual intractions in description of odd-odd deformed nuclei. Home: 2 Deodar Ln Alamo CA 94507 Office: Lawrence Livermore Nat Lab PO Box 808 L-3 Livermore CA 94551

STRUBLE, LESLIE JEANNE, civil engineer, educator; b. Olympia, Wash., May 30, 1947; d. Thomas Leighton and Eileen Claire (Lewis) Storey; m. Robert George Struble, Aug. 8, 1970; 1 child, Robert Thomas. BA in Chemistry, Pitzer Coll., 1970; MSCE, Purdue U., 1979, PhD in Civil Engring., 1987. Rsch. chemist CalMat Co., Colton, Calif., 1970-76; rsch. scientist Martin Marietta Corp., Balt., 1979-84; rsch. engr. Nat. Inst. Standards and Tech., Gaithersburg, Md., 1984-89; asst. prof. civil engring. U. Ill., Urbana, 1989—. Assoc. editor Jour. Am. Ceramic Soc., 1992—, Jour. Materials Civil Engring., 1990; contbr. articles to Jour. Am. Ceramic Soc., Cement and Concrete Rsch., other profl. publs. Chmn. bd. dirs. Relay (Md.) Children's Ctr., 1981. Mem. ASCE, ASTM (chmn. task group, subcoms.), Am. Ceramic Soc. (sec. cements div. 1983-84, chmn. 1987), Am. Concrete Inst., Materials Rsch. Soc. (symposium co-chair 1986, 92), Sigma Xi. Office: U Ill 205 N Mathews St Urbana IL 61801-2352

STRUCKHOFF, RONALD ROBERT, manufacturing engineer; b. Washington, Mo., Apr. 8, 1947; s. Virgil A. and Eileen A. (Dieckhaus) S.; m. Ava Jane Shortt, Jan. 16, 1965; children: Laura, Jeff, Gretchen, John. BS in Aeronautics, St. Louis U., 1992. Mem. field svc. staff Solar Turbines, Inc., San Diego, 1981-87; mfg. engr. McDonnell Douglas Aerospace, St. Louis, 1987—. Mem. AIAA, Soc. Mfg. Engrs., Soc. Automotive Engrs. Home: PO Box 1208 Lake Sherwood MO 63357

STRUECKER, GERHARD, analytical chemist; b. Dortmund, Federal Republic of Germany, Nov. 13, 1954; s. Hans-Walter and Martha (Schoof) S.; m. Sylvia Flachmann, Aug. 15, 1985; children: Juliane, Thomas. Diploma in chem. engring., Fachhochschule Münster, Burgsteinfurt, Germany, 1978; BS, Juniata Coll., Huntingdon, Pa., 1979. Tchr.

sci. Fachhochschule Münster, 1980-81; dept. chmn. Stadtwerke Hagen (Germany) AG, 1981—; mem. working com. on chemistry Arbeitsgemeinschaft Wasserwerke an der Ruhr, Gelsenkirchen, Germany, 1981—, chmn., 1993—; mem. working com. on gen. affairs Arbeitsgemeinschaft Trinkwassertalsperren, Wuppertal, Germany, 1986—. Contbr. articles to sci. jours. Mem. Am. Chem. Soc., N.Y. Acad. Scis. Lutheran. Avocations: photography, hiking, collecting stamps. Home: Soelder Strasse 118, D-44289 Dortmund 41 Northrhine-Westphalia, Germany Office: Stadtwerke Hagen AG, Hohenzollernstrasse 5-7, D-58095 Hagen 1 Northrhine Westphalia, Germany

STRUHL, KEVIN, molecular biologist, educator; b. N.Y.C., Sept. 2, 1952; s. Joseph and Harriet (Schachter) Struhl; m. Marjorie A. Oettinger, June 4, 1989. BS, MS, MIT, 1974; PhD, Stanford U., 1979; MA, Harvard U., 1989. Asst. prof. Harvard Med. Sch., Boston, 1982-86, assoc. prof., 1986-89, prof., 1989-91, Gaiser prof., 1991—; scientific adv. bd. Pharmagenics, Inc., Allendale, N.J., 1989—, Scriptech, Cambridge, Mass., 1993—. Jane Coffin Childs fellow, 1980; Searle scholar, 1983; recipient award in Microbiology Eli Lilly, 1990. Office: Harvard Med Sch Dept Biological Chemistry 240 Longwood Ave Boston MA 02115

STRYER, LUBERT, biochemist, educator; b. Tientsin, China, Mar. 2, 1938. B.S. with honors, U. Chgo., 1957; M.D. magna cum laude, Harvard U., 1961; DS (hon.), U. Chgo., 1992. Helen Hay Whitney fellow Harvard U., also Med. Research Council Lab., 1961-63; from asst. prof. to assoc. prof. biochemistry Stanford U., 1963-69; prof. molecular biophysics and biochemistry Yale U., 1969-76; Winzer prof. cell biology Stanford U. Sch. Medicine, 1976—, chmn. dept. structural biology, 1976-79; cons. NIH, NRC; pres., sci. dir. Affymax Rsch. Inst., Palo Alto, Calif., 1989-90; mem. sci. adv. bd. Jane Coffin Child's Fund Rsch. to Prevent Blindness, 1982-90; sci. adv. bd. Pew Scholars Profs. in Biomed. Scis. Mem. editorial bd.: Jour. Molecular Biology, 1968-72, Jour. Cell Biology, 1981—; assoc. editor: Annual Revs. Biophysics and Bioengineering, 1970-76. Recipient Am. Chem. Soc. award in biol. chemistry Eli Lilly & Co., 1970, Alcon award in vision Alcon Rsch. Inst., 1992. Fellow AAAS (Newcomb Cleveland prize 1991), Am. Acad. Arts and Scis.; mem. NAS, Am. Chem. Soc., Am. Soc. Biol. Chemists, Biophys. Soc., Phi Beta Kappa. Office: Stanford Sch Medicine Fairchild Ctr D133 Stanford CA 94305

STUART, CHARLES EDWARD, electrical engineer, oceanographer; b. Durham, N.C., Feb. 9, 1942; s. Charles Edward and Wilma Kelly Stuart; m. Margaret Ann Robinson, Jan. 9, 1982; children: Marjorie Kelly, Heather Alison. BSEE, Duke U., 1963. Engr. Westinghouse Electric Corp., Balt., 1963-65; sr. engr. Booz Allen Hamilton, Chevy Chase, Md., 1966-68; rsch. dir. B-K Dynamics Inc., Huntsville, Ala., 1969-78; oceanographer Office of Naval Rsch., Arlington, Va., 1979-84; dir. maritime system office Advanced Rsch. Projects Agy., Arlington, 1985—. Contbr. 12 papers on ocean acoustics to profl. jours. Mem. IEEE (sr., ad. com. 1991-93), Assn. Unmanned Vehicle Systems (trustee 1989-93), Acoustical Soc. Am. Methodist. Achievements include leading work in development of unmanned undersea vehicle technology. Home: Dept of Defense DARPA/UWO 4718 N 17th St Arlington VA 22207 Office: Advanced Rsch Projects Agy 3701 N Fairfax Dr Arlington VA 22203

STUART, JAMES DAVIES, analytical chemist, educator; b. Elizabeth, N.J., Sept. 30, 1941; s. Norman Fisher and Madeleine Davies (Harris) S.; m. Carol Ann Morrison, June 14, 1964; children: James Edward, Jean Ann. BS in Chemistry, Lafayette Coll., 1963; PhD in Chemistry, Lehigh U., 1969. Instr. Lafayette Coll., Easton, Pa., 1967-69; asst. prof. U. Conn., Storrs, 1969-75, assoc. prof., 1975—; dir. marine environ. lab. U. Conn., Avery Point, Conn., 1990-92; vis. lectr. U. Ga., Atlanta, 1976; vis. fellow Yale U., New Haven, 1983; mem. adv. bd. dept. health Water Supply Sect., State of Conn., Hartford, 1977-78; cons. IBM Instruments, Danbury, Conn., 1983-85, HNU Systems, Inc., Newton, Mass., 1991—; co-chair Town of Coventry (Conn.) Solid Waste Mgmt. Com. Contbr. over 40 articles to sci. jours. Co-prin. investigator methods for measuring sub-surface gasoline pollution EPA, 1987—. Mem. New Eng. Chromatography Coun. (exec. com. 1990—).

STUART, JAY WILLIAM, engineer; b. L.A., Aug. 26, 1924; s. Jay William and Mamie Marie (Pollock) S.; m. Nancy Giovinazzo, July 28, 1951; children: Tani Lynn Stuart Robertson, Joel Vanni Stuart. BSME, Caltech, 1946, MS in Aero. Engring., 1948, profl. degree in Aero. Engring., 1951. Registered profl. engr., Calif. Various engring. and sci. positions Calif., 1943-73; engring. specialist, hydro/aero. Aerojet Gen., Surface Effect Ship, Tacoma, Wash., 1973; engring. specialist Lockheed Missile Space/Surface Effect Ship, Sunnyvale, Calif., 1973-74; contract engr., prin. aero. Rohr Marine, Surface Effect Ship, San Diego, 1974-76; engring. specialist, heat exch. Aerojet Mfg. Co. AMCO, Fullerton, Calif., 1976-77; contract engr., aero. Hughes Missiles, AMRAM, Canoga Park, Calif., 1977-78; mem. tech. staff Jet Propulsion Lab., Pasadena, Calif., 1978-80; sr. spacecraft specialist TDRSS ABACUS Programming Corp./TRW, Redondo Beach, Calif., 1980-82; prin engr. Spacecom Contel divsn. TDRSS/TRW, Redondo Beach, 1982-86; contract engr., aero. sci. Rockwell Internat., Shuttle, Downey, Calif., 1986-88, contract engr., systems safety, 1989-90. Contbr. articles to profl. jours. Leader YMCA, Gardena, 1965-72. Scholarship Douglas Aircraft, 1947, 49. Mem. AIAA (mem. subcom. on ednl. tech. L.A. sect. 1969-71), NSPE (mem. skill conversion team dept. labor 1971-72, del. to Ministry of Machine Bldg., China 1986), Am. Def. Preparedness Assn., Heat Transfer and Fluid Mechanics Inst., Caltech Alumni Assn., U.S. Badminton Assn., Sigma Xi, Lambda Delta Lambda. Achievements include research in improved solution of the Falkner and Skan boundary-layer equation, low-drag specification of surface irregularities, complete spectra of the velocity fluctuations in the wake of a stalled aircraft; derivation of geophysical-Coriolis form of the Orr-Sommerfeld stability equation. Home: 17502 Valmeyer Ave Gardena CA 90248

STUART, SANDRA JOYCE, computer information scientist; b. Wheatland, Mo., Aug. 15, 1950; d. Asa Maxville and Inez Irene (Wilson) Friedley; m. John Kendall Stuart, Apr. 17, 1971; 1 child, Whitney Renee. Student, Cen. Mo. State U., 1968-69; AA (hon.), Johnson County Community Coll., 1980; BS in Bus. Adminstrn. cum laude, Avila Coll., 1992. Statis. asst. Fed. Crop Ins. Corp., Kansas City, Mo., 1978-83; mgr. Fed. Women's Program, Kansas City, 1979-80; mgmt. asst. Marine Corps Fin. Ctr., Kansas City, 1983-85, analyst computer systems, 1985-88; computer programmer analyst Corps. of Engrs., Kansas City, 1988-91; regional program mgr. FAA, Kansas City, 1991—. Author: The Samuel Walker History, 1983. Asst. supt. Sunday sch. Overland Park (Kans.) Christian Ch., 1979-80, supt., 1980-82. Mem. Wheatland High Sch. Alumni Assn. (pres. 1990-91). Avocations: needlework, genealogy, reading, traveling.

STUBBE, JOANNE, chemistry educator. Ellen Swallow prof. chemistry MIT, Cambridge. Arthur C. Cope scholar award Am. Chemistry Soc., 1993. Mem. NAS. Office: MIT Dept Chemistry 77 Massachusetts Ave Cambridge MA 02139

STUBBERUD, ALLEN ROGER, electrical engineering educator; b. Glendive, Mont., Aug. 14, 1934; s. Oscar Adolph and Alice Marie (LeBlanc) S.; m. May B. Tragus, Nov. 19, 1961; children: Peter A., Stephen C. B.S. in Elec. Engring, U. Idaho, 1956; M.S. in Engring, UCLA, 1958, Ph.D., 1962. From asst. prof. to assoc. prof. engring. UCLA, 1962-69; prof. elec. engring. U. Calif., Irvine, 1969—; assoc. dean engring. U. Calif., 1972-78, dean engring., 1978-83; chief scientist U.S. Air Force, 1983-85; dir. Electrical Communications and Systems Engring. div. NSF, 1987-88. Author: Analysis and Synthesis of Linear Time Variable Systems, 1964, (with others) Feedback and Control Systems, 2d edit., 1990; contbr. articles to profl. jours. Recipient Exceptional Civilian Svc. medal USAF, 1985, 90. Fellow IEEE (Centennial medal 1984), AIAA, AAAS; mem. Ops. Research Soc. Am., Sigma Xi, Sigma Tau, Tau Beta Pi, Eta Kappa Nu. Home: 19532 Sierra Soto Rd Irvine CA 92715-3841 Office: U Calif Dept Elec Engring and Computer Sci Irvine CA 92717

STUBBLEFIELD, JAMES IRVIN, physician, surgeon, army officer; b. Phila., Aug. 17, 1953; s. James Irvin Sr. and Geri (Harvey) S.; m. Linda Marie Simms, Aug. 12, 1978; children: Lindsay, Shannon. BSEE, MS in Bioengring., U. Pa., 1977; MD, Hahnemann U., Phila., 1982. Diplomate Am. Bd. Emergency Medicine, 1991. Mgr. energy engring. Norcross, Inc., Bryn Mawr, Pa., 1977-78; commd. 2d lt. U.S. Army, 1977, advanced through grades to lt. col., 1993; intern in gen. surgery Letterman Army Med. Ctr., San Francisco, 1982-83; flight surgeon, brigade surgeon 101st Airborne Div., Ft. Campbell, Ky., 1983-87; resident in emergency medicine Madigan Army Med. Ctr., Ft. Lewis, Wash., 1987-90; chief emergency med. svcs. Silas B. Hays Army Hosp., Ft. Ord, Calif., 1990-93, chief dept. emergency medicine and primary care, 1993—; flight surgeon attack helicopter battalion Operation Desert Storm, Persian Gulf, 1991. Fellow Am. Coll. Emergency Physicians; mem. AMA, U.S. Army Flight Surgeon Soc., Assn. Mil. Surgeons U.S., Tau Beta Pi, Eta Kappa Nu, Alpha Epsilon Delta. Roman Catholic. Avocations: jogging, tennis, table tennis, martial arts, billiards. Home: 22586 Oak Canyon Rd Salinas CA 93908-9606

STUBER, CHARLES WILLIAM, genetics educator, researcher; b. St. Michael, Nebr., Sept. 19, 1931; s. Harvey John and Minnie Augusta (Wilks) S.; m. Marilyn Martha Cook, May 28, 1953; 1 child, Charles William Jr. BS, U. Nebr., 1952, MS, 1961; PhD, N.C. State U., 1965. Vet., agrl. instr. Broken Bow (Nebr.) High Sch., 1956-59; research asst. U. Nebr., Lincoln, 1959-61; research geneticist Agrl. Research Service, USDA, Raleigh, N.C., 1962-75, supervisory research geneticist, research leader, 1975—; prof. genetics & crop sci. N.C. State U., Raleigh, 1975—. Assoc. editor Crop Sci. Jour., 1979-82, tech. editor, 1984-86, editor, 1987-89; author, co-author over 150 articles to jours. and chpts. to books. Chmn. coun. on ministries and numerous offices Highland United Meth Ch., Raleigh. Lt. USN, 1952-56. Named Outstanding Scientist of Yr., USDA-ARS, 1989. Fellow Am. Soc. Agronomy, Crop Sci. Soc. Am. (editor-in-chief 1987-91, pres. 1992-93); mem. AAAS, Genetics Soc. Am., Am. Genetic Assn. (sec. 1984-86), Sigma Xi, Phi Kappa Phi. Avocations: windsurfing, water skiing, sailing. Home: 1800 Manuel St Raleigh NC 27612-5510 Office: USDA-ARS NC State U Dept Genetics PO Box 7614 Raleigh NC 27695

STUCK, ROGER DEAN, electrical engineering educator; b. Ventura, Calif., Nov. 6, 1924; s. William Henry and Marian Grace (Ready) S.; m. Opal Christine Phillips, July 25, 1948; children: Dean, Phyllis, Sandra. BSEE, Calif. Inst. Tech., 1947; MSEE, N.C. State U., 1957. Elec. engr. Warren Wilson Coll., Swannanoa, N.C., 1947—, instr. elec. engring. physics, 1948-69, dean students, 1969-72, instr. physics, elec. engr., 1972-86. Author: (charts) The Periodic Table of Physical Concepts, 1977, The Periodic Table of Physical Concepts with Economic Concepts, 1980; (book) The Periodic Table of Physical Concepts Book of Definitions, 1980. Lt. (j.g.) USNR, 1942-46. Mem. Sigma Xi. Republican. Presbyterian. Achievements include identification of gravitational inductance and capacitance and splendor (MV13). Home: 65 Green Forest Rd Swannanoa NC 28778-2246

STUCKERT, GREGORY KENT, mechanical engineer, researcher; b. Calgary, Alta., Can., Dec. 11, 1963; s. Ronald Frederick Stuckert and Marilyn Joyce (Goldsmith) McKevitt. Student, U. Calgary, Can., 1981-82; BS in Aerospace Engring., Ariz. State U., 1985, MSc, 1987, PhD, 1991. Engr. Gen. Dynamics Space Systems Div., San Diego, 1990-91; rsch. scientist DynaFlow, Inc., Columbus, Ohio, 1991—. Mem. AIAA, ASME, Soc. Automotive Engrs. Home: 43 Forest Ridge Dr Worthington OH 43235 Office: DynaFlow Inc 3040 Riverside Dr # 109 Columbus OH 43221

STUDEBAKER, JOHN MILTON, utilities engineer, consultant, educator; b. Springfield, Ohio, Mar. 31, 1935; s. Frank Milton and Monaruth (Beatty) S.; m. Virginia Ann Van Pelt, Mar. 12, 1960; 1 child, Jacqueline Ann Allcorn. PhD, Columbia Pacific U., 1984. Cert. plant engr. Indsl engr. Internat. Harvest Co., 1957-60, supr. indsl. engring., 1960-66, gen. supr. body assembly, 1967-68, mgr. indsl. engring., 1968-70; mgr. manufacturing engring. Lamb Electric Co., 1970-72, Cascade Corp., 1972-76; engring. mgr. Bundy Tubing Corp., Winchester and Cynthia, Ky., 1978-84; utility cons. engr., 1989—; cons. engr. numerous orgns.; instr. numerous univs. including Boston U., Clemson U., Cornell U., Harvard U., Duquesne U., U. Ala., U. Ill., L.I. U., U. Wis., Ga. State U., James Madison U., Tex. Tech. U., Calif. State U., Pacific Luth. U., Fairleigh Dickinson U., San Francisco State U.; instr. Am. Mgmt. Assn., Rochester Inst. Tech., Ctr. for Profl. Advancement. Author: Slashing Utility Costs Handbook, 1992, Natural Gas Purchasing Handbook, 1993. Mem. Nat. Soc. Profl. Engrs., Am. Inst. Plant Engrs., Assn. Energy Engrs. (instr.), Doctorate Assn. N.Y. Educators. Republican. Home: 225 Al Fan Ct Winchester KY 40391

STUDER, JAMES EDWARD, geological engineer; b. Aurora, Colo., Sept. 1, 1961; s. Fredrick Ernest and Patricia Dora (McWilliams) S.; m. Anita Louise Palmer, Apr. 19, 1986. BS in Geol. Engring., U. Mo., Rolla, 1984, MS in Geol. Engring., 1985. Registered profl. engr., Kans., Fla., Tex. Engring. aide engring. div. pub. works City of Kansas City, Mo., 1981-83; civil engr. tech. U.S. Army Corps Engrs., Kansas City, 1984; staff engr. Woodward-Clyde Cons., St. Louis, 1985; staff to asst. project engr. Woodward-Clyde Cons., Overland Park, Kans., 1986-89, project engr., 1989-90; grad. teaching asst. U. Mo., Rolla, 1985; sr. project engr. Coastal Remediation Co., Norman, Okla., 1990-92; program dir. Hall Southwest Corp., Austin, Tex., 1992-93; S.W. region program mgr. Envirogen, Inc., Austin, 1993—; lectr. grad. sch. seminars, 1987-89, Nat. Seminar on RCRA Corrective Actions, 1990. Mem. environ. adv. bd. City of Round Rock, Tex., 1993. Mem. ASCE (assoc.), Assn. Ground Water Scientists and Engrs., Assn. Engring. Geologists (assoc.), Sigma Gamma Epsilon (W.A. Tarr award 1984). Roman Catholic. Achievements include discovery of rare mineral occurrence known as quartz pseudomorphed after barite, previously unrecorded cave in Missouri; research in application of numerical modeling technique to simulate pressure relief well network operation, innovative destruction technologies for restoration of hazardous waste sites. Office: Envirogen Inc 1106 Clayton Ln Ste 542W Austin TX 78723

STUDIER, FREDERICK WILLIAM, biophysicist; b. Waverly, Iowa, May 26, 1936; s. Frederick William and Maudine Modelle (Shoesmith) S.; m. Susan Sarah Cook, Jan. 26, 1962; children: Frederick, Carol. B.S., Yale U., 1958; Ph.D., Calif. Inst. Tech., 1963. Postdoctoral fellow dept. biochemistry Stanford U. Med. Sch., 1962-64; asst. biophysicist dept. biology Brookhaven Nat. Lab., Upton, N.Y., 1964-67; asso. biophysicist Brookhaven Nat. Lab., 1967-70, biophysicist, 1970-74, sr. biophysicist, 1974—, chair dept. biology, 1990—; adj. prof. biochemistry SUNY, Stony Brook. Recipient Ernest Orlando Lawrence Meml. award U.S. Dept. Energy, 1977. Mem. NAS, AAAS, Am. Acad. Arts and Scis., Am. Soc. for Biochemistry and Molecular Biology, Am. Soc. Microbiology, Biophys. Soc., Am. Soc. Virology. Achievements include research on nucleic acid physical chemistry, molecular genetics bacteriophage T7. Office: Biology Dept Brookhaven Nat Lab Upton NY 11973

STUDNESS, CHARLES MICHAEL, economist; b. Mpls., Nov. 2, 1935; s. Leo C. and Alma (Mehus) S.; m. Harriet Leah Katz, Oct. 27, 1968; children: Erica, Lisa, Roy. BA, U. Minn., 1957, MA, 1958; PhD in Econs., Columbia U., 1963. Lectr. CCNY, 1961-64, U. Minn., Mpls., 1964-65; economist Fed. Res. Bank N.Y., N.Y.C., 1965-67, N.Y. Stock Exchange, N.Y.C., 1967-68, Eastern Airlines, N.Y.C., 1968-70, Baker Weeks, N.Y.C., 1970-76, E.F. Hutton, N.Y.C., 1976-79; pres. Studness Rsch., Manhasset, N.Y., 1979—; lectr. Baruch Coll., N.Y.C., 1968-74; Contbg. editor Public Utilities Fortnightly, 1990—. Columnist, Pub. Utilities Fortnightly, 1979—.

STUELAND, DEAN THEODORE, emergency physician; b. Viroqua, Wis., June 24, 1950; s. Theodore Andrew and Hazel Thelma (Oftedahl) S.; m. Marlene Ann McClurg, Dec. 30, 1972; children: Jeffrey, Michael, Nancy, Kevin. BSEE, U. Wis., 1972, MSEE, 1973, MD, 1977. Diplomate Am. Bd. Internal Medicine, Am. Bd. Geriatric Medicine, Am. Bd. Emergency Medicine; cert. in addictions medicine, cert. med. rev. officer. Resident Marshfield (Wis.) Clinic, 1977-80, emergency physician, dir. emergency svc., 1981-93; emergency physician Riverview Hosp., Wisconsin Rapids, Wis., 1980-81; med. dir. Nat. Farm Medicine Ctr., Marshfield, 1986—, alcohol and other drug abuse unit St. Joseph's Hosp., Marshfield, 1988—; exec. com. Marshfield Clinic, 1989-91, treas., 1993; ACLS state affiliate faculty Am. Heart Assn., 1984—, nat. faculty, 1992—. Contbr. articles to profl. jours. Charter mem., pres. Hewitt (Wis.) Jaycees, 1984; bd. dirs. Northwood County chpt. ARC, 1988—. Fellow ACP, Am. Coll. Emergency Physicians (bd. dirs. Wis. chpt. 1984-90, v.p. 1990-91, pres. 1991-92, counselor 1993—), Am. Coll. Preventive Medicine; mem. Biomed. Engring. Soc. (sr. mem.), Am. Soc. Addictions Medicine. Mem. Missionary Alliance Ch. (bd. govs., treas. 1991—). Office: Marshfield Clinic 1000 N Oak Ave Marshfield WI 54449-5703

STUEMPFLE, ARTHUR KARL, physical science manager; b. Williamsport, Pa., Jan. 5, 1940; s. Arthur Carl and Jeanette Esther (Jacobs) S.; m. Linda Jean Campbell, Mar. 30, 1961; children: Jeffrey, Karl. BS in Physics, Drexel U., 1962; MS in Physics, Johns Hopkins Univ., 1971. Chief, test/measurement br. U.S. Army Chem. Sch., Ft. McClellan, Ala., 1962-64; physicist Edgewood Arsenal, Aberdeen Proving Ground, Md., 1964-71; rsch. physicist Chem. Systems Lab., Aberdeen Proving Ground, 1971-78; chief operational sci. br., rsch. dir. Chem. Rsch., Devel. and Engring. Ctr., Aberdeen Proving Ground, 1978-85, chief, physics div. rsch. dir., 1985-92; chief test methodology and program integration, rsch. & tech. dir. Edgewood Rsch., Devel., & Engring. Ctr., Aberdeen Proving Ground, 1992—; U.S. mem. aerosol tech. The Tech. Cooperation Program, Washington, 1982—; cons. phys. property subcom. Nat. Spray Drift Task Force, Wilmington Del., 1990. Contbr. articles to profl. jours.; patentee in field. Organizing mem. Edgewood Meadows Civic and Improvement Assn., 1990; lay reader, cantor, choir mem. Lord of Life Luth. Ch., Edgewood, 1979—; bus. mgr. LC Sewing Sch., Edgewood, 1971-75. Recipient R&D Achievment award U.S. Army, Washington, 1977, Meritorious Civilian Svc. award Dept. Army, Washington, 1986, AMC Spl. Features award Army Materiel Command, 1989; W.H. Walker Tech. Leadership award Chem. Rsch., Devel. and Engring. Ctr., 1987, Internat. Tech. Devel. award U.S./U.K./Can. Dept. Def., 1990. Mem. Am. Assn. Aerosol Rsch., Am. Def. Preparedness Assn., Sigma Xi (treas. admissions com. 1973-75). Republican. Achievements include research on aerosol and spray dissemination and characterization, on chemical warfare and chemical biological defense technology, mathematical modeling and operations research analyses, on aerosol transport, deposition and environmental fate, on evaporation and persistency, on powder technology, on simulants and simulation, on NBC survivability technology, on management of basic and applied research. Home: 2300 Perry Ave Edgewood MD 21040-2808 Office: Edgewood Rsch Devel and Engring Ctr Attn SCBRD-RT Aberdeen Proving Ground MD 21010-5423

STUFFLE, LINDA ROBERTSON, electrical engineer; b. Hibbing, Minn., Apr. 4, 1949; d. James Hamilton and Gertrude Lavina (Hansen) Robertson; m. Roger Ellis Laub, June 1, 1974 (wid. May 1984); m. Roy Eugene Stuffle, Oct. 3, 1985; 1 child, Robert Eugene. BSEE, Mich. Tech. U., 1971, MS in Bus. Adminstrn., 1978; MSEE, Purdue U., 1983. Registered profl. engr., Mich., Idaho. Relay engr. Minn. Power Co., Duluth, 1969-72; project engr. Modern Constructors, Inc., Duluth, 1972-73; instr. Mich. Tech. U., Houghton, 1973-75, asst. prof., 1975-83; teaching assoc. U. Mo., Rolla, 1983-88; sr. design engr. U.S. Elec. Motors, St. Louis, 1988-90; ptnr., sr. engr. Stuffle Engring. and Co., Cons. Svcs., Pocatello, Idaho, 1990—; G*POP fellowship evaluator Dept. Edn., Washington, 1979-80. Assoc. editor: IEEE Potentials, 1989-91; author: Solution Manual for Electromagnetic and Electromechanical Machines, 1985; co-author: (proceedings) Midwest Symposium on Circuits and Systems, 1987-92, IEEE Industrial Applications Society, 1989, (transactions) IEEE Power Delivery, 1990; co-editor: Proceedings of the 31st Midwest Symposium on Circuits and Systems. Mem. IEEE (sr.), Soc. Women Engring. (sr., exec. com. N.Y.C. chpt. 1980-84), Tau Beta Pi, Eta Kappa Nu. Presbyterian. Home: 1005 Park Ln Pocatello ID 83201

STUHL, OSKAR PAUL, organic chemist; b. Wilhelmshaven, Fed. Republic Germany, Dec. 23, 1949; s. Johannes Alexander and Johanna Wilhelmine (Hoelling) S. Dipl. Chem., U. Duesseldorf, 1976, Dr.rer.nat., 1978. Tutor, Institut fuer Organische Chemie, U. Dusseldorf, 1975-76, sci. assoc., 1976-79; mgr. product devel. Drugofa GmbH, Cologne, Fed. Republic Germany, 1980; mgr. sci. rels. R.J. Reynolds Tobacco GmbH, Cologne, 1981-88, mgr. sci. svcs., 1989—; cons. in field. Mem. editorial bd. Beitraege zur Tabakforschung Internat. Mem. Dusseldorf Museums Verein, Verein der Freunde des Hetjens-Museums, Verein der Freunde and Foerderer der U. Dusseldorf, Verein der Freunde des Stadtmuseums Dusseldorf, Met. Mus. Art (N.Y.C.), Friends Royal Acad. Arts, London, Art Soc. of Rheinlande and Westfalen, Gesellschaft der Freunde der Kunstammlung NRW; Gesellschaft der Freunde und Foerderer der Univ. Dusseldorf; Zuercher Kunstgesellschaft, Foerderverein NRW-Stiftung, Forum fuer Film (Duesseldorf), Deutsch-Japanische-Gesellschaft, Verein zur Foerderung Deutsch-Japanischer Beziehungen. Mem. Gesellschaft Deutscher Chemiker, Gesellschaft Deutscher Naturforscher und Aerzte, Max-Planck-Gesellschaft, Deutsche Gesellschaft fuer Arbeits hygiene, Am. Chem. Soc. (including various divs.), Chem. Soc. Japan, N.Y. Acad. of Scis., Royal Soc. Chemistry, Am. Pharm. Assn., Acad. Pharm. Rsch. and Sci., AAAS, Internat. Union Pure and Applied Chemistry, Am. Soc. Pharmacognosy, Fedn. Internat. Pharmaceutic, Christlich Demokratische Union, CDU-Wirtschaftsvereinigung. Roman Catholic. Clubs: Vereinigung AC Dusseldorf; PCL (London); KDStV Burgundia-Leipzig (Zu Dusseldorf) im CV, Golf Club Hilden. Contbr. articles to profl. jours.; patentee in field. Home: An der Thomaskirche 23, D-40470 Dusseldorf Germany Office: RJ Reynolds Tobacco GmbH, Maria-Ablass Platz 15, D 50668 Cologne 1, Germany

STUKEL, THERESE ANNE, biostatistician, educator; b. Ottawa, Ont., Can., May 18, 1952; came to U.S., 1983; d. Anton and Maria (Turk) Stukel; m. Edwin Augustus Gailits, July 16, 1983; children: André, Nicola. BSc, U. Ottawa, 1973; DEA, U. Paris, 1975; PhD, U. Toronto, 1983. Asst. prof. stats. Old. Dominion U., Norfolk, Va., 1983-84; from asst. prof. to assoc. prof. biostats. Dartmouth Med. Sch., Hanover, N.H., 1984—. Contbr. articles to Jour. Am. Statis. Assn., New Eng. Jour. Medicine. Nat. Cancer Inst. rsch. grantee, 1989-93. Mem. Sigma Xi. Office: Dartmouth Med Sch HB7927 Hanover NH 03755-3861

STULL, DANIEL RICHARD, research thermochemist, educator, consultant; b. Columbus, Ohio, May 28, 1911; s. Lucius Walter and Irene Mabel (Haldeman) S.; m. Ruth Louise Keck, Sept. 26, 1936 (dec. 1982); children: Louise Irene Stull Hassman, Richard Walter; m. Mary Morton Lowe, Apr. 28, 1984. BS in Chemistry, Math., Baldwin-Wallace Coll., 1933; PhD in Chemistry, Johns Hopkins U., 1937. Asst. prof. chemistry East Carolina U., Greenville, N.C., 1937-40; rsch. crew leader Dow Chem. Co., Midland, Mich., 1940-50, rsch. tech. expert, 1950-60, dir. thermal lab., 1960-69, rsch. scientist, 1969-76, cons., 1976-77; ret., 1976; cons. in field; rsch. adv. com. Mfg. Chemists Assn., Washington, 1958-65, Nat. Rsch. Coun. rev. bd. Nat. Bur. Standards, Washington, 1959-70; rsch. mem., commr. Internat. Union Pure and Applied Chemistry, 1963-72; printer Hobby Print Shop, Alembic Press, 1954-92. Author: Fundamentals of Fire and Explosion, 1976; author, editor books Joint Army Navy Airforce Rocket Propulsion Group, 1958-76; co-author: Chemical Thermodynamics of Organic Compounds, 1948-69; contbr. to more than 70 sci. rsch. publs. Fin. chmn. 1st United Meth. Ch., Midland, 1948-58; mem. Cosmos Club, Washington, 1967-72; mem. ch. choir various communities, 1928-90. Recipient Hugh Huffman Meml. award Calorimetry Conf., Ames, Iowa, 1965, Alumni Merit award Baldwin-Wallace Coll., Berea, Ohio, 1968, Book award Rsch. Soc. Am., 1969. Fellow Am. Inst. Chemists; mem. AAAS, Am. Chem. Soc. (local sect. award 1980), Sigma Xi. Achievements include extensive compilation of vapor pressure data, development of automatic strip chart recorder for platinum resistance thermometry, 1st automatic recording low temperature calorimeter, computer program to calculate thermodynamic functions, thermal equilibria. Home: 1113 W Park Dr Midland MI 48640-4250

STULL, DONALD LEROY, architect; b. Springfield, Ohio, May 16, 1937; s. Robert Stull and Ruth Branson; m. Patricia Ann Ryder, Dec. 29, 1959 (div. Dec. 1985); children: Cydney Lynn, Robert Branson, Gia Virginia. BArch, Ohio State U., 1961; MArch, Harvard U., 1962. Registered architect, Calif., Conn., Fla., Ky., Maine, Md., Mass., Mich., Mo., N.H., N.J., N.Y., R.I., Tenn., Tex., Va., D.C. Pres. Stull Assocs., Inc., Boston, 1966-84; pres. Stull and Lee, Inc., Boston, 1984—; mem. Loeb fellowship com. Harvard Grad. Sch. Design, Cambridge, 1969-80; mem. adv. bd. Boston Archtl. Ctr., 1972-80, Mus. Nat. Ctr. of Afro-Am. Artists, Boston, 1978—, Ohio State U. Sch. Architecture, 1980—; design prof. Harvard Grad. Sch. Design, 1974-81; mem. vis. design studio, Rice University, Houston, Tex., spring 1993; bd. dirs. Mus. of Afro-Am. History, Boston, 1975-85; mem. vis. com. Yale Sch. Art and Architecture, New Haven, Conn., 1972-76, William Henry Bishop chair Yale Sch. Architecture, 1975; mem. nat. presdl. design award jury Nat. Endowment for Arts, 1984, 88. Trustee Shaw U., 1973-75, Boston Found. for Architecture; mem. Design Adv. Panel, Balt.,

1976-80; chmn. Mass. Art Commn., Boston, 1978-80; commr. Boston Art Commn., 1980-92; mem. Design Adv. Group, Cambridge, 1980-90; commr. Boston Civic Design Commn., 1987—; adv. com. Suffolk Sch. Bus. Mgmt., 1989—; bd. dirs. Historic Boston, 1990—. Recipient Presdl. Design award Nat. Endowment for Arts, 1988; named one of Outstanding Young Men Boston, 1969, Outstanding Young Men Am., 1970, Centennial Yr. Outstanding Alumnus Ohio State U., 1970. Fellow AIA (nat. design com. 1972-84); mem. Boston Soc. Architects (bd. dirs. 1969, AIA Regional Design awards 1975, 80, 89), Mass. Soc. Architects. Office: Stull and Lee Inc Ste 1100 38 Chauncy St Boston MA 02111

STUMPF, BERNHARD JOSEF, physicist; b. Neustadt der Weinstrasse, Rhineland, Germany, Sept. 21, 1948; came to U.S., 1981; s. Josef and Katharina (Cervinka) S. Diploma physics, Saarland U., Saarbrucken, West Germany, 1975, Dr.rer.nat., 1981. Rsch. asst. physics dept. Saarland U., Saarbrucken, 1976-81; rsch. assoc. Joint Inst. Lab. Astrophysics, U. Colo., Boulder, 1981-84; instr. physics, physics dept. NYU, N.Y.C., 1984-86, asst. rsch. scientist Atomic Beams Lab., 1984-85, assoc. rsch. scientist Atomic Beams Lab., 1985-86; vis. assoc. prof. physics dept. U. Windsor (Ont., Can.), 1986-88; assoc. prof. physics dept. U. Idaho, Moscow, 1988—; chmn. Conf. on Atomic and Molecular Collisions in Excited States, Moscow, 1990. Contbr. articles to profl. jours. German Sci. Found. postdoctoral fellow U. Colo., Boulder, 1981-83; recipient Rsch. Opportunity award NSF, U. Idaho, 1990-91. Mem. AAAS, AAUP, German Phys. Soc., Am. Phys. Soc., N.Y. Acad. Scis. Office: U Idaho Dept Physics Moscow ID 83843

STUMPF, DAVID MICHAEL, food scientist; b. Lancaster, Pa., Apr. 3, 1953; s. Harry Martin and Katherine (Nemith) S.; m. Rochelle Marie Halsell, Nov. 26, 1975 (div. Jan. 20, 1987); children: Andrea Renee, Jenelle Elaine. BA, Penn State U., 1975. Rsch. scientist Hershey (Pa.) Foods Corp., 1976—. Author: (trade jour.) The Mobilometer, 1984, (newsletter) Confectionery Rheology, 1986. Fellow Royal Microscopical Soc.; mem. Inst. Food Technologists. Achievements include patents in field. Office: Hershey Foods Corp Box 805 Hershey PA 17033

STUMPF, PAUL KARL, former biochemistry educator; b. N.Y.C., N.Y., Feb. 23, 1919; s. Karl and Annette (Schreyer) S.; married, June 1947; children: Ann Carol, Kathryn Lee, Margaret Ruth, David Karl, Richard Frederic. AB, Harvard U., 1941; PhD, Columbia U., 1945. Instr. biochemistry U. Mich., Ann Arbor, 1946-48; from asst. prof. to prof. U. Calif., Berkeley, 1948-58; prof. U. Calif., Davis, 1958-84, prof. emeritus, 1984—; chief scientist Competitive Rsch. Grants Office USDA, Washington, 1988-91; cons. Palm Oil Rsch. Inst., Kuala Lumpur, Malaysia, 1982-92; sci. adv. bd. Calgene, Inc., Davis, Calif., 1990—; sci. adv. panel Md. Biotech. Inst., 1990-92. Co-Author: Outlines of Enzyme Chemistry, 1955, Outlines of Biochemistry, 5th ed. 1987; co-editor-in-chief Biochemistry of Plants, 1980; exec. editor Archives of Biochemistry/Biophysics, 1965-88; contbr. over 250 articles to profl. jours. Mem. planning commn. City of Davis, 1966-68. Guggenheim fellow, 1962, 69; recipient Lipid Chemistry award Am. Oil Chemists Soc., 1974, Sr. Scientist award Alexander von Humboldt Found., 1976, Superior Svc. Group award USDA, 1992. Fellow Linnean Soc. London; mem. NAS, Royal Danish Acad. Scis., Am. Soc. Plant Physiologists (pres. 1979-80, chmn. bd. trustees 1986-90, Stephen Hales award 1974, Charles Reid Barnes Life Membership award 1992), Yolo Fliers Country Club (Woodland, Calif.). Avocation: golf. Home: 764 Elmwood Dr Davis CA 95616-3517 Office: Univ of Calif Molecular/Cellular Biology Davis CA 95616

STUMPF, WALDO EDMUND, nuclear energy industry executive; b. Vryheid, Natal, South Africa, June 29, 1942; s. Edmund and Agnes (Röttcher) S.; m. Hettie Stumpf, June 12, 1965; 3 children. BScMetE, U. Pretoria, South Africa, 1964; PhD in Metallurgy, U. Sheffield, Eng., 1968. Registered metall. engr. Engr. Atomic Energy Corp. SA Ltd., Pretoria, 1968-87, sr. gen. mgr., 1987-89, CEO, 1990—; prof. extraord. dept. materials sci. and metall. engring. U. Pretoria, 1989—. Contbr. 25 sci. and tech. articles to profl. jours. Recipient Brunton medal U. Sheffield, 1968. Mem. South African Inst. Mining and Metallalurgy, South African Akademies Kuns en Wetenskap (assoc.), Fellowship of Excellence, Pretoria Toast Masters Club (pres. 1989). Avocations: hiking, reading, gardening. Home: 102 Van Wouw St, Groenkloof, Pretoria 0181, South Africa Office: Atomic Energy Corp SA Ltd, Pelindaba PO Box 582, Pretoria 0001, South Africa*

STUNKARD, ALBERT JAMES, physician, educator; b. N.Y.C., Feb. 7, 1922; s. Horace Wesley and Frances (Klank) S. BS, Yale U., 1943; MD, Columbia U., 1945; MD (hon.), U. Edinburgh, 1992. Intern in medicine Mass. Gen. Hosp., Boston, 1945-46; resident physician psychiatry Johns Hopkins Hosp., 1948-51, rsch. fellow psychiatry, 1951-52; rsch. fellow medicine Columbia U. Svc., Goldwater Meml. Hosp., N.Y.C., 1952-53; Commonwealth rsch. fellow, then asst. prof. medicine Cornell U. Med. Coll., 1953-57; mem. faculty U. Pa., 1957-73, 76—, prof. psychiatry, 1962-73, 76—, Kenneth Appel prof. psychiatry, 1968-73, chmn. dept., 1962-73; prof. psychiatry Med. Sch., Stanford U., 1973-76. Contbr. articles on psychol., physiol., sociol. and genetic aspects of obesity to profl. jours. Capt. M.C., AUS, 1946-48. Ctr. for Advanced Study in Behavioral Scis. fellow, 1971-72. Mem. Inst. Medicine of NAS, Am. Assn. of Chmn. of Depts. of Psychiatry (past pres.), Acad. Behavioral Medicine Rsch. (past pres.), Am. Psychosomatic Soc. (past pres.), Assn. Rsch. in Nervous and Mental Diseases (past pres.), Soc. Behavioral Medicine (past pres.). Achievements include contributions to the behavioral treatment of obesity and to understanding of sociological, psychological and genetic contributions to the disorder. Office: U Pa Sch Medicine Dept Psychiatry 3600 Market St Philadelphia PA 19104

STURDEVANT, WAYNE ALAN, engineering manager; b. Portland, Oreg., Apr. 3, 1946; s. Hervey Grable Sturdevant and Georgia Eileen (Rawls) Bright; m. Helen F. Radbury, Sept. 24, 1976; children: Wayne Alan Jr., Stephen Thomas, John Howard; children from previous marriage: Brian Alan, Daniel Robert. BS in Edn., So. Ill. U., 1980. With USAF, 1964-83, supt. of the Air Force's on the job tng. adv. svc., 1978-82; chief USAF On-Job-Tng., 1982-85; ret., 1985; lead engr. instl. tech. McDonnell Douglas Corp., St. Louis, 1985-88; sr. systems analyst Southeastern Computer Cons., Inc., Austin, 1988—; developed advanced concepts in occupational edn. and computer based tng. design. Contbr. articles on mgmt. and tng. innovations in the work place to profl. jours. Bishop LDS Ch., San Antonio, 1983-84, high councilor, Austin, 1986-90, mem. stake presidency, 1990—; commr. Boy Scouts Am., 1986—. Recognized for leadership in multi-nat. programs; recipient Citation of Honor, Air Force Assn., 1980; named Internat. Man of the Yr. Internat. Biog. Centre, Cambridge, Eng., 1992. Republican. Avocations: reading, camping, raquet sports. Home: 9214 Independence Loop Austin TX 78748-6312

STURGE, MICHAEL DUDLEY, physicist; b. Bristol, Eng., May 25, 1931; came to U.S., 1961, naturalized 1991; s. Paul Dudley and Rachel (Graham) S.; m. Mary Balk, Aug. 21, 1956; children: David Mark, Thomas Graham, Peter Daniel, Benedict Paul. BA in Engring. and Physics, Gonville and Caius Coll., Cambridge, Eng., 1952; PhD in Physics, Cambridge U., Eng., 1957. Mem. staff Mullard Rsch. Lab. (now Philips), Redhill, Eng., 1956-58; sr. rsch. fellow Royal Radar Establishment, Malvern, Eng., 1958-61; mem. tech. staff Bell Labs., Murray Hill, N.J., 1961-83, Bellcore, Red Bank, N.J., 1984-86; prof. dept. physics Dartmouth Coll., Hanover, N.H., 1986—; rsch. assoc. Stanford U., 1965, U. B.C., Vancouver, Can., 1969; vis. prof. Technion, Haifa, Israel, 1972, 76, 81, 85, Williams Coll., Williamstown, Mass., 1982, 84, Trinity Coll., Dublin, 1989, 93, U. Fourier, Grenoble, France, 1989, 91; exch. scientist Philips Rsch. Lab., Eindhoven, The Netherlands, 1973-74. Contbr. over 100 papers in solid state physics to profl. publs.; co-editor: Excitons, 1982; editor Jour. of Luminescence, 1984-90. Fellow Am. Phys. Soc. Office: Dartmouth Coll Dept Physics Wilder Lab Hanover NH 03755-3528

STURGES, LEROY D., engineering educator; b. Slayton, Minn., Mar. 21, 1945; s. Russell M. and M. Irene (Nelsen) S.; m. Sue A. Davis, Feb. 14, 1969; children: Jeffrey L., Jason L. B in Aero Engring., U. Minn., 1967, PhD, 1977. Engr., analyst Fgn. Tech. Div. USAF, Wright-Patt AFB, Ohio, 1968-72; assoc. prof. Iowa State U., Ames, 1977--. Author: Engineering Mechanics: Statics, 1993, Engineering Mechanics: Dynamics, 1993; contbr. articles to profl. jours. Capt. USAF 1967-72. Mem. Am. Soc. Engring. Edn., Soc. Rheology, Am. Acad. Mechanics, Tau Beta Pi, Sigma Gamma Tau, Sigma Xi. Office: Iowa State U 2019 Black Engring Bldg Ames IA 50011

STURGES, SIDNEY JAMES, pharmacist, educator, investment and development company executive; b. Kansas City, Mo., Sept. 29, 1936; s. Sidney Alexander and Lenore Caroline (Lemley) S.; m. Martha Grace Leonard, Nov. 29, 1957 (div. 1979); 1 child, Grace Caroline; m. Gloria June Kitch, Sept. 17, 1983. BS in Pharmacy, U. Mo., 1957, post grad.; MBA in Pharmacy Adminstrn., U. Kans., 1980; PhD in Bus. Adminstrn., Pacific Western U., 1980; cert. in Gerentology, Avila Coll., 1986. Registered pharmacist, Mo., Kans.; registered nursing home adminstr., Mo.; cert. vocat. tchr., Mo. Pharmacist, mgr. Crown Drugs, Kansas City, Mo., 1957-60; pharmacist, owner Sav-On-Drugs and Pharmacy, Kansas City, 1960-62; ptnr. Sam's Bargain Town Drugs, Raytown, Mo., 1961-62; pharmacist, owner Sturges Drugs DBA Barnard Pharmacy, Independence, Mo., 1962—; pres., owner Sturges Med. Corp., Independence, Mo., 1967-1977, Sturgess Investment Corp., Independence, 1967-1978, Sturwood Investment Corp., Independence, 1968—, Sturges Agri-Bus. Co., Independence, 1977—, Sturges Devel. Co., 1984—; bd. dirs. Comprehensive Mental Health Corp., Truman Med. Ctr., 1992; instr. pharmacology Penn Valley Community Coll., 1976-84; instr., lectr. various clubs and groups. Contbr. articles to profl. jours. Bd. dirs. Independence House, 1981-83; mem. Criminal Justice Adv. Commn., Independence, 1982—. Recipient Outstanding award Kans. City Alcohol and Drug Abuse Council, 1982. Mem. Mo. Sheriffs Assn., Mo. Pharm. Assn. (pharmacy dr. 1981, Pharmacists Against Drug Abuse award 1989), Mo. Found. Pharm. Care, U. Mo. Alumni Assn. Home and Office: Sturges Co 16805 E Cogan Rd Ste B Independence MO 64055-2815

STURM, RICHARD E., occupational physician; b. Ann Arbor, Mich., Mar. 4, 1951; m. Marci Lynn Pechauer, July 10, 1987; children: Jennifer, Caryn. MD, Mich. State U., 1978; MPH, U. Ill., 1980. Diplomate Am. Bd. Internal Medicine, Am. Bd. Preventive Medicine and Occupational Medicine. Intern then resident Cook County Hosp., Chgo., 1978-82; clin. physician Addison (Ill.) Med. Ctr., 1982-84, Lakeland Med. Group, Libertyville, Ill., 1984-85; med. program dir. Parkside Occupational Health Ctr., Niles and Elk Grove, Ill., 1985-92, No. Ill. Med. Ctr., McHenry, Ill., 1992—; dist. med. advisor U.S. Dept. of Labor Workers' Compensation, Chgo., 1981-84; attending emergency room physician Cook County Hosp., 1982-84, voluntary attending, 1982—; adj. instr. U. Ill. Coll. Medicine, 1987—. Achievements include research in designing medical surveillance programs for specific job exposures, applications of personal computer in occupational medical programs and trend forecasting. Office: No Ill Med Ctr 4201 Medical Ctr Dr McHenry IL 60050

STURROCK, PETER ANDREW, space science and astrophysics educator; b. South Stifford, Essex, England, Mar. 20, 1924; came to U.S., 1949; s. Albert Edward and Mabel Minnie (Payne) S.; m. Marilyn Fern Stenson, June 29, 1963; children: Deirdre, Colin; 1 child from previous marriage, Myra. BA, Cambridge (Eng.) U., 1945, MA, 1948, PhD, 1951. Scientist Telecommunications Rsch. Establishment, Malvern, Eng., 1943-46, Nat. Bur. Standards, Washington, 1949-50, Atomic Energy Rsch. Establishment, Harwell, 1951-53; fellow St. John's Coll., Cambridge U., 1952-55; rsch. assoc. Stanford (Calif.) U., 1955-61, prof. applied physics and engring. sci., then prof. space sci. and astrophysics, dept. applied physics, 1961—, dir. for plasma rsch., 1964-74, 80-83; dep. dir. Ctr. for Space Sci. and Astrophysics, 1983-92, dir., 1992—. Author: Static and Dynamic Electron Optics, 1955, Plasma Physics, 1993; editor: Plasma Astrophysics, 1967, Solar Flares, 1980, Physics of the Sun, vols. I, II, III, 1986. Recipient Gravity prize Gravity Found., 1967, Hale prize Am. Astron. Soc., 1986, Henryk Arctowski medal NAS, 1990, Space Sci. award AIAA, 1992; European Ctr. for Nuclear Rsch. fellow, 1957-58. Fellow AAAS, Royal Astron. Soc., Internat. Astron. Union; mem. Soc. for Sci. Exploration (pres. 1982—). Office: Stanford U ERL 306 Stanford CA 94305

STUSEK, ANTON, mechanical engineer, researcher; b. Trbovlje, Slovenia, Slovenian, Jan. 14, 1932; s. Friderik and Antonia S.; m. Cecilijr Abram; children: Andrej, Natasha. Student, Engring. Coll., 1949; BSME, U. Ljubljana, Slovenia, 1959; postgrad., U. Mo., 1963; MS in Mech. Engring., U. Zagreb, Croatia, 1971. Designer Mil. Tech. Inst., Belgrad, Yugoslavia, 1949-52; researcher, project engr. Shipbuilding Inst., Zagreb, 1959-73; lectr. dept. mech. engring. U. Ljubljana, 1973—. Author: Hydraulics and Pneumatics, 1969. Lt. col. Yugoslavian Navy, 1949-73. Decorated Medal with Silver Swords, Medal with Golden Swords, Silver medal Zayreb. Mem. ASME, Slovenian Soc. Engrs. and Techs., Soc. for Devel. and Promoting of Hydraulics and Pneumatics Aachen. Roman Catholic. Achievements include innovations of an axial piston pump, EH servovalve, EH steering systems for ships, hydraulic alternating current generator; research project in fluid power in Slovenia. Home: Ellerjeva 45, 61000 Ljubljana Slovenia Office: UL Fakulteta za strojnistvo, Askerceva 6, 61000 Ljubljana Slovenia, Slovenia

STUTES, ANTHONY WAYNE, mechanical engineer; b. Sulphur, La., Aug. 12, 1963; s. Willie Ray and Jeanette (Broussard) S.; m. Lisa Marie Clarey, May 25, 1991. BSME, La. Tech. U., 1987. Registered profl. engr., S.C. Surveyor Daniel Constrn. Co., Lake Charles, La., 1983; security officer La. State Racing Commn., Vinton, 1983; pipe fitter helper Himont USA, Lake Charles, 1984-85; ind. painter Ruston, La., 1985; draftsman MSE, Inc., Shreveport, La., 1986, project engr., 1987-88; project engr. CRS Sirrine Engrs., Inc., Greenville, S.C., 1988—. Mem. Shreveport Lions, 1988-89. Mem. ASME (vice chmn., sec. coll. rels. 1987-93), NSPE (vice chmn., sec. career guidance com., pub. rels. com.), S.C. Soc. Profl. Engrs. (Young Engr. of Yr. 1992), Am. Soc. Plumbing Engrs. (v.p. tech. com., v.p. legis. com., newsletter editor), Toastmasters (sec.-treas., v.p. membership). Roman Catholic. Office: CRS Sirrine Engrs Inc 1041 E Butler Rd Greenville SC 29607

STUTING, HANS HELMUTH, chemist, researcher, consultant; b. Clifton, N.J., Feb. 19, 1958; s. Hans Arthur Ewald and Berta (Flintrop) S.; m. Melissa Beth Rose, June 16, 1990. BS, Syracuse U., 1980; MS, U. Mass., 1983; PhD, Northeastern U., 1990. Mem. staff Waters Assocs., Milford, Mass., 1983-84, L.D.C. Analytical, Riviera Beach, Fla., 1984-86; dir. R&D Roche Biomed. Labs., Raritan, N.J., 1990—; pres., prin. Hans H. Stuting, tech. cons., Boston, Long Valley, N.J., 1987—. Contbr. articles to Jour. Chromatography, Analytical Chemistry. Achievements include development of coupling gradient HPLC with low angle laser light scattering detection for biopolymer molecular weight determination; chiral HPLC seperations and quantitations at the low nglmL level for drug metabolism and clinical trials protocols in addition to final pharmaceutical product monitoring. Home: 105 Wehrli Rd Long Valley NJ 07853 Office: Roche Biomed Labs subs Hoffmann-La Roche 69 First Ave Raritan NJ 08869

STUTMAN, LEONARD JAY, research scientist, cardiologist; b. Boston, Apr. 8, 1928; s. Herbert Hyman and Nellie (Wiener) S.; BS, MIT, 1948; MA, Boston U., 1949; MD, U. Rochester, 1953; m. Jeanne Ann Soblen, Dec. 23, 1951; children: Peter, David, Marc, Robin. Intern, resident medicine Bellevue Hosp., 1953-57; chief, med. services br. WPAFB, Dayton, Ohio, 1957-59; spl. advanced research fellow NIH, Nat. Heart Inst. 1959-61; instr. in clin. medicine N.Y. U. Coll. Medicine, 1956-61, asst. prof. pathology, 1961-65; assoc. prof. clin. medicine N.Y. Med. Coll., 1980—; head coagulation research lab. St. Vincent's Hosp. and Med. Center, N.Y., 1965—; attending physician St. Vincent's Hosp.; sr. attending physician medicine, sr. cardiologist Nyack (N.Y.) Hosp.; med. dir. Presdl. Life Ins. Co., Nyack, Urbaine Life Reinsurance Co., Tarrytown, N.Y.; bd. dirs. Metriplex, Inc. Cambridge, Mass., 1992—. Contbr. articles to profl. jours. Dir. cardiac epidemiology study Ford Found. Vera Inst.; mem. Internat. Com. on Thrombosis and Hemostasis. Capt. USAF, 1957-59. Fellow Am. Coll. Cardiology, ACP, N.Y. Acad. Medicine; mem. Am. Soc. Hematology, N.Y. Med. Soc., Sigma Xi. Home: 250 Townline Rd West Nyack NY 10994-2824 Office: 153 W 11th St New York NY 10011

STYGAR, MICHAEL DAVID, civil engineer; b. Norwich, Conn., Apr. 21, 1947; s. Michael P. and Mary M. (Wojtkiewicz) S.; m. Kathleen Maslanka, Apr. 26, 1980; children: Seth Andrew, Shem, Krystynka, Danyka. BSCE magna cum laude, Tufts U., 1969; MS in Constrn. Adminstrn. magna cum laude, George Washington U., 1976. Registered profl. engr., Va. Sr. constrn. ops. engr. Fed. Hwy. Adminstrn., Arlington, Va., 1971-79; sr. project mgr. Reinforced Earth, Arlington, 1979-82, Ralph M. Parsons Internat., Abu Dhabi, United Arab Emirates, 1982-84; sr. design engr. JMT Inc., Fairfax, Va., 1984; airport engr. FAA, Washington, 1984-85; dir. program mgmt. NAVFACENGCOM-PACDIV, Houston, 1985-87; chief engr. NAVFACENGCOM, Cherry Point, N.C., 1987-88; sr. project mgr., engr. Mass. Port Authority, Boston, 1988—. Mem. planning bd. Town Norfolk, Mass., 1991. With U.S. Army, 1969-71. Mem. NSPE, Mass. Soc. Civil Engrs., Boston Soc. Civil Engrs., Airport Assn. Coun. Internat., Am. Assn. Airport Execs. Roman Catholic. Office: Mass Port Authority 10 Park Plz Boston MA 02116

SU, DONGZHUANG, computer science educator, university official; b. Xiamen, Fujian, China, Dec. 6, 1932; s. Yuzhong and Lizhen (Huang) S.; m. Ningyu Ma, Jan. 20, 1961; children: Bin, Yanchun. BSc, Harbin Inst. Tech., People's Republic China, 1956. Head rsch. group Inst. Computing Tech., Acad. Sinica, Beijing, 1956-59; dir., lectr. Harbin Inst. Tech., 1959-65; dir., assoc. prof. N.W. Telecommunications Engring. Inst., Xi'an, People's Republic China, 1966-84; prof. computer sci., v.p. Beijing Info. Tech. Inst., 1984—; project mgr. North China Inst. Computer Tech., Beijing, 1974-76; dir. UN Devel. Program project Chinese Info. Processing Rsch. Ctr., Beijing, 1986—; exec. dir. Beijing Artificial Intelligence Cen. Lab., 1988-90, dir., 1990—; mem. com. sci. and tech. Ministry Machinery and Elctronics Industry, Beijing, 1989—. Author: Computer Organization and Devices, 1961, Computer Achitecture, 1981 (Nat. prize 1986). Mem. Chinese Info. Processing Soc. (v.p. 1991—), Assn. for Computing Machinery, IEEE Computer Soc., Chinese Inst. Electronics (sr.). Avocations: music, swimming. Office: Beijing Info Tech Inst, Weizikeng outside Desheng, Beijing 100101, China

SU, KENDALL LING-CHIAO, engineering educator; b. Fujian, China, July 10, 1926; came to U.S., 1948; s. Ru-chen and Sui-hsiong (Wang) S.; m. Jennifer Gee-tsone Chang, Sept. 10, 1960; children: Adrienne, Jonathan. BEE, Xiamen U., Peoples Republic China, 1947; MEE, Ga. Inst. Tech, 1949; PhD, Ga. Inst. Tech., 1954. Jr. engr. Taiwan Power Co., Taipei, Republic China, 1947-48; asst. prof. Ga. Inst. Tech, Atlanta, 1954-59, assoc. prof., 1959-65, prof., 1965-70, Regents prof., 1970—; mem. tech. staff Bell Labs., Murray Hill, N.J., 1957. Author: Active Network Synthesis, 1965, Time-Domain Synthesis of Linear Networks, 1969, Fundamentals of Circuits, Electronics, and Signal Analysis, 1978, Handbook of Tables for Elliptic-Function Filters, 1990, Fundamentals of Circuit Analysis, 1993; mem sci. adv. com. Newton Graphic Sci. mag., 1987—. Fellow IEEE; mem. Chinese Lang. Computer Soc., Phi Kappa Phi, Sigma Xi (pres. Ga. Inst. Tech chpt. 1968-69, 72-73, Faculty Research award 1957), Eta Kappa Nu. Methodist. Office: Ga Inst of Tech Sch of Elec Engring Atlanta GA 30332-0250

SU, QICHANG, physicist; b. Tianjin, China, July 27, 1962; s. Guangjun and Linfeng (Wang) S.; m. Jingyuan He. MA, U. Rochester, 1986, PhD, 1991. Rsch. assoc. Max-Planck Inst. for Quantum Optics, Munich, 1991-92; instr., fellow U. Rochester, N.Y., 1992—. Contbr. articles to profl. jours. Achievements include rsch. on dynamical properties of simple quantum systems under the influence of very intense perturbation, various strong-field processes such as multiphoton ionization, above-threshold ionization and high order harmonic generation; pioneering concept of dynamical atomic stabilization effects, ionization supression and electron localization, large degree of stabilization in finite-frequency finite-intensity short-pulse laser systems. Office: U Rochester Dept of Physics/Astronomy Rochester NY 14627

SU, TSUNG-CHOW JOE, engineering educator; b. Taipei, Taiwan, Republic of China, July 9, 1947; came to U.S., 1969; s. Chin-shui and Chenling (Shih) S.; m. Hui-Fang Angie Huang, Dec. 26, 1976; children: Julius Tsu-Li, Jonathan Tsu-Wei, Judith Tsu-Te, Jessica Tsu-Yun. BS, Nat. Taiwan U., 1968; MS in Aeronautics, Calif. Inst. Technology, 1970, AE, 1973; EngScD, Columbia U., 1974. Registered profl. engr., Fla., Tex. Rsch. teaching asst. Calif. Inst. Technology, Pasadena, 197-72; rsch. asst. Columbia U., N.Y.C., 1972-73; naval architect John J. McMullen Assoc., Inc., N.Y.C., 1974-75; asst. prof. civil engring. Tex. A&M U., College Station, Tex., 1976-82; assoc. prof. ocean engring. Fla. Atlantic U., Boca Raton, 1982-87, prof. ocean engring, 1987-92, prof. mech. engring., 1992—; prin. investigator clean room rsch., Fla. Atlantic U. Robotics Ctr., Boca Raton, 1986—. Contbr. over 60 articles and papers to profl. jours.; assoc. editor Jour. Engring. Mechs. Coord. Calif. Tech. Alumni Fund, South Fla. area, 1987-88. 2d lt. Chinese Army, 1968-69. Grantee in field. Fellow AIAA (assoc.); mem. ASME, ASCE (chmn. fluids com.), Soc. Mfg. Engrs., Calif. Tech. Alumni Assn., Royal Palm Improvement Assn. Home: 2150 Areca Palm Rd Boca Raton FL 33432-7969 Office: Fla Atlantic U Dept Mech Engring Boca Raton FL 33431

SU, YALI, chemist; b. Taiyuan, Shanxi, China, Mar. 13, 1963; came to U.S. 1987.; d. Xi and Jingrong (Dong) S.; m. Yong Liang, July 18, 1992. BS, Peking U., 1984, MS, 1987. Rsch. asst. U. Md., College Park, 1987-89, U. Notre Dame, Notre Dame, Ind., 1989--. Reilly fellow, 1992. Mem. Am. Chem. Soc., Internat. Soc. for the Study of the Origin of Life. Achievements include modification of the traditional mechanism of the gamma radiolysis of ethanol. Office: U Notre Dame Radiation Lab Notre Dame IN 46556

SU, YAOXI, fluid mechanics educator; b. Xiamen, Peoples Republic of China, July 28, 1934; s. Zhiping and Suying (Zhuang) S.; m. Peiqin Gao, July 2, 1960; children: Su Lei, Su Mang, Su Ting. BS, Xian Aero. Inst., 1957; MS, Northwestern Poly. U., Xian, Peoples Republic of China, 1964. Lectr. Northwestern Poly. U., 1964-84, assoc. prof., 1985-91, prof., 1991—. Contbr. articles to profl. jours. Recipient Nat. Award of Sci. and Tech. Progress, 1987. Mem. AIAA, Chinese Assn. Aerodynamics. Achievements include research into wind tunnel sidewall interference and boundary layer control, 3D wind tunnel contraction design and fish locomotion, computation of viscous/inviscid flow interaction. Office: Northwestern Poly U, PO Box 114, Xi'an 710072, China

SU, YUH-LONG, chemistry educator, researcher; b. Tainan, Republic of China, May 12, 1952; s. Yao-Chung and Hsien-fong (Chiang) Su; m. Jin-hua Huang, June 1, 1980; 1 child, Doris Mon-Yi. BS, Nat. Taiwan U., Taipei, 1976; PhD, Ohio State U., 1985. Rsch. assoc. Princeton (N.J.) U., 1985-86; assoc. prof. Nat. Taiwan U., Taipei, 1986-91, prof., 1991—; dir. student activities Nat. Taiwan U., Taipei, 1987-89; cons. The Youth Group, Taipei, 1987-90. Contbr. articles to profl. jours. Mem. (congressman) The Nat. Assembly, Taipei, 1991—. Served to 2d lt. Taiwanese Army, 1976-78. Recipient Outstanding Rsch. award The Nat. Sci. Coun., Taipei, 1991. Mem. Am. Chem. Soc., Soc. Applied Spectroscopy, Chinese Chem. Soc. Mem. Kuomingtan Party. Achievements include development of synthesis and characterization of metalloporphyrins of unusual oxidation states. Home: 131, Sect 1, Chung Hua Rd, 930 Taitung Taiwan Office: National Taiwan Univ, 1, Sect 4, Roosevelt Rd, 107 Taipei Taiwan

SUAMI, TETSUO, chemistry educator; b. Nagoya, Japan, Nov. 22, 1920; s. Masaaki and Kimie Suami; children: Takao, Hiroo. BS, Keio U., Yokohama, Japan, 1944, PhD, 1957. Instr. dept. applied chemistry Keio U., Yokohama, 1945-46, lectr., 1946-47, asst. prof., 1947-62, prof., 1962-85, prof. emeritus, 1985—; prof., head dept. chemistry Meisei U., Tokyo, 1985—; postdoctoral researcher dept. chemistry U. Pa., Phila., 1957-58, U. Ill., Urbana, 1958-60; vis. prof. Technische Hochschule, Darmstadt, Germany, 1966, dept. chemistry U. London, 1979. Co-editor: Aminocyclitol Antibiotics, 1979; author: Adr. Carboydrate Chemistry and Biochemistry, 1990; mem. editorial bd. Jour. Carbohydrate Chemistry, 1982. Recipient award Hattori Found., Tokyo, 1969. Mem. Chem. Soc. Japan (dir. 1969, award 1983), Am. Chem. Soc., Synthetic Organic Chemistry Assn. (v.p. 1976). Achievements include synthetic studies on carba-sugars (pseudo-sugars), total synthesis of validamycins, total synthesis of tunicamycins, molecular mechanisms of sweet taste (sweet amino acids, sugars and aspartame). Office: Meisei U, 2-1-1 Hodokubo Hino, Tokyo 191, Japan

SUAREZ, GEORGE MICHAEL, urologist; b. Havana, Cuba, Apr. 21, 1955; came to U.S., 1955; s. Miguel Angel and Elena (Sanchez) S. BA, Heideberg U., 1976; MD, U. Dominica, Portsmouth, 1980, Rutgers U., 1980. Diplomate Am. Bd. Urology; lic. physician, Ind., Fla., La. Intern straight gen. surgery Columbus-Cuneo-Cabrini Med. Ctr., Northwestern U. Med. Sch., Chgo., 1980-81; resident gen. surgery Columbus-Cuneo-Cabrini Med.

Ctr., Northwestern U. Med. Sch., 1981-82; urology rsch. fellow Tulane U. Sch. Medicine and Delta Regional Primate Ctr., New Orleans, 1982-83; resident, chief resident urology Tulane U. Sch. Medicine, New Orleans, 1983-87; attending urologist, dir. urodynamics lab. spinal cord unit VA Med. Ctr., Miami, Fla., 1987-90; attending urologist U. Miami Hosp. and Clinics, 1987-90, dir. Urodynamics Lab., Jackson Meml. Med. Ctr., 1987-90, dir. urology rehab. rsch. program Bantle Rehab. Rsch. Ctr., 1987—; attending urologist Jackson Meml. Hosp., Miami, 1987—; asst. prof. dept. urology U. Miami Sch. Medicine, 1987-90; cons. Sylvester Comprehensive Cancer Ctr., Miami, 1987—, Childrens Med. Svcs., Miami, 1987—, Avalon Technologies, Indpls., Mentor Corp., Santa Barbara, Calif., Cook Urol., Spence, Ind., Teknar Ultrasound, Inc., Santa Barbara, Schering Labs. N.J., Rorer Pharms., Ft. Washington, Pa.; attending urologist Doctors Hosp., Bat. Hosp., Childrens Hosp., South Miami Hosp., Larkin Hosp., Mercy Hosp., Victoria Hosp., West Gables Hosp., Cedars Med. Ctr., Kidney Stone Ctr. Contbr. articles to profl. jours. Founder, pres. For the Love of Life Found. Recipient Urology Rsch. award Touro Infirmary Hosp., New Orleans, 1983-84, award of excellence Video Urology, 1989. Mem. ACS, AMA, Am. Acad. Pediatrics, Am. Fertility Soc., Am. Med. Polit. Action Com., Am. Soc. Andrology,. Am. Urol. Assn., Colegio Interam. de Medicos y Cirujanos, Am. Confederation Urology, Cuban Am. Urol. Soc., Dade County Med. Assn., European Urologic Soc., Fla. Med. Assn., Fla. Urol. Soc., Internat. Continence Soc., Greater Miami Urol. Soc., N.Y. Acad. Scis., So. Med. Assn., World Med. Assn., Urodynamics Soc., Surg. Aid to Children of the World, Internat. Soc. Urology. Office: Miami Urologic Inst 7051 SW 62d Ave Miami FL 33143

SUAREZ, LUIS EDGARDO, mechanical engineering educator; b. Jujuy, Argentina, May 14, 1957; came to U.S., 1983; s. Luciano and Maria Mercedes (Colche) S. MS in Engring. Mechanics, Va. Poly Inst., 1984, PhD in Engring. Mechanics, 1986. Jr. engr. Atomic Energy Commn. Argentina, Cordoba, 1981; instr. part-time dept. of structures U. Cordoba, 1981-82, asst. prof. grad. programs, 1987-89; rsch. asst. engring. sci. and mechanics Va. Poly. Inst., Blacksburg, 1983-86, asst. prof. engr. sci. and mechanics dept., 1986-87; asst. prof. gen. engring. dept. U. P.R., Mayaguez, 1989-91, assoc. prof. gen. engring. dept., 1991—; proposal reviewer U.S. Army Rsch. Office, Mayaguez, 1991; paper reviewer Jour. Vibration and Acoustics, Jour. Engring. for Industry, Jour. Engring. Structures, Mayaguez, 1992; panelist to select scholarships Battelle, Raleigh, 1991-92. Co-author: Multinational Seismic Design Codes, Handbook, 1992; contbr. articles to profl. jours. Cunningham fellow Va. Poly. Inst., 1986; grantee U.S. Army Rsch. Office, 1990, Nat. Ctr. for Earthquake Engring. Rsch., 1992-93, NSF, 1993, NASA/Langley, 1993. Mem. ASCE, ASME, AIAA, Am. Soc. Engring. Edn., Am. Acad. Mechanics, Sigma Xi (Rsch. award Va. Tech. chpt. 1987). Roman Catholic. Achievements include development of methods for seismic analysis of mechanical equipment that are used in industry, several methods for dynamic analysis of large structural systems. Office: Univ Puerto Rico Gen Engr Dept Stefani Bldg Mayaguez PR 00680

SUAREZ QUIAN, CARLOS ANDRÉS, biology educator; b. La Habana, Habana, Cuba, Jan. 25, 1953; s. Andrés and Hortensia (Quian) S.; m. Kathryn C. Creel, July 24, 1982; children: Harrison C., Benjamin C. BS, U. N.C., 1975; PhD, Harvard U., 1983. Rsch. fellow NIH, Bethesda, Md., 1983-87; assoc. prof. Georgetown Med. Sch., Washington, 1987—; mem. editorial bd. Assn. Reproductive Tech., 1990-91; ad hoc reviewer NIH, Bethesda, 1993. Contbr. articles to profl. jours. Mem. Am. Soc. Cell Biology, Am. Soc. Andrology, Soc. for Study of Reproduction, Sigma Xi. Democrat. Roman Catholic. Achievements include over 31 publs. in peer reviewed jours. Office: Georgetown Med Ctr Dept Anatomy 3900 Reservoir Rd NW Washington DC 20007

SUBA, STEVEN ANTONIO, obstetrician/gynecologist; b. Columbia, Mo., July 4, 1957; s. Antonio Ronquillo and Sylvia Marie (Karl) S.; m. Brenda Charlene Crosby, Aug. 9, 1986; 1 child, Bethany Caroline. BA in Biology, St. Mary's U., San Antonio, 1979; MD, Tex. Tech U., 1984. Diplomate Am. Bd. Ob.-Gyn. Resident ob-gyn. Tex. Tech. U., Lubbock, Tex., 1984-87; chief resident ob-gyn. John Peter Smith Hosp., Ft. Worth, 1987-88; pvt. practice ob-gyn. Ft. Worth, 1988—. Fellow Am. Coll. Ob.-Gyn., mem. AMA, Tex. Med. Assn. Office: 6100 Harris Pky Ste 250 Fort Worth TX 76132-4107

SUBASIC, CHRISTINE ANN, architectural engineer; b. Erie, Pa., June 30, 1966; d. William Frederick and Joan Dorothy (Hoehn) Callista; m. Shawn Peter Subasic, June 3, 1989. B of Archtl. Engring. with distinction, Pa. State U., 1989. Engr. in tng. Jr. design engr. Cagley & Assocs., Rockville, Md., 1988, design engr., 1989-91; staff engr. Brick Inst. Am., Reston, Va., 1991—; mem. Passive Solar Industries Coun., Washington, 1992—. Contbr. articles to jours., papers to proceedings. Vol. Students Engaged in Engring., Washington, 1990-91. Scholar Pa. State U., 1981, 84-88, Beyers Panhellenic scholar Pa. State Panhellenic Coun., 1987. Mem. ASCE (assoc.), Am. Soc. for Testing and Materials, Tau Beta Pi, Phi Mu. Office: Brick Inst Am 11490 Commerce Park Dr Reston VA 22091

SUBBUSWAMY, MUTHUSWAMY, environmental engineer, researcher, consultant; b. Ramnad, India, June 5, 1938; came to U.S., 1980; s. Subbuswamy and Lakshmi S.; m. Karpagam, Sept. 14, 1969; children: Saishree, Padmalakshmi. BS, U. Madras, India, 1961; MS, U. Madras, 1972, U. Miami, Fla., 1983; PhD, U. Miami, 1990. Civil environ. engr. Pub. and Mcpl. Works, Madras, 1963-73; asst. prof. U. Madras, 1973-78; scientist Nat. Environ. Engr. Rsch. Inst., Madras, 1978-81; teaching and rsch. U. Miami, Fla., 1981-89; rsch. assoc. Fla. Atlantic U., Boca Raton, Fla., 1989-91; cons., environ. engr. Woodward-Clyde Cons., Tallahassee, Fla., 1991—; mem. Am. Soc. Civil Engrs., Miami, Fla., 1993—. Contbr. articles to profl. jours. Recipient Fellowship award U. Miami, Fla., 1983; grantee Dept. Edn., Fla. State, 1989. Mem. Am. Soc. Indsl. Engrs. Achievements include research in aquaculture tratment for waste water; development of rotor mechanism using bamboo blades for waste water treatment; design of weight lowering device for manual materials handling tasks, effect of plastic/solid wastes in anaerobic digestion. Home: 2300 B High Rd Tallahassee FL 32303 Office: Woodward-Clyde Consultants 3676 Hartsfield Rd Tallahassee FL 32303

SUBIRANA, JUAN ANTONIO, polymer chemist, educator; b. Barcelona, Spain, July 14, 1936; s. Juan B. and Maria (Torrent) S.; m. Mercedes Vilanova (div.); children: Miriam, Brian; m. J. Lourdes Campos, 1983; children: Marc, Lluis. PhD in Chemistry, U. Madrid, 1960, PhD in Chem. Engring., 1963. Postdoctoral fellow Harvard U., Cambridge, Mass., 1961-63; prof. chem. engring. U. Politecnica Catalunya, Barcelona, 1966—. Recipient Narcis Monturiol medal Catalunya Govt., 1986. Office: Dept Ingenieria Quimica-Etseib, Diagonal 647, E-08028 Barcelona Spain

SUBRAMANIAN, CHELAKARA SURYANARAYANAN, mechanical and aerospace engineering educator; b. Bombay, Dec. 24, 1950; came to U.S., 1988; s. Chelakara V. and Janaky (Subramanian) Suryanarayanan; m. Prabha Seshadri, Jan. 24, 1979; 1 child, Ajay. BE in Mech. Engring., Bangalore U., 1973; ME in Aerospace Engring., Indian Inst. Sci., 1975; PhD in Mech. Engring., U. Newcastle, 1983. Registered profl. engr., U.K.; internat. profl. engr. Sr. rsch. fellow Indian Inst. Sci., Bangalore, 1975-77; rsch. fellow, teaching asst. U. Newcastle, New South Wales, Australia, 1977-81; rsch. asst. Imperial Coll. Sci. and Tech., London, 1982-86; sr. rsch. engr. British Maritime Tech. Ltd., Newcastle, U.K., 1986-88; adj. rsch. prof. Naval Postgrad. Sch., Monterey, Calif., 1988-91; assoc. prof. Fla. Inst. Tech., Melbourne, 1992—; NSF, interanat. jour. reviewer NSF/ASME Transactions, 1988—; cons. Columbia Mech. Constrn., San Francisco, 1990—, IBM Corp., Endicott, N.Y., 1990—. Contbr. chpts. to books and articles to profl. jours. Co-ordinating mem. Indian Acads. in Am., N.Y., 1991-92. Recipient Sir Banco Gold medal Bangalore (India) U., 1973, Postgrad. fellowship Indian Inst. Sci., Bangalore, 1973-75, Rsch. fellowship U. Newcastle, New South Wales, 1977-81, Postdoctoral Rsch. fellowship Sci. and Engring. Rsch., London, 1982-86, NASA/Inst. for Computer Applications in Sci. and Engring. Faculty fellowship, Langley Rsch. Ctr., Va., 1992. Mem. AIAA, ASME, Am. Phys. Soc., Aero. Soc. India (assoc. mem.), Soc. Engrs. U.K. (corp. mem.). Achievements include rsch. in computer detection of large scale organized motions in a turbulent boundary layer has only limited validity. Home: Lockmar Estates 581 Vermount Rd NE Palm Bay FL 32907 Office: Fla Tech Dept Mech and Aerospace Engring 150 W University Blvd Melbourne FL 32901

SUCHORA, DANIEL HENRY, mechanical engineering educator, consultant; b. Youngstown, Ohio, Dec. 2, 1945; s. Stanley and Stella (Tocicki) S.; m. Patricia Ann Sanders, Sept. 2, 1968; children: Kevin, Sherri, Matthew. B of Engring., Youngstown State U., 1968, MS, 1970; PhD, Case Western Res. U., 1973. Registered profl. engr., Ohio. Project engr. Ajax Magnethermic Corp., Warren, Ohio, 1968-70; chmn. engring. tech. Ctrl. Ohio Tech. Coll., Newark, 1973-75; prof. mech. engring. Youngstown State U., 1975—; cons. Simmers Engring. and Crane Co., Canfield, Ohio, 1976—, Dexter Pultrusions Corp., Aurora, Ohio, 1976—. Bd. dirs. Boardman (Ohio) Community Baseball, 1992—; exec. v.p. Boardman Little League, 1983-91. Recipient fellowship NSF, 1970-73, Roemer prize in mech. engring. Henry Roemer Found., 1968. Mem. ASME (exec. com. 1975—), Sigma Xi (treas. 1991—). Roman Catholic. Home: 617 Oakridge Dr Boardman OH 44512 Office: Youngstown State Univ Dept Mech Engring Youngstown OH 44555

SUCHOW, LAWRENCE, chemistry educator, researcher, consultant; b. N.Y.C., June 24, 1923; s. Arthur David and Gussie (Rosen) S.; m. Rosalyn Hirsch, Jan. 21, 1968. BS, CCNY, 1943; PhD, Poly. Inst. of Bklyn., 1951. Analytical chemist Aluminum Co. of Am., Maspeth, N.Y., 1943-44; rsch. chemist Baker & Co. Inc., Newark, 1944, 46-47, Manhattan Project, Oak Ridge, Tenn., 1945-46, U.S. Army Signal Corps Engring. Labs., Fort Monmouth, N.J., 1950-54, Francis Earle Labs., Peekskill, N.Y., 1954-58; sr. rsch. chemist Westinghouse Elec. Corp., Bloomfield, N.J., 1958-60; mem. rsch. staff IBM T.J. Watson Rsch. Ctr., Yorktown Heights, N.Y., 1960-64; prof. chemistry N.J. Inst. of Tech., Newark, 1964-91, prof. emeritus, 1991—; mem. faculty Imperial Coll. Sci. and Tech., London, 1974. Author 1 chpt. in book; contbr. articles to profl. jours. With U.S. Army, 1944-46. Grantee (3) NSF, 1967-74, (2) N.J. Commn. on Sci. and Tech., 1988-89. Fellow N.Y. Acad. Sci.; mem. Am. Chem. Soc., Sigma Xi, Phi Lambda Upsilon. Achievements include rsch. on high-temperature inorganic reactions, physical properties of solids, X-ray crystallography, crystal growth, semiconductors, thin films, phosphors, nacreous pigments, rare earths, high Tc superconductors. Home: 6 Horizon Rd Fort Lee NJ 07024 Office: NJ Inst Tech Chemistry Div Newark NJ 07102

SUCHY, SUSANNE N., nursing educator; b. Windsor, Ont., Can., Sept. 20, 1945; d. Hartley Joseph and Helen Viola (Derrick) King; m. Richard Andrew Suchy, June 24, 1967; children: Helen Marie, Hartley Andrew, Michael Derrick. Diploma, St. Joseph Sch. Nursing, Flint, Mich., 1966; BSN, Wayne State U., 1969, MSN, 1971. RN, Mich. Afternoon supr., staff nurse operator and recovery rm. St. John Hosp., Detroit, 1966-70; nursing instr. Henry Ford Community Coll., Dearborn, Mich., 1972—, on leave 1988-90; coord. Dyad Project Harper Hosp., Detroit, 1990-92, clin. nurse specialist/oncology case mgr., 1988—; mem. Detroit Demonstration Site Team for defining and differentiating ADN/BSN competencies, 1983-87. Contbr. articles to profl. jours. Past bd. dirs., pres. St. Pius Sch. Mem. ANA, AACH, N.Am. Nursing Diagnosis Assn. (bylaws com. 1991-93, bylaws dir. 1993—), Mich. Nursing Diagnosis Assn. (pres. 1987-90, elected by-law chairperson 1991-92, treas. 1993—), Nat. League Nursing, Detroit Dist. Nurses Assn. (chmn. nominating com., legis. com.), Oncology Nursing Soc. (presenter abstract ann. conf. 1991, poster presentations ann. conf. 1991, 92, govt. relations chair 1992—), Daus. of Isabella (internat. dir., past regent, state treas.), Wayne State U. Alumni Assn., Sigma Theta Tau (nominating com. 1991-93). Roman Catholic. Home: 12666 Irene St Southgate MI 48195-1765 Office: Henry Ford Community Coll 5101 Evergreen Rd Dearborn MI 48128-1495

SUDAN, RAVINDRA NATH, electrical engineer, physicist, educator; b. Chineni, Kashmir, India, June 8, 1931; came to U.S., 1958, naturalized, 1971; s. Brahm Nath and Shanti Devi (Mehta) S.; m. Dipali Ray, July 3, 1959; children: Rajani, Ranjeet. B.A. with first class honors, U. Punjab, 1948; diploma, Indian Inst. Sci., 1952, Imperial Coll., London, 1953; Ph.D., U. London, 1955. Engr., Brit. Thomson-Houston Co., Rugby, Eng, 1955-57; Engr. Imperial Chem. Industries, Calcutta, India, 1957-58; research asso. Cornell U., Ithaca, N.Y., 1958-59; asst. prof. elec. engring. Cornell U., 1959-63, asso. prof., 1963-68, prof., 1968-75, IBM prof. engring., 1975—, dir. Lab. Plasma Studies, 1975-85, dep. dir. Cornell Theory Ctr., 1985-87; cons. Lawrence Livermore Lab., Los Alamos Sci. Lab., Sci. Applications Inc., Physics Internat. Co.; vis. research asso. Stanford U., summer 1963; cons. U.K. Atomic Energy Authority, Culham Lab., summer 1965; vis. scientist Internat. Center Theoretical Physics, Trieste, Italy, 1965-66, summers 1970, 73, Plasma Physics Lab. Princeton U., 1966-67, spring 1989, Inst. for Advanced Study, Princeton, N.J., spring 1975; head theoretical plasma physics group U.S. Naval Research Lab., 1970-71, sci. adviser to dir., 1974-75; chmn. Ann. Conf. on Theoretical Aspects of Controlled Fusion, 1975, 2d Internat. Conf. on High Power Electron and Ion Beam Research and Tech., 1977. Mem. editorial bd. Physics of Fluids, 1973-76, Comments on Plasma Physics, 1973—, Nuclear Fusion, 1976-84, Physics Reports, 1990—; co-editor Handbook of Plasma Physics; contbr. over 200 articles to sci. jours. Fellow IEEE, AAAS, Am. Phys. Soc. (Maxwell prize 1989), Nat. Rsch. Coun. (chmn. Plasma Sci. com. 1992-94). Achievements include patents (with S. Humphries, Jr) intense ion beam generator. Office: Cornell Univ 369 Upson Hall Ithaca NY 14853

SUDHAKARAN, GUBBI RAMARAO, physicist, educator; b. Sathnur, Karnataka, India, May 29, 1949; came to U.S., 1978; s. Gubbi Ramarao and Chandramma S.; m. Pushpa V. Doddaballapur, July 12, 1983; children: Sunil, Shaan. MS, Cen. Coll., Bangalore, 1972; PhD, U. Idaho, 1982. Robert A. Welch Found. postdoctoral fellow Tex. Tech. U., Lubbock, 1982-83; asst. prof. U. Idaho, Moscow, 1983-88; assoc. prof. physics SUNY, Oswego, 1988—. Contbr. articles to profl. publs. Mem. Am. Phys. Soc., Sigma Xi. Home: Box 146 Castle Dr Oswego NY 13126 Office: SUNY Dept Physics Oswego NY 13126

SUDIJONO, JOHN LEONARD, physicist; b. Medan, Indonesia, Aug. 25, 1966; s. Alexander and Monica (Megawati) S. BSc in Physics and Math., U. Tex., 1988; MSc in Physics, U. Mich., 1991, PhD in Physics, 1993. Rsch. asst. physics dept. U. Mich., Ann Arbor, 1988—. Named M. J. Reiger Physics scholar, 1985-87, Hughes Physics scholar, 1987-88; recipient AVS student prize award for Best Rsch., 1993. Mem. IEEE, Materials Rsch. Soc., N.Y. Acad. Sci., Am. Phys. Soc., Optical Soc. Am., Phi Beta Kappa, Phi Kappa Phi. Achievements include research in scanning force and scanning tunneling microscopes; in surface science with molecular beam epitaxy on compound semiconductors. Office: U Mich Physics Dept 1049 Randall Lab Ann Arbor MI 48109-1120

SUESS, JAMES FRANCIS, clinical psychologist; b. Evanston, Ill., Aug. 8, 1950; s. James Francis and Rae Love (Miller) S.; m. Linda Grace Powell, July 31, 1976; 1 child, Misty Lynne. BS, U. So. Miss., 1974, MS, 1978, PhD, 1982. Lic. psychologist, N.Y.; diplomate Am. Bd. Profl. Psychology, Am. Bd. Med. Psychotherapists. Assoc. psychologist State of Miss., Ellisville, 1978-80; clin. psychologist SUNY Med. Sch./Erie County Med. Ctr., Buffalo, 1982-84, supervisiong clin. psychologist, 1984-87, asst. dir., 1987—; dir. practica SUNY Med. Sch., 1982-90, faculty counsel, 1988—; cons. Buffalo Dept. Social Svcs., 1985—; mem. speakers bur. Erie Alliance for Mentally Ill, 1986—; vis. prof. U. Guadalajara Sch. Medicine, 1985—. Author: Annotated Bibliography of Sex Roles, 1972, Personality Disorder and Self Psychology, 1991; contbr. articles to refereed jours. including Perceptual and Motor Skills, Jour. Clin. and Consulting Psychology, Am. Annals of Deaf. With USAR, 1969-76. Fellow Am. Orthopsychiat. Assn., Soc. Personality Assessment; mem. Am. Psychol. Assn. Home: 3348 Staley Rd Grand Island NY 14072-2021 Office: Erie County Med Ctr 462 Grider St Buffalo NY 14215-3021

SUFIT, ROBERT LOUIS, neurologist, educator; b. Washington, July 16, 1950; s. Herbert and Alice Elizabeth (Rotzsch) S.; m. Diana Dills, Apr. 17, 1976 (div. Dec. 1988); children: Ben, Jessica, Alexandra. BA, Johns Hopkins U., 1972, MA, 1972; MD, U. Va., 1976. Diplomate Am. Bd. Psychiatry and Neurology, Am. Bd. Electrodiagnostic Medicine. House officer U. Pitts., 1977-80; post doctoral fellow U. Colo., Denver, 1980-82; asst. prof. U. Wis., Madison, 1982—. Contbr. 40 articles to profl. jours. and chpts. to books. Grantee NIH. Fellow Am. Acad. Neurology. Home: 564 Toepfer Ave Madison WI 53711-1666 Office: U Wis 600 Highland Ave Madison WI 53792-0001*

SUFKA, KENNETH JOSEPH, psychology educator; b. Mason City, Iowa, Apr. 15, 1960; s. Lawrence Reuben and Rose Elizabeth (Wojtanowicz) S.; m. Alice Eileen Von Wald, Dec. 19, 1987. BS, Iowa State U., 1986, MS, 1988, PhD, 1990. Postdoctoral fellow in pharmacology Drake U., Des Moines, 1990-91; instr. physiology/pharmacy U. Osteo. Medicine and Health Scis., Des Moines, 1990-91; temp. asst. prof. psychology Iowa State U., Ames, 1991-92; asst. prof. psychology U. Miss., Oxford, 1992—; reviewer, referee Physiology and Behavior, San Antonio, 1990—, Pharmacology Biochemistry and Behavior, San Antonio, 1990—. Contbr. articles to profl. jours. Faculty advisor U. Miss. chpt. Habitat for Humanity, Oxford, 1992. Mem. Am. Psychology Soc., Soc. for Neurosci., Internat. Behavioral Neurosci. Soc. Democrat. Roman Catholic. Office: U Miss Dept Psychology University MS 38677

SUGA, HIROSHI, chemistry educator; b. Kyoto, Japan, Feb. 28, 1930; s. Kazutaka and Sekae Suga; m. Hiroko Hayashi, Apr. 28, 1959; children: Kiyoshi, Takeshi, Yasushi, Eiko. BS, Osaka (Japan) U., 1953, DSc, 1960. Rsch. assoc. Osaka U., 1958-63, lectr., 1963-67, assoc. prof., 1967-79, prof., 1979-93, prof. emeritus, 1993—; vis. prof. Zhejiang U., Hangzhou, People's Republic of China, 1987; dir. chem. thermodynamics lab. Osaka U., 1984-89, dir., microcalorimetry rsch. ctr., 1989. Ministry of Edn. fellow, 1969; Brit. Coun. grantee, 1989-91; recipient Meml. award U.S. Calorimetry Conf., 1992. Mem. N.Y. Acad. Scis. (com. chem. thermodynamics 1982—, nat. com. chemistry 1988—), Soc. Calorimetry and Thermal Analysis (Japan) (pres. 1991—). Home: 17-39 Minoo 4, Minoo 562, Japan Office: Kinki URsch Inst Sci & Tech, Kowakae, Higashi Osaka 577, Japan

SUGAWARA, ISAMU, molecular pathology educator; b. Furenai, Hokkaido, Japan, June 25, 1949; s. Toichi and Yoshi (Veda) S.; m. Ryoko Takagi, May 18, 1980; children: Yu, Sachiko. MD, Hokkaido U., Sapporo, 1976; PhD, Tokyo U., 1980. Intern, 1976-80; with faculty of medicine Tokyo U., 1980; resident, 1980-83; lectr. Nara Med. U., 1983-85; sr. rsch. scientist Hoechst Japan Rsch. and Devel. Labs., Tokyo, 1985-87; lectr. Inst. Med. Sci. Tokyo U., 1987-90; assoc. prof. Saitama Med. Sch., Kawagoe, 1990—; nonregular lectr. Nara Med. U., 1986—; rsch. assoc. Karolinska Inst., Stockholm, 1980-82, U. Alta., Edmonton, Can., 1982-83. Author: English Paper Writing, 1991 (Cancer Rsch. award 1990), Molecular Pathology, 1991. Translator Japan Translation Fedn., Tokyo, 1990. Named Cancer Rsch. mem. Ministry of Edn., Tokyo, 1991. Mem. Japan Pathol. Soc., Am. Assn. Cancer Rsch., Am. Assn. Immunologists, Japan Cancer Assn., Japan Immunol. Soc., Internat. Immunopharm. Soc. Buddhist. Avocations: golf, classical music, reading, mountain climbing, baseball. Home: 1-3-8 Kurihara, Niiza, Saitama 352, Japan Office: Saitama Med Sch, 1981 Kamoda, Tsujido Kawagoe, Saitama 350, Japan

SUGAWARA, TAMIO, chemist; b. Shihoro, Hokkaido, Japan, Jan. 2, 1948; s. Hiroshi and Reiko (Nishioka) S.; m. Noriko Hashimoto, Feb. 24, 1972; children: Takeshi, Kumiko. B of Pharmacology, Kyoto Coll. of Sch., 1970; PhD, Osaka U., 1975. Rschr. Shionogi Rsch. Lab., Osaka, 1975-87, group leader, 1991—; group leader Drug Delivery System Inst., Noda, 1988-91; rsch. assoc. U. Montreal, Can., 1982-84. Contbr. articles to Carbohydrate Rsch. Home: 2-9-12 Mukogaoka, Sanda Sanda Hyogo 669-13, Japan Office: 5-12-4 Sagisu, Osaka 553, Japan

SUGERMAN, ABRAHAM ARTHUR, psychiatrist; b. Dublin, Ireland, Jan. 20, 1929; came to U.S., 1958, naturalized, 1963; s. Hyman and Anne (Goldstone) S.; m. Ruth Nerissa Alexander, June 5, 1960; children: Jeremy, Michael, Adam, Rebecca. B.A., Trinity Coll., Dublin, 1950; M.B., B. Chir., B.A.O., Trinity Coll., 1952; D.Sc., SUNY, Bklyn., 1962. Diplomate: Am. Bd. Psychiatry and Neurology. House officer Meath Hosp., Dublin, 1952-53, St. Nicholas Hosp., London, 1953-54; sr. house physician Brook Gen. Hosp., London, 1954; registrar in psychiatry Kingsway Hosp. Derby and Kings Coll. Med. Sch., Newcastle, Eng., 1955-58; clin. psychiatrist Trenton (N.J.) Psychiat. Hosp., 1958-59; cons. psychiatry, 1964-80; research fellow Downstate Med. Center, Bklyn., 1959-61; chief investigative psychiatry sect. N.J. Bur. Research, Princeton, 1961-73; cons. research, assoc. psychiatrist Carrier Clinic, Belle Mead, N.J., 1968-72, 78-90; dir. outpatient services Carrier Clinic, 1972-74, 77-78, med. dir., 1974-77; dir. research Carrier Found., Belle Mead, 1972-79; med. dir. addiction recovery services Community Mental Health Center, U. Medicine and Dentistry of N.J., Piscataway, 1990—; cons. psychiatry Med. Center, Princeton, N.J., 1972—; clin. assoc. prof. psychiatry Rutgers Med. Sch., New Brunswick, N.J., 1972-78; clin. prof. Robert Wood Johnson Med. Sch. (formerly Rutgers Med. Sch.), 1978—; vis. prof. Rutgers Center for Alcohol Studies, 1977-83, Hahnemann Med. Coll., Phila., 1978—; contbg. faculty Grad. Sch. Applied and Profl. Psychology, Rutgers U., 1974-78. Editor: (with Ralph E. Tarter) Alcoholism: Interdisciplinary Approaches to an Enduring Problem, 1976, Expanding Dimensions of Consciousness, 1978; contbr. articles to profl. jours. Bd. dirs. N.J. Mental Health Research and Devel. Fund, Princeton, 1968-74; v.p. Jewish Family Service, Trenton, 1972-78; 1st v.p. Trenton Hebrew Acad., 1972-75. Fellow AAAS, Am. Psychiat. Assn., Am. Coll. Neuropsychopharmacology, Am. Coll. Clin. Pharmacology, Am. Coll. Psychiatrists, Royal Coll. Psychiatrists; mem. AMA, Soc. Biol. Psychiatry, Assn. Research Nervous and Mental Diseases, Eastern Psychol. Assn., Am. Med. EEG Assn., Am. Soc. Addiction Medicine. Home: 125 Roxboro Rd Lawrenceville NJ 08648-3998 Office: Community Mental Health Ctr Piscataway NJ 08855

SUGIKI, SHIGEMI, ophthalmologist, educator; b. Wailuku, Hawaii, May 12, 1936; s. Sentaro and Kameno (Matoba) S.; AB, Washington U., St. Louis, 1957, M.D., 1961; m. Bernice T. Murakami, Dec. 28, 1958; children: Kevin S., Boyd R. Intern St. Luke's Hosp., St. Louis, 1961-62, resident ophthalmology, Washington U., St. Louis, 1962-65; chmn. dept. ophthalmology Straub Clinic, Honolulu, 1965-70, Queen's Med. Ctr., Honolulu, 1970-73, 80-83, 88-90, 93—; assoc. clin. prof. ophthalmology Sch. Medicine, U. Hawaii, 1973—. Served to maj. M.C., AUS, 1968-70. Decorated Hawaiian NG Commendation medal, 1968. Fellow ACS; mem. Am., Hawaii med. assns., Honolulu County Med. Soc., Am. Acad. Ophthalmology, Contact Lens Assn. Opthalmologists, Pacific Coast Oto-Ophthal. Soc., Pan-Pacific Surg. Assn., Am. Soc. Cataract and Refractive Surgery, Am. Glaucoma Soc., Internat. Assn. Ocular Surgeons, Am. Soc. Contemporary Ophthalmology, Washington U. Eye Alumni Assn., Hawaii Ophthal. Soc., Rsch. To Prevent Blindness. Home: 2398 Aina Lani Pl Honolulu HI 96822-2024 Office: 1380 Lusitana St Ste 714 Honolulu HI 96813-2449

SUGIYAMA, KAZUNORI, music producer; b. Tokyo, Aug. 18, 1950; came to U.S., 1976; s. Hiroshi and Michiko (Maeda) S.; m. Emi Fukui, Aug. 11, 1981. BS, Waseda U., 1974, postgrad., 1974-75; MA, Boston U., 1977. Jr. adminstrv. officer Japanese Mission to UN, N.Y.C., 1978-88; rep. N.Y. Toshiba EMI Records, Jazz Div., Tokyo, 1990—; rep. U.S. U.S. D.I.W. Records, Tokyo, 1991—; corr. Jazz Life, Tokyo, 1980-88; columnist OCS News, N.Y.C., 1982-90. Recording engr. (album) Bud and Bird/Gil Evans, 1988 (Grammy); producer V/Ralph Paterson, 1990 (Jazz Album of Yr.); co-producer The Nurturer/Geri Allen, 1991 (2d pl. Jazz Album of Yr.), Big Band & Quartet/David Murray, 1992 (Best Prodn. Jazz Album of Yr.); translator Autobiography of Miles Davis, 1989. Mem. Nat. Acad. Recording Arts and Scis. Avocations: tennis, travel. Office: 93 Mercer St 3W New York NY 10012

SUHRBIER, KLAUS RUDOLF, hydrodynamicist, naval architect; b. Gnoien, Germany, Sept. 12, 1930; arrived in U.K., 1966; s. Ulrich Julius and Dora Auguste (Elsaesser) S.; m. Inge Ursula Koepke, Oct. 1, 1955; children: Andreas, Karin. Dipl.Ing., U. Rostock, Germany, 1955. Chartered engr. Hydrodynamicist Institut fuer Schiffbau, Berlin, 1955-60, Versuchsanstalt fuer Binnenschiffbau e.V., Duisburg, Fed. Republic of Germany, 1960-63; sci. officer Inst. fuer Schiffbau U. Hamburg, Fed. Republic of Germany, 1963-66; sr. hydrodynamicist Vosper Ltd., Portsmouth, Eng., 1966; chief hydrodynamicist Vosper Thornycroft (U.K.) Ltd., Portsmouth, Eng., 1966-92; cons. ship hydrodynamics, 1992—; mem. cavitation com. 16th and 17th Internat. Towing Tank Conf., Leningrad, USSR, 1978-81, Gothenburg, Sweden, 1981-84; chmn. cavitation com. 18th Internat. Towing Tank Conf., Kobe, Japan, 1984-87; chmn. high speed marine vehicles com. Internat. Towing Tank Conf., Madrid, 1987-90. Co-author: (book) Dhows to Deltas, 1971; inventor reduction of cavitation erosion, 1974, 92; contbr. numerous articles and papers to profl. jours. and procs. Fellow Royal Soc. Naval

Architects; mem. Soc. Naval Architects and Marine Engrs., Schiffbautechnische Ges. e.V. Avocations: sailing, skiing, history. Home and Office: 30A Beach Rd, Emsworth, Hampshire PO10 7HR, England

SUIT, D. JAMES, civil engineer; b. Lewellen, Nebr., Aug. 29, 1951; s. D.W. (Donald William) and Ellen Leona (Green) S.; m. Catherine Lorraine Poppinga, Aug. 30, 1975. BCE, U. Nebr., 1974. Design engr. USDA Soil Conservation Svc., Lincoln, Nebr., 1974-75; civil engr. USDA Soil Conservation Svc., Lincoln, 1975-77; project engr. USDA Soil Conservation Svc., Scottsbluff, Nebr., 1977-84; asst. state constrn. engr. USDA Soil Conservation Svc., Bismarck, N.D., 1984-89; state conservation engr. USDA Soil Conservation Svc., Bozeman, Mont., 1989—. Mem. Jaycees, Scottsbluff, 1977-84. Recipient Eagle Scout award Boy Scouts Am., Oshkosh, Nebr., 1969. Mem. ASCE, Mont. Land Improvement Contractors Assn. Roman Catholic. Achievements include organizing a 475 member group into a structured scheduling process for constrn. activities in Mont., N.D., organizing tng. aids for all constrn. agys. in N.D., Mont. Home: 1220 S Black Bozeman MT 59715 Office: USDA Soil Conservation Svc 10 E Babcock St Bozeman MT 59715

SUITER, JOHN WILLIAM, industrial engineering consultant; b. Pasadena, Calif., Feb. 16, 1926; s. John Walter and Ethel May (Acton) S.; B.S. in Aero. Sci., Embry Riddle U., 1964; m. Joyce England, Dec. 3, 1952; children—Steven A., Carol A. Cons. indsl. engr., Boynton Beach, Fla., 1955—. Instr. U. S.C. Tech. Edn. Center, Charleston, 1967-69. Served as pilot USAF, 1944-46. Registered profl. engr., Fla. Mem. Am. Inst. Indsl. Engrs., Soc. Mfg. Engrs. (sr.), Computer and Automated Systems Assn., Methods-Time Measurement Assn. (assoc.), Soc. Quality Control. Home: PO Box 5262 Englewood FL 34224-0262

SUKANEK, PETER CHARLES, chemical engineering educator, researcher; b. N.Y.C., Sept. 15, 1947; s. Charles Anthony and Ruth Helena (Barry) S.; m. Kathleen Louise Lambert, Sept. 28, 1969; children: Stephen, Kevin, Jennifer. BChE, Manhattan Coll., 1968; MS, U. Mass., 1968, PhD, 1972. Project engr. USAF Rocket Propulsion Labs., Edwards, Calif., 1972-76; assoc. prof. Clarkson U., Potsdam, N.Y., 1976-90; staff engr. Philips Rsch. Labs., Eindhoven, Netherlands, 1986; prof., chair chem. engring. U. Miss., University, 1991—; cons. IBM, Burlington, Vt., 1982-90. Contbr. articles to profl. jours. Capt. USAF, 1972-76. Mem. Am. Inst. Chem. Engrs., Soc. for Rheology, Electrochem. Soc., Rotary Internat. Office: U Miss Dept Chem Engring University MS 38677

SUKOV, RICHARD JOEL, radiologist; b. Mpls., Nov. 13, 1944; s. Marvin and Annette Sukov; Susan Judith Grossman, Aug. 11, 1968; children: Stacy Faye, Jessica Erin. BA, BS, U. Minn., 1967, MD, 1970; student, U. Calif.-Berkeley, 1962-64. Diplomate Am. Bd. Radiology; lic. physician, Minn., D.C., Calif. Intern pediatrics U. Minn., Mpls., 1970-71; resident radiology UCLA Ctr. for Health Sci., 1973-76; fellow in ultrasound and computed tomography UCLA, 1976-77; staff radiologist Centinela Hosp. Med. Ctr., Inglewood, Calif., 1977-85, Daniel Freeman Marina Hosp. Med. Ctr., Marina del Rey, Calif., 1980—; dir. radiology Daniel Freeman Marina Hosp. Med. Ctr., Inglewood, Calif., 1988-90; asst. clin. prof. radiology UCLA Ctr. for Health Scis., 1977-83; adv. bd. Aerobics and Fitness Assn. Am., 1983—. Contbr. articles to profl. jours. Vol. Venice Family Clinic, 1985—. Lt. comdr. USPHS, 1970-72. U. Minn. fellow, 1964-65, 66, 70. Mem. AMA, Royal Soc. Medicine, Soc. Radiologists in Ultrasound (charter), Minn. Med. Alumni Assn., Los Angeles County Med. Assn., Calif. Med. Assn., Radiol. Soc. N.Am., L.A. Radiol. Soc. (continuing edn. com. 1990—, treas.), L.A. Ultrasound Soc., Am. Coll. Radiology. Avocations: skiing, jogging, travel. Office: Inglewood Radiology 323 N Prairie Ave Ste 160 Inglewood CA 90301-4597

SULKIN, STEPHEN DAVID, marine biology educator; b. Topeka, Aug. 14, 1944; s. Norman Manuel and Dorothy (Fine) S.; m. Shelly Sinclair Turpin, June 6, 1970; children: Kimberly, Tracy, Matthew. AB, Miami U., Oxford, Ohio, 1966; MS, Duke U., 1971, PhD, 1971. Rsch. asst. prof. Chesapeake Biol. Lab. U. Md., Solomons, 1971-76; asst. prof., head Horn Point Environ. Lab. U. Md., Cambridge, 1976-82, prof., 1984-85; prof., dir. Shanna Point Marine Ctr. Western Wash. U., Bellingham, 1985—. Editor Estuarian Coastal Shelf Sci., 1990—; contbr. articles to sci. publs. Recipient Olscamp Outstanding Rsch. award WWR, 1990. Mem. AAAS, Estuarian Rsch. Fedn. Home: 3605 Illinois Ln Bellingham WA 98226-4364 Office: Shannon Point Marine Ctr 1900 Shannon Point Rd Anacortes WA 98221-4042

SULLA, NANCY, computer coordinator, consultant, business owner; b. Phila., Sept. 15, 1955; d. Vincent F. and Margaret (Ether) S. BA in Elem. Edn., Fairleigh Dickinson U., 1977; MA in Computer Sci., Montclair (N.J.) State Coll., 1988. Cert. elem. edn. and data processing tchr., N.J., cert. supervisory. Tchr. 4th grade Ridgefield Park (N.J.) Bd. of Edn., 1977-78; tchr. maths. Ridgefield (N.J.) Bd. of Edn., 1978-81; tech. support mgr. Computer Scis. Corp., Lyndhurst, N.J., 1981-83; computer coord. Mahwah (N.J.) Bd. of Edn., 1983—; pres. Compuco, Inc., Ramsey, N.J., 1987—; developer Real Estate Sales Tracking System; co-developer LogoWriter Activities for Readers. N.J. State Dept. of Edn. grantee, 1986. Mem. N.Y. State Assn. for Computers and Tech. in Edn., N.J. Assn. for Ednl. Tech., Network for Action in Microcomputer Edn. Mem. Christian Ch. Avocations: music, cycling, tennis. Home: 93 Grove St Ramsey NJ 07446 Office: Mahwah Bd Edn Ridge Rd Mahwah NJ 07430-2021

SULLENTRUP, MICHAEL GERARD, structural engineer, consultant; b. Washington, Mo., May 15, 1958, s. William J. III and Ruth M. (Eckelkamp) S.; m. Glenda S. Brinker, Aug. 2, 1980; children: Jeremy M., Jane M. BSCE, U. Mo., 1980, MSCE, 1981. Rsch. asst. U. Mo., Columbia, 1979-80, rsch. specialist, 1980-81, teaching asst., 1981; quality control insp. Quinn Concrete Co., Marshall, Mo., 1980; engr. McDonnell Aircraft Co., St. Louis, 1982-85, sr. engr., 1987-89, lead engr., 1990—, recruiter, 1986—, product quality integrator, 1993. Contbr. articles to profl. jours.; author personal computer software. Tchr. Roman Cath. Ch., Hazelwood, Mo., 1986—; leader Boy Scouts Am., Hazelwood, 1989—. Mem. AIAA (sr.), ASTM, Digital Equipment Corp. User Soc., KC (officer, bd. dirs. Hazelwood, 1984—). Achievements include expertise in development and maintenance of relational database management systems, and in automation of fatigue and fracture mechanics analysis methods. Home: 7494 Naples Dr Hazelwood MO 63042 Office: McDonnell Douglas Aerospace MC 2702681 PO Box 516 Saint Louis MO 63166

SULLIVAN, JAMES, consultant; s. James E. and Kathern (Bilms) S.; m. Iris Rodriguez, June 11, 1965. BS in Indsl. Tech., Tenn. Tech. U., 1967; AS in Engring. Tech., Gratham Sch. Engring., 1976; MS, Clayton U., 1977; BT in Mfg. Engring., Elizabethtown Coll., 1977; MA in Bus. Mgmt., Ctrl. Mich. U., 1978; BS in Elec. Tech., Empire State Coll., 1980; MA in Indsl. Mgmt., Cen. Mich. U., 1980, MA in Personal Mgmt., 1982, BSBA, 1982; BS in Logistics and CIM, Empire State Coll., 1991; PhD in Adminstrn., Inst. Profl. Studies, 1992. Author: A Church in Error, 1971; contbr. articles to profl. jours. Mem. Am. Inst. Plant Engring., Svc. Engring. Soc., Soc. Mfg. Engring, ASHREA, ALOA, Sigma Iota Epsilon. Home: 240 Jean Wells Dr Goose Creek SC 29445

SULLIVAN, JAMES HARGROVE, JR., mechanical engineer, utility engineer; b. Carrollton, Ga., May 19, 1962; s. James Hargrove and Elaine (Scott) S.; m. Theresa Johnson, Jan. 26, 1985; children: Sean Paterick, Kayla Marie. BSME, So. Coll. Tech., 1985. Registered profl. engr., Ga.; cert. energy mgr.; chargered indsl. gas cons. Inst. Gas. Tech. Project engr. Cleveland Consol., Inc., Atlanta, 1985-87; HVAC (heating, ventilation and air conditioning) engr. Atlanta Gas Light Co., 1987-90, asst. divsn. indsl. engring., 1991, dir. HVAC engring., 1991—. Mem. ASHRAE (chair tech. com.), Ga. Soc. Profl. Engrs. Republican. Methodist. Office: Atlanta Gas Light Co PO Box 4569 Atlanta GA 30302

SULLIVAN, JAY MICHAEL, medical educator; b. Brockton, Mass., Aug. 3, 1936; s. William Dennis and Wanda Nancy (Kelpsh) S.; m. Mary Suzanne Baxter, Dec. 30, 1964; children: Elizabeth, Suzanne, Christopher. B.S. cum laude, Georgetown U., 1958, M.D. magna cum laude, 1962. Diplomate Nat. Bd. Med. Examiners, Am. Bd. Internal Medicine. Med. intern Peter Bent Brigham Hosp., Boston, 1962-63; resident Peter Bent Brigham Hosp., 1963-64, 66-67, chief resident, 1969-70, fellow in cardiology, 1964-66, dir. hypertension unit, 1970-74; Nat. Heart Inst. fellow, 1984, Med. Found. research fellow, 1967; preceptorship in biol. chemistry Harvard U. Med. Sch., Boston, 1967-69; asst. prof. medicine Harvard U. Med. Sch., 1970-74; dir. med. services Boston Hosp. for Women, 1973-74; prof. medicine, chief div. cardiovascular diseases U. Tenn. Coll. Medicine, Memphis, 1974—; vice-chmn. dept. medicine U. Tenn. Coll. Medicine, 1982-85; mem. staff Regional Med. Ctr., Memphis, VA, Bapt. Meml. hosps., U. Tenn. Medical Center-Wm. F. Bowld Hosp., Le Bonheur Children's Hosp., Saint Jude Children's Rsch. Hosp.; fellow Council for High Blood Pressure Research; cons. Nat. Heart, Lung and Blood Inst., 1974—, VA, 1983—. Contbr. articles to sci. jours. Served with M.C., U.S. Army, 1963-70. Fellow ACP (bd. govs.), Am. Coll. Cardiology (pres. Tenn. chpt.), Coun. on Circulation of Am. Heart Assn.; mem. AAAS, Assn. Univ Cardiologists, Assn. Profs. of Cardiology, Internat. Soc. Hypertension, Am. Fedn. Clin. Rsch., Am. Heart Assn. (pres. chpt. 1982-83, v.p. affiliate), Racquet Club of Memphis, Sigma Xi, Alpha Omega Alpha, Alpha Sigma Nu. Roman Catholic. Home: 6077 Maiden Ln Memphis TN 38120-3104 Office: 951 Court Ave Rm 353D Memphis TN 38163

SULLIVAN, JOHN FALLON, JR., government official; b. Washington, Sept. 6, 1935; s. John Fallon and Thelma (Simmons) S.; m. Eliza Johnstone Buchanan, Dec. 27, 1958; children: John Fallon III, James S. BSCE, Va. Mil. Inst., 1958. Registered profl. engr., Ala., Wis., Miss. Asst. area engr. Fed. Hwy. Adminstrn., Montgomery, Ala., 1961-62; area engr. Fed. Hwy. Adminstrn., Atlanta, 1963-67; hwy. engr. Fed. Hwy. Adminstrn., Washington, 1967-69; dist. engr. Fed. Hwy. Adminstrn., Madison, Wis., 1970-74; asst. div. adminstr. Fed. Hwy. Adminstrn., Tallahassee, 1974-78; div. adminstr. Fed. Hwy. Adminstrn., Dover, Del., 1978-80, Jackson, Miss., 1981—. 1st lt. U.S. Army, 1959. Mem. ASCE. Methodist. Home: 105 Sandra Cove Clinton MS 39056 Office: Fed Hwy Adminstrn 666 North St Ste 105 Jackson MS 39202

SULLIVAN, JOHN LAWRENCE, III, psychiatrist; b. Scranton, Pa., Nov. 24, 1943; s. John Lawrence and Jane Marie (Hoppel) S.; m. Paula Ann DeRemer, Mar. 3, 1979; 1 child, John Bradley. BA, Duke U., 1965; MD, Johns Hopkins U., 1969. Diplomate Am. Bd. Psychiatry and Neurology. Asst. prof. psychiatry Duke U., Durham, N.C., 1973-78, assoc. prof. of psychiatry, 1978-81; dep. dir. med. rsch. Dept. Vet. Affairs Cen. Office, Washington, 1981-83; chmn. dept. psychiatry Richard L. Roudebush VA Med. Ctr., Indpls., 1985—; prof. psychiatry and of neurobiology Ind. U., Indpls., 1985—; mem. Nat. Coun. on Health Care Tech., Washington, 1981-82. Co-editor: Foundations of Biochemical Psychiatry, 1976; editor: Biomedical Psychiatric Therapeutics, 1984; contbr. articles to profl. jours. Sr. fellow Duke U. Ctr. for Study of Aging and Human Devel., Durham, 1975; recognition for Disting. Pub. Svc. Dept. of Vet. Affairs, Washington, 1984. Fellow Am. Psychiat. Assn.; mem. Soc. Biol. Psychiatry, AAAS, Acad. Psychosomatic Medicine. Achievements include research interests include the neurobiology of monoamine oxidase and medical psychiatry. Office: RL Roudebush VA Med Ctr 1481 W 10th St Indianapolis IN 46202

SULLIVAN, KATHRYN D., geologist, astronaut; b. Paterson, N.J., Oct. 3, 1951; d. Donald P. and Barbara K. Sullivan. BS in Earth Scis., U. Calif., Santa Cruz, 1973; PhD in Geology, Dalhousie U., Halifax, N.S., Can., 1978, Dr. (hon.), 1985. With NASA, 1978—, astronaut, 1979—, mission specialist flight STS-41G, 1984, mission specialist flight STS-31, 1990; adj. prof. Rice U., Houston, 1985; mem. Nat. Commn. on Space, 1985—; mem. exec. panel Chief of Naval Ops., 1988—; first Am. woman to perform extra-vehicular activity. Lt. comdr. USNR. Recipient Space Flight medal NASA, 1984, 90, Exceptional Svc. medal, 1985, 91, Nat. Air and Space Mus. trophy Smithsonian Instn., 1985, Haley Space Flight award AIAA, 1991. Mem. AIAA, Geol. Soc. Am., Am. Geophys. Union, Soc. Women Geographers, Explorers Club, Sierra Club. Address: NASA Johnson Space Ctr Astronaut Office Houston TX 77058

SULLIVAN, LAWRENCE JEROME, aerospace engineer; b. Fall River, Mass., Sept. 30, 1953; s. Frederick Jeremiah and Mary Octavia (Gallagher) S. BS in Aerospace Engring., Cath. U., Washington, 1975. Rsch. asst. U. Mass., Amherst, 1978-79; design and stress analyst Structural Composites Inc., Azusa, Calif., 1979-81; stress and field engr. Martin Marietta Corp., Vandenberg AFB, Calif., 1981-84; R & D engr. specialist Rohr Inc., Chula Vista, Calif., 1984—. Mem. AIAA, Soc. for Advancement of Material and Process Engring. Home: Apt # 236 601 Telegraph Canyon Rd Chula Vista CA 91910 Office: Rohr Inc PO Box 878 Chula Vista CA 91912

SULLIVAN, LOUIS WADE, former secretary health and human services, physician; b. Atlanta, Nov. 3, 1933; s. Walter Wade and Lubirda Elizabeth (Priester) S.; m. Eve Williamson, Sept. 30, 1955; children: Paul, Shanta, Halsted. B.S. magna cum laude, Morehouse Coll., Atlanta, 1954; M.D. cum laude, Boston U., 1958. Diplomate: Am. Bd. Internal Medicine. Intern N.Y. Hosp.-Cornell Med. Ctr., N.Y.C., 1958-59, resident in internal medicine, 1959-60; fellow in pathology Mass. Gen. Hosp., Boston, 1960-61; rsch. fellow Thorndike Meml. Lab. Harvard Med. Sch., Boston, 1961-63; instr. medicine Harvard Med. Sch., 1963-64; asst. prof. medicine N.J. Coll. Medicine, 1964-66; co-dir. hematology Boston U. Med. Ctr., 1966; assoc. prof. medicine Boston U., 1968-74; dir. hematology Boston City Hosp., 1973-75; also prof. medicine and physiology Boston U., 1974-75; dean Sch. Medicine, Morehouse Coll., Atlanta, 1975-89, pres., until 1989, 1993—; sec. Dept. of Health and Human Svcs., Washington, 1989-93; mem. sickle cell anemia adv. com. NIH, 1974-75; ad hoc panel on blood diseases Nat. Heart, Lung Blood Disease Bur., 1973, Nat. Adv. Rsch. Coun., 1977; mem. med. adv. bd. Nat. Leukemia Assn., 1968-70, chmn., 1970; researcher suppression of hematopoiesis by ethanol, pernicious anemia in childhood, folates in human nutrition. John Hay Whitney Found. Opportunity fellow, 1960-61; recipient Honor medal Am. Cancer Soc., 1991. Mem. Am. Soc. Hematology, Am. Soc. Clin. Investigation, Inst. Medicine, Phi Beta Kappa, Alpha Omega Alpha. Episcopalian. Office: Morehouse Sch Medicine Office of the Pres 720 Westview Dr SW Atlanta GA 30310

SULLIVAN, NEIL SAMUEL, physicist, researcher, educator; b. Wanganui, Wellington, N.Z., Jan. 18, 1942; came to U.S., 1983; s. Reynold Richard and Edna Mary (Alger) S.; m. Robyn Annette Dawson, Aug. 28, 1965; children: Raoul Samuel, Robert Alexander and David Charles (twins). BSc with 1st class honors, U. Otago, N.Z., 1964, MSc in Physics, 1965; PhD in Physics, Harvard U., 1972. Postdoctoral rsch. Centre d'Etudes Nucleaires, Saclay, France, 1972-74, rsch. physicist, 1974-82; prof. physics U. Fla., Gainesville, 1982—, chair physics dept., 1989—; co-prin. US Nat. High Magnetic Field Lab., 1991. Contbr. numerous articles on quantum solids and nuclear magnetism to profl. jours., 1971—. Recipient prix Saintour, College de France, Paris, 1978, prix LaCaze, Academie des Sciences, Paris, 1982; Fulbright exch. grantee, 1965; Frank Knox Meml. fellow Harvard U., Cambridge, Mass., 1965-67. Mem. AAAS, Am. Assn. Physics Tchrs., Inst. Physics, Societe Francaise de Physique, European Phys. Soc., Am. Phys. Soc. Current work: Investigation of fundamental properties of solid hydrogen and solid helium at very low temperatures; studies of molecular motions using nuclear magnetic resonance; orientational disorder in molecular crystals, cryogenic detectors for dark matter particles and other cosmological relics of big bang theory; discovery of quadrupole glass phase of solid hydrogen, anomalous nuclear spin-lattice relaxation of solid 3He at interfaces; development of NMR techniques to study molecular dynamics at very low temperatures, quantum diffusion in solid hydrogen; design of ultra-sensitive low-noise cryogenic UHF detectors. Subspecialties: Condensed matter physics; Low temperature physics. Home: 4244 NW 76th Ter Gainesville FL 32606-4132

SULLIVAN, PAUL ANDREW, research electrical engineer; b. Sterling, Colo., June 22, 1944; s. William Edward Sullivan and Annie Olivia (Croll) Mueller; m. Joan Leslie Rasmussen, May 22, 1965 (div. May 1971); children: Kimberly Lynne, Heather Paige; m. Gerry Wood, Aug. 1, 1980. BS in Elec. Engring., Colo. State U., 1966; MS in Elec. Engring., U. So. Calif., 1968, PhD in Materials Sci., 1975. Mem. tech. staff surface sci. Hughes Rsch. Labs., Malibu, Calif., 1967-71, head micropattern replication, 1971-77, assoc. program mgr. Defense very high speed integrated circuits program, 1978-80; guest scientist Max Planck Inst. Solid State Rsch., Stuttgart, Germany, 1977-78; mgr. advanced devel. NCR Microelectronics, Ft. Collins, Colo., 1980-84, dir. digital signal processing, 1984-86; dir. engring. personal computer div. NCR, Clemson, S.C., 1987-89; dir. systems engring. CAD program Micrelectronics and Computer Tech. Corp., Austin, Tex., 1989-91; dept. head electronic packaging rsch. AT&T Bell Labs., Murray Hill, N.J., 1992—. Contbr. articles to profl. jours. Recipient Industrial Rsch. and Devel. 100 award Rsch. & Devel. Found., Chgo., 1985. Mem. IEEE (sr.), Assn. Computing Machinery, Sigma Xi, Eta Kappa Nu, Phi Kappa Phi. Achievements include patents for Apparatus Synchronizing an Opaque Video Tape with a Video Display, Strip Exposure Apparatus for Nucleation Medium, Alignment System and Method with Micromovement Stage, Hard X-Ray and Fluorescent X-Ray Detection of Alignment Marks for Precision Mask Alignment, Method and Apparatus for Mask-to-Wafer Gap Control in X-Ray Lithography, Process for Channeling Ion Beams, Method of Making CMOS by Twin-Tub Process Integrated with a Vertical Bipolar Transistor, Process for Fabricating a Bipolar Transistor with a Thin Base and an Abrupt Base-Collector Junction, Use of Selectively Deposited Tungsten for Contact Formation and Shunting Metallization, High Density, Low Power, Merged Vertical Fuse/Bipolar Transistor Device and Method of Fabrication. Office: AT&T Bell Labs 600 Mountain Ave Murray Hill NJ 07974

SULLIVAN, ROBERT SCOTT, architect; b. Alexandria, La., Sept. 8, 1955; s. Robert Wallace and Harriette Henri (Fedric) S. BA cum laude, Tulane U., 1979, BArch, 1979. Registered architect, N.Y., Calif.; cert. Nat. Coun. of Archtl. Registrations Bds. Staff architect Cavitt, McKnight, Weymouth, Inc., Houston, 1979-81, Hardy, Holzman, Pfeiffer Assocs., N.Y.C., 1981-83; ptnr. Sullivan, Briggs Assocs., N.Y.C., 1983-86; project architect Butler, Rogers, Baskett, N.Y.C., 1985-86; prin. R. Scott Sullivan AIA, Berkeley, Calif., 1986-89; ptnr. Talbott Sullivan Architects, Albany, Calif., 1989—; cons. Neometry Graphics, N.Y.C., 1983-86, dir, 1986—; dir. Middleton/Sullivan Inc., Alexandria, 1981—. Works include specific design projects at N.Y. Hist. Soc. exhibit Grand Cen. Terminal, N.Y.C., 1982, The Houston Sch. of Performing Visual Arts, 1980, The Pingry Sch., Bernards Twp., N.J., 1982, Arts Ctr. at Oak Knoll Sch., Summit, N.J., 1986. Vestry member St. Mark's Episc. Ch., Berkeley, 1988-89; bd. dirs. The Parsonage, Episcopal Diocese Calif., 1992—; cons. Commn. Accessibility, Episcopal Diocese Calif., 1991—. Mem. AIA, Calif. Council Architects, Archtl. League N.Y.C., Nat. Trust for Hist. Preservation, Victorian Soc. in Am., Tau Sigma Delta. Democrat. Episcopalian. Home: 1060 Sterling Ave Berkeley CA 94708 Office: 1323 Solano Ave Albany CA 94706

SULTZER, BARNET MARTIN, microbiology and immunology educator; b. Union City, N.J., Mar. 24, 1929; s. Moses Joseph and Florence Gertrude (Fischer) S.; m. Judith Ray Moreinis, Aug. 26, 1956; 1 child, Steven Bennett. BS, Rutgers U., 1950; MS, Mich. State U., 1951, PhD, 1958. Rsch. assoc. Princeton (N.J.) Labs., Inc., 1958-64; from asst. prof. to prof. microbiology SUNY, Bklyn., 1964—, interim chmn. dept. microbiology, 1980-82; vis. scientist Karolinska Inst., Stockholm, Sweden, 1971-72; vis. prof. Pasteur Inst., Paris, 1979-80. Assoc. editor Jour. of Immunology, 1983-86; contbr. book chpts. and over 50 articles to profl. jours. on microbiology and immunology; mem. editorial bd. Infection and Immunity, 1980-84. Pres. Tenants Assn. Gateway Pla., Manhattan, N.Y., 1990-92; mem. Community Bd. # 1, Manhattan, 1989—. 1st lt. USMC, 1952-55. Pres.'s fellow Am. Soc. Microbiology, 1957; grantee USPHS, NIH, Office of Naval Rsch., 1967—. Mem. AAAS, Am. Soc. Microbiology, Am. Assn. Immunologists, N.Y. Acad. Sci., Harvey Soc., Internat. Endotoxin Soc., Reticuloendothelial Soc., Sigma Xi. Achievements include patent for chemical detoxification of endotoxins and discovery of the genetic basis for mammalian responses to endotoxins including immunological and pathophysiological effects; developed first commercial immunological pregnancy test. Office: SUNY Health Sci Ctr 450 Clarkson Ave Brooklyn NY 11203-2098

SUMMERS, JAMES DONALD, agricultural engineering consultant; b. Hannibal, Mo., Apr. 29, 1955; s. Ralph Donald and Margaret Virginia (Howald) S.; m. Marcia Jean Dowding, Aug. 4, 1979; children: Andrew James. BS in Agrl. Engring., U. Mo., 1977, MS in Agrl. Engring., 1978; PhD in Engring., U. Nebr., 1983. Registered profl. engr., Nebr. Grad. rsch. asst. U. Mo., Columbia, 1977-78; asst. instr. U. Nebr., Lincoln, 1978-80, grad. rsch. asst., 1980-81; consulting engr. R.L. Large and Assocs., Inc., Lincoln, 1981-82; asst. prof. Okla. State U., Stillwater, 1982-86, assoc. prof., 1986-87; consulting engr. R.L. Large and Assocs., Inc., Lincoln, 1987—; mem. adv. com. Nat. Tchr. Certification, Lincoln, 1992. Contbg. author: (chpt.) Automotive Engineering and Litigation, Vol. III, 1989; contbr. articles to profl. jours. Trustee United Meth. Ch., Palmyra, Nebr., 1988—; tutor Adult Basic Edn., Palmyra, 1989—; mem. Dist. OR-1, Bd. Edn., Palmyra, 1988—. Mem. NSPE, Nat. Acad. Forensic Engrs., Am. Soc. Agrl. Engrs., Soc. Automotive Engrs. Office: RL Large and Assocs Inc 4333S 48th St Lincoln NE 68516

SUMMERS, MAX (DUANNE), entomologist, scientist, educator; b. Wilmington, Ohio, June 5, 1938; s. John William Summers and Helen Jane (Rolfe) Summers Kantner; children: Mark William Keith Dwayne; m. Sharon Braunagel, Dec. 28, 1991. AB magna cum laude, Wilmington Coll., 1962; PhD, Purdue U., 1968. Asst. prof. U. Tex., Austin, 1969-73, assoc. prof., 1973-75; prof. entomology Tex. A&M U., College Station, 1977-83, Disting. prof., 1983—, chair agrl. biotech., dir. Ctr. for Advanced Invertebrate Molecular Scis., Inst. Bioscis. and Tech., 1988—; vis. prof. U. Calif., Berkeley, 1976. Editor Virology Jour., 1983—; exec. editor Protein Expression and Purification; contbr. over 200 articles to profl. jours.; patentee baculovirus expression for vector system. Recipient J.V. Osmun Profl. Achievement award, 1988, President's award of honor Tex. A&M U., 1988, Alumni award Wilmington Coll., 1988; Alumni award Purdue U., 1989, Disting. Alumni award, 1992. Fellow AAAS, Am. Acad. Microbiology; mem. NAS, Am. Soc. for Virology (councilor 1982-85, pres. 1991-92), Am. Soc. Microbiology (lectr. Found. for Microbiology 1986-87), Soc. for Invertebrate Pathology, Genetics Soc. Am., Internat. Com. on Taxonomy of Viruses (exec. com., chair invertebrate virus subcom. 1988-93), Am. Soc. for Biochemistry and Molecular Biology, Entomol. Soc. of Am., Am. Acad. Microbiology, Sigma Xi. Home: 1908 Streamside Way Bryan TX 77801-2715 Office: Tex A&M U Dept Entomology 324 Minnie Belle Heep College Station TX 77843-2475

SUMNER, MALCOM EDWARD, agronomist, educator; b. June 7, 1933. BSc in Agriculture, Chemistry and Soil Sci., U. Natal, South Africa, 1954, MSc in Agriculture, Soil Physics cum laude, 1958; PhD in Soil Chemistry, U. Oxford, Eng., 1961. Sr. lectr. dept. soil sci. U. Natal, 1955-57, assoc. prof. dept. soil sci., 1961-71, prof., head dept. soil sci. and agrometeorology, 1971-77; prof. dept. agronomy U. Ga., Athens, 1977-91, regents' prof., coord. environ. soil sci. program dept. crop and soil scis., 1991—; hon. prof. dept. agrl. and environ. sci. U. Newcastle-Upon-Tyne, Eng., 1989—; cons. Brit. Commonwealth Devel. Corp., 1967-82, South African Chamber of Mines, 1969-76, FAO-UNDP Program, India, 1986-87, Food and Fertilizer Tech. Ctr., ASPAC, Taiwan, 1986, Combustion Chems. Corp., 1987-90, Masstock Inc., 1988, AGRILAB, 1987-93, Standard Fruit Co., Honduras, 1992, Australian Ctr. Internat. Agrl. Rsch., 1992. Author: (with B. Ulrich) Soil Acidity, 1991, (with B.A. Stewart) Soil Crusting: Physical and Chemical Processes, 1992; author: (with others) Interactions at the Soil Colloid-Soil Solution Interface, 1989, Ecological Indicators, 1992, Sustainable Land Management in the Tropics: What Role for Soil Science?, 1993, Innovative Management of Subsoil Problems, 1993, MacMillan Encyclopedia of Chemistry, 1993; contbr. 184 articles to profl. jours. Recipient Excellence in Rsch. award Ag Alumni, 1989, Sr. Frederick McMaster fellowship, 1992. Fellow Soil Sci. Soc. Am. (Rsch. award 1991), Am. Soc. Agronomy (Werner L. Nelson award 1991, Soil Sci. award 1991). Office: Univ of Georgia Athens GA 30601*

SUMRELL, GENE, research chemist; b. Apache, Ariz., Oct. 7, 1919; s. Joe B. and Dixie (Hughes) S. BA, Eastern N.Mex. U., 1942; BS, U. N.Mex., 1947, MS, 1948; PhD, U. Calif., Berkeley, 1951. Asst. prof. chemistry Eastern N.Mex. U., 1951-53; sr. rsch. chemist J. T. Baker Chem. Co., Phillipsburg, N.J., 1953-58; sr. organic chemist Southwest Rsch. Inst., San Antonio, 1958-59; project leader Food Machinery & Chem. Corp., Balt., 1959-61; rsch. sect. leader El Paso Natural Gas Products Co. (Tex.), 1961-64; project leader So. utilization research and devel. div. U.S. Dept. Agr., New Orleans, 1964-67, investigations head, 1967-73, rsch. leader Oil Seed and Food Lab., So. Regional Rsch. Ctr., 1973-84, collaborator, 1984— . Contbr. numerous papers to profl. jours. Served from pvt. to staff sgt. AUS, 1942-46.

Mem. AAAS, Am. Chem. Soc., N.Y. Acad. Scis., Am. Inst. Chemists, Am. Oil Chemists Soc., Am. Assn. Textile Chemists and Colorists, Rsch. Soc. Am., Phi Kappa Phi, Sigma Xi. Achievements include patents in field. Home: PO Box 24037 New Orleans LA 70184-4037 Office: 1100 Robert E Lee Blvd New Orleans LA 70179

SUN, BENEDICT CHING-SAN, engineering educator, consultant; b. Nanking, People's Republic of China, Nov. 5, 1934; came to U.S., 1955; s. Kuang-Yu Sun and Ta (Chen) Chiang; m. Alice Kau-Hwa Mao, Sept. 18, 1965; children: Christina, David, Eileen. BSME, Nat. Taiwan U., Taipei, 1955; MSME, Kans. State Coll., 1959; PhD in Theoretical and Applied Mechs., U. Ill., 1967. Tool engr. Boeing Airplane Co., Renton, Wash., 1959-60; jr. engr. IBM, San Jose, Calif., 1960-63; asst. prof. mech. engring. N.J. Inst. Tech., Newark, 1967-70, assoc. prof., 1970-90, assoc. prof. engring. tech., 1990—; cons. Stone & Webster Engring., Inc., Boston, N.Y.C., 1970-73, 79-80; prin. engr., cons. Ebasco Svcs., Inc., N.Y.C., 1980—. Contbr. articles to profl. jours. Mem. ASME, Am. Soc. Engring. Edn. Roman Catholic. Home: 17 Sunset Dr Whippany NJ 07981-1626 Office: NJ Inst Tech 323 Martin Luther King Jr Blvd Newark NJ 07102-1824

SUN, EMILY M., economics educator; m. Siao Fang Sun, June 23, 1951; children: Patricia Viane, Caroline Marie, Diana Kate. MA, U. Mich., 1950, PhD, 1957. Prof. econs. Northland Coll., Ashland, Wis., 1957-64; assoc. prof. econs. Manhattan Coll., Riverdale, N.Y., 1964-79, prof. econs., 1979—; cons. Maritime Adminstrn., U.S. Dept. Commerce, 1969-70. Contbr. articles to profl. jours. Rosenthal grantee, 1980; recipient Trustees award, Manhattan Coll., 1987, Bonus et Fidelis medal, 1989. Mem. Am. Econ. Assn., Acad. Internat. Bus. Office: Manhattan Coll Manhattan College Pky Bronx NY 10471-3913

SUN, HOMER KO, electrical engineer; b. Taipei, Dec. 26, 1961; came to the U.S., 1970; s. Hubert K. and Helen (Lee) S. BEE, SUNY, Stony Brook. System engr. Grumman Aircraft Systems, Bethpage, N.Y., 1984—. Buddhist. Office: Grumman Aircraft Systems E02-78 Grumman Blvd Calverton NY 11933

SUN, JING, electrical engineering educator; b. Hefei, Anhui, Peoples Republic of China, Apr. 7, 1961; came to U.S., 1984; d. Hansheng Sun and Xingzhen Xu; m. Bingchang Xu, July 4, 1986; 1 child, Alexander M. Xu. BS in Elec. Engring., U. Sci. and Tech. of China, Heifei, 1982, MS, 1984; PhD, U. So. Calif., 1989. Lectr. U. Sci. and Tech. of China, 1982-84; teaching asst. U. So. Calif., L.A., 1985-87, rsch. asst., 1988-89; asst. prof. elec. engring. Wayne State U., Detroit, 1989—; cons., lectr. Ford Motor Co., Dearborn, Mich., 1991—. Contbr. articles to profl. jours. Mem. IEEE, Detroit Soc. Engrs., Sigma Xi. Office: Wayne State U 5050 Anthony Wayne Dr Detroit MI 48202

SUN, KEUN JENN, physicist; b. Keelung, Taiwan, China, Mar. 5, 1949; came to U.S., 1977; s. Ching-Chuan and Tsai-Hsia (Chiou) S.; m. Hsi Wei, Dec. 19, 1976; children: Bor Jen, Phillip B.R. MS, U. Akron, 1979; PhD, U. Wis., Milw., 1986. Rsch. assoc. Nat. Rsch. Coun., Hampton, Va., 1986-88; rsch. scientist Coll. William and Mary, Williamsburg, Va., 1988—. Co-author: Physical Acoustics, Vol. XX, 1992; contbr. articles to profl. jours. Mem. IEEE, AAAS, Am. Phys. Soc., Material Rsch. Soc. Achievements include findings of spin-phonon interaction in magnetic superconductors, ultrasonic relaxation attenuation in magnetic superconductors, acoustic waveguide sensors for epoxy curing, and stuctural flaws detection with acoustic plate waves. Office: NASA Langley Rsch Ctr MS 231 Hampton VA 23185

SUN, LI-TEH, economics educator; b. Hong Kong, Dec. 5, 1939; s. Beh-Yu and Ruey-Jeng (Wang) S.; m. Ping Zhong, June 1, 1991. BA in Econs., Chung Hsing U., Taipei, Taiwan, 1962; MS in Econs., Okla. State U., 1968, PhD in Econs., 1972. Rsch. assoc. U. Mont., Missoula, 1969-70; lectr. Humboldt State U., Arcata, Calif., 1972-75; acad. resource specialist Chancelor's Office Calif. State U. and Colls., Long Beach, Calif., 1975-77; assoc. prof., prof. econs. Nat. Chung Hsing U., Taipei, 1977-81, chair dept. pub. fin., 1978-82; prof. econs. Moorhead (Minn.) State U., 1982—; coord. China programs Moorhead State U., 1987-89. Contbr. articles to profl. jours. Mem. adv. bd. Centre of Humanomics, 1985—. Named Prof. of Yr., Humboldt State U., Arcata, Calif., 1974. Mem. Moorhead Cen. Lions (newsletter editor 1990—, bd. dirs. 1983—, pres. 1987-88, Lion of Yr. 1984, 85, 90, 91). Avocations: table tennis, biking, travel. Home: 823 23d Ave S Moorhead MN 56560 Office: Moorhead State U Dept Econs Moorhead MN 56563

SUN, SIAO FANG, chemistry educator; b. Shaoshing, China, Feb. 19, 1922; came to U.S., 1949; s. Yuan and Yu C. Sun; m. M. Emily Chao, June 23, 1951; children: Patricia Vian, Caroline Marie, Diana Kate. MA, U. Utah, 1950; MS, Loyola U., 1956; PhD, U. Chgo., 1958, U. Ill., 1962. Prof. math. Northland Coll., Ashland, Wis., 1960-64; asst. prof. chemistry St. John's U., Jamaica, N.Y., 1964-70, assoc. prof. chemistry, 1970-75, prof. chemistry, 1975-92, adj. prof., 1992—; visiting scientist Nat. Ctr. Sci. Rsch., Strasbourg and Mendon-Bellevue, France, 1975-78, Carlsberg Lab., Copenhagen, 1981; staff scientist Max Planck Inst. Biophysical Chemistry, Gottingen, Germany, 1976. Contbr. articles to profl. jours. Office: St John's Univ Dept Chemistry Jamaica NY 11439

SUN, TUNG-TIEN, medical science educator; b. Chung King, Szechuan, People's Republic of China, Feb. 20, 1947; s. Chung-Yu and Wen (Lin) S.; m. Brenda Shih-Ying Bao, Aug. 14, 1971; children: I-Hsing, I-Fong. BS in Agrl. Chemistry, Nat. Taiwan U., Taipei, 1967; PhD in Biochemistry, U. Calif., Davis, 1974. Rsch. assoc. dept. biology MIT, Cambridge, 1974-78; asst. prof. depts. dermatology, cell biology and anatomy Johns Hopkins Med. Sch., 1978-81, assoc. prof. depts. cell biology and anatomy, dermatology, ophthalmology, 1981-82; assoc. prof. depts. dermatology and pharmacology NYU Med. Sch., N.Y.C., 1982-86, prof. depts. dermatology and pharmacology, 1986-90, Rudolf L. Baer prof. depts. dermatology and pharmacology, 1990—, assoc. dir. Skin Disease Rsch. Ctr., 1989—; mem. cell biology study sect. NIH, 1984-88. Mem. editorial bd. Differentiation, 1984—, Epithelial Cell Biology, 1990—; assoc. editor Jour. Investigative Dermatology, 1990—. Recipient Career Devel. award Nat. Eye Inst., 1978-82, Monique Neill-Caulier Career Scientist award, 1984-89; named Angus lectr. U. Toronto Med. Sch., 1986, Pinkus lectr. Am. Acad. Dermatopathologists, 1986, Liu lectr. Stanford Med. Sch., 1987, Susan Swerling lectr. Harvard Med. Sch., 1991. Fellow AAAS: mem. Am. Soc. Biol. Chemists, Am. Soc. for Cell Biology, Internat. Soc. Differentiation (bd. dirs. 1985-88), Soc. Investigative Dermatology (Montagna lectr. 1989), Assn. Rsch. in Vision Sci. and Ophthalmology. Achievements include research on keratins as markers for defining different pathways and stages of epithelial differentiation, on identification of stem cells in epidermis, corneal epithelium and hair follicle, and identification of uroplakins as novel urothelial differentiation markers. Office: NYU Med Sch Dept Dermatology 560 1st Ave New York NY 10016-6402

SUN, YANYI, research scientist; b. Beijing, June 14, 1944; came to U.S., 1986; d. Yufeng Sun and YiQing Li; m. Rirong Cheng, Sept. 30, 1972; children: Cheng. BS, U. Sci. and Tech. of China, 1967. Engr. Yan Shan Petrochemical Plant, Beijing, 1968-80; instr. Beijing (China) Indsl. U., 1980-83; lectr. Ctrl. Radio and TV U. of China, Beijing, 1984-86; rschr. TCSUH, U. Houston, Tex., 1986—. Contbr. articles to Phys. Rev. Letters, Nature, Physica C. Mem. Am. Phys. Soc. Office: U Houston Tex Ctr Superconducting Houston TX 77004

SUN, ZONGJIAN, physicist, researcher; b. Chongging, Sichuan, China, Mar. 15, 1945; came to U.S., 1988; s. Baolian and Zhongfen (Zhang) S.; m. Haiwen Dai, Jan. 4, 1982; 1 child, Runjia Sun. BS in Physics, Ea. Normal U., Shanghai, 1978; MS in Physics, Tongji U., Shanghai, 1981. Asst. prof. Pohl Inst., Tongji U., Shanghai, 1982-85, assoc. prof., 1985-87; guest scientist Nat. Inst. Stds. and Techs., Boulder, Colo.; assoc. researcher U. Colo., Boulder, 1989-90; rsch. scientist POC, Torrance, Calif., 1990—. Achievements include first use of prism-couplers to measure fiber scalar mode propagation constants in 1984 (about 10 months before other scientists); also in 1989 invention of world's smallest prism (0.005"). Office: POC 20600 Gramercy Pl Ste 103 Torrance CA 90501

SUNDARARAJAN, NARASIMHAN, electrical engineering educator; s. V. Narasimhan and N. Rengarayaki; m. Saraswathi Sundararajan, June 1, 1972; children: Sowmya, Sripriya. BE with honors, U. Madras, India, 1966; M Tech., Indian Inst. Tech., Madras, 1968; PhD, U. Ill., 1971. Grad. rsch. asst. U. Ill., Urbana, 1969-72; head control and guidance analysis sect. Indian Space Rsch. Orgn., Trivandrum, India, 1972-75; head mission analysis group Indian Space Rsch. Orgn., Trivandrum, 1976-81, head launch vehicle systems div., 1981; dep. dir. advanced tech. planning office Indian Space Rsch. Orgn., Bangalore, India, 1986-88; dir. launch vehicle design group Indian Space Rsch. Orgn., Trivandrum, 1988-91; assoc. prof. Sch. Elec. and Electronics Engring., Nanyang Technol. U., Singapore, 1991—; lectr. various instns. Editor: Jour. Indian Rocket Soc.; contbr. more than 45 articles to profl. jours ; reviewer for various jours. NRC fellow, 1975-76, 1981-83. Fellow AIAA (assoc.); mem. IEEE (sr.), Control System Soc., Robotics and Automation Soc., Acoustic Speech and Signal Processing Soc. Achievements include first to use lattice filters for identification and control of large space structures; rsch. in multivariable robust controller design, control and guidance of aircrafts, spacecrafts, missiles and launch vehicles. Home: 58 Nanyang Ter, 2263 Singapore Singapore Office: Nanyang Technol U, Sch Elec and Electronic Sup, 2263 Singapore Singapore

SUNDARESAN, P. RAMNATHAN, research chemist, consultant; b. Madras, India, Aug. 11, 1930; came to U.S., 1961; s. Peruvemba A. and Saraswathi Subramanian Ramnathan; m. Gloria Marquez Sundaresan, Dec. 23, 1970; children: Sita, Ramesh. BS, U. Banaras, Banaras, India, 1950, MS, 1953; PhD, Indian Inst. Sci., Bangalore, 1958. Rsch. assoc. Radiocarbon Lab., U. Ill., Urbana, 1961-62, Dept. Nutrition and Food Sci., MIT, Cambridge, Mass., 1962-64; vis. scientist, rsch. assoc. U.S. Army Rsch. Inst. Environ. Medicine, Natick, Mass., 1964-66, rsch. biochemist, 1966-68; chief, Lipids Lab. Rsch. Inst. St. Joseph Hosp., Lancaster, Pa., 1968-77; rsch. chemist FDA, Washington DC, 1977—; cons. Millersville (Pa.) State Coll., 1972-77, VA Med. Ctr., Washington DC, 1973-77, 1984—; panel mem. Source Evaluation Group. Contbr. articles to nat. and internat. profl. jours. Coun. Sci. and Indsl. Rsch. sr. rsch. fellowship, New Delhi, India, 1959-61; NIH rsch. grantee, 1970-77. Fellow Am. Inst. Chemists; mem. Am. Inst. Nutrition, Am. Soc. Biochemistry and Molecular Biology, Biochemical Soc. (U.K.), Am. Coll. Toxicology, Sigma Xi. Office: FDA MOD I 8301 Muirkirk Rd Laurel MD 20708

SUNDBERG, JOHAN EMIL FREDRIK, communications educator; b. Stockholm, Mar. 25, 1936; s. Halvar G.F. and Margit (Hammarberg) S.; m. Agneta Hagerman, 1982; 1 child, Susanna; m. Ulla E.M. Ahlestem, 1983; children: Martin, Erik. Organist, Upsala, 1957, PhD, 1966. Guest rsch. dept. speech communication Royal Inst. Tech., Stockholm, 1962-66, rschr. dept. speech comm., 1967-79, prof. head. music group, dept. speech comm., 1979—; adv. group Inst. Rsch. and Coordination Acoustics/Music, Paris, 1978—. Author: Science of the Singing Voice, 1987, Science of Musical Sound, 1991; editor 12 books; contbr. articles to profl. jours. Pres. Stockholm Bach Choir, 1973-79. Fellow Acoustical Soc. Am.; mem. Swedish Acoustical Soc. (pres. 1976-81), Royal Swedish Acad. Music. Office: Speech Comm/Music Acoustics, Box 70014, S-10044 Stockholm Sweden

SUNDERLIN, CHARLES EUGENE, consultant; b. Reliance, S.D., Sept. 28, 1911; s. Glen Eugene and Frances (Smith) S; m. Sylvia Alice Sweetman, July 8, 1936; children: Ann Elizabeth, Mary Cornelia, Katherine Patricia, William Dana. AB, U. Mont., 1933; BA, MA (Rhodes scholar), Oxford U., Eng., 1935; PhD, U. Rochester, 1939. Instr. chemistry Union Coll., 1938-41; instr., asst. prof. U.S. Naval Acad., Annapolis, Md., 1941-43, 45-46; sci. liason officer U.S. Office Naval Rsch., London, 1946-47, dep. sci. dir., 1948-49, sci. dir., 1949-51; dep. dir. NSF, Washington, 1951-57; spl. asst. to pres. Nat. Acad. Scis., Washington, 1965-69; v.p., sec. Rockefeller U., N.Y.C., 1969-76; spl. asst. Nat. Sci. Bd., Washington, 1976-78; exec. sec., staff dir. Com. on 10th Nat. Sci. Bd. Report, Washington, 1976-78; U.S. del. 6th and 7th Gen. Assemblies Internat Coun. Sci. Unions, Amsterdam, 1952, Oslo, 1955; mem. working party on Establishment of Internat. Adv. Com. on Sci. Rsch., Paris, 1953; Meeting of Dirs. Nat. Rsch. Ctrs., Milan, 1955, Symposium on Orgn. and Adminstrn., Applied Rsch., Vienna, 1956; mem. Com. Experts on Scientists' Rights, Paris, 1953, Nat. Acad. Scis. Workshop on Indsl. Devel. Taiwan, Rep. of China, 1968; chmn. AIAA/ASME 9th Structures, Structural Dynamics and Materials Conf., 1968; treas., bd. dirs. Engrs. and Scientists Com., Inc., People to People Program. Lt. USNR, 1943-45. Fellow AAAS, Chem. Soc. (London); mem. AIAA, Am. Chem Soc., Faraday Soc., Royal Instn. Gt. Britain, Soc. Chem. Industry, Wadham Assn. U.S., United Oxford and Cambridge Univ. Club (London), Internat. Club (Washington), Sigma Alpha Epsilon. Episcopalian. Home: 3036 P St NW Washington DC 20007 Other: 137 E Main St Cambridge NY 12816

SUNDERMAN, DUANE NEUMAN, chemist, research institute executive; b. Wadsworth, Ohio, July 14, 1928; s. Richard Benjamin and Carolyn (Neuman) S.; m. Joan Catherine Hoffman, Jan. 31, 1953; children: David, Christine, Richard. BA, U. Mich., 1949, MS, 1954, PhD in Chemistry, 1956. Researcher Battelle Meml. Inst., Columbus, Ohio, 1956-59; mgr. Battelle Meml. Inst., Columbus, 1959-69, assoc. dir., 1969-79, dir. internat. programs, 1979-84; sr. v.p. Midwest Rsch. Inst., Kansas City, Mo., 1984-90, exec. v.p., 1990—; exec. v.p. Midwest Rsch. Inst., Golden, Colo., 1990—; dir. Nat. Renewable Energy Lab., Golden, Colo. Contbr. numerous articles to profl. jours. Bd. dirs. Mid-Ohio chpt. ARC, 1982-83, U. Kansas City, 1985-90, Mo. Corp. for Sci. and Tech., Jefferson City, 1986-90. Mem. AAAS, Am. Chem. Soc., Am. Mgmt. Assn. Republican. Presbyterian. Avocation: computers. Office: Nat Renewable Energy Lab 1617 Cole Blvd Golden CO 80401-3305

SUNDERMAN, F(REDERICK) WILLIAM, JR., toxicologist, educator, pathologist; b. Phila., June 23, 1931; s. Frederick William and Clara Louise (Baily) S.; m. Carolyn Reynolds, Aug. 24, 1963; children: Frederick W. III, Elizabeth R., Emily L. BS, Emory U., 1952; MD, Thomas Jefferson U., 1955. Diplomate Am. Bd. Pathology. Intern, resident Jefferson Hosp., Phila., 1955-58; clin. chemist NIH, Bethesda, Md., 1960; assoc. in medicine Thomas Jefferson U., Phila., 1961-64; assoc. prof. pathology U. Fla. Med. Sch., Gainesville, 1964-68; prof., head lab. med. dept. U. Conn. Med. Sch., Farmington, 1968—, prof. toxicology, 1988—. Co-editor 17 books and monographs; contbr. 317 papers to sci. jours. Lt. USNR, 1958-60. Named Scientist of Yr. Assn. Clin. Scientists; recipient Ames award Am. Assn. Clin. Chemistry, 1978, Nickel Rsch. award Internat. Union of Pure and Applied Chemistry, 1992. Achievements include rsch. in toxicity and carcinogenicity of metals, especially nickel. Office: U Conn Health Ctr 263 Farmington Ave Farmington CT 06030-2225

SUNDICK, ROBERT IRA, anthropologist, educator; b. Bklyn., May 8, 1944; s. Philip G. and Thelma (Lechner) S.; m. Carol A. Strongin, June 11, 1967; children: Scott A., Jennifer L. BA, SUNY, Buffalo, 1966; MA, U. Toronto, Can., 1967, PhD, 1972. Diplomate Am. Bd. Forensic Anthropology. Prof. anthropology Western Mich. U., Kalamazoo, 1969—, also dept. chmn., 1988—; bd. dirs. Am. Bd. Forensic Anthropology; forensic anthropologist, 1980—. Author: (chpt.) Role of the Expert Witness, 1984; contbr. articles to profl. jours. Fellow Am. Acad. Forensic Scis. (chair, sec. 1988-90), Am. Anthrop. Assn.; mem. Am. Assn. Phys. Anthropology, Paleopathology Assn., Phi Kappa Phi. Home: 1010 Dobbin Dr Kalamazoo MI 49006 Office: Western Mich Univ Anthropology Dept Kalamazoo MI 49008-5032

SUNDLER, FRANK ESKIL GEORG, histology educator, biomedical scientist; b. Hjärsås, Sweden, Apr. 5, 1943; s. Georg and Gulli (Sundell) S.; m. Kristina Larsson, Aug. 6, 1971; children: Martin, Linda, Emil, Frida. Cand. med., Lund (Sweden) U., 1965, PhD, MD, 1973. Amanuens dept. histology Lund U., 1965-71, rsch. asst., 1971-77, asst. prof., 1979-85, assoc. prof., 1986—, chmn. dept. med. cell rsch., 1987—, cons. dept. surgery, 1980-83. Asst. editor jour. Regulatory Peptides, 1984—; sect. editor jour. Acta Physiol. Scand., 1992—; mng. editor jour. Anat. Embryology, 1993—; cooperating editor Cell Tissue Res., 1993—; contbr. articles to profl. publs. Mem. AAAS, N.Y. Acad. Scis., Swedish Endocrine Soc., Swedish Royal Physiographic Soc., Neuropeptide Club. Home: Vinkelvägen 4, 24010 Dalby Sweden Office: Lund U Dept Med Cell Rsch, Biskopsgatan 5, 22362 Lund Sweden

SUNDNES, GUNNAR, science foundation director, educator; b. Trondheim, Norway, Feb. 21, 1926; s. Ole Sigvard and Gudrun (Folstad) S.; m. Laila Baglo, July 5, 1952; children—Gunnar, Lars Orjan, Marianne. Cand., U. Oslo, 1954; Ph.D., U. Bergen, 1971. Scientist, Inst. Marine Resarch, Bergen, 1953-56, Zoophysiol. U. Oslo, 1956-58; sr. scientist Inst. Marine Research, Bergen, 1958-72; dir. Mus. Sci. U. Trondheim, 1972-80, prof. marine biology, 1972-94; mem. Am. Fisheries Soc., Marine Biol. Assn. U.K., Royal Norwegian Soc. Sci. and Letters (exec. dir., sec. gen.). Home: Veimesterstien 21, 7022 Trondheim Norway Office: DKNVS The Found, E Skakkes Gt 47B, 7013 Trondheim Norway

SUNG, DAE DONG, chemistry educator; b. Sichunmyon, Kyung Nam, Republic of Korea, June 17, 1945; s. Cha Sung and Soon Agh (Ha) S.; m. Buyng Hee Yoon, Apr. 13, 1975; children: Myo Ya, Yun Duck. BS, Dong-A U., Pusan, Republic of Korea, 1969, MS, 1977, DSc, 1981. Lectr. Pusan Nat. U., 1977-78, Kyung Nam Tech. Coll., Pusan, 1978-79; lectr. Dong-A U., 1979-81, from asst. prof. to assoc. prof., 1981-90, prof. chemistry, 1990—, head lab. basic scis., 1982-83, head Basic Sci. Inst., 1989-91; postdoctoral researcher Princeton U., N.J., 1983. Contbr. articles to profl. jours. Grantee Republic of Korea Sci. and Engring. Found., 1984, Ministry of Edn., Republic of Korea, 1987. Fellow Eng. Royal Soc. (rschr.); mem. Korean Chem. Soc. (author bull. 1985, 87, 90), Am. Chem. Soc. Avocation: tennis. Office: Dong-A U, Dept Chemistry, Saha-Gu Pusan 604-714, Republic of Korea

SUNG, KUO-LI PAUL, bioengineering educator. MA in Biology, Coll. William and Mary, 1975; MS in Physiology, Columbia U., 1977; PhD in Physiology, Rutgers-Columbia U., 1982; PhD in Bioengineering (hon.), Chonging U., China, 1993. Teaching assist. dept. physiology Taiwan U., 1970-72; rsch. asst. dept. biology Coll. Willam and Mary, 1972-74; rsch. asst. dept. physiology Coll. Physicians and Surgeons, Columbia U., 1974-77, rsch. worker dept. physiology, 1977-81, staff assoc. sci. dept. physiology and cellular biophysics, 1982; lectr. divsn. of circulatory physiology and biophysics dept. of physiology and cellular biophysics, Coll. Physicians and Surgeons Columbia U., 1986, 87; lectr. Inst. Biomedical Sci. Academia Sinica, 1987; assoc. rsch sci. dept. physiology and cellular biophysics Coll. Physicians and Surgeons, Columbia U., 1982-88; organizer and instr. Cell Biophysics Workshop Academia Sinica and Nat. Sci. Coun., Taiwan, China, 1987; assoc. rsch. bioengineer III, lectr. dept. applied mechanics and engring. scis.-bioengineering U. Calif., 1988-92, assoc. prof. of orthopaedic dept., Sch. Medicine, 1992—, assoc. prof. of bioengineering dept. applied mechanics and engring. scis.-bioengineering, 1992—; lectr. bioengineering ctr. Chongqing U., China, 1993; full mem. cancer ctr. U. Calif., San Diego, 1991, Inst. for Biomedical Engring., 1991—; organizer Cellular Adhesion: Signaling and Molecular Regulation Am. Physiol. Soc., 1994, main speaker Cell Biophysics Workshop Academia Sinica and Nat. Sci. Coun., China, 1987, Cellular Adhesion Workshop, West China of Med. Scis., China, 1993. Author various publs. Recipient New Investigator Rsch. award NIH, 1984-87, Best Jour. Paper award ASME, 1989, Chancellor award U. Calif., San Diego, 1988-89, The Whitaker Found. award, 1990, Melville medal ASME, 1990, Lamport award Biomedical Engring. Soc., 1992; Dr. Yat-Sen Sun Fellow Taiwan, China, 1967; Walter Russell Scholar, 1980-82. Mem. AAAS, Am. Physiol. Soc., N.Am. Soc. of Biorheology, Internat. Soc. of Biorheology, Biomedical Engring. Soc., Microcirculatory Soc., Sigma Xi. Achievements include research in influence of tumor suppressor genes on tumor cell metastasis, biophysical properties and molecular organization of cell membranes, healing mechanism of human ligament cells, adhesion between osteoblast and biomaterials, biophysical properties of blood cells and endothelial cells in inflammatory reponse, energy balance and molecular mechanisms of cell-cell interactions in immune response, intracellular ions, intracellular transmition and cell activation. Office: Univ of California Bioengineering Division R-012 La Jolla CA 92093*

SUNG, MING, chemical engineer; b. Shanghai, China, Nov. 17, 1947; came to U.S., 1968; s. Paul V. and Lin Y. (Tai) S.; m. Lily L. Chu, Sept. 3, 1977; children: Victor, Dominic. BS, Colo. Sch. Mines, Golden, 1971, MS, 1972; postgrad., U. Fla., 1972-74. Registered profl. engr., Mich. Process/project engr. Monsanto Co., St. Louis, 1974-78; process engr. BASF Corp., Wyandotte, Mich., 1978-80; head office process engr. Shell Devel. Co., Houston, 1980—; cons. tech. advisor ERDA/Dept. Energy, Washngton, 1976-78;' chief judge Chem. Processing Vaaler Award, Chgo., 1992; cons., lectr. chem./petrochm. industry, China, 1985, 87, 89; vol. tchr. N.W. Chinese Lang. Inst. Contbr. articles to profl. jours. Vol. Houston Internat. Festival, 1987-92; sec. Asian Am. Voter League, Houston, 1992. Mem. Am. Inst. Chem Engrs. (fin. com.), Chinese Am. Petroleum Assn. (petrochem. co-chair, bd. dirs.), Nat. Assn. Chinese Ams. (pres. 1988-90, bd. dirs.). Home: 15315 Quiet Creek Dr Houston TX 77095 Office: Shell Devel Co 3333 Hwy 65 Houston TX 77082-3101

SUNG, YUN-CHEN, computer graphics software developer. BS, Calif. Inst. Tech.; MS, UCLA. Formerly with Digital Prodns., Whitney/Demos Prodns.; now software developer MetroLight Studios, L.A. Credits include contbr. to opening animation sequence (film) Labyrinth. Office: MetroLight Studios Ste 400 5724 W 3rd St Los Angeles CA 90036-3078

SUNTHARALINGAM, NAGALINGAM, radiation therapy educator; b. Jaffna, Ceylon, June 18, 1933; married; 3 children. BSc, U. Ceylon, 1955; MS, U. Wis., 1966, PhD in Radiol. Scis., 1967. Asst. lectr. physics U. Ceylon, 1955-58; from instr. radiol. physics to assoc. prof. radiology Thomas Jefferson U. Med. Coll., Phila., 1962-72, prof. radiology, 1972—; vis. lect. grad. sch. medicine U. Pa., 1967-72, cons. dept. physics, 1968-72; cons. WHO, 1972. Recipient William D. Coolidge award Am. Assn. Physicists in Medicine, 1992. Fellow Am. Coll. Radiology, Am. Coll. Med. Physics (chmn. bd. dirs. 1988); mem. Am. Assn. Phys. Medicine (pres. 1983), Health Physics Soc. Achievements include research in radiation dosimetry, thermoluminescence dosimetry, and clinical dosimetry. Office: Thomas Jefferson University 11th & Walnut Sts Philadelphia PA 19107*

SUONINEN, EERO JUHANI, materials science educator; b. Viipuri, Finland, May 14, 1929; s. Herman and Helvi Irene (Rosendal) S.; m. Pirkko Kaarina Kovanen, June 28, 1958 (dec. Aug. 1986); children: Liisa Maria, Mikko Juhani, Marja; m. Vuokko Valvikki Koskela, Sept. 22, 1990. Diploma in engring., Finnish Inst. Tech., Helsinki, 1952; MS, MIT, 1954, PhD, 1957. Rsch. asst. MIT, Cambridge, 1954-57; rsch. worker Norelco, Mt. Vernon, N.Y., 1957; chief physicist Outokumpu Co., Helsinki, 1958-61; rsch. assoc. U. Ariz., Tucson, 1961-62; prof. tech. physics U. Oulu, Finland, 1962-69; prof. materials sci. U. Turku, Finland, 1969—; vis. prof. U. Conn., Storrs, 1972-73, Oreg. State U., Corvallis, 1983-84. Contbr. over 100 articles to tech. jours. U.S. Loan to Finland Found.-Fulbright grantee, 1953-54. Mem. Finnish Acad. Tech. Scis., Acad. Finland (natural sci. rsch. coun. 1986-91), Sigma Xi. Lutheran. Home: Rykmentintie 45 Apt 22, SF-20880 Turku Finland Office: U Turku Lab Materials Sci, Itäinen Pitkäkatu 1, SF-20520 Turku Finland

SUPLICKI, JOHN CLARKE, chemical and metallurgical engineer, administrator; b. Paterson, N.J., Apr. 23, 1950; s. Joseph Stanley and Margaret Elizabeth (Clarke) S.; m. Susan Laurel Kunkler, Nov. 8, 1975; children: Tara, Anne, Joseph, Robert, Thomas. BSChE, Clarkson Coll., 1972, MSChE, 1974. Application engr. Ecodyne Indsl. Waste Treatment, Union, N.J., 1974-77; project engr. Borden, Plant City, Fla., 1977-81; metallurgist Gardiner, Ft. Meade, Fla., 1981-83, Basso Chem., Jacksonville, Fla., 1983-85; process dev. engr. Haber, Towaco, N.J., 1985-86; mgr. mgmt. info. systems Forum for Scientific Excellence, Lake Hopatcong, N.J., 1986—. Author: (reference book) Cross Reference Index of Hazardous Chemicals, 1990, co-author: (reference books) List of Lists of Worldwide Hazardous Chemicals, 1990, Index of Hazardous Contents of Commercial Products, 1990, Compendium of Hazardous Chemicals in Schools and Colleges, 1990. Mem. AIChE, AME-SME, Phi Kappa Phi. Republican. Mem. LDS Ch. Achievements include air flotation cell patent used for mineral seperation greatly increased recovery, grade and size of particle flotation; set up massive database of hazardous chemicals and controls of commercial products. Home: Box 137 Glasser NJ 07837

SUPPES, PATRICK, statistics, education, philosophy and psychology educator; b. Tulsa, Mar. 17, 1922; s. George Biddle and Ann (Costello) S.; m. Joan Farmer, Apr. 16, 1946 (div. 1970); children: Patricia, Deborah, John

Biddle; m. Joan Sieber, Mar. 29, 1970 (div. 1973); m. Christine Johnson, May 26, 1979; children: Alexandra Christine, Michael Patrick. B.S., U. Chgo., 1943; Ph.D. (Wendell T. Bush fellow), Columbia U., 1950; LL.D., U. Nijmegen, Netherlands, 1979; Dr. honoris causa, Académie de Paris, U. Paris V, 1982. Instr., Stanford U., 1950-52, asst. prof., 1952-55, assoc. prof., 1955-59, prof. philosophy, statistics, edn. and psychology, 1959—; founder, chief exec. officer Computer Curriculum Corp., 1967-90. Author: Introduction to Logic, 1957, Axiomatic Set Theory, 1960, Sets and Numbers, books 1-6, 1966, Studies in the Methodology and Foundations of Science, 1969, A Probabilistic Theory of Causality, 1970, Logique du Probable, 1981, Probabilistic Metaphysics, 1984, Estudios de Filosofía y Metodologí de la Ciencia, 1988, Language for Humans and Robots, 1991, Models and Methods in the Philosophy of Science, 1993; (with Davidson and Siegel) Decision Making, 1957, (with Richard C. Atkinson) Markov Learning Models for Multiperson Interactions, 1960, (with Shirley Hill) First Course in Mathematical Logic, 1964, (with Edward J. Crothers) Experiments on Second-Language Learning, 1967, (with Max Jerman and Dow Brian) Computer-assisted Instruction, 1965-66, Stanford Arithmetic Program, 1968, (with D. Krantz, R.D. Luce and A. Tversky) Foundations of Measurement, Vol. 1, 1971, (with M. Morningstar) Computer-Assisted Instruction at Stanford, 1966-68, 1972, (with B. Searle and J. Friend) The Radio Mathematics Project: Nicaragua, 1974-75, 1976 (with D. Krantz, R.D. Luce and A. Tversky) Foundations of Measurement, Vol. 2, 1989, Vol. 3, 1990. Served to capt. USAAF, 1942-46. Recipient Nicholas Murray Butler Silver medal Columbia, 1965, Disting. Sci. Contbr. award Am. Psychol. Assn., 1972, Tchrs. Coll. medal for disting. service, 1978, Nat. medal Sci. NSF, 1990; Center for Advanced Study Behavioral Scis. fellow, 1955-56; NSF fellow, 1957-58. Fellow AAAS, Am. Psychol. Assn., Am. Acad. Arts and Scis.; mem. NAS, Math Assn. Am., Psychometric Soc., Am. Philos. Assn., Am. Philos. Soc., Assn. Symbolic Logic, Am. Math Soc., Académie Internationale de Philosophie des Scis. (titular), Nat. Acad. Edn. (pres. 1973-77), Am. Psychol. Assn., Internat. Inst. Philosophy, Finnish Acad. Sci. and Letters, Internat. Union History and Philosophy of Sci. (div. logic, methodology and philosophy of sci., pres. 1975-79), Am. Ednl. Research Assn. (pres. 1973-74), Croatian Acad. Scis. (corr.), Russian Acad. Edn. (fgn.), Norwegian Acad. Sci. and Letters (fgn.), European Acad. Scis. and Arts, Chilean Acad. Scis., Sigma Xi.

SUPPES, TRISHA, neuroscientist; b. Palo Alto, Calif., June 1, 1951; m. James M. Prosser. BA in Human Biology, Stanford U., 1973; PhD, UCLA, 1980; MD, Dartmouth Med. Sch., 1987. Undergrad. rsch. asst. dept. biology Stanford (Calif.) U., 1971, 72-73, rsch. asst. Inst. for Math. Studies in Social Scis., 1973-74; grad. student dept. anatomy UCLA, 1974-80, doctoral candidate dept. biology, 1977-80; postdoctoral fellow dept. neurology Stanford (Calif.) U., 1981-83; rsch. assoc. neurochem. pharmacology Med. Rsch. at Addenbrooke's Hosp., Cambridge, Eng., 1984; resident in adult psychiatry McLean Hosp., Belmont, Mass., 1987-91; rsch. assoc. dept. neurology Children's Hosp., Boston, 1990-91; clin. fellow psychiatry Harvard Med. Sch., Boston, 1987-92, fellow in neuroscience, 1990-92; asst. prof. dept. psychiatry U. Tex. Southwestern Med. Ctr., Dallas, 1992—; faculty gross anatomy UCLA Sch. of Medicine, Dept. of Anatomy, 1975-76, faculty microscopic anatomy, 1976, faculty dept. biology, 1977; faculty human nervous system and behavior Harvard Med. Sch., Boston, 1991. Author: Home Based Education: Needs and Technological Opportunities, 1976, The Effects of Tetraethylammonium on the Electrophysiological Properties of Sympathetic Neurons Isolated from the Superior Cervical Ganglia of Neonatal Rats Grown in Cell Culture, 1980; contbr. numerous articles to profl. jours. Recipient Chancellor's Patent Fund grant UCLA, 1980, McCormick fellowship Stanford U. Sch. Medicine, 1981-83, Summer Rsch. fellowship Scottish Rite Found. for Schizophrenic Rsch., 1984, Neurosci. fellowship in Psychiatry McLean and Mass. Gen. Hosps., 1990-92, Dista Travel Fellowship award Soc. Biol. Psychiatry, 1992, Nat. Alliance Rsch. Schizophrenia and Depression award, 1993. Achievements include discovery of neurons added to ctrl. nervous system of the crayfish after birth; discovery of seconds long Calcium-dependent ionic current in mammalion sympathetic neurons and quantification of time to relapse after lithium discontinuation in bipolar manic-depressive illness. Office: U Tex Southwestern Med Ctr PO Box 1 #600 5959 Harry Hines Blvd Dallas TX 75235-9101

SURESH, BANGALORE ANANTHASWAMI, information systems educator; b. Secunderabad, India, Dec. 10, 1951; came to U.S. 1984; s. B.S. Ananthaswami and B.A. Suma (Leela) Rao; m. Pushpa, July 10, 1980; children: Deepika S., Preetika S. B in Tech., Indian Inst. Tech., Madras, 1974; diploma, Nat. Inst. Tng. in Indsl. Engring., Bombay, 1976; PhD, Syracuse U., 1993. Corp. planning officer HMT Ltd., Bangalore, India, 1976-79; indsl. engr. Binny Ltd., Bangalore, India, 1979-80, acting factory mgr., 1980-81; asst. prodn. mgr. Brooke Bond India Ltd., Coimbatore, India, 1981-84; teaching asst. Syracuse U., 1984-88, univ. fellow, 1988-89; asst. prof. MIS Ind. U.-Purdue U., Ft. Wayne, 1989-91; asst. prof. info. systems N.J. Inst. Tech., Newark, 1991—; reviewer Jour. Systems Integration, 1992, Internat. Conf. Info. Systems, 1992, Ann. Meeting Decision Scis. Inst. Contbr. articles to profl. jours.; reviewer Jour. Systems Integration, 1992, others. Syracuse U. fellow, 1988-89. Mem. Ops. Rsch. Soc. Am. (session chmn. 30th ann. meeting 1990), Assn. for Computing Machinery, Decision Scis. Inst., Inst. Mgmt. Sci. (session chmn. 30th ann. meeting 1990, joint nat. meeting 1992, discussant 31st ann. meeting 1991), Spl. Interest Group Bus. Data Processing (rsch. grantee 1986), Spl. Interest Group Artificial Intelligence, Lions (charter sec. 1980), Leo Club (founder, sec. 1973). Hindu. Avocations: tennis, racquetball, music, photography, travel. Home: 588 Lathrop Ave Boonton NJ 07005-2247 Office: NJ Inst Tech University Heights Newark NJ 07102

SURFACE, STEPHEN WALTER, water treatment chemist, environmental protection specialist; b. Dayton, Ohio, Feb. 25, 1943; s. Lorin Wilfred and Virginia (Marsh) S.; m. Suzanne MacDonald, Aug. 29, 1964 (div.); 1 child, Jennifer Nalani; m. Sinfrosa Garay, Sept. 16, 1978; children: Maria Lourdes, Stephanie Alcantara. BS, Otterbein Coll., 1965; MA, U. So. Calif., 1970; postgrad., U. Hawaii, 1971. Tchr. Hawaii State Dept. Edn., Honolulu, 1970-71; staff chemist Del Monte Corp., Honolulu, 1971; head chemist USNPearl Harbor, Honolulu, 1971-76; staff chemist USN Pearl Harbor, Honolulu, 1976-90; chief office installation svcs., environ. protection Def. Logistics Agy., Camp Smith, Hawaii, 1990—. Contbr. articles to profl. jours. Recipient DuPont Teaching award, U. So. Calif., 1966. Fellow Am. Inst. Chemists; mem. Am. Chem. Soc., Am. Water Works Assn., Am. Def. Preparedness Assn., N.Y. Acad. Scis., Sigma Xeta, Phi Lambda Upsilon. Democrat. Methodist. Avocations: traveling, artifact collecting, landscaping. Home: 94-1139 Noheaiki St Waipahu HI 96797 Office: Def Logistics Agy DPAC-W Camp Smith HI 96861-4110

SURGI, ELIZABETH BENSON, veterinarian; b. New Orleans, June 11, 1955; d. Andrew Ernest Jr. and Mary Elizabeth (Steinlage) Benson; m. Marion Rene Surgi, May 22, 1981; children: Reneé Elizabeth, Sara Elizabeth. BS in Med. TEch., U. New Orleans, 1977; DVM, La. State U., 1984. Assoc. veterinarian West Park Vet. Svcs., Houma, La., 1984, Animal Emergency Svc., Schaumburg, Ill., 1984-85; staff veterinarian Anti-Cruelty Soc., Chgo., 1985-86; assoc. veterinarian Terry Animal Hosp., Wilmette, Ill., 1986-89; owner, chief veterinarian Sauganash Animal Hosp., Chgo., 1989—. Bd. dirs. Edgebrook-Sauganash unit Am. Cancer Soc., 1990—. Mem. Am. Vet. Med. Assn., Am. Vet. Dental Soc., Ill. Acad. Vet. Medicine (pres. 1992), Assn. Avian Veterinarians, Ill. State Vet. Med. Assn., Chgo. Vet. Med. Assn. Republican. Episcopalian. Office: Sauganash Animal Hosp 4054 W Peterson Ave Chicago IL 60646-6071

SURGI, MARION RENE, chemist; b. New Orleans, Dec. 19, 1956; s. George Edward and Barbara Ruth (Pearce) S.; m. Elizabeth Benson, May 22, 1981; children: Reneé Elizabeth, Sara Elizabeth. BS, U. New Orleans, 1979; PhD, La. State U., 1981-84. Gen mgr. Superior Amusement Co., New Orleans, 1975-80; rsch. assoc. U. New Orleans, 1980-81; grad. asst. La. State U., Baton Rouge, 1981-84; sr. rsch. chemist Signal Cos., Des Plaines, Ill., 1984-86; rsch. specialist AlliedSignal Corp, Des Plaines, 1986-90, rsch. mgr. in math. simulation sci., 1990—; bd. dirs. SAH, P.C., Chgo. Contbr. articles to profl. jours. Patentee in field. Recipient Merck award, 1981, others. Mem. Am. Chem. Soc., Phi Lambda Upsilon. Republican. Episcopalian. Avocations: bicycle racing, real estate renovation, business and real estate acquisitions. Home: 503 Oakdale Ave Glencoe IL 60022 Office: Allied-Signal 50 E Algonquin Rd Des Plaines IL 60016-6102

SURH, DAE SUK, engineering consulting company executive; b. Seoul, Republic of Korea, Nov. 15, 1945; s. Jung Ik and Young Sook (Lee) S.; m. Theresa Lee, Aug. 16, 1969; children: Gene, Joanne. BS, U. Mich., 1969; MS, Purdue U., 1975, PhD, 1979. Ops. rsch. mgr. LTV Steel, Cleve., 1979-83; cons. World Bank, Republic of Korea, 1983-84, Pohang Steel Co., Republic of Korea, 1984-85, Samsung Electronics, Republic of Korea, 1989-90; pres. Atworth Engring. Co., Seoul, 1986—; adj. prof. Daeku U., Republic of Korea, 1985-86; lectr. Korea Productivity Ctr., Seoul, 1991, High Profit Mgmt. Cons., Seoul, 1991; advisor CIM Korea '91, Seoul, 1991. Contbr. articles to profl. jours. Sgt. Korean Armed Forces, 1970-72. Recipient Lectr.'s award Korean Indsl. Engrs. Soc., 1984, Nat. Engring. Achievement award Min. Sci. and Tech., 1993. Mem. Inst. Indsl. Engrs. (sr.), Korean Chpt. Indsl. Engrs. (pres. 1991—), Korean Simulation Soc. (v.p. 1991—), Inst. Mgmt. Sci. Avocations: golf, tennis, music. Office: Dong Il CIM Engring, 944-1 Daechi Dong Kangnamku, Seoul Republic of Korea

SURIANELLO, FRANK DOMENIC, civil engineer; b. Buffalo, Apr. 12, 1961; s. Domenic and Columba (Fulfaro) S.; m. Catherine Denise Juhre, Oct. 6, 1990; 1 child, Domenic Juhre. BS in Civil Engring., Valparaiso U., 1979-83; postgrad. in bus. administrn., Canisus Coll., 1988-93. Registered profl. engr., N.Y. Pres. Surianello Constrn., Buffalo, 1983—. Mem. ASCE, Am. Concrete Inst. Home: 31 Shalamar Ct Amherst NY 14068 Office: Surianello Constrn 635 Wyoming Ave Buffalo NY 14215

SURMAN, OWEN STANLEY, psychiatrist; b. Boston, Apr. 21, 1943; s. Aaron Harry and Edith Anne (Silver) S.; m. Lezlie Anne Humber, July 19, 1969; children: Craig Bruce Hackett, Kathleen Bridget Lezlie. BSc with honors, McGill U., 1964, MD, CM, 1968. Diplomate Am. Bd. Psychiatry and Neurology. Intern in internal medicine Balt. City Hosp., 1968-69; clin. fellow in medicine Johns Hopkins U., Balt., 1968-69; resident in psychiatry Mass. Gen. Hosp., Boston, 1969-72; clin. fellow in psychiatry Harvard Med. Sch., Boston, 1969-72; clin. asst. in psychiatry Mass. Gen. Hosp., Boston, 1975-76, asst. in psychiatry, 1977-80, asst. psychiatrist, 1980-86, assoc. psychiatrist, 1986-89, psychiatrist, 1990—; instr. psychiatry Harvard U. Med. Sch., Boston, 1975-80, asst. prof., 1980-90, assoc. prof., 1990—; mem., psychiat. cons. Boston Ctr. Heart Transplant; mem. ethics com. Mass. Ctr. Organ Transplantation, 1988—; mem. subcom. Human Studies, Mass. Gen. Hosp., 1982—; mem. Inst. for Study of Smoking Behavior and Policy, John F. Kennedy Sch. Govt., 1982-89. Mem. editorial bd. Jour. Geriatric Psychiatry and Neurology, 1988—; contbr. articles, letters and book revs. to med. jours., chpts. to books. Bd. dirs. Unitarian-Universalist Area Ch., Sherborn, Mass., 1983-86, 93—; advancement officer troop 1 Boy Scouts Am., Sherborn, 1983-91. Lt. comdr. M.C., USNR, 1972-75. Milton Fund grantee, Upjohn Corp. grantee, Burroughs Wellcome Co. grantee, Eli Lily Corp. grantee, 1989. Fellow Am. Psychiat. Assn.; mem. AAAS, Mass. Med. Soc., N.Y. Acad. Scis., Am. Acad. Psychosomatic Medicine, Hastings Ctr. (assoc.), Johns Hopkins Med. and Surg. Soc., Libr. of Boston Athenaeum, Ford Hall Forum. Republican. Avocation: creative writing. Office: Mass Gen Hosp Wang ACC 815 15 Parkman St Boston MA 02114

SURYAVANSHI, O. P. S., corporate executive; b. Basti, India, Oct. 15, 1961; s. S.R.M. Singh and Kamla Singh. BA, Lucknow U., 1983; dip. profl. photography, London Sch., 1985. Chmn. Momco India, Gonda, 1988—; chmn. Omotel's India Pvt. Ltd., journalist, beaurochief, u.p. Editor, author: Prolonged Production, 1988. Fellow Planetary Soc.; mem. Master Photographer Assn., Profl. Photographer of Am., Inc., Royal Overseas League, Bus. Ptnrs. Club (bd. dirs. 1988), Execs. Club (London), Lions. Avocations: adventure works, social activity, reading, photography, worldwide travel. Home: 172 Civil Line, Gonda 271 001, India Office: Omotel's India Pvt Ltd, 1221 Munshi-Gumj, Barabanki 225 001, India

SUSI, ENRICHETTA, chemist, researcher; b. Chieti, Italy, June 7, 1941; d. Lorenzo and Egiziaca (Ricci) S.; m. Raffaele De Maria, June 25, 1965 (div. Dec. 1991); 1 child, Andrea. Grad., U. Bologna, Italy, 1964. Fellow U. Bologna, 1965-69; staff rschr. Nat. Rsch. Coun., Bologna, 1970—; mem. European Material Rsch. Soc. Network, Brussels, 1989—. Co-author: Scientific Authority, Women Authority, 1992; contbr. articles to profl. jours. 1st sec. Trade Union, 1988-92. Mem. Ipazia Woman Scientists Community. Office: CNR-Lamel Inst, Via Gobetti 99, 40100 Bologna Italy

SUSKIND, SIGMUND RICHARD, microbiology educator; b. N.Y.C., June 19, 1926; s. Seymour and Nina Phillips S.; m. Ann Parker, July 1, 1951; children: Richard, Mark, Steven. A.B., NYU, 1948; Ph.D., Yale U., 1954. Research asst. biology div. Oak Ridge Nat. Lab., 1948-50; USPHS fellow NYU Med. Sch., N.Y.C., 1954-56; mem. faculty Johns Hopkins U., Balt., 1956—; prof. biology Johns Hopkins U., 1965—, Univ. prof., 1983—, Univ. ombudsman, 1988-91, dean grad. and undergrad. studies, 1971-78, dean Sch. Arts and Scis., 1978-83; head molecular biology sect. NSF, 1970-71; cons. NIH, 1966-70, Coun. Grad. Schs., Mid States Assn. Colls. and Secondary Schs., 1973—, NSF, 1986; vis. scientist Weizmann Inst. of Sci., Israel, 1985; trustee Balt. Hebrew U., 1985-93; mem. adv. bd. La. Geriatric Ctr., 1990—. Author: (with P.E. Hartman) Gene Action, 1964, 69, (with P.E. Hartman and T. Wright) Principles of Genetics Laboratory Manual, 1965; editor: (with P.E. Hartman) Foundations of Modern Genetics series, 1964, 69; mem. sci. editorial bd. Johns Hopkins U. Press, 1973-76, 88-91. With USNR, 1944-46. NIH grantee, 1957-76. Fellow AAAS; mem. Am. Soc. Microbiology, Genetics Soc. Am., Am. Assn. Immunology, Am. Soc. Biol. Chemistry and Molecular Biology, Coun. Grad Schs., Assn. Grad. Schs., Northeastern Assn. Grad. Schs. (exec. com. 1975-76, pres. 1977-78). Research in microbial biochemical genetics and immunogenetics. Office: Johns Hopkins U Dept Biology and McCollum-Pratt Inst 34th and Charles Sts Baltimore MD 21218

SUSON, DANIEL JEFFREY, physicist; b. Oxnard, Calif., Aug. 19, 1962; s. Morris A. and Roslyn Etta (Menachof) S.; m. Carla Lee Phillips, Dec. 28, 1986; children: Rachel Louise, Avram Oscar. BA, U. Colo., 1984; MS, U. Tex. at Dallas, Richardson, 1986, PhD, 1989. Postdoctoral fellow Waseda U., Tokyo, 1989-90; programmer U. Tex. Southwestern Med. Ctr., Dallas, 1990-91; asst. prof. Tex. A&I U., Kingsville, 1991—; adj. prof. Dallas County C.C., 1991. Contbr. articles to profl. jours. Mem. Am. Phys. Soc. Home: 402 W Lee Kingsville TX 78363 Office: Tex A&I U Physics Dept CB 175 Kingsville TX 78363

SUSSMAN, DEBORAH EVELYN, designer, company executive; b. N.Y.C., May 26, 1931; d. Irving and Ruth (Golomb) S.; m. Paul Prejza, June 28, 1972. Student Bard Coll., 1948-50, Inst. Design, Chgo., 1950-53, Black Mountain Coll., 1950, Hochschule für Gestaltung Ulm, Fed. Republic Germany, 1957-58. Art dir. Office of Charles and Ray Eames, Venice, Calif., 1953-57, 61-67; graphic designer Galeries Lafayette, Paris, 1959-60; prin. Deborah Sussman and Co., Santa Monica, Calif., 1968-80; founder, pres. Sussman-Prejza and Co., Inc., Santa Monica, 1980-90, Culver City, Calif., 1990—; speaker, lectr. UCLA Sch. Architecture, Archtl. League N.Y.C., Smithsonian Inst., Stanford Conf. on Design, Am. Inst. Graphic Arts Nat. Conf. at MIT, Design Mgmt. Inst. Conf., Mass.; spl. guest Internat. Design Conf., Aspen, Colo., Fulbright lectr., India, 1976; speaker NEA Adv. Coun., 1985, Internat. Coun. Shopping Ctrs., 1986, USIA Design in Am. seminar, Budapest, Hungary, 1988, participant exhbn., Moscow, 1989, Walker Art Ctr., Mpls., 1989. Mem. editorial adv. bd. Arts and Architecture Mag., 1981-85, Calif. Mag., Architecture Calif. Fulbright grantee Hochschule für Gestaltung Ulm, 1957-58; recipient numerous awards AIA Nat. Inst. Honors, 1985, 88, Am. Inst. Graphic Arts, Calif. Coun. AIA, Communications Arts Soc., L.A. County Bd. Suprs., Vesta award Women's Bldg. L.A. Fellow Soc. Environ. Graphic Design; mem. AIA (hon.), Am. Inst. Graphic Arts (bd. dirs. 1982-85, founder L.A. chpt., chmn., 1983-84, numerous awards), L.A. Art Dirs. Club (bd. dirs., numerous awards), Alliance Graphique Internat. (elect. mem.), Architects, Designers and Planners Social Responsibility, Calif. Women in Environ. Design (adv. bd.), Trusteeship (affiliate Internat. Women's Forum), SEGD. Democrat. Jewish. Avocation: photography. Office: Sussman/Prejza & Co Inc 3960 Ince Blvd Culver City CA 90232-2635

SUSSMAN, GERALD JAY, electrical engineering educator; b. Bklyn., Feb. 8, 1947; s. Murry and Ethel (Kaplan) S.; m. Julie Esther Mazel, June 18, 1969. SB in Math, MIT, 1968, PhD, 1973. From asst. prof. to prof. MIT, Cambridge, 1973-91, Matsushita prof. elec. engring., 1991—. Author: A Computational Model of Skill Acquisition, 1974, Structure and Interpretation of Computer Programs, 1985; contbr. numerous articles to profl. jours. Recipient Karl Karlstom Edn. award ACM, 1991, MIT Convocation Program award, 1992. Fellow Am. Assn. for Artificial Intelligence, mem. Am. Astron. Soc., Am. Watchmakers Inst. Achievements include design of numerous spl. computers for scientific and engring. applications, including the Digital Orrery (now in Smithsonian Mus.) which was used to determine that the motion of Pluto is chaotic. Office: MIT Dept Elec Engring 77 Massachusetts Ave Cambridge MA 02139

SUSSMAN, KARL EDGAR, physician; b. Balt., May 29, 1929; s. Abram Alan and Sadye Deborah (Silverman) S.; married, 1955; children: Paula Barbara Sussman, Ann Laurie Sussman. BA, Johns Hopkins U., 1951; MD, U. Md., 1955. Diplomate Am. Bd. Internal Medicine. Intern Barnes Hosp., St. Louis, 1955-56; resident U. Colo. Health Scis. Ctr., Denver, 1958-61; asst. prof. medicine U. Colo. Med. Sch., Denver, 1963-68, assoc. prof. medicine, 1968-73, prof. medicine, 1973—; chief of medicine VA Med. Ctr., Denver, 1972-75, clin. investigator, 1975-80, chief med. rsch., 1984—; regional editor Diabetes Rsch. and Clin. Practice, Amsterdam, Netherlands, 1990—. Author/editor: Juvenile Type Diabetes and its Complications, 1971, Clinical Guide to Diabetes Mellitus, 1987. Fellow ACP; mem. Endocrine Soc., Western Assn. of Physicians. Achievements include rsch. on insulin secretion and action. Office: VA Med Ctr 1055 Clermont Denver CO 80220

SUSUMU, KAMATA, organic and medicinal chemist; b. Kurume, Fukuoka, Japan, Sept. 16, 1937. MSc, U. Tokyo, 1963, PhD, 1966. Rschr. Shionogi Rsch. Labs., Shionogi & Co., Ltd., Osaka, Japan, 1966—, gen. mgr. medicinal chemistry, 1990—; rsch. assoc. U. Alb., Can., 1974-75; rsch. fellow Calif. Inst. of Tech., 1975-76. Contbr. articles to profl. jours. Mem. Am. Chem. Soc., Pharm. Soc. of Japan, Chem. Soc. Japan, Soc. of Synthetic Organic Chemistry (Japan). Achievements include patents for medicinal drugs. Home: 1-18-14 Hikarigaoka, Takarazuka-shi 665, Japan Office: 5-12-4 Sagisu Fukushima-ku, Osaka 553, Japan

SUTHERLAND, C. A., metallurgical engineer. Recipient H.T. Airey award Can. Inst. Mining and Metallurgy, 1991. Office: care Xerox Tower Ste 1210, 3400 de Maisonneuve Blvd W, Montreal, PQ Canada H3Z 3B8*

SUTHERLAND, GEORGE LESLIE, retired chemical company executive; b. Dallas, Aug. 13, 1922; s. Leslie and Madge Alice (Henderson) S.; m. Mary Gail Hamilton, Sept. 9, 1961 (dec. Mar. 1984); children: Janet Leslie, Gail Irene, Elizabeth Hamilton; m. Carol Brenda Kaplan, Feb. 19, 1986. BA, U. Tex., Austin, 1943, MA, 1947, PhD, 1950. With Am. Cyanamid Co., various locations, 1951-87; asst. dir. research and devel. Princeton, N.J., 1969-70, dir. research and devel., agr. div., 1970-73; v.p. med. research and devel. Pearl River, N.Y., 1973-86, dir. med. research div., 1978-86, dir. chem. research div., 1980-81; v.p. corp. research tech. Pearl River, 1986-87. Served with USN, 1944-46. Mem. Assn. Research Dirs. (pres. 1975-76), AAAS, Am. Chem. Soc. Home: 42 Sky Meadow Rd Suffern NY 10901-2519

SUTHERLAND, JOHN BRUCE, IV, microbiologist; b. Tampa, Fla., Nov. 9, 1945; s. John Bruce and Lois (Larner) S.; m. Fatemeh Rafii, Dec. 30, 1982. AB in Biol. Sci., Stanford U., 1967; MS in Botany, U. Wis., 1973; PhD in Plant Pathology, Wash. State U., 1978. Postdoctoral fellow U. Idaho, Moscow, 1978-81; asst. prof. biol. sci. Tex. Tech. U., Lubbock, 1981-83; rsch. scientist Mich. Tech. U., Houghton, 1984-86; vis. assoc. prof. bacteriology U. Idaho, Moscow, 1986-88; microbiologist FDA, Jefferson, Ark., 1988—. Mem. editorial bd. Applied and Environ. Microbiology, 1984—, contbr. articles to Mycologia. Vol. Peace Corps, Ethiopia, 1967-69. Mem. Am. Soc. for Microbiology, Mycological Soc. Am., Am. Phytopathol. Soc., Sigma Xi. Methodist. Achievements include research in metabolism of aromatic compounds by fungi and actinomycetes. Home: 810 Green Oak Ln White Hall AR 71602 Office: Nat Ctr Toxicol Rsch FDA Jefferson AR 72079

SUTHERLAND, JOHN CLARK, physicist, researcher; b. N.Y.C.; s. William Benjamin and Julia (Corless) S.; m. Betsy Blake Middleton, Aug. 28, 1965. BS with highest honors, Ga. Inst. Tech., 1962, MS, 1964, PhD, 1967. Postdoctoral researcher Univ. Calif., Berkeley, 1969-72; asst. prof. Coll. Medicine Univ. Calif., Irvine, 1973-76, assoc. prof., 1976-77; biophysicist Brookhaven Nat. Lab., Upton, N.Y., 1977-88, sr. biophysicist, 1988—; mem. biophysics adv. panel Nat. Sci. Found., Washington, 1982-83; adj. prof. Dept. Physiology Sch. Medicine SUNY, Stonybrook, 1993—. Assoc. editor: Photochemistry and Photobiology, 1981-84; editorial bd. Electrophoresis, 1990—; contbr. 120 articles to profl. jours. Capt. U.S. Army, 1967-69. NDEA fellow Ga. Inst. Tech., Atlanta, 1962-65, Oak Ridge Grad. fellow Oak Ridge (Tenn.) Nat. Lab., 1965-67, P.H.S. Postdoctoral fellow NIH, 1969-72; recipient Rsch. Career Devel. award Nat. Cancer Inst., 1976-81; IR-100 award R & D Mag., 1987. Fellow Am. Phys. Soc.; mem. Biophysical Soc., Am. Soc. for Photobiology, European Soc. for Photobiology. Office: Biology Dept Brookhaven Nat Lab Upton NY 11973

SUTIN, NORMAN, chemistry educator, scientist; b. Ceres, Republic of South Africa; came to U.S., 1956; s. Louis and Clara (Goldberg) S.; m. Bonita Sakowski, June 29, 1958; children: Lewis Anthony, Cara Ruth. B.Sc., U. Cape Town (S. Africa), 1948, M.Sc., 1950; Ph.D., Cambridge U. (Eng.), 1953. Research fellow Durham U. (Eng.), 1954-55; research assoc. Brookhaven Nat. Lab., Upton, N.Y., 1956-57, assoc. chemist, 1958-61; chemist Brookhaven Nat. Lab., 1961-66, sr. chemist, 1966—, dept. chmn., 1988—; affiliate Rockefeller U., N.Y.C., 1958-62; vis. fellow Weizmann Inst., Rehovoth, Israel, 1965; vis. prof. SUNY-Stony Brook, 1968, Columbia U., N.Y.C., 1968-69, Tel Aviv U., Israel, 1973-74, U. Calif.-Irvine, 1977, U. Tex. Austin, 1979. Editor: Comments on Inorganic Chemistry Jour., 1980-87; mem. editorial bd. Jour. Am. Chem. Soc., 1985-89, Inorganic Chem., 1986-89, Jour. Phys. Chem., 1987-92; contbr. articles to profl. jours. Mem. NAS, Am. Acad. Arts and Scis., Am. Chem. Soc. (recipient award for disting. svc. in advancement of inorganic chemistry 1983). Office: Brookhaven Nat Lab Dept of Chemistry Upton NY 11973

SUTLIFF, KIMBERLY ANN, psychologist; b. Harrisburg, Pa., Mar. 25, 1964; d. Gregory Leo and Carlene (Samuels) S. BS in Psychology, Juniata Coll., 1986; MA in Psychology, Conn. Coll., 1988. Psychologist, therapist Green Ridge Counseling Ctr., Williamsport, Pa., 1989—. Mem. Harrisburg Area Psychol. Assn., Pa. Psychol. Assn., Am. Psychol. Assn. Democrat. Home: 219 S 14th St Lewisburg PA 17837 Office: Green Ridge Counseling Ctr 829 W 4th St Williamsport PA 17701

SUTNICK, ALTON IVAN, medical school dean, educator, researcher, physician; b. Trenton, N.J., July 6, 1928; s. Michael and Rose (Horwitz) S.; m. Mona Reidenberg, Aug. 17, 1958; children: Amy, Gary. A.B., U. Pa., 1950, M.D., 1954; postgrad. studies in biomed. math., Drexel Inst. Tech., 1961-62; postgrad. studies in biometrics, Temple U., 1969-70. Diplomate Am. Bd. Internal Medicine. Rotating intern Hosp. U. Pa., 1954-55, resident in anesthesiology, 1955-56, resident in medicine, 1956, USPHS postdoctoral research fellow, 1956-57; asst. instr. anesthesiology, then asst. instr. medicine U. Pa. Sch. Medicine, 1955-57; resident in medicine Wishard Meml. Hosp., Indpls., 1957-58; chief resident in medicine Wishard Meml. Hosp., 1960-61; resident instr. medicine Ind. U. Sch. Medicine, Indpls., 1957-58; USPHS postdoctoral research fellow Temple U. Hosp., 1961-63; instr., then asso. in medicine Temple U. Sch. Medicine, 1962-65; mem. faculty U. Pa. Sch. Medicine, 1965-75, assoc. prof. medicine, 1971-75; clin. asst. physician Pa. Hosp., 1966-71; research physician, then assoc. dir. Inst. Cancer Research, Phila., 1965-75; vis. prof. medicine Med. Coll. Pa., Phila., 1971-74; prof. medicine Med. Coll. Pa., 1975—, dean, 1975-89, sr. v.p., 1976-89; v.p. Ednl. Commn. Fgn. Med. Grads, 1989—; dir. clin. devel. Am. Oncologic Hosp., Phila., 1973-75; attending physician Phila. VA Hosp., 1967—, Hosp. Med. Coll. Pa., 1971—; cons. in field. Mem. U.S. nat. com. Internat. Union Against Cancer, 1969-72; mem. Nat. Conf. Cancer Prevention and Detection, 1973, Nat. Workshop Profl. Edn. in High Blood Pressure, 1973, Nat. Cancer Control Planning Conf., 1973; vice chmn. Gov. Pa. Task Force Cancer Control, 1974-76, chmn. com. cancer detection, 1974-76; mem. health research adv. bd. State of Pa., 1976-78; mem. diagnostic research adv. group Nat. Cancer Inst., 1974-78; chmn. coordinating com., comprehensive cancer center program Fox Chase Cancer Center, U. Pa. Cancer Center, 1975; cons. WHO, Govt. of India, 1979, Govt. of Indonesia, 1980, entire S.E. Asia region, 1981, U. Zimbabwe, 1989, Ministry of Health of Poland, 1992, Israel

Sci. Coun., 1992, U. Autonomade Guadalajara, Mex., 1993, Generalitat de Catalunya, Spain, 1993; mem. Nat. Conf. on Med. Edn. Author numerous articles in field.; Asst. editor: Annals Internal Medicine, 1972-75; editorial bd. other med. jours. Bd. dirs. Phila. Coun. Internat. Visitors, 1972-77, Israel Cancer Rsch. Fund, 1975—, Am. Assocs. Ben Gurion U., 1986—, Internat. Med. Scholars Program, 1988-89, Sight Savers Internat., 1988-91; trustee Ednl. Commn. Fgn. Med. Grads., 1987-89; adv. commn. Internat. Participation Phila. '76, 1973-76. Capt. M.C. AUS, 1958-60. Recipient Arnold and Marie Schwartz award in medicine AMA, 1976, Torch of Learning award Am. Friends of Hebrew U., 1981, medal Ben Gurion U. of Negev, Israel, 1985, medal U. Cath. de Lille, France, 1987, medal U. Belgrade, Yugoslavia, 1988, Founder's award and medal Med. Coll. Pa., 1989, St. Thomas Aquinas award Santo Tomas U. Med. Alumni Assn., The Philippines, 1989, medal Kiev Med. Inst., Ukraine, 1991, Benjamin Albagli medal Inst. de Pos-Graduacao Medica Carlos Chagas, Brazil, 1993. Fellow ACP, Coll. Physicians Phila. (censor 1977-86 , councillor 1977-86); mem. Am. Fedn. Clin. Research (pres. Temple U. chpt. 1964-65), Am. Assn. Cancer Research, Am. Soc. Clin. Oncology, Am. Dermatoglyphics Assn., Assn. Am. Cancer Insts., Assn. Am. Med. Colls., Northeast Consortium on Med. Edn. (treas. 1983-89, chmn. 1986-87), Council of Deans of Pvt. Free-Standing Med. Schs. (co-founder, nat. chmn. 1983-85), Pa. Council Deans (chmn. 1987-89), Am. Cancer Soc. (vice chmn. service com. Phila. div. 1974-76, bd. dirs. 1974-80, chmn. awards com. 1976), Am. Lung Assn., AMA, AAAS, Am. Heart Assn., NAFSA-Assn. Internat. Educators, Pan Am. Med. Assn., Phila. Coop. Cancer Assn., N.Y. Acad. Scis., Pa. Heart Assn., Heart Assn. Southeastern Pa., Pa. Med. Soc., Phila. County Med. Soc. (chmn. com. internat. med. affairs 196), Pa. Lung Assn., Phila. Assn. for Clin. Trials (bd. dirs. 1980-81), Health Systems Agy. Southeastern Pa. (gov. bd., exec. com. 1983-87, sec. 1985-87), Am. Assn. Ben Gurion U. (bd. dirs. 1986—), Soc. des Medecins Militaires Français, Internat. Med. Sch. Affiliates Consortium (co-founder, vice chmn. 1985-87), Phi Beta Kappa, Sigma Xi, Alpha Omega Alpha (councillor 1963-65). Discovered assn. of hepatitis B surface antigen with hepatitis; performed 1st studies of pulmonary Surfactant in adult human lung disease; developed cancer screening system based on risk status; pioneer in describing non-A non-B hepatitis, pioneer in showing relationship of body iron stores to cancer susceptibility and life expectancy; organized first symposium on problems of foreign medical graduates; coined word "ergasteric" for lab.-contracted disease; responsible for advances in assessment of clinical competence. Home: 2135 St James St Philadelphia PA 19103-4804

SUTPHEN, ROBERT RAY, mechanical engineer, manager; b. Yakima, Wash., Oct. 22, 1950; s. Robert Forest and Patricia Ann (Bowlby) S.; m. Sandra J. Woelk, Sept. 26, 1972 (div. 1981); children: Elizabeth Ann, Robecca Ray; m. Patricia B. Fendall, June 24, 1989; 1 child, Christopher Michael. BSME, Wash. State U., 1973. Plant mgr. Yakima (Wash.) Machine & Fdry., 1973-75; irrigation sys. design engr. Orchard-Rite Ltd., Yakima, 1975-76, mech. equip. design engr., 1976-91, product liability and safety mgr., 1991—; cons. Gleed Water Assn., 1984-85. Inventor in field. Fire capt. and e.m.t. Gleed Fire Dept., 1979— (Fireman of the Yr. 1980, 87). Republican. Baptist. Avocations: antique car restoration, computer hardware assembly, hiking, biking. Home: 441 Pleasant Valley Rd Yakima WA 98908-9697 Office: Orchard-Rite Ltd PO Box 9308 Yakima WA 98909-0308

SUTTER, CARL CLIFFORD, civil engineer; b. Waukegan, Ill., Mar. 7, 1957; s. Clifford George and Bertha Martina (Teer) S.; m. Elizabeth Ramlet, June 28, 1980; children: Rachel, Katherine. BS, U. Wis., 1981. Registered profl. engr., Wis. Resident engr. Ill. Dept. Transp., Dixon, 1980-84; staff engr. City of Fond du Lac, Wis., 1984-87; sr. project mgr. McMahon Assocs., Inc., Menasha, Wis., 1987—. Mem. ASCE (engring. mgmt. at project level 1992—, pres. Fox River Valley br. 1980-91, dir. Wis. sect. 1991—), Wis. Soc. Profl. Engrs. Home: 1831 S Bouten St Appleton WI 54915 Office: McMahon Assocs Inc 1377 Midway Rd PO Box 405 Menasha WI 54952-0405

SUTTER, JOSEPH F., aeronautical engineer, consultant, retired airline company executive; b. Seattle, Wash., Mar. 21, 1921; m. Nancy Ann French, June 14, 1943. B.A., U. Wash., 1943. Various engring. positions Boeing Comml. Airplane Co., Seattle, 1946-65, dir. engring. for Boeing 747, 1965-71, v.p., gen. mgr. 747 div., 1971-74, v.p. program ops., 1974-76, v.p. ops. and product devel., 1976-81, exec. v.p., 1981-86, cons., 1986-87; cons. Boeing Comml. Airplane Co., 1987—; chmn. aerospace safety adv. panel NASA, 1986; mem. Challenger Accident Commn., 1986. Served to lt. j.g. USN, 1943-45. Recipient Master Design award Product Engring. mag., 1965, Franklin W. Kolk Air Transp. Progress award Soc. Aero. Aerospace Coun., 1980, Elmer A. Sperry award, 1980, Nuts & Bolts award Transport Assn., 1983, Nat. Medal Tech., U.S. Pres. Reagan, 1985, Sir Kingsford Smith award Royal Aero. Soc. in Sydney, 1980, Wright Bros. Meml. Trophy, 1986; Joseph F. Sutter professorship established in his honor at U. Wash., Boeing Co., 1992. Hon. fellow AIAA (Daniel Guggenheim award 1990), Royal Aero. Soc. Gt. Brit.; mem. Internat. Fedn. Airworthiness (pres. 1989). Office: Boeing Comml Airplane Co PO Box 3707 Mail Stop 13-43 Seattle WA 98124

SUTTERBY, LARRY QUENTIN, internist; b. North Kansas City, Mo., Sept. 11, 1950; s. John Albert and Wilma Elizabeth (Henry) S.; m. Luciana Rises Magpuri, July 5, 1980; children: Leah Lourdes, Liza Bernadette. BA in Chemistry, William Jewell Coll., 1972; MD, U. Mo., Kans. City, 1976. Diplomate Am. Bd. Internal Medicine. Resident in internal medicine Mt. Sinai Hosp., Chgo., 1976-79; physician Mojave Desert Health Svc., Barstow, Calif., 1979-86; pvt. practice Barstow, 1986—; med. dir. Rimrock Villa Convalescent Hosp., Barstow, 1986-89, Mojave Valley Hospice, 1983—. Recipient Loving Care award Vis. Nurse Assn. Inland Counties, 1988. Mem. AMA, Calif. Med. Assn., San Bernadino County Med. Soc., Am. Soc. Internal Medicine, Am. Geriatric Soc., Acad. Hospice Physicians, Soc. Gen. Internal Medicine, Physicians Who Care, Am. Numismatic Assn., Combined Orgns. Numismatic Error Collectors Am. Democrat. Roman Catholic. Avocations: astronomy. Office: 209 N 2d Ave Barstow CA 92311

SUTTON, GEORGE W., aerospace company executive. BME, Cornell U.; PhD, Calif. Inst. Tech. Dir. Kaman Aerospace Corp., Tucson; v.p. Jaycor, San Diego, Helionetics, San Diego, AVCO, Everett, Mass.; scientific adv. USAF, Washington; mgr. GE, King of Prussia, Pa.; with Aero Thermo Tech., Arlington, Va. Author: Engineering Aspects of MHD, 1964, Engineering Magnetohydrodynamics, 1965, Direct Energy Conversion, 1967; editor-in-chief AIAA Jour., 1968—; contbr. papers to refereed tech. jours. Fellow AAAS, AIAA (Thermophysics award 1980, Svc. award 1988). Achievements include pioneering development of ablation heat protection for hypersonic flight (inorganic reinforced plastic) and successful flight test; conceptual design of high-power laser; design of high-energy laser mounted in military vehicle, kinetic energy interceptors. Office: PO Box 15627 Arlington VA 22215-0627

SUTTON, HARRY ELDON, geneticist, educator; b. Cameron, Tex., Mar. 5, 1927; s. Grant Edwin and Myrtle Dovie (Fowler) S.; m. Beverly Earlene Jewell, July 7, 1962; children: Susan Elaine, Caroline Virginia. B.S. in Chemistry, U. Tex., Austin, 1948, M.A., 1949; Ph.D. in Biochemistry, U. Tex., 1953. Biologist U. Mich., 1952-56, instr., 1956-57, asst. prof. human genetics, 1957-60; asso. prof. zoology U. Tex., Austin, 1960-64; prof. U. Tex., 1964—, chmn. dept. zoology, 1970-73, asso. dean Grad. Sch., 1967-70, 73-75, v.p. for research, 1975-79; mem. adv. council Nat. Inst. Environ. Health Scis., 1968-72, council sci. advs., 1972-76; mem. various coms. Nat. Acad. Scis.-NRC; cons. in field; bd. dirs. Associated Univs. for Research in Astronomy, 1975-79, Argonne Univs. Assn., 1975-79, Univ. Corp. for Atmospheric Research, 1975-79, Associated Western Univs., 1978-79. Author: Genes, Enzymes, and Inherited Disease, 1961, An Introduction to Human Genetics, 1988, Genetics: A Human Concern, 1985; editor: First Macy Conference on Genetics, 1960, Mutagenic Effects of Environmental Contaminants, 1972, Am. Jour. Human Genetics, 1964-69. Trustee S.W. Tex. Corp. Public Broadcasting, 1977-80, sec., 1979-80; bd. dirs. Ballet Austin, 1978-84; mem. Austin Arts Commn., 1991—. Served with U.S. Army, 1945-46. Mem. AAAS, Am. Soc. Human Genetics (dir. 1961-69, pres. 1979), Genetics Soc. Am., Am. Soc. Biochem. and Molecular Biology, Am. Chem. Soc., Tex. Genetics Soc. (pres. 1979), Environ. Mutagen Soc., Am. Genetic Assn. Club: Headliners (Austin). Achievements include research and publications in human genetics. Home: 1103 Gaston Ave Austin TX 78703-2507 Office: U Tex Dept Zoology 528 Patterson Laboratories Austin TX 78712

SUYDAM, PETER R., clinical engineer, consultant; b. Jersey City, Apr. 1, 1945; s. Stedman Mills and Winifred M. (Murphy) S.; m. Patricia Cunniff, Feb. 2, 1970 (dec. 1976); m. Jaimy Slifka, Feb. 11, 1978; children—Rycken Stedman, Stephen Michael. Student in engring. Rensselaer Poly. Inst.; student in pre-medicine, psychology, U. Rochester; B.S. in Bio-Engring., U. Ill.-Chgo., 1975. Cert. clin. engr.; cert. health care safety profl. Dir. clin. engring. Rush-Presbyn.-St. Luke's Med. Ctr., Chgo., 1975-81; pres. Syzygy, Inc., Chgo., 1978-81; lead auditor quality assurance Callaway Nuclear Power Plant, Union Elec. Co., St. Louis, 1981-84; sr. cons. Ellerbe Assocs., Inc., Mpls., 1984-86; div. mgr. CH Health Technologies, Inc., St. Louis, 1986—; project mgr. Landmark Contract Mgmt., Inc., 1988-89; dir. healthcare tech. planning The Cannon Corp., 1989-91; tech. advisor New V.I.P. Hosp., Riyadh, Saudi Arabia, 1991-92; pres. Analytic Systems Co., St. Louis, 1991—; staff cons. Joint Commn. on Accreditation for Hosps., Chgo., 1978-81; mem. tech. com. Safe Use of Electricity in Patient Care Areas of Health Care Facilities; mem. Bd. Examiners for Clin. Engring. Cert., 1980-85; com. mem. Midwest Med. Group Standards, Chgo. Hosp. Council, 1976-81. Contbr. articles to profl. jours. Served with USN, 1967-73. Mem. Assn. Advancement Med. Instrumentation (elec. safety com. 1980—), AAAS, IEEE (chpt. chmn. group on engring. in medicine and biology), Instrument Soc. Am., Am. Nat. Standards Inst., Am. Hosp. Assn., Nat. Fire Protection Assn. (health care, elec. and engring. sects.), Am. Soc. Hosp. Engrs., Am. Soc. Quality Control. Current work: Biotechnology applications in medicine and industry; quality assurance-all fields. Subspecialties: Biomedical engineering; Clinical engineering. Office: Tech Assessment PO Box 13577 Saint Louis MO 63138

SUZUKI, AKIRA, chemistry educator; b. Mukawa, Hokkaido, Japan, Sept. 12, 1930; s. Sadasuke and Na-e Suzuki; m. Yohko Iwahori, Oct. 12, 1957; children: Eriko, Rikako. BS in Chemistry, Hokkaido U., Sapporo, Japan, 1954, MSc in Chemistry, 1956, PhD in Chemistry, 1959. Rsch. asst. dept. chemistry Hokkaido U., 1959-61, assoc. prof. synthetic chemistry, 1961-73, prof. applied chemistry, 1973—; postdoctoral Purdue U., West Lafayette, Ind., 1963-65; vis. prof. applied chemistry Tohoku U., Sendai, Japan, 1982-83, dept. chemistry, 1989-90; vis. prof. indsl. chemistry Tokyo Inst. Tech., 1987-88, dept. chemistry,1 988-89; vis. prof. chemistry Osaka (Japan) U., 1987-88, U. Coll. Swansea, U.K., 1988; vis. prof. indsl. chemistry Tokyo U., 1989-90; Japanese rep. com. mem. Internat. Meeting on Boron Chemistry (IUPAC), Lodnon, 1987—. Contbr. over 290 articles to chemistry jours. Pres. Oh-Asa S-Town Assn., Hokkaido, 1991-92. Recipient Chem. Soc. Japan award, 1989, W.W. Lectureship, Eastman Kodak Co., Rochester, N.Y., 1986; Testimonial, Korean Chem. Soc., Seoul, Korea, 1987; Japan Ministry of Edn. grantee Tokyo, 1985-92. Mem. Am. Chem. Soc., Chem. Soc. Japan (v.p. 1992-93), Soc. Synthetic Organic Chemistry Japan. Avocations: reading, traveling. Home: Oh-Asa Naka-Machi 11-3, Ebetsu-shi Hokkaido 069, Japan Office: Hokkaido U, Dept Applied Chemistry, Sapporo Hokkaido 060, Japan

SUZUKI, FUJIO, immunologist, educator, researcher; b. Shibayama, Japan, June 25, 1946; came to U.S., 1980; s. Takeshi and Kimie S.; m. Katsuko Eda, Oct. 4, 1969; children: Emi, Sumihiro. BA in English Lit., Tohoku-Gakuin U., Sendai, Japan, 1968; PhD in Bacteriology, Sch. Medicine Tohoku U., Sendai, 1975. Postdoctoral fellow U. Tex. Med. Br., Galveston, 1980-82, asst. prof., 1982-84, assoc. prof., 1987-91, prof., 1991—; assoc. prof. Kumamoto (Japan) U., 1984-87; mem. sci. staff Shriners Burns Inst., Galveston, 1987—. Contbr. articles to profl. jours. Pres. Japanese Alumni Assn. Galveston, 1990—. Recipient Nohagi Rsch. award Tohoku U., 1972, James W. McLaughlin award U. Tex. Med. Br., 1980; Shriners N.Am. grantee U. Tex. Med. Br., 1987 (2), 1992. Mem. Am. Soc. Microbiology, Am. Assn. Cancer Rsch., Internat. Soc. Antiviral Rsch., N.Y. Acad. Scis. Republican. Buddhist. Achievements include discovery that suppressor T cells and suppressor macrophages are generated by the stimulation of thermal injury, that interferon-gamma is produced by the administration of various immunomodulators into man and animals, others. Home: 7714 Chantilly Cir Galveston TX 77551 Office: U Tex Med Br H-82 Clay Hall Galveston TX 77555

SUZUKI, JON BYRON, periodontist, educator; b. San Antonio, July 22, 1947; s. George K. and Ruby (Kanaya) S. BA in Biology, Ill. Wesleyan U., 1968; PhD magna cum laude in Microbiology, Ill. Inst. Tech., 1971; DDS magna cum laude, Loyola U., 1978. Med. technologist Ill. Masonic Hosp. and Med. Ctr., Chgo., 1966-67; instr. lab. in histology and parasitology Ill. Wesleyan U., Bloomington, 1967-68; med. technologist Augustana Hosp., Chgo., 1968-69; rsch. assoc., instr. microbiology Ill. Inst. Tech., Chgo., 1968-71; clin. rsch. assoc. U. Chgo. Hosps., 1970-71; clin. microbiologist St. Luke's Hosp. Ctr., Columbia Coll., Physicians and Surgeons, N.Y.C., 1971-73; assoc. med. dir. Paramed. Tng. and Registry, Vancouver, B.C., Can., 1973-74; dir. clin. labs. Registry of Hawaii, 1973-74; chmn. clin. labs. edn. Kapiolani Community Coll., U. Hawaii, Honolulu, 1974; lectr. periodontics, oral pathology Loyola U. Med. Ctr., Maywood, Ill., 1974-90; lectr. stomatology Northwestern U. Dental Sch., Chgo., 1982—; NIH rsch. fellow depts. pathology and periodontics Ctr. for Rsch. in Oral Biology, U. Wash.-Seattle, 1978-80; prof. dept. periodontics and microbiology U. Md. Coll. Dental Surgery, Balt., 1980-90; assoc. prof. div. dentistry and oral and maxillofacial surgery The Johns Hopkins Med. Inst., Balt., 1982—; practice dentistry specializing in periodontics Balt., Pitts.; dean Sch. Dental Medicine, U. Pitts.; cons. Dentsply Internat., York, Pa., U.S. Army, Walter Reed Med. Ctr., Washington, USN, Nat. Naval Med. Command, Bethesda, The Nutra Sweet Co., Deerfield, Ill.; cons. Food and Drug Adminstrn., Rockville, Md.; mem. Oral Biology/medicine study sect. NIH, Bethesda, 1985-90; vis. scientist to Moscow State U., USSR, 1972, NASA, Houston, 1976—; lectr. Internat. Congress Allergology, Tokyo, 1973; lab. dir. Hawaii Dept. Health. Author: Clinical Laboratory Methods for the Medical Assistant, 1974; mem. editorial bd. Am. Health Mag.; contbr. articles on research in microbiology, immunology and dentistry to sci. jours. Instr. water safety ARC, Honolulu, 1973—. Recipient Pres.'s medallion Loyola U., Chgo., 1977; named Alumnus of Yr., Wesleyan U., 1977. Fellow Acad. Dentistry Internat., Am Coll. Dentists, Internat. Coll. Dentists, Am. Coll. Stomatognathic Surgeons; mem. AAAS, ADA, AAUP, Am. Acad. Periodontology (diplomate), Am. Inst. Biol. Scis., Internat. Soc. Biophysics, Internat. Soc. Endocrinologists, Ill. Acad. Sci. (chmn. microbiology session of 65th ann. meeting 1972), Am. Internat. Assn. Dental Rsch. (pres. Md. chpt.), Am. Acad. Microbiology (diplomate), Am. Bd. Microbiology (examiner), N.Y. Acad. Scis., Sigma Xi, Omicron Kappa Upsilon (nat. pres.), Beta, Beta. Home: 3501 Terrace St Pittsburgh PA 15261-0001 Office: U Pitts Sch Dental Medicine Dean's Office Pittsburgh PA 15261

SUZUKI, KUNIHIKO, biomedical educator, researcher; b. Tokyo, Japan, Feb. 5, 1932; came to U.S., 1960; s. Nobuo and Teiko (Suzuki) S.; m. Kinuko Ikeda, Dec. 20, 1960; 1 child, Jun. BA in History and Philosophy of Sci., Tokyo U., 1955, MD, 1959; MA (hon.), U. Pa., 1971. Diplomate Nat. Bd. Med. Licensure Japan. Rotating intern USAF Hosp. Tachikawa, Tokyo, Japan, 1959-60; asst. resident in neurology Bronx (N.Y.) Mcpl. Hosp. Ctr.-Albert Einstein Coll. Medicine, 1960-61, resident in neurology, 1961-62, clin. fellow in neurology, 1962-64; instr. in neurology Albert Einstein Coll. Medicine, Bronx, 1964, asst. prof., 1965-68; assoc. prof. U. Pa. Sch. Medicine, Phila., 1969-71, prof. neurology and pediatrics, 1971-72; prof. neurology Albert Einstein Coll. Medicine, 1972-86, prof. neurosci., 1974-86; prof. neurology and psychiatry, faculty curriculum in neurobiology U. N.C. Sch. Medicine, Chapel Hill, 1986—, dir. Brain and Devel. Rsch. Ctr., 1986—; staff dept. neuropsychiatry Tokyo U. Faculty Medicine, 1960, U. Pa. Inst. Neurol. Scis., 1969-72; attending physician Bronx Mcpl. Hosp. Ctr., 1976-86, Hosp. Albert Einstein Coll. Medicine, 1977-86; vis. prof. fellowship Japan Soc. for Promotion Sci., 1980, Yamada Sci. Found., 1981; mem. neurology B study sect. NIH, 1971-75, guest scientist, 1984-85, program com. mental retardation and devel. disabilities, 1989-92; mem. basic neurosci. task force Nat. Inst. Neurol. and Communicative Disorders and Stroke, 1978, adv. panel directions and opportunities for future research, 1983, bd. sci. counselors, NIH, 1980-84; mem. adv. com. on fellowships Nat. Multiple Sclerosis Soc., 1974-77; jury St. Vincent Internat. award for Med. Sci., 1979; mem. adv. com. Eunice Kennedy Shriver Ctr., Waltham, Mass., 1974-84; med. adv. bd. Children's Assn. for Research on Mucolipidosis Type IV, 1983—; mem. U.S. Nat. Com. for Internat. Brain Research Orgn., 1985-89. Editor: Ganglioside Structure and Function, 1984; chief editor Jour. Neurochemistry, 1977-82, dep. chief editor, 1975-77; mem. editorial bd. Jour. Neuropathology and Exptl. Neurology, 1981-83, Neurosci., 1975—, Molecular Chem. Neuropathology, 1983—, Neurochem. Research, 1985—, Metabolic Brain Disease, 1985-87, Molecular Brain Research, 1985—, Jour. Molecular Neurosci., 1987—, Developmental Neurosci., 1987—, Jour. Neurosci. Rsch., 1993—; contbr. articles to profl. jours. Mem. Nat. Adv. Commn. on Multiple Sclerosis, 1973-74; mem. med. adv. bd. United Leukodystrophy Found., 1982-86, Nat. Tay-Sachs and Allied Diseases Assn., 1971—, Canavan Found., 1992—. Recipient A. Weil award Am. Assn. Neuropathologists, 1970, M. Moore award Am. Assn. Neuropathologists, 1975, Jacob K. Javits Neurosci. Investigator award NIH, 1985, 92, Humboldt Sr. Rsch. award Humboldt Found., 1990. Mem. Am. Soc. for Neurochemistry (pres. 1985-87, coun. 1973-77, 87-91), Internat. Soc. for Neurochemistry (coun. 1987-89, treas. 1989-93, pres. 1993—), Soc. for Neurosci., Am. Soc. Biochemistry and Molecular Biology, Am. Acad. Neurology, NAS, AAAS, Japanese Med. Soc. Am. (Disting. Scientist award 1985), Japanese Neurochem. Soc., Internat. Brain Rsch. Orgn., Am. Soc. Human Genetics, Japan Soc. Inherited Metabolic Disease (hon.). Club: University (Durham, N.C.). Avocations: piano, photography, bird watching, skiing. Office: U NC Chapel Hill Brain & Devel Rsch Ctr Campus Box 7250 Chapel Hill NC 27599-7250

SUZUKI, NOBUTAKA, chemistry educator; b. Nishio, Aichi, Japan, Nov. 8, 1942; s. Kihachiro and Masayo (Miwa) S.; m. Fumiko Sato, Mar. 21, 1971; children: Mina, Kumi. B of Chemistry, Nagoya U., Japan, 1966, D of Chemistry, 1972. Asst. prof. dept. chemistry Mie U., Tsu, Japan, 1971-88, assoc. prof., 1988; sr. rschr., group leader Biophoton project JRDC, Sendai, Japan, 1988-90; assoc. prof. Shimonoseki (Japan) Nat. U. Fisheries, 1990-92, prof., 1993—. Author: Natural Products Chemistry, 1975, 2d rev. edit., 1983, Bioluminescence of Chemiluminescence, Current Status, 1991, Oxygen Radicals, 1992; editor (book) The Roles of Oxygen in Chemistry and Biochemistry, 1988, (book/tape) Scientific English in Fisheries, 1992. Grantee Naito Meml. Found., 1977, Tokai Sci. Rsch. Found., 1986, Argl. Biological Chemistry Japan, 1990, Kiei-Kai Sci. Rsch. Found., 1991—, Skylark Rsch. Found., 1992. Mem. Am. Chem Soc., Am. Soc. for Photobiology, Argl. Biological Soc. Japan, Chem. Soc. Japan. Office: Nat U Fisheries, Yoshimi, Shimonoseki Yamaguchi 75965, Japan

SUZUKI, TAIRA, physics educator; b. Kawasaki, Japan, Dec. 3, 1918; s. Heijiro and Ishi S.; m. Kayo Fujita, Mar. 18, 1945; children: Satoshi, Jun, Ryo. MA, Hokkaido U., Sapporo, 1945; PhD, Hokkaido U., 1958. Assoc. prof. Tohoku U., Sendai, 1949-59; prof. Tokyo U., Inst. Solid State Physics, 1959-79, dir., 1968-73; prof. dept. applied physics Sci. U. Tokyo, 1979-88, prof. dept. material sci., 1988—; emeritus prof. Tokyo U., 1979—; scholar Brit. Council, Bristol U., Eng., 1955-56; asst. prof. U. Ill., Urbana, 1956-58; chmn. council Inst. Materials Research, Tohoku U., Sendai, 1986—; chmn. Kaya Conf. steering com., Japan Council of Promotion Sci., 1988—; mem. organizing com. of internat. conf. Fundamentals of Fracture, 1981—. Author: Dislocation Dynamics, 1991; author/editor: Science of Precious Metals, 1991; contbr. articles to profl. jours. Recipient 30th Gold Medal, Japan Inst. Metals, 1985; Honda prize, Meml. Soc. of Prof. Kotaro Honda, 1986. Mem. Phys. Soc. Japan, Japan Inst. Metals (hon.). Home: Hisamoto 1-17-1-517 Takatsu, Kawasaki 213, Japan Office: Dept Matl Sci and Tech, Sci Univ Tokyo, Noda Yamazaki 278, Japan

SUZUKI, TETSUYA, biochemistry educator; b. Hamamatsu, Shizuoka, Japan, Oct. 5, 1942; s. Kenji and Nobue (Takagi) S.; m. Machiko Suzuki, May 12, 1973; 1 child, Shingo. BS in Agr., Shizuoka U., 1966; MS in Agr., Kyoto U., 1968, D Agr., 1975. Rsch. asst. Rsch. Inst. Food Sci., Kyoto (Japan) U., Uji, 1968-75, rsch. assoc., 1975-80, instr., 1982-87, lectr., 1988; rsch. assoc. Okla. Med. Rsch. Found., Oklahoma City, 1980-81; assoc. prof. faculty fisheries Hokkaido U., Hakodate, Japan, 1988—; cons. Hakodate High Tech. Innovation Ctr., 1989-90; vol. Coll. & Univ. Partnership Program, Memphis. Vol. N.Mex. State Govt., Santa Fe, 1985-86. Sci. rsch. grantee Ministry Edn., Japan, 1978-80, 83-85, 86-88, 89-90. Mem. AAAS, Am. Chem. Soc., Coun. Agrl. Sci. and Tech., Soc. Free Radical Rsch., Japanese Biomed. Soc., Japan Soc. for Biosci., Biotech., Agrochemistry, N.Y. Acad. Scis. Lutheran. Achievements include research on physiologically functional phospholipid, functionality of water in biological systems, reactive oxygen stress to animals. Home: 1-102-43 Honcho, Hokkaido, Hakodate 040, Japan Office: Hokkaido U Faculty Fisheries, Minato, Hakodate 041, Japan

SUZUKI, YUICHIRO JUSTIN, biomedical scientist; b. Tokyo, Apr. 24, 1962. MS, Med. Coll. Va., 1989, PhD, 1991. Postdoctoral fellow U. Calif., Berkeley, 1991—. Contbr. articles to Free Radical Biology and Medicine, Biochemistry, IEEE Proceedings, Am. Jour. Physiology. A.D. Williams fellow Med. Coll. Va., Richmond, 1986, Grad. fellow 1989, Postdoctoral fellow Am. Heart Assn., Calif., 1992. Mem. Internat. Soc. for Free Radical Rsch., Biophysical Soc., Bay Area Oxygen Club (co-organizer 1991—), Sigma Xi. Achievements include modified technique for the measurement of sulfhydryl groups oxidized by reactive oxygen; the use of network thermodynamic modeling in free radical rsch.;inhibition of Ca-ATPase of vascular smooth muscle sarcoplasmic reticulum by superoxide radicals; presence of xanthine oxidase in smooth muscle; thioctic acid and dihydrolipoic acid are novel antioxidants; specifity of superoxide among homologous proteins; cigarette smoke exposure increases the cell water organization and membrane order of cultured T cells; alpha-lipoic acid prevents glucose-induced protein structural modifications and tumor necrosis factor-induced nuclear factor kappa B activation; discovery that superoxide stimulates inositol trisphosphate-induced calcium release. Office: Univ Calif 251 Life Scis Addition Berkeley CA 94720

SVAASAND, LARS OTHAR, electronics researcher; b. Oslo, Feb. 3, 1938; married; 4 children. MSc, Norwegian Inst. Tech., Trondheim, 1961, PhD, 1976. Rsch. scientist div. radar and electronics Norwegian Def. Rsch. Establishment, Kjeller, 1961-65; rsch. scientist div. theoretical electronics U. Trondheim, 1966-69, asst. prof., 1971-74, assoc. prof. phys. electronics, 1974-81, head dept. elec. engring. and computer sci., 1984-87; rsch. scientist Electronics Rsch. Labs., ELAB, Trondheim, 1969-71, head electro optics rsch. group, 1972-76; sci. advisor Continental Shelf and Petroleum Tech. Rsch. Inst., Iku, Norway, 1984—; prof. phys. electronics Norwegian Inst. Tech., U. Trondheim, 1982—; vis. and cons. prof. U. So. Calif. Sch. of Medicine, L.A., 1989—, U. Calif., Irvine, 1990—; bd. dirs. Norwegian Inst. Tech., 1990—, SINTEF Rsch. Group, Norway. Author books; co-editor: Lasers in Medical Science, 1985. Fellow Am. Soc. Laser Medicine and Surgery; mem. Am. Phys. Soc., Am. Soc. Photobiology, Internat. Soc. Optical Engring., Inst. Elec. Engrs., Norwegian Acad. Tech. Scis., N.Y. Acad. Scis., Royal Norwegian Soc. Scis. Office: U Trondheim Inst Tech, O.S. Bragstads plass 4,, Trondheim N-7034, Norway

SVE, CHARLES, mechanical engineer; b. Pana, Ill., Feb. 21, 1940; s. Erling and Mae Silvey (Priest) S.; m. Ruth Vivian Goodwin, Sept. 15, 1962; children: Charles Harold, Jennifer Ruth. BSCE, MIT, 1962, MSCE, 1963; PhD, Northwestern U., Evanston, Ill., 1968. Sr. engr. Space and Info. Systems, N.Am. Aviation, Downey, Calif., 1963-64, Missile Systems Div., Wilmington, Mass., 1964-66; sr. scientist The Aerospace Corp., El Segundo, Calif., 1968—. Mem. ASME, Am. Acad. of Mechanics. Office: Aerospace Corp PO Box 92957 Los Angeles CA 90009

SVEBAK, SVEN EGIL, psychology educator; b. Verdal, Middle Norway, Norway, Dec. 17, 1941; s. Hans Georg and Dagny (Strand) S.; m. Randi Myrseth, Apr. 2, 1971; children: Annette, Teresa. Grad. in Psychology, U. Oslo (Norway), 1970; D in Philosophy, U. Bergen, Norway, 1982. Lic. Psychologist. Instr. of psychology U. Oslo, 1967-68; rsch. asst. U. Bergen, 1968-70, asst. prof., 1970-76, assoc. prof. to prof., 1976—; vis. prof. Queens U. of Belfast, No. Ireland, 1987; adj. prof. U. Utah, 1990; prof. of medicine, U. of Trondheim, Norway, 1993—; lectr. of Behavioral Medicine Physiotherapy Coll. of Bergen, 1977-92; cons. in Archtl. Design CUBUS A-S, 1970—, Inst. Psychol. Counseling, Bergen, 1988—; co-organizer 3d Internat. Conf. Reversal Theory, Amsterdam, 1987; organizer 6th Internat. Conf. Reversal Theory, Bergen, 1993. Editor: Psychological Service Armed Forces, 1986; assoc. editor Internat. Jour. of Psychophysiology, 1983—; contbr. articles to profl. jours. Recipient numerous research grants, 1972—. Mem. Soc. for Psychophysiological Research, Psychophysiology Soc., Internat.

Orgn. of Psychophysiology, Reversal Theory Soc., Internat. Soc. for the Study of Individual Differences, Internat. Soc. Humor Studies. Avocations: boating, fishing, wildlife, jogging, skiing. Office: U Bergen Dept Biol & Med Psychology, Arstadveien 21, N-5009 Bergen Norway

SVENSON, ERNEST OLANDER, psychiatrist, psychoanalyst; b. Duluth, Minn., Oct. 16, 1923; s. Ernest G. and Mabel A. (Benson) S.; m. Raquel Lefevre, 1954 (div. 1965); children: Ernest E., Stuart K.; m. Shirley Zupancic, 1982. BS, Wayne State U., 1948, MD, 1952; BA, Augustana Coll., 1948. Diplomate Am. Bd. Psychiatry and Neurology. Intern Gorgas Hosp., C.Z., 1952-53, staff physician, 1953-54; resident Charity Hosp., New Orleans, 1954-57; psychoanalytic trainee New Orleans Psychoanalytic Inst., 1958-62; pvt. practice, 1958—; assoc. prof. psychiatry La. State U., 1962-80, clin. prof. psychiatry, 1980—; chmn. dept. psychiatry Touro Infirmary, 1973-76; mem. Gov.'s Adv. Com., Mental Health, 1971-72; cons. S.E. La. State Hosp., 1958-61; sr. vis. physician, Charity Hosp., 1965—; clin. prof. psychiatry Tulane U. Med. Sch., 1983—. Bd. dirs. New Orleans Area/Bayou River Health Systems Agy., 1978-82, exec. com., 1979-82, chmn. project rev. com., 1978-79; mem. Area Health Planning, 1971-82. Lt. (j.g.) USNR, 1942-46. Fellow Am. Psychiatric Assn.; mem. AAAS, Am. Psychoanalytic Assn. (com. new tng. facilities 1988-92), Internat. Psychoanalytic Assn., La. Med. Assn., Am. Coll. Psychiatrists, New Orleans Mental Health Assn. (bd. dirs. 1975-77), New Orleans Area Psychiatry Assn. (pres. 1969-70), La. Psychiatric Assn. (pres. 1969-71), New Orleans Psychoanalytic Soc. (pres. 1971-73), New Orleans Psychoanalytic Inst. (sec.-treas. 1975, tng. supr. analyst 1972—, pres. 1983-85, chmn. edn. com. 1983-85), N.Y. Acad. Scis., Alpha Omega Alpha. Home: 123 Walnut St Apt 1001 New Orleans LA 70118-4846 Office: 1301 Antonine St New Orleans LA 70115-3685

SVETIC, RALPH E., electrical engineer, mathematician; b. Chgo., Aug. 20, 1948; s. Ralph F. and Patricia M. (Lord) S.; m. Lynn D. Zapolski, Feb. 13, 1987. BS in Chemistry (with distinction), U. Ill., Chgo., 1970; MS in Engring., U. Wis., Milw., 1989. Chemist Nalco Chem., Chgo., 1971-76, Mobil Oil Corp., Milw., 1977-81; engr. Fiatron Systems Inc., Milw., 1981-86, Lachat Instruments, Milw., 1986, Biochem Internat. Inc., Waukesha, Wis., 1986-88; sr. engr. BCI Internat., Waukesha, 1989—. Mem. IEEE. Achievements include patent for food process antifoam; development of discrete squarewave transform. Home: Box 538 Milwaukee WI 53201 Office: BCI Internat W238 N1650 Rockwood Dr Waukesha WI 53188

SVIDERSKY, VLADIMIR LEONIDOVICH, neurophysiologist; b. Leningrad, Russia, Sept. 19, 1931; s. Leonid and Anastasia (Galkina) S.; m. Galina Evgenjevna Sviderskaya, July, 23, 1960; children: Elena Vladimirovna Sviderskaya. MD, Military Medical Acad., Russia, 1956; PhD in Biological Sci. (hon.), Pavpov Inst., Russia, 1970; Academician (hon.), USSR Acad. Sci., Russia, 1987. Cert. Physiologist, Russia. Sci. worker Sechenov Inst. Evolutionary Physiology and Biochemistry, Leningrad, Russia, 1957-67, head, dept. neurophysiology of invertebrates, 1968—, dir., 1982—. Author: Neurophysiology of Insect Flight, 1973, Basics of Insect Neurophysiology, 1980 (Orbeli award 1981), Locomotion of Insects, 1988; editor in chief: Journal Evolutionary Biochemistry and Physiology, 1988. Recipient State Premium award USSR, 1987. Mem. The Russian Physiol. Soc., St. Petersburg Sci. Ctr., St. Petersburg Physiol. Soc. Orthodox Russian. Achievements include the principles of motor control of insect locomotion; the study of comparative physiology of motor control in invertebrates and vertebrates. Office: Sechenov Inst of Evolutional, Prospekt Morisa Toreza 44, 194223 Saint Petersburg Russia

SVIKLA, ALIUS JULIUS, pharmacist; b. Merbeck, Germany, Jan. 12, 1947; came to U.S., 1949; s. Julius and Brone (Maksimavich) S. BS in Pharmacy, Northeastern U., Boston, 1973. Pharmacist Osco Drug, Cambridge, Mass., 1973-75; profl. sales rep. Pfizer Labs., N.Y.C., 1976-77; pharmacist, mgr. CVS Pharmacy, South Dennis, Mass., 1977—; drug abuse cons. Healthcare Assn., Boston, First Group of Boston, 1979-87; liason Kaunas Med. Acad., Lithuania. Served with USMC, 1965-68. Decorated Purple Heart. Mem. Am. Pharm. Assn., Internat. Pharm. Fedn., Mass. Pub. Health Assn., Mass. State Pharm. Assn., Lithuanian Am. Pharm. Assn., Mil. Order of Purple Heart (life), Fleet Res. Assn. (Cape Cod chpt. H.O.G.). Republican. Roman Catholic. Avocations: golf, sailing, tennis, fitness. Home: 9 Seagrove Rd South Dennis MA 02660 Office: CVS Pharmacy PO Box 715 Rte 134 Patriots Sq South Dennis MA 02660

SVOBODA, GORDON HOWARD, pharmacognosist, consultant; b. Racine, Wis., Oct. 29, 1922; s. Louis Joseph and Selma Alma Lena (Neumann) S.; m. Marjorie Ellen Huber, Aug. 20, 1945 (div. Oct. 1973); children: Carla Ellen, Sandra Louise, Karen Sue. BS, U. Wis., 1944, PhD, 1949. Part-time instr. U. Wis., Madison, 1945-49; asst. prof. pharmacy U. Kans., Lawrence, 1949-50; rsch. assoc. Eli Lilly & Co., Indpls., 1950-78; pvt. practice Indpls., 1979—; rsch. coord. Natural Resources, Inc., Indpls., 1969-73. Mem. editorial adv. bd. Jour. Pharm. Scis., 1965-70 (Ebert prize, Washington, 1967), Jour. Natural Products, 1966-84; author 8 books; contbr. over 69 articles to profl. publs. Recipient Achievement award Am. Pharm. Assn. Found., Washington, 1963, Hon. Citation U. Wis., Madison, 1963. Mem. Am. Soc. Pharmacognosy (hon., pres. 1963-64), Sigma Xi, Rho Chi (awardee 1964), Phi Eta Sigma, Phi Lambda Upsilon, Phi Kappa Phi. Lutheran. Achievements include 6 patents for isolation of medicinals from plants; discovery of leurocristine (vincristine), the drug for inducing complete bone marrow remission in acute lymphocytic leukemia of childhood; discovery of 36 alkaloids and verification presence of 5 others from the Madagascan periwinkle; isolation of acronycine (acronine), an alkaloid from the Australian scrub ash, possessing the broadest experimental antitumor spectrum of any known compound. Home and Office: 3918 Rue Renoir Indianapolis IN 46220-5618

SVRLUGA, RICHARD CHARLES, science and technology executive; b. Berwyn, Ill., Feb. 6, 1949; s. William J. and Ruth E. (Crowell) S.; m. Donna M. Hanson, Aug. 11, 1978; children: Sara M. Wachter, Krista A. Wachtel. BA in Math., U. Dubuque, 1971; MS in Edn., Ind. U., 1973; MBA, Boston U., 1978. Asst. to v.p. overseas program Boston. U., Heidelberg, Fed. Republic Germany, 1978-81; asst. dean Coll. Liberal Arts and grad. sch. Boston U., 1981-85; cofounder, exec. v.p., dir. Summit Tech., Inc., Waltham, Mass., 1985-90; pres., COO Seragen, Inc., Hopkinton, Mass., 1988—; co-founder, chmn. Clearflow, Inc., Boulder, Colo., 1992—; dir. Greenpages, Inc., Portsmouth, N.H., 1992—. Vol. Voluntary Action Ctr., Boston, 1975, Jr. Achievement, Newton, Mass., 1990, United Way of Metrowest Mass., 1991; trustee Arthritis Found. Mass. chpt., 1992—. Office: Seragen Inc 97 South St Hopkinton MA 01748-2204

SWAFFORD, STEPHEN SCOTT, electrical engineer; b. Greenville, S.C., July 25, 1962; s. Harold Ernest and Carol (Lollis) S.; m. Catherine Handley, Dec. 15, 1984; children: Matthew Stephen, Michael Edward. BS, Clemson U., 1984. Engr. in tng. Application engr. Square D Co., Columbia, 1985-88; elec. engr. Cryovac Div. Sryovac Div. W.R. Grace Co., Duncan, S.C., 1988-91, electrical systems engr., 1991--; advisor Wilson Vocat. Sch., Columbia, 1986-88. pres. Easley First United Meth. Men's Club, 1990--. Home: 700 James Rd Easley SC 29642 Office: Cryovac div WR Grace PO Box 464 Duncan SC 29334

SWAIM, JOHN FRANKLIN, physician, health care executive; b. Bloomingdale, Ind., Dec. 24, 1935; s. Max DeBaun and Edna Marie (Whitely) S.; m. Joan Dooley, Sept. 19, 1957 (div. Apr. 1979); children: John Franklin, Parke Allen, Pamela Ann; m. Peggy Lou Sankey, May 30, 1979; one child, Anne-Marie. BS cum laude, Ind. State U., 1959; MD, Ind. U., Indpls., 1963. Diplomate Am. Bd. Family Practice. Med. dir. Parke Clinic, Rockville, Ind., 1969—; pres. Parke Investments Inc., Rockville, 1972—, Vermillion Health Care Corp., Clinton, Ind., 1977—; bd. dirs. Parke State Bank, Rockville. Author: One Year and Eternity, 1978; also contbr. articles to profl. jours. Coroner, Parke County, Ind., 1972-82. Served to capt. USAF, 1963-67, Vietnam. Decorated Bronze Star. Mem. Am. Acad. Family Physicians, AMA, Ind. State Med. Assn. (dist. pres. 1986—), Midwest Fin. Assn. Republican. Club: Hoosier Assocs. (Indpls.). Lodges: Elks, Masons, Shriners. Avocations: reading and investing. Home and Office: Parke Clinic 503 Anderson St Rockville IN 47872-1008

SWAIMAN, KENNETH FRED, pediatric neurologist, educator; b. St. Paul, Nov. 19, 1931; s. Lester J. and Shirley (Ryan) S.; m. Phyllis Kammerman Sher, Oct. 1985; children: Lisa, Jerrold, Barbara, Dana. B.A. magna cum laude, U. Minn., 1952, B.S., 1953, M.D., 1955; postgrad., 1956-58; postgrad. (fellow pediatric neurology), Nat. Inst. Neurologic Diseases and Blindness, 1960-63. Diplomate: Am. Bd. Psychiatry and Neurology, Am. Bd. Pediatrics. Intern Mpls. Gen. Hosp., 1955-56; resident pediatrics U. Minn., 1956-58, neurology, 1960-63; postgrad. fellow pediatric neurology Nat. Inst. Neurologic Diseases and Blindness, 1960-63; asst. prof. pediatrics, neurology U. Minn. Med. Sch., Mpls., 1963-66; asso. prof. Nat. Inst. Neurologic Diseases and Blindness, 1966-69; prof., dir. pediatric neurology U. Minn. Med. Sch., 1969—, exec. officer, dept. neurology, 1977—, mem. internship adv. council exec. faculty, 1966-70; cons. pediatric neurology Hennepin County Gen. Hosp., Mpls., St. Paul-Ramsey Hosp., St. Paul Children's Hosp., Mpls. Children's Hosp.; vis. prof. Beijing U. Med. Sch., 1989. Author: (with Francis S. Wright) Neuromuscular Diseases in Infancy and Childhood, 1969, Pediatric Neuromuscular Diseases, 1979, (with Stephen Ashwal) Pediatric Neurology Case Studies, 1978, 2d edit., 1984, Pediatric Neurology: Practice and Principles, 1989; editor: (with John A. Anderson) Phenylketonuria and Allied Metabolic Diseases, 1966, (with Francis S. Wright) Practice Pediatric Neurology, 1975, 2d edit., 1982; mem. editorial bd.: Annals of Neurology, 1977-83, Neurology Update, 1977-82, Pediatric Update, 1977-85, Brain and Devel. (Jour. Japanese Soc. Child Neurology), 1980—, Neuropediatrics (Stuttgart), 1982-92; editor-in-chief: Pediatric Neurology, 1984—; contbr. articles to sci. jours. Chmn. Minn. Gov.'s Bd. for Handicapped, Exceptional and Gifted Children, 1972-76; mem. human devel. study sect. NIH, 1976-79, guest worker, 1978-81. Served to capt. M.C. U.S. Army, 1958-60. Fellow Am. Acad. Pediatrics, Am. Acad. Neurology (rep. to nat. council Nat. Soc. Med. Research); mem. Soc. Pediatric Research, Central Soc. Clin. Research, Central Soc. Neurol. Research, Internat. Soc. Neurochemistry, Am. Neurol. Assn., Minn. Neurol. Soc., AAAS, Midwest Pediatric Soc., Am. Soc. Neurochemistry, Child Neurology Soc. (1st pres. 1972-73, Hower award 1981, chmn. internat. affairs com., 1991—, mem. long range planning com. 1991—), Internat. Assn. Child Neurologists (exec. com. 1975-79), Profs. of Child Neurology (1st pres. 1978-80, mem. nominating com. 1986—), Japanese Child Neurology Soc. (Segawa award 1986, mem. nominating com. 1986—, chair internat. affairs com. 1991—, mem. long range planning com. 1991—), Soc. de Psiquiatria y Neurologia de la Infancia y Adolescencia, Phi Beta Kappa, Sigma Xi. Home: 420 Delaware St SE Minneapolis MN 55455-0374 Office: U Minn Med Sch Dept Pediatric Neurology Minneapolis MN 55455

SWAIN, EDWARD BALCOM, environmental research scientist; b. Stamford, Conn., Aug. 6, 1952; s. Rodney Towle and Florence (Brown) S.; m. Mary Ethel Keirstead, Dec. 26, 1981; children: John, Daniel. BA, Carleton Coll., 1974; PhD, U. Minn., 1984. Rsch. assoc. U. Minn., Mpls., 1984-88; rsch. scientist Minn. Pollution Control Agy., St. Paul, 1988—. Contbr. articles to profl. jours. Recipient NSF fellowship. Mem. AAAS, Ecol. Soc. of Am., Internat. Soc. for Limnology, Sigma Xi. Achievements include elucidation of the relationship between mercury air emissions and bioaccumulation in fish. Office: Minn Pollution Control Agy 520 Lafayette Rd Saint Paul MN 55155

SWAIN, ROBERT VICTOR, environmental engineer; b. Dillwyn, Va., Feb. 2, 1940; s. Frank Robert and Virginia Odell (Doss) S.; m. Norma Foley, Aug. 28, 1965 (div. Aug. 1987); 1 child, John Victor; m. Sandra Dawn Huffman Jones, July 23, 1988; children: Susan, Robert, Debra. BS, Carson-Newman Coll., 1962, Va. Tech., 1964; MS, Va. Tech., 1967. Civil engr. C.E., Huntington, W.Va., 1964; san. engr. Gannett Fleming, Camp Hill, Pa., 1965-72, Glace & Glace Inc., Harrisburg, Pa., 1972-82, Glace & Radcliffe, Inc., Winter Park, Fla., 1982-84, Post Buckley Schuh & Jernnigan, Orlando, Fla., 1984-86, Malcolm Pernie, Inc., Orlando, 1986-88, Engring. Sci., Inc., 1988, James M. Montgomery, Inc., Pasadena, Calif., 1988—. Mem. ASCE, Water Environ. Fedn., Am. Water Work Assn., Toastmasters Internat. Republican. Baptist. Home: 526 Park Rose Ave Monrovia CA 91016 Office: J M Montgomery Inc 250 N Madison Ave Pasadena CA 91109

SWAINE, ROBERT LESLIE, JR., chemist; b. Melrose, Mass., Aug. 27, 1950; s. Robert Leslie and Barbara Elizabeth (Allen) S.; m. Susan Ellen Hayes, June 2, 1973; children: Sarah Elizabeth, Kelly Ann, Samantha Lynne. BS, Northeastern U., Boston, 1973; MS, Rutgers U., 1975. Flavor chemist Florasynth, N.Y.C., 1974-76, Sherwin-Williams, Danbury, Conn., 1976-78; tech. mgr. Thomas J. Lipton, Allendale, N.J., 1978-83; sect. head Procter & Gamble Co., Cin., 1983—, rsch. fellow, 1993. Patentee fruit juice with diet beverage, comml. manufacture of citrus juices. Recipient Medallion award Internat. Fedn. Essential Oils and Aroma Trades; rsch. fellow Procter & Gamble, 1993. Fellow Am. Inst. Chemists; mem. Am. Chem. Soc., Soc. Flavor Chemists, Inst. Food Technologists. Office: Procter & Gamble Co 6210 Center Hill Ave Cincinnati OH 45224-1797

SWALES, JOHN E. (TED), retired horticulturalist; b. Kaleden, B.C., Can., 1924. BS in Agriculture, Ont. Agrl. Coll., U. Toronto, 1947. With B.C. Dept. Agriculture, 1948-49; district horticulturalist Nelson, B.C., 1948-53, Creston, B.C., 1958-63, Penticton, B.C., 1963-75; apple specialist Agriculture Can. Rsch. Sta., Summerland, B.C., 1975-79; head field svc. Okanagan Similkameen Cooperative Growers' Assn., Oliver, B.C., 1979-90. Recipient award of merit B.C. Fruit Growers Assn., 1989, Disting. Extension Specialist award Internat. Dwarf Fruit Tree Assn., 1989, Spl. Recognition award Can. Soc. Horticultural Sci., 1990; named Agrologist-of-Yr. B.C. Inst. Agrologists, 1986. AIC Fellowship award Agrl. Inst. Can., 1992. Office: 184 Highway 97 S1, C4 RR#1, Kaleden, BC Canada V0H 1K0*

SWALLEY, ROBERT FARRELL, structural engineer, consultant; b. Ponca City, Okla., June 1, 1930; s. Robert Arthur and Jeannette Dean (Edwards) S.; m. Mary Jo Durham, Oct. 18, 1965; children: Arthur Gentry, Susanne Evelyn. BS with distinction, U.S. Naval Acad., 1952; BSCE, U. Mo., 1958; MS, Stanford U., 1959. Registered profl. engr., structural engr. Sr. rsch. engr. USN Civil Engring. Lab., Port Hueneme, Calif., 1959-63; structural engr. Benham Blair & Affiliates, Oklahoma City, Okla., 1967-69; sr. project engr. AMF Inc., Advanced System Lab., Santa Barbara, Calif., 1969-73; structural engr. Penfield & Smith Engrs. Inc., Santa Barbara, 1973-77; structural engr., prin. engr., CEO Swalley Engring. Inc., Santa Barbara, 1978-93; cons., structural engr. Santa Barbara, 1993—. Contbr. tech. reports on Small Buried Arches, Design of a Cast in Place Personnel Shelter, Behavior of Buried Model Arch Structures, Loadings on Drydock Gates from Nuclear Explosions. Mem. ASCE (pres. 1976-77, v.p. 1983-84), NSPE (pres. 1978-84), Structural Engrs. Assn. Calif. Achievements include patent for Ventilator Blast Closure. Home and Office: 504 Consuelo Dr Santa Barbara CA 93110-1118

SWALLOW, BRENT MURRAY, agricultural economist, researcher; b. Maryfield, Can., Dec. 9, 1958; s. Gordon Henry and Arnetta Caroline (Hammond) S.; m. Kimberly Anne Smith, June 3, 1989. BSA, U. Saskatchewan, Can., 1981, MSc, 1983; MA, U. Wis., Madison, 1989, PhD, 1991. Rsch. assoc. Va. Poly. Inst., Blacksburg, Va., 1983-84; rsch. fellow Nat. U. Lesotho, 1984-87; grad. asst. U. Wis., Madison, 1987-91; agrl. econ. Internat. Livestock Ctr. for Africa, Nairobi, Kenya, 1991—; temporary advisor World Health Orgn., Geneva, Switzerland, 1992. Author: Marketing in Agricultural Development, 1992. Treas. Grad. Student Assn., U. Saskatchewan, 1982-83, Community Help Orgn., Roma, Lesotho, 1985-86; v.p. Taylor-Hibbard Club, U. Wis., 1988-89; student rep. Dept. Grad. Com., U. Wis., 1989-90. Recipient Dollie Hantleman scholarship U. Saskatchewan, Can., 1981-83, Saskatchewan Agrl. Rsch. Found. scholarship, U. Saskatchewan, 1981-82. Mem. Am. Agrl. Econ. Assn., Internat. Soc. for Ecological Econ., Internat. Assn. for Study of Common Property, Gamma Sigma Delta. Avocations: travel, reading, tennis, basketball. Home: PO Box 46847, Nairobi Kenya Office: Internat Livestock Ctr for Africa, PO Box 46847, Nairobi Kenya

SWALM, THOMAS STERLING, aerospace executive, retired military officer; b. San Diego, Sept. 28, 1931; s. Calvin D. and Margaret A. (Rynning) S.; m. Charlene La Vern Garner, June 26, 1954; children: Edward Steven, Lori Ann. BS, U. Oreg., 1954; MS in Pub. Adminstrn., George Washington U., 1964; grad., Air Command and Staff Coll., 1964, Nat. War Coll., 1974. Commd. USAF, 1954, advanced through grades to maj. gen., 1982; instr. fighter-interceptor weapons sch. USAF, Tyndall AFB, Fla., 1956; pilot 434th Fighter-Day Squadron USAF, George AFB, Calif., 1957-58; engring. test pilot and flight examiner 50th Tactical Fighter Wing, 10th Tactical Fighter Squadron USAF, Toul-Rosieres AFB, France, and Hahn AFB, Fed. Republic Germany, 1958-61; hdqrs. 12th USAF, Waco, Tex., 1961-64; instr. pilot, flight examiner 4453d Combat Crew Tng. Wing USAF, Davis-Monthan AFB, Ariz., 1965-66; flight comdr. 12th Tactical Fighter Wing USAF, Cam Ranh Bay AFB, Republic Vietnam, 1966-67; comdr. air-to-air flight instr. and chief R&D/OT&E sect. USAF Fighter Weapons Sch., Nellis AFB, Nev., 1967-70; comdr., leader Thunderbirds USAF, 1970-73; chief fighter attack directorate USAF, Kirtland AFB, N.Mex., 1974-75, dep. dir. test and evaluation, 1975-76; from vice comdr. to comdr. 8th Tactical Fighter Wing USAF, Kunsan AFB, Republic of Korea, 1976-78; comdr. 3d Tactical Fighter Wing USAF, Clark AFB, Philippines, 1978-79; comdr. 57th Fighter Weapons Wing, comdr. fighter weapons sch. USAF, Nellis AFB, Nev., 1979-80; comdr. 833d air div. USAF, Holloman AFB, N.Mex., 1980-81; comdr. tactical air warfare ctr. USAF, Eglin AFB, Fla., 1981-86; ret. USAF, 1986; pres. T. Swalm and Assocs., Ft. Walton Beach, Fla., 1986-91; v.p. Melbourne Systems Div. Grumman Corp., 1991—; v.p. Applications Group Internat., Inc., Atlanta, 1986-89; mem. sci. adv. bd. USAF, 1993—. Mem. editorial bd. Jour. Electronic Def., 1983-86; contbr. articles to profl. jours. Hon. chmn. Heart Assn., Las Vegas, Nev., 1972; exec. dir. Boy Scouts Am., Las Vegas and Alamagordo, N.Mex., 1970-81; chmn. AFA Scholarship Found., 1989-91; active Fla. Govs. Coun. for TQM, 1992-93; bd. dirs. Jr. Achievement, Cen. Fla., 1992-93. Decorated D.S.M., Legion of Merit with two oak leaf clusters, DFC, Air medal with 14 oak leaf clusters, Vietnam Service medal with three service stars, Republic Vietnam Campaign medal; recipient R.V. Jones Trophy Electronic Security Command, 1984. Mem. Air Force Assn., (exec. advisor, Jerome Waterman award 1985, Jimmy Doolittle Fellow 1986), Thunderbirds Pilots Assn., Old Mission Beach Athletic Club (founder), Assn. Old Crows (editorial bd., R.V. Jones trophy 1984), Order of Daedalians (flight capt.), Sigma Nu. Republican. Presbyterian. Avocations: golf, tennis, racquetball, sailing. Office: Grumman Melbourne Systems PO Box 9650 Melbourne FL 32902-9650

SWAMY, DEEPAK NANJUNDA, electrical engineer; b. Bangalore, India, July 1, 1965; s. Shimoga G. N. and Vani N. Swamy; m. Deepa Sunder, May 7, 1993. BSEE, Regional Engring. Coll., 1987; MBA, U. R.I., 1990. Ind. cons. South Kingstown, R.I., 1989; sr. analyst KMI Corp., Newport, R.I., 1990—. Author: (market study) Subscriber Loop to the Year 2001, 1991, European Markets for Fiberoptics, 1991, Fiber-in-the-Loop Markets in Europe, 1993; contbr. jours. and papers to profl. pubs. Computer instr. Literacy Vols./Homeless Shelter, Newport, 1991, 92; vol. Newport Hosp., 1990. Mem. IEEE, Am. Mensa, Beta Gamma Sigma, Phi Kappa Phi. Achievements include conducting strategic analyses and planning for world's top telecommunications companies. Home: 21 Princeton St Newport RI 02840 Office: KMI Corp 31 Bridge St Newport RI 02840

SWAN, CHARLES WESLEY, psychoneuroimmunologist; b. Hartsville, S.C., Mar. 11, 1942; s. Alexander and Anna Josephine (Toney) S.; m. Sharon Louise Johnson, Aug. 9, 1969; children: Simone Yvette, Charles Jr. PhD, Ohio Christian Coll., 1972; postgrad., U. Tex., Houston, 1991. Diplomate Am. Bd. Neuropsychology and psychotherapy. Clin. asst. Perkins State Hosp., Jessup, Md., 1965-69; pers. dir. Hotel Corp. Am., Balt., 1969-71; pvt. practice Swan Assocs., Poconos, Pa., 1971—; exec. dir. United Neighborhood Svcs., Scranton, Pa., 1971-72; state liaison officer HEW, Washington, 1972-85; pres., chief exec. officer Swan Assocs., Tampa, Fla., 1985—; presdl. policy advisor White House, Washington, 1991-92; cons. White House, Washington, 1988-92; lectr. health symposiums, 1972-92. Commr. Boy Scouts Am., Dallas, 1984-85; mentor Kellogg's Trust/Urban League, Tampa, 1990; Rep. Nat. Com. Presidential Club, Washington, 1991, Senatorial Inner Circle, 1993, Nat. Leadership Team, Washington, 1991. Recipient Silver Key award Nat. Mental Health Assn., Washington, 1969. Mem. AAAS, Black Psychologists (chmn. 1982-83, Disting. Svc. award 1990), Am. Assn. World Health, Nat. Libr. Medicine, N.Y. Acad. Sci., Am. Air Mus. in Britian (founding mem.). Achievements include research in the brain processing info./neuro-peptides, DNA, transmitters, hidden motion; man will discover that there is speed faster than the speed of light; genome will be the new big bang. Home: 16127 Ancroft Ct Tampa FL 33647-1041

SWAN, KENNETH CARL, physician, surgeon; b. Kansas City, Mo., Jan. 1, 1912; s. Carl E. and Blanche (Peters) S.; m. Virginia Grone, Feb. 5, 1938; children: Steven Carl, Kenneth, Susan. A.B., U. Oreg., 1933, M.D., 1936. Diplomate: Am. Bd. Ophthalmology (chmn. 1960-61). Intern U. Wis., 1936-37; resident in ophthalmology State U. Iowa, 1937-40; practice medicine specializing in ophthalmology Portland, Oreg., 1945—; staff Good Samaritan Hosp.; asst. prof. ophthalmology State U. Iowa, Iowa City, 1941-44; asso. prof. U. Oreg. Med. Sch., Portland, 1944-45, prof. and head dept. ophthalmology, 1945-78; Chmn. sensory diseases study sect. NIH; mem. adv. council Nat. Eye Inst.; also adv. council Nat. Inst. Neurol. Diseases and Blindness. Contbr. articles on ophthalmic subjects to med. pubs. Recipient Proctor Rsch. medal, 1953; Disting. Svc. award U. Oreg., 1963; Meritorious Achievement award U. Oreg. Med. Sch., 1968; Howe Ophthalmology medal, 1977; Aubrey Watzek Pioneer award Lewis and Clark Coll., 1979, Disting. Alumnus award Oreg. Health Scis. U. Alumni Assn., 1988, Disting. Svc. award, 1988; named Oreg. Scientist of Yr. Oreg. Mus. Sci. and Industry, 1959. Mem. Assn. Research in Ophthalmology, Am. Acad. Ophthalmology (v.p. 1978, historian), Soc. Exptl. Biology and Medicine, AAAS, AMA, Am. Ophthal. Soc. (Howe medal for distinguished service 1977), Oreg. Med. Soc., Sigma Xi, Sigma Chi (Significant Sig award 1977). Home: 4645 SW Fairview Blvd Portland OR 97221-2624 Office: Oreg Health Scis U Portland OR 97201

SWANBERG, CHRISTOPHER GERARD, environmental engineer; b. N.Y.C., Jan. 31, 1938; s. Clifford Duane and Margaret Mary (Spillane) S.; m. Sherrie Joy Shuler, Mar. 3, 1979; children: Daniel Christopher, Sarah Elizabeth. BS in Environ. Engring., Western Ky. U., 1981; MBA, U. Tulsa, 1991. Environ. engr. Atlantic Richfield (ARCO), Louisville, 1981-82; environ. project engr. Atlantic Richfield (ARCO), Russellville, Ky., 1982-84; coord. environ. affairs Atlantic Richfield (ARCO), Independence, Kans., 1984-86; sr. environ. engr. Atlantic Richfield (ARCO), Anaheim, Calif., 1986-88; mgr. environ. engring. Atlantic Richfield (ARCO), L.A., 1988-90; v.p. Heritage Environ. Svc., Inc., Tulsa, 1990-91; sr. v.p. Separation and Recovery Systems, Inc., Irvine, Calif., 1991—. Contbr. articles to profl. publs. Asst. commr. Am. Youth Soccer Orgn., Laguna Niguel, Calif., 1992; bd. dirs. Independence Day Care, 1985; advisor Jr. Achievement, Louisville, 1984. Achievements include development of thermal disorption technology to meet EPA land disposal restrictions for petroleum refining wastes, EPA accepted physical and chemical standards for solidification and stabilzation of organic wastes, first water based coatings for aluminum foil packaging industry to reduce air pollution; designer aluminum industry's zero process discharge rolling mill complex. Home: 6 Aleria Laguna Niguel CA 92677 Office: Separation & Recovery Syst 1762 McGaw Ave Irvine CA 92714

SWANBORG, ROBERT HARRY, immunology educator; b. Bklyn., Aug. 27, 1938; s. Harry Eric and Ruth Mildred (Wahlstrom) S.; m. Nancy Kay Phelan, Sept. 3, 1966 (div. 1986); children: Kirsten Ann, Eric Robert; m. Janice Fay Olstyn, Aug. 8, 1992. BS, Wagner Coll., Staten Island, N.Y., 1960; MS, L.I. U., 1962; PhD, SUNY, 1965. Instr. to assoc. prof. Wayne State U. Med. Sch., Detroit, 1966-75, prof. immunology, 1977—; vis. scientist Wenner-Gren Inst., Stockholm, 1975-76; mem. fellowship adv. com. Nat. Multiple Sclerosis Soc., N.Y.C., 1986-90. Contbr. over 100 articles to profl. jours., chpts. to books. NIH grantee, 1966—, Nat. Multiple Sclerosis Soc. grantee, 1970—; recipient Javits Neurosci. Investigator award NIH, 1992—. Mem. Am. Assn. Immunologists, Am. Soc. Investigative Pathology, Am. Soc. Neurochemistry, Internat. Soc. Neuroimmunology. Lutheran. Office: Wayne State U Med Sch 540 E Canfield Detroit MI 48201

SWANEY, CYNTHIA ANN, medical computer service sales executive, business consultant; b. Garfield Heights, Ohio, Feb. 25, 1959; d. Peter John and Juanita Catherine (Crowle) Christ; m. C. Keith Swaney, Aug. 4, 1984; children: Jason Scott, Samantha Jean. Grad. high sch., Pepper Pike, Ohio. With Park View Fed. S&L, Cleve., 1975-79; customer svc., teller, trainer Park View Fed. S&L, 1977-79; exec. sec., ops. mgr. Majestic Steel Svc., Solon, Ohio, 1979-84; v.p. adminstrn. Datashare Corp., Chagrin Falls, Ohio, 1984—; cons. Stenciler's Emporium, Hudson, Ohio, 1988—, Deep Springs Trout Club, Chardon, Ohio, 1988—, Hiram House Camp, Moreland Hills, Ohio, 1990—; numerous med. offices, 1986—. Trustee Hiram House Camp,

1991—. Avocations: reading, water sports. Office: Datashare Corp PO Box 743 Chagrin Falls OH 44022-0743

SWANK, ANNETTE MARIE, computer software designer; b. Lynn, Mass., Nov. 9, 1953; d. Roland Paterson and Rita Mary (Edwards) S. BSEE and Computer Sci., Vanderbilt U., 1975; postgrad., Pa. State U., 1992—. Lead programmer GE, Phila., 1975-80; system analyst SEI Corp., Wayne, Pa., 1980-82; mgr., designer Premier Systems, Inc., Wayne, Pa., 1982-85, dir., 1985-88, tech. advisor, 1988-90, tech. architect, 1990-92; tech. architect Funds Assocs. Ltd., Wayne, 1992—. Designer: (programming lang. and data dictionary) Vision, 1985. Treas. Master Singers, Plymouth Meeting, Pa., 1987-88. Mem. Assn. for Computing Machinery, Gamma Phi Beta (com. chmn. alumna Phila. 1986-87). Avocations: singing, dancing, bowling, bridge, wine tasting. Home: 136 Pinecrest Ln King Of Prussia PA 19406 Office: Funds Assocs Ltd 440 E Swedesford Rd Wayne PA 19087

SWANK, ROY LAVER, physician, educator, inventor; b. Camas, Wash., Mar. 5, 1909; s. Wilmer and Hannah Jane (Laver) S.; m. Eulalia F. Shively, Sept. 14, 1936 (dec.); children: Robert L., Susan Jane (Mrs. Joel Keizer) Stephen (dec.); m. Betty Harris, May 23, 1987. Student, U. Wash., 1926-30; M.D., Northwestern U., 1935; Ph.D., 1935. House officer, resident Peter Bent Brigham Hosp., Boston, 1936-39; fellow pathology Harvard Med. Sch., 1938-39; prof. neurology Oreg. Med. Sch., 1954-74, prof. emeritus, 1974—, also former head div. neurology; pres. Pioneer Filters. Served to maj. M.C. AUS, 1942-46. Recipient Oreg. Gov.'s award for research in multiple sclerosis, 1966. Mem. Am. Physiol. Soc., Am. Neurol. Assn., European Microcirculation Soc., Sigma Xi. Achievements include invented micro embolic filter; research of phys. chem. changes in blood after fat meals and during surg. shock, platelet-leukocyte aggregation in stored blood and in hypotensive shock; low-fat diet in multiple sclerosis; research of phys. chem. changes in multiple sclerosis (plasma proteins); importance of plasma proteins in multiple sclerosis. Home: 789 SW Summit View Dr Portland OR 97225-6185

SWANSON, HELEN ANNE, psychology educator; b. Youngstown, Ohio, Nov. 9, 1957; d. Charles Michael and Nancy Anne (Coman) Koutsourais; m. James David Swanson, July 21, 984; children: Marissa Helene, Elise Helena. MS, Ohio U., 1982, PhD, 1985. Asst. prof. psychology U. Wis. Ctr.-Manitowoc County, 1985-88; asst. prof. psychology U. Wis.-Stout, Menomonie, 1988-92, assoc. prof. psychology, 1992—; pre-sch. lead tchr. Sunrise Day Ctr. and Pre-sch., White Bear Lake, Minn., 1984-85; workshop presenter Head Start, Manitowoc, 1986, Mentor Program for Teen Pregnancy Prevention, Manitowoc, 1988, Wis. Assn. Restitution Programs, Brookfield, 1988. Contbr. articles to Bull. of Psychonomic Soc., Child Study Jour., Jour. Conflict Resolution. Bd. dirs. Big Bros./Big Sisters of Manitowoc County, 1986-88; adv. com. Juvenile Restitution Project, Manitowoc, 1986-88; task force Victim-Offender Mediation Program, Manitowoc, 1987. Rsch. grantee Ohio U., 1983, U. Wis., 1988, 89, 90, 91, 92, 93. Mem. Midwestern Psychol. Assn., Am. Ednl. Rsch. Assn., Nat. Sci. Tchrs. Assn., Wis. Coun. on Human Concerns. Democrat. Eastern Orthodox. Office: U Wis-Stout Dept Psychology Menomonie WI 54751

SWANSON, LEE RICHARD, computer security executive; b. Mpls., Apr. 21, 1957; s. Donald Jerome and Wildie (Greenwood) S.; m. Amy Jane Shutkin, Jan. 1, 1980 (div. Apr. 1991). BS, U. Minn., 1983. Owner, prin. Environ. Landforms, Inc., Minnetonka, Minn., 1974-80; v.p. Blomfield-Swanson, Inc., Mpls., 1981-85; contractor Citicorp Card Acceptance Svcs., Seattle, 1986-88; pres., mktg. dir., cons. Room Svcs. Computers, Bellevue, Wash., 1988-91; exec. v.p. First Step Computer Consultants, Inc., Novato, Calif., 1991—. Libertarian. Avocations: fishing, sea kayaking, sailboarding, sailing. Office: First Step Computer Consultants Inc 60 Galli Dr Ste T Novato CA 94949

SWANSON, RICHARD WILLIAM, statistician; b. Rockford, Ill., July 26, 1934; s. Richard and Erma Marie (Herman) S.; m. Laura Yoko Arai, Dec. 30, 1970. BS, Iowa State U., 1958, MS, 1964. Ops. analyst Stanford Rsch. Inst., Monterey, Calif., 1958-62; statistician ARINC Rsch. Corp., Washington, 1964-65; sr. scientist Booz-Allen Applied Rsch., Vietnam, 1965-67, L.A., 1967-68; sr. ops. analyst Control Data Corp., Honolulu, 1968-70; mgmt. cons., Honolulu, 1970-73; exec. v.p. SEQUEL Corp., Honolulu, 1973-75; bus. cons. Hawaii Dept. Planning and Econ. Devel., Honolulu, 1975-77, tax rsch. and planning officer Dept. Taxation, 1977-82; ops. rsch. analyst U.S. Govt., 1982-89; shipyard statisician U.S. Govt., 1989—. Served with AUS, 1954-56. Mem. Hawaiian Acad. Sci., Sigma Xi. Home: 583 Kamoku St Apt 3505 Honolulu HI 96826-5240 Office: Pearl Harbor Naval Shipyard PO Box 400 Honolulu HI 96860-5350

SWANSON, STEVEN CLIFFORD, clinical psychologist; b. Toronto, Ont., Can., May 15, 1964; came to U.S., 1987; s. Swan Roy and Mary Kathryn (Horner) S. BSc with high distinction, U. Toronto, 1987; MA, Calif. Sch. Profl. Psychology, Fresno, 1989, PhD, 1992. Teaching asst. Erindale Coll. U. Toronto, Mississauga, Ont., Can., 1986, rsch. asst., lab. supr. Erindale Coll., 1986-87; psychological trainee Baird Elem. Sch., Fresno, 1988, San Luis Obispo (Calif.) Community Mental Health Ctr., 1988-89, Calif. Sch. Profl. Psychology-Fresno Psychol. Svc. Ctr., 1989-90; sch. psychology Ctrl. Unified Sch. Dist., Fresno, 1989-90; predoctoral intern Camarillo (Calif.) State Hosp., 1990-91; postdoctoral intern Calif. Sch. of Profl. Psychology-Fresno Psychol. Svc. Ctr., 1992—; therapist Friendship Home, Fresno, 1992—; acad. counselor Erindale Coll., U. Toronto, 1986-87. Recipient Rita K. Teetzel In-Course scholarship, 1986, Mrs. Lois Spigel Book prize, 1986. Mem. APA, San Joaquin Psychol. Assn., Calif. State Psychol. Assn. Office: CSPP Psychol Svc Ctr 1260 M St Fresno CA 93721

SWARTZ, MORTON NORMAN, medical educator; b. Boston, Nov. 11, 1923; s. Jacob H. and Janet (Heller) W.; m. Cesia Rosenberg, Sept. 18, 1956; children: Mark David, Caroline Joan. BA, Harvard Coll., 1945; MD, Harvard U., 1947; MD (hon.), U. Geneva, Switzerland, 1988. Diplomate Am. Bd. Internal Medicine (subsplty. exam. com. 1971-76, bd. govs. 1979-85). Med. intern and resident Mass. Gen. Hosp., Boston, 1947-50, chief resident in medicine, 1953-54; USPHS postdoctoral rsch. fellow Johns Hopkins U., McCollum-Pratt Inst. Enzymology, Balt., 1954-56; chief infectious disease unit Mass. Gen. Hosp., Boston, 1956-90, chief James Jackson Firm, dept. medicine, 1990—; assoc. prof. medicine Harvard Med. Sch., Boston, 1967-73, prof., 1973—; vis. assoc. prof. biochemistry, Stanford Med. Sch., Palo Alto, Calif., 1969-70. Author: (with others) Osteomyelitis, 1971; editor: Current Clinical Topics in Infectious Diseases, 1980—; assoc. editor New Eng. Jour. Medicine, 1981—; contbr. articles to profl. jours. 1st lt. U.S. Army, 1950-52. Sir MacFarlane Burnett lectr. Australasian Soc. Infectious Disease, 1981. Fellow ACP (master 1988, Disting. Tchr. award 1989); mem. Am. Soc. Biochemistry and Molecular Biology, Am. Soc. for Clin. Investigation, Assn. Am. Physicians, Infectious Diseases Soc. Am. (Bristol award 1984, Feldman award 1989), Inst. Medicine. Jewish. Avocations: biology, bird watching. Home: 54 Shaw Rd Chestnut Hill MA 02167-3122 Office: Mass Gen Hosp 32 Fruit St Boston MA 02114-2698

SWARTZLANDER, EARL EUGENE, JR., engineering educator, former electronics company executive; b. San Antonio, Feb. 1, 1945; s. Earl Eugene and Jane (Nicholas) S.; m. Joan Vickery, June 9, 1968. BSEE, Purdue U., 1967; MSEE, U. Colo., 1969; PhD, U. So. Calif., 1972. Registered profl. engr., Ala., Calif., Colo., Tex. Devel. engr. Ball Bros. Rsch. Corp., Boulder, Colo., 1967-69; Hughes fellow, mem. tech. staff Hughes Aircraft Co., Culver City, Calif., 1969-73; mem. rsch. staff Tech. Svc. Co., Santa Monica, Calif., 1973-74; chief engr. Geophys. Systems Corp., Pasadena, Calif., 1974-75, staff engr. to sr. staff engr., 1975-79, project mgr., 1979-84, lab. mgr., 1985-87; dir. ind. R&D TRW Inc., Redondo Beach, Calif., 1987-90; Schlumberger Centennial prof. engring. dept. elec. and computer engring. U. Tex., Austin, 1990—; hon. conf. chair 3rd Internat Conf. Parallel and Distributed Systems, Taiwan, 1993; hon. chair 3d Internat. Conf. on Parallel and Distributed Systems, Taiwan, 1992; gen. chair 11th Internat. Symposium Computer Arithmetic, 1992, Internat. Conf. Application Specific Array Processors, 1990, Iternat. Conf. Wafer Scale Integration, 1989. Author: VLSI Signal Processing Systems, 1986; editor: Computer Design Development, 1976, Systolic Signal Processing Systems, 1987, Wafer Scale Integration, 1989, Computer Arithmetic Vol. 1 and 2, 1990; editor-in-chief Jour. of VLSI Signal Processing, 1989—, IEEE Transactions on Computers, 1991—, editor 1982-86, IEEE Transactions on Parallel and Distributed Systems, 1989-90; Hardware Area Editor ACM Computing Reviews, 1985—; assoc. editor: IEEE Jour. Solid-State Circuits, 1984-88; contbr. over 75 articles to profl. jours. and tech. conf. proc. Bd. dirs. Casiano Estates Homeowners Assn., Bel Air, Calif., 1976-78, pres. 1978-80; bd. dirs. Benedict Hills Estates Homeowners Assn., Beverly Hills, Calif., 1984—, pres. 1990—. Recipient Disting. engring. Alumnus award Prudue U., West Lafayette, Ind., 1989, named Outstanding Elec. Engr., 1992, Knight Imperial Russian Order of Saint John of Jerusalem, Knights of Malfa, 1993. Fellow IEEE; mem. IEEE Computer Soc. (bd. govs. 1987-91), IEEE Signal Proc. Soc. (Adcom. 1992—), IEEE Solid-State Cirs. Coun. (sec. 1992—), Eta Kappa Nu, Sigma Tau, Omicron Delta Kappa.

SWEARINGEN, DAVID CLARKE, general practice physician; b. Shreveport, La., Apr. 23, 1942; m. Marion Joan Adams; children: David, Joy. BS, Centenary Coll., 1963; MD, La. State U., 1967. Intern Confederate Meml. Med. Ctr., Shreveport, 1967-68, resident in ophthalmology, 1968-71; staff ophthalmologist U.S. Naval Hosp., Memphis, 1971-73; pvt. practice in ophthalmology Shreveport, 1973-78; jr. officer of deck USS Halsey CG23, 1979; comdr. med. corps, head dept. ophthalmology U.S. Naval Hosp., Jacksonville, Fla., 1981-84; med. dir. Bio Blood Components, Shreveport, 1985-88; dir. med. svcs. Cen. La. State Hosp., Pineville, La., 1988—; pres. Shreveport (La.) Eye and Ear Soc., 1973-74; med. cons. Cenla Chem. Dependency Coun., Pineville, 1989—, Work Tng. Facility, Pineville, 1989—. Bassoonist Cenla Symphonic Band, Pineville, La., 1988—; mem. Jacksonville Fla. Concert Choral, 1978-81; vestry mem. St. Michaels Ch., Pineville. Mem. Am. Legion. Republican. Epsicopalian. Home: 10 Azalea Rd Pineville LA 71360 Office: Cen LA State Hosp 242 W Shamrock Pineville LA 71360

SWEENEY, BRYAN PHILIP, structural and geotechnical engineer; b. Boston, May 1, 1954; s. James H. and Rita Rose (Gilligan) S. BSCE, U. Notre Dame, 1976; MSCE, Stanford U., 1982, PhD, 1987. Registered profl. engr., Mass. Designer Parsons, Brinckerhoff, Quade & Douglas, Boston, 1977-82; sr. engr. Haley & Aldrich, Cambridge, Mass., 1987—. Contbr. articles to jours. ASCE, ASTM, EERI Conf., U.S. Dot Report. Mem. com. Town of Hignham Soid Waste. Mem. ASCE (mem. earth retaining structures com. 1990—), Boston Soc. Civil Engrs. (mem. com. constrn. group 1991—), Deep Found. Inst. (mem. eviron. com. 1992—), Chi Epsilon (pres.). Achievements include research in excavations and contamination. Office: Haley & Aldrich 58 Charles St Cambridge MA 02141

SWEENEY, LUCY GRAHAM, psychologist; b. Davenport, Iowa, Nov. 14, 1946; d. B. Graham and Dorothy (Lawson) S.; m. Richard N. Tiedemann, Dec. 2, 1978 (div. 1989); 1 child, Susan Lee. BA with honors, U. Denver, 1968; MA in Devel. Psychology, Columbia U., 1977; PsyD, Rutgers U., 1990. Cert. family therapist. Profl. actress, 1968-73; dir. therapeutic play and recreation program St. Luke's Med. Ctr., N.Y.C., 1973-78; child life coord. St. Francis Hosp., Hartford, Conn., 1978-80; program cons. Child and Family Svcs., Torrington, Conn., 1980-81; clinician Resolve Community Counseling Ctr., Scotch Plains, N.J., 1981-84; staff psychologist women's inpatient unit Lyons (N.J.) VA Med. Ctr., 1990; psychologist, team leader women's treatment program Fair Oaks Hosp., Summit, N.J., 1990-92; cons. Kessler Inst. for Rehab., East Orange, N.J., 1992—, Resolve Community Counseling Ctr., Scotch Plains, N.J., 1992—; staff psychologist Richard Hall Community Mental Health Ctr., Bridgewater, N.J., , 1990. Contbr. articles to profl. jours. Recipient John Weyandt award for Outstanding Student in Theatre U. Denver, 1968. Mem. APA, N.J. Psychol. Assn., Am. Anorexia/Bulimia Assn., Phi Theta Kappa. Home: 21 Harwich Ct Scotch Plains NJ 07076-3165

SWEENEY, THOMAS LEONARD, chemical engineering educator, researcher; b. Cleve., Dec. 12, 1936; s. Patrick and Anne (Morrin) S.; m. Beverly Marie Starks, Dec. 30, 1961; children: Patrick E., Thomas J., Michael S., Kevin E. BS, Case Inst. Tech., 1958, MS, 1960, PhD, 1962; JD, Capital U., Columbus, Ohio, 1974. Bar: Ohio 1974, U.S. Supreme Ct. 1978. Registered profl. engr., Ohio. Asst. prof. then assoc. prof. chem. engring. The Ohio State U., Columbus, 1963-73, prof., 1973—, assoc. v.p. rsch., 1982—, acting v.p. rsch. and grad., 1989—; pres. The Ohio State U. Research Found., Columbus, 1988—; Mem. Ohio Hazardous Waste Facility Bd., Columbus, 1984—; cons. numerous orgns. Editor: Hazardous Waste Management, 1982, Management of Hazardous and Toxic Waste, 1985; contbr. articles to profl. jours. Mem. Am. Inst. Chem. Engrs. (exec. com., cen. Ohio sect., 1970-72), Am. Chem. Soc., Am. Soc. for Engring. Edn. (chmn. environ. engring. div., 1973-74). Roman Catholic. Office: Ohio State U 1960 Kenny Rd Columbus OH 43210-1063

SWEENY, CHARLES DAVID, chemist; b. Freeport, Pa., May 22, 1936; s. Charles A. and Ruth (Beale) S.; m. Barbara K. Scheid, Feb. 12, 1977. BS, Pa. State U., 1960; PhD, Calif. Coast U., 1990. Technician Alcoa Rsch. Labs., New Kensington, Pa., 1954-58; spectroscopist Am. Color & Chem. Corp., Lock Haven, Pa., 1963-66, supr. analyt. labs., 1966-78; mgr. quality control Am. Color and Chem., 1978-81; pres. CDS Labs., Inc., 1982—. Scoutmaster, W. Branch coun. Boy Scouts Am., 1966-76, counselor, 1976—, chmn. adminstrv. bd., 1975—; Mem. emergency planning com., econ. devel. Clinton County Planning Commn. Served with AUS, 1960-63. Fellow Am. Inst. Chemists; mem. Am. Chem. Soc., Inter Soc. Color Coun. (chmn. strength of colorants-dyes com. 1975-78), Am. Assn. Textile Chemists and Colorists (chmn. weathering com. 1978-81, chmn. color measurement com. 1984-87), Pa. State U. Sci. Alumni Assn. (pres. bd. dirs. 1985—). United Methodist. Contbr. articles to profl. jours. Patentee effluent treatment. Office: Mt Vernon St Lock Haven PA 17745

SWEET, RITA GENEVIEVE, civil engineer; b. Norristown, Pa., Nov. 7, 1959; d. John William and Genevieve Rita (Koehler) S.; m. Robert A. Bellitto, Apr. 4, 1992. BS in Biology summa cum laude, Bucknell U., 1981; MS in Civil Engring., Stanford U., 1983. Lic. profl. engr., Va. Staff engr. Speitel Assocs. Environ. Engrs., Mt. Laurel, N.J., 1983-85, No. Va. Planning Dist. Commission, Annandale, 1985-86; civil engr. Virginia Beach (Va.) Utilities Dept., 1987—. Mem. ASCE, Am. Water Works Assn., Phi Beta Kappa. Achievements include assisting in obtaining fed. permits for $200 million water supply project, 1988-93. Office: Dept Pub Utilities Mcpl Ctr Virginia Beach VA 23456

SWEETSER, THEODORE HIGGINS, mathematician; b. Winnipeg, Manitoba, Can., Nov. 4, 1948; s. Theodore Higgins and Amelia Sweetser; m. Jane Lee Valentine, June 25, 1977; 1 child, Catherine Elizabeth. BA, Boston Coll., 1970; PhD, U. Calif.-San Diego, La Jolla, 1979. Tech. staff Jet Propulsion Lab., Pasadena, Calif., 1979—. Contbr. articles to Jour. Astro. Scis., Jour. Optimization Theory and Applications, Am. Jour. Phys. Medicine, Advances in Astro. Scis. Recipient Group Achievement award NASA, 1983, 92. Mem. AIAA, Am. Astro. Soc., Am. Math. Soc. Republican. Roman Catholic. Home: 1721 Las Lunas St Pasadena CA 91106 Office: Jet Propulsion Lab 4800 Oak Grove Dr Pasadena CA 91109

SWERDLOW, HAROLD, biomedical engineer; b. Bklyn., May 8, 1957; s. Melvin Jerome and Norma Lee (Steinhauer) S. BA in Physics and Math., U. Calif., Santa Cruz, 1979; PhD in Bioengring., U. Utah, 1991. Engr. in tng. Rsch. assoc. U. Calif., San Francisco, 1979-85, U. N.Mex., Albuquerque, 1986-87; grad. rsch. fellow U. Utah, Salt Lake City, 1987-91, postdoctoral fellow, 1991—. Contbr. articles to Electrophoresis, Analytical Chemistry, Nucleic Acids Rsch., Jour. Biol. Chemistry, Math. Tchrs. Jour. Recipient award for scientific excellence Bausch & Lomb, 1975; NSF fellow, 1988-91. Home: 3400 Emigration Canyon Salt Lake City UT 84108 Office: U Utah 6160 Eccles Genetics Bldg Salt Lake City UT 84112

SWERDLOW, MARTIN ABRAHAM, physician, pathologist; b. Chgo., July 7, 1923; s. Sol Hyman and Rose (Lasky) S.; m. Marion Levin, May 19, 1945; children—Steven Howard, Gary Bruce. Student, Herzl Jr. Coll., 1941-42; B.S., U. Ill., 1945; M.D., U. Ill., Chgo., 1947. Diplomate: Am. Bd. Pathology. Intern Michael Reese Hosp. and Med. Center, Chgo., 1947-48; resident Michael Reese Hosp. and Med. Center, 1948-50, 51-52, mem. staff, 1974—, chmn. dept. pathology, v.p. acad. affairs, 1974-90; pathologist Menorah Med. Center, Kansas City, Mo., 1954-57; asst. prof., pathologist U. Ill. Coll. Medicine, Chgo., 1957-59, assoc. prof., 1959-60, clin. assoc. prof., 1960-64, clin. prof., 1964-66, prof., pathologist, 1966-72, assoc. dean, prof. pathology, 1970-72; prof. pathology, chmn. U. Mo., Kansas City, 1972-74; prof. pathology U. Chgo., 1975-89; Geever prof., head pathology U. Ill., 1989-93, prof. emeritus, 1993—; mem. com. standards Chgo. Health Systems Agy., 1976—. Served with M.C. U.S. Army, 1944-45, 50-54. Recipient Alumnus of Yr. award U. Ill. Coll. Medicine, 1973; Instructorship award U. Ill., 1960, 65, 68, 71, 72. Mem. Chgo. Pathology Soc. (pres. 1980—), Am. Soc. Clin. Pathologists, Coll. Am. Pathologists, Internat. Acad. Pathology, Am. Acad. Dermatology, Am. Soc. Dermatopathology, Inst. Medicine, AMA. Jewish. Office: Univ IL Coll Medicine 1819 W Polk St Chicago IL 60612

SWERN, FREDERIC LEE, engineering educator; b. N.Y.C., Sept. 9, 1947; s. Reuben Swern and Anne Lillian Goldberg; m. Gayle Regina Unger, Dec. 25, 1969; children: Lauren, Michael. BEE, City Coll., N.Y.C., 1969; MSEE, Newark Coll. Engr., 1974; PhD Engr. Sci., N.J. Inst. Tech., 1981. Registered profl. engr., N.J. Engr. Navigation and Control divsn. Bendix, Teterboro, N.J., 1969-72, UNIVAC, divsn. Sperry Rand, Morris Plains, N.J., 1972-73; sr. systems analyst Chubb & Son Inc., Short Hills, N.J., 1973-78; sr. engr. Flight Systems divsn. Bendix, Teterboro, N.J., 1978-84; cons. Swerlin Assoc., 1981—; assoc. prof. engring Stevens Inst. Tech., Hoboken, N.J., 1984—; researcher, cons., Westinghouse Elevator, Morristown, N.J., 1987, NASA Langley Rsch. Center, Hampton, Va., 1979-82, 1985-90, Hyatt Clark Industries, Clark, N.J., 1984-85, U.S. Army Armament Rsch. and Devel. Ctr., 1985-92, Gen. Motors Corp., Warren, Mich., 1983-84; reviewer Prentice-Hall, Englewood Cliffs, N.J.; cons. Gov. Commn. on Sci. and Tech., 1985-86. Contbr. to profl. jours. Mem. Inst. Elec. Electronic Engrs., Am. Soc. Mech. Engrs., Assn. Computing Machinery, Am. Inst. Aeronautics and Astronautics, Sigma Xi, Pi Tau Sigma. Achievements include Automatic Flight Control System Patent using instrument landing system information and including inertial filtering means for reduced ILS noise; Aircraft Control System Patent using inertial signals; Improved Flare Control Patent for transport aircraft. Home: 53 Village Rd Florham Park NJ 07932 Office: Stevens Institute of Technology Mechanical Engring Lab Castle Point Hoboken NJ 07030

SWETLIK, WILLIAM PHILIP, orthodontist; b. Manitowoc, Wis., Jan. 31, 1950; s. Leonard Alvin and Lillian Julia (Knipp) S.; m. Cheryl Jean Klein, June 30, 1973 (div.); children: Alison Elizabeth, Lindsey Ann, Adam William Swetlik. Student, Luther Coll., Decorah, Iowa, 1968-70; DDS, Marquette U., 1974; MS in Dentistry, St. Louis U., 1977. Diplomate Am. Bd. Orthodontics. Resident in gen. dentistry USPHS, Norfolk, Va., 1974-75; practice dentistry specializing in orthodontics Green Bay, Wis., 1977—; instr. oral pathology NE Wis. Tech. Coll., Green Bay, 1979-86. Author: (with others) Orthodontic Headgear, 1977. Mem. Prevention Walking Club, Family Crisis Ctr. of Green Bay. Served as lt. USPHS, 1974-75. Fellow Coll. Diplomates Am. Bd. Orthodontics; mem. ADA, Am. Assn. Orthodontists, Wis. Dental Assn. (Continuing Edn. award 1986), Wis. Soc. Orthodontists, Orthodontic Edn. and Research Found., Brown Door Kewaunee Dental Soc. (program chmn. 1985-86, sec., treas. 1986-87, v.p. 1987-88, pres. 1988-89), St. Louis U. Orthodontic Alumni Assn. (pres. 1988-89), Acad. Gen. Dentistry, Violet Club of Am. Roman Catholic. Avocations: racquetball, skiing, jogging, raising violets, recording equipment. Home: 2160 Green Leaf Rd DePere WI 54115-8621 Office: 2654 S Oneida St Green Bay WI 54304-5392

SWETS, JOHN ARTHUR, psychologist, scientist; b. Grand Rapids, Mich., June 19, 1928; s. John A. and Sara Henrietta (Heyns) S.; m. Maxine Ruth Crawford, July 16, 1949; children—Stephen Arthur, Joel Brian. B.A., U. Mich., Ann Arbor, 1950, M.A., 1953, Ph.D., 1954. Instr. psychology U. Mich., Ann Arbor, 1954-56; asst. prof. psychology M.I.T., Cambridge, 1956-60, assoc. prof. psychology, 1960-63; v.p. Bolt Beranek & Newman Inc., Cambridge, 1964-69, sr. v.p., 1969-74, gen. mgr. research, devel. and cons., dir., 1971-74; chief scientist BBN Labs., Cambridge, 1975—; lectr. dept. clin. epidemiology Harvard Med. Sch., 1985-88, dept. health care policy, 1988—; mem. corp. Edn. Devel. Ctr., Newton, Mass., 1971-75; vis. research fellow Philips Labs., Netherlands, 1958; Regents' prof. U. Calif., 1969; advisor vision com., com. on hearing and bioacoustics NAS-NRC, 1960—; mem. Commn. on Behavioral Social Scis. and Edn., NRC, 1988-92, vice chair, 1992-93, chmn., 1993—; advisor, cons., lectr. numerous govtl. and profl. orgns. in science. Co-author: (with D.M. Green) Signal Detection Theory and Psychophysics, 1966; (with R.M. Pickett) Evaluation of Diagnostic Systems: Methods From Signal Detection Theory, 1982; editor: Signal Detection and Recognition by Human Observers, 1964; (with L.L. Elliott) Psychology and the Handicapped Child, 1974; (with D. Druckman) Enhancing Human Performance, 1988; contbr. articles to profl. jours.; mem. editorial bd. Medical Decision Making, 1980-85, Psychological Science, 1989—. Past mem. numerous civic orgns.; mem. corp. Winchester Hosp., Mass., 1981-84. Fellow AAAS (coun. 1986-89), APA (Disting. Sci. Contbn. award 1990), Am. Acad. Arts and Scis., Acoustical. Soc. Am. (exec coun. 1968-71), Soc. Exptl. Psychologists (chmn. 1986, exec. com. 1986-89, Howard Crosby Warren medal 1985), Am. Psychol. Soc.; mem. NAS (Troland award com. 1991, chmn. 1992), Psychonomic Soc., Psychometric Soc., Soc. Math. Psychology, Evaluation Rsch. Soc., Sigma Xi, Sigma Alpha Epsilon, Winchester Country Club, Cosmos Club. Congregationalist (moderator). Office: Bolt Beranek & Newman Inc 10 Moulton St Cambridge MA 02138-1191

SWETT, SUSAN, chemical engineer; b. Newark, N.J., Sept. 25, 1962; d. Stephen F. and Annette (Palazzolo) S. BSChE, Stevens Inst. Tech., Hoboken, N.J., 1984. Registered profl. engr., N.J. Sr. staff engr. environ. div. T&M Assocs., Middletown, N.J., 1984—. Mem. Am. Inst. Chem. Engrs., ASCE, Soc. Women Engrs. Home: 12 Louis St Old Bridge NJ 08857-2235 Office: T&M Assocs 11 Tindall Rd Middletown NJ 07748-2717

SWIATEK, KENNETH ROBERT, neuroscientist; b. Chgo., Dec. 30, 1935. BS, North Ctrl. Coll., 1958; PhD in Biol. Sci., U. Ill., 1965. Rsch. assoc. biol. chemistry dept. pediatrics coll. medicine U. Ill., 1965-68; rsch. scientist Ill. Inst. Devel. Disabilities, 1968-70, adminstrv. rsch. scientist, 1970-75, dir. rsch., 1975-76, dir., 1976-80. Grantee NIH, 1971-74. Mem. AAAS, Am. Chem. Soc., Am. Assn. Mental Deficiency, Soc. Devel. Biology, Sigma Xi. Achievements include research in growth of nervous system with special emphasis on development of carbohydrate, ketones and amino acid metabolism in fetal and newborn brain tissue as affected by pain-relieving drugs of labor and delivery. Home: 5029 Central Ave Western Springs IL 60558*

SWIDEN, LADELL RAY, research center director; b. Sioux Falls, S.D., June 17, 1938; s. Alick and Mildred Elizabeth (Larson) S.; m. Phyllis Lorriane Enga, Sept. 10, 1961; children: David, Daniel, Shari. BSEE, S.D. State U., 1961; MBA, U. S.D., 1982. Registered profl. engr., S.D., Minn. Instrument engr. Honeywell, Mpls., 1962-67; v.p. sales Swiden Appliance and Furniture, Sioux Falls, 1967-68; engring. mgr. Raven Industries, Inc., Sioux Falls, 1968-84; v.p. engring. Beta Raven Inc., St. Louis, 1984-85; pres. Delta Systems, Inc., St. Louis, 1985-86; acting dir. Engring. and Environ. Rsch. Ctr. S.D. State U., Brookings, 1986—, dir. univ./industry tech. svc., 1988—. Patentee in field. Chmn. Indsl. Devel. Com., Brookings, 1989-90, vice-chair, 1988-89; chmn. bldg. com. Ascension Luth. Ch., 1988-90. Mem. NSPE, Instrument Soc. Am., Aircraft Owners and Pilots Assn., Exptl. Aircraft Assn., Am. Bonanza Soc., S.D. Engring. Soc. (pres. N.E. chpt. 1991-92), Rotary, Elks. Avocations: flying, travel. Home: 105 Heather Ln Brookings SD 57006-4123 Office: SD State U PO Box 2220 Brookings SD 57007-7011

SWIFT, DAVID LESLIE, environmental health educator, consultant; b. Chgo., Aug. 7, 1935; s. Harry Leslie and Rhoda Louise (Lawson) S.; m. Suzanne Neild Haller, June 6, 1959; children: Charles, Kevin, Austin, Christopher. BS, Purdue U., 1957; SM, Mass. Inst. Tech., 1959; PhD, Johns Hopkins U., 1963. Asst. chem. engr. Argonne (Ill.) Nat. Lab., 1963-65; postdoctoral fellow London Sch. Hygiene & Tropical Medicine, 1965-66; from asst. to assoc. prof. environ. health sch. hygiene Johns Hopkins U., Balt., 1966-78, prof. sch. hygiene, 1978—; vis. prof. Danish Nat. Inst. Occupational Health, Copenhagen, 1986-87; Thomson vis. lectr. London Sch. Hygiene, 1988; cons. Internat. Commn. Radiation Protection, mem. lung task force Nat. Coun. Radiation Protection. Mem. editorial bd. Jour. Aerosol Sci.; contbr. chpts. to various books and monographs, articles to profl. jours. Fellow Am. Sci. Affiliation; mem. Am. Indsl. Hygiene Assn., Am. Conf. Govtl. Indsl. Hygienists, Am Assn. Aerosol Rsch., Gesellschaft für

Aerosolforschung, TAu Beta Pi, Delta Omega. Home: 1020 Litchfield Rd Baltimore MD 21239

SWIFT, GERALD ALLAN, aeronautical engineer; b. Arkansas City, Kans., Nov. 18, 1964; s. Arlen Lee Swift and Karen Louise Shively Briggs; m. Lori Jean Briney, May 31, 1986. BS, U. Kans., 1987; MS, Air Force Inst. Tech., 1991. Commd. 2d lt. USAF, 1987, advanced through grades to capt., 1991; aircraft design engr. Wright Labs., Tech. Exploration Directorate, Wright-Paterson AFB, Ohio, 1988-90; aircraft stability and control engr. 46 Test Wing, Compatibility Engring., Eglin AFB, Fla., 1991—. Mem. AIAA (Team Aircraft Design Competition award 1986-87), Soc. Am. Mil. Engrs., Aircraft Owners and Pilots Assn. Home: 136 Midland Ct Niceville FL 32578 Office: USAF 46 Test Wing/EAEL Eglin A F B FL 32542

SWIFT, JOHN LIONEL, civil engineer; b. Rochester, N.Y., Jan. 19, 1947; s. Stanley Melven and Ethel Eleanor (Hyde) S.; m. Martha Aida Hernandez, July 19, 1969; children: John Paul, Mark Stuart. BSCE, Heald Engring. Coll., 1969. Registered profl. engr., Ohio, Ala., La., Fla., Tex., Ga., registered profl. land surveyor, Ohio. Chief dep. county engring. Geauga County Engr., Chardon, Ohio, 1977-81; sr. design engr. Rust Internat., Birmingham, Ala., 1981-82; civil project mgr. Guillot-Vogt Assocs. Inc., Metairie, La., 1982-84; dept. head civil engring. Morphy, Marofsky & Masson, New Orleans, 1984-85; sr. project mgr. RS & H, Inc., Roswell, Ga., 1985-86; v.p. Foxworth, Swift & Assocs., Inc., Marietta, Ga., 1986-91; project mgr. Entech, Inc., Marietta, 1991—. Mgr. Colt Baseball Team, Marietta, 1988-92; mem. Lake County Regional Transit Authority, Painesville, Ohio, 1976-79; county rep. Ohio Soc. Profl. Engrs., Mentor, 1976-81; mem. YMCA, bd. dirs., 1978-81; sustaining mem. LWV, 1977-81. Fellow ASCE (airfield pavement com. 1991—, bd. dirs. Ga. chpt. 1990—); mem. Ga. County Safety Coun. (bd. dirs. 1977-81), Kiwanis (bd. dirs. 1978-81). Republican. Roman Catholic. Home: 2679 Arbor Spring Way Marietta GA 30066 Office: Entech Inc Ste B-5 540 Powder Springs SE Marietta GA 30064

SWIFT, KERRY MICHAEL, physical chemist; b. Alamosa, Colo., June 22, 1957; s. Marlyn Dean and Marjorie (Roberts) S.; m. Leticia Diane Muñoz, July 23, 1977; children: Camille, Candice. MS in Chemistry, Colo. State U., 1981; BSEE, Air Force Inst. Tech., 1983. Comd. 2d lt. USAF, 1981, advanced through grades to capt., 1988; sci. officer USAF, Albuquerque, 1981-88; new products scientist Milton Roy, Urbana, Ill., 1989-91; software mgr. Cytomation, Ft. Collins, Colo., 1991-92; v.p. sales and mktg. Chromex, Albuquerque, 1992-93; phys. biochemist pharm. products divsn. Abbott Labs., Abbott Park, Ill., 1993—. Co-contbr. articles to Jour. Chem. Physics, Applied Optics, Proc SPIE, Jour. Fluorescence. Colo. grad. fellow Colo. State U., 1980. Achievements include co-development of multiharmonic Fourier transform fluorescence lifetime instrument. Office: Abbott Labs Pharm Dept Structural Biology Pharm Discovery Rsch Abbott Park IL 60064-3500

SWIGER, ROY RAYMOND, biomedical scientist; b. Castro Valley, Calif., June 16, 1967; s. Roy Earl and Jessenda Helen (Valente) S. BS, Calif. State U., Hayward, 1990, cert. in biotech., 1992, postgrad. Asst. microbiologist Ab-Tox Lab., Pleasanton, Calif., 1989; criminal technologist DNA crime lab. Dept. Justice, Berkeley, Calif., 1991-92; biomed. scientist Lawrence Livermore (Calif.) Nat. Lab., 1992—; sr. rsch. asst. Calif. State U., Hayward, 1989-90. Pres. Pioneer Heights Student Assn., Calif. State U., Hayward, 1988-90. Mem. Genetic and Environ. Toxicology Assn., Sigma Xi (assoc.). Republican. Roman Catholic. Home: 21239 Gary Dr # 314 Castro Valley CA 94546 Office: Lawrence Livermore Nat Lab Biology abd Biotechnology Rsch Program Bldg 361 Livermore CA 94550

SWILLER, RANDOLPH JACOB, internist; b. N.Y.C., Jan. 21, 1946; s. Abraham Irving and Helen (Emmer) S.; m. Florence Tena Davis, Sept. 3, 1967; children: Jeremy Adam, Rebecca Susan, Steven Eric. BA in Biology cum laude, Hofstra U., 1968; MD, Chgo. Med. Sch., 1972. Diplomate Am. Bd. Psychiatry and Neurology. Intern Long Island Jewish-Hillside Med. Ctr., New Hyde Park, N.Y., 1972-73; psychiatric resident SUNY Downstate Med. Ctr., Bklyn., 1973-76; asst. attending psychiatrist Maimonides Med. Ctr., Bklyn., 1976-78; medical resident, mem. med. ethics com. Jewish Hosp. Med. Ctr. of Bklyn., 1978-80; fellow in hematology North Shore U. Hosp., Manhasset, N.Y., 1980-81; attending physician medicine/hematology U. Community Hosp., Tamarac, Fla., 1982—; attending physician in internal medicine Fla. Med. Ctr., Lauderdale Lakes, 1982—, mem. med. utilization rev. com., 1986—, mem. credentials and qualifications com., 1990—; attending physician in internal medicine Coral Springs (Fla.) Med. Ctr., 1987—, mem. med. utilization rev. com., 1987-89. Mem. ACP, AMA, Am. Soc. Internal Medicine, Fla. Med. Assn., Broward County Med. Assn. Democrat. Jewish. Achievements include research in disseminated intravascular coagulation in obstetrical practice, angioimmunoblatic lymphadenopathy syndrome. Avocation: piano. Office: 7710 NW 71st Ct Ste 304 Fort Lauderdale FL 33321-2932

SWINDEN, H. SCOTT, earth scientist. Recipient Barlow medal Can. Inst. Mining and Metallurgy, 1992. Office: care Can Inst Mining Metallurg, 3400 de Maisonneuve Blvd W, Montreal, PQ Canada H3Z 3B8*

SWINEHART, FREDERIC MELVIN, chemical engineer; b. Midland, Mich., May 23, 1939; s. Richard Woods and Francis Lucille (Bellinger) S.; m. Carole Jean Markeson, June 14, 1965; children: Karen Margaret, Anneke Christine, Karl Frederic. MSChemE, U. Mich., 1963, MS Engring. in Info. and Control, 1965, PhD in Chem. Engring., 1966. Process devel. leader Dow Chem. Co., Midland, Mich., 1966-70, area supt., supt. mfg. ion exch., 1970-76; tech. mgr. pharm. mfg. Dow Chem. Co., Milan, Italy, 1976-80; area supt. saran films Dow Chem. Co., Midland, 1980-84, mgr. rsch. computing svcs., 1984-91; process info. cons. Dow Corning Corp., Midland, 1991-93; exec. cons. Digital Equiptment Corp., 1993—. Office: Digital Equiptment Corp 5301 No Eastman Rd Midland MI 48640

SWINGLE, HOMER DALE, horticulturist, educator; b. Hixson, Tenn., Nov. 5, 1916; s. Edward Everett and Sarah Elizabeth (Rogers) S.; m. Gladys F. Wells, Dec. 21, 1942 (dec. June 1961); 1 child, Janet Faye Swingle Sciscioli; m. Ella Margaret Porterfield, Dec. 19, 1962. BS, U. Tenn., 1939; MS, Ohio State U., 1948; PhD, La. State U., 1966. Tchr. vocat. agr. Spring City (Tenn.) High Sch., 1939-46; hort. specialist U. Tenn., Crossville, 1946-47; from asst. prof. to prof. horticulture U. Tenn., Knoxville, 1948-79, prof. emeritus non-credit program, 1979—; cons. in plant and water rels. Oak Ridge (Tenn.) Nat. Lab., 1971-75; chmn. collaborators Vegetable Breeding Lab. USDA, Charleston, S.C., 1975. Cons. editor: Growing Vegetables and Herbs, 1984; contbr. 49 articles to sci. jours. Fellow Am. Soc. Hort. Sci. (chmn. so. sect. 1970-71); mem. Lions (life), Gamma Sigma Delta (pres. U. Tenn. chpt. 1977, teaching award of merit 1978-79). Republican. Methodist. Home: 3831 Maloney Rd Knoxville TN 37920-2823

SWINK, LAURENCE NIM, chemist, consultant; b. Enid, Okla., Oct. 24, 1934; s. Lyle Nim and Zelia Alice (Murphy) S.; m. Barbara Jean Trumbull, Sept. 1960 (div. 1978); 1 child, Steven Marshall. BA in Chemistry, U. Wichita, 1957; MSc, Iowa State, 1959; PhD in Chemistry, Brown U., 1969. Lab. mgr. Tex. Instruments, Inc., Dallas, 1966-78; v.p. rsch. Amorphous Materials, Inc., Garland, Tex., 1978-80; advanced systems mgr. Xerox Corp., Dallas, 1981-82; tech. dir. Multi-Plate Co., Inc., Dallas, 1982-85; sr. project mgr. Siemens Transmission Systems, Tempe, Ariz., 1986-87; sr. tech. cons. Quari Electronics, Chandler, Ariz., 1987-89; pres. AZ/TEC Consulting, Tempe, 1989—. Contbr. articles to profl. jours. Capt. USAF, 1960-63. Mem. Am. Crystallographic Assn. Achievements include crystal structures, materials, microanalysis and process techniques; contributor to materials and process patents. Home and Office: 1617 E Julie Dr Tempe AZ 85283

SWINT, JOSEPH ELLIS, computer systems analyst; b. Columbus, Ga., Jan. 3, 1952; s. James Alvin and Lydia (Storey) S.; m. Marian Patrick, Oct. 12, 1991. AS, Southern Tech. Inst., 1972; postgrad., U. Southern Miss., 1978-84. Technician T.E.T.R.A. Systems, Marietta, Ga., 1973-74; field engr. Inforex Inc., Atlanta, 1974-76; assoc. engr. Ingalls Ship Bldg., Pascagoula, Miss., 1976-83, systems analyst, 1983—. Roman Catholic. Home: 11100 Pecan Rd Pascagoula MS 39581 Office: Chevron USA Hwy 611 S PO Box 1300 Pascagoula MS 39581

SWITZER, JON REX, architect; b. Shelbyville, Ill., Aug. 22, 1937; s. John Woodrow and Ida Marie (Vadalabene) S.; m. Judith Ann Heinlein, July 7, 1962; 1 child, Jeffrey Eric. Student, U. Ill., 1955-58; BS, Millikin U., 1972; MA, Sangamon State U., 1981. Registered architect Ill., Mo., Ohio, Colo.; registered interior designer, Ill. Architect Warren & Van Praag, Inc., Decatur, Ill., 1970-72; prin. Decatur, 1972-81, Bloomington, Ill., 1981-83; architect Hilfinger, Asbury, Cufaude, Abels, Bloomington, 1983-84; ptnr. Riddle/Switzer, Ltd., Bloomington, 1984-86; with bldg., design and constrn. div. State Farm Ins. Cos., Bloomington, 1986-89; architect The Riddle Group, Bloomington, 1989-91; prin. J. Rex Switzer, Architect, Bloomington, 1991—. Served with U.S. Army, 1958-61. Mem. AIA (pres. Bloomington chpt. 1983, Decatur chpt. 1976, v.p. Ill. chpt. 1986-87, sec. 1985, treas. 1984), Am. Archtl. Found., Chgo. Architecture Found., Nat. Trust Hist. Preservation, Frank Lloyd Found., Decatur C. of C. (merit citation 1974, merit award 1979), Masons (32d degree). Republican. Presbyterian. Avocations: swimming, hunting, fishing, reading, drawing. Home: 9 Mary Ellen Way Bloomington IL 61701-2014 Office: Ste 6A 2412 E Washington St Ste 6A Bloomington IL 61704

SWITZER, TERENCE LEE, civil engineer; b. Allentown, Pa., Aug. 12, 1957; s. Jethro Joseph and Mary Margaret (Bennes) S.; m. Karen Lynn Sames, Oct. 1, 1982; children: Jennifer Lynn, Heather Ann, Kristin Lee, Justin Kyle. Assoc. in Surveying, Pa. State U., Wilkes Barre, 1978; BS in Water Resources Engring., Pa. State U., Middletown, 1980. Registered profl. engr., Va. Pollution control engr. Va. State Water Control Bd., Virginia Beach, 1980-84, supr. engr., 1984-85; civil engr./planner Naval Air Sta. Oceana, Virginia Beach, 1985-87, architect and engr. project mgr., 1987-88, engring. program mgr., 1988-90, dir. engring., planning and energy mgmt., 1990—. Home: 732 Willow Oak Dr Chesapeake VA 23320 Office: Naval Air Sta Oceana Bldg 820 Staff Civil Engr Office Virginia Beach VA 23460-5120

SWOPE, ROBERT J., physical science laboratory administrator; b. Zanesville, Ohio, July 20, 1936; s. Wendell B. and Sarah A. (Russell) S.; m. Mary M. Gerber, Apr. 30, 1961; 1 child, Erin G. BA, Ohio Wesleyan U., 1958. Chemist Pacific Resins, Newark, Ohio, 1960-63; chemist Aerospace Guidance and Metrology Ctr., Newark, 1963-85, chief chemist, lab. chief phys. sci., 1985—. Named signficant contbns. Charles Stark Draper Lab., Inc., Boston, 1989. Fellow Am. Inst. Chemists; mem. Am. Chem. Soc., Coblentz Soc. Achievements include research in contamination control, aerospace material characterization, aerospace fluids. Office: Aerospace Guidance and Metrology Ctr MAEL Newark Air Force Base Newark OH 43057-5149

SWYSTUN-RIVES, BOHDANA ALEXANDRA, dentist; b. Kopychynci, Ukraine, Jan. 31, 1925; came to U.S., 1951; d. Peter and Maria (Ottawa) Swystun; m. John Rives, June 20, 1952 (div. 1960); 1 child, Peter A. DMD, Ludwig Maximillians Universitat, Munich, 1951; DDS, NYU, 1960. Dentist Dr. Joseph Matriss, East Rutherford, N.J., 1960-61; gen. practice dentistry Clifton, N.J., 1961—. Vol. dentist Felician Sisters Orphanage, Lodi, N.J., 1982—; mem. Presdl. Task Force, Washington. Mem. ADA (award for commitment to professionalism and health), Ukrainian Med. Assn., Ukrainian Nat. Assn., Ukrainian Inst. Am., Clifton-Pasaic (N.J.) C. of C. Republican. Ukrainian Catholic. Avocations: reading, fgn. langs., walking, gold jewelry. Office: 1 Portland Ave Clifton NJ 07011-2347

SYBLIK, DETLEV ADOLF, computer engineer; b. Mildenberg, Templin, Fed. Republic Germany, July 9, 1943; s. Adolf and Elsbeth Elfriede (Laux) S.; m. Judit Galambos, Aug. 3, 1974; children: Rita, Claudius. Degree in engring., Hohere Technische Bundeslehr-Anstalt, Vienna, 1962; BA in Econs., U. Vienna, 1970. Engr. M-V-T, Stockerau, Austria, 1962-63; systems engr. IBM Roece Comecon, Austria, 1970-77, IBM Austria, Vienna, 1977-83; project staff IBM Videotex/CEPT, Frankfurt, Fed. Republic Germany, 1983-86; adv. systems engr. IBM Germany, Stuttgart, 1986—; cons. Austrian PTT, Vienna, 1983-86. Author: Videotex, 1983. With Austrian mil., 1963-64. Mem. TSV Sports Club, Diving Club. Roman Catholic. Avocations: travel, reading, hunting, music, sports. Home: Spaichinger Weg 4, D-71229 Leonberg Germany

SYED, IBRAHIM BIJLI, medical physicist, educator, theologist; b. Bellary, India, Mar. 16, 1939; came to U.S., 1969, naturalized, 1975; s. Ahmed Bijli and Mumtaz Begum (Maniyar) S.; m. Sajida Shariff, Nov. 29, 1964; children: Mubin, Zafrin. BS with honors, Veerasaiva Coll., Bellary, U. Mysore, 1960; MS with honors and distinction, Central Coll., Bangalore, U. Mysore, 1962; diploma Radiol. Physics and Hosp. Physics, U. Bombay, 1964; DSc, Johns Hopkins U., 1972; PhD (hon.), Marquis Giuseppe Scicluna Internat. U., Malta, 1985. Cert. hazard control officer, 1980, internat. health care safety profl., 1980; Diplomate: Am. Bd. Radiology, Am. Bd. Health Physics. Lectr. physics Veerasaiva Coll., Bellary, U. Mysore, 1962-63; med. physicist, radiation safety officer Victoria Hosp., India, 1964-67, Bowring and Lady Curz on Hosp. and Post-grad Med. Rsch. Inst., Bangalore, India, 1964-67; cons. med. physicist, radiation safety officer ministry of Health, Govt. of Karnataka, India, 1964-67, Bangalore Nursing Home, India, 1964-67; med. physicist, radiation safety officer. Halifax (N.S., Can.) Infirmary, 1967-69; dir. med. physics, radiation safety officer Baystate Med. Ctr, Springfield, Mass., 1973-79, assoc. prof. Springfield Tech. Community Coll., also adj. prof. radiology Holyoke Community Coll. (Mass.), 1973-79; asst. clin. prof. nuclear medicine U. Conn. Sch. Medicine, Farmington, 1975-79; cons. med. physicist Mercy Hosp., Springfield, also Wing Meml. Hosp., Palmer, Mass., 1973-79; med. physicist, radiation safety officer VA Med. Ctr., Louisville, 1979—, exec. officer radiation safety com., 1979—; prof. medicine (med. physics, nuclear cardiology, endocrinology, metabolism and radionuclide studies) and nuclear medicine, U. Louisville Sch. Medicine, 1979—, dir. nuclear med. scis., 1980—; vis. prof. Bangalore U., 1987-88, Gulbarga U., India, 1987-88; vis. scientist Bhabha Atomic Rsch. Ctr., Bombay, India; course dir. licensing for nuclear cardiologists U. Louisville, 1980—; mem. admissions com. nuclear medicine program U. Louisville, 1980—; guest, relief examiner Am. Bd. Radiology, 1991; mem. panel of examiners Am. Bd. Health Physics; PhD thesis examiner U. Delhi, Internat. Inst. for Advanced Study, Clayton, Mo., 1985—; faculty mem. Med. Physicists of India Ann. Meeting, 1987; Internat. Atomic Energy Agy. tech. expert in nuclear medicine on mission to People's Republic of Bangladesh, 1986; founder, pres. Islamic Rsch. Found. for Advancement of Knowledge, Louisville, 1988—; cons. Coun. Sci. and Indsl. Rsch., Govt. India, 1980—, cons. Am. Coun. Sci and Health, 1980—, cons. gastoenterology and urology div. FDA, HHS, 1988—, cons. radiopharmaceutical div., 1989—; cons. Govt. of India in nuclear medicine, diagnostic radiol. physics, therapeutic radiol. physics, and radiation safety, 1992; cons. radiological and medical nuclear physics Govt. of India, UN Devel. Program, 1992. Author: Radiation Safety for Allied Health Professionals, Radiation Safety Manual, 1979; contbg. editor Jour. of Islamic Food and Nutrition Coun. of Am., 1986—, health and sci. column Muslim Jour., 1989—; freelance writer AL'FURQAN Internat., Norcross, Ga., 1990, Message Internat., Jamaica, N.Y., 1990; editor: Science and Technology for the Developing World, 1988; mem. editorial bd. Jour. Islamic Med. Assn., 1981—; regular contbr. President's Page; contbr. over 100 articles to sci. jours.; manuscript reviewer for Sci. and Med. Jours., 1973. Pres. Springfield Islamic Ctr., 1973-79, India Assn., Louisville, 1980-81; v.p. Islamic Cultural Assn., Louisville, 1979-80, trustee, 1980—, vice chmn. bd. trustees, 1980-84, chmn. bd. trustees, 1984-86; vice chmn., bd. trustees Islamic Cultural Assn. of Louisville, Inc., 1987—; ordained minister for Islamic marriages, 1983—; vol. Muslim Chaplain to Ky. State Reformatory, LaGrange, Ky., 1989—, to VA Med. Ctr., Louisville, 1990—, Luther Luckett Correctional Instn., LaGrange, Ky., St. Mary's Correctional Instn., Lebanon, Ky., 1990—; khatibs; legal advisor Islamic Cultural Assn. Louisville, 1986-87; notary pub. Commonwealth of Ky., 1983—; mem. speaker's bur. Louisville C. of C., 1980—, U. of Louisville, 1980—; chmn. Ky. state nat. alumni schs. com. Johns Hopkins U.; judge Ky. State Sci. Fair, 1985—; dir. Ctr. for Qur'an and Sci. Studies, 1988—; trustee India Community Found. Louisville, 1980—, chmn. bd., 1984—; trustee of Karnataka India Community Found. Louisville, Inc., 1988—; bd. dirs. Child Guidance Clinic, Springfield, 1973-79, Heritage Corp., Louisville, 1981—, others; active Am. Cancer Soc., Heart Fund; vol. Muslim chaplain Ky. State Regormatory, La Grange, , Dept. of Vets. Affairs Med. Ctr., Louisville, 1990—. Recipient Disting. Community Service award India Community Found., 1982; WHO fellow, Govt. India scholar Bhabha Atomic Research Center, Bombay, 1963-64; USPHS fellow Johns Hopkins U., 1969-72. Fellow Inst. Physics (U.K.), Am. Inst. Chemists, Royal Soc. Health, Am. Coll. Radiology, Internat. Acad. Med. Physics; mem. Am. Coll. Nuclear Medicine, Health Physics Soc., Am. Assn. Physicists in Medicine, Soc. Nuclear Medicine (faculty mem. ann. meeting 1987), Nat. Assn. Ams. of Asian Indian Descent (chmn. state pub. relations com. 1982—), Islamic Soc. N.Am., Islamic Soc. Balt. (founding mem.), Islamic Cultural Ctr., Louisville, Islamic Assn. Maritime Provinces Can., Halifax, N.S. (asst. sec. 1967-69), Health Physics Soc. (chmn. med. health physics com. 1989—, affirmative action com. 1984—), Am. Assn. Physicists in Medicine (mem. biol. effects com.), Assn. Muslim Scientists and Engr. N. Am. (program chm. annual conf. 1987, treas. 1987-88, sec. 1988—), AAUP, Soc. Nuclear Medicine India (life, faculty mem. ann. meeting 1987), Assn. Med. Physicists India (life), Med. and Biol. Physics div. of Can. Assn. Physicists, Hosp. Physicists Assn., N.Y. Acad. Scis., Islamic Assn. Maritime Provinces of Can., Ky. Med. Assn., Jefferson County Med. Soc. (assoc.), Sigma Xi. Islamic. Home: 7102 Shefford Ln Louisville KY 40242-2853 Office: 800 Zorn Ave Louisville KY 40206-1433

SYED, MOINUDDIN, electrical engineer; b. Jaipur, Rajhastan, India, Feb. 1, 1947; came to U.S., 1969; s. Masihuddin and Aisha Bibi (Ali) S.; m. Nasim Afroz Mustfai, Jan. 7, 1972; children: Mohsin, Ahson, Mona, Hasan. BEE, U. Karachi, Pakistan, 1968; MSEE, Tulane U., 1971. Registered profl. engr., Ont., Can. Sci. staff mem. Bell No. Rsch., Ottawa, Can., 1973-79; staff engr. Ont. region Bell Can., Toronto, 1979-80; sr. planning engr. Contel Calif., Victorville, 1980-82, supervising engr., 1982-85, 88—; staff engr. Contel Corp., Bakersfield, Calif., 1985-88; mem. ICEP com. Bell No. Rsch., Ottawa, 1973-80, Bell Can., Toronto, 1985-86; quality task force Contel Corp., Bakersfield, 1985-86; project mgr. Contel Calif., Victorsville, 1988-89, chmn. spl. task force on fiber to home project, 1993—; featured speaker USTA Conv., Portland, Oreg., 1977; participant continuing edn. programs Chapman Coll., Golden Gate U., Victor Valley Coll.; teaching and rsch. asst. Tulane U., 1969-73. Co-author: Characterization of Electrical Environment, 1976; contbr. articles to jours. Juror State Ct. System, Victorville, 1990; vol. Desert Knolls Elem. Sch., Apple Valley, Calif., 1991-93. Scholar Fauji Found., 1962-64, Ministry Edn., 1964-68. Mem. IEEE. Achievements include rsch. in elec. protection of telecom network, telecom protection for power stations, and use of fiber optics technology in telecommunications. Office: Contel Calif 16071 Mojave Dr Victorville CA 92392

SYKES, RICHARD BROOK, microbiologist; b. Eng., Aug. 7, 1942; married; 2 children. BS, Paddington Coll., 1965; MS, London U., 1968; PhD, Bristol U., 1972. Head antibiotics rsch. unit Glaxo Rsch. Labs., 1972-77; dir. dept. microbiology Squibb Inst. Med. Rsch., 1979-83, v.p. infectious and metabolic diseases, 1983-86; chief exec. Claxo Group Rsch., U.K., 1986-87; group R & D dir. Glaxo Holdings, 1987—; chmn., chief exec. Glaxo Group Rsch. Ltd.; pres. Glaxo Inc. Rsch. Inst. Mem. Brit. Soc. Antimicrobial Chemother. Office: Glaxo Research Inst 5 Moore Dr Research Triangle Park NC 27709*

SYLLA, RICHARD EUGENE, economics educator; b. Harvey, Ill., Jan. 16, 1940; s. Benedict Andrew and Mary Gladys (Curran) S.; m. Edith Anne Dudley, June 22, 1963; children: Anne Curran, Margaret Dudley. BA, Harvard U., 1962, MA, 1965, PhD, 1969. Prof. econs. and bus. N.C. State U., Raleigh, 1968-90; Henry Kaufman prof. history fin. insts. and markets NYU, N.Y.C., 1990—, prof. econs., 1990—; cons. Citibank NA, N.Y.C., 1979-82, Chase Manhattan Bank, N.Y.C., 1983-85; vis. prof. U. Pa., Phila., 1983, U. N.C., Chapel Hill, 1988. Author: The American Capital Market, 1975; co-author: Evolution of the American Economy, 1980, 2d edit., 1993, A History of Interest Rates, 1991; co-editor: Pattterns of European Industrialization, 1991; editor Jour. Econ. History, 1978-84. Study fellow NEH, 1975-76; rsch. grantee NSF, 1985-94. Mem. Am. Econ. Assn., Econ. History Assn. (v.p. 1987-88, trustee 1978-88, Arthur H. Cole prize 1970), Am. Finance Assn., Bus. History Conf. (trustee 1991—), So. Econ. Assn. (v.p. 1981-82). Avocations: golf, hiking, fishing, stamp collecting, arts. Home: 110 Bleecker St Apt 23D New York NY 10012-2106 Office: NYU 44 W 4th St New York NY 10012-1126

SYMES, CLIFFORD E., telecommunications engineer; b. Cardiff, Wales, Nov. 11, 1929; came to U.S., 1969; s. Clifford and Ivy (Edwards) S.; m. Beryl Quick, June 6, 1951 (div. Dec. 1989); children: Dilwyn Edward, Gareth David; m. Margarita Inocenta Nunez Palomino), Nov. 27, 1992. BSc in Engring., U. London, 1949; M in Engring. Tech., Bart Inst. Tech., London, 1963. Registered profl. engr., Ont., Eng., Ill. Mem. sci. staff North Electric, Bramalea, Ont., Can., 1964-69; project mgr. Automatic Electric, Chgo., 1969-74; dir. engring. G.I. Clare Div., Chgo., 1974-76, Standex, Cin., 1976-78; sales mgr. Richco, Chgo., 1978-84; cons. Mega Internat., San Diego, 1984—; cons. Battelle Cabs, Columbus, Ohio, 1971, Armco Steel, Middletown, Ohio, 1972. Contbr. articles to profl. jours. Asst. commr. Boy Scouts Am., Galion, Ohio, 1972. Sgt. Royal Air Force, 1946-49. Mem. Royal Air Force Assn. (life). Republican. Methodist. Achievements include research in magnetic applications in telephone switching. Home and Office: 12383 Calle Albara # 8 El Cajon CA 92019

SYMES, LAWRENCE RICHARD, university dean, computer science educator; b. Ottawa, Ont., Can., Aug. 3, 1942; s. Oliver Lawrence and Maybell Melita Blanche (Gilliard) S.; m. Evelyn Jean Hewett, Apr. 3, 1964; children—Calvin Richard, Michelle Louise, Erin Kathleen. B.A., U. Sask., Saskatoon, Can., 1963, postgrad. in math., 1964; M.S., Purdue U., 1966, Ph.D., 1969. Asst. prof. Purdue U., West Lafayette, Ind., 1969-70; assoc. prof. computer sci. U. Regina, Sask., Can., 1970-74, prof., 1974—, dir. computer ctr., 1970-75, head dept. computer sci., 1972-81, dean of sci., 1982-92; invited lectr. Xian Jiaotong U., 1983, Shandong Acad. Sci., People's Republic of China, 1987; bd. dirs. Hosp. System Study Group, Saskatoon, 1978—, chmn. bd., 1980-83; dir. SSTA Computer Services, Regina; mem. adv. coun. Can./Sask. Advanced Tech. Agreement, 1985-87; mem. Sask. Agrl. Rsch. Found. Bd., 1987-88; mem. steering com. IBM/Sask. Agreement, 1990-92. Contbr. articles to profl. jours. Can. Fed. Govt. grantee, 1977-84. Mem. Assn. Computing Machinery, Can. Info. Processing (pres. 1979-80), IEEE Computer Soc., Sask. ADA Assn. (bd. dirs. 1990—). Office: U Regina, Faculty Sci, Regina, SK Canada S4S 0A2

SYMINGTON, JANEY STUDT, cell and molecular biologist; b. St. Louis, June 29, 1928; d. Sidney Melchior and Jane Belle (Sante) Studt; m. Stuart Symington Jr., June 21, 1949; children: Anne Wadsworth Symington Rhodes, William Stuart Symington, Sidney Studt Symington, John Sante Symington. BA, Vassar, 1950; PhD, Harvard/Radcliffe, 1959. Rsch. assoc. dept. biology Washington U., St. Louis, 1958-65, asst. prof. rsch. dept. biology, 1965-72; HEW-USPHS spl. trainee cell biology and tissue culture Dept. Microbiology & Immunology, Washington U. Sch. Medicine, St. Louis, 1973-74, virology rsch. assoc., 1974-77; rsch. assoc. Inst. for Molecular Virology St. Louis U. Sch. Medicine, 1978-79, asst. rsch. prof., 1980-85, assoc. rsch. prof., 1985—; guest curator genetic engring./biotechnology St. Louis Sci. Ctr., 1989—; bd. dirs., v.p. Acad. Sci. St. Louis, 1979—; mem. sci. and engring. com. Regional Commerce and Growth Assn. St. Louis, 1988—. Contbr. articles to profl. jours. Past class fund chmn. Vassar; past founding mem. exec. bd. Women's Soc. Washington U.; past bd. dirs. theatre project co. Seven Coll. Nat. scholar, 1946-50. Mem. AAAS, Am. Inst. Biol. Scis., Am. Soc. for Cell Biology, Am. Soc. for Microbiology, Am. Soc. Plant Physiologists, Botanical Soc. Am., Cen. States Election Microscopy Soc., N.Y. Acad. Scis., Phi Beta Kappa, Sigma Xi. Episcopalian. Achievements include discovery of nodes of stability in RNA of tobacco mosaic virus; design of system for detection and analysis of proteins by transfer from 2D gels and immunoprobing; elucidation of aspects of the replication of tobacco mosaic virus and adenovirus; preparation and characterization of antibodies to viral regulatory and transforming proteins using synthetic peptides and use of these peptides to study activity of viral genes in human cells. Avocations: golf, theater, opera, photography, nature study. Home: 745 Cella Rd Saint Louis MO 63124-1611 Office: St Louis U Sch Medicine Inst for Molecular Virology Saint Louis MO 63110

SYMONS, GEORGE EDGAR, environmental engineering editor; b. Danville, Ill., Apr. 20, 1903; s. Martin Van Meter Matthew and Dora Alice (Martz) S.; m. Virginia Thompson, July 16, 1926; 1 child, James Martin. BS with honors in Chemistry, U. Ill., 1928, MS in San. Engring. Chemistry, 1930, PhD in San. Engring. Chemistry, 1932. Diplomate Am. Acad. Environ. Engrs.; registered profl. engr., N.Y. Various positions as asst. engr.

and chemist, 1920-35; chief chemist Buffalo Sewer Authority, 1936-43; assoc. editor, mng. editor Water and Sewage Works Scranton Pub. Co., N.Y.C., 1943-51; self employed cons. Larchmont, N.Y., 1951-64; editor Water & Wastes Engring. Reuben H. Donnelly Corp., N.Y.C., 1964-70; mgr. spl. projects Malcolm Pirnie, Inc., White Plains, N.Y., 1970-78; pvt. practice engring. editor, cons. Larchmont, 1978-84; engring. editor Palm Beach, Fla., 1984—. Author tech. books and manuals; editor engring. mags.; contbr. over 300 articles to profl. jours.; editor 29 manuals for U.S. Navy Bureau Yards & Docks, 4 manuals on maintenance and operation, 2 technical books. Fellow APHA, ASCE, NSPE, Am. Acad. Environ. Engrs., Am. Chem. Soc., Am. Inst. Chem. Engrs., Am. Pub. Works Assn., Am. Water Works Assn. (hon., mem. exec. com. 1968-77, pres. 1973-74), Inter-Am. Assn. San. Engrs., Water Pollution Control Fedn. (hon., mem. name change com. 1957-59, 90, 91, Emerson medal 1969), New Eng. Water Works Assn., Sigma Xi, Phi Lambda Upsilon, Alpha Chi Sigma, Phi Kappa Sigma. Achievements include engring. assignments in rsch., teaching, water and wastewater plant operation and design. Home and Office: 300 S Ocean Blvd 3-H Palm Beach FL 33480

SYMONS, TIMOTHY JAMES MCNEIL, physicist; b. Southborough, Kent, Eng., Aug. 4, 1951; came to U.S., 1977; s. Henry McNeil and Catherine Muriel (Rees) S.; m. Syndi Beth Master, Mar. 1, 1987; children: Henry Benjamin, Daniel Robert. BA, Oxford (Eng.) U., 1972, MA, DPhil, 1976. Rsch. fellow Sci. Rsch. Coun., Eng., 1976-77; postdoctoral fellow Lawrence Berkeley (Calif.) Lab., 1977-79, div. fellow, 1979-84, sr. physicist, 1984—, dir. nuclear sci. div., 1985—; vis. scientist Max-Planck Inst., Heidelberg, Germany, 1980-81; mem. U.S. nuclear physics del. to USSR, 1986; mem. program adv. coms. Gesellschaft fur schwerionen forschung, 1987-88, Continuous Electron Beam Accelerator Facility, 1989-92, Brookhaven Nat. Lab., 1991-92; mem. policy com. Relativstic Heavy Ion Collider, 1988—. Contbr. over 70 articles to sci. jours. Fellow Am. Phys. Soc. (chmn. Bonner prize com. 1990). Avocations: music, gardening, travel. Office: Lawrence Berkeley Lab 1 Cyclotron Rd Berkeley CA 94720

SYNEK, M., physics educator, researcher; b. Prague, Czechoslovakia, Sept. 18, 1930; came to U.S., 1958, naturalized 1963; s. Frantisek and Anna (Kokrment) S.; children: Mary Rose, Thomas Robert. Indsl. chemist Tech. Sch., Prague, 1946-50; cert. in liberal arts, Prague, 1951; MS in Physics with distinction, Charles U., Prague, 1956; PhD in Physics, U. Chgo., 1963. Analytical chemist Indsl. Medicine Inst., Prague, 1950-51; rsch. physicist Acad. of Scis., Prague, 1956-58; from asst. to assoc. prof. De Paul U., Chgo., 1962-67; prof. Tex. Christian U., Ft. Worth, 1967-71; lectr., researcher U. Tex.-Austin, 1971-75; tenured faculty U. Tex.-San Antonio, 1975—; sci. advisor Tex. Edn. Agy., Austin, 1971-73, U. Tex., 1971-73; advisor Student Physics Soc., active numerous univ. coms. Researcher in laser-crystal energy efficiency, laser fusion, space lasers, approximate estimate of the extra-terrestial intelligence probability, nuclear age requiring free elections. Contbr. articles to sci. jours. Campaigner United Way, San Antonio, 1975—. Judge Alamo Sci. Fair. Rsch. grantee Robert A. Welch Found., 1968-71, 76-83, 93—. Fellow AAAS, Am. Phys. Soc. (life), Tex. Acad. Sci., Am. Inst. Chemists; mem. AAUP, Am. Assn. Physics Tchrs., Am. Mus. Natural History, Internat. Platform Assn., N.Y. Acad. Scis., Tex. Faculty Assn., Disabled Am. Vets., Commdrs. Club, Am. Chem. Soc., Czechoslovak Nat. Council (dist. sec. Chgo. 1961-63), Czechoslovak Soc. Arts and Scis. Am., San Antonio Astronomical Assn. (mem. world affairs coun.), Bexar County Czech Heritage Soc., Sigma Xi (life), Sigma Pi Sigma. Roman Catholic.

SYNGE, RICHARD LAURENCE MILLINGTON, biochemist; b. Liverpool, Eng., Oct. 28, 1914; s. Laurence M. and Katharine (Swan) S.; m. Ann Stephen, 1943; 3 sons, 4 daus. Ph.D.; ed., Winchester Coll., Trinity Coll., Cambridge (Eng.) U.; D.Sc. (hon.), U. East Anglia, 1977, U. Aberdeen, 1987; Ph.D. (hon.), U. Uppsala, 1980. Biochemist Wool Industries Research Assn., Leeds, Eng., 1941-43; staff biochemist Lister Inst. Preventive Medicine, London, 1943-48; head dept. protein chemistry Rowett Research Inst., Aberdeen, Scotland, 1948-67; biochemist Food Research Inst., Norwich, Eng., 1967-76; hon. prof. Sch. Biol. Scis., U. East Anglia, 1968-84; vis. biochemist Ruakura Animal Research Sta., Hamilton, New Zealand, 1958-59. Mem. editorial bd. Biochem. Jour., 1949-55. Recipient (with A.J.P. Martin) Nobel prize for chemistry, 1952; John Price Wetherill medal Franklin Inst., 1959; named hon. fellow Trinity Coll., Cambridge U., 1972. Fellow Royal Soc. Chemistry, Royal Soc.; hon. mem. Royal Irish Acad., Royal Soc. New Zealand, Am. Soc. Biol. Chemists. Home: 19 Meadow Rise Rd, Norwich NR2 3QE, England

SYROMIATNIKOV, VLADIMIR SERGEEVICH, aerospace engineer; b. Archangelsk, Russia, Jan. 7, 1933; s. Sergey Arkadievich and Ekaterina Yakovlevna (Ivanova) S.; m. Svetlana Ilinichna Chumakova, July 14, 1959; children: Anton, Ekaterina. Diploma in engring., Bauman Poly., Moscow, 1956; degree in math., Moscow U., 1961; Candidate of Scis. in Engring., Machine Building Inst., Moscow, 1968; ScD, Tsniimash Inst., Kaliningrad, Russia, 1979. Mem. engring. staff NPO Energia, Kaliningrad, Moscow, Russia, 1956—, dept. chief, 1974-84, chief of br., 1984—; prof. MLTI, Mytischy, Moscow, 1982—; v.p., tech. dir. Space Regatta Consortium, Kaliningrad, 1989—; pres. Aquilon Firm, Kaliningrad, 1990—; participant in Sputnik and Yuri Gagarin's Vostok spacecraft devel.; active in Soyuz, Salyut, Mir and Progress Spacecraft design; chief designer of Russian docking system; working group chmn. of Apollo-Soyuz Test Project, 1971-75; tech. dir. of first Solar Sail Spacecraft, 1993, Space Shuttle-Mir docking system, 1993—. Author: Docking Systems for Spacecraft, 1984, Manned Spacecraft, 1984; editor: Reliability Engineering and Risk Assessment, 1984; contbr. chpts. to books. Recipient Lenin prize, Moscow, 1976. Fellow AIAA (assoc.); mem. Internat. Astronautic Acad., Russian Acad. Scis. Achievements include over 100 patents in fields of electromechanics; design of large deployable structures, spacecraft concepts and systems, docking systems for Soyuz, Progress and Salyut spacecrafts. Home: Apg 81, Pavla Korchagina 14, 129278 Moscow Russia

SYVANEN, MICHAEL, geneticist, educator; b. Vancouver, Wash., Dec. 27, 1943; s. Carl R. Syvanen and Helen T. Auhonen; m. Sue E. Greenwald, 1978. BS, U. Wash., 1966; PhD, U. Calif., Berkeley, 1972. From asst. prof. to assoc. prof. Harvard U. Med. Sch., Boston, 1976-86; prof. Sch. Medicine U. Calif., Davis, 1987—. Author: Bacteria, Plasmids and Phage, 1984; contbr. 50 articles and papers to profl. jours., 3 book revs. Mem. Am. Soc. Microbiology. Democrat. Achievements include devel. of theory of role of horizontal gene transfer in biol. evolution; contbn. to tech. devels. in biotech. (gene cloning). Office: U Calif Sch Medicine Dept Microbiology Davis CA 95616

SZAL, GRACE ROWAN, research scientist; b. Amsterdam, N.Y., July 17, 1962; d. David Anderson and Helen Marie (Bursese) R.; m. Timothy James Szal, Dec. 31, 1988; 1 child, Matthew David. BA, Siena Coll., 1983; MA, Rutgers U., 1985, PhD, 1988. Rsch. asst., instr. Rutgers U., New Brunswick, N.J., 1983-88; postdoctoral rsch. fellow U. Pa., Phila., 1988-89, Tex. Coll. Osteo. Medicine, Fort Worth, 1989-90; postdoctoral rsch. fellow Tex. Christian U., Fort Worth, 1990-91, assoc. rsch. scientist, 1991—. Contbr. articles to profl. jours. Recipient Nat. Rsch. Svc. award Nat. Inst. Drug Abuse, 1988-91. Mem. Am. Psychol. Soc., Soc. for Neurosci. Office: Tex Christian U PO Box 32880 Fort Worth TX 76129

SZASZ, ANDRAS ISTVÁN, physicist, educator, researcher; b. Budapest, Hungary, Nov. 4, 1947; s. István Szasz and Maria Rozsa; m. Susan Szasz-Csih, Aug. 8, 1971; children: Oliver, Nora. MS, Eötvös U., Budapest, 1972, PhD, 1974; CSc, St. Petersburg U., Russia, 1983. Asst. prof. Eötvös U., Budapest, 1974-84, assoc. prof., 1984—; head metal lab. Eötvös U., Budapest, 1983-87, head surface physics lab., 1987—; vis. prof. Strathclyde U., Glasgow, Scotland, 1987—. Author 3 books; contbr. over 200 sci. papers to profl. jours.; patentee. Mem. Eötvös Physics Soc., European Physics Soc., Am. Surface Electroplating Soc., Am. Elec. Chem. Soc. Avocation: travel. Home: Szent István krt 20, H-1137 Budapest Hungary Office: Eötvös U, Muzeum krt 6-8, H-1088 Budapest Hungary

SZASZ, THOMAS STEPHEN, psychiatrist, educator, writer; b. Budapest, Hungary, Apr. 15, 1920; came to U.S., 1938, naturalized, 1944; s. Julius and Lily (Wellisch) S.; m. Rosine Loshkajian, Oct. 19, 1951 (div. 1970); children: Margot Szasz Peters, Susan Marie. A.B., U. Cin., 1941, M.D., 1944; D.Sc. (hon.), Allegheny Coll., 1975, U. Francisco Marroquin, Guatemala, 1979. Diplomate: Nat. Bd. Med. Examiners, Am. Bd. Psychiatry and Neurology. Intern 4th Med. Service Harvard, Boston City Hosp., 1944-45; asst. resident medicine Cin. Gen. Hosp., 1945-46, asst. clinician internal medicine div. out-patient dispensary, 1946; asst. resident psychiatry U. Chgo. Clinics, 1946-47; tng. research fellow Inst. Psychoanalysis, Chgo., 1947-48; research asst. Inst. Psychoanalysis, 1949-50, staff mem., 1951-56; practice medicine, specializing in psychiatry, psychoanalysis Chgo., 1949-54, Bethesda, Md., 1954-56, Syracuse, N.Y., 1956—; prof. psychiatry SUNY Health Sci. Ctr., Syracuse, 1956-90, prof. psychiatry emeritus, 1990—; vis. prof. dept. psychiatry U. Wis., Madison, 1962, Marquette U. Sch. Medicine, Milw., 1968, U. N.Mex., 1981; holder numerous lectureships, including C.P. Snow lectr. Ithaca Coll., 1970; E.S. Meyer Meml. lectr. U. Queensland Med. Sch.; Lambie-Dew orator Sydney U., 1977; Mem. nat. adv. com. bd. Tort and Med. Yearbook; cons. com. mental hygiene N.Y. State Bar Assn.; mem. research adv. panel Inst. Study Drug Addiction; adv. bd. Corp. Econ. Edn., 1977—. Author: Pain and Pleasure, 1957, The Myth of Mental Illness, 1961, Law, Liberty and Psychiatry, 1963, Psychiatric Justice, 1965, The Ethics of Psychoanalysis, 1965, Ideology and Insanity, 1970, The Manufacture of Madness, 1970, The Second Sin, 1973, Ceremonial Chemistry, 1974, Heresies, 1976, Karl Kraus and the Soul-Doctors, 1976, Schizophrenia: The Sacred Symbol of Psychiatry, 1976, Psychiatric Slavery, 1977, The Theology of Medicine, 1977, The Myth of Psychotherapy, 1978, Sex by Prescription, 1980, The Therapeutic State, 1984, Insanity: The Idea and its Consequences, 1987, The Untamed Tongue: A Dissenting Dictionary, 1990, Our Right to Drugs: The Case for a Free Market, 1992, A Lexicon of Lunacy, 1993, Cruel Compassion, 1994; editor: The Age of Madness, 1973; cons. editor of psychiatry and psychology: Stedman's Medical Dictionary, 22d edit, 1973; contbg. editor: Reason, 1974—, Libertarian Rev., 1986—; mem. editorial bd. Psychoanalytic Rev, 1965—, Jour. Contemporary Psychotherapy, 1968—, Law and Human Behavior, 1977—, Jour. Libertarian Studies, 1977—, Children and Youth Services Rev, 1978—, Am. Jour. Forensic Psychiatry, 1980—, Free Inquiry, 1980—. Served to comdr., M.C. USNR, 1954-56. Recipient Stella Feiss Hofheimer award U. Cin., 1944, Holmes-Munsterberg award Internat. Acad. Forensic Psychology, 1969; Wisdom award honor, 1970; Acad. prize Institutum atque Academia Auctorum Internationalis, Andorra, 1972; Distinguished Service award Am. Inst. Pub. Service, 1974; Martin Buber award Midway Counseling Center, 1974, Thomas S. Szasz award Ctr. Ind. Thought , 1990, Alfred R. Lindesmith award for achievement in field of scholarship and writing Drug Policy Found., 1991; others; named Humanist of Year Am. Humanist Assn., 1973; Hon. fellow Postgrad. Center for Mental Health, 1961, Mencken award, 1981, Humanist Laureate, 1984, Statue of Liberty-Ellis Island Found. Archives Roster, 1986. Life fellow Am. Psychiat. Assn., Am. Psychoanalytic Assn., Internat. and Western N.Y. psychoanalytic socs. Home: 4739 Limberlost Ln Manlius NY 13104-1405 Office: 750 E Adams St Syracuse NY 13210-2306

SZEBEHELY, VICTOR G., aeronautical engineer; b. Budapest, Hungary, Aug. 10, 1921; s. Victor and Vilma (Stockl) S.; m. Jo Betsy Lewallen, May 21, 1970; 1 dau., Julia. M.E., U. Budapest, 1943, Ph.D. in Engring, 1945; Dr. (hon.), Eotvos U. Budapest, 1991. Asst. prof. U. Budapest, 1945-47; research asso. State U. Pa., 1947-48; asso. prof. Va. Poly. Inst., 1948-53; research asso. Model Basin, U.S. Navy, 1953-57; research mgr. Gen. Electric Co., 1957-62; asso. prof. astronomy Yale U., 1962-68; prof. aerospace enging. U. Tex., Austin, 1968—; chmn. dept. U. Tex., 1977-81, R.B. Curran Centennial chair in engring., 1983—; cons. NASA-Johnson Space Center, U.S. Air Force Space Command, Lawrence Berkeley Lab., U. Calif. Author 18 books; contbr. over 200 articles on space research, celestial mechanics and ship dynamics to profl. jours. Knighted by Queen Juliana of Netherlands, 1956. Fellow AIAA, AAAS; mem. Am. Astron. Soc. (Brouwer award div. dynamical astronomy 1977), Internat. Astron. Union (pres. commn. on celestial mechanics), NAE, European Acad. Arts, Scis., Lit. Home: 2501 Jarratt Ave Austin TX 78703-2432 Office: U Tex Dept Aerospace Engring and Engring Mechanics Austin TX 78712

SZEKELY, JULIAN, materials engineering educator; b. Budapest, Hungary, Nov. 23, 1934; came to U.S., 1966, naturalized, 1975; s. Gyula and Ilona (Nemeth) S.; m. Elizabeth Joy Pearn, Mar. 2, 1963; children: Richard J., Martin T., Rebecca J., Mathew T., David A. B.Sc., Imperial Coll., London, 1959; Ph.D., D.I.C., 1961; D.Sc., D.I.C., Eng., 1972. Lectr. metallurgy Imperial Coll., 1962-66; asso. prof. chem. engring. State U. N.Y. at Buffalo, 1966-68, prof., 1968-76; dir. Center for Process Metallurgy, 1970-76; prof. materials engring. Mass. Inst. Tech., Cambridge, 1976—; cons. to govt. and industry. Author: (with N.J. Themelis) Rate Phenomena in Process Metallurgy, 1971, (with W.H. Ray) Process Optimization, 1973, (with J.W. Evans and H.Y. Sohn) Gas-Solid Reactions, 1976, Fluid Flow Aspects of Metals Processing, 1979; (with J.W. Evans and J.K. Brimocombe) Mathematical Modelling of Metals Processing Operations; Editor: Ironmaking Technology, 1972, The Steel Industry and the Environment, The Steel Industry and The Energy Crisis, 1975, The Future of the World's Steel Industry, 1976, Alternative Energy Sources for the Steel Industry, 1977; Contbr. articles to profl. jours. Recipient Jr. Moulton medal Brit. Inst. Chem. Engrs., 1964; Extractive Metallurgy Div. Sci. award Am. Inst. Mining and Metall. Engrs., 1973; also Mathewson Gold medal, 1973; Howe Meml. lectr., 1979, Extractive Metall. lectr., 1987 ; Sir George Beilby Gold medal Brit. Inst. Chem. Engrs.-Soc. Chem. Industry-Inst. Metals, 1973; Curtis McGraw research award Am. Soc. Engring. Edn., 1974; Profl. Progress award Am. Inst. Chem. Engrs., 1974; Charles H. Jennings Meml. award Am. Welding Soc., 1983; John Simon Guggenheim fellow, 1974. Mem. Nat. Acad. Engring. Office: MIT Dept Materials Sci & Engring Rm 4-140 Cambridge MA 02139

SZENTPÁLY, LÁSZLÓ VON, chemistry educator; b. Budapest, Hungary, Apr. 4, 1942; s. László Von and Irene (Eloed) S.; m. Ratna Ghosh, Oct. 29, 1986. MA, U. Basel, Switzerland, 1966, PhD, 1969; DSc, U. Stuttgart, Germany, 1985. Postdoctoral fellow Phillips U., Marburg, Germany, 1969-72, asst. prof., 1972-78; sr. rsch. fellow U. Stuttgart, 1979-85, 86; vis. scientist U. Tex., El Paso, 1985; prof. U. De Guanajuato, Mex., 1987-89; assoc. prof. U. W.I., Kingston, Jamaica, 1989—; cons. Hewlett Packard Germany, Boeblingen, 1985-86. Co-author: The Molecular Basis of Cancer, 1985, Polynuclear Atomic Compounds, 1987, Recent Advances in Chemistry and Molecular Biology of Cancer, 1992. Fellow World Assn. of Theoretical Organic Chemists; mem. Am. Chem. Soc., Internat. Soc. Math. Chemistry. Achievements include rsch. on valence states interaction model of chemical bonding (VSI model). Office: Chemistry Dept, U WI, Mona Campus, Kingston Jamaica

SZERSZEN, JEDRZEJ BOGUMIL (ANDREW SZERSZEN), plant pathologist, educator; b. Warsaw, July 10, 1946; came to U.S., 1984; s. Bogumil Wincenty and Irena Romualda (Dynska) S.; m. Ingeborg Claudia Schulze, July 26, 1986 (div. Sept. 1992). BSc, Warsaw Agrl. U., 1968, MS, 1969; PhD, Nat. Acad. Scis., Warsaw, 1976. Plant pathologist Horticultural Plant Breeding Experiment Sta., Zielonki, Poland, 1969-70; rsch. asst. Mich. State U., East Lansing, 1970-71; asst. prof. Inst. of Ecology, Nat. Acad. Scis., Warsaw, 1977-85; postdoctoral fellow dept. plant pathology Tex. A&M U., College Station, 1985-86, rsch. assoc. dept. plant pathology, 1987-91; asst. prof. Mich. State U., East Lansing, 1991—. Contbr. chpts. to books and articles to profl. jours. Recipient Govtl. award for rsch. work Govt. of Poland, 1980. Mem. Am. Phytopathological Soc., Am. Peanut Rsch. and Edn. Assn. (Best Paper plant pathology sect. annual meeting 1988), N.Y. Acad. Scis., Sigma Xi, Gamma Sigma Delta. Roman Catholic. Achievements include discovery of pathogenesis-related proteins in cotyledons of peanut plants upon infection with fungi Aspergillus; isolation and cloning of gene that encodes the enzyme that controls levels of the plant growth hormone indole-3-acetic acid (IAA) by converting IAA into an inactive form 1-0-IAGlucose, thus making possible the control of auxin-promoted growth in transgenic plants by anti-sense RNA. Home: 1204 University Village East Lansing MI 48823 Office: Mich State Univ Dept Botany/Plant Pathology East Lansing MI 48824

SZEWCZYK, MARTIN JOSEPH, chemical engineer; b. Passaic, N.J., Jan. 24, 1954; s. Joseph Walter and Margaret (Balint) S.; m. Karen Mae Demolli, Oct. 3, 1981. BSChE, Newark Coll. Engring., 1976. Project engr. Diamond Shamrock Corp., Cleve., 1976-79; project mgr. plastics div. Gen. Electric, Selkirk, N.Y., 1979-84; project mgr. CPC Internat. Best Foods, Union, N.J., 1984-88, Huls Am. Inc., Piscataway, N.J., 1988—. Mem. AIChE (chmn. 1983-84), Cen. Jersey Collie Club. Office: Huls Am Inc PO Box 365 Piscataway NJ 08855

SZEWCZYK, PAWEL, research institute administrator, educator; b. Rydultowy, Katowice, Poland, Sept. 10, 1942; s. Pawel and Regina (Ciezkowski) S.; m. Eugenia Urszula Placzek, July 25, 1965; children: Judyta Karolina, Iwona Julia. MS, Jagiellonian U., Krakow, Poland, 1965; Candidate of Sci., Silesian U., Katowice, Poland, 1971; DSc, Curie U., Lublin, Poland, 1988. Head lab. Chem. Works Oswiecim, Poland, 1965-71, Polish Acad. Scis., Zabrze, 1971-80; dep. dir. Inst. Plastics and Paint Industry, Gliwice, Poland, 1980-90, dir. gen., 1990; assoc. prof. phys. and theoretical chemistry Inst. Plastics and Paints, 1989—; sec. bd. dirs. Chemi-Erg, Gliwice, 1991—; chmn. gen. coun. Polish Rsch Insts., 1992—. Author: Calibration of Gel Chromatography, 1987; mem. editorial bd. Polimery, 1991—; editor 2 jours. on resins, pigments and paints, 1984-90; also articles. Recipient sci. award Govt. of Poland, 1988; fellow U.S. Dept. State, 1968-69. Mem. Polish Acad. Scis. (Commn. Polymer Analysis 1972, Commn. Polymers 1992—), Internat. Union Pure and Applied Chemistry (sec. East European subgroup 1983-91, mem. Working Party), Polish Accreditation and Quality Assn. (mem. presidium), Club Rsch. Labs. (chmn. program coun. 1993—). Avocations: house and garden, reading, music, travel. Home: Konstytucji 3 Maja 39, 43-190 Mikolow Katowice, Poland Office: Inst Plastics and Paints, Chorzowska 50, 44-101 Gliwice Katowice, Poland

SZKLENSKI, THEODORE PAUL, electrical engineer; b. North East, Pa., June 12, 1959; s. Walter and Irene Szklenski. BSEE, Pa. State U., 1981. Registered profl. engr., Pa. Engr. level 1 Pa. Power & Light, Allentown, Pa., 1981-83; engr. level 2 Pa. Power & Light, Scranton, Pa., 1983-88; sr. engr. Pa. Power & Light, Allentown, 1988—. Mem. IEEE, Soc. Mfg. Engrs., Eta Kappa Nu, Tau Beta Pi. Achievements include research in gas cooling. Office: PP&L 2 N 9th St A9-4 Allentown PA 18101

SZMANDA, CHARLES RAYMOND, chemist; b. Antigo, Wis.; s. Raymond J. and Maxine Ann (Orgeman) S.; m. Margaret Lee Pritzl, Aug. 28, 1971; 1 child, Benjamin. BS in Chemistry, Loyola U., 1973; PhD in Phys. and Inorganic Chemistry, U. Wis., 1979. Mem. tech. staff AT&T Bell Labs., Murray Hill, N.J., 1979-81, Bell Labs. Device Devel. Lab., Allentown, Pa., 1981-85; sr. rsch. specialist Monsanto Co., St. Louis, 1985-86; dir. quality and new product devel. Aspect Systems, St. Louis, 1986-89; mgr. advance product devel., prin. scientist Shipley Co., Marlboro, Mass., 1989—. Mem. AAAS, Am. Chem. Soc., Am. Phys. Soc., Sigma Xi. Office: Shipley Co 455 Forest St Marlborough MA 01752

SZTRIK, JÁNOS, mathematics educator, researcher; b. Békéscsaba, Békés, Hungary, Sept. 20, 1953; s. Endre and Erzsébet (Juhász) S.; m. Rita Rigó, Dec. 18, 1982; children: Attila, Katinka. PhD, U. Debrecen, Hungary, 1981. Asst. U. Debrecen, 1978-80, jr. researcher, 1980-85, researcher, 1985-90, sr. researcher, 1990-91, assoc. prof., 1991—, cons. dept. psychology, 1985—; adj. faculty U. Econs., Budapest, 1980—; cons. Tech. U. Budapest, 1987—; leader rsch. group on performance evaluation and reliability U. Debrecen; lectr. in field. Co-author : (with Sztrik and Rigó) How to get Easier with Computer Science, 1991; contbr. articles to profl. jours. Recipient Pro Universitate award U. Debrecen, 1978. Mem. Am. Math. Soc., Bolyai Math. Soc., London Math. Soc., Neumann Computer Soc. Roman Catholic. Avocations: soccer, tennis, skiing. Home: Darabos 12, 5/27 Debrecen Hungary 4026 Office: U Debrecen, 4010 Debrecen Hungary

SZUHAJ, BERNARD FRANCIS, food research director; b. Lilly, Pa., Nov. 27, 1942; s. Theodore and Rose Dorothy (Karmen) S.; m. Carole Ann Brady, Dec. 26, 1964; children: Matthew, Timothy, Bernard. BS, Pa. State U., 1964, MS, 1966, PhD, 1969. Grad. asst. Pa. State U., Univ. Park, Pa., 1964-66; research asst. Pa. State U., Univ. Park, 1966-68; scientist Cen. Soya Co., Inc., Fort Wayne, Ind., 1968-73; research dir. Cen. Soya Co., Inc., Fort Wayne, 1973-84, dir. food research, 1984—; v.p. Am. Oil Chemists'Soc. Found., Ill., 1988—; bd. dirs. POS Pilot Plant Corp., Saskatoon, Canada, 1987—. Patentee in field; co-editor Lecithins, 1985; editor: (lecithins) Sources Manufacture & Uses, 1989. Mem. Am. Oil Chemists' Soc. (bd. dirs. 1989—), Am. Chem. Soc., Inst. Food Technologists, Inst. Shortening & Edible Oils, Sigma Xi. Democrat. Roman Catholic. Office: Cen Soya Co Inc PO Box 1400 Fort Wayne IN 46801-1400

SZUKICS, JAMES CHARLES, mechanical engineer; b. Dec. 1, 1957. BSME with highest honors, Rutgers U., 1980, MBA, 1989. Registered profl. engr., N.J. Engring. project mgr. Howmedica div. Pfizer Inc., Rutherford, N.J., 1985—. Contbr. article to profl. jour. NSF grantee, 1979. Mem. ASME, N.J. Soc. Profl. Engrs. Home: 3502 Cricket Cir Edison NJ 08820 Office: Howmedica Div Pfizer Inc 359 Veterans Blvd Rutherford NJ 07070

SZULC, ROMAN WŁADYSŁAW, physician; b. Poznan, Poland, Oct. 2, 1935; s. Witold Alfons and Michalina (Wróblewska) S.; m. Danuta Katarzyna Zielewicz, June 18, 1959; 1 child, Katarzyna Chmielecka. MD, K. Marcinkowski Acad. Medicine, Poznan, 1957. Intern Acad. Hosp. nr. 2, Poznan, 1959; resident U. Amsterdam, 1967; with Acad. Hosp., Poznan, 1973—, asst. Ear Nose and Throat Clinic, 1958-67; asst. prof. Inst. Anesthesiology Acad. Medicine, Poznan, 1966-78, assoc. prof. Inst. Anesthesiology, 1978-87, prof., dep. dir. Inst. Anesthesiology, 1987—; sec. Commn. Polish Acad. Scis., Warsaw, 1987-91, mem. experimental therapy com. 1991—. Capt. Polish Armed Forces, 1961-63. Mem. European Acad. Anaesthesiology, Poznanskie Towarzystwo Przyjaciól Nauk. Roman Catholic. Avocations: contemporary history, tourism. Home: Powidzka 19, 61-039 Poznan Poland Office: Intensive Therapy Clinic, Dluga 1/2, 61-848 Poznan Poland

SZYMANSKI, JOHN JAMES, physicist, educator; b. Buffalo, Sept. 10, 1960; s. Herman A. and Alice I. Rospond) S.; m. Christina Kasprzyk, Aug. 21, 1982; children: Lillian G., Clara T. BS, Carnegie Mellon U., 1981, MS, 1983, PhD, 1987. Postdoctoral fellow Los Alamos (N.Mex.) Nat. Lab., 1987-90; asst. prof. physics Ind. U., Bloomington, 1990—. Mem. Am. Phys. Soc., Sigma Xi. Home: 3816 Laura Way Bloomington IN 47401 Office: Ind U Cyclotron Facility 2401 Milo Sampson Ln Bloomington IN 47405

SZYMONIK, PETER TED, computer systems coordinator; b. Boleslawiec, Jelena Gora, Jelena, Poland, Nov. 13, 1963; came to U.S., 1964; s. Jan and Genowefa (Bielak) S.; m. Stephanie Christine Sans, Feb. 14, 1991. BA in Modern World History, Internat. Polit. Sci., U. Conn., 1988. Systems coord. Cummings & Lockwood, Hartford, 1989—; system operator GE Info. Network, Rockville, Md., 1986—; cons. Sacred Hoop of Am. Resource Exch., Greenwich, 1988-89. Editor and pub. mags. Simulations Online, 1991, GEnie Games RT NewsLetter, 1991; contbg. editor Strategy Plus mag., 1991—. Democrat. Avocations: conflict simulations, hist. analysis, internat. relations. Home: 161 Woodbury Cir Middletown CT 06457-5650 Office: Cummings & Lockwood Cityplace I Hartford CT 06103

TABAKOFF, BORIS, pharmacologist educator; b. Tien-Tsin, China, Sept. 27, 1942; s. Isaak and Bertha (Neidental) T.; m. Emelia Johnson. BA, U. Colo., 1966, PhD, 1970. Prof. U. Ill., Chgo., 1974-84; dir. Alcohol & Drug Abuse Rsch. and Tng. Ctr., Chgo., 1980-84; sci. dir. Nat. Inst. on Alcohol Abuse & Alcoholism, Bethesda, Md., 1984-90; vis. prof. U. Ill., Rockford, 1984—, Med. Coll. of Va., Richmond, 1989—; faculty fellow Inst. for Behavioral Genetics, Boulder, Colo., 1991—; prof., chmn. U. Colo. Sch. of Medicine, Denver, 1990—; sci. advisor Vt. Alcohol Rsch. Ctr., Colchester, 1988—; cons. in field. Author: Neuropharmacology of Ethanol, New Approaches, 1991, Comprehensive Textbook of Substance Abuse, 1992; contbr. articles to profl. jours. Acting dep. dir. Nat. Inst. on Alcohol Abuse and Alcoholism, Rockville, Md., 1985-86. Recipient Pres. Rank Meritorious Exec. award, 1989, Meritorious award to Disting. Alumnus, Chgo. Med. Sch., 1990, ADAMHA Adminstrv. award for Pub. Svc., 1986, RSA award for Sci. Excellence in Rsch., Am. Rsch. Soc. on Alcoholism, 1988, Jellinek Meml. award for Major and Continuing Contbns. to Alcohol Rsch., 1988. Mem. Rsch. Soc. on Alcholism (pres. 1983-85), Internat. Soc. for Biomed. Rsch. on Alcholism (pres. 1986-90), Am. Coll. of Neuropsychopharmacology, Am. Soc. for Pharmacology & Exptl. Therapeutics. Achievements include patents in Composition and Method for Reducing Blood Acetaldehyde Levels; Identification of Individuals Predisposed Toward Alcohol Abuse; Antimotion Sickness Apparatus; Treatment of Alcohol Withdrawal Symptoms. Office: U of Colorado Sch of Medicine 4200 E Ninth Ave Denver CO 80262

TABATA, LYLE MIKIO, mechanical engineer; b. Lahaina, Hawaii, June 28, 1956; s. Teruo and Marilyn (Tagomori) T.; m. Vianne Strom, Aug. 18, 1984; children: Benson, Mikio, Michael. BS in Mech. Engring. Tech., Bradley U., Peoria, Ill., 1978. Mech. engring. supr. Oahu Sugar Co., Ltd., Waipahu, Hawaii, 1979-80; process engr. Lihue Plantation Co., Ltd., Lihue, 1982-83, processing supt., 1983-85, chief engr., 1985-91, factory supt., 1991-92, mgr. factory opts., 1992—. Home: 4578 Hoomana Rd Lihue HI 96766 Office: Lihue Plantation Co Ltd 2970 Kele St Lihue HI 96766

TABATA, YUKIO, engineering researcher; b. Maizuru, Kyoto, Japan, Sept. 29, 1948; s. Denji and Kimie (Yamazoe) T.; m. Masayo Tsuneyama, Oct. 10, 1974; children: Kayoko, Kentaro. BS in Biology, Shizuoka U., Japan, 1971; MS in Physics, Kanazawa U., Japan, 1974. With devel. dept. Ricoh Co., Ltd., Tokyo, 1974-79; with rsch. ctr. Ricoh Reprographic Tech. (later Ricoh Imaging Tech.), Numazu, Shizuoka, Japan, 1979-86; assoc. rsch. and devel. engr. Ricoh Imaging Tech., Numazu, 1986-91; assoc. rsch. and devel engr. Ricoh RS Divsn., Numazu, 1991-92; assoc. rsch. and devel engr. 11th sect. rsch. and devel. ctr. Ricoh Chem. Products R&D, Numazu, 1992-93, assoc. rsch. and devel. engr. 13th sect., 1993—; part-time lectr. electronics Shizuoka Prefecture Sch. of Tech. at Numazu, 1989-90. Patentee electrophotographic processes; inventor electric-ink-transfer printing process; developer toner charge distbn. analyzer. Co-founder Shizuoka U. Equestrian Club, 1970; mem. Good Will Guide, Tokyo, 1986—. Mem. Physical Soc. Japan, Inst. Image Electronics Engrs. Japan, Alumni Assn. Shizuoka U. Equestrian Club (chmn. 1975-83). Avocation: tennis. Home: 49-8 Kamo, Mishima, Shizuoka 411, Japan Office: Ricoh's Numazu R & D Ctr, 146-1 Nishisawada, Numazu, Shizuoka 410, Japan

TABATABAI, M. ALI, agronomist; b. Karbala, Iraq, Feb. 25, 1934. BS, U. Baghdad, 1958; MS, Okla. State U., 1960; PhD in Soil Chemistry, Iowa State U., 1965. Rsch. assoc. soil biochemistry Iowa State U., Ames, 1966-72, from asst. prof. to assoc. prof., 1972-78, prof. soil chemistry and biochemistry, 1978—; cons. Electric Power Rsch. Inst., Palo Alto, 1978-83. Mem. AAAS, Am. Soc. Agronomy (Soil Sci. award 1992), Am. Chem. Soc., Am. Soc. Microbiology. Achievements include research in soil enzymology and chemistry of sulfur, nitrogen and phosphorus in soils, nutrient cycling in the environment. Office: Iowa St Univ of Science Ames IA 50011*

TABAU, ROBERT LOUIS, rheumatologist, researcher; b. Marseille, France, May 10, 1928; s. Victor and Valentine Tabau; m. Mireille Thonney de Blonay, Sept. 18, 1962; children: Laurence, Valerie, Herve. Grad., Faculty Pharmacy Marseille, 1950, D of Pharmacology, 1952; BS, U. Aix-Marseille, France, 1950; MD, Faculty Medicine Marseille, 1959, M of Human Biology, 1960, diploma in human biology rsch. Cert. specialist in rheumatology, med. biology, thermal, clinatic and nuclear medicine, homéopathie and acupuncture. Chief doctor Univ. Med. Clinic, Lausanne, Switzerland, 1961-62; pvt. practice in rheumatology, thermal and climatic medicine Aix Les Bains, France, 1962—; rsch. worker, then rsch. supr. Nat. Inst. for Health and Med. Rsch., France, 1965; dir. rsch. ctr. in osteoarticulatory pathology Nat. Inst. for Health and Med. Rsch., Marseille, 1965-79; asst. cen. lab. Hosp. de Conception, Marseille, 1952-56, head lab. of functional axplorations, 1958-65; asst. radiobiology lab. Ctr. de Lutte contre Cancer, Marseille, 1953-57; med. cons. Hosps. in Marseille Ctr. Rheumatology, 1970-75; instr., researcher and med. counselor Auvergne Thermale, Rhone Alpes Thermal. Author: Applied Radiations and Isotopes, 1957, Cesium 137 in Téléthé rapie, 1963, Goutte and Lithiase Urique, 1964, L'osteoporose, 1964, La Polyarthrite Rhumatoide, 1965; reporter various med. confs. and symposia. Named Chevalier for work on insecticides and pesticides Nat. Inst. Agronomic Rsch., 1966, Chevalier for svcs. to Edn. Govt. of France, 1980, Officier, 1990, Chevalier Nat. Order of Merit, 1980, Officier, 1986. Mem. French Chem. Soc., Marseille Soc. Pharmacies, Soc. Biology, Soc. Functional Medicine (hon.), Lyonaise Group Med. Studies, Internat. Ctr. Auricular Medicine and Acupuncture, Circle of Rheumatologists, French Soc. Clin. and Biol. Rsch. (chmn. 1980—), Portuguese Inst. Rheumatology (hon.), Rotary (diplomate). Avocations: skiing, tennis, golf. Home and Office: Le Chambord 3 Roche du Roi, 73100 Aix Les Bains France

TABBA, MOHAMMAD MYASSAR, civil engineer, manager, educator; b. Damascus, Syria, Jan. 25, 1946; s. Baha-Eddin and Hayat (Arafe) T.; m. Noha Dakkak, July 24, 1973; children: Omar, Rima, Sheriff. B.Sc. in Engring., Damascus U., 1968; M.Engring., McGill U., 1972, Ph.D., 1979. Registered profl. engr., Que. Structural engr. M.Backler & Assocs., Montreal, Que., Can., 1972-73; project engr. Alcan Aluminum, Montreal, 1973-75; geotech. project engr. Lavalin, Montreal, 1978-81; geotech. specialist Hydro-Que., Montreal, 1981-83; chief engr. Al-Issa Cons. Engrs., Riyadh, Saudi Arabia, 1983-85; dir. projects SHARACO, Riyadh, 1985-88; expert Riyadh Devel. Authority, 1989—; prof. Inst. of Pub. Adminstrn., Riyadh; aux. prof. dept. civil engring. McGill U., 1981-83. Contbr. articles to profl. jours. Nat. Rsch. Coun. Can. grantee, 1970-72, 75-78. Mem. ASCE, Am. Concrete Inst., ASTM, Que. Order Engrs. Office: Riyadh Devel Authority, PO Box 495, Riyadh 11411, Saudi Arabia

TABENKIN, ALEXANDER NATHAN, metrologist; b. Moscow, Apr. 19, 1933; came to U.S., 1977; s. Nathan Lev and Lyubov Akim (Isakovich) T.; m. Faina Michail Turok, Jan. 8, 1965; children: Boris, Lev. MSME, Machine-Tool Engring. Inst., Moscow, 1956. Sr. rsch. engr. Nat. R & D Ctr. Machine-Tool Industry, Moscow, 1963-77; product mgr., mktg. mgr. instrument Fed. Products Co., Providence, 1977—; cons. Morton Thiokol Co.; speaker various confs. and meetings, 1968—. Author: Instruments for the Measurements of Roundness, Straightness and Flatness, 1970; contbr. over 30 articles to profl. jours. Bd. dirs. Jewish Fedn. R.I., Providence, 1982-87. Recipient Disting. Labor medal USSR Govt., 1976. Mem. ASME (standards com. 1980—), All Union State Standard (standards com. 1973-77), Internat. Standards Orgn. (U.S. del. to ISO tech. com. TC-57 Metrology and Properties of Surfaces 1990—). Achievements include patent for methods of geometry characterization of roller bearings; development of families of geometry and surface finish gauges. Home: 52 Top St Providence RI 02906-2939 Office: Fed Products Corp 1144 Eddy St Providence RI 02905-4545

TABIBI, S. ESMAIL, pharmaceutical researcher, educator; b. Khoy, Iran, May 26, 1945; came to U.S., 1978; s. S. Ebrahim and Sharifeh Tabibi; m. Shahnaz Rahaie, Mar. 28, 1975; children: Shahrzad, Shirazeh, Shabnam. PharmD, U. Tabriz, Iran, 1969; PhD, U. Md., Balt., 1982. Lab. scientist I biochemistry dept. U. Md., 1979-82; vis. fellow Nat. Cancer Inst., Bethesda, Md., 1982-83; rsch. assoc. Roxane Labs., Inc., Columbus, Ohio, 1983-86; dir. pharm. R & D, H.G. Pars Pharm. Lab., Inc., Cambridge, Mass., 1986; dir. pharm. R & D MediControl Corp., Newton, Mass., 1986-89; v.p. R & D Medicontrol Corp., Newton, Mass., 1989-90; v.p. R & D, mem. sci. adv. bd. Micro Vesicular Systems, Inc., Nashua, N.H., 1990-92; assoc. prof. pharms. dept. U. R.I., Kingston, 1992—; adj. assoc. prof. Mass. Coll. Pharmacy, Boston, 1989—; mem. sci. adv. bd. Cell Rsch. Corp., Newton, 1989—. Contbr. articles to sci. jours., chpt. to book. Vice chmn. PTO, Chelmsford, Mass., 1991. Mem. Am. Pharm. Assn., Am. Assn. Pharm. Scientists, Controlled Release Soc., Rho Chi. Achievements include patents on oral cavity and dental microemulsion products, hexamethylmelamine containing parenteral emulsions, lipid-vesicles having an alkyd wall forming material; patents pending on liponomal gel products, heat-dehydrated emulsion composition, heat-dehydrated liponomal compositions, wan emulsion as liquid control release drug delivery. Office: Nat Cancer Inst-NIH Pharm Resources Br 6130 Executive Pla N Rockville MD 20852

TABLER, WILLIAM BENJAMIN, architect; b. Momence, Ill., Oct. 28, 1914; s. Clyde Lyeth and Frances Beatrice (Ridley) T.; m. Phyllis May Baker, June 12, 1937; children: William, Judith. B.S. cum laude, Harvard U., 1936, B.Arch., 1939, M.Arch., 1939. Architect specializing in hotels; prin. works include Hilton hotels in N.Y.C., Bklyn., Dallas, Pitts., San Francisco, Toronto, Rye Town, N.Y., Long Branch and Woodcliff Lake, N.J., Washington and Izmir, Turkey; Conrad Internat. Istanbul, Turkey; Intercontinental hotels in Lahore, Rawapindi, Jamaica, Ras Al Khaimah, Jeddah, Nairobi, Lusaka, Dacca, Amman, Karachi and Jerusalem; Marriott Phila., Sheraton Universal City, New Orleans, Brussels and Sheraton Centre, Toronto; Meridien hotels in Colombo, Sri Lanka, Cairo and Heliopolis, Egypt and Jakarta, Indonesia, Othon Palace in Rio and Bahia; Registry in Bloomington and Scottsdale; Grand Kempinski, Dallas; Hosts of Houston and Tampa; Sonesta Bermuda; Radisson Duluth, Lough Key, Ireland; New Otani L.A., Chosen, Korea; Stouffers, Chgo. and St. Louis; Bonaventure Montreal; Hanover, Woodstock and Princeton Inns; 15 Hospitality Motor Inns; also Harper and Stony Brook Coll. Dormitories; many others; mem. bldg. constrn. adv. council N.Y.C. Bldg. Dept., 1967—. Bd. dirs. Manhattan Eye, Ear and Throat Hosp., Community Hosp., Glen Cove. Served as lt. USNR, 1943-46, PTO. Recipient Horatio Alger award Am. Schs. and Colls. Assn., 1958; 1st prize for excellence in design Internat. Hotel, Queens C. of C., N.Y., 1958; Producers Council award, 1967. Fellow AIA (nat. chmn. bldg. codes com., pres. N.Y. chpt. 1967-68), ASCE; mem. Royal Inst. Brit. Architects, Bldg. Research Inst., N.Y. Bldg. Congress, NYU Hotel and Restaurant Soc., Am. Nat. Standards Inst. (exec. com. constrn. standards bd.), Nat. Fire Protection Assn. (chmn. sect. com. on residential occupancies, com. on safety to life), Ave. of Americas Assn. (bd. dirs.). Club: Harvard (bd. mgrs., exec. com., chmn. house com. N.Y.C.). Home: 44 Wolver Hollow Rd Glen Head NY 11545-2808 Office: 333 7th Ave New York NY 10001-5004

TABOR, DAVID, physics educator; b. London, Oct. 23, 1913; s. Charles and Rebecca (Weinstein) T.; m. Hannalene Stillschweig, Mar. 14, 1943; children: Daniel Charles, Michael. BS with honors, Imperial Coll., London, 1934; PhD, U. Cambridge, Eng., 1939, ScD, 1956; ScD (hon.), U. Bath, Eng., 1985. Rsch. asst. Commonwealth Sci. and Indsl. Rsch. Orgn., Melbourne, Victoria, Australia, 1940-45, acting head tribophysics sect., 1945; asst. dir. rsch. Cavendish Lab., U. Cambridge, 1946-64, reader physics, 1964-73, prof., 1973-81, head dept. physics and chemistry of solids, 1969-81, prof. emeritus, 1981—; Donald Julius Groen lectr. Instn. Mech. Engrs., London, 1991; vis. prof. Imperial Coll., 1981-89. Author: (textbook) Gases, Liquids and Solids, 1969, rev. edit., 1979, 91. Recipient Gold Medal of Tribology, Instn. Mech. Engrs., 1972, Nat. award Am. Soc. Lubrication Engrs., 1965, Mayo D Hersey award ASME, 1974. Fellow Royal Soc. London (Royal medal 1992), Inst. Physics (London, Guthrie medal 1975), Gonville and Caius Coll. (Cambridge). Office: U Cambridge Cavendish Lab, Madingley Rd, Cambridge CB3 0HE, England

TABOR, EDWARD, physician, researcher; b. Washington, Apr. 30, 1947; married; 4 children. BA, Harvard U., 1969; MD, Columbia U., 1973. Intern and resident Columbia-Presbyn. Med. Ctr., N.Y.C., 1973-75; rsch. investigator Bur. Biologics, Bethesda, Md., 1975-83; dir. divsn. anti-infective drug products FDA, Rockville, Md., 1983-88; assoc. dir. for biol. carcinogenesis Nat. Cancer Inst./NIH, Rockville, 1988—. Contbr. articles to more than 200 publs. Capt. USPHS, 1975—. Achievements include research in hepatitis virus, hepatocellular carcinoma. Office: NIH 9000 Rockville Pike Bethesda MD 20892

TABOR, JOHN MALCOLM, genetic engineer; b. Harrisburg, Pa., May 30, 1952; s. Benjamin Luther and Nancy Lee (Smith) T.; m. Deborah Ann, May 24, 1975 (div. 1987); 1 child, Lindsey Smith. BS, Elizabethtown Coll., 1974; PhD, Kans. State U., 1978. Postdoctoral fellow Roche Inst., Nutley, N.Y., 1978-80; vis. scientist, instr. MIT, Cambridge, 1980-82; sr. scientist Bristol-Myers Squibb Co., Syracuse, N.Y., 1982-83, dept. head, 1983-85, asst. dir., 1985-90; assoc. dir. Bristol-Myers Squibb Co., 5, N.Y., 1990-92, dir., 1992--. Editor: Genetic Engineering Technology in Industrial Pharmacy: Principles and Applications. Mem. Sigma Xi. Office: Bristol Myers Co PO Box 4755 Syracuse NY 13221

TABOR, THEODORE EMMETT, chemical company research executive; b. Great Falls, Mont., Dec. 28, 1940; s. John Edward and Alviva Lillian (Thorsen) T.; m. Jacqueline Lou Hart, Aug. 5, 1959; children: Lori, John, Lexi. BA, U. Mont., 1962; PhD, Kansas State U., 1967. Various research and devel. positions Dow Chem. Co., Midland, Mich., 1967-81, mgr. coop. research, 1981—; co. alt. rep. to Coun. for Chem. Rsch., 1982—; co. rep. Indsl. Rsch. Inst., U. Rsch. Rels. Dirs. Network, 1991—, Am. Chem. Soc., Com. Corp. Assocs., 1993—; program mgr. The Dow Chem. Co. Found., 1989—. Mem. AAAS, Am. Chem. Soc., Soc. Rsch. Adminstrn., Nat. Coun. U. Rsch. Adminstrn. (assoc.). Mem. United Ch. Home: 2712 Mt Vernon Dr Midland MI 48642 Office: Dow Chem Co 1776 Building Midland MI 48674

TABRISKY, PHYLLIS PAGE, physiatrist, educator; b. Newton, Mass., Aug. 28, 1930; d. Joseph Westley and Alice Florence (Wainwright) Page; m. Joseph Tabrisky, Apr. 23, 1955; children: Joseph Page, Elizabeth Ann, William Page. BS, Douglass Coll., 1952; MD, Tufts U., 1956. Cert. phys. medicine and rehab. Intern U. Ill. Hosp., Chgo., 1956-57; phys. medicine and rehab. residency U. Colo. Sch. Medicine, Denver, 1958-60; gen. med. officer dept. pediatric and medicine Coco Solo Hosp., Panama Canal Zone, 1961-62; staff physician dept. pediatrics Ft. Hood (Tex.) Army Hosp., 1963; instr. dept. rehab. medicine Boston (Mass.) U. Sch. Medicine, 1964-66; asst. prof. phys. medicine and rehab. U. Colo. Sch. Medicine, Denver, 1966-68; staff physician VA Med. Ctr., Long Beach, Calif., 1968-71; acting chief phys. medicine and rehab. VA Med. Ctr., Long Beach, 1971-73, asst. chief rehab. med. svcs., 1973-91, chief rehab. med. svc., 1992—; asst. clin. prof. phys. medicine and rehab. U. Calif. Coll. Medicine, Irvine, 1970-75, assoc. clin. prof., 1975-80, prof., 1980—, vice chair dept. phys. medicine and rehab., 1985—, dir. residency tng., 1982—. Fellow Am. Acad. Phys. Medicine and Rehab.; mem. Am. Congress Rehab. Medicine, Assn. Acad. Physiatrists, Alpha Omega Alpha. Republican. Episcopalian. Avocation: U.S. history. Office: VA Med Ctr 5901 E 7th St Long Beach CA 90822

TACHIBANA, AKITOMO, chemistry educator; b. Kyoto, Japan, Sept. 9, 1951; s. Sei and Taeko (Sonoyama) T.; m. Yumi Nagase, May 30, 1983; children; Keiichiro, Masako. B of Engring., Kyoto U., 1974, MS, 1976, PhD, 1979. Editorial bd. Jour. Math. Chemistry, Basel, Switzerland, 1991—. Mem. Inst. Fundamental Chemistry (sci. coord. 1988—). Home: 489-107 Takashiro, Shiga-cho, Shiga-gun 520-05, Japan Office: Kyoto U Faculty Engring, Sakyo-ku, Kyoto 606, Japan

TACHIBANA, TAKESHI, mechanical engineer, educator; b. Osaka, Japan, Oct. 20, 1951; s. Keiichi and Akiko (Yanagimoto) T.; m. Fumi Yasunaka, May 30, 1986; children: Yasuki, Misato. BS, U. Tokyo, 1978; MS, Stanford U., 1980, Engr. degree, 1981; D Engring., U. Tokyo, 1985; M Engring., Stanford U., 1981; PhD, U. Tokyo, 1985. Asst. prof. engring. Kyushu Inst. Tech., Kitakyushu, Japan, 1985-87, assoc. prof. mech. engring., 1987—. Grantee Tanigawa Found., Tokyo, 1986, Sumitomo Found., Tokyo, 1992. Mem. AIAA, Combustion Inst., Airship Assn. Buddhist. Office: Kyushu Inst Tech, 1-1 Sensuicho Tobata, Kitakyushu 804, Japan

TACHIWAKI, TOKUMATSU, chemistry educator; b. Kyoto, Japan, Oct. 27, 1938; s. Sensuke and Yasu (Kishimoto) T.; m. Teruko Otsubo, May 25, 1965; children: Kenji, Yasushi, Yuuko. B. Tech., Doshisha U., Kyoto, 1961, M. Tech., 1963; D Tech., Osaka (Japan) U., 1992. Asst. prof. dept. chem. engring. Doshisha U., Kyoto, 1963-83, lectr. prof. dept. chem. engring., 1983-88, assoc. prof. dept. chem. engring., 1988—. Contbr. articles to profl. jours. Mem. Am. Inst. Chem. Engrs. Avocations: golf, gardening, fishing. Home: 11-41 Tenjinyama Miyamaki, Tanabe-cho 610-03 Kyoto, Japan Office: Doshisha U Dept Chem, Engring Imadegawa-Karasuma, Kyoto 602, Japan

TACKER, WILLIS ARNOLD, JR, academic administrator, medical educator, researcher; b. Tyler, Tex., May 24, 1942; s. Willis Arnold and Willie Mae (Massey) T.; m. Martha J. McClelland, Mar. 18, 1967; children: Sarah Mae, Betsy Jane, Katherine Ann. BS, Baylor U., 1964, MD, PhD, 1970. Lic. physician, Ind., Ohio, Alaska, Tex. Intern Mayo Grad. Sch. Medicine Mayo Clinic, Rochester, Minn., 1970-71; pvt. practice Prudhoe Bay, Alaska, 1971; instr. dept. physiology Baylor Coll. Medicine, Houston, 1971-73, asst. prof. dept. physiology, 1973-74; clin. prof. family medicine Ind. U. Sch. Medicine, West Lafayette, Ind., 1981—; vis. asst. prof. Biomed. Engring. Ctr. Purdue U., West Lafayette, 1974-76, assoc. prof. Biomed. Engring. Ctr. and Sch. Vet. Medicine, 1976-79, assoc. dir. William A. Hillenbrand Biomed. Engring. Ctr., 1980—, prof. Biomed. Engring. Ctr. and Sch. Vet. Medicine, 1979—, acting dir. William A. Hillenbrand Biomed. Engring. Ctr., 1991—; asst. dir. divsn. sponsored programs Purdue Rsch. Found., 1990—; vis. rsch. fellow Sch. Aerospace Medicine, Brooks AFB, San Antonio, 1982; with Corp. Sci. and Tech., State of Ind., 1985-88; presenter, cons. in field. Author: Some Advice on Getting Grants, 1991; co-author: Electrical Defibrillation, 1980; author: (with others) Handbook of Engineering and Medicine and Biology, 1980, Implantable Sensors for Closed-Loop Prosthetic Systems, 1985, Encyclopedia of Medical Devices and Instrumentation, 1988; contbr. numerous articles to profl. jours. Chmn. bd. dirs. Assn. Advancement Med. Instrumentation Found., Arlington, Va., 1987—. Mem. Am. Heart Assn. (bd. dirs. Ind. affiliate 1975-81, med. edn. com. 1975-81, pub. health edn. com. 1975-81, chmn. ad hoc com. CPR tng. for physicians 1976-77, rsch. review com. 1988-90), Am. Physiol. Soc., Ind. State Med. Assn., Tippecanoe County Med. Soc., Assn. Advancement Med. Instrumentation (chmn. various coms., bd. dirs. 1981-84, pres. 1985-86), Am. Men and Women Sci., Alpha Epsilon Delta, Beta Beta Beta, Soc. Sigma Xi. Achievements include research in biomedical engineering, cardiovascular physiology, medical education, emergency cardiovascular care, motor evoked potentials, skeletal muscle ventricle; patents for an apparatus and method for measurement and control of blood pressure, electrode system and method for implantable defibrillators, pressure mapping system with capacitive measuring pad. Office: Purdue U 1293 A A Potter Bldg # 204 West Lafayette IN 47907-1293

TAENZER, JON CHARLES, scientist, electronics engineering consultant; b. Chgo., Nov. 10, 1942; s. Roderick Bendix and Marcella Ida (Galle) T.; m. Anita Jeanette Sanner, Aug. 10, 1969; 1 child, Bryce Jon. BSEE with honors, Purdue U., 1964; MSEE, Stanford U., 1966, PhD in Elec. Engring., 1971. Rsch. engr. Magnaflux Corp., Chgo., 1962-64; electronics engr. Hewlett Packard Corp., Palo Alto, Calif., 1965; rsch. assoc. Stanford (Calif.) U., 1966-71; rsch. fellow Stanford Rsch. Inst., Menlo Park, Calif., 1972, rsch. engr., 1973-77; head engring. Diasonics, Inc., Sunnyvale, Calif., 1978-79; staff scientist SRI Internat., Menlo Park, 1979-83; sr. staff scientist Adept Tech., Inc., San Jose, Calif., 1983-89; mem. tech. staff Next Computer, Inc., Redwood City, Calif., 1989-91; sr. staff scientist, mgr. magnetic products ReSound Corp., Redwood City, 1992—; pres., prin. Taenzer Cons., Palo Alto; cons. numerous cos. Author over 65 publs. and papers. Participant Civic Recycling Group, Palo Alto, Calif., 1985—. Recipient Courtier award Fowler House, 1963, 64. Mem. IEEE, Tau Beta Pi, Eta Kappa Nu. Achievements include 16 U.S. patents, rsch. in electronics, psychophysics, ultrasonics, robotics. Office: ReSound Corp 220 Saginaw Dr Redwood City CA 94063

TAFLOVE, ALLEN, electrical engineer, educator, researcher, consultant; b. Chgo., June 14, 1949; s. Harry and Leah (Natovich) T.; m. Sylvia Hinda Friedman, Nov. 6, 1977; children: Michael Lee, Nathan Brent. BS with highest distinction, Northwestern U., 1971, MS, 1972, PhD, 1975. Assoc. engr. IIT Rsch. Inst., Chgo., 1975-78, rsch. engr., 1978-81, sr. engr., 1981-84; assoc. prof. Northwestern U., Evanston, Ill., 1984-88, prof., 1988—; cons. Electric Power Rsch. Inst., Palo Alto, Calif., 1985-86, Lawrence Livermore Nat. Lab., 1985-87, Lockheed Missiles & Space Co., Sunnyvale, Calif., 1985-88, MRJ, Inc., Oakton, Va., 1987-90, U.S. Naval Rsch. Lab., Washington, 1988—, Cray Rsch., Inc., Eagan, Minn., 1991—, Village of Wilmette, Ill., 1991—, City of Wheaton, Ill., 1991—, B.C. Hydro, Vancouver, Can., 1991—, Commonwealth Edison, Chgo., 1992—, MIT Lincoln Lab., 1992—. Co-author: Computational Electromagnetics: Integral Equation Approach, 1993; contbr. over 50 articles to profl. jours.; holder 10 patents. Recipient Adviser to Yr. award Northwestern U., 1991; Cabell fellow Northwestern U., 1975; rsch. grantee USAF, Electric Power Rsch. Inst., Lawrence Livermore Nat. Lab., NSF, Office Naval Rsch., Gen. Dynamics Corp., Northrop Corp., Lockheed Corp., Sci. Applications, Inc., Cray Rsch., Inc., Northwestern Meml. Hosp., NASA Ames Ctr., NASA Lewis Ctr., 1977. Fellow IEEE (Best Paper award 1983); mem. AAAS, IEEE Antennas and Propagation Soc. (Disting. nat. lect. 1990-91, chmn. tech. program com. Internat. Symposium 1992), Electromagnetics Acad., Internat. Union Radio Sci. (commn. B and K), N.Y. Acad. Scis., Sigma Xi, Eta Kappa Nu, Tau Beta Pi. Avocation: amateur radio. Office: Northwestern U Dept Elec Engring & Computer Sci 2145 Sheridan Rd Evanston IL 60208-3118

TAGGART, AUSTIN DALE, II, chemist; b. Odessa, Tex., Jan. 6, 1952; s. Austin Dale and Evylinn (Abercrombie) T.; m. Darren L. Rosenbaum, 1991; stepchildren: Robert G., John W.; 1 child, Patrick E. MS in Chemistry, U. Houston, 1978; BS in Edn. cum laude, Tex. Tech U., 1975; Cert. of Completion, Odessa Jr. Coll., 1972. Teaching fellow U. Houston, University Park, 1975-78; research fellow U. Houston, 1978; chemist Shell Chem. Co., Deer Park, Tex., 1978-81; sr. process chemist Shell Chem. Co., 1988—; process chemist Shell Oil Co., Deer Park, 1981-88; instr. chemistry Houston Community Coll., 1985, instr. math, 1986; chemistry edn. cons. Scientiarvm, Houston, 1987—; mem. adj. faculty U. Houston, 1976-78. Author poems; contbr. articles to profl. jours. Solicitor, United Way Campaign, 1987-90. Robert A. Welch Found. fellow, 1975-78. Mem. Am. Chem. Soc., Am. Inst. Chemists, N.Y. Acad. Scis., Tex. Tech U. Ex-Students Assn., U. Houston Ex-Students Assn., Alpha Chi Sigma, Phi Theta Kappa. Independent. Lutheran. Avocations: computers, music, literature, drama, racquetball. Home: 810 Seamaster Dr Houston TX 77062-5104 Office: Shell Chem Co PO Box 100 Deer Park TX 77536-0100

TAGGART, G. BRUCE, professional society administrator; b. Phila., Apr. 8, 1942; s. Robert Henry Taggart and Rachael Elizabeth Burtt. BS in Physics, Coll. William and Mary, 1964; postgrad. in engineering mechanics, U. Pa., 1964-65; PhD in Physics, Temple U., 1971. Instr. dept. physics Drexel U., Phila., 1970; asst. prof. dept. physics Va. Commonwealth U., Richmond, 1971-77, assoc. prof., 1977-82, prof., 1982-83; from mgr. materials sci. tech. to prin. staff mem., phys. scis. tech. divsn. BDM Internat., McLean, Va., 1983-90; program dir. materials theory, divsn. materials rsch. NSF, Washington, 1990—; vis. asst. prof. dept. physics Temple U., Phila., 1970-71; rsch. assoc. with theory group Oak Ridge (Tenn.) Nat. Lab., 1974; vis. prof. dept. theoretical physics Oxford (Eng.) U., 1978, Fed. U. Pernambuco, Recife, Brazil, 1980; vis. assoc. prof. dept. physics U. Ill., Urbana, 1978-79; guest worker with statis. physics group thermophysics divsn. Nat. Inst. Standards and Tech., Gaithersburg, Md., 1978-88; lectr. dept. physics and astronomy U. Md., College Park, 1989-90; vis. scientist divsn. materials rsch. on leave from BDM Internat., NSF, 1989-90; presenter in field. Referee: Phys. Rev., Physics Letters, Jour. of the Physics and Chemistry of Solids, Acad. Press, DARPA, NSF, contbr. numerous articles to profl. jours. Scholar Coll. William and Mary, 1964; Ford fellow U. Pa., 1964-65; NSF summer fellow and Univ. fellow Temple U., 1971. Mem. AAAS, Am. Phys. Soc. (condensed matter physics divsn., materials physics divsn., high polymer physics divsn.), Materials Rsch. Soc., Sigma Pi Sigma. Achievements include research in condensed matter physics, materials science and statistical mechanics. Office: Nat Sci Found Math and Phys Scis 1800 G St NW Washington DC 20550

TAGOE, CHRISTOPHER CECIL, chemical engineer; b. Kumasi, Ghana, Feb. 19, 1958; came to U.S., 1977; s. Clement Erasmus and Comfort Morkor (Martey) T.; m. Deborah Ann Meredith, July 9, 1983; 1 child, Andrew. BS in Chem. Engring., U. Lowell, 1981; MS in Chem. Engring., Northeastern U., Boston, 1983. Registered profl. engr., Tex. Engr. E.I. DuPont, Houston, 1984-86, area engr., 1986-87; area engr. Cain Chem., Houston, 1988-89, sr. engr., 1988-89; lab. supr. Occidental Chem., Houston, 1989-91, process engring. mgr., 1991—; lab instr. Northeastern U., Boston, 1981-83, part-time mem. fac., 1983-84; mem. adv. bd. chem. engring. dept. Prairie View (Tex.) A&M U., 1992—. Mem. editorial adv. bd. Chem. Processing Mag., Chgo., 1990-91, mem. adv. bd. Technology for Tomorrow section, 1990-91. Mem. Am. Inst. Chem. Engrs., Omega Chi Epsilon. Methodist. Achievements include research on coal-water mixtures, coal slurry fuels preparation, olefins cracking and ethylene plant feed stock quality. Office: Occidental Chem 5761 Underwood Rd Pasadena TX 77507

TAHARA, EIICHI, pathologist, educator; b. Hiroshima, Japan, July 19, 1936; s. Sadako Tahara; m. Yoshie Shimamoto; children: Hidetoshi, Makoto, Eiji. MD, Hiroshima U., 1963, D in Med. Scis., 1968. Asst. dept. pathology Hiroshima U. Sch. Medicine, 1968-72, asst. prof., 1972-77, assoc. prof., 1977-78, prof., chmn., 1978—; councilor Hiroshima U., 1985-87; chief div. anatomical pathology Hiroshima U. Hosp., 1986—. Author: Current Encyclopedia of Pathology, 1984, Digestive Disease Pathology, 1988; contbr. articles to profl. jours. Grantee Found. for Promotion Cancer Rsch., 1991, Princess Takamatsu Cancer Rsch. Fund, 1990, Ministry Health and Welfare, 1989-91. Mem. Internat. Soc. Differentiation (bd. dirs. 1990—), Japanese Soc. Pathology (bd. dirs. 1990—), Japanese Cancer Assn. (councilor), Japanese Rsch. Soc. for Gastric Cancer (sec.), Rotary. Buddhist. Office: Hiroshima U Sch Medicine, 1-2-3 Kasumi Minami-ku, Hiroshima 734, Japan

TAHILRAMANI, SHAM ATMARAM, mechanical engineer; b. Mirpurkhas, India, Jan. 7, 1945; came to Sierra Leone, 1970; s. Atmaram Jiwatram and Sundri (Atmaram) T.; m. Rajni Sham, June 14, 1980; children: Rakesh, Rina, Rishi. Mech. Engr., Lukadhirji Engring. Coll., Morbi, Gujarat State, India, 1965. Sect. head Govt. Vehicle Pool Sect., Ahmedabad, Gujarat, 1965-70; gen. mgr. J. Lall's, Atmaram Group Co., Freetown, Sierra Leone, 1970-74; group dir. J. Lall's, Atmaram & Sons, Freetown, 1974-78; mng. dir. Atmaram Group West Africa, Freetown, 1978-80; mng. dir., dir. for internat. ops. Atmaram Group Internat., Freetown, 1980-84; chmn. bd. dirs. Atmaram Group Worldwide, 1984—. Mem. Instn. Mech. Engrs. London (assoc.), Rotary Club Freetown, Golf Club Freetown, Acqua Club Freetown, Freetown Bi Centenary Lodge. Mem. Internat. Soc. Krishna Conciousness. Avocations: boating, swimming, golf, snooker. Office: Atmaram & Sons, 4 Kissy Rd PO Box 512, Freetown Sierra Leone

TAI, DAR FU, chemist; b. Taiwan, Republic of China, Oct. 21, 1954; s. Yu Lin and Tsui-O (Wu) T.; m. Shu Chuan Wang, May 9, 1983. BS in Chemistry, Nat. Taiwan U., 1977; PhD in Chemistry, U. Pitts., 1985. Postdoctoral assoc. MIT, Cambridge, 1985-87; assoc. rsch. fellow Dept. Ctr. for Biotech., Taipei, 1988-90; assoc. prof. Soochow U., Taipei, 1989-90, Dai Yeh Inst. of Tech., Chang Hwa, Taiwan, 1990—; patent examiner Nat. Bur. of Standard, Taipei, 1988—; cons. Syn-Tech Chem. and Pharm. Co., Ltd., Tainan, 1990—, Kao Ching Chuan Soy Sauce Co., Taichung, Taiwan, 1990—. Contbr. articles to profl. jours. including Tetrahedron Letter, Jour. Am. Chem. Soc., and Biotech. Letter. Mem. Am. Chem. Soc., Chinese Inst. of Food Sci. and Tech. Achievements include exploration of interconversion of regulovasines, synthesis of opitcal active lysergic acid from tryptophan; synthesis of Olivin from Threonine; synthesis of unnatural peptides via subtilisin in organic solvent. Home: 15 Dah-Yung St 6F Yuan Lin, Chang Hwa Republic of China Office: Dai Yeh Inst of Tech, 112 Shan-Jeau Rd Dah-Tsuen, Chang-Hwa Taiwan

TAI, FRANK, aerospace engineer; b. Omaha, Apr. 10, 1955; s. Shou Nan and May (Chuang) T.; m. Lorraine Mae Fesq, May 14, 1988. BSME, U. Calif., Berkeley, 1977; MS in Automatic Controls Engring., MIT, 1979. Design engr. satellite attitude control systems Ball Aerospace, Boulder, Colo., 1979-84; mgr. satellite attitude control systems TRW, Redondo Beach, Calif., 1984-88; mgr. engring. Microcosm, Inc., Torrance, Calif., 1988-89; pres., engring. cons., founder Tech. Advancements, Inc., Playa del Rey, Calif., 1989—. Contbr. articles to profl. jours. Mem. AIAA, Am. Astronautical Soc., Sigma Xi, Tau Beta Pi, Pi Tau Sigma. Home: 6738 Esplanade Playa Del Rey CA 90293-7525 Office: Tech Advancements Inc 6738 Esplanade # 300 Playa Del Rey CA 90293-7525

TAICHMAN, NORTON STANLEY, pathology educator; b. Toronto, Ont., Can., May 27, 1936; s. Louis and Frances (Kline) T.; m. Louise Sheffer, June 1, 1958; children: Russell, Susan, Darren, Leslie, Audrey. DDS, U. Toronto, 1961; Diploma in Periodontics, Harvard U., 1964; PhD, U. Toronto, 1967; MSc (hon.), U. Pa., 1972. Asst. prof. U. Toronto, 1967-69, assoc. prof., 1969-72; prof., chmn. pathology dept. sch. dental medicine U. Pa., Phila., 1972—. Recipient Birnberg award Columbia U., 1987, Disting. Alumnus award Harvard U., 1988. Mem. Internat. Assn. Dental Rsch. (Rsch. Basic Sci. award 1985), Am. Soc. Microbiology, Soc. for Leukocyte Biology. Office: U Pa Dept Dental Pathology 4010 Locust St Philadelphia PA 19104-6002*

TAIGANIDES, E. PAUL, agricultural-environmental engineer, consultant; b. Polymylos, Macedonia, Greece, Oct. 6, 1934; s. Pavlos Theodorou and Sophia ((Elezidou) T.; m. Maro Taiganides, Dec. 25, 1961; children: Paul Anthony, Tasos E., Katerina. BS in Agri. Engring., U. Maine, 1957; MS in Soil and Water Engring., Iowa State U., 1961, D of Environ. Engring., 1963. Cert. engr., Iowa, Colo. Rsch. assoc., asst. prof. Iowa State U., Ames, 1957-65; prof. Ohio State U., Columbus, 1965-75; mgr., chief tech. adviser UN, FAO, Singapore, Singapore, 1975-84, mgr., chief engr., 1984-85; mgr., chief tech. adviser UN, FAO, Kuala Lumpur, Malaysia, 1985-87; mgr., owner EPT Cons., Columbus, 1987—; cons. EPD/Hong Kong, 1988-92, WHO, UN, Denmark, Poland, Czechoslovakia, 1972-75, Internat. Devel. Rsch. Ctr., Can., China, Asian, 1984-89, NAE, Thailand, 1990, FAO, Malaysia, Foxley & Co., Nu-Tek Foods; environ. advisor to Bertam Devel. Corp., Kuala Lampur, Malaysia, 1992—; waste cons. to U.S. Feed Grains Coun., Taiwan, Malaysia, 1992, Venezuela, 1993; pres. Fan Engring., (US) Inc., 1991—, Red Hill Farms, Ohio, 1992—. CRW Author: (video) Waste Resources Recycle, 1985, Pig Waste Treatment and Recycle, 1992; editor: Animal Wastes, 1977; co-editor Agricultural Wastes/ Biological Wastes, 1979; contbr. articles to profl. jours. Bd. govs., v.p. Singapore Am. Sch., Singapore, 1978-83; clergy-leity congress Greek Orthodox Ch., Houston, 1974. Recipient rsch. awards EPA, 1971-75, Water Resources Inst., 1968-73; rsch. grantee UNDP, FAO, IDRC, GTZ, Asean, 1975-88. Fellow Am. Soc. Agrl. Engrs. (chmn. dept., A.W. Farral award 1974), Am. Assn. Environ. Engrs. (diplomate); mem. Am. Soc. Engring. Edn. (div. chmn.), Singapore Lawn Tennis Assn. (v.p. 1980-84), Am. Club (mgmt. com. 1980-85), Sigma Xi. Greek Orthodox. Avocations: tennis, classical music, folk dancing. Home and Office: 1800 Willow Forge Dr Columbus OH 43220-4414

TAIMUTY, SAMUEL ISAAC, physicist; b. West Newton, Pa., Dec. 20, 1917; s. Elias and Samia (Hawatt) T.; BS, Carnegie Inst. Tech., 1940; PhD, U. So. Calif., 1951; m. Betty Jo Travis, Sept. 12, 1953 (dec.); children: Matthew, Martha; m. Rosalie Richards, Apr. 3, 1976. Physicist, U.S. Naval Shipyard, Phila. and Long Beach, Calif., 1942-46; rsch. asst. U. So. Calif., 1947-51; sr. physicist U.S. Naval Radiol. Def. Lab., 1950-52, SRI Internat., Menlo Park, Calif., 1952-72; sr. staff engr. Lockheed Missiles & Space Co., Sunnyvale, Calif., 1972-89; cons. physicist, 1971—. Mem. Am. Phys. Soc., Sigma Xi. Episcopalian. Contbr. articles to sci. publs. Patentee in field. Home: 3346 Kenneth Dr Palo Alto CA 94303-4217

TAIRA, FRANCES SNOW, nurse educator; b. Glasgow, Scotland, Feb. 27, 1935; came to U.S., 1959, naturalized, 1964; d. Thomas and Isabel (McDonald) Snow; m. Albert Taira, June 20, 1962; children: Albert, Deborah, Paul. B.S.N., U. Ill., 1974, M.S.N., 1976; Ed.D., No. Ill. U., 1980. Staff nurse various hosps., 1959-73; instr. nursing Triton Coll., 1976-81; asst. prof. nursing Loyola U., Chgo., 1981—; coord. Learning Resource Ctr., supr. local area computer network Loyola U., Chgo., 1989—. Mem. Chgo. Dept. Disability and Aging Providers Coun. Mem. ANA, Ill. Nurses Assn. (bd. dir.), U. Ill. Gen. Alumni Assn. (bd. dirs.), Sigma Theta Tau, Phi Delta Kappa. Roman Catholic. Author: Aging: A Guide for the Family, 1983, Home Nursing: Basic Rehabilitation Care of Adults, 1986, Independence: Building Upon the Strengths pf Aging People, 1988; contbr. articles to profl. jours.; contbg. author Saunders Rev. for NCLEX-RN, 1990. Home: 404 Atwater Ave Elmhurst IL 60126-3613 Office: Loyola U Lake Shore Campus 6525 N Sheridan Rd Chicago IL 60626-5311

TAIT, CARLETON DREW, geochemist; b. Holyoke, Mass., Dec. 19, 1957; s. Bruce Mossman and Diane (Carleton) T. BA, Williams Coll., Williamstown, Mass., 1980; MA, Washington U., St. Louis, 1982; PhD, Washingotn U., St. Louis, 1984. Postdoctoral N.C. State U., Raleigh, 1984-87; postdoctoral Los Alamos (N.Mex.) Nat. Lab., 1987-89, staff mem., 1989—. Referee jours. Inorganic Chemistry, Geochemistry Cosmochemistry, Jour. Phys. Chemistry, Am. Chem. Soc. Mem. Am. Chem. Soc., Am. Geophys. Union, Optical Soc. Am., Geochem. Soc. Achievements include research in optical spectroscopies for speciation studies in aqueous solutions and in fluid inclusions, speciation of platinum group elements and radionuclides (Pu, Np) to help determine migration potential. Home: 115 Valley Dr Santa Fe NM 87501 Office: Los Alamos Nat Lab Mailstop J514 Los Alamos NM 87545

TAKACH, PETER EDWARD, process engineer; b. Queens, N.Y., July 27, 1954; s. Oliver Leslie and Norma Rose (Paradiso) T.; m. Patricia Anne Travers, Feb. 18, 1989. BS in Chemistry, Washington Coll., 1976; MS in Chemistry, St. John's U., 1985; MBA in Fin., Adelphi U., 1990. Process R & D group leader PCK Tech., Melville, N.Y., 1978-84; mgr. engring. Multiwire Div. Kollmorgen Corp., Hicksville, N.Y., 1984-90; product and process R & D supr. Advanced Interconnection Tech., Islip, N.Y., 1990-91, mgr. engring., 1991—; security cons. BK Sweeney's Inc., Garden City, N.Y., 1989—; seminar presenter in field. Mem. Garden City Centennials, 1971—. Recipient PCK Tech. fellowship, 1981-83, Frederic W. Scheneble Tech. award Kollmorgen Corp., 1987. Mem. MENSA. Achievements include patents in Fully Additive Plating Process; Improved Electroplating Process; Mechanism of Photocyclization of Several Phenylimedazoles. Home: 796 Garden Dr Franklin Square NY 11010 Office: Advanced Interconnection 181 Freeman Ave Islip NY 11751

TAKAGI, HIDEAKI, computer scientist, mathematician; b. Mihara-cho, Hyogo-ken, Japan, Mar. 23, 1950; s. Yoshio and Kiyo (Kashu) T.; m. Yoko Aida, Dec. 6, 1974; children: Mayu, Takuma, Hayato. Diploma, U. Tokyo, 1972, MS, 1974; PhD, UCLA, 1983. Systems engr. IBM Japan, Ltd., Tokyo, 1974-79; researcher IBM Japan Sci. Inst., Tokyo, 1983-84, mgr. communication networks, 1984-85; mgr. distributed systems IBM Tokyo Rsch. Lab., 1985-87, mgr. founds. systems, 1987-90, mgr. project planning, 1990-93; cons. researcher, 1993—; prof. U. Tsukuba, 1993—. Author: Analysis of Polling Systems, 1986, Queueing Analysis, 1991; editor: Stochastic Analysis of Computer and Communication Systems, 1990; jour. editor IEEE/ACM Transactions on Networking Communications Soc., N.Y., 1986—, Performance Evaluation, 1984—, Queueing Systems, 1988—. Mem. IEEE (sr.), Ops. Rsch. Soc. Am., Assn. for Computing Machinery. Avocation: world trotting. Home: 747-3 Serizawa, Chigasaki-shi, Kanagawa 253, Japan Office: U Tsukuba-Inst Socio Econ Planning, 1-1-1 Tennoudai, Tsukuba Ibaraki 305, Japan

TAKAGI, SHIGERU, chemistry educator; b. Ueki, Kumamoto, Japan, Jan. 13, 1956; s. Masayuki and Sei (Terada) T. BS, U. Tokyo, 1979, MS, 1981, DSc, 1985. Asst. prof. Nagoya (Japan) Inst. Tech., 1988—, assoc. prof., 1992—. Co-author: Jikken-kagaku kouza 17, 1991. Mem. Japan Chem. Assn., Japan Soc. Analytical Chemistry, Am. Chem. Soc. Office: Nagoya Inst Tech Dept Chem, Gokiso-cho Showa-ku, Nagoya Aichi 466, Japan

TAKAGI, SHINJI, economist, educator; b. Kumamoto, Japan, Sept. 2, 1953; s. Shigesuke and Kyoko (Sasahara) T.; m. Sue Lynn Bergmark, June 10, 1980; children: Kenta Benjamin, Emi Rebecca, Naomi Elizabeth, Koji Michael. BA, Swarthmore Coll., 1976; MA, Brigham Young U., 1979; PhD, U. Rochester, 1983. Econ. Internat. Monetary Fund, Washington, 1983-87, 89-90, Bank of Japan, Tokyo, 1987-89; assoc. prof. U. Osaka, Japan, 1990—; sr. economist Ministry of Fin., Tokyo, 1992—; lectr. Johns Hopkins U., Washington, 1989-90; rsch. assoc. Fed. Res. Bank San Francisco, 1990—. Author: Exchange Rate Fluctuation and the International Monetary System, 1989, Introduction to International Monetary Economics, 1992; editor: Japanese Capital Markets, 1993; contbr. numerous articles to profl. jours. Grew Found. scholar, 1972-76; grantee Suntory Found., 1991-92. Mem. Am. Econ. Assn., Japan Assn. Econs. and Econometrics. Mormon. Office: U Osaka Dept Econs, 1-1 Machikaneyama, Toyonaka Osaka 560, Japan

TAKAHASHI, FUMIAKI, research mechanical engineer; b. Tokyo, July 9, 1950; came to U.S., 1981; s. Hideo and Fumiko (Kojima) T.; m. Mamiko Niimoto, Sept. 25, 1982; children: Marina, Reina. B Engring., Keio U., Tokyo, 1973, M Engring., 1975, D Engring., 1982. Lectr. Keio U., Tokyo, 1980-81; mem. rsch. staff Princeton (N.J.) U., 1981-88; rsch. engr. U. Dayton, Ohio, 1988—; cons. The BOC Group, Murray Hill, N.J., 1988, MBR Rsch., Inc., Princeton, 1986-88; mem. adv. bd. PneuMotor, Inc., Dayton, 1989-90. Contbg. author: Hydrogen Energy Progress, 1980, Alternative Energy Sources III, 1983, Recent Advances in the Aerospace Sciences, 1985. Predoctoral fellow Japan Scholarship Found., 1975-78. Mem. ASME, AIAA (Best Paper award Mini-Syposium 1992), The Combustion Inst. (program subcom. 1991-93). Achievements include research in stability mechanisms of jet diffusion flames, sooting correllations for hydrocarbon fuels, disruptive burning mechanisms of slurry fuel droplets. Office: U Dayton 300 College Park Ave Dayton OH 45469-0140

TAKAHASHI, IICHIRO, economics educator; b. Niigata, Japan, Mar. 14, 1922; s. Ryushichi and Maka (Hatakeyama) T.; m. Kyoko Nakahara, May 5, 1951; children: Ryutaro, Tetsuro. BA, Keio U., Tokyo, 1947; MS, Cornell U., 1957; PhD, Keio U., 1973. Researcher Nat. Rsch. Inst. of Agrl. Econs., Tokyo, 1947-54; sect. head Kyushu br. Nat. Rsch. Inst. of Agrl. Econs., Fukuoka, Japan, 1954-62; sect. head Tokyo head office Nat. Rsch. Inst. of Agrl. Econs., 1962-72, dept. head Tokyo head office, 1972-73; prof. Kyushu U., Fukuoka, 1973-85, Fukuoka U., 1985-92, Kumamoto U. of Commerce, 1992—; com. mem. Ministry Edn., Tokyo, 1977-79, Ministry Agrl. Forestry Fish, Tokyo, 1984-86, Ministry Internat. Trade and Industry, Tokyo, 1981-91. Editor: Bibliography on Agricultural Marketing, 1964, Import System and Domestic Marketing of Agricultural Products, 1982; author: Industrial Organization of Meat Industry, 1972, Marketing of Agricultural Products, 1985. Pres. Kyushu Agrl. Econs. Assn., Fukuoka, 1979-81. 2d lt. Japanese Army, 1963-65. Avocations: Utai, walking, travel. Home: 503 24-28 Hirao 3 chome, Chuo-ku, Fukuoka 810, Japan

TAKAHASHI, KAZUKO, organic chemistry researcher, educator; b. Kanra-machi, Kanra-gun, Gunma-ken, Japan, Sept. 20, 1935; d. Hamao and Sawa (Chigira) T. BS, Saitama U., Urawa, Japan, 1958; DSc, Tohoku U., Sendai, Japan, 1964. Rsch. assoc. Japan Atomic Energy Inst., Tokai, Japan, 1958-59; rsch. assoc. Tohoku U. Faculty of Sci., Sendai, 1964-89, lectr. chemistry, 1990—; lectr. various profl. confs.; mem. internat. exch. com. Tohoku U., Sendai, 1990—, fgn. student counselor, 1991—. Contbr. articles to profl. jours. Grantee Ministry of Edn., Sci., Tokyo, 1986, 91, Hayashi Meml. Found. for Female Nat. Scientists, Tokyo, 1992. Mem. Am. Chem. Soc., Japan Chem. Soc., Synthetic Organic Chemists Japan. Achievements include patents for synthesis of novel organic redox systems, electroconductors, new near-infrared dyes, optical storage media, nonlinear optical materials and novel electrochromic materials. Home: Kano Taihaku, 204 Nomura Bldg 3-14-10, Sendai 980, Japan Office: Tohoku Univ Dept of Chem, Faculty of Sci, Sendai 980, Japan

TAKAHASHI, KEIICHI, zoology educator; b. Yokkaichi, Japan, May 31, 1931; s. Shozo and Toshi (Imamura) T.; m. Mihoko Terada, Sept. 26, 1957; children: Michiko, Yoshiki. BSc, U. Tokyo, 1953, MSc, 1955, PhD, 1960. Instr. U. Tokyo, 1956-68, assoc. prof., 1968-73, prof., 1973-92, emeritus prof., 1992—; prof. Internat. Christian U., Tokyo, 1992—; rsch. fellow Bedford Coll., U. London, 1960-62; dir. Misaki Marine Biol. Sta., U. Tokyo, 1988—. Contbr. numerous articles to profl. jours. Mem. Japanese Soc. for Comparative Physiology and Biochemistry (pres. 1990—), Japan Soc. for Biol. Scis. in Space (v.p. 1987-92), Japan Soc. for Biol. Scis. Edn. (v.p. 1984—), Japan Soc. for Cell Biology (councillor 1983—), Zool. Soc. Japan (councillor), Japan Soc. for Sci. Edn. (bd. dirs.), Inst. of Biology of U.K., Sci. Coun. of Japan. Mem. Christian Ch. Avocations: music, arts. Home: 14-31 Yochomachi, Shinjuku-Ku, Tokyo 162, Japan Office: Internat Christian U Dept Biology, 3-10-2 Osawa, Mitaka, Tokyo 181, Japan

TAKAHASHI, MASAYUKI, aquatic ecologist; b. Yokosuka, Kanagawa, Japan, July 6, 1942; s. Masao and Haruyo (Sugi) T.; m. Toshie Tanaka, Apr. 4, 1969; children: Mari, Emi. BS, Tokyo (Japan) Kyoiku U., 1965, MS, 1967, DSc, 1970. Postdoctoral fellow Fisheries Rsch. Bd. Can., Biol. Sta., Nanaimo, B.C., 1970-72; rsch. scientist Inst. Oceanography, U. B.C., Vancouver, 1972-77; assoc. prof. Inst. Biol. Scis., U. Tsukuba, Ibaraki, Japan, 1977-86, Botany Dept., U. Tokyo, 1986—. Author: Biological Oceanographic Processes, 1st edit. 1973, 2d edit. 1977, 3d edit. 1984, English for Science & Technology, 1984. Recipient First prize Nat. Coun. for Advancement of Sci. Edn., 1958, third award Nat. Sci. Fair Internat., 1959, Postdoctoral fellowship Nat. Rsch. Coun., 1970, 1st Ecology Biwa prize, Shiga Prefectural Govt., 1992, Gold medal of Oceanographic Soc. of Japan, 1992. Home: 672-162 Kojirohazama, Tsukuba 305, Japan

TAKAHASHI, TSUTOMU, chemist; b. Nagasaki, Japan, Aug. 29, 1949; s. Kunio and Kyoko (Kurogane) T.; m. Yuko Inoue, May 14, 1978; children: Sakuya, Hakaru, Azuma, Fusako. BS in Chemistry, Kyoto (Japan) U., 1972, MS in Chemistry, 1974; PhD in Chem. Engring., SUNY, Buffalo, 1981. Researcher Sumitomo Chem. Co., Ltd., Niihama, Ehime, Japan, 1974-87; rsch. assoc. Sumitomo Chem. Co., Ltd., Tsukuba, Ibaraki, Japan, 1988—; rsch. asst. SUNY, Buffalo, 1979-81. Contbr. articles to profl. publs., including Macromolecules, Rev. Phys. Chem. of Japan, Chem. Engring. Comm., Advances in Chemistry Series, others. Mem. Am. Chem. Soc. Achievements include patents in field. Office: 6 Kitahara, Tsukuba Ibaraki 300-34, Japan

TAKAKI, RYUJI, physics educator; b. Hiroshima, Japan, Oct. 7, 1940; s. Takeshi and Tatsu Takaki; m. Junko Naito, June 5, 1971; children: Daisuke, Shunsuke. B., U. Tokyo, 1963, M., 1965, D., 1969. Lectr. Tokyo Noko U., Fuchu, Tokyo, 1969-70; assoc. prof. Tokyo Noko U., 1970-83, prof., 1983—; pres. Soc. for Sci. on Form, Japan, 1988—; chief project U.S.-Japan Joint Rsch. Project, 1985-86. Author several books; editor-in-chief editing com. FORMA, 1988—; assoc. editor editing com. Fluid Dynamic Rsch., 1991—. Mem. Physical Soc. Japan, Japanese Soc. Biorheology, Japan Soc. Fluid Mechanics. Office: Tokyo Noko Univ, Fuchu Japan 183

TAKAO, HAMA, physiological chemistry educator; b. Tokyo, July 22, 1931; s. Kohzoh and Testuko (Tajime) H.; m. Kazue Inouye, May 16, 1959; children: Norio, Hiroko, Yukiko. BS, Osaka (Japan) U., 1954, PhD, 1960. Asst. Osaka U., 1959-64, asst. prof., 1964-68; prof. Kobe-Gakuin U., Kobe, Japan, 1968-83; dean faculty pharm. scis. Kobe-Gakuin U., 1983-92, pres., 1992—. Author: Text Book of Radiopharmacy, 1986, Clinical Chemistry, 1988; editor: The Functional Biochemistry, 1988; contbr. articles to profl. jours. Trustee Japan Assn. Pvt. Pharm. Sch., Tokyo, 1984—. Mem. Japan Soc. Pharmacists, Japan Pharm. Soc. (trustee 1986-87, councilor 1972—), Japan Biochem. Soc. (trustee 1979-80, councilor 1968—). Avocations: basketball, traveling, cameras. Office: Kobe Gakuin U, Arise Igawadani cho, Nishi ku, Kobe 651-21, Japan

TAKASAKI, ETSUJI, urology educator; b. Tokyo, Apr. 24, 1929; s. Kuranosuke and Fumi Takasaki; m. Sachiko Shinkai, Nov. 1, 1960; children: Satoshi, Masumi, Hiromi. MD, U. Tokyo, 1955, D. Med. Sci., 1960. Instr. urology U. Tokyo, 1960-62, asst. prof., 1962-67; chief. urol. svc. Musashino Red Cross Hosp., Tokyo, 1967-69, Komagome Met. Hosp., Tokyo, 1969-74; prof. urology Dokkyo U. Sch. Medicine, Tochigi, Japan, 1974—; lectr. U. Tokyo, 1969-74. Author: Urolithiasis, 1978; contbr. articles to Japanese Jour. Urology, 1960, Jour. Urology, 1986, Urologia Internat., 1989. Mem. Japanese Urol. Assn. (bd. dirs. 1963—), Japanese Soc. Andrology (bd. dirs. 1982—), Japanese Soc. Endourology and ESWL (bd. dirs. 1988—), Internat. Soc. Urology (Paris). Avocation: Kendo (Japanese fencing). Office: Dokkyo U, Sch Medicine Shimotsuga gun, Mibu machi 880, Tochigi Japan 321 02

TAKASAKI, YOSHITAKA, telecommunications scientist, electrical engineer; b. Hitachi, Ibaraki, Japan, Nov. 21, 1938; s. Yoshihisa and Hiroko (Otani) T.; m. Asako Komuro, Oct. 26, 1969; children: Yoshinori, Miyuki. B. Engring., U. Tokyo, 1962; PhD, Tokyo Inst. Tech., 1979. Asst. researcher cen. rsch. lab. Hitachi, Ltd., Kokubunji, Tokyo, Japan, 1962-64, assoc. researcher cen. rsch. lab., 1964-67, researcher cen. rsch. lab., 1967-74, sr. researcher cen. rsch. lab., 1974-86, chief researcher cen. rsch. lab., 1986—; coord. Hatachi Tech. Edn., Shibuya, Tokyo, Japan, 1987—; lectr. Takushoku U., Hachioji, Tokyo, Japan, 1990—, Toyo U., Kawagoye, Saitama, Japan, 1993—. Author: Communicatins Channels: Characterization and Behavior, 1976, Electronics and Communications Handbook, 1979, Optical Communications Handbook, 1982, Optical Communications Theory and Its Applications, 1988, Digital Transmission Design and Jitter Analyses, 1991; contbr. more than 60 articles to profl. jours. Auditor Kohitsuji Kindergarten, Tokorozawa, Saitama, 1991. Fellow IEEE (sr.); mem. N.Y. Acad. Scis., Inst. of Electronics, Info. and Communications Engrs., Inst. Elec. Engrs., Inst. TV Engrs. Achievements include development of theory of clock recovery for Digital Transmission; of generalized theory of Variable Equalizers; of line coding for Fiber Optics Transmissions; of upgrading strategies for broadband ISDN systems; over 50 patents. Home: 5-12-12 Naka-Arai, Tokorozawa Saitama 359, Japan Office: Hitachi Ltd Cen Rsch Lab, Higashi Koigakubo, Kokubunji Tokyo 185, Japan

TAKASHIO, MASACHIKA, biochemist; b. Tokyo, Mar. 26, 1946. BA in Agrl. Chemistry, Tokyo U., 1971, PhD, 1981. Biochemist Sapporo Breweries Ltd., Tokyo, 1971—; chief rep. Sapporo Breweries Ltd., N.Y.C., 1989-93; gen. mgr. beer tech. dept. Sapporo Breweries Ltd., Tokyo, 1993—; guest scientist Inst. Microbial Chemisty, Tokyo, 1976-81. Mem. Am. Assn. Pharm. Scientists, Am. Soc. Microbiology, N.Y. Acad. Sci. Achievements include discovery and industrialization of novel enzymes uricase and urease for diagnostic use; discovery of ribocitrin. Office: Sapporo Breweries Ltd, 7-10-1 Ginza, Chuo-ku Tokyo 104, Japan

TAKEDA, HARUNORI, physicist; b. Tokyo, Sept. 10, 1948; came to U.S., 1971; s. Masami and Mikiko T.; m. Young-Ja C., Dec. 25, 1976; children: Theodore, Miyuki. BS in Physics, Waseda U., Tokyo, 1971; PhD in Physics, U. Pa., 1978. Rsch. scientist KMS Fusion Inc., Ann Arbor, Mich., 1982-84; staff mem. Los Alamos (N.Mex.) Nat. Lab., 1985—. Author: Nuclear Instruments and Methods. Office: Los Alamos Nat Lab MS-H825 AT-7 Los Alamos NM 87545

TAKEDA, YASUHIRO, chemistry educator; b. Tokyo, Nov. 16, 1940; s. Masataro and Susao (Ishida) T.; m. Haruki Yamashita, June 1, 1972; children: Kotaro, Taijiro. BA in Pharm. Sci., U. Tokyo, 1964, PhD in Pharm. Sci., 1969. Pharmacy diplomate. Rsch. assoc. U. Tokyo, 1969-72, Princeton (N.J.) U., 1972-75; rsch. fellow Mitsubishi Inst. for Life Sci., Tokyo, 1975-76; asst. prof. Kobe Coll. for Women, Nishinomiya, Japan, 1976-84, prof., 1984—. Contbr. articles to profl. jours. Office: Kobe Coll, 4-1 Okadayama, Nishinomiya 662, Japan

TAKEDA, YOSHIYUKI, chemical company executive; b. Okayama, Japan, Jan. 23, 1933; m. Emiko Takeda, Jan. 12, 1967; children: Masafumi, Fumie. BS, Doshisha U., Kyoto, Japan, 1956; DSc, Kyoto U., 1964. Rschr. Cen. Rsch. Lab., Mitsubishi Kasei Corp., Tokyo, 1961-66, 69, mgr. R & D, 1970-72; mgr. R & D Kurosaki Plant Cen. Rsch. Lab., Mitsubishi Kasei Corp., Kitakyushu, Japan, 1972-77, sr. mgr., 1978-89; rsch. fellow Fed. Inst. Tech., Zurich, 1967-68; tech. liaison mgr. R & D Corp. Japan, Kurume, 1990-92; tech. mgr. chem. engring. dept. Shinryo Chem. Ltd., Kitakyushu, 1993—. 8 patents in field. Recipient Tech. award Chem. Industries Assn. Japan, Tokyo, 1985, Invention award Japan Inst. Invention and Innovation, Tokyo, 1989. Mem. Chem. Soc. Japan, Am. Chem. Soc. Avocations: music, collecting topographic materials, editing. Home: 5-71 Yoshida-Honmachi Sakyo, Kyoto J-606, Japan Office: Shinryo Chem Ltd, Yahatanishi-ku, 1-2 Kurosaki-Shiroishi, Kitakyushu J-806, Japan

TAKEI, TOSHIHISA, otolaryngologist; b. L.A., Apr. 19, 1931; s. Taketomi and Mitsue (Hagihara) T.; m. Emiko Kubota, Jan. 25, 1955; children: H. Thomas, T. Robert. BA, UCLA, 1954; MD, Boston U., 1962. Diplomate, Am. Bd. Otolaryngology. Intern L.A. County Harbor Gen. Hosp., 1962-63; resident in otolaryngology L.A. County/U. So. Calif. Med. Ctr., 1963-67; staff physician Covina (Calif.) Ear, Nose & Throat Med. Group, 1968—; asst. prof. Sch. Medicine, U. So. Calif., L.A., 1968—. 1st lt. U.S. Army, 1955-56, Korea. Fellow Am. Acad. Otolaryngology, Royal Soc. Medicine. Republican. Buddhist. Office: Covina ENT Med Group Inc 236 W College St Covina CA 91723-1902

TAKEMOTO, KIICHI, chemistry educator; b. Osaka, Japan, Feb. 14, 1930; s. Kenzo and Hatsue (Asakura) T.; m. Etsuko Tamaki, Apr. 26, 1959; children: Tatsuo, Mariko. B of Engring., Osaka U., Japan, 1953, DEng, 1962. Asst. fellow Osaka City U., Japan, 1953-59, lectr., 1959-64, assoc. prof., 1964-69; prof. Osaka U., Japan, 1969-93, Ryukoku U., 1993—. Author: Functional Polymers, 1987. Recipient Award for Chem. Soc., 1989, Award for Polymer Sci. Achievements, 1991. Mem. Am. Chem. Soc., Chem. Soc. Japan, Soc. Polymer Sci. Japan. Avocations: travel, railways, stamp collections. Home: Sakae 4-10-26, Takatsuki Osaka 569, Japan Office: Rykoku U Sci Technl Faculty, Seta, Otsu Shiga 520-21, Japan

TAKENAKA, TADASHI, electrical engineer, educator; b. Wakayama, Japan, July 31, 1946; s. Hideichi and Yone Takenaka; m. Tamiko Saji, Oct. 20, 1974; children: Mika, Maki. B in Engring., Shibaura Inst. Tech., Tokyo, 1970; M in Engring., U. Electro-Communications, Tokyo, 1973; D in Engring., Kyoto (Japan) U., 1985. Rsch. assoc. Sci. U. Tokyo, Noda, Chiba, Japan, 1973-88, asst. prof., 1988-91, assoc. prof., 1991—; rsch. assoc. Pa. State U., University Park, 1986-87; vis. researcher Nat. Inst. Rsch. in Inorganic Materials, Tsukuba, Japan, 1988—. Co-author: (in Japanese) Science of Ceramics, 2d edit., 1993; co-editor: Data of High-Tc Superconductivity, 1990. Mem. IEEE, Am. Ceramic Soc., Materials Rsch. Soc. Home: Nishicho 5-9, Chiba-ken, 277 Kashiwa Japan Office: Sci U Tokyo, Faculty Sci & Tech, Chiba, 278 Noda Japan

TAKEOKA, SHINJI, chemist; b. Suginami-Ku, Tokyo, Japan, Jan. 22, 1963; s. Fukutaro and Tazuko Takeoka. B Engring., Waseda U., Shinjyuku-Ku, Tokyo, 1986, M Engring., 1988, D Engring., 1991. Researcher Rsch. Inst. for Prodn. Devel., Kyoto, Japan, 1990—; rsch. assoc. Waseda U., 1991-93, asst. prof., 1993—. Contbr. articles to profl. jours. Recipient Mizuno award Waseda U., 1991; Japan Soc. for Promotion of Sci. fellow, 1990-, 91. Mem. Chem. Soc. Japan, Soc. Polymer Sci. Japan, Am. Chem. Soc. Buddhist. Avocation: Kendo, music, piano. Home: 4-27-26 Higashihatsutomi, Kamagaya-shi Chiba Chiba-ken 273-01, Japan Office: Waseda U Dept Polymer Chem, 3 Ohkubo, Shinjyuku-ku Tokyo 169, Japan

TAKEOKA, TSUNEYUKI, neurology educator; b. Singu, Japan, Mar. 1, 1944; s. Toyoichi and Kiyoko (Tashiro) T.; m. Machiko Kira, Jan. 17, 1980; children: Tomomichi, Tsunetoshi. MD, Keio U., Tokyo, 1969. Diplomate Japanese Bd. Neurology. Lectr. dept. neurology Keio U., 1973-78; asst. prof. neurology Tokai U., Kanagawa, Japan, 1978-92; assoc. prof. Tokai U., Kanagawa, 1992—; vis. prof. dept. neurology Reed Neurol. Rsch. Ctr., UCLA Sch. Medicine, 1990-91. Mem. Japanese Soc. Neurology (counselor 1985—), Japanese Cerebrovascular Disease Soc. (counselor 1989—). Home: 2-12-5-607 Kyonan, Musashino Tokyo 180, Japan Office: Tokai U Sch Medicine, Dept Neurology, Isehara Kanagawa 259-11, Japan

TAKES, PETER ARTHUR, immunologist; b. Albany, N.Y., May 1, 1957; s. Arthur Peter and Mary Nicholas (Marin) T.; m. Alexandra Kavourinos, Feb. 27, 1983; 1 child, Michael Kavourinos. BS, Clarkson U., 1979; PhD, Ind. State U., 1985. Rsch. immunologist Sigma Diagnostics, St. Louis, 1988-89; rsch. supr. Sigma Diagnostics, 1989-90, immunodiagnostics mgr., 1990-92; scientist, clin. studies coord. Sigma Immunochem., St. Louis, 1992—; instr. basic life support Am. Heart Assn., St. Louis, 1986-88; adj. prof. Webster U., 1993—. Author: Microwave Procedures Manual, 1990; contbr. articles to several profl. jours. Pres. Waterman Condo. Assn., St. Louis, 1987-90; basketball referee Mo. State High Sch. Athletic Assn., St. Louis, 1990—. U. fellow Ind. State U., 1979-84, Postdoctoral fellow Washington U. Sch. Medicine, St. Louis, 1985-87, Summer Rsch. fellow, AHEPA Cooley's Anemia Found., 1981. Mem. AAAS, Am. Assn. Clin. Chemistry, Nat. Soc. Histotechnology, N.Y. Acad. Scis., Assn. Med. Lab. Immunologists. Office: Sigma Immunochem PO Box 14508 Saint Louis MO 63178

TAKESHIMA, YOICHI, psychology educator; b. Nankoku-shi, Japan, Feb. 13, 1952; s. Kenjiro and Tsuyako Takeshima. BS, U. Tsukuba, Tsukuba-shi, Japan, 1979, MS, 1982. Lectr. Kochi Jr. Coll., Kochi-shi, Japan, 1983-84, 91—, Kochi Gakuen Jr. Coll., Kochi-shi, 1986-89; clin. psychologist Geiyoin Hosp., Aki-shi, Japan, 1984-85; rschr. Kochi Med. Sch., Nankoku-shi, 1985—. Mem. AAAS, Japanese Soc. Hypnosis, The Japanese Soc. Psychiatry & Neurology, Japanese Soc. Psychosomatic Medicine, The Japanese Soc. Social Psychology, The Japanese Group Dynamics Assn., The Japanese Soc. Humanistic Psychol., Japanese Psychol. Assn., Japan Assn. Applied Psychology, Soc. for Psychical Rsch., Japanese Soc. Ednl. Psychology, Japanese Soc. for Parapsychology. Avocations: films, theatre, music, correspondence, swimming. Home: 3-19 Ekimae-cho, Kochi-shi, Kochi-ken 780, Japan Office: Kochi Med Sch Dept Physiol, Okohcho, Kochi, Nankoku 781-51, Japan

TAKHTADZHYAN, ARMEN LEONOVICH, botanist; b. June 10, 1910; m. Alice. Prof., botany Leningrad U., Leningrad, USSR; research staff mem. Komarov Inst. of Botany, Saint Petersburg, USSR; dir. Komarov Inst. of Botany, 1976-86; guest taxonomical researcher New York Botanical Garden, Bronx, N.Y. Author of 45 books. Recipient Hero of Labor medal, USSR, 1990; foreign assoc. mem., US Nat. Acad. Sciences, 1971. Russian Acad. Sciences, 1972. Office: Komarov Institute of Botany, Ulitsa Popova 2, 197022 Saint Petersburg Russia

TAKINO, MASUICHI, physician; b. Santo-cho, Asago-gun Hyogo Prefecture, Japan, Apr. 14, 1904; s. Kichitaro and Yukiko (Toda) T.; m. Yoshiko Aoyagi, Nov. 17, 1930; 4 children. Student, Kyoto U., 1927, 1928-31; grad., Nat. U. Japan, 1929; MD, Kyoto U., 1934. Lectr. internal medicine Kyoto (Japan) U., 1939-41; dir. dept. pathology Osaka Prefectural Sengokuso Hosp. of Pulmonary Tb, 1941-45; dir. Dainippon Zoki Inst. for Med. Research, 1947-61, Nippon Zoki Inst. for Constitutional Diseases, 1962-80; adviser Nippon Inst. Constn. Diseases, Osaka, Japan, 1985—; adv. Nippon Zoki Pharm. Co. Ltd., 1968-80; practice medicine specializing in allergy and autonomic nervous system, Osaka. Author: A New Direction in Asthma Treatment, 1952; (with Yoshitada Takino) Allergy and Asthma, 1956; (with Y. Takino and Kunikazu Sugahara) Pathogenesis and Therapy of Bronchial Asthma with Special Reference to Organ Vagotonia, 1976; Allergy and Autonomic Nervous System, 1979; (with Y. Takino and Tatsuro Takino) Respiratory Hypersensitivity-Its Cholinergic Background and Treatment, 1989; contbr. rsch. articles to profl. jours. Fellow Am. Coll. Chest Physicians (emeritus); mem. Japan Soc. Allergy (hon. mem., award 1970), Japan Soc. Internal Medicine (counselor), Japan Soc. Endocrinology (counselor), Japan Soc. Neurovegetative Research (counselor, award 1988), Japan Soc. Angiocardiology (counselor) Japan Soc. Constn. (counselor), Internat. Soc. Biometeorology, Japan Christian Med. Assn. Achievements include the earliest demonstration of the pulmonary artery reflex through histological and electrophysiological evidence for baroreceptors in pulmonary arterial wall in animals and man; discovery of electrophysiological evidence of baroreceptors of the pulmonary and Ductus Botalli or Ligamentum arteriosum of the rabbit. Home: 1-3-7 Chiyogaoka, Nara 631, Japan Office: Shinmido Bldg, 5-19 Minami-Kyuhoji-Cho, Chuo-ku, Osaka 541, Japan

TAKIZAWA, AKIRA, engineering educator; b. Fukui, Japan, May 6, 1927; s. Tokujiro and Raku (Takinami) T.; m. Mutsuko Kyuno, Apr. 29, 1955; children: Noriko, Jo. B of Engring., Tokyo Inst. Tech., 1951, D of Engring., 1961. Researcher Mitsubishi Rayon Co. Ltd., Ootake, Hiroshima, Japan, 1951-61; asst. lectr. Tokyo Inst. Tech., 1961-67, assoc. prof., 1968; prof. Nagoya Inst. Tech., Nagoya, Aichi, Japan, 1968-91, prof. emeritus, 1991; adj. prof. polymer phys. chemistry Wayne State U., Detroit, 1964-66; vice-chmn. 3d Internat. Polymer Conf., Nagoya, 1990. Author: Handbook of Membrane, 1991; author, editor: Polymer Membrane, 1975, Polymer Structure and Separation, 1976; contbr. articles to profl. jours. Coun. mem. Ministry of Edn., 1980-82, Ministry Internat. Trade & Industry, 1989-91. Mem. Soc. Fiber Sci. and Tech. (standing dir. 1975-85, award of the soc. 1977, disting. svcs. to fiber sci. and tech. award 1992), Soc. Polymer Sci. (standing dir. 1970-90, Disting. Svcs. to Polymer Sci. 1988), Soc. Membrane Sci. (councilor 1976-91). Home: 3-22-6, Takamoridai, Kasugai, Aichi 487, Japan

TAKO, MASAKUNI, chemistry researcher and educator; b. Gushikawa, Okinawa, Japan, Mar. 19, 1947; s. Seiei and Humi Tako; m. Sachiko Tako, Sept. 24, 1978; children: Masatada, Tetsuko, Masaakira. BS, U. Ryukyus, Okinawa, Japan, 1970; MS, Kagoshima (Japan) U., 1972; PhD, Kyushu U., Fukuoka, Japan, 1975. Rschr. Kagoshima U., 1975-78; head quality control Hormel Co. Ltd., Okinawa, 1978-82; asst. prof. U. Ryukyus, Okinawa, 1982-88, assoc. prof., 1988—. Contbr. articles to profl. publs. Grantee Japanese Soc. Biosci. Biotech. Agrochemistry, 1986, Japanese Ministry Edn., 1987, 91, 92. Avocations: classical music, piano, igo, baseball. Office: U Ryukyus, 1 Senbaru, Nishihara Okinawa 903-01, Japan

TAL, JACOB, obstetrician/gynecologist; b. Iassi, Rumania, Oct. 9, 1945; came to U.S., 1976; s. Efraym and Esther (Bercovici) T.; m. Phyllis E., Nov. 7, 1974; children: Noa, Lara, Eric. Degree, Faculte Medicine, Lyon, France, 1969; MD, Sackler Sch. Medicine, Tel-Aviv, 1976. Diplomate Am. Bd. Ob-Gyn., 1982, recert. 1991. With Meml. City, Houston; resident ob-gyn. Albert Einstein Coll. Medicine, Bronx, N.Y., 1980; practice medicine specializing in ob-gyn. West Houston, and Katy, Tex., 1980-87; chief of staff Katy Community Hosp., 1987—. Fellow ACS, Am. Coll. Ob-Gyn, Tex. Assn. Ob-Gyn. Home: 618 Sandy Port Houston TX 77079 Office: Katy-W Houston Ob-Gyn Assn 5618 Medical Center Dr Katy TX 77494-6340

TALAMANTES, FRANK J., biology educator; b. July 8, 1943. BA in Biology, U. St. Thomas, Houston, 1966; MA, Sam Houston State U., 1970; PhD in Endocrinology, U. Calif., Berkeley, 1974. With faculty U. Calif., Santa Cruz, 1974—, prof. biology; mem. MARC study sect., human embryology and devel. study sect., NIH; bd. overseers Harvard Medical Sch.; exam. com. biochemistry, cell and molecular biology test Grad. Record Exam. Contbr. articles to profl. jours., chpts. to books; assoc. Endocrinology, 1985-87. Merit Rsch. grantee NIH. Fellow AAAS; mem. Soc. Advancement of Chicanos and Native Ams. in Science (pres. 1987—). Achievements include research in biochemistry and physiology of prolactin, growth hormone and placental lactogens. Office: Univ of Calif Santa Cruz Dept of Biology Santa Cruz CA 95064*

TALBOT, BERNARD, government medical research facility official, physician; b. N.Y.C., Oct. 6, 1937; s. Harry and Gertrude (Salkin) T.; m. Ane Katrine Larsen, June 2, 1963; children: Akia, Kamilla. B.A., Columbia U., 1958, M.D., 1962; Ph.D, MIT, 1967. NIH postdoctoral fellow MIT, 1962-69; NSF postdoctoral fellow U. Rome, 1969-70; commd. USPHS, 1975—, advanced through grades to med. dir.; med. officer Nat. Cancer Inst., Bethesda, Md., 1971-75; spl. asst. intramural affairs NIH, Bethesda, 1975-78; spl. asst. to dir. NIH, 1978-81; dep. dir. Nat. Inst. Allergy and Infectious Diseases, Bethesda, 1981-87, med. officer nat. ctr. for rsch. resources, 1987—. Contbr. articles on protein chemistry to profl. jours., chpts. on recombinant DNA guidelines to books. Recipient Commendation medal USPHS, 1977, Meritorious Service medal, 1984. Mem. Phi Beta Kappa. Office: NIH 9000 Rockville Pike Bethesda MD 20892-0001

TALBOT, FRANK HAMILTON, museum director, marine researcher; b. Pietermaritzburg, Natal, Republic of South Africa, Jan. 3, 1930; came to U.S., 1982; s. Ralph West and Willemina (Altmann) T.; m. Mabel Suzette Logeman, July 20, 1953; children: Helen Campbell, Richard Bill, Jonathan Charles, Neil Hamilton. BSc, U. Witwatersrand, South Africa, 1949; MSc, U. Cape Town, South Africa, 1951, PhD, 1959. Fisheries rsch. scientist Brit. Colonial Svc., Zanzibar, 1954-57; marine biologist South African Mus., Cape Town, 1958-59, asst. dir., 1960-63; curator fishes Australian Mus., Sydney, 1964-65, dir., 1965-74; prof. environ. studies MacQuarie U., Sydney, Australia, 1975-81; exec. dir. Calif. Acad. Scis., San Francisco, 1982-88; dir. Nat. Mus. Natural History, Smithsonian Inst., Washington, 1989—. Contbr. articles to sci. jours. Fellow AAAS, Calif. Acad. Sci., Royal Zool. Soc. (Australia); mem. Mus. Assn. Australia (pres. 1973-74), Australian Marine Scis. Assn. (pres. 1971-72), Explorers Club. Home: Apt 405 2737 Devonshire Pl NW Washington DC 20008-3453 Office: Nat Mus Natural History Smithsonian Instn Washington DC 20560

TALBOTT, GEORGE ROBERT, physicist, mathematician, educator; b. San Diego, Oct. 1, 1925; s. George Fletcher and Mary (Lanz) T.; BA with honors, UCLA, 1960; DSc, Ind. No. U., 1973. Physicist, mem. tech. staff Rockwell Internat. Co., Anaheim, Calif., 1960-85; mem. faculty thermodynamics Pacific States U., 1971-77, prof., 1972-80, chmn. dept. math. studies, 1973-80; lectr. computer sci. Calif. State U., Fullerton, 1979—; cons. physics, computer sci.; disting. guest lectr. Brunel U., London, 1974, 76; spl. guest Forschungsbibliothek, Hannover, W. Ger., 1979; assoc. editor KRONOS jour., Glassboro (N.J.) U., 1978—; chief computer scientist and ednl. videotape dir. Specialized Software, Wilmot, Wis., 1982—. With M.C., U.S. Army, 1956. Recipient Vis. Scholar's award Western Mich. U., 1979. Mem. Am. Soc. Med. Technologists, Am. Math. Soc., Math. Assn. Am., Am. Soc. Clin. Pathologists (lic. med. lab. technologist). Buddhist. Author: Electronic Thermodynamics, 1973; Philosophy and Unified Science, 1977, Computer Applications, 1989, Sir Arthur and Gravity, 1990, Fermat's Last Theorem, 1991; co-inventor burner. Home: 4031 E Charter Oak Dr Orange CA 92669-2611

TALBURT, JOHN RANDOLPH, computer science educator; b. Batesville, Ark., Sept. 21, 1945; s. Ernest Everett and Geneva Edith (Hudson) T.; m. Rebeca Bono, Dec. 9, 1989; children: Caryn Elaine, Alyssa Lenore, Edgardo Arturo. BS, Ark. State U., 1967; MS, U. Ark., 1969, PhD, 1971. Cert. in data processing, 1981. Assoc. prof. Columbus (Ga.) Coll., 1971-77; systems cons. Data Mgmt. Svcs., Batesville, Ark., 1977-79; data processing mgr. Carter, Mitchum & Co. CPAs, Batesville, Ark., 1979-81; pres. Profl. Computer Software, Inc., Batesville, Ark., 1981-83; prof. U. Ark., Little Rock, 1983—, chmn. dept. computer and info. sci., 1985-93; pres. Ark. Soc. for Computers and Info. Tech., Little Rock, 1989—; co-dir. Ctr. for Artificial Intelligence and Expert Systems, Fayetteville, Ark., 1989-91; rsch. fellow Nat. Ctr. for Toxicol. Rsch., Jefferson, Ark., 1989-91. Contbr. articles to profl. jours. Grantee NSF, 1991-93, 1989-90, AT&T Corp., 1991; recipient Applied Rsch. grant Acxiom Corp. and Ark. Sci. and Tech. Authority, 1990-91. Mem. ACM, Spl. Interest Group on Applied Computing (treas. 1990—), Spl. Interest Group on Systems Documentation. Office: U Ark Little Rock Computer Sci 2801 S University Ave Little Rock AR 72204

TALIMCIOGLU, NAZMI METE, civil and environmental engineer, educator; b. Bolu, Turkey, Dec. 9, 1961; came to U.S., 1983; s. Kemal Selcuk and Perihan (Kip) T.; m. Nurhan Toguc, Aug. 17, 1987. BS, Istanbul Tech. U., 1982; ME, Stevens Inst. Tech., 1985, PhD, 1991. Computer cons. computer ctr. Stevens Inst. Tech., Hoboken, N.J., 1986-88, rsch. asst. civil/environ. engring. dept., 1989-91, rsch. asst., prof. ctr. for environ. engring., 1991—. Contbr. articles to profl. jours. V.p. Stevens Turkish Students Assn., Hoboken, 1989. Recipient scholarship Turkish Govt., 1983-86; named Outstanding Young Man of Am., Outstanding Young Men of Am., 1989. Mem. ASCE (assoc.), N.Y. Acad. Sci., Turkish Architectural and Engring. Soc., Nat. Groundwater Assn., Sigma Xi. Achievements include development of a state-of-the-art computer model for evaluation of the impact of contaminated soils on groundwater quality. Office: Stevens Tech CEE James C Nicoll Environ Lab Hoboken NJ 07030

TALKINGTON, DEBORAH FRANCES, microbiologist; b. Thomson, Ga., Oct. 21, 1954; d. Paul Archilus and Mary Elizabeth (Mathews) T.; m. Michael Todd Fallon, Sept. 28, 1985; 1 child, Katherine Elizabeth. BS magna cum laude, U. Ga., 1976, MS, 1980, PhD, 1983. Postdoctoral fellow U. Ala., Birmingham, 1984-88, rsch. assoc., 1988-89; rsch. microbiologist Ctrs. for Disease Control, Atlanta, 1990—. Vol. Arthritis Info. Svc. Ala., Birmingham, 1988-89; sec. community outreach com. Westminster Presbyn. Ch., Gwinnett County, Ga., 1992—. Recipient Reed Rumsey Rsch. award Am. Assn. Avian Pathologists, 1983; Internat. Orgn. Mycoplasmologists fellow, 1984; NIH grantee, 1985, 86. Mem. Am. Soc. Mircobiology, Phi Kappa PHi, Gamma Sigma Delta, Sigma Xi. Achievements include research in mechanisms of bacterial pathogenesis, streptococcal, pneumococcal, and mycoplasma virulence factors. Office: Ctrs for Disease Control 1600 Clifton Rd Atlanta GA 30333

TALLAL, PAULA, psychologist; b. Austin, Tex., May 12, 1947; d. Joseph J. and Ruth Lois (Epstein) Tallal; m. Ian Nigel Creese, July 22, 1972. BA, NYU, 1969; PhD, Cambridge U., Eng., 1973. Cert. clin. psychologist, Calif. Asst. prof. psychology Johns Hopkins U., Balt., 1976-79; dir. rsch. Children's Hosp. and Health Ctr., San Diego, 1979-85; asst. prof. psychiatry U. Calif., San Diego, 1979-80, assoc. prof. psychiatry, 1980-86, prof. psychiatry, 1986-88; adj. prof. dept. neurosci. U. Medicine and Dentistry N.J./N.J. Med. Sch., Newark, 1989—; prof., co-dir. CMBN Rutgers U., Newark, 1988—. Co-editor: Psychoneuroendocrinology, 1990; co-author: Language, Speech and Reading Disorders in Children, 1988; contbr. articles to profl. jours. Rita Rudel/Lucy Moses Meml. Lecture, 1989; Disting. Young Scientist award, Md. Acad. Sci., 1976, Outstanding Creativity in Sci. Pres.'s award, Notre Dame Coll., Md., 1977. Mem. Soc. Neurosci., Acoustical Soc. Am., N.Y. Acad. Sci., Am. Psychol. Assn., Internat. Neuropsychology Soc., Acad. of Aphasia, Am. Speech-Lang.-Hearing Assn. (Cert. of Appreciation 1982). Office: Rutgers U 197 University Ave Newark NJ 07102-1814*

TALLENT, MARC ANDREW, clinical psychologist; b. Newport News, Va., Dec. 3, 1954; s. Norman and Shirley Dorothy (Rudman) T. BA, Columbia U., 1978; MA, Adelphi U., 1980, PhD, 1983. Lic. psychologist, N.Y. Pvt. practice N.Y.C., 1985—; psychoanalytic psychotherapist Ctr. for Modern Psychoanalytic Studies, N.Y.C., 1988—; supervising psychologist div. spl. edn. Bd. Edn. of N.Y.C., 1991—; cons. Big Bros./Big Sisters N.Y.C., 1988—, Task Force on AIDS N.Y. State Psychol. Assn., N.Y.C., 1989—. Editor: Modern Psychoanalysis, 1986—. Mem. APA, Nat. Assn. for Advancement of Psychoanalysis, N.Y. State Psychological Assn., Phi Beta Kappa. Office: Profl Ste B 51 5th Ave New York NY 10003-4320

TALLENT, NORMAN, psychologist; b. Springfield, Mass., Sept. 28, 1921; s. Louis and Sarah (Steinman) T.; m. Shirley Dorothy Rudman, Jan. 23, 1949; children: Marc Andrew, Robert David, Anne Louise. BS, U. Ill., 1946, MS, 1947; MA, Columbia U., 1949, PhD, 1954. Clin. psychologist Elgin (Ill.) State Hosp., 1951-53; clin. psychologist VA Hosp., Hampton, Va., 1954-59, Northampton, Mass., 1959-87; instr. George Washington U., Washington, 1955-59; adj. prof. U. Mass., Amherst, 1962-87. Author: Practice of Psychological Assessment, 1992, Psychological Report Writing, 1976, 4th edit., 1993, Report Writing in Special Education, 1980, Psychology of Adjustment, 1978, Psychology: Understanding Ourselves and Others, 1972, 2d edit., 1977, Psychological Perspectives on the Person, 1967, Clinical Psychological Consultation, 1963; contbr. over 30 articles to profl. jours. Sgt. U.S. Army, 1943-45, ETO. Fellow Am. Psychol. Assn., Soc. for Personality Assessment. Home and Office: 41 Hillside Rd Northampton MA 01060-2119

TALMAGE, DAVID WILSON, physician, microbiology and medical educator, former university administrator; b. Kwangju, Korea, Sept. 15, 1919; s. John Van Neste and Eliza (Emerson) T.; m. LaVeryn Marie Hunicke, June 23, 1944; children: Janet, Marilyn, David, Mark, Carol. Student, Maryville (Tenn.) Coll., 1937-38; B.S., Davidson (N.C.) Coll., 1941; M.D., Washington U., St. Louis, 1944. Intern Ga. Baptist Hosp., 1944-45; resident medicine Barnes Hosp., St. Louis, 1948-50; fellow medicine Barnes Hosp., 1950-51; asst. prof. pathology U. Pitts., 1951-52; asst. prof., then assoc. prof. medicine U. Chgo., 1952-59; prof. medicine U. Colo., 1959—, prof. microbiology, 1960-86, disting. prof., 1986—, chmn. dept., 1963-65, assoc. dean, 1966-68, dean, 1969-71; dir. Webb-Waring Lung Inst., 1973-83, assoc. dean for research, 1983-86; mem. nat. council Nat. Inst. Allergy and Infectious Diseases, NIH, 1963-66, 73-77. Author: (with John Cann) Chemistry of Immunity in Health and Disease; editor: Jour. Allergy, 1963-67, (with M. Samter) Immunological Diseases. Served with M.C. AUS, 1945-48. Markle scholar, 1955-60. Mem. NAS, Inst. Medicine, Am. Acad. Allergy (pres.), Am. Assn. Immunologists (pres.), Phi Beta Kappa, Alpha Omega Alpha. Office: U Colo Sch Medicine Denver CO 80262

TALWANI, MANIK, geophysicist, educator; b. Patiala, India, Aug. 22, 1933; came to U.S., 1954; s. Bir Sain and Saraswati (Khosla) T.; m. Anni Fittler, Apr. 3, 1958; children: Rajeev Manik, Indira, Sanjay. BSc with honors, Delhi U., India, 1951, MSc, 1953; PhD, Columbia U., 1959; PhD (hon.), Oslo U., 1981. From rsch. scientist to assoc. prof. Lamont-Doherty Geol. Obs., Columbia U., N.Y.C., 1959-70, dir. obs., 1971-81; prof. Columbia U., N.Y.C., 1970-82; dir. Ctr. for Crustal Studies Gulf R & D Co., Pitts., 1981-83; chief scientist exploration div. Gulf R & D Co., Houston, 1983-85; Schlumberger prof. geophysics Rice U., Houston, 1985—; cons. govt. Iceland, 1982—; dir. Geotech. Inst. Houston Advanced Rsch. Ctr., Woodlands, 1985—; bd. dirs. Mgmt. Techs. Inc., N.J., 1987-92. Co-author: Geophysical Atlas of the Norwegian Sea; editor 5 books on earth sci., Maurice Ewing Mmel. Symposium; co-editor: Geophysical Atlases of Indian, Atlantic and Pacific Oceans; contbr. over 100 papers to profl. jours. Recipient Krishnan award Indian Geophys. Union, 1964, James B. Macelwane award Am. Geophys. Union, 1967, Maurice Ewing award, 1981, Exceptional Sci. Achievement award NASA, 1973, Guggenheim award, 1974, Woollard award Geol. Soc. Am., 1986, Alfred Wegener medal European Union of Geoscis., 1993; Fulbright-Hays fellow, 1974; Stackler Disting. Lectr. U. Telaviv, 1987. Fellow AAAS, Am. Geophys. Union (James B. Macelwane award 1964, Maurice Ewing award, 1981), Geol. Soc. Am. (George P. Woollard award 1984); mem. Soc. Exptl. Geophysicists, Am. Assn. Petroleum Geologists, Norwegian Acad. Scis., Petroleum Club, Acad. Nat. Scis. Russian Fedn., Houston Geophys. Soc. (hon. mem. 1993), Sigma Xi. Home: 1111 Hermann Dr 10-D Houston TX 77004 Office: Rice U PO Box 1892 Houston TX 77251-1892

TAM, ALFRED YAT-CHEUNG, pediatrician, consultant; b. Hong Kong, Aug. 28, 1953; s. Fun and and Yun-Ha (Ko) T.; m. Rosanna Yick-Ming Wong, Sept. 15, 1979; children: Joyce Joy-Yee, Jonathan Joy-Man. MB, BS, U. Hong Kong, 1977; MRCP (U.K.), Royal Coll. Physicians, 1982. Med. officer Queen Mary Hosp., Hong Kong, 1978-79, med. officer pediatric dept., 1979-84; lectr. pediatrics dept. U. Hong Kong, 1984-89; cons. paediatrician Grantham Hosp., Hong Kong, 1989—; fellow U. Hong Kong Centre of Asian Studies, 1989—. Sch. mgr. Sunnyside Sch., Hong Kong, 1984-90; bd. dirs. Haven of Hope Christian Svcs., Hong Kong, 1986-92; chmn. Haven Hope Hosp., Hong Kong, 1990-92. Brit. Commonwealth scholar, 1982-83. Fellow Royal Coll. Physicians of Edinburgh, Hong Kong Coll. Pediatricians, Am. Coll. Chest Physicians. Avocation: badminton. Office: Rm 1213 Melbourne Plz, 33 Queen's Rd Ctrl, Hong Kong Hong Kong

TAMARIZ, JOAQUIN, chemist, educator; b. Mexico City, Apr. 20, 1950; s. Ernesto Eduardo and Maria Teresa (Mascarua) T.; m. Iris Kaufman, April 23, 1984; children: Marcos Joaquin, Sebastian Pascal. BS, MS, Nat. Univ., Mexico, 1975; PhD, U. Lausanne, Switzerland, 1983. Chemistry lectr. Nat. Univ. Mex., 1972-76; researcher U. Grenoble, France, 1977; asst. prof. Nat. Sch. Biol. Scis. National Polytechnic Inst., Mex., 1982-87; prof. Nat. Sch. Biol. Scis. National Polytechnic Inst., 1988-93. Contbr. articles to Helvetica Chim Acta, Tetrahedron Letters, Synth. Communication, Jour. Organic Chem., Jour. Chem. Rsch., Med. Chem. Rsch., Heterocycles, Synthesis. Named Nat. Researcher of Nat. Researcher System Mexican Ministry of Edn., 1984; recipient Prix of Rsch. Nat. Polytechnic Inst., 1991. Mem. Mexican Chem. Soc., Am. Chem. Soc. Home: Barranquilla 99 Lindavista, Mexico City 07300, Mexico Office: Escuela Scis Biol Dept Química IPN, Prol Carpio y Plan Ayala, Mexico City 11340, Mexico

TAMASHIRO, THOMAS KOYEI, retired electrical engineer; b. Paia, Maui, Hawaii, Aug. 4, 1926; s. Kokichi and Mashi (Shinyashiki) T.; m. Mary E. Oden; children: Cheryl M., Venita. BEE, Tri-State U., Angola, Ind., 1951, EE, 1964. Project engr. Jackson & Church Co., Saginaw, Mich., 1953-55; project mgr. Aerojet-Gen. Corp., Azusa, Calif., 1955-72; sr. staff engr. Aerojet Nuclear Corp., Idaho Falls, Idaho, 1972-75; project mgr. Allied Chem. Corp., Idaho Falls, 1976-79, Exxon Nuclear-Idaho, Idaho Falls, 1979-84; project mgr., sect. mgr. Westinghouse Idaho Nuclear Co., 1984-90, ret. 1990; lectr. in field. Mem. Am. Nuclear Soc. (Mgr. of Yr. 1986), Eagle Rock Amateur Radio Club, Shriners, Masons. Home: 2514 W Barberry Ln Rte 9 Idaho Falls ID 83402

TAMAYE, ELAINE E., coastal/ocean engineer; b. Maui, Hawaii, June 14, 1952; d. Yoichi and Lorraine Kawano. BS in Engring., U. Hawaii, 1974, MS in Ocean Engring., 1977. Civil/hydraulic engr. U.S. Army Corps Engrs., Fort Shaffer, Hawaii, 1977-83; v.p. Edward K. Nodd Assocs., Honolulu, 1983—. Mem. Soc. Naval Architects and Marine Engrs. (assoc.), Sigma Xi (assoc.). Office: Edward K Noda Assocs 615 Piikoi St Ste 1000 Honolulu HI 96814

TAMIMI, NASSER TAHER, educator, medical physicist; b. Hebron, Jordan, July 10, 1951; came to U.S., 1982; s. Taher Hamid and Malakeh Ahmad Sultan; m. Afaf Tamimi, July 10, 1979; children: Ramzi, Malak, Ramiz, Basim, Rula, Omar, Ahmad. MS, Ohio U., 1985; PhD in Nuclear Physics, Kent. State U., 1989; MS in Med. Physics, Wayne State U., 1993. Rsch. asst. Mid. East Tech. U., Ankara, Turkey, 1978-79; tchr. high sch. Dar-ul Aitam, Jerusalem, Jordan, 1978-79; teaching asst. Ohio U., Athens, 1982-85; rsch. asst. Kent (Ohio) State U., 1988, asst. prof., 1988—, rsch. assoc., 1988—. Contbr. articles to profl. jours. Mem. Am. Phys. Soc. (presenter 1988—), Am. Assn. Physicists in Medicine, Sigma Xi. Home: 775 Lessig Rd Akron OH 44312 Office: Wayne State U 3990 John Rd Detroit MI 48201

TAMIN, AZAIBI, molecular virologist, researcher; b. Muar, Johor, Malaysia, Mar. 9, 1959; came to U.S., 1984; s. Hj Tamin Sahandan and Komsinah Hassan; m. Zabedah Ismail, Sept. 1, 1984; 1 child, Adam Zulfaqar Azaibi. BS in Microbiology with honors, U. Leeds, Eng., 1983; PhD in Microbiology, Oregon State U., 1989. Tutor in microbiology Nat. U. Malaysia, 1983-84; teaching asst. Oregon State U., Corvallis, Oreg., 1985-86; rsch. asst. Oregon State U., Corvallis, 1986-88; rsch. fellow U.S. NRC & Ctrs. for Disease Control, Atlanta, 1989-91; rsch. scientist Ctrs for Disease Control Measles Lab., Atlanta, 1991—; guest tchr. science Dar-Un-Nur Sch., Atlanta, 1993—. Contbr. articles to Virology. Capt. Outward Bound Sch., Malaysia (Merit award), 1978; boy-sgt. Royal Military Coll., Malaysia, 1978; pres. Malaysian Student Union, U. Leeds, Eng., 1983. Recipient scholarships Ministry of Edn., Malaysia, 1978-83, Dept. Pub. Svc., 1984-89, N.L. Tartar fellowship U.S. NRC, CDC, 1989-91; Carter Ctr. fellow, 1993. Mem. AAAS, Am. Soc. Virology, Am. Soc. Microbiology, N.Y. Acad. Scis. Islamic. Achievements include first to report a mutation in the cis-acting element in a gene responsible for both temperature sensitivity and drug resistant phenotypes; molecular pathobiology and evolution studies of continental U.S. orthopoxuiruses; research in poxvirus expression system and immune response studies of measles virus infection. Office: Ctrs for Disease Control MS G17 Atlanta GA 30333

TAMIR, THEODOR, electrophysics researcher, educator; b. Bucharest, Roumania, Sept. 17, 1927; came to U.S., 1958, naturalized, 1968; s. Martin and Helena (Hart) Berman; m. Hadassah Cohen, Oct. 5, 1949; children: Jonathan, Yael. B.S. Technion, Israel Inst. Tech., 1953, Dipl. Ingenieur, 1954, M.S., 1958; Ph.D., Poly. Inst. Bklyn., 1962. Instr. Technion Israel Inst. Tech., Haifa, 1956-58; research staff Poly. Inst. N.Y., 1958-62, mem. faculty, 1962—, prof. electrophysics, 1969-92, Univ. prof., 1992—, head dept. elec. engring., 1974-79; sci. and engring. cons. to indsl. and govtl. labs. Editor, author: Integrated Optics, 1975 (transl. into Russian and Chinese), Guided Wave Optoelectronics, 1988 (transl. into Russian); co-editor: Springer Series in Optical Sciences, 1979—; contbr. chpts. to books, articles to profl. jours. Served with Israeli Army, 1947-49. Awarded Instn. Premium, 1964, Electronics Premium, 1967, Instn. Elec. Engrs., London; citation for disting. research Polytechnic chpt. Sigma Xi, 1978. Fellow IEEE, Instn. Elec. Engrs. (London), Optical Soc. Am.; mem. Internat. Union Radio Sci., Sigma Xi. Home: 981 E Lawn Dr Teaneck NJ 07666-6604 Office: 333 Jay St Brooklyn NY 11201-2990

TAMM, IGOR, biomedical scientist, educator; b. Tapa, Estonia, Apr. 27, 1922; s. Alexander and Olga Tamm; m. Olive Pitkin, May 9, 1953; children: Carol, Eric, Ellen. Student, Tartu U. Med. Faculty, Estonia, 1942-43; med. candid. exam., Karolinska Mediko-Kirurgiska Inst., Stockholm, 1945; MD cum laude, Yale U., 1947. Intern, asst. resident Grace-New Haven Community Hosp., 1947-49; asst. in medicine Yale U. Sch. Medicine, 1947-49; asst. and asst. physician The Rockefeller Inst., N.Y.C., 1949-53, assoc. and assoc. physician, 1953-56, assoc. prof., assoc. physician, 1956-58, assoc. prof. and physician, 1958-64; prof. and sr. physician The Rockefeller U., N.Y.C., 1964-86, Abby Rockefeller Mauzé prof. and sr. physician, 1986-92, emeritus, 1992—; assoc. mem. Commn. on Acute Respiratory Diseases, Armed Forces Epidemiol. Bd., 1961-73; mem. virology and rickettsiology study sect. NIH, 1964-68; mem. bd. sci. cons. Sloan-Kettering Inst. Cancer Research, 1966-75, vice chmn., 1971-72, chmn., 1972-73; mem. study panel for allergy and infectious diseases Health Research Council of City of N.Y., 1968-75; mem. Am. Cancer Soc. adv. com. on virology and cell biology; gen. chmn. task force on virology Nat. Inst. Allergy and Infectious Diseases. Contbr. 252 sci. papers on biology of viruses and cells; assoc. editor: Jour. of Immunology, 1957-59, Procs. of Soc. for Exptl. Biology and Medicine, 1963-66; adv. editor: The Jour. of Exptl. Medicine, 1971-81; hon. editorial bd. Biochem. Pharmacology, 1974-84; mem. editorial bd. Jour. of Interferon Rsch., 1980-88; editor symposium on viruses Am. Jour. of Medicine, 1965; editor (with F.L. Horsfall) Viral and Rickettsial Infections of Man, 4th edit., 1965. Recipient Alfred Benzon prize, 1967; Centennial lectr. U. Ill., 1968. Fellow AAAS, N.Y. Acad. Scis. (Sarah L. Poiley award 1977); mem. NAS, Am. Soc. Microbiology, Am. Assn. Immunologists, Soc. for Exptl. Biology and Medicine, Am. Soc. Clin. Investigation, Am. Acad. Microbiology, Am. Soc. Cell Biology, Soc. Gen. Microbiology, Assn. Am. Physicians, Deutsche Gesellschaft für Hygiene und Mikrobiologie (corr.), Am. Soc. Virology, Internat. Cell Cycle Soc., Internat. Soc. Interferon Research, Harvey Soc., Alpha Omega Alpha. Achievements include discovery and detailed characterization of the urinary (Tamm-Horsfall) glycoprotein; discovery of double-stranded RNA as the genetic material of reovirus; discovery of selective inhibitors of mRNA synthesis; discoveries regarding growth factor and cytokine action. Office: Rockefeller Univ Dept of Cell Physiology and Virology 1230 York Ave New York NY 10021-6341

TAMMEN, JAMES F., plant pathologist, educator; b. Sacramento, Feb. 27, 1925; m. Marilyn L. McDonald; 2 children. Student, Sacramento Coll., 1945, U. Calif., Davis; BS in Plant Sci. with honors, U. Calif., Berkeley, 1949, PhD, 1954. Jr. plant pathologist Calif. Bureau Plant Pathology, Riverside, Calif.; plant pathologist State Plant Bd. Fla., Gainesville, chief plant pathology lab.; from asst. prof to prof. dept. plant pathology Pa. State U., University Pk., 1956-65, dept. head, 1965-76, rsch. scientist advance hort. systems, 1989—, adj. prof., 1989—; dean Coll. Agriculture, Mpls., 1976-81; pres. Oglevee Assocs. Inc., 1981-86; dir., rsch. scientist tech. transfer office Inst. Food and Agrl. Scis. U. Fla., 1986-89; cons., speaker in field. With U.S. Army Corps, 1943-45, ETO. Recipient Rsch. award Am. Carnation Soc., Distinction award Internat. Geranium Conf. Fellow Am. Phytopathological Soc. (pres. 1972-75, editor jour., chmn. various coms., founder APS Found., spl. com. APS Endowment Fund, Disting. Svc. award 1993). Achievements include patent in field. Office: Pennsylvania State U Advanced Horticultural Systems University Park PA 16802*

TAMOR, STEPHEN, physicist; b. N.Y.C., Nov. 29, 1925; s. Abraham B. and Anna (Kaufman) T.; m. Jeanne Cohen, Jan. 23, 1949; children: Lynne, Kenneth L., Michael A., Sharon E. BS, CCNY, 1944; PhD, U. Rochester, 1950. Physicist Oak Ridge (Tenn.) Nat. Lab., 1950-52; physicist, postdoctoral fellow radiation lab. U. Calif., Berkeley, 1952-55; physicist GE, Schenectady, N.Y., 1955-66; physicist space scis. lab. GE, Valley Forge, Pa., 1966-71; physicist Sci. Applications Inst. Co., La Jolla, Calif., 1971-88; ind. cons. La Jolla, 1988—. Contbr. articles to Phys. Rev., Jour. Chem. Physics, other sci. publs. With USN, 1944-46. AEC fellow, 1949-50, John Simon Guggenheim Found. fellow, 1963-64. Fellow Am. Phys. Soc.

TAMRAT, BEFECADU, aeronautical engineer; b. Addis Ababa, Ethiopia, July 12, 1947; came to U.S., 1976; s. Tamrat Aberra and Yirgedu Reta; m. Assefash Kidane, July 5, 1980. BS, U. Ariz., 1969, MS, 1970. Sr. engr. sounding rocket project Space Vector Corp., Northridge, Calif., 1977-80; project engring. specialist Northrop Corp., Hawthorne, Calif., 1980-88; lead aeronautical engr. Rockwell Internat. NAAO, El Segundo, Calif., 1988—. Contbr. articles on fighter aircraft agility and performance to profl. publs. Mem. AIAA. Democrat. Achievements include introduction of new method of assessing fighter aircraft agility and performance. Home: 22334 Harbor Ridge Ln # 3 Torrance CA 90502 Office: Rockwell Internat 201 N Douglas St El Segundo CA 90009

TAMS, THOMAS WALTER, civil and structural engineer, land surveyor; b. Trenton, N.J., Oct. 30, 1921; s. Thomas Walter and Ada Bonnie (Harrold) T.; m. Catherine Ann Nicholson, Sept. 15, 1950; children: Catherine A., Patricia A. BSCE, Lafayette Coll., 1949. Registered profl. engr. and land surveyor, N.J., Pa., profl. planner, N.J. Bridge designer Reading R.R., Phila., 1950-55; dist. office engr. Portland Cement Assn., Chgo., 1955-68; project engr. Albert C. Jones Assocs., Mt. Holly, N.J., 1968-70; twp. engr. Princeton (N.J.) Twp., 1970-78; engr. Bensalem (Pa.) Twp., 1978-84, authority engr., 1984-92. 2d lt. USAF, 1941-45. Mem. Am. Legion, Kiwanis (chmn. agr. commn. 1962). Roman Catholic. Home: 20 Upton Ln Yardley PA 19067

TAMURA, IMAO, retired engineering educator; b. Wakayama-ken, Japan, Sept. 30, 1923; s. Kumataro and Yoshiko (Noguchi) T.; m. Yoshiko Otsuka, Apr. 28, 1953; children: Masaka, Miki. B of Engring. in Metallurgy, Osaka U., 1948, DEng, 1958. Asst. faculty engring. Osaka (Japan) U., 1953-61, assoc. prof. Inst. for Sci. and Indsl. Rsch., 1961-64; assoc. prof. faculty engring. Kyoto (Japan) U., 1964, prof. faculty engring., 1964-87, emeritus prof., 1987—; mgr., researcher Rsch. Inst. for Applied Sci., Kyoto, 1971—; tech. advisor Sumitomo Metal Industries Ltd., Osaka, 1987-90. Author: Strengthening and Toughening of Steels, 1969; co-author: Fundamentals of Metallic Materials, 1978, Thermomechanical Processing of High Strength Low Alloy Steels, 1988; co-author, editor: Processing of the International Conference on Thermomechanical Processing of Steels and Other Metals, 1988. Recipient Thermo-Tech. award Tanigawa Thermal Found., 1984, pub. svc. award Hattori Pub. Svc. Found., 1985, Murakami Meml. award Murakami Commemoration Found., 1989. Fellow Am. Soc. for Metals; mem. Am. Soc. for Metals Internat., Japan Inst. for Metals (hon., v.p. 1986-87, Disting. Svc. award 1968), Iron and Steel Inst. Japan (Henderson award 1980, Nishiyama award 1987), Japanese Inst. for Heat Treatment (v.p. 1987—, pres., award), Internat. Fedn. for Heat Treatment and Surface Engring. (v.p. 1951-53), Internat. Fedn. Heat Treatment (hon. v.p. 1992). Home: 142-16 Banjojiki, Fushimi-ku, Kyoto 612, Japan Office: Rsch Inst Applied Scis, 49 Tanaka Oi-cho Sakyo-ku, Kyoto 606, Japan

TAN, BOEN HIE, biochemist; b. Padangan, Java, Indonesia, Dec. 14, 1926; s. King Hoo and Bwan Nio (Oei) T. BS, U. Leyden, Holland, 1952, MS, 1955, ScD, 1962. Profl. nuclear medicine specialist. Fellow, asst. prof. U. Leyden, Holland, 1953-55, 62-64; fellow, rsch. assoc. Max Planck Inst., Gottingen, Germany, 1961-62, U. Minn., Mpls., 1955-61, 64-68, 1972-73; rsch. assoc. N.Y. Hosp., Cornell Med. Ctr., N.Y.C., 1968-72; rsch. assoc., prof. U. Groningen, Maastricht, Holland, 1973-81; rsch. assoc. U. South Ala., Mobile, 1982-92; analytical biochemist Ala. Dept. Environ. Mgmt., Montgomery, 1992—. Contbr. over 55 articles to profl. jours. Treas. "Aesculapius" Leyden U. Pharm. Student Assn., 1952-53. Mem. Nederlandse Vereniging voor Nucleaire Geneeskunde, Am. Assn. for Clin. Chemistry, FASEB, AAAS, Am. Chem. Soc. Achievements include research on sulfhydryl, disulfide groups in denatured, renatured proteins; purification, analysis, pharmacokinetic, pharmacological activities of new drugs; alpha-1-antitrypsin, plasma proteins, enzymes, inhibitors, fibrin formation lysis; vanadate-sulfhydryl complexes and PDE activities, DNA damage and repair; diabetes and the heart. Home: PO Box 230451 Montgomery AL 36123-0451 Office: Ala Dept Environ Mgmt 1890-A Dickinson Dr Montgomery AL 36109

TAN, CHAI TIAM, civil engineer, precast concrete company executive; b. Segamat, Malaysia, Apr. 22, 1943; s. Boon Cheong Tan and Leong Yam Tey; m. Sam Moi Pang, Sept. 15, 1968; childen: Tang Hee, Tang Kwee, Tang Yew. Diploma civil engring. first class, Tech. Coll., Kuala Lumpur, 1965; Part I, Instn. Civil Engrs., U.K., 1966, Part II, 1967. Tech. cadet Drainage and Irrigation Dept. Govt. Malaysia, 1965-67; engr. Drainage and Irrigation Dept. Govt. Malaysia, Johor, 1967-72; engring. mgr. Hume Industries Ltd., 1972-78; mgr., dir. Gen. Concrete SDN BHD, 1978-81; gen. mgr., dir. Associated Concrete Products SDN BHD, 1981—; dir. Concrete Products Mktg. S/B, ACP Mktg. SDN Bhd, Pembinaan Sri Depan Sdn Bhd, ACP (Trengganu) Sdn Bhd, Perniagaan Heeky Sdn Bhd. Mem. Instn. Civil Engrs. U.K., Instn. Engrs. Malaysia. Avocations: golf, swimming, jogging, reading, singing. Home: 27 Lorong Kampar, Off Jalan Landasan, Klang Selangor 41300, Malaysia Office: Assoc Conrete Products Sdn Bhd, Jalan Kilang, Petaling Jaya, Selangor 46050, Malaysia

TAN, JOHN K., chemical company executive, educator; b. Fukien, China, Sept. 23, 1934; s. E. Ching Lam and Goct Sia (Kua) T.; m. Lily Go Tan, Mar. 4, 1967; children: Carolyn G., Edward John G., Herbert Joseph G., Steven Julian G. BS in Chem. Engring., Mapua Inst. Tech., Manila, 1960; M.Engring., Yale U., 1964; postgrad. Stevens Inst. Tech., Hoboken, N.J., 1964-68. Chem. engr. Globe Paper Mills, Inc., Malabon, Rizal, Philippines, 1960-61; supr. Internat. Chem. Industries Inc., Guiguinto, Bulacan, Philippines, 1961-62; grad. rsch. asst. dept. chem. engring. Yale U., New Haven, 1962-64; sr. rsch. chem. engr. Rsch. Ctr., rsch. and devel. div. Lever Bros. Co., Unilever Ltd., U.S.A., Edgewater, H.J., 1964-65; prin. rsch. chem. engr., 1965-68; tech. mgr., tech. asst. to pres. Internat. Chem. Industries Inc., Guiguinto, Bulacan, Philippines, 1968-70; tech. cons. Econ. Devel. Found., Inc., Makati, Rizal, Philippines, 1968-70; asst. gen. mgr., tech. dir. Philippines Fermentation Indsl. Corp., Guiguinto, Bulacan, Philippines, 1970-73; prof. dept. dept. chemistry and chem. engring. Grad. Sch. Chem. Engring., Mapua Inst. Tech., Intramuros, Manila, Philippines, 1968-79; pres., exec. v.p. Lakeview Garments, Inc., Taguig, Metro Manila, 1976-86, v.p., gen. mgr. Premier Steam Laundry, Inc., Taguig, 1978-85; pres., gen. mgr., No. Chem. Sales Corp., Manila, Tex-Chem Mktg. Corp., Manila, 1972—; pres., chmn. bd. Bay Tank Yard, Inc., Manila, 1972—; exec. v.p., gen. mgr. Multi-Land Devel. Corp., Manila, 1972—; advisor Nat. Bd. Advisors, internat. div. Am. Biog. Inst., Raleigh, N.C. Pres., Philippine Assn. Chem. Suppliers, Inc., Makati, Metro-Manila, 1983-85, adv. 1986—. Trustee, Philippine Cultural High Sch., 1983—, pres. 1991—; pres. Mapua Inst. Tech. Chemistry and Chem. Engring. Alumni Assn., 1985-86; trustee Prof. Lauro A. Limuaco Meml. Found.; v.p. Filipino Chinese Amateur Athletic Fedn., 1988—; exec. dir. Fedn. Filipino-Chinese Cs. of C. and Industry, Inc., 1993—; pres. Philippine Cultural High Sch. Edn. Found., Inc., Manila, 1991—. Conn. State Water Resources Commn. fellow Yale U., 1962-64; United Aircraft scholar Yale U., 1963-64; Univ. scholar Mapua Inst. Tech., Manila, 1955-60, Silver medalist in chem. engring., 1960; Internat. Biog. Assn. fellow, Cambridge, Eng., 1983—; named Outstanding Alumnus, Mapua Inst. Tech. Mem. AICE, Am. Chem. Soc., N.Y. Acad. Scis., Yale Sci. and Engring. Assn., Sigma Xi, Eta Sigma Mu. Club: Yale of the Philippines. Home: Dasmarinas Village, 1973 Kasoy St, Makati Metro Manila, The Philippines Office: No Chem Sales Corp, 950 Soler St, Binondo Manila The Philippines

TAN, KIM LEONG, pediatrician, neonatologist, medical educator; b. Malaysia, Oct. 30, 1936; s. Chim Ean Tan and Siew Bo Yeoh, m. Kwok Yee Leng, June 15, 1963; children: Min-Li, Min-Ching, Wei-Liang. MBBS, U. Singapore, 1962; DCH, U. London, 1967. Fellow Royal Coll. Physicians, 1977, Fellow Royal Australasian Coll. Physicians, 1978. Med. officer Ministry of Health, Singapore, 1963-68; lectr. U. Singapore, 1968-71, sr. lectr., 1971-75, assoc. prof., 1976-79; prof. Nat. U. Singapore, 1980—; chief dept. neonatology Nat. U. Hosp. Singapore, 1990—; Mem. WHO Expert Adv. Panel on Maternal and Child Health, 1993-97, Nat. Sci. Com. on Hepatitis and Related Disorders, Singapore, 1985—; chmn. Expert Com. on Immunization Program, Singapore, 1988—; chmn. Chpts. of Pediatricians, Acad. of Medicine, Singapore, 1989-90. Co-editor Procs. of 1st Asia Oceana Congress of Perinatology, 1979; mem. editorial bd. Jour. of AMA (SEA), 1984—; designer phototherapy setup. Mem. panel of doctors Kim Seng Community Ctr. Night Clinic, Singapore, 1981—; mem. exec. com. Children's Aid Soc., Singapore, 1981-91. Mem. Singapore Pediatric Soc. (chmn. research fund 1975—, Haridas Meml. lectr. 1972, 77), Ob-gyn. Soc. Singapore (Benjamin Henry Sheares lectr. 1976), Brit. Med. Assn., Singapore Med. Assn., Acad. of Medicine. Home: 259 6th Ave, Dynasty Garden Ct 1, Singapore 1027, Singapore Office: Nat U Hospital Dept Neonatology, 5 Lower Kent Ridge Rd, Singapore 0511, Singapore

TAN, KOK TIN, aerospace engineer; b. Singapore, Oct. 3, 1958; s. Cheng Guan Tan and Siew Hong Ng; m. Lee Huang Seng, Oct. 18, 1986; children: Benjamin Chen Weijie, Elizabeth Chen Jiahui. M.Eng. in ME, Nat. U. Singapore, 1986; PhD in Aerospace Engring., U. Mich., 1991. Project leader Def. Sci. Orgn., Singapore, 1985-88, dep. program mgr., 1991-92, project chief engr., 1992—. Contbr. articles to profl. publs. Recipient Def. Tech. award Ministry of Def., Singapore, 1987. Mem. AIAA, IEEE. Achievements include invention of maximal output admissible sets, controllers design for linear system subject to state and control restraints, pressure prediction around cylinder oscillating in steady flow. Home: 02-30 Blk 317 Shunfu Rd, Singapore 2057, Singapore Office: Def Sci Orgn, 20 Science Park Dr, Singapore 0511, Singapore

TAN, TJIAUW-LING, educator, psychiatrist; b. Pemalang, Java, Indonesia, June 2, 1935; came to U.S., 1967; naturalized, 1972; s. Ping-Hoey and Liep-Nio (Liem) T.; m. Esther Joyce Kho, June 2, 1961; children: Paul Budiman, Robert Yuling, Alice Ayling. BS, U. Indonesia Faculty Medicine, 1957, MD, 1961; postgrad. U. Indonesia, Jakarta, 1961-65, U. Calif. at L.A., 1967-71, Pa. State U., 1971-72. Diplomate Am. Bd. Psychiatry and Neurology, Am. Bd. Gen. Psychiatry, Am. Bd. Geriatric Psychiatry. Lectr. psychiatry U. Indonesia, Jakarta, 1965-67; psychiat. cons. Central Gen. Hosp., Jakarta, 1965-67; postdoctoral fellow U. Calif. at L.A. Brain Rsch. Inst., 1967-69; asst. rsch. psychiatrist, dept. psychiatry Neuropsychiat. Inst. U. Calif., L.A., 1969-70; asst. prof. psychiatry Pa. State U., 1972-87; assoc. prof. psychiatry Pa. State U., 1987—; chief inpatient psychiatry Univ. Hosp. Milton S. Hershey Med. Ctr., 1972—, dir. Behavioral Medicine Clinic, co-dir. Biofeedback Lab., 1975—; cons. psychiatry Family and Children's Svc. Lebanon County, Lebanon, Pa., 1971-79, Bd. dirs. Retarded Children's Assn. Dauphin County, Inc., 1971-73. Fellow Am. Psychiat. Assn.; mem. Pa. Psychiat. Soc., Central Pa. Psychiat. Soc., Pa. Med. Assn., Dauphin County Med. Soc., Assn. Advancement Behavior Therapy, Assn. Applied Psychophysiology and Biofeedback, Soc. Behavioral Medicine, N.Y. Acad. Scis., AAAS, Assn. Psychophysiol. Study of Sleep, Am. Acad. Sleep Disorder Medicine, Am. Assn. for Geriatric Psychiatry, Am. Geriatric Soc. Contbr. articles to profl. jours. Home: 1478 Bradley Ave Hummelstown PA 17036-9143 Office: Pa State U Coll Medicine Dept Psychiatry 500 University Dr Hershey PA 17033-2360

TAN, YOKE SAN, mechanical engineer; b. Mentakab, Pahang, Malaysia, Nov. 22, 1955; s. Tan Ghee Kam and Lim Jit Kean; m. Ho Yong Ying; children: Tan Zhi Chuen, Tan Zhi Hui. B Mech. Engring., U. Auckland, New Zealand, 1979. Svc. engr. Scott & English (M) Sdn Bhd, Ipoh, Perak, Malaysia, 1979-80; sales exec. Scott & English (M) Sdn Bhd, Kuala Lumpur, Malaysia, 1980-83; sales mgr. Scott & English (M) Sdn Bhd, Petaling Jaya, Malaysia, 1983-85; br. mgr. Scott & English (M) Sdn Bhd, Ipoh, Perak, 1985-89, Johor Bahru, Johor Bahur, Johor, Malaysia, 1990—. Mem. Malaysian Inst. Mgmt. (assoc.), Instn. Engrs. Malaysia (grad.), Instn. Mech. Engrs. (grad.), Kelab Rekreasi Tentera Utara, Chateau Wine Club (v.p. 1991), Golf Club Di Raja Perak, Tanjong Puteri Golf and Country Club, Rotary (treas. Pasir Gudang club 1992-93, sec., pres.-elect 1993—). Office: Scott & English (M) Sdn Bhd, Lot 15 33/4 Milestone, Johor Bahru Malaysia

TANABE, YO, chemistry researcher; b. Tamano-shi, Okayama, Japan, Apr. 14, 1950; s. Yoshisada and Misao (Tsuboi) T.; m. Yurie Fukai, Feb. 28, 1982; children: Mariko, Shin. BS, U. Tokyo, 1974, MS, 1976, PhD, 1980. Lectr. Niimi Women's Coll., Niimi-shi, Okayama, Japan, 1980-85, asst. prof., 1985-93, head libr., 1986-93, prof., 1993—. Contbr. articles to profl. jours. Bd. dirs. Libr. Assn. of Okayama-prefecture, Okayama-shi, 1987-89. Mem. Libr. Conf. on Prefecture Jr. Coll. (pres. 1991-92), Japan Libr. Assn. (bd. dirs. 1991-92, v.p. jr. coll. meeting 1991-92). Buddhist. Avocations: piano, long-distance runnning, tennis, skiing, skating. Home: 364-18 Monde, Soja-shi 719-11, Japan Office: Niimi Women's Coll, 2-1263 Nishigata, Niimi-shi 718, Japan

TANACREDI, JOHN T(HOMAS), ecotoxicologist; b. Bklyn., Oct. 28, 1947; s. John J. and Josephine (Longo) T.; m. Julianne Scala, Apr. 29, 1972; children: Jeannine Lyn, Ryan Gerard. MS in Environ. Health Sci., CUNY, 1974; PhD in Environ. Health Engring., Poly. U., Bklyn., 1988. rsch. assoc. U. R.I., Grad. Sch. of Oceanography, Nat. Park Svc., Coop. Park Study Unit, Narragansett, R.I., 1988—, N.Y. Zool. Soc., N.Y. Aquarium, Bklyn, 1990—. Rsch. intern U.S. EPA, Edison, N.J., 1972-74; environ. specialist adminstr. U.S. Dept. Transp., N.Y.C., 1974-78; rsch. ecologist Nat. Park Svc., Bklyn., 1978-92; rsch. assoc. invert. dept. Am. Mus. of Natural History, 1992—; rsch. assoc. Univ. R.I.-Grad. Sch. Oceanography-Nat. Pk. Svc.-Cooperative Pk. Study Unit, Narragansett, R.I., 1988—, N.Y. Zool. Soc., N.Y. Aquarium, Bklyn., 1990—. Host radio program Focus on our Environment, WHPC-NY, 1988-91. Pres. Enviro-qual Assocs., Inc., 1977—. With USN, 1968-70. U.S. HEW fellow, 1972; named to Wall of Fame, Kingsborough C.C., CUNY, 1989. Fellow Explorers Club. Democrat. Achievements include research in managing public lands in the public interest, biodepuration in shellfish. Home: 91 Horton Ave Valley Stream NY 11581 Office: Nat Park Svc Gateway NRA-FBF Brooklyn NY 11234

TANAKA, NOBUYOSHI, consulting engineering; b. Takamatsu, Japan, May 22, 1934; s. Moriyoshi and Natsuko T.; m. Machiko Chisaka, Oct. 22, 1966; 1 child, Tomonari. BS, Tokyo Coll. Sci., 1959; M. Engring., Kyoto (Japan) U., 1961. Researcher Showadenko Chem. Ind. Ltd., Tokyo, 1961, Inst. Solid-State Phys. U. Tokyo, 1963-66; head of sect. Cen. Rsch. Lab. Showadenko Ltd., Tokyo, 1970-80; vice dir. Tech. Rsch. Lab. Showadenko Ltd., Tokyo, 1980-81; head of office N. Tanaka Cons. Engr.'s Office, Chiba, Japan, 1983-88, Yokohama, Japan, 1989—; tech. advisor Japan IMSL Co., Ltd., Tokyo, 1986—; committeeman Japan Rsch. Devel. Corp., Tokyo, 1983—, Japan Small Bus. Corp., Tokyo, 1987—. Author: Nuclear Magnetic Resonance, 1978, Computer Integrated Manufacturing, 1991; contbr. articles to profl. jours. including Polymer Solid States, Multivariate Analysis of Polymers, TD Use Workstation. Mem. Assn. Japan Cons. Engrs., Chem. Soc. Japan, Info. Processing Soc. Japan, Am. Chem. Soc. (affiliate). Achievements include patents for method of nuclearmagnetic relaxation time in high resolution NMR, for liquid crystal condenser for relay, for method of anti-mildew wrapping film, for thermal conductivity measurement apparatus. Home: No 1642 212 D7 Kamariya, Kanazawa ku, Yokohama 236, Japan Office: N Tanaka Cons Engrs, Technowave Bldg 8F, No 1 1 25 Sin Urashima, Yokohama 221, Japan

TANAKA, RICHARD I., computer products company executive; b. Sacramento, Dec. 17, 1928; s. G. and Kei Tanaka; m. Edith M. Arita, Aug. 21, 1950; children: Steven Richard, Jean Elizabeth, John Richard, Anne Mariko. B.S. with highest honors, U. Calif., Berkeley, 1950, M.S., 1951; Ph.D., Calif. Inst. Tech., 1958. Sr. research engr. N.Am. Aviation, Inc., 1951-54; mem. tech. staff Hughes Aircraft Co., 1954-57; dept. mgr., sr. mem. computer research Lockheed Missiles & Space Co., Palo Alto, Calif., 1957-65; sr. v.p. Calif. Computer Products Inc., Anaheim, Calif., 1966-77; pres. Internat. Tech. Resources Co., Tustin, Calif., 1977-80; pres., chief exec. officer Systonetics, Inc., Fullerton, Calif., 1980-86; pres. Lundy Electronics & Systems, Inc., Glen Head, N.Y., 1986-89; chmn., chief exec. officer, pres. Scan-Optics, Inc., East Hartford, Conn., 1989—; vis. prof. U. Calif., Berkeley, 1962. Author: Residue Arithmetic and Its Applications to Computer Technology, 1967. Hughes fellow Calif. Inst. Tech., 1955-57. Fellow IEEE (nat. chmn. computer soc. 1965-66, centennial medal); mem. Internat. Fedn. Info. Processing (pres. 1974-77, hon. life mem., U.S. del.), Am. Fedn. Info. Processing Socs. (pres. 1969-71, disting. service award 1983), Phi Beta Kappa, Tau Beta Pi, Eta Kappa Nu. Home: 19 Stratford Park Bloomfield CT 06002-2143 Office: Scan-Optics Inc 22 Prestige Park Cir East Hartford CT 06108-1917

TANAKA, TOSHIJIRO, physicist; b. Ooda, Shimane, Japan, July 12, 1944; s. Takeshiro and Masayo (Tanaka) T.; m. Hiroko Kodashiro, Apr. 27, 1983; children: Haruna, Hideki, Yoshiki. BS, Ibaragi U., Mito, Japan, 1967; MS, Kobe (Japan) U., 1969; PhD, Osaka (Japan) U., 1976. Lectr. Kagoshima (Japan) Prefectural Jr. Coll., 1977-78, assoc. prof., 1978-84, prof., 1984—; vis. scholar U. Calif., Santa Barbara, 1979-80. Mem. com. Kagoshima Indsl. Tech. Assn., 1990—. Recipient MBC prize Minami Nihon Broadcasting Co. Ltd., Kagoshima, 1990. Mem. Phys. Soc. Japan, Am. Phys. Soc., Physics Edn. Soc. Japan. Avocations: tennis, igo, Japanese folk music, dance. Home: 1437-181-8 Shimoishikicho, Kagoshima 890, Japan Office: Kagoshima Prefectural Jr Coll, 44 Shimoishikicho, Kagoshima 890, Japan

TANCIG, PETER, federal official; b. Freudenstadt, Germany, 1944; m. Simona Tancig, 1971; 1 child, Suna. BSc, MSc, PhD, Faculty Elec. Engring. and Computer Sci., Ljubjana. Vis. scientist MIT, Boston, 1980-82; chmn. sect. computational linguistics Slovenian Assn. for Applied Linguistics; head Lab. for Lang. and Speech Techs. dept. computer sci. and informatics Josef Stefan Inst., Ljubljana; prof. computer sci. Faculty Elec. Engring. and Computer Sci., Ljubljana; nat. coord. EC PHARE Program of Tech. Assistance; pres. of coun. ZS Greens of Slovenia; Minister of Sci. and Tech., 1990—; pres. Interministerial Coordination Fgn. Tech. Assistance, Nat. Comm. for UNESCO, Interministerial Comm. for Informatization Pub. Adminstrn. Mem. ACM, IEEE, AAAI, ESCA. Mem. ZS, the Green Party. Office: Ministry of Science & Technology, Cankarjeva 5, 61000 Ljubljana Slovenia*

TANDON, DON ASHOKA, electrical engineer; b. Muzaffarpur, Bihar, India, Oct. 23, 1944; came to U.S., 1971; s. B.N. and G. (Mehrotra) T.; m. Reena Burman, June 7, 1972; children: Anurag, Anita. MSEE, Ohio State U., 1973; MBA, Xavier U., 1977. Registered profl. engr., Ohio, Tex. Protection and comm. engr. Columbus (Ohio) and So. Ohio Electic Co., 1973-79; product planner Diamond Power Co., Lancaster, Ohio, 1979-82; sr. engr. Tex.-N.Mex. Power Co., Texas City, Tex., 1982—. Mem. IEEE, NSPE, Tex. Soc. Profl. Engrs., Kiwanis (Texas City chpt., bd. dirs. 1988-90, sec. 1990-91, v.p. 1991-92). Home: 2302 17th St N Texas City TX 77590 Office: Tex NMex Power Co PO Box 2190 Texas City TX 77592-2190

TANDON, RAJIV, psychiatrist, educator; b. Kanpur, India, Aug. 3, 1956; came to U.S., 1984; s. Bhagwan Sarup and Usha (Mehrotra) T.; m. Chanchal Nammi Vohra; children: Neeraj, Anisha. Student, St. Xavier's Coll., Bombay, India, 1974; BS, All India Inst., New Delhi, 1980; MD, Nat. Inst. of MH, India, 1983. Sr. resident Mental Health and Neuro-Scis., India, 1983-84; resident U. Mich. Hosps., Ann Arbor, 1984-87, attending psychiatrist, 1987—; dir. schizophrenia program U. Mich., Ann Arbor, 1987—, assoc. prof., 1993—; cons. Lenawee County Community Mental Health, Adrian, Mich., 1985—. Author: Biochemical Parameters of Mixed Affective States; Negative Schizophrenic Symptoms: Pathophysiology and Clinical Implications; contbr. more than 70 articles to profl. jours. Recipient Young Scientist's award Biennial Winter Workshop on Schizophrenia, 1990, 92, Travel award Am. Coll. Neuropsychopharmacology/Mead, 1990. Mem. Am. Psychiat. Assn. (Wisniewski Young Psychiatrist Rschr. award 1993), World Fedn. Mental Health, Soc. for Neurosci., N.Y. Acad. Scis., Soc. Biol.

Psychiatry, Mich. Psychiat. Soc. Democrat. Hindu. Office: U Mich Hosps 1500 E Medical Ctr Dr Ann Arbor MI 48109-0116

TANENBAUM, ANDREW STUART, computer scientist, educator, author; b. N.Y.C., Mar. 16, 1944; s. Julian and Esther (Brown) T.; m. Suzanne Alida Baart, Sept. 24, 1974; children: Barbara Laura, Marvin Eric. SB, MIT, 1965; PhD, U. Calif., Berkeley, 1971. Researcher Math. Ctr., Amsterdam, The Netherlands, 1971-73; prof. Vrije U., Amsterdam, 1973—; cons. Bell Labs., Murray Hill, N.J., 1980-83; lectr. various orgns., 1971—. Author: Operating Systems: Design and Implementation, 1987, Computer Networks, 1989, Structured Computer Organization, 1990, Modern Operating Systems, 1992; (software) The Minix Operating System, 1987, The Amoeba Distributed System, 1991; contbr. articles to profl. jours. Mem. IEEE, Assn. for Computing Machinery, Sigma Xi. Avocations: travel, cooking. Office: Vrije U, De Boelelaan 1081a, 1081 HV Amsterdam The Netherlands

TANG, CHAO, physicist; b. Nanchang, China, Oct. 2, 1958; came to U.S., 1981; s. Danlin Tang and Yansheng Cheng. BS, U. Sci. Tech. China, 1981; PhD, U. Chgo., 1986. Rsch. assoc. Brookhaven Nat. Lab., Upton, N.Y., 1986-88, Inst. Theoretical Physics, Santa Barbara, Calif., 1988-91; rsch. scientist NEC Rsch. Inst., Princeton, N.J., 1991—; referee Phys. Rev. and Phys. Rev. Letters, 1987—, Europhysics Letters, Jour. de Physique, 1992—. Contbr. articles to profl. jours. Recipient Telegdi prize, U. Chgo., 1982. Mem. Am. Phys. Soc., Materials Rsch. Soc., Overseas Chinese Physics Assn. Achievements include research on Schrödinger equation in one-dimensional quasiperiodic potentials; pattern formation in nonequilibrium systems; fractals and scaling; self-organized criticality; dynamics of earthquake faults. Home: 4 Dannys Way Cranbury NJ 08512 Office: NEC Rsch Inst 4 Independence Way Princeton NJ 08540

TANG, DAH-LAIN ALMON, automatic control engineer, researcher; b. Taiwan, 1955; came to U.S., 1979; s. I-Chung and Hung-Ju (Hsia) T.; m. Yuan-Huei Debby Tseng, Aug. 1979; children: Jonathan, Johanna. MS, Purdue U., 1980, PhD, 1984. Sr. rsch. engr. GM Rsch. Labs., Warren, Mich., 1984-89; staff rsch. engr. GM Rsch. Labs., Warren, 1989-93; staff engr. Transp. Electronics div. TRW, 1993—. Mem. IEEE, ASME (assoc.), Soc. Automotive Engrs., Sigma Xi. Achievements include patents on sensors and control methods for automotive engine controls.

TANG, IGNATIUS NING-BANG, chemical engineer; b. Nanking, China, July 7, 1933; s. Hung Ching and Wen Shan (Shih) T.; m. Carol Yu Lin, Oct. 30, 1962; children: Marion Mei-Ling, Alice Mei-Hua. BS in Chem. Engring., Nat. Taiwan U., 1955; MS in Chem. Engring., U. N.D., 1960; MS in Applied Math., SUNY, Stony Brook, 1975, PhD, 1982. Rsch. scientist Brookhaven Nat. Lab., Upton, N.Y., 1964—. Author/co-author over 60 sci. pubs. and book chpts. Mem. Am. Chem. Soc., Am. Assn. Aerosol Rsch., Soc. Applied Spectroscopy. Achievements include research in theoretical and experimental studies of molecular clustering about ions, gas-phase kinetics, phase transformation and growth of atmospheric aerosols, thermodynamics of concentrated electrolyte solutions. Home: Box 156 Shoreham NY 11786 Office: Brookhaven Nat Lab Bldg 815 Upton NY 11973

TANG, JINKE, physicist; b. Changchun, Jilin, China, Jan. 17, 1961; came to U.S., 1984; s. Dingxiang and Zhiyun (Jiang) T.; m. Hsiao-Po Cheng, Aug. 7, 1990; 1 child, Sean. PhD in Physics, Iowa State U., 1989, MS in Metallurgy, 1990. Asst. prof. U. New Orleans, 1990—. U. New Orleans Summer scholar, 1991. Mem. Am. Phys. Soc., Am. Soc. Metals. Office: U New Orleans Dept Physics New Orleans LA 70148

TANG, PUI FUN LOUISA, fragrance research administrator; b. Hong Kong, Sept. 8, 1956; d. Tien Shen and Lai Kwan (Sun) T.; m. Chi Hoo Terry Mak, May 22, 1988. BS in Chemistry, Hong Kong Bapt. Coll., 1979; PhD in Chemistry, No. Ill. U., 1984. Postdoctorate Inst. for Biomedical Rsch. U. Tex., Austin, 1984-86; rsch. assoc. Chinese U. Hong Kong, 1987-88; regional mgr. fragrance application rsch. and devel. Givaudan Roure Ltd., Hong Kong, 1988—. Contbr. articles to profl. jours. Hong Kong Bapt. Coll. scholar 1976-78, Academic award 1979; No. Ill. U. fellow, 1983. Mem. Am. Chem. Soc. Achievements include one patent. Office: Givaudan Roure Ltd, 10 Hysan Ave 16/F Sunning Plz, Hong Kong Hong Kong

TANG, QING, optical engineer; b. Sichuan, People's Republic of China, Nov. 25, 1964; s. Shi-Jin and Qi-Ze (Liu) T.; m. Hui Zhang, Jan. 2, 1992. BS in Physics, sichuan U., 1984; MSEE, U. Conn., 1990, PhD in Elec. Engring., 1993. Rsch. asst. Sichuan U., 1984-87, U. Conn., Storrs, 1990—; vis. scholar Darmstadt (Fed. Republic Germany) U. Tech., 1988-90. Contbr. articles to profl. jours.; patentee in field. European Econ. Communities scholar, 1988; recipient fellowship U. Conn., 1992. Mem. Optical Soc. Am., Internat. Soc. Opti-Electronics, Eta Kappa Nu. Achievements include invention of chirp-encoded joint transform correlator and phase visualization using BaTio3 phase conjugating mirror; research in optical computing, optical signal processing, holographic optical elements, and opto-electronics. Office: U Conn Elec Engring Dept U-157 Storrs Mansfield CT 06268

TANG, YOU-ZHI, chemist, researcher; b. Guang-zhou, Guangdong, China, Dec. 12, 1959; arrived in Can., 1982; s. Ming-sui Tang and Jin-Yao Huang. BEng with honors, Guangdong Inst. Tech., 1982; PhD, Dalhousie U., Halifax, N.S., Can., 1988. Chartered Chemist. Chemist Guangdong Environ. Protection Bur., Guangzhou, 1982; vis. scientist Environment Can., Toronto, Ont., 1987-89; project scientist Concord Environ., Toronto, 1989-90, sr. scientist, 1990—; oversea cons. South China Inst. for Environ. Scis., Guangzhou, 1991—; adj. assoc. prof. Guangdong Inst. Tech., Guangzhou, 1993—. Contbr. articles to profl. jours., chpt. to book. Killam scholar Killam Trust, 1983-86; Can. Govt. lab. vis. fellow Nat. Sci. & Engring. Rsch. Coun. Can., 1987-88. Mem. Air & Waste Mgmt. Assn., Assn. Chem. Profession of Ont. Achievements include invention of a series of novel detectors/sensors for chemical analysis; developed sampling methods for airborne pollutants. Office: Concord Environmental, 2 Tippett Rd, Toronto, ON Canada M3H 2V2

TANG, YU, structural engineer; b. Hsinchu, Taiwan, Republic of China, Nov. 8, 1954; came to U.S., 1982; s. Ming-chao and Pao-heng (Tong) T.; m. Olivia S. Shyr, Oct. 17, 1988; 1 child, Alan T. MS, Nat. Taiwan U., 1979; PhD, Rice U., 1986. Instr. Wang-Nan Inst. Tech., Chung-Li, Taiwan, 1981-86; from rsch. asst. to rsch. assoc. Rice U., Houston, 1982-89; structural engr. Bechtel Inc., Houston, 1989; asst. engr. Argonne (Ill.) Nat. Lab., 1990—. 2d lt. U.S. Army, 1979-81. Mem. ASCE (assoc., Raymond C. Reese Rsch. prize 1993), Sigma Xi. Office: Argonne Nat Lab RE/208 9700 S Cass Ave Argonne IL 60439

TANGARONE, BRUCE STEVEN, biochemist; b. Hartford, Conn., June 30, 1961; s. John Thomas and Ingrid (Schultz) T.; m. Lourdes Castillo, Sept. 4, 1988. BS, SUNY, Syracuse, MS. Biochemist Genetics Inst., Andover, Mass., 1988—. Mem. Am. Chem. Soc. Achievements include purification to homoeneity and characterization of a laminarinase from Trichoderma longibrachiatum. Home: 28 Longview Dr Hampstead NH 03841 Office: Genetics Inst 1 Burtt Rd Andover MA 01810

TANGCO, AMBROSIO FLORES, health administrator, surgeon, orthopedist; b. Manila, Mar. 20, 1912; s. Prudencio Tangco y Vargas and Maria Roque (Flores) T.; m. Gorgonia Ginete Dineros, June 20, 1947 (dec. Mar. 1979); children: Maria Angelita, Maria Lourdes, Ambrisio Felipe, Prudencio Roberto. MD, U. The Philippines, 1936; postgrad., Harvard U., 1946-48. Intern U. The Philippines, Los Banos, 1936; from instr. to assoc. prof. U. of The Philippines, Manila, 1948-62; from instr. to profl. lectr. sch. nursing U. of The Philippines, Laguna, 1939-62; bd. regents U. of The Philippines, Diliman, Quezon City, 1962-82; from resident to sr. resident in surgery The Philippine Gen. Hosp., 1936-48; faculty mem. St. Lukes' Hosp. Sch. Nursing, 1962-82; v.p. Manila Med. Svcs., Inc., Ermita, Manila, 1956—; med. dir. Manila Drs. Hosp., Ermita, Manila, 1979—; also bd. dirs. Manila Med. Svcs., Inc., Ermita, Manila; pvt. practice surgery Ermita, Manila, 1979—; chmn., bd. dirs., pres. Orthopedic Equipment, Inc.; chmn. bd. dirs., v.p. Blue Ribbon Philippines, Inc.; cons. Makati (The Philippines) Med. Ctr.; bd. dirs. Children's Internat. Summer Village; del. Third World Conf. on Med. Edn., Miami; del. Travelling Seminar on the Preparation Tchrs. for Med. Schs., USSR, Armenia, 1966; chmn. com. on med. svcs. Manila Drs. Hosp., chmn. com. on credentials, chmn. com. on graduate tng., chmn. dept. surgery. Co-author: (with others) U.S. Observation Tour Report by Leaders of Philippines Medical Education, 1962; contbr. articles to profl. jours. Co-organizer, coord. Outreach Programs, The Philippines, 1979—; past chmn., pres. bd. dirs. Eusebio S. Garcia and U. The Philippines Medicine '36 Found. Recipient Cert. Appreciation U. The Philippines Med. Alumni Soc., Inc., 1955, Diploma Honor The Philippine Med. Assn., 1955, Meritorious Svc. award U. The Philippines Coll. Medicine, Quezon City, 1966. Fellow ACS (gov. The Philippines 1971-76, founder, past pres.); mem. Internat. Soc. for Traumatic and Orthopedic Surgery (nat. del. 1975-89), Internat. Soc. for Traumatic and Orthopedic Surgery (nat. sec. 1975, nat. del. 1976-89), The Philippine Coll. Surgeons (past pres., past chmn. adv. com. to bd. regents, Cert. Appreciation 1961), The Philippine Orthopedic Assn. (co-founder, past pres.), Manila Med. Soc. (Cert. Appreciation 1955, 65), Philippine Med. Assn., U. The Philippines Alumni Assn. Avocations: golf, music, photography, gardening, travel. Home: 73 Jose Wright St, Metro Manila San Juan, The Philippines Office: Manila Doctors Hosp, 667 United Nations Ave, Manila 1000, The Philippines

TANGLER, JAMES LOUIS, aeronautical engineer; b. Cleve., May 25, 1940; s. Louis Thomas and Mary Elizabeth (Borzy) T.; m. Melissa Ann Winger, Sept. 6, 1969; children: Kyle, Hilary. BSME, U. Dayton, 1963; MS in Aero. Engring., Pa. State U., 1966. Project engr. Bell Helicopter Co., Hurst, Tex., 1969-78; project mgr. Rockwell Internat. Wind Program, Boulder, Colo., 1978-84; sr. scientist Nat. Renewable Energy Lab., Golden, Colo., 1984—. Author papers, articles in field. Capt. U.S. Army, 1966-69, Vietnam. Recipient 100 award R&D Mag., 1991, Outstanding Achievement award Nat. Renewable Energy Lab., 1990, Tech. Achievement award Am. Wind Energy Assn., 1990, award for excellence in tech. transfer Fed. Lab. Consortium, 1990, Dirs. award Solar Energy Rsch. Inst., 1988. Achievements include patent for twin vortex rotor blade tips, patent pending for thick wind turbine airfoils; finding for advanced wind turbine blades. Home: 1740 Deer Valley Rd Boulder CO 80303 Office: Nat Renewable Energy Lab 1617 Cole Blvd Golden CO 80401

TANI, SHOHEI, pharmacy educator; b. Komi, Ngano, Japan, Mar. 5, 1942; s. Seitetsu and Kazuko (Tani) T.; m. Masako Onzawa, July 23, 1980; 1 child, Shoko. BS, Sizuoka (Japan) Coll. Pharmacy, 1965; MS, Chiba (Japan) U., 1967; PHD, Kyoto (Japan) U., 1972. Pharm. diplomate. Rsch. fellow U. Hawaii, Honolulu, 1972-73; rsch. asst. Kobe (Japan) Gakuin U., 1974-75, asst. prof., 1975-76, assoc. prof., 1976-85, prof., 1985—; vis. fellow U. N.C., Chapel Hill, 1984-86. Contbr. articles to profl. jours. Mem. Pharm. Soc. Japan, Mass Spectroscopy Soc. Japan, Am. Chem. Soc., Am. Soc. Phamacognosy. Home: 6-23-17 Hontamon, Tarumi Ku, Kobe 655, Japan Office: Kobe Gakuin U 518 Arise, Ikawadani Nishi Ku, Kobe 651-21, Japan

TANIGAWA, KANZO, Japanese minister of science and technology. Two-term mem. Ho. Councillors, constituency Kochi (formerly one-term mem. Ho. Reps.); dir. sci. and tech. divsn. Liberal Democrat. Party; chair Com. on Agrl. Forestry and Fisheries Ho. Councillors, Local Adminstrn. Ho. Councillors; dir., state min. Gen. Sci. and Tech. Agy., 1991—. Office: Science & Technology, 2-2 Kasumigaseki, Chiyoda-ku Japan*

TANIGUCHI, TOKUSO, surgeon; b. Eleele, Kauai, Hawaii, June 26, 1915; s. Tokuichi and Sana (Omaye) T.; BA, U. Hawaii, 1941; MD, Tulane U., 1946; 1 son, Jan Tokuichi. Intern Knoxville (Tenn.) Gen. Hosp., 1946-47; resident in surgery St. Joseph Hosp., also Marquette Med. Sch., Milw., 1947-52; practice medicine, specializing in surgery, Hilo, Hawaii, 1955—; chief surgery Hilo Hosp.; teaching fellow Marquette Med. Sch., 1947-49; v.p., dir. Hawaii Hardware Co., Ltd. Capt. M.C., AUS, 1952-55. Diplomate Am. Bd. Surgery. Fellow Internat., Am. colls. surgeons; mem. Am., Hawaii med. assns., Hawaii County Med. Soc., Pan-Pacific Surg. Assn., Phi Kappa Phi. Contbr. articles in field to profl. jours. Patentee automated catheter. Home: 277 Kaiulani St Hilo HI 96720-2530

TANIUCHI, KIYOSHI, mechanical engineering educator; b. Kahoku-choo, Kami-gun, Kōchi-ken, Japan, Aug. 8, 1926; s. Takeshi and Toshi (Y-oshimoto) T.; m. Teiko Wakamatsu, Jan. 7, 1960; 1 child, Satoshi. Student, Kanto Gakuin Tech. Coll., Yokohama, Japan, 1946; BEng, Meiji U., Tokyo, 1952, DEng (hon.), 1980. Asst. Meiji U. Sch. Sci. and Tech., Kawasaki-shi, Japan, 1957-81, lectr. mech. engring., 1981-91, assoc. prof., 1991—; instr. engring. Shibaura Inst. Tech., Ohmiya-shi, Saitama-ken, Japan, 1974-90. Co-author 3 books on mech. engring.; also articles. Mem. ASTM, Soc. Exptl. Mechanics, Japanese Soc. Mech. Engrs. Avocations: photography, art appreciation. Home: 5-10-12 Wakamatsu, Sagamihara 229, Japan Office: Meiji U Sch Sci and Tech, 1-1-1 Higashimita, Tama-ku, Kawasaki 214, Japan

TANKELEVICH, ROMAN LVOVICH, computer scientist; b. Odessa, USSR, Apr. 25, 1941; came to U.S., 1990; s. Lev Faivel and Irina Rudolph (Novak) T.; m. Raye Ja. Sapiro, Jan. 28, 1961; children: Alex, Kate. BS in Physics, Moscow Phys. Engring. Inst., 1963, MS in Physics, 1965; PhD in Computer Sci., Moscow State U., 1968. Rscher., head rsch. lab. Moscow Rsch. Inst. for Computing Tech., 1965-74; prof. Moscow Tech. Inst., 1974-81; dir. rsch. lab. Moscow Comml. Inst., 1981-89; dir. R & D Micrel, Inc., Denver, 1990; v.p. R & D System 6, Inc., Denver, 1989—. Author: Simulation of Physical Fields, 1968, Analog Systems for Simulation, 1974, Microprocessor Systems, 1979; tech. writer, translator All-Union Inst. for Sci. and Tech. Info., Moscow, 1963-85; patentee in field. Mem. Scientists for Polit. Freedoms Com., Moscow, 1981-86. Recipient Gold Medal Int. Exhbn., Moscow, 1987. Mem. N.Y. Acad. Scis. Jewish. Avocations: hiking, theater, music. Office: System 6 Inc Ste 130 602 Park Point Dr Golden CO 80401

TANNER, CRAIG RICHARD, fire and explosion engineer; b. Buffalo, July 24, 1949; s. Chester William and Gloria Ann (Steffan) T.; m. Barbara Ann Schmitt, June 10, 1976; 1 child, Travis C.J. Student, SUNY, Buffalo, 1971-75, St. Petersburg Jr. Coll., Clearwater, Fla., 1986—, U. Md., 1991—. Plant supr./cons. Indsl. Motors, Inc., North Tonawanda, N.Y., 1968—; pres. Metal Reclaiming Plant, North Tonawanda, 1975-77; mgr., pres. Seneca Electric, Buffalo, 1976-81; pres. TRC Industries/Tanner Imports, Clearwater, 1985—, Tanner Agy., Clearwater, 1985—; fire investigator, pub. adjuster Nat. Fire Adjusters, Inc., Clearwater, 1986-90, Nat. Casualty and Fire Adjusters, Inc., Clearwater, 1990—. Mem. Elec. Apparatus Svcs. Assn., Nat. Assn. Investigative Specialists, Nat. Assn. Fire Investigators, Internat. Assn. Arson Investigators. Republican. Roman Catholic. Office: Nat Casualty/Fire Adjusters PO Box 10198 Clearwater FL 34617

TANNER, DAVID BURNHAM, physics educator; b. Norfolk, Va., Mar. 12, 1945; s. James Taylor and Nancy Burnham (Sheedy) T.; m. Marcia Haney, July 29, 1972; children: James Daniel, Michael Gregory. BA, U. Va., 1966, MS, 1967; PhD, Cornell U., 1972. Postdoctoral assoc. U. Pa., Phila., 1972-74; asst. prof. Ohio State U., Columbus, 1974-79, assoc. prof., 1979-82; prof. physics U. Fla., Gainesville, 1982—, chmn. dept. physics, 1986-89; cons. Xerox Webster (N.Y.) Rsch. Ctr., 1978—; mem. rev. panel NSF, Washington, 1989. Co-editor: Electric Transport & Optical Properties, 1978, ETOPIM2-Proceedings 2nd International Conference on Electricity, 1989; contbr. articles to profl. jours. Fellow Am. Phys. Soc. (Isakson prize com. 1988-91). Achievements include research in infrared properties of solids, especially high temperature superconductors, organic conductors, novel metals, axion search. Home: 3619 NW 38th St Gainesville FL 32605 Office: Univ Fla 215 Williamson Hall Gainesville FL 32611

TANNER, JOANNE ELIZABETH, psychologist, researcher; b. Tulsa, Apr. 27, 1944; d. Eugene Simpson T. and Ada Charlotte (Thomas) Jaquet; m. Clarence Cooper, June 14, 1965 (div. 1976); 1 child, Duncan; m. Charles Ernest, Apr. 24, 1987. MusB, Oberlin Coll., 1965; postgrad., U. Calif., Santa Cruz, 1989-91, U. St. Andrews, Scotland, 1992-93. Rsch. asst. Gorilla Found., Woodside, Calif., 1981-92; postgrad. teaching asst. in anthropology U. Calif., Santa Cruz, 1989-92; lectr. Gorilla Found., 1984-92; symphony violinst, various cities, 1965-89; pvt. violin tchr., Cupertino, Calif., 1980-92; instr. violin Met. State Coll., Denver, 1967-72. Contbr. articles to Jour. of Pragmatics, Primates, Gorilla. Achievements include research in gestural communication of zoo gorillas. Home: 841 Paradise Park Santa Cruz CA 95060-7009 Office: U Saint Andrews, Scottish Primate Rsch Group, Saint Andrews KY169JU, Scotland

TANSEL, IBRAHIM NUR, mechanical engineer, educator; b. Istanbul, Apr. 19, 1956; came to the U.S., 1980; s. Zekeriya and Fatma Tansel; m. Berrin Eroktem, Sept. 10, 1982; 1 child, Aylin. MS, Tech. U. Istanbul, 1980; PhD, U. Wis., 1986. Engr. in tng. Engr. Tech. U. Istanbul, 1978-80; teaching asst. U. Wis., Madison, 1984-86; asst. prof. Tufts U., Boston, 1986-90, Fla. Internat. U., Miami, 1990—; mem. rev. com. Internat. Jour. Machine Tools and Mfg., Birmingham, England, 1992—; reviewer Transactions of ASME, N.Y.C., 1990—. Contbr. articles to Jour. Engring. for Ind. Transactions ASME, Jour. Intelligent Mfg., Transactions of North Am. Metal Rsch. Inst., Internat. Jour. Machine Tools and Mfg. Recipient Outstanding Rsch. award Fla. Internat. U., 1993. Mem. Soc. Mfg. Engrs. (sr., Outstanding Young Mfg. Engr. 1992), North Am. Metal Rsch. Inst. (sr.), ASME. Achievements include pioneering use of neural networks for classification of manufacturing related signals, use of wavelet transformations and neural networks combined to evaluate tool condition in drilling and milling. Home: 742 Sevilla Ave Coral Gables FL 33134 Office: Fla Internat U Dept Mech En Miami FL 33199

TANT, MARTIN RAY, chemical engineer; b. Seneca, S.C., Apr. 10, 1953; s. Larry Ray and Doris Jo Anne (Alexander) T.; m. Martha Ann Hite, July 10, 1976. BS, Old Dominion U., 1975; MS, Va. Tech., 1979, PhD, 1986. Chemist Naval Mine Engring. Facility, Yorktown, Va., 1975-76; chem. engr. Naval Surface Weapons Ctr., Dahlgren, Va., 1979-82; sr. rsch. engr. Dow Chem. USA, Freeport, Tex., 1986-88; sr. rsch. chem. engr. Eastman Chem. Co., Kingsport, Tenn., 1988—. Co-editor Ionomers: Synthesis, Structure, Properties and Applications, 1993; contbr. over 40 publs. to books and profl. jours. Cunningham Dissertation fellow, 1985. Mem. AICE, Am. Chem. Soc., Soc. Plastics Engrs. (Engring., Properties and Structure div., bd. dirs. 1992—). Office: Eastman Chem Co PO Box 1972 Kingsport TN 37662

TANTALA, ALBERT MARTIN, civil engineer; b. Phila., Oct. 13, 1938; s. Peter and Olga (Lewiski) T.; m. Genevieve Mary Drejerski, Feb. 15, 1969; children: Peter John, Albert M. Jr., Michael Walter. BA, U. Pa., 1960, BCE, 1961; MCE, Villanova U., 1965; grad., U.S. Army Command and Gen. Staff Coll., 1973, Indsl. Coll. Armed Forces, 1974. Civil engr. Peter Tantala & Sons, Phila., 1960-65, Miller-McNichol Assocs., Phila., 1965; structural engr. Allstates Engring. Co., Phila., 1965-66; civil engr. Page Communications Engrs., Phila., 1966; cons. Tantala Assocs., Phila., 1966—; bd. dirs. 3d Fed. Savs., Phila., 1984—. Bd. dirs. Holy Family Coll., Phila., 1982—; pres. Bridesburg Civic Assn., Phila., 1979-80. Lt. col. U.S. Army, 1962-88. Recipient Carl T. Humphrey award Villanova U., 1991; named Man of Yr. Optimist Club, 1991. Fellow ASCE (sect. pres. 1986-87, Civil Engr. of Yr. Phila. sect. 1989); mem. Nat. Soc. Profl. Engrs., Pa. Soc. Profl. Engrs., Profl. Engrs. in Pvt. Practice, Soc. Am. Mil. Engrs. Office: 4903 Frankford Ave Philadelphia PA 19124-2693

TANZER, ANDREW ETHAN, mechanical engineer; b. N.Y.C., Aug. 2, 1958; s. Marvin Lawrence and Betsy (Chernoff) T.; m. Linda MacDonald, July 3, 1960; 1 child, Benjamin James. BS in Automotive Engring., Weber State Coll., 1981; MS in Tech. Mgmt., Worcester Poly. Inst., 1988. Researcher U. Mass., Amherst, 1981-83; engr. II Digital Equipment Corp., Shrewsbury, Mass., 1983-85, sr. engr., 1985-89; prin. engr. Digital Equipment Corp., Colorado Springs, Colo., 1989—. Mem. Pi Tau Sigma. Achievements include patent for multiple roller compact tape transport, patent pending for unique automotive carburator repair and inspection fixture. Office: Digital Equipment Corp CX01-1/P26 301 Rockrimmon Blvd Colorado Springs CO 80919

TAO, KAR-LING JAMES, physiologist, researcher; b. Canton, Guangtong, China, Aug. 20, 1941; came to U.S., 1966; s. Koon-Man and Hor-Yu (Hau) T.; m. Harriet Hoi-Yin Tao, June 7, 1969; children: Kane Hon-Yu, Nelson Nan-Yu. Diploma in biology, Chinese U., Hong Kong, 1964; MSc in Biology, Tuskegee Inst., 1968; PhD in Horticulture and Botany, U. Wis., 1971. Sci. tchr. Tsung Tsin Assn. High Sch., Hong Kong, 1965-66; rsch. fellow Carver Rsch. Found., Tuskegee, Ala., 1966-68; rsch. asst. U. Wis., Madison, 1968-71; rsch. assoc. Cornell U., Ithaca, N.Y., 1971-77; plant physiologist USDA, Beltsville, Md., 1977-83; genetic resources officer Food and Agrl. Orgn. Internat. Bd. for Plant Genetic Resources, Rome, 1984-88, rsch. officer Food and Agrl. Orgn., 1989-92; agrl. officer plant prodn. and protection divsn. Seed and Plant Genetic Resources Svc./Food and Agriculture Orgn. UN, Rome, 1992—; advisor Zhejiang Agr. U., Hanzhou, China, 1990—, Beijing Vegetable Rsch. Ctr., 1988—; rsch. prof. Botany Rsch. Inst., Beijing, 1989—, co-chmn. Symposium Seed Storage, Edinburgh, U.K., 1989; mem. organizing com. Internat. Conf. Seed Sci. and Tech., Hanzhou, 1989; escort to cotton del. Office Sci. Exch., Washington, 1980; hon. prof. Zhongshan U., 1991. Author: Seed Physiology and Standards, 1981, Gene Bank Management and Seed Vigor, 1986, Seed Vigor, 1991; contbr. over 45 articles to profl. jours. Recipient Cert. Tech. Recognition, Tech. Recognition Co., 1981. Mem. Am. Soc. Plant Physiology, Am. Biog. Inst. (Gold medal 1991), Crop Sci. Soc. Am., Chinese Acad. Sci. (Hon. Rsch. Prof. 1989), Rsch. Inst. Plant Prodn. (gold medal 1988), N.Y. Acad. Scis. Office: AGPS Food/Agriculture Orgn UN, Viale delle Terme di Caracalla, Rome 00100, Italy

TAO, MARIANO, biochemistry educator; b. Davao, Philippines, Mar. 3, 1938; came to U.S., 1963; s. Bong-Hua and Siu-Hua (Co) T.; m. Pearl Koh, June 3, 1967; children: Stephen, Kevin. BS in Chem. Engring., Cheng Kung U., Tainan, Taiwan, 1962; PhD in Biochemistry, U. Wash., 1967. Sr. fellow U. Wash., Seattle, 1967-68; guest investigator Rockefeller U., N.Y.C., 1968-70; asst. prof. U. Ill., Chgo., 1970-74, assoc. prof., 1974-78, prof. dept. biol. chemistry, 1978—, acting head, 1979-80; mem. biochemistry study sect. NIH, 1985-89; established investigator Am. Heart Assn., 1973-78. Mem. Am. Soc. Biochemistry and Molecular Biology, Am. Chem. Soc. Achievements include membrane abnormality and diseases I and II. Home: 1814 Gigi Ln Darien IL 60561-3547 Office: U Ill Chgo 1853 W Polk St Chicago IL 60612-4316

TAO, RONGJIA, physicist, educator; b. Shanghai, China, Jan. 28, 1947; came to U.S., 1979; s. Yun Tao and Xiao-Mei Zou; m. Weiying Duanmu, Dec. 22, 1976; children: Han, Jing. MA, Columbia U., 1980, PhD, 1982. Rsch. assoc. U. Wash., Seattle, 1982-84; rsch. fellow U. Cambridge, Eng., 1984; rsch. asst. prof. U. So. Calif., L.A., 1984-85; asst. prof. physics Northeastern U., Boston, 1985-89; asst. prof. physics So. Ill. U., Carbondale, 1989-91, assoc. prof. physics, 1991-92, prof. physics, 1993—; cons. UN Developing Program, N.Y. and China, 1992—; chair Internat. Conf. on Electrorheological Fluids, Carbondale, 1991, Feldkirch, Austria, 1993. Editor, author: Electrorheological Fluids, 1992; contbr. articles to profl. publs. Office of Naval Rsch. grantee, 1990, 92; recipient award Omni mag., 1987. Mem. Am. Phys. Soc. Achievements include discovery that electric-field induced solidification is the physical mechanism of electrorheological fluids, the crystalline structure of electrorheological fluids is a body-centered tetragonal lattice. Office: So Ill U Dept Physics Carbondale IL 62901

TAO, YONG-XIN, mechanical engineer, researcher; b. Shanghai, China, Mar. 5, 1954; arrived in Can., 1989; m. Wei Wang, Sept. 21, 1982; children: Yuan, Alisa. MSc, Tongji U., Shanghai, 1981; PhD, U. Mich., 1989. Lectr., then asst. prof. Tongji U., 1978-85; rsch. asst. U. Mich., Ann Arbor, 1986-89; postdoctoral fellow U. Sask., Saskatoon, Can., 1989-91, profl. rsch. assoc. dept. mech. engring., 1991—; presenter at nat. and internat. profl. confs. Contbr. to profl. publs. Recipient Nat. Sci. award Ministry Edn., China, 1985; grantee Can. Mortgage and Housing Corp., 1991—. Mem. ASME, ASHRAE, AIAA, ASTM. Achievements include Chinese patent for a water calorimeter for measurement of specific heat of liquids and solids. Office: Dept Mech Engring, Univ Sask, Saskatoon, SK Canada S7N 0W0

TAPLEY, BYRON DEAN, aerospace engineer, educator; b. Charleston, Miss., Jan. 16, 1933; s. Ebbie Byron and Myrtle (Myers) T.; m. Sophia Philen, Aug. 28, 1959; children: Mark Byron, Craig Philen. B.S., U. Tex., 1956, M.S., 1958, Ph.D., 1960. Registered profl. engr., Tex. Engr. Structural Mechanics Research Lab. U. Tex., Austin, 1954-58, instr. mech. engring., 1958, prof. aerospace engring. and engring. mechanics, 1960—, chmn. dept. aerospace engring. and engring. mechanics, 1966-77, Woolrich prof. engring., 1974-80, dir. Ctr. Space Research, 1983—, Clare Cockrell Williams chair in aerospace engring., 1984—; mem. adv. com. on guidance control and nav. NASA, 1966-67, com. on space rsch., panel I, 1974-76, chmn. region IV, engring. coun. on profl. devel., 1974-76; chmn. geodesy com. NRC, 1981-84, mem. aeros. and space engring. bd., 1984-86, mem. space sci. bd., chmn. com. on earth studies, 1988-91; bd. dirs. Tex. Space Grant Consortium. Editor: Celestial Mech. Jour, 1976-79; assoc. editor: Jour. Guidance and Control, 1978-79; assoc. editor: Geophys. Revs, 1979-81. Dir. Tex. Space Grant Consortium, 1990—. Recipient NASA Exceptional Sci. Achievement medal, 1983. Fellow AAAS (pres. engring. sect.), AIAA (chmn. com. on astrodynamics 1976-78, Mech. and Control of Flight award 1989), Am. Geophys. Union (pres. geodesy sect. 1984-86); mem. ASME, IEEE, NAE, Am. Acad. Mechanics, Am. Astronautics Soc. (pres. div. dynamic astronomy 1988-89), Soc. Engring. Sci., Am. Astron. Soc., Internat. Astron. Union, Sigma Xi, Pi Tau Sigma, Sigma Gamma Tau, Phi Kappa Phi, Tau Beta Pi. Home: 3100 Perry Ln Austin TX 78731-5327

TAPPAN, CLAY MCCONNELL, environmental engineer, consultant; b. Endicott, N.Y., Oct. 13, 1955; s. Richard Alworth and Florence Allene (Wooster) T.; m. Soraya Elizabeth Mora, Dec. 29, 1983; children: Safia, Christopher. BS in Environ. Engring., Fla. Inst. Tech., 1985, MS in Engring. Mgmt., 1992. Registered profl. engr., Fla. Design engr. Smith and Gillespie Engring., Inc., Jacksonville, Fla., 1985-87; project mgr. Camp Dresser and McKee, Inc., Sarasota, Fla., 1987—. With USN, 1975-81. Mem. NSPE, Fla. engring. Soc. (chpt. treas. 1992-93). Republican. Roman Catholic.

TAPPEINER, GERHARD, dermatologist, educator; b. Vienna, Austria, Mar. 23, 1947; s. Josef and Edith (Schlossnickel) T.; m. Irene Worda, Aug. 16, 1980; children: Stephanie, Sophie. MD, Vienna Med. Sch., 1971. Diplomate Am. Acad. Dermatology. Resident in pathology U. Vienna, 1971-74, assoc. prof. dermatology Med. Sch., 1982—, cons. dept. dermatology I, Med. Sch., 1981—; resident in dermatology and immunology U. Innsbruck, Austria, 1976-80; rsch. fellow Mayo Clinic and Mayo Found., Rochester, Minn., 1974-76. Contbr. chpts. to books; contbr. articles to sci. publs. Recipient several nat. sci. awards. Mem. Nat. Dermatologic and Immunlogic Soc., Internat. Dermatologic and Immunologic Soc., Am. Acad. Dermatology, Arbeitsgemeinschaft Dermatologische Forschung, Deutsche Dermatologische Gesselschaft, Deutsche Gesellschaft für Immunologie, European Soc. Dermatol. Rsch., Österreichische Gesellschaft für Allergologie und Immunologie, Österreichische Gesellschaft für Dermatologie und Venerologie, Societé Francaise de Dermatologie et de Syphiligraphie (corr.). Roman Catholic. Home: Josefstädter Strasse 29, A-1080 Vienna Austria Office: U Vienna Dept Dermatology, Währinger Gürtel 18-20, A-1090 Vienna Austria

TAPPERT, FREDERICK DRACH, physicist; b. Phila., Apr. 21, 1940; s. Theodor Gerhart and Helen Louise (Carson) T.; m. Sally Faith Feldman, Mar. 3, 1979; children: Andrew Scott, Peter Alan. BS in Engring. Sci., Pa. State U., 1962; PhD in Physics, Princeton U., 1967. Mem. tech. staff Bell Telephone Labs., Whippany, N.J., 1967-74; sr. rsch. scientist Courant Inst. Math. Sci. NYU, N.Y.C., 1974-78; prof. applied marine physics U. Miami, Fla., 1978—; cons. in field, 1974—. Contbr. over 50 articles to profl. jours. Grantee Office Naval Rsch., 1974—, NSF, 1980—. Fellow Acoustical Soc. Am.; mem. Am. Phys. Soc., Am. Geophys. Union, Soc. Indsl. Applied Math. Achievements include development of parabolic equation model and split-step Fourier algorithm for electromagnetic and ocean acoustic propagation. Home: 907 Jeronimo Dr Coral Gables FL 33146 Office: U Miami RSMAS 4600 Rickenbacker Causeway Miami FL 33149

TARA (TARA SINGH), research chemist; b. Kotdata, Punjab, India, June 11, 1921; came to U.S., 1966; naturalized, 1972; s. Nand and Isar (Kaur) Singh; m. Rani Surinder, Dec. 29, 1954; children: Nina, Roopinder, Sylvia, Sonya. BS with honors, Punjab U., 1944, MS with 1st class honors, 1946; AM, Harvard U., 1949, PhD, 1950. Post doctorate fellow with Prof. R.B. Woodward Harvard U., 1950-51; Post doctorate fellow NRC, Can., 1953-54; prof. chemistry govt. colls., Punjab, India, 1954-58; prin. govt. colls., India, 1958-64; rsch. and devel. chemist PEBOC Ltd., Northolt, Eng., 1964-65, Unilever Rsch. Lab., Isleworth, Eng., 1965-66, Aldrich Chem.Co., Milw., 1966-76, Polyscis., Inc., Warrington, Pa., 1976-88, Calbiochem, La Jolla, Calif., 1989—. Author many books; contbr. numerous articles to profl. jours. Mem. Am. Chem. Soc. Avocations: studies in comparative religions, meditation. Home: 4202 Appleton St San Diego CA 92117-1901 Office: Calbiochem Corp 10933 N Torrey Pines Rd La Jolla CA 92037-1080

TARAFDAR, SHANKAR PROSAD, astrophysicist, educator; b. Ranaghat, India, Nov. 10, 1938; s. Panchu Gopal and Narayani Tarafdar; m. Anima Patra, May 17, 1966; children: Shantanu, Ashis, Soumen. MSc, Calcutta (India) U., 1961; PhD, Bombay U., 1973. Physics trainee Atomic Energy Establishment, Bombay, 1962-63; rsch. assoc. Tata Inst. Fundamental Rsch., Bombay, 1963-70, fellow, 1970-78, reader, 1978-85, assoc. prof. astrophysics, 1985-93, prof. astrophysics, 1993—; rsch. asst. dept. applied math. and astronomy U. Cardiff, U.K., 1974-77; sr. rsch. assoc. Calif. Inst. Tech., 1981-83. Editor: Astrochemistry, 1987; contbr. articles to profl. publs. Fellow Royal Astron. Soc.; mem. Internat. Astron. Union, Am. Astron. Soc., Astron. Soc. India. Hindu. Home: Tifr Housing Colony, 1003 Bhaskara, Bombay 400 005, India Office: Tata Inst Fundamental Rsch, Homi Bhabha Rd, Bombay 400 005, India

TARANIK, JAMES VLADIMIR, geologist, educator; b. Los Angeles, Apr. 23, 1940; s. Vladimir James and Jeanette Downing (Smith) T.; m. Colleen Sue Glessner, Dec. 4, 1971; children: Debra Lynn, Danny Lee. B.Sc. in Geology, Stanford U., 1964; Ph.D., Colo. Sch. Mines, 1974. Chief remote sensing Iowa Geol. Survey, Iowa City, 1971-74; prin. remote sensing scientist Earth Resources Observation Systems Data Ctr., U.S. Geol. Survey, Sioux Falls, S.D., 1975-79; chief non-renewable resources br., resource observation div. Office of Space and Terrestrial Applications, NASA Hdqrs., Washington, 1979-82; dean mines Mackay Sch. Mines U. Nev., Reno, 1982-87, prof. of geology and geophysics, 1982—; pres. Desert Research Inst., Univ. and Community Coll. System Nev., 1987—; adj. prof. geology U. Iowa, 1971-79; vis. prof. civil engring. Iowa State U., 1972-74; adj. prof. earth sci. U. S.D., 1976-79; program scientist for space shuttle large format camera expt. for heat capacity mapping mission, liaison Geol. Scis. Bd., Nat. Acad. Scis., 1981-82; dir. NOAA Coop. Inst. Aerospace Sci. & Terrestrial Applications, 1986—; program dir. NASA Space Grant consortium Univ. and Community Coll. System Nev., Reno, 1991—; team mem. Shuttle Imaging Radar-B Sci. Team NASA, 1983-88, mem. space applications adv. com. 1986-88; chmn. remote sensing subcom. SAAC, 1986-88; chmn. working group on civil space commercialization U.S. Dept. Commerce, 1982-84, mem. civil operational remote sensing satellite com., 1983-84; bd. dirs. Newmont Gold Co.; mem. NASA Space Sci. and Applications adv. com., 1988—, Nat. Def. Exec. Res., 1986—, AF Studies Bd., Com. on Strategic Relocatable Targets, 1989-91; developer remote sensing program and remote sensing lab. for State of Iowa, ednl. program in remote sensing for Iowa univs. and U. Nev., Reno; Office Space and Terrestrial Applications program scientist for 2d space shuttle flight; terrestrial geol. applications program NASA, 1986—. Contbr. to profl. jours. Served with C.E. U.S. Army, 1965-67; mil. intelligence officer Res. Decorated Bronze Star medal; recipient Spl. Achievement award U.S. Geol. Survey, 1978, Exceptional Sci. Achievement medal NASA, 1982, NASA Group Achievement award Shuttle imaging radar, 1990, NASA Johnson Space Ctr. Group Achievement award for large format camera, 1985; NASA prin. investigator, 1973, 83-88; NDEA fellow, 1968-71. Fellow AAAS, Explorers Club, Geol. Soc. Am.; mem. Am. Assn. Petroleum Geologists (charter, mem. energy minerals divsn., divsn. environ. geoscis.), Soc. Mining Engrs., Am. Inst. Profl. Geologists (certified, pres. Nev. sect. 1985-87), AIAA (sr. mem.), Am. Astronautical Soc. (sr. mem.), Soc. Exploration Geophysicists, Geosci. and Remote Sensing Soc. of IEEE (bd. dirs., geosat com. 1983—), Am. Soc. Photogrammetry (certified), Am. Soc. Photogrammetry and Remote Sensing (dep. dir. for remote sensing cert.), Internat. Soc. Photogrammetry and Remote Sensing (pres. working group II/4 1976-80, working group VII-5 non-renewable resources 1980-88, working group VII-4 geology and mineral resources applications 1992—), Nev. Quality and Productivity Inst. (chmn. bd.), Nev. Industry Sci., Engring., and Tech. Task Force, Grand Canyon Visibility Commn. (pub. adv. com. 1992—), Sigma Xi, Phi Kappa Phi (life). Home: 3075 Susileen Dr Reno NV 89509-3855 also: 2108 calle De Espana Las Vegas NV 89120 Office: Univ and Community Coll Sys Nev Desert Rsch Inst Pres' Office Reno NV 89512

TARANTOLA, ALBERT, geophysicist, educator; b. Barcelona, Catalunya, Spain, June 15, 1949; s. Joan and Josefina (Pitarque) T.; m. Maria Zamora, Apr. 4, 1971. Docteur Specialite, U. Paris, 1976, Docteur d'Etat, 1981. Asst. U. Paris, 1979-81, maitre-asst., 1981-83; prof. geophysics Inst. Physique Globe, Paris, 1983—; dir. Troisieme Cycle Geophys. Interne, Paris, 1991—; chmn. com. on math. geophysics Internat. Union Geodesy and Geophysics, 1987. Author: Inverse Problem Theory, 1979; contbr. articles to profl. publs.; editor Jour. Geophys. Rsch., 1989—. Recipient A. D'Abbadie prize French Acad. Scis., 1988, Conrad Schlumberger award European Assn. Exploration Geophysicists, 1989. Office: IPG Paris, 4 Place Jussieu, F-75252 Paris Cedex 05, France

TARASIDIS, JAMIE BURNETTE, aerospace engineer; b. Griffin, Ga., Aug. 13, 1966; d. John W. and Shirley Ann (Hardman) Kelley; m. Gregory Tarasidis, June 30, 1990. B in Aerospace Engr., Ga. Tech., 1987, MS, 1988. Sr. aerospace engr. Planning Rsch. Corp., Hampton, Va., 1988-89; rsch. engr. Ga. Tech. Rsch. Inst., Atlanta, 1989-90, Ga. Tech. Sch. Mech. Engring., Atlanta, 1990-91; adj. instr. Parks Coll. St. Louis U., Cahokia, Ill., 1991—. Vol. ARC, St. Louis, 1992. Recipient Pres.' fellow Ga. Tech., Atlanta, 1987, Zonta Amelia Earhart fellowship, Chgo., 1987; named Outstanding Young Women Am., 1988. Mem. AIAA. Republican. Greek Orthodox. Achievements include developed computational method for calculating air data parameters for hypersonic vehicles. Office: Parks Coll St Louis Univ Ste 101 425 N New Ballas Rd Saint Louis MO 63141

TARBELL, DEAN STANLEY, chemistry educator; b. Hancock, N.H., Oct. 19, 1913; s. Sanford and Ethel (Millikan) T.; m. Ann Hoar Tracy, Aug. 15, 1942; children: William Sanford, Linda Tracy, Theodore Dean. A.B., Harvard U., 1934, M.A., 1935, Ph.D., 1937. Postdoctoral fellow U. Ill., 1937; mem. faculty U. Rochester, 1938—, successively instr., asst. prof., asso. prof., 1938-48, prof. chemistry, 1948-62, Charles Frederick Houghton prof. chemistry, 1960—, chmn. dept., 1964—; Disting. prof. chemistry Vanderbilt U., 1967—, Branscom disting. prof., 1975-76, disting. prof. emeritus, 1981—; Guggenheim fellow and vis. lectr. chemistry Stanford U., 1961-62; Fuson lectr., 1972; cons. USPHS, Army Q.M.C.; mem. various sci. adv. bds. to govt. agencies. Author: (with Ann T. Tarbell) Roger Adams, Scientist and Statesman, 1981, Essays on the History of Organic Chemistry in the United States, 1875-1955, 1986 , also papers on history of chemistry. Recipient Herty award Ga. Sect. Am. Chem. Soc., 1973 Dexter award Div. History of Chemistry Am. Chem. Soc., 1989; Guggenheim fellow, 1946-47. Mem. Nat. Acad. Sci., Am. Chem. Soc. (chmn. div. history of chemistry 1980-81), Chem. Soc. London, Am. Acad. Arts and Scis., History of Sci. Soc. Home: 6033 Sherwood Dr Nashville TN 37215-5734

TARDONA, DANIEL RICHARD, marine naturalist, park ranger, educator; b. Bklyn., Nov. 9, 1953; s. Felix Carmine and Patricia Ann (Tynan) T.; m. Jayne Hardwick, Apr. 28, 1990. BA, Monmouth Coll., 1976; MA, Cen. Mich. U., 1981. Cert. sch. psychologist, N.J. Pediat. psychologist Montclair (N.J.) State Coll., 1982-83; radiation/hazardous substances safety asst. Cytogen Corp., Princeton, N.J., 1984-85; park naturalist Frozen Head State Natural Area, Wartburg, Tenn., 1986; park ranger Cape Hatteras Nat. Seashore, Manteo, N.C., 1987; park ranger Great Smoky Mtns. Nat. Pk., Gatlinburg, Tenn., 1987-90, asst. dist. supr., 1990-92; site supr. Timucuan Ecological and Hist. Preserve, Jacksonville, Fla., 1992—. Mem. editorial review bd. Infant Mental Health Jour., 1982-85; guest reviewer Edn. and Treatment of Children, 1980, 81; contbr. articles to profl. jours. Mem. Am. Littoral Soc., 1992—, Jacksonville Maritime Mus. Soc., Jacksonville, 1992—, Fla. Audubon Soc., 1992—. Recipient Trustees scholar Rider Coll., 1983-84. Mem. Am. Shell Fisheries Assn., Soc. for Marine Mammalogy, Animal Behavior Soc., Wildlife Soc., Nat. Park Rangers Assn., Sigma Gamma Epsilon, Phi Delta Kappa, Psi Chi, Nat. Assn. Underwater Instrs. (openwater I scuaba diver). Achievements include research in marine animal behavior and evolution and burrowing rates of bivalves, predatory behavior, phylogenetic size increase in marine mammals. Home and Office: Timucuan Ecological and Hist Preserve 12713 Fort Caroline Rd Jacksonville FL 32225

TARDY, DANIEL LOUIS, mechanical engineer, education consultant; b. Rouen, France, Feb. 17, 1934; s. Lucien Aime and Marie Claudine (Brichaux) T.; m. Françoise Jeanne Lehuerou Kerisel, Mar. 7, 1958; children: Marie Noel, Olivier, Nicole, Heléne, Cecile. Engr., Ecole Poly., Paris, 1955, Genie Maritime, Paris, 1958; PhD, U. Nantes, France, 1964. Prof. Nat. Sch. Higher Mechanics, Nantes, 1962-73; mgr. Entreprise Travaux Publics de l'Ouest, Nantes, 1971—. Capt. French Navy, 1958-64. Home: 16 rue Alfred de Musset, 44100 Nantes France Office: ETPO, 3 place du Sanitat, 44100 Nantes France

TARHINI, KASSIM MOHAMAD, civil engineering educator; b. Beirut, Lebanon, Aug. 14, 1959; came to U.S., 1979; s. Mohamad Ali and Dia (Omais) T.; m. Kathleen D. McCloud, Aug. 28, 1984. BSCE, U. Toledo, 1981, MSCE, 1982, PhD in Civil Engring., 1988. Registered profl. engr., Ohio. Constrn. engr. Almabani Gen. Contractors, Al-Khobar, Saudi Arabia, 1983; vis. instr., part-time instr. U. Toledo, 1984-88; mgr. rsch. Nat. Concrete Masonry Assn., Herndon, Va., 1989-90; asst. prof. civil engring. Valparaiso (Ind.) U., 1990—. Author manual: Strength of Materials Laboratory, 1985—; contbr. articles to profl. jours. Mem. ASCE (v.p. N.W. br. 1992-93, Daniel W. Mead award 1988), Am. Soc. Engring. Edn., Soc. Exptl. Mechanics, Xigma Xi, Tau Beta Pi, Phi Kappa Phi, Pi Mu Epsilon. Achievements include development of wheel load distribution factors formula for highway bridges, baseline flexural bond strength of PCL mortar, hy-span precast box culverts. Home: 2803 Village Ln # 2B Valparaiso IN 46383 Office: Valparaiso U Dept Civil Engring Valparaiso IN 46383

TARIQ, ATHAR MOHAMMAD, soil microbiologist; b. Khanewal, Punjab, Pakistan, Nov. 1, 1951; came to U.S., 1986; s. Chaudhary Khushi Mohammad and Aiysha K. Mohammad; m. Rukhshanda Athar; children: Joveria, Ahsan, Aroba, Fiza, Zain. BSc, U. Punjab, Lahore, Pakistan, 1969; MSc, U. Karachi, 1972, MPhil, 1981. Rsch. fellow U. Karachi, Pakistan, 1972-77, rsch. officer, 1977-82; sci. officer Pakistan Agrl. Rsch. Coun., Islamabad, Pakistan, 1982-84, project incharge, 1984-86; sr. sci. officer Pakistan Agrl. Residents Coun., Islamabad, Pakistan, 1990—; project leader Biol. Nitrogen Fixation Program, 1982-86. Referee Jour. Bot., Plant Sci., Soil Sci., Agriculture; contbr. articles to profl. jours. Pres. Pakistan Agrl. Scientist Forum, Islamabad, 1985-86. Travelling grantee Australian Agrl. Devel. Bureau, 1980; Internat. scholar Utah State U., 1990. Mem. Pakistan Sci. Soc., Pakistan Bot. Soc. (joint sec. 1984-87), Pakistan Soc. Nematologists, Pakistan Soc. for Advancement of Sci., Soc. for Range Mgmt., Pakistan Soil Sci. Soc., Coun. on Soil Testing and Plant Analysis, Nitrogen Fixing Tree Assn. Muslim. Avocations: sight-seeing, hiking, hunting, internat. travelling. Home: 2 Civil Lines, Khanewal Punjab, Pakistan Office: Utah State U Dept Range Sci Logan UT 84321 also: Pakistan Agrl Rsch Coun, NARC Lab PO NARC, Islamabad 45500, Pakistan

TARNECKI, REMIGIUSZ LESZEK, neurophysiology educator, laboratory director; b. Vloclawek, Poland, Oct. 8, 1933; s. Antoni and Maria (Chojnacka) T.; m. Danuta Stepień, Aug. 22, 1957. BS, Salesian Gymnasium, Aleksandrów, Poland, 1950; MS, U. Łódź (Poland), 1956; PhD, Nencki Inst., Warsaw, 1963. Instr., neurophysiology U. Łódź, 1956; head lab. Nancki Inst., Warsaw, 1971—, assoc. prof., 1972-83, prof., head lab., 1983—; adj. educator Nencki Inst., 1964, guest prof. U. Pa., Phila., 1978. Contbr. articles to profl. jours.; patentee in field. Rsch. fellow Nencki Inst., 1957-58, Postdoctoral fellow, 1971; recipient Polish Acad. Sci. awards 1976, 83, 90. 2em. Biomed. Engring. Com., Polish Acad., Scientific Com. Nencki Inst., Soc. Neurosci., Polish Neurosci. Soc., Rodin Remediation Acad. Office: Nencki Inst, Pasteura 3, 02-093 Warsaw Poland

TARNOWSKI, KENNETH J., psychologist; b. Springfield, Mass., Aug. 16, 1954. BS in Psychology, U. Mass., 1979; PhD in Clin. Psychology, U. S.C., 1984. Lic. psychologist, Ohio. Asst. prof. pediatrics Ohio State U. Coll. Medicine & Columbus Children's Hosp., 1984-88; assoc. prof. pediatrics, psychiatry and psychology Case Western Res. U., Sch. Medicine, Cleve., 1988—, head div. child and adolescent devel., 1988—. Contbr. articles to profl. jours. Fellow APA, Am. Psychol. Soc.; mem. Assn. for Advancement Behavior Therapy, Sigma Xi. Achievements include rsch. in areas of child psychopathology, behavior therapy and behavioral pediatrics. Office: Case Western Res Univ Dept Pediatrics 2500 Metro Health Dr Cleveland OH 44109

TARPLEY, WILLIAM BEVERLY, JR., physics and chemistry consultant; b. Richmond, Va., Oct. 12, 1917; s. William B. and Sallie M. (Gatewood) T.; m. Nancy Tarpley, Aug. 10, 1938 (dec. Mar. 1980); 1 child, William B. III; m. Phyllis Malmquist, May 7, 1988. PhD, Columbia U., 1951. R & D chemist Schering Corp., Bloomfield, N.J., 1937-52; sr. phys. scientist U.S. Army Biol. Lab., Frederick, Md., 1952-55; v.p. R & D, Aeroprojects, Inc., West Chester, Pa., 1955-69; dir. materials dept. R & D Labs., Franklin Inst., Phila., 1969-71; v.p. R & D, Fluid Energy Equipment & Processing, Hatfield, N.J., 1971, Organic Recycling subs. UOP, West Chester, 1971-77, Energy & Minerals Rsch., Exton, Pa., 1977-85; ret., 1985; cons. in surface physics and chemistry, Downingtown, Pa., 1985—. Avocations: reading, photography, travel. Address: 29 Gunning Ln Woodmont North Downingtown PA 19335

TARR, CHARLES EDWIN, physicist, educator; b. Johnstown, Pa., Jan. 14, 1940; s. Charles Larned and Mary Katherine (Wright) T.; m. Bex Suzanne Harrell, Sept. 4, 1964 (div. Feb. 1977); m. Gudrun Kiefer, Nov. 18, 1977. B.S. in Physics (Morehead scholar 1957-61), U. N.C., Chapel Hill, 1961, Ph.D., 1966. Research assoc. U. N.C., Chapel Hill, 1966, U. Pitts., 1966-68; mem. faculty U. Maine, Orono, 1968—; assoc. prof. physics U. Maine, 1973-78, prof., 1978—, chmn. dept., 1977-79, assoc. dean Coll. Arts and Scis., 1979-81, acting dean Grad. Sch., 1981-87, acting v.p. research, 1984-87, dean Grad. Sch., 1987—; gast docent U. Groningen, Netherlands, 1975-76; cons. in field. Contbr. articles to profl. jours. NASA grantee, 1970-72; NSF grantee, 1972—. Mem. IEEE, Am. Phys. Soc., Assn. Computing Machinery, Northeastern Assn. Grad. Schs. (mem. at large 1988—, pres. elect 1990-91, pres. 1991—), Sigma Xi. Quaker. Home: 519 College Ave Orono ME 04473-1211 Office: Univ Maine Grad Sch Orono ME 04469

TARRO, GIULIO, virologist; b. Messina, Italy, July 9, 1938; s. Emanuele and Emanuela (Iannello) T. MD, U. Naples, 1962, postgrad. in nervous diseases, 1968, PhD in Virology, 1971; postgrad. in med. and biol. scis., Roman Acad., 1979; hon. degree, U. Pro Deo, Albany, N.Y., 1989, St. Theodora Acad., N.Y., 1991. Asst. in med. pathology Naples U., Italy, 1964-66; rsch. assoc. div. virology and cancer rsch. Children's Hosp., Cin., 1965-68; asst. prof. rsch. pediatrics U. Cin. Coll. Medicine, 1968-69; rsch. fellow Nat. Rsch. Coun., Naples, 1966-74, rsch. chief, 1974; prof. oncologic virology U. Naples Coll. Medicine, 1971-85; prof. microbiology and immunology Sch. Specialization, U. Naples, 1972—; div. chief virology D. Cotugno Hosp. for Infectious Diseases, Naples, 1973—; sr. scientist Nat. Cancer Inst. Frederick (Md.) Ctr., 1973; project dir. Nat. Cancer Inst., Bethesda, Md., 1971-75; edn. minister rep. Zool. Sta., Naples, 1975-79; cons. Italian Pharmacotherapic Inst., Rome, 1980—; pres. De Beaumont Bonelli Found. for Cancer Rsch., Naples, 1978—. Author: Virologia Oncologica, 1979 (award 1985), Patologia dell'AIDS, 1991; patentee in field. Pres. Sci. Cultural Com., Torre Annunziata, Italy, 1984, Tumor Prevention Assn., Rome, 1984; mem. acad. senate Constantinian U., Providence, 1990. Maj. Italian Navy, 1982-84. Decorated Comdr. Nat. Order of Merit, 1991; recipient Internat. Lenghi award Lincei Acad., 1969, Gold Microscope award Italian Health Minister, 1973, Knights of Humanity award Internat. Register of Chivalry, Malta, 1976, Gold medal of Culture, Pres. of Italian Republic, 1975, Culture award 1985. Fellow AAAS; mem. Am. Soc. Microbiology, N.Y. Acad. Scis., Internat. Assn. for Leukemias, Internat. League Doctors for Abolition of Vivasection (pres. 1992—), Italian Soc. Immuno-Oncology (v.p. 1975—, pres. 1990—), Lions (pres. Pompei chpt. 1987-89, vice gov. dist. 108y 1991-92, pres. to fight cancer 1992—). Roman Catholic. Achievements include discovery of RSV virus in infant deaths in Naples. Home: 286 Posillipo, 80123 Naples Italy Office: D Cotugno Hosp USL 41, 54 Quagliariello, 80131 Naples Italy

TARTAGLIA, JAMES JOSEPH, molecular virologist; b. Bethpage, N.Y., June 17, 1959; s. James Vincent and Rosina (Santilli) T.; m. Kathleen Elizabeth Reid, Mar. 19, 1988; children: Kristen Michelle DiBiase, Anne Marie, Michael James. BS, Bucknell U., 1981; PhD, Albany Med. Coll., 1986. Postdoctoral fellow Roche Inst. Molecular Biology, Nutley, N.J., 1985-87; scientist II N.Y. State Dept. Health, Albany, 1987-90; sr. scientist Virogenetics Corp., Troy, N.Y., 1990-91, asst. dir. rsch., 1991—; adj. asst. prof. Albany Med. Coll., 1989—, SUNY Sch. Pub. Health, Albany, 1991—. Co-author: Immunochemistry of Viruses II, 1990; contbr. articles to Jour. Virology, Proceedings of Nat. Acad. Sci., Lancet. Bucknell U. scholar, 1977-81. Roman Catholic. Achievements include development of poxvirus-based recombinant viruses as vaccine candidates for human and veterinary application; research in optimizing safety and efficacy of poxvirus vectors, developing vaccine candidates against rabies virus, feline leukemia virus, equine influenza virus, psuedorabies virus, human immunodeficiency virus, etc. Office: Virogenetics Corp 465 Jordan Rd Troy NY 12180

TARTAKOVSKIY, VLADIMIR ALEXANDROVICH, chemist, researcher; b. Moscow, Aug. 10, 1932; s. Aleksandr Arnoldovich and Maria Dmitrievna T.; m. Natalyia Y. Melanchenko, July 5, 1958; 1 child, Tatiana; m. Inna K. Kozlova, Aug. 29, 1980. Diploma in Chemistry, Moscow, 1955; ChD, Inst. Org. Chemistry Acad. Scis. USSR, Moscow, 1958. Jr. researcher Inst. Org. Chemistry, 1955; sr. researcher USSR Acad. Scis., 1962, head of lab., 1971, prof., 1974, dir., 1988; mem. sci. coun. sec. academic and indsl. chemical insts., Moscow, St. Petersburg, Chernogolovka, 1970-93. Mem. editorial bd. Izvestia RAN Zhournal, Obstsheji Khimii, Zhournal Prikladnoi Khimii, 1975-93; author over 200 scientific articles. Recipient Lenin prize USSR Coun. Ministers, 1976; named Corresponding Mem. of USSR Acad. of Scis., 1987, Academian, 1992. Mem. Moscow Club Scientists. Achievements include 100 inventions and 1 discovery in the field of large energy materials. Office: Zelinskiy Inst of Organic Chem, Leninskiy Prospekt 47, 117912 Moscow Russia

TARTE, RODRIGO, agriculture educator, researcher, consultant; b. Republic Panama, Dec. 31, 1936; s. Luis Antonio and Mercedes (Ponce) T.; m. Maria De Lourdes Rubio, Nov. 22, 1964 (div. 1974); children: Rodrigo, Maria De Lourdes, Fernando, Carlos Antonio; m. Ana Maria Cufre, Dec. 5, 1974; children: Andres, Martin. BS, U. P.R., Mayagüez, 1958; MS, U. Calif., Riverside, 1964; PhD, Cornell U., 1974. Horticulturist Inter-Am. Svc. for Cooperation in Agr., Republic Panama, 1958-62; prof. horticulture, nematology U. Panama, 1964-77; dir. rsch. Union Banana Exporting Countries, Republic Panama, 1977-81; dir. gen. Agrl. Rsch. Inst., Republic Panama, 1981-83, Tropical Agrl. Rsch. and Edn. Ctr., Costa Rica, 1984-92; exec. dir. Natura Found., Republic Panama, 1992—; regional coord. Internat. Meloidogyne Project, C.Am., 1976-79; expert nematology FAO, Chile, 1980-81; expert integrated pest mgmt. FAO/UN Environment Program, 1983-86; trustee Tropical Agrl. Internat. Ctr., Cali, Colombia, 1984-90; chmn. collaborators adv. com. Internat. Benchmark Sites Network for Agrotech. Transfer, 1985-87; leader strategic planning Tropical Agrl. Rsch. and Edn. Ctr.'s 10 Yr. Plan, 1987; sec. gen. Regional Cooperative Network for Rsch. and Edn. in Agrl. and Natural Resources, C.Am., 1990-91. Recipient Outstanding Leadership award Rutgers U., 1983. Mem. Orgn. Tropical Am. Nematologist (pres. 1975-76), Panamanian Commn. Scis. and Tech. (pres. 1993-94). Roman Catholic. Avocations: painting, drawing, photography. Home: PO Box 6288, Panama 5, Panama Office: PO Box 2190, Panama 1, Panama

TARTELL, ROBERT MORRIS, dentist; b. Bronx, N.Y., June 22, 1926; s. Julius and Ida (Saunders) T.; m. Lottie Haid Schachter, June 12, 1948; children: Ross Howard, Marc Sorrel, Adam Ethan. BA, N.Y.U., 1945, DDS, 1948. Lic. dentist, N.Y. V.p., dir. Medden, Inc., Valley Stream, N.Y., 1957-71; mng. ptnr. Profl. Investors, N.Y.C., 1957-91; pres., dir. Roberts Adv. Svc., Inc., N.Y.C., 1957-91; dir. postgrad. edn. Am. Soc. Study of Orthodontics, N.Y.C., 1971-72; mng. ptnr. RBT Co., Elmsford, N.Y., 1987—; v.p., bd. dirs. Sport World of Am. Inc., West Chester, N.Y., 1992—. Producer: Gilbert and Sullivan Light Opera Co., West Hempstead, N.Y., 1980—; co-copywrite: (Operetta) H.M.S. Pinafore in Yiddish, 1986. Pres. West Hempstead Sch. Commn. League, 1968-69; dir., founder West Hempstead Scholarship Fund, 1970-71. 1st lt. U.S. Army, 1952-54, Panama. Mem. Am. Dental Assn., 1st Dist. Dental Soc., Gen. Semantics Inst., ACLU, Mensa, Common Cause, Mason, Gallatin. Jewish. Home: 690 Hawthorne St West Hempstead NY 11552-3112 Office: 201 W Merrick Rd Valley Stream NY 11580-5595

TARULLI, FRANK JAMES, aerodynamicist; b. Bklyn., Nov. 4, 1950; s. Michael Nicholas and Anne (Piccininni) T. B Aero. Engring., Bklyn. Poly.

U., 1972, M Aero. Engring., 1974. Engr., aerodynamicist Sikorsky Aircraft Co., Stratford, Conn., 1973—. Mem. Tau Beta Pi, Sigma Gamma Tau. Roman Catholic. Home: 231 Bay 35 St Brooklyn NY 11214 Office: Sikorsky Aircraft N Main St Stratford CT 06497

TARY, JOHN JOSEPH, engineer; b. Salem, Ohio, Oct. 28, 1922; s. John and Mary Elizabeth (Toth) T.; m. Elizabeth Jane Keyes, May 18, 1957; children: Mary Jude, Elizabeth Jane, Ann Kathleen. BS, Tri-State Coll., 1943. Cert. engr. Radio engr. Sta. WFAH, Alliance, Ohio, 1946-48; test engr. Babcock & Wilcox, Alliance, 1948-52; field engr. Bendix Radio, Balt., 1952-56; electronics engr. Nat. Bur. of Standards, Boulder, Colo., 1956-79; field svc. engr. St. Regis Paper, Denver, 1979-81; telecom engr. Tri-State Generation and Transmission Assn., Inc., Denver, 1981-93; cons. Tary Assocs., 1993—; exhibits chmn. Utilities Telecomm Coun., Colorado Springs, 1990; registration chmn. Internat. Conf. on Communications, Denver, 1991; organizer 25 tech. confs. Co-author: A Guide for Telecommunication Specialists, 1962; contbr. articles to Electronic Mag. Bd. dirs. Colo. State Sci. Fair, Denver, 1976-79. With U.S. Army 1944-46. Recipient Avant Garde award Vehicular Tech. Soc., 1984, RAB Innovation award, 1992. Mem. IEEE (conf. chmn. region 5 1988—, chmn. Denver sect. 1975, editor RockIEEE Overlook 1993—, Centennial medal 1984), Armed Forces Comm. and Electronics Assn. (chpt. pres. 1974, awards chmn. 1974), Associated Pub.-Safety Communications Officers. Roman Catholic. Home and Office: 7739 Spring Dr Boulder CO 80303-5036

TARZWELL, CLARENCE MATTHEW, aquatic biologist; b. Deckerville, Mich., Sept. 29, 1907; s. Matthew and Jessie J. (Wilson) T.; m. Vera V. Paiter, Sept. 3, 1938; children: Diane Kay Tarzwell Siegmund, Barbara Ann Tarzwell Fahey, Thomas Neil. Student, Eastern Mich. U., 1925-29; A.B., U. Mich., 1930, M.S., 1932, Ph.D., 1936; Sc. D. (hon.), Baldwin Wallace Coll., 1967. Inst. Fisheries Rsch. Mich. Dept. Conservation rsch. fellow U. Mich., Ann Arbor, 1930-33; doctorate problem trout steam improvement supr. Mich. Emergency Conservation Work Civilian Conservation Corps Stream Improvement, Lansing, 1933-34; asst. aquatic biologist, E. Coastal and Intermountain Region U.S. Fish and Wildlife Svcs., Stream Improvement, Salt Lake City, 1934; asst. range examiner Region 3 U.S. Forest Svcs., Lake and Stream Improvement, Albuquerque, 1935-38; asso. biologist, chief field sta. TVA, Decatur, Ala., 1938-43; chief biology sect. USPHS, Savannah, Ga., 1944-48; sci. dir., chief aquatic biology sect. R.A. Taft San. Engring. Center, Cin., 1948-65; founding dir. Nat. Marine Water Quality Lab., West Kingston, R.I., 1965-72, Nat. Fresh Water Quality Lab., Duluth, Minn., 1964-67; sr. rsch. adviser EPA, Nat. Environ. Rsch. Ctr., Corvallis, Oreg., 1972-75; adj. prof. Coll. Resource Devel., U. R.I., Kingston, 1966-75; vis. prof. Coll. Resource Devel., U. R.I., 1976-79; mem. aquatic life adv. com. Ohio River Valley San. Commn., 1952-68, Nat. Acad. Sci.-NRC Com. on Pest Control-Wildlife Relationships (subcom. on research), 1960-63; mem. expert adv. panel environ. health WHO, Geneva, 1962-64; mem. panel fisheries experts FAO, UN, 1962-63; mem. pesticide com. Internat. Assn. Game, Fish and Conservation Commrs., 1960-61; Am. del. 1st and 2d internat. meetings sci. research into water pollution OECD, Paris, 1961, 62, chmn. internat. com. on long-term effects toxicants on aquatic life, 1962-65; mem. adv. com. control stream temperatures Pa. San Water Bd., 1959-62; adv. com. water quality criteria project Calif. Water Pollution Control Bd., 1961-63; chmn. sec. interior's Nat. Tech. Adv. Com. on Water Quality Requirements for Fish, Other Aquatic Life and Wildlife, 1967-68; chmn., mem. water pollution com. Water Pollution Control Fedn., 1955-75; chmn. com. for devel. standard methods for bioassays, 1958-75, com. for methods of sampling and analysis, 1965-74; mem. N.Am. Game policy com., 1971-73; mem. com. on power plant siting Nat. Acad. Engring.-Nat. Acad. Scis., 1971-72; pollution abatement com. Am. Fisheries Soc., 1952-73; mem. regional task force, southeastern New Eng. water and related land resources study New Eng. River Basins Com., 1972-75; cons. Kuwait Inst. Sci. Research, 1975; mem. R.I. State Wide Planning Council, 1978-82, Ecology Action for R.I.; advisor Save the Bay, Bottle Bill Coalition. Author 136 publs. aquatic biology, water pollution, water quality criteria and standards, tocicity, studies, pesticides, malaria control, rodent ectoparasite and typhus control. Recipient Conservationist of Year award State of Ohio, 1961; Am. Motors Profl. Conservationist award, 1962; Aldo Leopold medal Wildlife Soc., 1963; Meritorious Service medal USPHS, 1964; Disting. Career award EPA, 1973; Bronze medal EPA, 1974. Mem. Am. Fisheries Soc. (hon. life., chmn. pollution com. 1950-51, 54-61, mem. 1962-73, AFS award of excellence 1974), Am. Soc. Limnology and Oceanography (nat. adv. com. 1950-51, mem.-at-large 1959, bd. dirs. 1960-62), Societas Internationalis Limnologiae, Wildlife Soc. (hon. life), Am. Soc. Ichthyologists and Herpetologists (life), Am. Inst. Biol. Scis., USPHS Commd. Officers Assn. (pres. Cin. br. 1955-56), Stoic (pres. 1928), N.Am. Lake Mgmt. Soc., Phi Beta Kappa, Sigma Xi, Kappa Delta Pi, Phi Kappa Phi, Phi Sigma, Pi Kappa Delta. Home: 380B Post Rd Wakefield RI 02879-7508

TASHIRO, KOHJI, macromolecular scientist; b. Osaka, Japan, Sept. 24, 1950; s. Masaru and Tatsu (Kusakawa) T.; m. Sawako Matsui, Oct. 20, 1979; 1 child, Masashi. DSc, Osaka U., 1977. Spl. rschr. Japanese Orgn. for Promotion of Sci., 1978-81; asst. prof. Faculty of Sci., Osaka U., 1984-91, lectr., 1991—; Editorial bd. Current Polymer Rsch. (Japan), 1988—. Mem. Polymer Soc. Japan (Young Scientist scholarship 1986), Japan Soc. Chemistry, Japanese Soc. Fiber Sci. and Tech. (Sakurada Takeshi Meml. award 1989), Am. Chem. Soc., Am. Phys. Soc. Home: Satsuki-ga-Oka West 6-1-107, Suita 565, Japan Office: Faculty of Sci Osaka U, Machikane-yama 1-1, Tokonaka 560, Japan

TASKER, JOHN BAKER, veterinary medical educator, college dean; b. Concord, N.H., Aug. 28, 1933; s. John Baker and Catherine Mabel (Baker) T.; m. Grace Ellen Elliott, June 17, 1961; children—Sybil Alice, Sarah Catherine, Sophia Ethel. D.V.M., Cornell U., 1957, Ph.D., 1963. Instr. Cornell U., Ithaca, N.Y., 1960-61; from assoc. prof. to prof. Cornell U., 1967-78; from asst. prof. to assoc. prof. Colo. State U., Fort Collins, 1963-67; prof. vet. clinical pathology, assoc. dean La. State U., 1978-84; prof. vet. pathology, dean Coll. Vet. Medicine, Mich. State U., East Lansing, 1984—; cons. Ralston-Purina Co., St. Louis, 1978, Universidad Nacional P. Urena, Dominican Republic, 1980, U. Nebr., Lincoln, 1982-83. Editor: Veterinary Clinics of North America, 1976. Served to lst lt. U.S. Army, 1958-60. Recipient Outstanding Instr. award Colo. State U. Vet. Coll., 1967; Norden Teaching award Cornell U. Vet. Coll., 1977. Mem. AVMA, Mich. Vet. Med. Assn., Am. Coll. Vet. Pathologists (diplomate; examiner 1972-74), Am. Soc. Vet. Clin. Pathology (pres. 1971-72), Assn. Am. Vet. Med. Colls. (exec. com. 1986-91, pres. 1989-90). Avocations: reading; traveling. Home: 1136 Portage Path East Lansing MI 48823 Office: Mich State U Coll Vet Medicine G-100 Vet Med Ctr East Lansing MI 48824

TASSÉ, YVON ROMA, engineer; b. St. Gabriel de Brandon, Que., Can., Oct. 1, 1910; s. Victor L. and Amilda (Laurendeau) T.; m. Pauline Boyer, Nov. 11, 1935; children: Suzanne, Michel, Ghislaine, Denise, Lucile, Yves, Jacques. BA cum laude, Coll. Ste.-Marie, Montreal, Que., 1930; BS in Applied Civil Engring. with great distinction, Ecole Polytech., Montreal, Que., 1935. Apparatus engr. Can. Gen. Electric Co. Ltd., Montreal, 1937, Quebec, 1938-48; founder, ptnr. Tassé, Sarault & Assocs., Elec. and Mech. Cons. Engrs., Que., 1948-57; v.p., treas. Gen. Diesel Inc., Que., 1952-58; mng. dir. Indsl. and Trade Bur. of Met., Quebec, 1957-58; mem. Ho. Commons, 1958-62; appted. Parliamentary sec. to Minister of Pub. Works, 1959; ind. generalist, 1962—; bd. dirs. Fertek Inc., Montreal, Hopital de l'Enfant-Jesus, Que. Author: L'Electricité par le vent, 1935, La Navigation d'hiver sur le Saint-Laurent jusqu' á la cité de Québec. 1957, L'Engagement de l'ingénieur canadien, 1987. Successively dir., 2d v.p., 1st v.p., and pres. Que. Bd. Trade, 1950-54; chmn. bd. dirs. Hopital de l'Enfant-Jesus, 1974-86; corp. warden The Corp. of the Seven Wardens Inc./Soc. des Sept. gardiens inc., Can., 1986-91; mem. Nat. Adv. Coun. on Aging, Can., 1988-92. Fellow Can. Acad. Engring.; mem. Assn. Profl. Engrs. Que. and Ont., Assn. Cons. Engrs. Can., Am. Inst. Elec. Engrs., Am. Inst. Heating and Air Conditioning, Inst. Mgmt. Consultants Que., Can. C. of C. (mem. exec. com. 1954-57), Que. Holistic Med. Assn. (allied), Royal 22d Regt. Officers' Mess (hon.), Can. Corps. Commrs. (gov.), Cercle de la Garnison (Que.), Order Can. Conservative. Roman Catholic. Avocations: music, reading, fishing, tennis, curling. Office: 2052 du Bois-Joli, Sillery, PQ Canada G1T 1E1

TASSONE, BRUCE ANTHONY, chemical company executive; b. Phila., Sept. 8, 1960; s. Bruno Anthony and Julia A. (D'Alonzo) T. BSME, Univ. Pa., Phila., 1982; MBA with distinction, Univ. Pa., 1986. Asst. sales mgr. Gen. Electric, Schenectady, N.Y., 1982-84; dir., gen. mgr. Teleflex, King of Prussia, Pa., 1986—. bd. dirs. Friends' Assn. for Children. Nominee, Entrepreneur of Year, Del. Valley, Phila., 1989. Mem. Soc. of Mech. Engrs., Beta Gamma. Republican. Roman Catholic. Avocations: fin. mgmt, reading, sports activities. Home: 1722 Ridgeway Rd Havertown PA 19083

TATAR, JOHN JOSEPH, computer scientist; b. Highland Park, Ill., Jan. 7, 1951; s. Joseph and Elizabeth (Demeter) T.; m. Gloria Estella Vega, Dec. 18, 1976; children: Laura Maria, Steven David, Rebecca Elizabeth. BA, Marquette U., 1973; MPA, Am. U., 1977; postgrad., Ill. Inst. Tech., 1984-87. Project mgr. Washington Sci. Mktg., 1976-80; mgmt. analyst Argonne (Ill.) Nat. Lab., 1980-83, asst. scientist, 1983-89, asst. program mgr. computer sci., 1989—. Contbg. author: Energy Conserving Technology for Industry, 1986. Mem. ACM, IEEE (assoc.). Home: Honlacker Strasse 2, Erfenbach Germany 6750 Office: Argonne Nat Lab 9700 S Cass Ave Argonne IL 60439

TATARINOV, LEONID PETROVICH, science administrator, paleontologist; b. Tula, Russia, USSR, Nov. 12, 1926; s. Peter Lukitch and Anna Nikolaevna (Makarevskaya) T.; m. Irina P. Shamardina, July 12, 1948 (div. 1958); 1 child, Ksenia Leonidovna; m. Susanna Gurgenovna Bulat, Oct. 5, 1959; 1 child, Marina Leonidovna. DSc, State U., Moscow, 1969. Cert. zoologist, paleontologist. Aspirant State U., Moscow, 1949-52; redactor Publish Office "Mir", Moscow, 1953-54; scientist Paleontological Inst., Moscow, 1955-60, chef of lab., 1961-75, dir., 1975-92; vice sec. dept. gen. biology Russian Acad. Sci., Moscow, 1975—, councillor, 1992—; chef redactor Paleontological Jour., Moscow, 1978-86, Zool. Jour., Moscow, 1986—. Author: Theriodonts of USSR, 1974, Morphological Evolution of Theriodonts, 1976 (State award 1978), Essays on the Evolutionary Theory, 1987. Mem. Russian Acad. Scis. (academician 1981—), Linnean Soc. (London, fgn.). Avocations: history, music. Home: Leninsky Prospect 129-1, Moscow 117513, Russia Office: Paleontological Inst, Profsoyuznaya 123, Moscow 117321, Russia

TATE, JEFFREY L., biology institute administrator; b. Mpls., Dec. 18, 1957; s. Ronald L. and Betty M. (Nilson) T.; m. Tracey J. Benson, Sept. 9, 1989. BSc in Botany, U. Minn., 1980, PhD in Plant Physiology, 1985, postgrad., 1992—. Staff scientist Biol. Process Tech. Inst. U. Minn., St. Paul, 1985-87; rsch. assoc. Balt. Co. U. Md., 1987-89; spl. asst. to dir. Biol. Process Tech. Inst. U. Minn., 1989—. Dist. dir. Dem. Farmer Labor Party, Roseville, Minn., 1992—. Mem. Minn. Biotech. Assn. (sec./treas. 1991-93), Coun. Biotech. Ctrs. (chmn. 1993—). Lutheran. Avocation: gardening. Home: 2511 Lexington Ave N Saint Paul MN 55113 Office: U Minn Biol Process Tech Inst 1479 Gortner Ave Ste 240 Saint Paul MN 55108

TATE, JOHN EDWARD, chemical engineer; b. Nashville, Mar. 18, 1930; s. John C. and Beatrice (Clark) T.; m. Faye Dishner, Nov. 7, 1952; children: John E. Jr., Valerie Smith, Julie Duncan. BSChE, Vanderbilt U., 1952. Rsch. engr. Tenn. Eastman Co., Kingsport, 1952-53; sect. head polyester devel. Monsanto Co., Decatur, Pensacola, Ala., Fla., 1953-66; dir. process devel. Phillips Fibers Corp., Greenville, S.C., 1966-70; group mgr. engring. Am. Air Filter Co., Louisville, 1970-74; engring. dir. Texfi Industries, Greensboro, N.C., 1974-81; gen. mgr. Omega Yarns, Inc., Asheboro, N.C., 1981-82; pres. Southridge Corp., Greensboro, 1982—. Active County Environ. Adv. Bd., Guilford County, N.C., 1978-82; mem. environ. affairs com. N.C. Textile Mfrs. Assn., 1976-81, mem. energy conservation com., 1976-82. Mem. AIChE, Am. Assn. Colorists and Chemists, Soc. Plastics Engrs., Forest Oaks Country Club, Brotherhood St. Andrew. Republican. Episcopalian. Achievements include patents in field. Home: 1912 Milan Rd Greensboro NC 27410 Office: Southridge Corp PO Box 10904 Greensboro NC 27404

TATE, MANFORD BEN, guided missile scientist, investor; b. Okolona, Ark., Aug. 3, 1916; s. Ernest and Mabel (Burley) T.; m. Marjorie Belle Stone, Sept. 9, 1942; children: Howard Ernest, Virginia Louise Tate Smythe, Barbara Anne. Student, Cen. Coll. Liberal Arts, 1935-37; BSCE, U. Mo., 1940, MSCE, 1942; PhD in Theoretical and Applied Mechanics, Iowa State U., 1949. Grad. fellow, instr., asst. prof., assoc. prof. U. Mo., Columbia, 1940-51; lectr, engring. mechanics U. Va., Hampton, 1944-45; grad. fellow, instr. Iowa State U., Ames, 1946-47; engr. U.S. Naval Ordnance Lab., White Oak, Md., 1950-56; lectr., advisor U. Md., White Oak, 1950-56; guided missile rsch. scientist Johns Hopkins U. Applied Physics Lab., Laurel, Md., 1956-78, prof. missile structures, 1958-64; pres., gen. ptnr. Tate Ptnrs., Silver Spring, Md., 1968—; concrete inspector Mo. Hwy. Dept., 1940; vis. prof. Cath. U. Am., Washington, 1956-58; aeronautical rsch. Nat. Adv. Com. Aeronautics, Langley Field, Va., 1944-45; cons. in field. Bridge designer Bur. of Bridges, 1941; Crane designer Dravo Corp. Pittsburgh, Pa.; Author: (with others) Selected Problems in Materials Testing, 1943. Contbr. 400 articles to profl. jours. Cub scouts chmn., cubmaster Boy Scouts Am., 1956-61, chmn. rev. bd., 1962-65; Pta, civil defense chmn., 1961-63. With Mo. Nat. Guard, 1935-37; Air Corps Res., 1944-45. Mem. ASCE, ASTM, ASME, Am. Soc. Engring Edn., Am. Soc. Ordnance Engrs., Soc. Exptl. Stress Analysis, Sigma Xi, Phi Kappa Phi, Chi Epsilon. Democrat. Methodist. Achievements include patents on the Tomahawk Cruise Missile; derived theoretical solution for shock strength of the Fat Man atom bomb.

TATERA, JAMES FRANK, chemist, process analysis specialist; b. Milw., June 27, 1946; s. Harry Frank and Agnes Rose (Szymanowski) T.; m. Kaaren Marie Piekarski, Sept. 9, 1972; children: Patrick, Monica, David. BS in Chemistry, Math., U. Wis., Oshkosh, 1968; postgrad., U. Minn., 1968, 71-73; MBA, Cen. Mich. U., 1982. Cert. specialist in analytical tech. Teaching rsch. assoc. chemistry dept. U. Minn., Mpls., 1968, 71-73; analytical chemist Dow Corning Corp., Midland, Mich., 1973-76, scale up engr. new products commercialization, 1976-78, prodn. bldg. supt. prodn. dept., 1978-80; analytical systems specialist project and plant engring. Dow Corning Ltd., Barry, Wales, 1981-84; analytical systems supr. plant engring. Dow Corning Corp., Carrollton, Ky., 1984-85, analytical systems specialist plant engring. and maintenance, 1985-87, sr. analytical and control specialist project engring., 1988-90, sr. analytical systems specialist strategic change program, 1991—; session developer, panelist, paper presenter, workshop presenter internat. confs.; U.S. Nat. Com. del. to Internat. Electrotech. Commn. SC65D, Paris, 1993; presenter in field. Contbr. articles to profl. jours. 1st lt. arty. U.S. Army, 1969-71. Decorated Bronze Star, Bronze Star with oak leaf cluster. Mem. Am. Chem. Soc. (rep. Vol. in Public Outreach program), Instrument Soc. Am. (chmn. SP 76 standards com. 1991—, pres. N.E. Mich. sect. 1979-80, various sect. offices 1976-79), Air and Waste Mgmt. Assn., Elks, Am. Legion, VFW, Delta Sigma Phi, Phi Lambda Upsilon, Sigma Iota Epsilon. Roman Catholic. Home: 2038 Ridgewood Ln Madison IN 47250-2729 Office: Dow Corning Corp PO Box 310 Mail Stop 32 Carrollton KY 41008

TATEYAMA, ICHIRO, gynecologist; b. Fukuoka, Japan, Aug. 21, 1952; s. Chikayoshi and Katsuyo Tateyama; m. Hiroko Uchino, Nov. 2, 1980; children: Ryoko, Ayako, Yuko. MD, Kyoto U., 1978, MD, PhD, 1989. Diplomate Bd. Obs.-Gyn. Resident Faculty of Medicine Kyoto U., 1978-79, Niigata (Japan) Prefectural Hosp., 1979-82; asst. Fukui (Japan) Med. Sch., 1982-84; chief dr. Tomita Hosp., Kyoto, 1984-86; co-chief dr. Kurashiki (Japan) Cen. Hosp., 1986-88, Tenri (Japan) Hosp., 1988-89; chief dr. Kishiwada (Japan) City Hosp., 1989—. Contbr. articles to profl. jours. Mem. Japan Soc. Cancer Therapy, Japan Soc. Cancer Rsch., Japan Soc. Fertility and Sterility, Japan Soc. Obs.-Gyn., Japan soc. Clin. Electron Microscopy. Avocations: piano, fishing. Home: NR 13 4-27-1 Kamori-cho, Kishiwada 596, Japan Office: Kishiwada City Hosp, 4-27-1 Kamori-cho, Kishiwada 596, Japan

TATINA, ROBERT EDWARD, biology educator; b. Chgo., May 18, 1942; s. Edward and Gertrude (Dase) T.; m. Geraldyne Lynn Sansone, May 26, 1978; children: Thomas, Heather. BS, No. Ill. U., 1965; MA, So. Ill. U., 1972, PhD, 1982. Tchr. Trewyn Jr. High Sch., Peoria, Ill., 1965-66; teacher biology Evergreen Park (Ill.) Community High Sch., 1966-69; prof. biology Dakota Wesleyan U., Mitchell, S.D., 1975—. Mem. Bot. Soc. Am., Am. Assn. Biology Tchrs., Nat. Sci. Tchrs. Assn., S.D. Acad. Sci. Home: 704 S Montana Mitchell SD 57301 Office: Dakota Wesleyan U 1200 W University Ave Mitchell SD 57301

TATNALL, GEORGE JACOB, aeronautical engineer; b. Cin., Aug. 9, 1923; s. George Henry and Ida Mae (Hazelbaker) T.; m. G. Virginia Morgan, Feb. 5, 1949; children: Robert, William, Jeffrey, Thomas, Jane. BSME in Aeronautics, U. Pitts., 1949. Devel. engr. Naval Air Devel. Ctr., Warminster, Pa., 1949-57; supr. electro-mech. design br., supr. radome antenna sect. Naval Air Devel. Ctr., Warminster, 1957-62, 63-78; engring. group leader Corning (N.Y.) Glass Works, 1962-63; cons. Semcor, Inc., Warminster, 1979-81; cons. pvt. practice Warminster, 1982-84; cons. Veda, Inc., Warminster, 1989—; Tech. advisor Seventh Fleet USN, S.E. Asia, during Vietnam War, 1971. Contbr. to sci. papers presented at confs. (many also pub. in proceedings); author: (with others) chpt. in book Environmental Simulation and Test Data; author: manuals and documentation aircraft equipment. With US Army AF, 1943-45, China. Mem. AIAA (pres. U. Pitts. chpt. 1948), VFW, Flying Tiger U.S. 14th Air Force Assn. Achievements include patents for Speed Brake Retarding Mechanism for an Air Dropped Store, Air Dropped Minature Sonobuoy, Rotatable and Tiltable Radome with ind. scan and tilt antenna; developed low noise coupling through laminar boundary layer for acoustic homing missile, test facilities for supersonic rain erosion of aircraft materials. Home: 551 Walter Rd Warminster PA 18974

TATYREK, ALFRED FRANK, materials engineer, research chemist; b. Hillside, N.J., Jan. 23, 1930; s. Frank Peter and Frances (Luxa) T. BS, Seton Hall U., 1954; postgrad., Rutgers U., 1956-57. Rsch. chemist Bakelite div. Union Carbide, Bloomfield, N.J., 1953-58, U.S. Radium Corp., Morristown, N.J., 1959-62; analytical chemist insp. Chem. Procurement Dist. U.S. Army, N.Y.C., 1962-64; rsch. chemist Picatinny Arsenal U.S. Army, Dover, N.J., 1964-73; chem. materials engr. U.S. Army Armament Rsch., Devel. and Engring. Ctr., N.J., 1973—. Patentee pyrotechnic compositions, chemiluminescent compounds and processes, crank case oil vacuum purification system for internal combustion engines; lectr., contbr. articles on mountaineering expdns. and adventures in the great mountain ranges of N.Am., S.Am., and Africa to mags.; contbr. over 50 sci. and tech. reports. First aid instr. ARC, Essex County, N.J., 1969-82; chief first aid Maplewood (N.J.) CD 1971-91; patrol dir. Nat. Ski Patrol, Phoenicia, N.Y., 1978-84, sr. patroller So. N.Y. region, 1979—. Climbed Mt. Blanc, highest mountain peak in Europe; climbed to a summit of 19,730 on Mt. Kilimanjaro, highest mountain peak in Africa, 1972; leader on climbs of Matterhorn, Monte Rosa, Switzerland's highest peak; also participated in numerous mountain expdns. in U.S. and Can., including 3 first ascents in No. Cascades of Wash. (the S.E. ridge of Mt. Goode, Aug. 1963, Peak 7732 via the Snow Chute, Aug. 1964, the East ridge of Bear Mt., Aug. 1964). Mem. Nat. Soc. Inventors, Magician's Round Table, Alpine Club of Can., Appalachian Mountain Club, Sierra Club, Sigma Xi (pres. Picatinny chpt. 1974-75, 79-80, 85-86). Roman Catholic. Achievements include six patents in field. Home: 27 Orchard Rd Maplewood NJ 07040-1919 Office: US Army Armament Rsch Devel and Engring Ctr Dover NJ 07806-5000

TAUBE, ADAM A.S., biometrician, educator; b. Uppsala, Sweden, Jan. 12, 1932; s. Nils E. and Gurli (Westgren) T.; m. Liliane H. Åqvist, Jan. 4, 1962; children: Didrik, Amelie, Caroline, Barbara. Bachelor, Uppsala U., 1957, Master, 1962, PhD, 1969. Asst. dept. stats. Uppsala U., 1956—; sr. advisor UNESCO, Addis Ababa, 1972-75; lectr. dept. stats. Uppsala U., 1965-91, prof., 1991—; cons., researcher Swedish Cancer Fund, Uppsala. Contbr. articles to profl. jours. Avocations: classical music, classical guitar. Home: Thunbergsvägen 19, Uppsala S-75238, Sweden Office: Uppsala Univ, Dept Stats, Box 513, Uppsala S-75120, Sweden

TAUBE, HENRY, chemistry educator; b. Sask., Can., Nov. 30, 1915; came to U.S., 1937, naturalized, 1942; s. Samuel and Albertina (Tiledetski) T.; m. Mary Alice Wesche, Nov. 27, 1952; children: Linda, Marianna, Heinrich, Karl. B.S., U. Sask., 1935, M.S., 1937, LL.D., 1973; Ph.D., U. Calif., 1940; Ph.D. (hon.), Hebrew U. of Jerusalem, 1979; D.Sc. (hon.), U. Chgo., 1983, Poly. Inst., N.Y., 1984, SUNY, 1985, U. Guelph, 1987; D.Sc. honoris causa, Seton Hall U., 1988; Lajos Kossuth U. of Debrecen, Hungary, 1988; DSc, Northwestern U., 1990, Northwestern U., 1991. Instr. U. Calif., 1940-41; instr., asst. prof. Cornell U., 1941-46; faculty U. Chgo., 1946-62, prof., 1952-62, chmn. dept. chemistry, 1955-59; prof. chemistry Stanford U., 1962—, Marguerite Blake Wilbur prof., 1976, chmn. dept., 1971-74; Baker lectr. Cornell U., 1965. Hon. mem. Hungarian Acad., Scis., 1988. Recipient Harrison Howe award, 1961; Chandler medal Columbia U., 1964; F.P. Dwyer medal U. N.S.W., Australia, 1973; Nat. medal of Sci., 1976, 77; Allied Chem. award for Excellence in Grad. Teaching and Innovative Sci., 1979; Nobel Prize in Chemistry, 1983; Bailar medal U. Ill., 1983; Robert A. Welch Found. Award in chemistry, 1983; Disting. Achievement award Internat. Precious Metals Inst., 1986; Guggenheim fellow, 1949, 55. Fellow Royal Soc. Chemistry (hon.), Indian Chem. Soc. (hon.); mem. NAS (award in chem. scis. 1983), Am. Acad. Arts and Scis., Am. Chem. Soc. (Kirkwood award New Haven sect. 1965, award for nuclear applications in chemistry 1955, Nichols medal N.Y. sect. 1971, Willard Gibbs medal Chgo. sect. 1971, Disting. Svc. in Advancement Inorganic Chemistry award 1967, T.W. Richards medal NE sect. 1980, Monsanto Co. award in inorganic chemistry 1981, Linus Pauling award Puget Sound sect. 1981, Priestley medal 1985, Oesper award Cin. sect. 1986, G.M. Kosolapoff award Auburn sect. 1990), Royal Physiographical Soc. of Lund (fgn. mem.), Am. Philos. Soc., Finnish Acad. Sci. and Letters, Royal Danish Acad. Scis. and Letters, Coll. Chemists of Catalonia and Beleares (hon.), Can. Soc. Chemistry (hon.), Hungarian Acad. Scis. (hon. mem.), Royal Soc. (fgn. mem.), Brazilian Acad. Scis. (corr.), Engring. Acad. Japan (fgn. assoc.), Australian Acad. Scis. (corr.), Chem. Soc. Japan (hon. mem. 1993), Phi Beta Kappa, Sigma Xi, Phi Lambda Upsilon (hon.). Office: Stanford U Dept Chemistry Stanford CA 94305

TAUBES, CLIFFORD H., mathematician educator. Prof. math. Harvard U., Cambridge, Mass. Recipient Oswald Veblen Geometry prize Am. Math. Soc., 1991. Office: Harvard U Dept of Math Cambridge MA 02138*

TAUC, JAN, physics educator; b. Pardubice, Czechoslovakia, Apr. 15, 1922; came to U.S., 1969, naturalized, 1978; s. Jan and Josefa (Semonska) T.; m. Vera Koubelova, Oct. 18, 1947; children: Elena (Mrs. Milan Kokta), Jan. Ing.Dr. in Elec. Engring., Tech. U. Prague, 1949; RNDr., Charles U., 1956; Dr.Sc. in Physics, Czechoslovak Acad. Scis., 1956. Scientist microwave research Sci. and Tech. Research Inst., Tanvald and Prague, 1949-52; head semiconductor dept. Inst. Solid State Physics, Czechoslovak Acad. Scis., 1953-69; prof. exptl. physics Charles U., 1964-69, dir. Inst. Physics,, 1968-69; mem. tech. staff Bell Telephone Labs., Murray Hill, N.J., 1969-70; prof. engring. and physics Brown U., 1970-83, L. Herbert Ballou prof. engring. and physics, 1983-92, L. Herbert Ballou prof. emeritus, 1992—, dir. material research lab., 1983-88; dir. E. Fermi Summer Sch., Varenna, Italy, 1965; vis. prof. U. Paris, 1969, Stanford U., 1977, Max Planck Inst. Solid State Research, Stuttgart, Germany, 1982; UNESCO fellow, Harvard, 1961-62. Author: Photo and Thermoelectric Effects in Semiconductors, 1962, also numerous articles; editor: The Optical Properties of Solids, 1966, Amorphous and Liquid Semiconductors, 1974; co-editor: Solid State Communications, 1963-92. Recipient Nat. prize Czechoslovak Govt., 1955, 69; Sr. U.S. Scientist award Humboldt Found., 1981, Silver medal Union of Czechoslovak Mathematicians and Physicists, 1992. Fellow AAAS, Am. Phys. Soc. (Frank Isakson prize 1982, David Adler award 1988); mem. NAS, European Phys. Soc. (founding), Czechoslovak Acad. Scis. (corr. 1963-71, 90-91, fgn. 1991-92, Hlavka medal 1992). Office: Brown U Div Engring Providence RI 02912

TAVAGLIONE, DAVID, computer scientist, consultant; b. Pittston, Pa., May 10, 1930; s. Anthony and Nellie (Ambrose) T.; m. Mary Lou Wolf (div. Mar. 1969); children: James David, Jerry Kenneth; m. Eileen Kay Swanson, Aug. 8, 1970. BA in Math. and Physics, U. Tex., 1962. Programming supr. Ohio State U. Computer Ctr., Columbus; sr. systems analyst W.Va. Network for Ednl. Telecomputing, Morgantown. Editor, author: Acronyms and Abbreviations of Computer Technology and Telecommunications, 1992. Served with USAF, 1950-54, Korea. Methodist. Achievements include invention of implantable stimulator with extracorporeal control of rate, amplitude and duty cycle; invention of laser scanning microscope. Home: Rte 4 Box 149-C Morgantown WV 26505

TAVANO, FRANK, physicist, electrical engineer; b. Melrose, Mass., Apr. 11, 1962. BSEE magna cum laude, U. Mass., Lowell, 1984; postgrad., U. Fla., 1992. Registered profl. engr., Fla. Rsch. intern Raytheon Missile

Systems Lab., Bedford, Mass., 1983-84; mem. tech. staff Hughes Aircraft Co., Torrance, Calif., 1984-85; sr. elec. engr. Lockheed Corp., Orlando, Fla., 1985-86; sr. electronics engr. Schwartz Electro-Optics, Inc., Orlando, 1987, Wood-Ivey Systems Corp., Winter Park, Fla., 1987-89, Bombardier Corp., Orlando, 1989-92; physics teaching asst. U. Fla., Gainesville, 1992—. Achievements include design of microwave phased array interferometer, low noise RF/microwave receivers, GaAs laser-based rangefinders, varactor tuned bandpass filters, millimeter wave transmission data link, IMPATT diode power combiners, varactor tuned Gunn oscillators; participated in testing of Patriot radar system. Home: 3616 SW 19th St Gainesville FL 32608 Office: U Fla Dept of Physics Gainesville FL 32611

TAWATA, SHINKICHI, agricultural engineering educator; b. Ginowan, Okinawa, Japan, Aug. 28, 1949; s. Shinki and Yoshiko Tawata. M of Agrl., Kyushu U., 1974, PhD, 1978. Postdoctoral assoc. SUNY, Stony Brook, 1979, U. Calif., Berkeley, 1979-82; asst. prof. U. Ryukyus, Japan, 1982—; cons. Japan Internat. Corp. Agy., Japan, 1986, Agrl. and Fishery Ministry, Japan, 1986-91, Okinawa Prefecture, Japan, 1989—. Author: Pesticides and Alternatives, 1990; contbr. articles to profl. jours. Grantee Bank of Okinawa, 1991, Okinawa Prefecture, 1992. Achievements include patent in growth regulation method using mimosine, patent in simple reduction method of mimosine and its decomposed product from leucaena. Home: 45 Aichi, 901-22 Ginowan Japan Office: Coll of Agrl U Ryukyus, 1 Senbaru, Nishihara 903-01, Japan

TAY, ROGER YEW-SIOW, electromagnetic engineer; b. Johore, Malaysia, June 11, 1958; came to U.S., 1980; s. Soon Tay and Sam Chin. BSEE, U. Lowell, 1983, MSEE, 1985; postgrad. (acad. guest), Swiss Fed. Inst. Tech., Zurich, 1993. Teaching asst. U. Lowell, Mass., 1983-85, Va. Tech., Blacksburg, 1985-86; elec. engr. Motorola Electronic Pte. Ltd., Singapore, 1987-89, sr. elec. engr., 1989-90; sr. electromagnetic engr. Motorola Inc., Ft. Lauderdale, Fla., 1990-91, staff electromagnetic engr., 1991—. Mem. IEEE, N.Y. Acad. Scis. Achievements include patents pending for broadband end fed dipole antenna with a double resonant transformer, broadband sleeve for monopole antenna, modified printed folded dipole. Home: 1445 NW 129th Way Sunrise FL 33323

TAYLOR, ALAN HENRY, geography educator, ecological consultant; b. Berkeley, Calif., Oct. 26, 1954; s. Terrence Henry Machel and Margaret Marilyn (Brown) T.; m. Kristin Ann Brusila, Dec. 30, 1983; children: Kendra, Erik. BSc, Calif. State U., Hayward, 1977; MSc, Oreg. State U., 1979; PhD, U. Colo., 1987. Asst. prof. geography U. Md., Balt., 1987-90; asst. prof. geography Pa. State U., University Park, 1990-93, assoc. prof., 1993—; cons. ecologist World Wildlife Fund, Chengdu, People's Republic China, 1984-85, 90, 92. Author: Dynamics of Bamboo Forests in Giant Panda Habitat in Wolong Natural Reserve, China, 1993; contbr. articles to profl. jours. Office: Pa State U Dept Geography 302 Walker Bldg University Park PA 16802

TAYLOR, ANDREW CHRISTOPHER, electronics and electrical engineer; b. Chester, Cheshire, Eng., Dec. 30, 1960; s. Derek and Alice (Davis) T. BS in Engring. with honors, Imperial Coll., London, 1984. Registered profl. engr., England. Projects mgr. Physiol. Instrumentation, Whitland, Wales, 1984-87, Novametrix Med. Systems Inc., Wallingford, Conn., 1987-90; mgr. oximetry Ohmeda Monitoring, Louisville, Colo., 1990-92; dir. rsch. and devel. Mallinckrodt Med., Ann Arbor, Mich., 1992—. Achievements include patents for oximetry. Home: 2584 Walnut Rd Ann Arbor MI 48103 Office: Mallinckrodt 1590 Eisenhower Pl Ann Arbor MI 48108

TAYLOR, BARRY LLEWELLYN, microbiologist, educator; b. Sydney, Australia, May 7, 1937; came to U.S., 1967; s. Fredrick Llewelyn and Vera Lavina (Clarke) T.; m. Desmyrna Ruth Tolhurst, Jan. 4, 1961; children: Lyndon, Nerida, Darrin. BA, Avondale Coll., Cooranbong, New South Wales, 1959; BSc with honors, U. New South Wales, Sydney, 1966; PhD, Case Western Res. U., 1973; postgrad., U. Calif., Berkeley, 1973-75. Vis. postdoctoral fellow Australian Nat. U., Canberra, 1975-76; asst. prof. biochemistry Loma Linda (Calif.) U., 1976-78, assoc. prof. biochemistry, 1978-83, prof. biochemistry, 1983—, prof., chmn. dept. microbiology, 1988—, interim dir. Ctr. for Molecular Biology, 1989—. Contbr. articles to profl. publs. Rsch. grantee Am. Heart Assn., 1978-85, NIH, 1981—. Mem. Am. Soc. Microbiology, Am. Soc. Biochemistry and Molecular Biology. Office: Loma Linda U Dept Microbiology Loma Linda CA 92350

TAYLOR, BRENDA CAROL, computer specialist; b. Kansas City, Kans., July 27, 1956; d. Collis Billy and Lela V. (Caldwell) T. AS, Community Coll. in Kansas City, 1976; BS in Computer Sci., Kansas State U., 1978. Fin. ops. asst., programmer United Telecom Computer Group, Overland Park, Kans., 1979-81; asst. data processing mgr., systems analyst Truman Nat. Life Ins. Co., Kansas City, Mo., 1981-82; computer programmer analyst Boilermarkers' Nat. Funds Office, Kansas City, Kans., 1982-84, USDA/ASCS, Kansas City, Mo., 1984-88; computer systems programmer, computer specialist Nat. Computer Ctr./OIRM/USDA, Kansas City, Mo., 1988—. Mem. Federally Employed Women (v.p. Heart of Am. local chpt. 1992—), Fed. Women's Program Com. (vice chmn. 1988-92), Am. Bus. Women Assn., Blacks in Gov., Nat. Assn. Female Execs., Alpha Kappa Alpha (Sorority, Inc.), Phi Theta Kappa (hon. frat. 75-77, pres. 1977). Home: 2827 N 55th St Kansas City KS 66104

TAYLOR, CELIANNA I., information systems specialist; b. Youngstown, Ohio; d. Paul Thornton and Florence (Jacobs) Isley; m. (div.); children: Polly, Jerry, Jim. Bachelors degree, Denison U., 1939; MLS, Western Res. U., 1942. Worked in several pub. libs. and u. libs., 1939-50; head Libr. Cataloging Dept. Battelle Mem. Inst., Columbus, Ohio, 1951-53; head Personnel Office and assoc. prof. of Libr. Adminstrn. Ohio State U. Librs., Columbus, Ohio, 1954-65; coord. Info. Svcs. and Assoc. Prof. of Libr. Adminstrn. Nat. Ctr. for Research in Vocat. Edn., Ohio State U., Columbus, 1966-70; sr. research assoc., adminstrv. assoc., assoc. prof. libr. adminstrn. Dept. of Computer and Info. Sci. Ohio State U., Columbus, 1970-86, retired assoc. prof. emeritus Univ. Librs., 1986—; mem. Task Force on a Spl. Collections Database, Ohio State U. Librs., Columbus, 1988-89, comm. systems and recs. coord. Ohio State U. Retirees Assn., Columbus, 1992-93; cons. for several profl. orgns. including Ernst & Ernst CPA's and Oreg. State Systems of Higher Edn. 1962-82. Author: (with J. Magisos) book, Guide for State Voc-Tech Edn. Dissemination Systems 1971, (with A.E. Petrarca, and R.S. Kohn) book, Info. Interaction 1982; several articles for profl. jours.; designer: info. systems, CALL System, 1977-82, Channel 2000 Proj. Home Info. Svc., 1980-81, Continuing Education Info. Ctr., 1989-90, Human Resources (HUR) System, 1976-77,1979-82, DECOS, 1975-86, Computer-asst. libr. System, Optical Scan System, 1972-73, ERIC Clearinghouse for vocat. edn., 1966-70. Bd. dirs. Columbus Reg. Info. Svc., 1974-78, Community Info. Referral Svc., Inc. 1975-81; chmn. Subcom. on Design, Info. and Ref. Com. Columbus United Community Council, 1972-73; dir. Computer Utility for Pub. Info. Columbus, 1975-81. Mem. ALA, Am. Soc. for Info. Sci., Assn. Faculty and Profl. Women Ohio State U., Columbus Metro Club, Coun. for Ethics in Econs., Olympic Indoor Tennis Club. Avocations: bicycling, bird watching, folk dancing, gourmet cooking, tennis. Home: 3471 Greenbank Ct Columbus OH 43221-4724

TAYLOR, D. LANSING, cell biology educator; b. Balt., Dec. 26, 1946. BS, U. Md., 1968; PhD in Biology, SUNY, Albany, 1973. Fellow biophysics Marine Biol. Labs., 1973-74; asst. prof. biology Harvard U., 1974-78, assoc. prof., 1978-82; prof. biology Carnegie-Mellon U., Pitts., 1982—. Editor Jour. Cell Biology, 1981—, Jour. Cell Motility, 1981—. Mem. Am. Soc. Cell Biology, Biophys. Soc., N.Y. Acad. Sci. Achievements include research in molecular basis of amoeboid movements, utilizing biochemical, cell biological and biophysical approaches and fluorescence spectroscopy. Office: Carnegie Mellon U Dept Biomed Engring Program 4400 5th Ave Pittsburgh PA 15213*

TAYLOR, DEAN PERRON, biotechnologist, researcher; b. Cordova, Alaska, Nov. 19, 1947; s. Lewis Dean Taylor and Emilie Kathleen (Perron) St. Pierre; m. JoBess Hineline, July 15, 1972; children: Matthew Pendleton, Mark Lewis Dean. Student, U. Alaska, 1966-67; BS, U. Wash., 1970; PhD, U. Wis., 1976. Postdoctoral fellow Stanford (Calif.) U., 1976-79; assoc. sr. investigator in microbiology Smith Kline and French Labs., King of Prussia, Pa., 1979-81, sr. investigator microbiology, 1981-83, sr. investigator molecular genetics, 1983-85, asst. dir. molecular genetics, 1985-89; mgr. sci. projects Panlabs, Inc., Bothell, Wash., 1989-90, regional dir. biotechnology, 1990—; co-organizer UCLA symposium, Frisco, Colo., 1990; referee various publs. Contbr. articles, abstracts to profl. publs. Postdoctoral fellow Am. Cancer Soc., 1976-78; Giannini Found. postdoctoral fellow Bank of Am., San Francisco, 1979. Fellow Sigma Xi; mem. AAAS, Am. Soc. Microbiology, Soc. Indsl. Microbiology. Achievements include 2 patents, research in identification, cloning of genes for novel Streptomyces protein protease inhibitors, expression and secretion of active HIV binding protein by Streptomyces. Office: Panlabs Inc 11804 N Creek Pky S Bothell WA 98017-8805

TAYLOR, DOUGLAS FLOYD, mechanical engineer; b. Swanton, Ohio, Oct. 7, 1942; s. Floyd L. and Pauline (Zimmerman) T.; m. Karen Sue Smith, Dec. 23, 1967; children: Jeffery Douglas, Kenneth Floyd. AAS in Mech. Engring., U. Toledo, 1989. Svc. technician Ohio Gas Co., Swanton, 1969-77; maintenance Pilliod Cabinet Co., Swanton, 1977-89, corp. project engr., 1989-90; project mgr. Canberra Corp., Toledo, 1990—. Fireman, Swanton Vol. Fire Dept., 1964-83. With U.S. Army Reserve, 1964-70. Mem. ASME, Soc. Mfg. Engrs. Home: 110 Bassett Ave Swanton OH 43558 Office: Canberra Corp 3610 Holland-Sylvania Rd Toledo OH 43615

TAYLOR, EDWARD CURTIS, chemistry educator; b. Springfield, Mass., Aug. 3, 1923; s. Edward Curtis and Margaret Louise (Anderson) T.; m. Virginia Dion Crouse, June 29, 1946; children: Edward Newton, Susan Raines. Student, Hamilton Coll., 1942-44, DSc (hon.), 1969; AB, Cornell U., 1946, PhD, 1949. Postdoctoral fellow Nat. Acad. Scis., Zurich, Switzerland, 1949-50; DuPont postdoctoral fellow chemistry U. Ill., 1950-51, faculty, 1951-54, asst. prof. organic chemistry, 1952-54; faculty Princeton U., 1954—, prof. chemistry, 1964—, A. Barton Hepburn prof. organic chemistry, 1966—, chmn. dept. chemistry, 1974-79; vis. prof. Technische Hochschule, Stuttgart, Fed. Republic Germany, 1960, U. East Anglia, 1969, 71; Disting. vis. prof. U. Buffalo, 1968, U. Wyo., 1977; Backer lectr. U. Groningen, Holland, 1969; mem. chemistry adv. com. Office Sci. Research, USAF, 1962-73, Cancer Chemotherapy Nat. Service Ctr., 1958-62; cons. rsch. divs. Procter & Gamble, 1953-80, Eastman Kodak Co., 1965-83, Tenn. Eastman Co., 1968-83, Eli Lilly & Co., 1970—, Burroughs Wellcome Co., 1983—, E.I. duPont de Nemours & Co., 1986-90, Polaroid Corp., 1986—, Dow Elanco Co., 1989—, DuPont Merck Pharm. Co., 1990—. Author: (with McKillop) Chemistry of Cyclic Enaminonitriles and o-Aminonitriles, 1970, Principles of Heterocyclic Chemistry; film course, 1974; editor: (with Raphael, Wynberg) Advances in Organic Chemistry, vols. I-V, 1960-65, (with Wynberg) vol. VI, 1969, vols. VII-IX, 1970-79, (with W. Pfleiderer) Pteridine Chemistry, 1964, The Chemistry of Heterocyclic Compounds, 1968—, General Heterocyclic Chemistry, 1968— ; organic chemistry editorial adviser, John Wiley & Sons Inc., 1968—; mem. editorial adv. bd. Jour. Medicinal Chemistry, 1962-66, Jour. Organic Chemistry, 1971-75, Synthetic Communications, 1971—, Heterocycles, 1973—, Chem. Substructure Index, 1971—, Advances in Heterocyclic Chemistry, 1983—. Recipient rsch. awards Smith Kline and French Found., 1955, Hoffman-LaRoche Found., 1964-65, Ciba Found., 1971, Disting. Hamilton award, 1977, U.S. Sr. Scientist prize Alexander von Humboldt Found., 1983, Disting. Alumni medal Hamilton Coll., 1990, F. Gowland Hopkins medal, 1993; sr. faculty fellow Harvard U., 1959; Guggenheim fellow, 1979-80. Fellow N.Y. Acad. Scis., Am. Inst. Chemists; mem. Am. Chem. Soc. (award for creative work in synthetic organic chemistry, 1974, chmn. organic chemistry div. 1976-77), German Chem. Soc., Chem. Soc. London, Internat. Soc. Heterocyclic Chemistry (5th Internat. award 1989), Phi Beta Kappa, Sigma Xi, Phi Kappa Phi. Home: 288 Western Way Princeton NJ 08540-5337

TAYLOR, GAYLON DON, industrial engineering educator; b. Anchorage, Mar. 25, 1960; s. Gaylon Don and Rita Merle (Eudy) T.; m. Jo Ellen Gibson, June 6, 1987; 1 child, Daniel Alexander. BS in Indsl. Engring., U. Tex., Arlington, 1983, MS in Indsl. Engring., 1985; PhD, U. Mass., 1990. Mfg. engr. Tex. Instruments, Lewisville, Tex., 1983-86; process engr. Digital Equipment Corp., Enfield, Conn., 1987-89; asst. prof. U. Ark., Fayetteville, 1990—; pres. Don Taylor Cons. Svcs., Fayetteville, 1991—. Reviewer Internat. Jour. of Prodn. Rsch.; contbr. articles to profl. jours. Grantee NSF, 1992, Material Handling Rsch. Ctr., 1991-92, Ark. Sci. and Tech. Authority, 1991. Mem. IEEE (sr., U. Ark. faculty advisor 1992—), Sigma Xi, Tau Beta Pi, Alpha Pi Mu, Phi Eta Sigma, Pi Mu Epsilon, Alpha Chi, Omega Rho. Office: U Ark 4207 Bell Engineering Ctr Fayetteville AR 72701

TAYLOR, HENRY L., aerospace psychologist, educator; b. Tallahassee, Ala., Nov. 1, 1933; M. Mary Charlton Garrison, Nov. 23, 1933; children: Kenneth H., Gregory L., Barry C/. BA in Psychology, Auburn U., 1956, MS in Psychology, 1957; PhD in Psychology, Fla. State U., 1965. Lic. comml. pilot, flight instr.. Commd. 2d lt. USAF, 1960, advanced through ranks to col., navigator, radar observer schs., 1957-59; instr., radar observer 13th fighter interceptor squadron Glasgow AFB, Mont., 1959-62; staff scientist, dep. chief divsn. comparative psychology, 6571st Aeromed. Rsch. Lab. Holloman AFB, N.Mex., 1965-67; instr., navigator C-130-E 345th tactical airlift squadron CCK, Taiwan, 1967-69; program monitor human resources lab. Andrews AFB, Md., 1969-72; mil. asst. for tng. and pers. tech. Office Sec. Defense, Washington, 1972-78; commandant, acad. instr., fgn. officer sch. Air U. Maxwell AFB, 1978-90; acting head aviation rsch. libr. Inst. Aviation U. Ill., Urbana-Champaign, 1978-80, dir., prof., dir. comml. ops. Willard Airport Inst., Aviation, 1980—, prof. dept. psychology, 1981—, present interim head pilot tng. Inst. Aviation, 1993—; part-time instr. No. Mont. Coll., Havre, 1961, Alamogordo (N.Mex.) C.C., 1966, Prince Georges County C.C., Largo, Md., 1977, Inst. Aviation, U. Ill., Urbana-Champaign, 1988—; cons. Camp Rockmount, Black Mountain, N.C., 1969-87, 89—, Dept. Energy, 1980, Human Resources Libr., USAF, 1981, chmn. 1982, Office Naval Rsch., 1981, FAA, 1981-82, NAS, 1983-84, W & L Assocs., 1985-89, Sci. Applications Internat. Corp., McLean, Va., 1990-92, Inst. Defense Analyses, Alexandria, Va., 1985—, Galaxy Sci. Corp., Atlanta, 1990-92; joint adv. com. exec. group Lincoln Lab., Boston, 1971-72; drug rsch. coord. com. Dept. Defense, 1972-75; tech. adv. group FAA, 1976-78; bd. visitors 93d Bombardment Wing Tng. Program, 1979-80; chmn. rsch. adv. panel ops. tng. divsn. USAF Human Resources Lab., 1983-84, divsn. tng. systems, 1983, 84, rsch. adv. panel divsn. ops. tng. 1985; apptd. by gov. Bd. Aeronautical Advisors, State of Ill, 1985-92; mem. Ill. Task Force Compatible Land Use, 1986-91; Chanute AFB Reuse Aviation/Tourism com. 1989-91; mem. strategic oversight com. nat. Aviation Tng., Maintenance and Tech. Ctr., 1989-90; mem. com. Intermodal Trasp. Ctr., 1991—; mem. sci. adv. bd. USAF, 1993—. Editor: Human Factors, 1983-84, Human Computer Interaction, 1990; editorial bd. Military Psychology, 1988—; contbr. articles to profl. jours., chpts. to books. Scout master Boy Scouts Am., Black Mountain, 1960-62, cub scout com. Oxon Hill, Md., 1969-73; dir. Camp Ahoy, Black Mountain, 1969-76, family camp Camp Ridgecrest, Black Mountain, 1990;. Fellow APA (at large, divsn. 19, exec. com. divsn. applied exptl. and engring. psychologists 1983, 84-90, 91—, sec. 1984-88, apptd. acting treas. 1986, award com. 1987-89, chair 1990, mid-yr. symposium com. 1987—, chair nominations and election com. 1987-88, pres. elect 1987, pres. 1988, past pres. 1989-90, chair mem. com. 1991), Am. Psychological Soc., Human Factors Soc. (awards com.), Aerospace Human Factors Assn. (pres. protem 1990-91, pres. 1991-92, acting newsletter editor 1992, past pres. 1992-93, chair awards com. 1992-93); mem. AAAS, AIAA, Aerospace Med. Assn. (program com. 1986—, long range planning com. 1987—, chair ad hoc com. human factors 1987-88, chair aerospace human factors com. 1988-91, constitution and by-laws com. 1988-90, membership com. 1989—, nominations and election com. 1991—, Raymond F. Longacre award 1992), Assn. Aviation Psychologists, IMAGE Soc., Soc. Automotive Engrs., Ill. Pub. Airports Assn. (bd. dirs. 1980-90, 2d v.p. 1983-84, 1st v.p. 1983-84, chmn. internal affairs com. 1984-86, pres. 1986-88, past pres. 1988-90, chair membership com. 1988-92), Univ. Aviation Assn., Arnold Air Soc., Urbana-Champaign C. of C. (joint pub. transp. com. 1980-90, pub. transp. com. 1990—), Rotary, Scabbard and Blade, Sigma Xi, Psi Chi, Phi Delta Kappa. Home: 144 Oak Tree Rd Seymour: U of Illinois Willard Airport Inst of Aviation Urbana IL 61801

TAYLOR, HUGH PETTINGILL, JR., geologist, educator; b. Holbrook, Ariz., Dec. 27, 1932; s. Hugh Pettingill and Genevieve (Fillerup) T.; m. Candis E. Hoffman, 1982. B.S., Calif. Inst. Tech., 1954; A.M., Harvard U., 1955; Ph.D., Calif. Inst. Tech., 1959. Asst. prof. geochemistry Pa. State U., 1960-62; mem. faculty div. geol. and planetary scis. Calif. Inst. Tech., 1962—, now prof. geology, Robert P. Sharp prof., 1981; Crosby vis. prof. M.I.T., 1978; vis. prof. Stanford U., 1981; William Smith lectr. Geol. Soc. London, 1976; Hofmann lectr. Harvard U., 1980; Cloos lectr. Johns Hopkins U., 1986; with U.S. Geol. Survey, Saudi Arabia, 1980-81. Author: The Oxygen Isotope Geochemistry of Igneous Rocks, 1968, Stable Isotopes in High Temperature Geological Processes, 1986, Stable Isotope Geochemistry, 1991; assoc. editor Bull. Geol. Soc. Am, 1969-71, Geochimica Cosmochimica Acta, 1971-76; editor Chem. Geology, 1985-91. Recipient Penrose medal Geol. Soc. Am., 1993. Fellow Am. Geophys. Union, Mineral. Soc. Am. (councillor), Am. Acad. Scis., Nat. Acad. Scis. (Day Medal 1993); mem. Geochem. Soc. (councillor). Republican. Office: Calif Inst Tech Dept Geological & Planetary Pasadena CA 91125

TAYLOR, HUGH RINGLAND, ophthalmologist, educator; b. Melbourne, Victoria, Australia, Nov. 10, 1947; m. Elizabeth Mara Dax, Dec. 2, 1968; children: Katherine Isabel, Bartholomew, Edward, Phoebe. B Med. Sci., U. Melbourne, 1969, MB, BS, 1971, MD, 1978. Resident med. officer Austin Hosp., Melbourne, 1972-73; ophthalmic registrar Royal Victorian Eye and Ear Hosp., Melbourne, 1974-76; assoc. dir. Nat. Trach. and Eye Health Program, R.A.C.O., Sydney, Australia, 1976-77; postdoctoral fellow Wilmer Inst., Balt., 1977-79; asst. prof. Johns Hopkins U., Balt., 1980-83, assoc. prof., 1983-90, prof. ophthalmology, 1990; prof. ophthalmology U. Melbourne, 1990—; cons. WHO, Geneva, 1979-90, dir. Collaborative Ctr. Prevention Blindness; cons. Mectizan Expert Com., Atlanta, River Blindness Found., Houston, Edna McConnell Clark Found., N.Y.C., Fred Hollows Found., Sydney. Recipient award Rsch. to Prevent Blindness, Inc., 1988, Internat. Orgn. Against Trachoma, 1990. Fellow Royal Australian Coll. Surgeons; mem. Royal Australian Coll. Ophthalmologists, Internat. Agy. for Prevention of Blindness, Assn. for Rsch. in Vision and Ophthalmology (award 1981, 88). Home: 27 Kireep Rd, Balwyn Victoria 3103, Australia Office: U Melbourne Dept Ophthalmol, 32 Gisborne St, Melbourne Victoria 3002, Australia

TAYLOR, IRVING, mechanical engineer, consultant; b. Schenectady, N.Y., Oct. 25, 1912; s. John Bellamy and Marcia Estabrook (Jones) T.; m. Shirley Ann Milker, Dec. 22, 1943; children: Bronwen D., Marcia L., John I., Jerome E. BME, Cornell U., 1934. Registered profl. engr., N.Y., Mass., Calif. Test engr. Gen. Electric Co., Lynn, Mass., 1934-37; asst. mech. engr. M.W. Kellogg Co., N.Y.C., 1937-39; sect. head engring. dept. The Lummus Co., N.Y.C., 1939-57; research engr. Gilbert and Barker, West Springfield, Mass., 1957-58, Marquardt Corp., Ogden, Utah, 1958-60, Bechtel, Inc., San Francisco, 1960-77; cons. engr. Berkeley, Calif., 1977-91; adj. prof. Columbia U., 1950-60, NYU, 1950-60. Contbr. articles to profl. jours. Fellow ASME (life, Henry R. worthington medal 1990); mem. Pacific Energy Assn., Soaring Soc. Am. (life), Sigma Xi (assoc.). Unitarian. Avocations: sailplane soaring, skiing, lawn bowling. Home: 300 Deer Valley Rd Apt 2P San Rafael CA 94903-5514

TAYLOR, J(AMES) HERBERT, cell biology educator; b. Corsicana, Tex., Jan. 14, 1916; s. Charles Aaron and Delia May (McCain) T.; m. Shirley Catherine Hoover, May 1, 1946; children: Lynne Sue, Lucy Delia, Michael Wesley. B.S., So. Okla. State U., 1939; M.S., U. Okla., 1941; Ph.D., U. Va., 1944. Asst. prof. bacteriology and botany U. Okla., Norman, 1946-47; assoc. prof. botany U. Tenn., Knoxville, 1948-51; asst. prof. botany Columbia U., N.Y.C., 1951-54; assoc. prof. Columbia U., 1954-58, prof. cell biology, 1958-64; prof. biol. sci. Fla. State U., Tallahassee, 1964-83, Robert O. Lawton disting. prof. biol. sci., 1983-90, prof. emeritus, 1990—; assoc. dir. Inst. Molecular Biophysics, Fla. State U., 1970-79; dir. Inst. Molecular Biophysics, 1980-85; cons. Oak Ridge Nat. Lab., 1949-51; research collaborator Brookhaven Nat. Lab., 1951-58; nat. lectr. Sigma Xi Research Soc. Author: Molecular Genetics, Vol. 1, 1963, Vol. 2, 1965, Vol. 3, 1979; DNA Methylation and Cellular Differentiation, 1983; also papers on molecular genetics; contbr. over 100 articles in field to profl. jours. Pres. Unitarian Ch. Tallahassee, 1968-70. Served with M.C. U.S. Army, 1944-46, PTO. Recipient Meritorious Research award Mich. State U., 1960; Guggenheim fellow Calif. Inst. Tech., 1958-59. Mem. Nat. Acad. Scis., AAAS, Am. Inst. Biol. Sci., Am. Soc. Cell Biologists (pres. 1969-70), Biophysics Soc., Genetics Soc. Am. Democrat. Office: Fla State U Inst Molecular Biophysics Tallahassee FL 32306

TAYLOR, JAMES HUTCHINGS, chancellor; b. Hamilton, Ont., Can., Mar. 25, 1930; s. John Douglas and Mabel (Pugh) T.; m. Mary Cosh, Oct. 18, 1957; children: Andrew, Sarah, Katherine, Pegatha, James. BA, McMaster U., 1951, LLD (hon.), 1989; BA, Oxford U., 1953, MA, 1983. With Dept. External Afairs, 1953-55, Vietnam, 1955-56, India, 1956-58, U.S.S.R., 1967-70, France, 1973-76; asst. under-sec. of state Dept. External Afairs, 1977-80, dep. under-sec., 1980-82, under-sec. of state, 1985-89; amb., permanent rep. of Can. North Atlantic Council, 1982-85; amb. of Can. to Japan, 1989-93; personal rep. for Prime Minister to Paris Econ. Summit, 1989; chancellor McMaster U., 1992—. Lt. Royal Can. Inf.,1950. Club: Five Lakes. Home: 541 Manor Ave, Ottowa, ON Canada Office: McMaster U, 1280 Main St W, Hamilton, ON Canada L8S 4L8*

TAYLOR, JAMES KENNETH, biology educator, science teaching consultant; b. Fall River, Mass., July 28, 1929; s. James H. and Annie (Hembrough) T.; m. Barbara Louise Gray, Apr. 18, 1953; children: Sherry Anne, Keith Alan. BSE, Bridgewater (Mass.) State U., 1947-51; MA, Columbia U., 1951-54. Tchr. Fall River Pub. Schs., 1951-56; instr. biology Westfield (Mass.) State Coll., 1956-61, asst. prof., 1961-66, assoc. prof., 1966-86, prof., 1986—. Active Westfield Conservation Commn., 1963—, chmn. 1980—; mem. Rep. City Com., Westfield, 1990—. Teacher Tng. grantee (8 grants) NSF, Washington, 1970-75; recipient Conservationist of Yr. award Westfield Grange, 1988. Mem. Nat. Sci. Tchrs. Assn., Nat. Assn. Biology Tchrs. (annual conv. commr., Boston 1983), Ecol. Soc. Am., Mass. Sci. Suprs. Assn. (chmn. 1988-89). Republican. Methodist. Office: Dept Biology Westfield State Coll Westfield MA 01086

TAYLOR, JOHN JOSEPH, nuclear engineer; b. Hackensack, N.J., Feb. 27, 1922; s. John J.D. and Johanna F. (Thibideau) T.; m. Lorraine Crowley, Feb. 5, 1943; children: John B., Nancy M., Susan M. BA, St. John's U., Jamaica, N.Y., 1942, DSc (hon.), 1975; MS, U. Notre Dame, 1947. Mathematician Bendix Aviation Corp., Teterburo, N.J., 1946-47; engr. Kellex Corp., N.Y.C., 1947-50; v.p. water reactor div. Westinghouse Electric Corp., Pitts., 1950-81; v.p. nuclear power Electric Power Research Inst., Palo Alto, Calif., 1981-83; mem. adv. com. Oak Ridge (Tenn.) Nat. Lab., 1973-83, Brookhaven Nat. Lab., Upton, N.Y., 1986—, Argonne (Chgo.) Nat. Lab., 1980-86, bd. dirs.; cons. Office of Tech Assessment, Washington, 1975—. Co-author Reactor Shielding Manual, 1953, Naval Reactor Physics Manual; contbr. articles to profl. jours. Served to lt. (j.g.) USN, 1942-45. Recipient Order of Merit, Westinghouse Electric Corp., 1957, George Westinghouse Gold medal ASME, 1990. Fellow AAAS (bd. dirs.), NAE, Am. Nuclear Soc. (bd. dirs. Walter Zinn award 1993, bd. dirs.); mem. Am. Phys. Soc., Internat. Atomic Energy Agy. (internat. adv. group), Vienna, Austria, Cosmos Club, Washington. Republican. Roman Catholic. Home: 15 Oliver Ct Menlo Park CA 94025-6685 Office: Electric Power Research Inst PO Box 10412 3412 Hillview Ave Palo Alto CA 94303

TAYLOR, JON GUERRY, civil engineer; b. Columbia, S.C., Nov. 8, 1939; s. Edward Lee and Sadie Margaret (Session) T.; m. Faye Elizabeth Taylor, June 3, 1961; children: Jon Jr., Sally, Joy. BSCE, U. S.C., 1965, postgrad., 1966-67, 88-89. Registered profl. engr. S.C., N.C., Ga., Fla., Va., Mich. V.p. Lyles, Bisset, Carlisle and Wolff of S.C., Columbia, 1966-70, Lowell Dunn Co., Miami, Fla., 1970-72, Capeletti Bros., Miami, 1972-77; prin. engr. Wilbur Smith Assocs., Charleston, S.C., 1977-80; dir. engring. and constrn. Gator Dock and Marine, Sanford, Fla., 1980-84; pvt. practice engring. Charleston, 1984—; lectr. Sea Grant, Ft. Lauderdale, Fla., 1983, Sarasota. Fla., 1985, Internat. Marina Inst., Boston, 1987, Clearwater, Fla., 1990, Ft. Lauderdale, 1990, 93, States Orgn. for Boating Access, Charleston, S.C., 1988, others. Contbr. articles in marina design to mags. and tech. jours. Served with U.S. Army, 1958-59. Sea grantee various orgns. Mem. NSPE, ASCE, Internat. Marina Inst., Civil Engrs. Club Charleston (pres. 1988-89). Republican. Methodist. Home: 1063 Deleisseline Blvd Mount Pleasant SC 29464-9549 Office: PO Box 1082 Mount Pleasant SC 29465

TAYLOR, JOSEPH CHRISTOPHER, audio systems engineer; b. Polkton, N.C., May 7, 1959; s. James Guy and Estell Rose (Wright) T.; m. Selena Renée Ray, May 7, 1983; children: Courtney Desireé, Emily Ashton. BSEE, Clemson U., 1983. Rep. service Neotek West, Los Angeles, 1983-84; engr.

electronics AFPRO/TRW, Redondo Beach, Calif., 1983-84; engr. audio Premore Inc., North Hollywood, Calif., 1984-87; field engr. West Coast region Adams Smith, Burbank, Calif., 1987-88; co-owner Update Audio/Video Services, Clemson, S.C., 1983—. Baptist. Achievements include research in digital fiber optic factory automation work stations, AI robotic linkage determinants, factory interactive communication and visualization. Avocations: computers, snorkeling. Office: Update Audio/Video Svcs 201 Kelly Rd Clemson SC 29631-1638

TAYLOR, JOSEPH HOOTON, JR., radio astronomer, physicist; b. Phila., Mar. 29, 1941; s. Joseph Hooton and Sylvia Hathaway (Evans) T.; m. Marietta Bisson, Jan. 3, 1976; children: Jeffrey, Rebecca, Anne-Marie. B.A. in Physics, Haverford Coll., 1963; Ph.D. in Astronomy, Harvard U., 1968; D.Sc. (hon.), U. Chgo., 1985. Research fellow, lectr. Harvard U., 1968-69; asst. prof. astronomy U. Mass., Amherst, 1969-72; assoc. prof. U. Mass., 1973-77, prof., 1977-81; prof. physics Princeton U., 1980—. Author: Pulsars, 1977. Recipient Dannie Heineman prize in astrophysics Am. Inst. Physics/Am. Astron. Soc., 1980, Tomalla Found. prize in gravitation and cosmology, 1985, Magellanic Premium award Am. Philos. Soc., 1990, Wolf Prize in Physics, Wolf Found., 1992, Nobel Prize in Physics, Nobel Foundation, 1993; MacArthur fellow, 1981. Fellow Am. Acad. Arts and Scis., Am. Phys. Soc.; mem. NAS (Henry Draper medal 1985, John J. Carty medal Advancement Sci. 1991), Am. Philos. Soc., Am. Astron. Soc., Internat. Sci. Radio Union, Internat. Astron. Union. Mem. Soc. of Friends. Home: 272 Hartley Ave Princeton NJ 08540-5656 Office: Princeton U Physics Dept Princeton NJ 08544

TAYLOR, KENNETH DOUGLAS, finance and computer consultant, educator; b. Topeka, Nov. 21, 1942; s. Olin Orlando and Lola Louise (Conley) T.; AB, George Washington U., 1964, MS in Stats., 1966; MS in Computer Sci. SUNY, 1990, PhD in Math. Eurotech, 1992, (univ. fellow); student of Peter Hilton; postgrad., McGill U., 1974, Bowdoin Coll., U. Montreal; m. Joy Ellen Rice, May 25, 1973 (div. Nov. 1981). Registered rep./stockbroker. Sr. programmer C-E-I-R, Inc., 1963, 69; instr. Army Map Svc., 1964-65; student instr. McGill U., 1966-71; rsch. assoc. U. Va. Med. Sch., 1972; fin. and computer cons., Plymouth, N.Y., 1973-87; computer scientist USAF, 1989-90; broker Russell Hawkes Assoc./Linsco/Pvt. Ledger, 1993—; sec. Richmond (Va.) Computer Club, 1977. Summer grantee NSF, Can. Research Council. Mem. ASTM, Am. Math. Soc. Author papers in field. Home: PO Box 490 Chenango Bridge NY 13745 Office: Russell Hawkes Assoc 783 Chenango St Binghamton NY 13901

TAYLOR, KENT DOUGLAS, molecular biologist; b. Indpls., June 9, 1952; s. Kenneth Dean and M. Roberta (Read) T.; m. Debra Laraine Evans, July 29, 1978; children: Anna Danielle, Michael James Evans. PhD in Biology, U. So. Calif., 1982; MA in Religion, Pepperdine U., 1993. Microbiologist Dept. Vets. Affairs Med. Ctr. Devel. Biology Lab., Sepulveda, Calif., 1978—; asst. prof. Calif. State U., Northridge, 1990—. Author: Methods in Enzymology, 1993; contbr. articles to profl. jours. Lectr. Ventura (Calif.) Foursquare Ch., 1990—, Van Nuys (Calif.) Foursquare Ch., 1984-86. Recipient Performance award Dept. Vets. Affairs, 1985. Mem. AAAS, Greater L.A. Zoo Assn., San Diego Zool. Soc. Republican. Achievements include refinement of nucleic acid hybridization techniques to a sensitivity of 10 femtogram RNA and application to expression of genes in early mouse embryos; estimation that 8-10% of RNA is ribosomal, 25% is mitochondrial in blastocysts. Office: Calif State U 18111 Nordhoff St Northridge CA 91330

TAYLOR, LAURA MARIE, interior designer; b. Bremen, Ga.; d. Jack Seagraves and Harriet (Hardaway) T. Student, Parsons Sch. Design, 1979; BFA in Art and Interior Design, U. Ga., 1980; postgrad., Harvard U., 1990. Interior designer Fin. Bldg. Cons., Atlanta, 1980-81, Cooper Carry Interiors Group, Atlanta, 1981-84; freelance interior designer Atlanta, 1984-85; interior design coord. Trammel Crow Co., Atlanta, 1985-90; prin. Laura Taylor Interiors, Atlanta, 1990—. Interior designer Vignette Egleston Hosp. Festival of Trees, 1990, Egleston Hosp. Festival of Trees, 1991, 92; mem. High Mus. Art, Atlanta, 1980—, Spl. Friends of Decorative Arts and Young Careers; mem. com. Beaux Arts Ball-Carnival Auction, 1991; Atlanta Symphony Decorator's Show House, 1992. Published in Atlanta Homes & Lifestyles mag., 1992. Mem. Am. Soc. Interior Designers (archives chmn. 1988, chmn. ann. state meeting 1991, designer show ho. 1991, Project Design award 1st place healthcare 1993, Spl. Recognition award 1993, awards chmn. 1993), Internat. Fursnishings and Design Assn., Network Exec. Women in Hospitatlity Design, Nat. Coun. for Interior Design Qualification (cert.), Jr. League Atlanta. Methodist. Home and Office: 300 Fountain Oaks Ln Atlanta GA 30342

TAYLOR, LAWRENCE STANTON, software engineer; b. Oklahoma City, May 18, 1959; s. Stanton Adelbert and Eleanor Louise (McDaniel) T.; m. Susan Lynn Kaiser, May 25, 1985. BA in Integrated Sci.-Computer Studies, Northwestern U., 1981, MS in Computer Sci., 1983. Assoc. program analyst Harris Bank & Trust, Chgo., 1982-83; software engr. Gen. Dynamics, Ft. Worth, 1983-86, sr. software engr., 1986-88, software design specialist, 1988-91; software engr. D. Appleton Co., Inc., Irving, Tex., 1991—. Murphy fellow Northwestern U., 1981. Achievements include specializing in migration of software to new computers. Home: 3105 Huron Trail Lake Worth TX 76135 Office: D Appleton Co Inc Ste 1141 222 W Las Colinas Blvd Irving TX 75039

TAYLOR, LELAND ALAN, chemical and materials scientist, researcher; b. Springfield, Vt., Oct. 8, 1945; s. Wilberne Kincaid and Very Katherine (Prouty) T; m. Frances Ellen LaBay. Student, Am. Internat. Coll., 1967; BS in Broad Field Sci., Lyndon State Coll., 1971. Electrical lab. technician, lab. chemist WHV Weidmann Industries, St. Johnsbury, Vt., 1979—. Co-author: (reports) Advanced Concepts for Transformer Pressboard Dielectric Constant and Mechanical Strength, 1982, HVDC Converter Transformer Insulation, 1985. Mem. Am. Chem. Soc., Soc. Plastics Engrs., N.Y. Acad. Scis., Green Mountain Water Pollution Control Assn., The Smithsonian Assocs. Achievements include high heat resistant low dielectric constant oil-impregnatable electrical insulating board. Office: EHV Weidmann Industries Memorial Dr Box 903 Saint Johnsbury VT 05819

TAYLOR, LELAND HARRIS, JR., civil engineer; b. Oakland, Calif., Mar. 26, 1958; s. Leland Harris and Avice M. (Hatton) T.; m. Swee Choo, Apr. 30, 1991. BS in Marine Systems Engring., Tex. A&M U., 1981; MCE, U. Calif., Berkeley, 1989. Engr. in tng. Pipeline engr. Bechtel Petroleum, Houston, 1981-82; field engr. McDermott Internat., Dubai, United Arab Emirates, 1982-85; ops. engr. McDermott Internat., Singapore, 1986-89, sr. ops. engr., 1989—. Mem. Soc. Naval Architects and Marine Engrs. (assoc.). Achievements include development of dead weight induced tension lifts to assist in installation of subsea pipelines; design of multiple pile installation template for marine construction. Office: McDermott Internat SE Asia, 47 Shipyard Rd, Singapore 2262, Singapore

TAYLOR, LESLI ANN, pediatric surgery educator; b. N.Y.C., Mar. 2, 1953; d. Charles Vincent Taylor and Valene Patricia (Blake) Garfield. BFA, Boston U., 1975; MD, Johns Hopkins U., 1981. Diplomate Am. Bd. Surgery. Surg. resident Beth Israel Hosp., Boston, 1981-88; rsch. fellow Pediatric Rsch. Lab. Mass. Gen. Hosp., Boston, 1984-86; fellow pediatric surgery Children's Hosp. of Phila., Phila., 1988-90; asst. prof. pediatric surgery U. N.C., Chapel Hill, 1990—. Author: (booklet) Think Twice: The Medical Effects of Physical Punishment, 1985. Recipient Nat. Rsch. Svc. award NIH, 1984-86. Mem. AMA. Achievements include research on organ preservation for pediatric liver transplantation and short bowel syndrome.

TAYLOR, LYLE H., physicist; b. Paten, Iowa, Oct. 23, 1936; s. Hilles J. and Alma (Fredericks) T.; m. Jane Kay Easley, Sept. 27, 1959; children: Steven, Alan, Susan, Kevin, Brian. BS, Iowa State U., 1958; MS, N.Mex. State U., 1961; PhD, U. Kans., 1967. Sr. scientist White Sands (N.Mex.) Missile Range, 1958-61, MidWest Rsch. Inst., Kansas City, Mo., 1961-64; sr. scientist Westinghouse Sci. and Tech. Ctr., Pitts., 1967-84, fellow scientist, 1984—. Contbr. articles to profl. publs. Pres. Lions Club, Murrysville, Pa., 1978-79; commr. Pony Tail Softball League for Girls, Murrysville, 1974-77, Little League Baseball for Boys, Murrysville, 1970-71. Recipient George Westinghouse award for Outstanding Engring. Achievement, 1984, George Westinghouse Innovation award, 1990. Mem. Am. Phys. Soc. Presbyterian. Achievements include leadership of tech. team that demonstrated high power mid-IR laser; design of CO2 laser for pumping nonlinear harmonic generator crystals; co-invention of Spectral Imaging Threat Evaluation Sensor System. Home: 3317 Benden Dr Murrysville PA 15668 Office: Westinghouse Sci/Tech Ctr 1310 Beulah Rd Pittsburgh PA 15235

TAYLOR, MARK JESSE, military engineer; b. Phillipsburg, N.J., Feb. 5, 1957; s. Jesse Ireland Jr. and Eleanor Jane (Meeker) T.; m. Hilda Susan Valdivia, June 15, 1983; children: Alexandra, Monique, Elizabeth Jane. BSCE, Norwich U., 1979; MS in Mgmt., Lesley Coll., 1985. Engr. in tng. Commd. officer USAF, 1979, advanced through grades to maj., 1991; dep. missile crew comdr. USAF, 400 Strategic Missile Squadron, F.E. Warren AFB, Wyo., 1980-82; standardization evaluation mgr. USAF, 90 Strategic Missile Wing, F.E. Warren AFB, 1982-83; missile combat crew commander USAF, 400 Strategic Missile Squadron, F.E. Warren AFB, 1983-84; chief code handler tng. USAF, 90 Strategic Missile Wing, F.E. Warren AFB, 1984-85; ICBM flight test mgr. Ballistic Missile Orgn., Norton AFB, Calif., 1985-88; advanced tech. project mgr. Ballistic Missile Orgn., Norton AFB, 1988-89; dir. maintenance, engring. 1st Space Launch Squadron, Cape Canaveral AFS, Fla., 1989-93; mgr. Ballistic Missile Tech. Program HQAF-SPACECOM, Peterson AFB, Colo., 1993—. Presenter in field. V.p. Parent-Tchrs. Assn., San Bernardino, Calif., 1987-89. Mem. ASCE, Soc. Am. Mil. Engrs., Air Force Assn. Republican. Home: 2127 E Greenwich Circle Colorado Springs CO 80909 Office: AFSPACECOM/DRMM Peterson AFB CO

TAYLOR, MIKE ALLEN, computer systems analyst; b. Clinton, Ind., Feb. 6, 1949; s. Bruce Robert and Frieda (Eckle) T.; m. Terri Joy Thompson, Nov. 29, 1974; children: Krystyl Nancy Roundy III, Suzette Lynn. Grad., USAF Air Command & Staff Coll., 1969; BS, Ind. U., Indpls., 1992. Sr. system cons. Blue Cross and Blue Shield of Ind., Indpls., 1977-89, H.A.S., Indpls., 1989-90; systems analyst GTE Data Svcs., Ft. Wayne, Ind., 1990—. Author: (computer program) Silver Bullet, 1993. Mem., 1st lt. CAP, Santa Barbara, Calif., 1963-80; mem. pres. coun. Christian Heritage Coll., El Cajon, Calif., 1992-93. Recipient Gen. Carl A. Spaatz award CAP, Santa Barbara, 1969, Falcon award, 1970. Mem. Associated PC User Group (pres., founder 1985-89), Ctrl. Ind. Atari User Group (pres. 1982-84), Ind. State Hist. Soc., MENSA. Republican. Baptist. Home: 4920 Vance Ave Fort Wayne IN 46815 Office: GTE Data Svcs 6920 Pointe Inverness Way Fort Wayne IN 46804

TAYLOR, MORRIS ANTHONY, chemistry educator; b. St. Louis, July 10, 1922; s. Henry Clay Nathaniel and Georgia Lee Anna (Kenner) T.; m. Millie Betty Fudge, July 17, 1948 (dec. Jan. 1969); children: Carla Maria, Morris Jr.; m. Veonnia Joyce McDonald, Aug. 4, 1973; children: Dorcas Lynnea, Demetrius Sirrom. BS in Chemistry, St. Louis U., 1952. Rsch. chemist Universal Match Corp., Ferguson, Mo., 1952-54; mfg. chemist Sigma Chem., St. Louis, 1954; clin. chemist 5th Army Area Med. Lab., St. Louis, 1955-56; analytical chemist U.S. Dept. Agr.-Agrl. Rsch. Svc. Meat & Poultry Inspection, St. Louis, 1956-67; supervisory chemist U.S. Dept. Agr.-Food Safety and Quality Svc., St. Louis, 1967-76, chemist in charge, 1976-79; adj. prof. chemistry St. Louis Community Coll., 1981—; rating panel mem. Bd. CSC, St. Louis, 1969-79; reviewer Assn. Ofcl. Analytical Chemists, St. Louis, 1969-79; collaborator FDA Labs. on Analytical Methods, St. Louis, 1969-79. Bd. mem. Draft Bd. III, St. Louis, 1970-76. With U.S. Army, 1942-46. Fellow Am. Inst. Chemists; mem. Am. Chem. Soc., Internat. Union Pure and Applied Chemistry, St. Louis U. Alumni Chemists. Roman Catholic. Home: 10410 Monarch Dr Saint Louis MO 63136-5612 Office: St Louis Community Coll 5600 Oakland Ave Saint Louis MO 63110-1393

TAYLOR, PATRICIA ELSIE, epidemiologist; b. Ayr, Queensland, Australia, Mar. 20, 1929; d. Ernest Howard and Mayzie Lucy (Kwong) Lee; m. Kenneth Douglas Taylor, Oct. 1, 1960; 1 child, Douglas Craig. BS, U. Queensland, 1952, postgrad., 1954; PhD, U. Calif., Berkeley, 1964; LLD (hon.), St. Francis Xavier U., N.S., Can., 1981. Mem. rsch. staff Queensland Inst. Med. Rsch., Brisbane, 1949-58; assoc. in epidemiology Sch. Pub. Health U. Calif., Berkeley, 1958-60; grad. rsch. fellow Inst. Nutrition for Cen. Am. and Panama, Guatemala, 1960-63; rsch. fellow Child Rsch. Ctr. of Mich., Detroit, 1965-66; sr. rsch. assoc. London Sch. Hygiene and Tropical Medicine, 1967-71; rsch. scientist Dept. Nat. Health and Welfare, Ottawa, Ont., Can., 1972-78; sr. cons. in virology and sci. rsch. Iranian Nat. Blood Transfusion Svc., Pasteur Inst., Tehran, 1978-80; epidemiologist Lindsley F. Kimball Rsch. Inst., N.Y. Blood Ctr., N.Y.C., 1981—; prin. dancer Queensland Ballet Theatre, 1956-58; solo dancer Guatemalan Nat. Ballet, 1960-63; mem. various U.S. Fed. Adv. Panels for infectious disease and AIDS. Trustee Cathedral of St. John the Divine, N.Y., 1981. Named Woman of Yr., Can. Women's Club of N.Y., 1992; Paul Harris fellow, 1983, Internat. fellow AAUW, 1954-55; grantee Fulbright Found., 1954-55, Rockefeller Found., 1955. Mem. Royal Acad. Dancing, West Point Soc. N.Y. (hon.), Order of Can. (hon.). Home: 146 W 57th St # 61T New York NY 10019-3323 Office: NY Blood Ctr Lindsley F Kimball Rsch Inst 310 E 67th St New York NY 10021

TAYLOR, PAUL ALLEN, chemical engineer; b. Provo, Utah, Jan. 9, 1952; s. Frank Arvil and Phyllis (Christensen) T.; m. Jacqueline Lou Kracker, Sept. 3, 1977 (div. Feb. 1993). BS, Brigham Young U., 1974. Devel. engr. Oak Ridge (Tenn.) Y-12 Plant, 1974-88; devel. engr. Oak Ridge Nat. Lab., 1988—. Co-author several articles for profl. jours. Recipient Outstanding Engring. Achievement award Tenn. Soc. Profl. Engrs., 1986. Mem. NSPE (local mathcounts chmn. 1990-92), AICE (local membership chmn. 1992—), Water Environment Fedn. (nat. program com. 1992—), Vols. in Tech. Assistance. Achievements include developing an in-situ biol. denitrification process that was used to treat four waste ponds at the Oak Ridge Y-12 Plant, developing a recycle system that eliminates wastewater discharges from the plating shop at Oak Ridge. Office: Oak Ridge Nat Lab Bldg 3017 PO Box 2008 Oak Ridge TN 37831-6044

TAYLOR, PAUL DUANE, JR., chemical engineer; b. Middlesex, N.J., Dec. 2, 1969; s. Paul Duane and Anita Marie (Majikas) T. BSChemE, Rutgers U., 1992. Lab. technician GAF, Wayne, N.J., 1989; rsch. asst. dept. chem./biochem. engring. Rutgers U., Piscataway, N.J., 1989-90, rsch. assoc., 1990-92; mfg. engr. ISP, Texas City, Tex., 1992-93; sales engr. ISP, Houston, Tex., 1993—; mentor Rutgers Engring. Open House, Piscataway, 1990, 91. Mem. AAAS, Am. Inst. Chem. Engrs. (pres. 1991-92, 1st pl. paper competition 1990, capt. ambassador program 1992). Republican. Roman Catholic. Achievements include discovery of mechanism for preferential metabolic pathways of monoaromatic substituted hydrocarbon degradation under denitrifying conditions; assistance in pilot scale demonstration of pneumatic fracturing tech.-an innovative approach to soil/clay decontamination. Home: 11300 Regency Green # 2102 Cypress TX 77429 Office: ISP Spring Cypress Rd PO Box 1076 Cypress TX 77429

TAYLOR, RANDY STEVEN, hospital administrator; b. Bell, Calif., July 17, 1951; s. Jerry Joe and Viann Beverly (Borthick) T.; m. Rochelle Helen Forten, Feb. 10, 1978; children: Zachary Scott, Chelsea Leigh. BA, Columbia Pacific U., 1987, MA, 1991. Gen. contractor R. Taylor Constrn., Nuevo, Calif., 1978—; gen. supt. Del-Tec, Riverside, Calif., 1984-86; inspector of record Office State Health and Planning, Riverside, Calif., 1986-91; asst. adminstr., v.p. facilities Hemet (Calif.) Valley Med. Ctr., 1986-88; dir. facilities Riverside (Calif.) Community Hosp., 1988—; cons. Developmental at Hemet and Riverside Community Hosp. Departmental, Riverside (Calif.) Community Hosp., 1988—; gen. contractor Residential Comml. Constrn., Riverside County, 1980—; liaison State Fire Marshals Office, West Covina, Calif., 1986—, Office of Statewide Health Planning and Devel., Sacramento, Calif., 1986—. Co-chair United Way Campaign, Riverside Community Hosp., 1990; chair United Way Golf Tournament, Moreno Valley, 1990; hon. mem. Bob Dole Campaign, Washington, 1990. Recipient Cert. of Am. United Way, 1990. Fellow Am. Hosp. Assn.; mem. Exec. 2000 Club (2 Certs. of Appreciation 1990), Nat. assn. of Gen. Contractors, Hosp. Coun. of So. Calif., Canyon Crest Country Club, RIV Community Hosp. Golf League (pres. 1988). Republican. Avocations: golf, snow skiing, surfing. Office: Riverside Community Hosp 4445 Magnolia Ave Riverside CA 92501-4135

TAYLOR, RICHARD EDWARD, physicist, educator; b. Medicine Hat, Alta., Can., Nov. 2, 1929; came to U.S., 1952; s. Clarence Richard and Delia Alena (Brunsdale) T.; m. Rita Jean Bonneau, Aug. 25, 1951; 1 child, Norman Edward. B.S., U. Alta., 1950, M.S., 1952; Ph.D., Stanford U., 1962; Docteur honoris causa, U. Paris-Sud, 1980; DSc, U. Alta., 1991; LLD (hon.), U. Calgary, Alta., 1993. Boursier Lab. de l'Accelerateur Lineaire, Orsay, France, 1958-61; physicist Lawrence Berkeley Lab., Berkeley, Calif., 1961-62; staff mem. Stanford (Calif.) Linear Accelerator Ctr., 1962-68, assoc. dir., 1982-86, prof., 1968—. Fellow Guggenheim Found., 1971-72, von Humboldt Found., 1982; recipient Nobel prize in physics, 1990. Fellow AAAS, Am. Phys. Soc. (W.K.H. Panofsky prize div. particles and fields 1989), Royal Soc. Can.; mem. Am. Acad. Arts and Scis., Can. Assn. Physicists, Nat. Acad. Scis. (foreign assoc.). Office: Stanford Linear Accelerator Ctr PO Box 4349 Palo Alto CA 94309-4349

TAYLOR, ROBERT BONDS, instructional designer; b. Alamo, Tenn., Feb. 23, 1939; s. Spicer Elmo and Verginia Clerc (Davis) T.; m. Nancy Chipman, Oct. 26, 1964 (div. 1975); children: Robert Kent, Midge Chipman; m. Gailya Ann Mowe, Sept. 15, 1976; 1 child, John David Clines. BS in Indsl. Tech., Tenn. Technol. U., 1962; MS in Indsl. Edn., U. Tenn., 1965. Supr. Tenn. Dept. Vocat.-Tech. Edn., Nashville, 1965-71; v.p. Deltronics Inc., Nashville, 1971-75; pres. Tech. Engrs. Inc., Nashville, 1975-81; instrnl. technologist Resource Systems Internat., Nashville, 1981-86; pres. Tech. Resource Group, Inc., Nashville, 1986—; cons. Duracell, Inc., Cleveland, Tenn., 1987-89, Dow Chem. USA, Midland, Mich., 1988-92, E.I. Du Pont de Nemours, Chattanooga, 1987-89. Author: Taskmaster: Train the Trainer, 1986, Variable Frequency Drives, 1987, Mechanical Power Transmission, 1988. Sgt. U.S. Army, 1962-68. Mem. Soc. Applied Learning Technology, Am. Chem. Soc., Inst. Indsl. Engrs., Am. Soc. Quality Control. Achievements include design and development of a computer-generated system for quantifying and qualifying technical training needs currently used by 10 Fortune 100 companies. Office: Tech Resource Group Inc Ste 200 2200 21st Ave S Nashville TN 37212

TAYLOR, RONALD RUSSELL, engineer; b. Idaho Falls, Idaho, Aug. 21, 1945; s. Russell Doane and Amelia (Carpenter) T.; m. Patricia Jane Knutzen; m. children: Kenneth, Jeffrey, Michael, Kristopher. A of Chem., Ricks Coll., 1967; BSCE, U. Idaho, 1971. Registered profl. engr. Utah, Wash. Staff engr. U&I Sugar, Moses Lake, Wash., 1971-74; tech. asst. to gen. supt. U&I Sugar, Salt Lake City, 1974-77; computer implementation leader U&I Sugar, Moses Lake, 1977-78; energy engr. Boise Cascade, Salem, Oreg., 1978-80, yeast plant mgr., 1980-82; corp. engr. Darigold Inc., Seattle, 1982—; mem. Citizens Energy Adv. Com., Salem, 1979-82, Seattle, 1984-86. With U.S. Army, 1963-68, Vietnam. Mem. LDS Ch. Office: Darigold Inc 635 Elliott Ave W Seattle WA 98119

TAYLOR, ROY LEWIS, botanist; b. Olds, Alta., Can., Apr. 12, 1932; s. Martin Gilbert and Crystal (Thomas) T. B.Sc., Sir George Williams U., Montreal, Que., Can., 1957; Ph.D., U. Calif. at Berkeley, 1962. Pub. sch. tchr. Olds Sch. Div., 1949-52; jr. high sch. tchr. Calgary Sch. Bd., Alta., 1953-55; chief taxonomy sect., research br. Can. Agrl. Dept., Ottawa, Ont., 1962-68; dir. Bot. Garden, prof. botany, prof. plant scis. U. B.C., Vancouver, 1968-85; pres., chief exec. officer Chgo. Horticultural Soc., 1985—; dir. Chgo. Bot. Garden, Glencoe, Ill., 1985—; Pres. Western Bot. Services Ltd. Author: The Evolution of Canada's Flora, 1966, Flora of the Queen Charlotte Islands, Vols. I and II, 1968, Vascular Plants of British Columbia: A Descriptive Resource Inventory, 1977; The Rare Plants of British Columbia, 1985. Mem. State of Ill. Bd. Nat. Resources and Conservation, 1987—; trustee Nature Ill. Found., 1990—, Bot. Gardens Conservation Internat., 1990—. Recipient George White Medal of Honor, Mass. Hort. Soc., 1986. Fellow Linnean Soc. London; mem. Can. Bot. Assn. (pres. 1967-68), Biol. Coun. Can. (pres. 1973-74), Am. Assn. Mus. (accreditation com. 1980-85, chmn. 1985-91, chmn. ethics commn. 1991-93), Am. Assn. Bot. Gardens and Arboreta (pres. 1976, 77, Award of Merit 1987), Ottawa Valley Curling Assn. (pres. 1968-69), B.C. Soc. Landscape Architects (hon.), B.C. Bot. Garden (hon.), Gov.-Gen.'s Curling Club of Can. (life), Univ. Club. Office: Chgo Botanic Garden PO Box 400 Glencoe IL 60022-0400

TAYLOR, SAMUEL JAMES, mathematics educator; b. Carrickfergus, Northern Ireland, Dec. 13, 1929; came to U.S., 1984; s. Robert James and Janie (Catherwood) T.; married; children—Richard, Charles, Jonathan, Helen. B.Sc., Queen's U., Belfast, No. Ireland, 1950; Ph.D., Cambridge U., 1954. Bye fellow Peterhouse, Cambridge U., Eng., 1953-55; lectr. Birmingham U., Eng., 1955-62; prof. London U., 1962-75, Liverpool U., Eng., 1975-83; vis. prof. U. B.C., Vancouver, Can., 1983-84; Whyburn prof. math. U. Va., Charlottesville, 1984—, chmn. dept., 1986-89. Author: Introduction to Measure and Probability, 1966, Exploring Mathematical Thought, 1970, Introduction to Measure and Integration, 1973; editor: Decomposition of Probability Distributions, 1964. Procter vis. fellow Princeton U., N.J. 1952. Fellow Cambridge Philos. Soc., Inst. Math. Stats.; mem. London Math. Soc. (sec. 1962-65, 68-71, editor 1980-83), Am. Math. Soc. Presbyterian. Office: U Va Dept Math and Astronomy Bldg Charlottesville VA 22903-3145

TAYLOR, SCOTT THOMAS, microwave engineer; b. Greensboro, N.C., Feb. 6, 1950; s. Charles Jackson and Barbara Jean (Edinger) T. BS in Math., U. Md., 1977, BSEE, 1980. Staff engr. Allied-Signal, Towson, Md., 1972-91; prin. engr. Ideas, Inc., Columbia, Md., 1992—; staff cons. Anghel Labs., Inc., Rockaway, N.J., 1991. Home: 10350 Cavey Lane Woodstock MD 21163 Office: Ideas Inc 7120 Columbia Gateway Dr Columbia MD 21046

TAYLOR, STEVE LLOYD, food scientist, educator, consultant; b. Portland, Oreg., July 19, 1946; s. Lloyd Emerson and Frances Hattie (Hanson) T.; m. Susan Annette Kerns, June 23, 1973; children: Amanda, Andrew. BS, Oreg. State U., 1968, MS, 1969; PhD, U. Calif., Davis, 1973. Chief food toxicology lab. Letterman Army Inst. Res., San Francisco, 1975-78; asst. prof. food rsch. inst. U. Wis., Madison, 1978-83, assoc. prof., 1983-87; prof., head dept. food sci. & tech. U. Nebr., Lincoln, 1987—, dir. food processing ctr., 1987—; cons. in field, 1978—; mem. food and nutrition bd. NAS, 1991—, chair food chems. codex com.; mem. sci. adv. bd. Internat. Life Scis. Inst./Allergy & Immunology Inst.; mem. med. adv. bd. Allerx. Editor: Food Science & Technology Series; mem. editorial bd. Jour. Food Protection; contbr. over 120 articles to sci. publs. Fellow Inst. Food Technologists; mem. Am. Chem. Soc., Am. Acad. Allergy & Immunology, Am. Peanut Rsch. & Edn. Soc., Internat. Assn. Milk, Food & Environ. Sanitarians. Democrat. Presbyterian. Achievements include patents in field including Nisin as an antibotulinal agent in processed cheese spreads. Home: 941 Evergreen Dr Lincoln NE 68510 Office: U Nebr Food Processing Ctr Lincoln NE 68583-0919

TAYLOR, WASHINGTON THEOPHILUS, mathematics educator; b. Moblie, Ala., July 24, 1931; s. W. Howard and Iona (Fuller) T.; divorced; children: Tara G. Taylor Marshall, Theolya RhoMayne Taylor Cooley. BS, Ala. State U., 1953; MBS, U. Colo., 1960; EdD, Okla. State U., 1969. Tchr. Bd. of Sch. Commrs., Mobile, 1953-54, 56-59, 1985-86; prof. math. So. U., Baton Rouge, 1960-84; instr. math. U. South Ala., Moblie, 1986-87; assoc. prof. math. Ala. State U., Montgomery, 1987-92; instr. math. Bishop State Community Coll., Mobile, Ala., 1992—; cons. in field. Mem. Scotlandville Area Adv. Com., Baton Rouge, 1960-84, Ala. Dem. Com., Moblie, 1984—; bd. dirs. Com. to Elect St. Amant to Ho. or Reps., Baton Rouge, 1982, Citizens for Williams for City Coun., Prichard, Ala, 1984. Lt. col. USAR, ret. Fellow Internat. Paper Co., 1957, NSF, 1958-60, So. Fellowship Found., 1967-68; vis. fellow Ohio State U., 1972-73; Women's Federated Club scholar, 1949. Mem. NEA (life), Nat. Coun. Tchrs. Math., Ala. Mental Health Assn. (bd. dirs.), Montgomery Mental Health Assn. (bd. dirs. 1987), Res. Officers Assn. (life), Mobile Area Mardi Gras Assn., Phi Delta Kappa, Alpha Phi Alpha (life), Pi Mu Epsilon. Avocations: reading, fishing. Home: 2150 Barlow St Mobile AL 36617

TAYLOR, WILLIAM BROCKENBROUGH, engineer, consultant, management consultant; b. Norfolk, Va., Mar. 11, 1925; s. Lewis Jerome and Roberta Page (Newton) T.; m. Nancy Dare Aitcheson, June 12, 1945; children: William B. Jr., Anne P. Taylor Cregger, Paul K., Katherine C. Taylor Nace, David A. BS, U.S. Mil. Acad., West Point, N.Y., 1945; MS in Engring., Johns Hopkins U., 1951. Profl. engr., D.C., Va. Commd. 2d lt. U.S.

Army Corps of Engrs., 1945, advanced through grades to maj., 1953, retired, 1954; gen. engr. Army AEC Nuclear Power Program, Washington, Ft. Belvoir, Va., 1955-60; engring. mgr. Army Mapping R&D, Ft. Belvoir, 1960-62; aerospace engr. NASA, Washington, 1962-67; R&D mgr. staff Army Hdqrs., Washington, 1967-69; tech. dir. Army R&D Lab., Ft. Belvoir, 1969-73; chief R & D U.S. Army Corps of Engrs., Washington, 1973-77; prin. engr. Planning Rsch. Corp., McLean, Va., 1978-80; cons. energy and mgmt. Alexandria, Va., 1980—; mem. study com. Nat. Rsch. Coun., Washington, 1983; mem. constrn. mgmt. task force Grace Commn. on Cost. Control, Washington, 1982. Contbr. articles to The Mil. Engr. jour., Wash. Acad. Scis. jour., others. Participant various civic activities. Fellow Soc. Am. Mil. Engrs. (pres. 1974); mem. AIAA, Wash. Acad. Scis., Sigma Xi. Republican. Episcopalian. Achievements include program definition of NASA's Apollo Applications Program; designs for geothermal energy power plant for Dominica; 5 mil. nuclear power plants. Home and Office: 4001 Belle Rive Ter Alexandria VA 22309

TAYLOR, WILLIAM LOGAN, electrical engineer; b. St. Joseph, Mo., Nov. 27, 1947; s. William Robert T. and Doris Jean (Benson) Oldham; m. Roberta Jean Crockett, July 24, 1967; 1 child, Mitchel William. BEE, Memphis State U., 1972. Registered profl. engr., Tex. Master mechanic Armour and Co., Memphis, 1969-72; dist. supr. Memphis Light, Gas and Water, Memphis, 1972-79; assoc. gen. mgr. Brownsville (Tex.) Pub. Utility Bd., 1979-81; mgr. ops. Magic Valley Electric Cooperative, Mercedes, Tex., 1981-84; electric dir. City of Greenville, Tex., 1984-87; gen. mgr. Kerrville (Tex.) Pub. Utility Bd., 1987—; bd. dirs. Tex. Mcpl. Power Agy., Bryan, 1985-87, sec., 1986-87; mem. electricians licensing bd. City of Greenville, 1984-87. Co-author: Neutral Currents in Three Phase Systems, 1984. Bd. dirs. Kerr County Econ. Devel. Found., Kerrville, 1991-92, sec./treas. exec. com., 1991-92; active regional water com., Kerrville, Kerr County, 1990, City of Kerrville Resources and Land Fill Com., 1991-92. Mem. IEEE, Tex. Soc. Profl. Engrs., NSPE, NRA (life), Kerrville Area C. of C. (chmn. bldg. com. 1991-92, bd. dirs. 1988-92, chmn. govtl. affairs com. 1989-92), Assn. Wholesale Customers Lower Colo. River Authority (bd. dirs. 1988-92, v.p. 1989-92), Tex. Pub. Power Assn. (bd. dirs. 1985, 86, 88-92, sec./treas. 1991, v.p. 1992), Am. Pub. Power Assn. (legis. and resolutions com. 1991-92), Hill Country Gun Club, Rotary, Jaycees (sec. Whitehaven chpt. 1970-73, exec. v.p. 1981, state dir. 1972). Methodist. Home: 150 Roundabout Ln Kerrville TX 78028 Office: Kerrville Pub Utility Bd 2250 Memorial Blvd Kerrville TX 78028

TEAFORD, NORMAN BAKER, engineering manager; b. Columbus, Ohio, Aug. 25, 1944; s. Gale Emerson and Madonna E. (Baker) T.; m. Jo Ann L. Bertler, Apr. 1, 1969; children: Timothy Todd, Tamara Therese. BCE, Ohio State U., 1967; MBA, Xavier U., 1973. Registered profl. engr. Ohio, Pa. Indsl. engr. Columbia Gas of Ohio, Inc., Portsmouth, 1967-78; area mgr. Columbia Gas of Ohio, Inc., Jackson, 1978-83, Columbia Gas of Pa., Inc., Charleroi, 1983—. Mem. ASCE, NSPE, Rotary (past pres.). Home: 163 1/2 State St Charleroi PA 15026 Office: Columbia Gas of Pa Inc Arentzen Blvd Charleroi PA 15022

TEAGUE, EDGAR CLAYTON, physicist; b. Trion, Ga., Jan. 19, 1941; married; 4 children. BS in Physics, Ga. Inst. Tech., 1961; PhD in Physics, U. North Tex., 1978. Project leader Tex. Instruments, Inc., Dallas, 1964-67; project leader Nat. Bur. Standards (now Nat. Inst. Standards and Tech.), Gaithersburg, Md., 1972-78, group leader, 1978-86, group leader, precision engr., 1986—; adv. bd. N.C. State U. Precision Engring. Ctr., Raleigh, 1986-90; invited speaker on high accuracy dimensional metrology, design, constrn. and use of precision instrumentation for metal-vacuum-metal tunneling, measuring surface roughness, light scattering methods for on-line control of surface roughness, scanning probe microscopies, stylus instruments. Editor-in-chief: Nanotech. jour., 1989—; contbr. over 60 tech. publs. Recipient Silver medal U.S. Dept. Commerce, 1981. Mem. Am. Phys. Soc., Am. Soc. Precision Engring. (chmn. bd. dirs. 1987), Sigma Pi Sigma. Achievements include patent on method of forming semiconductor regions in epitaxial layers, patent on flexure hinges. Office: Nat Inst Standards/Tech Bldg 220 Rm A117 Gaithersburg MD 20899

TEAKLE, NEIL WILLIAM, engineering executive; b. Manchester, Lancashire, Eng., Dec. 22, 1949; s. Vaughan Bernard and Irene Dorothy (Roberts) T. Engring. Diploma, Preston Poly Tech., Lancashire, Eng., 1965; MSc, Calif. U., 1982, PhD, 1989. Tech. dir. Radnav Cons., Lancashire, Eng., 1977-90, Miltec R & D Svcs., London, 1990—, Communication Devices Ltd., Harrow, Eng.; cons., R&D in field of electronics; cons. Marconi Underwater Weapon Systems, Shorts Missile Systems, European Space Agy. Contbr. articles to profl. jours. Mem. IEEE (sr.), Acad. Scis. (N.Y.). Avocations: photography, scuba diving, Persian/Arabic culture, archery, pistol shooting. Home: 20 Balmoral Rd, Saint Annes-On-Sea FY8-1ER, England Office: Miltec R & D Svcs, PO Box 17 Clifton Dr, Saint Annes-On-Sea FY8-3SB, England

TEAL, EDWIN EARL, engineering physicist, consultant; b. Dallas, May 10, 1914; s. Olin Allison and Azelia Clyde (Kidd) T.; m. Ruby Brown, Aug. 10, 1939; children: Barbara Teal Berry, Marilyn Teal Roberts. BA, Baylor U., 1935; MS, U. Mich., 1937. Mem. staff computer-seismic dept. Texaco, Inc., Houston, 1938-40, magnetometer operator, 1940-42; physicist Navy Underwater Sound Lab., New London, Conn., 1942-45, Texaco Geophys. Lab., Houston, 1945-47; physicist, chief engr. equipment dept. Bariod div. N.L. Industries, Houston, 1947-70; asst. to mgr. mfg. McCullough div. N.L. Industries, Houston, 1970-79; pvt. practice cons. physicist Houston, 1979—. Pres. Indsl. Mgmt. Club, Houston, 1963-64; area dir. Braeswood Pl. Homeowners Assn., Houston, 1981-83; advisor Jr. Achievement of Houston. Marston scholar Brown U., 1935-36. Mem. IEEE (life, chmn. Houston sect. 1976), Acoustical Soc. Am. (life), Sigma Pi Sigma. Baptist. Achievements include patent on Process of Employing Frontal Analysis Chromatography in Well Logging.

TEAL, GILBERT EARLE, industrial engineer; b. Balt., Md., July 22, 1912; s. Cecil Armstrong and Margaret (Trimble) T.; m. Evangeline Maxine Piper, Apr. 4, 1947; children: Saundra Gail, Sue Anne, Gilbert Earle II. BCE, U. Md., 1937; postgrad., George Washington U., 1946; MA, NYU, 1947, M of Adminstrv. Engring., 1947, DSc in Engring., 1952, PhD, 1956; grad. U.S. Army Command and Gen. Staff Coll., 1944, Air War Coll., 1951, Indsl. Coll. Armed Forces, 1956. Registered profl. engr., Calif. Commd. 2d lt. U.S. Army, 1934, advanced through grades to lt. col, 1947; engr. Travelers Ins. Co., Washington and N.Y.C., 1937-39; indsl. relations engr. U.S. Rubber Co., Naugatuck, Conn., 1939-41; commd. lt. col. USAF, 1947, advanced through grades to col., 1960, ret., 1960; chief scientist v.p. Dunlap & Assocs., Darien, Conn., 1960-70; dean coll. Western Conn. State U., Danbury, 1970-75, v.p. acad. affairs, 1975-76, v.p. emeritus, 1976—; pres. Pub. Service Research Inst., Stamford, Conn., 1961-63, Mil. Personnel Assocs., Washington, 1968-71; instr., prof. aerospace sci. Purdue U., Lafayette, Ind., 1958-60; instr. U. Md., Fed. Republic of Germany and France, 1952-55, vis. lectr., 1974-75; research assoc. prof. edn. NYU, 1968-70, instr. Mgmt. Inst., 1966-67; instr. U. Conn., 1974-76; cons. faculty Charter Oak Coll., 1974-76, now emeritus faculty; pres. Gilbert E. Teal & Assocs., Stephens City, Va., 1960—; chief exec. officer Randolph-Macon Acad., Front Royal, Va., 1977-78; chmn. mgmt. faculty Shenandoah U., 1978-83, prof. emeritus, 1983—. Author: (with R. Fabrizio) Your Car and Safe Driving, 1962; editor 8 books; contbr. articles to profl. jours. Dir. Community Devel. Action Program, Newtown, 1969-70; mem. Newtown Bd. Edn., 1962-63; 1st v.p. Danbury ARC, 1975-76, pres., 1976-77; bd. mgrs. Conn. State PTA, 1974-76. Served with inf., U.S. Army, 1935-37. Decorated Bronze Star; recipient Founders' Day award NYU, 1957.Mem. Am. Psychol. Assn., Am. Soc. Safety Engrs., Am. Psychol Soc., Ret. Officers Assn., Air Force Assn., Arnold Air Soc., Vets. Safety, Am. Legion, Fla. C. of C., Greater Danbury C. of C. (bd. dirs. 1975-76), Scabbard and Blade, Sigma Xi, Phi Delta Kappa (emeritus), Psi Chi, Delta Mu Delta. Republican. Clubs: Army-Navy. Lodges: Masons, Shriners, Rotary. Home: 6022 Valley Pike Stephens City VA 22655-9701 Office: Shenandoah U Winchester VA 22601

TEBRINKE, KEVIN RICHARD, manufacturing engineer; b. Red Oak, Iowa, Nov. 6, 1958; s. Richard Donald and Carolyn Elane (Schenck) T.; m. Lori Nelson, May 20, 1981; children: Kaitlyn Elizabeth, Michael Stewart. BSChemE, Iowa State U., 1981. Project engr. Colgate Palmolive, Kansas City, Kans., 1982-89; sr. mfg. engr. Warner Lambert, Rockford, Ill., 1989—. Mem. internat. mgmt. coun. YMCA, Rockford, 1991—. Achievements include development of static resistance air seal for pneumatic conveying, leader of multi-functional bubblegum creative development team. Home: 5852 Thatcher Dr Rockford IL 61114-5539

TEDESCHI, HENRY, bioscience educator; b. Novara, Italy, Feb. 3, 1930; came to U.S., 1947; s. Edoardo Vittorio and Paola (Minerbi) T.; m. Terry Lorraine Kershner, Nov. 29, 1957; children: Alexander, Devorah, David. BS, U. Pitts., 1950; PhD, U. Chgo., 1955. Asst. prof. U. Chgo., 1957-60; asst. prof. U. Ill. Med. Sch., Chgo., 1960-65, assoc. prof., 1965; prof. biosci. SUNY, Albany, 1965—. Author: (textbooks) Cell PHysiology, Molecular Dynamics, 1974, 2d edit., 1993, also monograph in field. Office: SUNY 1400 Washington Ave Albany NY 12222

TEDLOCK, DENNIS, anthropology and literature educator; b. St. Joseph, Mo., June 19, 1939. BA, U. N.Mex., 1961; PhD in Anthropology, Tulane U., 1968. Asst. prof. anthropology Iowa State U., 1966-67; asst. prof. rhetoric U. Calif., Berkeley, 1967-69; research assoc. Sch. Am. Research, 1969-70; asst. prof. anthropology Bklyn. Coll., 1970-71; asst. prof. Yale U., 1972-73; assoc. Univ. prof., anthropology and religion Boston U., 1973-82, Univ. prof., anthropology and religion, 1982-87; James H. McNulty prof. dept. English SUNY, Buffalo, 1987—; vis. asst. prof. Wesleyan U., 1971-72; adj. prof. U. N.Mex., 1980-81; mem. Inst. for Advanced Study, 1986-87. Author: Finding the Center: Narrative Poetry of the Zuni Indians, 1972, The Spoken Word and the Work of Interpretation, 1983, Days from a Dream Almanac, 1990 (Victor Turner prize 1991), Breath on the Mirror: Mythic Voices and Visions of the Living Maya, 1993; co-editor: Teachings from the American Earth, 1975; also translator; contbr. articles to profl. jours. and lit. mags. Recipient PEN transl. prize for Popol Vuh: The Mayan Book of the Dawn of Life, 1986, Victor Turner prize for Days from a Dream Almanac, 1991; Guggenheim fellow, 1986. Office: SUNY Buffalo Dept English Buffalo NY 14260

TEECE, DAVID JOHN, economics and management educator; b. Marlborough, New Zealand, Sept. 2, 1948; came to U.S., 1971; s. Allan Teece. BA, U. Canterbury, Christchurch, N.Z., 1970, MA in Commerce, 1971; MA in Econs., U. Pa., 1973, PhD, 1975. Asst. prof. Stanford U., 1975-78, assoc. prof., 1978-82; prof. U. Calif., Berkeley, 1982—, dir. Ctr. Rsch. in Mgmt., 1983—. Author: Profiting from Innovation, 1986, and numerous other articles and books. Mem. Am. Econ. Assn. Office: U Calif 554 Barrows Hall Berkeley CA 94720

TEGLASI, HEDWIG, psychologist, educator; b. Debrecen, Hungary; d. Kalman and Irene (Rubinfeld) T.; m. Saul Golubcow, Dec. 19, 1971; children: Jordan, Jeremy. BA, Douglass Coll., 1969; PhD, Hofstra U., 1975. Asst. prof. Ind. U. of Pa., 1975-78; asst. prof. U. Md., College Park, 1978-83, assoc. prof., 1983—. Author: Clinical Use of Storytelling, 1993; contbr. articles to profl. jours. Mem. APA, Nat. Assn. Sch. Psychologists, Md. Sch. Psychology Assn. (exec. bd. 1991—). Home: 10941 Broad Green Terr Potomac MD 20854 Office: Univ Md Benjamin Bldg College Park MD 20742

TEGTMEIER, RONALD EUGENE, physician, surgeon; b. Omaha, Jan. 16, 1943; s. Harvey and Edna T.; children: Anne, Amy; m. Victoria Susan, June 28, 1985; children: Justina Becerra, Gregory Galvan, Mark Tegtmeier. AB, Dartmouth Coll., 1965; BMS, Dartmouth Med. Sch., 1966; MD, Harvard Med. Sch., 1968. Diplomate Am. Bd. Plastic Surgery. Internship in surgery U. Colo. Med. Ctr., Denver, 1968-69, residency in gen. surgery, 1969-70; plastic surgery preceptorship Kingston-upon-Hull, England, 1973; residency in plastic surgery U. Mexico, Albuquerque, 1974-76, fellowship, 1976; plastic surgeon pvt. practice Arvada, Colo., 1977—, Artistic Ctr. for Cosmetic Surgery, Golden, Colo., 1988—; pres. Clear Creek Valley Med. Soc., Lakewood, Colo., 1983-84; speaker of ho. Colo. Med. Soc., denver, 1985-87. Author: Aesthetica Tapes, 1988—; patentee in field; contbr. numerous papers and publs. to profl. jours. Named Outstanding Bus. Person, Arvada Jaycees, 1978; recipient Arvada Image award, 1981, Denver Post Gallery of Fame award, 1979. Mem. Am. Soc. Plastic and Reconstructive Surgeons, Am. Soc. for Aesthetic Plastic Surgery. Avocations: scuba, music, skiing, tennis, model trains, flying, aquariums. Office: Artistic Ctr Cosmetic Surgery 14062 Denver West Pky Bldg 52 Golden CO 80401-3121

TEICH, MALVIN CARL, electrical engineering educator; b. N.Y.C., May 4, 1939; s. Sidney R. and Loretta K. Teich. S.B. in Physics, MIT, 1961; M.S.E.E., Stanford U., 1962; Ph.D. in Quantum Electronics, Cornell U., 1966. Research scientist MIT Lincoln Lab., Lexington, Mass., 1966-67; prof. engring. sci. Columbia U., N.Y.C., 1967—, chmn. dept. elec. engring., 1978-80; mem. faculty applied physics dept. Columbia Radiation Lab., Ctr. Telecommunications Rsch.; mem. scientific bd. Inst. Physics, Czech Acad. Scis., Prague. Author: (with B.E.A. Saleh) Fundamentals of Photonics, 1991; dep. editor: Quantum Optics: Jour. European Optical Soc.; bd. editors: Jour. Visual Comm. and Image Representation, 1989-92; contbr. articles to profl. jours.; patentee in field. Recipient Citation Classic award Inst. for Sci. Info., 1981, Meml. Gold medal of Palacky U., Czech Republic, 1992; Guggenheim Meml. Found. fellow, 1973. Recipient Citation Classic award Inst. for Sci. Info., 1981, Meml. Gold Medal of Palachy U., Czech Republic, 1992; Guggenheim Meml. Found. fellow, 1973. Fellow AAAS, Optical Soc. Am. (Optics Letters editorial adv. panel 1977-79), IEEE (Browder J. Thompson Meml. prize 1969), Am. Phys. Soc.; mem. Acoustical Soc. Am., Assn. Rsch. in Otolaryngology, Sigma Xi, Tau Beta Pi. Office: Columbia U Dept Elec Engring New York NY 10027

TEICHOLZ, PAUL M., civil engineering educator, administrator; b. N.Y.C., May 24, 1937; s. David and Rose (Frankel) T.; m. Susan Swire, June 28, 1959; children: Marc, Nina, Leslie. BCE, Cornell U., 1959; MCE, Stanford U., 1960, PhD, 1963. Registered profl. civil. engr., Calif. Cons., ptnr. Jacobs Assocs., San Francisco, 1963-68; dir. mgmt. info. systems and strategic planning Guy F. Atkinson Co., South San Francisco, 1968-88; prof. civil. engring., dir. Ctr. for Integrated Facility Engring. Stanford (Calif.) U., 1988—; dir. devel. Coll. Prep. Sch., Oakland, Calif., 1985-90. Contbr. articles to profl. jours. Mem. ASCE (Constrn. Mgmt. Man of Yr. 1985), Optical Rsch. Soc. Am. Democrat. Jewish. Office: Stanford U Ctr Integrated Facility Engring Terman Engring Ctr Stanford CA 94305-4020

TEITSMA, ALBERT, physicist; b. Kuinre, Overijsel, Netherlands, May 21, 1943; came to Can. 1954; s. Teye Rinse and Arendje (Hazes) T. BSc, McMaster U., Hamilton, Ont., 1966, PhD, 1975. Postdoctoral fellow McMaster U., Hamilton, Ont., 1975-76; rsch. scientist U. Guelph, Ont., 1976-80, Queen's U., Kingston, Ont., 1980-81; physicist Trans Can. Pipelines Ltd., Scarborough, Ont., 1981-91; asst. mgr. R&D Pipetronix Ltd., Scarborough, Ont., 1991-92, mgr. R&D, 1992—. Contbr. articles to profl. jours. Nat. Rsch. Coun. scholar, 1968, Govt. of Ont. scholar, 1969-71, Dept. of Physics/McMaster U. scholar, 1972-75. Mem. AAAS, Am. Soc. Non-destructive Testing, Can. Assn. Rsch. in Non-destructive Testing, Can. Assn. Physicists, Planetary Soc., Met. Toronto Zool. Soc. Achievements include first measurement of the three body potential of Xenon; invention of a new method for dating geological formations (fission Xenon isotope dating); development of a high resolution magnetic flux leakage in-line pipeline inspection tool; measurement of magnetic field changes at stress points of ferromagnetic materials (pipelines, hoists, crane booms). Office: Pipetronix Ltd, 450 Midwest Rd, Scarborough, ON Canada M1P 3A9

TEIXEIRA DA CRUZ, ANTONIO, pharmaceutical company executive; b. Oporto, Portugal, Oct. 20, 1935; s. Antonio Júlio Jr. and Perfeita do Carmo (Teixeira) Cruz; m. Maria de Lourdes Amorim, Oct. 31, 1965; children: Maria Claudia, João. MD, U. Oporto, 1961; PhD in Biochemistry, U. Luanda, Angola, 1974. Intern in endocrinology U. Hosp., Oporto, 1962-63; asst. prof. Luanda U. Sch. Medicine, 1965-74; intern in endocrinology Karolinska Hosp., Stockholm, 1969; vis. researcher dept. biochemistry U. Stockholm, 1969-70; prof. biochemistry U. Lisbon, Portugal, 1974-76; sci. dir. Bayer-Portugal SA, Lisbon, 1976—. Contbr. articles to sci. jours. With M.C., Portugese Army, 1963-65. Grantee U. Luanda, 1969-70, 72. Avocations: reading, music, swimming. Office: Bayer Portugal SA, Rua da Quinta do Pinheiro, 1495 Lisbon Portugal

TEIXIER, ANNIE MIREILLE J., research scientist; b. Paris, Oct. 18, 1937; d. André Jean and Raymonde (Julien) T.; m. Michel Ruaudel, June 24, 1960; children: Valerie, Sophie, Pascaline. Lic. in sci. (hon.), U. Paris, 1958; DSc (hon.), Chimie Organique, 1963. Cert. sci. researcher. Rsch. scientist Alimentation Équilibrée Comentry, Paris, 1960-62; rsch. scientist Commissariat Energie Atomique, Saclay, France, 1963—, rsch. scientist dept. chemistry, 1963-68, rsch. scientist dept. physics, 1968-70, rsch. scientist dept. electronics, 1970-81, rsch. scientist dept. phys. chemistry, 1981-90, rsch. scientist dept. physics, 1990—. Contbr. articles to profl. jours.; patentee in field. Recipient Chevalier award Order Palmes Acad., 1991. Mem. Soc. Francaise Chimie. Roman Catholic. Achievements include research in field of supramolecular engineering in the solid state. Office: Commissariat Energie Atomique, Svc Chimie Moléculaire, 91191 Saclay Gif Sur Yvette, France

TEKELIOGLU, MERAL, physician, educator; b. Ermenek, Konya, Turkey, Dec. 23, 1936; d. Sefik and Zeynep Tekelioglu; grad. Ankara (Turkey) Med. Faculty, 1961, specialist degree in histology-embryology, 1964, Docent, 1969; prof., Faculty Medicine Hacettepe U., Ankara, 1975; m. Ziya Uysal, July 1961 (div. 1984); 1 child, Kaya. Asst. in histology-embryology Faculty Medicine, Ankara U., 1961-64, chief asst., 1964-69, docent dept. histology-embryology, Hacettepe U., 1969-75, prof., dir. dept. histology-embrology, Anka Med. Faculty, 1983—; cons. electron microscopy. Mem. Turkish Electron Microscopy Soc., European Soc. Anatomy, Royal Micros. Soc., Cytochem. Soc. of Oxford, European Pineal Study Group, Turkish Soc. Natural Protection, Clair Hall Cambridge (Eng.) U. (assoc.), European Soc. Human Reproduction and Fertility (Belgium), Am. Fertility Soc. Moslem. Author: (with others) The Cell: Fine Structure and Function, 1972, 74, 78, 82, Medical Embryology, 1984, General Medical Histology, 1989, Sobotta, Atlas of Histology, 3d edit. (Turkish version), 1990, articles on fine morphology of early pregnant and HCG and HMG treated human endometrium and human brain morphology of slow-growing viral infections, degenerative diseases of the human central nervous system, related topics. Home: Süslü Sok 1/4 Tandogan, Mebusevleri, 06580 Ankara Turkey Office: Ankara U Dept History-Embryology, Morphology Bldg Sihhiye, 06339 Ankara Turkey

TEKIPPE, RUDY JOSEPH, civil engineer; b. Decorah, Iowa, Dec. 2, 1943; s. Roman R. and Florence (Gesing) T.; m. Sheryl Ellen Bresnahan; children: Cynthia, Timothy, Michele, Ted. BSCE, Iowa State U., 1965, MSCE, 1966; PhD, U. Wis., 1969; AMP, Harvard U., 1992. Registered profl. engr., Calif., La., Utah. Cons. engr. L.P. Erdman, Decorah, 1963-65; engr., sr. engr., supr. J.M. Montgomery, Cons. Engr., Pasadena, Calif., 1969-74, v.p., 1974-83, sr. v.p., 1984—, dir. technology, 1992—; assoc. prof. Iowa State U., Ames, 1983-84. Co-author 3 tech. text and reference books; contbr. tech. papers to profl. publs. Recipient Disting. Svc. award U. Wis., Madison, 1992. Mem. Tau Beta Pi, Chi Epsilon, Phi Beta Phi, Sigma Nu. Achievements include design of water pollution control facilities including municipal and industrial wastewater management and treatment plants. Home: 2222 Kinneloa Ranch Rd Pasadena CA 91107 Office: Montgomery Watson Americas 300 N Lake St Ste 1200 Pasadena CA 91101

TEKKANAT, BORA, materials scientist; b. May 24, 1953; m. Kim K. Groesbeck, Dec. 21, 1983. BS, MS of Metall. Engring., Mid. East Tech. U., Ankara, Turkey, 1979; MS of Materials Engring., U. Mich., 1983, PhD of Materials Engring., 1987. Grad. teaching asst. & rsch. asst. materials sci. engring. U. Mich., Ann Arbor, 1982-87; sr. polymer scientist materials rsch. dept. Johnson Controls Inc., Milw., 1987-92, group leader materials rsch. Ctrl. Rsch. Lab., 1992—. Contbr. articles to J. Thermoplastic Composite Mater, Polymer Engring. and Sci., Jour. Am. Physics. Mem. Soc. for Advancement of Materials and Process Engring., Soc. for Plastic Engrs., Materials Rsch. Soc., Am. Soc. for Metals, Inst. Packaging Profls. Achievements include patents for composite substrate for bipolar electrode, method and apparatus for determining the environmental stress cracking resistance of plastic articles, inverted closures for beverage containers, polyester-polyoefin blends containing a functionalized elastomer, composite substrate for Zn-Br Electrodes. Research includes thermoplastics, environmental stress cracking resistance, and radiation-induced crosslinks in PE. Home: 1707-8A N Prospect Ave Milwaukee WI 53202 Office: Johnson Controls Inc 1701 W Civic Dr Milwaukee WI 53209

TELEGDI, VALENTINE LOUIS, physicist; b. Budapest, Hungary, Jan. 11, 1922; s. George and Ella (Csillag) T. MS in Chem. Engring., U. Lausanne, Switzerland, 1946; PhD, ETH, Zurich, Switzerland, 1950; Doctors (hon.), U. Louvain, Belgium, 1989, U. Budapest, Hungary, U. Chgo., 1991. Various prof. levels U. Chgo., 1951-72, Enrico Fermi disting. svc. prof., 1972-76; prof. ETH, Zurich, 1972-89; vis. prof. Calif. Inst. Tech., Pasadena, 1979—; visitor Cern, Geneva, Switzerland, 1976—, chmn. scientific policy com., 1978-83. Recipient Wolf prize in physics, 1991. Mem. Nat. Acad. Sci., Am. Acad. Arts and Scis., Accad. Dei Lincei, Accad. di Torino, Acad. Patavina. Avocations: traveling, jazz, gastronomy. Home: 2 ch Taverney, Geneva 1218, Switzerland Office: CERN, PPE Div, 23 Geneva 1211, Switzerland

TELESETSKY, WALTER, government official; b. Boston, Jan. 22, 1938; s. Keril and Nellie (Krelka) T.; m. Sharron-Dawn Lamp, July 15, 1961; children: Stephanie Ann, Anastasia Marie. BS in Mech. Engring., Northeastern U., 1960; MBA, U. Chgo., 1961. Mem. tech. staff The Mitre Corp., Bedford, Mass., 1962-68; sr. mem. tech. staff Data Dynamics, Inc., Washington, 1969; phys. scientist NOAA, Rockville, Md., 1970-71, U.S. Gate Project coord., 1972-74, dir. U.S. Global Weather Experiment Project Office, 1974, dir. Program Integration Office, 1975-77, dir. Programs and Tech. Devel. Office, 1977-79, dir. Programs and Internat. Activities Office, 1979-81; dep. assoc. dir. for tech. svcs., chief AFOS ops. div. Nat. Weather Svc., Silver Spring, Md., 1981-86, dir. Office of Systems Ops., 1986—; liaison to NAS coms. on atmospheric scis., geophysics studies and internat. environ. programs, 1975-81; U.S. coord. U.S./Japan Coop. Program in Natural Resources, 1980-88; chmn. U.S.-Japan Marine Resources and Engring. Coordination Com., 1980-88; U.S. del. governing coun. UN Environ. Program and World Meteorol. Orgn.; mem. commn. for Basic Systems World Meteorol. Orgn., 1988—. Contbr. articles to profl. pubs. Recipient Silver medal Dept. Commerce, 1975. Mem. AAAS, Am. Geophys. Union, Am. Meteorol. Soc. Home: 16 Eton Overlook Rockville MD 20850-3003 Office: 1325 E West Hwy Silver Spring MD 20910-3233

TELETZKE, GERALD HOWARD, environmental engineer; b. Beaver Dam, Wis., Mar. 22, 1928; s. Gerhard Charles and Helen Ida (Monr) T.; m. Elaine Mae Gloudeman, June 21, 1951; children: Gary, Barbara. BSCE, U. Wis., 1951, MSCE, 1952, PhD, 1956. Registered profl. engr., Ind., Ariz.; Diplomate Am. Assn. Environ. Engrs. Instr., rsch. asst. U. Wis., Madison, 1951-56; asst., assoc. prof. Purdue U., West Lafayette, Ind., 1956-59; rsch. dir. EIMCO Corp., Palatine, Ill., 1959-61; devel. mgr. Zimpro Inc., Rothschild, Wis., 1961-65, exec. v.p., 1965-68, pres., 1969-85; dir. Ariz. Dept. Environ. Quality, Phoenix, 1987-88; cons. LAW Engring., Phoenix, 1989-93; mem. mgmt. adv. group U.S. EPA, Washington, 1982-85; adv. group on selection of ctrs. of engring excellence NSF, Washington, 1983-86. Mem. Wausau, Wis. Sch. Bd., 1972-75; pres. Friends of D.C. Everest Sch. Dist., Wausau, 1984-85. With U.S. Army, 1945-49, maj. USAR, 1949-63. Mem. ASCE, Am. Inst. Chem. Engrs., Water Environ. Fedn., Am. Water Works Assn., Air and Waste Mgmt. Assn. Achievements include 11 patents in the field. Home: 9425 N 87th St Scottsdale AZ 05258

TELEVANTOS, JOHN YIANNAKIS, engineering executive; b. Famagusta, Cyprus, June 30, 1952; came to U.S., 1977; s. Michael J. and Julia (Mela) T.; m. Diane A. Pavlides, July 22, 1978; children: Maria, Michelle. ACGI, Imperial Coll., London, 1973; BS, U. London, 1973; DIC, Imperial Coll., 1977; PhD, U. London, 1977. Sr. engr. Union Carbide Corp., Sistersville, W. Va., 1977, project scientist, 1980; tech. mgr. Union Carbide Corp., Charleston, W. Va., 1981, assoc. dir., 1986, dir., 1989; dir. ARCO Chemical Co., Charleston, W. Va., 1990; assoc. editor J. Polymer Sci., 1987--. Contbr. articles to profl. jours. Dir. Voluntary Mgmt. Assistance Program, Charleston, 1989--. Avocations: astronomy, photography. Office: Arco Chem Co PO Box 38007 Charleston WV 25303

TELLEP, DANIEL MICHAEL, aerospace executive, mechanical engineer; b. Forest City, Pa., Nov. 20, 1931. B.S. in Mech. Engring. with highest honors, U. Calif., Berkeley, 1954, M.S., 1955; grad. Advanced Mgmt.

Program, Harvard U., 1971. With Lockheed Missiles & Space Co., 1955—, chief engr. missile systems div., 1969-75, v.p., asst. gen. mgr. advanced systems div., 1975-83, exec. v.p., 1983-84, pres., 1984—, 1986—; chmn., chief exec. officer Lockheed Corp., 1989—; cons. in field; bd. dirs. 1st Interstate Bancorp., SCECorp. Contbr. article to profl. jours. Bd. govs. Music Ctr. L.A. County, 1991—; mem. adv. bd. U. Clalif. Berkeley Sch. Engring.; mem. Calif. Bus. Roundtable, 1992—; nat. chmn. vol. com. U.S. Savs. Bond Campaign, 1993. Recipient Tower award San Jose State U., 1985, Aeronautics/Propulsion Laurels award Aviation Week and Space Tech., 1993. Fellow AIAA (Lawrence Sperry award 1964, Missile Systems award 1986);, Am. Astronautical Soc.; mem. NAE, Nat. Aero. Assn., Soc. Mfg. Engrs., Sigma Xi, Pi Tau Sigma, Tau Beta Pi. Office: Lockheed Corp 4500 Park Granada Calabasas CA 91399

TELLER, EDWARD, physicist; b. Budapest, Hungary, Jan. 15, 1908; naturalized, 1941; s. Max and Ilona (Deutch) T.; m. Augusta Harkanyi, Feb. 26, 1934; children: Paul, Susan Wendy. Student, Inst. Tech., Karlsruhe, Germany, 1926-28, U. Munich, 1928; Ph.D., U. Leipzig, Germany, 1930; D.Sc. (hon.), Yale U., 1954, U. Alaska, 1959, Fordham U., 1960, George Washington U., 1960, U. So. Calif., 1960, St. Louis U., 1960, Rochester Inst. Tech., 1962, PMC Colls., 1963, U. Detroit, 1964, Clemson U., 1966, Clarkson Coll., 1969; LL.D., Boston Coll., 1961, Seattle U., 1961, U. Cin., 1962, U. Pitts., 1963, Pepperdine U., 1974, U. Md. at Heidelberg, 1977; D.Sc., L.H.D., Mt. Mary Coll., 1964; Ph.D., Tel Aviv U., 1972; D.Natural Sci., DeLaSalle U., Manila, 1981; D. Med. Sci. (n.c.), Med. U. S.C., 1983. Research assoc. Leipzig, 1929-31, Goettingen, Germany, 1931-33; Rockefeller fellow Copenhagen, 1934; lectr. U. London, 1934-35; prof. physics George Washington U., Washington, 1935-41, Columbia, 1941-42; physicist U. Chgo., 1942-43, Manhattan Engr. Dist., 1942-46, Los Alamos Sci. Lab., 1943-46; prof. physics U. Chgo., 1946-52; prof. physics U. Calif., 1953-60, prof. physics-at-large, 1960-70, Univ. prof., 1970-75; Univ. prof. emeritus, chmn. dept. applied sci. U. Calif., Davis and Livermore, 1963-66; asst. dir. Los Alamos Sci. Lab., 1949-52; cons. Livermore br. U. Calif. Radiation Lab., 1952-53; asso. dir. Lawrence Livermore Lab., U. Calif., 1954-58, 60-75; dir. Lawrence Livermore Radiation Lab., U. Calif., 1958-60; now dir. emeritus, cons. Lawrence Livermore Nat. Lab., U. Calif., Manhattan Dist. of Columbia, 1942-46; also Metall. and Lab. of Argonne Nat. Lab., U. Chgo., 1942-43, 46-52, and Los Alamos, N.Mex., 1943-46; also Radiation Lab., Livermore, Calif., 1952-75; sr. research fellow Hoover Instn. War, Revolution and Peace, Stanford U., 1975—; mem. sci. adv. bd. USAF; bd. dirs. Assn. to the Unite the Democracies; past mem. gen. adv. com. AEC; former mem. Pres.'s Fgn. Intelligence Adv. Nat. Space Coun. Bd. Author: (with Francis Owen Rice) The Structure of Matter, 1949, (with A.L. Latter) Our Nuclear Future, 1958, (with Allen Brown) The Legacy of Hiroshima, 1962, The Reluctant Revolutionary, 1964, (with G.W. Johnson, W.K. Talley, G.H. Higgins) The Constructive Uses of Nuclear Explosives, 1968, (with Segre, Kaplan and Schiff) Great Men of Physics, 1969, The Miracle of Freedom, 1972, Energy: A Plan for Action, 1975, Nuclear Energy in the Developing World, 1977, Energy from Heaven and The Earth, 1979, The Pursuit of Simplicity, 1980, Better a Shield than a Sword, 1987, Conversations on the Dark Secrets of Physics, 1991. Past bd. dirs. Def. Intelligence Sch., Naval War Coll.; bd. dirs. Fed. Union, Hertz Found., Am. Friends of Tel Aviv U.; sponsor Atlantic Union, Atlantic Council U.S., Univ. Ctrs. for Rational Alternatives; mem. Com. to Unite Am., Inc.; bd. govs. Am. Acad. Achievement. Recipient Joseph Priestley Meml. award Dickinson Coll., 1957, Harrison medal Am. Ordnance Assn., 1955; Albert Einstein award, 1958; Gen. Donovan Meml. award, 1959; Midwest Research Inst. award, 1960; Research Inst. Am. Living History award, 1960; Golden Plate award Am. Acad. Achievement, 1961; Gold medal Am. Acad. Achievement, 1982; Thomas E. White and Enrico Fermi awards, 1962; Robins award of Am., 1963; Leslie R. Groves Gold medal, 1974; Harvey prize in sci. and tech. Technion Inst., 1975; Semmelweiss medal, 1977; Albert Einstein award Technion Inst., 1977; Henry T. Heald award Ill. Inst. Tech., 1978; Gold medal Am. Coll. Nuclear Medicine, 1980; A.C. Eringen award, 1980; named ARCS Man of Yr., 1980, Disting. Scientist, Nat. Sci. Devel. Bd., 1981; Paul Harris award Rotary Found., 1980; Disting. Scientist Phil-Am. Acad. Sci. and Engring., 1981; Lloyd Freeman Hunt Citizenship award, 1982; Nat. medal of Sci., 1983; Joseph Handleman prize, 1983, Sylvanus Thayer Medal, 1986; Shelby Cullom Davis award Ethics & Pub. Policy Assn., 1988; Presdl. Citizen medal Pres. Reagan, 1989; Ettore Majorana Erice Scíenza Per La Pace award, 1990; Order of Banner with Rubies of the Republic of Hungary, 1990. Fellow Am. Nuclear Soc., Am. Phys. Soc., Am. Acad. Arts and Scis., Hungarian Acad. Scis. (hon.); mem. Nat. Acad. Scis., Am. Geophys. Union, Soc. Engring. Scis., Internat. Platform Assn. Research on chem., molecular and nuclear physics, quantum mechanics, thermonuclear reactions, applications of nuclear energy, astrophysics, spectroscopy of polyatomic molecules, theory of atomic nuclei. Office: Hoover Instn Stanford CA 94305 also: PO Box 808 Livermore CA 94550

TELLEZ, GEORGE HENRY, safety professional, consultant; b. Bogotá, Colombia, June 24, 1951; came to U.S., 1954; s. Jorge Enrique and Nohemi (Rodriguez) T.; m. Nora Reyes, Aug. 24, 1974; children: Shantell, Sabrina. Student, CCNY, 1972-74, U. Cen. Fla., 1979-83. Cert. safety prof. Am. Bd. Cert. Safety Profls. Loss control rep. Hartford Ins. Group, Orlando, Fla., 1977-81, asst. loss control mgr., 1983-89; loss control cons. Comml. Union Ins. Co., Orlando, 1981-83; account engring. exec. Hartford Splty. Co., Dallas, 1989-91; loss control exec. Anco Ins. Houston, 1991—; instr. defensive driving Nat. Safety Coun., 1983-84. Chmn. health and safety Casselberry Elem. Sch. PTA, Orlando, 1986-87, coord. bicycle safety program, 1987; founding donor Lewisville (Tex.) Edn. Found., 1990-91. Mem. Am. Soc. Safety Engrs. (asst. regional v.p. region VIII 1986-87, v.p. Cen. Fla. chpt. 1985-86, pres. 1986-87, chmn. nominating com. 1988, chmn. welcome com. S.W. chpt. 1989-90, Pres. Club 1984-85, 86-88, 89-90, Pres. Cir. 1985-86, 91-92, Safety Profl. of Yr. award Cen. Fla. chpt. 1988). Avocations: collectibles, water sports, Stephen S. King fan, golf. Office: Anco Ins Houston 16000 Barkers Point Ln Houston TX 77079-4009

TEMAM, ROGER M., mathematician; b. Tunis, Tunisia, May 19, 1940; s. Ange M. and Elise (Ganem) T.; m. Claudette Cukorja, Aug. 21, 1962; children: David, Olivier, Emmanuel. Agregation Math., U. Paris, 1962, DSc, 1967. Asst. prof. math. U. Paris, 1960-67, prof., 1967—; prof. Ecole Polytechnique, Paris, 1968-85. Author: (with I. Ekeland) Convex Analysis and Variational Problems, 1974, Navier-Stokes Equations, 1977, Mathematical Problems in Plasticity, 1983, Infinite Dimensional Dynamical Systems in Mechanics and Physics, 1988, Integral Manifolds..., 1988; contbr. 190 articles to sci. jours.; editor Math. Model. and Num. Analysis, Physica D, assoc. editor other profl. jours. Recipient Grand Prix Joannidès, Acad. Sci. Paris, 1993. Mem. Am. Math. Soc., Soc. Indsl. and Applied Math., Soc. Math. Applications of Industry (founding pres. 1983-87).ng pres. 1983-87).

TEMANEL, BILLY ESTOQUE, agronomy research director, educator, consultant; b. Buenavista, Agusan, The Philippines, Jan. 11, 1958; s. Aniceto Toledo and Laureana Selga (Estoque) T.; m. Luz Cureg Talosig, May 3, 1984; children: Billson, BilleChristian. BS in Agr., Mindanao State U., Marawi City, The Philippines, 1979; MS in Agr., Isabela (Phillipines) State U., 1984. Irrigation supr. Cocoa Investors Inc., Digos, Davao Sur, The Philippines, 1980-81; sci. rsch. asst. I Isabela State U., 1982-87; bd. sec. Quirino State Coll., Diffun, The Philippines, 1987-88, instr. I, 1989-92; agr. advisor RP-German Community Forest Mgmt. & Agroforest Project, Nagtipunan, Quirino, The Philippines, 1993; provincial coord. Dept. Sci. and Tech., Diffun, Quirino, 1992—; coord. Linkages Spl. projects, Diffun, 1989—; project leader Banana R&D Project, Diffun, 1990—, peanut R&D project, Quirino, 1987—. Dir. CARD Found. Inc., Diffun, 1990—, v.p., 1991; mem. CVARRD Regional Tech. Working Group; v.p., bd. dirs. Diffun Water Dist, 1993—. Recipient Cert. Achievement, ICRISAT, 1989, Cert. Appreciation award Dept. Agr., 1990, United Meth. Ch., 1989, Dept. Sci. Tech., 1992, Dept. Agr., 1992, FFP-FAHP-FFPCC, 1992, Tech. Panel Agr. Edn., 1991, DOST-DECS, 1991, ICRISAT, 1992, Presdl. Cert. Recognition, 1991, Plaque of Recognition, Dept. Sci. Tech., 1992, 93, CUARRD, 1993. Mem. Soc. for Advancement Vegetable Industry, Philippines Assn. Rsch. Mgrs., Philippines Assn. Vocat. Educators, Internat. Network Banana Plantains, Cereals Legumes Asia Network, Crop Sci. Soc. Philippines, Low-external Input Sustainable Agr. Network, Conservation Farming Movement Inc., Forests, Trees People Programme Network, Katipunam Multipurpose Coop. (ctrl. com.), Masons. Avocations: playing lawn tennis, walking, discoing, going to movies. Office: Quirino State Coll, Diffun 3401, The Philippines

TEMES, CLIFFORD LAWRENCE, electrical engineer; b. Jersey City, Feb. 4, 1930; s. Julius Howard and Hannah (Hass) T.; m. Vivian Lorraine Newman, July 12, 1953; children: David, Lisa, Joel. BEE, Cooper Union, 1951; MSEE, Case Inst. tech., 1954; PhD, Poly. Inst. N.Y., 1965. Rsch. scientist Nat. Adv. Com. for Aeronautics, Cleve., 1951-54; lab. supr. Columbia U. Electroncs Rsch. Lab., N.Y.C., 1956-60; sr. project engr. Fed. Sci. Corp., N.Y.C., 1960-65; tech. staff Gen. Rsch. Corp., Santa Barbara, Calif., 1965-74; dept. staff Mitre Corp., McLean, Va., 1974-77; head search radar br. Naval Rsch. Lab., Washington, 1977—; cons. Columbia U. Electronics Rsch. Lab., 1960-65; reviewer Prentice Hall, 1968-69, IEEE, 1960-74. Contbr. articles to profl. jours. With U.S. Army, 1954-56. Cooper Union scholar, 1947; recipient Naval Rsch. Lab. Performance award, 1991. Mem. IEEE, Sigma Xi. Achievements include patent on fragment-tolerant transmission line; inventor two-beam scanning antenna requiring no rotary joints, color-coded plan position indicator.

TEMIN, HOWARD MARTIN, scientist, educator; b. Philadelphia, Pa., Dec. 10, 1934; s. Henry and Annette (Lehman) T.; m. Rayla Greenberg, May 27, 1962; children: Sarah Beth, Miriam Judith. BA, Swarthmore Coll., 1955, DSc (hon.), 1972; PhD, Calif. Inst. Tech., 1959; DSc (hon.), N.Y. Med. Coll., 1972, U. Pa., 1976, Hahnemann Med. Coll., 1976, Lawrence U., 1976, Temple U., 1979, Med. Coll. Wis., 1981, Colo. State U., 1987, PM Curie, Paris, 1988; U. Medicine Dentistry N.J., 1989; DSc (hon.), Med. and Dental Coll., N.J., 1989; D. honoris causa, U. Pierre et Marie Curie, Paris, 1988. Postdoctoral fellow Calif. Inst. Tech., 1959-60; asst. prof. oncology U. Wis., 1960-64, assoc. prof., 1964-69, prof., 1969—, Wis. Alumni Rsch. Found. prof. cancer rsch., 1971-80, Am. Cancer Soc. prof. viral oncology and cell biology, 1974—, H.P. Rusch prof. cancer rsch., 1980—, Steenbock prof. biol. scis., 1982—; mem. rsch. policy adv. com. U. Wis. Med. Sch., 1979-83; cons. Office Tech. Assessment Panel on Saccharin, 1977; mem. Internat. Com. Virus Nomenclature Study Group for RNA Tumor Viruses, 1973-75, subcoms. HTLV and AIDS viruses, 1985; mem. virology study sect. NIH, 1971-74, mem. dir.'s adv. com., 1979-83; cons. working group on human gene therapy NIH/RAC, 1984-89, mem. Nat. Cancer Adv. Bd., 1987—; mem. NAS/IOM Com. for a Nat. Strategy for AIDS, 1986-88, NAS/IOM AIDS activities oversight com., 1988-90—, steering com. biomed. rsch. Global Program on AIDS, 1989-90, mem. Global Commn. AIDS, 1991-92, ad hoc vaccine adv. panel NIAID, 1990-91, chmn. adv. com. genertic variation immuno deficiency viruses, vaccine br. div. AIDS, NIAID, NIH, 1988—, co-chmn. IOM Roundtable for Devel. of Drugs and Vaccines Against AIDS, 1992—; mem. WHO adv. coun. on HIV and AIDS, 1993; mem. NIAIO HIV rsch. and devel. vaccine working group; co-chair Viral Heteogencity Focus Group, 1992—; mem. NAS Report Review Com., 1988—; mem. fundamental rsch. panel Nat. Conf. on Health Rsch. Principles, 1978; sponsor Fedn. Am. Scientists, 1976—; mem. Waksman award com. Nat. Acad. Sci., 1976-81; mem. U.S. Steel award Com., 1980-83, 1990—, 1982; mem. sci. adv. bd. Coordinating Coun. Cancer Rsch., 1989—; Disting. lectr. Hermann U., Conn., 1988; Ochoa lectr. Internat. Congress Biochemistry, Prague, 1988; Muller lectr. Internat. Congress Gen., Toronto, 1988; Schultz lectr. Inst. for Cancer Rsch., 1989; Bitterman Meml. lectr., N.Y.C., 1984—; inaugural lectr. Md. Biotech. Inst., 1990; Abraham White lectr. George Washington U., 1990, Latta lectr. U. Nebr., 1991; bd. dirs. Found. for Adv. Cancer Studies, 1988—. Assoc. editor: Jour. Cellular Physiology, 1966-77, Cancer Research, 1971-74; exec. editor Molecular Carcinogenesis, 1987; mem. editorial bd.: Jour. Virology, 1971—, Intervirology, 1972-75, Proc. Nat. Acad. Scis, 1975-80, Archives of Virology, 1975-77, Ann. Rev. Gen., 1983, Molecular Biology and Evolution, 1983—, Oncogene Research, 1987—, Jour. AIDS, 1988—, Human Gene Therapy, 1989, In Vitro Rapid Communication Section, 1984—, AIDS and Human Retroviruses, 1990—. Co-recipient Warren Triennial prize Mass. Gen. Hosp., 1971, Gairdner Found. Internat. award, 1974, Nobel Prize in medicine, 1975; recipient Med. Soc. Wis. Spl. commendation, 1971, Papanicolaou Inst. PAP award, 1972, M.D. Anderson Hosp. and Tumor Inst. Bertner award, 1972, U.S. Steel Found. award in Molecular Biology, 1972, Theobald Smith Soc. Waksman award, 1972, Am. Chem. Soc. award in Enzyme Chemistry, 1973, Modern Medicine award for Distinguished Achievement, 1973, Harry Shay Meml. lectr. Fels Rsch. Inst., 1973; Griffuel prize Assn. Devel. Recherche Cancer, Villejuif, 1973, New Horizons lectr. award Radiol. Soc. N.Am., 1968, G.H.A. Clowes lectr. award Assn. Cancer Rsch., 1974, NIH Dyer lectr. award, 1974, Harvey lectr. award, 1974, Charlton lectr. award Tufts U. Med. Sch., 1976, Hoffman-LaRoche lectr. award Rutgers U., 1979, Yoder hon. lectr. award St. Joseph Hosp., Tacoma, 1983, Cetus lectr. award U. Calif., Berkeley, 1984; DuPont U., 1985, Japanese Found. for Promotion Cancer Rsch. lectr. award, 1985, Herz Meml. lectr. award Tel-Aviv U., 1985, Amoros. Meml. lectr. award U. West Indies, 1986, Albert Lasker award in basic med. sci., 1974, Lucy Wortham James award Soc. Surg. Oncologists, 1976, Alumni Disting. Svc. award Calif. Inst. Tech., 1976, Gruber award Am. Acad. Dermatology, 1981, Abraham White award,lectureship George Washington U., 1990; named to Cen. High Sch. Hall of Fame, Phila., 1976; recipient Pub. Health Service Research Career Devel. award Nat. Cancer Inst., 1964-74, 1st Hilldale award in Biolog. Sci. U. Wis., 1986, Braund Disting. vis. prof. award U. Tenn., 1987, Eisenstark lectr. award U. Mo., 1987, 1st Wilmot vis. prof. award U. Rochester, 1987, Sophie Moss lectr. award Hahnemann U., 1991, Latta lectr. award U. Nebr., 1991, Nat. Medal of Sci. NSF, 1992. Fellow AAAS, Am. Acad. Arts and Scis., Wis. Acad. Sci., Arts and Letters, Am. Soc. Microbiology; mem. NAS, Am. Soc. for Therapeutic Radiology and Oncology, Inst. Sci., Am. Philos. Soc., Tissue Culture Assn. (hon.), Royal Soc., Inst. Medicine. Office: U Wis McArdle Mem Lab 1400 University Ave Madison WI 53706-1506

TEMPERLEY, H. N. V., mathematician educator. Emeritus prof. dept. applied mathematics U. Wales, Cardiff, Eng. Recipient Rumford medal Royal Soc. London, Eng., 1992. Office: University of Wales, Dept of Physics, Cathay Park Cardiff CF1 3NS, England*

TEMPERO, KENNETH FLOYD, pharmaceutical company executive, physician, clinical pharmacologist; b. Morrisville, Vt., Sept. 30, 1939; s. Howard Everett and Lucile Lois (Lalouette) T.; m. Jeanne Marie Smith, Jan. 12, 1980; children: Suzelle Jeanne, Gavin Kenneth. BSc, U. Nebr., 1961; MSc, Northwestern U., 1964, PhD, 1966, MD, 1967; MBA, Fairleigh Dickenson U., 1981. Straight med. intern Cin. Gen. Hosp., 1967-68; resident in internal medicine U. Minn. Hosps., Mpls., 1968-69, 71-72; dir. clin. pharmacology Merck/MSDRL, Rahway, N.J., 1973-75, sr. dir. clin. pharmacology, 1975-78, exec. dir. clin. research, 1978-83; sr. v.p. clin. research and med. affairs G.D. Searle & Co., Skokie, Ill., 1983-87; chmn., chief exec. officer MGI PHARMA, Inc. (formerly Molecular Genetics, Inc.), Mpls., 1987—; instr. U. Minn., Mpls., 1972-73; vis. physician Rockefeller U., N.Y.C., 1974-85; assoc. prof. Drew Postgrad. Med. Sch., Los Angeles, 1985-87. Contbr. numerous articles to sci. jours. Treas., Villa Turicum Assn., Lake Forest, Ill., 1985. Served to maj. U.S. Army, 1969-71, Vietnam. Trainee NIH, Northwestern U., 1963-66; NIH spl. research fellow U. Minn., 1972-73. Mem. ACP, Am. Soc. Clin. Pharmacology and Therapeutics, Drug Info. Assn. Methodist. Avocations: canoeing; woodworking; swimming. Office: MGI PHARMA Inc 9900 Bren Rd E Ste 300E Hopkins MN 55343-9612

TEMPLE, LEE BRETT, architect; b. Balt., June 7, 1956. BArch, Cornell U., 1979. Cert. Nat. Coun. Archtl. Registration Bds. Gen. ptnr. Temple Gebelein Partnership, Ithaca, N.Y., 1981-91; prin. and sole propr. Lee Temple Architect AIA, 1985—; Architect AIA, Ithaca, N.Y. and Crestone, Colo., 1985—; vis. critic dept. architecture Cornell U., Ithaca, 1981; vis. prof. Hobart Coll., 1981-82; asst. prof. architecture Syracuse U., N.Y., 1982-87; prof. architecture Hobart Coll., 1992-93. Prin. works include Athena Residence, Chapelle Frontenac; author: Medieval Town Study, 1981. Chmn. social justice com. Cornell Cath. Community, Ithaca, 1989-92, trustee parish coun., 1989-90; mem. founding bd. dirs. Eco Village at Ithaca, 1991-93; mem. steering com. Tibetan Resettlement Project at Ithaca, 1991-92. Recipient 1st prize Storey Com. Compact House Competition, 1983; Eidlitz fellow dept. arch. Cornell U., 1979, 81. Mem. AIA (design excellence award, 1987, residential design award, 1987 cen. N.Y. chpt.), N.Y. State Assn. Architects, Cousteau Soc. Home and Office: 228 Sundown Overlook Crestone CO 81131

TEMPLER, DONALD IRVIN, psychologist, educator; b. Chgo., Aug. 26, 1938; s. Irvin Lennox and Berthe Elizabeth (Litwin) T. BA in Psychology, Ohio U., 1960; MA, Bowling Green U., 1961; PhD, U. Ky., 1967. Lic. psychologist, Ky., N.J., Calif. Dir. dept. psychology Western State Hosp., Hopkinsville, Ky., 1965-67; asst. prof. Western Ky. U., Bowling Green, 1967-68; chief psychologist Carrier Clinic, Belle Mead, N.J., 1969-73; dir. psychology Pleasant Grove Hosp., Louisville, Ky., 1973-74; chief psychologist Waterford Hosp., St. John's, Newfoundland, 1974-75; pvt. practice Louisville, 1975-78; faculty Calif. Sch. Profl. Psychology, Fresno, 1978—. Contbr. articles to Jour. Abnormal Psychology, Jour. Consulting and Clin. Psychology, Acta Neurologica Scandanavia, Archives Gen. Psychiatry, and Develop. Psychology, Am. Jour. Mental Deficiency, Perceptual and Motor Skills, Physiology and Behavior, Am. Psychologist, Jour. Clin. Psychology, British Jour. Psychiatry, Psychol. Documents, Internat. Jour. Addictions, Rsch. Communications in Chem. Pathology and Pharmacology, others; author 3 books; editorial bd. Omega; reviewer Addictive Behaviors, Internat. Jour. Aging and Human Devel., Jour. Abnormal Psychology, Jour. Rsch. in Personality, Perceptual and Motor Skills, others. Fellow Am. Psychol. Soc., Am. Psychol. Assn. Office: Calif Sch Profl Psychology 1350 M St Fresno CA 93721

TEMPLETON, GEORGE EARL, II, plant pathologist; b. Little Rock, June 27, 1931; s. George Earl and Gladys (Jones) T.; m. Bobbie Nell Moore, Aug. 31, 1958; children: George Earl, Gary Lee, Patricia Jan, Larry Ernest. BS, U. Ark., 1953, MS, 1954; PhD, U. Wis., 1958. Asst. prof. U. Ark., Fayetteville, 1958-62, assoc. prof., 1962-67, prof., 1967-85, univ. prof., 1985-91, disting. prof., 1991—. 1st lt. Chem. Corp, 1954-56. Recipient Award of Excellence, Weed Sci. Soc. of Am., 1973, John White award for Agrl. Rsch., 1979, Disting. Rsch. award Ark. Alumni Assn., 1987, Burlington No. award U. Ark., 1988, Superior Sr. award USDA, 1990, Ruth Allen award Am. Phytopathol Soc., 1991. Fellow Am. Phytopathol. Soc. (Ruth Allen award 1991); mem. Weed Sci. Soc. of Am., Mycological Soc. of Am., Brit. Mycological Soc., Ark. Acad. Sci. (sec. 1964-69, pres. 1971-72). Methodist. Achievements include development and EPA registration of the mycoherbicide COLLEGO. Home: 2310 Winwood Dr Fayetteville AR 72703 Office: U Ark Dept Plant Pathology Fayetteville AR 72701

TEMPLETON, JOHN MARKS, JR., pediatric surgeon; b. N.Y.C. Feb. 19, 1940; s. John Marks and Judith Dudley (Folk) T.; BA, Yale Coll., 1962; MD, Harvard U., 1968; m. Josephine J. Gargiulo, Aug. 2, 1970; children: Heather Erin, Jennifer Ann. Intern, Med. Coll. Va., Richmond, 1968-69, resident, 1969; assoc. prof. pediatric surgery U. Pa. and Children's Hosp. of Pa., 1986—, dir. trauma program, 1989—; chmn. bd. Templeton Growth Fund, Ltd. Assoc. editor: Textbook of Pediatric Emergencies, 1988. Chmn. health and safety, exec. bd. Phila. coun. Boy Scouts Am.; Ea. Coll., Nat. Recreation Found., Melmark Home; nat. bd. dirs., pres. Pa. div. Am. Trauma Soc. Served with M.C., USNR, 1975-77. Mem. ACS, AMA, Am. Pediatric Surg. Assn., Am. Acad. Pediatrics, Am. Assn. Surgery of Trauma, Easton Assn. Surgery of Trauma, Phila. Coll. Physicians, Union League, Merion Cricket Club. Republican. Evangelical. Office: 4 King St West, Toronto, ON Canada M5W 1M3

TEMPLETON, NANCY VALENTINE SMYTH, research scientist; b. N.Y.C., Feb. 14, 1952; d. Philip Eugene and Bertha Georgianna (Rooney) Smyth; m. David D. Roberts, Jan. 9, 1993; children: Lizette L., Renée C. Student, U. Kans., 1969-70; BS in Edn. magna cum laude, CUNY, 1974; MS in Genetics & Cell Biology, U. Conn., 1982; PhD in Molecular Biology & Biochemistry, Wesleyan U., 1988. Rsch. asst. II dept. molecular biophysics and biochemistry Kline Biology Tower, Yale U., New Haven, 1982-83; rsch. instr. dept. med. oncology Vincent T. Lombardi Cancer Rsch. Ctr., Ga. U. Med. Ctr., Washington, 1988-89; biotech. fellow lab. pathology Nat. Cancer Inst./NIH, Bethesda, Md., 1989-91; staff fellow molecular/ hematology br. Nat. Heart, Lung & Blood Inst./NIH, Bethesda, 1991—; teaching asst. U. Conn., 1980, 81, Wesleyan U., 1983, 84, 85; presenter Yale U., 1983, Basic Scis. Rsch. Group Children's Hosp. Rsch. Found., Cin., 1985, 87, Life Scis. Group Los Alamos (N.Mex.) Nat. Lab., 1987, dept. pediatrics and microbiology Children's Hosp. L.A., 1987, du Pont Nat. User's Meeting, Wilmington, Del., 1990, 42d Ann. Meeting of Vet. Pathologists, Orlando, Fla., 1991. Contbr. articles to Biotechniques, Diagnostic Molecular Pathology, Genomics, EMBO, PNAS, Cancer Rsch., Molecular Cell Biology, Intenat. Jour. Invert. Reproduction, also abstracts. Women's scholar 1972-73, scholar Conn. State, 1981-82. Mem. AAAS, Am. Soc. for Biochemistry and Molecular Biology, Am. Soc. for Microbiology, Am. Assn. Pathologists, Women in Cancer Rsch., Sigma Xi. Achievements include rsch. in polymerase chain reaction, cloning and characterization of human melanoma interstitial collagenase, structure and function of FBwc, effects of juvenile hormone analog on reproduction and devel. of Daphnia magna. Home: 6808 Persimmon Tree Rd Bethesda MD 20817 Office: NHLBI/NIH Molecular Hematology Br Bldg 10 Rm 7D18 Bethesda MD 20892

TEMSAMANI, JAMAL, molecular biologist, researcher; b. Tangier, Morocco, Apr. 18, 1960; came to U.S., 1989; s. Abdellah and Khadouj (Bouzid) T. MS in Biochemistry, Universite Scis. and Tech., Montpellier, France, 1985, PhD in molecular biology, 1989. Fellow Worcester Found., Shrewsbury, Mass., 1989-91; rsch. scientist Hydridon, Inc., Worcester, Mass., 1991-92, sr. rsch. scientist, 1992—; cons. Hybridon, Inc., 1990-91; reporter Joun. de Tanger, 1989—. Contbr. articles to profl. jours. Recipient Ligue Regionale contre le Cancer, 1987, Assn. pour la Recherche Contre le Cancer, ARC, 1989. Mem. N.Y. Acad. Scis., Am. Assn. Advancement Sci. Office: Hybridon Inc One Innovation Dr Worcester MA 01605

TENCZA, THOMAS MICHAEL, chemist; b. Wallington, N.J., July 8, 1932; s. Michael and Mary (Wojdyla) T.; m. Sylvia T. Tlusty, Apr. 25, 1959. AB, Columbia U., 1954; MS, Seton Hall U., 1964, PhD, 1966; MBA, Fairleigh Dickinson U., 1971. Rsch. chemist Botanical Drugs, S.B. Renick Co., 1957-60; with Brustol Myers Products, 1960—; sr. dir. product devel. Bristol Myers Products, Hillside, N.J., 1988-92; sr. dir. scientific affairs Bristol Myers Products, Hillside, 1992-93; sr. dir. external devel., techs. Bristol Myers Products, 1993—. With U.S. Army, 1957-59. Achievements include eight patents. Office: Bristol Myers 1350 Liberty Ave Hillside NJ 07207

TENG, CHUNG-CHU, ocean engineer, researcher; b. Taipei, Republic of China, May 10, 1955; came to U.S., 1980; s. Pao-Shih and Mien-Yuan (Wang) T.; m. Ching-Luan Yang, Sept. 28, 1985. MS, Oreg. State U., 1983, PhD, 1986. Registered profl. engr., La. Rsch. engr. Slotta Engring. Assocs., Inc., Corvallis, Oreg., 1986-88; sr. engr. Computer Scis. Corp., Stennis Space Center, Miss., 1988-90, Sverdrup Tech., Inc., Stennis Space Center, 1990; sr. ocean system engr., project mgr. Nat. Data Buoy Ctr., Stennis Space Center, 1990—; mem. tech. adv. com. Oreg. State U., Office Naval Rsch., Univ. Rsch. Initiative, 1988-91. Contbr. articles to Jour. Waterway, Port, Coastal and Ocean Engring., Jour. of Ocean Engring. Mem. ASCE, ASME, AIAA (marine systems and techs. tech. com. 1989—), Marine Tech. Soc. (treas. Gulf Coast sect. 1991). Achievements include research on hydrodynamics of marine structures, buoy systems and engineering, ocean waves, wave measurement systems, ocean engineering facilities. Office: Nat Data Buoy Ctr Bldg 1100 Stennis Space Center MS 39529

TENG, MAO-HUA, materials scientist, researcher; b. Taipei, Taiwan, Oct. 27, 1958; came to U.S., 1987; s. Jih-Ching and Hsiao-Mei (Shih) T.; m. Hsiao-Wei Chang, Jan. 25, 1987; 1 child, Lin-Chieh. MS, Nat. Taiwan U., 1984; PhD, Northwestern U., Evanston, Ill., 1992. Teaching asst. dept. geology Nat. Taiwan U., Taipei, 1981-84, PhD rsch. asst., 1986-87; rsch. asst. dept. materials sci. Northwestern U., Evanston, 1987-91, postdoctoral fellow, 1992—. 2d lt. Chinese Air Force, 1984-86. Mem. Am. Ceramic Soc., Am. Phys. Soc., Materials Rsch. Soc., Minerals, Metals and Materials Soc. Home: 818 Noyes St Apt 3C Evanston IL 60201-2853

TEN KATE, PETER CORNELIUS, civil engineer; b. Paterson, N.J., Oct. 10, 1948; s. James and Kathryn (De See) Ten K.; m. Beverly Jean Soodsma, Dec. 26, 1975; children: Peter Erik, Daniel James, Karen Beth, David Alan. BSCE, Rutgers U., 1970; M in Bus Mgmt., Fairleigh Dickensen U., 1982. Cert. profl. engr., profl. planner, N.J. Civil engr. Porter & Ripa Assocs., Morristown, N.J., 1972-75, Burns & Roe, Inc., Oradell, N.J., 1976-77; project supt. R.A. Hamilton Corp., Hackenssack, N.J., 1977—, treas., 1986, v.p., 1988, pres., 1990. Mem. Wyckoff Fire Dept., 1970—; mem. design rev. com. Twp. of Wyckoff, 1985—. 1st lt. U.S. Army, 1970-72. Mem. ASCE, NSPE, Ea. Christian Sch. Assn. (bd. dirs., pres. bd. 1989-90, chmn. capital bond dr.). Republican. Home: 268 Monroe Ave Wyckoff NJ 07481 Office: R A Hamilton Corp 409 S River St Hackensack NJ 07602

TENNEY, LEON WALTER, civil engineer; b. San Francisco, Aug. 18, 1949; s. Duane Paul and Harriet Lea (Davis) T.; m. Debra Jane Clapper, June 21, 1986; 1 child, Alexander Stuart. BS in Civil Engring., Va. Poly. Inst., 1972; grad., Command & Gen. Staff Coll., 1983; MA in History, George Mason U., 1992. Profl. engr., Va. Commd. 2d lt. U.S. Army, 1972, advanced through grades to maj., 1983; staff officer U.S. Army Engr. Sch., Ft. Belvoir, Va., 1979-81; NTC observer, contr. U.S. Army C.E., Ft. Irwin, Calif., 1981-84; comdr. 84th engring. co. U.S. Army C.E., Nurnberg, Fed. Republic Germany, 1984-85; DEH ops. officer U.S. Army C.E., Stuttgart, Fed. Republic Germany, 1986-87; gen. staff officer U.S. Army C.E.-The Pentagon, Washington, 1988-90; coll. instr. U.S. Army C.E.-The Pentagon, Ft. Leavenworth, Kans., 1990-92; ret. U.S. Army, 1992; project planner Higginbotham/Briggs & Assocs., Colorado Springs, Colo., 1992—. Author: Seven Days in 1862: Battles Around Richmond, 1993. Mem. ASCE, Soc. Am. Mil. Engrs. Presbyterian. Home: 1385 Masthead Way Monument CO 80132 Office: Higginbotham/Briggs 540 N Cascade Colorado Springs CO 80903

TENNEY, STEPHEN MARSH, physiologist, educator; b. Bloomington, Ill., Oct. 22, 1922; s. Harry Houser and Caroline (Marsh) T.; m. Carolyn Cartwright, Oct. 18, 1947; children: Joyce B., Karen M., Stephen M. AB, Dartmouth; MD, Cornell U.; ScD (hon.), U. Rochester. From instr. to assoc. prof. of medicine and physiology U. Rochester Sch. Medicine, 1951-56; prof. physiology Dartmouth Med. Sch., Hanover, N.H., 1956-74; dean Dartmouth Med. Sch., 1960-62, acting dean, 1966, 73, dir. med. scis., 1957-59, chmn. dept. physiology, 1956-77, Nathan Smith prof. physiology, 1974-88, Nathan Smith prof. emeritus, 1988—; med. dir. Parker B. Francis Found., 1975-83, exec. v.p., 1984-89; Chmn. physiology study sect. NIH, 1962-65; tng. com. Nat. Heart Inst., 1968-71; mem. exec. com. NRC; mem. physiology panel NIH study Office Sci. and Tech.; mem. regulatory biology panel NSF, 1971-75; chmn. bd. sci. counselors Nat. Heart and Lung Inst., 1974-78; chmn. Commn. Respiratory Physiology Internat. Union Physiol. Scis. Asso. editor: Jour. Applied Physiology, 1976—, Handbook of Physiology; notes editor: News in Physiol. Sci., 1989—; editorial bd.: Am. Jour. Physiology, Circulation Research, Physiol. Revs; Contbr. articles to sci. jours. Served with USNR, 1947-49; sr. med. officer Shanghai. Markle scholar in med. sci., 1954-59. Fellow Am. Acad. Arts and Scis., AAAS; mem. Inst. Medicine of Nat. Acad. Scis., Am. Physiol. Soc., Am. Soc. Clin. Investigation, N.Y. Acad. Scis., Gerontol. Soc., Am. Heart Assn., Assn. Am. Med. Colls., , Alpha Omega Alpha, Sigma Xi.

TENNYSON, RODERICK C., aerospace scientist; b. Toronto, Ont., Can., June 7, 1937; s. Clarence A. and Rosalie (Calderine) T.; m. Judith Grace Williams, June 17, 1961; children: Shân, Marc, Kristin. BA, U. Toronto, Ont., 1960, MA, 1961, PhD, 1965. Prof. inst. aerospace studies U. Toronto, 1974—, dir., 1985—, chmn. dept. engring. sci., 1982-85; selected as Can. experimenter on space shuttle flights; dir. adv. bd. Can. Airworthiness; dir ctr. excellence for Inst. Space & Terrestrial Sci.; chmn. Can. Found. for Internat. Space U.; cons. in field. Contbr. numerous articles to profl. jours., chpts. to books. Fellow Can. Aeronautic and Space Inst. Avocations: sailing, writing, recreations. Home: 104 McClure Dr, King City, ON Canada M3H 5T6 Office: Inst Aerospace Studies, 4925 Dufferin St, Downsview, ON Canada M3H 5T6*

TENOPYR, MARY LOUISE WELSH (MRS. JOSEPH TENOPYR), psychologist; b. Youngstown, Ohio, Oct. 18, 1929; d. Roy Henry and Olive (Donegan) Welsh; AB, Ohio U., 1951, MA, 1951; PhD, U. So. Calif., 1966; m. Joseph Tenopyr, Oct. 30, 1955. Psychometrist, Ohio U., Athens, 1951-52, also housemother Sigma Kappa; personnel technician to research psychologist USAF, 1953-55, Dayton, Ohio, 1952-53, Hempstead, N.Y.; indsl. research analyst to mgr. employee evaluation N.Am. Rockwell Corp., El Segundo, Calif., 1956-70; asso. prof. Calif. State Coll.-Los Angeles, 1966-70; assoc. research educationist UCLA, 1970-71; program dir. U.S. CSC, 1971-72; dir. selection and testing AT&T, N.Y.C., 1972—; lectr. U. So. Calif., Los Angeles, 1967-70; vice chmn. research com. Tech. Adv. Com. on Testing, Fair Employment Practice Commn. Calif., 1966-70; adviser on testing Office Fed. Contract Compliance, U.S. Dept. Labor, Washington, 1967-73. Pres., ASPA Found., 1985-87; mem. Army Sci. Bd. Fellow Am. Psychol. Assn. (bd. profl. affairs, edn. and training bd., mem. council reps., pres. div. indsl. organizational psychology); mem. Eastern Psychol. Assn., Am. Soc. Personnel Adminstrn. (bd. dirs. 1984-87), Nat. Acad. Sci. (coms. on ability testing, math. and sci. edn., panel on secondary edn.), Soc. Indsl. and Organizational Psychology (Recipient Profl. Practices award 1984), Nat. Council Measurement in Edn., Psychometric Soc., Met. N.Y. Assn. Applied Psychology, Am. Ednl. Research Assn., Sigma Xi, Sigma Kappa, Psi Chi, Alpha Lambda Delta, Kappa Phi. Editorial bd. Jour. Applied Psychology, 1972-87, Jour. Vocat. Behavior; assoc. editor Am. Psychologist; contbr. chpts. to books and articles to profl. jours. Home: 557 Lyme Rock Rd Bridgewater NJ 08807-1604 Office: 1 Speedwell Ave Morristown NJ 07962-1906

TENORIO, MANOEL FERNANDO DA MOTA, computer engineering educator; b. Maceio, Alagoas, Brazil, Oct. 24, 1957; came to U.S., 1981; s. Igor Souza and Maria Luiza (Maia) T.; m. Anne Elizabeth Grant, July 4, 1987; children: Joshua Emanuel, Jonathan Daniel, Katherine Anna. BSEE, Nat. Inst. Telecomm., Brazil, 1979; MSEE, Colo. State U., 1984; PhD, U. So. Calif., 1987. Teaching asst., instr. Nat. Inst. Telecomm., 1977-79; dir. R&D C.S. Systems and Components Ltda, Itajuba, MG, Brazil, 1979-80; rsch. asst. Colo. State U., Ft. Collins, 1981-82, U. So. Calif., L.A., 1982-85; lectr. elec. engring. UCLA, 1985-86, U. So. Calif., L.A., 1985-86; rsch. scientist Varian Rsch. Ctr., Palo Alto, Calif., 1986-87; asst. prof. elec. engring. Purdue U., West Lafayette, Ind., 1987—, coord. parallel distributed processing structures lab., 1988—; vis. researcher Ctr. for Computer Tech., Campinas, Sao Paulo, Brazil, 1982-85; lectr., panelist, presenter workshops, cons. in field; assoc. researcher Inst. Logic, Theory and Philosophy of Sci., ILTC, Rio de Janeiro, 1990—. Mem. editorial bd. sci. computing and automation Elsevier, 1989-90; contbr. articles, tech. reports to profl. publs., chpts. to books. Conselho Nacional de Pesquisa scholar, 1981-85, Coordenadoria de Aperfeicoamento de Pessoal de Ensino Superior scholar, 1979-80; grantee Davis Ross, 1988, 87-89, Knowles Electronics, 1987, Battelle, 1990, Army Rsch. Office, 1989-92, Dept. HHS, 1988—, GE Found., 1991-92. Mem. IEEE (neural network coun. com. 1991—), Assn. for Computing Machinery, Am. Assn. Artificial Intelligence, Internat. Neural Network Soc. Achievements include patents for parallel knowledge processing method for integrated circuit implementation, random-like information decoding system, dynamical system circuit for dense memory storage, communication system employing chaotic signals. Home: 1305 King Arthur Dr Lafayette IN 47905 Office: Purdue U Sch Elec Engring West Lafayette IN 47907

TEODORESCU, GEORGE, industrial designer, educator, consultant; b. Bucharest, Rumania, Jan. 27, 1947; arrived in Fed. Republic of Germany, 1980; MArch, Inst. Architecture, Bucharest, 1972. Architect Trade Show Enterprise IPA, Bucharest, Rumania, 1972-76; indsl. designer Ministry for consumer art MIU, Bucharest, 1976-77; design mgr. Ministry for Edn., Bucharest, 1977-80; devel. engr. Leybold-Heraeus, Cologne, Fed. Republic of Germany, 1981-86; indsl. designer Leybold A.G., Cologne, 1986-90, design. mgr., 1990-93; prof. Art Acad. Stuttgart, 1993—; prof. U. Wuppertal, 1991-93. Recipient Silver medal Eurodidac, 1986, Best Design award, Design Coun. NRW, Essen, Fed. Republic of Germany, 1988, Industrie Form Hannover, Fed. Republic of Germany, 1990, State award, 1991, Design award Design Coun. Stuttgart, Fed. Republic of Germany, 1989, 90. Mem. Indsl. Designer Soc. Am. (hon. mention 1989, Silver award 1990), Verein Deutscher Industrie Designer, Design Mgmt. Inst. Achievements include 6 patents for modular systems for physics (Germany), 1 patent for screen changing device for electron-welding (Germany), patents for wind tunnel experiment kit (Germany, France). Home: Tuernicherstr 3 ap7 et16, D-50968 Cologne Germany Office: Tesign Design Consultancy, Bonnerstr 498, D-50960 Cologne 51, Germany also: Art Acad Stuttgart, Am Weissenhof 1, 70191 Stuttgart Germany

TEPLITZKY, PHILIP HERMAN, computer consultant; b. Bklyn., Apr. 12, 1949; s. Moe and Blanch (Lewis) T.; m. Harriet Grossman, Aug. 8, 1982; 1 child, Ben. BA, Harpur Coll., Binghamton, N.Y., 1970; MS, Sch. Advanced Tech., Binghamton, 1977. Ptnr. Plagman Group, N.Y.C., 1981-85; sr. mgr. Price Waterhouse, N.Y.C., 1985-86; dir. Coopers & Lybrand, N.Y.C., 1988-92; dir. tech. SystemHouse, N.Y.C., 1992—; mem. industry adv. panel NYU, N.Y.C., 1991—. Mem. editorial adv. bd. Auerbach Data Security Jour., 1991. Adminstrv. v.p. Jewish Family Svc., Rockland County, 1992—. Mem. IEEE, Assn. Computing Machinery. Office: SYS-TEMHOUSE 885 3d Ave New York NY 10022

TERABE, SHIGERU, chemistry educator; b. Toyokawa, Aichi, Japan, Oct. 21, 1940; s. Yutaka and Cho (Tsujita) T.; m. Yoshie Matsuoka, May 3, 1967; children: Masaki, Fumitaka, Kyosuke. B of Engring., Kyoto U., Japan, 1963, M of Engring., 1965, DEng, 1973. Rsch. chemist Shionogi & Co. Ltd., Osaka, Japan, 1965-68; rsch. assoc. Ariz. State U., Tempe, 1975-77; asst. prof. Kyoto U., Japan, 1978-84, assoc. prof., 1984-90; vis. scientist Northeastern U., Boston, 1985; prof. Himeji Inst. Tech., Kamigori, Hyogo, Japan, 1990—. Author: How to Use Experimental Instruments in Chemistry, 1990, Micellar Electokinetic Chromotography, 1992; assoc. editor Jour. of Microcolumn Separations, 1991—; mem. adv. bd. Analytical Chemistry, 1990-92, Chromatographia, 1992—, mem. editorial bd. Analytical Methods and Instrumentation, 1993—, Jour. Chromatography, 1991—, Jour. Biomed. and Biophys. Methods, 1992—; contbr. 90 rsch., rev. articles to sci. jours. Mem. Chem. Soc. Japan, Japan Soc. Analytical Chemistry, Am. Chem. Soc. Achievements include invention of micellar electrokinetic chromotography. Home: 280-52 Harimada-cho, Moriyama Shiga 524, Japan Office: Himeji Inst Tech, Faculty of Sci, Kamigori Hyogo 678-12, Japan

TERADA, YOSHINAGA, economist; b. Niigata, Japan, June 27, 1919; s. Sakujiro and Koma (Shoda) T.; m. Etsuko Kasao, Apr. 8, 1964; 1 child, Wakana. B. in Agr., Mie Coll. Agr. and Forestry, Tsu, Japan, 1941; M. in Agr., Kyoto (Japan) U., 1944; D. in Econs., Meiji U., Tokyo, 1962. Asst. rschr. Ohara Inst. Agr., Kurashiki, Japan, 1944-47; from instr. to prof. Kwansei-gakuin U., Nishinomiya, Japan, 1947-57; prof. Kanto Jr. Coll., Tatebayashi, Japan, 1957-59, Meiji U. Faculty of Agr., Kawasaki, Japan, 1959-90; dean Div. Agrl. Econs., Kawasaki, 1964-70; lectr. Inst. of the Ministry Agr. and Forestry, Hachiooji, Japan, 1970-87, Mie U., Tsu, 1973-87; prof. grad. sch. Meiji U., Kawasaki, Japan, 1978-90; councillor Sch. Legal Person Meiji U., Tokyo, 1980-88, lectr. faculty agr. and grad. sch., 1990—, prof. emeritus, 1992—; cons. adminstrv. com. Yantai (People's Republic China) Econ. and Tech. Devel. Zone, 1989—; councilor Japanese Ednl. Assn., Tokyo, 1957. Author: The Development of an Economic Order, 1957, Capitalism and Agricultural Cooperative Society in Japan, 1957, Agricultural Cooperative Society in Japan, 1964. Fellow Agrl. Econ. Soc. Japan, Assn. for Regional Agrl. and Forestry Econs., Japan Econ. Policy Assn., Internat. Assn. Agrl. Economists, Econ. Rsch. Soc., Short Poetry Soc. "Araragi". Avocations: horticulture, Tanka (Japanese short poetry), horse-riding. Home: Teraodai 2-8-1 (16-402), Tama-ku Kawasaki-shi 214, Japan Office: Meiji U Faculty Agri, Higashimita 1-1-1 Tama-ku, Kawasaki 214, Japan

TERALANDUR, PARTHASARATHY KRISHNASWAMY, audiologist; b. India, Nov. 6, 1956; came to U.S., 1980; s. Krishnaswamy Iyengar R. and Padmamma K. (Krishnaswamy) T.; m. Gita Parthasarathy, Jan. 27, 1985; 1 child, Shilpa Parthasarathy. MSc in Audiology, U. Salford, Eng., 1980; PhD in Audiology, U. Tex., 1987. Speech pathologist and audiologist Lions Speech and Hearing Ctr., Bangalore, India, 1976-78; cons. audiologist Chgo. Osteo. Med. Ctr., 1980-82, Ill. Masonic Med. Ctr., Chgo., 1980-82; chief audiologist Easter Seal Rehab. Ctr., Elgin, Ill., 1981-82; asst. prof. So. Ill. U., Edwardsville, 1987—; mem. So. Ill. Univ. at Edwardsville Rsch. Com., 1991-92. Contbr. articles to profl. jours. Fellow Am. Auditory Soc.; mem. Am. Speech-Lang. and Hearing Assn., Acoustical Soc. Am. Achievements include research on the effects of repetition rate and phase on ABR in neonates and adult subjects; normal and multiple sclerosis subjects. Home: 373 W Glen Dr Glen Carbon IL 62034 Office: So Ill Univ at Edwardsville Audiology Dept Edwardsville IL 62026-1776

TERAO, TOSHIO, physician, educator; b. Shimizu, Japan, Jan. 18, 1930; s. Eiji and Mitsuko (Katagiri) T.; m. Setsuko Nishigaki, Nov. 13, 1961; children: Toshiya, Yasuo, Yoshio. Diploma U. Tokyo, 1953, M.D., 1960. Intern, Tokyo U. Hosp., 1953-54; sr. scientist Nat. Inst. Radiol. Sci., Chiba, Japan, 1963-67; research assoc. Mayo Clinic, Rochester, Minn., 1970-72; asst. U. Tokyo, 1972-77, lectr. in medicine 1977-79; prof. medicine Teikyo U., 1980-91; pres. Teikyo U. Med. Hosp., 1987-93, dean, 1993—. Author, editor in field. Mem. Am. Acad. Neurology, Japanese Soc. Internal Medicine, Japanese Soc. Neurology, Japanese Soc. Neuropathology, Japanese Soc. EEG and Electromyography, Japanese Soc. Psychiatry and Neurology, Japanese Soc. Cerebrovascular Disease, Sigma Xi. Office: Teikyo U, 2-11-1 Kaga, Itabashiku, Tokyo 173, Japan

TERAOKA, IWAO, polymer scientist; b. Toyama, Japan, July 25, 1958; came to U.S., 1989; s. Tetsuo and Setsuko (Kitani) T.; m. Sadae Sakai, Jan. 8, 1981; children: John, Erika. M Engring., U. Tokyo, 1984, D Engring., 1989. Rsch. assoc. dept. applied physics U. Tokyo, 1986-89; vis. scientist IBM Almaden Rsch. Ctr., San Jose, Calif., 1989-90; vis. scientist dept. polymer sci. U. Mass., Amherst, 1990-92; asst. prof. Polytech. U., Bklyn., 1993—. Contbr. articles to profl. jours. Mem. Am. Phys. Soc., Am. Chem. Soc. Achievements include research in molecular theory on rotational diffusion of entangled rodlike molecules, theory on glass transition of isotropic melt of rodlike molecules, separation of a bimodal polymer solution by use of porous glasses. Office: Poly U Dept Chemistry Brooklyn NY 11201

TERASAWA, MITITAKA, physics educator; b. Ina, Nagano, Japan, Mar. 23, 1937; s. Kanko and Satomi (Hara) T.; m. Emiko Kawasaki, Nov. 25, 1963; children: Takako, Hiroko, Nobutaka. BA, Kyoto U., Japan, 1960; DEng, U. Tokyo, 1974. Chief researcher Toshiba Cen. Rsch. Lab., Kawasaki, Kanagawa, Japan, 1974-78, Toshiba Nuclear Engring. Lab., Kawasaki, 1978-87; prof. physics Himeji (Japan) Inst. Tech., 1987—; vis. researcher Kans. State U., Manhattan, 1978-79; mem. cons. com. Japan Atomic Energy Rsch. Inst., Tokai, Ibaraki, Japan, 1985—; cons. com. Kansai Atomic Energy Conf., Osaka, Japan, 1990—. Author: Progress in Metal Physics and Physical Metallurgy, 1986, Spectrochemical Analysis with Ion Beam, 1987, Handbook of Nuclear Power Engring.; also articles. Mem. Am. Phys. Soc., Phys. Soc. Japan (editorial com. 1977-78), Atomic Energy Soc. Japan (editorial com. 1980-83, Acad. award 1983), Japan Soc. Applied Physics, Japan Inst. Metals, Japan Synchrotron Radiation Soc. Avocation: photography. Home: Kita-Shinzaike 3-1-20, Himeji, Hyogo 670, Japan Office: Himeji Inst Tech, Shosha 2167, Himeji Hyogo 671-22, Japan

TEREZAKIS, TERRY NICHOLAS, power/utility engineer; b. Hartford, Conn., June 28, 1961; s. Nicholas George and Pualine Anast (Terzis) T. AAS with high honor, SUNY, Delhi, 1987; B. Tech., Rochester Inst. Tech., 1989. Field engr. GE Power Generation, Schenectady, N.Y., 1989-90; staff engr. Dayton (Ohio) Power and Light Co., 1990—, engr. II, 1991-92; engr. III, 1993—; energy advisor SUNY, Delhi, 1986-87; energy auditor Rochester (N.Y.) Inst. Tech. Rsch. Corp., 1988-89; instr. energy SW Ohio Assn. Energy Engrs., Cin., 1991—; mem. speakers bur. Dayton Power and Light Co. Phi Theta Kappa scholar, 1988-89, Alumni scholar Rochester Inst. Tech., 1988-89. Mem. ASHRAE, Assn. Energy Engrs. (cert. energy mgr., Energy Engr. of Yr. award 1990, Energy Profl. Devel. award 1993), Nat. Assn. Power Engrs. Greek Orthodox. Office: Dayton Power and Light Co 9200 Chautauqua Rd Miamisburg OH 45342

TERHUNE, ROBERT WILLIAM, optics scientist; b. Detroit, Feb. 8, 1926; married; 2 children. BS, U. Mich., 1947, PhD in Physics, 1957; MA, Dartmouth Coll., 1948. Supr. digital computation and logic design sect. Willow Run Labs. U. Mich., 1951-54, rsch. physicist, 1954-59, mgr. Solid State Physics Lab., 1959-60; rsch. physicist Sci. Lab. Ford Motor Co., Dearborn, Mich., 1960-65, mgr. physics electronics dept., 1965-75, sr. staff scientist engring. and rsch. staff, 1976-87; sr. mem. tech. staff JPL Calif. Tech., 1988—; vis. scholar Stanford U., 1975-76. Editor Optics Let. Jour., 1977-83. Recipient Sci. and Engring. award Drexel Inst. Tech., 1964. Mem. IEEE, Optical Soc. Am. (editor jour. 1984-87, Frederic Ives medal 1992), Am. Phys. Soc. Achievements include research in quantum electronics, nonlinear optics, optical properties of solids and surfaces, molecular spectroscopy, advanced instrumentation. Office: 1460 Peyfair Estates Dr Pasadena CA 91103*

TERLIZZI, JAMES VINCENT, JR., secondary education educator; b. Malden, Mass., Nov. 4, 1940; s. James Vincent and Emily (Richie) T.; m. Marianne Ella Klaucke, May 2, 1965; children: James Vincent III, Eric Kurt. BS in Biology, Norwich U., 1962; MS in Biology, Northeastern U., Boston, 1970. Cert. tchr. biology, chemistry, gen. sci. and supervision. Sci. tchr. Kingsley Hall Sch., South Egremont, Mass., 1965-67, Seeglitz Sch., Peabody, Mass., 1967-69; chemistry tchr. Peabody (Mass.) High Sch., 1969-71; sci. dept. head Peabody (Mass.) Vets. Meml. High Sch., 1971—; workshop dir. Salem (Mass.) State Coll. Sci. Collaborative, 1986—. 1st lt. U.S. Army, 1963-65, Germany. Named Tchr. Exchange to England, Fulbright USIA, Washington, 1990-91; inducted into Mass. Hall of Fame for Sci. Educators, 1993. Mem. Mass. Assn. Sci. Suprs. (pres. 1986-87, Outstanding Sci. Educator 1990), North Shore Sci. Suprs. Assn. (sec.-treas. 1973—), Nat. Sci. Suprs. Assn. (regional dir. 1985-89, software coord. 1989-92, membership chmn. 1992—), Nat. Sci. Tchrs. Assn. (adminstr.-tchr. liaison com. 1990—). Avocations: orienteering, ice hockey, travel. Home: 12 Juniper St Ipswich MA 01938-1633 Office: Peabody Vets Meml High Sch 485 Lowell St Peabody MA 01960-1392

TERNYIK, STEPHEN, polytechnology researcher; b. Hamm, Ruhr, Germany, July 29, 1960; s. Istvan and Marianne (Hochgeladen) T.; m. Lisa Steven, Nov. 4, 1988; 1 child, Simon. MA, Tech. U., Berlin, 1986. Pvt. practice polytech. researcher, trainer Königswinter, Germany, 1986—; vis. rsch. fellow dept. precision machinery engring./rsch. lab. Tokyo U., 1993; cons. Adult Coll., 1987—. Author: Social Learning Processes, 1989. Mem. Am. Mgmt. Assn., Am. Inst. Indsl. Engrs. Mem. European People's Party. Avocations: gymnastics, barbecuing. Home: Rheinallee 20, D 533 Königswinter Germany

TER-POGOSSIAN, MICHEL MATHEW, radiation sciences educator; b. Berlin, Apr. 21, 1925; naturalized, 1954; s. Michel and Anna (Suratoff) T-P.; m. Ann Garrison Scott, Mar, 3, 1967. BA, U. Paris, 1943, postgrad., 1943-45; postgrad., Inst. Radium, Paris, 1945-46; MS, Washington U., 1948, PhD, 1950. Instr. radiation physics Washington U. Sch. Medicine, St. Louis, 1950-51, asst. prof., 1951-56, assoc. prof., 1956-61, prof., 1961-73, prof. biophysics in physiology, 1964—, prof. radiation scis., 1973—; adv. various Dept. Energy and NIH coms.; mem. diagnostic radiology and nuclear medicine study sect., 1979-81. Mem. editorial bd. Am. Jour. Roentgenology, Postgrad. Radiology, Jour. Computer Assisted Tomography, Jour. Nuclear Medicine, Jour. de Biophysique & Medicine Nucleaire; editor: IEEE Trans. on Med. Imaging, 1982-83; patentee in field of radiol. physics. Wendell Scott lectr. Washington U., 1973, Benedict Cassen lectr., 1976, David Gould lectr. Johns Hopkins U., 1977, R.S. Landauer Meml. lectr., 1981, Hans Hecht lectr. U. Chgo., 1981, 2d Ann. Soc. Nuclear Medicine lectr., 1985; recipient Paul C. Aebersold award Soc. Nuclear Medicine, 1777, Herrman L. Blumgart M.D. Pioneer award 1984, Georg Charles de Hevesy Nuclear Medicine Pioneer award Soc. Nuclear Medicine, 1985, Internat. award Gairdner Found., 1993. Fellow Am. Physics Soc., Am. Coll. Radiology (hon.), Acad. Sci. St. Louis; mem. Am. Nuclear Soc., Am. Radium Soc., Radiation Research Soc., Inst. Medicine, Nat. Acad. Sci., Radiol. Soc. N.Am. (New Horizons lectr. 1968). Republican. Home: 2 Brentmoor Park Saint Louis MO 63105-3003 Office: Mallinckrodt Inst 510 S Kingshighway Blvd Saint Louis MO 63110-1076

TERRADAS, JAUME, ecologist; b. Barcelona, Catalonia, Spain, Dec. 19, 1943; s. Antoni and Josefina (Serra) T.; m. Marina Mir, Oct. 26, 1966; children: Guillem, Berta. D. en Ciencias, U. Barcelona, 1973. Adj. prof. botany U. de Barcelona, 1968-69; adj. prof. botany U. Autonoma de Barcelona, Bellaterra, Spain, 1971-74, prof. agregado ecology, 1974-81, chmn. ecology, 1981—; mem. Com. MAB-Spain, Madrid, 1983—; dir. Ctr. de Recerca Ecològica i Aplicacions Forestals, Bellaterra, 1988—. Author: Ecologia Avui, 1971, 4th edit., 1981, Ecologia y Educación Ambiental, 1979, Barcelona: Ecologia D'Una Ciutat, 1986; editor: ERS Sistemes Naturals, 1989, Quercus ilex ecosystems: function, dynamics and management, 1992. Adv. mem. Consell de Benestar Social Municipality, Barcelona, 1989—. Mem. Ecol. Soc. Am., Brit. Ecol. Soc., Assn. Española de Ecología Terrestre (pres. 1989—), Inst. Catalana D'Historia Natural (v.p. 1977-78). Office: CREAF U Autonoma, 08193 Bellaterra Spain

TERREBONNE, ANNIE MARIE, medical technologist, educator, clinical laboratory scientist; b. Isola, Miss., Mar. 17, 1932; d. Tommy and Alpha (Whitfield) Patterson; m. Frank Paul Terrebonne, May 7, 1960. A.A., Co-Lin Jr. Coll., 1950; B.S., Miss. State U., 1952; grad. Knoxville Gen. Hosp. Sch. Med. Tech., 1953. Cert. Nat. Cert. Agy. Med. Lab. Personnel. Med., x-ray and EKG technician Layman-Saffold Clinic, Knoxville, Tenn., 1952-55; med. technologist in bacteriology St. Dominic's Hosp., Jackson, Miss., 1956-58; parasitologist Oschner's Clinic and Hosp., New Orleans, 1959-65; asst. supr., med. technologist II spl. hematology dept. U. Tex. Med. Br., Galveston, 1969-91 , supr. med. technologist III Lab, 1991—; mem. research and devel. staff, 1974—, instr. med. tech. students, 1981-86, instr. med. students, residents, and hematology fellows, 1987—. Contbr. articles to profl. jours. Mem. Nat. Certification Agy. for Med. Technologists, Am. Assn. Med. Technologists, Assn. Advancement of Sci., Galveston Dist. Soc. Med. Technologists, Tex. Soc. Med. Technologists, Am. Soc. Clin. Pathologists (cert.), Miss. State U. Alumni Assn. Democrat. Methodist. Clubs: Loyalty, Found. for Christian Living, Bayou Vista Recreation, Positive Thinkers'. Lodge: Order of Eastern Star, Grand 1894 Opera House. Home: 353 Ling Dr Hitchcock TX 77563-2601 Office: U Tex Med Br Spl Hematology Dept 424 Clin Sci Bldg Galveston TX 77550-2709

TERRELL, W(ILLIAM) GLENN, university president emeritus; b. Tallahassee, May 24, 1920; s. William Glenn and Esther (Collins) T.; m. Gail Strandberg Terrell; children by previous marriage: Francine Elizabeth, William Glenn III. BA, Davidson Coll., 1942, LLD (hon.), 1969; MS, Fla. State U., 1948; PhD, State U. Iowa, 1952; LLD (hon.), Gonzaga U., 1984, Seattle U., 1985. Instr., then asst. prof. Fla. State U., Tallahassee, 1948-55; asst. prof., then assoc. prof., chmn. dept. psychology U. Colo., Boulder, 1955-59, assoc., acting dean Coll Arts and Scis., 1959-63; prof. psychology, dean Coll. Liberal Arts and Scis., U. Ill. at Chgo. Circle, 1963-65, dean faculties, 1965-67; pres. Wash. State U., Pullman, 1967-85, pres. emeritus, 1985—; Pres. Nat. Assn. State Univs. and Land-Grant Colls., 1977-78; cons. The Pacific Inst., Seattle, 1987—. Contbr. articles to profl. jours. Served to capt. inf. U.S. Army, 1942-46, ETO. Recipient Disting. Alumnus award U. Iowa, 1985. Fellow APA, Soc. Rsch. in Child Devel.; mem. AAAS, Sigma Xi, Phi Kappa Phi. Avocations: golf, reading, traveling. Home: 2438 36th Ave W Seattle WA 98199 Office: The Pacific Inst 1011 Western Ave Seattle WA 98104-1040

TERRIS, SUSAN, physician, cardiologist; b. Morristown, N.J., Sept. 5, 1944; d. Albert and Virginia (Rinaldy) T. BA in History, U. Chgo., 1967, PhD in Biochemistry, 1975, MD, 1976. Diplomate Am. Bd. Internal Medicine, Am. Bd. Endocrinology and Metabolism, Am. Bd. Cardiovascular Disease. Resident in internal medicine Washington U., Barnes Hosp., St. Louis, 1976-78; fellow in endocrinology and metabolism U. Chgo., 1978-80, fellow cardiology, 1980-83; fellow cardiology U. Mich., Ann Arbor, 1983-85, instr. cardiology, 1985-86; head cardiac catheterization lab., head cardiology Westland (Mich.) Med. Ctr., 1985. Contbr. articles to Jour. Biol. Chemistry, Am. Jour. Physiology, Am. Jour. Cardiology, Jour. Clin. Investigation, other profl. publs. Grantee Juvenile Diabetes Found., 1978-80, NIH, 1978-79. Mem. AAAS, Am. Heart Assn., N.Y. Acad. Sci., Am. Women in Sci. Achievements include rsch. demonstrating dependence of intracellular degradation of insulin upon its prior receptor-mediated uptake by liver; studies of effects of various drugs on human circulatory system.

TERRY, BECKY FAYE, electrical engineer, executive; b. Dixon, Ill., Sept. 27, 1959; d. Kenneth August and Frances Helen (Olson) Dippel; m. John Forbes Terry, July 9, 1978; children: Jameson Clark, Dylan Jon. BSEE, George Washington U., 1987. Tech. writer Palisades Inst. Rsch. Svcs., Arlington, Va., 1985-89, dir. engring., 1989-90, exec. v.p., 1990—; sec. Adv. Group Electron Devices, Washington, 1985—; presenter at profl. confs. Author, editor tech. reports. Republican. Lutheran. Achievements include increasing industry and government awareness of decline of the microwave tube industry. Home: 2162 Belvedere Dr King George VA 22485 Office: Palisades Inst Rsch Svcs Ste 307 2011 Crystal Dr Arlington VA 22202

TERRY, CHARLES JAMES, metallurgical engineer; b. Milw., Dec. 12, 1949; s. Norbert Mathew and Olive Louise (Runte) T.; m. Karen Barbara Macejkovic, June 9, 1973; children: Jeanine, Suzanne, Kristen, Michael. BS in Metall. Engring., U. Wis., 1972. Registered profl. engr., Wis. Prep. plant engr. Midland Coal Co., Trivoli, Ill., 1973; metall. engr. ASARCO, El Paso, Tex., 1973-76; rsch. engr. Allis Chalmers, Milw., 1977-80; metall. engr. Kennametal, Inc., Fallon, Nev., 1981-86; spl. products mgr. Kennametal, Inc., Fallon, 1986—. Eucharistic min. St. Patrick's Ch., Fallon, 1990—. Mem. AIME (Metall. Soc.), Soc. Mining Engrs. (Reno chpt.). Republican. Roman Catholic. Achievements include patent on tungsten carbide process, others. Home: 3155 Alcorn Rd Fallon NV 89406

TERRY, ROBERT DAVIS, neuropathologist, educator; b. Hartford, Conn., Jan. 13, 1924; m. Patricia Ann Blech, June 27, 1952; 1 son, Nicolas Saul. AB, Williams Coll., 1946, DSc (hon.), 1991; M.D., Albany (N.Y.) Med. Coll., 1950. Diplomate: Am. Bd. Pathology, Am. Bd. Neuropathology. Postdoctoral tng. St. Francis Hosp., Hartford, 1950, Bellevue Hosp., N.Y.C., 1951, Montefiore Hosp., N.Y.C., 1952-53, 54-55, Inst. Recherches sur le Cancer, Paris, France, 1953-54; sr. postdoctoral fellow Inst. Recherches sur le Cancer, 1965-66; asst. pathologist Montefiore Hosp., 1955-59; assoc. prof. dept. pathology Einstein Coll. Medicine, Bronx, N.Y., 1959-64; prof. Einstein Coll. Medicine, 1964-84, acting chmn. dept. pathology, 1969-70, chmn., 1970-84; prof. depts. neuroscis. and pathology U. Calif.-San Diego, 1984—; mem. study sect. pathology NIH, 1964-68; study sects. Nat. Multiple Sclerosis Soc., 1964-72, 74-78; mem. bd. sci. counselors Nat. Inst. Neurol. and Communicative Disorders and Stroke, NIH, 1976-80, chmn., 1977-80; mem. nat. sci. council Huntington's Disease Assn., 1978-81; mem. med. and sci. adv. bd. Alzheimer Assn., 1978-88; mem. sci. adv. bd. Max Planck Inst., Munich, 1990—. Mem. editorial adv. bd. Jour. Neuropathology and Exptl. Neurology, 1963-83, 85-88, Lab. Investigation, 1967-77, Revue Neurologique, 1977-87, Annals of Neurology, 1978-82, Ultrastructural Pathology, 1978-86, Am. Jour. Pathology, 1985-89. Served with AUS, 1943-46. Recipient Potamkin prize for Alzheimer Rsch., 1988, Met. Life Found. award, 1991. Fellow AAAS; mem. Am. Assn. Neuropathologists (pres. 1969-70, Meritorious Contbn. award 1989), N.Y. Path. Soc. (v.p. 1969-70, pres. 1971-73), Am. Assn. Pathologists, Am. Neurol. Assn., Am. Acad. Neurologists. Research, publs. on Alzheimer's disease and Tay Sachs disease. Office: U Calif San Diego Dept Neuroscis La Jolla CA 92093

TERRY, STUART L(EE), plastics engineer; b. Chgo., Apr. 8, 1942; s. Gordon M. and Fredrica (Gordon) T.; m. Linda Jane Littenberg, Aug. 25, 1963 (div. 1974); m. Mary Ann Stames, Feb. 16, 1980; children: Robin D. Andrews, Mark R. Andrews, Marc L. Terry, Robin M. Terry. BSChemE, Cornell U., 1965, PhD, 1968; MS in Mgmt., Rensselear U., 1972. From sr. rsch. engr. to mgr. tech. acquisitions Monsanto Corp., Springfield, Mass., 1968-88; dir. tech. Sonoco Products Co., Hartsville, S.C., 1988—. Mem. Tech. Assn. of Pulp and Paper Industry, Soc. Plastics Engrs., Futures Soc. Office: Sonoco Products Co 1 North 2d St Hartsville SC 29550

TERVO, TIMO MARTTI, ophthalmologist; b. Helsinki, Finland, Mar. 9, 1950; s. Martti Vilho and Pirkko Kaarina (Tuhkanen) T.; m. Marja Kaarina Kesä, Aug. 13, 1977; children: Tomi, Markku. MB, U. Helsinki, 1972, MD, 1975, PhD, 1977. Asst. dept. anatomy U. Helsinki, 1972-79, resident dept. neurosurgery, 1980, resident dept. ophthalmology, 1980-83, asst. prof. dept. anatomy, 1978-83, asst. prof. dept. ophthalmology, 1983-89, sr. ophthalmologist, 1983-89, chief physician dept. ophthalmology, 1989—; physician Helsinki Naval Sta., Kirkkonummi, Finland, 1979-80; cons. Labsystems Co., Helsinki, 1986-90, Biohit Co., Helsinki, 1990-92. Lt. Finnish Navy, 1978-79. Mem. Assn. Rsch. in Vision and Ophthalmology, Internat. Soc. Eye Rsch. Avocations: cross country and downhill skiing, sailing, vehicles. Home: Jatasalmentie 9, 00830 Helsinki Finland Office: Dept Ophthalmology Hensinki, Haartmaninkatu 4C, 00290 Helsinki Finland

TERWILLIGER, JOSEPH DOUGLAS, statistical geneticist; b. N.Y.C., Aug. 20, 1965; s. James Isaac and Hazel Beatrice (Townsend) T. BA, Johns Hopkins U., 1987; BMus, Peabody Conservatory of Music, Balt., 1987; MA, Columbia U., 1989, M of Philosophy, 1991, PhD, 1993. Music theory asst. Peabody Conservatory of Music, 1986-87; tuba player Bklyn. Brass, S.I., N.Y., 1989—; statis. geneticist Columbia U., N.Y.C., 1987—. Tuba player orchestras, N.Y.C., 1987—; contbr. articles to profl. jours. Recipient 1st Prize Tuba Nat. Arts Club, 1990. Mem. Am. Fedn. of Musicians (local 802), Am. Soc. of Human Genetics, Am. Statis. Assn., Tubists Universal Brotherhood Assn. Home: Box 63 Wawarsing NY 12489 Office: Columbia U Box 58 722 W 168th St Unit 58 New York NY 10032

TERZIAN, KARNIG YERVANT, civil engineer; b. Khartoum, Sudan, July 4, 1928; came to U.S.; s. Yeznig and Marie T.; m. Helen S., Dec. 21, 1958. BCE, Am. U. Beirut, 1949; MCE, U. Pa., 1954. Assoc., L. T. Beck & Assocs., 1956-60; prin. Urban Engineers, Inc., Phila., 1960—, now sr. v.p., cons. major transp. projects in Pa., N.Y., N.J., Nigeria, Zaire. Bd. dirs. Armenian Sisters Acad., 1970-74. Mem. ASCE, ASTM, Prestressed Concrete Inst. Armenian Apostolic. Office: Urban Engrs Inc 300 N 3d St Philadelphia PA 19106

TESAR, DELBERT, machine systems and robotics educator, researcher, manufacturing consultant; b. Beaver Crossing, Nebr., Sept. 2, 1935; s. Louis and Clara (Capek) T.; m. Rogene Kresak, Feb. 1, 1957; children: Vim Lee, Aleta Anne, Landon Grady, Allison Jeanne. B.Sc. in Mech. Engring., U. Nebr., 1958, M.Sc., 1959; Ph.D., Ga. Tech. U., 1964. Assoc. prof. U. Fla., Gainesville, 1965-71, prof., 1972-83, grad. research prof., 1983-84, dir., founder Ctr. Intelligent Machines and Robotics, 1978-84; Curran chair in engring. U. Tex., Austin, 1985—; lectr. in field.; mem. rev. panel Nat. Bur. Standards, Gaithersburg, Md., 1982-88; mem. sci. adv. bd. to Air Force, 1982-86, numerous others; interactor with Russian Acad. Sci. on sci. and tech. Author: (with others) Cam System Design, 1975. Patentee in field; contbr. articles to profl. jours.; assoc. editor 3 computer and mfg. jours. Expert witness house sci. and tech. com. U.S. Ho. of Reps., 1978-84. Mem. Fla. Engring. Soc. (Outstanding Tech. Achievement award 1982), ASME (machine design award 1987). Avocations: antiques; art; travel. Home: 8005 Two Cove Dr Austin TX 78730-3125 Office: U Tex Dept Mech Engring ETC 4.146 Austin TX 78712

TESKE, RICHARD HENRY, veterinarian; b. Christiansburg, Va., July 22, 1939; s. August Frank and Peggy Marie (Macomber) T.; m. Mary Helen Webb, June 11, 1961; children: Helen Desiree, Mary Michele. BS, Va. Tech. U., 1962; DVM, U. Ga., 1965; MS, U. Fla., 1966. Diplomate Am. Bd. Vet. Toxicology. Asst. prof. U. Fla., Gainesville, 1967; dir. toxicology Hill Top Rsch., Inc., Miamiville, Ohio, 1967-70; chief pharmacology/toxicology br. Ctr. for Vet. Medicine FDA, Beltsville, Md., 1971-77, dep. dir. div. med. rsch. Ctr. for Vet. Medicine, 1977-78, dir. div. vet. med. rsch. Ctr. for Vet. Medicine, 1978-82; assoc. dir. for sci. Ctr. for Vet. Medicine FDA, Rockville, Md., 1982-85, dep. dir. Ctr. for Vet. Medicine, 1985—; mem. vet. med. adv. bd. panel U.S. Pharmacopeial Conv., Rockville, 1984—. Fellow Am. Acad. Vet. Pharms. and Therapeutics, Am. Acad. Vet. and Comparative Toxicology. Office: FDA Ctr for Vet Medicine 7500 Standish Pl Rockville MD 20855-2773

TESS, ROY WILLIAM HENRY, chemist; b. Chgo., Apr. 25, 1915; s. Reinhold W. and Augusta (Detl) T.; m. Marjorie Kohler, Feb. 19, 1944; children: Roxanne, Steven. BS in Chemistry, U. Ill., 1939; PhD, U. Minn., 1944. Rsch. chemist, group leader Shell Devel. Co., Emeryville, Calif., 1944—, rsch. supr., 1959-61, 63-66; rsch. supr. Royal Dutch/Shell Plastics Lab., Delft, The Netherlands, 1962-63; tech. planning supr. Shell Chem. Co., N.Y.C., 1967-70; tech. mgr. solvents Shell Chem. Co., Houston, 1970-77, cons., 1977-79; ind. cons. Fallbrook, Calif., 1979—; pres. Paint Rsch. Inst., Phila., 1973-76. Editor, organizer: Solvents Theory and Practice, 1973; (with others) Applied Polymer Science, 1975, Applied Polymer Science, 2d edit., 1985. Pres. Assn. Indsl. Scientists, Berkeley, Calif., 1948-50, Minerinda Property Owners Assn., Orinda, Calif., 1965-67, Houston Camellia Soc., 1973-74. Fellow Am. Inst. Chemists; mem. Nat. Paint and Coatings Assn. (air quality com. 1967-79), Fedn. Socs. Coatings Tech. (bd. dirs. 1973-76, Roon award 1957, Heckel award 1978), Am. Chem. Soc. (divsn. polymeric materials chmn. 1978, exec. com. 1977—, Disting. Svc. award 1993). Achievements include discovery, development and recommendation of uses for existing and potential petrochemical raw materials in coatings, resins and polymer industries; research in Epon resins, in high polymer latices based on acrylates, styrene, ethylene and vinyl esters, in drying oils and alkyds, and in solvent blends based on solution theory. Home and Office: 1615 Chandelle Ln Fallbrook CA 92028-1707

TESSIER, MARK A., electrical engineer; b. Northampton, Mass., June 7, 1959; s. Norbert Alexander and Sally Ann Tessier; m. Maria Tessier. BS in Elec. Power Engring., Rensselaer Poly Inst., Troy, N.Y., 1981; MS, Rensselaer Poly. Inst., 1982. Registered profl. engr., N.Y. Assoc. distbn. engr. Mass. Electric Co., Worcester, North Andover, 1982-85; sr. relay engr. New England Power Svc. Co., Westboro, Mass., 1985—. Home: 35 Colburn St Northboro MA 01532 Office: New England Electric 25 Research Dr Westborough MA 01582

TESSLER, STEVEN, ecologist; b. Phila., July 8, 1954; s. Harry Tessler and Ella Mae (DeHaas) Barry; m. Karen DiNenno, Oct. 16, 1987; 1 child, Julia. AS in Natural Sci., Delaware County C.C., Media, Pa., 1974; BS in Animal Sci., Pa. State U., 1976; MS in Entomology, Purdue U., 1979; PhD in Entomology/Ecology, Pa. State U., 1991. Asst. arachnid curator Frost Entomology Mus. Pa. State U., University Park, 1975-76; rsch. technician dept. biology Bryn Mawr (Pa.) Coll., 1979-80; edn. specialist Sea World, Inc., San Diego, 1981-83; rsch. asst. Dept. Anatomy Temple U. Med. Sch., Phila., 1983-84; rsch. assoc. Sch. Forest Res. Pa. State U., University Park, 1989-90; asst. prof. Dept. Biology Lock Haven (Pa.) U., 1990-92; cons. State Coll., 1992-93; ecologist, database mgr. Shenandoah Nat. Park, Luray, Va., 1993—. Contbr. articles to profl. jours. Recipient Outstanding Student award, Entomol. Soc. Pa., 1990. Mem. Ecol. Soc. of Am., N.Am. Benthol. Soc., Am. Arachnol. Soc., Sigma Xi, Gamma Sigma Delta. Achievements include rsch. on structure of terrestrial and aquatic invertebrate communities, graphic representation of multivariate data. Office: Div Nat Resources and Scis Shenandoah Nat Park Luray VA 22835

TESTA, ANTHONY CARMINE, chemistry educator; b. N.Y.C., Nov. 19, 1933; s. Antonio C. and Madeline (Fusco) T.; m. Helga L. Kittlinger, Nov. 10, 1962. BS in Chemistry, CCNY, 1955; PhD in Chemistry, Columbia U., 1961. Rsch. chemist Lever Bros. Rsch. Ctr., Edgewater, N.J., 1961-63; asst. prof. St. John's U., N.Y.C., 1963-68, assoc. prof., 1968-71, prof. chemistry, 1971—. Contbr. over 75 articles to profl. jours. With U.S. Army, 1955-57. Humboldt fellow, 1970-71. Mem. Am. Chem. Soc., Inter-Am. Photochem. Soc. Achievements include chemical research in photochemistry and spectroscopy. Office: St Johns Univ Dept Chemistry Jamaica NY 11439

TESTA, STEPHEN MICHAEL, geologist, consultant; b. Fitchburg, Mass., July 17, 1951; s. Guiseppe Alfredo and Angelina Mary (Pettito) T.; m. Lydia Mae Payne, July 26, 1986; 1 child, Brant Ethan Gage. AA, Los Angeles Valley Jr. Coll., Van Nuys, 1971; BS in Geology, Calif. State U., Northridge, 1976, MS in Geology, 1978. Registered geologist, Calif., Oreg.; cert. profl. geol. scientist., Idaho, Alaska; cert. engring. geologist, Calif.; registered environ. asessor, Calif. Engring. geologist R.T. Frankian & Assocs., Burbank, Calif., 1976-78, Bechtel, Norwalk, Calif., 1978-80, Converse Cons., Seattle, 1980-82; sr. hydrogeologist Ecology Environment, Seattle, 1982-83; sr. geologist Dames & Moore, Seattle, 1983-86; v.p. Engring. Enterprises, Long Beach, Calif., 1986-89; pres. Applied Environ. Svcs., San Juan Capastrano, Calif., 1990—. Author: Restoration of Petroleum Contaminated Aquifers, 1990, Principles of Technical Consulting and Project Management, 1991, Geologic Aspects of Hazardous Waste Management, 1993; editor: Geologic Field Guide to the Salton Basin, 1988, Environmental Concerns in the Petroleum Industry, 1989; contbr. over 60 articles to profl. jours., chpts. and preface to books. Mem. Am. Inst. Profl. Geologists (profl. devel. com. 1986, continuing edn. com., program chmn., 1988—, Presidential Cert. of Merit, 1987, nat. screening bd. mem. 1992-93, exec. bd. mem. 1993), Los Angeles Basin Geol. Soc. (pres. 1991-92), Geol. Soc. Am., Am. Assn. Petroleum Geologists Pacific Section (environ. com., co-chmn. 1993—), Am. Mineralogical Soc., South Coast Geol. Soc. AAAS, Assn. Ground Water Scientists and Engrs., Assn. Engring. Geologists, Assn. Mil. Engrs., Environ. Assessment Assn., Mineral. Soc. Can., Hazardous Materials Research Inst., Calif. Water Pollution Control Assn., Sigma Xi. Republican. Roman Catholic. Achievements include research in igneous and metamorphic petrology, asphalt chemistry; development of methods for subsurface characterization and remediation, proprietary processes for incorporation via recycling of contaminated soil and materials into cold-mix asphaltic products. Home: 31232 Belford Dr San Juan Capistrano CA 92675-1833 Office: Applied Environ Svcs 27282 Calle Arroyo San Juan Capistrano CA 92675

TESTER, LEONARD WAYNE, psychology educator; b. Nampa, Idaho, Aug. 21, 1933; s. Walter Vernon and Dora Dorothy (Peters) T. BTh, Kansas City Coll., Overland, Kansas, 1957; MA, Abilene Christian Coll. (now Abilene Christian U.), 1961; STB, Harvard U., 1969; EdM, Columbia U., 1971, EdD, 1976, MPhil, 1979, PhD, 1981. Lic. psychologist, N.Y. Pers. mgr. Boston Safe Deposit & Trust Co., 1966-69; sr. counselor, prof. clin. counseling, chair human rels. N.Y. Inst. Tech., Westbury, 1971—; pvt. practice N.Y.C., 1983—; columnist Korea Times, N.Y.C., 1989-90; cons., grad. asst. Columbia U. Bus. Sch. and Tchrs. Coll., N.Y.C., 1977-81. Contbr. articles to profl. jours.; presenter workshops in field. Exec. dir. Ho. of te Carpenter, Boston, 1967-68; bd. dirs. Pierre (S.D.) Coun. of Arts; bd. dirs. Counseling Ctr. Episcopal Ch., Great Neck, N.Y., Tech. Sch. in N.Y.C. William Wayne Jackson Honors scholarship Harvard Div. Sch. Fellow Am. Orthopsychiat. Assn.; mem. APA, N.Y. Soc. Clin. Psychologists, N.Y. Soc. Hypnosis and Psychotherapy, others. Home: PO Box 20107 New York NY 10023-1482 Office: NY Inst of Tech 1855 Broadway New York NY 10023-7692

TETHER, ANTHONY JOHN, aerospace executive; b. Middletown, N.Y., Nov. 28, 1941; s. John Arthur and Antoinette Rose (Gesualdo) T.; m. Nancy Engel Pierson, Dec. 26, 1963 (div. July 1971); 1 child, Jennifer; m. Carol Susan Dunbar, Mar. 3, 1973; 1 child, Michael. AAS, Orange County Community Coll., N.Y., 1961; BS, Rensselaer Poly Inst., 1963; MSEE, Stanford (Calif.) U., 1965, PhD, 1969. V.p., gen. mgr. Systems Control Inc., Palo Alto, Calif., 1968-78; dir. nat. intelligence Office Sec. of Def., Washington, 1978-82; dir. strategic tech. DARPA, Washington, 1982-86; corp. v.p. Ford Aerospace, Newport Beach, Calif., 1986-90, LORAL, Newport Beach, 1990-92; corp. v.p., sector mgr. Sci. Application Internat., Inc., San Diego, Calif., 1992—; chmn., bd. dirs. Condyne Tech., Inc., Orlando, Fla., 1990-92; dir. Orincon, La Jolla, Calif. Contbr. articles to profl. jours. Recipient Nat. Intelligence medal DCI, 1986, Civilian Meritorious medal U.S. Sec. Def., 1986. Mem. IEEE, Cosmos Club, Sigma Xi, Eta Kappa Nu, Tau Beta Pi. Avocations: ham radio, skiing. Home: 4518 Roxbury Rd Corona Del Mar CA 92625-3125

TEUSCHER, EBERHARD, pharmacist; b. Halle, Germany, Dec. 9, 1934; s. Wolfram and Maria (Henke) T.; m. Gisela Hempel, July 31, 1964; 1 child, Franka. Degree in apothecary, U. Halle, 1958, dr. rer. nat., 1960; dr. habil., U. Greifswald, Germany, 1965; lectr., asst. prof., U. Greifswald, 1966. Lic. pharmacist. Asst. dept. pharmacognosy U. Halle, 1959-61; head asst. dept. botany U. Greifswald, 1962-66, asst. prof. dept. pharmacy, 1968-71, prof., 1971—; dean faculty natural scis. U. Greifswald, 1975-90. Author: Pharmacognosy, 1970, 78, 83, 88, 90, Natural Toxins, 1988. Mem. Soc. Med. Plant Rsch., German Pharm. Soc. Office: U Greifswald, Jahnstrasse 15A, 17489 Greifswald Germany

TEW, E. JAMES, JR., electronics company executive; b. Dallas, July 7, 1933; s. Elmer James and Bessie Fay (Bennett) T.; children: Teresa Annette, Linda Diane, Brian James. Student Arlington State Jr. Coll., 1955-57; B.B.A. in Indsl. Mgmt., So. Meth. U., 1969; M.S. in Quality Systems, U. Dallas, 1972, M.B.A. in Mgmt., 1975, EdD in Adult Edn., Nova U., 1986, postdoctoral, 1986, 89. Registered profl. engr., Calif. Mgr. quality assurance ops. Tex. Instruments Inc., Dallas, 1957—, chmn. corp. metric implementation com., co-chmn. credit com. Texins Credit Union; adj. faculty Richland Coll., Mountain View Coll. Precinct chmn., election judge, del. several county and state convs.; bus. computer info. systems adv. bd. U. North Tex., bd. dirs. ctr. for quality and productivity U. North Tex.; bd. examiners Malcolm Baldrige Nat. Quality award, U.S. Dept. Commerce, Nat. Inst. Standard and Tech., 1988, 89, 90, 91. With U.S. Army, 1953-55. Decorated Army Commendation medal with oak leaf cluster. Fellow Am. Soc. Quality Control (cert. as quality and reliability engr., chmn. Dallas-Ft. Worth sect. 1974-75). Fellow U.S. Metric Assn. (cert., chmn. cert. bd. 1986-87); mem.. U.S.Res. Officers Assn., Dallas C. of C. (chmn. world mfg. com. 1974-77, chmn. spl. task force career edn. adv. bd. 1973-74), Mensa, Sigma Iota Epsilon, Phi Delta Kappa. Baptist. Clubs: Texins Rod and Gun (pres. 1969-70), Texins Flying, Masons (32 degree). Contbr. articles to profl. jours. Home: 10235 Mapleridge Dr Dallas TX 75238-2256 Office: PO Box 660246 MS 3107 Dallas TX 75266

TEWARI, ASHISH, aerospace scientist, consultant; b. Hardoi, India, Sept. 11, 1964; s. Kamaleshwar Sahai and Sudharma (Mishra) T. MS in Aerospace Engring., U. Mo., Rolla, 1988, PhD in Aerospace Engring., 1992. Grad. rsch. asst. McDonnell Aircraft Co., St. Louis, 1987-92; aerospace scientist Nat. Aero. Lab., Bangalore, India, 1992-93; asst. prof. aerospace engring. Indian Inst. Tech., Kanpur, India, 1993—; cons. Gulfstream Aerospace Corp., Savannah, Ga., 1990, Aero. Devel. Agy., Bangalore, India, 1993; pvt. pilot FAA. Contbr. articles of Jour. Aircraft. Rsch. grantee Indian Ministry Def. Aeros. R & D Bd., 1992. Mem. AIAA, Planetary Soc. Achievements include development of nonplanar doublet-point and acceleration-potential methods for supersonic unsteady aerodynamics, a consistent, nongradient optimization process for unsteady aerodynamics, new method for optimizing unsteady aerodynamics when repeated-poles occur; designed, fabricated and flight demonstrated a VTOL aircraft. Office: Indian Inst Tech, Dept Aerospace Engring, Kanpur 208016, India

TEWARI, KEWAL KRISHAN, chemist, researcher; b. Panjab, India, Jan. 1, 1940; came to U.S., 1974; s. S.K.D. and Leela Tiwari; m. Vijaya Sandhir, 1969; children: Muneesh, Asheesh. MS, Panjab U., Chandigarh, India, 1962, PhD, 1972. Lectr. Dayanand Anglo Vedic, Chandigarh, 1962-64, Govt. Coll. for Boys, Chandigarh, 1968-71; sci. officer Indian Inst. Petroleum, Dehradun, India, 1972-74; postdoctoral fellow Polymer Inst. U. Detroit, 1974-75; chemist Detroit Water and Sewer Dept., 1975-78, asst. lab. supr., 1979-84, lab. supr., 1984—; bd. dirs. ERT Testing Svcs., Inc., Detroit, 1988—. Mem. editorial bd. Gas Chromatography Lit., 1974; contbr. articles to profl. jours. Achievements include research in transport and themodynamic properties of small molecules. Office: Detroit Water & Sewage Dept 303 S Livernois Detroit MI 48209

TEWARI, KIRTI PRAKASH, biophysicist; b. Mirzapur, India, June 20, 1943; came to U.S., 1987; s. Sri Bhagwati Dayal and Kalawati (Dixit) T.; m. Manju Shukla, Feb. 12, 1969; children: Prabodh, Nitin. MSc in Physics, Agra (India) U., 1964; PhD in Biophysics, All India Inst. Med. Sci., New Delhi, India, 1975. Lectr. in physics PB Degree Coll., Gorakhpur U., India, 1964-65; jr. sci. asst. Inst. Rsch. and Devel. Orgn., Dehradun, India, 1965-66; rsch. fellow All India Inst. Med. Sci., 1966-68, sr. sci. officer, 1968-77; asst. prof. Indira Gandhi Med. Coll., Simla, India, 1977-81; assoc. prof. I.G. Med. Coll., Simla, India, 1981-87; fellow Indian Acad Med. Physics, New Delhi, 1983; rsch. assoc., scientist U. Tex. Med. Br., Galveston, 1987—; corefellow, prof. Dr. Y.S. Parmar Inst. Math. Scis., Simla, 1985-87; sec. All India Biomed. Engring. Conf., New Delhi, 1974, Parmar Inst. Math. Conf., 1987. Contbr. articles to profl. publs. Mem. Biophys. Soc., Biomed Engring. Soc. India (founding mem.). Hindu. Achievements include development of new computational technique for finding in vivo arterial health; co-development of a new photoelectric technique for estimating plasma (blood) coagulation time; co-observations on vesiculation and membrane redistribution processes in squid giant axon due to injury, mechanical and thermal stresses; research on Ca-channels in the cerebrovascular system. Home: 215 Market St # 103E Galveston TX 77550-5645 Office: U Md Med Sch Div Neurosurgery Dept Surgery & Neurosurgery Baltimore MD

TEWARSON, REGINALD PRABHAKAR, mathematics educator, consultant; b. Pauri, Garhwal, India, Nov. 17, 1930; came to U.S., 1957; s. Seth Narottam and Chand (Mani) T.; m. Hedi Thomann, July 1, 1960 (div. Nov. 1990); children: Anita Jasmine, Monique Shanti. MA, Agra (Ind.) U., 1952; PhD, Boston U., 1961. Lectr. Lucknow (Ind.) U., 1951-57; sr. mathematician Honeywell EDP, Wellesley Hills, Mass., 1960-64; leading prof. applied math. and stats. SUNY, Stony Brook, 1964—, leading prof. physiology and biophysics dept., 1964—; cons. NIH, Washington, 1971-74. Author: Sparse Matrices, 1973; mem. editorial bd. Applied Math. Letters, 1986—, Math. Computer Modeling, 1991—, Pan. Am. Math. Jour., 1991—; contbr. articles to profl. jours. Centenary scholar Govt. of India, 1946-50, Crusade scholar U.S. Coun. Chs., 1957-59; rsch. grantee NIH, 1973—, Air Force Office Sch. Rsch. Math. and Info. Scis., 1983-85, NSF, 1993—. Mem. Am. Math. Soc., Soc. Indsl. and Applied Math., Soc. for Math. Biology. Democrat. Achievements include pioneering research on sparse matrices based largely on own research; co-development of mathematical model of kidney concentrating mechanism, of computer model of neuronal function. Home: 22 Night Heron Dr Stony Brook NY 11790-1108 Office: SUNY Applied Math and Stats Dept Stony Brook NY 11794-3600

TEWHEY, JOHN DAVID, geochemist, environmental consultant; b. Lewiston, Maine, Feb. 14, 1943; s. William Patrick and Majorie (Peterson) T.; m. Gloria Nolin, June 19, 1965; children: Kathryn, Meridith, Allison. BS, Colby Coll., 1965; MS, U. S.C., 1968; PhD, Brown U., 1975. Cert. geologist, Maine, Calif. Project manager Lawrence Livermore (Calif.) Nat. Lab., 1974-81; div. leader E.C. Jordan Co., Portland, Maine, 1981-87; pres. Tewhey Assocs., Portland, Maine, 1987—; adj. faculty U. So. Maine, Gorham, 1987-89. Author (text) Hydrogeochemistry, 1987, (map folio) Geology of the Irmo Quadrance, S.C., 1975. Mem. Gorham Planning Bd., 1987—. Capt. USAF, 1967-71. Mem. Am. Inst. Hydrology (cert.), Am. Geophys. Union, Am. Chem. Soc., Rotary (2d v.p. 1992—). Roman Catholic. Home: 3 Valley View Dr Gorham ME 04038 Office: Tewhey Assocs 500 Southborough Dr South Portland ME 04106-6903

THACH, ROBERT EDWARDS, biology educator; b. Oklahoma City, Okla., Feb. 2, 1939; s. William Thomas and Mary Elizabeth (Edwards) T.; m. Carol Ann Schmidt, Sept. 23, 1959 (div. Aug. 1967); children: Catherine Anne, Robert Edwards Jr.; m. Sigrid Stumpp, Apr. 20, 1968; 1 child, Christopher Alexander. AB, Princeton (N.J.) U., 1961; PhD, Harvard U., 1964. Asst. prof. biochemistry and molecular biology Harvard U., Cambridge, Mass., 1966-69, assoc. prof. biochemistry and molecular biology, 1969-70; assoc. prof. biol. chemistry Wash. U., St. Louis, 1970-73, prof. biol. chemistry, 1973—, prof. biology, 1977—; chmn. biology, 1977-81, dean Grad. Sch. Arts and Scis., 1993—; institutional biosafety com. Monsanto Co., St. Louis, 1980-83. Mem. editorial bd. Jour. of Biol. Chemistry, 1984-89, Archives of Biochemistry and Biophysics, 1972-78; editor Enzyme, 1990-91; contbr. articles to profl. jours. including Sci., Nature, Cell, and Jour. of Biol. Chemistry. Fellowship Woodrow Wilson Found., 1961-62, NSF, 1962-64, John Simon Gugenheim Meml. Found., 1969; grantee NSF, NIH, 1970—, AAAS, 1992. Mem. Am. Soc. of Biol. Chemists, Am. Soc. for Virology. Achievements include discovery of initiation codon "AUG" for mRNA translation, role for GTP in protein synthesis initiation, translational repressor for ferritin synthesis; devel. of methods for RNA synthesis. Office: Washington U Dept Biology One Brookings Dr Saint Louis MO 63130

THACH, WILLIAM THOMAS, JR., neurobiology and neurology educator; b. Okla. City, Jan. 3, 1937; s. William Thomas and Mary Elizabeth T.; m. Emily Ransom Otis, June 30, 1963 (div. 1979); children: Sarah Brill, James Otis, William Thomas III. AB in Biology magna cum laude, Princeton U., 1959; MD cum laude, Harvard U., 1964. Diplomate Am. Bd. Psychiatry and Neurology (in Neurology). Intern Mass. Gen. Hosp., Boston, 1964-65, asst. residency, 1965-66; staff assoc. physiology sect. lab. clin. sci. NIMH, Bethesda, Md., 1966-69; neurology resident, clin. and rsch. fellow Mass. Gen. Hosp., 1969-71; from asst. prof. neurology to assoc. prof. neurology Yale U. Sch. Medicine, New Haven, Conn., 1971-75; assoc. prof. neurobiology and neurology dept. anatomy and neurobiology Washington U. Sch. Medicine, St. Louis, 1975-80, prof. neurobiology and neurology dept. anatomy and neurobiology, 1980—, chief divsn. neurorehab. dept. neurology, 1992—; acting dir. Irene Walter Johnson Rehab. Inst. Washington U. Sch. Medicine, 1989-91, dir., 1991-92; attending neurologist Barnes Hosp., med. dir. dept. rehab.; attending neurologist Jewish Hosp., St. Louis Regional Hosp.; bd. sci. counselors NINCDS, 1988-92; mem. NIH Study Sect. Neurology A, 1981-85. Assoc. editor Somatosenory and Motor Research; contbr. numerous articles to profl. jours. Fulbright grantee U. Melbourne, Australia, 1959-60; NIH grantee, 1971—. Mem. Am. Physiological Soc., Am. Acad. Neurology, Soc. Neurosci., Am. Neurological Assn., Phi Beta Kappa, Sigma Xi, Alpha Omega Alpha. Achievements include research on brain control of movement and motor learning, roles of the basal ganglia and the cerebellum in health and disease. Home: 7325 Kingsbury Blvd Saint Louis MO 63130 Office: Washington Univ Dept Anatomy & Neurobiol 660 S Euclid Ave Saint Louis MO 63110

THACKER, BEN HOWARD, civil engineer; b. Boone, Iowa, Dec. 5, 1960; s. Stanley E. and Gloria R. (Reitz) T.; m. Kimberly Ann MacDougall, May 26, 1984; children: Nicholas Ryan, Brett Michael. BCE, Iowa State U., 1983; MCE, U. Conn., 1987. Registered profl. engr., Tex. Rsch. engr. Gen. Dynamics, Groton, Conn., 1984-88; sr. rsch. engr. S.W. Rsch. Inst., San Antonio, 1988—. Contbr. articles to Internat. Jour. Fracture, Structures, Structural Dynamics and Materials, Computers and Structures. Mem. ASCE, AIAA. Republican. Home: 6215 Rue Sophie San Antonio TX 78238 Office: SW Rsch Inst 6220 Culebra Rd San Antonio TX 78228

THAGARD, NORMAN E., astronaut; b. Marianna, Fla., July 3, 1943; s. James E. Thagard and Mary F. Nicholson; m. Rex Kirby Johnson; children: Norman Gordon, James Robert, Daniel Cary. BS, Fla. State U., 1965, MS, 1966; M.D., U. Tex., 1977. Astronaut NASA, 1978—, mission specialist Challenger Flight 2, mission specialist Space Lab 3; mission specialist STS-30 Magellan; payload comdr. STS-42, Internat. Microgravity Lab. 1. Contbr. articles to profl. jours. Served with USMC, 1969-70, Vietnam. Decorated 11 Air medals, Navy Commednation medal with Combat V, Marine Corps E award, Vietnam Svc. medal, Vietnamese Cross of Gallantry with Palm. Mem. AIAA, Phi Kappa Phi. Avocations: classical music, electronic design. Office: Lyndon B Johnson Space Ctr NASA Houston TX 77058

THAKKER, ASHOK, aerospace engineering company executive; b. Bombay, Aug. 9, 1947; came to U.S., 1970; s. Bhagwandas P. and Champa (Lakhpatla) T.; m. Sarla A. Bhate, Dec. 30, 1975; 1 child, Amish. B of Engring., U. Bombay, 1970; MS, S.D. Tech. Inst., 1971; PhD, Va. Polytech. Inst. & State U., 1974; MBA, Fla. Inst. Tech., 1980. Registered profl. engr., Va., Fla.; cert. mfg. engr. Instr. Va. Poly. Inst., Blacksburg, 1972-74; engr. Alcoa Tech. Ctr., Alcoa Ctr., Pa., 1974-76, sr. engr., 1976-78; group leader Pratt and Whitney Aircraft, West Palm Beach, Fla., 1978-83; project mgr. U.S. govt. programs Rolls-Royce, Inc., Atlanta, 1983-85, mgr. materials rsch., 1985-86, sr. project mgr., 1986-88, sr. mgr. engring., 1988—. Contbr. numerous articles to profl. jours. and tech. reports to coms. Chmn. Internat. Student Assn., Blacksburg, 1972-73; advisor Jr. Achievement Program, Palm Beach, Fla., 1982-83. R & D Sethna Scholarship award, 1970; recipient Best Paper award AIAA-Atlanta Symposium, 1987. Mem. ASTM (reviewer tech. papers, composites, fatigue and fracture coms.), Am. Soc. Materials (chpt. program chmn. 1982, internat. chpt. membership com. 1989, vice chmn., then chmn. Atlanta chpt. 1989, 90, nat. devel. coun. 1990-91, nat. govt. and pub. affairs com. 1992-94, advisor, prfnl. as resource for instruction in sci. and math. student outreach program 1992—, tech. awareness adv. com., edit. bd. Materials & Process Jour. 1993—), Soc. Exptl. Mechanics (chpt. chmn. 1978, 81-82, chmn. fatigue com. 1984), Soc. Advancement Material and Process Engring., Soc. Mfg. Engrs. (com. mem., 1st vice chmn. Atlanta chpt. 1991-92, chmn. 1992-93, sr. internat. dir. 1992-93), Materials Rsch. Soc., Am. Ceramics Soc. (exec. com.), NASA Space Consortium (exec. com. 1992-93), Toastmasters, Lions, Sigma Xi, Vinings Club. Clubs: Toastmasters, Lions, Sigma Xi, Vinings (social com. 1992-93). Avocations: music, reading, traveling, golf, tennis. Home: 2690 Spencer Trail Marietta GA 30062 Office: Rolls-Royce Inc 2849 Paces Ferry Rd NW Atlanta GA 30339-3769

THAKOR, NITISH VYOMESH, biomedical engineering educator; b. Nagpur, India, Feb. 9, 1952; came to U.S., 1976; s. Vyomesh H. and Jayshree V. Thakor; m. Ruchira N. Thakor, Dec. 17, 1983; children: Mitali N., Milan N. B of Tech., Indian Inst. Tech., Bombay, 1974; PhD, U. Wis., 1981. Engr. Philips India Ltd., Bombay, 1974-76; rsch. asst. U. Wis., Madison, 1977-81; asst. prof. Northwestern U., Evanston, Ill., 1981-83; asst. prof. Johns Hopkins U., Balt., 1984-87, assoc. prof., 1987—; cons. Biomed. Instrumentation, 1984—. Assoc. editor Jour. Ambulatory Monitoring, Med. Design and Material, 1978—, IEEE Jour. Transactions on Biomed. Engring., 1988—, Jour. Biol. Systems, 1993—; contbr. more than 50 articles to profl. jours. Recipient Presdl. Young Investigator award, 1985, NIH Rsch. Career Devel. award, Centennial medal U. Wis. Sch. Engring.; Fulbright scholar, 1987; grantee NSF, NIH. Mem. IEEE (sr.), Sigma Xi. Achievements include development of micro-computer based ambulatory ECG monitor; techniques for arryhythmia detection in implantable pacemakers and defibrillators, neurological signal processing and monitoring instrumentation. Home: 7273 Steamerbell Row Columbia MD 21045-5229 Office: Johns Hopkins U Med Sch Biomed Engring Dept 720 Rutland Ave Baltimore MD 21205-2109

THALDEN, BARRY R., architect; b. Chgo., July 5, 1942; s. Joseph and Sibyl (Goodwin) Hechtenthal; m. Irene L. Mittleman, June 23, 1966 (div. 1989); 1 child, Stacy Elizabeth. BArch, U. Ill., 1965; M in Land Architecture, U. Mich., 1969. Landscape architect Hellmuth, Obata, Kassebaum, St. Louis, 1969-70; dir. landscape architecture PGAV Architects, St. Louis, 1970-71; pres. Thalden Corp (formerly Saunders-Thalden & Assocs. Inc.), St. Louis, 1971—. Prin. works include Rock Hill Park, 1975 (AIA award 1977), Wilson Residence, 1983 (AIA award), Nat. Bowling Hall of Fame, 1983 (St. Louis RCGA award 1984), Village Bogey Hills (Home Builders award 1985), St. Louis U. Campus Mall (St. L. ASLA award 1989), Horizon Casino Resort, Lake Tahoe, Nev., St. Louis Airport's Radisson Hotel. Bd. dirs. St. Louis Open Space Council, 1973-83; trustee United Hebrew Temple, St. Louis, 1984-88; apptd. Mo. Lands Architect Coun., 1990. Named Architect of Yr. Builder Architect mag., 1986. Fellow Am. Soc. Landscape Architects (nat. v.p. 1979-81, pres. St. Louis chpt. 1975, trustee 1976-79, nat. conv. chair 1991); mem. AIA, World Future Soc. (pres. St. Louis chpt. 1984—). Avocations: painting, gardening, tennis. Home: 8 Edgewater Ct Saint Louis MO 63105 Office: Thalden Corp 7777 Bonhomme Ave Ste 2200 Saint Louis MO 63105-1911

THALL, ARON DAVID, cell biologist; b. Kalamazoo, Oct. 11, 1960. MS, Calif. State U., 1985; PhD, U. Calif., San Francisco, 1990. Rsch. assoc. City of Hope Med. Ctr., Duarte, Calif., 1984-86; postdoctoral fellow dept. pathology U. Mich., Ann Arbor, 1991; assoc. Howard Hughes Med. Inst., Ann Arbor, 1991—. Contbr. articles to Biochemistry, Annals N.Y. Acad. Sci. Heyl scholar Kalamazoo Coll., 1978-79, Regents scholar U. Calif., 1987-88. Mem. AAAS. Office: Howard Hughes Med Inst Bldg MSRB II 1150 W Medical Ctr Dr Ann Arbor MI 48109-0650

THALMANN, DANIEL, computer science educator; b. Geneva, Jan. 26, 1946; s. Roger and Andree (Batard) T.; m. Nadia Magnenat, May 20, 1977; children: Melanie, Vanessa, Sabrina. Diploma, Univ., Geneva, 1970, Post-grad. Cert., 1972, PhD, 1977. Prof. Univ., Montreal, Can., 1977-88; vis. prof. Univ., Lincoln, Nebr., 1979; invited researcher CERN, Computer Graphics Group, Geneva, 1983-84; prof. Swiss Fed. Inst. of Tech., Lausanne, Switzerland, 1988—; co-dir. MIRA Lab, Univ. Montreal, Can., 1980-88; dir. Computer Graphics Lab., EPFL, Lausanne, 1988—; head Computer Sci. Dept., EPFL, Lausanne, 1991-93. Author, editor 20 books; co-editor-in-chief Jour. of Visualization and Computer Animation, 1990—. Mem. IEEE Computer Sci. Soc., Assn. for Computing Machinery, Eurographics, SIGGRAPH, Computer Graphics Soc., Eurographics Working Group on Animation and Simulation. Avocations: golf, tennis, skiing. Home: 25 ch Pres-de-la-Gradelle, Ch-1223 Cologny Switzerland Office: Swiss Fed Inst Tech, Computer Graphics Lab, CH-1015 Lausanne Switzerland

THAM, FOOK SIONG, crystallographer; b. Teluk Intan, Perak, Malaysia, Feb. 9, 1956; came to U.S., 1979; s. Siew Koh and Yuet Siew (Lee) T. BS, Elmira (N.Y.) Coll., 1982; MS, Rensselaer Poly. Inst., 1984, PhD, 1987. Physics lab. asst. Tar Coll., Kuala Lumpur, Malaysia, 1978-79; chemistry/math. tutor Elmira Coll., 1979-82; chemistry teaching asst. Rensselaer Poly. Inst., Troy, N.Y., 1982-83, radiochemist, 1982-87, x-ray crystallographer, 1987—; cons. asst. Zeta Internat., Latham, N.Y., 1983-84. Contbr. articles to profl. jours. Vol. Pupil Assistance in Learning, Elmira, 1981-82. Recipient fellowship Rensselaer Poly. Inst., 1982-87. Mem. Am. Crystallograhic Assn., Sigma Xi. Achievements include structural determination of organic, inorganic, organic-metallic compounds using single crystal x-ray diffraction techniques, anomalies in K beta/K alpha intensity ratios of elements CR to BR; investigation of minimum detection limit of radioactive iodine in liquid waste; development of portable X-ray spectrometer; surface density determination of old stamps by X-ray fluorescence techniques. Home: 304 8th St Troy NY 12180 Office: Rensselaer Poly Inst 110 8th St Troy NY 12180

THAMPI, MOHAN VARGHESE, environmental health and civil engineer; b. Kuching, Sarawak, Malaysia, Mar. 25, 1960; s. Padmanabha Ramachandran and Sosamma (Varghese) T. Gen. Cert. Edn., Cambridge U., 1976; B in Tech. with honors, Indian Inst. Tech., Kharagpur, India, 1983; MS in Engring., U. Tex., 1985; DSc (hon.), London Inst. Applied Rsch., 1992. Registered profl. engr., Tex., Fla., registered environ. mgr.; cert. safety tng. OSHA; cert. Nat. Coun. Examiners for Engrs. and Surveyors. Assoc. engr. Brown & Caldwell, Dallas, 1985-87; project mgr. Brown & Caldwell, Orlando, 1987-88, Stottler Stagg & Assocs., Cape Canaveral, Fla., 1988-91; sr. project engr. Chastain-Skillman, Inc., Lakeland, Fla., 1991-93; project mgr. Glace & Radcliffe, Inc., Winter Park, Fla., 1993—. Author: Ultraviolet Disinfection Studies in a Teflon-Tube Reactor, 1985; contbr. articles to profl. jours. Active Rep. Pres.'s Citizens Adv. Commn., 1992. Recipient Cert. of Cont. Profl. Devel. award Fla. Engring. Soc., 1992. Mem. NSPE, ASCE (assoc.), Internat. Assn. Water Pollution Rsch. and Control, Am. Mensa, Am. Water Works Assn., Water Pollution Control Fedn. (com. for preparing design practice manuals 1989—), Internat. Freelance Photographers Assn., Internat. Platform Assn., Am. Mgmt. Assn., Am. Smokers Alliance, Smithsonian Instn., Nat. Geog. Soc., Nat. Registry Environ. Professionals, U. Tex. Ex-Students Assn., Wine Soc. Am., Nat. Family Opinion, Internat. Deep Purple Appreciation Soc., Wilson Ctr. Assocs., I.I.T. Kharagpur Tech. Found., NASA Tech Briefs Reader Opinion Panel, Chemical Engring. Jour. Product Rsch. Panel, Plant Engring. Editorial Quality Panel, Kharagpur Tech. Alumni Found., Knight Order of Templars (Jerusalem). Mar Thoma Syrian Christian. Avocations: photography, music, travel, sports. Office: Glace & Radcliffe Inc PO Box 1180 Winter Park FL 32790

THANGARAJ, AYYAKANNU RAJ, mechanical engineering educator; b. Klangad, Tamil Nadu, India, Jan. 9, 1958; came to U.S., 1980; s. Kotiappan and Ponnammal (Thangiah) Ayyakannu; m. Umarani Jayaraman, Feb. 10, 1985; children: Siddharth Ayyakannu, Samyuktha. BTech, Indian Inst. Tech., Madras, 1980; M Engring., Carnegie Mellon U., 1982, PhD, 1985. Registered profl. engr., Mich. Rsch. asst. Carnegie Mellon U., Pitts., 1980-85; vis. asst. prof. mech. engring. Mich. Technol. U., Houghton, 1985-87, asst. prof., 1987-93, assoc. prof., 1993—; cons. Colding Internat. Corp., Ann Arbor, Mich., 1990-91, B.H. Barkalow, Inc., Newaygo, Mich., 1991—, Gagleard Addis & Imbrunone PC, Royal Oak, Mich., 1992—. Editor conf. symposium procs.; contbr. articles to profl. jours. Merit scholar Govt. of India, 1975, grad. scholar Speedsteel, Inc., Sweden, 1982; rsch. grantee Dow Chem. Co., 1988. Mem. ASME, Am. Soc. for Engring. Edn., Soc. Mfg. Engrs. (sr., cert., sci. com. 1989, Outstanding Young Mfg. Engr. award 1990), Sigma Xi. Home: 1006 E 9th Ave Houghton MI 49931-1506 Office: Mich Technol U ME-EM Dept 1400 Townsend Dr Houghton MI 49931-1295

THAPA, KHAGENDRA, survey engineering educator; b. Murtidhunga, Dhankuta, Nepal, Oct. 16, 1950; came to U.S., 1982; s. Ranadhoj and Krishna (Basnet) T.; m. Rajani Basnet, July 7, 1981; children: Samrat, Birat, Charisma. BSc, U. East London, 1978; MSCE, U. N.B., Fredericton, Can., 1980; MS, Ohio State U., 1985, PhD, 1987. Rsch. asst. U. N.B., Fredericton, 1978-80; researcher Geodetic Survey Can., Ottawa, 1980; lectr. Engring. Inst. Tribhuvan U., Kathmandu, 1980-82; teaching/rsch. assoc. Ohio State U., Columbus, 1982-87; assoc. prof. Ferris State U., Big Rapids, Mich., 1987-91, prof. Sch. Tech., 1991—; researcher The Ctr. for Mapping, Columbus, Ohio, 1990; cons. Technicians, Architects and Engrs. Consultancy, Kathmandu, 1980-82; chairperson Engring. Topographic Mapping Com. Am. Soc. Photogrammetry and Remote Sensing, Washington, 1991—. Grantee NSF, Washington, 1989, 91. Mem. Royal Instn. Chartered Surveyors (rsch. grant 1988-89), Am. Congress on Surveying and Mapping, Am. Soc. Photogrammetry, Inst. Navigation, Geodetic Sci. Club (v.p. 1985-86). Achievements include devising a new technique to find inconsistent observations and constraints in horizontal networks, using linear programming in optimal design of leveling networks, devising a new method of line generalization in computer cartography and digital mapping; worked on critical points detection and data compression using zero-crossings; analysed the accuracy of spatial data used in geographic information system. Home: 20825 Edewood Dr Big Rapids MI 49307 Office: Ferris State U Coll Tech Surveying Engring Program 915 Campus Dr Rm 312 Big Rapids MI 49307-2277

THAYER, RONDA RENEE BAYER, chemical engineer; b. Burlington, Iowa, Oct. 4, 1965; d. Ronald Robert Bayer and Meredith Ann (Gilbert) Osbon; m. Dennis Norman Thayer, May 22, 1992. BS in Chem. Engring., Purdue U., 1988. Entry engr. Goodyear Tire & Rubber Co., Akron, Ohio, 1988-90, devel. engr., 1990—. Post advisor Engring. Explorer Scouts, Boy Scouts Am., Akron, 1988-90, head advisor, 1990-93. Mem. Am. Inst. Chem. Engrs. (sr. mem.), Soc. Mfg. Engrs., Philanthropic Edn. Orgn. (v.p. 1991-92). Republican. Office: Goodyear Tire & Rubber Co PO Box 3531 D/464B Akron OH 44309-3531

THEEUWES, FELIX, physical chemist; b. Duffel, Belgium, May 25, 1937. Licentiaat physics, Cath. U. Louvain, 1961, DSc in Physics, 1966. Tchr. St. Vincent Sch., Westerlo, Belgium, 1961-64; rsch. assoc. chemistry U. Kans., 1966-68, asst. prof., 1968-70; rsch. scientist pharm. chemistry Alza Corp., Palo Alto, Calif., 1970-74, prin. scientist, 1974—, v.p. prodn., rsch. and devel., 1980—; Louis Busse lectr. dept. pharmacology U. Wis., 1981. Mem. Am. Chem. Soc., Acad. Pharm. Sci., N.Y. Acad. Sci. Achievements include research in osmosis, diffusion, solid state physics, cryogenics, high pressure, thermodynamics, pharmacology, pharmacokinetics, calorimetry. Office: Alza Corp 950 Page Mill Rd PO Box 10950 Palo Alto CA 94303*

THEILE, BURKHARD, physicist; b. Soemmerda, Thuringa, Germany, July 16, 1940; s. Hans U. and Rose E. (Kuszmink) T.; m. Dagmar Guenster, June 9, 1966 (div. Oct. 1978); children: Gudrun, Ulrike; m. Heide Luise Mueller, Oct. 6, 1983. Diplom-physiker, T.U., Braunschweig, Germany, 1968, Dr.rer.nat, 1971. Asst. prof. T.U., Braunschweig, 1970-75, assoc. prof., 1975-80; program mgr. Dornier System, Friedrichshafen, Germany, 1981-83, Dornier GmbH, Neuilly, France, 1983-85; exec. v.p. Dornier of N.Am., Arlington, Va., 1986-90; gen. mgr. STN System Technik Nord, Bremen, Germany, 1991—; advisor for space programs German Ministry for Sci., Bonn, Germany, 1974-79. Contbr. 36 articles to profl. jours. Mem. AIAA, Deutsche Gesellschaft fur Luft-u. Raumfahrt. Office: STN Systemtechnik Nord, GMBH, D-28199 Bremen Germany

THEISZ, ERWIN JAN, scientist; b. Velky Slavkov, Slovakia, Oct. 20, 1924; s. Jan and Julia Maria (Copus) T.; m. Lydia Margrete Kracht, Aug. 3, 1963; 1 chld, Christine. MSPE, Wiesbaden (Germany) State U., 1975; dipl. engr., Kassel (Germany) Coll., 1983. R&D chem. engr. U.S. Civil Svc. Commn. (GS12), Newark, 1960-65; chemist Sloan Kettering Cancer Inst., N.Y.C., 1964-65; chem. R&D mgr. Beech Nut Life Savers, Port Chester, N.Y., 1966-67; chemist quality control York Rsch., Stamford, Conn., 1968-71; chemistry mgr. Ducon Co., Mineola, N.Y., 1972; quality control chemist Air Pollution Industry, Englewood, N.Y., 1973; analytical chemist Riverside Engring., N.Y.C., 1974; analytical chemist R&D Westchester County Health Dept., White Plains, N.Y., 1975—. Contbr. numerous articles to profl. jours. Lutheran. Achievements include patent for a method for inhibiting the growth of tumorous tissue cells in mammals such as humans. Home: 127 Randolph Rd White Plains NY 10607-1518

THENEN, SHIRLEY WARNOCK, nutritional biochemistry educator; b. San Mateo, Calif., Nov. 19, 1935; d. Matthew and Elsie (Gartner) Warnock; m. Allan Thenen, June 20, 1962; children: Matthew, Anna. BS, U. Calif., Berkeley, 1957, PhD, 1970. Postdocotoral fellow Harvard U. Med. Sch., Boston, 1970-72; asst. prof. Harvard U. Sch. Pub. Health, Boston, 1972-80, assoc. prof., 1980-84; prof. U. Minn., St. Paul, 1984—; mem. grant rev. panel NIH, Bethesda, Md.; mem. rev. panel USDA, Washington, Fedn. Am. Socs. Exptl. Biology, Bethesda. Recipient Career Devel. award NIH, 1975-80. Mem. Am. Inst. Nutrition, N.Am. Mycol. Assn. Office: U Minn Dept Food Sci and Nutrition 1334 Eckles Ave Saint Paul MN 55108-1040

THEOBALD, JÜRGEN PETER, physics educator; b. Siegen-Weidenau, Germany, Nov. 6, 1933; s. Walter and Frieda (Dietrich) T.; m. Dörte Prühs, March 3, 1961; children: Dirk, Jörn, Marc. Diploma in Physics, U. Bonn, Fed. Republic of Germany, 1959; Dr of Natural Scis., U. Mainz, Fed. Republic of Germany, 1963. Rsch. fellow Max Planck Inst. für Chemie, Mainz, Fed. Republic of Germany, 1961-63; sci. referee Cen. Bur. for Nuclear Measurements, Geel, Belgium, 1964-73; prof. Inst. Kernphysik der Tech. Hochschule, Darmstadt, Fed. Republic of Germany, 1973—. Contbr. numerous articles profl. jours. Mem. European Physical Soc., German Physical Soc. Avocation: Jazz. Home: Graupnerweg 42, D-6100 Darmstadt Federal Republic of Germany Office: Technische Hochschule, Schlossgartenstrasse 9, Darmstadt Germany

THEODORE, ARES NICHOLAS, research scientist, educator, entrepreneur; b. Kalamata, Greece, Oct. 28, 1933; came to U.S., 1954; s. Nicholas A. and Angeliki (Myseros) Theodoracopulos; m. Peggy Salvarakis, Sept. 3, 196l; children: Nicholas A., Angie A. BA cum laude, Westminster Coll., Salt Lake City, 1958; MS, U. Utah, 196l; postgrad., Case Western Res. U., 1967-68. Asst. prof. chemistry Westminster Coll., 196l-64; sr. rsch. chemist Diamond Shamrock Corp., Cleve., 1964-69; rsch. scientist A, Ford Motor Co., Detroit, 1969-73, sr. rsch. scientist, 1973-84, prin. rsch. scientist, 1984—. Contbr. articles to profl. jours.; patentee in field (57). Mem. ch. bd. Holy Cross Greek Orthodox Ch., Farmington Hills, Mich., 1986-88; campaigner Farmington Hills Dem. Com., 1986-88; mem. bus. coun. Boston Dem. Com., 1988—, Nat. Dem. Com., 1988—. U. Utah fellow, 1958-6l, NSF fellow, 1964. Mem. Am. Chem. Soc. (treas. 1967), Ahepa (bd. govs. Dearborn, Mich. 1985-86). Avocations: swimming, golf. Home: 34974 Valley Forge Dr Farmington MI 48331-3210

THEROUX, STEVEN JOSEPH, biologist, educator; b. Providence, Apr. 15, 1961; s. Joseph Clement and Catherine Margaret (Evans) T.; m. Diane Theresa D'Aniello, May 31, 1986; children: Melissa D'Aniello, Katrice D'Aniello. BA, R.I. Coll., 1984; PhD, U. Mass., 1989. Rsch. assoc. Worcester Found. for Exptl. Biology, Shrewsbury, Mass., 1989-91, Howard Hughes Med. Inst. U. Mass. Med. Ctr., Worcester, 1991-92; asst. prof. Assumption Coll., Worcester, 1992—. Contbr. articles to Jour. of Bacteriology, Nuclear Acid Rsch., Jour. Biol. Chemistry, Molecular Endocrinology, FEBS LEtters. Recipient award Am. Inst. Chemistry, 1984. Mem. N.Y. Acad. Scis., Coun. Undergrad. Rsch., Amnesty Internat., Phi Kappa Phi. Roman Catholic. Achievements include research in mechanism of signal transduction by epidermal growth factor receptor. Office: Assumption Coll 500 Salisbury St Worcester MA 01615-0005

THEUER, CHARLES PHILIP, surgeon; b. Madison, Wis., Sept. 19, 1963; s. Richard C. and Carol Ann (Imbruce) T.; m. Mari Elizabeth Doty, May 27. 1989; 1 child, Christian Zachary. BS, Mass. Inst. Tech., 1985; MD, U. Calif., San Francisco, 1989. Resident in surgery Harbor-UCLA Med. Ctr., Torrance, 1989—; sr. asst. surgeon USPHS, Bethesda, Md., 1991—. Contbr. articles to profl. jours. Vol. big bro. NCAA, Cambridge, Mass., 1983-85; blood drive coord. ARC, Cambridge, 1981; vol. health worker dept. epidemiology U. Calif., San Francisco,1 985-89. Mem. AMA, AAAS, Alpha Omega Alpha. Home: 12208 Atherton Dr Silver Spring MD 20902 Office: NIH Bldg 37 Rm 4B27 9000 Rockville Pike Bethesda MD 20892

THEWS, GERHARD, physiology educator; b. Königsberg, Germany, July 22, 1926; m. Gisela Bahling, 1958; 4 children. Student, U. Kiel. Rsch. fellow U. Kiel, 1957-61, asst. prof., 1961-62, assoc. prof., 1962-63, prof., dir. physiol. inst., 1964—; dean faculty medicine U. Kiel, 1968-69; v.p. Acad. Sci. and Lit., 1977-85, pres., 1985—; pres. German Physiol. Soc., 1968-69, Internat. Soc. Oxygen Transport, 1973-75; mem. German Scientific Coun., 1970-72. Author: Human Anatomy, Physiology and Pathophysiology, 1985, Autonomic Functions in Human Physiology, 1985, Human Physiology, 1989. Recipient Wolfgang Heubner prize Berlin, 1961, Feldberg prize London, 1964, Carl Diem prize, 1964, Adolf Fick prize Würzburg, 1969, Ernst von Bergmann medal Germany, 1986. *

THIEL, DANIEL JOSEPH, physicist, researcher; b. Tulsa, Aug. 31, 1963; s. John Emery and Mary Agnes (Arnold) T. BS, Tex. A&M U., 1985; PhD, Cornell U., 1992. Rsch. asst. Tex. A&M U., College Station, 1983-85; grad. rsch. asst. Cornell U., Ithaca, N.Y., 1986-92, rsch. assoc., 1992—. Contbr. articles to sci. jours. Active ACLU, Amnesty Internat. Nat. Merit scholar, 1981-85; McMullen Grad. fellow Cornell U., 1985; Tng. grantee NIH, 1986-89. Mem. Am. Physical Soc., Soc. Photo-Optical Instrumentation Engrs., Phi Kappa Phi, Sigma Pi Sigma. Democrat. Achievements include patent for mounting of an X-ray concentrator; development of an optical element, the tapered capillary X-ray concentrator which focuses hard X-rays to smallest sizes ever achieved; construction of a scanning Xúray microscope with such an element; determined the first molecular excited-state structure. Home: 209 N Plain St Ithaca NY 14850 Office: Cornell U Dept Applied Physics 212 Clark Hall Ithaca NY 14853

THIEL, THOMAS JOSEPH, information scientist, consultant; b. Upper Sandusky, Ohio, Dec. 31, 1928; s. Otto Peter and Lillian Susan (Orians) T.; m. Alice Ellen Miller, June 18, 1955 (div. Dec. 1982); children: Susan Marie Schworer, Christine Ellen, Joseph Allen, John Andrew; m. Jean Karen Singer, Mar. 9, 1984. BS, Ohio State U., 1956, MS, 1959. Rsch. scientist USDA, Columbus, Ohio, 1957-63; rsch. adminstr. USDA, St. Paul, Minn., 1963-72, Peoria, Ill., 1972-84; div. mgr. Advanced Systems Devel., Alexandria, Va., 1984-91; pres. Info. and Image Mgmt. Techs., Alexandria, 1991—; cons. Office of Sec. of Def., U.S. Army, USN, 1984-93, NATO, Mons, Belgium, 1988, Inst. De La Vie, Washington and Paris, 1991; dir. lab. U.S. Dept. Def. Author: CD-ROM Mastering for Information and Image Management, 1990, Automated Indexing of Document Image Management Systems, Ency. of Libr. and Info. Sci.; contbr. articles to Jour. Soil Sci. Soc. Am. Procs., 1960-62; contbr. articles to profl. jours. Sgt. U.S. Army, 1950-52, Korea, 1992-93. Fellow Soil and Water Conservation Soc. (life, bd. dirs. 1981-84, tech. transfer futures task force 1993—), Am. Soc. Agronomy (cert., emeritus), Soil Sci. Soc. Am. (emeritus), Assn. for Info. and Image Mgmt. (co-chair CD-ROM task force 1990--, chair electronic image mgmt. SIG 1991-92, chair emerging tech. adv. group 1992-93), Soc. for Applied Learning Tech. Achievements include research in the application of optical disk, compact disk read only memory and multimedia systems to office technical and scientific information management and dissemination, and in water movement in agricultural fields. Home: 4541 Saucon Valley Ct Alexandria VA 22312-3163 Office: II-Tech PO Box 11868 Alexandria VA 22312-0868

THIERRY, ROBERT CHARLES, microbiologist; b. Bourtzwiller, France, Sept. 12, 1938; s. Emile G. and Lucie M. (Meyer) T.; m. Ruth A. Biehrer, Jan. 18, 1958; children: Isabelle, Michel, Christophe, Catherine, Pierre, Simon. Dr. Medicine, U. Strasbourg, 1968; cert. immunobiology, Inst. Pasteur, Paris, 1971; cert. microbiology, Inst. Pasteur, 1972. Asst. Med. High Sch., Strasbourg, France, 1970-73; chief of rsch. Med. High Sch., 1973-86, conf. master, 1986—. Mem. French Soc. Immunobiology, French Soc. Microbiology, Ctr. Democrates Sociaux. Roman Catholic. Office: Faculte Medecine, 3 rue Koeberle, F 67000 Strasbourg France

THIERS, EUGENE ANDRES, mineral economist, educator; b. Santiago, Chile, Aug. 25, 1941; came to U.S., 1962, naturalized, 1976; s. Eugenio A. and Elena (Lillo) T.; m. Marie H. Stuart, Dec. 23, 1965 (div. 1979); children: Ximena, Eugene, Alexander; m. Patricia Van Metre, Jan. 29, 1983. B.S., U. Chile, 1962; M.S., Columbia U., 1965, D.Eng.Sc., 1970. Mgr.-tech. Minbanco Corp., N.Y.C., 1966-70; dir. iron info. Battelle Inst., Columbus, Ohio, 1970-75; minerals economist SRI Internat., Menlo Park, Calif., 1975-79, dir.-minerals and metals, 1979-83, sr. cons., 1983-86; bus. mgr. inorganics, 1986—; vis. prof. mineral econs. Stanford U., Calif., 1983—; bd. dirs Small Mines Internat. Contbr. articles, chpts. to profl. publs. Campbell and Krumb fellow Columbia U., 1965-67. Fellow AAAS; mem. AIME (chmn. Columbus sect. of Ohio chpt. 1973-75, chmn. Bay Area chpt. of metall. sect. 1979-80), AIME (San Francisco sect.). Home: 426 27th Ave San Mateo CA 94403-2402 Office: SRI Internat 333 Ravenswood Ave Menlo Park CA 94025-3493

THIND, GURDARSHAN S., medical educator; b. Lyallpore, Punjab, India, Oct. 17, 1940; came to U.S., 1963; s. S. Manmohan Singh and Sardarni Ajaib Kaur (Jawanda) T.; m. Rajinder Kaur Sekhon, June 25, 1967; children: Gurpreet K., Gurbir S. MB, BS, Punjab U., 1962; grad. diploma, U. Pa., 1965, MSc in Cardiology, 1966. Intern Rajindra Hosp. of Govt. Med. Clg., Patiala, India, 1962; intern U. Pa., 1963-64, cardiology fellow, 1964-67, resident in internal medicine, 1968-70; assoc. instr. in cardiology U. Pa., Phila., 1964-69, assoc. in medicine, 1970, asst. prof., 1970-72; asst. prof. Washington U., St. Louis, 1972-76; assoc. prof. U. Louisville, 1976-85, prof., 1985—, dir. hypertension sect., 1976—; lectr. univs. and sci. meetings in U.S. and ffgn. countries, 1965—. Contbr. over 130 articles to med. jours. Gen. sec. Sikh Study Circle Louisville, 1967-87; cons. Washington Sikh Ctr., 1988—; coord. Sikh Youth Camps, Silver Spring, Md., 1984—. Fellow ACP (life mem.), Am. Coll. Cardiology; mem. Am. Physiol. Soc., Am. Soc. Hypertension (founding), numerous others. Republican. Achievements include research on trace elements in arterial hypertension, renovascular hypertension, clinical and experimental systemic hypertension and cardiovascular diseases. Home: 17603 Popedale Rd Louisville KY 40245-4350 Office: U Louisville Health Sci Ctr Dept Medicine ACB Louisville KY 40292

THIO, ALAN POOAN, chemist; b. Jakarta, Indonesia, Jan. 17, 1931; came to U.S., 1957; s. Teng-Giok and Kang-Nio (Oei) T.; m. Tatty H. Lim, Feb. 24, 1957; children: Amy, Nirmayati, Susanti. BSc, U. Indonesia, 1954, MSc, 1957; PhD, U. Ky., 1960. Assoc. prof. Bandung (Indonesia) Inst. Tech., 1965-67; assoc. Coll. Pharmacy U. Ky., Lexington, 1967-70, sr. rsch. analyst Regulatory Svcs., 1970-77, specialist Regulatory Svcs., 1977—. Abstractor Chem. Abstracts Svc., 1966—; contbr. (book) EPA's Manual of Chemical Methods for Pesticides, 1976, articles to profl. jours. Mem. Am. Chem. Soc. Achievements include setting up pesticide residue laboratory; researched synthesis of spiro-oxindoles, in-block gas-chromatography of pesticides, analysis of lysine in feed. Office: U Ky Regulatory Svcs Regulatory Svcs Bldg Lexington KY 40546-0275

THIRY, PAUL, architect; b. Nome, Alaska, Sept. 11, 1904; s. Hippolyte A. and Louise Marie (Schwaebele) T.; m. Mary Thomas, Oct. 26, 1940; children—Paul Albert, Pierre. Diploma, Ecole des Beaux Arts, Fontainebleau, France, 1927; A.B., U. Wash., 1928; D.F.A. (hon.), St. Martins Coll., 1970; D.Arts (hon.), Lewis & Clark Coll., 1979. Pvt. practice architecture, 1929—; mem. Thiry & Shay, 1935-39; co-architect war work U.S. Navy Advance Base Depot, Tacoma, Fed. Pub. Housing Adminstrn. projects; community centers and appurtenances Port Orchard, Wash., 1940-44; practice in Salt Lake City, Alaska, Seattle, 1945—; pres. Thiry Architects, Inc., 1971—; architect in residence Am. Acad. in Rome, 1969. Author: (with Richard Bennett and Henry Kamphoefner) Churches and Temples, 1953, (with Mary Thiry) Eskimo Artifacts Designed for Use, 1978; Contbr. articles to profl. jours.; Projects exhibited and illustrated in books and periodicals.; Prin. works include Am. Embassy residence, Santiago, Chile (awarded Diploma Clegio de Arquitectos de Chile 1965); comprehensive plan Seattle Center (awarded A.I.A. citation for excellence in community architecture 1966); Seattle Coliseum (A.I.S.I. awards for design and engring. 1965), (A.I.S.C. award of excellence 1965), Washington State Library (A1A-ALA award 1964), U.S. 4th Inf. (Ivy) Div. Monument, Utah Beach, France, 1969 (hon. citizen Sainte Marie du Mont), Am. Battle Monuments Commn. Meml., Utah Beach, 1984; cons.: comprehensive plan Libby Dam-Lake Koocanusa Project, Mont.; cons. Chief Joseph Dam Powerhouse, Wash.; architect for Visitors Center (C.E. distinguished design award 1970, Chief of Engrs. cert. of appreciation 1974); cons.: New Melones Dam/Reservoir Project, Calif., 1972, Reregulating dam and powerhouse, Libby, Mont., 1977—; others.; works include coll. bldgs., chs., museums, and, govt. projects; also designer fabrics; carpets, furniture. Commnr. City of Seattle Planning Commn., 1952-61, chmn., 1953-54; rep. on exec. bd. Puget Sound Regional Planning Council, 1954-57; chmn. Central Bus. Dist. Study, 1957-60; mem. exec. com. Wash. State Hosp. Adv. Council, 1953-57; architect in charge Century 21 Expn., Seattle, 1957-62; mem. Dept. Interior Historic Am. Bldg. Survey Bd., 1956-61, vice chmn., 1958-61; mem. Joint Com. Nat. Capital, 1961, exec. bd., 1962-72; mem. President's Council Redevel. Pennsylvania Av., Washington, 1962-65; cons. FHA, 1963-67; cons. architect Nat. Capitol, 1964—; mem. Nat. Capital Planning Commn., 1963-75, vice chmn., 1972-75; mem. arts and archtl. com. J. F. Kennedy Meml. Library, 1964; mem. Postmaster Gen.'s Council Research and Engring., 1968-70; mem. Peace Corps Adv. Council, 1982-84. Decorated officier d'Academie with palms France; recipient Paul Bunyan award Seattle C. of C., 1949, Outstanding Architect-Engr. award Am. Mil. Engrs., 1977; named Distinguished Citizen in the Arts Seattle, 1962, Constrn. Man of Year, 1963; Academician N.A.D.; named to Coll. of Architecture Hall of Honor U. Wash., 1987. Fellow A.I.A. (pres. Wash. chpt. 1951-53, preservation officer 1952-62, chancellor coll. fellows 1962-64, chmn. com. nat. capital 1960-64, chmn. com. hon. fellowships 1962-64, Seattle chpt. medal 1984); mem. Nat. Sculpture Soc. (hon., Herbert Adams Meml. medal 1974, Henry Hering medal 1976), Am. Inst. Planners (exec. bd. Pacific N.W. chpt. 1949-52, hon. life), Am. Planning Assn. (hon. life, N.W. chpt. citation 1983), Soc. Archtl. Historians (life mem., dir. 1967-70), Nat. Trust Historic Preservation, Am. Inst. Interior Designers (hon.), Liturgical Conf. (life), Seattle Art Mus. (life), Seattle Hist. Soc. (hon. life), Oreg. Cavemen (hon.), Delta Upsilon, Tau Sigma Delta, Scarab (hon.). Clubs: Cosmos (Washington); Century (N.Y.C.); Architectural League N.Y. Home: 919 109th Ave NE Bellevue WA 98004-4493 Office: 800 Columbia St Seattle WA 98104-1995

THODE, HENRY GEORGE, chemistry educator; b. Dundurn, Sask., Can., Sept. 10, 1910; s. Charles Herman and Zelma Ann (Jacoby) T.; m. Sadie Alicia Patrick, Feb. 1, 1935; children: John Charles, Henry Patrick, Richard Lee. BS, U. Sask., 1930, MS, 1932, LLD, 1958; PhD, U. Chgo., 1934, U. Regina, 1983; DSc (hon.), U. Toronto, 1955, U. B.C., 1960, Acadia U., 1960, U. Laval, 1963, Royal Mill Coll., 1964, McGill U., 1966, Queen's U., 1967, U. York, 1972, McMaster U., 1973. Asst. prof. chemistry McMaster U., Hamilton, Ont., Can., 1939-42, assoc. prof., 1942-44, prof., 1944-79, prof. emeritus, 1979—; dir. rsch. McMaster U., 1947-61, head of dept., 1948-52, v.p., 1957-61, pres., vice chancellor, 1961-72; mem. Nat. Rsch. Coun. of Can., 1943-45, 55-61, Def. Rsch. Bd. of Can., 1955-61; dir. Western N.Y. Nuclear Rsch. Ctr., 1965-73, Atomic Energy of Can., Ltd., 1966-81, Stelco Inc., 1969-85; rsch. cons. Atomic Energy of Can., 1952-65. Mem. editorial adv. bd. Jour. of Inorganic & Nuclear Chemistry; contbr. articles to profl. jours. Gov. Ont. Rsch. Found., 1955-82, Royal Botanic Gardens, 1961-73. Sr. Fgn. Scientist fellow NSF, 1970, Hon. Shell fellow, 1974, Sherman Fairchild Disting. scholar Caltech, 1977; recipient H.M. Tory medal Royal Soc. Can., 1959, Centenary medal, 1982, Sir William Dawson medal, 1989, Arthur L. Day medal Geol. Soc. Am., 1980, Order of Ont., 1989, Montreal medal Chem. Inst. of Can., 1993. Mem. Am. Chem. Soc., Chem. Inst. of Can. (pres. 1951-52, medal 1957), Sigma Xi, Gamma Alpha. Achievements include research in separation of isotopes, mass spectrometry, isotope abundances in terrestrial, meteoritic and lunar materials. Office: McMaster Univ, 1280 Main St W Nuclear Research, Hamilton, ON Canada L8S 4M1*

THOLEY, PAUL NIKOLAUS, psychology educator, physical education educator; b. St. Wendel, Fed. Republic Germany, Mar. 14, 1937; s. Paul and Elisabeth (Stier) T.; m. Viktoria Kobler; children: Torsten, Ellen. D degree, Johann Wolfgang Goethe U., Frankfurt, Fed. Republic Germany, 1973. Prof. psychology Johann Wolfgang Goethe U., 1974-82, 89—; prof. sports U. Braunschweig, Fed. Republic Germany, 1982-88. Editor, co-editor scientific jours., books; contbr. articles to profl. jours. Recipient Prize for Best Dissertation, Johann Wolfgang Goethe U., 1973. Mem. Internat. Assn. Consciousness Rsch. (bd. dirs. 1989-91), Assn. Dream Rsch., Lucidity Assn., Gestalt Theory (contbr. 1979—, bd. dirs., coeditor 1991—). Home: Altbachstrasse 18, 66629 Freisen/Oberkirchen SAAR Germany Office: Univ Frankfurt, Mertonstr 17, D-60325 Frankfurt Germany

THOM, ARLEEN KAYE, surgeon; b. Jamestown, N.D., July 31, 1957; d. Eugene Walter and Caroline June (Randall) T. BS in Elec. and Electronic Engring., N.D. State U., 1979; MD, Johns Hopkins U., 1983. Diplomate Am. Bd. Surgery. Intern, then resident in gen. surgery Robert Wood Johnson U. Hosp.- U. Medicine & Dentistry N.J., New Brunswick, 1983-85, 87-90; clin. assoc. NIH, Bethesda, Md., 1989—. NIH Cancer & Nutrition fellow U. Pa., 1985-87. Office: NIH Bldg 10 Rm 2B01 Bethesda MD 20892

THOMADAKIS, PANAGIOTIS EVANGELOS, computer company executive; b. Volos, Magnesia, Greece, Sept. 2, 1941; s. Evangelos and Eleftheria (Hatzioannou) T.; m. Marianna Vamvakousi; children: George, Erika. B in Physics and Elec., U. Athens, Greece, 1964. Cert. elec. engr. Electronic engr. IBM World Trade Corp., Athens, 1964-68; resident mgr. IBM World Trade Corp., Kuwait, 1965-66; from country mgr. to sales mgr. IBM World Trade Corp., Athens, 1967-73; asst. br. office mgr. IBM World Trade Corp., Chgo., 1973-75; sales mgr. IBM World Trade Corp., Athens, 1975-76; mng. dir. Am. Computer Systems Ltd. (ACS), Athens, 1977—; mem. Thessaliki Rsch. Co., 1991. Mem. Hellenic Red Cross Orgn., Athens, 1988, Athens Home for the Aged, 1987, New Democracy (Conservative Party), 1986. Recipient Outstanding MGM award Prime Computer, Inc., 1981, Achievement award, 1982, Decoration C' degree of St. Mark of the Patriarchie of Alexandrie 1990, Decoration B' degree of St. Mark of the Patriarchie of Alexandrie, 1991, Silver Cross of St. Mark of the Patriarchie of Alexandrie, 1991. Mem. Athenian Club, Glyfadas Golf Club. Mem. New Democracy Party. Greek Orthodox. Avocations: music, golf. Office: ACS ltd, 3A Skra Str, 17673 Kallithea Greece

THOMANN, GARY CALVIN, electrical engineer; b. Burlington, Iowa, July 23, 1942; s. Calvin F. and Helen Mae (Magel) T. BS, U. Kans., 1965, MS, 1967, PhD, 1970. Registered profl. engr., Kans. Prin. investigator NASA, Slidell, La., 1971-75; asst. prof. elec. engring. Wichita (Kans.) State U., 1975-81, assoc. prof. elec. engring., 1982-87; sr. engr. Power Techs., Inc., Schenectady, N.Y., 1987—. Contbr. articles to profl. publs. Mem. IEEE (sr.). Office: Power Tech Inc 1482 Erie Blvd Schenectady NY 12305

THOMAS, ABDELNOUR SIMON, software company executive; b. Kfarhoune, Lebanon, Oct. 25, 1913; came to U.S., 1920; s. Simon Thomas and Mary Sawaga-Thomas; m. Eva Maria Balling, Mar. 26, 1951; children: Robert F., David C., Paul J., Simon S., Mary E., Jonn T. BS, Holy Cross Coll., 1937; MEd, Boston U., 1939, PhD, 1950. Registered profl. engr., Mass. Prof. math. and math. stats. Boston Coll.; founder, dir. rsch. activities A. S. Thomas, Inc., 1955—; vis. prof. MIT. Contbr. Ency. Britanica, numerous. sci. publs. Lt. sgt. USNR, 1942-46, PTO. Recipient Forty-One for Freedom award Sec. USN, 1967. Roman Catholic. Achievements include co-development of theory of graded and hybrid absorber materials, theory of modulated surface wave antennas, re-entry vehicle antennas; characterization of radar cross section of metallic, coated metallic and nonmetallic bodies. Office: A S Thomas Inc 355 Providence Hwy Westwood MA 02090

THOMAS, ADRIAN WESLEY, laboratory director; b. Edgefield, S.C., June 23, 1939; s. Hasting Adrian and Nancy Azalena (Bridges) T.; m. Martha Elizabeth McAllister, July 12, 1964; children: Wesley Adrian, Andrea Elizabeth. BS in Agrl. Engring., Clemson U., 1962, MS in Agrl. Engring., 1965; PhD, Colo. State U., 1972. Rsch. scientist USDA-Agrl. Rsch. Svc., Tifton, Ga., 1965-69, Fort Collins, Colo., 1969-72; rsch. leader USDA-Agrl. Rsch. Svc., Walkinsville, Ga., 1972-89; lab. dir. USDA-Agrl. Rsch. Svc., Tifton, 1989—; mem. acad. faculty Colo. State U. Ft. Collins, 1969-72; acad. faculty U. Ga., Athens, 1973—, grad. faculty, 1988—. Contbr. agrl. rsch. articles to profl. jours. With U.S. Army, 1962-63. Mem. Am. Soc. Agrl. Engrs., Am. Soc. Agronomy, Soil and Water Conservation Soc. Am., Soil Sci. Soc. Am., Sigma Xi, Alpha Epsilon, Gamma Sigma Delta, Phi Kappa Phi. Lutheran. Avocations: reading, gardening, yard care, remodeling home, sports. Office: USDA Agrl Rsch Svc PO Box 946 Tifton GA 31793-0946

THOMAS, CHARLES ALLEN, JR., molecular biologist, educator; b. Dayton, Ohio, July 7, 1927; s. Charles Allen and Margaret Stoddard (Talbott) T.; m. Margaret M. Gay, July 7, 1951; children: Linda Carrick, Stephen Gay. AB, Princeton (N.J.) U., 1950; PhD, Harvard U., 1954. Rsch. scientist Eli Lilly Co., Indpls., 1954-55; NCR fellow U. Mich., Ann Arbor, 1955-56, prof. biophysics, 1956-57; prof. biophysics Johns Hopkins U., Balt., 1957-67; prof. biol. chemistry Med. Sch. Harvard U., Boston, 1967-78; chmn. dept. cellular biology Scripps Clinic & Rsch. Found., La Jolla, Calif., 1978-81; pres., dir. Helicon Found., San Diego, 1981—; founder, chief exec. officer The Syntro Corp., San Diego, 1981-82; pres., chief exec. officer Pantox Corp., San Diego, 1989—; mem. genetics study sect. NIH, 1968-72; mem. rsch. grants com. Am. Cancer Soc., 1972-76, 79-85. Mem. editorial bd. Virology, 1967-73, Jour. Molecular Biology, 1968-72, BioPhysics Jour., 1965-68, Chromosoma, 1969-79, Analytic Biochemistry, 1970-79, Biochim Biophys. ACTA, 1973-79, Plasmid, 1977—. With USNR, 1945-46. NRC fellow, 1965-66. Mem. AAAS, Am. Fedn. Biol. Chemists, Genetics Soc. Am., Am. Chem. Soc. Achievements include research in genetic and structural organization of chromosomes, in practical assessment of individual antioxidant defense system by analytical biochemistry. Home: 1640 El Paso Real La Jolla CA 92037-6304 Office: Helicon Foundation 4622 Santa Fe St San Diego CA 92109-1601

THOMAS, CHARLES EDWIN, chemist, researcher; b. Richmond, Va., June 29, 1949; s. Charles Edwin and Kathleen (Merchant) T.; m. Christy Lynn Watson, Aug. 21, 1976; children: Adam, Laura, Emily. BS in Chemistry, Va. Commonwealth U., 1974, MS in Chemistry, 1982. Scientist Philip Morris R & D, Richmond, Va., 1976-82, project leader, 1983-88, rsch. scientist, 1989—; adj. faculty Va. Commonwealth U., Richmond, 1976-80. Manuscrip rev. bd. Tobacco Rsch., 1988—; contbr. articles to profl. jours. Mem. Am. Chem. Soc. (chmn. Va. sect., alt. councilor 1990—, others), Coresta Task Force. Episcopalian. Achievements include patents for Microwave-Base Moisture Monitor, Novel Smoking Machine Design, Environmental Tobacco Smoke Monitor, Supercritical Fluid Chromatography of Tobacco Specific Compounds. Home: 13208 Forest Light Ct Richmond VA 23233 Office: Philip Morris R & D 4201 Commerce Rd Richmond VA 23234

THOMAS, DAVID BURTON, chemical company executive; b. Alsager, Cheshire, England, May 21, 1943; s. Reginald Burton and Rose Mary T. BA, Open U., 1978. Technician Shell Rsch. Ltd., Sittingbourne, Kent, 1963-67; rsch. asst. Massey U., N.Z., 1967-69; sr. rsch. technician, scientist SittingBourne Rsch. Centre, 1970—. Mem. Ch. of Eng. Avocations: reading, gardening, swimming. Home: 28 Bramblefield Ln, Sittingbourne ME102SX, England Office: Shell Rsch Ltd, Broad Oak Rd, Sittingbourne Kent, England ME9 8AG

THOMAS, DAVID WAYNE, engineer; b. Carthage, Tex., Dec. 5, 1951; s. S. David and Mary Ann (Lewis) T.; m. Margaret Ann Ward, Jan. 4, 1974; children: Mindy Ann, Heather Lee. BS, Tex. A&M U., 1975. Rsch. asst. Engring. Ext., College Station, Tex., 1973-75; engr. B,S&B, Houston, 1975-76; inspection engr. Aramco Svcs. Co., Houston, 1976—. Mem. ASME, Am. Welding Soc., Am. Soc. Metals. Baptist. Home: 25603 Pecan Valley Spring TX 77380 Office: Aramco Svcs Co 9009 W Loop S Houston TX 77096

THOMAS, DUNCAN CAMPBELL, biostatistics educator; b. Bryn Mawr, Pa., Dec. 6, 1945; s. R. David and Virginia (Campbell) T.; m. Nina Benedik, Aug. 23, 1969; children: Wilder, David, Dylan. BA, Haverford (Pa.) Coll., 1967; MS, Stanford (Calif.) U., 1969; PhD, McGill U., 1976. Asst. prof. McGill U., Montreal, Quebec, Canada, 1976-82, assoc. prof., 1982-83; assoc. prof. U. So. Calif., L.A., 1984-88, prof., 1988—; vis. fellow Biostats. Unit, Cambridge, Eng.,1 988-89; cons. fallout study U. Utah, Salt Lake City, 1986-91; mem. Nat. Acad. Sci. BEIR V Com., Washington, 1986-89, Calif. Sci. Adv. Panel Prop. 65, Sacramento, 1989-92, U.S. EPA Radiation Adv. Com., Washington, 1986. Guest editor Jour. Chronic Disease, 1985-86; author numerous papers and reports in field. Rsch. grantee NCI, 1986-92, 91-94, Calif. Tobacco Related Disease Program, 1990-91. Fellow Am. Coll. Epidemiology; mem. Soc. Epidemiol. Rsch., Biometric Soc. Achievements include development of statistical methods for epidemiology, radiation and cancer research, methods for genetic epidemiology and reproductive effects of exposure to pesticide malathion. Office: U So Calif Dept Preventive Medicine 1420 San Pablo St PMB B201 Los Angeles CA 90063

THOMAS, E(DWARD) DONNALL, physician, researcher; b. Mart, Tex., Mar. 15, 1920; married; 3 children. BA, U. Tex., 1941, MA, 1943; MD, Harvard U., 1946; MD (hon.), U. Cagliari, Sardinia, 1981, U. Verona, 1991, U. Parma, 1992. Lic. physician Mass., N.Y., Wash.; diplomate Am. Bd. Internal Medicine. Intern in medicine Peter Bent Brigham Hosp., Boston, 1946-47, rsch. fellow hematology, 1947-48; NRC postdoctoral fellow in medicine dept. biology MIT, Cambridge, 1950-51; chief med. resident, sr. asst. resident Peter Bent Brigham Hosp., 1951-53, hematologist, 1953-55; instr. medicine Harvard Med. Sch., Boston, 1953-55; rsch. assoc. Cancer Rsch. Found. Children's Med. Ctr., Boston, 1953-55; physician-in-chief Mary Imogene Bassett Hosp., Cooperstown, N.Y., 1955-63; assoc. clin. prof. medicine Coll. Physicians and Surgeons Columbia U., N.Y.C., 1955-63; attending physician U. Wash. Hosp., Seattle, 1963—; prof. medicine Sch. Medicine U. Wash., Seattle, 1963-90, head divsn. oncology Sch. Medicine, 1963-85, prof. emeritus medicine Sch. Medicine, 1990—; dir. med. oncology Fred Hutchinson Cancer Rsch. Ctr., Seattle, 1974-89, assoc. dir. clin. rsch. programs, 1982-89, mem., 1974—; attending physician Harborview Med. Ctr., Seattle, 1963—, VA Hosp., Seattle, 1963—, Providence Med. Ctr., Seattle, 1973—, Swedish Hosp., Seattle, 1975—; consulting physician Children's Orthopedic Hosp. and Med. Ctr., Seattle, 1963—; mem. hematology study sect. NIH, 1965-69; mem. bd. trustees and med. sci. adv. com. Leukemia Soc. Am., Inc., 1969-73; mem. clin. cancer investigation review com. Nat. Cancer Inst., 1970-74; 1st ann. Eugene C. Eppinger lectr. Peter Bent Brigham Hosp. and Harvard Med. Sch., 1974; Lilly lectr. Royal Coll. Physicians, London, 1977; Stratton lectr. Internation Soc. Hematology, 1982; Paul Aggeler lectr. U. Calif., San Francisco, 1982; 65th Mellon lectr. U. Pitts. Sch. Medicine, 1984; Stanley Wright Meml. lectr. Western Soc. Pediatric Rsch., 1985; Adolfo Ferrata lectr. Italian Soc. Hematology, Verona, Italy, 1991. Mem. editorial bd. Blood, 1962-75, 77-82, Transplantation, 1970-76, Proceedings of Soc. for Exptl. Biology and Medicine, 1974-81, Leukemia Rsch., 1977—, Hematological Oncology, 1982—, Jour. Clin. Immunology, 1982—, Am. Jour. Hematology, 1985—, Bone Marrow Transplantation, 1986—. With U.S. Army, 1948-50. Recipient A. Ross McIntyre award U. Nebr. Med. Ctr., 1975, Philip Levine award Am. Soc. Clin. Pathologists, 1979, Disting. Svc. in Basic Rsch. award Am. Cancer Soc., 1980, Kettering prize Gen. Motors Cancer Rsch. Found., 1981, Spl. Keynote Address award Am. Soc. Therapeutic Radiologists, 1981, Robert Roesler de Villiers award Leukemia Soc. Am., 1983, Karl Landsteiner Meml. award Am. Assn. Blood Banks, 1987, Terry Fox award Can., 1990, Internat. award Gairdner Found., 1990, N.Am. Med. Assn. Hong Kong prize, 1990, Nobel prize in medicine, 1990, Presdl. medal of sci. NSF, 1990,. Mem. NAS, Am. Assn. Cancer Rsch., Am. Assn. Physicians (Kober medal 1992), Am. Fedn. Clin. Rsch., Am. Soc. Clin. Oncology (David A. Karnoksky Meml. lectr. 1983), Am. Soc. Clin. Investigation, Am. Soc. Hematology (pres. 1987-88, Henry M. Stratton lectr. 1975), Internat. Soc. Exptl. Hematology, Internat. Soc. Hematology, Academie Royale de Medicine de Belgique (corresponding mem.), Swedish Soc. Hematology (hon.), Swiss Soc. Hematology, Royal Coll. Physicians and Surgeons Can. (hon.), Western Assn. Physicians, Soc. Exptl. Biology and Medicine, Transplantation Soc. Office: Fred Hutchinson Cancer Ct 1124 Columbia St Seattle WA 98104

THOMAS, EDWIN L., materials engineering educator. Prof., materials science MIT, Cambridge, Mass. Recipient High-Polymer Physics Prize, Am. Chemical Soc., 1991. Office: MIT Dept of Materials Science 77 Massachusetts Ave Cambridge MA 02139

THOMAS, GARETH, metallurgy educator; b. Maesteg, U.K., Aug. 9, 1932; came to U.S., 1960, naturalized, 1977; s. David Bassett and Edith May (Gregory) T.; m. Elizabeth Virginia Cawdry, Jan. 5, 1960; 1 son, Julian Guy David. B.Sc., U. Wales, 1952; Ph.D., Cambridge U., 1955, Sc.D., 1969. I.C.I. fellow Cambridge U., 1956-59; asst. prof. U. Calif., Berkeley, 1960-63; asso. prof. U. Calif., 1963-67, prof. metallurgy, 1967—, assoc. dean grad. div., 1968-69, asst. chancellor, acting vice chancellor for acad. affairs, 1969-72; sci. dir. Nat. Ctr. Electron Microscopy, 1982—; cons. to industry. Author: Transmission Electron Microscopy of Metals, 1962, Electron Micoscopy and Strength of Crystals, 1963, (with O. Johari) Stereographic Projection and Applications, 1969, Transmission Electron Microscopy of Materials, 1980; contbr. articles to profl. jours. Recipient Curtis McGraw Research award Am. Soc. Engring. Edn., 1966, E.O. Lawrence award Dept. Energy, 1978, I-R 100 award R & D mag., 1987, Henry Clifton Sorby award, Internat. Metallographic Soc., 1987; Guggenheim fellow, 1972. Fellow Am. Soc. Metals (Bradley Stoughton Young Tchrs. award 1965, Grossman Publ. award 1966), Am. Inst. Mining, Metall. and Petroleum Engrs.; mem. Electron Microscopy Soc. Am. (prize 1965, pres. 1976), Am. Phys. Soc., Nat. Acad. Scis., Nat. Acad. Engring., Brit. Inst. Metals (Rosenheim medal 1977), Internat. Fedn. Electron Microscopy Socs. (pres. 1986-90), Brit. Iron and Steel Inst. Club: Marylebone Cricket (Eng.). Patentee in field. Office: U Calif Dept Math Sci and Engring 284 Heanot Mining Bldg Berkeley CA 94720

THOMAS, GARLAND LEON, aerospace engineer; b. Topeka, Aug. 29, 1920; s. Jasper McKinley and Margaret (Hickman) T.; m. Kathleen Hickey, Aug. 29, 1948; children: John, Patricia, LiLi. BS, Drury Coll., 1942; PhD, U. Mo., 1954. Fellow engr. Westinghouse Electric Corp., Pitts., 1953-59; assoc. prof. Drury Coll., Springfield, Mo., 1959-66, Fla. Inst. Tech., Melbourne, 1966-71; contractor NASA, Kennedy Space Center, Fla., 1971-85, aerospace engr., 1985—. Contbr. to Jour. of Acoustical Soc., Remote Sensing for Planners. Lt. commdr. USNR, 1944-46. Mem. Am. Phys. Soc., Am. Nuclear Soc., Am. Assn. Physics Tchrs., Biophys. Soc. Achievements include 5 patents in nuclear reactor field. Home: 1208 River Dr # 102 Melbourne FL 32901-7370 Office: Kennedy Space Ctr RT-ENG-1 Orlando FL 32899

THOMAS, GREG HAMILTON, electronics engineer; b. Morgantown, W.Va., Oct. 15, 1959; s. Robert Hamilton and Dorothy (Barglof) T.; m. Amber Dawn Noel, Dec. 21, 1985; children: Rachel Kathryn, Morgan Grace. BSEE, U. Cin., 1984; MBA, Ball State U., Muncie, Ind., 1993. Electronics engr. Naval Weapons Support Ctr., Crane, Ind., 1984—. Scoutmaster Troop 484 Boy Scouts Am., Bloomfield, Ind., 1986-88; trustee Bloomfield Presbyn. Ch., 1990—. Mem. Am. Soc. Naval Engrs., U.S. Naval Inst. (assoc.). Presbyterian. Achievements include application of impedence matching concepts to Trident II missile linear ordnance test equipment to raise test signal fidelity to a new standard; Stinger missile and hawk missile system engineering for USMC. Office: Naval Weapons Support Ctr Bldg 2045 Code 4032 Crane IN 47522

THOMAS, HAROLD WILLIAM, avionics systems engineer, flight instructor; b. Cle Elum, Wash., Sept. 29, 1941; s. Albert John and Margaret Jenny (Micheletto) T.; children: Gregg Wallace, Lisa Michele. BS, U. Wash., 1964; M of Engring., U. Fla., 1968. Sci. programmer Aerojet Gen. Corp., Sacramento, Calif., 1964-65; systems analyst GE Co., Daytona Beach, Fla., 1965-69; systems engr. GE Co., Phoenix, 1969-70; sr. software engr. Sperry Flight Systems, Phoenix, 1970-77; sr. systems engr. Honeywell, Inc., Phoenix, 1977-80; engr. section head Sperry Flight Systems, Phoenix, 1980-87; free lance flight instr., 1981—; staff engr. Honeywell, Inc., Phoenix, 1987—; designated engring. rep. Fed. Aviation Adminstrn., Long Beach, 1987—. Mem. AIAA, SAE Internat. Internat. Soc. Air Safety Investigators, Am. Mensa Ltd. Achievements include patent for rotating round dial aircraft engine instruments, patent for dynamic approach display format with plan and profile views. Home: 2514 W Pershing Ave Phoenix AZ 85029 Office: Honeywell INc 21111 N 19th Ave Phoenix AZ 85027

THOMAS, JACK WARD, wildlife biologist; b. Fort Worth, Sept. 7, 1934; s. Scranton Boulware and Lillian Louise (List) T.; m. Farrar Margaret Schindler, June 29, 1957; children: Britt Ward, Scranton Gregory. BS, Tex. A&M U., 1957; MS, W.Va. U., 1969; PhD, U. Mass., 1972. Wildlife biologist Tex. Game & Fish Commn., Sonora, 1957-60, Tex. Parks & Wildlife Dept., Llano, 1961-66; research biologist U.S. Forest Svc., Morgantown, W.Va., 1966-69, Amherst, Mass., 1970-73, LaGrande, Oreg., 1974—. Author, editor: Wildlife Habitats in Managed Forests, 1979 (award The Wildlife Soc. 1980), Elk of North America, 1984 (award The Wildlife Soc. 1985); contbr. numerous articles to profl. jours. Served to lt. USAF, 1957. Recipient Conservation award Gulf Oil Corp., 1983, Earle A. Childs award Childs Found., 1984, Disting. Svc. award USDA, Disting. Citizen's award, E. Oreg. State Coll., Nat Wildlife Fedn. award for Sci., 1990, Disting. Achievement award Soc. for Cons. Biology, 1990, Giraffe award The Giraffe Project, 1990, Scientist of Yr. award Oreg. Acad. Sci., 1990, Disting. Svc. award Soc. Conservation Biology, 1991, Sci. Conservation award Nat. Wildlife Fedn., 1991, Chuck Yeager award Nat. Fish and Wildlife Found., 1992, Conservationist of Yr. award Oreg. Rivers Coun., 1992, Chief's Tech. Transfer award USDA, 1992, Tech. Transfer award Fed. Lab. Consortium, 1993. Fellow Soc. Am. Foresters; mem. The Wildlife Soc. (cert., hon., pres. 1977-78, Oreg. chpt. award 1980, Arthur Einarsen award 1981, Spl. Svcs. award 1984, Aldo Leopold Meml. medal 1990, Group Achievement award 1990), Am. Ornithol. Union, Am. Soc. Mammalogists, Lions, Elks. Avocations: hunting, fishing, white-water rafting. Office: US Forest Svc 1401 Gekeler Ln La Grande OR 97850-9802

THOMAS, JERRY ARTHUR, soil scientist; b. Logansport, Ind., Mar. 5, 1942; s. Purnal Kidd and Dorothy Helen (Smith) T.; m. Virginia Amy York, Oct. 17, 1964; 1 child, Charles Edward. BS in Agronomy, Purdue U., 1965; MS in Soil Fertility, Pa. State U., 1968. Libr. chemistry dept. Purdue Univ., West Lafayette, Ind., 1963, tech. aid USDA, ARS, agronomy dept., 1963-65; grad. teaching asst. agronomy dept. Pa. State U., University Park, 1965-67; soil scientist USDA Soil Conservation Svc., Indpls., 1967-85, Ind. State Bd. Health, Indpls., 1985—; com. mem. Ind. State 4-H Rabbit com., West Lafayette, 1976—. Author publs. in field including Classification of the Sloping Soils of the West Baden Group in Monroe County, Ind., 1978, Soil Survey of Monroe County, Ind., 1981, Soil Survey of Lawrence County, Ind., 1985, Availability of Conservation Tillage Planting Systems for Northwestern Ind., 1985. Asst. scoutmaster Boy Scouts Am., Rensselaer, Ind., 1982—. Recipient Innovative award Coun. State Govts., 1988. Mem. Am. Soc. Agronomy, Soil Sci. Soc. Am., Soil and Water Conservation Soc. Am., Ind. Acad. Sci., Ind. Assn. Profl. Soil Classifiers, Masons, Eastern Star. Presbyterian. Home: 301 Park Ave Rensselaer IN 47978-3037 Office: Div Sanitary Engring Indiana State Bd Health 1330 W Michigan St Indianapolis IN 46202-2874

THOMAS, JOAB LANGSTON, academic administrator, biology educator; b. Holt, Ala., Feb. 14, 1933; s. Ralph Cage and Chamintney Elizabeth (Stovall) T.; m. Marly A. Dukes, Dec. 22, 1954; children: Catherine, David, Jennifer, Frances. AB, Harvard U., 1955, MA, 1957, PhD, 1959; DSc (hon.), U. Ala., 1981; LLD (hon.), Stillman Coll., 1987. Cytotaxonomist Arnold Aboretum, Harvard, 1959-61; faculty U. Ala., University, 1961-76, prof. biology, 1966-76, 88-91, asst. dean Coll. Arts and Scis., 1964-65, 69, dean for student devel. Coll., 1969-74, v.p., 1974-76, dir. Herbarium, 1961-76, dir. Arboretum, 1964-65, 66-69; pres. U. Ala., Tuscaloosa, 1981-88; chancellor N.C. State U., Raleigh, 1976-81; pres. Pa. State U., University Park, 1990—; bd. dirs. Blount, Inc., First Ala. Bank, Tuscaloosa, Lukens, Inc.; intern acad. adminstrn. Am. Coun. on Edn., 1971. Author: A Monographic Study of the Cyrillaceae, 1960, Wildflowers of Alabama and Adjoining States, 1973, The Rising South, 1976, Poisonous Plants and Venomous Animals of Alabama and Adjoining States, 1990. Bd. dirs. Internat. Potato Ctr., 1977-83, chmn. 1982-83; bd. dirs. Internat. Svc. for Nat. Agrl. Rsch., 1985-91. Named Outstanding Prof. U. Ala., 1964-65, Ala. Acad. Honor, 1983, Citizen of Yr., Tuscaloosa, 1987. Mem. Golden Key, Phi Beta Kappa, Sigma Xi, Omicron Delta Kappa, Phi Kappa Phi. Office: Pa State U 201 Old Main University Park PA 16802-1589

THOMAS, JOHN KERRY, chemistry educator; b. Llanelli, Wales, May 16, 1934; came to U.S., 1960; s. Ronald W. and Rebecca (Johns) T.; m. June M. Critchley, Feb. 28, 1959; children: Delia, Roland, Roger. BS, U. Manchester, Eng., 1954, PhD, 1957, DSc, 1969. Rsch. assoc. Nat. Rsch. Coun. Can., Ottawa, 1957-58; sci. officer Atomic Energy. U.K., Harwell, Eng., 1958-60; rsch. assoc. Argonne (Ill.) Nat. Lab., 1960-70; prof. chemistry U. Notre Dame, Ind., 1970-82, Nieuwland prof. chemistry, 1982—. Author: Chemistry of Excitation at Interfaces, 1984; mem. editorial bd. Macromolecules Langmuir, Jour. of Colloid and Interface Soc. Fellow Royal Soc. Chemistry; mem. Am. Chem. Soc., Radiation Rsch. Soc. (editorial bd. jour., rsch. award 1974), Photochem. Soc. Home: 17704 Waxwing Ln South Bend IN 46635-1327 Office: Univ Notre Dame Dept Chemistry Notre Dame IN 46556

THOMAS, JOHN MELVIN, surgeon; b. Carmarthen, U.K., Apr. 26, 1933; came to U.S., 1958; s. Morgan and Margaret (Morgan) T.; m. Betty Ann Mayo, Nov. 3, 1958; children: James, Hugh, Pamela. MB, BChir, U. Coll. Wales, U. Edinburgh, 1958. Intern Robert Packer Hosp., Sayre, Pa., 1958-59, chief surg. resident, 1963, pres. med. staff, 1968; assoc. surgeon Guthrie Clinic Ltd., Sayre, 1963-69, chmn. dept. surgery, 1969-91; pres. bd. dirs. Guthrie Clinic Ltd., 1972-89; trustee Robert Packer Hosp.; chmn. exec. com. Guthrie Healthcare System, 1990-92; guest examiner Am. Bd. Surgery, 1979, 81, 85. Bd. dirs. Donald Guthrie Found. for Rsch., pres., 1983—; bd. dirs. Pa. Trauma System Found., 1984-90, pres., 1988, 89; chmn. licensure and accountability Gov.'s Conf., 1974; bd. dirs. Vol. Hosps. Am., 1993—. Mem. ACS (gov. 1985-91), AMA, Am. Group Practice Assn., Am. Soc. Parental and Enteral Nutrition, Pa. Med. Soc., Bradford County Med. Soc., Cen. N.Y. Surg. Soc., Internat. Soc. Surgery, Soc. Surgery Alimentary Tract, Ea. Vascular Soc., Shepard Hills Country Club, Moselem Springs Golf Club. Presbyterian. Home: 383 Lansing Station Rd Lansing NY 14882 Office: Guthrie Clinic Ltd Sayre PA 18840

THOMAS, JOHN MEURIG, chemist, scientist; b. Llanelli, Wales, Dec. 15, 1932; s. David John and Edyth Thomas; m. Margaret Edwards, 1959; 2 daughters. MA, U. Coll. of Swansea; PhD, Queen Mary Coll.; LLD (hon.), U. Wales, 1984; DLitt (hon.), CNAA, 1987; DSc (hon.), Jtexiot Walt Univ., 1989. Sci. officer UKAEA, 1957-58, asst. lectr., 1958-59, lectr., 1959-65, reader, 1965-69; prof., head dept. of chemistry UCW, Aberystwyth, 1969-78; prof., head dept. physical chemistry, fellow King's Coll. U. Cambridge, 1978-86; dir., resident prof. Davy Faraday Research Lab. Royal Inst. Great Britain, 1986—; Fullenian prof. chemistry, 1988—; vis. prof. Tech. U. Eindhoven, The Netherlands, 1962, Pa. State U., 1963, 67, Tech. U. Karlsruhe, Fed. Republic Germany, 1966, Weizmann Inst., Israel, 1969, U. Florence, Italy, 1972, Am. U. in Cairo, 1973, IBM Rsch. Ctr., San José, 1977, QMC, 1986—, Imperial Coll., London, 1986—, Acad. Sinica, Beijing, Inst. Ceramic Sci., Shanghai, 1986—; Winegard vis. prof. Guelph U., 1982, BBC Welsh Radio Annual lectr., 1978; Gerhardt Schmidt Meml. lectr. Weizmann Inst., 1979; Van't Hoff Meml. lectr. Ray Dutch Acad. Arts & Sci., 1988; Disting. vis. lectr., London U., 1980, Royal Soc.-Brit. Assn. lectr., 1980, Baker lectr. Cornell U., 1982-83, dist. lectr. in chem. U. Western Ont., 1983, Sloan vis. prof. Harvard U., 1983, Disting. vis. lectr. Tex. A&M U., 1984, Salters lectr. Royal Instn., 1985, 86, Disting. vis. lectr. U. Notre Dame, Ind., 1986, Schuit lectr. Inst. Catalysis, U. Del., 1986, Battista lectr. Clarkson U., 1987, Hund-Klemm lectr. Max Planck Ctr., Stuttgart, 1987, Kenneth Pitzer lectr. U. Calif., Berkeley, 1988, Hallinond lectr. Mineral Soc. Great Britain, 1989, Fanaday lectr. Royal Soc. Chemistry, 1989, Bahenian lectr. Royal Soc., 1990, Bruce-Pueller prize lectr. Royal Soc. Edinburgh, 1990; Hon. bencher, Gray's Inn, 1987; mem. Radioactive Waste Mgmt. Com., 1978-80; mem. chem. com., SRC, 1976-78; mem. main com. SERC, 1986—; mem. adv. com. Davy-Faraday Labs, Royal Instn., 1978-80; sci. adv. com. Sci. Alexandria, 1979—; bd. govs., Weizmann Inst., 1982—; chmn. Chemrawn, IUPAC, 1987—; trustee BM, 1987—, Sci. Mus., 1990—. Author: (with W.J. Thomas) Introduction to the Principles of Heterogeneous Catalysis, 1967; Pan edrychwyf ar y nefoedd, 1978; contbr. articles in field. Recipient numerous medals and awards. Avocations: ancient civilizations, bird watching, hill walking, Welsh literature. Office: Royal Instn, 21 Albemarle St, London W1X 4BS, England

THOMAS, LEO J., manufacturing company executive; b. Grand Rapids, Mich., Oct. 30, 1936; s. Leo John and Christal (Dietrich) T.; m. Joanne Juliani, Dec. 27, 1958; children: Christopher, Gregory, Cynthia, Jeffrey. Student, Coll. St. Thomas, 1954-56; BS, U. Minn., 1958; MS, U. Ill., 1960, PhD, 1961. Rsch. chemist Eastman Kodak Co., Rochester, N.Y., 1961-67, lab. head, 1967-70, asst. div. head color photography div., 1970-72, tech. asst. to dir., 1972-75, asst. dir., 1975-77, v.p., dir., 1977-78, sr. v.p., dir., 1978-84; sr. v.p., gen. mgr. Life Scis., Rochester, N.Y., 1985-88; vice-chmn., chmn. Sterling Drug Inc., N.Y.C., 1988-89; group v.p., gen. mgr. Health Group Eastman Kodak Co., Rochester, 1989-91, group v.p., pres. imaging, 1991—; bd. dirs. Rochester Telephone Corp., John Wiley and Sons Inc., N.Y.C., Eastman Kodak Co. Mem. AICE, AAAS, Am. Acad. Arts and Scis., Am. Inst. for Medicine and Bioengring., Nat. Acad. Engring., Rochester C. of C. Office: Eastman Kodak Co 343 State St Rochester NY 14650-0234

THOMAS, LEONA MARLENE, health information educator; b. Rock Springs, Wyo., Jan. 15, 1933; d. Leonard H. and Opal (Wright) Francis; m. Craig L. Thomas, Feb. 22, 1955; (div. Sept. 1978); children: Peter, Paul, Patrick, Alexis. BA, Govs. State U., 1982, MHS, 1986; cert. med. records adminstrn., U. Colo., 1954. Dir. med. records dept. Meml. Hosp. Sweetwater County, Rock Springs, Wyo., 1954-57; staff assoc. Am. Med. Records Assn., Chgo., 1972-77, asst. editor, 1979-81; statistician Westlake Hosp., Melrose Park, Ill., 1982-84; asst. prof. Chgo. State U., 1984—, faculty senator, 1989-90, acting dir. med. record adminstrn. program, 1991-92; chairperson Coll. Allied Health Pers., 1986-88; mem. rev. bd. network Newsletter of the Assembly on Edn. Co-pres. Ill. Dist. 60 PTA, Westmont; liaison Ill. Trauma Registry, 1991. Mem. Assembly on Edn., Am. Health Info. Mgmt. Assn., Am. Pub. Health Assn., Ill. Pub. Health Assn., Chgo. and Vicinity Med. Records Assn. (publicity com. 1989-90), Ill. Assn. Allied Health Profls., Gov.'s State Alumni Assn. Democrat. Methodist. Home: 6340 Americana Dr Apt 1101 Clarendon Hills IL 60514-2249 Office: Chgo State U Coll Nursin & Allied Health 95th at King Dr Chicago IL 60608

THOMAS, LEWIS JONES, JR., anesthesiology educator, biomedical researcher; b. Phila., Dec. 13, 1930; s. Lewis Jones and Margaretta Eleanore (Schmid) T.; m. Jane E. Priem, June 18, 1955; children—Lewis Jones III., Sarah Jane Thomas Snell. B.S. in Biology, Haverford Coll., 1953; M.D. cum laude, Washington U., St. Louis, 1957. Diplomate Am. Bd. Anesthesiology. Assoc. dir. Biomed. Computer Lab., Washington U. Sch. Med., St. Louis, 1972-75, dir., 1975—, assoc. prof. physiology and biophysics, 1974-84, assoc. prof. elec. engring., 1978—, assoc. prof. anesthesiology, biomed. engring., 1974—, assoc. prof. physiology, dept. cell biology and physiology, 1985—; assoc. prof. Inst. Biomed. Computing, Washington U., 1985-89, prof., 1989—, assoc. dir. Inst. for Biomed. Computing, 1983-91, 92—, acting dir., 1991-92; cons. Health Resources Admin., Rockville, Md., 1974-75, Nat. Ctr. Health Svc. Rsch., Washington, 1978-82; mem. biomed. rsch. tech. rev. com., div. rsch. resources NIH, 1988-92; cons. Diagnostic Radiology Coordinating Com. NIH Planning Subcom., 1990; Nat. Task Force NIH Stategic Plan, 1992; NIH Reviewers Reserve NCPR, 1992—. Contbr. articles to profl. jours. and books. Bd. dirs. Bd. Edn., University City, Mo., 1970, 72-73; v.p. Symphony Orch., University City, 1969-78, pres. 1978-91. Sr. asst. surg. USPHS, 1962-64. Recipient USPHS Rsch. Career Devel. award, 1966. Mem. Am. Physiol. Soc., AAAS, N.Y. Acad. Scis. Avocations: music performance, recreational computing. Office: Washington U Biomed Computer Lab 700 S Euclid Ave Saint Louis MO 63110-1085

THOMAS, MARLIN ULUESS, industrial engineering educator; b. Middlesboro, Ky., June 28, 1942; s. Elmer Vernon and Hellen Lavada (Banks) T.; m. Susan Kay Stoner, Jan. 18, 1963; children: Pamela Claire Thomas Davis, Martin Phillip. BSE, U. Mich., Dearborn, 1967; MSE, U. Mich., Ann Arbor, 1968, PhD, 1971. Registered profl. engr., Mich. Asst. and assoc. prof. dept. ops. rsch. Naval Postgrad. Sch., Monterey, Calif., 1971-75; assoc. prof. systems design dept. U. Wis., Milw., 1976-78; mgr. tech. planning and analysis vehicle quality-reliability Chrysler Corp., Detroit, 1978-79; prof. dept. indsl. engring. U. Mo., Columbia, 1979-82; prof. indsl. engring., chmn. dept. Cleve State U., 1982-88, acting dir. Advanced Mfg. Ctr., 1984-85; prof., chmn. indsl. engring. Lehigh U., Bethlehem, Pa., 1988-93, chmn. dept., 1988—; prof., head Sch. Indsl. Engring. Purdue U., West Lafayette, Ind., 1993—; program dir. NSF, Washington, 1987-88. Contbr. numerous articles on indsl. engring. and ops. rsch. to profl. jours. With USN, 1958-62. Named Outstanding Tchr., U. Mo. Coll. Engring., 1980, Coll. Man of Yr, Cleve. State U. Coll. Engring., 1985. Fellow Inst. Indsl. Engrs; mem. Ops. Rsch. Soc. Am., Am. Soc. for Engring. Edn., Am. Soc. Quality Control, Am. Statis. Assn., Soc. Am. Mil. Engrs., VFW. Office: Purdue U Sch Indsl Engring 1287 Grissom Hall West Lafayette IN 47907-1287

THOMAS, MARTIN LEWIS H., marine ecologist, educator; b. London, Feb. 9, 1935; s. John Breton H. and Evelyn Jessie Thomas; m. Mary Lou Harley, June 15, 1976; children: Calluna, Tiana. BSc, U. Dunelon, Newcastle, U.K., 1956; MSA, U. Toronto, Ont., 1963; PhD, Dalhousie U., Halifax, N.S., 1970. Ont. diploma in horticulture. Rsch. biologist Fisheries Rsch. Bd. Can., London, Ont., 1956-62; rsch. scientist Fisheries Rsch. Bd. Can., Ellerswe, P.E.I., 1962-70; asst. prof. U. New Brunswick, Saint John, 1971-74, assoc. prof., 1974-77, prof., 1977—; bd. dirs. Marine Rsch. Group U. New Brunswick, Saint John, 1970—. Author: Introducing the Sea, 1973, A Guide to the Ecology of Shoreline and Shallow-Water Marine Communities of Bermuda, 1992; editor: Marine and Coastal Systems, 1985. Sgt. U.K. mil., 1954-56. Hon. rsch. assoc. Bermuda Aquarium, Flates, 1992. Mem. Marine Biology Assn. U.K., Brit. Biol. Soc., N.Y. Acad. Sci. (hon. life mem.). Office: U New Brunswick, PO box 5050, Saint John, NB Canada E2L 4L5

THOMAS, MICHAEL EUGENE, electrical engineer, researcher, educator; b. Dayton, Ohio, Apr. 24, 1951; s. Eugene Franklin and Helen Philomena (Lechner) T.; m. Martha Jane Brammer, Mar. 19, 1983; children: Daniel James, Jane Michelle, Alissa Marie, Rebecca Christine, Joseph Michael. MSEE, Ohio State U., 1976, PhD, 1979. Rsch. assoc. electro sci. lab. Ohio State U., Columbus, 1973-79; prin. engr. Johns Hopkins U., Laurel, Md., 1979—; postdoctoral fellow Naval Postgrad. Sch., Monterey, Calif., 1982. Author: (chpt.) Handbook of Optical Constants of Solids, 1991; contbr. articles to Jour. Infrared Physics, Jour. Applied Optics, IEEE Jour. Transactions Geosci. and Remote Sensing; contbr. numerous articles to profl. jours. Janney fellow Johns Hopkin's U., 1988, Parsons fellow, 1991-92; recipient Walter Berl award Johns Hopkin's U., 1988. Mem. IEEE, Internat. Soc. for Optical Engrs., Optical Soc. Am. Achievements include development of mathematical models for describing electromagnetic propagation in the atmosphere of the earth, seawater, and optical window materials. Office: Johns Hopkins U Applied Physics Lab Johns Hopkins Rd Laurel MD 20723

THOMAS, PETER, biochemistry educator; b. Bridgend, Eng., Apr. 25, 1946; married, 1980; 2 children. BSc, U. Wales, 1967, PhD in Biochemistry, 1971. A. K. Fellow chemistry Inst. Cancer Rsch., London, 1971-79; sr. rsch. assoc. Mallory gastrointestinal rsch. lab. Mallory Inst. Pathology, Boston City Hosp., 1979-86, from assoc. medicine to prin. assoc. medicine, 1979-85, asst. prof. surg. biochemistry, 1985-90; assoc. rsch. mem. Cancer Rsch. Inst., New England Deaconess Hosp., Boston, 1985—; assoc. prof. surg. biochemistry Harvard Med. Sch., 1990—. Mem. Am. Soc. Biol. Chemistry, Am. Gastroenterol. Assn., Am. Assn. Study Liver Disease, Am. Assn. Cancer Rsch., Biochemistry Soc., Protein Soc. Achievements include research on the metabolism of glycoproteins especially carcinoembryonic antigen, on interactions between the Kupffer cell and hepatocyte in glycoprotein handling, on mechanisms of transfer of proteins from blood to bile, on tumor cell surfaces' relationship to development of metasteses especially from colorectal cancer. Office: New England Deaconess Hosp Lab Cancer Biology Dept Surgery 50 Binney St Boston MA 02115*

THOMAS, RAYMOND JEAN, computer consultant; b. LaTrinité, Porhoët, France, July 27, 1923; s. Vincent Louis and Marie Joseph (LeHyarric) T.; m. Monique Andrée Collignon, Aug. 2, 1954; children: Françoise Louise. Lic. Lettres, U. Paris, 1953, D Lettres, 1983. Secondary tchr. Ireland, Eng., France, 1942-53; comml. agt. Soc. des Fonderies de Pont a Mousson, Nancy, France, 1953-56; instr. Assumption Coll., Worcester, Mass., 1956-60, Northeastern U., Boston, 1960-61, Tufts U., Medford, Mass., 1961-63; asst. Toulouse (France) U., 1963-69; maitre de confs. Aix Marseille II Fac Luminy, Marseilles, France, 1969-88; computer cons. Peymians, Brignoles, France, 1988—; acoustical researcher Bolt Beranek & Newman, Cambridge, Mass., 1973-75, Centre Nat. d'Etudes et Telecommunications, Lannion and Paris, 1974-84, acoustical and pedagogical researcher, 1956—. Contbr. articles to profl. jours.; patentee in field. Cpl. French Forces, 1944-45. Achievements include research of effective use of natural languages by computers, entirely automated teaching. Home and Office: Peymians, 83170 Brignoles France

THOMAS, RICHARD DEAN, toxicologist, pathologist; b. Payson, Utah, Feb. 14, 1947; s. L. Dean and Irene (Sheffield) T.; m. Jacquelyn Thomas, June 5, 1970; children: Austin, Sterling, Joel, Carmen, Cameron. BS in Chemistry, Utah State U., 1971; PhD in Medicinal Chemistry, Colo. State U., 1974; postdoctoral student, George Washington U., 1979-80. Diplomate Am. Bd. Toxicology. Rsch. biochemist and toxicologist Poison Plants Div. and Sugar Industries Div. USDA, 1969-71; sr. metabolism scientist Biochemistry Dept. CIBA-GEIGY Corp., 1974-76; mgr., toxicologist Ctr. Occupational & Environ. Safety & Health SRI Internat., 1976-78; sr. toxicologist, mem. dept. staff Environ. Chemistry and Biology Dept. METREK Div. MITRE Corp., 1978-81; dir. Div. Biochem. Toxicology Borriston Labs., Inc., 1981-82; with NRC, Nat. Acad. Scis., 1982—, from sr. staff officer to Director Com. on Toxicology, currently Director Human Toxicology and Risk Assessment; adj. prof. Uniformed Service U. of Health Scis., F. Edward Herbert Sch.of Medicine, Bethesda, Md., Dept. Toxicology Sch. Medicine of U. Md., Balt.; cons. Washington met. area toxicology, health assessment, environ. studies; presenter numerous seminars, confs. nationwide on public health/environmental related issues. Contbr. numerous articles to profl. jours. Fellow Am. Inst. Chemists; mem. AAAS, Am. Coll. Toxicology (pres. 1993—, v.p. 1991-92, mem. finance com. 1991-92, chmn. edn. com. 1990-91), Am. Chem. Soc. (chem. safety com. 1978-84, exec. com. 1978-81, chmn. nominations and election com., 1978-81, various other coms.), Am. Public Health Assn., Am. Soc. Applied Spectroscopy, Am. Industrial Hygiene Assn., U.S. Acad. Pathology, N.Y. Acad. Scis., Soc. Occupational Environ. Health, Soc. Tech. Comms., Soc. Toxicology, Environ. Mutagen Soc., Environ. Health Inst., Genetic Toxicology Assn., Delta Phi Kappa. Office: Nat Acad Scis NRC 2101 Constitution Ave Washington DC 20418

THOMAS, ROBERT GLENN, biophysicist; b. Watertown, N.Y., Oct. 9, 1926; s. Glenn Raymond and Mildred A. (Van Horn) T.; m. Kathryn Ann Bourcy, Sept. 3, 1949 (div. 1965); children: Carol A., Glenn M., Paula E.; m. Lanny Rae Monkman, Nov. 21, 1982; 1 child, Steven A. BS, St. Lawrence U., 1949; PhD, U. Rochester, 1955. Asst. prof. U. Rochester, N.Y., 1955-61; dept. head Lovelace Found., Albuquerque, 1961-73; group leader Los Alamos (N.Mex.) Nat. Lab., 1973-84; program mgr. U.S. Dept. of Energy, Washington, 1984-90, Argonne (Ill.) Nat. Lab., 1990—; mem. task group Internat. Commn. Radiol. Protection, Oxford, Eng., 1978-79; cons. Stauffer Chem., Farmington, Conn., 1979-84; U.S. rep. Commn. on European Communities, Washington and Brussels, 1986-90. Author: (with others) Industrial Hygiene and Toxicology, 1978, 2d edit., 1984, 3d edit., 1991; assoc. editor jour. Health Physics, 1976-79, Radiation Rsch., 1976-82; contbr. 75 articles to profl. jours. Pres., lt. gov. Civitan Club, Los Alamos and Albuquerque, 1971-76; pres. Los Alamos Sheltered Workshop, 1977-84, No. N.Mex. United Way, Los Alamos, 1980-81. With USN, 1943-46, CBI. Recipient Disting. Svc. award Chinese Nationalist Govt., Taiwan, 1987; Atomic Energy Commn. fellow, 1950-51. Mem. Am. Indsl. Hygiene Assn., N.Y. Acad. Sci., Health Physics Soc., Radiation Rsch. Soc., Civitan Internat. (lt. gov. dist. 1982). Republican. Achievements include research in radiation biophysics field operative, demographic modeling for AIDS epidemic. Home: PO Box 279 Bigfork MT 59911-0279

THOMAS, ROBERT LEIGHTON, physicist, researcher; b. Dover-Foxcroft, Maine, Oct. 10, 1938; s. Tillson Davis and Ruth (Leighton) T.; m. Sandra Evenson, June 23, 1962; 1 child, Stephen Leighton. AB, Bowdoin Coll., 1960; PhD, Brown U., 1965. Rsch. assoc. Wayne State U., Detroit, 1965-66, from asst. to assoc. prof. physics, 1966-76, prof. physics, 1976—, dir. Inst. for Mfg. Rsch., 1986—; mem. organizing com. 5th Internat. Meeting on Photoacousics, Heidelberg, Fed. Republic Germany, 1987; chmn. Gordon Rsch. Conf. in Nondestructive Evaluation, 1991-92. Assoc. editor Rsch. in Nondestructive Evaluation; contbr. over 100 articles to sci. jours. Fellow Am. Phys. Soc.; mem. Mich. Tech. Coun. (bd. dirs. S.E. div. 1988—), Acad. Scholars, Sigma Xi (chpt. pres. 1990-91). Achievements include patents for Thermal Wave Imaging Apparatus, for Vector Lock-In Imaging System, and for Single Beam Interferometer. Office: Wayne State U Inst for Mfg Rsch Detroit MI 48202

THOMAS, STEVEN JOSEPH, structural engineer; b. Baraboo, Wis., Mar. 8, 1952; s. Stuart Edmund and Marion Martha (Klaman) T.; m. Jennifer Lee Hasper, Mar. 14, 1980; children: James Joseph, Shawna Lee. BSCE with honors, U. Wis., 1978. Lic. structural engr., civil engr. Design engr. Varco Pruden Bldgs., Turlock, Calif., 1978-81; staff structural engr. Varco Pruden Bldgs., Memphis, 1981-85, nat. accounts sales engr., 1985; sr. design engr., project mgr. Varco Pruden Bldgs., Evansville, Wis., 1985-88; dist. sales mgr. Vulcraft, Walnut Creek, Calif., 1988-89; mgr. engring. svcs. ASC Pacific, Inc., Tacoma, Wash., 1989—. Author (computer software) various structural design/analysis, 1985-92. Football coach YMCA, Tacoma, 1991. With USAF, 1972-74. Mem. ASTM, Structural Engrs. Assn. Wash. (bd. dirs. 1992—). Office: ASC Pacific Inc 2141 Milwaukee Way Tacoma WA 98421

THOMAS, TELFER LAWSON, chemist, researcher; b. Montreal, Que., Can., June 1, 1932; s. Telfer M. and Eleanor Baptist (Tatley) T.; m. Anna A. Thomas, Apr. 24, 1959; children: Stephen, Christopher, Sharon. BSc in Chemistry with honors, McGill U., 1953, PhD in Organic Chemistry, 1957. Rsch. chemist Imperial Oil Ltd., Sarnia, Ont., Can., 1957-59, GAF Corp., Rensselaer, N.Y., 1959-62; prin. investigator Fisons Pharms., Rochester, N.Y., 1962—. Contbr. more than 15 articles to profl. jours. Mem. IEEE, ACS. Achievements include 13 patents in field. Office: Fisons Pharms 755 Jefferson Rd Rochester NY 14623-3233

THOMAS, THRESIA K., biomedical researcher, biochemistry educator; b. Kilianthara, Kerala, India, Aug. 30, 1952; came to U.S., 1978; d. Scaria Kunnappalliel and Anna Chengalamparampil; m. Thekkumkattil John Thomas, May 1, 1977; children: Jay, Jose, John, MaryAnn, David. BS, Alphonsa Coll., Palai, Kerala, 1973; PhD, Indian Inst. Sci., Bangalore, India, 1978. Postdoctoral scholar U. Mich., Ann Arbor, 1978-80; rsch. assoc. U. Minn., Mpls., 1980-88; asst. prof. Robert Wood Johnson Med. Sch. U. Medicine and Dentistry N.J., Piscataway, 1988—; mem. NIH exptl. therapeutics study sect., 1992—. Contbr. articles to scholarly and profl. jours. Grantee Environ. and Occupational Health Sci. Inst., 1988, 89, U. Medicine and Dentistry N.J. Found., 1989, NIH, 1986, 89, 92. Mem. Am. Assn. Cancer Rsch., Am. Fedn. Clin. Rsch., Endocrine Soc., Biophys. Soc. Roman Catholic. Achievements include initiation of a new area of research on interaction of polyamines with steroid receptors; contribution to understanding of function of growth regulatory molecules such as steroid hormones, to development of polyamine biosynthetic inhibitors as therapeutic agents; contribution to the mechanism of action of environmental contaminant TCDD (dioxin). Home: 40 Caldwell Dr Princeton NJ 08540-2908 Office: UMDNJ Robert Wood Johnson Med Sch/Environ Occ Health 675 Hoes Ln Piscataway NJ 08854-5635

THOMAS KING, MCCUBBIN, JR., physicist; b. Balt., June 1, 1925; s. Thomas King and Isabella (McElmoile) McC.; m. Mary Ann Lamb, Jan. 28, 1951; children: Mary, Thomas, Ruth. BEE, U. Louisville, 1946; PhD in Physics, Johns Hopkins U., 1951. Rsch. assoc. MIT, Cambridge, Mass., 1955-57; from asst. prof. to prof. Pa. State U., University Park, 1957-90, prof. emeritus, 1990—. Contbr. articles to sci. jours. Lt. (j.g.) USN, 1945-46. Fulbright fellow, 1982. Mem. Optical Soc. Am. (dir.-at-large 1958-60). Democrat. Episcopalian. Home: 909 Willard Cir State College PA 16803 Office: Pa State U Davey Lab University Park PA 16802

THOMBORSON, CLARK DAVID (CLARK DAVID THOMPSON), computer scientist, educator; b. Mpls., Jan. 11, 1954; s. John Murray and Lorraine (Johnson) T.; m. Barbara Anne Borske, Aug. 19, 1983; children: Luke Anthony, Juliet Minerva. BS in Chemistry with honors, Stanford U., 1975, MS in Computer Sci., Computer Engring., 1975; PhD in Computer Sci., Carnegie Mellon U., 1980. Asst. prof. dept. elec. engring./computer sci. U. Calif., Berkeley, 1979-86; sci. worker Inst. Tech. Cybernetics, Bratislava, Czechoslovakia, 1984; prof. computer sci. U. Minn., Duluth, 1986—; rsch. intern Xerox, Palo Alto, Calif., 1978; acad. visitor IBM, Yorktown Heights, N.Y., 1991; vis. prof. dept. elec. engring. and computer sci. MIT, Cambridge, 1992-93. Contbr. articles to tech. publs. Treas. 1st Unitarian Ch., Duluth, 1989-92. Grantee NSF, 1981—. Mem. IEEE, Assn. Computing Machinery, Computer Profls. for Social Responsibility. Achievements include development of a model for theoretical analysis of VLSI circuit complexity, VLSI algorithms for sorting, binary addition, fourier transformation, design of computer programs for wire routing in gate arrays, data compression. Office: U Minn Dept Computer Sci 320 HH Duluth MN 55812

THOMPSON, ALLAN ROBERT, sales engineer; b. L.A., Jan. 23, 1932; s. Bert and Monica (Foley) T.; m. Mavis D. Craig, June 16, 1956; children: Gayle L. Geiger, Tracy A. Pierick, Craig A. BSEE, U. So. Calif., 1955. Product mgr. Spl. Products Instrument Div., Conrac Systems West, Duarte, Calif., 1960-65; dir. engring. Conrac Systems West, Duarte, Calif., 1979-81; br. engring. mgr. Interstate Electronics, Anaheim, Calif., 1966-70, mgr. engring., 1976-77; cons. engr. Huntington Beach, Calif., 1970-73, 77-79; dir. system product engr. Varian Data Machines, Irvine, Calif., 1973-76; v.p., sales engr. A-F Sales Engring., San Dimas, Calif., 1981—. Mem. West Orange County Spl. Edn. Adv. Bd., Huntington Beach, 1983-85, vice chmn., 1987-89; chmn. Mesa View Parent Adv. Com., Huntington Beach, 1978-82; mem. com., treas. troop 1450 Boy Scouts Am., 1982-93. Comdr. USNR, 1951-92, ret., 1992. Mem. IEEE, Res. Officers Assn., Naval Res. Assn., Ret. Offices Assn. Republican. Methodist. Achievements include design/development/supervision of numerous computer controlled, display, process and test systems; installations have included New York Stock Exchange, commodity and futures exchanges and several airports worldwide. Home:

6582 Jardines Dr Huntington Beach CA 92647 Office: A-F Sales Engring 569 Covina Blvd San Dimas CA 91773

THOMPSON, ANDREW ERNEST, mathematics educator; b. Springfield, Mass., Oct. 17, 1947; s. Richard Ernest and Virginia Laurie (Knight) T.; m. Anne Adams, Apr. 6, 1973; children: Stephanie Anne, Elizabeth Clare, Adam Richard. BA, Bridgewater State Coll., 1969; M in Math., Worcester Polytechnic Inst., 1988. Cert. secondary math., secondary adminstrn., Mass. Tchr. Whitman (Mass.) Sch. Dept., 1969—; curriculum cons. Bridgewater (Mass.) Public Schs., 1987-89; Horace Mann coord. Whitman Public Schs., 1986-88. Supt. Cen. Sq. Congl. Ch., Bridgewater, Mass., 1989-90, 92—; cubmaster Boy Scouts Am., Bridgewater, 1988-92, troop com. 1992—; pres. Whitman Edn. Assn., 1982, 84, 92—; mem. Bridgewater Sch. Com., 1992—. Recipient Harvard Practitioner Harvard U., 1988; NSF grantee, 1989-90. Mem. Nat. Coun. Tchrs. Math., Math. Assn. Am., Assn. Tchrs. Math. in Mass., ASCD. Office: Whitman-Hanson Regional Sch Dist Franklin St Whitman MA 02382

THOMPSON, ANN MARIE, neuroscientist, researcher, educator; b. Corry, Pa., Sept. 13, 1956; d. Thomas Richard and Irene Mary (Jansen) Kolaja; m. Glenn Clark, Jan. 2, 1986; children: Daniel, Ned. BS in Biology, SUNY, Fredonia, N.Y., 1978; PhD, Baylor Coll. Medicine, 1987. Sr. rsch. asst. Baylor Coll. Medicine, Houston, 1987-89; asst. prof. rsch. Health Sci. Ctr. Otorhinolargy, U. Okla., Oklahoma City, 1990—; adj. asst. prof. Health Scis. Ctr., Otorhinolargy, U. Okla., 1989-90, Health Sci. Ctr., Anatomy, 1990—. Reviewer Otolaryngology Head and Neck Surgery, 1990—; co-author: (with others) Basic and Applied Aspects of Vestibular Function, 1988; contbr. articles to profl. jours. Vol. St. Mary's Episcopal Sch., Edmond, Okla., 1992. Rsch. grantee NIH, 1992, Okla. Ctr. for Adv. of Sci. and Tech., 1991, Deafness Rsch. Found., 1988-92. Mem. AAAS, Assn. for Rsch. in Otolaryngology, Soc. for Neuroscience, N.Y. Acad. Scis., Sigma Xi. Office: U Okla Health Sci Ctr Dept ORL PO Box 26901 Oklahoma City OK 73190

THOMPSON, BENJAMIN, architect; b. St. Paul, Minn., July 3, 1918; s. Benjamin Casper and Lynne (Mudge) T.; m. Mary Okes, Dec. 4, 1942 (div. 1967); children: Deborah, Anthony, Marina, Nicholas, Benjamin; m. Jane Fiske McCullough, Sept. 13, 1969. B.F.A., Yale U., 1941. Registered architect, Mass., Minn., Ohio, R.I., Vt., N.Y., Maine, Md., Va., Ga. Founding partner Architects Collaborative, Inc., Cambridge, Mass., 1946-65; pres., chmn. bd. Design Research Internat., Inc., Cambridge, 1953-70; chmn. dept. architecture Harvard, 1963-68; pres. Benjamin Thompson & Assocs., Inc., Cambridge, 1966—; Mem. com. arts and architecture John F. Kennedy Meml. Library, 1964, Blue Ribbon Panel for Parcel 8, Boston Govt. Ctr., 1964; mem. vis. com. R.I. Sch. Design, 1964; juror Boston Archtl. Center Competition, 1964; bd. overseers for fine arts Brandeis U., 1964; vis. com. Boston Mus. Fine Arts, 1964; chmn. design adv. panel city of Cambridge NASA project, 1967-68; design adv. com. Boston Redevel. Authority, 1972-79. Prin. works include faculty office, classroom, Harvard Law Sch., edn. library, Harvard Grad. Sch., Kirkland (N.Y.) Coll., indsl. and fine arts bldg, Mass. State Coll, St. Paul Acad, Phillips Acad, Andover, Mass.; sci. bldg. dormitories, Williams Coll., Mt. Anthony Union High Sch, Bennington Vt.; music bldg., Amherst Coll., State U. N.Y. at Buffalo; important commns. include Intercontinental hotels, Abu Dhabi, Al Ain, Cairo, Cambridge elderly housing, Worcester family housing, Harvard married student housing, restoration of, Faneuil Hall Markets, Boston (Progressive Architecture award 1975, Design and Environment Mag. award 1976, Harleston Parker medal 1978, Nat. Honor award, Bartlett award 1978, Internat. Design award 1978, Mass. Gov.'s design award 1986), Harborplace, Balt. (Urban Land Inst. Spl. award 1986), South Street Seaport Restoration, New Fulton Market, Pier 17, N.Y.C. (Boston Soc. Architects Boston Exports award 1986, Am. Inst. Steel Constrn. Inc. award for excellence 1987), Old Post Office, Washington, Ordway Music Theatre, St. Paul (Masonry Inst. Honor award for excellence 1986), NYU Law Sch., IBM Hdqrs., Southbury, Conn., Bayside Market Place, Miami, Fla. (The Waterfront Ctr. Excellence on the Waterfront Honor award 1987, Fla. Assocs. AIA award 1989), Jacksonville Landing, Fla. (Fla. Assn. Housing and Urban Redevel. award 1988, NEFPZA 1988 design award) Century City Marketplace, Los Angeles, Union Station, Washington (Boston Soc. Architects award 1989, Mayor's Ann. Environ. Design award 1989, Nat. Trust for Hist. Preservation Honor award 1989), Custom House Docks, Dublin, Ireland, McWhirters Marketplace, Brisbane, Australia. Served as lt. USNR, 1942-45. Recipient New Eng. Architecture and Landscape award Boston Arts Festival for Acad. Quadrangle Brandeis U., 1963, Maine Arts award Colby Coll, dormitories, 1967, Pres.'s Medal of Honor Kirkland Coll., 1972, Spl. award for Alaska Capital Competition entry Am. Soc. Landscape Architects, 1978, Louis Sullivan award, 1985, Bard award, 1985, Domino's 30 award, 1988, Platinum Circle award, 1988. Mem. Boston Soc. Architects (Harleston Parker medal 1961, 71, 73, Neighborhood Housing award 1974, Honor award 1981, 84, 85, 86), Archtl. League N.Y. (Nat. Gold medal 1962), AIA (honor award merit 1963, 1st honor award 1964, Nat. firm award 1964, 87, New Eng. honor award 1966, 1970-72, 74, Library award 1968, 74, nat. honor award 1968, 70, 78, Archtl. firm award 1987, Gold medal 1992), Am. Assn. Sch. Adminstrs. (sch. design citation 1965). Office: Benjamin Thompson & Assocs 1 Story St Cambridge MA 02138-4973

THOMPSON, CARLA JO HORN, mathematics educator; b. Oklahoma City, Feb. 10, 1951; d. Hubert Henry and Gilleen Cora (Hall) Horn; m. Michael J. Thompson, Aug. 4, 1973 (div. 1986); 1 child, Emily Jane. BS, U. Tulsa, 1972, MTA, 1973, EdD, 1980; postgrad, Okla. State U., 1975-77. Tchr. math. Sapulpa (Okla.) pub. schs., 1973-79; research asst. U. Tulsa, 1979-80; asst. prof. math. and statistics Tulsa Jr. Coll., 1980—; research statistician Social & Edn. Research Assocs., Tulsa, 1987—; vis. adj. prof. U. Tulsa, 1989—, Univ. Ctr. at Tulsa, 1989—; reviewer Harcourt/Brace & Javonovich, Tulsa, 1988-89; cons. Little Brown Pubrs., Tulsa, 1984-86; evaluator Random House Pubrs., Tulsa, 1982-84. Contbr. articles to profl. jours.; author test manual: Basic Mathematics, 1989. Vol. Tulsa pub. schs., 1988—; co-leader Brownie troop Girl Scouts U.S., Tulsa, 1988— Named Tchr. of the Yr., Sapulpa Pub Schs., 1979. U. Tulsa grantee, 1980. Mem. Okla. Council Tchrs. Math. (coll. rep., exec. bd. 1983—), Am. Math. Assn. Two-Yr. Colls. (editorial bd. 1985—), Nat. Council Tchrs. Math., Phi Delta Kappa, Kappa Delta Pi. Democrat. Avocations: piano, dance, ballet, bicycling, flying. Office: Tulsa Jr Coll 3727 E Apache St Tulsa OK 74115-3151

THOMPSON, CLARK DAVID See THOMBORSON, CLARK DAVID

THOMPSON, DAVID ALFRED, industrial engineer; b. Chgo., Sept. 9, 1929; s. Clifford James and Christobel Eliza (Sawin) T.; children: Nancy, Brooke, Lynda, Diane, Kristy. B.M.E., U. Va., 1951; B.S. in Indsl. Engring, U. Fla., 1955, M.S. in Engring, 1956; Ph.D., Stanford U., 1961. Registered profl. engr., Calif; cert. profl. ergonomist. Research asst. U. Fla. Engring. and Industries Exptl. Sta., Gainesville, 1955-56; instr. indsl. engring. Stanford U., 1956-58, acting asst. prof., 1958-61, asst. prof., 1961-64, asso. prof., 1964-72, prof., 1972-83, prof., asso. chmn. dept. indsl. engring., 1972-73, prof. emeritus, 1983—; mem. clin. faculty occupational medicine U. Calif. Med. Sch., San Francisco, 1985—; pres., chief scientist Portola Assocs., Palo Alto, Calif., 1965—; prin. investigator NASA Ames Research Center, Moffatt Field, Calif., 1974-77; cons. Dept. of State, Fed. EEO Commn.; maj. U.S. and fgn. cos.; cons. emergency communications ctr. design Santa Clara County Criminal Justice Bd., 1974, Bay Area Rapid Transit Control Ctr., 1977. Govt. of Mex., 1978, Amadahl Corp., 1978-79, Kerr-McGee Corp., 1979, Chase Manhattan Bank, 1980, St. Regis Paper Co., 1980-82, Pacific Gas & Electric, 1983-85, Pacific Bell, 1984-86, 89-93, IBM, 1988-91, Hewlett-Packard, 1990-91, Reuter's News Svc., 1990-92; mem. com. for office computers Calif. OSHA. Dir., editor: documentary film Rapid Answers for Rapid Transit, Dept. Transp., 1974; mem. editorial adv. bd. Computers and Graphics, 1970-85; reviewer Indsl. Engring. and IEEE Transactions, 1972-86; contbr. articles to profl. jours. Served to lt. USNR, 1951-54. HEW grantee, 1967-70. Mem. Am. Inst. Indsl. Engrs., Human Factors Soc., IEEE, Am. Soc. Safety Engrs., World Safety Orgn., Internat. Assn. Indsl. Ergonomics and Safety Rsch. Home: 121 Peter Coutts Cir Palo Alto CA 94305-2519 Office: Stanford U Dept Indsl Engring Stanford CA 94305

THOMPSON, DAVID RUSSELL, agricultural engineering educator, academic dean; b. Cleve., Apr. 4, 1944; s. Dwight L. and Ella Caroline (Wolff) T.; m. Janet Ann Schall, Aug. 27, 1966; children: Devin Mathew, Colleen Michelle, Darin Michael. BS in Agrl. Engring., Purdue U., 1966, MS in Agrl. Engring., 1967; PhD in Agrl. Engring., Mich. State U., 1970. Asst. prof. agrl. engring., food sci. and nutrition depts. U. Minn., St. Paul, 1970-75, assoc. prof., 1975-81, prof., 1981-85; prof. agrl. engring., head dept. Okla. State U., Stillwater, 1985-91, assoc. dean Coll. Engring., Architecture and Tech., 1991—; engr. ops. dept. Green Giant Co., La Sueur, Minn., 1978-79; reviewer Colo. State U., CSRS, USDA, Ft. Collins, 1989, foods, feeds and prodn. cluster U. Mo., Columbia, 1989, dept. agrl. engring. Pa. State U., University Park, 1990, Tex. A&M U., College Station, 1992, also 5 other univs.; reviewer USDA, 1983; vis. scholar Va. Poly. Inst. and State U., Blacksburg, 1983. Author: The Influence of Materials Properties on the Freezing of Sweet Corn , 1984, Mathematical Model for Predicting Lysine and Methionine Losses During Thermal Processing of Fortified Foods; contbr. over 50 articles to sci. jours., including Jour. Food Sci. Fellow Am. Soc. Agrl. Engrs. (div. chmn. 1976-77, bd. dirs. 1981-84, 87-89, FIEI Young Researcher award 1983, President's citation 1989); mem. ASHRAE, Inst. Food Technologists (program com. 1982-85, state officer 1987-89), Am. Soc. Engring. Edn., Sigma Xi, Phi Kappa Phi, Tau Beta Pi, Alpha Epsilon, Phi Eta Sigma, Gamma Sigma Delta. Office: Okla State U Coll Engring Architecture & Tech 111 Engineering N Stillwater OK 74078-0522

THOMPSON, DAVID WALKER, astronautics company executive; b. Phila., Mar. 21, 1954; s. Robert H. and Nancy S. (Walker) T.; m. Catherine K. Ahulii, April 16, 1983. BS in Aeronautics and Astronautics, MIT, 1976; MS, Calif. Inst. Tech., 1977; MBA, Harvard U., 1981. Project engr. Jet Propulsion Lab., Pasadena, Calif., 1976; aerospace engr. NASA, Houston, 1977; project mgr. NASA, Huntsville, Ala., 1977-79; spl. asst. to pres. Hughes Aircraft Co., Los Angeles, 1981-82; pres., chief exec. officer Orbital Scis. Corp., Dulles, Va., 1982—; cons. Rockwell Internat., Thousand Oaks, Calif., 1980-81, Rand Corp., Santa Monica, Calif., 1982. Recipient Nat. award Space Found., Houston, 1981, Nat. Medal Tech. U.S Dept. Commerce Tech. Adminstrn., 1991; Nat. Air and Space Mus. Trophy, 1990; fellow Hertz Found., 1976, NSF fellow, 1976, Rockwell Internat. fellow, Harvard U. fellow, 1979; named Va. Industrial of Yr., 1991, Satellite Exec. of Yr., 1990. Fellow AIAA (assoc., Young Engr./Scientist Yr. award 1984), mem. Nat. Space Club. Office: Orbital Sciences Corp 21700 Atlantic Blvd Dulles VA 20166

THOMPSON, ELBERT ORSON, retired dentist, consultant; b. Salt Lake City, Aug. 31, 1910; s. Orson David and Lillian (Greenwood) T.; m. Gayle Larsen, Sept. 12, 1935; children: Ronald Elbert, Karen Thompson Toone, Edward David, Gay Lynne. Student, U. Utah, 1928-30, 33-35; DDS, Northwestern U., 1939; hon. degree, Am. Coll. Dentistry, Miami, Fla., 1958, Internat. Coll. Dentistry, San Francisco, 1962. Pvt. practice dentistry Salt Lake City, 1939-78; ret., 1978; inventor, developer and internat. lectr. postgrad./undergrad. courses various dental schs. and study groups, 1953-83; developer, tchr. Euthenics Dentistry Concept; cons. in field. Contbr. numerous dental articles to profl. jours. Life mem. Rep. Presdl. Task Force, Washington, 1985—. Recipient Merit Honor award U. Utah, 1985; named Dentist of the Yr. Utah Acad. Gen. Dentistry, 1991, Father of Modern Dentistry, 1991. Mem. ADA (life), Utah Dental Assn. (life, sec. 1948-49, Disting. Svc. award 1980), Salt Lake City Dental Soc. (life, pres. 1945-46), Utah Dental Hygiene Soc. (hon.), Am. Acad. Dental Practice Adminstrn. (life, pres. 1965-66), Internat. Coll. Dentists, Am. Coll. Dentists, Sons of Utah Pioneers (life), Dinorators Club (charter), Northwestern U. Alumni Assn. (Merit award 1961), Omicron Kappa Upsilon. Mormons. Avocations: temple work, golf, photography, computer, writing. Home: 3535 Hillside Ln Salt Lake City UT 84109-4099

THOMPSON, GEORGE ALBERT, geophysics educator; b. Swissvale, Pa., June 5, 1919; s. George Albert Sr. and Maude Alice (Harkness) T.; m. Anita Kimmell, July 20, 1944; children: Albert J., Dan A., David C. BS, Pa. State U., 1941; MS, MIT, 1942; PhD, Stanford U., 1949. Geologist, geophysicist U.S. Geol. Survey, Menlo Park, Calif., 1942-76; asst. prof. Stanford (Calif.) U., 1949-55, assoc. prof., 1955-60, prof. geophysics, 1960—, chmn. geophysics dept., 1967-86, chmn. geology dept., 1979-82, Otto N. Miller prof. earth scis., 1980-89, dean sch. earth scis., 1987-89; cons. adv. com. reactor safeguards Nuclear Regulation Commn., Washington, 1974—; mem. bd. earth sci. NRC, 1986-88, vice chmn. Yucca Mountain Hydrology-tectonics panel NRC, 1990-92; bd. dirs. Inc. Rsch. Inst. for Seismology, Washington, 1984-92, exec. com., 1990-92; mem. sr. external events rev. com. Lawrence Livermore Nat. Lab., 1989-93; mem. Coun. on Continental Sci. Drilling, 1990-93. Author over 100 research papers. With USNR, 1944-46. Recipient G.K. Gilbert award in seismic geology, 1964; NSF postdoctoral fellow, 1956-57; Guggenheim Found. fellow, 1963-64. Fellow AAAS, Geol. Soc. Am. (coun. mem. 1983-86, George P. Woollard award 1983), Am. Geophys. Union; mem. NAS, Seismol. Soc. Am., Soc. Exploration Geophysicists. Avocation: forestry. Home: 421 Adobe Pl Palo Alto CA 94306-4501 Office: Stanford U Geophysics Dept Stanford CA 94305-2215

THOMPSON, GUY THOMAS, safety engineer; b. Chattanooga, Dec. 29, 1942; s. Thomas Nelson and Dorothy Leona (Dobbs) T.; m. Joy Ann Gray, July 22, 1966 (div. 1978); children: Jeffrey Leighton, Lydia Ann; m. Vicki Lynn Brogdon, Dec. 6, 1979; 1 child, Laura Lynn. BA in Engring. Electronic, Park Coll., Parkville, Mo., 1976; MS in Indsl. Safety, Middle Tenn. State U., 1989. Factory rep. Modern Maid Appliances, Chattanooga, 1966-68; biomed. tech. USAF, 1968-77, commd. 2d. lt., 1977, advanced through grades to capt., 1981, ret., 1987; tng. and safety coord. Murfreesboro (Tenn.) Area Vocat. Tech. Sch., 1987—; cons. Tenn. Elec. Coop. Assn., Nashville, 1987—, Tenn. Mcpl. Elec. Power Assn., Brentwood, 1987-92; dir. safety and loss control Okla. Assn. Electric Coops. Author: Tech/Logistic Development Plans, 1978, Comprehensive Safety, 1988; copyright: Root Causes of Electric Contact Accidents and Electric Utility Safety Policy; inventor hyperbaric breathing apparatus. Coach Little League, Charleston, S.C., 1962-64, Girl's Softball League, Warner Robins, Ga., 1978-80, Girl's Softball, Tullahoma, Tenn., 1984-86. Mem. Am. Soc. Safety Engrs., Nat. Utility Tng and Safety Edn. Assn. (coord. conf. 1989), Nat. Safety Coun., Tenn. Safety Congress (exec. com.), Am. Tech. Edn. Assn., Tri-County Bowling Assn. (pres. 1982-88), Arnold Engring. & Devel. Ctr. Golf Club (coun. 1984-88), Am. Legion, Phi Kappa Phi. Republican. Methodist. Avocations: golf, boating, Indian artifacts, bowling. Home: 15089 N Oak Dr Choctaw OK 73020 Office: Okla Assn Electric Coops 2325 NE Expwy Box 11047 Oklahoma City OK 73136

THOMPSON, H. BRADFORD, chemist, educator; b. Detroit, Apr. 22, 1927; s. Herbert Bradford and Margaret Ann (Gilbert) T.; m. Jane Elizabeth Lang, June 13, 1949. BS, Olivet Coll., 1948; MA, Oberlin Coll., 1950; PhD, Mich. State U., 1953. Rsch. assoc. Mich. State U., East Lansing, 1953-55, Iowa State U. Ames Lab., 1963-65; sr. rsch. assoc. U. Mich., Ann Arbor, 1965-67; prof. U. Toledo, 1967-90; asst. prof. Gustavus Adolphus Coll., St. Peter, Minn., 1955-57, assoc. prof., 1957-63, scholar in residence, 1990—. Mem. AAAS, Am. Chem. Soc., Am. Phys. Soc., Assn. for Computing Machinery, Sigma Xi. Presbyterian. Home: 1604 Riverview Rd Saint Peter MN 56082 Office: Gustavus Adolphus Coll Saint Peter MN 56082

THOMPSON, H. BRIAN, telecommunications executive; b. Buffalo, Mar. 24, 1939; s. Hugh and Margaret (Motsco) T.; m. Mary Ann Selby; children: Christiana, Brandon. BS ChemE, U. Mass., 1960; MBA, Harvard Bus. Sch., 1968. Research engr. Kendall Co., Walpole, Mass., 1960-62, Monsanto Research Corp., Everett, Mass., 1966; cons. McKinsey & Co., Washington, 1968-77; sr. v.p. planning and mktg. Resource Scis. Corp., Tulsa, 1977-79; exec. v.p. Gelman Scis., Ann Arbor, Mich., 1979; pres. Subscription TV Am., Rockville, Md., 1979-81; sr. v.p. corp. devel. MCI Communications Corp., Washington, 1981-85; pres. mid-atlantic div. MCI Telecommunications, Arlington, Va., 1985-87; exec. v.p. MCI Communications Corp., Washington, 1987—. Patentee non-woven fiber products. Trustee Capitol Coll., Laurel, Md., 1986—; bd. dirs. Leadership Washington, 1986. Office: MCI Communications Corp 1133 19th St NW Washington DC 20036-3695

THOMPSON, HAYDN ASHLEY, electronics engineer; b. Grimsby, Humberside, U.K., Feb. 4, 1965; s. Haydn Alexander and Sandra (Warman) T. BSEE, U. Wales, Bangor, Eng., 1986, PhD, 1990. Chartered engr. Researcher Royal Aircraft Establishment, Bedford, U.K., 1986-87, Rolls Royce plc, Bristol, U.K., 1987-90; contractor European Space Agy., Bangor, 1990; sr. electronics engr. Dowty Aerospace and Def., Loudwater, U.K., 1990-91, Marconi Radar Systems Ltd., Chelmsford, U.K., 1991—; cons. Esprit DG IX, European Commn., Brussels, 1990—; presenter in field. Author: Parallel Processing for Jet Engine Control, 1990; contbr. articles to IEE Computing and Control Engring. Jour., Jour. Systems and Control Engring., Microprocessors and Microsystems. Mem. AIAA, IEEE, Internat. Fedn. Automatic Control (internat. aerospace control com.), Inst. Elec. Engrs., Royal Aero. Soc., Dept. Trade and Industry Safety Critical Systems Club, Occam User Group. Achievements include development of parallel processing for real-time fault-tolerant flight control systems, massively parallel real-time signal processing units for radar systems. Home: 18 Lakin Close, Essex CM2 6RU, England Office: Marconi Radar & Control Sys, Writtle Rd, Essex CM1 3BN, England

THOMPSON, HOWARD DOYLE, mechanical engineer; b. Cedar City, Utah, Apr. 17, 1934; s. William Howard and Ann (Esplin) T.; m. Patricia Ann Frei, Aug. 22, 1956; children: Tamra, Kimberly Smith, Shauna Bigham, Trisha Leah Weeks, Stephanie Yorgason. BSChemE, U. Utah, 1957; MS in Engring., Purdue U., 1962, PhD, 1965. Registered profl. engr., Ind. Asst. prof. mech. engring. Purdue U., West Lafayette, Ind., 1965-69, assoc. prof. mech. engring., 1969-74; assoc. rsch. scientist Pratt and Whitney Aircraft, East Hartford, Conn., 1969-70; sr. mech. engr. Arnold Engring. Devel. Ctr., Tullahoma, Tenn., 1980-81; vis. scientist Wright Labs., Wright Patterson AFB, Ohio, 1990-91; prof. mech. engring. Purdue U., West Lafayette, 1974—; bd. dirs. Laser Inst. of Am., Orlando, Fla., 1988—; cons. USAF, Dayton, Ohio, 1972—. Author: Laser Velocimetry and Particle Sizing, 1979. Lt. (j.g.) USN, 1957-60. Recipient Solberg Best Tchr. award Purdue U., 1982. Assoc. fellow AIAA; mem. ASME, Laser Inst. of Am., Sigma Xi, Tau Beta Pi. LDS. Home: 212 Knox Dr West Lafayette IN 47906 Office: Purdue U Sch Mech Engring West Lafayette IN 47907

THOMPSON, JAMES ROBERT, JR., federal space center executive; b. Greenville, S.C., Mar. 6, 1936; s. James Robert Sr. and Mildred (Morgan) T.; m. Susan Lynn Sproul, June 15, 1958 (div. Aug. 1985); children: James Robert III, Susan P. Thompson Rowe, Scott A.; m. Sherry Kay Gray, Apr. 11, 1989. BS in Aero. Engring., Ga. Tech. Inst., 1958; MSMechE, U. Fla., 1963; postgrad., U. Ala., 1967-69, PhD (hon.), 1988. Performance analysis engr. Pratt & Whitney, West Palm Beach, Fla., 1960-63; mech. engr., advanced propulsion sect., R&D Devel. Lab Marshall Space Flight Ctr., NASA, Huntsville, Ala., 1963-66, aerospace technologist liquid propulsion systems, 1966-68, chief space & nuclear engines sect., Propulsion & Vehicle Engring. Lab., R&D Directorate, 1968-70, chief man/systems integration br., Astro. Lab., Sci. & Engring. Directorate, 1970-74, mgr. main engine project, Shuttle Projects Office, 1974-82, assoc. dir. for engring., Sci. and Engring. Directorate, 1982-83; dir. of ctr. Marshall Space Flight Ctr., NASA, Ala., 1986—; spl. asst. to assoc. adminstr. for space flight NASA hdqrs., Washington, Mar.-June, 1986, cons., June-Sept., 1986; dep. dir. tech. ops., Plasma Physics Lab. Princeton (N.J.) U., Apr.-Sept., 1986. Author tech. papers in field. Co-patentee rocket engine nozzle skirt with transpiration cooling. Chmn. exhibits commn. Space & Rocket Ctr., Huntsville, Ala., 1987—; mem. Ala. Space and Sci. Exhibit Commn., Montgomery, Ala., 1987-89, Ala. Insdl. Coun. on Engring., Montgomery, 1987-89, Huntsville C. of C. Leadership 2000, 1987-89; mem. adv. coun. Princeton U., 1987-89. Lt. USNR, 1958-60. Recipient awards from NASA including Exceptional Svc. medal Skylab Program, 1973, Disting. Svc. medal Space Shuttle, 1981, Disting. Pub. Svc. medal, 1988, Disting. Svc. medal, 1988, others, Meritorious Sr. Govt. Exec. award Pres. U.S., 1982, 87, Disting. Svc. award Hunstville C. of C., 1988; one of recipients Meml. Trophy awarded to Shuttle Return to Flight Team Nat. Space Club, 1989, Holley medal, Am. Soc. Mechanical Engineers, 1991. Fellow AIAA (assoc.), Nat. Space Club (bd. govs. 1987-89), Sigma Xi, Sigma Nu (comdr. 1957-58). Methodist. Avocations: golf, bridge. Office: NASA HQ Code AD 400 Maryland Ave SW Washington DC 20546

THOMPSON, JOE FLOYD, aerospace engineer, researcher; b. Grenada, Miss., Apr. 13, 1939; s. Joe Floyd and Bernice Thompson; m. Emilie Kay Wilson, June 1, 1974; children: Mardi, Douglass. BS, Miss. State U., 1961, MS, 1963; PhD, Ga. Tech., 1971. Aerospace engr. NASA Marshall, Huntsville, Ala., 1963-64; prof. Miss. State U., Starkville, 1964-90, dir., 1990—; mem. tech. adv. bd. Army Rsch. Lab., Aberdeen, Md., 1993—; dir. computer code Nat. Grid Project, 1993. Author: Numerical Grid Generation, 1985, (computer code) Eagle Grid System, 1987; sr. assoc. editor Applied Math. and Computation, 1985—; assoc. editor Numerical Heat Transfer, 1989—; mem. editorial bd. Computational Fluid Dynamics Jour., 1993—. Recipient Commdr.'s award Army Waterways Exp. Sta., Vicksburg, Miss., 1992. Mem. IEEE, AIAA (Aerodynamics award 1992), SIAM. Presbyterian. Achievements include establishment of NSF Engineering Research Center; pioneering work in field of numerical grid generation. Home: Miss State U Computational Fluid Dynamics Box 255 Mississippi State MS 39762 Office: Mississippi State U Box 6176 Mississippi State MS 39762*

THOMPSON, JOHN EVELEIGH, horticulturist, educator; b. May 30, 1941. BSA, U. Toronto, 1963; PhD, U. Alta., Edmonton, Can., 1966. Postdoctoral rsch. fellow U. Birmingham, Eng., 1966-67; from asst. prof. to assoc. prof. U. Waterloo, Ont., Can., 1968-76, prof., 1976-87, chmn. dept. biology, 1980-86; prof., chmn. dept. hort. sci. U. Guelph, 1987-90; prof., dean of sci. U. Waterloo, 1990—; vis. prof. Weizmann Inst. Sci., Israel, 1980, John Innis Inst., Eng., 1987; Lady Davis vis. prof. Hebrew U., Israel, 1980; manuscript referee Archives Biochemistry and Biophysics, Plant Physiology, Planta, Physiologia Plantarum, Phytochemistry, Can. Jour. of Botany, Can. Jour. Biochemistry, Jour. Phycology, Physiol. Plant Pathology; grant reviewer NSERC, NSF, USDA; chmn. plant biology grant selection com. NSERC, 1983-84; mem., exec. CCUBC, 1983-86, CSPP, 1981-83; v.p. Gordon Rsch. Conf. on Plant Senecence, 1984; mem. plant biology grant selection com. NSERC, 1981-84, interdisciplinary grant selection com., 1984-85, strat. grant selection panel food and agriculture, 1991—, steacie fellow selection com., 1991, selection com. rsch. grants, 1991, 92; pres. Gordon Rsch. Conf. Plant Senescence, 1988; chmn. ecternal review com. dept. biology Simon Fraser U., 1992; reviewer Plant Biotech. Ctr. U. Toronto, 1990; cons. Philip Morris Ltd., Richmond, Va., 1985. Author: (with others) Scanning Electron Microscopy of Mammalian Reproduction, 1975, Biochemistry and Physiology of Protozoa, 1980, Physiology of Membrane Fluidity, vol. II, 1985, Handbook of the Biology of Aging, 2d edit., 1985, Plant Senescence, 1988, Senescence and Aging in Plants, 1988, Plant Physiology, vol. X, 1991; contbr. numerous articles to profl. jours. Fellow Royal Soc. Can. (new fellows selection com. 1990-93). Achievements include patent in control of senescence in fruits, vegetables and flowers; research in preparation of plasma membranes from amoebae, dose-response relationships of heavy metals and selected pesticides using biological indicator systems, measurments of sublethal and lethal effects of contaminants of an aquatic ecosystem; development of DNA-mediated transformation systems for plants. Office: University of Waterloo, Faculty of Science, Waterloo, ON Canada N2L 3G1

THOMPSON, JOHN JAMES, chemist, researcher, consultant; b. Washington, June 11, 1942; s. John James and Ann Elizabeth (Bridges) T.; m. Barbara Ann Clark, Aug. 21, 1971; children: John James Jr., David Clark. BS, Allegheny Coll., 1968. Rsch. tech. Allied Chem. Corp., Morristown, N.J., 1963-68; rsch. chemist I Allied Chem. Corp., Morristown, 1968-71; rsch. chemist II Allied Chem. Corp., Buffalo, N.Y., 1971-81; rsch. chemist III Allied Signal Inc., Buffalo, 1981—. Mem. Am. Chem. Soc., Am. Soc. for Mass Spectrometry. Republican. Episcopalian. Office: Allied Signal Inc 20 Peabody St Buffalo NY 14210

THOMPSON, JOHN N., ecology, evolutionary biology educator, researcher; b. Pitts., Nov. 15, 1951; s. John C. and Cecilia (Kravich) T.; m. Jill Fransmith, Aug. 18, 1973. BA, Washington & Jefferson Coll., 1973; PhD, U. Ill., 1977. Vis. assoc. prof. entemology U. Ill., Urbana, 1977-78; asst. prof. botany and zoology Wash. State U., Pullman, 1978-82, assoc. prof., 1982-87; vis. faculty Imperial Coll. Silwood Park, Ascot, England, 1985-86; prof. Wash. State U., Pullman, 1987—. Author: Interaction and Coevolution, 1982. Fullbright scholar CSIRO, Canberra, Australia, 1991-92. Fellow AAAS, Royal Entomol. Soc.; mem. Ecol. Soc. Am., Am. Soc. Naturalists, Soc. Study of Evolution (exec. coun. 1988-90). Office: U Wash Dept Botony & Zoology Pullman WA 99164

THOMPSON, LARRY FLACK, chemical company executive; b. Union City, Tenn., Aug. 31, 1944; s. Rufus Russell and Polly (Flack) T.; m. Joan

Bondurant, Aug. 30, 1964; children: Anthony Scott, Russell Allen. BS, Tenn. Tech. U., Cookeville, 1966; MS, Tenn. Tech. U., 1968; PhD, U. Mo., Rolla, 1970. Mem. tech. staff Bell Labs., Murray Hill, N.J., 1971-80; dept. head AT&T Bell Labs., Murray Hill, N.J., 1981—; Author: Introduction to Microlithography, 1985; patentee in field. Mem. NAE, Am. Chem. Soc. (bd. dirs. 1993—, Indsl. Chemistry award 1993, Roy W. Tess award 1993), Am. Inst. Chem. Engring. Avocations: gardening, hunting. Home: 1511 Long Hill Rd Millington NJ 07946-1813 Office: AT&T Bell Labs 6C302 Lithographic Materials Rsch 600 Mountain Ave New Providence NJ 07974-2010

THOMPSON, LARRY JOSEPH, veterinary toxicologist, consultant; b. Green Bay, Wis., Mar. 5, 1955; s. Bernard Thomas and Marcella (Gajeski) T.; m. Lisa Kay Delap, Aug. 20, 1983; children: Amanda Kay, Anthony James. BS, U. Wis., River Falls, 1977; DVM, U. Ill., 1983. Diplomate Am. Bd. Vet. Toxicology. Rsch. assoc. U. Ill. Coll. Vet. Medicine, Urbana, 1980-83, teaching assoc., 1984-90; staff veterinarian Mid-North Animal Hosp., Chgo., 1983-84; clin. toxicologist, dir. biosafety N.Y. State Coll. Vet. Medicine-Cornell U., Ithaca, 1990—. Contbg. author: Heavy Metal Toxicosis in Horses, 1991, Copper-Molybdenum in Cattle, 1993. Pres. East Ithaca Pre-Sch. Coop., 1992-93. Office: Diagnostic Lab NYSCVM PO Box 786 Upper Tower Rd Ithaca NY 14851

THOMPSON, LAWRENCE FRANKLIN, JR., computer corporation executive; b. Winchester, Tenn., Feb. 12, 1941; s. Lawrence Franklin and Mildred C. T.; m. Carol Lee Lufkin, Oct. 9, 1965; children: Jeffrey, Maureen. BS in Internat. Affairs, U.S. Air Force Acad., 1963. Enlisted USAF, 1958; cadet USAF Acad., 1959-63, commd. 2d lt., 1963, advanced through grades to capt., 1966, pilot, 1963-69, resigned, 1969; pres. Collectors Showcase, Orange, Calif., 1973-76; owner Lufkins (Limited Edition Collectibles), Mission Viejo, Calif., 1971—; chief exec. officer Computer City, Inc., Austin, Tex., 1979-91; pres., chief exec. officer Productivity Unltd., Inc., Austin, 1991—; v.p. ABC Computers, Inc., Austin, Tex., 1981-83; pres. Computer Craft of Austin, 1983-84. Decorated D.F.C., Silver Star. Office: Process Dynamics Internat Ste 150 9442 Capital of Tex Hwy Austin TX 78759

THOMPSON, LEIGH LASSITER, psychologist, educator; b. Houston, Jan. 13, 1960; d. Don Raines and Ann Janet (Visintin) Thompson; m. Robert Warner Weeks, June 20, 1992. BS, Northwestern U., 1982, PhD, 1988; MA, U. Calif., Santa Barbara, 1984. Asst. prof. psychology U. Wash., Seattle, 1988-92, assoc. prof., 1992—. Editorial bd. Organizational Behavior and Human Decision Processes, Internat. Jour. Conflict Mgmt., 1990—; assoc. editor Group Decision Making and Negotiations; contbr. articles to Psychol. Bull., Jour. Personality and Social Psychology, Jour. Exptl. Social Psychology, and others. Workshop leader Guardian Ad Litem, Seattle, 1990, 91. Recipient Young Investigator award NSF, 1991, Grad. Rsch. award Sigma Xi Found., 1987; grantee NSF, 1991, 89—, Nat. Inst. Dispute Resolution, 1987, APA, 1989; fellow Ctr. for Advanced Study Behavioral Sci. Mem. APA (S. Rains Wallace Dissertation award 1989), Am. Psychol. Soc., Acad. Mgmt. Achievements include discovery of "incompatibilty" error in negotiation; development of experimental and computer paradigms to study negotiation and conflict. Office: U Wash Psychology NI-25 Seattle WA 98195

THOMPSON, LEROY, JR., army reserve officer, radio engineer; b. Tulsa, July 7, 1913; s. LeRoy and Mary (McMurrain) T.; B.S. in Elec. Engring., Ala. Poly Inst., 1936; m. Ola Dell Tedder, Dec. 31, 1941; 1 son, Bartow McMurrain. Commd. 2d lt. U.S. Amy Res., 1935, advanced through grades to col., 1963; signal officer CCC, 1936-40; radio engr. Officer Hdqrs. 4th C A., 1941; with signal sect. Hdqrs. Western Def. command and 4th Army, San Francisco, 1942, comdg. officer 234th Signal Ops. Co., 1942; asst. chief, chief signal corps ROTC U. Calif., Berkeley, 1942-43; radio engring. officer O.C. SigO War Dept., Washington, 1943; radio engring officer Hdqrs. 3105th Signal Service Co. Hdqrs. CBI, New Delhi, 1944; signal officer Hdqrs. Northern Combat Area Command, Burma, 1944; signal officer Hdqrs. OSS Det 101, Burma, 1945; signal officer Hdqrs. OSS, China, 1945, radio engr., tech. liaison officer, Central Intelligence Group, CIA, 1945-50; chief radio br. Hdqrs. FEC, Tokyo, 1950-53, chief radio engring br. Signal C Plant Engring. Agy., 1953-55; radio cons. to asst. dir. def. research and engring. communications, 1960-62; ret., 1973; pvt. research and devel. on communication and related problems, 1963—; owner Thompson Research Exptl. Devel. Lab. Lic. profl. radio engr., Ga. Mem. IEEE (life sr.), Vet. Wireless Operators Assn., Am. Radio Relay League, Nat. Rifle Assn., Mil. Order World Wars, Res. Officers Assn., Am. Motorcycle Assn., Nat. Wildlife Fedn. Baptist. Home: 6450 Overlook Dr Alexandria VA 22312-1327

THOMPSON, LOIS JEAN HEIDKE ORE, industrial psychologist; b. Chgo., Feb. 22, 1933; d. Harold William and Ethel Rose (Neumann) Heidke; m. Henry Thomas Ore, Aug. 28, 1954 (div. May 1972); children: Christopher, Douglas; m. Joseph Lippard Thompson, Aug. 3, 1972; children: Scott, Les, Melanie. BA, Cornell Coll., Mt. Vernon, Iowa, 1955; MA, Idaho State U., 1964, EdD, 1981. Lic. psychologist, N.Mex. Tchr. pub. schs. various locations, 1956-67; tchr., instr. Idaho State U., Pocatello, 1967-72; employee/orgn. devel. specialist Los Alamos (N.Mex.) Nat. Lab., 1981-84, tng. specialist, 1984-89, sect. leader, 1989—; pvt. practice Los Alamos, 1988—; sec. Cornell Coll. Alumni Office, 1954-55, also other orgns.; bd. dirs. Parent Edn. Ctr., Idaho State U., 1980; counselor, Los Alamos, 1981-88. Editor newsletter LWV, Laramie, Wyo., 1957; contbr. articles to profl. jours. Pres. Newcomers Club, Pocatello, 1967, Faculty Womens Club, Pocatello, 1968; chmn. edn. com. AAUW, Pocatello, 1969. Mem. APA, AACD, N.Mex. Psychol. Assn. (bd. dirs. div. II, 1990, sec. 1988-90, chmn. 1990), N.Am. Soc. Adlerian Psychology, N.Mex. Soc. Adlerian Psychology (pres. 1990, treas. 1991-93), Soc. Indsl. and Orgnl. Psychology, Nat. Career Counseling Assn. Mem. LDS Ch. Avocations: racewalking, backpacking, skiing, tennis, biking. Home: 340 Aragon Ave Los Alamos NM 87544-3505 Office: Los Alamos Nat Lab MS M589 HRD-3 Los Alamos NM 87545

THOMPSON, LYLE EUGENE, electrical engineer; b. Pocatello, Idaho, May 16, 1956; s. Clyde Eugene and Doris (Pratt) T.; m. Barbara Mae Dickerson, Dec. 31, 1986. Grad. high sch. Sr. diagnostic engr. Calma/GE, Santa Clara, Calif., 1978-83; mem. tech. staff Telecommunications Tech., Inc., Milpitas, Calif., 1983-84; proprietor/cons. Lyle Thompson Cons., Fremont, Calif., 1984-87; sys. analyst Raynet Corp., Menlo Park, Calif., 1987-88; proprietor/cons. Lyle Thompson Cons., Hayward, Calif., 1988-89; mgr. sys. design Raylan Corp., Menlo Park, Calif., 1989-90; dir. system design Raylan Corp., Menlo Park, 1990-91; pvt. practice cons. San Lorenzo, Calif., 1991—; cons. in field. Patentee in field. Mem. ACM, IEEE. Avocations: skiing, music, role playing. Home: 664 Paseo Grande San Lorenzo CA 94580-2364

THOMPSON, MALCOLM FRANCIS, electrical engineer; b. Charleston, S.C., Sept. 2, 1921; s. Allen R. and Lydia (Brunson) T.; m. Ada Rose O'Quinn, Jan. 20, 1943 (dec. 1987); children: Rose Mary, Nancy Belle, Susan Elizabeth, Frances Josephine; m. Milena N. Winckler, June 22, 1989. BS, Ga. Inst. Tech., 1943, MS, 1947; postgrad., MIT, 1947-49. Instr. dept. elec. engring. MIT, 1947-49; rsch. engr. Autonetics Co., Anaheim, Calif., 1949-70; tech. dir. SRC div. Moxon, Inc., Irvine, Calif., 1970-73; engring. mgr., mgr. computers and armament controls. Northrop Aircraft Div., Hawthorne, Calif., 1973-87; ind. cons., 1987—. Patentee in field. Capt. AUS, 1943-46. Mem. IEEE, NRA, Nat. Geog. Soc., Am. Ordnance Assn., Eta Kappa Nu. Home and Office: 1602 Indus St Santa Ana CA 92707-5308

THOMPSON, MARK ALAN, environmental engineer; b. Ellwood City, Pa., Oct. 30, 1957. BSCE, Ind. Inst. Tech., 1979. Cert. hazardous materials mgr., Tenn. Jr. engr. Resource Cons., Inc., Ft. Collins, Colo., 1984-85; owner, gen. mgr. Pro-Wash, Inc., Aurora, Colo., 1985-90, A.P. Gen., Inc., Aurora, 1987-90; engr., project mgr. EcoTek, Inc., Erwin, Tenn., 1990—. Mem. Heritage Bapt. Ch., Johnson City, Tenn., 1991—; adult leader cub group Boy Scouts Am., Johnson City, 1992—. Capt. U.S. Army, 1979-83. Mem. Soc. Am. Mil. Engrs., Nat. Water Well Assn., Hazardous Materials Control Rsch. Inst. Office: EcoTek Inc 1219 Banner Hill Rd Erwin TN 37650

THOMPSON, MILTON ORVILLE, aeronautical engineer; b. Crookston, Minn., May 4, 1926; s. Peter and Alma Teresa (Evenson) T.; m. Therese Mary Beytebiere, June 25, 1949; children: Eric P., Milton Orville, Brett, Peter K., Kye C. BS in Engring, U. Wash., 1953; postgrad., U. So. Calif., 1956-59. Assoc. engr. Boeing Co., Seattle, 1953-55; rsch. pilot, X-15 rocket air plane pilot NASA, Edwards, Calif., 1956-66; asst. dir. rsch. NASA, 1967-72, dir. rsch. projects, 1972-74, chief engr., 1974-78; assoc. dir. NASA Dryden Flight Rsch. Ctr., 1978—. Author: At the Edge of Space, 1992; contbr. articles to tech. jours. Naval aviator USNR, 1944-49. Recipient Octave Chanute award Am. Inst. Aeros. and Astronautics, 1967, Disting. Service medal NASA, 1978, Exceptional Service medal, 1981, Outstanding Leadership medal, 1988; named Elder Statesman of Aviation Nat. Aero. Assn., 1990. Mem. Soc. Exptl. Test Pilots (Ivan C. Kinchloe award 1966), U.S. Govt. Sr. Exec. Svc. (charter). First to fly lifting body entry vehicle. Home: 1640 W Ave L-12 Lancaster CA 93534 Office: NASA-Dryden Flight Rsch Ctr PO Box 273 Edwards CA 93523-0273

THOMPSON, RAHMONA ANN, plant taxonomist; b. Oklahoma City, Okla., June 17, 1953; d. Raymond D. and Marilyn Frances (Strong) James; m. Ronald K. Thompson, Aug. 2, 1971. BS in Botany, Okla. U., 1978, MS in Botany, 1981; PhD in Botany, Okla. State U., 1988. Rsch. assoc. Okla. Biol. Survey, Norman, 1988-90; interim curator Robert Bebb Herbarium, Norman, 1990-91; asst. prof. East Cen. U., Ada, Okla., 1991—; taxon editor Flora of N.Am., St. Louis, 1990—; bd. dirs. Flora of Okla. Project, Stillwater. Contbr. articles to profl. jours. Bd. dirs. Environ. Control Adv. Bd., Norman, 1991—. Mem. Okla. Native Plant Soc. (bd. dirs. 1992—), Okla. Soc. for Electron Microscopy (pres.-elect 1992-93). Democrat. Achievements include rsch. on taxonomic problems in Poaceae, especially generic boundries in Paniceae, spikelet anatomy. Office: East Central U Biology Dept Ada OK 74820

THOMPSON, RICHARD EDWARD, biochemist; b. Wichita, Kans., Oct. 17, 1946; s. Richard Clayton and Joan Elizabeth (Messer) T.; m. Ramona Kay Bell, May 29, 1971; children: Frances Michelle, Stacy Kay. BS in Chemistry, Wichita State U., 1968, MS in Chemistry, 1969; PhD in Biochemistry, Okla. State U., 1974. Postdoctoral fellow NIH, 1972-74; rsch. assoc. U. Cin. Med. Ctr., 1974-77; asst. prof. North Tex. State U., Denton, 1977-83; biochemist Abbott Labs., Irving, Tex., 1983-86, North Chgo., Ill., 1986—. Contbr. articles to profl. jours. including Biophysical Jour., Jour. of Chromatography, Analytical ChimActa. NIH grantee, 1979, grantee R.A. Welch Found., 1978. Mem. Am. Heart Assn. (Tex. affiliate ctrl. rsch. review com. 1978-83, grantee 1978-80). Presbyterian. Achievements include medical diagnostics patent, computer assisted optimization of HPLC methods development, regulation of liver cholesterol synthesis in diabetics, development of cell culture process. Home: 1616 Pleasant Ct Libertyville IL 60048 Office: Abbott Labs Abbott Park North Chicago IL 60064

THOMPSON, ROBERT ALLAN, aerospace engineer; b. Cleve., June 10, 1937; s. Roy Henry and Viola Alverta (Nehls) T.; BSEE, Case Western Reserve U., 1958; postgrad. Cleve. State U., 1959, John Marshall Law Sch., 1970; PhD, Union Inst., 1979; m. Louise Alberta Saari, Nov. 27, 1970. Rsch. engr. Sohio Satellite Tracking Sta., Standard Oil Rsch. Lab., Cleve., 1958-63, acting dir., 1964-65; tchr. Cleve. Bd. Edn., 1958-65; dir. Warrensville Heights Planetarium and Space Sci. Program, 1964-65; tchr. spl. programs faculty Case Inst. Tech., 1965; dir. planning phase sci. div. Cleve. Supplementary Edn. Ctr., 1965-66; dir. James A. Lovell Regional Space Ctr., Milw., 1967-73; engring. and edn. cons., Chgo., 1973-78, Mystic, Conn., 1978—; pres., chmn. bd. Spatialworld Corp., 1982—; lectr. U. Wis., Milw., 1968-71; chmn. secondary math. curriculum com. Cleve. Pub. Schs., 1963-64; mem. Wis. Aerospace Edn. Com., 1968-71; sec. Friends of Space Center, 1968-75. Recipient Leadership award Kiwanis Key Club, 1961; Goodwin Watson Inst. doctoral fellow, 1978-79. Registered profl. engr., Ohio, Wis., Conn., R.I. Fellow Brit. Interplanetary Assn.; mem. IEEE (sr. mem., exec. com., chmn. membership com. Cleve. sect. 1965-66), AAAS, AIAA (chmn. Wis. sect. 1969-70, sr. mem. Conn. sect. council mem. 1984-85, disting. lectr. 1987-89), Cleve. Engring. Soc., Cleve. Astron. Soc. (mem. exec. com. 1966-67), Case Alumni Assn., Union Inst. Alumni Assn. Author: The New Egoshell-An Individualized Space Age Reality, 1980; co-author (with wife): Egoshell-Planetary Individualism Balanced Within Planetary Interdependence!, 1987; contbr. articles to encys. and profl. jours. Home: 45 Deer Ridge Rd Stonington CT 06378 Also: PO Box 624 Mystic CT 06355-0629 Office: PO Box 2001 Mystic CT 06355-0624

THOMPSON, ROBERT CAMPBELL, orthopaedic surgeon; b. N.Y.C., Mar. 23, 1938; s. Theodore Campbell and Cornelia (Tomlin) T.; m. Patricia Ann Hadley, Nov. 10, 1962; children: Suzanne Campbell, Brett Nicole. BS in Biology, Calif. Inst. Tech., 1960; MD, Johns Hopkins U., 1964. Diplomate Am. Bd. Orthopaedic Surgery. Intern, resident Union Meml. Hosp., Balt., 1965-66; resident and fellow in orthopaedic surgery John Hopkins Hosp., Balt., 1968-72; asst. prof. orthopaedic surgery and nuclear medicine Johns Hopkins U. Sch. Medicine, Balt., 1972-74; attending orthopaedic surgeon Meml. Hosp., Easton, Md., 1974—; mem. Gov.'s Commn. on Arthritis and Related Diseases, State of Md., 1986-92; comml. pilot and flight instr. Contbr. articles to profl. publs. Lt. comdr. USPHS, 1966-68. Mem. Md. Orthopaedic Assn., Flying Physicians Assn. (bd. dirs. 1992—), Assn. Computing Machinery. Achievements include patent for hearing testing machine. Office: 602 Dutchmans Ln Easton MD 21601

THOMPSON, ROBERT JAMES, physicist; b. Bayshore, N.Y., June 29, 1962; s. Arthur Montgomery and Lilian Florence (Norman) T. BS, Ga. Tech. Inst., 1984; postgrad., U. Tex., 1987-92. Sr. staff technician Bellcore, Navesink, N.J., 1984-87; rsch. asst. U. Tex., Austin, 1988-89, Calif. Tech. Inst., Pasadena, 1989—. Democrat. Achievements include research in optical domain studies of curity quantum electrodynamics; observed vacuum-Rabi splitting for a single atom. Home: 85 N Holliston # 10 Pasadena CA 91106 Office: Calif Tech Inst Mail Code 12-33 Pasadena CA 91106

THOMPSON, ROBERT W., theoretical physicist; b. Mpls., 1919. BS with highest distinction, U. Minn., 1941; PhD, MIT, 1948. With Los Alamos, 1943-45; asst. prof. Indiana U., 1948-53; prof. physics U. Chgo., 1953. Contbr. to profl. jours. Achievements include discovery of the Neutral Kaons, the Isotope PU 240, and execution of the P it assembly for the Nagasaki Bomb; patents include the Multiwire Proportional Counter. Home: 5648 Dorchester 1/W Chicago IL 60637

THOMPSON, STEPHEN ARTHUR, publishing executive; b. Englewood, N.J., Jan. 24, 1934; s. Stephen Gerard and Doris Lillian (Evans) T.; m. Joan Frances O'Connor, May 12, 1955 (div. 1978); children: Stephen Andrew, Craig Allen, David John; m. Sandra Rene Fingernut, May 27, 1979. BS, Ohio State U., 1961. Physicist Rocketdyne div. North Am. Aviation, Canoga Park, Calif., 1961-62, Marquardt Corp., Van Nuys, Calif., 1962-63; mem. tech. staff Hughes Rsch. Labs., Malibu, Calif., 1963-69; editor, with advt. sales dept. Chilton Co., L.A., 1969-77; sr. v.p. corp. mktg. Cahners Pub. Co., Newton, Mass., 1977—; founder Design News Engring. Edn. Found., Newton, 1991—. Author: Basketball for Boys, 1970; contbr. articles to Jour. Spacecraft/Rockets, 1966. Club leader YMCA, Canoga Park, 1963-78; active PTA, Canoga Park, 1961-80; bd. dirs. Chatsworth (Calif.) High Booster Club, 1972-80. 1st lt., jet fighter pilot USAF, 1952-58. Mem. Bus. Profl. Advt. Assn. (Golden Spike award 1980, 81, 82, 83), L.A. Mag. Reps. Assn. (life), Nat. Fluid Power Assn. Achievements include patents for ion source, system and method for ion implantation of semiconductors. Office: Cahners Pub Co 275 Washington St Newton MA 02158-1611

THOMPSON, TRAVIS, psychology educator, administrator, researcher; b. Mpls., July 20, 1937; s. William Raymond and Loretta (Travis) T.; m. Anna Leyens, June 12, 1970; children: Rebecca Lynn, Jennifer Eva, Andrea Laura, Peter Erich. BA, U. Minn., 1958, MA, PhD, 1961. Lic. psychologist. NSF postdoctoral fellow U. Md., College Park, 1961-63; asst. prof. U. Minn., Mpls., 1963-66, assoc. prof., 1966-69, prof., 1969-91, dir. inst. disabilities studies, 1987-91; prof., dir. John F. Kennedy Ctr. Vanderbilt U., Nashville, 1991—; mem. space com. Peabody Coll., Vanderbilt U., 1991—, search com. assoc. dean for rsch. sch. nursing, 1991—, chairs and ctr. dirs. com., 1991—; vis. fellow Cambridge (U.K.) U., 1968-69; vis. scientist Nat. Inst. Drug Abuse, Rockville, Md., 1979-80; mem. com. Nat. Inst. Child Health, Bethesda, Md. Hastings Ctr., Rights of Retarded, Hastings-on-Hudson, 1977-78; mem. instl. review bd. Minn. Dept. Human Svcs., 1986-89; mem. exec. com. 1990 planning com. Gatlinburg Conf. on Rsch. in Mental Retardation, 1988—, exec. com., 1992; mem. Human Devel. 3 Rsch. Review com. NIH, 1988-91; mem. adv. com. Inst. on Community Intergration, U. Minn., 1988-91, Rehab. Rsch. and Tng. Ctr in Adolescent Health, 1989-91; mem. local adv. com. Geriatric Rsch., Edn. and Clin. Ctr., VA Med. Ctr., Mpls., 1989-91; cons. in field, including North Star Rsch. and Devel. Inst., Mpls., 1963-65, Honeywell, Inc., Mpls., 1964-67, Faribault (Minn.) State Hosp., 1968-75, 87—, Cambridge (Minn.) State Hosp., 1972-74, Ctr. for Behavioral Therapy, Mpls., 1973-80, Clara Doerr Residence, Inc., Mpls., 1975-77, Ill. Sci. Adv. Com. on Mental Health and Devel. Disabilities, 1975-77, Minn. Dept. Pub. Welfare, 1978-80, Psychol. and Behavioral Cons., St. Paul., 1981-85, People, Inc., Mpls., 1985., NIH, 1986, Emerson Sch., 1986-87, Alternative Intervaention Cons., Inc., Mpls., 1990-91, U. Tex. Mental Scis. Inst., 1991—, Livonia (Mich.) Pub. Schs., 1992; invited speaker in field including Oxford (U.K.) U., 1968, Mario Negri Inst., Milan, Italy, 1969, York U., Toronto, Can., 1971, Uppsala U., Sweden, 1972, U. Auckland, New Zealand, 1974, , European Behaviour Therapy Assn. Meeting, Uppsala, 1977, Bermuda Coll., 1983, Maudsley Hosp., U. London, 1985, European Behavioral Pharmacology Soc. Meeting, Antwerp, Belgium, 1986, Amsterdan, The Netherlands, 1990, Rehab. Inst., Dalhousie U., Halifax, Nova Scotia, Can., 1989, U. Otago, New Zealand, 1992. Author: Behavioral Pharmacology, 1964, (with J.P. Grabowski) Reinforcement Schedules and Multi-operant Analysis, 1972; editor: Behavioral Model of Mental Retardation, 1972, 2nd edit., 1977, Saving Children at Risk, 1992, (with Peter B. Dews and James Barrett) Advances in Behavioral Pharmacology, numerous others; regional editor Pharmacology Biochemistry & Behavior, 1974-88; mem. editorial bd. Psychopharmacologia, 1967-78, Applied Rsch. in Mental Retardation, 1978-89, Mental Retardation, 1978-88, Analysis and Intervention in Devel. Diabilities, 1981-88, Jour. for Analysis and Modification of Behavior (Italy), 1982-87, The Behavior Analyst, 1984-90, Jour. Spl. Edn., 1992—, Exptl. and Clin. Psychopharmacology, 1992—; guest reviewer Jour. Exptl. Analysis of Behavior, 1965—, Jour. Applied Behavioral Analysis, 1968—, Behavior Therapy, 1972-80, Jour. Pharmacology and Exptl. Therapeutics, 1966-80, Psychopharmacology, 1978-88; author ednl. films. Predoctoral fellow USPHS, 1959-61. Fellow APA (divsns. exptl. analysis behavior, history psychology, pres. divsn. psychopharmacology 1974, divsn. mental retardation, 1990, com. n ethics in protection of human participants in rsch. bd. sci. affairs 1988-91, congl. testifier 1988, 89, 90, 92, Don Hake award 1990), Coll. on Problems Drug Abuse (charter); mem. Am. Acad. Mental Retardation (state chpt. award 1981), Am. Assn. Univ. Affilited Programs (exec. bd. 1989—), Assn. for Advancement Behavior Therapy (mem.-at-large 1971-72, pres. Minn. chpt. 1972-73, chair profl. affairs ethics com. 1973-81), Am. Colhopharmacology (sci. assoc. 1979-81), Am. Assn. for Behavioral Analysis (sustaining, program chair for behavioral pharmacology and toxicology 1984, 85, editorial adv. com. 1986-90), Am. Acad. Pediatrics (com. on DSM IV PC), Nat. Assn. for Dual Diagnosis, Behavioral Pharmacology Soc. (pres. 1972-73), European Behavioral Pharmacology Soc., Twin Cities Soc. for Children and Adults with Autism (hon., bd. profl. advisors 1982-91). Achievements include co-discovery of technique for screening abuse liability of new drugs. Office: Vanderbilt U John F Kennedy Ctr Box 40 GPC Nashville TN 37203

THOMPSON, WARREN S., dean, academic administrator; b. Utica, Miss., Aug. 19, 1929; s. John Edwin Thompson and Collise (Kitchens) Gibson; m. Marilyn Burney, July 18, 1953; children: Ruth, Lydia, John, Dee. BS, Auburn U., 1951, MS, 1955; PhD, N.C. State U., 1960. Asst. forester Miss. Agr. Experiment Sta., Mississippi State, 1953-54, assoc. dir., 1983—; grad. asst. Auburn (Ala.) U., 1954-55, N.C. State U., Raleigh, 1955-57; wood technologist Masonite Corp., 1957-59; asst. prof. La. State U., Baton Rouge, 1959-63, assoc. prof., 1963-64; dir., prof. Forest Products Lab., Mississippi State, 1964—; dean Sch. Forest Resources Miss. State U., Mississippi State, 1983—; mem. USDA Coop. Forestry Rsch. Coun., McIntire Stennis Coun. Contbr. articles to profl. jours. 1st lt. U.S. Army, 1951-53, Korea. Mem. Am. Wood Preservers Assn. (chmn. rsch. com.), Nat. Assn. Profl. Forestry Schs. and Colls. (chmn. nat. com., pres.), Miss. Forest Assn. (Meritorious Svc. award), Forest Products Rsch. Soc. (Disting. Svc. award mid-south region, co-chmn. Nat. Planning Group, regional bd. dirs.), Soc. Wood Sci. and Tech. (chmn. accreditation com., pres.), Nat. Assn. State Univs. and Land Grant Colls. (chmn. commn. on forestry), N.C. State U. Alumni Assn. (Disting. alumnus award). Republican. Avocations: reading, gardening, fishing. Office: Miss State U Sch Forest Resources 104 Dorman Hall PO Drawer FP Mississippi State MS 39762-9999

THOMPSON, WILLIAM BENBOW, JR., obstetrician/gynecologist, educator; b. Detroit, July 26, 1923; s. William Benbow and Ruth Wood (Locke) T.; m. Constance Carter, July 30, 1947 (div. Feb. 1958); 1 child, William Benbow IV; m. Jane Gilliland, Mar. 12, 1958; children: Reese Ellison, Belinda Day. AB, U. So. Calif., 1947, MD, 1951. Diplomate Am. Bd. Ob-Gyn. Resident Gallinger Mun. Hosp., Washington, 1952-53; resident George Washington U. Hosp., Washington, 1953-55; asst. ob-gyn. La. State U., 1955-56; asst. clinical prof. UCLA, 1957-64; assoc. prof. U. Calif.-Irvine Sch. Med., Orange, Calif., 1964-92; dir. gynecology U. Calif.-Irvine Sch. Med., 1977-92; prof. emeritus U. Calif.-Irvine Sch. Med., Orange, 1993—; vice chmn. ob-gyn. U. Calif.-Irvine Sch. Med., 1978-89; assoc. dean U. Calif.-Irvine Coll. Med., Irvine, 1969-73. Inventor: Thompson Retractor, 1976; Thompson Manipulator, 1977. Bd. dirs. Monarch Bay Assn. Laguna Niguel, Calif. 1969-77, Monarch Summitt II A ssn. 1981-83. With U.S. Army, 1942-44, PTO. Fellow ACS, Am. Coll. Ob-Gyn., L.A. Ob-Gyn. Soc. (life); mem. Orange County Gynecology and Obstetrics Soc. (hon.), Am. Soc. Law and Medicine, Capistrano Bay Yacht Club (commodore 1975), Internat. Order Blue Gavel. Avocation: boating. Office: UCI Med Ctr OB/GYN 101 City Blvd W Orange CA 92668-2901

THOMPSON, WILLIAM CHARLES, civil engineer; b. Wausau, Wis., Feb. 2, 1954; s. William Joseph and Bernice Lucille (Willert) T.; m. Jacalyn Leigh Luedtke, June 12, 1976. BS in Civil and Environ. Engring., U. Wis., 1977. Registered profl. engr., Nebr. Staff engr. Union Pacific R.R. Co., Omaha, 1978-80, resident engr., 1980-81; roadmaster Union Pacific R.R. Co., Spokane, Wash., 1981-82; asst. divsn. engr. Union Pacific R.R. Co., North Platte, Nebr., 1983-84; divsn. engr. Union Pacific R.R. Co., Kansas City, Mo., 1984-88; dir. methods and rsch. Union Pacific R.R. Co., Omaha, 1989—; asst. mgr. test ops. Assoc. of Am. R.R., Pueblo, Colo., 1982-83. Contbr. articles to profl. jours. Treas. pack 435 Boy Scouts Am., Omaha, 1990-91, asst. scoutmaster, 1991—. Mem. ASCE, Am. Rlwy. Engring. Assn. (chmn. com. 1987-91), Am. Roadmasters and Maintenance of Way Assn., Am. Bridge and Bldg. Assn., Transp. Rsch. Bd. (chmn. com. 1992—). Achievements include patent (with other) for track fastening device. Office: Union Pacific RR Co 1416 Dodge St MC 3300 Omaha NE 68179

THOMSON, CYNTHIA ANN, clinical nutrition research specialist; b. Princeton, N.J., Nov. 9, 1957; d. John Burnham Miner and Sally Ann (Tollerton) Thompson; m. Robert James Thomson, Jan. 12, 1985; children: Daniel James, Patrick Cary. BS, W.Va. U., 1980; MS, U. Ariz., 1987. Registered Dietitian; Cert. Nutrition Support Dietitian. Sr. clin. dietitian Kino Hosp., Tucson, 1980-86; clin. dietitian U. Med. Ctr., Tucson, 1986-90, chief clin. dietitian, 1990-91; program devel., cons. El Rio Health Ctr., Tucson, 1992—; program coord. nutrition in med. edn. U. Ariz., Tucson, 1991—; exec. bd. Ariz. Dietetic Assn., 1984—; expert reviewer Dietitians in Nutrition Support, Chgo., 1990—; del. Am. Dietetic Assn. Ariz., 1990—. Editor: Arizona Diet Manual, 1990. Vol. educator Tucson AIDS Project, 1990—. Named Young Dietitian of Yr., Ariz. Dietetic Assn., 1986. Mem. Am. Dietetic Assn., Am. Soc. Parenteral and Enteral Nutrition, N.Y. Acad. Scis., Soc. for Tchrs. of Family Medicine. Home: 3003 N Conestoga Ave Tucson AZ 85749 Office: Univ Arizona Coll of Medicine 2231 E Speedway Blvd Tucson AZ 85719

THOMSON, GRACE MARIE, nurse, minister; b. Pecos, Tex., Mar. 30, 1932; d. William McKinley and Elzora (Wilson) Olliff; m. Radford Chaplin, Nov. 3, 1952; children: Deborah C., William Earnest. Assoc. Applied Sci., Odessa Coll., 1965; extension student U. Pa. Sch. Nursing, U. Calif., Irvine, Golden West Coll. RN, Calif., Okla., Ariz., Md., Tex. Dir. nursing Grays Nursing Home, Odessa, Tex., 1965; supr. nursing Med. Hill, Oakland, Calif.; charge nurse pediatrics Med. Ctr., Odessa; dir. nursing Elmwood Extended Care, Berkeley, Calif.; surg. nurse Childrens Hosp., Berkeley; med./surg. charge nurse Merritt Hosp., Oakland, Calif.; adminstr. Grace and Assocs.; advocate for emotionally abused children; active Watchtower and Bible Tract Soc.; evangelist for Jehovah's Witnesses, 1954—.

THOMSON, JOHN ANSEL ARMSTRONG, biochemist; b. Detroit, Nov. 23, 1911; s. John Russell and Florence (Antisdel) T.; m. June Anna Mae Hummel, June 24, 1938; children: Sheryll Linn, Patrisha Diane, Robert Royce. AA, Pasadena (Calif.) City Coll., 1935; AB cum laude, U. So. Calif., 1957; BGS (hon.), Calif. Poly. State U., 1961; MA, PhD, Columbia Pacific U., 1978, 79; DA, Internat. Inst. Advanced Studies, Clayton, Mo., 1979. Cert. secondary tchr., Calif. Chemist J.A. Thomson Bio-Organic Chemist, L.A., 1938, Vitamin Inst. (formerly J.A. Thomson Bio-Organic Chemist), L.A. and North Hollywood, Calif., 1939—; vocat. edn. instr. U.S. War Manpower Commn., 1943-44; chmn. activities coun. World Coun. of Youth, L.A., 1932; pres. Coun. of Young Men's Divs. Athletic Commns., YMCA Pasadena area, 1931, chmn. exec. coun., 1932; dist. officer Boy Scouts Am., San Fernando Valley coun., 1954-64, del. to nat. conf., 1959, and others. Author: (booklets) Whose Are the Myths?, 1949, Open Eyes, Illegaliza Agency Abuses, 1968, Non-toxic Vitamins-hormones Answers to Environmental, Public Problems, 1972, Lobby Interest Goals to Sequester Nutrients Among Those Rarely Educated in Them, 1973, Support of Pressures to Homeostasis, Normality, 1990, Minimization of Toxics in Agriculture, 1991; contbr. articles to jours. Prog. leader United Meth. Men Quadrennial Conf., Lafayette, Ind., 1982, instr. United Methodist Ch. nat. seminar for profls., Nashville, Tenn., 1983, pres. United Meth. Men, 1979-80, mem. adminstv. bd., 1952—, chmn. commn. ch. and soc., 1986—, First United Meth. Ch., North Hollywood; mem. Rep. county ctrl. Com. L.A. County, 1941-50, chmn. 63d assembly dist., 1948-50, Rep. state ctrl. com., Calif. 1948-50. Recipient Sci. and Industry award San Francisco Internat. Expn., 1940, various scouting leadership awards Boy Scouts Am., Civic Svc. award State of Calif., 1949, others. Mem. AAAS (life), Am. Inst. Biol. Scis., Soc. Nutrition Edn., N.Y. Acad. Scis., Am. Horticultural Soc., Am. Chem. Soc. (So. Calif. sect.), Internat. Acad. Nutrition and Preventive Medicine, Western Gerontol. Assn., Am. Forestry Assn., Am. Assn. Nurserymen, Soc. Am. Florists and Ornamental Horticulturists, Internat. Soc. Hort. Sci., Nat. Coun. Improved Health, Nat. Recreation and Parks Assn., Nat. Landscape Assn., Nat. Nutritional Foods Assn. (Pioneer Svc. award 1970), Nat. Health Fedn., Sierra Club, Soc. Colonial Wars (life), Kiwanis (projects panelist internat. confs. 1987, 91), numerous others. Republican. Achievements include origination of a high proportion of known uses for horticultural hormones, with first products, many of them via solely-invented and produced Horms 4, 1 and 2, Superthrive, Cutstart and Seedyield, multiple vitamins-hormones, distributed world wide; development of highest known efficacies in plant activating reviving, transplanting, growing, perfecting, rooting and seed invigoration; creation of water-miscible multiple vitamins powder Auzon, for humans creation of over 100 other formula products. Office: Vitamin Inst PO Box 230 5411 Satsuma Ave North Hollywood CA 91603-0230

THOMSON, N. R., civil engineering educator. Prof., civil engring. U. Waterloo, Ont., Can. Recipient Horst Leiphole medal Can. Soc. Civil Engring., 1992. Office: U Waterloo, Dept Civil Engineering, Waterloo, ON Canada N2L 3G1*

THOMSON, STUART MCGUIRE, JR., science educator; b. Rocky Mount, N.C., May 14, 1945; s. Stuart McGuire and Sarah Stilly (McLean) T.; m. Betty Jean Klapp, Mar. 8, 1986. BA in Psychology, N.C. State U., 1969, BS in Zoology, 1972; MS in Biology, East Carolina U., 1986. Cert. tchr., N.C. Tchr. sci. John Graham High Sch., Warrenton, N.C., 1968-70, Chesterfield (S.C.) High Sch., 1970-71, Pungo Christian Acad., Belhaven, N.C., 1989-91, Beaufort County C.C., Washington, N.C., 1986—; disability determination specialist N.C. Dept. Human Resources, Raleigh, 1973-79; mgr. computer div. Thomson TV Co., Washington, 1979-84; various part-time positions East Carolina U., Greenville, N.C., 1985-89. Contbr. articles to Alcohol and Jour. of the Elisha Mitchell Sci. Soc. Mem. AAAS, N.C. Acad. Sci., N.Y. Acad. Scis., Washington (N.C.) Community Band, Sigma Xi (assoc.). Presbyterian. Achievements include invention of clinical test that uses one urine sample to indicate a predisposition towards alcoholism. Home: PO Box 401 Washington NC 27889-0401 Office: Beaufort County CC PO Box 1069 Washington NC 27889

THONNARD, ERNST, internist, researcher; b. Aachen, Germany, May 19, 1898; s. Jean and Anna (Schoddart) T.; m. Constanza Marotti, Jan. 24, 1932; children: Claudia, Ingrid, Norbert. MD, U. Frankfurt, Germany. Intern Mcpl. Hosp., Frankfurt-Hoechst, Germany, 1923, asst., 1924-26; extern Inst. of Tropical Diseases, Hamburg, Germany, 1926-27; dozent U. Berlin, Germany, 1942-48; asst. United Fruit Co. Hosp., Panama, 1928-29; asst. supt. United Fruit Co. Hosp., Santa Marta, Colombia, 1929-32; pvt. practice, rsch. on nutrition Barranguilla, Colombia, 1948-58; guest sci. St. Elizabeth Hosp., Washington, 1962—; dir. Inst. Nutrition Universidad del Atlantico, Barrianguilla, Colombia, 1956-58. Contbr. articles to profl. jours. NIH grantee, 1964. Mem. AMA, Sigma Xi. Roman Catholic. Home: 800 4th St SW Washington DC 20024 Office: St Elizabeth Hosp Martin Luther King Ave Washington DC 20032

THORARENSEN, ODDUR C.S., pharmacist; b. Reykjavik, Iceland, Apr. 26, 1925; s. Stefan and Ragnheidur (Hafstein) T.; m. Asta Baldvinsdottir, Oct. 24, 1948; (div. Aug. 1956); children: Stefan, Baldvin (Hafsteinn); m. Unnur Arny Long Thorarensen, Nov. 27, 1958; children: Ragnheidur Katrin, Sigridur Elin, Unnur Alma. Grad. diploma, Reyjavik Latin Sch., Iceland, 1945; examinatus Pharmaciae, U. Iceland, 1948; candidatus Pharmaciae, Royal Danish Coll. Pharmacy, Copenhagen, Denmark, 1949, 53, 54; student, P.C.P. & S., Phila., 1950. Proprietary Pharmacist, 1963. Asst. pharmacist Laugavegs Apotek, Reykjavik, Iceland, 1950-51, chief pharmacist, 1954-57, owner, mgr., 1963—; gen. mgr. Efnagerd Reykjavikur Ltd., Reykjavik, Iceland, 1957-69; mng. dir. Torence Ltd., Iceland, 1981—. Contbr. articles to profl. jours. Chmn. and diverse other positions, Independence Party, Local Orgn., Gardabaer, Reykjavik, Iceland, 1980-90; mem. Diverse Comnx., Min. Health, Reykjavik, Iceland, 1970—, Parish Bd. Gardar Ch., Gardabaer, Reykjavik, Iceland, 1972-84. Mem. Iceland C. of C. (bd. electors 1974—), Gimli Lodge, Masonic Nat. Grand Lodge Iceland, Iceland Pharmacist Union (sec. 1955-56, chmn. 1956-57), Iceland Proprietary Pharmacist Assn. (sec. 1973-74, 77-80), Reykjavik Proprietary Pharmacist Assn. (chmn. 1974-77, sec. 1977-78). Lutheran. Avocations: reading, jazz music, freemasonry, science fiction, space exploration, computers. Office: Laugavegs Apotek, 16 Laugavegi PO Box 477, 121 Reykjavik Iceland

THORBURN, JOHN THOMAS, III, marine engineer, consultant; b. Boston, May 23, 1920; s. John and Ellen Louise (Doddy) T.; m. Doris Louise Kenney, June 16, 1951; children: Debra, John IV. PhD in Marine Engring. and Naval Architecture, MIT, 1968. Pres. Thorburn Marine, Boston, 1943-92; pvt. practice cons. Boston, 1992—. Home and Office: 20 Allston St Boston MA 02124-2223

THORGEIRSSON, GUDMUNDUR, physician, cardiologist; b. Djupavik, Strandasysla, Iceland, Mar. 14, 1946; s. Thorgeir Gestsson and Asa Gundmundsdottir; m. Bryndis Sigurjonsdottir, July 20, 1968; children: Thorgeir, Sigurjon Arni, Hjalti, Bogi, Asa Bryndis. MD, U. Iceland, 1973; PhD, Case Western Res. U., 1978. Diplomate Am. Bd. Internal Medicine, Am. Bd. Cardiovascular Disease. Intern., resident path. medicine U. Hosps., Cleve., 1974-80, teaching fellow, 1980-82; cardiologist Landspitalinn U. Hosp., Reykjavik, Iceland, 1982—; assoc. prof. U. Iceland, 1985—; cons. Heart Prevention Clinic, Reykjavik, 1982—; bd. dirs. Icelandic Heart Assn., Reykjavik; chmn. Icelandic Nutrition Bd., 1990—. Editor Icelandic Med. Jour., 1983-91; contbr. articles to profl. jours. Icelandic Sci. Fund grantee, 1983-89. Fellow Am. Coll. Cardiology (assoc.), Soc. Scientarum Islandica; mem. Icelandic Med. Assn., Icelandic Cardiological Soc. (pres. 1985-87), ACP. Avocations: books, Icelandic medical history, skiing. Home: Klapparas 4, Reykjavik Iceland Office: Landspitalinn U Hosp, Reykjavik Iceland

THORGEIRSSON, SNORRI SVEINN, physician, pharmacologist; b. Iceland, Dec. 1, 1941; came to U.S., 1972, naturalized, 1980; d. Thorgeir Jonsson and Sigurlina Sigujonsdottir; M.D., U. Iceland, 1968; Ph.D., U. London, 1971; m. Unnur Thorgeirsson, Sept. 5, 1969; children—Sif, Christian. Intern, Univ. Hosp., Reykjavik, Iceland, 1968-69; registrar, research fellow dept. clin. pharmacology Royal Postgrad. Med. Sch., London, 1969-71; vis. fellow Lab. Chem. Pharmacology, Nat. Heart and Lung Inst., NIH, Bethesda, Md., 1972-73, vis. scientist sect. devel. pharmacology, Neonatal and Pediatric Medicine br. Nat. Inst. Child Health and Human Devel., 1974-75, chief sect. on molecular toxicology devel. pharmacology br., 1975-76; head biochem. pharmacology sect. Lab. Chem. Pharmacology, Nat. Cancer Inst., 1976-81, chief Lab. Carcinogen Metabolism, 1981—; mem. Chem. Selection Working Group, 1978—; mem. Com. on Occupational Carcinogenesis, 1979; mem. com. on amines Nat. Acad. Scis., 1979-80; co-chmn. Internat. Conf. on Carcinogenic and Mutagenic N-Substituted Aryl Compounds, NIH, Bethesda, 1979; preceptor Pharmacology Research Assoc. Program, Nat. Inst. Gen. Med. Scis., 1977—; mem. biol. response modifiers decision network com. Nat. Cancer Inst., 1980; lectr. in field. Mem. Am. Assn. Cancer Research, AAAS, Am. Soc. Exptl. Pharmacology and Exptl. Therapeutics, Am. Chem. Soc., N.Y. Acad. Scis., Environ. Mutagen Soc., Soc. Toxicology, European Assn. Cancer Research. Contbr. numerous articles, chpts. to profl. publs.; research in mechanisms of chem. carcinogenesis, control of differentiation in neo-plastic cells. Office: National Cancer Institute Cancer Etiology 9000 Rockville Pike, Bldg 31 Bethesda MD 20892

THORN, JAMES DOUGLAS, safety engineer; b. Tyler, Tex., May 20, 1959; s. Douglas Howard and Patricia Ann (Kolb) T.; m. Tarry Annette Wolfe, Aug. 2, 1991; children: Brandon, Alex. Student, U. of Mary, Manama, Bahrain, 1982, S.W. Tex. State U., 1984-86, La. State U., 1989, W.Va. Tech., 1991-92. Cert. EMT, basic trauma life support, advanced cardiac life support, hazardous materials ops., hazardous waste ops., AHA/CPR instr. 3d officer Jackson Marine S.A., Manama, 1981; constrn. foreman Brown & Root S.A., Manama, 1982-83; barge officer Rezayat/Brown & Root E.C., Manama, 1983-84; safety insp. Brown & Root U.S.A., Carson, Calif., 1987-88; sr. safety insp. Brown & Root U.S.A., Taft, La., 1988-89; project safety mgr. Brown & Root Braun, Institute, W.Va., 1989-93; mgr. safety and health Brown & Root Braun, Phila., 1993—; safety cons. Assn. Builders and Contractors, Charleston, W.Va., 1990-93, mem., 1989-93, chmn. safety seminar, 1991-93; drill monitor Kanawha Valley Emergency Preparedness Coun., South Charleston, W.Va., 1990-93. Youth counsellor Neon League, St. Albans, W.Va., 1991; den leader cub scouts Boy Scouts Am., 1991—. Mem. Am. Soc. Safety Engrs., Nat. Assn. EMTs, Team 911, Great Wall of Tex. Soc. Avocations: golf, snow skiing, scuba diving. Office: Brown & Root Braun PO Box 60597 Philadelphia PA 19145

THORN, ROBERT JEROME, chemist; b. Constantine, Mich., Dec. 1, 1914; s. Edward R. and Gladys R. (Hassinger) T.; m. Mary C. Craig, Aug. 15, 1942 (dec. July 1992); children: Craig E., Susan J. (dec.). BS, Alma Coll., 1938; PhD, U. Ill., 1942. Chemist U.S. Rubber Co., Detroit, 1942-46; chemist Argonne (Ill.) Nat. Lab., 1946-48, assoc. chemist, 1948-58, sr. chemist, 1958-84, rsch. assoc., 1984—; vis. prof. U. Kans., Lawrence, 1967-68; cons. Internat. Atomic Energy Agy., Vienna, Austria, 1961-65; chmn. Gordon Rsch Conf. on High Temperature Chemistry, 1970. Contbr. over 80 articles to profl. publs., 5 chpts. to books. Fellow AAAS; mem. Am. Chem. Soc. (emeritus), Sigma Xi. Achievements include rsch. on correlation of energy and entropy, nonstorichiometry in oxides, conductivity of normal state of oxide and organic superconductors. Home: 4622 Douglas Rd Downers Grove IL 60515 Office: Argonne Nat Lab 9700 S Cass Ave Argonne IL 60439

THORNDIKE, ROBERT MANN, psychology educator; b. Washington, Mar. 2, 1943; s. Robert Ladd and Dorothy Vernon (Mann) T.; m. Elva Stewart, Dec. 21, 1963; children: Tracy Kathleen Thorndike-Christ, Kristi Ann. BA, Wesleyan U., 1965; PhD, U. Minn., 1970. From asst. to assoc. prof. Western Wash. State Coll., Bellingham, 1970-77, prof., 1977—. Author: (with R.W. Brislin and W.J. Lonner) Cross-Cultural Research Methods, 1973, Correlational Procedures for Research, 1978, Data Collection and Analysis: Basic Statistics, 1982 (with D. Lohman) A Century of Ability Testing, 1990, (with G.K. Cunningham, R.L. Thorndike, E.P. Hagen) Measurement and Evaluation in Psychology and Education, 1991; contbr. chpts. to books: Measurement and Evaluation in Rehabilitation, 1976, 87, Foundations of Contemporary Psychology, 1979; author numerous revs. Mem. AAUP (exec. bd. dirs. western Wash. U. chpt. 1972-77, v.p. 1977-78, pres. 1978-82, treas. Wash. state conf. 1976-79, v.p. 1980-81, pres. 1981-83, sec. nat.assembly state confs. 1982-84, vice chair 1985-87), Am. Ednl. Rsch. Assn. (divsn. D - measurement and stats.), Psychometric Soc., Soc. Multivariate Experimental Psychology, Am. Psychol. Assn., Am. Psychol. Soc., Nat. Power Squadron (com. on instr. qualification 1982-86, teaching aids com. 1986-91), Bellingham Yacht Club. Episcopalian. Avocations: sailing, golf, tennis, military history. Office: Western Wash U Dept Psychology Bellingham WA 98225-9089

THORNE, KIP STEPHEN, physicist, educator; b. Logan, Utah, June 1, 1940; s. David Wynne and Alison (Comish) T.; m. Linda Jeanne Peterson, Sept. 12, 1960 (div. 1977); children: Kares Anne, Bret Carter; m. Carolee Joyce Winstein, July 7, 1984. B.S. in Physics, Calif. Inst. Tech., 1962; A.M. in Physics (Woodrow Wilson fellow, Danforth Found. fellow), Princeton U., 1963, Ph.D. in Physics (Danforth Found. fellow, NSF fellow), 1965, postgrad. (NSF postdoctoral fellow), 1965-66; D.Sc. (hon.), Ill. Coll., 1979; Dr.h.c., Moscow U., 1981. Research fellow Calif. Inst. Tech., 1966-67, assoc. prof. theoretical physics, 1967-70, prof., 1970—, William R. Kenan, Jr. prof., 1981-91, Feynman prof. theoretical physics, 1991—; Fulbright lectr., France, 1966; vis. assoc. prof. U. Chgo., 1968; vis. prof. Moscow U., 1969, 75, 78, 82, 83, 86, 88, 90; vis. sr. rsch. assoc. Cornell U., 1977, A.D. White prof.-at-large, 1968-92; adj. prof. U. Utah, 1971-92; mem. Internat. Com. on Gen. Relativity and Gravitation, 1971-80, Com. on U.S.-USSR Coop. in Physics, 1978-79, Space Sci. Bd., NASA, 1980-83. Co-author: Gravitation Theory and Gravitational Collapse, 1965, Gravitation, 1973, Black Holes: The Membrane Paradigm, 1986. Alfred P. Sloan Found. Research fellow, 1966-68; John Simon Guggenheim fellow, 1967; recipient Sci. Writing award in physics and astronomy Am. Inst. Physics-U.S. Steel Found., 1969. Mem. Nat. Acad. Scis., Am. Acad. Arts and Scis., Am. Astron. Soc., Am. Phys. Soc., Internat. Astron. Union, AAAS, Sigma Xi, Tau Beta Pi. Achievements include research in theoretical physics and astrophysics. Office: Calif Inst Tech 130-33 Pasadena CA 91125

THORNTON, EARL ARTHUR, mechanical and aerospace engineering educator; b. Portsmouth, Va., Oct. 7, 1936; s. John S. and Dora James (Weatherly) T.; m. Carrie Margaret Bailey, July 21, 1962; children: Laura Ann, Charles Steven. BS, Va. Poly. Inst., 1959, PhD, 1968; MS, U. Ill., 1961. Registered profl. engr., Va. Rsch. engr. Newport News Shipbldg., 1961-62, David Taylor Naval Ship R&D Ctr., Portsmouth, 1962-69; educator Old Dominion U., Norfolk, Va., 1969-87; vis. scholar U. Tex., Austin, 1987-89; vis. prof. Tex. A&M U., College Station, 1989; prof. mech. and aerospace engring. U. Va., Charlottesville, 1989—, dir. Light Thermal Structures Ctr., 1989—. Assoc. editor AIAA Jour. Spacecraft and Rockets; co-author: The Finite Element Method for Engineers, 1982; contbr. articles to profl. jours. Recipient NASA grants, NASA cash awards. Assoc. fellow AIAA; mem. Tau Beta Pi. Achievements include research in the finite element method for engrs., heat transfer and thermal stress in aerospace structures. Office: U Va Dept Mech Aerospace & Nuclear Engring Thornton Hall Charlottesville VA 22903

THORNTON, J. RONALD, technologist; b. Fayetteville, Tenn., Aug. 19, 1939; s. James Alanda and Thelma White (McGee) T.; m. Mary Beth Packard, June 14, 1964 (div. Apr. 1975); 1 child, Nancy Carole; m. Martha Klemann, Jan. 23, 1976 (div. Apr. 1982); 1 child, Trey; m. Bernice McKinney, Feb. 14, 1986; 1 child, Paul Leon. BS in Physics & Math., Berry Coll., 1961; MA in Physics, Wake Forest Coll., 1964; postgrad., U. Ala., 1965-66, Rollins Coll., 1970. Research physicist Brown Engring. Co., Huntsville, Ala., 1963-66; sr. staff engr. Martin Marietta Corp., Orlando, Fla., 1966-75; dep. dir. NASA, Washington, 1976-77; exec. asst. Congressman Louis Frey, Jr., Orlando, 1978; pres. Tens Tec, Inc., Orlando, 1978-79; dir. So. Tech. Applications Ctr. U. Fla., Gainesville, 1979—; mem. Light Wave Tech. Com., Fla. High Tech. and Indsl. Coun., Tallahassee, 1986—, NASA Tech. Transfer Exec. Com., Washington, 1987—, Javits Fellowship Bd., Washington, 1986-91, Gov.'s New Product Award Com., Tallahassee, 1988—, Fla. K-12 Math. Sci. and Computer Sci. Edn. Quality Improvement Adv. Coun., 1989-92, Fla. Sci. Edn. Improvement Adv. Com., 1991—. Pres. Orange County Young Rep. Club, Orlando, 1970-71; treas. Fla. Fedn. Young Reps., Orlando, 1971-72; chmn. Fla. Fedn. Young Reps., Orlando, 1972-74; pres. Gainesville Area Innovation Network, 1988-89. Named Engr. Exhibiting Tech. Excellence and Accomplishment cen. Fla. chpt. Fla. Engring. Soc., 1975, Achievement award NASA, 1977. Mem. IEEE, Tech. Transfer Soc., Nat. Assn. Mgmt. and Tech. Assitance Ctrs. (bd. dirs. 1988, pres. 1992), Gainesville Area C. of C. Republican. Avocations: music, travel, reading. Home: RR 2 Box 740 Newberry FL 32669-9627 Office: STAC 1 Progress Blvd Box 24 Alachua FL 32615

THORNTON, JOSEPH SCOTT, research institute executive, materials scientist; b. Sewickley, Pa., Feb. 6, 1936; s. Joseph Scott and Evelyn (Miller) T.; divorced; children: Joseph Scott III, Chris P. BSME, U. Tex., 1957, PhD, 1969; MSMetE, Carnegie Mellon U., 1962. Engr. Walworth Valve Co., Boston, 1958; metall. engr. Westinghouse Astronuclear Lab., Large, Pa., 1962-64; instr., teaching assoc. U. Tex., Austin, 1964-67; group leader Tracor Inc., Austin, 1967-69, dept. dir., 1973-75; dept. mgr. Horizons Rsch., Inc., Cleve., 1969-73; chmn., chief exec. officer Tex. Rsch. Internat., Inc. (formerly Tex. Rsch. Inst., Inc.), Austin, 1975—. Contbr. numerous tech. papers to profl. publs.; editor: WANL Materials Manual, 2 vols., 1964; patentee in field. Fellow Alcoa, Austin, 1964, RC Baker Found., 1967. Mem. ASME, ASTM, Am. Soc. Metals Internat. (exec. com. 1965-66), Adhesion Soc. Office: Tex Rsch Internat Inc 9063 Bee Caves Rd Austin TX 78733-6201

THORNTON, KATHRYN C., physicist, astronaut; b. Montgomery, Ala., Aug. 17, 1952; d. William C. and Elsie Cordell; m. Stephen T. thornton; children: Carol Elizabeth, Laura Lee, Susan Annette. BS in Physics, Auburn U., 1974; MS in Physics, U. Va., 1977, PhD, 1979. NATO postdoctoral fellow Max Planck Inst. Nuclear Physics, Heidelberg, Fed. Republic Germany, 1979-80; physicist U.S. Army Fgn. Sci. & Tech. Ctr., Charlottesville, Va., 1980-84; with NASA, 1984—; astronaut Lyndon B. Johnson Space Ctr. NASA, Houston, 1985—, mission specialist Space Shuttle Discovery flight STS-33, 1989; aboard maiden flight Space Shuttle Endeavor, 1992. Mem. AAAS, Am. Phys. Soc., Sigma Xi, Phi Kappa Phi. Address: NASA Johnson Space Ctr Astronaut Office Houston TX 77058

THORNTON, PETER BRITTIN, mechanical engineer; b. N.Y., Apr. 18, 1949. BSME, Columbia U., 1971, MBA, 1977. V.p. Stern & Stern Ind., Inc., N.Y.C., 1978-82; pres. Stern & Stern Ind. Inc., N.Y.C., 1982—. Lt. USN, 1971-75. Mem. ASTM (chmn. subcom. D13.20 on inflatable restraints). Achievements include patents for directional electrostatic dissipating fabric and method, electrostatic dissipating fabric, peel ply material, low permeability fabric, low permeability fabric for automotive airbags and method for making same.

THORNTON, STAFFORD EARL, civil engineering educator; b. Campbell County, Va., July 29, 1934; s. Carlton Wilson and Bessie (Thornton) Rives; m. Frances Carolyn Umberger, Sept. 23, 1955 (dec. Dec. 1960); 1 child, Suzanne Elizabeth Thornton Garrett; m. Josephine Edmonds Whittle, Jan. 19, 1963; children William Stafford, Rives Whittle. BCE, U. Va., 1959, MCE, 1962. Registered profl. engr. Va., W.Va. Instr. U. Va., Charlottesville, 1960-61, rsch. scientist, 1962-64; from asst. prof. to prof. W.Va. Inst. Tech., Montgomery, 1964—; mem. W.Va. Registration Bd. for Profl. Engrs., Charleston, 1980-90. City engr. City of Montgomery, 1972—. With U.S. Army, 1954-56. Mem. ASCE (dir., v.p. 1978-84, pres.-elect 1993—), United Engring. Trustees, Inc. (treas. 1991-93). Office: WVa Inst Tech Tech Assistance Ctr Montgomery WV 25136

THORNTON, THOMAS ELTON, psychologist, consultant; b. Stewartsville, Mo., June 16, 1924; s. John William Frank and Leota Emma (Odell) T.; m. Margaret Ester Black, Mar. 25, 1952 (div. 1978); children: Laura, Jennifer, Thomas, Michael; m. Marian Gottlieb Linoff, July 8, 1984. AB, U. Mo., 1952; BS, U. Miami, Fla., 1954; PhD, Northwestern U., 1958. Chief psychologist Lake County Mental Health Clinic, Waukegan, Ill., 1958-63; dir. tng. VA Hosp., Miami, Fla., 1963-80; chief psychologist VA Med. Ctr., Miami, 1981-84, cons., 1984—. Contbr. articles to profl. jours. With USN, 1943-46. Mem. AAAS, APA, Dade County Psychol. Assn. (bd. dirs. 1970-72), Fla. Psychol. Assn. (Outstanding Svc. to Psychology award 1985), Sigma Xi, Phi Beta Kappa. Home and Office: 7450 SW 140th Dr Miami FL 33158

THORP, JAMES HARRISON, III, aquatic ecologist; b. Kansas City, Mo., July 23, 1948; s. James Harrison Jr. and Delores Ray (Robinson) T.; m. Margaret Ellen Svoboda, Jan. 21, 1970; children: Sara Elizabeth, Zachary Ward. BA, U. Kans., 1970; MS, N.C. State U., 1973, PhD, 1975. Rsch. assoc. Savannah River Ecology Lab., Aiken, S.C., 1975-82; vis. assoc. prof. Cornell U., Ithaca, N.Y., 1982-85; dir. Calder Conservation and Ecology Ctr. Fordham U., Bronx, N.Y., 1985-88, assoc. prof. biology, 1985-88; dir. water resources lab., prof. biology U. Louisville, 1988—; environ. cons., 1982—; team mem. Regional Urban Design Assistance Team, Fargo, N.C., Moorhead, Minn., 1989. Editor: Ecology and Classification of North America Freshwater Invertebrates, 1991, Energy and Environmental Stress in Aquatic Systems, 1978; assoc. editor Freshwater Invertebrate Biology; contbr. articles to Oikos, Ecology. Mem. Mayor's Aquarium Study com., Louisville, 1989—. Mem. AAAS, Ecol. Soc. Am., N.Am. Bethological Soc., Soc. Internat. Limnology. Office: Biology Dept U Louisville Louisville KY 40292

THORPE, JACK VICTOR, energy consultant; b. Fall River, Mass., July 6, 1949; s. John V. and Belmira (Aguiar) T.; m. Patricia Ehrlich, Aug. 18, 1990. BS in Indsl. Engring., S.E. Mass. U., 1972; MBA, U. Balt., 1985. Supervisory analyst Computer Data Systems, Inc., Rockville, Md., 1979-81; sr. survey mgr. Vanguard Tech., Fairfax, Va., 1981-82; sr. rsch. analyst Applied Mgmt. Scis., Silver Spring, Md., 1982-85, 1986-87; dep. div. dir. Maxima Corp., Bethesda, Md., 1985; rsch. assoc. Sheladia Assocs., Gaithersburg, Md., 1987-88; program mgr. Z, Inc., Silver Spring, 1988—; program mgr. nuclear fuel projects Dept. Energy. Contbr. articles to profl. publs. Mem. Am. Nuclear Soc., Ops. Rsch. Soc. Am., Inst. Mgmt. Sci., Washington Ops. Rsch. Soc. (chmn. small bus. 1991-92), Delta Mu Delta, Sigma Iota Epsilon. Republican. Roman Catholic. Home: 5536 April Journey Columbia MD 21044 Office: Z Inc 8630 Fenton St Silver Spring MD 20910

THORSON, JAMES ALDEN, gerontologist, author, consultant; b. Chgo., Oct. 8, 1946; s. Louis Carlyle and Kathryn Gertrude (Christman) T.; m. Judy Rae Johnson, Jan. 22, 1966; children: Robert, Peter. MS, U. N.C., 1971; PhD, U. Ga., 1975. Asst. prof. dept. adult edn. U. Ga., Athens, 1971-77; Isaacson prof., chair dept. gerontology U. Nebr., Omaha, 1979—; expert witness in field; cons. U. Nebr. Med. Ctr., Omaha, 1987—. Author: Psychology of Aging, 1978, Spiritual Well-Being of the Elderly, 1980, Tough Guys Don't Dice, 1989, Are You Still Working at the Home?, 1992; contbr. articles to profl. jours. Recipient John Tyler Mauldin award Ga. Gerontology Soc., 1977. Fellow Gerontol. Soc. Am. Democrat. Presbyterian. Achievements include authoring scales to assess lethal behaviors, death anxiety, and sense of humor. Office: U Nebr at Omaha Dept Gerontology Omaha NE 68182-0202

THORUP, RICHARD MAXWELL, soil scientist; b. Salt Lake City, Dec. 20, 1930; s. James B. and Agnes L. (Maxwell) T.; m. Lois F. Williamson, June 21, 1957 (div. Mar. 1979); children: Teria, Terry, Troy; m. Evelyn F. McKinney, Jan. 26, 1980; children: Christine, Deanna, Margaret. AA, Chaffey Coll., Ont., Calif., 1950; BA, BYU, 1955; MS, N.C. State U., 1957; PhD, U. Calif., Davis, 1962. Agronomist Chevron Chem. Co., Phoenix, 1960-61; field agronomist Chevron Chem. Co., Fresno, Calif., 1962-66; regional agronomist Chevron Chem. Co., Ft. Madison, Iowa, 1967-74; nat. mgr. agronomy Chevron Chem. Co., San Francisco, 1975-82, chief agronomist, 1983-86; cons. Agtech, Provo, Utah, 1987—; adj. prof. BYU, Provo, 1987—; chmn. soil improvement com. Calif. Fertilizer Assn., 1980-82. Editor: Western Fertilizer Handbook, 1985, ORTHO Agronomy Handbook, 1984, Calif. Soil Testing Procedures, 1980; contbr. articles to profl. jours. Bd. dirs. Nat. Kidney Found. Utah, Provo, 1989-90. Mem. Am. Soc. Agronomy, ARCPACS (bd. dirs. 1978-81, chmn. agronomy sub board 1980-81), Soil Sci. Soc. Am. Republican. Mem. Mormon. Home and Office: AGTECH 1741 N 1500 East Provo UT 84604

THOULESS, DAVID JAMES, physicist, educator; b. Bearsden, Scotland, Sept. 21, 1934; came to U.S., 1979; s. Robert Henry and Priscilla (Gorton) T.; m. Margaret Elizabeth Scrase, July 26, 1958; children: Michael, Christopher, Helen. BA, U. Cambridge, Eng., 1955, ScD, 1985; PhD, Cornell U., 1958. Physicist Lawrence Berkeley Lab., Calif., 1958-59; rsch. fellow U. Birmingham, Eng., 1959-61, prof. math. physics, 1965-78; lectr., fellow Churchill Coll. U. Cambridge, Eng., 1961-65; prof. physics Queen's U., Kingston, Ont., Can., 1978; prof. applied sci. Yale U., New Haven, 1979-80;

prof. physics U. Wash., Seattle, 1980—. Author: Quantum Mechanics of Many Body Systems, 2d edit., 1972. Recipient Maxwell medal Inst. Physics, 1973, Holweck prize Soc. Francaise de Physique-Inst. Physics, 1980, Fritz London award for Low temperature physics, Fritz London Meml. Fund, 1984, Wolf prize in physics, 1990, Paul Dirac medal Inst. Physics, 1993; Edwin Uehling disting. scholar U. Wash., 1988—. Fellow Royal Soc., Am. Acad. Arts and Scis. Office: U Wash Dept Physics FM-15 Seattle WA 98195

THOVSON, BRETT LORIN, physicist; b. Spokane, Wash., Nov. 29, 1960; s. Virgil Dean and Carol Marie (Coyle) T. BS in Physics, Wash. State U., 1984. Radiol. control tech. Dept. of the Navy, Bremerton, Wash., 1986—. Mem. Am. Phys. Soc., Astron. Soc. of the Pacific. Home: 618 Sheridan Apt 29 Bremerton WA 98310

THRONE, JAMES EDWARD, entomologist; b. Calembrone, Italy, Sept. 10, 1954; came to U.S., 1956; s. Thomas and Varsenig (Haroian) T.; m. Aileen Frances Lynch, Aug. 7, 1977 (div. 1984); 1 child, Evan; m. Janet Helene Van Kirk, Feb. 1, 1985. MS in Entomoloy, Wash. State U., 1978; PhD in Entomology, Cornell U., 1983. R.J. Reynolds postdoctoral rsch. fellow N.C. State U., Raleigh, 1983-85; rsch. entomologist USDA Agrl. Rsch. Svc., Savannah, Ga., 1985—. Mem. Entomol. Soc. Am., Ga. Entomol. Soc., Ga. Acad. Sci., Audubon Soc. (local pres. 1991—), Sigma Xi (pres Savannah club 1990-91). Home: 120 Nassau Dr Savannah GA 31410 Office: USDA ARS 3401 Edwin St Savannah GA 31405

THROWER, KEITH JAMES, chemist; b. Norwich, United Kingdom, Feb. 11, 1941; s. James Edward and Ivy Doris (Sandle) T.; m. Susan Margaret Thurlow, Sept. 17, 1966; children: Bridgit, Andrew, Nicholas. BS with honors, Birmingham U., 1963; DPhil, Oxford U., 1966. Chartered chemist. Scientific officer agrl. rsch. coun. Inst. for Rsch. on Animal Diseases, Newbury, Berkshire, United Kingdom, 1966-69; sr. analyst Upjohn Ltd., Crawley, United Kingdom, 1969-71, mgr. internat. product R&D, 1971-82, dir. pharm. rsch. lab., 1982-88; dir. European discovery capability Upjohn Co., Brussels, 1988-90; pharm. cons. Norvicon Assocs., Brussels, 1991—. Author: Corticosteroids: Chemistry and Biochemistry, 1979. Fellow Royal Soc. Chemistry (sec. Belgium sect.); mem. Am. Chem. Soc., Licensing Execs. Soc., Drug Info. Assn., Soc. Med. Rsch., Assn. Brit. Pharm. Industry (chmn. grad. edn. 1986-88, bioavailability working group 1979-80, sci. com. 1980-88). Mem. Anglican Ch. Office: Norvicon Assocs, Ave F Roosevelt 102-BTE 13, 1330 Rixensart Belgium

THUE-HANSEN, VIDAR, physics educator; b. Jena, Germany, Jan. 2, 1944; s. Finn and Magnhild (Thue) H.; m. Ingebjørg Rudshaug, Aug. 20, 1968; children: Therese, Richard, Cecilie. Cand. real., U. Oslo, 1968. Rsch. asst. Agrl. U. Norway, As, 1968-75, assoc. prof., 1975-83, prof. physics, 1983—, chmn. dept. physics, 1980-86; bd. dirs. Agrl. U. Norway, 1984-89, 93—, chmn. edn. com., 1992—; chmn. Isotope Lab., 1990-92; vice chmn. Instrument Svc. A/S, As, 1991—. Contbr. articles to profl. jours. Mem. Norwegian Phys. Soc., Norwegian Geophys. Soc., European Geophys. Soc., Optical Soc. Am. Home: JA Hielms Gt 4, Moss Norway 1500 Office: Agrl U, Dept Agrl Engring, As Norway 1432

THUILLIER, RICHARD HOWARD, meteorologist; b. N.Y.C., N.Y., Apr. 3, 1936; s. Howard Joseph and Louise (Schilling) T.; m. Barbara Unger; children: Stephen, David, Lawrence, Daniel. BS in Physics, Fordham U., 1959; MS in Meteorology, NYU, 1963, postgrad., 1963-66. Cert. cons. meteorologist. Instr. SUNY, 1963-66; dir. of research Weather Engrs. of Panama Inc., Panama City, Rep. Panama, 1966-68; cons. Oakland, Calif., 1968—; meteorologist and chief of research and planning Bay Area Air Quality Mgmt. Dist., San Francisco, 1968-76; sr. research meteorologist SRI Internat., Menlo Park, Calif., 1976-80, Pacific Gas & Electric, San Francisco, 1980—; lectr. Hunter Coll., N.Y.C., 1965-66, U. Calif. , Berkeley, 1973-74, San Jose State U., 1984—. Served to capt. USAF, 1959-62. Mem. Am. Meteorol. Soc., (pres. Panama Canal Zone chpt. 1967-68, San Francisco Bay chpt. 1971-72, Outstanding Contributions to Advance of Applied Meteorolgy award 1993), Sigma Xi (hon.). Republican. Roman Catholic. Avocations: skiing, bowling, golf, music, art.

THUMS, CHARLES WILLIAM, designer, consultant; b. Manitowoc, Wis., Sept. 5, 1945; s. Earl Oscar and Helen Margaret (Rusch) T. B. in Arch., Ariz. State U., 1972. Ptnr., Grafic, Tempe, Ariz., 1967-70; founder, prin. I-Squared Environ. Cons., Tempe, Ariz., 1970-78; designer and cons. design morphology, procedural programming and algorithms, 1978—. Author: (with Jonathan Craig Thums) Tempe's Grand Hotel, 1973, The Rossen House, 1975; (with Daniel Peter Aiello) Shelter and Culture, 1976; contbg. author: Tombstone Planning Guide, 5 vols., 1974. Office: PO Box 3126 Tempe AZ 85280-3126

THUNING-ROBINSON, CLAIRE, oncologist; b. Cin., Nov. 17, 1945; married, 1984. BA, St. Mary-of-the-Woods Coll., 1967; MS, Nova U., 1977, PhD in Biology, 1982. Sr. rsch. assoc. St. Vincent Charity Hosp., 1969-74, rsch. assoc., 1974-87, assoc. dir., 1988-90; dir. Goodwin Inst. Cancer Rsch., Plantation, Fla., 1990—; dir. grad. studies Goodwin Inst. Cancer Rsch., 1981—. Mem. AAAS, Am. Soc. Clin. Pathology, Am. Assn. Cancer Rsch. Achievements include investigating the control of cancer using combined hyperthermia and chemotherapy; the use of oxygen immunosuppression in promoting xenogenic tumor growth and blocking autoimmune disease; research on properties of recombinant herpes virus strains.

THURM, ULRICH, zoology educator; b. Sorau, Lausitz, Germany, July 8, 1931; s. Walter and Gertrud (Steiner) T.; m. Adelheid Reeh, Aug. 22, 1964 (dec. Dec. 1972); children: Rüdiger, Agnes, Frauke; m. Barbara Althoff, Nov. 4, 1974; children: Henrike, Gundolf. Student zoology, physics and chemistry, U. Göttingen and U. Freiburg, Fed. Republic Germany, 1951-54, U. Würzburg, Fed. Republic Germany, 1954-58; D of Natural Scis., U. Würzburg, Fed. Republic Germany, 1962; student zoology, physics and chemistry, U. Munich, 1958-62. Rsch. assoc. Max-Planck-Inst. for Biol. Cybernetics, Tübingen, Fed. Republic Germany, 1962-70; prof. zoology U. Bochum, Fed. Republic Germany, 1971-74; prof., dir. dept. neurobiology U. Münster, Fed. Republic Germany, 1974—, dean Faculty Natural Scis., 1977-78, dean biol. scis., 1982-83. Contbr. numerous articles to profl. jours. Mem. Rheinisch-Westfälische Akademie Wissenschaften, Soc. for Phys. Biology (chmn. 1979-8l), also other zool., physiol. and biophys. socs. Lutheran. Avocations: music, philosophy. Office: U Münster Neurobiology Inst, Badestrasse 9, D-4400 Münster Germany

THURMAN, PAMELA JUMPER, research scientist; b. Maysville, Ark., Sept. 1, 1947; d. John Wayne Nichols Jumper and Georgia Hogner Holland; m. Gary Oliver Thurman, Sept. 5, 1971. BA in Psychology, Northeastern Okla. State U., 1982, MS in Counseling Psychology, 1983; PhD in Clin. Psychology, Okla. State U., 1990. Advocate/vol. Help-in-Crisis, Tahlequah, Okla., 1980-82; intern Grand Lake Mental health Ctr., Claremore, Okla., 1983; therapist Psychol. Svcs. Ctr., Okla. State U., Stillwater, 1984-85; rsch. asst. dept. pediatric oncology U. Okla. Health Sci. Ctr., Oklahoma City, 1983-84; rsch. assoc. dept. psychiatry and behavioral scis. Am. Indian Rsch. Project, U. Okla. Health Sci. Ctr., Oklahoma City, 1984-87; clin. intern Children's Med. Ctr., Tulsa, 1986-87; therapist United Clinics of Counseling, Tulsa, 1988-89; human svcs. dir. Indian health Care Resource Ctr., Tulsa, 1987-90; co-adminstr. Coun. Oak Rsch. and Evaluation, Inc., Loveland, Colo., 1987—; rsch. assoc. Tri-Ethnic Ctr. for Study Drug Abuse Prevention Western Behavioral Studies, Colo. State U., Ft. Collins, 1990—; adj. prof. psychology Northeastern Okla. Stte U., Tahlequah, 1986-90, Tulsa Jr. Coll., 1987-88; evaluation specialist Am. Indian Resource Ctr., Tulsa, 1987-88; lectr. in field; instl. review group Office of Substance Abuse Prevention, Washington, 1990-91, Office of Treatment Improvement, Washington, 1991, Nat. Inst. Drug Abuse, Washington, 1991; expert panel mem. Nat. IOmpaired Driving Rev. Com., Washington, 1991-92; evaluation team Colo. Impaired Driving Prevention program, Denver, 1991. Contbr. numerous articles to profl. jours. Mem. Okla. Youth Suicide Prevention Task Force, 1987-89; bd. dirs. Okla. Fedn. Parents, 1989; mem. Cherokee Nation Children's Commn., 1988-89; mem. Epidemiology Task Force, State of Okla., 1989-90, Women's Substance Abuse Task Force, 1989-90; bd. dirs. Tulsa Mental health Assn., 1990. Mem. APA. Home: 81 Humboldt Dr Livermore CO 80536 Office: Tri-Ethnic Ctr Preven Rsch Clark Bldg C-138-A Fort Collins CO 80523

THURMAN, WILLIAM GENTRY, medical research foundation executive, pediatric hematology and oncology physician, educator; b. Jacksonville, Fla., July 1, 1928; s. Horace Edward and Theodosia (Mitchell) T.; m. Peggy Lou Brown, Aug. 11, 1949 (div. 1978); children—Andrew E., Margaret Anne, Mary Allison; m. Gabrielle Anne Martin, Jan. 22, 1980; 1 step child, Stephanie Anne. B.S., U. N.C., 1949; M.S., Tulane U. Sch. Pub. Health, 1960; M.D.C.M., McGill U., Montreal,, 1954. Prof. pediatrics Cornell U. Sch. Medicine, N.Y.C., 1962-64; prof. pediatrics U. Va., Charlottesville, 1964-73; dean sch. medicine Tulane U., New Orleans, 1973-75; provost Health Scis. Ctr. U. Oklahoma, Oklahoma City, 1975-80; pres., chief exec. officer Okla. Med. Research Found., Oklahoma City, 1979—; sr. cons. pediatrics Surgeon Gen. USAF, Washington, 1964—; mem. Diet and Nutrition Study Com., Nat. Cancer Inst., Bethesda, Md., 1969—, Profl. Edn. Com., Am. Cancer Soc., N.Y.C., 1973—. Author: (with others) Bone Tumors in Children, 1963, Pediatric Malignant Disease, 1964; contbr. articles to profl. jours. Bd. dirs. United Way, Oklahoma City, 1977—, Community Found., Oklahoma City, 1975—, ARC, Oklahoma City, 1982—. Served with U.S. Army, 1944-46, ETO. Markle scholar, 1959-64. Fellow Am. Acad. Pediatrics (dir. 1969-72); mem. Am. Pediatric Soc., Soc. Pediatric Research (councillor 1971-75), Soc. Mil. Cons., Assn. Ind. Research Insts. (pres. 1985-87), Oklahoma City C. of C. (dir. 1979), Alpha Omega Alpha. Baptist. Club: Petroleum. Avocations: physical fitness; sailing. Home: 10001 N Nancy Oklahoma City OK 73131-5237

THURSTON, WILLIAM RICHARDSON, oil and gas industry executive, geologist; b. New Haven, Sept. 20, 1920; s. Edward S. and Florence (Holbrooke) T.; m. Ruth A. Nelson, Apr. 30, 1944 (div. 1966); children: Karin R., Amy R., Ruth A.; m. Beatrice Furnas, Sept. 11, 1971; children: Mark P., Stephen P., Douglas P., Jennifer P. AB in Geol. Sci. with honors, Harvard U., 1942. Registered profl. engr., Colo. Field geologist Sun Oil Co., Corpus Christi, Tex., 1946-47; asst. to div. geologist Sun Oil Co., Dallas, 1947-50; chief geologist The Kimbark Co., Denver, 1952-59; head exploration dept. Kimbark Exploration Co., Denver, 1959-66; co-owner Kimbark Exploration Ltd., Denver, 1966-67, Kimbark Assocs., Denver, 1967-76, Hardscrabble Assocs., Denver, 1976-80; pres. Weaselskin Corp., Durango, Colo., 1980—. Bd. dirs. Denver Bot. Gardens, 1972—, Crow Canyon Ctr. for Archaeology, Cortez, Colo., 1980-92. Comdr. USNR, World War II, Korea. Decorated D.F.C. (with 3 stars). Mem. Am. Assn. Petroleum Geologists, Denver Assn. Petroleum Landmen, Rocky Mountain Assn. Petroleum Geologists, Four Corners Geol. Soc. Republican. Avocations: photography, gardening, reading. Office: Weaselskin Corp 12995 US Hwy 550 Durango CO 81301

TIBURCIO, ASTROPHEL CASTILLO, polymer chemist; b. Manila, Philippines, Jan. 12, 1957; came to U.S., 1968; s. Albino Fulinara and Bessie (Castillo) T. BS in Chemistry cum laude, U. Pitts., 1979; MS in Chem. Engring., Lehigh U., 1983, PhD in Polymer Engring. and Sci., 1988. Rsch. chemist E.I. DuPont, Troy, Mich., 1986-88; devel. chemist GE Electromaterials, Coshocton, Ohio, 1988—. Contbr. articles to ACS Organic Coatings and Plastic Chemistry, Jour. Adhesion Sci. and Tech., Jour. Applied Polymer Sci. Alumni gift program leader, mem. corp. coun. U. Pitts., 1991—. Recipient Outstanding Leadership award Pitt/Oakland YMCA, 1979. Mem. Am. Chem. Soc., Sigma Xi. Achievements include patent for getek-glass-reinforced electrical laminate. Home: 1283C Pine View Trail Newark OH 43055 Office: Ge Electromaterials 1350 S 2d St Newark OH 43055

TICE, DAVID CHARLES, aerodynamics engineer; b. Edmonton, Alta., Can., Apr. 13, 1960; came to U.S., 1962; s. Colin Charles and Claire Yvonne (Foran) T.; m. Melissa A. Buck, May 8, 1993. BS in Aerospace Engring., U. So. Calif., 1982; postgrad., Rutgers U., 1993—. Aerodynamics/performance engr. Rockwell Internat./N.Am. Aircraft, El Segundo, Calif., 1982-90, Lockheed Engring. & Scis. Co., Hampton, Va., 1990-92. Contbr. sci. papers to AIAA. Mem. Peninsula Fine Arts Ctr., Newport News, Va., 1991—. Dean's fellow Rutgers U. Grad. Sch. Mgmt., 1992-94. Mem. AIAA, U. So. Calif. Young Alumni Club (v.p. 1988-90). Achievements include research in advanced subsonic, supersonic and hypersonic aircraft concepts.

TICHENOR, WELLINGTON SHELTON, physician; b. N.Y.C., Oct. 29, 1950; s. Frank Ceccarelli and Cornelia Tichenor. BSE cum laude, Princeton U., 1972; MD, N.Y. Med. Coll., 1975. Diplomate Am. Bd. Internal Medicine. Intern N.Y. Med. Coll.-Met. Hosp., 1975-76, resident, 1976-78; assoc. med. dir. Am. Internat. Group, N.Y.C., 1978-80; pvt. practice internal medicine Balt., 1980-81, N.Y.C., 1981—; attending physician Beth Israel Hosp. North, N.Y.C., 1981—. Contbr. articles to Postgrad. Medicine, Jour. Am Coll. Immunology. Mem. ACP, Am. Acad. Allergy and Immunology, Am. Coll. Allergy and Immunology, N.Y. State and County Med. Soc., N.Y. State Allergy Soc. Office: 642 Park Ave New York NY 10021

TIEDEMANN, ARTHUR RALPH, ecologist, researcher; b. Grand Junction, Colo., Mar. 21, 1938; s. Henry Frederick and Elizabeth Mildred (Montgomery) T.; m. Sharon Estelle Bjorsness, Aug. 19, 1962. MS, U. Ariz., 1966, PhD, 1970. Rsch. range scientist Pacific N.W. Rsch. Sta., Wenatchee, Wash., 1969-72, rsch. project leader, 1972-79; rsch. project leader Intermountain Rsch. Sta., Provo, Utah, 1979-83; chief rsch. ecologist Pacific N.W. Rsch. Sta., La Grande, Oreg., 1983—; chmn. Shrub Rsch. Consortium, Provo, Utah, 1983, sustainable forest and range mgt. com. Blue Mountains Natural Resources Inst., La Grande, 1992-93. Author, editor: Research and Management of Bitterbrush, 1983; editor: Biology of Atriplex and Related Chenopods, 1984. Recipient Superior Services award U.S. Dept. Agr., 1974, 90; named Outstanding Alumnus, U. Ariz. Sch. Renewable Natural Resources, 1992. Mem. Soc. Range Mgmt., N.W. Scientific (assoc. editor 1977-83), Soc. Am. Foresters (assoc. editor forest sci. 1980-92), Sigma Xi (pres. ea. Oreg. state coll. chpt. 1992-93). Republican. Lutheran. Achievements include research on fire and forest management effects on water quality, underground systems of gambel oak, nutrient losses, nutrient cycling, and sustainable resource productivity, influence of large ungulates on forest sucession and productivity, effects of livestock grazing on water quality. Office: Pacific N W Rsch Sta 1401 Gekeler La Grande OR 97850

TIEDEMANN, HEINZ, biochemist; b. Berlin, Feb. 16, 1923; s. Otto and Elizabeth (Schleuss) T.; m. Hildegard Waechter, Aug. 6, 1953; children: Karl-Heinz, Christiane. MD, Humboldt U., Berlin, 1949; Dr.rer.nat., Freie U. Berlin, 1952; dozent, Albert Ludwig U., Freiburg, Fed. Republic Germany, 1957. Rsch. asst. Max Planck Inst. Zellphysiologie und Medizinische Forschung, Berlin & Heidelberg, Fed. Republic Germany, 1949-54, Heiligenberg-Inst., Heiligenberg/Baden, 1954-63; fellow Carnegie Instn. dept. embryology, Balt., 1963-64; mem., dept. head Max Planck Inst. Meeresbiologie, Wilhelmshaven, 1965-67; prof. Freie Univ. Berlin, 1967—. Author: The Biochemistry of Animal Development II & III, 1967; Organizer, 1980; editor Roux Archives of Devel. Biology, 1966—, Internat. Jour. Devel. Biology, 1987—. Recipient Theodor Boveri award Physico-Medica Wüerzbûrg, 1991. Mem. Internat. Soc. Devel. Biologists, Gesellschaft Biologische Chemie, Gesellschaft Deutscher Chemiker. Home: Wildtalstr 51, 79108 Freiburg Germany Office: Freie U Berlin Dept Biochemistry, Arnimallee 22, 14195 Berlin 33, Germany

TIEDJE, JAMES MICHAEL, microbiology educator, ecologist; b. Newton, Iowa, Feb. 9, 1942; married, 1965; 3 children. BS, Iowa State U., 1964; MS, Cornell U., 1966, PhD in Soil Microbiology, 1968. From asst. prof. to prof. Mich. State U., 1968-78, prof. microbial ecology, 1978—; dir. sci. and tech. ctr. microbial ecology NSF, 1988—; vis. assoc. prof. U. Ga., 1974-75; cons. NSF, 1974-77; vis. prof. U. Calif., Berkeley, 1981-82; mem. biotech. sci. adv. com. GPA, 1986-89, chair sci. adv. coun., 1988-90. Editori Applied Microbiology, 1974—, editor-in-chief, 1980—. Fellow AAAS, Am. Soc. Agronomy (Soil Sci. award 1990), Internat. Inst. Biotech.; mem. Am. Soc. Microbiology (award in Applied and Environ. Microbiology 1992), Soil Sci. Soc. Am., Ecol. Soc. Am. Achievements include research in dentrification, microbial metabolism of organic pollutants, and molecular microbiol. ecology. Office: Michigan State U Microbial Ecology Ctr 540 Plant & Soil Scis Bldg East Lansing MI 48824-1325*

TIELENS, STEVEN ROBERT, information specialist; b. Orlando, Fla., Mar. 8, 1953; s. James D. and Nancy A. (Caffie) T.; m. Cynthia A. Bradley, June 4, 1976. AS, Jones County Jr. Coll., Ellisville, Miss., 1986. Elec. foreman IBEW Local 474, Memphis, 1976-83; telecommunications adminstr., asst. ops. mgr. Forrest Gen. Hosp., Hattiesburg, Miss., 1984-87; dir. info. svcs. JFK Med. Ctr., Atlantis, Fla., 1987-91; info. tech. cons. West Palm Beach, Fla., 1991; mgr. user support Racal-Datacom, Inc., Sunrise, Fla., 1992—. Jones County Jr. Coll. acad. scholar. Mem. Phi Theta Kappa, Phi Beta Lambda (pres., 1st place state competition in data processing I and II, 3rd place nat. in data processing II, 1986). Home: 16097 Rustic Rd West Palm Beach FL 33470 Office: Racal Datacom Inc 1601 N Harrison Pkwy Sunrise FL 33323-2899

TIEMAN, ROBERT SCOTT, chemist; b. Ft. Thomas, Ky., Nov. 4, 1963; s. Roger George and Margo Ann (Parks) T.; m. Dawn Lynn Hodge, June 20, 1987; 1 child, Robert Hayden. BS in Chemistry, No. Ky. U., 1987; PhD in Chemistry, U. Cin., 1991. Development chemist Fosroc Inc., Georgetown, Ky., 1991-92, product mgr., 1992—. Author: Oxygen Sensors Based on the Polymer PDMDAAC, 1991; contbr. articles to profl. jours. Mem. Am. Chem. Soc. Achievements include co-invention of solid-state electrochemical oxygen sensor, of electrochemical biosensor for oxygen. Home: 100 Redding Rd Georgetown KY 40324 Office: Fosroc Inc 150 Carley Ct Georgetown KY 40324

TIEN, CHANG-LIN, chancellor; b. Wuhan, China, July 24, 1935; came to U.S., 1956, naturalized, 1969; s. Yun Chien and Yun Di (Lee) T.; m. Di Hwa Liu, July 25, 1959; children: Norman Chihnan, Phyllis Chihping, Christine Chihyih. BS, Nat. Taiwan U., 1955; MME, U. Louisville, 1957; MA, PhD, Princeton U., 1959. Acting asst. prof. dept. mech. engring. U. Calif., Berkeley, 1959-60, asst. prof., 1960-64, assoc. prof., 1964-68, prof., 1968-89, 90—, A. Martin Berlin prof., 1987-89, 90—, dept. chmn., 1974-81, also vice chancellor for research, 1983-85; exec. vice chancellor U. Calif., Irvine, 1988-90; chancellor U. Calif., Berkeley, 1990—; tech. cons. Lockheed Missiles and Space Co., GE; trustee Princeton (N.J.) U., 1991—; bd. dirs. Wells Fargo Bank. Contbr. articles to profl jours. Guggenheim fellow, 1965; recipient Heat Transfer Meml. award, 1974, Larson Meml. award Am. Inst. Chem. Engrs, 1975. Fellow AAAS, ASME (Max Jakob Meml. award), AIAA (Thermophysics award 1977); mem. NAE. Office: U Calif Chancellor's Office Berkeley CA 94720

TIEN, ROBERT DERYANG, radiologist, educator; b. Taipei, Taiwan, Republic of China, July 5, 1956; came to U.S., 1982; BS, Nat. Taiwan U., MD, 1981; MPH, Harvard U., 1984. Intern Nat. Taiwan U. Hosp., Taipei, 1980-81; resident Baylor Coll. Medicine, Houston, 1984-88; instr. U. Calif., San Francisco, 1988-89; asst. prof. U. Calif. Sch. Medicine, San Diego, 1989-91; co-dir. MR and NR Duke U. Med. Ctr., Durham, N.C., 1991—, head sect. neuroradiology, 1991—, assoc. prof. radiology, 1991—; cons. radiologist Nat. Taiwan U. Sch. Medicine, Taipei, 1988—, GE Med. Systems, Asia, 1992—, Beijing Neurol. Inst., 1993—; hon. prof. Capital Med. Hosp., Beijing, 1993—. Mem. AMA, Western Neuroradiology Soc., Am. Head and Neck Radiology Soc., Am. Soc. Neuroradiologists (sr.), Radiol. Soc. N.Am. Office: Duke Univ Med Ctr Box 3808 Erwin Rd Durham NC 27710

TIERNAN, J(ANICE) CARTER MATHENEY, mathematics specialist, consultant; b. Knoxville, Tenn., July 4, 1962; d. James Albert Matheney and Sandra Carter (Williams) Schou; m. Timothy R.M. Tiernan, Sept. 5, 1992. BA in Computer Sci., U. Tenn., 1983; MS in Computer Sci., U. Tex., Richardson, 1987; PhD in Computer Sci., U. Tex., Arlington, 1992. Software design engr. Tex. Instruments Inc., Dallas, 1983-90; faculty assoc. U. Tex., 1990-92; math. specialist Project Seed, Dallas, 1992—. Contbg. author: Artificial Intelligence Applications in Manufacturing, 1992. Mem. Am. Assn. for Artificial Intelligence, Assn. for Computing Machinery, Sigma Xi, Tau Beta Pi. Home: 5526 Merrimac Ave Dallas TX 75206

TIERNAN, ROBERT JOSEPH, research and development engineer; b. Boston, Dec. 14, 1935; s. Harold Myles and Anna Marie (Purcell) T.; m. Lorette Annette Haley, June 21, 1959; children: Robert J., Ruth Anne, Daniel George, Paul Louis. AB, Boston Coll., 1957, MS, 1959; PhD, MIT, 1969. Cert. tchr. math. and sci., Mass. Solid state physicist Naval Rsch. Lab., Washington, 1959-62; atomic physicist Nat. Bur. Standards, Washington, 1962-63; engr. scientist Sylvania Rsch. Lab., Waltham, Mass., 1963-64; R&D engr. Raytheon, Waltham, 1969-74; high temperature physics and chemistry staff Argonne (Ill.) Nat. Labs., 1974-75; prin. R&D engr. GTG Sylvania Lighting Product, Danvers, Mass., 1978—. Contbr. articlest to Jour. Solid State Electronics, Jour. Applied Physics, Jour. Chem. Physics, Jour. Am. Ceramic Soc., Jour. Electrochem. Soc., Jour. Chem. and Physics, Jour. S.P.I.E., IEEE Jour. Mem. S.P.I.E., Am. Ceramic Soc., Sigma Xi. Roman Catholic. Home: 224 North St Stoneham MA 02180 Office: Osram Sylvania Lighting Products 100 Endicott St Danvers MA 01923

TIERNEY, ANN JANE, neuroscientist; b. Washington, Dec. 16, 1955; d. Brian and Theresa (O'Dowd) T.; m. James Irwin Wallace, July 2, 1989; children: Jasmine, Gemma. BA, Cornell U., 1978; PhD, U. Toronto, Ont., Can., 1985. Grad. teaching asst. U. Toronto, 1978-84; postdoctoral fellow marine program Boston U., 1985-87; rsch. assoc. Cornell U., Ithaca, N.Y., 1987-93; lectr. Ithaca Coll., 1991; asst. prof. Colgate U., 1993—. Contbr. articles to Am. Midland Naturalist, Can. Jour. Zoology, Animal Learning and Behavior, Biol. Bull., Jour. Neurophysiology. Fellow, U. Toronto, 1980, U.S. EPA, 1985, NIH, 1987; Ont. grad. scholar, 1983. Mem. AAAS, Am. Soc. Zoologists, Animal Behavior Soc. Achievements include research in evolution of nervous systems and behavior, physiology of neural circuits, and functioning of ion channels in nerve cells. Home: RR 1 96A Butternut Ln Hamilton NY 13346 Office: Colgate U Dept Psychology Neurosci Program Hamilton NY 13346

TIERNEY, MICHAEL JOHN, mathematics and computer science educator; b. St. Louis, Feb. 19, 1947; s. John Thomas and Alice Marie (Krieger) T.; m. Edith L. Echelmeyer, Nov. 21, 1975 (div. Sept. 1984); 1 child, John E.; m. Virginia Lee Christian, Apr. 6, 1985. BS, St. Louis U., 1969, MS, 1971, PhD, 1974. Prof. math. and actuarial sci. Maryville Coll., St. Louis, 1974-83; prof. math. and computer sci. Va. Mil. Inst., Lexington, 1983—. Mem. AAUP, AAAS, Am. Math. Soc., Math. Assn. Am., Soc. Indsl. and Applied Math., Sigma Xi. Presbyterian. Avocations: tennis, landscaping. Home: 819 Gwynne Ave Waynesboro VA 22980-3342 Office: Va Mil Inst Lexington VA 24450

TIETJEN, JAMES, research institute administrator; b. 1933. With RCA Corp., from 1963, David Sarnoff Rsch. Ctr., from 1987; sr. v.p., the exec. v.p., now pres., c.e.o. SRI Internat., 1990—; also bd. dirs. Office: SRI Internat 333 Ravenswood Ave Menlo Park CA 94025-3493*

TIETKE, WILHELM, gastroenterologist; b. Niengraben, Germany, Oct. 15, 1938; came to U.S., 1969, naturalized, 1979; s. Wilhelm and Frieda (Schmeding) T.; m. Imme Schmidt, Oct. 15, 1965; children: Cornelia, Isabel. MD, U Goettingen (West Germany), 1968. Diplomate Am. Bd. Internat. Medicine, Am. Bd. Gastroenterology. Intern Edward W. Sparrow Hosp., Lansing, Mich., 1970; resident in internal medicine Henry Ford Hosp., Detroit, 1971-73; fellow in gastroenterology, 1973-75; practice medicine specializing in gastroenterology, Huntsville, Ala., 1975—; mem. vol. faculty, cons. U. Ala., Huntsville, 1976; clin. assoc. prof. internal medicine, 1979—; v.p. Huntsville Gastroenterology Assocs., P.C., 1979—. Fellow Coll. Gastroenterology; mem. AMA, Ala. Med. Soc., Am. Coll. Physicians, Am. Soc. Gastrointestinal Endoscopy. Lutheran. Lodge: Rotary. Home: 2707 Westminster Way Huntsville AL 35801 Office: 119 Longwood Dr Huntsville AL 35801-4205 also: PO Box 2169 Huntsville AL 35804-2169

TIETZE, LUTZ FRIEDJAN, chemist, educator; b. Berlin, Mar. 14, 1942; s. Friedrich and Hete-Irene (Kruse) T.; m. Karin Krautschneider; children: Martin, Maja, Andrea, Julia. Diploma, U. Kiel, 1966, PhD, 1968; habil. for Organic Chemistry, U. Münster, 1975. Rsch. assoc. MIT, 1969-71; lectr. U. Münster, 1971-76; prof. U. Dortmund, 1977-78; full prof. and inst. dir. U. Göttingen, Göttingen, 1978—; dean and prodean U. Göttingen, 1983-87, 91—; vis. prof. U. Wis., 1982; mem. Bd. of Faculties of Chemistry, Germany. Author: Reactions and Syntheses, 1981, (translated into Japanese 1984, English 1989), 2d edit., 1991, Basic Course in Organic Chemistry, 1993; contbr. more than 170 articles to profl. jours.; patentee in field. Recipient Karl-Winnacker award, Hoechst AG, Germany, 1976, Lit. prize, Fonds der Chem. Industry, Germany, 1982. Fellow Royal Soc. Chemists; mem. GdCh, Am. Chem. Soc., Chem. Soc. Argentina (h.c.), Academia Scientiarum Göttingen. Home: Stumpfe Eiche 73, D-37077 Göttingen Federal Republic of Germany Office: U Göttingen Inst Organic Chemistry, Tammannstrasse 2, D-37077 Göttingen Germany

TIFFANY, STEPHEN THOMAS, psychologist; b. Gowanda, N.Y., Mar. 13, 1954; s. Lester E. and Mary Elenore (Crowe) T.; m. Kristine E. Zimmer, Dec. 30, 1977; 1 child, Patrick. MS, U. Wis., 1980, PhD, 1984. Lic. psychologist, Ind. Clin. intern Dept. Psychiatry, U. Wis., Madison, 1983-84; asst. prof. psychology Purdue U., West Lafayette, Ind., 1984-90, assoc. prof., 1990—; vis. scientist Nat. Inst. on Drug Abuse, Balt., 1991. Contbr. articles to profl. jours. Grantee Nat. Inst. on Drug Abuse, 1987, 89, 92, Am. Cancer Soc., 1989, 91, 93; recipient Disting. Sci. award Am. Psychol. Assn., 1993. Mem. AAAS, Am. Psychol. Soc., Assn. for advancement of Behavior Therapy. Achievements include research on role of learning in drug tolerance, measurement of drug craving and theoretical conceptualizations of the function of drug craving in drug abuse. Home: 615 Meridian St West Lafayette IN 47906 Office: Purdue Univ Dept Psychology West Lafayette IN 47907

TIFFT, WILLIAM GRANT, astronomer; b. Derby, Conn., Apr. 5, 1932; s. William Charles and Marguerite Howe (Hubbell) T.; m. Carol Ruth Nordquist, June 1, 1957 (div. July 1964); children: Jennifer, William John; m. Janet Ann Lindner Homewood, June 2, 1965; 1 child, Amy, stepchildren: Patricia, Susan, Hollis. AB, Harvard Coll., 1954; PhD, Calif. Inst. Tech., 1958. Nat. sci. postdoctoral Australian Nat. Univ., Canberra, 1958-60; rsch. assoc. Vanderbilt U., Nashville, 1960-61; astronomer Lowell Obs., Flagstaff, Ariz., 1961-64; assoc. prof. U. Ariz., Tucson, 1964-73, prof., 1973—. Joint author: Revised New General Catalog, 1973; contbr. over 100 articles to profl. jours. NSF Predoctoral fellow, 1954-58; NSF Postdoctoral fellow, 1958-60; grantee NASA, NSF, ONR. Fellow Am. Astron. Soc.; mem. Internat. Astronomers Union. Achievements include discovery of redshift quantification and correlations relating to it, including variability mapping of large scale supercluster structure--first to detect voids. Office: U of Arizona Dept of Astronomy Tucson AZ 85721

TIGHE, THOMAS JAMES GASSON, JR., healthcare executive; b. Malden, Mass., July 11, 1946; s. Thomas J. G. and Barbara (Buckland) T.; m. Carolyn Payne, Mar. 29, 1969; children: Jessica, Chelsea, Alexandra. BA, Bates Coll., 1968; MSc, Columbia U., 1970; MPH, Johns Hopkins U., 1973. Adminstr. asst. Boston U. Med. Ctr. U. Hosp., 1970-71, asst. adminstr., 1971-72; asst. dir. Mary Imogene Bassett Hosp., Cooperstown, N.Y., 1973-80, Cen. Maine Med. Ctr., Lewiston, 1980-86; exec. v.p. Cen. Maine Healthcare Corp., Lewiston, 1986—; lectr. U. Maine, Augusta, 1987-88, St. Joseph's Coll., Windham, Maine, 1990—; corporator Androscoggin Savs. Bank, Lewiston, 1990—. Bd. dirs. Maine Acting Co., Lewiston, 1986-88, Auburn (Maine) Pub. Libr., 1987-90, LA Arts, Lewiston, 1988—. Fellow Am. Coll. Healthcare Execs. (bd. govs. 1993—, coun. regents 1986-92); mem. Am. Hosp. Assn., Soc. Healthcare Planning, Mktg., Maine Hosp. Assn. (bd. dirs. 1990—). Office: Cen Maine Healthcare Corp 364 Main St Lewiston ME 04240-7072

TIGHE-MOORE, BARBARA JEANNE, electronics executive; b. Wadsworth, Ohio, Jan. 12, 1961; d. Norton Raymond and Laura Alida (Frank) T.; m. Derek William Moore, June 26, 1982. AS in Electronic Engring. summa cum laude, Hocking Tech. Coll., 1981; AS in Electronic Data Processing magna cum laude, Sinclair Coll., 1986; BA Honors Coll. magna cum laude, Kent State U., 1988. Lic. amateur radio operator. Tech. writer Computer Dept. Sinclair Coll., Dayton, Ohio, 1983; project mgr. O'Neil & Assocs., Dayton, 1983-84; biomed., bio-acoustic real-time flight simulation tempest developer Systems Rsch. Labs., Dayton, 1984-86; owner, pres. Lida Ray Techs., Dayton, Ohio, 1978—; computer specialist Kent State U. Press, 1987-88; mgmt. analyst Electronic Warfare Frontier Engring. Inc., 1988-89; supr. small computer tech. svcs. Frontier Engring., Inc., 1989-90, project engr., 1990-92; ptnr. MKCC, Dayton, 1990—; sr. program mgr. C.E.T.A., Dayton, 1992—; ptnr. SDCC, Dayton, 1992—; pres. Lida Ray Techs., Dayton; mem. graphics steering com., mem. sanctioned UNIX software adv. team Aero. Systems Divsn.; program chair IEEE Internat. Wireless LAN Conf.; mem. Engring. Application Support Environ. Security Working Group; bd. dirs. MKCC, SDCC; speaker Wireless '93, Calgary, Alta., Nat. Aeronautics & Electronics Conf., Dayton, Ohio, 1993. Editor: Graphics Directions, 1990-91; pub. SDCC Cleaning Times. Counselor Kwam's Kinder Kamp; tchr. Bible Sch.; cook Meals on Wheels; organizer/cook funeral Svcs. Dinners. Recipient Vol. Citizen award Wadsworth C. of C, 1979, Ohio Essayist award, 1979, Virginia Perryman award, 1979, Disting. Leadership award, 1990, 91. Mem. IEEE (former treas.), Computer Soc. of IEEE (sec. 1991-92, vice chmn. 1992—), Engring. Mgmt. Soc. of IEEE, Tech. and Soc. of IEEE, Data Processing Mgmt. Assn., Assn. Computer Machinery, Assn. Internat. Students Econs. & Commerce (pres. 1986-87), Internat. Film Soc. (pres. 1986-88), Armed Forces Communications and Electronics Assn. (judge sci. fair we. dist. 1992—), Equestrian Team (point rider 1987-88), Fencing Club, Phi Theta Kappa, Mortar Bd., Omnicron Delta Kappa, Beta Gamma Sigma. Avocations: travel, investing, equestrian show jumping, soccer. Home: 3125 Glen Rock Rd Dayton OH 45420-1900

TIGNER, MAURICE, physicist, educator. Attended, Webb Inst. of Naval Architecture, 1954-56; BS in Physics, Rensselaer Polytechnic Inst., 1958; PhD in Exptl. Physics and Elec. Engring., Cornell U., 1962. Prof. dept. physics Cornell U., Ithaca, N.Y.; rsch. assoc. dept. physics Cornell U., 1963-68, sr. rsch. assoc. and dir. ops., accelerator facilities dept. physics, 1968-77; vis. scientist Deutsches Elec. Synchotron, Hamburg, Germany, 1972-73; prof. physics and staff mem., lab. nuclear studies Cornell U., 1977—; dir. SSC Cen. Design Group Univs. Rsch. Assn., 1984-89; vis. prof. dept. physics, U. Calif., 1987-89; cons. Bechtel National, Inc., 1989—; mem. RHIC Machine Com., BNL, 1992—; mem. dirs. adv. com. for accelerator and fusion rsch. divsn., Lawrence Berkeley Lab., 1992—; chmn. steering com. accelerator test facility, BNL, 1992—. Recipient Ernest Orlando Lawrence Meml. award U.S. Dept. Energy, Washington, 1990. Fellow AAAS, Am. Phys. Soc., Am. Acad. Arts and Scis.; mem. Nat. Acad. Sci. Office: Cornell Univ Dept of Physics Ithaca NY 14853

TIJSSEN, PETER H. T., molecular virology educator, researcher; b. Elden, Gelderland, The Netherlands, May 1, 1944; arrived in Can., 1973; s. Andries P. J. and An M. (Thuis) T.; m. Trics C. S. van der Slikke, Dec. 18, 1975; children: Andrew, Janice. Diploma in engring., Hort. Coll., Utrecht, The Netherlands, 1966; Doctorate, U. Wageningen, The Netherlands, 1972; PhD, U. Montreal, Que., Can., 1980. Head biochemistry sect. U. Montreal, 1975-85; assoc. prof. U. Que., Ville de Laval, Can., 1985-89, prof., 1989—; mem. Internat. Com. Taxonomy of Viruses. Author: Practice and Theory of Enzyme Immunoassays, 1985, Japanese edition, 1990; author, editor: Handbook of Parvoviruses, 1990, Hybridization with Nucleic Acid Probes, vols. I and II, 1993; editor Jour. Immunoassay; contbr. articles to profl. jours. Mem. Am. Soc. Microbiology. Achievements include research in densovirus, bluecomb virus, and parvoviruses. Home: 76 Winston Cir, Pointe-Claire, PQ Canada H9S 4X6 Office: Inst Armand Frappier, 531 Blvd des Prairies, Ville de Laval, PQ Canada H7V 1B7

TIJUNELIS, DONATAS, engineering executive; b. Kaunas, Lithuania, Nov. 6, 1935; came to U.S. 1949.; s. Juojas and Alicija (Gustaitis) T.; m. Judre Paliokas, July 7, 1962; children: Daina, Rasa, Aras, Nida. BS in Chem. Engring., Purdue U., 1958; MS in Chem. Engring., Northwestern U., 1960; DBA in Mgmt., Nova U., 1986. Registered profl. engr. Project leader Union Carbide Corp., Chgo., 1959-68; rsch. engr. Continental Can Co., Chgo., 1962; assoc. dir. Borg Warner Rsch. Ctr., Des Plaines, Ill., 1968-87; pres. DKT Engring., Buffalo Grove, Ill., 1987-88; div. mgr. USG Corp., Chgo., 1988; v.p. in rsch. and devel. VISKASE Corp., Chgo., 1988—; pres. Rsch. Dis. Assn. Chgo., 1991-92; mem. Ill. Engring. REgistration Bd., Springfeld, 1992--, Indsl. Rsch. Inst., Washington, 1992--. Editor: Manufacturing High Technology, 1987; co-author: Research and Development on a Minimum Budget, 1979; contbr. articles to profl. jours. Named Outstanding Faculty Lake Forest Grad. Sch., 1988. Mem. Soc. Plastic Engrs., Tau Beta Pi. Roman Catholic. Achievements include 14 patents. Home: 419 Weidner Rd Buffalo Grove IL 60089 Office: VISKASE Corp 6855 W 65th St Chicago IL 60638

TILBROOK, BRONTE DAVID, research scientist; b. Minlaton, Australia, July 6, 1955; s. Bill Tillbrook and Linda Isabel Whitbread; m. Tracey Marie Lincoln, Jan. 9, 1993. MS, U. Hawaii, 1982, PhD, 1992. Rsch. scientist Commonwealth Sci. & Indsl. Rsch. Marine Labs., Hobart, Tasmania, Australia, 1990—; fellow Coop. Rsch. Ctr. for Antarctic and So. Ocean Equipment, Hobart, 1993—. Recipient Newcomb Cleveland prize AAAS, 1991-92. Mem. Am. Geophys. Union, Am. Soc. Limnology and Oceanography, Geochem. Soc. Office: CSIRO Marine Labs, Castray Esplanade, Hobart Tasmania, Australia 7001

TILGHMAN, SHIRLEY MARIE, biology educator. PhD in biochemistry, Temple U., 1975. Prof. molecular biology Princeton (N.J.) U., 1986—, Howard A. Prior prof. in life scis.; investigator Howard Hughes Med. Inst. Office: Princeton U Dept Molecular Biology Princeton NJ 08544

TILL, JAMES EDGAR, scientist; b. Lloydminster, Sask., Can., Aug. 25, 1931; s. William and Gertrude Ruth (Isaac) T.; m. Marion Joyce Sinclair, June 6, 1959; children: David William, Karen Sinclair, Susan Elizabeth. BA, U. Sask., 1952, MA, 1954; PhD, Yale U., 1957. Postdoctoral fellow Connaught Med. Research Labs., Toronto, Ont., Can., 1956-57; mem. physics div. Ont. Cancer Inst., Toronto, 1957-67, div. biol. rsch., 1967-89, div. head, 1969-82, with div. epidemiology and stats., 1989—; assoc. dean U. Toronto, 1981-84, Univ. prof., 1984—. Contbr. articles on biophysics, cell biology, cancer research and decision sci. to sci. jours. Recipient Gairdner Found. Internat. award, 1969. Fellow Royal Soc. Can.; mem. Can. Bioethics Soc. Home: 182 Briar Hill Ave, Toronto, ON Canada M4R 1H9 Office: 500 Sherbourne St, Toronto, ON Canada M4X 1K9

TILLERY, BILL W., physics educator; b. Muskogee, Okla., Sept. 15, 1938; s. William Earnest and Bessie C. (Smith) Freeman; m. Patricia Weeks Northrop, Aug. 1, 1981; 1 child, Elizabeth Fielding; children by previous marriage: Tonya Lynn, Lisa Gail. BS, Northeastern U., 1960; MA, U. No. Colo., 1965, EdD, 1967. Tchr. Guthrie Pub. Schs., Okla., 1960-62; tchr. Jefferson County schs., Colo., 1962-64; teaching asst. U. No. Colo., 1965-67; asst. prof. Fla. State U., 1967-69; assoc. prof. U. Wyo., 1969-73, dir. sci. and math. teaching ctr., 1969-73; assoc. prof. dept. physics Ariz. State U., Tempe, 1973-75, prof., 1976—; cons. in field. Author: (with Ploutz) Basic Physical Science, 1964; (with Sund and Trowbridge) Elementary Science Activities, 1967, Elementary Biological Science, 1970, Elementary Physical Science, 1970, Elementary Earth Science, 1970, Investigate and Discover, 1975; Space, Time, Energy and Matter: Activity Books, 1976; (with Bartholomew) Heath Earth Science, 1984; (with Bartholomew and Gary) Heath Earth Science Activities, 1984, 2d edit. 1987, Heath Earth Science Teacher Resource Book, 1987, Heath Earth Science Laboratory Activity, 1987, Physical Science, 1991, 2d edit. 1993, Physical Science Laboratory Manual, 1991, 2d edit. 1993, Physical Science Instructor's Manual, 1991, 2d edit. 1993, Physical Science Laboratory Manual Instructor's Manual, 1991, 2d edit. 1993, (with Grant) Physical Science Student Study Guide, 1991, 2d edit. (with Claassen) 1993, Introduction to Physics and Chemistry: Foundations of Physical Science, 1992, Laboratory Manual in Conceptual Physics, 1992, Physics, 1993, Chemistry, 1993, Astronomy, 1993, Earth Science, 1993; editor: Ariz. Sci. Tchrs. Jour., 1975-85, Ariz. Energy Edn., 1978-84. Fellow AAAS; mem. Nat. Sci. Tchrs. Assn., Ariz. Sci. Tchrs. Assn., Assn. Edn. of Tchrs. in Sci., Nat. Assn. Research in Sci. Teaching. Republican. Episcopalian. Home: 8986 S Forest Ave Tempe AZ 85284-3142 Office: Ariz State U Dept Physics Tempe AZ 85287-1504

TILLEY, SHERMAINE ANN, molecular immunologist, educator; b. Shawnee, Okla., Feb. 22, 1952; d. Cecil Fern and Zona Emma (Evans) T. BA in Chemistry summa cum laude, Okla. City U., 1973; PhD in Biochemistry, Johns Hopkins U., 1980. Rsch. assoc. Albert Einstein Coll. Medicine, Bronx, 1980-85; rsch. asst. prof. NYU Sch. Medicine, N.Y.C., 1985—; asst. mem. Pub. Health Rsch. Inst., N.Y.C., 1985—; ad hoc reviewer SBIR grants NIH, 1989—; sec., staff coun. adv. com. Pub. Health Rsch. Inst., N.Y.C., 1990—, bd. dirs., 1993—. Contbr. articles to profl. jours. Recipient Letzeiser medal, Okla. City U., 1973; Nat. Arthritis Found. fellow, 1982-85; Life and Health Ins. Med. Rsch. Fund grantee, 1986-89; NIH grantee, 1988—. Mem. AAAS, Am. Assn. Immunologists, Am. Soc. Human Genetics. Achievements include demonstration of synergistic neutralization of HIV-1 by human monoclonal antibodies against the V3 loop and CD4 binding site of gpl20; patent in field. Office: Pub Health Rsch Inst 455 1st Ave Rm 1133 New York NY 10016-9102

TILLOTSON, DWIGHT KEITH, biologist; b. Wichita, Kans., Aug. 31, 1951; s. Dwight Sheldon and Barbara Nadine (Means) T.; m. Roxanne Laree Chartier, June 9, 1979; children: Vanessa, Michael, Kristin, John, Benjamin. BS in Biology, Kans. State U., 1973; MS in Sci. Edn., SUNY, Albany, 1975; MS in Entomology, Kans. State U., 1988. Sci. tchr. Unified Sch. Dist. 457, Garden City, Kans., 1975-84; biologist U.S. Army Corps of Engrs., Galveston, Tex., 1989-90; environ. resource specialist U.S. Army Corps of Engrs., Kearney, Nebr., 1990—. Contbr. articles to profl. jours. Mem. Sigma Xi. Republican. Presbyterian. Office: US Army Corps Engrs 1430 Central Ave Kearney NE 68847

TILSON, DOROTHY RUTH, word processing executive; b. Bloomsburg, Pa., Mar. 24, 1918; d. Roy Earl and Mary Etta (Masteller) Derr; m. Irving Tilson, Sept. 1949. BS, Bloomsburg U., 1940. Tchr. Madison Consol. Sch., Jerseytown, Pa., 1940-42; gage checker Phila. Ordinance Gage Lab., 1942-43; tabulating asst. Remington Rand, Phila., N.Y.C., 1943-46; copy writer Sears Roebuck, Phila., N.Y.C., 1946-48; statis. asst. Ford Internat., N.Y.C., 1949-56; word processing adminstrv. asst. Coopers & Lybrand, N.Y.C., 1956-91. Life mem. Rep. Senatorial Inner Circle, Washington, 1987—. Mem. Am. Movement for World Govt. (sec. 1991—), N.Y. Theosophical Soc. (libr. 1967—), UN Assn.-USA (mem. global policy project which includes internat. econ. governance). Home: 435 W 119th St 9G New York NY 10027-7142 Office: Coopers & Lybrand 1301 Ave Of The Americas New York NY 10019-6022

TILTON, JAMES CHARLES, computer engineer; b. Burlington, Wis., July 1, 1953; s. Charles Edwin and Mary Jean (Robinson) T.; m. Hac Hua, June 12, 1982; children: Man, Thoa, Phuong, Nancy. BA, Rice U., 1976, MEE, 1976; MS, U. Ariz., 1978; PhD, Purdue U., 1981. Sr. engr. Computer Scis. Corp., Silver Spring, Md., 1982-83, Sci. Applications Rsch., Riverdale, Md., 1983-85; computer engr. NASA Goddard Space Flight Ctr., Greenbelt, Md., 1985—. Editor workshop procs.; contbr. articles to profl. jours. Mem. ch. coun. The Greenbelt Community Ch., 1991-92. Mem. IEEE (geosci. and remote sensing soc. adminstrv. com. 1992—), Internat. Assn. for Pattern Recognition (chair tech. com. 7 1990-92). Mem. United Ch. of Christ. Office: NASA Goddard Space Flight Ctr Mail Code 930 1 Greenbelt MD 20771

TILTON, JAMES JOSEPH, research librarian; b. Sioux City, Iowa, Oct. 1, 1942; s. Jesse Verner and Antoinette Marie (Seek) T.; m. Doreen Carole Buck, Apr. 7, 1973; 1 child, James Joseph. BA in History, U. Va., 1968; MLS, U. Md., 1970. Libr. Defense Tech. Info. Ctr., Alexandria, Va., 1970-72; sr. reference libr. U.S. Dept. Housing & Urban Devel., Washington, 1972-78; program mgr. Bibliographical Ctr. for Rsch., Denver, 1978-80; mktg. mgr. Sorities Group, Inc., Springfield, Va., 1980-81; mktg. mgr. southeast region N.Y. Times Info. Svcs., Arlington, Va., 1982-83; libr., asst. libr. Inst. Def. Analyses, Alexandria, 1984-90; sr. libr. IIT Rsch. Inst., Lanham, Md., 1990—; cons. in field. V.p. Civic Assn., Annandale, Va., 1976-77, co-block capt., Springfield, 1990-92; vol. Fairfax County Pub. Schs., Springfield, 1980-88. With USAF, 1962-66. Independent. Achievements include planning of first DOD automated information center at DTIC; intra-automated database services as a research tool at HUD library; establishment of the Strategic Defense Initiative library at the Inst. Defense Analyses. Home: 6214 Greeley Blvd Springfield VA 22152 Office: BMD Tech Info Ctr Ste 708 1755 Jefferson Davis Hwy Arlington VA 22202

TILTON, ROBERT DAYMOND, chemical engineer; b. Summit, N.J., Sept. 2, 1964; s. Robert Joseph and Eleanor Svenborg (Nelson) T. BChE, U. Del., 1986; MS, Stanford U., 1987, PhD, 1991. Vis. scientist Inst. Surface Chemistry and Royal Inst. Tech., Stockholm, Sweden, 1991-92; Du Pont asst. prof. Carnegie Mellon U., Pitts., 1992—. Recipient Outstanding Young Faculty award E.I. Du Pont de Nemours & Co., 1992. Mem. Am. Inst. Chem. Engrs., Am. Chem. Soc. (Victor K. La Mer award 1993), Sigma Xi. Achievements include research in surface diffusion and intermolecular forces in protein absorption, polymer absorption phenomena. Office: Carnegie Mellon University 5000 Forbes Ave Pittsburgh PA 15213

TIMAR, TIBOR, chemist; b. Endrod, Bekes, Hungary, July 3, 1953; s. Beno and Gizella (Nemet) T.; m. Iren Nacsa, Mar. 27, 1982; 1 child, David. Degree in Chemistry with honors, U. Debrecen, Hungary, 1977, PhD, 1980, postdoctoral, 1992. Chemist Alkaloida Chem. Co., Tiszavasvari, Hungary, 1977-78, rsch. chemist, 1979-88, head of rsch., 1988—. Inventor in field; contbr. articles to profl. jours. Welch fellow Tex. A&M U., 1990-92. Mem. Am. Chem. Soc. Home: Kabay Janos 17, H-4440 Tiszavasvari Hungary Office: Alkaloida Chem Co, Kabay Janos 29, H-4440 Tiszavasvari Hungary

TIMASHEFF, SERGE NICHOLAS, chemist, educator; b. Paris, Apr. 7, 1926; came to U.S., 1939, naturalized, 1944; s. Nicholas S. and Tatiana (Rouzsky) T.; m. Marina J. Gorbunoff, 1953; 1 child, Marina S. BS, Fordham U., 1946, MS, 1947, PhD, 1951; D (hon.), U. Marseille-Aix, France, 1990. Instr. Fordham U., 1947-50; postdoctoral fellow Calif. Inst. Tech., 1951; rsch. fellow Yale U., 1951-55; head phys. chemistry investigation Ea. Regional Rsch. Lab. U.S. Dept. Agr., Wyndmoor, Pa., 1955-66, head Pioneering Rsch. Lab. Phys. Biochemistry, 1966-73; prof. dept. biochemistry Brandeis U., Waltham, Mass., 1966—; adj. prof. Drexel Inst. Tech., Phila., 1962-63; vis. prof. U. Paris, 1972-73, 82-83, 86-87, Universite Techniques de Lille, 1980; mem. BBCA Study Sect. NIH, 1968-72; mem. Fordham U. Coun., 1967—; sr. postdoctoral rsch. fellow NSF, 1959-61. Editor: (monograph series) Biological Macromolecules, 1966-75; volume editor Methods in Enzymology, 1972-86; mem. editorial bd. Archives Biochemistry and Biophysics, 1966-69, exec. editor, 1970-86; mem. editorial bd. Jour. Biol. Chemistry, 1972-78, Biochemistry, 1978-85, Biochim. Biophys. Acta, 1974-81, Archives Biochemistry and Biophysics, 1966-69, exec. editor, 1970-86. Recipient Arthur S. Flemming award Jr. C. of C. Washington, 1964, Disting. Svc. award U.S. Dept. Agr., 1965, Sci. Achievement award Fordham U., 1967, Frances Stone Burns award Am. Cancer Soc. Mass. div., 1974, Humboldt Rsch. award, 1993; Guggenheim fellow, 1972-73, sr. internat. Fogarty fellow, 1986-87. Fellow AAAS; mem. Am. Chem. Soc. (nat. award in chemistry of milk 1963, Phila. sect. award for creative rsch. 1966), Am. Soc. Biol. Chemistry, Biophys. Soc., Sigma Xi, Alpha Chi Sigma, Phi Beta Kappa. Achievements include rsch. on stabilization of proteins by solvent components, self-assembly of cellular organelles, mechanism of interaction of tubulin and microtubules with anti-cancer drugs, thermodynamics of the solution interactions of proteins with protein, biological and pharmacological ligands and protein structure forming and breaking agents. Office: Brandeis U Grad Dept Biochemistry 415 South St Waltham MA 02254-9110

TIMM, GARY EVERETT, science administrator, chemist; b. Sioux City, Iowa, Nov. 23, 1943; s. Everett Leroy and Jeanne (Anderson) T.; m. Phyllis Joan Eldred, July 10, 1971; children: Sean, Heather, Ryan. BS, La. State U., 1966; MS, U. Minn., 1971; MA, Humphrey Inst., 1973. Tech. advisor Mobile Source Air Pollution Control, Washington, 1973-78; policy analyst U.S. Dept. Energy, Washington, 1978-79; sect. chief Testing Rules Devel. Br. U.S. EPA, Washington, 1979-83, chief Test Rules Devel. Br., 1983-88, acting dir. existing chem. assessment, 1988-89, chief Chem. Testing Br., 1990-92; sr. environ. scientist Chem. Control Div. U.S. Dept. Energy, Washington, 1992—. Pres. Stuart Ridge Civic Assn., Herndon, Va., 1980-81; mem. airport adv. com. Fairfax County, Fairfax, Va., 1989—. Mem. Am. Chem. Soc. (lectr. 1989—). Avocations: music, travel. Home: 12025 Cheviot Dr Herndon VA 22070 Office: US EPA Chem Testing Br 401 M St SW E415D Washington DC 20460

TIMM, WALTER CLEMENT, chemical engineer, researcher; b. Phila., Sept. 28, 1931; s. Walter Clement Jr. and Geraldine Sarah (Hansford) T. BSChemE, U. Del., 1957. Process engr. Hyden Neport Chem. Co., Fords, N.J., 1957-58; plant chemist Flintkote Inc., Watertown, Mass., 1958-66; rsch. chemist Johns Manville (N.J.) Corp., 1967-69; rsch. engr. GAF Corp., Vails Gate, N.Y., 1969-81; sr. rsch. engr. Tarkett Inc., Vails Gate, N.Y., 1991—; mem. tech. affairs com. Resilient Floor Covering Inst., Rockville, Md., 1989-91. With U.S. Army, 1954-56. Mem. ASTM (com. for test methods 1990—), Soc. Plastics Engrs. Republican. Baptist. Achievements include patent for cross linking of plasticized PVC; inventor of 6 U.S. patents and 4 foreign patents. Office: Tarkett Inc Rt 94 Vails Gate NY 12584-9999

TIMMER, DAVID HART, civil engineer; b. Omaha, May 30, 1953; s. Donald Hendrik and Imogene Agnes (Hart) T.; m. Sharon Kathryn Fussnecker, May 11, 1974; children: Donald H., Douglas H., Karen M., Gerrit J. BA in History and Edn., Eastern Ky. U., 1975; BSCE, Ohio State U., 1976. Registered profl. engr., Ohio. Drafter survey crew Richland Engring. Ltd., Mansfield, Ohio, 1969-76, bridge inspector, 1976, civil engr., 1976-78, structural engr., 1978-80, design engr., 1980-88, chief bridge insp., 1984-88, project engr., 1989—. CCD instr. Resurrection Parish, Lexington, Ohio, 1986-88; coun. pres., asst. Webbs leader Cub Scout Pack 126, Bellville, Ohio, 1988-89, tiger cub leader, 1987-88. Mem. NSPE, ASCE. Republican.

TIMMERMAN, ROBERT WILSON, engineering executive; b. Abington, Pa., July 27, 1944; s. Clarence Arthur and Mildred Wilson (Slack) T.; m. Nancy Jean Spinka, Sept. 28, 1974; children: Robert Jr., Elizabeth Jane. M in Engring., Cornell U., 1966, BS, 1965; postgrad., Northwestern U., Evanston, Ill., 1971-72, U. Pa., 1972-74. Project engr. Monsanto Co., Springfield, Mass., 1966-68; staff engr. Stone and Webster Engring. Corp., Boston, 1968-71, United Engrs. and Constructor, Boston, 1974-75; sr. engr. R.W. Beck and Assoc., Wellesley, Mass., 1975-77; prin. R.W. Timmerman and Assoc., Boston, 1977—. Patentee in field; contbr. articles to profl. jours. Mem. ASME (sect. chmn. 1979-89, nat. com. pub. affairs, 1981), ASHRAE, Mass. Engrs. Coun. (treas. 1984-86). Presbyterian. Avocations: theater lighting. Home and Office: 25 Upton St Boston MA 02118-1609

TIN, KAM CHUNG, industrial chemist, educator; b. Hong Kong, Nov. 28, 1943; s. Tak Kwan and San Chun (Pul) T.; m. Vivian Ying Ding, Aug. 20, 1971; children: Dinah, David, Denise. BSc, Chinese U. Hong Kong, 1966; BSc with honors, Hong Kong U., 1967; PhD, U. Ottawa, Ont., Can., 1972. Rsch. assoc. Queen's U., Ont., Can., 1972-73; rsch. scientist, then indsl. chemist Can. Packers, Inc., Toronto, Ont., 1973-78, prodn. mgr. antibiotics, 1978-81, tech. supr. pharm. group, 1981-85; mgr. R&D Hands Fireworks, Inc., Ottawa, Ont., 85-87; tech. dir. Astra Can. Ltd., Ont., 87-90; sr. lectr. dept. applied sci. City Poly. Hong Kong, 1990—; gen. mgr., cons. Pharm. and Chem. Tech. Ctr., Hong Kong, 1992—; indsl. expert NATO Indsl. Adv. Group, Can., 1989-90. Contbr. articles on chlorination of sulfoxides, industrial processes for pharms., low-toxicity smoke compositions to profl. publs. Bd. dirs. Guelph (Can.) Dist. Multicultural Ctr., Inc., 1989-90; v.p. Guelph Chinese Cultural Assn., 1989-90. Mem. Am. Chem. Soc., Hong Kong Chem. Soc. (coun. 1991—, vice chmn. 1993—), Internat. Pyrotechnic Soc., N.Y. Acad. Scis. Achievements include U.S. patent, Can. patent for antiviral agents, development of industrial processes for semi-synthetic penicillins. Office: City Poly Hong Kong Appl Sc, 83 Tat Chee Ave, Kowloon Hong Kong

TINDELL, RUNYON HOWARD, aerospace engineer; b. N.Y.C., Mar. 26, 1933; s. Edward and Suzanne (Louis) T.; m. Carolyn May Werner, Apr. 29, 1961; children: Edward, Nancy. BSME, Bklyn. Poly. Inst., 1958. Engr. Republic Aviation Corp., Farmingdale, N.Y., 1958-64; prin. engr. Grumman Aerospace Corp., Bethpage, N.Y., 1964—; mem. Naval Aeroballistics com., Washington, 1980-82; presenter tech. papers at symposia. Contbr. articles on aircraft design to profl. publs. Fellow AIAA (assoc., air br. propulsion com. 1980—, Outstanding Contbn. award 1987); mem. Am. Soc. Automotive Engrs. Achievements include patent on flight control augmentation inlet device, invention of multi-blowing inlet lip, development of F-14 air inlet system, development of propulsion system sizing methodology. Office: Grumman Aircraft Systems Mail Stop B21-35 Bethpage NY 11714

TING, CHEN-HANSON, software engineer; b. Kun-Ming, Yun-Nan, China, Aug. 29, 1939; came to U.S., 1975; s. Chen-Wei and Ya-Hgoo (Hwang) T.; m. Kuei-Chiao Kan, July 29, 1969; children: Pei-Tao, Pei-Te, Pei-Chung. BS, Nat. Taiwan U., 1961; MS, U. Chgo., 1963, PhD, 1965. Rsch. fellow Calif. Inst. Tech., Pasadena, 1965-67; vis. assoc. prof. Nat. Taiwan U., Taipei, 1967-69; prof. Chung-Cheng Inst. Tech., Taoyuan, Taiwan, 1969-75; staff scientist Lockheed Missile and Space Co., Palo Alto, Calif., 1976-88; software engr. Maxtor Corp., San Jose, Calif., 1988-89; staff software engr. Applied Biosystems, Inc., Foster City, Calif., 1989—; v.p.

Forth Interest Group, San Jose, 1990-91. Author: Forth Notebook, Vol. 1, 1985, Vol. 2, 1990, FPC Manual, 1990, Forth Programming Language, 1980-92; editor newsletter More on Forth Engines, 1989-92. Recipient Figgy award Forth Interest Group, 1986. Democrat. Achievements include design and program of the high energy real time (radiographic) inspection system; design and implementation of Forth operating system; patent pending for catalyst microbiology workstation program. Office: Applied Biosystems Inc 850 Lincoln Centre Dr Foster City CA 94404

TING, PAUL CHENG TUNG, research scientist, inventor; b. Hopei, Republic of China, May 26, 1936; came to U.S., 1968; s. Wen and Pei-Feng Ting; m. Paule Myhre Ting (div. 1988); children: Benjamin Chang Li, Arlene Pei-Feng. BS in Aerospace Engring., Boston U., 1971, MS in Aerospace Engring., 1972; PhD, Va. Poly. Inst. and State U., 1975. Rsch. assoc. U. N.H., 1975-76; mem. tech. staff Space div. Rockwell Internat., 1976-88; prin. engr. Lockheed Engring. and Scis. Co., Houston, 1988-92; pres. Thermal Scis. Co., Houston, 1992—; founding mem. Internat. Hypersonic Rsch. Inst., Grand Forks, N.D., 1988—. Reviewer Aviation Week, Space Tech. Rsch. Adv., 1990-91; contbr. articles to nat. and internat. profl. jours. Mem. AIAA (reviewer Jour. Thermophysics and Heat Transfer 1986, 89). Achievements include patent pending for Method and Apparatus for Disposal/Recovery of Orbiting Space Debris (U.S., Europe, Russia, Ukraine, Kazistan, Japan, China), for method of anti/de-icing of runway and aircraft; development of thermodynamic energy balance equations for space shuttle orbiter gas compartment during ascent and entry; rsch. on physics, fluid dynamics, heat transfer and propulsion. Office: Thermal Scis Co 831 Fern Springs Ct Houston TX 77062-2196

TING, ROBERT YEN-YING, physicist; b. Kwei-yang, China, Mar. 8, 1942; came to U.S. 1965; s. Chi-yung and Shou-feng (Yang) T.; m. Teresa Yen-chun Ting, June 3, 1967; children: Paul H., Peggy Y. BS, Nat. Taiwan U., 1964; MS, MIT, 1967; PhD, U. Calif., San Diego, 1971. Rsch. engr. U.S. Naval Rsch. Lab., Washington, 1971-77, supervisory engr., 1977-80; supervisory physicist U.S. Naval Rsch. Lab., Orlando, Fla., 1980—; prof. George Washington U., 1972-80. Contbr. over 100 articles in rheology, polymer and acoustics to profl. jours. Fellow Acoustical Soc. Am.; mem. Am. chem. Soc., Am. Ceramics Soc., Am. Inst. Chem. Engrs. Office: US Naval Rsch Lab PO Box 568337 Orlando FL 32856-8337

TING, SAMUEL CHAO CHUNG, physicist, educator; b. Ann Arbor, Mich., Jan. 27, 1936; s. Kuan H. and Jeanne (Wong) T.; m. Susan Carol Marks, Apr. 28, 1985; children: Jeanne Min, Amy Min, Christopher M. BS in Engring, U. Mich., 1959, MS, 1960, PhD, 1962, ScD (hon.), 1978; ScD (hon.), Chinese U. Hong Kong, 1987, U. Bologna, Italy, 1988, Columbia U., 1990, Moscow State U., 1991. Ford Found. fellow CERN (European Orgn. Nuclear Research), Geneva, 1963; instr. physics Columbia U., 1964, asst. prof., 1965-67; group leader Deutsches Elektronen-Synchrotron, Hamburg, W.Ger., 1966; assoc. prof. physics MIT, Cambridge, 1967-68, prof., 1969—; Thomas Dudley Cabot Inst. prof. M.I.T., 1977—; program cons. Div. Particles and Fields, Am. Phys. Soc., 1970; hon. prof. Beijing Normal Coll., 1987, Jiatong U., Shanghai, 1987, U. Bologna, Italy, 1988. Assoc. editor: Nuclear Physics B, 1970; contbr. articles in field to profl. jours.; editorial bd.: Nuclear Instruments and Methods, Mathematical Modeling; advisor Jour. Modern Physics A. Recipient Nobel prize in Physics, 1976, De Gasperi prize in Sci., Italian Republic, 1988, Ernest Orlando Lawrence award U.S. Govt., 1976, Eringen medal Soc. Engring. Sci., 1977, Gold medal in Sci. City of Brescia, Italy, 1988, Golden Leopard award Town of Taormina, 1988; Am. Acad. Sci. and Arts fellow, 1975. Mem. NAS; fgn. mem. Pakistani Acad. Sci., Academia Sinica, Soviet Acad. Sci. USSR. Office: MIT Dept Physics 51 Vassar St Cambridge MA 02139-4308

TINKER, H(AROLD) BURNHAM, chemical company executive; b. St. Louis, May 16, 1939; s. H(arold) Burnham and Emily (Barnicle) T.; m. Barbara Ann Lydon, Feb. 20, 1965; children: Michael B., Mary K., Ann E. BS in Chemistry, St. Louis U., 1961; MS in Chemistry, U. Chgo., 1964, PhD in Chemistry, 1966. Sr. research chemist Monsanto, St. Louis, 1966-69, research specialist, 1969-73, research group leader, 1973-77, research mgr., 1977-81; tech. dir. Mooney Chems., Inc., Cleve., 1981-90, v.p. rsch. and devel., 1991—. Patentee in field; contbr. article to profl. jours. Mem. Am. Chem. Soc. (chmn. bd. St. Louis sect. 1978-79), Cleve. Assn. Rsch. Dirs. (v.p. 1989, pres. 1990, bd. dirs. 1991). Roman Catholic. Avocation: computers. Home: 2889 Manchester Rd Cleveland OH 44122-2570 Office: O M Group Mooney Chems Inc 2301 Scranton Rd Cleveland OH 44113-4395

TINNERINO, NATALE FRANCIS, electronics engineer; b. N.Y.C., Jan. 22, 1939; s. Frank and Anne (Miliotto) T.; m. Helga Cohnen, Jan. 23, 1971; children: Matthew, Stephanie, Patricia. BSEE, Manhattan Coll., 1961; MSEE, NYU, 1964. Electronics engr. Bendix Corp., Teleboro, N.J., 1961-64; sr. electronics engr. Norden divsn. United Techs., Norwalk, Conn., 1967-73; program mgr. Perkin Elmer Corp., Danbury, Conn., 1973-84; co-founder, chief engr. Visionetics Corp., Brookfield, Conn., 1984-89; dir. R & D Assembly Techs. divsn. Gen. Signal Corp., Horsham, Pa., 1989—. Author: Automatic Optical Inspection for Printed Circuits, 1989, Automatic Optical Inspection for Semiconductor-Assembly Equipment, 1992; editor Automatic Die Alignment System for Semiconductor Assembly Equipment, 1992; contbr. papers to profl. jours. Scholar Bishop Loughlin High Sch., 1953. Mem. IEEE. Roman Catholic. Achievements include patent for unique digital servo control system, automatic optical inspection system for printed circuit bds.; patents pending for unique TV/radar video noise reduction technique, U.S. currency front to back registration optical inspection system, automatic electro-optical die precising system for semiconductor manufacturing. Home: 4795 Essex Dr Doylestown PA 18901

TINOCO, IGNACIO, JR., chemist, educator; b. El Paso, Tex., Nov. 22, 1930; s. Ignacio and Laurencia (Aizpuru) T.; 1 child. B.S., U. N.Mex., 1951, D.Sc., 1972; Ph.D., U. Wis., 1954; postgrad., Yale, 1954-56. Mem. faculty U. Calif.-Berkeley, 1956—, prof. chemistry, 1966—; Rosser-Rivera disting. lectr. in biol. chemistry Calif. State U.-Los Angeles and U. Calif.-Riverside, 1983. Guggenheim fellow, 1964. Mem. Am. Chem. Soc. (Calif. award 1965), Am. Biophysics Soc., Am. Phys. Soc., Nat. Acad. Sci. Office: Univ of Calif Dept of Chemistry Berkeley CA 94720

TINTO, JOSEPH VINCENT, electrical engineer; b. N.Y.C., Aug. 23, 1957; s. Joseph Mario and Harriet Joyce (Murthum) T.; m. Sarah Elizabeth Smith, June 25, 1983; children: Elizabeth, David. BSEE, Maritime Coll., 1979. 3d engr. U.S. Steel Great Lakes Fleet, Duluth, Minn., 1979-80, Farrell Lines Shipping, N.Y.C., 1980-81; project engr. Allied Colloids Inc., Fairfield, N.J., 1981-83; mgr. engring. svcs. U.S. ops. Allied Colloids Inc., Suffolk, Va., 1983-91, mgr. engring. svcs. N.Am. ops., 1991—. Mem. IEEE, Water Environ. Fedn. Republican. Roman Catholic. Office: Allied Colloids 2301 Wilroy Rd Suffolk VA 23439

TIO, KEK-KIONG, mechanical engineering researcher; b. Medan, Indonesia, Aug. 20, 1960; came to U.S., 1978; s. Tio Siong-Khiang Teh Peh-Eng. BSME, U. So. Calif., 1983, MSME, 1985, PhD, 1990. Research engr. dept. applied mech. and engring. sci. U. Calif., La Jolla, 1991—. Contbr. articles to Jour. Heat Transfer, Internat. Jour. Heat and Mass Transfer, Jour. Thermophysics and Heat Transfer, Physics of Fluids A, Jour. Fluid Mechanics. Mem. ASME, Am. Phys. Soc., Phi Kappa Phi. Achievements include development of analytical method to solve linear problems with spatially periodic (and multiply connected) mixed boundary conditions; applications include spray cooling, heat transfer accross the interface of two solids in partial mechanical contact. Office: Univ Calif Dept Applied Mechanics & Engring Scis La Jolla CA 92093-0411

TIPIRNENI, TIRUMALA RAO, metallurgical engineer; b. Gudivada, India, 1948; came to U.S., 1973; s. Subrahmanyam and Vasumathi (Bobba) T.; m. Pavani Rathnam, Indupuganti, Mar. 1, 1978; children: Renuka, Anita, Vijay Srinivas. BS in Chem., Physics and Math., Andhra U., India, 1966; B Metall. Engring., Nagpur U., India, 1971; MS in Metallurgy, Stevens Inst. Tech., Hoboken, N.J., 1975. Chief metallurgist Structure Probe, Inc., Metuchen, N.J., 1975-78; lab. dir. Consolidated Testing Labs, Inc., New Hyde Park, N.Y., 1978-85; pres. L.I. Testing Labs, Inc., North Babylon, N.Y., 1985—. Pres. Telugu Literary and Cultural Assn., N.Y., 1978-79. Mem. ASM Internat. (chmn. L.I. chpt. 1986-87, 93-94), ASTM, The Metall. Soc., Am. Welding Soc., Soc. for Advancement of Materials and Process Engring., Soc. Automotive Engrs., Nat. Assn. Corrosion Engrs., Am. Soc. Quality Control, Soc. Plastics Engrs. Office: Long Island Testing Labs 243A Wyandanch Ave North Babylon NY 11704-1501

TIPP, KAREN LYNN WAGNER, school psychologist; b. Chgo., Feb. 15, 1947; d. Harry and Sarah (Damask) Wagner; m. Michael Harvey, Dec. 30, 1973; children: Brenda Alyse, Brandon Philip. BA in Gen. High Sch. Edn., Roosevelt U., 1971; B of Jewish Studies, Spertus Coll., 1973, cert. in sch. psychology, 1981, MA in Jewish Studies, 1993; MA in Jewish Studies, Nat. Louis U., 1993, MS, 1974. Cert. psychologist, Ill.; nat. cert. sch. psychologist. Tchr. Niles Twp. High Sch., Skokie, Ill., 1971-72; mgr. travel agy. Chgo., 1983-85; tchr. spl. edn. No. Cook County, Ill., 1972-90; tchr. Hebrew Chgo. Bd. Jewish Edn., 1969-90, interim prin. religious sch., 1989-90; sch. psychologist Chgo. Pub. Schs., 1990—; ind. ednl. therapist, Chgo., 1973-91; contract psychologist, N.W. Suburban Chgo., 1981-90; cons. learning disabled Chgo. Bd. Jewish Edn., 1983-90. Pres. Truman Coll. Coun., 1993—, City Coll. Chgo., 1985-87, nom. chair, 1990—; exec. sec. North Town Community Coun., Chgo., 1984-86, pres., 1989-91, v.p., 1991-93; pres. dist. 2 coun. Chgo. Bd. Edn., 1987-89, spl. ed. chair, 1990-92; exec. sec. North Town Civic League, 1978-80, pres., 1981-84; mem. coop. extension youth coun. U. Ill., sec., 1985-91, exec. coun., 1990-91; charter mem. Hild Culture Ctr., membership chair; beat rep. Chgo. Police Dept.; vice-chair Head Start, Salvation Army; corresponding sec. Day Care Ctr. Bd.-EVanston, 1978-81, Rogers Park Mental Health Coun.; vice chair Dewey Day Care Evanston, 1972-75, Rogers Park Montessori Sch., 1979-80; youth chair Indian Boundary Playground Bldg., 1986-89; mem. steering com. Rogers Park Centennial, 1991-93. Master Tchr. grantee Jewish Bd. Edn., Chgo., 1981-89, 20 Yr. award, 1990; recipient Community Leadership award Dept. Human Svcs., Chgo., 1985, North Town-Dorothy LeRoy Community Svc. award, 1992. Fellow Am. Orthopsychiat. Assn.; mem. NASP, Coun. Exceptional Children (liaison 1972-86), Ill. Psychol. Assn. (Sch. Psychologist of Yr. 1991), Family Resource Ctr. on Disabilities (spl. edn. com. 1990—), Profls. in Learning Disabilties (legis. chair 1987—), Children with Attention Deficit Disorder, Ill. Sch. Psychologists Assn. (Practitioner of Yr. 1991), Chgo. Assn. Sch. Psychologists, Greater Uptown Youth Network, Family Resource Handicapped (spl. edn. com.), Ill. 4-H Found. Avocations: education, animals, playground building, activities with youth and community. Home: 6730 N Maplewood Ave Chicago IL 60645-4620

TIPPUR, HAREESH V., mechanical engineer; b. Bangalore, India, May 9, 1958; came to U.S., 1983; s. Venkatanarasaiah and Premaleela Tippur; m. Jayshree A. Mysore, 1988; 1 child, Ankita H. M of Engring., Indian Inst. Sci., Bangalore, 1982; PhD, SUNY, Stony Brook, 1988. GALCIT rsch. fellow Calif. Inst. Tech., Pasadena, 1988-90; asst. prof. Auburn (Ala.) U., 1990—. Contbr. articles to Applied Optics, Exptl. Mechs., Internat. Jour. Fracture, Jour. of Mechs. and Physics of Solids, Optics and Lasers in Engring. Mem. ASME (assoc.), Am. Soc. Engring. Edn. (Southeastern Region Outstanding Rsch. award 1993), Am. Acad. Mechs., Soc. Exptl. Mechs. (Hetényi award 1992). Achievements include devel. of optical interferometric techniques to study fracture mechs. of materials. Office: Auburn U 201 Ross Hall Auburn AL 36849

TIPTON, JON PAUL, allergist; b. Lynchburg, Ohio, Nov. 8, 1934; s. Paul Alvin and Jeanette (Palmer) T.; m. Martha J. Johnson, Dec. 29, 1968; children: Nicole Ann, Paula Michelle. BS, Ohio U., 1956; MD, Ohio State U., 1960. Resident internal medicine Ohio State U. Hosps., Columbus, 1964-66; fellow in allergy and pulmonary disease Duke U. Med. Ctr., Columbus, 1963-64, 66-67; pvt. practice medicine specializing in allergies Athens, Ohio, 1967-74; pvt. practice medicine specializing in allergy Marietta, Ohio, 1974—; dir. cardio respiratory therapy Marietta Meml. Hosp., 1983, med. dir. pulmonary rehab. program, chief of medicine; cons. Ohio U. Hudson Health Ctr., 1967—, Marietta Coll. Health Ctr., 1974—, United Mine Workers of Am. Funds, 1984—. Vol. Marietta Rep. Hdqrs., 1978—; mem. choir St. Luke's Luth. Ch., Marietta, 1983—. Served to capt. USAF, 1961-63. Mem. Am. Acad. Allergy, Ohio State Med. Assn., Wash. County Med. Soc., Parkersburg Acad. Medicine. Republican. Methodist. Avocations: yardwork, piano, attending plays, football, children. Home: 101 Meadow Ln Marietta OH 45750-1345 Office: 100 Front St Marietta OH 45750-3142

TIPTON, KENNETH WARREN, agricultural administrator, researcher; b. Belleville, Ill., Nov. 14, 1932; s. Roscoe Roy and Martha Pearl (Davis) T.; m. Barbara Adds, Mar. 2, 1957; children: Kenneth Warren Jr., Nancy Tipton O'Neal. BS, La. State U., 1955, MS, 1959; PhD, Miss. State U., 1969. Asst. prof. Agrl. Ctr., La. State U., Baton Rouge, 1959-70, assoc. prof., 1970-75, prof., 1975—; supt. Red River Rsch. Sta., La. Agrl. Expt. Sta. Agrl. Ctr., La. State U., Bossier City, 1975-79; assoc. dir. La. Agrl. Expt. Sta. Agrl. Ctr., La. State U., Baton Rouge, 1979-89, dir. La. Agrl. Expt. Sta., vice chancellor, 1989—; mem. com. nine USDA/Coop. State Rsch. Svc., 1986-88; Expt. State Com. Orgn. Policy, 1988-91. Contbr. articles to Agronomy Jour., Jour. Econ. Entomology, Grain Sorgham Conf. Coach baseball program Am. Legion, 1969-74; scoutmaster Boy Scouts Am., Baton Rouge, 1970-75. Capt. USAF, 1955-58. Mem. Am. Soc. Agronomy, Crop Sci. Soc. Am., Coun. Agrl. Sci. Tech. Achievements include research on inheritance of fiber traits in cotton, resistance of grain sorghum hybrids to bird damage, tannin content of grain sorghum and effects of phosphorus on growth of sorghum. Home: 732 Baird Dr Baton Rouge LA 70808-5916

TIPTON, THOMAS WESLEY, flight engineer; b. Okmulgee, Okla., May 12, 1952; s. John Melvin and Norma Dean (Boyd) T.; children: Samuel Lawrence, Stacy Lynn; m. Evelyn Marie Harzinski, Sept. 17, 1988; 1 child, Jonathan Clark. Student, William Penn Coll., Oskaloosa, Iowa, 1970-72. Enlisted U.S. Navy, 1972; div supr. Fighter Squadron 101, Naval Air Sta., Oceana, Va., 1973-76; quality assurance officer Naval Air Sta., Keflavik, Iceland, 1976-79; drill instr. Recruit Tng. Ctr., Great Lakes, Ill., 1980-83; flight engr., evaluator Patrol Squadron 26, Brunswick, Maine, 1983-89; flight engr., test and evaluation Force Warfare Naval Air Warfare Ctr. Aircraft Div., Patuxent River, Md., 1989-93. Mem. Nat. Rifle Assn. (life), Smithsonian Air and Space Mus., Friends of the Kennedy Ctr., Nat. Geographic Soc., Nat. Wildlife Fund. Avocations: reading, woodworking, wildlife conservation, golf. Home: 436 Council Bluffs Lusby MD 20657 Office: Force Warfare Directorate Naval Air Warfare Ctr Patuxent River MD 20670

TIRAS, HERBERT GERALD, engineering executive; b. Houston, Aug. 11, 1924; s. Samuel Louis and Rose (Seibel) T.; m. Aileen Wilkenfeld, Dec. 14, 1955; children—Sheryle, Leslie. Student, Tex. A. and M. U., 1941-42; attended, Houston U., 1942-65, student, Nat. Defence U., 1986. Registered profl. engr., Calif. Cert. mfg. engr. in gen. mfg.; robotics; mfg. mgmt; gen. mgmt. Engr., Reed Roller Bit, Houston, 1942-60; pres. Tex. Truss, Houston, 1960-77; chief exec. officer Omnico, Houston, 1977—; Nat. Defense exec. res. resources officer, Region VI Fed. Emergency Mgmt. Agy., 1982—. Served to 1st lt. CAP, 1954-61. Mem. Machine Vision Assn., Nat. Defense U. Found., Am. Assn. Artificial Intelligence, Soc. Mfg. Engrs., Robot Inst. Am., Robotics Internat., Marine Tech. Soc., Coll. and Univ. Mfg. Ednl. Council (nat. dir.), Assn. of the Indsl. Coll. of the Armed Forces. Lodge: Masons, Shriners. Home: 9703 Runnymeade Dr Houston TX 77096-4219 Office: PO Box 2872 Houston TX 77252-2872

TIRKEL, ANATOL ZYGMUNT, physicist; b. Krakow, Poland, Aug. 30, 1949; s. Alfred and Helena (Better) T. BS with honors, Monash U., 1971, PhD, 1975. Rsch. scientist Varian Techtron, Melbourne, Victoria, Australia, 1975; lectr. RMIT, Applied Physics, Melbourne, 1976; rsch. engr. Martin Marietta Rsch. Labs., Balt., 1977-78; engring. supr. Hughes Aircraft Co., Torrance, Calif., 1978-80; sect. head TRW, Redondo Beach, Calif., 1980-81; lectr. RMIT, Communication Engring., 1982-84; sr. lectr. RMIT, Applied Physics, 1985-86; dir. Aranda Applied Rsch., Melbourne, 1987-89, Sci. Tech., Melbourne, 1990—; com. mem. ASIA, 1987-90, Victoria, Australia, Swinburne Course Advisory, Victoria, 1989-91, RMIT Course Advisory, 1983-86. Contbr. articles to profl. jours.; inventor in field. Mem. IEEE (sr. mem.), AIP, Australian Inst. Physics, Australian Scientific Industries. Avocations: travel, sports, photography, music.

TIRRE, WILLIAM CHARLES, research psychologist; b. St. Louis, Oct. 22, 1952; s. Charles F. and Edna Marie (Rademeyer) T.; m. Karla Jane Schuette, Mar. 26, 1983; 1 child, Matthew. BA in Psychology, St. Louis U., 1975; MS in Psychology, Ill. State U., 1977; PhD in Ednl. Psychology, U. Ill., 1983. Rsch. asst. U. Ill., Champaign, Ill., 1977-81; rsch. assoc. Inst. for Personality and Ability Testing, Savoy, Ill., 1981-82; rsch. psychologist USAF Armstrong Lab., Brooks AFB, Tex., 1982—. Cons. editor Jour. of Genetic Psychology, 1991—, Genetic, Social and Gen. Psychology Monographs, 1991—; contbr. articles to profl. jours. Sun. sch. tchr. MacArthur Park Luth. Ch., San Antonio, 1988—. Recipient Air Force Office of Sci. Rsch. Star Team award, 1992, Chief Scientist's Team award, 1989. Mem. Am. Psychol. Soc., Internat. Soc. for Study of Individual Differences. Achievements include research on individual differences in cognitive abilities; development of a comprehensive computer-administered aptitude battery for use in selection and classification of Air Force personnel; others. Office: Armstrong Lab 7909 Lindbergh Brooks AFB TX 78235-5352

TISDALE, PATRICK DAVID, retired pediatrician; b. Fayetteville, N.C.; s. Henry Edward and Mary Elisabeth (McCarthy) T.; m. Jacqueline Lee May, Apr. 15, 1955 (dec. 1969); children: Patrick David Jr., Daniel Joseph Quinn, Sean Christopher, James Shannon, Neal Cameron, Xuan L., Lien M., Mai L., Thuvan E., Kim; m. Anna Eleanor Asher, Feb. 28, 1993; 1 stepson, Hugh Warren. BS in Engring., U.S. Mil. Acad., 1946-50; postgrad., George Wash. U., 1953-54; MD, Georgetown Med. Sch., 1954-58; cert., Command and Gen. Staff Sch., Fort Leavenworth, Kans., 1968-69. Diplomate Am. Acad. Pediatrics. Commd. 2d. lt. U.S. Army, 1950, advanced through grades to col., 1973; demolitions expert U.S. Army Corps Engrs., 1950-53, comdr., chief of svc., 1954-73; comdr. 1st Med. Battalion, Vietnam, 1967-68; retired U.S. Army, 1973; pvt. practice Columbus, Ga., 1973-82; assoc. prof. Emory U., 1973-82; contract physician Wash., 1982-84, Arabia, 1984-87, Wash., 1987-88; pediatrican Mountain Comprehensive Health Care, Ky., 1988-91. Decorated Legion of Merit, Bronze Stars (3), Air medals (2), Honor Medals 1st class and 3d class (Vietnamese); recipient Presdl. Citation, Kop Chai Lai award Thomas A. Dooley Found., 1970, Franz Sichell award LaSalle Mil. Acad., 1972, William Randolph Hearst medals, 1945, 46. Mem. Am. Acad. Pediatrics, Assn. Grads. West Point, Nat. Geographic Soc. Democrat. Roman Catholic. Avocations: scuba diving, white water rafting, hiking, gardening, writing.

TISON, DAVID LAWRENCE, microbiologist; b. Whittier, Calif., Sept. 26, 1952; s. George L. and Phyllis M. T.; m. Regina Ann Malinowski, Sept. 22, 1979. BS, U. Puget Sound, 1974; MS, U. Idaho, 1976; PhD, Rensselaer Poly. Inst., 1980. Diplomate Am. Bd. Med. Microbiology. Rsch. asst. dept. bacteriology U. Idaho, Moscow, 1974-76; rsch. asst. dept. biology Rensselaer Poly. Inst., Troy, N.Y., 1976-77, teaching asst. dept. biology, 1977-78; grad. rsch. fellow Savannah River Lab. Oak Ridge Assoc. Univs., Aiken, S.C., 1978-80; rsch. assoc. Oreg. State U., Corvallis, 1980-82; fellow clin. microbiology U. Tex. Med. Br., Galveston, 1982-84; clin. microbiologist MultiCare Med. Ctr., Tacoma, 1984—; cons., lectr., presenter in field. Reviewer, contbr. articles to profl. publs., chpts. to books. Mem. adv. bd. Nature Ctr. at Snake Lake, Tacoma, 1987—. Fellow Am. Acad. Microbiology; mem. APHA, Am. Soc. Microbiology, Am. Soc. Clin. Pathologists, Am. Soc. Med. Tech., Sigma Xi. Office: MultiCare Med Ctr PO Box 5299 Tacoma WA 98415-0299

TISSER, CLIFFORD ROY, electrical engineer; b. N.Y.C., Dec. 30, 1946; s. Israel Solomon and Evelyn (Sander) T.; m. Rita L. Gehring, Jan. 26, 1973; children: David M., Daniel J. BEE cum laude, CCNY, 1967; MEE, U. Wis., Milw., 1980. Registered profl. engr., Wis. Field svc. engr. Allis-Chalmers Corp., Milw., 1969-73, customer svc. engr., 1973-76; sr. application engr. Siemens-Allis, Inc., Milw., 1976-80; project mgr. Entech Engring., Inc., Elm Grove, Wis., 1980-81; pres. Power Core Engring., Inc., Milw., 1981—; lectr. U. Wis. Extension, Madison, 1980-82, Milw. Sch. Engring., 1981-84; speaker, developer courses Hughes Inst. Continuing Edn., 1982—. Mem. phys. plant and ops. com. Shorewood (Wis.) Sch. Dist., 1983—. Mem. IEEE, NSPE. Office: Power Core Engring Inc MEC 223 2821 N 4th St Milwaukee WI 53212-2300

TITTMAN, JAY, physicist, consultant; b. Bayonne, N.J., Dec. 28, 1922; s. Martin David and Julia (Rosenblum) T.; m. Eleanor Gelber, June 4, 1944; children: Carol S., Nancy Tittman Keefe, Barbara Tittman Markusson. AB, Drew U., 1944; MA, Columbia U., 1948, PhD, 1951. Physicist Schlumberger-Doll Rsch. Ctr., Ridgefield, Conn., 1951-54, sect. head nuclear physics, 1954-67, dept. head physics rsch., 1967-72; dept. head engring. physics Schlumberger Well Svcs., Houston, 1972-78, dir. devel. engring., 1978-81; sci. cons. Schlumberger-Doll Rsch. Ctr., Ridgefield, 1981-83; cons. Jay Tittman Tech. Consulting Svcs., Danbury, Conn., 1983—; mem. nuclear logging subcom. Am. Petroleum Inst., Dallas, 1973-80, organizing com. 4th Small Accelerator Conf., Denton, Tex., 1976, organizing com. Conf. on Physics in Petroleum Industry, Austin, 1981; adj. prof. phys. sci. New Eng. Inst., Ridgefield, 1970-72. Author: Geophysical Well Logging, 1986; contbr. articles to profl. jours. Bd. dirs. Assn. Religious Communities, Danbury, 1969-72, 82—, Jewish Home for Elderly of Fairfield County, 1986—; pres. United Jewish Ctr. of Danbury, 1983-84. Lt. (j.g.) USNR, 1944-46, PTO. Fellow Am. Phys. Soc.; mem. AAAS, Soc. Petroleum Engrs., Soc. Profl. Well Log Analysts (Gold Medal 1993), Sigma Xi. Jewish. Achievements include 15 patents in nuclear well logging. Home and Office: 11 Tanglewood Dr Danbury CT 06811

TIWARI, SUBHASH RAMADHAR, chemical engineer; b. Ahmedabad, Gujarat, India, May 15, 1949; came to U.S., 1970; s. Ramadhar and Bachidevi T.; m. Carol L. Cramer; children: Shalimar, Julie, Stephanie, Jessica. MS in Chemistry, Gujarat U., 1970; BS in Chem. Engring., U. Tulsa, 1972. Plant engr. Monogram Mills, Ahmedabd, India, 1979-80; factory control engr. Firestone, Okla. City, 1980-81; chem. engr. Tech. Rsch. and Devel., Okla. City, 1981-83; reverse engring. program mgr. USAF Tinker AFB, Okla. City, 1983-88; sr. aerospace engr. U.S. Army ASAT Program Office, Huntsville, Ala., 1988—. Contbr. articles to profl. jours. Recipient Dr. Bhabha Rsch. scholarship Gov. of India, Ahmedabad, 1970, Commander's award Dept. of Air Force, 1985, 86, Hq Commanders award, Dept. Air Force, 1987, Sec. of Air Force Award, 1987, Sec. of Defense award, 1988, Tinker AFB; Cert. of Achievement Dept. Army, 1990, commendation Dept. of Army, Huntsville Ala., 1990, '91, '92; invited mem. Presidential Coun. for Tech. Exchange with Soviet Union, 1992. Achievements include an R.E. program that saved $144M; modification of mfg. at Firestone saved 1.5M Yr.; mem. of design team for one of largest NDI facilities in world. Home: 1216 Brandywine Ln Decatur AL 35601 Office: USASSDC-ASAT PO Box 1500 Huntsville AL 35807

TKACH, ROBERT WILLIAM, physicist; b. Lorain, Ohio, Jan. 26, 1954; s. Joseph John and Helen Jeanette (Ryan) T.; m. Ann Christine Von Lehmen, Aug. 17, 1979; 1 child, Nicholas Ryan. BSBA, U. Cin., 1976; PhD of Physics, Cornell U., 1982. Rsch. assoc. Cornell U., Ithaca, N.Y., 1982-84; staff AT&T Bell Labs., Holmdel, N.J., 1984—. Contbr. articles to profl. jours. Mem. IEEE (assoc. editor Jour. of Lightwave Tech. 1992-95), Am. Phys. Soc., Optical Soc. Am., World Shorinji Kempo Orgn. Achievements include patents in field. Home: 27 Westwood Rd Little Silver NJ 07739-1754 Office: AT&T Bell Labs 791 Holmdel-Keyport Rd Holmdel NJ 07733-0400

TOBE, STEPHEN SOLOMON, zoology educator; b. Niagara-on-the-Lake, Ont., Can., Oct. 11, 1944; s. John Harold and Rose T. (Bolter) T.; m. Martha Reller. BSc, Queen's U., Kingston, Ont., 1967; MSc, York U., Toronto, Ont., 1969; PhD, McGill U., Montreal, Que., Can., 1972. Rsch. fellow U. Sussex, Eng., 1972-74; asst. prof. U. Toronto, 1974-78, assoc. prof., 1974-78, prof., 1982—, assoc. dean scis., faculty arts and sci., 1988-93; vis. prof. U. Calif., Berkeley, 1981, Nat. U. Singapore, 1987, 1993—, U. Hawaii, 1988; mem. animal biology grant selection com. Natural Scis. and Engring. Rsch. Coun. Can., 1986-89, chair, 1988-89; lectr. Internat. Congress Entomology, Vancouver, B.C., Can., 1988; cons. in hydroponics. Editor Insect Biochemistry, 1987; mem. editorial bd. Jour. Insect Physiology, 1980—, Physiol. Entomology, 1985—, Life Scis. Advances, 1987—; contbr. numerous articles to profl. jours., book chpts. Recipient Pickford medal in comparative endocrinology, 1993; E.W.R. Steacie fellow Natural Scis. and Engring. Rsch. Coun. Can., 1982-84. Fellow Royal Soc. Can., Royal Entomol. Soc.; mem. AAAS, Can. Soc. Zoologists, Am. Soc. Zoologists, Entomol. Soc. Can. (C. Gordon Hewitt award 1982, Gold medal 1990), Entomol. Soc. Am., Soc. Exptl. Biology, Soc. for Neurosci., Sigma Xi. Avocations: amateur radio, gardening, hydroponics. Home: 467 Soudan

Ave, Toronto, ON Canada M4S 1X1 Office: U Toronto Dept Zoology, 25 Harbord St, Toronto, ON Canada M5S 1A1

TOBERT, JONATHAN ANDREW, clinical pharmacologist; b. London, Eng., Dec. 10, 1945; came to U.S., 1975; s. Gerald and Alexandra Tobert; m. Michelle Iris Sternfeld, Sept. 25, 1983; children: Emily, Gwen. MB, BChir, U. Cambridge, Eng., 1970; PhD, U. London, Eng., 1975. Teaching fellow U. London, 1971-75; postdoctoral fellow Harvard U., Cambridge, Mass., 1975-76; assoc. dir. Merck Sharp & Dohme Rsch. Labs., Rahway, N.J., 1976-79, dir., 1979-82; sr. physician Merck Sharpe & Dohme Rsch. Labs., Rahway, N.J., 1982-86, sr. dir., 1986-88, exec. dir., 1988-91, disting. sr. physician clin. rsch., 1992—. Contbr. articles to profl. jours. Lalor Found. fellow, 1975. Mem. Am. Heart Assn. (Arteriosclerosis Coun.). Achievements include development of lovastatin, the first HMG-CoA reductase inhibitor, which has revolutionized the treatment of hypercholesterolemia. Office: Merck Sharp & Dohme Rsch PO Box 2000 Rahway NJ 07065-0900

TOBET, STUART ALLEN, neurobiologist; b. Bklyn., Feb. 7, 1956; s. Martin and Roslyn (Katz) T.; m. Joan Caluda King, Apr. 9, 1979. BS, Tulane U., 1978; PhD, MIT, 1985. Postdoctoral fellow Shriver Ctr., Waltham, Mass., 1985-88, asst. scientist, 1989—; asst. prof. dept. neurology Harvard Med. Sch., Boston, 1990—. Contbr. articles to profl. jours. and chpts. to books. Recipient Whitaker fellowship, 1982-84, NIMH Postdoctoral fellowship, 1984-85; grantee NSF, 1990, W.F. Milton Fund, 1992. Mem. AAAS, Soc. for Neurosci., Endocrine Soc., Soc. for Study of Reproduction. Achievements include discovery of a region in ferret brain that differentiates under the influence of prenatal estrogen; identification of molecular and morphological sites of action for gonadal steroids during brain development. Office: Shriver Ctr 200 Trapelo Rd Waltham MA 02254

TOBIN, CALVIN JAY, architect; b. Boston, Feb. 15, 1927; s. David and Bertha (Tanfield) T.; m. Joan Hope Fink, July 15, 1951; children—Michael Alan, Nancy Ann. B.Arch., U. Mich., 1949. Designer, draftsman Arlen & Lowenfish (architects), N.Y.C., 1949-51; with Samuel Arlen, N.Y.C., 1951-53, Skidmore, Owings & Merrill, N.Y.C., 1953; architect Loebl, Schlossman & Bennett (architects), Chgo., 1953-57, v.p., 1953-57; v.p. Loebl Schlossman & Hackl, 1957—; Chmn. Jewish United Fund Bldg. Trades Div., 1969; chmn. AIA and Chgo. Hosp. Council Com. of Hosp. Architecture, 1968-76. Archtl. works include Michael Reese Hosp. and Med. Ctr., 1954—, Prairie Shores Apt. Urban Redevel., 1957-62, Louis A. Weiss Meml. Hosp., Chgo., Chgo. State Hosp., Central Community Hosp., Chgo., Gottlieb Meml. Hosp., Melrose Park, Ill., West Suburban Hosp., Oak Park, Ill., Thorek Hosp. and Med. Ctr., Chgo., Water Tower Pl., Chgo., Christ Hosp., Oak Lawn, Greater Balt. Med. Ctr., Shriners Hosp. for Crippled Children Chgo., Hinsdale Hosp., Hinsdale, Ill., South Chgo. Community Hosp., Chgo., Mt. Sinai Med. Ctr., Chgo., also numerous apt., comml. and community bldgs. Chmn. Highland Park (Ill.) Appearance Rev. Commn., 1972-73; mem. Highland Park Plan Commn., 1973-79; mem. Highland Park City Coun., 1974-89, mayor pro-tem, 1979-89; mem. Highland Park Environ. Control Commn., 1979-84, Highland Park Hist. Preservation Commn., 1982-89; bd. dirs. Highland Park Hist. Soc., Young Men's Jewish Coun., 1953-67, pres., 1967; bd. dirs. Jewish Community Ctrs. Chgo., 1973-78, bd. dirs., 1989—; Ill. Coun. Against Handgun Violence, 1989—; trustee Ravinia Festival Assn., 1990—. With USNR, 1945-46. Fellow AIA (2d v.p. Chgo. chpt.); mem. U. Mich. Alumni Soc. Coll. Architecture and Urban Planning (bd. govs. 1989—), U. Mich. Alumni Assn. (bd. dirs. 1990—, v.p. 1993-94), Pi Lambda Phi. Jewish. Clubs: Standard, Ravinia Green Country Club. Home: 814 Dean Ave Highland Park IL 60035-4749 Office: Loebl Schlossman & Hackl 130 E Randolph St Chicago IL 60601-6206

TODARO, GEORGE JOSEPH, pathologist; b. N.Y.C., July 1, 1937; s. George J. and Antoinette (Piccinni) T.; m. Jane Lehv, Aug. 12, 1962; children: Wendy C., Thomas M., Anthony A. B.S., Swarthmore Coll., 1958; M.D., N.Y. U., 1963. Intern NYU Sch. Medicine, N.Y.C., 1963-64, fellow in pathology, 1964-65, asst. prof. pathology, 1965-67; staff assoc. Viral Carcinogenesis br. Nat. Cancer Inst., Bethesda, Md., 1967-70, head molecular biology sect., 1969-70; chief Viral Carcinogenesis br. Nat. Cancer Inst. (Lab. Viral Carcinogenesis), 1970-83; sci. dir., pres. Oncogen, Seattle, 1987-90; sr. v.p. exploratory biomed. rsch. Bristol-Myers Squibb Pharm. Rsch. Inst., 1990—; faculty mem. Genetics Program, George Washington U.; affiliate prof. pathology U. Wash., Seattle, 1983—. Editor: Cancer Research, 1973-86, Archives of Virology, 1976—, Jour. Biol. Chemistry, 1979—; contbr. articles to profl. jours. Served as med. officer USPHS, 1967-69. Recipient Borden Undergrad. Research award, 1963, USPHS Career Devel. award, 1967, HEW Superior Service award, 1971, Gustav Stern award for virology, 1972, Parke-Davis award in exptl. pathology, 1975; Walter Hubert lectr. Brit. Cancer Soc., 1977. Mem. Nat. Acad. Scis., Am. Soc. Microbiology, Am. Assn. Cancer Research, Soc. Exptl. Biology and Medicine, Am. Soc. Biol. Chemists, Am. Soc. Clin. Investigation. Home: 1940 15th Ave E Seattle WA 98112-2829 Office: Oncogen 3005 1st Ave Seattle WA 98121-1035

TODD, ALEXANDER ROBERTUS (BARON TODD OF TRUMPINGTON), chemistry educator; b. Glasgow, Scotland, Oct. 2, 1907; s. Alexander and Jane (Lowrie) T.; m. Alison Sarah Dale, Jan. 30, 1937 (dec. 1987); children: Alexander Henry, Helen Todd Brown, Hilary Alison. B.Sc. (Carnegie Research scholar 1928-29), U. Glasgow, 1928, D.Sc., 1938; Dr.phil.nat., U. Frankfurt am Main, 1931; D.Phil., Oxford U., Eng., 1933; M.A., U. Cambridge, Eng., 1944; LL.D. (hon.), univs. of Glasgow, Melbourne, Edinburg, Cal., Manchester, Hokkaido, Melbourne, Edinburh, Cal., Manchester, Hokkaido; Dr.rer.nat. (hon.), U. Kiel; D.Litt. (hon.), U. Sydney; D.Sc. (hon.), univs. of London, Exeter, Warwick, Sheffield, Liverpool, Oxford, Leicester, Durham, Eng., Univ. of Wales, U. Madrid, Spain, U. Aligarh, India, U. Strasbourg, France, Harvard U., Yale U.,U. Mich., U. Paris, U. Adelaide, Australia, U. Strathclyde, Scotland, Australian Nat. U., U. Cambridge, U. Philippines, Tufts U., Chinese U. Hong Kong, Hong Kong U. Mem. staff Lister Inst. Preventive Medicine, London, 1936-38; reader biochemistry U. London, 1937-38; prof., dir. chem. labs. U. Manchester, Eng., 1938-44; prof. organic chemistry U. Cambridge, Eng., 1944-71, fellow Christ's Coll.,, 1944—, master, 1963-78; chancellor U. Strathclyde, 1963-91; dir. Fisons Ltd., London, 1963-78, Nat. Rsch. Devel. Corp., London, 1968-76; vis. prof. Calif. Inst. Tech., 1938, U. Chgo., 1948, U. Sydney, 1950, MIT, Cambridge, 1954, U. Calif., 1957, Tex. Christian U., 1980; chmn. adv. Coun. Sci. Policy, 1952-64, Royal Commn. Med. Edn., 1965-68. Contbr. articles to profl. jours. Chmn. Nuffield Found., London, 1936-80; chmn. govs. United Cambridge Hosps., 1969-74; chmn. trustees Croucher Found., Hong Kong, 1980-87, pres., 1988—. Created knight, 1954, baron (life peer), 1962; Order of Merit (U.K.); Order Rising Sun (Japan); recipient Nobel prize for chemistry, 1957; Pour le Merite (W. Germany), 1966; Lomonosov medal U.S.S.R. Acad. Sci., 1978; medals various chem. socs., sci. orgns., including Royal Copley medals Royal Soc., 1949; named master Salter's Co., 1961. Fellow Royal Soc. (pres. 1975-80), Australian Chem. Inst. (hon.), Manchester Coll. Tech. (hon.), Royal Soc. Edinburgh (hon.), Royal Coll. Physicians London (hon.), Royal Coll. Physicians, Surg. Glasgow; mem. AAAS, NAS, French Chem. Soc. (hon.), German Chem. Soc. (hon.), Spanish Chem. Soc. (hon.), Belgian Chem. Soc. (hon.), Swiss Chem. Soc. (hon.), Japanese Chem. Soc. (hon.), Australian Acad. Sci., Austrian Acad. Sci., Ghana Acad. Sci., Polish Acad. Sci., Russian Acad. Sci., Acad. Natural Philosophy Halle (Germany), Am. Philos. Soc., N.Y. Acad. Sci., Chem. Soc. (pres. 1960-62), Internat. Union Pure and Applied Chemistry (pres. 1963-65), Soc. Chem. Industry. Office: 9 Parker St, Cambridge England

TODD, BETH ANN, mechanical engineer, educator; b. Greensburg, Pa., Aug. 31, 1959; d. Daniel Webster and Ellen Hattie (Rugh) T. BS in Engring. Sci., Pa. State U., 1981; MS in Applied Mechanics, U. Va., 1986, PhD in Mech. and Aerospace Engring., 1992. Assoc. engr. Westinghouse Corp., West Mifflin, Pa., 1981-83; rsch. asst. Rehab. Engring. Ctr. U. Va., Charlottesville, 1983-90; instr. GMI Engring. and Mgmt. Inst., Flint, Mich., 1990-92; asst. prof. Univ. Ala., Tuscaloosa, 1992—. Vice pres. St. Mark Luth. Ch., Charlottesville, 1988, pres., 1989. Mem. ASME (chpt. bd. dirs. 1991-92), NSPE, Soc. Women Engrs., Resna, Sigma Xi, Tau Beta Pi, Pi Tau Sigma. Office: U Ala Dept Engring Mechanics PO Box 870278 Tuscaloosa AL 35487-0278

TODD, DONALD FREDERICK, geologist. Recipient Michel T. Halbouty Human Needs award Am. Assn. Petroleum Geologists, 1992. Office: Constellation Group 1700 Broadway Ste 420 Denver CO 80290-0101*

TODD, JAMES S., surgeon, educator, medical association administrator; b. Hyannis, Mass., 1931. Intern Presbyn. Hosp., N.Y.C., 1957-58, resident in surgery, 1959-63; resident in surgery Delafield Hosp., N.Y.C., 1963; resident ob-gyn. Sloane Womens Hosp., N.Y.C., 1958-59; resident Valley Hosp., Ridgewood, N.J.; clin. asst. prof. surgery U. Medicine and Dentistry N.J., Newark; pres. AMA. Office: American Medical Assocation 515 N State St Chicago IL 60610-4320*

TODD, RICHARD HENRY, retired physican, investor; b. Pottstown, Pa., Mar. 25, 1906; s. John Henry and Effie (Davis) T.; m. Lydia Carey Dick, Feb. 1, 1930; children: Richard Henry Jr., John Andrew. AB, Johns Hopkins U., 1929, MD, 1933. instr. Georgetown U. sect. Children's Hosp., Washington, 1961. Med. resident St. Lukes Hosp., N.Y.C., 1933-35; gen. practioner Hancock, Md., 1935-37, Middletown, Md., 1937-40; pediatric resident Children's Hosp., Washington, 1940-42; sch. physician St. Alban's Sch. for Boys, Washington, 1942-72; med. dir. Gallaudet Coll., Washington, 1942-72; pvt. practice Pediatric and Allergy Washington, 1942-80; clin. assoc. prof. Georgetown U. Med. Sch., Washington, 1942—. Co-author: Allergy in Relation to Pediatrics, 1951, The Allergic Child, 1963; contbr. articles to profl. jours. 2nd lt. inf. ROTC, 1929-39. Mem. AMA (life), Am. Assn. Pediatrics (life), Am. Acad. Allergy and Immunology, Am. Assn. Allergies (life), D.C. Med. Soc. (life), St. Andrew's Soc. D.C., Soc. of Cinn., Chevy Chase Country Club. Avocation: investments. Home: 4000 Cathedral Ave NW # 304B Washington DC 20016-5249

TODD, ROGER HAROLD, metallurgical engineer, failure analyst; b. Denver, Apr. 22, 1928; s. Harry Macintosh and May Alcestus (Phillips) T.; m. Dona Lou Seaton, Sept. 2, 1951; children: Brian R., Elaine L., Craig S., Kevin B. BSMetE, Colo. Sch. Mines, 1954; MSMetE, U. Idaho, 1961. Registered profl. engr., N.Y. Engr. Hanford Atomic Products Operation-GE, Richland, Wash., 1954-64; sr. scientist Battelle-Northwest, Richland, Wash., 1965-70; sr. engr. WADCO-Westinghouse, Richland, Wash., 1970-72; quality assurance engr. Niagara Mohawk Power Corp., Syracuse, N.Y., 1972-90, sr. materials engr., 1991—. Elder, Presbyn. Ch., Baldwinsville, N.Y., 1991. Lt. comdr. USNR, 1946-76. Mem. Am. Soc. Metals (chmn. 1981-82), ASME, Am. Soc. Nondestructuve Testing, Am. Welding Soc., Sigma Xi. Office: Niagara Mohawk Power Corp 300 Erie Blvd W Syracuse NY 13202

TODD, TERRY RAY, physicist; b. DeKalb, Ill., Oct. 9, 1947; s. Cedric W. and Minerva A. (Olesen) T.; m. Carol A. Dierker, July 15, 1978; children: Laura, Quentin, Nathan. BS in Math., No. Ill. U., 1969; MS in Physics, Pa. State U., 1972, PhD, 1976. Postdoctoral fellow Nat. Bur. Standards, Gaithersburg, Md., 1976-78; chief optical engr. Spectra Physics, Bedford, Mass., 1978-80; staff physicist Exxon Rsch. & Engring. Co., Florham Park, N.J., 1980-91; pres. T.R. Todd Enterprises Inc., Budd Lake, N.J., 1991-92; system engring. mgr. UOP/Guided Wave, El Dorado Hills, Calif., 1992—. Contbr. articles to profl. jours. Office: UOP/Guided Wave 5190 Golden Foothill Pkwy El Dorado Hills CA 95762

TODD, ZANE GREY, utility executive; b. Hanson, Ky., Feb. 3, 1924; s. Marshall Elvin and Kate (McCormick) T.; m. Marysnow Stone, Feb. 8, 1950 (dec. 1983); m. Frances Z. Anderson, Jan. 6, 1984. Student, Evansville Coll., 1947-49; BS summa cum laude, Purdue U., 1951, DEng (hon.), 1979; postgrad., U. Mich., 1965; DHL, U. Indpls., 1993. Fingerprint classifier FBI, 1942-43; electric system planning engr. Indpls. Power & Light Co., 1951-56, spl. assignments supr., 1956-60, head elec. system planning, 1960-65, head substation design div., 1965-68, head distbn. engring. dept., 1968-70, asst. to v.p., 1970-72, v.p., 1972-74, exec. v.p., 1974-75, pres., 1975-81, chmn., chief exec. officer, 1976-89, dir., chmn. exec. com., 1989—, chief exec. officer, 1981-89; chmn., pres. IPALCO Enterprises, Inc., Indpls., 1983-89, dir., chmn. exec. com., 1989—; chmn. bd., chief exec. officer Mid-Am. Capital Resources, Inc. subs. IPALCO Enterprises, Inc., Indpls., 1984-89, also bd. dirs., 1984—; gen. mgr. Mooresville Pub. Svc. Co., Inc., Ind., 1956-60; bd. dirs. Nat. City Bank Ind. (formerly Mchts. Nat. Corp.), Am. States Ins. Co.; hon. dir. 500 Festival Assocs., Inc., pres. 1987. Originator probability analysis of power system reliability; contbr. articles to tech. jours. and mags. Past pres. adv. bd. St. Vincent Hosp.; bd. dirs. Commn. for Downtown, YMCA Found., Crime Stoppers Cen. Ind., Corp. Community Coun.; past chmn., bd. trustees Ind. Cen. U. (now U. Indpls.); bd. govs. Associated Colls. of Ind.; Nat. and Greater Indpls. adv. bds. Salvation Army; mem. adv. bd. Clowes Hall. Sgt. AUS, 1943-47. Disting. Engring. Alumnus Purdue U., 1976; named Outstanding Elec. Engr. Purdue U., 1992, Knight of Malta, Order of St. John of Jerusalem, 1986. Fellow IEEE (past chmn. power system engring. com.); mem. ASME, NSPE, Power Engring. Soc., Ind. Fiscal Policy Inst. (bd. govs.), Ind. C. of C., Indpls. C. of C., Mooresville C. of C. (past pres.), PGA Nat. Country Club, Ulen Country Club, Columbia Club, Indpls. Athletic Club (past bd. dirs.), Meridian Hills Country Club (past bd. dirs.), Skyline Club (bd. govs.), Newcomen Soc. (chmn. Ind.), Rotary, Lions (past pres.), Eta Kappa Nu, Tau Beta Pi. Home: 7645 Randue Ct Indianapolis IN 46278-1565 Office: Indpls Power & Light Co PO Box 1595 Indianapolis IN 46206-1595

TODD OF TRUMPINGTON, BARON See TODD, ALEXANDER ROBERTUS

TOEKES, BARNA, chemical engineer, polymers consultant; b. Budapest, Hungary, Oct. 12, 1923; came to U.S., 1949; s. Barna and Jusztina (Szatmári) Tökés; m. Ida Maria Kálmán, Aug. 24, 1948 (div. 1966); m. Georgianna D. Doyle, Aug. 26, 1967; 1 child, C. Justin. BS in Engring., UCLA, 1955, postgrad., 1955-57. Rsch. engr. Stauffer Chem. Co., Richmond, Calif., 1955-60; sr. rsch. engr. Rexall Chem. Co., Paramus, N.J., 1960-69, Holyoke, Mass., 1960-69; plant mgr., gen. mgr. Southern Petrochemicals, Channelview, Tex., 1969-73; mgr. process engring. and devel. Polysar Resins, Inc., Leominster, Mass., 1973-77; system engring. mgr. Sperry Rsch. Ctr., Sudbury, Mass., 1978-82; prin. process engr. C. F. Braun & Co., Alhambra, Calif., 1982-85; cons. engr. Dart Container Corp., Mason, Mich., 1987—; cons. B. Toekes Cons., Baytown, Tex., 1977-78, Mason, Mich., 1985—. Contbg. author: Aromatic Fluorine Compounds, 1962, Fire Safety Aspects of Polymeric Materials, 1978; editor, co-author: Organic Working Fluid Properties, 1982, System Component Compatibility, 1982; contbr. articles to profl. jours. Mem. AICE, Am. Chem. Soc., Soc. Plastics Engrs. Achievements include 6 U.S. and 19 foreign patents in field of precesses for polymerization and devolatilization of polyvinyl aromatic compounds, equipment for polymerization processes. Home: 1908 Hagadorn Rd Mason MI 48854 Office: Dart Container Corp 432 Hogsback Rd Mason MI 48854

TOEPPE, WILLIAM JOSEPH, JR., retired aerospace engineer; b. Buffton, Ohio, Feb. 27, 1931; s. William Joseph Sr. and Ruth May (Hipple) T. BSEE, Rose-Hulman Inst. Tech., Terre Haute, Ind., 1953. Engr. Electronics divsn. Ralph M. Parsons Co., Pasadena, Calif., 1953-55; pvt. practice cons. Orange, Calif., 1961-62; engring. supr. Lockheed Electronics Co., City of Commerce, Calif., 1962-64; staff engr. Interstate Electronics Corp., Anaheim, Calif., 1957-61; engring. supr. Interstate Electronics Corp., Anaheim, 1964-89, ret., 1989. Author: Finding Your German Village, 1990, Gazetteers and Maps of France for Genealogical Research, 1990. Pres. Golden Circle Home Owners' Assn., Orange, 1989-93. With U.S. Army, 1953-57. Mem. Ohio Geneal. Soc. (life), Geneal. Soc. Pa., So. Calif. Geneal. Soc. Avocations: genealogy, music. Home: 700 E Taft Ave Unit 19 Orange CA 92665-4400

TOFT, JÜRGEN HERBERT, orthopedic surgeon; b. Berlin, May 4, 1943; s. Herbert and Erna (Milczinsky) T.; m. Waltraud Toft, Dec. 27, 1968; children: Philip, Felix, Frederike, Maurizio. Grad., Humboldt U., Berlin, 1971; MD, Saarland U., Homburg, Germany, 1976. Intern County Hosp., Biberach, Germany, 1972-75; resident Orthopedic Clinic, Tailfingen, Tübingen, 1975-76; orthopedic surgeon Ingolstadt, Golstadt, Germany, 1976-80, Remscheid, Germany, 1980-84; cons. pvt. practice Germany, 1976—; chief, knee specialist Ambulatory Surg. Ctr., Munich, Germany, 1985—; pres., founding mem. Wissenschaftliche Gesellschaft für Arthroskopische Chirurgie, Munich, 1984—. With inf. German Armed Forces, 1962-65. Mem. Internat. Knee Soc., Internat. Arthroscopy Assn., Deutsche Gesellschaft für Orthopädie und Traumatology. Avocations: mountaineering, langs., travel. Home: Neumühlhausen 14, 85664 Neumühlhausen Germany Office: Ambulantes Operationszentrm, Effnerstrasse 38, 81925 Munich Germany

TOGERSON, JOHN DENNIS, computer software company executive; b. Newcastle, England, July 2, 1939; arrived in Can., 1949; s. John Marius and Margaret (McLaughlin) T.; m. Donna Elizabeth Jones, Oct. 3, 1964 (div. 1972); children: Denise, Brenda, Judson; m. Patricia Willis, May 5, 1984. BME, GM Inst., Flint, Mich., 1961; MBA, York U., Toronto, Ont., 1971. Sr. prodn. engr. GM of Can., Oshawa, Ont., 1961-69; with sales, investment banking Cochran Murray, Toronto, 1969-72; pres. Unitec, Inc., Denver, 1972-79, All Seasons Properties, Denver, 1979-81, Resort Computer Corp., Denver, 1981—; mng. dir. VCC London (subs. of Resort Computer Corp.), 1992; bd. dirs. VCC London (sub. of 1st Nat. Bank U.K.), London, 1989—, mng. dir., 1992; pres., bd. dirs. Resort Mgmt. Corp., Dillon, Colo., 1980-81; presenter Assn. of Resort Developers Nat. Conv., 1993. Avocations: mountain biking, ice hockey. Office: Resort Computer Corp 2801 Youngfield St Ste 300 Golden CO 80401-2266

TÖGLHOFER, WOLFGANG, medical director, researcher; b. Judenburg, Austria, July 16, 1959; s. Franz and Dorothea (Muhr) T.; m. Kristine Roth, Apr. 6, 1991. MD, Med. U. Graz, 1984, gen. practitioner and oncology degree, 1988. Pediatric oncologist Ped. Univ., Giessen, Germany, 1985; resident in oncology Med. Univ., Graz, 1986-88; pediatric oncologist St. Anna Univ. Hosp., Vienna, 1989; med. dir. Roussel Uclaf, Vienna, 1990, Hoffmann La Roche, Vienna, 1991-93; internat. med. mgr. Hoffmann La Roche, Basel, Switzerland, 1993—. With Austrian mil., 1985-86. Mem. AAAS, World Med. Assn., Internat. Brain Club, World Future Soc., N.Y. Acad. Scis., German Mgmt. Club, Brit. Assn. Pharm. Physicians, German Assn. Pharm. Physicians, UK-Cytokine Club. Roman Catholic. Avocations: reading, music, painting, sailing. Office: Hoffmann La Roche, Grenzacherstrasse Bau 74, CH-4002 Basel Switzerland

TOIVANEN, PAAVO UURAS, immunologist, microbiologist, educator; b. Tuupovaara, Finland, June 14, 1937; s. Vilho Pekka and Suoma Helena (Silvennoinen) T.; m. Auli Marjaana Pirilä, Nov 3, 1961; children: Laura, Otto, Pekka, Hannes, Suoma. BS, Turku (Finland) U., 1958, MD, 1962, D Med. Scis., 1966. Asst. physician Dept. Med. Microbiology Turku U., 1961-69, docent, 1968-69, assoc. prof., 1969-78, prof. bacteriology, serology, 1978—; mem. Basel Inst. Immunology, 1979-80. Editor: Avian Immunology, Basis and Practice, 1987, Reactive Arthritis, 1988. Mem. Finnish Soc. Immunology (pres. 1979-82), Am. Assn. Immunologists, Am. Soc. Microbiology, Transplantation Soc., Finnish Soc. Pathology, Finnish Soc. Hematology, Scandinavian Soc. Immunology (treas. 1978-82). Lutheran. Office: Turku U Dept Med Microbiolo, Kiinamyllynkatu 13, SF-20520 Turku Finland

TOIVONEN, HANNU TAPIO, control engineering educator; b. Turku, Finland, Oct. 7, 1952; s. Nils Åke and Hellevi (Haapajärvi) T. MSc, Åbo Akademi, Turku, 1976, PhD, 1981. Rsch. fellow Acad. Finland, Turku, 1978-81; lectr. Åbo Akademi (Swedish U. of Åbo), Turku, 1984-88, assoc. prof., 1988—. Contbr. articles to profl. jours. The Royal Norwegian Coun. for Sci. and Indsl. Rsch. fellow, 1982. Mem. IEEE, N.Y. Acad. Sci., Assn. for Computing Machinery. Office: Åbo Akademi, Swedish U Åbo, Biskopsgatan 8, 20500 Turku Finland

TOKAR, JOHN MICHAEL, oceanographer, ocean engineer; b. Bayonne, N.J., Sept. 13, 1951; s. Thomas and Frances (Fayder) Wargo; m. Virginia De Bellis, Oct. 10, 1976; children: Melina, Laurel. BA in Chemistry, Jersey City State Coll., 1973; MS in Engring., Cath. U., 1989. Chem. oceanographer Atlantic Oceanographic Lab. NOAA, Miami, Fla., 1977-82, ops. officer ship rsch., 1982-83; project engr. Nat. Ocean Svc. engring. staff NOAA, Rockville, Md., 1983-86, quality assurance coord. Office of Oceanography, 1986-88, program mgr. Nat. Ocean Svc., Office of Ocean Svcs., 1988-89, liaison for oceanography Office of Corps. Ops., 1989-91; commdg. officer Ship FERREL NOAA, Norfolk, Va., 1991-93; chief Damage Assessment Ctr. Ops., 1993—. Author: (chpt.) Geotechnical Engineering of Ocean Waste Disposal, 1990; contbr. articles to profl. jours. Mem. AAAS, IEEE, Am. Geophys. Union, Marine Tech. Soc. Achievements include development of fiber optic chemical sensors for ocean applications. Home: 4935 Tall Oaks Dr Monrovia MD 21770 Office: Damage Assessment Ctr 1305 East-West Hwy Silver Spring MD 20910

TOKERUD, ROBERT EUGENE, electrical engineer; b. Great Falls, Mont., Aug. 30, 1936; s. Fred Eugene Tokerud and Helen A. (Tadevich) Thomas; m. Marlys Jo Bordeleau, 1957; children: Pamela, Torri, Marc, Camille, Corinne, David. BSEE, U. Calif., Berkeley, 1959; cert. Inst. Mgmt., Northwestern U., 1975. Sr. project engr. Sperry Utah Co., Salt Lake City, 1959-65; mgr. infosystems Lockheed Electronics Co., Houston, 1965-69; mgr. earth resources Lockheed Engring. and Sci. Co., Houston, 1969-74, asst. dir. sci. and applications, 1974-79, dir. bus. devel., 1980-87; life sci. program mgr. Lockheed Engring. and Sci. Co., Washington, 1987-89; pres. Lockheed Support Systems, Inc., Arlington, Tex., 1989—; v.p. Lockheed Corp., 1993—. Author conf. procs., other profl. publs. Bd. dirs. El Lago (Tex.) Water and Waste Mgmt. Dist., 1974. Mem. AIAA, Aerospace Med. Assn., Air Force Assn., Army Aviation Assn. Am. Office: Lockheed Support Systems 1600 E Pioneer Pky Arlington TX 76010-6541

TOKHEIM, ROBERT EDWARD, physicist; b. Eastport, Maine, Apr. 25, 1936; s. Edward George and Ruth Lillian (Koenig) T.; m. Diane Alice Green, July 1, 1962; children: Shirley Diane, William Robert, David Eric, Heidi Jean. BS, Calif. Inst. Tech., 1958, MS, 1959; Degree of Engr., Stanford U., 1962, PhD, 1965. Rsch. asst. Hansen Labs Physics Stanford (Calif.) U., 1960-65; microwave engr. Watkins-Johnson Co., Palo Alto, Calif., 1965-73, staff scientist, head ferrimagnetic R&D dept., 1966-69; physicist SRI Internat., Menlo Park, Calif., 1973—. Co-author: Tutorial Handbook on X-ray Effects on Materials and Structures, 1992; contbr. articles to IEEE and Am. Phys. Soc. Jour., Jour. Applied Phys. on shock compression and others. Mem. IEEE (sr. mem.), Am. Phys. Soc., Toastmasters Internat., Tau Beta Pi, Sigma Xi. Achievements include discovery of nonreciprocal line-coupled microwave ferrimagnetic filters, optimum thermal compensation axes in YIG and GaYIG ferrimagnetic spheres, shock wave equation-of-state computational models for porous materials. Home: 5 Trinity Ct Menlo Park CA 94025 Office: SRI International 333 Ravenswood Ave Menlo Park CA 94025

TOKOLY, MARY ANDREE, microbiologist; b. Manila, Dec. 4, 1940; (parents Am. citizens) d. Robert Francis Tokoly and Ruby Waunita (Shriner) Kaderli. BS, Tex. Woman's U., 1962, MS, 1964, PhD, 1974. Instr Victoria (Tex.) Coll., 1964-66; asst. prof. Kans. State Coll., Pittsburg, 1966-68, Kans. Newman Coll., Wichita, 1974-75; grad. teaching asst. Tex. Woman's U., Denton, 1968-74; microbiologist Nix Hosp., San Antonio, 1975-77, Met. Hosp., San Antonio, 1977—. Sec. Bexar County chpt. Czech Heritage Soc. Tex., San Antonio, 1988—. Robert A. Welch Found. grantee, 1971, 72. Mem. Am. Soc. Clin. Pathologists (registered microbiologist), Am. Soc. Microbiology, Tex. Soc. Microbiology, South Tex. Assn. Microbiology Profls., Am. Soc. Med. Tech., N.Y. Acad. Scis., AAUW, S.W. Assn. Clin. Microbiologists, Sigma Xi. Roman Catholic. Avocations: taxes, computers, Czech culture and traditions. Office: Met Hosp 1310 Mccullough Ave San Antonio TX 78212-5699

TOKUE, IKUO, chemist, researcher; b. Kohnosu, Japan, Nov. 12, 1947; m. Toshie Tsukada, Mar. 17, 1979; children: Yuliko, Ayako, Jun-ichi. BS, Yokohama Nat. U., 1970; DSc, U. Tokyo, 1975. Asst. faculty of sci. Niigata (Japan) U., 1975-81, lectr., 1981-84, assoc. prof., 1984—.

TOLAN, ROBERT WARREN, pediatric infectious disease specialist; b. Bowling Green, Ohio, Nov. 20, 1960; s. Robert Warren Tolan and Margaret Delores (Petter) Cardwell; m. Lenita Kay Newberg, May 15, 1983. BA, Ind. U., 1982, MA, 1983; MD, Washington U., St. Louis, 1987. Diplomate Nat. Bd. Med. Examiners, Am. Bd. Pediatrics. Resident in pediatrics Riley Hosp. for Children, Indpls., 1987-90; fellow in infectious diseases St. Louis Children's Hosp., 1990—. Co-author: Fever of Unknown Origin in Children, 1991; contbr. articles to Clin. Infectious Diseases, Pediatric Infectious Diseases Jour., Infection and Immunity, Jour. Clin. Microbiology. Nat. Merit scholar Pitts. Plate Glass, 1978; Am. Med. Sch. Pediatrics fellow, 1990. Fellow Am. Acad. Pediatrics; mem. AMA, Am. Soc. Microbiology, Infectious Diseases Soc. Am., Pediatric Infectious Diseases Soc., Soc. for Preservation and Encouragement of Barbershop Quartet Singing in Am. Democrat. Episcoplian. Achievements include part in the cloning of an outer membrane protein from Haemophilus influenzae type b which is being developed as a vaccine candidate; reviews of surgical management of pediatric endocarditis and of toxic shock syndrome and influenza; description of systemic pseudomalignant form of cat-scratch disease in normal children, the cloning of an outer membrane protein from Haemophilus influenzae type b, the lack of epidemiologic utility of analysis of lipopolysaccharide from the same organism. Office: St Louis Children's Hosp Washington U Sch Medicine 660 S Euclid Ave # 8230 Saint Louis MO 63110-1093

TOLEDO-PEREYRA, LUIS HORACIO, transplant surgeon, researcher, educator; b. Nogales, Ariz., Oct. 19, 1943; s. Jose Horacio and Elia (Pereyra) Toledo; m. Marjean May Gilbert, Mar. 21, 1974; children: Alexander Horacio, Suzanne Elizabeth. BS magna cum laude, Regis Coll., 1960; MD, Nat. U. Mex., 1967, MS in Internal Medicine, 1970; PhD in Surgery, U. Minn., 1976, Ph.D. in Hist. Medicine, U. Minn., 1984. Rotating intern Hosp. Juarez, Nat. U. Mex., 1966; resident in internal medicine Instituto Nacional de la Nutricion, Nat. U. Mex., 1968, 70; resident in gen. surgery U. Minn., 1970-76; resident in thoracic and cardiovascular surgery U. Chgo., 1976-77; dir. surg. research Henry Ford Hosp., Detroit, 1977-79, co-dir. transplantation, 1977-79; chief transplantation, dir. surg. research Mt. Carmel Mercy Hosp., Detroit, 1979-89; chief transplantation, dir. rsch. Borgess Med. Ctr., Kalamazoo, Mich., 1990—; instr. biochemistry and internal medicine Nat. U. Mex., 1963, 68; adj. prof. Sch. Health Sci. Mercy Coll., Detroit, 1983-91, history Western Mich. U., 1990—, biol. sci., 1991—; prof. surgery Nat. U. Mex., 1990—. Recipient Outstanding Achievement award U. Mex., 1961, 64, 67; Resident Research award Assn. Acad. Surgery, 1974; Cecil Lehman Mayer Research award Am. Coll. Chest Physicians, 1975. Mem. AMA, Transplantation Soc., Assn. Acad. Surgery, Am. Soc. Transplant Surgery, Soc. Organ Sharing, Am. Soc. Nephrology, Am. Assn. Immunologists, Am. Physiol. Soc., Soc. Exptl. Biology and Medicine, Am. Soc. Artificial Organs, European Dialysis and Transplantation Assn., Am. Diabetes Assn., European Soc. Study of Diabetes, No. Am. Soc. Dialysis Transplantation, Pan Am. Soc. Dialysis Transplantation. Roman Catholic. Club: Grosse Pointe Yacht. Guest editor various med. and transplant jours.; mem. editorial bd. Dialysis and Transplantation, 1979—, Rsch. in Surgery, 1991—, Cirugia Iberoamericana, 1992—, clin. adv. bd. Transpl. Proc., 1993—, Medico Interamericano, 1993—; assoc. editor Transplantology, 1990—; contbr. over 400 articles on organ preservation, transplantation, other surg. and med. related areas, and the history of medicine to profl. jours.

TOLER, JAMES C., electrical engineer; b. Carthage, Ark., Jan. 31, 1936. BSEE, U. Ark., 1957; MSEE, Ga. Inst. Tech., 1970. Prin. rsch. engr. Ga. Inst. Tech., Atlanta, 1966—, dir. bioengring. ctr., 1984—, dir. ctr. rehab. tech., 1987—. Fellow IEEE; mem. Bioelectromagnetics Soc. Achievements include research on interaction of electromagnetic waves with biological systems. Office: Ga Tech Rsch Inst Bioengring Ctr 225 North Ave NW Atlanta GA 30332-0001*

TOLETE-VELCEK, FRANCISCA AGATEP, pediatric surgeon, surgery educator; b. Santo Domingo, Ilocos Sur, The Philippines, Mar. 16, 1943; came to U.S., 1966; d. Celedonio Alvarez Tolete and Cristeta (Lopez) Agatep; m. Damir Velcek; children: John, Jennifer. BS, U. Philippines, 1962, MD, 1966. Diplomate Am. Bd. Surgery. Intern U. Philippines, Manila, 1965-66; straigt surg. intern St. Clare's Hosp., N.Y.C., 1966-67, resident, 1967-71; surg. asst. St. Barnabas Hosp., N.Y.C., 1971-72; rsch. fellow in pediatric surgery SUNY Health Sci. Ctr., Bklyn., 1972-73, pediatric surg. chief res., 1973-75; asst. instr. surgery SUNY Health Sci. Ctr., Bkyln., 1972-75, asst. prof., 1975-81, assoc. prof., 1981-91; prof. SUNY Health Sci. Ctr., Bklyn., 1991—; chief pediatric surgery L.I. Coll. Hosp., Bklyn.; attending staff mem. St. Vincent's Med. Ctr., S.I., N.Y., Meth. Hosp., Bklyn., St. Vincent's Hosp. & Med. Ctr., N.Y.C., Beth Israel Hosp., N.Y.C., Lenox Hill Hosp., N.Y.C. Contbr. articles to med. jours. Lt. col. M.C., USAR, 1987—. Named Most Outstanding Alumnus, U. Philippines, 1987. Fellow ACS, Am. Acad. Pediatrics; mem. Am. Pediatric Surg. Assn., Am. Assn. for Surgery of Trauma, Soc. Internat. De Chirurgie, John Madden Surg. Soc., N.Y. Surg. Soc., Sigma Xi, Phi Sigma, Phi Kappa Phi. Roman Catholic. Home: 471 Summit St Englewood Cliffs NJ 07632 Office: Health Sci Ctr at Bklyn 965 5th Ave New York NY 10021-1709

TOLIVER, LEE, mechanical engineer; b. Wildhorse, Okla., Oct. 3, 1921; s. Clinton Leslie and Mary (O'Neall) T.; m. Barbara Anne O'Reilly, Jan. 24, 1942; children: Margaret Anne, Michael Edward. BSME, U. Okla., 1942. Registered profl. engr., Ohio. Engr. Douglas Aircraft Co., Santa Monica, Calif., 1942, Oklahoma City, 1942-44; engr. Los Alamos (N.Mex.) Sci. Lab., 1946; instr. mech. engring. Ohio State U., Columbus, 1946-47; engr. Sandia Nat. Labs., Albuquerque, 1947-82; instr. computer sci. and math. U. N.Mex., Valencia County, 1982-84; number theory researcher Belen, N.Mex., 1982—. Co-author: (computer manuals with G. Carli, A.F. Schkade, L. Toliver) Experiences with an Intelligent Remote Batch Terminal, 1972, (with C.R. Borgman, T.I. Ristine) Transmitting Data from the PDP-10 to Precision Graphics, 1973, (with C.R. Borgman, T.I. Ristine) Data Transmission-PDP-10/Sykes/Precision Graphics, 1975. With Manhattan Project (Atomic Bomb) U.S. Army, 1944-46. Mem. Math. Assn. Am., Am. Math. Soc. Achievements include development of 28 computer programs with manuals. Home: 206 Howell St Belen NM 87002-6225

TOLL, JOHN SAMPSON, association administrator, former university administrator, physics educator; b. Denver, Oct. 25, 1923; s. Oliver Wolcott and Merle d'Aubigne (Sampson) T.; m. Deborah Ann Taintor, Oct. 24, 1970; children: Dacia Merle Sampson, Caroline Taintor. BS with highest honors, Yale U., 1944; AM, Princeton U., 1948, PhD, 1952; DSc (hon.), U. Md., 1973, U. Wroclaw, Poland, 1975; LLD (hon.), Adelphi U., 1978; hon. doctorate, Fudan U., Peoples Republic China, 1987; LHD (hon.), SUNY, Stony Brook, 1990; LLD (hon.), U. Md., Eastern Shore, 1993. Mng. editor, acting chmn. Yale Sci. mag., 1943-44; with Princeton U., 1946-49, Proctor fellow, 1948-49; Friends of Elementary Particle Theory Research grantee for study in France, 1950; theoretical physicist Los Alamos Sci. Lab., 1950-51; staff mem., assoc. dir. Project Matterhorn, Forrestal Rsch. Ctr., Princeton U., 1951-53; prof., chmn. physics and astronomy U. Md., 1953-65; pres., prof. physics SUNY, Stony Brook, 1965-78; pres., prof. physics U. Md., 1978-88, chancellor, 1988-89, chancellor emeritus, prof. physics, 1989—; pres. Univs. Rsch. Assn., Washington, 1989—; 1st dir. State U. N.Y. Chancellor's Panel on Univ. Purposes, 1970; physics cons. to editorial staff Nat. Sci. Tchrs. Assn., 1957-61; U.S. del., head sci. secretariat Internat. Conf. on High Energy Physics, 1960; mem-at-large U.S. Nat. Com. for Internat. Union of Pure and Applied Physics, 1960-63; chmn. rsch. adv. com. on electrophysics to NASA, 1961-65; mem. gov. Md. Sci. Resources Adv. bd., 1963-65; mem., also chmn. NSF adv. panel for physics, 1964-67; mem. N.Y. Gov.'s Adv. Com. Atomic Energy, 1966-70; mem. commn. plans and objectives higher edn. Am. Council Edn., 1966-69; mem. Hall of Records Commn., 1979; bd. dirs. Washington/Balt. Regional Assn., 1980-89; mem., chmn. adv. coun. Princeton Plasma Physics Lab, 1979-85; mem. Adv. Coun. of Pres.'s, Assn. of Governing Bds., 1980-1988, So. Regional Edn. Bd., 1980-90; mem. exec. com. Nat. Assn. State Univs. and Land Grant Colls., 1980-88; bd. dirs. Def. Systems Mgmt. Coll., 1982-88; mem. univ. programs panel of energy rsch. bd. Dept. Energy, 1982-83; mem. Washington-Balt. Regional Assn., 1983-84, Ctr. for the Study of the Presidency, 1983-84; mem. SBHE Adv. Com., 1983-89, Md. Gov.'s Chesapeake Bay Council, 1985; mem. resource com. State Trade Policy Council Gov.'s High Tech Roundtable Md. Dept. Econ. Devel., 1986-89; marine div. chmn. NASULGC, 1986; bd. dirs. Am. Council on Edn., 1986-89, Math. Scis. Edn. Bd. Nat. Acad. Scis., 1991-93; bd. trustees Aspen Inst. for Humanities, 1987-89; mem. Commn. on Higher Edn. Middle States Assn. Colls. and Schs., 1987; chmn. adv. panel on tech. risks and opportunities for U.S. energy supply and demand U.S. Office Tech. Assessment, 1987; chmn. adv. panel on internat. collaboration in def. tech., U.S. Office Tech. Assessment, 1989-91, Sea Grant Review Panel U.S. Dept. Commerce, 1992—; vis. prof. Nordic Inst. Theoretical Physics, Niels Bohr Inst., Denmark, U. Lund, Sweden, 1975-76. Contbr. articles to sci. jours. Recipient Benjamin Barge prize in math. Yale U., 1943, George Beckwith medal for Proficiency in Astronomy, 1944, Outstanding citizen award City of Denver, 1958, Outstanding Tchr. award U. Md. Men's League, 1965, Nat. Golden Plate award Am. Acad. Achievement, 1968, Copernicus award Govt. of Poland, 1973, Stony Brook Found. award for disting. contbns. to edn., 1979, Disting. Svc. award State of Md., 1981; named Washingtonian of Yr., 1985; John Simon Guggenheim Meml. Found. fellow Inst. Theoretical Physics U. Copenhagen, U. Lund, Sweden, 1958-59. Fellow Am. Phys. Soc., Washington Acad. Scis.; mem. Am. Assn. Physics Tchrs., Fedn. Am. Scientists (chmn. 1961-62), Philos. Soc. Washington, N.Y. Acad. Scis., Assn. Higher Edn., Nat. Sci. Tchrs. Assn., Yale Engring. Assn., NAACP (life), Phi Beta Kappa, Sigma Xi (Sci. Achievement award 1965), Phi Kappa Phi (Disting. Mem. 1990), Sigma Pi Sigma, Omicron Delta Kappa (hon.). Club: Cosmos. Achievements incl research on elementary particle theory; scattering. Office: Univs Rsch Assn 1111 19th St NW Ste 400 Washington DC 20036-3603 also: U Md Dept Physics College Park MD 20742-4111

TOLLE, GLEN CONRAD, mechanical engineer; b. San Antonio, May 20, 1939; s. Harold Albert and Marguerite Theresa (Conrad) T.; m. Rosemary Agnes Mattingly, June 1, 1963; children: Glen Conrad Jr., Michael John, Brian Patrick, Sherri Christine. BSME, U. Tex., 1962; MSME, Tex. Tech. U., 1963; PhD, U. Houston, 1970. Registered professional engineer, Tex. Sr. engr. Tex. Instruments, Stafford, 1970-71; engr. Tracor, Inc., Austin, Tex., 1971-72; asst. prof. U. Tex., Austin, 1972; assoc. prof. Tex. A&M U., College Station, 1972-79; sr. rsch. assoc. Mobil Oil Corp., Dallas, 1979-90; cons. engr. Tolle Engring., Inc., Plano, 1975—. Capt. U.S. Army, 1966-68, Vietnam. Mem. Sigma Xi, Pi Tau Sigma. Achievements include a method to remove drilled cuttings from wellbores. Office: Tolle Engring Inc Box 866911 Plano TX 75023

TOLLEFSEN, GERALD ELMER, chemical engineer; b. Chgo., Dec. 4, 1942; s. Thor N. and Florence Elenore (Roth) T.; m. Beverly Gayle Sura, May 21, 1978; children: Gayle, Sandra, Michael. BSCE, Mich. Tech. U., 1967. Mem. tech. staff Sinclair Rsch., Harvey, Ill., 1967-69, Arco Petroleum Products, Harvey, 1969-85; v.p. tech. svc. Calumet Lubricants Co., Princeton, La., 1985—. Mem. AICE, Am. Soc. Quality Control. Achievements include patents for lubricating oil processing, catalyst loading techniques, hydrogenation. Home: 9855 Holamy Ln Shreveport LA 71106 Office: Calumet Lubricants Co 10234 Hwy 157 Princeton LA 71067

TOLLESTRUP, ALVIN VIRGIL, physicist; b. Los Angeles, Mar. 22, 1924; s. Albert Virgil and Maureen (Petersen) T.; m. Alice Hatch, Feb. 26, 1945 (div. Nov. 1970); children: Kristine, Kurt, Eric, Carl; m. Janine Cukay, Oct. 11, 1986. BS, U. Utah, 1944; PhD, Calif. Inst. Tech., 1950. Mem. faculty Calif. Inst. Tech., Pasadena, 1950-77, prof. physics, 1968-77; scientist Fermi Nat. Lab., Batavia, Ill., 1977—, head collider detector facility. Co-developer superconducting magnets for Tevatron, Fermi Lab. Served to lt. (j.g.) USN, 1944-46. NSF fellow; Disting. Alumni award Calif. Inst. Tech., 1993. Fellow Am. Phys. Soc. (R.R. Wilson prize 1989, Nat. medal for tech. 1989); mem. AAAS. Democrat. Home: 29W 254 Renouf Dr Warrenville IL 60555 Office: Fermi Nat Lab PO Box 500 Batavia IL 60510-5000

TOLLIVER, GERALD ARTHUR, research psychologist, behavioral counselor; b. Charles City, Iowa, Nov. 17, 1935; s. Hillard A. and S. Fern T.; m. Barbara Joan Dieks, May 17, 1959 (div. 1987); children: Steven, Sara, Doug; m. Laura Lee karp, May 31, 1992. MA, George Washington U., 1965; PhD, U. Coll. London, 1974. Rsch. psychologist Walter Reed Army Inst. Rsch., Washington, 1960-65, Med. Rsch. Coun., London, 1965-69, NASA U. Calif., Berkeley, Calif., 1969-71; behavioral coun. Seattle Pub. Schs., 1973-78; rsch. psychologist Dept. Labor and Industries, Seattle, 1979-81; vocat. rehab. counselor pvt. practice, Seattle, 1981-86; rsch. scientist Alcohol and Drug Abuse Inst. U. Wash., Seattle, 1981-86. City councilman Winslow (Wash.) City Govt., 1980-84. Home: 12258 12th Ave NW Seattle WA 98177

TOLLNER, ERNEST WILLIAM, agricultural engineering educator, agricultural radiology consultant; b. Maysville, Ky., July 14, 1949; s. Ernest Edward and Ruby Geneva (Henderson) T.; m. Caren Gayle Crane, Sept. 27, 1987. BS, U. Ky., 1972; PhD, Auburn (Ala.) U., 1981. Rsch. specialist U. Ky., Lexington, 1972-74, rsch. engr., 1974-76; teaching asst. Tex. A&M U., College Station, 1976-77; rsch. specialist Auburn U., 1977-80; asst. prof. U. Ga., Griffin, 1980-85, assoc. prof., 1985-90, prof., 1990—; cons. Masstock Dairies, Montezuma, Ga., 1988, Twiggs Corner Condominium Assn., Peachtree City, Ga., 1986-92, Daucet Nature Trails, Griffin, Ga., 1989. Treas. Condominium Assn., Peachtree City, 1988-91. Mem. Am. Soc. Agrl. Engrs. (chair soil dynamics and other related coms.), Am. Soc. Agronomy. Achievements include patent for Cone Penetrometer; patent pending for Soil Core Sampler; first to use an X-Ray Tomographic Scanner devoted solely to agricultural research tasks. Home: 1010 Rogers Rd Bogart GA 30622 Office: U Ga Biology and Agrl Engring Dept Driftmeir Engring Ctr Athens GA 30602

TOLONEN, RISTO MARKUS, systems software engineer; b. Tampere, Hame, Finland, May 12, 1965; came to U.S., 1986; s. Osmo Otto and Kristi Marjatta (Hieranen) T.; m. Tina Maria Jokihaara, July 11, 1981; children: Stephanie Ann, Christopher Markus. BS in Computer Sci., Weber State U., 1991. Jr. software engr. Kiva Corp., Salt Lake City, 1988-90, software engr., 1990—; cons. in field. Sgt. Finnish Army, 1984-85. Mem. ACT.

TOLPADI, ANIL KUMAR, mechanical engineer, researcher; b. Shimoga, India, Feb. 28, 1960; s. Shivananda and Padma (Rao) T.; m. Anjana Shastri, Feb. 26, 1990; 1 child, Anagha. B of Tech. in Mech. Engring., Indian Inst. Tech., Kharagpur, India, 1981; MSME, Iowa State U., 1983; PhD in Mech. Engring., U. Minn., 1987. Rsch. scientist Sci. Rsch. Assocs., Glastonbury, Conn., 1987-89; rsch. staff Gen. Electric R & D Ctr., Schenectady, N.Y., 1989—; presenter various orgn. confs. Contbr. articles to Jour. of Heat Transfer, Internat. Jour. of Heat and Mass Transfer, Numerical Heat Transfer., Jour. Fluids Engring. Mem. ASME, AIAA, Phi Kappa Phi. Home: 2 Melissa Ct Albany NY 12205 Office: Gen Electric R & D Ctr PO Box 8 MS K1-ES206 Schenectady NY 12301

TOMA, JOSEPH S., defense analyst, retired military officer; b. Lawrence, Mass., June 2, 1930; s. Shaker J. and Angelina Toma; m. Sue Ann Hinds, June 14, 1952; children: Rebecca, Robert, Clifford. BS in Aerospace Engring., Air Force Inst. Tech., 1959; MA in Internat. Rels., Salve Regina U., 1986. Comml. pilot rating. Enlisted USAF, 1949, advanced through grades to lt. col., pilot, 1952-59, 64-66; nuclear weapons analyst USAF, Albuquerque, 1959-64; aerospace engring. hdqrs. USAF, Washington, 1966-71; retired USAF, 1971; ops. analyst Dept. of Defense, Arlington, Va., 1971-73; spl. asst. Joint Chiefs of Staff Pentagon, Washington, 1973—; lectr. Mil. War Colls., Harvard U. Kennedy Sch. Gov., 1987-91. Co-author: The First Information War, 1992, Control of Joint Forces, 1989. Decorated Air medal Korean Svc., 1953. Mem. Armed Forces Communications Electronics Assn., Aircraft Owners and Pilots Assn., Order of Daedalians. Achievements include development of aerospace system concepts plans and doctrine and findings in classified reports on weapon effects. Home: 3034 Sleepy Hollow Rd Falls Church VA 22042 Office: Joint Staff J6A Pentagon Washington DC 20318-6000

TOMASI, THOMAS B., cell biologist, administrator; b. May 24, 1927; s. Thomas B. and Ivis (Ratazzi) T.; children—Barbara, Theodore, Anne. A.B., Dartmouth Coll., Hanover, N.H., 1950; M.D., U. Vt., Burlington, 1954; Ph.D., Rockefeller U., 1965. Intern, resident, chief resident Columbia Presbyn. Hosp., N.Y.C., 1954-58, instr. medicine, 1958-60; prof., chmn. div. exptl. medicine U. Vt., Burlington, 1960-65; prof. medicine, dir. immunology SUNY, Buffalo, 1965-73; prof., chmn. immunology dept. Mayo Med. Sch., Rochester, Minn., 1973-81; dir. Cancer Ctr., Disting. Univ. prof., chmn. dept. cell biology U. N. Mex., Albuquerque, 1981-86; pres., CEO Roswell Park Cancer Inst., Buffalo, 1986—, chmn. dept. molecular medicine. Author: The Immune System of Secretions, 1976; contbr. over 200 articles to profl. jours. Served with USN, 1945-46. Mem. Am. Soc. Cell Biology, Am. Assn. Immunologists, Am. Assn. Cancer Research, Am. Soc. Clin. Investigation, Am. Fedn. Clin. Research, Assn. Am. Physicians. Roman Catholic. Avocations: skiing, tennis, hunting, fishing, gardening. Home: 7980 E Quaker St Orchard Park NY 14127-2017 Office: Roswell Park Cancer Inst Elm and Carlton Sts Buffalo NY 14263

TOMASOVIC, STEPHEN PETER, radiobiologist, educator; b. Bend, Oreg., Jan. 5, 1947; s. Peter Alexander and Barbara Ann (Scott) T.; m. Barbara Jean Davis, Aug. 8, 1970. MS, Oreg. State U., 1973; PhD, Colo. State U., 1977. Asst. prof. dept. radiology U. Utah, Salt Lake City, 1979-80; from asst. prof. to assoc. prof. dept tumor biology M. D. Anderson Cancer Ctr., Houston, 1980-92, sect. chief, 1988—, prof., 1992—; prof. grad. sch. biomed. scis. U. Tex. Health Sci. Ctr., Houston, 1992—; chmn. faculty senate, pres.'s exec bd. dirs. M. D. Anderson Cancer Ctr., Houston, 1991-93; vice-chmn. biotech. adv. com. North Harris Montgomery C.C., Kingwood, Tex., 1991—. Assoc. editor Internat. Jour. Radiation Oncology, Biology and Physics, 1989—, Cancer Detection and Protection, 1992—; co-editor Cancer Bull., 1993—. Staff sgt. U.S. Army, 1969-71, Vietnam. Mem. Am. Soc. Cell Biology, Am. Assn. Cancer Rsch., Radiation Rsch. Soc., N.Am. Hyperthermia Soc. (councilor 1990-92), Phi Sigma, Sigma Xi. Office: M D Anderson Cancer Ctr Box 108 1515 Holcombe Blvd Houston TX 77030

TOMASSINI, JOANNE ELIZABETH, virologist, researcher; b. Endicott, N.Y., Oct. 16, 1952; d. Emanuel Victor and Frances Mary (Mistretta) T.; m. Robert John Lynch, Oct. 6, 1984; children: Matthew Christopher, Jennifer Elizabeth. BS in Biology, Marywood Coll., 1974; MS in Biochemistry, Cornell U., 1979; PhD in Microbiology, U. Pa., 1986. Rsch. technician Cornell U., Ithaca, N.Y., 1974-78; staff biochemist Merck Sharp & Dohme Rsch. Lab., West Point, Pa., 1979-80, rsch. biochemist, 1980-86, sr. rsch. biochemist, 1986-89, rsch. fellow, 1989—; adj. faculty U. Pa., Phila., 1991—. Contbr. articles to Jour. Virology and Proceedings of Nat. Acad. Sci. Mem. AAAS, Am. Soc. Virology, Internat. Soc. Antiviral Rsch., Merck Employees for Excellence in Edn. Achievements include patents for Hepatitis A Subunit Antigen, Monoclonal Antibody Blocks Rhinoviruses Attachment; research in isolation of receptor protein for human rhinoviruses cloning, characterization of receptor for rhinoviruses, influenza virus antivirals. Office: Merck Rsch Labs WP26B 1114 West Point PA 19486

TOMASZKIEWICZ, FRANCIS XAVIER, imaging technology educator; b. Chgo., Oct. 11, 1946; s. Francis Joseph and Stephanie Louise (Chodniewicz) T.; m. Laurie Ann Weber; children: Sean, Kelly, Christopher, Patrick, Michael (dec.), Jamie (dec.); 1 stepchild, Steven. BA, U. Ill., Chgo., 1972; MA, Northeastern Ill. U., 1976. Lic. secondary sch. educator; cert. spl. art instr. Tchr. art Pub. Sch. Dist. # 126, Alsip, Ill., 1972-74; tchr. art Hadley Jr. High Sch., Glen Ellyn, Ill., 1974-86, tech. educator, 1986—; cons. curriculum for tech. literacy Ill. State Bd. Edn., 1989—. Contbr. articles to profl. jours.; author monograph on laser imaging, 1992. Named Tchr. of Yr. Glen Ellyn Jaycees, 1976, Those Who Excel Program, 1991; recipient commendation Ill. State Bd. Edn., 1989. Achievements include design and implementation of electronic imaging tech. course in mid. sch. curriculum and at grad. level. Office: Hadley Imaging Tech 1701 S Vine Ave Park Ridge IL 60068

TOMAZI, GEORGE DONALD, electrical engineer; b. St. Louis, Dec. 27, 1935; s. George and Sophia (Bogovich) T.; m. Lois Marie Partenheimer, Feb. 1, 1958; children: Keith, Kent. BSEE, U. Mo., Rolla, 1958, Profl. EE (hon.), 1970; MBA, St. Louis U., 1965, MSEE, 1971. Registered profl. engr., Mo., Ill., Wash., Ohio, Calif., Va. Project engr. Union Electric Co., 1958-66; dir. corp. planning Gen. Steel Industries, 1966-70; exec. v.p. St. Louis Research Council, 1970-74; exec. v.p. Hercules Constrn. Co., St. Louis, 1974-75; dir. design and constrn. div. Mallinckrodt, Inc., St. Louis, 1975—. Author: P-Science: The Role of Science in Society, 1972, The Link of Science and Religion, 1973. Active Nat. Kidney Found.; bd. dirs. U. Mo. Devel. Council, St. Louis Artists Coalition; elder Luth. Ch.; mem. adv. com. grad. sch. U. Mo., Columbia, mem. pres's. role and scope commn. Served with U.S. Army, 1959-61. Recipient award Acad. Elec. Engrs., U. Mo., Rolla. Mem. NSPE, IEEE (chmn. state govt. activities com. 1990—), Japan-Am. Soc., AAAS, Am. Inst. Chem. Engrs., Profl. Engrs. in Industry, Mo. Soc. Profl. Engrs. (Profl. Engr. in Industry 1989), Profl. Engrs. and Land Surveyors (chmn. Mo. bd. for architects 1989—), Am. Def. Preparedness Assn., U. Mo. Alumni Assn. (bd. dirs. 1972-78), Engrs. Club (pres. 1985-86), Mo. Athletic Club, Rotary, Sigma Pi. Office: Mallinckrodt Splty Chem Co 12723 Stoneridge Dr Florissant MO 63033

TOMBAUGH, CLYDE WILLIAM, astronomer, educator; b. Streator, Ill., Feb. 4, 1906; s. Muron D. and Adella Pearl (Chritton) T.; m. Patricia Irene Edson, June 7, 1934; children: Annette Roberta, Alden Clyde. AB, U. Kans., 1936, MA, 1939; DSc (hon.), Ariz. State U., 1960. Asst. Lowell Obs., Flagstaff, Ariz., 1929, asst. astronomer, 1938; instr. sci. Ariz. State Coll., Flagstaff, 1943-45; vis. asst. prof. astronomy UCLA, 1945-46; astronomer Aberdeen Ballistics Labs. Annex/White Sands Missile Range, Las Cruces, N.Mex., 1946—, chief optical measurement sect., 1948, chief research and evaluation br. planning dept. Flight Determination div., 1948-53, chief investigator search for natural satellites project, 1953-58, planetary astrophys. researcher, 1958—; research assoc. prof. astronomy N.Mex. State U., 1955-59, prof., 1965-73, prof. emeritus, 1973—, with planetary astrophysics research program, 1959—; discoverer planet Pluto, 1930, 1 globular star cluster, 1932, 5 galactic star clusters, variable stars, asteroids, clusters of galaxies; extensive search for distant planets and natural earth's satellites, studies in apparent distbn. extragalactic galaxies, geol. studies Mars' and Moon's surface features, prodn. telescope mirrors; mem. expdn. extension satellite research project, Quito, Ecuador, 1956-58; lectr. in field. Author: Out of the Darkness: the Planet Pluto, 1980; contbr. articles to profl. jours. Edward Emory Slosson scholar in sci. U. Kans., 1932-36; recipient Jackson-Guilt medal and gift Royal Astron. Soc. Eng., 1931, Fairbanks award Soc. Photog. Instrument Engrs., 1968, Bruce Blair award, 1965, Disting. Svc. citation U. Kans., 1966, Rittenhouse award, Phila., 1990, Golden Plate award Am. Acad. Achievement, 1991; named to White Sand Missile Range Hall of Fame, 1980, Internat. Space Hall of Fame; Clyde Tombaugh Scholars Endowment Fund established in his honor at N.Mex. State U., 1987. Fellow Soc. for Research on Meteorites, AIAA; mem. Am. Astron. Soc., Internat. Astron. Union, Astron. Soc. Pacific, Sigma Xi. Mem. Unitarian Ch. Avocations: grinding telescope mirrors, designing small telescopes. Home: PO Box 306 Mesilla Park NM 88047-0306

TOMEK, WILLIAM GOODRICH, agricultural economist; b. Table Rock, Nebr., Sept. 20, 1932; s. John and Ruth Genevieve (Goodrich) T. B.S., U. Nebr., 1956, M.A., 1957; Ph.D., U. Minn., 1961. Asst. prof. Cornell U., Ithaca, N.Y., 1961-66, NSF fellow, 1965, assoc. prof. agrl. econs., 1966-70, prof., 1970—, chmn. dept. agrl. econs., 1988-93; vis. econ. USDA, 1978-79; vis. fellow Stanford U., 1968-69, U. New Eng., Australia, 1988. Author: Agricultural Product Prices, 1990; editor: Am. Jour. Agrl. Econs., 1975-77; co-editor: Review of Futures Markets, 1993—; mem. editorial bd. Jour. Futures Markets, 1992—; contbr. articles to profl. jours. Served with U.S. Army, 1953-55. Recipient Earl Combs Jr. award Chgo. Bd. Trade Found. Mem. Am. Agrl. Econs. Assn. (pres. 1985-86), Am. Econ. Assn., Econometric Soc., Northeastern Agrl. Econs. Assn., Am. Agrl. Econs. Assn. (awards 1981, 89, fellow), Gamma Sigma Delta. Democrat. Methodist. Office: Cornell U Warren Hall Ithaca NY 14853-7801

TOMITA, ETSUJI, computer science educator; b. Gifu, Japan, Nov. 9, 1942; s. Yoshiyuki and Kiyoko T.; m. Mikiko, May 25, 1975; 1 child, Yoichi. B of Engring., Tokyo Inst. Tech., Japan, 1966, M of Engring., 1968, DEng, 1971. Rsch. assoc. Tokyo Inst. Tech., Tokyo, Japan, 1971-76; assoc. prof. U. of Electro-Communications, Chofu, Tokyo, Japan, 1976-86, prof., 1986—; dept. chmn. U. Electro-Communications, Tokyo, 1987, 92; group chmn. editorial bd. Info Processing Soc. of Japan, 1980-84. Mem. editorial bd. IECE Japan, 1980-84; author: Information Processing, 1983, Automata and Formal Languages, 1992; contbr. articles to profl. jours. Recipient Yonezawa award Inst. Electronics and Communication Engrs. of Japan. Mem. Assn. for Computing Machinery, European Assn. for Theoretical Computer Sci., Inst. Electronics, Info. & Communication Engrs. of Japan, Info. Processing Soc. of Japan, Japanese Neural Network Soc. Avocations: classical music. Home: 6-5-33 Kajigaya Takatsu-ku, Kawasaki 213, Japan Office: U Electro-Communications, 1-5-1 Chofugaoka, Chofu 182, Japan

TOMIZUKA, MASAYOSHI, mechanical engineering educator, researcher; b. Tokyo, Mar. 31, 1946; came to U.S., 1970; s. Makoto and Shizuko (Nagatome) T.; m. Miwako Tomizawa, Sept. 5, 1971; children: Lica, Yumi. MS, Keio U., Japan, 1970; PhD, MIT, 1974. Rsch. assoc. Keio U., 1974; asst. prof. U. Calif., Berkeley, 1974-80, assoc. prof., 1980-86, prof., 1986—. Assoc. editor: IFAC Automatica, 1993—; contbr. more than 100 articles to profl. jours. NSF grantee, 1976-78, 81-83, 86-89, 93—, State of Calif. grantee, 1984-86, 88-93. Fellow ASME (chmn. dynamic systems and control divsn. 1986-87, tech. editor Jour. Dynamic Systems Measurement and Control 1988—); mem. IEEE (assoc. editor Control Systems mag. 1986-88), Soc. Mfg. Engrs. (sr. mem.). Office: U Calif Dept Mech Engring Berkeley CA 94720

TOMLINSON, BRUCE LLOYD, biology educator, researcher; b. Toronto, Ont., Can., Dec. 15, 1950; s. Wilbur Harvey and Betty Joan (Greenslade) T.; m. Donna Elaine Massie, June 18, 1977. MSc, U. Waterloo, Waterloo, Ont., 1978, PhD, 1983. Postdoctoral fellow Ohio State U., Columbus, 1983-84, rsch. assoc., 1984-88; asst. prof. SUNY, Freedonia, 1988-93; assoc. prof. and chair dept. biology SUNY, Fredonia, 1993—. Contbr. articles to profl. jours. Ont. Grad. scholar, Ont. Gov.; Rsch. grantee Cottrell Rsch. Found., 1989. Mem. AAAS, Soc. Devel. Biology, Sigma Xi. Office: SUNY Dept Biology Jewett Hall Fredonia NY 14063

TOMLINSON, G. RICHARD, industrial engineer; b. Clarinda, Iowa, Mar. 20, 1942; s. George Richard Tomlinson and Nina Kathern (Koboldt) Davis; m. Paula Louise Matthaei, Dec. 31, 1978; children: Sean Richard, Tracy Diane. AS, Friends U., 1976, BS, BA in Indsl. Engring., 1978. Assoc. indsl. engr., group capacity analyst Cessna Aircraft Co., Wichita, Kans., 1965-70; staff analyst McCall Pattern Co., Manhattan, Kans., 1971-73; program adminstr. NCR Acctg. Computer, Wichita, 1973-76; assoc. engr. Beech Aircraft Corp., Wichita, 1976-84; sr. indsl. engr. Martin Marietta Aerospace, Balt., 1984-88; continuation engr., indsl. tech. mgr. Nat. Computer Systems, Iowa City, Iowa, 1988—; instr. emergency svc. USAF, 1980-81. Contbr. articles to various publs. Sr. v.p. Kans. Jaycees, 1972-73; lead bass Iowa City Choralaires, 1991, 92, 93, dir. social events. Lt. col. USAF aux., 1960—. Named Jaycee of Month Manhattan Jaycees, 1971; recipient Gil Rob Wilson aerospace award USAF aux., 1977, Paul E. Garver leadership award, 1978, Grover Leoning achievement award, 1979, Kans. Wing Life Saving medal, 1977. Mem. Soc. Automotive Engrs. (membership vice-chair Balt. chpt. 1986-88, vice chair resources Mississippi Valley chpt. 1992—). Methodist. Achievements include development of ion gun for clean air system, automatic booklet sealing machine, wing integral, single engine radar. Home: Box 488 509 Al Ruby Cir Hills IA 52235

TOMLINSON, GARY EARL, museum curator; b. Yorktown, Ind., June 13, 1951; s. Arthur Earl and J. Irene (Hickman) T.; m. Suzanne Marie Naessens, Dec. 28, 1974; 1 child, James Ronald Earl. BS, Ball State U., 1973; MA, Mich. State U., 1974. Cert. secondary physics and math. tchr., Mich. Assoc. curator Pub. Mus. Grand Rapids, Mich., 1976—; Spl. Program for Instnl. Competency in Astronomy agent Harvard/Smithsonian Ctr. Astrophysics, 1990—; coord. astronomy day Astron. League, Washington, 1983—; bd. dirs. Coalition for Excellence in Sci. and Math., Grand Rapids, Western Mich. Interactive Sci. Ctr., Grand Rapids. Author and editor: Astronomy Day Handbook, 1988, 2d edit., 1989; editor: Anthology of Astronomical Poetry, 1984, Planetarium Bibliography, 1992; also articles. Crew chief Monogalia County Emergency Med. Svcs., Morgantown, W.Va., 1975-76; advisor Gran Rapids Pub. Sch.'s Spectrum, 1978-80. Recipient Edmund Sci. award, N.J., 1980, Spectrum award Pub. Rels. Soc. Am., 1986, 93, G.R. Wright Svc. award Astron. League, Nat. Svc. award Western Amateur Astronomers, 1992; Hoosier scholar Gov. Ind., 1969. Mem. Am. Astron. Soc., Great Lakes Planetarium Assn. (bd. dirs. 1984—, pres. 1986-87, fellow 1986), Internat. Planetarium Soc. (rep. 1990-92), Sigma Pi Sigma, Sigma Zeta. Office: Pub Mus Grand Rapids 54 Jefferson Ave SE Grand Rapids MI 49503

TOMLINSON, ROGER W., geographer. Recipient CAG Svc. to the Profession of Geography award Can. Assn. Geographers, 1991. Office: Tomlinson Assocs Ltd, 17 Kippewa Dr, Ottawa, ON Canada K1S 3G3*

TOMLINSON, THOMAS KING, petroleum engineer; b. Houston, Oct. 6, 1934; s. Ira Bowman Tomlinson and Margaret Louise (King) Tomlinson Hochmuth; m. Jessamy Warner, June 19, 1954 (div. May 1980); children: Jane, King, Marshall, Curtis. B in Petroleum Engring., Tex. A&M U., 1957. Prodn. engr. Conoco, Inc., Houston, 1957-63; pres. Thomas K. Tomlinson, Inc., El Campo, Tex., 1963—. Capt. USAFR, 1957-58. Mem. Soc. Petroleum Engrs. Office: PO Box 466 El Campo TX 77437

TOMPA, GARY STEVEN, systems technology director, material scientist; b. Trenton, N.J., Feb. 9, 1958; s. Julius W. and Olga M. (Klein) T.; m. Deborah R. Tompa; children: Sharon Jessica, Rachael Elizabeth, Gary Stephen. BS in physics, chemistry, math. and biology with thesis and high honors, Stevens Inst. Tech., 1980, MS in Physics, 1982, PhD, 1986. Post doctoral rsch. fellow Stevens Inst. Tech., Hoboken, N.J., 1986-87; staff scientist EMCORE, 1987-88, II-IV project leader, 1988-89, project mgr., 1990-91, adminstr. tech. contracts, dir. engring., 1991-92, dir. systems tech., 1992—; mem. Indsl. Adv. Bd. Peer Review Panel, State of N.J. Commn. on Sci. and Tech., 1988. Contbr. over 30 reviewed articles to profl. jours. Recipient Alfred E. Meyer 1st in Physics award, 1980; Garden State fellow, 1980-84; Stanley Rsch. fellow 1980-86. Achievements include patents pending in field. Home: 681-H Dover Ct Somerville NJ 08876 Office: EMCORE Corp 35 Elizabeth Ave Somerset NJ 08873

TOMPKINS, LAURIE, biologist, educator; b. N.Y.C., Mar. 29, 1950; d. Bruce and Jean (Murray) T.; m. Lawrence Neil Yager, Jan. 6, 1990. BA, Swarthmore Coll., 1972; PhD, Princeton U., 1977. Postdoctoral fellow Brandeis U., Waltham, Mass., 1977-80; asst. prof. Temple U., Phila., 1981-86, assoc. prof., 1986-92, prof., 1992—; instr. Marine Biol. Lab., Woods HOle, Mass., 1985-88. Contbr. articles to profl. jours. Rsch. grantee NIH, NSF, Washington, 1981, 84, 87. Fellow AAAS; mem. Genetics Soc. Am., Soc. for Study of Evolution, others. Democrat.

TON, DAO-RONG, mathematics educator; b. Ho County, Anhui, People's Republic of China, Jan. 8, 1940; s. You-Bing Ton and Liang-Xian Sai; m. Kai-Yao He; 1 child, Ling Tong. BS, South Anhui U., Wuhu, People's Republic of China, 1962. With Dept. Sci. and Tech. of Anui Province, People's Republic of China, 1972-77; lectr. U. Sci. and Tech. China, Hefei, People's Republic of China, 1977-87; assoc. prof. math. Hohai U., Nanjing, People's Republic of China, 1988-90, prof., 1990—; vis. scholar Bowling Green (Ohio) State U., 1987-88; reviewer Math. Revs., 1989—. Contbr. articles to profl. jours. Mem. Am. Math. Soc., Math. Assn. Am., Math. Soc. People's Republic China, Soc. Ordered Algebra People's Republic China. Mohammedan. Home: 5-303 3 Xikang Rd, Nanjing Jiangsu 210024, China

TØNDERING, CLAUS, software engineer; b. Copenhagen, Denmark, Aug. 2, 1953; s. Trygve and Lise (Petersen) T.; m. Trine Miland Nielsen, Aug. 12, 1978; children: Maria, Nina, Rebekka, Michala. MSc, Tech. U. Denmark, Copenhagen, 1978. Systems programmer Søren T. Lyngsø, Copenhagen, 1978-80; sr. systems software engr. Dansk Data Elektronik, Copenhagen, 1980-89; systems analyst Dansk Data Elektronik, Palmerston North, New Zealand, 1990; chief tech. strategist Dansk Data Elektronik, Copenhagen, 1991-92, mgr. product mktg., 1992-93; systems engr. Olicom, Copenhagen, 1993—; tech. expert, ISO, 1987-89. Mem. N.Y. Acad. Scis. Lutheran. Home: Skovvaenget 16, DK-2800 Lyngby Denmark Office: Olicom A/S, Nybrovej 114, DK 2800 Lyngby Denmark

TONEGAWA, SUSUMU, biology educator; b. Nagoya, Japan, Sept. 5, 1939; came to U.S., 1963; s. Tsutomu and Miyoko (Masuko) T.; m. Mayumi Yoshinari, Sept. 28, 1985; children: Hidde, Hanna, Satto. BS, Kyoto U., Japan, 1963; PhD, U. Calif., San Diego, 1968. Rsch. asst. U. Calif., San Diego, 1963-64, teaching asst., 1964-68; mem. Basel (Switzerland) Inst. Immunology, 1971-81; prof. biology MIT, Cambridge, 1981—. Editorial bd. Jour. Molecular and Cellular Immunology. Decorated Order of Culture, Emperor of Japan; recipient Cloetta prize, 1978, Avery Landsteiner prize Gesselschaft für Immunologie, 1981, Louisa Gross Horwitz prize Columbia U., 1982, award Gardiner Found. Internat., Toronto, Ont., Can., 1983, Robert Koch Found. prize, Bonn., Fed. Republic Germany, 1986, co-recipient Albert Lasker Med. Rsch. award, 1987, Nobel prize in Physiology or Medicine, 1987; named Person with Cultural Merit Japanese Govt., 1983. Mem. NAS (fgn. assoc.), Am. Assn. Immunologists (hon.), Scandinavian Soc. Immunology (hon.). Office: MIT 77 Massachusetts Ave Cambridge MA 02139-4307

TONER, WALTER JOSEPH, JR., transportation engineer, financial consultant; b. Rutherford, N.J., July 22, 1921; s. Walter Joseph Toner and Rhea Virginia Carell; m. Barbara Jean Francis, Sept. 11, 1943; children: Sherry Francis, Walter J. III. Student, Wesleyan U., 1938-39; BS in Engring., U.S. Naval Acad., 1942. Profl. engr. Engring. mgr. Bethlehem Steel Corp., Boston, 1946-75; nuclear cons. Stone & Webster, Boston, 1975-77; v.p. project devel. & mgmt. Sverdrup Corp., Boston, 1977-86; cons., v.p., T.Y. Lin Internat., Boston, 1986—; v.p. Performance Index, Inc., 1991—. Lt. USN, 1942-46. Fellow Soc. Am. Mil. Engrs. (nat. trustee, v.p. 1978-85, life); mem. ASCE (life), DAV, Am. Consulting Engrs. Coun., Am. Railway Engrs. Assn., Boston Soc. Civil Engrs., Transp. Rsch. Bd., U.S. Naval Acad. Alumni Assn. (life, nat. trustee), U.S. Naval Inst. (life), Wardroom Club, Bass River Yacht Club (chair race com.), Wrinkle Point Beach Assn., Masons. Episcopalian.

TONG, MARY POWDERLY, mathematician educator, retired; b. N.Y.C., May 24, 1924; d. William Joseph and Katherine Colwell Powderly; m. Hing Tong, Aug. 19, 1956; children: Christopher, Mary Elizabeth, William, Jane Frances, James. BA, St. Joseph's Coll., 1950; MA, Columbia U., 1951, PhD, 1969. Instr. math. St. Joseph's Coll., Bklyn., 1951-54, Columbia Univ., N.Y.C., 1954-60; asst. prof. math. Univ. Conn., Storrs, 1960-66; assoc. prof. math. Fairfield (Conn.) Univ., 1966-70; prof. math. William Paterson Coll., Wayne, N.J., 1970-81. Contbr. articles to profl. jours. Trustee South Bergen Mental Health Ctr., Lyndhurst, N.J., 1988—. Recipient fellowship NSF, Washington, 1959-60. Mem. Am. Math. Soc., Math. Assn. Am., Am. Phys. Soc., N.Y. Acad. Scis., Delta Epsilon Sigma. Roman Catholic. Home: 725 Cooper Ave Oradell NJ 07649-2334

TONG, YIT CHOW, research scientist; b. Hong Kong, Aug. 19, 1948; s. Ling Hon and Ah Lan (Mak) T.; m. Kim Kheng Toh, Dec. 19, 1976; 1 child, Simon Yi-Hon. B of Engring., U. Melbourne, Australia, 1972, PhD, 1977. Rsch. officer U. Melbourne, Australia, 1976-78, sr. rsch. officer, 1979-82, rsch. fellow, 1983-84, sr. rsch. fellow, 1985-87, prin. rsch. fellow, 1988-92; assoc. prof. Nanyuang Tech. U., Singapore, Singapore, 1992—. Patentee in speech processors and prosthesis; editor: Cochlear Prostheses, 1990. Recipient First Class honors in Mech. Engring., U. Melbourne, 1972. Mem. Acoustical Soc. Am., Australian Physiol. & Pharmacol. Soc. Avocations: electronics, reading, tennis, table tennis. Home: 1 Charles St, Kew 3101, Australia Office: Nanyang Tech U, Sch EEE Nanyang Ave, Singapore 2263, Singapore

TONN, BRUCE EDWARD, social scientist, researcher; b. Chgo., Feb. 3, 1955; s. Ralph Bruce Tonn and Doris F. Astrin Evans; M. Diana Kathleen Bettencourt, Sept. 8, 1979; children: Jenna, Christopher, Shara. Mem. rsch. staff Oak Ridge (Tenn.) Nat. Lab., 1983-88, group leader, 1988—; adv. bd. Electric Power Rsch. Inst., Palo Alto, Calif., 1990-92. Contbr. articles to profl. jours. Grantee U.S. Census Bur., Suitland, Md., 1989, Bur. Labor Statistics, Washington, 1991, Electric Power Rsch. Inst., 1991, U.S. Forest Svc., Ft. Collins, Colo., 1992, EPA, Washington, 1992. Mem. Am. Planning Assn., Am. Assn. Artificial Intelligence, Social Sci. Computing Assn. (pres. elect 1992-93), Soc. Risk Analysis. Achievements include research in philosophy of long term planning, court of generations for U.S., inherent uncertainty, methods for synthesizing evidence, present costs for cleaning up hazardous waste in U.S., measure of information in a lower probability, generalization of Jeffrey conditionalization, representing risk perceptions using knowledge-based techniques. Office: Oak Ridge Nat Lab PO Box 2008 Oak Ridge TN 37831-6207

TONN, ROBERT JAMES, entomologist; b. Watertown, Wis., June 23, 1927; s. Harry James and Elise (Foogman) T.; m. Noemi C. Tonn; children: Sigrid M., Monica E. BS, Colo. State U., 1949, MS, 1950; MPH, Okla. Med. Sch., 1963; PhD, Okla. State U., 1959. Rsch. assoc La. State U., Costa Rica/New Orleans, 1961-63; dir. Taunton Field Sta., Taunton, Mass., 1963-65; chief PMO unit WHO, various locations, 1965-87; adj. prof. of parasitology U. Tex.-El Paso, 1988—; cons. USAID/VBC, 1987—. Contbr. numerous articles to profl. jours. Mem. Am. Soc. Tropical Medicine, Soc. Vector Ecology (pres. 1984), Am. Mosquito Control Assn., U.S./ Mex. Border Health Assn., Royal Soc. Tropical Medicine and Hygiene, Masons. Congregationalist. Home: RR 3 Box 505 Park Rapids MN 56470-9363

TOOHIG, TIMOTHY E., physicist; b. Lawrence, Mass., Feb. 17, 1928; s. Timothy Michael and Catherine Marie (Walsh) T. BS in Physics, Boston Coll., 1947; MS in Optics, U. Rochester, 1953; PhD in Physics, Johns Hopkins U., 1962; Licentiate in Philosophy, Weston Coll., 1957. Ordained priest Soc. Jesus, 1965. Rsch. asst. Johns Hopkins U., 1957-62, rsch. assoc., 1962-63; physicist 200 Be V project Lawrence Berkeley Lab., 1964-66; physicist Brookhaven Nat. Lab., 1967-70; physicist, various positions Fermi Nat. Accelerator Lab., 1970—; physicist superconducting supercollider Ctrl. Design Group, 1984-89, dep. head conventional systems divsn.; physicist SSC Lab., 1989—, dep. head conventional construction divsn.; chaplain Seton Hall High Sch., Patchogue, N.Y., 1967-70, St. Joseph High Sch., Westchester, Ill., 1971-89. Sgt. U.S. Army, 1946-47. Mem. Am. Phys. Soc. Roman Catholic. Achievements include discovery of eta meson. Measurement of pi meson and k meson form factors. Demonstration of charged particle steering by crystals. Demonstration of the existence of channeling radiation from crystals. Home: 12345 Inwood Dallas TX 75244 Office: SSCL CCD 2550 Beckleymeade MS2014 Dallas TX 75237-3997

TOOLE, JAMES FRANCIS, medical educator; b. Atlanta, Mar. 22, 1925; s. Walter O. Brien and Helen (Whitehurst) T.; m. Patricia Anne Wooldridge, Oct. 25, 1952; children: William, Anne, James, Douglas Sean. B.A., Princeton U., 1947; M.D., Cornell U., 1949; LL.B., LaSalle Extension U., 1962. Intern, then resident internal medicine and neurology U. Pa. Hosp., Nat. Hosp., London, Eng., 1953-58; mem. faculty U. Pa. Sch. Medicine, 1958-62; prof. neurology, chmn. dept. Bowman Gray Sch. Medicine Wake Forest U., 1962-83; vis. prof. neuroscis. U. Calif. at San Diego, 1969-70; vis. scholar Oxford U., 1989; mem. Nat. Bd. Med. Examiners, 1970-76; mem. task force arteriosclerosis Nat. Heart Lung & Blood Inst., 1970-81; chmn. 6th and 7th Princeton confs. cerebrovascular diseases; cons. epidemiology WHO, Japan, 1971, 73, 93, USSR, 1968, Ivory Coast, 1977, Japan, 1993; mem. Lasker Awards com., 1976-77; chmn. neuropharmocologic drugs com. FDA, 1979; cons. NASA, 1966. Author: Cerebrovascular Diseases, 4th edit., 1990; editor: Current Concepts in Cerebrovascular Disease, 1969-73, Jour. Neurol. Sci., 1990—; mem. editorial bd. Annals Internal Medicine, 1968-75, Stroke, 1972-91, Jour. AMA, 1975-77, Ann. Neurology, 1980-86, Jour. Neurol. Soc., 1990; mem. editorial bd. Jour. of Neurology, 1985-89, editor, 1990—. Pres. N.C. Heart Assn., 1976-77. Served with AUS, 1950-51; flight surgeon USNR, 1951-53. Decorated Bronze Star with V, Combat Med. badge. Fellow ACP, AAAS (life); mem. AMA, Am. Clin. and Climatol. Assn., Am. Heart Assn. (chmn. com. ethics 1970-75), Am. Physiol. Soc., Am. Neurol. Assn. (sec.-treas. 1978-82, pres. 1984-85, archivist, historian 1988—), World Fedn. Neurology (sec.-treas. 1982-89, mgmt. com. 1990—), Am. Acad. Neurology, Am. Soc. Neuroimaging (pres. 1992—), Internat. Stroke Soc. (exec. com. 1989—, program chair 1992), Assn. Brit. Neurologists (hon.), German Neurol Soc. (hon.), Austrian Soc. Neurology (hon.), Irish Nevrol Assn. (hon.), 1992, Nat. Stroke Assn. (bd. dirs. 1993—). Home: 1836 Virginia Rd Winston Salem NC 27104-2316

TOOMBS, RUSS WILLIAM, laboratory director; b. Troy, N.Y., July 11, 1951; s. George John and Olive Catherine (Blodgett) T.; m. Patrice Ann De Paul, Aug. 19, 1972; children: David Christopher, Mark Patrick. BS, Cornell U., 1973. Environ. scientist Wapora, Inc., Washington, 1973-74; bacteriologist N.Y. State Dept. Health, Wadsworth Ctr. for Labs. and Rsch., Albany, N.Y., 1974-76, sr. bacteriologist, 1976-78, assoc. bacteriologist, 1978-86, dir. ops., 1986-90, assoc. dir., 1990—. Contbr. articles to profl. jours. Mem. Saratoga Performing Arts Ctr., Am. Soc. for Microbiology. Roman Catholic. Home: 65 Huntleigh Dr Albany NY 12211-1175 Office: Wadsworth Ctr Labs and Rsch Empire State Pla Albany NY 12201-0509

TOOMRE, ALAR, mathematics educator, theoretical astronomer; b. Rakvere, Estonia, Feb. 5, 1937; came to U.S., 1949, naturalized, 1955; s. Elmar and Linda (Aghen) T.; m. Joyce Stetson, June 15, 1958; children—Lars, Erik, Anya. B.S. in Aero. Engring., B.S. in Physics, MIT, 1957; Ph.D. in Fluid Mechanics, U. Manchester, Eng., 1960. C.L.E. Moore instr. math. dept. MIT, Cambridge, 1960-62; asst. prof. applied math. MIT, 1963-65, assoc. prof., 1965-70, prof., 1970—; fellow Inst. for Advanced Study, Princeton, N.J., 1962-63. Contbr. articles to profl. jours. Guggenheim fellow, 1969-70, MacArthur fellow, 1984-89; Fairchild scholar, 1975, Marshall scholar, 1957-60. Fellow AAAS; mem. Am. Astron. Soc. (Dirk Brouwer award 1993), Internat. Astron. Union, Am. Acad. Arts and Scis., Nat. Acad. Scis. Office: MIT Room 2-371 77 Massachusetts Ave Cambridge MA 02139-4307

TOP, FRANKLIN HENRY, JR., physician, researcher; b. Detroit, Mar. 1, 1936; s. Franklin Henry Sr. and Mary (Madden) T.; m. Lois Elizabeth Fritzell, Sept. 23, 1961; children: Franklin H. III, Brian N., Andrew M. BS, Yale U., 1957, MD cum laude, 1961. Diplomate Am. Bd. Pediatrics. Intern, resident, infectious diseases fellow U. Minn. Hosps., Mpls., 1961-66; commd. officer U.S. Army, advanced through grades to col.; med. officer, dept. virus diseases Walter Reed Army Inst. Research, Washington, 1966-70, chief dept. virus diseases, 1973-76, dir. div. communicable diseases and immunology, 1976-79, dep. dir., 1979-81, dir. and comdt., 1983-87; chief dept. virology Seato Med. Research Lab., Bangkok, 1970-73; comdr. U.S.A. Med. Research Inst. of Chem. Def., Aberdeen Proving Ground, Md., 1981-83; ret. U.S. Army, 1987; sr. v.p. Praxis Biologics Inc., Rochester, N.Y., 1987-88; exec. v.p., med. dir. MedImmune, Inc., Gaithersburg, Md., 1988—. Contbr. over 40 articles to med. jours. Decorated Legion of Merit with 2 oak leaf clusters. Fellow Am. Acad. Pediatrics, Infectious Diseases Soc. Am.; mem. AMA, Am. Assn. Immunologists, Alpha Omega Alpha. Avocation: ornithology. Office: MedImmune Inc 35 W Watkins Mill Rd Gaithersburg MD 20878-4024

TOPAL, MICHAEL DAVID, biochemistry educator; b. London, Feb. 18, 1945; came to U.S., 1946; s. Lawrence Herbert and Jean Lupin T.; m. Maria Lilly Phillips, Sept. 7, 1969; children: Daniel Jacob, Aaron Joseph. BA, Adelphi U., 1967; PhD, NYU, 1972. Post doctoral fellow Princeton U., 1972-77; asst. prof. U. N.C., Chapel Hill, 1977-84, assoc. prof., 1984-90, prof., 1990—. Contbr. articles to Biochem., Jour. Biol. Chemistry, Nature, PNAS. Fellow Nat. Cancer Inst., Princeton U., 1974-77, Scholar award, Leukemia Soc. Am., 1984-89. Office: Univ NC Med Sch Lineberger Cancer Ctr Chapel Hill NC 27599-7295

TOPEL, DAVID GLEN, college dean, animal science educator; b. Lake Mills, Wis., Oct. 24, 1937; s. Walter C. and Mable (Strauss) T.; m. Jacqueline May Richardson, Nov. 21, 1964. BS, U. Wis., 1960; MS, Kans. State U., 1962; PhD, Mich. State U., 1965. Asst. prof. Iowa State U., Ames, 1965-67, assoc. prof., 1967-73, prof. dept. animal sci., 1973-88, head. dept., 1979-88, dean, Coll. Agr., 1988—; dir. Iowa Agr. Expt. Sta., Ames, 1988—, MidAm. Internat. Agrl. Consortium, Lincoln, Nebr., 1988—. Author: The Pork Industry - Problems and Progress, 1968. Secretariat World Food Prize, Iowa State U., Ames, 1991—. Recipient Outstanding Educator award Nat. Assn. Meat Purveyors, 1989. Mem. Am. Soc. Animal Sci. (sec. So. sect. 1988, rsch. award 1979), Inst. Food Tech., Am. Meat Sci. Assn. (bd. dirs. 1977), N.Y. Acad. Sci. Presbyterian. Avocations: fishing, golf. Home: 2630 Meadow Glen Rd Ames IA 50010-8239

TOPPEL, BERT JACK, reactor physicist; b. Chgo., July 2, 1926; s. Jacob S. and Sonia (Feldman) T.; m. Leona C. Weiss, June 11, 1950; children: Alison, Leslie, Debra. BS, Ill. Inst. Tech., 1948, MS, 1950, PhD, 1952. Assoc. physicst Brookhaven Nat. Lab., Upton, N.Y., 1952-56; sr. physicist Argonne (Ill.) Nat. Lab., 1956—. With USN, 1944-46. Mem. Am. Nuclear Soc., Am. Phys. Soc. Achievements include development of the Argonne Reactor Computation (ARC) System; Mc2 Code, REBUS-3 Code. Office: Argonne Nat Lab 9700 S Cass Ave Argonne IL 60439

TORANZOS, GARY ANTONIO, microbiology educator; b. Cochabamba, Bolivia, Apr. 5, 1958; came to U.S., 1976; s. Alberto and Emma (Soria) T.; m. Teresa Maria Candelas, Dec. 3, 1988. MS, U. Ariz., 1983, PhD, 1985. Postdoctoral assoc. U. Fla., Gainesville, 1985-86; postdoctoral assoc. U. P.R., Rio Piedras, 1986-88, asst. prof., 1988—; cons. in field; adv. bd. Industry-Univ. Rsch. Ctr., Rio Piedras, 1989—. Contbr. articles to profl. jours. and chpt. to book. Mem. AAAS, Am. Soc. for Microbiology, P.R. Soc. for Microbiology (councilar 1990-91), Am. Water Works Assn. (councilor 1990—). Office: U Puerto Rico Dept Biology Rio Piedras PR 00931

TORGET, ARNE O., electrical engineer; b. Cathlamet, Wash., Oct. 10, 1916; s. John B. and Anna J. (Olson) T.; m. Dorothy M. Lackie, Aug. 30, 1941; children: Kathleen, James, Thomas. BSEE, U. Wash., 1940. Registered profl. elec. engr., Calif. Design engr. Boeing, Seattle, 1940-41, asst. group engr., 1941-46; design specialist N.Am. Aviation, L.A., 1946-50, 60-64, elec. supr., 1950-55; design specialist Rocketdyne N.Am. Aviation, Canoga Park, Calif., 1955-60; design specialist Space Div. Rockwell Internat., Downey, Calif., 1964-79; commr. Wahkiakum Count Pub. Utility, Cathlamet, 1985—; Bd. dirs. Wash. Pub. Utility Dist. Utility Systems, Seattle, 1985—, Wash. Pub. Power Supply System, Richland, 1987—, Wash. Pub. Utility Dist. Assn., Seattle, 1985—. Mem. AAAS, IEEE, Elks. Republican. Roman Catholic. Home: 166 E Sunny Sands Rd Cathlamet WA 98612-9708 Office: Wahkiakum County Pub Utilities Dist 45 Riv St Cathlamet WA 98612

TORI, MOTOO, chemist, educator; b. Yatomi-cho, Japan, June 25, 1948; s. Naoji and Tomoko Tori; m. Kumiko Toda, May 5, 1975; children: Kanako, Utako, Takao. BS, Shizuoka U., 1972; DSc, U. Tokyo, 1977. Lectr. U. Saitama, Urawa, 1982-83; assoc. prof. Tokushima (Japan) Bunri U., 1983—. Author: Studies in Natural Products Chemistry, 1988, 2D NMR Workbook, 1992; translator: Nuclear Magnetic Resonance, 1988, Structure Elucidation by Modern NMR, 1990, Organic Chemistry, 1987. Home: Hachiman-cho Shingai 131-3, Tokushima 770, Japan Office: Tokushima Bunri U, Yamashiro-cho, Tokushima 770, Japan

TORIGIAN, PUZANT CROSSLEY, pharmaceutical company executive, clinical research pharmacist; b. Istanbul, Turkey, Sept. 21, 1922; s. John and Shakeh (Yaver) T.; BS in Pharm., Columbia U., 1949; postgrad. N.Y.U.; D.Sc. (hon.), U. Ea. Fla., 1971; m. Joanne Curatolo, Mar. 28, 1971; children: Christine, John, Michael. Mem. staff Torigian Labs. Inc., Queens Village, N.Y., 1939-50, sales mgr., 1955-58, pres., 1976-84; staff, Lloyd Chemists, Jamaica, N.Y., 1951-53; pres. Crossley Pharms. Corp., Queens Village, 1953-55; asst. to pres. Marvin R. Thompson Inc., Stamford, Conn., pres., 1958-59; ops. mgr. J.B. Williams Co., N.Y.C., 1959-60; pres. Tobison Personnel Agy., N.Y.C., 1960-64; pres. Bravo Smokes Inc., Hereford, Tex., 1964-74; plant mgr. Sterling Drugs Internat., Kuala Lumpur, Malaysia, 1974-76; pres. Challenger Industries, Fort Lee, N.J., 1978-87; investigational drug specialist Clin. Rsch. Pharm. Bronx (N.Y.) VA Med. Ctr., 1987—; pres. Anoush Parfumerie, Ft. Lee, 1978—; mng. dir. Found. for Alternative Research, Inc., 1985—; pres., chmn. bd. dirs. Safer Smokes, Inc., Fort Lee, 1986—; sci. dir., v.p. Pharmakon USA, 1992—, also bd. dirs.; treas. Am. Armenian Med. Philanthropic Fund, Inc.; bd. dirs. Constinople Am. Relief Soc. Served with USNR, 1942-46, PTO. Decorated knight comdr. Order St. George of Corinthia. Mem. Columbia U. Coll. Pharmacy Alumni Assn. (pres. 1963-64, editor Graduate 1956-63, Lion award, Disting. Service award), Fedn. Alumni Assns. Columbia U. (trustee 1960-64), Am. Pharm. Assn., Am. Soc. Parenteral and Enteric Nutrition, Parenteral Drug Assn., N.Y. Acad. Scis., Internat. Soc. Pharm. Engrs. (charter), AAAS, Kappa Psi (nat. historian 1961-63). Mem. Armenian Apostolic Ch. Author: How to Give Up Tobacco, 1971; 4 patents in field. Home: 2 Horizon Rd Fort Lee NJ 07024-6525 Office: Pharmakon USA 1 State St Plz New York NY 10004

TORII, SHUKO, psychology educator; b. Toyohashi, Aichi-ken, Japan, Apr. 5, 1930; m. Toshiko Mochizuki, June 19, 1975. BA, U. Tokyo, 1954, MA, 1956, PhD, 1964. Rsch. asst. Tokyo Inst. Tech., 1959-61, U. Tokyo, 1961-65; assoc. prof. Tokyo U. Agr. Tech., 1965-70; rsch. assoc. U. Mich., Ann Arbor, 1966-68; rsch. assoc. Inst. Molecular Biophysics Fla. State U., 1968-69; from assoc. prof. to prof. psychology U. Tokyo, 1970-91; prof. U. Sacred Heart, Tokyo, 1991—. Author: The World of Vision, 1979, Psychology of Vision, 1982, Visual Perception in the Congenitally or Early Blinded after Surgery, 1992; author, editor: Perception, 1983, Visually Handicapped and Technology of Sensory Substitution, 1984, The Visually Handicapped and Their Cognitive Activity, 1993. Mem. Japanese Psychonomic Soc. (assoc. editor 1981-88, editor 1988—), Japanese Psychol. Assn. (mem. editorial com. 1987-89, editor 1989-92), Optical Soc. Am. Home: 2-

17-26-204 Takada, Toshima-ku, Tokyo 171, Japan Office: U Sacred Heart, 4-3-1 Hiroo, Shibuya-ku Tokyo 150, Japan

TORII, SIGERU, chemistry educator; b. Ohmi-Hachiman, Shiga, Japan, June 25, 1932; s. Hiroko Torii, June 21, 1961; m. Hiroko Fujii; children: Azusa, Yukari, Katsura. B. Engring., Kyoto (Japan) Tech. U., 1956; M. Engring., Kyoto U., 1959, D. Engring., 1962. Assoc. prof. Okayama (Japan) U., 1963-70, prof., 1970—, dean faculty engring.; researcher Harvard U., Cambridge, Mass., 1970-71. Mem. Soc. Synthetic Organic Chemistry Japan (bd. dirs., Synthetic Organic Chemistry Japan award 1982, v.p. 1991—), Chem. Soc. of Japan (bd. dirs., award 1987), Rotary. Home: 4-18 Sany-odanchi-4c, Sanyo-cho, Akaiwa-Gun, Okayama 709-80, Japan Office: Okayama U Faculty Engring, 3-1-1 Tsushima-Naka, Okayama 700, Japan

TORII, TETSUYA, retired science educator; b. Takao, Taiwan, May 14, 1918; s. Nobuhei Torii and Masako Suzuki; m. Noriko; children: Tohru, Mari Tanaka, Nobuya. MS, Univ. Tokyo, 1943, DSc (hon.), 1956. Chem. lectr. Kanagawa U., Yokohama, Japan, 1952-55; assoc. prof. chem. U. Chiba, Japan, 1955-62, prof. chem., 1962-63; prof. chem. Chiba Inst. Tech., 1963-93; exec. sec. Japan Polar Rsch. Assn., Tokyo, 1964—; dir. Japan Chem. Analysis Ctr., Chiba, 1980-90. Editor: pictorial book Antarctica 1970 (adapted to put into the time capsule); discoverer: new mineral in Antarctica antarcticite 1965. Officer Japanese Navy, 1943-45. Recipient Silver Cup, Prime Minister of Japan, 1962, citation Ministry of Edn., Sci. and Culture, 1977, prize Geochem. Rsch. Assn., 1977. Mem. Am. Geophys. Union, Geochem. Soc. Japan, Limnological Soc. (U.S.A.), Balneological Soc. Japan (pres. 1988), Explorer Club (U.S.A.). Avocation: mountaineering. Home: 4 18 18 Ogikubo, Suginami-ku, Tokyo 167, Japan Office: 2 3 4 Hirakawa-cho, Chiyodaku, Tokyo 102, Japan

TORNETTA, FRANK JOSEPH, anesthesiologist, educator, consultant; b. Norristown, Pa., Jan. 22, 1916; s. Joseph F. and Maria (Ciaccio) T.; m. Edith Galullo, Nov. 21, 1941 (dec. 1952); m. Norma Zollers, July 16, 1957; children: Frank Jr., David A., Mark A. BS, Ursinus Coll., 1938; MA, U. Pa., 1940; PhD, NYU, 1943; MD, Hahnemann Med. Coll., 1946. Diplomate Am. Bd. Anesthesiology, 1953. Instr. U. Md., College Park, 1940, Hofstra Coll., Hempstead, N.Y., 1941; teaching fellow NYU, N.Y.C., 1941-43; asst. instr. Med. Sch. U. Pa., Phila., 1949-50; dir. dept. anesthesiology. dir. Sch. Anesthesia, founder Montgomery Hosp. Med. Ctr., Norristown, Pa., 1950-91; clin. assoc. prof. Sch. Medicine Temple U., Phila., 1985-91; lectr. Grad. Sch. St. Joseph's U., Phila., 1987-91. Contbr. articles to profl. jours. Chmn. task force Montgomery County Health Dept., Norristown, 1989-91; active Valley Forge chpt. Boy Scouts Am., Norristown, 1982. Lt. USN, 1943-50. Fellow Am. Coll. Chest Physicians, Am. Coll. Anesthesiologists, Coll. Physicians Phila.; mem. Pa. Soc. Anesthesiologists (pres. 1970), Montgomery County Med. Soc. (pres. 1969), Montgomery Hosp. Med. Staff Assn. (pres. 1960), Hahnemann Med. Coll. Alumni Assn. (v.p. 1982), KC. Republican. Roman Catholic. Home: 307 Anthony Dr Plymouth Meeting PA 19462-1109 Office: Montgomery Hosp Med Ctr 1300 Powell St Norristown PA 19401-3324

TORNO, LAURENT JEAN, JR., architect; b. St. Louis, June 13, 1936; s. Laurent Jean and Jane (Crawford) T.; m. Elizabeth Gareschè; children: Elizabeth G., Laurent J. III, Martha C., Jean-Paul. BArch, Washington U., 1960. Registered architect, Mo. Draughtsman Smith S. Entzeroth and William B. Ittner, St. Louis, 1956-60; designer Hellmuth, Obata and Kassabaum, St. Louis, 1960-62; project architect Smith and Entzeroth Architects, St. Louis, 1962-64; assoc. ptnr. Theodore C. Christner and Assocs., St. Louis, 1965-66; prin. ptnr., v.p. Berger-Field-Torno-Hurley, St. Louis, 1966-74; prin. L.J. Torno Jr. and Assocs., St. Louis, 1975—; affiliate asst. prof. Sch. Architecture Washington U., 1970-76; bd. dirs. Nat. Paraplegia Found., St. Louis, 1973-76; exec. com. Gov.'s Com. on Employment of Handicapped, St. Louis, 1974-76. Grantee Nat. Endowment for Arts, 1975. Mem. AIA, St. Louis Home Builders Assn. (Honor award 1970, 71, 72, 73, 74), St. Louis CSoc. Cummunication Arts (Honor award 1970, 74), Landmarks Assn. St. Louis (bd. dirs. 1964-66). Office: LJ Torno Jr and Assocs 7916 Kingsbury St Saint Louis MO 63105

TORRANCE, ELLIS PAUL, psychologist, educator; b. Milledgeville, Ga., Oct. 8, 1915; s. Ellis Watson and Jimmie Pearl (Ennis) T.; m. Jessie Pansy Nigh, Nov. 25, 1959 (dec. Nov. 1988). B.A., Mercer U., 1940; M.A., U. Minn., 1944; Ph.D., U. Mich., 1951. Tchr. Midway Vocational High Sch., Milledgeville, 1936-37; tchr., counselor Ga. Mil. Coll., 1937-40, prin., 1941-44; counselor student counseling bur. U. Minn., 1945; counselor counseling bur. Kans. State Coll., 1946-48, dir., 1949-51; dir. survival research field unit Stead AFB, Nev., 1951-57; dir. Bur. Ednl. Research, 1958-64; prof. ednl. psychology U. Minn., 1958-66; chmn., prof. dept. ednl. psychology U. Ga., 1966-78, Alumni Found. disting. prof., 1974-85; Alumni Found. disting. prof. emeritus, 1985—; advisor Torrance Ctr. for Creative Studies, U. Ga. Author: Torrance Tests of Creative Thinking, Thinking Creatively in Action and Movement, Style of Learning and Thinking, and Sounds and Images; contbr. articles to jours., mags., books. Trustee Creative Edn. Found.; founder Nat. Future Problem Solving Problem and Bowl, Nat. Scenario Writing Contest. Torrance Ctr. for Creative Studies, Torrance Creative Scholars, and Mentors Network established in his honor at U. Ga. Fellow Am. Psychol. Assn.; mem. Am. Ednl. Rsch. Assn., Am Soc. Group Psychotherapy and Psychodrama, Creative Edn. Leadership Coun., Nat. Assn. Gifted Children, Am. Creativity Assn., Phi Delta Kappa. Baptist. Home: 183 Cherokee Ave Athens GA 30606-4305

TORRE, BENNETT PATRICK, computer scientist, consultant; b. Newark, Feb. 20, 1962; s. Benvenuto Peter and Paula Marie (Caprio) T. BA in Communications, BS in Computer Sci, William Paterson U., 1985. Cert. FCC Gen. Class Amateur Radio. Programmer ADP, Roseland, N.J., 1986-89; cons. Omni-Tech Cons. Svcs., Edison, N.J., 1989—. Mem. DeMolay. Home: 19 Gregory Terr Belleville NJ 07109

TORRES, ANTONIO, civil engineer; b. N.Y.C., Jan. 20, 1956; s. Antonio Sr. and Andrea (Mercado) T.; m. Francisca Rivera, June 8, 1979; children: Armando L., Carlos A. BCE, U. Puerto Rico, 1979. Registered profl. engr., Ga. Assoc. engr. Boeing Comml. Airplane Co., Seattle, 1979-80; civil engr. USDA Forest Svc., Bedford, Ind., 1980-81; civil engr. U.S. Army Corps Engrs., Washington, 1981-83, Dharhan, Saudi Arabia, 1983-84, Savannah, Ga., 1984-88, Huntsville, Ala., 1988—. Bd. dirs. Georgetown Community Assn., Savannah, 1988. Mem. ASCE, Soc. Mil. Engrs. Home: 2223 Magna Carta Pl SW Huntsville AL 35803-2173 Office: U S Army Corps Engrs PO Box 1600 Huntsville AL 35807-4301

TORRES, GUIDO ADOLFO, water treatment company executive; b. Esmeraldas, Ecuador, Aug. 29, 1938; s. Carlos M. and Nora I. (Andrade) T.; m. Lupe N. Duran, Aug. 29, 1964; children: Guido, Alex, Juan Jose. B of Chem. Engring, Cen. U. Ecuador, 1967. Chief of study group Indsl. Devel. Ctr. Ecuador, Quito, 1965-69, tech. asst. to exec. dir., 1969-70; researcher Frakes Water Treatment Plant, Luverne, Minn., 1970-71; researcher Andean Water Treatment Soc. Anonima, Quito, 1972-84, gen. mgr., 1984-85, pres., 1985—. Author: Industrial Water Treatment; contbr. articles on water treatment to jours.; developer water treatment chems., equipment. Fellow Chem. Engring Coll. Ecuador (pres. 1976); mem. Am. Inst. Chem. Engrs., Am. Water Works Assn., N.Y. Acad. Scis., Quito Indsl. Chamber, Quito C. of C., Quito Small Industry Chamber. Club: Castillo Amaguana. Avocations: chess, jogging, soccer. Home: 261 Miravalle, Quito Pichincha, Ecuador Office: Andean Water Treatment Soc, El Batan 250 PO Box 17-01-3297, Quito Pichincha, Ecuador

TORRES, ISRAEL, oral and maxillofacial surgeon; b. El Paso, Tex., Sept. 5, 1934; s. Francisco Mendoza and Manuela (Gallardo) T.; m. Karen Marie Hensley, Aug. 22, 1970; children—Michael, George Stanley, Dianna. B.S., Tex. Western Coll., 1958; D.D.S., U. Tex.-Houston, 1963; postgrad. Health Sci. Ctr., Houston, 1963-66; diploma (hon.) XX Reunion de Provincia, Juarez, Chihuahua, Mexico, 1970, Ateneo Odontologica Mexicano, Valle de Bravo, Mexico, 1973, Colegio de Cirujanos Mexicanos, Juarez, 1975, 84. Cert. instr. Advanced Cardiac Life Support, 1986—; diplomate Am. Bd. Oral and Maxillofacial Surgery. Resident in oral surgery Methodist Hosp., Houston, 1963-64, Ben Taub Hosp., Houston, 1964-65, Hermann Hosp., Houston, 1965-66; practice dentistry specializing in oral and maxillofacial surgery, El Paso, 1966—; mem. staff Sun Towers Hosp., chief oral and maxillofacial surgery, 1975, 76, 77, 93; gov. appointee Tex. State Bd. Dental Examiners, 1993—; instr. pathology El Paso Community Coll., 1975-76; lectr. in field. Author, editor: Magnificent Obsession: In Quest of High Mountain Game, 1990; contbg. author Asian Hunter, 1989; contbr. articles to Jour. Oral/Maxillofacial Surgery. Bd. dirs. Am. Cancer Soc., El Paso, 1971-73, El Paso Cancer Treatment Ctr., 1971-74; bd. dirs. West Tex. Health Systems Agy., 1980-82, chmn., 1981-82; med. adv. com. W. Tex. Council of Regional Health. Recipient Bowie Exes award, El Paso, 1982, Mexican Consul Gen. Award of Appreciation, 1986, Mexican Social Security Service Award of Appreciation, 1986. Fellow Am. Coll. Oral and Maxillofacial Surgeons, Am. Assn. Oral and Maxillofacial Surgeons, Internat. Assn. Oral and Maxillofacial Surgeons, Southwest Soc. Oral and Maxillofacial Surgeons, Acad. Internat. Dental Studies, Pan Am. Med. Assn., EPSDT (dental adv. and rev. com. 1985-86), Tex. Soc. Oral and Maxillofacial Surgeons, The Explorers Club; mem. El Paso Dist. Dental Soc. (pres. 1979), U. Tex. Dental Br. alumni Assn. (life), Nat. Rifle Assn. (life), Tex. Rifle Assn. (life), Am. Found. for N.Am. Wild Sheep, Internat. Sheep Hunting Assn. Republican. Roman Catholic. Clubs: Anthony Rod and Gun, Grand Slam (life). Avocations: high mountain sheep hunting; outdoor activities. Home: 416 Lindbergh Ave # El El Paso TX 79932-2118 Office: 1201 E Schuster Ave Bldg 4A El Paso TX 79902-4642

TORRES, MANUEL, aerospace engineer; b. Bklyn., Feb. 13, 1963; s. Manuel and Irma (Delgado) T.; m. Dorothy Eleanor Tallman, Apr. 25, 1987. BS in Aerospace Engring., Poly. Inst. N.Y., 1985; MS in Aerospace Engring., Boston U., 1990; postgrad., Pa. State U., 1991. Wind tunnel tech. Grumman Aerospace Corp., Bethpage, N.Y., 1984-85; aerodynamic design engr. Textron Def. Sys., Wilmington, Mass., 1985-90; aerodynamic and aerophysics design engr. GE-Aerospace/Hypersonics, Phila., 1990-93; project engr. Martin Marietta Astro Space, Valley Forge, Pa., 1993—. Author, co-author various tech. reports. Mem. ASME, AIAA (vice chmn. for Greater Phila. 1992-93, chmn. 1993—, mem. nat. com.), Soc. Naval Architects and Marine Engrs., Soaring Soc. AM. Avocations: tennis, boating, fishing, skiing, music, art, swimming, gliding.

TORRES, RIGO ROMUALDO, electrical engineer; b. Villalba, P.R., Apr. 8, 1963; s. Angel and Adela (Cruz) T. BSEE, U. P.R., 1983; MSEE, George Washington U., 1988. Lab. instr. U. P.R., Mayaguez, 1983; electronics engr. Naval Surface Warfare Ctr., Silver Spring, Md., 1984-91; sr. ASW analyst Sci. Applications Internat. Corp., McLean, Va., 1991—; cons. on mine warfare Sci. Applications Internat., 1991—, cons. on personal computers, 1991—; contracting officers tech. rep. Naval Surface Warfare Ctr., 1989. Author tech. manuals: Weapon Specification - MK597 A1 Board, 1984, Weapon Specification - MK597 A2 Board, 1985, Weapon Specification - RECO Device, 1990. Counselor Sch. Engring., U. P.R., 1980-81. Recipient High Quality Performance Commendation, Dept. Navy, 1985, 86, Spl. Achievement award 1985, 88, 90, Sustained Superior Performance award, 1987. Mem. Acoustical Soc. Am. (assoc.), Opus Dei (cooperator). Achievements include test and evaluation of mine electronics; development of signal processing software for low probability of intercept sonar and shallow water active sonar; development of ASW mine avoidance software models. Home: 9327 Steeple Ct Laurel MD 20723 Office: Sci Applications Internat 1710 Goodridge Dr Mc Lean VA 22102

TORRES-AYBAR, FRANCISCO GUALBERTO, medical educator; b. San Juan, P.R., July 12, 1934; s. Francisco and Maria (Aybar) Torres; m. Elga Arroyo; children: Elga, JoAnn Marie. BS, U. P.R., 1956; MD, U. Barcelona, Spain, 1963. Diplomate Am. Bd. Pediatrics. Chief pediatric cardiology Ponce (P.R.) Dist. Hosp., 1970-91, chmn. dept. pediatrics, 1971-91, med. dir., 1980, prog. dir. pediatric tng. prog., 1970-91; prof., chmn. dept. pediatrics Cath. Univ. Sch. of Medicine, Ponce, 1978-80; chmn., dept. pediatrics Damas Hosp., Ponce, 1985—; prof., chmn. dept. pediatrics Ponce Sch. of Medicine, 1980—. Editorial bd.: Sci.-Ciencia jour., Ponce, 1975—; contbr. articles to profl. jours. Fellow Am. Acad. Pediatrics, Am. Coll. Physicians, Am. Coll. Cardiology, Am. Coll. Internat. Physicians, British Royal Soc. Health, InterAm. Coll. Physicians and Surgeons; mem. AAUP, Phi Delta Kappa. Republican. Roman Catholic. Home: A26 Jacaranda Ponce PR 00731 Office: 13 Calle Mayor Ponce PR 00731-5025

TORRES MEDINA, EMILIO, oncologist, consultant; b. Mexico, Aug. 8, 1934; s. Manuel Torres and Juana Medina; m. Luisa Torres Dec. 6, 1966; children: Patricia, Ana Luisa, Veronica, Jesus Manuel. MD, U. Mex., 1958-64; internist, Hosp. of Nutrition, Mexico, 1985-87; oncologist, Hosp. of Oncology, Mexico, 1987-90. Chief dept. radiation Clin. of Parque, Chih, Mexico, 1970-80, chief of med. edn., 1974-78; chief dept. radiation Central Oncology, Juarez, Mexico, 1981-92; chief. of med. edn. I.S.S.S.T.E., Juarez, Mexico, 1984-85; oncologist I.S.S.S.T.E., Juarez, 1981-85; oncology prof. Uach Y Uacj, Juarez, Mexico, 1981-92; mem. med. coun. Pensiones Civiles del Edo., Juarez, Chih., 1985-90; med. dir. Pensionesl Civiles del Edo., Juarez, Chih., 1982-88, Electronic Diagnosis of Juarez, 1981-92. Author: Policitemia on Rats by Hemolizates, 1965, Alkil beta-d Glicosias on Human Lymphs, 1970. Advisor Juarez Cancer Soc., 1986-92. Mem. Soc. Mex. Est. Oncol. Capitulo Chihuahua (pres. 1990-92), Soc. Medicina Interna Cd. Juarez (pres. 1992). Office: Centro Oncologico del Norte, Plutarco Elias Calles 1235, 32350 Juarez Mexico

TORRES-OLIVENCIA, NOEL R., aeronautical engineer; b. Arecibo, P.R., Dec. 30, 1966; s. Rosendo Torres-Rosa and Gladys B. Olivencia-Badillo. BS, Poly. U. N.Y., 1990; MS, Rensselaer Poly. Inst., 1991. Engr. in tng. Allison Gas Turbine subs. GM, Indpls., 1991, assoc. engr., 1993—; engring assoc. Pratt & Whitney subs. United Techs., East Hartford, Conn., 1992-93. Mem. NSPE, AIAA, Soc. Hispanic Profl. Engrs. Roman Catholic. Office: Allison Gas Turbine PO Box 420 S 10 Indianapolis IN 46206

TORREY, HENRY CUTLER, physicist; b. Yonkers, N.Y., Apr. 4, 1911; s. John Cutler and Mabel (Kelso) T.; m. Helen Post Hubert, Sept. 11, 1937; children: John Cutler, Meriel Torrey Goodwin. BSc, U. Vt., 1932, DSc (hon.), 1965; PhD, Columbia U., 1937. Instr. physics Pa. State U., State College, 1937-41, asst. prof., 1941-42; mem. staff radiation lab. MIT, Cambridge, 1942-46; assoc. prof. Rutgers U., New Brunswick, N.J., 1946-48, prof., 1948-76, dean grad. sch., 1965-74, prof. emeritus, 1976—; cons. Chevron Rsch. Corp., La Habra Calif., 1952-65; trustee, mem. organizing com. Univs. Space Rsch. Assn., 1968-70; bd. govs. Lunar Sci. Inst., 1969-70. Co-author: Crystal Rectifiers, 1948; mem. editorial bd. Rev. Sci. Instruments, 1963-65. Guggenheim Found. fellow, Paris, 1964-65. Fellow Am. Phys. Soc. Achievements include patents in Oil Well Logging; co-discovery of nuclear magnetic resonance. Home: 1050 George St Apt 12D New Brunswick NJ 08901-1019

TORRIERI, DON JOSEPH, electronics engineer, mathematician, researcher; b. Balt., Nov. 19, 1942; s. Peter and Mary Torrieri; m. Nancy Karen Weir, Jan. 27, 1971; children: Karen Marisa, Peter. BS, MIT, 1964; MS, Poly. U., Farmingdale, N.Y., 1966, U. Md.; PhD, U. Md., 1971. Electronics engr. Naval Rsch. Lab., Washington, 1971-77; mathematician Dept. of the Army, Adelphi, Md., 1977—; part-time faculty George Washington U., Washington, 1988—, Johns Hopkins U., Balt. Author: Principles of Military Communication Systems, 1981, Principles of Secure Communication Systems,1985, 2d edit., 1992; contbr. chpt. to book, articles to profl. jours. Coach boys soccer and basketball Calverton Recreation Coun., Silver Spring, Md., 1990—; coach girls softball Montgomery County, Md., 1993—. Mem. IEEE (sr.), Sigma Xi. Achievements include authorship of a textbook that is the standard in the field. Home: 2204 Hidden Valley Ln Silver Spring MD 20904 Office: Dept Army 2800 Powder Mill Rd Adelphi MD 20783

TORTORA, ROBERT D., mathematician; b. Youngstown, Ohio, Aug. 20, 1946. BS in Math. cum laude, Youngstown State U., 1968; MS in Math. Stats., Cath. U. Am., 1972; PhD in Probability and Stats., Bowling Green State U., 1975. Mathematician Nat. Security Agy., 1968-69; various positions ending in dir. rsch. and applications, Nat. Agrl. Stats. Svcs. USDA, 1975-90; chief statis. rsch. divsn. Census Bur., Washington, 1990-92, assoc. dir. statis. design, methodology, standards, 1992—; chair subcom. role of telephone data collection in fed. surveys Fed. Com. Statis. Methodology, 1980—; co-chair com. disclosure risk analysis Office Mgmt. and Budget; mem. stats. industrial advisory com. Ohio State U., 1990—; adv. com. George Mason Univ. Ctr. computational Stats., 1991—. Contbr. articles to profl. jours. 1st lt. U.S. Army, 1969-72. Mem. Am. Statis. Assn. (mem., chair govt. subcom. mem. com. 1986-91, ad hoc com. Office Sci. Affairs 1987-88, 91—, assoc. editor Jour. Ofcl. Stats. 1988—, program chair survey methods rsch 1992, organizer various meetings), Wash. Statis. Soc. (chmn. agr. 1980-81), Internat. Assn. Survey Statisticians. Office: Bureau of the Census Statistical Design Federal Center Washington DC 20233*

TOSK, JEFFREY MORTON, biophysicist, educator; b. N.Y.C., Nov. 26, 1951; s. Philip and Beatrice Tosk; m. Tambrey Anne Carlin, June 15, 1986. BA, SUNY, New Paltz, 1974; MA, Loma Linda (Calif.) U., 1986, PhD, 1989. Sr. rsch. chemist Naylor Dana Inst., Valhalla, N.Y., 1974-79; cons., 1979—; postdoctoral fellow Loma Linda U. Sch. Medicine, 1989-90, asst. rsch. prof. psychiatry, 1990—, asst. prof. neurology and physiology, 1992—, dir. Parkinson's Rsch. Ctr., 1992—; rsch. Neurobiologist Pettis VA Med. Ctr., Loma Linda, 1990—; dir. Parkinson's Rsch. Ctr., Loma Linda, 1992—. Contbr. articles to profl. jours., chpts. to books. Recipient Travel awards NSF, 1987, 90, Raymond Sarber Fellowship Am. Soc. Microbiology, 1988; NSF rsch. grantee, 1990. Mem. Am. Assn. Cancer Rsch., Biophys. Soc., Soc. Neurosci., Sigma Xi (grantee 1986). Achievements include patent for chemical sensor; noted also for studies of noncovalent interactions in biophys. systems. Office: Pettis VA Med Ctr Rsch Svc 151 11201 Benton St Loma Linda CA 92357

TOTH, DANNY ANDREW, pharmacist; b. Chattanooga, Tenn., Sept. 17, 1946; s. Andrew Frank and Billie Dove (Kemp) T.; m. Sharon Annell Burns, Apr. 8, 1982; children: John Nathan, Justin, Paul, Nathan Adam, Meg. AA, Hiwassee Coll., 1966; BS in Pharmacy, Mercer So. Sch. Pharmacy, 1970, postgrad., 1972-73. Registered pharmacist. Agt. Ga. Drugs and Narcotics Agy., 1970-73; dir. dept. of pharmacy Newton County Hosp., Covington, Ga., 1974-80; owner Holmes Pharmacy, LaGrange, Ga., 1981—; pres. Pharmacists United for the Future, La Grange, Ga., 1988—; cons. Public Health Dept.; mem. Dist. 4 AIDS Task Force, La Grange, 1989—, Ga. Dept. Med. Assistance Formulary Com., Atlanta, 1989—; chmn. Drug Utilization Rev. for West Point Pepperel, 1988—. Recipient Leadership award McKesson Drug Co., 1991, Pharmacy Leadership award Nat. Assn. Retail Druggists, 1991, Ruth Duncan Humanitarian award, 1989, Merck, Sharp and Dohme award for Outstanding Achievement in Pharmacy, 1991-92, Bristol Myers-Squibb Pres. award, 1991-92. Mem. Ga. Soc. Hosp. Pharmacists, LaGrange Child Care Coun., Nat. Assn. Retail Druggists (steering com. 3d party prescription plans 1992, vice-chmn. 1993), Am. Pharm. Assn., Ga. Pharm. Assn. (bd. dirs., 1984-87, 2d v.p., 1988, 1st v.p., 1989, pres.-elect, 1990, pres. 1991—), West Cen. Ga. Pharm. Assn. (pres. 1989, 90), Abbots Pharmaceutical Pharmacy Advisory Coun., 1992—, Masons. Avocations: hunting, fishing, golf. Home: PO Box 972 La Grange GA 30241-0972

TOTH, JAMES JOSEPH, chemical engineer; b. Cleve., Nov. 22, 1956; s. James A. and Doris (Rietz) T.; m. Joan Elizabeth Clark, June 24, 1989. BS, Carnegie Mellon, 1979; MS, Ohio State U., 1982. Cons. MLI, Cleve., 1982-85; chem. engr. UNISYS, Bristol, Tenn., 1985-87; sr. engr. Digital Equipment Corp., Greenville, S.C., 1987-90, prin. engr., 1991—; tech. adv. bd. Nat. Ctr. for Mfg. Sci., Ann Arbor, Mich., 1991. Contbr. articles to profl. jours. including Inst. of Printed Cirs. Tech. Jour., 1992, Material Rsch. soc., 1991 and Jour. of Inst. of Cir. Tech., 1991. Vice-pres. Friends of Internats., Greenville, 1990—. Std. Oil fellowship Dept. of Chem. Engring., Ohio State U., 1980. Mem. AICE, IEEE, Toastmaster Internat. (competent toastmaster 1992). Roman Catholic. Achievements include patents on metallized and plated laminates. Home: 604 Bear Dr #2 Greenville SC 29605

TOTH, JAMES MICHAEL, electronics engineer; b. Cleve., Apr. 13, 1943; s. Albert Andrew and Helen (Pap) T. BEE, Case Inst. Tech., 1965. Rsch. engr. Republic Steel Rsch., Cleve., 1965-75, sr. rsch. engr., 1975-85; prin. engr. Textron Materials Tech. Ctr., Euclid, Ohio, 1985-90; sr. staff engr. LTV Steel Tech. Ctr., Independence, Ohio, 1990—; bd. dirs., exec. v.p. Barclay Precision Toys Inc., Cleve., 1986-92. Contbr. articles to profl. jours. Pres. North Ohio Windstar Environ. Group, Lyndhurst, Ohio, 1990—; bd. chmn. Unity of Greater Cleve., Mayfield Heights, Ohio, 1989-90. Mem. Am. Soc. for NDT, Windstar Found. Achievements include patents on Ultrasonic Inspection Systems; Eddy Current Inspecting Equipment; Optical Inspection Method; invention of Equisonic Ultrasonic Search Unit. Office: LTV Steel Tech Ctr 6801 Brecksville Rd Independence OH 44131

TOTH, KAROLY CHARLES, electrical engineer, researcher; b. Budapest, Hungary, Sept. 16, 1954; came to U.S., 1987; s. Karoly and Ilona (Kozma) T.; m. Susan Korda, Sept. 30, 1978; children: Benedek, Borbala. Diploma in Elec. Engring., Tech. U., Budapest, 1977, diploma in Devel. and Rsch., 1979, tech. doctorate in Electrical Engring., 1980. Project and team leader Medicor Works, Budapest, 1980-82; adj. prof. Tech. U., Budapest, 1980-82, asst. prof., 1982-86, sr. asst. prof., 1986-90; rsch. assoc. Ohio State U., Columbus, 1987—; mem. com. Digital Imaging Work Group, Budapest, 1986-87. Mem. editorial bd. Hungarian Electronics Monthly, Budapest, 1986-87. Recipient Duane C. Brown award Ohio State U., 1990. Mem. IEEE, Am. Soc. Photogrammetry and Remote Sensing. Achievements include patents in field of image displaying and processing procedures. Home: 2926 Welsford Rd Columbus OH 43221 Office: Ohio State Univ Geodetic Sci and Surv 1958 Neil Ave Columbus OH 43210-1247

TOTH, LOUIS MCKENNA, chemist; b. Lexington, Ky., Aug. 27, 1941; s. Louis Andrew and Catherine C. (McKenna) T ; m. Marie Louise Sciortino, June 30, 1962; children: William Joseph, Michael Andrew, Mary Catherine. BS in Chemistry & Physics, La. State U., 1963; PhD in Chemistry, U. Calif., Berkeley, 1967. Rsch. staff mem. Oak Ridge (Tenn.) Nat. Lab., 1967-80, sr. rsch. staff mem., 1980—. Editor: The TMI Accident Diagnosis and Prognosis, 1986. Mem. Am. Chem. Soc. Achievements include research in molten salt chemistry, spectroscopy, nuclear fuel cycle chemistry, actinide chemistry, hydrolysis behavior. Office: Oak Ridge Nat Lab PO Box 2008 MS 6268 Oak Ridge TN 37831

TOTTEN, ARTHUR IRVING, JR., retired metals company executive, consultant; b. Laurel, Del., Mar. 15, 1906; s. Arthur Irving and Lena Meade (Fowler) T.; m. Margaret Ross, Nov. 10, 1934 (dec. Mar. 1988); children: Margaret Totten Peters, Fitz-Randolph Fowler, Eleanor Totten Shumaker. BS in Chemistry, Union Coll., Schenectady, 1928. Rsch. chemist E.I. du Pont de Nemours & Co., Inc., Parlin, N.J., 1928-32; prodn. supt. E.I. du Pont de Nemours & Co., Inc., Parlin, 1932-37; rsch. chemist E.I. du Pont de Nemours & Co., Inc., Phila., 1937-42; dir. packaging rsch. Reynolds Metals Co., Richmond, Va., 1946-71; exec. v.p. rsch. Reynolds Rsch. Corp., Richmond, 1966-71; ret., 1971; cons. Nat. Inventors Coun., Washington, 1940-46; presenter in field, U.S., Can., Eng., France; pres. Rsch. & Devel. Assocs., N.Y.C., 1965-67, chmn. bd., 1967—. Contbr. numerous articles on packaging and paints to profl. jours. Pres. River Rd. Citizens Assn., 1949; head comml. div. Richmond chpt. ARC, 1954, bd. dirs., 1955-58; trustee Va. Inst. Sci. Rsch., 1970—; v.p. Adult Devel. Ctr., Richmond, 1988—; pres. Westham Green Citizens Assn., 1982; sr. warden Episc. Ch., 1952, 60, 66. Lt. Col. CWS, U.S. Army, 1942-46. Recipient cert. of appreciation U.S. Army Natick Lab., 1971, Humanitarian award Adult Devel. Ctr., 1992. Mem. Am. Inst. Chemists (life, past chpt. pres.), Rsch. and Devel. Assn. (life, past pres., cert. of appreciation, 1971), Packaging Inst. (pres. 1971, Profl. award 1968), N.Am. Packaging Inst. (pres. 1971), World Packaging Inst. (v.p. 1971), Inst. Food Tech. (Indsl. Achievement award 1978), Packaging Edn. Found. (Hall of Fame 1973), Country Club Va., Richmond Engrs. Club (past bd. dirs.), Soixante Plus, Masons, Sigma Phi. Achievements include patents for camouflage coatings for aircraft and other military equipment in World War II; for caustic soluble inks for beer labels. Home: 300 N Ridge Rd Apt 43 Richmond VA 23229-7452

TOTTEN, GARY ALLEN, spectroscopist; b. Terre Haute, Ind., Aug. 18, 1949; s. Ernest Warren and Jane Ann (Jenks) T.; m. Diana Christine Wythe, May 2, 1970; children: Warren, Jason, Kathryn. BS in Biology and Chemistry, Ind. State U., 1978. Fourier transform infrared technician Internat. Minerals & Chem. Corp., Terre Haute, 1980-83; fourier transform infrared spectroscopist Pitman-Moore Inc., Terre Haute, 1983-90; tech. specialist Cummins Diesel Co., Columbus, Ind., 1990—; guest lectr. sci. sch. corps. Vigo, Clay, Vermillion and Sullivan counties, Ind., 1983-90. Contbr. articles to sci. jours. Scoutmaster troop 7 Boy Scouts Am., Terre Haute, 1985-90; pres. Vigo County Youth Soccer Assn., Terre Haute, 1985; pres., bd. bd. dirs. Children's Sci. and Tech. Mus., Terre Haute, 1988-89; precinct committeeman Terre Haute Rep. Com., 1989. With USN, 1970-74. Mem. Am. Chem. Soc., Soc. for Applied Spectroscopy, Sigma Xi. Office: Cummins Diesel Co 1900 McKinley MC 50183 Columbus IN 47201

TOTTEN, VENITA LAVERNE, chemist; b. Ft. Worth, Tex., Jan. 5, 1963; d. Darrell L. and Dorothy (Stubblefield) T. BA, La. Tech. U., 1985; PhD, Baylor U., 1990. Rsch. fellow Baylor U., Waco, Tex., 1986-90; sr. analytical and quality specialist Corn Products CPC Internat., Argo, Ill., 1990–. Mem. Humane Soc. Recipient Electroplates and Surface Finishers Soc. award, 1988. Mem. Am. Chem. Soc., Am. Soc. Cereal Chemists, Chgo. Chromatography Discussion Group, Iota Sigma Pi. Office: CPC Internat Box 345 Argo IL 60501

TOU, JULIUS T., electrical and computer engineering educator; b. Shanghai, China, Aug. 18, 1926; came to U.S., 1949; married; children: Albert, Fred, Ivan, Sylvia. BS, Chiao-Tung U., Shanghai, 1947; MS, Harvard U., 1950; DEng, Yale U., 1952; hon. degree, Jiao Tong U., Shanghai, 1986. Asst. prof. U. Pa., Phila., 1954-57; assoc. prof. Purdue U., Lafayette, Ind., 1957-61, vis. prof., 1961-62; dir. computer sci. Northwestern U., Evanston, Ill., 1962-64, prof., 1961-64; dir. Batelle Meml. Inst., Columbus, Ohio, 1964-67; grad. rsch. prof. U. Fla., Gainesville, 1967—, dir. Ctr. for Info. Rsch., 1971—; adj. prof. Ohio State U., Columbus, 1964-67. Author: Digital and Sampled-data Control Systems, 1959, Modern Control Theory, 1964, Optimum Digital Control, 1963, Pattern Recognition Principles, 1973; editor: Software Engineering, 1969, Information Systems, 1972, Computer-based Automation, 1984; editor Internat. Jour. Computer Info. Scis., 1972—. Fellow IEEE; mem. Academia Sinica. Office: U Fla Ctr Info Rsch 314 CSE Bldg Gainesville FL 32611

TOUFFAIRE, PIERRE JULIEN, physician; b. Orange, Vaucluse, France, Aug. 22, 1933; s. Rene C. and Pierrette G. (Vaubourg) T.; m. Josette Leontine Lotti, Nov. 25, 1961; children: Michel, Jean. Student, Ancien Extern Hosp., Marseille, France, 1957; M.D., U. Marseille, 1963. Practice medicine, Saint-Maximin, France, 1961—. Roman Catholic. Home: Allée des Aubepines, 83470 Saint-Maximin France Office: 4500 Point Lookout Rd Orlando FL 32808-1734

TOUHILL, C. JOSEPH, environmental engineer; b. Newark, Aug. 27, 1938; s. Charles J. and Caroline A (Lesaius) T.; m. Helen Elizabeth O'Malley, June 11, 1960; children: Gregory Joseph, Stephen Mark, Christopher Alan, Kathleen Elizabeth. BCE, Rensselaer Poly. Inst., 1960, PhD in Environ. Engring., 1964; SM, MIT, 1961. Diplomate Am. Acad. Environ. Engrs. Mgr. water and land resources dept. Battelle Meml. Inst., Richland, Wash., 1964-71; pres. Baker/TSA Inc., Pitts., 1977-90; group sr. v.p. ICF Kaiser Engrs. Inc., Pitts., 1990—; cons. various engring. firms, Washington and Pitts., 1971-77; trustee Am. Acad. Environ. Engrs., Annapolis, 1971-77, 83-86. Co-author: Hazardous Materials Spills Handbook, 1982, Hazardous Waste Management Engineering, 1987; editor: Resource Management in the Great Lakes Basin, 1971; mem. editorial bd. Environ. Progress Jour., 1979—. Bd. dirs. Suburban Gen. Hosp., Pitts., 1986—, Pennwood Savs. Assn., Pitts., 1991; vice chmn. Franklin Park (Pa.) Authority, 1977—. Fellow ASCE; mem. Am. Inst. Chem. Engrs. (chmn. environ. engring. div. 1977), Am. Chem. Soc. (editorial adv. bd. 1975-77). Office: ICF Kaiser Engrs Inc 2206 Almanack Ct Pittsburgh PA 15237-1502

TOURTILLOTT, ELEANOR ALICE, nurse, educational consultant; b. North Hampton, N.H., Mar. 28, 1909; d. Herbert Shaw and Sarah (Fife) T. Diploma Melrose Hosp. Sch. Nursing, Melrose, Mass., 1930; BS, Columbia U., 1948, MA, 1949; edn. specialist Wayne State U., 1962. RN. Gen. pvt. duty nurse, Melrose, Mass., 1930-35; obstet. supr. Samaritan Hosp., Troy, N.Y., 1935-36, Meml. Hosp., Niagara Falls, N.Y., 1937-38, Lawrence Meml. Hosp., New London, Conn., 1939-42, New Eng. Hosp. for Women and Children, Boston, 1942-43; dir. H. W. Smith Sch. Practical Nursing, Syracuse, N.Y., 1949-53; founder, dir. assoc. degree nursing program Henry Ford Community Coll., Dearborn, Mich., 1953-74; dir. pioneering use of learning techs. via mixed media USPHS, 1966-71; prin. cons., initial coord. Wayne State U. Coll. Nursing, Detroit, 1975-78; cons. curriculum design, modular devel., instructional media Tourtillott Cons., Inc., Dearborn, Mich., 1974—; condr. numerous workshops on curriculum design, instructional media at various colls., 1966—; mem. Mich. Bd. Nursing, 1966-73, chmn., 1970-72, mem. rev. com. for constrn. nurse tng. facilities, div. nursing USPHS, 1967-70, mem. nat. adv. coun. on nurse tng., Dept. Health Edn. and Welfare, 1972-76. Author: Commitment-A Lost Characteristic, 1982; contbg. co-author: Patient Assessment-History and Physical Examination, 1977-81; contbr. chpts., articles, speeches to profl. publs. Served to capt. Nurse Corps, U.S. Army, 1943-48; ETO. Recipient Disting. Alumnae award Tchrs. Coll. Columbia U., 1974, Spl. tribute 77th Legislature Mich., 1974, Disting. Alumnae award Wayne State U., 1975, Disting. Service award Henry Ford Community Coll., 1982. Mem. ANA, Nat. League Nursing (chmn. steering com. dept. assoc. degree programs 1965-67, bd. dirs. 1965-67, 71-73, mem. assembly constituent leagues 1971-73, council assoc. degree programs citation 1974), Mich. League for Nursing (pres. 1969-71), Mich. Acad. Sci., Arts and Letters, Am. Legion, Tchrs. Coll. Alumnae Assn., Wayne State U. Alumnae Assn., Phi Lambda Theta, Kappa Delta Pi.

TOUSEY, RICHARD, physicist; b. Somerville, Mass., May 18, 1908; s. Coleman and Adella Richards (Hill) T.; m. Ruth Lowe, June 29, 1932; 1 dau., Joanna. A.B., Tufts U., 1928, Sc.D. (hon.), 1961; A.M., Harvard, 1929; Ph.D., 1933. Instr. physics Harvard, 1933-36, tutor div. phys. scis., 1934-36; research instr. Tufts U., 1936-41; physicist U.S. Naval Research Lab. optics div., 1941-58, head instrument sect., 1942-45, head micron waves br., 1945-58, head rocket spectroscopy br., atmosphere and astrophysics div., 1958-67, space sci. div., 1967-78, cons., 1978—; Mem. com. vision Armed Forces-NRC, 1944—; line spectra of elements com. NRC, 1960-72; mem. Rocket and Satellite Research Panel, 1958—; mem. astronomy subcom. space sci. steering com. NASA, 1960-62, mem. solar physics subcom., 1969-71; prin. investigator expts. including Skylab; mem. com. aeronomy Internat. Union Geodesy and Geophysics, 1958—; U.S. nat. com. Internat. Commn. Optics, 1960-66; mem. sci. steering com. Project Vanguard, 1956-58; mem. adv. com. to office sci. personnel Nat. Acad. Scis.-NRC, 1969-72. Contbr. articles to sci. jours. and books. Bayard Cutting fellow Harvard, 1931-33, 35-36; recipient Meritorious Civilian Service award U.S. Navy, 1945; E.O. Hulburt award Naval Research Labs., 1958; Progress medal photog. Soc. Am., 1959; Prix Ancel Soc. Francaise de Photographie, 1962; Henry Draper medal Nat. Acad. Scis., 1963; Navy award for distinguished achievement in sci, 1963; Eddington medal, 1964; NASA medal for exceptional sci. achievement, 1974; George Darwin lectr. Royal Astron. Soc., 1963. Fellow Am. Acad. Arts and Scis., Am. Phys. Soc., Optical Soc. Am. (dir. 1953-57, Frederic Ives medal 1960), Am. Geophys. Union; mem. Internat. Acad. Astronautics, Nat. Acad. Scis., Am. Astron. Soc. (v.p. 1964-66, Henry Norris Russell lectr. 1966, George Ellery Hale award 1992), Soc. Applied Spectroscopy, AAAS, Am. Geophys. Union, Philos. Soc. Washington, Internat. Astron. Union, Nuttall Ornithol. Club, Audubon Naturalists Soc., Phi Beta Kappa, Sigma Xi, Theta Delta Chi. Home: 10450 Lottsford Rd Apt 231 Bowie MD 20721-2742 Office: US Naval Research Lab Washington DC 20375

TOUYZ, STEPHEN WILLIAM, clinical psychologist, educator; b. Cape Town, Republic of South Africa, Aug. 29, 1950; s. Harry and Tilly (Woolfowitz) T.; m. Rennette Dawn Elk, Jan. 18, 1976; children: Justin Lawrence, Lauren Marissa. B.S., U. Cape Town, 1972, Ph.D., 1976; B.S. with honors, U. Witwatersrand, 1974. Tutor U. Whitewatersand, 1974; sr. tutor U. Cape Town, 1974-75; sr. research asst. Groote Schuur Hosp., Cape Town, 1974-75, intern clin. psychologist, 1976-78; staff psychologist Royal Prince Alfred Hosp., Sydney, Australia, 1978-80; clin. lectr. U. Sydney, 1979-91, clin. assoc. prof. psychiatry, 1991—; head clin. psychology unit Royal Prince Alfred Hosp., Sydney, 1980-88; cons. psychologist anorexia nervosa unit Northside Clinic, Sydney, 1979-84, Royal Prince Alfred Hosp., 1983-88, Lynton Pvt. Hosp., 1984—; head dept. medical psychology Westmead Hosp, 1988—; hon. assoc. dept. psychology U. Sydney, 1979—; hon. clin. assoc. dept. psychology Macquarie U., 1990—; mem. Westmead Hosps. Sci. Coun.; cons. practice guidelines com. for eating disorders Am. Psychiat. Assn., 1990-93. Author: Grune and Stratton, 1984, Williams and Wilkins, 1985. Rsch. grantee South African Coun. for Sci. and Indsl. Research, 1973-76, Nat. Health and Med. Rsch. Coun. Australia, 1983-87, Ramaciotti Found., 1983-84, Elli Lilly 1991-93. Fellow Internat. Coll. Psychosomatic Medicine (v.p. Australia and New Zealand chpt. 1993—); mem. Australian Psychol. Soc., Australian Behaviour Modification Assn., Australian Soc., Psychiat. Research, Am. Psychol. Assn. (affiliate), Internat. Acad. Sci., Australian Sleep Rsch. Assn., Australian Soc. Study Brain Impairment, Inc. Office: Westmead Hosp, Dept Med Psychology, Westmead 2145 New South Wales, Australia

TOVE, SAMUEL B., biochemistry educator; b. Balt., July 29, 1921; s. Max George and Sylvia (Gotthelf) T.; m. Shirley Ruth Weston, July 22, 1945; children: Michael, Nancy, Deborah. BS, Cornell U., 1943; PhD, U. Wis., 1950. Asst. prof. N.C. State U., Raleigh, 1950-55, assoc. prof., 1955-60, prof., 1960-75, William Neal Reynolds prof., 1975—. Fellow AAAS, Am. Inst. of Nutrition. Office: NC State U Dept of Biochemistry Raleigh NC 27695-7622

TOWER, ALTON G., JR., pharmacist; b. Buffalo, N.Y., Jan. 16, 1927; m. Nan R. Spinner, Aug. 15, 1953; children: Adrienne, Michele, Renee. BS in Pharmacy, U. Buffalo, 1953. Registered pharmacist. Pharmacist Woldmans Drug Store, Buffalo, 1946-53; med. svc. rep. Strasenburgh Lab., Rochester, N.Y., 1953-66; pharmacist, mgr. Eckerd Drugs, Clearwater, Fla., 1966—. Dir. Am. Cancer Soc. Pinellas County, Fla., 1976—, pres., 1988-89 (Life Saver award 1988), Pinellas Pharmacist Soc. Dir. Com. Affairs; charter mem. Smoke Free Class of 2000, Pinellas County, 1988—. Recipient Vol. of Yr. award Am. Cancer Soc. Pinellas County, 1987, James Beal Award Pharmacist of the Yr. Fla. Pharm. Assn., 1992. Mem. Am. Pharm. Assn., Fla. Pharmacy Assn. (bd. dirs. 1981-85, speaker ho. of dels. 1986, R.Q. Richards award 1989, Bowl of Hygeia award 1990, Sid Simkowitz Involvement award 1991), Pinellas County Pharmacy Soc. (life; dir. 1968-73, 78-81, 89-91, pres. 1973, 88, Pharmacist of Yr. 1973). Avocations: gardening, hiking, travel. Office: Eckerds #2332 Pinellas Park FL 34666

TOWNES, CHARLES HARD, physics educator; b. Greenville, S.C., July 28, 1915; s. Henry Keith and Ellen Sumter (Hard) T.; m. Frances H. Brown, May 4, 1941; children: Linda Lewis, Ellen Screven, Carla Keith, Holly Robinson. B.A., B.S., Furman U., 1935; M.A., Duke U., 1937; Ph.D., Calif. Inst. Tech., 1939. Mem. tech. staff Bell Telephone Lab., 1939-47; assoc. prof. physics Columbia U., 1948-50, prof. physics, 1950-61; exec. dir. Columbia Radiation Lab., 1950-52, chmn. physics dept., 1952-55; provost and prof. physics MIT, 1961-66, Inst. prof., 1966-67; v.p., dir. research Inst. Def. Analyses, Washington, 1959-61; prof. physics U. Calif., Berkeley, 1967-86, prof. physics emeritus, 1986—; Guggenheim fellow, 1955-56; Fulbright lectr. U. Paris, 1955-56, U. Tokyo, 1956; lectr., 1955, 60; dir. Enrico Fermi Internat. Sch. Physics, 1963; Richtmeyer lectr. Am. Phys. Soc., 1959; Scott lectr. U. Cambridge, 1963; Centennial lectr. U. Toronto, 1967; Lincoln lectr., 1972-73, Halley lectr., 1976, Krishman lectr., 1992, Nishina lectr., 1992; dir. Gen. Motors Corp., 1973-86; mem. Pres.'s Sci. Adv. Com., 1966-69, vice chmn., 1967-69; chmn. sci. and tech. adv. com. for manned space flight NASA, 1964-69; mem. Pres.'s Com. on Sci. and Tech., 1976; researcher on nuclear and molecular structure, quantum electronics, interstellar molecules, radio and infrared astrophysics. Author: (with A.L. Schawlow) Microwave Spectroscopy, 1955; author, co-editor: Quantum Electronics, 1960, Quantum Electronics and Coherent Light, 1964; editorial bd.; Rev. Sci. Instruments, 1950-52, Phys. Rev., 1951-53, Jour. Molecular Spectroscopy, 1957-60, Procs. Nat. Acad. Scis., 1978-84; contbr. articles to sci. publs.; patentee masers and lasers. Trustee Calif. Inst. Tech., Carnegie Instn. of Washington, Grad. Theol. Union, Calif. Acad. Scis.; mem. corp. Woods Hole Oceanographic Instn. Decorated officier Légion d'Honneur (France); recipient numerous hon. degrees and awards including Nobel prize for physics, 1964; Stuart Ballantine medal Franklin Inst., 1959, 62; Thomas Young medal and prize Inst. Physics and Phys. Soc., Eng., 1963; Disting. Public Service medal NASA, 1969; Wilhelm Exner award Austria, 1970; Niels Bohr Internat. Gold medal, 1979; Nat. Sci. medal, 1983, Berkeley citation U. Calif., 1986; named to Nat. Inventors Hall of Fame, 1976, Engring. and Sci. Hall of Fame, 1983; recipient Common Wealth award, 1993. Fellow IEEE (life medal of honor 1967), Am. Phys. Soc. (pres. 1967, Plyler prize 1977), Optical Soc. Am. (hon., Mees medal 1968), Indian Nat. Sci. Acad., Calif. Acad. Scis.; mem. NAS (coun. 1969-72, 78-81, chmn. space sci. bd. 1970-73, Comstock award 1959), Am. Philos. Soc., Am. Astron. Soc., Am. Acad. Arts and Scis., Royal Soc. (fgn.), Pontifical Acad. Scis., Max-Planck Inst. for Physics and Astrophysics (fgn.). Office: U Calif Dept Physics Berkeley CA 94720

TOWNSEND, ARTHUR SIMEON, manufacturing engineer; b. Jacksonville, Fla., Oct. 27, 1950; s. Wallace Herbert and Eunice Evelyn (Stringer) T.; m. Janet Francine Livingston, Apr. 25, 1977; children: Matthew Curtis, Ethan Christopher. BS, MS in Indsl. Engring., U. South Fla., 1984. Registered profl. engr., N.C. Advanced mfg. engr. Ingersoll-Rand Co., Mocksville, N.C., 1984-89, sr. mfg. engr., 1989-92, supr. advanced mfg. engr., 1992—; contract engr. Gen. Electric Co., Largo, Fla., 1983-84; cons. computer integrated mfg. Precision Techs., Shelton, Conn., 1991—. With USN, 1975-81. Mem. Nat. Soc. Profl. Engrs., Soc. Mfg. Engrs. (treas. 1987), Inst. Indsl. Engrs. Achievements include research in application of statistical process control techniques to flexible manufacturing systems, applied metrology techniques in precision metal removal, computer integrated manufacturing and factory automation. Office: Ingersoll-Rand Co 501 Sanford Ave Mocksville NC 27028

TOWNSEND, DAVID JOHN, psychologist; b. Freeport, Ill., Nov. 7, 1946; s. Wilbur Howard and Pauline Amelia (Deabler) T.; m. Janis Barbara Lubawsky, Sept. 8, 1968; children: Maya, Michael. BA, U. Mich., 1968; PhD, Wayne State U., 1972. Asst. prof. Montclair State Coll., Upper Montclair, N.J., 1972-83, assoc. prof., 1983-88, prof., 1988—; psychology chair Montclair State Coll., 1993—. Author: Comprehending Oral and Written Language, 1987; reviewer Jour. Memory and Language, Memory and Cognition; contbr. articles to Jour. of Verbal Learning and Verbal Behvior, Lang. and Cognitive Processes, Cognition. Faculty scholar Montclair State Coll., 1992; NSF grantee, 1979-85. fellow Max Planck Inst. Psycholinguistics, Princeton U.; mem. Am. Psychol. Soc., Cognitive Sci. Soc., Psychonomic Soc. Office: Montclair State Coll Dept Psychology Upper Montclair NJ 07043

TOWNSEND, FRANK MARION, pathology educator; b. Stamford, Tex., Oct. 29, 1914; s. Frank M. and Beatrice (House) T.; m. Gerda Eberlein, 1940 (dec. div. 1944); 1 son, Frank M.; m. Ann Graf, Aug. 25, 1951; 1 son, Robert N. Student, San Antonio Coll., 1931-32, U. Tex., 1932-34; MD, Tulane U., 1938. Diplomate: Am. Bd. Pathology. Intern Polyclinic Hosp., N.Y.C., 1939-40; commd. 1st lt. M.C., U.S. Army, 1940, advanced through grades to lt. col., 1946; resident instr. pathology Washington U., 1945-47; trans. to USAF, 1949, advanced through grades to col., 1956; instr. pathology Coll. Medicine, U. Nebr., 1947-48; asso. pathologist Scott and White Clinic, Temple, Tex., 1948-49; asso. prof. pathology Med. Br. U. Tex., Galveston, 1949-59; dir. labs. USAF Hosp. (now Wilford Hall USAF Hosp.), Lackland AFB, Tex., 1950-54; cons. pathology Office of Surgeon Gen. Hdqrs. USAF, Washington, 1954-63, chief cons. group Office of Surgeon Gen. Hdqrs., 1954-55; dep. dir. Armed Forces Inst. Pathology, Washington, 1955-59; dir. Armed Forces Inst. Pathology, 1959-63; vice comdr. aerospace med. divsn. Air Force Systems Command, 1963-65; ret., 1965; practice medicine specializing in pathology San Antonio, 1965—; dir. labs. San Antonio State Chest Hosp.; consulting pathologist Tex. Dept. Health hosps., 1965-72; clin. prof. pathology U. Tex. Med. Sch., San Antonio, 1969-72; prof., chmn. dept. pathology Health Sci. Ctr. U. Tex. Med. Sch., 1972-86, emeritus chmn., prof., 1986—; cons. U. Tex. Cancer Center-M.D. Anderson Hosp., 1966-80, NASA, 1967-75; mem. adv. bd. cancer WHO, 1958-75; mem. Armed Forces Epidemiology Bd., 1983-91; bd. govs. Armed Forces Inst. Pathology, 1984—. Mem. editorial bd. Tex. Med. Jour., 1978-86; contbr. articles to med. jours. Mem. adv. coun. Civil War Centennial Commn., 1960-65; bd. dirs. Alamo Area Sci. Fair, 1967-73. Decorated D.S.M., Legion of Merit; recipient Founders medal Assn. Mil. Surgeons, 1961. Recipient Comdr.'s award Armed Forces Epidemiol. Bd., 1990; F.M. Townsend Chair of Pathology endowed in his honor by faculty of Dept. Pathology, U. Tex. Health Sci. Ctr., 1987. Fellow ACP, Coll. Am. Pathologists (edn. advisor on accreditation, commr. lab. accreditation South Cen. States region 1971-84), An. Soc. Clin. Pathologists (Ward Burdick award 1983), Aerospace Med. Assn. (H.G. Mosely award 1962); mem. AMA, AAAS, Tex. Med. Assn., Internat. Acad. Aviation and Space Medicine, Tex. Soc. Pathologists, Am. Assn. Pathologists, Internat. Acad. Pathology, Acad. Clin. Lab. Physicians and Scientists, Soc. Med. Cons. to Armed Forces, Torch Club. Home: PO Box 77 Harwood TX 78632-0077 Office: U Tex Health Sci Ctr Dept Pathology 7703 Floyd Curl Dr San Antonio TX 78284-7700

TOWNSEND, JAMES TARLTON, psychologist; b. Amarillo, Tex., July 9, 1939; s. Tarlton Byrd and Virgie Leona (McAlister) T.; m. June 5, 1959 (div. July 1967); children: Phillip J., Pamela J., Joan M.; m. Lisa Michelle Day, June 29, 1985; 1 child, Tarlise N. AB, Fresno State Coll., 1961; PhD, Stanford U., 1966. Asst. prof. dept. psychology U. Hawaii, Honolulu, 1966-68; asst. prof. dept. psychology Purdue U., West Lafayette, Ind., 1968-71, assoc. prof. dept. psychology, 1971-76, prof. dept. psychology, 1976-89; Rudy prof. psychology Ind. U., Bloomington, 1989—. Co-author: Stochastic Modeling of Elementary Psychological Processes, 1983; co-editor: (book series) Scientific Psychology Series for Erlbaum Associates, 1991—; author/co-author 21 book chpts. and reviews; editor Jour. Math. Psychology, 1985-90, mem. editorial bd., 1990—; contbr. over 37 articles to profl. jours. Recipient James McKeen Cattell award, 1992-93, Am. Coun. Learned Socs. award, 1992-93, Sr. Scientist fellowship Braunschweig U., 1976-77, numerous rsch. grants. Mem. Soc. for Math. Psychology (pres. 1984-89). Democrat. Achievements include development of mathematical meta-theories and implied methodologies for identifying mental architecture and subprocess connections through experimentally obtained response times and patterns of response probabilities; of mathematical models for human charcter recognition; of a hierarchical structure of stochastic dominance for the investigation of stochastic models of human behavior. Office: Ind U Psychology Dept Bloomington IN 47405

TOWNSEND, JAMES WILLIS, computer scientist; b. Evansville, Ind., Sept. 9, 1936; s. James Franklin and Elma Elizabeth (Galloway) T.; m. Leona Jean York, Apr. 20, 1958; 1 child, Eric Wayne. BS in Arts and Scis., Ball State U., 1962; PhD, Iowa State U., 1970. Rsch. technologist Neuromuscular div. Mead Johnson, Evansville, 1957-60; chief instr. Zoology dept. Iowa State U., Ames, 1965-67; asst. prof. Ind. State U., Evansville, 1967-72; cons. electron microscopy Mead Johnson Rsch. Ctr., Evansville, 1971-73; mgr. neurosci. Neurosci. Lab., Kans. State U., Manhattan, 1974-76; head electron microscopy Nat. Ctr. for Toxicology Rsch., Jefferson, Ark., 1976-82; dir. electron microscopy U. Ark. Med. Sci., Little Rock, 1982-87; dir. computer ops. Pathology Dept. U. Hosp., Little Rock, 1987—; workshop presenter Am. Soc. Clin. Pathology, 1980-81, Nat. Soc. Histotechnologists, 1984-88. With USAF, 1957. Contbr. articles to profl. jours.; reviewer Scanning Electron Microscopy, 1977-78. Nat. Def. fellowship NDEA, Iowa State U., 1962-65; recipient Chgo. Tribune award Chicago Tribune, 1955. Mem. Sigma Zeta, Sigma Xi. Baptist. Avocation: genealogy. Home: 4 Breeds Hill Ct Little Rock AR 72211-2514 Office: Univ Hosp Dept Pathology 4301 W Markham St Little Rock AR 72205-7101

TOWNSEND, JOHN WILLIAM, JR., physicist, retired federal aerospace agency executive; b. Washington, Mar. 19, 1924; s. John William and Elenore (Eby) T.; m. Mary Irene Lewis, Feb. 7, 1948; children: Bruce Alan, Nancy Dewitt, John William III, Megan Lewis. BA, Williams Coll., 1947, MA, 1949, ScD, 1961. With Naval Research Lab., 1949-55, br. head, 1955-58; with NASA, 1958-68, dep. dir. Goddard Space Flight Ctr., 1965-68; dep. adminstrn. Environmental Scis. Services Adminstrn., 1968-70; asso. adminstr. Nat. Oceanic and Atmospheric Adminstrn., 1970-77; pres. Fairchild Space and Electronics Co., 1977-82; v.p. Fairchild Industries, 1979-85; pres. Fairchild Space Co., 1983-85; sr. v.p. Fairchild Industries, 1985-87; chmn. bd. Am. Satellite Co., 1985, sr. v.p., exec. aerospace group, 1987, exec. v.p., 1987; dir. NASA Goddard Space Flight Ctr., 1987-90; ret., 1990; mem. U.S. Rocket, Satellite Rsch. Panel, 1950-60; chmn. space applications bd. NRC, 1985-87; bd. dirs., trustee Telos Corp., 1990-92; mem. adv. bd. Loral Corp., 1990-92; mem. coms. NRC, 1990—; bd. dirs CTA, Inc. Author numerous papers, reports in field. Pres. town council, Forest Heights, Md., 1951-55. Served with USAAF, 1943-46. Recipient Profl. Achievement award Engrs. and Architects Day, 1957; Meritorious Civilian Service award Navy Dept., 1957; Outstanding Leadership medal NASA, 1962; Distinguished Service medal, 1971, 90; recipient Arthur S. Fleming award Fed. Govt., 1963. Fellow AIAA, AAAS, Am. Meteorol. Soc.; mem. NAE (com. 1990—), Am. Phys. Soc., Am. Geophys. Union. (fin. com. 1991—), Internat. Astronautical Fedn. (mem., trustee internat., acad. astronautics), Sigma Xi. Home: 15810 Comus Rd Clarksburg MD 20871-9169

TOWNSEND, MILES AVERILL, aerospace and mechanical engineering educator; b. Buffalo, N.Y., Apr. 16, 1935; s. Francis Devere and Sylvia (Wolpa) T.; children: Kathleen Townsend Hastings, Melissa, Stephen, Joel, Philip. BA, Stanford U., 1955; BS MechE, U. Mich., 1958; advanced cert., U. Ill., 1963, MS in Theoretical and Applied Mechanics, 1967; PhD, U. Wis., 1971. Registered profl. engr., Ill., Wis., Tenn., Ont. Project engr. Sundstrand, Rockford, Ill., 1959-63, Twin Disc Inc., Rockford, 1963-65, 67-68; sr. engr. Westinghouse Electric Corp., Sunnyvale, Calif., 1965-67; instr., fellow U. Wis., Madison, 1968-71; assoc. prof. U. Toronto, Ont., Can., 1971-74; prof. mech. engring. Vanderbilt U., Nashville, 1974-81; Wilson prof. mech. and aerospace engring. U. Va., Charlottesville, 1982—, chmn. dept., 1982-91; ptnr., v.p. Endev Ltd., Can. and U.S., 1972—; cons. in field. Contbr. numerous articles to profl. jours.; 7 patents in field. Recipient numerous research grants and contracts. Fellow ASME (mem. coun. on engring., productivity com., tech. editor Jour. Mech. Design); mem. AAAS, N.Y. Acad. Scis., Sigma Xi, Phi Kappa Phi, Pi Tau Sigma. Avocations: running, reading, music. Home: 221 Harvest Dr Charlottesville VA 22903-4850 Office: U Va Dept Mech and Aerospace Engring Thornton Hall Charlottesville VA 22903-2442

TOWNSEND, PALMER WILSON, chemical engineer, consultant; b. Bronx, N.Y., Aug. 1, 1926; s. Atwood Halsey and Mildred Brower (Wilson) T.; m. Helen Anne Lydecker, Feb. 6, 1949; children: Janet M., Martha L., Andrew W., Amy E., Rebecca L. AB in Chemistry, Dartmouth Coll., 1947; BSChemE, Columbia U., 1947, MSChemE, 1948, PhDChemE, 1956. Instr. chem. engring. Columbia U., N.Y.C., 1948-53; sr. engr. pilot plant divsn. Air Reduction Co., Inc., Murray Hill, N.J. and N.Y.C., 1953-56, sect. head chem. engring. divsn., 1957-61, asst. dir., 1961-1964, mgr. exptl. engring., cen. engring. dept., 1964-66, asst. to group v.p., 1966-67, dir. comml. devel. Chem. and Plastics divsn., 1967-70; dir. comml. devel. Plastics divsn. Allied Chem. Corp., 1970-72; cons. Berkeley Heights, N.J., 1972—. Editor phys. sci. sect. Good Reading, 1956, 60, 64; contbr. articles to profl. jours. With USNR, 1944-46. DuPont fellow, 1951. Mem. AICE, Am. Chem. Soc., Soc. Plastic Engrs., Soc. Plastics Industries, Assn. Cons. Chemists and Chem. Engrs. (pres. 1991-92), Sigma Xi, Phi Lambda Upsilon, Theta Tau. Congregationalist. Achievements include patents for processes for trifluorethyl vinyl ether, trifluoro ethanol, cryogenic composition of matter, carbon dioxide snow; development of molding approach for the plastic spacers used in superconducting magnets for ISAbelle and the Superconducting Super Collider; establishment of rigid PVC molding materials in the Bell Telephone System; invention of improved processes for production of fluorochemical intermediates to anesthetic agents; research on plastics development and processes, chemicals and monomers, energy development, polyvinyl chloride resin production and market applications. Home and Office: 9 Bristol Ct Berkeley Heights NJ 07922-1306

TOWNSHEND, BRENT SCOTT, electrical engineer; b. Toronto, Ontario, Can., Nov. 24, 1959; s. John Barton and Gail Denise (Pointon) T.; m. Michèle Lamarre, Aug.4, 1990. BA in Sc., U. Toronto, Ontario, Can., 1982; PhD, MS, Stanford U., 1987. Engr. Mitel, Ottawa, Ontario, Can., 1982; rsch. assoc. Stanford (Calif.) U., 1982-87; mem. tech. staff Bell Labs. AT&T, Murray Hill, N.J., 1987-90; pres. Townshend Computer Tools, Montreal, Quebec, Can., 1990—; lectr. McGill U., Montreal, 1990—; cons. pvt. practice, Palo Alto, Calif., 1988-90. Author: (with others) Nonlinear Modelling and Forecasting, 1992. Recipient Natural Sciences & Engring. Rsch. Coun. Postgrad. award, Govt. Can., 1982-86, scholarship Spar Aerospace, Toronto, 1982, award Cockburn Inst. Design, Toronto, 1990. Mem. IEEE, Acoustical Soc. Am., Assn. for Computing Machinery, Sigma Xi. Achievements include development of techniques for direct stimulation of auditory nerve of deaf patients to restore hearing; created models of speech production using nonlinear dynamics. Home: 230 Sherbrooke St E Apt 502, Montreal, PQ Canada H2x 1E1 Office: Townshend Computer Tools, 10 Ontario W Ste 502, Montreal, PQ Canada

TOYAMA, TAKAHISA, oil company executive, engineer; b. Tokyo, Oct. 10, 1925; s. Riyoko Toyama, Mar. 16, 1954; 1 child, Reiko Tayama Wess. BS, Nihon U., Tokyo, 1950. Plant mgr. The Nisshin Oil Mills, Ltd., Yokohama, Japan, 1967-69; dir. The Nisshin Oil Mills, Ltd., Tokyo, 1979-84, mng. dir., 1984-88, standing adviser, 1988-93; bd. dirs. Kobayashi Pharmacy Industry Co., Ltd., Tokyo; project leader Palm Oil Rsch. Inst. of Malaysia, 1988, Bio Industry Devel. Ctr., Japan, 1983—; cons. chemical engring., 1993—. Mem. Japan Vegetable Food Protein Assn. (vice chmn. 1986—). Home: 471-2 Kawashima cho, Hodogayaku Kamihoshikawa, Garden Ct 718, 240 Yokohama Japan Office: Rsch Lab Nisshim Oil Mills, 1-3 Chiwaka cho, 221 Yokohama Japan

TOYODA, TADASHI, physicist; b. Chiba, Japan, June 18, 1949; s. Toshiyuki and Michi (Suzuki) T.; m. Hanako Aoki; children: Tetsu, Yukiko. BS, Nagoya (Japan) U., 1974; PhD, SUNY, Buffalo, 1978. Rsch. assoc. U. Tübingen, Germany, 1979-83; rsch. assoc. U. Alberta (Edmonton, Can.), 1983-85, asst. prof., 1985-88; assoc. prof. Nagoya Shoka Daigaku, Nagoya, 1988—. Author: New Con. Matt. Physics, 1990. Woodbarn fellow, SUNY, 1977-78, Alexander Von Humbolt fellow, Bonn, Germany, 1978-79. Mem. Japan Physical Soc., Am. Physical Soc. Home: 4-8-8 Nishi-Ikebukuro, Toshima-ku, Tokyo 171, Japan Office: Nagoya Shoka Daigaku, Sagamine, Nisshin-cho, Aichi Aichi-Ken, Japan

TOYOMURA, DENNIS TAKESHI, architect; b. Honolulu, July 6, 1926; s. Sansuke Fujimoto and Take (Sata) T.; m. Akiko Charlotte Nakamura, May 27, 1949; children—Wayne J., Gerald F., Amy J., Lyle D. BS in Archtl. Engring., Chgo. Tech. Coll., 1949; cert., U. Ill., Chgo., 1950, 53, 54; student, Ill. Inst. Tech., Chgo., 1953-54; cert., U. Hawaii-Dept. Def., Honolulu, 1966-67, 73. Lic. architect, Ill. 1954, Hawaii 1963; lic. real estate broker Ill., 1957. Designer, draftsman James M. Turner, Architect, Hammond, Ind., 1950-51, Wimberly and Cook, Architects, Honolulu, 1952, Gregg, Briggs & Foley, Architects, Chgo., 1952-54; architect Holabird, Root & Burgee, Architects, Chgo., 1954-55, Loebl, Schlossman & Bennett, Architects, Chgo., 1955-62; prin. Dennis T. Toyomura, AIA, Architect, Honolulu, 1963-83, Dennis T. Toyomura, FAIA, Architect, Honolulu, 1983—; fallout shelter analyst Dept. Def., 1967—; cert. analyst multi-distaster design, Dept. Def., 1973; cert. value engr. NavFacEngCom., Gen. Svc. Adminstrn., Environ. Protection Agy., Fed. Housing Agy., U.S. Corps. of Engrs., 1988; cons. Aloha Tower Devel. Corp., State of Hawaii Project Devel. Evaluation Team, 1989, archl. cons. 1991—; mem. Provost Selection Interview Com., Leeward Community Coll., U. Hawaii, 1987; design profl. conciliation panelist Dept. of Commerce and Consumer Affairs, State of Hawaii, 1983, 84; archtl. design examiner Nat. Council of Archtl. Registration Bd., State of Hawaii, 1974-78; cons. Honolulu Redevel. Agy., City and County of Honolulu, 1967-71; sec., dir. Maiko of Hawaii, Honolulu, 1972-74, Pacific Canal of Hawaii, 1972; mem. steering com. IX world conf. World Futures Studies Fedn., U. Hawaii, 1986; conf. organizer pub. forum 10th Hawaii Conf. in High Energy Physics, U. Hawaii, 1985, 60th Ann. Nat. Council of Archtl. Registration Bd. Conv., Maui, Hawaii, 1981, Hawaii State Bd. Conv. steering com.; del. Nat. Credential Com., NCARB, 1981, 125th Nat. AIA Conv., Honolulu, 1982, HS/AIA Conv. Com., steering com., budget and fin. chmn., treas., 1981-82, appointments State of Hawaii; mem. legis. adv. com. Hawaii State Legis.; vice chmn. Rsch. Corp. U. Hawaii, 1991-93, bd. dirs. 1986—; commr. Hawaii State Found. on Culture and the Arts, 1982-86, Gov.'s Com. on Hawaii Econ. Future, 1984; archtl. mem. Bd. Registration for Profl. Engrs., Architects, Land Surveyors and Landscape Architects, State of Hawaii, 1974-82, sec. 1980, vice chmn. 1981, chmn., 1982; mem. Nat. Coun. Engring. Examiners, 1974-82; mem. Nat. Coun. Archtl. Registration Bds., Western region del. 1974-82, nat. del. 1974-82. Editor, pub.: (directory) Japanese Companies Registered to do Business in Hawaii, 1991. Del. commr. state assembly Synod of Ill., United Presbyn. Ch. U.S.A., 1958, alt. del. commr. nat. gen. assembly, 1958, del. commr. L.A. Presbytery, 1965; mem. bd. session 2d Presbyn. Ch., Chgo., 1956-62, trustee, 1958-62; trustee 1st Presbyn. Ch., Honolulu, 1964-66, 69-72, sec., 1965, bd. sessions, 1964-72, 74-79; founding assoc. Hawaii Loa Coll., Kaneohe, 1964; mem. adv. commn. drafting tech. Leeward Community Coll., U. Hawaii, 1965—; bd. dirs. Lyon Arboretum Assn., U. Hawaii, 1976-77, treas.,1976. With AUS, 1945-46. Recipient Human Resources of U.S.A. award Am. Bicentennial Rsch. Inst., 1973, Outstanding Citizen Recognition award Cons. Engrs. Coun. Hawaii, 1975, cert. appreciation Gov. Hawaii, 1982, 86, 89, commendation Hawaii Ho. of Reps. and Senate, 1983, 90, 91, cert. appreciation Leeward Community Coll., U. Hawaii Adv. Com., 1971-86, 87—, medal for peace Albert Einstein Internat. Acad. Found., 1990, commendation resolution Honolulu, Japanese C. of C., 1991. Fellow AIA (Coll. Fellows 1983, bd. dirs. Hawaii Soc. 1973-74, treas. 1975, pres. Hawaii coun. 1990, Pres.'s Mahalo award 1981, Fellows medal 1983); mem. AAAS (life), ASTM, Am. Arbitration Assn. (mem. panel of arbitrators, 1983—), Acad. Polit. Sci. (life), Am. Acad. Polit. and Social Scis., N.Y. Acad. Scis., Chgo. Art Inst., Chgo. Natural History Mus., HonoAcad. Arts, Nat. Geog. Soc., Coun. Ednl. Facility Planners Internat. (bd. govs. N.W. region 1980-86, 89—), Bldg. Rsch. Inst. (adv. bd. of Nat. Acad. Sci.), Ill. Assn. Professions, Constrn. Specifications Inst. (charter mem. Hawaii), Constrn. Industry Legis. Orgn. (bd. dirs. 1973-81, 83—, treas. 1976-77, v.p. 1990-91, pres. 1991—), Japan-Am. Soc., Hawaii State C. of C. (bd. dirs. 1984-87), U. Hawaii Kokua O'Hui, O'Nahe Popo (bd. dirs. 1984—), Hawaii-Pacific Rim Soc. (bd. trustees 1988—, sec. 1991—), Malolo Mariners Club (purser 1964, skipper 19865, Hawaii), Alpha Lambda Rho, Kappa Sigma Kappa. Home: 2602 Manoa Rd Honolulu HI 96822-1703 Office: Dennis T Toyomura FAIA 1370 Kapiolani Blvd Honolulu HI 96814-3605

TRABIA, MOHAMED BAHAA ELDEEN, mechanical engineer; b. Alexandria, Egypt, Oct. 23, 1958; came to U.S., 1983; s. Mokhtar M. Trabia and Hoorya (El Fawal) T.; m. Salma El Safwani Trabia, Apr.15, 1983; children: Sarah S., Suzanne H. MS, U. Alexandria, 1983; PhD, Ariz. State U., 1987. Asst. prof. U. Nev., Las Vegas, 1987—. Contbr. articles to profl. jours. Mem. ASME, Am. Soc. Engring. Edn., Soc. Mfg. Engrs. Office: Univ Nevada Dept Mech Engring 4505 Maryland Pkwy Las Vegas NV 89154

TRABOLD, THOMAS AQUINAS, chemical engineer; b. Rochester, N.Y., Sept. 23, 1962; s. John Francis and Beatrice (Montanaro) T.; m. Nancy E. Hersam, Sept. 14, 1991. BS, Clarkson U., 1984, PhD, 1989. Postdoctoral researcher Argonne (Ill.) Nat. Lab., 1989; engr. GE Co., Schenectady, N.Y., 1989—. Contbr. articles to profl. jours. Mensa scholar, 1986. Mem. AICE, ASME. Achievements include development of novel instrumentation for two-phase flow based on laser Doppler velocimetry and high speed digital image analysis. Office: GE Co E7-135 PO Box 1072 Schenectady NY 12301

TRACESKI, FRANK THEODORE, materials engineer; b. Montague, Mass., May 5, 1953; s. Frank Edward and Victoria Carolyn (Krejmas) T.; m. Laurie Anne Lemay, May 1, 1990. BS in Chemistry, U. Mass., 1975; M Forensic Scis., George Washington U., 1980. Materials scientist/engr. U.S. Army Materials & Mechanics Rsch. Ctr., Watertown, Mass., 1979-89; materials engr., mgr. def. acquisition Office Asst. Sec. Def. (Prodn. and Logistics), Washington, 1989—; mem. chlorofluorocarbons adv. com. Dept. Def., 1990-91. Author: Specifications and Standards for Plastics and Composites, 1990; contbr. articles to profl. jours. and books. Recipient Outstanding Performance award Def. Standardization Program, Dept. Def., 1987. Mem. ASTM (com. D20 plastics award of recognition 1989), Soc. Advancement Material and Process Engring. Republican. Roman Catholic. Achievements include rsch. in advanced composites, def. acquisition and environ. Office: Prodn Resources Support Office 5203 Leesburg Pike Ste 1403 Falls Church VA 22041

TRACHT, ALLEN ERIC, electronics executive; b. Bethesda, Md., Aug. 14, 1957; s. Myron Edward and Diane Serena (Goldberg) T.; m. Donna June Carothers, Sept. 14, 1986; children: Michael, Diane, Daniel. BS in Physics and Elec. Engring., MIT, 1979; MSEE, Calif. Inst. Tech., 1980. Biomed. researcher Case Western Res. U., Cleve., 1980-85; exec. engr. IOtech. Inc., Cleve., 1985—; cons. engring. Keithley Instruments, Cleve., 1985. Contbr. articles to profl. jours. NIH grantee Case Western Res. U., 1982. Mem. IEEE, Assn. for Computing Machinery, Sigma Xi, Tau Beta Kappa, Eta Kappa Nu. Home: 3066 Scarborough Rd Cleveland OH 44118-4065

TRACY, RUSSELL PETER, biomedical researcher, educator; b. N.Y.C., Aug. 28, 1949; s. Arnold John and Madeline Cecilia (Bleier) T.; m. Paula Susan Babiarz, Aug. 28, 1976; children: Sarah, Patrick. BS, LeMoyne Coll., 1971; PhD, Syracuse U., 1978. Diplomate Am. Bd. Clin. Chemistry. Psotdoctoral fellow Mayo Clinic, Rochester, Minn., 1978-83; asst. prof. pathology dept., asst. dir. clin. chemistry U. Rochester, N.Y., 1983-84; asst. prof. depts. pathology and biochemistry U. Vt., Burlington, 1984-93; assoc. prof. U. Vt., 1993—; cons. Argonne (Ill.) Nat. Lab., 1980-83; v.p., part owner Haematologic Techs., Essex Junction, Vt., 1989—. Contbr. articles to Am. Jour. Cardiology, Annals of Epidemiology, Blood, Internat. Jour. Biochemistry, Calcified Tissue Internat., numerous others; mem. editorial bd. Jour. Applied and Theoretical Electrophoresis. Mem. Essex Econ. Devel. Commn., 1990—. Recipient Young Investigator award Acad. Clin. Lab. Physicians and Scientists, 1982, Outstanding Sci. Achievement by a Young Investigator, Am. Assn. Clin. Chemistry, 1984. Fellow Nat. Acad. Clin. Biochemistry; mem. Am. Assn. Clin. Chemistry, Am. Soc. Biochemistry and Molecular Biology. Achievements include copyright for software that scans densitometry of electrophoresis gels; rsch. on preparation of monoclonal antibodies from spots cut from two dimensional gels, in clin. chemistry. Home: 18 Skyline Dr Essex Junction VT 05452 Office: U Vt Dept Pathology Burlington VT 05405

TRAEGER, DONNA JEAN, health facility manager; b. Natrona Heights, Pa., Oct. 17, 1956; d. Andrew Thomas and Sophie Lucille (Mniecznikowski) Kovach; m. Francis John Traeger, July 3, 1982. BA/BS with high honors, Carlow Coll., 1978; MS, U. Pitts., 1983. Asst. sect. head clin. hematology and clin. microscopy Mercy Hosp. Pitts., 1978-85; lab mgr. hematology Med-Chek Labs., Pitts., 1985-91; supr. hematology Morristown (N.J.) Meml. Hosp., 1991—. Fund raiser March of Dimes, Pitts., 1991, Am. Heart Assn., Pitts., 1991. March of Dimes scholar, 1974. Mem. Am. Soc. Clin. Pathology (cert. hematology specialist, cert. gen. lab. medicine), Am. Soc. Med. Tech., Pa. Soc. Med. Tech., Clin. Lab. Mgmt. Assn., Polish Women's Alliance (scholar 1974), Allegheny County Women's Club (scholar 1974). Roman Catholic. Avocations: nautilus aerobics, bench aerobics, golf, reading. Home: PO Box 289 35 Main St Otisville NY 10963 Office: Morristown Meml Hosp Madison Ave Morristown NJ 07960

TRAGER, GARY ALAN, endocrinologist, diabetologist; b. N.Y.C., July 30, 1950; s. Jacob Morris and Elena (Tanzer) T.; m. Marie-Christine Nicole Lachal, Dec. 26, 1976; children: Ashley Audrey, Brendon Alden. BA in Biology and Anthropology, SUNY, Binghamton, 1972; MD, U. Cen. del Este, Dominican Republic, 1980. Subintern-rotating Jamaica (N.Y.) Hosp., 1979-80; intern and resident medicine Bklyn.-Cumberland Med. Ctr., 1980-83; fellow endocrinology SUNY, Stony Brook, 1983-85, clin. asst. instr., 1983-85; asst. attending Huntington (N.Y.) Hosp., 1985-90, assoc. attending, 1990—; mem. nutrition com. Huntington Hosp., 1987—, dir. diabetes club, 1985—. Mem. profl. edn. com. Am. Diabetes Assn., Long Island, Melville, N.Y., 1985—; mem. Am. Diabetic Assn. Fund, Long Island, N.Y., 1989—; ad hoc mem. Eaton's Neck Emergency Squad, Long Island, 1985-89. Mem. AMA, Am. Fertility Soc., Am. Diabetes Assn., Am. Soc. Internal Medicine, Am. Soc. Andrology, Am. Assn. Clin. Endocrinologists, Peripheral Neuropathy Inst., Clin. Endocrine Soc., An. Soc. Hypertension. Office: 158 E Main St Huntington NY 11743-2988

TRAHERN, CHARLES GARRETT, physicist; b. Clarksville, Tenn., July 30, 1953; s. Charles Anderson and Ann (Metcalf) T. BA, Rice U., 1975; MA, U. Rochester, 1977; PhD, Syracuse U., 1982. Postdoctoral fellow Syracuse (N.Y.) U., 1983-85, Rutgers U., Piscataway, N.J., 1985-86, Ohio State U., Columbus, 1987-89; scientist II Superconducting Super Collider Lab., Dallas, 1989—. Co-author: Lectures on Group Theory for Physicists, 1984. Mem. Am. Phys. Soc. (div. Particles and Fields). Office: Superconducting Supercollider Lab 2550 Beckleymeade MS 4011 Dallas TX 75237

TRAICOFF, ELLEN BRADEN, psychologist; b. Gary, Ind.; d. Charles Leonard and Blossom (Riggin) Braden; children: Ted, George, Anthony, Gerald, Phillip. BS, Ind. U., 1971, MS, 1974; grad., Family Inst. Chgo., 1978; D of Psychology, Forest Inst. Profl. Psychology, 1987. Edn. and family therapist Cath. Family Services, 1974-78; child and family specialist Porter Starke Services, Valparaiso, Ind., 1978-79; dir. family violence programming Southlake Ctr. Mental Health, Merrillville, Ind., 1979-83, dir. forensic services, 1983-84; cons. Lake County (Ind.) Dept. of Pub. Welfare, 1984-86; psychologist Walbash Valley Hosp., Lafayette, Ind., 1987-88, Mercy Hosp. and Med. Ctr., Chgo., 1988-89; pvt. practice Merrillville, Ind., 1988-91; Swanson Ctr., Michigan City, Ind., 1988-90; pvt. practice Indpls., 1991—; mem. staff Charter Barclay Hosp., Chgo., South Shore Hosp., Chgo., Univ. Hosp., Winona Hosp. Midwest Med. Ctr., Indpls., 1991—, Meth. Hosp., Indpls., 1993—. Contbr. articles to profl. jours. Del. White House Conf. on Families, 1980. Mem. Am. Assn. Marriage and Family Therapy (clin.), Family Inst. Chgo. Alumni Assn., Am. Psychol. Assn., Ind. Psychol. Assn., Ill. Psychol. Assn. Avocations: archeology, anthropology. Office: 10291 N Meridian St Ste 180 Indianapolis IN 46290

TRAINA, PAUL JOSEPH, environmental engineer; b. N.Y.C., Mar. 8, 1934; s. Peter and Mary (Panepinto) T.; m. Mary Ann Delehanty, Oct. 8, 1955; children: Peter F., Kenneth P., Jean Marie, Julie Ann, Marie L. BCE, Manhattan Coll., 1955; M in San. Engring., U. Mich., 1960. Registered profl. engr., Ga. Chief water resources USPHS, Atlanta, 1960-63, dep. dir. S.E. com. Water Pollution Control project, 1963-67; dir. tech. svcs. Fed. Water Pollution Control Adminstrn., Athens, Ga., 1967-70; dir. enforcement div. EPA, Atlanta, 1972-79, dir. water div., 1979-85; sr. cons. Camp Dresser McKee, Cambridge, Mass., 1985—. Mem. Am. Acad. Environ. Engring. (diplomate), Water Pollution Control Fedn., Air Waste Mgmt. Assn. Home: 2366 Woodcreek Ct Tucker GA 30084-3301

TRAINOR, PAUL VINCENT, weapon systems engineer, genealogy researcher; b. Elmira, N.Y., Dec. 26, 1948; s. Walter Francis and Margaret Camilla (McCarthy) T. BSEE, U. Rochester, 1970; MS in Indsl. Mgmt., Cen. Mo. State U., 1979. Commd. 2d lt. USAF, 1970, advanced through grades to maj., 1990; Titan II missile crew mem. 381st SMW, McConnell AFB, Kans., 1971-73; minuteman tech. engr. 90th SMW, F.E. Warren AFB, Wyo., 1974-78; ICBM field engr. SATAF, Whiteman AFB, Mo., 1978-79; project engr. BMO, Norton AFB, Calif., 1980-84; system safety engr. AFISC, Norton AFB, Calif., 1985-87; guidance system project officer BMO, Norton AFB, Calif., 1988-90; pvt. practice Moreno Valley, Calif., 1991—. Author Genealogy of Michael F. McCarthy Family, 1992; asst. editor (mil. handbook) Software System Safety Handbook, 1985. Vol. March Field Mus., March AFB, Calif., 1992. Republican. Roman Catholic. Home: PO Box 7854 Moreno Valley CA 92552

TRAMONTOZZI, LOUIS ROBERT, electrical engineer; b. Boston, Mar. 24, 1958; s. Fiore and Concetta (Antonellis) T. BSEE, U. Lowell, 1980; MSEE, U. Mass., 1984. Programmer Army Materials and Mechanics Rsch. Ctr., Watertown, Mass., 1975-77; assembler Omni-Spectra, Waltham, Mass., 1978; engr./sr. engr. Raytheon Co., Marlborough, Mass., 1980—. Recipient Jacob Ziskind fellow, 1980. Mem. IEEE, Middlesex Amateur Radio Club (pres. 1991-93), Eta Kappa Nu. Republican. Roman Catholic. Achievements include research on improved microwave synthesizer design; organized and increased efficiency of testing Air Force satellite communications equipment. Home: 48 Bridge St Watertown MA 02172 Office: Raytheon Co 1001 Boston Post Rd Marlborough MA 01752

TRAN, DEAN, opto-electronic devices engineer; b. Quang Ngai, Vietnam, Jan. 16, 1948; came to U.S., 1978; s. Thien Tran and Bung Thi Che; m. Hong Hoa Le, 1968; children: Hung Le, Hong Ha Le, Bao Quoc, Hien Thanh. BS in Physics, Chemistry, Saigon (Vietnam) U., 1970, MS in Chemistry, 1972; BSEE, Calif. State U., Long Beach, 1981. Prodn. engr. TRW, RF Devices Div., Lawndale, Calif., 1982-83, circuitry engr., 1983-84; R&D scientist TRW, Rsch. Staff Group, Redondo Beach, Calif., 1984-89; sr. staff engr. PCO, Inc., Chatsworth, Calif., 1989-91; tech. dir. Pach and Co., San Clemente, Calif., 1991—. Contbr. articles to Jour. Crystal Growth, Electronic Letter, other tech. publs. Mem. IEEE, So. Calif. Crystal Growth Assn., Soc. Photo-Optical Instrumentation Engrs., Laser and Electo-Optics Soc. Achievements include 9 invention disclosures, including NiCr etchant solution; preservation technique for high-vapor pressure materials at high temperature using in LPE crystal growth, submicron channel for pBC laser, stabilization and sppeding up CRT driver Amplifier up to 2 nanoseconds, 45 degree mirror on Silicon for hybrid semiconductor Laser Array package, Supercool heat exchanger for high power semiconductor Laser Array package, selective area planar growth on GaAs channel substrate by LEP for high efficiency solar cell, hybrid microfocus lens for semiconductor LED/LD devices, low threshold Visible semiconductor laser devices. Home: 9331 Coronet Ave Westminster CA 92683 Office: Pach and Co 941 Calle Negocio San Clemente CA 92672

TRAN, JOHAN-CHANH MINH, research scientist; b. Saigon, Vietnam, Oct. 13, 1958; came to U.S., 1975; s. Ho Van and Kim-Thoa (Nguyen) T. BS, UCLA, 1980; PhD, U. Calif., Riverside, 1988. Postdoctoral researcher U. Calif., San Diego, 1988-91; rsch. scientist div. nephrology Cedars-Sinai Med. Ctr., L.A., 1991—. Office: Cedars-Sinai Med Ctr Div Nephrology B-220 Los Angeles CA 90048

TRAN, JOHN KIM-SON TAN, chemical senses executive, research administrator; b. Quang-Binh, Vietnam, Oct. 4, 1945; Came to U.S., 1975.; s. Dong Tan Tran and Chieu Thi Nguyen;m. Ann Xuyen Thi, July 30, 1972; children: Joseph Quoc-Bao Tan, Michael Quoc-Binh Tan, Regina Thuy-Quyen Tan, John Quoc-An Tan. Baccalaureate degree, Nat. Exam. Bd., Saigon, Vietnam, 1966; student, U. Saigon, 1966-70; grad., Republic of Vietnam Army Acad., 1971; BBA, U. Pa., 1976-80, postgrad., 1981-83; MS in Polit. Sci. and Pub. Adminstrn., So. Ill. U., Edwardsville, 1989. Cert. rsch. adminstr. Journalist (TV, radio) Saigon bur. Tokyo Broadcasting System, 1968-75; tchr. English Cao-Nguyen Jr. Mil. Acad., Pleiku, Vietnam, 1971-75; bookkeeper, budget asst. U. Pa., Phila., 1976-81, bus. admins., 1981-84; bus. mgr., treas. Blackburn Coll., Carlinville, Ill., 1984-87; treas., adminstr. Monell Chem. Senses Ctr., Phila., 1987—, also sec. bd. dirs. Contbr. articles to profl. jours. Sec. gen. Young Christian Students Movement, Saigon, 1969-71; warrant officer, 1971-73; v.p. Vietnamese Cath. Community Archdiocese Phila., 1976-82; treas., bd. trustees Blackburn Coll., Carlinville, Ill., 1984-87. Lt. Republic of Vietnam Army (South), 1973-75. Mem. Nat. Coun. Univ. Rsch. Adminstrs., Inst. Mgmt. Accts., Soc. Rsch. Adminstrs., Assn. Ind. Rsch. Insts. Roman Catholic. Avocations: classical music, traveling. Home: 1346 East Ave Abington PA 19001-2445 Office: Monell Chem Senses Ctr 3500 Market St Philadelphia PA 19104-3308

TRAN, LONG TRIEU, industrial engineer; b. Saigon, Vietnam, Oct. 10, 1956; came to U.S., 1973; s. Nguyen Dinh and Thiet Thi (Nguyen) T.; m. Khanh Thi-Hong Phan, Aug. 3, 1988. BS in Mech. Engring. with honors, U. Kans., 1976; MS in Mech. Engring., MIT, 1980; MBA in Bus. Adminstrn. with honors, U. Louisville, 1993. Cert. quality engr.; cert. mfg. engr. Prodn. programming engr. GE, Cleve., 1980-81; advanced mfg. engr. GE, Louisville, 1981-82, quality systems engr., 1982-84, quality control engr., 1984-86, sr. quality info. equipment engr., 1986-89, sr. quality indsl. engr., 1990—; exec. advisor Jr. Achievement Inc., Louisville 1983-84. Mem. AAAS, Am. Soc. of Mech. Engrs., N.Y. Acad. Scis., Am. Soc. for Quality Control, Computer and Automated Systems Assn. (charter), Robot Inst. Am., Robotics Internat. (charter), Soc. Mfg. Engrs. (sr.), Instrument Soc. Am. (sr.), Am. Mgmt. Assn., Internat. Platform Assn., Indsl. Computing Soc. (founder), Sigma Xi, Pi Tau Sigma, Tau Beta Pi, Phi Kappa Phi. Republican. Achievements include research on grinding processes and material surface analysis. Home: 3423 Brookhollow Dr Louisville KY 40220 Office: GE AP 5-1NC Louisville KY 40225

TRAN, NANG TRI, electrical engineer, physicist; b. Binh Dinh, Vietnam, Jan. 2, 1948; came to U.S., 1979, naturalized, 1986; s. Cam Tran and Cuu Thi Nguyen; m. Thu-Huong Thi Tong, Oct. 14, 1982; children: Helen, Florence, Irene, Kenneth. BSEE, Kyushu Inst. Tech., Kitakyushu, Japan, 1973, MSEE, 1975; PhD in Materials Sci., U. of Osaka Prefecture, Sakai, Japan, 1978. Rsch. assoc. U. of Calif.-Irvine, 1979; engr., rsch. scientist Sharp Electronics, Irvine, 1979-80; sr. rsch. scientist Arco Solar Industries, Chatsworth, Calif., 1980-84; sr. rsch. specialist, group leader 3M Co., St. Paul, 1985—; cons., lectr. Japan industry mgmt. Contbr. articles to profl. jours; patentee in field. Mem. tech. com. various internat. confs. Scholarship fellow Vietnamese Govt., Japan, 1968-73, grad. scholarship fellow Rotary Internat., Japan, 1973-75, predoctoral fellow Japanese Govt., 1975-78. Mem. IEEE, Am. Vacuum Soc., Japan Soc. of Applied Physics. Achievements include research on thin film electroluminescent displays, amorphous silicon solar cells, image sensors, transparent conducting oxide films.

TRAN, PHUOC XUAN, mechanical engineer; b. Danang, Vietnam, Feb. 12, 1945; came to the U.S., 1981; s. Phung Xuan and Thanh Thi (Phan) T.; m. Thu Xuan Le; children: Vina T., Camly T., Prenn X. MS in Mech. Engring., Ga. Tech. U., 1974, PhD, 1985. Postdoctoral rsch. fellow U.S. Dept. Energy Morgantown (W.Va.) Energy Tech. Ctr., 1986-88; mech. engr. U.S. Dept. Energy Pitts. Energy Tech. Ctr., 1988—. Contbr. chpt. to book and articles to Internat. Heat and Mass Transfer, Internat. Jour. for Numerical Methods, Internat. Symposium on Combustion, Combustion and Flame. Mem. Combustion Inst., Sigma Xi. Office: US Dept Energy Pitts Tech PO Box 10940 MS-P4-340 Pittsburgh PA 15236

TRAN, TOAN VU, electronics engineer; b. Saigon, Vietnam, June 18, 1956; came to U.S., 1981; s. Kham The Tran and Bac Thi Vu; m. Mong Huyen Vu Nguyen, Oct. 27, 1992. AA, Fla. Community Coll., 1983; BS in Elec. Engring. with honors, U. Fla., 1986. Quality control technician Medfusion System, Inc., Duluth, Ga., 1988; instrumentation specialist Emory U., Atlanta, 1988-89; electronics engr. Dept. Def., Security Operational Test Site, Ft. McClellan, Ala., 1989—. Avocations: travel, tennis, guitar. Office: Dept Def Security Operational Test Site SMCAR FSN M Fort McClellan AL 36205

TRAN, TRI DUC, chemical engineer; b. Saigon, Vietnam, July 23, 1963; came to U.S. 1980; s. Khoa Van Tran and Tuyet Quan (Luu) T.; m. Kooi-Cheng Lo, Feb. 14, 1991; 1 child, Diem Samantha. BS, U. Calif., Berkeley, 1987; PhD, U. Wis., 1993. Co-op chem. engr. IBM, San Jose, 1986; rsch. asst. chem. engring. Lawrence Berkeley (Calif.) Lab., 1987, U. Wis., Madison, 1987—. Mem. Am. Chem. Soc., Electrochem. Soc., Am. Inst. Chem. Engrs., Sigma Xi. Home: 1317 Spring St #322 Madison WI 53715 Office: Univ of Wis Chem Engring Dept 1415 Johnson Dr Madison WI 53706

TRAN-CONG, TON, applied mathematician, researcher; b. Saigon, Vietnam, Jan. 16, 1953; arrived in Australia, 1971; s. Nhi and Chinh Tran-Cong; m. C. Nguyen, Feb. 22, 1980 (div. 1991); 1 child, Catherine. B in Aero. Engring., U. Sydney, Australia, 1976, M in Aero. Engring., 1978, PhD in Aero. Engring., 1980. Rsch. asst. U. Wollongong (Australia), 1980; rsch. officer BHP Co., Melbourne, 1980-81; rsch. assoc. U. NSW, Sydney, 1982; rsch. scientist Aero. Rsch. Lab., Melbourne, 1982-87, sr. rsch. scientist, 1987—. Contbg. editor Nhan-Quyen; contbr. articles to profl. jours; patentee in field. Mem. com. Vietnamese Community Assn. in Victoria, Melbourne, 1990. Mem. AIAA, Am. Math. Soc. Avocation: electronics. Home: 4 Langville Ct, East Malvern, Victoria 3145, Australia Office: Aero Rsch Lab, 506 Lorimer St, Melbourne 3207, Australia

TRAN-VIET, TU, cardiovascular and thoracic surgeon; b. Hue, Binh-Tri-Thien, Vietnam; Apr. 26, 1949; s. Yen Tran-Viet and Ngoc Hat Ho Thi; m. Thu-Huong Le, July 23, 1976; children—Chi, Thi. M.D., Med. Sch. of Tours, France, 1974. Intern, Centre Hospitalier, St. Brieuc, France, 1973-75; asst. Clinique du Val d'Or, St. Cloud, France, 1976-78; asst. Hopital Laennec, Paris, 1978-83, cardiovascular and thoracic surgeon, 1981—; vis. asst., 1981-83, asst. prof., 1984—. Author: Tetralogie de Fallot, 1981. Editor: (video jour.) Communication Medicale, 1983. Contbr. articles to profl. jours. Mem. Societe de Chirurgie Thoracique et Cardio Vasculaire de Langue Francaise, Societe Europenne de Cardiologie, Societe de Pathologie Exotique. Home: 87 b rue Georges Ducrocq, 57000 Metz France Office: Hopital-Clinique Claude Bernard, 97 Rue Claude Bernard, 57070 Metz France

TRAPANI, CATHERINE, special education and educational psychology educator; b. Bklyn., Nov. 23, 1952; d. Louis J. and Bernice (Blind) T. BA, Hofstra U., 1974, MS, 1975; PhD, U. Wis., 1986; postdoctoral student, Johns Hopkins U., 1987. Asst. to the dean New Coll. Hofstra U., Hempstead, N.Y., 1977-81; teaching asst. U. Wis., Madison, 1982-86; asst. prof. Ill. Benedictine Coll., Lisle, 1987-89, U. Chgo., 1989—; assoc. prof. clin. psychology, dir. Acad. and Social Skills Evaluation and Tutoring Svc., Chgo.; bd. dirs. Sonia Shankman Orthogenic Sch. Author: Transition Goals for Adolescents with Learning Disabilities, 1990—. Fed. Govt. Spl. Edn. fellow, 1975, E.B. Fred U. Wis. fellow, 1983-84. Mem. Am. Assn. Mental

Deficiency, Am. Edn. Rsch. Assn., Assn. for Learning Disabilities, Coun. for Learning Disabilities, Coun. for Exceptional Children, Ill. Orton Dyslexia Soc. (bd. dirs.), Orton Dyslexia Soc., Tourettes Syndrome Assn., Johns Hopkins Med. and Surg. Assn. Office: U Chgo 5841 S Maryland Ave Chicago IL 60637-1470

TRASATTI, SERGIO, chemistry educator; b. Fermo, AP, Italy, Mar. 13, 1937; s. Giovanni and Marcella (Conti) T.; m. Maria Paola Pacini, Dec. 2, 1961; children: Massimo, Stefano. D in Indsl. Chemistry, U. Milan, Italy, 1961. Lectr. U. Milan, 1962-66, from asst. prof. to prof., 1966-80, prof., 1980—. Author and editor: Electrodes of Conductive Metallic Oxides (2 vols.) 1980-81; contbr. 200 articles to sci. jours., 8 chpts. to many authors books. Recipient Miolati prize Assn. italiana di Chimicie Fisica, Padova, Italy 1975, Pergamon Gold medal, 1993; named B.A. Battista Disting. Lectr. Clarkson U., Potsdam, N.Y. Mem. Internat. Soc. Electrochemistry (pres. 1989-90), Polish Chem. Soc., Internat. Union of Pure and Applied Chemistry (chmn. eletro chem. commn. 1984-85). Office: U Milan Dept Phys Chemistry & Electrochemistry, Via Venezian 21, 20133 Milan Italy

TRAUB, PETER, biochemist; b. Stuttgart, Germany, June 27, 1935; came to U.S., 1965; m. Ulrike Elisabeth Drechsler, Dec. 20, 1963. PhD, U. Munich, 1963. Rsch. assoc. Max Planck Inst. Biochemistry, Munich, Germany, 1963-65; dir. Max Planck Inst. for Cell Biology, Ladenburg, Heidelberg, Germany, 1970—; rsch. assoc. dept. genetics U. Wis., Madison, 1965-68; rsch. assoc. dept. biology U. Calif., San Diego, 1968-70. Author: Intermediate Filaments, A Review, 1985. Home: Erlbrunnenweg 3, D-69259 Wilhelmsfeld Germany Office: Max-Planck Inst Cell Biol, Rosenhof, 68526 Ladenburg Germany

TRAVEN, KEVIN CHARLES, structural engineer; b. Lincoln Park, Mich., Mar. 4, 1959; s. Gary Lynn and Barbara Ann (Toporek) T.; m. Michelle Figueroa, Mar. 7, 1992. BSCE, Mich. Inst. Tech., 1983. Registered profl. engr., Fla. Civil engr. Misner Marine Constrn., Tampa, Fla., 1983-85, Wattman & Assocs., Orlando, Fla., 1985-87, Boyer Singleton Assocs., Orlando, 1987-88; structural engr. Hi-Tec Assocs., Orlando, 1988-90, Badger Design & Constructors, Tampa, 1990-91, Bechtel Savannah River, Inc., North Augusta, S.C., 1991—. Republican. Roman Catholic. Home: 2701 Ridgecrest Dr Augusta GA 30907-4816 Office: BSRI 601 Broad St Augusta GA 30901

TRAWICK, LAFAYETTE JAMES, JR., research scientist; b. Newark, Sept. 3, 1965; s. Lafayette James and Rose Eva (Smith) T. Cert., Univ. Medicine and Dentistry N.J., 1987; BS, Oakwood Coll., 1988. Rsch. teaching specialist Univ. Medicine and Dentistry N.J., Newark, 1989-91; microbiologist Pharmacia Diagnostics, Piscataway, N.J., 1991; assoc. scientist Ortho Diagnostic Systems, Raritan, N.J., 1991-93, Schering Plough Rsch. Inst., Kenilworth, N.J., 1993—; rschr. summer rsch. program Univ. Medicine and Dentistry N.J., 1987. Coord. Health Temperance Fair, Newark. Mem. Am. Mus. Natural History (assoc.), Liberty Sci. Ctr. (charter, vol. 1993—). Adventist. Home: 715 Ste D Elizabeth NJ 07208

TREADWELL, KENNETH MYRON, mechanical engineer, consultant; b. Cleve., May 5, 1923; s. Herbert Eugene and Flora Mae Belle (Robinson) T.; m. Sally Ann Skeel, Aug. 8, 1951; 1 child, Karen Ann. BS, U.S. Naval Acad., 1948, BSME, 1954; MSME, MIT, 1955. Advanced through grades to lt. USN, 1942-54; sr. engr. Westinghouse (Bettis Lab.), West Mifflin, Pa., 1955-59, mgr. reactor design, 1959-64, mgr. reactor analysis, 1964-68, mgr. reactor engring., 1968-72, mgr. fuel element devel., 1972-77, mgr. fuel element devel. and statistics, 1977-82; pres. Treadwell Cons., Pitts., 1983—; sr. cons. O'Donnell Assocs., Pitts., 1986—. Inventor reactor safety system, 1975. Mem. Whitehall Pa. Zoning Hearing Bd., Pitts., 1985-90. Mem. ASME, South Hills Country Club, Sigma Xi. Avocations: genealogy, rose culture, golf. Home and Office: 4983 Parkvue Dr Pittsburgh PA 15236-2053

TREBLE, FREDERICK CHRISTOPHER, electrical engineer, consultant; b. Aldershot, Hampshire, U.K., Feb. 20, 1916; s. Michael Frederick and Edith Ellen (Drysdale) T.; m. Elnora Jones, Mar. 9, 1940 (dec. Feb. 1991); children: Paul, Anne, John, Frances. BSc with honours, U. London, 1937. Chartered engr. Sci. officer Ministry of Aircraft Prodn., London, 1940-46; sr. sci. officer Ministry of Supply, London, 1946-57; sr. sci. officer Royal Aircraft Establishment, Farnborough, Eng., 1957-63, prin. sci. officer, 1963-77; consulting engr. photovoltaic solar energy Farnborough, 1977—; cons. PV stds. com. Internat. Electrotech. Commn., Geneva, 1983-90; sec. U.K. sect. Internat. Solar Energy Soc., 1980-90. Editor, contbg. author: Generating Electricity from the Sun, 1991; contbg. author: Energy-Present & Future Options, 1984; contbr. articles to conf. procs. Mem. Inst. Elec. Engrs., Solar Energy Soc. Achievements include early studies of effects of radiation damage in space solar cells; developed photovoltaic performance measurement techniques; developed large, lightweight deployable solar arrays for spacecraft. Home and Office: 43 Pierrefondes Ave, Farnborough GU14 8PA, England

TREE, DAVID ALAN, chemical engineering educator; b. Provo, Utah, Nov. 21, 1959; s. David Rees and Roberta (Johnson) T.; m. Donna Rae Jackson, Aug. 20, 1982; children: David, Steven, Brian. BS, Brigham Young U., 1984; MS, U. Ill., 1987, PhD, 1990. Rsch. assoc. U. Kassel, Germany, 1989-90; asst. prof. Okla. State U., Stillwater, 1990—. Contbr. articles to profl. jours. Com. chmn. Troop 822, Boy Scouts Am., Stillwater, 1990-92; high councilor local ch., Stillwater, 1992—. Rsch. grantee Okla. State U., 1990, Univ. Ctr. for Energy, Stillwater, 1991, 92; teaching grantee Ctr. for Effective Instruction, Stillwater, 1992. Mem. Am. Inst. Chem. Engrs. (sec., treas. Okla. section 1992), Soc. Rheology, Polymer Processing Soc. Achievements include quantitative measurement of flow-induced crystallization of high polymers. Office: Okla State Univ 423 Engring North Stillwater OK 74078

TREFETHEN, LLOYD MACGREGOR, engineering educator; b. Boston, Mar. 5, 1919; s. Francis Lord and Harriet (Nichols) T.; m. Florence Marion Newman, 1944; children—Gwyned, Lloyd Nicholas. B.S., Webb Inst. Naval Architecture, 1940; M.S., MIT, 1942; Ph.D., Cambridge U., Eng., 1950. Engr. Gen. Electric Co., Schenectady, 1940, 42-44; sci. cons. Office Naval Research, Am. Embassy, London, Eng., 1947-50; tech. aide to chief scientist Office Naval Research; exec. sec. Naval Research Adv. Com., Washington, 1951, Nat. Sci. Bd., 1951-54; tech. aide to dir. NSF, Washington, 1951-54; asst. prof. mech. engring. Harvard U., Cambridge, Mass., 1954-58; prof. mech. engring. Tufts U., Medford, Mass., 1958—, chmn. dept. mech. engring., 1958-69, 81; cons. Gen. Electric Co., 1955-88; NSF fellow Cambridge U., Eng., 1956; vis. prof. U. Sydney, Australia, 1964-65, 72, 79; vis. fellow Battelle Seattle Research Ctr., 1971; sr. vis. scientist Commonwealth Sci. and Indsl. Research Orgn., Melbourne, Australia, 1972; hon. research assoc. Harvard U., 1978; vis. prof. Stanford U., 1979; Russell Springer Prof. Mech. Engring., U. Calif., Berkeley, 1986. Served with USN, 1944-46. Recipient Golden Eagle award CINE Soc., 1967; honors Am. Film Festival, 1965, 66, Le Prix de Physique, 5th Internat. Sci. Film Festival, Lyon, 1968 (for films on surface tension phenomena). Fellow AAAS, ASME. Office: Tufts U Dept Mech Engring Medford MA 02155

TREFTS, ALBERT S., mechanical engineer, consultant; b. Cleve., July 26, 1929; s. George M. and Dorothy (Sharpe) T.; m. Joan Landenberger, June 20, 1952; children: Dorothy E., Albert S. Jr., William G., Deborah C., Elizabeth. BSME, Cornell U., 1953. Cert. engineer, N.Y., Ohio. Test eng. Morrison Steel Prod., Inc., Buffalo, 1955-56; sales eng. Worthington Corp., Buffalo, 1956-62; mech. eng. Linde Div. Union Carbide Corp., Tonawanda, N.Y., 1962-64, Davey McKee Corp., Cleve., 1964-82; cons. engr. various firms, Cleve., 1982—. Lt. USNR, 1953-55. Mem. Cleve. Eng. Soc., Cleve. Skating Club, Cleve. Playhouse Club, Western Res. Soc. (pres. 1989), Founders & Patriots, Sons & Daus. of Pilgrims (pres. 1986-88), New Eng. Soc. of Cleve. (pres. 1986, 88), Mayflower Soc., Barons of Magna Carta, Foudner of Hartford (Conn.), Founders of Newbury (Mass.), New Eng. Soc. of the Western Res., Cornel Club N.Y.C., First Families of Mass., Clearwater (Fla.) Country Club. Republican. Presbyterian. Avocations: genealogy, curling, golf, tennis. Home: PO Box 200527 Cleveland OH 44120-9527 also: 21101 Malvern Rd Shaker Heights OH 44120 Office: MK Ferguson 1 Erieview Plz Cleveland OH 44114-1715

TREICHLER, RAY, agricultural chemist; b. Rock Island, Ill., Sept. 10, 1907; s. Wallace and Pearl (Cushman) T.; B.S., M.S., Pa. State U., 1929; Ph.D., U. Ill., 1939; m. Kathryn Amelia Blakeley, June 13, 1942. Asst. state chemist Tex. Agrl. Expt. Sta., Tex. A&M Coll., College Station, 1929-40; chief, chemistry and biochemistry research Fish & Wildlife Service Labs., U.S. Dept. Interior, Laurel, Md., 1941-44; chief, biol. activities Office of Quartermaster Gen., U.S. Army, Washington, 1945-53; chief, toxic agents br. Rand D. Command, Army Chem. Center, Md., 1953-56, asst. to dir. med. research Chem. Warfare Labs., 1956-58; research adminstr. USAF, Bolling Field, Washington, 1958-68; tech. services mgr. H.D. Hudson Mfg. Co., Washington, 1968—. Fellow N.Y. Acad. Scis.; mem. Am. Chem. Soc., Entomol. Soc. Am., Am. Soc. Tropical Medicine and Hygiene, Am. Mosquito Assn., Am. Soc. Agrl. Engrs., ASTM, Sigma Xi, Gamma Sigma Delta. Club: Masons. Developed pesticide application equipment, prevention deterioration, chemistry and formulations pesticides, pesticide dissemination systems. Contbr. articles on vitamins, basal energy and endogenous nitrogen metabolism, nutrition, composition fishery products, toxic compounds, prevention material deterioration. Home: Apt 402 4740 Connecticut Ave NW Washington DC 20008-5632 Office: HD Hudson Mfg Co 1130 17th St NW Ste 500 Washington DC 20036-4604

TREIMAN, DAVID MURRAY, neurology educator; b. St. Louis, Oct. 13, 1940; s. Alfred Abraham and Dorothea Bader (Collins) T.; m. Lucy Ellen Jones, Apr. 9, 1967; children: Stephen Brant, Michael Andrew, Matthew Laurence, Daniel Robert. BA in Zoology, U. Calif., Berkeley, 1962; MD, Stanford U., 1967. Diplomate Am. Bd. Psychiatry and Neurology. Intern Duke U. Med. Ctr., Durham, N.C., 1967-68; resident in internal medicine, postdoctoral fellow Duke U., Durham, N.C., 1967-70, resident in neurology svc., postdoctoral fellow, 1970-73; chief neurology U.S. Navy Med. Ctr., Jacksonville, Fla., 1973-75, Yokosuka, Japan, 1975-78; from asst. to assoc. prof. neurology UCLA Sch. Med., 1978-90, prof. neurology, 1990—. Editor: Status Epilepticus, 1983, Neurobehavioral Problems in Epilepsy, 1991; contbr. articles to New England Jour. Medicine, Neurology, Epilepsy Rsch. Comdr. U.S. Navy, 1973-78. Grantee VA, NIH. Fellow Am. Acad. Neurology; mem. Am. Epilepsy Soc., Am. Neurol. Assn., Neurosci. Soc. Achievements include research in clinical and electrical characteristics of generalized convulsive status epilepticus. Office: UCLA School of Medicine Dept Neurology RNRC 710 Westwood Plz Los Angeles CA 90077

TREIMAN, SAM BARD, physics educator; b. Chgo., May 27, 1925; s. Abraham and Sarah (Bard) T.; m. Joan Little, Dec. 27, 1952; children—Rebecca, Katherine, Thomas. Student, Northwestern U., 1942-44; S.B., U. Chgo., 1948, S.M., 1949, Ph.D., 1952. Mem. faculty Princeton U., 1952—, instr., 1952-54, asst. prof., 1954-58, assoc. prof., 1958-63, prof. physics, 1963—, Eugene Higgins prof., 1976—, chmn. dept., 1981-87, chmn. univ. rsch. bd., 1988—. Author: (with M. Grossjean) Formal Scattering Theory, 1960, (with R. Jackiw and D.J. Gross) Current Algebra and Its Applications, 1972; Contbr. articles to profl. jours. Served with USNR, 1944-46. Recipient Oersted medal Am. Assn. Physics Tchrs. Mem. NAS, Am. Phys. Soc., Am. Acad. Arts and Scis. Home: 60 Mccosh Cir Princeton NJ 08540-5627

TREJO, LEONARD JOSEPH, psychologist; b. Mexico City, Feb. 24, 1955; came to U.S., 1972; s. Carlos and Margaret (Ryan) T.; m. Robin Lynn Boll, Aug. 7, 1976; children: Jonathan, Melina. BS in Psychology, U. Oreg., 1977; postgrad., U. Mich., 1977-78; MA in Psychology, U. Calif., San Diego, 1980, PhD in Psychology, 1982. Rsch. asst. psychobiology area dept. psychology U. Mich., Ann Arbor, 1977-78; rsch. and teaching asst. dept. psychology U. Calif., San Diego, 1978-82; sr. fellow depts. ophthalmology and biol. structure U. Wash., Seattle, 1982-84; instr. dept. psychology Point Loma Nazarene Coll., San Diego, 1988-91; rsch. psychologist Navy Pers. Rsch. and Devel. Ctr., San Diego, 1984—; lectr. dept. psychology U. Wash., Seattle, 1983. Contbr. articles to profl. jours., chpt. to book. Recipient NIH/Nat. Eye Inst. Nat. Rsch. Svc. award, 1982-84. Mem. AAAS, Am. Psychol. Soc. Achievements include development of the field of biopsychometric assessment, the application of psychophysiol. methods such as EEG, magnetoencephalography, and event-related potentials to human performance assessment; discovery of neural pathways which control the pupil's light reflex from eye to brain. Office: Navy Pers Rsch & Devel Ctr 53335 Ryne Rd San Diego CA 92152-7250

TREMAINE, SCOTT DUNCAN, astrophysicist; b. Toronto, Ont., Can., May 25, 1950; s. Vincent Joseph and Beatrice Delphine (Sharp) T. BSc, McMaster U., Hamilton, Ont., 1971; PhD, Princeton U., 1975. Postdoctoral fellow Calif. Inst. Tech., Pasadena, 1975-77; rsch. assoc. Inst. Astronomy, Cambridge, Eng., 1977-78; long-term mem. Inst. for Advanced Study, Princeton, N.J., 1978-81; assoc. prof. MIT, Cambridge, 1981-85; prof., dir. Canadian Inst. for Theoretical Astrophysics U. Toronto, 1985—. Author: Galactic Dynamics, 1987; contbr. articles to profl. jours. E.W.R. Steacie fellow Natural Scis. and Engring. Rsch. Coun., 1988; recipient H.B. Warner prize Am. Astron. Soc., 1983, Steacie prize, 1989, C.S. Beals award Canadian Astron. Soc., 1990, Rutherford medal Royal Soc. Can., 1990. Mem. Am. Acad. Arts and Scis. (fgn. hon.). Office: U Toronto-CITA McLennan Lab, 60 St George St, Toronto, ON Canada M5S 1A7

TRENARY, MICHAEL, chemistry educator; b. Los Angeles, Calif., July 8, 1956; s. Bernard Elroy and Jean Ann (Morris) T.; m. Wendy Greenhouse, June 10, 1984; children: Eleanor Jane, Russell Jack. BS, U. Calif., Berkeley, 1978; PhD, Mass. Inst. Tech., 1982. rsch. asst. prof. U. Pittsburgh, Pittsburgh, Penn., 1982-84; asst. prof. U. Ill., Chicago, IL, 1984-89, assc. prof., 1989-92; prof., 1992—. Contrib. articles to Jour. of Chem. Physics, Jour. Electron Spectroscopy, Surface Science, Chem. Phys. Letts. Recipient Dreyfus Tchr.-Scholar award Henry and Camille Dreyfus Found., 1989, U.Ill. Scholar award U. Ill. Found., 1990. Member Am. Chem. Soc., Am. Vacuum Soc., Am. Phys. Soc. Office: U Ill Dept Chemistry M/C111 Chicago IL 60680-4348

TRENCH, WILLIAM FREDERICK, mathematics educator; b. Trenton, N.J., July 31, 1931; s. George Daniel and Anna Elizabeth (Taylor) T.; m. Lucille Ann Marasco, Dec. 26, 1954 (div. Dec. 1978); children: Joseph William, Randolph Clifford, John Frederick, Gina Margaret; m. Beverly Joan Busenshut, Nov. 22, 1980. B.A. in Math., Lehigh U., 1953; A.M., U. Pa., 1955, Ph.D., 1958. Applied mathematician Moore Sch. Elec. Engring., U. Pa., 1953-56; with Gen. Electric Corp., Phila., 1956-57, Philco Corp., Phila., 1957-59, RCA, Moorestown, N.J., 1957-64; assoc. prof. math. Drexel U., Phila., 1964-67; prof. Drexel U., 1967-86; Andrew G. Cowles disting. prof. math. Trinity U., San Antonio, 1986—. Contbr. research articles in numerical analysis, ordinary differential equations, smoothing, prediction and spl. functions to profl. jours.; author: (with Bernard Kolman) Elementary Multivariable Calculus, 1971, Multivariable Calculus with Linear Algebra and Series, 1972; Advanced Calculus, 1978. Mem. Math. Assn. Am., Am. Math. Soc., Soc. Indsl. and Applied Math., Phi Beta Kappa, Eta Kappa Nu, Pi Mu Epsilon. Achievements include development of Trench's Algorithm for inversion of finite Toeplitz matrices, of fast algorithms for computing eigenvalues of structured matrices, of asymptotic theory of solutions of nonlinear functional differential equations under mild integral smallness conditions. Home: 211 W Rosewood Ave San Antonio TX 78212-2332 also: 413 Lake Drive W Divide CO 80814 Office: Dept Math Trinity U 215 Stadium Dr San Antonio TX 78212 also: 413 Lake Dr W Divide CO 80814

TRENCHER, GARY JOSEPH, nuclear engineer; b. Newark, Jan. 16, 1964; s. Enoch and Shirley Gwendolyn (Cook) T. BS in Engring. cum laude, N.J. Inst. Tech., 1990. Asst. rschr. Physics Dept., Rutgers U., Newark, 1989; assoc. engr. Pub. Svc. Elec. and Gas Co., Hancocks Bridge, N.J., 1990—. Conbr. articles to profl. jours. Mem. YJAD-Part of the Jewish Fedn. of Del. Community Involvement, Wilmington, 1992. Scholarship Inst. of Nuclear Power Ops., 1989-90. Mem. IEEE, Am. Soc. of Physics Students (pres. 1989-91). Achievements include 5 patents pending for processes on electrical and thermal mechanical devices. Home: 8 Sandalwood Dr #12 Newark DE 19713 Office: Nuclear Fuel/NDAB Endot Buttonwood Rd Hancocks Bridge NJ 08038

TRESCOTT, SARA LOU, water resources engineer; b. Frederick, Md., Nov. 17, 1954; d. Norton James and Mabel Elizabeth (Hall) T.; m. R. Jeffrey Franklin, Oct. 8, 1983. AA, Catonsville C.C., Balt., 1974; BA in Biol. Sci., U. Md., Balt., 1980. Sanitarian Md. Dept. Health & Mental Hygiene, Greenbelt, 1982; indsl. hygienist Md. Dept. Licensing & Regualtion, Balt., 1982-85; from water resources engr. to chief dredging div. Md. Dept. Natural Resources, Annapolis, 1985-92; chief navigation div. Md. Dept. Natural Resources, Stevensville, 1992—; chair adv. bd. EEO, Annapolis, 1990-92; tech. com. Nat. Mgmt. Info. Systems, Balt., 1983. Mem. ASCE, County Engrs. Assn. Md. Democrat. Achievements include research in beneficial uses of dredged material; development of technology for hydrographic surveying, providing Md. with an improved waterway transportation network. Home: PO Box 22 Woodbine MD 21797 Office: DNR Navigation div 305 Marine Academy Dr Stevensville MD 21666

TRETTER, JAMES RAY, pharmaceutical company executive; b. Boone, Iowa, June 7, 1933; s. Raymond J. and Freda E. (Ohge) T.; m. Neltje Van Loon; 1 child, Elsa. BS in Chemistry, Loras Coll., 1956; PhD in Chemistry, U. Calif., Berkeley, 1960. Chemist Pfizer, Inc., Groton, Conn., 1960-72, dir. med. chem. dept., 1972-74, dir. chem. process rsch., 1975-77, exec. dir. devel. rsch., 1977-80; v.p. R&D William H. Rorer, Inc., Ft. Washington, Pa., 1980-90; pres. Cen. Rsch. div. Rhone-Poulenc Rorer, Paris, 1990-92; pres. CEO Ixsys Inc., San Diego, 1992—. Avocation: sailing. Office: Ixsys Inc 3550 General Abromics Ct Ste L103 San Diego CA 92121 Office: Rhone-Poulenc Rorer Cen Rsch 500 Arcola Rd Collegeville PA 19426-0107

TREVES, JEAN-FRANÇOIS, mathematician educator; b. Brussels, Apr. 23, 1930; married, 1962; 2 children. Lic., U. Paris-Sorbonne, 1953, Dr(math), 1958. Asst. prof. math. U. Calif., Berkeley, 1958-60; assoc. prof. math. Yeshiva U., 1961-64; prof. Purdue U., 1964-71; prof. math. Rutgers U., New Brunswick, N.J., 1971—; vis. prof. Univ. Paris-Sorbonne, 1965-67, U. Paris, 1974-75. Mem. mission Orgn. Am. States, Brazil, 1961. Sloan fellow, 1960-64, Guggenheim fellow, 1974-75. Mem. Am. Math. Soc. (Chauvenet prize, 1971, Leroy P. Steele prize, 1991), Math. Soc. France. Achievements include the study of partial differential equations and functional analysis. Office: Rutgers Univ Dept of Math New Brunswick NJ 08903*

TREVISAN, MAURIZIO, epidemiologist, researcher; b. Naples, Italy, Jan. 31, 1952; came to U.S., 1979; s. Ilario and Bianca (Bruni) T.; m. Lisa Monagle, Dec. 22, 1983; children: Simona, Alessia, Stefan. MD magna cum laude, U. Naples, Italy, 1977; MS, SUNY, Buffalo, 1989. Cert. in medicine and surgery, Italy, 1977, diabetes and metabolic disease, Italy, 1980. Resident dept. internal medicine Med. Sch. U. Naples, 1977-79; rsch. fellow dept. community health and preventive medicine Med. Sch. Northwestern U., 1979-82; co-prin. investigator, dir. Cellular Ion Transport Lab Project Gubbio, U. Naples, 1983-85; asst. prof. dept. social and preventive medicine SUNY, Buffalo, 1985-88, clinical asst. prof. dept. family medicine, 1988-89, assoc. prof. dept. social and preventive medicine, 1988-92, clinical assoc. prof. nutrition program, 1989—, assoc. prof. dept. family medicine, 1989—, interim chair dept social and preventive medicine, 1991-92, prof. and chmn. dept. social and preventive medicine, 1993—; prin. investigator Women's Health Initiative WNY Vanguard Clin. Ctr., 1993—; vis. physican dept. physiology, Harvard Med. Sch., 1982; cons. inst. internal medicine and metabolic disease rsch. U. Naples; adj. asst. prof. dept community health and preventive medicine, Northwestern U. Med. Sch., 1987—. Fellow Am. Heart Assn. Coun. on Epidemiology. Recipient Rsch. Career Devel. award NIH, 1989—. Fellow, Internat. Soc. and Federation Cardiology. Achievements include population-based epidemiological investigation of ion transport abnormalities as risk factors for essential hypertension. Office: SUNY Buffalo Dept Social & Preventative Medicine 270 Farber Hall Buffalo NY 14214

TREYBIG, JAMES G., computer company executive; b. Clarendon, TX, 1940. BS, Rice U., 1963; MBA, Stanford, 1968. Mkgt. mgr. Hewlett-Packard Co., 1968-72; with Kleiner and Perkins, 1972-74; with Tandem Computer Inc., Cupertino, Calif., 1974—, now pres., chief exec. officer, dir. Office: Tandem Computers Inc 19333 Vallco Pky Cupertino CA 95014-2506

TRIBBLE, ALAN CHARLES, physicist; b. Little Rock, Aug. 11, 1961; s. George Alan and Barbara Jean (Stocks) T.; m. Elizabeth Ellen Gunion, July 30, 1988; 1 child, Matthew Alan. BS, U. Ark., 1983; MS, U. Iowa, 1986, PhD, 1988. Physicist Rockwell Internat., Seal Beach, Calif., 1988—. Grad. Student Rschr. Program fellowship NASA, 1987. Mem. AIAA, Am. Geophys. Union, Am. Phys. Soc. Achievements include rsch. observations of the plasma wake of the shuttle orbiter; devel. course and text on space environment effects.

TRICE, WILLIAM HENRY, paper company executive; b. Geneva, N.Y., Apr. 4, 1933; s. Clyde H. T.; m. Sandra Clayton, July 16, 1955; children—Russell, Amy. B.S. in Forestry, State U. N.Y., 1955; M.S., Inst. Paper Chemistry, Appleton, Wis., 1960, Ph.D., 1963. With Union Camp Corp., 1963—, tech. dir. bleached div., 1972-74; v.p., corp. tech. dir. research and devel. Union Camp Corp., Wayne, N.J., 1974-79; sr. v.p. tech. Union Camp Corp., 1979-85, exec. v.p., 1985—. Trustee, pres. Western Mich. U.-Paper Tech. Found., Syracuse Pulp and Paper Found. With USAF, 1955-57. Fellow TAAPI (bd. dirs. 1978-81), Indsl. Rsch. Inst. (alt. rep.), Inst. Paper Sci. and Tech. (trustee, exec. commn. alumni assn.). Home: 6 Hanover Rd Mountain Lakes NJ 07046-1004 Office: Union Camp Corp 1600 Valley Rd Wayne NJ 07470-2043

TRICKEL, NEAL EDWARD, chemist, chemical researcher; b. Monroe, Wis., June 26, 1954; s. Gorham Edward and Lilliam June (Roth) T. BS, U. Wis., Platteville, 1976. Cert. med. technologist, specialist in clin. chemistry. Clin. chemist Freeport (Ill.) Meml. Hosp., 1975—, histopathology chemist, 1985—; pres. Strommen-Trickel Industries, Inc., Monroe, Wis., 1986—; cons. U.S. Treasury Dept. Bur. Alcohol, Tobacco and Firearms, Milw., 1990; lectr. Lions Club, Monroe, 1980, Kiwanis Club, Freeport, 1988, Rotary Club, Lions Club, Freeport, 1989, Am. Def. Prepardness Assn. Small Arms Symposium (Internat.), Las Vegas, Nev., 1989. Inventor: patented noncorrosive tracer unit, 1986. Recipient scholarship, Rural Rehab. Assn., Platteville, Wis., 1975. Mem. Am. Chemical Soc., Am. Soc. Clin. Chemists, The Costeau Soc., Am. Defense Preparedness Assn. (corp. mem.). Republican. Achievements include patent in noncorrosive tracer unit for small arms ammunition; research in safer loading processes for small arms tracer ammunition. Home: 1911 10th St Monroe WI 53566-1832 Office: Freeport Meml Hosp 1045 W Stephenson St Freeport IL 61032-4899

TRICKEY, SAMUEL BALDWIN, physics educator, researcher, university administrator; b. Detroit, Nov. 28, 1940; s. Samuel Miller and Betty Irene (Baldwin) T.; m. Lydia Hernandez Dec. 28, 1962 (div. June 1981); children: Matthew J., Phillip J.; m. Cynthia Karle, Aug. 13, 1983. BA in Physics, Rice U., 1962; MS, Tex. A&M U., 1966, PhD in Theoretical Physics, 1968. Rsch. scientist Mason & Hanger-Silas Mason Corp., 1962-64; asst. prof. physics U. Fla., Gainesville, 1968-73, assoc. prof., 1973-77, prof. physics and chemistry, 1979—, dir. J.C. Slater Meml. Computing Lab., 1981-93, dir. Computer and Communications Resources Coll. Liberal Arts and Scis., 1986-90, exec. dir. info. techs. and svcs. Office of Provost, 1991—; prof. physics, chmn. physics and engring. physics Tex. Tech U., Lubbock, 1977-79; cons. Redstone Arsenal Ala., 1972-76; vis. rsch. scholar Mich. Tech. U., 1982-92; vis. scientist IBM Rsch. Ctr., San Jose, Calif., 1975-76; assoc. or dep. dir. Sanibel Symposia; cons. T div. Los Alamos Nat. Lab., 1984—, vis. scientist Max Planck Inst. für Astrophysik, Munich, 1985—. Exec. v.p. U. Fla. chpt. United Faculty of Fla., 1981-83. Named Tchr. of Yr., Coll. Arts and Scis., U. Fla., 1973-74. Fellow Am. Phys. Soc.; mem. Am. Assn. Physics Tchrs., Nat. R.R. Hist. Soc., Gulf Atlantic Yacht Club, San Juan 21 Class Assn., S.W. R.R. Hist. Soc., Phi Kappa Phi, Sigma Xi, Sigma Pi Sigma. Democrat. Presbyterian. Contbr. articles to profl. jours. Home: 723 NW 19th St Gainesville FL 32603-1102 Office: Quantum Theory Project Williamson Hall U Fla Gainesville FL 32611

TRICOLI, JAMES VINCENT, cancer genetics educator; b. Buffalo, Aug. 29, 1953; s. Vincent Peter and Theresa Magdeline (Siuda) T.; m. Margaret Wimmer, Aug. 16, 1975; 1 child, Lucas. BA, Canisius Coll., 1975; MA, SUNY, Buffalo, 1979, PhD, 1982. Postdoctoral researcher Roswell Park Cancer Inst., Buffalo, 1982-86; instr. Harvard Med. Sch., Boston, 1986-87; asst. prof. U. Cin. Coll. of Medicine, 1987—; Contbr. articles to profl. jours. including Biochemistry, Nature, Exp. Cell Rsch., Cancer Rsch., Genes Chromosomes and Cancer. Am. Cancer Soc. grantee, 1989, NIH Biomed. Rsch. grantee, 1989, Kidney Found. grantee, 1990, NIH/Nat. Cancer Inst. grantee, 1992. Mem. Am. Soc. Human Genetics, Soc. Basic Urol. Rsch.,

Planetary Soc., Sigma Xi. Achievements include research in purification of DNA Topoisomerase I, of mapping of the insulin-like growth factor I and II genes, of the characterization of IGF-I and -II genes in colon carcinoma, characterization of Y chromosomal gene expression in prostate carcinoma/tumor suppressor gene involvement in prostate cancer. Office: U Cin Coll Medicine 231 Bethesda Ave Cincinnati OH 45217

TRIDENTE, GIUSEPPE, immunologist educator; b. Triggiano, Puglia, Italy, Apr. 16, 1939; s. Nicola and Vincenza (Campobasso) T.; m. Sara Dal Dosso; children: Gaia, Ginerva. MD, Umiv. Bari, Italy, 1963, postgrad. in pathology, 1966; postgrad. in hematology, U. Padua, Italy, 1968. Med. Diplomate specialist. Jr. researcher, fellow Lady Tata Meml. Fund., U. Bari, 1965-68, Italian Minister Pub. Instrn., U. Bari, 1966; sr. rsch. fellow Dutch Nat. Rsch. coun., Risswisk, The Netherlands, 1968-69; asst. in pathology lab. Civil Hosp. and Univ., Padua, 1969-70; sr. scientist, asst. prof. transplant immunology & pathology Univ. Edmonton, Alberta, Can., 1971; prof. without tenure immunopathology U. Verona, Italy, 1971-75; prof. immunopathology U. Verona, 1975-82, prof., dir. immunology, 1983-90; dir. Inst. Immunology and Infectious Diseases, Univ. Verona, 1991—. Author: Immunology and Immunpathology, 1979; editor: Jour. of Immunolgy Rsch., 1989; contbr. over 200 publs. to med. and sci. jours. and books. Mem. Bd. Adminstrn. Univ. Verona, 1986-87; city hall coun. City of Verona, 1990-91; acad. senate U. Verona, 1991—. Avocations: sailing, skiing, restoring old furniture. Office: Inst Immunology & Infectious Diseases, Policlinico B Roma, 337100 Verona Veneto, Italy

TRIEGEL, ELLY KIRSTEN, geologist; b. Danbury, Conn., Dec. 10, 1947; d. Erich Valerian and Erika (Kirsten) T. BA magna cum laude, Western Conn. State Coll., 1971; MA, Harvard U., 1974; MS, U. Houston, 1977; PhD, U. Tenn., 1984. Cert. profl. ground water scientist, cert. profl. soil scientist; registered profl. geologist, N.C., Del., Fla. Rsch. geochemist Dresser-Magcobar, Houston, 1975-77; rsch. assoc. Oak Ridge (Tenn.) Nat. Lab., 1977-81; sr. project geologist Woodward-Clyde Cons., Plymouth Meeting, Pa., 1981-84; pres. Triegel & Assoc., Inc., Berwyn, Pa., 1984—; founder environ. cons. firm. Contbr. articles to profl. jours. Mem. Am. Soc. Testing and Materials, Assn. Ground Water Scientists and Engrs. (bd. dirs. 1991—), Geol. Soc. Am., Nat. Water Well Assn., Soil Sci. Soc. Am. Achievements include development of standard practice to sample soils and wastes for volatile organics; rsch. includes topics on chemistry of submerged soils. Office: Triegel & Assocs Inc 1235 Westlakes Dr Ste 320 Berwyn PA 19312

TRIFTSHÄUSER, WERNER, physics educator; b. Selb, Bavaria, Germany, Mar. 22, 1938; s. Hans and Lina (Ziegler) T.; m. Eva-Lucia Hrncirik, Nov. 13, 1981; children: Germar, Caroline, Natalie. Diploma in physics, Tech. U., Munich, 1962, D. in Natural Scis., 1965. Rsch. assoc. U. N.C., Chapel Hill, 1967-68, Queen's U., Kingston, Can., 1968-70; dir. rsch. group Nuclear Rsch. Ctr., Jülich, Fed. Republic Germany, 1970-75; prof. U. der Bundeswehr, München, 1975—. Mem. Am. Phys. Soc., Deutscher Hochschulverband, Internat. Adv. Com. on Positron Annihilation, N.Y. Acad. Scis. Home: Ringelnatzweg 3, 85521 Ottobrunn Germany Office: U der Bundeswehr München, Werner-Heisenberg-Weg 39, 85577 Neubiberg Germany

TRIGG, CLIFTON THOMAS, information systems consultant; b. Newark, Jan. 28, 1932; s. Thomas Clifton and Grace Louise (Gowen) T.; m. Mildred Kauffmann, Dec. 28, 1953 (div. 1958); m. M. E. DeWinter, June 9, 1963; children: Cathleen, Linda, Thomas. BS, Manmouth Coll., 1964; MBA, Pace U., 1967; postgrad., NYU, 1970. Phys. scientist U.S. Army Signal Corps. Engring. Lab., Ft. Monmouth, N.J., 1956-66; ops. rsch. analyst U.S. Army Electronics Command, Ft. Monmouth, 1966-87; cons. Middletown, N.J., 1987—. Lt. USAF, 1952-55. Princeton fellow Princeton U., 1982-83. Mem. IEEE, ACM, N.Y. Acad. Sci., Armed Forces Communication Electronics Assn. Achievements include patent for electrochemical. Home: 34 Lenape Trail Middletown NJ 07748 Office: Columbia Rsch Corp 788 Shewsbury Ave Tinton Falls NJ 07724

TRIGGER, KENNETH JAMES, manufacturing engineering educator; b. Carsonville, Mich., Sept. 6, 1910; married, 1939; 3 children. BS, Mich. State U., 1933, MS, 1935, ME, 1943. Asst. Mich. State Coll., 1933-34, instr. mech. engring., 1935-36; instr. mech. engring. Swarthmore (Pa.) Coll., 1937-38, Lehigh (Pa.) U., 1938-39; assoc. U. Ill., Urbana, 1939-40, from asst. prof. to prof., 1940-77, prof. mechanical, industrial engring.; cons. nuclear divsn. Union Carbide Corp., Continental Can. Co., Aeroprojects Inc., and Atlantic Richfield Co. Fellow Soc. Mfg. Engrs. (medal 1959), Am. Soc. Mech. Engrs. (Blackall award 1957, William T. Ennor Mfg. Tech. award 1992), Am. Soc. Metals; mem. Am. Soc. Engring. Edn. Achievements include research in metal cutting and machinability, physical metallurgy, cutting termperatures and temperature distribution in cutting of metals and mechanism of tool wear. Office: U Illinois Dept Mech Engr Indust Engr 140 1206 W Green St Urbana IL 61801*

TRIGIANO, ROBERT NICHOLAS, biotechnologist, educator; b. Johnstown, Pa., Nov. 30, 1953; s. Lucien Lewis and Betty Jean (Rice) T.; m. Margaret Kay Treadwell, Apr. 2, 1983; 1 child, Andrew Nicholas. BS, Juniata Coll., 1975; MS, Pa. State U., 1977; PhD, N.C. State U., 1983. Rsch. assoc., agronomist Green Giant Co., LeSueur, Minn., 1977-79; mushroom specialist Rol-Land Farms, Ltd., Blenheim, Ont., Can., 1979-80; postdoctoral assoc. U. Tenn., Knoxville, 1984-86, asst. prof., 1987-91, assoc. prof. dept. horticulture, 1991—. Assoc. editor Am. Soc. Hort. Sci., Arlington, Va., 1990-92, Plant Cell, Tissue and Organ Culture, 1992—. Grantee USDA, 1990-93. Mem. Am. Hort. Soc. (program chair 1993), Mycological Soc. Am., Internat. Assn. Plant Morphologists, Sigma Xi, Gamma Sigma Delta. Achievements include study in somatic embryogenesis in grasses and ornamental woody and flower crops, ultrastructure of host-pathogen relationships, DNA fingerprinting and physiology of plant pathogens. Home: 10629 Eagles View Dr Knoxville TN 37922 Office: Univ Tenn Ornamental Horticulture Knoxville TN 37901-1071

TRIGOBOFF, DANIEL HOWARD, psychologist; b. Bklyn., Jan. 21, 1953; s. Philip and Eileen G. (Dubin) T.; m. Eileen Hazel Ruff, Mar. 30, 1985. BA, SUNY, 1974; MA, U. Iowa, 1978, PhD, 1980. Lic. clin. psychologist. Clin. psychologist VA, Buffalo, 1980-87; program dir. Buffalo Psychiat. Ctr., 1988-89; program coord. Buffalo Gen. Hosp., 1989-92; adj. prof. SUNY, Buffalo, 1984—; clin. psychologist pvt. practice Amherst, N.Y., 1982—; cons. Mid-Erie Mental Health Svcs., Buffalo, 1980-89. Mem. adv. bd. Langston Hughes Inst., Buffalo, 1986. Mem. Am. Psychol. Assn. Jewish.

TRILLING, LEON, aeronautical engineering educator; b. Bialystok, Poland, July 15, 1924; came to U.S., 1940, naturalized, 1946; s. Oswald and Regina (Zakhejm) T.; m. Edna Yuval, Feb. 17, 1946; children: Alex R., Roger S. B.S., Calif. Inst. Tech., 1944, M.S., 1946, Ph.D., 1948. Research fellow Calif. Inst. Tech., 1948- 50; Fulbright scholar U. Paris, 1950-51, vis. prof., 1963-64; mem. faculty MIT, Cambridge, 1951—; prof. aeros. and astronautics MIT, 1962—; mem. Program in Sci. Tech. and Society, Engring. Edn. Mission to Soviet Union, 1958; vis. prof. Delft Tech. U., 1974-75; vis. prof. engring. Carleton Coll., 1987;. Pres. Met. Com. Ednl. Opportunity, 1967-70, Council for Understanding of Tech. in Human Affairs, 1984—. Guggenheim fellow, 1963-64. Fellow AAAS. Home: 180 Beacon St Boston MA 02116-1455 Office: MIT 77 Massachusetts Ave Cambridge MA 02139

TRIMBLE, JOHN LEONARD, sensor psychophysicist, biomedical engineer; b. Detroit, Feb. 27, 1944. BS, U. Ill., 1968; PhD in Physiology and Bioengineering, U. Ill. Med. Ctr., 1972. Rsch. assoc., asst. prof. Eye Rsch. Labs., U. Chgo., 1972-79, engring. dir., 1979-83; dir. rehab. R&D ctr. Hines (Ill.) VA Hosp., 1983—; pres. IMT, Inc., 1984—; vis. assoc. biology Calif. Inst. Tech., 1977-79; cons. Nat. Ctr. Health Care Tech., NIH, 1980—, office productivity, tech. and innovation and ctr. utilization fed. tech. U.S. Dept. Commerce, 1984—; clin. assoc. prof. dept. orthopedic surgery Loyola Med. Sch., 1983—; adj. assoc. prof. Pritzker Inst. Med. Engring., 1984—. Mem. AAAS, Inst. Elec. and Electronics Engrs., Soc. Neuroscience, Assn. Rsch. Vision and Ophthalmology, Sigma Xi. Achievements include research on neural modeling, biological signal processing, intelligent systems. Office: Hines VA Hosp Rehab R&D Ctr PO Box 20 Hines IL 60141-0020*

TRIMMER, BRENDA KAY, pharmacist; b. Carlisle, Pa., May 26, 1955; d. Harold Gleim and Mary Martha (McKinney) T. BS in Sci., Morehead State U., 1977; BSc in Pharmacy cum laude, Phila. Coll. Pharmacy and Sci., 1981; postgrad., Northwestern U., 1990. Lic. pharmacist, Pa., Ga., Tenn., Ala. Grad. intern Thomas Jefferson U. Hosp., Phila., 1981; dir. pharmacy Geriatric Pharmacy Systems, Chambersburg, Pa., 1981-83, Harrisburg, Pa., 1983-85; itinerant pharmacist Geriatric Pharmacy Systems, Pottsville, Pa., 1986-88; nutritional support pharmacist, cons. Polyclinic Med. Ctr., Harrisburg, 1985—; dir. pharmacokinetics Instnl. Pharmacy Cons., Griffin, Ga., 1988-93; cons. pharmacist Phila. region Pharmacy Corp. Am., 1993—; chmn. Acad. Long Term Care, Pa., 1987-89. Mem. Am. Diabetes Assn., First Ch. of Brethren. Named Outstanding Young Women of Am., 1981. Fellow Am. Soc. Cons. Pharmacists; mem. Am. Pharm. Assn., Am. Assn. Diabetes Educators, Am. Soc. Parenteral and Enteral Nutrition, Ga. Pharm. Assn., Ga. Soc. Parenteral and Enteral Nutrition, Pa. Pharm. Assn. (exec. coun. 1987-89), Ga. Acad. Cons. Pharmacists. Democrat. Avocation: travel. Home: 960 Innsbruck Dr Hummelstown PA 17036 Office: Instnl Pharmacy Cons 816 Everee Inn Rd Griffin GA 30223-4714

TRIMPE, MICHAEL ANTHONY, forensic scientist; b. Cin., Jan. 20, 1958; s. William Richard and Claire Ann (Peaker) T.; m. Joanne Itala DelliCarpini, Mar. 29, 1980; children: Anthony, Sean, Alexander. BS in Forensic Sci., Eastern Ky. U., 1979. Criminalist Hamilton County Coroner's Lab., Cin., 1980—; pres. Specialized County Arson Team, Cin., 1986-87. Contbr. articles to profl. jours. Mem. bd. edn. St. Aloysious on the Ohio, Cin., 1992, mem. marriage ministry, 1985—; bd. dirs. Greater Cin. Regional Fire and Arson Investigators Sem., 1988—. Recipient award for reducing arson Internat. Assn. of Arson Investigators, 1986. Fellow Am. Acad. Forensic Scis.; mem. Midwestern Assn. of Forensic Scis. (asron peer group mem. cert. com. 1990—), Internat. assn. of Forensic Scis., Miami View Golf Club (bd. dirs. 1992—). Office: Hamilton County Coroners Lab 3159 Eden Ave Cincinnati OH 45219

TRINKS, HAUKE GERHARD, physicist, researcher; b. Berlin, Germany, Feb. 19, 1943; s. Walter and Inge (Liese) T.; m. Maj Hötgen, May 6, 1969; children: Kerstin, Heike, Ole. PbD in Physics, U. Bonn, 1963. Researcher German French Rsch. Inst., France, 1969-71; devel. engr. Devel. Ctr. for Weapons, Meppens, Germany, 1972-74; prof. measurement technique U. German Armed Forces, Hamburg, Germany, 1975-81; prof. elec. engring. Tech. U. Hamburg, 1982-89; prof. engring. U. Tromsö, Norway, 1990—; dean engring. faculty U. Armed Forces, Hamburg, 1978-80, U. Tromsö, 1990—; v.p. Tech. U. Hamburg, 1987-89. Contbr. articles to profl. jours. Lt. German Army Paratoopers, 1963-64. Fellow AIAA (assoc.). Achievements include 10 patents: new type of fuse; instrumentation for satellites, for liquid gauging and for envionmental pollution measurement. Home: Hasenkamp 13, D2110 Bucholz Germany Office: Tech Univ Hamburg, Hamburger Schlosstrasse 20, D2100 Hamburg Germany

TRIPLETT, KELLY B., chemist; b. Cin.. BA, Northwestern U., 1968; PhD in Chemistry, U. Mich., 1974. Rsch. assoc. Mich. State U., 1974-76; rsch. chemist Stauffer Chem. Co., 1976-78; supr. Akzo Chem. Inc., Dobbs Ferry, N.Y., 1978-82, tech. mgr., 1982-84, bus. mgr., 1984-87, program mgr., bus. mgr., 1987-90, mgr. rsch. ctr., 1990—. Mem. Am. Ceramic Soc., Industry Rsch. Inst. Achievements include development of new research and development methodologies; research in new polymerization products and process, organometallics and transition metal chemistries; investigation of chemical routes to advanced ceramics. Office: Akzo Chemicals Inc 300 S Riverside Plz Chicago IL 60606*

TRIPLETT, WILLIAM CARRYL, physician, researcher; b. St. Marys, W.Va., May 9, 1915; s. Harry Carryl and Glenna Olive (Dotson) T.; m. Jan Dinsmoore, June 11, 1940 (div. 1961); children: William C. II, Jan Frances; m. Josephine Vann (div.); children: Harriett, Amber, Charles; m. Kathleen Quigley. BA, W.Va. U., 1936; MD, U. Md., 1940. Med. dir. Camp Wood (Tex.) Convalescent Ctr.; pvt. practice Corpus Christi, Tex., 1946-68, 72-88; dir. rsch. TRIAD Assocs. Inc., 1947—, ENA, 1968-92, Intercontinental Cardiac Rsch., 1984-92. Inventor and patentee in field. Bayfront adv. com. Corpus Christi, 1950-55; bay drilling com., Corpus Christi, 1954-56; environ. com., Tex., 1968-72. Capt. USCG Aux., 1942-46. Named Man of Yr. Camp Wood/Nucces Canyon C. of C., 1990; 18 awards as editor Costal Bd. Medicine, Corpus Christi. Mem. AMA, Tex. Med. Assn., Corpus Christi Yacht Club (comdr. 1953). Anglican. Avocations: hunting, fishing, yachting, hydroplane racing. Office: TRIAD Assocs Inc PO Box 517 Camp Wood TX 78833-0517

TRIPP, HERBERT ALAN, systems engineer; b. Rumford, Maine, Jan. 2, 1948; s. Herbert Baxter and Buelah M. (Beach) T.; m. Dorothy L., Oct. 21, 1983; 1 child, Heather Ann. BSCE magna cum laude, N.H. Coll., 1980. Tech. instr. Centronics Computer, Hudson, N.H., 1974-77; product support engr. Raytheon Data Systems, Norwood, Mass., 1977; tech. instr. Digital Equipment Co., Salem, N.H., 1977-81; product support engr. Digital Equipment Co., Washington, 1981-85, tech. program mgr., 1985-89, tech. devel. mgr., 1989—. With U.S. Army, 1968-71. Recipient Gold Key award, N.H. Coll., 1980. Mem. Nat. Space Club. Republican. Office: Digital Equipment Co 8301 Profdl Pl Landover MD 20785

TRIPP, LEONARD LEE, software engineer; b. L.A., Oct. 21, 1941; s. Leonard Henry and Allie Marie (Haws) T.; m. Celia Frank, Nov. 28, 1963; children: Valerie, Monica, Allyson, Justin, Mary Esther, Lee. BS, Brigham Young U., 1965, MS, 1967. Sci. programmer Boeing Co., Seattle, 1967-75, computer scientist, 1975-85, sr. computer scientist, 1985-91, assoc. tech. fellow, 1991—. Contbr. to profl. publs. Commr. King County Water Dist. 124, Federal Way, Wash., 1977-82, Lakehaven Sewer Dist., Federal Way, 1979-82, Federal Way Water and Sewer Bd., 1983-84. Recipient Silver Beaver award Boy Scouts Am., 1984. Mem. IEEE (vice-chair computer soc. std. activity bd. 1989-93, exec. vice-chair IEEE software engring. stds. com. 1993—, chair U.S. tech. adv. group to Internat. Orgn. for Stds./Internat. Electrotech. Commn./Joint Tech. Com. 1/SC7 software engring.), Assn. Computing Machinery, Math. Assn. Am., Boy Scouts Am. Achievements include participation in development of standard taxonomy of software engineering standards. Home: 28632 8th Pl S Federal Way WA 98003 Office: Boeing Comml Airplane PO Box 3707 MS 6Y-07 Seattle WA 98124

TRISCARI, JOSEPH, clinical research director; b. Termini, Italy, Apr. 17, 1945; s. Sebastiano and Francesca (Schifano) T.; m. Maria DiFabio, Apr. 30, 1966; children: Joseph M., Craig A., Erika. BS, Cornell U., 1971; MS, Fairleigh Dickenson U., 1975; MPhil, Columbia U., 1978, PhD, 1980. Teaching asst. Cornell U., Ithaca, N.Y., 1968-71; rsch. investigator Hoffman-LaRoche, Nutley, N.J., 1971-89; assoc. dir. Bristol-Myers Squibb, Princeton, N.J., 1989—. Mem. Bloomfield (N.J.) Bd. Health, 1987—. Mem. Am. Inst. Human Nutrition, Am. Soc. Clin. Pharmacology and Clin. Therapeutics, Am. Heart Assn., Am. Diabetes Assn., N.Y. Acad. Scis., Am. Assn. Advancement Sci., N. Am. Assn. Study Obesity. Office: Bristol-Myers Squibb PO Box 4000 Princeton NJ 08543-4000

TRIVEDI, JAY SANJAY, chemist; b. Ahmedabad, Gujarat, India, Jan. 2, 1961; came to U.S., 1982; s. Jagdishchandra P. and Manjula Trivedi; m. Swati L. Trivedi, May 28, 1988. BSc, St. Xavier's Coll., 1981; MS, Tuskegee U., 1987. Dir. Ambic Labs., Ahmedabad, 1979-82; lab. mgr. Toxicology Lab., Tuskegee (Ala.) U., 1986-88; sr. chemist Metpath Lab., Teterboro, N.J., 1988-89; sr. scientist Compuchem Labs., Research Triangle Park, N.C., 1989-90; scientist Abbott Labs., Abbott Park, Ill., 1990—. Leader, organizer Great Himalayan Expedition, India, 1979. Mem. Am. Chem. Soc., Am. Assn. Pharm. Scientists, Sigma Xi. Achievements include patent pending for the use of vitamin E as a penetration enhancer; research in area of pharm. compounds, their properties in solution. Office: Abbott Labs 1 Abbott Rd Abbott Park IL 60064-3500

TRIVELPIECE, ALVIN WILLIAM, physicist, corporate executive; b. Stockton, Calif., Mar. 15, 1931; s. Alvin Stevens and Mae (Hughes) T.; m. Shirley Ann Ross, Mar. 23, 1953; children: Craig Evan, Steve Edward, Keith Eric. B.S., Calif. Poly. Coll., San Luis Obispo, 1953; M.S., Calif. Inst. Tech., 1955, Ph.D., 1958. Fulbright scholar Delft (Netherlands) U., 1958-59; asst. prof., then assoc. prof. U. Calif. at Berkeley, 1959-66; prof. physics U. Md., 1966-76; on leave as asst. dir. for research div. controlled thermonuclear research AEC, Washington, 1973-75; v.p. Maxwell Labs. Inc., San Diego, 1976-78; corp. v.p. Sci. Applications, Inc., La Jolla, Calif., 1978-81; dir. Office of Energy Research, U.S. Dept. Energy, Washington, 1981-87; exec. officer AAAS, Washington, 1987-88; dir. Oak Ridge (Tenn.) Nat. Lab., 1988—; v.p. Martin Marietta Energy Systems, 1988—; head del. for joint NAS and Soviet Acad. Scis. meeting and conf. in the USSR on energy and global ecol. problems, 1989; chmn. math. scis. ednl. bd. NAS, 1990-93; chmn. coord. coun. for edn. NRC, 1991-93; bd. dirs. Bausch & Lomb, Inc., Rochester, N.Y.; active State of Tenn. Sci. and Tech. Adv. Commn., 1993—. Author: Slow Wave Propagation in Plasma Wave Guides, 1966, Principles of Plasma Physics, 1973; also articles. Named Disting. Alumnus Calif. Poly. State U., 1978, Disting. Alumnus Calif. Inst. Tech., Pasadena, Calif., 1987; recipient U.S. Sec. of Energy's Gold medal for Disting. Service, 1986; Guggenheim fellow, 1966. Fellow AAAS, IEEE, Am. Phys. Soc.; mem. AAUP, Nat. Acad. Engring., Am. Nuclear Soc., Am. Assn. Physics Tchrs., Capital Hill Club, Nat. Press Club, Sigma Xi. Achievements include patents in field. Home: 8 Rivers Run Way Oak Ridge TN 37830-9004 Office: Oak Ridge Nat Lab Office of Dir PO Box 2008 Oak Ridge TN 37831-2008

TRKULA, DAVID, biophysics educator; b. Patton Twp., Pa., Aug. 19, 1927; s. Milos and Dorothy (Macut) T.; m. Eleanor June Uhrrecht, Nov. 20, 1954; children: David Allen, Todd Max, Carol Lynn. BS, U. Pitts., 1949, MS, 1955, PhD, 1959. Instr. U. Pitts., 1957-59; asst. biophysicist M.D. Anderson Hosp. and Tumor Inst., Houston, 1959-61; physicist U.S. Army Biol. Labs., Ft. Detrick, Md., 1961-68; asst. prof. Baylor Coll. Medicine, Houston, 1968-85; adj. instr. Houston C.C., 1985—. With U.S. Army, 1945-46. Grantee Welch Found., Baylor Coll. Medicine, 1974-77, NIH, 1982-85. Mem. Sigma Xi. Democrat. Eastern Orthodox. Achievements include research in resistance to mutagenesis, infectivity of DNA, microbiol. aerosol stability, thymidine kinase isozymes and inactivation of viruses. Home and Office: 10603 Del Monte Dr Houston TX 77042

TROCKI, LINDA KATHERINE, geoscientist, natural resource economist; b. Erie, Pa., Oct. 7, 1952; d. Bernard Joseph and Catherine Frances (Manczka) T. BS in Geology with highest honors, N.Mex. Inst. Mining and Tech., 1976; MS in Geochemistry, Pa. State U., 1983, PhD in Mineral Econs., 1985. Staff mem. Los Alamos (N.Mex.) Nat. Lab., 1976-78, 83-90; geologist Internat. Atomic Energy Agy., Vienna, Austria, 1978-80; grad. rsch. asst. Los Alamos (N.Mex.) Nat. Lab., 1981-83, dep. group leader, 1990-92, dep. program dir., 1992—; com. mem. Global Found., Coral Gables, Fla., 1988—; mem. Strategic Lab. Coun., U.S. Dept. Energy, 1992—; mem. Chief of Naval Ops. Task force on Energy, Alexandria, Va., 1990-91. Contbr. to profl. publs. Pres. Vista Encantada Neighborhood Assn., Santa Fe, 1988-89. Fellow East West Ctr., Honolulu, 1988. Mem. AAAS, Internat. Assn. Energy Economists, Mineral Econs. and Mgmt. Soc. (pres.-elect). Office: Los Alamos Nat Lab MS F641 PO Box 1663 Los Alamos NM 87545-0001

TROST, BARRY MARTIN, chemist, educator; b. Philadelphia, Pa., June 13, 1941; s. Joseph and Esther T.; m. Susan Paula Shapiro, Nov. 25, 1967; children: Aaron David, Carey Daniel. B.A. cum laude, U. Pa., 1962; Ph.D., MIT, 1965. Mem. faculty U. Wis., Madison, 1965—; prof., chemistry U. Wis., 1969—, Evan P. and Marion Helfaer prof. chemistry, from 1976, Vilas research prof. chemistry; prof. chemistry Stanford U., 1987—, Tamaki prof. humanities and scis., 1990; cons. Merck, Sharp & Dohme, E.I. duPont de Nemours.; Chem. Soc. centenary lectr., 1982. Author: Problems in Spectroscopy, 1967, Sulfur Ylides, 1975; editor-in-chief Comprehensive Organic Synthesis, 1991—; editor: Structure and Reactivity Concepts in Organic Chemistry series, 1972—; assoc. editor: Jour. Am. Chem. Soc, 1974-80; editorial bd.: Organic Reactions Series, 1971—; editor-in-chief Comprehensive Organic Synthesis, 1991; contbr. numerous articles to profl. jours. Recipient Dreyfus Found. Tchr.-Scholar award, 1970, 77, Creative work in synthetic organic chemistry award, 1981, Baekeland medal, 1981, Alexander von Humboldt award, 1984, Guenther award, 1990, Janssen prize, 1990; named Chem. Pioneer, Am. Inst. Chemists, 1983; NSF fellow, 1963-65, Sloan Found. fellow, 1967-69, Am. Swiss Found. fellow, 1975—; Cope scholar, 1989. Mem. Am. Chem. Soc. (award in pure chemistry 1977), Nat. Acad. Scis., Am. Acad. Arts and Scis., AAAS, Chem. Soc. London. Office: Stanford U Dept Chemistry Stanford CA 94305

TROTT, KEITH DENNIS, electrical engineer, researcher; b. Boston, Nov. 17, 1952; s. Theodore Thompson Jr. and Catherine Eloise (Smith); div. 1982; children: Kerri Anne, Jennifer Marie; m. Natalie Jean Durbin, Mar. 4, 1983; 1 child, Christopher Michael. AS in Math., Clinton C.C., 1975; BS in Math., SUNY, Plattsburgh, 1977; MSEE, Syracuse U., 1981; PhD in Elec. Engring., Ohio State U., 1986. Enlisted USAF, 1972, commd. 2d lt., 1978, advance through grades to maj., 1989; avionics technician USAF, Plattsburgh AFB, 1973-78; test and instrumentation engr. Rome (N.Y.) Air Devel. Ctr., 1979-81; comm. electronic engr. TRI-TAC Joint Test Element, Ft. Huachuca, Ariz., 1981-83; doctoral student of USAF sponsored program Ohio State U., Columbus, 1983-86; radar cross sect. rsch. engr. Applied Electromagnetics div., Rom Lab., Hanscom AFB, Mass., 1986-91; sr. seeker tech. rsch. engr. Armament directorate, Wright Lab., Eglin AFB, Fla., 1991—; adj. asst. prof. Air Force Inst. Tech., Wright-Patterson AFB, Ohio, 1990—, math. instr. Daniel Webster Coll., Nashua, N.H., 1988-91. Contbr. articles to profl. jours. Musician Emerald Coast Community Band and White Sands Dance Band, Ft. Walton Beach, Fla., 1991—; bd. dirs. Merrimack (N.H.) Community Concert Band and Jazz Ensemble, musician, 1990-91; musician Griffiss (N.Y.) Community Band, 1979-81, dir.; pres. Lee (N.Y.) Little League, 1979-81. Decorated Meritorious Svc. medal, 1991, Commendation medal, 1981; recipient USAF R & D award, 1992, 3 Sci. Achievement awards, 1989-91, Rsch. Team of Yr. award Rome Lab., 1991. Mem. IEEE (sr., reviewer 1986—), Applied Comp Electromagnetics Soc., Sigma Xi, Eta Kappa Nu. Roman Catholic. Achievements include rsch. in electromagnetic scattering specializing in radar cross section physics, in bistatic radar, in picosecond pulse scattering physics, and in electromagnetic characterization of materials.

TROTTI, LISA ONORATO, psychology educator; b. Nyack, N.Y., Dec. 19, 1960; d. Francis Michael and Bruna Rose (Lenzovich) Onorato; m. Bryan Trotti. BA, Gettysburg Coll., 1982; MA, N.Mex. State U., 1984, PhD, 1989. Teaching asst. N.Mex. State U., Las Cruces, 1982-83, rsch. asst., 1983-87; preprofl. IBM/Human Factors, Boca Raton, Fla., summer 1983; asst. prof. Psychology Hartwick Coll., Oneonta, N.Y., 1987—. Author (with others): Human Computer Interaction, 1984, Empirical Studies of Programmers, 1986, Pathfinder Associative Networks: Studies in Knowledge Organization, 1990. Recipient Rsch. fellowship Hartwick Coll., 1990, rsch. grant, 1989, 90, 91. Mem. APA, Ea. Psychol. Assn., Software Psychology Soc., Human Factors Soc., Phi Kappa Phi, Psi Chi. Democrat. Office: Hartwick Coll Oneonta NY 13820

TROUBETZKOY, EUGENE SERGE, physicist; b. Clamart, France, Apr. 7, 1931; s. Serge E. and Marina N. (Gagarine) T.; m. Helen S. Kotchoubey, June 8, 1958; children: Irina, Serge, Natalie. Degree in sci., U. Paris, 1953; PhD, Columbia U., 1958. Sci. advisor United Nuclear Corp, N.Y.C., 1958-73; sr. rsch. assoc. Columbia U., N.Y.C., 1971-72, Math. Applications Group, N.Y.C., 1971-86; cons. various nat. labs., U.S., 1985—; dir. advanced rsch. and devel. Conceptual Graphic Images, N.Y.C., 1987—. Contbr. articles to profl. jours. Mem. Am. Phys. Soc., Assn. Russian Am. Scholars. Russian Orthodox. Office: CGI 100 Executive Blvd Ossining NY 10562

TROUT, MONROE EUGENE, hospital systems executive; b. Harrisburg, Pa., Apr. 5, 1931; s. David Michael and Florence Margaret (Kashner) T.; m. Sandra Louise Lemke, June 11, 1960; children: Monroe Eugene, Timothy William. AB, U. Pa., 1953, MD, 1957; LLB, Dickinson Sch. of Law, 1964, JD, 1969. Intern Great Lakes (Ill.) Naval Hosp., 1957-58; resident in internal medicine Portsmouth (Va.) Naval Hosp., 1959-61; chief med. dept. Harrisburg State Hosp., 1961-64; dir. drug regulatory affairs Pfizer, Inc., N.Y.C., 1964-68; v.p., med. dir. Winthrop Labs., N.Y.C., 1968-70; med. dir. Sterling Drug, Inc., N.Y.C., 1970-74, v.p., dir. med. affairs, 1974-78, sr. v.p., dir. med. affairs, bd. dirs., mem. exec. com., 1978-86; pres., chief exec. officer Am. Healthcare Systems, Inc., 1986—, chmn., 1987—, also bd. dirs.; chmn. bd. dirs. Am. Excess Ins. Ltd.; adj. assoc. prof. Bklyn. Coll. Pharmacy; spl. lectr. legal medicine, trustee Dickinson Sch. Law; trustee Ariz. State U. Sch. Health Adminstrn., 1988-91; mem. Sterling Winthrop Rsch. Bd., 1977-86; sec. Commn. on Med. Malpractice, HEW, 1971-73, cons., 1974; mem. Joint Commn. Prescription Drug Use, 1976-80; co-chmn. San Diego County

Health Commn. Mem. editorial bd. Hosp. Formulary Mgmt, 1969-79, Forensic Science, 1971—, Jour. Legal Medicine, 1973-79, Reg. Tox. and Pharmac, 1981-87, Medical Malpractice Prevention, 1985—; contbr. articles to profl. jours. Exec. com. White House mini conf. on aging, 1980; Republican dist. leader, New Canaan, Conn., 1966-68; mem. Town Council New Canaan, 1978-86 , vice chmn., 1985-86; bd. dirs. New Canaan Interchurch Svc. com., 1965-69, Athletes Kidney Found., Circle in Sq. Theater Inc., 1984-86, Nat. Com. for Quality Health Care, 1988-90, Friends Nat. Libr. Medicine, Criticare, Inc., West Co., Inc., Cytran, Inc., Cytyc, Inc., Gensia Inc., UCSD Found.; trustee Dickinson Sch. of Law, Cleve. Clin., 1971-87, U. Calif.-San Diego Thornton Hosp. and Med. Ctr., Albany Med. Coll., 1977-86, St. Vincent DePaul. Ctr. for the Homeless, 1987-90; trustee, vice chmn. Morehouse Med. Sch., 1980-89; assoc. trustee U. Pa.; bd. visitors U. Pa. Sch. Nursing; vice-chmn. Med. Commn. for Food and Shelter, Inc.; mem. Nat. Health Advisory Bd. AAA, N.Y. State Commn. Substance Abuse, 1978-80; chmn. bd. ACLM Found., 1983-87; chmn. Internat. B'nai B'rith Dinner, 1989. Served to lt. comdr. USNR, 1956-61. Recipient Alumni award of merit U. Pa., 1953, Disting. Alumni award Dickinson Sch. Law, 1989, Nat. Healthcare award Internat. B'nai B'rith, 1991; named to Hon. Order Ky. Cols., Tenn. Cols. Fellow Am. Coll. Legal Medicine (v.p., pres., bd. govs.); mem. AMA (Physicians Recognition awards 1969, 72, 76, 82, 85, 88, 92), Med. Execs. (pres. 1975-76), Delta Tau Delta. Lutheran. Office: 12730 High Bluff Dr Ste 300 San Diego CA 92130-2099

TROUT, THOMAS JAMES, agricultural engineer; b. Bluffton, Ohio, Mar. 30, 1949; s. Gerald Gene and Jeannette (Stauffer) T.; m. Vickie Lynn Traxler, June 22, 1976; children: Brian Traxler, Adam Traxler. BS, Case Western Res. U., 1972; MS, Colo. State U., 1974, PhD, 1979. Registered profl. engr., Idaho, Colo. Rsch. asst. prof. dept. agrl. engring. Colo. State U., Ft. Collins, 1979-82; agrl. engr. USDA AGrl. Rsch. Svc., Kimberly, Idaho, 1982—; cons. World Bank, Washington, USAID, Washington, 1979-81. Contbr. chpts. to books, articles to profl. pubis. Vice-chmn. Twinn Falls (Idaho) Parks and Recreation Commn., 1989-92. Mem. ASCE, Am. Soc. Agrl. Engrs. (com. chair 1989-92), U.S. Commn. Irrigation and Drainage. Office: USDA Agrl Rsch Svc 3793 N 3600 E Kimberly ID 83341

TROUTT, TIMOTHY RAY, mechanical engineer, educator; b. Cherokee, Okla., Aug. 28, 1950; s. Ray and Evelyn Joyce (Hadwiger) T.; m. Elizabeth Ann Bastion, Aug. 28, 1980; children: Amy, Amanda. BS in Physics, Okla. State U., 1972, MS in Mech. Engring., 1974, PhD, 1978. Rsch. assoc. U. So. Calif., L.A., 1978-80; asst. prof. Wash. State U., Pullman, 1980-86, assoc. prof. mech. engring., 1986—. Contbr. articles to profl. jours., chpts. to books. Mem. AIAA, Am. Phys. Soc., Phi Kappa Phi. Achievements include research on role of organized vortex structures in turbulent flow dispersion processes. Office: Wash State Univ Dept Mech Engring Pullman WA 99164-2920

TROWBRIDGE, JOHN PARKS, physician, nutritional medicine specialist, joint treatment specialist; b. Dinuba, Calif., Mar. 24, 1947; s. John Parks Sr. and Claire Dovie (Noroian) T.; m. Sabrina Lynne Williams, Aug. 12, 1987; children: Sharla Tyann, Lyndi Kendyll. AB in Biol. Scis., Stanford U., 1970; MD, Case Western Res. U., 1976; postgrad., Fla. Inst. Tech., 1983-85. Diplomate in Preventive Medicine, Am. Bd. Chelation Therapy, Nat. Bd. Med. Examiners. Intern in gen. surgery Mt. Zion Hosp. & Med. Ctr., San Francisco, 1976-77; resident in urol. surgery U. Tex. Health Sci. Ctr., Houston, 1977-78; pvt. med. practice Pain Relief Ctr. for Arthritis and Sports Injuries (formerly Ctr. for Health Enhancement), Humble, Tex., 1978—; chief corp. med. cons. Tex. Internat. Airlines, Houston, 1981-83; indsl. med. cons. to several heavy and light mfg. and svc. cos., Houston, 1979-84; immunology research asst. Stanford U. Med. Ctr., Stanford, Calif., 1967-70; night lab. supr. Kaiser Found. Hosp., Redwood City, Calif., 1971-72; advisor to bd. dirs. Am. Inst. Med. Preventics, Laguna Hills, Calif., 1988-90; featured lectr. profl. and civic orgns., U.S., 1983—. Co-author: The Yeast Syndrome, 1986, Chelation Therapy, 1985, Yeast Related Illnesses, 1987, 2d edit., 1990; contbr. Challenging Orthodoxy: America's Top Medical Preventives Speak Out, 1991; contbr. articles to profl. jours. Adv. bd. mem. Tex. Chamber Orchestra, Houston, 1979-80; med. dir. Humble unit Am. Cancer Soc., 1980-81; med. cons. personal fitness program Lake Houston YMCA, 1981-83. Nat. Merit scholar, 1965-69, Calif. State scholar, 1967-69; recipient Resolution of Commendation house of dels., 1974 Am. Podiatry Assn., Spl. Profl. Svc. Citation bd. trustees, 1976, Am. Podiatry Students Assn. Fellow Am. Assn. Orthopaedic Medicine, Am. Soc. for Laser Medicine and Surgery, Am. Coll. Advancement in Medicine (v.p. 1987-89, pres. elect 1989-91), Am. Acad. Neurol. and Orthopaedic Medicine and Surgery; mem. AMA, Am. Coll. Preventive Medicine, Am. Preventive Med. Assn. (charter, bd. dirs.), Am. Acad. Environ. Medicine, Am. Soc. Gen. Laser Surgery, Am. Acad. Thermology, Am. Assn. Nutritional Cons., Am. Soc. Life Extension Physicians (founding), Am. Assn. Physicians and Surgeons, Tex. Med. Assn., Harris County Med. Soc., Houston Acad. Medicine, Aerospace Med. Assn., N.Y. Acad. Scis., Internat. Acad. Bariatric Medicine, Med. Acad. Rheumatoid Disease, Huxley Inst. for Biosocial Rsch., Great Lakes Assn. Clin. Medicine (med. rsch. instnl. rev. bd.), Delta Chi. Avocations: private piloting, computer applications. Office: Pain Relief Ctr Arthritis and Sports Injuries 9816 Memoral Blvd Ste 205 Humble TX 77338

TROY, MARLEEN ABBIE, environmental engineer; b. Rome, N.Y., Apr. 23, 1957; d. Ephraim F. and Geraldine (Tankel) T. MS in Microbiology, U. R.I., 1982; BS in Biol. Sci., Drexel U., 1980, MSCE, 1986, PhD in Civil and Environ. Engring., 1989. Engr. in tng. Co-op. Phila. Water Dept., 1976-78; teaching asst., rsch. asst. U. R.I., Kingston, 1980-83; rsch. asst. Drexel U., Phila., 1983-89; project engr. OHM Remediation Svcs. Corp., Trenton, N.J., 1989—; project engr. site for biol. land treatment of diesel fuel contaminated soil. Contbr. chpts. to books Contaminated Soils-Diesel Fuel Contamination, 1992, Bioremediation-Field Experience, 1993; contbr. to Internat. Jour. Environ. and Pollution. Nat. Network Environ. Studies fellow U.S. Environ. Protection Agy., 1988; Steven Giegerich Meml. scholar Drexel U., 1988. Mem. ASCE (rev. Jour. Environ. Engrin. 1991—), Am. Soc. Microbiology. Office: OHM Remediation Svcs Corp 200 Horizon Ctr Blvd Trenton NJ 08691-1904

TROYER, ALVAH FORREST, seed corn company executive, plant breeder; b. LaFontaine, Ind., May 30, 1929; s. Alvah Forrest and Lottie (Waggoner) T.; m. Joyce Ann Wigner, Sept. 22, 1950; children: Anne, Barbara, Catherine, Daniel. B.S., Purdue U., 1954; M.S., U. Ill., 1956; Ph.D., U. Minn., 1964. Research assoc. U. Ill., Urbana, 1955-56; research fellow U. Minn., St. Paul, 1956-58; research sta. mgr. Pioneer Hi-Bred Internat., Inc., Mankato, Minn., 1958-65, research coordinator, 1965-77; dir. research and devel. Pfizer Genetics, St. Louis, 1977-81, v.p. and dir. research and devel., 1981-82; v.p. research and devel. DEKALB Plant Genetics, Ill., 1982-93; cons.Hybrid Seed Divsn. Cargill, Mpls., 1993—; researcher corn breeding, econ. botany, crop physiology, increasing genetic diversity, recent corn evolution. Contbr. articles to numerous publs.; developer of popular corn inbred lines and hybrids. Master sgt. U.S. Army, 1951-53, Korea. Recipient NCCPB Plant Breeding and Genetics award. Fellow AAAS, Am. Soc. Agronomy, Crop Sci. Soc. Am.; mem. Am. Genetic Assn., Genetic Soc. Am., N.Y. Acad. Sci., CAST, Sigma Xi, Gamma Sigma Delta, Alpha Zeta, Lambda Chi Alpha, Gamma Alpha, VFW. Methodist. Lodge: Masons. Home: 611 Joanne Ln De Kalb IL 60115-1862

TROYER, DERYL LEE, life sciences educator; b. York, Nebr., May 29, 1947; s. Glenn Titus and Aldene Ethel (Reeb) T.; m Joyce Arlene Larson, May 21, 1972; children: Travis Carl, Darcy Lea. DVM, Kans. State U., 1972, PhD, 1985. Assoc. Arlington Heights (Ill.) Animal Hosp., 1972-73; ptnr. Pawnee City (Nebr.) Animal Hosp., 1973-80; asst. prof. life scis. U. Ill., Urbana, 1985-86; asst. prof. Kans. State U., Manhattan, 1986-90, assoc. prof., 1991—. Contbr. articles to profl. jours. Office: Kans State U 228 Vet Med Scis Bldg Manhattan KS 66506

TROZZOLO, ANTHONY MARION, chemistry educator; b. Chgo., Jan. 11, 1930; s. Pasquale and Francesca (Vercillo) T.; m. Doris C. Stoffregen, Oct. 8, 1955; children: Thomas, Susan, Patricia, Michael, Lisa, Laura. BS, Ill. Inst. Tech., 1950; MS, U. Chgo., 1957, PhD, 1960. Asst. chemist Chgo. Midway Labs., 1952-53; assoc. chemist Armour Rsch. Found., Chgo., 1953-56; mem. tech. staff Bell Labs., Murray Hill, N.J., 1959-75; Charles L. Huisking prof. chemistry U. Notre Dame, 1975—, P.C. Reilly lectr., 1972, Hesburgh Alumni lectr., 1986, Disting. lectr. sci., 1986; vis. prof. Columbia U., N.Y.C., 1971, U. Colo., 1981, Katholieke U. Leuven, Belgium, 1983, Max Planck Inst. für Strahlenchemie, Mülheim/Ruhr, Fed. Republic Germany, 1990; vis. lectr. Academia Sinica, 1984, 85; Phillips lectr. U. Okla., 1971; C.L. Brown lectr. Rutgers U., 1975; Sigma Xi lectr. Bowling Green U., 1976, Abbott Labs., 1978; M. Faraday lectr. No. Ill. U., 1976; F.O. Butler lectr. S.D. State U., 1978; Chevron lectr. U. Nev., Reno, 1983; plenary lectr. various internat. confs.; founder, chmn. Gordon Conf. on Organic Photochemistry, 1964; trustee Gordon Rsch. Confs., 1988-92; cons. various chem. cos. Assoc. editor Jour. Am. Chem. Soc., 1975-76; editor Chem. Revs., 1977-84; editorial adv. bd. Accounts of Chem. Rsch., 1977-85; cons. editor Encyclopedia of Science and Technology, 1982-92; contbr. articles to profl. jours.; patentee in field. Fellow AEC, 1951, NSF, 1957-59. Fellow N.Y. Acad. Scis. (chmn. chem. scis. sect. 1969-70, Halpern award in Photochemistry 1980), AAAS, Am. Inst. Chemists; mem. Am. Chem. Soc. (Disting. Svc. award St. Joseph Valley sect. 1979, Coronado lectr. 1980, 93), AAUP, Sigma Xi. Roman Catholic. Home: 1329 E Washington St South Bend IN 46617-3340 Office: U Notre Dame Notre Dame IN 46556

TRPIS, MILAN, entomologist, scientist, educator; b. Mojsova Lucka, Slovakia, Dec. 20, 1930; came to U.S., 1971, naturalized, 1977; s. Gaspar and Anna (Sevcikova) T.; m. Ludmila Tonkovic, Dec. 15, 1956; children: Martin, Peter, Katarina. M.S., Comenius U., Bratislava, 1956; Ph.D., Charles U., Prague, 1960. Research asst. Slovak Acad. Sci., Bratislava, 1953-56; sci. asst. Slovak Acad. Sci., 1956-60, scientist, 1960-62, ind. scientist, 1962-69; ecologist-entomologist Wast Africa-Aedes Rsch. Unit, WHO, Dar es Salaam, Tanzania, 1969-71; asst. faculty fellow dept. biology U. Notre Dame, 1971-73, assoc. faculty fellow, 1973-74; assoc. prof. med. entomology Johns Hopkins U. Sch. Hygiene and Pub. Health, 1974-78, prof., 1978—, dir. labs. med. entomology; rsch. assoc. U. Ill., Urbana, 1966-67, Can. Dept. Agr., Lethbridge, Alta., 1967-68; dir. Biol. Rsch. Inst. Am., 1971-79; external dir. rsch. Liberiran Inst. Biomed. Rsch., 1981—; dir. AID project on transmission of river blindness in areas of Liberia and Sierra Leone; dir. WHO rsch. grant; tech. adv. com. AID Vector Biology and Control Project, 1986-91; dir. Johns Hopkins U./Fed. U. Tech. Akur Onchocerciasis Project in Nigeria, 1991—, Johns Hopkins U./Organisation de Coordination et de Cooperation pour la Lutte les Grandes Endemies-Pierre Richet Inst. Onchocerciasis Project, Bouaké, Ivory Coast, 1993—; dir. Johns Hopkins U./Pierre Richet Inst./ORSTOM onchocerciasis project in Ivory Coast, 1993—; trainer doctoral students, Africa, Asia, Cen. Am., 1979—. Editor: Jour. Biologia, 1956-71, Jour. Entomol. Problems, 1960-72; zool. sect.: Jour. Biol. Works, 1960-71; Contbr. articles to profl. jours. Dir. WHO project on prophylactic drugs for river blindness, Liberia, 1985-87. Recipient Slovak Acad. Sci., 1st prize for research project. Mem. AAUP, AAAS, Am. Inst. Biol. Soci., Am. Mosquito Control Assn., Am. Soc. Parasitologists, Helminthol. Soc. Washington, Am. Soc. Tropical Medicine and Hygiene, Entomol. Soc. Am., Am. Genetic Assn., Soc. of Vector Ecology, N.Y. Acad. Scis., Johns Hopkins U. Tropical Medicine Club, Smithsonian Assocs., Royal Soc. Tropical Medicine and Hygiene, Royal Entomol. Soc. of London, Sigma Xi, Delta Omega (Alpha chpt.). Home: 1504 Ivy Hill Rd Cockeysville MD 21030-1418 Office: Johns Hopkins U 615 N Wolfe St Baltimore MD 21205-2103

TRUAX, DENNIS DALE, civil engineering educator, consultant; b. Hagerstown, Md., July 25, 1953; s. Bernard James and Dorothy Hilda T.; m. Jeanie Ann Knable, Aug. 20, 1977. B.S. in Civil Engring., Va. Poly. Inst. and State U., 1976; M.S., Miss. State U., 1978, Ph.D., 1986. Registered profl. engr., Miss; cert. environ. engr. Asst. dep. constrn. mgr. Fairfax County, Va., 1972-74; design engr., Washington County, Md., 1976; instr. Miss. State U., Starkville, 1980-86, asst. prof. civil engring., 1986-91, assoc. prof., 1991—; environ. engring. cons. Lay leader Aldersgate United Meth. Ch., Starkville, 1982-85, chmn. pastor/parish relations, 1985-86, chmn. council on ministries, 1986-90, chmn. adminstrv. bd., 1990-92, chmn. fin. com., 1992—; adviser Triangle Fraternity, Starkville, Alumni Bd. Dirs. treas., 1989-93; bd. dirs. Meth. Student Ctr., Miss. State U., 1983-90, chmn. pastor/parish relations, 1984-86, v.p. bd., 1986, pres., 1987-89, treas., 1990-91; del. to Ann. Conf., Miss. Conf. United Meth. Ch., also vice chmn. com. on higher end.; active Starkville dist. lay council. Named Outstanding Young Man Am., U.S. Jaycees, 1983. Mem. ASCE (chair career guidance com. 1991-92, sec. Miss. sect. 1990-91, pres.-elect 1990-91, pres. 1991-92, advisor student chpt 1981-93), Am. Water Works Assn., Miss. Engring. Soc. (pres., pres. elect, region 3 v.p., bd. dirs., Tombigbee chpt. pres., chpt. pres.-elect), Nat. Soc. Profl. Engrs., Water Environ. Fedn. (rsch. com.), Sigma Xi (sec., pres.-elect, pres. Miss. State chpt.), Tau Beta Pi, Chi Epsilon. Democrat. Contbr. engring. articles to profl. jours. Home: 1054 Southgate Dr Starkville MS 39759-9673 Office: Miss State U PO Drawer CE Mississippi State MS 39762-5610

TRUDEL, MICHEL, virologist; b. Montreal, Que., Can., Feb. 5, 1944; married, 1967; 2 children. BA, U. Montreal, 1965, BSc, 1968, MSc, 1970; PhD in Cell Biology, U. Sherbrooke, 1973. Fellow cell membranes Nat. Cancer Inst., 1973-74, asst. prof., 1974-75; prof. virology Inst. Armand-Frappier, Laval, Que., 1975—; also dir. ctr. virology; grants, formation rschr. Ministry of Edn., Que., 1976-78, 88-91, Health and Welfare, Can., 1976-78, Nat. Rsch. Coun. Can., 1978-81, 84-93, Can. Med. Rsch. Coun., 1979-91, conseil de recherches en peche et agroalimentaire, Que., 1985-89. Mem. Am. Soc. Virologists, Internat. Soc. Antiviral Rsch., Can. Assn. Clin. Microbiology Infectious Diseases, Can. Soc. Microbiologists. Achievements include research on viral subunit vaccines and the implication of the physical form of the viral proteins that induce the immune response, neutralization epitopes rubella, respiratory syncytial virus, bovine herpes virus 1; research on synthetic peptides, recombinant vaccines. Office: Armand-Frappier Inst, 531 Blvd des Prairies Po 100, Laval, PQ Canada H7N 4Z3*

TRUEBLOOD, MARK, computer programmer, author; b. Cin., Feb. 23, 1948; s. William Oliver and Opal Lauretta (Hamilton) T.; m. Patricia Ann Bulman, May 16, 1981. AB-ScB in Physics, Brown U., 1971; MS in Astronomy, U. Md., 1983. Lab. technician Smithsonian Instn., Rockville, Md., 1972-74; mem. tech. staff Computer Scis. Corp., Silver Spring, Md., 1974-78; sr. computer programmer The Dilks Co., Herndon, Va., 1978-79; sr. computer analyst GE Space Div., Lanham, Md., 1979-81; program mgr. Ford Aerospace Corp., Seabrook, Md., 1981-90; sr. sci. programmer Nat. Optical Astronomy Obs., Tucson, 1990—; computer cons. pvt. practice, Potomac, Md., 1974-87; sci. dir. Winer Mobile Observatory, Sonoita, Ariz., 1983—. Co-author (with Russell M. Genet) Microcomputer Control of Telescopes, 1985; contbr. 16 articles on telescope control to profl. jours. Grantee Am. Astronomical Soc., 1987. Mem. Internat. Amateur Profl. Photoelectric Phometry Group, Am. Astron. Soc. (assoc.), Tucson Amateur Astronomy Assn. Democrat. Home: PO Box 797 Sonoita AZ 85637 Office: Nat Solar Obs PO Box 26732 Tucson AZ 85726-6732

TRUESDALE, GERALD LYNN, plastic and reconstructive surgeon; b. High Point, N.C., Aug. 3, 1949; s. Gonzales and Emma Dorothy (Allen) T.; m. Althea Ellen Sample, May 27, 1978; children: Gerard Lynn, Jessica Lynne. BS, Morehouse coll., 1971; MD, U. Chgo., 1975. Intern gen. surgery Emory U., 1975-78; resident gen. and plastic surgery Tulane Med. Ctr., 1978-82; pres. Greensboro (N.C.) Plastic Surg. Assocs. P.A., 1982—; bd. dirs. Greensboro Nat. Bank, 1989-92. Bd. dirs. N.C. A&T U. Found., Greensboro, 1988—, Natural Sci. Ctr., Greensboro, 1988—, Nat. Bank Greensboro; program chmn. Greensboro Men's Club, 1990; pres. Ea. Music Festival, 1992. Mem. AMA, NAACP (life, Greensboro chpt.), Greater Greensboro Med. Soc. (past pres.), Med. Splty. Jour. Club Greensboro, Beta Pi Phi. Home: 502 Staunton Dr Greensboro NC 27410-6071 Office: Greensboro Plastic Surg Assocs PA 901 N Elm St Greensboro NC 27401-1512

TRUHLAR, DONALD GENE, chemist, educator; b. Chicago, Ill., Feb. 27, 1944; s. John Joseph and Lucille Marie (Vancura) T.; m. Jane Teresa Gust, Aug. 28, 1965; children: Sara Elizabeth, Stephanie Marie. B.A. in Chemistry summa cum laude (scholar), St. Mary's Coll., Winona, Minn., 1965; Ph.D. in Chemistry, Calif. Inst. Tech., 1970. Asst. prof. chemistry and chem. physics U. Minn., Mpls, 1969-72; assoc. prof. U. Minn., 1972-76, prof., 1976-92; prof. Inst. Tech., 1993—; cons. Los Alamos Sci. Lab.; vis. fellow Joint Inst. for Lab. Astrophysics, 1975-76; sci. dir. Minn. Supercomputer Inst., 1987-88, dir., 1988—. Editor: Theoretica Chimica Acta, 1985—, Computer Phys. Comms., 1986—, Topics Physics Chem., 1992—, Understanding Chem. Reactivity; mem. editorial bd. Jour. Chemical Physics, 1978-80, Chem. Physics Letters, 1982—, Jour. Phys. Chemistry, 1985-87. Ruhland Walzer Meml. scholar, 1961-62; John Stauffer fellow, 1965-66, NDEA fellow, 1966-68, Alfred P. Sloan Found. fellow, 1973-77; grantee NSF, 1971—, NASA, 1987-92, U.S. Dept. of Energy, 1979—. Fellow Am. Phys. Soc.; mem. Am. Chem. Soc. (sec.-treas. theoretical chemistry sub divsn. 1980-89, councilor 1985-87, editor jour. 1984—). Achievements include research, numerous publications in field. Home: 5033 Thomas Ave S Minneapolis MN 55410-2240 Office: U Minn Minn Supercomputer Inst 1200 Washington Ave S Minneapolis MN 55455-0431

TRUJILLO, KEITH ARNOLD, psychopharmacologist; b. Marysville, Calif., Nov. 23, 1956; s. Filadelfio Emilio and Velma Arlene (Thompson) T.; m. Laurie Jo Fox, Oct. 8, 1988; children: Tara Hope Fox Trujillo, Aimee Grace Fox Trujillo. BA in Biology, Psychology, Chemistry, Calif. State U., Chico, 1979; PhD in Pharmacology and Toxicology, U. Calif., Irvine, 1985. Rsch. fellow Mental Health Rsch. Inst., U. Mich., Ann Arbor, 1986-90, sr. rsch. fellow, 1991, rsch. investigator, 1992—; adv. com. APA/NIMH Minority Fellowship Program, Washington, 1989—; liaison coord. Am. Psychol. Soc., Washington, 1988-92. Contbr. articles to profl. jours. NIMH postdoctoral traineeship, 1986; Nat. Inst. Drug Abuse fellow, 1987-90, grant, 1991; recipient Travel award Com. on Problems of Drug Dependence, 1989, 90, 92, Internat. Narcotics Rsch. Conf., 1989, 92. Mem. APA, AAAS, Am Psychol. Soc., Soc. for Neurosci., N.Y. Acad. Sci. Achievements include discovery that N-methyl-D-aspartate receptor antagonists inhibit development of opiate tolerance and physical dependence; discovery that drugs of abuse increase concentration of prodynorphin peptides in brain regions. Office: Univ of Mich Mental Health Rsch Inst 205 Zina Pitcher Pl Ann Arbor MI 48109-0720

TRULUCK, JAMES PAUL, JR., dentist, vintner; b. Florence, S.C., Feb. 6, 1933; s. James Paul and Catherine Lydia (Nesmith) TruL.; m. Kay Bowen (dec. Oct. 1981); children: James Paul III, David Bowen, Catherine Ann; m. Amelia Nickels Calhoun, Apr. 26, 1983; 1 child, George Calhoun. BS, Clemson (S.C.) U., 1954; DMD, U. Louisville, 1958. Pvt. practice Lake City, S.C., 1960—; founder, pres. TruLuck Vineyards & Winery, Lake City, 1976, Chateau TruLuck Natural Water Co., Lake City, 1990. Member bd. advisors Clemson U., 1978-84; mem. bd. visitors Coker Coll., Hartsville, S.C., 1978-84; pres., bd. dirs. Lions, Lake City, 1960-73; chmn. Greater Lake City Lake Commn., 1967-84. Capt. USAF, 1958-67. Recipient S.C. Bus. and Arts Partnership award S.C. State Arts Commn., 1988. Mem. ADA, Am. Assn. Vinters (bd. dirs. 1982-86), Am. Wine Soc. (nat. judge 1982-88), Am. Soc. Clin. Hypnosis (emeritus), Internat. Acad. Laser Dentistry (chartered), S.C. Dental Assn., Florence County Dental Assn., Soc. First Families of S.C. (exec. sec. 1991—), Descs. Colonial Govs. of Am., Descs. Magna Carta Barons Runnymede. Episcopalian. Avocations: genealogy, tennis, sailing, writing. Home: RR 3 Box 19 Lake City SC 29560-9309 Office: 125 Epps St Lake City SC 29560-2449

TRULY, RICHARD H., federal agency administrator; b. Fayette, Miss., Nov. 12, 1937; s. James B. Truly; m. Colleen Hanner; children: Richard, Michael, Daniel, Bennett, Lee Margaret. B.Aero. Engring., Ga. Inst. Tech., 1959. Commd. ensign U.S. Navy, 1959; advanced through grades to rear adm., assigned Fighter Squadron 33, served in U.S.S. Intrepid, served in U.S.S. Enterprise; astronaut Manned Orbiting Lab. Program USAF, 1965-69; astronaut NASA, from 1969, comdr. Columbia Flight 2, 1981; comdr. Columbia Flight 2 Challenger Flight 3, 1983; dir. Space Shuttle program, 1986-89; adminstr. NASA, 1989-92. Recipient Robert H. Goddard Astronautics award AIAA, 1990. Office: care NASA Office of the Adminstr 400 Maryland Ave SW Washington DC 20546

TRUNNELL, THOMAS NEWTON, dermatologist; b. Waterloo, Iowa, May 7, 1942; s. Thomas Lyle and Vivian (Dahl) T.; m. Patricia Rautiala, Aug. 2, 1974; children: Suzanne, Thomas, Sarah. AB cum laude, Princeton U., 1964; MD, U. Iowa, 1968. Diplomate Am. Bd. Dermatology. Intern U. So. Calif., L.A., 1969; resident NYU, 1972; pvt. practice dermatology Tampa, Fla., 1974—; asst. clin. prof. U. S. Fla., Tampa, 1975—. Contbr. articles to profl. jours. Maj. USAF, 1972-74. Mem. AMA, Am. Acad. Dermatology, Am. Assn. Dermatol. Surgeons, Fla. Med. Assn., Fla. Soc. for Dermatol. Surgeons (pres. 1993), Fla. Dermatol. Soc., Hillsborough County Med. Assn., Cutaneous Therapy Soc., Ducks Unltd. (organizing com. North Tampa chpt. 1988—). Republican. United Methodist. Avocations: fishing, hunting. Office: Thomas N Trunnell MD 13801 Bruce B Downs Blvd Tampa FL 33613-3911

TRURAN, WILLIAM R., electrical engineer; b. Franklin, N.J., Feb. 14, 1951; s. Wilfred Hardy and Stella Eva (Hall) T.; m. Virginia Lynn Johnson, Aug. 18, 1979; children: Michael, Wendy. BSEE, U. Tenn., 1972; MBA, Fairleigh Dickinson U., 1981; postgrad., Columbia U., NYU. Registered profl. engr., N.J., N.Y., Pa., Calif.; registered profl. planner, N.J. Design engr. Gordos Corp., Bloomfield, N.J., 1972-73; project engr. Edwards Engring., Pompton Plains, N.J., 1973-78; sr. engr. Apollo Tech., Whippany, N.J., 1978-81; elec. product mgr. Dodge-Newark, Fairfield, N.J., 1981—; pres. Trupower Engring., Sparta, N.J., 1984—; pres. T.E.C. Corp. of N.J., Sparta; cons. in field. Contbr. articles to profl. jours. Active foster child orgn. Christian Children's Fund. Mem. Nat. Soc. Profl. Engrs. (legis. action network, minuteman), Nat. Assn. Environ. Profls., Wilderness Soc., Sierra (foster child program). Episcopalian. Avocations: skiing, water skiing, triathlons, marathons, antique Corvettes. Home and Office: 37 Rainbow Trl Sparta NJ 07871-1724

TRUST, RONALD IRVING, organic chemist; b. Phila., May 23, 1947; s. Nathan and Dora (Lazow) T.; m. Arlene Eva Sherman, July 6, 1969; children: Phyllis, Paul. BS, Drexel U., 1969; PhD, Calif. Inst. Tech., 1974; MBA, Fairleigh Dickinson U., 1986. With Am. Cyanamid Co., Pearl River, N.Y., 1974—, asst. dir., global coord., 1990-92, asst. dir. regulatory affairs, 1992—; presenter on clinical supplies, drug accountability, clinical trial monitoring. Mem. AAAS, Drug. Info. Assn., Assocs. Clin. Pharmacology, N.Y. Acad. Scis. Achievements include patents on hypocholesteremic agents, antihypertensives, anxiolytic medicinal agents. Office: Am Cyanamid Co Middletown Rd Pearl River NY 10965

TRUTHAN, CHARLES EDWIN, physician; b. Cleve., Mar. 30, 1955; s. Jordan Alexander and Jean Marie (Knoll) T.; m. Joyce Lynn Miller, Dec. 4, 1982; children: Jennifer Ann, Patricia Jean. BA in Psychology, U. Toledo, 1979; DO, Ohio U., 1986. Diplomate Am. Bd. Osteo. Gen. Practitioners, Nat. Bd. Osteo. Med. Examiners. Instr. anatomy and physiology Hocking Tech. Coll., Nelsonville, Ohio, 1980-81; intern Brentwood Osteo. Hosp., Cleve., 1986-87; resident in family practice Davenport (Iowa) Med. Ctr., 1987-88; pvt. practice Wisconsin Dells, Wis., 1989—; chief med. svcs. Troop Med. Clinic #1, Ft. McCoy, Wis., 1992—; med. adviser Maple Heights (Ohio) Fire Dept., 1986-87; instr. ACLS Am. Heart Assn., Wis. Dells, 1985—; instr. EMT tng. MATC, Madison, Wis., 1989—, med. dir. basic trauma life support, 1990—; bd. dirs. Wis. Basic Trauma Life Support, 1992—. Firefighter, paramedic Willoughby Hills (Ohio) Fire Dept., 1973-84; firefighter Springfield Twp. Vol. Fire Dept., Toledo, 1977-79; EMT instr. Dept. Vocat. Edn., Trade and Indsl. Edn., Ohio, 1976-83; vol. APPAL Corps, Athens, Ohio, 1980; bd. dirs. S.E. Ohio Regional Coun. on Alcoholism, Athens, 1980-81; med. dir. Quad City Air Show, Davenport, 1988. Lt. comdr. USPHS, active duty with U.S. Coast Guard, 1988-89. res. 1989—. Mem. Am. Osteo. Assn., Wis. Assn. Osteo. Physicians and Surgeons (sec.-treas. 1991-92, pres.-elect 1992-93, pres. 1993-94), Am. Coll. Osteo. Physicians and Surgeons (pres. 1992-93), Res. Officers Assn. (life), Sauk County Med. Soc., Am. Acad. Family Practitioners, Assn. Mil. Osteo. Physicians and Surgeons, Assn. Mil. Surgeons U.S., Aircraft Owners and Pilots Assn., Exptl. Aircraft Assn. Avocations: flying, water sports, cross country skiing. Home: 1300 E Hiawatha Dr Wisconsin Dells WI 53965-9741 Office: Troop Med Clinic # 1 Bldg 1679 Fort McCoy WI 54656

TRUXAL, JOHN GROFF, electrical engineering educator; b. Lancaster, Pa., Feb. 19, 1924; s. Andrew Gehr and Leah Deldee (Groff) T.; m. Doris Teresa Mastrangelo, June 11, 1949; children—Brian Andrew, Carol Jean. A.B., Dartmouth, 1944; B.S., Mass. Inst. Tech., 1947, Sc.D., 1950; D.Eng. (hon.), Purdue U., 1964, Ind. Inst. Tech., 1971. Asso. prof. elec. engring. Purdue U., 1950-54; asso. prof. elec. engring. Poly. Inst. Bklyn., 1954- 57, prof., head dept., 1957-72, v.p. ednl. devel., 1961-72, dean engring., 1964-66, provost, 1966-68, acad. v.p., 1969-72; dean engring. State U. N.Y.,

Stony Brook, 1972-76; prof. engring. SUNY, 1976-91, Disting. Teaching prof. emeritus, 1991—; Cons. control engring. Author: Automatic Feedback Control System Synthesis, 1955, Introductory System Engineering, 1972, (with W.A. Lynch) Signals and Systems in Electrical Engineering, 1962; co-author: (with W.A. Lynch) The Man Made World, 1969, Man and His Technology, 1973, Technology: Handle With Care, 1975, The Age of Electronic Messages, 1991; editor: Control Engineers' Handbook, 1958. Recipient Rufus Oldenburger medal ASME, 1991. Fellow IEEE, AAAS; mem. NAE, Instrument Soc. Am. (pres. 1965), Am. Soc. Engring. Edn., Phi Beta Kappa, Sigma Xi, Tau Beta Pi, Eta Kappa Nu, Phi Kappa Psi. Home: 8 Avon Ct Dix Hills NY 11746 Office: SUNY Coll Engring Stony Brook NY 11794

TRYON, JOHN GRIGGS, electrical engineer educator; b. Washington, Dec. 18, 1920; s. Frederick Gale and Ruth (Wilson) T.; m. Helen Muenscher, June 21, 1948; children: Ralph W., Peter R. B in Physics, U. Minn., 1941; PhD in Engring. Physics, Cornell U., 1952. Profl. Engr., Alaska. Tech. staff Bell Telephone Lab., Whippany, N.J., 1951-58; prof. U. Alaska, College, 1958-69, Tuskegee (Ala.) Inst., 1969-75; prof. U. Nev., Las Vegas, 1975-87, prof. emeritus, 1987—. Capt. U.S. Army, 1942-46. Mem. Inst. Elec. and Electronic Engrs., Americal Phys. Soc., Am. Slar Energy Soc., Sigma Xi. Home: 631 Ave I Boulder City NV 89005-2729

TRYTEK, LINDA FAYE, microbiologist; b. Orlando, Fla., Oct. 23, 1947; d. J.C. and Blanche (Collins) Crews; m. Ray Trytek, Feb. 23, 1982; children: Tamara, Chris. Degree in child psychology, U. Cen. Fla., 1988. Cert. indsl. hygienist. Microbiologist Lyphomed Pharm., Orlando, 1979-88; microbiology supr. Orlando Labs., 1988-90; pres., lab dir. Tri-Tech Labs., Orlando, 1990—; mem. tech. svc. bd. Fla. Bottled Water, Orlando, 1988—. Campaign mgr. Kim Sheppard Campaign, Orlando, 1992. Mem. Fla. Soc. Microbiology, Nat. Geographic Soc. Democrat. Baptist. Office: Tri-Tech Labs Inc 1319 Renee Ave Orlando FL 32825

TSAI, CHI-TAY, mechanical engineering educator; b. Fuchien, Republic of China, Feb. 1, 1956; came to U.S., 1980; s. Ching-Sen and Hsiu-Jen (Huang) T.; m. Yuan-Ying Chao, Aug. 17, 1981; children: I-Hao, Lawrence. MS in Structural Mechanics, Rensselaer Poly. Inst., Troy, N.Y., 1981; PhD in Engring. Mechanics, U. Ky., 1985. Rsch. assoc. U. Ky., Lexington, 1986-88; rsch. scientist Air Force Inst. Tech., Dayton, Ohio, 1988-90; asst. prof. mech. engring. Fla. Atlantic U., Boca Raton, 1990—. Contbr. articles to profl. jours. NASA summer faculty fellow, Cleve., 1991; rsch. grantee NASA, Cleve., 1992, Air Force, Dayton, 1992. Mem. ASME (assoc.), Am. Acad. Mechanics, Am. Soc. Engring. Edn. Achievements include development of the first numerical model for the calculation of the quantity of dislocation density in semiconductor crystals grown from the melt; research in computational modeling of materials processing (crystal growth, high-speed machining, metal forming, etc.); micromechanics of defects in crystals; constitutive equation; viscoplasticity; mechanics of composite materials. Home: 21817 Linwood Way Boca Raton FL 33433 Office: Fla Atlantic U Dept Mech Engring 500 NW 20th St Boca Raton FL 33431

TSAI, MAVIS, clinical psychologist; b. Kowloon, Hong Kong, Sept. 30, 1954; came to U.S. 1966; d. Edwin Fang-Chin and Emily (Tseng) Tsai; m. Robert Joseph Kohlenberg, June 22, 1980; 1 child, Jeremy Tsai Kohlenberg. BA magna cum laude, UCLA, 1976; PhD, U. Wash., 1982. Undergrad. teaching asst. UCLA, 1975-76; teaching asst., predoctoral instr. U. Wash., Seattle, 1977-79; predoctoral lectr., psychology fellow Langley Parker Psychiat. Inst., San Francisco, 1980-81; predoctoral lectr. U. Wash., Seattle, 1981-82, ext. lectr., 1982-88; clin. psychologist in pvt. practice Seattle, 1982—; clin. supr. grad. students U. Wash., Seattle, 1989—. Co-author: Functional Analytic Psychotherapy, 1991; contbr. articles to profl. jours. Bd. dirs. Asian Counseling and Referral Svc., Seattle, 1984-85, Gifted Women's Conf., U. Wash., 1986-87. Calif. State scholar, 1972-76; recipient APA Minority fellowship, 1977-82, Danforth fellowship honorable mention, 1977. Mem. APA, Wash. Psychol. Assn., Asian Am. Psychol. Assn., Phi Beta Kappa. Avocations: guitar, mountain climbing, photography, tennis, woodworking. Office: 3245 Fairview Ave E Ste 303 Seattle WA 98102-3053

TSAI, TI-DAO, electrophysiologist; b. Shanghai, China, Dec. 17, 1936; s. Tong-Yun Cai and Cui-Ying Ma; m. Li-min Xiong, June 20, 1967; 1 child, Li Cai. Student, Moscow U., USSR, 1961. From asst. to assoc. prof. Shanghai Inst. of Physiology, Academia Sinica, Shanghai, China, 1961--; electrophysiologist Upjohn Co., Kalamazoo, Mich., 1990--; vis. scientist U. of Tex. Med. Branch, Galveston, 1981-82, U. Calgary, Alberta, Canada, 1987-88, U. Alberta, Edmonton, Alberta, Canada, 1989-90, Beckman Rsch. Inst., Duarte, Calif., 1990. Mem. editorial bd. Acta Physiologica Sinica, Shanghai, 1985-87; contbr. articles to profl. jours. Mem. Biophysical Soc., N.Y. Acad. Sci. Achievements include research on molecular and cellular electrophysiology. Office: Upjohn Co 301 Henrietta St Kalamazoo MI 49007-4940

TSAI, TOM CHUNGHU, chemical engineer; b. Kaohsiung, Taiwan, Oct. 24, 1948; came to U.S., 1971; s. Joyce Chionhwa Pai, Dec. 17, 1974; children: Wayne, Jimmy Payne. BS in Chem. Engring., Nat. Taiwan U., Taipei, 1970; MS in Chem. Engring., Purdue U., 1973, PhD in Chem. Engring., 1975. Registered profl. engr., Tex. Sr. process engr. CE-Lummus Co., Bloomfield, N.J., 1975-80; sr. engr. Bechtel Petroleum Inc., Houston, 1980-83; cons. engr. TDS Assocs., Houston, 1983-88; process engring. assoc. Dow Chem. Co., Freeport, Tex., 1988—. Co-author, contbr.: Ethylene-Keystone to the Petrochemical Industry, 1980, Kirk-Othmer Encyclopedia of Chemical Technology, 1980, Pyrolysis: Theory and Industrial Practice, 1983, Encyclopedia of Chemical Processing and Design, 1990, Unit Operations Handbook, 1992; contbr. articles to profl. jours. 2nd lt. R.O.C. Army, 1970-71, Taiwan. Mem. Am. Inst. Chem. Engrs., Assn. Am. Chinese Profls. (div. chmn. 1988-89). Achievements include rsch. in flare system design by microcomputer, yield correlations for AGO cracking, sizing a vertical separator by microcomputer, surface reactions in pyrolysis units, tech. improvement in heater design for olefins prodn.

TSAI, WEN-YING, sculptor, painter, engineer; b. Xiamen, Fujian, China, Oct. 13, 1928; came to U.S., 1950, naturalized, 1962; s. Chen-Dak and Ching-Miau (Chen) T.; m. Pei-De Chang, Aug. 7, 1968; children: Lun-Yi and Ming Yi (twins). Student, Ta Tung U., 1947-49; BS in Mech. Engring., U. Mich., 1953; postgrad., Art Students League N.Y., 1953-57, Faculty Polit. and Social Sci., New Sch., 1956-58. cons. engr., 1953-63; project mgr. Cosentini Assocs., 1962-63; project engr. Guy B. Panero, Engrs., 1956-60. Creator cybernetic sculpture based on prin. harmonic motion, stroboscopic effects; one-man shows include, Ruth Sherman Gallery, N.Y.C., 1961, Amel Gallery, N.Y.C., 1964, 65, Howard Wise Gallery, N.Y.C., 1968, Kaiser Wilhelm Mus. Haus Lange, Krefeld, Germany, 1970, Hayden Gallery of MIT, Cambridge, Ont. Sci. Centre, Toronto, Can., 1971, Corcoran Gallery Art, 1972, Denise René Gallery, 1972, 73, Musée d'Art Contemporain, Montreal, 1973, Museo de Arte Contemporáneo, Caracas, 1975, Wildenstein Art Center, Houston, 1978, Museo de Bellas Artes, Caracas, 1978, Hong Kong Mus. Art, 1979, Isetan Mus. Art, Tokyo, 1980, Galerie Denise René, Paris, 1983, Nat. Mus. History, Taipei, Taiwan, 1989, Taiwan Mus. of Art, Taichung, 1990; represented maj. internat. exhbns., also numerous group exhbns., in permanent collections, Centre Georges Pompidou, Paris; Tate Gallery, London, Albright-Knox Gallery, Buffalo Mus., Addison Gallery Am. Art, Andover, Mass., Museo de Arte Contemporáneo, Caracas, Museo de Bellas Artes, Caracas, Whitney Mus., Chrysler Art Mus., Orlando Sci. Ctr., MIT, Hayden Gallery, Kaiser Wilhelm Mus., Mus. Modern Art, Israel Mus., Jerusalem, Artware, Kunst und Elektronik, Honnover-Messe, Great Exploration-The Hands on Mus., Taiwan Mus. Art, Saibu Gas Mus., Nagoya City Mus., Mus. fü Holographie, Janagawa Sci. Pk., Hong Kong Sci. Mus., others; commd. works include: fountain at Land Mark, Hong Kong, 1980, , water sculpture at Shell Tower, Singapore, 1982, cybernetic upward falling fountains (2), Paris; creator spatial dynamic hydro-cybernetic systems for 42d Internat. Exhbn. Art-La Biennale di Venezia, 1986, Digital Visions-Computers and Art, Everson Mus. of Art, 1987, Contemporary Arts Ctr. Cin., 1987, IBM Gallery of Sci. and Art, N.Y.C., 1988, Phenomena Art Expo, Fukuoka, Japan, 1989, Artec '91, Wonderland of Sci.-Art Kanagawa Internat. Art Sci. Exhbn., Kawasaki, Japan, 1989, Nagoya, Japan, 1989, Vienna Messe-Wiener Festwochen, 1989, Kanagawa Internat. Art & Sci. Exhbn., Kawasaki, Japan, 1989, Artec Grandprix, Internat. Biennale in Nagoya, Japan, 1991 (Artec Grand Prix winner); featured: Art for Tomorrow-The 21st Century, CBS-TV, 1969, Video Variation, WGBH-TV, 1971, Science and Art, Japan TV Man Union, 1982, Art and Sci.-Innovation, Sta. WNET-TV, 1988, The World of Wen-Ying Tsai, Taiwan Pub. TV, 1991. John Hay Whitney fellow, 1963; MacDowell fellow, 1965; fellow Center Advanced Visual Studies, MIT, 1969, 70. Inventor upward falling fountain, computer mural, multiple light computer array, utilizing environ. feedback control system.

TSAKAS, SPYROS CHRISTOS, genetics educator; b. Parga, Epirus, Greece, Nov. 3, 1941; s. Christos Anastasios and Ioanna Alexandro (Tselios) T.; m. Kathleen Marie Flynn, Sept. 15, 1979; children: Joanne Emily, Spyridon Edward. BS in Animal Improvement, Agrl. U., Athens, Greece, 1965, M in Animal Improvement, 1966, PhD in Genetics, 1968; PhD in Population Genetics, U. Edinburgh, Scotland, 1981. Asst. prof. Agrl. U., Athens, Greece, 1971-73, assoc. prof., 1973-84, prof., 1984—; postdoctoral researcher U. Edinburgh, Scotland, 1976-80; faculty researcher Nat. Sci. Rsch. Ctr., France, 1983-84, U. Calif., Irvine, 1987-88; v.p. Faculty of Biology and Biotechnology, Athens, 1989-91, pres., 1991—; cons. LAVIPHARM, Athens, 1988—, VIKI Meat Processing Plant, Arta, Greece, 1990—, Ference Cosmetic Co., Athens, 1992; rep. of Greece, European Network on Farm Animal Genetics: Classical, Molecular and Quantitative, Edinburgh, 1991—. Contbr. numerous articles to profl. jours. Elected advisor Mayor, Parga, Greece, 1990. Lt. Army, 1969-71, Greece. Recipient UN F.A.O. scholarship, 1976-77, Scholarship, Onassis Found., 1979, NSF grant, 1983-84. Mem. N.Y. Acad. Sci., Genetics Soc. Am., U. Edinburgh Grads. Assn. & Alumni, Biol. Soc. of Greece, Zool. Soc. of Greece, Entomol. Soc. Greece, Onassis Found. Orthodox Greek. Avocations: inventing enzymatic product processes, music, swimming, football. Home: Pl Messologiou 2 Pangrati, 11634 Athens Greece Office: Agrl U Dept of Genetics, Votanicos 11855, Athens Greece

TSANG, JAMES CHEN-HSIANG, physicist; b. N.Y.C., June 1, 1946; s. Chih and See Yee (Chow) Tsang. BS, MS, MIT, 1968, PhD, 1973. Rsch. staff T.J. Watson Rsch. Ctr. IBM, Yorktown Heights, N.Y., 1973—. Fellow Am. Phys. Soc. Office: IBM T J Watson Rsch Ctr PO Box 218 Yorktown Heights NY 10592

TSAO, JOHN CHUR, materials engineer, government regulator; b. Taichung, Taiwan, People's Republic of China, June 8, 1952; came to U.S., 1966; s. Chi K. and Lily (Chen) T.; m. Wan-Qi Ting, Jan. 20, 1989; 1 child, Emily M. BSME, U. Md., 1974, MS, 1976. Registered profl. engr., Md. Stress analyst Bechtel Power Corp., Gaithersuburg, Md., 1974-79; mech. engr. Dept. of Navy, Indianhead, Md., 1979-80; mech. engr. U.S. Nuclear Regulatory Commn., Washington, 1980-81, reliability analyst, 1981-85, materials engr., 1985—. Vol. tutor U.S. Nuclear Regulatory Commn., 1985—; judge Mongomery County Sci. Fair, Gaithersburg, 1986—. Mem. ASME. Democrat. Presbyterian. Achievements include development of reactor vessel material data base. Office: US Nuclear Regulatory Commn Washington DC 20555

TSAPATSARIS, NICHOLAS, civil engineer, consultant; b. E. Orange, N.J., July 12, 1965; s. Leonidas Nicholas and Anna (Apostolakos) T. BS, Worcester Polytech. Inst., 1986, MS, 1987; MS, MIT, 1992. Registered profl. engr., N.Y., Mass., N.J., European Econ. Community. Project engr. Chazen Engring. & Land Surveying Co., Poughkeepsie, N.Y., 1987-89, Rhode & Soyka Consulting Engrs., Poughkeepsie, 1989-91; pvt. practice consulting engr. Ridgewood, N.J., 1991—. Mem. ASCE, Prestressed Concrete Inst. Greek Orthodox. Achievements include development of a knowledge-based expert system for preliminary selection of structural systems for tall buildings. Home: 20 Wilsey Sq Ridgewood NJ 07450

TSATSARONIS, GEORGE, mechanical engineering educator, researcher; b. Thessaloniki, Greece, Sept. 22, 1949; came to U.S., 1982; s. Asterios and Chrysoula (Ioannidou) T. Diploma in mech. engring., Nat. Tech. U., Athens, Greece, 1972; MBA, Tech. U., Aachen, Fed. Republic of Germany, 1976, PhD in Mech. Engring., 1977, Habilitation in Thermoeconomics, 1985. From rsch. assoc. to lectr. to sr. staff mem. Inst. of Thermodynamics, Aachen, 1972-82; rsch. prof. Energy & Environ. Engring. Ctr. Desert Rsch. Inst., Reno, Nev., 1982-86; prof. Ctr. for Electric Power Tenn. Tech. U., Cookeville, 1986—; reviewer in field. Assoc. editor Energy, The Internat. Jour., 1986—, Jour. Energy Resources Tech., 1988—; co-editor 12 bound vols.; contbr. numerous articles to profl. jours. Recipient Borchers award Tech. U. Aachen, 1977. Mem. ASME (chmn. systems analysis com. 1987-90, mem. exec. com. advanced energy systems divsn. 1993—), AAUP, Am. Inst. Chem. Engrs., Am. Soc. Engring. Edn., Greek Soc. Engrs., German Assn. Univ. Profs., Tenn. Acad. Scis., Sigma Xi. Democrat. Greek Orthodox. Office: Tenn Tech U PO Box 5032 Cookeville TN 38505

TSCHEUSCHNER, RALF DIETRICH, theoretical physicist; b. Hamburg, Fed. Republic Germany, Mar. 31, 1956; s. Dietrich Ernst and Ingeburg Lucie (Kirsch) T. Diploma in physics, U. Hamburg, 1978, PhD in Physics, 1987. Pub. rels. asst. Deutsches Elektronen-Synchrotron, Hamburg, 1978-80; lectr. physics and math. U. Hamburg, 1978-88; researcher, lectr. applied math. and telecommunications Bundeswehr U. Hamburg-Wandsbek, 1981-84; researcher and lectr. theoretical physics U. Hamburg, 1985—, researcher and lectr. musicology/computer music, 1988—; co-founder Schindler-Tscheuschner Music Rsch., Hamburg, 1991; vis. scientist U. Chgo., 1991. Asst. Emergency Svcs., Hamburg, 1975-85. Talent scholar Studienstiftung des Deutschen Volkes, Bonn, Fed. Republic Germany, 1974-81. Mem. Am. Phys. Soc., N.Y. Acad. Sci., German Phys. Soc., The Inter-Soc. for Electronic Arts, Amateur Radio Club German Fed. Mail, Friends of Studienstiftung des Deutschen Volkes. Avocations: aerobics, keyboard playing. Office: I Inst fuer Theor Physik, Jungiusstr 9, D-20355 Hamburg Germany

TSE, PO YIN, electrical engineer; b. Hong Kong, May 4, 1959; came to the U.S., 1980; s. Tin Sang T. and Hung (Ying) Kam; m. Pamela Elizabeth Goblick, March 7, 1987; 1 child, Melissa On. BEE, Northeastern U., 1984; postgrad., Worcester Poly. U. Component engr. Wang Labs., Lowell, Mass., 1984-86; aerospace vehicle engr. Textron Def. System, Wilmington, Mass., 1986-88; sr. component engr. Proteon Inc., Westboro, Mass., 1988—; chmn. Table Tennis Jour., Mass., 1988—. Mem. SPIE. Home: 155 Crescent St Shrewsbury MA 01545 Office: Proteon Inc 9 Technology Dr Westborough MA 01581

TSEKANOVSKIĬ, EDUARD RUVIMOVICH, mathematician, educator; b. Odessa, Ukraine, USSR, Mar. 15, 1937; s. Ruvim Solomonovich and Fanya Davidovna (Urlang) T.; m. Chelombytko Larisa Yakovlevna, June 29, 1977; 1 child, Vladislav Eduardovich. B in Math., Pedagogical Inst., Odessa, 1959; PhD, Low-Temperature Inst., Kharkov, USSR, 1964, DSc, 1970. Asst. prof. Inst. Radioelectronics, Kharkov, 1962-65; asst. prof., assoc. prof., head math. dept. Donetsk (USSR) State U., 1965-73, assoc. prof., 1973—; vis. prof. Leipzig U., Germany, 1990, Jagellonian U., Krakow, Poland, 1989; participant internat. confs. Timishoara, Romania, 1988, Lyptovsky, Czechoslovakia, 1989, Sapporo, Kobe, Japan, 1991. Author: Generalized functions Method in the Extension Theory of Unbounded Operators, 1973; contbr. articles to profl. jours. PhD degree com. mem. Inst. Applied Math. and Mechanics of Ukrainian Acad., Donetsk, 1975—. Mem. Am. Math. Soc. Avocations: classical music, reading. Office: SUNY Math Dept 106 Diefendorf Hall Buffalo NY 14214

TSENG, CHIA-JENG, computer engineer; b. Hsinchu, Taiwan, Republic of China, Mar. 3, 1949; came to the U.S., 1978; s. Kuo-Bang and Shiow-Taur (Liu) T.; m. Suh-Mei Lu, May 11, 1978; children: Daniel Hsueh-Pu, Albert Hsueh-Li. BS, Nat. Cheng Kung U., 1971; PhD, Carnegie-Mellon U., 1984. Rsch. engr. Telecommunication Labs., Chung-Li, Taiwan, 1975-77; instr. Nat. Tsing-Hua U., Hsinchu, 1977-78; mem. tech. staff AT&T Bell Labs., Murray Hill, N.J., 1984—. Contbr. articles to IEEE Transactions on Computer-Aided Design and various conf. proceedings. Mem. IEEE (sr.), Sigma Xi. Achievements include research in formal methods for high-level synthesis of digital systems including design capture and optimization techniques for data-path synthesis, control-path synthesis and pipeline synthesis. Office: AT&T Bell Labs 600 Mountain Ave Murray Hill NJ 07974

TSENG, CHRISTOPHER KUO-HOU, health scientist, administrator, chemist; b. Amoy, Fukien, China, Dec. 15, 1946; came to U.S., 1976; s. Ren-Jien and Sen-Yoon (Kuo) T.; m. Lucy You-Hua Kuan, Feb. 11, 1971; children: Jiun-Woei, Victor. MS, SUNY, Stony Brook, 1979, PhD, 1982. Rsch. assoc. SUNY, Stony Brook, 1982-83; vis. fellow Nat. Cancer Inst., Bethesda, Md., 1984-86, sr. staff fellow, 1986-88; chemist U.S. FDA, Rockville, Md., 1988-89; program officer Nat. Inst. Allergy and Infectious Diseases, Bethesda, 1989—. Reviewer Nucleosides and Nucleotides, 1989—, Internat. Soc. Antiviral Rsch., 1989—; contbr. articles to profl. jours. Mem. Am. Chem. Soc., Am. Soc. Microbiology, Internat. Soc. Antiviral Rsch. Achievements include patents in antiviral and anticancer cyclopentenyl cytosine; in 3-deazaneplanocin, intermediates for it and antiviral composition and method of treatment using it; patent pending for lipophilic, aminohydrolase-activated prodrugs; for acid stable dideoxynucleosides active against the cytopathic effects of HIV. Office: NIH Rm 3A23 6003 Executive Blvd Bethesda MD 20892

TSENG, HSIUNG SCOTT, engineer; b. Hualien, Taiwan, Republic of China, Sept. 18, 1948; came to U.S., 1979; s. Ron C. and Kwei May (Hsu) T.; m. Su T. Hsu, Mar. 30, 1973; 1 child, Kevin. MS, Osaka (Japan) U., 1979; PhD, U. Utah, 1984. Sr. R&D engr. Mobil Chem. Co. (TAITA), Kaohsiung, Republic of China, 1973-76; sr. R&D assoc., group leader B.F. Goodrich Co., Avon Lake, Ohio, 1985—. Author: Polymers for Information Storage, 1990; contbr. articles to Polymer, Jour. Polymer Sci./Physics Edn. Vice chmn. Taiwanese Student Assn., Osaka, 1978. Japanese Govt. scholar, 1976-78. Mem. Am. Chem. Soc., Soc. Plastic Engineers. Achievements include patent for Thermoplastic Polyurethanes having Improved Binder Properties for Magnetic Media Recording; several patents pending. Home: 1881 Holdens Arbor Run Cleveland OH 44145-2004 Office: BF Goodrich Co Walker & Moore Rds Avon Lake OH 44012

TSENG, TIEN-JIUNN, physics educator; b. Hankow, Hu Pei, China, Nov. 21, 1938; s. Tong-yuan and Yue-ying (Tao) T.; m. So-Lan Lem Tseng, Aug. 14, 1971; children: Lin-con, Lin-haw. BSc, Chung Yuan Christian U., Taiwan, Republic of China, 1962; PhD, U. N.B., Frederiction, Can., 1973. Assoc. prof. Chung Yuan Christin U., Chung Li, 1973-76, dean of students, 1975-78, prof., 1976—, dir. R&D, 1980-81, dean of sci., 1980-83; dir. libr. Chung Yuan Christian U., Chung Li, 1987-89, dir. chaplain's office, 1989-92, dean of students, 1992—; vis. scientist McGill U., Montreal, Que., Can., 1978-80, vis. prof., 1983-84; councillor in natural scis. Nat. Sci. Coun., Taipei, Republic of China, 1990—. Editorial bd. mem. of dictionary/mechanics Nat. Inst. for Compilation & Translation, Taipei, 1989—; contbr. articles to profl. jours. Chancillor (group) Kuo-Ming Tang polit. party, Chung Li; chmn. com. of elders and deacons/ch., Taipei, 1988—; elder The Holy Word Ch., Taipei, 1989—; mem. bd. China Evang. Mission, Taipei, 1989—. Recipient annual award/rsch. works Nat. Sci. Coun., 1984-92. Mem. Phys. Soc. of Republic of China (bd. dirs. 1984-90, 92—), N.Y. Acad. Scis., Am. Physics Soc., Phys. Edn. Soc., Sigma Xi, Phi Tao Phi. Avocations: ping-pong, reading. Office: Chung Yuan Christian U, Dept Physics, Chung-Li 320, Taiwan

TSENOGLOU, CHRISTOS, chemical engineer; b. Athens, Greece, July 26, 1954; came to U.S., 1978; s. Ioannis and Angela (Kahramanos) T. Diploma, Nat. Tech. U., Athens, 1978; MS, Northwestern U., 1981, PhD, 1985. Asst. prof. Stevens Inst. Tech., Hoboken, N.J., 1985-92, assoc. rsch. prof., 1992—; cons. Johnson and Johnson Dental Products, East Windsor, N.J., 1986—, Werner and Pfleiderer, Ramsey, N.J., 1987 —, Polymer Processing Inst., Hoboken, 1991—, Nat. Starch and Chem. Co., Bridgewater, N.J., 1992—; chmn. polymer rheology symposium 23rd Am. Chem. Soc. Mid. Atlantic Regional Meeting, Madison, N.J., 1990. Author: (with others) New Trends in Physics and Physical Chemistry of Polymers, 1989; contbr. articles to Jour. of Polymer Sci., Physics Edn., Macromolecules, Jour. of Rheology, Rheologica Acta, Polymer Engring. and Sci. Grantee Office Naval Rsch., 1986-89, NSF, 1993—. Mem. Am. Inst. Chem. Engrs., Am. Chem. Soc., Am. Physical Soc., Soc. of Rheology, Brit. Soc. of Rheology, N.Y. Acad. Scis. (vice-chair elect polymer sci. div. 1991-92). Achievements include development of molecular blending laws for polymer mixtures and composites, general rubber elasticity theory; study on the effects of fractal aggregation in suspension rheology; research on rheology of polymer melts, solutions, blends and suspensions, structure/property relationships in polymer fluids and elastomers, non-Newtonian fluid mechanics in polymer processing and mixing, molecular theories of viscoelasticity and diffusion in polymers. Home: 730 Hudson St Apt 22 Hoboken NJ 07030-5935 Office: Stevens Inst Tech Dept Chemistry & Chem Engring Hoboken NJ 07030

TSIEN, ROGER YONCHIEN, chemist and cell biologist; b. N.Y.C., Feb. 1, 1952; s. Hsue Chu and Yi Ying (Li) T.; m. Wendy M. Globe, July 30, 1982. AB summa cum laude in Chemistry and Physics, Harvard Coll., 1972; PhD in Physiology, U. Cambridge, 1977. Rsch. asst. U. Cambridge, Eng., 1975-78; asst. prof. Dept. Physiology-Anatomy U. Calif., Berkeley, 1981-85, assoc. prof., 1985-87, prof., 1987-89; prof. Pharmacology and Chemistry, investigator Howard Hughes Med. Inst. U. Calif., San Diego, 1989—; T.Y. Shen vis. prof. Medicinal Chem., MIT, 1991. Mem. editorial bds. Jour. Biological Chemistry, Molecular Biology of the Cell; contbr. chpts. to books, articles to profl. jours. Marshall scholar British Govt., 1972-75; Comyns Berkeley Rsch. fellow Gonville & Caius Coll., Cambridge, 1977-81; recipient Lamport prize N.Y. Acad. Scis., 1986, Javits Neuroscience Investigator award Nat. Inst. Neurological Disorders and Stroke, 1989—, Young Scientist award Passano Found., 1991, W. Alden Spencer award in Neurobiology Columbia U., 1991, Bowditch lectureship Am. Physiological Soc., 1992; Searle scholar, 1983-86. Mem. Phi Beta Kappa. Achievements include patents for Fluorescent Indicator Dyes for Calcium Ions, Chelators Whose Affinity for Calcium is Decreased by Illumination, New Photosensitive Calcium Chelators, Fluorescent Indicator Dyes for Calcium, Working at Long Wavelengths, Fluorescent Indicator Dyes for Alkali Metal Cations, Chelators whose Affinity for Calcium Ion is Increased by Illumination; 4 patents pending. Office: U California Howard Hughes Medical Inst M-047 Cellular & Molecular Medicine La Jolla CA 92093

TSIN, ANDREW TSANG CHEUNG, biochemistry educator; b. Hong Kong, July 19, 1950; came to U.S., 1979; s. Sai Nin and Chai Pong (Tsang) T.; m. Wendy L. Wickstrom, Jan. 20, 1979; 1 child, Cathy Mei. BS in Biology, Dalhousie U., Halifax, N.S., Can., 1973; MS in Zoology, U. Alberta, Edmonton, Alta., Can., 1976, PhD in Zoology, 1979; postgrad., Baylor Coll. Medicine, 1979-81. Prof. biochemistry and physiology U. Tex., San Antonio, 1990—; adj. prof. ophthalmology U. Tex. Health Sci. Ctr., San Antonio, 1990—; cons. Vision R & D, Lubbock, Tex., 1988-89, Alcon Lab., Ft. Worth, 1989—, Technology, Inc., Dayton, Ohio, 1985-86; U. Tex. at San Antonio adminstrv. officer, Radiation and Laser Safety, 1985-92. Contbr. articles to profl. jours. Scientific advisor, cons. NIH, Bethesda, Md., 1987, 88, NSF, Washington, 1987, 91, 92, 93. Named postgrad. scholar Nation Rsch. Coun. of Can., 1977-78, postdoctoral fellow Med. Rsch. Coun. Can., 1979-82, Alta. Heritage Found. Med. Rsch. fellow, 1981-82. Mem. AAAS, Am. Soc. Biochemistry and Molecular Biology, Assn. for Rsch. in Vision and Ophthalmology, Am. Soc. for Zoologists, Soc. for Neurosci. Achievements include research in neurobiology of the retina, biochemistry of membrane proteins, metabolism of retinoids, comparative animal physiology, and environmental and evolutionary biology. Office: U Tex San Antonio Divsn Life Scis San Antonio TX 78249

TSONIS, PANAGIOTIS ANTONIOS, biologist, researcher; b. Pili, Viotia, Greece, May 13, 1953; came to U.S., 1983; s. Antonios Anastasios and Isidora (Katerinitsas) T.; m. Katia Del Rio, May 28, 1988; children: Isidora, Sol Antoinette. BS, Patras U., 1977; MS, Nagoya U., 1980, PhD, 1983. Researcher Scripps Clinic, La Jolla, Calif., 1983-84, La Jolla Cancer Rsch. Found., 1984-88; prof. Nat. U., San Diego, Calif., 1986-88, Ind. U., Indpls., 1988-89, U. Dayton, Ohio, 1989—; adv. bd. scientific jour., Athens, 1989—. Editor: Recent Trends in Development, 1991; contbr. articles to profl. jours. Recipient 1st award NIH, 1990—; arthritis investigator, Nat. Arthritis Found., 1987-90, investigator Nat. Kidney Found., 1992. Mem. Soc. Devel. Biology. Achievements include study of cancer-related aspects of regeneration. Resistance in carcinogenisis in salamanders; mechanisms of limb regeneration; specific gene expression during regeneration. Office: Univ Dayton 300 College Park Dayton OH 45469-2320

TSUCHIDA, EISHUN, chemistry educator; b. Tokyo, May 2, 1930; mm. Hideko Tsuchida, Oct. 1, 1955; 1 child, Emi. BS, Sci. U. Tokyo, 1955; DSc, Waseda U., 1960. Cert. macromolecular chemistry. Rsch. assoc. Tokyo Inst. Tech., 1960-63; assoc. prof. Waseda U., Tokyo, 1963-72, prof., 1973—. Editor in chief Polymer Jour., 1986—; editor Jour. Macromolecular Sci., 1980—, Polymers for Advanced Technologies, 1990—; editor, author (with others): Macromolecular Complexes, 1991; contbr. over 500 articles to internat. jours.; author, co-author over 30 books. Mem. Chem. Soc. Japan (v.p. 1991-93, award 1985, Okuma Prize 1986), Soc. Polymer Sci. Japan (v.p. 1986-91, award 1990) Am. Chem. Soc., Royal Soc. Chemistry. Buddhist. Home: 2-10-10 Seki-minami, Nerima Tokyo 177, Japan Office: Waseda U, Dept Polymer Chemistry, Tokyo 169, Japan

TSUCHIYA, YUTAKA, photonics engineer, researcher; b. Mie-Pref, Japan, Mar. 22, 1942; s. Shiro and Setsuko (Morimoto) T.; m. Kazuko Nishida, Apr. 19, 1969; 1 child, Yuri. B of Engring., Aichi Inst. Tech., Nagoya, Japan, 1965; PhD in Engring., U. Tokyo, 1985. From engr. to div. mgr. Hamamatsu (Japan) Photonics, 1965-90, dep. dir. Cen. Rsch. Lab., 1990—. Inventor photon counting imaging system, optical oscilloscope, ultrafast streak cameras (Okochi meml. prize 1991); patentee in field. Recipient Suzuki Meml. award Inst. TV Engrs. Japan, 1978, Award for the Paper Inst. TV Engrs., Japan, 1981, Laser award Laser Soc. Japan, 1986, Chunich prize Chunichi Newspaper, Nagoya, Japan, 1991, Takayanagi Memorial prize, 1993. Mem. IEEE, OSA, APS, The Soc. Photo-Optical Instrumentation Engrs. (bd. dirs.). Avocations: swimming, classical music, game of go. Home: 2-9-38 Johoku, Hamamatsu 432, Japan Office: Hamamatsu Photonics, 5000 Hirakuchi, Hamakita 434, Japan

TSUDA, KYOSUKE, organic chemist, science association administrator; b. Urawa, Saitama, Japan, Feb. 10, 1907; s. Sosuke and Fusa (Kobayashi) T.; m. Eiko Mio, Apr. 22, 1936; children: Takeshi, Kazuko, Osamu. B., U. Tokyo, 1929, D., 1936. Asst. faculty of pharm. sci. U. Tokyo, 1929-38, assoc. prof. faculty of pharm. sci., 1938-51; prof. faculty of pharm. sci. Kyushu U., Fukuoka, Japan, 1951-55; prf. Inst. Applied Microbiology U. Tokyo, 1955-67, prof. emeritus, 1967; pres. Kyoritsu Coll. Pharmacy, Tokyo, 1967-84; dir. Rsch. Found. for Pharm. Sci., Tokyo, 1980—; lectr. New Year's Lecture at the Imperial Ct., Japan, 1979; chmn. Cen. Pharm. Affairs Coun. of Japan, 1975-81. Regional editor Tetrahedron & Tetrahedron Letters, 1967-78. Recipient Asahi Cultural award, 1965; named to Order of Culture, Japanese Cabinet under Premier Suzuki, 1982. Mem. Pharm. Soc. Japan (pres. 1980—), Japan Acad. (award 1966). Buddhist. Home: 27-20 Higashinaka-cho, Urawa 336, Japan Office: Rsch Found for Pharm Sci, 2-12-15 Shibuya Shibuya-ku, Tokyo 150, Japan

TSUDA, TAKAO, chemistry educator; b. Takatsuki, Ika, Japan, Aug. 22, 1940; m. Chizuko Tomita, Mar. 18, 1970; children: Ryotaro, Kozue. BS, Nagoya (Japan) Inst. Tech., 1963; MS, Nagoya U., 1965, D in Engring., 1974. Rsch. assoc. Nagoya U., 1966-79; assoc. prof. chemistry Nagoya Inst. Tech., 1979—; vis. scholar Stanford U., 1989-90. Author: Chromatography, 1989; editor jour. Chromatography, 1991; mem. editorial adv. bd. Jour. Microcolumn Separation, 1991; patentee rectangular capillary for capillary zone electrophoresis. Recipient Tokai Kagaku Kai Sho award Soc. Tokai Chemistry, 1980; grantee Suzuken Mel. Found., 1983, Toyota Found., 1984, Ishida Found., 1985, Daikoo Found., 1986; German Acad. Exch. Svc. scholar U. Von Mainz, 1984. Mem. Soc. for Chromatographic Sci. (editor), Am. Chem. Soc., Japan Chem. Soc., Japan Soc. for Analytical Chemistry. Avocations: social dancing. Home: Mongi 1-26 Iwasaki, Nisshin Aichi 470-01, Japan Office: Nagoya Inst Tech, Gokiso Showa, Nagoya Aichi 466, Japan

TSUGE, SHIN, chemistry educator; b. Nakatsugawa, Gifu, Japan, Feb. 1, 1939; s. Juichi and Shigeru (Yamamoto) T.; m. Nobuko Tamano, Apr. 24, 1966; children: Ken, Ryo. BS, Nagoya U., 1962, MS, 1964, PhD, 1970. Rsch. assoc. Nagoya U., 1965-75, assoc. prof., 1975-81, prof., 1981—; rsch. assoc. U. N.C., Chapel Hill, 1971-73. Editor: Jour. High Resolution Chromatography, Heidelberg, 1989—; editorial bds. Fresenis Zeitschrift Analtische Chemie, Heiderberg, 1980—, Jour. Analytical and Applied Pyrolysis, Amsterdam, 1979—; author books on polymer characterization, 1977—; contbr. articles in field. Recipient Award for Younger Researcher, Japan Soc. Analytical Chemistry, Tokyo, 1971, award Soc. Polymer Sci., Tokyo, 1979, Divisonal award Chem. Soc. Japan, 1987. Avocations: music, wine tasting. Home: Kitahonjigahara 3-12, Aichi, Owari-ashahi 488, Japan Office: Nagoya U, Dept Applied Chemistry, Nagoya 464, Japan

TSUI, CHIA-CHI, electrical engineer, educator; b. Shanghai, Apr. 23, 1953; came to the U.S., 1979; s. Xue-Zhu and Zhao-Hui (Zhang) T.; m. Rui Li. BS in Computer Sci., Concordia U., Montreal, 1979; MEE, SUNY, Stony Brook, 1980, PhD, 1983. Farmer Long River State Farm, Hei-Long-Jiang, Peoples Republic of China, 1969-75; asst. prof. dept. elec. engring. Northeastern U., Boston, 1983-88; assoc. prof. CUNY Coll. Staten Island, 1988—. Contbr. over 50 articles to profl. jours. Mem. IEEE (sr.). Home: 743 Clove Rd Staten Island NY 10310-2737 Office: CUNY Coll Staten Island Dept Applied Scis Staten Island NY 10301

TSUJI, HARUO, orthopaedics educator; b. Karasuyama, Tochigi, Japan, Jan. 3, 1933; s. Sabroh and Hiroko Tsuji; m. Keiko Tsuji, Oct. 25, 1963; children: Yuko, Tomoaki, Makiko. MD, Chiba (Japan) U., 1958, PhD, 1963. Asst. prof. Chiba U., 1963-69, assoc. prof., 1969-78; prof. of orthopaedics, chmn. Toyama (Japan) Med./Pharm. U., 1978—, pres. univ. libr., 1979-81. Author: Comprehensive Atlas of Lumbar Spine Surgery, 1991; editor jours. Spine, 1987, Jour. Japanese Ortopaedics Assn., 1988, Spine and Spinal Cord, 1988. Grantee Japanese Ministry Edn. and Sci., 1977-81, 82-93, Uehara Meml. Found., 1989. Mem. Japan Spine Rsch. Soc. (bd. dirs. 1974—, pres. 1993), Japanese Orthopaedic Assn., Soc. Internat. Chirugie Orthopeadique et Traumatologie, Internat. Soc. for Study of Lumbar Spine. Avocations: photography, painting. Home: Anyobo 162-2, Toyama 930, Japan Office: Toyama Med & Pharm U, Sugitani 2630, Toyama 930-01, Japan

TSUJI, MORIYA, immunologist; b. Tokyo, Jan. 1, 1958; came to the U.S., 1987; m. Veronica Colomer-Gould, July 11, 1993. MD, Jikei U., 1983; PhD, U. Tokyo, 1987. Rsch. asst. NYU Med. Ctr., N.Y.C., 1987-90, instr., 1990-91, asst. prof., 1991—. Translator: Introducing Immunology, 1987; contbr. articles to profl. jours. Scholar Nippon Ikuei-Kai, 1983-87; recipient Instnl. award Am. Cancer Soc., 1993, NYU Whitehead Presdl. award, 1994. Mem. Am. Assn. Immunologists, AAAS, N.Y. Acad. Sci. Achievements include research in role of CD4 lymphocytes in malaria immunity, functional role of gamma delta t lymphocytes in infectious disease. Office: NYU Sch Medicine 341 E 25th St New York NY 10010

TSURUTA, YUTAKA, building materials researcher; b. Shibuya-ku, Tokyo, Japan, Mar. 8, 1936; s. Akira and Masako (Iba T.; m. Yasuko Kiyosada, Nov. 1, 1966; children: Akane, Hikaru. BEng, Waseda U., Tokyo, 1958, MEng, 1960. Mem. staff Taisei Corp., Tokyo, 1960-70, mem. sr. staff, 1970-79, dir. finishing materials team, 1979-83; dir. bldg. materials div. Taisei Corp., Yokohama, Japan, 1983—; staff mem. Japan Indal. Standards, 1965—. Author: Waterproofing for Roofs, 1975, Waterproofing for Indoors, 1976, Details of Roofing and Waterproofing, 1977; patentee open joint of curtain wall. Recipient award Japan Soc. for Finishings Tech., 1993. Mem. ASTM, Archtl. Inst. Japan (chmn. waterproofing rsch. com. 1990—). Office: Taisei Corp, 344-1 Nase-cho, Totsuka-ku, Yokohama 145, Japan

TSUSUE, AKIO, geology educator; b. Tokyo, Sept. 24, 1928; s. Soichi and Eiko (Hirose) T.; m. Chiseko Hamada, May 1, 1957; children: Yoichi, Mariko. BS, U. Tokyo, 1954, MS, 1956, DSc, 1960. Rsch. fellow U. Tokyo, 1960-70; assoc. prof. econ. geology Kumamoto (Japan) U., 1970-73, prof., 1973-90, 92—, dean and prof., 1990-92; postdoctoral fellow Columbia U., N.Y.C., 1960-6l, rsch. assoc., 1961-62; rsch. assoc. Princeton (N.J.) U., 1962-63; vis. prof. Pa. State U., University Park, 1981; leader overseas field rsch. Korean granitic rocks and assoc. ore deposits, 1979-81, 82-84, 85-87. Mem. Soc. Econ. Geologists, Soc. Mining Geologists Japan (trustee 1977-79, 86-88), Mineral. Soc. Japan, Geochem. Soc. Japan, Geol. Soc. Japan. Mem. Liberal Democratic Party. Buddhist. Avocation: floriculture. Home: Shinyashiki l-l-35, Kumamoto 862, Japan Office: Kumamoto U, Kurokami 2-39-l, Kumamoto 860, Japan

TSZTOO, DAVID FONG, civil engineer; b. Hollister, Calif., Oct. 13, 1952; s. John and Jean (Woo) T.; m. Evelyn Yang, July 31, 1982; children: Michaela Gabrielle, Shawn Michael. BS, Calif. Poly. State U., 1974; MS in Engring., U. Calif., 1976. Registered profl. civil engr., Calif. Engr., dispatcher Conlec Corp., Hollister, Calif., 1972-73; rsch. asst. U. Calif. Engring. Dept., Berkeley, 1975-76; jr. civil engr. Contra Costa County Pub. Works, Martinez, Calif., 1977-78, asst. civil engr., 1978-81, civil engr. III, 1981-83; assoc. civil engr. East Bay Mucpl. Util. Dist., Oakland, Calif., 1983-88; sr. civil engr. East Bay Mucpl. Util. Distbr., Oakland, Calif., 1988—; chpt. chmn. We. Coun. Engrs., Martinez, Calif., 1977-82; mem. Nat. Soc. Profl. Engrs., Martinez, Calif., 1977-82. Co-author: Energy Absorbing Devices in Structures, 1977, EQ Testing of Stepping Frame with Devices, 1977, Development of Energy-Absorbing Devices, 1978. Sponsor Sing & Bring Children's Club, Oakland, San Lorenzo, Calif., 1982-90; v.p. Sun Country Homeowners, Martinez, Calif., 1977-82. Recipient Presdl. Design Achievement award Nat. Endowment for the Arts, Washington, 1984. Mem. Am. Soc. Civil Engrs., Tau Beta Pi, Phi Kappa Phi. Republican. Baptist. Achievements include patent for application work, Conlec Corp., Hollister, Calif., 1973. Office: E Bay Mcpl Util Dist 375-11th St Oakland CA 94607

TU, CHARLES WUCHING, electrical and computer engineering educator; b. Taipei, Taiwan, Republic of China, Jan. 5, 1951; came to U.S., 1965; BSc, McGill U., 1971; PhD, Yale U., 1978. Lectr. physics dept. Yale U., New Haven, Conn., 1978-80; mem. tech. staff AT&T Bell Labs., Murray Hill, N.J., 1980-87, disting. MTS, 1987-88; assoc. prof. dept. elec. and computer engring. U. Calif., La Jolla, 1988-91, prof. dept. elec. and computer engring., 1991—; adv. bd. mem. U.S. Molecular Beam Epitaxy Workshop, 1990—; com. mem. Elec. Materials Conf., 1991—. Author: (with others) MBE and the Technology of Selectively Doped Heterostructure Transistors, 1985; editor conf. proceedings. Recipient award UN Devel. Program, 1991; grantee Office Naval Rsch., 1988—, NSF, 1991—. Mem. IEEE, Materials Rsch. Soc. (co-chair symposium 1989, 92), Am. Vacuum Soc. (exec. bd. 1989-91), The Minerals, Metals and Materials Soc. Achievements include patents for shallow impurity neutralization, gaseous etching process. Office: U Calif Dept Elec & Computer Engring La Jolla CA 92093-0407

TU, KING-NING, materials science and engineering educator; b. Canton, China, Dec. 30, 1937; came to U.S., 1962; s. Ying-Chiang Tu and San-Yuk Chen; m. Ching Chiao, Sept. 25, 1964; children: Olivia, Stephen. BSc, Nat. Taiwan U., 1960; MSc, Brown U., 1964; PhD, Harvard U., 1968. Rsch. staff mem. IBM Watson Rsch. Ctr., Yorktown Heights, N.Y., 1968-93, sr. mgr. thin film sci. dept., 1978-85, sr. mgr. materials sci. dept., 1985-87; prof. dept. materials sci. & engring. UCLA, 1993—. Co-author: (textbook) Electronic Thin Film Science, 1992; editor: Materials Chemistry & Physics, 1992—. Fellow Am. Phys. Soc., The Metall. Soc. (Applications to Practice award 1988), Churchill Coll. (U.K.). Achievements include 8 patents on thin film technology for microelectronics. Office: IBM T. J. Watson Research Ct PO Box 218 Yorktown Heights NY 10598

TUBBS, DAVID EUGENE, mechanical engineer, marketing professional; b. Springfield, Ill., Jan. 12, 1948; s. Eugene Lewellyn and Jacqueline Flo (Jones) T.; m. Linda Alyson Smith, Aug. 2, 1970; children: Corbin David, Cavan Scott. BSME, Ill. Inst. Tech., 1970; postgrad., Okla. State U., 1992. Registered profl. engr., Ill. Project engr. Sargent & Lundy, Chgo., 1970-82, bus. devel. mgr., 1982-83; mgr. power sales Yuba Heat Transfer Corp., Tulsa, 1983-85; with press products mktg. Nordam, Tulsa, 1985-86; dir. mktg. Brooks Aero. Svc. div. Nordam, Tulsa, 1986-91; mech. dept. mgr. The Benham Group, Tulsa, 1991-93; chief mech. engr. EDECO Engrs./Cons, Tulsa, 1993—. Mem. ASME, Am. Welding Soc., Ill. Inst. Tech. Alumni Assn. (bd. dirs. 1977-80), Delta Tau Delta, Pi Tau Sigma. Republican. Club: Toastmasters. Avocations: racquetball, USSF soccer referee. Home: 3317 W Quincy St Broken Arrow OK 74012-9034 Office: EDECO Engrs/Cons 4500 S Garnett Tulsa OK 74146

TUBURAN, ISIDRA BOMBEO, aquaculturist; b. El Salvador, The Philippines, May 15, 1955; d. Rosaleo Bajuyo and Anatolia (Buna) Bombeo; m. Reynaldo Torres Tuburan, May 9, 1982; children: Angelie Marie, Ray Oliver, Gwen Valorie. BSc in Marine Biology, Xavier U., Cagayan de Oro, The Philippines, 1976; MSc in Fisheries, U. of the Philippines, Iloilo, 1980. Rsch. asst. S.E. Asian Fisheries Devel. Ctr., Iloilo, 1978-80, rsch. assoc., 1980-88, assoc. scientist, 1989—; R & D technician Syarikat Yadeka, Sabah, Malaysia, 1986; lectr. Tech. and Livelihood Resource Ctr., The Philippines, 1987-88. Contbr. articles to profl. publs. Grantee S.E. Asian Fisheries Devel. Ctr., 1976-78, Internat. Found. for Sci., 1987—; recipient Spl. Tng. award Brit. Coun., 1982-83. Mem. Asian Fisheries Soc., Philippine Assn. Marine Sci. Roman Catholic. Avocations: reading, ball games, swimming. Home: Lot 18 Block 19 NHA II, Mandurriao 5000, The Philippines Office: SE Asian Fisheries Devel Ct, Tigbauan, Iloilo 5021, The Philippines

TUCCI, JAMES VINCENT, physicist; b. Hollis, N.Y., Feb. 13, 1939; s. Francis Vincent and Marguerite Tucci; m. Dorothy Helen Tucci, June 2, 1962; 1 child, Angela. BA, Hofstra U., 1962; MS, U. Mass., 1965, PhD, 1966. Instr. U. Bridgeport, Conn., 1966, asst. prof., 1967-69, assoc. prof., 1969-72, prof., 1972—, chair of physics, 1968-71, 71-89, dir. div. of sci. and math., 1989—; cons. for attys. in accident reconstrn. and optics. Active Pub. Schs., Bridgeport Sch. System. Mem. Optical Soc. Am. Office: U Bridgeport Bridgeport CT 06601

TUCCIO, SAM ANTHONY, aerospace executive, physicist; b. Rochester, N.Y., Jan. 15, 1939; s. Manuel Joseph and Phillis (Cannizzo) T.; m. Jenney Laprell Elvington, May 1, 1982; children: David Samuel, Karen Ann, Rebecca Jean, Ashley Lauren. BS, U. Rochester, 1965. Research physicist Eastman Kodak Co., Rochester, 1965-72; program mgr. Lawrence Livermore (Calif.) Labs., 1972-75; sr. physicist Allied Corp., Morristown, N.J., 1975-81; gen. mgr. Allied Laser Products Div., Westlake Village, Calif., 1981-84; sr. bus. mgr. Northrop Corp., Hawthorne, Calif., 1984-89; dir. space bus. Loral Corp., Pasadena, Calif., 1989-92; dir. bus. devel. ThermoTrex Corp., San Diego, 1992—. Patentee in field; contbr. numerous articles to profl. jours. Recipient IR 100 award Indsl. Rsch. Assn., 1971. Republican. Methodist. Home: 18035 Polvera Way San Diego CA 92128-1123 Office: ThermoTrex Corp 9550 Distribution Ave San Diego CA 92121-2306

TUCKER, BEVERLY SOWERS, information specialist; b. Trenton, N.J., Dec. 1, 1936; d. Eldon Jones and Verbeda Eleanor (Roberts) Sowers; m. Harvey Richard Tucker, Dec. 27, 1958 (div. Nov. 1983); children: Randall Richard, Brian Alan. BS in Chemistry with distinction, Purdue U., 1958; MS in Geology, No. Ill. U., 1985; MA in Library and Info. Sci., Rosary Coll., 1989. Asst. rsch. librarian CPC Internat., Argo, Ill., 1958-62; chem. patent searcher Chgo., 1962-66; info. specialist C. Berger & Co., Wheaton, Ill., 1986, Amoco Corp., Naperville, Ill., 1987—; faculty Coll. Du Page, Glen Ellyn, Ill., 1989—. Mem. Spl. Libraries Assn., Ill. Fedn. Women's Club (treas. 5th dist. 1979-81, Outstanding Jr. Clubwoman award 1979-80), Garden Club Council Wheaton (pres. 1981-82), Wheaton Jr. Woman's Club (pres. 1977-78, Single Parent scholar 1984), Gardens Etc. Club (pres. 1978-79), Alpha Lambda Delta, Delta Rho Kappa, Theta Sigma Phi, Alpha Chi Omega (grantee 1985). Republican. Presbyterian. Avocations: bridge, needlework, gourmet cooking. Home: 1507 Paula Ave Wheaton IL 60187-6135 Office: Amoco Corp PO Box 3083 Warrenville Rd and Mill St Naperville IL 60566

TUCKER, KATHERINE LOUISE, nutritional epidemiologist, educator; b. Balt., Feb. 2, 1955; d. Richard Kennedy and Jeanne Margaret (Crowley) Tucker; m. Luis Miguel Falcon, June 6, 1987; 1 child, Alex Tucker. BSc, U. Conn., 1978; PhD, Cornell U., 1986. Program mgr. Cornell Nutritional Surveillance Program, Ithaca, N.Y., 1986; asst. prof. nutritional epidemiology McGill U., Ste. Anne de Bellevue, Que., 1986-89, Tufts U., Boston, 1989—; cons. in field. Editorial bd. Jour. Ecology of Food and Nutrition, 1992—; contbr. articles to profl. jours.; author monograph: Advances in Nutritional Surveillance, 1989. Andrew D. White fellow, 1981-83. Mem. Am. Inst. Nutrition. Office: Tufts Univ USDA Human Nutrition Rsch 711 Washington St Boston MA 02111

TUCKER, RANDOLPH WADSWORTH, engineering executive; b. Highland Pk., Ill., Dec. 3, 1949; s. Thomas Keith and Nancy Ellen (Jung) T.; m. Jean Marjorie Zenk, June 30, 1973 (div. April 1991); 1 child, Nicholas Randolph; m. Lori Kaye Hicks, June 21, 1991. BS in Fire Protection Engring., Ill. Inst. Tech., 1972; M in Mgmt., Northwestern U., 1979. Registered profl. engr., Ill., Tex., Fla., La., Ga. With Ins. Svcs. Office of Ill., Chgo., 1972-74, bldg. insp., fire protection cons., 1972-74; with Rolf Jensen & Assocs., Inc., Deerfield, Ill., 1974—, cons. engr., 1974-77, mktg. mgr., 1977-81, mgr. Houston office, 1981-83, v.p. engring., mgr. Houston, 1983-89, v.p., tech. officer for Atlanta, Houston, N.Y.C., and Washington offices, 1989-90, sr. v.p., 1990—; mem. adv. coun. Tex. State Fire Marshal, Austin, 1983-91; Dept. of Justice/Nat. Inst. Corrections cons. to Tex. Commn. on Jail Stds., 1993—. Editorial advisor Rusting Publs., N.Y.C., 1981—, Cahners Pub., 1993—; author articles in field. V.p. Juvenile Fire Setters Program, Houston, 1982-84. Named one of Outstanding Young Men Am., U.S. Jaycees, 1981. Mem. Soc. Fire Protection Engrs. (chmn. nat. qualifications bd. 1985, pres. Houston chpt. 1983-84), Soc. Mktg. Profl. Svcs. (pres. Houston chpt. 1985, nat. pres. 1989-90), Nat. Fire Protection Assn., Profl. Svcs. Mgmt. Assn., Internat. Conf. Bldg. Ofcls., So. Bldg. Code Cong. Internat., Inc., Bldg. Ofcls. and Code Adminstrs. Internat., Bldg. Ofcls. Assn. Tex., AIA (profl. affiliate), Tex. Soc. Architects (profl.affiliate), Houston C. of C. (vice chmn. fire protection com. 1983, govt. rels. com. 1984—), Aircraft Owners and Pilots Assn., Waller Country Club. Republican. Episcopalian. Avocations: flying, golf. Office: Rolf Jensen & Assoc Inc 13831 Northwest Freeway Ste 330 Houston TX 77040-5215

TUCKER, RICHARD DOUGLAS, computer engineer, consultant; b. Lowell, Mass., Jan. 14, 1962; s. Richard Leon and Gwendoline Sylvia (Coates) T.; m. Diane D. Mitchell, June 26, 1988 (div.). Degree in computer electronics, Wentworth Tech., Waltham, Mass., 1988. Owner Quantum Inc., Tewksbury, Mass., 1985-88; field engr. NYNEX Bus. Ctrs., Boston, 1989-91; systems engr. Sears Bus. Ctrs., Burlington, Mass., 1991-93, Glasgal Comm., Newton, Mass., 1993—. Mem. Internat. Soc. Cert. Electronics Technicians, Cert. NetWare Profls. Assn. Home: 106 W Bacon St Plainville MA 02762 Office: Glasgal Comm Inc 44 Mechanic St Newton Upper Falls MA 02164

TUCKER, RICHARD LEE, civil engineer, educator; b. Wichita Falls, Tex., July 19, 1935; s. Floyd Alfred and Zula Florence (Morris) T.; m. Shirley Sue Tucker, Sept. 1, 1956; children: Brian Alfred, Karen Leigh. BCE, U. Tex., 1958, MCE, 1960, PhD in Civil Engring., 1963. Registered profl. engr., Tex. Instr. civil engring. U. Tex., 1960-62; from asst. prof. to prof. U. Tex., Arlington, 1962-74, assoc. dean engring., 1963-74; v.p Luther Hill & Assoc., Inc., Dallas, 1974-76; C.T. Wells prof. project mgmt. U. Tex., Austin, from 1976, dir. Constrn. Industry Inst., from 1983, dir. Constrn. Engring. and Project Mgmt. Program; pres. Tucker and Tucker Cons., Inc., Austin, 1976—. Contbr. numerous articles and papers to profl. jours. Recipient Erwin C. Perry award, 1978, Faculty Excellence award, 1986, Joe J. King Profl. Engring. Achievement award Coll. Engring. U. Tex., Austin, 1990; named Outstanding Young Engr., Tex. Soc. Profl. Engrs., 1965, Outstanding Young Man, City of Arlington, 1967, Michael Scott Endowed Rsch. fellow Inst. for Constructive Capitalism, 1990-91, Ronald Reagan award for Individual Initiative, Construction Industry Inst., 1991 (1st recipient), Construction Engring. Construction Educator award, NSPE, 1993. Fellow ASCE (R.L. Peurifoy award 1986, Thomas Fitch Rowland prize 1987, Tex. sect. award of honor 1990); mem. NSPE (Constrn. Engring. Educator award Profl. Engrs. in Constrn. 1993), NRC, Soc. Am. Mil. Engrs., The Moles (hon.). Baptist. Office: U Tex at Austin Dept Civil Engring 3208 Red River St Ste 300 Austin TX 78712

TUCKER, ROBERT ARNOLD, electrical engineer; b. Atlanta, Jan. 30, 1941; s. Hugh Dorsey and Mary Ella (Dobbs) T.; m. Judy Elizabeth Henley, Sept. 18, 1964 (div. Sept. 13, 1979); children: Paige Elizabeth, Priscilla Elaine, Matthew Arnold; m. Joan Janet Ashton, Nov. 18, 1983; children: Beth Bennett, Russell E. McKenna III. BEE, Ga. Inst. Tech., 1964. Registered profl. engr., Ala., Fla., Ga., Tenn. Engr. in tng. Patterson & Dewar Engrs. Inc., 1964; substation design engr. Ga. Power Co., 1964-65, comml. sales engr., sr. engr., 1965-67, 72-73; dist. sales engr. Sylvania Elec. Products, Inc., 1967-71; regional sales mgr. The J.H. Spaulding Co., 1971-72; gen. sales mgr. Perimeter Lighting, Inc., 1973-74; utility mgmt. engr. Colonial Stores, Inc., 1975-76; prin. R. Arnold Tucker & Assocs., P.C., 1977-82; dist. sales engr. GTE Products Corp., 1982-86, mgr. tech. programs, 1986-90, engr. regional lighting/energy mgmt., 1990-93; comml. engr. Osram Sylvania Inc., Atlanta, 1993—. Contbr. articles to profl. jours. Mem. Kiwanis. Mem. NSPE, Illuminating Engring. Soc., Assn. Energy Engrs. Republican. Baptist. Home: 381 Shelton Woods Ct Stone Mountain GA 30088 Office: Osram Sylvania Inc 5169 Pelican Dr Atlanta GA 30349

TUCKER, ROBERT KEITH, environmental scientist, research administrator; b. Santa Ana, Calif., Feb. 22, 1936; s. Lloyd Levi and Margaret Corinne (Skiles) T.; m. Sharon Penney Langs, 1961 (div. 1974); children: Katherine Penney, Lee Ann; m. Joan Cook Luckhardt, Oct. 9, 1977. BA in Biochemistry, U. Calif., Berkeley, 1963; MA in Marine Biology, Humbolt State U., 1967; PhD in Zoology, Physiological Ecology, Duke U., 1971. Rsch. biologist Sandy Hook Lab. NOAA Nat. Marine Fisheries Svc., Highlands, N.J., 1971-77; rsch. scientist N.J. Dept. Environ. Protection and Energy, Trenton, 1977-78, mgr. water and biota unit Office Cancer and Toxic Substances Rsch., 1978-80, asst. dir. Office Cancer and Toxic Substance Rsch., 1980-83, dep. dir. Office Sci. and Rsch., 1983-86, dir. Divsn. Sci. and Rsch., 1986—; chmn. environ. task force Gov.'s Coun. Prevention Mental Retardation and Devel. Disabilities, 1984-86, Interagy. Task Force Prevention Lead Poisoning, 1986—; mem. bd. trustees N.J. Marine Scis. Consortium, 1987—; mem. N.J. Commn. Cancer Rsch., 1984—; mem. adv. bd. Environ. and Occupational Health Scis. Inst. U. Medicine and Dentistry N.J., 1987—, N.J. Comprehensive Cancer Ctr., 1991—, Greater N.Y. Bight Marine Rsch., 1992—; mem. external adv. bd. Nat. Inst. Occupational Safety and Health, 1990-92; mem. review panel estuarine rsch. northeast ctr. NOAA, Nat. Marine Fisheries Svc., 1989; organizer, dir. session 100th anniversary meeting Am. Soc. Testing and Materials, 1984; external com. mem. for PhD candidates Ruthers U., CUNY, Hunter Coll., U. Medicine and Dentistry N.J., 1972—. Contbr. over 50 articles to profl. jours. Fellow USPHS, 1967-70, recipient Merit citation NOAA, 1975, Outstanding Scientist award N.J. DEP, 1980. Mem. AAAS, Am. Chem. Soc. (organizer, dir. sessions on environ. chemistry Mid-Atlantic regional meeting 1979), Internat. Soc. Environ. Epidemiology, Soc. Occupational and Environ. Health, Crustacean Soc. Democrat. Unitarian. Achievements include reasearch in lead poisoning, problems dealing with petroleum contaminated soils, toxics in drinking water and groundwater, effects of toxic contaminants on marine resources. Office: NJ Dept Environ Protection & Energy Divsn Sci & Rsch CN409 Trenton NJ 08625

TUCKER, RODNEY STUART, electrical and electronic engineering educator, consultant; b. Melbourne, Australia, Mar. 14, 1948; s. Jack Wilbur and Irene Nason (Wilkinson) T.; m. Gretel Susan Lamont, May 10, 1978; childen: Fiona, Adrian. B of Engring., U. Melbourne, 1969, PhD, 1975. Lectr. U. Melbourne, Victoria, 1972-73; rsch. fellow U. Calif., Berkeley, 1975-76, Cornell U., Ithaca, N.Y., 1976-77; prin. rsch. scientist Plessey Co., Eng., 1977-78; sr. lectr. U. Queensland, Australia, 1978-82; mem. tech. staff AT&T Bell Labs., Holmdel, N.J., 1982-90; prof., head dept. elec. and electronic engring. U. Melbourne, Victoria, 1990—; assoc. dir. Australian Photonics Cooperative Rsch. Cntr., 1992—; cons. Harris Corp., Palo Alto, Calif., 1978-1980, Compact Engring., Patterson, N.J., 1990—; mem. mgmt. Australian Telecommunications and Electronics Rsch. Bd., Sydney, Australia, 1991—. Patentee in field of photonics; contbr. over 200 articles to profl. jours. Recipient Harkness fellowship Commonwealth Fund of N.Y., 1975-79. Fellow IEEE (editor Transactions on Microwave Theory and Techniques, N.Y.C., 1988-90), Inst. of Engrs. (Fisk Prize, 1970). Office: Univ Melbourne, Dept Elec & Electronic Engring, Parkville 3052, Australia Home: Hawthorn, 35 Grove Rd, 3122 Victoria Australia

TUCKER, ROY ANTHONY, electro-optical instrumentation engineer, consultant; b. Jackson, Miss., Dec. 11, 1951; s. Roy Anthony and Marjorie Faye (Human) T. BS in Physics, Memphis State U., 1978; MS in Sci. Instrumentation, U. Calif., Santa Barbara, 1981. Planetarium tech. Memphis (Tenn.) Mus. Planetarium, 1976-78; engring. tech. Kitt Peak Nat. Obs., Tucson, 1979; electro-optical engr. Multiple Mirror Telescope Obs., Tucson, 1981-83; rsch. engr., dept. physiology Univ. Ariz., Tucson, 1988-92; electro-optical engr. Applied Tech. Assocs., Inc., Albuquerque, N.Mex., 1992—; cons. Roy Tucker and Assocs., Vancouver, Wash., 1983-84, Tucson, 1984—. Sgt. USAF, 1972-76. Mem. AAAS, Am. Astron. Soc. Home: 5600 Gibson Blvd SE Apt 233 Albuquerque NM 87108

TUCKER, SHIRLEY LOIS COTTER, botany educator, researcher; b. St. Paul, Apr. 4, 1927; d. Ralph U. and Myra C. (Knutson) Cotter; m. Kenneth W. Tucker, Aug. 22, 1953. B.A., U. Minn., 1949, M.S., 1951; Ph.D., U. Calif.-Davis, 1956. Asst. prof. botany La. State U., Baton Rouge, 1967-71, assoc. prof., 1971-76, prof., 1976-82, Boyd prof., 1982—. Co-editor: Aspects of Floral Development, 1988; contbr. more than 90 articles on plant devel. to profl. jours. Fellow Linnean Soc., London, 1975—; Fulbright fellow Eng., 1952-53. Mem. Bot. Soc. Am. (v.p. 1979, program chmn. 1975-78, pres.-elect 1986-87, pres. 1987-88, Merit award 1989), Am. Bryological and Lichenological Soc., Brit. Lichenological Soc., Am. Inst. Biol. Scis., Am. Soc. Plant Taxonomists, Sigma Xi, Phi Beta Kappa. Home: 1022 Baird Dr Baton Rouge LA 70808-5922 Office: La State U Dept Botany Baton Rouge LA 70803

TUCKER, WILLIAM GENE, environmental engineer; b. Lebanon, N.H., Mar. 28, 1942; s. William S. Tucker and Lily A. (King) Clark; m. Elizabeth A. Glessing, Mar. 19, 1966 (div. July 1988); children: William A., Lisa B.; m. Virginia Massey, Aug. 28, 1992. BS in Chem. Engring., RPI, 1964; MS in Environ. Engring., U. Kans., 1967; PhD, U. Washington, 1975. Profl. engr., Ohio. Chem. process design engr. Allied Chem. Corp., Syracuse, N.Y., 1964-66; commd. officer U.S. Pub. Health Svc., Cin., 1967-70; rsch. mgr. U.S. EPA, N.C., 1972-90; vis. fellow J.B. Pierce Lab./ Yale U., New Haven, Conn., 1990-91; dep. dir. Pollution Control div. AEERL U.S. EPA, 1992--; edit. advisor Indoor Air Internat. Jour. Indoor Air Quality and Climate, 1990—; chmn. SSPC62 Com. for Revision Ventilation Standard, ASHRAE, Atlanta, 1991—. Editor: Sources of Indoor Air Contaminants, 1992. Mem. Am. Soc. Heating, Refrigerating, Air Conditioning Engrs., Internat. Soc. Indoor Air Quality and Climate, Internat. Acad. Indoor Air Scis. Office: USEPA MD-54 Research Triangle Park NC 27709

TUCKSON, REED V., university president. Pres. Charles R. Drew U., L.A. Office: Charles R Drew U Office of President 1621 E 120th St Los Angeles CA 90059

TUDOR, GREGORY SCOTT, land information system analyst; b. Seattle, May 6, 1965; s. John Mack and Dora Lou (Blair) T.; m. Thea Nanette Mounts, Mar. 24, 1990; 1 chld, Erin Moira. BA in Physics and Russian, Colgate U., 1987; MS Engring. in Land Surveying, U. Wash., 1989. Land survey technician U.S. Bur. Land Mgmt., Anchorage, Alaska, 1986-87; teaching asst. in civil engring. U. Wash., Seattle, 1988-89; land survey party chief Charles J. Shank, PLS, Bainbridge Island, Wash., 1989-90; land info. system analyst Wash. State Dept. Natural Resources, Olympia, 1990—; presenter at confs. in field. Mem. ASCE (assoc.), Am. Congress on Surveying and Mapping, Am. Soc. for Photogrammetry and Remote Sensing, Mountaineers, Tau Beta Pi. Achievements include rectification of aerial photographs using digital elevation models; research in global positioning system for geographic information systems, measurement based land information systems, least squares network adjustments. Office: Wash State Dept Natural Resources PO Box 47020 Olympia WA 98504-7020

TUDOR, JOHN JULIAN, microbiologist; b. Knoxville, Tenn., Sept. 5, 1945; s. John Julian and Ida Louise (Vanderpool) T.; m. Patricia Agnes Hudson, Jan. 4, 1969; children: Stephen Matthew Tudor, Michael Ryan Tudor. BS, U. Ky., 1967, PhD, 1977. Microbiology lectr. U. Ky., Lexington, Ky., 1968-71; biology instr. Winthrop Coll., Rock Hill, S.C., 1972-74; biology asst. prof. St. Joseph's U., Phila., 1977-84, biology assoc. prof., 1984-89; microbiology vis. prof. Temple U., Phila., 1988-89; biology prof. St. Joseph's U., Phila., 1989—. Contbr. articles to profl. jours. Fellow U.S. Pub. Health Svc., Washington U., St. Louis, 1971-72; recipient Sherago Excellence award Ky-Tenn Branch ASM, 1976; grantee Nat. Inst. Gen. Med. Sci., St. Joseph's U., Phila., 1989-92, Nat. Inst. Human Genome Rsch., St. Joseph's U., Phila., 1992—. Mem. Am. Soc. Microbiology, Coun. on Undergrad. Rsch., Sigma Xi. Office: St Josephs University 5600 City Ave Philadelphia PA 19131

TUFTE, ERLING ARDEN, civil engineer; b. Dallas, Jan. 19, 1946; s. Elmer Theodore and Agnes Irene (Gunderson) T.; m. Elizabeth Margit Thoreson, Nov. 29, 1968; children: John Erl, Kari Elizabeth. BSCE, U. N.D., 1969; MS in Indsl. and Mech. Engring., Mont. State U., 1982; PhD in Engring., N.D. State U., 1986. Registered profl. engr., Minn., Mont., N.D. Asst. city engr. City of Moorhead, Minn., 1972-74; city engr. City of Jamestown, N.D., 1974-78, City of Helena, Mont., 1978-80; cons. grad. students Mont. State U., Bozeman, 1980-82; instr. grad. students N.D. State U., Fargo, 1982-86; dir. pub. works City of Great Falls, Mont., 1986—; com. mem. Legislative Solid Waste Mgmt. Com., Helena, 1991—. Co-author, editor: Snow and Ice Control Plan, 1989, 91. Mem. City/County Environ. Protection Com., Mont.; mem. Govt. Infrastructure Assessment Task Force N.D. Lt. USN, 1969-72. Recipient WWTP of Yr. award EPA, 1991. Mem. ASCE, Am. Pub. Works Assn. (treas. Rocky Mountain chpt. 1992—, Man of Yr. award 1991, Top Ten Pub. Works Leaders 1992 award), Great Falls C. of C., Am. Water Works Assn., Masons. Achievements include development and presentation of courses for Rural Tech. Assistance Program; engineering numerous projects for redevel. of Helena downtown area; planning a variety of urban renewal project in ctrl. bus. dist. of Moorhead, Minn. Office: City of Great Falls Pub Works 1025 25th Ave NE Great Falls MT 59404

TUKEY, HAROLD BRADFORD, JR., horticulture educator; b. Geneva, N.Y., May 29, 1934; s. Harold Bradford and Ruth (Schweigert) T.; m. Helen Dunbar Parker, June 25, 1955; children: Ruth Thurbon, Carol Tukey Schwartz, Harold Bradford. B.S., Mich. State U., 1955, M.S., 1956, Ph.D., 1958. Research asst. South Haven Expt. Sta., Mich., 1955; AEC grad. research asst. Mich. State U., 1955-58; NSF fellow Calif. Inst. Tech, 1958-59; asst. prof. dept. floriculture and ornamental horticulture Cornell U., Ithaca, N.Y., 1959-64, assoc. prof., 1964-70, prof., 1970-80; prof. urban horticulture U. Wash., Seattle, 1980—, dir. Arboreta, 1980-92, dir. Ctr. Urban Horticulture, 1980-92; cons. Internat. Bonsai mag., Electric Power Rsch. Inst., P.R. Nuclear Ctr., 1965-66; lectr. in field; mem. adv. com. Seattle-U. Wash. Arboretum and Bot. Garden, 1980-92, vice chmn., 1982, chmn., 1986-87; vis. scholar U. Nebr., 1982; vis. prof. U. Calif.-Davis, 1973; mem. various coms. Nat. Acad. Scis.-NRC; bd. dirs. Arbor Fund Bloedel Res., 1980-92, pres., 1983-84. Mem. editorial bd. Jour. Environ. Horticulture, Arboretum Bull. Mem. nat. adv. com. USDA, 1990—; pres. Ithaca PTA; troop advisor Boy Scouts Am., Ithaca. Lt. U.S. Army, 1958. Recipient B.Y. Morrison award USDA, 1987; NSF fellow, 1958-59; named to Lansing (Mich.) Sports Hall of Fame, 1987; grantee NSF, 1962, 75, Bot. Soc. Am., 1964; hon. dr. Portuguese Soc. Hort., 1985. Fellow Am. Soc. Hort. Sci. (dir. 1970-71); mem. Internat. Soc. Hort. Sci. (U.S. del. to council 1971-90, chmn. commn. for amateur horticulture 1974-83, exec. com. 1974-90, v.p. 1978-82, pres. 1982-86, past pres. 1986-90, chmn. commn. Urban Horticulture 1990—), Internat. Plant Propagators Soc. (eastern region dir. 1969-71, v.p. 1972, pres. 1973, internat. pres. 1976), Am. Hort. Soc. (dir. 1972-81, exec. com. 1974-81, v.p. 1978-80, citation of merit 1981), Royal Hort. Soc. (London) (v.p. hon. 1993—), Bot. Soc. Am., N.W. Horticulture Soc. (dir. 1980-92), Arboretum Found. (dir. 1980-92), Rotary, Sigma Xi, Alpha Zeta, Phi Kappa Phi, Pi Alpha Xi, Xi Sigma Pi. Presbyterian. Home: 3300 E St Andrews Way Seattle WA 98112-3750 Office: U Wash Ctr for Urban Horticulture GF-15 Seattle WA 98195

TULENKO, MARIA JOSEFINA, pharmacist; b. Caguas, P.R., Dec. 26, 1930; d. Rafael and Pura (Mitja) Gandara; m. Alan George Tulenko, June 23, 1956; children: Christine Alane, Eric Alan. BS in Chemistry magna cum laude, U. P.R., 1953, BS in Pharmacy magna cum laude, 1953; MBA magna cum laude, Fairleigh Dickinson U., 1973. Registered pharmacist, N.J., P.R. Mfg. chemist Syntex Corp., Hato Rey, P.R., 1953-54; analytical chemist Hoffmann La Roche, Inc., Nutley, N.J., 1954-57, supr. quality control labs., 1962, mgr. quality control labs. and quality assurance, 1966, dir. quality control labs. and quality assurance, 1975-85; asst. v.p. sci. affairs Zenith Pharm. Inc., Northvale, N.J., 1985-89; v.p. regulatory affairs and quality mgmt. systems MOVA Pharm. Corp., Caguas, 1989—; mem. bd. pharmacy edn. program Rutgers U., New Brunswick. Vice pres. Tribute to Women in Industry, North Bergen, N.J., 1978; mem. exec. bd. Edn. Opportunity Fund, Bloomfield (N.J.) Coll. Recipient Tribute to Women in Industry award YWCA, 1977. Mem. Am. Pharm. Assn., Am. Soc. for Quality Control, Tribute to Women in Industry Mgmt. Forum (v.p. programs 1978), Sigma Xi. Republican. Roman Catholic. Home: 620 Grove St Clifton NJ 07013-3843 Office: MOVA Pharm Corp PO Box 8639 Caguas PR 00726-8639

TULLY, JOHN CHARLES, research chemical physicist; b. N.Y.C., May 17, 1942; s. Harry V. and Pauline (Fischer) T.; m. Mary Ellen Thomsen, Jan. 23, 1971; children: John Thomsen, Elizabeth Anne, Stephen Thomsen. BS, Yale U., 1964; PhD, U. Chgo., 1968. NSF postdoctoral fellow U. Colo. and Yale U., 1968-70; mem. tech. staff AT&T Bell Labs., Murray Hill, N.J., 1970-82, disting. mem. tech. staff, 1982-85, head phys. chemistry rsch. dept., 1985-90, head materials chem. rsch. dept., 1990—; vis. prof. Princeton (N.J.) U., 1981-82, Harvard U., Cambridge, Mass., 1991. Contbr. articles to sci. jours.; author, prodr. movie Dynamics of Gas-Surfact Interactions, 1979. NSF predoctoral fellow, 1965-68. Fellow AAAS, Am. Phys. Soc. (chem. physics exec. com. 1983-86); mem. Am. Chem. Soc. (chmn. theoretical chemistry subdiv. 1991-92, phys. chemistry div. 1993-94), Sigma Xi. Achievements include patent on Method and Apparatus for Surface Characterization Utilizing Radiation from Desorbed Particles; fundamental theoretical contributions towards atomic leval understanding of chemical reaction dynamics. Office: AT&T Bell Labs ID346 600 Mountain Ave Murray Hill NJ 07974

TULVE, NICOLLE SUZANNE, researcher; b. Newburgh, N.Y., July 16, 1970; d. Nicholas A. Jr. and Sibylle M. (Sage) T. BS, Oswego (N.Y.) State U., 1992. Rsch. aide Rsch. Ctr. at Oswego, 1990-92; rsch. asst. Wadsworth Ctr. Labs. Rsch., Albany, N.Y., 1992—. Mem. Sigma Xi. Republican. Roman Catholic. Home: 129 W 4th St Oswego NY 13126

TUMAY, MEHMET TANER, geotechnical consultant, educator, researcher administrator; b. Ankara, Turkey, Feb. 2, 1937; came to U.S., 1959; s. Bedrettin and Muhterem (Uybadin) T.; m. Karen Nuttycombe, June 15, 1962; children—Peri, Suna. B.S. in Civil Engring., Robert Coll. Sch. Engring. (Turkey), 1959; M.C.E., U. Va., 1961; postgrad. UCLA, 1963-64; Ph.D., Tech. U. Istanbul (Turkey), 1971; Fugro-Cesco postdoctoral research fellow U. Fla., Gainesville, 1975-76. Instr. civil engring. U. Va., Charlottesville, 1961-62; asst. prof. civil engring. U. Louisville, 1962-63; teaching fellow UCLA, 1963-64; asst. prof. civil engring. Robert Coll. Sch. Engring., Istanbul, 1966-71; assoc. prof. dept. civil engring. Bogazici U., Istanbul, 1971-75; assoc. prof. then prof. civil engring., coord. geotech. engring. La. State U., Baton Rouge, 1976—; adv. prof. U. Vicosa, Minas Gerais, Brazil, 1991—, Tongji U., Shanghai, People's Republic of China, 1991—; dir. Geomechanics Program NSF, Washington, 1990—; maitre de conferences Ecole Nationale des Ponts et Chaussees, Paris, 1980—; geotech. cons. Sauti, Spa, Cons. Engrs., Italy, 1969-72, SOFRETU-RATP, Paris, 1972-73, D.E.A., Cons. Engrs., Istanbul, 1974-75, BOTEK, Ltd., Istanbul, 1975—, Senler-Campbell Assos., Louisville, 1979—, Fugro Gulf-Geogulf, Houston, 1980—; cons. UN Devel. Program, 1982-84, 87; cons. in field. Contbr. articles to profl. jours. AID scholar, 1975-76; lic. civil engr., La., Turkish Chamber of Civil Engring; French Ministry External Relations scholar, 1982. Fellow ASCE; mem. Am. Soc. Engring. Edn., ASTM, La. Engring. Soc., Turkish Soil Mechanics Group (charter), Turkish Chamber Civil Engrs., Internat. Soc. Soil Mechanics and Found. Engring., Sigma Xi, Chi Epsilon, Tau Beta Pi. Home: 1915 W Magna Carta Pl Baton Rouge LA 70815-5521 Office: La State U Dept Civil Engring Baton Rouge LA 70803

TUN, MAUNG MYINT THEIN, analytical chemist; b. Sagaing, Burma, Mar. 3, 1950; s. U Phone Myint and Daw Khin Yi. B. in Chem. Engring., Rangoon (Burma) Inst. Tech., 1973; diploma indsl. mgmt., Stott's Corr. Coll., Melbourne, Australia, 1989; MS, Western Pacific U., 1990. Cert. chem. engring. Rsch. asst. Ctrl. Rsch. Orgn., Yangon, 1973-76; process engr. Paper & Chem. Industries Corp., Yangon, Yeni, 1977-83; chem. engr. Rice Bran Oil Industry, Yangon, 1984-88; prin. chemist Nauru Phosphate Corp., Nauru, 1989—. Friend World-wide Fund for Nature Internat., Gland, Switzerland, 1990; activitist Consumers Protection Group, Nauru, 1990, Myanmar, 1990. Mem. ASTM, British Standards Soc., Inst. Supervisory Mgmt., Tech. Assn. of the Pulp & Paper Industry, Assn. Official Analytical Chemists. Avocations: reading, writing, gardening, walking, tennis. Home: 9-101 Ward 7, Shwe Pyi Tha Myo Thit, Yangon Burma Office: Nauru Phosphate Corp, Nauru Nauru Address: PO Box 219, Nauru Nauru

TUNAC, JOSEFINO BALLESTEROS, biotechnology administrator; b. Vintar, Philippines, Nov. 11, 1942; came to U.S., 1966; s. Ireneo and Corazon (Ballesteros) T.; m. Corrie D., Sept. 7, 1968; children: Brooks, Corinna. MS, S.D. State U., 1968; PhD, Rutgers U., 1975. Rsch. microbiology Merck, Sharp & Dohme, Rahway, N.J., 1974-77; rsch. dir. Warner Lambert subs. Parke-Davis, Ann Arbor, Mich., 1977-86; pres., founder Fermical, Inc., Detroit, 1987—; cons. Nat. Cancer Inst., Bethesda, Md., 1985-86, Met. Ctr. for High Tech., Detroit, 1986-87, Mycosearch, Inc., Chapel Hill, N.C., 1985-88. Contbr. articles to Jour. of Fermentation & Bioengring., Jour. of Antibiotics, Drugs of the Future, Antimicrobial Agt. & Chemotherapy, Jour. Nat. Cancer Inst. Nat. Cancer Inst. grantee, 1987. Mem. Am. Assn. Cancer Rsch., Am. Soc. Microbiology, Soc. Indsl. Microbiology. Achievements include discovery of new antibacterial (epithienamycins), antifungal (hydroheptin), and anticancer (chloropentostatin, veractamycin, elactocin, fostriecin, others) agents; developement of fermentation process of athelmintic agent (avermectin); manufacture of antiviral agents (vidarabine, pentostatin); invention of a novel culture flask (Tunair flask system) and a high-aeration capacity bioreactor (Airmentor). Home: 5284 Collington Dr Troy MI 48098-2442 Office: Fermical Inc 2727 2d Ave Detroit MI 48201

TUNG, HSIEN-HSIN, chemical engineer. MS, Northwestern Univ., 1981, PhD, 1985. Rsch. fellow Merck Co., Inc., Rahway, N.J., 1986—. Contbr. articles to profl. jours. Recipient Speakers List award Merck Co., Inc. 1989—. Mem. AICE. Achievements include patents in field. Office: Merck Co Inc P O Box 2000 Rahway NJ 07065

TUNNELL, CLIDA DIANE, air transportation specialist; b. Durham, N.C., Nov. 20, 1946; d. Kermit Wilbur and Roberta (Brantley) T. BS cum laude, Atlantic Christian Coll., 1968; pvt. pilot rating, instr. rating, Air Care, Inc., 1971, 83. Cert. tchr. Tchr. Colegio Karl C. Parrish, Barranquilla, Colombia, S. Am., 1968-69, Nash County Schs., Nashville, N.C., 1969-86; ground sch. instr. Nash. Tech. Coll., Nashville, 1984-85; specialist, technician Am. Airlines, Dallas-Ft. Worth Airport, Tex., 1987—, A300 lead developer in flight tng. program devel., 1988-89, with flight ops. procedures flight ops. tech., 1990—, F100-fleet specialist flight ops. tech., 1992—; ednl. cons., Euless, Tex., 1989—. State Tchrs. Scholar N.C., 1964-68, Bus. and Profl. Women Scholar, 1980-81. Mem. 99, Internat. Orgn. Women Pilots (various offices), AMR Mgmt. Club. Avocations: flying, painting, writing, traveling. Home: PO Box 234 Euless TX 76039-0234

TUNNICLIFF, DAVID GEORGE, civil engineer; b. Ord, Nebr., Sept. 18, 1931; s. George Thomas and Ada Ellen (Ward) T.; m. Elaine Jean Interrante, Oct. 17, 1959 (div.); children: Martha Allison Tunnicliff Loeb, Vivian Jean; m. Joan Elizabeth Duchesneau, Oct. 26, 1975. BS, U. Nebr., 1954; MS, Cornell U., 1958; PhD, U. Mich., 1972. Registered profl. engr., Nebr., Mass. Engr. Nebr. Dept. Rds., Lincoln, 1954-60; asst. prof., then assoc. prof. Wayne State U., Detroit, 1960-67; chief tech. svcs. Warren Bros. Co., Cambridge, Mass., 1967-79; prin., cons. engr. D.G. Tunnicliff, Cons. Engr., Omaha, 1979—. Contbr. to profl. publs. Rep. precinct del., Detroit, 1965-66. With U.S. Army, 1955-56. Mem. ASTM (chair subcom. 1973—), ASCE, Assn. Asphalt Paving Technologists (bd. dirs. 1976-78), Transp. Rsch. Bd. (com. chair 1983-89). Mem. Evangel. Covenant Ch. Home: 9624 Larimore Ave Omaha NE 68134-3038 Office: DG Tunnicliff Cons Engr 9624 Larimore Ave Omaha NE 68134-3038

TURBEVILLE, ROBERT MORRIS, engineering executive; b. Cleve., May 2, 1951; s. Wilfred and Patricia Alice (Lamb) T.; m. Lisa Edelman, Apr. 2, 1977; children: Adam, Dennis, Diana. Student, Drew U., London, 1971-72; BA in History, W. Va. Wesleyan, 1973. Mgmt. trainee U.S. Steel, Pitts., 1973-74, foreman, 1975-79; mgr. standard products Heyl & Patterson, Inc., Pitts., 1979-83, sales mgr. to gen. mgr., 1983-88, v.p., 1988-91, pres., 1991—; dir. Heyl & Patterson, Inc., Pitts., 1988—; chmn. Bridge & Crane Inspection, Inc., Pitts., 1990—. Asst. leader Cub Scouts Am., Pitts., 1991—; coach Mt. Lebanon Soccer Assn., Pitts., 1990—. Mem. Coal Prep. Adv. Bd. (co-chmn. 1985-89), AIME, Process Equipment Mfrs. Assn. Republican. Methodist. Avocations: car restoration, travel, music. Office: Heyl & Patterson Inc Box 36 Pittsburgh PA 15230

TURCHI, PETER JOHN, aerospace and electrical engineer, educator, scientist; b. N.Y.C., Dec. 30, 1946; s. Charles Orlando and Fay Florence (Breglia) T.; m. Judith Ann Radogna, June 13, 1967; children: Janita Nicole, Rebecca Lenore. BSE in Aerospace and Mech. Sci./Physics, Princeton U., 1967, MA, 1969, PhD, 1970. Rsch. assoc. Plasma Propulsion Lab., Princeton U. (N.J.), 1963-70; plasma physicist Air Force Weapons Lab., Kirtland AFB, N.Mex., 1970-72; rsch. physicist Naval Research Lab., Washington, 1972-77, chief Plasma Tech. br., 1977-80; scientist R&D Assocs., Arlington, Va., 1980-81; dir. RDA Washington Rsch. Lab., Alexandria, Va., 1981-89; prof. Aero. and Astronautical Engring. Dept. Ohio State U., Columbus, 1989—; chmn. Megagauss Inst., Inc., Alexandria, 1979-89; chmn. mech. and aero. engring. adv. coun. Princeton U., 1988-92, mem. engring sch. adv. coun., 1988-92, dean's leadership coun., 1992—; resident collateral faculty Ohio Aerospace Inst., 1989—; lab cons. Los Alamos (N.Mex.) Nat. Lab., 1989—; interagy. sr. staff scientist USAF Phillips Lab., Kirtland AFB, N.Mex, 1990—; lectr. George Washington U., 1987-89, Air Force Pulsed Power Lecture Program, 1979-81; cons. on pulsed power tech.; chmn. 2d Internat. Conf. on Megagauss Fields, Arlington, 1979; chmn. Spl. Conf. on Prime-Power for High Energy Space Systems, Norfolk, Va., 1982; mem. organizing com. Megagauss Magnetic Field Confs. Pres., Collingwood Civic Assn. (Va.), 1980-81; rep. Mt. Vernon Council, Mt. Vernon Dist., Fairfax County, Va.; pres. Pulsed Power Conf. Inc., Albuquerque, 1985-87. Served to 1st lt. USAF, 1970-72. Recipient AIAA Nat. Student award, 1967, Invention award U.S. Air Force, 1972; Research Publ. award Naval Rsch. Lab., 1976, U.S. Navy and Air Force invention awards, 1978-83; NSF grad. fellow, 1967-70. . Mem. Am. Phys. Soc., AIAA (internat. chmn. 18th, 19th, 21st and 22d electric propulsion confs. 1985-91, mem. tech. com. plasmadynamics and lasers, 1983-86, mem. elec. propulsion com. 1987—, chmn., 1991—), IEEE (sr., tech. chmn. 5th and gen. chmn. 6th pulsed power confs. 1985-87, mem. plasma sci. and applications exec. com. 1987-89), Am. Soc. Engring. Edn., Planetary Soc., Sigma Xi, Tau Beta Pi. Clubs: Princeton Campus, Va. Ki Soc., Albuquerque Aikido Soc. Editor: Megagauss Physics and Tech., 1980; assoc. editor Jour. Propulsion and Power, 1990—; contbr. numerous articles in field to profl. jours.; patentee in field; research on electromagnetic implosion soft x-ray source, high energy x-ray generation by ultrahigh speed plasma flows, plasma flow switch for magnetic energy delivery above 10 megamperes; stabilized liner implosion system for controlled thermonuclear fusion. Office: Ohio State U 328 Bolz Hall 2036 Neil Ave Columbus OH 43210-1276

TURKEL, SOLOMON HENRY, physicist; b. N.Y.C., Sept. 8, 1911; s. Bernard and Rebecca (Kirschenbaum) T.; m. Mildred Cohen, May 7, 1941; children: Susan Turkel Reininger, Patricia Turkel Landsberg. BS, CCNY, 1932, MS, 1933. Instr. high sch. N.Y., 1941-42; physicist Argonne Nat. Lab., 1944-46, U. Chicago, 1944-46; prin. physicist dir. tech. info. Nuclear Engine Propulsion Aircraft, Oak Ridge, Tenn., 1946-49; mem. tech. staff Ops. Rsch. Office of Johns Hopkins U., Washington, 1948-56; systems analyst Northrop Corp., Hawthorne, Calif., 1956-59; sr. engr. GE Co., Phila., 1959-61; staff scientist Hughes Aircraft Co., Culver City, Calif., 1961-63, Aerospace Corp., San Bernardino, Calif., 1963-66, Rockwell Internat., Downey, Calif., 1966-76; ret. Contbr. articles to profl. jours. Bd. dirs. Nat. Conf. Christians and Jews, Orange County, Calif., 1976—, Anti-Defamation League of B'nai B'rith, 1976—; pres. So. Calif. Coun. B'nai B'rith, L.A., 1976-84. With U.S. Army, 1943-46. Recipient UN Medal, 1952. Fellow AAAS. Democrat. Jewish. Achievements include discovery and identification of many nuclear isotopes and radioactive elements. Home: 18325 Tamarind St Fountain Valley CA 92708-5829

TURKEVICH, ANTHONY LEONID, chemist, educator; b. N.Y.C., July 23, 1916; s. Leonid Jerome and Anna (Chervinsky) T.; m. Ireene Podlesak, Sept. 20, 1948; children: Leonid, Darya. B.A., Dartmouth Coll., 1937, D.Sc., 1971; Ph.D., Princeton U., 1940. Research assoc. spectroscopy physics dept. U. Chgo., 1940-41; asst. prof., research on nuclear transformations Enrico Fermi Inst. and chemistry dept., 1946-48, assoc. prof., 1948-53, prof., 1953-86, James Franck prof. chemistry, 1965-70, Distinguished Ser. prof., 1970-86, prof. emeritus, 1986; war research Manhattan Project, Columbia U., 1942-43, U. Chgo., 1943-45, Los Alamos Sci. Lab., 1945-46; Participant test first nuclear bomb, Alamagordo, N.Mex., 1945, in theoretical work on and test of thermonuclear reactions, 1945—, chem. analysis of moon, 1967—; cons. to AEC Labs.; fellow Los Alamos Sci. Lab., 1972—. Del. Geneva Conf. on Nuclear Test Suspension, 1958, 59. Recipient E.O. Lawrence Meml. award AEC, 1962; Atoms for Peace award, 1969. Fellow Am. Phys. Soc.; mem. N.Y. Acad. Sci. (Pregel award 1988), AAAS, Am. Chem. Soc. (nuclear applications award 1972), Am. Acad. Arts and Scis. Mem. Russian Orthodox Greek Cath. Ch. Clubs: Quadrangle, Cosmos. Home: 175 Briarwood Loop Oak Brook IL 60521-8713 Office: U Chicago Dept Chemistry 5640 S Ellis Ave Chicago IL 60637-1467

TURKLE, SHERRY, sociologist, psychologist, educator; b. N.Y.C., June 18, 1948; d. Milton and Harriet (Bonowitz) T.; m. Ralph Willard, Aug. 16, 1987; 1 child, Rebecca Ellen Turkle Willard. AB, Radcliffe Coll., 1970; PhD, Harvard U., 1976; PhD (hon.), Claremont Coll., 1990. Lic. psychologist, Mass. Asst. prof. sociology MIT, Cambridge, Mass., 1976-80, assoc. prof. sociology, 1980-91, prof. sociology, 1991—, mem. Lab. for Computer Sci., 1976—. Author: Psychoanalytic Politics: Freud's French Revolution, 1978, 2d edit., 1992, The Second Self: Computers and the Human Spirit, 1984. Fellow Guggenheim Found., 1981-82, Rockefeller Found., 1980-81; named Woman of Yr., Ms. Mag., 1984. Fellow AAAS. Office: MIT Rm E51-201C 70 Memorial Dr Cambridge MA 02139

TURKO, ALEXANDER ANTHONY, biology educator; b. Bridgeport, Conn., Aug. 19, 1943; s. Alexander I. and Elizabeth K. (Kulcsar) T.; m. Nancy Bally Hoinacky, Dec. 30, 1967; children: Michelle Lynn, Mark A. BA, So. Conn. State U., 1965, MS, 1967, postgrad., 1976. Assoc. prof. So. Conn. State U., New Haven, 1965—. Mem. AAUP. Home: 11 Birchwood Ln Monroe CT 06468-1025

TURMAN, GEORGE, former lieutenant governor; b. Missoula, Mont., June 25, 1928; s. George Fugett and Corinne (McDonald) T.; m. Kathleen Hager, Mar. 1951; children—Marcia, Linda, George Douglas, John, Laura. B.A., U. Mont., 1951. Various positions Fed. Res. Bank of San Francisco, 1954-64; mayor City of Missoula, 1970-72; mem. Mont. Ho. of Reps. from (Dist. 18), 1973-74; Mont. Pub. Service commr. (Dist. 5), 1975-80; lt. gov. State of Mont., 1981-88, resigned; apptd. Pacific NW Electric Power & Conservation Council, 1988; pres. Nat. Ctr. for Appropriate Tech., Butte, Mont., 1989—. Served with U.S. Army, 1951-53. Decorated Combat Inf. badge. Home: 1525 Gerald Ave Missoula MT 59801

TURNBULL, GORDON KEITH, metal company executive, metallurgical engineer; b. Cleve., Nov. 10, 1935; s. Gordon Gideon and Florence May (Felton) T.; m. Sally Ann Ewing, June 15, 1957; children: Kenneth Scott, Stephen James, Lynne Ann, June Patricia, James Robert. BS in Metall. Engring., Case Western Res. U., 1957, MS in Metall. Engring., 1959, PhD in Metall. Engring., 1962. Engr. then sr. engr. casting and forgings div., Cleve. Research ALCOA, 1962-67, sr. metallurgist quality assurance, Cleve. Forge Plant, 1967-68; group leader fabricating metallurgy div. ALCOA Tech. Ctr., Pitts., 1968-71, sect. head, mgr. ingot casting div., 1971-78, mgr. fabricating metallurgy div., 1978-79, asst. dir. finishes engring. properties and design, 1979-80; mgr. bus. planning services, corp. planning dept. ALCOA, Pitts., 1980-82, dir. tech. planning, 1982-86, v.p. tech. planning, 1986-91, exec. v.p. strategic analysis/planning and info., 1991—. Patentee method of not compacting titanium powder. Governing bd. Allegheny Ctr. Christian & Missionary Alliance Ch., Pitts. Mem. AIME, ASM, Nat. Acad. Engring. (rsch. bd.), Am. Foundrymen's Soc., Sigma Xi. Avocation: hockey. Home: 550 Fairview Rd Pittsburgh PA 15238-1745 Office: Aluminum Co Am Rm 2902 1501 Alcoa Bldg Pittsburgh PA 15219

TURNER, ANTHONY PETER FRANCIS, biotechnologist, educator; b. London, June 5, 1950; s. Thomas F. W. and Juliet M. (Frasca) T.; m. Elizabeth Caroline Bellamy, May 23, 1984; children: Ellen Katherine, Daniel Anthony John. Postgrad. cert. in edn., Christchurch Coll., 1973; MSc, U. Kent, Canterbury, Eng., 1977; PhD, Portsmouth U., Eng., 1980. Lectr. South Kent Coll. Tech., Folkstone, 1974-76; rsch. asst. U. Kent, 1976-77, rsch. fellow, 1980-81; rsch. asst. Portsmouth U., 1977-80; sr. fellow Cranfield (Eng.) Inst. Tech., 1981-89, prof., 1989—, head biotech. ctr., 1992—; bus. mgr. Cranfield Biotech. Ltd., 1989-91; project leader European Concerted

Action, Brussels, 1988-94; vis. prof. Tokyo Inst. Tech., 1989, U. Florence, 1993; presenter in field. Acad. editor Elsevier Applied Sci., London, 1986—; editor: Biosensors: Fundamentals and Applications, 1987, 89, Advanced in Biosensors, 1991, 92; editor jour. Biosensors and Bioelectronics, 1986—; contbr. over 200 papers to profl. publs. Recipient Energy prize Brit. Petroleum, 1982, Best Paper award Eurosensors, 1991, Pers. Investigation award Royal Soc., London, 1982; sr. fellow Brit. Diabetic Assn., 1982. Mem. European Biosensor Group (founder, chmn. 1987), European Soc. for Engring. and Medicine, Romanian Soc. Clin. Engring. and Medicine (hon.), U.K. Sensor Group (exec. 1991). Achievements include 19 patents in elucidation and application of mediated electrochemistry in biosensors; co-developer biosensor for blood glucose. Office: Cranfield Inst Tech, Cranfield Biotech Ctr, Cranfield MK430AL, England

TURNER, BILLIE LEE, botanist, educator; b. Yoakum, Tex., Feb. 22, 1925; s. James Madison and Julia Irene (Harper) T.; m. Virginia Ruth Mathis, Sept. 27, 1944 (div. Feb. 1968); children: Billie Lee, Matt Warnock; m. Pauline Henderson, Oct. 22, 1969 (div. Jan. 1975); m. Gayle Langford, Apr. 18, 1980; children (adopted)—Roy P., Robert L. B.S., Sul Ross State Coll., 1949; M.S., So. Meth. U., 1950; Ph.D., Wash. State U., 1953. Teaching asst. botany dept. Wash. State U., 1951-53; instr. botany dept. U. Tex., Austin, 1953; asst. prof. U. Tex., 1954-58, asso. prof., 1958-61, prof., 1961—, now S.F. Blake prof. botany, chmn., 1967-75, dir. Plant Resources Ctr., 1957—; Asso. investigator ecol. study vegetation of, Africa, U. Ariz., Office Naval Research, 1956-57; vis. prof. U. Mont., summers 1971, 73, U. Mass., 1974. Author: Vegetational Changes in Africa Over a Third of a Century, 1959, Leguminosae of Texas, 1960, Biochemical Systematics, 1963, Chemotaxonomy of Leguminosae, 1972, Biology and Chemistry of Compositae, 1977, Plant Chemosystematics, 1984; Asso. editor: Southwestern Naturalist, 1959—. Served to 1st lt. USAAF, 1943-47. NSF postdoctoral fellow U. Liverpool, 1965-66. Mem. Bot. Soc. Am. (sec. 1958-59, 60-64, v.p. 1969), Tex. Acad. Sci., Southwestern Assn. Naturalists (pres. 1967, gov.), Am. Soc. Plant Taxonomists, Internat. Assn. Plant Taxonomists, Soc. Study Evolution, Phi Beta Kappa, Sigma Xi. Office: U Tex Plant Resources Ctr Main Bldg 228 Austin TX 78712

TURNER, CARL JEANE, international business consultant, electronics engineer; b. Sevierville, Tenn., July 27, 1933; s. Kenneth Albert and Lenna Faye (Christopher) T.; m. Flossie Pearl Ingram, Dec. 11, 1954; children: Marcia, Kenneth, Theresa, Christopher, Robin. BEd, BSEE, MBA, Columbia Pacific U., PhD, 1983. With Civil Air Patrol, Fla. Wing., 1947-50, Fla. Air Nat. Guard, 1948-50, USAF, 1950-72, aviator, electronics engring., AEW & C, tactical reconnaissance, electronic warfare, spl. ops., Korea and Vietnam, advanced through grades to chief master sgt., ret. 1972; field engring., internat. mktg. Itek Corp., 1972-77, 78-81, sr. engr./analyst, chief instr. E-Systems, Inc., Greenville, Tex., 1977-78; gen. mgr. Optical Systems div. Itek Internat. Corp., Athens, Greece, 1978-79, gen. mgr. German programs joint venture Itek Internat. Corp./AEG Telefunken AG, Ulm, Germany, 1979-81, mgr. program planning and control, internat. ops. Applied Tech. div., Sunnyvale, Calif., 1981; mgr. export mktg. GTE Corp., Govt. Systems Group, Western Div., Mountain View, Calif., 1981-83; internat. sales mgr. Probe Systems, Inc., Sunnyvale, Calif., 1983-84; dir. internat. mktg. Gen. Instrument Corp., Def. Systems Group, Hicksville, N.Y., 1984-90; founder, pres., chief exec. officer Intermanagement Tech. Co., Longview, Tex., 1990—; bd. dirs. Am. Air Mus. in Britain, div. Imperial War Mus. Author, editor electronic warfare mgmt. courses and internat. bus. books. Named to Order of Seasoned Weasels; recipient George Washington Honor medal Freedoms Found., 1965, Presdl. Achievement award, 1982, Pres.'s Medal of Merit, 1982. Mem. IEEE, Assn. Old Crows, Nat. Assn. Profls. Office: 400 W Terrace Dr Longview TX 75601-3823

TURNER, JOHN ANDREW, economist; b. Chgo., July 9, 1949; s. Henry Andrew and Mary Margaret (Tilton) T.; m. Kathleen King Peery, June 21, 1975; 1 child, Sarah. BA, Pomona Coll., Claremont, Calif., 1971; MA, Stanford U., 1972; PhD, U. Chgo., 1977. Rsch. econ. SSA, Washington, 1976-80, U.S. Dept. Labor, Washington, 1980—; cons. OECD, Paris, 1989; chmn. Internat. Pension Conf., U.S. Dept. Labor, Washington, 1990—; adj. prof. George Washington U., 1994—. Author: Pension Policy for a Mobile Labor Force, 1993; editor: Trends in Pensions, 1989 (transl. into Japanese 1991), Pension Policy: An International Perspective, 1991, Trends in Health Benefits, 1993. Fulbright scholar Institut de Recherches Economiques et Sociales, France, 1994. Mem. Am. Econ. Assn. Methodist. Avocation: tennis. Home: 3713 Chesapeake St NW Washington DC 20016-1813 Office: US Dept Labor 200 Constitution Ave NW Washington DC 20210-0002

TURNER, JOHN FREELAND, foundation administrator, former federal agency administrator, former state senator; b. Jackson, Wyo., Mar. 3, 1942; s. John Charles and Mary Louise (Mapes) T.; m. Mary Kay Brady, 1969; children: John Francis, Kathy Mapes, Mark Freeland. BS in Biology, U. Notre Dame, 1964; postgrad., U. Innsbruck, 1964-65, U. Utah, 1965-66; MS in Ecology, U. Mich., 1968. Rancher, outfitter Triangle X Ranch, Moose, Wyo.; chmn. bd. dirs. Bank of Jackson Hole; photo-journalist; state senator from Sublette State of Wyo., Teton County, 1974-89; mem. Wyo. Ho. of Reps. Teton County, 1970-74; pres. Wyo. Senate, 1987-89; chmn. legis., minerals bus. and econ. devel. com. Teton County, Wyo., 1987-89; dir. Fish and Wildlife Svc. Dept. Interior, Washington, 1989-93; v.p. sustained devel. Conservation Fund, Arlington, Va., 1993—; vice chmn. Sec. of Interior's Nat. Parks Adv. Bd.; statewide coordinating Task Force U. Wyo., exec. commn. State Reps., adv. council Coll. Agriculture U. Wyo.; mem. Teton Sci. Sch. Bd., Nat. Wetland Forum, 1983, 87; mem. exec. com. Council of State Govt.; chmn. Pride in Jackson Hole Campaign, 1986; bd. dirs. Wyo. Waterfowl Trust; chmn. steering com. of UN Convention on Wetlands of Internat. Importance, 1990—; head U.S. delegation to Convention. on Internat. Trade Endangered Species. Author: The Magnificent Bald Eagle: Our National Bird, 1971. Mem. Western River Guides Assn., Jackson Hole Guides and Outfitters. Named Citizen of Yr. County of Teton, 1984; recipient Nat. Conservation Achievement award Nat. Wildlife Fedn., 1984, Sheldon Coleman Great Outdoors award, 1990, Pres.'s Pub. Svc. award The Nature Conservatory, 1990, Stewardship award Audobon Soc., 1992, Nat. Wetland Achievement award Ducks Unlimited, 1993. Republican. Roman Catholic.

TURNER, JOHN SIDNEY, JR., otolaryngologist, educator; b. Bainbridge, Ga., July 25, 1930; s. John Sidney and Rose Lee (Rogers) T.; m. Betty Jane Tigner, June 5, 1955; children: Elizabeth, Rebecca, Jan Marie. BS, Emory U., 1952, MD, 1955. Diplomate Am. Bd. Otolaryngology. Intern U. Va. Hosp., 1955-56; resident in otolaryngology Duke U. Med. Ctr., 1958-61; prof. otolaryngology Emory U., Atlanta, 1961—, chmn. dept., 1961—; ear specialist, chief otolaryngology Emory Clinic, 1961—; area cons. in field U.S. 3d Army, 1962-69; assoc. dir. heart disease control program Fla. Bd. Health, 1956-58; Ga. state chmn. Deafness Rsch. Found., 1968—; v.p. Clifton Casualty Ins. Co., Atlanta. Mem. internat. editorial bd. Drugs jour., 1982—, Ethicals in Med. Progress, 1982—, Dialogue jour., 1988—; mem. editorial bd. Otolaryngology—Head and Neck Surgery, 1991; contbr. chpts. to books, articles to profl. jours. With USPHS, 1956-58. Recipient Appreciation award Children of Fulton County and Fulton County Health Dept., 1975. Mem. AMA, So. Med. Assn. (chmn. otolaryngology sect. 1974, cert. of appreciation 1974), Ga. Soc. Otolaryngology (pres. 1973), Med. Assn. Ga., Med. Assn. Atlanta, Am. Acad. Otolaryngology, Triological Soc. (v.p., chmn. so. sect. 1991—), Am. Acad. Otolaryngic Allergy, Assn. Acad. Depts. Otolaryngology, Optimist (pres. Atlanta 1975), Alpha Omega Alpha. Democrat. Methodist. Home: 1388 Council Bluff Dr NE Atlanta GA 30345-4132 Office: Emory U Dept Otolaryngology 1365 Clifton Rd NE Atlanta GA 30307-1013

TURNER, LEAF, physics researcher; b. Bklyn., Mar. 23, 1943; s. Max and Florence Estelle (Tanenbaum) T.; m. Ruby Ann Sherman, Sept. 11, 1966; children: Alyssa Wendy, Lara Dawn, Ari Mark, Rima Jacqueline. AB, Cornell U., 1963; MS, U. Wis., 1964, PhD, 1969. Inst. postdoctoral fellow Weizmann Inst. of Sci., Rehovot, Israel, 1969; symposium fellow Niels Bohr Inst., Copenhagen, Denmark, 1969; postdoctoral fellow dept. of physics U. Toronto, Ontario, Can., 1969-71; asst. scientist space sci. and engring. ctr. U. Wis., Madison, 1971-72, project assoc. dept. physics, 1972-74, vis. prof. physics, 1984; tech. staff mem. Los Alamos (N.Mex.) Nat. Lab., 1974—; assoc. editor Physics of Fluids, 1985-87; mem. adv. com. Inst. for Fusion Studies, Austin, Tex., 1982-85; mem. internal adv. com. Ctr. for Nonlinear Studies, Los Alamos Nat. Lab. Contbr. articles to profl. jours. Physics instr. Los Alamos High Sch., 1985-91. Mem. Am. Phys. Soc., Phi Beta Kappa, Sigma Xi. Achievements include include devel. of Rayleigh Gans-Born Light Scattering by Ensembles of Randomly Oriented Anisotropic Particles; devel. of theory of turbulent magnetic relaxation of plasma with topological field constraints and others. Office: Los Alamos Nat Lab T-15 MS B217 PO Box 1663 Los Alamos NM 87545

TURNER, MALCOLM ELIJAH, biomathematician, educator; b. Atlanta, May 27, 1929; s. Malcolm Elijah and Margaret (Parker) T.; m. Ann Clay Bowers, Sept. 16, 1948; children: Malcolm Elijah IV, Allison Ann, Clay Shumate, Margaret Jean; m. Rachel Patricia Farmer, Feb. 1, 1968; children: Aleta van Riper, Leila Samantha, Alexis St. John, Walter McCamy. Student, Emory U., 1947-48; B.A., Duke U., 1952; M.Exptl. Stats., N.C. State U., 1955, Ph.D., 1959. Analytical statistician Communicable Disease Center, USPHS, Atlanta, 1953; rsch. assoc. U. Cin., 1955, asst. prof., 1955-58; asst. statistician N.C. State U., Raleigh, 1957-58; assoc. prof. Med. Coll. Va., Richmond, 1958-63, chmn. div. biometry, 1959-63; prof., chmn. dept. statistics and biometry Emory U., Atlanta, 1963-69; chmn. dept. biomath., prof. biostats. and biomath. U. Ala., Birmingham, 1970-82, prof. biostats. and biomath., 1982—; instr. summers Yale U., 1966, U. Calif. at Berkeley, 1971, Vanderbilt U., 1975; prof. U. Kans., 1968-69; vis. prof. Atlanta U., 1969; cons. to industry. Mem. editorial bd. So. Med. Jour., 1990—; contbr. articles to profl. jours. Fellow Ala. Acad. Sci., Am. Statis. Assn. (hon.), AAAS (hon.); mem. AAUP, AMA (affiliate), Biometrics Soc. (mng. editor Biometrics 1962-69), Soc. for Indsl. and Applied Math., Mensa, Sigma Xi, Phi Kappa Phi, Phi Delta Theta, Phi Sigma. Home: 1734 Tecumseh Trail Pelham AL 35124-1012

TURNER, RAYMOND EDWARD, chemistry educator, researcher; b. Portsmouth, Va., Dec. 13, 1948; s. Vernon and Kate Alicia (Ely) T.; m. Merlene Jeanette Blackett, Aug. 12, 1972 (div. June 1982); 1 child, Ebony Elysia; m. Margaret Elizabeth Alleyne, May 25, 1985. BS in Chemistry, Bklyn. Coll., 1974; MS, Fordham U., 1982; MS, PhD, Polytech U., Bklyn., 1986. Postdoctoral fellow Sch. of Pub. Health Harvard U., Boston, 1987-88; prof. maths. & chemistry Roxbury Community Coll., Boston, 1987—; rsch. assoc. in anatomy and cellular biology Tufts U., Boston, 1989—. Author: (textbook) Developmental Concepts in Science, 1991. Capt. MSC, USAR, 1982—. Mem. AAAS, Am. Chem. Soc., Sigma Xi. Methodist. Achievements include research on the solution properties of hyaluronic acid oligosaccharides, effects of hyaluronan molecular chain size on chondrogenesis in chick embryonic limb buds, hyaluronan binding protein and cell interaction in chondrogenesis. Office: Tufts U Dept Anatomy Cellular Biology 136 Harrison Ave Boston MA 02111-1800

TURNER, WILLIAM BENJAMIN, electrical engineer; b. Bklyn., Sept. 23, 1929; s. Jacob Joshua and Mollie (Klein) T. BEE, CCNY, 1955; MBA, NYU, 1964; DD (hon.), UCLA, 1978. Cert. tchr., N.Y. Chief engr. Esan Electronic Labs., Fla. and N.Y., 1969—; cons. in field, 1965—. Author: Theology - The Quintessence of Science, 1981, Nothing and Non-Existence, 1986, Hyper Light Speed Technology, 1992. Sgt. U.S. Army, 1951-53, Korea. Decorated Bronze Star. Mem. Mensa, Boynton Beach C. of C. Achievements include invention of world's fastest computers, advanced concepts in time theory, development of multi-dimensional geometry theory of the universe. Home and Office: 1916-B Palmland Dr Boynton Beach FL 33436

TURNER, WILLIAM OLIVER, III, civil engineer; b. Tacoma, Wash., Sept. 15, 1957; s. William Oliver Jr. and Betty Katherine (McKay) T.; m. Colette Talley, Aug. 8, 1981; 1 child, Andrew McKay. BSCE, Va. Poly. Inst. and State U., 1980; MBA, U. N.C., 1992. Registered profl. engr., N.C. Engr. in tng. N.C. Dept. Transp., Raleigh, N.C., 1980-82; assoc. engr. Carolina Power and Light Co., Raleigh, 1982-83, engr., 1983-86, sr. engr., 1986—. Active White Plains United Meth. Ch., Cary, N.C., 1987-89. Mem. ASCE, NSPE. Republican.

TURNER, WILLIAM RICHARD, retired aeronautical engineer, consultant; b. Pulaski, Va., Aug. 11, 1937; s. Richard Kunkle and Georgia Virginia (Shinault) T.; m. Shirley Jean Stoots, Dec. 10, 1958 (div. Aug. 1969); children: Pamela Jean, Ricky Alan; m. Delonda Gail Fanin, Mar. 23, 1970; children: Anita Jo, William Ballard; 1 adopted child, Shauna Lynn. BS in Aero. Engring., Va. Poly. Inst., 1965, MS in Aero. Engring., 1972. Registered profl. engr., Pa., Va., Calif., N.J., Mass., Del.; cert. flight test engr. FAA-Designated Engring. Rep., cert. commercial flight instr. Engr. space vehicle group Hercules Inc./Thiokol, 1965-74; pres. Turner Engring. Svcs. Co., Aberdeen, Md., 1974-80; engr. Balt. Gas and Electric, 1980-82; v.p. R.A.M.. Engrs., Ellicott City, Md., 1982-84; CAA/FAA comp. quality assurance engr. Brit. Aerospace Inc., Herndon, Va., 1984-88; sr. flt. test engr. USN, NATTC, Pax River, Md., 1988; mgr. quality engr. Hughes FIT. Simulations Ops., Herndon, 1988-90; ret., 1990; aeronaut. cons. Winchester, Va., 1990—; mem. rsch. adv. panel Aviation Week and Space Tech., N.Y.C., 1987—. Dir. youth Loudoun Valley Ch. of God, Purcellville, Va., 1992. Col. USMC, 1954-64, 68-71, with USMCR, 1954-84, ret., Korea, Vietnam. Mem. Soc. Flight Test Engrs. (chmn. 1988), Assn. Naval Aviation, U.S. Naval Inst., Am. Soc. Quality Control (Disting. Svc. award 1975), NATC, Tailhook Assn. Achievements include development of integration standard central air data computer F14, A6, A4, F4, S3 aircraft, of two stop 2 start rocket capability motor, of reliabilty-refueling system KC-135 A/C and Zuni Rocket advanced ignition system and baggage pod Jetstream 31 Aircraft (heavy and light load flight testing), also worked on design and static testing of same; design retro-rocket motor for moon lander. Home: 2338 Roosevelt Blvd Winchester VA 22601

TURNEY, STEPHEN ZACHARY, cardiothoracic surgeon, educator; b. Akron, Ohio, Sept. 28, 1935; s. Stephen T. and Olga M. (Zachar) T.; m. Carolyn Garney, June 25, 1960; children: Stephen G., Lisa S., Theodore Z., Katherine E. MD, Georgetown U., 1959. Diplomate Am. Bd. Thoracic Surgery, Am. Bd. Surgery. Intern Boston City Hosp., 1959-60; resident gen. surgery Cleve. Met. Gen. Hosp., 1960-66; resident thoracic surgery U. Hosp., Ann Arbor, Mich., 1967-69; asst. prof. surgery U. Md., Balt., 1970-74, assoc. prof. surgery, 1974—. Author 7 book chpts.; editor: Management of Cardiothoracic Trauma, 1990; contbr. 80 articles to profl. jours. Capt. U.S. Army, 1961-63. Fellow ACS; mem. Am. Assn. Thoracic Surgery. Roman Catholic. Achievements include patents for Blood Pressure Manifold, automated Respiratory Gas Monitoring System; development of first multibed automated respiratory gas monitoring system. Office: U Md Sch Medicine 22 S Greene St Baltimore MD 21201

TURNQUIST, PAUL KENNETH, agricultural engineer, educator; b. Lindsborg, Kans., Jan. 3, 1935; s. Leonard Otto and Myrtle Edith (Ryding) T.; m. Peggy Ann James, Dec. 22, 1962; children: Todd, Scott, Greg. BS Agrl. Engring., Kans. State U., 1957; MS in agrl. engring., Okla. State U., 1961, PhD agrl. engring., 1965. Registered profl. engr., Okla. Rsch. engr. Caterpillar Tractor Co., Peoria, Ill., 1957; instr., asst. prof. Okla. State U., Stillwater, 1958-62; assoc. prof., prof. S.D. State U., Brookings, 1964-76; prof., dept. head Auburn (Ala.) U., 1977—; mem. ABET Engring. Accreditation Commn., 1992—. Co-author: Tractors & Their Power Units, 1989; contbr. articles to profl. jours. Fellow Am. Soc. Agrl. Engrs. (life, trustee found. 1990-93, bd. dirs. edn. com. 1992—); mem. Am. Soc. for Engring. Edn., coun. Forest Engrs., Sigma Xi, NSPE. Methodist. Home: 1216 Nixon Dr Auburn AL 36830-6302

TURRO, NICHOLAS JOHN, chemistry educator; b. Middletown, Conn., May 18, 1938; s. Nicholas John and Philomena (Russo) T.; m. Sandra Jean Misenti, Aug. 6, 1960; children: Cynthia Suzanne, Claire Melinda. BA, Wesleyan U., 1960, DSc (hon.), 1984; PhD, Calif. Inst. Tech., 1963. Instr. Columbia U., N.Y.C., 1964-65, asst. prof., 1965-67, assoc. prof., 1967-69, prof. chemistry, 1969—, William P. Schweitzer prof. chemistry, 1982—, chmn. chemistry dept., 1981-84; Cons. E.I. duPont de Nemours and Co., Inc. Author: Molecular Photochemistry, 1965, (with G.S. Hammond, J.N. Pitts, D.H. Valentine) Survey of Photochemistry, vol. 1, 1968, vol. 2, 1970, vol. 3, 1971, (with A.A. Lamola) Energy Transfer and Organic Photochemistry, 1971, Modern Molecular Photochemistry, 1978; mem. editorial bd. Langmuir Macromolecules, Ency. of Phys. Sci. and Tech., Jour. of Reactive Intermediates. NSF fellow; Alfred P. Sloan Found. fellow.; Recipient Eastman Kodak award for excellence in grad. research, award for pure chemistry, 1973; U.S. Dept. Energy E.O. Lawrence award, 1983; Urey award Columbia U., 1983, Porter Medal award European Photchem. Assn., Japanese Photochem. Soc., Inter-Am. Photochem. Soc., 1994; Guggenheim fellow Oxford U., 1985; Fairchild scholar Calif. Inst. Tech., 1984-85. Mem. NAS, Am. Chem. Soc. (editorial bd. jour. 1984—, Harrison Howe award, Rochester, N.Y. sect. 1986, Arthur C. Cope award 1986), Am. Acad. Arts and Scis., Am. Chem. Soc. (Fresenius award 1973, award for pure chemistry 1974, James Flack Norris award 1987), Chem. Soc. (London), N.Y. Acad. Scis. (Freda and Gregory Halpern award in photochemistry 1977), Inter-Am. Photochemistry Soc. (photochemistry award 1991, European Photo-Chem. Assn.; Phi Beta Kappa, Sigma Xi. Home: 125 Downey Dr Tenafly NJ 07670-3005 Office: Columbia U 116th St and Broadway New York NY 10027

TURZILLO, ADELE MARIE, reproductive physiologist, researcher; b. Lackawanna, N.Y., Feb. 12, 1963; d. Rose Marie Eagan; m. Michael Edward Dwyer, May 30, 1992. BA, Cornell U., 1984, PhD, 1992. Sr. spl. technician Brigham Women's Hosp., Boston, 1985-86; grad. rsch. asst. Cornell U., Ithaca, N.Y., 1986-92; grad. teaching asst. Cornell U., Ithaca, 1986-92; postdoctoral fellow Colo. State U., 1992—. Contbr. articles to Jour. Reproductive Fertility, Domestic Animal Endocrinology. Named Outstanding Teaching Asst., Cornell U., Ithaca, 1988. Mem. Am. Soc. Animal Sci., Soc. for Study of Reproduction, Soc. for Study of Fertility, Sigma Xi. Democrat. Roman Catholic. Office: Colo State U Dept Physiology Fort Collins CO 80523

TUSTISON, RANDAL WAYNE, materials scientist; b. Fort Wayne, Ind., Dec. 4, 1947; s. Robert Eldon and Anna Marilla (Shreve) T.; m. Kathleen Elizabeth Nelson, Jan. 4, 1975; children: Eric Robert, Anna Elizabeth. BS in Physics, Purdue U., 1970; MSMetE, U. Ill., 1972, PhDMetE, 1976. Rsch. assoc. MIT, Cambridge, 1976-78; sr. scientist Raytheon Co., Lexington, Mass., 1978-81, prin. scientist, 1982-91; asst. mgr. advanced materials, 1991—, Raytheon Co., Lexington, Mass.; physicist 3M Co., St. Paul, 1981-82; instr. Northeastern U., Boston, 1983-86; U.S. del. 4th NATO Workshop on Passive IR Optical Materials, 1993. Contbr. articles to profl. jours. Mem. Am. Diabetes Assn., Juvenille Diabetes Found. Mem. Am. Vacuum Soc. (program com. 1986-91, chmn. vacuum tech. div. 1990-91), Am. Phys. soc., Soc. of Photo-Optical Instrumentation Engrs., Sigma Pi Sigma. Achievements include 10 U.S. patents, 3 fgn. patents in the field of high durability, infrared transparent materials and coatings. Office: Raytheon Co 131 Spring St Lexington MA 02173

TUTT, CHARLES LEAMING, JR., educational administrator, former mechanical engineering educator; b. Coronado, Calif., Jan. 26, 1911; s. Charles Leaming and Eleanor (Armit) T.; m. Pauline Barbara Shaffer, Aug. 16, 1933 (dec. Aug. 1981); children: Charles Leaming IV, William Bullard; m. Mildred Dailey LeMieux, Aug. 7, 1982; stepchildren: Linda Dailey LeMieux, Leslie Evans LeMieux. BSE, Princeton U., 1933, ME, 1934; D in Engring., Norwich U., 1967. Student engr. Buick Motor div. GM, Flint, Mich., 1934-36; engr. chassis unit sect. Buick Motor div. GM, 1936-38, spl. assignment engr., 1938-40; asst. prof. mech. engring. Princeton U., 1940-46; staff asst. ASME, N.Y.C., 1940-44; asso. editor Product Engring. mag. McGraw-Hill Pub. Co., N.Y.C., 1944-46; asst. to pres. Gen. Motors Inst., Flint, 1946-50; adminstrv. chmn. Gen. Motors Inst., 1950-60, dean engring., 1960-69, dean acad. affairs, 1969-75; pres. Sunnyrest Sanitarium, Colorado Springs, Colo., 1982—, chmn., 1989—. Contbr. articles to profl. jours. Mem. adv. com. Sloan Mus., Flint, 1965-82; trustee Norwich U., Northfield, Vt., 1963-76; bd. dirs. Engring. Found., N.Y.C., 1963-75, chmn., 1967-73; v.p. Friends of Pike Peak Libr. Dist., 1985-88, pres., 1986-88; mem. adv. bd. Pikes Peak Community Coll., 1986—. Fellow ASME (life, v.p. 1964-66, pres. 1975-76); mem. Soc. Mfg. Engrs. (dir. 1972-78), Am. Soc. Engring. Edn., Soc. Automotive Engrs., Colo. Soc. Profl. Engrs., Engrs. Coun. for Profl. Devel. (dir. 1975-80), Am. Soc. Metals, Soc. of Cin. in State of Va., Sigma Xi, Delta Tau Delta, Tau Beta Pi. Clubs.: Flint City, University (Flint), Wigwam (Deckers, Colo.), Princeton (N.Y.C.), Cooking, Cheyenne Mountain Country (Colorado Springs), Broadmoor Golf. Home: 20 Loma Linda Dr Colorado Springs CO 80906-4313

TUTTLE, DAVID BAUMAN, data processing executive; b. N.Y.C., Oct. 25, 1948; s. John Bauman and Charlotte (Root) T.; m. Mildred Suzanne Lamb, May 5, 1973 (div. May 1978); m. Nancy Viola Caraber, Mar. 14, 1981; children: Jason David, John Paul. Student, MIT, 1966-69. Assoc., sr. assoc. programer IBM Cambridge (Mass.) Sci. Ctr., 1968-71; staff programmer IBM VM/370 Devel., Burlington, Mass., 1971-76; sr. prin. S/W engr. Digital Equipment Corp., Maynard, Mass., 1976-78; mgr. Cambridge Telecom/GTE Telenet, Burlington, 1978-81; sr. scientist GTE Telenet, Burlington, 1981-84, chief scientist, 1984-85; sr. tech. cons. Prime Computer, Inc., Framingham, Mass., 1985-86; prin. tech. cons. Prime Computer, Inc., Framingham, 1986-89; sr. rsch. engr. Ungermann-Bass Inc., Andover, Mass., 1990-91; chief engr. Ungermann-Bass, Inc., Andover, Mass., 1991-93; strategy forum del. Corp. for Open Systems, McLean, Va., 1986-89, architecture com. mem., 1989, strategy forum nominating com., 1986-87; patent rev. com. Prime Computer, Inc., 1985-89. Co-author and editor: 3270 Display System Protocol, 1981, 83, Hotline BSC Access Method, 1970. Donor mem. Smithsonian Inst., Washington, 1980—. Mem. Assn. Computing Machinery, IEEE (computer soc.), Nat. Space Soc. (life mem.), The Cousteau Soc., USS Constitution Mus. Assn., Black and Blues of Killington (treas. 1986-89), Mandala Folk Dance Ensemble (dancer 1970-73). Republican. Presbyterian. Avocations: Duplicate Bridge (life master, Am. Contract Bridge League, 1983), alpine skiing. Home: 27 Heather Dr Reading MA 01867-3961

TUTUMLUER, EROL, civil engineer; b. Istanbul, Turkey, Apr. 28, 1967; came to U.S., 1989; s. Ali Kemal and Sabahat (Demirel) T. BS in Civil Engring., Bogazici U., Istanbul, 1989; MS in Civil Engring., Duke U., 1991, Ga. Inst. Tech., 1993; postgrad., Ga. Inst. Tech., 1993—. Rsch. asst. Duke U., Durham, N.C., 1989-91, rsch. asst., teaching asst. Ga. Inst. Tech., Atlanta, 1991—. Reviewer ASTM jour., 1991—. Office Ga. Tech. Geotech. Soc., Atlanta, 1991-92. Mem. ASCE, Ga. Tech Geotechnical Soc. (pres. 1993—). Moslem. Achievements include research in engring. sci. and mechanics; numerical and analytical methods in geomechanics. Home: 1185 Collier Rd Apt 31-D Atlanta GA 30318 Office: Ga Tech Sta PO Box 50467 Atlanta GA 30332

TWARDOWSKI, THOMAS EDWARD, JR., development chemist; b. Chgo., Dec. 2, 1964; s. Thomas Edward and Colleen R. (Moore) T. BSChE, U. Ill., 1986, MS, 1988, PhD, 1992. Grad. rsch. asst. U. Ill., Champaign, 1986-92; devel. chemist Nat. Starch & Chem. Co., Bridgewater, N.J., 1992—; vis. scientist Riso Nat. Lab., Denmark, 1988. Author: Some Molecular Contributions and Curing. . ., 1992; contbr. articles to Jour. Composite Materials, Macromolecules, Polymer Bull., others. Reader, monitor Rec. for the Blind, 1990—; vol. Am. Chem. Soc., 1992; mem. Pal program YWCA, Champaign, 1988. Mem. Am. Chem. Soc., Am. Inst. Chem. Engrs., Am. Phys. Soc., Alpha Chi Sigma. Achievements include characterization of contributions of entanglements in rubber elasticity; isolation of polymer backbone stiffness contribution to physical properties; simulation of thick graphite/epoxy composite laminates (computer modeling). Office: Nat Starch and Chem Co 10 Finderne Ave Bridgewater NJ 08807

TWEED, PAUL BASSET, chemical engineer, explosives and suicidology consultant; b. Zvinigorodka, Russia, Sept. 22, 1913; came to U.S., 1914; s. Jacob and Ida (Basset) T.; m. Mildred Emma Haycock, June 15, 1942 (dec. Dec. 1985); children: Bradford, Joel. BS, Rensselaer Poly. Inst., 1934, MS, 1935; MA, N.Y. State Coll. Tchrs., 1937. Chemist Am. Hard Rubber Co., Butler, N.J., 1936-40; chem. engr. Picatinny Arsenal, Dover, N.J., 1940-62; cons. engr. Avco Corp., Wilmington, Mass., 1962-66; engr. Martin Marietta Aerospace Co., Orlando, Fla., 1966-76; bus. mgr. We Care Inc., Orlando, Fla., 1976-84; instr. Henry George Sch. Social Sci., Newark, N.J., 1940-60; chmn. explosives subcom. Ordnance Engring. Handbook Series, Durham, N.C., 1955-62; cons. to loading sect. Am. Ordnance Assn., Washington, 1957-62; mem. Mutual Weapons Devel. Team, Washington, 1958, 60, Am. Inst. Chemists, Washington, Nat. Accreditation Appeals Bd., 1971-76; coord. course in explosive ordnance, Martin Marietta, 1969-70; mem. adv. bd. Lake Holden Water, 1974-84; guest lectr. on explosive trains Franklin Inst., Phila., 1970; coord. sci. lectures Orlando Sci. Ctr., 1989. Author, patentee in field explosives; author in field of suicidology. Mem. Fla. Cares,

Daytona Beach, 1987—, bd. dirs., 1990—; mem. Widowed Persons Svc., Orlando, 1989—, asst. coord. pot luck dinners, 1992-93, coord. holiday dinners, 1992—, Outreach vol., 1992—; mem. Fla. Consumer Action Coun. for Mental Health, 1991—, chmn. budget com. dist. 7, 1992—, steering com. dist. 7, 1992—, gen. com. dist. 7, 1992—, teleconference for drop-in ctr., 1992—; v.p., bd. dirs. Pathways Drop-In Ctr., 1993; mem. Alcohol, Drug Abuse and Mental Health subcoms. for Long Term Acute, Children, Elderly, Prevention, Substance Abuse and Forensic, 1990-93; sec. alliance First Unitarian Ch. Orlando, 1993, coord. Inner Reaches, 1992-93, mem. Universalist Svc. Com., 1989—. Mem. Am. Assn. Suicidology (mem. vol. com. 1987—, individual cert. com. 1987—), Fla. Assn. Crisis Svcs. (chmn. stats. com. 1987-88), Am. Assn. Ret. Persons, Hemlock Soc. Cen. Fla. (bd. dirs. 1988—), Hemlock Soc. Ctrl. Fla. (treas. 1988-89, 93, bd. dirs. 1988—), Books 'n Stuff, Moola Gala, Sigma Xi (pres. Winter Park Drop-in Ctr. 1991—).

TWEEDY, ROBERT HUGH, equipment company executive; b. Mt. Pleasant, Iowa, Mar. 24, 1928; s. Robert and Olatha (Miller) T.; B.S. in Agrl. Engring., Iowa State U., 1952; m. Genevieve Strauss, Aug. 15, 1969; children—Bruce, Mark; 1 stepdau., Mary Ellen Francis. Sr. engr. John Deere Waterloo Tractor Works, Waterloo, Iowa, 1953-64; mktg. rep. U.S. Steel Corp., Pitts., 1964-68; mgr. product planning agrl. equipment div. Allis-Chalmers Corp., Milw., 1969-76, mgr. strategic bus. planning Agrl. Equipment Co., 1976-85; mgr. strategic bus. planning Deutz-Allis Corp., 1985-89; project mgr. AGCO Corp., Batavia, Ill., 1989—; chmn. agrl. research com. Farm and Indsl. Equipment Inst., Chgo., 1974-76, mem. safety policy adv. com., 1972-89; mem. farm conf. Nat. Safety Council, Chgo., 1973-89; mem. industry sector adv. com. No. 16, U.S. Dept. Commerce, 1982-85; bd. dirs. C.V. Riley Meml. Found. Recipient citation in engring. Iowa State U., 1983. Fellow Am. Soc. Agrl. Engrs. (v.p. 1974-78, pres. 1981-82, gen. chmn. hdqrs. bldg. project 1968-70; chmn. Found. Trustees 1983-88, Wis. Engr. of Year award 1980, McCormick-Case Gold medal 1989); mem. Soc. Automotive Engrs., Masons. Patentee in field. Home: 1340 Bonnie Ln Brookfield WI 53045 Office: AGCO Corp 1500 N Raddant Rd Batavia IL 60510

TWIDALE, C(HARLES) R(OWLAND), geomorphologist, educator; b. Lincolnshire, Eng., Apr. 5, 1930; s. George Wilfred and Gladys May (West) T.; m. Kathleen Mary Gargini, Apr. 21, 1956; children: Nicholas, Richard Jonathan, Amanda Elizabeth. Ed. Wintringham Grammar Sch., Grimsby; BSc, U. Bristol, 1951, MSc, 1953, DSc, 1977; PhD, McGill U., 1957; D. Honoris Causa, U. Complutense, Madrid, 1991. Rsch. officer, div. land rsch. Commonwealth Sci. and Indsl. Rsch. Orgn., Canberra, 1952-57; mem. faculty dept. geography, geology, U. Adelaide, 1958—; vis. prof. geology and geophysics U. Calif., Berkeley, 1971; vis. prof. geology U. Tex., Austin, 1979. Nuffield Commonwealth bursary, 1965; NSF sr. fgn. scientist fellow, 1965-66, also vis. prof. Rensselaer Poly. Inst., Troy, N.Y.; mem. engring. and earth scis. panel Australian Rsch. Coun., 1992—. Author: Geomorphology, 1968, Structural Landforms, 1971, Analysis of Landforms, 1976, Granite Landforms, 1982; contbr. articles to profl. jours. Recipient Mueller medal Australian and New Zealand Assn. Advancement Sci., 1993. Fellow Royal Soc. South Australia (pres., 1975-76, Verco medal 1977), Geol. Soc. Am.; mem. Geol. Soc. Australia. Home: 7 Brecon Rd, Aldgate, Aldgate South Australia, 5154, Australia Office: Univ Adelaide, Geology & Geophysics Dept, Adelaide South Australia, 5000, Australia

TWIN, PETER JOHN, physics educator; b. London, July 26, 1937; m. Jean Leatherland, 1963; 2 children. Student, U. Liverpool. Lectr. U. Liverpool, Eng., 1964, sr. lectr., prof. experimental physics, 1988—; head Nuclear Structural Facility Daresbury Lab., Cheshire, 1983-88. Contbr. articles to profl. jours. Recipient John Price Wetherill medal Franklin Inst., 1991, Tom W. Bonner prize in nuclear physics Am. Phys. Soc., 1991. Office: U of Liverpool-Dept of Physics, POB 147, Liverpool L69 3BX, England*

TWINING, LINDA CAROL, biologist, educator; b. Paterson, N.J., July 8, 1952; d. Elmer Robert and Mildred Ruth (Weeks) T.; m. Charles Frederick Gerdes, Aug. 9, 1980; children: Andrew Micah Twining, Emilie Ruth Twining. BA, William Paterson Coll., 1972; MS, Rutgers U., 1975; PhD, U. Ill., 1982. Teaching asst. Rutgers U., New Brunswick, N.J., 1972-74, U. Ill., Champaign, 1974-81; asst. prof. biology Northeast Mo. State U., Kirksville, 1981-88; assoc. prof. Northwest Mo. State U., Kirksville, 1988—; reviewer biology textbooks for several publishers; corr. Sci.-by-Mail network, 1991-92. Participant Ann. Midwestern Conf. of Parasitologists; mem. choir, bell choir 1st Presbyn. Ch., Kirksville, 1983-92, tchr. Sun. sch., 1985—, elder, chair Christian edn. com., 1989-92. Mem. AAAS, Am. Soc. Microbiology, Am. Soc. Parasitologists, Assn. Women in Sci., Mo. Acad. Sci., Audubon Soc. Democrat. Office: Northeast Mo State U Divsn Sci Kirksville MO 63501

TWIST-RUDOLPH, DONNA JOY, neurophysiology and psychology researcher; b. Cape May, N.J., Dec. 3, 1955; d. Donald and Mary Ann (Johnson) Twist; m. Daniel Jay Rudolph, Jan. 10, 1981; children: Andrew, Adam. BS, Boston U., 1978; MA, SUNY, Stony Brook, 1984; PhD, SUNY, 1986. Licensed phys. therapist, N.Y., Conn. Teaching asst., dept. phys. therapy N.Y. U., 1980; teaching asst., dept. psychology SUNY, Stony Brook, 1982-83, teaching asst., dept. grad. psychology, 1984; intern, dept. rehab. medicine N.Y. U. Med. Ctr., Rusk Inst. Rehab. Medicine, N.Y.C., 1984-86, postdoctoral fellow, rsch. scientist, 1986-87; dir. rsch. and edn., chief of phys. therapy Norwalk (Conn.) Hosp., 1987—; state bd. examiner N.Y. State Phys. Therapy Licensing Exam. Profl. Svcs., Albany, 1986—; adj. asst. prof. Mt. Sinai Sch. Med., N.Y.C., 1990—. Contbr. articles profl. jours. Named Outstanding Young Woman Am., 1981, 83; grantee Easter Seal Rsch. Found., 1985-87, Rehab. Svcs. Adminstr. Dept. Edn., 1991; recipient Therapeutic Techs. Ins. award, 1989. Mem. Am. Phys. Therapy Assn., Am. Congress Rehab. Medicine, N.Y. Acad. Scis. Home: 381 Hemlock Rd Fairfield CT 06430-1857 Office: Norwalk Hosp 24 Stevens St Norwalk CT 06850-3894

TWOHEY, MICHAEL BRIAN, fishery biologist; b. Grand Rapids, Mich., Feb. 1, 1955; s. Edward Lewis and Margaret (Cook) T.; m. Claire Patten Shefferly, Sept. 20, 1991; 1 child, Kersten Fitzgerald. BS in Earth Sci., Edn., No. Mich. U., 1977, MA in Zoology, 1989. Fishery biologist U.S. Fish and Wildlife Svc., Marquette, Mich., 1990—. Trustee Marquette County Solid Waste Mgmt. Authority, 1989—; bd. dirs. Recycle! Marquette, 1988—; mem. Marquette County Solid Waste Planning Com. Mem. Sigma Xi, Phi Kappa Phi. Office: US Fish and Wildlife Svc 1924 Industrial Pky Marquette MI 49855

TYAGI, SOM DEV, physicist, educator; b. New Delhi, July 26, 1947; came to U.S. 1968, naturalized, 1979; s. Chander Bal and Vidya (Hanso) T.; m. Karen Lee Greenhalgh, June 29, 1974. B.Sc. with honors, Delhi U., 1967, M.Sc., 1969; M.S., Lowell U., 1972; Ph.D., Brigham Young U., 1976. Postdoctoral assoc. Drexel U., Phila., 1976-79, asst. prof. physics, 1979-84, assoc. prof., 1984—. Contbr. articles to profl. jours. Grantee U.S. Navy, U.S. Army, EPA, 1980—. Mem. Soc. Physics Students, Am. Phys. Soc., Sigma Xi, Sigma Pi Sigma. Home: 26 W Salisbury Dr Wilmington DE 19809-3416 Office: Drexel U Dept Physics 32nd and Chestnut Sts Philadelphia PA 19104

TYCKO, DANIEL H., physicist; b. L.A., Nov. 14, 1927; s. Aaron M. Tycko and Sonia (Sadicoff) Brown; m. Milicent Germansky, July 31, 1952; children: Benjamin, Robert, Jonathan. BA in Physics, UCLA, 1950; PhD in Physics, Columbia U., 1957. Sr. rsch. assoc. physics dept. Nevis Labs., Columbia U., N.Y.C., 1957-66; assoc. prof. physics Rutgers U., New Brunswick, N.J., 1966-67; assoc. prof. computer sci. SUNY, Stony Brook, 1967-70, prof., 1970-80; prin. scientist Technicon Instruments Corp., Tarrytown, N.Y., 1980-82, cons., 1982-89; cons. DHT Assocs., Stony Brook, 1989—; vis. scientist Centre d'Etude Nucleaires, Saclay, France, 1961-62; guest assoc. physicist Brookhaven Nat. Lab., Upton, N.Y., 1962-68; cons. Technicon Instruments Corp., Tarrytown, 1970-80. Contbr. articles to profl. jours. With U.S. Army, 1946-47. R&D grantee NSF, SUNY, Stony Brook, 1969-71. Jewish. Achievements include patents for methods for measuring physical properties of blood cells; research in pattern recognition to analyze multidimensional data from flow cytometers. Office: DHT Assocs PO Box 1033 Stony Brook NY 11790

TYCZKOWSKA, KRYSTYNA LISZEWSKA, chemist; b. Kazimierz Dolny, Poland, Feb. 13, 1946; came to U.S., 1981; d. Bogdan and Zofia (Plisiecka) Liszewski; m. Julius Kazimierz, June 12, 1971; 1 child, Marta Marianna. MS in Organic Chemistry, U. Maria Curie, Lublin, Poland, 1971; PhD in Tech. Sci., Tech. U., Lodz, Poland, 1981. From rsch. asst. to supr. analytical lab. Cen. Lab. Feed Industry, Lublin, Poland, 1971-81; vis. lectr. poultry sci. dept. N.C. State U., Raleigh, 1981-84, from analytical chemist to rsch. assoc. Coll. Vet. Medicine, 1984-92, sr. rsch. assoc. Coll. Vet. Medicine, 1992—. Contbr. articles to profl. jours. Mem. Assn. Ofcl. Analytical Chemists, Am. Soc. Mass Spectrometry. Republican. Roman Catholic. Achievements include developing analytical methods for drug residue in tissue and milk, neutrophil drug metabolism in elderly. Office: NC State U Coll Vet Medicine 4700 Hillsborough St Raleigh NC 27606

TYE-MURRAY, NANCY, research scientist; b. Bittburgh, Germany, Feb. 11, 1955; parents U.S. citizens; d. Joe B. and Janelle (Bowen) Tye; m. David John Murray, May 15, 1983; children: Ellen, Aubrey. BS, Tex. Christian U., 1977; MA, U. Iowa, 1979, PhD, 1984. Cert. clin. competency-audiology. Rsch. asst. dept. psychology U. Iowa, Iowa City, 1983-85; asst. rsch. scientist U. Iowa Hosps., Iowa City, 1985-90, assoc. rsch. scientist, 1990—. Author, editor: Cochlear Implants and Children: A Handbook, 1992; author: Communications Training for Children and Teenagers, 1992; contbr. articles to Jour. of the Acoustical Soc. Am., Jour. Speech and Hearing Rsch., Ear and Hearing. Prin. investigator Easter Seal Rsch. Found., 1986-89, Deafness Rsch. Found., 1987-89, Children's Miracle Network, 1989-90; investigator NIH/NINCDS Program Project, 1990—. Recipient editors award Volta Rev., 1992. Mem. Am. Speech-Lang.-Hearing Assn. (rev. bd. 1989-92, com. on noise exposure 1990-92), Acoustical Soc. Am., Acad. Rehab. Audiology (monograph editor 1992—), Alexander Graham Bell Assn. (rev. bd. 1992—). Achievements include rsch. in speech sci., aural rehab. and cochlear implants; developed four laser videodisc speech reading tng. programs which are used by cochlear implant recipients. Office: Dept Otolaryngology Univ Iowa Hosps Iowa City IA 52242

TYGRET, JAMES WILLIAM, civil engineer; b. Lynwood, Calif., Apr. 22, 1952; s. James Morris and Betty Jean (Watkins) T.; m. Dana Boswell Jordan, May 30, 1975; children: Samuel William, Daniel James. BS in Bioengring., Tex. A&M U., 1974. Registered profl. engr., Iowa, Ariz. Cons. engr. French Reneker Assocs., Fairfield, Iowa, 1976-82; dep. contract mgr. USAF, Tucson, Ariz., 1982-83; project engr. U.S. C.E., Tucson, 1983-84; chief constrn. mgmt. in Europe U.S. C.E., Frankfurt, Fed. Republic Germany, 1984-88, resident engr., 1988-89; resident engr. U.S. C.E., Incirlik, Turkey, 1989-91; dep. dir. engring. and housing U.S. Army, Ft. Carson, Colo., 1991—. Adult leader Am. Cub Scouts, Frankfurt, 1988-89, Incirlik, 1989-90; pres. PTSO, Incirlik, 1990-91. Mem. ASCE (pres. Deutschland Internat. Group 1987-89), Soc. Am. Mil. Engrs. (bd. dirs. Colorado Springs chpt. 1992). Home: 19280 Top of the Moor W Monument CO 80132 Office: US Army Dir Engring and Housing Fort Carson CO 80913

TYLER, CARL WALTER, JR., physician, health research administrator; b. Washington, Aug. 22, 1933; s. Carl Walter and Elva Louise (Harlan) T.; m. Elma Hermione Matthias, June 23, 1956 (dec. Dec. 1991); children—Virginia Louise, Laureen, Jeffrey Alan, Cynthia Kay. A.B., Oberlin Coll., 1955; M.D., Case-Western Res. U., 1959. Diplomate Am. Bd. Ob-Gyn. Rotating intern Univ. Hosps. of Cleve., 1959-60, resident in ob-gyn, 1960-64; med. officer USPHS, 1964; obstetrician-gynecologist USPHS Indian Health Service, Tahlequah, Okla., 1964-66; epidemic intelligence service officer Bur. Epidemiology, Ctrs. for Disease Control, Atlanta, 1966-67; dir. family planning evaluation div. Bur. Epidemiology, Ctrs. for Disease Control, 1967-80, asst. dir. for sci., 1980-82, acting dir. Ctr. for Health Promotion and Edn., 1982, dir. epidemiology program office, 1982-88, med. epidemiologist Office of Dir., 1988-90, asst. dir. for acad. programs, pub. health practice program office, 1990—; clin. asst. prof. ob-gyn Emory U. Sch. Medicine, Atlanta, 1966-80, clin. assoc. prof., 1980—, also clin. assoc. prof. preventive medicine and community health, adj. assoc. prof. sociology Coll. Arts and Scis., 1977-90; adj. assoc. prof. pub. health Sch. Pub. Health, 1990—; clin. prof. pub. health and community medicine Morehouse Sch. Medicine, Atlanta, 1990—; mem. Nat. Sleep Disorders Rsch. Commn., 1990—; mem. adv. com. on oral contraception WHO, Geneva, 1974-77, mem. adv. com. maternal and child health, 1982-88; lectr. in field. Editor: (monograph) Venereal Infections; assoc. editor: Maxcy-Rosenau Textbook of Public Health and Preventive Medicine, 13th edit., 1992; contbr. articles to profl. jours. Chmn. Dekalb County Schs. com. on instruction programs, subcom. on health, phys. edn. and safety, (Ga.), 1967-68; active Ga. State Soccer Coaches Assn., Atlanta, 1973-79, DeKalb County YMCA. Josiah Macy Found. fellow, 1956-58; NIH grantee, 1961-64; recipient Superior Service award, 1974, Meritorious Service medal USPHS, 1984, Disting. Service medal, 1988; Carl S. Shultz Population award APHA, 1976, medal of Excellence Ctrs. for Disease Control, 1984. Fellow Am. Coll. Ob-Gyn (chmn. community health com. 1974-77), Am. Coll. Preventive Medicine, Am. Coll. Epidemiol.; mem. Am. Epidemiologic Soc., Internat. Epidemiologic Assn., Assn. Tchrs. Preventive Medicine (bd. dirs. 1988-89), Am. Pub. Health Assn. (governing council 1976-78), Assn. Planned Parenthood Profls., Population Assn. Am., Sierra Club. Club: Briarcliff Woods Beach (Dekalb County, Ga.). Avocations: photography; camping. Office: HHS Ctrs for Disease Control Mailstop E-42 1600 Clifton Rd NE Atlanta GA 30329-4046

TYLER, EWEN WILLIAM JOHN, retired mining company executive, consulting geologist; b. Sheffield, Eng., Aug. 24, 1928; arrive in Australia, 1940; s. William Harold and Ethel (Matthew) T.; m. Aldyth Dorothy Watts, March 3, 1951; children: Brett, Jane, Timothy. B.Sc. with honours, U. Western Australia, Perth, 1949. Geologist, Geita Gold Mining Co., Ltd. (Tanzania), 1949-59; dir. Tanganyika Holdings Ltd., London, 1959-69, Melbourne, Australia, 1969-75; dir. Tanaust Proprietary Ltd., Melbourne, 1975-78, Ashton Mining Ltd., Melbourne, 1978-90; chmn. bd. dirs. Australian Diamond Exploration, Melbourne. Recipient Clunics Ross Nat. Sci. & Tech. award, 1992. Fellow Geol. Soc. Australia, Australasian Inst. Mining and Metallurgy; mem. Inst. Mining and Metallurgy, Order of Australia. Anglican. Clubs: Melbourne, Royal Automobile of Victoria (Melbourne).

TYLER, SETH, zoology educator; b. Chgo., Feb. 26, 1949; s. Lloyd P. and PHyllis M. T.; m. Mary S. Tyler, Aug. 31, 1970. BA with distinction, Swarthmore Coll., 1970; PhD, U. N.C., 1975. Killam fellow Dalhousie U., Halifax, Nova Scotia, 1975-76; asst. prof. U. Maine, Orono, 1976-82, assoc. prof., 1982-90, prof. dept. zoology, 1990—; assoc. prof. Wash. State U., Pullman, 1980; Panel mem. NSF, 1983-84. Editor: Advances in Biology of Turbellarians, 1986, Turbellarian Biology, 1991; contbr. 38 articles to profl. jours. Mem. Am. Soc. Zoologists (chair membership com. 1990), Am. Microscopical Soc. (mem. exec. com. 1983-86, 91—, bd. reviewers 1979-91, assoc. editor 1992—), Sigma Xi, Phi Beta Kappa. Achievements include documenting evolutionary significance of ultrastructural character, significance of functional morphology at ultrastructural level. Office: U Maine Dept Zoology 5751 Murray Hall Orono ME 04469-5751

TYNDALL, TERRY SCOTT, electrical engineer, biomedical engineer; b. Glendale, Calif., Aug. 21, 1962; s. Samuel Patrick and Elizabeth Jean (Sims) T.; m. Chris Anne Reyes, Apr. 29, 1989 (div. Mar. 1991). AS in Laser Physics, Pasadena City Coll., 1983; BSEE, Calif. Poly. Inst., 1988, postgrad., 1992—. Jr. electronic engr. Jet Propulsion Lab., Pasadena, Calif., 1984-88; electronic engr. Jet Propulsion Lab., Pasadena, 1988—. Mem. NSPE. Home: 5236 Hartwick St Los Angeles CA 90041

TYNER, C. FRED, federal agency administrator; b. Milw.; m. Lee Tyner; children: Michael, Rachel, Elizabeth. BA in Philosophy, Harvard U., 1963; MD, Case Western Reserve, 1967. Resident in neurology U. Wash. Med. Sch., Seattle, 1967-70; rsch. neuro-physiologist Walter Reed Army Inst. Rsch., Washington, 1970-75, chief dept. med. neurosciences, 1978-81, dir. divsn. neuropsychiatry, 1981-87, dir., comdt., 1987-92, dep. comdr. U.S. Army Med. Rsch. and Devel., 1992—. Decorated Legion of Merit. Office: Walter Reed Army Inst Rsch Office of Director Washington DC 20307-5100*

TYREE, LEWIS, JR., retired compressed gas company executive, inventor, technical consultant; b. Lexington, Va., July 25, 1922; s. Lewis Sr. and Winifred (West) T.; m. Dorothy A. Hinchcliff, Aug. 21, 1948; children: Elizabeth Hinchcliff, Lewis III, Dorothy Scott. Student, Washington & Lee U., 1939-40; BS, MIT, 1947. Cryogenic engr. Joy Mfg. Co., Michigan City, Ind., 1947-49; v.p. Hinchcliff Motor Service, Chgo., 1949-53; cons. engr. Cryogenic Products, Chgo., 1953-76, Liquid Carbonic Corp., Chgo., 1960-76; exec. v.p. Liquid Carbonic Industries, Chgo., 1976-87; bd. dirs. Liquid Carbonic Industries, Chgo., Worldwide Cryogenics (MVE), New Prague, Minn. Patentee in cryogenics. Served to 1st lt. U.S. Army, 1943-46, PTO. Mem. Soc. Cin., ASME, Am. Soc. Heating, Refrigeration, and Air Conditioning Engring., Hinsdale Golf Club, Lexington Golf and Country Club. Republican. Episcopalian. Home: Mulberry Hill Liberty Hall Rd Lexington VA 24450-1703

TYRL, PAUL, mathematics educator, researcher, consultant; b. Prague, Czechoslovakia, Dec. 24, 1951; came to U.S., 1970, naturalized, 1978; s. Vladimir Tyrl and Marta (Kocian) Kocian. BA with honors, Jersey City State Coll., 1977, MA, 1980; EdD Rutgers U., 1987. Cert. tchr. secondary edn., higher edn., N.J. Quality controller Agfa-Perutz, Munich, W. Ger., 1969-70; technician AT&T, Kearny, N.J., 1970-73; acquisition librarian Jersey City State Coll., 1973-74, post office supr., 1974-76, dir. math lab., instr. math. 1976-80; instr. math. Hudson County Community Coll. (N.J.), 1980-82, assoc. prof., coordinator math., 1982-84; prof., chmn. math., curriculum dir., acad. coord. Coll. New Rochelle (N.Y.), 1984—; researcher Rutgers U., New Brunswick, N.J., 1980—; cons. Jersey City Bd. Edn., N.J., 1982—. Contbr. articles to profl. jours. Recipient Commemorative medal of Honor, 1986, Cultural Doctorate award in Philosophy of Edn. World U., 1988. Mem. Nat. Council Tchrs. Math. (reviewer and referee), N.Y. Acad. Scis., Am. Ednl. Research Assn., Math. Assn. Am., Am. Math. Assn. Two-Yr. Colls., Assn. Supervision Curriculum Devel., Am. Mus. Natural History, Nat. Geog. Soc., Nat. Wildlife Fedn., Smithsonian Instn. Roman Catholic. Achievements include research in mathematics anxiety and mathematics problem solving. Office: New Rochelle Coll Sch New Resources 125 Barclay St New York NY 10007-2179

TYRRELL, ALBERT RAY, government liaison for industry; b. Indpls., Sept. 21, 1919; s. Laurence Ray and Nina Atherton (Mobley) T.; m. Sussie Fredrika Petersen, Oct. 4, 1954; children: Nina Fresanne, Robert Warren. AA, Reedley Jr. Coll., Calif., 1939; BA, Harvard U., 1948. With Lockheed Aircraft, Burbank, Calif., 1948-50; rep. Lockheed Aircraft, Dayton, Ohio, 1950-53; ops. mgr. Ind. Mil. Air Transport Assn., Washington, 1953-54; cons. Construciones Aeronauticas, Madrid, 1954-55; pres. Teleprompter of Washington, 1955-56; v.p. Teleprompter Corp., N.Y.C., 1956-57; v.p. mktg. Internat. Atlas Div. Atlas Corp., Oakland, Calif., 1958-63; chmn. mgmt. com. Global Assocs., Oakland, Calif., 1962-64; bd. dirs. Global Assocs., 1964-84; v.p. for congl. relations Atlas Corp., Washington; industry rep. Am. Mining Congress, Washington, 1964-84; founder joint venture Global Assocs., 1964-84. Apptd. mem. Def. Industry Adv. Coun., 1960. Col. USAF, 1940-50. Decorated DFC with 2 oak leaf clusters, Air medal with 3 oak leaf clusters, others. Mem. Harvard Club N.Y.C., Harvard Club Washington, Congl. Country Club, University Club, Masons, Quiet Birdmen. Baptist. Address: 3001 Veazey Ter NW # 407 Washington DC 20008

TZAGOURNIS, MANUEL, physician, educator, university dean and official; b. Youngstown, Ohio; came to Oct. 20, 1934,; s. Adam and Argiro T.; m. Madeline Jean Kalos, Aug. 30, 1958; children: Adam, Alice, Ellen, Jack George. B.S., Ohio State U., 1956, M.D., 1960, M.S., 1967. Intern Phila. Gen. Hosp., 1960-61; resident Ohio State U., Columbus, 1961-63, chief med. resident, 1966-67, instr., 1967-68, asst. prof., 1968-70, assoc. prof., 1970-74, prof., 1974—, asst. dean Sch. Medicine, 1973-75, assoc. dean, med. dirs. hosps., 1975-80, v.p. health services and dean of medicine, 1981—; gen. practice medicine Columbus, 1967—; mem. staff Ohio State U. Hosps.; mem. Coalition for Cost Effective Health Services Edn. and Research Group State of Ohio, 1983. Contbg. author: textbook Endocrinology, 1974, Clinical Diabetes: Modern Management, 1980; co-author: Diabetes Mellitus, 1983, 88. Citation Ohio State Senate Resolution No. 984, 1989. Capt. U.S. Army, 1962-64. Recipient Homeric Order of Ahepa Cleve. chpt., 1976, Phys. of Yr. award Hellenic Med. Soc. N.Y., 1989; citations Ohio Stae Senate and Ho. of Reps., 1975, 83. Mem. Franklin County Acad. Medicine, Assn. Am. Med. Colls., Deans' Council. Mem. Greek Orthodox Ch. Home: 1589 Stanford Rd Columbus OH 43212 Office: Ohio State U Coll Medicine 200 Meiling Hall 370 W 9th Ave Columbus OH 43210-1238

UBAN, STEPHEN ALAN, mechanical engineer; b. Waterloo, Iowa, May 10, 1950; children: William, Jacob. BS in Mech. Engring., Iowa State U., 1973. Registered profl. mech. engr., Oreg. Sr. rsch. engr. Smith & Loveless, Ecodyne, Lenexa, Kans., 1973-77; engring. mgr. Neptune Microfloc, Corvallis, Oreg., 1977-83, dir. R&D, 1983-85; with spl. projects dept. Johnson Screens, St. Paul, 1985-89; dir. R&D Johnson Filtration Systems, St. Paul, 1989-91, 1991—; dir. R&D Wheelabrator Engineered Systems, Inc., St. Paul, 1991—; com. mem. EPA, Washington, 1987. Mem. Product Devel. and Mgmt. Assn. (Minn. chpt. pres. 1991-93), Internat. Product Devel. and Mgmt. Assn. (v.p. new svcs. 1993), Am. Water Works Assn., Am. Mgmt. Assn., Greenpeace, Nat. Resource Def. Coun. Achievements include patents for microfiltration, underdrains, refinery flow distributor, dissolved air flotation. Office: Wheelabrator Engineered Systems Inc PO Box 64118 Saint Paul MN 55164

UBUKA, TOSHIHIKO, biochemistry educator; b. Kagaminocho, Okayama, Japan, Jan. 31, 1934; s. Yoshio and Shigeko (Hashimoto) U.; m. Satoko Iwamiya, Oct. 18, 1960; children: Takayoshi, Hiromi, Atsue. MD, Okayama U., 1959, PhD, 1964. With Okayama U., 1964-73, asst. prof., 1973-80, assoc. prof. Med. Sch., 1980-81, prof. Med. Sch., 1981—; rsch. assoc. Med. Coll. Cornell U., N.Y.C., 1968-71. Co-author: Methods in Enzymology, vol. 143, 1987; editor Acta Med Okayama, 1980—, Physiol Chem Phys and Med NMR, 1982—, Amino Acids, 1991—; chief editor Acta Med Okayama, 1987-90. Fellow Japanese Biochem. Soc., Japanese Soc. Nutrition and Food Sci.; mem. AAAS, N.Y. Acad. Scis., Internat. Soc. Amino Acid Rsch., Soc. Study Inborn Errors Metabolism. Buddhism. Achievements include research in sulfur biochemistry, sulfur nutrition, cysteine metabolism in mammals, protein modification with mixed disulfides, inborn errors of cysteine metabolism. Office: Okayama U Med Sch, 2-5-1 Shikatacho, Okayama 700, Japan

UCHIDA, HIDEO, engineer; b. Feb. 24, 1919. BS in Engring., Tokyo U. Prof. (hon.), dept. engring. Tokyo U., Japan. Mem. Nuclear Saftey Commn. Office: Japan Nuclear Safety Comm, 2-2-1 Kasumigaseki, Chiyoda-ku Japan*

UCHIYAMA, SHOICHI, mechanical engineer; b. Tokyo, Aug. 1, 1927; m. Teruko Shimizu Uchiyama, Nov. 12, 1962; children: Richard Junichi, Robert Hiroaki. BS, Chiba U., 1953; MS, UCLA, 1957, PhD, 1963. Asst. rsch. engr. UCLA, 1957-63; project engr. Aerospace Rsch. Assocs., Inc., W. Covina, Calif., 1963-65; mem. tech. staff N. Am. Rockwell, Downey, Calif., 1965-71; gen. mgr. NKK, Kawasaki, Japan, 1971-86; tech. counselor NKTEKS, Tokyo, 1986-93; tech. advisor Kawawa Internat. Patent Office, Tokyo, 1993—; coms. mem. High Temp. Com., Japan, 1973-77, Fluid Analysis Inst., 1975-78, ISES Com., Japan, 1975-80; lectr. Musashi Inst. Tech., Tokyo, 1987—, Tōin U. Yokohama, Japan, 1990—. Contbr. articles to profl. jours. Home: 5-9-2 Hiyoshidai, Togane Japan

UCHUPI, ELAZAR, geologist, researcher; b. N.Y.C., Oct. 31, 1928; s. Alfonso and Carmen (Urbizu) U. BS, CCNY, 1952; MS, U. So. Calif., 1954, PhD, 1962. Rsch. asst. U. So. Calif., L.A., 1955-62; rsch. asst. Woods Hole (Mass.) Oceanographic Inst., 1962-64, assoc. scientist, 1964-79, sr. scientist, 1979—; mem. Gulf of Mex. panel Joint Oceanographic Instns. Deep Earth Sampling, 1972-74, Sci. Com. for Oceanic Rsch. Working Group 41, 1973-74, steering com. U.S. Oceanographic Office Relief Map World's Oceans, Joint Oceanographic Instns. Site Survey Panel, 1978-85; J. Seward Johnson chair in oceanography W.H.O.I., 1989-93; compiler geological maps Ocean Margin Drilling. Mem. editorial staff Offshore Mag., 1972-74, Marine Geology, 1971-75. Recipient Cert. Recognition, Nat. Assn. Geology Tchrs., Inc. and its Crustal Evolution Edn. Project, 1979, Medal, Offshore Mag. Editorial Adv. Bd., 1974, Frances P. Shepard award, 1991. Fellow Geol. Soc. Am.; mem. Am. Assn. Petroleum Geologists (editorial staff Bulletin 1972-78, Cert. Recognition 1969), Am. Soc. Limnology and Oceanography, Archeol. Inst. Am., Sociedad Geologica de Espana, Soc. Econ. Paleontologists and Mineralogists, Sigma Xi. Achievements include research in seismic reflection, magnetic and gravity profiles of the eastern Atlantic continental margin and adjacent deep seafloor, suspended matter and other properties of surface waters of the northeastern Atlantic Ocean, the continental margin off western Africa: Angola to Sierra Leone, Senegal to Portugal, sediments of 3 bays of Baja, Calif.: Sebastian Viscaino, San Cristobal and Todos Santos, characteristics of sediments of the mainland shelf of southern Calif., submarine geology of the Santa Rosa-Cortes Ridge, sediments on the continental margin off eastern U.S., the continental slope between San Francisco and Cedrow Island, Mex., sediments of the Palos Verdes shelf, sediments and topography of Kane Basin, statistical parameters of Cape Cod Beach and eolian sands, basins of Gulf of Mex., structure of Georges Bank, and the continental margin of the Atlantic coast of the U.S., topography and structure of Northeast Channel, Gulf of Mex., and Cashes Ledge, Gulf of Maine, distribution and geologic structure of Triassic rocks in the Bay of Fundy and the northeastern part of the Gulf of Maine, Microrelief of the continental margin south of Cape Lookout, N.C., shallow structure of the Straits of Fla., sub-surface morphology of L.I., Block Island, Rhode Island sounds, and Buzzards Bay, bathymetry of the Gulf of Mex., slumping on the continental margin southeast of L.I., N.Y., woody debris on the mainland shelf off Ventura, southern Calif., the continental margin south of Cape Hatteras, N.C., the Atlantic continental shelf and slope of the U.S., geological structure of the continental margin off Gulf Coast of the U.S., and more. Office: Woods Hole Oceanographic Inst Dept Geology & Geophysics Woods Hole MA 02543

UDA, ROBERT TAKEO, aerospace engineer; b. Honolulu, Aug. 1, 1942; s. Masao and Irene Kuualoha (Waipa) U.; m. Karen Elizabeth Rowland, June 8, 1968; children: Atom Richard, Marc Edward, Heather Ann. BS in Aerospace Engring., U. Okla., 1966; MS in Astronautics, Air Force Inst. Tech., Wright-Patterson AFB, Ohio, 1968; BS in Gen. Bus., Univ. State of N.Y., Albany, 1988; MBA, U. La Verne, 1993. Commd. 2d lt. USAF, 1966, advanced through grades to capt., resigned, 1974; prin. engr. Planning Rsch. Corp., Kennedy Space Center, Fla., 1974-77; sr. preliminary design engr. Hamilton Std. divsn. United Techs. Corp., Windsor Locks, Conn., 1977-78; project engr. TRW, Inc. Def. & Space Systems Group, Redondo Beach, Calif., 1978-79; gen. mgr., product line mgr., program mgr. HR Textron, Inc., Valencia, Calif., 1979-83; v.p., gen. mgr. North Am. Mfg. Corp., Spanish Fork, Utah, 1983; chmn., pres., CEO Apollo Systems Tech., Inc., Canyon Country, Calif., 1983-87; mgr. advanced programs, project mgr., sr. engr. specialist Rockwell Internat. Corp., Downey, Calif., 1987—; adj. faculty Nat. U., San Diego, 1987. Prepared, pub. over 150 documents, reports, papers, publs. Mormon stake high counselor, 1984-88, stake clk., 1988-89, bishop Canyon Country, Calif., 1989—. Named Jaycee of Yr. Hawthorne, Calif., 1970-71, Sunnyvale, Calif., 1969-70. Fellow Brit. Interplanetary Soc.; assoc. fellow AIAA (chmn. missile systems engring. std. com. 1985-89, mem. stds. tech. coun. 1986—, space launch vehicles com. on stds. 1991—); mem. Am. Astronautical Soc., Air Force Assn., Regents Coll. Alumni Assn., Sigma Tau. Republican. Home: 19544 Delight St Canyon Country CA 91351 Office: Rockwell Internat Corp 2800 Westminster Blvd Seal Beach CA 90740

UDDIN, WAHEED, civil engineer; b. Karachi, Pakistan, Feb. 8, 1949; came to U.S., 1981; s. Hameed and Amjadi (Begum) U.; m. Rukhsana Tayyab, July 1, 1978; children: Omar W., Usman W. BSCE, U. Karachi, 1970; MS in Geotech. Engring., Asian Inst. Tech., Bangkok, 1975; PhD in Transp. Engring., U. Tex., 1984. Registered profl. engr., Tex. Lab. engr. Airport Devel. Agy., Ltd., Pakistan, 1971-73; materials engr. Netherlands Airport cons., Jeddah, Saudi Arabia, 1975-78; asst. rsch. engr. U. Petroleum and Minerals Rsch. Inst., Dhahran, Saudi Arabia, 1978-81; rsc. engr. Austin (Tex.) Rsch. Engrs., Inc., 1984-87; pavement/materials engr. Tex. R&D Found., Riverdale, Md., 1987-89; UN pavement expert UNCHS/Dubai Municipality, Dubai, 1989-91; asst. prof. U. Miss., University, 1991—; infrastructure cons. Engring. Mgmt. Applications, Inc., Silver Spring, Md., 1992—; liaison officer for Saudi Arabia and UAE Asian Geotech. Info. Centre, Bangkok, 1976-81. Contbr. over 45 articles to profl. jours. Chmn. adult lectures Muslim Community Ctr., Silver Spring, 1989. M of Engring. scholarship U.K. Govt., 1973-75. Mem. ASCE, ASTM, Internat. Soc. Asphalt Pavements (founder), U.K. Instn. Civil Engrs. (assoc.), Chi Epsilon. Achievements include patent for highway pavement nondestructive testing and analysis methodology, road user cost and benefit analysis software. Office: U Miss Dept Civil Engring University MS 38677

UDEINYA, IROKA JOSEPH, pharmacologist, researcher; b. Mgbowo, Enugu, Nigeria, Aug. 15, 1953; came to U.S., 1973; s. Udeinya Onuoha and Udumma (Omelu) Udeinya; m. Christie Nnena Anih, Aug. 30, 1976; children: Onuoha, Onyeka, Toochi, Omasirichi, Chinasa, Chineche. BA in Biology, Brandeis U., 1976; PhD in Pharmacology, West Va. U., 1979. Rsch. fellow NIH, Bethesda, Md., 1979-83; lectr. U. Nigeria Coll. Medicine, Enugu, 1983-87; sr. rsch. scientist Walter Reed Army Med. Ctr., Washington, 1987-89; assoc. prof. pharmocology, anasthesiology Howard U. Coll. Medicine, Washington, 1989—; cons. WHO, Geneva, 1980-81, U. Nigeria Teaching Hosp., Enugu, 1985-87. Contbr. articles to profl. jours. Recipient Tropical Disease rsch. grant WHO, 1983, rsch. grant U. Nigeria Senate, Nsukka, 1985, biotechnology rsch. fellowship Rockefella Found., 1984, NAS rsch. fellow, 1988. Mem. AAAS, Am. Soc. Tropical Medicine & Hygiene, Tissue Culture Assn., Sigma Xi. Achievements include establishment of in vitro method for studies of attachment of malaria infected erythrocytes to endothelial cells; demonstration of ability of immune sera to block and reverse attachment of infected enthrocytes to endothelial cells; establishment of continous-culture human endothelial cells without viral transformation. Home: 422 N Horners Ln Rockville MD 20850-1644 Office: Howard Univ Coll Medicine Dept Pharmacology Washington DC 20059

UDLER, DMITRY, physicist, educator; b. Kharkov, Ukraine, USSR, Dec. 15, 1954; came to U.S., 1989; s. Grigory and Meri (Tatiyevskaya) U. PhD, Inst. for Solid State Physics, Chernogolovka, USSR, 1988. Rsch. scientist Inst. for Solid State Physics, 1988-89, rsch. asst. prof. physics Northwestern U., Evanston, Ill., 1990—. Contbr. articles to profl. publs. Mem. Am. Phys. Soc., Materials Rsch. Soc., Metals Soc. Office: Northwestern U 2225 Sheridan Rd Evanston IL 60208-3108

UDVARDY, MIKLOS DEZSO FERENC, biology educator; b. Debrecen, Hungary, Mar. 23, 1919; came to U.S., 1958, naturalized, 1965; m. Maud Emilie Bjorklund, Feb. 11, 1951; children: Beatrice, M. Andrew, Monica Lilly. PhD, U. Debrecen, 1942, PhD honoris causa, 1990. Asst. biologist Hungarian Inst. Ornithology, 1942-45; rsch. assoc. in biology Hungarian Acad. Sci., 1945-48; rsch. fellow in zoology U. Helsinki, 1948-49, U. Uppsala, Sweden, 1949-50; asst. curator Swedish Mus. Natural History, 1951; vis. lectr. ecology U. Toronto, Ont., Can., 1951-52; lectr. zoology U. B.C., Vancouver, Can., 1952-53, asst. prof., 1953-59, assoc. prof., 1959-66; prof. biol. sci. Calif. State U., Sacramento, 1966-83, prof. emeritus, 1983—; asst. scientist Fisheries Rsch. Bd. Can., 1952-55; vis.. prof. U. Hawaii, 1958-59, U. Bonn., 1970-71; vis. spl. lectr. UCLA, 1963-64; Fulbright lectr. U. Honduras, 1971-72; summer lectr. U. Pa., 1958, U. Tex., 1961, U. Calif., Santa Barbara, 1964, U. No. Ariz., 1969; mem. internat. protecting bd. Biol. Sta. Wilhelminberg, Austria; mem. Point Reyes Bird Obs. Mem. Am. Ornithol. Union, Cooper Ornithol. Soc., Nat. Audubon Soc., Wilson Ornithol. Soc., Brit. Ornithol. Union, Assn. German Ornithology, Am. Assn. Geographers, Finnish Ornithol. Soc. (corr.), Spanish Ornithol. Soc., Swedish Ornithol. Soc. (life), Hungarian Acad. Scis. (ext.). Achievements include research on distributional and ecological zoogeography, animal ecology and behavior, ornithology, biology and distribution of Appendicularia. Office: Calif State U Dept Biol Scis 6000 J St Sacramento CA 95819

UDWADIA, FIRDAUS ERACH, engineering educator, consultant; b. Bombay, Aug. 28, 1947; came to U.S., 1968.; s. Erach Rustam and Perin P. (Lentin) U.; m. Farida Gagrat, Jan. 6, 1977; children: Shanaira, Zubin. BS, Indian Inst. Tech., Bombay, 1968; MS, Calif. Inst. Tech., 1969, PhD, 1972; MBA, U. So. Calif., 1985. Mem. faculty Calif. Inst. Tech., Pasadena, 1972-74; asst. prof. engring. U. So. Calif., Los Angeles, 1974-77, assoc. prof., 1977-83, prof. mech. engring., civil engring. and bus. adminstrn., 1983-86; prof. engring, bus. adminstrn. U. So. Calif., 1986—; also bd. dirs. Structural Identification Computing Facility U. So. Calif.; cons. Jet Propulsion Lab., Pasadena, Calif., 1978—, Argonne Nat. Lab., Chgo., 1982-83, Air Force Rocket Lab., Edwards AFB, Calif., 1984—. Assoc. editor: Applied Mathematics and Computation, Jour. Optimization Theory and Applications, Jour. Franklin Inst.; mem. adv. bd. Internat. Jour. Tech. Forecasting and Social Change; contbr. articles to profl. jours. Bd. dirs. Crisis Mgmt. Ctr., U. So. Calif. NSF grantee, 1976—; recipient Golden Poet award, 1990. Mem. AIAA, ASCE, Am. Acad. Mechanics, Soc. Indsl. and Applied Math., Seismological Soc. Am., Sigma Xi (Earthquake Engring. Research Inst., 1971, 74, 84). Avocations: writing poetry, piano, chess. Home: 2100 S Santa Anita Ave Arcadia CA 91006-4611 Office: U So Calif 430K Olin Hall University Park Los Angeles CA 90089-1453

UEBLEIS, ANDREAS MICHAEL, engineer; b. Wels, Austria, Jan. 20, 1963; s. Heinrich and Brigitte (Gratz) U. Grad. chem. engring., HTBLA, Wels, 1982; diploma engr. in materials sci., Montan U., Leoben, Austria, 1991. Jr. asst. Montan U. Inst. for Chemistry, Leoben, 1985-86; asst. engr. Montan U. Inst. Metals Sci., Leoben, 1986-90; chief engr. HILTI AG Corp. Rsch., Schaan, Liechtenstein, 1990—. Contbr. sci. articles to profl. jours. Mem. Am. Vacuum Soc., Soc. Petroleum Engrs., Deutsche Gesellschaft för Materialwissenschaften, Verein der Leobener Werkstoffwissenschafter, Soc. Suisse de Traitement de Surface. Roman Catholic. Avocations: music, tennis, squash, soccer, skiing. Home: Egelseestr 12 A, A 6800 Feldkirch Austria Office: HILTI AG, Corp Rsch IWO, FL9494 Schaan Liechtenstein

UEDA, EINOSUKE, physician; b. Osaka, Japan, Feb. 9, 1933; s. Kenzo and Sadako U.; m. Sachiko Ueda, Oct. 21, 1962; children: Yuko, Makoto. MD, Osaka (Japan) U., 1957, PhD, 1962. Intern Nat. Osaka Hosp., 1958; assoc. prof. internal medicine Ehime U. Med. Sch., Matuyama, Japan, 1974-80; v.p. Nat. Toneyama Hosp., Toyonaka, Japan, 1980-92; pres. Kinki-chu Nat. Hosp., 1992—; lectr. Osaka U. Med. Sch., 1980—, Hyogo Med. Coll., Nishinomiya, Japan, 1980—. Mem. N.Y. Acad. Sic., Sakai East Rotary Club. Home: 2-34 Suzuhara-cho, Itami, Hyogo 664, Japan Office: Kinkichuo Nat Hosp, 1180 Nagasone-cho, Sakai, Osaka 591, Japan

UFIMTSEV, PYOTR YAKOVLEVICH, radio engineer, educator; b. Altai region, Russia, USSR, July 8, 1931; came to U.S., 1990; s. Yakov Fedorovich and Vasilisa Vasil'evna (Toropchina) U.; m. Vera M. Umnova, 1958 (div. 1968); 1 child, Galina; m. Tatiana Vladimirovna Sinelschikova, May 3, 1986; children: Ivan, Vladimir. Grad., Odessa State U., USSR; PhD, Cen. Rsch. Inst. of Radio Industry, Moscow, 1959; ScD, St. Petersburg State U., Russia, 1970. Engr., sr. engr., sr. scientist Cen. Rsch. Inst. of Radio Industry, Moscow, 1954-73; sr. scientist Radio Engring. & Electronics Acad. Scis., Moscow, 1973-90; vis. prof., adj. prof. UCLA, 1990—; mem. Sci. Bd. of Radio Waves, Acad. Scis., Moscow, 1960-90. Author: Method of Edge Waves in the Physical Theory of Diffraction, 1962; contbr. over 80 articles to profl. jours. Recipient USSR State Prize, Moscow, 1990, Leroy Randle Grumman medal for outstanding sci. achievement, N.Y.C., 1991. Mem. AIAA, IEEE, Electromagnetics Acad. (U.S.), A.S. Popov Sci. Tech. Soc. Radio Engring., Electronics & Telecommunication (Russia). Achievements include origination of the Physical Theory of Diffraction, used for design of American stealth aircrafts and ships; for radar-cross-section calculation, and antenna design. Home: 715 Gayley Ave Apt 503 Los Angeles CA 90024-2489 Office: UCLA Dept Elec Engring 405 Hilgard Ave Los Angeles CA 90024-1594

UGWU, MARTIN CORNELIUS, pharmacist; b. Enugu, Anambra, Nigeria, Aug. 22, 1956; came to U.S., 1978; s. Nneji and Maria Uchenwa (Igwesi) U.; m. Renee Mashell Momon, June 30, 1990; 1 child, Martin Cornelius Jr. AA/AS in civil engring., Brevard Community Coll., 1980; BS in chemistry, Grambling State U., 1982; PharmD, Fla. A&M U., 1986. Registered clin. pharmacist Dept. Profl. Regulation, Fla. Pharmacist Rite Aid Pharmacy, Miami, Fla., 1986-87, mgr., 1989—; clin. pharmacist Mercy Hosp., Miami, 1987-88; mem. pharmacy and therapeutic com. Palmetto Gen. Hosp., Miami, 1986. Named one of Outstanding Young Men of Am., 1986. Mem. Am. Pharm. Assn., Am. Soc. Hosp. Pharmacists, Am. Soc. Parenteral and Enteral Nutrition, Fla. Pharmacy Assn. Roman Catholic. Home: 19051 NW 78 Pl Miami FL 33015 Office: Rite Aid Pharmacy 744 Lincoln Road Mall Miami FL 33139-2872

UHDE, GEORGE IRVIN, physician; b. Richmond, Ind., Mar. 20, 1912; s. Walter Richard and Anna Margaret (Hoopes) U.; m. Maurine Elizabeth Whitley, July 27, 1935; children—Saundra Uhde Seelig, Thomas Whitley, Michael, Janice. M.D., Duke U., 1936. Diplomate: Am. Bd. Otolaryngology. Intern Reading (Pa.) Hosp., 1936-37, resident in medicine, 1937-38; resident in otolaryngology Balt. Eye, Ear, Nose and Throat Hosp., 1938-40, U. Oreg. Med. Sch., Portland, 1945-47; practice medicine specializing in otolaryngology Louisville, 1948—; asst. prof. otolaryngology U. Louisville Med. Sch., 1945-62, prof. surgery (otolaryngology), head dept., 1963-92, prof. emeritus, 1992—, dir. otolaryngology services, 1963—; mem. staffs Meth., Norton's-Children's, Jewish, St. Joseph's, St. Anthony's, St. Mary and Elizabeth's hosps.; cons. Ky. Surg. Tb Hosp., Hazlewood, VA Hosp., Louisville, U. Louisville Speech and Hearing Center. Author 4 books.; Contbr. articles to profl. jours. Bd. dirs. Easter Seal Speech and Hearing Ctr. Lt. col. M.C. U.S. Army, 1940-45, ETO, Gen. Isenhower staff, 1943-45. Recipient Disting. Service award U. Louisville, 1972. Fellow A.C.S., Am. Acad. Ophthalmology and Otolaryngology, So. Med. Soc.; mem. N.Y. Acad. Scis., Am. Coll. Allergists, Am. Acad. Facial Plastic and Reconstructive Surgery, AAAS, Assn. U. Otolaryngologists, AAUP, Assn. Mil. Surgeons U.S., Am. Laryngol., Rhinol. and Otol. Soc., Am. Audiology Soc., Soc. Clin. Ecology, Am. Soc. Otolaryngology Allergy, Centurian Otol. Research Soc. (Ky. rep.), Am. Council Otolaryngology (Ky. rep. 1968—), Hoopes Quaker Found., SAR (life), Gen. Soc. Colonial Wars (hereditary mem.), Alpha Kappa Kappa. Democrat. Methodist. Clubs: Filson, Big Spring Country, Jefferson. Home: 708 Circle Hill Rd Louisville KY 40207-3627 Office: Med Towers Louisville KY 40202

UHL, CHARLES HARRISON, botanist, plant cytologist; b. Schenectady, N.Y., May 28, 1918; s. Harrison Carl and Florence Anna (Haupt) U.; m. Natalie Browning Whitford, Aug. 15, 1945; children: Natalie Jean, Mary Catherine, Charles H. Jr., Elizabeth Whitford. AB, Emory U., 1939, MS, 1941; PhD, Cornell U., 1947. From asst. prof. to prof. Emeritus Cornell U., Ithaca, N.Y., 1947—; Emerito La Sociedad Mexicana De Cactologia, A.C., Mexico City, 1976. Contbr. articles to profl. jours. Lt. USNR, 1942-45. Fellow Cactus and Succulent Soc. Am.; mem. AAAS, Bot. Soc. Am., Am. Soc. Taxonomists. Office: Cornell Univ Plant Biology Ithaca NY 14853-5908

UHLENBECK, KAREN KESKULLA, mathematician, educator; b. Cleve., Aug. 24, 1942; d. Arnold Edward and Carolyn Elizabeth (Windeler) Keskulla; m. Olke Cornelis, June 12, 1965 (div.). B.S. in Math., U. Mich., 1964; Ph.D. in Math., Brandeis U., 1968. Instr. math. MIT, Cambridge, 1968-69; lectr. U. Calif.-Berkeley, 1969-71; asst. prof., then assoc. prof. U. Ill., Urbana, 1971-76; assoc. prof., then prof. U. Ill., Chgo., 1977-83; prof. U. Chgo., 1983-88; Sid W. Richardson Found. Regents' Chair in Math. U. Tex., 1988—; speaker plenary address Internat. Congress Maths., 1990; mem. com. women on sci. and enging. NRC, 1992—. Author: Instantons and Four Manifolds, 1984. Contbr. articles to profl. jours. NSF Grad. fellow, 1964-68, Sloan Found. fellow, 1974-76, MacArthur Found. fellow, 1983-88; recipient Alumni Achievement award Brandeis U., 1988. Mem. AAAS. Nat. Acad. Scis., Alumni Assn. U. Mich. (Alumnae of Yr. 1984), Am. Math. Soc., Assn. Women in Math., Phi Beta Kappa. Avocations: gardening, canoeing, hiking. Office: U Tex Dept Math Austin TX 78712

UHM, DAN, process engineer; b. Seoul, Korea, Jan. 23, 1964; came to U.S. 1965; s. Do Sung Uhm and Eun Sook Park; m. Sang Hee Lee, Sept. 10, 1989. BS in Ceramic Engring., U. Wash., 1986; postgrad., Furman U., Greenville, S.C., 1989-91, Clemson (S.C.) U., 1991-92, Va. Inst. Tech., 1993—, Radford U., 1993—. With CIA, Washington, 1985-87; process engr. Kyocera Corp., Vancouver, Wash., 1987-89, Kemet Electronics Corp., Fountain Inn, S.C., 1989-92, Alcatel Telecom., Roanoke, Va., 1992—; cons. Korean Capacitor Industry, 1987-92; cons. tech. contracts Tongkook Electronics, Samryoong Corp., Sangyong Corp. Author tng. manual/text: Statistical Process Control, 1988, Quality Circles, 1989. Mem. Am. Ceramic Soc., Nat. Inst. Ceramic Engrs., Internat. Soc. Hybrid Microelectronics, Am. Soc. Quality Control. Achievements include replacement of 111 trichloroethane in the capacitor industry; development of unique molding process for polymer and epoxy materials, process for alloying of silver/palladium electrodes in ceramic chip capacitors. Home: 1005 Chestnut Mountain Dr Vinton VA 24179 Office: Alcatel Telecom 7635 Plantation Rd Roanoke VA 24019

UIJTDEHAAGE, SEBASTIAN HENDRICUS J., psychophysiology researcher; b. Rijswijk, Netherlands, Oct. 2, 1959; came to U.S. 1987; s. Josephus Leonardus Uijtdehaage and Theresia Maria Asberg Ruisch. BS, Free U., Amsterdam, 1984, MS, 1988; PhD, Pa. State U., 1991. Instr. Pa. State U., State College, 1989; rschr. psychophysiology UCLA, 1991—, instr., 1992—. Contbr. articles to profl. jours. Pres. Nat. Student Chamber Orch., Amsterdam, 1984; mgr. Symphony Orch. Free U., Amsterdam, 1982. Mem. Am. Psychol. Soc., Soc. Psychophysiol. Rsch. Achievements include research on effects of eating on motion sickness; eating decreases motion sickness susceptibility. Office: UCLA NPI 760 Westwood Plaza Los Angeles CA 90024

ULABY, FAWWAZ TAYSSIR, electrical engineering and computer science educator, research center administrator; b. Damascus, Syria, Feb. 4, 1943; came to U.S. 1964; s. Tayssir Kamel and Makram (Ard) U.; m. Mary Ann Hammond, Aug. 28, 1968; children: Neda, Aziza, Laith. BS in Physics, Am. U. Beirut, 1964; MSEE, U. Tex., 1966, PhDEE, 1968. Asst. prof. elec. and computer engring. U. Kans., Lawrence, 1968-71, assoc. prof., 1971-76, prof., 1976-84; prof. elec. engring. and computer sci. U. Mich., Ann Arbor, 1984—, dir. NASA Ctr. for Space Terahertz Tech., 1988—. Author: Microwave Remote Sensing, Vol. 1, 1981, Vol. 2, 1982, Vol. 3, 1986, Radar Polarimetry, 1990. Recipient Kuwait prize in applied scis. Govt. of Kuwait, 1987, NASA Group Achievement award, 1990. Fellow IEEE (gen. chmn. internat. symposium 1981-91, Disting. Achievement award 1983, Centennial medal 1984; mem. IEEE Trans. Soc. (editor Geosci. and Remote Sensing jour. 1981-84), IEEE Geosci. and Remote Sensing Soc. (pres. internat. symposium 1979-81), Internat. Union Radio Sci. Avocations: flying kites, racketball. Office: U Mich 3228 EECS 1301 Beal Ann Arbor MI 48109

ULAN, MARTIN SYLVESTER, retired hospital administrator, health services consultant; b. Wilkes Barre, Pa., May 12, 1912; s. John Albert and Elizabeth (Marcinak) U.; m. Gladys Cecilia Olsen, Oct. 29, 1938; children: Martin Olsen, Mardys Cecilia Brooke Leeper. BS in Pharmacy, Phila. Coll. Pharmacy and Sci., 1934, MS in Biology, 1936; cert., Columbia U., 1950, U. Chgo., 1952. Diplomate Am. Bd. of Pharmacy. Pharmacology asst. Phila. Coll. Pharmacy and Sci., 1934-39; research asst. LaWall & Harrisson Lab., Phila., 1934-39; chief pharmacist White Haven (Pa.) Sanatorium, 1939; asst. prof., chmn. dept. pharmacology Rutgers U. Coll. Pharmacy, Newark, N.J., 1940-50; asst. adminstr. Hackensack (N.J.) Hosp., 1950-54, adminstr., 1954-72; adminstr. York (Maine) Hosp., 1972-78; private cons. York, Maine, 1978—; cons. drugs and chems. Bd. of Econ. Warfare, Washington, 1942-45; chmn. Pharmacy team Unitarian Svc. Com., Munich, 1949; lectr. Columbia U., N.Y., 1961-72; bd. dirs. Creative Work Systems Rehab. Ctr., Saco, Maine. Pres. York Health Found., 1981-84; mem. Bd. Selectmen, York, 1986-89, 90-93, vice chmn., 1986-88, chmn., 1988-89, 90-91; trustee York Hosp.; mem. Pres.'s Com. on Mental Retardation, Washington, 1986—, Maine Adv. Com. on Mental Retardation, 1989—; bd. advisors New England U., 1992—; bd. dirs. York County Parent Awareness, 1992-93, York County Community Action, 1992—. Recipient Paul Harris fellowship. Fellow Am. Coll. Health Care Execs., Am. Coll. Apothecaries; mem. Am. Pharm. Assn., Am. Hosp. Assn. (ho. of dels. 1963-69), N.J. Pharm. Assn., Bergen County Pharm. Assn., Rotary, Rho Chi. Republican. Roman Catholic. Avocations: golf, gardening. Home: 73 Agamenticus Ave Cape Neddick ME 03902-7110

ULLMAN, EDWIN FISHER, research chemist; b. Chgo., July 19, 1930; s. Harold P. and Jane F. Ullman; m. Elizabeth J. Finlay, June 26, 1954; children—Becky L., Linda J. B.A., Reed Coll., Portland, Oreg., 1952; M.A., Harvard U., 1954, Ph.D., 1956. Research chemist Lederle Labs., Am. Cyanamid, Pearl River, N.Y., 1955-60; group leader central research div. Am. Cyanamid, Stamford, Conn., 1960-66; sci. dir. Synvar Research Inst., Palo Alto, Calif., 1966-70; v.p., dir. research Syva Co., Palo Alto, 1970—. Edit. bd.: Jour. Organic Chemistry, 1969-74, Jour. Immunoassay, 1979—, Jour. Clin. Lab., Analysis, 1986-87; contbr. articles to sci. jours. Patentee in field. NSF predoctoral fellow, 1952-53; U.S. Rubber Co. fellow, 1954-55. Recipient Clin. Ligand Assay Soc. Mallinckrodt award, 1981, Can. Soc. Clin. Chemists Health Group award, 1982, Inventor of Yr. award Peninsula Patent Law Assn., 1987. Fellow AAAS; mem. Am. Chem. Soc., Am. Assn. Clin. Chemists (Van Skyke award N.Y. sect. 1984, No. Calif. sect. award 1991), Am. Soc. Biol. Chemists, Clin. Ligand Assay Soc., Phi Beta Kappa. Office: Syva Co PO Box 10058 900 Arastradero Rd Palo Alto CA 94303

ULLMANN, EDWARD HANS, systems analyst; b. Bayshore, N.Y., Sept. 16, 1967; s. Edward and Helene (List) U. BS in Computer Sci., Alfred U., 1989. Systems analyst United Techs. Norden Systems, Melville, N.Y., 1989—; mem. C.A.L.S. com. Norden Systems, 1992—. Fireman Copiague (N.Y.) Fire Dept., 1990—. Republican. Lutheran. Home: 17 East Gate Copiaque NY 11726 Office: Norden Systems 75 Maxess Rd # 3 Melville NY 11747-3182

ULLOM, LAWRENCE CHARLES, JR., engineer; b. Wheeling, W.Va., Sept. 5, 1964; s. Lawrence Charles Sr. and Gertrude Virginia (Michel) U.; m. Deborah Lynn Ball, Mar. 25, 1989. AS in Electronics Engring., W.Va. No. C.C., Wheeling, 1984; BEE, W.Va. Inst. Tech., 1988. Simulation engr. Aircraft div. Naval Air Warfare Ctr., Patuxent River, Md., 1988—; cons. engr. LDG Electronics, St. Leonard, Md., 1991—. Contbr. articles to profl. jours. Office: Naval Air Warfare Ctr Aircraft Divsn Bldg 2035 Patuxent River MD 20670

ULRICH, ALFRED DANIEL, III, chemist; b. Pitts., Sept. 16, 1961; s. Alfred Daniel Jr. and Bernice Ann (Dreschel) U.; m. Sandra Elaine Triponey (div. Nov. 1986); 1 child, Douglas Scott; m. Tanya Renee Parker, Jan. 7, 1989; children: Danielle Nicole, Kenneth Parker. BS in Chemistry, U. Pitts., 1983, postgrad., 1983-84; MBA, U. N. Fla., 1991. Intern chemist U.S. Dept. Energy, Pitts., 1982-83; health physics technician Montefiore Hosp., Pitts., 1983-85; chemist PCR Group, Inc., Jacksonville, Fla., 1985-92, bus. devel. specialist, mktg. coord., 1992—; mem. interface task force High Temple Workshop, Dayton, Ohio, 1989-91. Contbr. articles to profl. jours.; patentee in field. Bd. dirs. Jacksonville Campus Ministries, 1986-87; little league coach Southside Youth Athletic Assn., Jacksonville, 1989-92, Mandarin Athletic Assn., 1993—; com. mem. Southside United Meth. Ch., Jacksonville, 1988-91, pres. discovery Sunday Sch. class, 1991. Named Outstanding Young Man of Am., Outstanding Young Men of Am., 1989. Mem. Soc. Plastic Engrs. (tech. paper rev. com. mem. 1989—, tech. session chmn. 1990-91), Soc. for Advancement of Materials and Process Engring., Am. Concrete Inst., Health Physics Soc. Republican. Avocations: fishing, golf, weight training, camping. Home: 10351 Belmont Stakes Ct Jacksonville FL 32257 Office: PCR Inc 8570 Phillips Hwy Ste 101 Jacksonville FL 32256-8273

ULRICH, JOHN AUGUST, microbiology educator; b. St. Paul, May 15, 1915; s. Robert Ernst and Mary Agnes (Farrell) U.; m. Mary Margaret Nash, June 6, 1940 (dec. May 1985); children: Jean Anne, John Joseph, Robert Charles, Karl James, Mary Ellen, Lenore Alice; m. Mary Matkovich, July 19, 1986. BS, St. Thomas Coll., 1938; PhD, U. Minn., 1947. Instr. De La Salle High Sch., Mpls., 1938-41; rsch. asst. U. Minn., Mpls., 1941-45, 49, Hormel Inst., U. Minn., Austin, 1945-49; instr. Mayo Clinic, U. Minn., Rochester, 1949-55; asst. prof. Mayo Found., U. Minn., Rochester, 1955-66; assoc. prof. U. Minn., Mpls., 1966-69; prof. U. N.Mex., Albuquerque, 1969-82, prof. emeritus, 1982—; chmn. Bacteriology & Mycology Study Sect., NIH, Washington, 1961-64, Communicable Diseases Study Sect., Atlanta, 1968-69; rsch. chmn. in field. Chmn. Zumbry Valley Exec. Bd., Boy Scouts Am., Rochester, 1953-55; mem. Gamehaven Exec. Bd., Boy Scouts Am., Rochester, 1952-62, Dem. Com., Olmsted County, Minn., 1964-69. Recipient Silver Beaver award Boy Scouts Am., 1962, Bishop's award Winona Diocese, 1962, Katahli award U. N.Mex., 1980. Mem. Am. Soc. Microbiology, Am. Chem. Soc., Elks, Am. Bd. Med. Mycology, Am. Acad. Microbiology. Democrat. Roman Catholic. Achievements include discoveries in food preservation; survival of microorganisms at low temperatures; urinary amino acid excretions in variety of disease states; post-operative wound infections; bacterial skin populations; hospital epidemiology. Home: 3807 Columbia Dr Longmont CO 80503-2117

ULRICH, JOHN ROSS GERALD, aerospace engineer; b. Kalispell, Mont., Nov. 25, 1929; s. Alva Austin and Hattie Lenora (Kingston) U.; m. Virginia Jean Breinholt, June 19, 1954; children: Virginia, John, Annette, Lenora, James. BS in Engring., Northrup U., 1952. Registered profl. engr., Colo. Engr. Lockheed Aircraft, Burbank, Calif., 1952-54, Radioplane, Van Nuys,

Calif., 1956-57; sect. head Aerojet-Gen., Sacramento, 1957-65; chmn. of SCUT Martin Marietta, Denver, 1965-92. Lectr. Arapahoe County Pub. Schs. With U.S. Army, 1954-56. Mem. AIAA. Achievements include invention of skyline technique for the solution to simultaneous equations, Photographic Strain Measurement Technique. Home: 3435 E Arapahoe Littleton CO 80122

ULSHEN, MARTIN HOWARD, pediatric gastroenterologist, researcher; b. N.Y.C., Mar. 5, 1944; s. Lawrence F. and Dorothy (Cohen) U.; divorced; children: Sarah Powell, Daniel; m. Sue Ellen McRae, Dec. 17, 1988. BA, U. Rochester, 1965, MD, 1969. Diplomate Am. Bd. Pediatrics, Am. Bd. Gastroenterology. Intern U. N.C., 1969-70; resident U. Colo., 1972-74, fellowship in pediatric gastroenterology, 1974-75; fellowship in pediatric gastroenterology Childrens Hosp., Boston, 1975-77; prof. U. N.C., Chapel Hill, 1977—. Assoc. editor Jour. Pediatrics; med. editor Pediatric Gastroenterology, Am. Bd. Pediatrics; contbr. articles to profl. jours. With USPHS, 1970-72. Office: U NC Dept Pediatrics CB # 7220 Chapel Hill NC 27599-7220

UMEKI, SHIGENOBU, physician, researcher; b. Arita, Wakayama, Japan, Sept. 23, 1951; s. Saichi and Katsuyo (Okamoto) U.; m. Yasuko Toshida, Feb. 11, 1980; children: Kazunori, Hirochika. Student, Gifu (Japan) Univ., 1971, postgrad., 1979. Jr. resident Gifu Univ. Sch. Med., 1977-79, asst. biochemistry, 1983-84; sr. resident internal medicine Kawasaki Med. Sch., Kurashiki, Japan, 1984-90, asst. prof., 1990-93; dir. internal medicine dept. Kumeda Hosp., Osaka, Japan, 1989-91, 93—. Fellow Japanese Coll. Physicians (award 1986), Internat. Coll. Angiology (award 1987), Am. Coll. Chest Physicians (award 1989); mem. Japanese Soc. Internal Medicine, Japanese Biochem. Soc., AMA. Avocations: travel, tennis, reading, painting. Home: 2981-1 Obu-Cho Kishiwada, Osaka 596, Japan Office: Kumeda Hosp, 2944 Obu-Cho Kishiwada, Osaka 596, Japan

UMMINGER, BRUCE LYNN, government official, scientist, educator; b. Dayton, Ohio, Apr. 10, 1941; s. Frederick William and Elnora Mae (Waltemathe) U.; m. Judith Lackey Bryant, Dec. 17, 1966; children: Alison Grace, April Lynn. BS magna cum laude with honors in biology, Yale U., 1963, MS, 1966, MPhil, 1968, PhD, 1969; postgrad., U. Calif., Berkeley, 1963-64; cert. univ. adminstrv./mgmt. tng. program, U. Cin., 1975; cert., Fed. Exec. Inst., 1984. Asst. prof. dept. biol. scis. U. Cin., 1969-73, assoc. prof. dept. biol. scis., 1973-75, acting head dept. biol. scis., 1973-75, prof. dept. biol. scis., 1975-81, dir. grad. affairs, 1978-79; program dir. regulatory biology program NSF, Washington, 1979-84, dep. dir. cellular bioscis. div., 1984-89, mem. sr. exec. svc., 1984—, acting div. dir., 1985-87, 88-89, div. dir. cellular bioscis. div., 1989-91, div. dir. integrative biology and neurosci. div., 1991—; sr. advisor on health policy Office of Internat. Health Policy Dept. State, Washington, 1988; exec. sec. Nat. Sci. Bd. Com. on Ctrs. and Individual Investigator Awards, 1986-88; mem. NSF rev. panel Exptl. Program to Stimulate Competitive Rsch, 1989, Rsch. Improvement in Minority Instns., 1986, 87, U.S.-India Coop. Rsch. Program, 1981-82, U.S.-India Exchange of Scholars Program, 1979-81; vice chmn. biotech. rsch. subcom. Fed. Coord. Coun. on Sci. Engring. and Tech. Office Sci. and Tech. Policy, 1991—; mem. group nat. experts on safety in biotech., OECD, 1988-89; mem. sr. exec. panel Exec. Potential Program, Office Pers. Mgmt., 1988-89; mem. space shuttle proposal rev. panel in life scis. NASA, 1978, rsch. assocs. in space biology award com., 1985-91, chmn. cell and devel. biology discipline working group, space biology program, 1990-91, chmn. gravitational biology panel, NASA Specialized Ctrs. Rsch. and Tng., 1990, mem. exec. steering com. in life scis., 1991, mem. gravitational biology facility sci. working group, 1992—, mem. NASA neurolab. steering com., 1993—; mem. panel study biol. diversity, Bd. Sci. and Tech. Internat. Devel. NRC, 1989; mem. adv. screening com. in life scis. Council for Internat. Exchange of Scholars, 1978-81; liaison rep. nat. heart, lung and blood adv. council NIH, 1979-87, nat. adv. child health and human devel. coun., 1990—, recombinant DNA adv. com., 1988; liaison rep. agrl. biotech. rsch. adv. com. USDA, 1989—, mem. interagy. rsch. animal com., 1984-88; interagy. working group internat. biotech., 1988—. Author book chpts. and contbr. articles to profl. jours.; assoc. editor Jour. Exptl. Zoology, 1977-79; editorial adv. bd. Gen. and Comparative Endocrinology, 1982. Mem. world mission com. Ch. of the Redeemer, New Haven, 1967-68; Sun. sch. steering com. Calvary Episcopal Ch., Cin., 1972-73, sr. acolyte, 1972-77, adult edn. com., 1975-76; adv. com. mem. Wakefield High Sch., 1991-92; sci. adv. com. Arlington Pub. Schs., 1987-92, PTA exec. bd., 1991-92, adv. coun. on instrn., 1991-92; adv. bd. mem. Campbell Comml. Coll., Cin., 1977-79. Recipient George Rieveschl, Jr. Rsch. award U. Cin., 1973, Outstanding Performance award and Sustained Superior Performance NSF, 1981, spl. achievement award 1985, Sr. Exec. Svc. Performance award 1986, 87, 90, 91, Presdl. Rank Meritorious Exec. award 1992; U. Cin. Grad. Sch. fellow 1977—, NSF fellow 1964; rsch. grantee NSF 1971-79. Fellow AAAS (coun. 1980-83, 89-90, mem. program com. for 1989 annual meeting 1988, chairperson-elect sect. G.-Biol. Scis. 1987-88, chairperson 1988-89, ret. 1989-90, mem. steering group sect. com. G 1987-90), N.Y. Acad. Scis.; mem. Am. Soc. Zoologists (sec., mem. exec. com. 1979-81, chmn. nominating com. 1981, sec. div. of comparative physiology and biochemistry 1976-77, chmn. Congl. Sci. Fellow Program com. 1986-89, mem. 1991-93), Am. Physiol. Soc. (program adv. com. 1978-81, program exec. com. 1983-86, mem. steering com., comparative physiology sect. 1978-81, sec. Am. Physiol. Soc.-Am. Soc. Zoologists Task Force on Comparative Physiology 1977-78), Am. Inst.is. (chmn. selection com. congl. sci. fellow in zool. scis. 1987, mem. congl. fellow liaisons com. 1991), Am. Soc. for Gravitational and Space Biology, Sr. Execs. Assn., Assn. of Yale Alumni (del. 1990-93), Mory's Assn., Yale Club (Washington), Masons (32 degree), K.T., Shriners, Sigma Xi (Disting. Rsch. award U. Cin. chpt. 1973, pres. U. Cin. chpt. 1977-79). Presbyterian. Achievements include development of science policy in cellular biosciences, integrative biology, neuroscience, and biotechnology; research in low temperature biology, in comparative physiology, endocrinology and biochemistry of fish, and in. Home: 4087B S Four Mile Run Dr Arlington VA 22204-5604 Office: NSF Div Integrative Biology and Neuroscience 1800 G St NW Washington DC 20550-0002

UMPLEBY, STUART ANSPACH, management science educator; b. Tulsa, Mar. 5, 1944; s. Joseph Gray Umpleby and Mary Carolyn (Woerheide) Cresap; m. Gertraud Maria Zangl, Mar. 7, 1986; children: Oliver Gray, Nicholas Anspach. BS in Engring., U. Ill., 1967, BA in Polit. Sci., 1967, MA in Polit. Sci., 1969; PhD in Communications, 1975. Engr. Westinghouse Electric Corp., Pitts., 1966, Machinen Fabrik Froriep, Dusseldorf, Germany, 1967; instr. U. Ill., Urbana, 1968-70; prof. George Washington U., Washington, 1975—; cons. U.S. Agy. Internat. Devel., Washington, 1979, IBM Intertrade, Vienna, 1990, Bled, Yugoslavia, 1990; lectr. Hitachi Ltd., Tokyo, 1970, Inst. for Systems Studies, Soviet Acad. of Scis., Moscow, 1983, 87, 91, Union of Scientists, Sofia, Bulgaria, 1988; guest scholar U. Pa., Phila., 1983, Internat. Inst. for Applied Systems Analysis, Vienna, 1984; guest prof. U. Vienna, 1990. Author: (with others) Adequate Modeling of Systems, 1983, Power, Autonomy, Utopia: New Approaches toward Complex Systems, 1986, Managers and National Culture, 1993, A Science of Goal Formulation: American and Soviet Discussions of Cybernetics and Systems Theory, 1991, also editor (with Vadim N. Sadovsky), Cybernetics of National Development, 1991, also editor (with Robert Trappl); contbr. articles to Cybernetics and Systems, Futures, Population and Environ., Systems Practice, Telecommunications Policy, Soc., Bus. and Soc. Rev., Ekistics, Policy Scis., Jour. Aesthetic Edn., others. Vol. human devel. projects Inst. of Cultural Affairs, Washington, 1976—. Rsch. grantee NSF, U. Ill., 1972-73, George Washington U., Washington, 1977-80, C. F. Kettering Found., U. Ill., 1973. Mem. Am. Soc. for Cybernetics (pres. 1980-82) Austrian Soc. for Cybernetic Studies (assoc. editor jour. 1990—). Office: George Washington U Dept Mgmt Sci Washington DC 20052

UMRIGAR, DARA NARIMAN, civil engineer; b. Bombay, Aug. 28, 1953; came to U.S., 1978; s. Nariman Darabshaw Umrigar and Villoo (Sorabji) Italia; m. Vira Pesi Raimalwalla, May 22, 1982; 1 child, Eric Dara. BSCE, U. Bombay, 1977; MSCE, Prairie View (Tex.) A&M U., 1988. Registered profl. engr., Tex. Grad. engr. trainee Engring. Constrn. Corp. Ltd., Bombay, 1977-78, engr., 1980; civil engring. aide Ardaman and Assocs., Inc., Tallahassee, 1979; grad. staff engr. Law Engring., Inc., Houston, 1980-89; engr. City of Houston, 1989—. Vol. Zoarastrian Assn. Houston, 1980—. Mem. ASCE, Sigma Xi, Tau Beta Pi. Achievements include research in evaluating flexural strength of Dramix steel fiber reinforced concrete. Home: 11743 Cliveden Dr Houston TX 77066-4620 Office: City of Houston 1100 Louisiana Ste 1100 Houston TX 77002

UMSCHEID, LUDWIG JOSEPH, government computer specialist; b. N.Y.C., Dec. 30, 1937; s. Ludwig and Bertha (Gruber) U.; m. Carole Patricia Fisher, Apr. 22, 1967. BS, Fordham U., 1959; MS, Adelphi U., 1963, NYU, 1967. Computer programmer Svc. Bur. Corp., IBM, N.Y.C., 1962; systems analyst Computer Applications Inc., N.Y.C., 1963-71; rsch. meteorologist, maj. systems specialist Environ. Rsch. Labs., NOAA, U.S. Dept. Commerce, Princeton, N.J., 1971—. Contbr. articles to profl. jours. Bd. dirs. Stony Brook Millstone Watershed Assn., Princeton; co-founder, v.p. Friends of Hopewell Valley Open Space, Pennington, N.J.; mem. Pennington Planning Bd.; chair Pennington Environ. Commn. 1st lt. USAF, 1959-62. Mem. IEEE, IEEE Computer Soc., Am. Meteorol. Soc., Assn. Computer Machinery, Assn. N.J. Environ. Commns., N.Y. Acad. Scis. Achievements include pioneering use of Fast Fourier Transforms for polar filtering in numerical weather prediction. Office: US Dept Commerce GFDL/NOAA PO Box 308 Princeton NJ 08542

UMSTADTER, KARL ROBERT, engineering physicist; b. Point Pleasant, N.J., Aug. 29, 1970; s. Grant Alan and Barbara Ann (Simons) U. BS in Engring. Physics, Rensselaer Poly. Inst., 1991, MS, 1993, postgrad., 1993—. Rsch. asst., grad. rsch. supr., grad. rsch. asst., grad. senator Rensselaer Poly. Inst., Troy, N.Y., 1989—. Contbr. articles to IEEE Design and Test Mag., others. Mem. IEEE, Am. Nuclear Soc., Pi Kappa Alpha, Sigma Xi, Alpha Nu Sigma. Republican. Roman Catholic. Achievements include patent for laser-based noncontact test system. Home: 2335A 15th St Troy NY 12180 Office: Rensselaer Poly Inst Design & Mfg Inst DMI-CII 9015 Troy NY 12180-3590

UNCKEL, PER, minister of education; b. Finspång, Ôstergôtland, Sweden. Educator Uppsala U.; chair Swedish Young Moderates, 1971-76; spokesman on energy questions Moderate Party, 1978-82; leader Nat. Campaign, 1980; party spokesman on Edn. and Sci. Referendum on Nuclear Power, 1982-86; sec. Moderate Party, 1986—, minister Edn., 1991—. Office: Ministry of Education, Mynttorget 1, 103 33 Stockholm Sweden*

UNDERHILL, ROBERT ALAN, consumer products company executive; b. Columbus, Ohio, June 9, 1944; s. Robert Alan and Grace Ruth (Smith) U.; m. Lynn Louise Stentz, Oct. 18, 1963; children: Robert Alan III, Richard Louis. Student, Case Western Res. U., 1962-64, Ohio State U., 1965. With tech. svc. dept. Gen. Tire & Rubber Co., Akron, Ohio, 1966-69; quality control engr. Edmont-Wilson Co., Canton, Ohio, 1969-70; mgr. quality assurance Pharmaseal Labs., Massillon, Ohio, 1970-72; mgr. R&D Internat. Playtex Corp., Paramus, N.J., 1972-78; mgr. R&D Kimberly-Clark Corp., Neenah, Wis., 1978-80, dir. R&D, 1980-83, v.p. R&D, 1983-93, sr. v.p. R&D, 1993—; bd. dirs. Appleton (Wis.) Med. Ctr., Outagamie County (Wis.) Chpt. Am. Red Cross. Patentee (U.S. and fgn.) med. device. Mem. exec. bd. Bay Lakes coun. Boy Scouts Am., 1988-92. Mem. Pi Delta Epsilon. Republican. Avocations: stock market investment analysis, travel. Home: 2193-C Sunrise Dr Appleton WI 54914-8766 Office: Kimberly-Clark Corp 2100 Winchester Rd Neenah WI 54956 also: 1400 Holcomb Bridge Rd Roswell GA 30076-2199

UNDERWOOD, ARTHUR LOUIS, JR., chemistry educator, researcher; b. Rochester, N.Y., May 18, 1924; s. Arthur Louis and Grace Ellen (Porter) U.; m. Elizabeth Knapp Emery, June 30, 1948 (dec. Jan. 1986); children: Paul William, Robert Emery, Susan Elizabeth. B.S. in Chemistry, U. Rochester, 1944, Ph.D. in Biochemistry, 1951. Research assoc. Atomic Energy Project, U. Rochester, 1946-51; research assoc. in chemistry MIT, Cambridge, 1951-52; asst. prof. chemistry Emory U., Atlanta, 1952-57, assoc. prof., 1957-62, prof., 1962—; research assoc. chemistry Cornell U., 1959-60; vis. prof. chemistry Mont. State U., summers, 1979-87. Co-author: Quantitative Analysis 6th edit., 1991; contbr. articles to profl. jours. Served with USN, 1944-46. Fellow AAAS; mem. Phi Beta Kappa, Sigma Xi. Republican. Methodist. Home: 1354 Springdale Rd NE Atlanta GA 30306-2419 Office: Emory U Dept Chemistry Atlanta GA 30322

UNDERWOOD, GEORGE ALFRED, mechanical engineer; b. Evanston, Ill., Dec. 14, 1924; s. George A. and Lillian (Johnsos) U.; m. Marian Larson, May 24, 1952; children: Paul G., Carol L., Joan E., Cheryl A. BSME, Ill. Inst. Tech., 1947. Registered profl. engr., Ill., Fla., Ky., Ga. Design engr. Internat. Harvester Co., Melrose Park, Ill., 1946-53; chief engr. K.J. Moore & Co., Chgo., 1953-56; chief machine design engr. Felt Products Mfg. Co., Skokie, Ill., 1956-65; owner Underwood Engring. Co., Skokie, Ill., 1965-68; chief engr. D.E. Miller & Assocs., Inc., Lincolnwood, Ill., 1968-72; spl. project engr. F.J. Littell Machine Inc., Chgo., 1972-77; pres., chmn. bd. Underwood & Assocs., Inc., North Ft. Myers, Fla., 1977—. Cub master, troop com. Boy Scouts Am., Skokie, 1953-77. Recipient Silver Beaver award Boy Scouts Am., 1963. Mem. ASME, NSPE, ASHRAE, Nat. Fire Protection Assn., Masons, Shriner. Republican. Methodist. Home: 503 NE 15th Ave Cape Coral FL 33909 Office: Underwood & Assocs Inc 390 Pondella Rd Apt 6 North Fort Myers FL 33903

UNGER, ISRAEL, dean, chemistry educator; b. Tarnow, Poland, Mar. 30, 1938; arrived in Can., 1951; s. David and Hinda (Fisch) U.; m. Marlene Parker, July 6, 1964; children: Sharon, Sheila. BSc, Sir George Williams, Montreal, Can., 1958; MSc, U. N.B., Can., 1960, PhD, 1963. Welsh post doctoral fellow U. Tex., 1963-65; from asst. prof. to prof. U. N.B., Fredericton, 1965-86, dean sci. dept., 1986—; coun. trustee Inst. for Rsch. on Public Policy, Ottawa, Ont. Can., 1981-86; chair adv. bd. sci. tech. info. Nat. Rsch. Coun., 1991; lectr. in field; chmn. various coms. U. N.B. Author: (with W.A. Noyes) Singlet and Triplet States: Benzene and Simple Aromatic Compounds, 1966; contbr. articles to profl. jours. Chief judge, N.B. Sci. Fair, 1984; judge Telesat Getaway Spl. Contest, Space Shuttle Atlantis, 1984; chairperson Dr. Everett Chalmers Hosp.- Univ. N.B. Liaison Com., 1988; active Prime Minister's Com. for Awards in Excellence in Edn., Ottawa, 1993. CIL fellow U. N.B., 1962-63; grantee NRC, 1965-78, NSERC, 1978-82, CAUT Rsch., 1983-85. Mem. Can. Assn. Univ. Tchrs. (pres. 1980-81), Sci. for Peace (nat. bd. dirs. 1982-85), B'nai Brith (pres. Fredericton chpt. 1981-83). Home: 66 Cameron Ct, Fredericton, NB Canada E3B 5A3 Office: Univ N B Faculty of Sci, PO Box 4400, Fredericton, NB Canada E3B 5A3

UNGERS, LESLIE JOSEPH, engineering executive; b. Painesville, Ohio, Sept. 17, 1951; s. R. Joseph and Lillian G. (Stampfel) U.; m. Priscilla Batsche, Nov. 5, 1977; children: Margeaux A., Joseph A. AB, Miami U., Oxford, Ohio, 1973; MS, U. Cin., 1984. Lectr. Lakeland C.C., Mentor, Ohio, 1974-75; mgr. indsl. hygiene PEI Assocs., Inc. div. IT Corp., Cin., 1975-87; adj. prof. U. Cin., 1980-81; pres. Ungers and Assocs., Inc., Cin., 1987—. Contbr. articles to Environ. Health Prospectives, Am. Indsl. Hygiene Assn. Jour., other profl. publs. Mem. hazardous waste com. Charterite Party, Cin., 1990. Fellow Acad. Kettering Fellows (sec.-treas. 1992—). Achievements include discovery that arsenic is released from semiconductor wafers following manufacturing presenting a health hazard to workers and contamination during production of integrated circuits. Home: 3422 Berry Ave Cincinnati OH 45208 Office: Ungers and Assocs Inc 1136 Saint Gregory St Cincinnati OH 45202-1724

UNVER, ERDAL ALI, research mechanical engineer; b. Nizip, Gaziantep, Turkey, Jan. 1, 1953; came to U.S., 1975; s. Ahmet Talat and Mukaddes (Seker) U.; m. Amira Victoria Margie Unver, Apr. 25, 1980; 1 child, Susan Ayse. BS, Tech. U. Istanbul, Turkey, 1975; MS, U. Calif., Berkeley, 1977; PhD, Lehigh U., 1981; MBA, George Washington U., 1991. Rsch. engr. U.S. Steel Rsch. Ctr., Monroeville, Pa., 1981-84, ENKA Constrn. Co. Rsch. Ctr., Istanbul, 1984-86; mech. engr. David Taylor Rsch. Ctr., Bethesda, Md., 1986-89; indsl. projects mgr. AT&T Bell Labs., Arlington, Va., 1989—. Contbr. articles to profl. jours. Mem. ASME, Acoustical Soc. Am. Avocations: travel, classical music, soccer. Home: 1515 Hugo Cir Silver Spring MD 20906 Office: AT&T Bell Labs 1919 S Eads St Ste 300 Arlington VA 22202

UPADHYAY, YOGENDRA NATH, physician, educator; b. Gorakhpur, India, Dec. 21, 1938; came to U.S., 1963; s. Murlidhar and Vansraji (Pande) U.; m. Cecile R. Yonish; children: Asha, Sameer, Sanjay. MB, BS, All India Inst. Med. Scis., New Delhi, 1962. Diplomate Am. Bd. Psychiatry and Neurology, Am. Bd. Pediatrics. Instr. in pediatrics Johns Hopkins U. Sch. Medicine, Balt., 1969-71; fellow in child psychiatry Johns Hopkins Hosp./Johns Hopkins U., Balt., 1971-72; resident, then sr. resident in psychiatry Albert Einstein Coll. Medicine/Bronx Mcpl. Hosp. Ctr., 1972-74, fellow in child psychiatry, 1974-75; chief, partial hosp. program for children, dept. psychiatry Brookdale Hosp., Bklyn., 1976-77; med. dir. West Nassau Mental Health Ctr., Franklin Sq., N.Y., 1977-80; asst. prof. clin. psychiatry SUNY, Stony Brook, 1978-92; dir. child and adolescent psychiatry Nassau County Med. Ctr., East Meadow, N.Y., 1980-92; sr. psychiatrist South Oaks Hosp., Amityville, N.Y., 1992—. Fellow Am. Psychiatric Assn. (cons. task force treatments psychiatric disorders 1989—), Am. Acad. Child and Adolescent Psychiatry. Office: 400 Sunrise Hwy Amityville NY 11701

UPHOFF, JOHN VINCENT, mechanical engineer; b. Deadwood, S.D., June 21, 1965; s. John Herman and Jan Marie (McCormick) U.; m. Deanna Lynn Wetrosky, Apr. 6, 1986; children: Matthew, Ryan, Nicholas. BSME, S.D. State U., 1988. Plant engr. Iowa Pub. Svc., Sioux City, 1988—. Mem. ASME (chmn. sect. 1990-92). Roman Catholic. Office: Iowa Public Svc 401 Douglas St Sioux City IA 51102

UPSHAW, TIMOTHY ALAN, chemical engineer; b. Paducah, Ky., Dec. 3, 1958; s. Edward Burrell and Emma Louise (Smith) U.; m. Claire Elizabeth Sutton, Sept. 18, 1982; children: Julie Ann, Bryan Alan, Amy Rebekah. BS, U. Ky., 1980. Registered profl. engr., Tenn., Ky. Chem. engr. Eastman Kodak, Kingsport, Tenn., 1980-87, sr. chem. engr., 1987-92; sr. process engr. Internat. Specialty Products, Calvert City, Ky., 1992—; adv. panel mem. Dow Jones & Co., Syosset, N.Y., 1992—; lectr. in field. Active Abundant Living Chrisitan Fellowship, Kingsport, 1988-92. Named Ky. Col. Commonwealth of Ky., 1975, Duke of Paducah City of Paducah, 1975. Mem. Wall St. Club, Omega Chi Epsilon, Tau Beta Pi. Achievements include research on conceptual design work for largest polethylene terephthalate plant in the world, lead design on 2d generation acetic anhydride plant chemicals from coal. Home: 1166 S Friendship Rd Paducah KY 42003 Office: ISP Chems PO Box 37 Calvert City KY 42029

UPTON, ARTHUR CANFIELD, retired experimental pathologist; b. Ann Arbor, Mich., Feb. 27, 1923; s. Herbert Hawkes and Ellen (Canfield) U.; m. Elizabeth Bache Perry, Mar. 1, 1946; children: Rebecca A., Melissa P., Bradley C. Grad., Phillips Acad., Andover, Mass., 1941; BA, U. Mich., 1944, MD, 1946. Intern Univ. Hosp., Ann Arbor, 1947; resident Univ. Hosp., 1948-49; instr. pathology U. Mich. Med. Sch., 1950-51; pathologist Oak Ridge Nat. Lab., 1951-54, chief pathology-physiology sect., 1954-69; prof. pathology SUNY Med. Sch. at Stony Brook, 1969-77, chmn. dept. pathology, 1969-70, dean Sch. Basic Health Scis., 1970-75; dir. Nat. Cancer Inst., Bethesda, Md., 1977-79; prof., chmn. dept. environ. medicine NYU Med. Sch., N.Y.C., 1980-92; emeritus, 1993—; prof. emeritus NYU Med. Sch., N.Y.C., 1993—; prof. pathology Sch. Medicine, U. N.Mex., Albuquerque, 1991—; attending pathologist Brookhaven Nat. Lab., 1969-77; dir. Inst. Environ. Medicine, Med. Sch., NYU, 1980-91; mem. various coms. nat. and internat. orgns.; lectr. in field. Assoc. editor: Cancer Research; mem. editorial bd.: Internat. Union Against Cancer. Served with AUS, 1943-46. Recipient Ernest Orlando Lawrence award for atomic field, 1965, Comfort-Crookshank award for cancer rsch. Inst. Med., NAS, 1979, Claude M. Fuess award 1980, Sarah L. Poilley award for pub. health, 1983, CHUMS Physician of Yr. award 1985, Basic Cell Rsch. in Cytology Lectureship award 1985, Fred W. Stewart award, 1986, Ramazzini award, 1986, Lovelace Med. Found. award, 1993; Sigma Xi nat. lectr., 1989-91. Fellow N.Y. Acad. Sci.; mem. Am. Assn. Pathologists and Bacteriologists, Internat. Acad. Pathology, Inst. Medicine Nat. Acad. Sci., Radiation Rsch. Soc. (councilor 1963-64, pres. 1965-66), Internat. Assn. Radiation Rsch. (pres. 1983-87), Am. Assn. Cancer Rsch. (pres. 1963-64), Am. Soc. Exptl. Pathology (pres. 1967-68), AAAS, Gerontol. Soc., Sci. Rsch. Soc. Am., Soc. Exptl. Biology and Medicine, Peruvian Oncology Soc. (hon.), Japan Cancer Assn. (hon.), N.Y. State Health Rsch. Coun. (chmn. 1982-90), Internat. Assn. for Radiation Rsch. (pres. 1983-87), Assn. Univ. Environ. Health Sci. Ctrs. (pres. 1982-90), Phi Beta Kappa, Phi Gamma Delta, Alpha Omega Alpha, Nu Sigma Nu, Sigma Xi. Achievements include research on pathology of radiation injury and endocrine glands, on cancer, on carcinogenesis, on experimental leukemia, on aging. Home: 3 Washington Square Village New York NY 10012 Office: Inst Environ Medicine NYU Med Ctr 550 First Ave New York NY 10016*

URBAN, CATHLEEN ANDREA, client technical support consultant; b. Elizabeth, N.J., June 7, 1947; d. Emil Martin and Susan (Rahoche) Cupec; m. Walter Robert Urban, Nov. 5, 1966; children: Karen Louise, Kimberly Ann. Student, Rutgers U., 1965-66, 91—; AS in Computer Info. Systems, Raritan Valley Community Coll., North Branch, N.J., 1990, AAS in Computer Programming, 1990. Office mgr. K-Mart Corp., Somerville, N.J., 1987-90; software developer Bell Communications Rsch., Piscataway, N.J., 1990-93, client tech. support cons., 1993—. Leader Somerset County 4-H Program, Bridgewater, 1978-87. Mem. Internat. Platform Assn., Golden Key Nat. Honor Soc., Phi Theta Kappa. Roman Catholic. Avocations: science fiction, reading, showing Siberian Huskies, candle making. Home: 570 Amwell Rd Neshanic Station NJ 08853-3404 Office: Bell Comm Rsch 444 Hoes Ln Piscataway NJ 08854-4104

URBAN, MAREK WOJCIECH, chemist educator, consultant; b. Krakow, Poland, Oct. 3, 1952; came to U.S., 1979; s. Ludwik and Wieslawa (Domon) U.; m. Katarzyna Maria, Apr. 21, 1979. BS in Materials Sci., U. Mining and Metall., Krakow, Poland, 1979; MS in Chemistry, Marquette U., 1981; PhD in Chemistry, Mich. Tech. U., 1984. Postdoctoral rsch. assoc. Case We. Res. U., Cleve., 1984-86; asst. prof. polymers and coatings N.D. State U., Fargo, 1986-89, assoc. prof. polymers and coatings, 1989—; lectr. Gordon Rsch. Confs. Author: Vibrational Spectroscopy of Molecules and Macromolecules on Surfaces, 1993; editor: Structure-Property Relations in Polymers; Spectroscopy and Performance, 1993; contbr. articles to profl. jours., chpts. to books. Mem. Am. Chem. Soc. (series editor on polymer surfaces and interfaces, polymer chemistry and polymeric materials div., sci. and engring. div., chair Internat. Symposium on Spectroscopy of Polymers 1991, Internat. Symposium on Hyphenated Methods in Polymer Chanod 1993), Am. Phys. Soc., Soc. Applied Spectroscopy (Minn. chpt. chairperson-elect 1987-88, chairperson 1988-89), Northwestern Soc. for Coatings Tech. Achievements include patents for "Rheo-Photoacoustic FT-IR Spectroscopic Methods and Apparatus" and "Surface Treatment of Silicone Rubber". Office: ND State U Dept Polymers & Coatings Fargo ND 58105

URENA-ALEXIADES, JOSE LUIS, electrical engineer; b. Madrid, Spain, Sept. 5, 1949; s. Jose L. and Maria (Alexiades Christodulakis) Urena y Pon. MSEE, U. Madrid, Spain, 1976; MS in Computer Science, UCLA, 1978. Rsch. asst. UCLA, 1978; systems analyst Honeywell Info. Systems, L.A., 1978-80; mem. tech. staff Jet Propulsion Lab., Pasadena, Calif., 1980-91; exec. dir. Empresa Nacional de Innovacion S.A., L.A., 1991—. Contbr. various articles to profl. jours. Two times recipient NASA Group Achievement award. Mem. IEEE, IEEE Computer Soc., IEEE Communications Soc., Assn. for Computer Machinery, World Federalist Assn., Spanish Profl. Am. Inc. Roman Catholic. Avocations: active photographer, Master's swimming. Home: 904 Dickson St Marina Dl Rey CA 90292-5513 Office: Empresa Nacional Innovacion SA 2049 Century Park E Ste 2770 Los Angeles CA 90067-3202

URICHECK-HOLZAPFEL, MARYANNE, pharmacist; b. Phila., Sept. 23, 1960; d. Joseph Francis and Norma Joan (Miller) Uricheck; m. J. Kirk Holzapfel, Sept. 26, 1992. BS in Pharmacy, Phila. Coll. Pharmacy, 1983. Pharmacy tech. St. Francis Hosp., Wilmington, Del., 1980; pharmacy intern Wilmington Med. Ctr., 1980-83, Abbott Labs., Chgo., 1982; pharmacist-in-charge Rite Aid, Wilmington, 1983-88; staff pharmacist, cons. Happy Harry's Inst. Pharmacy, Newark, Del., 1988—. Mem. Am. Soc. Cons. Pharmacists, Am. Pharm. Assn., Nat. Coun. State Pharm. Assn. Execs., Am. Soc. Assn. Execs., Del. State Soc. Hosp. Pharmacists, Del. Pharm. Soc. (exec. dir. 1990—). Roman Catholic. Avocations: gardening, dancing, reading, swimming. Home: 169 E Green Valley Cir Newark DE 19711-6792 Office: 111 Ruthar Dr Newark DE 19711-8016

URIST, MARSHALL RAYMOND, orthopedic surgeon, researcher; b. Chgo., June 11, 1914; s. Irwin and Minna (Vision) U.; m. Alice Elizabeth Pfund, Aug. 16, 1941; children—Marshall McLean, Nancy Scott Urist

Miller, John Baxter. B.A., U. Mich., 1936; M.S., U. Chgo., 1937; M.D., Johns Hopkins U., 1941; M.D. (hon.), U. Lund, Malmo, Sweden, 1977. Diplomate Am. Bd. Orthopedic Surgery. Practice medicine specializing in orthopedic surgery Los Angeles, 1948—; mem. staff U.S. VA Hosp. Wadsworth, Los Angeles, 1948-69; adj. assoc. prof. surgery Sch. Medicine, UCLA Med. Ctr., 1954-69, prof. orthopedic surgery, 1969—; dir. bone research lab. UCLA Med. Ctr., 1950—; cons. to surgeon gen. U.S. Army Com. on Trauma, D.C., 1963-71, U.S. Navy, D.C., 1964; keynote speaker Gordon Conf. Bone Growth and Regulation, 1988. Author: Bone: Fundamentals of Physiology of Bone, 1968, Fundamental & Clinical Bone Physiology, 1980; editor-in-chief Clin. Orthopedics and Related Research; contbr. 325 articles to med. jours. Served to lt. col. U.S. Army, 1943-46, ETO. Recipient Sir Henry Wellcome award Assn. Mil. Surgeons of U.S., 1947, Kappa Delta awards Am. Acad. Orthopedic Surgeons, Chgo., 1950, 81, Claude Bernard medal U. Montreal, Que., Can., 1962, Dallas Phemister lectureship U. Chgo., 1978; Disting. Service medal Am. Assn. Tissue Banks, 1988, Gold medal for sci. achievement Orthopedic Soc. Spain, 1988, Outstanding Rsch. award U.S. Army R&D Command, 1989, 91; Guggenheim Found. fellow, N.Y.C., 1972, Bristol Meyers Sqibb Zimmer award, 1993; honoree Conf. on Bioactive Factors, U. Tex., San Antonio, 1988. Fellow ACS, Royal Coll. Surgeons (Edinburgh); mem. AAAS, Assn. Bone and Joint Surgeons (pres. 1967-68, editor-in-chief jour. Clin. Orthopedic Rsch. 1966—), Hip Soc. (founding, pres. 1978-79), Am. Orthopedic Assn. (hon.). Republican. Avocation: agriculture. Discovered bone morphogenetic protein in 1965. Home: 796 Amalfi Dr Pacific Palisades CA 90272-4508 Office: 1033 Gayley Ave Los Angeles CA 90024-3417

URQUHART, ANDREW WILLARD, metallurgist, engineering executive; b. Burlington, Vt., Aug. 24, 1939; s. John Wardrop and Dorothy Helen (Hefflon) U.; m. Carolyn Fay Powell, Mar. 9, 1963; children—Marion, Dorothy. A.B., Dartmouth Coll., 1961, M.S., 1964, Ph.D., 1971. Engr. Div. of Naval Reactors, AEC, 1962-67, Creare, Inc., Hanover, N.H., 1967-68; metallurgist Gen. Electric Corp Research and Devel., Schenectady, 1971-75, br. mgr., 1975-84; v.p. research, devel. and engring. Lanxide Corp., Newark, Del., 1984-89, sr. v.p. tech., 1989-93, pres., CEO Lanxide Electronic Components, L.P., 1993—. Contbr. articles to profl. jours. Patentee in field. Served to lt. USN, 1962-67. Mem. ASM Internat. (chmn. Ea. N.Y. sect. 1978-79), Am. Ceramic Soc., Phi Beta Kappa, Sigma Xi. Avocations: outdoor and family activities. Home: 48 Bridleshire Rd Newark DE 19711-2454 Office: Lanxide Electronic Components LP PO Box 6077 1300 Marrows Rd Newark DE 19714-6077

URQUHART, SALLY ANN, environmental scientist, chemist; b. Omaha, June 8, 1946; d. Howard E. and Mary Josephine (Johnson) Lee; m. Henry O. Urquhart, July 31, 1968; children: Mary L., Andrew L. BS in Chemistry, U. Tex., Arlington, 1968; MS in Environ. Scis., U. Tex., Dallas, 1986. Registered environ. mgr.; lic. asbestos mgmt. planner, Tex. Rsch. asst. U. Tex. Dallas, Richardson, 1980-82; substitute sci. tchr. Dallas Ind. Sch. Dist., 1982; high sch. sci. tchr. Allen (Tex.) Ind. Sch. Dist., 1983-87; hazardous materials specialist Dallas Area Rapid Transit, 1987-90, environ. compliance officer, 1990-91, 1990-93. Pres. Beacon Sunday Sch. Spring Valley United Meth. Ch., Dallas, 1987, adminstrv. bd. dirs., 1989, com. status and role of women, 1992. Scholar Richardson (Tex.) br. AAUW, 1980. Mem. Am. Inst. Chemists, Am. Chem. Soc., Am. Soc. Safety Engrs., Am. Indsl. Hygiene Assn., Am. Conf. Govtl. Indsl. Hygienists (assoc.), Nat. Registry Environ. Profls., Soc. Tex. Environ. Profls. Avocations: jewelry design, counted cross stitching. Home: 310 Sallie Cir Richardson TX 75081-4229 Office: Dallas Area Rapid Transit PO Box 660163 Dallas TX 75266-7207

URSIC, SREBRENKA, information technology researcher, manager, consultant; b. Zagreb, Croatia, Mar. 4, 1947; s. Ivo Didolic and Katarina (Bilic) Radulovic Ursic; m. Ignac Lovrek, Nov. 24, 1971 (div. 1975); m. Damir Schenauer Vuk, Jan. 20, 1990. BSEE, U. Zagreb, 1970, MSEE, 1973. Product and software devel. specialist RIZ Semicondrs., Zagreb, 1970-80, asst. gen. mgr., 1980-82, bipolar integrated cir. researcher, 1982-84; application specific integrated cir. design group leader Rade Končar Inst., Zagreb, 1984-90; tech. cons., mgr. Sistemprojekt, Zagreb, 1990—; mem. working group for microelectronics UNESCO, Belgium, 1984; conf. chmn. Yugoslav Microelectronics Conf., 1988; del. Croatian Nat. Sci. Coun., Zagreb, 1982-84; invited speaker at profl. confs. Contbr. articles to profl. jours. Fellow MIDEM (exec. com. 1974-89); mem. IEEE, The Planetary Soc. Avocations: music, hatha yoga, holistic recuperation methods, tennis. Home: Kruziceva 4, 41000 Zagreb Croatia

URSO, CHARLES JOSEPH, physician; b. Bronx, N.Y., Jan. 3, 1942; s. Carl Peter and Anna L. (Jewidowicz) U.; m. Maria Regina Perret, Sept. 24, 1972; chidren: Charles M., Maria L., Laura A., Michael C. BS in Biology, L.I. U., 1963; MD, U. Bologna, Italy, 1970. Diplomate Am. Bd. Preventive Medicine and Occupational Medicine. Resident in internal medicine Misericordia/Fordham Hosp., Bronx, 1970-74; fellow in internal medicine St. Clare's Hosp., N.Y.C., 1974-75; staff physician N.Y. Telephone Co., N.Y.C., 1975-76, area med. dir., 1976-88, med. dir. environ. health, 1988-91, med. dir. health and fitness, 1991—; cons. internal medicine Louis Lasky Med. Ctr., Lindenhurst, N.Y., 1985—. Gov., treas. Flower Hill (N.Y.) Assn. Civic Action Com., 1982-90; sch. bd. adv. com. Manhasset (N.Y.) Sch. Dist., 1988-91; brigade surgeon Manhasset-Lakeville and Plandome Fire Depts., 1990-92. Fellow Am. Coll. Preventive Medicine; mem. Am. Coll. Sports Medicine, N.Y. State Med. Soc., N.Y. County Med. Soc., Nat. Fire Protection Assn. Home: 29 Mason Dr Manhasset NY 11030 Office: NY Telephone Co Rm 2549 1095 Ave of Americas New York NY 10036

USCHEEK, DAVID PETROVICH, chemist; b. University Heights, Ohio, July 9, 1937; s. Peter Ivanovich and Marie (Ocasek) U. BS, Case Western Res. U., 1959. Chemist The Glidden Co., Cleve., 1963-67, Mobil Chem. Co., Cleve., 1967-71, Limbacher Coatings, Cleve., 1971-72, Continental Products, Euclid, Ohio, 1972-80, Body Bros. Paint Corp., Bedford, Ohio, 1980-83, Harrison Paint Corp., Canton, Ohio, 1983-88, Akron (Ohio) Paint and Varnish, 1988—, cons. The Analyst, Chardon, Ohio, 1991—. Mem. Am. Chem. Soc., Internat. Union of Pure and Applied Chemists. Achievements include rsch. on EPA compliant waterborne and high solids coatings with abnormally low volatile organic compound content, high performance corrosion inhibitive water-based primers and topcoats for industrial applications. Home: 8602 Auburn Rd Chardon OH 44024 Office: Akron Paint and Varnish Inc 1390 Firestone Pkwy Akron OH 44301

USHE, ZWELONKE IAN, development engineer; s. Samson Chiwara and Daisy (Sibanda) U. BS, SUNY, Syracuse, 1986; BS in Chem. Engring., SUNY, Buffalo, 1988. Quality engr. 3M Co., St. Paul, Minn., 1988-90; devel. engr. Lydall Manning, Troy, N.Y., 1990--. Active Big Bros., Troy, 1990-92. Mem. AIChE, AFS, TAPPI. Office: Lydall Manning PO Box 328 Troy NY 12181

USHER, JOHN MARK, industrial engineering educator. BS in Chem. Engring., U. Fla., 1981; MSChemE, La. State U., 1983, MS in Indsl. Engring., 1986, PhD in Indstrl. Engring., 1989. Registered profl. engr., Miss. Process engr. Tex. Instruments, Inc., Dallas, 1981-82; rsch. asst. chem. engring. La. State U., Baton Rouge, 1982-85, rsch. asst., 1986-89; asst. prof. engring. Miss. State U., Starkville, 1989—; tech. assoc. Miss. State U. Ctr. for Robotics and Artificial Intelligence, Starkville, 1990—; tech. dir. Miss. State U./IBM Computer Integrated Mfg. Ctr., Starkville, 1991—. Mem. Inst. Indsl. Engring., Soc. Mfg. Engrs., Am. Assn. Artifical Intelligence. Achievements include research on computers in industry. Office: Miss State U PO Box U 101 McCain Mississippi State MS 39762

USHER, PETER DENIS, astronomy educator; b. Bloemfontein, S. Africa, Oct. 27, 1935; s. William Arthur and Clarinda Emily (Lawrenson) U.; m. Claire Bradford Pierce, Nov. 18, 1961 (div. 1979); 1 child, Robert Wetherbee. MSc, U. Orange Free State, Bloemfontein, S. Africa, 1959; PhD, Harvard U., 1965. Postdoctoral fellow Harvard-Smithsonian, Cambridge, Mass., 1966-67; rsch. scientist Am. Sci. and Engring., Cambridge, 1967-68; asst. prof. Pa. State U., University Park, 1968-73, assoc. prof., 1973-86, prof., 1987—. Contbr. articles to profl. jours.; editor: Astronomy Quarterly, 1982-89. Recipient awards/grants NASA, 1982-83, 83-89, NSF, 1976-77, 79-81, 83-84. Mem. Internat. Astron. Union, Royal Astron. Soc., Am. Astron. Soc. Achievements include first enunciation of the prin. of azimuth steerability for large transit telescopes; ind. discovery of new Poincare-Lighthill perturbation theory for ordinary differential equations and conditions for applicability, quantification of quasar variability and role in the Hubble diagram; variability of Cygnus X-2; introduction of the Q Statistic; the US survey for faint blue objects; the cataclysmic variable DV UMa. Office: Pa State Univ University Park PA 16802

USMAN, NASSIM, research chemist and biochemist; b. Montreal, June 18, 1959; came to U.S., 1986; s. Mohd and Hannelore (Grohmann) U. BSc, McGill U., 1982, PhD, 1987. Natural scis. and engring. rsch. coun. postdoctoral fellow MIT, Cambridge, Mass., 1987-89, NIH Fogarty Internat. rsch. fellow, 1989-90, postdoctoral assoc., 1990-92; sr. rsch. scientist Ribozyme Pharms. Inc., Boulder, Colo., 1992—; cons. Millipore, Bedford, Mass., 1987-92, Chem Genes Corp., Waltham, Mass., 1987-92. Referee Jour. Am. Chem. Soc., 1990—; contbr. articles to profl. jours. Nat. scis. and engring. rsch. coun. doctoral fellow, 1983-86. Mem. AAAS, Am. Chem. Soc. Achievements include development of the first practical chem. synthesis of RNA; patents in the field of RNA enzymes, or ribozymes, and molecular recognition of nucleic acids. Office: Ribozyme Pharms Inc 2950 Wilderness Pl Boulder CO 80301

USUKI, SATOSHI, physician, educator; b. Ehime, Japan, July 2, 1944; s. Wataru and Chieko (Doi) U.; m. Yoshie Inage, Aug. 8, 1974; children: Hiromune, Yoshimune, Chiemi. MD, Yokohama (Japan) City U., 1965, PhD, 1965; DSc, U. Tokyo, 1978, MD, 1981. Ednl. asst. U. Tokyo, 1971-78; prof. U. Tsukuba, Japan, 1978—; fellow, prof. of distinction Inst. Advanced Rsch. in Asian Sci. and Medicine, WHO, Hempstead, N.Y., 1990—; lectr. Huddinge U. Hosp., Karolinska Inst., Sweden, 1992. Promoter East Asian Econ. Caucus; trustee Minamiuwakai; mem. Japan-North Am. Med. Exch. Found., 1987, Internat. Human Resources Inst. Network. Recipient Japan-China Med. Assn. award, 1988, Li Shi Zhen Outstanding Manuscript award Hong Kong Inst. Promotion Chinese Culture, 1993. Fellow Inst. Advanced Rsch.; mem. AAAS, N.Y. Acad. Sci., Inter-Am. Sci. Hypertension, Fallopius Internat. Soc., Am. Inst. Ultrasound in Medicine, Internat. Soc. Gynecol. Endocrinology, Am. Assn. Gynecol. Laparoscopists, Inst. Growth Sci. Assn. (assoc., Japan), Am. Roentgen Ray Soc., Internat. Soc. Amino Acid Rsch., Am. Soc. Hypertension, Internat. Study Group for Steroid Hormones, Internat. Soc. Cardiovascular Pharmacotherapy, Internat. Menopause Soc., Internat. Soc. Infectious Diseases, Internat. Soc. Outer Space Law. Avocations: Japanese fencing, reading, travel, driving, fishing. Home: Kurakake 725-3, Tsukuba Ibaragi 305, Japan Office: U Tsukuba Inst Clin Medicine, Tennodai 1-1-1, Tsukuba Ibaragi 305, Japan

UTRACKI, L. ADAM, polymer engineer; b. Poland, Aug. 1, 1931; came to Can., 1968; Degree chem. engring., Poly. U., Poland, 1953, Dr.Eng., 1960; privat habil. dozent, Polish Acad. Scis., 1963. Rschr. Polish Acad. Scis., Lodz, Poland, 1956-65; postdoctoral fellow U. So. Calif., L.A., 1960-62; vis. prof. Case Western Res. U., Cleve., 1967-68; sr. rschr. Gulf Oil Can., Montreal, 1968-71; vis. scientist McGill U., Montreal, 1971-73; group leader CIL Inc. Explosives Lab., Montreal, 1973-80; sr. rsch. officer Nat. Rsch. Coun. Can./Indsl. Materials Inst., Montreal, 1980—. Author: Polymer Alloys and Blends, 1989, 91; editor: Progress in Polymer Processing, 1989, 90, 91; co-editor: Multiphase Polymers, 1989, Polymer Rheology and Processing, 1990, Current Topic in Polymer Science, 1987; editor Polymer Netwroks and Blends, Toronto, 1990—; editorial bd. Internat. Polymer Processing, Munich, 1986—, Advanced Polymer Tech., N.Y., 1984-89; contbr. 17 book chpts. and over 300 articles to profl. jours. Recipient Scientific Acheivement award Nat. Rsch. Coun., 1990. Mem. Can. Rheology Group (pres. 1989-90), Polymer Processing Soc. (pres. 1987-89), Soc. Plastics Engrs., VAMAS-Polymer Blends (chmn. 1989—). Achievements include 5 patents (internat.). Office: Nat Rsch Coun Can, 75 De Mortagne, Boucherville, PQ Canada J4B 6Y4

UTTS, JESSICA MARIE, statistics educator; b. Niagara Falls, N.Y., Oct. 13, 1951; d. Richard C. and Patricia (Highberger) U. BA, SUNY, Binghamton, 1973; MA, Pa. State U., 1975, PhD, 1978. Asst. prof. stats. U. Calif., Davis, 1978-84, assoc. prof., 1984-93; prof., 1993—. Recipient Disting. Teaching award U. Calif., Davis, 1985. Fellow AAAS, Inst. Math. Stats. (treas. 1988—), Am. Statis. Assn. Office: U Calif Divsn Stats Davis CA 95616

UWUJAREN, GILBERT PATRICK, economist, consultant; b. Oza Agbor, Bendel, Nigeria, May 6, 1945; came to U.S., 1985; s. Jacob Aghahowa and Victoria (Lasila) Uwujaren; m. Ngozi Buzugbe, Aug. 25, 1973; children: Jane, Janice, Jacob, Jo-Anne, Joseph, Jarune. BSc, U. Ibadan, Nigeria, 1971; MA, Columbia U., 1975, MPhil, 1977, PhD, 1977. Asst. lectr. U. Ife, Ibadan, 1972-73; sr. lectr. U. Ife, Ile-Ife, Nigeria, 1977-85; economist World Bank, Washington, 1985-89, cons., 1989-92; pres. Econ. Devel. Assocs., Burke, Va., 1993—. Contbr. articles to profl. jours. Recipient German Acad. award Govt. Fed. Republic Germany, Ibadan, 1970-71, Rockefeller award Rockefeller Found., Ibadan, 1971-73. Mem. Am. Econ. Assn. Home: 5801 Shana Pl Burke VA 22015-3663

UYEHARA, CATHERINE FAY TAKAKO (YAMAUCHI), physiologist, educator, pharmacologist; b. Honolulu, Dec. 20, 1959; d. Thomas Takashi and Eiko (Haraguchi) Uyehara; m. Alan Hisao Yamauchi, Feb. 17, 1990. BS, Yale U., 1981; PhD in Physiology, U. Hawaii, Honolulu, 1987. Postdoctoral fellow SmithKline Beecham Pharms., King of Prussia, Pa., 1987-89; asst. prof. in pediatrics U. Hawaii John Burns Sch. Medicine, Honolulu; rsch. pharmacologist Kapiolani Med. Ctr. for Women and Children, Honolulu, 1989—; statis. cons. dept. clin. investigation Tripler Army Med. Ctr., Honolulu, 1984-87, 89—, chief rsch. pharmacology sect., 1991—. Co-inventor method for preserving renal function in cases of rhabdomyolysis; contbr. articles to profl. jours. Mem. Am. Fedn. for Clin. Rch., Western Soc. Pediatric Rsch., Am. Physiol. Soc., N.Y. Acad. Scis., Soc. of Uniformed Endocrinologists. Democrat. Mem. Christian Ch. Avocations: swimming, diving, crafts, horticulture, music. Office: Tripler Army Med Ctr Dept Clin Investigation HSHK-CI Honolulu HI 96859-5000

UYEMOTO, JERRY KAZUMITSU, plant pathologist, educator; b. Fresno, Calif., May 27, 1939; married, 1965; 1 child. BS in Agronomy, U. Calif., Davis, 1962, MS in Plant Pathology, 1964, PhD in Plant Pathology, 1968. Lab. tech. U. Calif., Davis, 1963-67; from asst. to assoc. prof. virology N.Y. State Agrl. Expt. Sta., Cornell U., 1968-77; prof. Kansas State U., Manhattan, 1977-81; sr. staff scientist Advanced Genetic Scis., 1982-84; vis. scientist U. Calif., Davis, 1984-86, rsch. plant pathology, USDA Agrl. Rsch. Svc., 1986—. Recipient Lee M. Hutchins award Am. Phytopath. Soc., 1993. Mem. Assn. Applied Biologists, Am. Phytopath Soc.. Achievements include research on a variety of crop plants; research contributions were also made on virus diseases of apple and annual crop plants; ELISA protocols tested and/or established for serological indexing of all Prunus tree sources used for scion buds and seeds. Office: UC-Davis Dept of Plant Pathology USDA Agricultural Research Svc Davis CA 95616*

UYTTENBROECK, FRANS JOSEPH, gynecologic oncologist; b. Lier, Antwerp, Belgium, July 11, 1921; s. Joseph Uyttenbroeck and Augusta Verstreken; m. Elizabeth Andrea Switters, Oct. 24, 1946; children: Lia, Anna Maria, Godeliva, Maria Magdalena, Lutgardis. Student, Coll. St. Gummarus, Lier, 1939; MD, U. Louvain (Belgium), 1946; PhD, U. Amsterdam (the Netherlands), 1952. Prof. gynecology and obstetrics U. Antwerp, 1972-88, prof. emeritus, 1988—. cons. gynecologic oncology U. Antwerp. Named Comdr. Order of Leopold, Grand Officer of the Crown Order, Grand Officer of the Order of Leopold. Mem. Royal acad. Medicine of Belgium, Acad. of Surgery (Paris), Acad. Nationale de Medicine, Academia Europaea, N.Y. Acad. Scis., Soc. of Pelvic Surgeons, Soc. of Gynecologic Ongologists, ISVVD, Internat. Soc. Gynecologic Oncologists (founding mem. 1985, assoc.). Avocation: traveling. Home: 12 Ave Jan Van Rijswijck, 2018 Antwerp Belgium

VACHHER, PREHLAD SINGH, psychiatrist; b. Rawalpindi, Punjab, Pakistan, Nov. 30, 1933; came to U.S., 1960; s. Thakar Singh and Harbans Kaur (Ghai) V.; m. Margaret Mary Begley, Oct. 9, 1963; children: Paul, Sheila, Mary Ann, Eileen, Mark. Grad., Khalsa Coll., India, 1950; MD, Panjab U., Amritsar, India, 1956. Diplomate Am. Bd. Psychiatry. Staff N.J. State Hosp., Trenton, 1965-66, Wayne County Gen. Hosp., Eloise, Mich., 1966-68; pvt. practice Livonia, Mich., 1966-75, Woodstock, Va., 1992—; pres. Vachher Psychiat. Ctr., P.C., Livonia, 1975-91; dir. community psychiatry Northville (Mich.) State Hosp., 1968-71; cons. staff Kingswood Hosp., Ferndale, Mich., 1966—, Annapolis Hosp., Wayne, 1967-88, St. Joseph Mercy Hosp., Ann Arbor, 1971—; westland staff Margaret Montgomery Hosp., 1988-91; bd. dirs. Oakland Rental Housing Assn., 1990-91; med. dir. mental health unit Shenandoan County Meml. Hosp., Woodstock, Va., 1991—. Mem. Am. Psychiat. Assn., Va. Psychiat. Soc., Sikh Physicians in Mich. (bd. dirs. 1987), Canton C. of C. (pres. 1975), Sikh Bus. Profl. Coun. (pres. 1988—), Rotary (Canton and Plymouth, Mich., Woodstock). Office: Doctors Bldg # 105 Woodstock VA 22664

VAFEADES, PETER, mechanical engineering educator; b. Detroit, Dec. 26, 1957; s. George and Agnes-Athena (Kellargi) V.; m. Katherine Dakie, July 2, 1980. BS, U. Detroit, 1982; MS, Mich. State U., 1983, Mich. State U., 1984; PhD, Mich. State U., 1987. Lectr. U. Minn., Mpls., 1987-89; asst. prof. mech. engring. Trinity U., San Antonio, Tex., 1989—. Author: (software) Pdelie; Automatic Symbolic Solution of Partial Differential Equations, 1991; contbr. articles to profl. jours. Mem. AAUP, Am. Soc. Engring. Edn., Am. Math. Soc., Soc. Indsl. and Applied Math., Soc. Automotive Engrs. Office: Trinity U 715 Stadium Dr San Antonio TX 78212-3104

VAFOPOULOU, XANTHE, biologist; b. Thessaloniki, Greece, Aug. 22, 1949; d. Konstantin-Nikiforos Vafopoulos and Anna Gega. BA in Biology, U. Thessaloniki, 1972; MA in Biology, Bridgewater State Coll., 1978; PhD in Biology, U. Conn., 1980. Postdoctoral fellow U. Conn., Storrs, 1980-81, asst. prof. in residence, 1981-84, rsch. assoc., 1984-86; rsch. assoc. York U., North York, Ont., Can., 1987—. Contbr. numerous articles to profl. jours. Mem. Am. Soc. Zoologists, Can. Soc. Zoologists, Internat. Fedn. Comparative Endocrinological Socs., European Soc. Comparative Endocrinology. Greek Orthodox. Office: York U, 4700 Keele St, North York, ON Canada M3J 1P3

VAGUE, JEAN MARIE, physician; b. Draguignan, France, Nov. 25, 1911; s. Victor Francois and Marie (Voiron) V.; m. Denise MArie Jouve, Sept. 3, 1936; children: Philippe, Thierry, Irene (Mrs. Claude Juhan), Maurice. Baccalaureat, Cath. Coll., Aix en Provence, France, 1928; MD, Marseilles (France) U., 1935. Intern, Hotel Dieu Conception, Marseilles, 1930, resident, 1932-39; practice medicine specializing in endocrinology, Marseilles, 1943—; assoc. prof. Marseilles U., 1946-57, prof., clinic endocrinology, 1957—. Dir. Ctr. Alimentary Hygiene and Prophylaxis Nutrition Diseases Nat. Rys. Mediterranean region, 1958—; expert chronic degenerative diseases (diabetes) WHO, 1962—. Served to lt. French Army, 1939-40. Decorated Cross Legion Honor, Acad. Palms, knight pub. health, knight mil. merit, War Cross. Mem. Endocrine Soc. U.S., Am. Diabetes Assn., Royal Soc. Medicine (London), European Assn. for Study Diabetes, Spanish, Italian, French (past pres.) socs. endocrinology, French Acad. Medicine, Spanish Acad. Medicine, Italian Acad. Medicine, Belgian Acad. Medicine, French Lang. Diabetes Assn. (past pres.). Author: Human Sexual Differentiation, 1953, Notions of Endocrinology, 1965, Obesities, 1991, Dawn on Iaboc's Ford, 1993, History of Man, History of Men, 1993, others. Achievements include first identification of the metabolic and vascular complications of android obesity and their mechanism; research in demonstration of diabetogenic and atherogenic power of obesity with topographic distbn. fat in upper part of body, evolution of android diabetogenic obesity from 1st stage of efficacious hyperinsulinism to less efficacious hyperinsulinism and hypoinsulinism-neuro-germinal degeneration, degenerative lesions of germinal epithelium and nervous system. Home: 411 Ave du Prado E-6, 13008 Marseilles France Office: Clinique Endocrinologique, Hopital U de la Timone, Blvd Jean-Moulin, 13385 Marseilles France

VAHABZADEH, FARZANEH, food scientist, educator, researcher; b. Tehran, Iran, Feb. 22, 1951; d. Salman and Robabeh V. BS in Chemistry, U. Shahid Beheshti, Iran, 1974; MS in Nutrition, Food Sci., Utah State U., 1979-82, PhD, 1983-86. Chemist in pollution and chem. lab. Tehran Regioanl Water Org. Ministry of Energy, 1975-77; rsch. asst. nutrition and food sci. dept. Utah State U., Logan, 1979-86; asst. prof. Tehran Ploytech., 1986—; adv. com. mem., cons. Iranian Rsch. Org. for Sci. and Tech., Tehran, 1990—; student competition organizor Iranian Food Sci. and Tech. Assn., Teharn, 1991, 92; referee several sci. jours., Iran. Translator book on enzymetic analysis; author in field. Mem. Inst. Food Technologists (cert.), Am. Chem. Soc., Iranian Food Sci. and Tech. Assn. Achievements include research on mechanism of action of mineral chelators on clostridium botulinium growth in cured meat products, mechanism of pink color defect in turkey products, biotechnology and whey treatments; established two teaching laboratories (food quality control, food microbiology) in this dept. Office: Amir Kabir U Techs, Hafez Ave No 424, Tehran Iran

VAHAVIOLOS, SOTIRIOS JOHN, electrical engineer, scientist, corporate executive; b. Mistra, Greece, Apr. 16, 1946; s. John Apostolos and Athanasia (Pavlakos) V.; m. Aspasia Felice Nessas, June 1, 1969; children: Athanasia, Athena, Kristy. BSEE, Fairleigh Dickinson U., 1970; MSEE, Columbia U., 1972, M in Philosophy, 1975, PhDEE, 1976. Mem. tech. staff Bell Telephone Labs., Princeton, N.J., 1970-75, supr., 1975-76, dept. head, 1976-78; founder, pres., chief exec. officer Phys. Acoustics Corp., Princeton, 1978—; adviser Greece Ministry Def., Athens, 1986-88; bd. dirs. Orthosonics, Inc., N.Y.C. Contbr. more than 100 papers to profl. publs. 13 U.S. patents, 7 fgn. patentsin field. Bd. dirs. Holy Cross Greek Orthodox Sch. Theology, Boston, 1989—; adv. bd. Trenton State Coll., N.J., 1983—; chmn. Princeton sect. United Fund, 1976-78. Recipient Spartan Merit award Spartan World Soc., 1987, Entrepreneur of Yr. award Arthur Young/Inc. Mag., N.J., 1989. Fellow IEEE (Centennial medal award 1984), Am. Soc. Nondestructive Testing (bus. and fin. com. 1984-87, 88—, bd. dirs. 1985, sec. 1989, treas. 1990, v.p. 1991, pres. 1992, editor handbook on Acoustic Emission, 1988), Acoustic Emission Working Group; mem. ASTM, IEEE Indsl. Electronics Soc. (sr. mem. adminstrv. com. 1988, founder, v.p. conf. 1974-78, 2d prize Student Paper Contest 1970, Outstanding Young Engr. award 1984, editor Trans. on Indsl. Electronics 1976-82), N.Y. Acad. Scis. Independent. Greek Orthodox. Avocations: bird hunting, soccer, technical writing, gardening. Home: 7 Ridgeview Rd Princeton NJ 08540-7601 Office: Phys Acoustics Corp PO Box 3135 Princeton NJ 08543-3135

VAHEY, DAVID WILLIAM, measurement scientist, physicist; b. Youngstown, Ohio, Nov. 21, 1944; s. William Henry and Mildred (Herr) V.; m. Linda Sue Begley, Aug. 6, 1977; children: Brian, Michael. SB in Physics, MIT, 1966; MS in Elec. Engring., Calif. Inst. Tech., Pasadena, 1967, PhD, 1973. Rsch. scientist Battelle Meml. Inst., Columbus, Ohio, 1974-81; prin. physicist Accuray Corp. (now ABB), Columbus, 1981-88; sr. R&D scientist Internat. Paper, Tuxedo, N.Y., 1988—. Contbr. articles to profl. jours. Recipient Top 100 award IR&D Mag., Chgo., 1982. Mem. TAPPI (developer of ednl. module on light /optics 1991), Am. Forest and Paper Assn. (task team on postal automation issues). Achievements include measurement device for edges of intraocular lenses; online measurement of paper ultrasonic properties; prediction of paper curl tendency using light reflection. Office: Internat Paper Long Meadow Rd Tuxedo Park NY 10987

VAIDYA, KIRIT RAMESHCHANDRA, anesthesiologist, physician; b. Sihor, India, Feb. 20, 1937; came to U.S., 1971; s. Rameshchandra Harilal Vaidya and Kanta Bachubhai Mulani; m. Rashmi Kirit Vaidya; children: Kaushal, Sujal. BSc, Gujrat U., India, 1959; MB BS, Karnatak U., India, 1965. Intern St. Joseph Hosp., Providence, 1971-72; resident in anesthesiology R.I. Hosp., Providence, 1973, Boston City Hosp., 1974-76; clin. instr. anesthesiology Boston U. Sch. Medicine, 1977-79; staff anesthesiologist Bridgeport (Conn.) Hosp., 1979—; asst. clin. prof. anesthesiology U. Conn. Med. Ctr., Farmington, 1987—. Mem. Am. Assn. Physicians from India, Conn. Assn. Physicians from India, Fairfield County Med. Assn., Conn. State Soc. Medicine, Conn. State Soc. Anesthesiologists, Am. Soc. Anesthesiologists, Internat. Anesthesiology Assn. Hindu. Home: 54 Quail Trl Trumbull CT 06611-5259 Office: Bridgeport Anesthesia Assocs 965 White Plains Rd Ste 301 Trumbull CT 06611-4566

VAIL, PETER ROBBINS, geologist; b. N.Y.C., Jan. 13, 1930; s. Donald Bain and Eleanor (Robbins) V.; m. Carolyn Flesher, Sept. 15, 1956; children: Andrea, Susan, Timothy Edward. A.B., Dartmouth Coll., 1952; M.S., Northwestern U., 1955, Ph.D. (Shell fellow), 1959. Asst. geologist U.S. Geol. Survey, Spokane, Wash., 1952-56, Evanston, Ill.; research geologist Carter Oil Co., Tulsa, 1956-58, Jersey Prodn. Research Co., Tulsa, 1958-62; sr. research geologist Jersey Prodn. Research Co., 1962-65; sr. research specialist Exxon Prodn. Research Co., Houston, 1965-66; research assoc. Exxon Prodn. Research Co., 1966, research supr., 1966-70, sr. research assoc., 1970-72, sr. research adv., 1972-75, research scientist, 1975-81, sr. research scientist, 1981—; mem. Consortium for Continental Reflection Profiling Site Selection Com., 1974-81; mem. internat. subcom. on stratigraphic classification Internat. Union Geol. Seismology com. on Stratigraphy, 1976—; mem. ocean sci. bd. U.S. Nat. Acad. Scis., 1979-82; mem. Joint Oceanography Insts. Deep Earth Sampling Passive Margin Panel, 1978-84; William Smith lectr. Geol. Soc. London, 1978; Bullerwell lectr. U.K. Geophys. Soc., 1983; mem. Am. Petroleum Inst., 1978—; mem. Eur. Assn. Exploration Geophysicists, 1983. Contbr. to profl. jours. articles on seismic stratigraphy, global changes of sea level, tectonics. Recipient Offshore Tech. Conf. Individual Disting. Achievement Award, 1983, William H. Twenhofel medal Soc. Sedimentary Geology, 1992. Fellow Geol. Soc. Am. (research grants com. 1977-79, chmn. 1979, councilor 1979-82), AAAS; mem. Am. Assn. Petroleum Geologists (disting. lectr. 1975-76, marine geology com. 1977, research com. 1978-84, co-recipient Pres.'s award for best pub. paper 1979, co-recipient Matson award for author of best paper 1980 ann. conv. 1981), Soc. Exploration Geophysicists (Virgil Kaufmann Gold medal for advancement of science of geophys. exploration 1976, Disting. lectr. 1985, hon. mem. 1985—), Soc. Econ. Paleontologists and Mineralogists (Disting. speaker 1984 ann. meeting luncheon), Houston Geophys. Soc. (hon. mem.), Houston Geol. Soc. (Best Paper award 1982-83), Mayflower Soc., Sigma Xi. Home: 3745 Del Monte Dr Houston TX 77019-3017 Office: Exxon Production Research Co Box 2189 Houston TX 77001 also: Rice University PO Box 1892 6100 South Main Houston TX 77251

VAILAS, ARTHUR C., biomechanics educator. BS in Exercise Physiology, U. N.H., 1973; PhD in Exercise Physiology, U. Iowa, 1979, postgrad., 1979-82. Asst. prof. dept. kinesiology UCLA, 1982-88, assoc. prof., 1988; assoc. prof. U. Wis., Madison, 1988-91, prof., 1991—, dir. biodynamics lab., 1988—; session chair Gordon Rsch. Conf.; mem. life sci. adv. com. NASA, life sci. del. to COSMOS 1887, 2044 Missions; com. chair Musculoskeletal Implementation Group EDO and Countermeasure Sci. Plan; SLS-1 scientist. Contbr. 60 articles to profl. jours. Recipient Rsch. Svc. award NIH, Outstanding Sci. award USSR-NASA; named Disting. Scientist, CSU. Mem. Nat. Rsch. Coun. (mem. NASA com.). Office: U Wis Biodynamics Lab 2000 Observatory Dr Madison WI 53706*

VAISHNAVA, PREM PRAKASH, engineering educator; b. Jodhpur, India, Oct. 10, 1942; came to U.S., 1980; s. Hari Ram and Shanti devi (Tilawat) V.; m. Manju L. Sharma, Feb. 19, 1972; children: Sanjay, Ajay, Prashant. MSc, Jodhpur U., 1965, PhD, 1976. Asst. prof. Jodhpur U., 1965-78; rsch. assoc. Heriot-Watt U., Edinburgh, Scotland, 1978-80; asst. prof. W.Va. U., Morgantown, 1980-83; assoc prof. No. Ill. U., DeKalb, 1983-86; assoc. prof. GMI Engring. and Mgmt. Inst., Flint, Mich., 1986-91, prof., 1991—. Contbr. over 50 articles to profl. jours. Mem. Am. Phys. Soc., Material Rsch. Soc. Achievements include rsch. on high temperature superconductivity. Home: 607 Fremont Flint MI 48504

VAITUKAITIS, JUDITH LOUISE, medical research administrator; b. Hartford, Conn., Aug. 19, 1940; d. Albert George and Julia Joan (Vaznikaitis) V. BS, Tufts U., 1962; MD, Boston U., 1966. Investigator, med. officer reproductive rsch. Nat. Inst. Child Health and Human Devel., NIH, Bethesda, Md., 1971-74; assoc. dir. clin. rsch. Nat. Ctr. Rsch. Resources NIH, Bethesda, Md., 1986-91, dir. gen. clin. rsch. ctr., 1986-91, dep. dir. extramural rsch., 1991; acting dir. Nat. Ctr. Rsch. Resources NIH, Bethesda, 1991-92, dir., 1992—; from assoc. prof. to prof. medicine Sch. Medicine Boston U., 1974-86, assoc. prof. physiology, 1975-80, assoc. prof. ob-gyn., 1977-80, program. dir. gen. clin. rsch. ctr., 1977-86, prof. physiology, 1980-86; head sect. endocrinology and metabolism Boston City Hosp., 1974-86. Mem. editorial bd. Jour. Clin. Endocrin. and Metabolism, 1973-80, Proc. Soc. Exptl. Biol. and Medicine, 1978-87, Endocrine Rsch., 1984-88. Author: Clinical Reproductive Neuroendocrinology, 1982; contbr. articles to profl. jours. Recipient Disting. Alumna award Sch. Medicine, Boston U., 1983, Mallincrodt award for Inv. Rsch. Clin. Radiossay Soc., 1980. Mem. Am. Fedn. Clin. Rsch., Endocrine Soc., Am. Soc. Clin. Rsch., Soc. Exptl. Biology and Medicine, Assn. Am. Physicians, Soc. for Study of Reproduction, Am. Soc. Andrology. Office: Nat Ctr Rsch Resources NIH 9000 Rockville Pike Bethesda MD 20892

VAJDA, GYÖRGY, electrical engineer; b. Budapest, Hungary, June 18, 1927. Grad., Tech. U., Budapest, 1949, PhD, 1957, DSc, 1964. Asst. lectr. Tech. U., 1949-50, titular univ. prof., 1968—; sr. advisor Hungarian Acad. Scis., Budapest, 1950-52; asst. dir. Measurement Inst., Budapest, 1952-57; asst. dir. Inst. Electric Power Rsch., Budapest, 1957-62, gen. mgr., 1969; head dept. Ministry of Heavy Industry, 1962-69; vice chmn. Hungarian Atomic Energy Com., 1979. Author: Deterioration of Isolations, 1964, Electric Fields in Isolations, 1970, Design of Electrical Isolations, 1968, Energy Policy, 1980. Dep. Minister Culture and Edn., 1985—. Recipient State prize, 1975, Gold award People's Republic, 1967, Silver award, 1967, Liberation award, 1970, Zipernovsky prize, 1960; named to Order of Labour, 1977. Mem. Hungarian Acad. Scis. (Presidium 1985—), Assn. Hungarian Electrotechnicians (chmn. 1970), Assn. Energy (Conf. Internat. des Grands Réseaux Électriques, World Energy Conf.). Office: Magyar Tudományos Akadémia, Roosevelt-tér 9, 1051 Budapest Hungary*

VAKAKIS, ALEXANDER F., mechanical engineering educator; b. Samos, Greece, Dec. 16, 1961; came to U.S., 1987; s. Fotios and Anna (Prindezi) V.; m. Sotiria Koloutsou, Aug. 16, 1987. MSc, Imperial Coll., 1986; PhD, Calif. Inst. Tech., 1990. Researcher, tchr. Calif. Inst. Tech., Pasadena, 1987-90; asst. prof. U. Ill., Urbana, 1990—. Contbr. articles to Jour. Applied Mechanics, Jour. Sound and Vibration, Internat. Jour. Non-Linear Mechanics, Nonlinear Dynamics, SIAM Jour. Applied Math. Recipient NSF Rsch. Initiation award, 1992, Josephine De'Karman fellowship De'Karman Fellowship Trust, 1990, Powell Inst. fellowship Calif. Inst. Tech., 1990. Mem. ASME, AIAA, Soc. Exptl. Mechanics, Soc. for Indsl. and Applied Math., Am. Acad. Mechanics. Office: Univ Ill Dept Mechanical Indsl Engr 1206 W Green St Urbana IL 61801

VAKALO, EMMANUEL-GEORGE, architecture and planning educator, researcher; b. Athens, Greece, May 10, 1946; came to U.S., 1965; s. George Constantine and Eleni (Stavrinou) V.; m. Kathleen Leitgabel, July 20, 1974. BArch, Cornell U., 1969, MArch, 1972, M in Regional Planning, 1977; PhD, U. Mich., 1985. Instr. U. Mich., Ann Arbor, 1975-79, asst. prof., 1979-91, assoc. prof., 1991—; cons. JP Industries, Ann Arbor, 1987-88; guest prof. Tech. U. Wien, U. Okla., Carnegie-Mellon U., La. State U., U. Notre Dame, Ryerson Poly. Inst., Calif. State Poly. U., Pomona. Author: Visual Studies, 1983, Visual Syntax: Function and Production of Forms, 1988. Mem. Am. Planning Assn., Environ. Design Rsch. Assn., Nat. Inst. Archtl. Edn., Inst. Math. Geography. Avocations: photography, drawing, reading, sailing. Office: U Mich Coll Arch and Urban Planning 2000 Bonisteel Dr Ann Arbor MI 48109-2069

VAKIRTZIS, ADAMANTIOS MONTOS, systems analyst; b. Karpasi, Limnos, Greece, June 7, 1964; came to U.S. 1967.; s. Peter and Grammatiki (Alateras) V.; m. Frances Elaine Dalupan, Feb. 10, 1990. AS, Dutchess Community Coll., Poughkeepsie, 1984; BA, SUNY, Oswego, 1986. Computer programmer Donald Grenier Assocs., Mahopac, N.Y., 1986-87; systems analyst IPCO Corp., White Plains, N.Y., 1987-90, Cent. Hudson Gas and Electric, Poughkeepsie, 1990—. Mem. IEEE Computer Soc., Am. Hellenic Ednl. Progressive Assn., Greek Orthodox Young Adult League (v.p. 1990-92). Democrat. Greek Orthodox. Home: 29 Helen Ct Beacon NY 12508 Office: Cent Hudson Gas and Ele 284 South Ave Poughkeepsie NY 12601

VALBERG, LESLIE STEPHEN, medical educator, physician, researcher; b. Churchbridge, Sask., Can., June 3, 1930; s. John Stephen and Rose (Vikfusson) V.; m. Barbara Valberg, Sept. 14, 1954; children: John, Stephanie, Bill. M.D., C.M., Queens U., 1954, M.S.C., 1958. Cert. internal medicine specialist Royal Coll. Physicians and Surgeons of Can. Asst. prof. Queens U. Kingston, Ont., Can., from 1960—, prof., until 1975; prof. medicine U. Western Ont., London, 1975—, chmn. dept., 1975-85; dean faculty of medicine U. Western Ont., 1985-92; cons. Univ. Hosp., 1975—. Rsch. Assoc. Med. Council of Can., 1960-65. Fellow Royal Coll. Physicians and Surgeons; mem. Am. Gastroent. Assn., Med. Rsch. Coun. Can. (v.p. 1980-82). Home: 1496 Stoneybrook Crescent, London, ON Canada N5X 1C5 Office: U Western Ont, London, ON Canada N6A 5C1

VALCAVI, UMBERTO, chemistry educator; b. Desenzano, Italy, Sept. 15, 1928; s. Giuseppe and Rosa (Bianchini) V.; m. Ines Scrocca, Sept. 17, 1960; children: Rosella, Giampaolo. Grad. Indsl. Chemistry, U. Milan, Italy, 1955; Libero docente in organic chemistry, State U. Italy, 1965. Rschr. Farmitalia S.P.A., Milan, 1955-59; dir. rsch. Istituto Biochimico Italiano SPA, 1959-76, sci. dir., 1977-80, gen. mgr., 1981-86, pres., exec. com., 1986-89; prof. organic chemistry U. Milan, 1964—; bd. dirs. Pas-SCI SPA; mem. sci. bd. Lorenzini Found., 1977—. Author: Advanced Organic Chemistry, 1980, Applied Organic Chemistry, 1983, Bioorganic Chemistry, 1992; contbr. 100 articles to Internat. Jour. Organic Chemistry and Internat. Jour. Pharm. Chemistry; holder 26 patents. Recipient Gran Cross Knight, Italian Rep., 1980, Order of Merit, Commdr. Italian Rep., 1982. Fellow N.Y. Acad. Sci.; mem. AAAS, Am. Chem. Soc., Italian Soc. Pharm. Sics. (bd. dirs. 1982—), Rotary. Avocations: pre-history, origin of life. Home: 21 Viale Biancamaria, Milan 20122, Italy Office: Dipartimento di Chimica Orgn, 21 Via Venezian, Milan Italy

VALCOURT, BERNARD, Canadian government official, lawyer; b. St. Quentin de Restigouche, N.B., Can., Feb. 18, 1952; s. Bertin and Gerladine (Allain) V.; children: Annie, Edith. Student, U. N.B. Lawyer; mem. Ho. of Commons, 1984; formerly parliamentary sec. Ministry State for Sci. and Tech. and Ministry Revenue; former minister state for small bus. and tourism, from 1986, minister state for Indian affairs and no. devel., 1987-89; mem. Privy Council, 1986—; former minister consumer and corp. affairs, from 1989, minister fisheries and oceans, 1990-91, minister employment and immigration, 1991-93; minister Human Resources and Labour Can., Ottawa, Ont., 1993—. Mem. ABA, N.B. Bar Assn. Progressive Conservative. Roman Catholic. Office: Human Resources & Labor Can, 140 Promenade du Portage, Ottawa-Hull, ON Canada K1A 0J9

VALDATA, PATRICIA, English language educator, aviation writer; b. New Brunswick, N.J., Oct. 16, 1952; d. William Rudolph and Ethel Ann (Kovacs) V.; m. Robert W. Schreiber, Apr. 21, 1979. BA, Rutgers U., 1974, BA with honors, 1977; MFA, Goddard Coll., 1991. Supr. AT&T Long Lines, Bedminster, N.J., 1979-81; mgr. info. svcs. N.J. Ednl. Computer Network, Edison, 1981-83; tech. writer on contract to Bell Comm. Rsch., Piscataway, N.J., 1983-86; v.p. Computer System Design & Mgmt. Inc., Annandale, N.J., 1986-88; adj. instr. Del. Tech. and C.C., Wilmington, 1989-90, Cecil C.C., Elkton, Md., 1990—; adj. instr. dept. English, U. Del., Newark, 1991—; presenter 3d Nat. Women in Aviation Conf., 1992. Columnist Wilmington News Jour., 1992; contbr. numerous articles to Soaring mag., Women in Aviation mag. Mem. Women Soaring Pilots Assn. (co-founder, sec. 1986—), Soaring Soc. Am., 1-26 Assn. (set Md. altitude record 1991), Aircraft Owners and Pilots Assn., Exptl. Aircraft Assn., Atlantic Soaring Club (co-founder, pres. 1992). Office: U Del English Dept Memorial Hall Newark DE 19716

VALDEMARIN, LIVIO, computer peripherals company executive; b. Gorizia, Italy, Jan. 27, 1944; s. Antonio and Laura (Dibert) V.; m. Claudia Segatti, Oct. 24, 1970; children: Paolo, Marco. Maturita', Liceo Scientifico, Gorizia, 1962; D. in Electronics, U. di Trieste, Trieste, Italy, 1972. Cert. in engring., math. tchr. High sch. tchr. Instituto Tecnico Commerciale Fermi, Gorizia, 1971-86; pres. L-3 Elettronica Commerciale, Gorizia, 1972-76, Elcom, Gorizia, 1977—; treas. European Desktop Pub. Group, 1988. Capt. Italian mil., 1970-71. Mem. Rotary. Avocation: travel by camper. Office: Elcom SRL, via degli Arcadi 2, Gorizia Italy

VALENCIA, JAIME ALFONSO, chemical engineer; b. Arequipa, Peru, Apr. 2, 1952; came to the U.S., 1970; s. Luis A. and Julia (Chavez) V.; m. Cecelia Ann Ingram, Nov. 21, 1987; 1 child, Julia Nicole. BSChE, U. Md., 1974; DSc in Chem. Engring., MIT, 1978. Registered profl. engr., Tex. Mem. cons. staff Arthur D. Little, Inc., Cambridge, Mass., 1978-82; group leader Exxon Prodn. Rsch. Co., Houston, 1982-86; dir. tech. devel. Novatec, Inc., Houston, 1986-89; v.p. Novatec Prodn. Systems, Inc., Houston, 1988-91; prin. NOVI Enterprises, Houston, 1986—; cons. Exxon Prodn. Rsch. Co., Houston, 1990-92. Contbr.: Flue Gas Desulfurization, 1982; contbr. articles to Internat. Jour. Heat and Mass Transfer, Hydrocarbon Processing; author: AIChE Symposium Series, 1977. Mem. AIChE, MIT Alumni Club South Tex. (bd. dirs. 1990—), Houston Inventors Assn., Tau Beta Pi (pres. Houston alumnus chpt. 1985-90). Roman Catholic. Achievements include patents for cryogenic processing; invention of CFZ process for cryogenic separations; development of Novatec process for cleaning oil contaminated cuttings from drilling operations; basic principles for Ryan-Holmes technology for processing natural gas. Home: 9830 Stableway Dr Houston TX 77065 Office: NOVI Enterprises PO Box 22807 Houston TX 77227

VALENCIA, ROGELIO PASCO, electronics engineer; b. Paombong, Bulacan, The Philippines, Mar. 18, 1939; came to U.S., 1959; s. Silvino Carlos and Basilia Galang (Pasco) V.; m. Amelia Almendariz Gomez, May 31, 1965; children: Zenaida Leticia, Lucinda Amelia, Rogelio Pasco II. Student mech. engring., Mapua Inst. Tech., Manila, 1955-59; student English and math., Coll. William and Mary, 1963-64, numerous USCG tng. schs. Enlisted man USCG, 1959, advanced through grades to chief warrant officer; with USCG cutter Rush, Vietnam, 1970-71; sr. tech. officer USCG Loran Sta., Hokkaido, Japan, 1977-78; exec. officer USCG Loran Sta., Dana, Ind., 1978-79; ret., 1979; computer analyst Wyman & Gordon Co., Danville, Ill., 1979-80; precision measurement electronics lab. technician USAF, Rantoul, Ill., 1980-88, digital computer engr., 1988—. Designer synchronous Loran clock, field telephone monitor. Avocations: masonry, electronics, woodworking, camping. Home: 1303 Bradford Cir Saint Joseph IL 61873-9625 Office: USAF Civil Engring Rantoul IL 61868-5260

VALENTINE-THON, ELIZABETH ANNE, biologist; b. Worcester, Mass., Nov. 11, 1948; d. Lillian Elizabeth (Aalto) Valentine; divorced; 1 child, Michael. BA, Anna Maria Coll., 1971; MEd, Boston U., Heidelberg, Germany, 1974; MS, U. Wis., 1976; PhD, U. Essen, Germany, 1985. Lab technician Mason Rsch. Inst., Worcester, 1968-72, German Cancer Rsch. Ctr., Heidelberg, 1972-74; rsch. scientist Inst. Human Genetics, Essen, 1977-86, State Hygiene Inst., Bremen, Germany, 1987-90; lab. supr. Schiwara Practice Lab. Medicine, Bremen, 1990—; lectr. U. Bremen, 1988-90, Sch. Med. Tech., Bremen, 1990—. Contbr. articles to profl. jours. Rsch. grantee German Nat. Rsch. Agy., Toenjes-Vagt-Agy. Mem. AAAS, Am. Soc. Microbiology, German Soc. Virology, Internat. Com. of Jackson Lab., European Group for Rapid Viral Diagnosis, European Soc. Clin. Microbiology and Infectious Diseases, Alpha Mu Gamma, Delta Epsilon Sigma. Roman Catholic. Avocation: tennis. Home: Beim Kronskamp 12, D-28355 Bremen 33, Germany Office: Schiwara Practice Lab Medicine, Haferwende 12, 28357 Bremen Germany

VALENTINI, JAMES JOSEPH, chemistry educator; b. Martins Ferry, Ohio, Mar. 20, 1950; s. Julio Celeste and Dorothy Mary (Orell) m. Pamela Susan Burgess, Jan. 17, 1981; children: Evan David, Colin Burgess. BS, U. Pitts., 1972; MS, U. Chgo., 1973; PhD, U. Calif., Berkeley, 1976. Postdoctoral fellow Harvard U., Boston, 1977-78; rsch. fellow Los Alamos Nat. Lab., 1978-80, staff member, 1980-82, assoc. group leader, 1982-83, dep. group leader, 1983-84; assoc. prof. Univ. Calif., Irvine, 1984-87, prof., 1987-90; prof. Columbia Univ., N.Y.C., 1990—; cons. E.I. DuPont, 1987—; mem. rev. panel NASA, 1988; mem. Brookhaven Nat. Lab. Chemistry Dept. Rev., 1988, Air Force Office of Sci. Rsch. Chem. Scis. Rev. Panel, 1988-89; chmn. Gordon Rsch. Conf. on Atomic and Molecular Interactions, 1992; organizer SPIE Meeting on Laser Applications to Chem. Dynamics, 1987. Author: (with others) Spectrometric Techniques, 1985, Applications of Laser Spectroscopy, 1986, Gas Phase Bimolecular Reactions, 1989; contbr. articles to Jour. Chem. Physics, Jour. Phys. Chemistry, Chem. Physics Letters, Israel Jour. Chemistry. NSF fellow, 1972-75. Fellow Am. Phys. Soc.; mem. AAAS, AAUP, Am. Chem. Soc. (chmn. div. phys. chemistry, 1992—), Phi Beta Kappa, Sigma Xi. Office: Columbia U Dept Chemistry 116th St and Broadway New York NY 10027

VALENTINI, JOSE ESTEBAN, scientist, educator, chemical engineering director; b. Cordoba, Argentina, May 10, 1950; came to U.S., 1980; s. Cesar Santiago and Irma Judith (Drehock) V.; m. Silvia Dolores Castillo, Apr. 16, 1979; children: Bryan Esteban, Romina Natalia. MS, U. Buenos Aires, 1973; DSc, U. La Plata, Argentina, 1977; MS, U. Pitts., 1981; postgrad., Duke U., 1981-84. Postdoctoral fellow Duke U., Durham, N.C., 1981-84; prof. NYU Med. Ctr., N.Y.C., 1984—; sr. scientist E.I. DuPont, Brevard, N.C., 1984—; dir., scientist surface rheology group Rutgers U., N.J., 1985—; mem. adv. bd. dept. chem. engring. Rutgers U., N.J., 1986-91. Contbr. articles to profl. jours. Leader ARC, Argentina, 1973-74; mem. Rotary, Argentina, 1975, Jaycees, N.C., 1984-85. NIH grantee, 1982, 84; Ministry of Edn. scholar Argentina, 1979; recipient Presdl. award Pres. of Argentina, 1968. Mem. N.Y. Acad. Sci., Soc. of Spectroscopy, Am. Chem. Soc., Sigma Xi. Achievements include patents in coating technology; research in surface science, coating, cancer research, toxicology and rheology. Office: EI DuPont PO Box 267 Brevard NC 28712-0267

VALENTINO, JOSEPH VINCENT, engineering consultant; b. Port Chester, N.Y., July 2, 1930; s. Joseph Paul and Mary Sparkey (Caviola) V.; m. Mary Ann Mottola, May 21, 1954; children: Thomas, Jodi. Grad. high sch., Port Chester. Pres. Reliable Chimney and Stack Corp., Mamaroneck, N.Y., 1953-76; owner, pres. Joseph V. Valentino Cons. and Engring., Mamaroneck, 1976-88, Cogchair Thermo Tech., White Plains, N.Y., 1988—. sec. Dads' Club Ryeneck Schs., Mamaroneck, 1970, participant Student Aid Fund, 1973; bldg. com. Martin Luther King Ctr., Mamaroneck, 1980. Mem. N.Y. Acad. Scis. Achievements include patents in field. Home: 8 Saxonwood Park Dr White Plains NY 10605

VALENTINUZZI, MAX EUGENE, bioengineering and physiology educator, researcher; b. Buenos Aires, Argentina, Feb. 24, 1932; s. Maximo and Emma L. (Mazzulli) V.; m. Nilda Pontorno, May 16, 1957; children: Debora Fabiana, Veronica Sandra. BA, Colegio Nacional Buenos Aires, 1950; Degree in Elec. Engring., U. Buenos Aires, 1951-56; PhD in Physiology and Biophysics, Baylor Coll. Medicine, Houston, 1969. Technician Casalis Srl, Buenos Aires, 1954-55; elec. engr. Transradio Internat., Buenos Aires, 1955-60; rsch. assoc. Emory U., Atlanta, 1960-62; assoc. prof. U. del Sur, Bahia Blanca, Argentina, 1963-66; asst. prof. Baylor Coll. Medicine, Houston, 1969-73; prof. U. Tucuman, Argentina, 1973—; career investigator CONICET, Tucuman, 1977—; dir. Inst. Superior Investigaciones Biologicas, Tucuman, 1981-86. Contbr. articles to profl. jours. Recipient Nightingale prize Internat. Fedn. Med. and Biol. Engring., London, 1973, Houssay prize Argentine Soc. Biology, Buenos Aires, 1981, Catalina B. de Baron prize CORDIC, Buenos Aires, 1983-84, Golden Route prize Soc. Dist. Diarios, Buenos Aires, 1984. Mem. IEEE (sr.), Am. Physiol. Soc., Bioengring. Soc. Argentina, Biol. Engring. Soc. London, Acad. Nat Ingenieria, Acad. Ciencias Medicas Cordoba. Avocation: music. Office: U Tucuman, C C 28 Suc 2, 4000 Tucuman Argentina

VALIANT, LESLIE GABRIEL, computer scientist; b. Mar. 28, 1949; s. Leslie and Eva Julia (Ujlaki) V.; m. Gayle Lynne Dyckoff, 1977; children—Paul A., Gregory J. M.A., Kings Coll., Cambridge, U.K., 1970; D.I.C., Imperial Coll., London, 1971; Ph.D., U. Warwick, U.K., 1974. Vis. asst. prof. Carnegie-Mellon U., Pitts., 1973-74; lectr. U. Leeds, Eng., 1974-76; lectr., reader U. Edinburgh, Scotland, 1977-82; vis. prof. Harvard U., 1982, Gordon McKay prof. computer sci. and applied math., 1982—. Guggenheim fellow, 1985-86; recipient Nevanlinna prize Internat. Math. Union, 1986. Fellow Royal Soc., Am. Assn. for Artificial Intelligence. Office: Harvard U 33 Oxford St Cambridge MA 02138-2901

VALIEV, KAMIL AKHMETOVICH, engineering educator; b. Kazan, Tatarstan, Russia, Jan. 15, 1931; married, 1957; 2 children. Candidate of sci. in Physics, Kazan State Univ., Russia, 1958, DSc (hon.), 1963. Sr. educator, researcher in physics Kazan Pedagogical Inst., Kazan, Tatarstan, Russia, 1957-64; dir., Indsl. Inst. for Molecular Electronics, Moscow, Russia, 1965-77; lab. dir. Inst. Gen. Physics, Moscow, Russia, 1978-88; dir. Inst. Physics and Tech., Moscow, Russia, 1989—. Author Physics of Submicron Lithography, 1992; editor Microelectronics, 1988—. Recipient Lenin prize USSR, 1971, 72, 85. Fellow Third World Acad. Sci.; mem. Russian Acad. Sci. (correspondent). Achievements include patents and inventions in microelectronic technology. Home: Zelinsky 38 building 8 flat 85, Moscow Russia

VALLORT, RONALD PETER, mechanical engineer; b. Chgo., Nov. 14, 1942; s. Peter John and Ann Carrie (Mazur) V.; m. Carolyn Ruth Blum, June 11, 1966; children: Joseph Ronald, James Kurt. BSME, U. Ill., 1964, MSME, 1965. Registered profl. engr., 26 states. Design engr. Swift and Co., Chgo., 1965-68; dir. design, project mgr., constrn. engr. Globe Engring. Co., Chgo., 1968-83; v.p., dir. facility design A. Epstein and Sons Internat., Chgo., 1983-92; chief engring. officer The Haskell Co., Jacksonville, Fla., 1992—. Contbr. articles to profl. jours. Coord. funds athletic and booster com. St. Joseph Ch., Downers Grove, Ill., 1980-82, chmn. facilities and maintenance com., 1982-92; mem. Downers Grove Flood Control Bd., 1985-92. Mem. ASHRAE (pres. Chgo. chpt. 1977-78, chmn. meat, fish and poultry com. 1977-78, R in ASHRAE chmn. 1979-80, chmn. rsch. and tech. com. 1987-88, bd. dirs. 1992—), Internat. Inst. Ammonia Refrigeration (chmn. rsch. com. 1991—, bd. dirs. 1993—), Internat. Inst. Refrigeration (D-1 commn. 1990—), Nat. Fire Protection Assn. Achievements include design of soybean processing plant, dry sausage plant, United Air Lines Terminals, O'Hare Airport, Carnation ice cream plant, Meat Sci. Lab., Iowa State U. Home: 12536 Marsh Creek Dr Ponte Vedra Beach FL 32082 Office: The Haskell Co 111 Riverside Ave Jacksonville FL 32202

VALOSKI, MICHAEL PETER, industrial hygienist; b. Sewickley, Pa., Oct. 22, 1951; s. Peter and Julia (Hopta) V.; m. Alesia Marie DiIanni, Sept. 22, 1984; 1 child, Michael Peter Valoski II. BS cum laude, Ind. U. of Pa., 1973; MS in Hygiene, U. Pitts., 1977. Environ. health specialist Allegheny County Health Dept., Pitts., 1973-78; acoustician Environ. Acoustic Rsch. Co., Bridgeville, Pa., 1977-78; indsl. hygienist Mine Safety and Health Adminstrn., Pitts., 1978—. Contbr. articles to profl. publs. Mem. Am. Indsl. Hygiene Assn., Am. Conf. of Govtl. Indsl. Hygienists, Acoustical Soc. Am. (assoc.). Office: Mine Safety and Health Adminstrn 4800 Forbes Ave Pittsburgh PA 15213

VALRAND, CARLOS BRUNO, aerospace engineer; b. Havana, Cuba, Jan. 31, 1943; s. Carlos Heriberto Valrand and Maxima Justa Drake. B.Aerospace Engring., Ga. Inst. Tech., 1965. Engr. Boeing Aircraft Co., Seattle, 1965-67; aerodynamicist Lockheed-Ga. Co., Atlanta, 1967-70; mem. profl. staff TRW Systems Group, Houston, 1970-74; sect. head TRW Systems Group, Huntsville, Ala., 1974-76; staff engr. Fed. Systems div. IBM, Houston, 1976-83, adv. engr., 1983-89, sr. engr., 1989—; dir. Bithian Empire Mercantile Co., Houston. Contbr. articles to profl. jours.; editor: BEM Automobile Buyer's Guide, 1982. Mem. Rep. Nat. Com., Washington, 1984—. Mem. AIAA, AAAS, Toastmasters (v.p. edn. 1992—, pres. 1990, Competent Toastmaster 1988). Roman Catholic. Achievements include development of PERCAM reconfigurable air defense engagement simulator; space shuttle dynamic simulations; space station fault management knowledge based system. Home: 16419 Havenhurst Houston TX 77059

VALSARAJ, KALLIAT THAZHATHUVEETIL, chemical engineering educator; b. Tellichery, Kerala, India, Oct. 2, 1957; came to U.S., 1980; s. Mundayat B. Nambiar and Kalliat T. Bhanumathy; m. Nisha Valsaraj, Dec. 24, 1990; 1 child, Viveca. MS, Indian Inst. Tech., Madras, India, 1980; PhD, Vanderbilt U., 1983. Affiliate faculty U. Ark., Fayetteville, 1983-86; sr. rsch. assoc. Hazardous Waste Rsch. Ctr. La. State U., Baton Rouge, 1986-90, asst. prof., 1990—; mem. panel directions in separations NSF, 1989-90; cons. Balsam Engr. Cons., Salem, N.H., 1990-91; presenter in field. Contbr. over 50 articles to profl. jours. Grantee Dept. of Def., 1986-89, NSF, 1989, 92-95, EPA, 1989-91, 90-92. Mem. Am. Chem. Soc., Am. Inst. Chem. Engrs., Nat. Geographic Soc., Indian Chem. Soc. Achievements include patent for innovative groundwater treatment. Home: 2023 General Beauregard Ave Baton Rouge LA 70810 Office: La State U Dept Chem Engring Baton Rouge LA 70803

VÁMOS, GEORGE A., mechanical engineer, consultant; b. Budapest, Hungary, July 7, 1912; s. Soma and Rose (Bauer) V.; m. Helene Szarvas, Oct. 16, 1938; 1 child, Eva. Diploma Mech. Engring., Tech. U., Budapest, 1934, D Tech., 1956. Dept. mgr. Neményi Paper Mills, Csepel, Hungary, 1934-45; chief paper dept. Ministry of Industry, Budapest, 1945; dir. Paper Rsch. Inst., Budapest, 1949-73; dir. gen., prof. Coll. Tech. for Light Industry, Budapest, 1972-82; cons. Paper Rsch. Inst., Budapest, 1985—;

editor-in-chief Papiripar Jour., Budapest, 1957—; mem. bd. Nat. Com. for Tech. Devel., Budapest, 1962-78; mem. coun. Hungarian Fedn. Tech. & Sci. Soc., 1949—; lectr. in field. Author: Papiripari Kézikönyv, 1962, 2nd edit., 1980, Papiripari ABC, 1964, 2nd edit., 1984, Papirgyártasi Technólogia, 1981; contbr. 150 articles to profl. jours. Recipient Kossuth prize Pres. of the Republic, 1957; Golden Badge, Eucepa Coun., 1987. Mem. Tech. Assn. Paper & Printing Industry (hon. pres. 1983—), Liaison Europeenne Pour La Cellulose et Papier, Assn. Hungarian Sci. and Indsl. Journalists (pres. 1990—), others. Achievements include research in straw and short fibre pulps for paper making, drying, international cooperation in the paper industry and research. Home: Alkotás U 13, H 1123 Budapest Hungary Office: Tech Assn Paper & Printing Industry, Fö U 68, H 1371 Budapest Hungary POB 433

VAN ALLEN, JAMES ALFRED, physicist, educator; b. Mt. Pleasant, Iowa, Sept. 7, 1914; s. Alfred Morris and Alma E. (Olney) Van A.; m. Abigail Fithian Halsey, Oct. 13, 1945; children: Cynthia Schaffner, Margot Cairns, Sarah Trimble, Thomas, Peter. BS, Iowa Wesleyan Coll., 1935; MS, U. Iowa, 1936, PhD, 1939; ScD (hon.), Iowa Wesleyan Coll., 1951, Grinnell Coll., 1957, Coe Coll., 1958, Cornell Coll., Mt. Vernon, Iowa, 1959, U. Dubuque, 1960, U. Mich., 1961, Northwestern U., 1961, Ill. Coll., 1963, Butler U., 1966, Boston Coll., 1966, Southampton Coll., 1967, Augustana Coll., 1969, St. Ambrose Coll., 1982, U. Bridgeport, 1987. Research fellow, physicist dept. terrestial magnetism Carnegie Instn., Washington, 1939-42; physicist, group and unit supr. applied physics lab. Johns Hopkins U., Balt., 1942, 46-50; organizer, leader sci. expdns. study cosmic radiation Peru, 1949, Gulf of Alaska, 1950, Arctic, 1952, 57, Antarctic, 1957; Carver prof. physics, head dept. U. Iowa, Iowa City, 1951-85; Regents fellow Smithsonian Instn., 1981; rsch. assoc. Princeton, N.J., 1953-54; mem. devel. group radio proximity fuze Nat. Def. Rsch. Coun., OSRD; pioneer high altitude rsch. with rockets, satellites and space probes. Author: Origins of Magnetospheric Physics, 1983, First to Jupiter, Saturn and Beyond, 1981; contbg. author: Physics and Medicine of Upper Atmosphere, 1952, Rocket Exploration of the Upper Atmosphere; editor: Scientific Uses of Earth Satellites, 1956; acting editor Jour. Geophys. Rsch.-Space Physics, 1991-92; contbr. numerous articles to profl. jours. Lt. comdr. USNR, 1942-46, ordnance and gunnery specialist, combat observer. Recipient Physics award Washington Acad. Sci., 1949, Space Flight award Am. Astronautical Soc., 1958, Louis W. Hill Space Transp. award Inst. Aero. Scis., 1959, Elliot Cresson medal Franklin Inst., 1961, Golden Omega award Elec. Insulation Conf., 1963, Iowa Broadcasters Assn. award, 1964, Fellows award of merit Am. Cons. Engrs. Coun., 1978, Nat. Medal of Sci., 1987, Vannevar Bush award NSF, 1991; named comdr. Order du Merit Pour la Recherche et l'Invention, 1964; Guggenheim Found. rsch. fellow, 1951. Fellow Am. Rocket Soc. (C.N. Hickman medal for devel. Aerobee rocket 1949), IEEE, Am. Phys. Soc., Am. Geophys. Union (pres. 1982-84, John A. Fleming award 1963, 64, William Bowie medal); mem. NAS, AAAS (Abelson prize 1986), Iowa Acad. Sci., Internat. Acad. Astronautics (founding), Am. Philos. Soc., Am. Astron. Soc., Royal Astron. Soc. U.K. (gold medal 1978), Royal Swedish Acad. Sci. (Crafoord prize 1989), Am. Acad. Arts and Scis., Cosmos Club (Washington), Sigma Xi (Proctor prize 1987), Gamma Alpha. Presbyterian. Achievements include discovery of radiation belts around earth. Office: Univ Iowa 701 Van Allen Hall Iowa City IA 52242-1403

VAN ARSDEL, EUGENE PARR, tree pathologist, consultant meteorologist; b. Emaus, Pa., Dec. 4, 1925; s. William Campbell and Mabel Elizabeth (Hedde) Van A.; m. Rose Price, Aug. 23, 1948 (div. Aug. 1991); children: Jonathan Eugene, Elizabeth Rose. BS in Forestry, Purdue U., 1947; MS, U. Wis., 1952, PhD, 1954. Plant pathologist Lake States Forest Expt. Sta. U. Wis., Madison, 1956-59, plant pathologist, 1959-62; prin. plant pathologist, project leader No. Conifer Disease Rsch., No. Ctrl. Forest Expt. Sta. U. Minn., St. Paul, 1962-68, prof. plant pathology, 1967-68; assoc. prof. plant sci. Tex. A&M U., College Station, 1968-80; plant pathologist, forester Profl. Tree Svc., Inc., Bryan, Tex., 1981—; vis. prof. Yale U., New Haven, 1965-66. Assoc. editor Ecol. Soc. Am. Jour., 1968-71; contbr. chpts. to books, articles to Am. Meteorol. Soc., Am. Soc. Foresters, Am. Phytopath. Soc., others. An organizer, mem. Brazos Chorale, Bryan, Tex., 1970—. Grantee NSF, 1962-63, Tex. Peanut Producers Bd., 1972, USDA Forest Svc., 1973-77, Mrs. Lyndon B. Johnson, 1977-82, I.S.A. Rsch. Trust, 1986. Fellow AAAS; mem. Soc. Am. Foresters, Internat. Soc. Arboriculture, Am. Phytopath Soc. (emeritus). Achievements include development of effective treatment for wilt diseases in oaks of Texas; study of climatic.microclimatic relationship of the spread of white pine blister rust; research in long-distance transport of fungous spores. Office: Profl Tree Svc Inc 2112 Cavitt Dr Bryan TX 77801

VAN ARSDEL, WILLIAM CAMPBELL, III, pharmacologist; b. Indpls., June 27, 1920; s. William C. and Mabel (Hedde) Van A. BS, Oreg. State Coll., 1949, MS, 1951, PhD, 1959; MS, U. Oreg., 1954. Teaching fellow dept. zoology Oreg. State Coll., Corvallis, 1954-59, rsch. fellow dept. animal husbandry, 1954-59, jr. animal physiologist, 1959-60, asst. in animal physiology, 1960-63; pharmacologist FDA, Washington, 1963—. Contbr. articles to Jour. Am. Pharm. Assn., Oreg. State Coll. Agrl. Experiment Sta. Bull., The Physiologist, Am. Heart Jour., Brit. Vet. Jour., IEEE Transaction on Biomed. Engring., Sci. Tchr., Strength and Health, others. Mem. AAAS, Am. Chem. Soc., Am. Coll. Clin. Pharm., Am. Mus. Natural History, Japan-Am. Soc. Washington, N.Y. Acad. Sci., Washington Acad. Scis., South Pacific Underwater Medicine Soc., Undersea and Hyperbaric Med. Soc., Inc., Soc. Exptl. Biology and Medicine (Washington sect., corres. sec. 1977-80, 82-92, pres. 1981-82), Oreg. State Soc. Washington, Internat. Oceanographic Found., Ind. Soc. Washington, U. Oreg. Med. Sch. Alumni Assn., Toastmasters Internat., Nat. Rifle Assn., Sierra Club, Audubon Soc., Sigma Xi, others. Achievements include research in comparative embryology, pharmacology of intestinal motility, electrocardiology of domestic animals, ichthyology, electrocardiograms of small animals. Home: 1000 6th St SW # 301 Washington DC 20024 Office: FDA 5600 Fishers Ln Rockville MD 20857

VAN ASSCHE, FRANS JAN MAURITS, economics educator; b. Mechelen, Belgium, Nov. 26, 1948; s. Lodewijk Maurits and Mathilda (Van Hemelryck) Van A.; m. Bernice Ryckewaert, Jan. 3, 1975 (div. Aug. 1991). Handels-en Bedrijfs Econ. Ingenieur, U. Leuven, Belgium, 1974. Rsch. asst. U. Leuven, 1975-77; researcher Data Mgmt. Rsch. Lab., Control-Data, Brussels, 1977-86; prin. cons. James Martin Assocs., Brussels, 1986-89, v.p., 1989-91; mng. dir. James Martin & Co. Belgium, Brussels, 1991-93; mem. editorial bd. Info. Systems Jour. Author: Information System Methodologies, 1988; contbr. articles to profl. publs. Chmn. Stichting Ideale Samenleving Leuven, 1989. Mem. Contactgroep Beleidsinformatica Leuven. Avocations: vedic science, music, hiking. Home: Doornstraat 14, 3370 Boutersem Belgium Office: James Martin & Co, Ambachtenlaan 38, 3001 Heverlee Belgium

VAN BOOVEN, JUDY LEE, data processing manager; b. Kansas City, Mo., Oct. 26, 1952; d. Gene Warren and Jane Lewis (Wallace) Pulley; m. Cecil Carlin Van Booven, Aug. 19, 1972; children: Walter Matthew, Leia Christine, Kelly Diane, Matthew Carlin. Student, Cen. Mo. State U., Warrensburg, 1970-72; AAS, Penn Valley Community Coll., Kansas City, Mo., 1981; BS, William Jewell Coll., 1988. Bookkeeper Century Mills, Wilmington, N.C., 1972-73; acctg. clk. Forest Siding, North Kansas City, Mo., 1974-75, computer operator, 1975-77, programmer/ops. mgr., 1977-80; programmer/analyst Western Water Mgmt., Inc., North Kansas City, Mo., 1980-88, data processing mgr., 1989—; freelance programmer Modern Window Co., North Kansas City, 1978-80, Forest Lumber, Oklahoma City, 1975-80, Kay-Dee Systems, North Kansas City, 1980-82; systems cons. Forest Siding, 1980-82. Author 1st place essay, North Kansas City Centennial, 1988. Founder, chmn. Children's Book Drive, Kansas City, 1983—; mem. parent adv. com. Gracemor Sch., Kansas City, 1988-89, bd. dirs. PTA, 1989—, co-chmn. newsletter, 1989-91, 1st v.p., 1991-92, treas., 1992-93, sec., 1993—; asst. soccer coach Sherwood Soccer Club, GU10, 1991—, soccer coach GU8, 1991—. Recipient scholarship Bus. and Profl. Women's Assn., 1970, Bd. Regents, Mo., 1970; Mgr. of Yr. award Western Water Mgmt., 1992. Mem. ABWA (treas. 1983-84, sec. 1985-87, named Woman of the Yr. 1986), NAFE, Phi Theta Kappa. Baptist. Avocations: reading, hiking, camping, archaeology, geology. Home: 5143 N Richmond Ave Kansas City MO 64119-4063 Office: Western Water Mgmt Inc 1345 Taney St Kansas City MO 64116-4414

VAN BRUGGEN, ARIENA HENDRIKA CORNELIA, plant pathologist; b. Delfzyl, The Netherlands, Dec. 7, 1949; came to U.S., 1980; d. Adriaan and Nelly (Meerwaldt) Van B. MS, Agr. U., Wageningen, The Netherlands, 1976; PhD, Cornell U., 1985. Assoc. expert Food and Agriculture Orgn. of UN, Ethiopia, 1976-80; grad. asst. Cornell U., Ithaca, N.Y., 1980-84; postdoctoral Boyce Thompson Inst., Ithaca, N.Y., 1984-86; asst. prof. U. Calif., Davis, 1986-92, assoc. prof., 1992—. Contbr. articles to profl. jours. Recipient Ciba Geigy award Am. Phytopathology Soc.; Eriksson Gold medal Agr. U., Sweden, 1993; USDA grantee, 1988-92. Achievements include discovery of a new plant pathogen, rhizomonas suberifaciens. Office: U Calif Dept Plant Pathology Davis CA 95616

VAN BURKLEO, BILL BEN, osteopath, emergency physician; b. Tulsa, Nov. 21, 1942; s. Walter Russell and Joan Vera (Brimm) Van B.; m. Paula Mae Brinkley, Mar. 5, 1965 (div. Feb. 1974); children: Baron, Kristy and Kelly (twins). BS, U. Tulsa, 1965; DO, Okla. State U., 1981. Diplomate Nat. Bd. Osteo. Examiners. Defensive back, quarterback, punter Can. Football League, Ottawa, Calgary, 1966-73; dir. sports and spl. events Tulsa Cable TV, 1974-78; rotating intern Corpus Christi (Tex.) Osteo. Hosp., 1981-82; family physician Antlers (Okla.) Med. Clinic, 1982-90, Colbert (Okla.) Med. Clinic, 1989-90; dir. dept. emergency Valley View Regional Hosp., Ada, Okla., 1990—; regional med. dir. Okla. Spectrum Emergency Care, Inc.; mem. clin. faculty Coll. Osteo. Medicine, Okla. State U. Author newspaper column, several computer programs. Mem. Rep. Senatorial Inner Cir., Washington, 1990-91; affiliate faculty Am. Heart Assn. Named to Alltime Greats of Okla., Jim Thorpe Award Com., 1975. Mem. Am. Osteo. Assn., Am. Coll. Gen. Practitioners, Okla. Osteo. Assn., S.W. Okla. Osteo. Assn. (pres. 1990-91). Avocations: tennis, flying, sailing. Home: PO Box 2740 Ada OK 74821-2740

VAN CAMPEN, DARRELL ROBERT, chemist; b. Two Buttes, Colo., July 15, 1935; s. Robert Lewis and Pauline (Comer) Van C.; m. Orlene Crone, Sept. 8, 1958 (div. 1976); children: Anthony, Bryan; m. Judith Ann Gorsky, June 27, 1978; 1 child, John. BS, Colo. State U., 1957; MS, N.C. State U., 1960, PhD, 1962. Postdoctoral fellow Cornell U., Ithaca, N.Y., 1962-63; rsch. chemist USDA ARS, Ithaca, 1964-80, lab. dir., 1980—. Contbr. articles and revs. to profl. jours. and chpts. to books. NIH fellow, 1962. Mem. Sigma Xi, Alpha Zeta, Phi Kappa Phi. Avocations: golf, gardening, numismatics. Home: 117 Simsbury Dr Ithaca NY 14850-1728 Office: USDA ARS Plant Soil & Nutrition Lab Tower Road Ithaca NY 14853

VAN CLEEF, JABEZ LINDSAY, marketing professional; b. Cooperstown, N.Y., Nov. 19, 1948; s. John Henry and Persis (Hathaway) V.; m. Martha Millard, May 1, 1983 (div. 1993); children: Jane, Lucy. BA, Cornell U., 1970. Mgr. Willard Gallery, N.Y.C., 1976-78; mgr., prin. Writing Specialists, Atlanta, 1979-82; account exec. A.B. Isacson Assocs., N.Y.C., 1982-85; account supr. Gilbert, Whitney & Johns, Whippany, N.J., 1985-86; from mktg. mgr. to dir. mktg. Hosokawa Micron Internat., N.Y.C., Summit, N.J., 1986—. Sec. editorial bd. KONA, 1989—; co-author: Fundamentals of Powder Technology, 1991; contbr. articles to Am. Scientist, Chem. Engring., Rsch. and Devel. Active Montclair (N.J.) Art Mus., 1991—. Mem. Materials Rsch. Soc., N.Y. Acad. Scis., Harmonium Classical Choral Soc. (bd. dirs. 1991—), Garden State Quilter's Guild. Episcopalian. Home: 7089 N River Rd New Hope PA 18938 Office: Hosokawa Micron Internat 10 Chatham Rd Summit NJ 07901

VAN COTT, HAROLD PORTER, human factors professional; b. Schenectady, Nov. 16, 1925; s. Harrison Horton and Edith (Porter) Van C.; m. Madeleine P. Bouvier, Oct. 8, 1953; children: Laurent, Jeanne Marie, Anne. BA in Psychology, U. Rochester, 1948; MA in Psychology, U. N.C., 1952, PhD in Psychology, 1954. Dir. rsch. Am. Insts. Rsch., Washington, 1964-69; mng. editor APA, Washington, 1969-75; div. chief Nat. Bur. Stds., Rockville, Md., 1975-80; chief scientist Biotech. Inc., Arlington, Va., 1980-81; v.p. Essex Corp., Alexandria, Va., 1982-85; prin. staff officer NRC, Washington, 1985-92; cons. Van Cott Assocs., Bethesda, Md., 1992—; cons. Idaho Nat. Engring. Lab., Idaho Falls, 1992—, Planning Rsch. Corp., Reston, Va., 1992—, NRC, Washington, 1992—. Editor: Human Engineering Guide to Equipment Design, 1972; assoc. editor Jour. Human Factors Soc. With U.S. Army, 1945-46. Fellow AAAS, APA (pres. div. 21 1983-84), Human Factors Soc., Washington Acad. Sci. Presbyterian. Home and Office: Van Cott Assocs 8300 Still Spring Ct Bethesda MD 20817

VANDAM, LEROY DAVID, anesthesiologist, educator; b. N.Y.C., Jan. 19, 1914; s. Albert Herman and Esther Henrietta (Cahan) V.; m. Regina Phyllis Rutherford, Nov. 30, 1939; children: Albert Rutherford, Samuel Whiting. PhB magna cum laude, Brown U., 1934; MD, NYU, 1938; MA, Harvard U., 1967. Diplomate Am. Bd. Anesthesiology. Resident surgeon Beth Israel Hosp., Boston, 1942-43; fellow in surgery Johns Hopkins Hosp., Balt., 1945-47; asst. prof. anesthesia U. Pa. Med. Sch., Phila., 1951-54; instr. in anesthesia WHO, Copenhagen, 1953; surgeon Peter Bent Prigham Hosp., Boston, 1954-69; prof. of anaesthesia Harvard Med. Sch., Boston, 1967-80, prof. emeritus, 1980—; mem. com. on revision U.S. Pharmacopoeia, Washington, 1970-75. Editor=in-chief Jour. Anesthesiology, 1962-70; mem. editorial bd. New Eng. Jour. Medicine, 1976-80; author texts: (with others) Introduction to Anesthesia, 1967—, To Make the Patient Ready for Anesthesia, 1980, 2d edit., 1984, The Genesis of Contemporary Anesthesia, 1982; contbr. over 250 articles, revs. to profl. publs., chpts. to books. 1st lt. Med. Corps, AUS, 1943-73. Fellow Am. Coll. Cardiology; mem. Assn. Univ. Anesthesiologists (pres. 1964-65, Disting. Svc. award 1967), Halsted Surg. Soc., Aesculapian Club, Phi Beta Kappa, Sigma Xi, Alpha Omega Alpha. Home: 10 Longwood Dr 268 Westwood MA 02090 Office: Brigham & Women's Hosp 75 Francis St Boston MA 02115

VANDEBERG, JOHN THOMAS, chemical company executive. BA in Chemistry, Carroll Coll., Helena, Mont., 1962; MS in Chemistry, Loyola U. Chgo., 1966, PhD in Chemistry, 1969; M in Mmgmt., Northwestern U., 1988. Sect. leader analytical chemistry Desoto Inc., 1969-73, mgr. rsch. svcs., 1973-78, mgr. polymer devel., 1978-84, dir. polymer devel, fiber optic materials, new ventures rsch., 1984-89, dir. tech. radiation curable products, 1989-90; v.p. tech., founding mem. DSM Desotech, Inc., Elgin, Ill., 1990—. Author: Infrared Spectroscopy-Its Use in the Coatings Industry, 1969, An Infrared Spectroscopy Atlas for the Coatings Industry, 1979; contbr. over 15 articles to profl. jours. Achievements include patents in radiation curable coating compositions applied to fluorine-treated surfaces, light weight concrete and cementitious masonry products, new radiation curable technology-maleate-vinyl ether free radical polymerzations; research in plastic bottle for packaging nuts in food industry. Home: 415 W Oakwood Dr Barrington IL 60010

VAN DE KAR, LOUIS DAVID, pharmacologist, educator; b. Amsterdam, The Netherlands, Apr. 15, 1947; came to U.S. 1975; s. A. and M. (Barend) Van De K.; m. S.L. Schmitt, Oct. 28, 1978. MS, U. Amsterdam, 1974; PhD, U. Iowa, 1978. Postdoctoral fellow U. Calif., San Francisco, 1979-81; asst. prof. Loyola U. of Chgo., Maywood, Ill., 1981-87, assoc. prof., 1987-92, prof. pharmacology, 1992—. Fulbright fellow, 1975; NIH rsch. svc. awardee, 1980; NIMH grantee, 1990, NIDA grantee, 1989. Office: Loyola Univ of Chgo Dept Pharmacology 2160 S 1st Ave Maywood IL 60153

VAN DEMARK, ROBERT EUGENE, SR., orthopedic surgeon; b. Alexandria, S.D., Nov. 14, 1913; s. Walter Eugene and Esther Ruth (Marble) Van D.; m. Bertie Thompson, Dec. 28, 1940; children: Ruth Elaine, Robert, Richard. B.S., U. S.D., 1936; A.B., Sioux Falls (S.D.) Coll., 1937; M.B., Northwestern U., 1938, M.D., 1939; M.S. in Orthopedic Surgery, U. Minn., 1943. Diplomate Am. Bd. Orthopedic Surgery; cert. addl. qualifications surgery of the hand. Intern Passavant Meml. Hosp., Chgo., 1938-39; fellow in orthopedic surgery Mayo Found., 1939-43; 1st asst. orthopedic surgery Mayo Clinic, 1942-43; orthopedic surgeon Sioux Falls, 1946—; attending orthopedic surgeon McKennan Hosp., pres. med. staff, 1954, 70; attending orthopedic surgeon Sioux Valley Hosp., pres. staff, 1951-52; clin. prof. orthopedic surgery U. S.D., Sioux Falls, 1953—, adj. prof orthopedic anatomy, 1983—; med. dir. Crippled Children's Hosp. and Sch., 1956-84; chief hand surgery clinic VA Hosp., Sioux Falls., 1978-90; bd. dir. S.D. Blue Shield, 1976-88. Editor: S.D. Jour. Medicine; Contbr. articles to med. jours. Bd. dirs. S.D. Found. for Med. Care, 1976-83; hon. chmn. S.D. Lung Assn., 1982. Maj. U.S. Army, 1943-46. Recipient citation for outstanding svc. Pres.'s Commn. for Employment Physically Handicapped, 1960; Svc. to Mankind award Sertoma Internat., 1963; award for dedicated svcs. to handicapped S.D. Easter Seal Soc., 1969; Robins award for outstanding community svc., 1971; Humanitarian Svc. award United Cerebral Palsy, 1976; Alumni Achievement award U. S.D., 1977; Disting. Citizen award S.D. Press Assn., 1978; U. S.D. Med. Sch. Faculty Recognition award, 1980; outstanding contbns. to Handicapped Children award S.D. State Dept. Health, 1985, Community Svc. Health award Sioux Valley Hosp. Found., 1991; named Humanitarian of Yr. S.D. Human Svcs. Forum, 1987. Fellow ACS (first pres. S.D. chpt. 1952-53); mem. Am. Acad. Orthopedic Surgery, Am. Soc. Surgery Hand, Am. Assn. Hand Surgery (cert.), Am. Coll. Sports Medicine, Mid-Am. Orthopedic Assn., Am. Soc. Surgery Hand, Am. Acad. Cerebral Palsy, Clin. Orthopedic Soc., Doctors Mayo Soc., S.D. Med. Assn. (pres. 1974-75, Disting. Svc. award 1987, Spl. Presdl. award 1991), Sioux Falls Dist. Med. Soc., SAR, 500 First Families, Optimists, Minnehaha Country Club, Sigma Xi, Alpha Omega Alpha, Phi Chi. Lutheran. Home: 2803 Ridgeview Way Sioux Falls SD 57105-4243 Office: 1301 S 9th Ave Sioux Falls SD 57105-1042

VAN DEN AKKER, JOHANNES ARCHIBALD, physicist; b. L.A., Dec. 5, 1904; s. John and Mabel (Freebairn) Van den A.; m. Adelaide H. Carrier, June 20, 1930 (dec. Jan. 1955); 1 child, Valerie; m. Carmen L. Haberman, June 9, 1958 (dec. Mar. 1989); m. Margaret Koller, Jan. 20, 1990. BS in Physics/Engring., Calif. Inst. Tech., 1926, PhD in Physics, 1931. Instr. Washington U., St. Louis, 1930-34; prof. physics Inst. Paper Chemistry, Appleton, Wis., 1935-70; cons. Am. Can Co., Neenah, Wis., 1971-82, James River Corp., Neenah, 1982-85, Appleton, Wis., 1985—; lectr. short courses TAPPI, Atlanta. Co-author 10 books; contbr. Encyclopedia of Physics, 1st and 2d edit. and articles to physics and tech. jours. Sr. Fulbright scholar U. Manchester (Eng.), 1961-62; recipient Gold medal TAPPI, 1968. Fellow AAAS, Am. Phys. Soc., TAPPI; mem. Am. Inst. Physics, Optical Soc. Am., Am. Assn. Physics Tchrs., Sigma Xi, Tau Beta Pi, Phi Gamma Delta. Achievements include research on spatial distribution of x-ray photo-electrons, ultraviolet spectrophotometers, analog computers, methods for measuring centroid wavelengths of broad band/filter spectrophotometers, methods of measuring absolute reflectance of diffuse surfaces, designs of special instrumentation for study of optical, physical and mechanical properties of paper, and development of theories for these properties. Home: 1101 E Glendale Ave Appleton WI 54911-3144

VANDENBERG, EDWIN JAMES, chemist, educator; b. Hawthorne, N.J., Sept. 13, 1918; s. Albert J. and Alida C. (Westerhoff) V.; m. Mildred Elizabeth Wright, Sept. 9, 1950; children: David James, Jean Elizabeth. M.E. with distinction, Stevens Inst. Tech., 1939, Dr.Engring. (hon.), 1965. Research chemist Hercules Inc. Research Ctr., Wilmington, Del., 1939-44, asst. shift supr. Sunflower Ordnance Works, Kans., 1944-45, research chemist Research Ctr., Wilmington, 1945-57, sr. research chemist, 1958-64, research assoc., 1965-77, sr. research assoc., 1978-82; adj. prof. chemistry Ariz. State U., Tempe, 1983-91, rsch. prof. chemistry, 1992—. Author: Polyethers, 1975; Coordination Polymerization, 1983; Contemporary Topics in Polymer Science V, 1984, Catalysis in Polymer Synthesis, 1992. Patentee in field. Mem. adv. bd. Jour. Polymer Sci., 1967—, Macromolecules, 1979-81; chmn. Gordon Rsch. Conf. on Polymers, 1978. Mem. Am. Chem. Soc. (councillor Del. sect. 1974-81, chmn. 1976, chmn. div. polychemistry 1979, coordinator indsl. sponsors 1982—, Del. sect. award 1965, Nat. 1979, Indsl. Rsch. 100 award 1965, Polymer Chemistry award 1981, Exceptional Service award 1983, Applied Polymer Sci. award 1991, Charles Goodyear medal, 1991, Herman F. Mark award 1992). Home: 16223 E Inca Ave Fountain Hls AZ 85268-4518 Office: Ariz State U Dept Chemistry and Biochemistry Tempe AZ 85287-1604

VANDENBERG, JOHN DONALD, entomologist; b. Benton Harbor, Mich., Jan. 24, 1954; s. Robert Landis and Madelaine Louise (Westendorf) V.; m. Alice C. L. Churchill, Oct. 8, 1983. B.S. with Honors, U. Mich., 1975; M.S., U. Maine, 1977; Ph.D., Oreg. State U., 1982. Grad. rsch. asst. U. Maine, Orono, 1975-77; grad. teaching asst. Oreg. State U., Corvallis, 1977-78, grad. rsch. asst., 1978-82; postdoctoral assoc. Boyce Thompson Inst., Ithaca, N.Y., 1982-83; rsch. entomologist Agrl. Rsch. Svc., U.S. Dept. Agr., Beltsville, Md., 1983-87, rsch. leader Agrl. Rsch. Svc., Logan, Utah, 1987—; acting asst. dir. Midwest area Agrl. Rsch Svc., Peoria, Ill., 1991; equal employment opportunity counsellor, 1985-87. Contbr. articles to profl. jours. Mem. Soc. for Invertebrate Pathology, Entomol. Soc. Am., Am. Soc. for Microbiology, Sigma Xi (sec. 1989-90, chpt. pres. 1991). Avocations: singing, guitar, softball, gardening. Office: Utah State U USDA-ARS Bee Biology & Systematics Lab Logan UT 84322-5310

VAN DEN BERGH, SIDNEY, astronomer; b. Wassenaar, Netherlands, May 20, 1929; emigrated to U.S., 1948; s. Sidney J. and Mieke (van den Berg) vandenB.; m. Paulette Brown; children by previous marriage: Peter, Mieke, Sabine. Student, Leiden (The Netherlands) U., 1947-48; A.B., Princeton U., 1950; M.Sc., Ohio State U., 1952; Dr. rer. nat., Goettingen U., 1956. Asst. prof. Perkins Obs., Ohio State U., Columbus, 1956-58; research assoc. Mt. Wilson Obs., Palomar Obs., Pasadena, Calif., 1968-69; prof. astronomy David Dunlap Obs., U. Toronto, Ont., Can., 1958-77; dir. Dominion Astrophys. Obs., Victoria, B.C., 1977-86; prin. rsch. officer NRC Can., 1977—. Fellow Royal Soc. London, Royal Soc. Can.; mem. Am., Royal Astron. Soc. (assoc.), Canadian Astronomy Soc. (sr. v.p. 1988-90, pres. 1990-92). Home: 418 Lands End Rd, Sidney, BC Canada V8L 5L9

VAN DEN BOOM, WAYNE JEROME, industrial engineer; b. Bay City, Mich., Nov. 20, 1953; s. Raymond Francis and Iva Jean (Coupie) Van Den Boom; m. Esperanza Hope Hernandez, Nov. 21, 1981; children: Sean David, Kristine Ashley. B in Indsl. Engring., GM Inst., 1976; MA, Central Mich. U., 1980. Supr. prodn. Chevrolet Co., Bay City, Mich., 1976-78, indsl. engr., 1978-80; engr. material handling Chevrolet Co., Adrian, Mich., 1980-83; sr. project engr. packaging CPC Hdqrs., Warren, Mich., 1983-89, sr. project engr. vehicle program support, 1989-91; mgr. Outside Supplier and Allied Packaging Group, Warren, Mich., 1991-92, NAO Container Engring./Gen Assembly Group, Warren, 1992—. Avocations: gardening, golf, camping, photography. Home: 46347 Franks Ln Shelby Township MI 48315-5309 Office: NAO Containerization Gen Motors Corp MC 1604-14 31 Judson St Pontiac MI 48342-2230

VANDEN BOUT, PAUL ADRIAN, astronomer, physicist, educator; b. Grand Rapids, Mich., June 16, 1939; s. Adrian and Cornelia (Peterson) Vanden B.; m. Rachel Ann Eggebeen, Sept. 1, 1961; children—Thomas Adrian, David Anton. A.B., Calvin Coll., 1961; Ph.D., U. Calif.-Berkeley, 1966. Postdoctoral fellow U. Calif., Berkeley, 1966-67; postdoctoral fellow Columbia U., N.Y.C., 1967-68; instr. Columbia U., 1968-69, asst. prof., 1969-70; asst. prof. U. Tex., Austin, 1970-74; assoc. prof. U. Tex., 1974-79, prof., 1979-84; dir. Nat. Radio Astronomy Obs., Charlottesville, Va., 1985—; cons. NSF, NASA. Fellow Fulbright Found., Heidelberg, Fed. Republic Germany, 1961-62, Leiden, Netherlands, 1977. Fellow AAAS, Am. Phys. Soc.; mem. Am. Astron. Soc., Internat. Astron. Union, Internat. Radio Sci. Union. Office: Nat Radio Astronomy Obs 520 Edgemont Rd Charlottesville VA 22903

VANDENBURGH, CHRIS ALLAN, electrical engineer; b. Gary, Ind., Oct. 22, 1955; s. George William and Phyllis Jean VanDenburgh. BSEE, Rose Hulnan Inst. Tech., 1978. Regional insp. region III, team leader spl. inspection br. office U.S. Nuclear Regulatory Commn., Rockville, Md., 1987-89, sect. chief vendor inspection br. office, 1989-90, dep. dir. office of enforcement, 1990-93, chief reactor projects br. region V, 1993—. Roman Catholic. Home: 8 Rainbow Circle Danville CA 94526 Office: Nuclear Regulatory Commn Region V Office 1450 Maria Ln Walnut Creek CA 94596

VAN DEN HERIK, HENDRIK JACOB, computer science and artificial intelligence educator; b. Rotterdam, The Netherlands, Oct. 8, 1947; s. Arie and Jantje (Schreur) Van Den H.; m. Maria Aletta Raaphorst, Sept. 24, 1971; children: Seada Nada, Larissa Jasmijn, Kirsten Jelena. MSc in Applied Math. cum laude, Vrije U., Amsterdam, 1974; PhD in Tech. Scis., Delft U. of Tech., 1983. Asst. prof. Vrije U., Amsterdam, 1974-75; asst. prof. Delft (The Netherlands) Univ. Tech., 1975-84, assoc. prof., 1985-87; prof. U. Limburg, Maastricht, The Netherlands, 1987—, Leiden (The Netherlands) U., 1988—; vis. prof. McGill U., Montreal, Can., 1984; cons. Westmount, Delft, 1985-88; bd. dirs. Riks Maastricht, 1990—; sci. dir. SKBS, 1991—. Co-author 7 books; editor 5 books; contbr. numerous articles to profl. jours.

Mem. Dutch Assn. for Artificial Intelligence (founder, pres. 1990—), JURIX (founder, pres. 1988—), Dutch Assn. Computer Chess (founder, bd. dirs. 1980-82), Internat. Computer Chess Assn. (editor-in-chief 1983—), Am. Assn. for Artificial Intelligence, IEEE, Assn. for Computing Machinery. Avocations: computer chess, games, jogging, music, theater. Home: Pasteurlaan 4, Pijnacker 2641 ZE, The Netherlands Office: U Limburg, Kapoenstraat 2, Maastricht 6211 KW, The Netherlands

VAN DEN MUYZENBERG, LAURENS, engineer; b. Wageningen, Netherlands, June 28, 1933; s. Erwin Woldemar Botho and Nelly Wilhelmina Maria (Bogtman) v.d.M.; children: Louise, Jergen. MS, U. Delft, 1957. Rsch. asst. H.B. Maynard Inc., U.S., 1957-58, pres., 1977-87; indsl. engr. Kennecott Copper Corp., Chile, 1959-61; cons. Maynard Sweden, 1962-69, tech. dir., 1970-72, mng. dir., 1975-76; mng. dir. Maynard France, 1973-74, H.B. Maynard Ltd., United Kingdom, 1987-91; founder Muyzenberg Mgmt. Cons., Ltd., London, 1991—; cons. Thomson CSF-Signal, Holland, 1992, Hunting Engring., Ltd., Eng., 1990, Astra, Sweden, 1992, Calorgas, Eng., 1991, TNO, Holland, 1992. Author: Project Management, (indsl. engring. handbook) 1990; contbr. articles to profl. jours. Home: 1 Queens Ter, Windsor SL4 2AR, England Office: Muyzenberg Mgmt Cons Ltd, 11 Charles St, London W1X 7HB, England

VANDERBEEK, FREDERICK HALLET, JR., chemical engineer; b. Hackensack, N.J., Jan. 11, 1951; s. Frederick Hallet and Rosamond Marie (LeBlanc) V. B of Engring., Stevens Inst. Tech., 1973, M of Engring., 1983. Engr. I Combustion Engring., Inc., Windsor, Conn., 1973-75; design engr. Universal Oil Products, Inc., Darien, Conn., 1976-78; chem. engr. Gilbert Commonwealth, Inc., Reading, Pa., 1978-79; design engr. Wilputte Corp., 1980-81; engr. II Pub. Svc. Electric and Gas Co., Newark, N.J., 1984-89; instrumentation and controls engr. United Engrs. and Constructors, Phila., 1989-91; applications engr. Fischer & Porter Co., Warminster, Pa., 1991-92; control systems engr. Honeywell, Inc., Ft. Washington, Pa., 1992—. Mem. AICE, Inst. Soc. Am. Democrat. Roman Catholic. Home: 365 Newtown Rd #C25 Warminster PA 18974

VANDERFORD, FRANK JOSIRE, physicist, computer scientist consultant; b. Moose Lake, Minn., Oct. 19, 1921; s. William and Mary (Flaa) V.; m. Eleanor Marie Gibis, Feb. 8, 1945; children: Constance, Gail, Deborah. Cert. in Electronics, U. Minn., 1952. teacher's credential in elec. engring. and electronics. Electrical foreman Great No. Railroads, 1946-55; engr. Remington Rand Univac Computer Div., St. Paul, 1955-59, asst. field supr., co-designer transistor computer, 1956-59; field supr. Collins Radio Co., Newport Beach, Calif., 1959-69; engr. Control Data Corp., Santa Ana, Calif., 1969-71; staff engr. Hughes Aircraft Co., Fullerton, Calif., 1971—; physicist Rsch. and Devel. Dept., Newport Beach. Served with USN, 1940-46. Recipient Philippine Def. medal, 1942. Mem. Nat. Rep. Senatorial Com., U.S. Senatorial Club, Rep. Presdl. Task Force (charter), World War II Submarine Vets., Nat. Rifle Assn. Achievements include development of automatic computerized system for programming and testing "Proms" and "Pals" (Memory and Logic I C Devices); projects include: Goddard Satellite Tracking System, Apollo space tracking and timing system, numerous communication systems. Avocations: mini horse rancher, music, sports car restoration.

VAN DE RIET, JAMES LEE, environmental engineer; b. Norfolk, Va., Sept. 20, 1942; s. Peter Herbert and Ima Lee (Baker) Van de R.; m. Ruth Kleppe, Aug. 8, 1964; children: Robert Glen, Deborah Ruth, Peter James. BCE, Va. Poly. Inst., 1965, MS, 1966. Field engr. Buck Seifert & Jost, Norfolk, 1965-66, resident engr., 1967-70; office mgr. Buck Seifert & Jost, Virginia Beach, Va., 1970-72; sr. v.p. and dir. Buck Seifert & Jost, Inc., Virginia Beach, 1973—, bd. dirs.; appointed to adv. com. Va. Sewage Regulations. Mem. ASCE, NSPE, Am. Water Works Assn., Water Environ. Fedn. (bd. dirs. 1989-92, Bedel, 1989, Enslow-Hedgepeth 1992), Am. Acad. Environ. Engrs. Mem. Assembly of God Ch. Home: 1608 Arnold Cir Virginia Beach VA 23454-1607 Office: Buck Seifert & Jost Inc 760 Lynnhaven Pkwy Ste 140 Virginia Beach VA 23454

VAN DER KROEF, JUSTUS MARIA, political science educator; b. Djakarta, Indonesia, Oct. 30, 1925; came to U.S., 1942, naturalized, 1952; s. Hendrikus Leonardus and Maria Wilhelmina (van Lokven) van der K.; m. Orell Joan Ellison, Mar. 25, 1955 (dec.); children: Adrian Hendrick, Sri Orell. B.A., Millsaps Coll., 1944; M.A., U. N.C., 1947; Ph.D., Columbia U., 1953. Asst. prof. fgn. studies Mich. State U., 1948-55; Charles Dana prof., coord. dept. polit. sci. and sociology U. Bridgeport, Conn., 1956-92, prof. emeritus, 1992—; vis. prof. Nanyang U., Singapore, U. Philippines, Quezon City, Vidyodaya U., Sri Lanka Colombo; dir. Am.-Asian Ednl. Exchange, 1969—; chmn. editorial bd. Communications Research Services, Inc., Greenwich, Conn., 1971-80; mem. internat. adv. bd. Union Trust Bank, Stamford, Conn., 1974-88, adv. bd., 1988—; mem. nat. acad. adv. council Charles Edison Meml. Youth Fund; bd. dirs. WUBC-TV, Bridgeport, Conn., 1978-80. Author: Indonesia in the Modern World, 2 vols., 1954-56, Indonesian Social Evolution. Some Psychological Considerations, 1958, The Communist Party of Indonesia: Its History, Program and Tactics, 1965, Communism in Malaysia and Singapore, 1967, Indonesia Since Sukarno, 1971, The Lives of SEATO, 1976, Communism in Southeast Asia, 1980, Kampuchea: The Endless Tug of War, 1982, Aquino's Philippines. The Deepening Security Crisis, 1988, Territorial Claims in the South China Sea, 1992; mem. editorial bd. World Affairs, 1975—, Jour. Asian Affairs, 1975—, Asian Affairs, 1980—, Asian Profile, 1983—, Jour. of Govt. and Adminstrn., 1985—, Jour. of Econ. and Internat. Relations, 1987—, Asian Affairs Jour. (Karachi), 1992—; mng. editor: Asian Thought and Society, 1986—; book rev. editor: Asian Thought and Soc, 1976-85. Mem. City Charter Revision Com. City of Bridgeport, 1983-86, 90-92. Served with Royal Netherlands Marine Corps, 1944-45. Sr. fellow Research Inst. Communist Affairs, Columbia U., 1965-66, Rockefeller Found.; fellow U. Queensland, Brisbane, Australia, 1968-69; research fellow Inst. Strategic Studies, Islamabad, Pakistan, 1982—; research fellow Mellon Research Found., 1983, 90; research fellow Internat. Ctr. Asian Studies, Hong Kong, 1983—. Mem. Univ. Profs. Acad. Order (nat. pres. 1970-71), Pi Gamma Mu, Phi Alpha Theta, Lambda Chi Alpha, Alpha Sigma Lambda Phi Sigma Iota. Home: 165 Linden Ave Bridgeport CT 06604-5730

VANDERLINDE, WILLIAM EDWARD, materials engineer; b. Syracuse, N.Y., Oct. 12, 1959; s. Raymond Edward and Ruth (Hansen) V. BS in Physics, U. Va., 1981; PhD in Materials Engring., Cornell U., 1987. Sr. engr. U.S. Dept. Def., Columbia, Md., 1987—; vis. prof. of physics U. Md., Balt., 1989-90. Contbr. articles to Jour. Vacuum Sci. Tech. B., Materials Rsch. Soc. Symposium Proceedings, Loki Sci. Mag. IBM grad. fellow, 1983. Mem. Am. Phys. Soc., Am. Vacuum Soc. Office: Microelectronics Rsch Lab 9231 Rumsey Rd Columbia MD 21045-1924

VAN DER MEER, JAN, hematologist; b. Leeuwarden, Friesland, The Netherlands, Sept. 19, 1950; s. Jacob and Margaretha Helena Maria (Wolters) van der M.; m. Pyteke Goslinga, July 6, 1972; children: Jacob Sebastiaan, Paul Jan, Robert Louis. MD, U. Groningen, The Netherlands, 1972. Head of coagulation lab. U. Hosp. Groningen, The Netherlands, 1976-78, head div. of haemostasis thrombosis and rheology, 1983—, head haemophilia ctr. Groningen, 1983—, head of lab. for haemostasis thrombosis and rheology, 1983—, head sect. thrombosis and thrombolysis Thorax Ctr., 1989—, gen. dir. ctr. for clin. drug rsch., 1990—; bd. dirs. Cent. Haematology Labs. U. Hosp. Groningen, The Netherlands; coord. Dutch Rsch. Group for Coronary Bypass Surgery, Interuniversity Cardiologic Inst. of The Netherlands, Utrecht, 1985—; mem. Dutch adv. bd. for Treatment of Haemophilia, The Netherlands, 1986-87; mem. cardiovascular rsch. group Thorax Ctr., U. Hosp. Groningen, 1988—; adv. bd. angiology, 1988—; mem. scientific adv. bd. Fedn. of Dutch Thrombosis Svcs., The Hague, 1990—; scientific adv. bd. Dutch Thrombosis Found., 1990—. Contbr. numerous articles to profl. jours. Mem. Dutch Soc. on Internal Medicine, Dutch Soc. on Haematology, Dutch Soc. on Haemophilia Treatment, Benelux Soc. on Microcirculation (bd. dirs.), Dutch Soc. on Thrombosis and Haemostasis, Internat. Soc. on Thrombosis and Haemostasis. Avocation: painting. Office: Univ Hosp, Oostersingel 59, 9713 EZ Groningen The Netherlands

VAN DER MEER, SIMON, accelerator physicist; b. The Hague, The Netherlands, Nov. 24, 1925; s. Pieter and Jetske (Groeneveld) van der M.; m. Catharina M. Koopman, Apr. 26, 1966; children: Esther, Mathijs. Engring. degree in physics, Poly. U., Delft, The Netherlands, 1952; Dr. (hon.), U. Geneva, 1983, U. Amsterdam, The Netherlands, 1984, U. Genoa, Italy, 1985. Research engr. Philips Physics Lab., Eihdhoven, The Netherlands, 1952-55; sr. engr. CERN European Orgn. Nuclear Research, Geneva, 1956-90; ret., 1990. Co-recipient Nobel prize for physics, 1984. Mem. Royal Netherlands Acad. Scis. (corr.), (fgn. hon.) AAAS.

VAN DER MEIJ, GOVERT PIETER, physics educator; b. Haarlem, The Netherlands, July 1, 1951; s. Govert and Phyllis (Hart) Van der M.; m. Florina Johanna Burger, Mar. 15, 1979; children: Maarten Arthur, Wouter Gesinus Stanley. DSc in Math. and Sci., U. Leiden, 1984. Lectr. physics Inst. Petroleum and Gastech., Den Helder, The Netherlands, 1985—; former guest scientist Netherlands Energy Rsch. Foun. Avocations: badminton, sailing. Home: Doorzwin 1128, Den Helder 1788 KB, The Netherlands Office: Inst Petroleum and Gas Tech, Molenplein 1, Den Helder 1781 BZ, The Netherlands

VAN DER MEULEN, JOSEPH PIERRE, neurologist; b. Boston, Aug. 22, 1929; s. Edward Lawrence and Sarah Jane (Robertson) VanDer M.; m. Ann Irene Yadeno, June 18, 1960; children—Elisabeth, Suzanne, Janet. A.B., Boston Coll., 1950; M.D., Boston U., 1954. Diplomate: Am. Bd. Psychiatry and Neurology. Intern Cornell Med. div. Bellevue Hosp., N.Y.C., 1954-55; resident Cornell Med. div. Bellevue Hosp., 1955-56; resident Harvard U., Boston City Hosp., 1958-60, instr., fellow, 1962-66; assoc. Case Western Res. U., Cleve., 1966-67; asst. prof. Case Western Res. U., 1967-69, assoc. prof. neurology and biomed. engring., 1969-71; prof. neurology U. So. Calif., L.A., 1971—; also dir. dept. neurology Los Angeles County/U. So. Calif. Med. Center; chmn. dept. U. So. Calif., 1971-78, v.p. for health affairs, 1977—, dean Sch. Medicine, 1985-86, dir. Allied Health Scis., 1991—; vis. prof. Autonomous U. Guadalajara, Mex., 1974; pres. Norris Cancer Hosp. and Research Inst., 1983—. Contbr. articles to profl. jours. Mem. med. adv. bd. Calif. chpt. Myasthenia Gravis Found., 1971-75, chmn., 1974-75, 77-78; med. adv. bd. Amyotrophic Lateral Sclerosis Found., Calif., 1973-75, chmn., 1974-75; mem. Com. to Combat Huntington's Disease, 1973—; bd. dirs. Calif. Hosp. Med. Ctr., Good Hope Med. Found., Doheny Eye Hosp., House Ear Inst., L.A. Hosp. Good Samaritan, Children's Hosp. of L.A., Barlow Respiratory Hosp., USC U. Hosp., chmn., 1991—; bd. govs. Thomas Aquinas Coll.; bd. dirs. Assn. Acad. Health Ctrs., chmn., 1991-92; pres. Scott Newman Ctr., 1987-89; pres., bd. dirs. Kenneth Norris Cancer Hosp & Rsch. Inst. Served to lt. M.C. USNR, 1956-58. Nobel Inst. fellow Karolinska Inst., Stockholm, 1960-62; NIH grantee, 1968-71. Mem. AMA, Am. Neurol. Assn., Am. Acad. Neurology, L.A. Soc. Neurology and Psychiatry (pres. 1977-78), L.A. Med. Assn., Mass. Med. Soc., Ohio Med. Soc., Calif. Med. Soc., L.A. Acad. Medicine, Alpha Omega Alpha (councillor 1992—), Phi Kappa Phi. Home: 39 Club View Ln Palos Verdes Peninsula CA 90274-4208 Office: U So Calif 1985 Zonal Ave Los Angeles CA 90033-1058

VANDER MOLEN, JACK JACOBUS, engineering executive, consultant; b. Assen, Drenthe, Netherlands, May 28, 1916; came to U.S., 1947, naturalized 1952; s. Evert Moll and Victorina Sweelssen; m. Ina Mary Auerbach, 1946. ME, M.T.S., Haarlem, 1940; postgrad. computer programming, Ariz. Tech., 1982. Draftsman, designer Fokker Aircraft, Amsterdam, Netherlands, 1939-40; asst. plant mgr. Bruynzeel's Deuren Fabriek, Zaandam, Netherlands, 1941-44; civilian mgr. Allied Hdqrs. Rest Ctrs., Maastricht, Amsterdam, Netherlands, 1944-45; cen. staff tech. efficiency and orgn. Philips Radio, Eindhoven, Holland, 1945-47; indsl. engr. N.Am. Philips, Dobbs Ferry, N.Y., 1947-48; supr. methods and standards Otis Elevator Co., Yonkers, N.Y., 1948-51; staff engr., material handling and distbn., cons. Drake, Startzman, Sheahan & Barclay, N.Y.C., 1951-55; mgr. material handling engring. Crane Co., Chgo., 1955-60; assoc., cons. A.T. Kearney & Co., Inc., Chgo., 1960-67; pres., cons. J.J. Vander Molen & Co., Internat., Oak Park, Ill. and Sun City, Ariz., 1967—. Conceptual developer of plants, warehouses and terminals, computerized conversion of inventory into space requirements for food chains. With Dutch resistance, U.S. and Can. Armed Forces, 1940-45. Mem. ASME, Internat. Materials Handling Soc. (nat. dir., past pres. Chgo. chpt). Avocations: swimming, biking, music, reading, walking. Home: 10629 W Willowcreek Cir Sun City AZ 85373-1345 Office: PO Box 1656 Sun City AZ 85372-1656

VAN DER PLOEG, LEONARDUS HARKE THERESIA, molecular biologist; b. Breda, The Netherlands, Sept. 17, 1954; came to U.S., 1984; s. Antoine Gerard and Corry J.M. Van der P.; 1 child, Sasja Tse. MS, U. Amsterdam, The Netherlands, 1976, PhD, 1984. Rsch. scientist Netherlands Cancer Inst., Amsterdam, 1981-84; asst. prof. Columbia U., N.Y.C., 1984-87, assoc. prof., 1987—; dir. dept. genetics and molecular biology Merck Rsch. Labs., Rahway, N.J., 1991—; study sect. NIH, Washington, 1990—; mem. editorial bd. Molecular Cell Biology, 1991, Molecular Biochem. Parasitology, 1990. Contbr. articles to profl. jours. Recipient Searle scholarship, 1985, Nabisco award, 1987, Hirschl award, 1985, 1987 award Dutch Soc. Biochemistry, 1987. Mem. ASM. Achievements include development protozoal gentics and an improved understanding of host-parasite interaction. Office: Merck Rsch Lab PO Box 2000 Rahway NJ 07065

VANDER VELDE, WALLACE EARL, aeronautical and astronautical educator; b. Jamestown, Mich., June 4, 1929; s. Peter Nelson and Janet (Keizer) Vander V.; m. Winifred Helen Bunai, Aug. 29, 1954; children—Susan Jane, Peter Russell. B.S. in Aero Engring, Purdue U., 1951; Sc.D., Mass. Inst. Tech., 1956. Dir. applications engring. GPS Instrument Co., Inc., Newton, Mass., 1956-57; mem. faculty Mass. Inst. Tech., 1957—, prof. aero. and astronautics, 1965—; Cons. to industry, 1958—. Author: Flight Vehicle Control Systems, Part VII of Space Navigation, Guidance and Control, 1966, (with Arthur Gelb) Multiple-Input Describing Functions, 1968; also papers. Served to 1st lt. USAF, 1951-53. Recipient Edn. award Am. Automatic Control Coun., 1988. Fellow AIAA; mem. IEEE. Home: 50 High St Winchester MA 01890-3314 Office: MIT Cambridge MA 02139

VANDERVERT, LARRY RAYMOND, psychologist, educator, writer; b. Spokane, Wash., Dec. 28, 1938; s. Curtis Clark and Edythe Marie (Peachey) V.; m. Betty Jean McKinney, Jan. 21, 1967; children: Kimberly Jean Balocco, Bryce Raymond. BA in Psychology, Ea. Wash. U., 1966, MSc in Psychology, 1967; PhD, Wash. State U., 1977. Coll. faculty Spokane (Wash.) Falls C.C., 1969—. Author: Introductory Psychology, 1992; editor: (nat. newsletter) Network, 1987-91; contbr. articles to Neurological Positivism. Author: Introductory Psychology, 1992; contbr. articles to Neurological Positivism. Fellow APA (editor newsletter Network 1987-91, named tchr. of yr. in two yr. coll. 1989); mem. Soc. for Chaos Theory in Psychology (pres. 1991, co-founder), Am. Nonlinear Systems (founder, pres. 1993). Achievements include developed brain based epistemology termed Neurological Positivism. Home: W 711 Waverly Pl Spokane WA 99205-3271

VANDERVOORT, KURT GEORGE, physicist; b. L.A., May 11, 1959; s. Frances Edward and Jewell Gertrude (Failer) V.; m. Edith maria Biegler, Sept. 6, 1986. BS in Physics, Humboldt State U., 1986; MS in Physics, U. Ill., Chgo., 1988, PhD in Physics, 1991. Fgn. fisheries biologist Nat. Marine Fisheries Svc., Seattle, 1984; fisheries biologist Oreg. Fish and Wildlife, Newport, Oreg., 1985; grad. fellow Argonne (Ill.) Nat. Lab., 1988-91, potdoctoral fellow, 1991—. Contbr. articles to profl. jours. Mem. Am. Phys. Soc., Sigma Xi. Achievements include rsch. of upper critical field measurements of high temp. superconductors; invention of low field superconductor quantum interference device, magnetometer system for characterization of high-temp superconductors. Home: 225A Washington # 2W Oak Park IL 60302

VANDERWALKER, DIANE MARY, materials scientist; b. Springfield, Mass., Nov. 1, 1955. BS, Boston Coll., 1977; PhD, MIT, 1981. NATO fellow, 1981-82; asst. prof. SUNY, Stony Brook, 1983-85; materials rsch. engr. Army Rsch. Lab. (formerly U.S. Army Materials Tech. Lab.), Watertown, Mass., 1986—; cons. IBM, Yorktown Heights, N.Y. Contbr. articles to profl. pubis. Mem. N.Y. Acad. Scis. Roman Catholic.

VANDER WIEL, KENNETH CARLTON, computer services company executive; b. Sheldon, Iowa, July 6, 1933; s. Sylvan Vander Wiel and Irene F. (Basie) Taylor; m. Loretta Marie Smith, Aug. 28, 1969; children: Gretchen G., Alison June, Joseph W., Carol Ann, Andrea, Beth L., David. BA, Bowling Green U., 1955. With mgmt. devel. Dayton (Ohio) Power & Light, 1955-68; cons. G.W. Young & Assocs., Dayton, 1968-69; pres. Datamac Corp., Dayton, 1970-84, Carlton Leasing Co., 1975—, The Carlton Systems Group, Dayton, 1984-89, Carlton Computer Systems, Dayton, 1989—; ceo Carlton Computer Systems, Inc., Tampa, Fla., 1993—. Chmn. Montgomery County Data Processing Task Force, Dayton, 1982. With USN, 1955-57. Recipient Commendations, Montgomery County Commn., Dayton, 1982, Dayton Execs. Club, 1986. Avocations: sailing, fishing, travel. Office: Carlton Computer Systems 1887 Southtown Blvd Dayton OH 45439-1931

VANDE STREEK, PENNY ROBILLARD, nuclear medicine physician, researcher, educator; b. Highland Park, Mich., May 23, 1953; d. Richard Charles Robillard and E. Louise (Gee) Armstrong; m. Harley Eugene Vande Streek, Oct. 22, 1977; children: Gregory, Elizabeth. BA, Spring Arbor Coll., 1977; DO, Univ. Osteo. Med./Health Scis., Des Moines, 1983. Diplomate Am. Bd. Internal Medicine, Am. Bd. Nuclear Medicine. Rsch. asst. Parke-Davis & Co., Detroit, 1977-79; anatomy and microbiology lab. asst. U. Osteo. Medicine, Des Moines, 1980-81; intern in family practice Scott AFB (Ill.) Med. Ctr., 1983-84; chief internal medicine Williams Hosp., Williams AFB, Ariz., 1987-88; resident in internal medicine Wilford Hall USAF Med. Ctr., San Antonio, 1984-87, nuclear medicine fellow, 1988-90, staff nuclear medicine, 1990-91, chief positron emission tomography imaging svc., 1991—; asst. prof. radiology U. Tex. Health Sci. Ctr., San Antonio, 1991—; reaching fellow Uniformed Svcs. U. Health Scis., Bethesda, Md., 1985-87; instr. Incarnate Word Coll., 1992, mem. adv. coun. nuclear medicine tech. program, 1991—; guest lectr. various hosps. Contbr. chpt. (with R.F. Carretta and F.L. Weiland): Nuclear Medicine Approaches to Musculoskeletal Disease: Current Status, 1993. Participant City Disaster Preparedness, Des Moines, 1980, Phoenix, 1988; provider ACLS, ARC, 1983, 84, 87, 88, 91, provider advanced trauma life support, 1989. Maj. USAF, 1979—, Operation Desert Storm. Decorated USAF Achievement medal; Mallinckrodt grantee, Cin., 1991; recipient Young Leader award Spring Arbor (Mich.) Coll., 1992. Mem. Soc. Nuclear Medicine, Am. Coll. Nuclear Physicians, Soc. Air Force Physicians, Assn. Mil. Osteo. Physicians and Surgeons, Am. Osteo. Assn., Sigma Sigma Phi. Avocations: swimming, travel, hiking, reading, golf. Office: PO Box 1328 Rocklin CA 95677

VAN DEVENDER, J. PACE, physical scientist; b. Jackson, Miss., Sept. 12, 1947; m. Nancy Jane Manning, 1971; 3 children. BA in Physics, Vanderbilt U., 1969; MA in Physics, Dartmouth Coll., 1971; PhD in Physics, U. London, 1974. Physicist diagnostics devel. Lawrence Livermore Lab., 1969; mem. tech. staff pulsed power rsch. and devel. Sandia Nat. Labs., 1974-78, divsn. supr. pulsed power rsch. divsn., 1978-82, dept. mgr. fusion rsch., 1982-84; dir. pulsed power scis. Sandia Nat. Labs., Albuquerque, 1984-93, dir. corp. comm., 1993, dir. Nat. Indsl. Alliances Ctr., 1993—; mem. rev. bd. Compact Ignition Tokomak, Sandia Labs. rep., 1988—. Mem. editorial bd. Laser and Particle Beams, 1987-90. Mem. bd. trust Vanderbilt U., 1969-73. With U.S. Army, 1969-71. Recipient Ernest Orlando Lawrence Meml. award U.S. Dept. Energy, 1991; named one of 100 Most Promising Scientists Under 40, Sci. Digest, 1984; Marshal scholar U. London, 1971-74. Fellow Am. Phys. Soc.; mem. NAS (mem. bd. naval studies 1990-93), Phi Beta Kappa, Omicron Delta Kappa, Sigma Xi. Home: 7604 Lamplighter Ln NE Albuquerque NM 87109 Office: Sandia Nat Lab PO Box 969 Livermore CA 94550*

VANDEVENDER, ROBERT LEE, II, nuclear engineering consultant; b. Muncie, Ind., Nov. 16, 1958; s. Robert Lee and Evelyn June (Matthews) V.; m. Laura Jo Longfellow, June 11, 1977 (div. July 1990); children: Holly Suzanne, Robert Lee III, Bryan Matthew; m. Deborah Ann Keiffer, Sept. 26, 1992. Grad., Naval Nuclear Power Sch., Orlando, Fla., Nuclear Power Tng. Unit, West Milton, N.Y., 1979, Naval Engring. Lab. Technician Sch., West Milton, N.Y., 1979; AS, Mohegan Community Coll., Norwich, Conn., 1983; grad., GE Thermodynamics, Heat Transfer and Fluid Flow Sch., 1986; cert. achievement, Joliet Jr. Coll., 1986; grad., GE Sr. Reactor Operator Sch., 1986. Lead engring. lab. technician, staff instr. Knolls Atomic Power Lab. U.S. Dept. Def., West Milton, 1979-81; sr. reactor operator, simulator instr. GE Nuclear Tng. Svcs., Morris, Ill., 1985-87; sr. lead engr., nuclear engring. cons. ABB Impell Corp., Melville, N.Y., 1987-91, Chgo., 1991-93; nuclear engring. cons. Megan Corp., Allentown, Pa., 1993—; adminstrv. coord., cons. Davis Besse Nuclear Power Sta., Toledo Edison, 1987-88; adminstrv. cons., supr. ops. support Niagara Mohawk Power Corp., Oswego, N.Y., 1988-90; lead engring. cons. design baseline compolitation Phila. Electric Co., King of Prussia, 1991; engring. cons. final safety analysis Zion (Ill.) Nuclear Sta., Commonwealth Edison Co., 1991, leader engr. dual unit outage procedures, 1991-93, engring. cons. fire protection surveillances, 1993—. Author: The Vandevender, Wilson, McAshlan, Silvers and Kimmel Families, 1990, (contbg. author with Robert Friedberg) Paper Money of the United States; contbr. articles to various publs. Vol. examiner FCC, Gettysburg, 1985—; merit badge counselor Boy Scouts Am., Muncie, 1983. With USN, 1977-85, Ind. Guard Res., 1975-76, USNR, 1976-77, 85-87. Recipient radiosport diploma for operating achievement Internat. Amateur Radio Union, 1984, 85. Mem. Am. Legion, Am. Nuclear Soc. (life), Am. Radio Relay League (life, asst. tech. coord. 1985-87), Internat. Platform Assn., Am. Numismatic Assn. (life), VFW (life), Masons, Odd Fellows (grand ruler Ind. 1975-76, pres. region X 1975-76, rep. to UN Pilgrimage for Youth 1976), NRA (life), Soc. Paper Money Collectors (life). Democrat. Avocations: genealogy, amateur radio, numismatics. Home: PO Box 2032 Muncie IN 47307 Office: Megan Corp PO Box 201 Gurnee IL 60031-0201

VAN DE WALLE, CHRIS GILBERT, physicist; b. Gent, Belgium, May 10, 1959; came to U.S., 1982; s. Gaston Eduard and Louise Armanda (De Moor) Van de W. Engr., Rijks U. Gent, 1982; MS in Elec. Engring., Stanford U., 1983, PhD, 1986. Vis. scientist IBM Watson Rsch. Ctr., Yorktown Heights, N.Y., 1986-88; sr. mem. rsch. staff Philips Labs., Briarcliff Manor, N.Y., 1988-91; mem. rsch. staff Xerox Palo Alto (Calif.) Rsch. Ctr., 1991—; adj. prof. Columbia U., N.Y.C., 1991; chair Trieste (Italy) Semicondr. Symposium, 1992. Editor: Wide-Band-Gap Semiconductors, 1993; contbr. chpts. to books, numerous articles to profl. jours. Recipient Isabella Van Portugal prize U. Gent, 1982, prize City of Gent, 1982; Belgian-Am. Ednl. Found. fellow, 1982-83. Mem. Am. Phys. Soc., Materials Rsch. Soc. (Grad. Student award 1985). Office: Xerox PARC 3333 Coyote Hill Rd Palo Alto CA 94304

VAN DE WATER, THOMAS ROGER, neuroscientist, educator; b. Oceanside, N.Y., Dec. 6, 1939; s. Lynn and Lenora (Winterson) Van De W.; m. Jeanette Adele Vilece, July 11, 1964; children: Ann Marie, Thomas Scott, Christopher Lynlee, Elizabeth Adele. AAS in Forestry, Paul Smith Coll., 1959; BS in Biology, Western Carolina U., 1961; MA in Biology, Hofstra U., 1965; PhD in Biology, NYU, 1976. Rsch. assoc. Med. Sch. Yale U., 1964-65; rsch. scientist Med. Sch. NYU, 1967-68; dir. devel. otobiology Albert Einstein Coll. Medicine, Bronx, N.Y., 1981—, prof. otolaryngology, 1987—, prof. neurosci., 1991—, dir. rsch., 1990—; lectr., author Nobel Symposium, 1985; acad. dir. N.Y. Acad. Medicine Otolaryngology basic sci. course, 1988—. Editor: Biology & Change in Otobiology, 1986, Genetics of Hearing Impairment, 1990, Handbook of Auditory Research, Clinical Aspects; contbr. over 50 sci. publs. Eucharistic min. Holy Redeemer Ch., Freeport, N.Y. With U.S. Army Med. Corps, 1962. Mem. Assn. for Rsch. in Otolaryngology (sec./treas. 1990-93), N.Y. Acad. Sci., Am. Otological Soc. (assoc.), Soc. Neurosci., Oto-Rhinolaryngologica Collegium Amitae Sacrum. Achievements include establishment of organotypic culture of mammalian inner ear; definition of neuron-sensory cell interactions during inner ear development, role of epithelial-mesenchymal tissue interactions in the developing inner ear, role of growth factors during inner ear embryogenesis in both sensory and non-sensory systems, regeneration of mammalian auditory haircells. Home: 262 Pennsylvania Ave Freeport NY 11520-1329 Office: Albert Einstein Coll Medicine 1410 Pelham Pky S Bronx NY 10461-1101

VAN DISHOECT, EDWINE, physicist educator. Prof. U. Leiden. Recipient Maria Goeppert Mayer award Am. Phys. Soc., N.Y., 1993. Office: Univ Leiden, Stationsweg 46 PO Box 9500, N-2300 Ra The Netherlands*

VANDIVER, PAMELA BOWREN, research scientist; b. Santa Monica, Calif., Jan. 12, 1946; d. Roy King and Patricia (Woolard) Evans; m. J. Kim Vandiver, Aug., 1968 (div. 1984); 1 child, Amy. BA in Humanities and Asian Studies, Scripps Coll., 1967; postgrad., U. Calif., Berkeley, 1968; MA in Art, Pacific Luth. U., 1971; MS in Ceramic Sci., MIT, 1983, PhD in Materials Sci. and Near Eastern Archeology, 1985. Instr. in glass and ceramics Mass. Coll. of Art, Boston, 1972; lectr. MIT, Cambridge, 1973-78, rsch. assoc., 1978-85; rsch. phys. scientist Conservation Analytical Lab., Washington, 1985-89; sr. scientist in ceramics C.A.L. Smithsonian Instn., Washington, 1989—; bd. dirs. Rolatape Corp., Spokane, Wash.; guest researcher Nat. Inst. Standards & Tech., Gaithersburg, Md., 1989-91. Co-author: Ceramic Masterpieces, 1986; co-editor: Materials Issues in Art and Archaeology, 1988, 2d edit., 1991, 3d edit., 1992; bd. editors Archeomaterials, 1986-93; contbr. numerous articles to profl. jours. Advisor Lexington (Mass.) Montessori Sch., 1980; camping leader Girl Scouts U.S., Alexandria, Va., 1985-88; sponsor mentorship program Thomas Jefferson High Sch. of Sci. and Tech., Alexandria, 1992. Recipient Disting. Alumna Achievement award Scripps Coll., 1993. Mem. AAAS, Am. Inst. Archeology, Soc. Am. Archeology, Internat. Inst. of Conservation, Soc. for History of Tech., Am. Ceramics Soc. (ancient ceramics com. 1978—), Materials Rsch. Soc. (guest editor bull. 1992), Am. Chem. Soc., Cosmos Club, Sigma Xi. Avocations: sailing, diving, soaring, swimming, photography. Office: Smithsonian Inst Conservation Analytical Lab Washington DC 20560

VAN DUSEN, HAROLD ALAN, JR., electrical engineer; b. Bellingham, Wash., Aug. 9, 1922; s. Harold Alan Van Dusen and Eva Mary Kinnie; m. Joyce Hermance, June 18, 1949; children: Randall Dean, Wendy Gay, Shari Lynn. BSEE, Purdue U., 1949. Registered profl. engr., Wis. Engr. Line Material Co./McGraw-Edison, South Milwaukee, Wis., 1949-61; sr. engr. McGraw-Edison, South Milwaukee, Wis., 1961-87; ret.; cons. Lighting Product Design, South Milwaukee, Wis., 1987—; vol. exec. Internat. Exec. Svc. Corp., Port Said, Egypt, 1990. Contbr. articles to profl. jours. Participant People to People, China, 1984, Roadway Lighting Forum, Europe, 1970. With USAAF, 1942-45. Fellow Illuminating Engring. Soc. Achievements include 17 patents; research on interaction of outdoor lighting equipment with environmental effects such as wind induced vibration, dirt depreciation, optical plastics applications, glassware breakage, etc. Home and Office: 1629 Drexel Blvd South Milwaukee WI 53172

VAN DUSEN, JAMES, cardiologist; b. Kearny, N.J., Feb. 21, 1952; s. James Devere and Eleanor Rose (DiSalvo) Van D.; m. Patricia Ellen Sales, June 23, 1973; children: Matthew, Heidi, Scott. BA in Chemistry, Rutgers U., 1973; DO, Coll. Osteo. Medicine/Surgery, Des Moines, 1976. Intern, then resident in medicine Meml. Gen. Hosp., Union, N.J., 1976-79; fellow in cardiology Overlook Hosp./Columbia U., N.Y.C., 1979-81; cardiologist Cardiology Diagnostic Assocs., New Providence, N.J., 1981-87; dir. internal medicine residency program Meml. Gen. Hosp., Union, 1984-87; founder Heart Health Inst., 1st outpatient cardiac catheterization lab. in no. Fla. Fellow Am. Coll. Osteo. Internists, Am. Coll. Cardiology; mem. Rotary (bd. dirs. 1989-91, Paul Harris fellow). Roman Catholic. Home: 5 Caribe Ct Palm Coast FL 32137-8957 Office: Heart Health Inst 4 Office Park Dr Palm Coast FL 32137-3808

VAN DUSEN, LANI MARIE, psychologist; b. Alexandria, Va., July 23, 1960; d. Arthur Ellsworth and Ann Marie (Brennan) Van D. BS magna cum laude, U. Ga., 1982, MS, 1985, PhD, 1988. Cert. secondary tchr., Ga. Tchr. Henry County Sch. System, McDonough, Ga., 1982-83; rsch. psychologist Metrica Inc., Bryan, Tex., 1988; assoc. prof. psychology U. Ga., Athens, 1988-89, chmn. Conf. for Behavioral Scis., 1987; asst. prof. psychology Utah State U., Logan, 1989-93, assoc. prof. psychology, 1993—; cons. Western Inst. for rsch. and Evaluation, Logan, 1990—; bd. dirs. Human Learrning Clinic, Logan, 1990—; reviewer William C. Brown Pubs., 1990, Dushkin Pub. Group Inc., 1990-91. Contbr. articles to profl. jours. Fellow Menninger Found.; mem. APA, Psychonomic Soc., Am. Ednl. Rsch. Assn., AAUP, ASCD. Republican. Avocations: hiking, tennis, skiing, knitting, swimming. Home: 435 East 1200 North Logan UT 84321 Office: Utah State U Dept Psychology UMN 2810 Logan UT 84322-2810

VAN DYK, MICHAEL ANTHONY, software designer; b. Moscow, Idaho, Sept. 14, 1950; s. Victor and Alice (LaFrance) Van D.; m. Lorna Veraldi, 1984; children: Nathan, Daniel. BS, U. Wash., 1973. Cons. Salomon Bros., Inc., 1980-82, Citicorp/Citibank, N.Y.C., 1982-85, Drexel, Burnham, Lambert, Inc., N.Y.C., 1987-88; ptnr., pres. Mintz, Pappas & Van Dyk, Inc., N.Y.C., 1982-83; adj. prof. Fordham U., N.Y.C., 1984. Designer of software including Trader Workstation, 1988, Trading System, 1985, Toddlerware, 1991. Vice chair N.Y. Assn. Computing Machinery, 1980-84. Home and Office: 20565 NE 6th Ct Miami FL 33179-2415

VAN DYK, ROBERT, health care center executive; b. Dumont, N.J., Jan. 18, 1953; s. Marvin Bernard and Gertrude (Fuhr) Van D.; m. Holly Jean Harper, June 9, 1973; children: Kristina, Reed. BS in Bus. Mgmt. cum laude, Fairleigh Dickinson U., 1975; MHA in Hosp. Adminstrn., George Washington U., 1978. Asst. adminstr. Van Dyk Nursing and Convalescent Homes, Ridgewood, Montclair, N.J., 1978-80; adminstr. The Bucktail Med. Ctr., Renovo, Pa.; asst. adminstr. Divine Providence Hosp., Williamsport, Pa., 1980-83; pres., chief exec. officer Christian Health Care Ctr., Wyckoff, N.J., 1983—; bd. dirs. Bergen-Passaic Hosp. Coun., Westwood, N.J., 1986—. Mem. Nat. Fire Protection Agy., president's adv. coun. Houghton Coll., campaign cabinet United Way Bergen County; past mem. fin. com., chmn. devel. com., mem. exec. com. and bd. Northeastern Bible Coll.; past mem. bd. dirs. exec. com., chmn. fin. com., vice chmn. sch. bd. Hawthorne Christian Acad.; bd. dirs. Columbia Savings Bank. Mem. Am. Health Care Assn. (v.p. 1990—, bd. dirs., exec. com., membership com., pub. rels. com. 1990—, state del. region II 1986—, non-proprietary com. 1986—), Bergen County Coalition Concerned Nursing Home Adminstrs. (bd. dirs. 1985—), Nat. Assn. Pvt. Psychiat. Hosps. (polit. action com. 1986—), N.J. Assn. Health Care Facilities (bd. dirs. 1984—, legis. com. 1983—, chmn. edn. com. 1983—), Am. Coll. Health Care Adminstrs., Am. Hosp. Assn., N.J. Hosp. Assn. (mem. govt. rels. com.), No. N.J. Health Planning Coun. Republican. Avocations: skiing, tennis, golf, building/renovations. Home: 32 Chestnut St Mahwah NJ 07430-3104 Office: Christian Health Care Ctr 301 Sicomac Ave Wyckoff NJ 07481-2194

VAN DYKE, JACOB, civil engineer; b. Utrecht, Holland, Aug. 19, 1933; s. J.L. and J.S. (Van Vreumingen) Van D.; m. Nel Renooy, Nov. 29, 1958; children: Tessa, Jeroen. BS in Civil Engring., U. Utrecht, 1954. Registered civil engr. 1963. Asst. in bus. Hollandsche Beton Group N.V., The Hague, 1956-60; chief exec. Civil Engring. dept. Hillen & Roosen, Amsterdam, 1960-70; mng. dir. Slokker Bouw Group bv, Huizen, 1970-80; bd. dirs. BAM Group N.V., Bunnik, 1980—; chmn. bd. S.A.O.B., 1979-87; chmn. supervisory bd. S.B.R., 1987—; pres. Progresbouw, 1986-89; pres. V.G. Bouw, 1989-92. 1st lt. Netherlands Army, 1954-56. Mem. Netherlands Assn. Rd. Bldg. (pres. 1966-70), Netherlands Assn. Bldg. Industry (bd. supervisory), Edibouw (pres. 1990—), Curatorium Assn. Sci. Edn. Bldg. Industries (bd. supervisory multitop b.v. Zaandam, bd. supervisory sendit b.v. Tilburg), Order of Oranje Hassau (officer). Avocations: sports, skiing, hockey, golf, swimming. Home: Jan Steenlaan 49, 3723 BT Bilthoven The Netherlands Office: BAM Group NV, Runnenburg 11 PO Box 114, 3980CC Bunnik The Netherlands

VANDYNE, BRUCE DEWITT, quality control executive; b. Savannah, N.Y., Apr. 12, 1932; m. Shirley Mae Knox, Oct. 5, 1952; children: Daniel, Kathy, Deborah. Student, N.Y. State U., 1953. Food inspector USDA, Rochester, N.Y., 1953-55; head of quality control lab. Curtice Bros., Mt. Morris, N.Y., 1956-58; quality control mgr. Comstock Foods, Red Creek, N.Y., 1958-65, Seneca Foods Corp., Dundee, N.Y., 1966—. Republican. Methodist. Office: Seneca Foods Corp 74 Seneca St Dundee NY 14837

VANE, JOHN ROBERT, pharmacologist; b. Worcestershire, Eng., Mar. 29, 1927; s. Maurice and Frances Florence V.; m. Elizabeth Daphne Page, Apr. 4, 1948; children: Nicola, Miranda. BSc in Chemistry, U. Birmingham, 1946; MSc in Pharmacology, Oxford U., 1949, D Phil., 1953, DSc, 1970; MD (hon.), U. Cracow, Poland, 1977, Copernicus Acad. Medicine, Cracow; Hon. doctorate, Rene Descartes U., Paris, 1978; DSc (hon.), CUNY, 1980, Aberdeen U., 1983, N.Y. Med. Coll., Birmingham U., U. Surrey, 1984, Camerino U., Italy, 1984, Louvain, 1986, Buenos Aires, 1986; D honoris causa in Medicine and Surgery, U. Florence. Fellow Therapeutic Research Council, Oxford U., 1946-48; researcher worker Sheffield U., 1948-49, Nuffield Inst. Med. Research, Oxford U., 1949-51; Stothert research fellow Royal Soc., 1951-53; instr., then asst. prof. pharmacology Yale U. Med. Sch., 1953-55; mem. faculty Inst. Basic Med. Scis., Royal Coll. Surgeons Eng., 1955-73, prof. exptl. pharmacology, 1966-73; group research and devel. dir. Wellcome Found. Ltd., Beckenham, Kent, 1973-85; dir., chmn. William Harvey Research Inst. St. Bartholomew's Hosp. Med. Coll., London, 1986—; bd. dirs. Vanguard Medica Ltd., Sparta Pharms. Inc., U.S., Biofor Inc., U.S. Co-editor: Adrenergic Mechanisms, 1960; Prostaglandin Synthetase Inhibitors, 1974; Metabolic Functions of the Lung, Vol. 4, 1977; Handbook of Experimental Pharmacology, 1978; Prostacyclin, 1979; Interactions Between Platelets and Vessel Walls, 1981; contbr. numerous articles to profl. jours. Decorated knight bachelor; recipient Baly medal Royal Coll. Physicians; Albert Lasker Basic Med. Rsch. award; Peter Debye prize; Nuffield Gold medal; Ciba Geigy Drew medal Soc. for Endocrinology, 1981; Nobel prize in physiology or medicine, 1982; Galen Medal Worshipful Soc. Apothecaries, 1983; Louis Pasteur Found. prize, Santa Monica, Calif., 1984. Fellow ACP (hon.), Inst. Biology, Royal Soc. (Royal medal 1989), Brit. Pharm. Soc. (hon.), Royal Coll. Pathologists (hon.); mem. NAS (fgn. assoc.), Polish Pharm. Soc. (hon.), Physiol. Soc. (hon.), Royal Acad. Medicine Belgium, Royal Netherlands Acad. Arts and Scis., Polish Acad. Scis. (fgn.), Am. Acad. Arts and Scis. (fgn. hon.), Soc. Drug Research, Alpha Omega Alpha (hon.). Office: William Harvey Rsch Inst, St Bartholomew's Hosp Med Coll, London EC1M 6BQ, England

VANEGAS, HORACIO, neurobiology educator, director; b. Caracas, Venezuela, Sept. 3, 1939; s. Horacio and Carmen (Fischbach) V. B in Medicine and Surgery, Cen. U., Caracas, 1962; PhD, Yale U., 1968. Asst. prof. Instituto Venezolano de Investigaciones Cientificas, Caracas, 1969-77; assoc. prof. Instituto Venezolano Investigaciones Cientificas, Caracas, 1977-81, prof., 1981—; chair dept. biophysics and biochemistry Max-Planck Inst., Munich, 1984-86, dep. dir., 1986-88, dir., 1988—, guest researcher, 1975-76; dir. Latin Am. Ctr. for Biological Scis., 1987—; prof. Ctrl. U. Venezuela, Caracas, 1988—; vis. prof. U. Calif., San Francisco, 1982-83; pres. Quimbiotec, Inc., Venezuela, 1989—; bd. dirs. NRC, Venezuela, 1989—. Editor book; contbr. numerous articles to profl. jours. Dir. Centro Latinoamericano Ciencias Biologicas (UNESCO), Venezuela, 1987-90; v.p. U. Iberoamericana Postgrado, Salamanca, 1989—; chair Venezuelan Com. Fogarty Internat. Fellowship, 1988-90, Pew Latin Am. Fellowship, 1990—. Decorated Order Francisco Miranda (Venezuela), Order Andres Belio (Venezuela), Order Gabriela Mistral (Chile). Mem. Internat. Assn. for Study of Pain, Internat. Brain Rsch. Orgn., Soc. for Neurosci., European Neurosci. Assn., Yale Club of Venezuela. Office: Instituto Venezolano, Investigaciones Cientificas, Caracas 1020A, Venezuela*

VAN ENGELEN, DEBRA LYNN, chemistry educator; b. Burley, Idaho, Dec. 31, 1952; d. W. Dean and Eyvonne (Campbell) Van Engelen; m. John L. Crawford, Dec. 19, 1987; 1 child, Aaron C. Coghlan. BA, Washington U., 1974; PhD, Oreg. State U., 1987. Grad. fellow Oreg. State U., Corvallis, 1982-86; asst. prof. U. N.C., Asheville, 1986-92, assoc. prof. chemistry, 1992—; vis. prof. Emory U., Atlanta, summer 1992. Contbr. articles to profl. jours. NSF rsch. awardee, 1992, 93-96; grantee N.C. Bd. Sci. Tech., 1988-89, Blue Ridge Health Ctr., 1989. Mem. Am. Chem. Soc. (chpt. chair 1993, sec.-treas. 1991), Coun. on Undergrad. Rsch., Sigma Xi, Phi Kappa Phi, Phi Lambda Upsilon. Achievements include development of simultaneous absorbance and fluorescence gas-phase detection system for gas chromatography; research in monitoring pesticides in ground water. Office: Univ of NC Dept Chemistry Asheville NC 28804

VAN GEYT, HENRI LOUIS, architect; b. Brussels, Feb. 2, 1947; s. Jerome Charles and Marie Jose (Van Den Weghe) Van G.; m. Annie Claire Van Laere, Dec. 22, 1973; children: Celine, Adriaan. BArch, St. Lukes U., Brussels, 1972. Registered architect. Trainee Archiduk A & E, Leuven, Belgium, 1973-75, project architect, 1975-78, assoc., 1978—; cons. ACT, Brussels, 1990—. Pres. Jaycees, Leuven, 1988; expert adviser Comsn. European Communities, 1993—. Mem. Hospibel (bd. dirs., pres. 1990-93, hon. pres. 1993—). Avocations: tennis, classical music. Home: Kouter 15, 3060 Bertem Belgium Office: Archiduk A & E, Frederik Lintsstraat 45, 3000 Leuven Belgium

VAN GUNDY, SEYMOUR DEAN, nematologist, plant pathologist, educator; b. Toledo, Feb. 24, 1931; s. Robert C. and Margaret (Holloway) Van G.; m. Wilma C. Fanning, June 12, 1954; children: Sue Ann, Richard L. BA, Bowling Green State U., 1953; PhD, U. Wis., 1957. Asst. nematologist U. Calif., Riverside, 1957-63, assoc. prof., 1963-68, prof. nematology and plant pathology, 1968-73, assoc. dean rsch., 1968-70, vice chancellor rsch., 1970-72, chmn. dept. nematology, 1972-84, prof. nematology and plant pathology, assoc. dean rsch. Coll. Natural and Agrl. Scis., 1985-88, acting dean, 1986, interim dean, 1988-90, dean, 1990-93. Former mem. editorial bd. Rev. de Nematologie, Jour. Nematology and Plant Disease; contbr. numerous articles to profl. jours. NSF fellow, Australia, 1965-66; grantee Rockefeller Found., Cancer Res., NSF, USDA. Fellow AAAS, Am. Phytopathol. Soc., Soc. Nematologists (editor-in-chief 1968-72, v.p. 1972-73, pres. 1973-74). Home: 1188 Pastern Rd Riverside CA 92506-5619 Office: U Calif Dept Nematology Riverside CA 92521

VAN HEMMEN, HENDRIK FOKKO, vehicle engineer; b. Dordrecht, The Netherlands, Jan. 8, 1960; came to U.S., 1976; s. Henk and Gertruda Sabiena (Teffer) Van H.; m. Anne Crosley Forsyth, Apr. 5, 1986; children: Hendrik James, Hannah Gretchen. BS in Aerospace & Ocean Engring., Va. Polytech Inst. and State U, 1982. Registered profl. engr. N.Y. Engr. MAR Inc., Rockville, Md., 1980-81; engr., surveyor Am. Bur. Shipping, N.Y.C., 1982-84; chief engr. Johan Valentjn Inc., Newport, R.I., 1984-88; cons. Francis A. Martin & Ottaway Inc., N.Y.C., 1988—; cons. to underwriters, fin. institutions, attys., shipping cos. and vehicle mgrs. Designer various sail and power yachts, 1975—. Mem. AIAA (student engrs. coun. rep. 1981-82, Design Competition award 1982), Nat. Soc. Profl. Engrs., Nat. Acad. Forensic Engrs., Nat. Fire Protection Assn., Soc. Naval Architects and Marine Engrs. (N.Y. state papers chmn. 1992, vice chmn. student chpt. 1981-82, Nat. Student Paper award 1982), U.S. Naval Inst. (assoc.). Office: Francis A Martin & Ottaway 90 Washington St New York NY 10006

VAN HOFTEN, JAMES DOUGAL ADRIANUS, business executive, former astronaut; b. Fresno, Calif., June 11, 1944; s. Adriaan and Beverly (McCurdy) van H.; m. Vallarie Davis, May 31, 1975; children—Jennifer Lyn, Jamie Juliana, Victoria Jane. B.S., U. Calif.-Berkeley, 1966; M.S., Colo. State U., 1968, Ph.D., 1976. Asst. prof. U. Houston, 1976-78; astronaut NASA, Houston, 1978-86; sr. v.p., mgr. advanced systems line Bechtel Nat., Inc., San Francisco, 1986-93; project mgr. Hong Kong New Airport projects. Served with USN, 1969-74; lt. col. Air N.G. 1984—. Recipient Disting. Service award Colo. State U., 1984; Disting. Citizen award Fresno Council Boy Scouts Am., 1984; Disting. Achievement award Pi Kappa Alpha, 1984. Assoc. fellow AIAA; mem. ASCE (Aerospace Sci. and Tech. Application award 1984). Republican. Home: 46 Tai Tam Rd House C, Hong Kong Hong Kong Office: Internat Bechtel Inc, 8 Harbour Rd, Wanchai Hong Kong

VAN HOOSIER, GERALD LEONARD, veterinary science educator; b. Weatherford, Tex., June 4, 1934; s. Gerald L. Sr. and Louise (Ashcroft) Van H.; m. Marlene M., Aug. 2, 1959; children: Gunther, Paul. DVM, Tex. A&M, 1957. Diplomate Am. Coll. Lab. Animal Medicine. Postdoctoral fellow U. Calif., Berkeley, 1959-60, Wash. State U., Pullman, 1969-71; chief applied virology NIH, Bethesda, 1960-62; assoc. prof. Baylor U. Coll. Medicine, Houston, 1962-69; assoc. prof., dir. Wash. State U., Pullman, 1969-75; prof., chmn. U. Wash., Seattle, 1975—; cons. VA Med. Ctr., Seattle, Tacoma, Boise, Seattle Biomed. Rsch. Inst. Editor: Laboratory Hamsters, 1987; contbg. author: The Hamster, Clin. Chemistry of Lab. Animals, 1989; contbr. articles to profl. jours. With USPHS Res., 1959—. Recipient Disting. Alumni award Tex. A&M, 1991, Animal Investigation and Diagnosis Lab award NIH, 1979-91, Comparative Med. Tng. NIH, 1982-91. Mem. Am. Assn. Lab. Animal Sci. (pres. 1991-92, Charles A. Griffin award 1986). Office: U Wash Dept Comparative Med SB-42 Seattle WA 98195

VAN HORN, MARY RENEÉ, interior designer; b. Sidney, Ohio, Apr. 5, 1940; d. Jasper Joseph and Thelma Doris (Kelly) Hott; m. James F. Rex, Aug. 13, 1962 (div. Aug. 1978); children: Christine Anne Rex, Craig Alan Rex; m. W.E. Van Horn, Mar. 20, 1983. Cert. of tech. in interior design, San Jacinto, Pasadena, Tex., 1982, AAS in Interior Design, 1983. Cert. profl. interior designer. Designer Jerry's Decor, Houston, 1982, Fran Lyder's, Houston, 1982-83; designer Interiors By Reneé, Chgo., 1983-88, Memphis, 1988—. Mem. Am. Soc. Interior Designers (assoc.), Memphis Kennel Club, Mid South Terrier Club. Republican. Lutheran. Avocations: hobbies, cairn terriers, arabian horses. Home and Office: 8401 Hwy 70 Arlington TN 38002

VANHOUTEN, JACOB WESLEY, environmental project manager, consultant; b. Grand Rapids, Mich., July 24, 1958; s. Calvin Ross and Lois Mae (Sleight) VanH. AAS, Alpena (Mich.) Community Coll., 1978; BS, Ferris State Coll., 1980; MS, Cen. Mich. U., 1986. Registered environ. property assessor, environ. profl. Grad. teaching asst. Cen. Mich. U., Mt. Pleasant, 1982-85; water quality specialist TMI Environ. Svcs., Mt. Pleasant, 1986-88, project mgr., 1988—; adj. faculty Cen. Mich. U., 1986, Saginaw (Mich.) Valley State U., 1988. Contbr. articles to profl. jours. Recipient Beaver Island scholarship Cen. Mich. U., 1982. Mem. ASTM (tech. com.), Nat. Registry Environ. Profls., Nat. Environ. Tng. Assn., Mich. Soc. Environ. Profls., N.Am. Benthol. Soc., Am. Fisheries Soc. Republican. Avocations: hunting, fishing, camping, nature photography, writing. Office: TMI Environ Svcs 2600 Three Leaves Dr Mount Pleasant MI 48858-7913

VAN HOUTEN, ROBERT, nuclear engineer, consultant; b. Peoria, Ill., Oct. 2, 1923; s. Merle Burt and Leila Alice (Waltrip) Van H.; m. Dorothy May, Dec. 19, 1947; children: Karol, Merle, Merry, Debra. BSChemE, Washington U., St. Louis, 1947, PhD, 1950. Profl. engr., Ohio. Assoc. dir. chem. and metal. rsch. P.R. Mallory & Co., Indpls., 1956-58; prin. engr., mgr. nuclear materials GE Co., Evendale, Ohio, 1958-70; rsch. engr. Atomics Internat., Canoga Park, Calif., 1970-73; sr. nuclear engr. Office of Rsch.-Nuclear Regulatory Commn., Washington, 1974-89; tech. coord. Office of Sec.-Nuclear Regulatory Commn., Washington, 1990-92; adj. assoc. prof. nuclear engring. U. Cin., 1964-70; program mgr. TMI-2 Lower Head Exam OECD/NEA, Paris, 1988-89. Mem. Sigma Xi. Achievements include patent for Massive Metal Hydrides. Home: 20139 Laurel Hill Way Germantown MD 20874

VANIER, JACQUES, physicist; b. Dorion, Que., Can., Jan. 4, 1934; s. Henri and Emma (Boileau) V.; m. Lucie Beaudet, July 8, 1961; children: Lyne, Pierre. BA, U. Montreal, 1955, BSc, 1958; MSc, McGill U., 1960, PhD, 1963. Lectr. U. Montreal, 1961-63, McGill U., 1960-63; physicist Varian Assocs., Beverly, Mass., 1963-67, Hewlett Packard Co., Beverly, 1967; prof. elec. engring. U. Laval, Que., 1967-83; physicist Nat. Rsch. Coun., Ottawa, 1983—, head elec. and time standards, 1984-86, dir. Lab. Basic Standards, 1986-90, dir. gen. Inst. for Nat. Measurement Standards, 1990-93; cons. Communication Components Corp., Costa Mesa, Calif., 1974-76, EGG Co., Salem, Mass., 1979-82; chmn. com. A URSI. Author: Basic Theory of Lasers and Masers, 1971, The Quantum Physics of Atomic Frequency Standards, 1989; contbr. articles to profl. jours.; inventor nuclear quadrupole resonance thermometer. Fellow IEEE (Centennial medal 1984), Royal Soc. Can., Am. Phys. Soc.; mem. Can. Assn. Physicists. Office: 6149 Rivermill Cir, Orleans, ON Canada K1C 5N3

VAN LOPIK, JACK RICHARD, geologist, educator; b. Holland, Mich., Feb. 25, 1929; s. Guy M. and Minnie (Grunst) Van L.; 1 son, Charles Robert (dec.). B.S., Mich. State U., 1950; M.S., La. State U., 1953, Ph.D., 1955. Geologist, sect. chief, asst. chief, chief geology br. U.S. Army C.E., Waterways Expt. Sta., Vicksburg, Miss., 1954-61; chief engrs. environ. adv. bd. U.S. Army C.E., 1988—; chief area evaluation sect., tech. dir., mgr. Space and Environ. Sci. Programs, tech. requirements dir. geosciences ops. Tex. Instruments, Inc., Dallas, 1961-68; chmn. dept. marine sci. Tex. Instruments, Inc., 1968-74; prof. dept. marine sci., dir. sea grant devel., dean Center for Wetland Resources, La. State U., Baton Rouge, 1968—; chmn. Coastal Resources Directorate of U.S. Nat. Com. for Man and Biosphere, U.S. Nat. Commn. for UNESCO, 1975-82; dir. Gulf South Research Inst., 1974-89; mem. Nat. Adv. Com. Oceans and Atmosphere, 1978-84; mem. Lower Miss. River Waterway Safety Com. USCG 8th dist., 1983—; mem. adv. council Nat. Coastal Resources Research and Devel. Inst., 1985—; ofcl. del. XX Congreso Internacional, Mexico City, 1956, XII Gen Assembly Internat. Union Geodesy and Geophysics, Helsinki, 1960; chmn. panel on geography and land use Nat. Acad. Scis.-NRC, com. on remote sensing programs for earth resources surveys, 1969-77. Fellow Geol. Soc. Am., A.A.A.S.; mem. Am. Astronautical Soc. (dir. S.W. sect. 1967-68), Am. Soc. Photogrammetry (dir. 1969-72, chmn. photo interpretation com. 1960, 65, rep. earth scis. div. NRC 1968-71), Am. Geophys. Union, Am. Assn. Petroleum Geologists (acad. adv. com. 1973-78), Assn. Am. Geographers, Soc. Econ. Paleontologists and Mineralogists (mem. research com. 1962-65), Am. Mgmt. Assn., Soc. Research Adminstrs., Marine Tech. Soc., Am. Water Resources Assn., Soc. Am. Mil. Engrs., Sea Grant Assn. (exec. bd. 1972-74, 80-82, pres.-elect 1988-89, pres. 1989—), Nat. Ocean Industries Assn. (adv. council 1973—), Nat. Conf. Advancement Rsch., La. Partnership for Tech. and Innovation (bd. dirs. 1989—), Sigma Xi. Home: 9 Rue Sorbonne Baton Rouge LA 70808-4682 Office: La State U Office Sea Grant Devel Baton Rouge LA 70803

VAN MAERSSEN, OTTO L., aerospace engineer, consulting firm executive; b. Amsterdam, The Netherlands, Mar. 2, 1919; came to U.S., 1946; s. Adolph L. and Maria Wilhelmina (Edelmann) Van M.; m. Hortensia Maria Velasquez, Jan. 7, 1956; children: Maria, Patricia, Veronica, Otto, Robert. BS in Chem. Engring., U. Mo., Rolla, 1949. Registered profl. engr., Tex., Mo. Petroleum engr. Mobil Oil, Caracas, Venezuela, 1949-51; sr. reservoir engr. Gulf Oil, Ft. Worth and San Tome, Venezuela, 1952-59; acting dept. mgr. Sedco of Argentina, Comodoro Rivadavia, 1960-61; export planning engr. LTV Aerospace and Def., Dallas, 1962-69, R & D adminstr. ground transp. div., 1970-74, engr. specialist new bus. programs, 1975-80; mgr. cost and estimating San Franciso and Alaska, 1981-84; owner OLVM Cons. Engrs., Walnut Creek, Calif., 1984—; cons. LTV Aerospace and Def., Dallas, 1984—. Served with Brit. Army. Intelligence, 1945, Germany. Mem. SPE (sr.). Democrat. Roman Catholic. Clubs: Toastmasters (Dallas), (sec./treas. 1963-64), Pennywise (Dallas) (treas. 1964-67). Avocations: travel, photography. Home and Office: OLVM Cons Engrs 1619 Arbutus Dr Walnut Creek CA 94595-1705

VANN, JOSEPH MC ALPIN, nuclear engineer; b. Clinton, N.C., Dec. 30, 1937; s. Joseph Rose and Louise Myrtle (Beaver) V.; m. Edith Ausley, Apr. 1, 1961; 1 child, Natasha Vann Bottoms. BS in Math., N.C. State U., 1958; MEd in Math., U. N.C., Chapel Hill, 1961; M in Physics, East Carolina U., 1972; MS in Nuclear Engring., Va. Polytechnic Inst., 1973. Chmn. math. dept. Mt. Olive (N.C.) Coll., 1961-71; safety and licensing engr. Gen. Pub. Utility, Parsippany, N.J., 1973-76; nuclear engr. N.J. Dept. Environ. Protection, Trenton, 1976-81; sr. radiol. engr. Ebasco Co., N.Y.C., 1981-89; sr. engr. West Valley (N.Y.) Nuclear Svcs., 1990-92; emergency planner Princeton (N.J.) Plasma Physics Lab., 1992—; sr. scientist Gen. Physics Corp., Aiken, S.C., 1992. Contbr. articles to profl. jours. Mem. Am. Physical Soc., Sigma Xi, Sigma Pi Sigma, Pi Mu Epsilon. Home: 23 Hillview Ave Madison NJ 07940-1738 Office: Princeton Plasma Physics Lab Princeton NJ 28543

VANNICE, M. ALBERT, chemical engineering educator, researcher; b. Broken Bow, Nebr., Jan. 11, 1943; s. Duane M. and Eugenia R. (Farmer) V.; m. Bette Ann Clark, Jan. 2, 1971. BSChemE, Mich. State Univ., 1964; MS, Stanford Univ., 1966, PhD, 1970. Engr. Dow Chemical Co., Midland, Mich., 1966, Sun Oil Co., Marcus Hook, Pa., 1970; sr. rsch. engr. Esso Rsch. & Engr. Co., Linden, N.J., 1971-76; assoc. prof. Pa. State Univ., State Coll., 1976-80, prof., 1980—, disting. prof., 1991—; cons. Eastman Kodak Co., Rochester, N.Y., 1980—; adv. bd. Adsorption Sci. & Tech., 1982—. Editorial bd. Jour. of Catalysis, 1988—; contbr. articles to profl. jours. Recipient N.Y. Catalysis Soc. award, 1985, Profl. Progress award, 1986, P.H. Emmett award, 1987, Pa.-Cleve. Catalysis Soc., 1988, Humboldt Rsch. award, 1990. Mem. AICHE, ACS, N.Am. Catalysis Soc., MRS. Achievements include 9 patents; effects of Strong Metal-Support Interactions on catalytic behavior; studies of CO hydrogenation. Office: Pa State Univ 107 Fenske Lab University Park PA 16802

VAN OOTEGHEM, MARC MICHEL MARTIN, pharmacology educator; b. Aalst, Belgium, Nov. 11, 1927; s. Jozef Jan Marie and Maria Camilla Julia (De Meyer) Van O.; m. Monique Jacqueline Marie Rubbens, Sept. 25, 1957; children: Patrick, Martine, Isabelle. Grad. in pharmacy, U. Louvain, Belgium, 1954, grad. in indsl. pharmacy, 1955, PharmD, 1963. Indsl. pharmacist Labs. R.I.T., Genval, Belgium, 1966=67; asst. Swiss Fed. Inst.

Tech., Zurich, 1956-57; asst. U. Louvain, 1957-63, lectr., 1963-72; assoc. prof. pharmacy U. Antwerp, Belgium, 1972-75, prof., 1975—; mem. Belgian Nat. Pharmacopeia Commn., 1975—. Author: Biofarmacie der vaste toedieningsvormen, 1979, Ophthalmika, 1st edit., 1971, 4th edit., 1990; mem. editorial bd. Farmaceutisch Tijdschrift, 1973—, Pharm. Sci. Techniques, 1986—; contbr. over 120 articles. Decorated officer Order Leopold II, comdr. Order of Crown (Belgium). Mem. Internat. Fedn. Pharmaceutics, Arbeitsgemeinschaft fur pharmazeutische Verfahrenstechnik, Galencia, Internat. Soc. for Eye Rsch., Soc. Dakryology. Home: Heidestraat 5, B-3020 Herent Winksele, Belgium Office: U Antwerp, Universiteitsplein 1, B-2610 Wilrijk Belgium

VAN OS, NICO MARIA, research chemist; b. Hoorn, Netherlands, Feb. 9, 1944; s. Otto Hendricus Johannes and Petronella (Kuiper) Van O.; m. Johanna Maria Hendrika Schrier, June 4, 1986; children: Nicole, Suzanne, Ludo, Eva. PhD, U. Amsterdam, Netherlands, 1975. Rsch. chemist Shell Rsch., Amsterdam, 1975-90, sr. rsch. scientist, 1990—; scientific bd. Tenside-Surfactants-Detergents, 1991—. Co-author: Physico-Chem. Properties of Surfactants, 1993; referee Langmuir, 1990—, Jour. Colloid Interface Sci., 1990—; contbr. articles to profl. jours. Mem. Young Scientists Assn. (chmn. 1972—). Achievements include research on the physical chemistry of surfactants and industrial applications of surfactants. Office: Shell Rsch Lab, PO Box 3003, 1003 AA Amsterdam The Netherlands

VANRIPER, WILLIAM JOHN, computer graphics scientist, consultant; b. Catskill, N.Y., Aug. 30, 1951; s. Wilbur W. and Annie C. (Radeski) vanR. AB, Bard Coll., 1981; MS, Boston U., 1985. Chief scientist Cambridge (Mass.) Parallel Processors, 1985-87, Flamingo Graphics, Cambridge, 1987-90, Megaware, Woburn, Mass., 1990; v.p. engring. C2C, Indpls., 1991; pres. UVR, Cambridge, 1992—. Author: Encyclopedia of Graphics File Formats, 1993. Mem. IEEE, Assn. for Computing Machinery. Office: UVR PO Box 2684 Cambridge MA 02238

VAN ROMPUY, PAUL FRANS, economics educator; b. St. Kat Waver, Antwerp, Belgium, Dec. 18, 1940; children: Joeri, Heiko. MS in Econs., U. Ill., Belgium, 1966; PhD in Econs., U. Louvain, Belgium, 1970. From asst. prof. to prof. econs. U. Louvain, 1970—; rsch. dir. U. Leuven, 1970-75, chmn. dept. econs., 1982-85. Advisor, Vice-Prime Minister, Brussels, 1988-89; chmn. fin. com. High coun. Pub. Fin. Govt., 1989. Fullbright fellow, 1962. Mem. Am. Econ. Assn., European Econ. Assn. Home: Leeuwerikweg 37, B-3140 Keerbergen Belgium Office: U Louvain, E Van Evenstraat 2 B, B-3000 Louvain Belgium

VAN ROOSBROECK, WILLY WERNER, physicist; b. Antwerp, Belgium, Aug. 10, 1913; came to U.S., 1916; s. Gustave Leopold and Marie Joanna (DeGraef) van R.; m. Marjorie Anna Covert, Oct. 7, 1945 (dec. Feb. 1982). AB, Columbia U., 1934, MA in Physics, 1937. Mem. tech. staff Bell Lab., Murray Hill, N.J., 1937-78; cons. in field, 1978—. Contbr. 35 articles to profl. jours. Fellow Am. Phys. Soc.; mem. N.Y. Acad. Sci., Triple Nine Soc., Internat. Soc. Philos. Enquiry, Phi Beta Kappa. Achievements include 4 patents concerning semiconductor device principles and use of thermophoresis for efficient production of optical fibers; formulation and applications of basic differential equations governing transport and recombination of current carriers in semiconductors; theory of relaxation semiconductors, including mechanism of the fast recombination and explanation of Ovshinsky reversible threshold switching in diodes of amorphous semiconductor alloys. Home: 19 Whittredge Rd Summit NJ 07901-2824

VAN RYN, TED MATTHEUS, electrical engineer; b. The Hague, The Netherlands, July 1, 1948; arrived in Can., 1952; came to U.S., 1990; s. Adrianus and Janna Christina (Keymel) van R.; m. Ricki Ann Kennedy, Mar. 23, 1976 (div. 1980); children: Michael Allan, Christopher Allan, Sandra Ann; m. Judith Mavis Seltzer, Nov. 25, 1981; 1 child, Saul Marlon. BASc, U. Waterloo, Ont., Can., 1976. Registered profl. engr., Ont. Engr. Toronto (Ont.) Lab., IBM, 1976-90; bus. analyst programming systems line-of-bus. IBM, Somers, N.Y., 1990—. Mem. Assn. Profl. Engrs. Province Ont. Office: IBM PO Box 100 Rte 100 Somers NY 10000

VAN SANTEN, RUTGER ANTHONY, catalysis educator; b. Langedijk, Holland, Netherlands, May 28, 1945; s. Arie and Caroline (Teeling) VanS.; m. Marretje Carolina Schroten, Sept. 20, 1968; children: Marieke, Hanneke. Masters, U. Leiden, 1968, Phil degree, 1971. Postdoctoral fellow Stanford (Calif.) Rsch. Inst., 1971-72; rsch. chemist Shell Rsch., Netherlands, 1972-82, supr., 1984-87; supr. Shell Devel. Co., Tex., 1982-84; prof. Eindhoven U. Tech., Netherlands, 1988—; dir. Eindhoven U., Inst. Catalysis, Netherlands, 1989—; vis. scientist Free U. Amsterdam, 1977-78; prof. extraordinarius Eindhoven U. Tech., Netherlands, 1986-87. Author: Theoretical Heterogeneous Catalysis, 1991; editor: Fundamental Aspects of Heterogeneous Catalysis Studied by Particle Beams, 1991, Elementary Reaction Steps in Catalysis, 1993, Catalysis, An Integrated Course, 1993; contbr. articles to profl. jours. Recipient Golden medal Royal Dutch Chem. Soc., 1982; named to Chiapette lectureship, 1992, Schuit Catalysis lectureship, 1992. Mem. Royal Dutch Chem. Soc. (chmn. 1987-92), Dutch Catalysis Fedn. (chmn. 1988-91), Dutch Inst. Catalysis (dir. 1991—), Chem. Rsch. Coun. (vice chmn. 1989—). Achievements include contbn. to mechanistic understanding of important heterogeneous catalytic reactions; devel. of new concepts in hydrocarbon version catalysis; computational studies of metal- and zeolite catalysis. Office: Eindhoven U of Tech, Den Dolech 2 Postbus 513, 5612 AZ Eindhoven The Netherlands

VAN SCHILFGAARDE, JAN, agricultural engineer, government agricultural service administrator; b. The Hague, Netherlands, Feb. 7, 1929; came to U.S., 1946, naturalized, 1957; married; 3 children. B.S., Iowa State Coll., 1949, M.S., 1950, Ph.D. in Agrl. Engring. and Soil Physics, 1954. Instr., assoc. agrl. engr. Iowa State Coll., 1949-54; asst. prof. agrl. engring. N.C. State Coll., 1954-57, assoc. prof., 1957-62, prof., 1962-64; drainage engr. Agrl. Rsch. Svc. USDA, Raleigh, N.C., 1954-64; chief water mgmt. engr. soil and water conservation rsch. div. USDA, Beltsville, Md., 1964-67; assoc. dir. USDA, 1967-71, dir., 1971-72; dir. Salinity Lab. USDA, Riverside, Calif., 1972-84; dir. Mountain States Area Agrl. Rsch. Svc. USDA, Ft. Collins, Colo., 1984-86, assoc. dir. no. plains area Agrl. Rsch. Svc., 1986-91; assoc. dep. adminstr. for natural resources Agrl. Rsch. Svc. USDA, Beltsville, Md., 1991—; vis. prof. Ohio State U., 1962. Mem. ASCE, NAE, Am. Soc. Agrl. Engrs., Soil Sci. Soc., Soil Conservation Soc. Am. Office: Agrl Rsch Svc Nat Program Staff BARC-W 005 Beltsville MD 20705

VAN SCHUYVER, CONNIE JO, geophysicist; b. Tulsa, Oct. 10, 1951; d. Lloyd Lee and Mary Ellen (Scott) Parks; m. Larry Gene Van Schuyver, May 15, 1972 (div. 1982). Student, Okla. State U., Stillwater, 1969-71; BS, Tex A&M U., 1977; postgrad., Houston Community Coll., 1990, Univ. Houston, 1990—. Geophysicist Seiscom Delta, Inc., Houston, 1978-81; sr. geophysicist Champlin Petroleum Corp., Houston, 1981-84, BKW Seismic Processing (later ESP Earth Scis.), Houston, 1984-85, Geophys. Devel. Corp., Houston, 1987—; cons. CogniSeis Processing Svcs., Crawley, West Sussex, U.K., 1992—; paralegal analyst, Conoco, Inc., Houston, 1986. Mem. Soc. Exploration Geophysicists, Geophys. Soc. Houston, Sports Car Club of Am. Office: Stanley House Kelvin Way, Crawley W Sussex RH10 2SX, England

VAN SICKELS, MARTIN JOHN, chemical engineer; b. N.Y.C., July 21, 1942; s. Henry Martin and Veronica Bernadette (De Nier) Van S.; m. Marie Teresa Valenti, Jan. 28, 1968; children: Melissa, Michael. BChemE, CCNY, 1965; MChemE, NYU, 1966. Lic. profl. engrs., Tex., Ala. Process engr. Chem. Constrn. Corp., N.Y.C., 1966-67; process engr., sr. process engr. Haldor Topsoe, Inc., N.Y.C., 1967-71; program mgr., process mgr. J.F. Pritchard & Co., Kansas City, Mo., 1971-74; process mgr., mgr. process dept., mgr. chems./synthetic fuel The Rust Internat. Corp., Birmingham, Ala., 1974-83; v.p. proposals and projects Kellogg Rust Synfuels Inc., Houston, 1981-88; v.p. and dir. product tech., mktg. and lic. The M.W. Kellogg Co., Houston, 1988—, also dir. tech. mgmt.; bd. dirs. Materials Property Coun., N.Y.C., 1991—. Contbr. articles to profl. jours. Mem. AICE (Young Engr. of Yr. award 1977), AIA, Am. Chem. Soc., Lic. Execs. Soc. USA, Comml. Devel. Assn., Sigma Xi. Office: The M W Kellogg Co PO Box 4557 Houston TX 77210-4557

VANSUETENDAEL, NANCY JEAN, physicist; b. Plainfield, N.J., Dec. 13, 1957; d. Robert Frederick and Jean Beck Malone; m. Richard Lee Von Suetendael, Apr. 20, 1984. BA, Albright Coll., 1980; MA, Montclair State Coll., 1985; AS in Physics, Brookdale C.C., Lincroft, N.J., 1989; BS in Applied Physics, Math., Stockton State Coll. Cert. secondary edn. tchr. Engring. technician Sanders and Thomas, Inc., Jackson, N.J., 1983-85; quality assurance mgr. Orbiting Astron. Observatory, Inc., Lincroft, N.J., 1985-86; scientific researcher, writer Bell Comms. Rsch., Lincroft, 1986-87; software analyst, tester Concurrent Computer Corp., Lincroft, 1987-90; sr. ops. analyst, sr. mem. tech. staff, project mgr. Computer Resource Mgmt., Inc., Atlantic City, N.J., 1990—; tech. cons. FAA Task Force, Dallas-Ft. Worth Airport Capacity Office Design Team, Dallas-Ft. Worth and Atlantic City, 1991-92; field analyst, Windsor, Conn., 1989. Author: Software Quality Metrics, 1991, Parallel Processing Using E/SP, 1990, Stepping Through An E/SP Session Tutorials, 1990, others. Participant March of Dimes Walkathon, Freehold, N.J., 1972-76; del. Harlaxton Coll., Community Affairs, Grantham, Eng., 1978-79. Recipient grad. assistantship Montclair State Coll., Upper Montclair, 1980-81, Outstanding Tech. Publ. award, Soc. Tech. Communicators, 1990. Mem. Am. Inst. Physics, Barnegat Amateur Radio Club, FAA Flying Club, The Nature Conservancy, Stockton State Astronomy Club. Achievements include rsch. in laser applications, circuit analysis and design, wave theory and propagation, computer parallel processing techniques, unix operating system interactive programs, simulation modeling, databus architectures and airport capacity and design analyses. Home: 97 Water St Barnegat NJ 08005 Office: Computer Resource Mgmt Inc FAA Tech Ctr Bldg 270 Atlantic City Internat Air Atlantic City NJ 08405

VAN TAMELEN, EUGENE EARLE, chemist, educator; b. Zeeland, Mich., July 20, 1925; s. Gerrit and Henrietta (Vanden Bosch) van T.; m. Mary Ruth Houtman, June 16, 1951; children: Jane Elizabeth, Carey Catherine, Peter Gerrit. A.B., Hope Coll., 1947, D.Sc., 1970; M.S., Harvard, 1949, Ph.D., 1950; D.Sc., Bucknell U., 1970. Instr. U. Wis., 1950-52, from asst. to asso. prof., 1952-59, prof., 1959-61, Homer Adkins prof. chemistry, 1961-62; prof. chemistry Stanford, 1962-87, prof. emeritus chemistry, 1987—, chmn. dept., 1974-78; Am.-Swiss Found. lectr., 1964. Mem. editorial adv. bd.: Chem. and Engring. News, 1968-70, Synthesis, 1969-91, Accounts of Chem. Research, 1970-73; editor: Bioorganic Chemistry, 1971-82. Recipient A.T. Godfrey award, 1947; G. Haight traveling fellow, 1957; Guggenheim fellow, 1965, 73; Leo Hendrik Baekeland award, 1965; Prof. Extraordinarius Netherlands, 1967-73. Mem. NAS, Am. Chem. Soc. (Pure Chemistry award 1961, Creative Work in Synthetic Organic Chemistry award 1970), Am. Acad. Arts and Scis., English-Speaking Union (patron 1990—), Rolls Royce Owners Club, Churchill Club (bd. dirs. 1990-92, vice chmn. 1991-92), Los Altos Tomorrow (bd. dirs. 1991—). Home: 23570 Camino Hermoso Los Altos Hills CA 94024-6407 also: Moorings, PO Box 101, Saint Lucia West Indies

VAN TASSEL, ROGER ALAN, infrared phenomenologist, geophysicist; b. Orange, N.J., Oct. 19, 1936; s. William Francis and Marie (Simson) Van T. BA, Wesleyan U., Middleton, Conn., 1958; PhD, Northeastern U., 1972. Project scientist Geophysics Directorate Phillips Lab., Hanscom AFB, Mass., 1961-92, dir. optical environ. div., 1992—. Contbr. articles to profl. jours. Mem. Am. Geophys. Union. Unitarian-Universalist. Achievements include patent for hadamard spectrograph. Office: Phillips Lab GPO 29 Randolph Rd Hanscom AFB MA 01731-3010

VAN VELDHUIZEN, PETER JAY, oncologist, hematologist; b. Bellflower, Calif., Oct. 18, 1959; s. Peter and Grace (Van Surksum) Van V. BS in Chemistry, U. S.D., 1982, MD, 1986. Resident U. Kans. Med. Ctr., Kansas City, 1986-89, fellow, 1989—. Mem. AMA, ACP, Phi Beta Kappa, Alpha Omega Alpha. Home: Apt 1D 6885 W 51st Terr Mission KS 66202

VAN VLACK, LAWRENCE HALL, engineering educator; b. Atlantic, Iowa, July 21, 1920; s. Claude H. and Ruth (Stone) Van V.; m. Frances E. Runnells, June 27, 1943; children: Laura R., Bruce H. B.S. in Ceramic Engring., Iowa State U., 1942; Ph.D. in Geology, U. Chgo., 1950. Registered profl. engr., Mich. With U.S. Steel Co., 1942-53; mem. faculty U. Mich., 1953—, prof. materials sci. and engring., 1958—, chmn. dept., 1967-73; cons. in field. Author: Nature and Behavior of Engineering Materials, 1956, Elements of Materials Science, 2d edit., 1964, Physical Ceramics for Engineers, 1964, Materials Science for Engineers, 1970, Textbook of Materials Technology, 1973, Materials Science and Engineering, 1975, 6th edit., 1989, Aids for Introductory Materials Courses, 1977, Nickel Oxide, 1980, Materials for Engineering, 1982, Materials for Thermal Management, 1992. Served with USNR, 1945-46. Fellow Am. Soc. Metals (Sauveur lectr. 1979, Gold medal 1984, A.E. White disting. tchr. award 1985), Am. Ceramic Soc., AAAS; mem. AIME, Nat. Inst. Ceramic Engrs., N.Y. Acad. Scis., Sigma Xi, Tau Beta Pi, Alpha Chi Sigma, Alpha Sigma Mu (hon.). Methodist. Home: 2115 Nature Cove Ct Apt 309 Ann Arbor MI 48104-4987

VAN WALLENDAEL, LORI ROBINSON, psychology educator; b. Waukegan, Ill., Oct. 24, 1960; d. Bernard Albert and Eunice (Booth) Robinson; m. Shawn Edmund Van Wallendael, Aug. 30, 1986; 1 child, Heather Victoria. BA, MacMurray Coll., Jacksonville, Ill., 1982; MA, Northwestern U., 1985, PhD, 1986. Lectr. U. N.C., Charlotte, 1986-89, asst. prof. psychology, 1989—. Author: Instructor's Manual for Matlin's Psychology, 1992; textbook reviewer Brooks/Cole Pub. Co., Pacific Grove, Calif., 1989, McGraw-Hill Pub. Co., Cambridge, Mass., 1990; contbr. articles to profl. jours. Tech. vol., reading stage coord. Theatre Charlotte, 1989. Northwestern U. fellow, 1982-84; U. N.C.-Charlotte rsch. grantee, 1987, 89. Mem. Soc. for Judgment and Decision Making, Am. Psychol. Soc., Psychonomic Soc. (assoc.), Sigma Xi. Roman Catholic. Avocations: dollhouse and miniature building/collecting, theatrical costume design. Home: 5325 St Stephens Church Rd Gold Hill NC 28071-9438 Office: U NC-Charlotte Dept Psychology UNCC Sta Charlotte NC 28223

VAN WANING, WILLEM ERNST, computer scientist; b. Rotterdam, Zuid-Holland, The Netherlands, July 21, 1948; s. Willem Ernst and Maria Helena (Van den Blink) Van W.; m. Pamela Josephine Charlotte Harms, Aug. 9, 1979; children: Wouter Ernst, Marjoleine Annemieke. DSc, U. Leiden, The Netherlands, 1977. Dept. head Cen. Reken Inst., Leiden, 1977-83; scientist Ctr. for Math. and Computer Sci., Amsterdam, The Netherlands, 1983-85; asst. prof. U. Amsterdam, 1985-88; software engr., 1988-90; software architect Van Waning Software Architecture, Muiderberg, The Netherlands, 1990—; mgr. systems programming & networks SVB, 1991—; dir. Post Acad. Onderwys, Amsterdam, 1989. Inventor: Parallel Head Matching, 1987, Modules as 1st class values; author: (with others) Design Rules for Computer Integrated Manufacturing, 1986. Home and Office: Van Waning Software Archit, GH Breitnerlaan 5 Noord-Holland, 1399XD Muiderberg The Netherlands

VAN WINKLE, EDGAR WALLING, electrical engineer, computer consultant; b. Rutherford, N.J., Oct. 12, 1913; s. Winant and Jessie Walcott (Mucklow) Van W.; m. Jessie Stetler, Apr. 23, 1938 (dec. 1992); children: Barbara Van Winkle Clifton, Catrina Van Winkle Poindexter, Cornelia Van Winkle Schloss; m. Martha Polyé, May 22, 1993. B.E.E., Rutgers U., 1936; M.S. in Indsl. Engring., Columbia U., 1943, P.E. in Indsl. Engring., 1966. Registered profl. engr., N.J. Elec. engr. A.B. Dumont Labs. Passaic, N.J., 1943-48; chief engr. Facsimile Electronics, Passaic, 1948-52; cons. Bur. Ships, Washington, 1952; asst. sr. staff scientist Bendix Corp., Teterboro, N.J., 1952-67; sr. staff scientist Conrac Corp., West Caldwell, N.J., 1967-78; pres. Empac, Inc., Rutherford, N.J., 1979—. Contbr. articles, papers to profl. jours; patentee in field. Ruling elder Presbyterian Ch., Rutherford, 1984-91, chmn. endowment com., 1984—. Mem. IEEE (life, treas. artificial intelligence sect. North N.J. Chpt. 1982-84), Bendix Mgmt. Club (life), North N.J. Automatic Control Group (chmn. 1967-68), Met. Engring. Mgmt. (chmn. 1966-67), Mensa, Holland Soc., Green Pond Yacht Club (past commodore), Delta Phi. Republican. Club: Upper Montclair Country. Current work: Artificial intelligence and robotics. Subspecialty: Mathematical software.

VAN WINKLE, JON, chemical engineer; b. Hackensack, N.J., Oct. 29, 1930; s. John Rigby Gill and Margaret (Hulings) Van W.; children: Marcus, Sarah Caroline. BA cum laude, Amherst Coll., 1951; BS, MIT, 1953, MS, 1954; PhD, Rensselaer Poly. Inst., 1965; MBA, U. Mass., 1974. Registered profl. engr., N.Y., Ohio. Sr. chem. engr. Gen. Electric, various locations, 1970-82; project mgr., materials Holophane div. Manville, Newark, Ohio, 1982-87; mgr. tech. and engring. LCP Chem., Moundsville, W.Va., 1987-90; mgr., facilities supt. MCNC Ctr. for Microelectronics, Research Triangle Park, N.C., 1990—. Author articles on fuel cells, ultrapurification processes, corrosion and electrochem. phenomena. GE Found. fellow, 1962, NSF fellow, 1963, 64, Olin Matheson Co. fellow, 1965. Mem. AICE, Am. Chem. Soc., Sigma Xi (pres. 1992-93), Rotary (dir. 1979-82, sec. 1988-89, 91-92). Episcopalian. Achievements include patents in field. Home: 5804-11 Tattersall Dr Durham NC 27713-9026 Office: MCNC Ctr for Microelectronics PO Box 12889 3021 Cornwallis Dr Research Triangle Park NC 27709

VAN WISSEN, GERARDUS WILHELMUS JOHANNES MARIA, consulting engineering company executive; b. Voorburg, The Netherlands, Aug. 2, 1941; s. Johannes J.G. and Wilhelmina (Houweling) van W.; m. Marianne C.E. Struyk, July 6, 1965; children: Elise W.G., Wikke M., Annemarliese. Ir. Civil engr., Tech. U. Delft, The Netherlands, 1967. Asst. expert FAO, Rome, 1967-70; hydraulic engr. water master plan Rio Cai Agrar- und Hydrotechnik GmbH, Brazil, 1970-72; tech. dir. Ferrenafe irrigation scheme Agrar- und Hydrotechnik GmbH, Peru, 1972-74; team leader Tanga water master plan Agrar- und Hydrotechnik GmbH, Tanzania, 1974-76; regional dir. Africa dept. Essen, Fed. Republic Germany, 1976-78, sales mgr., 1978-82, mng. dir., 1982—. Mem. Pro Ruhrgebiet, Essen, 1984—. Mem. Royal Inst. Engrs., Verband unabhängig beratender Ingenieurfirmen (export com. 1982—), Industrie und Handelskammer (export com. 1983—), Deutscher Verband für Wasserwirtschaft und Kulturbau (export com. 1984—). Avocations: painting, fly fishing. Office: Agrar-und Hydrotechnik GmbH, Huyssenallee 66-68, D-4300 Essen 1, Germany

VANYO, EDWARD ALAN, environmental engineer; b. Kingston, Pa.. BS in Chem. Engring., Lafayette Coll., 1986. Site mgr. N.J. Dept. Environ. Protection, Trenton, 1986-88; project mgr. Chem. Waste Mgmt., Inc., Princeton, N.J., 1988-90; sr. environ. engr. Henkel Corp., Ambler, Pa., 1990—. Mem. AICE. Office: Henkel Corp 300 Brookside Ave Ambler PA 19002

VAN ZAK, DAVID BRUCE, psychologist; b. Santa Monica, Calif., Nov. 5, 1950; s. Martin and Anita (Cohen) Van Z.; m. Nina Weinstein, Feb. 15, 1991; 1 child, Joshua Jordan. PhD, U.S. Internat. U., 1984, postgrad., 1984—. Lic. psychologist, Calif., marriage, family and child therapist; cert. sex therapist. Psychologist Project Total Push, Watts, Calif., 1977-79; attending psychologist U. So. Calif. Dental Sch., L.A., 1989—; psychologist cons. Westwood Plz. Geriatric Facility, L.A., 1989—; clin. instr. Dept. Behavioral Dentistry, U. So. Calif., L.A., 1989—; pvt. practice L.A., 1988—; bd. dirs. Calif. Inst. for Socioanalysis, Long Beach, Calif., 1978-82; cons. Sign of the Dove Skilled Nursing, L.A., 1989-92, Calif. Assn. Marital and Family Therapists, 1980. Author: (with others) Behavioral Medicine, 1991; contbr. more than 20 articles on premenstrual syndrome and incontinence to profl. jours. Grantee Calif. Youth Authority, 1978-79, Calif. Dept. Health, 1982-83. Mem. APA, Calif. Psychol. Assn., L.A. Psychol. Assn., Internat. Psychosomatics Inst. (editor Behavioral Medicine Jour. 1993), Assn. for Applied Psychophysiology (chair job search com. 1985), biofeedback Soc. Calif. (bd. dirs. 1986-90, chair instrumentation 1986-90). Democrat. Jewish. Achievements include development of treatments for premenstrual syndrome, premenstrual affective syndrome, anal incontinence, urinary incontinence; rsch. on relaxation tng., stress mgmt., self-esteem, childhood autism, intelligence deterioration in old age and predictors of dementia. Published self-help tapes on hypnosis, relaxation, training and stress management. Developed stress management program for Japanese executives. Office: Acad Behavioral Medicine 11600 Wilshire Blvd Ste 10 West Los Angeles CA 90025

VAN ZUTPHEN, LAMBERTUS F.M. (BERT VAN ZUTPHEN), geneticist, educator; b. Gemert, The Netherlands, Oct. 27, 1941; s. Johannes and Petronella (Van den Berg) Van Z.; m. Anna Maria Rooymans; children: Yvonne, Esther. BS in Agrl. Engring., Agrl. Coll., Roermond, The Netherlands, 1961; M in Biology, U. Utrecht, The Netherlands, 1969, PhD, 1974. Biology tchr. Vitus Coll., Bussum, The Netherlands, 1968-69; from asst. prof. to assoc. prof. U. Utrecht, 1969-76, 77-83, prof., dept. head, 1983—, mem. ethical com., 1988—; nat. adv. com. Genetic Modification, 1991—; senats commn. Deutsche Forschungs Gemeinschaft, 1991—. Editor: Proefdieren en dierproeven, 1991, Principles of Laboratory Animal Science: A Contribution to the Humane Use and Care of Animals and to Experimental Results, 1993; mem. editorial bd. jours. JEANS, ATLA, Vet. Quar.; contbr. articles to profl. jours. Advisor HLO Coll., Utrecht, 1990—; vice-chmn. Nat. Adv. Com. on Animal Experimentation, Ryswyk, 1985—. Fellow Fogarty NIH 1976-77. Mem. Med. Com. Royal Netherlands Acad. Scis., Nat. Soc. Lab. Animal Sci. (pres. 1978-82), Netherlands Fedn. Lab. Animal Sci. Assn. (sec. 1984—), German Soc. Lab. Animal Sci. (v.p. 1987—). Avocations: badminton, walking, cycling. Home: Sanatoriumlaan 87, 3705 AN Zeist The Netherlands Office: U Utrecht, Yalelaan 2, Utrecht The Netherlands

VAN ZYTVELD, JOHN BOS, physicist, educator; b. Hammond, Ind., Nov. 12, 1940; s. Cornelius Jacob and Catherine (Bos) Van Z.; m. Carol Ann Vander Slik, Dec. 30, 1961; children: Jennifer Sue, Eric James, Sara Lynne. AB, Calvin Coll., 1962; MS, Mich. State U., 1964, PhD, 1967. Postdoctoral fellow physics U. Sheffield, Eng., 1967-68; from asst. prof. to prof. physics Calvin Coll., Grand Rapids, Mich., 1968—; sr. postdoctoral fellow in physics U. Leicester, Eng., 1974-75; sr. Fulbright lectr. physics Yarmouk U., Jordan, 1980-81; assoc., dir. solid state physics program NSF, Washington, 1983-85; sr. Fulbright rsch. fellow in physics U. Heraklion, Crete, Greece, 1990-91; vis. prof. physics Kyoto (Japan) U., 1991; chair sci. adv. com. U.S. Congress, 1972-78; cons. NSF, Washington, 1985—, Dept. Energy, Washington, 1988—; counselor physics coun. undergrad. rsch., 1986-90, pres., 1988-90. Author rsch. papers, rev. paper. Grantee Rsch. Corp., Tucson, Ariz., 1969-76, NSF, 1977—. Fellow Am. Sci. Affiliation; mem. AAAS, Am. Assn. Physics Tchrs., Am. Phys. Soc. Office: Calvin Coll Dept Physics 3201 Burton St SE Grand Rapids MI 49546

VARADARAJ, RAMESH, research chemist; b. Madras, India, Dec. 28, 1956; came to U.S., 1983; s. Kavaseri and Jayalakshmi Varadaraj; m. Shanti Subramaniam, Jan. 29, 1988; 1 child, Nevin. PhD, Ind. Inst. Sci., Bangalore, 1984. Postdoctoral fellow Georgetown U., Washington, 1984-86, Temple U., Phila., 1986-88; staff rsch. scientist Exxon Rsch. & Engring. Co., Clinton, N.J., 1988—. Contbr. over 50 articles to profl. jours. Achievements include patents for novel surfactants and additives, environmental hydrocarbon remediated technologies. Office: Exxon Rsch & Engring Co Rte 22 E Annandale NJ 08801

VARANDAS, ANTONIO JOAQUIM DE CAMPOS, chemistry educator; b. Mata Curia, Anadia, Portugal, Sept. 19, 1947; s. Alfredo Cerveira and Lusitana Batista (de Campos) V.;m. Edite Maria Dias Carvalheira, Feb. 26, 1972; children: Antonio Miguel, Pedro Luis. Diploma in chem. engring., U. Oporto, Portugal, 1971; PhD in Theoretical Chemistry, U. Sussex, Eng., 1976. Aux. prof. U. Coimbra, Portugal, 1977-82, assoc. prof., 1982-88, prof., 1988—, head chemistry dept., 1989-92, chmn. chem. sci. coun., 1992—; vis. rsch. scholar dept. chemistry Mpls. Supercomputer Inst., U. Minn., 1988. Co-author: Molecular Potential Energy Functions, 1984, Estrutura e Reactividade Molecular, 1986; editor, reviewer Portuguese Jour. of Chemistry, 1986; editor, reviewer Portuguese Jour. of Chemistry, 1985-92; contbr. over 90 articles to profl. jours. Lt. Portuguese Navy, 1978-79. NATO scholar, 1973-76, Brit. Coun. scholar, 1978; recipient Artur Malheiros prize for physics and chemistry Lisbon Acad. Scis., 1985. Mem. Am. Phys. Soc., European Phys. Soc. (bd. dirs. chem. physics sect. 1987—), Royal Soc. Chemistry, N.Y. Acad. Scis., Portuguese Chem. Soc. (v.p. Lisbon chpt. 1985-89, Ferreira da Silva prize 1991), Portuguese Phys. Soc. Roman Catholic. Home: R. Infanta D. Maria, 460,, 4E, 3000 Coimbra Portugal Office: U Coimbra, Dept de Quimica, 3049 Coimbra Portugal

VARGA, THOMAS, mechanical engineering educator; b. Szombathely, Hungary, Sept. 1, 1935; s. Emil and Carola (Klein) V.; m. Helga Caroline Kirnbauer, Sept. 8, 1962 (dec. Dec. 1984); m. Gerda Marie-Luise Hunold, July 30, 1986. Mech. engring. diploma, Tech. U., Vienna, Austria, 1960; D. in Sci. Tech., Swiss Fed. Inst. Tech., Zürich, Switzerland, 1966. Group leader Sulzer Bros. Ltd., Winterthur, Switzerland, 1966-78; lectr. Swiss Fed. Inst. Tech., Zürich, 1972—; full prof. U. Tech., Vienna, 1979—; expert Swiss

Fed. Nuclear Safety Inspectorate HSK), 1974—; head Inst. Resch. and Testing in Materials Technology, TVFA, 1979—. Author: Eisenwerkstoffe, 1972; contbr. numerous tech. articles to profl. jours. Mem. ASTM, Austrian Soc. for Materials Tech., Am. Soc. for Metals, German Welding Soc., German Soc. Materials Rsch. and Testing , Swiss Soc. Materials Tech. Lutheran. Avocations: literature, travelling, fossils. Office: TVFA Tech U Vienna, Karlsplatz 13, Vienna Austria

VARGAS, ROGER IRVIN, entomologist, ecologist; b. Long Beach, Calif., Jan. 10, 1947; s. Roger E. and Olga (Irvin) V.; m. Kathylyn Richardson, July 28, 1990; children: Noelani, Kela. BA in Zoology, U. Calif., Riverside, 1969; MS in Biology, San Diego State U., 1974; PhD in Entomology, U. Hawaii, 1979. Rsch. assoc. U. Hawaii, Honolulu, 1979-80; rsch. entomologist USDA Agr. Rsch. Svc., Honolulu, 1980—; USDA cons. on fruit flies, Mex., Cen. Am., Calif., 1980—. Contbr. articles on ecology, biol. control, behavior, pest mgmt. and mass-rearing of fruit flies to Jour. Econ. Entomology, Environ. Entomology, other profl. publs. Mem. AAAS, Entomol. Soc. Am., Hawaiian Entomol. Soc. Achievements include patent for Mediterranean fruit fly egg collection system, devel. of demographic and ecological data for Hawaiian fruit flies, devel. of mass-rearing techniques for sterile insect technique control of fruit flies. Office: USDA Tropical Fruit and Vegetables Rsch Lab PO Box 2280 Honolulu HI 96804

VARLEY, JOHN OWEN, engineer; b. Akron, Ohio, Feb. 8, 1963; s. John Francis and Marie Delores (Kirmes) V. BS in Mech. Engring., U. Akron, 1986. Plant engr. Goodyear Aerospace Systems Group, Akron, 1986-87; staff engr. The Glidden Co., Cleve., 1987—. Mem. Am. Soc. Heating Refrigerating and Air Condition Engrs. (mem. tech. com. 1992--), Am. Soc. Mech. Engrs. (dir. Akron chpt. 1987-88), Jaycees (exec. bd. mem. 1990-92). Office: The Glidden Co 925 Euclid Ave Cleveland OH 44115

VARLEY, REED BRIAN, fire protection engineer, consultant; b. Chgo., Aug. 24, 1931; s. Brian Jennings and Mildred (Reed) V.; m. Jewel Angela Woung, Nov. 7, 1992; children by previous marriage: Steven, Rebecca. BS in Fire Protection, Safety Engring., Ill. Inst. Tech., 1953, postgrad., 1958-65. Registered profl. engr., Fla., Ill., Colo., Calif. Engr. Mountain States Inspection Bur., Denver, 1953-62; assoc. engr. Ill. Inst. Tech. Rsch., Chgo., 1962-64; fire protection cons. Marsh and McClennan M&M, Chgo., 1964-75; group fire protection engr. Bland-Payne Australia Ltd., North Sydney, Australia, 1975-78; mgr., engr. Gage-Babcock and Assocs., Miami, 1978-86; prin. Varley-Campbell and Assocs., Miami, Fla., 1986—. Author tech. reports; contbr. articles to profl. publs. With U.S. Army, 1954-56. Mem. NSPE, Fla. Engring. Soc., Fla. Inst. Consulting Engrs., Soc. Fire Protection Engrs. (pres. Fla. chpt. 1983-85, treas. 1985-92). Office: Varley Campbell Assocs 1110 Brickell Ave Ste 430 Miami FL 33131

VARLOTTA, DAVID, anesthesiologist; b. Douglaston, N.Y., Jan. 9, 1958; s. Gerard Anthony and Isabel Natalie (Rich) V.; m. Barbara Lee Goldberg, May 26, 1983; children: Michele Leslie, Richard Max. BA in Chemistry, Emory U., 1975; DO, N.Y. Coll. Osteo. Medicine, 1983. Diplomate Am. Bd. Anesthesiologists. Intern Coney Island Hosp., Bklyn., 1983-84; resident internal medicine Maimonides Hosp., Bklyn., 1984-85; resident anesthesiology Montefiore Hosp., Bronx, N.Y., 1985-86; resident anesthesiology Cleve. Clinic Found., 1986-87, fellow cardiothoracic anesthesiology, 1987-88, fellow neuroanesthesia, 1988; asst. prof. U. South Fla. Coll. Medicine, Tampa, 1988-91, clin. asst. prof. anesthesiology, 1991—; staff anesthesiologist Univ. Community Hosp., Tampa, 1991—. Grantee Anaquest Pherms., Liberty Corner, N.J., 1989. Mem. AMA, Internat. Anesthesia Rsch. Soc., Am. Soc. Anesthesiologists, Soc. Cardiovascular Anesthesiologists (edn. com. 1989-93), Fla. Soc. Anesthesiologists (charter and bylaws com. 1991-92), Avila Golf and Country Club. Republican. Roman Catholic. Avocations: golf, travel, collecting sports memorabilia. Home: 5108 E Longboat Blvd Tampa FL 33615-4230 Office: U Community Hosp Dept Anesthesiology 3100 E Fletcher Ave Tampa FL 33613-4688

VARMA, ARVIND, chemical engineering educator, researcher; b. Ferozabad, India, Oct. 13, 1947; s. Hans Raj and Vijay L. (Jhanjhee) V.; m. Karen K. Guse, Aug. 7, 1971; children: Anita, Sophia. BS ChemE, Panjab U., 1966; MS ChemE, U. N.B., Fredericton, Can., 1968; PhD ChemE, U. Minn., 1972. Asst. prof. U. Minn., Mpls., 1972-73; sr. research engr. Union Carbide Corp., Tarrytown, N.Y., 1973-75; asst. prof. chem. engring. U. Notre Dame, Ind., 1975-77, assoc. prof., 1977-80, prof., 1980-88, Arthur J. Schmitt prof., 1988—, chmn. dept., 1983-88; vis. prof. U. Wis., Madison, fall 1981; Chevron vis. prof. Calif. Inst. Tech., Pasadena, spring 1982; vis. prof. Ind. Inst. Tech.-Kanpur, spring 1989, U. Cagliari, Italy, summer 1989, 92. Editor: (with others) The Mathematical Understanding of Chemical Engineering Systems, 1980, Chemical Reaction and Reactor Engineering, 1987; contbr. over 130 articles to profl. jours. Recipient Tchr. of Yr. award Coll. Engring., U. Notre Dame, 1991, Spl. Presdl. award U. Notre Dame, 1992, R.H. Wilhelm award AICE, 1993; Fulbright scholar; Indo-Am. fellow, 1988-89. Home: 1721 E Cedar St South Bend IN 46617-2535 Office: Dept Chem Engring U Notre Dame Notre Dame IN 46556

VARMA, BAIDYA NATH, sociologist, broadcaster; m. Savitri Devi. PhD, Columbia U., 1958. Radio broadcaster to India UN; Asian News Moderator Nat. Edn. TV Network, N.Y.C.; prof. emeritus sociology CUNY; prodr. radio dramas Voice of Am.; wrote, narrated over 200 documentary films, News of the Day; lectr. numerous univs. U.S., Can., Eng., India; chair Plenary Sessions World Congress of Sociology, Internat. Congress Anthrop. and Enthnological Scis; cons. Nat. Endowment Humanities, Ctr. Migration Studies, Dept. Energy, Wenner-Gren Found. Anthrop. Rsch. in U.S., Can. Coun., Indian Law Inst.; chair faculty seminars Columbia U.; presided Centenary Celebrations Indian Writers, N.Y.C.; vis. prof. Columbia U., other U.S., Indian Univs.; chair panel on religions and sexuality Parliament of World's Religions, 1993. Author: The Sociology and Politics of Development: A Theoretical Study, 1980, Social Science and Indian Society, 1985, New Directions in Theory and Methodology, 1993, Contemporary India (cert. of merit German Govt.), author, editor others; contbr. articles Ency. Americana, profl. jours.; edit. adv. nat., internat. sociol. jours. Assoc. trustee Wordsworth Trust; trustee Taraknath Das Found.; bd. scholars Buddhist Cultural Inst., U.S.; judge Permanent People's Tribunal Indsl. and Environ. Hazards and Human Rights, Rome; established Varma Found.; founding mems. Lincoln Ctr. for Performing Arts, N.Y.C.; chmn. bd. trustees Soc. for Restoration of Ancient Vidyadhams of India; trustee Internat. Found. for Vedic Edn., U.S. Sr. faculty fellow Am. Inst. Indian Studies, 1964-65, 1984-85; elected to Am. Film Inst.; guest fellow Oxford U., The Sorbonne, Inst. Advanced Study. Mem. N.Y. Acad. Scis., South Asian Sociols. (1st pres.), Soc. Indian Acads. in Am. (exec. com.), Global Orgn. People of Indian Origin (life). Home: 62 Belvedere Dr Yonkers NY 10705

VARMA, RAJ NARAYAN, nutrition educator, researcher; b. Shertallai, Kerala, India, Oct. 11, 1928; came to U.S., 1956; s. Rama and Amminikutty (Amma) V.; m. Valsamani Varma, May 24, 1956; children: Deepa, Usha. BSc, Banaras (India) U., 1948, MSc, 1951; MS, U. Calif., Davis, 1958, PhD, 1962. Registered dietitian; lic. dietitian. Asoc. prof. Fort Valley (Ga.) State Coll., 1969-73; asst. prof. Mont. State U., Bozeman, 1973-74; assoc. prof. Tuskegee (Ala.) U., 1974-76; asst. prof. Hunter Coll., CUNY, 1978-81; vis. assoc. prof. U. South Ala., Mobile, 1981-83; asst. prof. nutrition Youngstown (Ohio) State U., 1983-87, Disting. prof., 1987, assoc. prof., 1987—; rsch. cons. St. Vincent Charity Hosp., Cleve., 1986-89. Cancer Control Consortium of Ohio grantee, 1985. Mem. Am. Soc. for Parenteral and Enteral Nutrition, Am. Dietetic Assn., Nutrition Today Soc., Sigma Xi. Office: Youngstown State U 410 Wick Ave Youngstown OH 44555

VARMAVUORI, ANNELI, chemical society administrator, chemist; b. Helsinki, Finland, Apr. 7, 1939; d. Reino Kaarlo and Eva-Maija (Angervo) V. MSc in Chemistry, U. Helsinki (Finland), 1969; info. specialist, Helsinki U. Tech., 1974. Synthetic chemist Alko Ltd. Rsch. Lab., Helsinki, Finland, 1969; analytical chemist Orion Corp. Orion Pharm., Helsinki, Finland, 1969-72; sr. scientist, head of info. svc. Kemira Group Espoo Rsch. Ctr., Espoo, Finland, 1972-85; sec. gen. The Assn. of Finnish Chem. Socs., Helsinki, Finland, 1985—; pres. Finnish Chem. Congress, Helsinki, 1983-93; bd. dirs., v.p. Assn. Finnish Chem. Socs., Helsinki, 1978-85. Congress editor: 27th Internat. Congress of International Union of Pure and Applied Chemistry, 1979; editor: Kemistimatrikkeli-Kemistmatrikeln, 1986. Recipient Golden Badge of Merit Student Union of Chemistry, 1965, Pennant Chem. Soc. of No. Finland, 1989, Pennant of Merit Assn. Finnish Chem. Socs., 1990. Mem. Fedn. Finnish Chemists (bd. dirs., v.p. 1973-80, Badge of Merit 1982), Finnish Chem. Soc. (bd. dirs. 1975-91, pres. 1982, Pennant of Merit 1983), Finnish Soc. Chem. Engrs., Chem. Soc. Finland, Am. Chem. Soc., Assn. Franco-Finlandaise pour la Reserche Scientifique et Technique. Avocation: foreign languages. Office: Assn Finnish Chem Societies, Hietaniemenkatu 2, 00100 Helsinki Finland

VARMUS, HAROLD ELIOT, microbiologist, educator; b. Oceanside, N.Y., Dec. 18, 1939; s. Frank and Beatrice (Barasch) V.; m. Constance Louise Casey, Oct. 25, 1969; children: Jacob Carey, Christopher Isaac. AB, Amherst Coll., 1961, DSc (hon.), 1984; MA, Harvard U., 1962; MD, Columbia U., 1966. Lic. physician, Calif. Intern, resident Presbyterian Hosp., N.Y.C ., 1966-68; clin. assoc. NIH, Bethesda, Md., 1968-70; lectr. dept. microbiology U. Calif.-San Francisco, 1970-72; asst. prof. U. Calif., San Francisco, 1972-74; assoc. prof. U. Calif., San Franisco, 1974-79; prof. U. Calif., San Francisco, 1979—, Am. Cancer Soc. research prof., 1984-93; dir. National Institutes of Health, Bethesda, Md., 1993—; chmn. bd. on biology NRC. Editor: Molecular Biology of Tumor Viruses, 1982, 85; Readings in Tumor Virology, 1983; assoc. editor Genes and Development Jour., Cell Jour.; mem. editorial bd. Cancer Surveys. Named Calif. Acad. Sci. Scientist of Yr., 1982; co-recipient Lasker Found. award, 1982, Passano Found. award, 1983, Armand Hammer Cancer prize, 1984, GM Alfred Sloan award 1984, Shubitz Cancer prize, 1985, Nobel Prize in Physiology or Medicine, 1989. Mem. AAAS, NAS, Inst. Medicine of NAS, Am. Soc. Virology, Am. Soc. Microbiology, Am. Acad. Arts and Scis. Democrat. Home: 956 Ashbury St San Francisco CA 94117-4409 Office: National Institutes of Health 9000 Rockville Pike Bethesda MD 20892

VARNER, JOSEPH ELMER, biology educator, researcher; b. Nashport, Ohio, Oct. 7, 1921; s. George Ezra and Inez Charlotte (Gladden) V.; m. Carol Roberta Dewey, June, 1945 (div. 1971); children—Lee, Lynn, Karen, Beth; m. Jane Elanor Burton, June, 1976. B.Sc., Ohio State U., 1942, M.Sc., 1943, Ph.D., 1949; Docteur Honoris Causa (hon.), Nancy U., France, 1977. Chemist Owens Corning, Newark, 1943-44; analytical chemist Battelle Meml. Inst., Columbus, Ohio, 1946-47; from asst. prof. to prof. biology Ohio State U., Columbus, Ohio, 1950-61; postdoctoral Calif. Inst. Tech., Pasadena, 1953-54; NSF fellow Cambridge U., Eng., 1959-60; research scientist Martin-Marietta, Balt., 1961-65; prof. biochemistry Mich. State U., East Lansing, 1965-73; NSF fellow U. Wash., Seattle, 1971-72; prof. biology Washington U., St. Louis, 1973—, Am. Cancer Soc. scholar, 1980-81. Author, editor: Plant Biochemistry, 1965, 2d edit., 1976; contbr. articles on plant biology to profl. jours. Fellow AAAS; mem. NAS, Am. Soc. Plant Physiologists (pres. 1969, Stephen Hales Prize 1990), Soc. Developmental Biology (pres. 1986), Am. Acad. Arts and Sci. Avocation: soaring. Home: 7275 Kingsbury Blvd Saint Louis MO 63130-4141 Office: Washington U Dept Biology Saint Louis MO 63130

VARNEY, ROBERT NATHAN, retired physicist, researcher; b. San Francisco, Nov. 7, 1910; s. Frank Hastings Sr. and Emily Patricia (Rhine) V.; m. Astrid Margareta Riffolt, June 19, 1948; children: Nils Roberts, Natalie Rhine. AB with highest honors in Physics, U. Calif., Berkeley, 1931, MA, 1932, PhD, 1935; DSc (hon.), Leopold Franzens U., Innsbruck, Austria, 1983. Instr. NYU, 1936-38; asst. prof., assoc. prof., prof. Washington U., St. Louis, 1938-64; mem. rsch. lab. Bell Labs, Murray Hill, N.J., 1951-52; sr. mem. rsch. lab., sr. sci cons. Lockheed Missiles & Space Co., Palo Alto, Calif., 1964-75; guest prof. Leopold Franzens U., Innsbruck, 1977-78; mem. Mo. Gov.'s Sci. Adv. Com., St. Louis, 1960-64; mem. tech. staff Bell Telephone Labs., Murray Hill, N.J., 1951-52. Author: Engineering Physics, 1948; (with others) Methods of Experimental Physics, 1968, Introduction to ... Atmospheric Pollution, 1972; contbr. 82 articles to scholarly and profl. jours. Comdr. USNR, 1931-57. Fulbright fellow Leopold Franzens U., Innsbruck, 1971-72, 76-77, NSF sr. postdoctoral fellow Inst. Tech., Stockholm, 1958-59, NRC sr. postdoctoral fellow U.S. Army Ballistic Rsch. Lab., Aberdeen, Md., 1975-76; recipient Cross of Honor 1st Class Austrian Govt., 1981. Fellow Am. Phys. Soc.; mem. Am. Assn. Physics Tchrs., Phi Beta Kappa, Sigma Xi, Tau Beta Pi, Omicron Delta Kappa. Episcopalian. Achievements include research in electron swarms and atmospheric pollutants. Home: 4156 Maybell Way Palo Alto CA 94306-3820

VARNHAGEN, MELVIN JAY, mechanical engineer; b. N.Y.C., Mar. 13, 1944; s. Melvin Jay Varnhagen Sr. and Amy (Wertheimer) Hawkes. B Engring., Stevens Inst. Tech., 1966; MA in Indsl. Safety, NYU, 1973. Registered profl. engr., Mass., Vt., N.H.; cert. safety profl. Process engr. Pepperidge Farms, Inc., Norwalk, Conn., 1986-87; plant engr. Micrognosis, Inc., Danbury, Conn., 1987; process engr. Brake Systems Inc., Stratford, Conn., 1987; engring. specialist UNC Naval Products, Inc., Uncasville, Conn., 1987-88; plant engr. Spalding Sports Worldwide, Chicopee, Mass., 1989; product design engr. Kirby/Lester Electronics, Stamford, Conn., 1990; design engr. The Coca Cola Co., Atlanta, 1990; sr. engr. Brand Utility Svcs., Inc., Essex, Conn., 1991-92; mech. project mgr. Hi-Tech Extrusion Systems Inc., Westerly, R.I., 1992—; cons. MJV Engring. Co., Wilmington, Vt., 1976-86; advisor State of Vt. Bd. Profl. Engrs., 1981. Author: Status of Engineers in Today's Society, 1980. Water Safety instr. ARC, 1964. Recipient grant and stipend Nat. Inst. Occupational Safety and Health, NYU, 1972. Mem. NSPE, Vt. Soc. Profl. Engrs., Am. Soc. Home Inspectors, Profl. Singles Assn., Stevens Club. Achievements include advocate of professional registration of engineers in industry and honesty in military procurement; design of industrial and commercial equipment and products and the systems to manufacture them. Office: Hi-Tech Extrusion Systems PO Box 1995 Airport Dr E Westerly RI 02891

VARON, DAN, electrical engineer; b. Tel-Aviv, Israel, July 24, 1935; came to U.S., 1961; s. Reuven and Stephanie Varon; m. Judith Hilda Mansbach, Aug. 12, 1962; 2 children:. BSEE, The Technion, Haifa, Israel, 1957, Diplom Ingenieur, 1960; MSc, Polytechnic U., Bklyn., 1963; D of Engring. Sci., NYU, 1963. Engr. Israeli Air Force, Israel, 1957-61; rsch. fellow Polytechnic U., Bklyn., 1961-63; teaching asst. NYU, N.Y.C., 1963-65; mem. tech. staff Bell Telephone Labs., Whippany, N.J., 1965-69; mgr. Tymshare (Dial Data, Inc.), Newton, Mass., 1969-71; prin. engr. Raytheon Co., Marlborough, Mass., 1971—. Contbr. articles to Bell System Tech. Jour., IEEE, Radio Sci., Jour. Air Traffic Control Assn. Mem. IEEE (mem. tech. program com. 1970), N.Y. Acad. Scis., Assn. Computing Machinery.

VARSANYI-NAGY, MARIA, biochemist; b. Budapest, Hungary, Mar. 5, 1942; came to U.S., 1987; d. Bela and Iren (Csontos) N.; m. Janos Varsanyi, Jan. 23, 1965; 1 child, Eva. MA in Biochemistry, Tech. U. Budapest, 1970, PhD in Biochemistry, 1977; DSc in Med. Sci., Hungarian Acad. Sci., Budapest, 1991. Head lipid rsch. lab. Semmelweis Med. Sch., Budapest, 1970-78, head isotope rsch. lab., 1978-87, 89-90; vis. researcher dept. medicine U. Calif., Irvine, 1987-89; dir. diabetes rsch. lab. Children's Hosp. Orange County, Orange, Calif., 1990—; cons. immunology rsch. program Semmelweis U., Budapest, 1980—; lectr. in field. Contbr. articles, abstracts in field of diabetes, immunology and metabolism to sci. publs. Mem. Am. Diabetes Assn., European Assn. for Study of Diabetes, Am. Fedn. Clin. Rsch., Hungarian Diabetes Assn., Hungarian Immunology Assn., Hungarian Biochemistry Assn. Achievements include research in Type I diabetes as an autoimmune disease caused by activated macrophage product interleukin-1B induced free radical generation and its prevention by inhibiting macrophage activation. Office: Childrens Hosp Orange County 455 South Main St Orange CA 92668

VARSHAVSKY, ALEXANDER JACOB, molecular biologist; b. Moscow, Nov. 8, 1946; came to U.S., 1977; s. Jacob M. Varshavsky and Mary B. (Zeitlin) V.; m. Vera Bingham, Aug. 30, 1990; children: Roman Bingham, Victoria. BS in Chemistry, Moscow State U., 1970; PhD in Biochemistry, Inst. of Molecular Biology, Moscow, 1973. Asst. prof. Dept. Biology, MIT, Cambridge, 1977-80, assoc. prof., 1980-86, prof., 1986-92; H. Smits prof. cell biology Calif. Inst. Tech., Pasadena, 1992—. Contbr. over 100 articles to profl. jours. in the field of genetics and biochemistry. Mem. AAAS. Achievements include 8 patents; discoveries in the fields of DNA replication, chromosome structure and intracellular protein turnover. Office: Calif Inst Tech Divsn Biology Pasadena CA 91125

VARVAK, MARK, mathematician, researcher; b. Kiev, USSR, Feb. 13, 1939; came to U.S., 1987; s. Shlyoma and Anna (Berimskaya) V.; m. Nellie Albert, Feb. 1, 1973 (div. Oct. 1983); 1 child, Alexander. MS in Applied Math., Ukraine U., 1969; candidate of sci., Rsch. Inst. of Structures, Kiev, 1970. Sr. researcher Rsch. Inst. of Structures, Kiev, 1963-79; sr. engr. Constrn. Authority, Kiev, 1979-86; programmer Consulting Engring. Co., N.Y.C., 1988-91. Contbr. more than 40 articles to profl. jours. Mem. Soc. for Indsl. and Applied Math.

VARY, JAMES PATRICK, physics educator; b. Savanna, Ill., May 23, 1943; s. Willis L. and Ethice K. (McCabe) V.; m. Audrey Maria Zarba, June 11, 1966 (div. June 1989); children: William James, Brian Edward; m. Hildegard Maria Mummenthal, May, 18, 1991. BS, Boston Coll., 1965; MS, MPh, Yale U., 1968, PhD, 1970. Rsch. assoc. MIT, Cambridge, 1970-72; asst. physicist Brookhaven Nat. Lab., Upton, N.Y., 1972-74, assoc. physicist, 1974-75; asst. prof. physics Iowa State U., Ames, 1975-77, assoc. prof., 1977-81, prof., 1981—, acting dir. Internat. Inst. Theoretical and Applied Physics, 1993—; dir. nuclear theory program Ames Lab., 1977-82; vis. prof. Calif. Inst. Tech., 1986-87; Disting. vis. prof. Ohio State U., 1987-88; vis. prof. U. Washington, 1992; chmn. Gordon Rsch. Conf. Nuclear Physics, 1991; Contbr. over 135 articles to sci. jours. Editor of Conf. Proceedings of 2 Internat. Confs. in Physic, 1980-89. Recipient Alexander von Humboldt award for Sr. U.S. Scientists, 1992; Alexander von Humboldt fellow, 1979. Fellow Am. Phys. Soc., Union of Concerned Scientists, Am. Phys. Soc., Sigma Xi.

VASA, ROHITKUMAR BHUPATRAI, pediatrician, neonatologist; b. Rajula, Gujarat, India, July 26, 1947; came to U.S., 1973; s. Bhupatrai Jayantilal and Vijayalaxmi V.; m. Usha B. Shah, Feb. 26, 1970; children: Falguni, Monisha. MBBS, Med. Coll., Baroda, India, 1970, Diploma Child Health, 1971, MD in Pediatrics, 1973. Diplomate Am. Bd. Pediatrics, Sub-Bd. of Neonatal Perinatal Medicine, Am. Bd. Sports Medicine. Intern, then resident Beth Israel Hosp., N.Y.C., 1973-75; fellow in neonatology Bellvue Hosp. Med. Ctr., N.Y.C., 1975-77; attending pediatrician Gouverneur Hosp., N.Y.C., 1977-79; asst. chief pediatrics U.S. Army Hosp., Fort Campbell, Ky., 1979-81; dir. neonatology Mercy Hosp. and Med. Ctr., Chgo., 1981—. Contbr. articles to profl. jours. Fellow Am. Acad. Pediatrics; mem. AAAS, Soc. of Critical Care Medicine, Am. Coll. Sports Medicine, India Med. Assn. (sec. 1992, pres. elect 1993), Indian Acad. Pediatrics, Chgo. Pediatrics Soc., Physicians for Perinatal Care (sec. 1991-93). Hindu. Office: Mercy Hosp Dept Pediatrics Stevenson Expy at King Dr Chicago IL 60616

VASAK, DAVID JIŘI JAN, physicist, consultant; b. Prague, Bohemia, Czechoslovakia, Dec. 30, 1951; arrived in Fed. Republic Germany, 1968; s. Jan Z. and Edith L. (Wieluch) V.; m. Gudrun E. Giel, Nov. 19, 1982 (div. 1992); 1 child, Jan N. Diploma, U. Frankfurt, Fed. Republic Germany, 1979, PhD summa cum laude, 1985. Rsch., teaching assoc. U. Frankfurt Inst. für Theoretische Physik, 1979-84; guest scientist Ges. für Schwerionen-Forschung, Damstadt, Fed. Republic Germany, 1984-85; researcher U. Calif. Lawrence Berkeley Lab., 1985-86; sr. physicist NIS Ingenieur Ges. mbH, Hanau, Fed. Republic Germany, 1987-91; mgr. KPMG Nolan, Norton & Co., Frankfurt, 1991-93; ea. Europe rep. NIS, 1993. Contbr. articles to profl. jours. NATO rsch. fellow, 1985. Avocation: basketball. Home: Rhönstrasse 15, 63450 Hanau Germany Office: NIS Ingenieurgesellschaft mbH, Donaustr 23, 63452 Hanau Germany

VASILAKIS, ANDREW D., mechanical engineer; b. Somerville, Mass., Apr. 21, 1943; s. Dimitrios John and Penelope (Axiotakis) V.; m. Nancy E. Ellis, June 12, 1966; children: Kosta A., Elena C. BSME, Northeastern U., 1966; MSME, MIT, 1968. Engr. Dynatech Corp., Cambridge, Mass., 1967-69; sr. engr. Thermo Electron, Waltham, Mass., 1969-78; v.p. AMTI, Newton, Mas., 1978-84; mgr. engring. Tecogen, Waltham, 1984-89; sr. v.p. AMTI, Newton, 1989—. Mem. ASME, Assn. Energy Engrs. Achievements include patents for Low Pollution high Efficiency Burner; High Efficiency Water Heating System; Combustion Product Water Heater. Office: AMTI 151 California St Newton MA 02158

VASILIADES, JOHN, chemist; b. Kastoria, Greece, Jan. 20, 1945; came to the U.S., 1955; s. Evan and Alexandra (Sotiropoulos) V.; m. Anne E. Stephens, Apr. 16, 1977; children: Elizabeth, Evan. BA, Hunter Coll., 1967; PhD, U. Nebr., 1971. Instr. dept. chemistry U. Nebr., Lincoln, 1969-71; rsch. assoc. Purdue U., West Lafayette, Ind., 1971-72; resident, fellow dept. clin. pathology U. Ala., Birmingham, 1972-73, instr. sch. medicine dept. pathology, 1973-75, asst. prof., 1975-80; asst. prof. U. Mich., Ann Arbor, 1980-82; assoc. prof. dept. pathology Creighton U., Omaha, 1982-85; chief quality assurance program, expert witness Air Force Drug Testing Program, Brooks AFB, Tex., 1985-86; clin. assoc. prof. dept. pathology U. Tex. Health Scis. Ctr., San Antonio, 1986-87; lab. dir. human reproduction Pope Paul VI Inst., Omaha, 1988—; asst. prof. dept. chemistry U. Ala., Birmingham, 1973-80, dir. spl. chemistry and toxicology sect., 1973-80; assoc. dir. clin. chemistry, dir. rsch. dept. pathology U. Mich., 1980-82; dir. clin. chemistry and toxicology Creighton U./St. Joseph Hosp., 1982-85, toxicologist, 1983-85; assoc. prof. pharm. scis. sch. pharmacy and allied health professions Creighton U., 1983-85; adj. staff St. Joseph Hosp., 1983-85; lab. dir., toxicologist Toxicology and Clin/Chem Labs. Inc., Omaha, 1986—. Editor: Am. Assn. Drug Detection Labs. Newsletter, 1976-77; contbr. articles to Inorganic Chemistry, Analytical Chemistry, Ala. Jour. Med. Sci., Clin. Chemistry, Clin. Toxicology, Jour. Analytical Toxicology, Am. Jour. Clin. Pathology, Jour. Chromatography, Clin. Biochemistry, Advances in Neurotoxicology, Circulation. Cons. VA Hosp., Ann Arbor, Omaha. Mem. Am. Acad. Forensic Scientists, Am. Chem. Soc. (analytical sect.), Internat. Assn. Forensic Toxicologists, AAAS, Acad. Clin. Lab. Physicians and Scientists, Soc. Applied Spectroscopy, Am. Assn. Clin. Chemists, Am. Acad. Clin. Toxicology, N.Y. Acad. Sci., Nat. Acad. Forensic Scientists, Forensic Sci. Soc., Soc. Forensic Toxicologists, Sigma Xi. Home: 933 N Howard Fremont NE 68025 Office: Toxicology Clin/Chem Labs 7701 Pacific St Ste 300 Omaha NE 68114

VASQUEZ, RODOLFO ANTHONY, protein chemist, researcher; b. L.A., Dec. 30, 1954; s. Teofilo and Celia (Lopez) V.; m. Lucia Colin Maldonado, Mar. 25, 1978; children: Alicia, Stephen. BS in Biol. Scis., U. Calif., Davis, 1979. Rsch. assoc. Harbor-UCLA Med. Ctr., Torrance, 1979-80; quality assurance analyst Travenol Labs., L.A., 1980-81; sr. rsch. assoc. Hyland div. Baxter Healthcare Corp., L.A., 1981-87, supr. parenterals prodn., 1987-89, project mgr., 1989-91, program mgr., 1991—. Mem. Parenteral Drug Assn. Achievements include 2 U.S. patents (with others) for Method for Preparing Antihemophilic Factor by Cold Precipitation and for Improving Solubility of Recovered AHF Product, and for a Process for Purifiying Immune Serum Globulins, 1 European patent for Process for Preparing Immune Serum Globulins; U.S. patent pending for preparing stable intravenously-administrable immune globulin preparations. Office: Baxter Healthcare Corp Hyland Div 4501 Colorado Blvd Los Angeles CA 90039-1103

VASSILOPOULOU-SELLIN, RENA, medical educator. MD, A. Einstein, 1974. Assoc. prof. Univ. Tex., Houston, 1980—. Office: Anderson Cancer Ctr 1515 Holcombe Blvd #15 Houston TX 77030-4009

VASSY, DAVID LEON, JR., radiological physicist; b. Gaffney, S.C.. BS, Furman U., 1974; MS, Clemson U., 1978. Diplomate Am. Bd. Radiology and Therapeutic Radiol. Physics, Am. Bd. Med. Physics in Therapeutic Radiol. Physics. Radiol. physicist Mid-Atlantic Radiation Physics, Inc., Adelphi, Md., 1978-80, Spartanburg (S.C.) Regional Med. Ctr., 1980-84, Spartanburg Radiation Oncology, P.A., 1984—; mem. imaging adv. com. Sun Health, Inc., Charlotte, N.C., 1987—; cons. to numerous hosps. and pvt. physicians' offices. Contbr. articles to profl. publs. Mem. Am. Assn. Physicists in Medicine, Am. Soc. for Therapeutic Radiation Oncology, Health Physics Soc., Radiol. Soc. N.Am. Office: Spartanburg Radiol Oncology PO Box 4126 Spartanburg SC 29305-4126

VASUDÉVAN, ASURI KRISHNASWAMI, materials scientist; b. Mysore, Karnataka, India, May 26, 1943; came to U.S. 1967; m. Martha L. Farinacci, Sept. 12, 1981. MS, Stevens Inst. Tech., 1969; PhD, Case Western Res. U., 1974. Vis. scientist Wright-Patterson AFB, Dayton, Ohio, 1973-75; vis. asst. prof. So. Ill. U., Carbondale, 1975; vis. scholar Northwestern U., Evanston, Ill., 1975-77; staff scientist Aluminum Co. of Am., Pitts., 1978-88; program mgr. Office of Naval Rsch., Arlington, Va., 1988—. Author/editor: Aluminum Alloys, 1989, High Temperature Structural Silicides, 1992; editor: Monograms in Materials Science, 1991—; assoc. editor Materials Sci. &

Engring. Jour., 1984—; contbr. numerous articles to profl. jours. Recipient Tech. award Aluminum Co. of Am., 1984, 85, 87, R&D Award for 2090 Alloy, Indsl. Rsch., Chgo., 1986, Mgmt. award USN, 1989, 90, 91, 92. Mem. Am. Soc. for Materials, Transactions Metals Soc. Achievements include 5 patents on various aluminum alloy develolpments and processing. Home: 13241 Pleasant Glen Ct Herndon VA 22071-2344 Office: Office of Naval Rsch Code 4421 800 N Quincy St Arlington VA 22217-5660

VATHSAL, SRINIVASAN, electrical engineer, scientist; b. Tiruchira Palli, Tamil Nadu, India, Nov. 5, 1947; s. Rangasamy Srinivasan and Seshadri Jambakavalli; m. Srinivasa Raghavan Alamelu Vathsal, Feb. 5, 1975; children: V. Padmapriya, V. Srividya, V. Radha. BE, Thiagarajar Coll. Engring., Madurai, 1968; ME, Birla Inst Tech. and Sci., Pilani, 1970; PhD, Indian Inst. Sci., Bangalore, 1974. Scientist Vikram Sarabhai Space Ctr., Trivandrum, Kerala, 1974-78, VikramSarabhai Space Ctr., Trivandrum, Kerala, 1980-82; postdoctoral fellow DFVLR, Braunschweig, West Germany, 1978-80; prof. P.S.G. Coll. Tech., Coimbatore, 1982-84; sr. NRC-NASA rsch. scientist GSFC-NASA, Greenbelt, Md., 1984-86; prin. scientist, prof. Osmania U., Hyderabad, 1989-90; scientist DRDL, Hyderabad, 1989—, head post flight analysis directorate systems, 1992-93; cons. Project Prithvi/DRDL, Hyderabad, 1986-87; mem. CABS Projects, Bangalore, 1990-92; chmn. ECM Environment, Hyderabad, 1990-92. Contbr. articles to profl. jours. Mem. IEEE (sr.), AIAA, System Soc. India, Aero. Soc. India. Office: DRDL, Kanchanbagh, Hyderabad 500258, India

VAUGHAN, JOHN THOMAS, biology educator; b. St. Petersburg, Fla., Feb. 18, 1959; s. James Taylor and Peggy (Garrett) V.; m. Vicki Lynn Pittman, Aug. 24, 1980. BS in Biology, Appalachian State U., 1980; MS in Entomology, U. Fla., 1985; postgrad., Ohio State U., 1985—. Park naturalist South Mountains State Park, Connelly Springs, N.C., 1980; mgr. Longvue Motel, Boone, N.C., 1980-81; entomologist U. Fla./USDA, Gainesville, 1981-85; asst. prof. biology Daemen Coll., Amherst, N.Y., 1991—; environ. cons. SERPETEC, Quito, Ecuador, 1990; cons. Sci. Kit & Boreal Labs., Buffalo, 1992. Rsch. grant Tinker Found., 1990, Fla. Entomol. Soc., 1985; recipient Grad. Alumni Rsch. award Ohio State Grad. Sch., 1990. Mem. Soc. for the Study of Evolution, Soc. for the Study of Amphibians and Reptiles, Sigma Xi. Achievements include rsch. in evidence of male choice in Galapagos Lava Lizards (Tropidurus grayi). Office: Daemen Coll 4380 Main St Amherst NY 14226-3592

VAUGHAN, MARGARET EVELYN, psychologist, consultant; b. Mpls., Nov. 9, 1948; d. Robert Bergh and Evelyn (Glockner) Cedergren; m. William Vaughan Jr., July 30, 1981. BA, St. Cloud (Minn.) State U., 1972; MA, Western Mich. U., 1977, PhD, 1980. Lic. psychologist, Mass. Asst. prof. psychology Kalamazoo (Mich.) Coll., 1979-81; postdoctoral fellow Harvard U., Cambridge, Mass., 1981-82, rsch. assoc., 1982-83, rsch. assoc. Sch. of Bus., 1983-84; asst. prof. psychology Salem (Mass.) State Coll., 1984-88, assoc. prof. psychology, 1988—; cons. Shore Ednl. Collaborative, Medford, Mass., 1984—; bd. dirs. B.F. Skinner Found., Cambridge. Author: (with B.F. Skinner) Enjoy Old Age, 1983; editor-elect The Behavior Analyst Jour., 1991-93, editor, 1993—. Mem. APA, Assn. for Behavior Analysis, Phi Kappa Phi, Psi Chi, Alpha Lambda Delta. Office: Salem State Coll Dept Psychology Salem MA 01970

VAUGHAN, MARTHA, biochemist; b. Dodgeville, Wis., Aug. 4, 1926; d. John Anthony and Luciel (Ellingen) V.; m. Jack Orloff, Aug. 4, 1951 (dec. Dec. 1988); children: Jonathan Michael, David Geoffrey, Gregory Joshua. Ph.B., U. Chgo., 1944; M.D., Yale U., 1949. Intern New Haven Hosp., Conn., 1950-51; research fellow U. Pa., Phila., 1951-52; research fellow Nat. Heart Inst., Bethesda, Md., 1952-54, mem. research staff, 1954-68; head metabolism sect. Nat. Heart and Lung Inst., Bethesda, 1968-74; acting chief molecular disease br. Nat. Heart, Lung and Blood Inst., Bethesda, 1974-76, chief cell metabolism lab., 1974—; mem. metabolism study sect. NIH, 1965-68; mem. bd. sci. counselors Nat. Inst. Alcohol Abuse and Alcoholism, 1988-91; cons. editor Acad. Medicine. Editorial bd. Jour. Biol. Chemistry, 1971-76, 80-83, 88-90, assoc. editor, 1992—; editorial adv. bd. Molecular Pharmacology, 1972-80, Biochemistry,1 989—; editor Biochemistry and Biophysics Rsch. Communications, 1990-91; contbr. articles to profl. jours., chpts. to books. Bd. dirs. Found. Advanced Edn. in Scis., Inc., Bethesda, 1979-92, exec. com., 1980-92, treas., 1984-86, v.p., 1986-88, pres., 1988-90; mem. Yale U. Coun. com. med. affairs, New Haven, 1974-80. Recipient Meritorious Service medal HEW, 1974, Disting. Service medal HEW, 1979, Commd. Officer award USPHS, 1982. Mem. NAS, Am. Acad. Arts and Scis., Am. Soc. Biol. Chemists (chmn. pub. com. 1984-86), Assn. Am. Physicians, Am. Soc. Clin. Investigation. Home: 11608 W Hill Dr Rockville MD 20852-3751 Office: Nat Heart Lung & Blood Inst NIH Bldg 10 Rm 5N-307 Bethesda MD 20892

VAUGHAN, OTHA H., JR., aerospace engineer, research scientist; b. Anderson, S.C., July 1, 1929; s. Otha H. and Ethel (Mayfield) V.; m. Betty Frances McCoy; children: Thera Virginia, Leslie, Frances. BS in Mech. Engring, Clemson U., 1951, MS in Mech. Engring., 1959; postgrad., U. Tenn. Space Inst., Tullahoma, 1975-81, U. Ala., Huntsville, 1974-75. Registered profl. engr., Ala. Commd. 2nd lt. USAF, 1951, advanced through grades to lt. col., 1972, retired, 1979; rsch. engr. NASA Marshall Space Flight Ctr., Huntsville, Ala., 1980—. Contbr. over 60 articles to profl. jours. Charter Mem. Aviation Hall of Fame, Dayton, Ohio. Mem. AIAA (assoc. fellow), Nat. Soc. Profl. Engrs., Air Force Assn. (past v.p., Huntsville chpt., life mem.), Reserve Officers Assn. (past pres. Huntsville chpt., life mem.), Minute Man Soc. Ala., Antique Aircraft Assn. (life mem.), Mason, Shriner. Achievements include patent in Lunar Communications Receiver and Transmitter for Lunar Surface Missions; participation in design of rocket and space vehicle systems, research and development of Redstone, Jupiter, Juno, Saturn I, Saturn IB, and Saturn V, Skylab and Apollo program, and the space shuttle launch vehicle systems; research in environmental design criteria for lunar and planetary exploration vehicles, zero-g atmospheric cloud physics, and atmospheric electricity research. Home: 10102 Westleigh Dr SE Huntsville AL 35803-1647

VAUGHAN, STEPHEN OWENS, project chemist; b. Newton, N.C., Jan. 29, 1961; s. Thomas Owens and Betty Lee (McAlister) V.; m. Kathleen Fern Scherr, Apr. 16, 1988; 1 child, William Owens. BS in Chemistry, U. N.C., 1983; MA, Johns Hopkins U., 1985, PhD in Physical Chemistry, 1987. Rsch. analyt. chemist Rsch. Triangle Inst., Research Triangle Park, N.C., 1987-90; materials engr. ITT-Electro-Optical Products div., Roanoke, Va., 1990-91; mem. tech. staff ITT-Electro-Optical Prodn. div., Roanoke, Va., 1991-93; sr. project chemist Scott Splty. Gases, Plumsteadville, Pa., 1993—; presenter in field. Contbr. 7 articles to profl. jours. Chmn., host Cryolect Users Meeting, RTP, 1988. Mem. Am. Chem. Soc., Alpha Epsilon Delta. Achievements include patent pending for novel application of image intesifiers in chromatographic detection; research in instituted recycling programs, efforts to reduce chemical waste at point sources; current research involves the study of surface-gas interactions and mechanisms. Home: 201 Commons Way Doylestown PA 18901 Office: Scott Splty Gases 6141 Easton Rd Plumsteadville PA 18949-0310

VAUGHAN-KROEKER, NADINE, psychologist; b. Tampa, Fla., Aug. 30, 1947; d. Joseph Marcus and Velna Pearl (Jones) Williams; m. E. L. Vaughan III, 1966 (div. Aug. 1976); children: E. L. Vaughan, Heather Vaughan Oyarzun; m. Dennis Wayne Kroeker, Apr. 9, 1982; 1 child, Melanie Sage. BA in Criminal Justice, U. South Fla., 1974, MA with honors in Rehab. Counseling, 1975; PhD in Psychology, Saybrook Inst., 1990. Lic. mental health counselor, Fla., lic. clin. psychologist, Calif. Co-founder Women's Resource Ctr., Tampa, Fla., 1973-75; clin. dir. Oper. PAR Inc., Clearwater, Fla., 1975-76; exec. dir. Vocare Found., Oakland, Calif., 1976-78; community and organizational devel. specialist Calif. Employment Dept., Calif Rehab. Dept., Berkeley, Sacramento, 1978-82; cons., trainer N. Vaughan-Kroeker PhD Profl. Svcs., 1982—; part-time faculty psychology Sierra Coll., Rocklin, Calif., 1990—; exec. dir. Women's Resource Ctr. No. Calif., Nevada City, 1990-92; cons., trainer U.S. Forest Svc., Nevada City, 1990-91; chief devel. officer Delivery Club, Grass Valley, Calif., 1990—. Mem. Nevada County (Calif.) Task Force on Drug/Alcohol Abuse, 1990; bd. dirs. Foothill Theatre Co., Nevada City, 1988-89; vol. psychotherapist, bd. dirs. People with HIV, Grass Valley, 1989-92. Mem. Am. Psychol. Assn., Assn. Humanistic Psychology. Democrat. Avocations: theatrical performance (singing, acting, dance). Office: PO 331 Nevada City CA 95959

VAUGHN, EDDIE MICHAEL, mechanical engineer; b. Fairfield, Calif., Nov. 13, 1956; s. John Dennis and Velma May (Neeley) V.; m. Sharon Ann Jones, Nov. 19, 1982; children: Heather Renee, Jonathan Michael. BSME, U. Tenn., 1981. Registered profl. engr., Fla. Mech. engr. Fla. Power & Light Co., Juno Beach, 1981—. Mem. ASME (ops. and maintenance working group 1989—). Baptist. Home: 6125 Francis St Palm Beach Gardens FL 33418 Office: Fla Power & Light 700 Universe Blvd Juno Beach FL 33408

VAUGHT, RICHARD LOREN, urologist; b. Ind., Oct. 28, 1933; s. Loren Judson and Bernice Rose (Bridges) V.; widowed, July 1987; children: Megan, Niles, Barbara, Mary; m. Nancy Lee Gusa, Aug. 1992. AB in Anatomy and Physiology, Ind. U., 1955; MD, Ind. U., Indpls., 1958. Diplomate Am. Bd. Urology. Intern U.S. Naval Hosp., St. Albans, N.Y., 1958-59, resident in gen. surgery, 1959-60, resident in urology, 1960-63; spl. fellow Sloan Kettering Meml. Hosp. for Cancer and Allied Diseases, N.Y.C., 1962; pediatric urology observer Babies Hosp., Columbia-Presbyn. Med. Ctr., N.Y.C., 1962; head urology U.S. Naval Hosp., Beaufort, S.C., 1963-65; asst. chief urology, head pediatric urology U.S. Naval Hosp., San Diego, 1965-68; pvt. practice Plaza Urol., Sioux City; med. dir. dept. hyperbaric medicine St. Luke's Regional Med. Ctr., Sioux City, 1968—; pres., chmn. bd. dirs. Care Choices of Siouxland, Sioux City, 1987—. Organizer telecommunications system for deaf, Siouxland, 1983. Lt. comdr. USN, 1958-68. Fellow ACS, Internat. Soc. Cryosurgery, Am. Acd. Pediatrics; mem. Am. Urol. Assn., Soc. Pediatric Urology, Undersea and Hyperbaric Medicine Soc., Am. Soc. Laser Medicine and Surgery, Am. Lithotripsy Soc., Woodbury County Med. Soc. (pres.), Sertoma (Sertoman of Yr. 1983). Office: Plaza Urol PC 2800 Pierce St Ste 400 Sioux City IA 51104-3707

VAUX, HENRY JAMES, forest economist, educator; b. Bryn Mawr, Pa., Nov. 6, 1912; s. George and Mary (James) V.; m. Jean Macduff, Jan. 11, 1937; children: Henry J., Alice J. B.S., Haverford Coll., 1933, D.Sc. (hon.), 1985; M.S., U. Calif., 1935, Ph.D., 1948. Forest engr. Crown Willamete Paper Co., Portland, Oreg., 1936-37; instr. Sch. Forestry, Oreg. State Coll., Corvallis, 1937-42; assoc. economist La. Agrl. Expt. Sta., Baton Rouge, 1942-43; economist U.S. Forest Service, Berkeley, Calif. and Wash., 1946-48; lectr. Sch. Forestry, U. Calif.-Berkeley, 1948-50, assoc. prof., 1950-53, prof., 1953-78, prof. emeritus, 1978—, dean, 1955-65; asst. dir. Agrl. Expt. Sta. U. Calif., 1955-65, dir. Wildland Research Center, 1955-65; chmn. Calif. Bd. Forestry, 1976-83. Author tech. articles, bulls. Served from ensign to lt. (j.g.) USNR, 1943-46. Recipient Berkeley citation U. Calif., 1978, Disting. Service award Forestry Assn., 1986; hon. fellow Calif. Acad. Scis., 1986. Fellow AAAS, Soc. Am. Foresters (Gifford Pinchot medal 1983); mem. Forest History Soc. (dir. 1971-75), Sigma Xi, Xi Sigma Pi. Mem. Soc. of Friends. Home: 622 San Luis Rd Berkeley CA 94707-1726

VAWTER, WILLIAM SNYDER, computer software consultant; b. N.Y.C., Oct. 26, 1931; s. William Snyder and Flora Adeline (Woods) V.; m. June Elizabeth Lindahl Mahoney, June 22, 1957 (div. Sept. 1972); children: Jonathan, Timothy, Jamison; m. Donia Ibrahim Hamarneh, Dec. 26, 1976; 1 child, Rafic William. BA, Rutgers U., 1953; postgrad., Harvard Divinity Sch., 1957-58. 1st lt. U.S. Army Signal Corps, Fort Devens, Mass., 1953-55; computer analyst B.F. Goodrich Co., Watertown, Mass., 1956, Traffic Exec. Assn., Eastern R.R., N.Y.C., 1959-60; systems analyst Mangels Stores, N.Y.C., 1961, Gimbels, N.Y.C., 1962-63; mgr. environ. data applications Computer Applications, Inc., N.Y.C., 1963-64; systems analyst Fire Ins. Rsch. & Actuarial Assn., N.Y.C., 1965; v.p. Mgmt. Datavision, Inc., N.Y.C., 1965; pres. Mgmt. Data, Inc., Dover, Del., 1965—; cons. in field; with computerized reapportionment plan for subcom. N.Y. State Legislature, 1965; data base adminstr. NASA, Houston, 1976; computer mgr. Saudi Arabia N.G., Riyadh, 1980-81; computer cons. U.S. AID-funded Afghan Relief Project, Quetta, Pakisatan, 1990-91; logistics and computer cons. U.S. AID-funded Family Planning Project for Ministry of Health, Dhaka, Bangladesh, 1991-92; mem. MIS needs assessment team for Ministry of Health and Family Welfare for World Bank & Who, Dhaka, 1993. Patentee in field. Speaker Queens Village (N.Y.) Fair Housing Coun., 1967. Recipient Honor medal Freedoms Found., 1949, Tuition scholarship Rutgers U., 1949, N.Y. State Med. scholarship, 1955, Tuition scholarship, 1957. Mem. Assn. for Computing Machinery, Computer Soc. of IEEE, Royal Sch. Ch. Music, Am. Guild of Organists, Theta Chi Frat. Democrat. Episcopalian. Avocations: photography, yoga, teaching piano. Home: PO Box 232, Madaba Jordan Office: Mgmt Data Inc 15 E North St PO Box 899 Dover DE 19903-0899 also: GPO Box 3135, Dhaka 1000, Bangladesh

VAYNMAN, SEMYON, materials scientist; b. Odessa, USSR, Oct. 2, 1949; came to U.S., 1980; s. Kelman and Esther (Potashnik) V.; m. Dora Skladman, Nov. 18, 1977; children: Ethel, Alexander. MS in Chemistry, Odessa U., 1973; PhD in Materials Sci., Northwestern U., 1986. Rsch. scientist Rsch. Inst. Foundry Tech., Odessa, 1973-77, Rsch. Inst. Power Industry, Lvov, USSR, 1977-80, GARD, Niles, Ill., 1981-84; rsch. scientist, rsch. prof. Northwestern U., Evanston, Ill., 1986—; reviewer Jour. Electronic Packaging, 1989, IEEE publ., 1988—; mem. adv. bd. sci. com. for solder joints reliability Dept. Def., Washington, 1989-90. Contbr. sci. papers to profl. publs., chpts. to books. Mem. Am. Soc. Metals, Mineral, Metals and Materials Soc. (electronic packaging and interconnection materials com. 1989—). Achievements include 2 patents in foundry technology. Office: Northwestern U 1801 Maple Ave Evanston IL 60201

VAZ, NUNO ARTUR, physicist; b. Uige, Angola, Portugal, Nov. 23, 1951; came to U.S., 1976; s. Nuno and Ilda M. (Pedro) V.; m. Maria Joao, Apr. 16, 1977; children: Ana, Pedro, Bernardo. EE, Tech. U. Lisbon, Portugal, 1975; MA in Physics, Kent (Ohio) State U., 1977, PhD in Physics, 1980. Postdoctoral fellow CFMC, Lisbon, 1981-82; rsch. assoc. Kent State U., 1982-84; sr. rsch. scientist GM Rsch. Labs., Warren, Mich., 1984-86; staff rsch. scientist Techn. Leveraging Office, Warren, Mich., 1992—; rsch. scientist Technology Leveraging Office, 1992—. Author: (chpt.) Liquid Crystals, 1991; patentee polymer dispersed liquid crystals. Matsumae Internat. Found. fellow, 1983; Nato scholar, 1976-77, Fulbright Hays scholar, 1976-80. Mem. Internat. Soc. Magnetic Resonance, Am. Phys. Soc., Soc. Automotive Engrs., Sigma Xi. Roman Catholic. Office: GM Rsch Labs Physics Dept 30500 Mound Rd Warren MI 48090-9055

VEATCH, ROBERT MARLIN, medical ethics researcher, philosophy educator; b. Utica, N.Y., Jan. 22, 1939; s. Cecil Ross and Regina (Braddock) V.; m. Laurelyn Kay Lovett, June 17, 1961 (div. Oct. 1986); children: Paul Martin, Carlton Elliot; m. Ann Bender Pastore, May 23, 1987. B.S., Purdue U., 1961; M.S., U. Calif. at San Francisco, 1962; B.D., Harvard U., 1964, M.A., 1970, Ph.D., 1971. Teaching fellow Harvard U., 1968-70; research assoc. in medicine Coll. Physicians and Surgeons, Columbia U., 1971-72; assoc. for med. ethics Inst. of Society, Ethics and Life Scis., Hastings-on-Hudson, N.Y., 1970-75; sr. assoc. Inst. of Society, Ethics and Life Scis., 1975-79; prof. med. ethics Kennedy Inst. Ethics Georgetown U., 1979—, prof. philosophy, 1981—, dir., 1989—; adj. prof. depts. community and family medicine and ob/gyn, 1984—; mem. vis. faculty various colls. and univs.; mem. governing bd. Washington Regional Transplant Consortium, 1988—; bd. dirs. Hospice Care D.C., 1989, pres., 1993—; active United Network Organ Sharing Ethics Com., 1989—, Washington Regional Transplant Consortium Governing Bd., 1988—. Author: Value-Freedom in Science and Technology, 1976, Death, Dying, and the Biological Revolution, 1976, rev. edit., 1989, Case Studies in Medical Ethics, 1977; A Theory of Medical Ethics, 1981, The Foundations of Justice, 1987, The Patient as Partner, 1987; (with Sara T. Fry) Case Studies in Nursing Ethics, 1987, The Patient-Physician Relation: The Patient as Partner, Part 2, 1991, (with James T Rule) Ethical Questions in Dentistry, 1993; editor or co-editor: Bibliography of Society, Ethics and the Life Sciences, 1973, rev. edit., 1978, The Teaching of Medical Ethics, 1973, Death Inside Out, 1975, Ethics and Health Policy, 1976, Teaching of Bioethics, 1976, Population Policy and Ethics, 1977, Life Span: Values and Life Extending Technologies, 1979, Cases in Bioethics from the Hastings Center Report, 1982, Medical Ethics, 1989, Cross Cultural Perspectives in Medical Ethics, 1989, (with Edmund D. Pellegrino and John P. Langan) Ethics, Trust, and the Professions, 1991; assoc. editor Ency. of Bioethics; editorial bd. Jour. AMA, 1976-86, Jour. Medicine and Philosophy, 1980—, Harvard Theol. Rev., 1975—, Jour. Religious Ethics, 1981—; editorial adv. bd.: Forum on Medicine, 1977-81; contbg. editor Hosp. Physician, 1975-85, Am. Jour. of Hosp. Pharmacy, 1989—; sr. editor Kennedy Inst. of Ethics Jour., 1991—; contbr. articles to profl. jours. Mem. Soc. Christian Ethics. Home: 11200 Richland Grove Dr Great Falls VA 22066-1104 Office: Georgetown U Kennedy Inst of Ethics Washington DC 20057

VEDUNG, EVERT OSKAR, political science educator; b. Sveg, Jamtlands lan, Sweden, Apr. 9, 1938; s. Oskar Emanuel and Brita (Andreasson) V.; m. Siv Birgit Dahlstrom, June 5, 1963; children: Andreas Ove, Jonas Evert, Linus Oskar Emanuel. Fil dr, Uppsala U., 1971. Asst. prof. polit. sci. Uppsala (Sweden) U., 1969-74, assoc. prof., 1974-85; spl. researcher Social Sci. Coun., Uppsala, 1986-91; vis. prof. Kyung Hee U., Seoul, 1985, 88, 90, Vienna U., 1987, 89, 91, 93; cons. Am. Coun. for an Energy-Efficient Economy, Washington, 1982-83, German Energy Ministry, Bonn, 1989, AKF, Copenhagen, 1990-93. Author: Unionsdebatten 1905, 1971, Political Reasoning, 1982, Utvärdering Politik och Förvaltning, 1991, Statens Markpolitik, Kommunerna och Historiens Ironi, 1993; co-editor: Politics as Rational Action, 1980; contbr. articles on policy evaluation, implementation, nuclear energy, energy conservation, green parties and comparative methods to profl. publs. Evaluator Energy Conservation Commn., 1980-82, Council for Bldg. Research, 1984—, Gen. Acctg. Office, 1984-85; policy analyst Bd. for Spent Nuclear Fuel, 1986-87 (all Stockholm). Ford Found. fellow Harvard U., 1975-76; Jubilee fellow Aarhus (Denmark) U., 1979; Am.-Scandinavian Found. fellow U. Tex., Austin, 1982-83. Mem. Swedish Polit. Sci. Assn., Am. Polit. Sci. Assn., Am. Evaluation Assn. Avocations: fly fishing, environmental concerns, classical music. Home: Ymerg 14 1tr, S-753 25 Uppsala Sweden Office: Uppsala U, Dept Govt PO Box 514, S-751 20 Uppsala Sweden

VEENING, HANS, chemistry educator; b. Arnhem, The Netherlands, May 7, 1931; came to U.S., 1944; s. John Dirk and Cornelia J. (DeGoede) V.; m. Elizabeth I. Timmerman, Sept. 7, 1957. AB, Hope Coll., 1953; PhD, Purdue U., 1959. Instr. chemistry Bucknell U., Lewisburg, Pa., 1958-60, asst. prof. chemistry, 1960-67, assoc. prof. chemistry, 1967-72, prof. chemistry, 1972—, chmn. chemistry dept., 1986—, presdl. prof. chemistry, 1990—; vis. prof. U. Amsterdam, The Netherlands, 1966-67, Oak Ridge (Tenn.) Nat. Lab., 1972-73, Free U., Amsterdam, 1984-85. Contbr. 63 articles to profl. jours. NSF fellow, 1966; grantee NSF, NIH, Petroleum Rsch. Fund and Camille and Henry Dreyfus Found., 1966-92. Mem. Am. Chem. Soc. (chmn. 1970), Royal Dutch Chem. Soc., Phi Beta Kappa (hon.), Sigma Xi. Achievements includes first to report liquid chromatographic separation of metal complexes. Office: Bucknell U Dept Chemistry Lewisburg PA 17837

VEERARAGHAVAN, DHARMARAJ THARUVAI, materials scientist, researcher; b. Tenkasi, Tamilnadu, India, Jan. 4, 1965; came to U.S., 1987; s. T. V. and Jayalakshmi V. (Dharmaraj) V. BS, Bombay U., 1985; MS, Indian Inst. Tech., Bombay, 1987; PhD, N.Mex. Inst. Mining & Tech., 1990. Rsch. assoc. U. Ill., Urbana, 1990—; cons. Agri-Tech Industries, Champaign, Ill., 1992—. Contbr. articles to Applied Biochemistry & Biotechnology, Jour. Applied Polymer Sci., Polymer Sci. & Engring., Bulletin Am. Phys. Soc. Sponsor World Vision, Calif., 1989—. Recipient Merit Cum Means scholarship Indian Inst. Tech., Bombay, 1985-87, grad. fellowship N.Mex. Inst. Mining and Tech., Socorro, 1988-90. Mem. Am. Soc. Metals, Am. Chem. Soc., Am. Phys. Soc. Achievements include co-founder of Quantitative Characterization of Biodegradation of Plastics/Polymers-related materials. Home: 300 S Goodwin Ave Apt 215 Urbana IL 61801 Office: Univ Ill 1304 W Green St Urbana IL 61801

VEIGEL, JON MICHAEL, corporate professional; b. Mankato, Minn., Nov. 10, 1938; s. Walter Thomas and Thelma Geraldine (Lein) V.; m. Carol June Bradley, Aug. 10, 1962. BS, U. Washington, 1960; PhD, UCLA, 1965. Program mgr. Office of Tech. Assessment, U.S. Congress, Washington, 1974-75; div. mgr. Calif. Energy Commn., Sacramento, 1975-78; asst. dir. Solar Energy Rsch. Inst., Golden, Colo., 1978-81; pres. Alt. Energy Corp., Rsch. Triangle Park, N.C., 1981-88, Oak Ridge (Tenn.) Assoc. U., 1988—; bd. dirs. Am. Coun. Energy Efficient Economy, Washington, Pacific Internat. Ctr. for High Tech. Rsch., Honolulu; mem. energy engring. bd. Nat. Rsch. Coun. Contbr. articles to jours. Bd. dirs. Oak Ridge Community Found., also chmn.; trustee Maryville Coll., Mendeleyev U., Moscow, Russia. 1st lt. USAF, 1965-68. Mem. AAAS. Avocations: photography, flying. Office: Oak Ridge Assoc Univs PO Box 117 Oak Ridge TN 37831-0117

VEITH, MICHAEL, chemist; b. Görlitz, Germany, Nov. 9, 1944; s. Werner and Inge (Reichert) V.; m. Marie-Martine Hars, Oct. 1, 1971; children: Frederike, Sebastian, Charlotte, Emilie. Diploma in chemistry, U. Munich, Fed. Republic of Germany, 1969, PhD, in Natural Sci., 1971; Habilitation, U. Karlsruhe, Fed. Republic of Germany, 1977. Asst. U. Karlsruhe, 1971-77, lectr., 1977-79; prof. U. Braunschweig, Fed. Republic of Germany, 1979-84, U. Saarland, Saarbrücken, Fed. Republic of Germany, 1984—. Contbr. articles to profl. jours. and chapts. to books. Recipient chemistry award Acad. Wissenschaften, Göttingen, Fed. Republic of Germany, 1982, Leibniz award German Sci. Found., Bonn, 1991; Winnacker scholar, Frankfort, Fed. Republic of Germany, 1978, Heisenberg scholar, German Sci. Found., 1978. Office: U of Saarlandes, Stadtwald, D-6600 Saarbrücken Saar, Germany

VEITH, ROBERT WOODY, hematologist; b. New Orleans, May 20, 1952; s. Paul Frederick Veith and Mary Ellen (Morse) Bradshaw; m. Robin Claire Gautreau; children: Rebekka, Claire, Jacob Paul, Emma Ruth. BA, La. State U., 1973, BS, 1974, MD, 1978. Diplomate Am. Bd. Internal Medicine, Am. Bd. Hematology, Am. Bd. Oncology. Intern Meml. Hosp., Worcester, Mass., 1978-79; resident La. State U., New Orleans, 1979-81; rsch. fellow U. Wash., Seattle, 1981; asst. prof. La. State U., New Orleans, 1985-91, assoc. prof., 1991—. Mem. AAAS, Am. Soc. Hematology, Am. Soc. Clin. Oncology. Office: La State U Sch Medicine 1542 Tulane Ave New Orleans LA 70116

VEIZER, JÁN, geology educator; b. Pobedim, Slovakia, June 22, 1941; came to Can., 1973; s. Viktor and Brigita (Brandstetter) V.; m. Elena Ondrus, July 30, 1966; children: Robert, Andrew Douglas. Prom. Geol., Comenius U., Bratislava, Slovakia, 1964; RNDr, Comenius U., Bratislava, Slovak Republic, 1968; CSc, Slovak Acad. Sci., Bratislava, Slovakia, 1968; PhD, Australian Nat. U., Canberra, 1971. Asst. lectr. Comenius U., 1963-66; research scientist Slovak Acad. Sci., 1966-71; vis. asst. prof. UCLA, Los Angeles, 1972; vis. research scientist U. Göttingen, Fed. Republic Germany, 1972-73; research scientist U. Tübingen, Fed. Republic Germany, 1973; from asst. prof. to full prof. U. Ottawa (Can.), 1973—; prof. Ruhr U., Bochum, Fed. Republic of Germany, 1988—; cons. NASA, Houston, 1983-86; vis. prof. and scholar Northwestern U., Evanston, Ill., 1983-87; vis. fellow Australian Nat. U., 1979; vis. prof. U. Tübingen, 1974; Lady Davis professorship Hebrew U., Jerusalem, 1987. Contbr. articles to profl. jours., chpts. to books. Served to j.lt. Med., 1965-66, Czechoslovakia. Recipient W. Leibniz prize German Rsch. Found., 1992; named Rsch. Prof. Yr., 1987; Humboldt fellow, 1980, Killam Research fellow Can. Council, 1986-88. Fellow Royal Soc. Can. (Willet G. Miller medal 1991), Geol. Assn. Can. (Past-Pres. medal 1987), Geol. Soc. Am.; mem. Geochem. Soc. Am. Roman Catholic. Club: Ski. Avocations: reading, hiking, skiing, history. Office: U Ottawa Dept Geology, Derry/Rust Rsch Group, Ottawa, ON Canada K1N 6N5 also: Ruhr U Inst Geologie, Lehrstuhl Sedimentgeologie, 4630 Bochum Federal Republic of Germany

VELAGALETI, RANGA RAO, agronomist, environmental scientist; b. Gudem, Andhra, India, May 26, 1946; came to U.S., 1976; s. Narsimha Rao and Suvarchala V.; m. Poonam R. Velagaleti, Oct. 11, 1975; children: Parashant, Naina. BSc with honors, U. Delhi, India, 1965, MSc, 1967, PhD, 1971. Postdoctoral fellow Rothamstead Exptl. Sta., Harpenden, U.K., 1975-76, Charles F. Kettering Rsch. Lab., Yellow Springs, Ohio, 1976-77; coord. internat. programs Charles F. Kettering Rsch. Labs., Yellow Springs, Ohio, 1982-84; rsch. scientist Boyce Thompson Inst., Cornell U., Ithaca, N.Y., 1977-79; staff scientist Internat. Inst. Tropical Agr., Ibadan, Nigeria, 1979-81; prin. rsch. scientist Battelle Meml. Inst., Columbus, Ohio, 1984-87, sr. rsch. scientist, 1987-88, group leader, 1988-90, dept. mgr., 1991—. Contbr. papers to profl. publs. Mem. Am. Soc. Agronomy, Crop Sci. Soc. Am., Soil Sci. Soc. Am., Coun. Agrl. Sci. and Tech. Achievements include patent for autoclavable, reusable plant growth system and method. Office: Battelle Meml Inst 505 King Ave Columbus OH 43201-2693

VELARDO, JOSEPH THOMAS, molecular biology and endocrinology educator; b. Newark, Jan. 27, 1923; s. Michael Arthur and Antoinette (Iacullo) V.; m. Forresta M. Monica Power, Aug. 12, 1948 (dec. July 1976). AB, U. No. Colo., 1948; SM, Miami U., 1949; PhD, Harvard U., 1952. Rsch. fellow in biology and endocrinology Harvard U., Cambridge, Mass., 1952-53; rsch. assoc. in pathology, ob-gyn. and surgery Sch. Medicine Harvard U., Boston, 1953-55; asst. in surgery Peter Bent Brigham and Women's Hosp., Boston, 1954-55; asst. prof. anatomy and endocrinology Sch. Medicine, Yale U., New Haven, 1955-61; prof. anatomy, chmn. dept. N.Y. Med. Coll., N.Y.C., 1961-62; cons. N.Y. Fertility Inst., 1961-62; dir. Inst. for Study Human Reprodn., Cleve., 1962-67; prof. biology John Carroll U., Cleve., 1962-67; mem. rsch. and edn. divs. St. Ann Obstetric and Gynecologic Hosp., Cleve., 1962-67; head dept. rsch. St. Ann Hosp., Cleve., 1964-67; prof. anatomy Stritch Sch. Medicine Loyola U., Chgo., 1967-88, chmn. dept. anatomy Stritch Sch. of Medicine, 1967-73; cons. Internat. Basic and Biol.-Biomed. Curricula, Lombard, Ill., 1979—; mem. adv. coun. IBC Who's Who Reference Titles, Internat. Biog. Ctr., Cambridge, Eng., 1990—; hon. appointee rsch. bd. advisors Am. Biog. Inst., Raleigh, N.C., 1991—; course moderator laparoscopy Brazil-Israel Congress on Fertility and Sterility, and Brazil Soc. of Human Reproduction, Rio de Janeiro, 1973. Author, contbr. to: (with others) Histochemistry of Enzymes in the Female Genital System, 1963, The Ovary, 1963, The Ureter, 1967, rev. edit., 1981; editor, contbr. Endocrinology of Reproduction, 1958, Essentials of Human Reproduction, 1958; cons. editor, co-author: The Uterus, 1959; contbr. Trophoblast and Its Tumors, 1959, Hormonal Steroids, Biochemistry, Pharmacology and Therapeutics, 1964, Human Reproduction, 1973; co-editor, contbr.: Biology of Reproduction, Basic and Clinical Studies, 1973; contbr. articles to profl. jours. Apptd. U.S. del. to Vatican, 1964; charter mem. U.S. Rep. Presdl. Task Force, 1988—; rep. U.S. Senate Inner Circle, 1988—. With USAAF, 1943-45. Decorated Presdl. Unit citation with 2 oak leaf clusters; award recipient Lederle Med. Faculty Awards Com., 1955-58; named hon. citizen City of Sao Paulo, Brazil, 1972; U.S. del. to Vatican, 1964. Fellow AAAS, N.Y. Acad. Scis., Gerontol. Soc., Pacific Coast Fertility Soc. (hon.); mem. Am. Assn. Anatomists, Am. Soc. Zoologists, Am. Physiol. Soc., Endocrine Soc., Soc. Endocrinology (Gt. Britain), Soc. Exptl. Biology and Medicine, Am. Soc. Study Sterility (Rubin award 1954), Internat. Fertility Assn., Pan Am. Assn. Anatomy, Midwestern Soc. Anatomists (pres. 1973-74), Mexican Soc. Anatomy (hon.), Harvard Club, Sigma Xi, Kappa Delta Pi, Phi Sigma, Gamma Alpha, Alpha Epsilon Delta. Achievements include extensive original research and publications on the physiology and development of decidual tissue (experimental equivalent of the maternal portion of the placenta) in the rat; biological investigation of eighteen human adenohypophyses (anterior lobes of the human pituitary glands); induction of ovulation utilizing highly purified adenohypophyseal gonadotropic hormones in mammals; and the interaction of steroids in reproductive mechanisms. Office: 607 E Wilson Ave Lombard IL 60148-4062

VELASCO, JAMES, chemist, consultant; b. Limon, Mex., Oct. 11, 1919; came to U.S., 1920; s. Salvador and Manuela (Pelayo) V.; m. Anna Lucille Baber, Dec. 30, 1950; children: Jamie, Terence, Dean. BS, U. Colo., 1949. Specifier Cen. Sci. Co., Chgo., 1950-51; analytical chemist Reid Murdoch Food Corp., Chgo., 1951-52; analytical chemist USDA, Beltsville, Md., 1952-56, rsch. chemist, 1956-83; ret., 1983; Cons. Neotec Instrument Co., Silver Spring, Md., Pacific Sci., Silver Spring, NIR Systems, Silver Spring, U.S. FDA, Washington. Contbr. articles to sci. jours. Coach Boys' Club, Beltsville, 1964-68. 1st lt. U.S. Army, 1941-45, ETO. Democrat. Roman Catholic. Achievements include patent in glassware for chromatographic column analysis of neutral oil in crude oilseed extracts, development of apparatus to detect aflatoxin in oil seed meals, design of fluorometer for aflatoxin in florisil layer. Home: 4708 Cardinal Ave Beltsville MD 20705

VELASCO NEGUERUELA, ARTURO, biology educator; b. Madrid, Feb. 26, 1944; s. Arturo Velasco Arruebarrena and Trinidad (Negueruela) Gómez; m. Consuelo Medina Sanchez, Dec. 20, 1969; children: Velasco Medina Arturo, Velasco Medina Fernando. Grad. in Chemistry, Complutense U., Madrid, 1968, grad. in Pharmacy, 1974, PhD in Pharmacy, 1978. Registered pharmacist. Chemist Minas Tarna SA, Madrid, 1968-72; prof. Complutense U., 1972—; cons. Direccion Gen. Investigation Ciencia y Tecnologia, Madrid, 1990-91, Ministerio Edn. y Ciencia. Contbr. over 100 articles to profl. jours. Recipient Premio Abello award Real Academia de Farmacia, Madrid, 1983. Mem. Soc. Española de Bromatologia, Soc. Micologica Castellana, Real Sociedad Española de Historia Natural, Spanish Pharm. Soc. for Medicinal Plants (pres.). Roman Catholic. Achievements include research on natural plant products, primary aromatic plants and essential oils. Office: Complutense Univ, Dept Botany Faculty Biology, 28040 Madrid Spain

VELASQUEZ, PABLO, mining executive; b. Potosí, Bolivia, Mar. 8, 1962; came to U.S., 1978; s. Felipe and Bertha (Zambrana) V.; m. Candace Marie Connally, Jan. 18, 1986; children: Veronica, Sergio. BS in Metall. Engring., S.D. Sch. Mines, 1984, postgrad., 1984-85. Cons. Occidental Minera-Bolivia, Potosí, 1985—; lab. dir. Malapai Resources Co., Casper, Wyo., 1987-88, process engr., 1988-89, plant supt., 1989-90; plant-mine supt. Total Minerals Corp., Casper, 1990-92, prodn. mgr., 1993—; tech. group mem. Total Minerals Corp., Houston, 1991—. Mem. Soc. Mining Engring., Ctrl. Wyo. AIME. Achievements include research on vanadium recovery from in-situ leach liquors, ground water restoration at the Irigaray Mine, Wyo., elimination of selenium from ISL aquifers. Home: 211 Cummings Ave Buffalo WY 82834

VELASQUEZ PEREZ, JOSE R., cardiologist; b. Caripito, Monagas, Venezuela, Sept. 18, 1949; s. Rafael Agustin and Carmen (Perez) V.; m. Tomasa Argelia Gamero, Dec. 28, 1973; children: Jesus Jose, Jesus Rafael. Degree, Agustin R. Hernandez, Anaco, Venezuela, 1969; MD, U. de Oreinte, Bolivar, 1976; M in Cardiology, U.N.A.M., Mexico City, 1981. Pvt. practice Consultorio Cardiologico, Cumana, Venezuela, 1981—; chief cardiology Josefina de Figuera, Cumana, 1981—. Contbr. articles to profl. jours. Named to Order Antonio Jose de Sucre Govt. of Venezuela. Fellow Internat. Soc. Holter Monitoring (assoc.); mem. Cumana Heart Found. (pres. 1986—), Am. Soc. Echocardiography. Avocation: deep sea fishing. Home: Av Gran Mariscal # 152, 6101 Cumana Venezuela Office: Josefina de Figueroa, Apt # 027, 6101 Cumana Venezuela

VELK, ROBERT JAMES, psychologist; b. Chgo., Feb. 27, 1938; s. Jerry E. and Sylvia B. (Wladar) Vlk; m. Vera A. Kraml, Nov. 25, 1961; children—Robert Frank, Cheryl Anne. B.B.A., Northwestern U., 1963, M.B.A., 1968; M.A., Rutgers U., 1980, Ph.D., 1983. Asst. mgr. product decorations Meyercord Co., Carol Stream, Ill., 1959-65, nat. account mgr., 1965-68; assoc. Kepner Tregoe, Inc., Princeton, N.J., 1968-70, Western region mgr., 1970-72, dir. mktg. N.Am. ops., 1972-73; pres. Creative Leadership Inc., Princeton, 1973-83; pres. Cognitive Sci. Corp., Ft. Collins, Colo., 1983—; dir. mgmt. Devel. Ctr. Anderson Sch. Mgmt. U. N.Mex., 1991—. Author: Information and Imagination, 1978; Thinking About Thinking, 1978. Mem. Am. Psychol. Assn., Am. Soc. Tng. and Devel., Nat. Soc. Performance and Instrn., Cognitive Sci. Soc. Clubs: Christian Businessmen's Com. of Central Jersey (chmn. 1974-75), Gideon's. Office: U NM Anderson Grad Sch Mgmt 9440 Callaway Cir NE # NE Univ Of New Mexico NM 87131-1221

VELKOV, SIMEON HRISTOV, civil engineer; b. Pleven, Bulgaria, May 5, 1928; came to U.S., 1984; s. Hristo Dimitrov and Mara Simeonova (Zankova) V.; widowed; 1 child, Juliana. MS in Civil, Structural Engring., State Poly. U., Sofia, Bulgaria, 1952. Registered profl. engr., Bulgaria, Mass. With Ministry of Pub. Works, Sofia, 1953-69; tech. advisor Ministry of Interior, Addis Ababa, Ethiopia, 1969-73; dir. design control Nat. Water Supply Authority, Sofia, 1973-77; chief. civil engring. (design) Fed. Ministry of Housing and Environ., Lagos, Nigeria, 1977-84; prin. engr. Mass. Dept. Pub. Works, Boston, 1986—. Contbr. numerous articles to profl. jours. Office: Mass Dept Pub Works 10 Park Plz Boston MA 02116-3973

VELTMAN, ARIE TAEKE, electronic engineer; b. Eindhoven, Brabant, Netherlands, Nov. 6, 1957; s. Marinus-Jan and Greetje (Van Gorcum) V. IR, Tech. U., 1988. Scientist Amsterdam Univ., 1988-89, Netherlands Energy Rsch. Found., Petten, 1989—. Contbr. articles to profl. jours. Mem. Royal Dutch Inst. Engrs. Home: Drevelstraat 20, 1825 KK Alkmaar NH, The Netherlands Office: Netherl En Res Fnd, Dept DE PO Box 1, 1755 ZG Petten NH, The Netherlands

VELTMAN, MARTINUS J., physics educator. John D. MacArthur prof. physics U. Mich., Ann Arbor. Office: U Mich Dept of Physics Ann Arbor MI 48108

VENABLE, ROBERT ELLIS, crop scientist; b. Clovis, N.Mex., Sept. 30, 1952; s. Charles Edward and Evelylee (Harville) V.; m. Linda Sue Campbell, Oct. 9, 1976 (div. 1982); m. Melva Ivette Roman, Sept. 24, 1988; 1 child, Jonathan Shelby. BA in Biology, Hendrix Coll., 1974; MS in Natural Sci., U. Ark., 1976. Quality control mgr. Thompson-Hayward Chem./Platte, Greenville, Miss., 1975-82; scientist Velsicol Chem./Sandoz, Chgo., 1982-88; surp. formulations devel. Ecogen, Inc., Langhorne, Pa., 1988—. Mem. ASTM (pesticide com.). Achievements include research of crop protection on water dispersable extruded pellets. Office: Ecogen Inc 2005 Cabot Blvd W Langhorne PA 19047

VENEDAM, RICHARD JOSEPH, chemist, educator; b. Phoenix, Feb. 25, 1953; s. John Richard and Helen Alma (MacDonald) V.; m. Valeria I. Cappell, Mar. 29, 1981 (div. 1987). BS in Chemistry, No. Ariz. U., 1985, MA in Ednl. Adminstrn., 1989, EdD in Ednl. Leadership, 1993. Cert. tchr., Ariz., Alaska. Staff engr. C.R. Ward Corp., Phoenix, 1976-81; dir. Capitol Porjects Lower Kuskokwim Sch. Dist., Bethel, Alaska, 1981-84; tchr. jr. high level Havasupai Elem. Sch., Supai, Ariz., 1986-88; tribal engr. Havasupai Tribal Coun., Supai, 1986-91, cons. engr., 1991—; supr. instrumentation dept. chemistry No. Ariz. U., Flagstaff, 1988—; bd. dirs. Salt River Project Tchrs. Adv. Coun., Phoenix, 1988—,. Author: Alternative Energy Sources the Environment and You, 1992; contbr. articles to Jour. Electrochem. Soc., others. Bd. dirs. adv. com. on safety Alaska Bd. of Game, 1983-84; referee Am. Youth Soccer, Flagstaff, 1988—. With U.S. Army, 1972-75. Grantee Nat. Park Svc., 1991. Mem. Electrochem. Soc., Am. Chem. Soc., Sigma Xi, Phi Lambda Upsilon. Republican. Home: 1120 W Azure St Flagstaff AZ 86001 Office: No Ariz Univ Dept Chemistry Flagstaff AZ 86011

VENEKLASEN, PAUL SCHUELKE, physicist, acoustics consultant; b. Grand Haven, Mich., June 10, 1916; s. James Theodore and Ann Susan (Schuelke) V.; m. Elizabeth Busby; children: Lee Harris, Mark Schuelke. BS in Math., Physics with honors, Northwestern U., 1937, MS in Math., Physics, 1938; postgrad., UCLA, 1938-41. Registered profl. engr., Calif. Teaching asst., rsch. asst. UCLA, 1938-42, rsch. assoc. Nat. Def. Rsch. Com., 1942-43; rsch. assoc. div. phys. war rsch. Duke U., 1943; rsch. assoc. electro-acoustic lab. Harvard U., Cambridge, Mass., 1943-45; rsch. physicist Altec-Lansing Corp., 1946-50; dir., acoustical cons. Paul S. Veneklasen and Assocs., Inc., Santa Monica, Calif., 1947—; acoustical cons., coord. chorale Carmel Bach Festival, 1947-65; cons. design of auditoriums for performing arts; presenter at profl. confs. Contbr. articles to profl. pubs. Fellow Acoustical Soc. Am.; mem. AIAA, Nat. Coun. Acoustical Cons. (honoree 1992). Achievements include patents for sound shield for telephone transmitters; earphone sockets; earphone socket structure; ear defender; acoustic fabric testing instrument; shielded condenser microphone, shielded cable system for microphones, others. Office: Paul S Veneklasen & Assocs 1711 16th St Santa Monica CA 90404

VENET, CLAUDE HENRY, architect, acoustics consultant; b. Lyon, France, Aug. 10, 1946; came to U.S., 1981; s. René Joseph and Marcellé (Michel) V.; m. Blanca Eppenstein Portella, Feb. 12, 1981 (div. Feb. 1984); m. Mouna Bennani-Smires, Feb. 29, 1992. Dipl. Elec. Eng., ESTA, Rochefort, France, 1968; Lic. Physics, U. Paris, 1971; MArch, So. Calif. Inst. Architecture, 1986. Mgr. Ling Dynamics/Altec U.K., Royston, Eng., 1971-72; mng. dir. CVE Enterprises, London, 1972-75; sales dir. Macinnes/Amcron France, Paris, 1975-77; tech. dir. Audio Cons. Coordination, Rio de Janeiro, 1977-81; cons. Paramount (Sound) Films Corp., Glendale, Calif., 1981-82; pres. Architecture and Engring. in Acoustics, Belleville, France, 1986-91, Archicoustics Inc., Miami, Fla., 1991—. Vol. Architects Without Frontiers, Paris, 1990—. Cpl. French Air Force, 1962-67. Mem. AIA (assoc.), AAAS, Am. Inst. Physics, Acoustical Soc. Am., Audio Engring. Soc., Order French Architects, Chamber French Cons. Engrs. Achievements include design of computer-driven, polymorphic, multi-use theater with continuous variable acoustics/geometry, variable-shape, multi-acoustics concept in recording studio design. Office: Archicoustics Corp PO Box 971176 Miami FL 33197-1176

VENETSANOPOULOS, ANASTASIOS NICOLAOS, electrical engineer, educator; b. Athens, Greece, June 19, 1941; emigrated to Can., 1968; s. Nicolaos Anastasios and Elli (Papacondilis) V. Diploma, Athens Coll., 1960; diploma in elec. and mech. engring., Nat. Tech. U., Athens, 1965; MS, Yale U., 1966, MPhil, 1968, PhD, 1969. Registered profl. engr., Greece, Ont. Asst. in instrn. engring. and applied sci. Yale U., 1966-68, research asst., 1968-69; lectr. U. Toronto, Ont., Can., 1968-69; asst. prof. elec. engring. U. Toronto, 1970-73, assoc. prof., 1973-81, prof., 1981—, chmn. communications group dept. elec. engring., 1974-78, 81-86, assoc. chmn. elec. engring., 1978-79, mem. elec. engring. exec. com., 1974-79, 81-86, mem. elec. engring. curriculum com., 1972-79; acad. visitor Imperial Coll. Arts and Tech., U. London, 1979-80; vis. prof. Nat. Tech. U. Athens, spring 1979-80, Fed. U. of Technology, Switzerland, 1986-87, U. Florence, Italy, summer 1987; cons. elec. engring. Consociates Ltd.. Editor Can. Elec. Engring. Jour., 1981-83; contbr. over 400 articles to profl. jours; contbr. to 10 books. Mem. allocations and agy. relations com. United Community Fund, Toronto, 1971-74; pres. Hellenic-Can. Cultural Soc., 1972-75; sec. gen. Greek Community Met. Toronto, 1973-75. Fulbright travel grantee in U.S., 1965; Def. Research Bd. Can. grantee, 1972-75, UN grantee; NSF grantee; J.P. Bickell Found. grantee; Natural Scis. and Engring. Research Council of Can. Fellow IEEE (fin. chmn. internat. symposium on circuit theory 1973, tech. program chmn. internat. conf. communications 1978, 86, vice-chmn. Toronto sect. 1976-77, chmn. 1977-79, assoc. editor transactions on circuits and systems 1985-87, guest editor spl. issue Transactions on Circuits and Systems in Image Processing 1987), Engring. Inst. Can.; mem. Assn. Profl. Engrs. Ont., Assn. Profl. Elec. Engrs. Greece, Assn. Profl. Mech. Engrs. Greece, Can. Soc. Elec. Engring. (chmn. Toronto sect. 1975-77, nat. dir. 1976-88, pres. 1983-86), AAAS, Yale Sci. and Engring. Assn., N.Y. Acad. Scis., Tech. Chamber Greece, Am.-Hellenic Ednl. Progress Assn. (v.p. Toronto sect. 1973-75, pres. 1975-77), Intercultural Council (chmn. ednl. com. 1971-80, sr. v.p. 1977-80), Sigma Xi. Office: U Toronto Dept Elec Engring, Toronto, ON Canada M5S 1A4

VENEZIANO, PHILIP PAUL, biologist, educator; b. Boston, May 9, 1921; s. Paul and Teresa (Terranova) V.; m. Velma Dora Taylor, Aug. 3, 1943; children: Jan Ellen Veneziano Tilsen, Valerie Kay. MS in Zoology, U. Wyo., 1946; PhD in Biol. Scis., Northwestern U., 1964. Asst. chemist petroleum rsch. lab. U.S. Bur. Mines, Laramie, Wyo., 1947-50; chemist U.S. Geol. Survey, Dept. Interior, Austin, Tex., 1952-54; rsch. assoc. dept. biology Northwestern U., Evanston, Ill., 1954-61; asst. prof. zoology Ill. Inst. Tech., Chgo., 1962-65; prof. biology Wright Coll., Chgo., 1966-91, prof. emeritus, 1991—. Contbr. articles on embryology to sci. jours. With U.S. Army, 1939-40. Mem. Sigma Xi. Democrat. Methodist. Achievements include research on magnetic effects on embryo development. Home: 3133 Walden Ln Wilmette IL 60091

VENKATASUBRAMANIAN, RAMA, electrical engineer; b. Kodaikonal, India, May 25, 1961; came to U.S., 1983; s. R.V. Subramanian and K.S. Sulochana; m. Lalitha T. Subramanian, May 29, 1988; 1 child, Anusuya. PhD, Rensselaer Poly. Inst., 1989. Rsch. engr. Rsch. Triangle Inst., Research Triangle Park, N.C., 1989—. Recipient Allen B. Dumont prize for grad. achievement, 1989. Mem. IEEE, Material Rsch. Soc., Internat. Thermoelectric Soc. Achievements include first demonstration of luminescence from planar, quantized Ge structures; first proposed the concept of monolithically-interconnected superlattice-type structures for high-performance thermoelectric elements; development of Eutectic-Metal-Bonding of GaAs onto Si, microstimulator for visual prosthesis, highest efficiency for a GaAs solar cell on polycrystalline Ge; first demonstration of double tunnel junction GaInP2/GaAs cascade solar cell; highest reported minority carrier lifetime in heteroepitaxial GaAs on Si. Home: 104 Old Rockhampton Ln Cary NC 27513 Office: Research Triangle Inst 3040 Cornwallis Rd Research Triangle Park NC 27709

VENKATESAN, DORASWAMY, astrophysicist, physics educator; b. Coimbatore, India. BSc, Loyola Coll., India, 1943; MSc, Benares Hindu U., 1945; PhD in Cosmic Rays, Gujarat U., 1955. Lectr. physics K. P. Coll., Allahabad, India, 1947-48; lectr. Durbar Coll., Rewa, India, 1949; sr. rsch. asst. cosmic rays Phys. Rsch. Lab., Ahmedabad, India, 1949-56; fellow inst. electron physics Royal Inst. Tech., Stockholm, 1956-57; fellow Nat. Rsch. Coun. Can., 1957-60; rsch. assoc. space physics and astrophysics U. Iowa, 1960-63, asst. prof. physics, 1963-65, cons. high altitude balloon program, 1965; assoc. prof. physics U. Alta., 1965-69; prof. physics U. Calgary, Alta., 1969—. Fellow Brit. Inst. Physics; mem. Am. Geophysics Union, Can. Assn. Physicists, Can. Astronomy Soc. Achievements include research in solar terrestrial relations, in astrophysics involving studies of cosmic rays, radiation belts, ionospheric absorption, auroral x-rays, geomagnetism, solar activity, cosmic x-ray sources and interplanetary medium. Office: University of Calgary, 2500 Univ Dr NW, Calgary, AB Canada T2N 1N4*

VENKATESH, YELDUR PADMANABHA, biochemist, researcher; b. Kalale, Karnataka, India, Dec. 2, 1953; came to U.S., 1981; s. Padmanabha and Anasuya (Bai) Rao; m. Poornima Venkatesh, June 26, 1981; 1 child, Madhava. BS in Chemistry, Bangalore (India) U., 1970; MS in Biochemistry, Kasturba Med. Coll., Manipal, India, 1974; PhD in Biochemistry, Indian Inst. Sci., Bangalore, 1981. Lectr. Kasturba Med. Coll., 1974-75; postdoctoral fellow Washington U. Sch. Medicine, St. Louis, 1981-83, NIH trainee in immunology, 1984-85; rsch. assoc. Smith Kline & French Labs., King of Prussia, Pa., 1985-87; asst. rsch. scientist ImmunoGen, Inc., Cambridge, Mass., 1987-88, rsch. scientist, 1988-92, sr. rsch. scientist, 1992—. Author: Annals New York Academy Sciences, 1983; contbr. articles to profl. jours. including Internat. Jour. Peptide and Protein Rsch., Jour. Immunology, Molecular Immunology. U. Grants Commn. fellow Bangalore, 1975-77, 77-79. Mem. Am. Assn. Immunologists, N.Y. Acad. Scis., Protein Soc., Soc. for Complex Carbohydrates. Hindu. Achievements include research on protein structure-function relationships in bovine pancreatic ribonuclease A, human complement protein C3, ricin, and monoclonal antibodies, effect of chemical modification (deamidation) on protein folding, development of affinity chromatography media for separation of biologicals, development of novel cross-linkers for the preparation of antibody heteroconjugates, protocols for removal of toxic contaminants and minimizing aggregation in the proprietary toxin (blocked ricin) and immunotoxins, a new process for increasing the yield of blocked ricin used in the production of immunotoxins for cancer therapy. Home: 8 Rocky Nook Malden MA 02148-1560 Office: ImmunoGen Inc 148 Sidney St Cambridge MA 02139-4239

VENNE, LOUISE MARGUERITE, librarian; b. St. Jean, Que., Can., Jan. 30, 1944; d. Robert and Thérèse (Joyal) V. MSc in Botany, U. Montreal, Can., 1967, BSc in Libr. Sci., 1968. Cert. corp. bibliothecaires profl., Que. Ref. and collection devel. sci. and engring. libr. Laval U., Que., Can., 1968-73; phycologist sci. lab. U. Montreal, 1974-79; head libr. Indsl. Material bd. Can. Inst. Sci. and Tech. Info., Coun. Nat. Rsch., Montreal, 1979—. Co-author: Plantes Sauvages Printanieres, Plantes Sauvages des Villes et des champs 1 & 2, Plantes Sauvages Comestibles et au Menu, Plantes Sauvages des Lacs Tourbieres et Rivieres, Les Fougeres, Preles, Lycopodes, 1973-92. Home: 3753 Mercier, Montreal Quebec, Canada Office: Conseil Nat de Recherches, 75 Boul de Mortagne, Boucherville, PQ Canada J4B 6Y4

VENTER, RONALD DANIEL, mechanical engineering educator, researcher, administrator; b. East London, Republic of South Africa, Jan. 3, 1944; came to Can., 1974; s. Daniel Frans and Myrtle (Quirk) V.; m. Beryl Venter, July 17, 1975. BASc, U. Witwatersrand, Johannesburg, Republic of South Africa, 1966; M in Engring., McMaster U., Hamilton, Can., 1969, PhD, 1971. Head mechanics div. Indsl. Diamond div. DeBeers, Johanneburg, 1971-72, head high pressure systems Indsl. Diamonds div., 1972-74; asst. prof. dept. mech. engring. U. Toronto, 1975-78, prof., 1981—, assoc. chmn. dept. mech. engring., 1979-81, chmn., 1981-91, vice dean Faculty of Applied Sci. and Engring., 1993—, Wallace G. Chalmers prof. engring. design, 1993—; dir. Ont. Ctr. for Automotive Arts Tech., St. Catherine's, 1983-89, Ctr. for Hydrogen and Electrochem. Studies, Toronto, 1986—, Indsl. R&D Inst., Midland, Ont., 1992—. Co-author: Plane Strain Slip Line Fields, 1982; contbr. articles to profl. jours. Recipient Gold medal award Chamber of Mines South Africa, 1966. Mem. ASME, Soc. Mfg. Engrs., Assn. Profl. Engrs. Ont. Conservative. Roman Catholic. Home: 55 Thorncrest Rd, Islington, ON Canada W9A 1S8 Office: U of Toronto, 5 King's College Rd, Toronto, ON Canada M5S 1A4

VENTOSA, ANTONIO, microbiologist, educator; b. Cordoba, Spain, June 22, 1954; s. Jose and Maria Luisa (Ucero) V.; m. Cristina Cutillas, Mar. 20, 1981; children: Antonio, Jose Javier. B, Coll. Virgen del Carmen, Cordoba, 1970; grad. in pharmacy, U. Granada, Spain, 1976, PhD in Pharmacy (hon.), 1981. Asst. prof. U. Granada, Spain, 1977-82; asst. prof. U. Sevilla, Spain, 1982-84, assoc. prof., 1985—, vice dean faculty pharmacy, 1993—. Editor: Microbiologia Jour., 1985—, Systematic and Applied Microbiology Jour., 1991—; contbr. chpts. in books and articles to profl. jours. Mem. AAAS, Am. Soc. Microbiology, Spanish Soc. Microbiology, Soc. Gen. Microbiology, Soc. Applied Microbiology, Deutsche Assn. for Hygiene and Microbiology, Internat. Com. on Systematic Bacteriology (sec. subcom. on Taxonomy of Halobacteriaceae 1982—), World Fedn. Culture Collection (exec. bd. 1992—). Avocation: music. Office: U Sevilla Fac Pharmacy, Dept Microbiology, 41012 Seville Spain

VENTRICE, MARIE BUSCK, engineering educator; b. Allentown, Pa., Oct. 17, 1940; d. Poul Gunni and Edith Marie (Peterson) B.; m. Carl A. Centrice, Jan. 25, 1960; children: Ruth Esther, Carl Alfred Jr., James August. BS, Tenn. Tech. U., 1966; MS, Auburn U., 1968; PhD, Tenn. Tech. U., 1974. Registered profl. engr. Tenn. Instr. dept. engring. sci. Tenn. Tech. U., Cookeville, 1969-70, asst. prof. dept. mech. engring., 1974-79, assoc. prof., 1979-86, interim dir. Ctr. Electric Power, 1985-88, prof., 1986—, assoc. dean coll. engring., 1989—. Contbr. 36 tech. reports and papers. Mem. AAUP, ASME, Am. Soc. Engring. Edn., Mat. Soc. Prof. Engrs., Soc. Women Engrs. Achievements include development of an analog technique for the study of combustion instability. Home: 183 Paris St Cookeville TN 38501 Office: Tenn Tech U Coll Engring Cookeville TN 38505

VENTURA, OSCAR NESTOR, chemistry educator, researcher; b. Montevideo, Uruguay, Apr. 14, 1957; s. Oscar Lenin and Melba Raquel (Perez) V.; m. María Beatriz Romero, Sept. 26, 1983; children: Oscar Diego, Sebastian Gabriel. BS in Chemistry, U. Uruguay, 1980, MSc in Quantum Chemistry, 1982; PhD in Quantum Chemistry, U Uruguay, 1993. Rsch. asst. faculty chemistry U. Uruguay, Montevideo, 1980-82, asst. prof., 1982-86, assoc. prof., 1986-90, full-time prof., 1990—; dir. inst. chemistry Fac. Scis., Montevideo, 1990-92; vis. scientist Max-Planck Inst. Astrophysics, Garching, Germany, 1988, U. Barcelona, Spain, 1984-85; cons. Commn. European Communities, Brussels, 1989-90, Banco Interamericano de Desarrollo project Conicit-Venezuela, 1993; adv. bd. mem. Nat. Rsch. Group, Montevideo, 1992. Co-author: Nuevas Tendencias En Química Teórica, 1990, Computational Chemistry, Structure, Interactions and Reactivity, 1992; referee Jour. of Am. Chem. Soc., Jour. of Molecular Structure; contbr. articles to profl. jours. Coun. mem. U. Uruguay, 1988-90. Recipient Nat. Rsch. prize Nat. Rsch. Coun., Uruguay, 1981, Rotary Found. award Rotary Internat., U.S.A., 1984, Young Rschr. award Third World Acad. Scis., Italy, 1991; named EEC Rsch. fellow Commn. of European Commun, Brussels, 1988, Alexander von Humboldt fellow, Germany, 1993. Mem. Am. Chem. Soc., Internat. Assn. Theoretical Chem. Physics, Internat. Union Pure and Applied Chemistry, World Assn. Theoretical Chem. Physics, Polish Chem. Soc., Argentinian Soc. Phys. Chemistry. Social-democrat. Agnostic. Achievements include research in molecular modelling; computational study of chemical reactions, solvation, hydrogen-bonded clusters, quantum chemistry; research grants from UNESCO, Commission European Communities NSF, Swedish Agy. Research & Coop. Home: Joaquin Ñóñez 2771 apto 203, Montevideo 11000, Uruguay Office: Facultad de Química/Química Cuántica, Avda Gral Flores 2124, Montevideo 11800, Uruguay

VERA GARCIA, RAFAEL, food and nutrition biochemist; b. Santa Rosa, Misiones, Paraguay, May 6, 1939; s. Rafael Vera and Juana Bautista (Garciá León) V. Pharm. chemist degree, Nat. U. Asunción (Paraguay), 1961; postgrad., U. Rochester, 1963-65; D. in Biochemistry, Nat. U. Asunción (Paraguay), 1966. Asst. prof. radiobiology Faculty Chem. Scis., Nat. U. Asunción, 1971, assoc. prof. radiobiology, 1974; rsch. fellow on nutritional biochemistry NIRD, U. Reading (Eng.), 1974-75; rsch. fellow biology dept.

Ctr. Recherche Nucléare Saclay (France), 1978-79; head human nutrition program Nat. Inst. Tech. and Standards, Asunción, 1976-88; head food biochemistry and nutrition rsch. dept. U. Nac. Asunción, 1981—, prof. radioisotope methodolgy, 1983—; asst. prof. nutritional biochemistry Faculty Chem. Scis., U. Nat. Asunsión, 1990—, dir. rsch. dept., 1991—; rsch. fellow on nutritional biochemistry Dept. of Nutrition and Bromatology II, Faculty of Pharmacy U. Complutense, Madrid, 1984. Dir. Publicaciones FCQ, 1982—; contbr. articles to profl. jours. Named one of 12 Most Outstanding People of 1991, 1 de Marzo Radiobroadcasting System. Mem. Assn. Biochemists, Nuclear Medicine and Biology Soc., Nat. Com. Food Protection-Nat. Health Coun., Assn. Profs. of Faculty of Chem. Scis., Paraguayan Nutrition Soc., Spanish Bromatology Soc. (hon.), San Jose Sch. Alumni, Garden Club. Roman Catholic. Avocations: karate, music, martial arts, TV, soccer. Home: Mayor Fleitas 896, Asunción Paraguay Office: Faculto de Ciencias Quimicas-Una, PO Box Casilla # 1055, Asunción Paraguay

VERCELLI, DONATA, immunologist, educator. Prof. Med. Sch. Harvard U. Recipient Burroughs Wellcome Developing Investigator award Immunopharmacology of Allergic Diseases Am. Acad. Allergy and Immunology, 1991. Office: Harvard Medical School Childrens Hospital Cambridge MA 02138*

VERDERY, ROY BURTON, III, gerontologist, consultant; b. Bennington, Vt., June 15, 1947; s. Roy Burton and Charlotte (Gilbert) V.; m. Marlene Hannah Honek, June 22, 1979; stepchildren: Lori Larks, David Larks. PhD, U. Calif., Berkeley, 1975; MD, U. Miami, 1983. Diplomate Am. Bd. Internal Medicine, Geriatrics. Sr. fellow U. Wash., Seattle, 1976-79; asst. prof. rsch. U. Montreal, Can., 1979-81; med. resident Dartmouth-Hitchcock Med. Ctr., Hanover, N.H., 1983-85; med. staff fellow NIH-Nat. Inst. of Aging-Gerontology Rsch. Cntr., Balt., 1985-88; asst. prof. Johns Hopkins U., Balt., 1987-88, Wake Forest U., Winston-Salem, N.C., 1988-92; assoc. prof. U. Ariz., Tucson, 1992—. Recipient Rsch. Career award NIH, Nat. Inst. Aging, 1990. Fellow Am. Coll. Physicians, Am. Heart Assn.; mem. Am. Geriatric Soc. (New Investigator 1991), Gerontol. Soc. Am. Achievements include research in the role of energy metabolism, inflammation and lipoprotein metabolism in elderly people with failure to thrive and cachexia. Office: Ariz Ctr on Aging 1821 E Elm St Tucson AZ 85719

VERDESCA, ARTHUR SALVATORE, internist, corporate medical director; b. Cliffside Park, N.J., May 25, 1930; s. Cosimo Theodore and Giulia Elvira (DeLipsis) V.; m. Ann Edith Copping, June 24, 1961; children: Stephen, Julia, Edith. AB, Columbia U., 1951, MD, 1955. Diplomate Am. Bd. Internal Medicine. Intern St. Luke's Hosp., N.Y.C., 1955-56, resident, 1956-57, 59-60, fellow Nat. Heart Inst., 1960-61; staff physician Western Electric, N.Y.C., 1961-63, assoc. hdqrs. med. dir., 1963-65, hdqrs. med. dir., 1965-85; corp. med. dir. Am. Internat. Group, N.Y.C., 1985—. Author: Live, Work and Be Healthy, 1980. Capt. USAF, 1957-59. Fellow ACP, Am. Acad. Occupational Medicine; mem. N.Y. Occupational Med. Assn. (pres. 1979-80). Roman Catholic. Avocation: crossword puzzle construction. Home: 19 Randolph Dr Morristown NJ 07960-5319 Office: Am Internat Group Inc 70 Pine St New York NY 10270-0199

VERDONK, EDWARD DENNIS, physicist; b. Edmonton, Alta., Can., Aug. 31, 1961; came to U.S., 1986; s. Dennis and Doris (Blom) V. BSc with honors, U. Alta., 1986; MA, Wash. U., St. Louis, 1989, PhD, 1992. Rsch. asst. Dept. of Geophysics, U. Alta., Edmonton, 1986, Lab. for Ultrasonics, Wash. U., St. Louis, 1988-92; mem. tech. staff Hewlett-Packard Med. Lab., Palo Alto, Calif., 1993—. Contbr. articles to profl. jours. including Jour. of Clin. Investigation, Circulation, Jour. of the Acoustical Soc. of Am. Mem. Am. Soc. of Non-destructive Testing (bd. dirs. 1991-92). Office: Hewlett-Packard Co Bldg 26U-17 3500 Deer Creek Rd Palo Alto CA 94303

VERGA SHEGGI, ANNAMARIA, physicist; b. Florence, Italy, Sept. 30, 1929; d. Egidio and Annunziata (Niccoli) Verga; m. Scheggi, Apr. 21, 1954. M in Math. and Physics, U. Florence, 1953, PhD, 1968. From researcher to rsch. dir. Ist. Ricerca sulle Onde Elettromagnetiche, Florence, 1961-85, dir. Nat. Program Electrooptical Techs., 1988—; prof. U. Florence, 1970-72, U. Bari, Italy, 1985-86. Co-author 9 patents; contbr. 160 articles to profl. jours. Mem. Assn. Elettrotecnica ed Elettronica Italiana (pres. Optoelectronics Topical Group 1987-89), Consiglio Nazionale delle Ricerche/Union Radio Sci. Internat. (pres. Italian com. 1990—), Adv. Group for Aerospace R&D (mem. panel 1981-89), Internat. Optical Fiber Sensors Conf. (steering com.), Internat. Soroptimist Club. Office: Ist Ricerca sulle Onde Elettromagnetiche, Via Panciatichi 64, 50127 Florence Italy

VERHEYEN, MARCEL MATHIEU, homoeopathist, consultant; b. Stokkem, Belgium, Dec. 11, 1951; s. Rene and Maria (Peeters) V.; m. Nicole Stevens Verheyen, July 17, 1974; children: Renee, Rachel. MD, K.U.L Cath. U., Leuven, Belgium, 1979; student, Joszef Coll., Hasselt, Belgium, 1970. Medical Doctor. Editor Beter, 1983; cons. Enterprises for rsch. and devel. of natural med., The Netherlands, 1982; rscher. Homoepathy and Phytotherapy, Switzerland, 1988. Author: Homoepathy For the Whole Family, 1986; editor: Gezondheidsnieuws, 1983. Mem. Nederlandse Vereniging voor Fytotherapie, Belgische Vereniging vor Fytotherapie, European Sci. Coop. for Phytotherapy. Avocations: music, nature, sports. Home: 5 Troliebergplein, Leuven 3010, Belgium Office: Ctr for Natuurgeneeskunde, Prev Medicine/Homoeopathy, 1 Cortenstraat, 6211 HT Maastricht The Netherlands

VERILLON, FRANCIS CHARLES, chemist; b. Paris, France, Sept. 24, 1944; s. Bernard August and Jacqueline (Masbou) V.; m. Monique Francoise Geneau, Aug. 1, 1968; children; Sylvestre, Julien. MBA, Inst. Adminstrn. Entreprises, Paris, 1979; PhD in Analyt. Chemistry, U. Pierre et Marie Curie, Paris, 1977. Lab. mgr. Savcongo Soap Factory, Brazzaville, Congo, 1971-72; rsch. engr. Srti, Thomson group, Buc, France, 1973-78; product mgr. Gilson Med. Electronics, Villiers Le Bel, France, 1979—. Mem. editorial bd. Lab. Robotics and Automation, 1989-92; contbr. articles to profl. jours. Found. Amparo de Pesquisa grantee, 1969. Mem. Am. Chem. Soc., Groupement Pour l'Avancement Des Methodes Spectroscopiques d'Analyse, Groupe Francais De Bio-Chromatographie, Amnestry Internat. Achievements include research in innovations for automated chromatography at the laboratory scale; patents for Peparation of Biological Samples Prior to HPLC, Supercritical Fluid Chromatography with Automated Pressure Control (France). Home: 41 45 Rue De Domremy, 75013 Paris France Office: Gilson Med Electronics SA, 72 Rue Gambetta BP 45, 95400 Villiers Le Bel France

VERMA, DHIRENDRA, civil engineer; b. New Delhi, Dec. 26, 1961; came to U.S., 1984; s. Prem and Krishna (Srivastava) Narain. B of Tech., Indian Inst. Tech., 1984; MS, Case Western Res. U., 1988, PhD, 1990. Registered profl. engr., Mich. Sr. project engr. Altair Engring., Troy, Mich., 1989-93; product engr. Chrysler Corp., Auburn Hills, Mich., 1993—; cons. Chrysler Corp., GM, Ford, Dow Chems., Detroit, 1989—; mem. senate Case Western Res. U., 1988. Contbr. articles to Jour. Structural Divsn. Vol. Mich. Humane Soc., Auburn Hills, 1989—. Mem. ASCE (assoc., contbr. articles to jour.). Achievements include development of accurate numerical schemes for simulation of non-linear dynamic events, finite element technologies applied to automobile crashworthiness, stochastic process applications to engineering mechanics, structural reliability and application to civil engineering structures, development of new methodologies for predicting fatigue crack growth in metals. Office: Chrysler Corp CIMS 484-36-01 800 Chrysler Dr W Auburn Hills MI 48287

VERMA, RAM SAGAR, geneticist, educator, author, administrator; b. Barabanki, India, Mar. 3, 1946; came to the U.S., 1972; s. Gaya Prasad and Late Moonga (Devi) V.; m. Shakuntala Devi, May 4, 1962; children: Harendra K., Narendra K. BSc, Agra U., India, 1965, MSc in Quantitative Genetics, 1967; PhD in Cytogenetics, U. Western Ont., London, Ont., Can., 1972; diploma clinical cytogenetics, The Royal Coll. Pathologists, London, 1984. Diplomate The Royal Coll of Pathologists, London; lic. dir. clin. Cytogenetics, N.Y.C. and N.Y. state. Rsch. and teaching asst. dept. plant scis. U. Western London, Ont., Can., 1967-73; postdoctoral rsch. assoc. cytogenetics U. Colo. Dept. of Pediatrics, Denver, 1973-76; instr. to prof. human cytogenetics dept. of medicine Health Sci. Ctr. SUNY, Bklyn., 1976—, prof. dept. anatomy and cell biology, 1988—; chief cytogenetics div. hematology and cytogenetics Interfaith Med. Ctr. (formerly Jewish Hosp. and Med. Ctr. Bklyn.), 1980-86; chief div. genetics L.I. Coll. Hosp., Bklyn., 1986—; cons. WHO, Switzerland, 1982, Nat. Geog. Soc., Washington, 1982, Phototake, 1982-87; mem. cytogenetic adv. com. Prenatal Diagnosis Lab. N.Y.C. Dept. Health, 1978-90, Genetic Task Force N.Y. State, N.Y.C., 1976—; reviewer grants Nat. and Internat. Health Agys. and Socs.; lectr. colls., univs. and profl. assns. Author: Heterochromatin: Molecular and Structural Aspects, 1988, The Genome, 1990, (with A. Babu) Human Chromosomes: Manual of Basic Techniques, 1989; editor-in-chief: Advances in Genome Biology, 1989; contbr. of over 250 abstracts and presentations and over 250 articles to profl. publs. including Am. Jour. Ob.-Gyn., Blood, Jour. Med. Genetics, Japanese Jour. Human Genetics, Onclogy, Cytobios, Am. Jour. Human Genetics, Am. Jour. Clin. Oncology, Internat. Jour. Cancer, Chromosoma, Cytogenetics. Apptd. to Adv. Coun. to Asst. Commr. City of N.Y. Dept. Health, Bur. Lab. Svcs., 1988. Nat. Merit scholar Gov. India, 1964-67, 1965-67; rsch. scholar Nat. Rsch. Coun. Can. and U. Western Ont., 1967-72, also teaching assistantship, 1972-73; rsch. grantee N.Y. State Dept. Health, Albany, 1985, 85-86, Cancer Treatment Fund, Cornell Med. Coll., 1985-86, United Leukemia Fund, Cornell Med. Coll., 1985-86, 86-87, Nat. Cancer Inst. of Health, Md., 1985-86, 86-87, 97-88, 88-90, Nat. Cancer Inst., 1976-77, 77-78, 78-80. Fellow AAAS, Assn. Clin. Scientists, The Inst. of Biology, N.Y. Acad. Scis., N.Y. Acad. Medicine (assoc.); mem. Am. Assn. Clin. Rsch., Am. Fedn. Clin. Rsch., Am. Genetic Assn. (life), Am. Soc. Cell Biology, Am. Soc. Human Genetics (life), European Soc. Human Genetics, Fedn. Am. Scientists, Genetic Soc. Am., Genetic Soc. Can., Genetic Toxicology Assn., Internat. Assn. Human Biologists, Indian Soc. Human Genetics (life), Soc. Exptl. Biology and Medicine, The Royal Coll. of Pathologists, London. Achievements include research in differentiation of eukaryotic chromosomes with special interest on molecular aspects of structural organization of hetero-and euchromatin, cytological detection of cell damage using old and new classical methods of cytogenetics, application of animal models to understand the human genetic diseases, mechanisms of human cancer using DNA probes and blotting techniques, application of various banding techniques in basic and clinical cytogenetics, automation of human genome using computers. Home: 45-38 Springfield Blvd Bayside NY 11361 Office: The L I Coll Hosp Div of Genetics Brooklyn NY 11201

VERMEER, MARK ELLIS, project engineer; b. LeMars, Iowa, July 10, 1949; s. Elmer H. and Harriet L. (DeBoer) V.; m. Debra K. Van Aartsen, Jan. 8, 1972 (div. Dec. 1988); children: Julie, Robyn. BS in Edn., Northwestern U., 1972; BSCE with distinction, Iowa State U., 1976; MBA with honors, U. S.D., 1988. Registered profl. engr. Tchr. Northeast Hamilton High Sch., Blairsburg, Iowa, 1971-74; project engr. DeWild Grant Reckert, Sioux City, Iowa, 1976-81; project mgr. Younglove, Sioux City, Iowa, 1981-91; sr. project engr. Todd and Sargent, Inc., Ames, Iowa, 1991—. Mem. Leadership Sioux City, 1990; organizer Sioux City Revercade, 1990; campaign leader Iowa 6th Dist. Rep. Campaign, Sioux City, 1986. Mem. ASCE, NSPE, Tau Beta Pi. Home: 1315 Big Blustem Ct # B4 Ames IA 50010

VERNARELLI, MICHAEL JOSEPH, economics educator, consultant; b. Rochester, N.Y., Nov. 24, 1948; s. S. John and Angelica Dolores (Morabito) V.; m. Joan Ann Taylor, Oct. 4, 1975; children: Jacqueline Andrea, Laurel Aileen. BA in Econs., U. Mich., 1970; MA in Econs., SUNY, Binghamton, 1974, PhD in Econs., 1978. Account analyst Travelers Ins. Co., Rochester, 1970-71; prof. econs. Rochester Inst. Tech., 1976—, chmn. dept., 1987—; econs. cons. Rochester Downtown Devel. Corp., 1980; rsch. economist div. housing rsch. HUD, Washington, 1980-81, vis. scholar, 1980; pres., forensic economist Rochester Econ. Cons., 1983—; vis. prof. U.S. Bus. Sch. in Prague, 1992-93. Contbg. author: Federal Housing Policy and Desegregation, 1986. Mem. Brighton (N.Y.) Bd. Archtl. Rev., 1990-91, mem. planning bd., 1991—. Recipient Eisenhart award Rochester Inst. Tech., 1987; grantee SUNY, Binghamton, 1974. Mem. Am. Econ. Assn., Nat. Assn. Forensic Economists, Ea. Econ. Assn., Greater Rochester C. of C. (panel mem. bus. trends com. 1987—), Omicron Delta Epsilon. Roman Catholic. Avocation: golf. Home: 133 Esplanade Dr Rochester NY 14610-3325 Office: Rochester Inst Tech Rochester NY 14623-0887

VERNBERG, FRANK JOHN, marine and biological sciences educator; b. Fenton, Mich., Nov. 6, 1925; s. Sigurd A. and Edna (Anderson) V.; m. Winona M. Bortz, Sept. 7, 1945; children—Marcia Lynn, Eric Morrison, Amy Louise. A.B., DePauw U., 1949, M.A., 1950; Ph.D., Purdue U., 1951. Prof. zoology Duke Marine Lab., Beaufort, N.C., 1951-69; Belle W. Baruch prof. marine ecology, dir. Belle W. Baruch Coastal Research Inst., U. S.C., Columbia, 1969—, interim dean Coll. Sci. and Math., 1993—; vis. prof. U. Coll. West Indies, Jamaica, 1957-58, U. Sao Paulo, Brazil, 1965; program dir. exptl. analytical biogeography of sea Internat. Biol. Program, 1967-69; mem. com. manned orbital research lab. Am. Inst. Biol. Scis.-NASA, 1966-68; pres. Estuarine Research Fedn., 1975-77. Contbr. articles to profl. jours.; spl. editor: Am. Zoologist, 1963; mem. editorial bd.: Biol. Bull., 1977-80; editor: Jour. Exptl. Marine Biology and Ecology, 1978—. Served with USNR, 1944-46. Recipient W.S. Proctor award Sigma Xi, award Drug Sci. Found., 1987; Guggenheim fellow, 1957-58; Fulbright-Hayes fellow, 1965. Fellow AAAS, Am. Soc. Zoologists (sec.-treas. div. comparative physiologists 1959-61, sec.-treas. edn. com. 1960-62, mem. council 1965-67, pres. 1982), Southeastern Estuarine Research Soc. (pres. 1974-76), So. Assn. Marine Labs. (pres. 1993), Estuarine Research Fedn. (pres. 1975-77), S.C. Wildlife Fedn. (Constructionist of Yr. 1983). Office: U SC Belle W Baruch Coastal Rsch Inst Columbia SC 29208

VERNICK, ARNOLD SANDER, environmental engineer; b. N.Y.C., May 2, 1933; s. Joseph Leon and Beatrice (Carlin) V.; m. Lynne Beatrice Bowin, Sept. 16, 1962; children: Jeffrey Francis, Kenneth Charles. BS, Queens Coll., 1956; BS in Civil Engring., Columbia U., 1956; MS, NYU, 1970. Registered profl. engr. N.J., N.Y., Vt., N.H., Mass., Maine, R.I., Conn., Pa., Del., Md., Va., W. Va., Mich., Ill., N.C., Ohio, D.C. Project engr. Esso Standard Oil Co., Linden, N.J., 1956-62; civil engr. Alexander Potter Assocs., N.Y.C., 1962-63; dist. engr. Gulfstan Corp., Middlesex, N.J., 1963-64; sanitary engr. Chem. Constrn. Corp., N.Y.C., 1964-66, Gibbs & Hill, N.Y.C., 1966-68; mgr. environ. engring. Burns & Roe Indsl. Svcs. Corp., Oradell, N.J., 1968-88; v.p. Geraghty & Miller, Inc., Rochelle Park, N.J., 1988—. Author (with others) Handbook of Industrial Wastes Pretreatment, 1980; editor: Handbook of Wastewater Treatment Processes, 1981; contbr. articles to profl. jours. With AUS, 1958, comdr. USPHS Res. Corp. Fellow ASCE (chmn. exec. com. environ. engring. divsn. 1990-91); mem. Nat. Soc. Profl. Engrs., Am. Acad. Environ. Engrs. (diplomate), Am. Water Works Assn., Air & Waste Mgmt. Assn., Water Environ. Fedn., Tau Beta Pi. Achievements include directing projects in the delineation and remediation of ground water and soil contamination, hazardous and solid waste mgmt., water supply, wastewater treatment and pollution abatement. Home: 602 James Ln River Vale NJ 07675 Office: Geraghty & Miller Inc 201 W Passaic St Rochelle Park NJ 07662

VERNIKOS, JOAN, science association director; b. Alexandria, Egypt, May 9, 1934; came to U.S., 1960; d. Apostolos and Catherine (Manganari) V.; m. Constantine Danellis, June 10, 1960 (dec. Apr. 1971); children: Eftihia, George; m. Geoffrey Cyril Hazzan, Sept. 4, 1978. B Pharmacy, U. Alexandria, 1955; PhD, U. London, 1960. Asst. prof. Ohio State U. Med. Sch., Columbus, 1961-64; rsch. scientist NASA Ames Rsch. Ctr., Moffet Field, Calif., 1966-93; chief human studies NASA Ames Rsch. Ctr., Moffett Field, Calif., 1972-76, acting dep. dir. life sci., 1976; assoc. dir. space rsch. NASA Ames Rsch. Ctr., Moffet Field, Calif., 1987-93, chief life sci. divsn., 1988-93; dir. life and biomed. scis. and applications NASA HQ Code UL, Washington, 1993—; mem. pharmacology study sect. NIH, Bethesda, Md., 1974-78; cons. Adv. Group for Aerospace R&D, Cologne, Germany, 1991; mem. aerospace medicine adv. com. NASA, Washington, 1988-93; rsch. assoc. NAS, NASA Ames Rsch. Ctr., 1964-66. Author: Hormones and Behavior, 1972, Neuroregulators and Psychiatric Disorders, 1977, Selye's Guide to Stress Research, Vol. 1, 1980, Strategies for Mars, 1993, Stress: Neurochemical and Molecular Approaches, 1992; author, editor: Inactivity: The Physiology of Bedrest, 1986. Mem. adv. bd. Nat. Hispanic U., San Jose, Calif., 1991-93. Recipient medal for sci. achievement NASA, 1973. Fellow Aerospace Med. Assn. (Strughold award 1990); mem. Endocrine Soc., Am. Soc. Pharmacology and Exptl. Therapeutics, Soc. Neuroscience. Greek Orthodox. Achievements include two patents; research on hormone indexes of stress; understanding brain mechanisms regulating stress response and evidence for coping mechanisms; contributor to understanding physiological responses and adaptation to weightlessness of space flight, drug treatments for postflight orthostatic intolerance; minimal requirements for artificial gravity in prolonged space missions. Office: Life and Biomed Scis/Appls NASA HQ Code UL 300 E St SW Washington DC 20546

VERNON, JACK ALLEN, otolaryngology educator, laboratory administrator; b. Kingsport, Tenn., Apr. 6, 1922; s. John Allen and Mary Jane (Peters) Vernon Hefley; m. Betty Jane Dubon, Dec. 12, 1946 (div. 1972); children—Stephen Mark, Victoria Lynn; m. Mary Benson Meikle, Jan. 2, 1973. B.A. in Psychology, U.Va., 1948, M.A. in Psychology, 1950, Ph.D. in Psychology, 1952. Instr. psychology Princeton U., N.J., 1952-54, asst. prof., 1954-60, assoc. prof., 1960-64, prof., 1964-66; prof. otolaryngology Oreg. Health Sci. U., Portland, 1966—; also dir. Oreg. Hearing Rsch. Ctr. Oreg. Health Sci. U. Inventor in field. Author: Inside the Black Room, 1963. Adv. Office Civil Defense, Washington, 1961-62. Served to 2d lt. USAAF, 1943-44. Recipient Guest of Honor award 1st Internat. Tinnitus Seminar, 1979. Mem. Assn. Research Otolaryngology (pres. 1973-74), Am. Acad. Ophthalmology and Otolaryngology. Democrat. Lodge: Rotary. Avocations: woodworking; sailing; skiing; reading. Home: 17505 UU NW Sauvie Island Portland OR 97231 Office: Oreg Hearing Rsch Ctr 3181 SW Sam Jackson Park Rd Portland OR 97201*

VERNON, SIDNEY, physician, publisher, author; b. N.Y.C., Nov. 12, 1906; s. Hyman and Lillian (Zonenberg) V.; m. Rosalie Silverstein, (dec. Oct. 1983); children—Kenneth, Sheridan. B.S., CCNY, 1926; M.D., L.I. Coll. Hosp., 1930. Intern, Bellevue Hosp., N.Y.C., 1930-31; resident Backus Hosp., Norwich, Conn., 1931-32; gen. practice medicine, Willimantic, Conn., 1932-41, 1952—; commd. 1st lt. U.S. Army, 1941, advanced through grades to lt. col., 1946; med. officer U.S. Army, 1941-50; chief surgery Arrowhead, Two Harbor Hosp., Minn., 1950-51; chief surgeon Army Hosp., Waltham, Mass., 1947, Ft. Monmouth, N.J., 1948-49, Air Force Hosp., Hempstead, N.Y., 1950 Author: How to Understand People, 1982; Reach for Charisma, 1984. Contbr. 75 articles to profl. jours. Decorated Bronze Star. Recipient Hon. Laymans award Community Assn. Health Phys. Edn. and Recreation. Fellow ACS, Internat. Coll. Surgery; mem. Am. Coll. Sports Medicine, AMA, Am. Bd. Abdominal Surgery (bd. govs.), Willimantic C. of C. (chmn. community council phys. fitness 1958-61), Am. Pain Soc., Internat. Assn. Pain., Inflammation Club Upjohn Pharm. Jewish. Lodge: B'nai Brith (pres. 1939-40).

VERRET, DOUGLAS PETER, semiconductor engineer; b. Thibodaux, La., Mar. 1, 1947; s. Norman Peter and Kate Marie (Marcello) V.; m. Patricia Ellen Coogan, Mar. 1, 1975; children: Sybil Margaret, Laurence Francis. BS summa cum laude, Spring Hill Coll., 1969; MS in Physics, Purdue U., 1974; PhD in Physics, U. New Orleans, 1978. Physics tchr. Arch. Rummel High Sch., Metairie, La., 1973-74; asst. prof. physics Xavier U., New Orleans, 1974-79; process engr. digital products div. Tex. Instruments, Houston, 1979-82, bipolar bicmos devel. mgr. digital products divsn., 1991—, prodn. mgr. application specific products div., 1991—; dir. mfg. techniques Sematech, Austin, Tex., 1988-90; program chmn. Bipolar/Bicmos Crcts. and Tech. Meeting, 1993. Contbr. articles to profl. publs. Fellow NSF, 1970-74, UNCF, 1969, Tex. Instruments, 1985, 92—. Mem. IEEE (sr.), Am. Phys. Soc., Sigma Xi. Democrat. Roman Catholic. Achievements include 16 patents and 9 patents pending for double poly self-aligned bipolar transistor, bicmos (digital) fabrication method, self-aligned w-plugged vias, Si-Ge bipolar transistor fabrication method, Cu interconnects, others. Home: 13807 Baytree Dr Sugar Land TX 77478

VERRILLO, RONALD THOMAS, neuroscience educator, researcher; b. Hartford, Conn., July 31, 1927; s. Francesco Paul and Angela (Forte) V.; m. Violet Silverstein, June 3, 1950; children—Erica, Dan, Thomas. B.A., Syracuse U., 1952; Ph.D., U. Rochester, 1958. Asst. prof. Syracuse U., 1957-62, research assoc., 1959-63, research fellow, 1963-67, assoc. prof., 1967-74, prof., 1974—, assoc. dir. Inst. Sensory Research, 1980-84, dir., 1984—, dir. grad. neurosci. program, 1984—; advisor com. on hearing, bioacoustics and biomechanics NRC. Author: Adjustment to Visual Disability, 1961 (award 1962). Contbr. several chpts. to books, articles to profl. jours. Served with USN, 1945-46. Fellow Am. Found. for Blind, 1956, NATO, 1970; grantee NSF, 1969-72, 84-87, NIH, 1972—. Fellow Acoustical Soc. Am.; mem. Psychonomic Soc., Soc. for Neurosci. N.Y. Acad. Scis., Sigma Xi (research award 1982). Home: 312 Berkley Dr Syracuse NY 13210-3031 Office: Syracuse University Inst for Sensory Research Merrill Ln Syracuse NY 13244

VERRY, WILLIAM ROBERT, mathematics researcher; b. Portland, Oreg., July 11, 1933; s. William Richard and Maurine Houser (Braden) V.; m. Bette Lee Ronspiess, Nov. 20, 1955 (div. 1981); children: William David, Sandra Kay Verry Londregan, Steven Bruce, Kenneth Scott; m. Jean Elizabeth Morrison, Oct. 16, 1982; step-children: Lucinda Jean Hale, Christine Carol Hale Fortner, Martha Jean Johnson, Brian Kenneth Lackey, Robert Morrison Lackey. BA, Reed Coll., 1955; BS, Portland State U., 1957; MA, Fresno State U., 1960; PhD, Ohio State U.-Columbus, 1972. Instr. chemistry Reedley (Calif.) Coll., 1957-60; ops. research analyst Naval Weapons Center, China Lake, Calif., 1960-63; ordnance engr. Honeywell Ordnance, Hopkins, Minn., 1963-64; sr. scientist Litton Industries, St. Paul., 1964-67; project mgr. Tech. Ops., Inc., Alexandria, Va., 1967-70; research assoc. Ohio State U., Columbus, 1970-72; prin. engr. Computer Sci. Corp., Falls Church, Va., 1972-77; mem. tech. staff MITRE Corp., Albuquerque, 1977-85; C3 program dir., assoc. prof. math. sci. Clemson U., S.C., 1985-87; mgr. simulation and modeling Riverside Research Inst., Rosslyn, Va., 1987-91; mgr. Hillcrest Gardens, Livermore, Calif., 1992—. Founder, minister Christian Love Ctr. Mem. Ops. Research Soc. Am. Home and Office: 550 Hillcrest Ave Livermore CA 94550-3771

VERSTRAETE, MARY CLARE, biomedical engineering educator; b. Detroit, Jan. 7, 1960; d. Joseph Leon and Lois Patricia (Lynch) V. BS, Mich. State U., 1982, MS, 1984, PhD, 1988. Asst., then assoc. prof. dept. biomed. engring. The Univ. of Akron, Ohio, 1988—; mem. Institutional Rev. Bd. for the Protection of Human Subjects, U. Akron, 1990—. Reviewer Medicine & Science in Sports & Exercise jour., 1989—; contbr. articles to profl. jours. Mem. ASME (regional bd. 1984—), Am. Soc. Biomechanics, Biomed. Engring. Soc. (advisor student chpt.), Sigma Xi. Office: The Univ of Akron Dept Biomed Engring Akron OH 44325-0302

VERWEY, TIMOTHY ANDREW, structural engineer, consultant; b. Jacksonville, Fla., Aug. 26, 1964; s. John and Bernice Janett (Lane) V. BSCE, U. Cen. Fla., 1987. Engr. Structural Techs., Apopka, Fla., 1988-92, Camp Dresser & McKee Inc., Orlando, Fla., 1992—. Mem. ASCE (assoc.). Republican. Presbyterian. Home: 2427 Piedmont Lakes Blvd Apopka FL 32703 Office: Camp Dresser & McKee Inc 1950 Summit Park Dr Ste 300 Orlando FL 32810

VESELY, KAREL, chemist, educator; b. Brno, Czechoslovakia, Oct. 25, 1921; s. Vitezslav and Helena (Polckova) V.; m. Darja Novotna, Mar. 8, 1945; children: Petr, Tomas. Degree, U. Prague, Czechoslovakia, 1947, ScD, 1964. Rsch. chemist Inst. Macromolecular Chemistry, Brno, 1951-65, 70-89; prof. chemistry Masaryk U., Brno, 1965-70, 90—. Author: Polyreactions, 1955, Polymers, 1992; contbr. to profl. publs. Mem. Am. Chem. Soc., Czech Chem. Soc. (medal 1991). Achievements include 40 patents in stabilization of polymers, filled polymers. Home: Lerchova 42, Brno 60200, Czech Republic

VEST, CHARLES MARSTILLER, university administrator; b. Morgantown, W.Va., Sept. 9, 1941; s. Marvin Lewis and Winifred Louise (Buzzard) V.; m. Rebecca Ann McCue, June 8, 1963; children—Ann Kemper, John Andrew. BSME, W.Va. U., 1963; MS in Engring., U. Mich., 1964, PhD, 1967; DEng (hon.), Mich. Tech. U., 1992. Asst. prof., then assoc. prof. U. Mich., Ann Arbor, 1968-77, prof. mech. engring., 1977-90, assoc. dean acad. affairs Coll. Engring., 1981-86, dean Coll. Engring., 1986-89, provost, v.p. acad. affairs, 1989-90; pres. MIT, Cambridge, 1990—; vis. assoc. prof. Stanford U., Calif., 1974-75; dir. E.I. du Pont de Nemous and Co., 1993—. Author: Holographic Interferometry, 1979; assoc. editor Jour. Optical Soc. Am., 1982-83; contbr. articles to profl. jours. Trustee Wellesley Coll., Environ. Rsch. Inst. Mich. WGBH, Boston Mus. Sci., E.I. duPont de Nemours and Co., Woods Hole Oceanographic Inst., New Eng. Aquarium. Recipient Excellence in Rsch. award U. Mich., 1980, Disting. Svc. award,

1972, Disting. Visitor award U. La Plata, Argentina, 1978, Centennial medal Am. Soc. Engring. Edn., 1993. Fellow AAAS, NAE, Am. Acad. Arts and Scis., Optical Soc. Am.; mem. AWME, Nat. Acad. Engring., Sigma xi, Tau Beta Pi, Pi Tau Sigma. Presbyterian. Home: 111 Memorial Dr Cambridge MA 02142-1348 Office: MIT 77 Mass Ave Bldg 13 Rm 2090 Cambridge MA 02139-4307

VESTAL, TOMMY RAY, lawyer; b. Shreveport, La., Sept. 19, 1939; s. Louie Wallace and Margaret (Golden) V.; m. Patricia Marie Blackwell, Jan. 24, 1981; children: Virginia Ann Yancy, John Wallace Vestal, Douglas William Yancy. BSME, U. Houston, 1967, JD, 1970. Bar: Tex. 1970, U.S. Patent Office 1972, U.S. Ct. Appeals (D.C. cir.) 1975. Patent atty. Am. Enka Corp., Asheville, N.C., 1970-71, Akzona Inc., Asheville, 1971-84, Akzo Am., Inc., Asheville, 1985-86; sr. patent atty. Fibers div. BASF Corp., Enka, N.C., 1986-87, div. patent counsel, 1987-89, sr. patent counsel, 1989-90; pvt. practice law, 1990-91; dir. Geary Glast & Middleton, Dallas, 1992; ptnr. Falk, Vental & Fish, Dallas, 1992—. Mem. ABA, State Bar Tex., Am. Intellectual Property Law Assn., Carolina Patent, Trademark and Copyright Law Assn. (bd. dirs. 1983-85, 2d v.p. 1985-86, 1st v.p. 1986-87, pres. 1987-88), Asheville C. of C. (chmn. legal affairs com.), Phi Alpha Delta. Republican. Lutheran. Lodge: Kiwanis (pres. 1982). Avocations: golf, fishing, hiking. Home: 3109 Squireswood Dallas TX 75006 Office: 700 N Pearl Dallas TX 75201

VESTMAR, BRIGEL JOHANNES AHLMANN, information systems agency adviser, scientist; b. Aalborg, Denmark, July 26, 1937; s. Kai Brigel Ahlmann and Karly (Gregersen) V.; m. Jette Jensen, Dec. 28, 1961; children: Peter, Brigel. MSc, Tech. U. Copenhagen, 1963; Dr.Ir., Tech. U. Delft, 1975; postgrad., Inst. Methodes Direction Ent., Lausanne, Switzerland, 1975. Scientist Shape Tech. Ctr., The Hague, The Netherlands, 1965, sr. scientist, 1965-69, prin. scientist, 1969-80; asst. prin. tech. adviser NATO Communications and Info. Systems Agy., Brussels, 1980-92, tech. adv. users svcs. divsn., 1992—; rep. to several tech. NATO coms. and working groups. Contbr. numerous articles in field to profl. publs. Lt. Royal Danish Navy, 1963-65. Mem. Danish Soc. Chem., Civil, Electrical and Mech. Engrs., IEEE, Internat. L'Institut Pour L'Etude des Methodes de Direction de L'Entreprise Alumni Assn. Home: Madeliefjeslaan 22, Tervuren, 3080 Brussels Belgium Office: NACISA, Rue de Geneve 8, 1140 Brussels Belgium

VETRO, JAMES PAUL, electrical project engineer; b. Binghamton, N.Y., May 9, 1960; s. Dominic and Florence Anna (Herbon) V.; m. Carol Elaine Buettner, Aug. 14, 1982; children: Sarah Anne, Laura Christine, Amy Marie. AS in Engring. Sci., Broome C.C., Binghamton, N.Y., 1979; BS in Elec. Engring., SUNY, Buffalo, 1981; MS in Elec. Engring., Syracuse U., 1983. Registered profl. engr., Wis. Elec. design engr. link flight simulation div. Singer, Binghamton, 1981-84; sr. elec. project engr. GE Med. Systems, Milw., 1984—. Mem. AAAS. Roman Catholic. Home: 226 Meadowside Ct Pewaukee WI 53072 Office: GE Med Systems PO Box 414 W828 Milwaukee WI 53201

VEZERIDIS, MICHAEL PANAGIOTIS, surgeon, researcher, educator; b. Thessaloniki, Greece, Dec. 16, 1943; came to U.S., 1974; s. Panagiotis and Sofia (Avramidis) V.; m. Therese Mary Statz; children: Peter Statz, Alexander Michael. MD, U. Athens, 1967; MA (hon) ad eundem, Brown U., 1989. Diplomate Am. Bd. Surgery. Fellow surg. rsch. Harvard Med. Sch./Mass. Gen. Hosp., Boston, 1974-77; resident U. Mass., Worcester, 1977-80; fellow in surg. oncology Roswell Park Meml. Inst., Buffalo, 1980-81, Attending surgeon, 1981-82; staff surgeon VA Med. Ctr., Providence, 1982-84; asst. prof. surgery Brown U., Providence, 1982-88; chief surg. oncology VA Med. Ctr., Providence, 1984—, assoc. chief surgery, 1986—; cons. in surgery R.I. Hosp., Providence, 1987—; surg. oncologist Roger Williams Med. Ctr., Providence, 1989—; assoc. dir. div. surg. oncology Brown U., Providence, 1989—, assoc. prof. surgery, 1988—; chmn. profl. edn. com. R.I. div. Am. Cancer Soc., Providence, 1987-89, bd. dirs., 1987—, pres.-elect. 1989-91, pres., 1991—; vis. prof. U. Patras (Greece) Med. Sch., 1988; mem. sci. adv. com. Clin. Rsch. Ctr., Brown U., Providence, 1989-91. Contbr. articles to profl. jours. and chpts. in med. books. Mem. parish coun. Ch. of Annunciation, Cranston, R.I., 1985-91; v.p. Hellenic Cultural Soc. Southeastern New Eng., Providence, 1987-89. Merit Review Cancer Research grantee Vets. Adminstr., 1983-89; named Profl. Fed. Employee of Yr., R.I. Fed. Exec. Coun., 1987; decorated Commendation medal USN. Fellow ACS; mem. Soc. Surg. Oncology, Assn. for Acad. Surgery, Am. Soc. Clin. Oncology, N.Y. Acad. Scis. (life), Soc. for Surgery of the Alimentary Tract, Am. Assn. for Cancer Rsch., Collegium Internat. Chirurgiae Digestivae, New Eng. Cancer Soc., New Eng. Surg. Soc., Quidnessett Country Club. Greek Orthodox. Avocations: classical music, reading, fencing, tennis, squash, cross-country skiing. Home: 50 Limerock Dr East Greenwich RI 02818-1643 Office: Roger Williams Med Ctr 825 Chalkstone Ave Providence RI 02908-4728

VÉZINA, MONIQUE, Canadian government official; b. Rimouski, Qué., Canada, July 13. Mem. cabinet, minister external relations, mem. Parliament, Govt. of Canada, Ottawa, Ont., 1984-86, minister supply and services and receiver gen., 1986-87; minister of state for transport Govt. of Canada, 1987-88, minister of state for employment and immigration, minister of state for srs., 1988—. Chmn. parents com. Lower St. Lawrence Sch. Bd., 1964-77; bd. dirs., pres. Assoc. Family Orgns., Qué., Can., 1974-81; nat. pres. Dames Hélène de Champlain, 1976-79; pres. Fédn. des Caisses populaires Desjardins du Bas St-Laurent, 1976-84, Girardin-Vaillancourt Found., 1976-84; bd. dirs. Fédn. des Caisses populaires et d'économies Desjardins du Québec 1977-84, Société immobilière du Qué., 1984; mem. Conseil supérieur de l'éducation du Qué., 1978-82, chmn. secondary sch. bd., 1978-82; dep. bd. chmn. Régie de l'assurance automobile du Qué., 1978-81; chmn. bd. dirs. Institut cooperatif Desjardin, 1981-84,78-81. Office: House of Commons, Parliament Bldg Rm 442 N Ctr Block, Ottawa, ON Canada K1A 0A6

VEZIROGLU, TURHAN NEJAT, mechanical engineering educator, energy researcher; b. Istanbul, Turkey, Jan. 24, 1924; came to U.S., 1962; s. Abdul Kadir and Ferruh (Bilim) V.; m. Bengi Isikli, Mar. 17, 1961; children: Emre Alp, Oya Sureyya. A.C.G.I., City and Guilds Coll., London, 1946; B.Sc. with honors, U. London, 1947; D.I.C., Imperial Coll., London, 1948; Ph.D., U. London, 1951. Engring. apprentice Alfred Herbert Ltd., Coventry, U.K., 1945; project engr. Office of Soil Products, Ankara, Turkey, 1953-56; tech. dir. M.K.V. Constrn. Co, Istanbul, 1957-61; assoc. prof. mech. engring. U. Miami, Coral Gables, Fla., 1962-65; prof. U. Miami, Coral Fables, Fla., 1966—; dir. grad. studies mech. engring. U. Miami, Coral Gables, Fla., 1965-71, chmn. dept. mech. engring., 1971-75, assoc. dean research Coll. Engring., 1975-79, dir. Clean Energy Research Inst., 1974—; UNESCO cons., Paris; vis. prof. Middle East Tech. U., Ankara, 1969. Editor-in-chief: Internat. Jour. Hydrogen Energy, 1976—. Pres. Learning Disabilities Found., Miami, 1972-73, advisor, 1974-80. Recipient Turkish Presdl. sci. award Turkish Sci. and Tech. Research Found., 1975; named hon. prof. Xian Jiaotong U., China, 1982. Fellow AAAS, ASME, Instn. Mech. Engrs.; mem. Internat. Assn. Hydrogen Energy (pres. 1975), AIAA, Assn. Energy Engrs., Am. Nuclear Soc., Am. Soc. Engring. Edn., AAUP, Internat. Soc. Solar Energy, Systems Engring. Soc., Sigma Xi. Home: 4910 Biltmore Dr Miami FL 33146-1724 Office: U Miami Clean Energy Rsch Inst PO Box 248294 Miami FL 33124-8294

VIANA, THOMAS ARNOLD, computer scientist; b. Fall River, Mass., Nov. 17, 1951; s. manuel Lopes and Jennie Tillie (Sroka) V.; m. Colleen Judith Larrivee, Nov. 23, 1974; 1child, Andrew Thomas. BS in Math., Southeastern Mass. U. (now U. Mass. at Dartmouth), 1973, MS in electrical engring./ computer sci., 1979. Tech. staff programmer Logicon, Inc., Newport, R.I., 1974-76; computer specialist Naval Under water Systems Ctr., Newport, 1976-80, Naval Undersea Warfare Ctr., Newport, 1980-. Coach Bristol Youth Soccer Assn., R.I., 1989-90, Bristol Little League, 1990; aide Bristol 4th July Com., 1988-. Mem. Assn. Computing Machinery, Am. Assn. for Artifical Intelligence, Digital Equipment Computer Users Soc. (spl. interest group coms. 1980-86). Democrat. Roman Catholic. Office: Naval Underseas Warfare Ctr Code 221 Bldg 1171-1 Newport RI 02841-5047

VIANCO, PAUL THOMAS, metallurgist; b. Rochester, N.Y., Dec. 28, 1957; s. George William and Josephine Rose (Sardisco) V. BS in Physics, SUNY, 1980; MS in Mechanical and Aeronautical Engring., U. Rochester, 1981, PhD in Materials Sci., 1986. Sr. mem. tech. staff Sandia Nat. Labs., Alburquerque, 1987—. Mem. ASME, Am. Welding Soc. (chmn. subcom. 1992—), ASM Internat., The Metalurgical Soc., Sandia Skeet Club (treas.). Home: 4012 Shenandoah Pl NE Albuquerque NM 87111 Office: Sandia Nat Labs P O Box 5800 Dept 1831 Albuquerque NM 87185

VICE, CHARLES LOREN, electromechanical engineer; b. LaVerne, Okla., Jan. 2, 1921; s. Cyrus Christopher and Ethel Segwitch (Hoy) V.; m. Katherine Margaret Maxwell, July 14, 1949; children: Katherine Lorene, Charles Clark, Ann Marie. Cert., Oreg. State U., 1944, BSME, 1947; postgrad., U. So. Calif., 1948-55. Registered profl. engr., Calif. Mgr. magnetic head div. Gen. Instrument Corp., Hawthorne, Calif., 1959-62; sr. staff engr. magnetic head div. Ampex Corp., Redwood City, Calif., 1962-66; chief mech. engr. Collins Radio Corp., Newport Beach, Calif., 1967-69; pres. FerraFlux Corp., Santa Ana, Calif., 1970-78; sr. staff engr. McDonnell Douglas Computer Systems Co., Irvine, Calif., 1979-89, Santa Ana, Calif., 1989; ret. McDonnell Douglas Computer Systems Co., 1989; cons. Teac Corp. Japan, 1974-78, Otari Corp. Japan, 1975-77, Univac Corp., Salt Lake City, 1975-76, Crown Radio Corp. Japan, 1979-80, Sabor Corp. Japan, 1982, Empire Corp., Tokyo, 1987-89, DIGI SYS Corp., Fullerton, Calif., 1988-89, Puritan Bennett Aerosystems, El Segundo, Calif., 1989—. Patentee in field. Served with U.S. Army Engrs., 1943-46. Decorated Bronze Star. Mem. NSPE. Republican. Club: Toastmasters. Avocations: piano, singing. Home: 5902 E Bryce Ave Orange CA 92667-3305 Office: Precision Cons Inc 5902 E Bryce Ave Orange CA 92667-3305

VICEPS, KARLIS DAVID, solar residential designer; b. Syracuse, N.Y., Jan. 12, 1956; s. Karlis and Velta (Augskaps) V.; m. Catherine Elliott Hale, Aug. 9, 1985. BS with high distinction, Worcester Poly. Inst., 1978. Engr. in tng. Pickard & Anderson, Auburn, N.Y., 1978-79; carpenter Kaufman Construction, Inc., Taos, N.Mex., 1980-85; sole propr. Energyscapes, Taos, N.Mex., 1985—; chair adv. bd., v.p. Roadrunner Recyclers, Inc., Taos, 1991-93, vol. Contbr. article to Solar Today. Com. mem. Taos Clean and Beautiful Non-Motorized Pathways, 1993. Mem. Am. Solar Energy Soc. (mem. solar bldgs. bd. 1988), N.Mex. Solar Energy Assn., Renew Am., Union Concerned Scientists, Chi Epsilon. Achievements include development of passive solar domestic hot water and nightime heating from solar gain. Office: Energyscapes PO Box 2264 Taos NM 87571

VICK, AUSTIN LAFAYETTE, civil engineer; b. Cedervale, N.Mex., Jan. 28, 1929; s. Louis Lafayette and Mota Imon (Austin) V.; BSCE, N.Mex. State U., 1950, MSCE, 1961; m. Norine E. Melton, July 18, 1948; children: Larry A., Margaret J., David A. Commd. 2d lt. USAF, 1950, advanced through grades to capt., 1959, ret., 1970; ordnance engr. Ballistics Rsch. Lab., White Sands Proving Ground, Las Cruces, N.Mex., 1950-51, civil engr., 1951-55, gen. engr. White Sands Missile Range, 1957-73, phys. scientist adminstr., 1955-57, 73—; owner A.V. Constrn., Las Cruces, 1979—; realtor Campbell Agy., Las Cruces, 1979-84; cons. instrumentation systems, ops. maintenance and mgmt., 1984—; pres., treas. Survey Tech., Inc., 1985—; cons. in field, Las Cruces, 1984—. Mem. outstanding alumni awards com. N.Mex. State U., 1980. Recipient Outstanding Performance award Dept. Army, White Sands Missile Range, 1972, Spl. Act awards, 1967, 71, 75. Mem. Mil. Ops. Research Soc. (chmn. logistics group 1968-69), Am. Def. Preparedness Assn. (pres. 1970-72), Assn. U.S. Army (v.p. 1970-71), Am. Soc. Photogrametry, Am. Astronautical Soc. (sr. mem.), N.Mex. State U. Acad. Civil Engring. Contbr. articles to profl. jours. Home and Office: 4568 Spanish Dagger Las Cruces NM 88001-7643

VICK, JOHN, engineering executive; b. Feb. 12, 1933. B in Elec. Engring., Tex. A&M U., M in Elec. Engring. Mem. aerophysics group rsch. and engring. Gen. Dynamics, Ft. Worth, 1959-61, from sr. aerosystems engr. to project aerosystems engr. F-111 radar group, 1961-67, group engr. F-111 aerosystems project office, 1967-68, asst. project engr., 1968-73, aerosystems group engr. electronic fabrication ctr., 1973-75, chief, 1975-77, mgr., 1977-80, dir. support requirements and systems dept., 1980-88, divsn. v.p. mil. electronics, 1988-91, divsn. v.p. rsch. and engring., 1991—. Office: General Dynamics Corp Fort Worth Div General Dynamics Blvd Fort Worth TX 76101*

VICK, MARIE, retired health science educator; b. Saltillo, Tex., Jan. 22, 1922; d. Alphy Edgar and Mollie (Cowser) Pitts; m. Joe Edward Vick, Apr. 5, 1942; children: Mona Marie, Rex Edward. B.S., Tex. Woman's U., Denton, 1942, M.A., 1949. Tchr. Coahoma (Tex.) High Sch., 1942-43, Santa Rita Elem. Sch., San Angelo, Tex., 1943-45, Crozier Tech. High Sch., Dallas, 1946-47, Monroe Jr. High Sch., Omaha, 1947-48; instr. Tex. Woman's U., Denton, 1948-50; tchr. San Angelo (Tex.) Jr. High Sch., 1957-58, San Angelo (Tex.) Sr. High Sch., 1957-58, Harlingen Bonham Elem. Sch., 1958-59, Harlingen (Tex.) High Sch., 1959-62; prof. health sci. Coll. Edn. U. Houston, 1962-80. Author: A Collection of Dances for Children, 1970; Health Science in the Elementary School, 1979; contbr. articles to profl. jours.; artist in oil, watercolor and acrylic. Mem. exec. bd. Health Care Task Force of Walker County. Recipient Cert. of Achievement, Tex. Commn. Intercollegiate Athletics for Women, 1972, Research Service award Tex. Cancer Control Program, 1978-79, Plaudit award Nat. Dance Assn., 1982, Disting. Service award Pan Am. U., 1983, Service citation Am. Cancer Soc., Cert. of Appreciation, Tex. div. Am. Cancer Soc., 1980; Favorite Prof. honoree Cap and Gown Mortar Bd., U. Houston, 1974. Mem. AAHPERD (dance editor 1971-74), NEA, Am. Sch. Health Assn., So. Assn. Health, Phys. Edn. Coll. Women (sec. dance sect. 1970-73), Tex. State Tchrs. Assn. (sect. chmn. 1964-65), Tex. Assn. Health, Phys. Edn. and Recreation (chmn. dance sect. 1968-69), Tex. Assn. Coll. Tchrs., Tex. Women's U. Nat. Alumnae Assn. (life), Tex. Women's Pioneer Club, Am. Assn. Ret. Persons (chmn. legis. com. Huntsville chpt. 1988-90, bd. dirs., liaison person Walker County commrs. 1989-90, chmn. community svc. project Walker County Unpaved Rd. Survey 1989, mem. exec. bd. 1992-94), Nat. Ret. Tchrs. Assn. (legis. chmn. 1988-89), Tex. Assn. Ret. Tchrs., Property Owners Assn. (organizer, past pres.), U. Houston Assn. Ret. Profs., Huntsville Garden Club (treas.). Democrat. Methodist. Home: RR 6 Box 681A Huntsville TX 77340-9806

VICKERMAN, KEITH, biologist; b. Huddersfield, U.K., Mar. 21, 1933; s. Jack and Mabel (Dyson) V.; m. Moira Dutton, Sept. 16, 1961; 1 child, Louise Charlotte. BSc, U. Coll. London, 1955; PhD, London U., 1960, DSc, 1970. Wellcome rsch. fellow U. Coll. London, 1958-63, tropical rsch. fellow of the Royal Soc., 1963-68; reader in zoology U. Glasgow, Scotland, 1968-74; prof. zoology U. Glasgow, 1974-84, Regius prof. zoology, 1984—; cons. expert WHO Panel on Parasitic Diseases, 1973—. Author: (with F.E.G. Cox) The Protozoa, 1967; contbr. numerous articles to profl. jours. Fellow Royal Soc., Royal Soc. Edinburgh, U. Coll. London. Fellow Royal Soc. Tropical Medicine and Hygiene, Linnean Soc.; mem. Soc. Protozoologists (Brit. sect. pres. 1977-80). Avocations: sketching, gardening. Home: 16 Mirrlees Dr, Glasgow Scotland G12 0SH Office: Univ of Glasgow, Glasgow Scotland G12 8QQ

VICKERS, AMY, engineer. BA in Philosophy, NYU, 1980; MS in Engring. Scis., Dartmouth Coll., 1986. Project coord. N.Y.C. Dept. Environ. Protection, N.Y.C., 1981-83; freelance cons. USCG, N.Y.C., 1983-85; dir. N.Y.C. Coun. Community Environ. Protection, N.Y.C., 1986-87; project mgr. Mass. Water Resources Authority, Boston, 1987-89, Brown and Caldwell Engring., Boston, 1989-91; prin., owner Amy Vickers and Assocs., Boston, 1991—; mem. Am. Water Works Assn., Denver, 1987—, AWWA Water Conservation Com., Denver, New Eng. Bus. for Social Responsibilities, Boston, 1991—. Contbr. articles to profl. jours. Office: Amy Vickers and Assocs 100 Boylston St Ste #1015 Boston MA 02116-4610

VICKERS, JAMES HUDSON, veterinarian, research pathologist; b. Columbus, Ohio, Apr. 21, 1930; s. Carl James and Olga Elizabeth (Schaer) V.; m. Valerie Janet May, Apr. 5, 1964; 1 child, Dana Carlton. BS, Ohio State U., 1952, DVM, 1958; MS, U. Conn., 1966. Diplomate Am. Coll. Vet. Pathologists. Veterinarian Columbus Mcpl. Zoo, 1958-60; dir. pathology dept. Lederle Labs., Pearl River, N.Y., 1960-70; v.p., dir. rsch. Primelabs. Inc., Farmingdale, N.J., 1970-72; dir. spl. studies Johnson & Johnson, Washington Crossing, N.J., 1972-73; dir. divsn. veterinary svcs.Ctr. for Biologics FDA, Bethesda, Md., 1973—; rep. interagy. rsch. animal commn. FDA, Bethesda, 1974-89; cons. Gov. Arab Republic Egypt, Cairo, 1976-84, Paul Erlich Inst., Frankfurt, Fed. Republic Germany, 1977-78; chmn. com. on animalcare Ctr. for Biologics, Bethesda. Contbr. chpts. to books. Spokesman Urbana (Md.) Civic Assn., 1987. Capt. U.S. Army, 1952-54. Recipient Presdl. citation Pres. James Carter, 1980, Alumni Svc. award Ohio St. Coll. Vet. Medicine, 1987, FDA Commr.'s spl. citation, 1988. Mem. Am. Vet. Med. Assn., Soc. Toxicology, Internat. Acad. Pathology, Ohio State Vet. Med. Assn., Zane Grey's West Soc. (pres. 1987—), Westerners Internat. Avocations: fine arts, books. Office: Ctr for Biologics Rsch 8800 Rockville Pike Bethesda MD 20892-0001

VICKERY, EUGENE LIVINGSTONE, retired physician, writer; b. Fairmount, Ind., Nov. 27, 1913; s. Lee Otis and Grace (Hawkins) V.; BS with distinction, Northwestern U., 1935, MB, 1940, MD, 1941; m. Millie Margaret Cox, Dec. 21, 1941; children: Douglas Eugene, Constance Michelle, Anita Sue, Jon Livingstone. Intern Evanston (Ill.) Hosp., 1940-41; pvt. practice medicine, Lena, Ill., 1946-84; chmn. med. records com. Freeport Meml. Hosp., 1954-64, sec. staff, 1964-67, chairman credentials com., 1964-69, v.p. staff, 1967-69, chief staff, 1969-71, chmn. constn. and bylaws com., 1971-80; mem. staff St. Francis Hosp.; local surgeon Ill. Central R.R.; health officer Lena, 1948-84; mem. Stephenson County Bd. Health, 1966-75, v.p., 1969-75; mem. peer rev. policy com. No. Ill. Found. Med. Care. Mem. Lena Sch. Bd., 1951-54; mem. Lena Library Bd., 1958-62; med. dir. Civil Def., rural Stephenson County, Ill., 1961-70; mem. exec. bd. Blackhawk Area council Boy Scouts Am., recipient Silver Beaver award Nat. Coun., 1968, Distinguished Eagle award Nat. Council, 1977, mem. nat. coun., 1971-93; bd. dirs. Stephenson County unit Am. Cancer Soc. Served from 1st lt. to maj. AUS, 1941-46. Decorated Legion of Merit; recipient Lena Community Service award, 1972; named Ill. Family Physician of Yr., 1981, named Illinois Communicator of Year Ill. Woman's Press Assn., 1989; recipient Silver Wreath, Nat. Eagle Scout Assn., 1982. Mem. Stephenson County (pres.), Ill. (chmn. med.-legal council 1976-79) med. socs., AMA, Am. (mental health com. 1981), Ill. (chmn. bd. dirs., pres. 1979) acads. family physicians, Am. Numis. Assn., Nat. Rifle Assn., Arctic Inst. N.Am., Am. Legion, Phi Beta Kappa. Republican. Mem. Evang. Free Ch. Lion. Clubs: Apple Canyon, Masons (32 deg.), Shriners (Legion of Honor). Author: Dad Calls Me Jack; Adventures in Rhyme; Life Goes On, The Ramiluk Stories; author Weekly poetry column Vic's Verse. Arctic expeditions: Yukon River from its source, 1935, St. Lawrence Island, Bering Sea and N.W. Alaska, 1966, (with Richard E. Byrd Polar Ctr.) low altitude flight to North Geog. Pole, Baffin Island and Ungava Bay Eskimo Villages, 1970, N.W. Hudson's Bay Eskimo Villages, 1974, Cen. and Western Can. Arctic and Mackenzie Delta, 1990. Home and Office: 602 Oak St Lena IL 61048-9716

VICTOR, ANDREW CROST, physicist, consultant, small business owner; b. N.Y.C., Nov. 4, 1934; s. Joseph and Stella (Crost) V.; m. Dorothy Tresselt, Dec. 9, 1955; children: Lisa Ann, Jean Sylvia Victor Lindsteadt, Joseph Andrew. BA in Chemistry, Swarthmore (Pa.) Coll., 1956; MS in Physics, U. Md., 1961. Physicist Nat. Bur. Standards, Washington, 1956-62; physicist Naval Weapons Ctr., China Lake, Calif., 1962-67, br. head, 1967-80, program mgr., 1980-89; physicist, owner Victor Tech., Ridgecrest, Calif., 1990—; mem., chmn. exhaust plume subcom. Joint Army, Navy, NASA, Air Force Propulsion Group, Laurel, Md., 1964-89; mem., leader exhaust plume and propulsion hazards com. The Tech. Coop. Program, U.S., UK, Can., Australia, New Zealand, 1967-89; mem. plume working group Adv. Group Aerospace Rsch. and Devel./NATO, Brussels, 1987-90. Contbr. articles to profl. jours. Fellow AIAA (assoc.); mem. Am. Def. Preparedness Assn., System Safety Soc., Internat. Pyrotechnics Soc., Sigma Xi. Achievements include research in exhaust plume technology, explosives and warheads, insensitive munitions and hazards, rocket propulsion, rocket plume signatures and high temperature thermodynamics.

VIDA, STEPHEN ROBERT, environmental engineer; b. N.Y.C., Feb. 13, 1951; s. Stephen Robert and Anna (Leszkovics) V.; m. Louisa Kramer, June 29, 1974; children: Robert Marc, Kristine Michelle. B of Engring., Manhattan Coll., 1973, M of Environ. Engring., 1975; MBA, NYU, 1984. Registered profl. engr. Environ. engr. Havens and Emerson, Saddle Brook, N.J., 1973-75, U.S. EPA, N.Y.C., 1975-81; exec. dir. Two Bridges Sewerage Authority, Lincoln Park, N.J., 1981-82; cons. Coopers and Lybrand, N.Y.C., 1982-87; mgr. Ernst and Young, N.Y.C., 1987-91; environ. engr. U.S. EPA, N.Y.C., 1991—. Commr., coach Syosset (N.Y.) Soccer Club, 1988—. Recipient fellowship U.S. EPA, 1973. Mem. NSPE (advisor scholarship com. Nassau County 1975-82, N.Y. state level 1990-91), Tau Beta Pi, Chi Epsilon. Home: 19 Lilac Dr Syosset NY 11791

VIDIC, BRANISLAV, cell biologist; b. Mitrovica, Serbia, Yugoslavia, May 20, 1934; came to U.S. 1965; s. Jeyrem and Olga (Andric) V.; m. Ljubica Jovanovic, June, 1976 (div.); m. Holda Sanchez, Apr. 19, 1985; 1 child, Alexander. BA, KGH, Yugoslavia, 1952; ScD, U. Belgrade, Yugoslavia, 1959. Asst. prof. U. Novisad, Yugoslavia, 1960-62, U. Lausanne, Switzerland, 1962-65; assoc. prof. cell biology St. Louis U., 1965-71; prof. cell biology Georgetown U., Washington, 1971—; vis. prof. Walter Reed Hosp., Washington, 1973—, U. Nancy, Frances, 1992, U. Belgrade, 1987—, U. Berne, Switzerland, 1978-79. Author: Atlas of the Ear, 1970, Atlas of the Human Body, 1984, Dissection Manual, 1984; contbr. book: Ordan's Oral Histology, 1979; contbr. articles to profl. jours. Pres. We Care, Washington, 1988—. With Yugoslavian Army, 1959-60. Recipient Golden Apple in teaching Georgetown U., 1992; Serbian Acad. Scis. grantee, 1991, NIH grantee, 1974, Am. Heart Assn. grantee, 1981, Tobacco Rsch. grantee, 1975. Mem. AAAS, Am. Soc. Cell Biology, Am. Assn. Anatomists, N.Y. Acad. Scis., Serbian Acad. Scis. Achievements include research in transplacental effect of tobacco smoke on pulmonary elastic tissue; thermal disposition of phospholipid molecules in a bi-layer configuration. Office: Georgetown Univ Dept Anatomy & Cell Biology 3900 Reservoir Rd NW Washington DC 20007

VIDOVICH, DANKO VICTOR, neurosurgeon, researcher; b. Zagreb, Croatia, Dec. 29, 1958; came to U.S., 1991; s. Mladen and Zdenka (Radonichich) V. MD, Zagreb U., Croatia, 1982, MSc in Biology, 1990. Neurosurgeon Clin. Hosp. Sisters of Mercy, Zagreb, Croatia, 1986-91; sr. rschr. Allegheny Singer Rsch. Inst., Pitts., 1991—. Contbr. articles to profl. jours. Achievements include performing first laser assisted nerve anastomosis in humans; laser assisted embryonic tissue transplantation in spinal cord injury. Office: Allegheny Singer Rsch Inst 420 E North Ave Ste 302 Pittsburgh PA 15212

VIEILLARD-BARON, BERTRAND LOUIS, engineer, corporation official; b. Tunis, Tunisia, Aug. 31, 1940; s. Henri Marie and Antoinette (Penet) V-B.; grad. French Ecole Polytechnique, 1962, French Ecole du Génie Maritime, 1965; m. Béatrice Michaud, Aug. 4, 1964; children: Emmanuel, Loic, Anne, Hubert, Mayeule. Metall. engr. for nuclear reactor Indret, France, 1965-70, 70-71; dir. rsch. ctr. Le Creusot, France, 1971-75, corp. dir. rsch. and devel., Creusot-Loire, Paris, 1975-84; mng. dir. French Rsch. Inst. for Shipbldg., 1984-92; v.p. spl. corp. projects Framatome, 1992—; adviser French Adminstrn. Naval Techs. Mem. French Shipbuilders Assn. (bd. dirs. 1987), Institut Francais De Recherchec Pour L'Exploitation De La Mer (chmn. tech. com., com. rsch. and devel. in Eurepean shipbldg.), French Maritime Acad. Home: 33 de Lattre, 78150 Le Chesnay France Office: 47 Rue de Monceau, 75008 Paris France

VIENKEN, JOERG HANS, chemical engineer; b. Wittlich, Fed. Republic Germany, June 1, 1948; s. Walter and Ruth (Würtenberger) V.; m. Karin Bock; children: Hans, Peter, Claudia. Diploma Ing., Tech. U., Darmstadt, Fed. Republic Germany, 1975; D. Ing., Tech. U., Aachen, Fed. Republic Germany, 1980. Scientist Nuclear Rsch. Ctr. Jülich (Fed. Republic Germany), 1976-84, Inst. Biotechnology, Würzburg, Fed. Republic Germany, 1984-85, R&D Bus. Unit Membrana, Wuppertal, Fed. Republic Germany, 1985-88; head dept. Inst. Med. Membrane Application Akzo Wuppertal, 1988—. Contbr. chpts. to books. Fellow Vereinigung Allgem. Angew. Mikrobiologie, European Soc. Artificial Organs, Japanese Soc. Artificial Organs, Japanese Soc. Dialysis Therapy. Home: Hausfeld 79, 42399 Wuppertal Germany Office: Akzo Faser AG, Membrana Oehder Strasse 28, 42289 Wuppertal Germany

VIERRA, FRANK HUEY, environmental engineer, consultant; b. Downey, Calif., Oct. 21, 1946; s. Frank Vierra Jr. and Coralee (Huey) Butts; m. Janet C. Wagely, (div. 1981); m. Wanda F. Robertson, Mar. 6, 1982; children: Jonnie Barbour, Keith, Jesse Summers, Kris, Kory. Grad. high sch., Corona, Calif. Plant mgr. Cargill Poultry Products, California, Mo., 1977-80; maintenance supr., engring. coord., then protein plant supt. Nat. By-Products, Blackwater, Mo., 1980-85; protein plant supt. Pilgrim's Pride

Corp., Mt. Pleasant, Tex., 1985-87, mgr. wastewater plant, 1987-90; mgr. environ. engring., continuous improvement advisor Pilgrim's Pride Corp., Pittsburg, Tex., 1990—; owner, mgr. QC Svcs., Mt. Pleasant, 1987—. With USN, 1964-67, 69-71. Mem. Am. Inst. Plant Engrs., AM. Soc. for Quality Control, Tex. Water Environ. Assn., Water Environ. Fedn. Home: 1110 S Merritt Ave Mount Pleasant TX 75455 Office: Pilgrims Pride Corp 110 S Texas St Pittsburg TX 75646

VIESSMAN, WARREN, JR., academic dean, civil engineering educator, researcher; b. Balt., Nov. 9, 1930; s. Warren and Helen Adair (Berlinckee) V.; m. Gloria Marie Scheiner, May 11, 1953 (div. Apr. 1975); children: Wendy, Stephen, Suzanne, Michael, Thomas, Sandra; m. Elizabeth Gertrude Rothe, Aug. 8, 1980; children: Heather, Joshua. B in Engring., Johns Hopkins U., 1952, MS in Engring., 1958, DEng, 1961. Registered profl. engr., Md. Engr. W. H. Primrose & Assocs., Towson, Md., 1955-57; project engr. Johns Hopkins U., Balt., 1957-61; from asst. to assoc. prof. N.Mex. State U., Las Cruces, 1961-66; prof. U. Maine, Orono, 1966-68, U. Nebr., Lincoln, 1968-75; sr. specialist Libr. Congress, Washington, 1975-83; prof., chmn. U. Fla., Gainesville, 1983-90, assoc. dean for rsch. and grad. study, 1990-91, assoc. dean for acad. programs, 1991—; vis. scientist Am. Geophys. Union, 1970-71; Maurice Kremer lectr. U. Nebr., 1985; lectr. Harvard U. Water Policy Seminar, 1988, Wayne S. Nichols Meml. Fund Ohio State U., 1990; mem. steering com. on groundwater and energy U.S. Dept. Energy, 1979-80; mem. task group on fed. water rsch. U.S. Geol. Survey, 1985-87; mem. com. of the water sci. and tech. bd. NAS, 1986-90; mem. water resources working group Nat. Coun. on Pub. Works Improvement, 1987; chmn., chief of engrs. Environ. Adv. Bd., Washington, 1991—; chmn. solid and hazardous waste mgmt. adv. bd. State U. System Fla. Co-author: Water Supply and Pollution Control, 1993, Water Management: Technology and Institutions, 1984, Introduction to Hydrology, 1988; contbr. over 145 articles to profl. jours. Mem. Water Mgmt. Com., Gainesville, 1983-88, Fla. Environ. Efficiency Study Commn., 1986-88. 1st lt. U.S. Army C.E., 1952-54, Korea. Fellow ASCE (Julian Hinds award 1989), Am. Water Resources Assn. (nat. pres. 1991, Icko Iben award 1983), Univs. Coun. on Water Resources (pres. 1987), Sigma Xi, Tau Beta Pi. Avocations: scuba diving, woodworking. Office: U Fla Coll Engring 312 Weil Hall Gainesville FL 32611-2083

VIEZER, TIMOTHY WAYNE, economist; b. Cleve., Jan. 13, 1959; s. Lawrence Stephen and Elaine Pearl (Thompson) V.; m. Jody Claire Russell, Oct. 14, 1988; 1 child, Jessica Marlene. BBA, BA with honors, Kent State U., 1982, MA in Econs., 1987, MA, 1989. Adminstrv. asst. Arthur Andersen & Co., Cleve., 1985; corp. economist Centerior Energy Corp., Cleve., 1985-90; grad. teaching assoc. Ohio State U., Columbus, 1990-91; grad. rsch. assoc. Nat. Regulatory Rsch. Inst., Columbus, 1991—; asst. economist Huntington Nat. Bank, Columbus, 1991-92; instr. econs. Cuyahoga C.C., Parma, Ohio, 1989-90; lectr.. Baldwin-Wallace Coll., Berea, Ohio, 1989-90, Ohio Dominican Coll., Columbus, 1991, Capital U., 1991—, Columbus State C.C., 1992—, Otterbein Coll., 1993—. Co-author: The Soviet Occupation of Afghanistan, 1986. Named one of Outstanding Young Men of Am. U.S. Jaycees, 1981. Mem. Am. Econ. Assn., Nat. Assn. Bus. Economists, Assn. Iron and Steel Engrs. (assoc.), Internat. Assn. Bus. Forecasting (asst. sec. 1986, bd. dirs. 1987), Columbus Assn. Bus. Economists. Roman Catholic. Home: 6837 Welland St Dublin OH 43017-1482

VIGDOR, MARTIN GEORGE, psychologist; b. Bronx, N.Y., Jan. 14, 1939; s. Leo and Ida (Rosenblatt) V.; m. Lorraine Louise Retta, Mar. 31, 1971; 1 child, Neil Andrew. BA, CCNY, 1959; PhD, NYU, 1971. Lic. psychologist, N.Y., Conn. Psychologist intern VA, Bklyn., 1965-70; clin. psychologist Family Ct., N.Y.C., 1971-72; staff psychologist Cath. Charities Guidance Inst., Bronx, 1973-78, Jewish Child Care Assoc., N.Y., 1973-78, Pleasantville (N.Y.) Diagnostic Ctr., 1978-81; clin. psychologist Putnam Community Svcs., Inc., Carmel, N.Y., 1982-84; clin. dir. Green Chimneys Children's Svcs. Inc., Brewster, N.Y., 1981—; adj. prof. The Union Grad. Sch., Cin., 1987—; cons. Northeast Counseling Ctr., Katonah, N.Y., 1984—. Contbr. articles to profl. jours. Pres. Temple Beth Elohim, Brewster, 1985-87; v.p. Jewish Fedn., Danbury, Conn., 1990-92, pres., 1992—. Mem. APA, N.Y. Psychol. Assn., Conn. Psychol. Assn., Residential Treatment Facilty Coalition N.Y. (bd. dirs. 1988—), Com. Psychologists of Voluntary Child Care Agys., Rotary (pres. Brewster club 1988-89, dist. gov. rep. 1986-87, asst. to dist. gov. 1989-90, Paul Harris fellow 1989, Dean of coll. of pres. 1990-92). Jewish. Avocations: traveling, gardening, reading, walking, computer programming. Office: Green Chimneys Childrens Svcs Inc Putnam Lake Rd Brewster NY 10509-1113

VIGFUSSON, JOHANNES ORN, scientific officer; b. Akureyri, Iceland, Mar. 15, 1945; s. Vigfus Thorarinn Jonsson and Sigridur Huld Johannesdottir; m. Barbara Ruth Keller, Oct. 18, 1980; 1 child, Vanessa Stefania. Diploma in theoretical physics, U. Zurich, Switzerland, 1974, PhD, 1975; M in Piano, Conservatory Zurich, 1977. Asst. U. Zurich, 1974-75, researcher, 1975-83, 85-87; sci. officer Swiss Fed. Nuclear Safety Inspectorate, Würenlingen, 1987—; vis. asst. prof. CUNY, 1983-84; vis. fellow Princeton (N.J.) U., 1984-85; tchr. Reisehochschule Zurich, 1968-87; concert pianist, 1977—. Contbr. articles to profl. jours. Pres. Friends of Iceland Soc., 1970-72, 73-78. Recipient Landolt prize Conservatory Zurich, 1975. Mem. Am. Phys. Soc., Swiss Phys. Soc., Icelandic Phys. Soc., N.Y. Acad. Scis., Planetary Soc., Icelandic Club Switzerland (counsellor 1989-90). Avocations: skiing, sailing, history of culture and civilization.

VIGGERS, ROBERT FREDERICK, mechanical engineering educator; b. Tacoma, Wash., Jan. 18, 1923; s. Stuart Thomas and Lucile Dorthy (Cooley) V.; m. Edna Sivia Bonn, 1945 (div. 1971); children: Robert Stuart, Christine Ann, Rebbeca Lee. BSME, U. Wash., 1944; MSME, Oreg. State U., 1949. Instr. gen. engring. U. Wash., Seattle, 1946-48; instr. mech. engring. Oreg. State U., Corvallis, 1948-49; instr. mech. engring. Seattle U., 1949-51, asst. prof. mech. engring., 1951-56, assoc. prof. mech. engring., 1956-64, prof. mech. engring., 1964-90, prof. emeritus, 1990—; pres. R.F. Viggers, Inc., Profl. Engr., Seattle, 1949—; biomed. engring. cons. Hope Heart Inst., Seattle, 1959—. Author: Prosthetic Replacement of the Aoritic Valve, 1972, Coagulation, 1972. Leader Explorer Scouts, Seattle, 1958-72; chmn. hiking Camp Fire Girls, Seattle, 1965-75. With USN, 1944-46, PTO. Named Man of Yr., Interdisciplinary Biomed. Soc., 1972. Mem. ASME, Am. Soc. Engring. Edn., Sigma Xi. Democrat. Achievements include patents in a device for conditioning bunker "C" or residual fuel oil for use in railroad diesel engines; in an aortic heart valve prosthesis. Home: 1358 Windcrest Ln Anacortes WA 98221 Office: Seattle Univ 900 Broadway Seattle WA 98122

VIGLER, MILDRED SCEIFORD, retired chemist; b. North East, Pa., Sept. 6, 1914; d. William and May Elizabeth (Currie) Sceiford; m. Roland Elmer Vigler, Mar. 19, 1934 (div. May 1952). BA, Lake Erie Coll., Painesville, Ohio, 1935. Tchr. Ashtabula County Bd. Edn., Cork, Ohio, 1935-36, Geneva Twp., Ohio, 1936-38; chemist Interlake Iron Corp., Erie, Pa., 1942-45; analytical rsch. chemist Standard Oil Co. Ohio, Cleve., 1945-79, part-time researcher, 1981-84, ret., 1979. Contbr. articles to sci. jours. including SAS Jour., Analytical Chemistry, Applied Spectroscopy. Named Disting. Alumna, Lake Erie Coll., 1983. Mem. Am. Chem. Soc. (Ameritus award 1986), AAUW, Lake View Country Club. Episcopalian.

VIGMO, JOSEF, retired geriatrician; b. Reykjavík, Iceland, Nov. 12, 1922; arrived in Sweden, 1956; s. Olaf Johan Olsen-Vigmostad and Aline Josefine (Zachariassen) Hervik; m. Soffía Axelsdóttir, Jan. 24, 1953; children: Terje, Sylvi Aline. MD, U. Iceland, 1953; postgrad., U. Gothenburg, Sweden, 1960. Lic. in internal medicine, cardiology, geriatrics, Sweden. Asst. med. officer Sandträsks Tuberculosis Sanatorium, Sweden, 1953-54; resident in pulmonary diseases Sandträsks Tuberculosis Sanatorium, 1956-57; rotating intern White Meml. Hosp. and Clinic, Loma Linda U., L.A., 1954-55; resident in internal medicine Piteå County Hosp., Sweden, 1958-59, Kalix and Skene County Hosps., Norrköping Gen. Hosp., Sweden, 1961-65; sub-chief med. officer Hultafors Health Ctr., Sweden, 1960; sub-chief med. officer geriatric dept. Borås Gen. County Hosp., Sweden, 1966-77; chief med. officer geriatric dept. Borås Gen. County Hosp., 1977-87; ret., 1987; consulting cardiologist, Borås Gen. County Hosp., 1978—, lectr. Sch. Nursing, 1967—. Recipient Gold medal Älvsborg County Council, 1987. Mem. Swedish Med. Assn., Swedish Geriatrics Assn., Swedish Assn. Chief Med. Officers, South-Älvsborg County Assn. Chief Med. Officers. Lutheran. Avocations: linguistics, genealogy. Home: Båleröd, Sjövägen 1, S-452 97 Strömstad Sweden

VIJAYABHASKAR, RAJAGOPAL COIMBATORE, chemist, educator; b. Tadepaligudem, India, Nov. 2, 1966; came to the U.S., 1990; s. C.B. Rajagopal and C.S. Kusuma Bai. BS, MS, Madras U., 1989; postgrad., Fla. Atlantic U., 1990—. Teaching asst. dept. chemistry Fla. Atlantic U., Boca Raton, 1990—. Achievements include research of analytical chemistry and organic chemistry. Office: Fla Atlantic U Dept Chemistry 500 NW 20th St Boca Raton FL 33431

VIJH, ASHOK KUMAR, chemistry educator, researcher; b. Multan, India (now Pakistan), Mar. 15, 1938; s. Bishamber Nath and Prem Lata (Bahl) V.; m. Danielle Blais (div.); 1 child, Aldous Ian. BSc with honors, Panjab U., India, 1960, MSc with honors, 1961; PhD, Ottawa U., 1966; LLD (hon.), Concordia U., 1988; DSc (hon.), Waterloo U., 1993. Group leader Inst. Research Hydro-Quebec, 1969-74, program leader, 1975-81, maitre-de-research, 1973—; vis. prof., thesis dir. INRS-Energie, U. Que., 1970—; Xerox lectr. U. Montreal, 1991. Author: Electrochemistry of Metals and Semiconductors, 1973; editor: Oxides and Oxide Films; mem. editorial bd. Applied Physics Communications, Internat. Jour. Hydrogen Energy, Materials Chemistry and Physics, also several other jours.; contbr. over 250 articles to profl. jours. Decorated officer Order of Can.; chevalier de Ordre Nationale de Quebec; knight Order of Malta; recipient Urgel Archambault medal Assn. Canadienne pour Advancement Scis., 1984, Commemorative medal 125th Anniversary Confederation Can., 1992. Fellow IEEE, Royal Soc. Chemistry, Chem. Inst. Can. (Noranda lectr. 1979, Palladium medal 1990), Inst. Physics (U.K.), Royal Soc. Can. (dir. applied scis. and engring. div. 1990, Acad. Sci., Thomas W. Eadie medal 1989), Am. Phys. Soc., Third World Acad. Scis., Nat. Acad. Scis. India, Acad. Francophone d'Ingenieus (France; founding fellow 1993); mem. European Acad. Arts, Scis. and Humanities (academician), Electrochem. Soc. (Lash Miller award 1973), Can. Council (Izaak Walton Killam Meml. prize 1987), Materials Rsch. Soc. India (hon.). Office: Hydro-Quebec Research Inst, 1800 Montee Ste Julie, Varennes, PQ Canada J0L 2P0

VIKTIL, MARTIN, electrical engineer; b. Trondheim, Norway, Nov. 19, 1944; m. Micheline Marie Evrard, July 17, 1976; children: Thomas, Astrid. Grad., Naval Radio Sch., Bergen, Norway, 1963, Tech. Coll., Trondheim, 1972; MSc in Elec. Engring., Tech. U., Trondheim, 1977; bus. economist, U. Trade and Acctg., Norway, 1992. Radio officer Shipowner P. Smedvig, Venator, Norway, 1966-68; maintenance engr. RNCSIR/ESRO, Ny Ålesund,, Norway, 1972-74; rsch. engr. NDRE, Kjeller, Norway, 1977-85; sr. scientist ABB Corp., Nesbru, Norway, 1986—; vis. scientist Synertek Inc., Santa Clara, Calif., 1981-83; lectr. U. Oslo, Norway, 1986—, external examiner, 1988—; external examiner U. Trondheim, 1988—; project leader Nordic Boundary Scan Project, 1989-92. Contbr. articles to profl. jours. Leader Parents Orgn. for Primary Sch., Slemmestad, Norway, 1989-91. Recipient EKF award for test concepts towards the yr. 2000, 1990. Mem. Assn. Norwegian Chartered Engrs. Office: ABB Corp Rsch, Bergervegen 12, Nesbru N-1360, Norway

VILCEK, JAN TOMAS, medical educator; b. Bratislava, Czechoslovakia, June 17, 1933; came to U.S., 1965, naturalized, 1970.; s. Julius and Friderika (Fischer) V.; m. Marica F. Gerhath, July 28, 1962. M.D., Comenius U., Bratislava, 1957; C.Sc. (Ph.D.), Czechoslovakia Acad. Sci., Bratislava, 1962. Fellow Inst. Virology, Bratislava, 1957-62, head of lab., 1962-64; asst. prof. microbiology NYU Med. Ctr., N.Y.C., 1965-68, assoc. prof., 1968-73, prof., 1973—, head biol. response modifiers, 1983—; lectr. Chinese Acad. Med. Sci., Beijing, 1981, 83, Osaka U., 1987-88; chmn. nomenclature com. WHO, 1981-86, cons. biol. standardization com., 1982-88; mem. adv. com. Cancer Soc., 1981-87, chmn., 1983; expert French Ministry Health, 1983-88; mem. sci. adv. bd. Max Planck Inst., Munich, 1987—. Author: Interferon, 1969; editor in chief Jour. Archives of Virology, 1975-86; editor: Interferons and the Immune Systems, 1984, Tumor Necrosis Factor: Structure, Function and Mechanism of Action, 1991; mem. editorial bd. Virology, 1979-81, Archives of Virology, 1986-92, Infection and Immunity, 1983-85, Antiviral Research, 1984-88, Jour. Inteferon Rsch., 1980—, Jour. Immunological Methods, 1986—, Natural Immunity and Cell Growth Regulation, 1986-92, Jour. Immunology, 1987-89, Lymphokine Rsch., 1987—, Jour. Biol. Chemistry, 1988-90, ISI Atlas Sci., Immunology, 1988-89, Jour. Cellular Physiology, 1988—, Cytokine, 1989—, Biologicals, 1989—, Acta Virologica, 1991—, Cellular Immunology, 1993—; contbr. articles to profl. jours. Mem. rev. panel Israel Cancer Rsch. Fund, 1993—; mem. fellowship rev. com. Am. Heart Assn., 1992—. Recipient Rsch. Career Devel. award USPHS, 1968-73, Recognition award Japanese Inflammation Soc., 1989, Outstanding Investigator award Nat. Cancer Inst., NIH, 1991—; grantee USPHS, numerous other orgns. Mem. AAAS, Soc. Gen. Microbiology, Am. Soc. Microbiology, Am. Assn. Immunologists, Internat. Soc. Inteferon Rsch. Office: NYU Med Ctr 550 1st Ave New York NY 10016-6402

VILCHE, JORGE ROBERTO, physical chemistry educator; b. La Plata, Argentina, June 8, 1946; s. Roberto Marcos and Delia Olga (Falcon Bravo) V.; m. Maria Estela Testoni, Aug. 23, 1947; 1 child, Maria Viktoria. Chem. engr. degree, U. La Plata, 1968. Fellow Inst. of Electrochemistry U. Karlsruhe (Fed. Republic Germany), 1970-71; fellow INIFTA Inst. Phys. Chemistry U. La Plata, 1972-73; sr. materials engr. FATE Electronic R&D Div., Buenos Aires, 1974-76; rsch. scientist CONICET Nat. Rsch. Coun., Buenos Aires, 1976—; head corrosion sect. INIFTA Inst. Phys. Chemistry U. La Plata, 1976—, prof. phys. chemistry, 1976—; fellow von Humboldt Found., Karlsruhe, 1988-89; mem. adv. bd. Provincia Buenos Aires Rsch. Coun., La Plata, 1976-83, Nat. Inst. Indsl. Tech., Buenos Aires, 1990—. Contbr. articles to profl. jours., author tech. reports; patentee in field. Recipient Young Author Prize in Chem. Sci. and Tech., Argentina, 1984, Young Author Prize in Corrosion Sci., OAS, 1986. Mem. Electrochem. Soc., Internat. Soc. Electrochemistry, Nat. Assn. Corrosion Engrs. Roman Catholic. Home: Calle 57 No 898, 1900 La Plata Argentina Office: INIFTA, Sucursal 4 CC 16, 1900 La Plata Argentina

VILLA, ROBERTO RICCARDO, chemist; b. Milano, Italy, July 30, 1961; s. Aldo and Rosetta (Puricelli) V.; m. Giuliana Salani, Apr. 24, 1993. Degree in chemistry, U. Milano, 1986, PhD, 1989. Lab. mgr. Prassis-Sigma Tau, Milano, 1989-91; lab and nuclear magnetic resonance mgr. Magis Farmaceutici, Brescia, Italy, 1991—. Contbr. articles to Jour. Organic Chemistry, 1988, Tetrahedron, 1988, Tetrahedron Letters, Bull. Soc. Chem. Franc. Mem. Regional and Provincial Youthful Movement of Dem. Cristiana, Milano, 1980-83. With Quartermaster Corps., 1986-87, Bologna. Mem. Soc. Chimica Italiana, Am. Chem. Soc., World Wildlife Fund, Gruppo Sportivo San Michele. Roman Catholic. Achievements include patent on concerning synthesis of cardiovascular compounds (Italy), patents pending on cardiovascular compounds (Europe). Home: Via Bronzino 6, 20133 Milano Italy Office: Magis Farmaceutici, Via Cacciamali 36, 25125 Brescia Italy

VILLAFRANCA, JOSEPH J., biochemistry educator; b. Silver Creek, N.Y., Mar. 23, 1944; s. Joseph Nicholas and Mildred (Dolce) C.; children: Jennifer, June. BS, SUNY, Fredonia, 1965; PhD, Purdue U., 1969. Evan Pugh prof. chemistry Pa. State U., University Park, 1971—; cons. Monsanto Corp., St. Louis, 1985-89, Eastman Kodak Co., Rochester, N.Y., 1985-87. Author 132 sci. publs. Mem. Am. Chem. Soc. (councilor 1986-89), Am. Soc. for Biochemistry and Molecular Biology. Avocation: skiing. Office: Pa State U Dept Chemistry 152 Davey Lab University Park PA 16802-1009

VILLAIRE, WILLIAM LOUIS, chemical engineer; b. Bay City, Mich., May 28, 1967; s. Edward Carl and Susann Helen (Freiders) V. Student, Delta Coll., University Center, Mich., 1985-87; BS, Mich. State U., 1990. Funeral home employee Ambrose/Squires Funeral Chapels, Bay City, Mich., 1984-87; pro shop employee Bay City Country Club, 1983-86; lab technician, student co-op Dow Chemical Mich. Divsn., Midland, 1986-87; lab. worker Carcinogenisis Lab. Mich. State U., East Lansing, 1987-88, maintenance supr. McDonel Hall, 1988-90; chem. process engr. AC Rochester div. GM, Flint, Mich., 1990-93, shift engr. waste water treatment plant. Referee Greater Flint Hockey Referee Assn., 1990—. Roman Catholic. Office: GM Corp AC Rochester Divsn 1300 N Dort Hwy Flint MI 48556

VILLANI, DANIEL DEXTER, aerospace engineer; b. Lewiston, Maine, Aug. 31, 1947; s. Leonard and Helen Marie (McMorran) V.; m. Angela Mary Miller, Sept. 16, 1967; children: Constance Lynne, Adam Neil, Dorothy Ellen. BS in Physics with honors, Calif. Inst. Tech., 1969; PhD in Aerospace Engring., Princeton (N.J.) U., 1983. Reliability engr. Jet Propulsion Lab., Pasadena, 1969; engring. programmer Tex. Instruments, Princeton, 1973-74; with Hughes Aircraft Co., L.A., 1975—, mem. tech. staff, sr. scientist/engr., 1985—; dep. team chief Magellan Radar System Engring. Team, Jet Propulsion Lab., 1991-92; payload systems engr. GMS and GOES Meteorol. Satellites, Tokyo, Washington, L.A., 1982-84. Song-writer/guitarist Cephas Gospel/Rock Band, 1983-84. Choir dir., guitarist Holy Innocents' Cath. Ch., Long Beach, Calif., 1985—; vol. Youth Motivation Task Force, L.A., 1990-92. Recipient Pub. Svc. medal NSAS, 1992, Group Achievement award NASA, 1992. Mem. AIAA, Calif. Inst. Tech. Alumni Assn., Sigma Xi. Democrat. Roman Catholic. Achievements include patent for alignment system for encoders. Office: Hughes Aircraft S50/X331 PO Box 92919 Los Angeles CA 90009

VILLARREAL, CARLOS CASTANEDA, engineering executive; b. Brownsville, Tex., Nov. 9, 1924; s. Jesus Jose and Elisa L. (Castaneda) V.; m. Doris Ann Akers, Sept. 10, 1948; children: Timothy Hill, David Akers. B.S., U.S. Naval Acad., 1948; M.S., U.S. Navy Postgrad. Sch., 1950; LL.D. (hon.), St. Mary's U., 1972. Registered profl. engr. Commd. ensign U.S. Navy, 1948, advanced through grades to lt., 1956; comdg. officer U.S.S. Rhea, 1951, U.S.S. Osprey, 1952; comdr. Mine Div. 31, 1953; resigned, 1956; mgr. marine and indsl. operation Gen. Electric Co., 1956-66; v.p. mktg. and adminstrn. Marquardt Corp., 1966-69; head Urban Mass Transit Adminstrn., Dept. Transp., Washington, 1969-73; commr. Postal Rate Commn., 1973-79, vice chmn., 1975-79; v.p. Washington ops. Wilbur Smith and Assocs., 1979-84, sr. v.p., 1984-86, exec. v.p., 1987—, also bd. dirs.; lectr. in field; mem. industry sector adv. com. Dept. Commerce; mem. sect. 13 adv. com. Dept. Transp., 1983-86. Contbr. to profl. jours. Mem. devel. com. Wolftrap Farm Park for the Performing Arts, 1973-78; mem. council St. Elizabeth Ch., 1982-86, chmn. fin. com.; mem. bd. edn. St. Elizabeth Sch.; bd. dirs. Assoc. Catholic Charities, 1983-86; mem. fin. com. Cath. Charities, U.S.A. Decorated knight Sovereign Mil. Hospitaller Order St. John of Jerusalem of Rhodes and Malta, 1981; recipient award outstanding achievement Dept. Transp. Fellow ASCE, Am. Cons. Engrs. Coun. (vice chmn. internat. com.); mem. IEEE, NSPE (pres. D.C. soc. 1986-87, bd. dirs. 1988-91), Am. Pub. Transit Assn., Soc. Naval Architects and Marine Engrs., Soc. Am. Mil. Engrs., Am. Rds. and Transp. Builders Assn. (chmn. pub. transp. adv. coun.), Transp. Rsch. Bd., Washington Soc. Engrs., Internat. Bridge, Tunnel and Turnpike Assn., Intelligent Vehicle Hwy. System Am. (chmn. fin. com., bd. dirs.), Univ. Club, Army-Navy Club. Republican. Roman Catholic. Office: 2921 Telestar Ct Falls Church VA 22042

VILLARRUBIA, JOHN STEVEN, physicist; b. New Orleans, Jan. 3, 1957; s. Roger Eugene Jr. and Marie L. (Richard) V.; m. Maria Rachael Benischek, July 13, 1991; 1 child, Mark. BS, La. State U., 1979; MS, Cornell U., 1987, PhD, 1987. Vis. scientist IBM Watson Rsch., Yorktown Heights, N.Y., 1987-89; physicist Nat. Inst. Standards and Tech., Gaithersburg, Md., 1989—. NSF fellow, 1979-82. Mem. Am. Phys. Soc., Am. Vacuum Soc. (scholar 1981-82, Russell & Sigurd fellow 1983). Republican. Roman Catholic. Achievements include first atomic resolution images of Si(111)7x7 rest atom layer.

VILLAVERDE, ROBERTO, civil engineer; b. Chihuahua, Mex., Aug. 14, 1945; came to U.S., 1975; s. Jesus and Carolina (Lazo) V.; m. Phyllis Arlene Potocky, June 21, 1980; children: Derrick Anton, Carlina Jeanette. BS, U. Chihuahua, 1968; MS, Nat. U. Mex., Mexico City, 1971; PhD, U. Ill., 1980. Reg. civil engr., Calif. Asst. prof. San Diego State U., 1980-81; rsch. prof. nat. U. Mex., Mexico City, 1981-82; asst. prof. U. Calif., Irvine, 1983-88, assoc. prof., 1988--. Co-author: International Handbook of Seismic Resistance Design, 1993; contbr. articles to profl. jours. Recipient Faculty Devel. award San Diego State U., 1981, Career Devel. award U. Calif., 1986; OTCA scholar, 1971; Faculty Rsch. fellow U. Calif., 1985. Mem. Am. Soc. Civil Engrs., Am. Soc. for Engring. Edn., Earthquake Engring. Rsch. Inst., Seismological Soc. Am. Roman Catholic. Office: U Calif Civil Engring Dept Irvine CA 92717

VILLAX, IVAN EMERIC, chemical engineer, researcher; b. Magyaróvár, Hungary, Apr. 16, 1925; arrived in Portugal, 1952; s. Edmond Joseph and Marianne (Manninger) V.; m. Diane Houssemayne Du Boulay, Feb. 8, 1958; children: Peter, Guy, Sofia, Miguel. Grad. chem. engr., Tech. U., Budapest, Hungary, 1948, D Universitas (hon), 1992. Rsch. asst. Agrl. Rsch. Ctr., Clermont-Ferrand, France, 1949-52; rsch. dir. Inst. Pasteur, Lisbon, Portugal, 1952-61; pres. Hovione-Sociedade Quimica, SA, Lisbon, 1959—. Contbr. articles to profl. jours. Mem. Am. Chem. Soc., Am. Soc. for Microbiology, Grémio Literário, N.Y. Acad. Scis. Roman Catholic. Achievements include patents in preparation of new topical corticusterold esters, new homogeneous rhodium catalysts and their use in stereospecific hydrogenation. Office: Hovione-Sociedade Quimica SA, Sete Casas Loures, 2670 Lisbon Portugal

VILLCHUR, EDGAR, audiology research scientist; b. N.Y.C., May 28, 1917; s. Mark and Mariam (Vinograd) V.; m. Rosemary Mackay Shafer, Nov. 28, 1945; children: Miriam Villchur Berg, Mark. BS, CCNY, 1938, MSEd, 1940. Contbg. editor Audio jour., Mineola, N.Y., 1952-56; instr. electronics NYU, N.Y.C., 1952-57; pres., dir. rsch. Acoustic Rsch., Inc., Cambridge, Mass., 1954-67, Found. for Hearing Aid Rsch., Woodstock, N.Y., 1967—; vis. scientist Albert Einstein Coll. Medicine, N.Y.C., 1982-84, MIT, Cambridge, 1978-79; referee Jour. Acoustical Soc. Am., 1970—. Author: Handbook of Sound Reproduction, 1957, Reproduction of Sound, 1962; contbr. Besancon Ency. of Physics, 1985; contbr. articles to profl. jours. Capt. U.S. Army Air Corps, 1941-46, PTO. Decorated Bronze Star; named to Audio Hall of Fame, 1980; recipient Emile Berliner award Berliner Hist. Soc., 1974, Silver medal Audio Engring. Soc., 1972, Hi-Fi News award Hi-Fi News and Record Rev., London, 1984. Mem. Acoustical Soc. Am. (referee assn. jour. 1970—). Jewish. Achievements include development of acoustic suspension loudspeaker, dome high-frequency loudspeaker (on exhibit at Smithsonian Instn.), multi-channel compression/equalization for hearing aids. Office: Found Hearing Aid Rsch PO Box 306 Woodstock NY 12498

VINCENT, FREDERICK MICHAEL, neurologist, educational administrator; b. Detroit, Nov. 19, 1948; s. George S. and Alyce M. (Borkowski) V.; m. Patricia Lucille Cordes, Oct. 7, 1972; children: Frederick Michael, Joshua Peter, Melissa Anne. BS in Biology, Aquinas Coll., 1970; MD, Mich. State U., 1973. Diplomate Am. Bd. Psychiatry and Neurology, Am. Bd. Electrodiagnostic Medicine, Nat. Bd. Med. Examiners. Intern St. Luke's Hosp., Duluth, Minn., 1974-75; resident in neurology Dartmouth Med. Sch., Hanover, N.H., 1975-77, instr. dept. medicine, chief resident neurology, 1977-78; chief, neurology sect. Munson Med. Ctr., Traverse City, Mich., 1978-84; asst. clin. prof. medicine and pathology Mich. State U., East Lansing, 1978-84, chief sect. neurology Coll. Human Medicine, 1984-87; clin. prof. psychiatry and internal medicine Mich. State U., 1989—; clin. prof. medicine, 1990—; pvt. practice Neurology, Neuro-oncology and Electrodiagnostic Medicine, Lansing, Mich., 1987—; clin. and research fellow neuro-oncology Mass. Gen. Hosp., Boston, 1985; clin. Fellow in neurology Harvard Med. Sch., Boston, 1985; cons. med. asst. program Northwestern Mich. Coll., Traverse City, 1983-84; neurology cons. radio call-in show Sta. WKAR, East Lansing, 1984—, WCMU TV, 1987. Author: Neurology: Problems in Primary Care, 1987, 2d edit., 1993; contbr. articles to profl. jours. Fellow NSF, 1969, Nat. Multiple Sclerosis Soc., 1971. Fellow ACP, Am. Acad. Neurology, Am. Assn. Electrodiagnostic Medicine; mem. Am. Coll. Legal Medicine, Am. Acad. Clin. Neurophysiology, Am. Heart Assn., Am. Soc. Clin. Oncology, Am. EEG Soc., Am. Fedn. Clin. Rsch., Am. Soc. Neurol. Investigation, Am. Sleep Disorders Assn., Am. Epilepsy Soc., Soc. for NeuroSci., N.Y. Acad. Scis., Am. Soc. Law and Medicine, Am. Soc. for Neuro-Rehab., Movement Disorders Soc., Univ. Club, Alpha Omega Alpha. Roman Catholic. Office: 405 W Greenlawn Ave Ste 430 Lansing MI 48910-2889

VINCENT, JAMES LOUIS, biotechnology company executive; b. Johnstown, Pa., Dec. 15, 1939; s. Robert Clyde and Marietta Lucille (Kennedy) V.; m. Elizabeth M. Matthews, Aug. 19, 1961; children: Aimee Archelle, Christopher James. BSME, Duke U., 1961; MBA in Indsl. Mgmt., U. Pa., 1963. Mgr. Far East div. Tex. Instruments, Inc., Tokyo, 1970-72; pres. Tex.

Instrument Asia, Ltd., Tokyo, 1970-72; v.p. diagnostic ops., pres. diagnostics div. Abbott Labs., North Chgo., Ill., 1972-74, group v.p., bd. dirs., 1974-81, exec. v.p., COO, bd. dirs., 1979-81; corp. group v.p., pres. Allied Health and Sci. Products Co. Allied Corp., Morristown, N.J., 1982-85; chmn., chief exec. officer Biogen, Inc., 1985—; bd. dirs. Continental Bank Corp., Millipore Corp. Bd. dirs. dean's coun. Duke U.; chmn. exec. bd. Wharton Grad. Sch., U. Pa.; bd. dirs. Found. for the Nat. Tech. Recipien Young Exec. Achievement Young Execs. Club, Chgo., 1976, Disting. Alumni award Duke U., 1988, The Found. for Nat. Tech. medal. Mem. Biotech. Industry Orgn. (bd. dirs.), Econ. Club Chgo., Shoreacres Country Club, Algonquin Club, Boston Club, Chgo. Club, The Links (N.Y.C.). Republican. Presbyterian. Office: 14 Cambridge Ctr Cambridge MA 02142-1481

VINER, MARK WILLIAM, psychiatrist; b. Burbank, Calif., May 26, 1961; s. Mervyn J. and Marilyn (Grossman) V.; m. Kellie Ann Hachee, Jan. 20, 1990; children: Vincent Marc, Arielle Jaye. BA, U. So. Calif., 1985; MD, Spartan Health Sci. U., 1989. Intern in internal medicine Griffin Hosp., Yale U., Derby, Conn., 1990; PGY II, resident psychiatrist Norwich (Conn.) Hosp., 1991, chief resident, 1992—; PGY III, resident psychiatrist U. Conn. Health Ctr., Farmington, 1992—. Author: Der Tatsachenbericht, 1981. Edmondson fellow U. So. Calif., 1981. Mem. Am. Coll. Physicians (assoc.), Am. Psychiat. Assn., Conn. Psychiat. Soc. (co-chair residents com. 1992—), N.Y. Acad. Scis., Hartford County Med. Soc. Home: 269 New Britain Ave Unionville CT 06085 Office: U Conn Health Ctr 263 Farmington Ave Farmington CT 06032

VINGOE, FRANCIS JAMES, clinical psychologist; b. San Diego, Oct. 20, 1931; arrived in Wales, 1975; s. Alfred and Mary Ellen (James) V.; m. Dolores Marquerite Chevillard, Apr. 1957 (div. 1965); 1 child, Sylvie Lamorna; m. Grace Roberta Cameron, Apr. 15, 1966; children: Lisa Michelle, Wendy Sue, Michael Jan. BA with honors, Calif. State U., San Diego, 1956; MA, Calif. State U., San Francisco, 1960; PhD, U. Oreg., 1965. Diplomate in hypnosis Am. Bd. Psychol. Examiners; chartered clin. psychologist, criminological and legal psychologist. Teaching asst. U. Nebr., Lincoln, 1956-57; instr. math. Cogswell Poly. Coll., San Francisco, 1959-61; ednl. psychologist Bremerton (Wash.) Pub. Schs., 1961-63; instr. psychology Olympic Coll., Bremerton, 1961-63; lectr. Napa Coll., 1963-64; counselor, psychometrist, instr. U. Oreg., Eugene, 1964-65; rsch. assoc. State of Oreg., 1965; asst. prof. psychology Colo. State U., Ft. Collins, 1965-68; assoc. prof. psychology SUNY, Cortland, 1968-71; sr. lectr. clin. psychology U. Groningen, The Netherlands, 1971-75; prin. clin. psychologist Univ. Hosp. Wales, Cardiff, 1975-85; lectr. U. Wales Coll. Medicine, Cardiff, 1975—; cons. clin. psychologist Univ. Hosp. Wales, Cardiff, 1985—; cons. Poudre R-1 Schs., Ft. Collins, 1966-68, Tri County Head Start Programme, Torrington, Wyo., 1967-68; vis. lectr. Wells Coll., Aurora, N.Y., 1968-70. Author: Clinical Psychology and Medicine, 1981; cons. editor Internat. Jour. Clin. and Exptl. Hypnosis; assoc. editor Brit. Jour. Exptl. and Clin. Hypnosis, 1982, Jour. Contemporary Hypnosis, 1991; clin. editor Brit. Jour. Exptl. and Clin. Hypnosis, 1982; contbr. articles to profl. jours. With USN, 1949-53. Faculty Rsch. fellow SUNY, Cortland, 1970; Faculty Rsch. grantee Colo. State U., 1968. Fellow Brit. Psychol. Soc.; mem. Brit. Soc. Exptl. and Clin. Hypnosis (chmn. Wales and West of Eng. br. 1982-85), Brit. Soc. Clin. and Exptl. Hypnosis (chmn. 1985). Avocations: swimming, sailing, wind surfing, travelling, reading. Home: 87 Blackoak Rd Cyncoed, Cardiff CF2 6QW, Wales Office: Univ Hosp Wales Dept Clin, Psychology Heath Park, Cardiff CF4 4XW, Wales

VINICK, FREDRIC JAMES, chemist; b. Amsterdam, N.Y., June 18, 1947; s. Louis Bernard and Margarete B. (Schiller) V.; m. Carol Lynn Berns, June 28, 1970; children: Julie, Peter, Andrew. BA, Williams Coll., 1969; PhD, Yale U., 1973. Sr. scientist CIBA-Geigy, Summit, N.J., 1975-78; sr. rsch. scientist Pfizer, Groton, Conn., 1978-83, sr. rsch. investigator, 1983-84, project leader, 1984-87, asst. dir., 1987-91, dir. chemistry, 1991—. Sr. editor Ann. Reports in Medicinal Chemistry, 1988-90; editorial bd. Drug Design and Discovery jour., 1991—. NSF fellow, 1970, NIH fellow, 1973. Mem. Am. Chem. Soc., N.Y. Acad. Sci. Achievements include patents on synthesis of aspartame, navel phosphodiesterase inhibitors, discovery of first non-peptide substance P antagonists. Home: 55 Trumbull Rd Waterford CT 06385 Office: Pfizer Eastern Point Rd Groton CT 06340

VINING, CRONIN BEALS, physicist; b. Balt., Aug. 22, 1957; s. Theron Marcus and Rita Mary (Kirby) V.; m. Elizabeth Platt, Feb. 23, 1979. BS in Physics, Va. Polytech. Inst., 1978; PhD, Iowa State U., 1983. Sr. physicist GE, Phila., 1984-87; physicist Jet Propulsion Lab., Pasadena, Calif., 1987—; cons. Marlow Industries, Dallas, 1990-91. Contbr. articles to profl. jours. Mem. Internat. Thermoelectric Soc. (pres. 1994, conf. editor, 1990, co-editor 1991, adv. com. 1990—), Am. Phys. Soc., Materials Rsch. Soc., Mensa. Achievements include developing first self-consistent theoretical models for silicon-germanium thermoelectric alloys. Specializing in growth, characterization, and theory of thermoelectric materials for power generation and refrigeration. Home: 11738 Moorpark St # J Studio City CA 91604 Office: Jet Propulsion Lab 4800 Oak Grove Dr Pasadena CA 91109

VINOKUR, ROMAN YUDKOVICH, physicist, engineer; b. Bezshad, Ukraine, USSR, June 9, 1948; came to U.S., 1991; s. Yuda M. Vinokur and Polina M. Rosenblit; div.; 1 child, Ilya R. MS in Physics, Moscow Inst. Physics/Tech., 1972; PhD in Applied Acoustics, All Union Inst. Bldg. Physics, Moscow, 1986. Engr. Exptl. Inst. of Automatics, Redkino, USSR, 1972-75; physics rschr. Lab. Bldg. Acoustics, Moscow, 1975-91; physics rschr. engring. lab. Lasko Metal Products, Inc., Chester, Pa., 1991—; mem. sect. scientific coun. Moscow Inst. Bldg. Physics, 1988-91. Contbr. articles in physics and math (for students) to profl. jours. and popular mags. Lector, coun. mem. Moscow Ho. of Scientific and Engring. Propaganda, 1979-91. Recipient 2d prize for one of best works in field of sound insulation of walls and windows, Moscow Coun. of Soc. of Scientists and Engrs., 1979. Mem. Acoustical Soc. Am., ASME. Jewish. Achievements include development of a theory of sound insulation of windows, walls, floors, and partitions. Office: Lasko Metal Products Inc 820 Lincoln Ave West Chester PA 19380

VINZ, FRANK LOUIS, electrical engineer; b. Laredo, Tex., Jan. 5, 1932; s. Louis and Margaret Reeves (Schaer) V.; m. Mary Margaret Harlow, June 24, 1956; children: Laura Lee, Susan Elizabeth, Bradley Louis. BSEE, Tex. A&M U., 1953; MSEE, USAF Inst. Tech., 1963; postgrad. in Elec. Engring., U. Tenn., 1967. Project officer Air Force Armament Ctr., 1955-58; electronic engr. Army Ballistic Missile Agy., Redstone Arsenal, Ala., 1958-60; engring. supr. Marshall Space Flight Ctr. NASA, Huntsville, Ala., 1960-89; prin. engr. BDM Internat., Inc., Huntsville, 1989-91; sr. engr. Loral Aerosys divsn., Huntsville, 1991—. Contbr. articles to Jour. Inst. Navigation, SPIE Symposium on Intelligent Robots. Trustee, chmn. Presbyn. Ch., Huntsville, 1974-76, elder, 1959-63, 78-81, 83-85; commdr. Huntsville Power Squadron, 1984-85. Grad. sch. scholarship NASA-MSFC, U. Tenn., 1966-67. Mem. AIAA (com. for flight simulation), Huntsville Assn. Tech. Socs. (sec. 1992—). Achievements include patent pending for computer vision system for automatic docking of spacecraft. Home: 1006 San Ramon Ave SE Huntsville AL 35802 Office: Loral Aero Sys Divsn 620 Discovery Dr Huntsville AL 35806

VIOLET, WOODROW WILSON, JR., retired chiropractor; b. Columbus, Ohio, Sept. 19, 1937; s. Woodrow Wilson and Alice Katherine (Woods) V.; student Ventura Coll., 1961-62; grad. L.A. Coll. Chiropractic, 1966; m. Judith Jane Thatcher, June 15, 1963; children: Woodina Lonize, Leslie Alice. Pvt. practice chiropractic medicine, Santa Barbara, Calif., 1966-73, London, 1973-74, Carpinteria, Calif., 1974-84; past mem. coun. roentgenology Am. Chiropractic Assn. Former mem. Parker Chiropractic Rsch. Found., Ft. Worth; mem. Scripps Clinic Rsch. Coun. Served with USAF, 1955-63. Recipient award merit Calif. Chiropractic Colls., Inc., 1975, cert. of appreciation Nat. Chiropractic Antitrust Com., 1977. Mem. Nat. Geog. Soc., L.A. Coll. Chiropractic Alumni Assn., Delta Sigma. Patentee surg. instrument. Home: 2575 Bedford Ave Hemet CA 92545-5308

VIOLETTE, CAROL ANN, chemist, environmental compliance consultant; b. Westport, Conn., Jan. 12, 1956; d. Russell Andrew Smith, Oct. 24, 1987; 1 child, William Joseph. BS in Chemistry, U. Conn., 1985, PhD in Chemistry, 1990. Sr. lab. technician Delta Rubber Co., Danielson, Conn., 1985-86, regulatory compliance cons., 1986—; commr. Killingly (Conn.) Inland Wetlands and Watercourses Commn., 1991—. Contbr. articles to profl. jours. Recipient Conn. High Tech. award Conn. Bd. Higher Edn., 1988-89, 89-90. Mem. Nat. Fire Protection Assn., Air and Waste Mgmt. Assn., Am. Chem. Soc., Hazardous Materials Control Rsch. Inst., Environ. Policies Coun. Achievements include development of a waterborne vulcanizable adhesive to replace MEK based system; reduced MEK emissions to the air by 90% this year at one small manufacturing facility. Home: PO Box 681 Dayville CT 06241

VIOLETTE, GLENN PHILLIP, construction engineer; b. Hartford, Conn., Nov. 15, 1950; s. Reginald Joseph and Marielle Theresa (Bernier) B.; m. Susan Linda Begam, May 15, 1988. BSCE, Colo. State U., 1982. Registered profl. engr., Colo. Engring. aide Colo. State Hwy. Dept., Glenwood Springs, Colo., 1974-79, hwy. engr., 1980-82; hwy. engr. Colo. State Hwy. Dept., Loveland, Colo., 1979-80; project engr. Colo. State Hwy. Dept., Glenwood Canyon, Colo., 1983—; guest speaker in field. Contbg. editor, author, photographer publs. in field. Recipient scholarship Fed. Hwy Adminstrn., 1978. Mem. ASCE, Amnesty Internat., Nat. Rifle Assn., Siera Club, Audubon Soc., Nature Conservancy, World Wildlife Fund, Cousteau Soc., Chi Epsilon. Office: Colorado Dept Transp 202 Centennial St Box 1430 Glenwood Springs CO 81602

VIPULANANDAN, CUMARASWAMY, civil engineer, educator; b. Colombo, Sri Lanka, Jan. 7, 1956; s. Valliipuram and Maheswary Ammal Cumaraswamy; m. Giritha Vipulanandan, Aug. 5, 1986; 1 child: Geethanjali. MS, Northwestern U., 1981, PhD, 1984. Teaching asst. Northwestern U., Evanston, Ill., 1981-84; asst. prof. U. Houston, 1984-90, assoc. prof., 1990—. Contbr. articles to Jour. Materials in Civil Engring., Cement and Concrete Rsch., Material Rsch. Bull., Jour. Hazardous Materials, Hazardous Waste and Hazardous Materials, Jour. Applied Polymer Sci., Transp. Rsch. Record, Jour. Geotech. Engring., Jour. Am. Ceramic Soc., Materials Jour. American Concrete Inst., Polymer Engring. and Sci., Engring. Fracture Mechanics, Jour. Composite Materials. Mem. ASCE (assoc.), Materials Rsch. Soc., Am. Concrete Inst., Hazardous Materials Control Resources Inst. Achievements include development of new bonding tests to evaluate grouts and fracture properties of polymer concrete; developed superconducting composite; developed methods for treating contaminated soils. Office: U Houston 4800 Calhoun Houston TX 77204-4791

VIRGIL, SCOTT CHRISTOPHER, chemistry educator; b. San Mateo, Calif., Nov. 4, 1965; s. Leigh LaSeur and Dolores V. BS, Calif. Inst. tech., 1987; PhD, Harvard U., 1991. Asst. prof. chemistry MIT, Cambridge, 1992—. Contbr. articles to profl. jours. Camille and Henry Dreyfus Found. New Faculty awardee, 1991; David and Lucile Packard Found. fellow, 1992—. Achievements include notable findings concerning the molecular details of the cyclization step in sterol biosynthesis. Office: MIT Dept Chemistry 77 Massachusetts Ave Cambridge MA 02139

VISCO, FERDINAND JOSEPH, cardiologist, educator; b. Bklyn., July 8, 1941; s. Joseph Thomas and Susan (Baratta) V.; m. Laurie Judith Glass, Sept. 18, 1983; 1 child, Melissa; children by previous marriage, Ruth, Joseph, Jennifer. BS in Biology, Fairfield U., 1963; MD, U. Padua, Italy, 1969. Diplomate Am. Bd. Internal Medicine, specialty cardiovascular disease; lic. physician and surgeon, N.Y. Intern medicine Flushing (N.Y.) Hosp., 1969-70; jr. and sr. resident medicine Cath. Med. Ctr., Queens Hosp. Ctr., Jamaica, N.Y., 1970-72; fellow cardiology Nassau County Med. Ctr., East Meadow, N.Y., 1972-74; dir. medicine Freeport (N.Y.) Hosp., 1975; instr. medicine SUNY, Stony Brook, 1973-75; instr. medicine Albert Einstein Coll. Medicine, Bronx, N.Y., 1975-78, asst. prof. medicine, 1978—; assoc. cardiologist Bronx Lebanon Hosp., 1975—, attending physician, 1978—, dir. noninvasive cardiology lab., 1975—. Fellow Am. Coll. Cardiology; mem. ACP, Am. Soc. Echocardiography. Avocations: computers, ham radio, golf. Office: Bronx Lebanon Hosp 1650 Grand Concourse Bronx NY 10457

VISCUSO, SUSAN RICE, psychologist, researcher; b. Gresham, Oreg., Dec. 7, 1961; d. Robert Richardson and Elizabeth Lois (Theiring) Rice; m. Patrick Demetrios Viscuso, Aug. 18, 1985. BA, Cath. U. Am., 1984; MSc, Brown U., 1987, PhD, 1989. Fellow Army Rsch. Inst., Alexandria, Va., 1989; rsch. psychologist CIA, Langley, Va., 1989—. Author: (book chpt.) Representing Simple Arithmetic in Neural Networks, 1989; contbr. articles to profl. jours. Fellow NSF, 1985-88, NRC, 1988-89; recipient scholarship Elks Club, 1980. Mem. APS, Cognitive Sci. Soc., Sigma Xi, Phi Beta Kappa. Republican. Greek Orthodox. Achievements include researching mental models and comprehension of written instructions; how arithmetic facts are represented in human memory and simulations of simple arithmetic learning using neural network computer models. Office: CIA Washington DC 20505

VISEK, WILLARD JAMES, nutritionist, animal scientist, physician, educator; b. Sargent, Nebr., Sept. 19, 1922; s. James and Anna S. (Dworak) V.; m. Priscilla Flagg, Dec. 28, 1949; children: Dianna, Madeleine, Clayton Paul. B.Sc. with honors (Carl R. Gray scholar), U. Nebr., 1947; MSc (Smith fellow in agr.), Cornell U., 1949, Ph.D., 1951; M.D. (Peter Yost Fund scholar), U. Chgo., 1957; DSc (hon.), U. Nebr., 1980. Diplomate Nat. Bd. Med. Examiners, 1960. Grad. asst., lab. animal nutrition Cornell U., 1947-51; AEC postdoctoral fellow Oak Ridge, 1951-52; research assoc., 1952-53; research asst. pharmacology U. Chgo., 1953-57, asst. prof., 1957-61, assoc. prof., 1961-64; rotating med. intern U. Chgo. Clinics, 1957-58, 58-59, 59; prof. nutrition and comparative metabolism, dept. animal sci. Cornell U., Ithaca, N.Y., 1964-75; prof. clin. sci. (nutrition and metabolism) Coll. Medicine and dept. food sci. U. Ill. Coll. Agr., Urbana-Champaign, 1975—; prof. dept. internal medicine U. Ill. Coll. Medicine, Urbana-Champaign, 1986—; Brittingham vis. prof. U. Wis., Madison, 1982-83; Hogan meml. lectr. U. Mo., 1987; mem. subcom. dog nutrition, com. animal nutrition NRC-Nat. Acad. Sci., 1965-71; adv. council Inst. Lab. Animal Resources, NRC-Nat. Acad. Sci., 1966-69; sub-com. animal care facilities Survey Inst. Lab. Animal Resources, 1967-70; cons., lectr. in field; mem. sci. adv. com. diet and nutrition cancer program Nat. Cancer Inst., 1976—; mem. nutrition study sect. NIH, 1980-84; chmn. membership com. Am. Inst. Nutrition Am. Soc. Clin. Nutrition, 1978-79, 80-83, 85; cons. VA, NSF, indsl. orgns.; Wellcome vis. prof. in basic med. scis. Oreg. State U., 1991-92.; bd. sci. counselors USDA, 1989—. Mem. editorial bd. Jour. Nutrition, 1980-84, editor, 1990—; contbr. articles to profl. jours. Active local Boy Scouts Am. Served with AUS, 1943-46. Nat. Cancer Inst. spl. fellow M.I.T., research fellow Mass. Gen. Hosp., 1970-71; recipient Alumni award Neb. 4-H, 1967, Osborne and Mendel award, 1985, U. Ill. Coll. Medicine faculty merit award, 1988; sr. scholar U. Ill., 1988. Fellow AAAS, Am. Inst. Nutrition, Am. Soc. Animal Sci. (chmn. subcom. antimicrobials, mem. regulatory agency com. 1973-78); mem. Soc. Pharmacology and Exptl. Therapeutics, Am. Inst. Nutrition (council 1980-83, 85-86), Soc. Exptl. Biology and Medicine, Am. Soc. Clin. Nutrition, Am. Therapeutic Soc., Am. Gastroenterol. Assn., Am. Bd. Clin. Nutrition, Innocents Soc., Fedn. Am. Socs. Exptl. Biology (sci. steering group life scis. rsch. office, adv. com. 1986-92), Am. Bd. Nutrition (bd. dirs.), Nat. Dairy Coun. (rsch. adv. com. 1987-91, vis. prof. nutrition program 1981-92), Gamma Alpha (pres. 1948-49), Phi Kappa Phi (pres. 1981-82), Alpha Gamma Rho (pres. 1946-47), Gamma Sigma Delta. Presbyterian (elder). Lodge: Masons. Home: 1405 W William St Champaign IL 61821-4406 Office: U Ill 190 Med Sci Bldg 506 S Mathews St Urbana IL 61801

VISHNIAC, ETHAN TECUMSEH, astronomy educator; b. New Haven, Sept. 29, 1955; s. Wolf Vladimir and Helen Frances (Simpson) V.; m. Ilene Joy Busch, June 13, 1976; children: Cady Anne, Miriam Rachel. BS and BA summa cum laude, U. Rochester, 1976; MA, Harvard U., 1980, PhD, 1980. Rsch. assoc. Princeton (N.J.) U., 1980-82; lectr. U. Tex., Austin, 1982-84, asst. prof., 1984-88, assoc. prof., 1988—. Contbr. numerous articles to jours. Recipient Presdl. Young Investigator, 1985; Alfred Sloan fellow, 1986. Mem. Am. Astron. Soc. (Helen B. Warner prize 1990), Internat. Astron. Union, Am. Phys. Soc. Office: U of Tex Astronomy Dept Austin TX 78712

VISICH, KAREN MICHELLE, mechanical engineer; b. Milw., Aug. 25, 1964; d. Kenneth John Spara and June Mary (Kibble) Muszynski; m. Michael Visich, Oct. 7, 1989. BS in Mech. Engring., Milw. Sch. Engring., 1986; MS in Mech. Engring., Purdue U., 1988. Rsch. asst. Purdue U., West Lafayette, Ind., 1986-88; mem. tech. staff The Aerospace Corp., El Segundo, Calif., 1988-92; engr. Applied Measurement Systems Inc., New London, Conn., 1992—. Vol. Animal Welfare League, New London, 1992. Recipient Scholarship Johnson Found., Milw., 1983-85. Mem. AIAA, ASME. Achievements include research on aerodynamic detuning and aeroelasticity of a supersonic turbomachine rotor. Office: Applied Measurement Systems Inc 400 Bayonet St Ste 101 New London CT 06320

VISKANTA, RAYMOND, mechanical engineering educator; b. Lithuania, July 16, 1931; came to U.S., 1949, naturalized, 1955; s. Vincas and Genovaite (Vinickas) V.; m. Birute Barbara Barpsys, Oct. 13, 1956; children: Renata, Vitas, Tadas. BSME, U. Ill., 1955; MSME, Purdue U., 1956, PhD, 1960. Registered profl. engr., Ill. Asst. mech. engr. Argonne (Ill.) Nat. Lab., 1956-59, student rsch. assoc., 1959-60, assoc. mech. engr., 1960-62; assoc. prof. mech. engring. Purdue U., West Lafayette, Ind., 1962-66, prof. mech. engring., 1966-86, Gross disting. prof. engring., 1986—; guest prof. Tech. U. Munich, Germany, 1976-77, U. Karlsruhe, Germany, 1987; vis. prof. Tokyo Inst. Tech., 1983. Contbr. more than 300 tech. articles in profl. jours. Recipient Sr. U.S. Scientist award Alexander von Humboldt Found., 1975, Sr. Rsch. award Am. Soc. Engring. Edn., 1984, Nusselt-Reynolds prize, 1992; Japan Soc. for Promotion Sci. fellow, 1983. Fellow ASME (Heat Transfer Meml. award 1976, Max Jakob Meml. award 1986, Melville medal 1988), AIAA (Thermophysics award 1979); mem. AAUP, AAAS, NAE, Sigma Xi, Pi Tau Sigma, Tau Beta Pi. Home: 123 Pawnee Dr West Lafayette IN 47906-2167 Office: Purdue Univ 1288 Mechanical Engring Bldg West Lafayette IN 47907-1288

VISOTSKY, HAROLD MERYLE, psychiatrist, educator; b. Chgo., May 25, 1924; s. Joseph and Rose (Steinberg) V.; m. Gladys Mavrich, Dec. 18, 1955; children: Jeffrey, Robin. Student, Herzl Coll., Chgo., 1943-44, Baylor U., 1944-45, Sorbonne, 1945-46; B.S., U. Ill., 1947, M.D. magna cum laude, 1951. Intern Cin. Gen. Hosp., 1951-52; resident U. Ill., Ill. Research and Ednl. Hosp., also Neuropsychiat. Inst., Chgo., 1952-55; asst. prof. U. Ill. Coll. Medicine, 1957-61, assoc. prof. psychiatry, 1965-69, dir. psychiat. residency tng. and edn., 1955-59; prof., chmn. dept. psychiatry and behavioral scis. Northwestern U. Med. Sch., Chgo., 1969-91; dir. Psychiat. Inst., chmn. dept. psychiatry Northwestern Meml. Hosp., 1969-91; sr. attending physician Evanston (Ill.) Hosp.; Polio respiratory center psychiat. cons. Nat. Found. Infantile Paralysis, U. Ill., 1955-59; dir. mental health Chgo. Bd. Health div. mental health services, 1959-63; dir. Ill. Dept. Mental Health, 1962-69; examiner Am. Bd. Psychiatry and Neurology, 1964—; cons. Center Mental Health and Psychiat. Services, Am. Hosp. Assn., 1979—; mem. 1st U.S. mission on mental health to USSR State Dept. Mission, 1967; chmn. task force V Joint Commn. on Mental Health of Children, 1967-69; mem. adv. com. on community mental health service Nat. Inst. Mental Health, 1965-67; mem. profl. adv. bd. Jerusalem Mental Health Center; profl. adv. group Am. Health Services, Inc.; adv. com. Joint Commn. Accreditation Hosps., Council Psychiat. Facilities; mem. spl. panel mental illness for bd. dirs. ACLU; rector Lincoln Acad. of Ill. Faculty Social Service; mem. select com. psychiat. care and evaluation HEW; mem. faculty Practising Law Inst.; bd. overseers Spertus Coll. Judaica, Chgo., 1981-83. Contbr. articles to psychiat. jours., chpts. psychiat. textbooks. Trustee Erikson Inst. Early Edn., Ill. Hosp. Assn., Mental Health Law Project, Washington. Served with AUS, 1942-46. Decorated D.S.C., Purple Heart, Bronze Star; recipient Edward A. Strecker award Inst. Pa. Hosp., 1969; Med. Alumnus of Year award U. Ill., 1976; Disting. Service award Chgo. chpt. Anti-Defamation League B'nai B'rith, 1978. Fellow Am. Orthopsychiat. Assn. (dir. v.p. 1970-71, pres. 1976-77, Leadership in Community Health Programs award 1986), Am. Psychiat. Assn. (chmn. council on mental health services 1967-68, v.p. 1973-74, bd. trustees, 1973-83, sec. 1981-83, chmn. council nat. affairs 1975-78, com. on abuse of psychiatry and psychiatrists, 1980-84, chmn. council internat. affairs 1984-89, Adminstrv. Psychiatry award 1985, Simon Bolivar award 1982, Spl. Presdl. Commendation award 1988), AAAS, Am. Coll. Psychiatrists (charter, bd. regents 1976-79, v.p. 1980, pres. 1983-84, E.B. Bowis award 1981, Gold medal for contbns. to psychiatry 1988), Chgo. Inst. Medicine, Am. Coll. Mental Health Adminstrs. (founding fellow); mem. Am. Assn. Chmn. Dept. Psychiatry , Am. Assn. Social Psychiatry (v.p. 1976, pres.-elect 1987, pres. 1988-90), Council Med. Splty. Socs., AMA, Ill. Psychiat. Soc. (pres. 1965-66), Am. Coll. Psychoanalysts, Mental Health Law Project (bd. trustees 1973-79) World Assn. Social Psychiatry (councilor, exec. coun. 1984—), World Psychiatric Assn. (rev. com. 1985—, sec. of meetings 1990—), NAS, Inst. of Med. (commn. to develop methods useful for VA in estimating physician needs 1988—, expert rev. panel to set priorities for mental health 1989), Chgo. Consortium for Psychiatric Rsch. (chmn., bd. dirs. 1989—), Alpha Omega Alpha. Home: 2748 Lincolnwood Dr Evanston IL 60201-1229 Office: Northwestern Meml Hosp Superior St & Fairbanks Ct Chicago IL 60611

VISSER, MATTHEW JOSEPH, physicist; b. Wellington, New Zealand, Mar. 21, 1956; s. Herman Matthew and Cecilia Johanna (Regter) V. BS in Physics, BA in Math., BS/honors, Victoria U., New Zealand, 1976, 77, 78, MS in Math., MA in Physics, 1981; PhD in Physics, U. Calif., Berkeley, 1984. Postdoctoral fellow U. So. Calif., L.A., 1984-87, Los Alamos (N.Mex.) Nat. lab., 1987-89; postdoctoral fellow Washington U., St. Louis, 1989-91, asst. rsch. prof., 1991—. Contbr. articles to profl. jours. Co-recipient 2nd prize essay contest, Gravity Rsch. Found., 1986. Mem. Am. Phys. Soc. Achievements include contbn. to theoretical understanding of the fifth force and to limits on the existence of their putative effect, multiple contbns. to theory of traversable and quantum wormholes and hawking's chronology protection conjecture. Office: Washington U Dept Physics Saint Louis MO 63130-4899

VISSOL, THIERRY-LOUIS, senior economist, researcher; b. Chabanais, Charente, France, Mar. 3, 1951; s. Jacques-Marie and Claire-Octavie (Benoit-Guyod) V.; m. Carola Maggiulli; children: David, Andrea. Diplome Univ. Tech., U. Limoges (France), 1971, DEA Econ., 1976, Doctorat Econ., 1979. Chargé de cours (Econ.) I.U.T. Gestion, Limoges, France, 1975-80; asst. (Econ.) U. Sciences Economiques, Limoges, France, 1976-80; chargé de cours (History of Econ.) U. Econ., Limoges, France, 1981-82; adminstr. European Community Commn., Brussels, 1980-86, sr. adminstr., 1986-91, head of unit ECU, 1991—; chargé de cours Inst. Catholique des Hautes Etudes Commerciales, Brussels, 1986-91, dir. Centre Ecu et Prospective d'Integration Monetaire Europeenne, 1987—. Editor: (econ. jour.) De Pecunia, 1989—; contbr. numerous scientific articles to profl. jours. Avocations: history, art, music, travel. Office: European Community Commn, Rue de la Loi 200, 1049 Brussels Belgium

VISWANATH, DABIR SRIKANTIAH, chemical engineer; b. Bangalore, India, Aug. 5, 1934; s. Srikantiah and Kamalamma Viswanath; m. Pramila Viswanath, Jan. 5, 1967; 1 child, Arvind. BS, Mysore U., Bangalore, 1953; DIIS, Indian Inst. of Sci., Bangalore, 1956; MS, U. Rochester, 1960, PhD, 1962. Chemist Essen & Co., Bangalore, 1953-54; chem. engr. Sarabhai Chems., Bangalore, 1956-57; asst. prof. to prof., chmn. Indian Inst. of Sci., Barda, 1965-78; vis. prof. Tex. A&M, College Station, 1978-79; prof. U. Mo., Columbia, 1979-90, prof., chmn., 1990—. Co-author: Data Book on Viscosity of Liquids, 1989; contbr. articles to profl. jours. Pari Hargovandas fellowship Indian Inst. of Sci., Lever Bros. fellowship U. Rochester; recipient Halliburton Travel awards, Mo; grantee NSF, 1985, 1990-91, U. Wis., 1982-83, IBM Corp., 1988-91, USEPA-Kans. State, 1988-91, Waste Mgmt. of N.Am., 1992-93. Fellow Am. Inst. of Chemists; mem. AAAS, Am. Inst. of Chem. Engrs., Am. Chem. Soc., Columbia Rotary Club, Indian Inst. of Chem. Engrs., Catalyst Soc. of India, Am. Inst. of Chemistry. Achievements include rsch. in the chemistry of ordorant compounds, generalized thermodynamic properties of real gases-generalized compressibility charts, thermodynamic properties of CC14, supercritical extraction, hydrocarbon oxidation, thermophysical properties of gases, liquids and polymer composites. Home: 507 Onofrio Ct Columbia MO 65203 Office: U Mo Dept Chem Engring Columbia MO 65211

VISWANATHAN, RAMASWAMY, physician, educator; b. Coimbatore, India, Aug. 20, 1949; came to U.S., 1972; s. Thiruvalangadu and Bhavani Krishnamurthy Ramaswamy; m. Kusum Ramakrishna, June 15, 1980; children: Vikram, Vivek, Vidya. MB, BS, U. Madras, 1972; D of Med. Sci., SUNY, 1989. Diplomate Am. Bd. Psychiatry and Neurology, added qualifications in geriatric psychiatry, added qualifications in addiction psychiatry. Med. intern Bklyn.-Cumberland Med. Ctr., 1972-73; resident in internal medicine L.I. Jewish-Hillside Med. Ctr.-Queens Hosp. Ctr. Affiliation, N.Y.C., 1973-74; resident in psychiatry SUNY Health Sci. Ctr., Bklyn., 1974-77, fellow in psychosomatic medicine, 1976-78, fellow in research tng. in psychiatry, 1977-79, mem. staff, 1978—, clin. asst. prof. psychiatry, 1979-

87, instr. in medicine, 1979—, clin. assoc. prof., 1987-90, assoc. prof. clin. psychiatry, 1990—, assoc. dir. med.-psychiat. liaison svc., 1981-83, 84—, med. dir. Anxiety Disorders Clinic, 1982—, acting dir. med.-psychiat. liaison svc., 1983-84, mem. com. cancer edn. and preventive oncology, 1983-92, dir. course on life-threatening illness, dying and death, 1984-89, dir. med. interviewing course, 1985-89, dir. doctor-patient relationship course, 1989-90, dir. intro. to clin. medicine-human dimension course, 1990—, cons. AIDS unit, 1989—; with student evaluation and promotion com. SUNY Health Sci. Ctr., 1989—, with course dirs.' com., 1989—; mem. exec. com. Univ. Hosp., 1992—; cons. Bklyn. VA Med. Ctr., 1986-89; pvt. practice medicine specializing in psychiatry, psychosomatic medicine, behavior therapy, hypnosis and sex therapy Bklyn., 1978—. Contbr. articles to profl. jours. Curriculum coun. Herricks Pub. Sch. dist., 1992—. Fellow ACP, Am. Psychiat. Assn. (com. on consultation, liaison psychiatry & primary care edn., coun. on internat. affairs), Indian Psychiat. Soc. (life); mem. AMA, Bklyn. Psychiat. Soc. (councillor, pres. elect 1988-90, pres. 1990-92, chmn. com. on AIDS 1990-92, chmn. com. on consultation-liaison psychiatry 1990-92, chmn. disaster response com. 1991-92, chmn. legis. & pvt. practice com. 1992—, chmn. internat. affairs com. 1992—), Assn. Advancement Behavior Therapy, Acad. Psychosomatic Med., Soc. for Liaison Psychiatry (bd. dirs. 1986-90, 1993—, sec. 1987-88), Anxiety Disorders Assn. Am., N.Y. State Psychiat. Assn. (com. on govt. health programs, task force on practice guidelines, legis. com., chmn. edn. com. 1992—), Soc. Exploration Psychotherapy Integration, Am. Assn. Psychiatrists from India (founder, life, exec. com. 1979-85, 92—, sec. 1992—, Sci. and Svc. award 1990). Office: SUNY Health Sci Ctr 450 Clarkson Ave Box 127 Brooklyn NY 11203

VITA, JAMES PAUL, software engineer; b. Bronx, July 18, 1955; s. Salvatore J. and Mary (Macaluso) V.; m. Elizabeth A. Peterson, June 10, 1978. BA, NYU, 1976; MA, L.I. U., 1978; MS, CUNY, 1987. Applications programmer CUNY, N.Y.C., 1985-88; sr. assoc. programmer IBM, Poughkeepsie, 1988—; cons. Mental Health Assn. in Dutchess County, Poughkeepsie, 1989-91. Contbr. articles to profl. jours. Mem. IEEE, Computer Soc. of IEEE, Assn. for Computing Machinery, N.Y. Acad. Scis., Golden Key, Phi Beta Kappa. Home: 100 Brandy Ln Wappingers Falls NY 12590-6445 Office: IBM PO Box 390 Poughkeepsie NY 12602-0390

VITALI, JUAN A., nuclear engineer, consultant; b. Caracas, Venezuela, June 24, 1962; came to U.S., 1981; s. Bruno and Maria (Vitellozzi) V. BS in Nuclear Engring., U. Fla., 1984, M in Engring., 1987, PhD, 1992. Rsch. asst. U. Fla., Gainesville, 1985-90, 91-92; rsch. scientist RTS Labs., Inc., Alachua, Fla., 1990-91, Gen. Imaging Corp., Gainesville, 1992—; staff cons. High Tech. and Industry Coun., Office of the Gov., Tallahasee, Fla.,1991. Contbr. articles to profl. jours. Mem. ASME, Am. Nuclear Soc., Hispanic Alumni Assn. U. Fla. Democrat. Roman Catholic. Achievements include Nuclear Augmented Thrusters (patent pending), X-Ray Food Irradiator (patent pending). Office: Gen Imaging Corp PO Box 13455 Gainesville FL 32604

VITEK, RICHARD KENNETH, scientific instrument company executive; b. Chgo., Feb. 1, 1935; s. Martin and Mildred (Veverka) V.; m. Marilyn W. Young, June 23, 1956; children: Christine, Debra, Evelyn. AB, Albion Coll., 1956; MS, U. Mo., 1958. Rsch. chemist Allied Chem. Corp., Morristown, N.J., 1958-64; rsch. chemist AEC/Nat. Lead Co., Cin., 1957; sales mgr. Aldrich Chem. Co., Inc., Milw., 1964-66, dir. mktg., 1966-68; pres., chmn. Fotodyne Inc. and Variquest Techs., Inc., Heartland, Wis., 1968—. Bd. mem. various civic and indsl. orgns. Mem. Am. Chem. Soc., Council Ind. Mgrs., Wis. Acad. Scis., Arts and Letters, Ind. Bus. Assn. Wis., Wis. Bus. Assn. Independent Republican. Congregationalist. Contbr. numerous articles to profl. jours. and books. Achievements include 4 invention patents; discovery of Bismuth Dimethylglyoxime and F3NO. Office: 950 Walnut Ridge Dr Hartland WI 53029-9388

VITERBI, ANDREW JAMES, electrical engineering and computer science educator, business executive; b. Bergamo, Italy, Mar. 9, 1935; came to U.S., 1939, naturalized, 1945; s. Achille and Maria (Luria) V.; m. Erna Finci, June 15, 1958; children: Audrey, Alan, Alexander. SB, MIT, 1957, SM, 1957; PhD, U. So. Calif., 1962; DEng (honoris causa), U. Waterloo, 1990. Research group supr. C.I.T. Jet Propulsion Lab., 1957-63; mem. faculty Sch. Engring. and Applied Sci., UCLA, 1963-73, assoc. prof., 1965-69, prof., 1969-73; exec. v.p. Linkabit Corp., 1973-82; pres. M/A-Com Linkabit, Inc., 1982-84; chief scientist, sr. v.p. M/A-Com. Inc., 1985; prof. elec. engring. and computer sci. U. Calif., San Diego, 1985—; vice chmn., chief tech. officer Qualcomm Inc., 1985—; chmn. U.S. Commn. C, URSI, 1982-85; vis. com. dept. elec. engring. and computer sci. MIT, 1984—. Author: Principles of Coherent Communication, 1966, (with J.K. Omura) Principles of Digital Communication and Coding, 1979; bd. editors: Proc. IEEE, 1968-77; mem. bd. editors: Information and Control, 1967, Transactions on Info. Theory, 1972-75. Recipient award for valuable contbns. to telemetry, space electronics and telemetry group IRE, 1962, best original paper award Nat. Electronics Conf., 1963, outstanding papers award, info. theory group IEEE, 1968, Christopher Columbus Internat. Communications award, 1975, Aerospace Communication award AIAA, 1980, Outstanding Engring. Grad. award U. So. Calif., 1986; co-recipient Nat. Electronics Conf. C and C Found. award, 1992; Marconi Internat. fellow, 1990. Fellow IEEE (Alexander Graham Bell medal 1984, Shannon lectr. internat. symposium on info. theory 1990); mem. NAE. Office: Qualcomm Inc 10555 Sorrento Valley Rd San Diego CA 92121-1617

VITETTA, ELLEN S., microbiologist educator, immunologist. BA, Conn. Coll.; MS, NYU, 1966, PhD, 1968. Prof. microbiology Southwestern Med. Sch., U. Texas, Dallas, Tex.; dir. Cancer Immunobiology Ctr., U. Texas, Dallas, Tex.; Sheryle Simmons Patigian Disting. chair in cancer immunobiology Southwestern Med. Sch., U. Tex., Dallas; prof. dept. microbiology U. Tex., Southwestern Med. Ctr., Dallas; bd. sci. coun. NCI Cancer Treatment Bd., 1993; sci. adv. bd. Howard Hughes Med. Inst., 1992—; Kettering selection com. GM Cancer Rsch. Foun., 1987-88; task force NIAID in Immunology, 1989-90; mem. sci. bd. Ludwig Inst., 1983—. Editorial bd.: Advances in Host Defense Mechanisms, 1983—, Annual Review of Immunology, 1991—, Bioconjugate Chemistry, 1989-93, Cellular Immunology, 1984-93, Current Opinion in Immunology, 1992—, FASEB Journal, 1987—, International Journal of Onocology, 1992—, International Soc. Immunopharmacology, 1989—, Journal of Immunology, 1975-78, Molecular Immunology, 1978-93; Assoc. Editor: Cancer Research, 1986—; Immunochemistry Section Editor: Journal of Immunology, 1978-82; Co-Editor in Chief: Therapeutic Immunology, 1992—. Recipient Women's Excellence in Sci. award Fedn. Am. Soc. Expn. Biology, 1991, Taittinger Breast Cancer Rsch. award Komen Found., 1983, Pierce Immunotoxin award, 1988, NIH Merit award, 1987—, U. Tex. Southwestern Med. Sch. Faculty Teaching awards 1989, 91, 92, FASEB Excecllence in Sci. award, 1991, Abbot Clinical Immunology award Am. Soc. Microbiologists, 1992, Past State Pres. award Tex. Fed. Bus. Profl. Women's Club, 1993. Mem. Am. Assn. Immunologists (pres. 1993—), Internat. Soc. Immunopharmacology (councillor 1991—). Achievements include co-discovery of IL-4, development of immunotoxins and identification of IgD on murine B cells. Office: Univ of Texas Southwestern Medical Ct 5323 Harry Hines Blvd Dallas TX 75235*

VITULLI, WILLIAM FRANCIS, psychology educator; b. Bklyn., July 17, 1936; s. William S. and Sadie Rosaria (Stallone) V.; m. Betty Jean Sheubrooks, June 15, 1961; children: Paige Vitulli Baggett, Quinn Anthony, Sherik Denise. BA, U. Miami, 1961, MS, 1963, PhD, 1966. Lic. psychologist, Ala. Grad. asst. U. Miami, Coral Gables, Fla., 1961-65; asst. prof. psychology U. South Ala., Mobile, 1965-69, assoc. prof., 1969-75, prof., 1975—; v.p. Ala. Bd. Examiners in Psychology, Montgomery, 1982-84; rsch. cons. Drug Edn. Coun., Mobile, 1988-93. Mem. editorial bd. Jour. Sport Behavior, 1978—; contbr. articles to profl. jours. Mem. adv. bd. Contact Mobile, 1987—. Named Prof. of Quarter, Alpha Lambda Delta, 1977-78. Mem. APA, Southeastern Psychol. Assn., Ala. Psychol. Assn. (pres. 1975), Italian-Am. Cultural Soc. South Ala. (chair hist.-cultural com. 1982), Sigma Xi, Psi Chi (faculty adviser U. South Ala. chpt. 1972-80). Roman Catholic. Avocations: jogging, athletics research and analysis, fishing. Home: 2025 Maryknoll Ct Mobile AL 36695-3829 Office: U S Ala 307 University Blvd N Mobile AL 36688-0001

VIVERA, ARSENIO BONDOC, allergist; b. Cebu City, Philippines, Oct. 29, 1931; s. Arsenio R. and Ramona del Mar (Bondoc) V.; A.A., Cebu Coll., U. Philippines, 1950, M.D., 1954. Intern, Philippines Gen. Hosp., 1954-55; resident in medicine Beekman-Downtown Hosp., N.Y.C., 1955-57, Detroit Meml. Hosp., 1957-58; resident in allergy Robert A. Cooke Inst. Allergy, Roosevelt Hosp., N.Y.C., 1958-59, fellow in allergy, 1959-61; sr. cons. scientist Philippines Nat. Inst. Sci. and Tech., Manila, 1961-62; practice medicine specializing in allergy, N.Y.C.; chief allergy dept. attending physician N.Y. Polyclinic Med. Sch. and Health Center, 1972-77, adj. prof., 1972-77; clin. attending physician Robert A. Cooke Inst. Allergy, 1969—; asst. attending physician N.Y. Infirmary, N.Y.C., 1969—; chief allergy, attending St. Vincent's Hosp. and Med. Center, N.Y.C., 1977—. Diplomate Am. Bd. Allergy and Immunology. Fellow Am. Acad. Allergy, Am. Coll. Allergists, Am. Assn. Clin. Immunology and Allergy; mem. N.Y. Allergy Soc., AMA, Am. Assn. Cert. Allergists, N.Y. Acad. Scis., N.Y. Acad. Medicine, Am. Geriatric Soc., N.Y. State Med. Soc., N.Y. County Med. Socs. Office: 681 Lexington Ave Fl 5 New York NY 10022-2607

VIVES, STEPHEN PAUL, biology educator, fish biologist; b. Bartlesville, Okla., Oct. 13, 1958; s. Van Carl and Loretta May (Bloom) V.; m. Debra S. Thompson, Aug. 16, 1980; 1 child, Stephen Carl. BS, Okla. State U., 1980, MS, 1982; MS, U. Wis., 1986, PhD, 1988. Postdoctoral assoc. dept. zoology Ariz. State U., Tempe, 1988-90; asst. prof. dept. biology Ga. So. U., Statesboro, 1990—. Contbr. articles to Copeia, The Southwestern Naturalist, The Am. Midland Naturalist, Animal Behaviour. Guyer predoctoral fellow U. Wis., 1987; Ga. So. U. rsch. grantee, 1992, U.S. Fish and Wildlife Svc. cooperator, 1992. Mem. AAAS, Am. Soc. Ichthyologists and Herpetologists (Raney award 1987), Animal Behavior Soc., Desert Fishes Coun. Office: Ga So U Dept Biology LB-8042 Statesboro GA 30460

VIVIANI, GARY LEE, electrical engineer; b. McAllen, Tex., June 1, 1955; s. Alfred Paul Viviani and Madeleine Ann Yost; m. Deborah Ruth Claussen, Aug. 20, 1977; children: Sarah, Emily. BSEE, Purdue U., 1977, PhD in Elec. Engring., 1980. Registered profl. engr., Tex. Rsch. scientist E.I. DuPont de Nemours & Co., Wilmington, Del., 1980-82; Gulf States Utilities rsch. prof. Lamar U., Beaumont, Tex., 1982-88; prin. engr. Textron Def. Systems, Wilmington, Mass., 1988—; cons. Gulf States Utilities, Beaumont, 1982-88; advisor NSF, Washington, 1986. Referee jours.; contbr. over 30 articles to profl. jours. Named outstanding young engring. faculty, Gulf-Southwestern region, Am. Soc. for Engring. Edn., 1983; 4 yr. scholar USAF, 1973. Mem. IEEE (sr., vice chmn. Beaumont sect. 1986-87, referee jours.), AAAS, Soc. Indsl. and Applied Math. Lutheran. Achievements include patent for advanced cryptological device; invention of musical electronic lock, multi-stable device, first dynamical electronic device to achieve multiple stable equilibria; research in controllable multi-stable phenomena, aircraft landing and large scale computer-based control systems. Home: PO Box 454 Topsfield MA 01983-0654

VIVONA, DANIEL NICHOLAS, chemist; b. Chgo., Apr. 13, 1924; s. Daniel and Mary Rose (Lamonico) V.; student Chgo. City Coll., 1941-42, 46; BA, U. Maine, 1951; MS, Pa. State U., 1953; postgrad. Purdue U., 1953-56; m. Helen Mary Belanger, Sept. 14, 1950; 1 son, Daniel Maurice. Instr. chemistry Purdue U., Lafayette, Ind., 1955-56; with Minn. Mining and Mfg. Co., St. Paul, 1956-86, sr. chemist, 1969-79, info. scientist, 1979-81, quality assurance sr. chemist, 1981-86; cons., 1986—. With USAAF, 1943-45. Decorated Air medal with oak leaf clusters, DFC. Dow Corning fellow, 1952-53. Mem. Am. Chem. Soc., Phi Beta Kappa. Roman Catholic. Home: 3253 Kraft Cir N Lake Elmo MN 55042-9720 Office: Beta of Dan Vivona PO Box 128 Lake Elmo MN 55042-0128

VIZCAINO, HENRY P., mining engineer, consultant; b. Hurley, N.Mex., Aug. 28, 1918; s. Emilio D. and Petra (Perea) V.; m. Esther B. Lopez, Sept. 16, 1941; children: Maria Elena, Rick, Arthur, Carlos. BS in Engring., Nat. U., Mexico City, 1941; BS in Geology, U. N.Mex., 1954. Registered profl. engr. With Financiera Minera S.A., Mexico City, 1942-47; gen. mgr. Minas Mexicanas S.A., Torreon, Mex., 1947-51; exploration engr. Kerr McGee Corp., Okla., 1955-69; cons. Albuquerque, 1969-75, 84—; regional geologist Bendix Field Engring., Austin, Tex., 1976-79; staff geo-scientist Bendix Field Engring., Grand Junction, Colo., 1979-81; sr. geologist Hunt Oil Co., Dallas, 1981-84. Contbr. articles to profl. publs. Mem. Rotary, Elks. Republican. Congregationalist. Home and Office: 20 Canoncito Vista Rd Tijeras NM 87059-7833

VLACH, JEFFREY ALLEN, environmental engineer; b. Detroit, May 18, 1953; s. Robert Allen and Virginia Mae (Melton) V.; m. Diane Kay Daugherty, Oct. 27, 1984; children: Elizabeth Daugherty, Meredith Anna. BS, Purdue U., 1975. Cert. asbestos bldg. inspector, mgmt. planner. Environ. specialist D.E. McGillem and Assocs., Inc., Indpls., 1975-80, United Cons. Engrs., Inc., Indpls., 1980-88, Beam, Longest & Neff, Inc., Indpls., 1988—; asbestos bldg. inspector, mgmt. planner EPA, 1989. Conservation coord. Amos Butler chpt. Nat. Audubon Soc., Indpls., 1980-82. Named Eagle Scout Boy Scouts Am., 1969. Mem. ASCE (affiliate), Nat. Wildlife Fedn., Ind. Recycling Coalition, Nat. Assn. Environ. Profls. Office: Beam Longest & Neff Inc 8126 Castleton Rd Indianapolis IN 46250-2099

VLACHOPOULOS, JOHN, chemical engineering educator; b. Volos, Greece, Aug. 11, 1942; arrived in Can., 1968; s. Apostolos and Kleopatra (Anagnostou) V. Dipl. Ing, NATU, Athens, Greece, 1965; MS, Washington U., St. Louis, 1968, DSc, 1969. Registered profl. engr., Ont., Can. Asst. prof. chem. engring. McMaster U., Hamilton, Ont., 1968-74, assoc. prof. chem. engring., 1974-79, prof. chem. engring., 1979—, chmn. dept. chem. engring., 1985-88; sabbatical leave U. Stuttgart, Germany, 1975, CEMEF Ecole des Mines, Paris, 1981-82, 88-89; cons. in field, Can., U.S., Europe and Japan. Contbr. rsch. papers to profl. publs. Fellow Chem. Inst. Can.; mem. Am. Inst. Chem. Engring., Soc. Plastics Engrs., Soc. Rheology, Polymer Processing Soc. Greek Orthodox. Achievements include creation of commercial software packages used for simulation of plastics processing. Office: McMaster U Chem Engring Dpt, 1280 Main St W, Hamilton, ON Canada L8S 4L7

VLADAVSKY, LYUBOV, computer scientist, educator; b. Kiev, USSR, Dec. 13, 1957; came to U.S., 1980; d. Semyon and Frida (Schechtman) Shcherbakov; m. Boris Vladavsky, Jan. 5, 1977; 1 child, Valerie. BS in Computer Sci., U. Md., 1983; MS in Computer Sci., George Washington U., 1988. Programmer, analyst Western Union, McLean, Va., 1983-84; software engr. Advanced Tech. Systems, Vienna, Va., 1984-87; sr. computer scientist Logicon, Inc., Washington, 1987—; lectr. U. Denver, Arlington, Va., 1987-89, U. Md., College Park, 1989—. Contbr. papers to sci. publs. Mem. Parent Tchr. Student Assn., Thomas Jefferson Sch. Sci. and Tech., Va., 1991. Mem. ACM (Washington chpt.), Spl. Interest Group (Washington chpt. SIG Ada symposium program com. 1991), Internat. Rational Users Group (conf. chair 1988-90, chair 1990-92), Ada Semantic Interface Specification Working Group (publicity/meeting chair 1992—). Office: Logicon Inc 475 School St SW Washington DC 20024-2799

VLADIMIROV, VASILIY SERGEYEVICH, mathematician; b. Diaglevo, Leningrad, USSR, Jan. 9, 1923; s. Sergei and Maria (Sokolova) V.; m. Nina Ovsjannikova; children: Sergei, Michail. D of Math., Steklov Inst., Moscow, 1960; prof., Phys.-Tech. Inst., Moscow, 1965; academitian, Acad. Sc. of USSR, 1970. Jr. rsch. worker Steklov Inst. Math. Acad. Sci. USSR, Leningrad, 1948-50; sr. rsch. worker Steklov Inst. Math. Acad. Sci. USSR, Moscow, 1956-69, head dept. of math. phys., 1969, vice-dir., 1986-88, dir., 1988-93; head math. dept. All-union Rsch. Inst. Exptl. Physics, Arsamas-16, 1950-55; prof. Moscow Phys. Tech. Inst., 1964-87. Author: Mathematical Problems in One-Speed Theory of Transfer, 1961 (Liapount Gold medal 1971), Methods of Several Complex Variables, 1964, Equations of Mathematical Physics, 1967, Distributions in Mathematical Physics, 1976, Tauberian Therorum for Distributions, 1986. Soldier Air-Forces, 1941-45. Recipient Gold Star Sickle and Hammer Russian Govt., 1983, Lenin Orders, 1975, 83, Labour Red Banner Orders, 1967, 73, Order of the Great Pathriotic War II degree, 1985, State Prises, 1953, 87. Mem. Acad. Scis. of USSR, Saxsonian Acad. of Sci. (fgn.), Voievoding Acad. Sci. and Arts (fgn.), Soc. Math. and Physics of Chechoslovakia (hon.), Moscow Math. Soc., Internat. Assn. in Math. and Physics. Avocations: skiing, fishing. Office: Steklov Inst Math, Vavilov St 42, V-312 Moscow 117966 GSP-1, Russia*

VLIEGENTHART, JOHANNES FREDERIK G., educator bio-organic chemistry; b. Zuilen, Netherlands, Apr. 7, 1936; s. Frderik Johannes and Maria (Vervoort) V.; m. Magda Franciska Oosterhuis, Oct. 1, 1966; children: Marion, Maarten, Victor. Doctorandus, Univ. Utrecht, 1960, doctor sci, 1967; doctor h.c., Debrecen, Hungary, 1992, Lille, France, 1992. Prof. bio-organic chem., 1975—; rsch. dir. Byvoet Ctr. Biomolecular Rsch., Utrecht, Netherlands, 1988—. Editorial bd., mng. editor European Jour. Biochemistry, 1990—. Mem. Royal Netherlands Acad. Arts and Scis., Am. Soc. Biochemistry (hon.), Royal Swedish Acad. Scis. Home: Van Zyldreef 20, NL- 3981GX Bunnik The Netherlands Office: PO Box 80075, NL 3508 Te Utrecht The Netherlands

VOAS, SHARON JOYCE, environment and science reporter; b. Rapid City, S.D., June 6, 1955; d. J. Paul and Mary O. (Voas) Olinger; m. Terry Hertel, July 7, 1973 (div. 1977). BA summa cum laude, U. N.H., 1982; MS, Columbia U., 1988. State govt. reporter Lebanon (N.H.) Valley News, 1983; cts. and police reporter Concord (N.H.) Monitor, 1984-85; Hispanic/housing reporter Transcript-Telegram, Holyoke, Mass., 1986-87; sci./environment reporter Pitts. (Pa.) Post-Gazette, 1988—; Author: (newspaper series) Ten Years After Three-Mile Island, 1989, The New Zoos, 1989, Profit at Non-Profit Hospitals, 1991, To Fur of Not to Fur, 1992. Recipient 2d pl. N.E. Associated Press Newspaper Editors Assn., 1987; fellow Coun. for Advancement of Sci. Writing, 1987; recipient 1st pl. sci. writing award Am. Tentative Soc., 1988, Golden Quill 2d sci. reporting, 1990, 91, Matrix journalism award, 1993. Mem. AAAS, Investigative Reporters & Editors, Soc. Environ. Journalists, Phi Beta Kappa. Unitarian. Office: Pittsburgh Post-Gazette 50 Blvd of the Allies Pittsburgh PA 15218

VODONICK, EMIL J., engineer; b. Chisholm, Minn., Oct. 10, 1918; s. Joseph Leanard and Mary Ann (Furst) V.; m. Marion Winifred Smith, Feb. 4, 1944; children: E. John, David Smith, Mark Joseph, Virginia Susan, Laurie Ann, Mary Ellen. B in Civil Engring., U. Minn., 1942; MS, U. Calif., 1972. Cert. profl. engr. Calif., Ariz., Tex. Structural rsch. engr. various, 1942-58; staff asst. v.p. Sundstrand Machine Tool Co., Rockford, Ill., 1958-61; chief program mgr. Thiokol Chem. Corp., Brigham City, Utah, 1961-64; project engr. Autonetics Systems Div., Anaheim, Calif., 1964-70; environ. cons. NASA, Clear Lake, Tex., 1972-74; dir. environ. health Coconino County, Flagstaff, Ariz., 1974-76; engr. Air Resources Bd. Calif. State, El Monte, 1976-82; tech. v.p. Vodonick Assocs., Garden Grove, Calif., 1977--; instr. prof. Ala. Poly. Inst. Ext., Brimingham, 1948-49; instr. Rockford Coll., Ill., 1959-60; sr. engring. cons. SCI, Anaheim, Calif., 1979-80. Mem. County State Com., Brigham City, Utah, 1961-62; precinct capt. Garden Grove, 1988, 90. Republican. Roman Catholic. Achievements include development of the largest dynamic shaking systems in the U.S., development of down draft stove to reduce emissions, design and development to test aircraft using a free floating balanced loading. Home: 11091 Faye Ave Garden Grove CA 92640

VODYANOY, VITALY JACOB, biophysicist, educator; b. Kiev, Ukraine, USSR, June 2, 1941; came to U.S., 1979; s. Jacob and Vera (Reznik) V.; m. Galina Rubin, Apr. 22, 1967; 1 child, Valerie. MS in Physics, Moscow Physical Engring. Inst., 1964; PhD in Biophysics, Agrophysical Rsch. Inst., Leningrad, USSR, 1973. Asst. prof. Inst. of Semiconductors, Leningrad, USSR, 1965-72; assoc. prof. A.F. Ioffe Physicotech. U., Leningrad, 1972-78; sr. rsch. scientist NYU, 1979-82; rsch. assoc. U. Calif., Irvine, 1982-89; assoc. prof. Auburn (Ala.) U., 1989-93, prof., 1993—; ad hoc reviewer Nat. Science Found., Washington, 1985—. Author: (with others) Membrane Biophysics, 1971, Physics of Solid State and Neutron Scattering, 1974, Receptors Events and Transduction Mechanisms in Taste and Difaction, 1989, Molecular Electronics: Biosensors and Biocomputers, 1989, CNS Neurotransmitters and Neuromodulators, 1993; contbr. over 55 articles to profl. jours.; inventor device for film deposition. Recipient grants Nat. Sci. Found. U. Calif., 1982-85, 1985-88, U.S. Army Rsch. Office, 1985-88, U. Calif., 1986-88, U. Calif, 1988-92, U. Calif., Auburn U. Fellow Inst. for Biol. Detection Systems; mem. AAAS, Am. Assn. Nutritional Cons., The Am. Physical Soc., The Biophysical Soc.. Republican. Jewish. Avocation: medical herbs. Home: 541 Summertrees Dr Auburn AL 36830-6766 Office: Auburn U Coll of Vet Medicine 212 Green St Auburn AL 36849-6121

VOET, JUDITH GREENWALD, chemistry educator; b. N.Y.C., Mar. 10, 1941; d. Philip and Gertrude (Gevertzman) Greenwald; m. Donald Voet, Jan. 30, 1965; children: Wendy, Douglas. BS, Antioch U., 1963; PhD, Brandeis U., 1969. Postdoctoral fellow U. Pa., Phila., 1969-72; rsch. assoc. Haverford, Pa., 1972-74, Fox Chase Cancer Ctr., Phila., 1974-76; lectr. U. Pa., Phila., 1976-77, U. Del., Newark, 1977-78; assoc. prof. Swarthmore (Pa.) Coll., 1978—. Author: Biochemistry, 1990; contbr. articles to profl. jours. Office: Swarthmore Coll 500 College Ave Swarthmore PA 19081

VOGEL, EUGENIO EMILIO, physics educator; b. Temuco, Cautin, Chile, Jan. 4, 1946; s. Carlos Ernesto and Margarita (Matamala) V.; m. Berta Dominguez, May 17, 1967 (div. Nov. 1984); children: Tatiana, Natalia; m. Maria Angelica Osorio, Jan. 12, 1985; 1 child, Paulina. Licenciado, U. Concepcion, Chile, 1969; MA, Johns Hopkins U., Balt., 1972, PhD, 1975. Asst. prof. U. Andes, Merida, Venezuela, 1975-76; prof. U. Concepcion, Chile, 1976-81; vis. scholar Max Planck Inst., Stuttgart, Germany, 1979-80; assoc. mem. Centro Atomico Bariloche, Argentina, 1989—; prof. U. Frontera, Temuco, Chile, 1982—; bd. dirs. U. Frontera, Temuco, 1985-88, 90-92. Author: Origenes de la Mecanica Cuantica, 1986; contbr. numerous articles to profl. jours., presentations to scientific meetings. Cand. for Rector U. Frontera, Temuco, 1990. Ford Found. fellow, Concepcion, Chile, 1964-68, Gilman fellow Johns Hopkins U., Balt., 1970; recipient Sabbatical award Fundación Andes, Santiago, Chile, 1988-89, Rsch. award Fulbright Commn., Chile-U.S., 1991; Fondecyt grantee, Santiago, 1987—. Mem. Am. Phys. Soc. (life mem.), Chilean Phys. Soc. (bd. dirs. 1978-79, pres. 1991—). Avocations: photography, mountain walking. Office: U Frontera, Ave Francisco Salazar 01145, Temuco Cautin, Chile

VOGEL, GERHARD, HANS, pharmacologist, toxicologist; b. Bucarest, Roumania, Sept. 9, 1927; s. Eugen Georg and Emilie Katharina (Sturm) V.; m. Anna Theresia Zoller, Dec. 23, 1988. Pharmacist degree, U. Erlangen, 1951; physician degree, U. Tubingen, 1955; assoc. prof. degree, U. Marburg, 1967; honorary prof. degree, U. Frankfurt, 1979. Resident City Hosp., Heidenheim, Germany, 1956-57; senior scientist endocrinology lab. Dept. of Pharmacology, Hoechst AG, Frankfort, Germany, 1958-69, dir., 1967-78; dir. Pharma Rsch. Experimental Medicine, Hoechst AG, Frankfort, Germany, 1977-79, Pharma Preclinical Evaluation and Devel., Hoechst AG, Frankfort, Germany, 1980-88, Decision Bd. on Pharm. Devel., Hoechst AG, Frankfort, Germany, 1989-90; cons. Pharmaceutical and Medical Rsch. Devel., Hofheim, Germany, 1990—; mem. several scientific assns., Germany/USA, 1970—. Editor several books on workshop confs. and symposia; contbr. over 100 articles on biomechanics to profl. jours. Home: Mainzer Strasse 40, D 65719 Hofheim Germany Office: Hoechst AG, Bruening Strasse, D 65926 Frankfurt am Main Germany

VOGEL, H. VICTORIA, psychotherapist, educator. B.A., U. Md., 1968; M.A., NYU, 1970; M.A., NYU, 1975; M.Ed., Columbia U., 1982, postgrad., 1982—; cert., Am. Projective Drawing Inst., 1983. Art Therapist Childville, Bklyn., 1962-64; tchr., Montgomery County (Md.) Jr. High Sch., 1968-69; with High Sch. div. N.Y.C. Bd. Edn., 1970—, guidance counselor, instructor, psychotherapist in pvt. practice; clinical counseling cons. psychodiagnosis and devel. studies, The Modern School, 1984—; art/play therapist Hosp. Ctr. for Neuromuscular Disease and Devel. Disorders, 1987—; employment counselor-adminstr. N.Y. State Dept. Labor Concentrated Employment Program, 1971-72; intern psychotherapy and psychoanalysis psychiat. div. Cen. Islip Hosp., 1973-75; with Calif. Grad. Inst., L.A.; Columbia U. Tchrs. Coll., N.Y. intern psychol. counseling and rehab. N.J. Coll. Medicine, Newark, 1979. Mem. com. for spl. events NYU, 1989; participant clin. and artistic perspectives Am. Acad. Psychoanalysis Conf., 1990. Mem. APA, AAAS, Am. Psychol. Soc., Am. Orthopsychiat. Assn., Am. Soc. Group Psychotherapy & Psychodrama (publs. com. 1984—), Am. Counseling Assn., N.Y.C. Art Tchrs. Assn., Art/Play Therapy, Assn. Humanistic Psychology (exec. sec. 1981), Tchrs. Coll. Adminstrv. Women in Edn., Phi Delta Kappa (editor chpt. newsletter 1981-84, exec. sec. Columbia U. chpt. 1984—, chmn. nominating com. for chpt. officers 1986—, nominating com. 1991, pub. rels. exec. bd. dirs. 1991, rsch. rep. 1986—), Kappa Delta Pi. Author: The Never Ending Story of Alcohol, Drugs and Other Substance Abuse, 1992, Variant Sexual Behavior and the Aesthetic Modern Nudes, 1992, Psychological Science of Behavioral Interventions, 1993.

VOGEL, RONALD BRUCE, food products executive; b. Vancouver, Wash., Feb. 16, 1934; s. Joseph John and Thelma Mae (Karker) V.; m. Donita Dawn Schneider, Aug. 8, 1970 (dec. June 1974); 1 child, Cynthia Dawn; m. Karen Vogel, Feb. 14, 1992. BS in Chemistry, U. Wash., 1959. Glass maker Peuberthy Instrument Co., Seattle, 1959-60; lab. technician Gt. Western Malting Co., Vancouver, 1960-62, chief chemist, 1962-67, mgr. corp. quality control, 1967-72, mgr. customer svcs., 1972-77, v.p. customer svcs., 1977-79, v.p. sales, 1979-84, gen. mgr., 1984-89, pres., CEO, 1989—; chmn. bd. dirs. R&K Bus. Mgmt. Cons., Vancouver. Contbr. articles to profl. jours. Chmn. bd. dirs. Columbia Empire Jr. Achievement, Portland, Oreg., 1991-92. With U.S. Army, 1954-56. Recipient numerous awards. Mem. Master Brewers Assn. Am. Home: 30103 SE Shepherd Rd Washougal WA 98671

VOGELEY, CLYDE EICHER, JR., engineering educator, consultant; b. Pitts., Oct. 19, 1917; s. Clyde Eicher and Eva May (Reynolds) V.; m. Blanche Wormington Peters, Dec. 15, 1947; children: Eva Anne, Susan Elizabeth Steele. BFA in Art Edn., Carnegie Mellon U., 1940; BS in Engring. Physics, U. Pitts., 1944, PhD in Math., 1949. Rsch. engr. Westinghouse Rsch. Labs., East Pitts., Pa., 1944-54; adj. prof. math. U. Pitts., 1954-64; sr. scientist Bettis Atomic Power Lab., W. Mifflin, Pa., 1956-59; supr. tech. tng. Bettis Atomic Power Lab., W. Mifflin, 1959-71; mgr. Bettis Reactor Engring. Sch., W. Mifflin, 1971-77, dir., 1977-92; cons. U.S. Dept. Energy, Washington, 1992—; cons. Bettis Atomic Power Lab., W. Mifflin, 1954-56; U.S. Navy Nuclear Power Schs., Mare Island, Calif., Bainbridge, Md., 1959-69. Author: (grad. sch. course) Non-linear Differential Equations, 1954; (rev. text) Ordinary Differential Equations, Rev. edit. 5, Shock and Vibration Problems, Rev. Edit. 6, 1991. Pres., trustee Whitehall (Pa.) Pub. Libr., 1985. Recipient USN commendation, Naval Reactors Br., 1992. Mem. IEEE (life), Am. Phys. Soc., Associated Artists of Pitts., Pitts. Watercolor Soc., Sigma Pi Sigma, Sigma Tau, Sigm Xi. Presbyterian. Achievements include patents for Automatic Continuous Wave Radar Tracking System, Modulating Signals Passing Along Ridges Waveguides, Ridged Waveguide Matching Device, Method for Joining Several Ridged Waveguides, Antenna Feed Modulation Unit, others. Home: 185 Peach Dr Pittsburgh PA 15236

VOGELSTEIN, BERT, oncology educator. BS, U. Pa., 1970; MD, Johns Hopkins U. Rsch. assoc. Nat. Cancer Inst., 1976-78; prof. dept. oncology Johns Hopkins U. Sch. Medicine, Balt., 1978—; advisor Nat. Insts. Health Scientific Review Groups, Nat. Cancer Inst. Assoc. editor Genes, Chromosomes and Cancer; mem bd. reviewing editors Science Magazine; contbr. article to profl. jours. Recipient Gairdner Found. Internat. award Gairdner Found., 1992, ACS Medal Honor Am. Camcer Soc., 1992, Richard Lounsbery award Nat. Acad. Scis., 1993. Mem. NAS, Am. Acad. Arts Scis. Achievements include revolutionizing our understanding of complex genetic mutations that occur when an normal bowel epithelial cell is transformed into a malignant cell. Office: Johns Hopkins Univ Sch Medicine 720 Rutland Ave Baltimore MD 21205*

VOGL, ANNA KATHARINA, chemical engineer; b. Chgo., Apr. 1, 1967; d. Anton and Magdalena (Windt) V. B Engring., Lakehead U., Thunder Bay, Ont., Can., 1989. Automation engr. UOP, Riverside, Ill., 1989-91; simulation engr. UOP, Des Plaines, Ill., 1991—. Mem. Am. Inst. Chem. Engrs., Soc. Women Engrs., Instrument Soc. Am. Roman Catholic. Office: UOP 25 E Algonquin Rd Des Plaines IL 60017

VOGL, OTTO, polymer science and engineering educator; b. Traiskirchen, Austria, Nov. 6, 1927; came to U.S., 1953, naturalized, 1959; s. Franz and Leopoldine (Scholz) V.; m. Jane Cunningham, June 10, 1955; children: Eric, Yvonne. Ph.D., U. Vienna, 1950; Dr. rer. nat. h.c., U. Jena, 1983; Dr h.c., Poly. Inst., Iasi, Romania, 1992. Instr. U. Vienna, 1948-55; research asso. U. Mich., 1953-55, Princeton U., 1955-56; scientist E.I. Du Pont de Nemours & Co., Wilmington, Del., 1956-70; prof. polymer sci. and engring. U. Mass., 1970-83, prof. emeritus, 1983—; Herman F. Mark prof. polymer sci. Poly. U., Bklyn., 1983—; guest prof. Kyoto U., 1968, 80, Osaka U., 1968, Royal Inst. Stockholm, 1971, 87, U. Freiburg, Germany, 1973, U. Berlin, 1977, Strasbourg U., 1976, Tech. U. Dresden, 1982; guest Soviet Acad. Sci., 1973, Polish Acad. Sci., 1973, 75, Acad. Sci. Rumania, 1974, 76; cons. in field. Chmn. com. on macromolecular chemistry Nat. Acad. Sci. Author: Polyaldehydes, 1967, (with Furukawa) Polymerization of Heterocyclics, 1973, Ionic Polymerization, 1976, (with Simionescu) Radical Co and Graftpolymerization, 1978, (with Donaruma) Polymeric Drugs, 1978, (with Donaruma and Ottenbrite) Polymers in Biology and Medicine, 1980, (with Goldberg and Donaruma) Targeted Drugs, 1983, (with Immergut) Polymer Science in the Next Decade, 1987; contbr. articles to profl. jours. Recipient Fulbright award, 1976, Humboldt prize, 1977, Chemistry Pioneer award, 1985, Gold medal City of Vienna, Austria, 1986, Exner medal, 1987, Mark medal, 1989, Honor Ring, City of Traiskirchen, 1989; Japan Soc. Promotion of Sci. sr. fellow, 1980. Mem. Am. Chem. Soc. (chmn. div. polymer chemistry 1974, chmn. Conn. Valley sect. 1974, award applied polymer chemistry 1990), Am. Inst. Chemistry, AAAS, Austrian Chem. Soc., Japanese Soc. Polymer Sci. (life, award 1991), Sigma Xi, N.Y. Acad. Sci., Austrian Acad. Sci., Pacific Polymer Fedn. (pres.). Home: 349 Oxford Rd New Rochelle NY 10804-3324 Office: Poly Univ Six MetroTech Ctr Brooklyn NY 11201-2990

VOGT, ERICH WOLFGANG, physicist, university administrator; b. Steinbach, Man., Can., Nov. 12, 1929; s. Peter Andrew and Susanna (Reimer) V.; m. Barbara Mary Greenfield, Aug. 27, 1952; children: Edith Susan, Elizabeth Mary, David Eric, Jonathan Michael, Robert Jeremy. B.S., U. Man., 1951, M.S., 1952; Ph.D., Princeton U., 1955; D.Sc. (hon.), U. Man., 1982, Queen's U., 1984, Carleton U., 1988; LL.D. (hon.), U. Regina, 1986. Research officer Chalk River (Ont.) Nuclear Labs., 1956-65; prof. physics U. B.C., Vancouver, 1965—; Assoc. dir. TRIUMF Project, U. B.C., 1968-73, dir., 1981—, v.p. univ., 1975-81; chmn. Sci. Council B.C., 1978-80. Co-editor: Advances in Nuclear Physics, 1968—; Contbr. articles to profl. jours. Decorated officer Order of Can.; recipient Centennial medal of Can., 1967. Fellow Royal Soc. Can., Am. Phys. Soc.; mem. Canadian Assn. Physicists (past pres.). Office: Triumf, 4004 Wesbrook Mall, Vancouver, BC Canada V6T 2A3

VOGT, ROCHUS EUGEN, physicist, educator; b. Neckarelz, Germany, Dec. 21, 1929; came to U.S., 1953; s. Heinrich and Paula (Schaefer) V.; m. Micheline Alice Yvonne Bauduin, Sept. 6, 1958; children: Michele, Nicole. Student, U. Karlsruhe, Germany, 1950-52, U. Heidelberg, Germany, 1952-53; S.M., U. Chgo., 1957, Ph.D., 1961. Asst. prof. physics Calif. Inst. Tech., Pasadena, 1962-65, assoc. prof., 1965-70, prof., 1970—, R. Stanton Avery Disting. Service prof., 1982—, chmn. faculty, 1975-77, chief scientist Jet Propulsion Lab., 1977-78, chmn. div. physics, math. and astronomy, 1978-83; acting dir. Owens Valley Radio Obs., 1980-81; v.p. and provost Calif. Inst. Tech., Pasadena, 1983-87; vis. prof. physics MIT, 1988—; dir. Caltech/MIT Laser Interferometer Gravitational Wave Observatory Project, 1987—. Author: Cosmic Rays (in World Book Ency.), 1978, (with R.B. Leighton) Exercises in Introductory Physics, 1969; contbr. articles to profl. jours. Fulbright fellow, 1953-54; recipient Exceptional Sci. Achievement medal NASA, 1981, Profl. Achievement award U. Chgo. Alumni Assn., 1981. Fellow AAAS, A. Phys. Soc. Achievements include research in astrophysics and gravitation. Office: Calif Inst Tech 102-33 Pasadena CA 91125

VOIGT, HANS-DIETER, oil company executive, researcher, educator; b. Jüterbog, Germany, Oct. 23, 1941; s. Gustav and Helene (Atlas) V.; m. Edeltraut Lorenz, May 12, 1967; children: Kristin, Astrid. MS, Mining Acad., Freiberg, Germany, 1968, PhD, 1977. Group leader Erdöl-Erdgas Stendal, Germany, 1968-72; sect. leader Erdöl-Erdgas Gommern, Germany, 1973-90, project mgr., 1991; project mgr. VEGO OEL GmbH, Germany, 1992—; lectr. Mining Acad., 1979—; cons. Geothermie Neubrandenburg, Germany, 1987-90. Author: Geohydrodynamics, 1985, Heat and Mass Flow, 1992, Geothermics, 1991; also numerous articles; numerous patents in field. With German Army, 1962-63. Mem. German Sci. Soc. for Erdöl, Erdgas und Kohle, Soc. Petroleum Engrs. Avocations: reading, sports, travel. Home: B-Brecht-Strasse 14C, 39120 Magdeburg Germany Office: VEGO OEL GmbH, Madgeburger Chaussee, 39245 Gommern Germany

VOIGT, ROBERT G., numerical analyst; b. Olney, Ill., Dec. 21, 1939; s. Donald E. and Jean C. (Fishel) V.; m. Susan J. Strand, Aug. 25, 1962; children: Christine, Jennifer. BA, Wabash Coll., 1961; MS, Purdue U., 1963; PhD, U. Md., 1969. Mathematician Naval Ship R & D Ctr., Washington, 1962-69, 71-72; vis. prof. U. Md., College Park, 1969-71; asst. dir. Inst. for Computer Application in Sci. and Engring., Hampton, Va., 1973-83, assoc. dir., 1983-86, dir., 1986—; mem. tech. adv. bd. NSF, NASA, Dept. of Energy, and others. Co-author: Solution of Partial Differential Equations on Parallel and Vector Computers, 1985; co-editor 7 books; editor numerous jours.; contbr. 21 articles to profl. jours. Recipient Pub. Svc. award NASA, 1989. Mem. IEEE, Am. Math. Soc., Assn. for Computing Machinery, Soc. for Indsl. and Applied Math. (sec. 1987—). Office: Inst Computer Application NASA Langley Rsch Ctr Mail Stop 132 C Hampton VA 23665

VOIGTMAN, EDWARD GEORGE, JR., chemist educator; b. St. Louis, Dec. 26, 1949; s. Edward George Voigtman Sr. and Margaret Ann (Tracey) Dobsch; m. Janiece Lee Leach, May 31, 1975. BS, Rensselaer Polytech. Inst., 1972; PhD, U. Fla., 1979. Assoc. U. Fla., Gainesville, 1979-86; asst. prof. U. Mass., Amherst, 1986-92, assoc. prof., 1992—. Author: (simulation shareware) Voigt fx, 1992; mem. editorial working team Spectrochimica Acta Electronica, 1991—. Mem. Am. Chem. Soc., Soc. for Applied Spectroscopy. Office: U Mass LGRT-102 Chemistry Amherst MA 01003

VOISIN-LESTRINGANT, EMMANUELLE MARIE, pharmacologist, consultant; b. Rochefort, France, Jan. 1, 1956; came to U.S., 1986; d. Jacques Marie and Bernadette Andre (Gaudechon) Voisin; m. Pierre-Yves Lestringant, June 17, 1983; children: Adeline, Pauline. PhD, Univ. Paris XII, 1984. Researcher Inst. Gustave Roussy, Villejuif-Paris, 1979-85; preclin. coord. Servier Labs., Paris, 1985-86; researcher NIH, Bethesda, Md., 1986-88; pharmacologist FDA, Rockville, Md., 1988-90; dir. R & D Besins-Iscovesco U.S., Inc., Herndon, Va., 1990-91, v.p. R & D, 1991—; chmn. Com. on Reproductive Toxicology, FDA, Rockville, 1989-90, mem. Neurotoxicity Assessment Com., 1989-90, working group on Devel. of Guidance for Stereoisomers, 1989-90. Contbr. articles to Cancer Rsch., Regulatory Pharmacology & Toxicology, Food & Drug & Cosmetic Law jour. Named Fogarty Fellow, NIH, Bethesda, Md., 1987-88. Mem. Am. Coll. Clin. Pharmacology, N.Y. Acad. Scis., Drug Info. Assn., Regulatory Affairs Profl. Soc., Am. Assn. Pharm. Scientists. Achievements include major contributions in international drug development and registration, and of interspecies comparisons in drug development. Home: Quintiles Chelsea Pl 1007 Slater Rd Morrisville NC 27560-9745

VOJCAK, EDWARD DANIEL, metallurgist; b. Chgo., Mar. 15, 1960; s. Edward Donald and Joyce Denise (Dibiase) V. Student, U. Ill. Chgo. Circle, 1980-82; BS, UIC Chgo., 1983. Metallurgist Bliss & Laughlin Steel, Harvey, Ill., 1984-87, mgr. quality engring., 1989-92; supr. heat treat Brad Foote Gear Works, Cicero, Ill., 1987-89; plant metallurgist LaSalle Steel, Hammond, Ind., 1992—. Author symposia procs., 1992. Mem. ASTM (com. mem. 1984-89), Am. Soc. Metals, The Metall. Soc., Math. Assn. Am., U. Ill. Alumni Assn. Office: LaSalle Steel Co 1412 150th St Hammond IN 46327

VOJTA, PAUL ALAN, mathematics educator; b. Mpls., Sept. 30, 1957; s. Francis J. and Margaret L. V. B in Math., U. Minn., 1978; MA, Harvard U., 1980, PhD, 1983. Instr. Yale U., New Haven, 1983-86; postdoctoral fellow Math. Scis. Rsch. Inst., Berkeley, Calif., 1986-87; fellow Miller Inst. for Basic Rsch., Berkeley, 1987-89; assoc. prof. U. Calif., Berkeley, 1989-92, prof., 1992—; mem. Inst. for Advanced Study, Princeton, 1989-90. Author: Diophantine Approximations and Value Distribution Theory, 1987. Recipient perfect score Internat. Math. Olympiad, 1975. Mem. Am. Math. Soc. (Frank Nelson Cole Number Theory prize 1992), Math. Assn. Am., Phi Beta Kappa, Tau Beta Pi. Avocations: computer, skiing. Office: Univ Calif Dept of Math Berkeley CA 94720

VOJTECH, RICHARD JOSEPH, nuclear physicist; b. Havre de Grace, Md., Oct. 21, 1959; s. George Louis and Emily (Cerny) V.; m. Cynthia Dawn Turner, Dec. 17, 1983; 1 child, Richard Joseph Jr. BS in Physics, Loyola Coll., Balt., 1981; PhD in Nuclear Physics, U. Notre Dame, 1987; postgrad., SUNY, Stony Brook, 1987-88. Math. aide AMSAA-FEATD, U.S. Dept. Def., Aberdeen, Md., 1978-81; teaching asst. U. Notre Dame, Ind., 1981-83, rsch. asst., 1983-87; rsch. scientist SUNY, 1988-90; sr. scientist EG&G Energy Measurements, Washington, 1990-91, sci. specialist, 1992—. Contbr. articles to profl. jours. Presdl. and Md. Senatorial scholar Loyola Coll., 1977-81. Mem. Am. Phys. Soc., Am. Nuclear Soc., Health Physics Soc. (plenary). Achievements include project scientist for U.S. Dept. Energy Remote Sensing Laboratory. Office: EG&G Energy Measurements PO Box 380 Suitland MD 20752

VOLBERG, HERMAN WILLIAM, electronic engineer, consultant; b. Hilo, Hawaii, Apr. 6, 1925; s. Fred Joseph and Kathryn Thelma (Ludloff) V.; m. Louise Ethel Potter, Apr. 26, 1968; children: Michael, Lori. BSEE, U. Calif., Berkeley, 1949. Project engr. Naval Electronics Lab., San Diego, 1950-56; head solid state rsch. S.C. div. Gen. Dynamics, San Diego, 1956-60; founder Solidyne Solid State Instruments, La Jolla, Calif., 1958-60; founder, v.p. electronics div. Ametek/Straza, El Cajon, Calif., 1960-66; founder, cons. H.V. Cons., San Diego, 1966-69; sr. scientist Naval Ocean Systems Ctr., Oahu, Hawaii, 1970-77; chief scientist Integrated Scis. Corp., Santa Monica, Calif., 1978-80; founder, pres. Acoustic Systems Inc., Goleta, Calif., 1980-84; founder, pres. Invotron, Inc., Lafayette, Calif., Murray, Utah, 1984—; tech. dir. Reson, Inc., Santa Barbara, Calif., 1992—; lectr. solid state course UCLA and IBM, 1956-62; instr. Applied Tech. Inst., Columbia, Md., 1988—. Contbr. articles to IRE Bull., IEEE Ocean Electronics Symposium. Mem. adv. panels for advanced sonar systems and for high resolution sonars, USN, 1970-77. 1st lt. U.S. Army, 1944-47, ETO. Recipient award of merit Dept. Navy, 1973. Mem. IEEE, NRA, Planetary Soc., Assn. Old Crows, Masons, Elks. Achievements include patent for device for detecting and displaying the response of tissue to stimuli, high rate neutralizer (HIRAN), crane high-voltage sensing system. Avocations: flying, Judo black belt. Home and Office: 41 W 6830 S Salt Lake City UT 84107-7124

VOLDMAN, STEVEN HOWARD, electrical engineer; b. Rochester, N.Y., Sept. 8, 1957; s. Carl Jerome and Blossom (Passer) V.; m. Annie Curry Brown, July 1986; children: Aaron Samuel, Rachel Pesha. BS, U. Buffalo, 1979; MS, MIT, 1981, EE, 1982; MS in Engring. Physics, U. Vt., 1986, PhD, 1991; postgrad. resident study, IBM, 1988-91. Engring. asst. R.E. Ginna Nuclear plant Rochester Gas & Electric, N.Y., 1977, 78; rsch. assoc. MIT, Boston, 1979-81, rsch. assoc. high voltage rsch. lab., 1981-82; staff level engr. IBM, Burlington, Vt., 1982—, mem. 4-Mb DRAM devel. staff, 1985-88, mem. 16-Mb DRAM devel. staff, 1991-93, adv. engr., 1993—. Contbr. articles to Internat. Electron Device Meeting, Conf. on Elec. Insulation and Dielectric Phenomena, Transaction Elec. Devices, Computational Method in Elec. Engring., Numerical Analysis of Sem. Devices and Integrated Crcts., Device Rsch. Conf., Electrochem. Soc., Internat. Conf. on Microelectronic Test Structures, IEEE Transaction on Nuclear Sci., ECS Low Temperature Procs., Jour. Applied Physics, Jour. Electrostats. Discharge and Elec. Overstress Conf. Procs.; patentee in field. Mem. IEEE, AAAS, Vt. Inventor's Assn., Sigma Xi, Phi Eta Sigma, Tau Beta Pi. Democrat. Avocations: painting, photography, travel.

VOLKERT, MICHAEL RUDOLF, molecular geneticist; b. Jena, Germany, July 10, 1949; came to U.S., 1953; s. Rudolf Gerhard and Erika Wally (Hofmann) V.; m. Margaret Mary LeRoux, Aug. 21, 1971; children: Mark LeRoux, Emily LeRoux. BS, U. Wis., 1971; MS, Iowa State U., 1973; PhD, Rutgers U., 1977. Postdoctoral fellow U. Calif., Berkeley, 1977-80; sr. staff fellow NIH, Research Triangle Park, N.C., 1980-85; asst. prof. U. Mass. Med. Sch., Worcester, 1985-87, assoc. prof., 1987—. Mem. AAAS, Genetics Soc. of Am., Am. Soc. for Microbiology. Office: Dept Molecular Genetics and Microbiology 55 Lake Ave North Worcester MA 01655

VOLKMAN, ALVIN, pathologist, educator; b. Bklyn., June 10, 1926; s. Henry Phillip and Sarah Lucille (Silverstein) V.; m. Winifred Joan Grinnell, June 12, 1947 (div. Aug. 1967); children: Karl Frederick, Nicholas James, Rebecca Jane Evans, Margaret Rose Werrell, Deborah Ann Falls; m. Carol Ann Fishel, Jan. 26, 1973; 1 child, Natalie Fishel; 1 stepchild Jeffrey C. Moore. BS, Union Coll., 1947; MD, U. Buffalo, 1951; D.Philosophy, U. Oxford (Eng.), 1963. Diplomate Nat. Bd. Med. Examiners, Am. Bd. Pathology. Intern, Mt. Sinai Hosp., Cleve., 1951-52; research fellow dept. anatomy Western Res. U. Sch. Medicine, 1952-54; resident, then sr. resident, then asst. in pathology Peter Bent Brigham Hosp., Boston, 1956-60; asst. prof. pathology Columbia U. Coll. Physicians and Surgeons, 1960-66; asst. mem., then assoc. mem. Trudeau Inst., Saranac Lake, N.Y., 1966-67; prof. dept. pathology East Carolina U. Sch. Medicine, Greenville, N.C., 1977—, acting chmn. dept. pathology, 1989-90, assoc. dean for rsch. and grad. studies, 1989—; mem. NIH study sect. immunological scis., 1975-79, chmn., 1977-79. Served to lt. USNR, 1954-56. Am. Cancer Soc. scholar, 1961-63; Arth and Rheumat Found. fellow 1952-54. Mem. AAAS, Am. Assn. Pathologists, Am. Assn. Immunologists, Am. Soc. Hematology, Reticuloendothelial Soc., Am. Soc. Microbiologists, N.Y. Acad. Scis., Soc. Leukocyte Biology (hon. life). Contbr. articles to sci. jours. Office: East Carolina U Sch Medicine Brody Bldg Greenville NC 27858

VOLKMAN, DAVID J., immunology educator; b. Bklyn., Jan. 11, 1945; s. Clarence and Ruth (Fox) V.; m. Pamela Marian Bickerman, Jan. 29, 1967; children: Eric, Aaron Jon. BS in Math., Union Coll., 1966; PhD in Biochemistry, U. Rochester, 1971, MD with distinction, 1976. Diplomate Am. Bd. Internal Medicine, Am. Bd. Allergy and Immunology, Am. Bd. Diagnostic Lab. Immunology. Intern, 1976-77, resident internal medicine, 1977-78; rsch. assoc. Sloan-Kettering Inst., N.Y.C., 1971-72; resident medicine U. Pitts., 1976-78; clin. assoc. Nat. Inst. Allergy/Infectious Diseases, Bethesda, Md., 1978-82; sr. investigator NIAID, NIH, Bethesda, 1983-85; assoc. prof. medicine SUNY, Stony Brook, 1985—; mem. immunological scis. study sect. NIH, Bethesda, 1987-91; mem. sci. com. III Internat. AIDS Conf., Washington, 1987. Assoc. editor Jour. of Immunology, 1984-91; editorial bd. Clin. Immunology and Immunopathology, 1991—; contbr. articles to profl. jours. Sr. surgeon USPHS, 1978-85. Fellow Am. Coll. Physicians, Am. Acad. Allergy & Immunology. Jewish. Achievements include first cloning of human antigen-specific T lymphocytes; first application of antigen-toxin conjugates to abrogation of B cell reactivity; first demonstration of effect of HTLV-I infection on the function of specific T cell clones; research on human immune response to Lyme Disease. Home: 15 James Neck Rd Saint James NY 11780-9738 Office: SUNY Dept of Allergy Health Scis Ctr. T-16040 Stony Brook NY 11794

VOLPE, ERMINIO PETER, biologist, educator; b. N.Y.C., Apr. 7, 1927; s. Rocco and Rose (Ciano) V.; children—Laura Elizabeth, Lisa Lawton, John Peter. B.S., City Coll. N.Y., 1948; M.A., Columbia, 1949, Ph.D. (Newberry award 1952), 1952. Mem. faculty Newcomb Coll., 1952-60, prof. zoology, 1960-64; prof. biology Tulane U., New Orleans, 1964-81; W.R. Irby disting. prof. biology Tulane U., 1979-81, chmn. dept., 1964-66, 69-79, asso dean grad. sch., 1967-69; prof. basic med. scis. (genetics) Mercer U. Sch. Medicine, Macon, Ga., 1981—; cons. Nat. Commn. Undergrad. Edn. in Biol. Scis., 1964-71; mem. steering com. Biol. Scis. Curriculum Study, 1966-70; panelist NRC, 1967-70; mem. U.S. Nat. Commn. for UNESCO, 1968-72; regional lectr. Sigma Xi, 1970-72; lectr. Elderhostel, 1988—; chmn. Advanced Placement Test in Biology, Ednl. Testing Service, 1975-80. Author: (textbook) Understanding Evolution, 1985, Human Heredity and Birth Defects, 1971, Patterns and Experiments in Developmental Biology, 1973, Man, Nature, and Society, 1975, The Amphibian Embryo in Transplantation Immunity, 1980, Biology and Human Concerns, 1993, Patient in the Womb, 1984, Test-Tube Conception: A Blend of Love and Science, 1987; mem. editorial bd. jour. Copeia, 1962-63; asso. editor Jour. Exptl. Zoology, 1968-76, 84-85; editor (jour.) Am. Zoologist, 1975-80; contbr. articles to profl. jours. Served with USNR, 1945-46. Fellow AAAS; mem. Genetics Soc. Am., Am. Soc. Zoologists (pres. 1981), Am. Soc. Naturalists, Soc. Devel. Biology, Soc. Study Evolution, Am. Soc. for Cell Biology, Am. Soc. Human Genetics, Phi Beta Kappa (v.p. Tulane U. chpt. 1962), Sigma Xi (pres. Tulane U. chpt. 1964, faculty award 1972.). Office: Mercer Univ Sch Medicine PO Box 134 Macon GA 31207-0002

VOLPÉ, ROBERT, endocrinologist; b. Toronto, Ont., Can., Mar. 6, 1926; s. Aaron G. and Esther (Shulman) V.; m. Ruth Vera Pullan, Sept. 5, 1949; children: Catherine, Elizabeth, Peter, Edward, Rose Ellen. MD, U. Toronto, 1950. Intern U. Toronto, 1950-51, resident in medicine, 1951-52, 53-55, fellow in endocrinology, 1952-53, 55-57, sr. rsch. fellow dept. medicine, 1957-62, McPhedran fellow, 1957-65, asst. prof., 1962-68, assoc. prof., 1968-72, prof., 1972-92, prof. emeritus, 1992—, dir. div. endocrinology and metabolism, 1987-92, chmn. Centennial Com., 1987-88; attending staff St. Joseph's Hosp., Toronto, 1957-66; active staff Wellesley Hosp., Toronto, 1966—; dir. endocrinology rsch. lab. Wellesley Hosp., 1968—, physician-in-chief, 1974-87; trans-Atlantic vis. prof. Caledonia Endocrine Soc., 1985; Hashimoto Meml. lectr. Kyushu U., Fukuoka, Japan, 1992. Author: Systematic Endocrinology, 1973, 2d edit., 1979, Thyrotoxicosis, 1978, Auto-immunity and Endocrine Disease, 1985, Thyroid Function and Disease, 1989, Autoimmunity in Endocrine Disease, 1990; also over 260 rsch. articles on immunology of thyroid disease; editorial bd. Endocrine Pathology; past editorial bd. Jour. Clin. Endocrinology and Metabolism, Clin. Medicine, Clin. Endocrinology, Annals Internal Medicine. Served with Royal Can. Naval Vol. Res., 1943-45. Recipient Goldie medal for med. rsch. U. Toronto, 1971, Novo-Nordisk prize Irish Endocrine Soc., 1990; Med. Rsch. Coun. Can. grantee, 1955—. Fellow Royal Coll. Physicians Can. (coun. 1988—, chmn. annual meetings com. 1988—, sci. program com. 1988—, chmn. rsch. com. 1990—), Royal Coll. Physicians Edinburgh, Royal Soc. Medicine, ACP (gov. for Ont. 1978-83); mem. AAAS, Can. Soc. Endocrinology and Metabolism (past pres., Sandoz prize lectr. 1985, disting. svc. award 1990), Toronto Soc. Clin. Rsch. (Baxter prize lectr. 1984), Can. Soc. Clin. Investigation, Am. Thyroid Assn. (pres. 1980-81, disting. scientist award 1991), Assn. Am. Physicians, Endocrine Soc., Am. Fedn. Clin. Rsch., Can. Soc. Nuclear Medicine (Jamieson prize lectr. 1980), Can. Inst. Acad. Medicine, N.Y. Acad. Sci., European Thyroid Assn. (corr.), Latin Am. Thyroid Assn. (corr.), Soc. Endocrinology and Metabolism of Chile (hon.), Japan Endocrine Soc. (hon., gold medal 1986), Donalda Club, Alpine Ski Club (bd. dirs. 1987-89), U. Toronto Faculty Club. Home: 3 Daleberry Pl, Don Mills, ON Canada M3B 2A5 Office: Wellesley Hosp, Toronto, ON Canada M4Y 1J3

VOLPICELLI, FREDERICK GABRIEL, computer scientist; b. Bklyn., Nov. 9, 1946; s. Fred. G. and Irma J. (Carosa) V.; m. Elizabeth Anne Lamb, De. 27, 1969; children: Zachary X., Alicia M., Gabriel B. BA in Math., NYU, 1969; MS in Computer Sci., Pratt Inst. Engring. and Sci., 1973. Cert. computer profl. Internat. Assn. Computer Profls. Tchr. math. Aviation High Sch., Queens, N.Y., 1969-85; computer sci. lectr. Iona Coll., New Rochelle, N.Y., 1977-78; asst. prof. computer sci. Mercy Coll., Dobbs Ferry, N.Y., 1978-85; cons., owner Mikron Computer Cons., Harwich, Mass., 1985-; sr. product mgr. Quest Techs., 1993; v.p. ops. The Barrich Cos., Inc., Newport, R.I., 1991, Jerome J. Manning and Co., Inc., Boston, 1990-91; devel. officer, programmer Realty Works, Orleans, Mass., 1988-89; MLS supr. Cape Cod & Islands Bd. Realtors, 1992. Author: (computer program) Realty Works, 1988. Regents scholar, 1964-69. Mem. Assn. for Computing Machinery. Home: 3 Russell Dr Harwich MA 02645 Office: Mikron Computer Cons 3 Russell Dr Harwich MA 02645

VOLTMER, MICHAEL DALE, electric company executive; b. Des Moines, July 26, 1952; s. Robert D. and Kathy A. (Miller) V.; m. Joann H. Hove, Sept. 9, 1978; children: Gerad Frank, Anna Christine. B.S., Luther Coll., 1974. Founder, pres. Voltmer Electric Co., Decorah, Iowa, 1974—; Chmn. Winneshiek County Rep. Party, Decorah, 1982-83, fin. chmn., 1984—; pres. Sunflower Child Care Ctr. Inc., Decorah, 1979-80; mem. Sec. Planning and Zoning Commn., City of Decorah, 1983—, chmn. City Comprehensive Planning Com.; pres. Good Shephard Luth. Ch. Mem. Illuminating Engring. Soc., Nat. Fire Protection Assn., Oneota Golf and Country Club (pres. 1985), Oneota Valley Cultural Club (v.p.), Elks. Avocations: golf; racquetball. Home: 622 North St Decorah IA 52101-9806 Office: Voltmer Electric Inc 1826 St Hwy 9 Decorah IA 52101-9212

VOLTZ, STERLING ERNEST, physical chemist, researcher; b. Phila., Apr. 17, 1921; s. Harry John and Gertrude Irene (Derr) V.; m. Betty Morgan, Nov. 6, 1943; children: Sandra Elizabeth, Karen Lee. BA, Temple U., 1943, MA, 1947, PhD, 1952. Rsch. chemist Houdry Process Corp., Linwood, Pa., 1951-58; group leader Sun Oil Co., Marcus Hook, Pa., 1958-60; supervising engr. GE, Phila., 1960-62; cons. liaison scientist GE, Valley Forge, Pa., 1962-68; rsch. assoc. Mobil Rsch. & Devel. Corp., Paulsboro, N.J., 1968-80, adminstrv., 1980-86; pvt. practice Media, Pa., 1986—. Contbr. articles to Jour. Phys. Chem., Jour. Am. Chem. Soc., Jour. Organic Chemistry, Analytical Chemistry, Jour. Soc. Automotive Engrs., Jour. Chem. and Engring. Data, Jour. Am. Inst. Chem. Engrs. and others. Lt. (j.g.) USN, 1943-

46, ETO. Mem. AAAS, Am. Chem. Soc. (Phila. sect.), Catalysis Soc., Catalysis Club. Phila. (sec.-treas., chmn., dir. 1957-60), Am. Legion, Disabled Am. Vets., Sigma Xi. Achievements include 23 patents for Simulation of Catalytic Cracking Process, for Compatible Mixtures of Coal Liquids and Petroleum Based Fuels, for Reactivation of Automotive Exhaust Oxidation Catalyst, for Increasing Antiknock Value of Olefinic Gasoline, for Preparation of Aromatic Hydrocarbons, for Process for Dehydrocyclizing Heterocyclic Organic Compounds, for Alumina Stabilized by Thoria to Resist Alpha Alumina Formation, for Method of Treating Chromium Oxide, others; invention of plastic dry bag; co-development of commercial methanol-to-gasoline process, of fuel cell for space power applications, including first successful operation in space flight; development of catalysts and processes for petroleum and petrochemical conversions, of electronic apparatus to measure dielectric properties during oxidation reactions and establish reaction kinetics; establishment of relationship between catalytic properties, surface chemistry, and semiconductivity properties of metal oxide catalysts; research on catalytic systems for automotive emissions control including kinetic model of oxidation of carbon monoxide and hydrocarbons. Home: 6 E Glen Cir Media PA 19063-4712

VON BERG, ROBERT LEE, chemical engineer educator, nuclear engineer; b. Wheeling, W.Va., June 14, 1918; s. Leo and Ceynora Grey (Jenkins) Von B.; m. Kate Langley Hopkins, June 14, 1947; children: Eric, Gretchen, Karl, Karin. BS, W. Va. Univ., 1940, MS, 1941; ScD, M.I.T., 1944. Registered profl. engr., N.Y. Industrial engr. E.I. Du Pont de Nemours & Co., Wilmington, Del., 1944-46; prof. chem. engr., nuclear engr. Cornell Univ., Ithaca, N.Y., 1946-88; chem. engr. Dow Chem. Co., Midland, Mich., 1953-54; rsch. fellow Delft Tech. Inst., Delft, Netherlands, 1960-61; staff mem. Los Alamos (N.Mex.) Scientific Lab., 1967-68; vis. prof. Univ. Newcastle, Australia, 1974-75, Univ. Canterbury, Christchurch, New Zealand, 1982, 88, 92; prof. emeritus, 1988—; cons. Oak Ridge (Tenn.) Nat. Lab., 1950-51, Brookhaven (N.Y.) Nat. Lab., 1950-60, Los Alamos Scientific Lab., 1961-69, Dow Chem. Co., Midland, 1951-55, Bristol Labs., Syracuse, N.Y., 1962-66, Office of Saline Water, Washington, 1980-85. Contbr. articles to profl. jours. Mem. Am. Chem. Soc., Am. Inst. Chem. Engrs., Sigma Xi, Tau Beta Pi, Phi Lambda Upsilon. Presbyterian. Achievements include patent for the process for forming phenolformadehyde resinous condensates in continous tubular reactors. Home: 501 Hanshaw Rd Ithaca NY 14850 Office: Cornell U Sch Chem Engring Ithaca NY 14853

VON BERNUTH, ROBERT DEAN, agricultural engineering educator, consultant; b. Del Norte, Colo., Apr. 14, 1946; s. John Daniel and Bernice H. (Dunlap) von B.; m. Judy M. Wehrman, Dec. 27, 1969; children: Jeanie, Suzie. BSE, Colo. State U., 1968; MS, U. Idaho, 1970; MBA, Claremont (Calif.) Grad. Sch., 1980; PhD in Engring., U. Nebr., 1982. Registered profl. engr., Calif., Nebr. Agrl. product mgr. Rain Bird Sprinkler Mfg., Glendora, Calif., 1974-80; instr. agrl. engring. U. Nebr., Lincoln, 1980-82; from assoc. prof. to prof. U. Tenn., Knoxville, 1982-90; prof., chmn. Mich. State U., East Lansing, 1990—; v.p. Von-Sol Cons., Lincoln, 1980-82; prin. Von Bernuth Agrl. cons., Knoxville, East Lansing, 1982—. Patentee in field. With USNR, 1970-92, Vietnam. Decorated DFC (2); recipient Disting. Naval Grad. award USN Flight Program, Pensacola, Fla., 1970. Mem. ASCE, Am. Soc. Agrl. Engrs. (chair irrigation group), Irrigation Assn. (chair awards com.), Naval Reserve Assn. Avocations: flying, skiing, antique tractors. Office: Mich State U Agrl Engring 215 Farrall Hall East Lansing MI 48824

VONDRA, LAWRENCE STEVEN, aerospace engineer; b. Lincoln, Nebr., Nov. 20, 1963; s. Elmer Edward and Caroline (Mioni) V.; m. Deborah Sue Hickmott, June 13, 1987; children: Bradley Scott, Valerie Nicole. BS in Engring., Ariz. State U., 1989. Aerospace engr. Orbital Scis. Corp./Space Data Div., Chandler, Ariz., 1990—. Mem. AIAA. Baptist. Home: 2015 E University Dr #32 Tempe AZ 85281 Office: Orbital Scis Corp 3380 S Price Rd Chandler AZ 85248

VON ESCHEN, ROBERT LEROY, electrical engineer, consultant; b. Glasgow, Mont., Oct. 3, 1936; s. Leroy and Lillian Victoria (Eliason) Von E.; m. Carolyn Kay Frampton, Dec. 14, 1965; children: Eric Leroy, Marc Alfred. BSEE, Mont. State U., 1961. Registered profl. engr., Pa. Resident elec. engr. transmission and distbn. divs. Stanley Cons., Inc., 1962-63, cons. engr. fossil div., 1963-64, 66-68; cons. engr. hydro div. Stanley Cons., Inc., Monrovia, Liberia, 1965-66; cons. engr. fossil div. Gilbert Assocs., Inc., 1968-72, cons. engr. nuclear div., 1972-74, 77-84, project site mgr., 1974-77; cons. engr. nuclear div. United Energy Svcs. Corp., 1984-86; safety systems functional insp. nuclear div. United Energy Svcs. Corp., Marietta, Ga., 1986-91, mgmt. assessment, 1992; mgmt. assessment Mason & Hanger-Silas Mason Co., Inc., Amarillo, Tex., 1992-93, maintenance planning mgr., 1993—; tech. cons. World Bank, Monrovia, 1963; engring. cons. USN, Manila, 1967. Founder, dir. Madison (Ohio) Computer Soc., 1983-85; v.p., bd. dirs. N.E. coun. Boy Scouts Am., Painesville, 1983-85. Recipient Silver Beaver award Boy Scouts Am., 1981, 84. Mem. IEEE, NRA, NPSE, Nat. Assn. Ret. Persons, Soc. Am. Mil. Engrs., Am. Def. Preparedness Assn., Tex. Profl. Engrs. Soc., Masons (life), Elks, Scottish Rite (life), Shriners. Avocations: target and skeet shooting, drafting and electronics. Home: 3445 Gladstone Ln Amarillo TX 79121 Office: Mason & Hanger-Silas Mason Co Inc PO Box 30020 Amarillo TX 79177

VON FISCHER, GEORGE HERMAN, social psychologist, unified social systems scientist, management consultant, data processing executive; b. Cin., Oct. 24, 1935; s. George Henry and Dorothea Ann (Steffens) Von F.; m. Patricia L. Seward, June 21, 1961 (div. 1981); children: Gary L., Michael L. BBA, U. Cin., 1962, PhD, 1984; MA, U. Akron, 1968. CPA, Ohio. Auditor Arthur Young & Co., Cin., 1959-61; systems analyst AVCO Corp., Cin., 1961-63; exec. dir. long-range planning The Hoover Worldwide Corp., Canton, Ohio, 1964-71; instr. U. Cin., U. Akron, Kent State U., Edgecliff Coll., 1972-78; asst. prof. No. Ky. U., Highland Heights, 1978-80; dir. MIS Cin. Electronics, 1980-81; mgmt. cons. New Eng. Trade Adjustment Assistance Ctr., Boston, 1982; chief contractor ADP Evaluation Divsn., Def. Contract Mgmt. Dist. N.E., Boston, 1983—. Mem. corp. planning adv. com. Cleve. State U., 1968; mem. long-range planning adv. com. U. Cin., 1973-74; reader Recording for the Blind, Cambridge, Mass., 1992—. With U.S. Army, 1954-56. Named Best Actor of Yr. Ohio Com. Theater Orgn., 1968; fellow U. Cin., 1972-73. Mem. AAAS, Union of Concerned Scientists. Achievements include development of documentation and methodological procedures for government oversight of contractor ADP operations, management and facilitites; research in unified social science; facilitation in implementing total quality management in the federal government. Home: 570 Massachusetts Ave Boston MA 02118 Office: 395 Summer St Boston MA 02210

VON GOELER, EBERHARD, physics educator; b. Berlin, Feb. 22, 1930; s. Friedrich-Karl and Margarethe (von Knorre) von G.; m. Marleen D. Poole, Aug. 4, 1960; children: Friedel, John, Katherine. MS, U. Ill., 1955, PhD, 1961. Rsch. assoc. U. Ill., Urbana, 1960-61; rsch. scientist Desy, Hamburg, Fed. Republic Germany, 1961-63; asst. prof. physics Northeastern U., Boston, 1963-66, assoc. prof. physics, 1966-73, prof. physics, 1973—; vis. scientist Fermi Nat. Accelerator Lab., Batavia, Ill., 1971-72, Stanford Linear Accelerator Ctr., Palo Alto, Calif., 1978-79, U. Houston, 1986-87. Author, editor introductory physics lab. text; author, editor conf. procs. in field; contbr. articles to sci. jours. NSF grantee, 1966—. Mem. Am. Phys. Soc. Office: Northeastern U Dept Physics Boston MA 02115

VON HAARTMAN, HARRY ULF, international transportation company executive; b. Sundvall, Sweden, Mar. 19, 1942; s. Nils Erik and Ulla Margareta (Lavonius) von H.; m. Heidi Savolainen, Sept. 5, 1969; children: Harriet, Henri, Hans. MSc in Engring., Helsinki (Finland) U. Tech., 1967; BS in Econs., Helsinki U., 1971. Trainee Beton Monierbau, Braunschweig, Germany, 1965; researcher Oy Partek Ab, Parainen, 1966; engr. Erkki Juva oy, Helsinki, 1967; dep. mgr. Rakennustietosäätiö, Helsinki, 1967-70; sales mgr. Oy Lohja AB, Helsinki, 1970-72, R&D mgr., 1972-73, export mgr., 1974-77; dir. material adminstrn. Oy Lohja AB, Virkkala, Finland, 1977-82; pres. Oy Victor Ek ab, Helsinki, 1982—; vice chmn. Finnish Forw Assn., Helsinki, 1984—; mem. adminstrv. coun. Yrittäjäin Fennia, Helsinki, 1988—; chmn. Travek Travel Bur., 1982; mem. bd. Cargo Express, Helsinki, 1987-89, Freeport of Finland Ltd., 1991—; vice chmn. Employers Confedn., 1986—. Chief author: Lightweight Concrete, 1975. Lt. Finnish inf., 1961-62. Named to Order of Lions in Finland, 1st Class. Mem. Helsinki C. of C. (adminstrv. coun. 1989—). Liberal. Avocations: sports, bridge. Office: Oy Victor Ek Ab, Kanavaranta 9, 00160 Helsinki Finland

VON HEIMBURG, ROGER LYLE, surgeon; b. Chgo., Feb. 5, 1931; s. Franklin Dederick and Alice Julia (Zebuhr) von H.; m. Mary Ellen Janson, July 12, 1952; children: Mary Deborah, Donald Franklin. AB, Johns Hopkins U., 1951, MD, 1955; MS in Surgery, U. Minn., Rochester, 1964. Diplomate Am. Bd. Surgery. Intern Johns Hopkins Hosp., Balt., 1955-56; resident in surgery Mayo Clinic, Rochester, 1958-62, chief resident in surgery, 1962, asst. to staff in surgery, 1962-64; practice medicine specializing in surgery Green Bay, Wis., 1964—; staff St. Vincent Hosp., Green Bay, 1964—, Bellin Meml. Hosp., Green Bay, 1964—. Contbr. articles to profl. jours. Mem. State Bd. of Health Care Info., 1988-91; reapptd., 1991—. Lt. USNR, 1956-58. Fellow ACS; mem. State Med. Soc. Wis. (bd. dirs. 1980-89, vice-chmn. 1983-87, chmn. 1987-89, pres.-elect 1989-90, pres. 1990-91), Wis. Chpt. ACS (v.p. 1985-86, pres.-elect 1986-88, pres. 1988-90), Brown County Med. Soc. (pres. 1986), Wis. Surg. Soc. (coun. mem. 1987-90). Republican. Methodist. Avocations: piano, auto repair. Home: 344 Terraview Dr Green Bay WI 54301-1523 Office: Webster Clinic 900 S Webster Ave Green Bay WI 54301-3508

VON HILSHEIMER, GEORGE EDWIN, III, neuropsychologist; b. West Palm Beach, Fla., Aug. 15, 1934; s. George E. Jr. and Dorothy Sue (Bridges) Von H.; m. Catherine Jean Munson, Dec. 27, 1968 (div. Oct. 1987); children: Dana Germaine, George E. IV, Alexandra; m. Jonnie Mae Warner, June 29, 1991. BA, U. Miami, 1955; PhD, Saybrook Inst., 1977. Diplomate Acad. Psychosomatic Medicine, Am. Bd. Behavioral Medicine, Am. Acad. Pain Mgmt., Am. Bd. Cert. Managed Care Providers. Sr. minister Humanitas, N.Y.C., 1959-64; headmaster Summerlane Sch., North Branch, N.Y., 1964-69; supt. Green Valley Sch., Orange City, Fla., 1969-74; neuropsychologist Growth Insts., Maitland, Fla., 1974—; cons. Sci. Adv. Bd. EPA, Washington, 1974-84; chmn. Certification Bd., Internat. Coll. Environ. Medicine, 1991—; mem. Bd. Assn. Diagnostic Effective and Brief Therapy, dir. curriculum, 1993—. Author: How to Live With Your Special Child, 1970, Understanding Problems of Children, 1975, Allergy, Toxins and the LD Child, 1977, Psychobiology of Delinquents, 1978, Depression Is Not a Disease, 1989, Brief Therapy, 1993; editor Human Learning, Washington. Mem. spl. bd. Fla. Symphony Orch., 1992—. With U.S. Army, 1957-59. Fellow Royal Soc. Health, Internat. Coll. Applied Nutrition, Acad. Psychosomatic Medicine; mem. Toastmasters Club, Rotary, Phi Kappa Phi, Omicron Delta Kappa, Alpha Sigma Phi. Mem. Ch. of Brethren. Achievements include establishment of minor physical anomalies as significant predictor of physical and mental disease; demonstrated that treatment by neurofeedback significantly reduced criminal recidivism and that delinquency is a functio of physical disease. Home: 160 W Trotters Dr Maitland FL 32751-5736 Office: AAT 175 Lookout Pl # 1 Maitland FL 32751-5533

VON HIPPEL, FRANK NIELS, public and international affairs educator; b. Cambridge, Mass., Dec. 26, 1937; s. Arthur Robert and Dagmar (Franck) Von H.; m. Patricia Bardi, June, 1987; 1 child from previous marriage, Paul Thomas. S.B., MIT, 1959; Ph.D., Oxford U., 1962. Research assoc. U. Chgo., 1962-64; research assoc. Cornell U., Ithaca, N.Y., 1964-66; asst. prof. Stanford U., Calif., 1966-69; assoc. physicist Argonne Nat. Lab., Ill., 1970-73; research physicist Princeton U., N.J., 1974-83, prof. pub. and internat. affairs, 1983—; dir. Bull. of Atomic Scientists, Chgo., 1983-86, chmn. editorial bd., 1991—; mem. editorial bd. Science and Global Security, 1989—. Author: Advice and Dissent, 1974, Citizen Scientist, 1991; contbr. articles to profl. jours. Rhodes scholar, 1959-62; fellow MacArthur Found., 1993. Fellow AAAS (bd. dirs. 1987-88), Am. Phys. Soc. (Forum award 1977); mem. Fedn. Am. Scientists (chmn. 1979-84, Pub. Svc. award 1989), Fedn. Am. Scientists Fund (chmn. 1986—). Home: 3 University Way Princeton Junction NJ 08550-1617 Office: Princeton U Dept Pub Inter Affairs Princeton NJ 08544

VON KEHL, INGE, toxicologist-pharmacologist; b. Frankfurt, Main, Fed. Republic of Germany, Jan. 16, 1933; came to U.S., 1954; s. Karl and Käthe (Greiner) K.; divorced; 1 child, Timothy. BS in Chemistry, Biology, NE Mo. State U., 1959; MS, U. Idaho, 1963; PhD in Eurotech. Rsch., Southhampton U., 1988. Prof. Kirksville Mo. Coll. of Medicine & Surgery, 1963-85; prof. pharmacy U. San Marco, Lima, Peru, 1985-87; instr. chemistry U. SW La., Lafayette, 1987-88; rsch. assoc. U. Mo., Kansas City, 1988-90; quality assurance officer Heritage Found., Kansas City, 1990—. Editor: The Biology and Chemistry of Hydroxamic acids, 1982; contbr. articles to profl. jours.; editorial cons. Jour. Am. Osteopathic Assn., 1971-82; abstractor Am. Chem. Soc. Nat. Def. Act fellow U.S. Govt., 1961. Mem. Am. Chem. Soc., Am. Soc. Exptl. Pharmacology, Sigma Xi. Achievements include 3 patents on the synthesis of phenylpropyl hydroxamic acids. Home: 10012 Oakdell St Kansas City MO 64114-4145

VON KUTZLEBEN, SIEGFRIED EDWIN, engineering consultant; b. Veckerhagen, Germany, May 19, 1920; s. Erich Melchior and Katherina Helene (Klotz) von K.; m. Ursula Herta, July 21, 1915; children: Bernd E., Roy E., Werner E. BS, NYU, 1951. Naval disarmament control officer U.S./Brit. Mil. Govt. Germany, 1945-47; plant engr. Colgate-Palmolive-Peet, Jersey City, 1951; sr. sales engr. C.E. Lummus Nederland, The Hague, The Netherlands, 1951-58, mng. dir., 1958-71; exec. v.p. C.E. Lummus Co., Bloomfield, N.J., 1971-74; pres. C.E. Constrn. Internat., Bloomfield, 1974-81; group v.p. C.E. Lummus Group, Bloomfield, 1974-81; dir. Fluor Europe Ltd., London, 1981-82; mng. dir. Fluor Nederland B.V., Haarlem, The Netherlands, 1982-85; dir. Kaefer Techs., Houston, 1985-90, Bremtex Corp., Houston, 1990—; cons. to industry. Bd. deacons Am. Protestant Ch. The Hague, 1964-68; trustee N.J. Independent Colls., 1974-75, Suomi Coll., Hancock, Mich., 1978-81; pres. Am. C. of C. The Netherlands, 1970-71, 83-85. Decorated Officer Order Oranje-Nassau (Netherlands), 1971, Comdr. Order Lion of Finland (Finland), 1973. Mem. Am. Inst. Chem. Engrs. Lutheran. Avocations: swimming, sailing, hiking.

VON LINSOWE, MARINA DOROTHY, information systems consultant; b. Indpls., July 21, 1952; d. Carl Victor and Dorothy Mae (Quinn) von Linsowe; m. Clayton Albert Wilson IV, Aug. 11, 1990; 1 dau., Kira Christina von Linsowe. Student Am. River Coll., Portland State U. Verbal operator Credit Bur. Metro, San Jose, Calif. and Portland, Oreg., 1970-72; computer clk. Security Pacific Bank, San Jose, 1972-73; proof operator Crocker Bank, Seaside, Calif., 1973-74; proof supr. Great Western Bank, Portland, 1974-75; bookkeeper The Clothes Horse, Portland, 1976-78; computer operator Harsh Investment Co., Portland, 1978-79; data processing mgr. Portland Fish Co., 1979-81; data processing mgr. J & W Sci. Inc., Rancho Cordova, Calif., 1981-83; search and recruit specialist, data processing mgr. Re:Search Exec. Recruiters, Sacramento, Calif., 1983; sr. systems analyst Unisys Corp. (formerly Burroughs), 1983-91; sr. systems cons. FileNet Corp., Portland, Oreg., 1991-92; owner Optimal System Svcs., Portland, Oreg., 1992—; mfg. specialist, computer conversion cons., Portland. First violinist Am. River Orch. Recipient Bank of Am. Music award, 1970. Mem. NAFE, APICS, Am. Prodn. and Inventory Control Soc., Am. Mgrs. Assn., MENSA, Data Processing Mgmt. Assn. Republican. Lutheran. Home: 5902 SW Canby St Portland OR 97219-1264

VON OHAIN, HANS JOACHIM P., aerospace scientist; b. Dessau, Fed. Republic of Germany, Dec. 14, 1911; came to U.S., 1947; s. Wolf and Katherine L. (Nagel) von O.; m. Hanny Lemke, Nov. 26, 1949; children: Stephen, Christopher, Catherine, Stephanie. PhD in physics and aerodyn., U. Goettingen, Fed. Republic of Germany, 1935; DSc (hon.), U. W.Va., 1982. Head jet proplusion devel. div. Heinkel Aircraft Corp., Rostock, Fed. Republic of Germany, 1935-45; cons. U.S. Navy, Stuttgart, Fed. Republic of Germany, 1945-46; chief scientist USAF, Dayton, Ohio, 1947-79, with Aerospace Research Lab., 1963-75, with Propulsion Lab., 1975-79; now aerospace research cons. U. Dayton Research Inst. Multiple patents in field; contbr. article to profl. jours. Recipient R. Tom Sawyer award ASME, 1990. Fellow AIAA (hon., Goddard prize 1966, Daniel Guggenheim Medal award 1991); mem. NAE (Charles Stark Draper prize 1991). Club: Wings (N.Y.C.). Home: 3305 Nan Pablo Dr Melbourne FL 32934 Office: U Dayton Research Inst 300 College Park Ave Dayton OH 45469-0002

VON RIESEN, DANIEL D., chemistry educator; b. Beatrice, Nebr., Nov. 20, 1943; s. John H. and Rosemary (Brazelton) Von R.; m. Karen K. Stone (div.); m. Lois J. Puccio, Sept. 10, 1987. BA, Hastings Coll., 1965; PhD, U. Nebr., 1971. Asst. prof. Hamilton Coll., Clinton, N.Y., 1971; mem. faculty Roger Williams U., Bristol, R.I., 1972—, chmn. dept. chemistry, 1975—; mktg. specialist Isco Inc., Lincoln, Nebr., 1985-86. Mem. Am. Chem. Soc., New England Assn. Chemistry Tchrs. Office: Roger Williams U Bristol RI 02809

VON SCHULLER-GOETZBURG, VIKTORIN WOLFGANG, economist, consultant; b. Vienna, July 1, 1924; s. Viktorin Stefan and Paula Judith (Binder) von S-G.; M.S., U. Paris, 1948; Ph.D., U. Vienna, 1954; M.B.A., Vienna Bus. Sch., 1957. Asst. prof. Vienna Bus. Sch., 1954-58; internat. fellow Stanford Research Inst., Menlo Park, Calif., 1959-60; head econs. and mktg. research IBB, Vienna and Fiduciaire Internationale, Paris, 1960-67; with SRI Internat., Zurich, 1968-91, mgr. environ. and spl. studies Chem. Industries div. Europe, 1978-87, mgr. environ. and spl. studies Process Industries div., 1987-89, sr. cons., 1989-91; ind. cons., 1992—; cons. B.A.U., Dutch Ministry of Environment, EEC Commn., Euratom, I.A.R.C., UNEP. Smith-Mundt fellow, 1959. Mem. European Indsl. Mktg. Research Assn., European Chem. Mktg. Research Assn., Austrian Chem. Soc., German Chem. Soc., Austrian Assn. Graduated Economists. Clubs: Ancient Order St. George, Gesellschaft der Musikfreunde, Austrian Touring. Contbr. articles to profl. jours. Home: 3 Nibelungengasse, A-1010 Vienna Austria Office: 142 Allenmoosstrasse, CH8050 Zurich Switzerland

VON SEGESSER, LUDWIG KARL, cardiovascular surgeon; b. Lucerne, Switzerland, Mar. 15, 1952; s. Ludwig and Mathilde (Glutz) von S.; m. Marie Dinh, June 27, 1979; children: Ludwig, Jeanne. MD, U. Basel, Switzerland, 1979, PD, 1989. Cert. Surg. Bd. Switzerland, 1985. Resident Kantonsspital Obwalden, Sarnen, Switzerland, 1979; resident Univ. Hosp., Geneva, 1980-83, mem. staff cardiovascular surgery, 1983-85; fellow cardiovascular surgery Tex. Heart Inst., Houston, 1986; staff surgeon Clinic for Cardiovascular Surgery Univ. Hosp., Zurich, 1987—; Mem. European Bd. of Cardiovascular Perfusion, chmn. 1993—. Author: Arterial Grafting for Myocardial Revascularization, 1990; contbr. over 200 articles to profl. jours. Recipient Goetz-Preis award U. Zurich, 1991, award Swiss Cardiology Found., 1991. Fellow ACS; mem. Swiss Soc. Thoracic and Cardiovascular Surgery (sec. 1989—), Swiss Soc. Surgery, Swiss Soc. of Cardiology, German Soc. Thoracic and Cardiovascular Surgery, European Assn. Cardio-Thoracic Surgery, European Soc. Artificial Organs, Soc. Critical Care Medicine, Internat. Soc. Heart & Lung Transplantation, Internat. Soc. Surgery, The Soc. Thoracic Surgeons, Assn. Advancement Med. Instrn., Am. Soc. Artificial Internal Organs. Achievements include research on improvement of blood exposed surfaces in perfusion devices. Office: Univ Hosp Clin Cardiovascul Surg, Ramistrasse 100, Zurich CH-8091, Switzerland

VON SPRECHER, ANDREAS, chemist; b. Bern, Switzerland, Sept. 13, 1951; s. Hans and Bertha (Kuster) von S.; m. Sabine Schneider, Feb. 12, 1986; children: Raeto, Stephanie. PhD, U. Basel, 1978. Postdoctoral fellow U. Ariz., Tucson, 1978-79; rsch. fellow CIBA-GEIGY Ltd., Basel, 1979-87, sr. rsch. fellow, 1988-91, head chemistry rsch. allergy/asthma, 1991—. Contbr. articles to profl. jours. Mem. Neue Schweizerische Chemische Gesellschaft, Am. Chem. Soc. Achievements include patents in the field. Office: CIBA-GEIGY Ltd, K 136 4 81, CH 4002 Basel Switzerland

VON TAAFFE-ROSSMANN, COSIMA T., physician, writer, inventor; b. Kuklov, Slovakia, Czechoslovakia, Nov. 21, 1944; came to U.S., 1988; d. Theophil and Marianna Hajossy; m. Charles Boris Rossmann, Oct. 19, 1979; children: Nathalie Nissa Cora, Nadine Nicole. MD, Purkyne U., Brno, Czechoslovakia, 1967. Intern Valtice (Czechoslovakia) Gen. Hosp., 1967-68, resident ob-gyn, 1968-69; med. researcher Kidney Disease Inst., Albany, N.Y., 1970-71; resident internal medicine Valtice Gen. Hosp., 1972-73; gen. practice Nat. Health System, Czechoslovakia, 1973-74; pvt. practice West Germany, 1974-80; med. officer Baragwanath Hosp., Johannesburg, South Africa, 1984-85, Edendale Hosp., Pietermaritzburg, South Africa, 1985-86; pvt. practice Huntingburg, Ind., 1988—; med. researcher, 1966—. Contbr. articles on medicine to profl. jours.; inventor, patentee in field. Office: Medical Arts Plz Huntingburg IN 47542

VON TERSCH, FRANCES KNIGHT, analytical chemist; b. Akron, Ohio, Aug. 24, 1963; d. John Gibbons Jr. and Mary (Edson) Knight; m. Robert Lee Von Tersch, Dec. 7, 1991. BA, Agnes Scott Coll., Atlanta, 1985; PhD, U. Ga., 1990. Teaching asst. U. Ga., Athens, 1985-86, rsch. asst., 1986-90; sr. rsch. chemist Ciba-Geigy Corp., McIntosh, Ala., 1990—. Mem. AAAS, Am. Chem. Soc., Sigma Delta Epsilon. Republican. Presbyterian. Achievements include research in analytical chemistry. Office: Ciba-Geigy Corp McIntosh AL 36553

VON VICZAY, MARIKA (ILONA), naturopathic medical doctor, physician; b. Papa, Hungary, Sept. 30, 1935; came to U.S., 1957; d. Joseph Viczay and Matild Hegedüs; m. Howard C. Warren, Aug. 8, 1966 (div. 1969); 1 child, Lillian B. Warren. Masters degree, U. Budapest, Hungary, 1956; Doctor degree, NYU, 1972; MD (hon.), Gesundheltszentrum, Austria, 1992. Adminstr. Ancora (N.J.) Hosp., 1958-60; pres. Rsch. Inner Serenity, Chgo., 1972-85; dir. Isis Health and Rejuvenation Clinic, Asheville, N.C., 1985—; dep. gov. ABI Rsch. Assoc., Raleigh, N.C.; rschr. Bio-Oxidative Medicine, Dallas, 1990. Contbr. articles to profl. jours. Recipient Superior Achievement award Life Force Co., 1986, Answer to Cancer Rsch. award N.Y. Acad. Scis., 1986. Mem. MAYR-Artze Physicians Med. Assn., Am. Naturopathic Med. Assn., Acad. Hollistic Med. Assn., Occidental Inst. Rsch. Found. (assoc.). Achievements include research in the lymphatic system and treatment resolution "electro-lymphatic therapy"; in health restoration-age prevention through a complete detoxification system "the Isis method". Home: 1 Celia Pl Asheville NC 28801 Office: Isis Health Rejuvenation Clinic 16 Arlington Asheville NC 28801

VON WEIZSÄCKER, ERNST ULRICH, environmental scientist; b. Zürich, Switzerland, June 25, 1939; s. Carl Friedrich and Gundalena (Wille) Von W.; m. Christine Radtke; children: Jakob, Paula, Adam, Franz, Maria. Diploma in Physics, U. Hamburg, Fed. Republic of Germany, 1965; PhD in Biology, U. Freiburg, Fed. Republic of Germany, 1969. Fellow Protestant Interdisc Rsch. Inst., Heidelberg, Fed. Republic of Germany, 1969-72; prof. Biology U. Essen, Fed. Republic of Germany, 1972-75; pres. U. Kassel, Fed. Republic of Germany, 1975-80; dir. UN Ctr. for Sci. and Tech., N.Y.C., 1981-84, Inst. for European Environ. Policy, Bonn, 1984-91; pres. for climate, environ. and energy Wuppertal (Fed. Republic Germany) Inst., 1991—. Author, editor: Offene Systeme I, 1974, Erdpolitik, 1989. Recipient Pfaff prize Pfaff Found., 1977, Premio de Natura award, 1989. Mem. AAAS, German Zool. Soc., Club of Rome. Mem. Social Dem. Party. Lutheran. Avocation: chess. Office: Wuppertal Inst, Doeppersberg 19, 5600 Wuppertal Germany

VON WINKLE, WILLIAM ANTON, electrical engineer, educator; b. Bridgeport, Conn., Nov. 29, 1928; s. William Mathias and Lillian (Wiggleworth) Von W.; m. Arlene McDermott, July 24, 1950; children: Linda, Lee Ellen, Donna, Nancy, William, Karl, Patricia, Eric. B of Engring., Yale U., 1950, M of Engring., 1952; PhD, U. Calif., Berkeley, 1961. Engr. Navy Underwater Sound Lab., New London, Conn., 1952-66, dir. for rsch., 1966-89; chief scientist Old Ironsides, Inc., New London, 1989—; prof. math. and engring. U. New Haven, West Haven, Conn., 1952—, U. Conn., Storrs, 1953—. Editor 2 IEEE spl. issues, 1977, 84. Mem. Water and Pollution Commn., New London, 1990—. Recipient Adm. Martel award Nat. Security Industry Assn., 1988; USN Dept. grad. fellow, 1957-59. Fellow IEEE, Acoustical Soc. Am.; mem. Cosmos Club, Conn. Acad. of Sci. and Engring. (charter), Sigma Xi. Achievements include patents for sonar performance computer and deep integrated virtual array. Home: 105 Gardner Ave New London CT 06320 Office: WA Von Winkle 105 Gardner Ave New London CT 06320

VON ZUR MÜHLEN, ALEXANDER MEINHARD, physician, internal medicine educator; b. Riga, Latvia, May 13, 1936; arrived in Germany, 1939.; s. Alexander and Kira (Velitschkowski) von zur M.; m. Karen Berg, 1958 (div. 1977); children: Insa, Friederike, Patrick; m. Ulrike Warnecke, 1977; children: Constantin, Nicolas. Grad., Med. Sch. Freiburg, Fed. Republic Germany, 1963. Asst. Med. Sch. Göttingen, Fed. Republic Germany, 1965-74, sr. asst., 1974; prof. Internal Medicine Med. Sch. Hannover, Fed. Republic Germany, 1974—. Contbr. articles to profl. jours. Mem. German Soc. Endocrinology, German Soc. Internal Medicine. Office:

Med Hochschule Hannover, Konstanty-Gutschow Str 8, D-3000 Hannover Germany

VOO, LIMING M., biomedical engineer, researcher; b. Shanghai, China, Jan. 14, 1959; came to U.S., 1985; s. Da-Hong Wu and Yu-Hua Cao. BSME, Shanghai U. Sci. and Tech., 1982; MSME, U. Iowa, 1987. Mech. engr. Shanghai Special Machine Tools, 1982-85; teaching asst. U. Iowa, Iowa City, 1987-92, rsch. asst., 1985—; presenter in field. Vice chmn. Young Engrs. Adv. Com., Shanghai, 1983-85. Internat. Student scholar U. Iowa, 1985-87. Mem. ASME (student), Sigma Xi (assoc.). Office: U Iowa Dept Biomed Engring 1202 Engineering Bldg Iowa City IA 52242

VOOK, FREDERICK WERNER, electrical engineer; b. Syracuse, N.Y., Nov. 3, 1966; s. Richard Werner and Julia Elizabeth (Deskins) V. BS, Syracuse U., 1987; MSc, Ohio State U., 1989, PhD, 1992. Grad. rsch. asst. dept. elec. engring. Ohio State U., Columbus, 1987-92; lead engr. Motorola Wireless Data Group, Arlington Heights, Ill., 1992—. U. fellow Ohio State U., 1987-88, grad. fellow IBM, 1990-92. Mem. IEEE (reviewer 1991—), Sigma Xi, Tau Beta Pi.

VOORHEES, FRANK RAY, biology educator; b. Pekin, Ill., Dec. 8, 1935; s. Frank Holten and Rachel (Haig) V.; m. Ruth N. Stevenson, Aug. 15, 1958 (dec. Jan. 1990); children: Rachel, Christie; m. Suzanne K. Templeton, May 2, 1992. MS, U. Ill., 1968, PhD, 1969. Asst. prof. biology Knox Coll., Galesburg, Ill., 1969-75; prof. biology Cen. Mo. State U., Warrensburg, 1975—. Contbr. articles to profl. jours. Mem. Mo. Acad. Sci. (pres. 1992-93). Episcopalian. Office: Cen Mo State U Dept of Biology WCM 306 Warrensburg MO 64093

VOORSANGER, BARTHOLOMEW, architect; b. Detroit, Mar. 23, 1937; s. Jacob H. and Ethel A. (Arnstein) V.; m. Lisa Livingston, 1964; m. Catherine Hoover, Sept. 10, 1983; children—Roxanna Virginia (dec.), Matthew Ansley. A.B. cum laude, Princeton U., 1960; diplome, Fontainebleau, 1960; M.Arch., Harvard U., 1964. Assoc. Vincent Ponte, Montreal, Que., Can., 1964-67, I.M. Pei & Ptnrs., 1968-78; dir. I.M. Pei & Ptnrs., Iran, 1975-78; co-chmn. Voorsanger & Mills (Architects), N.Y.C., 1978-90; founder, prin. Voorsanger & Assocs., Architects, N.Y.C., 1990—; founder Taylor/Voorsanger Urban Designers, 1991; lectr. Bennington (Vt.) Coll., U. Pa., Columbia U., Harvard U.; guest critic, lectr. Yale U., Pratt Inst., CUNY, RISD, U. Cin., Syracuse U., U. Tex., Arlington. Exhbns. include: NYU, Archtl. Assn., London, Harvard Grad. Sch. Design, Vacant Lots Housing Study, N.Y., Deutsches Architeckur Mus., Frankfurt, Mus. Finnish Architecture, Avery Lib.Centennial Exhbn. Columbia Univ., Helsinki, Bklyn. Mus.; major projects include: Le Cygne Restaurant, Neiman houseboat, NYU Midtown Ctr., NYU Bus. Sch. Library, La Grandeur housing, NYU dormitories, Hostos Community Coll., N.Y.; finalist Bklyn. Mus. masterplan internat. competition, expansion and master plan Pierpont Morgan Libr., R. Krasnow Apt., N.Y.C.; fellow J. Pierpont Morgan Libr., N.Y. Mem. vis. com. R.I. Sch. Design, U. Tex.-Arlington; mem. N.Y. Historical Soc. (Architects Circle Steering Com.), chmn. bd. advisors Temple Hoyne Buell Ctr. Study Am. Architecture, Columbia U., 1989—; bd. dirs. Sir John Soane Mus. Found. 1st lt. U.S. Army, 1960-61. Recipient awards N.Y.C. chpt. AIA, AIA/Better Homes, Bard City Club, Interiors mag., Stone Inst., AIA/Libr., Lumen, Pratt Inst., NYU, N.Y.C. Art Commn. Fellow AIA (bd. dirs. N.Y.C. chpt. 1979-81, v.p. 1987, chmn. Brunner award com. 1978-80, Bard award pres.-elect N.Y.C. chpt. 1984, Nat. Honor award N.Y. State award); mem. Archtl. League N.Y.C. (bd. dirs.), Sir John Soane Mus. Found., N.Y. Found. for Architecture (bd. dirs.), Century Assn., River Club, Tuesday Evening Club (N.Y.C.), Wadawanuck Club. Home: 350 E 52nd St Apt 9G New York NY 10022-6739 Office: 246 W 38th St Fl 14 New York NY 10018-5886 also: 83 Main St Stonington CT 06378

VOOTS, TERRY LYNNE, technical sales engineer; b. Joliet, Ill., Nov. 15, 1956; s. David Eugene and Phyllis Jean (Dugan) V.; m. Gina Marie Munzert, Apr. 1, 1977; children: Eric J., Ryan J. AAS in Electronics Tech., Kishwaukee Coll., Malta, Ill., 1981, AS in Indsl. Tech., 1981. With GTE Automatic Electric, Genoa, Ill., 1981-83; tech. support engr. Electro Scientific Ind., Portland, Oreg., 1983-87; tech. sales engr. Photon Kinetics Inc., Beaverton, Oreg., 1987—. Precinct com. Rep. Party, Joliet, Ill., 1975. Mem. Audio Engring. Soc., TIA/EIA Telecom. Industry Assn. (voting). Lutheran. Achievements include designing a technic for photo elastic stress optic correction factor in optical fiber. Home: 1376 Brookcliff Dr Marietta GA 30062

VORA, MANHAR MORARJI, radiopharmaceutical scientist; b. Dumra, Kutch, India, Nov. 5, 1948; s. Morarji Damji and Premila Vora; m. Kalpana Vasanji Shah, Jan. 6, 1974; children: Shailey, Neil. MS, Western Ky. U., 1972; PhD, U. Ga., 1978. Rsch. assoc. Atlanta U. Ctr., 1978-80; sr. chemist Mt. Sinai Med. Ctr., Miami Beach, Fla., 1980-86; adj. prof. Med. Sch. U. Miami, Fla., 1982-86; acting chmn. radionuclide and cyclotron ops. King Faisal Specialist Hosp. and Rsch. Centre, Riyadh, Saudi Arabia, 1986—; rsch. proposal reviewer King Abdulaziz City for Sci. and Tech., Riyadh, 1990—; monograph reviewer U.S. Pharmacopiea, Rockville, Md., 1989—; cons. in field. Contbr. rsch. articles to internat. jours. Mem. Am. Chem. Soc. (sec. Miami chpt. 1982-83, chmn. 1983-84), Soc. Nuclear Medicine, Internat. Isotope Soc., Acad. Pharm. Scientists, Assn. Ofcl. Analytical Chemists, Rho Chi. Achievements include development of chromatographic methods for radiopharmaceuticals. Office: King Faisal Specialist Hosp, and Rsch Ctr, PO Box 3354 MBC-03, Riyadh 11211, Saudi Arabia

VORA, MANU KISHANDAS, chemical engineer, quality consultant; b. Bombay, India, Oct. 31, 1945; s. Kishandas Narandas and Shantaben K. (Valia) V.; m. Nila Narotamdas Kothari, June 16, 1974; children: Ashish, Anand. BSChemE, Banaras (India) Hindu U., 1968; MSChemE, Ill. Inst. Tech., Chgo., 1970, PhD in ChemE, 1975; MBA, Keller Grad. Sch. Mgmt., Chgo., 1985. Grad. asst. Ill. Inst. Tech., 1969-74; rsch. assoc. Inst. Gas Tech., Chgo., 1976-77, chem. engr., 1977-79, engring. supr., 1979-82; mem. tech. staff AT&T Bell Labs., Holmdel, N.J., 1983-84, Naperville, Ill., 1984—; quality cons. Naperville, 1989—, Milw., 1990—; speaker in field. Invited editor Internat. Petroleum Encyclopedia, 1980. Chmn. Save the Children Holiday Fund Drive, 1986—; trustee Avery Coonley Sch., Downers Grove, Ill., 1987-91; pres., dir. Blind Found. for India, Naperville, 1989—; dir. Nat. Ednl. Quality Initiatives, Inc., Milw., 1991—, fellow, 1992. Recipient Non-Supervisory AA award Affirmative Actions Adv. Com., 1987, Outstanding Contbn. award Asian Am. for Affirmative Actions, 1989, Disting. Svc. award Save The Children, 1990. Mem. Am. Soc. Quality Control (cert., standing rev. bd. 1988—, editorial rev. bd. 1989, nat. quality month regional planning com. 1989—, nat. cert. com. 1989—, chmn. cert. process improvements subcom. 1990—, Chgo. sect. exec. bd., sr. mem., vice chair section. affairs, 1993-94, Spl. award 1991, Century Club award 1992, Am. Mgmt. Assn. (assoc.), Chgo. Assn. of Tech. Socs. (Ann. Merit award 1992), Indian Sub-continent Cultural Club (pres. 1985—). Hindu. Avocations: reading, photography, travel, philanthropic activities. Home: 2S749 Theresa Ct Oak Brook IL 60521 Office: AT&T Bell Labs 2000 N Naperville Rd Naperville IL 60566-9397

VORBRUEGGEN, HELMUT FERDINAND, organic chemist; b. June 12, 1930; s. Josef and Ingeborg (Gindler) V.; m. Erika Krupke, July 23, 1965; 1 child, Gerd. PhD in Organic Chemistry, U. Göttingen, Germany, 1958. Rsch. assoc. Inst. Tech., Stockholm, 1958-59, Stanford (Calif.) U., 1959-63, Woodward Rsch. Inst., Basel, Switzerland, 1963-66; dept. head cen. rsch. Schering A.G., Berlin, 1966—; adj. prof. Inst. Tech., Berlin, 1973—. Contbr. 120 articles, revs. to profl. jours. Achievements include development of new synthetic methods of synthesis of nucleosides (e.g. anti-AIDS therapy); new standard synthetic reagents such as trimethylsilyltriflate. Home: Wilkestr 7, 13507 Berlin Fed Republic of Germany Office: Schering AG, D-13342 Berlin Germany

VORBURGER, THEODORE VINCENT, physicist, metrologist; b. N.Y.C., Aug. 30, 1944; s. Theodore Xavier and Mary (McBride) V.; m. Joanne Schuh, July 13, 1968; children: Judith, Jane. BS, Manhattan Coll., 1965; MS, Yale U., 1966, PhD, 1969. Rsch. asst. Yale U., New Haven, 1967-69; postdoctoral rsch. assoc. U. Del., Newark, 1969-71; asst. prof. Del. State Coll., Dover, 1971-72; physicist Nat. Bur. Standards, Gaithersburg, Md., 1972-87; supervisory physicist Nat. Inst. Standards and Tech., Gaithersburg, 1987—; tech. adviser Rapid Optics Fabrication Tech. Program Dept. Def., 1985-86; tech. cons. Chinese Univ. Devel. Project, Washington, 1986; chmn. com. on surface texture ANSI/ASME, 1986-93. Editor: Surface Metrology Collection in Optical Engring., 1985; contbr. articles to profl. publs. Cath. vol. Montgomery County Detention Ctr., Rockville, Md., 1983—. Recipient Silver medal Dept. Commerce, Washington, 1992. Mem. Am. Phys. Soc., Soc. Photo-optical Instrumentation Engrs. Republican. Roman Catholic. Achievements include calibrations in surface finish, development of sinusoidal roughness calibration blocks, research system to study light-scattering from rough surfaces. Office: Nat Inst Standards/Tech Gaithersburg MD 20899

VORHIES, MAHLON WESLEY, veterinary pathologist, educator; b. Fairfield, Iowa, June 26, 1937; s. Harold Wesley and Edith Mae (Bender) V.; m. Ilene Lanore Hoffman, Aug. 29, 1959; children—Susan Rae, Robert Wesley. D.V.M., Iowa State U., 1962; M.S., Mich. State U., 1967. Veterinarian in pvt. practice Riverside, Iowa, 1962-64; clin. instr. Mich. State U., East Lansing, 1964-67; vet. pathologist Iowa State U., Ames, 1962-72; vet. pathologist, dir., dept. head S.D. State U., Brookings, 1972-86; dir., dept. head Kans. State U. Coll. Vet. Medicine, Manhattan, 1986—; cons. NIH, Commonwealth Pa., Winrock Internat. Hdqrs., U.S. Dept. Agr., FAO-Fundagro/MIAC, LIFE, Quito, Ecuador, 1991. Contbr. articles to profl. jours. Trustee Brookings United Presbyn. Ch., 1975-78, elder, 1981-84; mem. adv. com. Pipestone Area Vocat. Tech. Inst., 1984. Mem. AVMA, Assn. Am. Vet. Med. Colls., S.D. Vet. Med. Assn. (Vet. of Yr. award 1979), Kans. Vet. Med. Assn., Am. Assn. Avian Pathologists, U.S. Animal Health Assn., Am. Assn. Vet. Lab. Diagnosticians (E.P. Pope award 1984, pres. 1977), North Central Conf. Vet. Lab. Diagnosticians (chmn. 1984), Commn. Vet Medicine, Phi Zeta, Gamma Sigma Delta. Clubs: Brookings Country (dir. 1983-86). Lodge: Shrine (chmn. 1984-85). Home: 2035 Rockhill Cir Manhattan KS 66502-3952 Office: Kans State U Coll Vet Medicine Manhattan KS 66506

VORNDAM, PAUL ERIC, analytical and organic chemist; b. Quincy, Ill., Aug. 17, 1947; s. Herbert John and Virginia A. (Bradney) V.; m. Patricia A. Kellerman, Jan. 28, 1971 (div. Nov. 1992); children: Jeffrey M., Jeremy M.; m. Margaret E. Van der Weerdt, Nov. 15, 1992. BS, Millikin U., 1969; MS, Ill. State U., 1971; PhD, U. Colo., 1987. Assoc. prof. chemistry USAF Acad., Colorado Springs, Colo., 1980-91; group leader Colo. Analytical R & D Corp., Colorado Springs, 1991—. Contbr. articles to Jour. Am. Chem. Soc., Jour. Organic Chemistry. Mem. AAAS, Am. Chem. Soc. (alt. councillor Colorado Springs 1991-92). Home: 418 Echo Ln Colorado Springs CO 80904 Office: Colo Analytical R & D Corp 4720 Forge Rd Ste 108 Colorado Springs CO 80907

VOSS, JÜRGEN, chemistry educator; b. Hamburg, Germany, Feb. 19, 1936; s. Adolph Bruno and Berta Frieda (Schüler) V.; m. Jutta Voss, Feb. 19, 1965; children: Eva Lotte, Sibylle Konstanze. D. in Natural Scis., U. Hamburg, 1965, Habilitation, 1972. Asst. U. Hamburg, 1964-73, lectr., 1973-77, prof., 1977—. Author: (with others) Houben-Weyl's Handbook, 1985, Comprehensive Organic Synthesis, 1991; contbr. 125 articles to sci. jours. Confidential agt. Friedrich-Ebert-Found., Bonn, Germany, 1980—. Mem. Gesellschaft Deutscher Chemiker, Am. Chem. Soc. Lutheran. Home: Hochestieg 34A, 22391 Hamburg Germany Office: U Hamburg, Martin Luther King Platz 6, 20146 Hamburg Germany

VOSS, PAUL JOSEPH, physicist; b. Chgo., Mar. 10, 1943; s. Paul Joseph and Irene Ester (Bergman) V.; divorced; children: Kalila Laurel Mage, Lisa Jean Voss. BS, Syracuse U., 1969; MS, Johns Hopkins U., Balt., 1972. From assoc. physicist to sr. physicist Johns Hopkins U. Applied Physics Lab., Laurel, Md., 1969—, facility mgr. guidance system evaluation lab., 1981—; chmn. AEGIS Scenario Cert. Com., Washington, 1987—. Avocations: skiing, windsurfing, shooting. Home: 5635B Harpers Farm Rd Columbia MD 21044-2002 Office: Johns Hopkins U Applied Physics Lab Johns Hopkins Rd Laurel MD 20723-1140

VOSS, STEVEN RONALD, environmental engineer; b. Mpls., Feb. 4, 1962. BS, U. Minn., 1988; MS, Clarkson U., 1992. Engr. C&S Engrs. Inc., Syracuse, N.Y., 1990-92; project engr. Conestoga-Rovers & Assocs., St. Paul, 1992—. Mott fellow C.S. Mott Found., 1988. Mem. ASCE, Internat. Assn. for Gt. Lakes Rsch., Am. Chem. Soc. Office: Conestoga-Rovers Assocs 1801 Old Hwy 8 Ste 114 Saint Paul MN 55112

VOSS, TERENCE J., human factors scientist, educator; b. Cin., June 29, 1942; s. Harold A. and Marguerite (Canavan) V.; m. Charmaine E. Wilson, Sept. 3, 1983. BA, SUNY, Geneseo, 1965; MA, Fla. Atlantic U., 1972; postgrad., U. Mont., 1973-78. Dept. dir., sr. staff scientist Essex Corp., Alexandria, VA., 1980-88; sr. human factors scientist ARD Corp., Columbia, Md., 1988-90; lead human factors scientist, fellow engr. Westinghouse Savannah River Co., Aiken, S.C., 1990—; cons. in field; mem. adj. faculty psychology dept. DePaul U., Chgo., 1990; human factors cons. U.S. Dept. Energy, 1990—. Contbr. articles to profl. jours. Named Citizen Amb., People to People Internat., 1985. Mem. AAAS, Am. Nuclear Soc., Human Factors Soc., Sci. Rsch. Soc. N.Am., Sigma Xi. Achievements include contributions resulting in improvements to nuclear facilities and procedures thereby reducing human error, facilitating human behavior and increasing operator and public safety. Home: 19 Shrewsbury Ln Aiken SC 29803-6299 Office: Westinghouse Savannah River Co Bldg 706-15C Aiken SC 29803

VOSS, WERNER, dermatologist; b. Hagen, Germany, Apr. 6, 1949; s. Karl and Herta (Koch) V.; m. Barbara Finke, June 16, 1973; children: Viola, Marcel. MD, Univ. Hosp., Muenster, 1973. Resident in dermatology Univ. Hosp., Muenster, 1973-78; pvt. practice Muenster, 1978—; pres. Dermatest Rsch. Co., Muenster, 1983—; pres. Med. Data Svc. Co., Muenster, 1987—. Home: Birkenweg 4, Münster Germany 4400 Office: Dermatest GmbH, Birkenweg 4, Münster Germany 4400

VOSS, WERNER KONRAD KARL, architect, engineer; b. Muenster, Fed. Republic Germany, Dec. 9, 1935; s. Reinhard and Maria (Ewers) V.; m. Sabine Gerda Franziska Glaser, Nov. 28, 1963; children: Katharina, Alexander, Franziska. Diploma in civil engring., U. Braunschweig, Braunschweig, Fed. Republic Germany, 1964, diploma in archtl. engring., 1968. Asst. U. Braunschweig, 1964-69, chief engr., 1969-79; prin. Werner Voss & Ptnrs., Braunschweig, 1969-80, Voss & Petersen, Braunschweig, 1989. Fellow Chamber of Architects. Office: Werner Voss & Ptnrs, Harzburger Str 13, D 3300 Braunschweig Germany

VOSSEL, RICHARD ALAN, systems engineer; b. Somerville, N.J., Oct. 28, 1953; s. Louis Fredrick and Dorothy Marie (Snyder) V.; m. Laura Josephine Knox, July 23, 1983; children: Joshua, Melina, Christina. BA in Math., Christopher Newport U., 1983. Elec. technician USAF, McChord AFB, Wash., 1972-76; programmer USAF, Langley AFB, Va., 1976-80; system programmer SDC, Langley AFB, Va., 1980-83; system engr. UNISYS, Newport News, Va., 1983-87; system engr. mgr. SYSCON Corp., Williamsburg, Va., 1987—. Scout leader Boy Scouts of Am. Recipient UNISYS Achievement award for Tech. Excellence, 1986. Achievements include devel. of knowledge in area of automatic identification and use in DOD activities. Home: 202 Terrys Run Yorktown VA 23693 Office: SYSCON Corp 309 McLaws CR Ste K Williamsburg VA 23185

VOSSLER, JOHN ALBERT, civil engineer; b. Newburgh, N.Y., Oct. 9, 1925; s. Vernon Martense and Frieda (Bachmann) V.; m. Betiejean Sleight Erts, Sept. 4, 1948; children: Karen Ann, Susan Jean. BS in Structural Engring., U. Mich., 1951. Registered profl. engr., N.Y., Conn. Structural engr. Marine div. Maxon Constrn. Co., Tell City, Ind., 1953-56; staff engr. Systems Mfg. div. IBM, Poughkeepsie, N.Y., 1956-68; project mgr. Real Estate & Constrn. div. IBM, Stamford, Conn., 1968-88; adv. engr. Gen. Systems div. IBM, Hopewell Junction, N.Y., 1988-89; dir. devel. Getter, Segner & Gironda, P.E., P.C., Valhalla, N.Y., 1989-93; prin. John A. Vossler Assocs., Danbury, Conn., 1988—. Bd. dirs. Lake Pl. Condo Assn., 1989—.With U.S. Army, 1944-46; PTO; 1st lt. USAF, 1951-53. Mem. ASCE, NSPE, Conn. Soc. Profl. Engrs., Conn. Engrs. in Pvt. Practice, Am. Radio Relay League. Home: 12-147 Boulevard Dr Danbury CT 06810 Office: John A Vossler Assocs 12-147 Boulevard Dr Danbury CT 06810

VOSSOUGHI, SHAPOUR, chemical and petroleum engineering educator; b. Siahkal, Gilan, Iran, June 25, 1945; s. Mirza Aghasi and Ghamar Talat (Farahpour) V.; m. Ziba Mani, Nov. 6, 1973; children: Anahita, Sarah. G-rad. diploma, McGill U., Montreal, Can., 1971; MSc., U. Alta., Edmonton, Can., 1973, PhD, 1976. Instr. Arya-Mehr U., Tehran, Iran, 1967-70; rsch. assoc. U. Kans., Lawrence, 1976-77, asst. scientist, 1977-78; sr. scientist Nat. Iranian Oil Co., Tehran, 1978-79; asst. scientist U. Kans., Lawrence, 1979-81, assoc. scientist, 1981-82, assoc. prof., 1982—; researcher Shell Rsch. Ctr., Rijswijk, the Netherlands, 1978-79; sabbatical researcher Elf Aquitaine, Pau, France, 1989-90; tech. presenter in field; cons. for UN, Nat. Iranian Oil Co. and U. of Petroleum Industry, Ahwaz, Iran. Contbr. 20 publs. to profl. jours. including Can. Jour. Chem. Engring., Jour. Can. Petroleum Tech., Soc. Petroleum Engrs. Jour., Jour. Thermal Anal., Indsl. Engring. Chem. Fundamentals, Trans. Soc. Petroleum Engrs., Thermochimica Acta, Chem. Engring. Commun., SPE Reservoir Engring., Jour. Petroleum Sci. and Engring. Faculty fellow EXXON Edn. Found., 1982-85; rsch. grantee Columbian Resources Inc., Topeka, 1989, U. Kans., 1983-89, Dept. of Energy, 1982-84, 92—, Core Labs., 1980-81. Mem. Soc. Petroleum Engrs., Am. Inst. Chem. Engrs., N.Am. Thermal Analysis Soc., Soc. of Rheology, Sigma Xi. Achievements include patent in field. Home: 1035 Lakecrest Lawrence KS 66049-0001 Office: U Kans 4006 Learned Hall CPE Lawrence KS 66045

VOUGHT, FRANKLIN KIPLING, pharmaceuticals chemist, researcher; b. Balt., Feb. 16, 1965; s. Franklin Joseph and Terri (Vought) Zunt. BA, U. Miami, 1990. Commd. seaman E-3 USCG, 1985, advanced through grades to petty officer 2d class, 1992; aviation survivalman USCG, San Juan, P.R., 1985-86; rescue swimmer USCG, Miami, Fla., 1986-92; chemistry lab. instr. U. Miami, Coral Gables, 1991-92, rsch. asst., mgr. lab. for water rsch., 1992—; R&D chemist Barker Norton Pharmacuetical, Miami, 1991—. Recipient Bowman Ashe scholarship U. Miami, 1989, Rsch. in Phys. Chemistry fellowship U. Miami, 1992. Mem. AAAS, U. Miami Alumnae. Episcopalian. Home: Apt 5 7515 SW 59th Ave Miami Lakes FL 33014 Office: Baker Norton Pharmaceutical 8800 NW 36th St Miami FL 33166

VOUTSAS, APOSTOLOS THEOHARIS, chemical engineer; b. Salonica, Macedonia, Greece, Mar. 12, 1966; came to U.S., 1989; s. Theoharis A. and Georgia Alice (Vermatou) V. BSc with honors, Aristotle U., Thessaloniki, Greece, 1989; MSc, Lehigh U., 1991. Cert. engr., Pa. Rsch. asst. Lehigh U., Bethlehem, Pa., 1989-92, rsch. engr., 1992—; summer intern Applied Materials, Santa Clara, Calif., 1993. Contbr. articles to profl. jours. Mem. Electrochem. Soc., Am. Inst. Chem. Engrs., Materials Rsch. Soc. Republican. Greek Orthodox. Achievements include pioneering rsch. in reduced-thermal-budget polycrystalline silicon deposition by Thermal Chem.-Vapor-Deposition techniques; exptl. investigation of the effect of key process parameters on the structure of silicon thin films obtained by LPCVD and UHV-CVD; formulation of theoretical models reflecting the experimental trends; made major application in area of Active Matrix LCDs. Office: Lehigh U Sherman Fairchild Ctr 161 Memorial St E Bethlehem PA 18015

VOYIADJIS, GEORGE ZINO, civil engineer, educator; b. Cairo, Egypt, Dec. 15, 1946; s. Zino Dimitri and Eleni (Mavridou) V.; m. Christina George Tziortzi, Nov. 4, 1978; children: Helena G., Andrew G. BSc with highest honors, Ain Shams U., Cairo, 1969; MSc, Calif. Inst. Tech., 1970; DSc, Columbia U., 1973. Reg. eng. in tng., 1981. Sr. stress analyst Nuclear Power Svcs., Inc., N.Y.C., 1973-75, EBASCO Svcs., Inc., N.Y.C., 1975; assoc. prof. U. Petroleum and Minerals, Dhahran, Saudi Arabia, 1975-80; prof. civil engring., acting assoc. dean Grad. Sch. La. State U., Baton Rouge, 1980—. Editor: Mechanics of Material Interfaces, 1986, Advances in the Theory of Plates and Shells, 1990, Microstructural Characterization in Constitutive Modeling of Metals and Granular Media, 1992, Damage in Composite Materials, 1993; contbr. numerous articles on mech. behavior of solids to profl. publs. Mem. ASCE, ASME, Am. Soc. Engring. Educators, Am. Acad. Mechanics, Soc. Engring. Sci., Sigma Xi. Democrat. Achievements include study of modeling mechanical behavior of metals and metal matrix composites; refined theory of plates and shells and numerical simulation of elasto-plastic contact problems. Home: 12718 N Oak Hills Pky Baton Rouge LA 70810 Office: La State Univ Dept Civil Engring Baton Rouge LA 70803

VOYSEST, OSWALDO, agronomist, researcher; b. Callao, Peru, Sept. 18, 1933; s. Alejandro and Constanza V.; m. Magda Diaz, Jan. 8, 1959; children: Magda, Oswaldo. MS, U. Calif., Davis, 1964; PhD, Mich. State U., 1970. Plant breeder bean program La Molina Agrl. Experiment Sta., Lima, Peru, 1959-69, head bean program, 1970-75; agronomist bean program CIAT (Internat. Ctr. for Tropical Agr.), Cali, Colombia, 1976—. Author: Bean Varieties in Latin America and Their Origin, 1983; editor: Common Beans: Research for Crop Improvement, 1991. Vice pres. Rotary, Maranga, Peru, 1969. Rockefeller Found. fellow, 1958, 61-63, 67-70. Mem. Am. Soc. Agronomy, Am. Soc. Crop Sci. Methodist. Achievements include research in breeding 5 bean varieties currently grown in the Peruvian coast. Office: CIAT, A A 6713, Cali Colombia

VRACHAS, CONSTANTINOS AGHISILAOU, aeronautical engineer; b. Jaffa, Palestine, Mar. 22, 1931; s. Aghisilaos Costas and Helen Nicola (Karatzas) V.; m. Maria Georghiou Papanastassiou, Dec. 31, 1955; children: Helen, Danae Cleopatra. Diploma in aero. engrin., Brit. Air U., 1954; BBA, Cyprus Coll., 1971. Chartered engr., Euro. engr. Apprentice Cyprus Airways, Nicosia, 1949-51, aircraft engr., 1954-56; insp. Arab Airways, Amman, Jordan, 1956-58, Aden Airways, 1958-60; sr. chief engr. Cyprus Airways, 1960-89; chief airworthiness surveyor Dept. Civil Aviation, Ministry of Communications and Works, Nicosia, 1990—; com. mem. Joint Airworthiness Regulations, Amsterdam, 1991—, Air Legislation, Nicosia, 1989—. Contbr. articles to profl. jours. Fellow Royal Aero. Soc., British Inst. Mgmt., Chartered Inst. Transp., AIAA (fellow). Home: 28 Irinis St, Flat 4 Strovolos, Nicosia 148, Cyprus Office: Ministry of Communications Works, Dept Civil Aviation, Nicosia Cyprus

VREELAND, ROBERT WILDER, electronics engineer; b. Glen Ridge, N.J., Mar. 4, 1923; s. Frederick King and Elizabeth Lenora (Wilder) V.; m. Jean Gay Fullerton, Jan. 21, 1967; 1 son, Robert Wilder. BS, U. Calif., Berkeley, 1947. Electronics engr. Litton Industries, San Carlos, Calif., 1948-55; sr. devel. electronics engr. U. Calif. Med. Ctr., San Francisco, 1955-89; ret.; cons. electrical engring; speaker 8th Internat. Symposium Biotelemetry, Dubrovnik, Yugoslavia, 1984, RF Expo, Anaheim, Calif., 1985, 86, 87. Contbr. articles to profl. jours., also to internat. meetings and symposiums; patentee in field. Recipient Chancellor's award U. Calif., San Francisco, 1979; cert. appreciation for 25 years' service U. Calif., San Francisco, 1980. Mem. Nat. Bd. Examiners Clin. Engring. (cert. clin. engr.), IEEE, Assn. Advancement Med. Instrumentation (bd. examiner), Am. Radio Relay League (pub. service award 1962). Home: 45 Maywood Dr San Francisco CA 94127-2007 Office: U Calif Med Ctr 4th and Parnassus Sts San Francisco CA 94143

VRENTAS, CHRISTINE MARY, chemical engineer, researcher; b. Chgo., June 16, 1953; d. John and Antoinette (Golonka) Jarzebski; m. James Spiro Vrentas, June 8, 1975; children: Catherine, Jennifer. BS, Ill. Inst. Tech., 1975; MS, Northwestern U., 1977, PhD, 1981. Asst. prof. Pa. State U., University Park, 1981-83, adj. asst. prof., 1983-87, adj. assoc. prof., 1987-90, adj. prof., 1990—. Contbr. articles to profl. jours. Chmn. com. PTO, State College, Pa., 1989-92. Mem. AICHE, Soc. Rheology, Tau Beta Pi, Phi Lambda Upsilon. Office: Pa State Univ Dept Chem Engring 119 Fenske Lab University Park PA 16802

VRIJENHOEK, ROBERT CHARLES, biologist; b. Rotterdam, Netherlands, Mar. 13, 1946; came to U.S., 1949; s. Adrian Maria and Anna Maria (Van Stratum) V.; m. Linda Cheryl Packard, Jan. 13, 1968; children: Michelle Anne, Eric Michael. BA, U. Mass., 1968; PhD, U. Conn., 1972. Asst. prof. So. Meth. U., Dallas, 1972-74; asst. prof. Rutgers U., New Brunswick, N.J., 1974-78, assoc. prof., 1978-83, prof. biology, 1983—; dir. Ctr. Theoretics and Applied Genetics, New Brunswick, N.J., 1988—. Editor-in-chief Evolution, 1987-90; assoc. editor Conservation Biology, 1987—; editorial bd. Jour. Heredity, 1993—; contbr. articles to sci. publs. Grantee NSF, 1974, 76, 79, 83, 86, 88, 89, 93. Fellow AAAS. Office: Rutgers Univ New Brunswick NJ 08903

VROMAN, GEORGINE MARIE, medical anthropologist; b. Padang, Sumatra, Indonesia, Mar. 4, 1921; came to U.S., 1947; d. Gerrit Samuel and Marie Elisabeth (Bickes) Sanders; m. Leo Vroman, Sept. 10, 1947; children: Geraldine E., Peggy Ann. MD, Rijks U., Utrecht, Netherlands, 1947; PhD in Anthropology, New Sch. Social Rsch., 1979. Rsch. asst. pathology St. Peter's Gen. Hosp., New Brunswick, N.J., 1947-50; adj. prof. Ramapo (N.J.) Coll., 1980-81; cons. Cognitive Rehab. Svc. Bellevue Hosp. Ctr., N.Y.C., 1981-83, cons. Geriatric Clinic, 1984—; assoc. Cognitive Rehab. Svcs., Sunnyside, N.Y., 1984—. Chief editor: Women at Work, 1988. Bd. dirs. Parents' Assns., Pub. Sch. 125, 209, 43, Manhattan, Bklyn., 1957-67; leader troops Girl Scouts U.S.A., Manhattan, Bklyn., 1959-69, asst. dist. commr., 1964-69. Fellow Assn. Applied Anthropology; mem. Am. Assn. Anthropology, Met. Med. Anthropology Assn., N.Y. Acad. Scis., N.Y. Neuropsychology Group. Achievements include research in cognition, lang. and consciousness, integrating theory and practice in clin. neuropsychology. Home: 2365 E 13th St Brooklyn NY 11229

VU, BIEN QUANG, mechanical engineer; b. Quang-Ngai, Vietnam, Sept. 24, 1951; came to U.S. 1975; s. Binh Quang and Huong Thi (Ngo) V.; m. Gabrielle Vuong Harrison, Apr. 6, 1985; children: Alexandra, Joceline, Sophianne. MS, U. Idaho, 1977; PhD, U. Colo., 1981. Mech. engr. Naval Ocean Systems Ctr., San Diego, 1981-84; sr. rsch. scientist LTV, Dallas, 1984-85; staff mem. MIT/Lincoln Labs., Lexington, 1985-86, System Planning Corp., Arlington, Va., 1986-91; sr. staff scientist Sci. Applications Internat. Corp., Arlington, Va., 1991—. Contbr. 15 articles to profl. jours. Named Best Rschr. of the Yr., System Planning Corp., 1990. Mem. AIAA (mem. tech. com. 1988-91). Achievements include research in areas of high velocity impact, composite materials, and development in areas of missile defense technology. Home: 9221 Dorothy Ln Springfield VA 22153 Office: SAIC 1700 N Moore Ste 1820 Arlington VA 22209

VU, CUNG, chemical engineer; b. Ha Dong, Vietnam, Jan. 1, 1951; came to U.S., 1981; s. Thieu Vu and Sam (Thi) Vuong; m. Mai-Chi Nguyen, Dec. 5, 1981; children: Michelle Lan-Chi, Meghan Xuan-Chi. BE in Chem. Engring., U. Sydney, Australia, 1975; PhDChemE, Monash U., Australia, 1979. Rsch. engr. ICI Australia Ltd., Melbourne, 1979-80; sr. rsch. engr., engring. assoc. Uniroyal Chem. Co., Middlebury, Conn., 1981-86; rsch. engr., sr. rsch. engr., engring. assoc. W.R. Grace & Co., Columbia, Md., 1986—. Contbr. articles to profl. jours. Colombo Plan scholar Australian Govt., 1970-74, Monash Grad. scholar Monash U., 1975-78. Mem. TAPPI, Am. Inst. Chem. Engrs., Am. Chem. Soc., Soc. for Advancement Material and Process Engring. (Uniroyal Tech. Achievement award, Grace Devel. award). Achievements include 10 patents for coatings for automobile and construction applications. Office: WR Grace & Co 7379 State Route 32 Columbia MD 21044-4098

VUKOVIC, DRAGO VUKO, electronics engineer; b. Dubrovnik, Croatia, Yugoslavia, Sept. 9, 1934; s. Vuko and Katica (Simunovic) V.; m. Marija Kakarigji, May 15, 1956; children: Katija, Snjezana, Sanja. Diploma engr., Faculty of Electrotehnics, Zagreb, 1961; MS, Faculty of Econs., Zagreb, 1977. Supervising engr. Elektrojug, Dubrovnik, 1961-62; lighting designer Arhitekt, Dubrovnik, 1963-64; designer, supr. Biro za izgradnju, Dubrovnik, 1964-71, mng. dir., 1971-74; leader architect team Atelier LAPAD, Dubrovnik, 1974—; tchr., head electrotechnical div. Nautical Coll., Dubrovnik, 1963-67; cons. engr. Dubrovnik, Cavtat, Dubrovnik, 1968-75; collaborator Faculty of Architecture, Zagreb, 1974—. Contbr. in field. Recipient Town Planning & Architecture award Yugoslav Competition Com., Titograd, 1978; Plaque Assn. Visual Artists and Applied Arts of Croatia, 1982. Mem. Assn. Visual Artists of Applied Arts of Croatia, Soc. Ind. Geodesists, Civil Engrs. and Architects, Radio Club (Dubrovnik), Am. Radio Relay League Inc, Am. Biog. Inst. Rsch. Assn. Office: Atelier LAPAD, L Rogovskog 4, 50000 Dubrovnik Croatia

VULIS, DIMITRI LVOVICH, computer consultancy executive; b. Leningrad, Russia, Dec. 29, 1964; came to U.S., 1979; s. Lev Klyukvin and Inna Vulis; m. Maryam Inzel, Mar. 30, 1988; 1 child, Daniel Benjamin. BA in Math. cum laude, CUNY, 1985, MA in Math., 1989, postgrad., 1989—. Pres. D&M Consulting Svcs., Inc., Forest Hills, N.Y., 1990—; cons. in math. Fordham U., N.Y.C., 1989—, Cornerstone Asset Mgmt., N.Y.C., 1989—, Possev-USA, N.Y.C., 1990—. Author: (computer program) Russian TeX, 1989. Mem. Am. Math. Soc., TeX Users' Group. Republican. Avocations: cryptography, golf. Office: D&M Cons Svc Inc 67-67 Burns St Forest Hills NY 11375

VULPETTI, GIOVANNI, physicist; b. Reggio, Calabria, Italy, Sept. 17, 1945; s. Attilio and Elvira (Guaiana) V.; m. Rosella Ritrovato, Aug. 3, 1978; children: Desiree, David. Diploma in Nuclear Physics, U. Rome, 1969, PhD in Physics, 1973. Fellow Nuclear Energy Lab., Casaccia/Rome, Italy, 1969-70; space researcher Univ. Rome, 1974-78; sr. scientist Telespazio Spa, Rome, 1979-84, mission analysis function head, 1985-89, advanced mission studies br. head, 1990, sci. asst. to gen. mgr.; mem. astrodynamics com. Internat. Astron. Fedn., Paris, 1987—; guest scientist Nat. Ctr. for Propulsion and Energy, Milan, 1971-73. Contbr. articles to profl. jours. and publs. Fellow Brit. Planetary Soc.; mem. AIAA (astrodynamics com. on standards 1991—), Internat. Acad. Astronautics (corr., co-chmn. I.S.E.C. 1990-91, small satellite subcom. 1991—, lunar devel. subcom. 1993). Achievements include development of unified design-oriented mathematical picture of the today-known rocket and non-rocket propulsion systems for high-performance space missions, detailed trajectory and engine design concepts for potential future very-low-mass launchers and space tugs using the matter-antimatter annihilation; research on impact of high mass-into-energy-conversion propulsion systems to optimization of high payload or high-final-speed deep-space trajectories, datailed analysis for use of Nb-Ti superconductors in space as source of field in magnetic sail propulsion systems, studies of optimal space trajectories of deep-space probes powered by staged/combined propulsions: rocket and solar sail, photon solar sail and solar-wind magnetic sail. Office: Telespazio Spa, Via Tiburtina 965, 00156 Rome Italy

VUORINEN RUPPI, SAKARI ANTERO, materials scientist; b. Juupajoki, Finland, June 5, 1948; s. Viljo Antero and Kerttu Inkeri (Peuramo) V.; m. Raili Maarit Numminen, July 31, 1982; 1 child, Robert. MSc, Helsinki (Finland) U. Tech., 1977; PhD, Tech. U. Denmark, Copenhagen, 1982. Rschr. Helsinki U. Tech., 1976; rsch. assoc., assoc. prof. Tech. U. Denmark, Copenhagen, 1978-84; sr. rschr. Acad. of Finland, Helsinki, 1984-87; adj. prof. Helsinki U. Tech., 1986—; sr. staff metallurgist Seco Tools AB, Fagersta, Sweden, 1988—; vis. scientist Uppsala (Sweden) U., 1987; rsch. engr.-cons. Matine, Helsinki, 1980-81. Contbr. articles on surface and interface studies of solid materials to profl. jours.; patentee in surface and coatings tech. Wihuri award Wihuri Found., 1980. Mem. AAAS, N.Y. Acad. Scis., Metall. Soc., Danish Soc. Chem., Civil, Elec. and Mech. Engrs. Office: Seco Tools AB, 737 82 Fagersta Sweden

VUŠKOVIĆ, LEPOSAVA, physicist, educator; b. Lešnica, Yugoslavia, Apr. 23, 1941; d. Djordje and Kristina (Obućina) Jovanović; m. Marko VušKović, Feb. 2, 1964 (div. Oct. 1982); children: Kristina, Ivo; m. Svetozar Popović, July 18, 1987; 1 stepchild, Ljubica Popović. Diploma in Phys. Chemistry, U. Belgrade, Yugoslavia, 1963, MS in Physics, 1968, PhD in Physics, 1972. Rsch. fellow Inst. Physics U. Belgrade, 1964-73, from rsch. scientist to head Atomic Physics Lab., 1973-78, dir. atomic laser and high energy physics div., 1981-85; assoc. prof. U. of Arts, Belgrade, 1973-85; assoc. rsch. prof. dept. physics NYU, N.Y.C., 1985-93; assoc. prof. dept. physics Old Dominion U., Norfolk, Va., 1993—; mem. gen. com. Internat. Conf. on Physics of Electronic and Atomic Collisions, 1977-81, mem. organizing com. VIII Conf., 1973, mem. organizing com. VII Symposium on Physics of Ionized Gases, Dubrovnik, Yugoslavia, 1976, 82. Author: (textbook) Physical Bases of Film, 1985, (with others) Metrology of Gaseous Pollutants, 1981, Investingation of Electron-Atom Laser Interactions, 1982. Mem. Am. Phys. Soc., Optical Soc. Am., European Phys. Soc., Yugoslav Phys. Soc. Office: Old Dominion U Physics Dept Norfolk VA 23529

WAALAND, IRVING THEODORE, retired aerospace design executive; b. Bklyn., July 2, 1927; s. Trygve and Marie Waaland; m. Helen Rita Katz, Apr. 7, 1961; children: Theodore, Neil, Elizabeth, Scott, Diane. BA magna cum laude, NYU Coll. Engring., 1953. Project engr. Grumman Corp., Bethpage, N.Y., 1953-74; v.p., chief designer Northrop Corp., Pico Rivera, Calif., 1974-93. Patentee in field. With USAF, 1946-48. Fellow AIAA (Aircraft Design award 1989, Aircraft Design cert. merit 1989, Wright Bros. lectr. in Aeronautics 1991); mem. NAE, Am. Def. Preparedness Assn. (Leslie E Simon award 1990). Home: 65 Rollingwood Dr Palos Verdes Peninsula CA 90274-2425

WAALER, BJARNE ARENTZ, physician, educator; b. Bergen, Norway, Apr. 18, 1925; s. Rolf and Gudrun Waaler; M.D., U. Oslo, 1950, D.M., 1959; m. Gudrun Arentz, Dec. 19, 1950; children: Hans Michael Arentz, Finn, Astrid. Various positions in Norwegian hosps., 1950-56; research fellow physiology lab. physiology Oxford (Eng.) U., 1958-61; prof. physiology U. Oslo Med. Faculty, 1962-92; dean med. faculty, 1974-77, rector, 1977-85; retired U. Oslo Med. Facility, 1992. Decorated comdr. Royal Order St. Olav. Mem. Norwegian Acad. Sci. and Letters (pres. 1990—), Physiol. Soc. Eng. Author papers in field. Office: U Oslo, Dept Physiology, Postbox 1103 Blindern, 0317 Oslo 3, Norway

WAALKES, MICHAEL PHILLIP, toxicologist; b. Balt., Aug. 6, 1953; s. T. Phillip and Frances Elizabeth (Brewster) W.; m. Michele Ann Conte, Nov. 12, 1976; children: Phillip Lee, Joseph Michael, Thomas Edward. BA, Hope Coll., 1975; PhD, W.Va. U., 1981. Postdoctoral fellow dept. pharmacology U. Kans. Med. Ctr., Kansas City, 1981-83; staff fellow Nat. Cancer Inst., Frederick, Md., 1983-88, rsch. pharmacologist, 1988-90, chief inorganic carcinogenesis sect., 1990—; adj. prof. toxicology U. Md. Environ. Toxicology Program, Balt., 1989—. Contbr. articles to profl. jours.; editor: Target Organ Toxicity Series: Carcinogenesis, 1992; editorial bd.: Toxicology and Applied Pharmacology, 1988—. Fellow Nat. Inst. Environ. Health Sci., 1981-83, NIH, 1977-81. Mem. Soc. Toxicology (Achievement award 1990), Am. Assn. for Cancer Rsch. Achievements include research in genetic susceptibility to metal carcinogenesis, target organ specificity. Home: 4220 Springview Ct Jefferson MD 21755 Office: Nat Cancer Inst FCRDC Bldg 538 Rm 205E Frederick MD 21702

WACHSMAN, HARVEY FREDERICK, neurosurgeon, lawyer; b. Bklyn., June 13, 1936; s. Ben and Mollie (Kugel) W.; m. Kathryn M. D'Agostino, Jan. 31, 1976; children: Dara Nicole, David Winston, Jacqueline Victoria, Lauren Elizabeth, Derek Charles, Ashley Max, Marea Lane, Melissa Roseanne. B.A., Tulane U., 1958; M.D., Chgo. Med. Sch., 1962; J.D., Bklyn. Law Sch., 1976. Bar: Conn. 1976, N.Y. 1977, Fla. 1977, D.C. 1978, U.S. Supreme Ct. 1980, Pa. 1984, Md. 1986, Tex. 1987. Diplomate Nat. Bd. Med. Examiners; cert. Am. Bd. Legal Medicine, Am. Bd. Profl. Liability Attys. (pres.); cert. civil trial advocate Nat. Bd. Trial Advocacy (trustee). Intern surgery Kings County Hosp. Ctr., Bklyn., 1962-63; resident in surgery Kingsbrook Med. Ctr., Bklyn., 1964-65; resident in neurol. surgery Emory U. Hosp., Atlanta, 1965-69; practice medicine specializing in neurosurgery Bridgeport, Conn., 1972-74; ptnr. firm Pegalis & Wachsman, Great Neck, N.Y., 1977—, Wachsman & Wachsman, Great Neck, 1976—; adj. prof. neurosurgery SUNY, Stony Brook; adj. prof. law St. John's U. Sch. Law.; adj. prof. Bklyn. Law Sch., U. South Fla. Coll. Medicine. Author: American Law of Medical Malpractice, Vol. I, 1980, 2nd edit., 1992, American Law of Medical Malpractice, Vol. II, 1981, American Law of Medical Malpractice, Vol. III, 1982, Cumulative Supplement to American Law of Medical Malpractice, 1981, 82, 83, 84, 85; mem. editorial bd.: Legal Aspects of Med. Practice, 1978-82. Fellow Am. Coll. Legal Medicine (mem. bd. govs. 1986, chmn. edn. com. 1983—, chmn. 1985 nat. meeting, New Orleans, chmn. 1988 nat. meeting, Va., bd. dirs. ACLM Found.), Am. Acad. Forensic Scis., Royal Soc. Medicine, Royal Soc. Arts (London), Royal Soc. Medicine (London), Roscoe Pound Found. of Assn. Trial Lawyers Am.; mem. ABA, Am. Soc. Law and Medicine, Congress Neurol. Surgeons, Assn. Trial Lawyers Am., Soc. Med. Jurisprudence (trustee), N.Y. Bar Assn., Conn. Bar Assn., Fla. Bar Assn., D.C. Bar Assn., N.Y. Acad. Scis., Assn. Trial Lawyers Am. (bd. govs.), N.Y. Trial Lawyers Assn., Conn. Trial Lawyers Assn., Fla. Acad. Trial Lawyers, Md. Trial Lawyers Assn., Tex. Trial Lawyers Assn., Pa. Trial Lawyers Assn., Nat. Bar Assn. (mem. com. on South Africa), Nassau County Bar Assn., Fairfield County Med. Soc., Nassau-Suffolk Trial Lawyers Assn. Club: Cosmos (Washington). Home: 55 Mill River Rd Oyster Bay NY 11771-2711 Office: 175 E Shore Rd Great Neck NY 11023-2430

WACHSPRESS, MELVIN HAROLD, electrical engineer, consultant; b. N.Y.C., Jan. 31, 1926; s. Sidney and Jean (Lichtenstein) W.; m. Ruth Lessner, Aug. 19, 1951; children: Eric, Deborah. BEE, The Cooper Union, 1949; MEE, NYU, 1954. Registered profl. engr., N.Y. Staff engr. to chief scientist Gen. Instrument Corp., Hicksville, N.Y., 1963-90; cons. pvt. practice, 1990—. Served in USN, 1944-46. Achievements include patent for Microwave Omnidirectional Antenna. Home: 67 Alexander Dr Syosset NY 11791

WACKER, WARREN ERNEST CLYDE, physician, educator; b. Bklyn., Feb. 29, 1924; s. John Frederick and Kitty Dora (Morrissey) W.; m. Ann Romeyn MacMillan, May 22, 1948; children: Margaret Morrissey, John Frederick. Student, Georgetown U., 1946-47; M.D., George Washington U., 1951; M.A. (hon.), Harvard, 1968. Intern George Washington U. Hosp., 1951-52, resident, 1952-53; resident Peter Bent Brigham Hosp., Boston, 1953-55; Nat. Found. Infantile Paralysis fellow, 1955-57; investigator Howard Hughes Med. Inst., Boston, 1957-68; mem. faculty Harvard Med. Sch., Boston, 1955—; assoc. prof. medicine Harvard Med. Sch., 1968-71, Henry K. Oliver prof. hygiene, 1971-89, prof. hygiene, 1989—, dir. univ. health services, 1971-89, acting master Mather House, 1974-75, acting master Kirkland House, 1975-76, master Cabot House, 1978-84; sr. med. cons. Risk Mgmt. Found. of the Harvard Med. Instns., Cambridge, 1992—; vis. scholar St. Mary's Hosp. Med. Sch., 1964; vis. prof. U. Tel Aviv, 1987; dir. Applied Mgmt. Systems, Burlington, Mass., Millipore Corp., Bedford, Mass. Author: Magnesium and Man, 1981; sec., editorial adv. bd.: Biochemistry, 1962-76; assoc. editor: Magnesium; contbr. articles to med. and sci. jours. Vestryman St. Paul's Episc. Ch., Brookline, Mass., 1965-68, 76-79, 91—; bd. dirs. Harvard Community Health Plan, Boston, 1973-84; mem. fin. com., 1984-86, mem. corp., 1986—; bd. dirs. Bishop Rhinelander Found., Cambridge, 1973-76, 78-84, Controlled Risk Ins. Co., 1976-78; pres. bd. overseers Peter Bent Brigham Hosp., Boston, 1979-84; trustee Brigham and Women's Hosp., Boston, 1984-89, Risk Mgmt. Found., 1979-92; mem. mgmt. bd., med. bd. MIT, 1985—; mem. corp. Mt. Auburn Hosp., Cambridge, 1986—; mem. adv. bd. hospitality program Episc. Diocese Mass., 1989—. Served to 1st lt. USAAF, 1942-45. Decorated Air medal, D.F.C., Liberation medal (Greece); named Disting. Alumnus, George Washington U., 1963; recipient Cert. of Merit, Soc. Magnesium Research, 1985. Mem. Am. Chem. Soc., Am. Soc. Biol. Chemistry, Am. Soc. Clin. Investigation, AMA, Mass. Med. Soc., A.C.P., Am. Coll. Health Assn. (pres. 1981, Boynton award 1986), Biochemistry Soc. (London), Am. Coll. Nutrition, Sigma Xi, Alpha Omega Alpha. Democrat. Episcopalian. Clubs: Harvard (Boston); Cosmos (Washington). Home: 91 Glen Rd Brookline MA 02146-7764 Office: Risk Mgmt Found 840 Memorial Dr Cambridge MA 02139

WADA, AKIYOSHI, physicist; b. Tokyo, June 28, 1929; s. Koroku and Haruko (Kikkawa) W.; m. Sachiko Naito, Mar. 16, 1958; children: Akihisa, Akihide. BS, U. Tokyo, 1952, PhD, 1959. Asst. prof. dept. physics Faculty of Sci. U. Tokyo, 1962-71, prof., 1971-90, dean Faculty of Sci., 1989-90, prof. emeritus, 1990—; dir. Sagami Chem. Rsch. Ctr., Sagamihara, Japan, 1986—; pres. Biophys. Soc. Japan, Tokyo, 1973-75; chmn. biophys. com. Sci. Coun. Japan, Tokyo, 1979-85; mem. coun. Human Frontier Sci. Program, Strasburg, France, 1990—, Sci. Coun. of Japan, Tokyo, 1991—. Author: d-Helix as a Macrodipole, 1976, Molten Globule State of Protein, 1984, DNA Sequencing Strategy, 1987, Computer in Biophysics. Recipient Shimpo prize Chem. Soc. of Japan, 1961, Matsunaga prize Matsunaga Found., 1971, Simadu prize Shimadu Found., 1983. Mem. Human Genome Orgn., Sci. and Tech. Agy., Tokyo Club. Home: 11-1-311 Akasaka 8, Minato-ku Tokyo 107, Japan Office: Sagami Chem Rsch Ctr, 4-1 Nishi-Ohnuma 4, Sagamihara Kanagawa 229, Japan

WADA, EITARO, biogeochemist; b. Tokyo, Sept. 21, 1939; s. Torakichi and Kachiko (Katoh) W.; m. Setsuko Tanabe, May 3, 1967; children: Tohru, Akane, Hiroshi. BS, Tokyo U. Edn., 1962, MS, 1964, PhD, 1967. Rsch. assoc. U. Tokyo, 1967-76; vis. scientist U. Tex., Port Aransas, 1974-75; chief Mitsubishi-Kasei Inst. Life Scis., Tokyo, 1976—, dir. social and environ. sci. dept., 1989-91; prof. Ctr. Ecol. Rsch. Kyoto U., 1991—; instr. Sch. Meteorology, Chiba, Japan, 1967-68. Mem. New Zealand Antarctic Expedition, 1980-81. Co-author: Marine Biochemistry, 1971; Heavy Nitrogen, 1980. Author/editor: Story of the Sea, 1984, Nitrogen in the Sea: Forms, Abundnces and Rat Processes, 1991. Author jour. Nature, 1981; assoc. editor Oceanographic Soc. Japan, Tokyo, 1974. Pres. Sengoku Children's Assn., Kawasaki, Kanagawa, Japan, 1984, Bd. Isotopenpraxis (Germany), 1987—. Recipient Okada price Oceanographic Soc. Japan, 1974. Mem. Geochem. Soc. Japan, Oceanographic Soc. Japan, Internat. Vereinigung fur Theoretische und Angewandte Limnology, Japanese Soc. Soil Sci. and Plant Nutrition. Buddhist. Avocation: travel. Home: 2-31-1 Suge-Sengoku, Tamaku Kawasaki 214, Japan Office: Kyoto U Ctr Ecol Rsch, 4-1-23 Shimosakamoto, Otsu Shiga 520-01, Japan

WADDELL, DAVID GARRETT, electrical engineer, consultant; b. Houston, Feb. 10, 1941; s. David Lewis and Portia Jane (Garrett) W.; m. Patricia Ann Gardner, July 2, 1977; 1 child, Sarah Gail. Student, Lamar U., 1962-65. Pres. Counterparts Cons. Inc., St. Petersburg, Fla., 1985—. With USAF, 1958-62. Mem. Internat. Motorsports Assn., Sports Car Club Am., Sportscar Vintage Racing Assn. Republican. Office: Counterparts Cons Inc 3773 Central Ave Ste A668 Saint Petersburg FL 33713-1800

WADDEN, RICHARD ALBERT, environmental science educator, consultant, researcher; b. Sioux City, Iowa, Oct. 3, 1936; s. Sylvester Francis and Hermina Lillian (Costello) W.; m. Angela Louise Trabert, Aug. 9, 1975; children—Angela Terese, Noah Albert, Nuiko Clare. Student, St. John's U., Collegeville, Minn., 1954-56; B.S. in Chem. Engring., Iowa State U., 1959; M.S. in Chem. Engring, N.C. State U., 1962; Ph.D. in Chem. and Environ. Engring., Northwestern U., 1972. Registered profl. engr., Ill.; cert. indsl. hygienist. Engr. Linde Co., Tonnawanda, N.Y., 1959-60, Humble Oil Co., Houston, 1962-65; instr. engring. Pahlavi U. Peace Corps, Shiraz, Iran, 1965-67; tech. adviser Ill. Pollution Control Bd., Chgo., 1971-72; asst. dir. Environ. Health Resource Ctr. Ill., Chgo., 1972-74; asst. prof. environ. and occupational health scis. Sch. Pub. Health U. Ill.-Chgo., 1972-75, assoc. prof., 1975-79, prof., 1979—, dir., 1984-86, 87-92; dir. Office Tech. Transfer U. Ill. Ctr. for Solid Waste Mgmt. and Resch., 1987-92; dir. indsl. hygiene program Ednl. Resource Ctr. Occupational Safety and Health U. Ill.-Chgo.; vis. scientist Nat. Inst. Environ. Studies, Japan, 1978-79, invited scientist, 1983, 84, 88; cons. air pollution control, health implications of energy devel., indoor air pollution. Author: Energy Utilization and Environmental Health, 1978, (with P.A. Scheff) Indoor Air Pollution, 1983, Engineering Design for Control of Workplace Hazards, 1987; contbr. numerous articles to profl. publs. Sr. Internat. fellow Fogarty Internat. Ctr.-NIH, 1978-79, 83; WHO fellow, 1984. Mem. Am. Chem. Soc., Am. Inst. Chem. Engrs., Am. Acad. Environ Engrs. (diplomate), Am. Acad. Indsl. Hygiene (diplomate), Air and Waste Mgmt. Assn., Am. Indsl. Hygiene Assn. Office: U Ill Sch Pub Health PO Box 6998 Chicago IL 60680-6998

WADDEN, THOMAS ANTONY, psychology educator; b. Richmond, Va., Sept. 3, 1952; s. Thomas Antony Jr. and Mary Lloyd (Cradock) W.; m. Jan Robin Linowitz, Nov. 11, 1984; children: David Joseph, Michael James. AB magna cum laude, Brown U., 1975; PhD, U. N.C., 1981. Psychology intern Boston VA Med. Ctr., 1980-81; instr. in psychology U. Pa. Sch. Medicine, Phila., 1981-82, asst. prof. psychology, 1982-87, assoc. prof. psychology, 1987-91; prof. psychology, dir. clin. tng. Syracuse (N.Y.) U., 1992—; clin. dir. Obesity Rsch. Group, U. Pa., Phila., 1983-91; dir. Ctr. for Health and Behavior, Syracuse U., 1992—. Assoc. editor Annals of Behavioral Medicine, 1990—, mem. editorial bd. Behavioral Therapy, Internat. Jour. Obesity, 1986—, Health Psychology, 1989-92, Jour. Consulting and Clin. Psychology; editor (with T.B. VanItallie) Treatment of the Seriously Obese Patient, 1992, (with A.J. Stunkard) Obesity: Theory and Therapy, 1993; contbr. chpts. to books; writer numerous sci. papers. Recipient Nat. Rsch. Svc. award Nat. Inst. Mental Health, 1983-85, Rsch. Scientist Devel. award, 1987-92. Mem. APA, Soc. Behavioral Medicine (bd. dirs. 1987-90), Assn. for Advancement of Behavior Therapy (New Rschr. award 1986), Acad. Behavioral Medicine, N.Y. Acad. Scis., Phi Beta Kappa, Sigma Xi. Democrat. Avocations: tennis, squash, symphonic music, guitar. Home: 8304 Glen Eagle Dr Manlius NY 13104 Office: Syracuse U 503 Huntington Hall Syracuse NY 13244

WADDINGTON, DAVID JAMES, chemistry educator; b. Edgware, Middlesex, U.K., May 27, 1932; s. Eric James and Marjorie Edith (Harding) W.; m. Isobel Hesketh, Aug. 17, 1957; children: Matthew James, Rupert John, Jessica Katharine. BSc, Imperial Coll., London, 1953, PhD, 1956. Head of sci. dept. Wellington Coll., Berkshire, 1960-64; sr. lectr. U. York, U.K., 1965-78, prof., 1978—. Co-author: Organic Chemistry Through Experiment, 4th edit., 1985, Modern Organic Chemistry, 4th edit., 1985, Chemistry: The Salters' Approach, 1990; editor: Chemical Education in the Seventies, 2d edit., 1982, Teaching School Chemistry, 1985, Education, Industry and Technology, 1987. Recipient Nyholm medal Royal Soc. of Chemistry, 1985, Brasted award Am. Chem. Soc., 1988. Mem. Internat. Coun. of Scientific Unions (chmn. com. on teaching of sci. 1989—). Home: Murton Hall, York United Kingdom Office: U York, York Y01 5DD, England

WADE, JAMES WILLIAM, aerospace engineer; b. Mpls., Feb. 15, 1964; s. Richard Gene and Maxine Louise (Dodge) W.; m. Katherine Louise Aune, Oct. 29, 1988; children: Andrew James, Alexandra Katherine. BA, Gustavus Adolphus Coll., St. Peter, Minn., 1986; MS, U. Ill., 1987; PhD, U. Colo., 1991. Cert. instrument flight instr. Ind. contractor Universal Energy Systems, Albuquerque, summer 1987; assoc. engr. Lockheed Engring & Scis., Houston, 1987-89; rsch. asst. U. Colo., Boulder, 1989-91; sr. engr. Lockheed Engring. & Scis., Houston, 1991-93; aerospace engr. NASA-Johnson Space Ctr., Houston, 1993—. Co-adminstr. ch. youth group Shepherd of the Hills, Boulder, 1990-91. Rsch. grantee Sigma Xi, St. Peter, Minn., 1985-86; Century XXI fellow U. Colo., Boulder, 1989-91. Mem. AIAA, Aircraft Owners and Pilots Assn., Mensa, Sigma Xi. Lutheran. Home: 1417 Chestnest Springs Houston TX 77062 Office: NASA-JSC Mail Code ET3 Houston TX 77058

WADE, MICHAEL JOHN, ecology and evolution educator, researcher; b. Evanston, Ill., Oct. 21, 1949; s. John Francis and Mary Loretta (Sweeney) W.; m. Debra Lynn, Sept. 4, 1987; children: Catherine, Megan, Travis. BA, Boston Coll., 1971; PhD, U. Chgo., 1975. Asst. prof. dept. biology U. Chgo., 1975-81, assoc. prof., 1981-86, prof. dept. ecology and evolution, 1986—, chmn. dept., 1991—; mem. predoctoral panel NSF, 1979-84, population biology panel, 1986-88, genetics study sect. NIH, 1988-91. Mem. Soc. for Study Evolution (v.p. 1985). Office: U Chgo 1101 E 57th St Chicago IL 60637-1573

WADE, ROBERT GLENN, engineering executive; b. Sturgeon, Mo., Nov. 21, 1933; s. Robert Clifford and Mildred Guinn (Bartee) W.; m. Geraldine Harris, Dec. 27, 1959; 1 child, Carolyn Ruth. BSCE, U. Mo., 1955. Registered profl. engr., Mo., Kans. Structural engr. Carter-Waters Corp., Kansas City, Mo., 1958-62; project mgr. Pfuhl & Stevson, Kansas City, 1962-76; prin. Stevson-Hall & Wade, Inc., Kansas City, 1976-82; pres. Structural Engring. Assocs., Inc., Kansas City, 1982-85, chmn., chief exec. officer, 1985—; mem. Mo. Bd. Architects, Engrs. and Land Surveyors, 1992—; mem. Midwest Concrete Industry Bd., pres., 1975-76. Contbg. author: Quality Assurance for Consulting Engineers, 1986. Com. mem. Downtown Coun., Kansas City, 1990. 1st lt. USAF, 1956-58. Recipient 1st merit award Midwest Concrete Industry Bd., 1976, award of excellence Am. Inst. Steel Constrn., 1982, Excellence in Design award Prestressed Concrete Inst., 1988. Fellow ASCE (pres. Kansas City sect. 1986-87, Leadership award 1987); mem. Am. Cons. Engrs. Coun. (firm rep., bd. dirs. 1987-88), Cons. Engrs. Coun. Mo. (firm rep., pres. 1986-87, Svc. award 1987). Avocation: golf. Office: Structural Engring Assocs 101 W 11th St Kansas City MO 64105-1805

WADE, RODGER GRANT, financial systems analyst; b. Littlefield, Tex., June 25, 1945; s. George and Jimmie Frank (Grant) W.; m. Karla Kay Morrison, Dec. 18, 1966 (div. 1974); children: Eric Shawn, Shannon Annelle, Shelby Elaine; m. Carol Ruth Manning, Mar. 28, 1981. BA in Sociology, Tex. Tech. U., 1971. Programmer First Nat. Bank, Lubbock, Tex., 1971-73, Nat. Sharedata Corp., Odessa, Tex., 1973; asst. dir. computing ctr. Odessa Community Coll., 1973-74; programmer/analyst Med. Sci. Ctr., Tex. Tech U., Lubbock, 1974-76; sys. mgr. Hosp. Info. Sys., Addison, Tex., 1976-78; programmer, analyst Harris Corp., Grapevine, Tex., 1978-80, Joy Petroleum, Waxahachie, Tex., 1980-82; owner R&C Bus. Sys./Requerdos de Santa Fe, N.Mex., 1982-84; fin. sys. analyst Los Alamos (N.Mex.) Tech. Assocs.,

1984—; owner El Rancho Herbs, Santa Fe, 1988-91, Wade Gallery, Santa Fe, 1990-91, Wade Systems, Santa Fe, 1992—. Vol. programmer Los Alamos Arts Coun., 1987-88; mem. regulations task force N.Mex. Gov.'s Health Policy Adv. Com.; vol. systems support Amigos Unidos of Taos, 1990—. Republican. Avocation: photography. Home: RR 5 Box 271H Santa Fe NM 87501-9805 Office: Los Alamos Tech Assocs 1650 Trinity Dr Los Alamos NM 87544-3065

WADE, STACY LYNN, computer specialist; b. Douglas, Ariz., Aug. 29, 1965; d. Gilbert Wayne and Georgia Beth (Scott) Caddell; m. Scott Alan Wade, Apr. 2, 1988; 1 child, Travis Ryan. BS, DeVry Inst. Tech., 1986. Programmer analyst Nev. Power Co., Las Vegas, 1986-89; computer specialist U.S. Dept. Energy, Las Vegas, 1989—. Republican. Baptist. Office: US Dept Energy 2753 S Highland Las Vegas NV 89109

WADE, THOMAS EDWARD, university research administrator, electrical engineering educator; b. Jacksonville, Fla., Sept. 14, 1943; s. Wilton Fred and Alice Lucyle (Hedge) W.; m. Ann Elizabeth Chitty, Aug. 6, 1966; children: Amy Renee, Nathan Thomas, Laura Ann. BSEE, U. Fla.-Gainesville, 1966, MSEE, 1968, PhD, 1974. Cert. Rsch. Adminstr., 1993. Interim asst. prof. U. Fla.-Gainesville, 1974-76; prof. elec. engring. Miss. State U., Starkville, 1976-85, state-wide dir. state widemicroelectronics research lab., Miss., 1978-85, assoc. dean., prof. electrical engring. U. South Fla., Tampa, 1985—, dir. Engring. Indsl. Experiment Sta., 1986-93, exec. dir. Ctrs. for Engring. Devel. and Rsch., 1985-90, mem. presdl. faculty adv. com. for rsch. and tech. devel., 1986-88, mem. fed. demonstration project com. for contracts and grants, 1986-88; mem. adv. bd. USF Exec. Fellows Program, 1987-91; chmn. evaluation task force applied rsch. grants program High Tech. and Industry Coun. State of Fla., 1988-90, vice chmn. microelectronics and materials subcommittee 1987—, mem. telecommunications subcom, 1988-89, chmn. legis. report com. FHTIC, 1989-90; mem. Tampa Bay Internat. Super Task Force, 1986-92, vice chmn. edn. com. 1988; dir. Fla. Ctr. for Microelectronics Design and Test, 1986-88; bd. dirs. NASA Ctr. Comml. Devel. of Space, Space Communications Center, Fla., 1990—; rev. panel govt.-univ.-industry rsch. round table for fed. demonstration project, NAS, 1988; solid state circuit specialist Applied Micro-Circuits Corp., San Diego, 1981-82; sr. scientist NASA Marshall Space Flight Ctr., Huntsville, Ala., 1983; scientist Trilogy Semiconductor Corp., Santa Clara, Calif., 1984; bd. dirs. NASA Ctrs. for the Commercial Devl. Space, Space Comm. Ctr., Fla.; cons. in field. Author: Polyimides for VLSI Applications, 1984, (U.S. Army handbook) Modern VLSI Circuit Fabrication Processess, 1984, Photosensitive Polyimides for VLSI Applications, 1986, VLSI Multilevel Interconnection State-of-the-Art Seminar, 1989—; contbr. 120 articles to profl. jours. Vol., United Fund, Miss. State U., 1983-85. Recipient Outstanding Engring. Teaching award Coll. Engring. U. Fla., 1976, Cert. of Recognition NASA (5 times), 1981-88, Outstanding Research award Sigma Xi, 1984, Outstanding Contbn. to Sci. and Tech. award Fla. Gov., 1989, 90. Mem. AAAS, NSPE, IEEE (sr. mem., guest editor periodical 1982, gen. chmn. Internat. VLSI Multilevel Interconnection Conf. annually 1984—, editor conf. proceedings 1985—, chmn. acad. affairs com. CHMT Soc. 1984-86, gen. chmn. univ./govt./industry microelectronics symposium, 1981, tech. program commn., 1991, bd. dirs. workshop on tungsten and other refractory metals 1987-90), Am. Soc. Engring. Edn. (gen. chmn. engring. research counc. ann. meeting 1987, chmn. engring. rsch. coun. adminstrv. com. 1987-90, chmn. coun., 1990-92, session chmn. ann. meeting 1990, 92, bd. dirs. 1990-92, mem. Nominations Com. 1992-93, mem. Long Range Planning Com. 1992-93, recipient Centennial Certificate 1992), World Future Soc., Internat. Soc. Hybrid Microelectronics, Assn. U.S. Army (bd. dirs. Suncoast chpt. 1991—), Soc. Photo Optical Instrumentation Engring., Univ. Faculty Senate Assn. of Miss. (organizer 1985), Am. Vacuum Soc., Am. Phys. Soc., Am. Electronics Assn., World Future Soc., Am. Inst. Physics, Nat. Coun. Univ. Rsch. Adminstrn., Soc. Rsch. Adminstrs. (external rels. com. for SRA 1988-91), Fla. Engring. Soc. (v.p. edn. com. 1988-89, pres. 1989-90, bd. dirs. 1989-90, Fla. engring. found. trustee 1989-90, ann. meeting steering com. 1989-90, Outstanding Svc. to the Profession award 1992), Soc. Am. Mil. Engring., Order of Engrs., 1991—, Tampa Bay Internat. Super Task Force (vice chmn. edn. com. 1988,) 1987-92, Sigma Xi (v.p. 1985), Tau Beta Pi (Fla. Alpha chpt. pres. 1969, 71, nat. outstanding chpt. award 1969, 71, faculty advisor Miss. Alpha chpt. 1977-85, faculty advisor Fla. Gamma chpt. 1986—), Eta Kappa Nu (pres. 1968), Sigma Tau, Omicron Delta Kappa, Soc. Am. Inventors,e Key (v.p. 1972, sec, 1971), Epsilon Lambda Chi (founder 1970, pres. 1971). Club: Rotary (Paul Harris Fellow 1987, perfect attendance award 1986—, chmn. com. on environ. issues 1990). Active First Bapt. Ch., Temple Terrace, Fla., vice-chmn. bd. deacons 1989-90, chmn. bd. deacons, 1990-91, chmn. pastor search com. 1990-91, vice chmn. long range plannning com., 1989-91. Avocations: collecting antique furniture, carpentry, restoring antique sports cars, basketball. Home: 5316 Witham Ct E Tampa FL 33647-1026

WADEY, BRIAN LEU, polymer engineer; b. Montreal, Que., Can., Apr. 30, 1941; came to U.S., 1982; s. Gerald Henry Wadey; m. Heather E. Henry, May 15, 1965; children: Allison, Matthew. B Applied Sci., St. Francis Xavier, Antigonish, Nova Scotia, 1967. Registered profl. engr., Va. Tech. sales engr. Union Carbide, Lachine, Que., 1967-72; bus. devel. mgr. Dow Chem., Montreal, 1972-77; tech. mgr. B.F. Goodrich, Kitchener, Ont., 1977-82; tech. dir. Pantasote, Passaic, N.J., 1982-85; N.Am. tech. mgr. BASF Corp., Parsippany, N.J., 1985—. Mem. ASTM, Soc. Plastics Engrs. (treas. 1974-77, vinyl inst. bd. 1987-89, health and safety chair 1990—), Soc. Automotive Engrs. (fogging com. 1987—). Achievements include research on the effect of polymeric additives on automotive interior trim. Office: BASF Corp 100 Cherry Hill Rd Parsippany NJ 07054

WADLINGTON, WALTER JAMES, legal educator; b. Biloxi, Miss., Jan. 17, 1931; s. Walter and Bernice (Taylor) W.; m. Ruth Miller Hardie, Aug. 20, 1955; children: Claire Hardie, Charlotte Taylor Griffith, Susan Miller, Derek Alan. A.B., Duke U., 1951; LL.B., Tulane U., 1954. Bar: La. 1954, Va. 1965. Pvt. practice New Orleans, 1954-55, 58-59; asst. prof. Tulane U., 1960-62; mem. faculty U. Va., 1962—, prof law, 1964—, James Madison prof., 1970—, prof. legal medicine, 1979—, Harrison Found. rsch. prof., 1990-92; tutor civil law U. Edinburgh, Scotland, 1959-60; vis. Tazewell Taylor prof. law Coll. William and Mary, spring 1986; program dir. Robert Wood Johnson Med. Malpractice Program, 1985-91; mem. adv. bd. clin. scholars program, 1989—. Author: Cases and Materials on Domestic Relations, 1970, 2d edit., 1990, Cases and Materials on Law and Medicine, 1980, (with Whitebread and Davis) Children in the Legal System, 1983; editor-in-chief Tulane U. Law Rev., 1953-54. Served to capt. AUS, 1955-58. Mem. Va. Bar Assn., Am. Law Inst., Inst. Medicine of NAS, Order of Coif. Home: 1620 Keith Valley Rd Charlottesville VA 22903 Office: U Va Sch Law Charlottesville VA 22903

WADMAN, WILLIAM WOOD, III, health physicist, consulting company executive, consultant; b. Oakland, Calif., Nov. 13, 1936; s. William Wood, Jr., and Lula Fay (Raisner) W.; children: Roxanne Alyce Wadman Hubbing, Raymond Alan (dec.), Theresa Hope Wadman Beaudreaux; m. Barbara Jean Wadman; stepchildren: Denise Ellen Varine Skrypkar, Brian Ronald Varine Skrypkar. M.A., U. Calif., Irvine, 1978. Radiation safety specialist, accelerator health physicist U. Calif. Lawrence Berkeley Lab., 1957-68; campus radiation safety officer U. Calif., Irvine, 1968-79; dir. ops., radiation safety officer Radiation Sterilizers, Inc., Tustin, Calif., 1979-80; prin., pres. Wm. Wadman & Assocs. Inc., 1980-87; pres. Intracoastal Marine Enterprises Ltd., Martinez, Calif., 1981-86; mem. team No. 1, health physics appraisal program NRC, 1980-81; cons. health physicist to industry; lectr. sch. social ecology, 1974-79, dept. community and environ. medicine U. Calif., Irvine, 1979-80, instr. in environ. health and safety, 1968-79, Orange Coast Coll., in radiation exposure reduction design engring. Iowa Electric Light & Power; trainer Mason & Hanger-Silas Mason Co., Los Alamos Nat. Lab.; instr. in medium energy cyclotron radiation safety UCLBL, lectr. in accelerator health physics, 1966, 67; curriculum developer in field. Active Cub Scouts; chief umpire Mission Viejo Little League, 1973. Served with USNR, 1955-63. Recipient award for profl. achievement U. Calif. Alumni Assn., 1972, Outstanding Performance award U. Calif., Irvine, 1973. Mem. Health Physics Soc. (treas. 1979-81, edittor proc. 11th symposium, pres. So. Calif. chpt. 1977, Professionalism award 1975), Internat. Radiation Protection Assn. (U.S. del. 4th Congress 1977, 8th Congress 1992), Am. Nuclear Soc., Am. Public Health Assn. (chmn. program 1978, chmn. radiol. health sect. 1979-80), Campus Radiation Safety Officers (chmn. 1975, editor proc. 5th conf. 1975), ASTM. Club: UCI Univ. (dir. 1976, sec. 1977, treas. 1978). Contbr. articles to tech. jours. Avocation: achievement include research in radiation protection and environmental sciences. Home: 3687 Red Cedar Way Lake Oswego OR 97035-3525 Office: 675 Fairview Dr Ste 246 Carson City NV 89701-5436

WADSÖ, B. INGEMAR, science educator; b. Varberg, Sweden, Apr. 13, 1930; m. Anne Marie Wadsö; children: Lars, Cecilia. PhD, Lund U., Sweden, 1962. Full prof. Lund (Sweden) U., 1981—. Author more than 200 sci. papers. Achievements include patents in field. Office: Lund U, Box 124, 22100 Lund Sweden

WADZINSKI, HENRY TEOFIL, physicist; b. Wilkes-Barre, Pa., Dec. 22, 1938; s. Teofil Gerald and Helen Telesfora (Kitlowski) W.; m. Pauline Anita Walker, June 11, 1966; children: Steven John, Thomas Luke. BA magna cum laude, Harvard Coll., 1960; MA, U. Calif., Berkeley, 1963; PhD, Johns Hopkins U., 1968. Physicist Lab. Aime-Cotton, Orsay, France, 1968-70; sr. analyst Smithsonian Astrophys. Observatory, Cambridge, Mass., 1970-78; physicist, analyst Bedford (Mass.) Rsch. Assocs., 1978-86; physicist Boston Coll., Chestnut Hill, Mass., 1986—; mem. sci. panel Project STAR Harvard-Smithsonian Ctr. for Astrophysics, 1988—, mem. adv. bd. Project Insight, 1990—. Active Boy Scouts Am., Arlington, Mass., 1980—; bd. dirs. Arlington Youth Visit Exch. Program, 1989—. Recipient Skylab Achievement award NASA, 1974, Skylab Group Achievement award, 1974. Mem. Am. Phys. Soc., Am. Geophys. Union, AAAS, N.Y. Acad. Scis., Sigma Xi. Roman Catholic. Home: 71 Menotomy Rd Arlington VA 02174-6111 Office: Boston Coll Inst Space Rsch Barry Pavilion 885 Centre St Newton MA 02159

WAELDE, LAWRENCE RICHARD, chemist; b. Teaneck, N.J., Dec. 27, 1951; s. Clinton Brewster and Eileen Florence (Kennedy) W.; m. Soledad Nelita Acedillo, May 24, 1975; children: Christine Ann, Richard Adams. BS, Fairleigh Dickinson U., 1976; postgrad., Syracuse U., 1969-72. Project leader Muralo Paints, Bayonne, N.J., 1974-79; lab. mgr. Lazon Paints, Fair Lawn, N.J., 1979-84; plant mgr. Stevens Paint Co., Yonkers, N.Y., 1984-86; mgr. powder coatings Troy Chem. Corp., Newark, 1986—. Mem. N.Y. Soc. Coating Tech. (tech. chmn. 1990—), Powder Coating Inst. (tech. mem.). Office: Troy Chem Corp 1 Ave L Newark NJ 07105

WAELSCH, SALOME GLUECKSOHN, geneticist, educator; b. Danzig, Germany, Oct. 6, 1907; came to U.S., 1933, naturalized, 1938; d. Ilya and Nadia Gluecksohn; m. Heinrich B. Waelsch, Jan. 8, 1943; children—Naomi Barbara, Peter Benedict. Student, U. Konigsberg, Germany, U. Berlin, 1927-28; Ph.D., U. Freiburg, Germany, 1932. Rsch. assoc. in genetics Columbia U., 1936-55; assoc. prof. anatomy Albert Einstein Coll. Medicine, 1955-58, prof., 1958-63, prof. genetics, 1963—, chmn. dept. genetics, 1963-76; mem. study sects. NIH. Contbr. numerous articles on devel. genetics. Recipient Nat. Medal of Sci., Nat. Sci. Found., 1993. Fellow AAAS, Am. Acad. Arts and Scis.; mem. NAS, N.Y. Acad. Scis. (hon. life), Am. Soc. Zoologists, Am. Assn. Anatomists, Genetics Soc., Soc. Devel. Biology, Am. Soc. Naturalists, Am. Soc. Human Genetics, Sigma Xi. Office: Albert Einstein Coll Med Dept Molecular Genetics 1300 Morris Park Ave Bronx NY 10461-1924

WAESCHE, R(ICHARD) H(ENLEY) WOODWARD, combustion research scientist; b. Balt., Dec. 20, 1930; s. J(oseph) Edward and Margaret Steuart (Woodward) W.; m. Lucy Spotswood White, June 29, 1957; children: Charles Russell, Ann Spotswood. BA, Williams Coll., 1952; postgrad. U. Ala., 1956-58; MA, Princeton U., 1962, PhD, 1965. Rsch. scientist Rohm & Haas Redstone div., Huntsville, Ala., 1954-59; rsch. asst. Princeton U., 1961-64; sr. rsch. scientist Rohm & Haas, Huntsville, 1964-66; sr. rsch. engr. United Tech. Rsch. Ctr., East Hartford, Conn., 1966-81; prin. scientist Atlantic Rsch. Corp., Gainesville, Va., 1981-93; sr. sci. Sci. Applications Internat. Corp., 1993—, Atlantic Rsch. Corp., 1992—, Battelle Meml. Inst., 1992—; cons. Goodyear Corp., 1959-60, Princeton U., 1965, NRC, 1985-86, NASA, 1987—, Def. Adv. Rsch. Project Agy., 1988—, Directed Techs, 1989—. Assoc. editor Jour. Spacecraft and Rockets, 1975-80, editor-in-chief, 1980-86, Jour. Propulsion and Power, 1986—; contbr. numerous articles to profl. jours.; mem. exec. bd. Dictionary of Modern Science and Technology; mem. exec. adv. bd. Encyclopedia of Physical Science and Technology. Chmn. Fine Arts Commn., Glastonbury, Conn., 1975-77. Served to cpl. U.S. Army, 1952-54. Recipient JANNAF Recognition award, 1988; Guggenheim fellow, 1959-61. Fellow AIAA (chmn. propellants and combustion tech. com. 1975-77, propulsion tech. group coord. 1979-81, tech. activities com. 1979—, publs. com. 1980—, inst. devel. com. 1988—, fin. com. 1988—, dir. propulsion & energy 1992—, bd. dirs. 1992—, Best Paper in Solid Rockets 1989); mem. Am. Phys. Soc., Combustion Inst., Evergreen Country Club (Haymarket, Va.), Sigma Xi. Episcopalian. Home: 4319 Banbury Dr Gainesville VA 22065-1122 Office: Sci Applications Internat Corp 1710 Goodridge Dr Mc Lean VA 22102

WAGENKNECHT, WALTER CHAPPELL, radiologist; b. Evanston, Ill., July 1, 1947; s. Edward Charles and Dorothy (Arnold) W.; m. Elizabeth Ann Martin, July 27, 1985; children: Brendan, James. AB in History, Boston U., 1969, ThM in Bibl. Studies, 1973, MD, 1979; BA in Biology, U. Mass., 1975. Diplomate Am. Bd. Radiology. Med. intern Faulkner Hosp., Boston, 1979-80; radiology resident St. Vincent Hosp., Worcester, Mass., 1980-83; fellow in pediatric radiology U. Mass. Med. Ctr., Worcester, 1983-84; radiologist South County Hosp., Wakefield, R.I., 1984-85, Heywood Meml. Hosp., Gardner, Mass., 1985—. Jacob Sleeper fellow Boston U., 1973-74. Mem. Am. Coll. Radiology, Am. Roentgen Ray Soc., Radiol. Soc. N.Am., Phi Beta Kappa. Home: 177 Leo Dr Gardner MA 01440 Office: Wachusett Radiology Inc 250 Green St Ste 203 Gardner MA 01440

WAGENSBERG, JORGE, physicist; b. Barcelona, Spain, Dec. 2, 1948; s. Isaac Wagensberg and Sara Lubinski. Degree Physics, U. Barcelona (Spain), 1971, PhD Physics, 1976. Asst. prof. physics U. Barcelona, 1972-76, prof., 1976—; sci. dir. Mus. Sci. of Fund "la Caixa", 1986-91, dir., 1991—; v.p. Institut d-Humanitats, 1986—; dir., creator Metatemas, 1983—, Materia Viva, 1990—; Author: Nosotros y la Ciencia, 1980, Ideas Sobre la Comple jidad del Mundo, 1985; contbr. articles to profl. jours. Mem. Math. Biology Soc., Fundación Conde de Barcelona. Home: Laforja 95, E-08021 Barcelona Spain Office: Museu de la Ciencia, Teodor Roviralta 55, E-08022 Barcelona Spain

WAGERS, WILLIAM DELBERT, JR., theoretical geneticist; b. Denver, Mar. 4, 1949; s. William Delbert and Christine Varah (Buchanan) W.; m. Susan Rebecca Hotchkiss, 1969 (div. June 1974); 1 child, Christian Michel. BBA, M in Liberal Arts, So. Meth. U., 1971; MS, U. North Tex., 1993. Systems analyst Rapidata, Inc., Fairfield, N.J., 1973-75; account mgr. Burroughs Machines, Ltd., London, 1975-77; sr. systems analyst Jung Systeme, Karlsruhe, 1977, Moore Bus. Systems, Denton, Tex., 1978-81; dir. systems devel. Scott Instruments Corp., Denton, 1981-86; pvt. practice author, cons. Denton, 1985-89; rschr. biol. scis. U. North Tex., Denton, 1992—; mem. allocations subcom. United Way, Denton, 1982; mem. Land Use Planning Com., Denton, 1982; mem. City Energy Subcom., Denton, 1983. Precinct chmn. Rep. Party, Denton, 1979-88; election judge Rep. Party, Denton, 1984-86; assemblyman U. North Tex. Student Assembly, Denton, 1992-93. Recipient ACE-Best Local Origination Newcast award Assn. for Cable Excellence, 1982. Mem. Tri-Beta. Democrat. Roman Catholic. Achievements include patent for Process of Man-Machine Interactive Educational Instruction Using Voice Response Verification; discovery of biological basis of mythological texts e.g. Bible, Cabbalah, known as Biocosmic Theory. Home: Ste 241 1407 Bernard Denton TX 76201 Office: Univ North Tex PO Box 5218 Denton TX 76205

WAGGONER, JOHN EDWARD, psychology educator; b. Harrisburg, Pa., Jan. 18, 1959; s. John William and Lenore Dorothy (Irrgang) W.; m. Dianne Marie Yarko, July 15, 1988. BA in Psychology, Shippensburg State Coll., 1981; MS, Pa. State U., 1985, PhD, 1989. Asst. prof. psychology Bloomsburg (Pa.) U., 1988—. Contbr. articles to profl. jours. Mem. APS, Jean Piaget Soc., Soc. for Rsch. in Child Devel. Achievements include research on grasping the meaning of metaphor, children's comprehension of emotional metaphor, interaction theories of metaphor, metaphor and context in the language of emotion. Home: PO Box 550 Millville PA 17846 Office: Bloomsburg U Psychology Dept 2124 McCormick Bloomsburg PA 17815

WAGGONER, LEE REYNOLDS, federal agency administrator; b. Uniontown, Pa., Oct. 20, 1940; s. Thomas Aubrey and Genevie (Ruse) W.; m. Beverly Petersen, June 12, 1965; children: D'Etta, Lee. BS in Zoology, Ohio U., 1962; BS in Pharmacy, W.Va. U., 1969; M in Forensic Sci., George Washington U., 1977. Diplomate Am. Bd. Forensic Document Examiners. Spl. agt. FBI, Washington, 1972—. Contbr. articles to Jour. Forensic Scis., FBI Law Enforcement Bull. Lt. comdr. USNR, 1962-73. Fellow Am. Acad. Forensic Scis. Home: 15308 Carrolton Rd Rockville MD 20853 Office: FBI 10th & Pennsylvania Ave NW Washington DC 20535

WAGGONER, SUSAN MARIE, electronics engineer; b. East Chicago, Ind., Sept. 1, 1952; d. Joseph John and Elizabeth (Monyok) Vasilak; m. Steven Richard Waggoner, July 31, 1976. AS, Ind. U., 1975, BA in Journalism, 1976, BS in Physics, 1982, M in Pub. Affairs, 1991. Engring. technician Naval Surface Warfare Ctr., Crane, Ind., 1978-82, electronics engr. test and measurement equipment, 1982-91, electronics engr. batteries, 1991—. Mem. Am. Soc. Naval Engrs., Fed. Mgrs. Assn., Federally Employed Women, Am. Rose Soc., Am. Hort. Soc., Indpls. Rose Soc., Theatre Circle of Ind. U., Sigma Pi Sigma. Home: RR 5 Box 387 Loogootee IN 47553-9337 Office: Naval Surface Warfare Ctr 300 Hwy 361 Crane IN 47522-5001

WAGMAN, GERALD HOWARD, library administrator; b. Newark, Mar. 4, 1926; s. David and Sophie (Milinsky) W.; B.S., Lehigh U., 1946; M.S., Va. Poly. Inst. and State U., 1947; m. Rhoda Kirschner, Dec. 9, 1948; children: Jan Donald, Neil Mark. Tech. research asst. Squibb Inst. for Med. Research, New Brunswick, N.J., 1947-49, research asst., 1954-57; mgr. Yankee Radio Corp., N.Y.C., 1950-54; assoc. biochemist Schering Corp. (now Schering-Plough Rsch. Inst.), Kenilworth, N.J., 1957-58, biochemist, 1958-65, sr. biochemist, 1966-68, sect. leader, 1969-70, mgr. antibiotics dept., 1970-74, assoc. dir. microbiol. scis.-antibiotics, 1974-76, assoc. dir. microbiol. scis. and head screening lab., 1977-78, dir. microbiol. strain lab., 1979-84, antibiotics isolation, 1984-85, dir. microbial products chem. screening, 1985-87, prin. scientist, 1987-89, mgr. libr. info. ctr., 1989-93; tech. writing cons., editor, 1993—; mem. adv. bd. Nat. Cert. Commn. in Chemistry and Chem. Engring., 1985-88. Coun. mem. Troop 23 Boy Scouts Am., 1964-66; communications officer East Brunswick Civil Def. and Disaster Control, 1966-71; mem. sci. adv. com. East Brunswick Bd. Edn., 1960-68; bd. dirs. Tamarack N. Homeowners Assn., 1983-84, 89-93, pres., 1989-93. Recipient Public Svc. award Am. Radio Relay League, 1965. Chartered chemist, Gt. Britain. Fellow Am. Inst. Chemists; mem. AAAS, ALA, Spl. Librs. Assn., Am. Chem. Soc., Am. Soc. Microbiology, Am. Inst. Biol. Scis., Soc. Indsl. Microbiology, N.Y., Acad. Sci., Soc. Applied Bacteriology (Gt. Britain), Royal Soc. Chemistry, Sigma Xi, Tau Delta Phi. Author: Chromatography of Antibiotics, 1973, rev. edit., 1984; mem. editorial bd. Antimicrobial Agents and Chemotherapy, 1971-74; co-editor: Isolation, Separation and Purification of Antibiotics, 1978, Natural Products Isolation, 1989; contbr. articles to profl. jours. and books. Patentee in field. Home: 17 Crommelin Ct East Brunswick NJ 08816-2406 Office: Schering-Plough Rsch Inst 2015 Galloping Hill Rd Kenilworth NJ 07033

WAGNER, BERNHARD RUPERT, computer scientist; b. Munich, July 24, 1951; s. Richard L. and Therese (Kratzer) W.; m. Petra J. Hoermann, Dec. 4, 1976; children: Eva, Hubert, Barbara. MS, U. Pierre et Marie Curie, Paris, 1976; Dipl. Math., Ludwig-Maximilians U., Munich, 1978; PhD in Computer Sci., ETH, Zurich, Switzerland, 1986. Software engr. Siemens AG, Munich, 1978-82; rsch. asst. Swiss Fed. Inst. Tech., Zurich, 1983-86; asst. prof. Brigham Young U., Provo, Utah, 1986-87; computer scientist Ciba-Geigy AG, Basel, Switzerland, 1987—; cons. Mettler AG, Greifensee, Switzerland, 1986; seminar developer Ciba-Geigy AG, Basel, 1987-90. Contbr. articles to profl. jours. Bd. dirs. Internationaler Bauorden, Worms, Germany, 1968—; mem. Freiburg-Madison Soc., Freiburg, 1989—. Mem. ACM, Spl. Interest Group for Operating Systems. Avocation: ski racing. Home: Tannenbergstr 4F, Freiburg D-79117, Germany

WAGNER, CARRUTH JOHN, physician; b. Omaha, Sept. 4, 1916; s. Emil Conrad and Mabel May (Knapp) W. A.B., Omaha U., 1938; B.Sc., U. Nebr., 1938, M.D., 1941, D.Sc., 1966. Diplomate: Am. Bd. Sugery, Am. Bd. Orthopaedic Surgery. Intern U.S. Marine Hosp., Seattle, 1941-42; resident gen. surgery and orthopaedic surgery USPHS hosps., Shriners Hosp., Phila., 1943-46; med. dir. USPHS, 1952-62; chief orthopaedic service USPHS Hosp., San Francisco, 1946-51, S.I., N.Y., 1951-55; health mblzn. USPHS Hosp., 1959-62; asst. surgeon gen. dep. chief div. hosps. UPHS, 1957-59; chief div. USPHS, 1962-65, USPHS (Indian Health), 1962-65; dir. Bur. Health Services, 1965-68; Washington rep. AMA, 1968-72; health services cons., 1972-79; dept. health services State of Calif., 1979—. Contbr. articles to med. jours. Served with USCGR, World War II. Recipient Pfizer award, 1962; Meritorious award Am. Acad. Gen. Practice, 1965; Distinguished Service medal, 1968. Fellow A.C.S. (bd. govs.), Am. Soc. Surgery Hand, Am. Assn. Surgery Trauma, Am. Geriatrics Soc., Am. Acad. Orthopaedic Surgeons; mem. Nat. Assn. Sanitarians, Am. Pub. Health Assn. Sanitarians, Am. Pub. Health Assn., Washington Orthopaedic Club, Am. Legion, Alpha Omega Alpha. Lutheran. Club: Mason (Shriner). Home: 6234 Silverton Way Carmichael CA 95608-0757 Office: PO Box 638 Carmichael CA 95609-0638

WAGNER, GLENN NORMAN, pathologist; b. Phila., Sept. 6, 1946; s. Harry William and Alice Thelma (Aspen) W.; m. Joan Cecilia Cappello, June 30, 1971; 1 stepson, Kenneth Alan Scott. AB, Temple U., 1969; DO, Phila. Coll., 1974. Diplomate Am. Bd. Pathology. Commd. ensign USN, 1970, advanced through grades to capt., 1988; intern Naval Regional Med. Ctr., San Diego, 1974-75, resident in pathology, 1975-79; fellowship in forensic pathology Armed Forces Inst. of Pathology, 1980-81; in aerospace med. tng. (fligh surgery) Naval Aeromed. Inst., Pensacola, Fla., 1981; police officer, detective and criminalist Phila. Police Dept., 1968-74; forensic pathologist Armed Forces Inst. Pathology, Washington, 1979-88, chief dep. med. examiner, 1988-90, asst. med. examiner, 1990-91, dep. dir., 1991—; forensic cons. Nat. Cancer Inst., Bethesda, Md., 1990—, clin. asst. prof. Uniform Svcs. U. Health Scis., Bethesda, 1984—. Author: Mass Disasters Identification, 1991, Aerospace Pathology, 1991, CAP Forensic Pathology Handbook, 1993. Asst. scoutmaster Boy Scouts Am., Olney, Md., 1992. Capt. USN, 1970—. Decorated Legion of Merit. Fellow Coll. Am. Pathologists (com. mem. 1989—), Am. Soc. Clin. Pathologists, Am. Acad. Forensic Sci., Aerospace Med. Assn.; mem. AMA, AAAS, Am. Osteo. Assn., Am. Chem. Soc., Am. Assn. Analytical Chemists, Nat. Assn. Med. Examiners. Republican. Achievements include national and international reputation as forensic pathologist in mass identification casualties, aerospace mishaps, human factor analyses, biodynamics, scene reconstructions. Office: Armed Forces Inst Pathology 6825 16th St NW Washington DC 20306

WAGNER, HENRI PAUL, aerospace engineer; b. Luxembourg City, Luxembourg, June 4, 1960; came to U.S., 1990; s. Gaston Pierre and Anne Maria (Berg) W. Candidate engr., U. Liege, Belgium, 1982; diploma in engring., U. Stuttgart, Fed. Republic Germany, 1989, postgrad., 1989—. Researcher in aerospace engring. U. Stuttgart, 1989—. Contbr. articles to profl. jours. Mem. AIAA, Deutsche Gesellschaft für Luft- und Raumfahrt. Achievements include research on gradient driven instabilities in stationary MPD truster flows. Office: U Stuttgart Inst for Space Systems, Pfaffenwaldring 31, D-7000 Stuttgart 80, Germany

WAGNER, HENRY NICHOLAS, JR., physician; b. Balt., May 12, 1927; s. Henry N. and Gertrude Loane W.; m. Anne Barrett Wagner, Feb., 1951; children—Henry N., Mary Randall, John Mark, Anne Elizabeth. A.B., Johns Hopkins U., 1948, M.D., 1952; D.Sc. (hon.), Washington Coll., Chestertown, Md., 1972, Free U., Brussels, 1985; M.D. (hon.), U. Gottingen, 1988. Chief med. resident Osler Med. Service, Johns Hopkins Hosp., Balt., 1958-59; asst. prof. medicine, radiology Johns Hopkins Med. Instns., 1959-64, asso. prof., 1964-65, prof., dir. divs. nuclear medicine and radiation health sci., 1965—. Author numerous books in field.; contbr. articles to med. jours. Served with USPHS, 1955-57. Recipient Georg von Hevesey medal, 1976. Fellow ACP; mem. Inst. Medicine of NAS, AMA (coun. sci. affairs, Sci. Achievement award 1991), Balt. City Med. Soc. (past pres.), World Fedn. Nuclear Medicine and Biology (past pres.), Am. Bd. Nuclear Medicine (founding mem.), Soc. Nuclear Medicine (past pres.), Am. Fedn. Clin. Research (past pres.), Research Socs. Council (past pres.), Assn. Am. Physicians, Am. Soc. Clin. Investigation, Phi Beta Kappa. Home: 5607 Wildwood Ln Baltimore MD 21209-4520 Office: 615 N Wolfe St Baltimore MD 21205-2103

WAGNER, JOEL H., aerospace engineer. BS in Mech. Engring., Iowa State U., 1976, MS, 1977; postgrad. in Mech. Engring., U. Conn. Registered profl. engr., Conn. With rsch. ctr. United Techs., Hartford, Conn., 1977—. Recipient Gas Turbine award ASME, 1991. Achievements include research in heat transfer in rotating systems, prediction of turbine component erosion, axial fan aerodynamics and noise, compressor and turbine aerodynamics, and effects of rotation on internal cooling of gas turbine blades. Office: United Techs Rsch Ctr 411 Silver Lane Hartford CT 06108*

WAGNER, JOHN PHILIP, safety engineering educator, science researcher; b. Trenton, N.J., Feb. 29, 1940; s. Joseph and Anna Wagner; m. Carol Anne Hammond, June 14, 1969; children: John Joseph (Jay), Timothy Andrew. BS in Chemistry, St. Joseph's U., 1961; MSChemE, Johns Hopkins U., 1964, PhDChemE, 1966. Registered prof. engr., Tex. Rsch. asst. chemistry Johns Hopkins U., Balt., 1961-62, rsch. fellow chem. engring., 1962-66; assoc. chemist Applied Physics Lab. Johns Hopkins U., Silver Spring, Md., summer 1962, sr. engr. Applied Physics, 1966-72; sr. rsch. scientist Factor Mut. Rsch. Corp., Norwood, Mass., 1972-73; rsch. supr. Gillette Rsch. Inst., 1973-78; staff engr., sr. staff engr. EXXON Rsch. and Engring. Co., 1978-83; assoc. prof. indsl. engring. Tex. A&M U., 1985-89, assoc. prof. nuclear engring., 1989—; assoc. dir. and rsch. engr. Food Protein Rsch. Devel. Ctr. Tex. Engring. Expt. Station., 1983-90; assoc. dir., rsch. engr. Engring. Bioscis. Rsch. Ctr Tex. A&M U., College Station, 1990—; cons. O'Melveny & Myers, L.A., 1987-88, Lawrence Livermore Nat. Lab., Exxon Co.-USA, Englehard Industries, Gillette Rsch. Inst., Liberty Mut., Champion Internat, John Deere. Mem. editorial adv. bd. Jour. Polymer-Plastics Tech. and Engring., 1987—, Indsl. Crops and Products, 1991—; co-guest editor Jour. Bioresources Tech., 1991; contbr. chpts. to books, articles to profl. jours.; patentee in field. Grantee USDA/DOD, 1984-93. Mem. Am. Oil Chemists Soc. (environ. com. 1985-86), Assn. for the Advancement of Indsl. Crops, Am. Chem. Soc., Am. Inst. Chem. Engrs., Am. Soc. Engring. Edn., Cath. Alumni Club Balt. (pres. 1968), Sigma Xi, Phi Lambda Upsilon. Avocations: philatelics, numismatics, sports. Office: Tex A&M U FM Box 183 College Station TX 77843-2146

WAGNER, JOSEPH EDWARD, veterinarian, educator; b. Dubuque, Iowa, July 29, 1938; s. Jacob Edward and Leona (Callahan) W.; m. Kay Rose (div. Apr. 1983); children: Lucinda, Pamela, Jennifer, Douglas. DVM, Iowa State U., 1963; MPH, Tulane U., 1964; PhD, U. Ill., 1967. Asst. prof. U. Kans. Med. Ctr., Kansas City, 1967-69; assoc. prof. U. Mo. Coll. Vet. Medicine, Columbia, 1969-72, prof. vet. medicine, 1972—; cons. Harlan Sprague Dawley, Indpls., 1984—. Author: The Biology and Medicine of Rabbits and Rodents, 1989. Recipient award of excellence in lab. animal medicine Charles River Found., Wilmington, Mass., 1986. Mem. AMVA, Am. Coll. Lab. Animal Medicine (pres. 1985-86), Am. Assn. Lab. Animal Scis. (pres. 1980-81). Office: U Mo-Coll of Veterinary Medicine Dept of Vet Pathology 1600 W Rollins Rd Columbia MO 65211-0001

WAGNER, KAREL, JR., nuclear engineer, educator; b. Prague, Čsr, Czech Republic, Feb. 15, 1931; s. Karel and Anna (Titti) W.; m. Sonja Vayrova, Sept., 24, 1963; children: Karel, Antonin. MEE, Tech. U., Plzeň, ČSSR, 1950-54, PhD Reactor Engring., 1968; M. Reactor Engring., Charles U., 1957; MBA, CEI U. Geneva, 1970. Testing engr. ŠKODA Elect. Machinery Plant, Plzeň, 1954-56; rsch. and devel. officer ŠKODA Nuclear Machinery Plant, Plzeň, 1957-68, head rsch. and devel. group, 1968-80, sci. officer, 1985-89, advisor to dir., 1993; head commissioning ŠKODA Nuclear Machinery Plant, J. Bohunice, ČSSR, 1981-85; chmn. CS Atomic Energy Commn., Prague, 1990-92; lector Tech. U., Plzeň, 1964-93; chmn. Complex State Rsch. and Devel. Program for LWR, Prague, 1969-73, State Rsch. and Devel. Program for Instrumentation and Control. LWR, 1965-80. Author: (textbook) Control of Nuclear Power, 1964, (screenplay) A New Way of Reactor Control, 1963; translator Instrumentation of NPP, 1965, System Logigue, 1971. Recipient Tech. Project award CS Min. Heavy Industry, 1960, State Sci. Prize CS Govt., 1963, Commissioning Nuclear Power Plant award CS Min. Energy, 1985. Mem. CS Sci. and Tech. Soc., CS Nuclear Soc., CS Auto-Moto Club. Achievements include patents for device for measuring the thickness of austenitic weld deposits on carbon steel wells (reactor pressure vessels), device for position measurement of reactor control rod drives. Home: Litohlavská, 337 01 Rokycany Czech Republic Office: ŠKODA Nuclear Machinery Co Ltd, Orlik 266, 316 06 Plzen Czech Republic

WAGNER, MARVIN, general and vascular surgeon, educator; b. Milw., Feb. 20, 1919; s. Benjamin and Ella (Drotman) W.; m. Shirley Semon; children: Terry, Jeffrey, Penny. MD, Marquette U., 1944, MS, 1951. Diplomate Am. Bd. Surgery. Intern Mt. Sinai Med. Ctr., Milw., 1944-45, jr. and sr. resident in surgery, 1945-46, 47-50; pvt. practice Milw., 1950—; mem. staff Columbia, Milw. Children's, Milwaukee County Gen., St. Joseph's, VA, Froedtery Meml. Luth. hosps., Good Samaritan Med. Ctr., Sinai-Samaritan Ctr.; chmn., chief dept. surgery St. Michael's Hosp., 1965-69, pres. med. staff, 1981-82; vascular cons. Trinity Meml. Hosp., Waukesha (Wis.) Meml. Hosp.; clin. prof. surgery, adj. prof. anatomy Med. Coll. Wis., Milw.; mem. occupational adv. com. Milw. Area Tech. Coll., 1982-83; lectr., condr. workshops, site visitor in field; also others. Author: (with T. Lawson) Segmental Anatomy: Applications to Clinical Medicine, 1982 (Most Outstanding Book in Health Scis. award Assn. Am. Pubs. 1982), Atlas of Chest Imaging; contbr. over 85 articles to med. jours. including Surgery, Wis. Med. Jour., Am. Jour. Obstet. Surg. Gynecology, Modern Medicine, AMA Archives Surgery, Marquette Med. Rev., Am. Jour. Gastroenterology, Surg. Gynecology and Obstetrics, Sci., Transplantation Bull., Angiology, Abdominal Surgery, Am. Jour. Surgery, Archives Surgery, Jour. AMA. Mem. United Way Corp., 1975-78; chmn. physicians div. United Fund, 1972, bd. dirs., 1973-76, chmn. profl. div., 1973, co-chmn. doctor's div., 1977; mem. agy. facilities rev. com. and steering com. Southeastern Wis. Health Systems Agy., 1976-77; mem. adlumni fund raising com. Marquette U., 1971-72; mem. fund rasing com. project 75, Med. Coll. Wis., 1975-76. Recipient Disting. Svc. award Med. Coll. Wis., 1980, Alumnus of Yr. award, 1985, citation Milw. County Bd. Supervisors, 1988; Marvin Wagner endowed chair in anatomy and cellular biology named in his honor, 1988-91; grantee Am. Heart Assn., 1957-59, Milw. Cancer Soc., 1959-60, Wis. Heart Assn., 1960, USPHS, 1960-62, 86-89, NIH, 1960-62, Taitel, 1961, 62, 64, 65, 66, 3M Corp., 1968, Med. Coll. Wis., 1972, Winters Rsch. Found., 1976-80, McMillan Pub. Co., 1979-82, Tisshberg Found., 1985. Fellow ACS (sci. exhibit award 1957, 70); mem. AAUP, AMA (Physician's recognition award 1980-85, 89), Am. Assn. Anatomists, Cen. Surg. Soc., Collegium Internat. Chirurglae Digestivae, Soc. for Surgery Alimentary Tract, Am. Assn. Clin. Anatomists, Milw. Acad. Medicine, Milw. Acad. Surgery (coun. 1973-76), N.Y. Acad. Scis., Western Surg. Assn., Wis. Heart Assn., Wis. Surg. Soc. (coun. 1973-76), Med. Soc. Milwaukee County (pres. 1975, President's citation 1975), Alpha Omega Alpha. Achievements include patent for spandex sutures and prosthesis patches. Office: Med Coll Wis Anatomy and Cellular Biology 8701 W Watertown Plank Rd Milwaukee WI 53226-4801

WAGNER, MARY EMMA, geologist educator, researcher; b. Wilmington, Del., June 20, 1927; d. John Mercer and Emma Welch (Davis) Mertz; m. Daniel Hobson Wagner, Sept. 10, 1949; children: David Hobson, Christopher Daniel, Thomas John, Elizabeth Ann. BA, Mt. Holyoke Coll., 1948; MS, Bryn Mawr Coll., 1966, PhD, 1972. Lectr. in geology U. Pa., Phila., 1972-93. Contbr. articles to Am. Jour. Sci., Bull. Geol. Soc. Am., Geol. Soc. Am. Centennial Vol., Jour. of Geology. Grantee NSF, Bryn Mawr Coll., 1977-79, U. Pa., 1984-87; Pew Found., U. Pa., 1990. Mem. Geol. Soc. Am. (bd. dirs. n.e. sect. 1984-86), Am. Geophys. Union, Phila. Geol. Soc. (pres. 1984-86). Home: 36 Laurel Cir Malvern PA 19355 Office: U Pa Dept Geology Philadelphia PA 19104-6316

WAGNER, M(AX) MICHAEL, computer scientist, educator; b. Muenster, Germany, Dec. 5, 1949; arrived in Australia, 1974; s. Max and Wilma (Schrader) W.; m. Patricia Ann Faury, Aug. 24, 1981; children: Rita, Ingmar, Ingrid. Dipl. Phys., U. Munich, 1973; PhD in Computer Sci., Australian Nat. U., 1979. Rsch. scientist Tech. U., Munich, 1979-81, Nixdorf AG, Paderborn, Fed. Republic Germany, 1983-84; lectr. Nat. U., Singapore, 1981-83; sr. lectr. U. Wollongong, Australia, 1984-86, U. New South Wales, Canberra, Australia, 1986—. Editor proc. 1st and 2d Australian Internat. Confs. on Speech Sci. and Tech., 1986, 88; contbr. articles to profl. publs. Mem. IEEE, Australian Speech Sci. and Tech. Assn. Inc. (found. chmn. 1988-92), East Timor Found. Inc. (chmn. 1989—). Home: 40 Froggatt St, Turner 2601, Australia Office: U New South Wales, Univ Coll/ADFA, Canberra 2600, Australia

WAGNER, NORMAN JOSEPH, III, chemical engineering educator; b. Phila., Aug. 7, 1962; s. Norman Joseph II and Gertrude Eugene (Mamrod) W.; m. Sabine Banerjee, Dec. 21, 1992. BS, Carnegie Mellon U., 1984; PhD, Princeton U., 1989. NATO/NSF postdoctoral fellow U. Konstanz, Germany, 1988-89; dir. postdoctoral Los Alamos (N.Mex.) Nat. Lab., 1989-90; asst. prof. U. Del., Newark, 1991-. Mem. Am. Inst. Chem. Engrs., Soc. Rheology, Am. Chem. Soc., Am. Soc. Engring. Edn. Office: U Del Dept Chem Engring Newark DE 19716

WAGNER, THOMAS ALFRED, chemical engineer; b. Noerdlingen, Fed. Republic Germany, May 31, 1953; s. Alfred and Agnes (Guenther) W. Diploma in engring., FH Aalen, Fed. Republic Germany, 1979. Process engr. Siemens AG, Munich, 1979-82, process engr. dry etching, 1982-86, mgr. process engring. and dry etching, 1986—, cons. of computer use and integration in the dept., 1986-88, mgr. structuring processes microsystems, 1989—. Avocation: computer graphics. Home: Hermanstr 19, D-86150 Augsburg Federal Republic of Germany Office: Siemens AG, ZPL1 TW 32, D-81739 Munich Germany

WAGNER, WARREN HERBERT, JR., educator, botanist; b. Washington, Aug. 29, 1920; s. Warren Herbert and Harriet Lavinia (Claflin) W.; m. Florence Signaigo, July 16, 1948; children: Warren Charles, Margaret Frances. A.B., U. Pa., 1942; Ph.D., U. Calif. at Berkeley, 1950; spl. student, Harvard, 1950-51. Instr. Harvard, summer 1951; vis. prof., 1991; faculty U. Mich. at Ann Arbor, 1951—, prof. botany, 1962—, curator pteridophytes, 1961—, dir. Bot. Gardens, 1966-71, chmn. dept. botany, 1975-77; spl. rsch. higher plants, origin and evolution ferns, groundplan/divergence methods accurate deduction phylogenetic relationships fossil and living plants, pteridophytes of Hawaii, 1962-65; prin. investigator project evolutionary characters ferns NSF, 1960—, monograph grapeferns, 1980—, pteridophytes of Hawaii, 1991—; chmn. Mich. Natural Areas Coun., 1958-59; mem. Smithsonian Coun., 1967-72, hon. mem., 1972—; cons. mem. Survival Svc. Commn., Internat. Union for Conservation of Nature and Natural Resources, 1971—; mem. nat. hist. coun. Nat. Mus., 1989—. Trustee Cranbrook Inst. Scis. Recipient Distinguished Faculty Achievement award U. Mich., 1975, Amoco Outstanding Tchr. award, 1980, Disting. Sr. Lectr. award U. Mich., 1986. Fellow AAAS (sec. sect. bot. scis., v.p. sect. 1968), Am. Acad. Arts and Scis.; mem. NAS, Am. Fern Soc. (sec. 1952-54, curator, librarian 1957-77, pres. 1970, 71, hon. mem. 1978), Am. Soc. Plant Taxonomists (coun. 1958-65, pres. 1966, Asa Gray award 1990), Soc. for Study Evolution (v.p. 1965-66, coun. mem. 1967-69, pres. 1972), Am. Soc. Naturalists, Internat. Assn. Pteridologists (v.p. 1981-87, pres. 1987—), Bot. Soc. Am. (pres. 1977, Merit award 1978, Henry Allan Gleason award 1992), Mich. Bot. Club (pres. 1967-71), Torrey Bot. Club, New Eng. Bot. Club, Internat. Soc. Plant Morphologists, Internat. Assn. Plant Taxonomy, Sigma Xi, Phi Kappa Tau. Home: 2111 Melrose Ave Ann Arbor MI 48104-4067

WAGNER, WILLIAM MICHAEL, data processing department administrator; b. Saratoga Springs, N.Y., May 6, 1949; s. Harold Wilbur and Alice Frieda (Stauffacher) W.; m. Barbara Lee Galarneault, Jan. 25, 1980; 1 child, Harold Galarneault Wagner. BA, Bradley U., 1971; MA, U. Tex., 1973. Tchr. Happy Grove High Sch., Hector's River, Jamaica, 1974-75; specialist software systems U. Tex., Austin, 1977-88, asst. dir., 1988—. Contbr. articles to profl. jours. Mem. Software AG User Group Internat. (ADABAS product rep. 1986-87, v.p. 1987-88, pres. 1989-91, adv. bd. 1990—), Computer Measurement Group, Littlefield Soc., Chancellor's Soc., Phi Eta Sigma, Phi Kappa Phi, Omicron Delta Kappa, Sigma Pi Sigma, Kappa Delta Rho. Republican. Avocations: squash, cycling, skiing, scuba diving, railroads, naval mil. history. Office: U Tex Box Q Univ Sta Austin TX 78713

WAGONER, ROBERT VERNON, astrophysicist, educator; b. Teaneck, N.J., Aug. 6, 1938; s. Robert Vernon and Marie Theresa (Clifford) W.; m. Lynne Ray Moses, Sept. 2, 1963 (div. Feb. 1986); children: Alexa Frances, Shannon Stephanie; m. Stephanie Nightingale, June 27, 1987. B.M.E., Cornell U., 1961; M.S., Stanford U., 1962, Ph.D., 1965. Research fellow in physics Calif. Inst. Tech., 1965-68, Sherman Fairchild Disting. scholar, 1976; asst. prof. astronomy Cornell U., 1968-71, asso. prof., 1971-73; asso. prof. physics Stanford U., 1973-77, prof., 1977—; George Ellery Hale disting. vis. prof. U. Chgo., 1978; mem. Com. on Space Astronomy and Astrophysics, 1979-82, theory study panel Space Sci. Bd., 1980-82, physics survey com. NRC, 1983-84; grant selection com. NSERC (Can.), 1990-93. Contbr. articles on theoretical astrophysics and gravitation to profl. publs., mags.; co-author Cosmic Horizons. Sloan Found. rsch. fellow, 1969-71; Guggenheim Meml. fellow, 1979; grantee NSF, 1973-90, NASA, 1982—. Fellow Am. Phys. Soc.; mem. Am. Astron. Soc., Internat. Astron. Union, Tau Beta Pi, Phi Kappa Phi. Patentee. Office: Stanford U Dept Physics Stanford CA 94305-4060

WAHL, MARTHA STOESSEL, mathematics educator; b. Ottumwa, Iowa, Mar. 9, 1916; d. Theodore A. and Anna Theresa (Coday) Stoessel; m. John Schempp Wahl, Dec. 27, 1943 (dec. Aug. 1982); children: Elizabeth A. O'Connor, Richard Carl, Patrick Theodore. Aa, Ottumwa Heights Coll., 1936, diploma; BA, U. Iowa, 1938; AM, Columbia U., 1942, 6th yr. diploma, 1959. Asst. prof. maths. Western Conn. State Coll., Danbury, 1959-78, assoc. prof. maths., 1978-85, full prof. math., 1985-86, ret., 1986. Author: I Can Count the Petals of a Flower, 1976, 2d edit., 1985 (Edn. Pres. award Picture Story 1977); contbr. articles to profl. jours. Participant meetings City of Ridgefield, Conn., 1955—; grantor John and Stacey Wahl scholarships Physics dept. U. Iowa, 1982—; participant elder hostels U. Iowa, 1979—. Mem. Nat. Coun. Tchrs. Maths. (life), AAUW (life), AAUP, DAR, Phi Delta Kappa (exec. bd., Excellence in Teaching award 1985), Delta Kappa Gamma. Demcorat. Roman Catholic. Avocations: attending elder hostels, travel, writing memoirs. Home: 1 Huckleberry Ln Ridgefield CT 06877-5705

WAIDELICH, DONALD LONG, electrical engineer, consultant; b. Allentown, Pa., May 3, 1915; s. John A. Sr. and Maisie Hamilton (Long) W.; m. Florence Emma Bennethum, June 6, 1939; 1 child, Ann Louise. BEE, Lehigh U., 1936, MS, 1938; PhD, Iowa State U., 1946. Registered profl. engr., Mo. Instr., asst. prof. electrical engring. U. Mo., Columbia, 1938-44, assoc. prof., prof., 1946-85, prof. emeritus, 1985—, assoc. head engring. experiment sta., 1955-60, chair dept., 1960-61; electrical engr. Naval Ordnance Lab., Silver Spring, Md., 1944-46; cons. Naval Electronics Lab., San Diego, 1948-50, Argonne (Ill.) Nat. Lab., 1950-60, Nat. Aeronautics & Space, Green Belt, Md., 1961-70, Hughes Aircraft Co., Santa Monica and El Segundo, Calif., 1970-88; Fulbright prof. Cairo U., 1950-52, U. New South Wales, Sydney, Australia, 1960-62. Author: (with G. Lago) Transients in Electrical Circuits, 1958. Com. mem. Civic and Univ. Retirees, Columbia, Mo., 1985—. Recipient Rsch. award Sigma Xi, 1977. Fellow IEEE (life, Excellence award 1985), AIEE, Inst. Radio Engrs.; mem. Am. Soc. Engring. Edn. (life), Nat. Soc. Profl. Engrs. (life). Episcopal. Achievements include research on rectifiers, electromagnetic testing of materials, microwave antennas, electrostatics, and magnetic fields. Home: 104 E Ridgeley Rd Columbia MO 65203 Office: U Mo Dept Elec & Comp Engring 25 W Engineering Bldg Columbia MO 65211

WAINBERG, ROBERT HOWARD, biology educator; b. Toronto, Ont., Can., Oct. 17, 1953; came to U.S., 1976; s. Jack and Helene (Leibowitz) W.; m. Kirsten Lis Mygil, May 9, 1983 (div. Apr. 1991). BSc in Zoology magna cum laude, U. Toronto, 1976; MS in Zoology, Ohio U., 1978; PhD in Zoology, U. Tenn., 1983. Rsch. asst. dept. zoology and microbiology Ohio U., Athens, 1976-78; rsch. asst. dept. zoology U. Tenn., Knoxville, 1978-79, biostats. cons., 1980-86, sr. instr. gen. biology program, 1983-86, mem. bd., biology coord., supr. ednl. advancement program, 1983-85, mem. bd., instr. premed. enrichment program, 1984-86; NRC-NIH postdoctoral fellow dept. human oncology U. Wis. Clin. Cancer Ctr., Madison, 1986-87, lab. supr. hyperthermia sect. dept. human oncology, 1987-88; asst. prof. biology Piedmont Coll., Demorest, Ga., 1988—; pre-nursing advisor coop. program at Nell Hodgson Woodruff Sch. Nursing, Emory U., Atlanta, 1990—. Co-author: Fundamentals of Biological Investigation for General Biology Students at the University of Tennessee, 1985, (reference book) Preoperational Assessment of Water Quality and Biological Resources of Chickamauga Reservoir, Watts Bar Nuclear Plant, 1973-1985, 1986; contbr. articles to Prostaglandins, Internat. Jour. Hyperthermia, Lipids. Guest lectr. and speaker various local orgns., including Boy Scouts Am., hosps., Rotary and Kiwanis clubs, 1988—; participant speakers bur. Piedmont Coll.-Habersham County (Ga.) Coop. Ednl. Program, 1989—. Recipient teaching merit award U. Tenn., 1986, teaching excellence and campus leadership award Sears-Roebuck Found. and Piedmont Coll., 1991; Hilton A. Smith grad. fellow U. Tenn., 1982-83. Mem. Ga. Acad. Scis., Sigma Xi, Phi Kappa Phi. Jewish. Achievements include research on using certain bacterial permeases as models for studying the combined modalities of hyperthermia with various drugs to overcome multi-drug resistance in cancer cells; research which has shown that composition of cell membranes of certain cancer lines are altered to ostensibly enhance their survival following a hyperthermic insult. Office: Piedmont Coll Dept Biology PO Box 10 Demorest GA 30535

WAINIONPAA, JOHN WILLIAM, systems engineer; b. Quincy, Mass., July 13, 1946; s. Frank Jacob and Jennie Sofia (Kaukola) W.; m. S. Linda Rapo, Oct. 18, 1969; children: Heidi Liisa, Erik David, Sinikka Lin. BSEE, U. N.Mex., 1972; MS in Aero. Engring., Naval Postgrad. Sch., 1981. Engr.-in-tng., Colo. Enlisted USN, 1968, commd. ens., 1972, advanced through grades to lt. comdr., 1982; flight instr. Tng. Squadron 27, Corpus Christi, Tex., 1973-75; aircraft, mission comdr. Patrol Squadron 49, Jacksonville, Fla., 1976-79; ops. officer Anti-Submarine Warfare Ops. Ctr., Kadena, Okinawa, Japan, 1982-84; launch and control systems officer Naval Space Command, Dahlgren, Va., 1984-86; naval space systems ops. officer U.S. Space Command, Colorado Springs, 1986-88; ret. USN, 1988; systems engr. CTA Inc., Colorado Springs, 1988-93, tng. coord., 1993—. Merit badge counselor Boy Scouts Am., Colorado Springs, 1986—. Mem. AIAA (sr.), IEEE, U.S. Naval Inst., Sigma Tau, Eta Kappa Nu. Avocations: music, skiing, volunteer church activities. Office: CTA Inc 7150 Campus Dr Ste 100 Colorado Springs CO 80920-6592

WAINRIGHT, SAM CHAPMAN, marine scientist, educator; b. Bridgeport, Conn., July 19, 1954; s. Ralph B. and Dorothy (Chapman) W.; m. Patricia D. Ottens, July 27, 1984. BA, Lycoming Coll., Williamsport, Pa., 1976; MS, Fla. Atlantic U., Boca Raton, 1982; PhD, U. Ga. Grad. rsch. asst. Inst. of Ecology, U. Ga., Athens, 1983-88, postdoctoral assoc., 1988; postdoctoral assoc. Marine Biology Lab., Woods Hole, Mass., 1989-91, Allan Hancock Found., U. So. Calif., L.A., 1991-92; asst. prof. marine biology Inst. Marine/Coastal Scis., Rutgers U., New Brunswick, N.J., 1992—; reviewer NSF, Washington, 1988—, Limnology and Oceanography, 1989—, Marine Biology, Berlin, Germany, 1990—, Estuaries, 1991—, Nat. Oceanic and Atmospheric Adminstrn., Washington, 1991—. Contbr. articles to Marine Biology, Fresh Water Biology, Marine Ecology Progress Series, Science. Mem. AAAS, Am. Soc. for Limnology and Oceanography, Am. Fisheries Soc., Northeastern Estuarine Rsch. Soc., Sigma Xi (grantee 1984). Achievements include research on The Georges Bank marine food web, the role of resuspended marine sediments in planktonic food webs, marine ecology using stable isotope mass spectrometry. Office: Rutgers U Inst Marine/Coastal Scis PO Box 231 New Brunswick NJ 08903-0231

WAINWRIGHT, STEPHEN A., zoology educator, design consultant; b. Indpls., Oct. 9, 1931; s. Guy A. and Jeannette (Harvey) W.; m. M. Ruth Palmer, July 25, 1956; children: Peter C., Ian P., Archer T., Jennifer S. BS in Zoology, Duke U., 1953; BA in Zoology, Cambridge U., England, 1958; PhD, U. Calif., Berkeley, 1962; MA, Cambridge U., 1963. Assoc. prof. zoology Duke U., Durham, N.C., 1964-76, prof., 1976-85, J.B. Duke prof., 1985—; adj. prof. design sch. N.C. State U., Raleigh, 1983—. Author: (with others) Mechanical Design in Organisms, 1976, Axis and Circumference, 1988. Served as cpl. U.S. Army, 1953-55. NSF postdoctoral fellow, 1962-64. Fellow AAAS; mem. Marine Biol. Assn. United Kingdom, Am. Soc. Zoologists (pres. 1988), Am. Soc. Biomechanics (pres. 1981). Avocation: sculpture. Office: Duke U Dept Zoology Durham NC 27706

WAIS DE BADGEN, IRENE RUT, limnologist; b. Buenos Aires, Argentina, May 24, 1957; d. Samuel and Beatriz (Leibovich) Wais; m. Javier Martin Badgen, June 12, 1980; children: Ezequiel Dario, Natalia Andrea, Ivan Matias. Bachelor's degree, Nat. Coll. Buenos Aires U., 1973, Lic. in Ciencias Biologicas, 1978; postgrad., Oreg. State U., 1982. Diplomate in biology and stream ecology. Lab. asst. Nat. Mus. Natural History and Nat. Inst. for Natural Scis. Rsch., Buenos Aires, 1976-78, asst. researcher, 1978-81, main researcher, 1981-87, head limnological dept., 1987—; cons. in limnology Hidronor S.A. , Buenos Aires, 1979-81, Entidad Binacional Yacyreta, Buenos Aires, 1989-90; vis. prof. U. Austral, Valdivia, Chile, 1983, Ctr. Investigac, Aguas, 1985, Fundacion Vida Silvestre, Olavarria, Argentina, 1987. Contbr. articles to profl. jours and chpts. to books. Coord. Internat. Rhithrobiologists, Buenos Aires, L.A., 1984—, Programme Environ. Edn., Buenos Aires, 1989—; pres. Ambientis Found., Buenos Aires, 1989—. Named One of Ten Outstanding Young Argentinans Jr. Chamber Internat., 1988. Fellow Orgn. Am. State of USA, Nat. Coun. for Scientific and Tech. Rsch. Avocations: embroadery, sports, reading, writing. Home: Anasco 1792, 1416 Buenos Aires Argentina Office: Argentine Mus Natural Scis, Av Angel Gallardo 470, 1405 Buenos Aires Argentina

WAIT, JAMES RICHARD, electrical engineering educator, scientist; b. Ottawa, Ont., Can., Jan. 23, 1924; came to U.S., 1955, naturalized, 1960; s. George Enoch and Doris Lillian (Browne) W.; m. Gertrude Laura Harriet, June 16, 1951; children: Laura, George. BASc, U. Toronto, Ont., 1948, MASc, 1949; Ph.D. in Electromagnetic Theory, 1949, 1951. Research engr. Newmont Exploration, Ltd., Jerome, Ariz., 1949-52; sect. leader Def. Research Telecommunications Establishment, Ottawa, 1952-55; scientist U.S. Dept. Commerce Labs., Boulder, Colo., 1955-80; adj. prof. elec. engring. U. Colo., Boulder, 1961-80; prof. elec. engring., geosci. U. Ariz., Tucson, 1980-88, Regents prof., 1988-90, emeritus Regents prof., 1990—; prin. Geo-Em Cons., Tucson, 1990—; fellow Coop. Inst. Research Environ. Scis., 1968-80; sr. scientist Office of Dir. Environ. Research Labs., Boulder, 1967-70, 72-80; vis. research fellow lab. electromagnetic theory U. Denmark, Copenhagen, 1961; vis. prof. Harvard, 1966-67, Catholic U., Rio de Janeiro, 1971; vis. prof. elec. engring. U. B.C., Vancouver, Can., 1987; mem.-at-large U.S. nat. com. Internat. Sci. Radio Union, 1963-65, 69-72, del. gen. assemblies, Boulder, 1957, London, 1960, Tokyo, 1963, Ottawa, 1969, Warsaw, Poland, 1972, Lima, Peru, 1975, Helsinki, Finland, 1978; sec. U.S. nat. com., 1976-78; Lansdowne lectr. U. Victoria, B.C., Can., 1992. Founder Jour. Radio Sci, 1959, editor, 1959-68; assoc. editor: Pure and Applied Geophysics, 1964-75, Geoexploration, Ludea, Sweden, 1983-91; co-editor internat. series monographs on electromagnetic waves Pergamon Press, 1961-73, Instn. Elec. Engrs, London, 1974—. Served with Canadian Army, 1942-45. Recipient Gold medal Dept. Commerce, 1958; Samuel Wesley Stratton award Nat. Bur. Standards, 1962; Arthur S. Flemming award Washington C. of C., 1964; Outstanding Publ. award Office Telecommunications, Washington, 1972; Rsch. and Achievement award Nat. Oceanic and Atmospheric Adminstrn., 1973; Van der Pol gold medal, 1978; Evans fellow Otago U., New Zealand, 1990. Fellow IEEE (mem. adminstrv. com. on antennas and propagation 1966-73, Harry Diamond award 1964, Centennial medal 1984, Disting. Achievement award geosci. and remote sensing 1985, Disting. Achievement award antennas and propagation soc. 1990, Heinrich Hertz medal 1992), AAAS, Instn. Elec. Engrs. (Gt. Britain), Sci. Research Soc. Am. (Boulder Scientist award 1960); mem. NAE. Home and Office: 2210 E Waverly St Tucson AZ 85719-3848

WAITE, LAWRENCE WESLEY, physician; b. Chgo., June 27, 1951; s. Paul J. and Margaret E. (Cresson) W.; m. Courtnay M. Snyder, Nov. 1, 1974; children: Colleen Alexis, Rebecca Maureen, Alexander Quin. BA, Drake U., 1972; DO, Coll. Osteo. Medicine and Surgery, Des Moines, 1975; MPH, U. Mich., 1981. Diplomate Nat. Bd. Osteo. Examiners. Intern Garden City Osteo. Hosp., Mich., 1975-76; practice gen. osteo. medicine, Garden City, 1979-82, Battle Creek, 1982—; assoc. clin. prof. Mich. State U. Coll. Osteo. Medicine, East Lansing, 1979—; dir. med. edn. Lakeview Gen. Osteo. Hosp., Battle Creek, Mich., 1983-87; cons. Nat. Bd. Examiners Osteo. Physicians and Surgeons, 1981-88; chief med. examiner Calhoun County, 1991—. Writer TV program Cross Currents Ecology, 1971; editor radio series Friendship Hour, 1971-72. Bd. dirs., instr. Hospice Support Services, Inc., Westland, Mich., 1981-86; mem. profl. adv. council Good Samaritan Hosp., Battle Creek, 1982-83; bd. dirs. Neighborhood Planning Council 11, Battle Creek, 1982-92; mem. population action council Population Inst., 1984—; exec. bd. officer Battle Creek Area Urban League, 1987-91; vestryman St. Thomas Episcopal Ch., 1990-93; leader Boy Scouts Am. Served to lt. comdr. USN, 1976-79. State of Iowa scholar, 1969. Mem. AMA, Aerospace Med. Assn., Nat. Eagle Scouts Assn. (life), Am. Osteo. Assn., S. Cen. Osteo. Assn. (officer, state del. 1983—), Am. Pub. Health Assn., Am. Acad. Osteopathy, Bermuda Hist. Soc. (life). Avocations: geography, medieval history, genealogy. Home: 140 S Lincoln Blvd Battle Creek MI 49015 Office: 3164 Capital Ave SW Battle Creek MI 49015-4108

WAKE, DAVID BURTON, biology educator, researcher; b. Webster, S.D., June 8, 1936; s. Thomas B. and Ina H. (Solem) W.; m. Marvalee Hendricks, June 23, 1962; 1 child, Thomas Andrew. B.A., Pacific Luth. U., 1958; M.S., U. So. Calif., 1960, Ph.D., 1964. Instr. anatomy and biology U. Chgo., 1964-66, asst. prof. anatomy and biology, 1966-69; assoc. prof. zoology U. Calif., Berkeley, 1969-72, prof., 1972-89, prof. integrative biology, 1989-91, John and Margaret Gompertz prof., 1991—; dir. Mus. Vertebrate Zoology U. Calif., Berkeley, 1971—. Author: Biology, 1979; co-editor: Functional Vertebrate Morphology, 1985, Complex Organismal Functions: Integration and Evolution in the Vertebrates, 1989. Mem. nat. bd. Nat. Mus. Natural History. Recipient Quantrell Teaching award U. Chgo., 1967, Outstanding Alumnus award Pacific Luth. U., 1979; grantee NSF, 1965—; Guggenheim fellow, 1982. Fellow AAAS, NRC (bd. biology 1986-92); mem. Internat. Union for Conservation of Nature and Natural Resources (chair task force on declining amphibian populations 1990-92), Am. Soc. Zoologists (pres. 1992), Am. Soc. Naturalists (pres. 1989), Am. Soc. Ichthyologists and Herpetologists (bd. govs.), Soc. Study Evolution (pres. 1983, editor 1979-81), Soc. Systematic Biology (coun. 1980-84), Herpetologist's League (Disting. Herpetologist 1984). Home: 999 Middlefield Rd Berkeley CA 94708-1509

WAKED, ROBERT JEAN, chemical engineering executive; b. Beirut, Lebanon, Nov. 7, 1957; came to U.S. 1976.; s. Jean and Linda (Fadel) W.; m. Kathleen Vaughn, Nov. 11, 1989. B in Chem. Engring., U. Dayton, 1979, MS in Chem Engring., 1981; MBA, Winthrop U., 1992. Rsch. and devel. chemist NCR Systemedia, Miamisburg, Ohio, 1979-81; devel. engr. Ciba-Geigy, McIntosh, Ala., 1982-84; prodn. engr. SCM Metal Products, N.C., 1985-88, project engr., 1988-89; process engr. Waste Chem, Rock Hill, S.C., 1989-91; mgr. process tech. Thermal KEM, Rock Hill, 1991—. Mem. adv. com. York Tech. Coll., Rockhill, 1991—. Recipient 2 Tech. awards SCM Orgn., 1988, 89. Mem. Am. Inst. Chem. Engr., MBA Assn. Office: Themalkem 2324 Vermesdale Rd Rock Hill SC 29731-2664

WAKELYN, PHILLIP JEFFREY, chemist, consultant; b. Akron, Ohio, Apr. 29, 1940; s. Arthur Thomas and Elsa Katherine (Koch) W. BS in Chemistry, Emory U., 1963; MS in Textile Chemistry, Ga. Inst. Tech., 1968; PhD in Textile Chemistry, U. Leeds, Eng., 1971. Cert. textile technologist, U.K. Rsch. chemist Dow Chem. Co., Williamsburg, Va., 1963-66; rsch. chemist Dow-Badische Co., Williamsburg, summers 1966-67; rsch. assoc. Tex. Tech. U. Textile Rsch. Ctr., 1971-73; lectr. Tex. Tech. U. Dept. Textile Engring., Lubbock, 1972-73; head chem. rsch. Tex. Tech. U. Textile Rsch. Ctr., Lubbock, 1971-73; mgr. environ. health, safety Nat. Cotton Coun. Am., Washington, 1973—; Nat. Adv. Com. Flammable Fabrics Act, Washington, 1976-78; adj. prof. Tex. Tech. U. Dept. Chem. Engring., 1974-78; cons., expert witness, 1971—. Editor: Washed Cotton, 1986, Proceedings Cotton Dust Research Conference, 1976-93; contbr. more than 100 articles to profl. jours. Anglo-Am. fellow Internat. Wool Secretariate, Leeds, 1968-71, Stribiling Found. fellow Ga. Inst. Tech., 1966-67; named to Am. Men and Women of Sci.; recipient Cert. of Appreciation USDA, 1975, CPSC, 1978, Disting. Svc. award Nat. Cotton Ginnen Assn., 1992. Fellow Am. Assn. Textil Chemists and Colorists, N.Y. Acad. Sci.; mem. Am. Chem. Soc., Am. Indsl. Hygiene Assn. (v.p. Mid South Sect. 1983-84), Textile Inst., Scientific Rsch. Soc., Am. Oil Chemists Soc., Sigma Xi. Achievements include testifying at fed. OSHA cotton dust hearings representing 5 different sectors of the cotton industry, at fed. OSHA hearings on formaldehyde, hazard communication and air contaminants and before 5 different state OSHAS. Office: Nat Cotton Coun Am 1521 New Hampshire Washington DC 20036

WAKEMAN, THOMAS GEORGE, mechanical engineer; b. Watertown, N.Y., June 28, 1945; s. George Lewis and Agnus Neomi (Kilburn) W.; m. Connie Anne Bailey; 1 child, Kenneth Loren. BSME, Ohio U., 1968; MS, U. Cin., 1971, PhD, 1982. Registered profl. engr., Ohio. Engr. Gen. Electric, Cin., 1968-78, engring. mgr., 1978—; adj. faculty U. Cin., 1983—; advisor dept. mech. engring. Ohio U., Athens, 1984—. Contbr. articles to profl. jours. Recipient Svc. award Ohio U., 1991. Mem. Soc. Automotive Engrs. Republican. Roman Catholic. Achievements include 31 patents received and 5 pending in the jet engine field. Home: 527 Ridge Ave Lawrenceburg IN 47025

WAKIMURA, YOSHITARO, economics educator; b. Dec. 6, 1900. Grad., Tokyo U., 1924. Auditor Nisshin Elect., Japan; vis. prof. economics Tokyo U., Japan; now pres. Nippon Gakushiin, Tokyo, Japan. Recipient medal of Honor with Blue Ribbon, 1963; 1st Cl. Order of the Sacred Treasure, 1971. Office: Nippon Gakushiin, 7-32 Ueno Park, Taito-ku Tokyo 110, Japan*

WAKNINE, SAMUEL, dental materials scientist, researcher, educator; b. Casablanca, Morocco, Apr. 26, 1959; came to U.S., 1973; s. David and Rebecca (Ouaknine) W. BS in Biol. Sci., U. Calif., Irvine, 1982; M Dental Materials Sci., NYU, 1984; PhD, U. Conn., 1991. Practical dentistry fellow Laguna Beach, Santa Ana, Calif., 1980-81; assoc. rsch. scientist dental splty. tech. group Am. Hosp. Corp., Irvine, 1981-82; dental materials cons. restorative composite div. Pentron Corp., Wallingford, Conn., 1982-83, mgr. R & D, 1983-84, dir. R & D, 1984-86, v.p. R & D, 1986-88; v.p. R & D and mfg. Jeneric/Pentron, Inc., Wallingford, 1988—; adj. asst. prof. dept. dental materials sci., div. restorative and prosthodontic scis. NYU Dental Ctr., N.Y.C., 1987—; adj. asst. prof. dept. gen. dentistry practice Case-Western Res. U., Cleve., 1991—; lectr., presenter seminars in field U.S. and worldwide. Contbr. more than 40 rsch. articles to dental jours., chpts. to biomed. texts; contbr. to ednl. tapes and audio/video. Liaison North African Jewish Assn., Calif., N.Y., 1973—. Grantee Nat. Inst. Dental Rsch., NIH, 1987-92. Mem. AAAS, Internat. Assn. for Dental Rsch. (chmn. sci. sessions 1984—), Am. Assn. for Dental Rsch., Biomed. Materials Rsch. Congress Soc., Acad. Dental Materials, Am. Soc. for Metals, Am. Chem. Soc., Am. Inst. Chemists, Conn. Electron Microscopy Soc., Internat. Scanning Electron Microscopy Soc., N.Y. 1st Dist. Dental Soc., Dental Materials Group, N.Y. Acad. Scis. Achievements include international patents in dental/bioprosthesis field, U.S., Germany, Can., Switzerland, Liechtenstein; development of over 50 bioprosthesis dental products in field of operative, restorative, reconstructive and adhesive dentistry of optimized hybrid dental composites and mechanisms of adhesion to tooth enamel, dentin and cementum of polymer composites, metal alloys and ceramic suprastructures and implants; characterization and optimization of biocompatible biomedical dental prosthetic devices/restorative materials. Home: 12C Limewood Ave Branford CT 06405-5303 Office: U Conn Health Ctr Sch Dental Medicine L-6100 Farmington CT 06032

WAKOFF, ROBERT, electrical engineer, consultant; b. N.Y.C., Feb. 8, 1925; s. Morton Meyer and Yetta (Kreisfeld) W.; m. Ruth Mackler, sept. 18, 1949; children: Michael Bruce, David Bradley, Brian Scott. BEE, CCNY, 1948. Test engr. Bendix Aviation Corp., Teterboro, N.J., 1948-50; missile coord. Bell Aircraft, Buffalo, 1950-52, missile elec. group, 1952-58, missile project engr., 1958-60; prelimmary design engr. Bell Aerospace Textron, Buffalo, 1960-65, proposal mgr., 1965-85, ret., 1985; proposal cons. Western N.Y. Tech. Devel. Ctr., Buffalo, 1985—. Contbr. articles to profl. jours. Pres. Temple Sinai, Buffalo, 1979-81; vol. cancer info. specialist Roswell Park Meml. Cancer Inst., Buffalo, 1986—. With USN, 1944-46. Mem. IEEE (sr.). Democrat. Jewish. Home and Office: 10 Hardt Ln Amherst NY 14226-2508

WALCH, MARIANNE, microbiologist; b. Annapolis, Md., Jan. 12, 1957; d. Anthony Jr. and Elizabeth Evalyn (Thompson) W. BS, U. Md., 1980; SM, Harvard U., 1981, PhD, 1986. Biol. aide U.S. Fish and Wildlife Svc., Laurel, Md., 1979-80; teaching fellow Harvard U., Cambridge, Mass., 1981-83, NSF predoctoral fellow, 1980-83, rsch. asst., 1983-86; rsch. assoc. Ctr. Marine Biotech., U. Md., Balt., 1986-88; rsch. asst. prof. Ctr. Marine Biotech., U. Md., College Park, 1988—; materials engr. Naval Surface Warfare Ctr., Silver Spring, Md., 1990—; exec. planning com. Nat. Study of Indicators and Pathogens in Shellfish, 1989-91; del. UNESCO Workshop on Marine Sci. Teaching, Paris, 1988; small bus. innovative rsch. program reviewer and tech. rep., 1991—; co-dir. NSWC Molecular Computing Program, 1991-92; adj. asst. prof. U. Md. U. Coll., 1993—. Author: (with others) Dahlem Conference: Structure of Biofilms, 1989, Encyclopedia of Microbiology, 1992; contbr. articles to profl. jours. Vol. NOW, Boston, 1982-84, Cambridge Women's Ctr. and Shelter, 1983-85; participant U. Md. Women's Forum, College Park, 1990-92. Mem. AAAS, Am. Soc. Microbiology (mem. newsletter editor com. on status of women in microbiology), Nat. Shellfisheries Assn., Nat. Assn. Corrosion Engrs. Democrat. Achievements include 2 patents for use of bacterial films and products to enhance set of oysters in hatcheries; rsch. on microbiologically influenced corrosion, biomagnification of pollutants in biofilms, bioremediation of hydrocarbons and metals, microbial cues for invertebrate settlement and metamorphosis, use of bacterial photopigments for molecular computing and holographic applications. Office: Ctr for Marine Biotech 600 E Lombard St Baltimore MD 21202

WALCZAK, ZBIGNIEW KAZIMIERZ, polymer science and engineering researcher; b. Gostyn, Poland, Mar. 27, 1932; came to U.S. 1964.; s. Ignacy P. and Stanislawa (Zychlewicz) W.; m. M. Krystyna Szymusik-Zygmuntowicz; 1 child, Agatha C. Engr. master, Poly. of Lodz, Poland, 1956. Tech. mgr. Coop. Chem. Industries, Lodz, 1956-59; chief chemist Persöner Concernen AB, Ystad, Sweden, 1960-62; rsch. chemist Shawinigan (Que.) Chems. Ltd., 1963-64, E.I. DuPont De Nemours Inc., Old Hickory, Tenn., 1964-67; sr. project leader Phillips Petroleum Co., Bartlesville, Okla., 1967-70, 1967-70; rsch. assoc. Inmont Corp., Clifton, N.J., 1973-82; rsch. fellow Kimberly-Clark Corp., Roswell, Ga., 1982—. Author: Formation of Synthetic Fibers, 1977, Thermal Bonding of Fibers, 1992, novel organic compounds; patentee synthetic fibers. Active Underground Boy Scouts as part of Polish Resistance Movement of Armia Krajowa, 1943-45. Home: 1350 Witham Dr Atlanta GA 30338-3337 Office: Kimberly-Clark Corp 1400 Holcomb Bridge Rd Roswell GA 30076-2190

WALD, FRANCINE JOY WEINTRAUB (MRS. BERNARD J. WALD), physicist, academic administrator; b. Bklyn., Jan. 13, 1938; d. Irving and Minnie (Reisig) Weintraub; student Bklyn. Coll., 1955-57; B.E.E., CCNY, 1960; M.S., Poly. Inst. Bklyn., 1962, Ph.D., 1969; m. Bernard J. Wald, Feb. 2, 1964; children—David Evan, Kevin Mitchell. Engr., Remington Rand Univac div. Sperry Rand Corp., Phila., 1960; instr. Poly. Inst. Bklyn., 1962-64, adj. research asso., 1969-70; lectr. N.Y. Community Coll., Bklyn., 1969, 70; instr. sci. Friends Sem., N.Y.C., 1975-76, chmn. dept. sci., 1976—. NDEA fellow, 1962-64. Mem. Am. Phys. Soc., Am. Assn. Physics Tchrs., Assn. Tchrs. in Ind. Schs., N.Y. Acad. Scis., Nat. Sci. Tchrs. Assn., AAAS, Sigma Xi, Tau Beta Pi, Eta Kappa Nu.

WALD, GEORGE, biochemist, educator; b. N.Y.C., N.Y., Nov. 18, 1906; s. Isaac and Ernestine (Rosenmann) W.; m. Frances Kingsley, May 15, 1931 (div.); children: Michael, David; m. Ruth Hubbard, 1958; children: Elijah, Deborah. B.S., NYU, 1927, D.Sc. (hon.), 1965; M.A., Columbia U., 1928, Ph.D., 1932; M.D. (hon.), U. Berne, 1957, U. Leon, Nicaragua, 1984; D.Sc. (hon.), Yale U., 1958, Wesleyan U., 1962, McGill U., 1966, Amherst Coll., 1968, U. Rennes, 1970, U. Utah, 1971, Gustavus Adolphus U., 1972, Columbia U., 1990; D.H.L. (hon.), Kalamazoo Coll., 1984. NRC fellow at Kaiser Wilhelm Inst. Berlin and Heidelberg, U. Zurich, U. Chgo., 1932-34; tutor biochem. scis. Harvard U., 1934-35, instr. biology, 1935-39, faculty instr., 1939-44, assoc. prof. biology, 1944-48, prof., 1948-77, Higgins prof. biology, 1968-77, prof. emeritus, 1977—; vis. prof. biochemistry U. Calif., Berkeley, summer 1956; Nat. Sigma Xi lectr., 1952; chmn. divisional com. biology and med. scis. NSF, 1954-56; Guggenheim fellow, 1963-64; Overseas fellow Churchill Coll., Cambridge U., 1963-64; participant U.S.-Japan Eminent Scholar Exchange, 1973; guest China Assn. Friendship with Fgn. Peoples, 1972; v.p. Permanent Peoples' Tribunal, Rome, 1980—. Co-author: General Education in a Free Society, 1945, Twenty Six Afternoons of Biology, 1962, 66, also sci. papers on vision and biochem. evolution. Recipient Eli Lilly prize Am. Chem. Soc., 1939, Lasker award, Am. Pub. Health Assn., 1953, Proctor medal Assn. Rsch. in Opthalmology, 1955, Rumford medal Am. Acad. Arts and Scis., 1959, Ives medal Optical Soc. Am., 1966, Paul Karrer medal in chemistry U. Zurich, 1967; co-recipient Nobel prize for physiology or medicine, 1967; T. Duckett Jones award Helen Hay Whitney Found., 1967; Bradford Washburn medal Boston Mus. Sci., 1968; Max Berg award, 1969, Priestley medal Dickinson Coll., 1970. Fellow Nat. Acad. Sci., Am. Acad. Arts and Scis., Am. Philos. Soc. Home: 21 Lakeview Ave Cambridge MA 02138-3325 Office: Harvard U Biol Labs Cambridge MA 02138

WALDEN, OMI GAIL, public affairs and government relations specialist; b. Alma, Ga., Dec. 25, 1945; d. Banner H. and Naomi (Thomas) Lee; A.B. in Journalism, U. Ga., 1967; exec. mktg. mgmt. program Stanford U., 1981; m. Ralph Edward Walden, Apr. 27, 1968. Asst. dir. pub. relations Ga. Ports Authority, Savannah, 1968-69; dir. pub. relations U.S. HUD Model Cities Program, Alma, 1970-73, citizens participation coordinator, 1970, dir. research and evaluation, 1971-72; fed. and state relations coordinator, policy advisor on energy and environ. issues Former Gov. Jimmy Carter and Gov. George Busbee, Atlanta, 1973-76; dir. Ga. Office Energy Resources, Atlanta, 1976-78; asst. sec. conservation and solar applications U.S. Dept. Energy, Washington, 1978-79, adv. to sec. for conservation and solar mktg., 1979-80; now pres. Omi Walden & Assocs., pub. affairs and govt. relations cons. firm; exec. dir. Nat. Energy Mgmt. Inst., Washington; gov. rep. to Pres. Intergovernmental Sci., Engring. and Technology Advisory Panel, 1978. Democrat. Baptist. Contbr. articles in field to profl. jours. Home: 829 Colony House E 145 15th St Atlanta GA 30361 Office: 1750 New York Ave NW Washington DC 20006-5301

WALDEN, ROBERT THOMAS, physicist, consultant; b. Paducah, Ky., Mar. 25, 1939; s. Charles Robert and Anna Catherine (Robertson) W.; m. Nellie Sue Clayton, June 9, 1962; children: Clayton Thomas, Alan Keith. BS, Murray State U., 1961, MS, 1968; PhD, Miss. State U., 1973. Tchr. math. Dongola (Ill.) High Sch., 1962-63; instr. physics Paducah C.C., 1963-68; asst. prof. Ky. Wesleyan Coll., Owensboro, 1968-70, chair sci. Miss. Gulf Coast C.C., Perkinston, 1973-87; staff scientist Nat. Ctr. Phys. Acoustics, U. Miss., University, 1987-91; prof. physics Mid-Continent Coll., Mayfield, Ky., 1990; cons. Walden Assocs., Paducah, 1990—; vis. prof. U. Miss., University, 1986-87; cons. Paducah Gaseous Diffusion Plant, 1992—; organizer, chair 1st nat. symposium on agroacoustics; organizer, chair Miss. Alliance Sci. Advancement. Contbr. articles to Jour. Molecular Spectroscopy, Jour. Miss. Acad. Sci. Dir., founder Stone County Jr. Basketball League, Wiggins, Miss., 1979-82; vice-chmn. Stone Coutny Hosp. Bd., Wiggins, 1983-85. Mem. Acoustical Soc. Am., Rotary (Rotarian of Yr. Wiggins unit 1981), Sigma Pi Sigma, Phi Kappa Phi. Baptist. Achievements include initiation of acoustic remote sensing of bark beetles. Home: 420 Hutchinson Ave Paducah KY 42003 Office: Walden Assocs 453 Hutchinson Ave Paducah KY 42003

WALDEN, W. THOMAS, lead software engineer, consultant; b. Tampa, Fla., Oct. 21, 1950; s. William T. Walden Sr. and Naida (Hohenthaner) Farmer; m. Shirley Hopper, Apr. 1, 1973 (div. Sept. 1976); children: Christine, Benjamin; m. Marie Diane McAvey, July 5, 1977; children: Joseph, Robert, Sean. BS in Computer Sci., Fla. Inst. Tech., 1985. Material control supr. Regency Co., Satellite Beach, Fla., 1974-76; sr. computer applications specialist Harris Corp., Melbourne, Fla., 1976-86; pres. Computer Sorcery, Palm Bay, Fla., 1986-87; systems analyst Q-Bit Corp., Palm Bay, Fla., 1987-89; systems engr. Computer Task Group, Orlando, Fla., 1989-90; lead software engr. Harris Corp., Palm Bay, Fla., 1990—. Author: (software) Printing Routines, 1989, Scatter and Gather with Clipper, 1991, Bit-Mapped Security, 1992, Downsizing to Clipper, 1993. Asst. dist. com. Brevard Youth Soccer League, Brevard County, Fla., 1987-90; pres. Palm Bay Youth Soccer, 1992—. Recipient Clipper Champion award Nantucket Corp., 1991, Dist. Com. award Fla. Youth Soccer Assn., 1990. Mem. Space Coast Clipper Developers Group (founding pres. 1991—). Republican. Baptist. Home: 393 Pipit St NE Palm Bay FL 32907

WALDFOGEL, LARUE VERL, electrical engineering executive; b. Anderson, Ind., Apr. 28, 1926; s. Charles and Lena Susan (Sieler) W.; m. Elizabeth Joan Buehrer, June 20, 1948;children: Roger Dean, Sandra Mae, Susan Kay. B Engring., U. Toledo, 1950. Registered profl. engr., Ohio, Mich., Ind.; master electrician. Elec. engr. Elec. Autolite Co., Toledo, Ohio, 1950-52; plant elec. engr. Bohn Aluminum Co., Adrian, Mich., 1952-54; v.p. engring. Lenawee Elec. Co., Adrian, 1954-61; elec. engr. Straight Engring. Co., Inc., Adel, Iowa, 1961-72, Stout Constrn. Co., Inc., Sylvania, Ohio, 1972-82, Midstates Terminals, Inc., Toledo, 1982; dir. elec. engring. Jones & Henry Engrs., Inc., Toledo, 1982—. Served in USN, 1944-46, Japan. Mem. AIEE, NSPE, Am. Legion, VFW, Lions. Home: 3668 Consear Rd Lambertville MI 48144 Office: Jones & Henry Engrs Inc 2000 W Central Ave Toledo OH 43606

WALDMANN, THOMAS ALEXANDER, medical research scientist, physician; b. N.Y.C., Sept. 21, 1930; s. Charles Elizabeth (Sipos) W.; m. Katharine Emory Spreng, Mar. 29, 1958; children—Richard Allen, Robert James, Carol Ann. A.B., U. Chgo., 1951; M.D., Harvard U., 1955. Diplomate Am. Bd. Allergy and Immunology. Intern Mass. Gen. Hosp., Boston, 1955-56; clin. assoc. Nat. Cancer Inst. NIH, Bethesda, Md., 1956-58, sr. investigator, 1958-68, head immunophysiology sect., 1968-73, chief metabolism br., 1971—; cons. WHO, 1975, 78; bd. dirs. Found. for Advanced Edn. in Scis., Bethesda, 1980—, treas., 1988-90, v.p., 1990—; William Dameshek vis. prof. Tufts U., 1983; Burroughs Welcome vis. prof. U. Calif., Irvine, 1984; mem. med. adv. bd. Howard Hughes Med. Inst., 1986-92. Author: Plasma Protein Metabolism, 1970; contbr. over 500 articles to profl. jours. Discoverer diseases intestinal lymphangiectasia and allergic enteropathy. Served with USPHS, 1956-58, 59-63, 75—. Recipient Henry M. Stratton medal Am. Hemotology Soc., 1977; named Man of Yr. Am. Leukemia Soc., 1980; recipient G. Burroughs Mider award NIH, 1980; Disting. Service medal Dept. Health and Human Services, 1983. Fellow Am. Acad. Allergy (Bela Schick award 1974, John M. Shelton award 1984, Lila Gruber prize 1986, Simon Shubitz prize 1987, CIBA-GEIGY Drew award 1987, Milken Family Med. Found. Disting. Basic Scientist prize, Artois Latour Internat. Rsch. prize 1991, Bristol-Myers Cancer prize 1992, Artois-Ballet-Latour prize); mem. NAS, Am. Acad. Arts and Scis., Inst. Medicine, Nat. Acad. Scis., Assn. Am. Physicians, Am. Soc. Clin. Investigation (mem. editorial bd. 1978-80, 83-88), Clin. Immunology Soc. (pres. 1988). Avocation: photography. Home: 3910 Rickover Rd Silver Spring MD 20902-2329 Office: Nat Cancer Inst NIH Bldg 31 9000 Rockville Pike Bethesda MD 20892

WALDOW, STEPHEN MICHAEL, radiation biologist, educational consultant; b. Buffalo, Aug. 19, 1959; s. Stephen Robert Waldow and Patricia Ann (Smith) Dowd; m. Kathleen Ann Donnelly, Dec. 29, 1984; children: Meghan Kerry, Patrick Vincent, Stephen Patrick, Daniel Robert. BS in Biology, Niagara U., 1981; PhD in Radiation Biology, SUNY, Buffalo, 1984. Assoc. rsch. coord. Wenske Laser Ctr., asst. dir. Ravenswood Hosp. Med. Ctr., Chgo., 1985-86; dir. div. radiation rsch. Cooper Hosp.-Univ. Med. Ctr., Camden, N.J., 1987-91; assoc. prof. dept. radiation oncology Temple Univ., Phila., 1991—; mem. com. Z136.3 Am. Nat. Standards Inst., N.Y.C., 1986-87; clin. asst. prof. dept. radiology U. Medicine and Dentistry N.J.-Robert Wood Johnson Med. Sch., Camden, 1987-91; edn. cons., mem. edn. adv. bd. Laser Ctrs. Am., Inc., Cin., 1988—; vis. asst. prof. dept. environ. sci. Rutgers U., New Brunswick, N.J., 1990-91. Contbr. articles to profl. jours. Scholar Niagara U., 1977-81; travel grantee Am. Soc. for Laser Medicine and Surgery, 1986. Mem. AAAS, Am. Soc. for Photobiology, Radiation Rsch. Soc. (travel grantee 1986), N.Y. Acad. Scis. Roman Catholic. Achievements include first to show synergistic interaction of microwave hyperthermia and photodynamic therapy in vivo; one of first to show efficacy of low power Nd:YAG laser-induced tumor hyperthermia in vivo. Home: 14 Country Hollow Cir Sicklerville NJ 08081-3905 Office: Spectrum Lectures Inc PO Box 393 Sicklerville NJ 08081-0393

WALDRON, KENNETH JOHN, mechanical engineering educator, researcher; b. Sydney, NSW, Australia, Feb. 11, 1943; came to U.S., 1965; s. Edward Walter and Maurine Florence (Barrett) W.; m. Manjula Bhushan, July 3, 1968; children: Andrew, Lalitha, Paul. BEngring., U. Sydney, 1964, M Engring. Sci., 1965; PhD, Stanford U., 1969. Registered profl. engr., Tex. Acting asst. prof. Stanford (Calif.) U., 1968-69; lectr., sr. lectr. U. NSW, Sydney, 1969-74; assoc. prof. U. Houston, 1974-79; assoc. prof. mech. engring. Ohio State U., Columbus, 1979-81, prof., 1981—, Nordholt prof., 1984—, chmn. dept. mech. engring., 1993—. Co-author: Machines That Walk, 1988; editor: Advanced Robotics, 1989; contbr. over 170 articles to profl. jours. and conf. procs. Fellow ASME (tech. editor Trans. Jour. Mech. Design 1988-92, Leonardo da Vinci award 1988, Mechanisms award 1990); mem. Soc. Automotive Engrs. (Ralph R. Teetor award 1977), Am. Soc. for Engring. Edn. Achievements include work on adaptive suspension vehicle project. Office: Ohio State U 206 W 18th St Columbus OH 43210

WALECKI, WOJCIECH JAN, physicist, engineer; b. Warsaw, Poland, Jan. 10, 1964; came to U.S., 1988; s. Jan Andrzej and Stanislawa M. (Prusak) W.; m. Anna Maria Klosowska; 1 child, Katherine. MSc, Warsaw U., 1988, Brown U., 1992. Teaching asst. Purdue U., West Lafayette, Ind., 1988-89; rsch. asst. Brown U., Providence, 1989-92; ind. cons. elec. engring., constrn., 1992; materials engr. MIM Corp., Providence, 1990-91. Mem. Am. Phys. Soc. Office: Brown Univ Dept Physics Box 1843 Providence RI 02912

WALI, MOHAN KISHEN, environmental science and natural resources educator; b. Kashmir, India, Mar. 1, 1937; came to U.S., 1969, naturalized, 1975; s. Jagan Nath and Somavati (Wattal) W.; m. Sarla Safaya, Sept. 25, 1960; children: Pamela, Promod. BS, U. Jammu and Kashmir, 1957; MS, U. Allahabad, India, 1960; PhD, U. B.C., Can., 1970. Lectr. S.P. Coll., Srinagar, Kashmir, 1963-65; rsch. fellow U. Copenhagen, 1965-66; grad. fellow U. B.C., 1967-69; asst. prof. biology U. N.D., Grand Forks, 1969-73, assoc. prof., 1973-79, prof., 1979-83, Hill rsch. prof., 1973, dir. Forest River Biology Area Field Sta., 1970-79, Project Reclamation, 1975-83, spl. asst. to univ. pres., 1977-82; staff ecologist Grand Forks Energy Rsch. Lab., U.S. Dept. Interior, 1974-75; prof. Coll. Environ. Sci. and Forestry, SUNY, Syracuse, 1983-89, dir. grad. program environ. sci., 1983-85; prof., Sch. Natural Resources, 1990—; dir., Sch. Nat. Resources, assoc. dean, Coll. Agr., 1990-93; vice chmn. N.D. Air Pollution Adv. Council, 1981-83; co-chair IV Internat. Congress on Ecology, 1986. Editor: Some Environmental Aspects of Strip-Mining in North Dakota, 1973, Prairie: A Multiple View, 1975, Practices and Problems of Land Reclamation in Western North America, 1975, Ecology and Coal Resource Development, 1979, Ecosystem Rehabilitation-Preamble to Sustainable Development, 1992; sr. editor Reclamation Rev., 1976-80, chief editor, 1980-81; chief editor Reclamation and Revegetation Rsch., 1982-87; contbr. articles to profl. jours. Recipient B.C. Gamble Disting. Teaching and Svc. award, 1977. Fellow AAAS, Nat. Acad. of Scis. India; mem. Ecol. Soc. Am. (chmn. sect. internat. activities 1980-84), Brit. Ecol. Soc., Can. Bot. Assn. (dir. ecology sect. 1976-79, v.p. 1982-83), Ohio Acad. Sci., Torrey Bot. Club, Am. Soc. Agronomy, Am. Inst. Biol. Sci. (gen. chmn. 34th ann. meeting), Internat. Assn. Ecology, Internat. Soc. Soil Sci., N.D. Acad. Sci. (chmn. editorial com. 1979-81), Sigma Xi (nat. lectr. 1983-85, pres. Ohio State chpt. 1993-94, pres. Syracuse chpt. 1984-85, Outstanding Rsch. award U. N.D. chpt. 1975). Office: Ohio State U Sch Natural Resources 2021 Coffey Rd Columbus OH 43210-1044

WALIA, SATISH KUMAR, microbiologist, educator; b. Fathepur, India, June 29, 1951; came to the U.S., 1980; s. Anand Sarup and Kamla Devi Walia; m. Venna Rani, Apr. 28, 1980; children: Sandeep Kumar, Sonia. MS, Panjab U., 1975; PhD, Mahrishi Dayanand U., 1980. Demonstrator Med. Coll., Rohtak, India, 1976-80; postdoctoral assoc. U. Fla., Gainesville, 1980-82, asst. instr., 1982-84; asst. prof. Oakland U., Rochester, Mich., 1984-90, assoc. prof., 1990—; mem. U.S. EPA Environ. Biology Rev. Panel, Washington, 1986; cons. Providence Hosp., Southfield, Mich., 1986-89, Nat. Ctr. Mfg. Sci., Ann Arbor, Mich., 1991, Enviroclean Inc., Troy, Mich., 1991-92. Contbr. articles to Plasmid, European Jour. Clin. Microbial Infectious Disease, Applied Environ. Microbiology, Virology, Gene. Grantee U.S. EPA, 1986, 90. Mem. Am. Soc. for Microbiology, AAAS, AAUP, Indian Microbiologist Assn. Am. (pres. 1992), SMI Inc. (pres. 1992), Sigma Xi. Achievements include development of microbial processes for degradation of chlorinated biphenyls; research using gene probes in detection of PCB degrading bacteria, construction of bacterial strains for production of specialty chemicals, metabolic pathways for degradation of PCBs. Home: 793 Dressler Ln Rochester Hills MI 48307 Office: Oakland U Dodge Hall Rochester MI 48309-4401

WALINSKY, PAUL, cardiology educator; b. Phila., June 21, 1940; s. Aaron and Bess (Kleiman) W.; m. Stephanie Sosenko, Nov. 27, 1971; children: Shira, Daniel. BA, Temple U., 1961; MD, U. Pa., 1965. Cert. Nat. Bd. Med. Examiners, Am. Bd. Internal Medicine Cardiovascular. Instr. medicine

Thomas Jefferson U., Phila., 1973-75, asst. prof. medicine, 1975-79, assoc. prof. medicine, 1979-82, prof. medicine, 1982—; cons. EP Technologies, Mountain View, Calif., 1991—, Baxter Edwards, Irving, Calif., 1988-91. Contbr. articles to profl. jours.; inventor method for high frequency ablation, percutaneous microwave catheter angioplasty. Capt. USAF, 1967-69. Fellow Am. Coll. Cardiology, ACP; mem. AMA, Pa. Med. Soc., Phila. County Med. Assn. Achievements include invention of perfusion balloon catheter, microwave aided balloon angioplasty with lumen measurement, intravascular ultrasonic imaging catheter and method for making same. Office: Thomas Jefferson U 111 S 11th St Philadelphia PA 19107

WALIZE, REUBEN THOMPSON, III, health research administrator; b. Williamsport, Pa., May 28, 1950; s. Reuben Thompson Jr. and Marion Marie (Smith) W.; m. Kathleen Anne Smith, Aug. 13, 1979; children: Heather, Amanda, Reuben IV. BS, Pa. State U., 1972; MPH magna cum laude, U. Tenn., 1975. Manpower planner Northcentral Pa. Area Health Edn. System, Danville, 1975-76, asst. dir., 1976, exec. dir., 1976-78; health mgr. Seda-Cog, Timberhaven, Pa., 1978; exec. asst. VA Med. Ctr., Erie, Pa., 1978-81; trainee VA Med. Ctr., Little Rock, 1981; adminstrv. officer rsch. svc. VA Med. Ctr., White River Junction, Vt., 1981-88; mgmt. analyst Dept. Vets. Affairs Med. Ctr., Roseburg, Oreg., 1988-90, health systems specialist, 1990-92; adminstrv. officer rsch. Vets. Affairs Med. Ctr. Am. Lake, Tacoma, 1992—; Mem. Pa. Coun. Health Profls., 1975-77, Ctrl. Pa. Health System Agy. Manpower Com., 1975-77; mem. Interagy. Coun. Geisinger Med. Ctr., Danville, 1976-78; liaison for rsch. Dartmouth Med. Sch., Hanover, N.H., 1981-88; cons. in field. Recipient Man of Achievement award Queens Coll., Eng., 1978, Student Am. Med. Assn. Found. award, 1975; 1st pl. Douglas County Lamb Cooking Contest, 1992. Mem. APHA, AAAS, N.Y. Acad. Scis., Soc. Rsch. Adminstrs., Pa. State Alumni Assn., Nat. Audubon Soc., Steamboaters, Nat. Wildlife Fedn., Record Catch Club, VIP Club. Avocations: fly fishing, fly tying, gardening, photography, gourmet cooking. Home: 1103 25th Ave SE Puyallup WA 98374-1362

WALKENBACH, RONALD JOSEPH, foundation executive, pharmacology educator; b. Hermann, Mo., Mar. 15, 1948; s. Walter Bernard and Lillian Ann (Ochsner) W.; m. Mary McLean Hodge, June 21, 1975 (div. July 1981); m. DeAnna Marie Roemer, Oct. 3, 1984. B.S. in Chemistry, Quincy Coll., Ill., 1970; Ph.D. in Pharmacology, U. Mo., Columbia, 1975. Postdoctoral fellow U. Va., Charlottesville, 1975-77; postdoctoral fellow U. Mo., Columbia, 1977-78; lab. investigator Eye Research Found., Columbia, 1978-81, exec. dir., 1981—. Contbr. research articles to profl. jours. Troop coordinator Boy Scouts Am., Columbia, 1985. Research grantee Nat. Eye Inst., NIH, 1978—. Mem. AAAS, Assn. Research in Vision and Ophthalmology. Roman Catholic. Lodge: Lions (2d v.p. 1985). Avocations: hunting, fishing, skiing, hiking, woodworking. Office: Eye Research Found Mo Inc 404 Portland St Columbia MO 65201-6506

WALKER, BRUCE EDWARD, anatomy educator; b. Montreal, Que., Can., June 17, 1926; s. Robinson Clarence and Dorothea Winston (Brown) W.; m. Lois Catherine McCuaig, June 26, 1948; children—Brian Ross, Dianne Heather, Donald Robert, Susan Lois. B.S., McGill U., 1947, M.S., 1952, Ph.D., 1954; M.D., U. Tex. at Galveston, 1966. Instr. anatomy McGill U., 1955-57; asst. prof. anatomy U. Tex. Med. Br., 1957-61, assoc. prof. anatomy, 1961-67; prof. Mich. State U., East Lansing, 1967—, chmn. dept., 1967-75. Contbr. articles to profl. jours. Mem. Am. Assn. Anatomists, Teratology Soc., Am. Assn. for Cancer Research. Office: Mich State U Anatomy Dept East Lansing MI 48824

WALKER, DANIEL JAY, biologist; b. Wadena, Minn., Sept. 13, 1960; s. Larry Lee and Janis Carolyn (Graham) W.; m. Sara Sue Kollack, Aug. 15, 1992. BS in Biology, U. N.D., 1984; MS, U. Wis., Eau Claire, 1988; PhD, Bowling Green State U., 1993. Grad. teaching asst. U. Wis., Eau Claire, 1984-87; grad. teaching fellow Bowling Green (Ohio) State U., 1988-89, 90-92, grad. rsch. fellow, 1989-90. Contbr. article to Jour. Parasitology. Mem. Am. Soc. Parasitology, Am. Soc. Tropical Medicine and Hygiene, Helminthological Soc. Wash., Am. Microscopy Soc., Sigma Xi. Republican. Home: 318 Rawson Apt L-6 Dundee MI 48131 Office: Bowling Green State U N College Dr Bowling Green OH 43403

WALKER, DUNCAN MOORE HENRY, electrical engineer; b. Lynwood, Calif., Dec. 4, 1956; s. Duncan Partrick and Margaret Theresa (Mueller) W.; m. Mary Patrick, Apr. 29, 1989. BS in Engring., Caltech, 1979; MS, Carnegie Mellon U., 1984, PhD, 1986. Engr. Hughes Aircraft Co., Culver City, Calif., 1977-78, Digital Equipment Corp., Hudson, Mass., 1979-81; rsch. engr. Carnegie Mellon U., Pitts., 1986-93; assoc. prof. Tex. A&M U., College Station, 1993—. Author: Yield Simulation for Integrated Circuits, 1987; gen. chair IEEE Internat. Workshop on Defect and Fault Tolerance in VLSI Systems, 1992. Mem. IEEE, Assn. for Computing Machinery, Sigma Xi. Achievements include devel. of techniques for statis. simulation of functional yield loss in integrated circuits. Office: Tex A&M U CS Dept College Station TX 77845

WALKER, EARL E., manufacturing executive; b. St. Louis, Mo., Feb. 21, 1921; s. Thomas T. and Ella Mary (Steggerman) W.; m. Myrtle Agnew, Sept. 27, 1942; children: Mary, Tom, Nancy, Peggy. Grad., Ranken Tech. Sch., 1941. Welder Curtis-Wright Aircraft Co., St. Louis, 1941-49; welder McDonnell-Douglas Aircraft Co., St. Louis, 1949-51, foreman, 1951-53; pres. S.N.W. Welding, St. Louis, 1952-53, Coeur Lane Mfg. Co., St. Louis, 1953-55, Carr Lane Mfg. Co., St. Louis, 1955—; bd. dirs. All Am. Products Co., Los Angeles, Am. Assn. Indsl. Mgmt. Chmn. Coop. Edn. Program, Scottish Rite Clinic for Childhood Lang. Disorders, U. Tex. Sch. Nursing; sec. of interior, Jefferson Nat. Meml. Commn., 1986—. Named Subcontractor of Yr. Small Bus. Adminstrn., Midwest Region VII, 1987; recipient Silver Platter award, Scottish Rite, Mo. Valley. Mem. Soc. Mfg. Engrs. (internat. dir., chmn. nat. exposition com., Nat. Businessman of Yr. 1975, Mgmt. Achievement award, Eli Whitney award 1991, Edn. Found. award for individual achievement 1991), Masons (33 degree). Avocation: golf. Office: Carr Lane Mgf Co 4200 Carr Lane Ct Saint Louis MO 63119-2196

WALKER, ESPER LAFAYETTE, JR., civil engineer; b. Decatur, Tex., Sept. 22, 1930; s. Esper Lafayette and Ruth (Mauldin) W.; B.S., Tex. A&M U., 1953; B.H.T., Yale U., 1958; m. Sara Lynn Dunlap, Oct. 2, 1955; children—William David, Annette Ruth. Design engr. Tex. Hwy. Dept., Austin, 1956-57; dir. Dept. Traffic Engring., High Point, N.C., 1958-63; v.p. Wilbur Smith & Assos., Houston, 1968-89, sr. v.p., 1989—. Pres. Meadow Wood PTA, 1976-77; chmn. Pack 902 com. Sam Houston council Boy Scouts Am., 1973-74, treas. Troop 904 com., 1976-80; treas. Stratford High Band, 1981-82, bd. dirs., 1980-82; baseball team mgr. Spring Br. Sports Assn., 1975-77; mem. adminstrv. bd. Meml. Drive Meth. Ch., 1971-77, 83-89, 90-93, bldg. com., 1974-82, fin. com., 1975-77, 83-85, trustee, 1983-89, chmn. bd. trustees, 1990-93. Served to 1st lt. C.E., AUS, 1953-56. Recipient Key Man award High Point Jaycees, 1962. Registered profl. engr., Tex., S.C., Colo., Ark., Wis., La., Okla., N.Mex., Wyo. Mem. Nat., Tex. socs. profl. engrs., High Point Jaycees (dir.), Tex. A&M U. Alumni Assn. (ctr. urban affairs council 1985-90, vice chmn. 1988), Houston C. of C. (chmn. transit com. 1975-79), Inst. Transp. Engrs. (pres. So. sect. 1963). Clubs: Galveston Country Club, Summit, Plaza. Home: 14216 Kellywood Ln Houston TX 77079-7410 Office: 908 Town And Country Blvd Ste 400 Houston TX 77024-2298

WALKER, FRANK BANGHART, pathologist; b. Detroit, June 14, 1931; s. Roger Venning and Helen Frances (Reade) W.; m. Phyllis Childs; children: Nancy Anne, David Carl, Roger Osborne, Mark Andrew. B.S., Union Coll., N.Y., 1951; M.D., Wayne State U., 1955, M.S., 1962. Diplomate Am. Bd. Pathology. Intern Detroit Meml. Hosp., 1955-56; resident Wayne State U. and affiliated hosps., Detroit, 1958-62; pathologist, 1962—; dir. labs. Detroit Meml. Hosp., 1984-87, Cottage Hosp., Grosse Pointe, Mich., 1984—; pathologist, dir. labs. Macomb Hosp Ctr. (formerly South Macomb Hosp.), Warren, Mich., 1966—, Jennings Meml. Hosp., Detroit, 1971-79, Alexander Blain Hosp., Detroit, 1971-85; ptnr. Langston, Walker & Assocs., profl. corp., Grosse Pointe, 1968—; instr. pathology Wayne State U. Med. Sch., Detroit, 1962-72, asst. clin. prof., 1972—. Pres. Mich. Assn. Blood Banks, 1969-70; mem. med. adv. com. ARC, 1972-83; mem. Mich. Higher Edn. Assistance Authority, 1975-77; trustee Alexander Blain Meml. Hosp., Detroit, 1974-83, Detroit-Macomb Hosp. Corp., 1975—; bd. dirs. Wayne State Fund, 1971-83. Capt. M.C., AUS, 1956-58. Recipient Disting. Svc. award Wayne State U. Med. Sch., 1990. Fellow Detroit Acad. Medicine; mem. AMA (coun. on long-range planning and devel. 1982-88, vice chmn. 1985-87, chmn. 1987-88, trustee 1988—), Am. Bd. Pathology (trustee 1982—, treas. 1984-91, v.p. 1991-92, pres. 1993—), Coll. Am. Pathologists (disting. svc. award 1989), Am. Soc. Clin. Pathologists (sec. 1971-77, pres. 1979-80, disting. svc. award 1989), Mich. Soc. Pathologists (pres. 1980-81), Wayne County Med. Soc. (pres. 1984-85, trustee 1986-91, chmn. 1990-91), Mich. Med. Soc. (bd. dirs. 1981-90, vice chmn. 1986-88, chmn. 1988-90), Am. Assn. Blood Banks, Mich. Assn. Blood Banks, Wayne State U. Alumni Assn. (bd. govs. 1968-71), Wayne State U. Med. Alumni Assn. (pres. 1969, trustee 1970-85, Disting. Alumni award 1974), Econ. Club Detroit, Detroit Athletic Club, Lochmoor Club, Mid-Am. Club, Phi Gamma Delta, Nu Sigma Nu, Alpha Omega Alpha. Republican. Episcopalian. Home and Office: 14004 Harbor Place Dr Saint Clair Shores MI 48080-1528

WALKER, GRAHAM CHARLES, biology educator; b. Boston, Feb. 8, 1948; s. Charles Bertram and Margaret Elizabeth (Biehn) W.; m. Janet Elizabeth Haliburton, May 30, 1970; 1 child, Gordon Andrew. BSc with honors, Carleton U., Ottawa, Ont., Can., 1970; PhD, U. Ill., 1974. Asst. prof. biology MIT, Cambridge, Mass., 1976-80, assoc. prof. biology, 1980-86; prof. biology MIT, Cambridge, 1986—. Editor-in-chief Jour. Bacteriology, 1991—, editor, 1985-91; editorial bd. Mutation Rsch., Amsterdam, Netherlands, 1982—; contbr. articles to Procs. NAS USA, Cell, Microbiology Rev. Housemaster McCormick Hall, MIT, Cambridge, 1986-92. Margaret MacVicar Faculty fellow MIT, 1992-2002, John Simon Guggenheim Meml. fellow, 1984, Woodrow Wilson fellow, 1970; recipient Rita Allen Career Devel. award, 1978-83. Mem. Am. Soc. for Microbiology, Genetics Soc. Am., Am. Chem. Soc., Environ. Mutagen Soc. Achievements include discovery of umuDC analogs on plasmid pKM101; demonstration of repetoire of SOS genes; demonstrated critical role of exopolysaccharide in nodule invation by Rhizobium meliloti; cloning and analysis of roles of UmuDC in UV mutagenesis; demonstrated RecA-mediated cleavage activates UmuD, second symbiotically active exopolysaccharide in R. meliloti, DnaK as a molecular thermometer, differential transcriptional activation by Ada protein. Office: MIT Biology Dept 77 Massachusetts Ave Cambridge MA 02139

WALKER, HARRELL LYNN, plant pathologist, botany educator, researcher; b. Minden, La., May 14, 1945; s. George Harrell Walker and Janice Ora (Nix) Smith; m. Diane Marie Reynolds, Sept. 6, 1975 (dec. Aug. 1988); children: Mark Thomas, Alan Reynolds. BS, La. Tech. U., 1966; MS, U. Ky., 1969, PhD, 1970. Biol. rsch. asst. U.S. Army Med. Lab., Ft. Meade, Md., 1970-72; postdoctoral rsch. asst. dept. fisheries Auburn (Ala.) U., 1972-74; plant pathologist Ala. Dept. Agriculture, Montgomery, 1974-75, asst. dir. plant industry divsn., 1975-76; rsch. scientist U.S. Dept. Agr./Agril. Rsch. Svc., Stoneville, Miss., 1976-84; dir. rsch. sta. Mycogen Corp., Ruston, La., 1984-87; prof. botany La. Tech. U., Ruston, 1987—. Editor: Biological Control of Weeds With Plant Pathogens, 1982; contbr. over 40 articles to profl. jours. With U.S. Army, 1970-72. Recipient Inventor's award U.S. Dept. Commerce, Washington, 1985; grantee U.S. Dept. Agr., 1981-83, La. Soybean Rsch. Bd., 1992-93. Mem. Am. Phytopathol. Soc., Weed Sci. Soc. Am., So. Weed Sci. Soc. Republican. Southern Baptist. Achievements include 7 patents related to biological control of weeds with plant pathogens, manipulation of microorganisms for control of plant diseases. Office: La Tech U Dept Biol Scis Ruston LA 71272

WALKER, JAMES CALVIN, computer software systems developer, politician; b. Egypt, Miss., Sept. 19, 1951; s. John Jasper Walker and Minnie Mary (Brooks) Holliday; m. Brenda Marie Caston, July 31, 1976; children: Jelani, Akilah, Tahirah, Rashida, Jamilah. BS in Computer Sci., Jackson (Miss.) State U., 1972; MS, Purdue U., 1975. User cons. Purdue U., Lafayette, Ind., 1972-75; mem. tech. staff Bell Telephone Labs., Whippany, N.J., 1975-83; supr. Bell Telephone Labs., Piscataway, N.J., 1983-84; dist. mgr. Bell Communications Rsch., Piscataway, 1984-90, div. mgr., 1990-91, exec. dir., 1991—. Mem. mcpl. planning com. N.J. State Youth Svcs. Commn., Treanton; liaison Gov.'s Alliance Against Drugs and Alcohol; bd. dirs. Somerset County Family Counseling, Bound Brook, N.J.; legis. aide N.J. State Legislature; councilman Franklin Twp., Somerset, N.J.; mem. cen. N.J. bd. dirs. NCCJ. Recipient Community Svc. award NAACP; named N.J. Black Achiever Newark YM/YWCA, Man of Yr., Nat. Assn. Negro Bus. and Profl. Women; NSF grantee. Mem. IEEE, Assn. Computing Machinery, Rotary (bd. dirs. 1985-87). Democrat. Baptist. Avocations: piano, gardening, singing, weight lifting. Home: 581 Waldorf St Somerset NJ 08873-2453 Office: Bellcore 444 Hoes Ln 4A-601 Piscataway NJ 08854

WALKER, JIMMIE KENT, mechanical engineer; b. Lawton, Okla., May 1, 1940; s. James K. and Ruth L. (Fleming) W.; m. Carol T. Walker, Mar. 22, 1962; (div. Aug. 1979); children: Mollie Walker Freeman, Keith M.; m. Joan F. Korth, Feb. 11, 1983; 1 stepson, Raymond Van Buskirk. BSME, Okla. U., 1963; MBA, U. S.D., 1991. Design engr. Cessna Aircraft, Hutchinson, Kans., 1965-68, 72-75; chief engr. Lyons (Kans.) Mfg. Corp., 1968-72; sales engr. Prince Mfg. Co., Sioux City, Iowa, 1975-79, mfg. mgr., 1979-83, dir. engring., 1983—. Pres. Hutchinson Jaycees, 1968; treas. Lyons Jaycees, 1970; mem. adv. bd. adult edn. Western Iowa Tech., Sioux City, 1980-90, Cosmo, 1992-93; judge Physics Olympics, 1992-93; bd. dirs. Jr. Achievement Sioux City, 1993—. Maj. U.S. Army, 1972. Mem. Am. Welding Soc., Am. Soc. Metals, Soc. Profl. Engrs., Tau Beta Pi, Sigma Tau, Beta Gamma Sigma. Roman Catholic. Achievements include patents for square wire, internal piston lock for hydraulic cylinders, design of high efficiency, low speed tractor pto pump, design of durable series/rephase implement carrier wheel cylinders. Home: 4816 Royal Ct Sioux City IA 51104-1128 Office: Prince Mfg Co PO Box 537 Sioux City IA 51102-0537

WALKER, JOHN MICHAEL, consulting environmental hydrogeologist; b. New Orleans, Apr. 9, 1942; s. William Jefferson and Frances Juanita (Nummy) W.; m. Terry Ann Leff, Aug. 22, 1969; children: Juli Christine, Erik Michael, Ariel Lindsey. BS in Geology, Old Dominion U., 1973; postgrad., Colo. State U., 1980. Cert. profl. geologist. Constrn. inspector U.S. Corps of Engrs., Arlington, Va., 1963-68; geologist U.S. Corps of Engrs., Norfolk, Va., 1970-73; field project engring. geologist U.S. Dept. Interior Bur. of Reclamation, Denver, 1974-78; project design engring. geologist U.S. Corps of Engrs., Savannah, Ga., 1978-81; sr. consulting engr. E.I. du Pont de Nemours & Co., Beaumont, Tex., 1981-91; ret., 1991; pres. Sci. Environ. Tng., Inc., Beaumont, Tex., 1989—; tng. com. Dupont Well Operators Network, 1988-89; cons. in field. Producer edn. films Geology, Injection Wells, 1989, Environ. Regulations, 1990. Planning com. Hazardous and Solid Waste Inst. U. Tex. Sch. of Law, Austin, 1988-89; mem. Environ. Health & Safety Law, 1990. Sgt. U.S. Army, 1968-70. Recipient Accomplishment award E.I. Dupont de Nemours & Co., Tenn., 1988, Bronze award for engring. excellence, Tex., 1990, Safety award Dept. of Interior, 1979, TELLY award for tng. film excellence, 1990, Environ. Excellence award Dupont Chems., 1992. Mem. Assn. Engring. Geology, Assn. Groundwater Scientists and Engrs., Underground Injection Practices Coun., The Platform Soc., MENSA. Republican. Episcopalian. Achievements include patent pending on "Dilithium Crystals," 1993; copyright on "Gifts of the Magi," 1993. Home and Office: Sci Environ Tng Inc 6795 Knollwood Dr Beaumont TX 77706-5408

WALKER, JOHN NEAL, agricultural engineering educator; b. Erie, Pa., Feb. 19, 1930; s. Gordon Durwood and Marie Katherine (Beck) W.; m. Betty Jane Bloeser, Aug. 14, 1954; children: David Thomas, Janet Ann. B.S. in Agrl. Engring, Pa. State U., M.S., 1958; Ph.D., Purdue U., 1961. Registered profl. engr., cons., Ky. Extension agrl. engr. Pa. State U., 1954-58; mem. faculty U. Ky., 1960—, prof. agrl. engring., 1966-74, chmn. dept., 1974-81, acting dir. Inst. Mining and Minerals Research, 1981-82, assoc. dean Coll. Agr., 1982-88, assoc. dean Coll. Engring., 1989—, dir. Ctr. for Robotics and Mfg. Systems, 1991—; Duggar lectr. Auburn (Ala.) U., 1976. Author articles in field, chpts. in books. Officer USNR, 1951-77. Named Gt. Tchr. U. Ky. Alumni Assn., 1974. Fellow Am. Soc. Agrl. Engrs. (exec. coun. 1984-87, 88-91, v.p. adminstrv. coun. 1984-87, pres. 1989-90, MBMA award 1974); mem. Order Ky. Cols. Office: U Ky S-129 Agrl Sci North 177 Anderson Hall Lexington KY 40506

WALKER, JOHN SCOTT, mechanical engineering educator; b. Washington, D.C., May 25, 1944. BS, Webb. Inst. Naval Architecture, 1966; Phd, Cornell U., 1970. Asst. prof. theoretical & applied mech. U. Ill., Urbana, 1971-75, assoc. prof., 1975-78, asst. dean coll. engring., 1980-81, prof. theoretical & applied mech., 1978-88, prof. mech. & indsl. engring., 1988—. Contbr. articles to profl. jours. Mem. ASME (Pi Tau Sigma Gold Medal award 1976), Am. Soc. Engring. Edn., Am. Nuclear Soc., Am. Acad. Mechanics. Achievements include authoring or co-authoring more than 112 publications in field of magnetohydrodynamics. Office: U Ill Dept Mech and Indsl Engring 1206 W Green St Urbana IL 61801

WALKER, LAWRENCE REDDEFORD, ecologist, educator; b. Ann Arbor, Mich., Sept. 11, 1951; m. Elizabeth Ann Powell. MS, U. Vt., 1977; PhD, U. Alaska, 1985. Postdoctoral assoc. Stanford (Calif.) U., 1985-87; asst. prof. U. P.R., Rio Piedras, 1988-91, U. Nev., Las Vegas, 1992—. Contbr. to profl. publs. Mem. Ecol. Soc. Am., Am. Inst. Biol. Scis., Assn. Tropical Biology. Achievements include research on nitrogen-fixing plants, thicket-forming species, mechanisms driving primary plant succession. Office: Univ Nev Las Vegas 4505 Maryland Pky Las Vegas NV 89154-4004

WALKER, RAYMOND JOHN, physicist; b. L.A., Oct. 26, 1942; s. Raymond Osmund and Marie Dorothy (Peterman) W. BS, San Diego State U., 1964; MS, UCLA, 1969, PhD, 1973. Rsch. assoc. U. Minn., Mpls., 1973-77; rsch. geophysicist Inst. Geophysics and Planetary Physics UCLA, 1977—; mgr. planetary plasma interactions node project scientist NASA Planetary Data System; mem. numerous coms. on space physics and the mgmt. of space physics data NRC and NASA. Contbr. articles to profl. jours. Mem. AAAS, Am. Geophys. Union (chair info. tech. com. 1990-92), Am. Astron. Soc. (div. Planetary Sci.). Achievements include research in magnetospheric physics, in planetary magnetospheres, in global magnetohydrodynamic simulation of solar wind-magnetosphere interaction, in data management, in magnetic field modeling. Home: 11053 Tennessee Ave Los Angeles CA 90064 Office: UCLA IGPP 405 Hilgard Ave Los Angeles CA 90024-1567

WALKER, RICHARD, JR., nephrologist, internist; b. Dayton, Ohio, Sept. 1, 1948; m. Madeleine Ann Walker. BS cum laude, Ohio State U., 1970, MD, 1973. Diplomate Nat. Bd. Med. Examiners, internal medicine, nephrology, critical care medicine Am. Bd. Internal Medicine. Intern medicine U. Tex. Southwestern, Dallas, 1973-74, resident internal medicine, 1974-76, fellow nephrology, 1976-78; staff nephrologist and internist Bay Med. Ctr., Panama City, Fla., 1978—, HCA Gulf Coast Hosp., Panama City, 1978—; med. dir. Panama City Artificial Kidney Ctr., 1978—. Mem. AMA, Fla. Med. Assn., Bay County Med. Soc., Am. soc. Nephrology, Internat. Soc. Nephrology, Fla. Soc. Nephrology, Renal Physicians Assn., Am. Soc. Internal Medicine, Fla. Soc. Internal Medicine, Nat. Kidney Found., Alpha Omega Alpha. Home: 320 Bunkers Cove Rd Panama City FL 32401-3912 Office: Nephrology Assocs PA 504 N Macarthur Ave Panama City FL 32401-3656

WALKER, ROBERT MOWBRAY, physicist, educator; b. Phila., Feb. 6, 1929; s. Robert and Margaret (Seivwright) W.; m. Alice J. Agedal, Sept. 2, 1951 (div. 1973); children: Eric, Mark; m. Ghislaine Crozaz, Aug. 24, 1973. B.S. in Physics, Union Coll., 1950, D.Sc., 1967; M.S., Yale U., 1951, Ph.D., 1954; Dr honoris causa, Université de Clermont-Ferrand, 1975. Physicist Gen. Electric Research Lab., Schenectady, 1954-62, 63-66; McDonnell prof. physics Washington U., St. Louis, 1966—; dir. McDonnell Center for Space Scis., 1975—; vis. prof. U. Paris, 1962-63; adj. prof. metallurgy Rensselaer Poly. Inst., 1958, adj. prof. physics, 1965-66; vis. prof. physics and geology Calif. Inst. Tech., 1972, Phys. Research Lab., Ahmedabad, India, 1981, Institut d'Astrophysique, Paris, 1981; nat. lectr. Sigma Xi, 1984-85; pres. Vols. for Internat. Tech. Assistance, 1960-62, 65-66, founder, 1960, bd. dirs., 1961—; mem. Lunar Sample Analysis Planning Team, 1968-70, Lunar Sample Rev. Bd., 1970-72; adv. com. Lunar Sci. Inst., 1972-75; mem. temporary nominating group in planetary scis. Nat. Acad. Scis., 1973-75, bd. on sci. and tech. for internat. devel., 1974-76, com. planetary and lunar exploration, 1977-80, mem. space sci. bd., 1979-82; bd. dirs. Univs. Space Research Assn., 1969-71; mem. organizing com. Com. on Space Research-Internat. Astron. Union, Marseille, France, 1984; mem. task force on sci. uses of space sta. Solar System Exploration Com., 1985-86; mem. Antarctic Meteorite Working Group, 1985-92; mem. NASA Planetary Geosci. Strategy Com., 1986-88; mem. European Sci. Found. Sci. Orgn. Com., Workshop on Analysis of Samples from Solar System Bodies, 1990; chmn. Antarctic Meteorite Working Group, 1990-92. Decorated Officier de l'Ordres des Palmes Academiques (France), 1993; recipient Disting. Svc. award Am. Nuclear Soc., 1964, Yale Engring. Assn. award for contbn. to basic and applied sci., 1966, Indsl. Rsch. awards, 1964, 65, Exceptional Sci. Achievement award NASA, 1970, E.O. Lawrence award AEC, 1971, Antarctic Svc. medal NSF, 1985; NSF fellow, 1962-63. Fellow AAAS, Am. Phys. Soc., Meteoritical Soc. (Leonard medal 1993), Am. Geophys. Union; mem. NAS (J. Lawrence Smith medal 1991), Am. Astron. Soc. Achievements include research and publs. on cosmic rays, nuclear physics, geophysics, radiation effects in solids, particularly devel. solid state track detectors and their application to geophysics and nuclear physics problems; discovery of fossil particle tracks in terrestrial and extra-terrestrial materials and fission track method of dating; application of phys. scis. to art and archaeology; lab. studies of interplanetary dust and interstellar grains in primitive meteorites. Home: 3 Romany Park Ln Saint Louis MO 63132-4211

WALKER, ROBERT WYMAN, environmental sciences educator; b. Arlington, Mass., Mar. 15, 1933; s. Kenneth C. and Clara A. (Wyman) W.; m. Muriel Lillian Parent, Sept. 8, 1959; 1 child, Jeffrey Warren. BS, U. Mass., 1955, MS, 1959; PhD, Mich. State U., 1963. Assoc. prof., dir. environ. scis. program U. Mass., Amherst, 1963—; vis. prof. U. Toulouse, France, 1972-73, 90, U. Otago, Dunedin, New Zealand, 1980. Sch. com. mem. Montague, Mass., 1968-71; town meeting mem., Montague, 1980-85; bd. of health Montague, 1985. With U.S. Army, 1954-57. Office: U Mass Environ Scis 112 Stockbridge Hall Amherst MA 01003

WALKER, ROGER GEOFFREY, geology educator, consultant; b. London, Mar. 26, 1939; s. Reginald Noel and Edith Annie (Wells) W.; m. Gay Parsons, Sept. 18, 1965; children—David John, Susan Elizabeth. B.A., Oxford U., Eng., 1961; D.Phil., Oxford U., 1964. NATO postdoctoral fellow Johns Hopkins U., Balt., 1964-66; geology faculty McMaster U., Hamilton, Ont., Can., 1966—; prof. McMaster U., 1973—; vis. scientist Marathon Oil Research Ctr., Littleton, Colo., 1973-74, Amoco Can., Calgary, Alta., 1982; vis. prof. Australian Nat. U., Canberra, 1981; tchr. 80 profl. short courses on various aspects of oil exploration in clastic reservoirs, Can., U.S., Brazil, Australia, Japan; mem. grant selection com. earth scis. sect. Nat. Scis. and Engring. Rsch. Coun. Can., 1981-84. Editor: Facies Models, 1979, 2d edit., 1984, 3d edit., 1992; contbr. over 120 articles to profl. jours. Recipient operating and strategic grants Nat. Scis. and Engring. Rsch. Coun. Can., 1966—. Fellow Royal Soc. Can.; mem. Geol. Assn. Can. (assoc. editor 1977-80, past president's medal 1975), Can. Soc. Petroleum Geologists (Link award 1983, R.J.W. Douglas Meml. medal 1990), Am. Assn. Petroleum Geologists (Disting. lectr. 1979-80), Soc. Econ. Paleontologists and Mineralogists (pres. eastern sect. 1975-76, coun. for mineralogy 1979-80, hon. mem. 1991, assoc. editor 1970-78), Can. Assn. Univ. Tchrs., Internat. Assn. Sedimentologists. Avocations: skiing, classical music, photography, model railroading. Home: 71 Robin Hood Dr, Dundas, ON Canada L9H 4G2 Office: McMaster U, Dept Geology, 1280 Main St W, Hamilton, ON Canada L8S 4M1

WALKER, SYDNEY, III, pharmacologist, psychiatric administrator; b. Chgo., Oct. 4, 1931. BA in Zoology, UCLA, 1953; MS in Physiology, U. So. Calif., 1956; postgrad., U. Calif., San Francisco, 1959-60; MD, Boston U., 1964. Diplomate Am. Bd. Psychiatry and Neurology (fellow). Profl. rugby player Am. All Stars, 1953; intern in surgery Presbyn. St. Luke's Hosp., Chgo., 1964-65; resident in neurosurgery U. Pitts. Sch. Medicine, 1965; resident, teaching fellow in psychiatry U. Pitts. Sch. Medicine/Western Psychiat. Inst., 1965-66; resident in psychiatry UCLA Neuropsychiatric Inst., 1966-68; resident in neurology L.A. County/U. So. Calif. Med. Ctr., 1968-70; pvt. practice neuropsychiatrist La Jolla, Calif., 1971-74; dir., diagnostic neuropsychiatrist So. Calif. Neuropsychiatric Inst., 1974—; founder, chmn. bd. dirs. Behavioral Neurology Internat., 1986—; invited lectr. and presenter in field. Editor-in-chief Neuropsychiatric Bulletin, 1976—. 1st lt. U.S. Army, 1954-55, Korea. Grantee NIH, 1964. Fellow Royal Soc. Medicine; mem. Am. Acad. Postgrad. Med. Edn. (chmn. bd. dirs. 1986—), Am. Acad. Neurology (assoc.), Am. Psychiat. Assn. (assoc.), Behavioral

Neurology Internat., Calif. Med. Assn., San Diego County Med. Assn., Soc. Biol. Psychiatry, World Fedn. Neurology. Office: So Callif Neuropsychiatric Inst 6794 La Jolla Blvd La Jolla CA 92037

WALKER, TIMOTHY JOHN, aerospace engineer; b. Hatton, Derby, Eng., Jan. 14, 1948; came to U.S., 1988; s. Thomas Archy and Dorothy Elsie (Goring) W.; m. Janis Elizabeth Dodkin, Aug. 9, 1969; children: Larissa Jayne, Simon Timothy. BSME, Bournemouth Coll. Tech., 1972. Apprentice Brit. Aerospace, Hurn, Eng., 1964-69, design engr., 1969-72, planning engr., 1973-79, prin. planning engr., 1980-84; sr. planning engr. C.F. Taylor, Christchurch, Eng., 1979-80; design engr. supr. GED Ltd., Borehamwood, Eng., 1984-88, The Dee Howard Co., San Antonio, 1988—. Mem. AIAA. Home: 3502 Forest Ln San Antonio TX 78247 Office: The Dee Howard Co PO Box 469001 San Antonio TX 78247

WALL, BRIAN RAYMOND, forest economist, policy analyst, consultant; b. Tacoma, Wash., Jan. 26, 1940; s. Raymond Perry and Mildred Beryl (Pickert) W.; m. Joan Marie Nero, Sept. 1, 1962 (div. Aug. 1990) children: Torden Erik, Kirsten Noel. BS, U. Wash., 1962; MF, Yale U., 1964. Forestry asst. Weyerhaeuser Timber Co., Klamath Falls, Oreg., 1960; inventory forester West Tacoma Newsprint, 1961-62; timber sale compliance forester Dept. Nat. Resources, Kelso, Wash., 1963; rsch. forest economist Pacific N.W. Rsch. Sta., USDA Forest Svc., Portland, Oreg., 1964-88, cons. 1989—; co-founder, bd. dirs. Cordero Youth Care Ctr., 1970-81; owner Brian R. Wall Images and Communications; cons. to govt. agys., Congress univs., industry; freelance photographer. Co-author: An Analysis of the Timber Situation in the United States, 1982; contbr. articles, reports to profl. publs., newspapers. Interviewed and cited by nat. and regional news media. Recipient Cert. of Merit U.S. Dept. Agr. Forest Service, 1982. Mem. Soc. Am. Foresters (chmn. Portland chpt. 1973, Forester of Yr. 1975), Conf. of Western Forest Economists Inc. (founder, bd. dirs. 1988-91, treas. 1982-87), Portland Photographic Forum, Common Cause, Nat. Audubon Soc., Zeta Psi. Home and Office: 16810 S Creekside Ct Oregon City OR 97045-9206

WALL, FREDERICK THEODORE, chemistry educator; b. Chisholm, Minn., Dec. 14, 1912; s. Peter and Fanny Maria (Rauhala) W.; m. Clara Elizabeth Vivian, June 5, 1940; children: Elizabeth Wall Ralston, Jane Vivian Wall-Meinike. B.Chemistry, U. Minn., 1933, Ph.D., 1937. Mem. faculty chemistry dept. U. Ill., 1937-64, dean grad. coll., 1955-63; prof., chmn. dept. chemistry U. Calif., Santa Barbara, 1964-66, vice chancellor rsch., 1965-66; vice chancellor grad. studies and research, prof. chemistry U. Calif. at San Diego, 1966-69; exec. dir. Am. Chem. Soc., Washington, 1969-72; prof. chemistry Rice U., Houston, 1972-78, San Diego State U., 1978-81, U. Calif., San Diego, 1982-91; Pres. Assn. Grad. Schs., 1961; trustee Inst. Def. Analyses, 1962-64; mem. governing bd. Nat. Acad. Scis.-NRC, 1963- 67. Author: Chemical Thermodynamics, 1958; editor: Jour. Phys. Chemistry, 1965-69; contbr. articles on quantum mechanics and statis. mechanics to sci. jours. Mem. Am. Chem. Soc. (Pure Chemistry award 1945, dir. 1962-64), Finnish Chem. Soc. (corr.), Am. Acad. Arts and Scis., Nat. Acad. Scis., Am. Phys. Soc., AAAS, Sigma Xi, Phi Kappa Phi, Phi Lambda Upsilon (hon.). Achievements include early work on Monte Carlo computer simulation of macromolecular configurations and of basic reaction probabilities. Home: 2468 Via Viesta La Jolla CA 92037-3935

WALL, JACQUELINE REMONDET, industrial psychologist, rehabilitation counselor; b. Paris, Dec. 25, 1958; came to U.S., 1959; d. Jack Whitney and Hazel Aline (Riley) Hargett; m. Mel Dennis Remondet, Aug. 5, 1977 (div. Mar. 1984); m. David Gordon Wall, Jan. 27, 1990; 1 child, Jeanette Renee. BA, Southeastern La. U., 1978; MA, U. Tulsa, 1982, PhD, 1989. Lic. profl. counselor, Okla. Program coord. Hillcrest Med. Ctr., Tulsa, 1982-88; coord. psychol. svcs. Rebound Inc.-Cane Creek Hosp., Martin, Tex., 1989-90; psychologist Sea Pines Rehab. Hosp., Melbourne, Fla., 1990; indsl. psychology intern Morris & Assocs., Jackson, Miss., 1990-91; ind. cons. indsl. psychology, 1991-92; postdoctoral fellow clin. respecialization program IIT, 1992—; instr. Tulsa Jr. Coll., 1989; rsch. asst. U. Tulsa, 1984-86, La. State U., Baton Rouge, 1980, Med. Sch., Tulane U., New Orleans, 1979-80; part-time instr. Wayne State U., 1991; presenter in field. Contbr. book chpt., articles to profl. jours. Recipient rsch. grant U. Tulsa, 1982. Mem. APA, Soc. for Indsl.-Orgnl. Psychology, Southeastern La. U. Thirteen Club, Sigma Xi, Psi Chi, Phi Kappa Phi, Phi Lambda Pi. Home: 2199 Anita St Grosse Pointe MI 48236-1429 Office: Ill Inst Tech Dept Psychology IIT Ctr Chicago IL 60616

WALL, SONJA ELOISE, nurse, administrator; b. Santa Cruz, Calif., Mar. 28, 1938; d. Ray Theothornton and Reva Mattie (Wingo) W.; m. Edward Gleason Holmes, Aug. 1959 (div. Jan. 1968); children: Deborah Lynn, Lance Edward; m. John Aspesi, Sept. 1969 (div. 1977); children: Sabrina Jean, Daniel John; m. Kenneth Talbot LaBoube, Nov. 1, 1978 (div. 1989); 1 child, Tiffany Amber. BA, San Jose Jr. Coll., 1959; BS, Madonna Coll., 1967; student, U. Mich., 1968-70. RN, Calif., Mich., Colo. Staff nurse Santa Clara Valley Med. Ctr., San Jose, Calif., 1959-67, U. Mich. Hosp., Ann Arbor, 1967-73, Porter and Swedish Med. Hosp., Denver, 1973-77, Laurel Grove Hosp., Castro Valley, Calif., 1977-79, Advent Hosp., Ukiah, Calif., 1984-86; motel owner LaBoube Enterprises, Fairfield, Point Arena, Willits, Calif., 1979—; staff nurse Northridge Hosp., L.A., 1986-87, Folsom State Prison, Calif., 1987; co-owner, mgr. nursing registry Around the Clock Nursing Svc., Ukiah, 1985—; staff nurse Kaiser Permanente Hosp., Sacramento, 1986-89, hospice nurse, 1990-93; home care nurse HSSI-Care Point, Placerville, Calif., 1992—; owner Royal Plantation Petites Miniature Horse Farm. Contbr. articles to various publs. Leader Coloma 4-H, 1987-91; mem. mounted divsn. El Dorado County Search and Rescue, 1991-93; docent Calif. Marshall Gold Discovery State Hist. Park, Coloma, Calif. Mem. AACN, NAFE, Soc. Critical Care Medicine, Am. Heart Assn. (CPR trainer, recipient awards), Calif. Bd. RNs, Calif. Nursing Rev., Calif. Critical Care Nurses, Soc. Critical Care Nurses, Am. Motel Assn. (beautification and remodeling award 1985), Nat. Hospice Nurses Assn., Soroptimist Internat. Calif., Am. Miniature Horse Assn. (winner nat. grand championship 1981-82, 83, 85, 89), DAR (Jobs Daus. hon. mem.), Cameron Park Country Club. Republican. Episcopalian. Avocations: pinto, paint, Thoroughbred and miniature horses, real estate devel., tennis, hiking, golf. Home and Office: Around the Clock Nursing Svc PO Box 559 Coloma CA 95613-0559

WALLACE, BONNIE ANN, biophysics educator, researcher; b. Greenwich, Conn., Aug. 10, 1951; d. Arthur Victor and Maryjane Ann W. BS in Chemistry, Rensselaer Poly. Inst., 1973; PhD in Molecular Biophysics and Biochemistry, Yale U., 1977. Postdoctoral rsch. fellow Harvard U., Boston, 1977-78; asst. prof. dept. biochemistry and molecular biophysics Columbia U., N.Y.C., 1979-86, assoc. prof. 1986; prof. dept. chemistry, dir. Ctr. for Biophysics Rensselaer Poly. Inst., 1987-92; reader in crystallography U. London, 1991—; vis. scientist MRC Lab. Molecular Biology, Cambridge, Eng., 1978; Fogarty sr. fellow Birkbeck Coll., U. London, 1990. Contbr. numerous articles to profl. jours. and books. Jane Coffin Childs fellow, 1977-79; recipient Irma T. Hirschl award, 1980-84; Camille and Henry Dreyfus tchr.-scholar, 1986; named Hot Young Scientist Fortune Mag., 1990. Mem. Aspen Ctr. for Physics Fellowship, 1986, Biophys. Soc. (nat. coun., Dayhoff award 1985), Am. Crystallographic Assn., Brit. Crystallographic Assn., Biochem. Soc. Britain (coun. mem. peptides and proteins group), Sigma Xi, Phi Lambda Upsilon. Office: Birkbeck Coll, Dept Crystallography, Univ London, London WC1E 7HX, England

WALLACE, DONALD JOHN, III, rancher, former pest control company executive; b. Houston, May 17, 1941; s. D.J. Jr. and Doris Jill (Gano) W.; m. Patricia Anne McShane, Sept. 3, 1964 (div. 1984); children: Donald John IV, Megan; m. Nena Jo Isenhower, June 1, 1985 (div. 1989); 1 child, Andrew. BBA in Mktg., Texas A&M U., 1963. Regional sales dir. Orkin Exterminating Co., Inc., Dallas, 1977-79, br. mgr., 1979-80, dist. mgr., 1980-83, comml. region mgr., 1983-85, regional sales dir., 1985-86; owner Omega Telex, Dallas, 1986-88; rancher Valley View, Tex., 1988—. Mem. Tex. Structural Pest Control Bd., Austin, 1983-84. Mem. Nat. Pest Control Assn., Tex. Pest Control Assn., Dallas Pest Control Assn. Republican. Roman Catholic. Avocations: hiking, fishing, hunting, skiing. Home: 1400 Trails End Valley View TX 76272-9530

WALLACE, F. BLAKE, aerospace executive, mechanical engineer; b. Phoenix, Az., Jan. 10, 1933. BMechE, Calif. Inst. Tech., 1955; MS in Engring., Ariz. State U., 1963, PhD in Engring., 1967. Preliminary design engr. Pratt & Whitney, East Hartford, Conn., 1955-59; chief engr. advanced tech. Garrett Corp., Phoenix, 1959-80; mgr. advanced plans and progress Aircaft Engring. and Aircraft Engine Groups GE, Evendale, Ohio, 1981-83; gen. mgr. Allison div. GM, Indpls., 1983—; v.p. GM, 1987—; mem. aero. adv. com. NASA, 1985—. Author numerous tech. papers. Fellow AIAA (chmn. air breathing propulsion tech. com. 1977-78, Air Breathing Propulsion award 1991), U.S. Advanced Ceramic Assn. (chmn. 1987-89). Office: GM Allison Div 2001 S Tibbs Ave # 420 Indianapolis IN 46241-4893

WALLACE, JAMES JR., engineering executive, researcher; b. New Brunswick, N.J., Oct. 6, 1932; s. James and Mary Angela (Devaney) W.; m. Nancy Elizabeth Vivian, June 14, 1958; children: Patrick, Kathleen, Megan, Anne, Michael. Bs, U.S. Merchant Marine Acad., 1954; MS, U. Notre Dame, 1959; PhD, Brown U., 1963. Rsch. engr. Bendix Aviation, South Bend, Ind., 1958-59; sr. staff scientist Avco Everett (Mass.) Rsch. Lab., 1963-74; pres. Far Field, Inc., Sudbury, Mass., 1974—; cons. Textron, Everett, 1973-88, Army Rsch. Office, Rsch. Triangle, N.C., 1984-87, Los Alamos (N.Mex.) Nat. Lab., 1986-91. Contbr. 30 articles to profl. jours. Active in local politics, Sudbury, 1963-83, church activities, Sudbury, 1963-80. Lt. USN, 1954-56. Mem. IEEE, AIAA, Internat. Soc. Optical Engring. (speaker 1986), Optical Soc. Am. (speaker 1973), N.Y. Acad. Scis. Democrat. Roman Catholic. Achievements include research in atmospheric propagation of high energy lasers; one of first to iniate nonlinear theory of atmospheric propagation (thermal blooming) and develop computational procedures for solving the problem. Home and Office: Far Field Inc 6 Thoreau Way Sudbury MA 01776

WALLACE, JANE HOUSE, geologist; b. Ft. Worth, Aug. 12, 1926; d. Fred Leroy and Helen Gould (Kixmiller) Wallace; A.B., Smith Coll., 1947, M.A., 1949; postgrad. Bryn Mawr Coll., 1949-52. Geologist, U.S. Geol. Survey, 1952—, chief Pub. Inquiries Offices, Washington, 1964-72, spl. asst. to dir., 1974—, dep. bur. ethics counselor, 1975—, Washington liaison Office of Dir., 1978—. Recipient Meritorious Service award Dept. Interior, 1971, Disting. Svc. award, 1976, Sec.'s Commendation, 1988, Smith Coll. medal, 1992. Fellow Geol. Socs. Am., Washington (treas. 1963-67); mem. Sigma Xi (asso.). Home: 3003 Van Ness St NW Washington DC 20008-4701 Office: Interior Bldg 19th and C Sts NW Washington DC 20006 also: US Geol Survey 103 National Ctr Reston VA 22092

WALLACE, JOAN S., psychologist; b. Chgo., Nov. 8, 1930; d. William Edouard and Esther (Fulks) Scott; m. John Wallace, June 12, 1954 (div. Mar. 1976); children—Mark, Eric, Victor; m. Maurice A. Dawkins, Oct. 14, 1979. A.B., Bradley U., 1952; M.S.W., Columbia U., 1954; postgrad., U. Chgo., 1965; Ph.D., Northwestern U., 1973; H.H.D. (hon.), U. Md., 1979; L.H.D. (hon.), Bowie State Coll., 1981; LLD (hon.), Ala. A&M U., 1990. Lic. social psychologist, social worker. Asst. prof., then assoc. prof. U. Ill.-Chgo., 1967-73; assoc. dean, prof. Howard U., Washington, 1973-76; v.p.-programs Nat. Urban League, N.Y.C., 1975-76; v.p. adminstrn. Morgan State U., Balt., 1976-77; asst. sec. adminstrn. USDA, Washington, 1977-81, adminstr. Office Internat. Cooperation and Devel., 1981-89; rep. to Trinidad and Tobago Inter Am. Inst. for Cooperation in Agr., USDA, 1989; speaker in field. Contbr. articles, chpts. to profl. publs. Chair Binat. Agrl. Research and Devel. Fund, 1987. Recipient Disting. Alumni award Bradley U., 1978, Meritorious award Delta Sigma Theta, 1978, award for leadership Lambda Kappa Mu, 1978, award for outstanding achievement and svc. to nation Capital Hill Kiwanis Club, 1978, Links Achievement award, 1979, Presdl. Rank for Meritorious Exec., 1980, NAFEO award, 1989, Community Svc. award Alpha Phi Alpha, 1987, Pres.' award for outstanding pub. svc. Fla. A&M U., 1990. Mem. APA, NASW, AAAS, Am. Consortium for Internat. Pub. Adminstrn. (exec. com., governing bd. 1987), Soc. Internat. Devel. (Washington chpt.), Sr. Exec. Assn., Soc. for Internat. Devel., White House Com. on Internat. Sci., Engring. and Tech., Internat. Sci. and Edn. Coun. (chmn. 1981-89), Am. Evaluation Assn., Consortium Internat. Higher Edn. (adv. com.), Caribbean Studies Soc., Caribbean Assn. of Agriculture Economists, Assn. Polit. Psychologists, Pi Gamma Mu. Episcopalian. Avocations: crafts, painting, collecting international art. Home: 6010 S Falls Circle Dr Fort Lauderdale FL 33319 Office: US Dept Agriculture Office of Sec Washington DC 20250

WALLACE, JOHN LAWRENCE, immunophysiologist, educator; b. Toronto, Ont., Can., Sept. 25, 1956; s. Richard Victor and Dora Kathleen (Walker) W.; m. Beth Colleen Chin, July 31, 1987; children: Alexandra Emily, Meredith Evelyn. BSc, Queen's U., Kingston, Ont., 1979, MSc, 1980; PhD, U. Toronto, 1983. Postdoctoral fellow Wellcome Rsch. Found., Beckenham, Eng., 1984-86; asst. prof. Queen's U., Kingston, Ont., 1986-89; assoc. prof. U. Calgary, Alta., 1989-93, prof., 1993—. Office: University of Calgary, 3330 Hospital Dr NW, Calgary, AB Canada T2N 1N4

WALLACE, JOHN MICHAEL, meteorology educator; b. Flushing, N.Y., Oct. 28, 1940; 3 children. BS, Webb Inst. Naval Architecture, 1962; PhD, MIT, 1966. From asst. to assoc. prof. U. Wash., Seattle, 1966-67, prof. atmos. sci., 1977—; dir. Joint Inst. Study Atmosphere and Ocean, 1980—; adj. assoc. prof. environ. studies U. Wash., 1973—. Mem. Am. Meteorol. Soc. (Meisinger award 1975, Carl-Gustav Rossby Rsch. medal 1993), Am. Geophys. Union (Macelwane award 1972). Achievements include rsch. in gen. circulation; tropical meteorology. Office: U Wash Dept Atmos Scis AK-40 Seattle WA 98195*

WALLACE, LOUISE MARGARET, critical care nurse; b. Norwich, Conn., June 15, 1942; d. Irving Clifford and Helen Lucille (Fain) Hayden; m. R.D. Wallace, Dec. 2, 1967; 1 child, Donald Orville. Grad., Joseph Lawrence Sch. Nursing, Conn., 1963; student, Miami-Dade (Fla.) Jr. Coll., 1966-67, Yavapai Coll., 1970. Nurse ICU and ob-gyn. dept. George Washington U. Hosp., Washington, 1964-65; nurse pediatrics dept. Jackson Meml. Hosp., Miami, Fla., 1965-66; nurse ICU Bapt. Hosp., Miami, 1966-67; nurse ICU and CCU N. Shore Hosp., Miami, 1967-71; nurse ICU and CCU VA Med. Ctr., Prescott, Ariz., 1971-84; nurse ICU and CCU VA Med. Ctr., Poplar Bluff, Mo., 1984-93, relief clin. coord., 1991-92, 93—; instr. nursing Miami-Dade Jr. Coll., 1968-69; instr. basic CPR, Prescott, 1975-81. Mem. Am. Assn. Critical Care Nurses, Am. Diabetes Assn. Avocations: needlework, knitting, traveling, dogs. Home: HC 1 Box 76 Grandin MO 63943-9602

WALLACE, ROBERT EARL, geologist; b. N.Y.C., July 16, 1916; s. Clarence Earl and Harriet (Wheeler) W.; m. Gertrude Kivela, Mar. 19, 1945; 1 child: Alan R. BS, Northwestern U., 1938; MS, Calif. Inst. Tech., 1940, PhD, 1946. Registered geologist, Calif., engring. geologist, Calif. Geologist U.S. Geol. Survey, various locations, 1942—; regional geologist U.S. Geol. Survey, Menlo Park, Calif., 1970-74; chief scientist Office of Earthquakes, Volcanoes and Engring. U.S. Geol. Survey, Menlo Park, 1974-87; asst. and assoc. prof. Wash. State Coll., Pullman, 1946-51; mem. adv. panel Nat. Earthquake Prediction Evaluation Coun., 1980-90, Stanford U. Sch. Earth Sci., 1972-82; mem. engring. criteria rev. bd. San Francisco Bay Conservation and Devel. Commn. Contbr. articles to profl. jours. Recipient Meritorious Service award U.S. Dept. Interior, 1978, Disting. Service award U.S. Dept. Interior, 1978; Japanese Indsl. Tech. Assn. fellow, 1984. Fellow AAAS, Geol. Soc. Am. (chair cordillidan sect. 1967-68), Earthquake Engring. Research Inst., mem. Seismol. Soc. Am. (medalist 1989). Avocations: bird watching, ham radio, water color painting. Office: US Geol Survey MS-977 345 Middlefield Rd Menlo Park CA 94025-3591

WALLACE, ROBERT EUGENE, II, rail transit systems executive; b. Winchester, Ind., Aug. 22, 1956; s. Robert E. and Martha Ellen (Flatter) W.; m. Katherine Jeannine Bezrutch, Aug. 2, 1979; 1 child, Brian Mitchell. BS, Ind. State U., 1982; MBA, Ind. Wesleyan U., 1990. Process engr. Twigg Corp., Martinsville, Ind., 1982-84; project engr. Stewart Warner, Indpls., 1984-85; quality mgr. Morris Machine Co. Inc., Indpls., 1985-88; engring. mgr. SMC Pneumatics, Inc., Indpls., 1988-90, project mgr., 1990-92; Global Mktg. Assocs., 1992—. Clk. Marion County Election Bd., Indpls., 1985-88; v.p. Countryside Homeowners Assn., Indpls., 1984-86. With USN, 1976-80. Mem. Soc. Mfg. Engrs., Am. Passenger Transit Assn. (assoc.), Epsilon Pi Tau. Home: 5250 Woodside Dr Indianapolis IN 46208 Office: Global Mktg Assocs 5250 Woodside Dr Ste 120 Indianapolis IN 46208

WALLACE, ROBERT LUTHER, II, engineer; b. Ronceverte, W.Va., Jan. 27, 1949; s. Robert Luther and Eloise Virginia (Houck) W.; m. Dorothy James, May 1970 (div. 1979); m. Lucy Alice Frazier, June 13, 1981; children: Sheena Rene, Lacey Christina. BS in Aerospace Engring., W.Va. U., 1970, MS in Indsl. Engring., 1973. Project engr. Naval Air Systems Commd., Washington, 1972-78, Air Force Logistics Command, Wright Patterson AFB, Ohio, 1978-81; engring. work leader Air Force Logistics Command, 1981-84; logistics mgmt. specialist Mil. Airlift Commd., Scott AFB, Ill., 1984-86; program mgr. Air Force Logistics Commd., Wright-Patterson AFB, 1986-91; chief sci., emgring. and mfg. mission requirements unit 645 Mission Support Squadron, Wright-Patterson AFB, Ohio, 1991—. Trustee Idle Hour Swim Club, Beavercreek, Ohio, 1990—; mem. Beavercreek Schs. Strategic Planning Com., 1990; asst. lay minister Peace Evang. Luth. Ch., Beavercreek, 1987—, asst. dir. evang., 1987-89, v.p. for ministries, 1992—. Mem. ASTD, Soc. Linguistics Engrs., Am. Inst. Aeronautics and Astronautics, Federal Mgrs. Assn., Inst. Indsl. Engrs. (v.p. 1983-84). Republican. Avocations: running, lifting weights. Home: 3755 Olde Willow Dr Beavercreek OH 45431-2469 Office: 645 MSSQ/MSUAS Ste 4 5215 Thurown St Wright Patterson AFB OH 45433-5544

WALLACE, TERRY CHARLES, SR., technical administrator, researcher, consultant; b. Phoenix, May 18, 1933; s. Terry Milton Wallace and Fair June (Hartman) Wallace Timberlake; m. Yvonne Jeannette Owens, May 21, 1955; children—Terry Charles, Randall James, Timothy Alan, Sheryl Lynn, Janice Marie. B.S., Ariz. State U., 1955; Ph.D., Iowa State U., 1958. Mem. staff Los Alamos Nat. Lab., 1958-71, dep. group leader, 1971-80, group leader, 1980-83, assoc. div. leader, 1983-89, tech. program coord., 1989-91, retired, 1991; ptnr. Stonewall Enterprises, Los Alamos, 1966-71. Contbr. chpts., numerous articles to profl. publs. Patentee in field. Fundraiser Los Alamos County Republican Party, N.Mex., 1983-84. Served to 1st lt. Chem. Corps, U.S. Army, 1959-61. Mem. Am. Chem. Soc., AAAS. Methodist. Home: 146 Monte Rey Dr S Los Alamos NM 87544-3826 Office: 146 Monte Rey Dr S Los Alamos NM 87544

WALLACE, TERRY CHARLES, JR., geophysicist, educator; b. Ames, Iowa, June 30, 1956; married; 1 child. BS in Geophysics, BS in Math, N.Mex. Inst. Mining Tech., 1978; MS, Calif. Inst. Tech., 1980, PhD in Geophysics, 1983. Asst. prof. dept. geoscis. U. Ariz., Tucson, 1983-88, curator Mineral Mus., 1985—, assoc. prof. seismology dept. geoscis., 1988—. Mem. AAAS, Seismology Soc. Am., Am. Geophysics Union (James E. Macelwane medal, 1993). Achievements include research in computational seismology, global crustal structure and explosion source physics. Office: U Arizona Dept Geosciences Tucson AZ 85721*

WALLEM, DANIEL RAY, metallurgical engineer; b. Seattle, Dec. 1, 1961; s. Oscar Eugene and Suzanne (McCluskey) W. BS in Metall. Engring. magna cum laude, U. Wash., 1984; MS, Ohio State U., 1986. Registered profl. engr., Wash. Factory support engr. mfg. R&D Boeing Corp., Auburn, Wash., 1987-90; chief nickel and cobalt alloy metall. engr. Boeing Materials Tech., Renton, Wash., 1990—; cons. Utilx Corp., Kent, Wash., 1992; propr. D. Wallem Co., Metall. Cons., Renton, 1992—; counselor for metallurgy Boy Scouts Am., Issaquah, Wash., 1989-92. Mem. Am. Soc. for Metals Internat., Tau Beta Pi. Office: Boeing Comml Airplane Group PO Box 3707 Mail Stop 3K-12 Seattle WA 98124

WALLER, JAMES EDWARD, JR., experimental psychologist, educator; b. Columbus, Ga., Aug. 23, 1961; s. James Edward Sr. and Billie Jean (Barbour) W.; m. Patricia Marie Pearson, Aug. 4, 1985; children: Brennan Martin, Hannah Marie. MS, U. Colo., 1985; PhD, U. Ky., 1988. Instr. U. Colo., Boulder, 1983-85, U. Ky., Lexington, 1985-88; asst. prof. Asbury Coll., Wilmore, Ky., 1988-89; asst. prof. Whitworth Coll., Spokane, Wash., 1989-92, assoc. prof., 1992—; vis. prof. Cath. U., Eichstaat, Fed. Republic of Germany, summer 1992, Tech. U., Berlin, summer 1990. Contbr. articles to profl. jours. Advisor Salvation Army. Mem. Am. Psychol. Soc., Coun. Tchrs. of Psychology (coord. western region 1992—), Western Psychol. Assn., Soc. for Computers in Psychology. Republican. Presbyterian. Achievements include research in social loafing; in psychohistory of holocaust; in value self-confrontation. Home: W 608 Falcon Ave Spokane WA 99218 Office: Whitworth Coll Dept of Psychology Spokane WA 99251-0706

WALLER, JOHN LOUIS, anesthesiology educator; b. Loma Linda, Calif., Dec. 1, 1944; s. Louis Clinton and Sue (Bruce) W.; m. Jo Lynn Marie Haas, Aug. 4, 1968; children: Kristina, Karla, David. BA, So. Coll., Collegedale, Tenn., 1967; MD, Loma Linda U., 1971. Diplomate Am. Bd. Anesthesiology. Intern Hartford (Conn.) Hosp., 1971-72; resident in anesthesiology Harvard U. Med. Sch.-Mass. Gen. Hosp., Boston, 1972-74, fellow, 1974-75; asst. prof. anesthesiology Emory U. Sch. Medicine, Atlanta, 1977-80, assoc. prof., 1980-86, prof., chmn. dept., 1986—; cons. Arrow Internat., Inc., Reading, Pa., 1988—; bd. dirs. Clifton Casualty Co., Colo.; mem. adv. com. on anesthetic and life support drugs FDA, Washington, 1986-92; numerous vis. professorships and lectures. Contbr. articles to med. jours. Maj. M.C., USAF, 1975-77. Recipient cert. of appreciation Office Sec. Def., 1983. Fellow Am. Coll. Anesthesiologists, Am. Coll. Chest Physicians; mem. AMA, Am. Soc. Anesthesiologists, Soc. Cardiovascular Anesthesiologists (pres. 1991-93), Internat. Anesthesia Rsch. Soc. (trustee 1984—), Assn. Univ. Anesthetists, Soc. Acad. Anesthesia Chairmen (councillor 1989—), Assn. Cardiac Anesthesiologists. Avocations: tennis, sailing, swimming. Office: Emory U Hosp Dept Anes 1364 Clifton Rd NE Atlanta GA 30322-1104

WALLER, NIELS GORDON, psychologist; b. N.Y.C., Apr. 13, 1957; s. Howard and JoDelle (Rundquist) W. ALM, Harvard U., 1989; PhD, U. Minn., 1990. Asst. prof. U. Calif., Davis, 1990—; dir. Calif. Twin Registry, Davis, 1990—. Author: (with others) Bronchial Asthma, 1992; contbr. articles to profl. jours. Recipient Thomas Small prize Harvard U., 1990. Mem. APA, Behavior Genetics Assn., Psychometric Soc., Am. Statis. Assn. Achievements include research on genetic influence on religious interests discovered using twins reared apart. Office: U Calif Dept Psychology Young Hall Davis CA 95616

WALLERSTEIN, LEIBERT BENET, economist; b. Bklyn., July 5, 1922; s. William Mark and Ray Leah (Goldberg) W.; m. Alice Stehle, Oct. 10, 1929; stepchildren: Nora Odendahl, Steven Odendahl. BA in Econs., U. N.Mex., 1950; MA in Econs., U. Minn., 1951; PhD in Econs., Social Scis., Columbia Pacific U., 1988. With U.S. Corps of Engrs. and Army Air Corps Matl. Command, Washington, 1943-46; Merchant Marine; with AUS, 1945-47, U.S. Navy Bur. ORD, 1954-55, U.S. Dept. Labor, 1956-67, HUD, 1967-69, U.S. DOT, 1961-80; faculty U.S. Merchant Marine Acad., Kings Point, 1977, U. Minn., U. Md., U. New Mex., Pentagon, Georgetown U., Ben Franklin U., Montgomery Coll., Montgomery County Adult Edn. Ctr.; cons. economist, 1980—; v.p. Reano Co., L.A. Contbr. articles to profl. jours. Vol. Shepard's Table, Washington, 1985—, Chevy Chase (Md.) grade schs., 1986, polit. pres. campaigns, 1980, 84, 92; v.p. Univ. Young Dems., U. N.Mex., Albuquerque, 1947-50, NAACP, 1947-50. Mem. Atlantic Econ. Soc., Am. Econ. Assn., Economists Club Washington (charter), Sr. Club (chmn. 1983—), Disabled Am. Vets., Am. Legion. Jewish. Avocations: fishing, hiking, reading, swimming, fgn. travel. Home and Office: 3505 Thornapple St Chevy Chase MD 20815-4014

WALLERSTEIN, ROBERT SOLOMON, psychiatrist; b. Berlin, Jan. 28, 1921; s. Lazar and Sarah (Guensberg) W.; m. Judith Hannah Saretsky, Jan. 26, 1947; children—Michael Jonathan, Nina Beth, Amy Lisa. B.A., Columbia, 1941, M.D., 1944; postgrad., Topeka Inst. Psychoanalysis, 1951-58. Assoc. dir., then dir. rsch. Menninger Found., Topeka, 1954-66; chief psychiatry Mt. Zion Hosp., San Francisco, 1966-78; tng. and supervising analyst San Francisco Psychoanalytic Inst., 1966—; clin. prof. U. Calif. Sch. Medicine, Langley-Porter Neuropsychiat. Inst., 1967-75, prof., chmn. dept. psychiatry, also dir. inst., 1975-85, prof. dept. psychiatry, 1985-91, prof. emeritus, 1991—; vis. prof. psychiatry La. State U. Sch. Medicine, also New Orleans Psychoanalytic Inst., 1972-73, Pahlavi U., Shiraz, Iran, 1977, Fed. U. Rio Grande do Sul, Porto Alegre, Brasil, 1980; mem., chmn. rsch. scientist career devel. com. NIMH, 1966-70; fellow Ctr. Advanced Study Behavioral Scis., Stanford, Calif., 1964-65, 81-82, Rockefeller Found. Study Ctr., Bellagio, Italy, 1992. Author 11 books and monographs; mem. editorial bd. 12 profl. jours.; contbr. articles to profl. jours. Served with AUS, 1946-48. Recipient Heinz Hartmann award N.Y. Psychoanalytic Inst., 1968, Disting. Alumnus award Menninger Sch. Psychiatry, 1972, J. Elliott Royer award U. Calif., San Francisco, 1973, Outstanding Achievement award No. Calif. Psychiat. Soc., 1987, Mt. Airy gold medal, 1990, Mary Singleton Sigourney award, 1991. Fellow ACP, Am. Psychiat. Assn., Am. Orthopsychiat. Assn.; mem. Am. Psychoanalytic Assn. (pres. 1971-72), Internat. Psychoanalytic Assn. (v.p. 1977-85, pres. 1985-89), Group for Advancement Psychiatry, Brit. Psycho-Analytical Soc. (hon.), Phi Beta Kappa, Alpha Omega Alpha. Home: 290 Beach Rd Belvedere Tiburon CA 94920-2472 Office: 655 Redwood Hwy Ste 261 Mill Valley CA 94941-3011

WALLIN, B(ENGT) GUNNAR, neurophysiology educator; b. Karlstad, Sweden, Oct. 26, 1936; s. Ragnar and Elsa (Källström) W.; m. Ann Marie Lindbom, Aug. 6, 1960; children: Maria, Johan, Staffan. MD, U. Uppsala, Sweden, 1963. Rsch. physiologist U. Calif., San Francsico, 1963-64; staff mem. dept. clin. neurophysiology U. Uppsala, Sweden, 1969-84; prof. clin. neurophysiology U. Göteborg, Sweden, 1984—, chmn. dept. clin. neuroscis., 1993—; vis. prof. Va. Commonwealth U., Richmond, 1980, Baker Med. Rsch. Inst., Melbourne, Australia, 1991. Contbr. sci. articles to profl. jours. Recipient Australian-European award Australian Govt., Canberra, 1988-89. Mem. Swedish Soc. Medicine (Alvarenga's prize 1988), Swedish Soc. Clin. Neurophysiology (chmn. 1988-90), Internat. Union Physiol. Scis. (mem. commn. on autonomic nervous system 1982-90). Office: U Göteborg, Sahlgren's Hosp, Dept Clin Neurophysiology, 41345 Göteborg Sweden

WALLING, CHEVES THOMSON, chemistry educator; b. Evanston, Ill., Feb. 28, 1916; s. Willoughby George and Frederika Christina (Haskell) W.; m. Jane Ann Wilson, Sept. 17, 1940; children—Hazel, Rosalind, Cheves, Janie, Barbara. A.B., Harvard, 1937; Ph.D., U. Chgo., 1939. Rsch. chemist E.I. duPont de Nemours, 1939-43, U.S. Rubber Co., 1943-49; tech. aide Office Sci. Research, Washington, 1945; sr. rsch. assoc. Lever Bros. Co., 1949-52; prof. chemistry Columbia U., N.Y.C., 1952-69; Disting. prof. chemistry U. Utah, Salt Lake City, 1970-91, prof. chemistry emeritus, 1991—. Author: Free Radicals in Solution, 1957; also numerous articles. Fellow AAAS; mem. Nat. Acad. Scis., Am. Acad. Arts and Scis., Am. Chem. Soc. (editor jour. 1975-81, James Flack Norris award 1970, Lubrizol award 1984). Home: Box 537 Jaffrey NH 03452

WALLIS, ROBERT JOE, pharmacist, retail executive; b. Lawton, Okla., Dec. 26, 1938; s. John L. and Bertha Leora (Blake) W.; m. Rubena Ann Hennessee, June 1, 1958; children: Jeffrey Allen, Joseph Robert, Justin Matthew. BS in Pharmacy, Okla. U., 1962. Pharmacist, mgr. Hyde Drug, Oklahoma City, 1962-77, v.p., mgr., 1977-82, pres., 1982—; mem. McKesson Drug Co. Small Chain Adv. Bd., San Francisco, 1986-87. Bd. mgmt. YMCA, 1989—. Mem. Oklahoma City Profl. Businessmen's Assn. (pres. 1962, 72), Nat. Assn. chain Drug Stores (sml. chain com. 1991—). Republican. Methodist. Avocations: golf, tennis, church activities. Office: Hyde Drug Inc 5108 N Shartel Ave Oklahoma City OK 73118-6094

WALLIS, W(ILSON) ALLEN, economist, educator, statistician; b. Phila., Nov. 5, 1912; s. Wilson Dallam and Grace Steele (Allen) W.; m. Anne Armstrong, Oct. 5, 1935; children: Nancy Wallis Ingling, Virginia Wallis Cates. AB, U. Minn., 1932, postgrad., 1932-33; postgrad. fellow, U. Chgo., 1933-35, Columbia U., 1935-36; DSc, Hobart and William Smith Colls., 1973; LLD, Roberts Wesleyan Coll., 1973, U. Rochester, 1984; LHD, Grove City Coll., 1975; D of Social Scis., Francisco Marroquin U., Guatemala, 1992. Economist Nat. Resources Com., 1935-37; instr. econs. Yale U., 1937-38; asst. to assoc. prof. econs. Stanford U., 1938-46; Carnegie rsch. assoc. Nat. Bur. Econ. Rsch., 1939-40, 41; dir. war rsch. Statis. Rsch. Group Columbia U., 1942-46; prof. stats. and econs. U. Chgo., 1946-62, chmn. dept. stats., 1949-57, dean Grad. Sch. Bus., 1956-62; pres. (title later chancellor) U. Rochester, N.Y., 1962-82; under sec. for econ. affairs U.S. Dept. State, Washington, 1982-89; resident scholar Am. Enterprise Inst., Washington, 1989—; staff Ford Found., 1953-54; fellow Ctr. for Advanced Study in Behavioral Scis., 1956-57. mem. math. div. NRC, 1958-60; bd. dir. Nat. Bur. Econ. Rsch., 1953-74; spl. asst. Pres. Eisenhower, 1959-61; pres. Nat. Commn. Study of Nursing and Nursing Edn., 1967-70; chmn. Commn. Presdl. Scholars, 1969-78; mem. Pres.'s Commn. on All-Vol. Armed Force, 1969-70; chmn. Pres.'s Commn. Fed. Stats., 1970-71; mem. Nat. Coun. Ednl. Rsch., 1973-75; chmn. Adv. Coun. Social Security, 1974-75; bd. dirs. Corp. Pub. Broadcasting, 1975-78, chmn., 1977-78. Author: (with others) Consumer Expenditures in the United States, 1939, A Significance Test for Time Series and Other Ordered Observations, 1941, Sequential Analysis of Statistical Data: Applications, 1945, Techniques of Statistical Analysis, 1947, Sampling Inspection, 1948, Acceptance Sampling, 1950, Statistics: A New Approach, 1956, The Nature of Statistics, 1962, Welfare Programs: An Economic Appraisal, 1968; An Overgoverned Society, 1976; co-compiler: The Ethics of Competition and Other Essays by Frank H. Knight, 1935; chmn. editorial adv. bd.: Internat. Ency. Social Scis., 1960-68; contbr. articles to profl. jours. Trustee Tax Found., 1961-82, chmn. bd., 1972-75, chmn. exec. com., 1975-78; bd. overseers Hoover Instn. War, Revolution and Peace, 1972-78; trustee Eisenhower Coll., 1969-79, Nat. Opinion Rsch. Ctr., 1957-62, 64-68, Com. Econ. Devel., 1965-71, Colgate Rochester Div. Sch., 1963-82, Ctr. Govtl. Rsch., Inc., 1962-82, Internat. Mus. Photography at George Eastman House, 1963-82, Robert A. Taft Inst. Govt., 1973-77, Ethics and Pub. Policy Ctr., 1980-82, 89—; mem. Com. on the Present Danger, 1980-82, 1989—; chmn. bd. overseers Ctr. Naval Analyses, 1967-82. Recipient Sec.'s Disting. Svc. award Dept. of State, Washington, 1988. Fellow Am. Soc. Quality Control, Inst. Math. Stats., Am. Statis. Assn. (editor Jour. 1950-59, pres. 1965), Am. Acad. Arts and Scis.; mem. Am. Econ. Assn. (exec. com. 1962-64), Rochester C. of C. (trustee 1963-68, 70-75), Mont Pelerin Soc. (treas. 1949-54), Washington Inst. of Fgn. Affairs, Phi Beta Kappa, Chi Phi, Beta Gamma Sigma. Clubs: Cosmos (Washington); Bohemian (San Francisco). Office: Am Enterprise Inst 1150 17th St NW Washington DC 20036-4603

WALLNAU, LARRY BROWNSTEIN, psychologist educator; b. New Haven, May 23, 1949; s. David Wolfe and Frances (Brownstein) W.; m. JoAnn T. McDermott, June 29, 1991; 1 child, Naomi Aviv. BA cum laude, U. New Haven, 1973; PhD, SUNY, Albany, 1977. Rsch. assoc. SUNY, Albany, 1977-79; asst. prof. psychology SUNY, Brockport, 1977-84, assoc. prof. psychology, 1984—; bd. dirs. Child and Adolescent Stress Mgmt. Inst., Brockport. Co-author: Statistics for the Behavioral Sciences, 1985, 88, 92, Essentials of Statistics for the Behavioral Sciences, 1991, also accompanying study guides, 1985, 91, 92. Mem. AAAS, APA (Teaching Psychology div.), Textbook Authors Assn., Am. Psychol. Soc. (charter mem.), Oak Orchard Yacht Club (rec. sec. 1993—). Achievements include development of techniques for undergraduate instruction of statistics and research methods; research in biopsychology and health psychology. Home: 29 South Ave Brockport NY 14420 Office: Dept Psychology SUNY Brockport NY 14420

WALLNER, FRANZ, engineer, educator; b. Berlin, Mar. 24, 1937; s. Franz and Kundry (Siewert) Wallner-Baste. Diploma in engring., Tech. U. Berlin, 1964, DEng, 1970. Rsch. fellow Tech. U. Berlin, 1964-69, prof. automobile engring., 1974—; rsch. fellow Westend Surg. Clinic, Westend Surg. Clinic, Free U. Berlin, 1969-75; pres. BMT Messtechnik GmbH, Berlin, 1976—. Contbr. numerous articles on automobile and artificial heart rsch. and bioengring. to profl. jours. Mem. Soc. Automotive Engrs., Verein Deutscher Ingenieure, Verein Deutscher Elektrotechniker, Internat. Ozone Assn., Reitclub Grunewald. Home and Office: Argentinische Allee 32a, D-14163 Berlin Zehlendorf, Germany

WALSCHOT, LEOPOLD GUSTAVE, conservator; b. Halle, Brabant, Belgium, May 21, 1936; s. Edward and Jeanne (Lesenne) W.; m. Lutgarde Elisabeth Ophalffens, July 10, 1961; 1 child, Eddy. MS in Geography, State U., Ghent, Belgium, 1958, DSc in Geography, 1967. Asst. Soil Survey Ctr., Ghent, 1958-62; gratis asst. Geol. Inst., State U., Ghent, 1958-62, asst., 1962-67, sr. asst., 1967-69, conservator libr. and mus., 1969—; lectr. geology Higher Inst. Architecture, Ghent, 1970-88; mng. dir. Nat. History and Med. Partnership, Ghent, 1976—; lectr. maritime geography Higher Nautical Sch., Antwerp, 1993—; scientific adviser Dirk Frimout Ctr., Poperinge, 1993—. Author: (book) Bibliography of Quaternary Geology of Belgium, 1962, (catalog) Library Geological Institute University of Ghent, 1963, (encyclopedia) Oosthoek's: several subjects, 1976-81; co-author: (book) Proceedings of 10th Salt-Water Intrusion Meeting, 1988; editor: (serial) Abstracts of Belgian Geology and Phys. Geography, 1969-76, (jour.) Natuurwetenschappelijk Tijdschrift, 1972—; co-operator: (serial) Internat. Geog. Bibliography, 1970—, (ency.) Academic American Encyclopedia, 1979; contbr. articles to profl. jours. Recipient medal of Merit City of Ghent, 1969, Knight of the Crown Order Belgian govt., Brussels, 1975, Knight of the Leopold Order, 1980, Civil First Class medal, 1984, Officer of the Crown order, 1985, Officer's Cross of the Leopold Order, 1988, Silver medal of Merit, Province of Brabant, Brussels, 1983; named Laureate interuniv. contest, Univ. Found., Brussels, 1959. Mem. European Assn. Sci. Editors, Belgian Soc. Geog. Studies, Belgian Soil Sci. Soc., Belgian Com. Engring. Geology, Royal Geol. and Mining Soc. of The Netherlands, Indian Soc. Soil Sci. (life), Soc. Ghent Geographers (sec. 1963—). Avocations: Town of Halle history, playing bombardon. Office: U Ghent Geol Inst, Krijgslaan 281, B-9000 Ghent East Flanders, Belgium

WALSH, CHRISTOPHER THOMAS, biochemist, department chairman; b. Boston, Feb. 16, 1944; married; 1 child. BA, Harvard U., 1965; PhD in Life Sci., Rockefeller U., 1970. Helen Hay Whitney Found. fellow Brandeis U., 1970-72; from asst. prof. to prof. chemistry and biology MIT, 1972-87, assoc. dir. Whitaker Coll. Mgmt., 1979-82, Uncas and Helen Whitaker prof., 1980-85, Karl Taylor Compton prof., 1985-87, chmn. chemistry dept., 1982-87; David Wesley Gaiser prof. Harvard U., 1987-91, chmn. dept. biol. chemistry and molecular pharmacology Med. Sch., 1987—, Hamilton Kuhn prof., 1991—; cons. Merck, Sharp & Dohme Rsch. Labs., 1975-81, Monsanto Corp. Res. Labs., 1980-81, Johnson & Johnson, 1982-83, Hoffman LaRoche, 1982—, Genzyme & Bioinfo Assocs., 1983—, Firmenich, S A, 1986-90, Enzymatics, 1988—, Biotage, 1989—; mem. panel rsch. grants study NSF, 1977-79, panel study sect. biochemistry NIH, 1978-82, gen. med. coun., NIH, 1983-85, Chemal Rsch. Group, WHO, 1984-86; co-chmn. Gordon Rsch. Conf. Enzymes, Coenzymes & Molecular Biology, 1978, Conf. Methanogenesis, 1984; chmn. study sect. biochemistry NIH, 1982—; lectr. in field. Alfred P. Sloan Found. fellow, 1975-77, Camille and Henry Dreyfus Tchr.-Scholar grantee, 1976-80; recipient Eli Lilly award, 1979. Mem. NAS, Inst. Medicine-NAS, Am. Acad. Arts & Sci., Am. Soc. Biol. Chemists, Am. Chem. Soc., Am. Soc. Microbiology. Achievements include research in enzymatic reation mechanisms, phosphoryl and pyrophosphoryl transfers, flavin-dependent enzymes, membrane biochemistry and mechanism of active transport. Office: Dana Farber Cancer Inst 44 Binney St Boston MA 02115*

WALSH, DANIEL STEPHEN, systems engineering consultant; b. Newark, Dec. 3, 1941; s. Vincent Daniel and Marie Frances (Fischl) W.; B.S. in Mech. Engring., N.J. Inst. Tech., 1963, M.S. in Systems Engring., 1974; M.B.A., N.Y.U., 1969, ScD, 1992; children: Lorraine, Kevin, Maureen. Staff systems analyst, indsl. engr. Merck & Co., Rahway, N.J., 1967-70; mgmt. cons. Delicia Inc., Elizabeth, N.J., 1970-71; mgr. spl. projects Supermarkets Gen. Corp., Woodbridge, N.J., 1971-83; systems engring. cons., 1983—; dir. sys. engring. Bell Comm. Rsch. Piscataway, N.J., 1983—; adj. prof. N.J. Inst. Tech., 1974—, Fairleigh Dickinson U.; cons.; condr. seminars profl. assns. and instns.; guest lectr. Rutgers U. Mem. planning bd., Clark, N.J., 1978-79. Served with USAF, 1963-67. Howard B. Begg fellow, 1974; registered profl. engr. Mem. Ops. Research Soc. Am., Inst. Mgmt. Sci., Inst. Indsl. Engrs., Assn. Computer Users, Sigma Pi, Alpha Pi Mu, Tau Beta Pi, Alpha Iota Delta. Roman Catholic. Contbr. articles to profl. jours. Home and Office: 155 Jupiter St Clark NJ 07066-2212

WALSH, GREGORY SHEEHAN, optical systems professional; b. Buffalo, Dec. 24, 1955; s. John Kevin and Ruth (Murphy) W.; m. Patricia DelGiudice, Apr. 8, 1976; children: James, Kevin. BBA, U. North Fla., 1989; cert. in submarine periscope design, Dept. of Navy, 1992. Optical systems specialist Naval Aviation Depot, Jacksonville, Fla., 1984-91, Trident Refit Facility, Kings Bay, Ga., 1991—; tech. adv. Naval Tech. Tng., Pensacola, Fla., 1992; bd. dirs. Strategic Bus. Plan Naval Aviation Depot, Jacksonville, 1989; cons. Small Bus. Adminstrn., Jacksonville, 1989. Coach Orange Park (Fla.) Soccer Assn., 1989—. With USN, 1974-78. Achievements include design of submarine periscope stadimeter sling, submarine optical periscope bushing, submarine optical periscope torque sleeve, submarine optical quick evacuation plug puller, aircraft optical hot mock-up; performed the first article acceptance test on the FA-18 aircraft optical systems. Home: 450 Sigsbee Ct Orange Park FL 32073-3409

WALSH, JANE ELLEN MCCANN, health care official; b. Uniontown, Pa., Jan. 16, 1941; d. Albert Benton and Dorothy Rose (Ruble) McCann; B.A., Hood Coll., 1963; postgrad. Northwestern U., 1964-65; M.A., Antioch Coll., 1978; m. John Daniel Walsh, June 8, 1973; stepchildren: Christopher, Mark, Jonathan, Jennifer. Research asst. Cert. Nat. Interviewers, Chgo., 1963-64; project dir. Assn. Am. Med. Colls., Evanston, Ill., 1965-68; systems analyst Research Found. Mental Hygiene, Inc., Orangeburg, N.Y., 1968-72; systems analyst Nat. Center Health Svcs. Research, Rockville, Md., 1972-73; research coord. U. Calif., Berkeley, 1973-75; coord. Pvt. Initiative in PSRO, San Francisco, 1975-76; cons., 1977-78; asso. dir. tech. svcs. Western Consortium for Public Health, Berkeley, 1978—; cons. pub. health info. systems, health svcs. evaluation, mental health info systems, data sources. Recipient Martha Schaeffer Shaw award, 1960. Mem. Am. Public Health Assn., Assn. Health Svcs. Research, Common Cause, Mus. Soc. San Francisco, Smithsonian Instn. Democrat. Presbyterian. Club: Highlands Country. Author: Introduction to Standard Mumps, 1978; developer automated system for storage and retrieval of clin. psychiat. data, 1969; designed and assisted in implementation of state-wide mgmt. info. system for indigent care program, 1986-92; research on multi-splty. group practices delivering primary care, procedures for conducting concurrent quality assurance. Home: 50 Schooner Hill Oakland CA 94618 Office: 2001 Addison St Ste 200 Berkeley CA 94704-1195

WALSH, JOHN HARLEY, medical educator; b. Jackson, Miss., Aug. 22, 1938; s. John Howard and Aimee Nugent (Shands) W.; m. Courtney Kathleen McFadden, June 12, 1963 (div. 1979); children: Courtney Shands (Mrs. Peter Phleger), John Harley Jr.; m. Mary Carol Territo, Feb. 4, 1989. BA, Vanderbilt U., 1959, MD, 1963. Diplomate Am. Bd. Internal Medicine. Resident N.Y. Hosp. Cornell Med. Ctr., N.Y.C., 1963-67; rsch. assoc. Bronx VA Hosp., N.Y., 1969-70; clin. investigator Wadsworth VA Hosp., L.A., 1971-73; asst. to assoc. prof. UCLA, 1970-78, prof. medicine, 1978—, dir. Integrated Gastroenterology Tng. Program, 1983-89, dir. Div. Gastroenterology, 1988—, Dorothy and Leonard Straus prof., 1989—; dep. dir. Ctr. Ulcer Rsch. and Edn., L.A., 1974-80, assoc. dir., 1980-87, dir. 1987—; adv. coun. mem. NIH, Bethesda, Md., 1982-85; mem. Nat. Digestive Disease Adv. Bd., 1987—. Assoc. editor: Gastroenterology Jour., 1976-86; mem. editorial bd. Am. Jour. Physiology, Jour. Clin. Endocrinology and Metabolism, Peptides Jour.; contbr. articles to profl. jours. Served with USPHS, 1967-69. Recipient Rorer award So. Calif. Soc. Gastroenterology, 1972, Western Gastroent. Rsch. prize Western Gut Club, 1983, Merit award NIH, 1987. Mem. Am. Gastroent. Assn., Am. Soc. Clin. Investigation, Assn. Am. Physicians, Endocrine Soc., Western Soc. Clin. Investigation (counselor 1978-81). Episcopalian. Avocations: tennis, golf, ballet, modern fiction. Home: 266 Dolores St San Francisco CA 94103-2203 Office: UCLA Sch Medicine Ctr for Health Scis Rm 44-146 10833 Le Conte Ave Los Angeles CA 90024-1602*

WALSH, KENNETH ALBERT, chemist; b. Yankton, S.D., May 23, 1922; s. Albert Lawrence and Edna (Slear) W.; m. Dorothy Jeanne Thompson, Dec. 22, 1944; children: Jeanne K., Kenneth Albert, David Bruce, Rhonda Jean, Leslie Gay. BA, Yankton Coll., 1942; PhD, Iowa State U., 1950. Asst. prof. chemistry Iowa State U., Ames, 1950-51; staff mem. Los Alamos Sci. Lab., 1951-57; supr. Internat. Minerals & Chem. Corp., Mulberry, Fla., 1957-60; mgr. Brush Beryllium Co., Elmore, Ohio, 1960-72; assoc. dir. tech. Brush Wellman Inc., Elmore, 1972-86; cons., patentee in field. Democratic precinct chmn., Los Alamos, 1956, Fremont, Ohio, 1980. Mem. AIME, Am. Chem. Soc. (sect. treas. 1956), Am. Soc. for Metals Internat., Toastmasters Internat., Theta Xi, Phi Lambda Upsilon. Methodist. Home: 2106 Kensington Dr Tyler TX 75703-2232

WALSH, SCOTT WESLEY, reproductive physiologist, researcher; b. Wauwatosa, Wis., July 23, 1947; s. Virgil C. and Harriet E. (Jacobson) W.; m. Cynthia Lea Sorenson, Oct. 10, 1981 (div. Mar. 1987). BS (with hon.), U. Wis., 1970, MS, 1972, PhD, 1975. Asst. prof. U. N.D. Sch. Medicine, Grand Forks, 1975-76; asst. to assoc. scientist Oreg. Primate Ctr., Beaverton, 1976-80; asst. prof. Oreg. Health Scis. U., Portland, 1978-80, Mich. State U., E. Lansing, 1980-85; assoc. prof. U. Tex. Health Sci. Ctr., Houston, 1985-90; prof. Med. Coll. Va., Richmond, 1990—; grant reviewer NIH, NICHD, Washington, 1988—. Grantee NIH, NICHD, 1983—; named Dean's Teaching Excellence list U. Tex. Health Sci. Ctr., Houston, 1988-89. Mem. Soc. for Gynecologic Investigation, Am. Physiol. Soc., Endocrine Soc., Soc. for Study of Reproduction, Phi Kappa Phi. Achievements include discovery of an imbalance of increased thromboxane, decreased protacyclin and increased lipid peroxides in placentas obtained from women with pregnancy-induced hypertension. Office: Medical Coll Virginia Dept OB/GYN Box 34 Richmond VA 23298-0034

WALSH, THOMAS JOSEPH, neuro-ophthalmologist; b. N.Y.C., Sept. 18, 1931; s. Thomas Joseph and Virginia (Hughes) W.; m. Sally Ann Maust, June 21, 1958; children—Thomas Raymond, Sara Ann, Mary Kelly, Kathleen Meghan. BA, Coll. Fordham, 1954; MD, Bowman Gray Med. Sch., 1958. Intern St. Vincent's Hosp., N.Y.C., 1958-59; resident ophthalmology Bowman Gray Med. Sch., Winston-Salem, N.C., 1961-64; fellow neuro-ophthalmology Bascom Palmer Eye Inst., Miami, Fla., 1964-65; practice medicine specializing in neuro-ophthalmology Stamford, Conn., 1965—; dir. neuro-ophthalmology service, asst. prof. ophthalmology and neurology Yale Sch. Medicine, New Haven, 1965-74; assoc. prof. Yale Sch. Medicine, 1974-79, prof., 1979—, also bd. permanent officers; dir. ophthalmology Stamford Hosp., 1978-83; mem. staff St. Joseph Hosp., Yale New Haven Hosp.; cons. to surgeon gen. army in neuro-ophthalmology Walter Reed Hosp., Washington, 1966—, VA Hosp., West Haven, 1965—, Silver Hill Found., New Canaan, Conn., 1974—; frequent lectr. various univs. Contbr. articles to various publs. Mem. adv. bd. Stamford Salvation Army, 1972—; mem. med. bd. Darien Nurses Assn., Conn., 1972—; surgeon Darien Fire Dept., 1969—. With AUS, 1959-61. Decorated Knight of Malta, 1983; Centennial fellow Johns Hopkins, 1976. Mem. AMA, Conn., Fairfield County med. socs., Acad. Ophthalmology, Oxford Ophthal. Congress, Acad. Neurology, Am. Assn. Neurol. Surgeons, Internat. Neuro-Ophthalmology Soc., Soc. Med. Cons. to Armed Forces, Cosmos Club (Washington), Darien County Club, Yale Club (N.Y.C), Lions, Army-Navy Club. Office: 1100 Bedford St Stamford CT 06905-5301

WALSH, THOMAS JOSEPH, neuroscientist, educator; b. Chgo., Sept. 12, 1952; s. Joseph Michael and Myra Lynn (Kalmar) W. BA in Psychology, No. Ill. U., 1975; PhD in Biopsychology, Syracuse U., 1980. Rsch. psychologust U.S. EPA, Research Triangle Park, N.C., 1980-82; staff fellow NIEHS, Research Triangle Park, N.C., 1982-84; rsch. assoc. U. N.C., Chapel Hill, 1984-86; assoc. prof. psychology Rutgers U., New Brunswick, N.J., 1986—; cons. Burroughs-Wellcome, Research Triangle Park, 1986-90. Contbr. articles to profl. jours. Grantee NIH, 1986—, NSF, 1991—, BRSG, 1986—. Mem. Soc. for Neurosci., Internat. Neurotoxicology Assn. (sci. organizing com. 1988—). Achievements include development of animal models of neurological diseases; Alzheimer's disease; neurotransmitter interactions and memory. Home: 3404 Wildwood Ct Monmouth Junction NJ 08852 Office: Rutgers Univ Dept Psychology New Brunswick NJ 08903

WALSHE, BRIAN FRANCIS, management consultant; b. White Plains, N.Y., June 8, 1958; s. Kevin D. and Anna G. (Touhy) W. BSCE, Northeastern U., 1981; MBA, U. Mich., 1986. Assoc. engr. Stone & Webster Engring. Corp., Boston, 1981-82, engr., 1982-84; assoc. Metzler and Assocs., Deerfield, Ill., 1986-87, sr. assoc., 1987-90, mgr., 1990—; guest lectr. in field; expert witness testimony in constrn. mgmt. Mem. Am. Soc. Cost Engrs., Am. Nuclear Soc., Inst. Mgmt. Cons., Fin. Mgmt. Assn., MENSA. Republican. Home: 1730 N Clark St # 2408 Chicago IL 60614

WALSH-MCGEHEE, MARTHA BOSSE, conservationist; d. Leon and Lenore (Carter) Bosse; m. Leo S. Walsh, Sept. 30, 1972 (div. Oct. 1982); m. Donald B. McGehee, Aug. 6, 1992. Student, U. Mo., 1966, Baker U., 1966-67, Marymount-Manhattan, 1980-82. Flight attendant TWA, N.Y.C., 1967-78; pres. Island Conservation Effort, 1988—; trustee Rare Ctr. for Tropical Bird Conservation, Phila., 1987-91. Ptnr. in conservation World Wildlife Fund, Washington, 1986—; assoc. World Resources Inst., Washington, 1987—; mem. St. Croix (V.I.) Landmarks Soc., 1985—, Sherman (Conn.) Hist. Soc., 1986, Sherman Libr. Assn., 1986. St. Croix Environ. Assn., 1987—; mem. Saba Conservation Found. Nature Conservancy. Mem. Caribbean Conservation Assn., St. Lucia Naturalists Soc., Cedam Internat., Soc. Caribbean Ornithology, Friends of Abaco Parrot. Republican. Avocations: reading, bird watching, horseback riding, scuba diving. Home: 90 Edgewater Dr Apt 901 Coral Gables FL 33133 also: Windwardside, Saba Netherlands Antilles

WALTER, JAMES FREDERIC, biochemical engineer; b. Arlington, Va., Dec. 3, 1956; s. John Emory and Barbara (Ward) W.; m. Paula Ann Hanchak, Apr. 19, 1983; children: Jonathan Anthony, James Alexander. BS, U. Va., 1979; PhD, U. Calif., Berkeley, 1983. Rsch. engr. W.R. Grace, Columbia, Md., 1983-85, sr. rsch. engr., 1985-88, mgr. biochem. engring., 1988—; lectr. U. Md., College Park, 1985-87. Contbr. chpts to books; contbr. articles to Chem. Engring. Jour., Phytopathology. Named Outstanding Young Leader Optimists, 1974. Mem. AIChE, Trigon (sec. 1974-75), Alpha Chi Sigma. Achievements include patents for storage of stable azadirachtin formulations, selective production of L-serine derivative isomers, a novel vermiculite formulation with fermentor biomass of biocontrol fungi to control soilborne pathogens, others. Office: WR Grace 7379 Rt 32 Columbia MD 20861

WALTER, PAUL HERMANN LAWRENCE, chemistry educator; b. Jersey City, N.Y., Sept. 22, 1934; s. Helmuth Justus and Adelaide C. J. (Twardy) W.; m. Grace Louise Carpenter, Aug. 25, 1956; children: Katherine Elizabeth Walter Bousquet, Marjorie Allison Walter Moran. BS, MIT, 1956; PhD, U. Kans., 1960. Rsch. scientist DuPont Cen. Rsch. Dept., Wilmington, Del., 1960-67; prof. chem. Skidmore Coll., Saratoga Springs, N.Y., 1967—, chair chemistry and physics, 1975-85. Translator: (book) Foundations of Crystal Chemistry, 1968; contbr. articles to sci. publs. Fellow Chem. Inst. Can.; mem. AAAS, AAUP (pres. 1984-86), Am. Chem. Soc. (bd. dirs. 1991—, chmn. 1993), Sociedad Quimica de Mexico (hon.). Democrat. Presbyterian. Achievements include patents in field. Home: 9 Walter Dr Saratoga Springs NY 12866-9233 Office: Skidmore Coll Dept Chemistry and Physics N Broadway Saratoga Springs NY 12866-1632

WALTERS, D. ERIC, biochemistry educator; b. Circleville, Ohio, Jan. 20, 1951; s. David E. and Sarah G. (Riley) W.; children: Abigail Lee, Matthew Steven. BS in Pharmacy, U. Wis., 1974; PhD in Medicinal Chemistry, U. Kans., 1978. Postdoctoral researcher Ind. U., Bloomington, 1978-79; rsch. scientist Kraft Foods, Glenview, Ill., 1979-82, Searle Pharm. Co., Skokie, Ill., 1982-85; group leader NutraSweet Co., Mount Prospect, Ill., 1985-91; assoc. prof. Chgo. Med. Sch., North Chicago, Ill., 1991—; adj. asst. prof. U. Ill., Chgo., 1987—; cons. Molecular Simulations Inc., Burlington, Mass., 1991—; mem. spl. study sects. NSF, Washington, 1983-84. Author: Opiates, 1986; editor: Sweeteners: Discovery, Molecular Design, Chemoreception, 1991; referee Jour. Medicinal Chemistry, Chem. Senses; contbr. articles to profl. jours. Mem. Am. Chem. Soc., Bread for the World, Woodstock Inst. for Sci. and Humanities, Sigma Xi. Lutheran. Achievements include 2 patents on new sweeteners, other patents pending; research on computational biochemistry and drug design. Office: Chgo Med Sch Dept Biol Chemistry 3333 Green Bay Rd North Chicago IL 60064

WALTERS, JOHN LINTON, electronics engineer, consultant; b. Washington, Mar. 8, 1924; s. Francis Marion Jr. and Roma (Crow) W.; m. Grace Elizabeth Piper, June 19, 1948; children: Richard Miller, Gretchen Elizabeth, Christopher Linton, John Michael, Kim Anne. BS, U.S. Naval Acad., 1944; SM, Harvard U., 1949; DrEng, Johns Hopkins U., 1959. Staff mem. Los Alamos (N.Mex.) Sci. Lab., 1949-52; rsch. assoc. Johns Hopkins U., Balt., 1952-59; assoc. elec. engr. Brookhaven Nat. Lab., Upton, N.Y., 1959-62; rsch. scientist Johns Hopkins U., Balt., 1962-70; electronics engr. Naval Rsch. Lab., Washington, 1970—; sci. advisor Comdr. 6th Fleet, Gaeta, Italy, 1979-80; lectr. dept. elec. engring. Johns Hopkins U., 1964-65. Lt. (j.g.) USN, 1944-47, PTO. Recipient commendation Dir. of Navy Labs., 1979. Mem. IEEE, Sigma Xi. Achievements include research on theoretical evaluations of electronic countermeasures, refinements to measurement techniques used particle accelerators, analysis of radar and jamming phenomena. Home: 212 Winchester Beach Dr Annapolis MD 21401 Office: Radar Divsn Naval Rsch Lab Washington DC 20375-5000

WALTERS, KENN DAVID, scientist, company executive; b. Birmingham, Eng., July 1, 1957; s. Kenneth Walters; m. Barbara Grabert, Aug. 9, 1985; 1 child, Natascha Ruth. BSc in Computer Sci., Highbury Tech. Sch., Portsmouth, Eng., 1973; MSc in Computer and Info. Systems, Pacific Western U.,

Calif., 1989; PhD in Computer Info. Sytems, Pacific Western U., 1991. Programmer Arinco, London, 1973-75; project leader I.C.L., London, 1976-78, Racal Electronics, Eng., 1978-80; chief designer Aramco, Saudi Arabia, 1980-81; cons. M.O.D., Eng., 1981-82; cons., project leader I.T.T., Fed. Republic Germany and U.S., 1982-86; dir. internat. Canaan Computers, U.S., 1986-87; div. mgr. Nixdorf Computers A.G., Munich, 1987-89; chief exec. officer Capricorn Systems Assocs., Munich, 1989-91, ESP Informatik GMbH, Munich, 1991—. Author: Programming Practices and Guidelines, 1985. Fellow Inst. Analysts and Programmers; mem. Data Processing Inst., Inst. for Data Processing Mgmt., N.Y. Acad. Scis. Avocations: shooting, reading, classic cars.

WALTERS, LORI ANTIONETTE, electrical engineer, consultant; b. Smyrna, Tenn., Jan. 30, 1967; d. Ronnie Edwin and Donna Gail (Tagliente) Brandon; m. Howard Corey Walters V, June 22, 1991. BEE, Vanderbilt U., 1989. Elec. engr. Consoer Townsend & Assocs., Inc., Nashville, 1989-90, Brown and Caldwell Consultants, Atlanta, 1990—. Mem. NSPE, Ga. Soc. Profl. Engrs. (Mathcounts Coord. Gwinnett County Chpt. 1992-93, state co-chair 1993-94). Roman Catholic. Office: Brown and Caldwell 53 Perimeter Center E Ste 500 Atlanta GA 30346-1905

WALTERS, MARTEN DOIG, chemical engineer; b. Coventry, Eng., Nov. 3, 1953; came to U.S., 1981; s. Charles Alfred and Betty Elizabeth (Bass) W.; m. Jacqueline Greenshaw, Dec. 23, 1983; children: David John, Michael Joseph, Robert James. BSc in Chem. Engring. with honors, U. Manchester, Eng., 1975. Chartered engr., U.K.; European engr., Europe. Process engr. Norsk Hydro Co., Avonmouth, Eng., 1975-77, Felixstowe, Eng., 1977-81; sr. process engr. Davy McKee Co., Lakeland, Fla., 1981-88, 88-90, IMC Fertilizer Co., Mulberry, Fla., 1988; process supr. Bromwell and Carrier Co., Lakeland, 1990—; cons. Jordanian Phosphate Mining Co., Amman, 1988-90, World Environment Ctr. for Waste Minimization Program, Lithuania, 1993. Contbr. to conf. procs., tech. publs. Mem. AICE, Instn. Chem. Engrs. U.K., Engring. coun. U.K., European Fedn. Nat. Engring. Assns. Achievements include design and start-up of fertilizer plants in Iraq, China, Morocco, Yugoslavia, U.K. and U.S. Home: 1211 Fairfax St S Lakeland FL 33813 Office: Bromwell and Carrier Co 5925 Imperial Pky Lakeland FL 33860

WALTERS, TRACY WAYNE, mechanical engineer; b. Long Beach, Calif., Apr. 4, 1963; s. Larry Eugene Walters and LaDonna Gay (Lollio) Behm; m. Kathryn Louise Hirlinger, Jan. 20, 1990; 1 child, Jared Larry. BSME, U. Calif., Santa Barbara, 1985, MSME, 1986. Rsch. asst. U. Calif., Santa Barbara, 1985-86; sr. engr. Gen. Dynamics Space Systems Div., San Diego, 1986-92; rsch. engr. Babcock & Wilcox Alliance (Ohio) Rsch. Ctr., 1992—. Contbr. articles to profl. jours. Mem. Canton 1st Ch. of Nazarene. Recipient grad. fellowship U. Calif., 1985—. Mem. ASME (assoc.), AIAA, Creation Rsch. Soc. Republican. Achievements include co-development of Atlas I and II as expendable launch vehicles; development of surge fluid transient analysis code. Home: 3917 Conrad Dr C-10 Spring Valley CA 92077 Office: Babcock and Wilcox Alliance Rsch Ctr 1562 Beeson St Alliance OH 44601

WALTERS, WILLIAM PLACE, aerospace engineer; b. Bay Shore, N.Y., Feb. 24, 1943; s. William Edward and Amanda Mae (Place) W.; m. Sharon Marie Coffey, Mar. 15, 1968; children: Kristin Marie, Jill Ann. BS, N.C. State U., 1964; MS, U. Ala., Huntsville, 1968; PhD, U. Ill., 1972. Sr. engr. Northrop Space Labs., Huntsville, 1964-69; rsch. assoc. Coord. Space Labs., Urbana, Ill., 1969-72; tech. staff mem. MIT Lincoln Labs., Lexington, Mass., 1972-74; chief engr. Stone & Webster, Boston, 1974-76; rsch. engr. Ballistics Rsch. Lab., Aberdeen, Md., 1976-92, Army Rsch. Lab., Aberdeen, 1992—; sr. cons. Comm. Rsch. Ctr., Ottawa, Ont., Can., 1972-73; lectr. Computational Mechanics, Balt., 1985-91; adj. prof. Drexel U., Phila., 1988—; presenter, cons., instr., researcher in field of ballistics. Author: Fundamentals of Shaped Charges, 1989, High Velocity Impact Dynamics, 1990; contbr. over 100 articles to profl. publs. Recipient award NASA-Marshall Space Flight Ctr., 1968. Fellow ASME, Ballistics Rsch. Lab.; mem. Am. Def. Preparedness Assn. (life), Am. Phys. Soc., Sigma Gamma Tau. Achievements include 7 patents in field. Office: Army Rsch Lab Aberdeen Proving Ground Aberdeen MD 21005

WALTHER, HERBERT, physicist, educator; b. Ludwigshafen, Fed. Republic Germany, Jan. 19, 1935; s. Philipp and Anna (Lorenz) W.; m. Margot Gröschel, July 27, 1962; children: Thomas, Ulrike. Diploma in physics, U. Heidelberg, 1960, PhD, 1962. Postdoctoral fellow U. Heidelberg, 1962-63; sci. asst. U. Hannover (Fed. Republic Germany), 1963-68, lectr., 1968-69; guest scientist lab. Aimé Cotton CNRS, Orsay, France, 1969; vis. fellow Joint Inst. for Lab. Astrophysics, U. Colo., 1970; prof. U. Bonn (Fed. Republic Germany), 1971; prof. physics U. Cologne (Fed. Republic Germany), 1971-75, U. Munich, 1975—; dir. Max-Planck-Institut fur Quantenoptik, Garching, Fed. Republic Germany, 1981—; mem. senate, coun. NSF, Fed. Republic Germany, 1978-84; mem. planning Max-Planck-Soc., 1982-86; chmn. commn. atomic and molecular physics Internat. Union Pure and Applied Physics, 1984-87; Stanley H. Klosk lectr. NYU, 1985; Loeb lectr. Harvard U., 1990. Author several books in field. Recipient Max Born prize Inst. of Physics, The German Phys. Soc., 1978, Einstein prize Indsl. and Univ. Rsch. Affiliates, 1988, Gaub medal Braunschweigische Wiss. Gesellschaft, 1989, Albert A. Michelson medal Franklin Inst., 1993, King Faisal Internat. Prize in Physics, Faisal Found., 1993. Fellow Optical Soc. Am. (publ. com. 1986, Charles Hard Townes award 1990); mem. European Phys. Soc. (div. chmn. 1987, chmn. quantum electronics div. 1987-89), German Phys. Soc. (bd. dirs. 1979-82), Roland Eötvös Phys. Soc. (hon.), Bavarian Acad. Scis., Akademie der Naturforscher Leopoldina, Am. Acad. of Arts and Scis. (foreign hon. mem. 1993). Home: Egenhoferstr 7a, 8000 München 60, Federal Republic Germany Office: Max-Planck Inst Quantenoptk, Ludwig Prandtl Str 10, 8046 Garching Germany

WALTHER, JOSEPH EDWARD, health facility administrator, retired physician; b. Indpls., Nov. 24, 1912; s. Joseph Edward and Winona (McCampbell) W.; m. Mary Margaret Ruddell, July 11, 1945 (dec. July 1983); children: Mary Ann Margolis, Karl, Joanne Landman, Suzanne Conran, Diane Paczesny, Kurt. BS, Ind. U., 1936, MD, 1936; postgrad., U. Chgo., Harvard U., U. Minn., 1945-47. Diplomate Nat. Bd. Med. Examiners, Am. Bd. Internal Medicine, Am. Bd. Gastroenterology. Intern Meth. Hosp. and St. Vincent Hosp. of Indpls., 1936-37; physician, surgeon U.S. Engrs./Pan Am. Airways, Midway Islands, 1937-38; chief resident, med. dir. Wilcox Meml. Hosp., Lihue, Kauai, 1938-40; internist, gastroenterologist Meml. Clinic Indpls., 1947-83, med. dir., pres., chief exec. officer, 1947—; founder, pres. Doctors' Offices Inc., Indpls., 1947—; founder, pres., chief exec. officer Winona Meml. Found. and Hosp. (now Walther Cancer Inst.), Indpls., 1956—; clinical asst. prof. medicine Ind. U. Sch. Medicine, Indpls., 1948-93, clin. asst. prof. emeritus, 1993—. Author: (with others) Current Therapy, 1965; mem. edit. rsch. bd. Postgrad. Medicine, 1982-83; contbr. articles to profl. jours. Bd. dirs. March of Dimes, Marion County div., 1962-66, Am. Cancer Soc., Ind. div., 1986—. Col. USAAF, 1941-47, PTO. Decorated Bronze Star, Silver Star, Air medal, Soldiers medal; recipient Clevenger award Ind. U., 1989, Disting. Alumnus award Ind. U. Sch. Medicine, 1990. Master Am. Coll. Gastroenterology (pres. 1970-71, Weiss award 1988); mem. AMA (del. 1970-86), Soc. Cons. to Armed Forces, Ind. Med. Assn., Marion County Med. Assn., Ind. U. Alumni Assn. (life), Hoosier Hundred (charter), Highland Golf and Country Club, Waikoloa Golf and Country Club (Hawaii), Indpls. Athletic Club, 702 Club. Republican. Home: 4266 N Pennsylvania St Indianapolis IN 46205-2613 Office: Walther Cancer Inst 3202 N Meridian St Indianapolis IN 46208-4646

WALTON, ALAN, oceanographer; b. Consett, County Durham, England, May 29, 1932; s. Albert and Mary Walton; m. Dorothy Foreman, Aug. 18, 1956; children: Deborah Anne, Karen Elizabeth, Susan Caryl. BS with honors, U. Durham, Eng., 1953, PhD, 1956; DSc, U. Glasgow, Scotland, 1971. Asst. sci. dir. Isotopes Inc., Westwood, N.J., 1959-64; sr. lectr. U. Glasgow, Scotland, 1965-70; head chem. oceanography Bedford Inst., Halifax, N.S., Can., 1970-78; dir. Internat. Lab. of Marine Radioactivity, Monaco, 1979-82, 86-90; spl. advisor Fisheries and Oceans, Ottawa, Ont., Can., 1982-84; dir. gen. Bayfield Lab., Burlington, Ont., 1984-85; dir. Ctr. Océanographique, Rimouski, P.Q., Can., 1990—; chmn. marine standards program N.R.C., Ottawa, 1972-80; vice-chmn. UN Expert Group Standards, Paris, 1992—. Contbr. over 7,100 articles on environ. sci. to jours. including Nature, Jour. Environ. Radioactivity, Radiocarbon. Fellow Royal Soc. Chem. (U.K.). Home: University of Quebec, 418 William Price, Rimouski, PQ Canada G5L 8W4 Office: Univ du Quebec, 310 Alleé des Ursulines, Rimouski, PQ Canada G5L 3A1

WALTON, CHARLES MICHAEL, civil engineering educator; b. Hickory, N.C., July 28, 1941; s. Charles O. and Virginia Ruth (Hart) W.; m. Betty Grey Hughes; children: Susan, Camila, Michael, Gantt. BS, Va. Mil. Inst., 1963; MCE, N.C. State U., 1969, PhD, 1971. Research asst. N.C. State U., Raleigh, 1967-71; transp. planning engr. N.C. Hwy. Commn., Raleigh, 1970-71; asst. prof. civil engring. U. Tex., Austin, 1971-76, assoc. prof., 1976-83, prof., 1983-87, Bess Harris Jones Centennial prof. natural resource policy studies, 1987-91, Paul D. and Betty Robertson Meek Centennial prof. engring., 1991-93, Ernest H. Cockrell Centennial chair engring., 1993—, chmn. dept. civil engring., 1988—; transp. cons., 1970—; assoc. dir. Ctr. for Transp. Rsch., U. Tex., 1980-88; chmn., exec. com. Transp. Rsch. Bd., NRC, 1991. Contbr. articles to profl. jours. Past chmn. Urban Transp. Commn., Austin. Fellow ASCE (Harland Bartholomew urban planning award 1987, Frank M. Masters transp. engring. award 1987, James Laurie prize 1992), Inst. Transp. Engrs.; mem. NSPE, NAE, Soc. Automotive Engrs., Urban Land Inst., Soc. Am. Mil. Engrs., Ops. Rsch. Soc. Am., Internat. Rd. Fedn. (bd. dirs.), Austin C. of C. (Leadership Austin Program). Democrat. Methodist. Home: 3404 River Rd Austin TX 78703-1031 Office: U Tex Dept Civil Engring Austin TX 78712

WALTON, ERNEST THOMAS SINTON, physicist; b. Dungarvan, County Waterford, Ireland, Oct. 6, 1903; s. J.A. Walton; m. Winifred Isabel Wilson, 1934; 4 children. Student, Meth. Coll., Belfast, Northern Ireland; MSc, Trinity Coll., Dublin, Ireland; PhD, Cambridge (Eng.) U.; DSc (hon.), Queen's U., Belfast, Gustavus Adolphus Coll., Minn., U. Ulster, Northern Ireland; PhD, Dublin City U. Erasmus Smith's prof. natural and exptl. philosophy Trinity Coll., Dublin, 1947-74, fellow emeritus, 1974—; hon. fellow Inst. Physics, London. Recipient Overseas Research scholar, 1927-30, Sr. Research award, dept. sci. and indsl. research, 1930-34, Clerk Maxwell scholar, 1932-34, Hughes medal, Royal Soc., 1938, Nobel prize for physics, 1951. Home: 26 St Kevin's Pk Dartry Rd, Dublin 6, Ireland Office: Trinity Coll, Dept Physics, Dublin Ireland

WALTON, HAROLD FREDERIC, retired chemistry educator; b. Tregony, Cornwall, Eng., Aug. 25, 1912; s. James and Martha Florinda Jane (Harris) W.; m. Sadie Goodman, June 17, 1938 (dec. Jan. 1987); children: James, Elizabeth Louise, Daniel Goodman. BA, U. Oxford, Eng., 1934, DPhil, 1937. Rsch. chemist Permutit Co., Birmingham, N.J., 1938-40; asst. prof. Northwestern U., Evanston, Ill., 1940-46; from asst. prof. to prof. chemistry U. Colo., Boulder, 1947-82, chmn. chemistry dept., 1962-66, prof. emeritus, 1982—; vis. prof. U. Trujillo, Peru, 1966-67, Fulbright-Hays program, 1970; vis. prof. Pedagogical Inst., Caracas, Venezuela, 1972. Author: Inorganic Preparations, 1948, Principles and Methods of Chemical Analysis, 1952, 2d edit., 1964, Elemenatry Quantitative Analysis, 1958, (with others) Ion Exchange in Analytical Chemistry, 1973, Ion Exchange in Analytical Chemistry, 1990, others; editor: Benchmark Papers in Ion-Exchange Chromatography, 1976. Recipient Dal Nogare award on Chromatography, Del. Valley Chromatography Forum, 1988; named hon. prof. U. San Marcos, Lima, Pery, U. Trujillo. Fellow AAAS; mem. Am. Chem. Soc. (chmn. Colo. sect. 1958), Chem. Soc. Peru (corr.). Home: PO Box 1837 Boulder CO 80306 Office: U Colo Campus Box 216 Boulder CO 80309-0216

WALTON, JEFFREY HOWARD, physicist; b. Honolulu, July 29, 1958; s. George Edward and Marilyn Ruth (Pearson) W.; m. Kathleen Clark Scharff, May 31, 1986. BS, U. Fla., 1980, MS, 1983; PhD, Coll. William and Mary, 1989. Postdoctoral researcher Naval Rsch. Lab., Washington, 1989-91, Fla. State U./Nat. High Magnetic Field Lab., Tallahassee, 1991—. Contbr. articles to profl. publs. NRC postdoctoral assoc., 1989. Mem. Am. Phys. Soc., Sigma Pi Sigma. Office: Fla State Univ Dept Chemistry PO Box 3006 Tallahassee FL 32306-3006

WALTON, KIMBERLY ANN, medical laboratory technician; b. Balt., June 16, 1965; d. Lucian Wayne and Arlene Catherine (Hopkins) W. AA in med. lab. tech., Essex Coll., 1985. Med. tech. Balt. Rh Typing Lab., 1985-91. Mem. AAAS, Am. Soc. Clin. Pathologists (cert.), Am. Soc. Med. Tech., Am. Assn. Blood Banks. Office: Baltimore Rh Typing Lab 400 W Franklin St Baltimore MD 21201-1833

WALTON, THOMAS CODY, chemical plastics engineer; b. Waterbury, Conn., June 8, 1956; s. James Praul and Mary Elizabeth (Cody) W. BS in Chem. Engring., Northeastern U., 1982; MS in Plastics Engring., U. Lowell, 1986. Tech. mgr. Lewcott Chems. and Plastics Corp., Millbury, Mass., 1982-85; group leader composites R&D Emerson and Cuming, Inc., Canton, Mass., 1985-87; founder, pres. Plastec Assocs., Inc., Pepperell, Mass., 1987—; program mgr. Foster Miller, Inc., Waltham, Mass., 1989-92; program mgr., sr. engr. Aspen Systems, Inc., Marlborough, Mass., 1992—. Contbr. articles to profl. publs. Del. Mass. Rep. party, 1990—. Recipient five SBIR rsch. grants. Mem. Soc. Advancement of Material and Process Engring (chmn., bd. dirs. Boston chpt. 1989, 92, 93), Soc. Plastics Engrs. (dir. advanced polymer composites divsn. 1989—). Roman Catholic. Achievements include work on solventless thermal spray painting technology, electron and x-ray composite curing technology for spacecraft components, resin transfer molding technology for thermoplastic matrix composites, racked and golf shaft composite technology, biodegradable plastics. Office: Aspen Systems Inc 184 Cedar Hill St Marlborough MA 01752

WALTZ, JOSEPH MCKENDREE, neurosurgeon, educator; b. Detroit, July 23, 1931; s. Ralph McKinley and Bertha (Seelye) W.; m. Janet Maureen Journey, June 26, 1954; children: Jeffrey McKinley, Mary Elaine, David Seelye, Stephen McKendree; m. Marilyn Liska, June 5, 1967; 1 child, Tristana McKendree. Student, U. Mich., 1950; B.S., U. Oreg., 1954, M.D., 1956. Diplomate Am. Bd. Neurol. Surgery. Surg. intern U. Mich. Hosp., 1956-57, gen. surg. resident, 1957-58, clin. instr. neurosurgery, 1960-63; neurosurg. assoc. St. Barnabas Hosp., N.Y.C., 1963—; assoc. dir. Inst. Neurosci., 1974—, dir. dept. neurol. surgery, 1977—; assoc. cons. in neurosurgery Englewood (N.J.) Hosp., 1964—; assoc. prof. neurosurgery NYU Med. Ctr., 1974—; asst. prof. dept. surgery (neurosurgery) N.Y. Coll. Osteo. Medicine, 1989—; bd. dirs. Neurol. Surgery Rsch. Found., 1978. Author papers on functional neurosurg. treatment of abnormal movement disorders cerebral palsy, others; cryothalamectomy-cryopulvinectomy and implantation brain pacemakers; chpt. in book on cryogenic surgery; contbr. chpts. to Cryogenic Surgery, Neurology, 1982, Advances in Neurology, 1983. Patentee 4-electrode quadrapolar computerized spinal cord stimulator. Mem. sci. adv. bd. Dystonia Med. Research Found., 1980—; trustee St. Barnabas Hosp., 1980—. Served to capt. M.C. AUS, 1958-60. Recipient bronze award Am. Congress Rehab. Medicine, 1967, World Community Svc. award Rotary. Mem. AMA, Am. Paralysis Assn., World Soc. Stereotactic and Functional Neurosurgery, Congress Neurol. Surgeons, Math. Assn. Am., Internat. Neural Network Soc., Soc. for Cryobiology, N.Y. State Med. Soc., Bronx County Med. Soc., N.Y. State Neurosurg Soc., Nat. Ski Patrol, Phi Beta Pi. Achievements include spl. rsch. on neurophysiology and treatment of epilepsy, basal ganglia disorders, abnormal movement disorders, cerebral palsy, also neurosurg. application stereotactic thalamic surgery and spinal cord stimulation. Home: Four B Island South 720 Milton Rd Rye NY 10580-3258 Office: St Barnabas Hosp 4422 3d Ave Bronx NY 10457

WALTZ, RONALD EDWARD, physicist; b. Indpls., Nov. 21, 1943; s. Eugene Edward and Evelyn (Griffith) W.; m. Candace Dumlao, July 24, 1967; children: Justin, Jonathan. BS, Purdue U., 1966; PhD, U. Chgo., 1970. Rsch. assoc. MIT, Boston, Mass., 1971-73; rsch. scientist Visidyne Inc., Burlington, Mass., 1973-75; sr. tech. advisor Gen. Atomics, San Deigo, 1975—; vis. prof. Nagoya U., Japan, 1986; adj. prof. physics Univ. Calif., San Diego, 1986—. Contbr. over 50 articles to profl. jours. NSF Postdoctoral fellow, 1970-71. Fellow Am. Phys. Soc. Office: Gen Atomics 13-303 PO Box 85608 San Diego CA 92186

WAMBOLDT, MARIANNE ZDEBLICK, psychiatrist; b. Chgo., Oct. 25, 1954; d. William Thomas and Mary (Demko) Zdeblick; m. Frederick Steven Wamboldt, Sept. 11, 1976; children: Krystyna Noel, Alexander Steven. BS in Biology magna cum laude, Marquette U., 1976; MD, U. Wis., 1980. Diplomate Am. Bd. Child Psychiatry, Am. Bd. Neurology and Psychiatry. Intern in psychiatry U. Wis. Hosps., Madison, 1981-82, resident in psychiatry, 1982-85, fellow in child psychiatry, 1984-85; fellow in clin. rsch. lab. clin. sci. sect. brain & behavior NIMH, Bethesda, Md., 1985-86; fellow in child psychiatry dept. psychiatry U. Colo. Health Scis. Ctr., Denver, 1989-90, asst. prof. psychiatry, 1990—; dir. psychophysiologic unit Nat. Jewish Ctr. for Immunology and Respiratory Medicine, Denver, 1990—, acting dir. pediatric psychiatry, 1991-92, dir. pediatric psychiatry, 1992—; behavioral neurophysiology rsch. adminstr. behavioral scis. rsch. br., NIMH, Rockville, Md., 1988-89; HLA transfusion coord. Milw. Blood Ctr., 1976-77; staff adolescent psychiatrist Psychiatric Inst. Montgomery County, Rockville, 1986-88; invited participant Rsch. Careers in Child Psychiatry Workshop, Stanford U., NIMH, 1984; mem. pediatric merger task force Nat. Jewish Ctr. for Immunology and Respiratory Medicine, 1990-92, quality assurance chair divsn. pediatric psychiatry, 1991—, mem. IRB com., 1991—, mem. ethics com., 1991—; mem. exec. com. child psychiatry U. Colo. Health Scis. Ctr., 1991—, mem. residency edn. com. child psychiatry, 1991—; presenter in field. Reviewer Contemporary Psychology, 1989—; assoc. editor Psychiatry, 1990—; ad hoc reviewer Family Process, 1991—; contbr. papers to Am. Jour. Psychiatry, Behavioral Neurosci., Developmental Psychobiology, Handbook of Anxiety Disorders, Am. Jour. Family Therapy, Contemporary Psychology. Scout leader Girls Scouts Am., Denver, 1992. Mem. Acad. Psychosomatic Medicine, Am. Acad. Child Psychiatry, Am. Family Therapy Assn., Am. Psychiatric Assn., Colo. Child and Adolescent Psychiatric Soc., Colo. Psychiatric Soc., Developmental Psychobiology Rsch. Group, Physicians for Social Responsibility, Soc. for Rsch. in Child Devel., Women in Medicine. Democrat. Roman Catholic. Achievements include rsch. in family factors in adolescent asthma outcome; in epidemiology of symptom perception in asthmatic children; in neurobiology of parental behaviors, suicide in youth, epidemiology and family interactional patterns of learning disabilities, epidemiology and prevention of respiratory virus transmission in the closed population. Office: Nat Jewish Ctr Immunology Dept Pediatrics 1400 Jackson St Denver CO 80206

WANG, ALLAN ZUWU, cell biologist; b. Shanghai, China, June 1, 1939; came to U.S., 1982; m. Qin-Yu Chen, Mar. 30, 1968; 1 child, George Qi. MD, Shanghai Med. U., 1962; PhD, Chinese Acad. Scis., 1966. Rsch. assoc. Chinese Acad. Sci., Shanghai, 1966-80; postdoctoral rsch. fellow Imperial Cancer Rsch. Fund, London, 1980-82; assoc. rsch. scientist U. Tex. System Cancer Ctr., Houston, 1982-83; assoc. prof. Chinese Acad. Scis., 1983-85; rsch. scientist Fox Chase Cancer Ctr., Phila., 1987-90; acting scientific dir. Western Pa. Hosp. Found. Rsch. Inst., Pitts., 1990-91, dir. rsch. dept. medicine, 1990—; vis. assoc. prof. Med. Coll. Pa., 1986-87; mem. rsch. bd. advisors Am. Biophysical Inst., N.C., 1989—, Internat. Biophysical Ctr., Cambridge, England, 1989—. Contbr. articles to profl. jours. Mem. AAAS, Am. Soc. Cell Biology, Am. Assn. Cancer Rsch., British Soc. Cell Biology, British Assn. Cancer Rsch. Achievements include pharmacological study of anti-cancer drug Methoxyl Sarcolysin; application of image processing in SEM investigation of experimental tumor matastasis, synovicyte chemotaxis and migration. Office: Western Pa Hosp Found 4800 Friendship Ave Pittsburgh PA 15224

WANG, BOR-JENQ, mechanical engineer; b. Taipei, Taiwan, Republic of China, Oct. 15, 1959; came to U.S., 1984; s. Wan-Peng and Shu-Mei (Chen) W. BS, Nat. Taiwan U., Taipei, Republic of China, 1982; MS, SUNY, Buffalo, 1985; PhD, MIT, 1990. Rsch. asst., teaching asst. dept. mech. engring. SUNY, Buffalo, 1984-85; rsch. asst. Lab. for Mfg. and Productivity at MIT, Cambridge, Mass., 1986-91; sr. rsch. engr. Packard Electric Div.-GM, Warren, Ohio, 1991—; asst. to Dr. Irving H. Shames for engring. textbook publ., 1985. Contbr. articles to profl. publs. Recipient Book Coupon award Nat. Taiwan U., 1982, Chiang Kai-Shek scholarship, 1982, Edison medal Edison Nature Sci. Found., 1978.

WANG, BOR-TSUEN, mechanical engineering educator; b. Taipei, Taiwan, Republic of China, Oct. 24, 1961; s. Shen-Teh and Hsin (Cheng) W.; m. Meng Jane Hwung, June 3, 1989; 1 child, Vinson. BS, Tamkang U., 1985; MS, Va. Poly. Inst. and State U., 1988, PhD, 1991. Assoc. prof. Nat. Pingtung Poly. Inst., Pingtun, Taiwan, China, 1991—. Contbr. articles to profl. jours. 2nd lt. Chinese Army, 1981-83, Taiwan. Mem. AIAA, ASME, Chinese Soc. Mech. Engring., Acoustical Soc. of Am., Engring. Soc. for Advanced Mobility Land Sea Air and Space. Buddhist. Achievements include rsch. on computer design and intelligent material structure systems. Office: Nat Pingtung Poly Inst, Pingtung 91207, Taiwan

WANG, CHAO-CHENG, mathematician, engineer; b. Peoples Republic of China, July 20, 1938; came to U.S., 1961; s. N.S. and V.T. Wang; m. Sophia C.L. Wang; children: Ferdinand, Edward. BS, Nat. Taiwan U., 1959; PhD, Johns Hopkins U., 1965. Registered profl. engr., Tex. Asst. prof. Johns Hopkins U., Balt., 1966-68, assoc. prof., 1968-69; prof. Rice U., Houston, 1968-79, Noah Harding prof., 1979—, chmn. math. sci. dept., 1983-89, chmn. mech. engring. and materials sci., 1991—. Author numerous books in field; contbr. articles to profl. jours. Named Disting. Young Scientist Md. Acad. Sci., 1968. Mem. ASME, Soc. Natural Philosophy (treas. 1985-86), Am. Acad. Mechs. Office: Rice Univ Dept Mech Engring and Materials Sci Houston TX 77251

WANG, CHIA PING, physicist, educator; came to U.S., 1963, naturalized; (parents Chinese citizens).; s. Guan Can and Tah (Lin) W. BS, U. London, 1950; MS, U. Malaya, 1951; PhD in Physics, U. Malaya and U. Cambridge, 1953; DSc in Physics, U. Singapore, 1972. Asst. lectr. U. Malaya, 1951-53; mem. faculty Nankai U., Tientsin, 1954-58, prof. physics, 1956-58, head electron physics div., 1955-58, mem. steering com. nuclear physics div., 1956-58, head electron physics Lanchow Atomic Project, 1958; mem. faculty Hong Kong U.; mem. faculty Chinese U., Hong Kong, 1958-63, prof. physics, 1959-63, acting head physics, math. depts., 1959; rsch. assoc. lab. nuclear studies Cornell U., Ithaca, N.Y., 1963-64; assoc. prof. space sci. and applied physics Cath. U. Am., Washington, 1964-68; assoc. prof. physics Case Inst. Tech. Case Western Res. U., Cleve., 1966-70; vis. scientist, vis. prof. U. Cambridge (Eng.), U. Leuven (Belgium), U.S. Naval Rsch. Labs., U. Md., MIT, 1970-75; rsch. physicist radiation lab. U.S. Army Natick (Mass.) R & D Command, 1975—; pioneer in fields of nuclear sub-structure (now often referred to as parton), nucleon sub-unit structure, multiparticle prodn., cosmic radiation, picosecond time to pulse-height conversion, thermal physics, lasers, microwaves. Contbg. author: Atomic Structure and Interactions of Ionizing Radiations with Matter in Preservation of Food by Ionizing Radiation, 1982; contbr. numerous articles to profl. jours. Recipient Outstanding Performance award Dept. Army, 1980, Quality Increase award, 1980. Mem. Am. Phys. Soc., Inst. Physics London, N.Y. Acad. Scis., AAAS, Sigma Xi. Home: 28 Hallett Hill Rd Weston MA 02193-1753 Office: US Army Natick Rsch and Devel Ctr Natick MA 01760

WANG, CHIH CHUN, chemist, scientific administrator; b. Beijing, Oct. 9, 1932; came to U.S.; s. W.H. and Hilda (Yang) W.; m. Lena Y. Liu, Dec. 1959 (dec. Dec. 1967); 1 child, Joyce; m. Betty R. Tung, Mar. 29, 1969; children: Francis, Bessie. BS in Chem. Engring., Nat. Taiwan U., 1955; MS in Chem. Engring., Kans. State U., 1959; PhD in Phys. Chemistry, Colo. State U., 1962; postdoctoral student, U. Kans., 1963. Mem. tech. staff RCA Labs. div. RCA Corp., Princeton, N.J., 1963-67, rsch. leader, 1967-73, fellow, 1973-88; v.p. Am. Lumi Corp., Cranbury, N.J., 1989-91, pres., 1991—; cons. NASA, Washington, 1978-79, Gen. Elec. Co., Milw., 1989-90; joint mng. dir. Phosphor Tech. Ltd., Essex, Eng., 1989-90. Author, editor: Heteroepitaxial Semiconductor Devices, 1978; author: Characterization of Semiconductors, 1979; contbr. numerous articles to profl. jours.; inventor, patentee more than 20 U.S. patents. Mem. ACS, Electrochem. Soc. for Info. Display (div. editor Jour. 1978-91), Am. Assn. in Crystal Growth, Sigma Pi Sigma, Sigma Xi. Home: 41 Maple Stream Rd East Windsor NJ 08520 Office: Am Lumi Corp 2525 Rt 130 Bldg B Cranbury NJ 08512

WANG, DAHONG, architect, consultant; b. Peking, People's Republic of China, July 6, 1919; s. Chung-hui and Soulian (Yang) W.; m. Meili Lin; children: Elaine, Shouchang, Daniel, Eean, Patrick Jay-one. MArch, Harvard U., 1943; MA, Cambridge U., 1950. Registered class A architect. Dir. D. Wang & Assocs., Taipei, Taiwan, 1952—; cons. Tourism Bur., Taipei, 1980—, Rapid Transit System, Taipei Mcpl. Govt., 1988—. Designer man's first moon landing Selene Monument, 1969; translator (into Chinese) The Picture of Dorian Gray. Mem. Shanghai City Planning Bd., 1946. Recipient Cambridge Archtl. Soc. award, 1937, GM Competition award,

1945. Mem. Taipei Architects' Assn., Cambridge Soc., Sino-Brit. Cultural and Econ. Assn. Avocations: writing, swimming, skiing, music, fencing. Office: Rm 1 10th Fl, 197 Chung Hsiao E Rd Sec 4, Taipei 105, Taiwan

WANG, FRANKLIN FU YEN, materials scientist, educator; b. Macau, China, Sept. 19, 1928; s. Kwan Hsien and Shu Tsiang (Hsiang) W.; m. Katharine Wang, Sept. 8, 1956; children—Jennifer, Alexander. B.A., Pomona Coll., 1951; M.S., U. Toledo, 1953; Ph.D., U. Ill., 1956. Dir. research Glascote Products, Cleve., 1956-58; research scientist A.O. Smith Corp., Milw., 1958-61; research staff mem. Sperry Rand Research Center, Sudbury, Mass., 1961-66; asso. prof. dept. materials sci. SUNY, Stony Brook, 1966-72; prof. SUNY, 1972—, chmn., 1971-74. Author: Introduction to Solid State Electronics, 1980; editor: Materials Processing: Theory and Practices, 1980; prin. editor: Materials Letters, 1982. Fellow Am. Inst. Chemists, Am. Ceramic Soc.; mem. IEEE, Am. Soc. Engring. Edn., Am. Chem. Soc., Assn. Crystal Growth, Am. Phys. Soc., Engring. Sci. Soc., Materials Research Soc. Office: NSF Math & Phys Scis 1800 G St NW Washington DC 20550

WANG, I-TUNG, atmospheric scientist; b. Peking, People's Republic of China, Feb. 16, 1933; came to U.S., 1958; s. Shen and Wei-Yun (Wen) W.; m. Amy Hung Kong; children: Cynthia P., Clifford T. BS in Physics, Nat. Taiwan U., 1955; MA in Physics, U. Toronto, 1957; PhD in Physics, Columbia U., 1965. Rsch. physicist Carnegie-Mellon U., Pitts., 1965-67, asst. prof., 1967-70; environ. systems engr. Argonne (Ill.) Nat. Lab., 1970-76; mem. tech. staff Environ. Monitoring and Svcs. Ctr. Rockwell Internat., Creve Coeur, Mo., 1976-80, Newbury Park, Calif., 1980-84; sr. scientist, combustion engr. Environ. Monitoring and Svcs. Inc., Newbury Park, Camarillo, 1984-88; sr. scientist ENSR Corp (formerly ERT), 1988; pres. EMA Co., Thosand Oaks, Calif., 1989—; tech. advisor Bur. of Environ. Protection, Republic of China, 1985; environ. cons. ABB Environ., 1989-92, ARCO, 1990-91, Du Pont (SAFER Systems Divsn.), 1992—. Contbr. papers to profl jours. First violin Conejo Symphony Orch., Thousand Oaks, Calif., 1981-83. Grantee Bureau of Environ. Protection, Taiwan, 1985. Mem. N.Y. Acad. of Scis., Air and Waste Mgmt. Assn., Sigma Xi. Avocations: violin and chamber music. Office: EMA Co Ste 435 2219 E Thousand Oaks Blvd Thousand Oaks CA 91362-2921

WANG, JAI-CHING, physics educator; b. Taipei, Taiwan, May 20, 1940; came to U.S., 1968; s. Jim-Bin and Tzu (Liou) W.; m. Shou-Mei Chen, May 14, 1972; 1 child, Amy. MST, U. Wis., Superior, 1969; MS, U. Mass., 1972, PhD, 1976. Tchr. Wan-Hou Girls' High Sch., Taipei, 1963-65; instr. Tam-Kang Coll. Arts & Scis., Taipei, 1965-68; asst. prof. Physics Ala. A&M U., Normal, 1976-82, assoc. prof. Physics, 1982—, interim chair Physics, 1990-91, asst. chair Physics, 1992—; summer faculty rsch. fellow NASA/ASEE, Huntsville, Ala., 1982-84, 92; physicist U.S. Army, Redstone, Ala., 1986. Author: Magnetic Properties of the Itinerant Femimagnets at Low Temperatures, 1977. Prin. Huntsville Chinese Sch., 1983—; pres. Huntsville Chinese Assn., 1980 (Achievement award 1987), advisor Soc. Physics Students, Ala. A&M U., 1982— (Appreciation award 1985). Recipient Disting. Svc. award Constitution Bicentennial Commn., Madison, Ala., 1987. Mem. Am. Physics Soc., Am. Assn. for Crystal Growth, Sigma Pi Sigma. Office: Ala A&M U PO Box 28 Normal AL 35762

WANG, JAMES CHUO, biochemistry and molecular biology educator; b. Kiangsu, China, Nov. 18, 1936; came to U.S., 1960; s. Chin and H.-L. (Shih) W.; m. Sophia Shu-lan Hwang, Dec. 23, 1961; children: Janice S., Jessica A. BS, Nat. Taiwan U., 1959; MA, U. S.D., 1961; PhD, U. Mo. Coll. Arts and Sci., 1964. Asst. instr. Nat. Taiwan U., Taipei, 1959-60; rsch. fellow in chemistry Calif. Inst. Tech., Pasadena, 1964-66; asst. prof. chemistry U. Calif. at Berkeley, 1966-69, assoc. prof., 1969-74, prof., 1974-77; prof. biochemistry and molecular biology Harvard U., Cambridge, Mass., 1977-88, Mallinckrodt prof. biochemistry and molecular biology, 1988—; Chancellor's Disting. lectr. U. Calif., Berkeley, 1984; mem. molecular biology study sect. NIH, 1988-91, chair, 1990-91; disting. faculty lectr. U. Tex., M.D. Anderson Cancer Ctr., 1989. Mem. editorial bd. Quar. Rev. Biophysics, 1988—. Guggenheim fellow Guggenheim Found., 1986-87; recipient Disting. Alumnus award U. Mo. Coll. Arts and Scis., 1991. Fellow Am. Acad. Arts and Scis.; mem. NAS (molecular biology award 1983), Third World Acad. Scis., Academia Sinica (Taipei). Office: Harvard U Dept Biochem & Mo Bio 7 Divinity Ave Cambridge MA 02138-2092

WANG, JIAN GUANG, engineering physicist; b. Qian-Xian, China, Nov. 15, 1944; s. Zhong-Zuo and Pei-lan Wang; m. Mei-li Chen, Apr. 27, 1975; 1 child, Andrew Yu. Diploma, Tsing-Hua U., 1968; MSE in Nuclear Engring., U. Mich., 1987, PhD in Nuclear Engring., 1989. Gen. mgr. Xian (China) Radio Co., 1969-78; dep. head rsch. group Inst. of High Energy Physics, Beijing, 1978-86; vis. scholar Rutherford-Appleton Lab., Oxford, Eng., 1981-82; vis. scientist Stanford (Calif.) Linear Accelerator Ctr., 1985-86; mem. rsch. faculty U. Md. Lab. for Plasma Rsch., College Park, 1989—. Contbr. articles to profl. jours. Mem. Am. Phys. Soc., IEEE. Achievements include design and development of radio electronic, nuclear electronic, and radiation detection instruments; high power microwave generation from mechanism of free electron cyclotron masers; electron beam transport and diagnostics, particle accelerators, beam physics. Office: U Md Lab for Plasma Rsch College Park MD 20742

WANG, JIAN-MING, research optics scientist; b. Zhaozhou, China, May 23, 1963; came to U.S., 1989; s. Xuchen and Shuqin (Zhang) W.; m. Shu-Yu Wang, Nov. 20, 1987; 1 child, Seraphina T. MS, Inst. Optics, Changchun, China, 1985; PhD with very high honors, U. Paris-Sud, 1989. Rsch. asst. scientist Inst. Optics, 1985; rsch. asst. U. Paris-Sud, 1986-89; rsch. assoc. U. So. Calif., L.A., 1989-92; rsch. scientist InterDigital Telecom., Port Washington, N.Y., 1992—. Contbr. articles to Applied Optics, Photonic Switching Proc. Mem. Optical Soc. Am., Optical Soc. France. Home: 3 Carla Ct Huntington Station NY 11746 Office: InterDigital Comm Corp 833 Northern Blvd Great Neck NY 11021

WANG, JIAN-SHENG, materials scientist; b. Taihe, Jiangxi, China, July 7, 1941; came to U.S., 1980; s. Shuo-Ru and Feng-Zi (Liu) W.; m. Hui-Jian Liu, Aug. 17, 1963; children: Yi-Bin, Yi-Min. MS, Northwestern Poly. U., China, 1963; PhD, Stanford U., 1985. Rsch. engr. Iron and Steel Rsch. Inst., Beijing, 1963-80; rsch. asst. Stanford (Calif.) U., 1982-85; sr. rsch. assoc. Harvard U., Cambridge, Mass., 1985—; vis. scholar Stanford U., 1980-82; vis. scientist Max-Planck Inst. für Eisenforschung, Düsseldorf, Germany, 1989-90. Co-author: Steels for Power Stations, 1982; contbr. articles to Jour. Materials Rsch., Acta Metallurgica et Materialia, Scripta Metallurgica et Materialia, Materials Scis. and Engring., New Metallic Materials, Acta-Scripta Metallurgica Proceedings Series, Sagamore Army Materials Research Conference Proceedings. Recipient Nat. Prize sci. and technol. innovation China Nat. Com. Sci. and Tech., Beijing, 1981, Honorable Documents, Ibid, Beijing, 1978. Mem. Materials Rsch. Soc., Am. Soc. Metals, Am. Physics Soc. Achievements include patents for a low-alloying heat-resistant steel with the creep property compariable to A1S1 317, chromium-maganese and maganese-aluminum type nickel free heat resistant steels; notable findings in the relation between embrittlement sensitivity and surface and interface segregations of impurities, effects of the mobility of segregant atoms, the directional dependence of ductile versus brittle response of the interface in bi-materials and bi-crystals, the effects of the loading phase angle on the fracture bahaviour of interfaces in bi-materials, the alloying principle of low alloying heat-resistant steels, the mechanism of low-carbon Bainite phase transformation and its creep-strengthening effect. Home: 52 Albion St Somerville MA 02143 Office: Harvard Univ 29 Oxford St Cambridge MA 02138

WANG, JI-PING, pharmacist, researcher; b. Tang-Shan, Hebei, China, Sept. 13, 1948; came to U.S., 1981; d. Shao-Hsun and Liang-Shou (Hu) W.; m. Si Luo, Feb. 16, 1978; 1 child, Lai Laura Luo. BS in Pharmacy, U. Wash., 1987, MS, 1992. X-ray technician Xin-Xian Hosp., Shanxi, People's Republic of China, 1972-73; pharmacist Drug Inspection Inst., Beijing, People's Republic of China, 1978-81; rsch. technician U. Wash., Seattle, 1985-87; pharmacist Luke's Pharmacy, Seattle, 1988-91; rsch. scientist U. Washington, Seattle, 1991—. Contbr. articles to Jour. of Chromatography, Pharm. Rsch., Clin. Pharmacology and Therapeutics. Recipient Scholastic Achievement award U. Wash., 1985, Merck award, 1986. Mem. Am. Pharm. Assn., Rho Chi (Achievement award 1985). Office: U Washington BG-20 Seattle WA 98195

WANG, JI-QING (CHI-CHING WONG), acoustician, educator; b. Shanghai, China, Dec. 9, 1929; m. Frances Ya-Sing Tsu Wang, May 11, 1958; children: Tony, Pearl. BA, Hangchow U., 1951. From asst. prof. to assoc. prof. Tongji U., Shanghai, 1954-78, prof., 1979—; head Archtl. Acoustics and Noise Control Group, Shanghai, 1978-90; dep. dir. Inst. Acoustics, Shanghai, 1979-90; chmn. Grad. Program Archtl. Sci., Shanghai, 1986—; pres. Acoustical Cons. Assocs., Shanghai, 1984—; bd. regents Tongji U., 1986—; chief acoustical cons. for most major auditoriums and theatres in China. Author: Sound Insulation and Noise Abatement in Architectural Design, 1959, Sound Systems in Auditorium Practice, 1980; editor, author Handbook of Architectural Acoustic Design, 1987; co-editor in chief Technical Acoustics, 1986—; contbr. articles to Jour. Acoustical Soc. Am., Acta Acustica, Chinese Jour. of Acoustics. Recipient 2d Place award Ministry of Planning and Construction, Beijing, 1989; grantee Nat. Sci. Found., Beijing, 1985-89, Nat. Acad. Grants, Beijing, 1982-84, Nat. Com. for Acoustical Standardization, Beijing, 1981-91. Mem. Acoustical Soc. Am., Acoustical Soc. Shanghai (pres. 1987-91), Acoustical Soc. China (exec. coun. 1988—), Nat. Com. for Acoustical Standardization (vice-chmn. subcom. bldg. acoustics 1981—). Achievements include establishment of several acoustical standards in China, noise criteria of impact to the community, room acoustical design criteria for auditoriums, improvement of sound insulation for lightweight walls. Home: 59 Fu-shing Rd W, Shanghai 200031, China also: 160 Marietta Dr San Francisco CA 94127 Office: Tongji U Inst Acoustics, 1239 Si-pin Rd, Shanghai 200092, China

WANG, JOSEPH JIONG, astrophysicist; b. Shanghai, China, Aug. 14, 1962; came to U.S., 1985; s. Kezhong and Rongyu (Wan) W. B in Engring., Tsinghua U., 1985; SM, MIT, 1988, PhD, 1991. Mem. tech. staff Jet Propulsion Lab., Calif. Inst. Tech., Pasadena, 1991—. Contbr. articles to profl. jours. including Phys. Fluids, Jour. Geophys. Res., Jour. Spacecraft Rockets, AIAA Jour., Geophys. Res. Letters, Rsch. Aeras spacecraft-plasma interactions, space plasma physics, parallel computing. Mem. Am. Geophys. Union, Am. Phys. Soc., Sigma Xi. Achievements include presentation of a complete theory on dynamic coupling between hypersonic plasma flow and large, high-voltage object on ion-plasma-time scale; discussion of the plasma wake formation and structure in the large-dimension/high voltage regime; development of 3D electromagnetic PIC codes for MIMD parallel computers. Office: Jet Propulsion Lab MS301-460 4800 Oak Grove Dr Pasadena CA 91109

WANG, KEVIN KA-WANG, pharmaceutical biochemist; b. Hong Kong, May 30, 1961; came to U.S., 1989; s. Howe and Alice (Chan) W.; m. Alice Ip; 1 child, Jonathan. BS in Biochemistry, U. Guelph, 1984; PhD in Pharm. Scis., U. B.C., 1989. Rsch. assoc. Wayne State U., Detroit, 1989-91; sr. scientist Parke-Davis Pharm. Rsch. Warner-Lambert Co., Ann Arbor, Mich., 1991-92, rsch. assoc., exec. mgr., group leader, 1992—; co-chmn. Biomed. Rsch. Coun. Symposiums, U. Mich., Ann Arbor, 1992; invited speaker Internat. Scientific Symposium, 1988—. Author: Principles of Medical Biology, 1993; contbr. chpts. to books and articles to profl. jours. Ontario scholar, 1980; U. Grad. fellow, U. B.C., 1987-88, Am. Heart Assn. Rsch. fellow, 1989-91. Mem. Can. Soc. Chem. Tech. (Merit award 1985), Chem. Inst. Can. (Silver medal 1984), N.Y. Acad. Sci., Soc. Neurosci. Achievements include development of selective calpain inhibitors (patent pending). Office: Parke-Davis/Warner-Lambert 2800 Plymouth Rd Ann Arbor MI 48105

WANG, KUO CHANG, aerospace engineering educator; b. Anhui, China, Nov. 9, 1924; m. Betty Shih-Pi Hua, June 15, 1958; children: Edward, Joyce, Philip. PhD, Renssaler Poly. Inst., 1961. Engr. Chinese Arsenals, 1948-53; rsch. scientist Balt. div. Martin Co., 1961-67, Rsch. Inst. for Advanced Studies, Martin-Marietta Corp., Balt., 1967-73; sr. rsch. scientist Martin Marietta Labs., Balt., 1973-79; prof. San Diego State U., 1980—. Contbr. articles to profl. jours. Rsch. grant Air Force Office of Scientific Rsch., 1970-85, Office of Naval Rsch., 1977-82, Army Rsch. Office, 1972-75. Mem. AIAA (tech. achievement award 1981), Sigma Xi, Kappa Phi. Achievements include discovery of open separation concept; pioneering rsch. in three-dimensional laminar boundary layer and separated flow structure, lifting-line theory of low-aspect-ratio wings, dissociation and thermal radiation effects in high temperature gas flow. Home: 7273 Alliance Ct San Diego CA 92119 Office: San Diego State U Dept Aerospace Engring San Diego CA 92182

WANG, KUO-KING, manufacturing engineer, educator. BSME, Nat. Ctrl. U., China, 1947; MSME, U. Wis., 1962, PhD in Mech. Engring., 1968. Sibley prof. mech. engring. Cornell U., Ithaca, N.Y.; founder, dir. Cornell Injection Molding Program, 1974—; cofounder Cornell Mfg. Engring. and Productivity Program, Advanced CAE Tech., Inc. Recipient Disting. Svc. citation U. Wis., 1990. Fellow ASME (Blackall Machine Tool and Gage award 1968, William T. Ennor Mfg. Tech. award 1991), Soc. Mfg. Engrs. (Frederick W. Taylor Rsch. medal 1987); mem. CIRP, ASM Internat., Nat. Acad. Engring., Am. Welding Soc. (Adams Meml. Membership award 1976), Soc. Plastic Engrs., Polymer Processing Soc. Achievements include pioneering research in friction welding and applications of solid modeling to CAD/CAM. Office: Cornell Univ Dept Mech/Aero Engring Upson Hall Ithaca NY 14853*

WANG, LIANG-GUO, research scientist; b. Foochow, People's Republic of China, Apr. 23, 1945; parents Chi-hsi Wang and Yunqing Chen; m. Shu-fen Zhang, Sept. 27, 1977; children: Zhijing, Zhijian. BS in Physics, Peking U., Beijing, 1969; MS in Physics, Ohio State U., 1983, PhD in Physics, 1986. Tech. mgr. and electronics engr. various cos., People's Republic of China, 1971-78; rsch. asst. Inst. of Academia Sinica, Beijing, 1978-80, U. Ky., Lexington, 1981, Ohio State U., Columbus, 1981-86, U. Va., Charlottesville, 1987-89; rsch. scientist Coll. of William and Mary, Williamsburg, Va., 1989—; cons. NASA Langley Rsch. Ctr., Hampton, Va., 1989—. Contbr. articles to profl. jours. Recipient Pub. Svc. medal NASA, 1992. Mem. Am. Phys. Soc., Am. Geophys. Union, Optical Soc. Am., Internat. Soc. for Optical Engring, Photonics Soc. of Chinese-Ams. Achievements include patents in field; research and development of highly sensitive frequency modulation laser spectrometers, advanced laser and passive radiometer instruments for airborne and spaceborne applications. Home: 4 Poulas Ct Hampton VA 23669-1863 Office: NASA Langley Rsch Ctr M/S 468 Hampton VA 23665

WANG, LIN, physicist, computer science educator, computer software consultant; b. Dandong, China, June 11, 1929; came to U.S., 196l, naturalized, 1972; s. Lu-Ting and Shou-Jean (Sun) W.; m. Ingrid Ling-Fen Tsow, July 8 1961; children: W. Larry, Ben. BS in Physics, Taiwan U., 1956; MS in Physics, Okla. State U., 1965, PhD in Physics, 1972. Mem. physics faculty Cheng Kung U., Tainan, China, 1957-6l; asst. prof. physics Southwestern Okla. State U., Weatherford, 1965-72; prof., chmn. physics dept. N.E. Coll. Arts and Sci., Maiduguri, Nigeria, 1973-75; mem. tech. staff Pacific Engring. Corp., Bellevue, Wash., 1976-78; sr. software engr., Far East cons. Electro-Sci. Industries, Inc., Portland, Oreg., 1979-82; mem. sr. computer sci. faculty South Seattle C.C., 1983—. Mem. Assn. for Computing Machinery, Am. Phys. Soc., AAUP. Avocations: classical music, world travel. Home: 9322 168th Pl NE Redmond WA 98052

WANG, LING DANNY, research scientist; b. Nanking, Republic of China, Jan. 5, 1948; came to U.S., 1973; s. Gibbs and Chin-Inn Hall; m. Denise King, 1981; children: Grace, Ray. BS, Fu-Jen U., Taipei, Taiwan, 1971; MS, U. Wis., 1974; PhD, U. Nev., 1982. Rsch. asst. Inst. of Zoology, Taipei, 1972-73; postdoctoral fellow U. Pa., Phila., 1982-85; rsch. assoc. Temple U., Phila., 1985-87; assoc. rsch. fellow Inst. of Biomed. Scis., Taipei, 1987—; coord. cardiovascular div. Inst. of Biomed. Scis., Taipei, 1989—; mem. various orgnl. scientific com., 1987—. Contbr. articles to profl. jours. including Biophys. Biochem. Rsch. Com., Thrombosis Rsch., Fibrinegen, Thombosis, Coagulation and Fibrinalaysis, Blood. With USMC, 1971-72. Recipient Svc. award NIH, 1982-85; fellowship Am. Heart Assn., 1986-87; grantee Nat. Sci. Coun., 1987—. Mem. Am. Chemistry Soc., Chinese Cell Biology Soc., Sigma Xi. Home: 200 Larchwood Rd Springfield PA 19064 Office: Inst of Biomed Scis, Academia Sinica, Taipei 11529, Taiwan

WANG, LIQIN, materials scientist; b. Hangzhou, China, Dec. 28, 1954; s. Zhushen and Jianqin (Tao) W.; m. Junhui Li, Jan. 28, 1982; 1 child, Xiaofan Wang. MS, Shanghai Jiaotong U., 1984; PhD, U. Md., 1992. Rsch. assoc. Xian Jiaotong U., Xian, China, 1984-87; coord. U. Md., College Park, 1990-92, rsch. scientist, 1992–; cons. Chi Assocs., Arlington, Va., 1990-92. Contbr. articles to profl. jours. Mem. ASM, Phi Kappa Phi. Achievements include one patent. Office: U Md Dept Materials and Nuclear College Park MD 20742

WANG, PAUL WEILY, physics educator; b. Kao-Hsiung, Taiwan, Republic of China, Nov. 4, 1951; came to U.S., 1979; s. Yao Wen Wang and Yue Hua Lo; m. Diana Chung-Chung Chow, June 9, 1979; children: Agnes J., Carol H., Alfred Z. PhD, SUNY, Albany, 1986. Rsch. asst. prof. Vanderbilt U., Nashville, 1986-90; asst. prof. U. Tex., El Paso, 1990—; cons. EOTec Inc., 1987-88. Contbr. articles to Jour. Applied Physics, Nuclear Instru. & Meth., Springer Series in Surface Scis., Jour. Non-Crys. Solids, Lasers. Fellow Inst. for the Study of Defects in Solids; mem. Am. Optics Soc., Am. Phys. Soc. Achievements include development of iron in silicon gettered by thermally grown silicon dioxide thin film, of hydrogen migrate toward defects in silicon during annealing; discovery that lifetimes of luminescence centers in silica are stimulated by particle bombardments, that defects introduced by gamma-ray radiation enhance the luminescence in silica, development of defects creation mechanism in silica by 5 and 50 eV photons, discovery of silver diffuses and precipitates thermally on the surface in ion exchange sodium calcium silicate glass. Home: 6890 Orizaba Ave El Paso TX 79912-2324 Office: U Tex Dept Physics El Paso TX 79968

WANG, PETER ZHENMING, physicist; b. Quanzhou, Fujian, People's Republic of China, Nov. 30, 1940; came to U.S., 1983; s. Guohua and Shunhua (Chen) W.; m. Grace Ruhui Xu, Mar. 14, 1967; children: Yili, Yile. MS, Qinghua U., Peking, People's Republic of China, 1964; postgrad., U. Tex., Dallas, 1983-84. Sr. engr. Particle Accelerator Inst., Shanghai, 1964-83; electronic engr. Benchmark Media Systems, Inc., Syracuse, N.Y., 1984-87; project engr. McGaw Inc., Carrollton, Tex., 1988—; physicist High Energy Physics Inst., Peking, 1978-79. Co-author: (book) Vacuum World, 1984. Tchr. Bible study, Plano, Tex., 1990. Baptist. Achievements include research and design of a variety of proton and electron accelerators for low energy nuclear physics experiments, industries and hospitals; design of 50 GEV proton synchrotron, design of audio distribution amplifiers and consoles for BBC, ABC, and other TV and radio stations; development and design of air bubble detector, pressure transducer and noise reduction solution for the microprocessor based infusion therapy instrument used in hospitals. Home: 1510 Chesterfield Dr Carrollton TX 75007-2847 Office: McGaw Inc 1601 Wallace St Carrollton TX 75006-6652

WANG, QUNZHEN, mechanical engineer; b. Qingzhou, China, June 21, 1965; came to U.S., 1987; s. Laisui and Yuxiang (Zhao) W.; m. Lin Tian, Aug. 24, 1990. MS, Mich. Tech. U., 1989; postgrad., Pa. State U., 1989—. Rsch. asst. Mich. Tech. U., Houghton, 1987-89, Pa. State U., State College, 1989—. Contbr. articles to profl. jours. Fellow Li Found., Inc., San Francisco, 1987, Chinese Govt., 1985. Mem. ASME, Am. Phys. Soc. Home: 504 Elm Rd State College PA 16801 Office: Pa State U B12 Hallowell Bldg State College PA 16802

WANG, RAN-HONG RAYMOND, optical engineer, scientist; b. Tai Chung, Taiwan, Republic of China, May 31, 1957; came to U.S., 1986; s. Tsing-Thu and Mi (Chaug) W.; m. Min-Shine Chow, Jan. 6, 1985; children: Patrick, Allen. BS, Soo Chow U., Taipei, Taiwan, 1979; MS, Tsing Hwa U., Hsingchu, Taiwan, 1981, U. Ariz., 1988; PhD, U. Ariz., 1990. 2d lt. Chinese Army, Taiwan, Republic of China, 1981-83; supr., asst. Melles Griot Ky., Hsingchu, Taiwan, Republic of China, 1983-86; grad. student U. Ariz., Tucson, 1986-90; sr. devel. scientist Melles Griot Inc., Irvine, Calif., 1990—. V.p. Chinese Student Assn., Tucson, 1988-89. Recipient Phi Tau Phi Honors award, 1981. Mem. Optical Soc. Am., Am. Mgmt. Assn., Internat. Soc. Optical Engring., Optical Engring. Soc. of Rep. of China, Photonics Soc. Chinese-Ams. Office: Melles Griot 1770 Kettering Irvine CA 92714-5670

WANG, RUQING, research physicist; b. Yunnan, China, Jan. 20, 1941; s. Wenxing and Daishi Wang; m. Qingmei Zhang, Mar. 21, 1973; children: Quan, Xu. BS in Physics, Yunnan U., Kunming, China, 1962; PhD in Physics, Kingston U., Eng., 1993. Rsch. asst. prof. Dalian (Liaoning) Inst. Chem. Physics, Chinese Acad. Scis., 1962-86; guest scientist Nat. Bur. Standards, Gaithersburg, Md., 1986-88, Nat. Phys. Lab., Teddington, Middlesex, Eng., 1988-91, Kingston U., Surrey, Eng., 1988-91; rsch. worker Nat. Rsch. Coun., Ottawa, Can., 1991-92; rsch. scientist Cross Cancer Inst., Edmonton, Alta., Can., 1993—. Contbr. articles to profl. jours. Mem. Chinese Phys. Soc., Chinese Optical Soc., Chinese Soc. Computational Physics, Am. Phys. Soc.

WANG, SEMYUNG, mechanical engineer, researcher; b. Seoul, Republic of Korea, Oct. 1, 1957; s. Hacksoo and Kiyeon (Park) W.; m. Kyunghee Shin, Sept. 19, 1982; 1 child, Eunice. BS, Hanyang U., Seoul, 1980; MS, Wayne State U., 1986; PhD, U. Iowa, 1991. Design engr. Daewoo Heavy Industries, Inchon, Korea, 1982-83; rsch. asst. Wayne State U., Detroit, 1984-87; rsch. asst. Ctr. for computer aided design U. Iowa, Iowa City, 1988-90, project leader Ctr. for computer aided design, 1991—; adj. asst. prof. mechanical engring. dept. U. Iowa, 1992—. Contbr. articles to profl. jours. Mem. AIAA, ASME, Soc. Automobile Engrs. Baptist. Office: Univ Iowa Ctr for Computer Aided Design Iowa City IA 52242

WANG, SHIH-LIANG, mechanical engineering educator, researcher; b. Taipei, Taiwan, May 27, 1955; s. Chin-chin and Yu-Hua (Wang) W.; m. Yue-Min Wang, May 16, 1981; children: Harris, Christine. BS in Mech. Engring., Nat. Tsing Hua U., Taiwan, 1977; PhD, Ohio State U., 1986. Registered profl. engr. Engr. Indsl. Tech. Rsch. Inst., Hsingchu, Taiwan, 1979-81; rsch. assoc. Ohio State U., Columbus, 1983-86; asst. prof. dept. mech. engring. N.C. A&T State U., Greensboro, 1986-91, assoc. prof., 1991—; vis. scholar Oak Ridge Nat. Lab., 1989; cons. Sterling Co., Greensboro, N.C., 1991. Contbr. articles to ASME Jour. Mech. Design, Internat. Jour. Robotics and Automation. 2d lt. Taiwan Army, 1977-79. DuPont fellow, 1990. Mem. ASME (sec. Carolina sect. 1991-92), Sigma Xi. Achievements include pioneering research on robots with linkage structures.

WANG, SHI-QING, physicist; b. Peking, China, Oct. 6, 1959; s. DaLai Wang and Yiyi Chen; married; children: Janice G., Alicia D. BS, Wuhan U., 1982; PhD, U. Chgo., 1987. Postdoctoral fellow UCLA, 1987-89; asst. prof. Case Western Res. U., Cleve., 1989—. Office: Case Western Res Univ 10900 Euclid Ave Cleveland OH 44106

WANG, SHU-SHAW (PETER WANG), educator; b. Taichung, Taiwan, Republic of China, May 31, 1953; came to U.S., 1984; s. Chi-Jen and Gui-Jan (Tswai) W.; m. Mei-Lien Su Wang, May 22, 1983; children: Elaine Ying, Joanne Ying. PhD in Elec. Engring., U. Tex., Arlington, 1991. Rsch. engr. Chung-Shin Inst. Sci. and Tech., Taiwan, 1979-84; faculty assoc. circuit theory U. Tex., Arlington, 1991—. Mem. IEEE, Optical Soc. Am., Sigma Xi (outstanding doctoral rsch. award N.C. 1992), Tau Beta Pi. Achievements include patent for optical guided-mode resonance filters. Office: U Tex at Arlington Room 518 Box 19016 Univ Of Texas At Arlington TX 76019

WANG, TAYLOR GUNJIN, science administrator, astronaut, educator; b. Shanghai, China, June 16, 1940; came to U.S., 1963; m. Beverly Fung, 1966; children: Kenneth, Eric. BS, UCLA, 1967, MS, 1968, PhD, 1971. Mgr. microgravity sci. and applications program Jet Propulsion Lab., Pasadena, Calif., 1972-88, cons., 1987-89; Space Shuttle astronaut-scientist NASA, 1983-85; Centennial prof., dir. Ctr. for Microgravity Rsch. and Applications Vanderbilt U., Nashville, 1988—. Contbr. over 150 articles to profl. jours.; inventor living cells encapsulation as a cure of hormone deficiency states in humans; over 20 patents in field. Fellow Acoustical Soc. Am.; mem. AIAA, Am. Phys. Soc., Assn. Space Explorers-USA (pres. 1988), Sigma Xi. Office: Vanderbilt U Sta B Box 6079 Nashville TN 37235

WANG, WEN-YING LEE, systems specialist; b. Taipei, Taiwan, China, Apr. 15, 1949; came to U.S. 1971; d. Tsao-pu and May-shien (Hsu) Lee; m. Chia-Lin J. Wang, Dec. 15, 1973; children: Joann Lee, Diane Lee. MS, U. Va., 1973, Villanova U., 1982. Chemist Weston Environ. Cons. Co., West Chester, Pa., 1980-82; specialist Sci. Computing div. DuPont Co., Wilmington, Del., 1982—; instr. Del. Tech. Coll., Wilmington, 1991-92. Office: E I DuPont de Nemours PO Box 80320 Wilmington DE 19880-0320

WANG, XUEMIN, biochemistry educator; b. Hanggang, Hubei, People's Republic of China, Jan. 13, 1958; came to U.S., 1982; s. Jianglong Wang and Linxin Han; m. Ling Zheng, May 18, 1991; 1 child, Melissa Ling. MSc, Ohio State U., 1984; PhD, U. Ky., 1987. Postdoctoral researcher La. State U., Baton Rouge, 1988-91; asst. prof. Kans. State U., Manhattan, 1991—. Contbr. articles to profl. jours. Mem. AAAS, Am. Soc. Plant Physiologists. Achievements include research in biochemical pathways and regulation of phospholipid metabolism in higher plants. Office: Kans State U Dept Biochemistry Willard Hall Manhattan KS 66506

WANG, YAO, engineering educator; b. Zhejiang, China, Dec. 30, 1962; d. Chang-Fa Wang and Yu-Ying Chen; m. Hai Sang, Feb. 14, 1986. MSEE, Tsinghua U., Beijing, 1985; PhD in Elec. Engring., U. Calif., Sant Barbara, 1990. Asst. prof. Polytechnic U., Bklyn., 1990—; cons. AT&T Bell Labs., Holmdel, N.J., 1992—; vis. lectr. U. Calif., Santa Barbara, 1990. Office: Polytechnic U 6 Metrotech Ctr Brooklyn NY 11201

WANG, YIBING, physicist; b. Hunan, China, Aug. 10, 1940; m. Yonge Shan, Aug. 30, 1969; children: Lian, Hong. BS, Harbin U., China, 1964; PhD, UCLA, 1983. Rsch. assoc. China Inst. Radiation Protection, Taiyuan, China, 1964-80, Dept. Radiol. Scis., U. Calif., Irvine, 1982-83; assoc. prof. China Inst. Radiation Protection, Taiyuan, 1983-85; prof. med. physics Beijing 514 Hosp., China, 1985—; med. physicist Wadsworth VA Hosp., L.A., 1981; dep. dir. dept. radiol. 514 Hosp., 1987-90vis. prof. Coll. Arts and Scis., Beijing U., 1987, Gen. Hosp. of PLa, Beijing, 1987-89; mem. Cons. Com. of PLA for Med. Equip., 1988—, Cons. of Profl. Title of China Yuan Wang Gen. Co., 1991—. Author: Medical Dosimetry, Handbook of Radiation Protection, 1992; contbr. articles to profl. jours.; editor Chinese Jour. Med. Imaging Tech., 1988—. The Spl. in Sci. and Tech. Ministry of Nuclear Industry, 1970, 4th in Sci. and Tech., 1979, 2nd in Sci. and Tech., 1986, 4th in Sci. and Tech., 1979, 3d in Sci. and Tech., 1987. Mem. Chinese Assn. Med. Imaging (bd. dirs. 1989—), Am. Assn. of Physicists in Medicine, Am. Inst. Physics. Achievements include pioneering set-up of internal dosimetry laboratory in China and development of related technology; involvement in development of DSA in U.S. and introduction of technology to China. Office: Beijing 514 Hosp, PO Box 1602, Beijing China 100101

WANG, YING ZHE, mechanical and optical engineer; b. Beijing, July 15, 1936; s. Ja Zheng and Ming Ru (Li) W.; m. Zhi Mei Zhang, Jan. 28, 1960; children: James X., Frank. MA, U. Tanjing, China, 1959. Engr. Beijing Opto-Elec. Rsch. Inst., 1975-82; rschr. Stanford (Calif.) U. Dept. Physics, 1982-83; engr. Coopervision Corp., Sunnyvale, Calif., 1984-88, Spectra-Physics Inc., Mountain View, Calif., 1988-89; sr. engr. Lexel Laser, Inc., Fremont, Calif., 1989—. Mem. SPIE, Nat. Optical Instrumentation Soc. (China). Home: 377 Creekside Dr Palo Alto CA 94306 Office: Lexel Laser Inc 48503 Milmont Dr Fremont CA 94538

WANG, ZHAO ZHONG, physics researcher; b. Shanghai, People's Republic China, Jan. 23, 1946; s. Zhu-Kang and Shu-zhen (Chen) W.; m. Ya-Xin Yu, Sept. 17, 1971; children: Zhi-Yuan, Bin-Yuan. BA, U. Sci. & Tech. of China, Beijing, 1968; PhD, U. Grenoble (France), 1985. Cert. solid state physicist. Lectr. U. Sci. & Tech. of China, Hefai, People's Republic China, 1978-79; maitre de conf. U. Grenoble, 1984-85; rsch. assoc. Princeton (N.J.) U., 1985-88, rsch. staff mem., 1988-90; dir. rsch. Lab. Microstructures and Microelectronics Centre National de la Recherche Scientifique, Paris, 1987—. Mem. Am. Phys. Soc., N.Y. Acad. Scis. Home: 40 Rue des Aulnes, 92330 Sceaux France Office: L2M/CNRS, 196 Ave H Ravera, 92220 Bagneux France

WANG, ZHI JIAN, aerospace engineer; b. Leiyang, Hunan, China, Oct. 3, 1964; came to U.S., 1991; s. Le Yao and Qiao Ying (Liu) W.; m. Xiao Jie Qu, Feb. 13, 1989; 1 child, Diane Dian Li. BSc, Changsha Inst. Tech., Hunan, 1985; PhD, U. Glasgow, Scotland, 1990. Rsch. fellow U. Glasgow, 1990-91; rsch. asst. Oxford (Eng.) U., 1991; rsch. engr. CFD Rsch. Corp., Huntsville, Ala., 1991—. Contbr. articles to profl. publs. Mem. AIAA. Achievements include research on high-resolution schemes in computational fluid dynamics. Office: CFD Rsch Corp 3325 D Triana Blvd Huntsville AL 35805

WANN, LAYMOND DOYLE, petroleum research scientist; b. Magazine, Ark., Apr. 25, 1924; s. Vernon Cecil and Emma (McCrary) W.; B.S. in Physics (Phi Eta Sigma scholar), Okla. State U., 1949, M.S., 1950; m. Betty Lou Brown, Nov. 6, 1948; children: Jacqueline, Lyndall Doyle. With Conoco Inc., Ponca City, Okla., 1951—, sr. research scientist, 1957-60, research group leader, 1960-81, assoc. research dir., 1981-84, staff scientist, 1984-85, ret., 1985; cons. in disciplines of phys. . Mem. Mcpl. Airport Bd., Ponca City. Served with AUS, 1942-46; ETO. Decorated Bronze Star. Mem. Am. Petroleum Inst. chmn. well logging subcom.), IEEE, Aircraft Owners and Pilots Assn., Seaplane Pilots Assn., VFW, Am. Legion, Phi Kappa Phi, Pi Mu Epsilon, Sigma Phi Sigma. Republican. Episcopalian (vestryman). Contbr. articles on elec. and radioactive well-logging, elec. design to profl. jours. Patentee in field. Home: 1501 Monument Rd Ponca City OK 74604-3522 Office: 1000 S Pine St Ponca City OK 74601-7501

WANZER, MARY KATHRYN, computer company executive, consultant; b. South Bend, Ind., Sept. 12, 1942; d. Cyril Joseph and Kathryn Alice (Dumke) Tlusty; m. Boyd Eugene Wanzer, May 30, 1964; children: Adam James, Christopher James. BS, Northland Coll., 1964; student, Am. U., Washington, 1972-73. Tchr. Montgomery Co. Md. Schs., Rockville, 1964-66; mathematician Johns Hopkins U., Silver Spring, Md., 1966-68; systems analyst ITT Fed. Elec. Corp., Kennedy Space Ctr., Fla., 1968-69; computer programmer Atlantic City (N.J.) Hosp., 1969-71; project leader Fairfax Hosp. Assn., Falls Church, Va., 1971-73; sr. systems analyst Xerox Corp., Leesburg, Va., 1973-76; software engr. E-Systems, Falls Church, Va., 1982-85; pres. Atlantic Office Svcs., Ltd., Bethany Beach, Del., 1988—; cons. Chesapeake Utilities, Dover, Del., 1990, Intervet., Millsboro, Del., 1990-92; MIS mgr. Thompson Pub. Group, Salisbury, Md., 1992-93; systems analyst Mountaire, Selbyville, Del., 1993—. Leader LaLeche League, Annandale, Va., 1980-83; v.p. No. Va. Hockey Club, Fairfax County, Va., 1986-87. Roman Catholic. Avocations: boating, swimming. Home: 941 Lake View Dr Bethany Beach DE 19930-9675 Office: The Office Answer 5 Starboard Ct # 1 Bethany Beach DE 19930-9679

WARBURTON, RALPH JOSEPH, architect, engineer, planner, educator; b. Kansas City, Mo., Sept. 5, 1935; s. Ralph Gray and Emma Frieda (Niemann) W.; m. Carol Ruth Hychka, June 14, 1958; children: John Geoffrey, Joy Frances. B.Arch., MIT, 1958; M.Arch., Yale U., 1959, M.C.P., 1960. Registered architect, Colo., Fla., N.J., Md., Ill., N.Y., Va., D.C.; registered profl. engr., Fla., N.J., N.Y.; registered community planner, Mich., N.J. With various archtl. planning and engring. firms Kansas City, Mo., 1952-55, Boston, 1956-58, N.Y.C., 1959-62, Chgo., 1962-64; chief planning Skidmore, Owings & Merrill, Chgo., 1964-66; spl. asst. for urban design HUD, Washington, 1966-72, cons., 1972-77; prof. architecture, archtl. engring. and planning U. Miami, Coral Gables, Fla., 1972—, chmn. dept. architecture, archtl. engring. and planning, 1972-75, assoc. dean engring. and environ. design, 1973-74; dir. grad. urban and regional planning program, 1973-75 81, 87—; advisor govt. Iran, 1970; advisor govt. France, 1973, govt. Ecuador, 1974, govt. Saudi Arabia, 1985; cons. in field, 1972—, lectr., critic design juror in field, 1965—; mem. chmn. Coral Gables Bd. Architects, 1980-82; mem. archtl. portfolio jury Am. Sch. and U., 1993. Assoc. author: Man-Made America: Chaos or Control, 1963; editor: New Concepts in Urban Transportation, 1968, Housing Systems Proposals for Operation Breakthrough, 1970, Focus on Furniture, 1971, National Community Art Competition, 1971, Defining Critical Environmental Areas, 1974; contbg. editor: Progressive Architecture, 1974-84; editorial adv. bd.: Jour. Am. Planning Assn., 1983-88, Planning for Higher Edn., 1986—, Urban Design and Preservation Quarterly, 1987—; contbr. over 100 articles to profl. jours.; mem. adv. panel Industrialization Forum Quar., 1969-79, archtl. portfolio jury Am. Sch. and Univ., 1993. Mem. Met. Housing and Planning Council, Chgo., 1965-67; mem. exec. com. Yale U. Arts Assn., 1965-70; pres. Yale U. Planning Alumni Assn., 1983—; mem. ednl. adv. com. Fla. Bd. Architecture, 1975. Recipient W.E. Parsons medal Yale U., 1960; recipient Spl. Achievement award HUD, 1972, commendation Fla. Bd. Architecture, 1974, Fla. Trust Historic Preservation award, 1983, Group Achievement award NASA, 1976; Skidmore, Owings & Merrill traveling fellow MIT, 1958; vis. fellow Inst. Architecture and Urban Studies, N.Y.C., 1972-74; NSF grantee, 1980-82. Fellow AIA (nat. housing. com. 1968-72, nat. regional devel. and natural resources com. 1974-75, nat. systems devel. com. 1972-73, nat. urban design com. 1968-73, bd. dirs. Fla. S. chpt. 1974-75), ASCE, Fla. Engring. Soc.; mem. Am. Inst. Cert. Planners (exec. com. dept. environ. planning 1973-74), Am. Soc. Engring. Edn. (chmn. archtl. engring. div. 1975-76), Nat. Soc. Archtl. Engrs. (founding profl.), Nat. Soc. Profl. Engrs., Nat. Sculpture Soc. (allied profl. mem.), Nat. Trust Hist. Preservation (principles and guidelines com. 1967), Am. Soc. Landscape Architects (hon., chmn. design awards jury 1971, 1972), Am. Planning Assn. (Fla. chpt. award excellence 1983), Internat. Fedn. Housing & Planning, Am. Soc. Interior Designers (hon.), Urban Land Inst. (assoc.), Omicron Delta Kappa, Sigma Xi, Tau Beta Pi. Club: Cosmos (Washington). Home: 6910 Veronese St Coral Gables FL 33146-3846 Office: 420 S Dixie Hwy Coral Gables FL 33146-2222 also: U Miami Sch Architecture Coral Gables FL 33124-5010

WARCHOL, MARK FRANCIS ANDREW, design engineer; b. New Kensington, Pa., Apr. 11, 1954; s. Frank and Patricia Marie (Snyder) W.; m. Stephanie Marie Synder, Dec. 30, 1982; children: Erin, Rachel. BSEET, Pa. State U., 1976; M in Material Sci., U. Va., 1990. Registered profl. engr., Pa. Designer Estey Organ Co., New Kensington, Pa., 1974, Essex Group, Kittanning, Pa., 1976-77, Compuguard, Pitts., 1977-79; sr. non-destructive evaluation engr. Alcoa, Alcoa Center, Pa., 1979—; mem. readers coun. Laser Focus World, Westford, Mass., 1988-89. Co-author: Field Demonstrations of Communication Systems for Disturbed Automation, Vol. 2, 1980. Mem. Am. Soc. for Nondestructive Test (cert.). Achievements include patents relating to the detection of soluable hydrogen in molten aluminum; development of automated ultrasonic inspection systems for flat rolled aerospace aluminum plate; co-development of first power line carrier communications system to integrate watt-hour meters, reverse vending machine to recycle aluminum cans, trademarked Telegas II instrument. Home: 1334 Taylor Ave New Kensington PA 15068-5533 Office: Aluminum Co Am 100 Tech Dr Alcoa Center PA 15069

WARD, CHARLES RAYMOND, systems engineer; b. Lansing, Mich., Oct. 23, 1949; s. George Merrill and Dorothy Irene (Hupp) W.; m. Sarah Hopkins Eddy, June 23, 1979; children: Katherine Emily, Rachel Elizabeth. BS in Math., Purdue U., 1971; MSEE, Naval Postgrad. Sch., 1977. Commd. ensign USN, 1971, advanced through grades to lt. commdr., served on USS Barbel, 1972-75, served on USS James Madison, 1978-81, served on USS Alabama, 1983-85; strategic navigation project mgr. Strategic Systems Programs, Arlington, Va., 1985-91; surveillance towed array sensor system, mgr. of plans and programs TRW Systems Div., McLean, Va., 1991—. Editor: Trident Navigation Standard Operating Procedures, 1991, Acoustic Warfare Operating Doctrine, 1992. Grounds com. chmn. Burke (Va.) United Meth. Ch., 1989—, worship com. chmn., 1993. Mem. IEEE, Eta Kappa Nu, Sigma Xi. Republican. Achievements include research in automatic depth and pitch control for a near surface submarine. Office: TRW Systems Division PO Box 10400 Fairfax VA 22031-0400

WARD, DEREK WILLIAM, lead systems analyst; b. Rockford, Ill., Aug. 2, 1956; s. Derek William Henry and Eileen Mary (Jacques) W.; m. Susan Marie Manley, Aug. 20, 1985 (div. 1990). BSBA in Computer Scis., Econs., U. Ill., 1978. Lead systems analyst LTV-Vought Missiles Div., Dallas, 1984-91, Loral-Vought Systems Div., Dallas, 1991—. Contbr. articles to profl. publs. Capt. U.S. Army, 1978-83. Mem. Shriners, Star In the East Lodge. Republican. Episcopalian. Achievements include analytical quantification of line-of-sight kenetic energy and non-line-of-sight chemical energy synergies, active countermeasures effects on existing/future anti-armor force structures and weapons. Home: Apt 1413 2636 Verandah Ln Arlington TX 76006 Office: Loral Vought Systems Div PO Box 650003 Mail Stop WT 52 Dallas TX 75265-0003

WARD, JAMES VERNON, biologist, educator; b. Mpls., Mar. 27, 1940; s. Vernon G. and Nela J. (Kemmer) W.; m. Janice Ann Walstrom, June 14, 1963. BS, U. Minn., 1963; MA, U. Denver, 1967; PhD, U. Colo., 1973. Biology tchr. Denver Pub. Schs., 1963-70; asst. prof. Colo. State U., Fort Collins, Colo., 1974-79, assoc. prof., 1979-83, prof., 1983—; lectr. in field. Editor: The Ecology of Regulated Streams, 1979; author: Aquatic Insect Ecology, 1992, Illustrated Guide to the Mountain Stream Insects of Colorado, 1992; contbr. over 100 sci. papers to profl. jours.; editor Regulated Rivers, 1987—. Mem. North Am. Benthological Soc. (pres. 1987-88), Ecol. Soc. of Am., Sigma Xi. Office: Colorado State U Department of Biology Fort Collins CO 80523

WARD, JOAN GAYE, psychologist; b. Englewood, N.J.; d. James A. and Eda D. (Mullan) W. BA, Miami U., 1956; MA, New Sch. for Social Rsch., 1965; PhD, NYU, 1973, cert. psychotherapy and psychoanalysis, 1981. Rsch. assoc. Mktg. Survey & Rsch. Corp., N.Y.C., 1962-67; counselor NYU, N.Y.C., 1967-68, instr., 1970-71; psychologist Bur. Child Guidance, Sch. Bd. Support Teams, Bronx, 1969—; field trainer Sch. Bd. Support Teams, Bronx, 1989—; practicum supr. Fordham U., N.Y.C., 1989-90, 91-92; internship supr. L.I. U., Dobbs Ferry, N.Y., 1990-91; pvt. practice psychotherapy N.Y.C., 1969—; adjunct asst. clinical prof. NYU, 1992-93. NYU rsch. fellow, 1967-68. Mem. APA, N.Y. State Psychol. Assn., Ea. Psychol. Assn., Nat. Assn. Sch. Psychologists, N.Y. Assn. Sch. Psychologists, Postdoctoral Soc. Avocations: sailing, cross country skiing. Home: 91 Schofield St City Island NY 10464-1533

WARD, JOHN WESLEY, pharmacologist; b. Martin, Tenn., Apr. 8, 1925; s. Charles Wesley and Sara Elizabeth (Little) W.; m. Martha Isabelle Hendley, Dec. 7, 1947; children: Judith Carol, Charles Wesley, Richard Little. A.A., George Washington U., 1948, B.S., 1950, M.S., 1955; Ph.D., Georgetown U., 1959. Research assoc. in pharmacology Hazleton Labs., Falls Church, Va., 1950-55, head dept. pharmacology Hazleton Labs., 1955-58, chief depts. biochemistry and pharmacology, 1958-59; with A. H. Robins Co., Richmond, Va., 1959—, dir. biol. research, 1978-80, dir. research, 1980—, v.p. research, 1982-89, v.p., gen. mgr. R & D div., 1989-90, ret.; lectr. in pharmacology Med. Coll. Va., 1960-64, adj. assoc. prof. pharmacology, 1982-90; guest lectr. Seminar on Good Lab. Practices, FDA, Washington, 1979, Chgo., 1979, San Francisco, 1979; apptd. expert pharmacologue toxicologue, France, 1986. Contbr. articles on pharmacology, toxicology and medicinal chemistry to profl. pubs. Served with USMC, 1943; Served with USN, 1944-46; Served with U.S. Army, 1944. Mem. AAAS, N.Y. Acad. Sci., Va. Acad. Sci., Am. Chem. Soc., Soc. Toxicology (charter), Am. Soc. Pharmacology and Exptl. Therapeutics, Internat. Soc. Regulatory Toxicology and Pharmacology (charter), Pharm. Mfrs. Assn. (chmn. animal care and use com. 1971-88), Am. Assn. for Accreditation Lab. Animal Care (chmn. bd. trustees 1976-80), Sigma Xi. Clubs: Willow Oaks (Richmond); Cosmos (Washington), Masons (Washington). Achievements include patents in field. Home: 10275 Cherokee Rd Richmond VA 23235-1107

WARD, LLEWELLYN O(RCUTT), III, oil producer; b. Oklahoma City, July 24, 1930; s. Llewellyn Orcutt II and Addie (Reisdorph) W.; m. Myra Beth Gungoll, Oct. 29, 1955: children: Casidy Ann, William Carlton. Student, Okla. Mil. Acad. Jr. Coll., 1948-50; BS, Okla. U., 1953; postgrad. Harvard U., 1986. Registered profl. engr., Okla. Dist. engr. Delhi-Taylor Oil Corp., Tulsa, 1955-56; ptnr. Ward-Gungoll Oil Investments, Enid, Okla., 1956—; owner L.O. Ward Oil Ops., Enid, 1963—; v.p. 1420 Lahoma Rd. Inc., Enid, 1967—, also bd. dirs.; mem. Okla. Gov.'s Adv. Coun. on Energy; rep. to Interstate Oil Compact Commn.; bd. dirs. Community Bank and Trust Co. Enid. Chmn. Indsl. Devel. Commn., Enid, 1968—; active YMCA; mem. bd. visitors Coll. Engring., U. Okla.; mem. adv. coun. Sch. Bus., trustee Phillips U., Enid, Univ. Bd., Pepperdine, Calif.; Okla. chmn. U.S. Olympic Com., 1986—; chmn. bd. Okla. Polit. Action Com., 1974—, Bass Hosp.; Rep. chmn. Garfield County, 1967-69; Rep. nat. committeeman from Okla.; bd. dirs. Enid Indsl. Devel. Found. Served with C.E., U.S. Army, 1953-55. Named to Order of Ky. Cols. Mem. Am. Inst. Mining and Metall. Engrs., Ind. Petroleum Assn. Am. (area v.p., bd. dirs.), Okla. Ind. Petroleum Assn. (pres., bd. dirs.), Nat. Petroleum Council, Enid C. of C. (v.p., then pres.), Alpha Tau Omega. Methodist. Clubs: Toastmasters (pres. Enid chpt. 1966), Am. Bus. (pres. 1964). Lodges: Masons, Shriners, Rotary (pres. Enid 1990-91). Home: 900 Brookside Dr Enid OK 73703-6941 Office: 502 S Fillmore St Enid OK 73703-5703

WARD, MILTON HAWKINS, mining company executive; b. Bessemer, Ala., Aug. 1, 1932; s. William Howard and Mae Ivy (Smith) W.; m. Sylvia Adele Randle, June 30, 1952; children: Jeffrey Randle, Lisa Adele. BS in Mining Engring., U. Ala., 1955, MS in Engring., 1981; MBA, U. N.Mex., 1974. Registered profl. engr., Tex., Ala. Supr., engr. San Manuel (Ariz.) Copper Corp., 1955-60; mine supt., divsn. supt., gen. supt. of mines, divsn. engr. Kerr-McGee Corp., Oklahoma City and Grants, N.Mex., 1960-66; gen. mgr. Homestake Mining Co., Grants, 1966-70; v.p. ops. Ranchers Exploration & Devel. Corp., Albuquerque, 1970-74; pres., bd. dirs. Freeport Minerals Co., N.Y.C., 1974-85; pres., COO Freeport-McMoRan, Inc., New Orleans, 1985-92, also bd. dirs.; chmn., pres., CEO Cyprus Minerals Co., Englewood, Colo., 1992—; bd. dirs. Mineral Info. Inst., Inc., Internat. Copper Assn., Adv. Coun. Internat. Investments; bd. dirs., mem. exec. com. Contbr. articles to profl. jours. Past trustee New Orleans Mus. Art, former pres.; trustee Children's Hosp., New Orleans, Tulane U. Bd. Adminstrs.; bd. dirs. Smithsonian Nat. Mus. Natural History, Nat. Mining Hall of Fame and Mus.; past mem. pres. coun. Contemporary Arts Ctr.; former adv. com. chmn. Tulane U. Bioenviron. Rsch. Ctr.; former adv. bd. bus. coun. Tulane U. Sch. Bus.; others. Recipient Daniel C. Jackling award Soc. Mining, Metallurgy and Exploration, 1992. Fellow Inst. Mining & Metallurgy (London); mem. Am. Mining Congress (chmn., dir. 1989-92, vice chmn., dir.), AIME (former sect. chmn., Disting. Mem. award, Saunders Gold medal), Am. Found. Phosphate Prodn. (bd. dirs.), Am. Australian Assn., Mining and Metall. Soc. Am. (pres. 1981-83, mem. exec. com.), Can. Inst. Mining and Mettal., Soc. Mining. Engrs. (Jackling award 1992), NAM (mem. natural resource com.), N.Mex. Mining Assn. (former chmn., dir.), Internat. Copper Assn. (bd. dirs.), Adv. Coun. Internat. Investments, Internat. Coun. Metals and Environ. (bd. dirs.), New Orleans City Club (bd. govs.), Met. Club Denver, Sky Club, Copper Club, Univ. Club, Mining Club (v.p., gov. 1979—), Petroleum Club (New Orleans), Sierra Club. Republican. Presbyterian. Clubs: City, New Orleans Country; Univ. (N.Y.C.). Avocations: tennis, enology, flying. Office: Cyprus Minerals Co 9100 E Mineral Circle Englewood CO 80155

WARD, ROBERT LEE, civil engineering educator; b. St. Louis, Dec. 29, 1948; s. Hubert Vere and Edna (Ashworth) W.; m. Karen Lucinda Gentry, June 6, 1972; children: Heather, Jennifer, Brian, Vanessa. BS, U. Mo., Rolla, 1971, MS, 1974; PhD, U. Ark., 1988. Registered profl. engr., Mo., Kans., Ohio. Project engr. Tex. Eastman Co., Longview, 1975-78; city engr. City of O'Fallon (Mo.), 1979-80; asst. prof. dept. engring. St. Louis Community Coll., Florissant, 1980-84; instr. civil engring. Kans. State U., Manhattan, 1984-86; rsch. asst. agrl. engring. U. Ark., Fayetteville, 1986-88; assoc. prof. civil engring. N.Mex. State U., Las Cruces, 1988-89; assoc. prof. civil engring. Ohio No. U., Ada, 1989—, Herbert F. Alter chair of engring. sci.; cons. Richard D. Irwin, Inc., Boston, 1992, Hull & Assocs., Toledo, 1990; sec. coun. Ohio No. U., 1990-93, mem. acad. affairs com., 1990-93. Mem. ASCE, Ohio Soc. Profl. Engrs., Am. Soc. Engring. Edn., Chi Epsilon, Phi Kappa Phi. Methodist. Home: 123 S Johnson St Ada OH 45810 Office: Ohio No Univ Dept Civil Engring Ada OH 45810

WARD, WANDA ELAINE, psychologist; b. Atlanta, June 27, 1954; d. Clifford R. Ward and Elaine (Phillips) Ward Jackson. BA in Psychology, Princeton U., 1976; PhD in Psychology, Stanford U., 1981. Asst. prof. psychology U. Okla., Norman, 1981-88, assoc. prof. psychology, 1988—, founding dir. Ctr. for Rsch. on Multi-Ethnic Edn., 1986—; vis. asst. prof. Ctr. for Study of Reading, U. Ill., Urbana, 1984-85; vis. scholar Ctr. for Social Orgn. of Schs., Johns Hopkins U., Balt., 1990; program dir. career access programs NSF, Washington, 1992—; mem. joint legis. task force on literacy Okla. State Senate, Oklahoma City, 1987-88, mem. legis. task force on minority tchr. recruitment, 1989-90; mem. Okla. Scholar-Leadership Enrichment Program Adv. Coun., Norman, 1990-92; presenter workshops, seminars, symposia in field. Sr. co-editor: Key Issues in Minority Education: Research Directions and Practical Implications, 1989; contbr. articles to profl. publs. Mem.-at-large exec. com. alumni coun. Princeton U., 1989-91; bd. dirs. Support Ctr. of Okla., Oklahoma City, 1990-92; mem. minority adv. bd. KOCO-TV 5, Oklahoma City, 1989-92. Ford Found. fellow, 1976-81; recipient A.C. Hamilton Tribute of Appreciation and Commendation, 12th ann. conf. Nat. Black Caucus of State Legislators, 1988. Mem. APA, Am. Ednl. Rsch. Assn., Assn. Black Psychologists, Western Psychol. Assn. Office: NSF 1800 G St NW Washington DC 20550

WARD, WILLIAM WADE, clinical immunologist; b. London, May 21, 1953; came to U.S., 1964; s. Robert G. and Rita Ann (McLaren) W.; m. Janet Beth Thompson, Oct. 10, 1977; children: Stephanie, Geoffrey, Kristen. MS, SUNY, Syracuse, 1982; MA, Cen. Mich. U., 1984. Enlisted USAF Hosp., 1972, advanced to maj., 1988; chief lab. svc. USAF Hosp, Wurtsmith AFB, Mich., 1977-80; dir. armed svcs. whole blood processing lab. McGuire AFB, N.J., 1982-85, Lackland AFB, Tex., 1988—; chief immunology Wilford Hall Med. Ctr. USAF Hosp., Lackland AFB, Tex., 1988—; predoctoral fellow Med. Coll. Va., Richmond, 1985-88; clin. asst. prof. med. tech. U. Tex. Health Sci. Ctr., San Antonio, 1991—; chief biomed. scientist USAF Biomed. Scis. Corps, Washington, 1992. Contbr. articles to profl. jours. Recipient Outstanding Clin. Scientist award, 1993. Mem. Am. Soc. Clin. Pathologists (cert. med. technologist, cert. blood bank specialist), Am. Soc. Microbiology, Assn. Med. Lab. Immunologists, Am. Assn. Blood Banks, Sigma Xi. Office: Wilford Hall Med Ctr PSLCI Ste 1 2200 Bergquist Dr Lackland AFB TX 78236-5300

WARDEN, GARY GEORGE, computer engineer; b. Balt., Sept. 2, 1951; s. R.C. and Carsey (Farley) W.; m. Diane Susan Trick, Feb. 15, 1975; children; Stephanie, Samuel, Rachel, Nathan. BS in Computer Engring., Wright State U., 1978; Diploma in Biblical Studies, Centerville (Ohio) Bible Coll., 1980. Project engr. Simulation Tech., Dayton, Ohio, 1978-80; chief engr. Systran Corp., Dayton, Ohio, 1980—. Contbr. articles to confs. Soccer coach Community Say Soccer, Tipp City, Ohio, 1986-90, Dayton Christian Mid. Sch., 1989-91. Sgt. USAF, 1970-74. Pastor Brethren in Christ Ch. Achievements include development of replicated memory networking technology. Office: Systran Corp 4126 Linden Ave Dayton OH 45432

WARDEN, GLENN DONALD, burn surgeon; b. Palo Alto, Calif., Jan. 25, 1943; s. John and Thelma Warden; m. Nori Katherine Bartschi, Mar. 30, 1971; children: Glenn David, Nori Lei, Emilie Nicole, Lianna Katherine. Student, U. Ariz., 1961-64; MD, U. Utah, 1968. Diplomate Am. Bd. Surgery; qualifications in surg. critical care; lic., Utah, Tex., Ohio. Intern in surgery dept. surgery U. Utah Med. Ctr., Salt Lake City, 1968-69, resident in gen. surgery dept. surgery, 1969-71, 74-76; fellow in renal transplantation - nephrology dept. medicine divsn. nephrology VA Hosp., Salt Lake City, 1970-71; dir. Intermountain Burn Ctr. U. Utah Sch. Medicine, Salt Lake City, 1977-85, dir. trauma divsn. dept. surgery, 1978-85, co-dir. surg. ICU, 1983-84, from instr. to asst. prof. to assoc. prof. dept. surgery, 1976-85, prof. dept. surgery, 1985; prof. dept. surgery, dir. divsn. burn surgery U. Cin. Med. Ctr., 1985—; chief of staff Cin. unit Shriners Burn Inst.; mem. exec. com. dept. surgery U. Cin. Med. Ctr., 1985, com. on reappointment, promotion and tenure, 1985—, chmn., 1991, dir. search com. oral surgery, 1991, adv. com. dept. phys. medicine & rehab., 1985-87, dir. search com., 1985-87, med. scis. scholars com., 1985-87, faculty forum exec. com., 1987-89, operating rm. subcom., 1985—, equipment control com., 1986—, trauma com., 1985—, spl. care units com., 1988—; critical care unit cons. emergency med. svcs. Salt Lake City Dept. Health, 1977-85, bd. dirs. emergency trng. coun., 1982-85, curricumum com., 1984-85; burn cons. Handicapped Children's Svcs., Utah State Dept. Health, 1981-85; mem. univ. senate U. Utah, 1984-85; chmn. I.V. subcom. U. Utah Med. Ctr., 1976-80, 82-85, oper. rm. exec. com., 1979-85, pharmacy and therapeutics com., 1979-85, infection ctrl. com., 1979-85, med. dir. disaster com., 1981-85, ambulatory svcs. com., 1982-85, emergency med. svcs. com., 1983-85, critical care com., 1984-85; contbr. (with others) various exhibits and posters ACS, U.S. Army Inst. Surg. Rsch., Brooke Army Med. Ctr., Am. Burn Assn.; lectures, presentations, symposiums ACS, Internat. Transplantation Soc., Am. Assn. Surgery of Trauma, Western Surg. Assn., Brooke Army Med. Ctr., Am. Burn Assn., Internat. Soc. for Burn Injuries, Assn. Acad. Surgery, Inst. Surg. Rsch., Intermountain Pediatric Soc., Nat. Inst. Gen. Med. Sci., Soc. Univ. Surgeons, Marion Labs., U. Cal., U. Texas S.W. Med. Ctr., Ky. Surg. Soc., Ohio Respiratory Care Soc., Ohio Burn Team Soc., Am. Soc. Parenteral and

Enteral Nutrition, Grady Meml. Hosp., Atlanta, U. Iowa Hosps., Miami Valley Hosp., Christ Hosp., Cin., Internat. Soc. Surgery, Ohio State U., Fedn. Am. Socs. for Exptl. Biology, Am. Assn. Tissue Banks, U.S. Army Inst. Surg. Rsch., Am. Soc. Plastic & Reconstructive Surg. Nurses, others. Mem. editorial bd. Jour. Burn Care and Rehab., 1982—; contbr. numerous chpts. to books including Burns of the Upper Extremities, 1976, Surgery of the Ambulatory Patient, 1980, Management of the Burned Patient, 1987, Manual of Excision, 1988, Immune Consequences of Trauma, Shock and Sepsis, 1989, Total Parenteral Nutrition, 1991, The Critical Care of the Burned Patient, 1992, Host Defense Dysfunction in Trauma, Shock and Sepsis, 1993; contbr. numerous articles to profl. jours. Maj. U.S. Army, 1971-74. Named Internat. Hon. prof. surgery Third Mil. Med. Coll., People's Republic of China; Mel Ott scholar U. Ariz., 1962, Paul Spaulding scholar U. Ariz., 1963, Gen. Resident scholar U. Ariz., 1963, Martha Bamburger scholar U. Utah Sch. Medicine, 1965. Fellow ACS (Utah chpt. 1980, sec.-treas. 1984); mem. Internat. Soc. for Burn Injuries, Internat. Soc. of Surgery, Internat. Burn Found. of U.S. (bd. dirs. 1992), Pan-Pacific Surg. Assn., Pan-Am. Med. Assn., Am. Burn Assn. (com. on orgn. and delivery of burn care 1976-80, com. edn. 1986, chmn. 1986-87, sec. 1987-90, 1st v.p. 1990-91, pres.-elect 1991-92, pres. 1992-93, bd. trustees 1993), Transplantation Soc., Soc. for Leukocyte Biology (formerly Reticuloendothelial Soc.. for Acad. Surgery, Utah Med. Assn. (del. 1979-81), Am. Assn. Tissue Banks (inter-regional exch. com. 1977-79, skin coun. 1989-91), Western Transplant Assn., Am. Soc. Transplant Surgeons, Am. Soc. Parenteral and Enteral Nutrition (edn. com. 1981—), Southwestern Surg. Congress, Salt Lake Surg. Soc., Utah Soc. Certified Surgeons (treas. 1981, pres. 1982), Am. Assn. Surgery of Trauma, Univ. Surg. Soc., Surg. Infection Soc. (chmn. fellowship com. 1986-88), Soc. Univ. Surgeons, Western Surg. Assn., Soc. Critical Care Medicine, Acad. Medicine Cin., Ohio State Med. Assn., Cin. Surg. Soc., Surg. Biology Club III, Ohio Burn Team Soc., U. Cin. Grad. Surg. Soc., Eastern Great Lakes Burn Study Group, Am. Surg. Assn., Ctrl. Surg. Assn., Wound Healing Soc., Am. Trauma Soc. (liaison bd. 1990). Office: Shriners Burn Inst 3229 Burnet Ave Cincinnati OH 45229

WARD-MCLEMORE, ETHEL, research geophysicist, mathematician; b. Sylvarena, Miss., Jan. 22, 1908; d. William Robert and Frances Virginia (Douglas) Ward; m. Robert Henry McLemore, June 30, 1935; 1 child, Mary Frances. BA, Miss. Woman's Coll., 1928; MA, U. N.C., 1929; postgrad., U. Chgo., 1931, Colo. Sch. Mines, 1941-42, So. Meth. U., 1962-64. Head math. dept. Miss. Jr. Coll., 1929-30; instr. chemistry, math. Miss. State Coll. for Women, 1930-32; rsch. mathematician Humble Oil & Refining Co., Houston, 1933-36; ind. geophys. rsch., Tex. and Colo., 1936-42, Ft. Worth, 1946—; geophysicist United Geophys. Co., Pasadena, Cal., 1942-46; tchr. chemistry, physics, Hockaday Sch., Dallas, 1958-59, tchr. math., 1959-60, tchr. chemistry, 1968-69; tchr. chemistry Ursuline Acad., Dallas, 1964-67; geophys. cons., Dallas, 1957—; with Eugene McDermott Libr., U. Tex., rsch. geophysicist. Author: China, 1983, Bibliography of the Publications of the Texas Academy of Science, 1929-87, 1989, The Academies of Science of Texas (1880-1987), 1989, also annotated bibliographies of sedimentary basins, 1981; contbr. articles to profl. jours. Mem. AAAS, Am. Math. Soc., Acads. of Sci. Tex., Math. Assn. Am., Am. Geophys. Union (40 yr. Mem. Rsch. Silver Pin award 1988), Seismol. Soc. Am., Soc. Exploration Geophysicists (50 yr. Gold cert. 1986, Hon. Membership award 1989, hon. life), Soc. Indsl. and Applied Math., Am. Chem. Soc., Inst. Math. Statis., Tex Acad. Sci. (Appreciation cert. 1985), Dallas Geophys. Soc. (hon. life 1986, Disting. Svc. award 1988), Sigma Xi. Home: 8600 Skyline Dr Apt 1107 Dallas TX 75243-4158 Office: U Tex Eugene McDermott Libr MC3-418 Box 830643 Richardson TX 75083

WARDWELL, JAMES CHARLES, computer and management consultant; b. Portchester, N.Y., July 2, 1952; s. Charles Marshall and Inez (Slate) W. BS in Physics, U. Bridgeport, 1974; MA in Physics, Kent State U., 1976; MS in Physics, Ga. Inst. Tech., 1979. Adminstrv. specialist office v.p. acad. affairs Ga. Inst. Tech., Atlanta, 1979-80; dir. tech. and adminstrv. svcs. dept. physics NYU, N.Y.C., 1982-86; dir. budget and computer systems Union Theol. Sem., N.Y.C., 1986-88; cons., 1988-89; CEO Trinity Unltd., Inc., Norwalk, Conn., 1989—. Recipient Founder's award Boy Scouts Am., 1989, Saint George award Protestant Episcopal Ch. of USA and Boy Scouts of Am., 1991. Mem. Am. Phys. Soc. (life), Order of the Arrow (vigil mem., fin. advisor 1989—, Founders award 1989).

WARE, BENJAMIN RAY, university administrator; b. Ponca City, Okla., Sept. 21, 1946; s. Clyde Elmer and Lois Aliene (Smith) W.; m. Sheridan Lee Welch, May 28, 1967 (div. 1976); 1 child, Winston Arthur; m. Claudia Borman, Dec. 21, 1979; children: Jeffrey Bright, Amelia Marie. BS in Chemistry, Okla. State U., 1968; PhD in Biophys. Chemistry, U. Ill., 1972. Asst. prof. chemistry Harvard U., Cambridge, Mass., 1972-75, assoc. prof., 1975-79; prof. and chmn. dept. chemistry Syracuse U., 1979-84, Kenan prof. sci., 1984-91, v.p. rsch., 1989-92, v.p. rsch., computing, 1992—. Contbr. articles to profl. jours. Grantee, NIH, NSF; Alfred P. Sloan fellow, 1976-80. Fellow AAAS; mem. Phi Beta Kappa, Phi Kappa Phi. Achievements include invention of electrophoretic light scattering; first to combine laser light scattering and fluorescence photobleaching recovery to distinguish mutual and tracer diffusion; first to apply laser Doppler velocimetry to protoplasmic streaming. Home: 2600 State Route 174 Marietta NY 13110-9619 Office: Syracuse Univ 3-014D Ctr Sci and Tech Syracuse NY 13244-1100

WARE, LAWRENCE LESLIE, JR., microbiologist; b. Montgomery, W.Va., Sept. 12, 1920; s. Lawrence Leslie Sr. and Dorothy (Gay) W.; m. Elizabeth Fisher, Jan. 27, 1946; children: David Allan, Kenneth Robert, Barbara Ann. BS, Roosevelt U., 1950. Microbiologist U. Chgo., 1946-51, U.S. Army Biol. Warfare Project, Frederick, Md., 1951-59; microbiologist, Indian Hosp. lab. dir. USPHS, Phoenix, 1959-65; microbiologist, med. info. officer U.S. Army Med. R & D. Command, Washington, 1966-86. Contbr. articles to profl. jours. Fellow Am. Acad. Microbiology, Sigma Xi (sec. 1978-83, pres. 1983-86). Home: 9224 Kristin Ln Fairfax VA 22032

WARFEL, CHRISTOPHER GEORGE, mechanical engineer, design consultant; b. Kingston, N.Y., July 27, 1958; s. John David and Dorothy Anne (Diemer) W. BS in Forestry Engring., SUNY, Syracuse, 1981; BS in Gen. Engring., Syracuse U., 1981; MSME, U. Mass., 1986. Registered profl. engr. Engr. intern Metro. Washington Coun. Govts., summer 1983, 84; staff engr. Pub. Svc. Co. of N.H., Manchester, 1986—. Contbr. articles to profl. jours. and newspapers. Asst. wrestling coach YMCA, Manchester, 1991, 92. Regent's scholar N.Y., 1977-78. Mem. Assn. Energy Engrs., Assn. Heating, Ventilation and Refrigeration Engrs., Cogeneration Inst. Achievements include research in renewable energy technology and energy grant analysis for non-profit and government institutions. Home: PO Box 1593 Manchester NH 03105-5000

WARFEL, JOHN HIATT, medical educator, retired; b. Marion, Ind., Mar. 3, 1916; s. Robert A. and Mary (Hiatt) W.; m. Marjorie Jane Wolfe, Oct. 28, 1942; children: Barbara, Susan, David. BS, Capital U., 1938; MS, Ohio State U., 1941; PhD, Case Western Res. U., 1948. Assoc. prof. anatomy SUNY, Buffalo, 1949-86. Author: The Extremities, 6th edit., 1993, The Head, Neck and Trunk, 6th edit., 1993. Lt. (s.g.) USNR, 1942-46. Mem. Am. Assn. Anatomists, Mason, Sigma Xi. Methodist.

WARING, JOHN ALFRED, research writer, lecturer, consultant; b. San Francisco, Dec. 30, 1913; s. John A. and Mary (Wheeler) W. Student pub. schs. Yachting, marine editor Chgo. Tribune, 1934-47; editor Kellogg Messenger, Kellogg Switchboard & Supply Co., Chgo., 1945-49; rsch. cons. Baxter Internat. Econ. Rsch. Bur., Inc., investment counselling, N.Y.C., 1951-52; rsch. asst. mktg. rsch. dept. Fuller, Smith & Ross, Inc., advt., N.Y.C., 1953; rsch. writer, cons. Twentieth Century Fund, N.Y.C., 1953-54; rsch. cons. Ford Motor Co., Dearborn, Mich., 1955; chief rschr. Internat. Fact Finding Inst., Lawrence Orgn., pub. rels. cons., N.Y.C., Washington, 1957-58; lectr. on energy, tech. and history World Power Conf., Montreal, 1958, First Energy Inst., Am. U., Washington, 1960, Nat. Archives, Washington, 1962, Smithsonian Mus. History and Tech., Washington, 1968; guest lectr. social responsibility in sci. U. Md., 1972, guest lectr. sci. and environment, 1974; lectr. Internat. Conf. on Energy and Humanity, Queen Mary Coll., U. London, 1972, World Energy Conf., Detroit, 1974, History of Sci. Soc., Norwalk, Conn., 1974; rsch. cons. PARM Project, Nat. Planning Assn., Washington, 1961-62; cons. Sci., Tech. and Fgn. Affairs Seminar, Fgn. Svc. Inst., Dept. State, 1965; vis. lectr. social implications of sci. for N. Am. U. Alta., Edmonton (Can.), 1981; lectr. Carnahan Conf. Harmonizing Tech. with Soc., U. Ky., Lexington, 1987; rsch. cons. Program of Policy Studies in Sci. and Tech., George Washington U., 1967-68; del. U.S. commn. UNESCO Conf., San Francisco, 1969; inaugural lectr. Future of Sci. and Soc. in Am. Seminar U.S. Civil Svc. Commn., Washington, 1970; editorial cons. Nat. Acad. Engring., Washington, 1971; rsch. cons., analyst Seminar Sch., Indsl. Coll. Armed Forces, Ft. Lesley J. McNair, Washington, 1958-74; researcher Office Plans and Programs, U.S. Army Med. Dept., Washington, 1975-76; asst. editor Def. Systems Mgmt. Rev. mag. Def. Systems Mgmt. Coll., Ft. Belvoir, Va., 1977-78; ret., 1978. Contbr. chpts. to books; compiler statis. tabulations. Mem. AAAS, Soc. History of Tech. (charter), History Sci. Soc., Technocracy, Washington Acad. Scis., Washington Soc. Engrs. (sec. 1977), Internat. Soc. Gen. Semantics, N.Y. Acad. Sci., D. C. Area Phi Beta Kappa Assn. Home: 1320 S George Mason Dr Arlington VA 22204-3851

WARNE, RONSON JOSEPH, mathematics educator; b. East Orange, N.J., June 14, 1930; s. Ronson Joseph and Mildred (Morton) W.; m. Gloria Jane La France, Oct. 24, 1950. BA, Columbia U., 1953; MS, NYU, 1955; PhD, U. Tenn., 1959. Teaching asst., instr. U. Tenn., Knoxville, 1955-59; asst. prof. Math. La. State U., New Orleans, 1959-63; assoc. prof. Math. Va. Polytech. Inst., Blacksburg, 1963-64; prof. Math. W.Va. U., Morgantown, 1964-70, U. Ala., Birmingham, 1970-89, King Fahd U. of Petroleum and Minerals, Dhahran, Saudi Arabia, 1989—. Contbr. articles to profl. jours. Oak Ridge Nat. Lab. fellow, 1960, Dryser fellow, U. Tenn., 1957; vis. rsch. scholar U. Calif., Berkeley, 1982. Mem. Am. Math. Soc. Avocations: body building, weightlifting, running. Office: KFUPM # 1564, Dhahran 31261, Saudi Arabia

WARNE, WILLIAM ELMO, irrigationist; b. nr. Seafield, Ind., Sept. 2, 1905; s. William Rufus and Nettie Jane (Williams) W.; m. Edith Margaret Peterson, July 9, 1929; children: Jane Ingrid (Mrs. David C. Beeder), William Robert, Margaret Edith (Mrs. John W. Monroe). AB, U. Calif., 1927; DEcons (hon.), Yonsei U., Seoul, 1959; LLD, Seoul Nat. U., 1959. Reporter San Francisco Bull. and Oakland (Calif.) Post-Enquirer, 1925-27; news editor Brawley (Calif.) News, 1927, Calexico (Calif.) Chronicle, 1927-28; editor, night mgr. L.A. bur. AP, 1928-31, corr. San Diego bur., 1931-33, Washington corr., 1933-35; editor, bur. reclamation Dept. Interior, 1935-37; on staff Third World Power Conf., 1936; assoc. to reviewing com. Nat. Resources Com. on preparation Drainage Basin Problems and Programs, 1936, mem. editorial com. for revision, 1937; chief of information Bur. Reclamation, 1937-42; co-dir. (with Harlan H. Barrows) Columbia Basin Joint Investigations, 1939-42; chief of staff, war prodn. drive WPB, 1942; asst. dir. div. power Dept. Interior, 1942-43, dept. dir. information, 1943; asst. commr. Bur. Reclamation, 1943-47; apptd. asst. sec. Dept. Interior, 1947, asst. sec. Water and Power Devel., 1950-51; U.S. minister charge tech. cooperation Iran, 1951-55, Brazil, 1955-56; U.S. minister and econ. coord. for Korea, 1956-59; dir. Cal. Dept. Fish and Game, 1959-60, Dept. Agr., 1960-61, Dept. Water Resources, 1961-67; v.p. water resources Devel. and Resources Corp., 1967-69; resources cons., 1969—; pres. Warne & Blanton Pubs. Inc., 1985-90, Warne Walnut Wrancho, Inc., 1979—; Disting. Practitioner in Residence Sch. Pub. Adminstrn., U. So. Calif. at Sacramento, 1976-78; adminstr. Resources Agy. of Calif., 1961-63; Chmn. Pres.'s Com. on San Diego Water Supply, 1944-46; chmn. Fed. Inter-Agy. River Basin Com., 1948, Fed. Com. on Alaskan Devel., 1948; pres. Group Health Assn., Inc., 1947-51; chmn. U.S. delegation 2d Inter-Am. Conf. Indian Life, Cuzco, Peru, 1949; U.S. del. 4th World Power Conf., London, Eng., 1950; mem. Calif. Water Pollution Control Bd., 1959-67; vice chmn. 1960-62; mem. water pollution control adv. bd. Dept. Health, Edn. and Welfare, 1962-65, cons., 1966-67; chmn. Calif. delegation Western States Water Council, 1965-67. Author: Mission for Peace-Point 4 in Iran, 1956, The Bureau of Reclamation, 1973, How the Colorado River Was Spent, 1975, The Need to Institutionalize Desalting, 1978; prin. author: The California Experience with Mass Transfers of Water over Long Distances, 1978; editor Geothermal Report, 1985-90. Served as 2d lt. O.R.C., 1927-37. Recipient Disting. Svc. award Dept. Interior, 1951, Disting. Pub. Svc. Honor award FOA, 1955, Order of Crown Shah of Iran, 1955, Outstanding Svc. citation UN Command Korea, 1959, Order of Indsl. Sv. Merit Bronze Star, Korea, 1991. Fellow Nat. Acad. Pub. Adminstrn. (sr., chmn. standing com. on environ. and resources mgmt. 1971-78); mem. Nat. Water Supply Improvement Assn. (pres. 1978-80, Lifetime Achievement award 1984), Internat. Desalination Assn. (founding mem., Lifetime Disting. Service award 1991), Soc. Profl. Journalists, Sutter Club, Nat. Press Club (Washington), Lambda Chi Alpha. Home and Office: 1570 Madrono Ave Palo Alto CA 94306-1015

WARNER, BARRY GREGORY, geographer; b. Cambridge, Ont., Can., July 20, 1955; s. Gregory O. and Alma (Jansen) W. B in Environ. Studies, U. Waterloo, 1978, MS, 1980; PhD, Simon Fraser U., Burnaby, Can., 1984. Rsch. asst. prof. U. Waterloo, Ont., 1985-89, rsch. assoc. prof., 1989-91, assoc. prof. geography, 1991—; interim dir. Wetlands Rsch. Inst., 1991—; vis. prof. U. Neuchatel, 1993. Editor: Methods in Quaternary Ecolgy, 1990; contbr. articles to profl. jours. Postdoctoral fellow Natural Scis. & Engring. Rsch. Coun. of Can., 1984-85, rsch. fellow, 1985-90. Fellow Geol. Assn. Can. Office: Univ of Waterloo, Dept Geography, Waterloo, ON Canada N2L 3G1

WARNER, JANET CLAIRE, software design engineer; b. Portland, Oreg., May 2, 1964; d. W. J. and Wendelyn A. (Twombly) W. Student, Clackamas Community Coll., 1982-85; BS in Computer Sci., U. Portland, 1987, MSEE, 1992. Systems asst. U. Portland, 1986-87, programmer Applied Rsch. Ctr., 1987; software design engr. Photon Kinetics, Inc., Beaverton, Oreg., 1987-92; ind. software cons., 1993—. Mem. IEEE, Assn. Computing Machinery (chmn. U. Portland chpt. 1986-87), Soc. Women Engrs. (treas. Oreg. sect. 1988-89), Eta Kappa Nu (treas. Theta Beta chpt. 1991-92), Portland Rose Soc. Avocations: drawing, photography, sailing, swimming, downhill skiing, gardening.

WARNER, LESLEY RAE, biology educator; b. Milton, New Zealand, July 21, 1940; arrived in Australia, 1970; d. Raymond John and Fay Catherine (Sanders) Wilson; m. Roger Joseph Smales (div. 1981); children: Alastair Grantly, Fiona Ruth; m. George Oliver Warner. BHSc, Otago U., Dunedin, New Zealand, 1961, BSc, 1963, MSc with honors, 1965; PhD, Adelaide U., Australia, 1975; grad. diploma in Edn., Adelaide Coll. Arts/Edn., 1981. Sr. lectr. Gippsland Inst., Churchill, Victoria, Australia, 1982-85; head dept. biology Capricornia Inst., Rockhampton, Australia, 1986-89; dean Sch. Sci. Univ. Coll. Central Queensland, Rockhampton, Victoria, Australia, 1989-90; head dept. biology Univ. Central Queensland, Rockhampton, Australia, 1991—. Author: Gippsland Flavours, 1988. Fellow Royal Soc. South Australia; mem. Wildlife Disease Assn., Australian Inst. Biology, Australian Soc. Parasitology, Zonta Internat. (Rockhampton pres. 1991-93). Mem. Uniting Church of Australia. Office: U Central Queensland, Rockhampton 4702, Australia

WARNER, RICHARD DAVID, research foundation executive; b. Batavia, N.Y., July 22, 1943; s. Wesley Elmer and Adelia May (Elwell) W.; m. Susan Connelley, June 6, 1964; children: Mindy Sue, Craig Mathew; m. Eula Lorraine Anderson, Mar. 1, 1974; 1 child, Richard David Jr. BS, Houghton Coll., 1965; MS, Pa. State U., 1967. Rschr. Eastman Kodak Co., Rochester, N.Y., 1967-72; dir. computer programs Rayne Internat., N.Y.C., 1972-73; dir. R & D Consol. Internat., Chgo., 1973-74, Nashville Electrographics Co., 1975-76; rsch. dir. Graphic Arts Tech. Found., Pitts., 1977—; mem. coun. Rsch. and Engring., Chadsford, Pa., 1989—; vice chmn. Internat. Assn. Rsch. Insts. for Graphic Arts Industries, Dramstadt, Fed. Republic Germany, 1990—. Contbr. articles to profl. publs. Achievements include invention of GATF's color communicator; patent for GATS's Dot Gain Scale-II; patent pending for GATF's aceutance guide. Office: Graphic Arts Tech Found 4615 Forbes Ave Pittsburgh PA 15213-3796

WARNER, ROLLIN MILES, JR., economics educator, financial planner, real estate broker; b. Evanston, Ill., Dec. 25, 1930; s. Rollin Miles Warner Sr. and Julia Herndon (Polk) Clarkson. BA, Yale U., 1953; cert. in law, Harvard U., 1956; MBA, Stanford U., 1960; cert. in edn. adminstrn., U. San Francisco, 1974. Asst. to v.p. fin. Stanford U., 1960-63; instr. history Town Sch., San Francisco, 1963-70; instr. econs. and math., dean Town Sch., 1975—; prin. Mt. Tamalpais, Ross, Calif., 1972-74; dir. devel. Katharine Branson Sch., Ross, 1974-75, instr. in econs. and math., computer-aided design and Mathematica; cons. Educators Collaborative, San Anselmo, Calif., 1983—, Nat. Ctr. for Fin. Edn., San Francisco, 1986—. Author: America, 1986, Europe, 1986, Africa, Asia, Russia, 1986, Greece, Rome, 1981, Free Enterprise at Work, 1986. Scoutmaster to dist. commr. Boy Scouts Am., San Francisco, 1956—. Recipient Silver Beaver award Boy Scouts Am., 1986. Mem. Am. Econs. Assn., Am. Mgmt. Assn., Math. Assn. Am., Inst. Cert. Fin. Planners, Manteca Bd. Realtors, Boston Computer Soc., Berkeley Macintosh User's Group, Real Estate Cert. Inst., Grolier Club N.Y., Univ. Club San Francisco, San Francisco Yacht Club (Belvedere, Calif.), Old Oundelian Club London. Office: Town Sch 2750 Jackson St San Francisco CA 94115-1195

WARNICK, WALTER LEE, mechanical engineer; b. Balt., May 31, 1947; s. Marvin Paul and Freda (Wilt) W.; m. Metta Ann Nichter, May 2, 1970; children: Ashlie Colleen, Leah Brooke. BS in Engring., Johns Hopkins U., 1969; PhD, U. Md., 1977. Registered profl. engr., Md. Engr. Westinghouse Electric Co., Linthicum, Md., 1969-71, U.S. Naval Rsch. Lab., Washington, 1971-77; engr. U.S. Dept. Energy, Washington, 1977-85, sr. exec., 1985—; Dept. Energy rep. Nat. Acid precipitation Assessment Program, Washington, 1981—. Author: Warnick Families of Western Maryland, 1988, Wilt Families of Western Maryland, 1991. Pres. citizen's adv. com. to Howard County Bd. of Edn., Ellicott City, Md., 1980-82. Mem. ASME, Sigma Xi. Office: US Dept Energy Washington DC 20585

WARNOCK, DAVID GENE, nephrologist, pharmacology educator; b. Parks, Ariz., Mar. 5, 1945. MD, U. Calif., San Francisco, 1970. Assoc. prof. medicine and pharmacology U. Calif., San Francisco, 1982—; chief nephrology sect. VA Med. Ctr., 1983—. Mem. AAAS, Am. Physiol. Soc., Am. Soc. Clin. Investigation, Am. Soc. Nephrology. Office: U Ala Nephrology Rsch & Tng Ctr UAB Sta Birmingham AL 35294*

WARREN, AMYE RICHELLE, psychologist, educator; b. Waycross, Ga., Jan. 22, 1958; d. John Roy and Vera Jo (Ballew) W.; m. David M. Leubecker (div.); m. Randall Travis Cagle, June 10, 1990. BS in Psychology, Ga. Tech. U., 1980, PhD, 1984. Asst. prof. U. Tenn., Chattanooga, 1984-87, assoc. prof., 1987—; vis. scholar U. Ariz., Tucson, 1990-91. Contbr. articles to profl. jours. and chpts. to books. Rsch. planning grantee NSF, 1990; named Outstanding Tchr. U. Tenn. Nat. Alumni Assn., 1987. Mem. Am. Psychol. Assn., Soc. for Rsch. in Child Devel., Am. Psychology-Law Soc. Democrat. Presbyterian. Office: U Tenn Dept Psychology 615 McCallie Ave Chattanooga TN 37403

WARREN, BARBARA KATHLEEN (SUE WARREN), wildlife biologist; b. Appleton, Wis., Oct. 3, 1943; d. Richard Grant and Beatrice Marie (Kath) Henika. Diploma, St. Luke's Sch. Nursing, San Francisco, 1965; AS in Forest Tech., Green River Community Coll., Auburn, Wash., 1976; BS in Wildlife Biology, U. Calif., Davis, 1990. RN, Calif.; cert. merit wildlife program devel. Nurse ICU, Ross (Calif.) Gen. Hosp., 1965-66; nurse emergency room CCU, St. Luke's Hosp., 1966-68; head nurse ICU and CCU, Valley Gen. Hosp., Auburn, 1971-76; forest technician Wash. Dept. Natural Resources, Husum, 1976-77; nurse emergency room ICU, Marshall Hosp., Placerville, Calif., 1977-78; forestry and wildlife biology technician U.S. Forest Svc., Pioneer, Calif., 1978-89, trainer for critical incident stress, Region 5, 1987—, dist. wildlife biologist, 1989-90, career advisor, 1990—; asst. forest wildlife biologist U.S. Forest Svc., 1990-91, asst. dist. ranger, 1991-92. Chmn. outdoor program com. Girl Scouts U.S.A., Sacramento, 1981-85, master planning cons., 1984-86; vol. ARC, Sacramento, 1985—, vol. disaster nurse, 1986—. With Nurse Corps, U.S. Army, 1968-71. Recipient award for outstanding svc. Girl Scouts U.S.A., 1987, Role Model of Yr. award, 1989; Sustained Superior Performance and Host of Yr. award Eldorado Nat. Forest, 1988, Regional Affirmative Action award U.S. Forest Svc., 1990, Outstanding Woman award YWCA, 1991. Mem. Wildlife Soc. Democrat. Avocations: music, photography, camping, crafts. Office: Dist Ranger Pineridge Rd PO Box 300 Shaver Lake CA 93664

WARREN, CHRISTOPHER CHARLES, electronics executive; b. Helena, Mont., July 27, 1949; s. William Louis and Myrtle Estelle (Moren) W.; m. Danette Marie Georgde, Apr. 21, 1972; 1 child, Jeffrey Scott. Grad. high sch., Helena, 1967. Electrician Supreme Electronics, Helena, 1972-81; v.p., svc. technician Capital Music Inc., Helena, 1981—; state exec. Amusement & Music Operators Assn. Coun. of Affiliated States, Chgo., 1990-92, bd. dirs., 1992—. Sgt. USAF, 1968-72, Vietnam. Mem. Internat. Flipper Pinball Assn. (sec., treas. 1992—), Mont. Coin Machine Operators State 8-Ball (chmn.), Valley Nat. 8 Ball Assn. (charter), Amusement and Music Operators Assn. (bd. dirs. 1992—), Ducks Unltd., Eagles, Moose, Rocky Mountain Elk Found. Avocations: photography, restoring old cars and trucks, hunting, fishing. Home: 8473 Green Meadow Dr Helena MT 59601-9379 Office: Capital Music Inc 3108 Broadwater Ave Helena MT 59601-9201

WARREN, D. ELAYNE, environmental sanitarian; b. Wynne, Ark., Nov. 18, 1949; d. Thelma Asilee Warren; m. Thomas G. Casey II, Feb. 20, 1990; 1 child, Adam Samuel. BA, St. John's Coll., 1973; postgrad., U. Md., 1984-86. Indexer N.Y.C., 1973-81; info. mgmt. specialist Hill and Assocs. Inc., Annapolis, Md., 1982-87; info. mgr. Inst. for Injury Reduction, Dunkirk, Md., 1988-90; registered sanitarian State of Md., Cumberland, 1990—. Contbr. articles to profl. jours. Vol. Mt. Savage Hist. Soc., 1991. Quaker. Home: PO Box 490 Mount Savage MD 21545-0490

WARREN, DONALD WILLIAM, physiology educator, dentistry educator; b. Bklyn., Mar. 22, 1935; s. Sol B. and Frances (Plotkin) W.; m. Priscilla Girardi, June 10, 1956; children: Donald W. Jr., Michael C. BS, U. N.C., 1956, DDS, 1959; MS, U. Pa., 1961, PhD, 1963; d Odontology honoris causa, U. Kuopio, Finland, 1991. Asst. prof. dentistry U. N.C., Chapel Hill, 1963-65, dir. Craniofacial Ctr., 1963—, assoc. prof., 1965-69, prof., 1969-80, chmn. dept. dental ecology, 1970-85, Kenan prof., 1980—, rsch. prof. otolaryngology, 1985—; cons. NIH, Bethesda, Md., 1967—, R. J. Reynolds-Nabisco, Winston-Salem, N.C., 1986—. Contbr. articles to profl. jours. Recipient Honor award Am. Cleft Palate Assn./Craniofacial Assn., 1992, O. Max Garner award U. N.C. Bd. Govs., 1993. Fellow AAAS, Internat. Coll. Dentists, Am. Speech and Hearing Lang. Assn., Internat. Assn. Dental Rsch., Acoustical Soc. Am., Am. Cleft Palate Assn. (pres. 1981-82, Disting. Svc. award 1984), Am. Cleft Palate Edn. Found. (pres. 1976-77). Avocations: horse related activities, running, farming. Home: PO Box 1356 Southern Pines NC 28388 Office: U NC Dental Rsch Ctr CB # 7455 Chapel Hill NC 27514

WARREN, JOAN LEIGH, pediatrician; b. St. Louis, Oct. 15, 1957; d. Harold Lee and Lorraine Jeanette (Hurley) W.; m. Frank Snipes, May 15, 1981 (div. 1989); 1 child, Mallory. BA, U. Mo., Kansas City, 1980, MD, 1982. Diplomate Am. Bd. Pediatrics. Resident in pediatrics Sacred Heart Children's Hosp., Pensacola, Fla., 1982-85; pvt. practice, St. Louis, 1985-87; dir. pediatric svcs. Barnes St. Peters (Mo.) Hosp., 1988—. Fellow Am. Acad. Pediatrics; mem. NAFE, St. Louis Pediatric Soc. Avocations: tennis, softball, camping. Office: Barnes St Peters Hosp 10 Hospital Dr Saint Peters MO 63376

WARREN, RICHARD M., experimental psychologist, educator; b. N.Y.C., Apr. 8, 1925; s. Morris and Rae (Greenberg) W.; m. Roslyn Pauker, Mar. 31, 1950. BS in Chemistry, CCNY, 1946; PhD in Organic Chemistry, N.Y. U., 1951. Flavor chemist Gen. Foods Co., Hoboken, N.J., 1951-53; rsch. assoc. psychology Brown U., Providence, 1954-56; Carnegie sr. rsch. fellow Coll. Medicine NYU, 1956-57; Carnegie sr. rsch. fellow Cambridge (Eng.) U., 1957-58, rsch. psychologist applied psychology Rsch. Unit, 1958-59; rsch. psychologist NIMH, Bethesda, Md., 1959-61; chmn. psychology Shimer Coll., Mt. Carroll, Ill., 1961-64; assoc. prof. psychology U. Wis., Milw., 1964-66, prof., 1966-73, rsch. prof., 1973-75, disting. prof., 1975—; vis. scientist Inst. Exptl. Psychology, Oxford (Eng.) U., 1969-70, 77-78. Author: (with Roslyn Warren) Helmholtz on Perception: Its Physiology and Development, 1968, Auditory Perception: A New Synthesis, 1982; contbr. articles on sensation and perception to profl. jours. Fellow APA, Am. Psychol. Soc.; mem. AAAS, Acoustical Soc. Am., Am. Chem. Soc., Am. Speech and Hearing Assn., Sigma Xi. Office: U Wis Dept Psychology Milwaukee WI 53201

WARREN, RICHARD WAYNE, obstetrician and gynecologist; b. Puxico, Mo., Nov. 26, 1935; s. Martin R. and Sarah E. (Crump) W.; m. Rosalie J. Franzola, Aug. 16, 1959; children: Lani Marie, Richard W., Paul D. BA, U. Calif., Berkeley, 1957; MD, Stanford U., 1961. Intern, Oakland (Calif.) Naval Hosp., 1961-62; resident in ob-gyn Stanford (Calif.) Med. Ctr., 1964-67; practice medicine specializing in ob-gyn, Mountain View, Calif., 1967—; mem. staff Stanford and El Camino hosps.; pres. Richard W. Warren M.D., Inc.; assoc. clin. prof. ob-gyn Stanford Sch. Medicine. Served with USN, 1961-64. Diplomate Am. Bd. Ob-Gyn. Fellow Am. Coll. Ob-Gyn; mem. AMA, Am. Fertility Soc., Am. Assn. Gynecologic Laparoscopists, Calif. Med. Assn., San Francisco Gynecol. Soc., Peninsula Gynecol. Soc., Assn. Profs. Gynecology and Obstetrics, Royal Soc. Medicine, Shufelt Gynecol. Soc. Santa Clara Valley. Contbr. articles to profl. jours. Home: 102 Atherton Ave Menlo Park CA 94027-4021 Office: 2500 Hospital Dr Mountain View CA 94040-4106

WARREN, TOMMY MELVIN, petroleum engineer; b. Tallassee, Ala., Aug. 26, 1951; s. Henry M. and Christene (Jones) W.; m. Carolyn Milam; children: Lisa, Wesley, Eric. BS in Mineral Engring., U. Ala., Tuscaloosa, 1973, MS in Mineral Engring., 1976. Ops. engr. Amoco Prodn. Co., Lafayette, La., 1973-75; rsch. engr. Amoco Prodn. Co., Tulsa, 1976-86, rsch. supr., 1986—; mem. steering com. Nat. Program for Advanced Drilling and excavation Techs., 1992. Contbr. articles to Jour. Petroleum Tech., Soc. Petroleum Engrs. Jour., SPE Drilling Engring. Recipient Rossiter Raymond award AIME, 1982. Mem. So. Petroleum Engrs. (chmn. tech. coverage drilling com. 1985, com. chmn. Drilling Engring. award 1984-86, tech. editor 1987-90, rev. chmn. 1990-92) Cedrick K. Ferguson medal 1985). Republican. Baptist. Achievements include patents for drilling tools and methods. Home: Rt 1 Box 130 10 Coweta OK 74429 Office: Amoco Prodn Co PO Box 3385 4502 E 41st St Tulsa OK 74102

WARRINGTON, ROBERT O'NEIL, JR., mechanical engineering educator and administrator, researcher; b. Sparta, Wis., Mar. 13, 1945; s. Robert O'Neil and Virginia (Johnson) W.; m. Anne Cabell, Aug. 14, 1968; children: Robert O'Neil III, Daniel Scott, Kristy Cabell. BS in Aerospace Engring., Va. Poly. Inst., 1967; MS in Mech. Engring., U. Tex.-El Paso, 1971; PhD in Mech. Engring., Mont. State U., 1975. Instr. U. Tex.-El Paso, 1971; instr. Mont. State U., Bozeman, 1975-76, asst. prof., 1976-80, assoc. prof., 1980-83; prof. mech. engring., dept. head La. Tech. U., Ruston, 1983—, track and field ofcl., 1983—; cons. Atlas Processing Co., Shreveport, La., 1984—. Author: Montana Energy Primer, 1978; contbr. articles to profl. jours. Coach Little League, Bozeman, 1978-83, Soccer League, Bozeman, 1981-83. Served with U.S. Army, 1968-70. Grantee NSF, Exxon Found., Hewlett-Packard Corp., C.E., Dept. Energy, 1975—, AT&T Found. Mem. ASME (K-8 com. 1983—), Am. Soc. for Engring Edn. (vice-chmn., program chmn. mech. engring. div.), Mont. Engring. Assocs. (sec.-treas. 1979—), Nat. Ski Patrol, Sigma Chi. Republican. Roman Catholic. Office: LA Tech Univ Mech Engring Dept PO Box 10348 Tech Sta Ruston LA 71272

WARSHAW, STEPHEN ISAAC, physicist; b. N.Y.C., Mar. 26, 1939; s. Montague Israel and Sadie (Walters) W.; m. Harriet Golden, Aug. 2, 1964; children: Anne, Michael. BSc, Poly. Inst. Bklyn., 1960; PhD, Johns Hopkins U., 1966. Postdoctoral rsch. assoc. dept. physics U. Ill., Urbana, 1966-68; sr. physicist U. Calif./Lawrence Livermore Nat. Lab., 1968—. Author jour. articles, lab. reports, computer programs. Mem. Am. Phys. Soc., Acoustical Soc. Am., Am. Geophys. Union. Home: 218 Canyon Creek Ct San Ramon CA 94583 Office: Lawrence Livermore Nat Lab L 298 PO Box 808 Livermore CA 94550

WARTH, JAMES ARTHUR, physician, researcher; b. N.Y.C., Apr. 30, 1942; s. Peter and Anne (Furgang) W.; m. Maria Archer Russell, May 3, 1969; children: David M., Andrew A. BS, Tufts U., 1963, MD, 1967. Diplomate Am. Bd. Internal Medicine, Am. Bd. Hematology, Am. Bd. Oncology. Hematologist Harvard Health Svcs. Harvard U., Cambridge, Mass., 1976-77, officer, 1976-77; attending hematologist Harper-Grace Hosps., Detroit, 1977-84; asst. prof. medicine Wayne State U., Detroit, 1977-84; rsch. scientist New Eng. Med. Ctr., Boston, 1984-86; attending hematologist-oncologist The Faulkner Hosp., Boston, 1986—; asst. prof. medicine Tufts U. Sch. Medicine, Boston, 1986—; course dir. phys. diagnosis Faulkner Hosp., Boston, 1993—; cons. NIH, Bethesda, Md., 1987, 80-83, Mas. Profl. Rev. Orgn., Waltham, 1991—; vis. prof. Yale U., New Haven, 1986; invited lectr. on oxidants and the erythrocyte SUNY, Syracuse, 1991, Tufts U.-NEMC, Boston, 1992; faculty advisor Sch. Medicine, Tufts, U., 1991—: appearance on NBC affiliate TV news program, Detroit, 1980; rsch. lectr. NIH, Tarrytown, N.Y., 1986. Contbg. author: (textbook) Hematologic Disorders in Maternal-Fetal Medicine, 1990; reviewer Am. Jour. Hematology, 1986, Jour. of Andrology, 1990-92; contbr. articles to profl. jours. Maj. U.S. Army, 1969-71. Rsch. grantee NIH, 1980-83, 83-86, spl. fellowship, 1974-76. Fellow Am. Coll. Physicians; mem. Am. Soc. Hematology, Am. Fedn. Clin. Rsch., Biomembranes in Sickle Cell Rsch. Group. Avocations: art, music, architecture, tennis. Office: Faulkner Hosp 1153 Centre St Rm 5950 Boston MA 02130

WARWICK, WARREN J., pediatrics educator; b. Racine, Wis., Jan. 27, 1928; married; 2 children. BA, St. Olaf Coll., 1950; MD, U. Minn., 1954. Med. fellow pediatrics U. Minn., Mpls., 1955-57, med. fellow specialist, 1959-60, from instr. to assoc. prof. pediatrics, 1960-78, prof. pediatrics, 1978—; mem. ctr. program com. Nat. Cystic Fibrosis Rsch. Found., 1964-66, chmn. med. care com., 1966-71, coop. study com., 1971-72; mem. exec. bd. sci.-med. com. Internat. Cystic Fibrosis (Mucoviscidosis) Assn., 1970-80; mem. nat. data registry com. Cystic Fibrosis Found., 1972-86; Annalisa Marzotto cystic fibrosis chmn. Cardiovascular Rsch. fellow Alpha Omega Phi, 1955-57, Rsch. fellow Am. Heart Assn., 1959-60; recipient Rsch. Career Devel. award USPHS, 1961-66. Achievements include research in pulmonary diseases, experimental pathology, immunology, cystic fibrosis. Office: U Minn Hosp Cystic Fibrosis Ctr 420 Delaware St SE Minneapolis MN 55455*

WASCH, ALLAN DENZEL, technical publication editor; b. Barberton, Ohio, Jan. 3, 1962; s. Denzel David and Irmgard Elisabeth (Herrmann) W. BS in Biology, Mich. State U., 1984. Editor ASM Internat., Materials Park, Ohio, 1985—. Mem. Lions. Democrat. Home: 641 W Glendale St Bedford OH 44146 Office: ASM Internat Materials Information Materials Park OH 44073

WASFIE, TARIK JAWAD, surgeon, educator; b. Baghdad, Iraq, July 1, 1946; m. Barina Y. Wasfie, Mar. 11, 1975; children: Giselle, Nissan. BS, Central U., Iraq, 1964; MD, Baghdad Med. Sch., 1970. Surg. rsch. assoc. Sinai Hosp. of Detroit/Wayne State U., 1981-85; clin. fellow Coll. Phys. & Surg., Columbia U., N.Y.C., 1985-91, postdoctoral rsch. scientist, 1987-91; attending surgeon Mich. State U./McLaren Hosp., Flint, 1991—. Contbr. articles to profl. jours. NIH grantee, 1984. Fellow ACS (assoc.); mem. AMA, Mich. State Med. Soc., Flint Acad. Surgeons, Am. Soc. Artificial Internal Organs, Internat. Soc. Artificial Organs, Soc. Am. Gast. Endoscopic Surgeons. Achievements include production of antiidiotypic antibodies and their role in transplant immunology; development of percutenous access device. Home: 1125 Kings Carriage Grand Blanc MI 48439

WASHBURN, BARBARA, cartography researcher; m. Henry Bradford Washburn Jr., Apr. 27, 1940; children: Dorothy Polk, Edward Hall, Elizabeth. Asst. to. Henry Bradford Washburn Jr. Recipient Centennial award Nat. Geog. Soc., 1988. Home: 220 Somerset St Belmont MA 02178-2011*

WASHBURN, H. BRADFORD, JR., museum administrator, cartographer, photographer; b. Cambridge, Mass., June 7, 1910; s. Henry Bradford and Edith (Hall) W.; m. Barbara Teel Polk, Apr. 27, 1940; children: Dorothy Polk, Edward Hall, Elizabeth Bradford. Grad., Groton Sch., 1929; A.B., Harvard U., 1933, A.M., 1960, D.H.L. (hon.), 1975; postgrad., Inst. Geog. Exploration, 1934-35; postgrad. hon. degrees; Ph.D., U. Alaska, 1951; D.Sc., Tufts U., 1957, Colby Coll., 1957, Northeastern U., 1958, U. Mass., 1972, Curry Coll., 1982; D.F.A., Suffolk U., 1965; D.H.L., Boston Coll., 1974; LL.D., Babson Coll., 1980. Instr. Inst. Geog. Exploration, Harvard U., 1935-42; dir. Mus. Sci., Boston, 1939-80, chmn. of the corp., 1980-85, hon. dir., 1985—; dir. Mountaineer in Alps, 1926-31; explorer Alaska Coast Range, 1930-40; served as leader numerous mountain, subarctic area explorations; cons. various govtl. agys. on Alaska and cold climate equipment; leader in spl. expdns. investigating high altitude cosmic rays, Alaska, 1947; rep. Nat. Geog. Soc., 17th Internat. Geog. Congress, 1952; leader Nat. Geog. mapping expdns. to, Grand Canyon, 1971-75; chmn. Mass. Com. Rhodes Scholars, 1959-64; chmn. arts and scis. com. UNESCO conf., Boston, 1961; mem. adv. com. John F. Kennedy Library, 1977; mem. vis. com. Internat. Mus. Photography, 1978; mem. U.S. Nat. Commn. for UNESCO, 1978; lectr. work of Yukon Expdn., Royal Geog. Soc., London, 1936-37, on mapping Grand Canyon, 1976; lectr. Mus. Imaging Tech., Bangkok, 1989, Royal Geog. Soc., London, on mapping Mt. Everest, 1990. Contbr. articles, photographs on Alaska, Alps, glaciers, and mountains to mags., books.; editor, pub. 1st large-scale map Mt. McKinley, Am. Acad. Arts and Scis.-Swiss Found. Alpine Rsch., Bern, 1960; mapped Mt. Kennedy for Nat. Geog. Soc., 1965, Grand Canyon, 1971-74, Muldrow Glacier (Mt. McKinley), 1977; editor new chart, Squam Lake, N.H., 1968, new Grand Canyon map for Nat. Geog. Soc., 1978, Bright Angel Trail map, 1981; photo-mapped Mt. Everest for Nat. Geog. Soc., 1984; dir., pub. large-scale map of Mt. Everest for Nat. Geog. Soc., 1984-88; project chief new 1:50,000 map of Mt. Everest for Nat. Geog. Soc. and Boston Sci. Mus., 1988; pub. Tourist Guide to Mt. McKinley, 1971, new map of Presdl. Range, N.H, 1989; completed new large-scale relief model Mt. Everest, 1990; one-man photographic shows Whyte Art Mus., Banff, Can., Internat. Mus. Photography, N.Y.C., Rochester, N.Y. Bd. overseers Harvard, 1955-61; trustee Smith Coll., 1962-68, Richard E. Byrd Found., 1979-84, Mt. Washington Obs., 1979—; mem. Task Force on Future Financing of Arts in Mass., 1978; hon. bd. dirs. Swiss Found. Alpine Research, 1984—. Recipient Royal Geog. Soc. Cuthbert Peek award for Alaska Exploration and Glacier Studies, 1938, Burr prize Nat. Geog. Soc., 1940, 65, Stratton prize Friends of Switzerland, 1970, Lantern award Rotary Club, Boston, 1978, New Englander of Yr. award New Eng. Coun., 1974, Gold Research medal Royal Scottish Geog. Soc. (with wife), 1979, Alexander Graham Bell award Nat. Geog. Soc., 1980, Disting. Grotonian award Groton Sch., 1979, Explorers medal Explorers Club, 1984, award for lifelong contbns. to cartography and surveying Engring. Socs. New Eng., 1985; named Bus. Statesman of Yr. Harvard Bus. Sch. Assn., Boston, 1970; named to Acad. Disting. Bostonians Boston C. of C., 1983; one of nine Photographic Masters, Boston U. Fellow Royal Geog. Soc. London, Harvard Travelers Club (Gold medal 1959), Nat. Geog. Soc. (with wife, Centennial award 1988), AAAS, Am. Acad. Arts and Scis., Am. Geog. Soc. (hon.), Calif. Acad. Sci., Assn. Sci. Tech. Centers (hon.), Arctic Inst. N.Am., U. Ariz. Ctr. for Creative Photography; mem. Explorers Club (hon. dir.), French Alpine (hon. mem. Groupe de Haute Montagne, Paris), Phi Beta Kappa (hon.), Chinese Assn. Sci. Expeditions (hon.). Clubs: Commercial, Harvard Varsity, St. Botolph (hon. life), Aero Club of New Eng. (hon.), Harvard Mountaineering (Boston) (hon., past pres.); American Alpine (.Y.C.) (hon.); Alpine (London) (hon.); Sierra of San Francisco (hon.); Mountaineers (Seattle) (hon.); Mountaineering of Alaska (hon.); hon. mem. several clubs. Leader 1st ascent Mt. Crillon, Alaska, 1934, Nat. Geog. Soc. Yukon Expdn., 1935; leader 1st aerial photog. exploration Mt. McKinley, 1936, ascended its summit, 1942, 47, 51; leader 1st aerial exploration St. Elias range, 1938; 1st ascents Mount Sanford and Mount Marcus Baker in Alaska, 1938, Mt. Lucania, Yukon, 1937, Mt. Bertha, Alaska, 1940, Mt. Hayes, Alaska, 1941; 1st ascent West side Mt. McKinley 1951; leader Nat. Geog. Soc. Mt. Everest mapping project, 1981-88; expdn. to S.E. Asia, guest Chinese Acad. Scis., met with King of Nepal, 1988; leader expdn. to Nepal, 1992; 1st laser-distance observation to summit of Mt. Everest, 1992; 50th trip to Alaska to open exhibit of own photos Anchorage Art Mus., 1993. Home: 220 Somerset St Belmont MA 02178-2011 Office: Science Park Boston MA 02114

WASHINGTON, ALLEN REED, chemist. BS in Biology, Northeast Okla. State U., 1976. Lic. water-sewer lab. operator, Okla. Operator waste treatment plant Ft. Howard Paper Co., Muskogee, Okla., 1978-80; plant chemist pollution control City of Muskogee, 1980-87, plant chemist water plant, 1987—. Mem. AFSCME (bd. dirs., negotiator), Am. Water Works Assn., Okla. Water and Pollution Control Assn., Water Environ. Fedn., U.S. Chess Fedn. Office: Muskogee Water Plant 3500 Port Pl Muskogee OK 74403

WASHINGTON, DAVID EARL, computer systems professional; b. Ft. Valley, Ga., Feb. 7, 1962; s. Earl Lewis and Charlotte Ruth (Chapman) W.; m. Diana Renee Turner, May 29, 1992. AS in Logistics Mgmt., Ga. Mil. Coll., 1992. Electronics technician Robins AFB, Warner Robins, Ga., 1985—, mgr. VAX cluster, 1990—. With USN, 1981-85. Recipient Civil Svc. Superior Performance award Robins AFB, 1990-92. Mem. DEC User Group. Methodist. Achievements include computer systems management for first paperless depot repair facility for Department of Defense; management of computer systems for low altitude navigation and targeting infrared for night (LANTIRN) weapon system. Home: 150 Wrights Mill Cir Warner Robins GA 31088 Office: Bldg 640 WR-ALC/LYPBD Robins AFB GA 31098

WASHINGTON, REGINALD LOUIS, pediatric cardiologist; b. Colorado Springs, Colo., Dec. 31, 1949; s. Lucius Louis and Brenette Y. (Wheeler) W.; m. Billye Faye Ned, Aug. 18, 1973; children: Danielle Larae, Reginald Quinn. BS in Zoology, Colo. State U., 1971; MD, U. Colo., 1975. Diplomate Nat. Bd. Med. Examiners, Am. Bd. Pediatrics, Pediatric Cardiology. Intern in pediatrics U. Colo. Med. Ctr., Denver, 1975-76, resident in pediatrics, 1976-78, chief resident, instr., 1978-79, fellow in pediatric cardiology, 1979-81, asst. prof. pediatrics, 1982-1988, assoc. prof. pediatrics, 1988-90, assoc. clin. prof. pediatrics, 1990—; staff cardiologist Children's Hosp., Denver, 1981-90; v.p. Rocky Mountain Pediatric Cardiology, Denver, 1990—; mem. admissions com. U. Colo. Sch. Medicine, Denver, 1985-89; bd. dirs. Children's Health Care Assn.. Cons. editor Your Patient and Fitness, 1989—. Adv. bd. dirs. Equitable Bank of Littleton, Colo., 1984-86; bd. dirs. Cen. City Opera, 1989—, Cleo Parker Robinson Dance Co., 1992—, Rocky Mountain Heart Fund for Children, 1984-89, Raindo Ironkids, 1989—; nat. bd. dirs. Am. Heart Assn., 1992—; bd. dirs. Nat. Coun. Patient Info. and Edn., 1992—, Children's Heart Allliance, 1993—. Named Salute Vol. of Yr. Big Sisters of Colo., 1990. Fellow Am. Acad. Pediatrics (cardiology subsect.), Am. Coll. Cardiology, Am. Heart Assn. (coun. on cardiovascular disease in the young, exec. com. 1988-91, nat. devel. program com. 1990-94, vol. of yr. 1989, pres. Colo. chpt. 1989-90, Torch of Hope 1987, Gold Heart award Colo. chpt. 1990, bd. dirs. Colo. chpt., exec. com. Colo. chpt. 1987—, grantee Colo. chpt. 1983-84, mem. editorial bd. Pediatric Exercise Scis. 1988—), Soc. Critical Care Medicine; mem. Am. Acad. Pediatrics/Perinatology, N.Am. Soc. Pediatric Exercise Medicine (pres.), Colo. Med. Soc. (chmn. sports medicine coun. 1993—), Denver C. of C. (gov.'s coun. for phys. fitness 1990-91, Leadership Denver 1990), Denver Athletic Club, Met. Club, Glenmoor Golf Club. Democrat. Roman Catholic. Avocations: skiing, fishing. Home: 7423 Berkeley Cir Castle Rock CO 80104-9278 Office: Rocky Mountain Pediatric Cardiology 1601 E 19th Ave Ste 5600 Denver CO 80218-1022

WASSELL, STEPHEN ROBERT, mathematics educator, researcher; b. Santa Monica, Calif., Jan. 17, 1963; s. Desmond Anthony and Catherine Ann (Stephens) W. BS in Architecture, U. Va., Charlottesville, 1984, PhD in Math., 1990. Programmer, analyst UNISYS, McLean, Va., 1984-85, graphics artist, 1986; tutor, Summer Transition Program U. Va., Charlottesville, 1987-88, teaching asst., 1986-90; asst. prof. math. Sweet Briar (Va.) Coll., 1990—; prof. of record Ctr. for the Liberal Arts, U. Va., 1991; vis. asst. prof. math., Charlottesville, Va., 1992; doctoral cons., Charlottesville, 1989-90; tutor, Charlottesville, 1987-90. Author: (with L.E. Thomas) Schrödinger Operators, 1992; contbr. chpt. to book. Awarded grad. assistantship U. Va., 1986-90; Gordon T. Whyburn fellow, 1985-86. Mem. Am. Math. Soc., Math. Assn. Am., Sigma Nu (Beta chpt. treas. 1985-86). Achievements include patents for solar powered lawnmover, for solar shed, for ear muffs. Home: RR 1 Box 196 North Garden VA 22959-9602 Office: Sweet Briar Coll Dept Math Scis Sweet Briar VA 24595

WASSERBURG, GERALD JOSEPH, geology and geophysics educator; b. New Brunswick, N.J., Mar. 25, 1927; s. Charles and Sarah (Levine) W.; m. Naomi Z. Orlick, Dec. 21, 1951; children: Charles David, Daniel Morris. Student, Rutgers U.; BS in Physics, U. Chgo., 1951, MSc in Geology, 1952, PhD, 1954, DSc (hon.), 1992; Dr. Hon. Causa, Brussels U., 1985, U. Paris, 1986; DSc (hon.), Ariz. State U., 1987. Research assoc. Inst. Nuclear Studies, U. Chgo., 1954-55; asst. prof. Calif. Inst. Tech., Pasadena, 1955-59, assoc. prof., 1959-62, prof. geology and geophysics, 1962-82, John D. MacArthur prof. geology and geophysics, 1982—; served on Juneau Ice Field Research Project, 1950; cons. Argonne Nat. Lab., Lamont, Ill., 1952-55; former mem. U.S. Nat. Com. for Geochem., com. for Planetary Exploration Study, NRC, adv. council Petroleum Research Fund, Am. Chem. Soc.; mem. lunar sample analysis planning team (LSAPT) Manned Spacecraft Ctr., NASA, Houston , 1968-71, chmn.,1970; lunar sample rev. bd. 1970-72; mem. Facilities Working Group LSAPT, Johnson Space Ctr., 1972-82; mem. sci. working panel for Apollo missions, Johnson Space Ctr., 1971-73; advisor NASA, 1968-88, physical scis. com., 1971-75, mem. lunar base steering com., 1984; chmn. com. for planetary and lunar exploration, mem. space sci. bd. NAS, 1975-78; chmn. div. Geol. and Planetary Scis., Calif. Inst. Tech., 1987-89; vis. prof. U. Kiel, Fed. Republic of Germany, 1960, Harvard U., 1962, U. Bern, Switzerland, 1966, Swiss Fed. Tech. Inst., 1967, Max Planck Inst., Mainz and Heidelberg, Fed. Republic of Germany, 1985; invited lectr., Vinton Hayes Sr. Fellow, Harvard U., 1980, Jaeger-Hales lectr., Australian Nat. U., 1980, Harold Jeffreys lectr. Royal Astron. Soc., 1981, Ernst Cloos lectr., Johns Hopkins U., 1984, H.L. Welsh Disting. lectr. U. Toronto, Can. 1986., Danz lectr. U. Washington, 1989. Green vis. prof. U. B.C. 1982; 60th Anniversary Symposium speaker, Hebrew U., Jerusalem, 1985. Recipient Group Achievement award, NASA, 1969, Exceptional Sci. Achievement award, NASA, 1970, Disting. Pub. Service medal, NASA, 1973, J.F. Kemp medal Columbia U., 1973, Profl. Achievement award U. Chgo. Alumni Assn., 1978, Disting. Pub. Service medal with cluster, NASA, 1978, Wollaston medal Geol. Soc. London, 1985, Sr. Scientist award, Alexander von Humboldt-Stiftung, 1985, Crafoord prize Royal Swedish Acad. Scis., 1986. Gold medal Royal Astron. Soc., 1991; named Hon. Fgn. Fellow European Union Geoscis., 1983, recipient Holmes medal 1987; Rgents fellow Smithsonian Inst. Fellow AAAS, Am. Geophysical Union (planetology sect., Harry H. Hess medal 1985), Geol. Soc. Am. (life, Arthur L. Day medal 1970), Meteoritical Soc. (pres. 1987-88, Leonard medal 1975); mem. Geochem. Soc. (Goldschmidt medal 1978), Nat. Acad. Scis. (Arthur L. Day prize and lectureship 1981, J. Lawrence Smith medal 1985), Norwegian Acad. Sci. and Letters, Am. Phil. Soc. Research interests include geochemistry and geophysics and the application of the methods of chemical physics to problems in the earth scis. Major researches have been the determination of the time scales of nucleosynthesis, connections between the interstellar medium and solar material, the time of the formation of the solar system, the chronology and evolution of the earth, moon and meteorites, the establishment of dating methods using long-lived natural radio-activities, the study of geologic and lunar processes using nuclear and isotopic effects as a tracer in nature, the origin of natural gases, and the application of thermodynamic methods to geologic systems. Office: Calif Inst of Tech Divsn Geol & Planetary Scis Pasadena CA 91125

WASSERMAN, GERALD STEWARD, psychobiology educator; b. Bklyn., Nov. 22, 1937; s. Julius and Bessie (Weissman) W.; m. Louise Janet Mund, June 17, 1962; children: Mark Daniel, Rachel Lynn. BA, NYU, N.Y.C., 1961; PhD, MIT, 1965. Rsch. asst. NYU-Bellevue Med. Ctr., N.Y.C., 1959-61; grad. asst. MIT, Cambridge, Mass., 1961-63, grad. fellow, 1963-65; postdoctoral fellow NIH, Bethesda, Md., 1965-67; asst. prof. U. Wis., Madison, 1967-70, assoc. prof., 1970-85; prof. psychobiology Purdue U., West Lafayette, Ind., 1975—. Author: Color Vision, 1978; mem. editorial bd. Color Rsch. and Application, 1977-80; adv. editor Contemporary Psychology, 1981-87; mem. editorial bd. Biol. Signals, 1990—; contbr. articles to profl. jours. Pres. Temple Israel, West Lafayette, Ind., 1984-85. Recipient First Prize Midwest region Johns Hopkins Nat. Search for Computing to Assist the Disabled, Chgo., 1991. Fellow Am. Psychol. Soc., Optical Soc. Am.; mem. Internat. Soc. Psychophysics, Internat. Brain Rsch. Orgn., Acoustical Soc. Am., Soc. Neurosci. Libertarian. Jewish. Achievements include patent for sensory prosthesis; promulgation of task dependence hypothesis of sensory coding, temporal summation as an index of mental timing; design of artificial receptor for use in nerve deafness. Office: Purdue U Dept Psychol Scis West Lafayette IN 47906-1364

WASSERMAN, HARRY HERSHAL, chemistry educator; b. Austin, Mass., Dec. 1, 1920; s. Maurice Leonard and Rebecca (Franks) W.; m. Elga Ruth Steinherz, Jan. 1, 1947; children: Daniel M., Diana R., Steven A. BS, MIT, 1941; MA, Harvard U., 1942, PhD, 1949; MA (hon.), Yale U., 1962. Rsch. asst. O.S.R.D. Penicillin Project, 1945; instr., asst. prof. dept. chemistry Yale U., New Haven, Conn., 1948-57, assoc. prof. dept. chemistry, 1957-62, prof. dept. chemistry, 1962-82, Eugene Higgins prof. chemistry, 1982—; cons. Union Carbide Corp., 1956-66, Sandoz Corp., 1966-74, SmithKline Beckman Corp., 1974-80, Ortho Pharmaceuticals, 1980—. Author numerous articles on organic chemistry in sci. jours. Capt. USAF, 1942-45. Recipient William Devane Tchr.-Scholar award Phi Beta Kappa, 1977, Catalyst Teaching award Chem. Mfrs. Assn., 1985, Outstanding Tchr. award Yale U., 1985, Aldrich Synthetic Chemistry award Am. Chem. Soc., 1987. Mem. NAS, Am. Acad. Scis., Conn. Acad. Scis. Avocation: water color painting. Home: 192 Bishop St New Haven CT 06511-3718 Office: Yale U Dept Chemistry 225 Prospect Ave East Haven CT 06512-1958

WASSERMAN, KAREN BOLING, clinical psychologist, nursing consultant; b. Olney, Ill., July 29, 1944; d. Kenneth G. and Betty Jean (Varner) Boling; m. James M. Wasserman, Apr. 14, 1965; children: Nicole C., Michael B. RN, Barnes Hosp. Sch. Nursing, St. Louis, 1965; BA, Antioch Coll., 1977; Dr. of Psychology, Wright State U., 1986. Lic. psychologist, Miss., Ohio, Ind.; RN, Miss., Mo., Ohio. Staff nurse various med. facilities, 1965-76; instr. practical nurse program Ind. Vocat. Tech. Coll., Richmond, 1976-77; staff, float nurse Good Samaritan Hosp., Dayton, Ohio, 1977-78; pub. health nurse coord. Bur. Alcoholism Svcs., Dayton, 1978-79; alcoholism counselor IV Bur. Alcoholism Svcs., Dayton, Ohio, 1979-82; practicum student Wright State U. Sch. Profl. Psychology, Dayton, 1983-85; psychology intern Balt. VAMC Consortium, 1985-86; clin. psychologist Dayton VAMC, 1987-89; co-owner Fairhaven Pvt. Mental Health Clinic, Biloxi, Miss., 1989—; clin. psychologist Gulf Oaks Hosp., Biloxi, 1989—; Sand Hill Hosp., Gulfport, Miss., 1993—; psychiatric nursing cons. Mercy Hosp., Omaha, Council Bluffs, Ia., 1987. Chmn. community svcs. Altrusa Internat., Biloxi, 1990— (treas. 1993-94); mem. Evangelism com. First United Meth. Ch., Gulfport, Miss., 1991-93; Friend of the Rainbow Warrior, Greenpeace, 1986-93. Recipient Alumnae award in Acads., Barnes Hosp. Sch. Nursing, 1965. Mem. APA, Ohio Psychol. Assn., Miss. Psychol. Assn. (continuing edn. com. 1990—). Avocations: architecture, gardening, travel, movies. Office: Fairhaven Pvt Mental Health 280 Rue Petit Bois Biloxi MS 39531

WASSERMAN, ROBERT HAROLD, biology educator; b. Schenectady, Feb. 11, 1926; s. Joseph and Sylvia (Rosenberg) W.; m. Marilyn Mintz, June 11, 1950; children: Diane Jean, Arlene Lee, Judith Rose. B.S., Cornell U., 1949, Ph.D., 1953; M.S., Mich. State U., 1951. Research assoc. AEC project U. Tenn., Oak Ridge, 1953-55; sr. scientist med. div. Oak Ridge Inst. Nuclear Studies, 1955-57; asso. prof. dept. phys. biology N.Y. State Vet. Coll., Cornell U., 1957-63, prof., 1963—, James Law prof. physiology, 1989—, acting head phys. biology dept., 1963-64, 71, 75-76, chmn. dept. / sect. physiology, 1983-87, mem. exec. com. div. biol. sci., 1983—; vis. fellow Inst. Biol. Chemistry, Copenhagen, 1964-65; chmn. Conf. on Calcium Transport, 1962; co-chmn. Conf. on Cell Mechanisms for Calcium Transfer and Homeostasis, 1970; mem. adv. bd. Vitamin D Symposia, 1976—; mem. adv. bd. Symposia Calcium-Binding Proteins, 1977-88, chmn., 1977; mem. food and nutrition bd. NRC; cons. NIH, Oak Ridge Inst. Nuclear Studies; mem. pub. affairs com. Fedn. Am. Socs. Exptl. Biology, 1974-77 ; chmn. com. MPI, NRC. Bd. editors: Calcified Tissue Research, 1977-80, Procs. Soc. Exptl. Biol. Medicine, 1970-76, Cornell Veterinarian, Jour. Nutrition; contbr.: articles to profl. jours. Served with U.S. Army, 1944-45. Recipient Mead Johnson award, 1969, Andre Lichtwitz prize INSERM, 1982; Guggenheim fellow, 1964-65, 72, W.F. Neuman award Am. Soc. Bone and Mineral Rsch., 1990; NSF-OECD fellow, 1964-65. Fellow Am. Inst. Nutrition, mem. Am. Physiol. Soc., Soc. Exptl. Biology and Medicine, AAAS, Nat. Acad. Scis., Sigma Xi, Phi Kappa Phi, Phi Zeta. Home: 207 Texas Ln Ithaca NY 14850-1758

WASSERMAN, STANLEY, statistician, educator; b. Louisville, Aug. 29, 1951; s. Irvin Levitch and Jeanne (Plattus) W.; m. Sarah Wilson, Feb. 3, 1974; children: Andrew Joseph, Eliot Miles. BS in Econs., U. Pa., 1973; PhD in Stats., Harvard U., 1977. Asst. prof. U. Minn., Mpls., 1977-82; assoc. prof. U. Ill., Urbana, 1982-88, prof. psychology, stats., sociology, 1988—; vis. researcher Columbia Univ., N.Y.C., 1978; cons., expert witness EEOC, Cleve., 1979-81; cons. V.A. Med. Ctr., Mpls., 1980-82, AT&T

Communications, Basking Ridge, N.J., 1988-90. Assoc. editor: Sociological Methodology, 1978-81, Psychometrika, 1988—, Am. Statistician, 1993—; guest editor: Sociol. Methods and Rsch., 1992. Treas. Montessori Sch. of Champaign-Urbana, Savoy, Ill., 1990-92. Grantee NSF, Washington, 1979-81, 84-89; postdoctoral fellow Social Sci. Rsch. Coun., N.Y.C., 1978. Fellow Am. Statis. Assn. (assoc. editor jour. 1987—); mem. AAAS, Psychometric Soc., Royal Statis. Soc., Classification Soc. N.Am. (sec., treas. 1993—). Achievements include reseach in applied statistics, categorical data analysis, social network analysis. Home: 1709 Pleasant St Urbana IL 61801-5830 Office: U Ill 603 E Daniel St Champaign IL 61820-6267

WATANABE, KOUICHI, pharmacologist, educator; b. Manchuria, Japan, Aug. 26, 1942; s. Tetsuya and Mine W.; children: Toshikazu, Yoshihiro, Motohiro. BS, Tokyo Coll. Pharmacy, 1966; MS, Osaka U., 1968; PhD, 1971; LPIBA, 1986; DSc (hon.) Internat. U. Found., 1987, World U. Roundtable, 1988. Vis. fellow reprodn. rsch. br. Nat. Inst. Child Health and Devel., NIH, Bethesda, Md., 1971-73; vis. scientist dept. pharmacology Coll. Medicine, Howard U., Washington, 1973-75, asst. prof., 1975-83; asst. prof. pharmacology U. Hawaii, 1983—; drug info. officer med. info. dept. Minophagen Pharm. Co., Tokyo, 1993—. Contbr. articles to sci. jours. Named Man of Yr., ABI, 1991, 93, IBC, 1991, 92; Am. Cancer Soc. grantee, 1980-81. Mem. Am. Soc. Pharmacology and Exptl. Therapeutics, N.Y. Acad. Scis. (inaugural mem.), Am. Soc. Hypertension (charter). Subspecialties: Chemotherapy; Molecular pharmacology. Current work: Mechanism of action of various antineoplastic agts. on calmodulin. Vinca alkaloids found to be calmodulin inhibitors. Suggested that amounts of calmodulin or its binding proteins may be endogenous regulators of antineoplastic action or transport of these drugs. Home: Rm 10, Aoba-S0, 3-7-17, Higashi-izumi Komae City Tokyo 201, Japan Office: Minophagen Pharm Co, New Shinsaka Bldg 4th Fl 8-10-22, Akasaka Minato-ku Tokyo 107, Japan

WATANABE, KYOICHI A(LOYSIUS), chemist, researcher, pharmacology educator; b. Amagasaki, Hyogo, Japan, Feb. 28, 1935; s. Yujiro Paul and Yoshiko Francisca (Hashimoto) W.; m. Kikyoko Agatha Suzuki, Nov. 22, 1962; children: Kanna, Kay, Kenneth, Kim, Kelly, Katherine. BA, Hokkaido U., 1958, PhD, 1963. Lectr. Sophia U., Tokyo, 1963; rsch. assoc. Sloan-Kettering Inst., N.Y.C., 1963-66, assoc., 1968-72, assoc. mem., 1972-81, mem., prof., 1981—; rsch. fellow U. Alta., Edmonton, Can., 1966-68; assoc. prof. Cornell U. Med. Coll., N.Y.C., 1972-81, prof. pharmacology, 1981—; mem. study sect. NIH, Washington, 1981-84. Achievements include total synthesis of nucleasic antibiotics, novel heterocycle ring transformation, C-nucleose chemistry; antiviral and anticancer nucleosides; intercalating agents; modified oligonucleotides. Office: Sloan Kettering Inst 1275 York Ave New York NY 10021

WATANABE, RICHARD MEGUMI, medical research assistant; b. San Fernando, Calif., Sept. 7, 1962; s. Takashi and Toshiko (Yamane) W. BS, U. So. Calif., L.A., 1986; MS, U. So. Calif., 1989, postgrad., 1989—. Lab. asst. U. So. Calif. Sch. Medicine, L.A., 1985-87; data entry clk. L.A. County/U. So. Calif. Med. Ctr. Women's Hosp., 1985-89; rsch. asst. U. So. Calif. Sch. Medicine, L.A., 1987—; statis. cons. U. So. Calif. Sch. Medicine, 1988, dir. kinetic core, 1992—. Recipient Pacific Coast Fertility Soc. 1st prize in-tng. award, 1990; NIH fellow, 1990. Mem. AAAS, Am. Diabetes Assn., Am. Physiol. Soc., European Assn. Study of Diabetes. Avocations: music woodworking. Office: 1333 San Pablo St MMR-620 Los Angeles CA 90033

WATANABE, TOSHIHARU, ecologist, educator; b. Kyoto, Japan, June 6, 1924; s. Seizo and Fusa Watanabe; m. Sumiko Isebo, Nov. 3, 1952; children: Ikuko, Naoki. DSc, Kyoto U., 1961. Prof. Nat. Kanazawa (Japan) U., 1972-75, Nat. Nara (Japan) Women's U., 1975-88; prof. Fgn. Studies Kansai U., Hirakata/Osaka, Japan, 1988—; owner Diamond Resort Hawaii Owner's Club, 1989—; pres. Inst. Sci. Rsch. to Hydrospherical Ecology, 1990—. Author: Encyclopedia of Environmental Control Technology, 1990; editor: Japanese Jour. Diatomology, 1985, Japanese Jour. Water Treatment Biology, 1971. Profl. mem. Ministry of Constrn., Kinki dist., 1980—; vice-chmn. Com. on environ. pollution, Nara Prefecture and City, 1982—, com. mem. Wakayama, 1980—. Avocations: Indian ink drawing, music. Home: Higashigawa-cho 518, Shinkyogoku St Nakagyo-ku, Kyoto 604, Japan Office: Kansai U Fgn Studies, Kitakatahoko-cho 16-1, Hirakata Osaka 573, Japan

WATANABE, YOSHIHITO, chemistry educator; b. Morioka, Iwate, Japan, May 30, 1953; s. Keiji and Teruko (Anetai) W.; m. Yumiko Iwabuch, Jan. 7, 1982; children: Haruka, Takuma. BS, Tohoku U., 1976; PhD, U. Tsukuba, 1982. Postdoctoral fellow U. Mich., Ann Arbor, 1982-84, asst. rsch. scientist, 1984-85; rsch. staff mem. Princeton U., Princeton, N.J., 1985-87; asst. prof. Keio U., Tokyo, 1987-89; sr. researcher Nat. Chem. Lab. for Industry, Tsukuba, Japan, 1989-90; assoc. prof. chemistry Kyoto U., 1990—. Author: Macintosh for Scientific Presentation, 1990; contbr. articles to profl. jours. Mem. Chem. Soc. Japan, Am. Chem. Soc. Avocations: tennis, skiing, baseball. Office: Kyoto U, Divsn Molecular Engring, Kyoto 606-01, Japan

WATERBORG, JAKOB HARM, biochemistry educator; b. Utrecht, The Netherlands, Dec. 31, 1948; came to U.S., 1980; s. Harm and Rutgerina C. (Wyburg) W.; m. Marie Antoinette Calon, Oct. 21, 1977; children: Harm, Wieteke, Linda. Doktoraal, U. Nymegen, The Netherlands, 1974, PhD, 1980. Postdoctoral rsch. asst. U. Calif., Davis, 1980-83; rsch. faculty U. Sussex, Brighton, Eng., 1983-84; rsch. assoc. U. Nev., Reno, 1984-87, rsch. specialist, 1987-88; asst. prof. U. Mo., Kansas City, 1988—. Contbr. articles to Plant Molecular Biology, Biochemistry, Plant Physiology, Jour. Biol. Chemistry, Biochimica Biophysica ACTA. Mem. AAAS, Am. Soc. Plant Physiologists, Internat. Soc. for Plant Molecular Biology, Am. Soc. for Biochemistry and Molecular Biology. Office: U Mo 5100 Rockhill Rd Kansas City MO 64110

WATERMAN, DANIEL, mathematician, educator; b. Bklyn., Oct. 24, 1927; s. Samuel and Anna (Robson) W.; m. Mudite Upesleja, Nov. 4, 1960; children—Erica, Susan, Scott. B.A., Bklyn. Coll., 1947; M.A., Johns Hopkins U., 1948; Ph.D., U. Chgo., 1954. Research assoc. Cowles Commn. Research in Econs., Chgo., 1951-52; instr. Purdue U., West Lafayette, Ind., 1953-55, asst. prof., 1955-59; asst. prof. U. Wis.-Milw., 1959-61; prof. Wayne State U., Detroit, 1961-69; prof. Syracuse (N.Y.) U., 1969—, chmn. math dept., 1988—; cons. Martin-Marietta, Denver, 1960-61; researcher in real and Fourier analysis. Editor: Classical Real Analysis, 1985: contbr. articles to profl. jours. Fulbright fellow U. Vienna, 1952-53. Mem. Math. Assn. Am., Am. Math. Soc. (council mem.-at-large 1975-78), Sigma Xi. Home: 116 Donridge Dr Syracuse NY 13214-2344 Office: Syracuse U Dept Math Syracuse NY 13244-1150

WATERS, DEAN A., engineering executive; b. Jersey City, May 2, 1936; s. Edward G. and Edna Mae (McCabe) W.; m. Jacquelyn R. Waters; children: Heather A., Geoffrey D. BE, Yale U., 1957, BS, 1958; MS, N.C. State U., 1960; PMD, Harvard U., 1978. Registered profl. engr., Tenn. Dept. head Union Carbide Corp., Oak Ridge, Tenn., 1972-77, dep. div. dir., 1977-80; div. dir. Union Carbide/Martin Marietta, Oak Ridge, 1980-92; program dir. Martin Marietta-Oak Ridge Nat. Lab., 1992—; advisor U. Tenn. Engring. Sch., Knoxville, 1989-92; bd. mem. PEAC, Knoxville, 1987-92. Recipient E.O. Lawrence award Dept. Energy, Washington, 1977. Mem. Nat. Soc. Profl. Engrs., Am. Soc. Engring. Mgrs. (chpt. pres. 1986-87), Tenn. Soc. Profl. Engrs., Rotary (pres. 1992-93, Paul Harris fellow). Achievements include 21 patents in field for gas centrifuge and liquid centrifuge. Office: Oak Ridge Nat Lab PO Box 2001 Oak Ridge TN 37831-7291

WATERS, ROBERT GEORGE, laser engineer; b. Bklyn., Dec. 27, 1948; s. Robert Wilburne and Jane Evelyn (Turner) W.; m. Barbara Elizabeth Nunziata, June 14, 1970; 1 child, Carl Anthony. BS in Physics, Stevens Inst. Tech., Hoboken, N.J., 1970; PhD in Physics, CUNY, N.Y.C., 1979. Assoc. scientist Exxon Enterprises Inc., Elmsford, N.Y., 1979-81; sr. staff scientist Airtron div. Litton Systems, Morris Plains, N.J., 1981-84; dir. R&D Opto-Electronics Ctr., McDonnell Douglas, Elmsford, N.Y., 1984—; mem. program com. Opto-Electronics/Laser Conf., L.A., 1990—, Conf. on Lasers and Electro-Optics, Balt., 1993—. 1st lt. USAF, 1971-73. Achievements include two patents in field; developments in high power semiconductor lasers, diode laser reliability innovations; contbns. to visible diode lasers.

WATERS, WILL ESTEL, horticulturist, researcher, educator; b. Smithtown, Ky., Sept. 19, 1931; m. Elizabeth Sumler; children: Marie, Alan, Sheila. BS in Agronomy, U. Ky., 1954, MS in Soil Physics, 1958; PhD in Plant Nutrition, Vegetable Crops, U. Fla., 1960. Asst. prof. floricultural rsch. Gulf Coast Experiment Sta., Bradenton, Fla., 1960-66, assoc. prof. ornamental horticulture, 1966-70; prof., horticulturist, head Ridge Ornamental Hort. Lab., Apopka, Fla., 1968-70; prof., dir. Gulf Coast Rsch. & Edn. Ctr., Bradenton, 1970—; designer staff, equipment and programs, supr. devel. Agrl. Rsch. & Edn. Ctr. at Apopka, 1967-70; dir. Agrl. Rsch. & Edn. Ctr. at Dover, Fla., 1971—; dir. Agrl. Rsch. & Edn. Ctr. at Immokalee, Fla., 1971-86; author, supr. 4 fed. grants Gulf Coast Rsch. & Edn. Ctr., established MS and PhD coop. tng. programs and periodic formal undergrad. courses. Contbr. over 275 sci., tech. and popular articles to profl. jours. Named Ky. Col., Gov. Wendell Ford, 1972; recipient Appreciation award Fla. Flower Assn., 1978; named to Manatee County Agrl. Hall of Fame, 1988. Fellow Am. Soc. Hort. Sci. (Alex Laurie award 1968); mem. Soc. Am. Florists, So. Weed Conf., Soc. Region Am. Soc. Hort. Sci. (Leadership and Adminstrn. award 1990), Fla. Soil and Crop Sci. Soc., Fla. State Hort. Soc. (Outstanding Rsch. Paper award ornamental sect. 1967, Presdl. Gold Medal award 1968, sectional v.p. 1969), Fla. Ornamental Growers Assn. (Rsch. & Edn. award 1986), Fla. Nurserymen and Growers Assn. (Disting. Svc. award Manasota chpt. 1977, Outstanding Industry Person award 1981), Fla. Farm Bureau, Fla. Foliage Assn., Fla. Strawberry Growers Assn., Interamerican Soc. Tropical Horticulture. Achievements include development of horticultural production and handling recommendations in plant nutrition, growing media, soil and water chemistry, herbicides, post harvest keeping quality, and cultivar evaluations of primary flower, foliage and vegetable crops. Office: Gulf Coast Rsch & Ednl Ctr 5007-60th St E Bradenton FL 34203

WATERS, WILLIAM FREDERICK, psychology educator; b. Dayton, Ohio, Apr. 25, 1943; s. Martin Edward and Augusta (Packer) W.; m. Harriet Lee, Apr. 8, 1978; children: Janice Beth, Jocelyn Marlowe, Jason Nathaniel. BA, Tulane U., 1964; MS, Case Western Res. U., 1966, PhD, 1969. Diplomate Am. Bd. Profl. Psychology, Am. Bd. Sleep Medicine. Prof. psychiatry Sch. Medicine U. Mo., Columbia, 1968-79, prof. psychology dept. psychology, 1970-79, dir. psychol. svcs. Mental Health Ctr., 1968-74; prof. psychology dept. psychology La. State U., Baton Rouge, 1979—, dir. clin. psychology tng. program, 1979-89; coord. sleep disorders program Ochsner Clinic of Baton Rouge, 1988—; coord. clin. neurosci. rsch. Pennington Biomed. Rsch. Ctr., Baton Rouge, 1992—; prin. investigator rsch. grant Alton Ochsner Med. Found., 1989, tng. grant NIMH, 1980-83, 83-86. Fellow APA, Am. Psychol. Soc., Am. Sleep Disorders Assn.; mem. Soc. Psychophysiol. Rsch. Achievements include research in extinction as the primary process in the efficacy of systematic desensitization treatment of phobias, in the development of a reliable and valid psychometric assessment of the physiological aspects of emotional response, in the below-zero habituation of the human orienting response; that habituation of the orienting response plays a basic role in human selective attention; and that individual differences in orienting response, stress, and negative emotions influence psychophysiological insomnia. Home: 7732-C Jefferson Pl Blvd Baton Rogue LA 70809 Office: La State U Dept Psychology Audubon Hall Baton Rouge LA 70803

WATKINS, CHARLES BOOKER, JR., mechanical engineering educator; b. Petersburg, Va., Nov. 20, 1942; s. Charles Booker and Haseltine Lucy (Thurston) W.; m. Judith Griffin; children: Michael, Steven. B.S. in Mech. Engring. cum laude, Howard U., 1964; M.S., U. N.Mex., 1966, Ph.D., 1970. Registered profl. engr., D.C. Mem. tech. staff Sandia Nat. Labs., Albuquerque, 1964-71; asst. prof. dept. mech. engring. Howard U., Washington, 1971-73; prof., chmn. dept. mech. engring. Howard U., 1973-86; Herbert G. Keyser prof. mech. engring. dean Sch. Engring. CCNY, 1986—; cons. U.S. Army, U.S. Navy, NSF, pvt. industries, 1984-85. Research grantee NSF, U.S. Navy, Nuclear Regulatory Commn.; research grantee Dept. Energy, NASA; recipient Ralph R. Teetor award Soc. Automotive Engrs., 1980; Sandia Labs. doctoral fellow; NDEA fellow. Fellow ASME; mem. AIAA (council), AAAS, Soc. Automotive Engrs., Am. Soc. Engring. Edn., Sigma Xi, Omega Psi Phi, Tau Beta Pi. Home: 171 Sherman Ave Teaneck NJ 07666-4121 Office: CCNY Sch Engring Convent Ave New York NY 10031

WATKINS, CHARLES REYNOLDS, medical equipment company executive; b. San Diego, Oct. 28, 1951; s. Charles R. and Edith A. (Muff) W.; children: Charles Devin, Gregory Michael. BS, Lewis and Clark Coll., 1974; postgrad., U. Portland, 1976. Internat. salesman Hyster Co., Portland, Oreg., 1975-80, Hinds Internat. Corp., Portland, 1980-83; mgr. internat. sales Wade Mfg. Co., Tualatin, Oreg., 1983-84; regional sales mgr. U.S. Surg., Inc., Norwalk, Conn., 1984-86; nat. sales mgr. NeuroCom Internat., Inc., Clackamas, Oreg., 1986-87; pres. Wave Form Systems, Inc., Portland, 1987—; bd. dirs. U.S. Internat., Inc., Portland, Clearfield Med., Minorax, Inc. Bd. dirs. Portland World Affairs Coun., 1980. Mem. Am. Soc. Laser Medicine and Surgery, Am. Fertility Soc., Am. Assn. Gynecol. Laparoscopists, Portland City Club. Republican. Avocations: flying, photography, travel. Office: Wave Form Systems Inc PO Box 3195 Portland OR 97208-3195

WATKINS, GEORGE M., surgeon, educator; b. Nashville, Mar. 23, 1935; s. George Miller and Brownie (Appleton) W.; m. Alison Shiverick Hall, June 25, 1966; children: Laura, Erica. BA, Vanderbilt U., 1957. Diplomate Am. Bd. Suergery. Assoc. prof. surgery U. South Fla., Tampa, 1975-79, vice chmn. dept. surgery, 1979-84, chief trauma, prof. surgery, 1979-87, chief surgery, 1979-85; dir. dept. surgery Easton (Pa.) Hosp., 1987-90; prof. surgery U. Ill. Coll. Medicine, Peoria, 1990—. Lt. USN, 1960-64. Mem. Am. Assn. Surgeons (treas. 1981-84), Am. Trauma Soc. (pres. 1986-88). Home: 223 W Hollyridge Peoria IL 61614 Office: U Ill Coll Medicine 420 NE Glen Oak Ave Ste 302 Peoria IL 61603

WATKINS, JAMES DAVID, government official, naval officer; b. Alhambra, Calif., Mar. 7, 1927; s. Edward Francis and Louise Whipple (Ward) W.; m. Sheila Jo McKinney, Aug. 19, 1950; children: Katherine Marie, Laura Jo, Charles Lancaster, Susan Elizabeth, James David, Edward Francis. BS, U.S. Naval Acad., 1949; MS, Naval Postgrad. Sch., 1958; LHD (hon.), Marymount Coll., 1982, N.Y. Med. Coll., 1988; DSc (hon.), Dowling Coll., 1983, U. Ala., 1991; LLD (hon.), Cath. U., 1985. Commd. ensign USN, 1949, advanced through grades to adm., 1979, comdg. officer U.S.S. Snook, 1964-66, exec. officer U.S.S. Long Beach, 1967-69; head submarine/nuclear power distbn. control br. Bur. Naval Pers., Dept. Navy, Washington, 1969-71, dir. enlisted pers. div., 1971-72; asst. chief of naval pers. for enlisted pers. control Bur. Naval Personnel, Dept. Navy, Washington, 1972-73; comdr. Cruiser-Destroyer Group 1 USN, 1973-75; dep. chief naval ops. manpower Navy Dept., Washington, 1975-78, chief of naval pers., 1975-78, chief Bur. Naval Pers., 1975-78; comdr. U.S. Sixth Fleet USN, 1978-79; vice chief naval ops. Navy Dept., 1979-81, comdr.-in-chief U.S. Pacific Fleet, 1981-82, chief naval ops., 1982-86; chmn. Presdl. Commn. on Human Immunodeficiency Virus Epidemic, 1987-88; sec. Dept. of Energy, Washington, 1989—. Decorated D.S.M. with 1 gold star, Legion of Merit with 2 gold stars, Bronze Star medal with Combat V; recipient Disting. Alumni award Naval Postgrad. Sch., 1958, Chairman's award Am. Assn. Engring. Socs., 1991. Mem. U.S. Naval Acad. Alumni Assn. Roman Catholic. Lodge: Knights of Malta. Office: Dept Energy 1000 Independence Ave SE Washington DC 20585-0001

WATKINS, WILLIAM H(ENRY), electrical engineer; b. Kingsport, Tenn., Oct. 29, 1929; s. Lewis Henry and Hazell (Duff) W.; m. Nell Rose Walters, Jan. 28, 1950; children: William David, Sharon Renee. Electronics degree, U.S. Army, 1953. Owner, mgr. Food Distributorship, Kingsport, Tenn., 1954-58, Watkins Stereo, Kingsport, 1959-81, Watkins Engring., Kingsport, 1981—; cons. Infinity, Inc., Chatsworth, Calif., 1975—. Patentee in field. With U.S. Army, 1952-54. Recipient Golden award SONY of Am., 1968, Volume award GE, 1968, Best Speaker Design award Stereophile Mag., 1984-85. Mem. Am. Legion. Methodist. Avocations: audio, music, crafts. Office: Watkins Engring 1019 E Center St Kingsport TN 37660-4990

WATSON, DAVID COLQUITT, electrical engineer, educator b. Linden, Tex., Feb. 9, 1936; s. Colvin Colquitt and Nelena Gertrude (Keasler) W.; m. Flora Janet Thayn, Nov. 10, 1959; children: Flora Janeen, Melanie Beth, Lorrie Gaylene, Cheralyn Gail, Nathan David, Amy Melissa, Brian Colvin. BSEE, U. Utah, 1964, PhD in Elec. Engring. (NASA fellow), 1968. Electronic technician Hercules Powder Co., Magna, Utah, 1961-62; rsch. fellow U. Utah, 1964-65, rsch. asst. microwave devices and phys. electronics lab., 1964-68; sr. mem. tech. staff ESL, Inc., Sunnyvale, Calif., 1968-78, head dept. Communications, 1969-70; sr. engring. specialist Probe Systems, Inc., Sunnyvale, 1978-79; sr. mem. tech. staff ARGO Systems, Inc., Sunnyvale, 1979-90; sr. mem. tech. staff GTE Govt. Systems Corp., Mountain View, Calif., 1990-91; sr. cons. Watson Cons. Svcs., 1991-92; sr. staff engr. ESL, Inc., 1992— ; mem. faculty U. Santa Clara, 1978-81, 1992—, San Jose State U., 1981—, Coll. Notre Dame, 1992—, Chapman U., 1993—. Contbr. articles to IEEE Transactions, 1965-78; co-inventor cyclotron-wave rectifier; inventor gradient descrambler. Served with USAF, 1956-60. Mem. IEEE, Phi Kappa Phi, Tau Beta Pi, Eta Kappa Nu. Mem. LDS Ch. Office: GTE Govt Systems Corp 100 Ferguson Dr Mountain View CA 94043-5294

WATSON, DAVID RAYMOND, chemist; b. Crossett, Ark., Dec. 27, 1933; s. William Pugh and Frances Elizabeth (Turner) W.; m. Mary Sue Parham, Mar. 27, 1954; children: Kirk David, Melanie Diane. BS in Chemistry, U. Ark., 1956. Analytical chemist Reynolds Metals Co., Bauxite, Ark., 1956-57, project dir., 1957-66, dir. ceramic rsch., 1966-71, dir. chem. rsch., 1971-84, mgr. alumina and ceramics lab., 1984—. Contbr. articles to profl. jours. Fellow Am. Ceramic Soc.; mem. Nat. Inst. Ceramic Engrs. Republican. Baptist. Achievements include patents on low-soda alumnina process, alumina preparation and product, method of molding alumina, alumina system, improvements relating to manufacturer of sintered alumina, hollow ceramic balls as auto catalyst supports, others. Home: 1527 Cedarhurst Benton AR 72015 Office: Reynolds Metals Co PO Box 97 Bauxite AR 72011

WATSON, DONALD CHARLES, cardiothoracic surgeon, educator; b. Fairfield, Ohio, Mar. 15, 1945; s. Donald Charles and Pricilla H. Watson; m. Susan Robertson Prince, June 23, 1973; children: Kea Huntington, Katherine Anne, Kirsten Prince. BA in Applied Sci. Lehigh U., 1968, BS in Mech. Engring., 1968; MS in Mech. Engring., Stanford U., 1969; MD, Duke U., 1972. Diplomate Am. Bd. Thoracic Surgery, Am. Bd. Surgery. Intern in surgery Stanford U. Med. Ctr., Calif., 1972-73, resident in cardiovascular surgery, 1973-74, resident in surgery, 1976-78, chief resident in heart transplant, 1978-79, chief resident in cardiovascular and gen. surgery, 1978-80; clin. assoc. surgery br. Nat. Heart and Lung Inst., 1974-76, acting sr. surgeon, 1976; assoc. cardiovascular surgeon dept. child health and devel. George Washington U., Washington, 1980-84, asst. prof. surgery, asst. prof. child health and devel., 1980-84, attending cardiovascular surgeon dept. child health and devel., 1984-89; assoc. prof. surgery, 1984-89; assoc. prof. pediatrics U. Tenn.-Memphis, 1984-90, prof. surgery, 1989—, prof. pediatrics, 1990—, chmn. cardiothoracic surgery, 1984—; mem. staff Le Bonheur children's Med. Ctr., Memphis, chmn. cardiothoracic surgery, 1984—; mem. staff William F. Bowld Med. Ctr., Memphis, Regional Med. ctr. at Memphis, Baptist Meml. Med. Ctr., Memphis; cons. in field; instr. advanced trauma life support; profl. cons., program reviewer HHS. Contbr. chpts., numerous articles, revs. to profl. publs. Served to lt. comdr. USPHS, 1974-76. Smith Kline & French fellow Lehigh U., 1967; NSF fellow Lehigh U., 1968; univ. interdepartmental scholar and univ. scholar Lehigh U., 1968. Fellow Am. Coll. Cardiology, Am. Coll. Chest Physicians (forum cardiovascular surgery, council critical care), Southeastern Surg. Congress, Am. Acad. Pediatrics (surgery sect.), ACS; mem. Assn. Surg. Edn., Am. Assn. Thoracic Surgery, Soc. Thoracic Surgeons, So. Thoracic Surg. Assn., Am. Thoracic Soc., Assn. Acad. Surgery, Internat. Soc. Heart Transplantation, Am. Fedn. Clin. Research, Found. Advanced Edn. in Scis., Andrew G. Morrow Soc., Council on Cardiovascular Surgery of Am. Heart Assn., Soc. Internat. di Chirig, AAAS, N.Y. Acad. Sci., AMA, NIH Alumni Assn., Stanford U. Med. Alumni Assn., Duke U. Med. Alumni Assn., Duke U. Alumni Assn., Stanford U. Alumni Assn., Lehigh U. Alumni Assn., Smithsonian Assocs., Sierra Club, U. Tenn. Pres.'s Club, LeBonheur Pres.'s Club, U.S. Yacht Racing Assn., Pilots Internat. Assn., Nat. Assn. Flight Instrs., Aircraft Owners and Pilots Assn., Order Ky. Cols., Phi Beta Kappa, Tau Beta Pi, Pi Tau Sigma, Phi Gamma Delta. Republican. Presbyterian. Club: Memphis Racquet. Avocations: sailing; racquet sports; flying. Office: U Tenn 956 Court Ave # 232 Memphis TN 38163-0001

WATSON, JAMES DEWEY, molecular biologist, educator; b. Chicago, Ill., Apr. 6, 1928; s. James Dewey and Jean (Mitchell) W.; m. Elizabeth Lewis, 1968; children: Rufus Robert, Duncan James. BS, U. Chgo., 1947, DSc, 1961; PhD, Ind. U., 1950, DSc, 1963; LLD, U. Notre Dame, 1965; DSc, L.I. U., 1970, Adelphi U., 1972, Brandeis U., 1973, Albert Einstein Coll. Medicine, 1974, Hofstra U., 1976, Harvard U., 1978, Rockefeller U., 1980, Clarkson Coll., 1981, SUNY, 1983; DSc (hon.), U. Buenos Aires, Argentina, 1986, Rutgers U., 1988, Bard Coll., 1991, U. Cambridge, 1993, Fairfield U., 1993, U. Stellenbosch, 1993. Rsch. fellow NRC, U. Copenhagen, 1950-51; Nat. Found. Infantile Paralysis fellow Cavendish Lab., Cambridge U., 1951-53; sr. rsch. fellow biology Calif. Inst. Tech., 1953-55; asst. prof. biology Harvard U., 1955-58, assoc. prof., 1958-61, prof., 1961-76; dir. Cold Spring Harbor Lab., 1968—; assoc. dir. Nat. Ctr. for Human Genome Rsch. NIH, 1988-89, dir. Nat. Ctr. for Human Genome Rsch., 1989-92. Author: Molceular Biology of the Gene, 1965, 4th edit., 1986, The Double Helix, 1968, (with John Tooze) The DNA Story, 1981, (with others) The Molecular Biology of the Cell, 1983, 2d edit., 1989, (with John Tooze and David Kurtz) Recombinant DNA, A Short Course, 1983, 2d edit., 1992. Named Hon. fellow Clare Coll., Cambridge U.; recipient (with F.H.C. Crick) John Collins Warren prize Mass. Gen. Hosp., 1959, Eli Lilly award in biochemistry Am. Chem. Soc., 1959, Albert Lasker prize Am. Pub. Health Assn., 1960, (with F.H.C. Crick) Rsch. Corp. prize, 1962, (with F.H.C. Crick and M.H.F. Wilkins) Nobel prize in medicine, 1962, Presdl. medal of freedom, 1977. Mem. NAS (Carty medal 1971), Am. Philos. Soc., Am. Assn. Cancer Rsch., Am. Acad. Arts and Scis., Am. Soc. Biol. Chemists, Royal Soc. (London), Acad. Scis. Russia, Danish Acad. Arts and Scis. Home: Bungtown Rd Cold Spring Harbor NY 11724-2209 Office: Cold Spring Harbor Lab PO Box 100 Cold Spring Harbor NY 11724-0100

WATSON, JAMES EDWIN, physicist; b. Mpls., Mar. 9, 1952; s. James William and Hazel (Lammers) W.; m. Debra Bernard, Oct. 10, 1981; children: Anna, Ellen, Tyler. BS in Engring. Physics, Ohio State U., 1974; PhD of Physics, U. Ill., 1981. Staff AT&T Bell Labs., Allentown, Pa., 1981-90; rsch. specialist Fiber Optics Lab. 3M Co., St. Paul, 1990—. Mem. Optical Soc. Am., Soc. Photo Instrumentation Engrs. Achievements include development of polarization-independent lithium niobate optical switch. Office: 3M Company 3M Ctr 260 5B 08 Saint Paul MN 55144

WATSON, JOYCE MARGARET, observatory librarian; b. Manchester, Eng., Aug. 9, 1922; d. Herbert Thomas and Margaret Helen (Cole) Eke; m. Thomas J. Rey, May 29, 1955 (div. Jan. 1965); children: Pamela H., Lilli J., Antonia M.; m. Alan P. Watson, Mar. 30, 1985. Head libr. Ferranti, Ltd., Hollinwood, Eng., 1948-52, EMI Ltd., Hayes, Eng., 1952-55; libr. U. Lowell (Mass.) Rsch. Found., 1964-69; head libr. Smithsonian Astrophys. Obs., Cambridge, 1969-86, info. specialist, 1991—; head libr. Harvard-Smithsonian Ctr. for Astrophysics, Cambridge, 1986-90; librarian emerita Smithsonian Astrophys Obs, 1990—; mem. Working Group for Modern Astron. Methodology, Strasbourg, France, 1987—; various working groups on astron. data NASA, Washington, 1988—. Contbr. articles, reports to profl. publs. Mem. Am. Astron. Soc., Astron. Soc. Pacific, Spl. Librs. Assn. (chair physics-astronomy-math. div. 1986-87). Home: 34 Gorham St Chelmsford MA 01824-2913 Office: Smithsonian Astrophys Obs 60 Garden St Cambridge MA 02138-1596

WATSON, MARY ELLEN, ophthalmic technologist; b. San Jose, Calif., Oct. 29, 1931; d. Fred Sidney and Emma Grace (Capps) Doney; m. Joseph Garrett Watson, May 11, 1950; children: Ted Joseph, Tom Fred, Pamela Kay Watson Niles. Cert. ophthalmic med. technologist and surg. asst. Ophthalmic technician Kent W. Christoferson, M.D., Eugene, 1965-80; ophthalmic technologist, surg. asst., adminstr. I. Howard Fine, M.D., Eugene, 1980—; course dir. Joint Commn. Allied Health Pers. in Ophthalmology, 1976--, lectr., mem. faculty, 1983--, skill evaluator and site coord., Eugene, 1988--; interant. instr. advanced surgical techniques. Contbr. articles to profl. jours. Recipient 5-Yr. Faculty award Joint Commn. for Allied Health Pers. in Ophthalmology, 1989. Mem. Allied Tech. Pers. in Ophthalmology, Internat. Women's Pilots Assn. Avocation: flying. Home: 2560 Chaucer Ct Eugene OR 97405-1217 Office: I Howard Fine MD 1550 Oak St Eugene OR 97401-7701

WATSON, RAYMOND COKE, JR., college president, engineering educator, mathematics educator; b. Anniston, Ala., Aug. 31, 1926. BS, Jacksonville State U.; MSE, U. Ala.; MS, U. Fla.; PhD in Engring. Sci., Calif. Coast U. Chief engr. Dixie Svc. Co., 1948-54; head dept. physics and engring. Jacksonville State U., 1954-60; v.p. engring. and rsch. Teledyne Brown Engring., 1960-70; dir. continuing edn., engring. and math. U. Ala., Huntsville, 1970-76; pres., prof. engring. and math. Southeastern Inst. Tech., Huntsville, 1976—; adj. assoc. prof. U. Ala., Huntsville, 1961-70; cons. various def. industries, 1970—. Mem. IEEE, AIAA, Optical Soc. Am., Ops. Rsch. Soc. Am., Inst. Mgmt. Sci., Internat. Soc. Optical Engrs., Inst. Indsl. Engrs. Achievements include research in defense systems, space systems and electro-optics, sensor technologies, space based defense. Office: Teledyne Brown Engineering PO Box 070007 300 Sparkman Dr NW Huntsville AL 35807*

WATSON, ROBERT BARDEN, physicist; b. Champaign, Ill., Apr. 14, 1914; s. Floyd Rowe and Estelle Jane (Barden) W.; m. Genevieve L. Carter, Oct. 11, 1941; children: Ann Barden, Roberta Gail, Douglas Carter. AB, U. Ill., 1934; MA, UCLA, 1936; PhD, Harvard U., 1941. Teaching asst. UCLA, 1935-36; teaching asst., instr. Harvard U., Cambridge, Mass., 1936-41, rsch. assoc., 1940-41, rsch. assoc., Underwater Sound Lab., 1941-45; from asst. to assoc. prof. physics U. Tex., Austin, 1945-60, rsch. assoc., 1946-60; physical scis. adminstr. U.S. Dept. of the Army, Washington, 1960-76; cons. archtl. acoustics, Austin, 1946-60; rep. to various nat. and internat. sci. groups U.S. Army, Washington, 1960-76. Contbr. articles to Jour. Acoustical Soc. Am., Jour. Applied Physics. Fellow AAAS (Sci. Freedom and Responsibility award 1993), Acoustical Soc. Am.; mem. IEEE, Optical Soc. Am., Assn. Am. Physics Tchrs., Wash. Acad. Sci., Sigma Xi, Sigma Pi Sigma. Achievements include 3 patents concerning recorder, radar and sonar.

WATSON, ROBERT TANNER, physical scientist; b. Columbus, Ohio, Sept. 25, 1922; s. Rolla Don and Gladys Margaret (Tanner) W.; m. Jean Mehlig, Oct. 7, 1944; children—Melinda Jean, Parke Tanner, John Mehlig, Todd Pennell, Kate Ann. B.A., DePauw U., 1943; Ph.D., MIT, 1951. With photo products dept. E. I. duPont de Nemours & Co., Inc., Parlin, N.J., 1951-55; with electron tube div. RCA, Marion, Ind., 1955-59; with Internat. Tel. & Tel. Corp., 1959-71, v.p., gen. mgr. indsl. labs. div., 1962-63, pres., 1963-68; dep. tech. dir., etc. Aerospace and Def. Group, 1968-71; gen. phys. scientist OT and NTIA U.S. Dept. Commerce, 1971-90; process studies program dir., science division, Mission to Planet Earth NASA, Washington, D.C., 1990—. Mem. Gov.'s Air Pollution Control Bd., 1966-70; Allen County chmn. Ind. Sesquicentennial, 1965; trustee Ind. Ednl. Services Found., 1968-71; bd. dirs. YMCA of Greater Ft. Wayne and Allen County, 1966-69. Served to lt. (j.g.) USNR 1943-46 combat, PTO; USNR, 1946-55. Mem. Am. Inst. Mgmt. (fellow pres.'s council 1970), Ft. Wayne C. of C. (dir. 1966-68), Allen County-Ft. Wayne Hist. Soc. (Old Fort com.), Optical Soc. Am., IEEE, Sigma Xi, Phi Eta Sigma, Beta Theta Pi. Research and engring. in x-ray diffraction, radioactive scattering, solid state energy levels, electron emission, magnetic and photog. media, electro-optical, Laser and telecommunication systems. Mgr. several engring. firsts. Patentee in field. Home: 1770 Lang Dr Crofton MD 21114-2145

WATSON, STEVEN EDWARD, family physician; b. Purcell, Okla., Nov. 9, 1952; s. Herman Edward and Freada Adelle (Stevens) W.; m. Jami Sue Mitchell, Sept. 16, 1988; children: Travis Edward, Ethan. BS in Biology, N.E. Okla. State U., 1975; student, U. Okla., 1976; DO, Okla. State U., 1979. Intern house physician Grand Prairie (Tex.) Community Hosp., 1979-80; emergency rm. physician, family physician Hillcrest Med. Ctr., Oklahoma City, 1980—; med. dir. Dept. Human Svcs., State of Okla., Oklahoma City, 1985-87; provider program free of charge Psychol. Counseling Svcs., Okla.; med. dir. Med. Assts. ProgramOkla. CIty Jr. Coll. Capt. Spl. Forces U.S. Army, 1970-72. Recipient Cert. of Appreciation, Dept. Human Svcs., State Okla., 1986; dedicated behavior medicine ctr. in his name for outstanding achievements and contbns. Hillcrest Health Ctr., 1991. Mem. Am. Legion, Am. Osteo. Assn., Am. Coll. Gen. Practitioners, Okla. Osteo. Assn., South Cen. Dist. Med. Soc. Avocations: former professional football player, electronics repair, community basketball, youth soccer. Home: 10617 W Country Dr Oklahoma City OK 73170-2415 Office: Watson Family Clinic 6501 S Western Ave Ste B Oklahoma City OK 73139-1705

WATSON, STUART LANSING, chemist; b. Bristol, Va., Jan. 30, 1948; s. Stuart Lansing Sr. and Anna Ruth (Harmon) W.; Martha Ann Moser, May 30, 1970; children: Douglas Stuart, David Hugh. BS, Carson-Newman Coll., 1969; PhD, U. Tenn., 1976. Chemist Union Carbide Corp., South Charleston, 1976-85; sr. v.p. E.R. Carpenter Co. Inc., Richmond, Va., 1985-- Capt. USAF, 1969-73.

WATT, (ARTHUR) DWIGHT, JR., computer programming and microcomputer specialist, educator; b. Washington, Jan. 25, 1955; s. Arthur Dwight and Myrtle Lorraine (Putnam) W.; m. Shari Elizabeth Gambrell, July 30, 1988. BA, Winthrop U., 1977, MBA, 1979; EdD, U. Ga., 1989. Cert. data processing, Inst. Certification of Computer Professionals; cert. instr. first aid, safety ARC. Data processing instr. York Tech. Coll., Rock Hill, S.C., 1977-78; computer ctr. asst. Winthrop Coll., Rock Hill, 1976-79; data processing instr. Brunswick (Ga.) Coll., 1979-80; system operator, asst. programmer Sea Island (Ga.) Co., The Cloister, 1981; pvt. practice data processing cons. Swainsboro, Ga., 1981—; computer programming/microcomputer specialist instr. Swainsboro Tech. Inst., 1981—; cons. and speaker in field; chmn. exec. bd. computer curricula Ga. Dept. Tech. and Adult Edn., 1990-92. Author: District Revenue Potential and Teachers Salaries in Georgia, 1989; co-author: District Property Wealth and Teachers Salaries in Georgia, 1990, Factors Influencing Teachers Salaries: An Examination of Alternative Models, 1991, Local Wealth and Teachers Salaries in Pennsylvania, 1992, School District Wealth and Teachers' Salaries in South Carolina, 1993. Chmn. Emanuel County chpt. ARC, Swainsboro, 1989-90, 92-93, bd. dirs., 1989—; pres. United Meth. Men, Swainsboro, 1984-86. Named Outstanding Young Citizen Swainsboro Jaycees, 1985, Postsecondary Tchr. of the Year Ga. Bus. Edn. Assn., 1985. Mem. Ga. Bus. Edn. Assn. (dist. bd. dirs. 1986), Ga. Vocat. Assn., Data Processing Mgmt. Assn., Swainsboro Jaycees (treas. 1984-89, pres. 1987-88), Ga. Jaycees (v.p. area C membership 1988-89, chaplain 1989-90, bd. dirs. region 6, 1990-91, chmn. state shooting edn. 1991-92, treas. S.E. chpt. fair 1993), U.S. Jaycees (nat. rep. shooting edn. program 1992—, Shooting Edn. State Program Mgr. of Yr. award 1992). Methodist. Home: PO Box 1637 Swainsboro GA 30401-4637 Office: Swainsboro Tech Inst 201 Kite Rd Swainsboro GA 30401-1898

WATT, JEFFREY XAVIER, mathematics sciences educator, researcher; b. Indpls., July 14, 1961; s. John Hayden and Beverly Ann (Schneider) W. BS in Geophysics, Mich. Technol. U., 1983; MS in Geology, Emory U., 1984; MS in Applied Math., Purdue U., 1986; PhD in Math., Ind. U., 1990. Lectr. Ind. U.-Purdue U., Indpls., 1988-90; asst. prof. math. scis. Purdue U. Sch. Sci., Indpls., 1990—; cons. Park Tutor High Sch., Indpls., 1990—. Eucharistic min. St. Albert the Gt. Ch., Mich., 1982; referee Ind. High Sch. Athletic Assn., 1983—. NSF rsch. grantee, 1992—. Mem. Indpls. Hockey Ofcls. Assn. (exec. bd. 1986—). Avocation: sports. Home: PO Box 2813 Indianapolis IN 46206-2813

WATT, NORMAN RAMSAY, chemistry educator, researcher; b. Providence, Oct. 29, 1928; s. Albert Byron and Marie Louise (Turcotte) W. ScB, Brown U., 1951; PhD, U. Conn., 1967. Teaching fellow NYU, N.Y.C., 1951-54; teaching asst. U. Conn., Storrs, 1954-55, asst. instr., 1955-59, instr., 1960-61, 64-67, asst. prof., 1967-82; instr. Danbury (Conn.) State Coll., 1959-60. Home: 41 Garden Hills Dr Cranston RI 02920

WATT, WILLIAM STEWART, physical chemist; b. Perth, Scotland, Feb. 25, 1937. BSc, U. St. Andrews, Scotland, 1959; PhD in Phys. Chemistry, U. Leeds, 1962. Fellow Cornell U., 1962-64; rsch. chemist Corneel Aeronautics Lab., Buffalo, 1964-71; head chem. laser sect. Naval Rsch. Lab., 1971-73, dep. head laser physics br., 1973-76, head laser physics br. optical sci. divsn., 1976-79; gen. mgr. wash ops. W. J. Schafer Assoc., Arlington, Va., 1979-80, v.p. program devel., 1980-90, sr. v.p., dir. programs, 1991—. Recipient J. B. Cohen Rsch. prize, 1962. Mem. IEEE (assoc. editor Jour. Quantum Electronics), Am. Phys. Soc., Combustion Inst., Sigma Xi. Achievements include research in laser physics and development, laser-induced chemistry, energy transfer and reaction rate measurements, optical diagnostics. Office: W J Schafer Assoc Inc Ste 800 1901 N Fort Myer Dr Arlington VA 22209*

WATTERS, THOMAS ROBERT, geologist, museum administrator; b. West Chester, Pa., Feb. 1, 1955; s. Frank Edward Sr. and Beatrice Josephine (Speirs) W.; m. Nancy Rae Tracey, June 18, 1983; children: James T. Samantha E., Adam T. BS in Earth Scis., West Chester U., 1977; MA in Geology, Bryn Mawr Coll., 1979; PhD in Geology, George Washington U., 1985. Rsch. fellow Am. Mus. Natural History, 1978-80; rsch. asst. dept. terrestrial magnetism Carnegie Instn. Washington, 1980-81; rsch. geologist Ctr. for Earth and Planetary Studies Smithsonian Instn., Washington, 1981-89, supervisory geologist Ctr. for Earth and Planetary Studies, Nat. Air and Space Mus., 1989—, acting. chmn. Ctr. for Earth and Planetary Studies, 1989-92, chmn. Ctr. for Earth and Planetary Studies, 1992—; lectr. in field. Contbr. over 100 abstracts and papers. Active Plum Point Elem. PTA, 1991—, mem. exec. bd. 1991-92; coach Calvert T-Ball, 1992—. William. P. Phillips Meml. scholar; grantee NASA, 1983—, 85-89, 86-88, 87-88, 87-90, 89— (two grants), 91—; recipient cert. award NASA, 1983, 86, 89, 91, 92. Mem. AAAS, Geol. Soc. Am. (mem. editorial bd. GEOLOGY 1993—), Am. Geophysical Union (Editor's citation 1992), Kappa Delta Phi. Achievements include research in terrestrial and planetary tectonics and tectonophysics; morphological and structural comparisons of tectonic features on the terrestrial planets and analogous features on Earth; geologic mapping of Mars. Office: Smithsonian Instn Nat Air & Space Museum Rm 3789 Ctr Earth & Planetary Studies Washington DC 20560

WATTS, HELEN CASWELL, civil engineer; b. Brunswick, Maine, July 28, 1958; d. Forrest and Frances Caswell; m. Austin Watts. BS in Civil Engring., U. N.H., 1980; cert. 5th yr. pulp and paper, U. Maine, 1983. Registered profl. engr., Maine. Constrn. engr., 1980-82, Oklahoma Dept. Transp., Okla. City, 1983-84; design engr. Structural Design Cons., Inc., Portland, Maine, 1985; facility engr. Bath (Maine) Iron Works, 1986—. Mem. ASCE (assoc.), Tech. Assn. Pulp and Paper Industry (assoc.). Achievements include permits and construction of 300 ton transporter roadway across wetlands and mitigation site. Office: Bath Iron Works Corp 700 Washington St Bath ME 04530-5001

WATTS, HELENA ROSELLE, military analyst; b. East Lynne, Mo., May 29, 1921; d. Elmer Wayne and Nellie Irene (Barrington) Long; m. Henry Millard Watts, June 14, 1940; children—Helena Roselle Watts Scott, Patricia Marie Watts Foble. B.A., Johns Hopkins U., 1952, postgrad., 1952-53. Assoc. engr., Westinghouse Corp., Balt., 1965-67; sr. analyst Merck, Sharp & Dohme, Westpoint, Pa., 1967-69; sr. engr. Bendix Radio div. Bendix Corp., Balt., 1970-72; sr. scientist Sci. Applications Internat. Corp., McLean, Va., 1975-84; mem. tech. staff The MITRE Corp., McLean, 1985—; adj. prof. Def. Intelligence Coll., Washington, 1984-85. Contbr. articles to tech. jours. Mem. IEEE, AAAS, AIAA, Nat. Mil. Intelligence Assn., U.S. Naval Inst., Navy League of U.S., Air Force Assn., Assn. Former Intelligence Officers, Assn. Old Crows, Mensa, N.Y. Acad. Sci. Republican. Roman Catholic. Avocations: photography; gardening; reading. Home: 4302 Roberts Ave Annandale VA 22003-3508 Office: MITRE Corp W965 7525 Colshire Dr Mc Lean VA 22102-7500

WATTS, KENNETH MICHAEL, computer scientist; b. Chgo., Feb. 24, 1964; s. James Herman and Ethel Edna (Whitsell) W.; m. Shirley Faye Hilton, July 14, 1984. Student, So. Ill. U., 1982-84. Student programmer So. Ill. U., Carbondale, 1982-83, student computer cons., 1983-84; computer mktg. rep. Tandy Corp./Radio Shack, Carbondale, 1982-85; retail store mgr. Tandy Corp./Radio Shack, Owensboro, Ky., 1985-86, sr. computer specialist, 1986-88; exec. v.p. Computer Links Assocs. Inc., Owensboro, 1988; sr. systems adminstr. Commonwealth Aluminum Corp., Lewisport, Ky., 1988—. Fellow Digital Equipment Computer Users Soc., Louisville FUSE Group. Republican. Methodist. Achievements include design of administrative information network; development of information processing downsizing strategy in accordance with industry standards with goal of client-server and end-user computing. Office: Commonwealth Aluminum Corp PO Box 480 Ky Hwy 1957 Owensboro KY 42351

WATTS, MALCOLM L., chemist; b. Bristol, Somerset, Eng., Oct. 23, 1937; came to U.S., 1979; s. Geoffrey H. and Lillian Florence (Pye) W.; m. Lovat Simpson, June 28, 1992; children: George, Isla, Elizabeth, Claire. BSc in Chemistry with honors, Bristol U., 1958, PhD in Chemistry, 1961. Plant supr. ICI, North Yorishire, Eng., 1961-64, plant start-up supr., 1964-67, sr. plant start-up supr., 1968-70, project mgr., 1971-76, sect. mgr., 1977-79; tech. dir. ICI, Houston, 1980-84; program mgr. ICI, Wilmington, Del., 1985—. Mem. IEEE, Am. Chem. Soc. Achievements include patent in use of methonolamine to prevent corrosion in cracker dilution steam systems. Home: 807 Lisadell Dr Kennett Square PA 19348 Office: ICI Americas Concord Pike Wilmington DE 19897

WATTS, MARVIN LEE, minerals company executive, chemist, educator; b. Portales, N.Mex., Apr. 6, 1932; s. William Ellis and Jewel Reata (Holder) W.; m. Mary Myrtle Kiker, July 25, 1952; children: Marvin Lee, Mark Dwight, Wesley Lyle. BS in Chemistry and Math., Ea. N.Mex. U., 1959, MS in Chemistry, 1960; postgrad. U. Okla., 1966, U. Kans., 1967. Analytical chemist Dow Chem. Co., Midland, Mich., 1960-62; instr. chemistry N.Mex. Mil. Inst., Roswell, 1962-65, asst. prof., 1965-67; chief chemist AMAX Chem. Corp., Carlsbad, N.Mex., 1967-78, gen. surface supt., 1978-84; pres. N.Mex. Salt and Minerals Corp., 1984—; chem. cons. Western Soils Lab., Roswell, 1962-67; instr. chemistry N.Mex. State U., Carlsbad, 1967—; owner, operator cattle ranch, Carlsbad and Loving, N.Mex., 1969—; bd. dirs. Mountain States Mutual Casualty Co., 1991; gen. mgr. Eddy Potash, Inc., 1987—; dir. Soil Conservation Svc.; mem. Roswell dist. adv. bd. Bur. Land Mgmt. Bd. dirs. Southeastern N.Mex. Regional Sci. Fair, 1966; mem. adv. bd. Roswell dist. Bur. Land Mgmt.; mem. Eddy County Fair Bd., 1976—, chmn., 1978, 82; mem. pub. sch. reform com.; chmn. higher edn. reform com.; mem. sponsor of N.Mex. Pub. Sch. Reform Act, bd. dirs. Carlsbad Regional Med. Ctr., 1976-78; pres. bd. Carlsbad Found., 1979-82; adv. bd. N.Mex. State U. at Carlsbad, 1976-80; vice chmn. bd. Guadalupe Med. Ctr.; bd. dirs. N.Mex. State U. Found.; state senator N.Mex. Legis., 1984-89. Mem. Rep. State Exec. com., 1972—; Rep. chmn. Eddy County (N.Mex.), 1970-74, 78-82. dirs. Conquistador coun. Boy Scouts Am., Regional Environ. Ednl. Rsch. and Improvement Orgn. Served with Mil. Police Corps, AUS, 1953-55; Germany. Recipient Albert K. Mitchell award as outstanding Rep. in N.Mex., 1976; hon. state farmer N.Mex. Future Farmers Am.; hon. mem. 4-H. Fellow N.Mex. Acad. Sci.; mem. Am. Chem. Soc. (chmn. subsect.), Western States Pub. Lands Coalition, Carlsbad C. of C. (dir. 1979-83), N.Mex. Mining Assn. (dir.), AIME (chmn. Carlsbad potash sect. 1975), Carlsbad Mental Health Assn., N.Mex. Inst. Mining and Tech. (adv. bd. mining dept.), Am. Angus Assn., Am. Quarter Horse Assn., N.Mex. Cattle Growers Assn. (bd. dirs. 1989—), Carlsbad Farm and Ranch Assn., Nat. Cattleman's Assn. Baptist. Kiwanis (Disting. lt. gov.), Elks. Home: PO Box 56 Carlsbad NM 88221-0056 Office: PO Box 31 Carlsbad NM 88221-5601

WATTS, MICHAEL ARTHUR, materials engineer; b. San Pedro, Calif., Nov. 18, 1955; s. Melvin A. Watts and Arlene P. Ault; m. Susan J. Reis, Dec., 1973 (div. Apr., 1983); m. Ann E. Winkelman, July 21, 1988; children: Arthur, Andy, Erin Winkelman, Heather Winkelman. BS in Materials Engring., N.Mex. Inst. Mining & Tech., 1988. Radar technician western svc. div. Raytheon, Fountain Valley, Calif., 1978-80; sr. engr. analyst Bournes Instruments, Riverside, Calif., 1980-82; mem. tech. staff Monolithic Microsystems, Santa Cruz, Calif., 1982-83; student rsch. asst. N.Mex. Inst. Mining and Tech., Socorro, 1983-88; corrosion engr. Arco Alaska, Achorage, 1989—. With USAF, 1974-78. Mem. Nat. Assn. Corrosion Engrs., Am. Soc. Metallurgy, Tau Beta Pi (pres. 1987-88). Achievements include patent in Non-contact Infared Soldering System, other patents pending. Home: 13670 Karen St Anchorage AK 99515 Office: Arco Alaska KFE/NSK 39 PO Box 196105 Anchorage AK 99519-6105

WATTSON, ROBERT KEAN, aeronautical engineer; b. Kansas City, Mo., Oct. 18, 1922; s. Robert Kean and Anna Laurie (Smith) W.; m. Velda Henrichs, Nov. 27, 1943 (dec. 1992); children: Robert K. III, Keith H., Elsie A., Vincent A., Meredith S., Bruce C. BSME, Okla. State U., 1946; SM in aeronautical engineering, MIT, 1948. Registered profl. engr., N.D., Ind. Instr. Mech. Engring. Okla. State U., 1946-47; assoc. prof. Mech. Engring. North Dakota State U., 1948-53; engr., section head Cessna Aircraft Co., 1953-56; chief engr., assoc. prof., head Engring. Rsch. Dept. U. Wichita, 1956-60; program mgr., mgr. V/STOL aerodynamics, new product aerodynamics The Boeing Co., Wichita, 1960-63; chief R & D, chief tech. staff, chief aerodynamics Lerjet Industries Inc., 1963-67; aerodynamics engr., adv. product engr. Beech Aircraft Corp., 1967-70; assoc. prof., assoc. chmn., prof., assoc. chmn., aeronautical engring. Tri-State U., 1970-78; engring. staff specialist Gates Learjet Corp., 1978-85; assoc. prof., head dept. mech. engring. tech. Oregon Inst. Tech., 1985-87; prof., assoc. chmn. aerospace engring. Tri-State U., 1987-93; prof. emeritus, 1993; cons. in field; mem. Adv. Performance Sect., United Aircraft Ltd., Longeuil, Quebec, 1972, Aerodynamics Propulsion Unit, Gates Learjet Corp., Wichita, ASEE/NASA summer faculty fellow Systems Design Project Group, Langley Rsch. Ctr., 1975, veh. ops. rsch. Langley Rsch Ctr. 1986, 1987; adj. prof. Okla. State U., 1980. Contbr. articles to profl. jours. Lt. Col. USAFR, 1965. Assoc. fellow Am. Inst. Aeronautics and Astronautics (assoc., sec. chmn., 1961, 76). Home: 5004 W Robinson Wichita KS 67212

WAUER, ROLAND HORST, biologist; b. Idaho Falls, Idaho, Mar. 22, 1934; s. Herman Horst and Adeline (Delaby) W.; m. Betty Joe Newman, Aug. 27, 1976; children: Rebecca, Katrina, Susan, Stephan. BS in Wildlife Mgmt., San Jose State U., 1957; MS in Biol. Sci., Sul Ross State U., 1970. Park naturalist Death Valley (Calif.) Nat. Monument, 1957-63, Zion (Utah) Nat. Pk., 1963-66; chief pk. naturalist Big Bend (Tex.) Nat. Pk., 1966-72; chief scientist S.W. Region Nat. Pk. Svc., Santa Fe, 1972-79; chief div. natural resources Nat. Pk. Svc., Washington, 1979-83; asst. supt. Great Smoky Mountains Nat. Pk., Gatlinburg, Tenn., 1983-86; resource specialist Caribbean Nat. Pk. Svc., St. Croix, 1986-89; ret., 1989—; mem. Nat. Acad. of Sci. Bd., 1989-91. Author: Naturalist's Mexico, 1992, Naturalist's Big Bend, 1980, Birds of Big Bend National Park, 1979, Birds of Zion National Park, 1965, Birds of the Eastern National Parks, U.S. and Canada, 1992; contbr. numerous articles to profl. jours. Recipient Spl. Achievement award Nat. Pk. Svc., 1974, 75, 88, Commendation, Sec. of Interior, 1980, 1st Francis Jacot award, 1988. Mem. George Wright Soc., Tex. Coastal Mgmt. Plan, Chihuahuan Desert Rsch. Inst. Democrat. Lutheran. Achievements include devel. and establishment of Nat. Pk. Svc. Resource Mgmt. Tng. Program. Home: 202 Padre Ln Victoria TX 77901

WAUGH, JOHN STEWART, chemist, educator; b. Willimantic, Conn., Apr. 25, 1929; s. Albert E. and Edith (Stewart) W.; married 1983; children: Alice Collier, Frederick Pierce. AB, Dartmouth Coll., 1949; PhD, Calif. Inst. Tech., 1953; ScD (hon.), Dartmouth Coll., 1989. Research fellow physics Calif. Inst. Tech., 1952-53; mem. faculty Mass. Inst. Tech., 1953—, prof. chemistry, 1962—, Albert Amos Noyes prof. chemistry, 1973-88, inst. prof., 1989—; vis. prof. U. Calif.-Berkeley, 1963-64; lectr. Robert Welch Found., 1968; Falk-Plaut lectr. Columbia U., 1973; DuPont lectr. U. S.C., 1974; Lucy Pickett lectr. Mt. Holyoke Coll., 1978; Reilly lectr. U. Notre Dame, 1978; Spedding lectr. Iowa State U., 1979; McElvain lectr. U. Wis., 1981; Vaughan lectr. Rocky Mountain Conf., 1981; G.N. Lewis meml. lectr. U. Calif., 1982; Dreyfus lectr. Dartmouth Coll., 1984; G.B. Kistiakowsky lectr. Harvard U., 1984; O.K. Rice lectr. U. N.C., Chapel Hill, 1986, Baker lectr. Cornell U., 1990; Smith lectr. Duke U., 1992; sr. fellow Alexander von Humboldt-Stiftung; also vis. prof. Max Planck Inst., Heidelberg, 1972; vis. scientist Harvard U., 1976; mem. chemistry adv. panel NSF, 1966-69, vice chmn., 1968-69; mem. rev. com. Argonne Nat. Lab., 1970-74; mem. sci. and edn. adv. com. Lawrence Berkeley Lab., 1980-86; exchange visitor USSR Acad. Scis., 1962, 75; mem. vis. com. Tufts U., 1966-69, Princeton, 1973-78; mem. fellowship com. Alfred P. Sloan Found., 1977-82; Joliot-Curie prof. École Supérieure de Physique et Chemie, Paris, 1985. Author: New NMR Methods in Solid State Physics, 1978; editor: Advances in Magnetic Resonance, 1965-87; assoc. editor: Jour. Chem. Physics, 1965-67, Spectrochimica Acta, 1964-78; mem. editorial bd. Chem. Revs., 1978-82, Jour. Magnetic Resonance, 1989—, Applied Magnetic Resonance, 1989—. Recipient Irving Langmuir award, 1976, Gold Pick Axe award, 1976, Pitts. award Spectroscopic Soc. Pitts., 1979, Wolf prize, 1984, Pauling medal, 1985, Calif. Inst. Tech. Disting. Alumnus award, 1987, Killian award, 1988, ISMAR prize, 1989, Richards medal, 1992; Sloan fellow, 1958-62; Guggenheim fellow, 1963-64, 72; Sherman Fairchild scholar Calif. Inst. Tech., 1989. Fellow Am. Acad. Arts and Scis., Am. Phys. Soc. (chmn. div. chem., physics 1983-84), AAAS; mem. NAS, Slovenian Acad. Sci. and Arts (fgn. corr.), Phi Beta Kappa, Sigma Xi. Office: MIT 77 Massachusetts Ave Cambridge MA 02139-4307

WAUGH, WILLIAM HOWARD, biomedical educator; b. N.Y.C., May 13, 1925; s. Richey Laughlin and Lyda Pearl (Leamer) W.; m. Eileen Loretta Garrigan, Oct. 4, 1952; children: Mark Howard, Kathleen Cary, William Peter. Student, Boston U., 1943, W.Va. U., 1944; MD, Tufts U., 1948, postgrad., 1949-50. Cardiovascular rsch. trainee Med. Coll. Ga., Augusta, 1954-55, asst. rsch. prof., 1955-60, assoc. medicine, 1957-60; assoc. prof. medicine U. Ky., Lexington, 1960-69, chair cardiovascular rsch., 1963-71, prof. medicine, 1969-71; prof. medicine and physiology East Carolina U., Greenville, N.C., 1971—; head renal sect. U. Ky. Coll., Lexington, 1960-68; chmn. dept. clin. scis. E.Carolina U., Greenville, 1971-75; chmn. E. Carolina U. Policy and Rev. Com. on Human Rsch., Greenville, N.C., 1971-90. Contbr. articles to profl. jours. With AUS, 1943-46; capt. USAF, 1952-54. Fellow Am. Coll. Physicians; mem. Am. Physiology Soc., Am. Heart Assn., Am. Soc. Nephrology, Am. Assn. for the Advancement Sci. Republican. Achievements include basic advances in excitation contraction coupling in vasc. smooth muscles; basic advances in autoregulation of renal blood flow and urine flow; adj. therapy in acute lung edema; noncovalent antisickling agents in sickle cell hemoglobinopathy. Home: 119 Oxford Rd Greenville NC 27858 Office: E Carolina U Sch Medicine Dept Physiol Greenville NC 27858

WAUTIER, JEAN LUC, hematologist; b. Colombelles, Calvados, France, Nov. 10, 1942; s. Marcel Jean and Léonie Benoite (Eliat) W.; m. Marie Paule Pepin, July 19, 1967; children: Jean Baptiste, Pauline. MD, U. Caen, France, 1971; degree in hematology, U. Paris, 1973, D of Biol. Scis., 1979. Asst. Hosp. St. Louis, Paris, 1972-75; cons. Hosp. Lariboisière, Paris, 1976-78, med. dir. Ctr. Transfusion, 1979—; sr. investigator Nat. Inst. Med. Research, Paris, 1980—; prof. hematology U. Paris VII, 1988—; chmn. AIDS com. Author: The Role of Platelets; co-author: Sang et Vaisseaux, 1987, sci. films; editor: Nouvelle Revue Francaise d'Hematologie, 1983—, Internat. Vascular News Letter, Clin. Hemorheology. Mem. exec. com. Assn. Personnels Sportifs Adminstrns. Paris, 1984. Recipient O. Lemonom price Paris Acad. Scis., 1983. Fellow Royal Soc. Medicine; mem. Internat. Soc. Hematology, French Soc. Hematology, Soc. Clin. Hemorheology, N.Y. Acad. Scis., Internat. Soc. Thrombosis Hemostasis, European Soc. Clin. Investigation, Convention Libérale Européenne et Sociale. Avocations: sailing, skiing, tennis, golf. Office: Hosp Lariboisiere Ctr Transfusion, 2 Rue Ambroise Pare, 75010 Paris France

WAXMAN, RONALD, computer engineer; b. Newark, Nov. 28, 1933; s. Benjamin and Rose (Lifson) W.; m. Pearl Latterman, June 19, 1955; children: David, Roberta, Benjamin. BSEE, N.J. Inst. Tech., 1955; MEE, Syracuse U., 1963. Engr. IBM, Poughkeepsie, N.Y., 1955-56, 58-64, East Fishkill, N.Y., 1964-70, Poughkeepsie and Kingston, N.Y., 1970-80; sr. engr. IBM, Manassas, Va., 1980-87; prin. scientist U. Va., Charlottesville, 1987—; IEEE rep. to Internat. Elec. Commn. U.S. tech. activities group for internat. design automation stds.; mem. steering com. very high speed integrated circuits hardware description lang. VHDL Users Group, 1987-91. Contbr. numerous articles to profl. jours. and tech. presentations. 1st lt. USAF, 1956-58. Fellow IEEE, IEEE Computer Soc. (bd. govs. 1989—, founder and chmn. design automation standards subcom. 1983-88, chmn. design automation tech. com. 1988-90, vice chmn. tech. activities bd. 1991-92, Disting. Vis. 1986-88, Meritorious Svc. cert. 1988, TAB Pioneer award 1989), Internat Fedn. Info. Processing (mem. working group hardware description langs., comp. soc. rep. to tech. com.), Assn. Computing Machinery (mem. SIGDA adv. bd. 1988-91), Eta Kappa Nu, Sigma Xi. Achievements include patents in field. Office: U Va Ctr for Semicustom Integrated Systems Dept EE Thornton Hall Charlottesville VA 22903

WAYBURN, LAURIE ANDREA, environmental and wildlife foundation administrator, conservationist; b. San Francisco, Sept. 27, 1954; d. Edgar A. and Cornelia (Elliott) W. BA, Harvard U., 1977. Program mgr. UN Environment Program, Nairobi, Kenya, 1978-81; program cons. UNESCO,

Paris and Montevideo, Uruguay, 1983-86; cons. BBC, London, 1986-87; exec. dir. Point Reyes Bird Obs., Stinson Beach, Calif., 1987—; speaker numerous confs., 1987—; coord. Cen. Calif. Coast Biosphere Res., 1988—; com. mem. U.S. Nat. Program Man and Biosphere, Washington, 1989—; bd. dirs. UN Assn. Panel, San Francisco, People for A GGNRA, San Francisco, 1989—; del. Internat. Coun. Bird Preservation, Washington, 1989—. Contbr. articles to profl. jours. Advisor Switzer Found., San Francisco, 1990—. Avocations: hiking, birding, music writing. Home: PO Box 272 Bolinas CA 94924-0272

WAYLAND, MARILYN TICKNOR, medical researcher, evaluator, educator; b. Detroit, Sept. 12, 1949; d. George Gary and Vera Virginia (Lux) Rieckhoff; m. William James Wayland, Sept. 22, 1979; children: Jessica, James. BA, Wayne State U., 1972, PhD, 1983. Project dir. Oakland County Emergency Med. Svcs., Pontiac, Mich., 1980-81, Wayne State U. Med. Sch., Detroit, 1981-84; dir. clin. rsch. St. John Hosp. Med. Ctr., Detroit, 1984—; biliary rsch. dir. Mich. Mobile Lithotripsy, Inc., Detroit, 1988—. Contbr. articles to Acad. Medicine, Michigan Hosps., Jour. Med. Edn., Med. Bull.; editor Med. Bull., 1984—. Grantee Mich. Health Care Edn. & Rsch. Found., Detroit, 1986, '89. Mem. Am. Assn. Med. Colls., Am. Pub. Health Assn., Am. Ednl. and Rsch. Assn., Mich. Assn. of Med. Educators. Office: St John Hosp & Med Ctr 22101 Moross Rd Grosse Pointe MI 48236-2172

WAYLAND, RUSSELL GIBSON, JR., geology consultant, retired government official; b. Treadwell, Alaska, Jan. 23, 1913; s. Russell Gibson and Fanchon (Borie) W.; m. Mary Mildred Brown, 1943 (div. 1964); children: Nancy, Paul R.; m. Virginia Bradford Phillis, Dec. 24, 1965. B.S., U. Wash., 1934; A.M., Harvard, 1937; M.S., U. Minn., 1935, Ph.D., 1939. Engr., geologist Homestake Mining Co., Lead, S.D., summers 1930-39; with U.S. Geol. Survey, 1939-42, 1952-80, chief conservation div., 1966-78; research phys. scientist Office of Dir., 1978-80; energy minerals cons., 1980—; Washington rep. Am. Inst. Profl. Geologists, 1982-88; commr. VA Oil and Gas Conservation Bd., 1982-90; with Army-Navy Munitions Bd., 1942-45, Office Mil. Govt. and Allied High Commn., Germany, 1945-52; instr. geology U. Minn., 1937-39. Author sci. bulls. in field. Served to lt. col. AUS, 1942-46. Decorated Army Commendation medal; recipient Distinguished Service award Dept. Interior. Mem. AIME, Mineral Soc. Am., Geol. Soc. Am., Am. Inst. Profl. Geologists, Soc. Econ. Geologists, Assn. Engring. Geologists, Cosmos Club, Sigma Xi, Tau Beta Pi, Phi Gamma Delta, Sigma Gamma Epsilon, Gamma Alpha, Phi Mu Alpha, Sinfonia. Episcopalian. Home and Office: 4660 35th St N Arlington VA 22207-4462

WAYLAND, SARAH CATHERINE, cognitive psychologist; b. Tucson, Nov. 22, 1963; d. James Robert and Susan Carrie Shier (Martz) W.; m. Alan Keith Thompson, May 19, 1986. BA, Rice U., 1985; PhD, Brandeis U., 1990. Rsch. asst. dept. psychology Rice U., Houston, 1983-85, Brandeis U., Waltham, Mass., 1986-90; rsch. asst. Aphasia Rsch. Ctr., Boston VA Med. Ctr., 1986; postdoctoral fellow psychology dept. Northeastern U., Boston, 1990—. Contbr. articles to Aphasiology, Jour. Applied Psycholinguistics, Jour. Gerontology: Psychol. Scis., Behavioral Rsch. Methods, Instrumentation and Computers. Brandeis fellow Brandeis U., 1985-90; grantee NIH, 1990-93. Mem. APA, Am. Psychol. Soc., Acoustical Soc. Am., Psychonomic Soc. Office: Northeastern U Dept Psychology 125 NI Boston MA 02115

WAYMAN, PATRICK ARTHUR, astronomer; b. Bromley, Eng., Oct. 8, 1927; came to Ireland, 1964; s. Lewis John and Mary (Palmer) W.; m. Mavis McIntyre Smith Gibson, June 19, 1954; children: Russell, Karen, Sheila. BA, Cambridge U., Eng., 1948, MA, 1952, PhD, 1952. Sr. sci. officer Royal Greenwich Obs., Sussex, Eng., 1952-62, head dept., 1962-64; sr. prof. astronomy sect. Dublin (Ireland) Inst. for Advanced Studies, 1964-92. Contbr. articles to profl. jours. Mem. Royal Astron. Soc. (assoc.), Royal Irish Acad., Internat. Astron. Union (gen. sec. 1979-82, editor publ. 1979-82). Anglican. Home: Glebe Cottage, Wicklow Ireland

WAYNE, LAWRENCE GERSHON, microbiologist, researcher; b. L.A., Mar. 11, 1926; married; 5 children. BS, UCLA, 1949, MA, 1950, PhD in Microbiology, 1952. Chief rsch. lab. VA Hosp., San Fernando, Calif., 1952-71; chief Tb rsch. lab. VA Hosp., Long Beach, Calif., 1971—; assoc. clin. prof. U. Calif., Calif. Coll. Medicine, Irvine, 1970—; mem. infectious disease rsch. program com. VA, 1961-64, pulmonary disease rsch. progres com., 1964; mem. lab. com. VA-Armed Forces Coop. Study Chemotherapy Tb, 1961-66; cons. Calif. Dept. Pub. Health, 1963-69; mem. adv. com. actinomycetes Bergey's Manual Trust, 1967—; mem. judicial com. Internat. Com. System Bacteria, 1973-86, 78—; mem. bacteria and mycology study sect. Nat. Inst. Allergy and Infectious Disease, 1971-74. Recipient Bergey award, 1988. Fellow Am. Acad. Microbiology; mem. Am. Soc. Microbiology (J. Roger Porter award 1991), Am. Thoracic Soc. (sec.-treas. 1972-74). Achievements include research in natural history and diagnositc techniques orf tuberculosis and fungus diseases, physiology and classification of mycobacteria. Office: Veterans Admin Hospital Tuberculosis Research Lab 5901 E Seventh St Long Beach CA 90822*

WAZNEH, LEILA HUSSEIN, organic chemist; b. Beirut, May 3, 1957; came to U.S., 1988; d. Hussein and Fatme (Kanso) W. BS in Chemistry, Lebanese U. Beirut, 1982, M Chemistry, 1984; PhD in Organic Chemistry, U. Rennes, France, 1987. Instr. sci. Irchad High Sch., Beirut, 1981-82; instr. chemistry Makassed Nursing Coll., Beirut, 1988; postdoctoral rsch. scientist Columbia U., N.Y.C., 1988-91, staff assoc. rsch. scientist, 1992—; participant first Women Internat. Leadership Program, Internat. House, N.Y.C., 1991-92. Contbr. articles to profl. publs. Grantee Kellogg Found., 1991-92. Mem. AAAS, Am. Chem. Soc. Muslim. Achievements include characterization of reactive thioaldehydes by mass spectrometry, synthesis of new class of compounds, thiocyanohydrins. Office: Columbia U Ste 1520 701 W 168th St New York NY 10032

WAZONTEK, STELLA CATHERINE, computer programmer, analyst, software engineer; b. Bethlehem, Pa., Feb. 17, 1961; d. Edward Walter and Stella Bernice (Stankus) W. BS in Computer Sci., Moravian Coll., 1983; MS in Computer Sci., N.J. Inst. Tech., 1989; postgrad., LaSalle U., 1991—. Software engr. RCA Aerospace GE, Moorestown, N.J., 1983-87; sr. programmer, analyst Paramax Corp., Warminster, Pa., 1987-93; sr. systems analyst Unisys Corp., Blue Bell, Pa., 1993—. Mem. IEEE, Assn. for Computing Machinery (treas. student chpt. 1982-83), Upsilon Pi Epsilon. Republican. Roman Catholic. Avocations: horseback riding, hiking, reading, softball. Home: 3927 Freemansburg Ave Bethlehem PA 18017

WEATHERBEE, CARL, retired chemistry educator, genealogist; b. Michigan City, Ind., Nov. 21, 1916; s. Walter and Rachel (Edwards) W.; m. Lucile Westwood, Sept. 1, 1950; children: Gordon Dean, Carleen, Linda, Cecilia, Tina. AB, Hanover Coll., 1940; MA, U. Ill., 1946; PhD, U. Utah, 1950. Asst. prof. chemistry Reed Coll., Portland, Oreg., 1950; asst. prof. chemistry U. Hawaii, Honolulu, 1950-52; prof. chemistry, chmn. chemistry dept. Millikin U., Decatur, Ill., 1952-82; prof. chemistry emeritus Millikin U., Decatur, 1982—; cons. Lincoln Labs., Decatur, 1957-59. Editor: Ill. State Genealogical Soc., 1978-81; contbr. many articles on chem., genealogical subjects to profl. jours., periodicals. With U.S. Army, 1941-45. Recipient chem. rsch. grants, 1950-70, Disting. Svc. award Ill. State Geneal. Soc., 1982, Alumni Achievement award Hanover Coll., 1992; Computer Ctr. in Scovill Sci. Hall at Millikin U. named in his honor, 1989. Fellow Am. Inst. Chemists; mem. Am. Chem. Soc. (pres. Decatur-Springfield subsect. 1968, mem. organic subcom. div. chem. edn. 1968-74), Ill. Assn. Chemistry Tchrs. (pres. 1955-56, 56-57), Ill. State Acad. Sci. (budget com. 1961-65, 68-71, chmn. jr. acad. com. 1965, 2d v.p. 1968, historian 1974-78, chmn. chemistry sect. 1965, assoc. editor trans. 1964-67), Decatur Genealogical Soc. (hon. life, 2d v.p. 1974-77), Am. Legion (life), Alpha Chi Sigma, Sigma Xi, Phi Kappa Phi (pres. Millikin chpt. 1957-58), Sigma Zeta (nat. editor Sigma Zetan 1957-59), Gamma Sigma Pi, Delta Epsilon. Republican. Mormon. Avocations: baseball, fishing, coin, stamp collecting, gardening. Home: 1360 W Macon St Decatur IL 62522-2704 Office: Millikin U Dept Chemistry 1184 W Main St Decatur IL 62522-2039

WEATHERFORD, GEORGE EDWARD, civil engineer; b. Oakdale, Tenn., Jan. 8, 1932; s. Walter Clyde and Kathleen (Hinds) W.; m. Martha Jeannette Beck, July 9, 1960; children: Kathleen Jeannette Weatherford-Hommeltoft, Elizabeth Lynn. BSCE, Ind. Inst. Tech., Fort Wayne, 1957; BS Engr. in Constrn., U. Mich., 1959; MSBA, St. Francis Coll., 1975. Registered profl. engr., Ind., Ga., Ohio, Minn., Iowa, S.C., Pa., Ky., Ill., Md., La., Tenn., Mich. Plant engr. Cen. Soya Co., Inc., Decatur, Ind., 1959; civil engr. Cen. Soya Co., Inc., Decatur, 1959-64; county hwy. engr. Allen County Ind. Govt., Ft. Wayne, 1964-66; sr. civil engr. Cen. Soya Co., Inc., Fort Wayne, 1966-69, engring. mgr., 1969-77, prin. engr., 1977—; ind. cons. 1964—. Author book chpts.; contbr. articles to profl. jours. Trustee Ft. Wayne YWCA, 1973-76. Sgt. USMC, 1950-54. Mem. ASCE (state treas. 1957), NSPE, Am. Concrete Inst., Am. Inst. Steel Constrn., Nat. Grain and Feed Assn. (fire and explosion rsch. and edn. com.), Ill. Assn. Structural Engrs., Grain Elevator and Processing Soc. (edn. programming com.). Republican. Mem. Disciples of Christ Ch. Home: 3617 Delray Dr Fort Wayne IN 46815-6012

WEATHERILL, NIGEL PETER, mathematician, researcher; b. Staincliffe, Yorkshire, Eng., Nov. 1, 1954; s. Ernest and Barbara (Smith) W.; m. Barbara Ann Hopkins, Sept. 4, 1976; children: George James, Laura Carys. BSc, U. Southampton, Eng., 1976, PhD, 1979. Dep. leader rsch. team Anglian Water Authority, Cambridge, Eng., 1980-81; sr. project supr. Aircraft Rsch. Assn., Bedford, Eng., 1981-87; rsch. staff mem. dept. mech. and aerospace engring. Princeton U., 1986; reader dept. civil engring. Univ. Coll. of Swansea, Wales, 1987—; vis. prof. NSF Engring. Rsch. Ctr., Miss., 1991—; cons. Aircraft Rsch. Assn., Bedford, 1987-90, Water Quality Mgmt., Cambridge, 1988—. Contbr. articles to Internat. Jour. Numerical Methods, Computer Math. with Applications, Aero. Jour. of the Royal Aero. Soc., others. Chmn. European Rsch. Community on Flow, Turbulence and Combustion, U.K.-South, 1988-92; mem. transonic aerodynamics com. Engring. Scis. Data Unit, 1988-92. Recipient Busk award Royal Aero. Soc. London, 1991, Wolfson Acad. award The Wolfson Trust, 1992. Fellow Inst. Math. and Its Applications; mem. AIAA, Chartered Mathematician of United Kingdom. Mem. Ch. of England. Achievements include early development work on multiblock grid generation, also on Delaunay triangulation; research on numerical simulation of inviscid transonic flow over a complete aircraft. Office: Univ Coll of Swansea, Dept Civil Engring, Swansea Wales SA2 8PP

WEATHINGTON, BILLY CHRISTOPHER, analytical chemist; b. Bossier City, La., Dec. 3, 1951; s. Billy and Christine (Lowman) W.; m. Tamara R.A. Horman, Aug. 23, 1973 (div. May 1975); m. Gwendolyn C. Adamson, May 19, 1979; 1 child, Leia C. BA, Auburn U., 1972, BS, 1979. Chemist U.S. Dept. Agriculture, Bettsville, Md., 1978-80; sr. rsch. chemist Midwest Rsch. Inst., Riyadh, Saudi Arabia, 1980-82; quality assurance mgr. Hittman Assocs. Inc., Columbia, Md., 1982-84, dir. mktg., 1984-86, tech. dir., 1986-88; lab. dir. RMC Environ. Svcs. Inc., Pottstown, Pa., 1988-89, v.p. analytical div., 1989—. Contbr. articles to profl. jours. Mem. ASTM, Am. Chem. Soc. Achievements include patent disclosure for destruction of simple and complex cyanides; research on the destruction of cyanide in wastewaters, on phytoxins in Rhizoctonia Solani, on the monitoring of contaminated soil for methyl mercury, on quality assurance in field laboratory, and on single blind versus double blind evaluations. Home: 1535 Kauffman Rd Pottstown PA 19464-2307 Office: RMC Environ Svcs Inc 88 Robinson St Pottstown PA 19464-6440

WEAVER, ARTHUR LAWRENCE, physician; b. Lincoln, Nebr., Sept. 3, 1936; s. Arthur J. and Harriet Elizabeth (Walt) W.; BS (Regents scholar) with distinction, U. Nebr., 1958; MD, Northwestern U., 1962; MS in Medicine, U. Minn., 1966; m. JoAnn Versemann, July 6, 1980; children: Arthur Jensen, Anne Christine. Intern U. Mich. Hosps., Ann Arbor, 1962-63; resident Mayo Grad. Sch. Medicine, Rochester, Minn., 1963-66; practice medicine specializing in rheumatology and internal medicine, Lincoln, 1968—; mem. staff Bryan Meml. Hosp., chmn. dept. rheumatology, 1976-78, 82-85, 89-91, vice-chief staff, 1984-87; mem. courtesy staff St. Elizabeths Hosp., Lincoln Gen. Hosp.; mem. cons. staff VA Hosp.; chmn. Juvenile Rheumatoid Arthritis Clinic, 1970-88; assoc. prof. dept. internal medicine U. Nebr., Omaha, 1976—; med. dir. Lincoln Benefit Life Ins. Co., Nebr., 1972-90; mem. exam. bd. Nat. Assn. Retail Druggists; mem. adv. com. Coop. Systematic Studies in Rheumatic Diseases III. Bd. dirs. Nebr. chpt. Arthritis Found., 1969—; mem. tech. cons. panel for rheumatology Harvard Resource Based Relative Value Study; trustee U. Nebr. Found., 1974—. Served to capt., M.C., U.S. Army, 1966-68. Recipient Outstanding Nebraskan award U. Nebr., 1958, also C.W. Boucher award; Philip S. Hench award Rheumatology, Mayo Grad. Sch. Medicine, 1966. Diplomate Am. Bd. Internal Medicine, Am. Bd. Rheumatology. Fellow ACP (Nebr. council 1983—), Am. Rheumatism Assn. (com. on rheumatologic practice 1983-87, pres.-elect Cen. region 1983-84, pres. Cen. region 1984-85); mem. AMA, Am. Coll. Rheumatology (sec. 1991—, 1st Paulding Phelps award, bd. dirs. 1985—, planning com. 1987—, exec. com. 1991—, sec. 1991-93, pres. rsch. and edn. found. 1991—, 2nd v.p. 1993—), Am. Soc. Internal Medicine (coord. com. physician payment svcs. 1988—), Nebr. Soc. Internal Medicine (Internist of Yr., 1988), Nebraska Rheumatism Assn., Nebr. Med. Assn., Lancaster County Med. Soc., Mayo Grad. Sch. Medicine Alumni Assn., Arthritis Health Professions Assn. (com. on practice 1984-87), Nat. Soc. Clin. Rheumatology (program chairperson 1986-87, 88, exec. com. 1987—), Midwest Cooperative Rheumatic Disease Study Group, (chmn. exec. com. 1986—), Arthritis Found. (profl. del.-at-large 1987-88, 89, 90, Nat. Vol. Svc. citation, 1988), Phi Beta Kappa, Sigma Xi, Alpha Omega Alpha, Pi Kappa Epsilon, Phi Rho Sigma. Republican. Presbyterian. Editorial bd. Nebr. Med. Jour., 1982—; contbr. articles to med. jours. Home: 4239 Calvert Pl Lincoln NE 68506-4252 Office: 2121 S 56th St Lincoln NE 68506

WEAVER, CAROLYN LESLIE, economist, public policy researcher; b. Washington, Jan. 20, 1952; d. Kenneth Faulkner and Margaret Mae (Taylor) Weaver; m. Robert John Mackay, Aug. 12, 1980; children: Taylor Cotesworth, Bennett Faulkner. BA, Mary Washington Coll., 1973; PhD in Econs., Va. Polytech. Inst. & State U., 1977. From instr. to asst. prof. econs. Tulane U., New Orleans., 1976-78; from asst. prof. to assoc. prof. and rsch. assoc. Ctr. for Pub. Choice, Va. Poly. Inst. and State U., Blacksburg, 1978-83; chief profl. staff mem. on social security U.S. Senate Com. on Fin., Washington, 1981-84; sr. rsch. fellow Hoover Inst., Stanford U., Calif., 1984-86; resident scholar, dir. social security & pension project Am. Enterprise Inst., Washington, 1987—; sr. advisor Nat. Commn. on Social Security Reform, 1982-83; cons. U.S. Senate Fin. Com., 1984, Social Security Adminstrn., 1984-85; mem. U.S. Disability Adv. Coun., 1987-88, U.S. Social Security Commrs. Disability Adv. Com., 1989, Social Security Pub. Trustees Working Group on Trust Fund Solvency, 1989-90, Acad. Bd. Advisors Americans for Generational Equity, 1986-92, Ind. Inst., 1986—, Retirement Policy Inst., 1989—; founding mem. Nat. Acad. Social Ins., 1988—. Author: The Sources and Dimensions of Crisis in Social Security: A First Step Toward Meaningful Reform, 1981, Crisis in Social Security: Economic and Political Origins, 1982; editor: Social Security's Looming Surpluses: Prospects and Implications, 1990, Disability and Work, 1991, Regulation mag., 1986-88; sr. editor Am. Enterprise mag., 1989—; contbr. numerous articles to profl. jours., editorials to bus. publs. Grad. fellow The Scaife Found, 1973-75, The Earhart Found., 1975-76; rsch. grantee NSF, Washington, 1979-81. Mem. Am. Econs. Assn., Nat. Acad. Social Ins. (founding mem.). Episcopalian. Office: Am Enterprise Inst 1150 17th St NW Washington DC 20036-4603

WEAVER, CHARLES LYNDELL, JR., architect; b. Canonsburg, Pa., July 5, 1945; s. Charles Lyndell and Georgia Lavelle (Gardner) W.; m. Ruth Marguerite Uxa, Feb. 27, 1982; children: Charles Lyndell III, John Francis. BArch, Pa. State U., 1969; cert. in assoc. studies U. Florence (Italy), 1968. Registered architect, Pa., Md., Mo., Va., Ky. With Celento & Edson, Canonsburg, Pa., part-time 1966-71; project architect Meyers & D'Aleo, Balt., 1971-76, corp. dir., v.p., 1974-76; ptnr. Borrow Assocs.-Developers, Balt., 1976-79, Crowley/Weaver Constrn. Mgmt., Balt., 1976-79; pvt. practice architecture, Balt., 1976-79; cons., project mgr. U. Md., College Park, 1979-80; corp. cons. architect Bank Bldg. & Equipment Corp., Am., St. Louis, 1980-83; dir. archtl. and engring. svcs. Ladue Bldg. & Engring. Inc., St. Louis, 1983-84; v.p., sec. Graphic Products Corp.; pres. CWCM Inc. Internat., 1987—; dir. K-12 Edn. Market Ctr. and sr. program mgr., Sverdrup Corp., 1989—; vis. Alpha Rho Chi lectr. Pa. State U., 1983; vis. lectr. Washington U. Lindenwood Coll., 1987; panel mem. Assn. Univ. Architects Conv., 1983. Project bus. cons. Jr. Achievement, 1982-85; mem. cluster com., advisor Explorer Program, 1982-85. Recipient 5 brochure and graphic awards Nat. Assn. Indsl. Artists, 1973; 1st award Profl. Builder/Am. Plywood Assn., 1974; Honor award, 2 articles Balt. chpt. AIA, 1974; Better Homes and Gardens award Sensible Growth, Nat. Assn. Home Builders, 1975; winner Ridgely's Delight Competition, Balt., 1976. Mem. ASCD, BBC Credit Union (bd. dirs. 1983-85), Vitruvius Alumni Assn., Penn State Alumni Assn., BOCA, NFPA, Am. Assn. Sch. Adminstrs., Coun. Ednl. Facilities Planners, Assn. Sch. Bus. Officials, Alpha Rho Chi (nat. treas. 1980-82, dir. nat. found. treas. 1989—). Home and Office: 1318 Shenandoah Ave Saint Louis MO 63104-4123

WEAVER, CHRISTOPHER S(COT), scientist; b. N.Y.C., Feb. 6, 1951; s. Richard B. and Mildred (Stier) W.; m. Constance Joan Bohon, Sept. 1, 1991. MS/MA, Wesleyan U., Middletown, Conn., 1975, PhD, 1977; ME, MIT, 1985. Registered profl. engr. Mgr. forecasting tech. Am. Broadcasting Co., N.Y.C., 1977-79; v.p. sci. and tech. Nat. Cable TV, Washington, 1979-81; chief engr. U.S. Congress, Washington, 1981-82; pres. Media Tech. Ltd., Rockville, Md., 1982—; cons. engr., lectr. FCC, Washington, 1979-82; cons. engr. Corp. for Pub. Broadcasting, 1979-85; advisor NSF, Washington, 1979-84; lectr. MIT, 1983-89; U.S. rep. to U.S. delegation to 20th Assembly of Internat. Union of Radio Sci. Contbr. articles to profl. jours. Benefactor Nat. Holocaust Mus., Washington, 1989—. Seaman U.S. Merchant Marines, 1965-68. Recipient Nat. Sci. award NSF, Washington, 1969, Sci. award, N.Y. State, 1969, Japan Conf. award Internat. Inst. Edn., N.Y.C., 1975. Mem. IEEE, Soc. Motion Picture Engrs. Jewish. Achievements include key design of Congl. Data Network. Office: Media Tech Ltd 1370 Piccard Dr Rockville MD 20850

WEAVER, CRAIG LEE, civil engineer; b. Somerset, Pa., June 7, 1954; s. Harold William and Blanche Arlene (Will) W.; m. Rae Ann Walker, May 14, 1977; children: Jennifer Marie, Ryan Elliott. BS in Civil Engring. Tech., U. Pitts., 1976; MBA, Ind. (Pa.) U., 1984. Registered profl. engr., Pa. Civil engr. Projec Inc., Pitts., 1976-77; project engr. L. Robert Kimball & Assoc., Ebensburg, Pa., 1977-80; dept. mgr. The EADS Group, Somerset, 1980—. Coun. pres. St. Paul's United Ch. of Christ, Somerset, 1989. Mem. NSPE (chpt. pres. 1986-87, Young Engr. of Yr. 1987, Engr. of Yr. 1993), ASCE, Am. Soc. Hwy. Engrs. (sect. pres. 1989-90), Profl. Engrs. in Pvt. Practice (state chair 1990-92). Republican. Office: The EADS Group PO Box 837 1065 Tayman Ave Somerset PA 15501-0837

WEAVER, JERRY REECE, management scientist, educator; b. Tuscaloosa, Ala., Apr. 30, 1946; s. Cecil and Janice Margaret (Reece) W.; m. Brenda Jo Alexander, May 30, 1970; children: Alexander Evan, Hilary Jane. MA, U. Ala., 1969; PhD, U. Tenn., 1979. Mathematician U.S. Dept. Navy, 1971-72; instr. U. Tenn., Knoxville, 1972-80; prof. mgmt. sci. U. Ala., Tuscaloosa, 1980—; cons. Ala. Productivity Ctr., Tuscaloosa, 1988—. Contbr. articles to profl. publs. Mem. Ops. Rsch. Soc. Am., Decision Scis. Inst. Methodist. Office: U Ala 348 Alston Hall Tuscaloosa AL 35487

WEAVER, KENNETH, gynecologist, researcher; b. Whitetop, Va., Dec. 4, 1933; s. Grover Cleveland and Violet Elaine (Baldwin) W.; children: Teresa Marie, Janice Eileen, Beverly Lynn, Pamela Jean, Cynthia Ann; m. Shelby Jean Davis, June 15, 1966. BA, U. N.C., 1957, MD, 1960. Diplomate Am. Bd. Ob-Gyn. Intern U.S. Pub. Health Svc., Boston, 1960-61; med. officer Cherokee (N.C.) Indian Hosp., 1961-64; gen. physician Haywood County Hosp., Waynesville, N.C., 1964-70; obstetrician, gynecologist Haywood County Hosp., Waynesville, 1974-77; resident U. Ark. Med. Ctr., Little Rock, 1970-74; asst. prof. U. Ark., Little Rock, 1977-78, acting chmn., dept. ob-gyn., 1978; pvt. practice Johnson City, Tenn., 1978—; mem. Gov. Com. on Cancer, Raleigh, N.C., 1970-71; dir. Maternity and Infant Care, Little Rock, 1977-78; assoc. prof. James H. Quillen Coll. Medicine, Johnson City, 1978-83. Contbr. articles to sci. jours. Fellow Am. Coll. Ob-Gyn., Am. Coll. Nutrition, Am. Assn. Gynecol. Laparoscopists, N.Y. Acad. Scis. Achievements include six patent devices having to do with laser surgery and other gynecol. and urol. uses; rsch. in magnesium and preeclampsia, in magnesium and migraine, in relationship between magnesium and blood platelet function. Home: 377 Tavern Hill Rd Jonesborough TN 37659-5026 Office: 1103 Jackson Blvd Jonesborough TN 37659

WEAVER, KENNETH NEWCOMER, geologist, state official; b. Lancaster, Pa., Jan. 16, 1927; s. A. Ross and Cora (Newcomer) W.; m. Mary Elizabeth Hoover, Sept. 9, 1950; children—Wendy Elaine, Matthew Owen. BS, Franklin and Marshall Coll., 1950; MA, Johns Hopkins U., 1952, PhD, 1954. Instr. geology Johns Hopkins, 1953- 54; ops. analyst Ops. Rsch. Office, Washington, 1954-56; chief geologist, then mgr. geology and quarry dept. Medusa Portland Cement Co., Wampum, Pa., 1956-63; dir., state geologist Md. Geol. Survey, Balt., 1963—; chmn. Md. Land Reclamation Com., 1967-76, 78—; Gov.'s rep. Interstate Oil Compact Commn., Interstate Mining Compact Commn.; mem. outer shelf adv. com. U.S. Dept. Interior; chmn. Md. Topographic Mapping Com.; mem. com. on surface mining and reclamation Nat. Acad. Scis., 1978, vice chmn. com. on disposal of excess spoil, 1980-81, mem. com. on geologic mapping, 1983, liaison mem. bd. earth scis., 1982-88, mem. com. on water resources rsch., 1989—, chmn. com. on abandoned minelands research priorities, 1987; mem. subcom. on mgmt. of maj. underground constrn. projects Nat. Acad. Engring.; mem. Md. Commn. on Artistic Property. Served with AUS, 1954-56. Fellow Geol. Soc. Am. (sec. N.E. sect. 1985—), AAAS; mem. Am. Inst. Mining Engrs., Am. Inst. Profl. Geologists (editor 1983-84, Ben H. Parker Meml. medal 1992), Am. Geol. Inst. (governing bd. 1973, exec. com. 1989-90), Am. Water Rsch. Assn., Geol. Soc. Washington, Assn. Am. State Geologists (pres. 1973), Sigma Xi. Republican. Presbyn. (elder). Club: Johns Hopkins (Balt.). Home: 14002 Jarrettsville Pike Phoenix MD 21131-1409 Office: Md Geol Survey 2300 St Paul St Baltimore MD 21218-5210

WEAVER, KERRY ALAN, construction engineer; b. Shamokin, Pa., Jan. 26, 1952; s. John Elwood and Arlene Betty (Wary) W.; m. Mary Rebecca Thompson, May 17, 1975 (div. Aug. 1990); m. Laura Ilene Shach, June 14, 1992. BS in Civil Engring., Pa. State U., 1973; postgrad., Loyola Coll. Engr. in tng. Project engr. The Whiting-Turner Contracting Co., Towson, Md., 1973—. Treas. Community Assn., Phoenix, Md., 1992. Mem. ASCE, Am. Concrete Inst., Constrn. Specification Inst. Republican. Office: Whiting-Turner Contracting 300 E Joppa Rd Towson MD 21286

WEAVER, MICHAEL ANTHONY, mining engineer, consultant; b. Pitts., Aug. 25, 1948; s. James S. and Ellen W.; m. Patricia L. Barry; children: Scott-Patrick, Dulany. BSCE, U. Pitts., 1976, MS in Mining Engring., 1979, MBA, 1980. Registered profl. engr. Asst. project engr. John T. Boyd Co., Pitts., 1973-75; rsch. mining engr. U.S. Bur. Mines, Pitts., 1975-79, mining engr., 1979-80; chief mining engr. Colo.-Ute Elec. Assn., Montrose, 1980-82; asst. prof. Pa. State U., State College, 1982-88; prin. mining engr., sr. mining engr. Indpls. Power & Light, 1989—; mem. fossil and synthetic fuel sect. Edison Electric Inst., Washington. Mem. Soc. Mining Engrs. Office: Indpls Power & Light PO Box 1595 Indianapolis IN 46206-1595

WEAVER, MICHAEL JOHN, chemist, educator; b. London, Mar. 30, 1947. BSc, U. London, 1968, PhD in Chemistry, DIC, 1972. Rsch. fellow in chemistry Calif. Inst. Tech., 1972-75; asst. prof. chemistry Mich. State U., 1975-80, assoc. prof., 1980—; with dept. chemistry Purdue U., Lafayette, Ind., 1982—. Mem. Am. Chem. Soc. Achievements include research in kinetics of electrode processes, structure of electrode-electrolyte interfaces, measurement of rapid electrochemical reaction rates, theories of electron transfer kinetics, chemistry of metal macrocyles. Office: Purdue Univ Dept Chemistry Lafayette IN 47907*

WEBB, GREGORY FRANK, retired aeronautical engineer; b. Bristol, Eng., Jan. 25, 1921; s. Llewellyn Frank and Catherine Ellen (Baker) W.; m. Barbara Cecil Lewis, Dec. 21, 1981. Higher nat. cert. aero. engring., Medway Tech. Coll., Gillingham, Kent, Eng., 1946. Apprentice, inspector, draughtsman Short Bros. (Rochester & Bedford) Ltd., Rochester, Eng., Swindon, Eng., 1936-47; chief draughtsman Navarro Aircraft Constrn. Co., Heston, Middlesex, Eng., 1947-48; project and devel. engr. Brit. S.Am., Brit. Overseas, Brit. European Airways Corps., London, 1948-57; sr. design engr. Sir W.G. Armstrong Whitworth Aircraft Ltd., Coventry, Warwicks, Eng., 1958-59; R&D engr. Pressed Steel-Fisher Ltd., Cowley, Oxford, Eng., 1960-68; project engr. Dredge & Marine Ltd., Penryn, Cornwall, Eng., 1971-77; tech. author, illustrator Hunting Engring. Ltd., Ampthill, Bedford, Eng., 1979-86; analyst engring. design Sci. Rsch. Coun., London, 1967-68; press officer Coun. Engring. Instns., West of Eng. Group, 1976-78. Editor, author (with others): Manuals for Marine Craft & Defence Equipment, 1971-86; contbr. articles to profl. jours. Corr. mem. Assn. Engring. & Shipbldg. Draughtsman, Rochester, 1945-47; founding-mem. Air-Britain, The Airship

Assn., London, 1970's, RAES Mgmt. Studies Group, London, 1960's; instr. Air Tng. Corps, 1974-77. Fellow AIAA (assoc., 40 Yr. cert. 1987); mem. Engring. Coun. (chartered engr.), Royal Aero. Soc. (hon. sec. Medway chpt. 1945-47, 50 Yr. cert. 1993), Inst. Mgmt. (assoc.). Conservative. Anglican. Achievements include 4 patents for vehicle engineering safety and processes; pioneering research on civil and military aircraft, seating, hover dredgers and vehicles; research in methods of industrial training and safety techniques and instruction, in analyses of engineering design, in methodology of innovation. Home: Wayside 24 King St, Mortimer Common, Reading RG7 3RS, England

WEBB, JOHN ALLEN, JR., engineering executive; b. Lakewood, Ohio, May 26, 1946; s. John Allen and Rhoda Lillian (Grumney) W.; m. Susan Henrietta Neff, Sept. 28, 1968; children: Amanda H., Heather A., Michelle L. BSEE, Ohio No. U., 1968; M (hon.), The Cleve. Clin. Found., 1972; MBA, Baldwin-Wallace Coll., 1981. Registered profl. engr. Ohio. Control systems rsch. engr. NASA Lewis Rsch. Ctr., Cleve., 1968-78, project engr., 1978-82, deputy project mgr., 1982-86, SCADA systems mgr., 1986-89, br. chief, 1989—. Contbr. tech. papers to pubs. Mem. IEEE, Tau Beta Pi. Republican. Lutheran. Achievements include patent for circuit for detecting initial systole and dicrotic notch. Home: 27010 Butternut Ridge Rd North Olmstead OH 44070 Office: NASA Lewis Rsch Ctr 21000 Brookpark Rd Cleveland OH 44135

WEBB, O(RVILLE) LYNN, physician, pharmacologist, educator; b. Tulsa, Aug. 29, 1931; s. Rufus Aclen and Berla Ophelia (Caudle) W.; m. Joan Liebenheim, June 1, 1954 (div. Jan. 1980); children—Kathryn, Gilbert, Benjamin; m. Jeanne P. Heath, Aug. 24, 1991. B.S., Okla. State U., 1953; M.S., U. Okla., 1961; Ph.D. in Pharmacology, U. Mo., 1966, M.D., 1968. Diplomate Nat. Bd. Med. Examiners, Am. Bd. Family Practice. Research assoc. in pharmacology U. Okla., 1959-61; research fellow NIH, 1962-66; instr. pharmacology U. Mo., Columbia, 1966-68, asst. prof., 1968-69; intern, U. Mo. Med. Center, 1968-69; family practice, New Castle, Ind., 1969-89, med. dir. VA Clinic, Lawton, Okla., 1989—; clin. assoc. prof. family medicine U. Okla. Sch. Medicine, 1989—; adj. assoc. prof. pharmacology U. Okla. Sch. Medicine, 1989—; mem. staff Henry County Meml. Hosp., New Castle, 1969-89; guest prof. pharmacy and pharmacology Butler U. Coll. Pharmacy, Indpls., 1970-75; owner, dir. Carthage Clinic, 1975-89; clin. assoc. prof. family medicine Ind. U., 1986-89; county physician, jail med. dir. Henry County, Ind., 1976-89. Bd. dirs. Lawton Philharmonic, 1990—. Recipient Cert. of merit in Pharmacol. and Clin. Med. Research, 1970; Med. Student Research Essay award Am. Acad. Neurology, 1968. Fellow Am. Acad. Family Physicians; mem. AMA (ann. award recognition 1975—), Ind. State Med. Assn., Am. Coll. Sports Medicine, AAAS, N.Y. Acad. Sci., Am. Soc. Contemporary Medicine and Surgery, Festival Chamber Music Soc. (bd. dirs. Indpls. 1981-87), Mensa, Sigma Xi. Clubs: Columbia, Skyline (Indpls.). Lodge: Elks. Author: (with Blissitt and Stanaszek, Lea and Febiger) Clinical Pharmacy Practice, 1972; contbr. articles to profl. jours. Office: VA Clinic Comanche County Hosp Lawton OK 73503

WEBB, PAUL, physicist. With Gen. Electric Can., Vaudreuil, Que. Recipient Outstanding Achievement in Indsl. and Applied Physics award Can. Assn. Physicists, 1991. Office: Gen Electric Canada Inc, Vaudreuil, PQ Canada J7V 8P7*

WEBB, RICHARD A., physicist; b. L.A., Sept. 10, 1946; married; 2 children. BA, U. Calif., Berkeley, 1968; MS, U. Calif., San Diego, 1970, PhD, 1973. Rsch. assoc. U. Calif., San Diego, 1973-75; from asst. to assoc. rsch. physicist Argonne Nat. Lab., 1975-78; mem., mgr. rsch. staff T.J. Watson Ctr. IBM, Yorktown Heights, 1978—. Recipient Simon Mem prize, 1989. Fellow Am. Physics Soc. (Oliver E. Buckley condensed Malter Physics prize 1992). Achievements include research in macroscopic quantum tunneling in Josephson junctions at low temperatures, investigations of the Aharonov-Bohm effect and universal periodic conductance fluctuations in very small semiconducting and normal metal rings, measurement and temperature, magnetic field and Femri Engery dependencies of the conduction process of very small Si MOSFET devices in both insulating and metallic regimes. Office: IBM T.J. Watson Rsch Ctr PO Box 218 Yorktown Heights NY 10598*

WEBB, STEVEN GARNETT, engineering educator; b. El Paso, Tex., June 8, 1958; s. Carl Robert and Earnestine (Garnett) W.; m. Dina Marie Duell, July 7, 1984; children: Brandon Christopher, Ryan Matthew, Sean Robert. B in Astro. Engring. and Math., USAF Acad., 1980; M in Aerospace Engring., Princeton (N.J.) U., 1981; PhD in Aero. Engring., Air Force Inst. Tech., 1988. Commd. 2d lt. USAF, 1980, advanced through grades to maj., 1991; project officer Dept. for Tech. L.A. Air Force Sta., 1981-83; program mgr. Air Force Space Tech. Ctr., Albuquerque, 1983-85; assoc. prof. Dept Engring. Mechanics, USAF Acad., 1988—. Contbr. articles to profl. jours. including Optics Letters, Vertica, Jour. Exptl. Mechanics. Explorer post advisor Boy Scouts Am., 1986-88; deacon Faith Presbyn. Ch., Colorado Springs, 1990-92. Guggenheim fellowship Guggenheim Assn., Princeton U., 1980. Mem. AIAA (structural dynamics tech. com. 1990-92, jour. reviewer 1990—), ASME (jour. reviewer 1990-92), Air Force Assn. Episcopalian. Office: Dept Engring Mechanics HQ USAFA/DFEM U S A F Academy CO 80840

WEBB, WATT WETMORE, physicist, educator; b. Kansas City, Mo., Aug. 27, 1927; s. Watt Jr. and Anna (Wetmore) W.; m. Page Chapman, Nov., 1950; children: Watt III, Spahr C., Bucknell C. BS, MIT, 1947, ScD, 1955. Rsch. engr., asst. dir. rsch. Union Carbide Metals Co., Niagara Falls, N.Y., 1947-52, 55-61; prof. applied physics Cornell U., Ithaca, N.Y., 1961—, dir. Sch. Applied and Engring. Physics, 1983-88; dir. NIH-NSF Resource Biophysical Imaging and Opto-electronics, 1988—, dir. Biophysics Program, 1991—; NIH scholar-in-residence Fogarty Internat. Ctr. for Advanced Study, 1988-92; mem. adv. panels Materials Adv. Bd., 1958-59, 63-64, NSF, 1974—; co-chair NAS panel on sci. interfaces and tech. applications, Physics Through the 90s, 1983-86. Mem. adv. com. Physics Today, 1991—; assoc. editor Phys. Rev. Letters, 1975-91; mem. editorial bd. Biophysics Jour., 1975-78, mem. publ. com., 1976-83; contbr. 200 articles to profl. jours. Guggenheim fellow, 1974-75. Fellow AAAS, Am. Phys. Soc. (chmn. 1988-89, exec. com. div. biol. physics 1975-77, Biol. Physics prize 1991), Am. Inst. Med. and Biol. Engrs. (founding 1992); mem. Nat. Acad. Engrin., Biophys. Soc. (mem. coun. 1972-75, 82-85), Am. Soc. Cell Biology, Am. Soc. Gen. Physiology, Internat. Soc. Optical Engring., Cornell Rsch. Found. (bd. dirs., exec. com. 1983—), Ithaca Yacht Club, Rochester Yacht Club, N.Y. Yacht Club. Achievements include patents in optical instruments, two photon laser microscopy, fluorescent probes, microcrystals, welding technology. Office: Cornell U Clark Hall Ithaca NY 14853

WEBBER, DAVID MICHAEL, civil engineer; b. Yuba City, Calif., Sept. 30, 1966; s. James Frank and Joan Pauline (Sarkes) W.; m. Melinda Ann Johnson, July 7, 1990. BSCE, U. Tenn., 1990, MSCE, 1991. Registered engr.-in-tng., Ill. Rsch. asst. Transp. Ctr., Knoxville, Tenn., 1990-91; project engr. Farnsworth & Wylie, Bloomington, Ill., 1991-93, David Volkert & Assocs., Mobile, Ala., 1993—. Vol. McLean County Rep. party, Bloomington, 1992. Mem. ASPE, Ill. Soc. Profl. Engrs., Inst. Transp. Engrs. (assoc.). Roman Catholic. Home: 3655 Old Shell Rd # 121 Mobile AL 36608 Office: David Volkert & Assocs PO Box 7434 Mobile AL 36670

WEBER, ALFONS, physicist; b. Dortmund, Germany, Oct. 8, 1927; s. Alexander and Ilona (Banda) W.; m. Jeannine K. Weber, Oct. 8, 1955; children: Karl, Louise, Paul. PhD, Ill. Inst. Tech., 1956. Instr. physics Ill. Inst. Tech., Chgo., 1953-56; from asst. prof. physics to prof. Fordham U., Bronx, N.Y., 1957-81, prof. physics and chemistry, 1976-81, chmn. dept. physics, 1964-70; rsch. physicist Nat. Inst. Standards and Tech., Gaithersburg, Md., 1977—, acting chief molecular spectroscopy div., 1980-81, chief molecular physics div., 1982—; with chem scis. divsn. U.S. Dept. Energy, 1991-92, chem. divsn. NSF, 1992—. Editor: Raman Spectroscopy of Gases and Liquids, 1979; Structure and Dynamics of Weakly Bound Molecular Complexes, 1987, Spectroscopy of the Earth's Atmosphere and Interstellar Medium, 1992; mem. editorial bd. Jour. of Raman Spectroscopy, Jour. Chem. and Phys. Reference Data. V.p. Union Free Dist. # 1 Sch. Bd., Eastchester, N.Y., 1970-73. Postdoctoral fellow NRC Can., U. Toronto, 1956-57. Fellow Am. Phys. Soc. (councillor 1987-91); mem. AAAS, Optical Soc. Am. Ellis R. Lippincott award 1991), Coblentz Soc. Office: Nat Inst Standards and Tech Molecular Physics Div Rm B 268 Bldg 221 Gaithersburg MD 20899

WEBER, ARTHUR PHINEAS, chemical engineer; b. Bklyn., Mar. 10, 1920; s. Irving and Bertha (Irgang) W.; m. Jean Betty Abelman, July 11, 1942; children: Sheldon Geoffrey, Diane Leslie. BChemE., CCNY, 1941. Registered profl. engr., N.Y., Ohio, Pa. Design engr. chem. equipment Hendrick Mfg. Co., Carbondale, Pa., 1941-42; exec. engr. chem. plant design Chemurgy Design Corp., N.Y.C., 1942-44; dir process devel and process engring, nuclear energy design The Kellex Corp., N.Y.C., 1944-49; tech. dir. mixing and chem. and ceramic machinery Internat. Engring. Inc., Dayton, Ohio, 1949-69; owner Arthur Phineas Weber, engrs., N.Y.C., 1951—; pres. A.P. Weber Co., Inc., N.Y.C., 1959—; tchr., instr. chem. engring. CCNY, 1942-44; adj. assoc. prof. chem. engring. NYU, 1951-54; adj. prof. chem. engring. Poly. Inst. Bklyn., 1955-67. Contbr. articles to profl. jours. Mem. spl. subcom. on glossary for nuclear sci. and engring. NCR. Recipient cert. of svc. for atomic bomb contbn., U.S. War Dept., 1945. Fellow AAAS; mem. AICE, NSPE, Am. Chem. Soc., N.Y. Acad. Scis., Am. Soc. Safety Engrs., N.Y. Soc. Architects, Unity of Nassau County Club (pres.), Old Westbury Golf and Country Club (pres.), Long Island Golf Assn. (pres.), Metro. Golf Assn. (pres.). Home: 1334 Surrey Ln Rockville Centre NY 11570 Office: 265 Sunrise Hwy Rockville Centre NY 11570

WEBER, CHARLES WALTON, chemistry educator; b. Phoenixville, Pa., Jan. 16, 1943; s. Harry Charles and Mildred (Cullum) W.; m. Cheryl Anne Knauer, May 13, 1966; 1 child, Candice. BS, Phila. Coll. Pharmacy and Sci., 1964; PhD, U. Pa., 1969. Instr. SUNY, Buffalo, 1969; from asst. to assoc. prof. Del. Valley Coll., Doylestown, Pa., 1969—; analytical chemistry cons. Delaware Valley Coll., Doylestown, Pa., 1980—. Author: (mag.) Delaware Valley Express, 1970—, editor, 1989—; contbr. article to book Greenberg's Guide to Lionel Trains, 1945-69, 1985. Mem. Train Collectors Assn. (pres. atlantic divsn. 1974-76). Republic. Lutheran. Office: Delaware Valley Coll Rte # 202 Doylestown PA 18901

WEBER, DAVID ALEXANDER, physicist; b. Lockport, N.Y., Mar. 6, 1939; s. Fred Leonard John and C. Gladys (Woodcock) W.; m. Sandra Jean Watson, Aug. 26, 1961; children: Sarah D. Beisheim, David A. Jr. BS, St. Lawrence U., 1956; PhD, U. Rochester, 1971. Rsch. asst. Sloan Kettering Cancer Inst., N.Y.C., 1961-68; fellow, asst., assoc. prof. U. Rochester, N.Y., 1968-87; sr. scientist, head nuclear medicine rsch. group Brookhaven Nat. Lab., Upton, N.Y., 1987—; prof. radiology SUNY, Stony Brook, 1988—. Contbr. articles to profl. jours. Fellow NIH 1978-79. Fellow Am. Coll. Nuclear Physicians; mem. Am. Assn. Physicists in Medicine (assoc. editor 1979-82, 88-93), Health Physics Soc. (chpt. pres. 1973-75), Soc. Nuclear Medicine (pres. computer coun. 1977-78, pres. instrumentation coun. 1986-87, chmn. MIRD com. 1988-94), Sigma Xi. Office: Brookhaven Nat Lab Bldg 490 Med Dept Upton NY 11973

WEBER, DAVID FREDERICK, genetics educator; b. Terre Haute, Ind., Nov. 18, 1939; s. Walter John and Marguerite Johanna (Anliker) W.; m. Darlene Marie Hohman, Aug. 17, 1963; children: Julilynne, Mark David. BA, Purdue U., 1961; MS, Ind. U., 1964, PhD, 1967. Prof. genetics Ill. State U., Normal, 1967—. Atomic Energy Commn. grantee, 1970-82, USDA grantee, 1985—. Home: 2115 E Taylor Bloomington IL 61701

WEBER, DAVID JOSEPH, biochemist, educator; b. Chestertown, Md., July 29, 1962; s. Robert Joseph and Sara Ann (Fallowfield) W.; m. Alice Koegel, Sept. 17, 1988. BS, Muhlenberg Coll., 1984; PhD, U. N.C., 1988. Postdoctoral fellow Johns Hopkins Sch. Medicine, Balt., 1988-92; asst. prof. U. Md. Med. Sch., Balt., 1992—; teaching asst. U. N.C., Chapel Hill, 1984-85, rsch. asst., 1985-88. Contbr. articles to Biochemistry, Jour. Biol. Chemistry. Recipient Nat. Rsch. Svc. award NIH, 1989-92, Young Investigators award Johns Hopkins Sch. Medicine, Balt., 1992. Mem. Am. Chem. Soc., Biophys. Soc., AAAS. Democrat. Achievements include discovery of how amino acids interact in wild type enzymes by comparing results from single and double mutants of staphylococcal nuclease; research in structure and location of DNA in staphylococcal nuclease. Home: 3813 Keswick Rd Baltimore MD 21211

WEBER, DONALD CHARLES, entomologist; b. Washington, Feb. 16, 1958; s. George and Isabelle (Pearson) W.; m. Ann Christine Reid, Aug. 13, 1983; 1 child, Sarah Reid. BA in Biology and Environ. Studies, Williams Coll., 1979; MS in Entomology, U. Calif., Berkeley, 1984; PhD in Entomology, U. Mass., 1992. Leader Cauliflower IPM Project U. Calif., Berkeley, 1983-84; orchard rsch. technologist/entomology U.S. Dept. Agr./Appalachian Fruit Rsch. Sta., Kearneysville, W.Va., 1984-86; specialist sweet corn integrated pest mgmt. U. Mass., Amherst, 1986-88; oberassistent applied entomology Swiss Fed. Inst. of Tech., Zurich, 1993—. Author book chpt. Colorado Potato Beetle, 1993; contbr. articles to profl. jours. Mem. Ashfield (Mass.) Conservation Commn., 1990-92; pres. Fernald Entomol. Club, Amherst, 1987-88, Appalachian Fruit Rsch. Sta. Employees Assn., Kearneysville, W.Va., 1986. Recipient Dwight Botanical prize Williams Coll., Williamstown, Mass., 1979, Univ. fellowships U. Mass., 1986-87, Mass. IPM Spl. Rsch. grant, Amherst, 1992; named Outstanding Teaching Asst., U. Calif., Berkeley, 1983, Switzer Environ. fellow, Switzer Found., Concord, N.H., 1990-91. Mem. Entomol. Soc. Am. (chmn. EB student com. 1991-92, John Henry Comstock award 1993), Coleopterists Soc., Sigma Xi, Phi Beta Kappa. Achievements include first to report and quantify enormous aggregation of Colo. beetle along woody field borders and use of this to develop new cultural controls; first documentation of positive relationship between weed abundance in sweet corn and European corn borer. Office: Swiss Federal Inst Tech, Clausiusstrasse 21, Ch-8092 Zurich Switzerland

WEBER, EICKE RICHARD, physicist; b. Muennerstadt, Fed. Republic Germany, Oct. 28, 1949; s. Martin and Irene (Kistner) W.; m. Magdalene Graff (div. 1983); m. Zuzanna Liliental, June 10, 1985. BS, U. Koeln, Fed. Republic of Germany, 1970, MS, 1973, PhD, 1976, Dr.Habil., 1983. Sci. asst. U. Koeln, 1976-82; rsch. asst. U. Lund, Sweden, 1982-83; asst. prof. Dept. Material Sci. U. Calif., Berkeley, 1983-87, assoc. prof., 1987-91, prof. materials sci., 1991—; prin. investigator Lawrence Berkeley Lab., 1984—; vis. prof. Tohoku U., Sendai, Japan, 1990; cons. in field; internat. fellow Inst. for Study of Defects in Solids, SUNY, Albany, 1978-79; chmn. numerous confs.; lectr. in field. Contbr. more than 180 articles to profl. jours.; editor: Defect Recognition and Image Processing in III-V Compounds, 1987, Imperfections in III-V Compounds, 1993; co-editor: Chemistry and Defects in Semiconductor Structures, 1989, others; series co-editor: Semiconductors and Semimetals, 1991—, Growth and Characterization of Semiconductor Materials, 1992—. Recipient IBM Faculty award, 1984; rsch. grantee Dept. of Energy, 1984—, Office Naval Rsch., 1985—, Air Force Office Sci. Rsch., 1988—, NASA, 1988-90, Nat. Renewable Energy Lab., 1992—. Mem. Am. Phys. Soc., Matls. Rsch. Soc. Achievements include first identification of point defects formed by dislocation motion in silicon; determination of the energy levels of antisite defects in GaAs, of 3d transition metal solubility and lattice site in silicon, of mechanism of internal gettering in silicon; research in defects formed in III/V thin films and interfaces; on lattice mismatched heteroepitaxial growth; in structure and electronic properties of metal GaAs heterostructures; in nature and electronic properties of defects in GaAs and related compounds; in transition metal gettering in silicon; polysilicon for photovoltaic applications; in field modulated microwave absorption of high-Tc superconductors; scanning tunneling microscopy of semiconductor thin films and interfaces; on electron paramagnetic resonance of defects in semiconductors. Office: U Calif Dept Materials Sci 272 Hearst Mining Bldg Berkeley CA 94720

WEBER, GEORG FRANZ, immunologist; b. Erlangen, Germany, July 7, 1962; came to U.S., 1989; s. Otto and Margret (Hartung) W.; m. Chitra Edwin, Sept. 21, 1991; 1 child, Ramona Sara. BS, Ohm-Gymnasium, Erlangen, Germany, 1981; MD, Julius Maximilians U., Wuerzburg, Germany, 1988, PhD, 1988. Rsch. assoc. U. S. Ala., Mobile, 1989; rsch. fellow dept. biochemistry and Dana-Farber Cancer Inst. Harvard U., Boston, 1990-91, rsch. fellow dept. pathology and Dana-Farber Cancer Inst., 1991-93, instr., 1993—. Contbr. articles to profl. jours. Mem. Amnesty Internat., N.Y.C., 1991—. Deutsche Forschungsgemeinschaft fellow, 1989-91. mem. AMA, Deutscher Aerzteverband, Oxygen Soc. Achievements include research into reactive oxygen species in medicine and 1991 link of enzyme defect to childhood seizures, theory of chess, biomechanics. Office: Dana-Farber Cancer Inst 44 Binney St Boston MA 02115

WEBER, GEORGE, oncology and pharmacology researcher, educator; b. Budapest, Hungary, Mar. 29, 1922; came to U.S., 1959; s. Salamon and Hajnalka (Arvai) W.; m. Catherine Elizabeth Forrest, June 30, 1958; children: Elizabeth Dolly Arvai, Julie Vibert Wallace, Jefferson James. BA, Queen's U., 1950, MD, 1952; MD (hon.), U. Chieti, Italy, 1979, Med. Faculty, Budapest, 1982, U. Leipzig, Fed. Republic of Germany, 1987, Tokushima (Japan) U., 1988; Kagawa (Japan) U., 1992. Rsch. assoc. Montreal Cancer Inst., 1953-59; prof. pharmacology Ind. U. Sch. Medicine, Indpls., 1959—; dir. Lab for Exptl. Oncology Sch. Medicine, Ind. U., Indpls., 1974—, prof. Lab. for Exptl. Oncology, 1974-90, disting. prof. Lab. for Exptl. Oncology, 1990—; chmn. study sect. USPHS, Washington, 1976-78; sci. adv. com. Am. Cancer Soc., N.Y.C., 1972-76, Atlanta, 1994—, Damon Runyon Fund, N.Y.C., 1971-76; mem. U.S. Nat. Com., Internat. Union Against Cancer, Washington, 1974-80, 90-94, NAS, Washington, 1974-80, 90-94. Editor: Advances in Enzyme Regulation, Vols. 1-34, 1962—; assoc. editor Jour. Cancer Rsch., 1969-80, 82-89. Recipient Alecce Prize for cancer rsch. Tiberine Acad., Rome, 1971, Best Prof. award Student AMA, Indpls., 1966, 68, G.F. Gallanti prize for enzymology Internat. Soc. Clin. Chemists, 1984, Outstanding Investigator award Nat. Cancer Inst., NIH, 1986-93. Mem. Am. Soc. for Pharmacology and Exptl. Therapeutics, Am. Assn. Cancer Rsch. (G.H.A. Clowes award 1982), Am. Physiol. Soc., Biochem. Soc., Russian Acad. Sci. (hon.), Hungarian Cancer Soc. (hon.), Hungarian Acad. Scis. (hon.), Acad. Scis. Bologna (Italy) (hon.). Home: 7307 Lakeside Dr Indianapolis IN 46278-1618 Office: Ind Sch Medicine Lab Exptl Oncology 702 Barnhill Dr Indianapolis IN 46202-5200

WEBER, JOHN BERTRAM, architect; b. Evanston, Ill., Oct. 15, 1930; s. Bertram Anton and Dorothea Hennecke (Brammer) W.; m. Sally Ann French; children: Suzanne French Weber Roulston, Jane Marie Weber McCarthy, Patricia Weber Blodgett, Nancy Brammer. AB in Architecture, Princeton U., 1953; postgrad., Ill. Inst. Tech., 1959. Registered architect. Field engr. United Constrn. Co., Riverdale, N.D., 1952; draftsman Bertram A. Weber Architect, Chgo., 1947- 53; architect Betram A. Weber Architect, Chgo., 1958-1973; field engr. Atkinson United Constrn. Co., Greenup and Ashland, Ky., 1956-58; ptnr., proprietor Weber & Weber Architects, Chgo., Northbrook and Winnetka, Ill., 1973—; Mem. Ill. Architecture Act Revision task force, 1982-89. Prin. works include Prestwick Country Club, the 3175 Commercial Ave. Bldg., Northbrook, med. office bldg. and additions to Bi-county hospital, Warren, Mich., additions and alterations to Detroit Osteopathic Hosp., addition to Duraclean Internat. Bldg., Deerfield, additions to The Admiral (a retirement home in Chgo.), and numerous pvt. residences, churches, comml., ednl., and recreational bldgs. Active Winnetka (Ill.) Community Caucus, 1965, 74, mayor's adv. com. on bldg. codes, Chgo., 1975-80; chmn. bldg. com. Winnetka Community House, 1977-81, Winnetka Zoning Bd. Appeals, 1983-88, chmn., 1987-88; deacon, elder Winnetka Presbyn. Ch. With USN, 1953-56. Fellow Ill. Soc. Architects (bd. dirs. 1969-84, 91—, pres. 1976-78); mem. Northbrook C. of C., Architects Club Chgo. (pres. 1981, bd. dirs. 1976-86), AIA (health facility com. Chgo. chpt. 1969-76), Ill. Architect-Engr. Coun. (chmn. 1981-82, del. 1976-87, 93—), Builders Club Chgo. (bd. dirs. 1966—, pres. 1973-74), Am. Legion, Old Willow Club (pres. 1982-83, bd. dirs. 1980-82), Mchts and Mfrs. Club, Dairymens Country Club. Home: 415 Berkeley Ave Winnetka IL 60093-2109 Office: Weber & Weber Architects 415 Berkeley Ave Winnetka IL 60093-2109

WEBER, LAVERN JOHN, marine life administrator, educator; b. Isabel, S.D., June 7, 1933; s. Jacob and Irene Rose (Bock) W.; m. Shirley Jean Carlson, June 19, 1959 (div. 1992); children: Timothy L., Peter J., Pamela C., Elizabeth T.; m. Patricia Rae Lewis, Oct. 17, 1992. AAS, Everett Jr. Coll., 1956; BA, Pacific Luth. U., 1958; MS, U. Wash., 1962, PhD, 1964. Instr. U. Wash., Seattle, 1964-67, asst. prof., 1967-69, acting state toxicologist, 1968-69; assoc. prof. Oreg. State U., Corvallis, 1969-75, prof., 1976—, asst. dean grad. sch., 1974-77; dir. Hatfield Marine Sci. Ctr. Oregon State U., Newport, 1977—, supt. Coastal Oreg. Marine Exptl. Sta., 1989—. Pres., trustee Newport Pub. Libr., 1991-92, Yaquina Bay Econ. Found., Newport, 1991-92; v.p. Oreg. Coast Aquarium, 1985—. Recipient Pres. award Newport Rotary, 1984-85. Mem. South Slough Mgmt. Commn., Am. Soc. Pharm. and Exptl. Therapy, West Pharm. Soc., Soc. Toxicology, Soc. Exptl. Biol. Med. (n.w. divsn., pres. 1978, 82, 87) , Pacific N.W. Assn. Toxicologists (chair 1985-86, coun. 1991-93), Western Assn. Marine Lab. (pres. 1993). Avocations: woodworking, reading, walking, scuba, gardening. Office: Oregon State Univ Hatfield Marine Sci Ctr Aquarium 2030 Marine Science Dr Newport OR 97365-5296

WEBER, LOWELL WYCKOFF, internist; b. Aberdeen, S.D., June 13, 1923; s. Lowell Henry and Grace Manzer (Wyckoff) W.; m. Lillian Marie Vassbotn, Sept. 1, 1951; children: Laurel Ann, Lynn Faye, Lowell II, Launcelot Granger. BS, U. N.D., 1945; MD, U. Ill., Chgo., 1946. Diplomate Am. Bd. Internal Medicine. Rotating intern St. Luke's Hosp., Chgo., 1946-47; resident in internal medicine Mpls. VA Hosp. and U. Minn. Hosp., Mpls., 1949-52; ptnr., owner Abbott Clinic, Mpls., 1952-92; retired, 1992. Capt. U.S. Army Med. Corps, 1947-49. Congregational. Home: 8691 Rich Rd Minneapolis MN 55437

WEBER, MICHAEL HOWARD, nuclear control operator; b. Provo, Utah, Sept. 9, 1960; s. Allen Howard and Bonnie Jilene (Hoggan) W.; m. Laura Jean Smith, May 19, 1990. AAS in Nuclear Tech., Aiken Tech. Coll., 1983; BS in Nuclear Sci., U. Md., 1992. Lic. sr. reactor operator. Aux. operator Carolina Power & Light Co., New Hill, N.C., 1983-88, control operator, 1988-92, sr. control operator, 1992—. Recipient scholarship Aiken County Homebuilders Assn., 1982. Mem. Am. Nuclear Soc. Republican. Lutheran. Home: 215 Maple Ln Fuquay-Varina NC 27526 Office: Carolina Power & Light Co PO Box 165 New Hill NC 27562

WEBER, THOMAS WILLIAM, chemical engineering educator; b. Orange, N.J., July 15, 1930; s. William A. and Dorothy (Negus) W.; m. Marianne S. Hartmann, June 4, 1966; children—Anne Louise, William Alois. B.Chem. Engring., Cornell U., 1953, Ph.D., 1963; M.S. in Chem. Engring., Newark Coll. Engring., 1958. Registered profl. engr., N.Y. Chem. engr. econs. and planning Esso Research & Engring., Linden, N.J., 1955-58; instr. Cornell U., 1961-62; asst. prof. SUNY-Buffalo, 1963-66, assoc. prof. chem. engring., 1966-82, prof., 1982—, assoc. chmn. dept., 1980-82, chmn. dept., 1982-89. Author: An Introduction to Process Dynamics and Control, 1973. Named Prof. of Yr., Tau Kappa Chi, 1965; recipient Chancellor's award for excellence in teaching, 1981, Tchr. of Yr. award Tau Beta Pi, 1982. Fellow Am. Inst. Chem. Engrs. (chmn. We. N.Y. sect. 1969-70, Profl. Achievement award We. N.Y. sect. 1978), Am. Soc. Engring. Edn. (chmn. instrumentation div. 1975-77, chmn. St. Lawrence sect. 1979-80, 92—, chmn. div. experimentation and lab.-oriented studies 1985-86, Outstanding Zone Campus Rep. award 1988, AT&T Found. award 1987-88), Tech. Socs. Coun. Niagara Frontier (sec. 1973-75, pres. 1975-76, treas. 1978—), Sigma Xi, Phi Kappa Phi, Tau Beta Pi, Theta Xi. Presbyterian. Club: Swedish of Buffalo (pres. 1974-76). Home: 52 Autumnview Rd Buffalo NY 14221

WEBLEY, PAUL ANTHONY, chemical engineer; b. Cape Town, South Africa, July 16, 1961; s. John Anthony and Maureen Elizabeth (Du Toit) W.; m. Catherine Rose Curry-Hyde, Dec. 7, 1985; children: Alec, Emma. BS in Chem. Engring., MIT, 1986, PhD in Chem. Engring., 1989. Asst. prof. MIT/Dow Chem., Midland, Mich., 1990-92; rsch. engr. Air Products & Chems., Inc., Allentown, Pa., 1992—; cons. Dow Chem. Co., Midland, 1990-92. Contbr. articles to Energy and Fuels, Indsl. Engring. Chem. Res., AIChE Jour., ACS Symposium Series. Mem. Am. Inst. Chem. Engring., Sigma Xi. Achievements include South African patent in a new pressure-compensating microflow calorimeter; patentee pending process to cool and dehumidify the feed to membrane or adsorption based N2-Generators. Home: 6413 Manzanita Dr Macungie PA 18062 Office: Air Products & Chemicals 7201 Hamilton Blvd Allentown PA 18062

WEBSTER, ALEXANDER JAMES, agrologist; b. St. Walburg, Sask., Can., Sept. 5, 1925; s. Andrew Oliver and Bella (Buick) W.; m. Margaret Jean Robinson, May 6, 1949; children—Craig Richard, Stuart Blaire, Brenda Lea. B.S.A., U. Sask., 1949; M.Ed., Colo. State U., 1953. Agrl. rep. Sask. Dept. Agr., 1949-55, asst. dir. agr. ext., Regina, Sask., 1955-65, dir. animal industry, 1965-67, dir. prodn. and mktg., 1967-72, asst. dep., 1972-78, acting dep., Regina, Sask. 1979, dep., dept. rural affairs, 1979-82, chmn. Land Bank Commn., 1982-83, mgr. Sask. Farm Purchase Program, 1983-84, exec. dir. land adminstrn., 1984-85; dir. internat. projects Agdevco (now O & T Agdevco), 1985—; president Exp Consulting, Ltd., 1984—. Councilor Village of Kannata Valley, Silton, Sask., 1982-84; head Four Man Agrl. Mission to China, 1980; leader Red Meat Trade Mission, 1971. Served with Royal Can. Navy, 1944-46. Recipient AIC Fellowship Award, Agricultural Inst. of Canada, 1991. Mem. Sask. Inst. Agrologists (pres.), Can. Inst. Agrologists, Can. Soc. Extension (pres.). Mem. United Ch. Can. Club: Speed Skating. Lodge: Rotary. Home: 3910 Hill Ave, Regina, SK Canada S4S 0X5 Office: Agdevco 1106 Chestemere Plaza, 2500 Victoria Ave, Regina, SK Canada

WEBSTER, EDWARD WILLIAM, medical physicist; b. London, Apr. 12, 1922; came to U.S., 1949, naturalized, 1957; s. Edward and Bertha Louisa (Cornish) W.; m. Dorothea Anne Wood, June 24, 1961; children: John Stein, Peter Wood, D. Anne, Edward Russell, Mark Vincent, Susan Victoria. BSc in Elec. Engring., U. London, 1943, PhD, 1946; postgrad., MIT, 1949-51, 65-66, Columbia U., 1966; AM (hon.), Harvard U., 1989. Diplomate: Am. Bd. Radiology in radiol. physics, Am. Bd. Health Physics. Research engr. English Electric Co., Stafford, Eng., 1945-49; travelling fellow lab. for nuclear sci. MIT, 1949-50, staff scientist, 1950-51; lectr. U. London, 1952-53; physicist Mass. Gen. Hosp., Boston, 1953—, chief radiol. scis. div., 1970—; prof. radiology Harvard U. Med. Sch., Boston, 1975—; prof. radiology div. health scis. and tech. Harvard-MIT, 1978-86; examiner Am. Bd. Radiology, 1958-84, chmn. physics com., 1966-76; cons. IAEA, Vienna, Austria, 1960-62; mem. Nat. Council on Radiation Protection, 1964-89, dir., 1981-88, hon. mem., 1989; mem. task group Internat. Commn. on Radiol. Protection, 1966-70; cons. WHO, Geneva, 1964, 67; mem. coms. Nat. Acad. Scis., Washington, 1962-68, 71-74, 77-80, 83-84, 85-86; mem. study sect. Nat. Inst. Med. Scis., Bethesda, Md., 1969-73, Environ. Control Adminstrn., Washington, 1969-72; mem. adv. com. U.S. Nuclear Regulatory Commn., Washington, 1971-93; Garland lectr. Calif. Radiol. Soc., 1980; Verstandig lectr. U. Tenn. Coll. Medicine, 1982; Williams lectr. Am. Assn. Physicists in Medicine, 1983; mem. adv. com. on environ. hazards VA, 1984—, U.S. del. UN Sci. Com. on Effects of Atomic Radiation, 1987—; lectr. Harvard Sch. Pub. Health, 1971-86; cons. Radiation Effects Rsch. Found., Hiroshima, Japan, 1988; Taylor lectr. Nat. Coun. on Radiation Protection, 1992. Author: A Basic Radioisotopes Course, 1959, Atlas of Radiation Dose Distributions, 1965, Radiation Safety Manual of MGH, 1965, Physics in Diagnostic Radiology, 1970; co-author: Instrumentation and Monitoring Methods for Radiation Protection, 1978, Low-level Radiation Effects, 1982; co-editor: Advances in Medical Physics, 1971, Biological Risks of Medical Irradiations, 1980; inventor composite shields against low energy X-rays, 1970. Robert Blair travelling fellow London County Council, 1949; USPHS fellow, 1965; NIH grantee, 1958-80. Fellow Health Physics Soc. (Landauer award 1985, Failla award 1989), Am. Coll. Radiology (commn. mem. 1963—, Gold medal 1991); mem. Am. Assn. Physicists in Medicine (dir. 1958-65, pres. 1963-64 Coolidge medal), Soc. Nuclear Medicine (trustee 1973-77), Radiol. Soc. N.Am. (v.p. 1977-78), New Eng. Roentgen Ray Soc. (hon. mem., exec. com. 1976-77), Radiation Research Soc., Sigma Xi (nat. lectr. 1988-89). Office: Mass Gen Hosp Fruit St Boston MA 02114-2620

WEBSTER, HENRY DEFOREST, experimental neuropathologist; b. N.Y.C., Apr. 22, 1927; s. Leslie Tillotson and Emily (deForest) W.; m. Marion Havas, June 12, 1951; children: Christopher, Henry, Sally, David, Steven. AB cum laude, Amherst Coll., 1948; MD, Harvard U., 1952. Intern Boston City Hosp., 1952-53, resident, 1953-54; resident in neurology Mass. Gen. Hosp., 1954-56, research fellow in neuropathology, 1956-59; prin. investigator NIH research grants for electron microscopic studies of peripheral neuropathy, 1959-69; mem. staffs Mass. Gen., Newton-Wellesley hosps.; instr. neurology Harvard Med. Sch., 1959-63, assoc. in neurology, 1963-66, asst. prof. neuropathology, 1966; assoc. prof. neurology U. Miami Sch. Medicine, 1966-69, prof., 1969; head sect. cellular neuropathology Nat. Inst. Neurol. Diseases and Stroke, Bethesda, Md., 1969—; assoc. chief Lab. of Neuropathology and Neuroanat. Scis., 1975-84; chief Lab. Exptl. Neuropathology, 1984—; disting. scientist, lectr. dept. anatomy Tulane U. Sch. Medicine, 1973; Royal Coll. lectr. Can. Assn. Neuropathologists, 1982; Saul Korey lectr. Am. Assn. Neuropathologists, 1992; chmn. Winter Conf. on Brain Rsch., 1985, 86; head neuropathology delegation to visit China in 1990, Citizen Amb. Program, People to People Internat.; mem. exec. com. rsch. group on neuromuscular disease World Fedn. Neurology, 1986-93. Author: (with A. Peters and S.L. Palay) The Fine Structure of the Nervous System, 1970, 76, 91; contbr. articles to sci. jours. Recipient Superior Service award USPHS, 1977, A. von Humboldt award Fed. Republic Germany, 1985; named hon. prof. Norman Bethune U. of Med. Scis., Chanchun, China, 1991. Mem. Am. Assn. Neuropathologists (v.p. 1976-77, pres. 1978-79, Weil award 1960), Internat. Soc. Neuropathology (councillor 1976-80, v.p. 1980-84, exec. com. 1980-84, 86—, pres. 1986-90), Internat. Congress Neuropathology (sec. gen. VIII 1978), Peripheral Nerve Study Group (exec. com. 1975-93, chmn. 1977 meeting), Am. Neurol. Assn., Am. Acad. Neurology, Am. Soc. Cell Biologists, Am. Assn. Anatomists, Soc. Neurosci., Washington Ctr. Photography, Ausable Club, Japanese Soc. Neuropathology (hon.), Ausable Club,. Phi Beta Kappa. Office: NIH Bldg 36 Rm 4A 29 Bethesda MD 20892

WEBSTER, JEFFERY NORMAN, technology policy analyst; b. Erie, Pa., Oct. 23, 1954; s. Norman A. and Betty B. (Bessetti) W.; m. Harriet Marie McGinley, Nov. 27, 1982; 1 child, Jessica Marie. BA, Pa. State U., 1980; MPA, U. So. Calif., 1985. Evaluator, technologist U.S. Gen. Acctg. Office, L.A., 1981—; treas. So. Calif. Space Bus. Roundtable of the World Space Found., L.A., 1991—. Co-author numerous technology assessment reports to the Congress, 1983—. Vol. program mgr. Union Sta. Found., Pasadena, Calif., 1985. Mem. AIAA, Rotary. Achievements include congl. testimony on tech. risks and scientific utility of NASA's space sta. design, congl. report on tech. risks of assembling and maintaining NASA's space sta., loss of irreproduceable space sci. data due to faulty archiving practices by NASA, improvements needed in NASA's spacecraft computer technology. Home: 1969 N Hill Ave Altadena CA 91001 Office: US Gen Acctg Office Ste 1010 350 S Figueroa Los Angeles CA 90071

WEBSTER, JOHN GOODWIN, biomedical engineering educator, researcher; b. Plainfield, N.J., May 27, 1932; s. Franklin Folger and Emily Sykes (Boody) W.; m. Nancy Egan, Dec. 27, 1954; children: Paul, Robin, Mark, Lark. BEE, Cornell U., 1953; MSEE, U. Rochester, 1965, PhD, 1967. Registered profl. engr., Wis. Engr. North American Aviation, Downey, Calif., 1954-55; engr. Boeing Airplane Co., Seattle, 1955-59, Radiation Inc., Melbourne, Fla., 1959-61; staff engr. Mitre Corp., Bedford, Mass., 1961-62, IBM Corp., Kingston, N.Y., 1962-63; asst. prof. elec. engring. U. Wis.-Madison, 1967-70, assoc. prof. elec. engring., 1970-73, prof. elec. and computer engring., 1973—; cons. Gen. Electric Co., 3M Co., Johnson & Johnson. Author: (with others) Medicine and Clinical Engineering, 1977, Sensors and Signal Conditioning, 1991; editor: Medical Instrumentation: Application and Design, 1978, 2d edit., 1992, Clinical Engineering: Principles and Practices, 1979, Design of Microcomputer-Based Medical Instrumentation, 1981, Therapeutic Medical Devices: Application and Design, 1982; Electronic Devices for Rehabilitation, 1985; Interfacing Sensors to the IBM-PC, 1988, Encyclopedia of Medical Devices and Instrumentation, 1988, Tactile Sensors for Robotics and Medicine, 1988, Electrical Impedance Tomography, 1990, Teaching Design in Electrical Engineering, 1990, Prevention of Pressure Sores, 1991. Recipient Rsch. Career Devel. award NIH, 1971-76; NIH fellow, 1963-67. Fellow IEEE, Am. Inst. Med. and Biol. Engring., Instrument Soc. Am. (Donald P. Eckman Edn. award 1974); mem. Biomed. Engring. Soc., Am. Soc. Engring. Edn. (Western Electric Fund award 1978), Assn. Advancement Med. Instrumentation. Office: Univ Wis Dept Elec and Computer Engring 1415 Johnson Dr Madison WI 53706-1691

WEBSTER, JOHN ROBERT, chemical engineer; b. Riverdale, Calif., May 5, 1916; s. John Hamilton and Elizabeth Mae (Smith) W.; m. Phyllis Fridlund, Aug. 21, 1943; children: Richard, Ann, Mark. AB, Fresno (Calif.) State U., 1939. Registered profl. engr., Calif. Chief chemist Lindsay (Calif.) Olive Growers, 1939-60, plant supt., 1961-66, tech. dir., 1967-81, sales engr. biomass-fired power plants, 1982—. Mem. Am. Chem. Soc., C. of C. Achievements include patents for pimiento gel ribbon for stuffing olives, frozen sliced ripe olives packaged without brine, process for canning Spanish style olives, ripe olive process, olive tank that stirs and oxidizes olives via compressed air. Home: 386 Alameda St Lindsay CA 93247

WEBSTER, LOIS SHAND, association executive; b. Springfield, Ill., Sept. 25, 1929; d. Richings James and C. Odell (Gilbert) S.; m. Terrance Ellis Webster, Feb. 12, 1954 (dec. July 1985); children: Terrance Richings, Bruce Douglas, Andrew Michael. BA, Millikin U., 1951; cert. in libr. tech., Coll. Du Page County, Glen Ellyn, Ill., 1974; postgrad. libr. sci., No. Ill. U., 1977-82. Exec. asst. Am. Nuclear Soc., La Grange Park, Ill., 1973—. Contbr. articles and book chpts. to profl. pubis. Field dir. Springfield coun. Girl Scouts U.S., 1951-54; libr. advisor Du Page County coun. Girl Scouts U.S., 1973-74. Recipient Octave J. Du Temple award Am. Nuclear Soc., 1989. Mem. Spl. Librs. Assn. (divsn. chmn. 1984-85, chmn. by-laws com. 1987-89, bd. dirs. 1989-92, sec. 1990-91, visioning com. 1992—), Coun. Engring. and Sci. Soc. Execs., Am. Soc. Assn. Execs., Met. Chgo. Libr. Assembly (bd. dirs. 1982-85). Avocations: travel, genealogy. Home: 560 Dorset Ave Glen Ellyn IL 60137-5703 Office: Am Nuclear Soc 555 N Kensington Ave La Grange Park IL 60525-5592

WEBSTER, MERLYN HUGH, JR., manufacturing engineer, information systems consultant; b. Beaver Falls, Pa., Nov. 7, 1946; s. Merlyn Hugh and Helen Ruth (Dillon) W.; m. Linda Jeanne Gundlach, June 14, 1969; children: Matthew Jason, Nathaniel Kevin. AA, Palomar Coll., San Marcos, Calif., 1975; BA, Chapman Coll., 1978. Registered profl. engr., Calif. Sr. cons., pres. WEB Internat. Corp., 1992—; mfg. analyst NCR Corp., Rancho Bernardo, Calif., 1968-72, indsl. engr., 1972-76, sr. indsl. engr., 1976-78; sr. project mgr. Tektronix, Beaverton, Oreg., 1978-83, corp. distbn. I.E. mgr., 1983-86; sr. info. systems cons. Intel Corp., Hillsboro, Oreg., 1986—; pres. WEB Internat. Corp., Tualatin, Oreg., 1992—; cons. material handling Intel Mfg., Puerto rico and Ireland, 1989-92; cons. info. systems M.I.S.I., N.Y.C., 1992-93. Chmn. United Way Hillsboro, Oreg., 1986. With USMC, 1964-68, Vietnam. Mem. NSPE, Inst. Indsl. Engrs. (cert.), Shelby Car Club Am. Republican. Home: 5200 SW Joshua St Tualatin OR 97062-9792 Office: WEB Internat Corp 5200 SW Joshua Tualatin OR 97062-9792

WEBSTER, OWEN WRIGHT, chemist; b. Devils Lake, N.D., Mar. 25, 1929; s. Daniel Milton and Maude May (Wright) W.; m. Lillian Brostek; children: Ellen, Anne, John, James, Mary. BS in Chemistry, N.D. U., 1951, DSc (hon.), 1986; PhD in Chemistry, Pa. State U., 1955. Research chemist E.I. DuPont De Nemours, Wilmington, Del., 1955-74, group leader, 1974-79, research supr., 1979-84, research leader, 1984—. Patentee in field; contbr. articles to profl. jours. Mem. AAAS, Am. Chem. Soc. (chmn. Del. sect. 1975-76, Excellence in Research award 1987, Applied Polymer Sci. award 1993), Sigma Xi. Republican. Roman Catholic. Avocations: chess, bridge, golf. Home: 2106 Navaro Rd Wilmington DE 19803-2310 Office: EI DuPont De Nemours Exptl Sta E 328 Wilmington DE 19880

WEBSTER, RAYMOND EARL, psychology educator, psychotherapist; b. Providence, Dec. 3, 1948; s. Earl Harold and Madeline (D'Antuono) W.; m. Angela Grenier, Jan. 31, 1984; children: Matthew Raymond, Patrick Gregory, Timothy Andrew. BA, R.I. Coll., 1971, MA, 1973; MS, Purdue U., 1976; PhD, U. Conn., 1978. Lic. psychologist, N.C.; cert. sch. psychologist III (supervising), spl. edn. tchr., N.C. Dir. pupil svcs. and spl. edn. Northeastern Area Regional Edn. Svcs., Wauregan, Conn., 1978-79; dir. alternative vocat. sch. Capital Region Edn. Coun., West Hartford, Conn., 1979-83; prof. psychology, dir. sch. psychology program East Carolina U., Greenville, N.C., 1983—; rsch. assoc. ednl. psychology U. Conn., Storrs, 1976-78; cons. Bolton (Conn.) Pub. Schs., 1976-78, Columbia (Conn.) Pub. Schs., 1976-78, N.C. Dept. Instrn., Raleigh, 1983—; speaker at profl. meetings. Guest reviewer Jour. Applied Behavior Analysis, 1975; mem. editoral bd. Psychology in Schs., 1987—; contbr. numerous chpts. to books and articles to profl. jours. Trustee N.C. Ctr. for Advancement of Teaching, Cullowhee, 1990—. Sgt. U.S. Army Spl. Forces N.G., 1969-75. Recipient spl. distinction award Conn. Assn. Sch. Psychologists, 1983. Mem. APA, Nat. Assn. Sch. Psychologists (cert., alt. del. 1985-86, spl. distinction in profl. devel. 1982, 83), Nat. Acad. Neuropsychology, Sigma Xi. Methodist. Avocations: running, bicycling, flower gardening. Home: 200 Williams St Greenville NC 27858-8712 Office: East Carolina U Rawl Bldg Greenville NC 27834-4353

WECHSBERG, MANFRED INGO, chemical researcher; b. Maehr-Ostrau, Czechoslovakia, May 1, 1940; s. Josef and Helene (Pawlas) Wechsberg. Degree in engring., Tech. U., Vienna, Austria, 1965, PhD in Chemistry, 1966. Asst. prof. Tech. U., 1966; vis. rschr. U. Washington, Seattle, 1967, Princeton (N.J.) U., 1968, U. Calif., Berkeley, 1969; chemist Bayer AG, Leverkusen, Fed. Republic Germany, 1970-76; head rsch. dept. Chemie Linz AG, Austria, 1979-88, head mktg. rsch., 1988—. Patentee in field. Bd. dirs. Soc. Austrian-German Culture, Linz, 1984-92, European Conf. Human Rights and Self Determination, Bern, Bonn, Linz. Mem. Austrian Chem. Soc. (bd. dirs. 1985-92), Verein der Leichtathleten des OTB OO, Athletics Club, Akademische Gildenschaft (pres. 1983-91). Avocations: mountaineering, skiing, music, literature, theatre. Home: Im Blumengrund 9/35, A-4060 Leonding Austria

WEDEMEYER, ERICH HANS, physicist; b. Osterode, Fed. Republic Germany, Aug. 16, 1927; s. Theodor and Ella (Osterwald) W.; m. Christel Seeringer, Aug. 18, 1956; 1 child, Sibylle. Diploma for physics, U. Göttingen, Fed. Republic Germany, 1956, D. Natural Scis., 1969. Rsch. physicist Aerodyn. Versuchsanstalt, Göttingen, 1956-61, Ballistic Rsch. Lab., Aberdeen, Md., 1962-66; head dept. Deutsche Forschungs-und Versuchsanstalt Luft und Raumfahrt, Göttingen, 1966-80, cons., 1990—, rsch. physicist, 1982-90; vis. prof. Von Karman Inst., Brussels, 1980-82. Contbr. articles to Ing. Arch., Jour. Fluid Mech., Jour. Aircraft; author: Liquid Filled Projectile Design, 1969. With German Air Force, 1943-45. Mem. AIAA. Evangelic Lutheran. Achievements include development of the theory of confined fluid motion; research on Hydrodynamic Stability; on Wind Tunnel Techniques; 3 patents on adaptive wall test sections in wind tunnels. Home: Am Weinberg 37, 37176 Norten Hardenberg Germany Office: DLR Göttingen, Bunsenstrasse 10, 37073 Göttingen Germany

WEDEPOHL, LEONHARD M., electrical engineering educator; b. Pretoria, Republic of South Africa, Jan. 26, 1933; s. Martin Willie and Liselotte B.M. (Franz) W.; m. Sylvia A.L. St. Jean; children: Martin, Graham. B.Sc. (Eng.), Rand U., 1953; Ph.D., U. Manchester, Eng., 1957. Registered profl. engr., B.C. Planning engr. Escom, Johannesburg, Republic of South Africa, 1957-61; mgr. L.M. Erricson, Pretoria, Republic of South Africa, 1961-62; sect. leader Reyrolle, Newcastle, Eng., 1962-64; prof., head dept. Manchester U., 1964-74; dean engring. U. Man., Winnipeg, Can., 1974-79; dean applied sci. U. B.C., Vancouver, Can., 1979-85, prof. elec. engring., 1985—; mem. Sci. Research Council, London, 1968-74; dir. Man. Hydro, Winipeg, 1965-69, B.C. Hydro, Vancouver, 1980-84, B.C. Sci. Council, 1981-84; v.p. Quantic Labs., Winnipeg, 1986; cons. Horizon Robotics, Saskatoon, 1986; chmn. implementation team Sci. Place Can., 1985; cons. CEPEL, Rio de Janeiro; adv. Man. High Voltage D.C. Rsch. Ctr. Contbr. articles to sci. jours.; patentee in field. Named Hon. Citizen City of Winnipeg, 1979. Fellow Instn. Elec. Engrs. (premium 1967); mem. Assn. Profl. Engrs. B.C. Avocations: music; cross-country skiing; hiking. Office: U BC Dept Elec Engring, 2324 Main Hall, Vancouver, BC Canada V6T 1W5

WEED, RONALD DE VERN, engineering consulting company executive; b. Indian Valley, Idaho, Sept. 1, 1931; s. David Clinton and Grace Elizabeth (Lavendar) W.; m. Doris Jean Hohener, Nov. 15, 1953; children: Geraldine Gayle, Thomas De Vern, Cheryl Ann. BSChemE, U. So. Calif., 1957; MS in Chem. Engring., U. Wash., 1962; LLB, La Salle U., Chgo., 1975; postgrad., Century U., Beverly Hills, Calif., 1979—. Registered profl. engr., Washington, Calif. Devel. engr. GE Co., Richland, Washington, 1957-65, Battelle N.W. Labs., Richland, 1965-68; oper. plant engr. NIPAK, Inc., Kerens, Tex., 1968-72; aux. systems task engr. Babcock & Wilcox Co., Lynchburg, Va., 1972-74; materials and welding engr. Bechtel Group Cos., San Francisco, 1974-85; cons. engr. Cygna Energy Svcs., Walnut Creek, Calif., 1985-91; with inter city fund Cygna Energy Svcs., Oakland, Calif., 1991—. Contbr. rsch. reports, papers and chpts. in books; patentee in field. With U.S. Army, 1951-53. Mem. Am. Inst. Chem. Engrs., Am. Welding Soc., Nat. Assn. Corrosion Engrs. (cert., sect. vice chmn. and chmn. 1962-68). Avocations: reading, photography, gardening. Home and Office: 74 Sharon St Pittsburg CA 94565-1527

WEEKES, TREVOR C., astrophysicist; b. Dublin, Ireland, May 21, 1940; came to U.S. 1966; s. Gerard and Florence (Murtagh) W.; m. Ann Katherine Owens, Sept. 30, 1964; children: Karina, Fiona, Lara. BSc, U. Coll. Dublin, 1962, PhD, 1966; DSc, Nat. U. Ireland, 1978. Lectr. Univ. Coll. Dublin, 1964-66; postdoctoral fellow NRC, Cambridge, Mass., 1966-67; astrophysicist Smithsonian Astrophys. Obs., 1967-92, sr. astrophysicist, 1992—; resident dir. Whipple Obs., Amado, Ariz., 1969-76; vis. prof. Royal Greenwich Obs., U.K., 1980-81. Mem. Am. Astron. Soc., Am. Phys. Soc., Royal Astron. Soc. Democrat. Roman Catholic. Office: Whipple Obs PO Box 97 Amado AZ 85645-0097

WEEKS, DAVID JAMISON, electrical engineer; b. Cookeville, Tenn., Dec. 10, 1950; s. Howard Elmer and Louise (Jamison) W.; m. Sharon Sue Guest, Oct. 27, 1979; children: Rebekah Anne, Stephen James, Jonathan David. BSEE, Tex. Tech. U., 1978; M of Adminstrn. Sci., U. Ala., 1984; MS in Engring., Southeastern Inst. Tech., 1986. Rsch. asst. elec. engring. laser lab. Tex. Tech. U., Lubbock, 1975-78; systems engr. E-Systems, Garland, Tex., 1978-79; computer software analyst Teledyne Brown Engring., Huntsville, Ala., 1979-81; engr. NASA, Marshall Space Flight Ctr., Ala., 1981—. Contbr. articles to profl. jours. Chmn., bd. dirs. Choose Life, Huntsville, 1991—; mem. Eagle Forum, Huntsville, 1990—. Staff sgt. USAF, 1969-73. Mem. Eta Kappu Nu (pres. Gamma Nu chpt. 1977-78). Home: 2207 Linde St Huntsville AL 35810-4357 Office: NASA Code EL54 Huntsville AL 35812

WEEKS, JOHN DAVID, chemistry and physical science educator; b. Birmingham, Ala., Oct. 11, 1943; s. Arthur Andrew Weeks and Grace Hicks (Ezell) Marquez; m. Kaja Parming, May 27, 1987; 1 child, Aili. AB, Harvard U., 1965; PhD, U. Chgo., 1969. Postdoctoral fellow U. Calif., San Diego, 1969-71, U. Cambridge, Eng., 1971-72; mem. tech. staff AT&T Bell Labs., Murray Hill, N.J., 1972-90; prof. Inst. Phys. Sci. and Tech., dept. chemistry U. Md., College Park, 1990—. Contbr. articles to profl. jours. Fellow Am. Phys. Soc.; mem. AAAS, Am. Chem. Soc. (Hildebrand award 1990). Home: 15301 Watergate Rd Silver Spring MD 20905-5779 Office: U Md Inst Phys Sci and Tech College Park MD 20742

WEEKS, JOHN RANDEL, IV, nuclear engineer, metallurgist; b. Orange, N.J., Oct. 30, 1927; s. John Randel and Marion (Heberton) W.; m. Barbara A. Brewster, July 16, 1951; children: Ann Brewster, John Randel V. MetE, Colo. Sch. Mines, 1949; MS, U. Utah, 1950, PhD in Metallurgy, 1953. From assoc. scientist to sr. scientist Brookhaven Nat. Lab., Upton, N.Y., 1953—; sr. metallurgist U.S. AEC, Bethesda, Md., 1972-74; adj. prof. Poly. Inst. N.Y., Bklyn., 1978-86. Editor: Corrosion by Liquid Metals, 1970; editor book, procs.: Environmental Degradation of Materials in Nuclear Power Systems-Water Reactors, 1986, 88; contbr. papers on nuclear materials, metallurgy to profl. publs. Organist, choir dir. 1st United Meth. Ch., Port Jefferson, N.Y. Recipient Richard Tucker award Materials Soc., 1990. Fellow Am. Soc. Metals (chpt. pres. 1962-63, 89-90), Am. Nuclear Soc. (chpt. pres. 1986-87). Achievements include research of behavior of materials in nuclear reactors. Home: 25 Acorn Ln Stony Brook NY 11790-2127 Office: Brookhaven Nat Lab Bldg 197 C Upton NY 11973

WEEKS, THOMAS J., chemist; b. Tarrytown, N.Y., Aug. 31, 1941. BA, Colgate U., 1963; PhD in Chemistry, U. Colo., 1967. Rsch. chemist, rsch. engring. ctr. Johns-Manville Corp., Manville, N.J., 1966-68; sr. rsch. chemist Union Carbide Corp., Tarrytown, 1970-72; project chemist Union Carbide Corp., 1972-73, sr. staff chemist, 1973-75, supr., 1975-78; group leader Ashland Chemical Corp., 1978-80, rsch. mgr., 1980-83, com. devel. mgr., 1983-86, dept. head R&D, 1987—. Mem. AAAS, Am. Chem. Soc., Catalysis Soc. Achievements include research in management of research and development on thermosetting polymer systems, waste management and safety research and development building administration, synthesis, testing and manufacturing of heterogeneous catalysts. Office: Rsch & Development Lab PO Box 2219 Columbus OH 43216-2219*

WEERTMAN, JOHANNES, materials science educator; b. Fairfield, Ala., May 11, 1925; s. Roelof and Christina (van Vlaardingen) W.; m. Julia Ann Randall, Feb. 10, 1950; children: Julia Ann, Bruce Randall. Student, Pa. State Coll., 1943-44; B.S., Carnegie Inst. Tech. (now Carnegie Mellon U.), 1948, D.Sc., 1951; postgrad., Ecole Normale Superieure, Paris, France, 1951-52. Solid State physicist U.S. Naval Research Lab., Washington, 1952-58; cons. U.S. Naval Research Lab., 1960-67; sci. liaison officer U.S. Office Naval Research, Am. Embassy, London, Eng., 1958-59; faculty Northwestern U., Evanston, Ill., 1959—; prof. materials sci. dept. Northwestern U., 1961-68, chmn. dept., 1964-68, prof. geol. scis. dept., 1963—, Walter P. Murphy prof. materials sci., 1968—; vis. prof. geophysics Calif. Inst. Tech., 1964, Scott Polar Research Inst., Cambridge (Eng.) U., 1970-71, Swiss Fed. Inst. Reactor Research, Switzerland, 1986; cons. U.S. Army Cold Regions Research and Engring. Lab., 1960-75, Oak Ridge Nat. Lab., 1963-67, Los Alamos Sci. Lab., 1967—; co-editor materials sci. books MacMillan Co., 1962-76. Author: (with Julia Weertman) Elementary Dislocation Theory, 1964, 2d edit. 1992; mem. editorial bd. Metal. Trans, 1967-75, Jour. Glaciology, 1972—; assoc. editor Jour. Geophys. Research, 1973-75; contbr. articles to profl. jours. Served with USMC, 1943-46. Honored with naming of Weertman Island in Antarctica.; Fulbright fellow, 1951-52; recipient Acta Metallurgica gold medal, 1980; Guggenheim fellow, 1970-71. Fellow Am. Soc. Metals, Am. Phys. Soc., Geol. Soc. Am., Am. Geophys. Union (Horton award 1972, AIME Mathewson Gold medal 1977); mem. AAAS, Nat. Acad. Eng., Am. Inst. Physics, Internat. Glaciological Soc. (Seligman Crystal award 1983), Arctic Inst., Am. Quaternary Assn., Explorers Club, Sigma Xi, Tau Beta Pi, Phi Kappa Phi, Alpha Sigma Mu, Pi Mu Epsilon. Club: Evanston Running. Home: 834 Lincoln St Evanston IL 60201-2405

WEG, JOHN GERARD, physician; b. N.Y.C., Feb. 16, 1934; s. Leonard and Pauline M. (Kanzleiter) W.; m. Mary Loretta Flynn, June 2, 1956; children: Diane Marie, Kathryn Mary, Carol Ann, Loretta Louise, Veronica Susanne, Michelle Celeste. BA cum laude, Coll. Holy Cross, Worcester, Mass., 1955; MD, N.Y. Med. Coll., 1959. Diplomate: Am. Bd. Internal Medicine. Commd. 2nd lt. USAF, 1958, advanced through grades to capt., 1967; intern Walter Reed Gen. Hosp., Washington, 1959-60; resident, then chief resident in internal medicine Wilford Hall USAF Hosp., Lackland AFB, Tex., 1960-64; chief pulmonary sect. Wilford Hall USAF Hosp., 1964-66, chief inhalation sect., 1964-66, chief pulmonary and infectious disease service, 1966-67; resigned, 1967; clin. dir. pulmonary disease div. Jefferson Davis Hosp., Houston, 1967-71; from asst. prof. to assoc. prof. medicine Baylor U. Coll. Medicine, Houston, 1967-71; assoc. prof. medicine U. Mich. Med. Sch. Univ. Hosp., Ann Arbor, 1971-74; prof. U. Mich. Med. Sch. Univ. Hosp., 1974—; physician-in-charge pulmonary div., 1971-81, physician-in-charge pulmonary and critical care med. div., 1981-85; cons. Ann Arbor VA, 1971—, Wayne County Gen. hosps., 1971-84; mem. adv. bd. Washtenaw County Health Dept., 1973—; mem. respiratory and nervous system panel, anesthesiology Sect. Nat. Ctr. Devices and Radiol. Health, FDA, 1983—, chmn., 1985-88. Contbr. med. jours., reviewer, mem. editorial bds. Decorated Air Force Commendation medal; travelling fellow Nat. Tb and Respiratory Disease Assn., 1971; recipient Aesculpaius award Tex. Med. Assn., 1971. Fellow Am. Coll. Chest Physicians (chmn. bd. govs. 1976-79, gov. Mich. 1975-79, chmn. membership com. 1976-79, prof.-in-residence 1972—, chmn. critical care coun. 1982-85), Am. Coll. Chest Physicians and Internat. Acad. Chest Physicians (exec. council 1976-82, pres. 1980-81), ACP (chmn. Mich. program com. 1974); mem. AAAS, Am. Fedn. Clin. Rsch., AMA, Am. Thoracic Soc. (sec.-treas. 1974-76), Am. Assn. Inhalation Therapy, Air Force Soc. Internists and Allied Specialists, Soc. Med. Consultants to Armed Forces, Internat. Union Against Tb, Mich. Thoracic Soc. (pres. 1976-78), Mich. Lung Assn. (dir., Bruce Douglas award 1981), Am. Lung Assn., Rsch. Club U. Mich., Assn. Advancement Med. Instrumentation, Central Soc. Clin. Rsch., Am. Bd. Internal Medicine (subsplty. com. on pulmonary disease 1980-86, critical care medicine test com. 1985-87, critical care medicine policy com. 1986-87), N.Y. Med. Coll. Alumni Assn. (medal of honor 1990), Alpha Omega Alpha. Home: 3060 Exmoor Rd Ann Arbor MI 48104-4132 Office: B I H 245 Box 0026 1500 E Medical Center Dr Ann Arbor MI 48109

WEGNER, GARY ALAN, astronomer; b. Seattle, Dec. 26, 1944; s. Herbert Edward and Melba Jean (Gardner) W.; m. Cynthia Kay Goodfellow, June 25, 1966; children: Josef, Kurt, Christian, Peter-Jûrgen, Emma. Student, Wash. State U., Pullman, 1963-65; BS, U. Ariz., 1967; PhD, U. Wash.,

Seattle, 1971. Fulbright fellow Mount Stromlo Obs., Camberra, A.C.T., 1971-72; departmental demonstrator in astrophysics Oxford U., Eng., 1972-75; sr. sci. rsch. officer South African Astron. Obs., Capetown, Republic of South Africa, 1975-78; Annie J. Cannon fellow U. Del., Newark, 1978-79; asst. prof. Pa. State U., State College, 1979-82; asst. prof. to assoc. prof. physics and astronomy Dartmouth Coll., Hanover, N.H., 1982-88, Margaret Anne and Edward Leede Disting. prof. physics and astronomy, 1988—; dir. Mich.-Dartmouth-MIT Obs., 1991—; vis. astronomer Cornell U., 1992. Editor: White Dwarfs, 1989; contbr. articles to jours. in field. Keeley fellow Wadham Coll., Oxford, 1992-93, vis. fellow in astrophysics Oxford U., 1992-93; recipient numerous grants NSF, NASA. Mem. Am. Astron. Soc., Internat. Astron. Union. Luthran. Office: Dartmouth Coll Dept Physics & Astronomy Wilder Lab Hanover NH 03755

WEGST, AUDREY V., physicist, consulting firm executive; b. Stamford, Conn., Sept. 5, 1934; d. Russell Grant and Vivian (Norman) Smith; m. Sept. 6, 1958 (div. 1972); children: Gregory, Andrew. BA, Mt. Holyoke, 1956; PhD, U. Kans., 1979. Cert. Am. Bd. Radiology, Am. Bd. Med. Physics. Instr. U. Mich., Ann Arbor, 1957-63; assoc. prof. U. Kans. Med. Ctr., Kansas City, 1971-83; physicist Int. Atomic Energy Agy., Vienna, Austria, 1983-85; v.p., CEO Diagnostic Tech. Cons., Kansas City, Mo., 1985—; com. mem. NCRP, Washington, 1985—, Dept. Energy, Washington, 1991—, numerous others. Contbr. 50 articles to profl. jours. Mem. Am. Assn. Med. Physics (various offices 1971—), Health Physics Soc., Soc. Nuclear Medicine. Home: 5420 Pawnee Fairway KS 66205 Office: Diagnostic Tech Cons 4747 Troost Kansas City MO 64110

WEHEBA, ABDULSALAM MOHAMAD, immunologist, allergist, consultant; b. Tanta, Gharbya, Arab Republic of Egypt, Apr. 15, 1930; s. Mohamad Moustafa Weheba and Anisa (Mahmoud) Omar; m. Aida Mohamad, Mar. 29, 1957 (dec. Nov. 1988); children: Sawsan, Safa, Sameeh, Sanaa, Sami; m. Zeynab Ahmad, Oct. 9, 1989. MB, BCh, Alexandria (Egypt) U., 1955. House officer Alexandria U. Hosps., 1955-56; primary care physician Ministry Health, Al-Ras, Saudi Arabia, 1956-60; pvt. practice, Riyadh, Saudi Arabia, 1960-75, 77—; researcher, fellow Immunology Inst., Acad. Medicine, Vienna, Austria, 1975-76; cons. Med. City Hosp., Alexandria, 1983—. Contbr. articles to med. jours. Fellow N.Y. Acad. Scis., Internat. Coll. Angiology, Semouha Sporting Club. Avocations: travel, photography, gardening. Home: Villa 59 Khalil Kayat, Roushdy, Alexandria 21311, Egypt Office: PO Box 211, Riyadh 11411, Saudi Arabia

WEHMEIER, HELGE H., chemical company executive. With Bayer AG, 1965; pres., CEO Agfa Corp., Ridgefield Park, N.J., 1989—, Miles, Inc., 1989—. Office: Bayer USA Inc 500 Grant St Pittsburgh PA 15219-2502 Office: Miles Inc Mobat Rd Pittsburgh PA 15205*

WEHNER, HENRY OTTO, III, pharmacist, consultant; b. Birmingham, Ala., Mar. 3, 1942; s. Henry O. Jr. and Carolyn (Kirkland) W.; m. Sammye Ruth Murphy, June 8, 1974 (div. July 1989). AA, Daytona Beach Community Coll., 1967; BS in Biology, North Ga. Coll., Dahlonega, 1971; BS in Pharmacy, U. Ga., 1978. Registered pharmacist, Fla., Ga.; cert. sci. tchr. grades 7-12, Ga. Tchr. biology Irwin County High Sch., Ocilla, Ga., 1971-75; extern Eckerd Drugs, Athens, Ga., 1977; intern/extern St. Mary's Hosp., Athens, 1977; pharmacy intern Button Gwinnett Hosp., Lawrenceville, Ga., 1978; co-owner, mgr. Hiawassee (Ga.) Pharmacy, 1978-79; staff pharmacist Dyal's Pharmacy, Daytona Beach, Fla., 1979, Little Drug Co., New Smyrna Beach, Fla., 1979-80; staff pharmacist, mgr. Super X Drugs, New Smyrna Beach, 1980-81; staff pharmacist Fish Meml. Hosp., New Smyrna Beach, 1981-92, Halifax Med. Ctr., Daytona Beach, Fla., 1992—. With USAF, 1961-65. Mem. Am. Pharm. Assn., Fla. Soc. Hosp. Pharm., Volusia County Pharm. Assn., Eastern Shores Soc. Hosp. Pharmacists (charter mem.), Phi Lambda Sigma, Phi Theta Kappa. Methodist. Avocations: painting, cycling, tennis. Office: Halifax Med Ctr PO Box 1350 303 N Clyde Morris Blvd Daytona Beach FL 32114

WEHRBERGER, KLAUS HERBERT, physics educator, research manager; b. Tauberbischofsheim, Fed. Republic Germany, May 8, 1959; s. Otmar and Gerda (Schäfer) W.; m. Dagmar G. Bollmann, Jan. 11, 1985; children: Christian, Elias. Diploma, U. Marburg, Fed. Republic Germany, 1984; PhD, Technische Hochschule Darmstadt, Fed. Republic Germany, 1988. Rsch. fellow Ind. U., Bloomington, 1989-90; rsch. asst. Technische Hochschule Darmstadt, 1986-89, 90-91, asst. prof. physics, 1991-92; program dir. for physics Deutsche Forschungsgemeinschaft, Bonn, Germany, 1992—. Contbr. articles to Phys. Rev., Phys. Rev. Letters, Phys. Letters, Nuclear Physics. Rsch. grantee Deutsche Forschungsgemeinschaft, 1989-90. Mem. Am. Phys. Soc., German Phys. Soc. Achievements include research on theoretical nuclear physics and relativistic models. Office: DFG, Kennedyallee 40, D-5300 Bonn 2, Germany

WEHRING, BERNARD WILLIAM, nuclear engineering educator; b. Monroe, Mich., Aug. 3, 1937; s. Bernard Albert and Alma Christina (Graf) W.; m. Margaret Mary Robinson, Sept. 5, 1959; children: Mary Ann, James, Susan, Barbara. B.S.E. in Physics, U. Mich., 1959, B.S.E. in Math, 1959; M.S. in Physics, U. Ill., 1961, Ph.D. in Nuclear Engring, 1966. Asst. prof. nuclear engring. U. Ill., Urbana, 1966-70, assoc. prof., 1970-77, prof., 1977-84, asst. dean engring., 1981-82; prof. nuclear engring. N.C. State U., Raleigh, 1984-89, dir. nuclear reactor program, 1984-89; prof. mech. engring. U. Tex., Austin, 1989—, dir. Nuclear Engring. Teaching Lab., 1989—; cons. Argonne and Los Alamos nat. labs.; mem. crosssect. evaluation working group Brookhaven Nat. Lab. Contbr. sects. to books, articles to profl. publs. AEC fellow, 1963-65; NSF grantee, 1968—. Fellow Am. Nuclear Soc.; mem. Am. Soc. Engring. Edn., Am. Nuclear Soc. (standards com.), Am. Phys. Soc., IEEE. Achievements include contributing in the generation of basic nuclear data and development of new instruments and experimental techniques. Home: 8907 Spring Lake Dr Austin TX 78750-2932 Office: U Tex Nuclear Engring Teaching Lab Balcones Rsch Ctr Austin TX 78712

WEI, GAOYUAN, chemistry educator; b. Chenxi, Hunan, China, June 26, 1961; s. Xiangfu and Fuhua (Tian) W.; m. Weiwei Luan. B in Polymer Engring., South China Inst. Tech., Guangzhou, People's Rep. of China, 1982; MSc in Chem. Engring., Cornell U., 1985; PhD in Chemistry, U. Wash., 1990. Rsch. assoc. chemistry dept. U. Wash., Seattle, 1990; rsch. assoc. Cavendish Lab. U. Cambridge, Eng., 1990-92; lectr. chemistry dept. Peking U., Beijing, 1992—. Mem. AAAS, Am. Chem. Soc., Am. Phys. Soc., Internat. Union Pure and Applied Chemistry, Inst. of Physics. Mem. Am. Chem. Soc., Am. Phys. Soc., Internat. Union Pure and Applied Chemistry, Inst. of Physics, N.Y. Acad. Scis. Avocations: writing, sight-seeing, jogging, hiking, badminton. Home: Yandongyuan Peking U 606 Bldg 7A, Beijing 100871, China Office: Peking U, Dept Chemistry, Beijing 100871, China

WEI, I-YUAN, research and development manager; b. Taipei, Taiwan, Republic of China, Nov. 1, 1940; came to U.S., 1967; s. Kun-Te and Kun (Lu) W.; m. Shirley Chen, Dec. 28, 1968; children: Jerray, Jiaying. BS, Nat. Taiwan Normal U., 1963; MS, Nat. Taiwan U., 1966; PhD, Tufts U., 1971. Rsch. assoc. U. N.C., Chapel Hill, 1972-75; scientist Sprague Electric Co., North Adams, Mass., 1976-77; mgr. R & D Republic Foil, Salisbury, N.C., 1977-81; sr. engr. AMP Inc., Harrisburg, Pa., 1981-86, mem. tech. staff, 1986-88, project mgr., 1988-90, mgr. plating rsch., 1990—. Contbr. articles to Jour. Chinese Chem. Soc., Jour. Chem. Physics, Jour. Am. Chem. Soc., Inorganic Chemistry, Chem. Phys. Lett., Jour. Magna. Reson. Mem. Am. Electroplaters and Surface Finishers Soc. (chmn. electronics finishing com. 1991—, chmn., meetings and symposia, 1993—, dir. tech. edn. 1993—), Electrochem. Soc., Ctrl. Pa. Chinese Assn. (pres. 1984, chmn. bd. 1985—). Achievements include rsch. in deuteron quadrupole coupling constants in nitrobenzenes, hydrogen-bonded systems of aluminum hydroxides and anodic oxide films by deutron magnetic resonance, palladium and palladium-nickel as contact materials; 1 patent. Home: 1046 W Areba Ave Hershey PA 17033 Office: AMP Inc PO Box 3608 Harrisburg PA 17105

WEI, JAMES, chemical engineering educator, academic dean; b. Macao, China, Aug. 14, 1930; came to U.S., 1949, naturalized, 1960; s. Hsiang-chen and Nuen (Kwok) W.; m. Virginia Hong, Nov. 4, 1956; children: Alexander, Christina, Natasha, Randolph (dec.). B.S. in Chem. Engring, Ga. Inst. Tech., 1952; M.S., Mass. Inst. Tech., 1954, Sc.D., 1955; grad., Advanced Mgmt. Program Harvard, 1969. Research engr. to research assoc. Mobil Oil, Paulsboro, N.J., 1956-62; sr. scientist Princeton, N.J., 1963-68; mgr. corp. planning N.Y.C., 1969-70; Allan P. Colburn prof. U. Del., Newark, 1971-77; Sherman Fairchild distinguished scholar Calif. Inst. Tech., 1977; Warren K. Lewis prof. MIT, Cambridge, 1977-91, head dept. chem. engring., 1977-88; Pomeroy and Betty Smith prof. chem. engring., dean Sch. Engring. and Applied Sci. Princeton (N.J.) U., 1991—; vis. prof. Princeton, 1962-63, Calif. Inst. Tech., 1965; cons. Mobil Oil Corp.; cons. com. on motor vehicle emissions Nat. Acad. Sci., 1972-74, 79-80; mem. sci. adv. bd. EPA, 1976-79; mem. Presdl. Pvt. Sector Survey Task Force on Dept. Energy, 1982-83. Bd. editors Chem. Tech, 1971-80, Chem. Engring. Communications, 1972—; cons. editor chem. engring. series, McGraw-Hill, 1964—; editor-in-chief: Advances in Chemical Engineering, 1980; Contbr. papers, monographs to profl. lit., The Structure of Chemical Processing Industries, 1979. Recipient Am. Acad. Achievement Golden Plate award, 1966. Mem. Am. Inst. Chem. Engrs. (dir. 1970-72, Inst. lectr. 1968, Profl. Progress award 1970, Walker award 1980, Lewis award 1985, v.p. 1987, pres. 1988, Founders award 1990), Am. Chem. Soc. (award in petroleum chemistry 1966), Nat. Acad. Engring. (nominating com. 1981, peer com. 1980-82, membership com. 1983-85), AAAS, Am. Acad. Arts and Scis., Academica Sinica of Taiwan, Sigma Xi. Home: 571 Lake Dr Princeton NJ 08540 Office: Princeton U Engring Quadrangle Princeton NJ 08544-5263

WEI, MUSHENG, mathematics educator; b. Danyang, Jiangsu, China, Jan. 8, 1948; s. Rongtao and Qiaoqing (Zhang) W.; m. Jine Sun, Jan. 28, 1979; 1 child, Chong. BS, Nanjing U., People's Republic of China, 1982; MS, Brown U., 1984, PhD, 1986. Postdoctoral fellow U. Minn., Mpls., 1986-87, Mich. State U., East Lansing, 1987-88, Ohio State U., Columbus, summer 1987, 88; assoc. prof. East China Normal U., Shanghai, People's Republic of China, 1988-92, prof., 1992—; group leader numerical math. East China Normal U., 1991—' vis. scholar Ohio State U., Columbus, 1991-92, McGill U., Montreal, Can., 1992. Contbr. articles to profl. jours. Recipient Sci. Progress 3rd award Nat. Edn. Com., Peoples Republic China, 1990. Mem. Am. Math. Soc., Shanghai Soc. Indsl. and Applied Maths. Office: East China Normal U Dept Math, 3663 Zhongshan Road, Shanghai 200062, China

WEI, ROBERT PEH-YING, mechanics educator; b. Nanking, China, Sept. 16, 1931. BSE, Princeton U., 1953, MSE, 1954, PhD in Mech. Engring., 1960. Instr. mech. engring. Princeton U., 1954-57, rsch. asst. aero. engring., 1958-59; assoc. technologist, fracture mechanics applied rsch. lab. U.S. Steel Corp., 1959-61, technologist, 1961-62, sr. rsch. engr., 1962-64, assoc. rsch. cons., 1964-66; assoc. prof. mechanics Lehigh U., Bethlehem, Pa., 1966-70, prof., 1970—. Mem. ASTM, Sigma Xi. Achievements include research in fracture mechanics, mechanics and metallurgical aspects of fatigue crack growth and stress corrosion cracking, experimental stress analysis. Office: Lehigh U Dept Mechanical Engring & Mech Packard Lab Bethlehem PA 18015*

WEI, VICTOR KEH, electrical engineer; b. Taipei, Taiwan, Jan. 13, 1954; s. Chih-Shek and Yin-Hsia (Huang) W.; m. Betty H. Fang, Oct. 10, 1988; 1 child, Francine. BSEE, Nat. Taiwan U., 1976; PhD in Elec. Engring., U. Hawaii, 1980. Mem. tech. staff Bell Telephone Labs., Murray Hill, N.J., 1980-83; mem. tech. staff Bellcore, Morristown, N.J., 1984-88, dist. mgr., 1988—. contbr. articles to profl. jours. Mem. IEEE (bd. govs. 1991—). Achievements include research in locally adaptive data compression schemes, gernalized hamming weights. Office: Bellcore Rm 26 339 445 South St Morristown NJ 07960-1910

WEI, YAU-HUEI, biomedical research scientist; b. Taichung, Taiwan, Republic of China, May 29, 1952; s. Yuing-Kwei and Jen (Wu) W.; m. Yeh-Jen Lin, Aug. 11, 1978; children: Li-Sing, Tien-Sing, Mu-Sing, Teng-Sing. BS, Nat. Taiwan U., Taipei, Republic of China, 1974; PhD in Biochemistry, SUNY, Albany, 1980. Postdoctoral fellow SUNY, Albany, 1980-81; assoc. prof. biochemistry Nat. Yang-Ming Med. Coll., Taipei, 1981-85, prof., chair dept. biochemistry, 1985-91, dean student affairs, 1989-91; researcher Clin. Rsch. Ctr. Taipei Vets. Gen. Hosp., 1984—; cons. Taichung Vets. Gen. Hosp., 1993—; dir. Instrumentation Ctr., Nat. Yang-Ming Med. Coll., 1985-88, Ctr. for Molecular and Cell Biology, Nat. Yang-Ming Med. Coll., Taipei, 1993—. Author: (book) New Experimental Biochemistry, 1986, Recent Advances in Molecular and Biochemical Research on Proteins, 1993; editor: (book) Molecular Mechanism of Alcohol, 1989, Advances in Biotechnology and Molecular Biology, 1990; exec. editor: (jour.) Sci. Monthly, Taipei, 1989—. Recipient Biomed. Rsch. award Ching-Ling Med. Found., Taipei, 1983, Disting. Rsch. award The Nat. Sci. Coun. R.O.C. Taipei, 1987-89, 89-91, 92-94, Disting. Teaching award Min. Edn., Taipei, 1989-90. Mem. AAAS, Fedn. Asian and Oceanian Biochemists (mem. coun.), Chinese Biochem. Soc. (pres. 1991-93, bd. editors jour. 1986—), Soc. for Chinese Bioscientists in Am., Chinese Physiol. Soc., Chinese Clin. Biochemistry Soc. Baptist. Avocations: ping-pong, basketball, traveling, stamp and coin collecting. Home: 155 Li-Long St, Sec 2 Shih-Pai, Taipei 112, Taiwan Office: Nat Yang-Ming Med Coll Dept Biochem, 155 Li-Long St Sec 2, Taipei 112, Taiwan

WEICHERT, DIETER HORST, seismologist, researcher; b. Breslau, Silesia, Germany, May 2, 1932; came to Can., 1954; s. Kurt Herman and Margarete Adelheid (Buresch) W.; m. Edith Struning (div. 1983), children: Thomas, Andreas. BASc, U. B.C., Vancouver, Can., 1961, PhD, 1965; MS, McMaster U., Hamilton, Ont., Can., 1963. Rsch. scientist earth physics br. Dominion Obs., Ottawa, Ont., 1965-78; rsch. scientist Pacific Geosci. Ctr., Sidney, B.C., 1978-81, head earthquake studies, 1981-88, assoc. dir., acting dir., 1988-90; rsch. scientist Geol. Survey Can., Sidney, B.C., 1990—; cons. U.K. Atomic Energy Authority, Blacknest, Eng., 1970; guest lectr. Fredericiana U., Karlsruhe, Germany, 1971; mem. Can. Earthquake Engring. Com., Ottawa, 1982—; supt. constrn. industry, Germany, 1951-57. Contbr. over 80 articles to profl. jours. Woodrow Wilson fellow, 1961; Inco fellow, 1962-64. Mem. Seismol. Soc. Am. Achievements include research on underground nuclear explosion detection and identification; seismic hazard and strong seismic ground motion. Office: Geol Survey Can, 9860 W Saanich Box 6000, Sidney, BC Canada V8L 4B2

WEICHSELBAUM, RALPH R., oncologist chairman. BS, U. Wis., 1967; MD, U. Ill., Chgo., 1971. Intern Alameda County Hosp., Oakland, Calif., 1971-72; resident in radiation therapy Harvard Med. Sch., Boston, 1972-75; assoc. prof. radiation therapy Harvard Med. Ctr., Boston, 1980-84; assoc. prof. dept. cancer biology Harvard Sch. Public Health, Boston, 1983-84; prof., chmn. dept. radiation and cellular omcology Pritzker Sch. Medicine U. Chgo., 1984—, Harold H. Hines Jr. prof., chmn., 1990; head Michael Reese/U. Chgo. Ctr. Radiation Therapy, 1984—. Contbr. articles to profl. jours. Office: Ctr for Radiation Therapy MC 0085 5841 S Maryland Ave Chicago IL 60637-1404

WEIDNER, DONALD J., geophysicist educator; b. Dayton, Ohio, Apr. 26, 1945; s. Virgil Raymond and Aletha Winifred Weidner; m. Deborah Mary Ray, April 13, 1968; children: Raymond V., Jennifer L. AB in Physics cum laude, Harvard, 1967; PhD in Geophysics, Mass. Inst. Tech., 1972. Asst prof. SUNY, Stony Brook, N.Y., 1972-77, assoc. prof., 1977-82, prof. geophysics, 1982—; dir. Mineral Physics Inst., SUNY, 1988—, Ctr. for High Pressure Rsch, SUNY, 1991—. Am. Geophysical Union fellow, 1981; recipient James B. Macelwane award Am. Geophysical Union, 1981. Achievements include building (with others) the high pressure facility at Stony Brook SUNY; large volume high pressure studies with synchrotron radiation; determining the equation of state of earth materials; phase stability fields of minerals and has pioneered the use of this system to determine the yield strength of these materials; design team leader for the large volume experiments that the GeoCars program is preparing for the Advanced Photon Source. Office: SUNY Sci & Tech Ctr High Pressure Rsch Dept Earth & Space Scis Stony Brook NY 11794

WEIGEL, HENRY DONALD, civil engineer; b. Somers Point, N.J., Dec. 14, 1964; s. Henry Frank and Joan Alma (Sampson) W. BSCE, Villanova U., 1987. Lic. ocean operator USCG, 1987. Capt., mate various sportfishing vessels, Ocean City, N.J., 1977-88; asst. engr. Adams, Rehmann & Heggan Assoc., Hammonton, N.J., 1987-88, project engr., 1990—. With USN, 1988-90. Mem. Nat. Assn. Profl. Engrs., Villanova U. Engring. Alumni Soc., Cape May County Party and Charter Boat Assn. Episcopalian. Home: 61 Tyler Rd Greenfield NJ 08230 Office: Adams Rehmann & Heggan Assocs 850 S White Horse Pike Hammonton NJ 08037

WEIGEL, OLLIE J., dentist, mayor; b. Guthrie County, Iowa, Sept. 29, 1922; s. Verne Noble and Ethel Rebecca (Johnson) W.; m. Mary Kathryn Finnegan, June 3, 1944; children: John, Marilyn, Larry, Susan. DDS, U. Iowa, 1951. Practice dentistry Ankeny, 1951—; mayor City of Ankeny, 1974—. Mem. Ankeny City Coun., 1966-73, Des Moines Area C.C. Found. Bd., 1993—, Des Moines Area Metro Forum, 1985-93. 2d lt. USAAF, 1943-45, ETO. Mem. ADA, Iowa Dental Assn., Des Moines Dist. Dental Assn., Ankeny C. of C. (pres., Outstanding Citizen 1976), Mid Iowa Assn. Local Govts., League of Iowa Municipalities (pres. 1976-77), Am. Legion, Lions. Republican. Methodist. Avocation: fishing. Home: 2506 NW 4th St Ankeny IA 50021-1002 Office: 306 SW Walnut St Ankeny IA 50021-3042

WEIGEL, PAUL HENRY, biochemistry educator, consultant; b. N.Y.C., Aug. 11, 1946; s. Helmut and Jeanne (Wakeman) W.; m. Nancy Shulman, June 15, 1968 (div. Dec. 1987); 1 child, Dana J.; m. Janet Oka, May 17, 1992. BA in Chemistry, Cornell U., 1968; MS in Biochemistry, Johns Hopkins U., Balt., 1969, PhD in Biochemistry, 1975. NIH postdoctoral fellow Johns Hopkins U., Balt., 1975-78; asst. prof. U. Tex. Med. Br., Galveston, Tex., 1978-82, assoc. prof., 1982-87; prof. biochemistry and cell biology U. Tex. Med. Br., Galveston, 1987—, vice chmn. dept. human biol. chemistry and genetics, 1990—, acting chmn. dept. human biology, chemistry and genetics, 1992-93; mem. NIH Pathobiochemistry Study Sect., Washington, 1985-87; cons. Teltech, Mpls., 1985—. Contbr. articles to profl. jours.; patentee in field. Treas. Bayou Chateau Neighborhood Assn., Dickinson, Tex., 1981-83, v.p., 1983-84, pres., 1984-86. With U.S. Army, 1969-71. Grantee NIH, 1979—, Office Naval Rsch., 1983-87, Tex. Biotech., 1989—; recipient Disting. Tchr. award U. Tex. Med. Br., 1989, Disting. Rsch. award, 1989. Mem. Am. Chem. Soc., Am. Soc. Cell Biology, Am. Soc. Biochemistry and Molecular Biology. Democrat. Lutheran. Avocations: raquetball, basketball card collecting, poetry, camping. Home: 1109 Hidden Oaks League City TX 77573 Office: U Tex Med Br Dept Human Biol Chemistry and Genetics Basic Sci Bldg Galveston TX 77555-0647

WEIGLE, ROBERT EDWARD, mechanical engineer, research director; b. Shiloh, Pa., Apr. 27, 1927; s. William Edgar and Hilda Geraldine (Fans) W.; m. Mona Jean Long, Aug. 13, 1949; 1 child, Geoffrey Robert. BCE in Structures, Rensselear Poly. Inst., 1951, MS in Mechanics, 1957, PhD in Mechanics, 1959. Registered profl. engr., N.Y., Pa. Assoc. rsch. scientist Rensselear Poly. Inst., Troy, N.Y., 1955-59; chief scientist Watervliet Arsenal, 1959-77; technical dir. U.S. Army and Armament R & D Command, 1977-82; dir. U.S. Army Rsch. Office, 1982-88; dir. phys. sci. lab. N.Mex. State U., Las Cruces, 1988—; tech. dir., then dir. Benet Weapons Lab., 1959-77; chmn. numerous DoD and Army coms. Contbr. articles to profl. jours. Recipient Meritorious Civilian Service award for cannon breech design U.S. Army, 1964, U.S. Army Materiel Command citation for engineering achievement in Vietnam, 1966, Presidential citation for development of cannon firing simulator, 1965; elected to Am. Acad. of Mechanics, 1972. Mem. NSPE, ASME, AAAS, Am. Soc. Testing Materials, Am. Def. Preparedness Assn. (Crozier prize 1985), Nat. Conf. Advancement of Rsch. (program com. 1987, exec. conf. com., host rep. NCAR-46 ann. conf. 1992), Army Sci. Bd. (chmn. rsch. and new initiatives group), Soc. Exptl. Mechanics, Tau Beta Pi, Chi Epsilon, Sigma Xi. Office: N Mex State U Physical Sci Lab PO Box 30002 Las Cruces NM 88003-8002

WEIGMANN, HANS-DIETRICH H., chemist; b. Rostock, Germany, Jan. 12, 1930; m. Christa Weigmann; 2 children, Stefanie, Jessica. Vordiplom. in Chemistry, U. Hamburg, Germany, 1954, Diplom. in Organic Chemistry, 1958; Dr. rer. nat. in Organic Chemistry, Tehcnische Hochschule, Aachen, Germany, 1960. Scientist German Wool Rsch. Inst., 1960-61; postdoctoral fellow Textile Rsch. Inst., Princeton, N.J., 1961-63, sr. scientist, 1963-67, assoc. dir. chem. rsch., 1967-70, assoc. dir. rsch., 1970—; com. tech. programs 4th Internat. Wool Textile Rsch. Conf., 1969-70; chmn. Gordon Rsch. Conf. Fiber Sci., 1975. Co-recipient Best Paper award Textile Chemist and Colorist, 1982, Lit. award Soc. Cosmetic Chemists, 1986. Mem. Am. Chem. Soc. (divsn. cellulose, wood and fiber chemistry, divsn. organic coatings and plastics chemistry, chmn. symposium, 1970, Inst. Textile Sci. Can., 1982), Am. Assn. Textile Chemists and Colorists (com. RA91 applied dyeing theory, exec. com. rsch., com. RA33, colorfastness to atmospheric contaminants, Internat. Dyeing Symposium, 1983, Millson award com., 1987, Olney Medal, 1990). Office: Textile Research Institute PO Box 625 Princeton NJ 08542*

WEIHS, DANIEL, engineering educator; b. Kweilin, China, Oct. 29, 1942; s. Hugo and Rozsi (Glass) W.; m. Nira Zaretsky, Feb. 14, 1967; children: Joseph, Daphne, Ory. BS in Aeronautical Engring., Technion U., Haifa, Israel, 1964, MS in Aeronautical Engring., 1968, PhD, 1971. Project engr. Israeli Air Force, 1964-67; sr. lectr. Technion Israel Inst. Tech., Haifa, 1973-78, assoc. prof., 1978-83, prof., 1983—, dean faculty aerospace engring., 1986-88; head Aeronautical Rsch. Ctr., Haifa, 1986-88; dir. Neaman Inst. for Advanced Studies, Haifa, 1990—; cons. in field, 1975—, Southwest Rsch. Inst., San Antonio, 1989-90; dir. Israel Aircraft Industries, Lod, 1987-90, Beitshemesh (Israel) Industries, 1982-84, Iseael Limnological and Oceanographic Corps.), 1993—; mem. exec. com. Israel Space Agy., Tel Aviv, 1987—; chmn. Israel Sci. Satellite Program, 1987-89; active Nat. Com. Space Rsch., 1991—; trustee Ben Gurion U., Beer Sheva, Israel, 1989—; mem. adv. com. Marine Sci. Inst., Haifa, 1990—; sr. rsch. assoc. Nat. Rsch. Coun., La Jolla, Calif., 1978-79, Mountain View, Calif., 1984-85. Capt. Israeli Air Force, 1960-67. Royal Soc. vis. fellow, 1971-72. Mem. Israel Soc. Aeronautics and Astronautics, Am. Soc. Zoologists, Soc. Underwater Tech. Avocations: music, art, stamp collecting. Office: Technion Israel Inst of Tech, Faculty Aerospace Engring, Haifa Israel 32000

WEIL, EDWARD DAVID, research scientist, chemist, educator; b. Phila., June 13, 1928; s. Irving E. and Minna M. (Stainbrook) W.; m. Barbara Joy Hummel, Sept. 11, 1952; children: David L., Claudia E. BS in Chemistry, U. Pa., 1950; PhD in Organic Chemistry, U. Ill., 1953; MBA, Pace U., 1982. Chemist, supt. Hooker Chem. Co., Niagara Falls, N.Y., 1950-65; supr., sr. scientist Stauffer Chem. Co., Dobbs Ferry, N.Y., 1965-86; ind. cons., patent agt., propr. Intertech. Svcs., 1986—; dir. exploratory rsch. Adelphi Rsch. Ctr., Garden City, 1986-87; rsch. prof., Poly. U., Bklyn., 1987—; . Contbr. articles to Kirk-Othmer Ency., Ency. Polymer Sci., Rsch. Mgmt. Jour., others. Recipient IR-100 award Indsl. Rsch. Mag. Achievements include over 210 patents for commercial pesticides, flame retardants, processes, agricultural chemicals, others. Mem. Am. Chem. Soc. (chmn. profl. rels. com. N.Y. sect. 1980—), Assn. Cons. Chemists and Chem. Engrs., Sigma Xi. Home and Office: 6 Amherst Dr Hastings On Hudson NY 10706-3302

WEIL, MAX HARRY, physician, medical educator, medical scientist; b. Baden, Switzerland, Feb. 9, 1927; came to U.S., 1937, naturalized, 1944; s. Marcel and Gretl (Winter) W.; m. Marianne Judith Posner, Apr. 1955; children: Susan Margot, Carol Juliet. AB, U. Mich., 1948; MD, SUNY, N.Y.C., 1952; PhD, U. Minn., 1957. Diplomate Am. Bd. Internal Medicine and Critical Care Medicine, Nat. Bd. Med. Examiners. Intern in internal medicine Cin. Gen. Hosp., 1952-53; resident U. Minn. Hosps., Heart Hosp., VA Hosp., Mpls., 1953-55; rsch. fellow U. Minn., Mpls., 1955-56; sr. fellow Nat. Heart Inst., Mayo Clinic, Rochester, Minn., 1956-57; chief cardiology City of Hope Med. Ctr., Duarte, Calif., 1957-59; asst. clin. prof. U. So. Calif. Sch. Medicine, L.A., 1957-59, asst. prof., 1959-63, assoc. prof., 1963-71, prof., 1971-81; disting. prof., chmn. dept. medicine, chief div. cardiology U. Health Scis., Chgo. Med. Sch., North Chicago, Ill., 1982-91, disting. univ. prof., 1992—; adj. prof. Northwestern Univ. Med. Sch., Chgo., 1992—; prof. clin. med. bioengring. U. So. Calif., 1972-81; adj. prof. medicine, 1981—, Northwestern U. Med. Sch., Chgo., 1992—; dir. Shock Rsch. Unit, 1961-81, Inst. Critical Care Medicine Ann. Symposium, 1963—, Ctr. Critically Ill, U. So. Calif., L.A., 1968-80, Inst. Critical Care Medicine, Palm Springs, Calif. and North Chicago, 1974—, also pres.; attending cardiologist children's div. Los Angeles County/U. So. Calif. Med. Ctr., 1958-65, attending physician, 1958-71, sr. attending cardiologist, 1968-73, sr. attending physician, 1971-81; vis. prof. anesthesiology/critical care medicine, U. Pitts., 1973—; clin. prof. UCLA, 1981—; numerous vis. professorships and lectureships; cons. in field; prin. researcher numerous grants and rsch. projects; Weil Internat. lectr. in critical care medicine U. So. Calif., L.A., 1987. Sect. editor Archives Internal Medicine, 1983-87; guest editor Am. Jour. Cardiology; mem. editorial bd. Am. Jour. Medicine, 1971-79, 81-85, Jour. Chest, 1980-85, Jour. Circulatory Shock, 1979—, Clin. Engring. Newsletter, 1980—, Methods of Info. in Medicine, 1977-91; editorial adv. bd. Emergency Medicine, 1978—, Issues in

Health Care Tech., 1983—; assoc. editor Critical Care Medicine, 1973-74, mem. editorial bd., 1973-88; editor-in-chief Acute Care, 1983-90; contbr. articles to profl. jours.; patentee in field. Pres. Temple Brotherhood, Wilshire Blvd. Temple, L.A., 1967-68; bd. dirs. Hollywood Presbyn. Med. Ctr., 1976-81, L.A. chpt. Met. Am. Heart Assn., 1962-67, Chgo. chpt. Met. Am. Heart Assn., 1982—. Served with U.S. Army, 1946-47. Recipient prize in internal medicine SUNY, 1952, Alumni medallion SUNY, 1970; Disting. Svc. award Soc. Critical Care Medicine, 1984; numerous rsch. grants, 1959—; named Disting. Alumni Lectr., 1967, Oscar Schwindetzky Meml. Lectr. Internat. Anesthesia Rsch. Soc., 1978; recipient Lawrence R. Medoff award Chgo. Med. Sch., 1987, Morris L. Parker Rsch. award, 1989; Lilly scholar, 1988-89. Master ACP; fellow Am. Coll. Cardiology (chmn. emergency cardiac care com. 1974-81), ACP, Am. Coll. Chest Physicians (coun. clin. cardiology, coun. critical care medicine), Am. Coll. Clin. Pharmacology, Am. Coll. Critical Care Medicine, Am. Heart Assn. (coun. circulation, coun. thrombosis, coun. cardiopulmonary diseases), N.Y. Acad. Sci., Chgo. Soc. of Internal Medicine; mem. AMA (sec. editor jour. 1969-72), IEEE, Ill. Med. Assn., Lake County Med. Assn., Los Angeles County Med. Assn., Am. Physiol. Soc., Am. Soc. Pharmacology and Exptl. Therapeutics, Am. Soc. Echocardiography, Am. Soc. Nephrology, Am. Trauma Soc. (founding mem.), Assn. Computing Machinery, Assn. Am. Med. Colls., Assn. Program Dirs. Internal Medicine, Cen. Soc. Clin. Rsch., Chgo. Cardiol. Group (sec.-treas. 1986—, chmn. 1988-90), Chgo. Soc. Internal Medicine, Lake County Heart Assn. (bd. govs. 1983-86), Intensive Care Soc. U.K., L.A. Soc. Internal Medicine, Soc. Exptl. Biology and Medicine, Western Soc. Clin. Rsch., Fedn. Am. Socs. Am. Soc. Parenteral and Enteral Nutrition, Nat. Acad. Practice (disting. practitioner), Skull and Dagger, Sigma Xi, Alpha Omega Alpha. Jewish. Avocations: swimming, tennis, photography, philosophy-economics. Office: U Health Scis Chgo Med Sch 3333 Green Bay Rd North Chicago IL 60064-3095

WEILAND, PETER LAWRENCE, military officer; b. Detroit, Nov. 23, 1956; s. Peter Lawrence and Jeanette Rose (Wochaski) W.; m. Elizabeth Anne Hester, June 9, 1984; children: Peter Carter, Benjamin Cooper. BS, U.S. Mil. Acad., 1979; MS, Rice U., 1990, PhD, 1991. Commd. 2d lt. U.S. Army, 1979, advanced through grades to maj., 1992; mem. Significant Interest Group-Artificial Intelligence, Hampton, Va., 1991—. Contbr. articles and papers to conferences. Advisor Phoebus High Sch. Robotics Program, Hampton, Va., 1991—; co-organizer 24Hour Marathon for Spl. Olympics, Hampton, 1992. Recipient Disting. Honor Grad. U.S. Army Ranger Sch., Ft. Benning, Ga., 1980, Meritorious Svc. medal U.S. Army, Karlsruhe, Germany, 1987. Mem. IEEE, Soc. Am. Mil. Engrs. Avocations: golf, tennis, sailing, computers, stamp collecting. Office: HQ TRADOC ATCD-B Fort Monroe VA 23651

WEILER, KURT WALTER, radio astronomer; b. Phoenix, Mar. 16, 1943; s. Henry Carl and Dorothy (Esser) w.; m. Geertje Stoelwinder, June 8, 1979; children: Corinn Nynke Yoon, Anil Erick Jivan, Sanna Femke Lee. BS, U. Ariz., 1964; PhD, Calif. Inst. Tech., 1970. Guest investigator Netherlands Found. for Radioastronomy, Groningen, 1970-74; sci. collaborator Inst. for Radioastronomy, Bologna, Italy, 1975-76; sr. scientist Max Planck Inst. for Radioastronomy, Bonn, W.Ger., 1976-79; program dir. NSF, Washington, 1979-85; head interferometric rsch. sect., project dir. optical interferometer project Naval Rsch. Lab, Washington, 1985—; Halley steering com. NASA, Washington, 1981-85. Author: WSRT Users Guide, 1973, 75; editor Radio Astronomy from Space, 1987; contbr. over 100 articles to profl. jours. and mags. Mem. Am. Astron. Soc., Royal Astron. Soc., Internat. Astron. Union, Internat. Sci. Radio Union, Nederlandse Astronomen Club, Jaguar Club, Nat. Capital Club. Home: 6232 Cockspur Dr Alexandria VA 22310-1504 Office: Naval Rsch Lab Code 1000 4555 Overlook Ave SW Washington DC 20375-5000

WEILER, PAUL CRONIN, law educator; b. Port Arthur, Ont., Can., Jan. 28, 1939; s. G. Bernard and Marcella (Cronin) W.; m. Florrie Darwin, 1988; children: Virginia, John, Kathryn, Charles. B.A. with honors, U. Toronto, 1960, M.A. with honors, 1961; LL.B., Osgoode Hall Law Sch., 1964; LLM, Harvard Law Sch., 1965; LL.D., U. Victoria, 1981. Bar: Ont. Prof. law Osgoode Hall Law Sch., 1965-72; chmn. Labour Relations Bd. B.C., 1973-78; Mackenzie King prof. Can. studies Harvard Law Sch., 1978-80, Henry J. Friendly prof. law, 1980—; Henry J. Friendly prof. of law, 1993—; prin. legal investigator Harvard U. Med. Practice Study Group; impartial umpire AFL-CIO; chief reporter Am. Law Inst. Tort Reform Project; cons. to U.S. Commn. on Comprehensive Health Care (Pepper Commn.); spl. counsel Govt. of Ont. Rev. of Workers' Compensation, 1980-88; mem. pub. rev. bd. United Auto Workers, chief counsel Pres.' commn. Future Worker-Mgmt. Rels., 1993—; panelist, US/Canada Free Trade Agreement Softwood Lumber Arbitration, 1992-93. Author: Labor Arbitration and Industrial Change, 1970; In the Last Resort: A Critical Study of the Supreme Court of Canada, 1974; (with others) Labor Relations Law in Canada, rev. edit. 1974; (with others) Studies in Sentencing in Canada, 1974; Reconcilable Differences: New Directions in Canadian Labour Law, 1980; Reforming Workers Compensation, 1980; MEGA Projects: The Collective Bargaining Dimensions, 1981; Protecting the Worker From Disability, 1983; Governing the Workplace: The Future of Labor and Employment Law, 1990; (with others) Patients, Doctors, and Lawyers: Medical Injury, Malpractice Litigation and Patient Compensation, 1990; Medical Malpractice on Trial, 1991; A Measure of Maleprachice, 1992; (with others) Cases, Materials and Problems on Sports & the Law, 1993; contbr. articles to profl. jours. Mem. Nat. Acad. Arbitrators, Nat. Acad. Social Ins., Nat. Acad. Sci., Inst. Medicine. Roman Catholic. Club: Cambridge Tennis (Cambridge, Mass.). Office: Harvard U Law Sch 1525 Massachusetts Ave Cambridge MA 02138

WEILING, FRANZ JOSEPH BERNARD, retired botany and biometry educator; b. Dülmen, Fed. Republic Germany, Sept. 20, 1909; s. Bernard Johann Theodor and Franziska Josefa Gertrudis (Wewers) W.; m. Elisabeth Dorothea Jungewelter, Nov. 9, 1945 (dec. Sept. 1985); children: Irmgard, Margret, Jürgen, Günter. Student Philosophy, Olinda, Brazil, 1929-31; student theology, Salvador, Brazil, 1931-34; D. in Biology, Physics, Math., U. Münster, Fed. Republic Germany, 1940; HHD (hon.), Villanova U., Pa., 1987. Adminstr. of an asst. place U. Münster, 1945-47; mem. agrl. faculty U. Bonn, Bonn, Fed. Republic Germany, 1947—; privat dozent U. Bonn, 1949—, prof. botany, 1957—, prof. botany and biometry, 1962—, emeritus, 1974—. Contbr. sci. papers to profl. jours. Sgt. Germany Army, 1941-45. Recipient Golden Mendel medal Czechoslovakian Acad. Sci., 1992. Mem. N.Y. Acad. Scis., Biometric Soc. (coun. 1969-71), Bernoulli Soc., Am. Statis. Assn., Intern Soc. History Sci., Am. Genetical Assn., German Genetical Assn., German Soc. Botany, German Assn. Applied Botany. Roman Catholic. Avocation: horticulture. Home: Zur Marterkapelle 65, D 53127 Bonn Lengsdorf, Germany

WEILL, HANS, physician, educator; b. Berlin, Aug. 31, 1933; came to U.S., 1939; s. Kurt and Gerda (Philipp) W.; m. Kathleen Burton, Apr. 3, 1958; children: Judith, Leslie, David. B.S., Tulane U., 1955, M.D., 1958. Diplomate: Am. Bd. Internal Medicine. Intern Mt. Sinai Hosp., N.Y.C., 1958-59; resident Tulane Med. Unit, Charity Hosp. La., New Orleans, 1959-60; chief resident Tulane Med. Unit, Charity Hosp. La., 1961-62, sr. vis. physician, 1972—; NIH research fellow dept. medicine and pulmonary lab. Sch. Medicine Tulane U., New Orleans, 1960-61; instr. medicine Sch. Medicine Tulane U., 1962-64, asst. prof. medicine, 1964-67, assoc. prof., 1967-71, prof. medicine, 1971—, Schlieder Found. prof. pulmonary medicine, 1985—; chief Environ. Medicine sect. Tulane Med. Center, 1980—; dir. univ. Ctr. for Bioenviron. Rsch., 1989-93; dir. interdisciplinary research group in occupational lung diseases Nat. Heart, Lung and Blood Inst., 1972-92, mem. nat. adv. council, 1986-90, chmn. pulmonary disease adv. com., 1982-84; active staff Tulane Med. Center Hosp., 1976—; program dir. Nat. Inst. for Environ. Health Sci., 1992—; cons. pulmonary diseases Touro Infirmary, New Orleans, 1962—; cons. NIH, Nat. Inst. Occupational Safety and Health, Occupational Safety and Health Adminstrn., USN, NAS, EPA; lectr., participant workshops and confs. profl. groups in U.S., France, Can., U.K.; dir. Nat. Inst. Environ. Health Sci Superfund. Basic Rsch. Program, 1992—. Mem. editorial bd. Am. Rev. of Respiratory Disease, 1980-85, CHEST, 1987-91; editor Respiratory Diseases Digest, 1981; guest editor Byssinosis conf. supplement, CHEST, 1981. Fellow Am. Acad. Allergy, Royal Soc. Medicine, ACP; mem. Am. Thoracic Soc. (pres. 1976), Am. Lung Assn. (bd. dirs. 1975-78), New Orleans Acad. Internal Medicine (sec., treas. 1973-75), Am. Coll. Chest Physicians (gov. for La. 1970-75), Am. Fedn. Clin. Research, So. Soc. Clin. Investigation, N.Y. Acad. Scis., Brit. Thoracic Assn., Internat. Epidemiol. Assn., Am. Heart Assn. (task force on environment and cardiovascular system 1978), Brit. Thoracic Soc., Phi Beta Kappa, Alpha Omega Alpha. Home: 333 Friedrichs Ave Metairie LA 70005-4518 Office: Tulane U Sch Medicine Sect Environ Medicine 1430 Tulane Ave New Orleans LA 70112-1210

WEIMAR, ROBERT ALDEN, environmental engineering executive; b. Winchester, Mass., May 15, 1950; s. Alden Walter and Thelma Louise (Simmons) W.; m. Lark Elizabeth Waldmann, June 18, 1972; children: Heather Joy, Maegan Elizabeth. BS, U. Mass., 1972. Assoc. engr. Camp Dresser & mcKee, Inc., Boston, 1972-86; v.p., pres. Normandean Engrs. Inc., Concord, N.H., 1986-89; v.p. Camp Dresser & McKee, Inc., Cambridge, Mass., 1989—. Chmn. Hampstead (N.H.) Solid Waste Com., 1977-90, Hampstead Bldg. Com., 1980-83; rep. Southeastern Regional Refuge Dist., Kingston, N.H., 1986-90. Mem. ASCE, Solid Waste Assn. N.Am. (pres. chpt. 1992—, v.p. Mass. chpt. 1991). Achievements include research in water supply. Home: 44 Stony Ridge Rd Hampstead NH 03841 Office: Camp Dresser & McKee Inc 10 Cambridge Ctr Cambridge MA 02142

WEIMER, PAUL K(ESSLER), electrical engineer; b. Wabash, Ind., Nov. 5, 1914; s. Claude W. and Eva V. (Kessler) W.; m. Katherine E. Mounce, July 18, 1942; children: Katherine Weimer Lasslob, Barbara Weimer Blackwell, Patricia Weimer Hess. A.B., Manchester Coll., 1936, D.Sc. (hon.), 1968; M.A. in Physics, U. Kans., 1938; Ph.D. in Physics, Ohio State U., 1942. Prof. phys. sci. Tabor Coll., 1937-39; research engr. David Sarnoff Research Center, RCA Labs., Princeton, N.J., 1942-65; fellow tech. staff David Sarnoff Research Center, RCA Labs., 1965-81, cons., 1981—. Recipient TV Broadcasters award, 1946, Zworykin TV prize Inst. Radio Engrs., 1959, Sarnoff award in sci. RCA, 1963, Kulturpreis award German Photographic Soc., 1986, Albert Rose Electronic Imager of Yr. award, 1987, Pioneer award N.J. Inventors Hall of Fame, 1991. Fellow IEEE (Outstanding Paper award Solid State Cirs. Conf. 1963, 65, Morris Liebmann prize 1966); mem. NAE. Holder 90 patents in field; active in initial devel. of TV camera tubes, solid state sensors and thin-film transistors. Home and Office: 112 Random Rd Princeton NJ 08540-4146

WEIN, ALAN JEROME, urologist, educator, researcher; b. Newark, Dec. 15, 1941; s. Isadore R. and Jeanette Frances (Abrams) W. A.B. cum laude, Princeton U., 1962; M.D., U. Pa., 1966. Diplomate Am. Bd. Urology (trustee). Intern mixed surgery Hosp. U. Pa., Phila., 1966-67, resident surgery, 1967-68; resident urology U. Pa., Phila., 1969-72, fellow Harrison Dept. Surg. Rsch. Urology Sch. Medicine, 1968-69, asst. instr. surgery Sch. Medicine, 1967-68, asst. instr. urology, 1969-71, instr., 1971-72, asst. prof., 1974-76, assoc. prof., 1976-83, prof., 1983—, asst. chief urology, 1974-79, dir. Urodynamic Evaluation Ctr., 1974—, chmn. div. urology, 1981—, chief urology, 1981—; dir. resident edn. com. div. urology Sch. Medicine U. Pa., 1976—, coord. program urologics oncology, 1976—; chief urology VA Hosp., Phila., 1974-82, attending urologist, 1982—; asst. surgeon Children's Hosp. Phila., 1974—; cons. CDC Coun. Incontinence, 1990—; assoc. surgeon Pa. Hosp., Phila., 1977—; attending urologist Grad. Hosp., Phila., 1980—. Author: (with D.M. Barett) Controversies in Neuro-Urology, 1984, Voiding Function and Dysfunction: A Logical and Practical Approach, 1988, (with A.R. Mundy and T.P. Stephenson) Urodynamics: Principles, Practice and Application, 1984, (with P.M. Hanno) A Clinical Manual of Urology, 1987, (with Hanno, Staskin, and Krane) Interstitial Cystitis, 1990; editorial bd. asst. Urol. Survey, 1978-81; editorial bd. cons. Investigative Urology, 1978-81; mem. editorial bd. World Jour. Urology, 1982—, Am. Urol. Assn. Update series, 1983—, Urol. Survery, 1987—, Internat. Jour. Impotence Rsch: Basic and Clin. Studies, 1989—, Urology, 1991—; ad hoc reviewer Cancer, 1985—; cons. editor Sexuality and Disability, 1985—; asst. editor Jour. Urology, 1980-89, ad hoc reviewer clin. sect., 1989—, mem. editorial bd. investigative sect., 1989—; assoc. editor Neurourology and Urodynamics, 1982—; contbr. over 510 articles and abstracts to profl. jours. Mem. coun. urology Nat. Kidney Found., Inc.; mem. lectrs. bur. Am. Cancer Soc., 1984—; mem. adv. panel Help for Incontinent People, 1987—; mem. adv. bd. Simon Found., 1987—; mem. med. adv. bd. Institial Cystitis Assn., 1987—; chmn. bladder health coun. Am. Found. Urologic Disease, 1990—; trustee Am. Bd. Urology, 1990—. Served to maj. MC, U.S. Army, 1972-74. Grantee VA, 1974-79, 79, 81, 81-84, 82-85, 85-88, 88-92, Eaton Labs., 1975-76, 78-80, McCabe Rsch. Fund, 1975-82, 87-88, Merrell Nat. Labs., 1979-82, 1980-82, Nat. Kidney Found., 1980-81, NIH, 1980-83, 83-88, 84-87, 87—, Roche Labs., 1981, Smith Kline and French Labs., 1982, 86-88, Eli Lilly Labs., 1986-88, 91, Found. Interstitial Cystitis, 1986-87, 87-88, Sterling Drug Co., 1991. Fellow ACS; mem. AAAS, AMA (cons. com. drug evaluation 1977—), Am. Acad. Clin. Neurophysiology, Am. Assn. Surgery of Trauma, Am. Assn. Clin. Urologists, Am. Assn. Genito-Urinary Surgeons , Am. Fertility Soc., Am. Inst. Ultrasound in Medicine, Am. Soc. Pharmacology and Exptl. Therpeutics, Am. Soc. Andrology, Am. Soc. Clin. Oncology, Am. Urol. Assn. (chmn. home study courses 1982—, rsch. com. 1985—, editorial com mid-Atlantic sect. 1988—), Assn. Acad. Surgery, Can. Urol. Assn., Clin. Soc. Genito-Urinary Surgeons , Ea. Cooperative Oncologic Group, Endourol. Soc., Internat. Continence Soc., Nat. Assn. VA Physicians, N.Y. Acad. Scis., Coll. Physicians Phila., Del. Valley Soc. Transplant Surgeons, John Morgan Soc., Pa. Med. Assn., Pa. Oncologic Soc., Phila. Acad. Surgery, Phila. County Med. Soc., Phila. Profl. Standards Rev. Orgn., Phila. Urologic Soc. (pres. 1990-91), Ravdin-Rhoads Surg. Soc., Urol. Assn. Pa., Radiation Therapy Oncology Group (genitourinary working com. 1980—), Royal Soc. Medicine, Soc. Internat. d'Urologie, Soc. Basic Urologic Rsch., Soc. Sex Therapy and Rsch., Soc. Govt. Svc. Urologists, Soc. Pelvic Surgeons, Soc. Univ. Surgeons, Soc. Univ. Urologists, Soc. Urologic Oncology, Univ. Urologic Forum, Urodynamics Soc. (exec. com. 1980—), Uroespondence Club, Sigma Xi, Alpha Omega Alpha. Home: 609 Robinson Ln Haverford PA 19041-1921 Office: Hosp U Pa 3400 Spruce St-5 Silverstein Philadelphia PA 19104

WEINACHT, RICHARD JAY, mathematician; b. Union City, N.J., Dec. 10, 1931; s. Richard Jacob and Margaret Theresa (Rattiger) W.; Bernice Theresa Benson, Oct. 8, 1955; chdlren: Paul, John, Karen Anne, Richard Joseph, Joseph Benson, Judith Mary. BSCE cum laude, U. Notre Dame, 1953; MSCE, Columbia U., 1955; PhD in Math., U. Md., 1962. Postdoctoral fellow Courant Inst., NYU, 1962-63; asst. prof. math. U. Del., Newark, 1962-66, assoc. prof. math., 1966-74, prof. math., 1974—; vis. assoc. prof. Renseelaer Poly. Inst., Troy, N.Y., 1969-70; vis. prof. Technische Hochschule Darmstadt, Germany, 1976-77, U. Rome, 1984; NSF postdoctoral fellow NSF, 1962-63; Fulbright fellow Fulbright Found., 1976-77; DAAD Fgn. scholar, 1981; fellow U. Del., 1990-91. Contbr. articles to profl. jours. Office: U Del Dept Math Newark DE 19716

WEINBERG, ALVIN MARTIN, physicist; b. Chgo., Apr. 20, 1915; s. J.L. and Emma (Levinson) W.; m. Margaret Despres, June 14, 1940 (dec. 1969); children: David, Richard; m. Gene K. DePersio, Sept. 20, 1974. A.B., U. Chgo., 1935, A.M., 1936, Ph.D., 1939; LL.D., U. Chattanooga, Alfred U.; D.Sc., U. Pacific, Denison U., Kenyon Coll., Worcester Poly. Inst., U. Rochester, Stevens Inst. Tech., Butler U., U. Louisville, Tulane U. Research assoc. math. biophysics U. Chgo., 1939-41, Metall. Lab., 1941-45; joined Oak Ridge Nat. Lab., 1945, dir. physics div., 1947-48, research dir. lab., 1948-55, dir. lab., 1955-74; dir. office Energy Research and Devel., Fed. Energy Office, 1974; dir. Inst. Energy Analysis, Oak Ridge, 1975-85; disting. fellow Oak Ridge Associated Univs., 1985—; mem. Pres.'s Sci. Adv. Com., 1960-62, Pres.'s Medal of Sci. Com. Author: Reflections on Big Science, 1967, (with E.P. Wigner) Physical Theory of Neutron Chain Reactors: Continuing the Nuclear Dialogue, 1985, Nuclear Reactions: Science and Trans-Science, 1992; co-author: The Second Nuclear Era, 1985; co-editor: The Nuclear Connection, 1985, Strategic Defenses and Arms Control, 1987; editor: Eugene Wigner's Collected Works on Nuclear Energy, 1992. Recipient Atoms for Peace award, 1960, E.O. Lawrence award, 1960, U. Chgo. Alumni medal, 1966, Heinrich Hertz award, 1975, N.Y. Acad. Scis. award, 1976, Enrico Fermi award, 1980, Harvey prize, 1982, Eugene Wigner award in reactor physics, 1992. Mem. Nat. Acad. Scis. (applied sci. sect.), Am. Nuclear Soc. (pres. 1959-60), Nat. Acad. Engring., Am. Acad. Arts and Scis., Am. Philos. Soc., Royal Netherlands Acad. Sci. (fgn. asso.). Home: 111 Moylan Ln Oak Ridge TN 37830-5351 Office: Oak Ridge Associated Univs PO Box 117 Oak Ridge TN 37830

WEINBERG, ROBERT ALLAN, biochemist, educator; b. Pitts., Nov. 11, 1942; s. Fritz E. and Lore (Reichhardt) W.; m. Amy Shulman, Nov. 19, 1976; children—Aron, Leah Rosa. S.B., MIT, 1964, Ph.D, 1969; Ph.D. (hon.), Northwestern U., 1984. Instr. Stillman Coll., Tuscaloosa, Ala., 1965-66; research fellow Weizmann Inst., Rehovoth, Israel, 1969-70, Salk Inst., LaJolla, Calif., 1970-72; from asst. prof. to prof. biochemistry MIT, Cambridge, 1973—; mem. Whitehead Inst., Cambridge, 1984—. Contbr. articles to profl. jours. Recipient Bristol Myers award, 1984, Armand Hammer award, 1984, Gairdner Found. Internat. award, 1992. Mem. Nat. Acad. Sci. (sci. award 1984). Avocations: genealogy, house building. Office: Whitehead Inst 9 Cambridge Ctr Cambridge MA 02142-1479

WEINBERG, SAMUEL, pediatric dermatologist; b. N.Y.C., Jan. 12, 1926; s. Harry and Rose (Stecher) W.; m. Pearl Oksner, Dec. 12, 1948; children: Ronald Andrew, Robin Ann. MB, Chgo. Med. Sch., 1947, MD, 1948. Clin. asst. prof. dermatology to prof. dermatology Med. Ctr. NYU, N.Y.C., 1961—. Author: Color Atlas of Pediatric Dermatology, 1975, 2d rev. edit., 1990. Capt. USAF, 1951-53. Fellow Am. Acad. Pediatrics, ACP, Am. Acad. Dermatology (chmn. pediatric dermatology sect. 1978-81, com. dermatolog. subspecialties 1984-86, task force pediatric dermatology 1981-84); mem. Soc. Pediatric Dermatology (charter, pres. 1980-81). Office: NYU Med Ctr 530 1st Ave New York NY 10016-6402

WEINBERG, STEVEN, physics educator; b. N.Y.C., NY, May 3, 1933; s. Fred and Eva (Israel) W.; m. Louise Goldwasser, July 6, 1954; 1 child, Elizabeth. BA, Cornell U., 1954; postgrad., Copenhagen Inst. Theoretical Physics, 1954-55; PhD, Princeton U., 1957; AM (hon.), Harvard U., 1973; ScD (hon.), Knox Coll., 1978, U. Chgo., 1978, U. Rochester, 1979, Yale U., 1979, CUNY, 1980, Clark U., 1982, Dartmouth Coll., 1984, Columbia U., 1990, U. Salamanca, 1992, U. Padua, 1992; PhD (hon.), Weizmann Inst., 1985; DLitt (hon.), Washington Coll., 1985. Rsch. assoc., instr. Columbia U., 1957-59; rsch. physicist Lawrence Radiation Lab., Berkeley, Calif., 1959-60; mem. faculty U. Calif., Berkeley, 1960-69, prof. physics, 1964-69; vis. prof. MIT, 1967-69, prof. physics, 1969-73; Higgins prof. physics Harvard U., 1973-83; sr. scientist Smithsonian Astrophys. Lab., 1973-83; Josey prof. sci. U. Tex., Austin, 1982—; sr. cons. Smithsonian Astrophys. Obs., 1983—; cons. Inst. Def. Analyses, Washington, 1960-73, ACDA, 1973; Sloan fellow, 1961-65; chair in physics Coll. de France, 1971; mem. Pres.'s Com. on Nat. Medal of Sci., 1979-82, Coun. of Scholars, Library of Congress, 1983-85; sr. adv. La Jolla Inst.; mem. Com. on Internat. Security and Arms Control, NRC, 1981, Bd. on Physics & Astronomy, 1989-90; dir. Jerusalem Winter Sch. Theoretical Physics, 1983—; mem. adv. coun. Tex. Superconducting Supercollider High Energy Rsch. Facility, 1987; Loeb lectr. in physics Harvard U., 1966-67, Morris Loeb vis. prof. physics, 1983—; Richtmeyer lectr., 1974; Scott lectr. Cavendish Lab., 1975; Silliman lectr. Yale U., 1977; Lauritsen Meml. lectr. Calif. Inst. Tech., 1979; Bethe lectr. Cornell U., 1979; de Shalit lectr. Weizman Inst., 1979; Cherwell-Simon lectr. Oxford U., 1983; Bampton lectr. Columbia U., 1983; Einstein lectr. Israel Acad. Arts and Scis., 1984; Hilldale lectr. U. Wis., 1985; Clark lectr. U. Tex., Dallas, 1986; Dirac lectr. U. Cambridge, 1986; Klein lectr. U. Stockholm, 1989; Sloan fellow, 1961-65; mem. Supercollider Sci. Policy Com., 1989—. Author: Gravitation and Cosmology: Principles and Application of the General Theory of Relativity, 1972, The First Three Minutes: A Modern View of the Origin of the Universe, 1977, The Discovery of Subatomic Particles, 1982; co-author (with R. Feynman) Elementary Particles and the Laws of Physics, 1987, Dreams of a Final Theory, 1992; rsch. and publs. on elementary particles, quantum field theory, cosmology; co-editor Cambridge U. Press, monographs on math. physics; mem. adv. bd. Issues in Sci. and Tech., 1984-87; mem. sci. book com. Sloan Found, 1985—; mem. editorial bd. Jour. Math. Physics, 1986-88; mem. bd. editors Daedalus, 1990—; mem. bd. assoc. editors Nuclear Physics B. Bd. advisors Santa Barbara Inst. Theoretical Physics, 1983-86; bd. overseers SSC Accelerator, 1984-86; bd. dirs. Headliners Found., 1993—. Recipient J. Robert Oppenheimer meml. prize, 1973, Dannie Heineman prize in math. physics, 1977, Am. Inst. Physics-U.S. Steel Found. sci. writing award, 1977, Nobel prize in physics, 1979, Elliott Cresson medal Franklin Inst., 1979, Madison medal Princeton U., 1991, Nat. Medal of Sci. NSF, 1991. Mem. Am. Acad. Arts and Scis. (past councilor), Am. Phys. Soc. (past councilor at large, panel on faculty positions com. on status of women in physics), NAS (supercollider site evaluation com. 1987-88), Einstein Archives (adv. bd. 1988—), Internat. Astron. Union, Coun. Fgn. Rels., Am. Philos. Soc., Royal Soc. London (fgn. mem.), Am. Mediaeval Acad., History of Sci. Soc., Philos. Soc. Tex., Phi Beta Kappa. Clubs: Saturday (Boston); Headliners, Tuesday (Austin); Cambridge Sci. Soc.

WEINBERGER, FRANK, information systems advisor; b. Chgo., Sept. 18, 1926; s. Rudolph and Elaine (Kellner) W.; m. Beatrice Natalie Fixler, June 27, 1953; children: Alan J., Bruce I. BSEE, Ill. Inst. of Tech., 1951; MBA, Northwestern U., Evanston, 1959. Registered profl. engr., Ill, Calif. Engr. Admiral Corp., Chgo., 1951-53; sr. engr. Cook Rsch., Chgo., 1953-59; mem. tech. staff Rockwell Internat., Downey, Calif., 1959-80, info. systems advisor, 1980—. Pres. Temple Israel, Long Beach, Calif., 1985-87, bd. dirs. 1973-85. With USN, 1944-46. Mem. Assn. for Computer Machinery. Democrat. Jewish. Avocation: microcomputers. Home: 3231 Yellowtail Dr Los Alamitos CA 90720 Office: Rockwell Internat 12214 Lakewood Blvd Downey CA 90241

WEINBERGER, MYRON HILMAR, medical educator; b. Cin., Sept. 21, 1937; s. Samuel and Helen Eleanor (Price) W.; m. Myrna M. Rosenberg, June 12, 1960; children: Howard David, Steven Neal, Debra Ellen. BS, Ind. U., Bloomington, 1959, MD, 1963. Intern Ind. U. Med. Ctr., Indpls., 1963-64, resident in internal medicine, 1964-66, asst. prof. medicine, 1969-73, assoc. prof., 1973-76, prof., 1976—, dir. Hypertension Research Ctr., 1981—; USPHS trainee in endocrinology and metabolism Stanford U. Med. Ctr., Calif., 1966-68, USPHS spl. fellow in hypertension, 1968-69. Contbr. articles to profl. jours. Fellow ACP, Am. Coll. Cardiology, Am. Coll. Nutrition, Am. Soc. for Clin. Pharmacology and Therapeutics; mem. AAAS, Am. Fedn. Clin. Research, AMA, Am. Heart Assn., Am. Soc. Nephrology, Internat. Soc. Nephrology, Central Soc. Clin. Research, Endocrine Soc., Internat. Soc. Hypertension, Soc. for Exptl. Biology and Medicine. Home: 135 Bow Ln Indianapolis IN 46220-1023 Office: Ind U Hypertension Research Ctr 541 Clinical Dr Indianapolis IN 46202-5111

WEINBRENNER, GEORGE RYAN, aeronautical engineer; b. Detroit, June 10, 1917; s. George Penbrook and Helen Mercedes (Ryan) W.; BS, M.I.T., 1940, MS, 1941; AMP, Harvard U., 1966; m. Billie Marjorie Elwood, May 2, 1955. Commd. 2d lt. USAAF, 1939, advanced through grades to col., 1949; def. attaché Am. embassy, Prague, Czechoslavakia, 1958-61; dep. chief staff intelligence Air Force Systems Command, Washington, 1962-68; comdr. fgn. tech. div. U.S. Air Force, Wright-Patterson AFB, Ohio, 1968-74; comdr. Brooks AFB, Tex., 1974-75; ret., 1975; exec. v.p. B.C. Wills & Co., Inc., Reno, Nev., 1975-84; lectr. Sch. Aerospace Medicine Brooks AFB, Tex., 1975-84; chmn. bd. Hispaño-Technica S.A. Inc., San Antonio, 1977—; adv. dir. Plaza Nat. Bank, San Antonio; cons. Def. Dept., 1981, Dept. Air Force, 1975-84. Decorated D.S.M., Legion of Merit, Bronze Star, Air medal, Purple Heart; Ordre National du Merite, Medaille de la Resistance, Croix de Guerre (France). Fellow AIAA (asso.); mem. San Antonio C. of C., World Affairs Council, Air Force Assn. (exec. sec. Tex. 1976-82), Assn. Former Intelligence Officers (nat. dir.), Air Force Hist. Found. (dir.), U.S. Strategic Inst., Nat. Mil. Intelligence Assn., Tex. Aerospace & Nat. Def. Tech. Devel. Coun., Am. Astronautical Soc., Aerospace Ednl. Found. (trustee), Disabled Am. Vets. (life), Mil. Order World Wars, Am. Legion, Assn. Old Crows, Kappa Sigma. Roman Catholic. Clubs: Army-Navy (Washington). Home: 7400 Crestway Dr Apt 903 San Antonio TX 78239-3094 Office: PO Box 18484 San Antonio TX 78218-0484

WEINER, ANDREW MARC, electrical engineering educator, laser researcher; b. Boston, July 25, 1958; s. Jason and Geraldine Hannah (Aronson) W.; m. Brenda Joyce Garland, Apr. 1, 1989. SB in Elec. Engring., MIT, 1979, SM, 1981, ScD, 1984. Mem. tech. staff Bellcore, Red Bank, N.J., 1984-89, dist. mgr., 1989-92; prof. elec. engring. Purdue U., West Lafayette, Ind., 1992—; assoc. editor IEEE Jour. Quantum Electronics, 1988—; adv. editor Optics Letters, 1989—. Fannie and John Hertz Found. grad. fellow, 1979-84. Fellow Optical Soc. Am. (tech. coun. 1988-91, Adolph Lomb award 1990); mem. IEEE (sr.; Traveling Lectr. award Lasers and Electro-optics Soc. 1988-89). Avocations include invention of techniques for manipulating the shapes of ultrashort laser pulses; pioneering studies of ultrafast nonlinear optics. Office: Purdue U Sch Elec Engring West Lafayette IN 47907-1285

WEINER, GEORGE JAY, internist; b. Plainview, N.Y., Mar. 1, 1956; m. Teresa Emily Wilhelm, July 30, 1983; children: Aaron, Miriam, Nathan. BA, Johns Hopkins U., 1978; MD, Ohio State U., 1981. Resident in internal medicine Med. Coll. Ohio, Toledo, 1981-85; hematology/oncology fellow U. Mich., Ann Arbor, 1985-89; asst. prof. medicine U. Iowa, Iowa City, 1989—. Achievements include prin. investigation of monoclonal antibody therapy of malignancy. Office: Univ of Iowa C32GH Iowa City IA 52242

WEINER, HAROLD M., radiologist; b. Phila., Apr. 30, 1937; s. Louis A. and Anna (Becker) W.; divorced; children: Lori, Julie. BA summa cum laude, Temple U., 1958, MD with hons., 1962. Diplomate Am. Bd. Radiology, Nat. Bd. Med. Examiners. Intern Polyclinic Hosp., Harrisburg, Pa., 1962-63; resident in pathology VA Hosp., Phila., 1965, resident in radiology, 1966-68; staff radiologist Sacred Heart Hosp., Norristown, Pa., 1969-73, chief radiologist, 1973—; chmn. infectious disease com., 1976-77, patient care com., 1977-80, med. audit com., 1981-87; sec.-treas. med. staff, 1982-85. Presenter numerous papers med. meetings. Served to capt. M.C. USAF, 1963-65. Decorated Legion of Honor Chapel of Four Chaplains. Mem. AMA, Am. Heart Assn. (coun. on cardiovascular radiology), Pa. Radiologic Soc., Pa. Med. Soc., Montgomery County Med. Soc. (chmn. continuing edn. com. 1980-84, bd. dirs. 1983-89, v.p. 1983-84, pres. 1985-86, del. to PA Med. Soc., 1979—, chair nominationg com. 1988-89), Phila. Roentgen Ray Soc., Am. Coll. Radiology (alt. councilor 1983-85), N.Y. Acad. Sci., Soc. for Magnetic Resonance Imaging, Radiolog. Soc. N.Am., Alpha Omega Alpha. Avocations: photography, chess, art, bridge, music. Office: Sacred Heart Hosp 1430 Dekalb St Norristown PA 19401-3406

WEINER, MELVIN MILTON, electrical engineer; b. Boston, Dec. 5, 1933; s. William Wolf and Kate (Berkowitz) W.; m. Sandra Roseman, Aug. 16, 1964; children: Steven William, Robert Jay. BS, MIT, 1956, MS, 1956. Coop. student Philco Corp., Phila., 1953-56; project engr. Chu Assocs., Littleton, Mass., 1956-59; cons. elec. engr. Brookline, Mass., 1959-76; sr. engr. EG&G Inc., Boston, 1963-65; sr. staff engr. Honeywell Radiation Ctr., Boston, 1965-67; staff engr. Am. Sci. and Engring. Inc., Cambridge, Mass., 1967-68; prin. rsch. engr. Avco Everett (Mass.) Rsch. Lab., 1971-78; mem. tech. staff Mitre Corp., Bedford, Mass., 1978—. Founder, chmn. Motor Vehicle Safety Group, 1962—. Mem. IEEE, Am. Phys. Soc., Solar Energy Soc., Optical Soc. Am., Sigma Xi, Eta Kappa Nu (chpt. pres. 1965-68, nat. dir. 1969-71). Achievements include books and patents in field; research in electromagnetic wave propagation, antennas, communication and radar systems. Home: 56 Marcellus Dr Newton Center MA 02159

WEINER, MYRA LEE, toxicologist; b. Princeton, N.J., Sept. 26, 1944; d. Lawrence and Mollie (Doctor) Lee; m. Paul Harvey Weiner, June 15, 1969; children: Lawrence, Rebecca, Sarah. BA, Stern Coll. for Women, 1966; PhD, SUNY, Bklyn., 1971. Diplomate Am. Bd. Toxicology. Postdoctoral fellow Rsch. Inst. Pharm. Scis., U. Miss. Sch. Pharmacy, Oxford, 1971-72, rsch. assoc., 1972-73; adj. prof. Mercer County C.C., Trenton, N.J., 1975-77; toxicologist FMC Corp., Princeton, N.J., 1980-89, sr. rsch. toxicologist, 1989-91, mgr. toxicology programs, 1991—; mem. safety com. Internat. Pharm. Excipients Counsel, 1991—, hydrogen peroxide task force European Chem. Industry Ecology and Toxicology Ctr., Brussels, 1990-92, tributyl phosphate task force Synthetic Organic Chems. Mfrs. Assn., Washington, 1989—. Author: (chpt.) Lithium in Biology and Medicine, 1991, Advances in Chitin and Chitosan, 1992; co-author monograph, 1992; contbr. articles to profl. jours. Mem. AAAS, Soc. Toxicology, Genetic Toxicology Soc., Chem. Mfrs. Assn. (chmn. aryl phosphates toxicology subcom. 1992—), Sigma Xi. Office: FMC Corp PO Box 8 Princeton NJ 08543

WEINER, NORMAN, pharmacology educator; b. Rochester, N.Y., July 13, 1928; m. Diana Elaine Weiner, 1955; children: Steven, David, Jeffrey, Gareth, Eric. BS, U. Mich., 1949; MD, Harvard U., 1953. Diplomate Am. Bd. Med. Examiners. Intern 2d and 4th Harvard Med. Svc., Boston City Hosp., 1953-54; rsch. med. officer USAF, 1954-56; instr. dept. pharmacology-biochemistry Sch. of Aviation Medicine, San Antonio, 1954-56; from instr. to asst. prof. Harvard Med. Sch., Boston, 1956-67; prof. pharmacology U. Colo. Health Sci. Ctr., Denver, 1967—; chmn. dept. pharmacology U. Colo. Health Sci., Denver, 1967-87; interim dean U. Colo. Sch. Medicine, 1983-84; div. v.p. Abbott Labs., Abbott Park, Ill., 1985-87. ecipient rsch. career devel. award USPHS, 1963, award Kaiser Permanente, 1974, 81; spl. fellow USPHS, London, 1961-62. . Spl. fellow USPHS, London, 1961-62; recipient USPHS Rsch. Career Devel. award, 1963, Kaiser Permanente award, 1974, 81, Allan D. Bass lectureship Vanderbilt U. Sch. Medicine, 1983, Pfizer lectureship Tex. Coll. Osteopathic Medicine, Ft. Worth, 1985; disting. Volwiler Rsch. fellow Abbott Labs., 1988; named disting. prof U. Colo., 1989. Mem. AAAS, Am. Soc. for Pharmacology and Exptl. Therapeutics (Otto Krayer award 1985), N.Y. Acad. Scis., Assn. Med. Sch. Pharmacology, Am. Soc. Neurochemistry, Western Pharmacology Soc., Am. Coll. Neuropsychopharmacology, Soc. Neurosci., Biochem. Soc., Internat. Brain Rsch. Orgn., Internat. Soc. Neurochemistry, Rsch. Soc. on Alcoholism, Phi Beta Kappa, Sigma Xi, Alpha Omega Alpha, Phi Eta Sigma, Phi Lambda Upsilon, Phi Kappa Phi. Office: U Colo Health Sci Ctr Pharmacology Dept 4200 E 9th Ave Rm C236 Denver CO 80261-0001

WEINER, RICHARD LENARD, hospital administrator, educator, pediatrician; b. N.Y.C., May 23, 1951; s. Irving and Martha E. (Pell) W. AB in Biology, NYU, 1972; MD, Albert Einstein Coll. Medicine, 1975. Diplomate Am. Bd. Pediatrics. Instr. in pediatrics Albert Einstein Coll. of Medicine, Bronx, 1978-80; pediatrician New Rochelle (N.Y.) Hosp. Med. Ctr., 1978-80; asst. dir. pediatrics Hosp. of Albert Einstein Coll. of Medicine, Bronx, 1980-86; assoc. dir. peditarics Einstein-Weiler Hosp. MMC, Bronx, 1986—; dir. pediatric evaluation unit Einstein-Weiler Hosp., Bronx, 1990—; coord. pediatric med. edn. New Rochelle Hosp., 1978-80; chmn. pvt. practice governance coun. dept. pediatrics Albert Einstein Coll. of Medicine, 1988-91. Mem. editorial adv. bd. Primary Care/Emergency Decisions, 1984-88; contbr. articles to Jour. Pediatrics, Pediatrics, Pediatric Infectious Diseases, Emergency Decisions. Recipient Physician's Recognition award AMA, 1982, 86. Mem. Phi Beta Kappa, Beta Lambda Sigma. Jewish. Office: Einstein Hosp 1825 Eastchester Rd Bronx NY 10461-2301

WEINER, WILLIAM JERROLD, neurologist, educator; b. Chgo., June 28, 1945; s. Leonard and Maxine Adrianne (Rappaport) W.; m. Susan Rosenband, Sept. 10, 1967; children: Monica, Miriam. BS, U. Ill., Urbana, 1966; MD, U. Ill., Chgo., 1969. Diplomate Am. Bd. Psychiatry and Neurology. Intern Presbyn. St. Luke's Med. Ctr., Chgo., 1969-70, neurology resident, 1971-73; asst. prof. neurology U. Chgo., 1975-77; assoc. prof. dept. neurol. scis., dept. pharmacology Rush U., Chgo., 1977-83; prof. neurology U. Miami, Fla., 1983—; dir. Movement Disorders Ctr. U. Miami Sch. Medicine, 1983—. Author: A Textbook of Clinical Neuropharmacology, 1981, Movement Disorders: A Comprehensive Survey, 1989; editor: Neurology for the Non-Neurologist, 1981, 2d edit., 1989, Emergent and Urgent Neurology, 1992, Drug-Induced Movement Disorders, 1992; contbr. numerous articles to profl. publs. Lt. comdr. USN, 1973-75. Fellow Am. Acad. Neurology; mem. AAAS, Movement Disorders Soc., Am. Neurol. Assn., Phi Beta Kappa, Alpha Omega Alpha. Achievements include research and publication on movement disorders including Parkinson's disease, Huntington's disease, Tourette's syndrome, dystonia and neuropharmacology. Office: Univ Miami Sch Medicine 1501 NW 9th Ave Miami FL 33136

WEINGARTEN, JOSEPH LEONARD, aerospace engineer; b. N.Y.C., June 5, 1944; s. Herman H. and Irene Jane (Binzer) W.; 1 child, Toby. B of Mech. Engring., NYU, 1966; student, Air War Coll., 1976. Chief engr. air transportability Test Loading Agy., Wright-Patterson AFB, Ohio, 1972-74; project engr. dept. engring. USAF, Wright-Patterson AFB, 1966-72, sr. project engr. dept. engring., 1974-76, planning and project engr. dept. engring., 1976-81, chief mgmt. ops. dept. engring., 1981-83, sr. tech. planner dept. engring., 1983-92; tech. asst. DCS engring. and tech. mgmt. Air Force Material Command, Wright-Patterson AFB, 1992-93; founder Huffman Wright Inst., 1993—; CEO Weingarten Gallery, Dayton, Ohio, 1967—; sec., treas., bd. dirs. Ohio Designer Craftsmen, Columbus; sec. Ohio Designer Craftsmen Enterprise, Columbus, 1982—; chmn. continuing edn. design dept. Affiliate Socs. Coun., Dayton, 1971-74, chmn. edn. coord. com. Kettering Inst., Wright State U., 1974-76, chmn. scientist and engr. awards panel, 1990-91, mem., 1992—. Contbr. articles on systems engring. to Aeronautical Systems div. Mech. Engring. Jour. (1st place award nat. contest 1970), Aerospace Industries Assn., Proceedings Fourth Intersoc. Conf. on Transp., Air Force Systems Command, USAF Spl. Purpose Report, Gems and Minerals, Ceramics Monthly, The Crafts Report, Macintosh Software. Scoutmaster Troop 81 Boy Scouts Am., Kettering, Ohio, 1985-91, com. mem., 1991—, dist. chmn. Sequioa Dist. Miami Valley Coun., 1991—; pres. Friends of Montessori Sch. South Dayton, 1978—. Capt. USAF, 1967-71. Named Eagle Scout Boy Scouts Am., 1962; recipient Disting. Eagle award Boy Scouts Am., 1992. Mem. AIAA (sr. mem., air transport systems tech. com. 1976-78, 80-82, Lawrence Sperry award 1977), ASME (sr. mem.), Am. Nat. Standards Inst. (materials handling 5 com. 1968-70), Soc. Automotive Engrs. (aircraft ground support equiment com. 1969-75). Achievements include patents for expendable air cargo pallet, mail container, collapsible air cargo container, process for reinforcing extruded articles, process for large scale extrusions, air flotation cargo handling system, integral aircraft barrier net, load distributive cargo platform, laminated plastic packaging material, computer printer paper support, and investment casting mold base; developments include 3g cargo restraint criteria used worldwide on aircraft/spacecraft/shuttles, rope extraction system for C-5A, system for large scale structural plastics extruxions, advanced planning documents for Air Force, report in new type of DOD procurement system; other achievements include the design and creation of jewelry sold in museums and retail stores.

WEINGAST, MARVIN, laboratory director; b. Bklyn., Jan. 1, 1943; s. Abe and Rose (Altein) W. BS, L.I. U., 1967, MS, 1971; postgrad., Poly. Inst., 1967-68. Analytic and pollution chemist Amerada Hess Corp., Pt. Reading, N.J., 1969-73; asst. lab. dir. Chem. Constrn., North Brunswick, N.J., 1973-74; dir. Indsl. Hygiene Lab. Nat. Starch and Chemical, Bridgewater, N.J., 1974—; grant com. mem. Ctr. for Hazardous and Toxic Substance Mgmt., Newark, 1988—; mem. Sourland Regional Citizens Planning Coun., Neshanic, N.J., 1989—. Contbr. to book: Small Business Programs, 1980; contbr. articles to profl. jours. Recipient Chemistry Dept. award L.I. U., 1967, Teaching fellowship Poly. Inst., 1967, L.I. U., 1968. Mem. MENSA, Am. Chem. Soc., Am. Conf. Chem. Labeling, Soc. Toxicology. Achievements include development of improved system for identification of hazardous chemicals; organization of first global monitoring of indsl. workers to hazardous workplace chemicals. Office: Nat Starch & Chem Co 10 Finderne Ave Bridgewater NJ 08807

WEINGOLD, HARRIS D., aerospace engineer. BS in Aerospace Engring., NYU, 1960, MS in Aerospace Engring., 1961. With sci. analysis group Pratt & Whitney, Hartford, Conn., 1965-74, with compressor tech. and rsch. group, 1974—. Mem. ASME (Gas Turbine award 1990), AIAA. Office: Pratt & Whitney 400 Main St Hartford CT 06108*

WEINLANDER, MAX MARTIN, retired psychologist; b. Ann Arbor, Mich., Sept. 9, 1917; s. Paul and Emma Carol (Lindemann) W.; BA, Ea. Mich. Coll., 1940; MA, U. Mich., 1942, PhD, 1955; M.A., Wayne U., 1951; m. Albertina Adelheit Abrams, June 4, 1946; children: Bruce, Annette. Psychometrist, VA Hosp., Dearborn, Mich., 1947-51; sr. staff psychologist Ohio Div. Corrections, London, 1954-55; lectr. Dayton and Piqua Centers, Miami U., Oxford, Ohio, 1955-62; chief clin. psychologist Child Guidance Clinic, Springfield, Ohio, 1956-61, acting dir., 1961-65; clin. psychologist VA Center, Dayton, Ohio, 1964-79; cons. Ohio Divsn. Mental Hygiene; summer guest prof. Miami U., 1957, 58, Wittenberg U., 1958; adj. prof. Wright State U., Dayton, 1975-76; cons. State Ohio Bur. Vocat. Rehab., Oesterlen Home Emotionally Disturbed Children. Pres. Clark County Mental Health Assn., 1960, Clark County Health and Welfare Club, 1961; mem. Community Welfare Coun. Clark County, 1964; chmn. Comprehensive Mental Health Planning Com. Clark County, 1964; trustee United Appeals Fund, 1960. Mem. citizens adv. coun. Columbus Psychiat. Inst., Ohio State U. Served as sgt. AUS, 1942-46. Fellow Ohio Psychol. Assn. (chmn. com. on utilization of pscyhologists; treas., exec. bd. 1968-71); mem. Am. Psychol. Assn., Ohio Psychol Assn., Mich. Psychol. Assn., DAV, U. Mich. Pres. Club, Pi Kappa Delta, Pi Gamma Mu, Phi Delta Kappa. Republican. Lutheran. Lodge: Kiwanis. Contbr. 18 articles to psychology jours. Home: 17185 Valley Dr Big Rapids MI 49307-9523

WEINRICH, JAMES DONALD, psychobiologist, educator; b. Cleve., July 2, 1950; s. Albert James and Helen (Lautz) W. AB, Princeton U., 1972; PhD, Harvard U., 1976. Postdoctoral fellow, then instr. Johns Hopkins U., Balt., 1980-82; rsch. assoc., then asst. rsch. prof. psychiatry Boston U., 1983-87; asst. rsch. psychobiologist, project mgr. U. Calif., San Diego, 1987-89, asst. rsch. psychobiologist, ctr. mgr., 1989-91, sr. investigator sexology, 1991—; bd. dirs. Found. Sci. Study of Sexuality, Mt. Vernon, Iowa. Author: Sexual Landscapes, 1987; co-editor: Homosexuality: Social, Psychological and Biological Issues, 1982, Homosexuality: Research Implications for Public Policy, 1991. Mem. Internat. Acad. Sex Rsch., Soc. for Sci. Study of Sex (Hugo Beigel award 1987), Am. Coll. Sexologists (cert.), Phi Beta Kappa. Avocations: computers, photography. Office: Univ Calif San Diego 2760 5th Ave Rm 200 San Diego CA 92103-6325

WEINRICH, STANLEY DAVID, chemical engineer; b. Bklyn., Aug. 21, 1945; s. Nathan and Freda Rebecca (Lew) W.; m. Arlene Gail Goldfarb, Oct. 4, 1980; children: Neal Fredric, Lauren Jamie. BS, The Cooper Union, 1966; PhD, Princeton U., 1971; MBA, Pace U., 1981. Rsch. engr. E.I. duPont de Nemours & Co., Wilmington, Del., 1971-72; asst. prof. Worcester (Mass.) Poly. Inst., 1972-76; project mgr. Exxon Rsch. & Engring. Co., Florham Park, N.J., 1976-77; mgr. energy tech. Internat. Paper Co., N.Y.C., 1977-85; mgr. tech. svcs. Internat. Paper Co., Pineville, La., 1985-87; dir. statis. quality control Internat. Paper Co., Memphis, 1987-89, mgr. mfg. tech. svcs., 1989-91; dir. environ. svcs. compliance Internat. Paper Co., Binghamton, N.Y., 1991-92; mgr. environ. health, safety Internat. Paper Co., Memphis, 1992—. Mem. Sigma Xi, Tau Beta Pi.

WEINSAFT, PAUL PHINEAS, retired physician, administrator; b. Zbaraz, Austria, July 18, 1908; came to U.S., 1941; s. Lipa and Basia (Landesberg) W.; m. Rachel Rosenfeld, Mar. 21, 1931. BS, Sorbonne, Paris, 1931, MD, U. Paris Faculty of Medicine, 1935. Diplomate Am. Bd. Internal Medicine. Asst. surgeon Soldiers' Home, Chelsea, Mass., 1941-45; physician pvt. practice Winthrop, Mass., 1945-51; staff physician VA Hosp., Martinsburg, W.Va., 1952-55; asst. dir. profl. svc. VA Hosp., N.Y.C., 1955-61; med. dir. Bklyn. Hebrew Home & Hosp., 1961-64; chief geriatric sect. Coney Island Hosp., Bklyn., 1964-67; med. dir. Met. Jewish Geriatric Ctr., Bklyn., 1967-79; ret., 1979—; cons. in field. Sci. and med. editor The Jewish Forward, N.Y.C.; contbr. numerous articles to profl. jours. Fellow ACP, N.Y. Acad. Medicine; mem. AMA, N.Y. State and County Med. Soc., Mass. Med. Soc., Am. Coll. Cardiology, N.Y. Acad. Scis. Democrat. Jewish. Avocation: linguistics. Home: 205 W 54th St New York NY 10019-5518

WEINSTEIN, DAVID IRA, industrial chemist; b. Phila., Oct. 7, 1953; s. Sidney Weinstein and Naomi June (Altman) Habib; m. Mindy Baylinson, Oct. 31, 1982; 1 child, Seth Andrew. BS, Villanova U., 1975, PhD, 1982. Asst. prof. Rosemont (Pa.) Coll., 1983-84; postdoctoral fellow Army R & D Ctr. NRC, Dover, N.J., 1984-85; group leader Nalco Chem. Co., Naperville, Ill., 1985—. Contbr. articles to profl. jours. Tech. program coord. explorer program Boy Scouts Am., Naperville, 1989—. Mem. ASTM (chmn. task group 1986-91), TAPPI. Achievements include development of thermogravimetric methods to study the burn-off of rolling oils and predict their behavior in anneal furnaces; research in thermal stability of polynitrobishomocubanes; used thermogravimetry to elucidate the deuterium kinetic isotope effect in the thermal decomposition of trinitro-triazacyclooctane and trinitro-triazacyclohexane. Office: Nalco Chem Co 1 Nalco Ctr Naperville IL 60563-1198

WEINSTEIN, GEORGE WILLIAM, ophthalmologist, educator; b. East Orange, N.J., Jan. 26, 1935; s. Henry J. and Irma C. (Klein) W.; m. Sheila Valerie Wohlreich, June 20, 1957; children: Bruce David, Elizabeth Joyce, Rachel Andrea. AB, U. Pa., 1955; MD, SUNY, Bklyn., 1959. Diplomate Am. Bd. Ophthalmology (bd. dirs. 1981—). Intern then resident in ophthalmology Kings County Hosp., Bklyn., 1959-63; asst. prof. ophthalmology Johns Hopkins U., Balt., 1967-70; head ophthalmology dept. U. Tex., San Antonio, 1970-80; prof., Jane McDermott Shott chmn. W.Va. U., Morgantown, 1980—. Author: Key Facts in Ophthalmology, 1984; editor: Open Angle Glaucoma, 1986; editor Ophthalmic Surgery jour., 1971-81, Current Opinion in Ophthalmology jour., 1988—; contbr. articles to profl. jours. Served to lt. comdr. USPHS, 1963-65. Sr. Internat. fellow Fogarty Internat. Ctr. NIH, 1987. Mem. ACS (bd. govs. 1983-85, bd. regents 1987-92), Assn. Univ. Profs. of Ophthalmology (pres. 1986-87), Am. Acad. Ophthalmology (bd. dirs. 1980-92, chmn. long range planning com. 1986-89, pub. and profl. sec. 1983-89, pres.-elect 1990, pres. 1991, Honor award, Sr. Honor award), Alpha Omega Alpha (faculty 1987), Am. Coll. Surgeons (bd. govs. 1972-75, bd. regents 1988-92), Am. Ophthalmology Soc. (coun. 1992—). Jewish. Avocations: music, photography, tennis, basketball. Home: 28 Lakeview Dr Morgantown WV 26505-9275 Office: W Va U Coll Medicine Dept Ophthalmology Morgantown WV 26506

WEINSTEIN, HERBERT, chemical engineer, educator; b. Bklyn., Mar. 10, 1933; s. Abraham and Pauline (Feldman) W.; m. Judith Cooper, Apr. 6, 1957; children: Michael Howard, Edward Marc, Ellen Rachel. B.Engring. in Chem. Engring, Coll. City N.Y., 1955; M.S. in Chem. Engring, Purdue U., 1957; Ph.D., Case Inst. Tech., 1963. Staff mem. Los Alamos Sci. Lab., 1956-58; research engr. NASA Lewis Research Center, Cleve., 1959-63; asst. prof. chem. engring. Ill. Inst. Tech., 1963-66, assoc. prof., 1966-72, prof., 1972-77; dir. Center for Biomed. Engring., 1973-77; prof. CUNY, 1977—; Herbert G. Kayser prof. of chem. engring., 1987—; vis. rsch. assoc., mem. Med. Rsch. Inst. Michael Reese Hosp. and Med. Ctr., Chgo., 1965-77; vis. prof. mech. engring. Technion-Israel Inst. Tech., 1972-73; vis. prof. biomed. engring. Rush Med. Coll., Chgo., 1973-76; summer prof. Exxon Rsch. and Engring. Co., annually, 1981-92; Lady Davis vis. prof. Technion-Israel Inst. Tech., 1985; cons. to industry, rsch. labs. Mem. Am. Inst. Chem. Engrs., Sigma Xi. Jewish. Research and publs. on fluidization, chem. reactor engring., fluid mechanics, biomed. engring. Office: CUNY Dept Chem Engring New York NY 10031

WEINSTEIN, I. BERNARD, physician; b. Madison, Wis., Sept. 9, 1930; s. Max and Frieda (Blachman) W.; m. Joan Anker, Dec. 21, 1952; children: Tamara, Claudia, Matthew. BS, U. Wis., 1952, MD, 1955, hon. degree, 1992. Intern, then resident in medicine Montefiore Hosp., N.Y.C., 1955-57; clin. asso. NIH, 1957-59; research asso. Harvard U. Med. Sch., 1959-60, MIT, 1960-61; mem. faculty Columbia U. Coll. Phys. and Surg., 1961—, prof. medicine, pub. health and genetics, 1973—; dir. Columbia-Presbyn. Cancer Ctr., 1985—; attending physician Presbyn. Hosp., N.Y.C.; cons. Nat. Cancer Inst., Internat. Agy. Rsch. Cancer. Author rsch. papers on biochemistry, molecular biology, cancer biology, chem. carcinogenesis. Served with USPHS, 1957-59. Recipient Meltzer medal Soc. Exptl. Biology and Medicine, Alumni award U. Wis. Med. Sch., 1986, Conte award Environ. Health Inst., 1990; European Molecular Biology Orgn. fellow Imperial Cancer Research Fund, London, 1970-71. Mem. Inst. Medicine NAS, Am. Soc. Microbiology, Am. Soc. Clin. Investigation, Am. Soc. Biol. Chemists, Am. Soc. Cell Biology, Internat. Soc. Quantum Biology, Am. Assn. Cancer Rsch. (pres. 1990, Clowes award 1987), Phi Beta Kappa, Sigma Xi, Alpha Omega Alpha. Jewish. Home: 249 Chestnut St Englewood NJ 07631-3135 Office: Columbia-Presbyn Cancer Ctr 701 W 168th St New York NY 10032-2704

WEINSTEIN, LEONARD HARLAN, institute program director; b. Springfield, Mass., Apr. 11, 1926; s. Barney Willard Weinstein and Ida Pauline (Feinberg) Weinstein Clark; m. Sylvia Jane Sherman, Oct. 15, 1950; children: Beth Rachel, David Harold (dec.). BS, Pa. State U., 1949; MS, U. Mass., 1950; PhD, Rutgers U., 1953. Postdoctoral fellow Rutgers U., New Brunswick, N.J., 1953-55; plant physiologist Boyce Thompson Inst., Yonkers, N.Y., 1955-63; program dir. Boyce Thompson Inst., Ithaca, N.Y., 1963-91, bd. dirs., 1976—; dir. ecosystem rsch. ctr. Cornell U., Ithaca, 1988-90; bd. dirs. Boyce Thompson Southwestern Arboretum, Superior, Ariz.; adj. prof. dept. natural resources Cornell U., Ithaca, 1979—. Contbr. articles (150) to profl. jours. and chpts. to books. Mem. sci. adv. bd. EPA, Washington, 1988-91; mem. com. natural resources NASULGS, 1986—. Grantee NIH, NSF, HEW, Am. Cancer Soc., NASA, EPA, DOE, USDA. Mem. AAAS, Am. Soc. Plant Physiologists, Soc. Environ. Toxicologists and Chemists, Sigma Xi, Pi Alpha Xi, Gamma Sigma Delta. Home: 608 Cayuga Heights Rd Ithaca NY 14850-1424 Office: Cornell U 125 Boyce Thompson Inst Tower Rd Ithaca NY 14853

WEINSTEIN, LEONARD MURREY, aerospace engineer, consultant, researcher; b. Louisville, Aug. 7, 1940; s. Max Weinstein and Shirley (Aronovitz) Zeimer; m. Runell Edna Alford, Feb. 10, 1967; 1 stepchild, Sandra Loretta Raven. BS, Fla. State U., 1962; MS, George Washington U., 1972, ScD, 1981. Aerospace engr. NASA, Hampton, Va., 1962-84, group leader, 1984-91, sr. rsch. scientist, 1991—; conf. presenter in field. Contbr. over 50 articles to profl. jours., including AIAA Jour., Jour. Radiology. Recipient spl. achievement award NASA Langley Rsch. Ctr., 1970, 79, 89, Space Act award, 1988; IR-100 award Instrument and Rsch. mag., 1987; numerous others. Mem. AIAA (sr.). Achievements include patents for liquid thickness gage, ice detector. Home: 13 Burke Ave Newport News VA 23601 Office: NASA 1A E Reid St MS 170 Hampton VA 23681

WEINSTEIN, MARIE PASTORE, psychologist; b. N.Y.C., Oct. 3, 1940; d. Edward and Sarah (Mancuso) Pastore; children: Arielle Rebecca, Damon Alexander. BA in Polit. Sci. and Lit., Ind. U.; MS in Psychology, L.I. U.; PhD in Ednl. Psychology, CCNY, 1986. Cert. sch. psychologist; lic. psychologist, N.Y. Sch. psychologist evaluation unit Bd. Edn., N.Y.C., 1977-78; dir., adminstr. learning ctr. Guidance Ctr. Flatbush, Bklyn., 1978-82; clin. team coord./psychologist Lorge Upper and Lower Sch., N.Y.C., 1982-85; psychologist devel. disabilities ctr. Roosevelt Hosp., N.Y.C., 1985-87; chief psychologist Blueberry Treatment Ctrs., Bklyn., 1987-89; cons. psychologist Ctr. for Children & Families, St. Albans, N.Y., 1989—; cons. psychologist United Cerebral Palsy Hearst Presch., Bklyn., 1988-89, Charles Drew Day Ctr., Queens Village, N.Y., 1982-85, Warbasse Nursery Sch., Bklyn., 1981-85; adj. asst. prof. Baruch Coll. CUNY, 1989; pvt. practice, Bklyn.; rsch. cons. Children's TV Workshop, N.Y.C., 1979; clin. cons. Bedford Stuyvesant Mental Health Ctr., Bklyn., 1990; cons. dist. 2 N.Y.C. Bd. Edn., 1988; guest lectr. Met. Hosp. Dept. Psychiatry, N.Y.C., 1988; edn. cons. Lit. Vols. N.Y., 1974-76. Contbg. author to children's ency., 1970. Mem. Am. Psychol. Assn., Internat. Congress on Child Abuse and Neglect, Am. Orthopsychiat. Assn. (program com.), Manhattan Fedn. Child and Adolescent Svcs. Office: 26 Court St Ste 2112 Brooklyn NY 11242

WEINSTEIN, NORMAN JACOB, chemical engineer, consultant; b. Rochester, N.Y., Dec. 31, 1929; s. Sol. and Anne (Trapunsky) W.; m. Ann Francine Keiles, June 30, 1957; children: Maury S., Aaron S., Kenneth B. BChemE, Syracuse U., 1951, MChemE, 1953; PhD, Oreg. State Coll., 1956. Registered profl. engr. Chem. engr. ESSO Rsch. Engring. Co., Linden, N.J., 1956-60; sr. engr. ESSO Rsch. Engring. Co., Baton Rouge, 1960-65; engring. assoc. ESSO Rsch. Engring. Co., Florham Park, N.J., 1965-66; asst. dir. engring. and devel. Princeton Chem. Rsch. Inc., 1966-67, dir. engring. and devel., 1967-69; pres. Recon Systems Inc., Raritan, N.J., 1969—. Author: Thermal Processing of Municipal Solid Waste for Resource and Energy Recovery, 1976; contbr. numerous articles to profl. jours. Mem. AICE (chmn. air sect., environ. div.), Am. Chem. Soc. ASTM, N.Y. Acad. Scis., Assn. Cons. Chemists and Chem. Engrs. Democrat. Achievements include 8 patents in chemical processing; expert in waste oil technology. Home: 1175 Canal Rd RD 1 Princeton NJ 08540 Office: Recon Systems Inc 5 Johnson Dr Box 130 Raritan NJ 08869-0130

WEINSTEIN, ROY, physics educator, researcher; b. N.Y.C., Apr. 21, 1927; s. Harry and Lillian (Ehrenberg) W.; m. Janet E. Spiller, Mar. 26, 1954; children: Lee Davis, Sara Lynn. B.S., MIT, 1951, Ph.D., 1954; Sc.D. (hon.), Lycoming Coll., 1981. Research asst. Mass. Inst. Tech., 1951-54, asst. prof., 1956-59; asst. prof. Brandeis U., Waltham, Mass., 1954-56; assoc. prof. Northeastern U., Boston, 1960-63, prof. physics, 1963-82, exec. officer, chmn. grad. div. of physics dept., 1967-69, chmn. physics dept., 1974-81; spokesman MAC Detector Stanford U., 1981-82; dean Coll. Natural Scis. and Math. U. Houston, 1982-88; prof. physics, 1982—; dir. Inst. Beam Particle Dynamics U. Houston, 1985—; assoc. dir., spokesman Tex. Ctr. for Superconductivity, 1987-89; vis. scholar and physicist Stanford (Calif.) U., 1966-67, 81-82; bd. dirs. Perception Tech., Inc., Winchester, Mass., Omniwave Inc., Gloucester, Mass., Wincom Inc., Woburn, Mass.; cons. Visidyne Inc., Burlington, Mass., Houston Area, Rsch. Ctr, Stanford U., Harvard U., Cambridge, Mass., Cambridge Electron Accelerator, mem. adv. com., 1967-69; mem. portfolio evaluation com. Houston Ptnrs., 1990—; chmn. bd. dirs. Xytron Corp., 1986-91; dir., mem. exec. com. Houston Area Rsch. Ctr., 1984-87; 3d ann. faculty lectr. Northeastern U., 1966; chmn.

organizing com. 4th ann. Internat. Conf. on Meson Spectroscopy, 1974, chmn. program com. 5th ann., 1977, mem. organizing com. 6th ann., 1980, 83; chmn. mgmt. group Tex. Accelerator Ctr., Woodlands, Tex., 1985-90; chmn. Tex. High Energy Physicists, 1989-91; keynote speaker MIT Alumni series, 1988; permanent mem. exec. com. Large Vol. Detector (Underground Neutrino Telescope, Italy). Author: Atomic Physics, 1964, Nuclear Physics, 1964, Interactions of Radiation and Matter, 1964; editor: Nuclear Reactor Theory, 1964, Nuclear Materials, 1964; editor procs.: 5th Internat. Conf. on Mesons, 1977; contbr. numerous articles to profl. jours. Mem. Lexington (Mass.) Town Meeting, 1973-76, 77-84; vice chmn. Lexington Coun. on Aging, 1977-83. With USNR, 1945-46. NSF fellow Bohr Inst., Copenhagen, 1959-60, Stanford U., 1969-70, Guggenheim fellow Harvard U., 1970-71; NSF grantee, 1961, 63, 65, 66, 68, 70—; recipient Tex. Rsch. awards, 1986-87, 90—, U.S. Dept. Energy awards, 1974, 77, 87—, NASA award, 1990—, Founders award World Congress on Superconductivity, 1988, Elec. Power Rsch. Inst. award, 1990, 91, 92. Fellow Am. Phys. Soc.; mem. Am. Assn. Physics Tchrs., N.Y. Acad. Scis., Masons, Sigma Xi, Phi Kappa Phi (chpt. pres. 1977-79, Nat. Triennial Disting. Scholar prize 1980-83), Pi Lambda Phi. Democrat. Unitarian. Achievements include measurement of fine structure of positronium; first measurement of rho meson coupling to gamma rays, of phi meson decay to two muons; discovery of non-applicability of Lorentz contraction to length measured by a single observer, disproof of splitting of A2 meson; achievement of highest magnetic field for any permanent magnet, in YBa2Cu3O7. Home: 4368 Fiesta Ln Houston TX 77004-6603 Office: U Houston Coll Natural Sci and Math Houston TX 77204-5506

WEINSTEIN, WILLIAM STEVEN, technical engineer; b. Newark, Aug. 28, 1947; s. Abraham Weinstein and Hilda (Bushman) Blonde; m. Ellen Faith Weinberger, Dec. 24, 1972; children: Lenard Scott, John Ryan. Engr. Precision Indsl. Design, Newark, 1967-68; enlisted USAF, 1968, tech. sgt., technician, 1968-82, resigned, 1982; engr. MPCS Video Industries, N.Y.C., 1982-83; technician CBS, N.Y.C., 1983—. Mem. N.Y. Acad. Scis. Home: 255 Greenbriar Cir Tobyhanna PA 18466-3008

WEINSTOCK, GEORGE MATTHEW, biology educator, researcher; b. Chgo., Feb. 6, 1949. BS, U. Mich., 1970; PhD, MIT, 1977. Staff scientist Nat. Cancer Inst.-Frederick (Md.) Cancer Rsch. Facility, 1980-83, sr. scientist, 1983-84; assoc. prof. U. Tex. Med. Sch., Houston, 1984-90, prof., 1990—; instr. Internat. Ctr. for Gen. Engring., Trieste, Italy, 1990—, Cold Spring Harbor (N.Y.) Lab., 1986-90. Recipient Jane Coffin Childs fellowship Bank of Am.-Giannini Found., 1977-79, 79-80, Outstanding Tchr. award Grad. Sch. of Biomed. Sci., Houston, 1987. Mem. AAAS, Genetics Soc. Am., Am. Soc. for Biochemistry and Molecular Biology, Am. Soc. for Microbiology. Achievements include patents for system for expressing foreign genes in bacteria, vaccine against bovine Pasteurella haemolytica. Office: U Tex Med Sch Dept Biochemistry 6431 Fannin Houston TX 77030

WEINSWIG, SHEPARD ARNOLD, optical engineer; b. Boston, June 12, 1936; s. Irving and Eve (Popovsky) W.; m. Lorraine Bowers, Dec. 21, 1968; children: Deborah Lora, Mark B. BA, Harvard U., 1959; MS, U. Ariz., 1974. Physicist ECA, Cambridge, Mass., 1961-65; spectroscopist Tufts Dental Sch., Boston, 1965-66; physicist Valpey Corp., Holliston, Mass., 1966-68; optical engr. Tex. Instruments, Dallas, 1971-75, Lockheed, Sunnyvale, Calif., 1975-76, ITT, Ft. Wayne, Ind., 1976—. Contbr. articles to profl. jours. Founder, 1st chmn. adv. bd. on disabled students Ft. Wayne Community Schs., 1968-71. With U.S. Army, 1957. Mem. Optical Soc. Am., Soc. Photo/Optical Instrumentation Engrs. (co-chmn. lens conf. ann. mtg. 1991-93). Jewish. Achievements include design of 7 different operational earth orbiting instruments used to measure and track weather conditions for NOAA. Office: ITT PO Box 3700 1919 W Cook Rd Fort Wayne IN 46819

WEINTRAUB, HAROLD M., geneticist; b. Newark, June 2, 1945; married; 2 children. BA, Harvard U., 1967; PhD in Cell Differentiation, U. Pa., 1971, MD, 1973. From asst. to assoc prof. biochem. scis. Princeton (N.J.) U., 1973-77; rschr. dept genetics Fred Hutchinson Cancer Rsch. Ctr., Seattle, 1978—; mem. study sect. molecular biology NIH; investigator Howard Hughes Med. Inst. Assoc. editor Jour. Cell Biology; asst. editor Science & Cell; contbr. articles to over 180 profl. jours. Scholar Rita Allen Found., 1976-81. Mem. NAS (Richard Lounsberry award, 1991), Am. Acad. Arts and Sci. Achievements include research in structure, function and replication of eukaryotic chromosomes, cell transformation by avian tumor viruses, gene regulation and development. Office: F Hutchinson Cancer Rsch Ctr Dept Genetics 1124 Columbia St Seattle WA 98104-2092*

WEINTRAUB, JOSEPH, computer company executive; b. N.Y.C., May 15, 1954; m. Valerie Weintraub; children: Sharon, Anna. BA, CUNY, 1968; PhD, Harvard, 1972; postgrad., NYU. Programmer Great Am. Ins. Co., 1968-72, Time Inc., 1972-73; systems devel. mgr. Abraham & Straus, 1973-80; project mgr., lecturer, instr. Pace U., 1980-81; with Interpublic Group of Companies, 1981-87; dir. advanced technology Thinking Software Inc., Flushing, N.Y., 1987—. Author: Exploring Careers in Computing, 1990. Recipient Loebner prize Cambridge Ctr. for Behavioral Studies, 1992. Mem. Mensa, Am. Assn. Artificial Intelligence. Achievements include writing of first program to pass the Turing Test, first speaking expert system; patent for AC/TV. Office: Thinking Software 46-16 65th Pl Flushing NY 11377

WEIR, D. ROBERT, metallurgical engineer, engineering executive. Sr. v.p. tech. Sherritt Gordon Ltd., Ft. Saskatchewan, Alta., Can. Recipient H.T. Airey award Can. Inst. Mining and Metallurgy,1990. Office: care Can Inst Mining Metallurg, 3400 de Maisonneuve Blvd W, Montreal, PQ Canada H3Z 3B8*

WEIR, EDWARD KENNETH, cardiologist; b. Belfast, No. Ireland, Jan. 7, 1943; came to U.S. 1973; s. Thomas Kenneth and Violet Hilda (ffrench) W.; m. Elizabeth Vincent Pearman, May 29, 1971; children: Fergus G., Conor K. BA, U. Oxford, U.K., 1964; MA, BM, BCh, U. Oxford, 1967, DM, 1976. Diplomate Am. Bd. Internal Medicine. Sr. house physician Nuffield Dept. Medicine, Radcliffe Infirmary, Oxford, 1970-71; registrar in cardiology Groote Schuur Hosp., Cape Town, South Africa, 1971-73; postdoctoral rsch. fellow U. Colo., Denver, 1973-75; cons. pediatric cardiologist U. Cape Town Med. Sch., 1975-76; cons. cardiologist U. Natal Med. Sch., Durban, South Africa, 1976-77; assoc. prof. medicine U. Minn., Mpls., 1978-85, prof. medicine, 1985—; staff physician Va. Med. Ctr., Mpls., 1978—; dir. Grover Confs. on Pulmonary Circulation, 1984-92. Co-editor: Pulmonary Hypertension, 1984, The Pulmonary Circulation in Health and Disease, 1987, Pulmonary Vascular Physiology and Pathophysiology, 1989, The Diagnosis and Treatment of Pulmonary Hypertension, 1992, The Pulmonary Circulation and Gas Exchange, 1993. Fulbright scholar, 1973-75; Sr. Internat. Fogarty fellow, 1993. Fellow Am. Coll. Cardiology, Royal Coll. Physicians London; mem. Am. Heart Assn. (Minn. affiliate bd. dirs. 1989-93, Nat. Cardiopulmonary Coun. membership and nominating coms. 1992—), Pulmonary Circulation Found. (treas. 1985—). Office: VA Med Ctr 111C 1 Veteran's Dr Minneapolis MN 55391

WEIS, SERGE, neuropathologist; b. Petange, Luxembourg, Aug. 11, 1960; s. Albert and Louise (Lacroix) W. MD, Med. Sch. Vienna, Austria, 1988. Rsch. asst. Inst. for Anatomy, Vienna, 1983-85, Boltzmann Inst. for Clin. Neurobiology, Vienna, 1985-87; rsch. fellow Inst. for Anatomy, Lübeck, Germany, 1988-89, Inst. Neuropathology, Munich, 1989—; chief Morphometry Lab., Munich, 1989—. Author: The Human Brain, 1992; co-author: Computer Graphics, 1991; editor: HIV-Infection of the Central Nervous System; contbr. articles to profl. jours. Mem. Internat. Soc. for Stereology, Soc. for Quantitative Morphology, Quantitative Diagnostics in Pathology. Socialist Party. Avocations: sailing, photography, music, reading. Home: Elisabethstrasse 85, Munich Germany D-8000 Office: Inst for Neuropathology, Thalkirchnerstrasse 36, Munich Germany D-8000

WEISBIN, CHARLES RICHARD, nuclear engineer; b. Bklyn., Jan. 4, 1944; s. Alma (Schwartz) Lovitt; m. Alison Norma Weisbin, June 20, 1964; children: Daniel Mark, Amy Gayle. MS in Nuclear Engring., Columbia U., 1965, DSc in Nuclear Engring., 1969. Group leader Oak Ridge (Tenn.) Nat. Lab., 1977-80, section head, 1980-89, dir. Ctr. for Engring. Systems Advanced Rsch., 1982-89, dir. robotics and intelligence systems, 1986-89; mgr. telerobotics tech. Jet Propulsion Lab., Pasadena, Calif., 1989-93, mgr. robotic systems and advanced computer tech. sect., 1989-93; mgr. rover and telerobotic tech., 1993—; mem. joint tech. panel on robotics DOD Joint Dirs. Labs., 1986-89; assoc. prof. computer sci. U. Tenn., Knoxville, 1984-89; program chmn. 2nd Internat. Conf. on Artificial Intelligence, IEEE Computer Soc., 1985; co-chmn. U.S. NASA Telerobotics Working Group; chmn. robotics and telepresence Space Tech. Interagency Group, 1992—. Author: Sensitivity and Uncertainty Analysis of Reactor Performance Parameters, 1982; mem. editorial bd. Applied Intelligence, 1990—; contbr. articles to profl. jours. Recipient NASA Exceptional Svc. medal, 1993. Mem. IEEE (Cert. Appreciation 1987), Am. Nuclear Soc. (program chmn. 1977-79), Robotics and Automation Soc., Sigma Chi, Tau Beta Pi. Republican. Jewish. Achievements include initiation of robotics and intelligent systems at Oak Ridge; research on sensitivity analysis, on non-destructive assay of spent nuclear fuel, supervised inspection, and emergency response robotics. Home: 775 Starlight Heights Dr La Canada Flintridge CA 91011 Office: Jet Propulsion Lab 4800 Oak Grove Dr Pasadena CA 91109-8099

WEISEL, GEORGE FERDINAND, retired zoology educator; b. Missoula, Mont., Mar. 21, 1915; s. George Ferdinand and Thula (Toole) W.; m. Maxine George, 1941 (div. 1945); 1 child, Anna Afton; m. Angela McCormick, Jan. 14, 1950; children: Thula, George. BA, U. Mont., 1941, MA, 1942; PhD, UCLA, 1948. Rsch. asst. U. Mich., Ann Arbor, 1943; rsch. asst. U. Mont., Missoula, 1942, prof. zoology, 1948-87; rsch. asst. Scripps Inst. Oceanography, La Jolla, Calif., 1946-47. Author: Men and Trade on Northwest Frontier, 1955; contbr. articles to profl. jours. Pres. W. Mont. Fish and Game, Missoula, 1963; dir. legal and edn1. fund Mont. Wildlife Assn., 1970-73; bd. dirs. Scientists for Pub. Info., St. Louis, 1960s. Lt. (j.g.) USN, 1943-46, PTO. Mem. Am. Soc. Ichthyologists, Am. Soc. Zoology, Nat. Resources Def. Coun., Sigma Xi. Democrat. Home: 615 Pattee Canyon Missoula MT 59803

WEISMAN, JOEL, nuclear engineering educator, engineering consultant; b. N.Y.C., July 15, 1928; s. Abraham and Ethel (Marcus) W.; m. Bernice Newman, Feb. 6, 1955; 1 child, Jay (dec.). B.Ch.E., CCNY, 1948; M.S., Columbia U., 1949; Ph.D., U. Pitts, 1968. Registered profl. engr., N.Y., Ohio. Plant engr. Etched Products, N.Y.C., 1950-51; from jr. engr. to assoc. engr. Brookhaven Nat. Lab., Upton, N.Y., 1951-54; from engr. to fellow engr. Westinghouse Nuclear Energy Systems, Pitts., 1954-59, from fellow engr. to mgr. thermal and hydraulic analysis, 1960-68; sr. engr. Nuclear Devel. Assocs., White Plains, N.Y., 1959-60; assoc. prof. nuclear engring. U. Cin., 1968-72, prof. nuclear engring., 1972—, dir. nuclear engring. program, 1977-86, dir. lab. basic and applied nuclear research, 1984—. Co-author: Thermal Analysis of Pressurized Water Reactors, 1970, 2d edit., 1979, Introduction to Optimization Theory, 1973, Modern Power Plant Engineering, 1985; editor: Elements of Nuclear Reactor Design, 1977, 2d edit., 1983; contbr. tech. articles to profl. jours.; patentee in field. Mem. Cin. Environ. Adv. Council, 1976-78; mem. Cin. Asian Art Soc., 1977—, v.p., 1980-82, pres., 1982-84; mem. exec. bd. Air Pollution League Greater Cin., 1980-90. Sr. NATO fellow, Winfrith Lab., U.K. Atomic Energy Authority, 1972; sr. fellow Argonne Nat. Lab., Ill., 1982; NSF research grantee, 1974-78, 82-85, 86-89; recipient Dean's award U. Cin. Coll. Engring., 1987. Fellow Am. Nuclear Soc. (v.p. Pitts. sect. 1957-58, mem. exec. com. thermal-hydraulics div. 1989-92); mem. Am. Inst. Chem. Engrs., Sigma Xi. Democrat. Jewish. Avocation: Japanese art. Home: 3419 Manor Hill Dr Cincinnati OH 45220-1522 Office: U Cin Dept Mech Ind & Nuclear Engring Cincinnati OH 45221

WEISS, ARMAND BERL, economist, association management executive; b. Richmond, Va., Apr. 2, 1931; s. Maurice Herbert and Henrietta (Shapiro) W.; BS in Econs., Wharton Sch. Fin., U. Pa., 1953, MBA, 1954; D.B.A., George Washington U., 1971; m. Judith Bernstein, May 18, 1957; children: Jo Ann Michele, Rhett Louis. Cert. assn. exec. Officer, U.S. Navy, 1954-65; spl. asst. to auditor gen. Dept. Navy, 1964-65; sr. economist Center for Naval Analyses, Arlington, Va., 1965-68; project dir. Logistics Mgmt. Inst., Washington, 1968-74; dir. systems integration Fed. Energy Adminstrn., Washington, 1974-76; sr. economist Nat. Comm. Supplies and Shortages, 1976-77; tech. asst. to v.p. System Planning Corp., 1977-78; chmn. bd., pres., chief exec. officer Assns. Internat., Inc., 1978—; chmn. bd. dirs., chief fin. officer RAIL Digital Corp., 1988-91; v.p., treas. Tech. Frontiers, Inc., 1978-80; sr. v.p. Weiss Pub. Co., Inc., Richmond, Va., 1960—; v.p. Condo News Internat., Inc., 1981; v.p., bd. dirs. Leaders Digest Inc., 1987-88; sec., bd. dirs. Mgmt. Svcs. Internat. Inc., 1987-88; adj. prof. Am. U., 1979-80, 89-90; vis. lectr. George Washington U., 1971; assoc. prof. George Mason U., 1984; treas. Fairfax County (Va.) Dem. com. 1992—, assisted Pres. Clinton, v.p. Gore transition at White House, 1993, pres. Washington Mgnt. and Business Assn., 1993—; chmn. U.S. del., session chmn. NATO Symposium on Cost-Benefit Analysis, The Hague, Netherlands, 1969, NATO Conf. on Operational Rsch. in Indsl. Systems, St. Louis, France, 1970; pres. Nat. Council Assns. Policy Scis., 1971-77; chmn. adv. group Def. Econ. Adv. Council Dept. Def., 1970-74; resident asso. Smithsonian Instn., 1973—; expert cons. Dept. State, GAO; undercover agt. FBI, 3 yrs. Del. Pres.'s Mid-Century White House Conf. on Children and Youth, 1950; scoutmaster Japan, U.S., leader World Jamborees, France, Can., U.S., 1945-61; Eagle scout, 1947; U.S. del. Internat. Conf. on Ops. Rsch., Dublin, Ireland, 1972; organizing com. Internat. Cost-Effectiveness Symposium, Washington, 1970; speaker Internat. Conf. Inst. Mgmt. Scis., Tel Aviv, 1973, del., Mexico City, 1967. Mem. bus. com. Nat. Symphony Orch., 1968-70, Washington Performing Arts Soc., 1974-88; bus. mgr. Nat. Lyric Opera Co., 1983—; mem. mktg. com. Fairfax Symphony Orch., 1984-91; bd. dirs. Mc Lean (Va.) Orch., 1992—; exec. com. Mid Atlantic council Union Am. Hebrew Congregations, 1970-79, treas., 1974-79, mem. nat. MUM com., 1974-79; mem. dist. com. Boy Scouts Am., 1972-75; bd. dirs. Nat. Council Career Women, 1975-79; Va. Acad. Scis., 1991—. Recipient Silver medal 50-yard free style and half mile swimming meet No. Va. Sr. Olympics, 1990. Fellow AAAS, Washington Acad. Scis. (gov. 1981—, v.p. 1987-88, pres.-elect 1989-90, gov. 1981-92, pres. 1990-91, past pres. 1991-92); mem. Ops. Research Soc. Am. (chmn. meetings com. 1969-71; chmn. cost-effectiveness sect. 1969-70), Washington Ops. Research/Mgmt. Sci. Council (editor newsletter 1969—, sec. 1971-72, pres. 1973-74, trustee 1975-77, bus. mgr. 1976—), Internat. Inst. Strategic Studies (London), Am. Soc. Assn. Execs. (membership com. 1981-82, cert.), Inst. for Mgmt. Sci., Am. Econ. Assn., Wharton Grad. Sch. Alumni Assn. (exec. com. 1970-73), Am. Acad. Polit. and Social Sci., Nat. Eagle Scout Assn., Am. Legion, Navy League of the U.S., Greater Wash. Soc. Assn. Execs., Fairfax County C. of C., Vienna, Va. C. of C., Alumni Assn. George Washington U. (governing bd. 1974-82, chmn. univ. publs. com. 1976-78, Alumni Service award 1980), Alumni Assn. George Washington U. Sch. Govt. and Bus. Adminstrn. (exec. v.p. 1977-78, pres. 1978-79), George Washington U. Doctoral Assn. (sr. v.p. 1968-69), Nat. Assn. Acad. Sci. (del. 1991-93). Jewish. (pres. temple 1970-72). Club: Wharton Sch. Washington (sec. 1967-69, pres. 1969-70, exec. dir. 1987—, Joseph Wharton award 1991). Co-editor: Systems Analysis for Social Problems, 1970, The Relevance of Economic Analysis to Decision Making in the Department of Defense, 197ard More Effective Public Programs: The Role of Analysis and Evaluation, 1975. Editor: Cost-Effectiveness Newsletter, 1966-70, Operations Research/Systems Analysis Today, 1971-73, Operation Research/Mgmt. Sci. Today, 1974-87; Feedback, 1969—, Condo World, 1981; assoc. editor Ops. Research, 1971-75; publisher: IEEE Scanner, 1983-89, Spl. and Individual Needs Tech. (SAINT) Newsletter, 1987-88, Jour. Parametrics, 1984-88. Home: 6516 Truman Ln Falls Church VA 22043-1821

WEISS, GEORGE ARTHUR, orthodontist; b. Bklyn., Feb. 1, 1921; s. Nathan L. and Ida (Rosenthal) W.; m. Jacqueline Hellermann, Jan. 28, 1945; children: Ellen Joy Weiss Finberg, Leslie Donna Weiss Schoenfeld. BA, Bklyn. Coll., 1941; DDS, Columbia U., 1944, cert. Orthodontics, 1954. Diplomate Am. Bd. of Orthodontics. Dir. dentistry Dental Clinic Southen Japan, Kobe, 1946; chief dentistry Olmstead AFB, Middleburg, Pa., 1947; pvt. practice Orthodontics Bayside, N.Y., 1947—; chief orthodontics Community Svc. Soc., N.Y.C., 1954-57; dir. orthodontics Jamaica (N.Y.) Hosp., 1977—; cons. State Aid Orthodontic Program for Handicapped Children, Suffolk County, N.Y., 1987—. Founder, Oakland Gardens Jewish Ctr., Bayside, N.Y., 1949; Bayside Oaks Jewish Ctr., Bayside, 1954. Major U.S. Army, 1944-48, Japan. Recipient Chemistry award, Am. Inst. Chem., N.Y.C., 1941. Fellow Am. Coll. Dentistry; mem. Am. Bd. Orthodontists, Am. Assn. Orthodontists, Northeastern Soc. Orthodontists, Dental Soc. State of N.Y. (bd. govs. 1976-86, mem. coun. on ins. 1965-75), Queens County Dental Soc. (trustee 1960—, chief adminstr. retirement fund, 1964-72, pres. 1975, Disting. Svc. award, 1988), Old Westbury Golf and Country Club, Alpha Omega Dental Soc. Avocations: golf, finance, insurance. Home and Office: 5901 Springfield Blvd Flushing NY 11364-1996

WEISS, GEORGE HERBERT, mathematician, consultant; b. N.Y.C., Feb. 19, 1930; s. Morris and Violet (Mayer) W.; m. Delia Esther Orgel, Dec. 20, 1961; children: Miriam Judith, Alan Keith, Daniel Mordechai. BA, Columbia U., 1951; MA, U. Md., 1953, PhD, 1958. Physicist USN, White Oak, Md., 1951-61; asst. prof. U. Md., College Park, 1959-63; fellow Rockefeller U., N.Y.C., 1963-64, Weizmann Inst., Rehovot, Israel, 1958-59; mathematician NIH, Bethesda, Md., 1964—; cons. GM, IBM, GE. Author: Lattice Dynamics in the Harmonic Approximation, 1963, 2d edit., 1971, The Master Equation in Chemical Physics, 1977, Aspects of the Random Walk, 1993. With U.S. Army, 1954-56. Recipient Disting. Svc. in Math. award Washington Acad. Sci., 1967, Disting. Svc. award NIH, 1970. Avocation: photography. Office: NIH Bethesda MD 20892

WEISS, HERBERT KLEMM, aeronautical engineer; b. Lawrence, Mass., June 22, 1917; s. Herbert Julius and Louise (Klemm) W.; m. Ethel Celesta Giltner, May 14, 1945 (dec.); children: Janet Elaine, Jack Klemm (dec.). B.S., MIT, 1937, M.S., 1938. Engr. U.S. Army Arty. Bds., Ft. Monroe, Va, 1938-42, Camp Davis, N.C., 1942-44, Ft. Bliss, Tex., 1944-46; chief WPN Systems Lab., Ballistic Research Labs., Aberdeen Proving Grounds, Md, 1946-53; chief WPN systems analysis dept. Northrop Aircraft Corp., 1953-58; mgr. advanced systems devel. mil. systems planning aeronutronic div. Ford Motor Co., Newport Beach, Calif., 1958-61; group dir., plans devel. and analysis Aerospace Corp., El Segundo, Calif., 1961-65; sr. scientist Litton Industries, Van Nuys, Calif., 1965-82; cons. mil. systems analysis, 1982—; Mem. Sci. Adv. Bd. USAF, 1959-63, sci. adv. panel U.S. Army, 1965-74, sci. adv. commn. Army Ball Research Labs., 1973-77; advisor Pres.'s Commn. Law Enforcement and Adminstrn. Justice, 1966; cons. Office Dir. Def., Research and Engring., 1954-64. Contbr. articles to profl. jours. Patentee in field. Recipient Commendation for meritorious civilian service USAF, 1964, cert. appreciation U.S. Army, 1976. Fellow AAAS, AIAA (assoc.); mem. IEEE, Ops. Research Soc. Am. Republican. Presbyterian. Club: Cosmos. Home: PO Box 2668 Palos Verdes Peninsula CA 90274-8668

WEISS, JOSEPH, physician; b. Kosice, Czechoslovakia, June 23, 1913; came to U.S., 1949; s. Abraham Adolf and Johanna (Nagy) W.; m. Eva Farkas, Apr. 24, 1944; 1 child, Julia. MD, Charles U., Prague, Czechoslovakia, 1947. Diplomate Am. Bd. Family Practice. Clin. asst. Charles U., 1947-48; resident in medicine Lebanon Hosp., Bronx, N.Y., 1949-51; clinician N.Y.C. Dept. Health, 1954-85; clin. instr. medicine N.Y. Med. Coll., N.Y.C. and Valhalla, N.Y., 1973-90; bd. dirs. La Guardia-Health Ins. Plan Greater N.Y., 1979-85, chmn. peer rev. com., 1985—; pres. Semmelweis Sci. Soc., N.Y.C., 1973-74; pres. Am.-Hungarian Med. Assn., N.Y.C., 1973-74. Recipient Presdl. Scroll Semmelweis Sci. Soc., 1974, medal Internat. Conv. Pathophysiology, Prague, 1975, Internat. Symposium Cardiomyopathies, Bratislava, Czechoslovakia, 1985. Fellow InterAm. Coll. Physicians and Surgeons, Am. Geriatric Soc., Am. Acad. Family Practice; mem. AMA, N.Y. State Med. Soc., Queens County Med. Soc., N.Y. Acad. Scis. Home: 69-33 170th St Flushing NY 11365-3309

WEISS, MAX TIBOR, aerospace company executive; b. Hajduananas, Hungary, Dec. 29, 1922; came to U.S., 1929, naturalized, 1936; s. Samuel and Anna (Hornstein) W.; m. Melitta Newman, June 28, 1953; children: Samuel Harvey, Herschel William, David Nathaniel, Deborah Beth. BEE, CCNY, 1943; MS, MIT, 1947, PhD, 1950. Rsch. assoc. MIT, 1946-50; mem. tech. staff Bell Tel. Labs., Holmdel, N.J., 1950-59; assoc. head applied physics lab. Hughes Aircraft Co., Culver City, Calif., 1959-60; dir. electronics rsch. lab. The Aerospace Corp., L.A., 1961-63, gen. mgr. labs. div., 1963-67, gen. mgr. electronics and optics div., 1968-78, v.p., gen. mgr. lab. ops., 1978-81, v.p. engring. group, 1981-86; v.p. tech. and electronics system group Northrop Corp., L.A., 1986-91; v.p., gen. mgr. electronics systems div. Northrop Corp., Hawthorne, Calif., 1991—; asst. mgr. engring. ops. TRW Systems, Redondo Beach, Calif., 1967-68; mem. sci. adv. bd. USAF. Contbr. articles to physics and electronics jours.; patentee in electronics and communications. With USNR, 1944-45. Fellow Am. Phys. Soc., IEEE, AIAA, AAAS; mem. NAE, Sigma Xi. Office: Northrop Corp 2301 W 120th St Hawthorne CA 90251-5032

WEISS, MICHAEL JAMES, chemistry educator; b. N.Y.C., May 29, 1941; s. Irving and Florence (James) W.; m. Myra Lee Landau, June 10, 1967; 1 child, Amy Merril Weiss Musikar. BS in Chemistry, Bklyn. Coll., 1964; MS in Chemistry, Adelphi U., 1966; PhD in Biochemistry, N.Y. Med. Coll., 1971. Supr. clin. chemistry Queens Gen. Hosp., N.Y.C., 1971-77; dir. biochemistry Kingsborough Med. Ctr., N.Y.C., 1977-78; lab. dir. Nassau (N.Y.) Diagnostic Lab., 1978-86; chemistry educator N.Y. Bd. Edn., 1986—; lab. cons., N.Y., 1986—. Author: (chpt.) The Antigenic Nature of Mamalian Cell Membranes, 1976; contbr. articles to profl. jours. Maj. U.S. Army, 1978—. Mem. AAAS, N.Y. Acad. Sci., Armed Forces Soc. Med. Lab. Scientists. Achievements include research in oneofetal antigens.

WEISS, RICHARD GERALD, chemist educator; b. Akron, Ohio, Nov. 13, 1942; s. William and Lillie (Goldstein) W.; m. Jeanne Ann Clasquin, Feb. 26, 1966; children: Margaret, Linnea, David. BS, Brown U., 1965; MS, U. Conn., 1967, PhD, 1969. Fellow Calif. Inst. Tech., Pasadena, 1969-71; vis. prof. U. de São Paulo, Brazil, 1971-74; asst., assoc. prof. Georgetown U., Washington, 1974-86, prof., 1986—; cons. World Bank, Washington, 1983-91, Fuisz Technologies, Chantilly, Va., 1988—; mem. discipline adv. com. Fulbright Commn., Washington, 1992—. Fellow NIH 1969-71, NAS 1971-74. Office: Georgetown U Dept Chemistry Washington DC 20057

WEISS, ROBERT FRANKLIN, podiatrist; b. Bridgeport, Conn., May 2, 1946; s. Murray Harold and Sara (Kramer) W.; m. Kathy Barbara Herstein, June 6, 1971; children: Lauren Jennifer, Scott Heath. Student, Norwalk (Conn.) Community Coll, 1965-67, U. Bridgeport, 1967; DPM, Ohio Coll. Podiatric Medicine, 1971. Resident James C. Giuffré Med. Ctr., Phila., 1971-72; chief dept. podiatry St. Joseph Med. Ctr., Stamford, Conn., 1985—; mem. dept. podiatry Stamford Hosp., 1987—, Norwalk Hosp., 1987—; police surgeon Conn. State Police, Meriden, 1985-91; sports medicine cons. IBM, White Plains, N.Y., 1985-91. Author: Archives of Podiatric Medicine & Foot Surgery, 1978, 2d edit., 1979; mem. editorial bd. Conn. Runner mag., 1991—; syndicated columnist Sports Medicine & Health, 1980-91; inventor foot support walking and running systems. Mem. trial adv. com. U.S. Olympic Marathon, Buffalo, 1984, Liberty Park, N.J., 1988. Fellow Am. Acad. Podiatric Sports Medicine (chmn. credentials and exam. com. 1985-87, Robert Barnes Disting. Svc. award 1988); mem. Am. Coll. Ft. Surgeons (assoc.), Am. Coll. Sports Medicine, Am. Med. Athletic Assn., Am. Podiatric Med. Assn., Conn. Podiatric Med. Assn. (peer rev. com. 1984-86), Fairfield County Podiatric Med. Assn. Avocations: fishing, skiing. Home: 350 Barrack Hill Rd Ridgefield CT 06877-3031 Office: Running Doctor Inc 800 Post Rd Darien CT 06820-4622

WEISS, ROBERT M., urologist, educator; b. N.Y.C., Jan. 13, 1936; s. David and Laura W.; m. Ilana Shemer, May 20, 1973; children—Erik Daniel, Dana Alexandra. B.S. magna cum laude, Franklin and Marshall Coll., Lancaster, Pa., 1957; M.D., SUNY, Bklyn., 1960; M.A. (hon.), Yale U., 1976. Diplomate: Am. Bd. Urology, Nat. Bd. Med. Examiners. Intern Cornell Med. Div., Bellevue Hosp., N.Y.C., 1960-61; resident in gen. surgery Beth Israel Hosp., N.Y.C., 1961-62; resident in urology Squier Urol. Clinic, Presbyn. Hosp., N.Y.C., 1963-64, 65-67; vs. fellow Columbia U. Coll. Physicians and Surgeons, N.Y.C., 1964-65, adj. assoc. prof. pharmacology, 1975-77, adj. prof. pharmacology, 1977—; mem. faculty Yale U. M.ed. Sch., New Haven, 1967—, prof. urology, 1976-88, prof., chief sect. of urology, 1988—; attending urology Yale-New Haven Hosp., New Haven, 1967-88, head sect. of urology, 1988—; cons. West Haven VA Hosp., Waterbury (Conn.) Hosp. Contbr. articles to med. publs. Served with USAR, 1962-63. Fellow ACS, Am. Acad. Pediatrics; mem. Am. Assn. Genito-Urinary Surgeons, Am. Physiol. Soc., Soc. Gen. Physiologists, Assn. Univ. Urologists, Soc. Pediatric Urology, Am. Urol. Assn., Am. Soc. Clin. Pharmacology and Therapeutics, Internat. Urodynamics Soc., AAAS, Internat. Soc. Dynamics of Upper Urinary Tract, Clin. Soc. Genito-Urinary Surgeons, Phi Beta Kappa, Sigma Xi. Office: 333 Cedar St New Haven CT 06510-3289

WEISS, ROBERT MICHAEL, dentist; b. Bklyn., June 5, 1940; s. Henry and Rena (Bluth) W.; (Trustees scholar) L.I. U., 1958-61; DDS, N.Y.U., 1965; postdoctoral cert. LD Pankey Inst. for Advanced Dental Edn., 1979; m. Irene Marilyn Sternick, June 30, 1962; children: Lori Ann, Julie Lynn, Karen Michelle. Pvt. practice dentistry, Avon, Conn., 1967—, pres. Avon Dental Group, P.C., 1972—; nat. cons. Conn. Gen. Ins. Co. for ins. coverage for Gen. Electric Co., 1980—; cons. CNA Ins. Co., 1988—; bd. dirs. Sentinel Bank. Chmn. Children's Dental Health Week, Hartford County, 1971; chmn. Jewish Adult Edn., West Hartford, Conn., 1986-87; trustee Temple Beth Israel, 1983—. Served to capt. USAF, 1965-67. Fellow Acad. Gen. Dentistry; mem. Am. Soc. Preventive Dentistry (pres. Conn. chpt.), Am. Dental Assn., Hartford Dental Soc., Conn. State Dental Assn. House Dels. (del. 1992—), Chronic Fatigue Immune Dysfunction Syndrome (Conn. bd. dirs. 1992—), So. New Eng. Assn. Practice Adminstrn., Starnard Beach Assn. (pres. 1984-86). Avon Jr. C. of C. (pres. 1971-72), Alpha Omega, Sigma Alpha Mu. Mason. Home: 74 Ferncliff Dr West Hartford CT 06117-1014 Office: 20 W Avon Rd Avon CT 06001

WEISS, SAMUEL ABRAHAM, psychologist, psychoanalyst; b. N.Y.C., May 13, 1923; s. Kasiel and Sophie (Schachter) W.; m. Alice Langer, May 20, 1958; children: Benjamin Z., Naomi E., Susan J. BA, Yeshiva U., 1944; MA, NYU, 1948, PhD, 1957. Diplomate Am. Bd. Profl. Psychology, Nat. Register of Health Svcs.; cert. in psychoanalytic psychotherapy, psychoanalysis. Intern Bellevue Psychiat. Hosp., N.Y.C., 1955-56; assoc. rsch. scientist NYU Med. Ctr., N.Y.C., 1956-59, rsch. scientist, 1959-68, assoc. dir. amputee psychology rsch., 1958-66; assoc. prof. psychology Yeshiva U., N.Y.C., 1961-71; psychol. cons. Stern Coll. for Women, Yeshiva U., N.Y.C., 1960-71; psychologist/psychotherapist/psychoanalyst in pvt. practice N.Y.C.; cons. N.Y. State Div. Vocat. Rehab., 1958-73. Contbr. articles to profl. jours. Fellow AAAS (Rosette award 1991), APA (editorial cons. rehab. psychology 1972-80). Jewish. Home: 80-40 Lefferts Blvd Kew Garden NY 11415 Office: 7 Park Ave Ste 66 New York NY 10016-4330

WEISS, SCOTT JEFFREY, civil engineer; b. Boston, Mar. 6, 1968; s. Irving and Charlotte Shirley (Kwatcher) W.; m. Meredith Eileen O'Brien, Oct. 31, 1992. BSCE, U. Mass., 1990, MS in Civil Engring., 1993. Registered engr.-in-tng., Mass. Asst. civil engr. D.L. Bean, Inc., land cons., surveyors, engrs., Westfield, Mass., 1989-91; grad. rsch. asst. civil engring. dept. U. Mass., Amherst, 1991; asst. engr. II, Fay, Spofford & Thorndike, Inc., engrs., Lexington, Mass., 1991—; instr., civil engring. U. Mass., Lowell, 1993—. Dept. Transp. fellow U. Mass., 1990-91. Mem. ASCE (assoc.), Boston Soc. Civil Engring. (younger mem. com., jour. editorial bd.), Tau Beta Pi, Chi Epsilon. Office: Fay Spofford & Thorndike 191 Spring St Lexington MA 02173

WEISS, STANLEY ALAN, mining, chemicals and refractory company executive; b. Phila., Dec. 21, 1926; s. Walter Joseph and Anne Betty (Lubin) W.; m. Lisa Popper, May 23, 1958; children: Lori Christina, Anthony Walter. Student, Georgetown U., 1950-51; fellow, Ctr. for Internat. Affairs, Cambridge, Mass., 1977-78. Founder Minera La Mundial, SA, San Luis Potosi, Mexico, 1954-56, Manganeso Mexicano SA, Mexico City, 1957-61, Mercurio Internacional SA, Mexico City, 1957-61; founder Flux, SA, Mexico City, 1960-82, Sao Paulo, Brazil, 1964-74; founder, chmn. bd., chief exec. officer Am. Minerals, Inc., El Paso, Tex., 1960-91; co-founder, exec. v.p. Ralstan Trading & Devel. Corp., 1968-91, chmn. bd. dirs., 1991; chmn. bd. dirs. Am. Premier, Inc., King of Prussia, Pa., 1991—. Author: Manganese: The Other Uses, 1976; contbr. articles on nat. security issues. Founder, chmn. Bus. Execs. for Nat. Security, Washington, 1982—; bd. dirs. New Am. Schs. Devel. Corp., Washington; active The Coun. for Excellence in Govt. With U.S. Army, 1944-46, ATO. Mem. Am. Bus. Conf., Am. Ditchley Found., Internat. Inst. Strategic Studies (Eng.), World Econ. Forum (Switzerland), Garrick Club, Queens Club (London), The Royal Instn. of G.B. (London). Jewish. Avocations: tennis, skiing. Home: 2126 Connecticut Ave NW # 5 Washington DC 20008-1729 Office: American Premier Inc Ste 330 1615 L St NW Washington DC 20036

WEISSKOPF, VICTOR FREDERICK, physicist; b. Vienna, Austria, Sept. 19, 1908; came to U.S., 1937, naturalized, 1942; s. Emil and Martha (Gut) W.; m. Ellen Tvede, Sept. 5, 1934 (dec. Aug. 1989); children: Thomas Emil, Karen Louise; m. Duscha Schmid, May 1991. Ph.D., U. Goettingen, Germany, 1931. Rsch. assoc. U. Copenhagen, Denmark, 1932-34, Inst. of Tech., Zürich, Switzerland, 1934-37; asst. prof. physics U. Rochester, N.Y., 1937-43; with Manhattan Project, Los Alamos, N.M., 1943-46; prof. physics Mass. Inst. Tech., 1946-60; dir. gen. European Orgn. for Nuclear Research, Geneva, Switzerland, 1961-65; Inst. prof. Mass. Inst. Tech., 1965—; chmn. high energy physics adv. panel AEC, 1967-73. Author: (with J. Blatt) Theoretical Nuclear Physics, 1952, Knowledge and Wonder, 1962, Physics in the Twentieth Century, 1972; (with K. Gottfried) Concepts of Particle Physics, 1984; The Privilege of Being a Physicist, 1989; The Joy of Insight: Passions of a Physicist, 1991; also articles on nuclear physics, quantum theory, radiation theory, etc. in science jours. Recipient Max Planck medal Germany, 1956, Hi Majorana award, 1970, G. Gamov award, 1971, Boris Pregel award, 1971, Prix Mondial Cino del Duca France, 1972, L. Boltzmann prize Austria, 1977, Nat. Medal of Sci. U.S., 1980, Wolf prize, Jerusalem, 1982, Enrico Fermi award U. S. Dept. Energy, 1988, Pub. Welfare medal Nat. Acad. Scis., 1991, Compton medal for statesmanship in sci. Am. Inst. Physics, 1992. Fellow Am. Phys. Soc. (pres. 1960, Forum award 1991); mem. Nat. Acad. Scis., Am. Acad. Arts and Scis. (pres. 1975-79), French Academie des Scis. (corr.), Austrian Acad. Sci. (corr.), Danish Acad. Sci. (corr.), Bavarian Acad. Sci. (corr.), Scottish Acad. Sci. (corr.), Soviet Acad. Sci. (corr.), Pontifical Acad. Sci., Spanish Acad. Sci. (corr.), German Acad. Sci. Home: 20 Bartlett Ter Newton MA 02159-2314

WEISSMANN, HEIDI SEITELBLUM, radiologist, educator; b. N.Y.C., Feb. 4, 1951; d. Louis and June (Joseph) Seitel Bloom; m. Murray H. Weissmann, June 16, 1973; 1 dau., Lauren Erica. BS in Chemistry magna cum laude, Bklyn. Coll., CUNY, 1970, MD, Mt. Sinai Sch. Medicine, N.Y.C., 1974. Diplomate Nat. Bd. Med. Examiners. Intern Montefiore Med. Ctr. Bronx, N.Y., 1974-75, resident in diagnostic radiology, 1975-78; fellow in computerized transaxial tomography and ultrasonography N.Y. Hosp.-Cornell U. Med. Ctr., N.Y.C., N.Y., 1978-79; instr. in radiology and nuclear medicine Albert Einstein Coll. Medicine, Montefiore Med. Ctr., Bronx, N.Y., 1979-80; asst. prof. radiology and nuclear medicine Albert Einstein Coll. Medicine and Montefiore Med. Ctr., Bronx, N.Y., 1980-84, assoc. prof. nuclear medicine, 1984—, assoc. prof. radiology, 1986—; adj. attending physician Montefiore Med. Ctr., 1979-87; chmn. Nuclear Medicine Grand Rounds: Greater N.Y., 1980-87; physician coord. Nuclear Medicine Technologist In-Service Tng. Program, 1982-86; cons. NIH, 1984-86, NIH Diagnostic Radiology, 1985-86. Assoc. editor Nuclear Medicine Ann., 5 vols., 1979-84, editor, 5 vols., 1985—; contbr. chpts. to books, articles to jours.; reviewer Jour. of Radiology, 1981—, mem. editorial adv. bd., 1985-86, assoc. editor, 1986—; reviewer. Jour. of Nuclear Medicine, 1981—, Am. Jour. of Roentgenology, 1986—, Gastroenterology, 1986—, Western Jour. of Medicine, 1985—; contbr. audiovisual programs and films. Recipient Saul Horowitz, Jr., Meml. award (Disting. Alumnus award), Mt. Sinai Sch. Medicine, 1980, Pres.' award, Am. Roentgen Ray Soc., 1979, Berta Rubinstein, M.D., Resident award, 1978, others. Mem. Radiol. Soc. N.Am. (mem. subcom. for nuclear medicine of program com., 1981, 82, 83, chmn. 1984, 85, 86), Soc. Nuclear Medicine (trustee 1983-87, 88—, sec.-treas. Correlative Imaging Council 1979-82, exec. bd. 1982-84, pres. 1984-86, exec. bd. 1986—, mem. acad. council 1980—, task force on interrelationship between nuclear medicine and nuclear magnetic resonance 1983-85 , gov. Greater N.Y. chpt. 1983-85, treas., 1985-86, 86-87, 2d ann. Tetalman award of Edn. and Research Found. 1982, mem., vice chmn. coms. and subcoms.), Soc. Gastrointestinal Radiologists, Am. Inst. Ultrasound in Medicine, N.Y. Acad. Scis., Assoc. Alumni Mt. Sinai Med. Ctr., Nuclear Radiology Club (chmn. 1983—). Phi Beta Kappa.

WEISS-WUNDER, LINDA TERESA, neuroscience research consultant; b. Passaic, N.J., June 15, 1960; d. Werner Emil and Rose-Marie Edith (Müller) Weiss; m. Richard John Wunder, Apr. 29, 1989; children: Katie Alison Wunder, Eric Joseph Wunder. BA, SUNY, Binghamton, 1982; PHD, Med. Coll. Pa., 1991. Neurosci. rsch. cons. U. Va., Charlottesville, 1992—. Contbr. articles to profl. jours. Co-dir. Community Presch. Orgn., Lake Monticello, Palmyra, Va., 1992-93. Mem. N.Y. Acad. Scis., Soc. for Neurosci. Home: 318 Jefferson Dr W Palmyra VA 22963

WEISWEILER, PETER, physician, educator; b. Jerxheim, Germany, Feb. 8, 1945; m. Helga Marth, Feb. 17, 1975; children: Silke, Katja. MD, U. Munich, 1973. Prof. medicine, head metabolic rsch. U. Munich, 1988—. Contbr. articles to profl. jours. Mem. European Soc. Clin. Investigation, Am. Heart Assn., others. Office: U Munich Metabolic Rsch, PO Box 152229 Holzstrasse 25, 80052 Munich Germany

WEISZ, ADRIAN, chemist; b. Cluj, Romania, Sept. 28, 1949; came to U.S., 1989; s. George and Berta (Hirschfeld) W.; m. Nancy Joyce Schatz, Mar. 30, 1986; children: Daniel, Noah, Ilana. BSc, Technion Israel Inst. Tech., 1973, MSc, 1979, DSc, 1984. Lady Davis fellow dept. chemistry Technion-Israel Inst. Tech., Haifa, 1987-89; sr. staff fellow Ctr. Food Safety and Applied Nutrition FDA, Washington, 1989—; vis. scientist Mario Negri Inst. Pharmacol. Rsch., Milan, 1979-80; vis. fellow lab. clin. sci. NIMH-NIH, Bethesda, Md., 1984-87. Contbr. articles to Jour. Am. Chem. Soc., Jour. Organic Chemistry, other profl. publs. Recipient Wolf Found. award, Herzlia, Israel, 1983. Mem. Am. Chem. Soc., Am. Soc. Mass Spectrometry, Israel Soc. Mass Spectrometry. Achievements include development of a countercurrent chromatography technique for separation of multigram mixtures of organic acids. Office: FDA HFS-126 200 C St SW Washington DC 20204

WEISZ, PAUL B(URG), physicist, chemical engineer; b. Pilsen, Czechoslovakia, July 2, 1919; naturalized, 1946; s. Alexander and Amalia (Sulc) W.; m. Rhoda A.M. Burg, Sept. 4, 1943; children: Ingrid B., P. Randall. Student, Tech. U. Berlin, 1938-39; BS, Auburn U., 1940; ScD, Swiss Fed. Inst. Tech., Zurich, 1965, ScD (hon.), 1980. Research physicist Bartol Research Found., Swarthmore, Pa., 1940-46; Research physicist Mobil Oil Corp. (formerly Socony Mobil Oil Corp.), 1958-61, sr. scientist, 1961-69, mgr. process research sect., 1967-69; mgr. Central Research Lab. Mobil Research & Devel. Corp., Princeton, N.J., 1969-82, sr. scientist and sci. adv., 1982-84; Disting. prof. chem. and bio-engring. sci. U. Pa., 1984-90, prof. emeritus, 1990—; adj. prof. Pa. State U., 1992—; cons. rsch. and tech. strategy, 1984—; vis. prof. Princeton U., 1974-76, mem. adv. council dept. chem. engring., 1973-78; mem. adv. and resource council Princeton U. Sch. Engring., 1974-78; chmn. center policy bd. Center for Catalytic Sci. and Tech., U. Del., 1977-81; mem. energy research adv. bd. U.S. Dept. Energy, 1985-90. Editor: Advances in Catalysis, 1956—; editorial bd. Jour. Catalysis, 1962-83, Chem. Engring. Communications, 1972-78; monthly columnist Sci. of the Possible, Chemtech, 1980-83; contbr. numerous articles to sci. jours.; holder 80 patents. Recipient ann. award Catalysis Club Phila., 1973, Lavoisier medal Société Chimique de France, 1983, Nat. Medal of Tech. U.S. Dept. Commerce Tech. Adminstrn., 1992. Fellow Am. Phys. Soc., Am. Inst. Chemists (Chem. Pioneer award 1974); mem. AICE (R.H. Wilhelm award 1978), Am. Chem. Soc. (sci. award South Jersey sect. 1963, E.V. Murphree award 1972, Leo Friend award 1977, chemistry of contemporary tech. problems award 1986, Carothers award 1987), N.Y. Acad. Scis., Nat. Acad. Engring., Soc. Chem. Industry (perkin medal 1986, Nat. medal of tech. 1992), Nassau Club (Princeton). Quaker. Office: Univ Pa Dept Chem Engring Philadelphia PA 19104

WEISZ, REUBEN R., neurology educator; b. Timisoara, Romania, June 11, 1946; s. Eugene and Yolanda W.; m. Jonina Schwartz, May 25, 1975; children: Ori, Elan. MD magna cum laude, U. Tel Aviv, Israel, 1973. Diplomate Am. Bd. Neurology. Asst. prof. Ind. U., Indpls., 1979-80; asst. prof. Chgo. Med. Sch., North Chgo., 1980-82, assoc. prof., 1982—; asst. chief dept. neurology North Chgo. VA Med. Ctr., 1980-91, chief dept. neurology, 1991. Author chpt. in book, several papers. Fellow Am. Acad. Neurology. Office: 2645 Washington Ste 400 Waukegan IL 60085

WEITHMAN, ALLAN STEPHEN, fisheries biologist; b. Tiffin, Ohio, Jan. 6, 1953; s. Allan Stephen and Bonnie Mae (Smith) W.; m. Angela Mary Daniels, May 11, 1974; children: Michael Robert, Scott Edward. MS in Fisheries, U. Mo., 1975, PhD, 1978. Cert. fisheries scientist. Fisheries rsch. biologist Mo. Dept. of Conservation, Columbia, 1978-84; environ. svcs. supr. Mo. Dept. of Conservation, Jefferson City, Mo., 1984—. Assoc. editor Transactions of the Am. Fisheries Soc., 1990-91; author: (with others) Inland Fisheries Management in North America, 1992; contbr. articles to profl. jours. including Fisheries and Creel and Angler Survey Symposium. Mem. Am. Fisheries Soc., Sigma Xi. Roman Catholic. Home: 2712 Mallard Ct Columbia MO 65201 Office: Mo Dept of Conservation PO Box 180 Jefferson City MO 65102

WEITZMAN, STANLEY HOWARD, ichthyologist; b. Mill Valley, Calif., Mar. 16, 1927; s. John Howard and Iva May (Hager) W.; m. Marilyn Jean Sohner, Feb. 7, 1948; children: Earl David, Anna Lisa. AB, U. Calif., Berkeley, 1951, MA, 1953; PhD, Stanford U., 1960. Sr. lab. technician U. Calif., Berkeley, 1951-56; instr. dept. anatomy Stanford U. Sch. Medicine, Palo Alto, Calif., 1956-62; assoc. curator fishes Smithsonian Instn., Washington, 1963-67, curator fishes, 1967—; editorial bd. Jour. Copeia, 1964-68. Contbr. over 200 articles to profl. jours. Grantee Smithsonian Instn., 1966—, NSF, 1965-68. Mem. Am. Soc. Ichthyologists & Herpetologists (life, treas. 1973-78, bd. govs. 1970-78, 80-85, Robert H. Gibbs Jr. Meml. award 1991), Soc. Brasileira de Ictiologia, Soc. Brasileira de Zoologia, Sigma Xi. Achievements include research in the evolution and classification of recent teleost fishes especially South Am. fishes and oceanic fishes, the biogeography and systematics of South Am. freshwater fishes. Office: Smithsonian Institution Museum Fishes Div Washington DC 20560

WEITZMANN, CARL JOSEPH, biochemist; b. Columbus, Ohio, Mar. 28, 1955; s. Walter Raul and Margaret (Noe) W.; m. Charlotte Ann Ellis, Jan. 7, 1984; children: Eric Gordon, Peter James. BA in Chemistry, Harper Coll., 1977; PhD in Biochemistry, U. Pa., 1989. Staff scientist Pall Corp., Glen Cove, N.Y., 1991—. Contbr. articles to profl. jours. Mem. AAAS. Achievements include production of large-subunit ribosomal RNA in vitro and methods of studying and isolating methylating enzymes and pseudouridine synthetases. Office: Pall Corp 30 Sea Cliff Ave Glen Cove NY 11542

WEKSLER, BABETTE BARBASH, hematologist; b. N.Y.C., Jan. 18, 1937; d. Philip and Roslyn (Weichsel) Barbash; m. Marc E. Weksler, 1958; children: David, Jennifer. BA, Swarthmore Coll., 1958; MD, Columbia U., 1963. Intern Bronx (N.Y.) Mcpl. Hosp. Ctr., 1963-64; resident in internal medicine Georgetown U. Hosp., Washington, 1965-67; prof. medicine Cornell U. Med. Coll., N.Y.C., 1981—; attending physician N.Y. Hosp., N.Y.C., 1981—. Editorial bd.: Am. Jour. Medicine, Stroke, Atherosclerosis; contbr. articles to med. jours. Mem. LWV, Am. Soc. Investigative Pathology, Am. Physiology Soc., Am. Heart Assn. (mem. exec. com., thrombosis coun.), Am. Soc. Hematology (chmn. edn. com. 1991—), Am. Soc. Clin. Investigation, Assn. Am. Physicians. Office: Cornell Univ Med Ctr Medicine-Hematology 1300 York Ave New York NY 10021-4805

WELBER, IRWIN, research laboratory executive; b. 1924; married. BS, Union Coll., 1948; MEE, Rensselaer Poly. Inst., 1950. With AT&T Bell Labs., Murray Hill, N.J., 1950-85, v.p. transmission systems, 1981; pres. Sandia Nat. Labs., Albuquerque, 1986-89, also bd. dirs.; corp. adv. bd. AMP Corp., 1990—.

WELCH, ASHLEY JAMES, engineering educator; b. Ft. Worth, May 3, 1933; married, 1952; 3 children. BS, Tex. Tech U., 1955; MS, So. Meth. U., 1959; PhD in Elec. Engring., Rice U., 1964. Aerophys. engr. Gen. Dynamics, Ft. Worth, 1957-60; instr. elec. engring. Rice U., 1960-64, from asst. to assoc. prof., 1964-74, dir. engring. computing facility, 1970-71, dir. biomed. engring. program, 1971-75; prof. elec. and biomed. engring. U. Tex., Austin, 1975—, Marion E. Forsman Centennial prof. engring., 1985—. Fellow IEEE, Am. Soc. Lasers Surg. Medicine (bd. dirs. 1989-92). Research in laser-tissue interaction, application of lasers in medicine. Office: Univ of Texas at Austin Dept of Elec and Computer Engring Austin TX 78712

WELCH, GEORGE OSMAN, electrical engineer; b. Houston, Aug. 16, 1948; s. George and Lucille (Polasek) W.; m. Darla True, June 22, 1974; children: Trey, Truley, Trevor. MEE, Rice U., 1971; postgrad., Baylor U. Registered profl. engr., Tex. Analyst Lockheed Co., Houston, 1972-73, 75-77, Tex. Instruments, Houston, 1979; sr. analyst Abbott Med. Electronics, Houston, 1979-81; cons. Control Applications, Houston, 1980-86; pres. Exec. Software Systems, Friendswood, Tex., 1986—. Charter mem. Rep. Presdl. Task Force, Washington,1 991-92. Grantee George R. Brown Found., 1971-72. Mem. Instrument Soc. Am. Home: 405 Chester Dr Friendswood TX 77546 Office: Exec Software Systems PO Box 828 Friendswood TX 77546

WELCH, PHILIP BURLAND, electronics company executive; b. Portland, Maine, Oct. 15, 1931; s. Philip Gerald Welch and Clara Jenny (Berry) Hawxwell; m. Sheila May Preston, May 19, 1960; children: Jahna Holly Welch Roth, Victoria Preston Welch Johnsen. Student, Berklee Coll., 1955-58. Nat. sales mgr. Akai Am. Ltd., Anaheim, Calif., 1970-73, BSR, USA, Blaupunkt, N.Y., 1973-76; nat. sales and mktg. mgr. Philips High Fidelity Labs, Ft. Wayne, Ind., 1976-79; dir. mktg. Pioneer Electronics, Moonachie, N.J., 1979-82; pres. Schneider N.Am. Ltd., Dayton, N.J., 1982-83; v.p. Lyons Assn., Indpls., 1986-88; pres. Nat. Electric Mktg. Co., Jacksonville, Fla., 1975—, Hemisphere Enterprises Corp., Jacksonville, 1988—; cons. ContraTech Corp., Portland, Oreg., 1986-87, Kukje Internat., N.Y.C., 1986, FCI Inc., Jersey City, N.J., 1985, others; cons. Multiform Products, Inc., Jacksonville, 1989-90, gen. mgr., 1990—; established Akai and Philips audio brand in U.S. Contbr. articles to profl. jours. With USAF, 1950-54. Named Man of Decade Audio/Video Cons. USA, 1982, Man of Yr. Soc. of Audio Cons., 1974. Republican. Avocations: flying, golf. Office: Phil Welch Enterprises 1099155 San Jose Blvd Jacksonville FL 32223-7229

WELCH, WILLIAM JOHN, astronomer, educator; b. Chester, Pa., Jan. 17, 1934; s. William Taylor and Ruth (van Leuven) W.; m. Jill C. Tartar, July 4, 1980; children by previous marriage—Eric, Leslie, Jeanette. B.S., Stanford U., 1955; M.S., U. Calif., Berkeley, 1958, Ph.D., 1960; Hon. Dr., Universite de Bordeaux, 1979. Asst. prof. elec. engring. U. Calif., Berkeley, 1960-65; asso. prof. U. Calif., 1965-69, prof. astronomy and elec. engring., 1969—; dir. Radio Astronomy Lab., 1972. Contbr. numerous articles on radio astronomy and related fields to profl. jours. Bd. trustees Asso. Univs., Inc.; mem. Arecibo Adv. Bd. Grantee in radio astronomy NSF; Grantee in radio astronomy NASA. Mem. Internat. Astron. Union, Internat. Union Radio Sci., AAAS, Am. Astron. Soc. Home: 2727 Shasta Rd Berkeley CA 94708-1923 Office: U Calif-Berkeley Radio Astronomy Lab 601 Campbell Hall Berkeley CA 94720

WELCHER, RONNIE DEAN, waste management and environmental services executive; b. Okemah, Okla., Dec. 9, 1946; s. William J.D. and Ruth Arlena (Reed) W.; Joyce Laverne Goodmon; children: Shelly, Kris, Brad. A in Adminstrv. Mgmt., Tulsa Jr. Coll., 1992—. Customer svc. rep. City of Tulsa, 1968-72, wastewater plant operator, 1972-76, lead operator III, 1976—; pres., owner Enviroserv of Okla., Sapulpa, 1976—; mem. Pub. Works Safety Com., Tulsa, 1990-92; approved instr. Okla. State Dept. Health, Oklahoma City, 1980. Bd. dirs. First Ch. of God, Sapulpa, 1984-86. With U.S. Army, 1966-68. Mem. Okla. Water Pollution Control Assn., Water Environ. Fedn., Double A Club of Okla. Home: 3110 Frontier Rd Sapulpa OK 74066 Office: Enviroserv of Okla 3110 Frontier Rd Sapulpa OK 74066

WELDON, WILLIAM FORREST, electrical and mechanical engineer, educator; b. San Marcos, Tex., Jan. 12, 1945; s. Forrest Jackson and Rubie Mae (Wilson) W.; m. Morey Sheppard McGonigle, July 30, 1968; children: William Embree, Seth Forrest. BS in Engring. Sci., Trinity U., San Antonio, 1967; MSME, U. Tex., 1970. Registered profl. engr., Tex. Engr. Cameron Iron Works, Houston, 1967-68; project engr. Glastron Boat Co., Austin, Tex., 1970-72; chief engr. Nalle Plastics Co., Austin, 1972-73; rsch. engr. U. Tex., Austin, 1973-77, tech. dir. Ctr. Electromechanics, 1977-85, dir. Ctr. Electromechanics, 1985-93, prof., 1985-93; mem. permanent com. Symposium on Electromagnetic Launch Tech., 1978—; cons. numerous cos. and govts., 1973—. Contbr. over 271 articles to profl. publs. Bd. dirs. Water Control & Improvement Dist. No. 10, Travis County, Tex., 1984—. Recipient Peter Mark medal Electromagnetic Launch Symposium, 1986, IR 100 award Indsl. Rsch. mag., 1983. Fellow Am. Soc. Mech. Engrs.; mem. IEEE (sr.), Nat. Soc. Profl. Engrs. Achievements include 28 patents for rotating electrical machines, pulsed power, and electromagnetic propulsion. Office: U Tex Ctr Electromechanics 10100 Burnet Rd Bldg 133 Austin TX 78758

WELLER, SOL WILLIAM, chemical engineering educator; b. Detroit, July 27, 1918; s. Ira and Bessie (Wieselthier) W.; m. Miriam Damick, June 11, 1943; children—Judith, Susan, Robert, Ira. B.S., Wayne State U., 1938; Ph.D., U. Chgo., 1941. Asst. chief coal hydrogenation U.S. Bur. Mines, Pitts., 1945-50; head fundamental rsch. Houdry Process Corp., Linwood, Pa., 1950-58; mgr. propulsion rsch. Ford Aeronutronic Co., Newport Beach, Calif., 1958-61; dir. chem. lab. and materials rsch. lab. Philco-Ford Co., Newport Beach, 1961-65; prof. chem. engring. SUNY-Buffalo, 1965—; emeritus, 1989; C.C. Furnas prof. SUNY-Buffalo, 1983—; vis. fellow Oxford U., 1989. Author numerous sci. papers, book chpts., ency. entries. Fulbright lectr. Madrid, 1975, Istanbul, 1980. Mem. Am. Chem. Soc. (chmn. Orange County sect. 1964, H.H. Storch award 1981, E.V. Murphree award 1982, Schoellkopf medal 1984, Dean's award 1991), ASTM (founder com. D32 on catalysts), Am. Inst. Chem. Engrs. (chmn. catalysis subcom. 1972-73). Achievements include patents in field. Office: SUNY Buffalo 305 Furnas Hall Buffalo NY 14260

WELLER, THOMAS HUCKLE, physician, emeritus educator; b. Ann Arbor, Mich., June 15, 1915; s. Carl V. and Elsie A. (Huckle) W.; m. Kathleen R. Fahey, Aug. 18, 1945; children: Peter Fahey, Nancy Kathleen, Robert Andrew, Janet Louise. A.B., U. Mich., 1936, M.S., 1937, LL.D. (hon.), 1956; M.D., Harvard, 1940; Sc.D., Gustavus Adolphus U., 1975, U. Mass., 1985; L.H.D., Lowell U., 1977. Diplomate Am. Bd. Pediatrics. Teaching fellow bacteriology Harvard Med. Sch., 1940-41, research fellow tropical medicine, pediatrics, 1947-48, instr. comparative pathology, tropical medicine, 1948-49, asst. prof. tropical pub. health Sch. Pub. Health, 1949-50, assoc. prof., 1950-54, Richard Pearson Strong prof. tropical pub. health, 1954-85, prof. emeritus, 1985—, head dept., 1954-81; intern bacteriology and pathology Children's Hosp., Boston, 1941; intern medicine Children's Hosp., 1942, asst. resident medicine, 1946, asst. dir. research div. infectious diseases, 1949-55; mem. commn. parasitic diseases Armed Forces Epidemiol. Bd., 1953-72, dir., 1953-59; charge parasitology, bacteriology, virology sections Antilles Dept. Med. Lab., P.R. Author sci. papers. Served to maj. M.C. AUS, 1942-46. Recipient E. Mead Johnson award for devel. tissue culture procedures in study virus diseases Am. Acad. Pediatrics, 1953, Kimble Methodology award, 1954, Nobel prize in physiology and medicine, 1954, George Ledlie prize, 1963, Weinstein Cerebral Palsy award, 1973, Stern Symposium honoree, 1972, Bristol award Infectious Diseases Soc. Am., 1980, Gold medal and diploma of honor U. Costa Rica, 1984, First Sci. Achievement award VZV Rsch. Found., 1993. Fellow Am. Acad. Arts and Scis.; mem. Harvey Soc., AMA, Am. Soc. Parasitologists, Am., Royal socs. tropical medicine and hygiene, Am. Pub. Health Assn., AAAS, Am. Epidemiological Soc., Nat. Acad. Scis., Am. Pediatric Soc., Assn. Am. Physicians, Soc. Exptl. Biology and Medicine, Am. Assn. Immunologists. Soc. Pediatric Research, Phi Beta Kappa., Sigma Xi, Alpha Omega Alpha. Home and Office: 56 Winding River Rd Needham MA 02192-1025

WELLING, LARRY WAYNE, pathologist, educator, physiologist; b. Kansas City, Mo., Oct. 22, 1940; s. Fredrick Joseph and Florence Mary (Hake) W.; m. Louise Watson, Dec. 28, 1967; 1 child, Matthew. AB in Chemistry, Rockhurst Coll., 1961; MD, U. Kans., 1965, PhD in Physiology, 1972. Diplomate Am. Bd. Pathology, Anatomic Pathology. Intern and resident in pathology U. Kans. Med. Ctr., Kansas City, 1965-68; asst. pathologist Armed Forces Inst. Pathology, Washington, 1968-70; rsch. fellow dept. pathology U. Kans. Med. Ctr., 1970-72; rsch. assoc. VA Med. Ctr., Kansas City, 1972-75, clin. investigator, 1975-80, staff pathologist, 1980—; asst. prof. U. Kans. Med. Ctr., 1972-77, assoc. prof., 1977-82, prof. pathology, 1982—, prof. physiology, 1984—. Author rsch. publs. in area of renal physiology; contbr. articles, revs. to profl. publs. Capt. U.S. Army, 1968-70. Grantee NIH, 1973-88, Am. Heart Assn., 1975-92. Mem. Am. Physiol. Soc., Am. Soc. Invest. Pathology, Am. Soc. Nephrology, Internat. Soc. Nephrology, Sigma Xi. Democrat. Achievements include research in relationship between histologic structure and transport function in renal tubules. Office: VA Med Ctr Rsch Svc 4801 Linwood Blvd Kansas City MO 64128

WELLNER, ROBERT BRIAN, physiologist; b. Syracuse, N.Y., Sept. 1, 1948; s. Robert Frank and Flora (Grome) W.; m. Dana Marie Barber, Aug. 5, 1972; children: Jeffrey Christopher, Elizabeth Mackenzie, Eric Douglas. BS in Biochemistry, SUNY, Syracuse, 1973, PhD in Biochemistry, 1980. Rsch. assoc. F. Hebert Sch. Medicine, Bethesda, Md., 1980-85; sr. staff fellow NIH, Bethesda, 1985-90; rsch. physiologist U.S. Army Med. Rsch. Inst. Infectious Diseases, Frederick, Md., 1990—; mem. vis. scholars selection com. NIH, Bethesda, 1988-90, peer grant reviewer, 1990—; speaker in field. Contbr. articles to Am. Jour. Physiology, Jour. Cellular Physiology, Jour. Biol. Chemistry, chpt. to book: Biology of the Salivary Glands, 1991. Mem. Am. Soc. Biochemistry and Molecular Biology, Am. Soc. Cell Biology (symposium speaker 1986), N.Y. Acad. Scis. Democrat. Achievements include elucidation of mechanisms involved in the autonomic regulation of epithelial transport events in human salivary glands. Office: US Army Med Rsch Inst Infectious Diseases Fort Detrick Frederick MD 21702-5011

WELLS, DAVID JOHN, university official, mechanical engineer; b. Ithaca, N.Y., Jan. 4, 1949; s. Arthur John and Dorothy Helen (Edwards) W.; m. Jane Baran, July 10, 1971; children: Jacob David, Abbe Grace, Anastasia Catherine. BS in Interdisciplinary Engring. and Mgmt., Clarkson U., 1972, MSME, 1980, PhD in Engring. Sci., 1985. Lic. profl. engr., Conn., Wyo. Planning engr. Newport News (Va.) Shipbuilding, 1973-76, Stone & Webster Engring., Boston, 1976-78; instr., counselor Clarkson U., Potsdam, N.Y., 1978-81, dir., 1986—86; project mgr., mgr. Combustion Engring., Windsor, Conn., 1981-86; dir. engring. and mgmt. program, mem. adminstrv. coun. Clarkson U., Potsdam, N.Y., 1986—, mem. exec. com., 1986—; chair, dean admission search com., 1992—; cons. in field; cons. Excellence in Edn. Action Plan, Potsdam Pub. Schs. Contbr. articles to profl. jours. Bd. dirs. Windsor Pub. Schs., 1985-86. Mem. ASME, IEEE (editorial bd. pres.), Engring. Mgmt. Soc., 21st Century Ltd. (edn. com.). Home: Rte 1 Bagdad Rd Box 368A Potsdam NY 13676 Office: Clarkson U Engring and Mgmt Program Potsdam NY 13699

WELLS, HERSCHEL JAMES, physician, former hospital administrator; b. Kirkland, Ark., Feb. 23, 1924; s. Alymer James and Martha Thelma (Cross) W.; m. Carmen Ruth Williams, Aug. 5, 1946; children: Judith Alliece Wells Jarecki, Pamela Elliece Wells McKinven, Joanne Olivia Wells Bennett. Student, Emory U., 1941-42, U. Ark., 1942-43; MD, U. Tenn., 1946. Rotating intern, then resident internal medicine Wayne County Gen. Hosp. (and Infirmary), Eloise, Mich., 1946-50; dir. infirmary div. Wayne County Gen. Hosp. (and Infirmary), 1955-65, gen. supt., 1965-74; dir. Wayne County Gen. Hosp. (Walter P. Reuther Meml. Long Term Care Facility), 1974-78; rev. physician DDS, SSA, Traverse City, Mich., 1978—. Served to maj. M.C. AUS, 1948-55. Mem. AMA, Mich. Med. Soc., Am. Fedn. Clin. Rsch., Masons (32 deg.), Alpha Kappa Kappa, Pi Kappa Alpha. Home and Office: 9651 N3 Rd Copemish MI 49625

WELLS, JAMES DAVID, JR., military officer; b. Carbondale, Pa., Mar. 30, 1969; s. James David and Muriel (Torch) W. BS, U.S. Mil. Acad., 1992. Commd. 2lt. U.S. Army, 1992; platoon leader U.S. Army Corps Engrs., Ft. Hood, Tex., 1992--. Mem. Am. Soc. Civil Engrs., Soc. Mil. Engrs. Republican. Roman Catholic. Home: 813 Railroad St 407 Twin Creek Dr # 14M Killeen TX 76543

WELLS, JAMES DOUGLAS, chemical engineer; b. Seattle, Apr. 24, 1953; s. John Douglas Wells and Betty Lo (Swafford) Garrard; m. Carol Denise King, Mar. 19, 1982; children: Rachel Lynn, Maegen Christine. AS, C.C. Allegheny County, Pitts., 1977; BSChemE, Cornell U., 1979. Engr. Dow Chem. USA, Midland, Mich., 1979-84, sr. engr., 1984-89, mfg. specialist, 1989-92, sr. mfg. specialist, 1992—; cons. Dowl Elanco Solids Processing Task Force, Midland, 1990—. V.p. Midland County Humane Soc., 1980-83. With U.S. Army, 1971-73. Republican. Office: Dow Chem USA 1000 Building Midland MI 48667

WELLS, JON BARRETT, engineer; b. Sewickley, Pa., Oct. 21, 1937; s. Calvin and Martha Barrett (Byrnes) W.; m. Nancy Lou LaFrance, Nov. 18, 1967; children: James Jonathan, Tiffany Lynn. BSEE, Calif. Poly U., 1961. Various positions Bell & Howell Co. Datatape Div., Pasadena, Calif., 1961-73; chief engr. Bell & Howell Co. Datatape Div., 1973-75; engring. mgr. Bell & Howell Co. Datatape Div., Baldwin Park, Calif., 1975-87, Lundy Fin. Systems, Rancho Cucamonga, Calif., 1987—; pres. Datatape Fed. Credit Union, Pasadena, 1978-93; v.p. Recognition Tech. Users Assn., Boston, 1987—; sec. Am. Nat. Standard Inst. X9B6, Washington, 1989-92. Patentee in field; contbr. articles to publs. Sec., founder Pasadena Neighborhood Housing Svcs., Pasadena, 1976-80. Mem. IEEE, Pasadena IBM Personal Computer Users Group, U.S. Power Squadrons, Aircraft Owners and Pilots Assn., Internat. Underwater Explorers Soc. Republican. Avocations: flying, boating, scuba diving, computers. Home: 2058 E Maverick La Verne CA 91750-2211 Office: Lundy Fin Systems 9431 Hyssop Dr Rancho Cucamonga CA 91730-6107

WELLS, LIONELLE DUDLEY, psychiatrist; b. Winnsboro, S.C., Nov. 22, 1921; s. Lionelle Dudley and Mary Wells; m. Mildred Wohltman, June 28, 1945 (dec. 1986); children: Lucia, Lionelle, John, Diane; m. Eilene Bromfield, Sept. 23, 1989. BS, U. S.C., 1943; MD, Med. U. S.C., 1945; grad., Boston Psychoanalytic Inst., 1960. Diplomate Am. Bd. Psychiatry and Neurology; lic. physician, S.C., Mass.; cert. in psychoanalysis. Intern Met. Hosp., N.Y.C., 1945-46; psychiatry resident VA Hosp., North Little Rock, Ark., 1948-50; asst. resident in Psychiatry Graylyn, Bowman-Gray Sch. Medicine, Winston-Salem, 1950-51; instr. psychiatry U. Ark., 1949-51, Mass. Gen. Hosp./Harvard Med. Sch., Boston, 1955-69; clin. instr. psychiatry Harvard Med. Sch., Boston, 1969-78; lectr. psychiatry Boston U. Sch. Medicine, 1977—; asst. clin. prof. psychiatry Harvard Med. Sch., 1978-93; lectr. psychiatry Tufts U. Med. Sch., Boston, Mass., 1981—; cons. staff Newton-Wellesley Hosp., Newton, Mass., 1983—; assoc psychiatrist Mass. Gen. Hosp., Boston, 1975-82, psychiatrist, 1982—; courtesy staff Waltham Weston Hosp. and Med. Ctr., 1977—; cons. Edith Nourse Rogers Meml. VA Med. Ctr., Bedford, Mass., 1966—; cons. in psychiatry VA Outpatient Clinic, Boston, 1959—, others in past; chmn. bd., chief exec. officer Bay State Health Care, 1984-91; nominating com. Am. Managed Care and Rev. Assn., 1988-89, others. Contbr. articles to profl. jours. Recipient Robert Wilson award, Med. U. S.C., 1943, 44. Fellow Am. Coll. Physician Execs., Am. Psychiat. Assn. (life); Mem. AMA, Am. Psychoanalytic Assn., Am. Geriatric Soc., Mass. Psychiat. Soc., Mass. Med. Soc., Boston Psychoanalytic Soc. and Inst., Boston Soc. for Gerontologic Psychiatry (membership chmn. and dir. 1974-76). Home and Office: 73 Rolling Ln Weston MA 02193-2474

WELSH, ELIZABETH ANN, immunologist, research scientist; b. Eureka, Calif., June 16, 1960; d. James Francis Welsh and Marie Louise (Roy) Kelleher-Roy. AB, U. Calif., Berkeley, 1983; PhD, U. Tex., Houston, 1990. Staff rsch. assoc. U. Calif., San Francisco, 1983-85; predoctoral rsch. fellow M.D. Anderson Cancer Ctr., U. Tex., Houston, 1985-90; postdoctoral rsch. fellow U. Tex. Med. Sch., Houston, 1990-91, Stanford (Calif.) U. Sch. Medicine, 1991—. Contbr. articles to profl. jours. Recipient R.E. Smith fellowship U. Tex. M.D. Anderson Cancer Ctr., 1986-87, Rosalie B. Hite Cancer Rsch. fellowship, 1987-90, Bristol Myers Squibb Rsch. fellowship Dermatology Found., 1991, Roche Labs. Rsch. fellowship, 1992, Walter V. and Idun Berry Rsch. fellowship Stanford U., 1992-95. Mem. AAAS, Am. Assn. Immunologists, Soc. Investigative Dermatology.

WELTER, LEE ORRIN, anesthesiologist, biomedical engineer; b. Wyandotte, Mich., Feb. 17, 1943; s. Leo Joseph Patrick and Mary Geraldine (Richards) W.; m. Patricia Ann Kinney, Dec. 26, 1964 (div. June 1983); children: Leanne Elizabeth Welter Folsom, Beth Ann Welter Belknap, Richard Patrick, Justin Kinney; m. Lynda Joyce Wong, Feb. 14,1985; stepchildren: Marcus Joseph Laven, Monique Ming Laven. MS in Bioengring., U. Mich., 1968, MD, 1974. Diplomate Am. Bd. Anesthesiology. Systems analyst Wyandotte Gen. Hosp., 1969-71; clin. engr. P.J. Pollard Assocs., Warren, Mich., 1972; house officer anesthesiology U. Mich. Med. Sch., Ann Arbor, 1974-77, instr. anesthesiology, 1977-79; anesthesiologist Sacramento, 1979-86; vis. instr. anesthesiology U. Mich. Med. Sch., Ann Arbor, 1986; anesthesiologist Chico, Calif., 1986—; med. advisor BoMed Med. Mfg. Ltd., Irvine, Calif., 1980-84. Bd. dirs. Chico Tennis Assn., 1992—. Mem. Am. Soc. Anesthesiologists, Human Factors Soc., Soc. Tech. in Anesthesia, Assn. for Advancement of Med. Instrumentation. Home: PO Box 8540 Chico CA 95927-8540

WELTMAN, JOEL KENNETH, immunologist; b. N.Y.C., May 22, 1933; s. Charles and Frances (Seasonwein) W.; m. K. Reulla Avatichi, June 26, 1956; children: Alica C., Orlee R. BA, NYU, 1954; MD, SUNY, Bklyn., 1958; PhD, U. Colo., 1963; MA, Brown U., Providence, 1972. Diplomate Am. Bd. Allergy and Immunology. Clin. assoc. prof. medicine Brown U., Providence, 1979—; pres. Itox, Inc., Providence, 1988—. Fellow Am. Soc. Biol. Chemists; mem. New Eng. Soc. Allergy (councilman 1992—). Achievements include patent for screening antibodies. Office: 300 Hanover St Fall River MA 02720

WELTY, JAMES R., mechanical engineer, educator; b. Garden City, Kansas, Oct. 23, 1933. BS, Oreg. State U., 1954, MS, 1959, PhD in Chem. Engrng., 1962. Test engr. Pratt & Whitney Aircraft, 1954; instr. mech. engring. Oreg. State U., 1958-61, from asst. prof. to assoc. prof., 1962-67, prof. mech. engring., 1967—, head dept. mech. engring., 1970—; rsch. engr. U.S. Bur. Mines, Oreg., 1962-64; mem. tech. staff Bell Tel. Labs., Pa., 1964; vis. prof. Thayer Sch. Engring. Dartmouth Coll., 1967. Recipient Edwin F. Church medal Am. Soc. of Mech. Engrs., 1990, rsch. grants U.S. Environ. Protection Agy., 1968-71, U.S. AEC, U.S. Dept. Energy, NSF, 1969—. Mem. Am. Soc. Mech. Engrs. Achievements include research in heat transfer, natural convection in liquid metals, non Newtonian fluids in natural and forced flows, numerical modeling of thermal plumes, fluidized bed heat transfer. Office: Oregon St Univ College of Engineering Mechanical Eng Dept Corvallis OR 97311

WELTY, KENNETH HARRY, civil engineer; b. Spirit Lake, Iowa, July 16, 1933; s. Kenneth Bertram and Josephine Louise (Tott) W.; m. Patricia Julienne Fremming, June 22, 1958; chidlren: David Keith, Michael Kent, Lisa Ann Welty Pagliocchini. BS in Civil Engring., U. Ariz., 1961. Highway engr. trainee U.S. Bur. Pub. Rds., Washington, 1961-64; highway rsch. engr. U.S. Dept. Trans. Fed. Highway Adminstrn., Washington, 1964-67, highway engr. Office Planning, 1967--. Leader Boy Scouts Am., Springfield, Va., 1970-82; watch coord. Fairfax County Neighborhood Watch, Springfield, 1990--. With U.S. Army, 1955-57. Mem. Tau Beta Pi. Methodist. Home: 6423 Cabell Ct Springfield VA 22150-1326

WELZIG, WERNER, philologist; b. Vienna, Austria, Aug. 13, 1935. PhD in Philosophy, 1958. Asst. Vienna Univ., 1958, habil., 1963, prof. German lang., lit., 1968—; guest. prof. Univ. Southern Calif., 1969—. Author Beispielhafte Figuren. Tor. Abenteurer u. Einsiedler bei Grimmelshausen, 1963, Der deutsche Roman im 20. Jhdt., 1967, Der Typus der dt. Balladenauth, 1977. Mem. Austrian Acad. Scis. Office: Austrian Acad Sci, Dr Ignaz Seipel-Pl 2, 1010 Vienna Austria*

WEN, CHENG PAUL, electrical engineer; b. Canton, China, July 3, 1933; came to U.S., 1952; s. Sau Paul and Chukling Tu Man; m. Laura T. Chan, Aug. 23, 1958; children Theresa A., Lisa A. BSEE, U. Mich., 1956, MSEE, 1957, PhD in Elec. Engring., 1963. Mem. of tech. staff RCA Labs., Princeton, N.J., 1963-74; group leader Rockwell Internat., Thousand Oaks, Calif., 1974-76, Dallas, 1976-78; sect. head Rockwell Internat., Anaheim, Calif., 1978-82; dept. mgr. Hughes Aircraft, Torrance, Calif., 1982-90; chief scientist Hughes Aircraft, 1990—; adj. instr. Calif. State U., Northlake, 1986-90. Contbr. 30 tech. papers to profl. meetings and jours. V.p. Conejo Swimming Assn., Thousand Oaks, Calif., 1976, Dallas Swim Club, 1977. Mem. IEEE (sr.), Am. Phys. Soc. Roman Catholic. Achievements include 20 U.S. patents including the invention of Coplanar Wave Guides, 1969. Office: Hughes Aircraft Co 3100 W Lomita Blvd Torrance CA 90509

WENDHOLT, NORMAN WILLIAM, civil engineer; b. Huntingburg, Ind., Jan. 25, 1958; s. Henry Charles and Edwina (Elmer) W.; m. Wilma Joan Morlan, Nov. 20, 1982; children: Brent Alan, Jill Ann, Beth Lynn. BS, U. So. Ind., 1980. Registered profl. engr., Ind. Project engr. Biagi and Assocs., Evansville, Ind., 1980-88; project civil engr. Kramer Group, Evansville, 1988-90; civil engring. mgr. John Brown, Inc., Newburgh, Ind., 1990; civil engr. Wilderman and Assocs., Evansville, 1990-91; hwy engr. County of DuBois, Jasper, Ind., 1991—. Mem. ASCE (chpt. v.p. 1990-92), NSPE, Ind. Assn. County Engrs. Home: 6025 S 500 E Saint Anthony IN 47575 Office: DuBois County Hwy Dept 1066 S State Rd 162 Jasper IN 47546

WENDT, HANS WERNER, life scientist, educator; b. Berlin, July 25, 1923; came to U.S., 1953; naturalized, 1967; s. Hans O. and Alice (Creutzburg) W.; m. Martha A. Linger, Dec. 23, 1956 (div. 1979); children: Alexander, Christopher, Sandra; m. Judith A. Hammer, June 25, 1988. MSc, U. Hamburg, Germany, 1949; PhD in Psychopharmcology, U. Marburg, Germany, 1953. Psychology diploma, Germany. Rsch. asst. U. Marburg, 1949-53; rsch. assoc. Wesleyan U. and Office Naval Rsch., Middletown, Conn., 1952-53; asst. prof., field dir. internat. project U. Mainz, Germany, 1955-59; engring. psychologist to prin. human factors scientist Link Aviation, Apollo Simulator Systems, Binghamton, N.Y., 1959-61; assoc. to prof. psychology Valparaiso (Ind.) U., 1961-68; prof. psychology Macalester Coll., St. Paul, 1968—; rsch. fellow chronobiology/geomedicine U. Minn., Mpls., 1980—; vis. scholar Harvard U., 1953-54; various cons. and reviewer, 1961—; hon. prof. sci. U. Marburg, Germany, 1971—. Contbr. 80 articles to profl. jours., chpts. to books. Recipient Disting. Sr. Scientist award, Alexander von Humboldt Found., 1976. Mem. Internat. Soc. Biometeorology, Internat. Soc. Chronobiology, Bioelectromagnetics Soc., Soc. Sci. Exploration, N.Y. Acad. Scis., others. Home: 2180 Lower Saint Dennis Rd Saint Paul MN 55116-2831 Office: Macalester Coll 1600 Grand Ave Saint Paul MN 55105-1899

WENGERD, SHERMAN ALEXANDER, geologist, educator; b. Millersburg, Ohio, Feb. 17, 1915; s. Allen Stephen and Elizabeth (Miller) W.; m. Florence Margaret Mather, June 12, 1940; children: Anne Marie Wengerd Riffey, Timothy Mather (dec.), Diana Elizabeth Wengerd Roach, Stephanie Katherine Wengerd Allen. AB, Coll. Wooster, 1936; MA, Harvard U., 1938, PhD, 1947. Registered profl. engr., N.Mex.; profl. geologist; lic. pilot, FAA. Geophysicist, Shell Oil Co., 1937; mining geologist, Ramshorn, Idaho, 1938; Austin teaching fellow Harvard U., 1938-40; rsch. petroleum geologist Shell, Mid-continent, 1940-42, 45-47; prof. geology U. N.Mex., 1947-76; ret., 1976; disting. prof. petroleum geology, 1982; rsch. geologist, 1947—, Petroleum Ind., 1976—; past co-owner Pub. Lands Exploration, Inc., Corona and Capitan Oil Cos.; ltd. ptnr. Rio Petro Oil Co., Dallas. Col. aide-de-camp staff Gov. State N.Mex., 1992. Served to lt. comdr. USNR, 1942-45; capt. Res., ret. Recipient Disting. Alumnus citation Coll. Wooster, 1979. Author chpts. in textbooks and encys., articles in geol. bulls., newsletters. Fellow Geol. Soc. Am., Explorers Club of N.Y., Ret. Officers Assn.; mem. Nat. Assn. Scholars, Four Corners Geol. Soc. (hon. life mem., pres. 1953), N.Mex. Geol. Soc. (hon. life mem. 1972—), Albuquerque Geol. Soc. (hon. life mem. 1989—), Am. Assn. Petroleum Geologists (hon. life, nat. editor 1957-59, pres. 1971-72, chmn. adv. council 1972-73, Presdl. award 1948, Sydney Powers Memorial medal 1992), Am. Petroleum Inst. (acad. mem., exploration com. 1970-72), Am. Inst. Profl. Geologists (state sect. pres. 1970, nat. editor 1965-66), Soc. Econ. Paleontologists and Mineralogists (dir. found. 1982-86), Aircraft Owners and Pilots Assn., Nat. Aero. Assn. (life), OX5 Aviation Pioneers (life), Silver Wings Flying Fraternity (life), Thomas L. Popejoy Soc., Naval Res. Assn. (life), Assn. Naval Aviation, Am. Legion (life), VFW (life), U. N.Mex. 21 Club, Sigma Xi, Sigma Gamma Epsilon, Phi Kappa Phi. Home: 1040 Stanford Dr NE Albuquerque NM 87106-3720

WENNER, EDWARD JAMES, III, microelectronics executive, facility/safety engineer; b. Covington, Va., Sept. 19, 1956; s. Edward James Jr. and Doris J. (Curtis) W.; m. Linda Christine Guptill, Dec. 31, 1982; children: Edward James IV, Cassandra Lea. BS in Indsl. Engring., U. Ark., 1978. Registered profl. engr., Tex. Indsl. engr. Rockwell Internat. (MED), Newport Beach, Calif., 1978-79; facilities engr. Xerox Corp. (Microelectronics Ctr.), El Segundo, Calif., 1979-88; process engr. microelectronics CRS Sirrine Engrs., Inc., Greenville, S.C., 1988-90; facility mgr. Microelectronics and Engring. Rsch. Bldg. U. Tex., Austin, 1990—. Mem. Semiconductor Safety Assn. Office: U Tex at Austin Code 78600 BRC Microelectronics Engring Rsch Austin TX 78712

WENNERSTROM, ARTHUR JOHN, aeronautical engineer; b. N.Y.C., Jan. 11, 1935; s. Albert Eugene and Adele (Trebus) W.; m. Bonita Gay Westenberg, Sept. 6, 1969 (div. Jan. 1989); children: Bjorn Erik, Erika Lindsay; m. Vicki Lynn Merrick, Feb. 17, 1990. BS in Mech. Engring., Duke U., 1956; MS in Aero. Engring., MIT, 1958; DSc of Tech., Swiss Fedn. Inst. Tech., Zurich, 1965. Sr. engr. Aircraft Armaments, Inc., Cockeysville, Md., 1958-59; rsch. engr. Sulzer Bros., Ltd., Winterthur, Switzerland, 1960-62; project engr. No. Rsch. and Engring. Corp., Cambridge, Mass., 1965-67; rsch. leader Air Force Aerospace Rsch. Lab., Dayton, Ohio, 1967-75, Air Force Aero Propulsion Lab., Dayton, 1975-91; dir. NATO Adv. Group for Aerospace R & D, Paris, 1991-94; mem. tech. adv. com., von Karman Inst. for Fluid Dynamics, Rhode-St-Genese, Belgium, 1988-94, bd. dirs.; lectr. in field. Contbr. articles to profl. jours. 1st lt., USAF, 1962-65. Recipient Cliff Garrett Turbo Machinery award Soc. Automotive Engrs., 1986; named Fed. Profl. Employee of Yr. Dayton C. of C., 1975; fellow Air Force Wright Aeronautical Labs., 1987. Fellow AIAA (assoc. editor 1980-82, Air Breaking Propulsion award 1979), ASME (chmn. turbomachinery com., gas turbine divsn. 1973-75, mem. exec. com. 1977-82, chmn. 1980-81, program chmn. internat. gas turbine conf. 1976, Beijing internat. gas turbine symposium 1985, mem. nat. nominating com. 1985-87, mem. TOPC bd. on rsch. 1985-88, mem. at large energy conversion group 1986-88, mem. bd. comms. 1989-91, editor Jour. Engring. for Gas Turbines and Power 1983-88, chmn. bd. editors 1989-91, founder, editor Jour. Turbomachinery 1986-88, R. Tom Sawyer award 1993). Achievements include introduction of wide-chord integrally-bladed fan, introduction of swept blading into mil. aircraft turbine engines; 5 patents in field. Home: 58 Blvd Emile Augier, Paris 75016, France Office: Adv Group for Aerospace, 7 Rue Ancelle, Neuilly sur Seine 92200, France

WENTRUP, CURT, chemist, educator; b. Holtug, Denmark, July 11, 1942; s. Peder Wernemann and Marta (Christiansen) Madsen; m. Edeline Marie Byrne, Mar. 27, 1975; children: Jens Killian, Neal Torben, Cormac Erik. Degree, U. Copenhagen, 1966, Dr. scient., 1977; PhD, Australian Nat. U., 1969. Chartered chemist. Rsch. asst. U. de Lausanne, Switzerland, 1969-71, charge de cours, 1971-73, pvt.-docent, 1973-76, maitre-asst., 1973-76; prof. U. Marburg, Germany, 1976-85; prof., head organic chemistry sect. U. Queensland, Brisbane, Australia, 1985—. Author: Reaktive Zwischenstufen, 1978, Reactive Molecules, 1984; contbr. 150 articles to Jour. Am. Chem. Soc., Jour. Organic Chemistry, Jour. Phys. Chemistry, Acta Chemica Scandinavica, Helvetica Chimica Acta, Angewandte Chemie. Peter and Emma Thomson scholar, 1964-66; Alexander von Humboldt fellow, 1969; rcipient Van't Hoff award Royal Netherlands Chem. Soc., 1973; grantee numerous orgns. Fellow Royal Soc. Chemistry, Royal Australian Chem. Inst.; mem. Am. Chem. Soc., Deutsche Chemische Gesellschaft. Achievements include research in reactive intermediates, in particular developments of flash vacuum pyrolysis/matrix isolation. Office: U Queensland, Dept Chemistry, Brisbane 4072, Australia

WENTS, DORIS ROBERTA, psychologist; b. L.A., Aug. 26, 1944; d. John Henry and Julia (Cole) W. BA, UCLA, 1966; MA, San Francisco State U., 1968; postgrad., Calif. State U., L.A., 1989-90, Claremont (Calif.) Grad. Sch., 1990—. Lic. ednl. psychologist, credentialed sch. psychologist, Calif. Sch. psychologist Diagnostic Sch. for Neurologically Handicapped Children, L.A., 1969-86; pvt. practice Monterey Park, Calif., 1986—; instr. Calif. State U., L.A., 1977. Co-author: Southern California Ordinal Scales of Development, 1977. Mem. Western Psychol. Assn., L.A. World Affairs Coun., L.A. Conservancy, Zeta Tau Alpha (officer Santa Monica alumnae chpt. 1970—, Cert. of Merit 1979), Sigma Xi. Avocations: travel, watersports, theatre, bridge, photography. Office: Claremont Grad Sch Dept Psychology Claremont CA 91711

WENTZLER, THOMAS H., chemical company executive; b. Saginaw, Mich., May 14, 1947; s. John McConnell and Betty Elaine (Hynan) W.; m. Kathleen Kissell, Sept. 13, 1969; children: Paul, Rudy. BS in Mineral Processing Engring., Pa. State U., 1969, MS in Mineral Processing, 1971; MBA, U. Mich., 1973. From rsch. analyst to market mgr. Dow Chem. Co., Houston, Midland, Mich., 1973-81; v.p. Tetra Resources, The Woodlands, Tex., 1981-83, pres., 1983-86; sr. v.p. Tetra Technologies, The Woodlands, 1987-93, sec., 1982—; vice chmn. Tetra Technologies, The Woodlands, 1987—, 1987-93. Mem. Am. Inst. Chem. Engrs., Am. Inst. Mining Engrs., Sigma Xi, Phi Kappa Phi. Roman Catholic. Home: 28 Autumn Cres The Woodlands TX 77381 Office: Tetra Technologies 25025 I-45 N Spring TX 77380-2176

WEPMAN, BARRY JAY, psychologist; b. Pittsfield, Mass., Mar. 12, 1944; s. William and Frances (Winetsky) W.; m. Molly W. Donovan, Apr. 8, 1989; children: Joshua M., Noah E. BS, Tufts U., 1965, DMD, 1970; PhD, U. Houston, 1973. Teaching asst. U. Houston, 1971-73; asst. prof. NYU, 1973-74; asst. prof. U. Medicine and Dentistry N.J., Newark, 1974-78, assoc. prof., 1978-80, clin. assoc. prof., 1980-82; adj. faculty Rutgers U., Piscataway, N.J., 1976-79; cons. psychologist Grace Counseling Ctr., Madison, N.J., 1988-91. Contbr. articles to profl. jours. and chpts. to books; mem. editorial bd. Voices: The Art and Science of Psychotherapy, 1988—, The Psychotherapy Patient, 1986—. NIDR fellow NIH, 1970-73. Mem. Am. Psychol. Assn., Am. Acad. Psychotherapy (exec. coun. 1985-88). Office: 3254 Jones Ct NW Washington DC 20007

WERBACH, MELVYN ROY, physician, writer; b. Tarzana, Calif., Nov. 11, 1940; s. Samuel and Martha (Robbins) W.; m. Gail Beth Leibsohn, June 20, 1967; children: Kevin, Adam. BA, Columbia Coll., N.Y.C., 1962; MD, Tufts U., Boston, 1966. Diplomate Am. Bd. Psychiatry and Neurology. Intern VA Hosp., Bklyn., 1966-67; resident in psychiatry Cedars-Sinai Med. Ctr., L.A., 1969-71; chmn. dept. mental health Ross-Loos Med. Group, L.A., 1972-75; dir. psychol. svcs., clin. biofeedback UCLA Hosp. and Clinics, 1976-80; pres. Third Line Press, 1986—; asst. clin. prof. Sch. Medicine, UCLA, 1978—; mem. nutritional adv. bd. Cancer Treatment Ctrs. Am., 1989—. Author: Third Line Medicine, 1986, Nutritional Influences on Illness, 1987, 2d edit., 1993, Nutritional Influences on Mental Illness, 1991, Healing Through Nutrition, 1993; Nutritional Medicine,, Eng., 1990—; mem. editorial bd. Jour. of Nutritional Medicine, 1993—; mem. med. adv. bd. Let's Live Mag., 1989—; contbr. articles to med. jours. Mem. Biofeedback Soc. Calif. (life, pres. 1977, Cert. Honor 1985).

WERBOS, PAUL JOHN, neural research director; b. Darby, Pa., Sept. 4, 1947; s. Walter Joseph and Margaret Mary (Donohue) W.; m. Lily Fountain, July 13, 1979; children: Elizabeth, Alexander, Maia. BA magna cum laude, Harvard U., 1967; MSc, London Sch. Econs., 1968; MA, Harvard U., 1969, PhD, 1974. Rsch. assoc. MIT, Cambridge, Mass., 1973-75; asst. prof. U. Md., College Park, 1975-78; math. statistician U.S. Census Bur., Suitland, Md., 1978-79; energy analyst U.S. Dept. Energy, Washington, 1979-88, 89; program dir. NSF, Washington, 1988, 89—. Author: The Roots of Backpropagation: From Ordered Derivatives to Neural Network & Political Forecasting, 1993; contbr. chpt. to Handbook of Intelligent Control, 1992. Regional dir., Washington rep. L-5 Soc. (merged with Nat. Space Soc.), Washington, 1980s. Mem. Internat. Neural Network Soc. (pres. 1991-92, sec. 1990). Quaker-Universalist. Achievements include patent pending for elastic fuzzy logic and associated adaptation techniques; devised theory of intelligence; created alternative formulation of quantum theory. Home: 8411 48th Ave College Park MD 20740 Office: NSF Rm 1151 1776 G St NW Washington DC 20550

WERMIEL, JARED SAM, mechanical engineer; b. Washington, May 15, 1949; s. Nathan and Ruth (Glass) W.; m. Carolyn Rose Wermiel, Aug. 24, 1975; children: Joel, Beth. BSChemE, Drexel U., 1972. Registered profl. engr., Md. Mech. engr. Bechtel Power Corp., Gaithersburg, Md., 1972-78; aux. system engr. U.S. Nuclear Regulatory Commn., Bethesda, Md., 1978-82; sect. chief U.S. Nuclear Regulatory Commn., Bethesda, Rockville, Md., 1982-90; br. chief U.S. Nuclear Regulatory Commn., Rockville, Md., 1990—. Mem. AIChE. Home: 16635 Shea Ln Gaithersburg MD 20877 Office: US Nuclear Regulatory Commn OWFN 8H3 Washington DC 20555

WERNER, ANDREW JOSEPH, physician, endocrinologist, musicologist; b. Budapest, Hungary, June 5, 1936; came to U.S., 1956; s. Steven and Clara (Gutfreund) W.; m. Elaine Audrey Friedenn; 1 child, Andrea Lisa. MD, Med. Coll. of Va., 1962. Intern Kings County Hosp. Downstate Med. Ctr., Bklyn., 1962-63; resident in internal medicine N.Y. Med. Coll. Flower and 5th Ave. Hosps., N.Y.C., 1963-65; NIH fellow in endocrinology Mt. Sinai Hosp., N.Y.C., 1965-66; attending physician Mt. Sinai Med. Ctr., N.Y.C., 1966—; mem. professorial faculty Mt. Sinai Sch. of Medicine, N.Y.C., 1966—; cons. in endocrinology Hosp. for Joint Diseases-Orthopedic Inst., N.Y.C., 1979—. Author: Wolfgang Amadeus Mozart, Summa Summarum,

1990; co-author: Malignant Tumors of the Thyroid: Clinical Concepts and Controversies, 1992. Patron Met. Opera Assn. Recipient Festungs Medallion State of Salzburg, Austria, 1989. Fellow N.Y. Acad. Medicine; mem. Am. Diabetes Assn., Endocrine Soc., N.Y. Acad. Scis., Am. Assn. Clin. Endocrinologists, The Philharmonic-Symphony Soc. of N.Y., Am. Inst. for Verdi Studies, Internat. Stiftung Mozarteum-Salzburg (Austria), Internat. Salzburg Assn. (pres. N.Y.C. and Salzburg 1989—). Office: 1112 Park Ave New York NY 10128-1235

WERNER, FRANK DAVID, aeronautical engineer; b. Junction City, Kans., Mar. 14, 1922; s. Joseph and Elizabeth Gertrude (Pearce) W.; m. Alice Eva Martel, June 1, 1946; children: JoAnn, David, Robert. BS, Kans. State U., 1943; MS, U. Minn., 1948, PhD, 1955. Registered profl. engr., Wyo. Jr. physicist Applied Physics Lab., Johns Hopkins U., Silver Spring, Md., 1943-47; scientist, lectr. U. Minn., Mpls., 1947-56; pres., dir. Rosemount Engring. Co. (now Rosemount, Inc.), Mpls., 1956-68, Origin Inc., Jackson, Wyo., 1968—; bd. chmn., dir. Tech Line Corp., Jackson, 1987—; bd. dirs. various corps. Contbr. articles to profl. jours. Dir., pres. bd. dirs. Am. Rehab. Found., Mpls., 1968-71; bd. dirs. Interstudy, Mpls., 1971, Jackson Hole Group, Inc., Jackson, 1992, Teton Village (Wyo.) Svc. and Improvement Dist., 1989-92; fire chief Teton Village Vol. Fire Dept., 1972-82. Mem. Am. Phys. Soc., Am. Instrument Soc., Teton Pines County Club, Sigma Xi. Achievements include 70 patents in the field of measuring instruments, mainly for aerospace measurements of temperature, pressure, fluid flow. Home: Box 70 Teton Village WY 83025 Office: Origin Inc Star Rte Box 9 Jackson WY 83001

WERNER, MARK HENRY, neurologist, researcher; b. Louisville, June 10, 1954; s. Stanley Gerald and Sara (Berolzheimer) W.; m. Yana Serita Banks, June 2, 1985; children: Adam, Isaac. BA, U. N.C., 1975; MD, Bowman Gray, 1981. Medicine resident New Hanover Hosp., Wilmington, N.C., 1981-82; neurology resident Vanderbilt Med. Ctr., Nashville, 1982-85; neuro-oncology fellow Duke Med. Ctr., Durham, N.C., 1985-87; rsch. assoc. Duke Med. Ctr., Durham, 1987-91; asst. prof. U. South Fla., Tampa, 1991—. Contbr. articles to profl. jours. Dem. vol. Durham, 1990. Fellow Nat. Cancer Ctr., 1986-87, Assn. Brain Tumor Rsch., 1986-88, 89-91, Am. Heart Assn., 1989-91. Mem. AAAS, Am. Acad. Neurology, Phi Beta Kappa. Jewish. Achievements include first to publish presence of EGF receptors in neurons of human nervous system, that diacyglycerol could overcome aspirin inhibition of platelet aggregation; rsch. in immunoreactive EGF receptors in human gliomas, diacyglycerol accumulation in platelets, multiphasic diacylglycerol generation in platelets, differentiation of glioma cells. Office: H Lee Moffit Cancer Ctr Neurology Svc 12902 Magnolia Dr Tampa FL 33612-9497

WERNER, ROY ANTHONY, aerospace executive; b. Alexandria, Va., June 30, 1944; s. William Frederick and Mary Audrey (Barksdale) W.; m. Paula Ann Privett, June 8, 1969; children: Kelly Rene, Brent Alastair. BA, U. Cen. Fla., 1970; MPhil, Oxford U., 1973; MBA, Claremont (Calif.) Grad. Sch., 1986. Reporter St. Petersburg (Fla.) Times, 1968-69; assoc. dir. White House Conf. on Youth, 1970-71; exec. sec. Oxford Strategic Studies Group, 1971-73; internat. officer Fed. Energy Adminstrn., 1973-74; mem. legis. staff U.S. Senate, Washington, 1974-79; prin. dept. asst. Sec. of The Army, Washington, 1979-81; dir. policy rsch. Northrop Corp., L.A., 1982-83, spl. asst. to sr. v.p., mktg. to mgr. program planning and analysis electronics system divsn., 1989; chmn. U.S. delegation/polit. com. Atlantic Treaty Assn. Meeting, Brussels, 1989, mem. U.S. delegation, Paris, 1990, others; staff dir. East Asian and Pacific Affairs subcom. U.S. Sentate Fgn. Rels. Com., 1977-79; mem. Atlantic Coun. of U.S., 1985—; speaker Pacific Parliamentary Caucus, numerous acad. confs. in U.S. and East Asia. Editorial bd. Global Affairs, 1982-86; contbr. numerous articles to profl. jours./publs. Pres. Irvine (Calif.) Boys and Girls Club, 1990-92, v.p., 1989-90, bd. dirs., 1987—; treas. Irvine Temporary Housing, 1988-91, bd. dirs., 1986-91; chmn. fin. com. Outreach Univ. United Meth. Ch., Irvine, 1989-90, others; corp. sec. Irvine Housing Opportunities, 1988-89, bd. dirs. Harbor Area Boys and Girls Clubs, 1991-94. Maj. USAR, 1963-81. Recipient Disting. Alumnus award U. Ctrl. Fla. Alumni Assn., Orlando, 1982, Outstanding Civilian Svc. medal Dept. of the Army, 1981, Atlantic Coun. of the U.S. Sr. Rsch. Fellow, 1988-89. Mem. Am. Fgn. Svc. Assn., Am. Def. Preparedness Com. Democrat. Methodist. Home: 28 Fox Hill Irvine CA 92714-5493

WERNICK, JUSTIN, podiatrist, educator; b. N.Y.C., Feb. 26, 1936; s. Charles and Ethel (Crown) W.; m. Susan Schoenfeld, Oct. 16, 1960; children: Elissa, Peter. D Podiatric Medicine, N.Y. Coll. Podiatric Medicine, N.Y.C., 1959. Diplomate Am. Bd. Podiatric Orthopedics. Pvt. practice Seaford, N.Y., 1960-78; co-founder, exec. v.p. Langer Biomechanics Group, Inc., Deer Park, N.Y., 1969—; prof. orthopedics N.Y. Coll. Podiatric Medicine, 1969—; mem. adv. bd. Rockport Shoe Co., Marlboro, Mass., 1988-92; lectr. in field, U.S. and abroad. Co-author: A Practical Manual for a Basic Approach to Biomechanics, 1972; guest editor Jour. Current Podiatric Medicine, 1989; editorial adv. Podiatry Tracts. Fellow Am. Coll. Foot Orthopedics, Am. Acad. Podiatric Sports Medicine; mem. Am. Podiatric Med. Assn., N.Y. State Podiatric Med. Assn. (Podiatrist of Yr. award 1976), Nat. Acad. Practice in Podiatry (Disting. Practitioner award 1985). Republican. Jewish. Avocations: photography, skiing, traveling, golf. Home: 96 5th Ave Apt 6J New York NY 10011-7611 Office: Langer Biomechanics Group 450 Commack Rd Deer Park NY 11729

WERT, CHARLES ALLEN, metallurgical and mining engineering educator; b. Battle Creek, Iowa, Dec. 31, 1919; s. John Henry and Anna (Spotts) W.; m. Lucille Vivian Mathena, Sept. 5, 1943; children: John Arthur, Sara Ann. B.A., Morningside Coll., Sioux City, 1941; M.S., State U. Iowa, 1943, Ph.D., 1948. Mem. staff Radiation Lab., Mass. Inst. Tech., 1943-45; instr. physics U. Chgo., 1948-50; mem. faculty U. Ill. at Urbana, 1950—, prof., 1955, head dept. metall. and mining engring., 1967-86, prof. emeritus, 1989; cons. to industry. Author: Physics of Metals, 1970, Opportunities in Materials Science and Engineering, 1977; also articles.; Cons. editor, McGraw Hill Book Co. Recipient sr. scientist award von Humboldt-Stiftung. Fellow Am. Phys. Soc., Am. Soc. Metals, AAAS, AIME; mem. Sigma Xi. Home: 1708 W Green St Champaign IL 61821-3796 Office: U Ill Metallurgy & Mining Bldg Urbana IL 61801

WERTENBERGER, STEVEN BRUCE, laser applications engineer; b. Caldwell, Idaho, May 30, 1953; s. Ralph Lee and Pamela Naden (Roedel) W.; m. Judith Ann Ludwig, May 19, 1979; children: Brooke Nicole, Brittany Leigh. B Mech. Engring., GM Inst., Flint, Mich., 1977. Engr. assembly div. GM, Van Nuys, Calif., 1972-79; sr. engr. truck and bus. GM, Shreveport, La., 1979-83; spot welding engr. GM Fanuc Robotics, Troy, Mich., 1983-84; mgr. spot welding GM Fanuc Robotics, Auburn Hills, Mich., 1984-88, mgr. laser applications, 1988—. Home: 3436 Edmunton Dr Rochester Hills MI 48306

WERTHEIMER, DAVID ELIOT, medical facility administrator, cardiologist; b. N.Y.C., Dec. 21, 1955; s. Norman and Marlene (Ratzen) W.; m. Joyce Corman, Jan. 31, 1978; children: Adam, Mitchell, Jessica. BA in Biology, CUNY, Queens, 1976; MD, Loyola U., 1979. Med. dir. dept. cardiology HCA Med. Ctr., Port St. Lucie, Fla., 1984—; dir. invasive cardiology Heart Inst. Port St. Lucie, 1987-92; chief med. officer Fla. dept. Profl. Regulation, 1991. Mem. Fla. Bd. Medicine, 1990. Fellow Am. Coll. Cardiology, Am. Coll. Chest Physicians, Am. Coll. Angiology, Coun. on Geriatric Cardiology (founder), Soc. for Cardiac Angiography. Office: Heart Inst Port St Lucie 1700 SE Hillmoor Dr Port Saint Lucie FL 34952-7544

WERZBERGER, ALAN, pediatrician; b. Toronto, Ont., Can., Dec. 4, 1954; came to U.S., 1985; s. Bernard and Clara (Hilman) W.; m. Sabina Fischman, June 18, 1978; children: Samuel, Moshe, Yehuda, Jacob Joseph, Mayer, Joel, Susan, Henry Werzberger. MD, U. Toronto, 1981. Intern, resident Hosp. for Sick Children, Toronto, Ont., 1981-85; Pvt. practice Monroe, N.Y., 1985—; assoc. attending Dept. Pediatrics Good Samaritan Hosp., Suffern, N.Y., 1985—, St. Agnes Hosp., White Plains, N.Y., 1992—; clin. asst., prof. pediatrics N.Y. Med. Coll., Valhalla, 1991—; dir. Kiryas Joel Inst. Medicine, Monroe, 1991—. Fellow Am. Acad. Pediatrics; mem. Med. Group Mgmt. Assn. Achievements include publication of the first demonstration of efficacy of a vaccine against Hepatitis A. Office: RD5 Box 157 Forest Rd Monroe NY 10950

WESCOTT, ROGER WILLIAMS, anthropology educator; b. Phila., Apr. 28, 1925; s. Ralph Wesley and Marion (Sturges-Jones) W.; m. Hilja J. Brigadier, Apr. 11, 1964; children: Walter, Wayne. Grad., Phillips Exeter Acad., 1942; B.A. summa cum laude, Princeton U., 1945, M.A., 1947, Ph.D., 1948; M.Litt., Oxford U., 1953. Asst. prof. history and human relations Boston U. and Mass. Inst. Tech., 1953-57; assoc. prof. English and social sci., also dir. African lang. program Mich. State U., 1957-62; prof. anthropology and history So. Conn. State Coll., 1962-66; prof., chmn. anthropology and linguistics Drew U., Madison, N.J., 1966—; Presdl. prof. Colo. Sch. Mines, 1980-81; first holder endowed Chair of Excellence in Humanities U. Tenn., 1988-89; fgn. lang. cons. U.S. Office Edn., 1961; pres. Sch. Living, Brookville, Ohio, 1962-65; exec. dir. Inst. Exploratory Edn., N.Y.C., 1963-66; Korzybski lectr. Inst. Gen. Semantics, N.Y.C., 1976; forensic linguist N.J. State Cts., 1982-83; host Other Views, N.J. Cable TV, Trenton, 1985-87. Author: A Comparative Grammar of Albanian, 1955, Introductory Ibo, 1961, A Bini Grammar, 1963, An Outline of Anthropology, 1965, The Divine Animal, 1969, Language Origins, 1974, Visions, 1975, Sound and Sense, 1980, Language Families, 1986, Getting It Together, 1990; also poems and articles; host, program dir. Other Views, N.J. Cable TV, 1985-87. Rhodes scholar, 1948-50; Ford fellow, 1955-56; Am. Council Learned Socs. scholar, 1951-52. Fellow Am. Anthrop. Assn., AAAS, African Studies Assn.; mem. Assn. for Poetry Therapy, Internat. Soc. Comparative Study of Civilizations (co-founder, pres. 1992—), Linguistic Assn. Can. and U.S. (pres. 1976-77), Internat. Linguistic Assn., Com. for the Future, Soc. for Hist. Rsch. (v.p.), Internat. Orgn. for Unification of Terminological Neologisms (1st v.p), Phi Beta Kappa. Home: 16A Heritage Crest Southbury CT 06488-1370

WESEMAEL, FRANÇOIS, physics educator. Prof. physics U. Montreal, Quebec, Canada. Recipient Rutherford Meml. medal Royal Soc. Can., Ottawa, Ontario, 1992. Office: Univ of Montreal, PO Box 6128 St A, Montreal, PQ Canada H3C 3J7*

WESEMANN, WOLFGANG, biochemistry educator; b. Essen, Germany, Apr. 1, 1931. Diploma in chemistry, U. Bonn, 1958, PhD, 1961; Habilitation, U. Marburg, 1968. Sci. asst. Inst. Chemistry, U. Bonn, Fed. Republic Germany, 1959-62, U. Nijmegen, The Netherlands, 1962-65; curator Inst. Biochemistry, U. Marburg, Fed. Republic Germany, 1965-68, prof. biochemistry, 1971—; head dept. neurochemistry U. Marburg (Fed. Republic Germany), 1991—; mng. dir. Inst. Physiol. Chemistry, 1991—. Contbr. over 100 articles to profl. jours. Mem. Collegium Internationale Neuro-Psychopharmacologicum, Deutsche Gesellschaft fur Biologische Chemie, Gesellschaft Deutscher Chemiker, Internat. Soc. for Neurochemistry, European Soc. for Neurochemistry. Home: Friedhofstrasse 25 a, D-3550 Marburg Germany Office: U Marburg Inst Physiol Chem, Hans-Meerwein-Strasse, D-3550 Marburg Germany

WESLEY, IRENE VARELAS, research microbiologist; b. L.A., Nov. 6, 1943; d. Manuel Menard Varelas and Rosario Balcorta Latini; m. Ronald Douglas Wesley, Dec. 14, 1968; children: Nicole Amelia, Carlos. BA, UCLA, 1965, DPH, 1973; MA, U. Calif., Irvine, 1967. Assoc. prof. Metro. State Coll., Denver, 1975-80, U. Colo., Denver, 1977-80; microbiologist Plum Island Lab., USDA-ARS, Greenport, N.Y., 1980-83, Nat. Animal Disease Ctr., USDA-ARS, Ames, Iowa, 1985—. Contbr. articles to profl. jours. Danforth Found. winner for outstanding undergrad. teaching, 1975-80. Mem. Am. Soc. Microbiology, Internat. Assn. Milk, Food, Environ. Sanitarians, Sigma Xi. Democrat. Roman Catholic. Achievements include research in DNA probe for Campylobacter Fetus (patent). Home: 4719 Dover Dr Ames IA 50010 Office: USDA-ARS Nat Animal Disease Ctr 2300 Dayton Rd Ames IA 50010

WESLEY, JAMES PAUL, theoretical physicist, lecturer, consultant; b. St. Louis, July 28, 1921; s. Edgar Bruce and Nanny Fay (Medford) W.; m. Margaret Ellen Martin, June 1943 (div. 1952); 1 child, Martin Medford; m. Dorothy Ree Casey, Aug. 1952 (div. 1963); children: William Casey, David Douglas, Gina Teresa; m. Michele Dudek, June 1963 (div. 1967); 1 child, Sherry Allison; m. Gabriele Beate Modest, July 30, 1975; children: Carl-Eric, Julia Ann, Benjamin Fredrik. BA, U. Minn., 1943; MA, UCLA, 1949, PhD, 1952; postdoctoral, Scripps Inst. Oceanography, La Jolla, Calif., 1951-52. Various positions, 1943-50; prof. physics U. Idaho, Moscow, 1953-56; rsch. geophysicist Newmont Exploration, Ltd., Jerome, Ariz., 1955-56; rsch. physicist Lawrence Radiation Lab. U. Calif., 1956-61; NIH rsch. fellow, fellow Ctr. Advanced Study Behavioral Sci., Stanford, Calif., 1961-62; rsch. physicist U. Denver, 1962-63, Melpar, Inc., Falls Church, Va., 1963, Roland F. Beers, Inc., Alexandria, Va., 1964; assoc. prof. physics U. Mo., Rolla, 1964-74; lectr., cons. on quantum theory, space-time physics, others Berlin and Blumberg, Germany, 1974—. Author: Ecophysics, 1974, Causal Quantum Theory, 1983, Advanced Fundamental Physics, 1991; co-editor: Proc. Internat. Conf. on Space-Time Absoluteness, 1982, Proc. Conf. on Foundations of Mathematics and Physics, 1991; editor: Progress in Space-Time Physics, 1987; contbr. over 92 articles on fundamental physics, geophysics and wave propagation to profl. jours. Nat. Bur. Standards rsch. fellow, 1950. Mem. Am. Phys. Soc., AAAS, European Phys. Soc., Deutschen Physics Gesellschaft. Unitarian. Avocations: piano, stamp collecting, hiking. Home and Office: Weiherdammstrasse 24, 78176 Blumberg Germany

WESLEY, STEPHEN BURTON, energy services executive; b. Louisville, July 13, 1949; s. Leon and Montie C. (Burton) W.; m. Kun Wanna Jarusin, May 22, 1972; 1 child, Thomas Jayson. AA, Somerset (Ky.) Coll., 1969; student, Community Coll. of Air Force, Maxwell, AFB, 1970-77; AA, Watterson Coll., 1977; student, U. Louisville, 1978-80. Cert. energy mgr., lighting efficiency profl. Electronics tech. Kegco, Somerset, 1972-74; instrument tech. Ky. Air Nat. Guard, Louisville, 1974-78; application engr. Johnson Controls, Inc., Louisville, 1978-81, sales engr., 1981-88, energy svcs. mgr., 1988—; adv. bd. Ivy Tech Vocat. Sch., Jeffersonville, Ind., 1988-90. Inventor pitot tube removal tool. Lay dir. Walk to Emmaus, Elizabethtown, Ky., 1989. Sgt. USAF, 1969-73. Mem. Assn. Energy Engrs. Baptist. Avocations: fishing, reading, church work, investments, genealogy. Home: 333 Chalet Dr Lebanon Junction KY 40150-8200 Office: Johnson Controls Inc 1808 Cargo Ct Louisville KY 40299-1980

WESOLOWSKI, DAVID JUDE, geochemist; b. Canonsburg, Pa., May 12, 1954; s. John Stanley and Esther Katherin (Steifvater) W.; m. Mary Ann Ruffing, May 9, 1975; children: James John, Anne Marie, Steven David, Peter Michael. BS in Geology, U. Pitts., 1976; PhD in Geochemistry/Mineralogy, Pa. State U., 1984. Asst. geologist U.S. Bur. Mines, Pitts., 1974-76; exploration geologist U.S. Steel Corp., Virginia, Minn., 1976-77; rsch. staff Oak Ridge (Tenn.) Nat. Lab., 1983-89, geochemistry group leader, 1989—; adj. assoc. prof. U. Tenn., Knoxville, 1990—. Assoc. editor Geochimica et Cosmochimica Acta, Columbus, Ohio, 1992—; contbr. articles to sci. publs. Eugene P. Wigner fellow, 1983-85, Philips Petroleum Co. fellow, 1980-82; numerous rsch. grants U.S. Dept. Energy, 1983-93. Mem. Geochem. Soc., Geol. Soc. Am., Am. Geophys. Union, Soc. Econ. Geologists. Achievements include research in aqueous and stable isotope geochemistry; speciation of metals and mineral solubilities in high temperature brines, effect of composition and temperature on isotope partitioning in geothermal brines. Office: Oak Ridge Nat Lab PO Box 2008 Oak Ridge TN 37831-6110

WESS, JULIUS, nuclear scientist. Researcher Max Planck Inst., Munchen, Germany. Recipient Eugene P. Wigner award, Am. Nuclear Soc., 1992. Office: Max Planck Inst fur Physik, Fohringer Ring 6 Postfach 401212, D-8000 Munich 40, Germany*

WESSEL, JAMES KENNETH, engineering executive; b. Batavia, Ohio, Aug. 18, 1939; s. Louis Henry and Esther Alberta (Baumann) W.; m. Helen Jean Esz, Aug. 6, 1960; children: Jeffrey Allen, Mark Adam. BSChemE, U. Cin., 1962. Product engr. Dow Corning Corp., Elizabethtown, Ky., 1962-68; supr. tech. svc. & devel. Dow Corning Corp., Midland, Mich., 1969-71, econ. evaluator, 1971-72, mgr. govt. mktg., 1972-74, mgr. automotive R & D, 1974-77, mgr. contract R & D, 1977-79, chmn. product mgmt. group, 1977-79; dir. R & D Dow Corning Wright Corp., Memphis, 1979-85; dir. coop. R & D Dow Corning Corp., Washington, 1985-90, dir. fed. bus. devel., 1990—. Mem. ASM (sec. govt. affairs com.), AIAA, Fedn. Material Socs. (trustee 1990—), Am. Ceramic Soc. Presbyterian. Achievements include patents for Methods of Curing Elastomers, Methods of Dewatering Coal. Office: Dow Corning Corp 1800 M St NW Ste 325 South Washington DC 20036

WESSEL, MORRIS ARTHUR, pediatrician; b. Providence, Nov. 1, 1917; s. Morris Jacob and Bessie (Bloom) W.; m. Irmgard Rosenzweig, June 1, 1952; children: David, Bruce, Paul, Lois. BA, Johns Hopkins U., 1939; MD, Yale U., 1943. Diplomate Am. Bd. Pediatrics. Intern Babies Hosp., N.Y.C., 1943-44; asst. dir. pediatric outpatient clinic Yale New Haven (Conn.) Hosp., 1951-52, dir. pediatric outpatient clinic, 1952-57; staff pediatrician, collaboration project Yale U. Sch. Medicine, 1957-62, instr. in pediatrics, 1950-53, clin. asst. prof., 1963-61, clin. assoc. prof. of pediatrics, 1961-75, clin. prof. pediatrics, 1975—; consulting pediatrician Clifford Beers Child Guidance Clinic, 1967—; bd. dirs. Clifford Beers Guidance Clinic, New Haven, 1950-55, Women's Health Svc., New Haven, 1992—, Child Welfare League, N.Y.C. Author: Parents Book on Raising a Healthy Child, 1987. Maj. AUS, 1944-47, ETO. Mem. New Haven Country Med. Soc., Conn. State Med. Soc., Am. Acad. Pediatrics, Soc. Adolescent Medicine.

WESSEL, PAUL, geology and geophysics educator; b. Sarpsborg, Norway, Aug. 31, 1959; came to U.S., 1985; s. Jan and Ingrid (Aspestrand) W.; m. Jill Kathleen Mahoney, Oct. 7, 1990. BS in Geology, U. Oslo, 1982, MS in Geology & Geophysics, 1984; PhD in Geology & Geophysics, Columbia U., 1990. Grad. rsch. asst. Columbia U., Palisades, N.Y., 1985-89; postdoctoral rsch. fellow Hawaii Inst. Geophysics, Honolulu, 1989-91; asst. prof. Sch. of Ocean & Earth Sci. Tech., U. Hawaii, Honolulu, 1991—. Author computer software, also articles. Rsch. fellow Royal Norwegian Coun. for Sci. & Indsl. Rsch., Oslo, 1985-87. Mem. Am. Geophys. Union, Hawaii Ctr. for Volcanology, Sigma Xi. Achievements include research on lithospheric geodynamics, analysis of earth's gravity field and isostasy, and the thermomechanical evolution of the oceanic lithosphere; developed software systems relevant to field. Office: SOEST-U Hawaii 2525 Correa Rd Honolulu HI 96822

WESSELLS, NORMAN KEITH, biologist, educator, university administrator; b. Jersey City, May 11, 1932; s. Norman Wesley and Grace Mahan Wessells; m. Catherine Pyne Briggs; children: Christopher, Stephen, Philip, Colin, Elizabeth. B.S., Yale U., 1954, Ph.D., 1960. Asst. prof. biology Stanford (Calif.) U., 1962-65, asso. prof., 1965-70, prof., 1971—, chmn. biol. sci., 1972-78; acting dir. Hopkins Marine Sta., 1972-75, asso. dean humanities and scis., 1977-81, dean, 1981-88; prof. biology, provost, v.p. acad. affairs U. Oreg., Eugene, 1988—. Author: (with F. Wilt) Methods in Developmental Biology, 1965, Vertebrates: Adaptations, 1970, Vertebrates: A Laboratory Text, 1976, 81, Tissue Interactions and Development, 1977, Vertebrates; Adaptations; Vertebrates: Physiology, 1979, (with S. Subtelny) The Cell Surface, 1980, (with J. Hopson) Biology, 1988, (with Hopson) Essentials of Biology, 1990. Served with USNR, 1954-56. Am. Cancer Soc. postdoctoral fellow, 1960-62; Am. Cancer Soc. scholar cancer research, 1966-69; Guggenheim fellow, 1976-77. Mem. Soc. Devel. Biology (pres. 1979-80), Am. Soc. Zoologist. Office: U Oreg Office of Provost Johnson Hall Eugene OR 97403

WESSELSKI, CLARENCE J., aerospace engineer. Aerospace technician Lockheed Engring., Houston. Recipient Engr. of the Yr. award AIAA, 1990. Office: Aerospace Technician Lockheed engineering MS B14 Houston TX 77058*

WESSINGER, WILLIAM DAVID, pharmacologist; b. Honolulu, Nov. 8, 1951; s. William D. and Virginia L. Wessinger; m. Larua L. Fike; children: Eliza, Kathleen, Marinna. BS, Rutgers U., 1973; PhD, Med. Coll. Va., 1983. Instr. U. Ark. for Med. Scis., Little Rock, 1985-86, asst. prof., 1986-92, assoc. prof., 1992—. Contbr. articles to profl. jours. Grantee Nat. Inst. on Drug Abuse. Mem. Am. Soc. Pharmacology and Exptl. Therapeutics, Behavioral Pharmacology Soc., Behavioral Toxicology Soc., Soc. for Neurosci. (pres. Ark. chpt. 1992—), Coll. on Problems of Drug Dependence. Achievements include research in drug dependence, stimulus properties of drugs, drug self-administration, drugs of abuse, and abused drug testing. Office: U Ark for Med Scis Dept Pharmacology Toxicology 4301 W Markham St Little Rock AR 72205

WESSJOHANN, LUDGER ALOISIUS, chemistry educator, consultant; b. Melle, Germany, May 19, 1961; s. H.H. and A. Wessjohann; m. Dagmar S. Dronsek, May 16, 1991. Diploma, U. Hamburg, Germany, 1987, PhD, 1990. Rsch. fellow U. Oslo, Norway, 1987-88; postdoctoral fellow Stanford (Calif.) U., 1990-91; guest prof. GTZ/U. Federal de Santa Maria, Brazil, 1990, 93; asst. prof. U. Munich, Germany, 1992—; cons. in field. Contbr. articles to profl. jours. Recipient Talented Student scholarship Studienstiftung Dt. Volk, 1984-87, Promotions stipend, 1988; named Royal Norwegian Postdoctoral fellow, 1987, Feodor Lynen fellow, 1990-91. Fellow Am. Chem. Soc., Gesellschaft Deutscher Chemiker, Am. Orchid Soc., Deutsche Orchideen Gesellschaft. Office: U Munich Inst Organic Chemistry, Karl str 23, D-80333 Munich Germany

WEST, CHARLES DAVID, chemistry educator; b. Riverside, Calif., July 25, 1937; s. Charle Cecil and Maurine (Kaylor) W.; m. Julia MacLaren, Dec. 20, 1964; children: Edward, Charles, Elizabeth. BA, Pomona U., 1959; PhD, MIT, 1965. Scientist Bechman Inst., Inc., Fullerton, Calif., 1964-67; prof. chemistry Occidental Coll., L.A., 1967—, chair dept. chemistry, 1983-86. Author: Quantitative Analysis, 1987; contbr. articles to profl. publs. Mem. Am. Chem. Soc. (chair So. Calif. chpt. 1991-92), Soc. for Applied Spectroscopy (chair). Achievements include first to link photoelectric detection to a plasma, first in U.S. to use ultrasonic atomization for chemical analysis. Office: Occidental Coll 1600 Campus Rd Los Angeles CA 90041

WEST, DENNIS PAUL, pharmacologist, pharmacist, educator; b. Moline, Ill., July 16, 1944; s. Paul Melvin and Edith Mae (Farmer) W.; m. Lee Ellen Larson, Mar. 16, 1968; children: James Paul, Patricia Lee. BS in Pharmacy, U. Ill., Chgo., 1967; MS in Pharmacy, U. Iowa, 1974; PhD in Pharmacology, Pacific We. U., 1992. Cert. pharmacist, Ill., Iowa, Calif. Instr. U. Iowa, Iowa City, 1969-74; clin. asst. prof. U. Ill., Chgo., 1975-77, asst. prof., 1978-80, assoc. prof., 1981-89, prof., 1990—; v.p. GenDerm Corp., Lincolnshire, Ill., 1991—; mem. dermatology panel U.S. Pharmacopeia, Bethesda, Md., 1981—. Author: (with E. Abel) The Integumentary System, 1987; contbr. articles to profl. and sci. jours., chpts. to texts. Hon. Rsch. fellow and hon. lectr. U. London, 1985-86. Fellow Am. Coll. Clin. Pharmacy; mem. AAAS, Am. Acad. Dermatology (affiliate), Am. Soc. Hosp. Pharmacists, Am. Coll. Clin. Pharmacology, Am. Assn. Pharm. Scientists, Am. Pharm. Assn., Am. Soc. Clin. Pharmacology and Therapeutics, Soc. for Investigative Dermatology, Licensing Execs. Soc., Ill. Coun. Hosp. Pharmacists, Ill. Coll. Clin. Pharmacy, N.Y. Acad. Scis., No. Ill. Soc. Hosp. Pharmacists, Chgo. Dermatological Soc. Achievements include research in clinical pharmacology, dermatology. Office: GenDerm Corp 600 Knightsbridge Pkwy Lincolnshire IL 60069

WEST, EARLE HUDDLESTON, communications company professional; b. Nashville, Sept. 13, 1955; s. Earle H. and Tommie West; m. Diane M. Matthews, Oct. 7, 1978; children: Benjamin, Lauren, Sarah, Nathan. BS in Math., Harding U., 1976; MSEE, Ga. Tech. U., 1978; MBA, St. Edwards U., 1981. Engr. Semiconductor div. Motorola, Austin, 1978-82; mgr. AT&T Bell Labs., Holmdel, N.J., 1982—. Deacon Freehold (N.J.) Ch. of Christ, 1989. Mem. Ch. of Christ. Home: 32 Georgian Bay Dr Morganville NJ 07751 Office: AT&T 101 Crawfords Corner Rd Holmdel NJ 07733

WEST, JACK HENRY, petroleum geologist; b. Washington, Apr. 7, 1934; s. John Henry and Zola Faye (West) Pigg; m. Bonnie Lou Ruger, Apr. 1, 1961; children: Trent John, Todd Kenneth. BS in Geology, U. Oreg., 1957, MS, 1961. Cert. petroleum geologist. Geologist Texaco Inc., L.A and Bakersfield, Calif., 1961-72; asst. dir. devel. geologist Texaco Inc., L.A., 1972-78; geologist Oxy Petroleum Inc., Bakersfield, 1978-80, div. geologist, 1980-83; exploitation mgr. Oxy U.S.A. Inc./Cities Svc. Oil and Gas, Bakersfield, 1983-89; sr. petroleum advisor WZI Inc., Bakersfield, 1990—. Active Beyond War, Bakersfield, 1983-90. Mem. Am. Assn. Petroleum Geologists (pres. Pacific sect. 1988-89, adv. coun. 1992—), Soc. Petroleum Engrs., San Joaquin Geol. Soc. (pres. 1984-85), Alfa Romeo Owners Club. Republican. Methodist. Avocations: music, sports cars.

WEST, JOHN THOMAS, surgeon; b. Live Oak, Fla., June 23, 1924; s. James Whitaker and Lelah Eulalia (Moore) W.; m. Ruth Marit Blakely, June 18, 1948; children: Phyllis Ann, Rebecca Ruth, James Carl, Jeffrey Moore,

Paul Blakely. BS, U. Mich., 1946; MD, Vanderbilt U., 1951. Diplomate Am. Bd. Surgery. Commd. officer USPHS, 1951, advanced through grades to capt., 1963; rotating intern USPHS Hosp., Seattle, 1951-52; chief surgery USPHS Alaska Native Hosp., Anchorage, 1957-60, resident gen. surgery, 1954-57; chief surgery USPHS Hosp., Seattle, 1963-69, USPHS Indian Hosp., Phoenix, 1969-71; sr. investigator surg. br. Nat. Cancer Inst., USPHS, Bethesda, Md., 1960-63; ret., 1971; clin. assoc. prof. Tex. Tech U., Lubbock, 1974-77; pvt. practice, La Grange, Ga., 1971-74, 77—; mem. active staff West Ga. Med. Ctr., La Grange, 1971-74, 77—. Bd. dirs. Ga. divsn. Am. Cancer Soc., 1972-77, 77-92. Recipient Meritorious Svc. medal USPHS, 1968. Fellow ACS, Soc. Surg. Oncology. Presbyterian. Achievements include report of facilitation of major hepatic resection by an innovation in the surgical exposure of the liver. Home: 134 Hickory Ln La Grange GA 30240-8622 Office: West Ga Oncology Assocs PC 301 Medical Dr Ste 503 La Grange GA 30240-4156

WEST, JOHNNY CARL, aeronautical engineer; b. Knox City, Tex., Sept. 25, 1951; s. Carl Wendell and Vida Marie (Leach) W.; m. Deborah Lynn Carney, Aug. 25, 1973; 1 child, Trisha Marie. BSME, U. Ark., 1974; MS in Systems Mgmt., St. Mary's U., 1978; MS in Aeronautical Engring., Air Force Inst. Tech., 1982; postgrad., Def. Systems Mgmt. Coll., 1992. Advanced through grades to lt. col. USAF, 1990; F100 engine propulsion engr. San Antonio ALC, Kelly AFB, Tex., 1974-78; program mgr. NATO JP-8 conv. HQ AFLC, Wright Patterson AFB, Ohio, 1978-81; X-29 flight controls flight test engr. USAF, Edwards AFB, Calif., 1983-85; chief spl. projects br. 6520 Test Group, Edwards AFB, Calif., 1985-86; chief R&D officer HQ AFMPC, Randolph AFB, Tex., 1987-90; dep. dir. test NASP Joint Program Office, Wright Patterson AFB, 1990-92; team lead, curtailment B-2 Program Office Wright Patterson AFB, 1992—. Sun. sch. tchr. Ch. of the Nazarene, Beavercreek, Ohio, 1991-92. Mem. AIAA, SAE. Home: 3087 Blue Green Dr Dayton OH 45431-2685

WEST, MARK DAVID, software engineer; b. Hanover, N.H., Nov. 18, 1964; s. David John and Elizabeth Ann (Roberts) W. BS, Bloomsburg U., 1986. Programmer, analyst Shared Med. Systems, Inc., Malvern, Pa., 1986-90; cons. Integrated Systems Cons. Group, Wayne, Pa., 1990—, internal tng. coord., 1991—. Mem. IEEE, Assn. Computing Machinery, Digital Equipment Corp. Users Soc. Democrat. Congregationalist. Office: Integrated Systems Cons Ste 200 575 Swedesford Rd E Wayne PA 19087

WEST, MICHAEL HOWARD, dentist; b. New Orleans, Sept. 11, 1952; s. Howard Thomas and Helen Katherine (Nolen) W.; m. Betty Namie, Aug. 3, 1973 (div.); 1 child, Nolan Joseph; m. Berbara Lynn Barlow, Apr. 8, 1989; children: Christopher, Nicole, Katherine. BS, U. So. Miss., 1975; DDS, La. State U., 1977. Diplomate Am. Bd. Odontology. Coroner pro tem Forrest County, Miss., 1986, dep. med. examination investigator, 1986—; pvt. practice, 1981—; clin. asst. prof. Sch. Dentistry La. State U., 1990—, U. Miss., Jackson, 1990—; diplomate People's Republic of China forensic sci. exch. Contbr. articles to profl. jours. Capt. USAF, 1973-81. Grantee Am. Bd. Forensic Odontology, 1991, Nat. Inst. Justice, Dept. Justice, 1992; recipient Reidar Sognnaes award Am. Soc. Forensic Odontology, 1982, Letter of Commendation, City of Kinner, 1982. Fellow Am. Acad. Forensic Science; mem. Internat. Assn. for Identification. Achievements include pioneer work in UV and alternative light photography of wound on human skin, conformation of single bullet theory in assassination of John F. Kennedy, first bitemark convictions in Miss., Ga. and Ark. Home: 130 NW Circle Hattiesburg MS 39401 Office: Plaza Dental Ctr 4600 W Hardy St Ste 11 Hattiesburg MS 39404

WEST, PHILIP WILLIAM, chemistry educator; b. Crookston, Minn., Apr. 12, 1913; s. William Leonard and Anne (Thompson) W.; m. Tenney Constance Johnson, July 5, 1935 (dec. Feb. 1964); children: Dorothy West/ Farwell, Linda West Gueho (dec.), Patty West Elstrott; m. Foymae S. Kelso, July 1, 1964. B.S., U. N.D., 1935, M.S., 1936, D.Sc. (hon.), 1958; Ph.D., State U. Iowa, 1939; postgrad., Rio de Janeiro, 1946. Chemist N.D. Geol. Survey, 1935-36; research asst. san. chemistry U. Iowa, 1936-37; asst. chemist Iowa Dept. Health, 1937-40; research microchemist Econ. Lab., Inc., St. Paul, 1940; faculty La. State U., 1940-80, prof. chemistry, 1951-80, Boyd prof., 1953-80, emeritus, 1980—, chmn. ann. symposium modern methods of analytical chemistry, 1948-65, dir. Inst. for Environmental Scis.,, 1967-80; co-founder, chmn. bd. West-Paine Labs. Inc., Baton Rouge, 1980-93; O. M. Smith lectr. Okla. State U., 1955; vis. prof. U. Colo., 1963, Rand Afrikaans U., 1980; adj. prof. EPA, 1969-80; founder Kem-Tech. Labs., Inc., Baton Rouge, 1954, chmn. bd., 1965-74; co-founder West-Paine Labs., Inc., Baton Rouge, 1978, pres., 1978-93, chmn. bd., lab dir., 1990; mem. 1st working party sci. com. on problems of environment, 1971-74; pres. analytical sect. Internat. Union Pure and Applied Chemistry, 1965-69, mem. sect. indsl. hygiene and toxicology, 1971-73, mem. air quality sect., 1973-75; mem. tech. adv. com. La. Air Pollution Control Com., 1979—; mem. Gov.'s Task Force Environ. Health, 1983-85; mem. sci. adv. bd. EPA, 1983-84; cons. WHO; tech. expert Nat. Bur. Standards Nat. Vol. Lab. Accreditation Program, 1988—; chmn. bd., CEO West & Assoc., Inc., 1992—. Author: Chemical Calculations, 1948, (with Vick) Qualitative Analysis and Analytical Chemical Separations, 2d edit., 1959, (with Bustin) Experience Approach to Experimental Chemistry, 1975; editor: (with Hamilton) Science of the Total Environment, 1973-78, (with Macdonald) Analytica Chimica Acta, 1959-78, Reagents and Reaction for Qualitative Inorganic Analysis; co-editor: Analytical Chemistry, 1963; asst. editor: Mikrochemica Acta, 1952-78, Michrochem. Jour, 1957-75; adv. bd.: Analytical Chemistry, 1959-60; publ. bd.: Jour. Chem. Edn, 1954-57; contbr. articles to profl. jours. Recipient Honor Scroll award La. sect. Am. Inst. Chemistry, 1972. Fellow AAAS; mem. Am. Chem. Soc. (Southwest award 1954, Charles E. Coates award 1967, Analytical Chemistry award 1974, award for Creative Advances in Environ. Sci. and Tech.), La. Acad. Sci., Air Pollution Control Assn., Am. Indsl. Hygiene Assn., Austrian Microchem. Soc. (hon.), Soc. of Analysts Eng. (hon.), Internat. Union Pure and Applied Chemistry (pres. commn. I, pres. analytical div.), Japan Soc. for Analytical Chemistry (hon.), La. Cancer and Health Found., Sigma Xi, Phi Lambda Upsilon, Phi Kappa Phi, Alpha Epsilon Delta, Alpha Chi Sigma, Tau Kappa Epsilon. Office: West-Paine Labs Inc 7979 G S R I Ave Baton Rouge LA 70820-7499

WEST, ROBERT CULBERTSON, chemistry educator; b. Glen Ridge, N.J., Mar. 18, 1928; s. Robert C. and Constance (MacKinnon) W.; children: David Russell, Arthur Scott, Derek. B.A., Cornell U., 1950; A.M., Harvard U., 1952, Ph.D., 1954. Asst. prof. Lehigh U., 1954-56; mem. faculty U. Wis.-Madison, 1956—, prof. chemistry, 1963—, Eugene G. Rochow prof., 1980; adj. prof. So. Oreg. State Coll., 1984—; indsl. and govt. cons., 1961—; Fulbright lectr. Kyoto and Osaka U., 1964-65; vis. prof. U. Würzburg, 1968-69, Haile Selassie I U., 1972, U. Calif.-Santa Cruz, 1977, U. Utah, 1981, Inst. Chem. Physics Chinese Acad. Sci., 1984, Justus Liebigs U., Giessen, Fed. Republic Germany, U. Estadual de Campinas, Brazil, 1989; Abbott lectr. U. N.D., 1964, Seydel-Wooley lectr. Ga. Inst. Tech., 1970, Sun Oil lectr. Ohio U., 1971, Edgar C. Britton lectr. Dow, Midland, Mich., 1971, Jean Day Meml. lectr. Rutgers U., 1973; Japan Soc. for Promotion Sci. vis. prof. Tohoku U., 1976, Gunma U., 1987; Lady Davis vis. prof. Hebrew U., 1979; Cecil and Ida Green honors prof. Tex. Christian U., 1983; Karcher lectr. U. Okla., 1986; Broberg lectr. N.D. State U., 1986; Xerox lectr. U. B.C., 1986, McGregory lectr. Colgate U., 1988; Lady Davis vis. prof. Technion Israel Inst. Tech., 1990; Humboldt prof. Tech. U. Munich, Fed. Republic Germany, 1990; George W. Watt lectr. U. Tex., Austin, 1992; vis. prof. U. Estadual de Campinas, Brazil, 1993; Dozor vis. fellow Ben Gurion U. of the Negev, Israel, 1993. Co-editor: Advances in Organometallic Chemistry, Vols. I-XXXVI, 1964—, Organometallic Chemistry--A Monograph Series, 1968—; contbr. articles to profl. jours. Pres. Madison Community Sch., 1970-81; founder, bd. dirs. Women's Med. Fund, 1971—; nat. bd. dirs. Zero Population Growth, 1980-86; bd. dirs., v.p. Protect Abortion Rights Inc., 1980; lay minister Prairie Unitarian Universalist Soc., 1982. Recipient F.S. Kipping award, 1970; Outstanding Sci. Innovator award Sci. Digest, 1985; Chem. Pioneering award Am. Inst. Chemists, 1988; Wacker Silicon prize, 1989; Humboldt U.S. Scientist award, 1990. Mem. Am. Chem. Soc., Chem. Soc. (London), Japan Chem. Soc., AAAS, Wis. Acad. Sci. Home: 305 Nautilus Dr Madison WI 53705-4333

WEST, ROBERT MACLELLAN, science education consultant; b. Appleton, Wis., Sept. 1, 1942; s. Clarence John and Elizabeth Ophelia (Moore) W.; m. Jean Sydow, June 19, 1965; 1 child, Christopher. BA, Lawrence Coll., 1963; SM, U. Chgo., 1964, PhD, 1968. Rsch. assoc. Princeton (N.J.) U., 1968-69; asst. prof. Adelphi U., Garden City, N.Y., 1969-74; curator of geology Milw. Pub. Mus., 1974-83; dir. Carnegie Mus. Natural History, Pitts., 1983-87, Cranbrook Inst. Sci., Bloomfield Hills, Mich., 1987-91; prin. RMW Sci. Action, Washington, 1992—; CEO Informal Sci., Inc., Washington, 1993—; adj. prof. U. Wis., Milw., 1974-83; com. mem. Indo-U.S. Subcom., 1990—. Contbr. articles to profl. jours. Bd. dirs. Friends of the New Zoo, Pitts., 1984-87; treas. East Mich. Environ. Action Coun., Birmingham, Mich., 1987-92. Recipient Arnold Guyot prize Nat. Geographic Soc., 1982; named Man of Yr. in Sci. by Vectors Pitts., 1988; NSF fellow, 1968; NSF rsch. grantee, 1970-80, Nat. Geographic Soc. rsch. grantee, 1973, 76, 77, 79, 80, 82. Mem. Nat. Coun. Sci. Edn. (bd. dirs. 1984-88, 92—), Nepal Natural History Soc. (advisor 1992—), Soc. Vertebrate Paleontology, Geol. Soc. Am., Paleontology Soc., Soc. Study Evolution, Am. Soc. Mammalogists, Am. Assn. Mus., Rotary Club Bethesda-Chevy Chase. Avocations: history, sports. Office: Informal Sci Inc PO Box 42328 Washington DC 20015

WEST, WILLIAM WARD, aerospace engineer; b. San Antonio, Dec. 2, 1963; s. William Gene and Shirley Lee (Bunker) W.; m. Karen Joan Horvath, June 16, 1990. BS in Aerospace Engring., St. Louis U., 1988; postgrad. in Space Sci., U. Houston, 1988—. Flight dynamics engr. Rockwell Space Ops. Co., Houston, 1988-90, with SMS Control/Propulsion Inst., 1990—. Active Rockwell Speaker's Bur., Houston. Mem. AIAA, British Interplanetary Soc. Achievements include participation in training of space shuttle flight crews.

WESTBROOK, FRED EMERSON, agronomist, educator; b. Arlington, Tenn., Dec. 19, 1916; s. Atha Moden and Albertha (Graham) W.; m. Virginia Gray, July 8, 1944; children: Anita Westbrook McClendon, Fred Emerson Jr. BS, Tenn. State U., 1946, MS, 1947; PhD, Mich. State U., 1954. Tchr. vocat. agr. Robertson County High Sch., Springfield, Tenn., 1947-48; instr. agronomy Tenn. State U., Nashville, 1948-57, head dept. plant sci., 1957-72; extension agronomist USDA Extension Svc., Washington, 1973-78, extension agronomy program leader, 1982-89; priorities analyst plant sci. USDA Sci. and Edn. Adminstrn., Beltsville, Md., 1978-82; forestry commr. Tenn. State Forestry, Nashville, 1989—; chmn. working group on Africanized honey bee USDA, 1988-89; cons. Tenn. State U. Sch. Agr., Nashville, 1989—. Contbr. articles to profl. jours. Sgt. U.S. Army, 1941-45, ETO. Rockefeller Found. fellow, 1953-54. Mem. Am. Soc. Agronomy, Soil and Water Conservation Soc. (Tenn. sec., treas. 1972). Achievements include discovery of ladino clover (trifolium repens) to be a biennial rather than a perennial. Home and Office: 3509 Albion St Nashville TN 37209-2553

WESTBROOK, SUSAN ELIZABETH, horticulturist; b. Canton, Ohio, Sept. 27, 1939; d. Walter Simon and Rosella Hunt Tolley; m. Edward D. Westbrook, July 2, 1966 (div. 1980); 1 child, Tyler Hunt. Student, Smithdeal-Massey, Richmond, Va., 1958-59; student in Spanish, U. Honduras, 1960; student biology/geology, Mary Washington Coll., 1960, 72, 73; student hort., Prince Georges Community Coll., 1987-88. Farm owner Spotsylvania, Va., 1972-83; office mgr. Tolley Investments, Inc., Fredericksburg, Va., 1980-83; real estate agt. Cooper Realty, Fredericksburg, Va., 1981-83; salesperson Meadows Farms Nursery, Chantilly, Va., 1986-93. Author booklets: Japanese Maples, 1990, Fruit Trees, 1989; author radio format: Gardening in Virginia, 1960; co-author computer program: Plantscape, 1990. Sec. Rep. Party, Spotsylvania, 1972-83, Elko County, Nev., 1968; judge Bd. Elections, Spotsylvania, 1980-83, cand. bd. suprs., 1979. Mem. Nat. Wildlife Fedn., Md. Nurserymen's Assn., Friends of the Nat. Arboretum. Avocations: travel, gardening, plant research and identification. Home: 6110 S Virginia Lane Dahlgren VA 22448

WESTBY, TIMOTHY SCOTT, oil company research engineer; b. Fargo, N.D., Apr. 16, 1957; s. Joseph Arlo and Dorothy Mae (Nye) W.; m. Holli Leigh Huber, Mar. 17, 1987; 1 child, Katherine Elizabeth. SBChemE, MIT, 1979; PhDChemE, U. Tex., 1984; postgrad., U. Houston Law Ctr., 1990—. Researcher Energy Lab., MIT, Cambridge, 1976-79; rsch. asst. U. Tex., Austin, 1979-84, teaching asst., 1981-83; assoc. rsch. engr. Shell Devel. Co., Houston, 1984-87, rsch. engr., 1987-91, sr. rsch. engr., 1991—; mem. adv. com. Ohio Combustion Rsch., Columbus, 1985-90, Pa. Coal Rsch. Coop., University Station, 1986-89. Contbr. articles to profl. jours.; patentee method for in situ coal drilling. Campaigner United Way, Houston, 1989-91. Scholar MIT, 1975-79; fellow U.S. dept. Energy, 1979-82, Getty Oil Co., 1983-84. Mem. ASTM (com. D-5, 1989—), ASME (advisor rsch. com. on corrosion and deposits from flue gases 1988—), Phi Delta Phi. Avocations: gardening, racquet sports, Italian food, old sports cars, skiing. Office: Shell Devel Co 3333 Hwy 6 S Houston TX 77082-3101

WESTERDAHL, JOHN BRIAN, nutritionist, health educator; b. Tucson, Dec. 3, 1954; s. Jay E. and Margaret (Meyer) W.; m. Doris Mui Lian Tan, Nov. 18, 1989. AA, Orange Coast Coll., 1977; BS, Pacific Union Coll., 1979; MPH, Loma Linda U., 1981. Registered dietitian. Chartered herbalist. Nutritionist, health educator Castle Med. Ctr., Kailua, Hawaii, 1981-84, health promotion coord., 1984-87, asst. dir. health promotion, 1987-88, dir. health promotion, 1988-89; dir. nutrition and health rsch. Health Sci., Santa Barbara, Calif., 1989-90; sr. nutritionist Shaklee Corp., San Francisco, 1990—; talk show host Nutrition and You, Sta. KGU Radio, Honolulu, 1983-89; nutrition com. mem. Hawaii div. Am. Heart Assn., Honolulu, 1984-87; mem. nutrition study group Govs. Conf. Health Promotion and Disease Prevention for Hawaii, 1985. Mem. AAAS, Am. Coll. Sports Medicine, Am. Dietetic Assn., Am. Nutritionists Assn., Am. Coll. Nutrition, Soc. for Nutrition Edn., Nat. Wellness Assn., Nutrition Today Soc., Am. Soc. Pharmacognosy, Inst. Food Technologists, Hawaii Nutrition Coun. (v.p. 1983-86,m pres.-elect 1988-89, pres. 1989), Hawaii Dietetic Assn., Calif. Dietetic Assn., N.Y. Acad. Scis., Seventh-day Adventist Dietetic Assn., several other profl. assns. Republican. Seventh-Day Adventist. Avocations: swimming, scuba diving. Office: Shaklee Corp 444 Market St San Francisco CA 94111-5325

WESTERHOUT, GART, retired astronomer; b. The Hague, The Netherlands, June 15, 1927; came to U.S., 1962, naturalized, 1969; s. Gerrit and Magdalena (Foppe) W.; m. Judith Mary Monaghan, Nov. 14, 1956; children: Magda C., Gart T., Brigit M., Julian C., Anthony K. Drs., Leiden U., Netherlands, 1954, Ph.D., 1958. Asst. Leiden U. Observatory, 1952-56, sci. officer, 1956-59, chief sci. officer, 1959-62; prof., dir. astronomy U. Md., 1962-73, chmn. div. math. and phys. scis. and engring., 1972-73, prof. astronomy, 1973-77; sci. dir. U.S. Naval Observatory, Washington, 1977-93; vis. astronomer Max Planck Inst. Radio Astronomy, Bonn, Germany, 1973-74; Vice chmn. div. phys. sci. NRC, 1969-73, mem. com. on radio frequencies, 1971-82; trustee Asso. Univ. Inc., 1971-74; mem. Inter Union Commn. on Allocation of Frequencies, 1974-82; mem. sci. council Stellar Data Center, Strasbourg, France, 1978-84, chmn., 1981; chmn. working group on astrometry, astronomy survey com. Nat. Acad. Scis., 1979-81; mem. Arecibo adv. bd. Nat. Astronomy and Ionosphere Center, 1977-80, chmn., 1979-80. Contbr. on radio astronomy, spiral structure of our galaxy and astrometry to profl. jours. Recipient citation for teaching excellence Washington Acad. Scis., 1972; U.S. Sr. Scientist award Alexander von Humboldt Stiftung, Ger., 1973; NATO fellow, 1959. Mem. Internat. Astron. Union (chmn. working group on astron. data 1985-91), Internat. Sci. Radio Union (pres. Commn. on radio astronomy 1975-78), Am. Astron. Soc. (councilor 1975-78, v.p. 1985-87), Royal Astron. Soc., Sigma Xi. Roman Catholic. Home: 811 W 38th St Baltimore MD 21211

WESTERMAN, PHILIP WILLIAM, biomedical researcher, medical educator; b. Brisbane, Australia, June 16, 1945; came to U.S., 1971; s. Stephen Edward and Jeanne Minta (Parish) W.; m. Janice Mary Carson, Jan. 10, 1970; children: Natasha Ann, Karen Lynn. BSc, U. Sydney, Australia, 1967, PhD, 1971. Post-doctoral fellow Case Western Res. U., Cleve., 1971-73; rsch. assoc. Australian Nat. U., Canberra, 1974-75, Calif. Inst. Tech., Pasadena, 1975-76; asst. prof. Northeastern Ohio U. Coll. Medicine, Rootstown, 1976-82, assoc. prof., 1982-88, prof., 1988—; mem. Liquid Crystal Inst., Kent (Ohio) State U., 1980—, adj. prof. chemistry, 1982—, full grad. faculty mem. Sch. Biomedical Scis., 1981—. Contbr. over 80 articles to profl. jours. Treas. Am. Field Svc. (exchange student orgn.), Kent, 1990-92. Recipient Student Teaching award Northeastern Ohio U. Coll. Medicine, 1979, 82, 89, Rsch. award NIH, 1980-87, Rsch. Challenge award Bd. Regents State of Ohio, 1988-91, Grant-in-Aid, Am. Heart Assn., 1992. Mem. Am. Chem. Soc., Biophys. Soc., Am. Soc. Biochemistry and Molecular Biology, Internat. Liquid Crystal Soc., N.Y. Acad. Scis., Sigma Xi (pres. Kent State U. chpt. 1990-92). Home: 454 Dansel St Kent OH 44240 Office: Northeastern Ohio Univ Coll Medicine PO Box 95 Rootstown OH 44272

WESTERN, ARTHUR BOYD, JR., physics educator; b. Detroit, Feb. 29, 1944; s. Arthur Boyd and Eileen Rita (Mulaney) W.; m. Jonnee Lynn Givens, Dec. 28, 1966; children: Douglas Arthur, Sara Lynn, Teresa Lynn. BS in Physics and Math., Rollins Coll., 1965; PhD in Physics, Mont. State U., 1976. Asst. prof. Mont. Coll. Mineral Sci., Butte, 1977-81, assoc. prof., head dept. physics and geophys. engring., 1981-86; assoc. prof. Rose-Hulman Inst. Tech., Terre Haute, Ind., 1986-88, prof., 1988—, head dept. physics and applied optics, 1991—; staff scientist Mountain States Energy, Butte, 1983-88. Editor: Physics Exam File, Vol. I and II, 1986; contbr. articles to profl. jours. Pres. Lost Creek PTA, Terre Haute, 1989-90. Mem. Am. Assn. Physics Tchrs., Soc. Photo-Optic Instrumentation Engrs., Soc. Experimental Mechanics. Achievements include first experimental verification of tricritical point in potassium dihydrogen phosphate (KDP), and originator (with W. P. Crummett) of sonic ranger for physics instruction in kinematics. Office: Rose-Hulman Inst Tech 5500 Wabash Ave Terre Haute IN 47803

WESTFALL, LINDA LOUISE, software engineer, metrics analyst; b. Milw., Jan. 12, 1954; d. Jay Nelson and Alice May (Moss) Vogelsong; m. Robert Lee Westfall, Aug. 24, 1974. BS, Carnegie-Mellon U., 1976; MBA, U. Tex., Dallas, 1990. Math. tchr. Kings Park (N.Y.) Jr. High Sch., 1976-77; software engr. Gen. Dynamics, Ft. Worth, 1978; programmer, analyst Electrospace Systems, Richardson, Tex., 1978-81; software engr. Near Space, Inc., Plano, Tex., 1981; mgr. prodn. software Control Systems Internat., Carrollton, Tex., 1981-87; software engr., metrics analyst DSC Communications Corp., Plano, Tex., 1987—, quality action team facilitator, 1990—. Vol. Better Breathing Club, Plano Hosp., 1987-88. Career facilitation grantee U. Tex., Arlington, 1978. Mem. IEEE, Am. Soc. for Quality Control, Assn. for Software Engring. Excellence (membership chair 1991-92, chmn. 1992-93), Nat. Mgmt. Assn. (instr. supervisory and mgmt. skills course 1991-93), Am. Bus. Women's Assn., Inst. Cert. Profl. Mgrs., Toastmasters Internat. (exec. v.p. 1990, ed. v.p. 1992, pres. 1992-93, competent toastmaster, Table Topics Area Contest winner 1990, 91, division 92, Evaluation Contest Area winner 1991), Alpha Xi Delta (pres. Plano Alums chpt. dir. 1985-88, Diamond Quill Scholarship award 1972—). Avocations: camping, hiking, reading, crafts. Home: 2632 Pin Oak Ln Plano TX 75075-3117 Office: DSC Communications 1000 Coit Rd MS 171 Plano TX 75075

WESTHEIMER, FRANK HENRY, chemist, educator; b. Balt., Jan. 15, 1912; s. Henry Ferdinand and Carrie (Burgunder) W.; m. Jeanne Friedmann, Aug. 31, 1937; children: Ruth Susan, Ellen. AB, Dartmouth Coll., 1932, ScD (hon.), 1961; MA, Harvard U., 1933, PhD, 1935; ScD (hon.), U. Chgo., 1973, U. Cin., 1976, Tufts U., 1978, U. N.C., 1983, Bard Coll., 1983, Weizmann Inst., 1987, U. Ill. at Chgo., 1988. Rsch. assoc. U. Chgo., 1936-37, asst. prof., 1941-44, assoc. prof., 1946-48, prof. chemistry, 1948-53; vis. prof. Harvard U., 1953-54, prof. chemistry, 1954-82, sr. prof., 1982-83, prof. emeritus, 1983—, chmn. dept., 1959-62; Overseas fellow Churchill Coll., U. Cambridge, Eng., 1962-63; mem. Pres.'s Sci. Adv. Com., 1967-70; research supr. Explosives Research Lab., Nat. Def. Research Com., 1944-45; chmn. com. survey chemistry Nat. Acad. Scis., 1964-65. Assoc. editor Jour. Chem. Physics, 1942-44, 52-54; editorial bd. Jour. Am. Chem. Soc, 1960-69, Procs. Nat. Acad. Scis., 1983-89; contbr. articles to profl. jours. Recipient Naval Ordnance Development award, 1946, Army-Navy cert. of appreciation, 1946, James Flack Norris award in phys.-organic chemistry, 1970, Willard Gibbs medal, 1970, Theodore W. Richards medal, 1976; award in chem. scis. Nat. Acad. Sci., 1980, Richard Kokes award, 1980, Charles Frederick Chandler medal, 1980, Rosenstiel award, 1981, Nichols medal, 1982, Robert A. Welch award, 1982, Ingold medal, 1983, Cope award, 1982, Nat. Medal of Sci., 1986, Paracelsus medal, 1988, Priestley medal, 1988, Repligen award, 1992; fellow Columbia U. NRC, 1935-36, Guggenheim Found., 1962-63, Fulbright-Hays Found., 1974. Mem. Nat. Acad. Sci. (council 1971-75, 76-79), Am. Philos. Soc. (council 1981-84), Am. Acad. Arts and Scis. (sec. 1985-90), Royal Soc. (fgn. mem.). Home: 3 Berkeley St Cambridge MA 02138-3409

WESTLOCK, JEANNINE MARIE, health care consultant; b. Pitts., Mar. 14, 1959; d. Russell and Shirley Anita (Meredith) W. BS in Applied Math., U. Pitts., 1981. Cert. profl. ins. woman, health ins. assoc. Ch. organist St. Luke Roman Cath. Ch., Carnegie, Pa., 1981-92; actuarial analyst Blue Cross & Blue Shield Mut. of Ohio, Cleve., 1984-89, health benefits mgmt. specialist, 1989-92, sr. health cons., 1992—; Active campaign state sen. Gary Suhadolnik, Cleve., 1988—; organist and vocalist St. Elizabeth Ann Seton parish, Carnegie, Pa., 1992—. Active polit. campaign State Sen. Gary Suhadolnik, Cleve., 1988—. Mem. Nat. Assn. Ins. Women (local exec. bd. 1986-90, local pres. 1989-90, state dir., 1990-91, Achievement award 1988, Nat. Rookie of the Yr. 1987, State Ins. Women of the Yr. 1988), Ins. Women of Cleve. (Ins. Woman of the Yr. 1986), Cleve. Women's City Club, Nat. Assn. Health Underwriters, NAFE, Toastmasters (local pres. 1988), Alpha Delta Pi. Roman Catholic. Avocations: travel, sports/fitness, music, needlework. Home: 12900 Lake Ave Apt 1703 Cleveland OH 44107-1556

WESTMAN, WESLEY CHARLES, psychologist; b. Schuyler, Nebr., Feb. 19, 1936; s. Meade Levi and Violet Edra (Abbott) W.; m. Noelene R. Westman, Dec. 27, 1957; children: Mark Phillip, Charlene Ann. BA, U. Va., 1961; MS, U. Wis., 1962, PhD, 1965; MS in Mgmt., Fla. Internat. U., 1976. Project dir. Stateride Planning Project, Hartford, Conn., 1966-68; pvt. practice West Hartford, 1967-71; exec. dir. Methadone Maintenance Treatment Program Hartford Dispensary, 1971; psychologist alcohol/drug dependency program VA Med. Ctr., Miami, 1971-72, 87-90, chief alcohol/drug dependency program, 1972-87; clin. dir. Grant St. Partnership, New Haven, 1991-92; cons. VA Central Office, Washington, 1972-85. Author: The Drug Epidemic, 1970, Bask Factos on Drug Abuse, 1972; contbr. articles to profl. jours. Sgt. USMC, 1955-58. Mem. Am. Psychol. Assn. Democrat. Congregationalist. Home: 7701 SW 176th St Miami FL 33157

WESTPHAL, ANDREW JONATHAN, physicist, researcher; b. Tulsa, June 7, 1961; s. James Adolph and Jean (Wimbish) W. BA, Rice U., 1984; MSAA, U. Wash., 1986; PhD, U. Calif., Berkeley, 1992. Programmer Worley Engring., London, 1979-80; staff engr. Flight Safety Internat., Tulsa, 1980-82, 84; summer scientist Jet Propulsion Lab., Pasadena, Calif., 1983; teaching asst. U. Calif., Berkeley, 1986-87, postdoctoral fellow, 1992—; exec. sec. small-class explorer rev. panel NASA, Balt., 1989. Contbr. articles to Nuclear Instruments and Methods, Phys. Rev., Nature. Mem. Am. Phys. Soc. Achievements include pioneer work in measurement of cross sections for electron capture and stripping by ultrarelativistic ions, cosmic-ray experiment on first Antarctic circum-polar balloon flight. Home: 355 Colusa Ave Kensington CA 94707 Office: Univ Calif Berkeley Dept Physics Berkeley CA 94720

WESTPHAL, KLAUS WILHELM, university museum director; b. Berlin, Mar. 20, 1939; came to U.S., 1969; s. Wilhelm Heinrich and Irmgard (Herize) W.; m. Margaret Elisabeth Dorothea Wagner, May 16, 1969; children: Barbara, Marianne, Christine. BS in Geology, Eberhard-Karls Universität, Germany, 1960, MS, 1964, PhD in Paleontology, 1969. Dir. geology mus. U. Wis. Madison, 1969—; bd. dirs. natural history coun. U. Wis. Madison, 1973—, Friends of Geology Mus., Inc., 1977—; nat. speaker on paleontology Outreach, 1977—; instr. paleontology U. Wis., 1977—; leader expeditions fossil vertebrates including dinosaurs, 1977—. Participant various tchr.-tng. projects Wis. Pub. Schs. Lutheran. Home: 3709 High Rd Middleton WI 53562 Office: U Wis Geology Mus 1215 W Dayton St Madison WI 53706

WETHERILL, GEORGE WEST, geophysicist, planetary scientist; b. Phila., Aug. 12, 1925; s. George West and Leah Victoria (Hardwick) W.; m. Phyllis May Steiss, June 17, 1950; children: Rachel, George, Sarah. Ph.B., U. Chgo., 1948, S.B. in Physics, 1949, S.M., 1951, Ph.D. in Physics, 1953. Mem. staff dept. terrestrial magnetism Carnegie Inst., Washington, 1953-60; prof. geophysics and geology UCLA, 1960-75, chmn. dept. planetary and space sci., 1968-72; dir. dept. terrestrial magnetism Carnegie Inst., Washington, 1975-91, mem. sci. staff, 1991—; cons. NASA, NSF, Nat. Acad. Sci. Editor Ann. Rev. of Earth and Planetary Sci.; assoc. editor, Meteoritics, Icarus; contbr. articles to profl. jours. With USN, 1943-46. Recipient G.K. Gilbert award Geol. Soc. Am., 1984, Profl. Achievement Citation U. Chgo.

Alumni Assn., 1985. Fellow Am. Acad. Arts and Scis., Am. Geophys. Union (pres. planetology sect. 1970-72, recipient H.H. Hess medal, 1991), Meteoritical Soc. (v.p. 1971-74, 81-83, pres. 1983-85, Leonard medal 1981); mem. NAS, Geochem. Soc. (v.p. 1973-74, pres. 1974-75), Internat. Assn. Geochem. and Cosmochemistry (pres. 1977-80), Internat. Astron. Union, Am. Astron. Soc. Div. Planetary Scis. and Div. Dynamic Astronomy (G.P. Kuiper prize 1986), Religious Soc. Free Quakers, Internat. Soc. for Study of Origin of Life. Episcopalian. Office: Carnegie Inst 5241 Broad Branch Rd NW Washington DC 20015-1395

WETLAUFER, DONALD BURTON, biochemist, educator; b. New Berlin, N.Y., Apr. 4, 1925; s George C. and Olga (Kirckhoff) W.; m. Lucille D. Croce, May 5, 1950; children—Lise, Eric. B.S. in Chemistry, U. Wis., Madison, 1946, M.S. in Biochemistry, 1952, Ph.D., 1954. Chemist Argonne (Ill.) Nat. Lab., 1944, 46-47, Bjorksten Lab., Madison, 1948-50; Carlsberg Lab., Copenhagen, 1955-56; research asso. Harvard U., 1956-61, tutor biochem. sci., 1958-61; asst. prof. biochemistry Ind. U. Med. Sch., 1961-62; asso. prof., then prof. biochemistry U. Minn. Med. Sch., 1962-75; DuPont prof. chemistry U. Del., Newark, 1975—, chmn. dept., 1975-85; vis. investigator Max Planck Inst. Ernahrungsphy., 1974-78; mem. fellowship rev. com. NATO, 1970; cons. Nat. Inst. Gen. Med. Sci., 1964—, NSF, 1980—. Author research papers in field of protein biochemistry, protein folding and high performance protein purification; indsl. cons. NSF predoctoral fellow, 1952-54; Nat. Found. Infantile Paralysis postdoctoral fellow, 1955-56; Am. Heart Assn. postdoctoral fellow, 1956-58; grantee USPHS, 1961—; grantee NATO, 1974-77; grantee NSF, 1977—; grantee AEC, 1962; recipient Career Devel. award USPHS, 1961-62. Mem. AAAS, Am. Chem. Soc. (councilor, alt. councilor divsn. biol. chemistry 1975-87), Am. Soc. Biochemistry and Molecular Biology, The Protein Soc., Chromatography Forum, Phi Beta Kappa. Office: U Del Dept Chemistry & Biochemistry Newark DE 19716

WETSCH, JOHN ROBERT, computer systems administrator; b. Dickinson, N.D., Aug. 27, 1959; s. Joseph John and Florence Mae (Edwards) W.; m. Laura Jean Johnson, Aug. 29, 1981; children: Julie Elizabeth, Katherine Anne, John Michael. BS, U. State of N.Y.-Regents Coll., Albany, 1984; MA, Antioch U., 1989. Radiation physics instr. Grand Forks (N.D.) Clinic, 1983-85; sr. programmer Planning Rsch. Corp., Cavalier Air Force Sta., N.D., 1987-91; pres. Dakota Sci. Inc., Langdon, N.D., 1988—; instr. U. N.D.-Lake Region, Devils Lake, 1988-91; systems adminstr. U.S. Courts Fine Ctr., Raleigh, N.C., 1991—; cons. on Wave Obs.-- N.D. Proposal, Gov.'s Office, Bismarck, 1991; founder, developer Dakota Sci., Inc., Langdon, 1988—. Author: (with others) COMPUTE!'s 2nd Book of Amiga, 1988; contbr. articles to COMPUTE! Jour. of Progressive Computing, 1987. Mem. coll. scholarship selection com. Cavalier Air Force Sta., N.D., 1990; program coord. Lake Region Outreach, U. N.D., Cavalier Air Force Sta., 1988-91; pres. Zeta Rho chpt. Pi Kappa Alpha, Grand Forks, 1981. SMITS scholar N.D. Acad. Sci., 1990; Larimore-Mathews scholar U. N.D., Grand Forks, 1978, N.D. Acad. Sci. scholar, 1978; recipient Westinghouse Sci. Talent Search award, 1978. Mem. Am. Assn. for Advancement Sci., Regents Coll. Degrees (grad. resource network), IEEE Computer Soc. Republican. Roman Catholic. Achievements include microcomputer short range weather forecasting algorithm; model of the motion of freely falling bodies in 3 dimensions as an elliptic paraboloid; study in astronomy and culture, and astronomy's impact on devel. on Western civilization. Home: 5069 Tall Pines Ct Raleigh NC 27609-4662 Office: US Courts Fine Ctr 310 New Bern Ave Raleigh NC 27611-9998

WETTE, EDUARD WILHELM, mathematician; b. Radevormwald, Germany, Feb. 4, 1925; s. Eduard and Anna Auguste (Finkensieper) W.; self-taught; m. Anna Maria Elisabeth Mohrhauer, Dec. 28, 1950; 1 dau., Adelheid Margaretha Elisabeth. Ind. rschr.; mem. adv. council Internat. Logic Rev., Bologna, Italy, 1975. Served with 405 Regt., 1944-45. Recipient prize Bonn U., 1968. Mem. Assn. Symbolic Logic, Am. Math. Soc., Math. Assn. Am., History of Sci. Soc., Soc. History of Tech., N.Y. Acad. Scis., Centro Superiore di Logica e Scienze Comparate. Author articles on finite contradictions within pure arithmetic; anti-relativistic morphometric representation of the motions' totality; electronic ether-curvatures; intrafinite recursive world calculus. Home: 14 Blumenstrasse, D42477 Radevormwald Germany Office: PO Box 4115, Uckerath, D53767 Hennef Germany

WETZEL, JOHN PAUL, structural dynamics engineer; b. Milw., Apr. 4, 1963; s. Max O. and Lucille S. (Strietzel) W.; m. Kathy L. Zimmerman, May 31, 1986; children: Ryan Louis, Erika Leigh. BS in Engring. Mechanics, U. Wis., 1985, MS in Engring. Mechanics, 1986. Assoc. staff mem. BDM Internat., Inc., Albuquerque, 1986-91; structural dynamics engr. Grumman Space Sta. Integration Div., Reston, Va., 1991—. Kurt F. Wendt scholar, 1985. Mem. AIAA, ASCE (tech. co-chair engring. constrn. and ops. in space 1987-88, 88-90, vice-chair; vice-chair dynamics and controls com., Aerospace Scis. and Tech. Applications award 1990). Office: Grumman Space Sta Integration Div 1760 Business Center Dr Reston VA 22090

WEY, JONG SHINN, research laboratory executive; b. Kaohsiung, Taiwan, Oct. 26, 1944; came to U.S. 1968; s. Tan-Ding and Chao (Lee) W.; m. Hseh-Yi Su, 1965; children: John, Nancy. BS in Chem. Engring., Nat. Taiwan U., 1967; PhD, Clarkson U., 1973. Sr. rsch. chemist Eastman Kodak Co., Rochester, N.Y., 1973-80, rsch. assoc., 1980-84, lab. head, 1984-91, sr. lab. head, 1991—. Co-author: Preparation and Properties of Solid State Materials, 1981, Handbook of Industrial Crystallization, 1992. Recipient Jour. award Photographic Sci. and Engring. Fellow Am. Inst. Chem. Engrs.; mem. Soc. Photographic Scientists and Engrs. Achievements include patents in field; published papers on crystallization, precipitation, nucleation, growth, size distribution contro and photographic emulsion. Office: Eastman Kodak Co Rsch Labs Bldg 59 Rochester NY 14650-1736

WEYANT, JOHN PETER, engineering economic systems educator; b. Bklyn., July 23, 1947; s. John Pershing and Elizabeth Anna (Siwik) W.; m. Kathleen Favor, June 3, 1984; children: Christopher Favor, Melissa Elizabeth. BS, Rensselaer Poly. Inst., 1969, MS, 1970; MS, Rensselaer Poly. Inst., 1971; PhD, U. Calif., Berkeley, 1976. Staff mem. Rand Corp., Santa Monica, Calif., 1975-76; postdoctoral fellow Kennedy Sch. Govt., Harvard U., Cambridge, Mass., 1976-77; rsch. assoc. dpet. ops. rsch. Stanford (Calif.) U., 1977-80, sr. rsch. assoc. dept. ops. rsch., 1980-84, assoc. prof. dept. engring. econ. systems, 1984-89, prof. dept. engring. econ. systems, 1989—; chmn. rev. panels on acid deposition rsch. EPA, 1983-87; mem. Office Tech. Assessment Adv. Bd. on Gas Supply, 1984-85, NAS Rev. Com. on Gas Rsch. Inst., 1988-89; founder with others Energy Modeling Forum, Stanford U., 1976-77, bd. dirs., 1984—. Editor: Handbook on Energy Economy Modeling, 1991; contbr. articles to profl. jours. including Omega, Mgmt. Sci., Am. Econ. Review, Econ. Perspectives Ops. Rsch. Grantee to support energy modeling forum U.S. Dept. Energy, 1984—, U.S. EPA, 1990—, Electric Power Rsch. Inst., Palo Alto, Calif., 1984—. Mem. Internat. Assn. for Energy Econs. (v.p. 1988-90), Ops. Rsch. Soc. Am., Assn. for Energy Engrs., Am. Econ. Assn. Home: 861 Allardice Way Stanford CA 94305-1050 Office: Stanford U Dept Engring Econ Systems Stanford CA 94305

WEYENBERG, DONALD RICHARD, chemist; b. Glenvil, Nebr., July 11, 1930; s. Clyde H. and Elva I. (Hlavaty) W.; m. Barbara Ann Oppenheim, Dec. 26, 1955; children: Ann Louise, Thomas Richard. B.S. in Chemistry, U. Nebr., 1951; Ph.D., Pa. State U., 1958; P.M.D. Program, Harvard U., 1968. Research chemist Dow Corning Corp., Midland, Mich., 1951-65, research mgr., 1965-68, dir. corp. devel., 1968-69, dir. silicone research, 1969-71, bus. mgr., 1971-76, dir. research, 1976-79, v.p. research and devel., 1979-86, chief sci. and sr. v.p. research and devel., 1987-93, sci. emeritus, 1993—; bd. dirs. Dow Corning, Dendritech; Hurd lectr. Northwestern U., 1992—. Bd. editors: Organometallics Jour., 1980—; contbr. articles to sci. jours., chpts. to books; patentee silicone materials. Bd. visitors Memphis State U., 1981—; mem. indsl. bd. advisors U. Nebr., 1988. Named Alumni fellow Pa. State U., 1988. Mem. Am. Chem. Soc. (chmn. Midland sect. 1967 Outstanding Achievement in Promotion Chem. Scis. award Midland sect.), Indsl. Research Inst., N.Y. Acad. Scis., Sigma Xi. Lodge: Rotary. Home: 4601 Arbor Dr Midland MI 48640-2644 Office: Dow Corning Corp PO Box 994 2200 W Salzburg Rd Midland MI 48686

WEYLER, WALTER EUGEN, JR., process engineer; b. Boston, Feb. 20, 1963; s. Walter Eugen and Nancy Prudence (Haines) W.; m. Kimberly Ann Weber, Apr. 6, 1991; 1 child, Jessica. BSChe, Northwestern Univ., 1985. Plant engr. Valspar Corp., Azusa, Calif., 1986-87, prod. engr., 1987; project engr. Valspar Corp., Wheeling, Ill., 1987-88; process engr. H.B. Fuller, Mpls., 1988-89, acting plant mgr., 1989-90, sr. process engr., 1990—. Author: Powder Coatings Mag., 1991. Dispasments com. United Way Mpls., 1991—. Mem. AICE (pres. 1985), Soc. Mfg. Engrs. (sr., pres. 1991). Achievements include engineered and supervised construction of Valspar's first electronically controlled semi-automated latex paint manufacturing plant. Home: 3527 47th Ave S Minneapolis MN 55406 Office: H B Fuller Co 3200 Labore Rd Vadrais Heights MN 55110

WEYMANN, RAY J., astronomy educator; b. Los Angeles, Dec. 2, 1934; s. August Charles and Lucile (Rausch) W.; m. Barbara Lee McDermott, June 16, 1956; children—Lynn Elizabeth, Catherine Ann, Steven Christopher. B.S., Calif. Inst. Tech., Pasadena, 1956; Ph.D., Princeton U., 1959. Postdoctoral fellow Calif. Inst. Tech., Pasadena, 1959-61; asst. prof. U. Ariz., Tucson, 1961-63, assoc. prof., 1964-67, prof. astronomy, from 1967, dir. Steward Obs., 1970-75; prof. astronomy U. Ariz., Tuscon, 1975-86; asst. prof. UCLA, 1963-64; dir. Mt. Wilson and Las Campanas Observatories, from 1986. Contbr. articles to profl. jours. Mem. Am. Astron. Soc. (v.p. 1981-83), Astron. Soc. Pacific (pres. 1973-75), Am. Acad. Arts and Sci., Nat. Acad. Scis. Office: Carnegie Inst Washington Mt Wilson & Las Campanas Obs 813 Santa Barbara St Pasadena CA 91101-1232

WHALEN, BARBARA RHOADS, mechanical engineer, consultant; b. Washington, Feb. 10, 1960; d. Robert H. and Janet (Miller) Rhoads; m. Timothy John Whalen, June. 18, 1983; children: Elizabeth Anne, Thomas Christopher. BA, BS in mech. engring., Bucknell Univ., 1983. Registered profl. engr. Project engr. Korda Nemeth Engring., Columbus, Ohio, 1983-87; project mgr., group mgr. Mueller Assocs., Inc., Baltimore, Md., 1987-92; construction mgr. Bon Secours Hosp., Baltimore, 1992—. Mem. Am. Soc. Heating Refrigerating and Air Conditioning Engrs., Am. Soc. Hosp. Engrs., Chesapeake Area Soc. Hosp. Engrs. Office: Bon Secours Hosp 2000 W Baltimore St Baltimore MD 21223

WHALEN, THOMAS EARL, psychology educator; b. Toledo, June 26, 1938; s. T. Mylo and Alice E. (Tallman) W.; m. Carolyn Margaret Lapham, Dec. 24, 1960; children: Jennifer Susan, Holly Elizabeth. BA, UCLA, 1960; MA, San Diego State U., 1967; PhD, U. Conn., 1970. Cert. secondary tchr., Calif. Secondary tchr. San Diego City Schs., 1964-68; rsch. assoc. Southwest Regional Lab., Inglewood, Calif., 1969; prof. Calif. State U., Hayward, Calif., 1970—; ednl. psychology dept. chair 1977-79, assoc. dean sch. edn. 1987-89, Calif. State U. Hayward; rsch. cons. Evaluation Assocs. San Francisco Bay Area Schs., 1971-88, Lawrence Livermore Nat. Lab., Livermore, Calif., 1982-83. Author: (text book) Ten Steps to Behavioral Research, 1989; contbr. articles to profl. jours. Lt. USN, 1960-63. Recipient U.S. Office of Edn. fellowship, U. Conn., 1968-70, post doctoral scholarship Am. Edn. Rsch. Assn., U. Iowa, 1972. Mem. Am. Ednl. Rsch. Assn., APA, Calif. Ednl. Rsch. Assn. (bd. dirs. 1982-84), Bay Area Coun. on Measurement and Evaluation in Edn. (pres. 1976-77), United Profs. of Calif. (exec. bd. Calif. State U. Hayward 1975-76). Avocations: golf, travel, gardening. Home: 325 Conway Dr Danville CA 94526-5511 Office: Calif State U 25800 Carlos Bee Blvd Hayward CA 94542-3001

WHALEY, MAX WELDON, chemist; b. Duncan, Okla., Aug. 21, 1947; s. Julius Weldon and Zula (Lomax) W.; m. Jerita Wynne Southern, June 5, 1971; children: Jason, Melissa, Stephen. BS, Okla. State U., 1969. Lab. technician physiol. sci. dept. Okla. State U., Stillwater, 1970-74, lab. mgr. physiol. sci. dept., 1974-76; analytical toxicologist Okla. Animal Disease Diagnostic Lab., Okla. State U., Stillwater, 1979—. Coach Stillwater Parks and Recreation Baseball, 1990-91. Mem. Assn. Ofcl. Analytical Chemists Internat. Mem. Ch. of Christ. Home: Rt 3 Box 29 Stillwater OK 74075 Office: Okla State U Okla Animal Disease Diagnostic Lab Stillwater OK 74078

WHANG, KYU-YOUNG, computer science educator, director; b. Pusan, Korea, Mar. 2, 1951; s. Sam. H. and Young H. (Kim) W.; m. Jung H. Song, Nov. 12, 1977; 1 child, Euijong. BS summa cum laude, Seoul Nat. U., 1973; MS, Korea Advanced Inst. Sci. and Tech., 1975, Stanford U., 1982; PhD, Stanford U., 1984. Sr. engr. Agy. for Def. Devel., Daejeon, Republic of Korea, 1975-78; system programmer Almaden (Calif.) Rsch. Lab. IBM, 1981; rsch. staff mem. IBM Watson Lab., Yorktown, 1983-91; assoc. prof. Korea Advanced Inst. Sci. and Tech., 1990—; dir. DB&KE Lab. Ctr. for Artificial Intelligence Rsch., Daejeon, 1990—; invited Teach-the-Tchrs. lectr. Korea Info. Sci. Soc., Seoul, 1990; cons. Sam Sung Co., Ki-Hung, Republic of Korea, 1991, Hewlett-Packard Labs., Palo Alto, Calif., 1993; conf. co-chair Far East Workshop Geog. Info. Sys., Singapore, 1993. Editor The VLDB Jour., Distributed and Parallel Databases; contbr. numerous articles to profl. jours. Mem. IEEE (sr., program vice chair Internat. Conf. on Data Engring. 1990-92, program co-chair 1989, assoc. editor Data Engring. Bull. 1990—, Cert. of Appreciation 1989, 90, 92), Assn. for Computing Machinery, Korean Info. Sci. Soc. (chair spl. interest group databases 1992—). Avocations: golf, swimming, skiing. Home: 209-903 Hyundai 2d Apt, Kae-Po Dong, Seoul Republic of Korea Office: Korea Advanced Inst Sci and Tech, 373-1 Koo-Sung Dong, Daejeon Republic of Korea

WHANG, SUNG H., metallurgical science educator; b. Suhbyuck, Bongwha, South Korea, Feb. 17, 1936; came to U.S., 1972; s. Ki Chae and Myung Mae (Kim) W.; m. Okki Lim, Jan. 20, 1961; children: Hewon, Jun-Ho, Jun-Shik, Jun-Sun. BS, Seoul (Korea) Nat. U., 1962; D. Engring. and Sci., Columbia U., 1978. Lectr. Seoul U., 1969-72; staff scientist Northeastern U., Boston, 1979-81, sr. scientist, 1981-85; assoc. prof. Poly. U., Bklyn., 1985-91, prof., 1991—; mem. bd. rev. Metall. Transactions A., Pitts., 1989—, Metallurgical Soc., Pitts., 1989—. Editor: High Temperature Superconducting Compoung I, TMS, 1989, High Temperature Aluminides and Intermetallics, TMS, 1990, High Temperature Superconducting Compound II, TMS, 1990, High Temperature Superconducting Compound III, TMS, 1991, Jour. Material Sci. & Engring., 1992, vols. 132, 133, 1992; contbr. articles to profl. jours. Mem. Metall. Soc. of AIME, Am. Soc. Metals, Materials Rsch. Soc., Sigma Xi. Achievements include patents in field. Office: Poly U Six Metrotech Ctr Brooklyn NY 11201

WHARTON, DANNY CARROLL, zoo biologist; b. Ontario, Oreg., Mar. 13, 1947; s. Carroll Curtis and Norma (Grigg) W.; m. Marilyn Christine Hoyt, Sept. 22, 1973; children: Amanda, Catherine, Margaret, Arcadio. BA in Psychology, Coll. Idaho, 1969; MA in Indsl. Adminstrn., Sch. for Internat. Tng., 1975; PhD in Biology, Fordham U., 1990. Rsch. assoc. Foresta Inst., Carson City, Nev., 1973-74; curatorial asst. Woodland Park Zool. Garden, Seattle, 1974-79; asst. curator N.Y. Zool. Soc./The Wildlife Conservation Soc., Bronx, 1979-85; assoc. curator N.Y. Zool. Soc., Bronx, 1985-89, curator, 1989—; advisor Internat. Snow Leopard Trust, Seattle, 1986; mem. US-USSR Environ. Agreement of U.S. Fish and Wildlife Svc., 1983. Contbr. articles to profl. jours. Vol. U.S. Peace Corps., Ecuador, 1969-71. Fulbright scholar, U. Münster, Fed. Republic Germany, 1976-77. Fellow Am. Assn. Zool. Parks and Aquariums (chmn. gorilla species survival plan 1992—, chmn. snow leopard species survival plan 1986—, co-chmn. marsupial and monotieme taxon adv. group 1990—); mem. Soc. for Conservation Biology, Internat. Union for Conservation of Nature/Species Survival Commn. (mem. captive breeding specialist group). Office: Wildlife Conservation Soc/ NY Zool Soc 185th St and Southern Blvd Bronx NY 10460

WHARTON, RALPH NATHANIEL, psychiatrist, educator; b. Boston, June 15, 1932; s. Nathaniel Philip and Deeda (Levine) W.; AB cum laude, Harvard U., 1953; MD, Columbia U., 1957, degree in psychoanalysis, 1970; children: Naida, Philip, Laura. Intern, Cornell div. Bellevue Hosp., N.Y.C., 1957-58; resident Columbia-Presbyn. Med. Center, N.Y.C., 1961-64; practice medicine, specializing in psychiatry and psychopharmacology, N.Y.C., 1964—; assoc. psychiatry Coll. Physicians and Surgeons, N.Y.C., 1964-69, asst. prof. clin. psychiatry, 1969-72, assoc. prof., 1972-83, prof., 1984—; sr. research psychiatrist N.Y. State Psychiat. Inst., N.Y.C., 1964-70; assoc. attending in psychiatry Columbia-Presbyn. Hosp., 1970—, ex-officio mem. bd. trustees, pres. Soc. Practitioners Columbia-Presbyn. Med. Center, 1980-82; attending psychiatrist Columbia-Presbyterian Med. Ctr., 1984—. Mem. bd. regent, chmn. Wharton Fund for Brain Rsch. Served with M.C., U.S. Army, 1958-61. Named one of Best Drs., N.Y. mag. Fellow N.Y. Acad. Medicine, Am. Psychiat. Assn., Am. Coll. Psychoanalysts (bd. regents 1989—); mem. AMA (mem. legis. action com.), Soc. Biol. Psychiatry, Royal Soc. Medicine, Lotos Club, Salon de Virtuosi (founding bd. mem.), Harvard Club. Author numerous publs. in field. Office: Columbia Presbyn Med Ctr 1070 Park Ave Ste 1D New York NY 10128-1000

WHARTON, THOMAS WILLIAM, health administration executive; b. St. Louis, Nov. 20, 1943; s. Thomas William and Elaine Margaret (Bassett) w.; divorced; children: Thomas William, Christopher John. BSc in Econs., U. Mo., 1967; M in Health Adminstrn., U. Ottawa, Ont., Can., 1978. Asst. to exec. dir. Ottawa Civic Hosp., 1978-80; exec. dir. Cariboo Meml. Hosp., Williams Lake, B.C., Can., 1980-83; dir. clinic and rehab. services Workers' Compensation Bd., Vancouver, B.C., 1983-89; pres. Gold Canyon, Lansdowne & Bradner Resources Corps., 1989—, Diagnostic and Health Cons., Vancouver, 1989—; ptnr., dir. Lynn Valley Med. Ctr., North Vancouver, B.C., 1993. Recipient Founder award Cariboo Musical Soc., 1983; named Lord of the Manors of Wharton and Kirkby Stephen (Eng.), 1991. Mem. Can. Coll. Health Service Execs., Am. Coll. Health Execs., Am. Acad. Med. Adminstrs., Health Adminstrs. Assn. B.C., U. Ottawa Health Service Alumni Assn. (pres. 1983-84). Avocations: swimming, music, art.

WHARTON, WILLIAM POLK, consulting psychologist, retired educator; b. Hopkinsville, Ky.; s. William Polk and Rowena Evelyn (Wall) W.; m. Lillian Marie Andersen, Mar. 11, 1944; 1 child, Christine Evelyn Wharton Leonard. BA, Yale U., 1934; MA, Tchrs. Coll., 1949; PhD, Columbia U., 1952. Diplomate Am. Bd. Profl. Psychology, Am. Bd. Psychotherapy; lic. psychologist, Pa. Rsch. advt. promotion, advt. sales Esquire Inc., N.Y.C., 1934-40; dir. counseling, prof. edn., counseling psychologist Allegheny Coll., Meadville, Pa., 1952-74, emeritus dir. and prof. edn., 1974—; prof., dir. The Ednl. Guidance Clinic, Meadville, 1958-74; pvt. practice cons. psychologist Meadville, 1974—; cons. U.S. Army Edn. Ctr., Ft. Meade, Md., 1960-61; rsch. adv. coun. Ednl. Devel. Ctr., Berea, Ohio, 1971-72; cons. to pres. Alliance Coll., Cambridge Springs, Md., 1975-76. Mem. editorial bd. Psychotherapy, 1966-68; reviewer Jour. Coll. Student Personnel, 1984-88; contbr. articles to profl. jours. Chmn. MH/MR Bd. Crawford County Pa., Meadville, 1970-73; com. chmn., Drug and Alcohol Coun. Crawford County Pa., Meadville, 1973-76; ethics com. chmn. North West Pa. Psychol. Assn., 1975-78; del. Pa. Mental Health Assn. Crawford County, 1978-79. Capt. U.S. Army, 1941-46; lt. col. USAR, 1960—. Psychotherapy Research Group vis. fellow, 1961-62; Romiett Stevens scholar, 1951. Fellow Pa. Psychol. Assn.; mem. Am. Psychol. Assn. (Disting. Contbn. award 1985), Am. Assn. for Counseling and Devel., Nat. Vocat. Guidance Assn., Pa. Coll. Personnel Assn. (chmn. 1956-57), Phi Beta Kappa. Phi Delta Kappa, Kappa Beta Pi, Pi Gamma Mu. Home and Office: 415 N Main St Meadville PA 16335-1510

WHEALTON, JOHN H., physicist; b. Bklyn., Apr. 27, 1943; s. Daniel J. and Isabelle K. (Baines) W.; m. Katharine M. Owens, Aug. 1, 1972; children: Karl, Linda, Thomas. BS, Lowell U., 1966; MS, U. Del., 1968, PhD, 1971. Rsch. assoc. Brown U., Providence, 1971-73, U. Colo., Boulder, 1973-75; staff scientist Oak Ridge (Tenn) Nat. Lab., 1975—; bd. editors Rev. Sci. Instruments. Contbr. over 250 articles to profl. jours.; patentee in field. Mem. Bd. Edn., Oak Ridge, 1990. Fellow Am. Phys. Soc.; mem. Nat. Sci. Tchrs. Assn., Am. Assn. Physics Tchrs., Math. Soc. Am., Nat. Sch. Bd. Assn., Nat. Coun. Tchrs. Math., Tenn. Inventors Soc., The Inventors Forum Martin Marietta Energy Systems Inc., Tenn. Sch. Bd. Assn. Avocations: symphony, community band, clarinet, model trains. Home: 185 Outer Dr Oak Ridge TN 37830-5364 Office: Oak Ridge Nat Lab Fusion Energy Div Oak Ridge TN 37831-8071

WHEAT, MYRON WILLIAM, JR., cardiothoracic surgeon; b. Sapulpa, Okla., Mar. 24, 1924; s. Myron William and Mary Lee (Hudiburg) W.; m. Erlene Adele Plank, June 12, 1949 (div. June 1970); children: Penelope Louise, Myron William III, Pamela Lynn, Douglas Plank; m. Carol Ann Karmgard, June 18, 1970; 1 child, Christopher West. AB, Washington U., St. Louis, 1949; MD cum laude, Washington U., 1951. Diplomate Am. Bd. Surgery, Am. Bd. Thoracic Surgery. Instr., clin. fellow Washington U., St. Louis, 1956-58; asst. prof. surgery U. Fla., Gainesville, 1958-65, prof. surgery, 1965-72; dir. profl. svcs., chief clin. physician U. Fla. Shands Teaching Hosp., Gainesville, 1968-72; prof. surgery, dir. thoracic and cardiothoracic surgery U. Louisville Sch. Medicine, 1972-75; cardiothoracic surgeon Cardiac Surg. Assocs., P.A., St. Petersburg, Fla., 1975-91, Clearwater, Fla., 1978—. Author (with others) 14 books; contbr. articles to profl. jours. 1st lt. USAF, 1943-46, ETO. Named First Howard W. Lillenthal Meml. lectr. Mt. Sinai Hosp., 1963. Fellow Am. Coll. Cardiology (chmn. bd. govs. 1968-69), Am. Coll. Surgeons (gov.); mem. Am. Surg. Assn., Am. Assn. for Thoracic Surgery, So. Surg. Assn., So. Thoracic Surg. Assn., Soc. Thoracic Surgeons, Soc. Thoracic Surgeons Great Britain and Ireland, Alpha Omega Alpha. Republican. Avocation: field trials-bird dogs. Home: 1772 Long Bow Ln Clearwater FL 34624-6402 Office: Myron W Wheat Jr MD 1260 S Greenwood Ave Ste E Clearwater FL 34616-4195

WHEELER, ANNEMARIE RUTH, biochemist; b. Eugene, Oreg., Nov. 26, 1959; d. David Keith and Nolene (Wade) W. BA in Biology and Chemistry, Whitman Coll., 1982; PhD Biochemistry, Colo. State U., 1987. Lab. teaching asst. dept. Chemistry Whitman Coll., Walla Walla, Wash., 1978-81, lab. teaching asst. dept. Biology, 1980-81; undergraduate rsch. participant dept. Chemistry Argonne (Ill.) Nat. Lab., 1981; grad. teaching asst. dept. Biochemistry Colo. State U., Ft. Collins, 1981-82, grad. rsch. asst. dept. Biochemistry, 1982-87; postdoctoral fellow dept. Biochemistry U. Minn., St. Paul, 1987-89; sr. scientist Beckman Instruments, Inc., Brea, Calif., 1989--. Contbr. articles to profl. jours. Treas. Puente Hills Dojo of Aikido Yoshinkai, Hacienda Heights, Calif., 1990--; founding mem. Aikido Yoshinkai Minn., Mpls., 1990; mem., vol. Friends of the Fullerton Arboretum, Calif., 1991--. Recipient Outstanding Community Svc. award Vol. Ctr. Greater Orange County, 1992, Signet Table award Whitman Coll., 1979-80, Cert. Appreciation Aikido Yoshokai N.Am., 1991, Cert. Appreciation Linda Vista Elem. Sch., 1992, 93; Colo. Grad. fellow Colo. State U., 1982-83. Mem. SPIE (co-chair symposium), Am. Chem. Soc. (pub. outreach participant 1991--), Grad. Women in Sci. (pres. L.A. chpt. 1992-93), Am. Assn. for the Advancement Sci. Achievements include the characterization of intrinsic fluorescence of RNA polymerase and of poly deoxyethenoadenine; demonstrated ion concentration and type dependence and DNA sequence and form dependence of specific transcription by both holo and core enzymes of RNA polymerase. Home: 17094 Los Angeles St Yorba Linda CA 92686 Office: Beckman Instruments Inc PO Box 8000 200 S Kraemer Blvd Brea CA 92622-8000

WHEELER, CARL, mathematics educator; b. Geneva, N.Y., Sept. 7, 1932. AB, Hamilton Coll., 1953; MA in Liberal Studies, Wesleyan U., 1970. Tchr., adminstr. Mid-Pacific Inst., Honolulu, 1956-58, 61-85, chair maths. dept., tchr., 1987—; admissions officer Hamilton Coll., Clinton, N.Y., 1958-61; tchr. Punahou Sch., Honolulu, 1985-87; mem. Math. Edn. Leadership Network, Washington, 1990—; teaching fellow Ctr. for Excellence in Edn., San Diego, 1990; cons. coll. board, Educational Testing Svcs. With U.S. Army, 1954-56. Woodrow Wilson Found. fellow, 1985; grantee NSF; Tandy Tech. scholar, 1991; recipient Presdl. award, 1990. Mem. Nat. Coun. Tchrs. Math., Hawaii Coun. Tchrs. Math. (v.p. 1986-88), Oahu Maths. League (pres. 1990-93, cons. coll. bd. 1993), Math. Assn. Am., Hawaii State Math. Coalition, Coun. of Presdl. Awardees in Math. Office: Mid Pacific Inst 2445 Kaala St Honolulu HI 96822-2299

WHEELER, CLARENCE JOSEPH, JR., physician; b. Dallas, Sept. 25, 1917; s. Clarence Joseph Sr. and Sadie Alice (McKinney) W.; m. Alice Mary Freels, Dec. 6, 1942 (dec.); m. Marie McLendon May, Dec. 30, 1983; children: Stephen Freels, C.J. III, Robert McKinney, THomas Michael, David Ritchey. BS in math., S. Meth. U., 1941, BA in Psychology, 1946; MD, John Hopkins U., 1950. Diplomate Am. Bd. Surgery; cert. ACLS, advanced trauma life support. Intern John Hopkins Hosp., Balti., 1950-51; resident in surgery Barnes Hosp., St. Louis, 1951-54; fellow thoracic surgery U. Wis. Hosp., Madison, 1954-56, instr. surgery, 1955-56; attending surgeon Welborne Clinic Baptist Hosp., Evansville, Ind., 1956-57; mem. consulting staff Tex. Children's Hosp., 1957-70; courtesy and consulting staffs Pasadena Hosp., Spring Br. Hosp., others, Houston, 1957-70; mem. active staff Hermann Hosp., Houston, 1957-70, St. Luke's Hosp., Houston, 1957-70, Meth. Hosp., Houston, 1957-70, St. Joseph's Hosp., Houston, 1957-70, Meml. Hosp., Houston, 1957-70, Ben Taub Gen. City/County Hosp., Houston, 1957-70, Diagnostic Hosp., &, 1957-70; attending surgeon Lindley Hosp., Duncan, Okla., 1970-71; sr. attending, chief surgery Gordon Hosp.,

Lewisburg, Tenn., 1971-73; chief thoracic surgery Lewisburg Community Hosp., 1973-75; mem. active med. staff, med. dir. Carver Family Health Clinic, 1975-82; dir. emergency dept. Meth. Med. Ctr. Ill., Peoria, 1975-82; mem. staff Contract Emergency Med. Care, Houston and Dallas, 1982-88; med. dir. substance abuse unit Terrell (Tex.) State Hosp., 1988-90; med. dir. Schick-Shadel Hosp., Dallas-Ft. Worth, 1991—; med. dir., chief of staff Schick-Shadel Hosp., Ft. Worth, 1991-93; med. dir. Skillman Med. Ctr., Dallas, 1993—; instr. surgery U. Wis. Med. Sch., 1955-56; clin. instr. surgery Baylor Coll. Medicine, Houston, 1959-70; lectr. surgery U. Tex. Postgrad. Sch., Houston, 1957-70; clin. asst. prof. surgery U. Ill. Sch. Medicine, Peoria, 1977-82. Treas. Samuel Clark Red Sch. PTA, Houston, 1959-61; bd. dirs. Salvation Army Boys Club, Houston; mem. Am. Mus. of Nat. History, Met. Mus. Art, Smithsonian Inst., Dallas Symphony Assn., Dallas Opera Soc., Dallas Theatre Ctr., Theatre Three Assn. Capt. USMCR, 1942-45, PTO. Decorated DFC with three stars, Air medal with four stars, Pacific Combat Theatre Ribbon with three stars, Purple Heart, Vietnamese Medal of Health (1st class), Vietnamese Medal Social Welfare, Navy Commendation medal, Presdl. Unit citation medal, Meritorious Bronze Star. Fellow ACS, Am. Coll. Angiology, Am. Coll. Chest Physicians, Royal Soc. Medicine, Internat. Coll. Surgeons, Am. Coll. Gastroentology, Southeastern Surg. Congress, Internat. Assn. Proctologists; mem. AAAS, Am., Tex., So., Indsl. Med. Assns., Harris County, St. Louis, Marshall County (pres.) med. socs., Am., Tex. thoracic socs., Nat. TB Assn., Am. Assn. History Medicine, Am. Soc. Contemporary Medicine and Surgery, Am. Soc. Addiction Medicine (cert.), Am., Tex., Houston heart assns., John Hopkins Med. and Surg. Soc., Southwestern Surg. Congress, Tex. Anti-TB Assn., Houston Gastroenterological Soc., Houston Surg. Soc., Postgrad. Med. Assembly S. Tex., Am. Cancer Soc., Am. Soc. Abdominal Surgeons, Marine Corps Officer's Assn., Naval Res. Officer's Assn., Nat. Geog. Soc., Greater Dallas Res. Officers Assn., Mil. Order of the World Wars, Navy League, Internat. Club, Sierra Club, Rotary, Kappa Sigma, Phi Eta Sigma, Kappa Mu Epsilon, Psi Chi. Episcopalian. Address: 4007 Cochran Heights Ct Dallas TX 75220-5032

WHEELER, DONALD KEITH, community and economic development specialist; b. Miami, Fla., Oct. 30, 1960; s. Clifford Keith and Elizabeth Grace (Bonilla) W. AA in Broadcasting, Miami-Dade Community Coll., 1981; BA in Polit. Sci., U. Fla., 1983; M of Pub. and Internat. Affairs, U. Pitts., 1985. Planning intern Am. Cancer Soc., Miami, 1985; assoc. planning and systems devel. Area Agy. on Aging, Miami, 1985-86; rsch. assoc. Fla. Internat. U., North Miami, 1986-87, adminstrv. coord., vis. instr., 1987-88; planner, grantwriter SER Jobs for Progress, Miami, 1988-90; tribal planner Miccosukee Tribe of Indians, Dade County, Fla., 1990—; mem. Miccosukee Tribal Community Rev. Bd., Dade County, 1990. Advisor Am. Cancer Soc., Miami, 1980-83. Recipient Outstanding Svc. award Am. Cancer Soc., 1981, Outstanding Community Svc. award The Miami Herald, 1979. Mem. Acad. Polit. Sci., Am. Soc. Pub. Adminstrn., Ptnrs. of the Ams., Toastmasters Internat. (pres. 1989), Sigma Phi Epsilon. Baptist. Avocations: reading, tennis, swimming. Home: 4554 SW 128th Pl Miami FL 33175-4614 Office: Miccosukee Tribe of Indians PO Box 440021 Miami FL 33144-0021

WHEELER, GEORGE CHARLES, consulting company executive; b. Balt., Oct. 9, 1923; s. George Charles and Julia Elizabeth (Watrous) W.; m. Dorothy W. Whittemore, Sept. 13, 1947 (div. 1977); children: Scott, Craig, Mark, Matthew, Tracy, Bruce; m. Clare Frances Weiner, Jan. 21, 1978. BS in Metall. Engring., Lehigh U., 1944. Various engring. and supervisory positions GE, Mass. and N.Y., 1944-62; mgr. materials, welding and nondestructive testing Knolls Atomic Power Lab., G.E., Schenectady, N.Y., 1962-68, mgr. nondestructive testing, 1968-85; pres., chief exec. officer Wheeler Nondestructive Testing, Inc., Schenectady, 1985—; cons. UN, N.Y.C., 1985—, Internat. Atomic Energy Agy., Vienna, Austria, 1985—, numerous others; guest lectr. Rensselaer Poly. Inst., Troy, N.Y., Union Coll., Schenectady, 1978-87; mem. math. sci. and tech. com. Schenectady County Community Coll., 1978-85, adj. prof., 1987—. Author: Guide to Developing Certification Exams, 1992; contbg. editor Jour. of ASNT. Troop leader Schenectady area Boy Scouts Am., 1957-64. Recipient GE Power Systems Engring. award. Mem. ASTM (com. internat. stds., com. nondestructive testing), NRA (life), Am. Soc. Nondestructive Testing (hon. life, bd. dirs. 1976-85, pres. 1983-84), Brit. Inst. Nondestructive Testing, Am. Soc. Metals, Nature Conservancy (life), Am. Welding Soc. Avocations: mountaineering, flying, gunning, photography. Office: Wheeler Nondestructive Testing Inc 29 Front St Schenectady NY 12305-1301

WHEELER, GEORGE LAWRENCE, chemist, educator; b. Rockville Center, N.Y., June 16, 1944; s. George Weton and Dorothy (Wolf) W.; m. Marcia Wheeler; children: David, Jean. PhD, U. Md., 1973. Postdoctoral fellow Yale U., New Haven, Conn., 1973-75; NIH tng. fellow Yale Med. Sch., New Haven, 1975-77, rsch. affiliate, 1990—; from asst. prof. to prof. U. New Haven, 1977-81, Jacob Buckman prof., 1983—; staff mem. Los Alamos (N.Mex.) Nat. Lab., 1981-83; mem. internal adv. bd. Ctr. for Non-Linear Studies Los Alamos Nat. Lab., 1981-83; chmn. dept. chemistry and chem. engring. U. New Haven, 1983-88. Contbr. articles to Nature, Proceedings Nat. Acad. Sci. Grantee NSF, 1980, U.S. Dept. Energy, 1981, Conn. Dept. Environ. Protection, 1991. Mem. Am. Chem. Soc., AAAS, N.Y. Acad. Sci., Am. Crystallographic Assn. Achievements include discovery of G protein involved in biochemistry of vision. Office: U New Haven Dept Chemistry 300 Orange Ave West Haven CT 06516

WHEELER, HAROLD ALDEN, retired radio engineer; b. St. Paul, May 10, 1903; s. William Archie Wheeler and Harriet Maria Alden; m. Ruth Gregory, Aug. 25, 1926 (dec. Feb. 1986); children: Dorothy, Caroline, Alden Gregory. BS in Physics, George Washington U., 1925, DSc (hon.), 1972; DEngring. (hon.), Stevens Inst. Tech., 1978, Polytechnic U., 1992. Engr. Hazeltine Service Corp., N.Y.C. and Bayside, N.Y., 1929-39; v.p., chief cons. engr. Hazeltine Service Corp., Little Neck, N.Y., 1940-45; cons. radio physicist Great Neck, N.Y., 1946-59; pres. Wheeler Labs Inc., Great Neck, 1947-68; dir. Hazeltine Corp., Little Neck, 1959-70, v.p., 1959-65, chmn., 1965-70; dir. Hazeltine Corp., Greenlawn, N.Y., 1971-83, chmn., 1971-77, chmn. emeritus, 1977-87, chief scientist, 1971-87; cons. Office of Sec. of Def., Washington, 1950-53; mem. Def. Sci. Bd., Washington, 1961-64. About 180 patents in field, including diode automatic volume control, 1932; author: Wheeler Monographs, Vol. 1, 1953, Hazeltine the Professor, 1978, Early Days of Wheeler and Hazeltine Corporation, 1982, Hazeltine Corporation in World War II, 1993; numerous papers in, procs. and transactions of IRE and IEEE. Recipient Modern Pioneer award Nat. Assn. Mfrs., 1940. Fellow IEEE (medal of honor 1964), IRE (Morris Liebmann prize 1940), Radio Club Am. (Armstrong medal 1964); mem. Nat. Acad. Engring., Inst. Elec. Engrs. (U.K.), Sigma Xi, Tau Beta Pi, Gamma Alpha. Republican. Unitarian. Home: 4900 Telegraph Rd Apt 523 Ventura CA 93003-4125

WHEELER, JOHN CRAIG, astrophysicist, writer; b. Glendale, Calif., Apr. 5, 1943; s. G.L. and Peggy Wheeler; m. Hsueh Lie, Oct. 29, 1967; children: Diek Winters, J. Robinson. BS in Physics, MIT, 1965; PhD in Physics, U. Colo., 1969. Asst. prof. astronomy Harvard U., Cambridge, Mass., 1971-74; assoc. prof. U. Tex., Austin, 1974-80, prof., 1980—, Samuel T. and Fern Yanagisawa Regents prof. astronomy, 1985—, chmn. astronomy dept., 1986-90; vis. fellow Joint Inst. Lab. Astrophysics, Boulder, Colo., 1978-79, Japan Soc. for Promotion of Sci., 1983; 1st vis. prof. Assn. Univs. for Rsch. in Astronomy, 1990; mem. exec. com. astrophysics workshop Aspen (Colo.) Ctr. for Physics, Tex. Symposium on Relativistic Astrophysics. Author: The Krone Experiment, 1986. Recipient undergrad. teaching award Coll. Natural Scis., U. Tex., 1984, teaching award Bd. Visitors, Dept. Astronomy, U. Tex., 1986; Fulbright fellow, Italy, 1991. Mem. Internat. Astron. Union, Am. Astron. Soc., Sigma Xi. Avocations: running, writing, reading. Office: U Tex Dept Astronomy Austin TX 78712

WHEELER, THOMAS JAY, biochemist; b. Schenectady, N.Y., Apr. 12, 1951; s. Dean Bussman and Alice Effie (Smith) W.; m. Valerie Jean Ives, June 17, 1973; children: Jay Ives, Eric Smith. SB, MIT, 1973; PhD in Biochemistry, Brandeis U., 1979. Postdoctoral fellow Sect. of Biochemistry, Cornell U., Ithaca, N.Y., 1978-81; asst. prof. Dept. Biochemistry, U. Louisville, 1981-90, assoc. prof., 1990—. Editorial bd. Biochimica et Biophysica Acta, Amsterdam, 1989—; contbr. articles to profl. jours. including Jour. Biol. Chemistry, Ann. Rev. Physiol., Biochemistry. Rsch. grants Juvenile Diabetes Found., 1986-88, 89-92, Ky. affiliate Am. Heart Assn., 1982-83, 86-88, NIH, 1982-85; postdoctoral fellowship NIH, 1978-81. Mem. AAAS, Am. Soc. for Biochemistry and Molecular Biology, Ky. Acad. of Sci., Ky. Assn. of Sci. Educators and Skeptics, Nat. Ctr. for Sci. Edn., Sigma Xi, Phi Beta Kappa. Achievements include research on kinetics and regulation of glucose transport, reconstitution of transport proteins in liposomes. Office: U Louisville Dept Biochemistry Health Scis Ctr Louisville KY 40292

WHEELER, WESLEY DREER, marine engineer, naval architect, consultant; b. N.Y.C., Aug. 3, 1933; s. Wesley Lunn and Rosalie (Smith) W.; m. Dolores Janes-Wheeler, May 27, 1989; children: Wesley P., Jonathan H., Deborah B. BS in Mech. Engring., Worcester Poly. Inst., 1954; MSE in Naval Architecture and Marine Engring., U. Mich., 1958. Naval architect Am. Bulk Carriers, N.Y.C., 1966-68; port engr. Am. Trade and Prodn. Co., N.Y.C., 1968-69; pres. Techmarine, Inc., N.Y.C., 1969-71; asesor tecnico Astilleros Espanoles SA, Cadiz, Spain, 1971-72; tech. dir. Am. Bulk Carriers, N.Y.C., 1972-74; pres. Wesley D. Wheeler Assoc., Ltd., N.Y.C., 1974-83; v.p. J.J. Henry Co. Inc., N.Y.C., 1983; pres. Wheeler Assocs., N.Y.C., 1983—; Am. rep. Blohm & Voss, Astilleros Españoles, Sud Marine, Dakar Marine, Shanghai Shipyard, Tandanoor. Elder and trustee Fifth Ave. Presbyn. Ch., N.Y.C., sec.; dir. Sutton Area Community, N.Y.C., Soc. Naval Architects and Marine Engrs., N.Y.C., 1990-92, treas.; chmn. Soc. Marine Cons., N.Y.C., 1987-92. Mem. Maritime Assn. Port of N.Y./N.J., Soc. Marine Port Engrs. N.Y., Soc. Maritime Arbitrators Inc., Royal Inst. Naval Architects, Inst. Marine Engrs., Asociacion Ingenieros Navales Madrid. Presbyterian. Avocations: boating, skiing, tennis, golf, travel. Home: 60 Sutton Pl S New York NY 10022-4168

WHEELER, WILLIAM CRAWFORD, agricultural engineer, educator; b. Maysville, Ga., Feb. 19, 1914; s. James D. and Pearl (Chandler) W.; B.S. in Agrl. Engring., U. Ga., 1940; student U. Tenn., 1941-42; M.S. in Agrl. Engring., Va. Polytech. Inst., 1951; postgrad. Mich State U., 1954-55; m. Annie Adams, May 23, 1942; children—Betty Catherine, James David. Supr. farm Ga. Vocational and Trade Sch., Monroe, 1932-36, instr. farm mechanics, 1940-41; jr. engr. U.S. Dept. Agr. Soil Conservation Service, Monroe; instr. agrl. engrng. U. Tenn., 1941-42, asso. prof. agrl. engring., 1946-53, specializing in rural electrification, 1946—; prof. agrl. engring. U. Conn., 1953-72, head dept., 1953-69, prof. emeritus, 1972—; sabbatical leave for waste disposal study in Europe and Can., 1969; conducted summer course in farm mechanics U. Ark., Fayetteville, 1948; partner Lynwood Devels., Storrs, Conn. Mem. N.E. Regional Research Techn. Com. on Poultry Housing, 1953-65, Com. on Mechanization of Forage Crops, 1954-58; com. on Mechanization of Fruit and Vegetable Harvest, 1959-71, Com. on Improvement Efficiency in Harvesting Apples, 1966-71; chmn. projects com. Conn. Electrification Council, 1953-66; mem. farm electric service com. N.E. Council, 1953-72; mem. water resources Inst. U. Conn., 1964-69. Served as maj. 329th inf. 83d Div., U.S. Army, 1942-46. Decorated Bronze Star, Purple Heart. Registered engr., Tenn. Fellow Am. Soc. Agrl. Engrs. (chmn. North Atlantic sect. 1962); mem. Am. Soc. Engring. Edn., Nat. Soc. Profl. Engrs., Alpha Zeta, Scabbard and Blade, Gridiron, Aghon, Gamma Sigma Delta. Conglist. (chmn. council 1964). Author articles in field. Home: 1116 Linkside Dr Atlantic Beach FL 32233-4387

WHEELOCK, KENNETH STEVEN, chemist; b. Kansas City, Mo., Sept. 18, 1943; s. Kenneth Lewis and Clara Mae (Hanenkratt) W.; m. Mary Corinne Percy, June 30, 1972; children: Michael Steven, Celeste Marie. BSc, U. Mo., Kansas City, 1965; PhD, Tulane U., New Orleans, 1970. Registered patent agent. Chemist Exxon Rsch. & Devel. Labs., Baton Rouge, La., 1969-72, rsch. chemist, 1972-77, staff chemist, 1977-83, sr. staff chemist, 1983-86; assoc. prof. physics La. State U., Baton Rouge, 1987; sr. rsch. chemist Phillips Petroleum Co., Bartlesville, Okla., 1987-91; chmn. Prakti Katalysts, Bartlesville, Okla., 1992-93; patent agt. GE Plastics, Pittsfield, Mass., 1993—; cons. dept. chemistry Tulane U., New Orleans, 1970-75. Advisor Jr. Achievement, Baton Rouge, 1971; sec. Baton Rouge Orchid Soc., 1983, Bartlesville Gifted and Talented, 1989. NDEA trainee, Tulane U., New Orleans, 1965-67, NASA fellow, 1967-69. Fellow Am. Inst. Chemists (profl. rels. com. 1991, 92, patents com. 1992); mem. AAAS, Am. Chem. Soc. (program chmn. petroleum div. 1976-77), Licensing Execs. Soc., Assn. Univ. Tech. Mgrs., N.Y. Acad. Sci., Sigma Xi. Episcopalian. Achievements include 20 patents; preparation and determination of crystal structure of (211) phase of 123 superconductors; invention of randomly cross-linked smectites, of high surface area supported perovskite catalysts and method for preparation; selective auto exhaust catalysts; theory of finely divided metals; bonding model for zerovalent acetylene and olefin complexes; fluidized catalytic cracking catalysts. Office: GE Plastics One Plastics Ave Pittsfield MA 01201

WHEELOCK, SCOTT A., physicist; b. Wyandotte, Mich., Jan. 2, 1967; s. Gary J. and MaryAnn Wheelock. BS, Saginaw Valley State U., 1989. Devel. physicist Hemlock (Mich.) Semiconductor Corp., 1988—, corp. laser safety officer, 1990—. Mem. Am. Phys. Soc., Nat. Honor Soc. Achievements include patent for float zone processing of particulate silicon. Office: Hemlock Semiconductor Corp 12334 Geddes Rd Hemlock MI 48626

WHEELON, ALBERT DEWELL, physicist; b. Moline, Ill., Jan. 18, 1929; s. Orville Albert and Alice Geltz (Dewell) W.; m. Nancy Helen Hermanson, Feb. 28, 1953 (dec. May 1980); children—Elizabeth Anne, Cynthia Helen; m. Cicely J. Evans, Feb. 4, 1984. B.Sc., Stanford U., 1949; Ph.D., Mass. Inst. Tech., 1952. Teaching fellow, then rsch. assoc. physics MIT, Boston, 1949-52; with Douglas Aircraft Co., 1952-53, Ramo-Wooldridge Corp., 1953-62; dep. dir. sci. and tech. CIA, Washington, 1962-66; with Hughes Aircraft Co., L.A., 1966-88, chmn., chief exec. officer, 1987-88; vis. prof. MIT, 1989; mem. Def. Sci. Bd., 1968-76, Pres.'s Fgn. Intelligence, 1983-88, presdl. commn. on space shuttle Challenger accident, 1986; trustee Calif. Inst. Tech., Aerospace Corp., Rand Corp. Author 30 papers on radiowave propagation and guidance systems. Fellow IEEE, AIAA (Von Karman medal 1986); mem. NAE, Am. Phys. Soc., Coun. on Fgn. Rels., Sigma Chi. Republican. Episcopalian. Address: 181 Sheffield Dr Montecito CA 93108

WHELAHAN, YVETTE ANN, nursing administrator, consultant; b. Detroit, May 17, 1943; d. David William and Margaret (Korte) Humberstone; m. John Thomas, May 15, 1965; children: John Thomas, Elizabeth Lee, Katherine Theresa. Diploma in nursing, St. Mary's Med. Ctr., 1963-65; cert. in mgmt., Maryville (Tenn.) Coll., 1987. Cert. operating rm. nurse; RN, Fla. Pvt. scrub nurse Eye, Ear, Nose & Throat Clinic, Atlanta, 1965-68; ophthalmology clinic coord. St. Marys' Med. Ctr., Knoxville, Tenn., 1968-88; dir. nursing Eye Ctr. of Fla., Ft. Myers, 1988-90; dir. surg. svcs. East Point Hosp., Lehigh Acres, Fla., 1990—. V.p. T. Ball League, Ft. City Recreation Ctr., Knoxville, 1978, St. Joseph's PTA, Knoxville, 1982, sec.; chairperson St. Jude Walk-A-Thon 1980-82, Memphis, 1980-82; bd. dirs. Teen Ctr. of Knoxville, 1980—. Mem. Assn. Oper. Rm. Nurses (sec. 1980-82, bd. dirs. chmn. 4 yr. com. program, chmn. membership com., pres. 1993-94), Am. Soc. Ophthalmic Registered Nurses (faculty cons. San Francisco Congress 1987—), Fla. Orgn. Nurse Execs. (bd. dirs.), Shellcoast Assn. Oper. Rm. Nurses (treas. 1991—), Fla. Assn. Oper. Rm. Nurses (nom. com. 1993-94). Republican. Roman Catholic. Avocations: walking, swimming, sewing, cooking. Home: 6220 Augusta Dr Apt 417 Fort Myers FL 33907-5750 Office: East Point Hosp 1500 Lee Blvd Lehigh Acres FL 33936-4835

WHELAN, ELIZABETH ANN MURPHY, epidemiologist; b. N.Y.C., Dec. 4, 1943; d. Joseph and Marion (Barrett) Murphy; m. Stephen T. Whelan, Apr. 3, 1971; 1 child, Christine B. BA, Conn. Coll., 1965; MPH, Yale U., 1967; MS, Harvard U., 1968, D.S., 1971. Coordinator County study Planned Parenthood, 1971-72; research assoc. Harvard Sch. Pub. Health, Boston, 1975-80; exec. dir. Am. Council Sci. and Health, N.Y.C., 1980-92, pres., 1992—; mem. com. on pesticides and toxics EPA; mem. U.S. Com. of Vital Stats., HHS; mem. Nat. Adv. Com. on Meat and Poultry Inspection, USDA. Author: Panic in the Pantry, 1975, 92, A Baby?...Maybe, 1975, Preventing Cancer, 1978, The Nutrition Hoax, 1983, A Smoking Gun, 1984, Toxic Terror, 1984, 93, Balanced Nutrition, 1989; contbr. articles to profl. jours. Bd. dirs. Food and Drug Law Inst., Nat. Agrl. Legal Fund, Media Inst., N.Y. divsn. Am. Cancer Soc. Recipient Medal Disting. Achievement award Conn. Coll., 1979, Am. Pub. Health Assn. Environ. award, 1992. Mem. APHA (Early Career award 1982, Homer Calver award 1992), Am. Inst. Nutrition, Am. Med. Writers Assn. (Walter Alvarez award 1986), U.S. Com. Vital Stats. Office: Am Council Sci and Health 2d Fl 1995 Broadway New York NY 10023

WHELAN, JOSEPH L., neurologist; b. Chisholm, Minn., Aug. 13, 1917; s. James Gorman and Johanna (Quilty) W.; m. Gloria Ann Rewoldt, June 12, 1948; children: Joe, Jennifer. Student, Hibbing Jr. Coll., 1935-38; BS, U. Minn., 1940, MB, 1942, MD, 1943. Diplomate Am. Bd. Psychiatry and Neurology. Intern Detroit Receiving Hosp., 1942-43; fellow neurology U. Pa. Hosp., Phila., 1946-47; resident neurology U. Minn. Hosps., Mpls., 1947-49; chief neurology svc. VA Hosp., Mpls., 1949; spl. fellow electroencephalography Mayo Clinic, Rochester, Minn., 1951; practice medicine specializing in neurology Detroit, 1949-73, Petoskey and Gaylord, Mich., 1973—; asst. prof. Wayne State U., 1957-63; chief neurology svcs. Grace Hosp., St. John's Hosp., Bon Secour Hosp., Detroit; cons. neurologist No. Mich. Hosps., Charlevoix Area Hosp., Community Meml. Hosp., Cheboygan, Ostego Meml. Hosp., Gaylord; instr. Med. Sch., U. Minn., 1949; cons. USPHS, Detroit Bd. Edn. Contbr. articles to profl. jours. Founder, mem. ad hoc Com. to Force Lawyers Out of Govt.; chmn. Reagan-Bush U.S. Presdl. Campaign, Kalkaska County, Mich., 1980; chmn. Jack Kemp U.S. Presdl. Campaign, Kalkaska County, 1988. Capt. AUS, 1943-46. Fellow Am. Acad. Neurology (treas. 1955-57), Am. Electroencephalography Soc.; mem. AMA, AAAS, Assn. Rsch. Nervous and Mental Diseases, Soc. Clin. Neurologists, Mich. Neurol. Assn. (sec.-treas. 1967-76, Disting. Physician award 1988), Mich. med. Assn., No. Mich. Med. Soc., N.Y. Acad. Scis., Grosse Pointe (Mich.) Club. Republican. Roman Catholic. Address: Oxbow 9797 N Twin Lake Rd NE Mancelona MI 49659

WHELEN, ANDREW CHRISTIAN, microbiologist; b. El Paso, Tex., July 17, 1959; s. Henry James and Frances Annette (Lasiter); m. Jaclyn Kay, Sept. 21, 1991. BS, S.D. State U., 1981; PhD, U. N.D., 1985. Grad. asst. dept. microbiology S.D. State U., Brookings, 1980-82, U. S.D., Vermillion, S.D., 1982-83; from grad. asst. to sr. grad. asst. dept. microbiology U. N.D., Grand Forks, 1983-85; commd. 2d lt. U.S. Army, 1982, advanced through grades to maj., 1993; chief virology Letterman Army Med. Ctr., San Francisco, 1986-88; chief microbiology Landstuhl (Germany) Army Regional Med. Ctr., 1989-93; adj. faculty San Francisco State U., 1987-88, U. Md., European div., 1989-90, Uniformed Svcs. U. Health Scis., European div., 1991-93. Pres. Jr. Officers Assn., Landstuhl, 1990-91; commr. Installation Softball League, Landstuhl, 1990-91; vol. Spl. Olympics, San Antonio, 1988. Decorated 2 Meritorious Svc. medals; postdoctoral fellow in Microbiology, Mayo Clinic, Rochester, Minn., 1993—. Mem. Am. Soc. Microbiology, Am. Soc. Parasitologists, German-Am. Med. Soc., Sigma Xi. Home: 1811 Robinson Ave El Paso TX 79902 Office: Mayo Clinic Dept Lab Sci Pathology Clin Microbiology 470 Hilton Bldg Rochester MN 55905

WHICKER, FLOYD WARD, biology educator, ecologist; b. Cedar City, Utah, July 24, 1937; married; 3 children. BS, Colo. State U., 1962, PhD in Radiation Biology, 1965. From asst. to assoc. prof. Colo. State U., Ft. Collins, 1965-80, prof. radiation biology, 1980—; cons. We. Radiation Cons., Inc., Rockwell Internat. Recipient Ernest Orlando Lawrence Meml. award U.S. Dept. Energy, 1990. Fellow AAAS; mem. Ecol. Soc. Am., Wildlife Soc., Health Physics Soc., Sigma Xi (hon. scientist). Achievements include research in radiation ecology, radionuclide behavior in natural ecosystems and radiation effects on plant and animal populations. Office: Colorado St Univ Colorado State University CO 80523*

WHIGHAM, MARK ANTHONY, computer scientist; b. Mobile, Ala., Jan. 14, 1959; s. Tommie Lee Sr. and Callie Mae (Molette) W. BS in Computer Sci., Ala. A&M U., 1983, MS in Computer Sci., 1990. Computer programmer U.S. Army Corps of Engrs., Huntsville, Ala., 1985-88; programmer analyst, coord. acad. computing Ala. A&M U., Normal, Ala., 1988-89; programmer analyst II, DEC systems coord., instr. part-time computer sci. dept. Ala. A&M U., 1989-91; systems engr. Advanced Bus. Cons. Inc.-La. div. Dow Chem. Co., 1991—; owner Whigham's Computer Cons., 1990—; systems engr. DOW Chem. Co., Plaquemine, La., 1991—; part-time instr. computer sci. dept. Ala. A&M U., 1989-91. Active Huntsville Interdenominational Ministerial Fellowship, Huntsville, 1984. Mem. Ala. Coun. for Computer Edn., Assn. for Computing Machinery, Huntsville Jaycees, Nat. Soc. Black Engrs., So. Poetry Assn., Nat. Arts Soc., Internat. Black Writers and Artists Assn., Optimists, Sigma Tau Epsilon, Alpha Phi Omega. Baptist. Avocations: chess, skating, reading, playing piano. Home: PO Box 3032 Huntsville AL 35810 Office: Dow Chem Co La Divsn-Computer Svcs Dept PO Box 150 Plaquemine LA 70765-0150

WHINERY, MICHAEL ALBERT, physician; b. Watsford, Waysford, Eng., June 30, 1951; s. Leo Howard and Doris Eileene Watsford and Alma Piper; m. Tatjana Dunnebier, 1976 (dec. Jan. 1981); m. Judy Renee Wright, Apr. 30, 1983; children: Rhiannon Daire Eilien, Terron Rae Lee. BS, Okla. U., 1976; D of Osteopathy, Okla. State U., 1980. Bd. cert. physician in gen. practice. Intern Hillcrest Health Ctr., Oklahoma City, Okla., 1980-81; with McLoud Clinic, McLoud, Okla.; house physician McLoud Nursing Ctr., 1988—; med. examiner Pottawatomie County Health, McLoud, 1983—. Author: Poetic Voices of America, 1991. Mem. Presdl. Order Merit Nat. Repub. Senatorial Com., Washington, 1991, Presdl. Task Force, 1983—, Senatorial Commn. Repub. Senatorial Inner Circle, Washington, 1991. With USMC, Vietnam. Recipient Acknowledgment of Outstanding Contbn. in Clin. Rsch. award SANDOZ Labs., 1992. Mem. Am. Legion, C. of C., Jr. C. of C., U.S. Senatorial Club (preferred mem.), U.S. Congressional Act Bd. (state advisor 1990-91). Baptist. Avocations: fishing, music, composing. Office: McLoud Clinic PO Box 713 107 S Main McLoud OK 74851

WHINNERY, JOHN ROY, electrical engineering educator; b. Read, Colo., July 26, 1916; s. Ralph V. and Edith Mable (Bent) W.; m. Patricia Barry, Sept. 17, 1944; children—Carol Joanne, Catherine, Barbara. B.S. in Elec. Engring, U. Calif. at Berkeley, 1937, Ph.D., 1948. With GE, 1937-46; part-time lectr. Union Coll., Schenectady, 1945-46; asso. prof. elec. engring. U. Calif., Berkeley, 1946-52, prof., vice chmn. div. elec. engring., 1952-56, chmn., 1956-59, dean Coll. Engring., 1959-63, prof. elec. engring., 1963-80, Univ. prof. Coll. Engring., 1980—; vis. mem. tech. staff. Bell Telephone Labs., 1963-64; research sci. electron tubes Hughes Aircraft Co., Culver City, 1951-52; disting. lectr. IEEE Microwave Theory and Technique Soc., 1989-92. Author: (with Simon Ramo) Fields and Waves in Modern Radio, 1944, 2d edit. (with Ramo and Van Duzer), 1985, (with D. O. Pederson and J. J. Studer) Introduction to Electronic Systems, Circuits and Devices; also tech. articles. Chmn. Commn. Engring. Edn., 1966-68; mem. sci. and tech. com. Manned Space Flight, NASA, 1963-69; mem. Pres.'s Com. on Nat. Sci. Medal, 1970-73, 79-80; standing com. controlled thermonuclear research AEC, 1970-73. Recipient Lamme medal Am. Soc. Engring. Edn., 1975, Centennial medal, 1993, Engring. Alumni award U. Calif.-Berkeley, 1980, Nat. Medal of Sci. NSF, 1992; named to Hall of Fame Modesto High Sch. (Calif.), 1983, ASEE Hall of Fame, 1993.; Guggenheim fellow, 1959. Fellow IRE (bd. dirs. 1956-59), IEEE (life, bd. dirs. 1969-71, sec. 1971, Edn. medal 1967, Centennial medal 1984, Medal of Honor 1985), Optical Soc. Am., Am. Acad. Arts and Scis.; mem. NAS, NAE (Founders award 1986), IEEE Microwave Theory and Techniques Soc. (Microwave Career award 1977), Phi Beta Kappa, Sigma Xi, Tau Beta Pi, Eta Kappa Nu. Congregationalist. Home: One Daphne Ct Orinda CA 94563 Office: U Calif Dept Electrical Engineering Berkeley CA 94720

WHIPPLE, ANDREW POWELL, biology educator; b. Columbus, Ohio, Feb. 12, 1949; s. Quentin Powell and Joan Pierce (Armstrong) W.; m. You-Ying Wang, June 23, 1973; children: Joan Chuan-Lee, Kyle Wang, Elizabeth Chuan-Fei, Daniel Wang. BS in Microbiology, Ohio State U., Columbus, 1971; MS in Biology, SUNY, Albany, 1974, PhD in Cell Biology, 1979. Rsch. asst. SUNY, Albany, 1975-78, univ. fellow, instr. biology, 1979; postdoctoral rsch. fellow Dana Farber Cancer Inst. Harvard Med. Sch., Boston, 1979-81; prof. biology Montreat (N.C.)-Anderson Coll., 1981-84, Taylor U., Upland, Ind., 1984—; vis. prof. Biology Dept. Tunghai U., Taichung, Taiwan, 1992-93; cons. James River Corp., Neenah, Wis., 1985, Agro-K, Mpls., 1986-88; summer faculty rsch. fellow USAF, Wright-Patterson AFB, Ohio, 1987-88. Contbr. articles to profl. jours. Mem. Am. Soc. Cell Biology, Am. Sci. Affiliation, Tissue Culture Assn., Soc. Chinese Bioscientists in Am. Republican. Presbyterian. Achievements include patent (with other) for Method for Treating Waste Fluid with Bacteria; adaptation of human lymphoblastoid cells to serum-free culture; description of response of transformed cells to growth factor stimulation as compared to their normal, non-transformed counterparts. Home: PO Box 448 Upland IN 46989-0448 Office: Taylor U Biology Dept Upland IN 46989

WHISTLER, ROY LESTER, chemist, educator, industrialist; b. Morgantown, W.Va., Mar. 3, 1912; s. Park H. and Cloe (Martin) W.; m. Leila Anna Barbara Kaufman, Sept. 6, 1935; 1 child, William Harris. B.S., Heidelberg Coll., 1934, D.Sc. (hon.), 1957; M.S., Ohio State U., 1935; Ph.D., Iowa State U., 1938; D.Litt. (hon.), St. Thomas Inst., 1982; D.Agr., Purdue U., 1985. Instr. chemistry Iowa State U., 1935-38; research fellow Bur. Standards, 1938-40; sect. leader dept. agr. No. Regional Rsch. Lab., 1940-46; prof. biochemistry Purdue U., 1946-76, Hillenbrand distinguished prof., asst. dept. head, 1974-82; Hillenbrand disting. prof. emeritus Purdue U., Lafayette, Ind., 1982—; chmn. Inst. Agrl. Utilization Research, 1961-75; vis. lectr. U. Witwatersrand, 1961, 65, 77, 85, Czechoslovakia and Hungary, 1968, 85, Japan, 1969, Taiwan, 1970, Argentina, 1971, New Zealand, Australia, 1967, 74; vis. lectr. Acad. Sci., France, 1975, Vladivostock Acad. Sci., 1976, Brazil, 1977, Egypt, 1979; lectr. Bradley Polytech. Inst., 1941-42, People's Repulic China, 1985; adv. Whistler Ctr. for Carbohydrate Chemistry; indsl. cons. Dir. USAir, Pfanstiehl Lab., Inc., Sperti Drug Products, Greenwich Pharm. Inc.; mem. NRC sub-com. nomenclature biochemistry; pres. Lafayette (Ind.) Applied Chemistry. Author: Polysaccharide Chemistry, 1953, Industrial Gums, 1959, 2d rev. edit., 1976, 3d rev. edit., 1992; rev. edit.: Methods of Carbohydrate Chemistry, series, 1962—; co-author: Guar, 1979; editor: Starch-Chemistry and Technology, 2 vols., 1965, 67, rev. edit., 1984; editorial bd. Jour. Carbohydrate Research, 1960-91, Starchs Chemistry and Technology, 1985; bd. advisors: Advances in Carbohydrate Chemistry, 1950—, Organic Preparations and Procedures Internat., 1970—, Jour. Carbo-Nucleosides-Nucleotides, 1973-77, Starke, Starch, 1979—; contbr. 500 articles to profl. jours. Recipient Sigma Xi Rsch. award Purdue U., 1953, German Saare medal, 1974, Sterling Henricks award USDA, 1991; Roy L. Whistler internat. award in carbohydrates named in his honor; Whistler Ctr. for Carbohydrate Rsch., Purdue U., Named in his honor. Fellow AAAS, Am. Chem. Soc. (chmn. Purdue sect. 1949-50, carbohydrate div. 1951, cellular div. 1962, nat. councilor 1953-87, bd. dirs. 5th dist. 1955-58, chmn. com. edn. and students, chmn. sub-com. polysaccharide nomenclature, symposium dedicated in his honor 1979, hon. fellow award cellulose div. 1983, Hudson award 1960, Anselme Payen award 1967, Starch award Japanese 1967, Carl Lucas Alsburg award 1970, Spencer award 1970, 75, Disting. Svc. award 1983, named one of 10 outstanding chemists Chgo. sect. 1948), Am. Inst. Chemists (pres. 1982-83, Gold Medal 1992), Am. Assn. Cereal Chemists (pres. 1978, Thomas Burr Osborne award 1974), Internat. Carbohydrate Union (pres. 1972-74); mem. Lafayette Applied Chemistry (pres. 1970—), Argentine Chem. Soc. (life), Sigma Xi (pres. Purdue sect. 1957-59, nat. exec. com. 1958-62, hon. life mem. 1983—), Phi Lambda Upsilon. Lodge: Rotary (pres. 1966).

WHITCOMB, JAMES HALL, geophysicist, foundation administrator; b. Sterling, Colo., Dec. 10, 1940; s. Clay Thane and Julia Melvina W.; m. Sandra Lynn McMurdo, July 13, 1965 (div. 1978); m. Teresa R. Idoni, Feb. 3, 1989. Geophysics engring. degree, Colo. Sch. of Mines, 1962; MS in Oceanography, Geophysics, Oreg. State U., 1964; PhD in Geophysics, Calif. Inst. Tech., 1973. Grad. rsch. asst. dept. oceanography Oreg. State U., Corvallis, 1962-64; geophysicist ctr. astrogeology U.S. Geol. Survey, Flagstaff, Ariz., 1964-66; Fullbright-Hayes program rsch. fellow seismol. inst. U. Uppsala, Sweden, 1966-67; grad. rsch. asst. seismol. lab. Calif. Inst. Tech., Pasadena, 1967-73, sr. rsch. fellow seismol. lab., 1973-79; assoc. prof. attendant rank dept. geol. scis. U. Colo., Boulder, 1979-82, fellow Coop. Inst. Rsch. in Environ. Scis., 1979-84; v.p. technical applications and mktg. ISTAC, Inc., Pasadena, 1984-88; program dir. seismology NSF, Washington, 1989—; expert witness U.S. Ho. Reps. Com. on Sci. and Tech., 1977; mem. geodynamics rev. bd. Jet Propulsion Lab., 1980-82, com. on geodesy Nat. Acad. Scis., 1982-85; pres. Boulder Systems, Inc., Pasadena, 1987-88. Scholar State of Colo., 1958-62, Mobil Oil Co., 1960; recipient Outstanding Achievement award U.S. Geol. Survey, 1964; fellow Sweden-Am. Found., 1966. Mem. AAAS, Am. Geophysical Union, Seismol. Soc. Am., Soc. Exploration Geophysicists (scholar 1963), Tau Beta Pi, Phi Kappa Phi, Sigma Xi. Office: Nat Sci Found Geosciences 1800 G St NW Washington DC 20550

WHITE, BERTRAM MILTON, chemicals executive; b. Boston, Nov. 17, 1923; s. Samuel Louis and Jennie Anne (Cohen) W.; m. Bernice Hannah Ginns; children: Mark Alan, Leland Jeffrey. BS, Lowell Inst. Tech., Cambridge, Mass., 1943. Product mgr. Philipps Bros. Chems. Inc., Holbrook, Mass., 1952-65, Sobin Chems. Inc., South Boston, 1965-69; pres. Solvent Chems. Co., Inc., Malden, Mass., 1969-73; v.p. I.C.C. Chems. Inc., N.Y.C., 1973-80; sr. v.p. Asoma Chems. Inc., Boston, 1980-83, Laporte Chems. USA, Hackensack, N.J., 1983-84; pres. Gen. Plastics and Chems. Co., Natick, Mass., 1984-91, GFI Chems. Inc., Sudbury, Mass., 1991—; bd. dirs. Sudexco N.V., Brussels, Recochem Inc., Montreal, Que., Can.; treas. U.S. Antimony Sales Corp.; pres. Tech. Exporters of Am., Miami, Fla. Served with Corps of Engring. U.S Army, 1943-46, ETO. Decorated Purple Heart. Mem. Drug Chem. and Allied Trades Assn., New Eng. Chemists Club, N.Y.C. Chemists Club, Salesmen's Assn. of Am. Chem. Industry. Jewish. Avocations: tennis, golf, boating. Office: GFI Chems Inc 111 Boston Post Rd Sudbury MA 01776

WHITE, DAVID CLEAVELAND, microbial ecologist, environmental toxicologist; b. Moline, Ill., May 18, 1929; s. Frederick Berryhill and Dorothy (Cleaveland) W.; m. Sandra Jean Shoults, July 7, 1957; children: Winifred Shoults, Christopher Cleaveland, Andrew Berryhill. AB magna cum laude, Dartmouth Coll., 1951; MD, Tufts U., 1955; PhD, Rockefeller U., 1962. Rotating intern Hosp. of U. Pa., 1955-56; asst. prof., assoc. prof., then prof. biochemistry U. Ky., Lexington, 1962-72; prof. biol. sci. Fla. State U., Tallahassee, 1972-86; disting. scientist U. Tenn./Oak Ridge Nat. Lab., Knoxville, 1986—; prof. microbiology, ecology U. Tenn., Knoxville, 1986—; prin. investigator Oak Ridge (Tenn.) Nat. Lab., 1986—; mem. adv. com. Ctr. Theol. Inquiry, Princeton (N.J.) U., 1986-91, dir. Ctr. for Environ. Biotech., 1991—, Inst. Applied Microbiology, Knoxville, 1986-91; mem. sci. adv. panel Mich. State Ctr. Microbial Ecology, Lansing, 1989—, Mont. State Ctr. for Biofilm Engring., Bozeman, Mont., 1991—; dir. Microbial Insights, Inc., Knoxville; Welcome vis. prof. U. Okla., Norman, 1984-85; speaker at profl. confs. Author: Sex, Drugs and Pollution, 1983, 2d edit., 1985; founding editor-in-chief Jour. Microbiol. Methods, 1985—; author 340 refereed sci. publs. Lt. M.C. USN, 1956-58. Recipient P.R. Edwards award S.E. br. Am. Soc. Microbiology, 1981, Proctor & Gamble Applied and Environ. Microbiology award Am. Soc. Microbiology, 1993, Applied and Environ. Microbiol. award ASM, 1993, Antarctic Svc. medal USN/NSF, 1984, Sci. and Tech. Achievement award U.S. EPA, 1987. Presbyterian. Achievements include discovery of signature biomarker technique for microbial biomass, community structure and nutritional status from environmental samples, microbial ecology of deep subsurface, tropical and antarctic sediments, microbial biofilms in microbial influenced corrosion, biosensors environmental biotechnology. Office: Ctr for Environ Biotech 10515 Research Dr Ste 300 Knoxville TN 37932-2575

WHITE, DORIS GNAUCK, science educator, biochemical and biophysics researcher; b. Milw., Dec. 24, 1926; d. Paul Benjamin and Johanna (Syring) Gnauck; m. Donald Lawrence White Sr., Oct. 9, 1954 (div. Jan. 1986); children: Stanley, Dean, Victor, Donald Lawrence Jr. BS with honors, U. Wis., 1947, MS, 1949, PhD, 1956. Cert. tchr., Wis. Tchr. agr. U.S. Army Disciplinary Barracks, Milw., 1946-50; chairperson dept. sci. Waunakee (Wis.) High Sch., 1950-51; 4-H leader extension div. USDA, Wis., N.J., 1950—; tchr. prof. U. Wis. Lab. High Sch., Madison, 1951-56; grad. teaching asst. health, rural, adult edn. U. Wis., Madison, 1951-56; prof. sci. edn., curriculum and instrn. William Paterson Coll., Wayne, N.J., 1957—; sci. teaching specialist Frankford (N.J.) Twp. Schs., 1962; steering com. N.J. Sci. Conv., 1977—, coll. liaison N.J. Sci. Suprs., 1979—; sr. faculty and grand marshall William Paterson Coll., 1992—; Eisenhower grant participant Belleville (N.J.) Pub. Schs. 1992—; participant N.J. Sci. and Math Coalition NSF grant, 1993—. Mem. nat. sci. tchrs. manuscript rev. panel Jour. Coll. Sci. Teaching, 1991—. Active 4-H Club Leadership, Morristown, N.J., 1968— (N.J. Alumni award 1991), Geraldine Rockefeller Dodge Found. Animal Shelter, Madison, N.J., 1968—, St. Hubert's Giralda; transporter of clothes and food for poor and homeless of Paterson, N.J., 1971—; lic. blood tester for salmonella/fowl typhoid U.S. Dept. Agr., 1987—; mem. panel on curriculum improvement N.J. Commr. Edn., 1990-91, program com. N.J. Sci. Conv., 1978—, sex equality com. N.J. Dept. Edn., 1987, sci. core proficiencies panel, N.J. Sci. Coalition, 1989-90; judge presdl. candidates for N.J. schs. N.J. Dept. Edn., Trenton, 1985-88; judge sci. fairs Carteret, N.J., 1991, N.J. Sci. Olympiad, 1993; seer Morristown, 1986-92, Haledon, N.J., 1957-60; trustee N.J. Sci. Suprs. Assn., 1980—. Recipient Educator award Am. Cancer Soc., 1967; Dyes Rsch. grantee William Paterson Coll. Alumni Found., 1989-90; grantee NSF. Fellow N.J. Sci. Tchrs. Assn. (exec. bd. 1978—, indsl. liaison com. 1990-92); mem. Am. Poultry Assn. (life, lic. judge 1948—), Am. Chem. Soc., Am. Minor Breeds Assn., Nat. Sci. Tchrs. Assn., Nat. Sci. Suprs. Assn., N.J. Acad. Sci. (chair sci. edn. divsn. 1990—, liaison to sci. tchrs. 1992—), N.J. Sci. Suprs. Assn. (exec. bd. 1979—, pres. 1981, Outstanding Sci. Supr. award 1986, President's award 1992), N.J. Physics Tchrs. Assn., N.J. Sci. and Tech. Assn. Republican. Methodist. Achievements include development of chromosome mapping of fowl genes, two new breeds of squash which are now commercial varieties, penguin-like ducks; research in poultry genetics, genetics of Cucurbitaceae, ultrasound treatment of plant seeds as related to seed germination and plant growth, laser holographic interferometry to measure plant growth occuring in seconds, measurement in angstroms of light, growth regulators in plants, cat eye genetics, catatonic effects in fowl, herpes virus research with fowl; experiments with leaf berms for highway sound barriers, stuffing leaf berms with waste tires for highway median barriers, stuffing leaf berms with waste paper for highway sound barriers; experiments with tom turkey head ornaments colors of red, white and blue as indicators of contentment or stress; design of ocean wave and tidal pumps to transport glacial melt; design of portalbe methane gas generator; cereal grains genetics research. Home: 7 Beaver Brook Rd Annandale NJ 08801-9405 Office: William Paterson Coll 408 Raubinger Hall 300 Pompton Rd Wayne NJ 07470-2103

WHITE, EMIL HENRY, chemistry educator; b. Akron, Ohio, Aug. 17, 1926. B.S., U. Akron, 1947; M.S., Purdue U., 1948, Ph.D. in Chemistry, 1950. Fellow U. Chgo., 1950-51; fellow Harvard U., 1951-52; instr. organic chemistry Yale U., New Haven, 1952-56; from asst. prof. to assoc. prof. Johns Hopkins U., Balt., 1957-64, prof. organic chemistry, 1964-80, D. Mead Johnson prof. chemistry, 1980—. Recipient Md. Chemist of Yr. award, 1980; Guggenheim fellow, 1958-59, NIH sr. fellow, 1965-66, 72-73. Home: 4400 Green Valley Rd Union Bridge MD 21791-9030 Office: Johns Hopkins U Dept Chemistry Baltimore MD 21218

WHITE, EUGENE THOMAS, III, nuclear project engineer; b. Chesapeake, Va., Nov. 1, 1956; s. Eugene T. Jr. and Carolyn Beulah (Parker) W.; m. Celia Ann Walston, May 19, 1979; children: Gregory, Jordan, Aaron. Cert. nuclear propulsion plant operator, USN, 1973; Assoc. in Nuclear Engring. Tech., Thomas Edison State Coll. Cert. engr.-in-tng., Va.; qualified in submarines. Nuclear engr. Norfolk Naval Shipyard, Portsmouth, Va., 1981—. Sunday sch. tchr., communion steward Hickory (Va.) United Meth. Ch., 1988—. With USN, 1975-81. Democrat. Office: Norfolk Naval Shipyard Nuclear Engring Dept Chesapeake VA 23709

WHITE, EUGENE VADEN, pharmacist; b. Cape Charles, Va., Aug. 13, 1924; s. Paul Randolph and Louise (Townsend) W.; m. Laura Juanita LaFontaine, Aug. 28, 1948; children: Lynda Sue, Patricia Louise. BS in Pharmacy, Med. Coll. Va., 1950; PharM (hon.), Phila. Coll. Pharmacy and Sci., 1966. Pharmacist McKim & Huffman Drug Store, Luray, Va., 1950, Miller's Drug Store, Winchester, Va., 1950-53; pharmacist, ptnr. Shiner's Drug Store, Front Royal, Va., 1953-56; pharmacist, owner Eugene V. White, Pharmacist, P.C., Berryville, Va., 1956—; Sturmer lectr. Phila. Coll. Pharmacy and Sci., 1979; Lubin vis. prof. U. Tenn. Sch. Pharmacy, Memphis, 1974; mem. bd. visitors Sch. Pharmacy, U. Pitts., 1969. Author: The Office-Based Family Pharmacist, 1978; created first office practice in community pharmacy, 1960, developed patient medication profile record, 1960. 2d lt. USAAC, 1943-45. Recipient Nat. Leadership award Phi Lambda Sigma, 1979, Outstanding Pharmacy Alumnus award Med. Coll. Va. Sch. Pharmacy Alumni Assn., 1989. Fellow Am. Coll. Apothecaries (J. Leon Lascoff award 1973); mem. Am. Pharm Assn. (Daniel B. Smith award 1965, Remington Honor medal 1978), Va. Pharm. Assn. (Pharmacist of Yr. award 1966, Outstanding Pharmacist award 1992). Methodist. Avocations: reading, woodworking, computer. Office: 1 W Main St Berryville VA 22611-1340

WHITE, FRANK M., mechanical engineer, educator. B in Mech. Engring., Ga. Inst. Tech., 1954, PhD in Mech. Engring., 1959; S.M. in Mech. Engring., MIT, 1956. With aerospace engring. dept. Ga. Tech.; prof. mech. and ocean engring. U. R.I., Kingston, 1967—; cofounder dept. ocean engring. U. R.I. Author: Viscous Fluid Flow, 1974, Fluid Mechanics, 1979, Heat Transfer, 1984, Heat and Mass Transfer, 1987; contbr. articles to Encyclopedia of Science & Technology and over 80 tech. papers and reports to profl. jours.; assoc. editor Jour. Fluids Engring., 1974-77, tech. editor, 1979-90. Recipient Westinghouse award Am. Soc. Engring. Edn., 1970, Disting. Alumnus award Ga. Tech., 1990. Fellow ASME (chmn. bd. editors, mem. publs. com., Bd. Comms., Lewis F. Moody award 1973, Fluids Engring. award 1991); mem. AIAA, Sigma Xi. Office: U RI Prof Mech & Ocean Engring Kingston RI 02881*

WHITE, GEORGE EDWARD, pedodontist; b. Jamestown, N.Y., July 31, 1941; s. Gordon Ennis and Margaret (Appleyard) W. AB, Colgate U., 1963; DDS, SUNY, Buffalo, 1967; PhD, MIT, 1973; DBA, Century U., 1982. Intern, then resident Children's Hosp., Buffalo, 1967-69; prof., chmn. dept. pediatric dentistry Tufts U. Sch. Dental Medicine, Boston, 1973—; chief dept. oral pediatrics New Eng. Med. Center Hosp., Boston, 1973-80; pvt. practice pedodontics, Boston, 1974—; lectr. MIT, 1975-80; cons. Abcor, Inc.; nat., internat. lectr. Nat. Inst. Dental Rsch. grantee, 1973—. Author: Dental Caries: A Multifactorial Disease, 1975, To Stand Alone, 1979; co-author: Maxillofacial Orthopedics: For the Growing Child, 1983; founder, editor-in-chief Jour. Pedodontics, 1976; editor: Clin. Oral Pediatrics, 1979; founder, editor-in-chief Mastering Pediatric Dentistry, 1993; contbr. articles to profl. jours. Fellow Am. Acad. Pediatric Dentistry, Acad. Gen. Denistry, Internat. Coll. Dentistry; mem. Am. Assn. Functional Orthodontist, Northeast Craniomandibular Soc., Platform Soc., Fedn. Dentaire Internationale, Sigma Xi, Omicron Kappa Upsilon. Office: Tufts U Sch Dental Medicine Dept Pediatric Dentistry 1 Kneeland St Boston MA 02111-1527

WHITE, JOHN AUSTIN, JR., engineering educator, dean, consultant; b. Portland, Ark., Dec. 5, 1939; s. John Austin and Ella Mae (McDermott) W.; m. Mary Elizabeth Quarles, Apr. 13, 1963; children: Kimberly Elizabeth White Brakmann, John Austin III. BS in Indsl. Engring., U. Ark., 1962; MS in Indsl. Engring., Va. Poly. Inst., 1966; PhD, Ohio State U., 1969; PhD (hon.), Cath. U. of Leuven, Belgium, 1985, George Washington U., 1991. Registered profl. engr., Va. Indsl. engr. Tenn. Eastman Co., Kingsport, 1961-63, Ethyl Corp., Baton Rouge, 1965; instr. Va. Poly. Inst. and State U., Blacksburg, 1963-66, asst. prof., 1970-72, assoc. prof., 1972-75; teaching assoc. Ohio State U., Columbus, 1966-70; assoc. prof. Ga. Inst. Tech., Atlanta, 1975-77, prof., 1977-84, Regents' prof., 1984—, Gwaltney prof., 1988—, dean engring., 1991—; asst. dir. engring. NSF, 1988-91; founder, chmn. SysteCon Inc., Duluth, Ga., 1977-84; exec. cons. Coopers & Lybrand, N.Y.C., 1984-93; mem. mfg. studies bd. NRC, Washington, 1986-88; bd. dirs. CAPS Logistics, Russell Corp. Co-author: Facility Layout and Location: An Analytical Approach, 1974 (Inst. Indsl. Engrs. Book of Yr. award 1974), 2d edit., 1991, Analysis of Queueing Systems, 1975, Principles of Engineering Economic Analysis, 3d edit., 1989, Capital Investment Decision Analysis for Management and Engineering, 1980, Facilities Planning, 1984 (Inst. Indsl. Engrs. Book of Yr. award 1984); editor: Production Handbook, 1987; co-editor: Progress in Materials Handling and Logistics, Vol. 1, 1989; also numerous articles to profl. jours., chpts. to books and handbooks in field, conf. procs. Recipient Outstanding Tchr. award Ga. Inst. Tech., 1982, Disting. Alumnus award Ohio State U. Coll. Engring., 1984, Reed-Apple award Material Handling Edn. Found., 1985, Disting. Svc. award NSF, 1991. Fellow Am. Inst. Indsl. Engrs (pres. 1983-84, facilities planning and design award 1980, Outstanding Indsl. Engr. award-region III 1974, region IV 1984, Albert G. Holzman Disting. Educator award 1988, Outstanding Pub. award, 1988, David F. Baker Disting. Rsch. award 1990), Am. Assn. Engring. Socs. (bd. govs., chmn. 1986, Kenneth Andrew Roe award 1989); mem. Nat. Acad. Engring., Ark. Acad. Indsl. Engring., Am. Soc. Engring. Edn., Coun. Logistics Mgmt., Internat. Material Mgmt. Soc. (Material Mgr. of Yr. 1989), Soc. Mfg. Engrs. (Mfg. Educator award 1990), Nat. Soc. Profl. Engrs., Ops. Rsch. Soc. Am. (hon.), Sigma Xi, Alpha Pi Mu, Omicron Delta Kappa, Phi Kappa Phi, Tau Beta Pi, Omega Rho. Baptist. Avocations: reading, golf, writing. Office: Ga Inst Tech Coll Engring Office of Dean Atlanta GA 30332

WHITE, KENNETH WILLIAM, electrical engineering consultant; b. Middletown, Ohio, Jan. 28, 1947; s. William A. Jr. and Dorothy (Deardorff) W.; m. Judith Flower, Sept. 6, 1969 (div. July 1987); children: Charles A., Shawn K.; m. Marty Gabriel, Dec. 2, 1989. BSEE, Cornell U., 1969, MEEE, 1970. Grad. asst. Cornell U., Ithaca, N.Y., 1969; elec. engr. GE, Schenectady, N.Y., 1969; group leader Procter & Gamble Co., Cin., 1970-84; dir. Ctr. Excellence, R.J.R. Nabisco and R.J. Reynolds Tobacco Co., Winston-Salem, N.C., 1984-89; founder, pres. Visual*Sense*Systems, Ithaca, 1989—; chmn. VISION '90 Conf., Detroit; numerous presentations on machine vision to various profl. socs. Exec. producer: (video) The World of Machine Vision, 1992; contbr. numerous articles to profl. jours. Ruling elder 1st Presbyn. Ch. Glendale, Cin., 1980-82; bd. dirs. Cornell Exec. Coun. SW Ohio, Cin., 1981-84. 1st lt. U.S. Army, 1969-72. Mem. IEEE, Soc. Photo-Optical Instrumentation Engrs., Machine Vision Assn. of Soc. Mfg. Engrs. (charter, chmn. consumable goods com. 1980-88, cert. mfg. engr. in machine vision, chmn. bd. dirs. 1987-92), Cornell Soc. of Engrs. (bd. dirs. 1991—), Instrument Soc. Am. (liaison food and pharm. industry div. 1986-89). Achievements include patents in Package Inspection System; Automated Cargo Loading System; method and apparatus for detecting the deposition of and adhesive on a travelling WEB, component inspection apparatus and method. Office: Visual Sense Systems 314 Meadow Wood Ter Ithaca NY 14850-9470

WHITE, LARRY KEITH, electrical engineer; b. Lumberton, N.C., Aug. 3, 1948; s. Jack David and Edith (Cagle) W.; m. Margaret Amy Brearey, Aug. 4, 1973; children: Jeffrey Louis, Amy Kristen. BSE, N.C. State U., 1970; MBA in Fin., U. N.C., Wilmington, 1990. Registered profl. engr. Jr. engr. Carolina Power and Light Co., Wilmington, 1970-72, div. engr. mgr., 1984-86, div. engr. supr., 1989—; sr. engr. Carolina Power and Light Co., Jacksonville, N.C., 1972-81; dist. engr. Carolina Power and Light Co., Jacksonville, 1981-84, divsn. engring. supr., 1989-93, corp. transp. mgr., 1993—. Course developer Tng. Course Effective Decision Making, 1982. Chmn. Jacksonville Planning Bd., 1980-84; mem. Wilmington Citizens Adv. Budget Com., 1992, Planning Commn., 1993—. Recipient Govs. award Edn. N.C. Bus. Com. for Edn., 1986. Mem. AMA, Wilmington C. of C. (co-chair ops. Edn. Coun.), Nat. Soc. Profl. Engrs., N.C. Soc. Engrs., Rotary (chair bd. dirs. dist. 7730 found. 1992—, pres. Wilmington chpt. 1989-90, bd. dirs. 1985—, bd. dirs. Jacksonville chpt. 1982-84), N.C. State Univ. Student Aid Assn., N.C. State U. Alumni Assn., U. N.C. Wilmington Alumni Assn. (bd. dirs. MBA chpt. 1990—). Home: 406 John Mosby Dr Wilmington NC 28412 Office: Carolina Power & Light PO Box 1110 Wilmington NC 28402-1110

WHITE, MICHAEL ELIAS, aerospace engineer; b. Cheverly, Md., Sept. 19, 1958; s. Elias and Grace (Gebran) W.; m. Cathy Darlene Utt, Apr. 30, 1988; children: Michael Elias II, Athena Marie. BS in Aerospace Engring., U. Md., 1980, MS in Aerospace Engring., 1981. Assoc. engr. Johns Hopkins U. Applied Physics Lab., Laurel, Md., 1981-86; engr., sr. staff Johns Hopkins U. Applied Physics Lab., Laurel, 1986-91, engr., prin. staff, 1991—, dep. program mgr. nat. aero. plane, 1988-90, program mgr. nat. aero. plane, 1990—, asst. group supr., 1990—; mem. adv. bd. on fellows and profs. Applied Physics Labs., Johns Hopkins U., Laurel, 1991-93. Contbr. articles to AIAA Jour. Propulsion and Power, APL Tech. Digest, APL Tech. Rev. Coach Lanham (Md.) Boys and Girls Club, 1976-81. Recipient Gene Zara award Nat. Aero. Plane-Joint Program Office, Dayton, Ohio, 1988; named Engr. of Yr. under age 36, D.C. Coun. Engrs./Archtl. Socs., Washington, 1992. Mem. AIAA (named Outstanding Young Engr./Scientist Nat. Capital Sect. 1990), Sigma Gamma Tau. Republican. Antiochian Orthodox Christian. Achievements include successful application of tangential mass addition to control of shockwave/boundary layer interactions in a scramjet inlet; design and testing scramjet inlet for dual combuster ramjet engine. Office: Johns Hopkins Univ Applied Physics Lab Johns Hopkins Rd Laurel MD 20723

WHITE, MICHAEL ERNEST, animal scientist; b. Ames, Iowa, Feb. 5, 1958; s. Donald Benjamin and Jean (Grove) W.; m. Susan Kay Swanson, June 19, 1982; children: Sarah Elizabeth, Eric Michael. BS, U. Minn., 1980, PhD, 1986. Lab. technician trainee U. Minn., St. Paul, 1977-81, rsch. asst., 1981-84, doctoral fellow, 1984-86; asst. prof. animal sci. Ohio State U., Columbus, 1986-91; asst. prof. animal sci. U. Minn., St. Paul, 1992-93, assoc. prof., 1993—; reviewer grant proposals U.S. Dept. agr., NSF, NIH, 1987—. Reviewer Jour. of Animal Sci., Jour. of Nutrition, Transgenic Rsch., 1987—; contbr. articles to profl. jours. Cantor St. Judes Cath. Ch., 1980-82. USDA grantee, 1986-94; U. Minn. grad. scholar, 1982. Mem. AAAS, Am. Soc. Animal Sci. (chmn. Midwestern sect. com. on growth, devel., muscle biology and meat sci. 1990-91), Ohio Acad. Sci., Sigma Xi, Phi Kappa Phi (award for acad. excellence 1980—), Gamma Sigma Delta. Roman Catholic. Achievements include research on cellular and molecular role of the insulin-like growth factors and their binding proteins in growth and development. Office: U Minn 1354 Eckles Ave Saint Paul MN 55108-6160

WHITE, NORMAN ARTHUR, engineer, corporate executive, educator; b. Hetton-le-Hole, Durham, Eng., Apr. 11, 1922; s. Charles Brewster and Lillian Sarah (Finch) W.; m. Joyce Marjorie Rogers, Dec. 16, 1944 (dec. July 1982); children: Howard Russell, Lorraine Avril; m. Marjorie Iris Rushton, May 14, 1983. BS in Engring., U. London, 1949, PhD in Econs. U. London, 1973; MS, U. Philippines, 1955; grad. Advanced Mgmt. Program, Harvard U., 1968. Chartered engr., U.K., Eur., Ing., EC. Rsch. engr. and product devel. mgr. Royal Dutch/Shell Group, 1945-64, gen. mgr. spl. product div., 1964-68, chief exec. new enterprises div., 1968-72, chmn./dir. Shell Oil and Mining operating cos., 1963-72; dir., chmn. exec. com. Tanks Oil & Gas Ltd., London, 1974-85; prin. exec. Norman White Assocs., 1972—; spl. adv. Hambros Bank Ltd., London, 1972-76; corp. adv. Placer Dome Ltd., Vancouver, B.C., Can., 1973-78; oil advisor Tanks Consol. Investments plc, Nassau, The Bahamas, 1974-85; dir. Environ. Resources Ltd., 1973-87; mem. acad. adv. bd. engring. U. London, 1976-85; dep. chmn. Strategy Internat., Ltd., 1976-82; vice chmn. Transat Energy Inc., Washington, 1977-90; chmn. KBC Advanced Tech., 1979-90, Am. Oil Field Systems plc, 1980-85; dep. chmn. Brit. Can. Resources Ltd., Calgary, 1980-83; chmn. Ocean Thermal Energy Conversion Systems Ltd., 1982—, Tesel-Gearhart plc, 1983-85, Process Automation & Computer Systems Ltd, 1985—, Kelt Energy plc, 1986-87, Andaman Resources plc, 1986-90; corp. advisor Kennedy & Donkin Group, 1986—; chmn. Delta Media Solutions Ltd., 1989-92; bd. dirs. Henley Centre for Forecasting, 1974-92, dep. chmn., 1974-87; bd. dirs. COM-TEK Resources Inc., Denver, 1987-93; dir. Proscyon Ptnrs. Ltd., 1992—; vis. prof. Manchester Bus. Sch., 1971-89, Henley-The Mgmt. Coll., Henley-on-Thames, Eng., 1976-90, The City U., London, 1989—; vis. lectr. Royal Coll. Def. Studies, 1981-84; mem. Brit. nat. com. World Energy Conf., 1977-88, mem. conservation commn., 1979-87; treas. World Petrol. Congs., 1983-91; chmn. Brit. nat. com. World Petrol. Congs., 1987—, v.p., 1991—. Author: (with others) Financing the International Petroleum Industry, 1978, Oil Substitution - World Outlook to 2020, 1983, Handbook of Engineering Management, 1989; contbr. articles to profl. jours. Mem. Senate U. London, 1974-87, adv. coun. Inst. U.S. Studies, London, 1984-92; chmn. Joint Bd. for Engring. Mgmt., 1989-93; mem. governing bd. King Edward VI's Royal Grammar Sch., Guildford, 1976—; mem. parliamentary and sci. com. House of Commons, 1977-83, 87-92; chmn. transnat. satellite Edn. Ctr. Dept. of Ednl. Studies U. Surrey, 1991—, bus. advisor, 1989—. Fellow Inst. Mech. Engrs. (mem. council 1981-85, 87-91, chmn. engring. mgmt. div. 1981-85), Instn. Mining and Metallurgy, Inst. Energy, Inst. Petroleum (mem. council 1975-81, v.p. 1978-81), Royal Soc. Arts, Inst. Mgmt., Royal Aero. Soc., Royal Inst. Internat. Affairs, Royal Inst.; mem. Am. Soc. Petroleum Engrs., Can. Inst. Mining, Freeman City of London, Liveryman, Worshipful Co. Engrs., Worshipful Co. Spectacle Makers, Guild World Traders, LSE Club, Athaenaeum Club, Harvard Bus. Club, St. George's House (Windsor Castle, assoc.). Home: Green Ridges, 6 Downside Rd, Guildford Surrey GU4 8PH, England Office: 9 Park House, 123-125 Harley St, London W1N 1HE, England

WHITE, ROBERT MARSHALL, physicist, government official, educator; b. Reading, Pa., Oct. 2, 1938; s. Carl M. and Miriam E. White; m. Sara Tolles; children: Victoria, Jonathan. B.S. in Physics, MIT, 1960; Ph.D., Stanford U., 1964. Vis. scientist Osaka U., Japan, 1963; NSF postdoctoral fellow U. Calif., Berkeley, 1965-66; asst. prof. physics Stanford U., 1966-70; NSF postdoctoral fellow, Cambridge, Eng., 1970-71; mgr. solid state research area Xerox PARC, 1971-78, mgr. storage technology, 1978-83, prin. scientist, 1983-84; v.p. research and tech. Control Data Corp. Data Storage

Products Group, Mpls., 1984-86, chief tech. officer, v.p. research and engring, 1986-89; v.p., dir. advanced computer techs. Microelectronics & Computer Tech. Corp., Austin, Tex., 1989-90; under-sec. of commerce for tech., Dept. Commerce, Washington, 1990—; lectr. dept. applied physics Stanford U., 1971-81; vis. scientist Ecole Polytechnique, Paris, 1976-78, U. Pernambuco, Brazil, 1978; cons. prof. applied physics, prin. investigator Magnetic Thin Film Program, Stanford U., 1982—; adj. prof. dept. physics U. Minn., 1987—; guest Chinese Acad. Scis., 1982. Author: Quantum Theory of Magnetism, 1970 (Russian transl., 1972, Polish transl., 1979); Long Range Order in Solids, 1979 (Russian transl., 1982); Quantum Theory of Magnetism, 1983; Introduction to Magnetic Recording, 1985. Contbr. articles to profl. jours. Bd. advisors Inst. Tech. U. Minn., 1987; mem. State Minn. Com. on Sci. and Tech. Research and Devel., 1987-90; mem. adv. bd. U. Ill. Coll. Engring. Recipient Alexander von Humboldt Prize, Fed. Republic of Germany, 1981, Tyler Environ. Achievement prize U. So. Calif. 1992. Fellow AAAS, IEEE (disting. lectr. Magnetics Soc., chmn. nat. materials, mem. editorial bd. SPECTRUM, adv. bd. com. on magnetic materials 1984), Am. Phys. Soc.; mem. NAE, NRC (commn. material sci. and engring., nat. steering com. advanced steady state neutron source, mem. panel radiation research, vice chmn. IUPAP commn. on magnetism), Conf. Magnetism and Magnetic Materials (adv. com. 1976-78, 80-95, program com. 1973-75, chmn. 1981, intermag conf. 1991), Internat. Conf. Magnetism (program chmn. 1985); mem. Panel on Advanced Computing of the Japanese Tech. Evaluation Ctr.; mem. Nat. Adv. Com. on Semiconductors, 1990-92, Mfg. Forum, 1991, Nat.Critical Techs. Panel, 1990-91. Office: Dept Commerce 14th Constitution Ave NW Washington DC 20230-0001

WHITE, ROBERT STEPHEN, physics educator; b. Ellsworth, Kans., Dec. 28, 1920; s. Byron F. and Sebina (Leighty) W.; m. Freda Marie Bridgewater, Aug. 30, 1942; children: Nancy Lynn, Margaret Diane, John Stephen, David Bruce. AB, Southwestern Coll., 1942, DSc hon., 1971; MS, U. Ill., 1943; PhD, U. Calif., Berkeley, 1951. Physicist Lawrence Radiation Lab., Berkeley, Livermore, Calif., 1948-61; head dept. particles and fields Space Physics Lab. Aerospace Corp., El Segundo, Calif., 1962-67; physics prof. U. Calif., Riverside, 1967-92, dir. Inst. Geophysics and Planetary Physics, 1967-92, chmn. dept. physics, 1970-73; prof. physics emeritus Rsch. Physicist Inst. Geophysics and Planetary Physics, Riverside, 1992—; lectr. U. Calif., Berkeley, 1953-54, 57-59. Author: Space Physics, 1970; contbr. articles to profl. jours. Officer USNR, 1944-46. Sr. Postdoctoral fellow NSF, 1961-62; grantee NASA, NSF, USAF, numerous others. Fellow Am. Phys. Soc. (exec. com. 1972-74); mem. AAAS, AAUP, Am. Geophys. Union, Am. Astron. Soc. Republican. Methodist. Home: 5225 Austin Rd Santa Barbara CA 93111 Office: U Calif Inst Geophysics & Planetary Physics Riverside CA 92521

WHITE, SCOTT RAY, engineering educator; b. Kansas City, Mo., Feb. 14, 1963; s. Raymond William and Connie Kay (Givens) W.; m. Karen Christine Penny, May 19, 1984; children: Amy Jessica, Evan Scott. BS, U. Mo., Rolla, 1985; MS, Washington U., 1987; PhD, Pa. State U., 1990. Prodn. engr. AT&T Tech. Systems, Kansas City, 1985; grad. student fellow NASA Langley (Va.) Rsch. Ctr., 1985-87; summer fellow McDonnell Douglas Rsch. Labs., St. Louis, 1987; instr. Engring. Graphics Pa. State U., State Coll., 1988; rsch. asst. Composites Mfg. Tech. Ctr. Pa. State U., State Coll., 1987-90; instr. Engring. Sci. and Mechanics Pa. State U., State Coll., 1989-90; asst. prof. Aero. and Astro. Engring. U. Ill., Urbana, 1990--. Mem. Am. Inst. Aero. and Astro. (faculty advisor 1990--), Soc. for the Advancement of Materials and Processing (faculty advisor 1991--), Am. Soc. Mechanical Engrs., Am. Soc. Composites. Achievements include the development of continuous manufacturing techniques for composite materials; first comprehensive mechanics of processing model for composites manufacturing. Home: 2510 Southwood Dr Champaign IL 61821 Office: U Ill 104 S Wright St Urbana IL 61801

WHITE, SUSIE MAE, school psychologist; b. Madison, Fla., Mar. 5, 1914; d. John Anderson and Lucy (Crawford) Williams; m. Daniel Elijah White, Oct. 20, 1958 (dec. Sept. 29, 1968). BS, Fla. Meml. Coll., St. Augustine, 1948; MEd, U. Md., 1953; postgrad., Mich. State U., 1955, Santa Fe Community Coll., 1988. Elem. tchr. Grove Park (Fla.) Elem. Sch., 1943; tchr. Douglas High Sch., High Springs, Fla., 1944-55; sch. psychologist Alachua County Sch. Bd., Gainesville, Fla., 1956-69; coord. social svcs. Alachua County Sch. Bd., Gainesville, 1970; owner, dir. Mother Dear's Child Care Ctr., Gainesville, 1989—. Del. Bapt. World Alliance, Bapt. Conv. Fla., Tokyo, 1970; state dir. leadership Fla. Bapt. Gen. Conv., 1971-85. Recipient Cert. of Appreciation Fla. State Dept. Edn., Tallahassee, 1971, Appreciation for Disting. Svc. award Fla. Gen. Bapt. Conv., Miami, 1979. Mem. Nat. Ret. Tchrs. Assn., Alachua County Tchrs. Assn., Fla. Meml. Coll. Nat. Alumni Assn., AAUW, Heroines of Jerico, Masons. Democrat. Avocations: gardening, speaking, working with police on crime prevention. Office: Child Care Ctr 811 NW 4th Pl Gainesville FL 32601-5049

WHITE, THOMAS DAVID, biology educator. BA in Biology, San Jose State U., 1971, MA in Biology, 1975; PhD in Biology, UCLA, 1985. Lab. instr. human anatomy and physiology U. Santa Clara, Calif., 1975-76; lectr. West Valley C.C., Cupertino, Calif., 1977; teaching fellow UCLA, 1978-85, vis. asst. prof. biology, 1986-89, lectr. biology, 1990-91; instr. Ctr. for Advancement of Academically Talented Youth Johns Hopkins U., summer 1986; asst. prof. biology SUNY, Buffalo, 1991—; staff zoologist Hovenweep Nat. Monument Archeol. Survey, summer 1976; faculty rsch. mem. Savannah River Ecology Lab., 1988—. Reviewer Copeia, Am. Zoologist; contbr. articles to profl. publs.; author reports, abstracts, presentations in field. Grantee NSF, 1975, U.S. Dept. Energy, 1992, UCLA, 1983, 84, N.Y. Dept. Environ. Conservation and Conservation Fund of U.S. Fish and Wildlife Svc., 1992—; Fulbright-Hays scholar, 1982. Mem. AAAS, Am. Soc. Zoologists, Soc. Neurosci., Am. Soc. Mammalogists, Internat. Soc. Neuroethology, Internat. Congress of Vertebrate Morphology. Home: 304 Joe McCarth Dr # 6 Amherst NY 14228 Office: SUNY Buffalo Biology Dept 1300 Elmwood Ave Buffalo NY 14222

WHITE, VIRGINIA, chemistry educator; b. Cardiff, Wales, June 17, 1939; came to U.S., 1968; d. Stanley Herbert and Thelma Alice (Hansen) Preston; m. Brian White, Feb. 19, 1936 (div.); children: Kathleen Sally, David William. BSc magna cum laude, U. Wales, 1961; MA in Chemistry, Smith Coll., 1978. Tchr. chemistry Highfields Bilateral Sch., Wolverampton, Eng., 1960-61, Newport (Wales) High Sch., 1961-66; geochemist Nova Scotia Rsch. Found., Can., 1966-68; lab. instr. chemistry Smith Coll., Northampton, Mass., 1970-77, lab. dir., instr. chemistry, 1977—; chair Necuse Lab. Devel., 1988-90. Author: the Outermost Island, 1985, Light and Matter, 1990; contbr. articles to profl. jours. Fellow Sigma Xi; mem. Smith Coll. Amunae Soc. Achievements include research in medicinal bush medicines. Office: Smith Coll Dept Chemistry Northampton MA 01063

WHITE, WILLIAM, research physicist; b. Millbrook, Ont., Can., Apr. 1, 1928; s. William and Pearl Emma (Lamb) W.; m. Francoise L. Babin, May 28, 1960; children: William Gregory, Gregory Scott, Eric Lachlan. BS in Engring. Physics, Queen's U., Kingston, Ont., 1950; PhD in Nuclear Physics, McGill U., 1961. Dir. rsch. Nuclear Chgo. and G.D. Searle, Des Plaines, Ill., 1962-79, Siemens Gammasonics, Inc., Hoffman Estates, Ill., 1981-93; prin. White R&D Cons., Cary, Ill., 1979—. Contbr. articles on sci. and rsch. mgmt. to profl. jours. Mem. Soc. Nuclear Medicine. Avocations: fishing, history of French people in Am. Home: 277 Little Stonegate Rd Cary IL 60013-2504

WHITE, W(ILLIAM) ARTHUR, geologist; b. Sumner, Ill., Dec. 9, 1916; s. Millard Otto and Joy Olive (Atkins) W.; m. Alma Evelyn Simonton McCullough, June 21, 1941. B.S., U. Ill., 1940, M.S., 1947, Ph.D., 1955. With Ill. Geol. Survey, Urbana, 1943-79, geologist, 1955-58, head clay resources and clay mineral tech. sect., 1958-72, geologist, 1972-79, geologist emeritus, 1979—; pvt. cons. geologist Urbana, 1979-88; prof. geology Fed. U. Rio Grande do Sul, Brazil, 1970. Contbr. articles to profl. jours. Fellow Geol. Soc. Am., Mineral. Soc. Am., AAAS; mem. Internat. Clay Mineral Soc., Am. Clay Mineral Soc., Ill. Acad. Sci., Mus. Natural History (assoc.), Nat. Geog. Soc., Am. Chem. Soc., Colloid Chem. Soc., Soc. Econ. Petrologists and Mineralogists, Inter-Am. Soc., Farm Bur., Am. Assn. Retired Persons, Order of United Comml. Travelers Am., U. Ill. Alumni Assn., Sigma Xi. Home: 603 E Colorado Ave Urbana IL 61801-5923

WHITE, WILLIAM BLAINE, geochemist, educator; b. Huntingdon, Pa., Jan. 5, 1934; s. William Bruce and Eleanor Mae (Barr) W.; m. Elizabeth Loczi, Mar. 27, 1959; children—Nikki Elizabeth White McCurry, William Brion (dec.). B.S., Juniata Coll., 1954; Ph.D., Pa. State U., 1962. Research asso. Mellon Inst., Pitts., 1954-58; asst. prof. Pa. State U., University Park, 1963-67; asso. prof. Pa. State U., 1967-72, prof. geochemistry, 1972—, chmn. grad. program in materials, 1990—. Assoc. editor The American Mineralogist, 1972-75, Materials Rsch. Bull., 1979—, Jour. Am. Ceramic Soc., 1985—, Water Resources Bull., 1992—; earth scis. editor Nat. Speleological Soc. Bull., 1964—; author: Geomorphology and Hydrology of Karst Terrains, 1988, (with Elizabeth L. White) Karst Hydrology: Concepts from the Mammoth Cave Area, 1989, (with M. Susan Barger) Daguerreotype: Nineteenth-century Technology and Modern Science, 1991; contbr. articles to profl. jours. Home: 542 Glenn Rd State College PA 16803-3472 Office: Pa State U Materials Rsch Lab University Park PA 16802

WHITEHEAD, NELSON PETER, foreign service officer; b. Washington, Sept. 12, 1960; s. Edwin Nelson and Marguerite (Janko) W. Degree, U. Grenoble, France, 1980; BA, Washington and Lee U., 1984, BS, 1984; M Engring., U. Va., 1991. Dir. testing def. meteorol. satellite program Westinghouse Space Div., Balt., 1985-86; joined Fgn. Svc. U.S. Dept. State, 1986; engr. U.S. Dept. State, Washington, 1986—. Mem. IEEE, Internat. Soc. Optical Engring., Lasers and Electro-Optics Soc. Presbyterian.

WHITEHURST, BROOKS MORRIS, chemical engineer; b. Reading, Pa., Apr. 9, 1930; s. David Brooks and Bessie Ann (Lowry) W.; B.S., Va. Poly. Inst. and State U., 1951; m. Carolyn Sue Boyer, July 4, 1951; children: Garnett, Anita, Robert. Sr. process asst. Am. Enka Corp., Lowland, Tenn., 1951-56; sr. process devel. engr. Va.-Carolina Chem. Corp., Richmond, Va., 1956-63; project engr. Texaco Inc., Richmond, 1963-66; mgr. engring. services Texasgulf, Inc., Aurora, N.C., 1967-80, mgr. spl. projects and long range planning, 1980-81; pres. Whitehurst Assocs., Inc., New Bern, N.C., 1981—; instr., lectr., cons. alternative sources of energy community colls. and univs.; presenter paper Solar World Forum, Brighton, Eng., 1981. Co-chmn. N.C. state supt. task force on secondary edn., 1974—; mem. N.C. Personnel Commn. for Public Sch. Employees; mem. N.C. state adv. com. on trade and indsl. edn., 1971-77; chmn. Gov.'s Task Force Vols. in the Workplace, 1981; chmn. State Adv. Council Career Edn., 1977—; gov.'s liaison for edn. and bus., 1978-79. Registered profl. engr., N.C. Recipient commendation Pres. U.S., 1981. Mem. Am. Inst. Chem. Engrs., Am. Inst. Chemists (dir. 1980-84, cert.), N.C. Inst. Chemists (pres. 1975-77), Nat. Soc. Profl. Engrs., N.C. Soc. Profl. Engrs., Royal Soc. Chemistry. Achievements include patents, and current work on biodegradable chelate systems, muncipal yard waste disposal, micronutrients for agriculture, waste rubber recycling. Home: 1983 Hoods Creek Rd New Bern NC 28564-9103 Office: PO Box 3335 New Bern NC 28564-3335

WHITELEY, JAMES MORRIS, retired aerospace engineer; b. Bangs, Tex., Jan. 27, 1927; s. Charles David and Ruby May (Snead) W. BS, Daniel Baker Coll., 1951; postgrad., U. Va., 1951-52, So. Meth. U., 1954-57. Rsch. scientist Nat. Adv. Com. for Aeronautics, Hampton, Va., 1951-52; engring. specialist Gen. Dynamics, Ft. Worth, 1952-91; ret., 1991. With USN, 1945-46, PTO. Assoc. fellow AIAA; mem. AARP, The Air Force Assn., Am. Legion. Home: 1207 Roaring Springs Rd Fort Worth TX 76114

WHITEMAN, WAYNE EDWARD, army officer; b. Hudson, N.Y., Dec. 23, 1957; s. Edward Earl and Lynnette Elnor (Richards) W.; m. Catherine Ann Crawford, May 31, 1986; children: Ashley Marie, Elizabeth Ann. BS, U.S. Mil. Acad., 1979; MSCE, MIT, 1987. Registered profl. engr., Va. Commd. 2d lt. U.S. Army, 1979, advanced through grades to maj., 1990; platoon leader, staff constrn. engr. 76th Engr. Bn. U.S. Army, Ft. Meade, Md., 1979-83; co. comdr. 2nd Engr. Bn. U.S. Army, Munson, South Korea, 1983-85; asst. prof. dept. civil and mech. engring. U.S. Mil. Acad., West Point, N.Y., 1987-90; engr. ops. officer 41st Engr. Bn. U.S. Army, Fort Drum, N.Y., 1991—; rsch. assoc. Los Alamos (N.Mex.) Nat. Lab., summer 1989, 90. Mem. ASCE, Soc. Am. Mil. Engrs., Sigma Xi, Phi Kappa Phi. Republican. Lutheran. Home: Pond Hill Rd Box 39 Chatham NY 12037 Office: 41st Engr Bn Fort Drum NY 13602

WHITENER, PHILIP CHARLES, aeronautical engineer, consultant; b. Keokuk, Iowa, July 9, 1920; s. Henry Carroll and Katherine Ethel (Graham) W.; m. Joy Carrie Page, Oct. 9, 1943; children: David A., Barbara C., Wendy R., Dixie K. BSME, U. N.Mex., Albuquerque, 1941. Ordained to elder Presbyn. Ch., 1956. Engr. Boeing Airplane Co., Seattle, 1941-47, supr. wind tunnel model design, 1947-57, project engr. B-52 flight test, 1957-62, engring. mgr. Fresh I hydrofoil, 1962-65, configurator supersonic transport, 1965-70, with preliminary design advanced concepts, 1970-83, ret., 1983; pres., chief engr. Alpha-Dyne Corp., Bainbridge Island, Wash., 1983—. Inventee in field. Organizer Trinity Ch., Burien, Wash., 1962, Highline Reformed Presbyn., Burien, 1970, Liberty Bay Presbyn., Poulsbo, Wash., 1978; pres. Whitener Family Found., Bainbridge Island, 1979; dir. Mcpl. League of Bainbridge, 1993—. Republican. Avocations: designing, computers, boating. Home: 5955 Battle Point Dr NE Bainbridge Is WA 98110-3407

WHITESIDE, THERESA LISTOWSKI, immunologist, educator; b. Katowice, Poland, Mar. 10, 1939; came to U.S., 1959; d. Edward A. Listowski and Valerie H. (Grembowiec) Price; m. G.A. Whiteside Jr., 1961 (div. 1971); 1 child, George A. III; m. Thomas H. Nimick Jr., May 1, 1982. MA in Biology, Columbia U., 1964, PhD in Microbiology, 1967. Diplomate Am. Bd. Med. Lab. Immunology. Postdoctoral fellow NYU Sch.Medicine, N.Y.C., 1967-69, rsch. scientist, 1969-71; spl. NIH fellow Columbia U. Coll. Physicians and Surgeons, N.Y.C., 1971-73; asst. prof. Sch. Medicine U. Pitts., 1973-79, assoc. prof., 1979-89, prof. pathology, 1989—; dir. immunologic monitoring and diagnostic lab. Pitts. Cancer Inst., 1986—; mem. Am. Bd. Med. Lab. Immunology, Washington,1 982-88; mem. instl. grant rev. bd. ACS, Atlanta, 1990—; pres. Assn. Med. Lab. Immunologists, Chester, Va., 1992. Contbr. sci. articles to profl. jours., chpts. to books. Bd. dirs. Kidney Found. Western Pa., Pitts., 1979-88, Salvation Army, Pitts., 1983-88, Pa. chpt. Lupus Found., Pitts., 1988—. Quincy WArd Boese fellow, 1964; NIH spl. fellow, 1971-73; Fogarty sr. internat. fellow, 1984. Fellow Am. Acad. Microbiology; mem. Am. Assn. Immunologists, Am. Assn. Cancer Rsch., Am. Assn. Pathologists, Clin. Immunology Soc. Office: Pitts Cancer Inst W1041 BST DeSoto at O'Hara St Pittsburgh PA 15213

WHITESIDES, GEORGE MCCLELLAND, chemistry educator; b. Louisville, Kty., Aug. 3, 1939; m. Barbara Breasted; children: George Thomas, Benjamin Haile. AB, Harvard U., 1960; PhD, Calif. Inst. Tech., 1964. Asst. prof. dept. chemistry MIT, Cambridge, 1963-69, assoc. prof., 1969-71, prof., 1971-75, Arthur C. Cope prof., 1975-80, Haslam and Dewey prof., 1980-82; prof. dept. chemistry Harvard U., Cambridge, 1982-86, Mallinckrodt prof., 1986—. Recipient Pure Chemistry award Am. Chem. Soc., 1975, Harrison Howe award Rochester sect., 1979, Remsen award Am. Chem. Soc., 1983, Arthur C. Cope Scholar award Am. Chem. Soc., 1989, Disting. Alumni award Calif. Inst. Tech., 1980; Alfred P. Sloan fellow, 1968. Fellow AAAS; mem. NAS, Am. Acad. Arts and Scis. Office: Harvard U Dept of Chemistry 12 Oxford St Cambridge MA 02138-2900

WHITFIELD, GARY HUGH, research director; b. Widnes, U.K., Dec. 6, 1951; arrived in Can., 1957; s. Fred Sparks and Marie (Doyle) W.; children: Megan, Brian. BSc, U. Guelph, Ont., 1975, MSc, 1977; PhD, Mich. State U., 1981. Biosystem analyst Environment Can., Sault Ste. Marie, Ont., 1981-82; rsch. scientist Agr. Can., Lethbridge, Alta., 1982-86; rsch. scientist Agr. Can., Harrow, Ont., 1986-90, sect. head, 1990-91, asst. dir., 1991-92; dir. Agr. Can., Delhi, Ont., 1992—. Contbr. articles to profl. jours., chpt. to book. Mem. Entomol. Soc. Am. Achievements include original research in biology and control of insect pests of corn, greenhouse crops, sugar beets and soybeans. Office: The Delhi Rsch Station, PO Box 186, Delhi, ON Canada N4B 2W9

WHITFIELD, GRAHAM FRANK, orthopedic surgeon; b. Cheam, Surrey, Eng., Feb. 8, 1942; came to U.S., 1969, naturalized, 1975; s. Reginald Frank and Marjorie Joyce (Bennett) W. BSc, King's Coll., U. London, 1963, PhD, Queen Mary Coll., U. London, 1969; MD, N.Y. Med. Coll., 1976. Rsch. scientist Unilever Rsch. Lab., Eng., 1963-66; postdoctoral fellow dept. chemistry Temple U., 1969-71, instr., 1971-72, asst. prof., 1972-73; resident in surgery N.Y. Med. Coll. Affiliated Hosps., N.Y.C., 1976-78, resident in orthopedics, 1978-79, sr. resident in orthopedic surgery, 1979-80, chief resident, 1980-81; attending orthopedic surgeon Good Samaritan Hosp., West Palm Beach, Fla., 1981-87, JFK Med. Ctr., Lake Worth, Fla., 1981—. Recipient N.Y. Med. Coll. Surg. Soc. award, 1976. Fellow Internat. Coll. Surgeons; mem. AMA, Fla. Med. Assn., Palm Beach County Med. Soc., Royal Inst. Chemistry (Eng.), So. Orthopedic Assn., Fla. Orthopedic Soc., Sigma Xi. Clubs: Beach (Palm Beach), Colette; Brit. Schs. and Univs., Soc. Sons of St. George (N.Y.C.); Govs. of Palm Beaches (West Palm Beach), Explorers' Club (N.Y.C.). Lodge: Rotary. Author: (with Joseph Cohn and Louis Del Guercio) Critical Care Readings, 1981; editorial bd., contbg. editor Hosp. Physician, 1978-82; cons. editor Physician Asst. and Health Practitioner, 1979-82; orthopedic cons. Conv. Reporter, 1980-82; assoc. editor in chief Critical Care Monitor, 1980-82; edit. bd. Complications in Orthopedics, 1986—; practice panel cons. in orthopedic surgery Complications in Surgery, 1982—. Office: 1870 Forest Hill Blvd West Palm Beach FL 33406-8901

WHITFORD, PHILIP CLASON, biology educator; b. Milw., Apr. 18, 1951; s. Philip Burton and Kathryn Dana (Reuber) W.; m. Karen Rae Firnrohr, Sept. 27, 1975; 1 child, Kirstina Rae. MS, U. Wis., Stevens Point, 1976; PhD, U. Wis., Milw., 1987. Lectr. U. Wis., Milw., 1980-86; asst. prof. biology dept. Winona (Minn.) State U., 1986-92, U. Wis., Whitewater, 1992-93, Capital U., Columbus, Ohio, 1993—; anatomy text reviewer Benjamin Cummings, Menlo Park, Calif., 1984-85; reviewer biol. papers Wis. Acad. Sci. and Letters, Madison, 1980-86; mem. planning com. Midwest Animal Behavior Meetings, U. Wis., Milw., 1980-81, Internat. Symposium for Wildlife Mgmt. on Pvt. Lands, Milw., 1979-80. Contbr. articles to profl. jours. Mem. Wildlife Soc., Minn. Acad. Sci., Sigma Xi. Achievements include first person to quantify vocal communication of Canada geese. Home: Box 316 Rt 2 Montello WI 53949 Office: Capital U Biology Dept Biology Dept Columbus OH 43209

WHITLOCK, EDWARD MADISON, JR., civil engineer; b. Farmville, Va., Oct. 30, 1933; s. Edward Madison and Mattie Leigh (Fretwell) W.; married, 1975 (div.); children: Edward Madison III, Janis Loretta Brown; m. Jacqueline Powell. BSCE, Va. Mil. Inst., 1955. Registered profl. engr. 17 states, Can., Australia. Structural designer Va. Steel Co., Inc., Richmond, 1958-60; prin. assoc. Wilbur Smith Assocs., Columbia, S.C./Australia, 1960-65; v.p. Wilbur Smith Assocs., N.Y.C., 1965-70; sr. exec. v.p. Wilbur Smith Assocs., New Haven, 1970-85; pres., CEO Whitlock Assocs., Columbia, S.C., 1985—; advisor NSF, 1973-74. Contbr. articles to profl. jours.; author monograph: ENO Transportation Foundation, 1982. Bd. dirs. Rideshare, South Cen. Conn., New Haven, 1984-85; pres. POC Horizontal Property Regime, Isle of Palms, S.C., 1991—; mem. Skyland Homeowners, Columbia, 1991—; bd. cons. ENO Transp. Found., 1980-83. Mem. ASCE (chmn. hwy. exec. com. 1981-82, pres., founder Midlands br. 1961-63, chmn. nat. transp. policy com. 1980-82, Hwy. Div. award 1985). Home: 188 Castle Rd Columbia SC 29210

WHITLOCK, LAURA ALICE, research scientist; b. Birmingham, Ala., Feb. 23, 1959; d. Richard Gordon and Virginia Irene (Rowell) W. BS with honors, Southwestern at Memphis, 1981; PhD, U. Fla., 1989. Rsch. asst. Los Alamos (N.Mex.) Nat. Lab., 1984-88, collaborator space astronomy and astrophysics, 1989-92; rsch. scientist Nichols Rsch. Corp., Huntsville, Ala., 1989-92, Inst. for Space Sci. and Tech., Gainesville, Fla., 1990—, Univ. Space Rsch. Assoc. Goddard Space Flight Ctr., Greenbelt, Md., 1992—. Contbr. articles to Astrophys. Jour., Astronomy and Astrophysics. Faculty scholar Southwestern at Memphis, 1977-81; faculty fellow U. Fla., 1981. Mem. Am. Astron. Soc., Optical Soc. Am., Assn. Women in Sci., Soc. Photo-Optical Instr. Engrs., Sigma Pi Sigma. Achievements include reorganization of the 10-year, all-sky Vela 5B X-ray satellite database for study of time variability in cosmic X-ray sources; prediction of recurrences of X-ray transient 4U0115+63 which have subsequently been observed; development of system-level Monte Carlo simulation of IR sensor performance operating in a high-radiation environment. Office: NASA-GSFC Code 668 1 Greenbelt MD 20771

WHITLON, DONNA SUE, neuroscientist, researcher; b. Flushing, N.Y., Oct. 24, 1952; d. Leonard E. and Eileen S. W.; m. Jack M. Rozental, Jan. 6, 1980; two children. BS, Mich. State U., 1974; PhD, U. Wis., 1979. Rsch. asst., NIH predoctoral trainee U. Wis. Dept. Biochemistry, Madison, 1974-79; biol. lab. scientist Manufactaras Arriol, Santo Domingo, Dominican Republic, 1980; high sch. sci. tchr. Carol Morgan Sch., Santo Domingo, Dominican Republic, 1981; postdoctoral fellow McArdle Lab. for Cancer Rsch. U. Wis., Madison, 1981-84, asst. scientist, 1985-91, assoc. scientist Waisman Cntr., 1991—. Contbr. articles to profl. jours. Mem. Assn. for Rsch. in Otolaryngology, Soc. for Neuroscience. Office: U Wis Room 636 Waisman Ctr 1500 Highland Ave Madison WI

WHITMAN, ALEXANDER H., JR., civil engineer, consultant; b. Boston, May 23, 1944; s. Alexander H. and Sylvia (Choate) W.; m. Laura Grinnell Koehne, Aug. 30, 1969; children: Nell, Alexandra. BS, Lehigh U., 1972; MBA, Rutgers U., 1983. Registered profl. engr., N.Y., Pa., Wash., Alaska. Engr. Gannett Fleming, Harrisburg, Pa., 1972-86; mgr. design svcs. Ecology and Environment, Inc., Buffalo, 1986-91; mgr. western div. engring. Ecology and Environment, Inc., Seattle, 1991—. With USN, 1964-67. Mem. NSPE. Republican. Episcopalian. Office: Ecology and Environment Inc 999 3rd Ave Ste 1500 Seattle WA 98104

WHITMYER, ROBERT WAYNE, soil scientist, consultant, researcher; b. Elkhart, Ind., Feb. 5, 1957; s. Wayne Ellsworth and Janet Sue (Housour) W.; m. Mary Kathleen Cory, June 7, 1980; children: Sydney Michelle, Kellie Mairin, Steffani Marie, Emily Claire. Student, Mpls. Coll. Art and Design, 1975-76; BS in Natural Resources cum laude, Ball State U., 1980; MS in Soil Sci., U. Minn., St. Paul, 1984. Cert. individual sewage treatment system inspector, designer, site evaluator, Minn., soil tester, morphological evaluator, Wis., profl. soil scientist. Rsch. asst. Soil Physics Lab. U. Minn., St. Paul, 1980-84; soil scientist Hakanson Anderson Assocs., Anoka, Minn., 1984-85; code enforcement officer, sewage treatment operator County of Washington, Stillwater, Minn., 1985-90; soil scientist Owen Ayres & Assocs., Madison, Wis., 1990—. Contbr. articles to profl. jours. Member com. Rice Creek Watershed Dist., Arden Hills, Minn., 1989, Shoreview (Minn.) Environ. Quality Com., 1988-90. Mem. Am. Registry Cert. Profl. in Agronomy, Crops and Soils, Soil Sci. Soc. Am., Soil Water Conservation Soc. Am., Wis. Onsite Waste Disposal Assn., Gamma Theta Upsilon (Iota Omega chpt.). Achievements include development of menu based computer program for monitoring and reporting all aspects of small community wastewater collection/treatment system operation, of formula for evaluating actual water flow velocity in soil pores; research on and identification of on site sewage treatment systems effective in removal of nitrogen. Office: Ayres Assoc 2445 Darwin Rd Madison WI 53704-3116

WHITNER, JANE MARVIN, scientific applications programmer; b. Oakland, Calif., Aug. 29, 1935; d. Chauncey Hill and Alice Belle (Cromwell) Whitner. BA in Biol. Sci., San Jose State U., 1958; MA in Biostatistics, U. Calif., Berkeley, 1960. EDP programmer San Mateo County EDP Ctr., Redwood City, Calif., 1962-65; sci. programmer Lockheed Missiles & Space Co., Sunnyvale, Calif., 1967-68, Stanford U. Med. Ctr., 1969-73; sci. sys. programmer Physics Internat. Co., San Leandro, Calif., 1980-84; bioanalyst Syntex Rsch. Corp., Palo Alto, Calif., 1985—. Mem. ACM, Astron. Soc. Pacific, Smithsonian Instn., U. Calif. Alumni Assn., Commonwealth Club of Calif. Avocations: astronomy, chess, writing, poetry, art collecting.

WHITNEY, GLAYDE DENNIS, psychologist, educator, geneticist; b. Sidney, Mont., Apr. 25, 1939; s. Russell Taylor and Althea May (Zuber) W.; m. Yvonne Marie Miels, June 20, 1965 (div. 1990); children: Scott, Timothy. BA cum laude, U. Minn., 1961, PhD, 1966. Postdoctoral fellow Inst. Behavior Genetics U. (Boulder) Colo., 1969-70; asst. prof. psychology Fla. State U., Tallahassee, 1970—; cons. NIH and NSF, Washington, 1976—; adv. group mem. Colo. Alcohol Rsch. Ctr., Boulder, 1986—. Assoc. editor Jour. Behavior Genetics, 1981-84; contbr. over 100 articles to profl. jours. Capt. USAF, 1966-69. Recipient Claude Pepper award Nat. Inst. Deafness and Communication Disorders, 1990; grantee NIH and NSF, 1970—; fellow NIMH, 1963. Mem. AAAS, Assn. Chemoreception Scis., Behavior Genetics Assn. (treas. 1978-81), NRA, Phi Beta Kappa. Republican. Roman Catholic. Office: Fla State U Psychology Dept R-54 Tallahassee FL 32306-1051

WHITNEY, ROBERT A., JR., veterinarian, government public health executive; b. Oklahoma City, July 27, 1935; m. Elizabeth Teeple, June 29, 1986; children: Stacy, Tara, Mark, Laura. BS, Okla. State U., Stillwater, 1958, DVM, 1959; MS in Pharmacology, Ohio State U., Columbus, 1965. Lic. veterinarian. Chief vet. resources br. NIH, Bethesda, Md., 1972-85, acting dir. Div. Rsch. Svcs., 1984-85, dir. Div. Rsch. Svcs., 1985-90, acting dir. Rsch. Res., 1988-90; dir. Nat. Ctr. Rsch. Res., Bethesda, Md., 1990-92, dir. office of animal care and use, 1986-92; chief vet. officer USPHS, 1986-90, dep. surgeon gen., 1992—; cons. Pan. Am. Health Orgn., 1992—. Co-author: Laboratory Primate Handbook, 1973. Assoc. editor Jour. of Med. Primatology, 1975—, Lab. Animal Sci., 1976-82; cons. editor Am. Jour. Primatology, 1980—. Lt. col. U.S. Army. 1959-71. Decorated DSM, 1987; recipient USPHS Commendation medal, 1976, Meritorious Svc. medal, 1984, Surgeon Gen.'s Exemplary Svc. medal, 1988; NIH Director's award 1984, Disting. Alumnus award Ohio State U., 1986, Disting. Grad. Recognition award Okla. State U., 1989, The Charles River Found. prize, 1989, Karl F. Meyer award Am. Vet. Epidemiol. Soc., 1989. Mem. AVMA, Am. Assn. Lab. Animal Sci. (pres. 1977-78), Am. Assn. Lab. Animal Practitioners, Am. Coll. Lab. Animal Medicine (bd. dirs. 1968-69, pres. 1992), Am. Soc. Primatologists. Home: 5213 Bangor Dr Kensington MD 20895-1104 Office: Office of Surgeon Gen Parklawn Bldg Rm 18-67 5600 Fishers Ln Rockville MD 20857

WHITNEY, WILLIAM PERCY, II, materials scientist; b. Springfield, Ill., Aug. 19, 1940; s. William Percy and Dorothy (Verrier) W.; m. Linda Elizabeth Tapp, Dec. 1, 1962; children: Elizabeth Ann Amundson, William Percy III. Degree in metallurg. engring., Colo. Sch. Mines, 1961; PhD in Ceramic Sci., Pa. State U., 1970. Chemist Corning (N.Y.) Glass Works, 1961-70, sr. chemist, 1970-74, sr. rsch. scientist, 1974-78, supr. X-ray analysis, 1978-82, mgr. instrumental analysis, 1982-88; sr. staff scientist Corning, Inc., 1988—; mem. standards com. ANSI, Washington, 1989—. Contbr. chpt. to Standard Methods of Chem. Analysis, 6th edit., 1966; contbr. articles to profl. jours. Scoutmaster Boy Scouts Am., 1985-86. Mem. Rotary (pres. Big Flats, N.Y. chpt. 1976-87, 86-87, Paul Harris fellow 1988). Achievements include patents for elec. heating units, furnace and method with sensor. Office: Corning Inc HP-AB-00-1 Corning NY 14831

WHITSEL, RICHARD HARRY, biologist; b. Denver, Feb. 23, 1931; s. Richard Elstun and Edith Muriel (Harry) W.; children by previous marriages: Russell David, Robert Alan, Michael Dale, Steven Deane. BA, U. Calif., Berkeley, 1954; MA, San Jose State Coll., 1962. Sr. rsch. biologist San Mateo County Mosquito Abatement Dist., Burlingame, Calif., 1959-72; environ. program mgr., chief of watershed mgmt. Calif. Regional Water Quality Control Bd., Oakland, 1972—; mem. grad. faculty water resource mgmt. U. San Francisco, 1987-89. Served with Med. Service Corps, U.S. Army, 1954-56. Mem. Entomol. Soc. Am., Entomol. Soc. Wash., Am. Mosquito Control Assn., Calif. Alumni Assn., The Benjamin Ide Wheeler Soc., Nat. Parks and Conservation Assn. (life), Sierra Club. Democrat. Episcopalian. Contbr. articles to profl. jours. Home: 4331 Blenheim Way Concord CA 94521-4258 Office: Calif Regional Water Quality Control Bd 2101 Webster St Oakland CA 94612-3027

WHITTAKER, PETER ANTHONY, biology educator; b. Cranwell, England, Sept. 2, 1939; s. William Leonard and Anne (Chapman) W.; m. Margaret Lynne Bowering, Aug. 6, 1966; children: Graham (dec.), Robert, Janet, Elizabeth. BSc with honors, Liverpool U., 1961; PhD, Leicester U., 1964. Lectr. in botany Hull (Eng.) U., 1964-68; lectr. in biochemistry Sussex U., Brighton, Eng., 1968-78; prof. biology, head dept. Maynooth (Ireland) Coll., 1978—; chmn. Royal Irish Acad., Nat. Commn. for Microbiology, Dublin, 1986—. Author 5 books in field of cell biology and evolution, also rsch. papers. Fellow Inst. Biology U.K.; mem. Inst. Biology of Ireland (pres. 1992—), Royal Zool. Soc. Ireland (v.p. 1992—). Roman Catholic. Avocations: walking, camping, touring, photography, classical music. Home: Mariaville, Maynooth Kildare, Ireland Office: Maynooth Coll, Biol Dept, Maynooth Kildare, Ireland

WHITTEMORE, PAUL BAXTER, psychologist; b. Framingham, Mass., Apr. 11, 1948; s. Harry Ballou and Margaret (Brown) W. BA in Religion, Eastern Nazarene Coll., 1970; MDiv., Nazerne Theol. Sem., 1973; MA in Theology, Vanderbilt U., 1975, PhD in Theology, 1978; PhD in Clin. Psychology, U. Tenn., 1987. Lic. psychologist. Asst. prof. philosophy and edn. Trevecca Nazarene Coll., Nashville, Tenn., 1973-76; asst. prof. philosophy and theology Point Loma Coll., San Diego, Calif., 1976-80; asst. prof. philosophy Middle Tenn. State U., Murfreesboro, 1980-83; clin. psychology intern. LAC/U. So. Calif. Med. Ctr., L.A., 1986-87; coord. behavior health ctr. Calif. Med. Ctr., L.A., 1987-88; clin. asst. prof. family medicine sch. medicine U. So. Calif., L.A., 1988—; pvt. practice psychologist Newport Beach, Calif., 1989--; mem. behavioral sci. faculty Glendale Advantist Family Practice Residency Program, Glendale, Calif., 1989-90; inpatient group therapist Ingleside hosp., Rosemead, Calif., 1990-92; founder, pres. The Date Coach, 1992—. Contbr. articles to profl. jours. Recipient Andrew W. Mellon Postdoctoral Faculty Devel. award Vanderbilt U., 1981. Mem. Am. Acad. Religion, Am. Philosophical Assn., Am. Assn. Univ. Prof. (chpt. v.p. 1982-83), Am. Psychological Assn. Achievements include discovery of link between phenylthiocarbamide tasting and depression. Office: 3901 MacArthur Blvd #200 Newport Beach CA 92660

WHITTEN, CHARLES ARTHUR, geodetic consultant; b. Redfield, S.D., Oct. 2, 1909; s. Herbert William and Mabel (Hales) W.; m. Brena Evelyn Uber, July 7, 1933; children: William Barclay, David Hart, John Charles. Student, U. Mich., summer 1929; BA, Carthage Coll., 1930, DSc (hon.), 1965, DSc (hon.), U. New Brunswick, Can., 1974, D of Engring. (hon.), Karlsruhe (Germany) U., 1975. Mathematician U.S. Coast and Geodetic Survey, Washington, 1930-46, chief triangulation br., 1946-62, chief electronic computing divsn., 1962, dep. dir. Office Phys. Scis., 1963-65, chief rsch. group, 1965-68, chief geodesist, 1968-72; geodetic cons. Silver Spring, Md., 1972—. Editor: Contemporary Geodesy, 1959; contbr. articles to profl. jours. Recipient silver and gold medals Dept. Commerce, 1959, 72, Levallois medal Internat. Assn. Geodesy, 1979, Mercherikov medal, 1987. Fellow Am. Geophys. Union (gen. sec. 1967-74, William Bowie medal 1980, Charles Whitten medal 1985); mem. AAAS, Internat. Union of Geodesy and Geophysics (fin. com. 1963-79, pres. 1975-79), Am. Congress on Surveying and Mapping, Cosmos Club (Washington). Lutheran. Achievements include award named in honor by the American Geophysical Union in 1985 to be given bi-annually in recognition of basic research in geodynamics. Home: 9606 Sutherland Rd Silver Spring MD 20901

WHITTEN, DAVID GEORGE, chemistry educator, researcher; b. Washington, Jan. 25, 1938; s. David Guy and Miriam Deland (Price) W.; m. Jo Wright, July 9, 1960; children: Jenifer Marie, Guy David. A.B., Johns Hopkins U., 1959; M.A., John Hopkins U., 1961, Ph.D., 1963. Asst. prof. chemistry U. N.C., Chapel Hill, 1966-70, assoc. prof., 1970-73, prof., 1973-80, M.A. Smith prof., 1980-83; C.E. Kenneth Mees prof. U. Rochester, N.Y., 1983—, chair dept. chemistry, 1988-91; dir. Ctr. for Photoinduced charge transfer, 1989—; mem. adv. com. for chemistry NSF; cons. Eastman Kodak Co.; Rochester, N.Y. Alfred P. Sloan fellow, 1970; John van Geuns fellow, 1973; recipient special U.S. scientist award Alexander von Humboldt Found., 1975; Japan Soc. for Promotion of Sci. fellow, 1982. Mem. AAAS, Am. Chem. Soc. (award in colloid and surface chemistry 1992), Internat. Union of Pure and Applied Chemistry (commn. on photochemistry), Interam. Photochem. Soc. (pres.). Democrat. Episcopalian. Home: 72 Canterbury Rd Rochester NY 14607-3405 Office: U Rochester Dept Chemistry 404 Hutchinson Hall Rochester NY 14627

WHITTEN, JERRY LYNN, chemistry educator; b. Bartow, Fla., Aug. 13, 1937; s. John Graves and Dorothy Iola (Jordan) W.; m. Mary Hill (div. Sept. 1977); 1 child, Jerrard John; m. Adela Chrzeszczyk, June 21, 1980; 1 child, Christina. BS in Chemistry, Ga. Inst. Tech., 1960, PhD, 1964. Cert. chemist. Rsch. assoc. to instr. Princeton (N.J.) U., 1963-65; asst. prof. chemistry Mich. State U., East Lansing, 1965-67; asst. prof. chemistry SUNY, Stony Brook, 1967-68, assoc. prof., 1968-73, prof., 1973-89, chmn. chemistry dept., 1985-89; prof. chemistry, dean Coll. Phys. and Math. Scis. N.C. State U., Raleigh, 1989—; vis. prof. Centre Européen de Calcul Atomique et Moléculaire, Orsay, France, 1974-75, Univ. Bonn and Wuppertal, Fed. Republic Germany, 1979-80, Eidgenossische Technische Hochschule, Zurich, Switzerland, 1984. Contbr. over 100 articles to profl. jours. Recipient Alexander von Humboldt U.S. Sr. Scientist award, 1979; grantee Petroleum Rsch. Fund, 1966-67, 74-76, 77-81, NSF, 1967-72, U.S. Dept. Energy, 1977—; SDIO/ONR grantee, 1991—; Alfred P. Sloan fellow, 1969-71. Mem. AAAS, Am. Phys. Soc., Am. Chem. Soc., N.Y. Acad. Scis., Sigma Xi. Democrat. Episcopalian. Avocations: boating, tennis, skiing. Office: NC State U Coll Dept Phy and Math Scis Box 8201 Raleigh NC 27695-8201

WHITTINGHAM, M(ICHAEL) STANLEY, chemist; b. Nottingham, Eng., Dec. 22, 1941; came to U.S., 1968, naturalized, 1980; s. William Stanley and Dorothy Mary (Findlay) W.; B.A. in Chemistry, Oxford U., 1964, M.A. (Gas Council scholar 1964-67), D.Phil., 1968; m. Georgina Judith Andai, Mar. 23, 1969; children: Jenniffer Judith, Michael Stanley. Rsch. asso., head solid state electrochemistry group Materials Center, Stanford U., 1968-72; mem. staff Exxon Research Co., Linden, N.J., 1972—, group head solid state chem. physics, 1975-78, dir. solid state scis., 1978-80, mgr. chem. engring. tech., 1980-84; dir. phys. scis. Schlumberger Co., Ridgefield, Conn., 1984-88; prof. chemistry, dir. The Inst. for Materials Rsch., SUNY, 1988—; cons., lectr. in field. Fellow JSPS U. Tokyo; mem. Electrochem. Soc. (Young Author award 1971), Am. Chem. Soc. (chmn. solid state sect. 1987, chmn. Binghamton sect. 1991), Am. Phys. Soc., N.Y. Electrochem. Soc. (chmn. 1980-81). Author, editor papers in field; author 5 books. Achievements include patents in field; reversible (rechargeable) lithium batteries and methods for making intercalation batteries; method for making T1S2 mixed material cathodes, high briteness luminescent displays. Home: 396 Meeker Rd Vestal NY 13850-3230 Office: SUNY Dept Chemistry Binghamton NY 13902-6000

WHITTLE, FRANK, aeronautical engineer; b. Coventry, Eng., June 1, 1907; married; 2 children. Grad., RAF Coll., Cranwell, 1928, RAF Officer's Sch. Engring, 1934, Cambridge U., 1936; numerous hon. degrees. Co-founder Power Jets Ltd., 1936; Served to Air Commodore RAF, ret., 1948; tech. advisor Brit. Overseas Airway Corp., Eng., 1948-52; tech. adviser Shell Group, Eng., 1953-57; cons. Rolls-Royce, Eng., 1961-70; mem. faculty US Naval Acad., from 1977. Recipient, Gold Medal, Royal Aeronautical Soc., 1944, James Watt Internat. Gold Medal, 1977, Daniel Guggenheim Gold Medal, 1946, Gold Medal, Fed. Aero. Internat., 1951, Franklin Medal, 1956, Goddard Award, 1965, (with Hans von Ohain) Charles Stark Draper Prize, NAE, 1991, for invention of the jet engine; Knighted by King George VI, 1948; recipient Order of Merit, Queen Elizabeth II, 1986. Fellow, Royal Soc., (U.K.) Fellowship Engrings.; Am. Acad. Arts and Scis. (fgn. mem.), NAE (fgn. assoc., Charles Stark Draper prize 1992).

WHITTLE, PHILIP RODGER, chemistry educator, crime laboratory director; b. Russell Springs, Ky., July 11, 1943; s. Garnet T. and Agnes R. (Rexroat) W.; m. Donna Kay Hollatz, June 18, 1967; children: Bruce, Brian. BS in Chemistry, U. Ky., 1965; PhD in Organic Chemistry, Iowa State U., 1969. Postdoctoral fellow U. Colo., Boulder, 1969-70; asst. prof. Mo. Southern State Coll., Joplin, 1970-74, assoc prof., 1974-79, prof., 1979—, dir. Regional Crime Lab., 1972—; cons. Chem Staat Labs., Neosho, Mo., 1976-88; profl. adv. com. Oak Hill Hosp., Joplin, 1977—. Contbr. articles to profl. jours. Speaker to civic orgns.; planning com. Diamond (Mo.) Sch., 1985-88. Fellow Am. Acad. Forensic Sci.; mem. Midwestern Assn. Forensic Scientists, Am. Soc. Crime Lab. Dirs. (rsch. adv. com., charter mem.), Mo. Assn. Crime Lab. Dirs. (pres. 1980, 1990-91, charter mem.), Am. Chem. Soc. (sec.-treas. 1972-73), Phi Beta Kappa, Phi Lambda Upsilon, Sigma Xi. Republican. Lutheran. Achievements include co-design of ground source heat pump, numerous improvements in forensic laboratory techniques. Home: Rt 8 Box 312 Joplin MO 64804 Office: MSSC Regional Crime Lab 3950 Newman Rd Joplin MO 64801

WIATR, CHRISTOPHER L., microbiologist; b. Chgo., Jan. 5, 1948; s. Joseph Thomas and Beatrice Harriet (Kaminski) Wiatr; m. Jeanne Lynn Malecki, Oct. 20, 1978; children: Kelli Jean, Christopher Joseph, Kaycee Lynn, Kirby Ann, Nicholas Aloysius. BS, Ill. Benedictine Coll., 1969; MS, IIT, 1974; PhD, U. Ill., Chgo., 1985. Cert. tchr. Tchr., coach St. Rita High Sch., Chgo., 1969-74; rsch. microbiologist Swift & Co./Esmark/Beatrice Foods, Chgo., 1974-75, lab. mgr., 1975-76, tech. dir. rsch. and quality assurance, 1976-79; sr. microbiologist Nalco Chem. Co. Water and Waste Treatment R & D, Naperville, Ill., 1985-87, sr. rsch. microbiologist, 1988, group leader, 1989-91; group leader Pulp & Paper Chems. R & D, Naperville, Ill., 1991—; reviewer Nat. Assn. Corrosion Engrs., Houston, 1989—. Co-author: (book chpt.) Food Preservation by Irradiation, 1978; contbr. articles to profl. jours. Com. Maplebrook I Swim Club, Naperville, 1990-92; Eagle scout, merit badge counselor Boy Scouts Am., 1963—. Named Researcher of Yr., Nalco Chem. Co., 1987. Mem. TAPPI (microbiology and microbial tech. com. 1993—), Am. Chem. Soc., Nat. Assn. Corrosion Engrs., Soc. Indsl. Microbiology, Am. Soc. for Microbiology, Sigma Xi (pres. Nalco chpt. 1991-92). Roman Catholic. Achievements include 4 U.S. patents and 2 European patent applications; enzyme applications for controlling microbial slime on equipment surfaces; nontoxic biocontrol; recognition as expert on industrial biofilms. Office: Nalco Chem Co R & D One Nalco Ctr Naperville IL 60563-1198

WIBERG, KENNETH BERLE, chemist, educator; b. Bklyn., Sept. 22, 1927; s. Dan and Solveig Berle W.; m. Marguerite Koch, Mar. 18, 1951; children—Patricia, Robert, William. B.S., Mass. Inst. Tech., 1948; Ph.D., Columbia, 1950. Mem. faculty U. Wash., 1950-62, prof. chemistry, 1958-62; prof. chemistry Yale U., 1962—, Whitehead prof., 1968-90, Eugene Higgins prof., 1990—, chmn. chem. dept., 1968-71; Boomer lectr. U. Alta., Can., 1959; Mem. chemistry advisory panel Air Force Office Sci. Research, 1960-66, NSF, 1965-68, chmn., 1967-68. Author: Laboratory Technique in Organic Chemistry, 1960, Interpretation of NMR Spectra, 1962, Physical Organic Chemistry, 1964, Computer Programming for Chemists, 1966, also articles.; Editor: Oxidation in Organic Chemistry, 1966, Sigma M.O. Theory, 1970; mem. editorial bd.: Organic Syntheses, 1963-71, Jour. Organic Chemistry, 1968-72, J. Computer Chem., 1982—, Tetrahedron, 1981—, Chemical Revs., 1986—; bd. editors: Jour. Am. Chem. Soc, 1969-72. Recipient James Flack Norris award, 1973, Arthur C. Cope award 1988; Sloan Found. fellow, 1958-62; Guggenheim fellow, 1961-62; Cope scholar, 1986. Mem. Nat. Acad. Scis., Am. Acad. Arts and Scis., Am. Chem. Soc. (exec. com. organic div. 1961-63, Calif. sect. award 1963, Linus Pauling award 1992). Home: 160 Carmalt Rd Hamden CT 06517-1904 Office: Yale U Chemistry Dept New Haven CT 06520

WICHA, MAX S., oncologist, educator; b. N.Y.C., Mar. 24, 1949; m. Sheila Crowley; children: Jason, Allyson. BS in Biology summa cum laude with honors, SUNY, Stony Brook, 1970; MD, Stanford U., 1974. Diplomate Am. Bd. Internal Medicine; lic. physician, Mich., Ill. Intern in internal medicine U. Chgo. Hosps. and Clinics, 1974-75, jr., sr. resident in internal medicine, 1975-77; rsch. assoc. lab. pathophysiology Nat. Cancer Inst./NIH, Bethesda, Md., 1977-79, fellow in clin. oncology, 1978-80, investigator lab. pathophysiology, 1979-80; asst. prof. internal medicine divsn. hematology and oncology U. Mich., Ann Arbor, 1980-83, assoc. prof., 1983-88, prof., 1988—, mem. tumor metastasis, extracellular matrix, reproductive endocrinology programs, 1982—, dir. divsn. hematology and oncology, dir. Simpson Meml. Rsch. Inst., 1984-93, mem. program in cellular and molecular biology, 1984—, dir. Comprehensive Cancer Ctr., 1987—; mem. cancer rsch. com. U. Mich., 1981—, sci. adv. bd. dental rsch. inst., 1983—, dean's adv. coun., 1988—, reproductive endocrinology selection com., breast care ctr. exec. com., 1988—, exec. dir.'s adv. coun., 1992—, chair instl. rev. com. gene therapy program project., 1992—, dean's adv. com. Howard Hughes Med. Inst., 1992—, strategic planning policy and organizational com. health scis. info. tech. and networking, 1992—; vis. prof. Mich. State U., 1985, Harvard U., Boston, 1986, Wash. State U., 1986, Boston U., 1986, Wayne State U./Harper Grace Hosps., 1987, U. Ill., 1987, Med. Coll. Wis., 1987., U. Chgo., 1987, Eppley Inst. for Rsch. in Cancer and Allied Diseases, Omaha, 1988, U. Nebr., Omaha, 1988, U. Minn./Minn. VA Hosp., 1988, MD Anderson Cancer Ctr., Houston., Mt. Sinai Med. Ctr., N.Y.C., Am. Cancer Soc., Kalamazoo, 1989, Gainesville, Fla., 1990, Orlando, Fla., 1990, Pezcoller Symposium, Rovereto, Italy, 1990, Prince Henry's Hosp., Melbourne, Australia, 1990, Northwestern U. Med. Ctr., Chgo., 1990, Meml. Sloan-Kettering Cancer Ctr., N.Y.C., 1990, Tex. S.W. U., Dallas, 1990, Mich. State U., 1991; lectr. U. Mich., 1990; mem. NIH Site Visit team U. Calif. Cancer Rsch. Lab., Berkeley, 1985; ad hoc mem. cell biology and physiology study sect. NIH, 1985, 86, study sect., Bethesda, 1991; mem. NCI Site Visit team Norris Cotton Dartmouth Cancer Ctr, 1989, Howard U., Wash., 1989, Howard U. Parent Com., 1989, MD Anderson Cancer Ctr., Houston, 1992; sci. advisor U. Colo. Cancer Ctr., Denver, 1990, Samuel Waxman Cancer Rsch. Found., Mt. Sinai Med. Ctr., N.Y.C., 1988-93; mem. NCI Adv. Panel, Bethesda, 1991; mem. sci. adv. com. U. Tex. San Antonio Cancer Ctr., U. Miami Sylvester Cancer Ctr., Mich. State U., East Lansing, Norris-Cotton Cancer Ctr., Dartmouth-Hitchcock Med. Ctr., Hanover, N.H., Mich. Cancer Found., Detroit, V. T. Lombardi Cancer Rsch. Ctr., Georgetown U., Washington, 1992—, MD Anderson Cancer Ctr., Houston, 1992—; mem. extramural sci. adv. bd. UCI Clin. Cancer Ctr., U. Calif. Irvine, Orange, 1992—; mem. NCI SPORE in Prostate Cancer Study Sect., 1992; chair NCI Cancer Ctr. Support Rev. Com., 1993; NCI Site Visit chair Jefferson Cancer Ctr., Phila., 1992, Worcester (Mass.) Cancer Found., 1993, Duke U. Cancer Ctr., Durham, N.C., 1993, Lineberger Comprehensive Ctr., Chapel Hill, N.C., 1993; mem. NCI Comprehensive Cancer Ctrs. Review, 1993; cons. Warner Lambert Co., 1980—. Assoc. editor: Molecular and Cellular Differentiation, 1993; co-editor: The Hematopoietic Microenvironment, 1993; mem. editorial bd. Blood, Molecular and Cellular Differentiation, Jour. Lab. and Clin. Medicine, Cancer Rsch., 1993—, Oncology, Cancer Prevention Internat.; reviewer Nature, Science, Proceedings of NAS, Jour. Clin. Investigation, Jour. Cell Biology, Exptl. Cell Rsch., Exptl. Hematology, Cancer., Clin. and Exptl. Metastasis, Jour. Nat. Cancer Inst., Tissue & Cell, Am. Inst. Biol. Scis., Am. Jour. Pathology, Jour Immunology, Jour. Med. Scis., NSF, Oncology Rsch., Lab. Investigation, Breast Cancer Rsch. and Treatment; contbr. over to profl. jours., chpts. to books.; invited lectr. in field. With USPHS, 1977-80. Recipient NSF RSch. award SUNY, 1969, Eli Luke and David Jacob Rsch. award Stanford U. Sch. Medicine, 1974, Upjohn Achievement Excellence in Medicine award, Outstanding Med. Resident award U. Chgo. Hosps., 1977, Jerome Conn Excellence in Rsch. award, 1983; grantee NIH, 1991—, 93—, Am. Cancer Soc., 1992—, Suntory Rsch. Inst., 1992-93. Mem. AAASN, Am. Assn. for Cancer Rsch. (state legis. com. 1992—, finance com. 1992—), Am. Fedn. for Clin. Rsch. (selections com. midwest sect. 1986—, comm. com., 1986—, awards com., 1986—), Am. Soc. for Cell Biology, Am. Soc. Hematology (com. on publs. 1991-93), Assn. Am. Cancer Insts. (bd. dirs. 1993—), Am. Soc. for Clin. Investigation, Am. Soc. Clin. Oncology (award selection com. 1992—), Ctrl. Soc. for Clin. Rsch., Mich. Soc. Hematology and Oncology, Southwest Oncology Group, Assn. Community Cancer Ctrs. Achievements include patents for antibodies to human mammary cell growth inhibitor and methods of production and use, human mammary cell growth inhibitor and methods of production and use; research in regulation of cell growth and differentiation, molecular mechanisms of tumor metastasis. Office: U Mich Simpson Meml Inst for Med Rsch 102 Observatory St Ann Arbor MI 48109-2020

WICKHAM, M(ARVIN) GARY, optometry educator; b. Ft. Morgan, Colo., Dec. 23, 1942; s. Marvin Gilbert W. and Dorothy Mae (Frazell) West; m. Irene Mary Wilhelm, Mar. 20, 1965. BS, Colo. State U., 1964, MS, 1967; PhD, Wash. State U., 1972. Rsch. physiologist VA, Gainesville, Fla., 1971-74; asst. prof. U. Fla., Gainesville, 1972-74; rsch. physiologist VA, San Diego, 1974-79; asst. rsch. biologist morphology of the eye U. Calif., San Diego, 1974-79; assoc. prof. histology, ocular anatomy and physiology Northeastern State U., Tahlequah, Okla., 1979-85; prof. histology, gen. biology Northeastern State U., Tahlequah, 1986-88, prof. optometry, histology, human genetics, gen. biology, 1988—; pres. Northeastern State U. Faculty Assn., 1989-91. Contbr. articles profl. jours., Tulsa Tribune, 1987—. Recipient Glaucoma Studies grant VA, 1975, Core Em Facility grant, 1977-79, Focal Argon Laser Lesions grant, 1979; Morphology of Mammal Eyes grant NIH, 1980, Computer-Based Image Analysis grantee Nat. Eye Inst., 1990. Mem. AAAS, Am. Soc. Zoologists, Am. Inst. Biological Scis., Assn. for Rsch. in Vision and Ophthalmology. Achievements include co-development of first argon laser treatment for glaucoma; first co-documentation of movement of silicone away from clinically implanted breast prosthesis devices using EDXA; co-application of vitreous carbon as a knife-making material; confirmation of generality of occurrence of encapsulated receptors inside cetacean eyes. Office: Northeastern State U Coll Optometry Tahlequah OK 74464

WICKHAM-ST. GERMAIN, MARGARET EDNA, mass spectrometrist; b. Kansas City, Mo., June 7, 1956; d. Ronald Lee and Mary Ann (Nicholas) Wickham; m. Christopher Newman St. Germain, June 11, 1988; 1 child, Mark Anthony. BS in Chemistry, St. Mary Coll., Leavenworth, Kans., 1978; student, U. Mo., 1979-80. Lab. technician VA Hosp., Kansas City, 1977; jr. chemist Midwest Rsch. Inst., Kansas City, 1978-80, jr. mass spectrometrist, 1980-81, asst. mass spectrometrist, 1981-85, assoc. mass spectrometrist, 1985-86, mass spectrometrist, 1986-90, sr. mass spectrometrist, 1990—. Co-author: Priority Pollutants, 1983. Active Mid-Continent coun. Girl Scouts U.S., 1962—, St. Bernadette's Ch., Kansas City, 1986—. Mem. Am. Chem. Soc. (mem. nat. younger chemist com. 1988-91, chair memberships com. Kansas City chpt. 1979-80, sec. 1984, chair-elect 1985, chair 1986, past chair 1987, chair chemistry conf. 1989—, founding mem. regional younger chemist com. 1990—, Chemagro Essay award 1977, 78), Am. Soc. for Mass Spectrometry, Soc. for Applied Spectroscopy, Kappa Gamma Pi (Excellence award 1978), Delta Epsilon Sigma (Excellence award 1978). Avocations: medicinal herbs, children's songs, gardening. Office: Midwest Rsch Inst 425 Volker Blvd Kansas City MO 64110-2299

WICKLAND, DIANE ELIZBETH, ecologist; b. Palmyra, Wis., Jan. 21, 1951; d. Anton Anthony and Dorthy Mae (Morris) W.; m. Paul Donis, July 4, 1987. BA, U. Wis., 1972; PhD, U. N.C., 1983. Postdoctoral fellow Jet Propulsion Lab., Pasadena, Calif., 1983-85, mem. tech. staff, 1985-87; program mgr. NASA, Washington, 1987-90, acting br. chief, 1990—. Author: (chpts.) Modern Ecology: Basic & Applied, 1991, Heavy Metal Tolerance in Plants, 1990, Theory & Applications: Optical Remote Sensing, 1989. Recipient Performance award NASA, 1988. Mem. Assn. for Women in Sci., Ecol. Soc. Am., Bot. Soc. Am., Am. Inst. for Biol. Scis. Office: NASA Code SEPO3 Washington DC 20546

WICKMAN, HERBERT HOLLIS, physical chemist, condensed matter physicist; b. Omaha, Sept. 30, 1936. AB, Mcpl. U. Omaha, 1959; PhD, U. Calif., Berkeley, 1965. Mem. tech. staff Bell Telephone Labs., Murray Hill, N.J., 1964-71; prof. phys. chemistry Oreg. State U., Corvallis, 1971-87; program dir. condensed matter physics NSF, Washington, 1987—. Contbr. over 60 articles to profl. jours. Fellow Am. Phys. Soc.

WICKRAMANAYAKE, GODAGE BANDULA, environmental engineer; b. Sri Lanka, Mar. 9, 1953; came to U.S., 1980; s. Godage Collin and Maginawathie W. (Dissanayake) W.; ,/ Sandhya Rukmal Karunaratne, Dec. 1, 1983; children: Nilesh Chamaka, Shenali Danica. BSc in Civil Engring., U. Sri Lanka, Peradeniya, 1976; M Engring. in Environ. Engring., Asian Inst. Tech., Bangkok, Thailand, 1979; PhD in Civil Engring., Ohio State U., 1984. Registered profl. engr., Ohio, Pa. Civil engr. Irrigation Dept., Sri Lanka, 1976-77; grad. fellow Asian Inst. Tech., Bangkok, 1977-79, rsch. engr., 1979-80; rsch. assoc. Ohio State U., Columbus, 1980-84; rsch. scientist Battelle Meml. Inst., Columbus, 1984-90; sr. cons. Environ. Internat. Corp., Princeton, N.J., 1990—. Contbr. articles to profl. jours., conf. procs., tech. books, environ. engring. handbook. Recipient Royal Danish award Danish Govt., 1977. Mem. ASCE. Buddhist. Achievements include patent for apparatus for treatment of soils contaminated with organic pollutants; patent pending for method for treatment of soils contaminated with organic pollutants. Home: 4 Woodland Ct West Windsor NJ 08512 Office: Environ Internat Corp 210 Carnegie Ctr Princeton NJ 08540

WICKS, FRANK EUGENE, mechanical engineer, educator; b. Watertown, N.Y., Mar. 5, 1939; s. Rollo Eugene and Hazel Tallman (Jerome) W.; m. Virginia Dawn Garilli, July 18, 1964; children: Frank, Linda. B Marine Engring., SUNY, Bronx, 1961; MSEE, Union Coll., Schenectady, N.Y., 1966; PhD, Rensselaer Poly. Inst., 1976. Registered profl. engr., N.Y. Engr. GE, Schenectady, 1961-75; nuclear supr. Rensselaer Poly. Inst., Troy, N.Y., 1976-87; prof. mech. engring. Union Coll., Schenectady, 1988—; const. Nat. Inst. Standards and Tech., 1987—. Contbr. articles to tech. and popular publs. Rep. candidate for U.S. Congress, 23d Dist., N.Y., 1980, 82, 84. Named in Patents of Week feature N.Y. Times, 1987, 89, 91. Mem. IEEE (sr.), ASME (Energy Resources Tech. award 1987), Am. Nuclear Soc., N.Y. Acad. Scis., Experimental Aircraft Assn., Soc. Automotive Engrs., Sigma Xi, Tau Beta Pi, Pi Tau Sigma. Achievements include patents for gas air conditioner, device for generating electricity from a gas furnace, use of diesel engine cycles heat to power turbine. Office: Union Coll Dept Mech Engring Schenectady NY 12308

WICKS, MOYE, III, chemical engineer, researcher; b. Houston, Aug. 18, 1932; s. Frank Melville and Gladys Evelyn (Paine) W.; m. Shirley Ann Kane, Feb. 1, 1958; children: Moye IV, John, Thomas, Paul, Mary Elizabeth. BS in Math., U. Houston, 1955, BSChemE, 1955, MSChemE, 1958, PhDChemE, 1965. Process designer, math. engr. Phillips Petroleum, Bartlesville, Okla., 1957-59; rsch. engr. Shell Devel., Houston, 1965-67, supr. fluid mechanics, 1968-75; tech. mgr. Shell Oil, Houston, 1976-78, dept. mgr., 1979-81; corp. sec. Shell Cos. Found., Houston, 1981-82; rsch. advisor Shell Devel. Co., Houston, 1983—; co-dir. AGA-API Project NX28, on Multiphase flow, Houston, 1960-65. Author: (with others) Advances in Solid-Liquid Flow in Pipes, 1970; co-author: Modern Chemical Engineering I, 1963; contbr. articles to profl. jours. NSF fellow, 1961-65. Fellow Am. Inst. Chem. Engrs. (exec. com. South Tex. sect. 1989, chmn. Houston 1991, chmn. tech. com. design inst. multiphase processing 1973-79, Best Fundamental Paper 1965). Roman Catholic. Achievements include 30 U.S. and international patents.

WICKSTROM, CARL WEBSTER, analytical chemist; b. Munising, Mich., Dec. 3, 1944; s. George Berger and Edith (Webster) W.; m. Rosemary Else Buckley Wickstrom, June 17, 1973; 1 child, Natica Else Buckley Wickstrom. BS in Chemistry, U. Mich., Ann Arbor, 1966; PhD in Phys. Chemistry, MIT, 1970. Tech. specialist U.S. Army, Natick Rsch. Lab., Natick, Mass., 1970-72; scientist Polaroid, Waltham, Mass., 1972-82; cons. Wakefield, Mass., 1982-83; dir. quality assurance Chem Design, Fitchburg, Mass., 1983-90, dir. analytical methods, 1990—. Author: (book) Websters of Haverhill, Mass., 1986; contbr. to Jour. Molecular Spectroscopy. Environ. lectr. high sch. chemistry classes, Wakefield, Natick and Fitchburg, Mass., 1991. With USA, 1970-72. Mem. Masonic Lodge (jr. deacon), Royal Arch Masons, Birka Lodge of VASA, U. Mich. Alumni Assn. Republican. Episcopalian. Home: 30 Strathmore Rd Wakefield MA 01880-1420 Office: Chem Design 99 Development Rd Fitchburg MA 01420-6000

WICKWIRE, PATRICIA JOANNE NELLOR, psychologist, educator; b. Sioux City, Iowa; d. William McKinley and Clara Rose (Pautsch) Nellor; BA cum laude, U. No. Iowa, 1951; MA, U. Iowa, 1959; PhD, U. Tex., Austin, 1971; postgrad. U. So. Calif., UCLA, Calif. State U., Long Beach, 1951-66; m. Robert James Wickwire, Sept. 7, 1957; 1 son, William James. Tchr., Ricketts Ind. Schs., Iowa, 1946-48; tchr., counselor Waverly-Shell Rock Ind. Schs., Iowa, 1951-55; reading cons., head dormitory counselor U. Iowa, Iowa City, 1955-57; tchr., sch. psychologist, adminstr. S. Bay Union High Sch. Dist., Redondo Beach, Calif., 1962-82, dir. student svcs. and spl. edn.; cons. mgmt. and edn.; pres. Nellor Wickwire Group, 1981—; mem. exec. bd. Calif. Interagency Mental Health Coun., 1968-72, Beach Cities Symphony Assn., 1970-82; chmn. Friends of Dominguez Hills (Calif.), 1981-85. Lic. ednl. psychologist, marriage, family and child counselor, Calif. Mem. AAUW (exec. bd., chpt. pres. 1962-72), L.A. County Dirs. Pupil Svcs. (chmn. 1974-79), L.A. County Personnel and Guidance Assn. (pres. 1977-78), Assn. Calif. Sch. Adminstrs. (dir. 1977-81), L.A. County SW Bd. Dist. Adminstrs. for Spl. Edn. (chmn. 1976-81), Calif. Assn. Sch. Psychologists (bd. dirs. 1981-83), Am. Psychol. Assn., Am. Assn. Sch. Adminstrs., Calif. Assn. for Measurement and Evaluation in Guidance (dir. 1981, pres. 1984-85), Am. Assn. Counseling and Devel. (chmn. Coun. Newsletter Editors 1989-91, mem. com. on women 1989-92), Assn. Measurement and Eval. in Guidance (Western regional editor 1985-87, conv. chair 1986, editor 1987-90, exec. bd. dirs. 1987-91), Calif. Assn. Counseling and Devel. (exec. bd. 1984—, pres. 1988-89, jour. editor 1990—), Internat. Career Assn. Network (chair 1985—), Pi Lambda Theta, Alpha Phi Gamma, Psi Chi, Kappa Delta Pi, Sigma Alpha Iota. Contbr. articles in field to profl. jours. Home and Office: 2900 Amby Pl Hermosa Beach CA 90254

WIDEBURG, NORMAN EARL, biochemist; b. Chgo., Mar. 8, 1933; s. Norman Otto and Edna May (Jensen) W.; m. Joy Ruth Kelly, Sept. 20, 1958; children:: Laura A., Mary J., Sandra K., Douglas N. BS, Ill. Inst. Tech., 1954; MS, U. Wis., 1956. Biochemist Abbott Labs., Abbott Park, Ill., 1958—. Mem. Am. Chem. Soc., Am. Soc. for Microbiology. Office: Abbott Labs Dept 47D Bldg Ap-9A Abbott Park IL 60064

WIDEMAN, GILDER LEVAUGH, obstetrician/gynecologist; b. Kansas, Ala., May 25, 1927; s. Grant Holcomb and Minnie Daly (Hoket) W.; m. Sara Frances De BardeLeben, Sept. 17, 1949; children: Elizabeth, Anne, Margaret, Katherine, John. BS, Birmingham So. Coll., 1953; MD, U. Ala. Med. Sch., 1956. Diplomate Nat. Bd. Med. Examiners, Am. Bd. Obstetrics and Gynecology; lic. MD, Ala. Internship St. Vincent's Hosp., Birmingham, Ala., 1956-57; resident in obstetrics and gynecology U. Hosp., Birmingham, Ala., 1957-60; asst. prof. ob-gyn. U. Hosp. and Hillman Clinic, Birmingham, Ala.; asst. clin. prof. dept. medicine U. Ala. Med. Ctr., Birmingham; staff mem. AMI Brookwood Hosp. Contbr. numerous articles to profl. jours. Recipient Outstanding Resident's award Med. Coll. Ala., A.O.A., 1960. Mem. AMA, Am. Coll. Ob-Gyn. Soc., Am. Soc. for Coposcopy, Am. Fertility Soc., Am. Soc. Clin. Pharmacology and Therapeutics, Am. Gynecol. Soc. for Study of Breast Cancer, Am. Coll. Clin. Pharmacology, Royal Soc. Medicine, Intenat. Med. Assn.-Guadalajara, Mex., Med. Assn. of State of Ala., Jefferson County Med. Soc., Birmingham Acad. Medicine, Ala. Ob-Gyn. Soc., Birmingham Ob-Gyn. Soc., U. Ala. Sch. of Medicine Alumni Assn. Office: Brookwood Ob/Gyn Clinic PC 2006 Brookwood Med Ctr Birmingham AL 35209

WIDMER, CHARLES GLENN, dentist, researcher; b. Daytona Beach, Fla., Jan. 8, 1955; s. Ernest Clyde and Martha Elizabeth (Hunter) W.; m. Alyson Lynn Byrd, Jul. 11, 1981; children: Kathryn Michelle, Elizabeth Ann. BS, Emory Univ., 1977, DDS, 1981; MS, SUNY, 1983. Asst. prof. Sch. Dentistry Emory Univ., Atlanta, 1983-91; assoc. prof. Coll. Dentistry Univ. Fla., Gainesville, 1991—. Contbr. articles to profl. jours. Contbr. Atlanta Zoo, 1985-91; sec., treas., v.p., pres. Neuroscience Group of Internat. Assn. Dental Rsch., Washington, 1989-93; sec., treas., v.p., pres. Assn. Univ. TMD and Orofacial Pair Programs, 1990—, NIH reviewer Nat. Inst. Health, Washington, 1988-89. Recipient Utilization of TEPs in Dentistry grant NIH, 1986-92, Orgn. and Function of Human Muscles NIH, 1991—, Methods for Conducting TMJ Imaging using MRI NIH, 1988-89, Rsch. Career Devel. award NIH, 1991-96. Mem. N.Y. Acad. Sci., Am. Dental Assn., Soc. Neuroscience, Internat. Brain Rsch. Orgn., Internat. Assn. Dental Rsch. Office: Univ Fla Dept Oral and Maxillofacial Surgery Box 100416 JHMHSC Gainesville FL 32610

WIDOM, BENJAMIN, chemistry educator; b. Newark, Oct. 13, 1927; s. Morris and Rebecca (Hertz) W.; m. Joanne McCurdy, Dec. 21, 1953; children: Jonathan, Michael, Elisabeth. A.B., Columbia U., 1949; Ph.D., Cornell U., 1953; DSc (hon.), U. Chgo., 1991. Research assoc. U. N.C., Chapel Hill, 1952-54; instr. chemistry Cornell U., Ithaca, N.Y., 1954-55; asst. prof. Cornell U., 1955-59, assoc. prof., 1959-63, prof., 1963-83, Goldwin Smith prof., 1983—; van der Waals prof. U. Amsterdam, 1972; vis. prof. Harvard U., 1975; IBM vis. prof. Oxford U. (Eng.), 1978; Lorentz prof. U. Leiden, 1985; vis. prof. Katholieke Universiteit Leuven, 1988. Author: (with J.S. Rowlinson) Molecular Theory of Capillarity, 1982. Served with U.S. Army, 1946-47. Recipient Clark Disting. Teaching award Cornell U., 1973; recipient Boris Pregel award for chem. physics rsch. N.Y. Acad. Scis., 1976, Dickson prize for sci. Carnegie-Mellon U., 1986, Hirschfelder prize in theoretical chemistry U. Wis., 1991. Fellow Am. Phys. Soc., Am. Acad. Arts and Scis.; mem. NAS, Am. Philos. Soc., Am. Chem. Soc. (Langmuir award in chem. physics 1982, Hildebrand award in theoretical and exptl. chem. of liquids 1992), N.Y. Acad. Scis. Home: 204 The Pky Ithaca NY 14850-2247 Office: Cornell U Chemistry Dept Ithaca NY 14853

WIECHEL, JOHN F., mechanical engineer; b. Toledo, Aug. 18, 1951; s. Robert G. and Mary (Kohn) W.; m. Patricia A. Azbell, June 21, 1975; children: Nathan, Matthew, Ann Marie. BSME, Purdue U., 1973, MSME, 1975; PhD, Ohio State, 1983. Mechanical engr. Texaco, Inc., Houston, 1975-80; sr. rsch. engr. Ohio State U., Columbus, 1980-84; engring. cons. SEA, Inc., Columbus, 1984—; chmn. Internat. Standards Assn., Geneva, 1988—. Contbr. articles to profl. jours. Methodist. Achievements include patents for oil recovery methods and neonatal jaundice measurement. Home: 170 Abbot Ave Worthington OH 43085 Office: SEA Inc 7349 Worthington-Galena Rd Columbus OH 43085

WIED, GEORGE LUDWIG, physician; b. Carlsbad, Czechoslovakia, Feb. 7, 1921; came to U.S., 1953, naturalized, 1960; s. Ernst George and Anna (Travnicek) W.; m. Daga M. Graaz, Mar. 19, 1949 (dec. Aug. 1977); m. Kayoko Y. Yamauchi, Nov. 1, 1990. M.D., Charles U., Prague, 1945. Intern County Hosp., Carlsbad, Czechoslovakia, 1945; intern U. Chgo. Hosps., 1955; resident in ob-gyn U. Munich, Fed. Republic Germany, 1946-48; practice medicine specializing in ob-gyn West Berlin, 1948-53; asst. ob-gyn Free U., West Berlin, 1948-52; assoc. chmn. dept. ob-gyn Moabit Hosp., Free U., West Berlin, 1953; asst. prof., dir. cytology U. Chgo., 1954-59, assoc. prof., 1959-65, prof., 1965-91, mem. bd. adult edn., 1964-68, prof. pathology, 1967-91, Blum-Riese prof. ob-gyn, 1968-91, acting chmn. dept. ob-gyn, 1974-75. Contbr. articles to profl. jours.; editor-in-chief Jour. Reproductive Medicine, Acta Cytologica, Analytical and Quantitative Cytology, Clinical Cytology; editor: Introduction to Quantitative Cytochemistry, Automated Cell Identification and Cell Sorting. Hon. dir. Chgo. Cancer Prevention Ctr., 1959-83; chmn. jury Maurice Goldblatt Cytology award, 1963-92. Recipient Cert. of Merit, U.S. Surgeon Gen., 1952, Maurice Goldblatt Cytology award, 1961, George N. Papanicolaou Cytology award, 1970. Mem. Am. Soc. Cytology (pres. 1965-66), Mex. Soc. Cytology (hon.), Spanish Soc. Cytology (hon.), Brazilian Soc. Cytology (fgn. corr.), Indian Acad. Cytology (hon.), Latin-Am. Soc. Cytology (hon.), Japanese Soc. Cytology (hon.), German Soc. Cytology (hon.), Internat. Acad. Cytology (pres. 1977-80), Central Soc. Clin. Research, Chgo. Path. Soc., Chgo. Gynecol. Soc. (hon.). Am. Soc. Cell Biology, German Soc. Ob-Gyn. Bavarian Soc. Ob-Gyn. German Soc. Endocrinology. Swedish Soc. Medicine (hon.), Sigma Xi. Home and Office: 1640 E 50th St Chicago IL 60615-3161

WIEDE, WILLIAM, JR., chemical engineer, consultant; b. Chgo., July 11, 1954. MSChemE, Purdue U., 1979, PhD in Chem. Engring., 1984. Staff rsch. engr. Texaco, Inc., Houston, 1984-86, USG Rsch. Ctr., Libertyville, Ill., 1986-88, Amoco Chem. Co., Chgo., 1988—. Contbr. articles to profl. jours. Dave Ross fellow Purdue U., 1977. Mem. AICE, ASME, Am. Assn. Cost Engrs. Achievements include expertise in cost estimating engineering and economic analysis; design of computer-aided cost engineering database and spreadsheet estimating tools for the oil and chemical process industry. Home: 441 S Illinois Ave Villa Park IL 60181-2959

WIEDL, SHEILA COLLEEN, biologist; b. Buffalo, Feb. 19, 1950; d. Frank George and Corinne Ruth (Nuskay) W.; m. Warren J. Kramer, May 8, 1993; one child, Colleen Bryce. B.S., Daemen Coll., 1972; M.S., U. Notre Dame, 1974; Ph.D., SUNY-Buffalo, 1986. Instr., Holy Cross Jr. Coll., South Bend, Ind., 1973-74; research technician SUNY, Buffalo, 1975-78; entomol. asst. N.Y. State Health Dept., 1979-80; entomol. intern Ohio Dept. Health, 1981; prof. natural scis. Trocaire Coll., Buffalo, 1974-85; postdoctoral scientist Am. Cyanamid, Lederle Labs., Pearl River, N.Y., 1985-86, clin. research assoc., 1986-89; assoc. mgr. CIBA Consumer Pharms., Edison, N.J., 1989-90; assoc. dir. clin. projects AKZO/Organon Inc., West Orange, N.J., 1990—; adj. prof. Ramapo Coll. of N.J., 1988—, Rockland Community Coll., 1989—. Mem. N.Y. State Assn. Two-Year Colls., Assn. Gnotobiotics, N.Y. State Archeol. Assn. Roman Catholic. Club: Notre Dame Alumni. Contbr. articles to profl. jours. Home: 11 Oakland Dr White House Station NJ 08889 Office: AKZO/Organon Inc 375 Mt Pleasant Ave West Orange NJ 07052-2798

WIEGAND, GORDON WILLIAM, flow cytometrist, electro-optic consultant; b. Balt., Dec. 7, 1948; s. William F. and Lillian F. Wiegand; m. Marie L. Martinez, July 8, 1977; 1 child, Anthony L. BA in Biology, Metro. State U., 1980. Flow cytometrist U. Colo., Denver, 1980-82; rsch. assoc. Coulter Electronics, Hialeah, Fla., 1982-85; biologist Nat. Inst. Environ. Health Sci., Durham, N.C., 1985-89; sr. rsch. assoc. Coulter Corp., Hialeah, 1987-89; rsch. assoc. Johns Hopkins Sch. Medicine, Balt., 1989—; cons. U. Miami, Fla., 1991. With USMC, 1968-72, Vietnam. Mem. AAAS, Am. Indian Sci. and Engring. Soc., Soc. Analytical Cytology. Achievements include research in measurement of cellular RNA, DNA and protein multi-emission laser microfluorometry. Home: 308 Robin Hood Rd Havre De Grace MD 21078

WIEGAND, STANLEY BYRON, engineer; b. Santa Monica, Calif., May 18, 1955; s. Frank Byron and Jean Wiegand. Student, Calif. State U., Northridge, 1986—. Engr. Metro-Goldwyn-Mayer, Culver City, Calif., 1977-88, Lorimar Telepictures, Culver City, 1989-90, Sony/Columbia Pictures, Culver City, 1991—. Mem. tech. subcom. Acad. Motion Picture Arts and Scis., Beverly Hills, Calif., 1979-80. Mem. U. So. Calif. Cinema/TV Alumni Assn. Home: 1828 Midvale Ave # 3 Los Angeles CA 90025 Office: Sony/Columbia Studios 10202 W Washington Blvd Culver City CA 90232

WIEGAND, THOMAS EDWARD VON, psychophysicist, consultant; b. Ridgewood, N.Y., Nov. 29, 1960; s. Edward Charles and Geraldine Jane (Karle) W. AB, Franklin and Marshall Coll., 1987; BA, Columbia U., 1987, MA, 1990, MPhil, 1991, PhD, 1993. Pres. SEI, Lancaster, Pa., 1980-82; rschr. psychophysics lab. Columbia U., N.Y.C., 1987-91, rschr. vision lab., 1991-93; rschr. rsch. lab. electronics MIT, 1993—; cons. Spectrum Physics Cons., Acton, Mass. Author: Multiple Moduli and Payoff Functions in Psychophysical Scaling, 1991; contbr. articles to profl. jours. Mem. IEEE, AAAS, Assn. for Rsch. in Vision and Ophthalmology, N.Y. Acad. Scis., Sigma Pi Sigma, Sigma Xi. Achievements include design of system for scaling experiments, computational model of light adaptation dynamics; research in virtual environments and teleoperators. Office: MIT Rsch Lab Electronics 50 Vassar St # 36-755 Cambridge MA 02139

WIEGHARDT, KARL ERNST, chemistry educator; b. Göttingen, Germany, July 25, 1942; s. Karl Eugen and Elisabeth (Klinkenborg) W.; m. Gertraud Willfahrt, Apr. 26, 1941; 1 child, Jan. PhD, U. Heidelberg, 1969. Lectr. U. Heidelberg, Fed. Republic Germany, 1974-75; assoc. prof. U. Hannover, Fed. Republic Germany, 1975-80; prof. Ruhr-U., Bochum, Fed. Republic Germany, 1980—. Mem. Am. Chem. Soc., Royal Soc. Chemistry, Gesellschaft Deutscher Chemiker. Office: Ruhr-U, D-4630 Bochum Germany

WIEGMAN, LENORE HO, chemist; b. Shanghai, Republic of China, Aug. 18, 1931; came to U.S., 1951; d. Molin and Wan (Chuck) Ho; m. Hans Wiegman, Mar. 17, 1961; 1 child, Elkan Douglas. BS, Mich. State U., 1955, MS, 1958; PhD, U. Amsterdam, 1979. Rsch. asst. U. Wash., Seattle, 1960-61; rsch. chemist U. Columbia, N.Y.C., 1961-62, Am. Cyanamid, Princeton, N.J., 1962-65, AKZO Chem. Co., Amsterdam, 1969-71, U. Amsterdam, 1971—; mem. Netherland Corrosion Centrum, Holland, 1984—; mem. discussion group electo. Chem., Holland, 1974—. Contbr. articles to profl. jours. including Jour. of Dental Rsch. and Jour. of Dentistry. Mem. Am. Chem. Soc., Women's Internat. Network, Sigma Delta Epsilon, Iota Sigma Pi, Pi Mu Epsilon. Achievements include a patent for a method of manufacturing amonium polyphosphates, to increase the expansion of plaster of paris for casting purposes and to improve plaster of paris. Home: Bosplaat 22, 1025AT Amsterdam The Netherlands

WIELAND, DAVID HENRY, aerospace structural engineer; b. Moline, Ill., July 7, 1960; s. Robert Joseph and Geraldine Louise (McIntire) W.; m. Ida Cruz, Sept. 20, 1986; children: Seth, Cory. BS in Aero. Engring., U. Ill., 1985. Rsch. engr. Southwest Rsch. Inst., San Antonio, 1989-92, sr. rsch. engr., 1992—. Author procs. profl. confs. Capt. USAF, 1985-89. Mem. AIAA (structure tech. com. 1991, editor newsletter). Achievements include research on prediction of fatigue life of cold worked fastener holes using fracture mechanics. Office: Southwest Rsch Inst 6200 Culebra Rd San Antonio TX 78228

WIELICZKA, DAVID MICHAEL, physics educator, researcher; b. Kansas City, Kans., Oct. 31, 1955; s. Joseph Raymond and Katherine (Super) W.; m. Mary Louise Gast, Sept. 12, 1981; children: Amy, Brian, Kristen. BA, Benedictine Coll., 1977; PhD, Iowa State U., 1982. Rsch. fellow Ames (Iowa) Lab., 1982-84; asst. prof. U. Mo., Kansas City, 1984-91, assoc. prof., 1991—; cons. ThermoDynamics, Inc., Merriam, Kans., 1987—; mem. adv. bd. Kansas City Mus., 1989—. Author: (chpt.) Optical Constants of Materials II, 1990; contbr. articles to profl. jours. Recipient Rsch. grants, NIH, 1990, 92, Chem. Rsch. and Devel. Ctr., 1985, 89. Mem. Am. Phys. Soc., Am. Assn. Physics Tchrs., Optical Soc. Am. Office: Univ Mo Kansas City 1110 E 48th Kansas City MO 64110

WIEMAN, CARL E., physics educator; b. Corvallis, Oreg., Mar. 26, 1951; m. Sarah Gilbert. BS, MIT, 1973; PhD, Stanford U., 1977. Asst. rsch. physicist dept. physics U. Mich., Ann Arbor, 1977-79, asst. prof. physics, 1979-84; assoc. prof. physics U. Colo., Boulder, 1984-87, prof. physics, 1987—; fellow Joint Inst. for Lab. Astrophysics, Boulder, 1985—; Loeb lectr. Harvard U., 1990-91; Rosenthal Meml. lectr. Yale U., Columbia U., 1988. Fellow Guggenheim Found., 1990-91. Fellow Am. Phys. Soc.; mem. Optical Soc. Am. Office: U Colo JILA Boulder CO 80309-0440

WIEMANN, MARION RUSSELL, JR., biologist, microscopist; b. Chesterton, Ind., Sept. 7, 1929; s. Marion Russell and Verda (Peek) W.; 1 child from previous marriage, Tamara Lee (Mrs. Donald D. Kelley). BS, Ind. U., 1959; PhD (hon.), World U. Roundtable, 1991. Histo-rsch. technician U. Chgo., 1959, rsch. asst., 1959-62, rsch. technician, 1962-64; tchr. sci. Westchester Twp. Sch., Chesterton, Ind., 1964-66; with U. Chgo., 1965-79, sr. rsch. technician, 1967-70, rsch. technologist, 1970-79; prin. Marion Wiemann & Assocs., cons. R&D , Chesterton, Ind., 1979-89. Author: Tooth Decay, Its Cause and Prevention Through Controlled Soil Composition, 1985, The Mechanism of Tooth Decay, 1985; contbr. articles to profl. jours. and newspapers. Vice-chmn. The Duneland 4th of July Com., 1987-91. With USN, 1951-53. Recipient Disting. Tech. Communicator award Soc. for Tech. Communication, 1974, Internat. Order Merit, 1991; ennobled Royal Coll. Heraldry, Australia, 1991; named Sagamore of the Wabash Gov. Ind., 1985; McCrone Rsch. Inst. scholar, 1968. Fellow World Lit. Acad.; mem. Internat. Soc. Soil Sci., Internat. Graphoanalysis Soc., VFW (charter mem., bd. dirs., post judge adv. 1986-91, apptd. post adj. 1986-91, Cross of Malta 1986), Govs. Club. Address: PO Box 532 Chesterton IN 46304

WIEMER-SUMNER, ANNE-MARIE, psychotherapist, educational administrator; b. Ger., Mar. 3, 1938; came to U.S., 1949, naturalized, 1956; d. Franz and Margaret (Neubacher) Wiemer; BA, Hunter Coll., 1963; MA, N.Y. U., 1965; PhD Union Inst., 1989; cert. Psychoanalytic Individual and Group Therapy, Washington Square Inst. Psychotherapy, 1975, 76; m. Eric Eden Sumner, May 24, 1974 (dec.); children: Erika, Trevor. Adminstrv. asst., counselor, asst. chmn. admissions N.Y. U., N.Y.C., 1956-69; asst. dean student Hunter Coll., N.Y.C., 1969-71; dean students Cooper Union Advancement Art and Sci., N.Y.C., 1971—; supr. Washington Sq. Inst. for Psychotherapy, N.Y.C., 1977-81; pvt. practice psychotherapy, N.Y.C. Trustee Grace Ch. Sch., 1985-91, Greenwich Village Soc. for Historic Preservation, 1991—; pres. Washington Sq. Assn., 1987—. Mem. Coun. Psychoanalytic Psychotherapists, Am. Psychol. Assn., Am. Group Psychotherapy Assn., Am. Orthopsychiat. Assn., Internat. Assn. Group Psychotherapy, Nat. Accreditation Assn. and Am. Exam. Bd. Psychoanalysis, N.Y. State Assn. Practicing Psychotherapists, Coll. Placement Council, Eastern Coll. Personnel Officers, Megantic Fish & Game Corp. Home: 7-13 Washington Sq N New York NY 10003 Office: Cooper Union Cooper Sq New York NY 10003-7102

WIENERT, JAMES WILBUR, chemical engineer; b. Cherokee, Iowa, Nov. 1, 1945; s. Wilbur Louis and Winola (Wiese) W.; m. Betty Ann Lorenzen, Feb. 21, 1971; children: Kyle, Eric, Scott. BSChemE, Iowa State U., 1968. Prodn. engr. Chemplex Co., Clinton, Iowa, 1972-79, ethylene product supt., 1979-83; prin. process engr. Enron Chem. Co., Clinton, 1983-86; tech. engr. Texaco Chem. Co., Port Arthur, Tex., 1986-88, light olefins prod. supt., 1988-89; sci. prin. process engr. Quantum Chem., LaPorte, Tex., 1989-91; ethylene product supt. Quantum Chem., LaPorte, 1991—. Contbr. articles to profl. jours. With U.S. Army, 1970-71. United Methodist. Home: 732 Texas Ave Port Neches TX 77651-4306 Office: Quantum Chem 1515 Miller Cut Off Rd La Porte TX 77571

WIENKER, CURTIS WAKEFIELD, physical anthropologist, educator; b. Seattle, Feb. 3, 1945; s. Curtis Howard and Ruth (Daniels) W.; m. Cherie Kolbeck, Sept. 19, 1966 (div. 1987); 1 child, Heather. MA, U. Ariz., 1972, PhD, 1975. From instr. to prof. anthropology U. South Fla., Tampa, 1972—, assoc. dean coll. social and behavioral scis., 1988-90, assoc. dean Coll. Arts and Scis., 1990-93. Contbr. articles to sci. jours. Mem. Friends of Mus. of Sci. and Industry, Tampa,1 973—; trustee Nat. Kidney Found. Fla., Tampa, 1985-88. Chautauqua fellow NSF, 1974-75; grantee NSF, 1977-78, Fla. Regional Med. Program, 1975-76. Fellow Human Biology Coun., Am. Acad. Forensic Scis.; mem. Am. Assn. Phys. Anthropologists, Sigma Xi. Office: Univ South Fla Dept Anthropology 4202 E Fowler Ave Tampa FL 33620

WIENS, DARRELL JOHN, biologist, educator; b. Hutchinson, Kans., Feb. 9, 1950; s. Menno Simon and Mary Esther (Kroeker) W.; m. Arleen Louise Cook, May 18, 1975; children: Eric Cook-Wiens, Galen Cook-Wiens. BA, Bethel Coll., 1972; MA, U. Kans., 1975; PhD, Kans. State U., 1982. Instr. in biology Bethel Coll., North Newton, Kans., 1975-77; faculty fellow Fresno (Calif.) Pacific Coll., 1983-84; postdoctoral fellow Stanford (Calif.) Med. Sch., 1984-86; postdoctoral researcher Ohio Agrl. R&D Devel. Ctr., Wooster, 1986-88; asst. prof. biology U. No. Iowa, Cedar Falls, 1988—; adj. asst. prof. Kans. State U., Manhattan, 1992; adj. prof. Inst. Poly. Nacional, Mexico City, 1982-83; adv. bd. Allen Coll. Nursing, Waterloo, Iowa, 1991—; textbook reviewer Mosby-Yearbook, Inc., St. Louis, 1991. Contbr.: (chpt.) Cellular and Molecular Biology of Experimental Mammary Cancer, 1986, Progress in Developmental Biology, 1986; contbr. articles to European Jour. Cell Biology, Differentiation, Devel. Biology, Jour. Dairy Sci. Grantee Am. Heart Assn., 1990-92. Mem. Am. Soc. for Cell Biology, Fedn. Am. Soc. Exptl. Biology, AAAS, Iowa Acad. Scis., Sigma Xi. Mennonite. Achievements include research in early heart development in chick embryo, including actin transitions, extracellular matrix molecules, and acne drug isotretinoin, differentiation of mammary gland epithelial cells in co-culture with adipolytes, molecular biology of milk protein expression in murine and bovine mammary glands. Home: 2304 Tremont Cedar Falls IA 50613 Office: U No Iowa Dept Biology Cedar Falls IA 50614

WIERBOWSKI, CYNTHIA ANN, quality systems manager; b. Wilkes-Barre, Pa., July 9, 1957; d. Bernard John and Edna Elaine (Reeves) W.; 1 child, Michael John Macheska II. BA in Biology, East Stroudsburg U., 1979; MS in Biochemistry, U. Scranton, 1985. Quality control technician Akzo Salt Inc., Clarks Summit, Pa., 1979-81, chemist, 1981-85, rsch. and tech. chemist, 1985-88, quality assurance specialist, 1988-90, quality systems mgr., 1990—. Recipient Presdl. Bonus award for Outstanding Svc., Akzo Salt Inc., 1990. Mem. Am. Chem. Soc.

WIERMAN, JOHN CHARLES, mathematician, educator; b. Prosser, Wash., June 30, 1949; s. John Nathaniel and Edith Elizabeth (Ashley) W.; m. Susan Shelley Graupmann, Aug. 13, 1971; 1 child, Adam Christopher. BS in Math., U. Wash., 1971, PhD in Math., 1976. Asst. prof. math. U. Minn., Mpls., 1976-81; asst. prof. Johns Hopkins U., Balt., 1981-82, assoc. prof., 1982-87, prof., 1987—, chmn. math. scis. dept., 1988—; sr. rsch. fellow Inst. Applied Math. and its Applications, Mpls., 1987-88. Co-author: First-Passage Percolation on the Square Lattice, 1978; contbr. articles to profl. jours. Grad. fellow NSF, 1971-74; NSF rsch. grantee, 1976—. Fellow Inst. Math. Stats.; mem. Bernoulli Soc., Am. Math. Soc., Am. Statis. Assn., Math. Assn. Am., AAAS, Phi Beta Kappa. Office: Johns Hopkins U Dept of Scis. Math 34th & Charles Sts Baltimore MD 21218

WIERNIK, PETER HARRIS, oncologist, educator; b. Crocket, Tex., June 16, 1939; s. Harris and Molly (Emmerman) W.; m. Roberta Joan Fuller, Sept. 6, 1961; children: Julie Anne, Lisa Britt, Peter Harrison. B.A. with distinction, U. Va., 1961, M.D., 1965; Dr. h.c., U. of Republic, Montevideo, Uruguay, 1982. Diplomate Am. Bd. Internal Medicine (mem. sub-bd. med. oncology 1981-88). Intern Cleve. Met. Gen. Hosp., 1965-66, resident, 1969-70; resident Osler Svc. Johns Hopkins Hosp., Balt., 1970-71; sr. asst. surgeon USPHS, 1966, advanced through grades to med. dir., 1976; sr. staff assoc. Balt. Cancer Rsch. Ctr., 1966-71, chief sect. med. oncology, 1971-76, chief clin. oncology br., 1976-82, dir., 1976-82; assoc. dir. div. cancer treatment Nat. Cancer Inst., 1976-82; asst. prof. medicine U. Md. Sch. Medicine, Balt., 1971-74, asso. prof., 1974-76, prof., 1976-82; Gutman prof., chmn. dept. oncology Montefiore Med. Ctr., 1982—; head div. med. oncology Albert Einstein Coll., 1982—; assoc. dir. for clin. rsch.Albert Einstein Cancer Ctr., 1982—; cons. hematology and med. oncology Union Meml. Hosp., Greater Balt. Med. Center, Franklin Sq. Hosp.; bd. dirs. Balt. City unit Am. Cancer Soc., 1971-78, chmn. patient care com., 1972-75, mem. profl. edn. and grants com. N.Y.C. div., 1983-90, mem. nat. clin. fellowship com., 1984—; mem.

med. adv. com. Nat. Leukemia Assn., 1976-88, chmn. med. adv. com., 1989—; chmn. adult leukemia com. Cancer and Leukemia Group B, 1976-82; prin. investigator Eastern Coop. Oncology Group, 1982—; chmn. gynecol. oncology com., 1986-88, chmn. leukemia com., 1988—; sci. cons. Vt. Regional Cancer Ctr., 1987—. Editor: Controversies in Oncology, 1982, Supportive Care of the Cancer Patient, 1983, Neoplastic Diseases of the Blood, 1985, 2d edit., 1991; assoc. editor Medical Oncology and Tumor Pharmacotherapy, 1987-91, sr. editor, 1991—; co-editor: Year Book of Hematology, 1986—, Handbook of Hematologic and Oncologic Emergencies, 1988; N.Am. editor Jour. Cancer Research and Clin. Oncology, 1986-89; mem. editorial bd. Cancer Treatment Reports, 1972-76, Leukemia Research, 1976-86, 91—, Leukemia, 1986—, Cancer Clin. Trials, 1977—, Hosp. Practice, 1979—, Jour. Clin. Oncology, 1989-91; sect. editor antineoplastic drugs Jour. Clin. Pharmacology, 1985—; co-editor Am. Jour. Med. Scis., 1976-81; also articles, chpts. in books. Mem. editorial bd. PDQ Nat. Cancer Inst., 1987—. Recipient Z Soc. award U. Va., 1961, Byrd S. Leavell Hematology award U. Va. Sch. Medicine, 1965. Fellow AAAS, ACP, Am. Coll. Clin. Pharmacology, Internat. Soc. Hematology, Royal Soc. Medicine (London); mem. Am. Soc. Clin. Investigation, Am. Soc. Clin. Oncology (chmn. edn. and tng. com. 1976-79, 84, subcom. on clin. investigation 1980-82, program com. 1990, pub. issues com., 1990—), Am. Assn. Cancer Rsch., Am. Soc. Hematology, Am. Fedn. Clin. Rsch., Am. Acad. Clin. Toxicology, Internat. Soc. Exptl. Hematology, N.Y. Acad. Sci., Am. Soc. Hosp. Pharmacy, Am. Soc. Clin. Pharmacology and Therapeutics, Am. Radium Soc. (program com. 1987—, exec. com. 1988—, publ. com. 1988-92, sec. 1990-91, pres.-elect, 1992-93; pres. 1993—), Polish Oncology Soc. (hon.), Harvey Soc., Uruguan Hematology Soc. (hon.), Phi Beta Kappa (assoc. 1991—), Sigma Xi, Alpha Omega Alpha, Phi Sigma (award 1961). Home: 43 Longview Ln Chappaqua NY 10514-1304 Office: Montefiore Med Ctr 111 E 210th St Bronx NY 10467-2490

WIERSEMA, HAROLD LEROY, aerospace engineer; b. Erie, Ill., Sept. 17, 1919; s. Clarence John and Tena (Griede) W.; m. Joanne Kearney, Mar. 19, 1955; children: Roger Kent, Marilyn Tena. BS, U. Ill., 1949. Aerospace engr. Space Div. Rockwell Internat., Downey, Calif., 1953-78; sr. spl. engr. Boeing Mil. Airplane Co., Wichita, Kans., 1978-86; aerospace engr., avionics cons. Long Beach, Calif., 1986—. Com. chmn. Boy Scouts Am., Lynwood, Calif., 1968-70; pres. Compton (Calif.) Pacific Little League, 1966-69; deacon Presbyn. Ch., Southgate, Calif., 1965-70. Col. USAF, World War II, Korea. Decorated D.F.C., Air medal (5); recipient Mach Buster award N.Am. Aviation, Edwards AFB, 1963, Order of Arrow, Boy Scouts Am., 1967. Mem. IEEE (life), U.S. Air Force Assn., 388th Bomb Group/8th Air Force, Nat. Geog. Soc., UCLA Alumni Assn., Shriners. Republican. Presbyterian. Home: 5451 Jonesboro Way Buena Park CA 90621-1615

WIERZBICKI, JACEK GABRIEL, physicist, researcher; b. Lódz, Poland, Oct. 27, 1948; came to U.S., 1986; s. Gabriel Wiktor and Jadwiga Krystyna (Skarzynska) W.; m. Grazyna Maria Chawrona, Aug. 31, 1974; children: Grazyna, Przemystaw, Danuta, Kinga. MS in Physics, U. Lódz, 1971, MS in Math., 1973, PhD in Physics, 1981. Researcher U. Lódz, 1971-75; reseacher Joint Inst. for Nuclear Rsch., Dubna, USSR, 1975-79; med. physicist Oncological Ctr., Lódz, 1980-83; lectr. Fed. U. Tech., Bauchi, Nigeria, 1983-86; rsch. fellow Ohio U., Athens, 1986-88; asst. prof. U. Ky., Lexington, 1988-92; assoc. prof. Wayne State U., Detroit, 1993—; Russian translator Am. Inst. Physics, N.Y., 1987. Contbr. over 60 articles to sci. jours. Mem. Am. Phys. Soc., Am. Assn. Physicists in Medicine, Radiation Rsch. Soc., Sigma Xi. Roman Catholic. Achievements include rsch. in interactions of neutrons with light nuclei, in devel. of radium treatment in gynecology and radiation protection in the hospital, in interactions of 24 MeV neutrons with nuclei and optical model analysis, in use of Cf-252 in therapy of cancer, clinical radiobiology. Home: 3422 Shakespeare Dr Troy MI 48098 Office: Wayne State U Radiation Oncology Ctr 3990 John R St Detroit MI 48201

WIESEL, TORSTEN NILS, neurobiologist, educator; b. Upsala, Sweden, June 3, 1924; came to U.S., 1955; s. Fritz Samuel and Anna-Lisa Elisabet (Bentzer) W.; 1 dau., Sara Elisabet. MD, Karolinska Inst., Stockholm, 1954; D Medicine (hon.), Karolinska Inst, Stockholm, 1989; AM (hon.), Harvard U., 1967; ScD (hon.), NYU, 1987, U. Bergen, 1987. Instr. physiology Karolinska Inst., 1954-55; asst. dept. child psychiatry Karolinska Hosp., 1954-55; fellow in ophthalmology Johns Hopkins U., 1955-58, asst. prof. ophthalmic physiology, 1958-59; assoc. in neurophysiology and neuropharmacology Harvard U. Med. Sch., Boston, 1959-60; asst. prof. neurophysiology and neuropharmacology Harvard U. Med. Sch., 1960-64, asst. prof. neurophysiology, dept. psychiatry, 1964-67, prof. physiology, 1967-68, prof. neurobiology, 1968-74, Robert Winthrop prof. neurobiology, 1974-83, chmn. dept. neurobiology, 1973-82; Vincent and Brooke Astor prof. neurobiology, head lab. Rockefeller U., N.Y.C., 1983—, pres., 1992—; Ferrier lectr. Royal Soc. London, 1972; NIH lectr., 1975; Grass lectr. Soc. Neurosci., 1976; lectr. Coll. de France, 1977; Hitchcock prof. U. Calif.-Berkeley, 1980; Sharpey-Schafer lectr. Phys. Soc. London; George Cotzias lectr. Am. Acad. Neurology, 1983. Contbr. numerous articles to profl. jours. Recipient Jules Stein award Trustees for Prevention of Blindness, 1971, Lewis S. Rosenstiel prize Brandeis U., 1972, Friedenwald award Assn. Rsch. in Vision and Ophthalmology, 1975, Karl Spencer Lashley prize Am. Philos. Soc., 1977, Louisa Gross Horwitz prize Columbia U., 1978, Dickson prize U. Pitts., 1979, Nobel prize in physiology and medicine, 1981, W.H. Helmerich III award 1989. Mem. Am. Physiol. Soc., Am. Philos. Soc., AAAS, Am. Acad. Arts and Scis., Nat. Acad. Arts and Scis., Swedish Physiol. Soc., Soc. Neurosci. (pres. 1978-79), Royal Soc. (fgn. mem.), Physiol. Soc. (Eng.) (hon. mem.). Office: Rockefeller U Pres Office 1230 York Ave New York NY 10021

WIESENBERG, RUSSEL JOHN, statistician; b. Kaukauna, Wis., Apr. 9, 1924; s. Emil Martin and Josephine (Appelbaker) W.; m. Jacqueline Leonardi, Nov. 23; children: James Wynne, Deborann Donna. BS, U. Wis., 1951, postgrad. Cornell U., 1960-61, U. Mich., 1969, George Washington U., 1976. Analyst, Gen. Electric Co., West Lynn, Mass., 1951-56; specialist Internat. Gen. Electric Co., Rio de Janeiro, Brazil, 1956-59; statistician Gen. Motors Corp., Lockport, N.Y., 1959-65, sr. statistician, Harrison Radiator div., 1965-78, sr. reliability engr., 1978-82, sr. reliability statistician, 1982-87 . Auditor, Community Chest Fund, 1952-55; umpire Little League Baseball, 1962-65; committeeman Buffalo Area council Boy Scouts Am., 1962—, Cub Scout committeeman, 1962-64, Webelos cubmaster, 1963-64; mem. Nat. Congress Parents and Tchrs., 1963—; heart fund Vol. Heart Assn., 1968; tournament dir. Am. Legion Baseball, 1975; vol. United Way campaign, 1983, nat. telethon March of Dimes, 1983-84. Served with AUS, 1943-46. Decorated Bronze Star. Mem. AAAS, Am. Statis. Assn., Nat. Register Sci. and Tech. Pers., U. Wis. Alumni Assn., Artus, Internat. Platform Assn., Phi Kappa Phi. Lutheran. com.). Contbr. articles to profl. jours. Home: 14 Norman Pl Buffalo NY 14226-4233

WIESENDANGER, J(OHN) ULRICH, consulting engineer; b. Yonkers, N.Y., July 12, 1913; s. Ulrich and Grace (Percival) W.; m. Georgia Fuller, Jan. 21, 1954; children: John U. Jr., Peter S. CE, Cornell U., 1935; LLB, Fordham U., 1941. Registered profl. engr., N.Y., Mass., Vt., N.H., Maine: registered land surveyor, Maine. Structural engr. Elwyn E. Seelye and Co., N.Y.C., 1936-41; prin. civil engr. Tuttle, Seelye, Place and Raymond, N.Y.C., 1941-46, Seelye, Stevenson, Value and Knecht, N.Y.C. and Maine, 1946-51; prin. J.U. Wiesendanger, Cons. Engr., East Winthrop, Maine, 1951—. Author, editor: Data Books for Civil Engineers. Mem. ASCE (life), Am. Acad. Environ. Engrs. (life), Am. Cons. Engrs. Coun., Am. Inst. Steel Constrn., Am. Concrete Inst. Home and Office: PO Box 8 East Winthrop ME 04343

WIESNER, JEROME BERT, engineering educator, researcher; b. Detroit, May 30, 1915; s. Joseph and Ida (Friedman) W.; m. Laya Wainger, Sept. 1, 1940; children: Stephen Jay, Zachary Kurt, Elizabeth Ann, Joshua A. B.S., U. Mich., 1937, M.S., 1938, Ph.D., 1950. Assoc. dir. U. Mich. Broadcasting Service, 1937-40; chief engr. Acoustical Record Lab., Library of Congress, 1940-42; staff MIT Radiation Lab., Cambridge, Mass., 1942-45, U. of Calif. Los Alamos Lab., 1945-46; mem. faculty MIT, Cambridge, 1946-71, dir. research lab. of electronics, 1952-61, head dept. elec. engring., 1959-60, dean of sci., 1964-66, provost, 1966-71, pres., 1971-80, inst. researcher and prof., 1980—; spl. asst. to Pres. on sci. and tech., 1961-64; chmn. Pres.'s Sci. Adv. Com., 1961-64; chmn. tech. assessment adv. coun. Office Tech. Assessment, U.S. Congress, 1976-79; bd. dirs. Cons. for Mgmt., Inc., The Faxon Co., Magnascreen, Rothko Chapel. Author: Where Science and Politics Meet, 1965, ABM—An Evaluation, 1969. Bd. govs. Weizman Inst. Sci.; trustee Woods Hole Oceanographic Inst., Kennedy Meml. Trust; bd. overseers Harvard U., 1987-93; bd. dirs. MacArthur Found.; life mem. MIT Corp., Cambridge, Carnegie Commn. Recipient Vannevar Bush award NSF, 1992. Fellow IEEE, Am. Acad. Arts and Scis.; mem. NAS (Pub. Welfare medal 1993), Am. Philos. Soc., AAUP, Am. Geophys. Union, Acoustical Soc. Am., Nat. Acad. Engring., MIT Corp. (life), Sigma Xi, Phi Kappa Phi, Eta Kappa Nu, Tau Beta Pi. Home: 61 Shattuck Rd Watertown MA 02172-1310 Office: MIT 20 Ames St # 15207 Cambridge MA 02142-1308

WIESNER, ROBERT, aeronautical engineer; b. Phila., Mar. 21, 1936; s. Manuel and Sara (Kahan) W.; m. Libby Goldstein, June 28, 1959; children: Renae, Elyse, David. BSME, Drexel U., 1959, MS in Aero. Engring., 1963, MS in Engring. Mgmt., 1970. Registered profl. engr., Pa. Performance aero. unit chief Helicopters Div. Boeing Co., Phila., 1968-72; tech. model Model 179 Helicopters divsn. Boeing Corp., Phila., 1972-73, mgr. aero. dept., 1973-79, HXM project engr., 1979-82, project engr. Model 360, 1982-90, project mgr. Model 360, 1990-91, mgr. advanced systems, 1991—; bd. dirs. Pa. State U. Indsl. and Profl. Adv. Coun., University Park, 1986-90, 92—. Pres. Marple Summit Civic Assn., Delaware County, Pa., 1974, Marple Newtown Band Assn., Delaware County, 1981. 2nd lt. U.S. Army, 1959-60. Mem. AIAA, Army Aviation Assn. Am., Am. Helicopter Soc. (chpt. pres. 1989-90, regional v.p. 1992—, Howard Hughes award to Boeing Helicopter Model 360 team 1988). Achievements include patent for structure for cooling tail rotor gearbox. Home: 711 Winchester Rd Broomall PA 19008 Office: Boeing Def and Space Group P O Box 16858 Philadelphia PA 19142

WIEST, JOHN ANDREW, dentist; b. Cheverly, Md., Apr. 13, 1946; s. Louis Madox and Martha Elizabeth (Wilson) W.; m. Billie Ann O'Hern, Aug. 1968 (div. June 1981); children: Laurie Carrolin, Courtney Elizabeth; m. Dawn Michelle McCain, Oct. 13, 1984. Student, Bridgewater Coll., 1964-65; BSc, U. Md., 1965-68, postgrad., 1968-69, DDS summa cum laude, 1975. Lic. dentist, Fla. Med. technician USPH Hosp., Balt., 1972-75, Balt. County Hosp., 1973-75; assoc. dentist Richard R. Powell DDS, Tampa, Fla., 1975-79; ptnr. Richard R. Powell DDS, Tampa, 1979-88; pvt. practice Tampa, 1988—; cons. DentAll of Fla., Tampa, 1978-79. With U.S. Army, 1969-71, Vietnam. Mem. ADA, Acad. Gen. Dentistry, Fla. Dental Assn., West Coast Dental Assn., Hillsborough County Dental Assn., Psi Omega, Dental Frat. Alumno Assn., U. Md. Dental Sch. Alumni Assn., Elks, Tampa Handball Group (recruitment/devel. chmn.). Avocations: handball, running, fishing, gardening. Office: 315 W Busch Blvd Tampa FL 33612-7829

WIEWIOROWSKI, TADEUSZ KAROL, research chemist, consultant; b. Sopot, Poland, Nov. 3, 1935; came to U.S., 1957; s. Karol I. and Gertrude V. (Pluzkiewicz) W.; m. Mathilde Adele Emke, Nov. 22, 1962; children: Thaddeus Charles, Mathilde Wiewiorowski Young. BS in Chemistry, Loyola U. South, 1959; PhD in Chemistry, Tulane U., 1967. Chemist Freeport Sulphur Co., New Orleans, 1959-60, rsch. chemist, 1960-65, sr. rsch. chemist, 1965-69; asst. mgr. R&D Freeport Minerals Co., New Orleans, 1969-78, mgr. R&D, 1979-81, dir. R&D, 1981-86; v.p. R&D Freeport McMoran Inc., New Orleans, 1986-90; cons. New Orleans, 1991—; Editor: Sulfur Research Trends, 1972; contbr. various publs., book chpts., 1966-70; co-chmn. Mardis Gras Symposium on Sulfur Chemistry, 1971. Serves on La. Stimulus for Excellence in Rsch. com., Baton Rouge, 1988—. Mem. Am. Chem. Soc. (chmn. La. sect. 1971). Achievements include 25 patents, including 6 in sulfur technology, 5 relating to hydrometallurical leaching of nickel and cobalt, 8 relating to phosphoric acid technology, 2 related to uranium hydrometallurgy, 4 related to hydrometallurgy of gold. Home and Office: 2620 Danbury Dr New Orleans LA 70131

WIGGINS, GLENN B., entomologist; b. Toronto, Ontario, Canada, Jan. 29, 1927. BA, U. Toronto, 1949, MA, 1950, PhD, 1958. Asst. biologist Fisheries Rsch. Bd. Can., 1950-51, asst. curator dept. entomology, 1952-61, from asst. curator in charge to curator in charge, 1961-76; curator dept. entomology Royal Ontario Mus., 1964—; prof. zoology U. Toronto, 1968—; vis. prof. U. Minn., 1970, 72, 74, U. Mont., 1981. Recipient Entomological Soc. of Can. Gold Medal award, Entomological Soc. Can., 1992. Mem. Entomological Soc. of Am., N.Am. Benthol Soc. Achievements include research in systematic entomology especially trichoptera, aquatic entomology, biology of temporary pools, domiciliary invertebrates, evolution. Office: Royal Ontario Museum, Toronto, ON Canada M5S 2C6*

WIGGINS, ROGER C., internist, educator; b. Tetbury, Eng., May 26, 1945. BA, Cambridge U., Eng., 1968; BChir, Middlesex Hosp. Med. Sch., London, 1971, MB, MA, 1972. House physician dept. medicine The Middlesex Hosp., London, 1971-72; house surgeon Ipswich (Eng.) and East Suffolk Hosps., 1972; sr. house officer Hammersmith Hosp., The Middlesex Hosp., Brompton Hosp., London, 1972-74; rsch. registrar The Middlesex Hosp. Med. Sch., London, 1975-76; postdoctoral fellow Scripps Clinic and Rsch. Found., La Jolla, Calif., 1976-78, rsch. assoc., 1978-79, asst. mem. 1, 1979-81; asst. prof. U. Mich., Ann Arbor, 1981-84, assoc. prof., 1984-90, prof., 1990—, chief nephrology, 1988—, dir. O'Brien Renal Ctr., 1988—, dir. NIH Nephrology Tng. Program, 1988—; lectr., speaker in field. Author chpts. to books; assoc. editor: Jour. Am. Soc. Nephrology, Clin. Sci.; contbr. articles to profl. jours. First Broderip scholar, 1971, Harold Boldero scholar, 1971, James McIntosh scholar, 1971, The Berkeley fellow Gonville and Caius Coll., 1976; recipient Leopold Hudson prize, 1971, The William Henry Rean prize, 1971, Disting. Rsch. Jerome W. Conn award, 1984. Fellow Royal Coll. Physicians (U.K.); mem. Am. Assn. Pathologists, Am. Assn. Immunologists, Am. Soc. Nephrology (mem. subcom. on curriculum and specialty devel. 1990-92, mem. abstract rev. team 1991, mem. program com. 1992), Am. Fedn. Clin. Rsch., Am. Soc. Clin. Investigation, Ctrl. Soc. Am. Fedn. Clin. Rsch. Home: 3142 Park Ridge Dr Ann Arbor MI 48103 Office: U Mich Nephrology Div 3914 Taubman Ctr Ann Arbor MI 48109-0364

WIGHTMAN, ARTHUR STRONG, physicist, educator; b. Rochester, N.Y., Mar. 30, 1922; s. Eugene Pinckney and Edith Victoria (Stephenson) W.; m. Anna-Greta Larsson, Apr. 28, 1945 (dec. Feb. 11, 1976); 1 dau., Robin Letitia; m. Ludmila Popova, Jan. 14, 1977. B.A., Yale, 1942; Ph.D., Princeton, 1949; D.Sc., Swiss Fed. Inst. Tech., Zurich, 1968, Göttingen U., 1987. Instr. physics Yale, 1943-44; from instr. to asso. prof. physics Princeton, 1949-60, prof. math. physics, 1960-92; prof. emeritus, 1992—; Thomas D. Jones prof. math. physics Princeton, 1971—; vis. prof. Sorbonne, 1957, École Polytechnique, 1977-78. Served to lt. (j.g.) USNR, 1944-46. NRC postdoctoral fellow Inst. Teoretisk Fysik, Copenhagen, Denmark, 1951-52; NSF sr. postdoctoral fellow, 1956-57; recipient Dannie Heineman prize math. physics, 1969. Fellow Am. Acad. Arts and Scis., Royal Acad. Arts; mem. Nat. Acad. Scis., Am. Math. Soc., Am. Phys. Soc., AAAS, Fedn. Am. Scientists. Office: Princeton U 350 Jadwin Hall Princeton NJ 08544

WIGLER, MICHAEL H., molecular biologist; b. N.Y.C., Sept. 3, 1947. BA,, Princeton U., Princeton, 1970; MMS,, Rutgers U., Rutgers, 1972; Ph.D. in microbiology,, Columbia U., Columbia, 1978. With Cold Spring Harbor Lab., head Mammalian Cell Genetics Sect., 1978-; adj. prof. Columbia Univ. Dept. of Genetics, N.Y.C., 1988-. Recipient Pfizer Biomed. Rsch. award, 1986, G.H.A. Clowes Meml. award for cancer rsch. Am. Assn. for Cancer Rsch., 1991; Am. Bus. Cancer Found. grantee. Mem. NAS. Office: Cold Spring Harbor Lab PO Box 100 1 Bungtown Rd Cold Spring Harbor NY 11724

WIGNARAJAH, KANAPATHIPILLAI, plant physiologist, researcher, educator; b. Batticaloa, Sri Lanka, Dec. 26, 1944; came to U.S., 1988; s. Sinnathoamby and Nagaratnam (Nallathamby) K.; m. Asha Vasanti Ramcharan, Aug. 2, 1984; children: Avisha Nia, Amira Tari. BS in Botany, U. Ceylon, Colombo, Sri Lanka, 1969; PhD in Plant Physiology, U. Liverpool, Eng., 1974. Asst. lectr. in botany U. Ceylon, Sri Lanka, 1969-71; rsch. assoc. agronomy dept. U. Western Australia, 1974-75; lectr. U. Malawi, Africa, 1975-76, U. West Indies, Trinidad, 1976-84; sr. lectr. U. Guyana, S. Am., 1985-86; rsch assoc. U. Wales, Bangor, United Kingdom, 1986-87; rsch. assoc. Ctr. Nat. de la Recherche Sci., Montpellier, France, 1987-88, U. Okla., Norman, 1988-89, U. Calif., Santa Cruz, 1989-90; plant scientist The Bionetics Corp., Moffett Field, Calif., 1990—; cons. Nat. Inst. for Sci. and Tech., Georgetown, Guyana, 1985-86; Inter-Am. Inst. for Coop. in Agrl., 1985-86; reviewer for Tropical Agrl., 1980-86, Oecologia Plantarium, 1986-87, Environ. and Exptl. Botany, 1990—, Grant Proposals to NASA, 1991—. Contbr. articles to profl. jours. Sec. Ceylon Nat. Hist. Soc., Sri Lanka, 1969-71. Recipient Wheat Bd. Rsch. grant, Australian Reserve Bank, Nedlands, 1974, Swedish Guest scholarship The Swedish Inst., Stockholm, 1980, King Gustav Lectr. medal, U. Stockholm, 1980, Yamani Found. U. fellowship, U. Wales, 1986, European Econ. Commn., Centre Nat. de la Recherche Scientifique fellowship Montpelier, France, 1987. Fellow Indian Chem. Soc.; mem. Scandinavian Soc. Plant Physiologists, Phytochem. Soc. Europe. Hindu. Achievements include major findings on adaptation of plants to environmental stresses, such as salinity, waterlogging and anaerobiosis. Office: The Bionetics Corp NASA Ames Rsch Ctr Moffett Field CA 94035-1000

WIGNER, EUGENE PAUL, physicist, educator; b. Budapest, Hungary, Nov. 17, 1902; came to U.S., 1930, naturalized, 1937; s. Anthony and Elisabeth (Einhorn) W.; m. Amelia Z. Frank, Dec. 23, 1936 (dec. 1937); m. Mary Annette Wheeler, June 4, 1941 (dec. Nov. 1977); m. Eileen C.P. Hamilton, Dec. 29, 1979. Chem. Engr. and Dr. Engring., Technische Hochschule, Berlin, 1925; hon. D.Sc., U. Wis., 1949, Washington U., 1950, Case Inst. Tech., 1956, U. Chgo., 1957, Colby Coll., 1959, U. Pa., 1961, Thiel Coll., 1964, U. Notre Dame, 1965; D.Sc. (hon.), Technische Universität Berlin, 1966, Swarthmore Coll., 1966, Université de Louvain, Belgium, 1967; Dr.Jr., U. Alta., 1957; L.H.D. (hon.), Yeshiva U., 1963; hon. degrees, U. Liège, 1967, U. Ill., 1968, Seton Hall U., 1969, Cath. U., 1969, Rockefeller U., 1970, Israel Inst. Tech., 1973, Lowell U., 1976, Princeton U., 1976, U. Tex., 1978, Clarkson Coll., 1979, Allegheny Coll., 1979, Gustav Adolphus Coll., 1981, Stevens Inst. Tech., 1982, SUNY, 1982, La. State U., 1985. Asst. Technische Hochschule, Berlin, 1926-27, asst. prof., 1928-33; asst. U. Göttingen, 1927-28; lectr. Princeton U., 1930, part-time prof. math. physics, 1931-36; prof. physics U. Wis., 1936-38; Thomas D. Jones prof. theoretical physics Princeton U., 1938-71; on leave of absence, 1942-45; with Metall. Lab., U. Chgo., 1946-47; as dir. research and devel. Clinton Labs.; dir. CD Rsch. Project, Oak Ridge, 1964-65; Lorentz lectr. Inst. Lorentz, Leiden, 1957; cons. prof. La. State U., 1971-85, ret., 1985; mem. gen. adv. com. AEC, 1952-57, 59-64; mem. math. panel NRC, 1952-54; physics panel NSF, 1953-56; vis. com. Nat. Bur. Standards, 1947-51; mem. adv. bd. Fed. Emergency Mgmt. Agy., 1982-91. Author: (with L. Eisenbud) Nuclear Structure, 1958, The Physical Theory of Neutron Chain Reactors (with A.M. Weinberg), 1958, Group Theory and its Applications to the Quantum Mechanics of Atomic Spectra, 1931, English translation, 1959, Symmetries and Reflections, 1967, Survival and the Bomb, 1969. Decorated medal of Merit, 1946, Order of Banner of Republic of Hungary, Rubies, 1990; recipient Franklin medal Franklin Inst., 1950, citation N.J. Tchrs. Assn., 1951, Enrico Fermi award AEC, 1958, Atoms for Peace award, 1960, Max Planck medal German Phys. Soc., 1961, Nobel prize for physics, 1963, George Washington award Am. Hungarian Studies Found., 1964, Semmelweiss medal Am. Hungarian Med. Assn., 1965, Nat. Sci. medal, 1969, Pfizer award, 1971, Albert Einstein award, 1972, Golden Plate medal Am. Acad. Achievement, 1974, Disting. Achievement award La. State U., 1977, Wigner medal, 1978, Founders medal Internat. Cultural Found., 1982, Medal of the Hungarian Central Rsch. Inst., Medal of the Autonomous Univ. Barcelona, Am. Preparedness award, 1985, Lord Found. award, 1989; named Nuclear Pioneer, Soc. Nuclear Medicine, 1977, Colonel Gov. of La., 1983. Mem. AAAS, Royal Soc. Eng. (fgn.), Royal Netherlands Acad. Sci. and Letters, Am. Nuclear Soc. (first recipient Eugene P. Wigner award 1990), Am. Phys. Soc. (v.p. 1955, pres. 1956), Am. Math. Soc., Am. Assn. Physics Tchrs., Am. Acad. Arts and Scis., Am. Philos. Soc., Nat. Acad. Scis., N.Y. Acad. Scis. (hon. life mem.), Austrian Acad. Scis., German Phys Soc., Franklin Inst., Acad. Sci., Gottingen, Germany (corr.), Hungarian Acad. Sci. (hon.), Austrian Acad. Scis. (hon.), Hungarian L. Eötvös Phys. Soc. (hon.), Sigma Xi. Office: Princeton U Jadwin Hall Princeton NJ 08540

WIIG, KARL MARTIN, management scientist, systems engineer; b. Karasjok, Finnmark, Norway, Feb. 8, 1934; came to U.S., 1957; s. Alf Kristian and Margarethe (Soylann) W.; m. Elisabeth Hemmersam Nielsen, June 10, 1958; children: Charlotte Elisabeth, Erik Daniel. BS, Case Inst. Tech., 1959, MS, 1964. Researcher Chr. Michelsen Inst., Bergen, Norway, 1960-64; systems engr. GE, Cleve., 1964-66; mgr. systems engring. Dundee (Mich.) Cement Co., 1966-70; chmn. of the bd. Abacus Alpha, Inc., Newton, Mass., 1980-81; mgr. systems and policy analysis Arthur D. Little, Inc., Cambridge, Mass., 1970-80, dir. artificial intelligence, 1981-87; ptnr. Coopers & Lybrand, Dallas, 1987-89, The Wiig Group, Arlington, Tex., 1989—; presenter in field. Author: Expert Systems: A Manager's Guide, 1990, The Economics of Offshore Oil and Gas Supplies, 1977, (publs.) Managing Knowledge: Executive Perspectives, 1989, Knowledge-Based Systems and Issues of Integration, 1988, Management of Knowledge: A New Opportunity, 1988; contbr. articles to Mgmt. Rev., Asset-Based Finance Jour., The Bankers Mag., Computer, Annales de AICA. With Norwegian Army, 1953-54. Mem. Internat. Assn. Knowledg Engrs. (trustee 1990—). Achievements include patent in variable ratio power steering. Home and Office: The Wiig Group 7101 Lake Powell Dr Arlington TX 76016-3517

WIIK, BJÖRN H., physicist researcher, director. Staff physicist Deutsches Elektronen-Synchrotron (DESY), Hamburg, Germany, 1972-76; sr. scientist DESY, 1976-81; prof., physics U. Hamburg, Hamburg, Germany, 1981-93; dir.gen. DESY, 1993—, co-dir. HERA project. Office: Univ Hamburg Dept Physics, Edmund Siemers Allee 1, S-2000 Hamburg 13, Germany

WIJAYA, ANDI, clinical laboratory executive, clinical chemistry educator; b. Klaten, Central Java, Indonesia, July 2, 1936; s. Yantik and Kwan Eng (Sie) W.; m. Mariani Nursanti, Feb. 5, 1935; 1 child, Rini Mariani. MS in Pharmacy, Bandung Inst. Tech., Indonesia, 1962; PhD in Clin. Chemistry, U. Munster, West Germany, 1976; MBA, Kennedy Western U., U.S.A., 1986. Cert. clin. chemist. Researcher Pharm. Industry, Solo, Indonesia, 1963-66; dir. Pharmacist, Solo, Indonesia, 1966-68; asst. prof. Atmajaya U., Solo, Indonesia, 1968-73; exec. dir. Prodia Clin. Labs., Bandung, Indonesia, 1975—; assoc. prof. clin. chemistry Bandung Inst. Tech., 1980—, Pajajaran U., Bandung, Indonesia, 1983—; cons. Directory Lab. Services, Indonesian Ministry of Health, 1980-85. Mem. Indonesian Assn. Clin. Chemistry (founder, exec. bd. mem., Clin. Chemistry award 1986), Am. Assn. Clin. Chemistry, Clin. Ligand Assay Soc., Clin. Lab. Mgmt. Assn., N.Y. Acad. Scis., Internat. Soc. Clin. Enzymology; fellow Nat. Acad. Clin. Biochemistry. Christian. Avocations: orchids, ferns. Office: Prodia Clin Labs, Wastukencana 38, 40116 Bandung Indonesia

WIJESUNDERA, VISHAKA, civil engineer, scientist; b. Galle, Sri Lanka, Apr. 2, 1962; came to U.S., 1989; d. Wimal Premasiri Widanapathirana and Dacy Wickrama Guneratne; m. H.A.R.V. Wijesundera, Apr. 4, 1988. BS in Civil Engring., U. Moratuwa, Sri Lanka, 1987; MS in Civil Engring., U. New Orleans, 1991. Instr. engring. U. Moratuwa, 1987; demonstrator for engring. Open U., Sri Lanka, 1987-89; grad. rsch. asst. dept. civil engring. U. New Orleans, 1990-91, grad. rsch. asst. dept. urban and pub. affairs, 1992—. Mem. ASCE, Am. Water Resources Assn., Sigma Xi (assoc., cert. of recognition, 1992). Home: U New Orleans Box 917 New Orleans LA 70148

WIJEYSUNDERA, NIHAL EKANAYAKE, mechanical engineering educator, industrial consultant; b. Kandy, Sri Lanka, July 14, 1943; arrived in Singapore, 1980; s. Tikiri Banada and Manike (Ekanayake) W.; m. Kamani Chrisanthi Kobbekaduwa, June 8, 1972; children: Duminda Nalaka, Harindra Channa. BS in Engring., U. Ceylon, Peradeniya, Sri Lanka, 1967; MS, U. Birmingham, Eng., 1969, PhD, 1972. Asst. lectr. U. Ceylon, Peradeniya, 1967-72, lectr., 1972-76, sr. lectr., 1976-78; asst. prof. mech. engring. Drexel U., 1978-80; sr. lectr. Nat. U. Singapore, 1980-86, coord. postgrad. program, mech. engring. dept., 1984—, assoc. prof., 1986—; indsl. cons. thermal engring., 1980—; prin. investigator funded projects, 1980—; external moderator for exams. U. Sri Lanka, Inst. Engrs. Sri Lanka, 1980—; reviewer rsch. proposal NSF, U.S.A., 1990. Reviewer tech. papers for profl. publs., 1980—; contbr. over 70 papers to profl. jours. and conf. proceedings (numerous citations). Mem. ASME, ASHRAE (bd. dirs. Singapore chpt. 1991—, coun. mem. 1991—), Internat. Solar Energy Soc., Inst. Engrs. Singapore (sec. mech. and elect. tech. com. 1987-88). Avocations: running, swimming, reading, music. Office: Nat U, Dept Mech Engring, Kent Ridge 0511, Singapore

WIKARSKI, NANCY SUSAN, information technology executive; b. Chgo., Jan. 26, 1954; d. Walter Alexander and Emily Regina (Wejnerowski) W.; m.

Michael F. Maciekowich, Dec. 5, 1976 (div. Feb. 1985). BA, Loyola U., Chgo., 1976, MA, 1978; PhD, U. Chgo., 1990. Paralegal Winston & Strawn, Chgo., 1978-79; real estate analyst Continental Bank, Chgo., 1979-84, systems analyst, 1984-88, ops. officer, 1988-89, automation cons., 1989-92; systems mgr. Sears Mortgage Banking Group, Vernon Hills, Ill., 1992—. Author: German Expressionist Film, 1990. Fellow U. Chgo., 1987-90. Mem. NAFE, Am. Mensa, Chgo. Computer Soc., Alpha Sigma Nu. Avocation: music. Office: Sears Mortgage Banking Gp 440 N Fairway Dr Vernon Hills IL 60061

WIKER, STEVEN FORRESTER, industrial engineering educator; b. Alhambra, Calif., Sept. 29, 1952; s. Bruce Forrester and Joan (Centers) W.; m. Jody Louise Wiker, Jan. 24, 1976; children: Douglas Forrester, James McCallum. BS in Physiology, U. Calif., Davis, 1975; MS in Biol. Scis., Washington U., 1981; MS in Indsl. Engring., U. Mich., 1982, PhD in Indsl. Engring., 1986. Rsch. project officer USCG, Washington, 1976-79; rsch. asst. U. Mich., Ann Arbor, 1979-86; rsch. engr. Naval Ocean Systems Ctr. Lab., Kailua, Hawaii, 1986-87, Naval Ocean Systems Ctr., San Diego, 1987-88; asst. prof. indsl. engring. U. Wis., Madison, 1988-93, head indsl. ergonomics rsch. lab., 1989-93; assoc. prof.dept. environ. health U. Wash., Seattle, 1993; faculty indsl. engring. Dept. Environ. Health, U. Wash., Seattle, 1993—; sr. rsch. engr. James Miller Engring., Inc., Ann Arbor, 1981-88; dirs. telerobotics lab. Wis. Ctr. for Space Automation and Robotics, 1991-93. Contbr. articles to profl. jours. Comdr. USCGR, 1976—. Recipient Achievement medal USCG, 1988, 92, Humanitarian Svc. medal, 1993; fellow Ford Motor Co., Detroit, 1983-86; grantee Nat. Inst. Occupational Safety and Health, Washington, 1979-84, NASA, Ctrs. for Disease Control. Mem. Am. Soc. Safety Engrs., Inst. Indsl. Engrs., Internat. Soc. Biomechs., Aerospace Med. Soc., Human Factors Soc., Res. Officers Assn., N.Y. Acad. Scis., Sigma Xi, Alpha Pi Mu. Avocations: pvt. pilot, amateur photography. Office: Univ of Wash SC-84 Dept Environ Health Seattle WA 98195

WIKLUND, K. LARS C., anesthesiologist, scientist, educator; b. Uppsala, Sweden, Oct. 10, 1943; s. Knut and Elisabeth (Rabenius) W.; m. Ulla B. Anderson; children: Per K.E., Clara M.E. MD, Uppsala U., 1969, PhD, 1975. Clin. researcher dept. anesthesiology Univ. Hosp. Uppsala, 1969-75, asst. prof., then assoc. prof. neuroanesthesia, 1975-86, prof., chmn. dept. anesthesiology and intensive care, 1986—. Contbr. articles to med. jours. Recipient Hon. prize in acute medicine Oslo U., 1991. Fellow Royal Coll. Anesthetists; mem. Swedish Soc. Anesthesia and Intensive Care (bd. dirs. 1985-91), European Acad. Anesthesiology, Am. Soc. Anesthesiologists. Home: Sveavägen 2, S 75236 Uppsala Sweden

WIKMAN, GEORG KARL, institution administrator; b. Helsingborg, Sweden, Aug. 15, 1943; s. Folke Karl and Hanna Linnea (Pramberg) W.; matriculation exam. Ulricehamn Coll., 1963; BSc in Math., U. Gothenburg, 1968, MSc, 1970, MSc in Physics, 1972. Lectr. dept. math. U. Gothenburg, 1965-66; lectr. Uddevalla Coll., 1970-72; lectr. dept. theoretical physics U. Lund, 1973-74; dir. Swedish Herbal Inst., Gotenburg, 1975—. Mem. Am. Soc. Pharmacognosy, Inst. for Advanced Rsch. in Asian Sci. and Medicine, Inst. Noetic Scis., Gesellschaft für Arzneipflanzenforschung. Office: Swedish Herbal Inst, Viktoriagatan 15, 411 25 Gothenburg Sweden

WIKTOR, PETER JAN, engineer; b. Astrida, Rwanda, Oct. 12, 1956; came to U.S., 1960; s. Tadeusze Jan and Anna (Krzyzanowska) W.; m. Deirdre Ruth Meldrum, Aug. 19, 1989. BS, U. Pa., 1978; MS, Rensselaer Polytech., 1984; PhD, Stanford U., 1992. Engr. McDonnell Douglas, Long Beach, Calif., 1978-80; design engr. Hughes Helicopter, L.A., 1982; rsch. asst. Rensselaer Polytech. Inst., Troy, N.Y., 1982-84; engr. GM Rsch. Lab., Detroit, 1983, Jet Propulsion Lab., Pasadena, Calif., 1984-87; rsch. asst. Stanford (Calif.) U., 1987-92. Contbr. articles to profl. jours. Mem. Am. Soc. Mech. Engrs., Am. Astron. Soc., Sigma Xi. Achievements include briefing of NASA on reactionless precision pointing actuator, rotating transformer equations, and thruster systems for liquid helium cooled spacecraft. Home: 3226 NE 87th St Seattle WA 98115

WIKTOROWICZ, JOHN EDWARD, research biochemist; b. Nairobi, Kenya, Dec. 23, 1949; came to U.S., 1951; s. Janusz Stanislaw and Cristina Bronislawa (Dziedzic) W.; m. Michelle Zgonina, Feb. 12, 1972; children: Alexis, Sloane, Conner. BS, Ill. Inst. Tech., 1974; PhD, U. Tex. Med. Br., Galveston, 1978. Teaching and rsch. asst. U. Tex. Med. Br., 1975-78; rsch. fellow Calif. Inst. Tech., Pasadena, 1978-81; assoc. rsch. fellow Scripps Cinic and Rsch. Found., La Jolla, Calif., 1981-82; asst. prof. Va. Poly. Inst. and State U., Blacksburg, 1982-89; R&D chemist Applied Biosystems Inc., Foster City, Calif., 1989-90, supr. R&D, 1990-91, group leader R&D, 1991—; cons. Granada Television, Ltd., Manchester, Eng., 1980, Automated Dynamics Corp., Laguna Hills, Calif., 1981-85, Am. Alaskan Malamute Breeders Assn., 1986-89. Contbr. to profl. publs. Rep. sch. site coun. Santa Teresa Elem. Sch., San Jose, Calif., 1991—. With USNR, 1971-77. Damon Runyon rsch. fellow Damon Runyon/Walter Winchell Cancer Rsch. Fund, 1978-80; recipient Outstanding Rsch. award Nat. March of Dimes, 1977. Mem. Protein Soc., Am. Soc. Biochemistry and Molecular Biology. Achievements include patents for flow-rate controlled surface charged coating for capillary electrophoresis, and capillary electrophoresis method with polymer tube coatins. Office: Applied Biosystem Inc 850 Lincoln Centre Dr Foster City CA 94404

WILBER, JOHN FRANKLIN, endocrinologist, educator; b. Bronxville, N.Y., Oct. 16, 1935; s. Franklin Morrow Wilber and Edith (Shephard) Smith; m. Joan Eddleman, Jan. 18, 1974; children: Margaret, Jennie; children by previous marriage: Douglas, Damon. BA, Amherst Coll., 1957; MD, Harvard U., 1961. Diplomate Am. Bd. Internal Medicine. Intern Peter Bent Brigham Hosp., Boston, 1960-62, resident, 1962-63; resident medicine Barnes Hosp., St. Louis, 1965-66; from asst. prof. to prof. medicine Northwestern U., Chgo., 1968-74; chief sect. endocrinology, acting head dept. medicine La. State U. Med. Ctr., New Orleans, 1974-89; prof., head div. endocrinology U. Md. Med. Ctr., Balt., 1989—. Contbr. chpts. to textbooks, articles to peer-reviewed jours. Vestryman St. David's Ch., Balt., 1992—. Col. USPHS, 1965-68. Grantee NIH, VA. Mem. Am. Thyroid Assn. (pres. 1990-91), Endocrine Soc. (program chmn. 1991-92), Assn. Am. Physicians, Am. Soc. Clin. Investigators, Md. Club, Phi Beta Kappa, Alpha Omega Alpha. Democrat. Episcopalian. Achievements include development of first human thyrotropin immunoassay, discovery of new functions for thyroid-releasing hormone, cloning of human prepro TRH gene, patent in field. Home: 3704 N Charles St Baltimore MD 21218 Office: Univ Md Med Ctr 22 S Greene St Baltimore MD 21201

WILBUR, KARL MILTON, zoologist, educator; b. Binghamton, N.Y., May 7, 1912; married, 1946; 2 children. B.A., Ohio State U., 1935, M.A., 1936; Ph.D., U. Pa., 1940. Asst. zoologist Ohio State U., 1935-36, instr. zoology, 1941-42; instr. U. Pa., 1939-40; Rockefeller fellow NYU, 1940-41; asst. prof. physiology Dalhousie U. Med. Sch., 1942-44; assoc. prof. Duke U., Durham, N.C., 1946-50, prof., from 1950, now James B. Duke emeritus prof. zoology; physiologist AEC, 1952-53; vis. prof. U. de Sao Paulo, 1965, OAS, Guayaquil, Ecuador, 1973; program advisor Ford Found., Caracas, Venezuela, 1967-69; adv. panel com. on biology ORNL, 1950-56; mem. vis. com. Woods Hole Oceanographic Inst., 1966-67; cons. Ford Found., 1969-70, Nat. Coun. Marine Resources and Engring. Devel., 1969, Fgn. Area Fellowship Program, 1969; mem. sci. rev. com. U. N.C., 1970-80; mem. adv. com. Coun. for Internat. Exch. of Scholars, 1977-80. Recipient dedication of Mechanisms of Biomineralization in Animals and Plants Biomineralization Symposium, 1980, Mechanisms of Calcification in Biol. Systems Symposium, 1983. Mem. Am. Physiol. Soc., Am. Soc. Naturalists, Soc. Gen. Physiology, Am. Soc. Zoologists (chmn. div. comparative physiology and biochemistry 1967-79), Phi Beta Kappa. Office: Duke U Dept Zoology 239 Biol Sci Durham NC 27706

WILBUR, PAUL JAMES, mechanical engineering educator; b. Ogden, Utah, Nov. 8, 1937; s. Earl Burton and Ada (James) W.; m. Twyla Beck Wilbur, June 8, 1960; children: Wendy Lee, Dagny Ann. BS, U. Utah, 1960; PhD, Princeton U., 1968. Registered profl. engr., Colo. Nuclear power engr. U.S. Atomic Energy Commn., Washington, 1960-64; prof. of mech. engring. Colo. State U., Fort Collins, 1968—; bd. dirs., sec. Ion Tech Inc., Fort Collins; cons., researcher NASA, Washington, 1970—. Author: Solar Cooling, 1977; contbr. numerous articles to profl. jours. Lt. USN, 1960-64. Mem. ASME (pres. local sect.), AIAA (tech. com., jour. editor). Home: 1500 Teakwood Ct Fort Collins CO 80525-1954 Office: Colo State U Mech Engring Dept Fort Collins CO 80523

WILBUR, ROBERT LYNCH, botanist, educator; b. Annapolis, Md., July 4, 1925; s. Ralph Sydney and Elizabeth Ellen (Lynch) W.; m. Jeanne Marie Doucette; children: Martha, Ralph, Ellen, Mark, Margaret, Lenore. BS, Duke U., 1946, MA, 1947; PhD, U. Mich., 1952. Asst. prof. botany U. Ga., Athens, 1952-53, N.C. State U., Raleigh, 1953-57; asst. prof. botany Duke U., Durham, N.C., 1957-63, assoc. prof., 1963-69, prof. dept. botany, systematic botanist, 1970—. Roman Catholic. Home: 2613 Stuart Dr Durham NC 27707-2835 Office: Duke U Dept of Botany Durham NC 27707

WILCHINS, SIDNEY A., gynecologist; b. Paterson, N.J., Feb. 2, 1940; s. Philip Aaron and Esther (Blake) W.; m. Carole Diane Brill, June 23, 1963, (div. Mar. 1985); children: Joan Helen, Edward Victor; m. Estelle Angel, Mar. 15, 1985; children: Jacqueline, Susan. BA in Biol. Scis., Rutgers U., 1961; MD, Georgetown U., 1965. Diplomate Am. Bd. Ob Gyn. Clin. instr. N.J. Med. Sch., Newark, N.J., 1971-73; clin. asst. prof. N.J. Med. Sch., Newark, 1973-78, clin. assoc. prof., 1978—; adj. rsch. prof. N.J. Inst. Tech., Newark, 1978—; assoc. dir. Pilgrim Med. Ctr., Montclair, N.J., 1982—; med. dir. Ultrasound Diagnostic Sch., Union, N.J., 1989-91, N.J. Menopause Found., 1992—; gynecol. cons. Organon/Akzo, 1991—. Author, editor: Cryosurgery and Medicine, 1990; contbr. articles to profl. jours. Lt. USNR, 1965-69. Fellow Am. Coll. Ob-Gyn., Am. Coll. Surgeons, N.Y. Acad. Medicine; mem. N.Y. Acad. Scis., Forensic Soc. Ob-gyn., Colonia Country Club. Achievements include patent pending on Intraperitoneal Hyperthermia Device, pregnancy conducto for labor software copyright; application of chaost level to analysis of labor physiology. Home: 154 Devon Rd Colonia NJ 07067-3205 Office: 14 E Westfield Ave Roselle Park NJ 87204

WILCOX, CHARLES JULIAN, geneticist, educator; b. Harrisburg, Pa., Mar. 28, 1930; s. Charles John and Gertrude May (Hill) W.; m. Eileen Louise Armstrong, Aug. 27, 1955; children: Marsha Lou, Douglas Edward. BS, U. Vt., 1950; MS, Rutgers U., 1955, PhD, 1959. Registered profl. animal scientist. Dairy farm owner, operator Charlotte, Vt., 1955-56; prof. U. Fla., Gainesville, 1959—; cons. in internat. animal agrl. various orgns., Great Britain, France, Sudan, Can., Mex., El Salvador, Ecuador, Brazil, Bolivia, Peru, Columbia, Venezuela, Dominican Republic, Sweden, Norway, 1965—. Author: (with others) Animal Agriculture, 1973, 2d edit., 1980, Improvement of Milk Production in Tropics, 1990; editor: Large Dairy Herd Management, 1978, 93. 1st Lt. U.S. Army, 1951-53, Korea. Recipient Disting. Svc. award Fla. Purebred Dairy Assn., 1985, Jr. Faculty award Gamma Sigma Delta, 1968, Sr. faculty award Gamma Sigma Delta, 1984, Internat. award for disting. svc. to agrl. Gamma Sigma Delta, 1987, 3 Campaign ribbons U.S. Army, Combat Infantry badge U.S. Army. Mem. Am. Dairy Sci. Assn., Am. Soc. Animal Sci., Brazilian Soc. Genetics (editorial bd 1979—), Am. Registry Profs. Animal Sci. (examining bd. 1987—), Fla. Holstein Assn. (pres. 1979), Fla. Guernsey Cattle Club (pres. 1974-76). Republican. Avocations: spectator sports (baseball, football, basketball, tennis). Office: Univ Fla Dairy Sci Dept Gainesville FL 32611

WILCOX, PAUL HORNE, academic administrator, researcher; b. N.Y.C., Oct. 2, 1950; s. Richard Leon and Madge Muncie (Horne) W.; m. Elizabeth Winston Wyman, Aug. 24, 1985. BA, Bennington (Vt.) Coll., 1979. Coord. Aspen (Colo.) Inst. for Humanistic Studies, 1979-81; pres. Aspen Internat. Assocs., 1981-84; mgr. Nordstrom Inc., L.A., 1984-86; pres. human engring. The Pacific Inst., Seattle, 1986-87; dir. ops. Tellsyn Group, L.A., 1987-88; chmn. Inst. for Mgmt. Studies, Seattle, 1988—; mem. grad. rev. bd. Antioch U., Seattle, 1990—; book reviewer Synapsia The Brit. Brain Jour., Marlow, Buckinghamshire, Eng., 1990—; advisor N.W. Ethics Inst., Seattle, 1990—; founding educator Gstaad Colloquium The Lastis Found., Gstaad, Switzerland, 1988; mem. adv. bd. Lumatron Corp., 1992—. Fundraiser United Way, L.A., 1986; trustee United Ch. of Christ, Seattle, 1989—, chmn. bd. trustees, 1991—; bd. dirs. Aspen Pub. Radio Sta. NPR, 1982; vol. Wash. Lit., 1990—; bd. dirs. Wash. State Lit. Hotline, 1992—; active World Affairs Coun., 1993—; chair Statewide Lit. Outreach, 1993. Rsch. fellow U. Pa., 1991. Mem. Brain Club (bd. dirs. Marlow-Buckinghamshire chpts. 1990—), Beaver Bay Club, Hillsboro Club. Avocations: skiing, hiking, music. Office: Inst for Mgmt Studies Ste 410 200 First Ave West Seattle WA 98119

WILCOX, RANO ROGER, psychology educator; b. Niagara Falls, N.Y., July 6, 1946; s. Howard Clinton and Phyllis Hope (Stevens) W.; m. Karen Lesley Thompson, Apr. 25, 1986; 1 child, Quinn Alexander. BA, U. Calif., Santa Barbara, 1968, MA in Math., 1976, PhD in Ednl. Psychology, 1976. Sr. rsch. assoc. UCLA, 1976-81; prof. psychology U. So. Calif., L.A., 1981—. Author: New Statistical Procedures for Social Sciences, 1987, Statistics for Social Sciences, 1993; assoc. editor Psychometrika; mem. editorial bd. 3 jours.; also over 100 articles. Recipient T.L. Saaty award Am. Jour. Math. & Mgmt. Scis., 1984. Fellow Am. Psychol. Soc., Royal Statis. Soc.; mem. Psychometric Soc., Am. Statis. Assn., Inst. Math. Stats., Biometric Soc., Am. Ednl. Rsch. Assn. Achievements include research on improved methods for comparing groups and measuring achievement; resistant measures of correlation and regression; substantial gains in power when testing hypotheses. Office: U So Calif Dept Psychology Los Angeles CA 90089-1061

WILCOX, RICHARD CECIL, information systems executive; b. Houston, Aug. 5, 1959; s. Cecil Taylor and Katherine (Keeble) W.; m. Connie Lee Clifton, Apr. 8, 1984; children: Barry Alvin, Ashley Elizabeth, Lindsay Catherine, Richard Clifton. BA in Behavioral Sci., Houston Bapt. U., 1982; MS in Mgmt., Sam Houston State U., Huntsville, Tex., 1987. Ordained minister to Bapt. Ch., 1979. Cons. Internat. Communication Mgmt., Houston, 1982-85, gen. mgr., 1985-86; pres., chief exec. officer Integrated Brokerage Co., Houston, 1986-88; asst. v.p., mgr. Tex. Commerce Bank Nat. Asn., Houston, 1988-90; mgr., admstr. info. systems Ernst and Young, Houston, 1990—; voting mem. communications steering com., Fin. Interchange, Inc., Houston, 1988-91. Lay counselor, dept. dir., Fist Bapt. Ch., Houston, 1984-88, Kingsland Bapt. Ch., Katy Tex., 1988-92. Mem. S.W. Communications Assn. (bd. dirs. legis. affairs com. 1988-89, bd. dirs., vice chmn. 1989-90), Houston Mus. Fine Arts. Republican. Avocations: classical literature, archaeology, music. Home: 4 Hartwick Ct Conroe TX 77304-1301 Office: Ernst and Young 1221 Mckinney St Ste 2400 Houston TX 77010-2007

WILCOX, RICHARD HOAG, information scientist; b. Wooster, Ohio, Sept. 23, 1927; s. Raymond Boorman and Hazel (Hoag) W.; m. Jean Balderston, May 13, 1950; children: Linda, Kathryn. BS in Elec. Engring., Lafayette Coll., Easton, Pa., 1951, Elec. Engr., 1955; M Engring. Adminstrn., George Washington U., Washington, 1964. Enlisted USN, 1945-47; commd. ensign USNR, 1951, advanced through grades to comdr., 1969; electronic scientist, ops. rsch. analyst U.S. Naval Rsch. Lab., Washington, 1951-58; ops. rsch. analyst, electronic engr. Office Naval Rsch., Washington, 1958-62, head info. systems br., acting dir. math. scis. div., 1961-68; head resource evaluation div. Exec. Office of Pres., Office Emergency Preparedness, Washington, 1968-69, head info. systems div., 1969-74; chief mil. affairs div. U.S. Arms Control and Disarmament Agy., Washington, 1974-75, chief arms transfer div., 1975-78, sr. scientist, 1978—; commr., dir. Commn. on Profls. in Sci. and Tech., Washington, 1967-92; asst. dir. computers and comm. Fed. Emergency Mgmt. Agy., Washington, 1981; vis. scholar Ctr. for Strategic and Internat. Studies, Georgetown U., Washington, 1983-84. Co-editor: Redundancy Techniques for Computing Systems, 1962, Computer and Information Sciences, 1964, Research Program Effectiveness, 1965; contbr. articles to profl. jours, chpts. to books. Recipient Superior Civiliam Svc. award Office Naval Rsch., 1966, Citation Pres. U.S., 1973, Superior Honor award U.S. Arms Control and Disarmament Agy., 1993. Mem. AAAS, George Washington U. Club, Sigma Xi, Tau Beta Pi. Achievements include patent for microwave multiplier device; devising info.-theoretic measure of randomness of human performance; mgmt. devel. and operation of first operational computer conferencing system; creation a variety of novel info. systems. Home: 5702 Cedar Bluff Pl Temple Hills MD 20748 Office: US Arms Control & Disarmament 320 21st St NW Washington DC 20451

WILCOX, ROGER CLARK, psychologist, researcher; b. Zanesville, Ohio, Apr. 1, 1934; s. Clark Lewis and Mildred Adelaide (O'Hara) W.; m. Joy Ann Barr, Nov. 2, 1956; children: Beth Hartigan, Wells Lewis, Judd O'Hara. BA, Ohio State U., 1959, MA, 1960; PhD, U. Tenn., 1968. Lic. psychologist, Ohio. Prin. Roger Wilcox, PhD, Inc., Zanesville, Ohio, 1976—; dir. adminstrn. Comprehensive Mental Health, Zanesville, 1974-76; prof. Ohio U., Zanesville, 1970-74; dept. chmn. Calif. Poly., St. Luis Obispo, 1968-70, Wilberforce (Ohio) U., 1966-68; indsl. cons. Visual Info. Inst., Xenia, Ohio, 1966-90; adminstrv. cons. Jefferson County Mental Health, Steubenville, Ohio, 1977-82. Author: Psychological Consequences of Being a Black American, 1971; contbr. more than 15 articles on verbal learning and verbal behavior to profl. jours. Budget chmn. United Way of Am., Zanesville, Muskingum County, 1970-73; co-founder Ohio Assn. Mental Health Dirs., Columbus, 1976. with U.S. Army, 1953-56. Fellow USPHS, 1960, U.S. Office Edn., 1968. Mem. APA, Am. Arbitration Assn., Ohio Psychol. Assn. Democrat. Home: 1054 Terrace Ct Zanesville OH 43701 Office: Roger Wilcox PhD Inc PO Box 8078 Zanesville OH 43702-8078

WILCOX, WALTER MARK, elementary particle physicist, educator; b. Chgo., Aug. 13, 1954; s. Marion Walter and Alice Kathryn (Lynch) W. BS with highest honors, So. Meth. U., 1975; PhD, UCLA, 1981. Assoc. researcher R&D Assocs., Marina Del Rey, Calif., 1979-81; vis. asst. prof. Okla. State U., Stillwter, 1981-83; rsch. assoc. TRIUMF, Vancouver, B.C., Can., 1983-85, U. Ky., Lexington, 1985-86; asst. prof. Baylor U., Waco, Tex., 1986-93, assoc. prof., 1993—. Contbr. articles to Phys. Review, Physics Letters, Nuclear Physics, 1983—. Bd. dirs. Waco (Tex.) Habitat for Humanity, 1991. Grantee: NSF, Washington, 1992. Mem. AAAS, Am. Phys. Soc., Phi Beta Kappa. Methodist. Achievements include first calculation of lattice hadron form factors (with Richard Woloshyn); innovation and application of lattice gauge techniques to particle phenomenology. Office: Baylor U Dept Physics Waco TX 76798-7316

WILCOX, WAYNE F., plant pathologist, educator, researcher; b. Newman, Calif., July 1, 1950; s. Donald Franklin and Sara Jane W.; m. Linda Theresa Pembroke, Nov. 16, 1969; children: Miranda Corrina, Holly Carolyn. BS in Plant Pathology (Pomology), U. Calif., Davis, 1977, MS in Plant Pathology, 1978, PhD, 1982. Asst. extension prof. U. Ky., 1982-84; asst. prof. Cornell U., N.Y. State Agrl. Expt. Sta., Geneva, 1984-90, assoc. prof., 1990—; dept. extension leader Cornell U., state liaison rep. Nat. Impact Assessment Program, mem. statewide fruit competitiveness and profitability com., IPM commodity com. fruit; rschr. in field. Contbr. numerous articles to profl. jours. Recipient CIBA-GEIGY award Am. Phytopathological Soc., 1993. Mem. Am. Phytopathological Soc. (sec. deciduous tree fruit disease workers 1984-85, chmn. 1985-86, editor jours. 1985-92, extension com. 1988-90), Am. Soc. Hort. Sci. Achievements include patents in biological control of phytophthora by trichoderma and by gliocladium; notable findings include elucidation of role of phytophthora species as casual agents of root rot of fruit crops, particularly their interaction with soil water status as it effects pathogenesis. Office: Cornell U NY State Agrl Expt Sta Dept Plant Pathology Geneva NY 14456

WILCZEK, ELMAR ULRICH, aviation professional; b. Dietenheim, Germany, Sept. 30, 1948; s. Egon Karl and Ursula Cäcilie (Rehmet) W. Diplomate in engring., Tech. U., Braunschweig, Germany, 1976; Dr.-Ing., Tech. U., Aachen, Germany, 1984. Researcher Tech. Univ, Aachen, 1976-85; project mgr. Dornier Luftfahrt GmbH, Friedrichshafen, Germany, 1986—, naval aviation profl., 1989—; lectr. seaplanes Fachhochschule, Aachen, 1986—; govt. cons. Fed. Naval Supreme Ct., Hamburg, Germany, 1989—. Contbr. articles to profl. jours., publs. 2d lt. Fed. German Army, 1967-70, lt. comdr. res. German Naval Air Arm, 1986—. Mem. AIAA, U.S. Naval Inst. (life), Deutsche Gesellschaft für Luft-und Raumfahrt, Gemeinschaft E-Stellen Travemünde, Assn. Naval Aviation (life). Achievements include promotion of the advanced amphibious aircraft with high seaworthiness for environ. protection of the seas, the wing-in-ground effect craft for sea transportation and new hydrofoil systems for boats and seaplanes. Office: Dornier Luftfahrt GmbH, D-88039 Friedrichshafen Germany

WILCZEK, FRANK ANTHONY, physics educator; b. Mineola, N.Y., May 15, 1951; s. Frank John and Mary Rose (Cona) W.; m. Elizabeth Jordan Devine, July 3, 1973; children: Amity, Mira. BS in Math., U. Chgo., 1970; MA in Math., Princeton U., 1971, PhD in Physics, 1973. Instr. Princeton (N.J.) U., 1973-74, asst. prof., 1974-76, assoc. prof., 1978-79, prof., 1980-81; prof. Inst. for Theoretic Physics, Santa Barbara, Calif., 1981-88, Inst. for Advanced Study, Princeton, 1989—; vis. fellow Inst. for Advanced Study, Princeton, 1977-78; vis. prof. Harvard U., 1987-88. Author: Longing for the Harmonies, 1988, Fractional Statistics and Anyon Superconductivity, 1990; contbr. articles to profl. jours. Recipient J.J. Sakurai prize Am. Phys. Soc., 1986; A.P. Sloan fellow, 1975-77, MacArthur fellow, 1982-87, Huttenback prof. U. Calif., Santa Barbara, 1984-88. Mem. NAS, AAS. Avocations: chess, music, logic puzzles. Home: 112 Mercer St Princeton NJ 08540-6827 Office: Inst Advanced Study Dept Natural Scis Olden Ln Princeton NJ 08540-4920

WILCZYNSKI, RYSZARD LESLAW, economist, educator; b. Cieszyn, Poland, Jan. 6, 1949; s. Stanislaw and Maria (Hruby) W.; m. Maria Jolanta Longawa, Jan. 26, 1980; 1 child, Dorota. MA in Econs., Warsaw Sch. Econs., 1972, PhD in Econs., 1977. Lectr. dept. econs. Warsaw Sch. Econs., 1972-92, prof. econs., dep. dir. Inst. Developing Economies, 1992-93; fin. counselor Embassy of Rep. Poland, Washington, 1993—; advisor Ministry of Fin., Warsaw, 1989—. Author: Education and Economic Development in Less Advanced Countries, 1989; contbr. articles to profl. jours. Rsch. fellow Alexanders von Humboldt Found., 1987-88, Friedrich Ebert Found., 1984. Mem. Soc. Humboldtiana Polonorum. Avocations: jazz music, movies, sports. Home: Marco Polo 4/53, 02-776 Warsaw Poland Office: Warsaw Sch Econs, Al Niepodlegeosci 162, 02-554 Warsaw Poland

WILD, HANS JOCHEN, systems engineering executive; b. Leipzig, Sachsen, Germany, Apr. 19, 1935; s. Hans Bruno and Annelise (Maurer) W.; m. Ute Brigitta Eberle, Mar. 4, 1961; children: Barbara, Anne, Hans. Diploma, Bonn U., Germany, 1958. Asst. researcher Bonn U., Germany, 1958-62; systems engr. IBM Germany, Stuttgart, 1962-66; dist. systems engring. mgr. IBM Germany, Essen, Duesseldorf, 1967—. Co-author: Endogene Process Systematik, 1964; contbr. articles to profl. jours. Mem. Internat. Neural Soc. Avocations: neural network research, recreational computing, jogging, 19th century impressionists. Home: Im Heinental 61, 72218 Wildberg Germany

WILDER, DAVID GOULD, orthopaedic biomechanics researcher; b. Neptune, N.J., July 24, 1952; s. Charles Moulton Gould and Christine Bayard (Clark) W.; m. Kathleen Mary Hill, July 11, 1981; children: Alison Bell, Braden Ames. MSME, U. Vt., 1978, PhD in Mech. Engring., 1985. Registered profl. engr., Vt. Grad. rsch. asst. orthopaedics dept. U. Vt., Burlington, 1974-81, rsch. asst., lab. tech. IV, orthopaedics dept., 1981-85, rsch. asst. prof. orthopaedics dept., 1985-91, rsch. assoc. prof. orthopaedics dept., 1991—; cons. Rehab. Techs., Inc., Burlington, 1989—; external reviewer Can. Med. Rsch. Coun., Toronto, 1987—. Contbr. articles to Jour. Occupational Medicine, Ergonomics, Bull. Hosp. Joint Diseases, Jour. Biomechanics, Jour. Biomech. Engring. Recipient Rsch. award Am. Back Soc., 1987, Volvo award, 1980, Vienna award, 1992; grantee NIH, Nat. Inst. Occupational Safety and Health, Orthopedic Rsch. and Edn. Found., Nat. Inst. Disability and Rehab. Rsch., Whitaker Found., 1983—. Mem. ASME (local exec. 1981-88), Am. Soc. Biomechanics (meeting chmn. 1988-89), Internat. Soc. Study Lumbar Spine (site selection com. 1990—, Volvo award 1980), Orthopaedic Rsch. Soc., Tau Beta Pi, Sigma Xi. Achievements include research on lower back pain and driving vehicles. Office: U Vt Rehab Stafford 4 Burlington VT 05405-0068

WILDER, JAMES ROBBINS, mechanical engineer; b. Sumter, S.C., Sept. 28, 1956; s. Arthur Harrison and Josephine (Williams) W.; m. Becky Lynn Jones, Jan. 17; children: Bobbie Jo, Katie Lynn. AA, U. S.C., 1979; BS in Math., Coll. of Charleston, 1979; BSME, Clemson U., 1982. Registered profl. engr., S.C., Ga. Master The Patterson Sch., Lenoir, N.C., 1979-80; engring. coop. Duke Power Co., Seneca, S.C., 1980-82; assoc. engr. Santee Cooper-S.C. Pub. Svc. Authority, Georgetown and Moncks Corner, S.C., 1982-85; design engr. Med. Coll. Ga., Augusta, 1985-87; sr. engr. C.T. Main Cons. Engrs., Aiken, S.C. and Charlotte, N.C., 1987-88; staff engr. E.I.

duPont de Nemours, Inc., Aiken, 1988-89; discipline engring. mgr. Westinghouse Savannah River Co., Aiken, 1989-90, Bechtel Savannah River, Inc., Aiken, 1990-91; chief engr. UGR Architects, Augusta, Ga., 1991-92; process engr. Monsanto, Augusta, Ga., 1992—. Assoc. Wycliffe Bible Translators, 1990. Recipient First Nat. Search award Johns Hopkins U., 1980. Mem. ASME. Pentecostal. Avocations: aviation, golf, sailing, horseback riding, nusic. Home: 410 Whaley Pond Rd Graniteville SC 29829 Office: Bechtel Savannah River Inc 703-45A Aiken SC 29808-0001

WILDT, DANIEL RAY, physicist; b. Boonville, Ind., May 22, 1956; s. Arthur Curtis and Betty Jean (Buster) W.; m. Janice Louise Laird, July 18, 1981; children: Michael Ryan, Blake Alan. BS, Murray State U., 1978; MS, Calif. State U., 1985. Test/software engr. Rockwell Internat., Canoga Park, Calif., 1978-80; laser physicist Rockwell Internat., Canoga Park, 1980-85; sr. scientist W.J. Schafer Assocs., Calabasas, Calif., 1985-91; program mgr. space based laser Dept. Def./Ballistic Missile Def. Orgn., Washington, 1991—. Mem. AIAA, Am. Def. Preparedness Assn.

WILES, ANDREW J., mathematican, educator; b. England, 1952; married; children. BS in mathematics, Oxford U, England; Ph.D. in mathematics, Cambridge U, England. Lecturer Inst Advanced Stud, Princeton, NJ; asst./assoc. prof. mathematics Harvard U, Cambridge, MA; prof. mathematics Princeton U, Princeton, NJ, 1982—. Achievements include discovery of solution for Pierre de Fermat's last theory of 1637. Office: Princeton U Dept Mathematics Princeton NJ 08544*

WILETS, LAWRENCE, physics educator; b. Oconomowoc, Wis., Jan. 4, 1927; s. Edward and Sophia (Finger) W.; m. Dulcy Elaine Margoles, Dec. 21, 1947; children—Ileen Sue, Edward E., James D.; m. Vivian C. Wolf, Feb. 8, 1976. B.S., U. Wis., 1948; M.A., Princeton, 1950, Ph.D., 1952. Research asso. Project Matterhorn, Princeton, N.J., 1951-53, U. Calif. Radiation Lab., Livermore, 1953; NSF postdoctoral fellow Inst. Theoretical Physics, Copenhagen, Denmark, 1953-55; staff mem. Los Alamos Sci. Lab., 1955-58; mem. Inst. Advanced Study, Princeton, 1957-58; mem. faculty U. Wash., Seattle, 1958—; prof. physics U. Wash., 1962—; cons. to pvt. and govt. labs.; vis. prof. Princeton, 1969, Calif. Inst. Tech., 1971. Author: Theories of Nuclear Fission, 1964, Nontopological Solitons, 1989, also over 165 articles. Del. Dem. Nat. Conv., 1968. NSF sr. fellow Weizmann Inst. Sci., Rehovot, Israel, 1961-62; Nordita prof. and Guggenheim fellow Lund (Sweden) U., also Weizmann Inst., 1976—; Sir Thomas Lyle rsch. fellow U. Melbourne, Australia, 1989; recipient Alexander von Humboldt sr. U.S. scientist award, 1983. Fellow Am. Phys. Soc., AAAS; mem. Fedn. Am. Scientists, AAUP (pres. chpt. 1969-70, 73-75, pres. state conf. 1975-76), Phi Beta Kappa, Sigma Xi. Club: Explorers. Research on theory of nuclear structure and reactions, nuclear fission, atomic structure, atomic collisions, many body problems, subnuclear structure and elementary particles. Office: U Washington Dept Physics FM 15 Seattle WA 98195

WILEY, DALE STEPHEN, mechanical engineer; b. Summit, N.J., Apr. 11, 1954; s. James Hundley and Francelle (Adams) W.; m. Debra Alaine Moeller, May 19, 1979; children: Joshua Hundley, Heather Alaine, Jennifer Rose. BSME, Milw. Sch. of Engring., 1977; MBA, U. Wis., Oshkosh, 1985. Registered profl. engr., Wis. Mech. engr. Gilbert Commonwealth Assocs., Inc., Jackson, Mich., 1977-79, Wis. Pub. Svc. Corp., Green Bay, Wis., 1979—. Chmn. Oconto (Wis.) Harbor Commn., 1981-87; mem. Green Bay Plan Commn., 1990—. Mem. ASME. Home: 2561 S Trillium Circle Green Bay WI 54313 Office: Wis Pub Svc PO Box 19002 600 N Adams St Green Bay WI 54301-9002

WILEY, DON CRAIG, biochemistry and biophysics educator; b. Akron, Ohio, Oct. 21, 1944; s. William Childs and Phyllis Rita (Norton) W.; m. Katrin Valgeirsdottir; 1 child, William Valgeir; children from previous marriage: Kristen D., Craig S. BS in Physics and Chemistry, Tufts U., 1966; PhD in Biophysics, Harvard U., 1971. Asst. prof. dept. biochemistry and molecular biology Harvard U., Cambridge, Mass., 1971-75, assoc. prof., 1975-79, prof. biochemistry and biophysics, 1979—, investigator Howard Hughes Med. Inst., 1987—; mem. biophys. chemistry study sect. NIH, 1981-85; Shipley Symposium lectr. Harvard Med. Sch., 1985, Peter A. Leermakers Symposium lectr. Wesleyan U., 1986, K.F. Meyer lectr. U. Calif., San Francisco, 1986, John T. Edsall lectr. Harvard U., 1987, Washburn lectr. Boston Mus. Sci., 1987, Harvey lectr. N.Y. Acad. Sci., 1988, XVI Linus Pauling lectr. Stanford U., 1989; rsch. assoc. in medicine Children's Hosp., Boston, 1990—. Contbr. numerous articles to profl. jours. Recipient Ledlie prize Harvard U., 1982, Louisa Gross Horwitz prize Columbia U., 1990, William B. Coley award Cancer Rsch. Inst., 1992, V.D. Mattia award, 1992, Passano Found. Laureate award, 1993, Emil von Behring prize, 1993; European Molecular Biology fellow, 1976. Mem. NAS (lectr. 1988), AAAS, Am. Acad. Arts and Scis., Am. Chem. Soc. (Nichol's Disting. Symposium lectr. N.E. sect. 1988), Am. Crystallographic Assn., Am. Soc. for Chemistry and Molecular Biology, Am. Soc. for Virology, Biophys. Soc. (Nat. lectr. 1989), Protein Soc. Office: Harvard U Dept Biochemistry & Molecular Biology 7 Divinity Ave Cambridge MA 02138-2092 also: Children's Hosp Lab of Molecular Medicine 320 Longwood Ave Boston MA 02115

WILEY, JASON LARUE, JR., neurosurgeon; b. Canandaigua, N.Y., Dec. 2, 1917; s. Jason LaRue and Eva Althea (Moore) W.; m. Alma Williams, Jan. 4, 1944 (div. Feb. 1956); children: Robert W., Richard L.; m. Ann Valentine Gerrish, Apr. 14, 1956 (div. July 1979); children: Martha V., Pamela M., Catherine A. Student, Antioch Coll., 1934-37; MD, Harvard U., 1941. Diplomate Am. Bd. Surgery, Am. Bd. Neurol. Surgery. Intern Kings County Hosp., Bklyn., 1941-42; asst. resident surgery Ellis Hosp., Schenectady, N.Y., 1948-49, from asst. to assoc. resident surgery Rochester (N.Y.) Gen. Hosp., 1949-51; from asst. to assoc. to chief resident neurosurgery Yale U. and Hartford Hosp., New Haven and Hartford Conn., 1951-54; practice medicine specializing in neurosurgery Kansa City, Mo., 1954-56, Rochester, 1956—; chief neurosurgery Rochester Gen. Hosp., 1959-71, emeritus neurosurgeon, 1989—; clin. asst. prof. neurosurgery U. Rochester, 1961-88. Mem. Bd. for Profl. Med. Conduct, N.Y. State Dept. Health, Albany, N.Y., 1985—. Served to lt. comdr. USN, 1942-47, PTO. Mem. Med. Soc. County Monroe, Med. Soc. State N.Y., N.Y. State Neurosurg. Soc. (bd. dirs. 1972-77), Congress Neurol. Surgeons, Am. Assn. Neurol. Surgeons. Republican. Club: Canandaigua Yacht. Avocations: sailing, skiing, fishing, genealogy. Office: 1445 Portland Ave Rochester NY 14621-3008

WILEY, JOHN EDWIN, cytogeneticist; b. Roanoke, Va., Mar. 2, 1951; s. James Edwin and Marie Rita (Cassell) W. BA, U. N.C., Greensboro, 1973, MA, 1976; PhD, N.C. State U., 1981. Diplomate Am. Bd. Med. Genetics-Clin. Cytogenetics. Biomed. researcher St. Paul's Coll., Lawrenceville, Va., 1981-82; postdoctoral trainee U. Wis., Madison, 1982-84; mem. faculty East Carolina U. Sch. Medicine, Greenville, N.C., 1984—. Contbr. articles to profl. jours. Biomed. rsch. support grantee United Way, Greenville, 1986-87, USPHS, Washington,1987-90. Mem. AAAS, Am. Soc. Human Genetics, Am. Soc. Zoologists, Am. Soc. Ichthyologists and Herpetologists. Democrat. Achievements include observation that certain genes on frog chromosomes seem to move frequently around, that chromosome constitution in many breast cancer tumors seems normal, that in some patients with ring X chromosomes the ring may not be turned off, that the addition of tumor promoting agents helps white blood cells in many vertebrates to divide, and that DNA sequences on ends of frog chromosomes are the same as those on the ends of human chromosomes. Home: 102 Hunters Ln Greenville NC 27834-8829 Office: East Carolina U Sch Medicine Moye Blvd Greenville NC 27858-4300

WILEY, SAMUEL KAY, chemical engineer; b. Chanute, Kans., Mar. 10, 1944; s. Ira S. and Edith B. (Rodgers) W.; m. Susan W. Wells, Feb. 4, 1967; 1 child, Sara K. BSChemE, U. Tex., 1967; MBA, U. Houston, 1974. Registered profl. engr., Tex., La. Pollution control supr. Monsanto Co., Texas City, 1967-68; regional sales mgr. John Zink Co., Houston, 1968-72; v.p. John Zink Co., Tulsa, 1982-85; prin. Resource Mgmt. Cons., Houston, 1972-82; pres. McGill Environ., Tulsa, 1985-90; dir. IT Corp., Tulsa, 1990-92; pres. Encor, Baton Rouge, 1992—; bd. dirs. Tulsa Authority for Recovery of Energy. Contbr. articles to profl. jours. Mem. AICE, Am. Petroleum Inst., Am. Waste Mgmt. Assn. Achievements include research on incineration of hazardous waste, hydrocarbon emissions from loading, controlling vapor losses, toxic waste incinerators. Office: Encor 12021 Lakeland Park Blvd Baton Rouge LA 70809

WILEY, WILLIAM RODNEY, microbiologist, administrator; b. Oxford, Miss., Sept. 5, 1931; s. William Russell and Edna Alberta (Threlkeld) W.; m. Myrtle Louise Smith, Nov. 10, 1952; 1 child: Johari. B.S., Tougaloo Coll., Miss., 1954; M.S., U. Ill., Urbana, 1960; Ph.D., Wash. State U., Pullman, 1965. Instr. electronics and radar repair Keesler AFB-U.S. Air Force, 1956-58; Rockefeller Found. fellow U. Ill., 1958-59; research assoc. Wash. State U., Pullman, 1960-65; research scientist dept. biology Battelle-Pacific N.W. Labs., 1965-69, mgr. cellular and molecular biology sect. dept. biology, 1969-72, inst. coordinator, life scis. program, assoc. mgr. dept. biology, 1972-74, mgr. dept. biology, 1974-79, dir. research, 1979-84; sr. v.p., dir. Pacific N.W. div. Battelle Meml. Inst., Richland, Wash., 1984—; adj. assoc. prof. microbiology Wash. State U., Pullman, 1968—; found. assoc. Pacific Sci. Ctr., Seattle, 1989—; bd. dirs. Sta. KCTS Channel 9, Seattle, 1990—; trustee Oreg. Grad. Inst. Sci. and Tech., 1990—; cons. and lectr. in field. Contbr. chpts. to books, articles to profl. jours. Co-author book in microbiology. Bd. dirs. Wash. Tech. Ctr., 1984-88, sci. adv. panel Wash. Tech. Ctr., 1984-88, Fed. Res. Bank of San Francisco (Seattle branch) 1991—; mem. adv. com. U. Wash. Sch. Medicine, 1976-79; trustee Gonzaga U., 1981-89, bd. regents, 1968-81; bd. dirs. MESA program U. Wash., Seattle, 1984-90, United Way of Benton & Franklin Counties, Wash., 1984—, Tri-City Indsl. Devel. Council, 1984-92; mem. Wash. Council Tech. Advancement, 1984-85; bd. dirs. Forward Wash., The Voice for Statewide Econ. Vitality, 1984—, N.W. Coll. and Univ. Assn. for Sci., 1985—; mem. Tri-City Univ. Ctr. Citizens Adv. Council, 1985—; apptd. Wash. State Higher Edn. Coordinating Bd., 1986-89, Wash. State U. Found., 1986-89; mem. Wash. State U. bd. Regents, 1989—; bd. dirs. Washington Roundtable, 1989—, Goodwill Games, 1989-90; mem. adv. coun. Mont. State Sci. and Tech., 1990-91; mem. bd. overseers Whitman Coll., 1990—; mem. external adv. bd. Clark Atlanta U., 1991—; mem. Cen. Wash. U. Inst. for Sci. and Society Bd. of Advisors, 1991—, Engring. exec. com. Southern U., Baton Rouge, La, 1992—; mem. Coun. Govt. Univ. Industry Rsch. Roundtable, 1993—; bd. trustees Fred Hutchinson Cancer Rsch. Ctr., 1992—; engring. exec. com. Southern U., 1992—. With U.S. Army, 1954-56. Mem. Am. Soc. Biol. Chemists, Am. Soc. Microbiology, AAAS, Soc. Exptl. Biology and Medicine, Sigma Xi. Office: Battelle Meml Inst Pacific NW Divsn Battelle Blvd Richland WA 99352

WILGEN, FRANCIS JOSEPH, mechanical engineer; b. Melrose, Minn., June 26, 1945; s. Ben and Marie Sabella (Billmeyer) W. BS in Agrl. Engring., U. Minn., 1967, MS in Agrl. Engring., 1968; MSME, U. Wis., 1975, PhD in Mech. Engr., 1977. Engring. coord. Am. Peace Corp., Kabul, Afghanistan, 1968-70; assoc. prin. engr. Gen. Mills, Inc., Mpls., 1977—. With U.S. Army, 1970-72. Mem. Instrument Soc. Am. Achievements include elucidation of the role of drying conditions in affecting the drying and structural properties of cereal foods; determination of the effects of flexible appendages on the whirl stability of rotating shafts. Home: 7200 York Ave S # 320 Edina MN 55435 Office: Gen Mills Inc 9000 Plymouth Ave N Minneapolis MN 55427

WILGOCKI, MICHAL, electrochemist, chemistry educator; b. Chorzow, Silesia, Poland, Sept. 18, 1947; s. Waclaw and Czeslawa Helena (Bloch) W. M of Chemistry, High Pedagogical Sch., Opole, Poland, 1971; D of Chemistry, U. Wroclaw, 1976. vis. scientist U. Copenhagen, 1981. Author: Polarographic Determination of Stability Constants, 1982 (award 1983), Interpretation of Stability Constants Obtained from Electrometric Data for Zinc Triad Cation-Ethanediamine Complexes in Aqueous Solution, The Paradox of Complex Equivalence, Their Acid Properties and Chelation Effects, 1993; co-author: Low Temperature Electrochemistry and Spectroelectrochemistry, 1993; contbr. articles to profl. jours. Mem. Solidarity, Wroclaw, 1980. Recipient awards Minister of Sci., Edn. and Technics of Poland, Warsaw, 1977, 83. Fellow Polish Chem. Soc. Roman Catholic. Home: Klodnicka 45/8, PL54217 Wroclaw Poland Office: Inst of Chemistry/Univ, F Joliot-Curie 14, PL50383 Wroclaw Poland

WILHELM, DALLAS EUGENE, JR., biology educator; b. Sterling, Kans., Sept. 5, 1942; s. Dallas Eugene Sr. and Genieve (Blair) W. BS, Ft. Hays State U., 1964; PhD, Tex. Tech U., 1977. Lab. instr. Ft. Hays Kans. State Coll., 1964-66; instr. biology Baker U., Baldwin City, Kans., 1966-68; park naturalist Davis Mountains Tex. State Park, Ft. Davis, 1972; wildlife biologist U.S. Fish & Wildlife Svc., Denver, 1974; part-time instr. Tex. Tech. U., Lubbock, 1972-76; asst. prof., chair sci. div. Lincoln Meml. U., Harrogate, Tenn., 1976-79; assoc. prof., chmn. biology Hastings (Nebr.) Coll., 1979—; contbr. Nebr. Natural Heritage Program, Lincoln, 1987—; chairperson biology dept. evaluation team, Kearney State Coll., 1986; com. mem. chiropractic physicians tech. com., Nebr. Dept. Health, 1988; com. mem. nurse anaesthetist tech. com. nebr. Dept. Health, 1989. Lt. USN, 1967-71. Mem. Am. Soc. of Mammalogists, Southwestern Assn. of Naturalists, Nebr. Acad. Sci., Nat. Assn. of Advisors for the Health Professions, Sigma Xi. Office: Hastings Coll 7th & Turner Ave Hastings NE 68901

WILHELM, HARLEY A., mechanical engineer; b. Iowa, Aug. 5, 1900. BA, Drake U., 1923; PhD, Iowa State U., 1931. Emeritus prof. Iowa State U., 1970—; mem atomic Energy Prog. Iowa State U., 1942, Ames Lab. Dept. Energy. Recipient Gold medal Am. Soc. Mech. Engrs., 1990, ASME medal, 1990. Home: 513 Hayward Ave Ames IA 50010*

WILHELMS, DON EDWARD, geologist; b. L.A., July 5, 1930; s. William Leslie and Allene Marie (Schmitt) W. BA, Pomona Coll., 1952; MA, UCLA, 1958, PhD, 1963. Geologist U.S. Geol. Survey, Menlo Park, Calif., 1962-86; ret., 1986—. Contbr. chpt.: Geology of the Terrestrial Planets, 1984, Planetary Mapping, 1990; author: The Geologic History of the Moon, 1987, To A Rocky Moon: A Geologist's History of Lunar Exploration, 1993; contbr. articles to profl. jours. With U.S. Army, 1953-55. Fellow AAAS, Am. Geophys. Union, Geol. Soc. Am. (G.K. Gilbert award 1988). Achievements include synthesis of stratigraphy and geologic history of the Moon, development of methods of lunar and planetary geologic mapping.

WILKE, GUNTHER, chemistry educator; b. Heidelberg, Germany, Feb. 23, 1925; s. Ernest W.; m. Dagmar Kind. Prof. chemistry Max-Planck Inst. fûr Kohlenforschung, Bochum, Germany, 1963—, dean dept. chemistry, 1968-69; dir. MPI for Coal Rsch., Mûlheim/Ruhr, Germany, 1969—. Fellow Am. Chem. Soc. (Willard Gibbs medal 1991); mem. Rhein-Westf. Acad. of Sci., German Acad. of Scientists, Royal Dutch Acad. of Sci. (fgn.). Office: Max Planck Inst fûr Kohlenforschung, Kaiser-Wilhelm Platz 1, D-4330 Mûlhein Germany*

WILKERSON, WILLIAM EDWARD, JR., civil engineer; b. Monroe, La., Nov. 9, 1946; s. W.E. Wilkerson. BSCE, La. Tech., 1969. Registered profl. engr., La., W.Va., Ala., Miss., Fla., Ill., Ohio, S.C., Tenn, Wis., Tex., Ga. Hwy. engr. W.Va. Dept. of Hwy., Charleston, 1969-75; sr. v.p. BCM Engrs., Inc., Mobile, Ala., 1975—. Mem. NSPE, ASCE, ASTM (com. E-50 environ. assessment), Am. Water Works Assn., Ala. Soc. Profl. Engrs. (pres.), Ala. Assn. Water Pollution Control, Ala. Water and Pollution Control Assn., La. Soc. Profl. Engrs., Mobile Area Coun. Engrs., Mobile C. of C., NAt. Asbestos Coun., Steel Structures Painting Coun., Constrn. Specification Inst. Achievements include research in construction services, controlling trihalomehtanes in drinking water, asbestos abatement. Office: BCM Engrs Inc PO Box 1784 Mobile AL 36633

WILKES-GIBBS, DEANNA LYNN, psychologist, educator; b. Libby, Mont., Oct. 1, 1955; d. Robert James and Grace Mary (Travis) Wilkes; m. Jeffrey Shane Reiter, Mar. 19, 1989; 1 child, Benjamin Louis Reiter. BA, U. Calif., San Diego, 1981; PhD, Stanford U., 1986. Rsch. asst. Dept. Psychology U. Calif., San Diego, 1979-80, Yale U., New Haven, Conn., 1980-81, Stanford (Calif.) U., 1980-82; asst. prof. Wesleyan U., Middletown, Conn., 1986—, chair cognitive sci. program, 1987-90. Consulting reviewer Jour. Memory and Lang., Cognitive Sci., Memory and Cognition, Lang., Child Devel.; contbr. articles to profl. jours. Grantee Ford Found. 1988, 92. Mem. Am. Psychol. Soc., Psychonomic Soc. (assoc.), Soc. Text and Discourse, Cognitive Sci. Soc. Achievements include research on the processes and products of knowledge coordination through lang.; relations of beliefs, intentions, and actions in collective behavior; socially-shared cognition. Office: Wesleyan U Dept Psychology Judd Hall Middletown CT 06459-0408

WILKINS, CORNELIUS KENDALL, chemist, researcher; b. Cleve., July 2, 1938; s. Cornelius Kendall and Addie (Williams) W.; m. Anna Johanna Honnef, Sept. 1, 1965 (div. 1980); children: Cassandra, Tal; m. Marianne Hoegh, Jan 13, 1981; children: Katrine, Marc. BS in Chemistry, Case Inst. Tech., 1960; PhD in Organic Chemistry, Ohio State U., 1964. Rsch. scientist Unilever Rsch., Vlaardingen, Holand, 1968-73; rsch. assoc. dept. botany U. B.C., Vancouver, Can., 1973-77; rsch. scientist Insternat. Centre Insect Physiology and Ecology, Nairobi, Kenya, 1977-78; assoc. prof. Northeastern Ill. U., Chgo., 1979-80; devel. chemist Drubin As, Copenhagen, Denmark, 1981-82, N. Foss Electric Als, Hillerod, Denmark, 1982-85; adj. Tech. U. Denmark, Lyngby, 1986-89, rsch. assoc., 1989-91; cons. Nat. Ins. Occupational Health, Copenhagen, 1991-92, rsch. assoc., 1992—. Recipient 1st prize Nordic Gastech. Ctr., Oslo, 1990; NSF fellow, 1960. Achievements include research in separation and structure determination of plant polyphenolics identification of vegetable and spice volatiles and their use in quality classification. Home: Borgager 62, 2620 Albertslund Denmark Office: Nat Inst Occupational Health, Lersø Parkallé 105, 2100 Copenhagen Denmark

WILKINS, DANIEL R., nuclear engineer, nuclear energy industry executive. BS in Engring. Sci., Case Inst. Tech.; SM, ScD in Nuclear Engring., MIT. Former head advanced boiling water reactor program GE, gen. mgr. dept. nuclear power systems engring., now gen. mgr. dept. nuclear svcs. and projects; Registered profl. nuclear engr., Calif. Contbr. over 30 tech. papers to sci. jours. Recipient George Westinghouse Gold medal ASME, 1992. Mem. Am. Nuclear Soc. (Mark Mills award). Office: General Electric Co Nuclear Energy 175 Curtner Ave MC835 San Jose CA 95125*

WILKINS, MAURICE HUGH FREDERICK, biophysicist; b. Pongaroa, New Zealand, Dec, 15, 1916; s. Edgar Henry and Eveline (Whittaker) W.; m. Patricia Ann Chidgey, Mar. 12, 1959; children: Sarah Fenella, George Hugh, Emily Lucy Una, William Henry. Ph.D., St. John's Coll., Cambridge, 1940; LL.D., U. Glasgow, 1972. Research with Manhattan Project, U. Calif., Berkeley, 1944; lectr. St. Andrews U., 1945; mem. faculty Kings Coll., London, 1946—, dep. dir. biophysics unit Med. Research Council,, 1955-70, dir. biophysics unit, 1970-72, dir. neurobiology unit, 1972-74, prof. molecular biology, 1962-70, prof. biophysics, 1970-81, also dir. MRC cell biophysics unit (formerly Med. Research Council neurobiology unit), 1974-80. Decorated comdr. Brit. Empire; recipient Albert Lasker award Am. Pub. Health Assn., 1960, Nobel prize for physiology and medicine (with F.H.C. Crick and J.D. Watson), 1962; fellow King's Coll., 1973—. Fellow Royal Soc., 1959; mem. Brit. Biophys. Soc. (past chmn.), Am. Soc. Biol. Chemists (hon.), Brit. Soc. for Social Responsibility in Sci. (pres. 1969), Am. Acad. Arts and Scis. (fgn. hon.). Research, publs. on structure of nerve membranes and X-ray diffraction analysis of structure of DNA; devel. of electron trap theory of phosphorescence and thermo-luminescence; light microscopy techniques for cyto-chem. research, including use of interference microscope for dry mass determination in cells.

WILKINS, TRACY DALE, microbiologist, educator; b. Sparkman, Ark., July 25, 1943; s. James Edward and Lena Belle (Wilcox) W. BS, U. Ark., 1965; PhD, U. Tex., 1969. Postdoctoral U. Ky. Med. Sch., Lexington, 1969-71; asst. prof. Va. Poly. Inst. State U., Blacksburg, 1972-75, assoc. prof., 1972-75, prof., 1980-85, head dept. anaerobic microbiology, 1985-93; dir. Ctr. for Biotechnology Va. Poly. Inst. & State U., Blacksburg, Va., 1993—; pres. TechLab., Inc., 1990—. Contbr. articles to profl. jours.; patentee in field. NIH grantee, 1975—, Nat. Cancer Inst. grantee, 1979—. Mem. Am. Soc. Microbiology, Soc. Intestinal Microecology and Disease (pres. 1989-91). Avocations: woodworking, horses, hunting, fishing. Office: Va Poly Inst & State U Ctr for Biotechnology Anaerobe Lab Complex Blacksburg VA 24061

WILKINSON, CLIFFORD STEVEN, civil engineer; b. Orange, N.J., Mar. 18, 1953; s. Clifford James and Elizabeth Adelade (Fairbanks) W.; m. Judith Anne Simon, Oct. 6, 1979; children: Brian Steven, David James, Caitlin Elizabeth. BCE, N.J. Inst. Tech., 1975, MCE, 1986. Registered profl. engr., N.J. Sr. assoc. Killam Assocs., Millburn, N.J., 1975-92, Paulus Sokolowski and Sartor, Warren, N.J., 1992—. Coach little league, youth soccer, Bridgewater, N.J., 1991-93; com. mem. North Br. Reformed Ch., Bridgewater, 1991-93. Mem. NSPE, N.J. Soc. Profl. Engrs., Water Environ. Fedn., Water Pollution Control Assn. Republican. Home: 520 Rolling Hills Rd Bridgewater NJ 08807

WILKINSON, GEOFFREY, chemist, educator; b. Todmorden, Eng., July 14, 1921; s. Henry and Ruth (Crowther) W.; m. Lise Schou, July 17, 1951; children: Anne Marie, Pernille Jane. BSc, Imperial Coll., London, 1941, PhD, 1946; DSc (hon.), U. Edinburgh, U. Granada, 1977, Columbia U., 1979, U. Bath, 1980, Essex U., 1989. With NRC, Can., 1943-46; staff Radiation Lab. U. Calif., Berkeley, 1946-50; mem. faculty MIT, 1950-51; faculty chemistry dept. Harvard U., 1951-55; faculty Imperial Coll. Sci., Tech. and Medicine, U. London, 1955-88; prof. emeritus Imperial Coll. Sci. and Tech., U. London, 1988; Falk-Plaut vis. lectr. Columbia, 1961; Arthur D. Little vis. prof. MIT, 1967; Hawkins Meml. lectr. U. Chgo., 1968; 1st Mond lectr. Royal Soc. Chemistry, 1981; Chini lectr. Italian Chem. Soc., 1981; Tovborg Jensen lectr. U. Copenhagen, 1992. Author: (with F.A. Cotton) Advanced Inorganic Chemistry: A Comprehensive Text, 5th edit., 1988; Basic Inorganic Chemistry, 2d edit., 1987. John Simon Guggenheim fellow, 1954; recipient award inorganic chemistry Am. Chem. Soc., 1966, Centennial Fgn. fellow, 1976, Royal Soc. Chem. Transition medal chemistry, 1972; Lavoisier medal French Chem. Soc., 1968; Nobel prize in chemistry, 1973; Hiroshima U. medal, 1978; Royal medal, 1981; Galileo medal U. Pisa, 1983; Longstaff medal, 1987; Royal Soc. Chem. 1st Polyhedron prize, 1989; hon. fellow Inst. Tech. U. Manchester, Eng., 1989; Messel medal Soc. Chem. Industry, 1990. Fellow Royal Soc.; fgn. mem. Royal Danish Acad. Sci., Am. Acad. Arts and Scis., Nat. Acad. Scis., Spanish Sci. Research Council. Office: Imperial Coll Sci Tech & Med, Dept Chemistry, London SW7 2AY, England

WILKINSON, GERALD STEWART, zoology educator; b. San Francisco, Jan. 17, 1955; s. Wilbur and Stella Rose (Stratman) W.; m. Cynthia Ann Pearson, May 4, 1985; 1 child, Sara Michelle Pearson. BS in Zoology, U. Calif., Davis, 1977; PhD in Biology, U. Calif.-San Diego, La Jolla, 1984. Teaching asst. U. Calif.-San Diego, La Jolla, 1978-83; NATO postdoctoral fellow U. Sussex, Eng., 1984, U. Edinburgh, Scotland, 1985; NIMH postdoctoral fellow U. Colo., Boulder, 1985-87; asst. prof. dept. zoology U. Md., College Park, 1987-92, assoc. prof., 1992—. Adv. editor Behavioral Ecology and Sociobiology. Recipient Searle scholar award Chgo. Community Trust, U. Md., 1988; Lilly teaching fellow Lilly Found., U. Md., 1989. Mem. Internat. Soc. Behavioral Ecology, Animal Behavior Soc., Soc. for Study of Evolution. Democrat. Achievements include notable finding of reciprocal food sharing in vampire bats; rsch. in coop. behavior in mammals, quantitative genetic studies of behavior. Office: U Md Dept Zoology College Park MD 20742

WILKINSON, PETER MAURICE, physician consultant, pharmacologist; b. Manchester, Eng., Aug. 6, 1941; s. Clarence Rallison and Doris (Hacking) W.; m. Barbara Modsley, Apr. 3, 1965; children: Ian, Mark, Elizabeth, Suzanne. MB ChB, Manchester U., 1964, MSc, 1973. Physician Christie Hosp. Nat. Health Svc. Trust, Manchester, 1975—; mem. Mancester lymphoma group Christie Hosp. NHS Trust, Manchester, 1976—; lectr. in pharmacology Manchester U., 1975—; lectr. in medicine Harvard Med. Sch., Boston, 1976-77; vis. prof. med. oncology Sidney Farber Cancer Inst., Boston, 1976-77; mem. steering com. EORTC Pharmacokinetics and Metabolism Group, 1970-81; mem. Med. Rsch. Com. Working Party on testicular tumours, 1979—, on bladder cancer, 1979-86. Co-editor jour. Cancer Chemotherapy Pharmacology; contbr. numerous articles to profl. jours. Bradley Meml. Surgical scholar, 1964. Fellow Royal Coll. Physicians; mem. Brit. Assn. Cancer Rsch. (mem. com. 1979-82), Brit. Soc. Clin. Pharmacology, Brit. Assn. Cancer Physicians, Am. Soc. Clin. Oncology, N.Y. Acad. Scis. Avocations: golf, music. Home: 12 Ramsdale Rd, Bramhall, Stockport Cheshire SK7 2PZ, England Office: Christie Hosp NHS Trust, Wilmslow Rd Withington, Manchester M20 9BX, England

WILKINSON, R. L., agriculturalist. Recipient AIC Fellowship award Agrl. Inst. Can., 1992. Home: 1971 Casa Marcia Crescent, Victoria, BC Canada V8N 2X5*

WILKINSON, RONALD EUGENE, engineer; b. Citronelle, Ala., Dec. 7, 1945; s. Emmett Eugene and Faye Cecile (Tanner) W.; m. Peggy Lee Hergesheimer, Dec. 30, 1966; children: Rhonda Denise, Damon Eugene. BS in Aero. Engring., Auburn U., 1968. Design engr. Pratt & Whitney, West Palm Beach, Fla., 1969-72; with Aircraft Products Teledyne Continental Motors, Mobile, Ala., 1972—, devel. engr., project engr., sr. project engr., mgr. advanced programs to program mgr., dir. bus. devel., 1990, dir. engring., 1990-91, v.p. engring., 1991—. Recipient SAE Manly award 1987 Best Tech. Paper on Aero Engines, SAE Excellence in Presentation award, 1987, 89, Teledyne PAcific Group Excellence in Mgmt. award, 1987; named Outstanding Aerospace Engring. Alumni, Auburn U., 1986. Mem. AAIA, Soc. Automotive Engrs., Assn. Unmanned Vehicle Systems, Am. Mgmt. Assn. Republican. Methodist. Achievements include patent for liquid cooled engine which powered the voyager aircraft on its nonstop, non-refueled flight around the world; devel. and design of the engine and turbochargers which powered the Boeing Condor-an experimental high-altitude long-endurance aircraft which set a world altitude record of 66980 feet for piston engine-powered aircraft; holder six patents relating to aircraft piston engine technology. Home: 6804 Hunters Ct Mobile AL 36695-2705 Office: Aircraft Products Teledyne Continental Motors PO Box 90 Mobile AL 36601

WILKINSON, RONALD STERNE, science administrator and historian; b. Chgo., Feb. 16, 1934; s. Maurice Sterne and Florence Marie (Colby) W.; m. Karen Ensinger, June 14, 1969 (div. 1976). BA, Mich. State U., 1960, PhD, 1969. Chemist Berry Bros., Detroit, 1955-57; mem. faculty Mich. State U., East Lansing, 1960-70; sci. specialist Libr. of Congress, Washington, 1970-90, sr. sci. specialist, 1990—; assoc. in bibliography Am. Mus. Nat. History, N.Y.C., 1976-82; trustee William T. Hornaday Conservation Trust, LaJolla, Calif., 1989—. Author: John Winthrop, Jr. and the Origins of American Chemistry, 1969, Benjamin Wilkes, The British Aurelian, 1982; editor-in-chief The Mich. Entomologist (later The Great Lakes Entomologist), 1966-71; contbr. more than 150 articles to sci. and history of sci. publs. Ryder scholar Mich. State U., U. London, 1960, Woodrow Wilson (Found.) fellow Harvard U., 1960-61, Fulbright scholar Univ. Coll., London, 1965-66. Fellow Linnean Soc. London, Geol. Soc. London, Royal Entomol. Soc. London; mem. Grolier Club (N.Y.C., asst. editor 1979-82). Democrat. Home: 228 9th St NE Washington DC 20002-6110 Office: Libr of Congress Washington DC 20540

WILKINSON, TODD THOMAS, project engineer; b. Mitchell, S.D., June 23, 1959; s. Ralph W. and Margret Jean (Kent) W. BSME, S.D. Sch. Mines and Tech., 1981; MS in Applied Mech., U. Cin., 1984. Registered profl. engr., Ohio. Pilot Singleton/Shea Spray Svc., Pierre, S.D., 1977-80; program engr. GE Aircraft Engines, Cin., 1981-85; field rep. naval air propulsion ctr. GE Aircraft Engines, Trenton, N.J., 1985-86; field rep. Grumman F-14 flight test GE Aircraft Engines, Calverton, N.Y., 1986-88; project engr. MATV flight demonstration GE Aircraft Engines, Cin., 1989—. Tech. illustrator: (book) The Instrument Pilot Handbook, 1980. Mem. AIAA. Roman Catholic. Achievements include co-invention on vectoring ring support system; comml. rated pilot.

WILKINSON, WILLIAM KUNKEL, psychologist, educator; b. San Luis Obispo, Calif., Nov. 20, 1959; s. James Sydney and Jane (Kunkel) W.; m. Mary Patricia McCarthy, Aug. 18, 1984; 1 child, Kyle Patrick. BA with distinction in psychology, San Diego State U., 1982; MA, EdD, No. Ariz. U., 1989. Cert. sch. psychologist K-12, Ariz.; lic. psychologist, N.Mex. Sch. psychol. trainee Raskob Learning Inst., Oakland, Calif., 1986; grad. rsch. asst. Ctr. for Excellence in Edn., Flagstaff, Ariz., 1987-88, grad. teaching asst., 1987-88; psychol. cons. Peach Springs/Grand Canyon Sch. Dist., Ariz., 1986-88; pre-doctoral intern Johns Hopkins U., Balt., 1988-89; psychologist Las Cruces (N.Mex.) Pub. Schs., 1991-92; asst. prof. psychology N.Mex. State U., Las Cruces, 1989—. Guest editor: Research in Higher Education, 1990—, Jour. Applied Behavior Analysis, 1988—, Test Critique in Buros' Mental Measurements Yearbook, 1990; contbr. numerous articles to profl. jours.; first author: Research for Human Service Professionals, 1993. Grantee Ctr. for Excellence in Edn., 1988. Mem. APA (dissertation rsch. award 1988), Am. Ednl. Rsch. Assn., Nat. Assn. Sch. Psychologists, Psychol. Soc. Ireland, Phi Kappa Phi. Achievements include research on individual differences in epistemological reasoning, research on self-report measure combining 6 different models of knowledge style in hope of distilling meaningful modes of epistemological perspectives. Office: 3591 Venus St Las Cruces NM 88003-0001

WILKNISS, PETER E., foundation administrator, researcher; b. Berlin, Germany, Sept. 28, 1934; U.S. citizen.; s. Fritz and Else (Stueber) W.; m. Edith P. Koester, May 25, 1963; children: Peter F., Sandra M. MS in Chemistry, Tech. U., Munich, Ger., 1958, PhD in Radio and Nuclear Chemistry, 1961. Rsch. chemist, radiological protection officer U.S. Naval Ordnance Sta., 1961-64, head nuclear chemistry branch, 1964-66; rsch. oceanographer U.S. Naval Rsch. Lab., 1966-70, head chemical oceanography branch, 1970-75; mgr. Nat. Ctr. Atmospheric Rsch. Program NSF, Washington, 1975-76, mgr. Internat. Phase of Ocean Drilling/Ocean Sediment Coring Program, 1976-80, mgr. Ocean Drilling Project Team, AAEO Directorate, 1980, dir. divsn. Ocean Drilling Programs, 1980-81, sr. sci. assoc. Office of Dir., 1981-82, dep. asst. dir. Sci, Tech., Internat. Affairs Directorate, 1982-84, dir. divsn. Polar Programs, 1984-93, sr. sci. assoc. Geoscis. Directorate, 1993—; liaison mem. NRC, NAS, Marine Bd., 1978-81, Polar Rsch. Bd., 1984-93; mem. atmospheric chemistry and radioactivity com. Am. Meteorological Soc., 1975-78; mem. interagy. com. atmospheric scis., 1975-76, space station adv. com., NASA, 1988-93. Contbr. 61 articles to sci., tech. jours., USN reports; over 100 formal presentations nat., internat. sci. confs., symposia, meetings; participant 16 nat., internat. workshops. Presdl. citation AIA, 1993; Wilkniss mountain Antarctic named in his honor Sec. Interior, U.S. Bd. Geographic Names, 1992. Mem. AAAS, Am. Geophysical Union, Antarctican Soc., Sigma Xi. Episcopalian. Avocations: soccer, swimming, skiing. Home: 8814 Stockton Pkwy Alexandria VA 22308 Office: National Science Foundation 1800 G St NW Rm 620 Washington DC 20550

WILKS, ALAN DELBERT, chemical research and technology executive, researcher; b. Liberal, Kans., Sept. 4, 1943; s. Delbert Elvado and Mabel Ida (Howell) W.; m. Irvana Sue Keagy, June 11, 1967; 1 child, Jolin Rai. BS in Chemistry, U. Kans., 1965; PhD in Analytical Chemistry, U. Iowa, 1970. Chemist UOP Rsch. Ctr., Des Plaines, Ill., 1969-76, group leader catalysis div., 1976-77, mgr. catalysis rsch., 1977-84; dir. phys. chemistry and surface sci. Allied-Signal Inc., Des Plaines, 1984-88, dir. materials sci. EMS sector, 1988-89, v.p. chems. and process tech., 1989—; cons. Las Alamos (N.Mex.) Nat. Lab., 1990—. Contbr. articles to profl. jours.; holder 5 patents. Mem. Am. Chem. Soc., Am. Vacuum Soc., Am. Ceramic Soc., N.Y. Acad. Sci., Chgo. Catalyst Club, Alpha Chi Sigma. Home: 1201 W Cleven Mount Prospect IL 60056 Office: Allied-Signal Inc 50 E Algonquin Rd Des Plaines IL 60017*

WILKS, RONALD, engineer; b. Killamarsh, Eng., Feb. 24, 1930; came to U.S., 1957; s. Fred and Jane (Nichols) W.; m. Dorothy Robinson, Jan. 23, 1954; children: Helen Margaret, Kathryn Julia, Jeremy Peter, Timothy Duncan, Heather Jane. MSc, Coll. of Tech., Kingston-upon-Hull, Eng., 1954. Systems integration engr. GE Manned Space Systems Divsn., King of Prussia, Pa., 1960-66; divsn. chief engr. Day & Zimmermann, Inc., Phila., 1966-77; internat. aviation cons. Berwyn, Pa., 1971-77; mgr. bus. devel. Lavelle Aircraft Co., Newtown, Pa., 1977-81; sr. bus. devel. specialist Lockheed Aero. Systems Co., Marietta, Ga., 1981-90; owner Wilks & Assocs., Marietta, 1990—. Co-author: (tech. publ.) Motion Picture Techniques and Equipment--As Applied to Manned Space Simulation, 1965. Chmn. Competitive Swimming YMCA, Berwyn, 1973-76. Fellow AIAA (assoc.); mem. Brit. Assn. Aviation Consultants, Am. Solar Energy Assn. Achievements include patent for External Fuel Tank for Supersonic Aircraft. Office: Wilks & Assocs PO Box 70663 Marietta GA 30007

WILL, EDWARD EDMUND, oil company executive; b. Elizabeth, N.J., July 4, 1952; m. Patricia Susan Granat. BA, Reed Coll., 1975; MBA, Harvard U., Boston, 1980; dipl., U. Paris, 1972. Mgr. strategic planning NL Shaffer/NL Industries, Houston, 1980-82; v.p. mktg. and internat. ops. NL Acme Tool Inc./NL Industries, Inc., Houston, 1982-87; v.p. internat. ops. Western Petroleum Svcs./Western Co. of N.Am., Houston, 1987-88; v.p. Western Co. N.Am., 1988; pres. Western Petroleum Svcs., 1988—; adj. prof bus. policy U. Houston, 1981-82. Contbr. articles to profl. jours. Mem. Phi Beta Kappa. Office: Western Co NAm 515 Post Oak Blvd Houston TX 77027-9407

WILL, THOMAS ERIC, psychologist; b. Feb. 28, 1949. BA in Edn., Concordia Coll., St. Paul, 1972; MPH, U. Minn., 1980, postgrad. in epidemiology, 1980-81; PsyD with distinction, Forest Inst. Profl. Psychology, Des Plaines, Ill., 1986. Lic. psychologist. Rsch. asst. dept. epidemiology Nat. Cancer Inst. U. Minn., Mpls., 1980-81; mem. crisis intervention staff Forest Psychiat. Hosp., Des Plaines, 1982-83; psychiat. researcher Ill. State Psychiat. Inst., Chgo., 1983-84; psychometrician Evaluation Ctr./U. Chgo. Med. Ctr., 1983-84; health psychologist Group Health, Inc. Mental Health Ctr., Mpls., 1986—; assoc. core faculty Minn. Sch. Profl. Psychology, Mpls., 1988-93; adj. asst. prof. Forest Inst. Profl. Psychology, Des Plaines, 1983-85; regional faculty The Fielding Inst., Santa Barbara, Calif., 1991—. Contbr. articles to profl. jours. Recipient grants, Group Health, Inc., 1988. Mem. APA, Nat. Acad. Neuropsychology, Nat. Register of Health Svc. Providers in Psychology, Psi Chi. Office: Group Health Inc 2701 University Ave SE Minneapolis MN 55414

WILLANDER, LARS MAGNUS, physics educator; b. Varberg, Halland, Sweden, July 2, 1948; s. Nils Ture and Inez Dagmar (Gustafsson) W.; m. Kerstin Marianne Larsson, Aug. 10, 1974; children: Johan, Josefine, Johanna. MSc in Physics, Lund (Sweden) U., 1974; cert. tchr.'s competence, Malmö (Sweden) U., 1974; MSc in Engring. Physics, Uppsala (Sweden) U., 1976; BA in Econs., Stockholm U., 1977; Dr.Sc., Royal Inst. Tech., Stockholm, 1984; docent competence, Linköping (Sweden) U., 1988. Rsch. engr. Philips Corp., Stockholm, 1976-80; rsch. asst. Royal Inst. Tech., 1980-84; specialist Nobel Industries, Stockholm, 1984-85; assoc. prof. Linköping (Sweden) U., 1985—. contbr. over 150 sci. articles to profl. publs. Mem. IEEE, Am. Phys. Soc., N.Y. Acad. Scis. Achievements include research in semiconductor devices and semiconductor physics. Home: Karl Dahlgrensgatan 6, S-582 28 Linköping Sweden Office: Linköping U, Dept Physics, S-581 83 Linköping Sweden

WILLARD, SCOTT THOMAS, reproductive physiologist; b. Marysville, Ohio, May 3, 1968; s. James Thomas and Florence Marie (Hiser) W. BS in Animal and Vet. Sci., U. R.I., 1990; postgrad., Tex. A&M U., 1990—. Vet. technician Washington County Vet. Hosp., Kingston, R.I., 1989-90; rsch. asst. Tex. A&M U., College Station, 1990—. Contbr. to profl. publs. Mem. AAAS, Am. Soc. Animal Sci., Soc. for Study of Reproduction, Am. Assn. Zool. Parks and Aquariums, Am. Assn. Zoo Keepers (book rev. com. 1991—). Achievements include research on domestic and exotic animal reproduction with emphasis on deer reproduction, ultrasonography and reproductive toxicology. Home: RD 9 287 Whiteacre Dr Bethlehem PA 18015 Office: Tex A&M U Dept Animal Sci Kleberg Ctr College Station TX 77843

WILLEKE, GENE E., environmental engineer, educator; b. Dola, Ohio, Sept. 5, 1934; s. Gale Charles and Mildred Elizabeth (Marling) W.; m. Carol Ann Blomquist, Sept. 8, 1962; children: Andrew Gerhard, Jonathan Christopher. AB in Math., Ohio No. U., 1956, BS in Civil Engring., 1957; MS in Civil Engring., Stanford U., 1960, PhD in Civil Engring., 1969. Registered profl. engr., Ohio. Civil engr. Madison Engring. Assn., Mansfield, Ohio, 1957; hydraulic rsch. engr. U.S. Bur. Pub. Rds., Washington, 1960-62; supr. sanitary engring. USPHS, Chgo., 1962-65; acting asst. prof. Stanford (Calif.) U., 1968-70; prof. Ga. Inst. Tech., Atlanta, 1970-77; dir. and prof. Inst. of Environ. Sci., Oxford, Ohio, 1977—; cons. in field. Treas. Boy Scouts Am., Oxford, 1988-90. With U.S. Army, 1957-59. Hwy. Rsch. Bd. awardee, 1971. Mem. ASCE, Nat. Assn. Environ. Profls., Nat. Soc. Profl. Engrs., Am. Geophys. Union, Am. Meteorol. Soc., Assn. Am. Geographers, Am. Soc. Engring. Edn. Democrat. Lutheran. Home: 1007 Cedar Dr Oxford OH 45056-2416 Office: IES Miami Univ 102 Boyd Hall Oxford OH 45056

WILLETT, KENNETH G., software engineer; b. The Dalles, Oreg., Dec. 25, 1952; s. Glen Owen and Lydia Emily (Smith) W.; m. Theresa Marie Laskowski, July 8, 1978; children: Catherine, Claire, Christopher, Colin. BA in Math/Physics, Whitman Coll., 1974; MS in Computer Sci., Wash. State U., 1981. Software engr. Tektronix, Beaverton, Oreg., 1974-80, Four Phase Systems, Portland, 1980-81, Mentor Graphics Corp., Wilsonville, Oreg., 1981—; chief architect Cad Framework Initiative, Austin, 1991—. Mem. Phi Beta Kappa. Democrat. Roman Catholic. Office: Mentor Graphics Corp 8005 SW Boeckman Rd Wilsonville OR 97077

WILLETT, PETER, information science researcher; b. London, Apr. 20, 1953; s. David William and Marigold Patricia (Illing) W.; m. Marie Therese Gannon, July 15, 1978; 1 child, Barbara Jane. MA, Oxford U., 1975; PhD, Sheffield U., 1979. Lectr. U. Sheffield, 1979-85, sr. lectr., 1985-88, reader, 1988-91, prof., 1991—. Editorial bd. Jour. of Chemical Information and Computer Sciences, Jour. of Am. Soc. for Info. Sci.; author: Similarity and Clustering in Chemical Information Systems, 1987, Parallel Database Processing, 1990, Three-Dimensional Chemical Structure Handling, 1991; contbr. over 200 articles to profl. publs. Recipient numerous rsch. awards including Skolnick award Am. Chem. Soc., 1993. Fellow Inst. of Info. Scientists; mem. British Computer Soc. Achievements include rsch. in documental retrieval and chemical structure handling, particularly in areas of document clustering, chemical similarity and three-dimensional chemical databases. Office: U Sheffield, Western Bank, Sheffield S10 2TN, England

WILLEY, EDWARD NORBURN, physician; b. Ann Arbor, Mich., June 14, 1933; s. Norman LeRoy and Mary Helen (Norburn) W.; m. Joan Bryant, Jan. 9, 1965 (div. Jul. 1992); children: Dawn, Linda, Karen. AB, Univ. Mich., 1954, MD, 1958. Diplomate Am. Bd. Pathology. Internship Duke U., Durham, N.C., 1958-59; residency Univ. Mich., Ann Arbor, 1959-63; pathologist Duval Medical Ctr., Jacksonville, Fla., 1963-67, Palms of Pasadena Hosp., St. Petersburg, Fla., 1967-85; consulting physician pvt. practice, St. Petersburg, Fla., 1985—. Recipient Crapo C. Smith scholar Univ. Mich., 1951-58, James B. Angell scholar, 1954. Fellow Coll. Am. Pathologists, Am. Acad. Forensic Sci., Royal Soc. Medicine; mem. AMA. Home: 7869 9th Ave S Saint Petersburg FL 33707-2730 Office: 6727 1st Ave S # 204 Saint Petersburg FL 33707-1341

WILLEY, GORDON RANDOLPH, retired anthropologist, archaeologist, educator; b. Chariton, Ia., Mar. 7, 1913; s. Frank and Agnes Caroline (Wilson) W.; m. Katharine W. Whaley, Sept. 17, 1938; children—Alexandra, Winston. A.B., U. Ariz., 1935, A.M., 1936, Litt.D. (hon.), 1981; Ph.D., Columbia U., 1942; A.M. honoris causa, Harvard U., 1950; Litt.D. honoris causa, Cambridge U., 1977, U. N.Mex., 1984. Archaeol. asst. Nat. Park Service, Macon, Ga., 1936-38; archaeologist La. State U., 1938-39; archaeol. field supr. Peru, 1941-42; instr. anthropology Columbia U., 1942-43; anthropologist Bur. Am. Ethnology, Smithsonian Instn., 1943-50; Bowditch prof. archaeology Harvard U., 1950-83, sr. prof. anthropology, 1983-87, chmn. dept. anthropology, 1954-57; vis. prof. Am. archaeology Cambridge (Eng.) U., 1962-63; mem. expdns. to Peru, Panama, 1941-52, Brit. Honduras, 1953-56, Guatemala, 1958, 60, 62, 64, 65, 66, 67, 68, Nicaragua, 1959, 61, Honduras, 1973, 75-77. Author: Excavations in the Chancay Valley, Peru, 1943, Archaeology of the Florida Gulf Coast, 1949, Prehistoric Settlement Patterns in the Viru Valley, Peru, 1953, Introduction to American Archaeology, 2 vols, 1966-71, The Artifacts of Altar de Sacrificios, 1972, Excavations of Altarde Sacrificious, Guatemala: Summary and Conclusions, 1973, Das Alte Amerika, 1974, The Artifacts of Seibal, Guatemala, 1978, Essays in Maya Archaeology, 1987, Portraits in American Archaeology, 1988, New World Archaeology and Culture History, 1990, Excavations at Seibal: Summary and Conclusions, 1990; co-author: Early Ancon and Early Supe Cultures, 1954, The Monagrillo Culture of Panama, 1954, Method and Theory in American Archaeology, 1958, Prehistoric Maya Settlements in the Belize Valley, 1965, The Ruins of Altar de Sacrificios, Department of Peten, Guatemala: An Introduction, 1969, the Maya Collapse: An Appraisal, 1973, A History of American Archaeology, 1974, 3d edit., 1993, The Origins of Maya Civilization, 1977, Lowland Maya Settlement Patterns: A Summary View, 1981, the Copan Residential Zone, 1993; co-editor: Courses Toward Urban Life, 1962, Precolumbian Archaeology, 1980, A Consideration of the Early Classic Period in the Maya Lowlands, 1985; editor: Prehistoric Settlement Patterns of the New World, 1956, Archaeological Researches in Retrospect, 1974. Overseas fellow Churchill Coll., Cambridge U., 1968-69; decorated Order of Quetzal Guatemala; recipient Viking Fund medal, 1953; Gold medal Archaeol. Inst. Am., 1973; Alfred V. Kidder medal for achievement in Am. Archaeology, 1974; Huxley medal Royal Anthrop. Inst., London, 1979; Walker prize Boston Mus. Sci., 1981; Drexel medal for archaeology Univ. Mus., Phila., 1981, Golden PLate award Am. Acad. Achievement, 1987. Fellow Am. Anthrop. Assn. (pres. 1961), Am. Acad. Arts and Sci., London Soc. Antiquaries, Soc. Am. Archaeology (pres. 1968, Disting. Svc. award 1980); mem. Nat. Acad. Sci., Am. Philos. Soc., Royal Anthrop. Inst. Gt. Britain and Ireland, Cosmos Club (Washington), Tavern Club (Boston), Phi Beta Kappa; corr. mem. Brit. Acad.

WILLEY, JAMES CAMPBELL, internist, educator; b. Toledo, Apr. 12, 1952; s. John Douglas and Marilynn (Miller) W.; m. Elisabeth Barras James, Oct. 24, 1981; 1 child, Marilynn M.J. BA, U. Vt., 1974; MD, Med. Coll. Ohio, 1978. Diplomate Am. Bd. Internal Medicine. Intern VA Hosp., Washington, 1978-79, resident in internal medicine, 1979-81; med. staff fellow Nat. Cancer Inst., Bethesda, 1981-84, biotech. fellow, 1986-89, expert, 1989-90; sr. staff fellow Nat. Heart, Lung, Blood Inst., Bethesda, 1984-86; asst. prof. U. Rochester (N.Y.) Sch. Medicine, 1990—. Contbr. articles to Jour. Cell Physiology, Cancer Rsch., other sci. publs. Mem. AAAS, Am. Assn. Cancer Rsch., Am. Soc. Human Genetics, Alpha Omega Alpha. Congregationalist. Achievements include research on role of cyclic AMP induction in control of cell proliferation, in vitro malignant transformation of human epithelial cells, evidence for bronchial epithelial cell tumor suppressor gene on short arm of chromosome 17. Home: 77 Crawford St Rochester NY 14620 Office: Univ Rochester 535 Elmwood Ave Rochester NY 14620

WILLIAMS, ALLAN NATHANIEL, agricultural economist; b. Georgetown, Guayana, Oct. 4, 1946; arrived in Trinidad and Tobago, 1981; s. Clement Allan and Lilian Princess (Greene) W.; m. Mavis Nathalie Christian, Aug. 1, 1981; 1 child, Lindiwe. BA, Miami U., Oxford, Ohio, 1968, MA, 1969; MA, Cornell U., 1972, PhD, 1976. Instr. Miami U., 1969-70, 72-73; cons. UN Devel. Program, Kingston, Jamaica, 1974, UN Environ. Program, Mexico City, 1977; cons. Caribbean environ. project UN Environ. Program, Port of Spain, Trinidad and Tobago, 1978; dir. Assn. for Caribbean Transformation Ltd., Port of Spain, 1978—; developer methodologies to incorporate environ. costs into devel. projects for Ctr. Human Ecology and Health, Pan Am. Health Orgn., Mexico City, 1979; compiler methodologies to bridge data gaps in agrl. stats.; creator mgmt. info. system for planning agrl. rsch. in East Africa, Inter-Govtl. Authority for Drought and Devel.; adminstr. programs of tech. and fin. assistance to small farmers and coops. in organizing prodn. and mktg. in Eastern Caribbean. Contbr. articles to profl. jours.; co-author commputer software package Agrl. Info. System, 1984-88; designer database system Nat. Agrl. Rsch. Info. System, 1990. Fellow Inst. for Internat. Edn., 1965-68, Robert McNamara fellow World Bank, 1983. Mem. Trinidad and Tobago Econs. Assn. (exec. mem. 1988-90), Internat. Third World Legal Studies Assn., Internat. Assn. Econs. Self Mgmt. (nt. corr. 1981-83), OI Com. Internat. (founding). Avocations: swimming, photography, home gardening. Office: ACT Ltd, 3 Pelham St Belmont, Port of Spain Trinidad and Tobago

WILLIAMS, ANN MEAGHER, hospital administrator; b. Hull, Mass., May 28, 1929; d. James Francis Meagher and Dorothy Frances (Antone) Mullins; m. Joseph Arthur Williams, May 15, 1954; children: James G., Mara A., A. Scott (dec.), Gordon M., Mark J., Antoinette M., Andrea M. BS, Chestnut Hill Coll., 1950; MS, Boston Coll., 1952. Radioisotope biologist Air Force Cambridge Rsch. Ctr., Bedford, Mass., 1952-55; asst. mgr. Roxbury Businessmen's Exch., Boston, 1956-66; owner, operator Chatterlane, Osterville, Mass., 1961-66; realtor James E. Murphy Inc., Hyannis, Mass., 1968-77; dir. community affairs Cape Cod Hosp., Hyannis, 1977—. Bd. dirs. Community Coun., Mid Cape, Mass., 1977-88, Cape Cod Mental Health Assn., 1977-82, Ctr. for Individual and Family Svcs., Mid Cape, 1982-87, Am. Cancer Soc., Mid Cape, 1981—; mem. sch. com. Cape Cod Regional Tech. High Sch., 1978—, United Way of Cape Cod, 1988-89; chmn. fin. com. City of Barnstable, Mass., 1969-77. Named Woman of Yr. Bus./Profl. Women's Club, 1982; recipient Cert. Appreciation Am. Cancer Soc., 1983, 88, Pres. Recognition award United Way Cape Cod, 1989. Mem. Am. Soc. Hosp. Mktg. and Pub. Rels., New Eng. Hosp. Pub. Rels. Mktg. Assn., Southeastern Mass. Hosp. Pub. Rels. Assn., Nat. Assn. Hosp. Devel., Chestnut Hill Coll. Alumnae Assn., Rotary Club of Osterville (bd. dirs. 1992—), Hyannis Area C. of C. (bd. dirs. 1993—). Roman Catholic. Avocation: community theater. Home: 8 E Bay Rd Osterville MA 02655-1909 Office: Cape Cod Hosp 27 Park St Hyannis MA 02601-3276

WILLIAMS, ANTHONY M., chemist; b. Orangeburg, S.C., Oct. 18, 1950; s. Clifton M. and Louella (Summers) W.; m. Julia Edith Scott, June 15, 1991; children: Dwayne, Sandra; children from previous marriage: Talya, Anthony. BS, Claflin Coll., 1972. Chemist AT&T Bell Labs., Murray Hills, N.J., 1974-91; chemist Union Camp Corp., Eastover, S.C., 1991—, safety specialist, 1982—. Contbr. articles to profl. jours. Mem. S.C. Lab. Mgmt. Soc. Home: J4 Glenfield Apts Orangeburg SC 29115 Office: Union Camp Corp PO Box B Hwy 601 Eastover SC 29044

WILLIAMS, ARTHUR E., federal agency administrator; b. Watertown, N.Y.; m. Carol Waite; children: Scott, Christina, Cheryl. BA in Math., St. Lawrence U., Canton, n.Y., 1960; BSCE, Rensselaer Poly. Inst.; M in Civil Engring./Econ. Planning, Stanford U.; grad., U.S. Army Comd. and Gen. Staff Coll., U.S. Naval War Coll. Registered profl. engr., Minn. Comdr. various assigments U.S. Army Corps Engrs.; chief engrs., comdr. U.S. Army Corps Engrs., Washington, 1992—. Decorated Bronze Star, Legion of Merit. Office: Dept of the Army Chief of Engineers 20 Masschusetts Ave NW Washington DC 20314*

WILLIAMS, BRIAN JAMES, geotechnical engineer; b. St. Paul, Mar. 16, 1959; s. James Harold and Louise Alice (McDonough) W. BS in Geol. Engring., S.D. Sch. Mines and Tech., 1981, MSCE, 1987. Registered profl. engr., Wash.; registered profl. geologist, Idaho. Cons. geologist/engr. Geo-Tech Resources, Helena, Mont., 1981-85; geotech. engr. No. Engring. and Testing, Inc., Gt. Falls, Mont., 1985-89, Chen-Northern, Inc., Pasco, Wash., 1989-92, Shannon & Wilson, Inc., Kennewick, Wash., 1992—. Mem. ASTM (tech. subcom. 1991—), ASCE (assoc.). Achievements include research, design and testing of state-of-the-art RCRA Impoundments used to store mixed chemical and radioactive wastes. Office: Shannon & Wilson Inc 1354 Grandridge Blvd Kennewick WA 99336

WILLIAMS, BRUCE WARREN, psychologist; b. Coronado, Calif., Feb. 23, 1948; s. E. Royce and Camilla (Forde) W.; m. Joanna Matthes Pyper, Nov. 28, 1970 (div. 1979); m. Patricia Gwynn Head, July 11, 1981. BA, Stanford U., 1970; PhD, U. Calif., Santa Barbara, 1979. Lic. psychologist, D.C., Calif. Asst. prof. San Diego State U., 1978-79; chief psychologist Forest Haven, Laurel, Md., 1979-85; dir. spl. programs D.C. Village, Washington, 1985-87; clin. psychologist Mental Retardation/Developmental Disabilities Adminstrn., Washington, 1987-90; cons. psychologist St. Elizabeth Hosp., Washington, 1989-90; pvt. practice San Diego and L.A., 1990—; psychology cons. South Cen. L.A. Regional Ctr., 1990—; chmn. human rights com. St. John's Child Devel. Ctr., Washington, 1985-90. Lt. USN, 1970-74. Mem. APA, Assn. for Behavior Analysis, Am. Assn. Mental Retardation (chair So. Calif. chpt. 1993—, sec. D.C. chpt. 1988-90). Home: 147 W Acacia Ave #153 Glendale CA 91204 Office: South Cen LA Reg Ctr Devel 2160 W Adams Blvd Los Angeles CA 90018

WILLIAMS, CHARLES MURRAY, computer information systems educator, consultant; b. Ft. Bliss, Tex., Dec. 26, 1931; s. Robert Parvin and Barbara (Murray) W.; m. Stanley Bright, Dec. 31, 1956; children: Margaret Allen Williams Becker, Robert Parvin, Mary Linton Williams Bondurant. BS, Va. Mil. Inst., 1953; MS, Stanford U., 1964; PhD, U. Tex., 1967. Physicist USAF, Kirtland AFB, N.Mex., 1956-58; staff mem. Sandia Labs., Albuquerque, 1958-62; programmer analyst Control Data Corp., Palo Alto, Calif., 1962-63; mathematician Panoramic Rsch., Inc., Palo Alto, 1963-64; mem. tech. staff Thomas Bede Found., Los Altos, Calif., 1964-65; rsch. scientist assoc. U. Tex., Austin, 1965-67; asst. prof. Computer Sci. Pa. State U., State College, 1967-72, Va. Poly. Inst. and State U., Blacksburg, 1972-75; assoc. prof. Computer Info. Systems Ga. State U., Atlanta, 1975-83, prof. Computer Info. Systems, 1983—; cons. Visicon Inc., State Coll., 1970-72, Broomall (Pa.) Industries Inc., 1973-79, Bausch & Lomb Inc., Rochester, N.Y. 1981, BellSOUTH Media Techs. Inc., Atlanta, 1987-90; mem. tech. staff Bell Labs., Whippany, N.J., 1979; textbook reviewer various publs.

including Harper & Row, Prentice Hall, Simon & Schuster, 1976—. Contbr. articles to profl. computer graphics and image processing publs. Bd. dirs. Ga. Striders, 1993—. 1st lt. USAF, 1954-56. Recipient Silver medals in 1500-meter and 3000-meter runs 60-64 age div. Athletic Congress Nat. Masters Indoor Track and Field Championships, 1992, Gold medals in 5000-meter and 10,000-meter runs, Bronze medal in 1500-meter run, 1992, Gold medals in 1500-meter and 3000-meter runs, 1993; ranked 3d nationally Running Times, 1992. Mem. Nat. Computer Graphics Assn. (hon., Ga. bd. dirs. 1979-85, bd. dirs. Ga. chpt. 1985-89, sec. 1988-89), Computer Graphics Pioneers, Upsilon Pi Epsilon, Omicron Delta Kappa. Republican. Episcopalian. Avocation: competitive road running (recipient numerous gold and silver medals). Home: 316 Argonne Dr NW Atlanta GA 30305-2814 Office: Ga State U University Pla Atlanta GA 30303

WILLIAMS, CHARLES OLIVER, JR., dentist; b. Greensboro, N.C., Sept. 1, 1939; s. Charles Oliver and Amy (Groves) W.; m. Nancy Jane McIntosh, July 31, 1965; children: Cristin McIntosh, Charles Oliver III. Student U. N.C., 1957-61; D.M.D., U. Louisville, 1966. Practice dentistry, Winnsboro, S.C., 1966—; assoc. mem. staff Fairfield Meml. Hosp., Winnsboro. Editor-in-chief S.C. Dental Jour., 1970-83. State publs. chmn. Nat. Children's Dental Health Week, Fairfield County, 1967; vice chmn. bd. dirs. Richard Winn Acad.; bd. visitors Med. U. S.C., Charleston; chmn. profl. div. United Fund, Fairfield County, 1967-68; mem. Fairfield County Mental Health Assn., 1970—; bd. dirs. Fairfield County coun. Boy Scouts Am., First Union Bank, Mt. Zion Soc.; pres. Winnsboro Cotillion. Recipient Dental Editors award Ohio State U., ADA, 1970. Mem. S.C. Central Dist. Dental Assn. (pres.; past pres. peer rev. bd.), S.C. Dental Assn. (past bd. govs.), Mid Carolina Dental Study Acad. (pres.), AAAS, ADA, Nat. Rehab. Assn., Fairfield Jr. C. of C. (past v.p.), Am. Soc. Dentistry for Children, Southeastern Acad. Prosthodontics, Am. Assn. Dental Editors, Mid Carolina Dental Study Acad., Fedn. Dentaire Internat., Royal Soc. Health (Gt. Britain), Order Ky. Cols., Delta Sigma Delta (Nat. Lit. Excellence award 1966). Past sr. warden Episcopal Ch. Office: 204 E Washington St Winnsboro SC 29180

WILLIAMS, CHARLES WILMOT, engineer, consultant; b. Detroit, Feb. 3, 1916; s. Walter Henry and Frances (Churchill) W.; m. Mary Helen Wendel, Oct. 19, 1940; children: James Wendel, Robert Churchill. BS in Engring., Princeton U., 1938; MS in Automotive Engring., Chrysler Inst., 1940. Registered profl. engr., Mich. Student engr. Chrysler Corp., Detroit, 1938-40, project engr., 1940-41, mgr. aircraft engine testing, 1942-45, asst. chief engr., 1946-51, mfg. dir. missile divsn., 1951-61; rsch. dir. Fed. Mogul Corp., Detroit, 1961-81; ind. cons. Birmingham, Mich., 1981—. Pres. Sr. Mens Club Birmingham, 1991, sec., v.p., 1988-90, chmn. 1993. Fellow AIAA (assoc.); mem. SAE (life, vice chmn. 1946-51), Engring. Soc. Detroit. Episcopalian. Achievements include patents on rolling element bearing. Home: 1824 Pine St Birmingham MI 48009

WILLIAMS, CURT ALAN, mechanical engineer; b. Dallas, Nov. 4, 1953; s. Arthur Berri and Betty Lou (Calvert) W.; m. Cynthia Ann Corliss, Sept. 25, 1976; children: Christopher Aaron, Craig Allen. BSME, U. Tex., 1976. Registered prof. engr., Tex. Engring. coop. Vought Aeros., Dallas, 1973-75; tech. svc. engr. nordel EPDM process E.I. duPont de Nemours and Co., Beaumont, Tex., 1976-79; tech. svc. engr. dacron yarn rsch. E.I. duPont de Nemours and Co., Old Hickory, Tenn., 1979-81; project engr. supr. photo systems div. E.I. duPont de Nemours and Co., Rochester, N.Y., 1981-84; product engring. supr. connector systems E.I. duPont de Nemours and Co., Camp Hill, Pa., 1984-85, tech. svcs. supr. connector systems, 1985-86; gold recovery engr. E.I. duPont de Nemours and Co., Emigsville, Pa., 1986; rsch. assoc. Electronics Tech. Ctr. E.I. duPont de Nemours and Co., Research Triangle Park, N.C., 1986-88, facilities and safety cons., 1988-91; rsch. assoc. duPont Beaumont Works E.I. duPont de Nemours and Co., Inc., Beamont, Tex., 1991-92, tech. assoc. duPont Beaumont Works, 1992—. Co-patentee method for eluting absorbed golf from carbon. Mem. ASME, Am. Soc. Safety Engrs., Am. Soc. Indsl. Security. Avocations: amateur radio, water skiing. Home: 10605 Dunhill Ter Raleigh NC 27615-1438 Office: E I du-Pont de Nemours & Co Hwy 347 S Beaumont TX 77704

WILLIAMS, DAMON S., civil engineer. BSCE, U. Ill., 1978; BS in Physics, Roosevelt U., 1969. Registered profl. engr., Ariz., Calif., Nev. Sr. engr. John Carollo Engrs., Phoenix, 1978-87; pres., prin. engr. Damon S. Williams Assocs., Phoenix, 1987—; mem. Joint U.S./Soviet Delegation on Environ. Preservation and Restoration, Lake Baikal, Russia, 1990, on A Comprehensive Land Use Policy and Allocation Program for the Soviet Portion of the Lake Baikal Watershed, 1991; mem. ind. cons. delegation of water quality, St. Petersburg, Russia, 1992; mem., chmn. adv. com. Operator Cert. for State of Ariz., 1984-88. Contbr. articles to profl. jours. mem. ASME, NSPE, Ariz. Water and Pollution Control Assn., Am. Water Works Assn., Water Environment Fedn. Office: Damon S Williams Assocs 645 E Missouri Ave Ste 270 Phoenix AZ 85012-1369

WILLIAMS, DAVID ALLAN, dentist, educator; b. Dayton, Ohio, June 30, 1949; s. Robert Eugene and Mary Ellen (Moore) W.; divorced. BS, Mich. State U., 1971; DDS, Case Western Res. U., 1975. Clin. assoc. prof. Northwestern U. Dental Sch., Chgo., 1978—; gen. practice dentistry Northbrook, Ill., 1979—, Chgo., 1980—. Bd. dirs. United Way of Northbrook. With USN, 1975-78. Armed Forces Health Profls. Scholar, 1972-75. Fellow Acad. Gen. Dentistry; mem. ADA (commn. on dental accreditation), Ill. Dental Soc., Chgo. Dental Soc., Acad. Operative Dentistry, Northbrook Rotary, Delta Tau Delta. Lutheran. Avocations: sailing, skiing, cycling. Office: 666 Dundee Rd Ste 801 Northbrook IL 60062-2734

WILLIAMS, DAVID BERNARD, educator; b. Leeds, Yorkshire, Eng., July 25, 1949; came to U.S. 1976, naturalized 1985; s. Joseph Edward and Catherine (Tull) W.; m. Margaretha Johanna Louwers, June 12, 1976; children—Matthew John, Bryn Joseph, Stephen David. B.A., Cambridge U., Eng., 1970, M.A., 1973, Ph.D., 1974. Chartered engr., U.K. Postdoctoral research fellow Cambridge U., Eng., 1974-76; asst. prof. Lehigh U., Bethlehem, Pa., 1976-79, assoc. prof., 1979-82, prof. materials sci. engring., 1982—; steering com. Argonne Nat. Labs Electron Microscope Group, Ill., 1984—. assoc. editor Jour. Microscopy, Oxford, Eng., 1984-92, editor, 1992—, Jour. Electron Microscopy Technique, N.Y.C., 1983—. Author: Practical Analytical Electron Microscopy in Materials Science, 1984. Recipient Burton medal Electron Microscopy Soc. Am., 1984; Robinson award Lehigh U., 1979, William J. Priestly Disting. professorship, 1984. Fellow Inst. Metals, Royal Microscopical Soc.; mem. Metall. Soc. of AIME, Electron Microscopy Soc. Am., Microbeam Analysis Soc. (dir., pres. 1991-92). Roman Catholic. Avocations: rugby; running; piano; reading. Office: Lehigh U Dept Material Sci and Engring Whitaker Lab 5 E Packer Ave Bethlehem PA 18015-3195

WILLIAMS, DAVID VANDERGRIFT, organizational psychologist; b. Balt., Feb. 5, 1943; s. Laurence Leighton and Mary Duke (Warfield) W.; m. Diane M. Gayeski, Aug. 23, 1980; 1 child, Evan David Williams. BA, Gettysburg (Pa.) Coll., 1965; MA, Temple U., 1967; PhD, U. Pa., 1971. Asst. prof. psychology Ithaca (N.Y.) Coll., 1970-75, assoc. prof. psychology, 1975—; ptnr. OmniCom Assocs., Ithaca, 1979—; Cons. and speaker in field. Co-author: Interactive Media, 1985, (computer based tng.), interactive video software, 1979—; contbr. to books and articles to profl. jours. Bd. dirs. McCormick Ctr., Brooktondale, N.Y., 1988—, Ctr. for Religion, Ethics and Social Policy, Cornell U., 1975-77, Eco-Justice Task Force, Ithaca, 1975-78; trustee Montessori Sch. Ithaca, 1993—; advisor Tau Kappa Epsilon, Ithaca, 1993—. Rsch. fellow U.S. Office of Edn., 1967-70; recipient various grants. Mem. ASTD, APA, Nat. Soc. for Performance and Instrn., Am. Correctional Assn., Am. Montessori Soc., Welsh Nat. Gymanfa Ganu Assn., Ithaca Yacht Club, Tau Kappa Epsilon. Avocations: singing, sailing, sign language, cooking, travel. Office: OmniCom Assocs 407 Coddington Rd Ithaca NY 14850

WILLIAMS, DENNIS THOMAS, civil engineer; b. Washington, Aug. 22, 1925; s. Dennis Thomas and Margaret Madelene (Henley) W.; m. Jane Elizabeth Fisher, Aug. 16, 1926; children: Roy Thomas, Laurence James, Laura Josephine, Eric. ME, Pratt Inst., Bklyn., 1959; postgrad., Howard U., Washington, 1943-46. Civil engr. N.Y.C. Dept. Pub. Works, 1952-58; supr. Bur. Water Pollution Control, N.Y.C., 1959-77; elec. engr. N.Y.C. Bur. Environ. Protective Agy., N.Y.C., 1978-87; tech. advisor North River Community Environ. Rev. Bd., N.Y.C., 1986—; cons. engr. N.Y. State D.E.C., 1991—. Author screen plays: Campaign, 1988, Napoleon, 1988. Fellow Profl. Engrs. Alumni Assn.; mem. Water Pollution Control Fedn., Artists for Mental Health Assn., Greenpeace, Greater Paterson C. of C., Kiwanis. Democrat. Mormon. Avocation: travel. Home: 284 Wall Ave Paterson NJ 07504-1014

WILLIAMS, DONALD JOHN, research physicist; b. Fitchburg, Mass., Dec. 25, 1933; s. Toivo John and Ina (Kokkinen) W.; m. Priscilla Mary Gagnon, July 4, 1953; children: Steven John, Craig Mitchell, Eino Stenroos. B.S., Yale U., 1955, M.S., 1958, Ph.D., 1962. Sr. staff physicist Applied Physics Lab., Johns Hopkins U., 1961-65; head particle physics br. Goddard Space Flight Center, NASA, 1965-70; dir. Space Environ. Lab., NOAA, Boulder, Colo., 1970-82; prin. investigator Energetic Particles expt. NASA Galileo Mission, 1977—; prin. staff physicist Johns Hopkins U. Applied Physics Lab., 1982-89, dir. Milton S. Eisenhower Rsch. Ctr., 1990—; mem. nat. and internat. sci. planning coms.; chmn. NAS com. on solar-terrestrial rsch., 1989—. Author: (with L.R. Lyons) Quantitative Aspects of Magnetospheric Physics, 1983; assoc. editor: Jour. Geophys. Research, 1967-69, Revs. of Geophysics and Space Research, 1984-86; editor: (with G.D. Mead) Physics of the Magnetosphere, 1969, Physics of Solar-Planetary Environments, 1976; mem. editorial bd.: Space Sci. Revs., 1975-85; contbr. articles to profl. jours. Served to lt. USAF, 1955-57. Recipient Sci. Research award, 1974; Disting. Authorship award, 1976, 85. Fellow Am. Geophys. Soc.; mem. Am. Phys. Soc., Internat. Assn. Geomagnetism and Aeronomy (pres. 1991—), Sigma Xi. Home: 14870 Triadelphia Rd Glenelg MD 21737-9408

WILLIAMS, EARL GEORGE, acoustics scientist; b. Cleve., Jan. 2, 1945; s. Jack Harry and Mary Helen (Theodore) W.; m. Virginia Handy, May 14, 1977; children: Elizabeth, Ned Daniel. BS, U. Pa., 1967; Masters, Harvard U., 1968; PhD, Pa. State U., 1979. Scientist Naval Air Devel. Ctr., Warminster, Pa., 1968-71; tchr. The Putney (Vt.) Sch., 1971-74; freelance cellist Vt. Symphony, Burlington, 1974-75; postdoctoral Pa. State U., State College, Pa., 1979-82; rsch. physicist Naval Rsch. Lab., Washington, 1982—. Contbr. articles to profl. jours. Recipient Berman Rsch. Publication award Naval Rsch. Lab., 1987, 90. Fellow Acoustical Soc. Am. (Alumni award 1990); mem. Tau Beta Pi. Achievements include two patents. Home: 4507 Twinbrook Rd Fairfax VA 22032 Office: Naval Rsch Lab Code 5137 Washington DC 20375

WILLIAMS, EDDIE R., university administrator; b. Chgo., Jan. 6, 1945; s. Eddie R. Anna Maude (Jones) W.; m. Shirley Ann King, May 31, 1966; children: Karen Lynn, Craig Dewitt, Evan Jonathan. BA in Math., Ottawa U., 1966; PhD in Complex Variables, Columbia U., 1971. Asst. to v.p. of acad. affairs San Diego State U., 1977; assoc. dir. operating budgets No. Ill. U., DeKalb, 1978-83, dep. dir. budge and planning, 1983, dir. budget and planning, 1983-84, asst. v.p. adminstrv. affairs, 1984-85, assoc. prof. math., 1978—, v.p. fin. and planning, 1985—; mem. pres.'s exec. cabinet No. Ill. U., DeKalb, 1985—, mem. univ. coun., 1985—; rep. Bd. Regents Facilities Com., Springfield, Ill., 1985—; speaker in field; cons. Winnebago County Opportunity In Indsl. Careers, Rockford, Ill., 1974-75. Co-author: Fundamentals of Mathematics. Asst. pastor South Park Bapt. Ch., Chgo., 1970—; tchr. Stateville Prison, Ill., 1973-74. Capt. USNR, 1980—. Decorated Navy Commendation Medal for Meritorious Achievement; grantee AV math. lab. U.S. Govt., 1974. Fellow Am. Coun. Edn.; mem. Nat. Assn. Math., Nat. Naval Officer Assn., Sigma Xi. Office: No Ill U Divsn Fin and Planning De Kalb IL 60115

WILLIAMS, EDWARD F(OSTER), III, environmental engineer; b. N.Y.C., Jan. 3, 1935; s. E. Foster Jr. and Ida Frances (Richards) W.; m. Sue Carol Osenbaugh, June 5, 1960; children: Cecile Elizabeth, Alexander Harmon. BS in Engring., Auburn U., 1956; MA in History, Memphis State U., 1974. Registered profl. engr., Tenn. Engr. Buckeye Cellulose Corp. (subs. of Procter & Gamble), Memphis, 1957, process safety engr., 1960; resident constrn. engr. Buckeye Cellulose Corp. (subs. of Procter & Gamble), Perry, Fla., 1960-61; staff engr. Buckeye Cellulose Corp. (subs. of Procter & Gamble), Memphis, 1961-70; chief engr., v.p. Enviro-trol, Inc., Memphis, 1970-73; v.p., then pres. Ramcon Environ. Corp., Memphis, 1973-80; pres. E.F. Williams & Assocs., Inc., Memphis, 1980—; bd. dirs. Mobile Process Tech. Inc., Memphis; v.p. Environ. Testing and Consulting, Inc., Memphis, 1985—. Author: Fustest with the Mostest, 1968, Early Memphis and Its River Rivals, 1969; editor Environ. Control News for So. Industry, 1971—. State rep. Tenn. Gen. Assembly, 1970-78; mem. Shelby County Bd. Commrs., Memphis, 1978—, chmn., 1987-88, 90-92; del. Rep. Nat. Conv., 1988, 92; vice chmn. Memphis-Shelby Local Emergency Planning Com., 1986—. Capt. USAF, 1957-60. Named Tenn. Water Conservationist of the Yr. Tenn. Conservation League, 1973, Tenn. Legis. Conservationist of the Yr. Nat. Wildlife Fedn., 1974, Memphis Outstanding Engr. Memphis Joint Engrs. Coun., 1980. Mem. NSPE, ASME, Am. Acad. Environ. Engrs. (diplomate), Water Pollution Control Fedn., Am. Indsl. Hygiene Assn. (chpt. pres.), Am. Soc. Safety Engrs., Air Pollution Control Assn., Engrs. Club Memphis (bd. dirs. 1979-80), Rotary, C. of C., Tenn. Hist. Soc. (v.p. 1972)., Tenn. Hist. Commn. (vice chmn. 1987—), West Tenn. Hist. Soc. (pres. 1983-85). Republican. Presbyterian. Avocation: history. Home: 148 Perkins Ext Memphis TN 38117-3127 Office: EF Williams & Assocs 751 E Brookhaven Cir Memphis TN 38117-4549 also: Box 241813 Memphis TN 38124

WILLIAMS, FORMAN ARTHUR, combustion theorist, engineering science educator; b. New Brunswick, N.J., Jan. 12, 1934; s. Forman J. and Alice (Pooley) W.; m. Elsie Vivian Kara, June 15, 1955 (div. 1978); children: F. Gary, Glen A., Nancy L., Susan D., Michael S., Michelle K.; m. Elizabeth Acevedo, Aug. 19, 1978. BSE, Princeton U., 1955; PhD, Calif. Inst. Tech., 1958. Asst. prof. Harvard U., Cambridge, Mass., 1958-64; prof. U. Calif.-San Diego, 1964-81; Robert H. Goddard prof. Princeton U., N.J., 1981-88; prof. dept. applied mechs. and engring. scis. U. Calif., San Diego, 1988—. Author: Combustion Theory, 1965, 2d edit., 1985; contbr. articles to profl. jours. Fellow NSF, 1962; fellow Guggenheim Found., 1970; recipient U.S. Sr. Scientist award Alexander von Humboldt Found., 1982, Silver medal Combustion Inst., 1978, Bernard Lewis Gold medal Combustion Inst., 1990. Fellow AIAA; mem. Am. Phys. Soc., Combustion Inst., Soc. for Indsl. and Applied Math., Nat. Acad. Engring., Nat. Acad. Engring. Mex. (fgn. corr. mem.), Sigma Xi. Home: 8002 La Jolla Shores Dr La Jolla CA 92037-3230 Office: U Calif San Diego Ctr Energy & Combustion Rsch 9500 Gilman Dr La Jolla CA 92093

WILLIAMS, FRANCIS LEON, engineering executive; b. McGill, Nev., Sept. 19, 1918; s. Leon Alfred and Mazie Arabella (Blanchard) W.; m. Ailsa Bailey, Oct. 1944 (div.); children: Rhonda, Grapham, Alison; m. Marita I. Furry, Feb. 23, 1974. Student, Calif. Inst. Tech., 1940-41, UCLA, 1946-47, Am. TV Labs., 1948; BME, Sydney U., Australia, 1952; postgrad., San Jose State Coll., 1958-60, Foothill Coll., 1961, Regional Vocat. Ctr., San Jose, Calif., 1962, Alexander Hamilton Inst., 1971-72, Lane Community Coll., 1978-85. Project engr., prodn. supr. Crompton, Parkinson, Australia Pty., Ltd., Sydney, 1949-50; field and sales engr. Perkins Australia Pty., Ltd., Sydney, 1951-54; chief mech. engr. Vicon Corp., San Carlos, Calif., 1955-60; design engr., group leader Lockheed Missiles and Space Co., Sunnyvale, Calif., 1960-70; prin. Astro-Tech Cons. Co., Los Altos, Calif., 1971-72; mech. designer Morvue and Morden Machines, Portland, Oreg., 1973-74; sr. mech. design engr. Chip-N-Saw div. Can-Car of Can., Eugene, Oreg., 1974-75; sales mgr. Indsl. Constrn. Co., Eugene, 1975-76, gen. mgr., 1977-78; ops. mgr. Steel Structures, Eugene, 1976-77; mech. design and project engr. Carothers Co., Eugene, 1978-80; chief engr. Bio Solar and Woodex Corps., Eugene and Brownsville, Oreg., 1980-83; cons. and design engr. Am. Fabricators, Woodburn, Oreg., 1983-84; design engr., draftsman Peterson Pacific Corp., Pleasant Hill, Oreg., 1984-85, Jensen Drilling Co., Glenwood, Oreg., 1985; design engr. Judco & Ball Flight Dryers, Inc., Harbor City, Calif., 1985-86; sr. v.p. The Richelsen Co., also cons.; chief engr. Peterson Pacific Corp., Eugene, 1984-93, mgr. new product devel. R&D, 1993—; also cons.; advisor solid waste recovery County Bd. Commr.'s Office, Eugene, 1984-85. Contbr. articles to profl. jours.; patentee in field. Chmn. bldg. and grounds Westminster Presbyn. Ch., Eugene, 1984-86. Served with USAF, 1941-45. Democrat. Lodge: Elks. Avocation: writing. Home: 2324 Lillian St Eugene OR 97401-4916

WILLIAMS, FREDERICK WALLACE, research chemist; b. Cumberland, Md., Sept. 24, 1939; s. Richard and Erna (Cole) W.; m. Lois Anne Taylor, June 26, 1964; children: Jennie Linn, William Frederick, Mary Beth. BS, U. Ala., 1961, MS, 1963, PhD, 1965. Postdoctoral researcher Naval Rsch. Lab., Washington, 1965-66, rsch. chemist, 1966-73, head combustion sect., 1973—, tech. dir. USS Shadwell, 1988—; pres. Washington Chromatography Discussion Group, 1975-76; chmn. Eastern States Combustion Inst., 1985-87. Author over 450 papers, reports and presentations. Pres. Forest Knolls Citizen Assn., Ft. Washington, Md., 1980-90, Ft. Washington Recreation Assn., 1985-87. Mem. Am. Chem. Soc., Combustion Inst., Washington Chromatography Disc. Gr. Achievements include research on analytic air purifier for use with gas chromatography; co-discovery of transition flame; development of personal air sampler; establishment of 200 and 10,000 ft3 confined space fire test facility. Office: Naval Rsch Lab Code 6183 Washington DC 20375-5000

WILLIAMS, GARY MURRAY, medical researcher, pathology educator; b. Regina, Sask., Can., May 7, 1940; s. Murray Austin and Selma Ruby (Domstad) W.; m. Christine Julia Lundberg; children: Walter, Jeffrey, Ingrid. BA, Washington and Jefferson Coll., 1963; MD, U. Pitts, 1967. Diplomate Am. Bd. Pathology, Am. Bd. Toxicology. Assoc. prof. pathology Temple U., Phila., 1971-75; mem. Fels Rsch. Inst., Phila., 1971-75; rsch. prof. N.Y. Med. Coll., Valhalla, 1975—; dir. med. scis., chief. pathology and toxicology div. Am. Health Found., Valhalla, 1975—; mem. toxicology study sect. NIH, Bethesda, Md., 1985-87; mem. working groups Internat. Agy. Rsch. on Cancer, Lyon, France, 1986, 87, 89, 91. Editor: Sweeteners: Health Effects, 1988; co-editor: Cellular Systems for Toxicity Testing, 1983; editor-in-chief Cell Biology and Toxicology, 1984—, Antioxidants: Chemical, Physiol., Nutritional and Toxicol. Aspects, 1993; author over 340 sci. papers. Lt. comdr. USPHS, 1969-71. Recipient Sheard-Sanford award Am. Soc. Clin. Pathologists U. Pitts., 1967. Mem. Am. Assn. Cancer Rsch., Am. Assn. Pathologists, Soc. Toxicology (Arnold J. Lehman award 1982), Luth. Brotherhood Fraternal Soc. (dist. pres. 1988-90), Phi Beta Kappa, Alpha Omega Alpha. Home: 8 Elm Rd Scarsdale NY 10583-1410 Office: Am Health Found 1 Dana Rd Valhalla NY 10595-1549

WILLIAMS, GEORGE CHRISTOPHER, biologist, ecology and evolution educator; b. Charlotte, N.C., May 12, 1926; s. George Felix and Margaret (Steuart) W.; m. Doris Lee Calhoun, Jan. 25, 1951; children: Jacques, Sibyl, Judith, Phoebe. AB, U. Calif., Berkeley, 1949; PhD, UCLA, 1955. Instr. and asst. prof. Mich. State U., East Lansing, 1955-60; assoc. prof. dept. ecology and evolution SUNY, Stony Brook, 1960-66, prof., 1966-90; adj. prof. Queens U., Kingston, Ont., Can., 1980—. Author: Adaptation and Natural Selection, 1966, Sex and Evolution, 1975, Natural Selection: Domains, Levels and Challenges, 1992; co-author: Evolution and Ethics, 1989; editor Quar. Rev. Biology, SUNY, 1965—. With U.S. Army, 1944-46. Recipient Eminent Ecologist award Ecol. Soc. Am., 1989, Daniel Giraud Elliot medal Nat. Acad. Sci., 1992; fellow Ctr. Adv. Study Behavioral Sci., Stanford, 1981-82, Guggenheim Found., 1988-89. Fellow AAAS, NAS, Soc. Study Evolution (v.p. 1973, pres. 1989), Am. Soc. Ichthyologists and Herpetologists, Am. Soc. Naturalists (editor 1974-79), Icelandic Natural History Soc. Office: SUNY Ecology & Evolution Stony Brook NY 11794

WILLIAMS, HEATHER, neuroethologist, educator; b. Spokane, Wash., July 27, 1955; d. James Edward and Maria Greig (Davenport) W.; m. Patrick David Dunlavey, Oct. 19, 1986; 1 child, Maria Greig. AB, Bowdoin Coll., 1977; PhD, Rockefeller U., 1985. Watson fellow Hebrew U., Eilat, Israel, 1977-78; asst. prof. Rockefeller Univ. Field Rsch. Ctr., Millbrook, N.Y., 1986-88, Williams Coll., Williamstown, Mass., 1988—. Contbr. articles to Sci., Animal Behaviour, Jour. Neurbiology, Proc. NAS, others. Mem. course setting team U.S. Orienteering/World Championships, Harriman State Park, N.Y., 1993. MacArthur Found. fellow, 1993. Mem. Soc. for Neurosci., Internat. Soc. for Neuroethology, Animal Behavior Soc., Sigma Xi, Phi Beta Kappa. Office: Williams College Dept Biology Williamstown MA 01267

WILLIAMS, HUGH ALEXANDER, JR., mechanical engineer, consultant; b. Spencer, N.C., Aug. 18, 1926; s. Hugh Alexander and Mattie Blanche (Megginson) W.; BS in Mech. Engring., N.C. State U., 1948, MS in Diesel Engring. (Norfolk So. R.R. fellow), 1950; postgrad. Ill. Benedictine Coll. Inst. Mgmt., 1980; m. Ruth Ann Gray, Feb. 21, 1950; children: David Gray, Martha Blanche Williams Heidengren. Jr. engr.-field service engr. Baldwin-Lima Hamilton Corp., Hamilton, Ohio, 1950-52, project engr., 1953-55; project engr. Electro-Motive div. Gen. Motors Corp., La Grange, Ill., 1955-58, sr. project engr., 1958-63, supr. product devel. engine design sect., 1963-86, staff engr., advanced mech. tech., 1986-87. Trustee Downers Grove (Ill.) San. Dist., 1965-92, pres., 1974-91, v.p., 1991-92; pres. Ill. Assn. San. Dists., 1976-77, bd. dirs., 1977-89; mem. statewide policy adv. com. Ill. EPA, 1977-79; mem. DuPage County Intergovtl. Task Force Com., 1988-92; elder Presbyn. Ch. Served with USAAC, 1945. Registered profl. engr., Ill. Recipient Trustee Svc. award Ill. Assn. San. Dists., 1986, Citizens award Downers Grove Evening chpt. Kiwanis, 1991. Fellow ASME (chmn. honors and awards com. 1993—, Diesel and Gas Engine Power Div. Speaker awards 1968, 84, Div. citation 1977, Internal Combustion Engine award 1987, exec. com. Internal Combustion Engine div. 1981-87, 88-92, chmn. 1985-86, sec. 1988-92); mem. Soc. Automotive Engrs. (life), ASME (chmn. Soichiro Honda medal com. 1987-92), Ill. Assn. Wastewater Agys. (Outstanding Mem. award 1990, hon. mem. 1992), Lions, Masons (32 degree), Sigma Pi. Republican. Editor: So. Engr., 1947-48; contbr. articles to profl. jours. Patentee in field. Home: 2108 Weybridge Dr Raleigh NC 27615-5562

WILLIAMS, IRVING LAURENCE, physics educator; b. Newport, R.I., Dec. 3, 1935; s. Leroy Payton and Alberta Helen (Troy) W.; m. Carrie Mae Graves, Aug. 26, 1967; children: Cheryl Anita, Carla Chantrase. EdB, R.I. Coll., 1957; MA in Teaching, Brown U., 1962; PhD, NYU, 1975. Cert. teaching, R.I. Classroom tchr. Newport (R.I.) Sch. Dept., 1962-63; prof. physics Morgan State U., Balt., 1963-67; prof. physics Nassau Community Coll., Garden City, N.Y., 1967—, asst. to pres., 1980-85; adj. prof. Hofstra U., Hempstead, N.Y., 1980-87; dist. clk. Roosevelt (N.Y.) Sch. Bd., 1989-91. Co-author: (lab. workbook) Meterology Lab. Exercises, 1975, 76. Treas. Econ. Opportunity Commn., Nassau County, N.Y., 1984; trustee Grace Lutheran Ch., Malverne, N.Y., 1987, Roosevelt Bd. Edn., 1988; active Roosevelt Rep. Club, 1989; mem. sch. bd. Grace Lutheran Sch., Malverne, 1991. With U.S. Army, 1957-60. Recipient Chancellor's award SUNY, 1975, Citzen's award EOC Nassau County, Hempstead, 1987, Roosevelt Educator's award, Roosevelt Coun., 1989; NSF Weather Svc. grantee, Washington, 1989. Mem. AAUP, Nat. Sci. Tchrs. Assn., Am. Assn. Physics Tchrs., Soc. Coll. Sci. Tchrs., N.Y. Acad. Sci., Am. Assn. Higher Edn., N.Y. Assn. Two Yr. Colls. Republican. Avocation: piano. Home: 220 Beechwood Ave Roosevelt NY 11575-1634 Office: Nassau C C Physical Sciences Dept Garden City NY 11530

WILLIAMS, JACK MARVIN, chemist; b. Delta, Colo., Sept. 26, 1938; s. John Davis and Ruth Emma (Gallup) W.; m. Joan Marlene Davis, Mar. 7, 1958; 3 children. B.S. with honors, Lewis and Clark Coll., 1960; M.S., Wash. State U., 1964, Ph.D., 1966. Postdoctoral fellow Argonne (Ill.), Nat. Lab., 1966-68, asst. chemist, 1968-70, assoc. chemist, 1970-72, chemist, 1972-77, sr. chemist, group leader, 1977—; vis. guest prof. U. Mo., Columbia, 1980, 81, 82, U. Copenhagen, 1980, 83, 85; chair Gordon Rsch. Conf. (Inorganic Chemistry), 1980. Bd. editors Inorganic Chemistry, 1979—, assoc. editor, 1982—. Crown-Zellerbach scholar, 1959-60; NDEA fellow, 1960-63; recipient Disting. Performance at Argonne Nat. Labs. award U. Chgo., 1987, Centennial Disting. Alumni award Wash. State U., 1990. Mem. Am. Crystallographic Assn., Am. Chem. Soc. (treas. inorganic div. 1982-84), Am. Phys. Soc., AAAS. Office: Chemistry Div 9700 Cass Ave Lemont IL 60439-4801

WILLIAMS, JACK RAYMOND, civil engineer; b. Barberton, Ohio, Mar. 14, 1923; s. Charles Baird and Mary (Dean) W.; m. Mary Berneice Jones, Mar. 5, 1947 (dec.); children: Jacqueline Rae, Drew Alan; m. Betty Ruth Scholfield, Nov. 9, 1990. Student Colo. Sch. Mines, 1942-43, Purdue U., 1944-45; BS, U. Colo., 1946. Gravity and seismograph engr. Carter Oil Co., Western U.S. and Venezuela, 1946-50; with Rock Island R.R., Chgo., 1950-80, structural designer, asst. to engr. bridges, asst. engr. bridges, 1950-63, engr. bridges system, 1963-80; sr. bridge engr. Thomas K. Dyer Inc., 1980-82; v.p. Alfred Benesch & Co., 1982—. Served with USMCR, 1943-45. Fellow ASCE; mem. Am. Concrete Inst., Am. Ry. Bridge and Bldg. Assn. (past pres.), Am. Ry. Engring. Assn. (past chmn. com. 8, Concrete and

Foundations). Home: 293 Minocqua St Park Forest IL 60466-1942 Office: 233 N Michigan Ave Chicago IL 60604

WILLIAMS, JAMES RICHARD, human factors engineering psychologist; b. Chgo., Apr. 16, 1932; s. James Henry and Margaret Lucille (Keefer) W.; m. Jonetta Rae Gilbert, Dec. 19, 1959; children: Janise Rebecca, Jason Richard. BS in Psychology, Purdue U., 1958, MS in Human Factors/Indsl. Psychology, 1960; PhD in Edn., NYU, 1971. Technical asst. Sci. Rsch. Assocs., Chgo., 1960-61; sr. systems cons. System Devel. Corp., Paramus, N.J., 1961-64; human factors engr. Kollsman Instrument Corp., Elmhurst, N.Y., 1964-66; project mgr. System Devel. Corp., Paramus, 1966-69; supr. tng. and standards Bell Labs., Piscataway, N.J., 1969-74; dist. mgr. AT&T, Basking Ridge, N.J., 1975-80; mem. technical staff Bell Labs., Piscataway, 1981-83; learning technologist Bell Communications Rsch., Piscataway, 1984—; cons. NYU, N.Y.C., 1968-70; chair U.S. Tech. Adv. Group, to internat. orgn. for standardization in ergonomics of human system interaction, 1988—. Editor: International Standards for Menu Dialogues and Command Dialogues with Computer Systems. Cub master Boy Scouts, Watchung, N.Y., 1973-74, asst. scout master, 1975-78. With USAF, 1951-55. Mem. Am. Psychol. Soc. (charter), Assn. for Computing Machinery (spl. interest group on computer-human interaction), Human Factors and Ergonomics Soc. (cert. in profl. ergonomics, rep. 1986—), Delta Rho Kappa, Kappa Delta Pi. Avocations: winemaking-co-owner Del Vista Vinyards, astronomy, body bldg. Home: 137 County Rd 513 Frenchtown NJ 08825-9705 Office: Bell Communications Rsch 6 Corporate Pl Piscataway NJ 08854-4120

WILLIAMS, JAMES SAMUEL, JR., chemist; b. Lackawanna, N.Y., Dec. 27, 1943; s. James Samuel Sr. and Adele S. (Thomas) W.; m. Maureen Smith, Oct. 27, 1967 (div. Aug. 1970); 1 child, Colleen Buccitelli. AS, Erie C.C., 1966. Chief lab. technician Lucidol, Tonawanda, N.Y., 1966-70; lab. dir. Buffalo Sewer Authority, 1972—. Mem. Am. Chem. Soc., Am. Water Works Assn., Water Environment Fedn., Am. Diabetes Assn., World Wildlife Fund, N.Y. State Assn. Approved Environ. Labs., Buffalo Zool. Soc. Democrat. Roman Catholic. Office: Buffalo Sewer Authority Foot of West Ferry St Buffalo NY 14213

WILLIAMS, JIMMIE LEWIS, research chemist; b. Indianola, Miss., June 3, 1953; s. West and Lorene (Mayfield) W. BS in Chemistry, Jackson (Miss.) State U., 1975; MS in Chemistry, Yale U., 1977; PhD in Inorganic Chemistry, U. Calif., Riverside, 1983. Sr. rsch. scientist Corning (N.Y.) Inc., 1983-89, rsch. assoc. in chemistry, 1989—, project leader ceramic rsch., 1990—; speaker in field. Contbr. articles to profl. jours. and conf. procs. Coach sport programs, Corning, 1983—; vol. Big Bro. Program, Corning, 1983-85. Named Outstanding Young Men in Am., 1980, 83. Fellow Am. Inst. Chemists; mem. Am. Chem. Soc., Am. Ceramic Soc., Air and Waste Mgmt. Assn., Soc. Automotive Engrs., NACCP (v.p. Elmira, N.Y. br. 1991—), Sigma Xi (pres. local chpt. 1990-91). Achievements include patents in field; research on automotive and industrial emissions control. Office: Corning Inc SP-DV-1-9 Corning NY 14831

WILLIAMS, JOHN HOWARD, architect, retired; b. Littleton, N.C., Feb. 7; s. Ruffin Hampton and Emma Maude (Fitts) W.; m. Thelma Lorena McGuffin, June 9, 1944; children: Juan McGuffin and Jeffrey Howard. BS in Archtl. Engring. with honors, N.C. Agrl. and Tech. U., 1942; Diploma/Architects Rev., U. Va./No. Extension, Falls Church, 1958; postgrad., U. Minn., 1965. Registered architect. Draftsman, architect, asst. chief Tech. Svc. Section/Div. Hosp., Pub. Health Svc., Washington, 1947-67; chief facilities planning br., Fed. Health Programs Svcs. HHS, Washington, 1967-73, chief facilities planning and constrn., Indian Health Svc., 1973-84; cons. health care facilities. Co-dir. hosp. systems studies, 1971, bed determination methodology, 1979; dir. health facilities planning manuels, 1980, cost estimating system, 1981. Vestryman Calvary Episcopal Ch., Washington, 1957; mem. Brotherhood of St. Andrew, Washington. Recipient Merrick medal for excellence in mechanic arts N.C. Agrl. and Tech. U., Greensboro, 1942, Outstanding Svc. and Dedication award Indian Health Svc., Health and Human Svcs.; established John H. Williams scholarship in archtl. engring. N.C. A&T State U., 1980. Mem. AIA (emeritus), Acad. of Architecture for Health, Forum for Health Care Planning, Nat. Soc. Archtl. Engrs., Alpha Kappa Mu, Kappa Alpha Psi. Avocations: tutoring, travel, golf, reading, TV.

WILLIAMS, JUDITH RENEE, architect; b. Orlando, Fla., Jan. 8, 1968; d. Joseph Jerry and Natalie (Smigelski) W. AA, U. Fla., 1989, B Design in Architecture, 1991; MS in Architecture, Carnegie Mellon U., 1993. Med. receptionist Dr. J.J. Williams, Franklin, N.C., 1981-83; chaplain's sec. Mt. Pisgah Acad., Candler, N.C., 1983-86; personal sec., asst. archtl. libr. Andrews U., Berrien Springs, Mich., 1986-87; teaching asst. Carnegie Mellon U., Pitts., 1991-93; architect pvt. practice, Cocoa Beach, Fla., 1993—. Scholar Andrews U., 1986-87. Adventist. Avocations: tae kwon do, scuba diving, swimming, tennis, racquetball. Home: 3609 N Atlantic Ave Apt 306C Cocoa Beach FL 32931-3412

WILLIAMS, KARL MORGAN, instrumentation engineer, researcher; b. Lincoln, Nebr., Nov. 5, 1958; s. Morgan Glen and Julia Ella (Bishop) W.; m. Linda Jan Lock, Aug. 13, 1983; children: Violet, Ethan, Eric, Eli. BS, Tex. A&M U., 1984, MS, 1989. Design engr. O.I. Corp., College Station, Tex., 1985-87; rsch. engr. Phyto Resource Rsch., College Station, 1987-89; product engr. O.I. Corp., College Station, 1989—. Contbr. articles to profl. jours. Mem. Am. Chem. Soc., Am. Soc. for Gravitational and Space Biology. Republican. Methodist. Achievements include three patents in a permeation pH adjustment system for nitrogen sensitive electrolytic conductivity detector for gas chromatography; a sweep gas method for keeping photo ionization detector lamp windows clean; a method for mounting two detectors onto the same port in gas chromatography; and three patents pending. Home: 1009 Edgewood Bryan TX 77802 Office: O I Corp PO Box 9010 College Station TX 77841

WILLIAMS, KOURT DURELL, operations analyst; b. L.A., June 15, 1962; s. Wayne R. and Jo P. (McFrazier) W.; m. Monica L. Rentie, Dec. 19, 1987; children: Andrise Monique, Kimberli Antoinette. BA, UCLA, 1985; MBA, Nat. U., San Diego, 1989. Econ. analyst Lockheed Corp., Burbank, Calif., 1984-86, ops. analyst advanced devel. programs, 1986-87; ops. analyst Rockwell Space Systems Div., Downey, Calif., 1987—; mem. Industry Environ. Task Force, 1985-86. Chmn. Black Edn. Commn., L.A. Unified Schs., 1990-92; chmn. pro team Black Leadership Coalition on Edn., L.A. County. Mem. NAACP (life), AIAA, Nat. Mgmt. Assn., Nat. Notary Assn., Kappa Alpha Psi (life, grad. advisor).

WILLIAMS, LUTHER STEWARD, biologist, federal agency administrator; b. Sawyerville, Ala., Aug. 19, 1940; s. Roosevelt and Mattie B. (Wallace) W.; m. Constance Marie Marion, Aug. 23, 1963; children: Mark Steward, Monique Marie. B.A. magna cum laude, Miles Coll., 1961; M.S., Atlanta U., 1963; Ph.D., Purdue U., 1968, D.Sc. (hon.), 1987. NSF lab. asst. Spelman Coll., 1961-62; NSF lab. asst. Atlanta U., 1962-63, instr. biology, faculty research grantee, 1963-64, asst. prof. biology, 1969-70, prof. biology, 1984-87, pres., 1984-87; grad. teaching asst. Purdue U., West Lafayette, Ind., 1964-65, grad. research asst., 1965-66, asst. prof. biology, 1970-73, assoc. prof., 1973-79, prof., 1979-80, NIH Career Devel. awardee, 1971-75, asst. provost, 1976-80; dean Grad. Sch., prof. biology Washington U., St. Louis, 1980-83; v.p. acad. affairs, dean Grad. Sch. U. Colo., Boulder, 1983-84; Am. Cancer Soc. postdoctoral fellow SUNY-Stony Brook, 1968-69; assoc. prof. biology MIT, 1973-74; spl. asst. to dir. Nat. Inst. Gen. Med. Scis., NIH, Bethesda, Md., 1987-88; dep. dir. Nat. Inst. Gen. Med. Scis. NIH, Bethesda, Md., 1988-89; sr. sci. advisor to dir. NSF, Washington, 1989-90, asst. dir. for edn. and human resources, 1990—; chmn. rev. com. MARC Program, Nat. Inst. Gen. Med. Scis., NIH, 1972-76; grant reviewer NIH, 1971-73, 76, NSF, 1973, 76-80, Med. Research Council of N.Z., 1976; mem. life scis. screening com. recombinant DNA adv. com. HEW, 1979-81; mem. nat. adv. gen. med. sci. council NIH, 1980-85; mem. adv. com. Office Tech. Assessment, Washington, 1984-87; chmn. fellowship adv. com. NRC Ford Found., 1984-85; mem.-at-large Grad. Record Exam. Bd., 1981-85, chmn. minority grad. edn. com., 1983-85; mem. health, safety and environ. affairs. com. Nat. Labs., U. Calif., 1981-87; mem. adv. panel Office Tech. Assessment, U.S. Congress, 1985-86; mem. fed. task force on women, minorities and the handicapped in sci. and tech., 1987-91; mem. adv. panel to dir. sci. and tech. ctrs. devel. NSF, 1987-88; mem. nat. adv. com. White House Initiative on Historically Black Colls. and Univs. on Sci. and Tech., 1986-89; numerous other adv. bds. and coms. Contbr. sci. articles to profl. jours. Vice chmn. bd. advisors Atlanta Neighborhood Justice Ctr., 1984-87; bd. dirs. Met. Atlanta United Way, 1986-87, Butler St. YMCA, Atlanta, 1985-87; trustee Atlanta Zool. Assn., 1985-87, Miles Coll., 1984-87, Atlanta U., 1984-87, 90—; mem. nominating com. DANA Found. NIH predoctoral fellow Purdue U., 1966-68. Mem. Am. Soc. Microbiology, Am. Chem. Soc., Am. Soc. Biol. Chemists (mem. ednl. affairs com. 1979-82, com. on equal opportunities for minorities 1972-84), AAAS, N.Y. Acad. Scis. Home: 11608 Split Rail Ct Rockville MD 20852-4423 Office: NSF 1800 G St NW Washington DC 20550-0002

WILLIAMS, M. WRIGHT, psychologist, educator; b. Houston, May 10, 1949; s. Marvin Wright and Mary Katherine (Lacey) W.; m. Brenda Eileen Wobig, Apr. 6, 1985; 1 child, Christopher Wright. BA, U. Tex., 1971; MS, Fla. State U., 1976, PhD, 1978. Lic. psychologist, Tex. Staff psychologist VA Med. Ctr., Houston, 1979—; pvt. practice Houston, 1980—; asst. prof. Baylor U. Coll. of Medicine, Houston, 1980—; adj. clin. asst. prof. U. Houston, 1982—. Contbr. articles to profl. jours. USPHS/NIMH fellow Fla. State U., 1973-76; recipient Cert. of Recognition, DAV, 1980. Mem. APA (cert. of recognition 1981), Am. Group Psychotherapy Assn., Am. Assn. Correctional Psychologists, Southeastern Psychol. Assn., Tex. Psychol. Assn., Houston Psychol. Assn. (exec. com. 1984-93, sec.-treas. 1986-87). Presbyterian. Avocation: jogging. Home: 4126 Falkirk Ln Houston TX 77025-2909 Office: VA Med Ctr Psychology Svc 2002 Holcombe Blvd Houston TX 77030-4211

WILLIAMS, MARK EDWARD, geriatrician; b. Charlottesville, Va., Feb. 2, 1950; s. William Lee and Merlyn (Carlton) W.; m. Jane Clark Williams, Aug. 11, 1973; children: John, James. AB, U. N.C., 1972, MD, 1976. Diplomate Am. Bd. Internal Medicine and Geriatric Medicine. Instr. medicine U. N.C. Sch. Medicine, Chapel Hill, 1979-80, med. dir., nurse practitioner evening clinic, 1979-80, asst. prof. medicine, 1985-88, assoc. prof., dir. program on aging, 1988—; fellow, instr. medicine (geriatrics) Monroe Community Hosp. U. Rochester (N.Y.) Sch. Medicine, 1980-81, sr. instr. medicine Monroe Community and Strong Meml. hosps., 1981-82, asst. prof., 1982-84, dir. Geriatric Consultative Svc. Monroe Community Hosp., 1983-84; co-dir. Geriatric Edn. Ctr. U. N.C. Sch. Pub. Health, Chapel Hill, 1986-88; mem. study sects. and panels Nat. Inst. on Aging Monitoring Bd., 1983-89; mem. consensus devel. conf. panel NIH, 1991; mem. Adv. Panel for Office Tech. Assessment, Washington, 1987-89; cons. pvt. and govt. founds. Assoc. editor: Am. Geriatrics Soc. Geriatrics Rev. Syllabus, 1992; mem. bd. Am. Geriatrics Soc., 1987—; author comml. computer software biostats., music, children's edn.; contbr. articles to profl. jours., chpts. to books; reviewer med. jours. John Motley Morehead scholar U. N.C., Chapel Hill, 1972-76, Robert Wood Johnson clin. scholar, 1978-80; fellow Salzburg (Austria) Seminar, 1983; grantee Robert Wood Johnson Found., U. Rochester, 1983-86, Nat. Int. Aging, U. Rochester, 1983-88, Bur. Health Professions, Health Resources and Svcs. Adminstrn., U. N.C., Chapel Hill, 1985-88, 88-92, John and Mary Markle Found., U. N.C., 1989, Robert Wood Johnson Found., U. N.C., Chapel Hill, 1987-91, Nat. Inst. Aging, U. N.C., Chapel Hill, 1990—. Mem. ACP, AAAS, Am. Geriatrics Soc. (bd. dirs. 1991—), Soc. Gen. Internal Medicine, Am. Assn. Pub. Health, Gerontol. Soc. Am., Am. Fedn. Clin. Rsch., Assn. Anthropology and Gerontology, N.C. Med. Soc. (mem. aging com. 1987—), Nat. Coun. on Aging, INc., Royal Soc. Medicine, Sherlock Holmes Soc. London, Assn. Acad. Geriatric Medicine Programs. Achievements include creation of timed manual performance test-a well validated protocol that predicts future needs for care; first report of randomized clin. trial in ambulatory geriatric assessment. Office: U NC Sch Medicine Program on Aging 141 MacNider Bldg CB # 7550 Chapel Hill NC 27599-7550

WILLIAMS, MARSHA KAY, data processing executive; b. Norman, Okla., Oct. 26, 1963; d. Charles Michael and Marilyn Louise (Bauman) Williams; m. Dale Lee Carabetta, Dec. 13, 1981. Student, Metro. State Coll., Denver. Data processing supr. Rose Mfg. Co., Englewood, Colo., 1981-84, Mile High Equip. Co., Denver, 1984-88; sr. tech. systems analyst Ohmeda Monitoring Systems, Louisville, Colo., 1988—. Mem. Bus. and Profl. Women's Assn. (Young Careerist 1991). Home: 3302 W 127th Ave Broomfield CO 80020-5822 Office: Ohmeda Monitoring Systems 1315 W Century Dr Louisville CO 80027-9560

WILLIAMS, MARSHA RHEA, computer scientist, educator, researcher, consultant; b. Memphis, Aug. 4, 1948; d. James Edward and Velma Lee (Jenkins) W.; BS, Beloit Coll., 1969; MS in Physics, U. Mich., 1971; MS in Systems and Info. Sci., Vanderbilt U., 1976, PhD in Computer Sci., 1982. Cert. data processing (CDP). Engring. coop. student Lockheed Missiles & Space Co., Sunnyvale, Calif., 1967-68; asst. transmission engr. Ind. Bell Telephone Co., Indpls., 1971-72; systems analyst, instr. physics Memphis State U., 1972-74; computer-assisted instrn. project programmer Fisk U., 1974-76; mem. tech. staff Hughes Rsch. Labs., Malibu, Calif., 1976-78; assoc. systems engr. IBM, Nashville, 1978-80; rsch. and teaching asst. Vanderbilt U., Nashville, 1980-82, spl. asst. to dean Grad. Sch., spring 1981, minority engr. advisor, 1975-76; cons. computer-assisted instrn. project Meharry Med. Coll., Nashville, summer 1982; assoc. prof. computer sci. Tenn. State U., Nashville, 1982-83, 84-90, full, tenured prof., 1990—, univ. marshal, 1992—; assoc. prof. U. Miss., Oxford, 1983-84, faculty senator; assoc. program dir. Applications of Advanced Techs. Sci. and Engring. Edn., NSF, 1987-88, apptd. USRA Sci. and Engring. Edn. Coun., Advanced Design Program, 1992—; cons. on minority scientists and engrs. Univ. Space Rsch. Assn., Washington, 1988; vis. scientist CSNET-Minority Instn. Networking Project Bolt, Beranek & Newman, Cambridge, Mass., 1989; mem. tech. staff Bell Communications Rsch., Red Bank, N.J., 1990; presenter papers profl. meetings. Editor-in-chief newspaper Pilgrim Emanuel Bapt. Ch., 1975-76; adv. Chi Rho Youth Fellowship, Temple Bapt. Ch., 1975-81, adv. com. Golden Outreach Sr. Citizens Fellowship, 1979-80, 86-87, 89—, Women's Day speaker, 1979, 81, Ebeneezer MIssionary Bapt. Ch., 1993; adviser Nat. Soc. Black Engring. Students, 1983-84; founder, coord. Tenn. State U. Assn. for Excellence in Computer Sci., Math. and Physics (AE-COMP), 1986-87, coord. Tech. Opportunites Fair, 1986, 87; dir. Tenn. State U. Minorities in Sci., Engring. & Tech. Rsch. Project-MISET, 1989—. Recipient Disting. Instr. award 1984; grantee Digital Equipment Corp., 1989-92; faculty rsch. grantee Tenn. State U., 1993. Mem. NAACP (nat. judge ACT-SO sci. olympics 1992), Assn. Computing Machinery, IEEE Computer Soc., Am. Soc. Engring. Mgmt., Data Processing Mgmt. Assn. (edn. chmn., bd. dirs. 1986), Tenn. Acad. Sci. Achievements include founding Assn. for Excellence in Computer Sci., Math., and Physics, research in database, network and human-computer interfacing for broadening minority participation in science, engineering and technology. Home: PO Box 270093 Nashville TN 37227-0093 Office: Tenn State U Dept Physics Math & Computer Sci 3500 Merritt Blvd Nashville TN 37209-1561

WILLIAMS, MAURICE, clinical psychologist; b. Jackson, Miss., Feb. 18, 1952; s. Dan and Barbara Jean (Cole) W.; m. Thomasine McNair, Oct. 23, 1982; children: Elliott Maurice, Wesley Thomas. BS, Jackson State U., 1974; MS, U. Ga., 1976, PhD, 1978. Lic. clin. psychologist, Ala. Lectr. U. Ga., Athens, 1977-78; asst. prof. Tougaloo (Miss.) Coll., 1979-82, acting dept. chair, 1980-82; staff psychologist East La. State Hosp., Jackson, 1982-88; psychologist II Bryce Hosp., Tuscaloosa, Ala., 1988—; adj. prof. Jackson State U., 1978-79. Mem. APA, Southeastern Psychol. Assn., La. Psychol. Assn., Chi Psi. Office: Bryce Hosp 200 University Blvd Tuscaloosa AL 35401-1294

WILLIAMS, MAX LEA, JR., engineer, educator; b. Aspinwall, Pa., Feb. 22, 1922; s. Max Lea and Marguerite (Scott) W.; m. Melba Cameron, July 26, 1967; children: Gregory S., Christine C. Williams Mann, Richard L. BS in Mech. Engring, Carnegie Inst. Tech., 1942; MS, Calif. Inst. Tech., Aero. Engr. and PhD in Aeronautics, 1950. Registered profl. engr. Calif., La., Pa., Utah. Mem. faculty, successively lectr., asst. prof., assoc. prof., prof. aeronautics Calif. Inst. Tech., 1948-66; dean, prof. engring. U. Utah, 1965-73, disting. prof. engring., 1973; dean, prof. engring. U. Pitts., 1973-85, dean emeritus, disting. service prof. engring., 1985—, disting. svc. prof. engring. emeritus, 1990—; Gen. Lew Allen disting. vis. prof. aero. Air Force Inst. Tech., 1985-87; rsch. scholar von Humbolt Found., 1992-93; mem. exec. com. Internat. Congress Fracture, ed. Internat. Jour. Fracture, 1965—; mem. sci. adv. bd. USAF, 1985-89, ad hoc mem., 1989—; sci. advisor Acquisition Logistic Ctr. USAF, 1987-88; sci. dir. NATO Adv. Study Inst., Italy, 1967; mem. biomaterials advisory commn. Nat. Inst. Dental Research, 1967-70; mem. chem. rocket advisory com. NASA, 1968-73; mem. engring. advisory com. NSF, 1969-72; mem. nat. materials adv. bd. NRC, chmn. Council on Materials Structures and Design, 1975-77; assoc. mem. Def. Sci. Bd., 1975-76, 82-85; chmn. structural mechanics com. Interagy. Task Force OSTP, coop. automotive res. program, 1979-80; cons., lectr. in field; cons. Dept. Def., 1970-93, Dept. State, 1973-81; dir. MPC Corp., Pitts., 1974-87; bd. dirs. U.S. World Energy Conf., Washington, 1977-83; pres. Utah Engring. Devel. Found., 1969-79; founding chmn. BCR Nat. Lab., U. Pitts., 1983-85; adviser Regional Indsl. Devel. Corp., Pitts., 1973-83, Ams. for Energy Independence, Washington, 1976-82; mem. nat. engr. adv. bd. Mercer U., 1988—. Editor in chief Internat. Jour. Fracture, 1965—; cons. editor Ency. Solid State, 1969-79; contbr. numerous articles to profl. jours. Served to capt. USAAF, 1943-46. NSF sr. postdoctoral fellow Imperial Coll., London, 1971-72; recipient Adhesion Research award ASTM, 1975, Solid Rocket Tech. Achievement award AIAA, 1988, Merit Civilian Svc. citation USAF, 1989; von Humboldt Found. Rsch. scholar, 1993—. Fellow Soc. Exptl. Mechanics; assoc. fellow Am. Inst. Aero. and Astronautics; mem. AAAS, Utah Acad. Scis., N.Y. Acad. Scis., Am. Soc. Engring. Edn., Am. Chem. Soc., Internat. Congress on Fracture (exec. com. 1965—, hon. fellow 1989), Soc. Rheology, Sigma Xi (nat. lectr. 1972), Theta Tau, Tau Beta Pi. Clubs: Cosmos (Washington); University (Pitts.). Home: 1236 Murrayhill Ave Pittsburgh PA 15217-1217

WILLIAMS, MORGAN LEWIS, pharmacist; b. Summerville, S.C., Feb. 28, 1948; s. Carroll S. and Venie Aline (Pound) W. BS in Pharmacy, U. S.C., 1971; Dipl., Niles Bryant Sch., Sacramento, 1982, Am. Sch. Piano Tuning, Morgan Hill, Calif., 1979. Lic. pharmacist, S.C., N.C.; registered tuner technician. Pharmacist VA Hosp., Columbia, S.C., 1972-75, Med. Ctr. Pharmacy, Florence, S.C., 1975-78, Revco Drugs, Cheraw, S.C., 1978—; organist Mt. Hebron United Meth. Ch., W. Columbia, S.C., 1967-76, Latta (S.C.) United Meth. Ch., 1977-81, Swansea (S.C.) United Meth. Ch., 1983-86, St. David's Episcopal Ch., Cheraw, 1989—; piano technician, Florence, Columbia, Cheraw, 1978—; inst. piano tech. Cheraw, 1991. Asst. editor Lodge Newsletter, Trestle Board, 1985-89. Mem. Piano Technicians Guild (editor 1991—), Am. Guild Organists, S.C. Pharm. Assn. (pres. 1981-82), Optimist Club, Masons (master 1989, musician 1990—). Republican. Episcopalian. Avocations: bowling, playing piano, physical fitness, singing. Home: 151 Tim Hickey Cir Cheraw SC 29520-7620 Office: Revco Drugs 932 Chesterfield Hwy Cheraw SC 29520-7008

WILLIAMS, NORRIS HAGAN, JR., biologist, educator, curator; b. Birmingham, Ala., Mar. 31, 1943; s. Norris Hagan Sr. and Ernestyne Edna (Brown) W.; m. Nancy Jane Fraser, June 26, 1970; children: Matthew Ian, Luke Fraser. BS, U. Ala., 1964, MS, 1967; PhD, U. Miami, 1971. Asst. prof. Fla. State U., Tallahassee, 1973-78, assoc. prof., 1978-81; assoc. curator U. Fla., Gainesville, 1981-83, curator, 1983—, dept. chmn., 1985—. Co-author: Orchid Genera of Costa Rica, 1986, Identification Manual for Wetland Plant Species of Florida, 1987; author: (chpt.) Orchid Biology II, 1982, Handbook of Experimental Pollination Biology, 1983; contbr. article to Biol. Bull., 1983. Mem. Assn. for Tropical Biology, Bot. Soc. Am., Soc. for Study Evolution, Linnean Soc. of London. Democrat. Office: U Fla Fla Mus Natural History Gainesville FL 32611

WILLIAMS, OLIVER ALAN, JR., computer scientist; b. Phila., Feb. 4, 1948; s. Oliver Alan Sr. and Matilda (Minnick) W. BS in Psychology, Temple U., 1968; MS in Computer Sci., Hartford Grad. Ctr., 1977; PhD in Computer Info. Sci., Rensselear Polytech. Inst., 1980. System programmer Hartford-ITT Ins. Co., 1970-74; software analyst Chesebrough Pond, Inc., Greenwich, Conn., 1976-78; system analyst Diversified Tech. Cons., Inc., Waterbury, Conn., 1978-80; auditor security analyst ISQ Assocs., Inc., Waterbury, 1980—. Lt. (j.g.) USNR. Office: ISQ Assocs Inc PO Box 9281 Waterbury CT 06724

WILLIAMS, PETER CHARLES, engineer; b. Chgo., July 14, 1949; s. Maurice Jaquetot and Betty Jane (Bath) W.; m. Ann Hazard Sawyer, Aug. 28, 1971; children: Marion, Joan, Joseph. BS in Chem. Engring., Tufts U., 1972, MSME, 1972. Registered profl. engr., Ohio. Environ. cons. Eli Bulba, Inc., Cambridge, Mass., 1973-74; ind. engring. cons. Gates Mills, Ohio, 1975-76; project engr. Whitey Co., Highland Heights, Ohio, 1976-83, chief engr., 1983-91; chief engr. Crawford Fitting Co., Solon, Ohio, 1992—. Mem. ASME, Sigma Xi, Tau Beta Pi. Achievements include 28 U.S. patents and numerous foreign patents. Office: Crawford Fitting Co 29500 Solon Rd Solon OH 44139

WILLIAMS, PETER MACLELLAN, nuclear engineer; b. N.Y.C., Aug. 30, 1931; s. Gilbert Harris and Evelyn (Buss) W.; m. Lois Crane, Oct. 6, 1956; children: Jane, Gilbert, Katherine, Anne, Louise, Robert. B in Chem. Engring., Cornell U., 1954; MS in Nuclear Engring., MIT, 1957; PhD in Nuclear Engring., U. Md., 1971. Engr. DuPont Savannah River, Aiken, S.C., 1954-55; task engr. AGN, San Ramon, Calif., 1957-60; project mgr. Am. Machine & Fdry., Greenwich, Conn., 1960-62; research staff Princeton U., N.J., 1962-67; sr. project mgr., specialist in high temperature gas cooled reactors U.S. Nuclear Regulatory Commn., Washington, 1967-91; dir. div. high temperature gas cooled reactors U.S. Dept. of Energy, Washington, 1991—; mem. Chernobyl Tracking Team, 1986; U.S. del. to gas-cooled reactors working group, Internat. Atomic Energy Agy., 1991; steering com. mem. U.S.-Japan Implementing Agreement on gas-cooled reactors, 1991. Contbr. articles to profl. jours.; author various reports. Scoutmaster Boy Scouts Am., Potomac, Md., 1972, cubmaster, 1983-86; pres. PTA Winston Churchill High Sch., Potomac, 1981. Assoc. fellow AIAA; mem. Am. Nuclear Soc., Sigma Xi. Democrat. Unitarian. Achievements include patent for liquid core nuclear rocket. Home: 9418 Thrush Ln Potomac MD 20854-3991 Office: US Dept Energy Adv Reactors Div HTGRS NE-451 Mail Stop Washington DC 20585

WILLIAMS, PHILIP COPELAIN, gynecologist, obstetrician; b. Vicksburg, Miss., Dec. 9, 1917; s. John Oliver and Eva (Copelain) W.; B.S. magna cum laude, Morehouse Coll., 1937; M.D., U. Ill., 1941; m. Constance Shielda Rhetta, May 29, 1943; children—Philip, Susan Carol, Paul Rhetta. Intern, Cook County Hosp., Chgo., 1942-43, resident in ob-gyn, 1946-48; resident in gynecology U. Ill., 1948-49; practice medicine specializing in ob-gyn, Chgo., 1949—; mem. staff St. Joseph Hosp., Ill. Masonic Hosp., Cook County Hosp., McGaw Hosp.; clin. prof. Med. Sch. Northwestern U., Chgo. Bd. dirs. Am. Cancer Soc. Chgo. unit and Ill. div. Served with U.S. Army, 1943-45. Recipient Civic award Loyola U., 1970; Edwin S. Hamilton Interstate Teaching award, 1984; diplomate Am. Bd. Ob-Gyn, Fellow ACS, Internat. Coll. Surgeons; mem. AMA, Chgo., Ill. med. socs., AMA, Chgo. Gynecol. Soc. (treas. 1975-78, pres. 1980-81), Am. Fertility Soc., Inst. Medicine, N.Y. Acad. Scis., AAAS. Presbyn. Clubs: Barclay, Carlton, Plaza. Contbr. articles to profl. jours. Home: 1040 N Lake Shore Dr Chicago IL 60611-1165 Office: 200 E 75th St Chicago IL 60619-2299

WILLIAMS, QUENTIN CHRISTOPHER, geophysicist; b. Wilmington, Del., Jan. 1, 1964; s. Ferd Elton and Anne Katherine (Lindberg) W.; m. Elise Barbara Knittle, Dec. 19, 1987; 1 child, Byron Frederick. AB, Princeton U., 1983; PhD, U. Calif., Berkeley, 1988. Rsch. geophysicist Inst. of Tectonics, U. Calif., Santa Cruz, 1988-91; asst. prof. dept. earth sci. U. Calif., Santa Cruz, 1991—. Contbr. articles to profl. jours. Presdl. Faculty fellow, 1993—. Mem. Am. Geophys. Union, Am. Phys. Soc. Office: U Calif Santa Cruz Dept Earth Sciences Santa Cruz CA 95064

WILLIAMS, REDFORD BROWN, medical educator; b. Raleigh, N.C., Dec. 14, 1940; s. Redford Brown Sr. and Annie Virginia (Betts) W.; m. Virginia Carter Parrott, August 9, 1940; children: Jennifer Betts, Lloyd Carter. AB, Harvard U., 1963; MD, Yale U., 1967. Diplomate Am. Bd. Internal Medicine. Intern, then resident Yale-New Haven Med. Ctr., 1967-70; sr. surgeon USPHS, Bethesda, Md., 1970-72; asst. prof. Duke U. Med. Ctr., Durham, N.C., 1972, prof. psychiatry, 1977—, prof. psychology, 1990—, dir. behavioral medicine rsch. ctr., 1985—; cons. NIH rev. coms., Bethesda, 1977—. Author: The Trusting Heart, 1989, Anger Kills, 1993; contbr. articles to profl. jours. Dir. N.C. Heart Assn., Chapel Hill, 1980-83. Recipient Rsch. Scientist award NIMH, 1974—; NIH grantee, 1986—. Fellow Soc. Behavioral Medicine (pres. 1984-85, Upjohn Disting. Scientist award 1992); mem. Am. Psychosomatic Soc. (bd. dirs. 1978-81, pres. 1992—). Unitarian Universalist. Avocation: tennis. Office: Duke U Med Ctr Box 3708 Durham NC 27710

WILLIAMS, RICHARD, JR., animal scientist; b. Centerville, Miss., Jan. 31, 1947; s. Richard Sr. and Effie Etherine Williams; m. Voletta Ann Polk, Mar. 8, 1981; children: Sharon, TeAndrea, Orland, DeVeron. BS, Alcorn State U., 1970; MS, Miss. State U., Starkville, 1976; postgrad., Wayne State U., U. Wash., U. Minn. Cert. tchr. in agrl. edn., math. and sci. Forestry specialist Oreg. State U., Portland, 1966-67; instr., animal scientist Alcorn State U., Lorman, Miss., 1970—, instr. coop. ext., 1989—. Contbr. to Jour. Animal Sci., Miss. Acad. of Sci. Rsch. Symposiums; co-author 6 MAFES publs. Deacon Holly Grove M.B. Ch., 1989—; active Future Farmers Am. Mem. Kiwanis (pres. 1989, v.p. 1993, judging team 4-H club 1991, Outstanding Kiwanian 1991), Omega Psi Phi (pres. 1989-91, Omega Man of Yr. 1991), Alpha Tau Alpha, Gamma Sigma Delta, Masons. Democrat. Baptist. Home: Acorn State University Alcorn State U PO Box 1027 Lorman MS 39096 Office: Alcorn State U Agrl Rsch Program PO Box 330 Lorman MS 39096

WILLIAMS, ROBERT CARLTON, electrical engineer; b. Copperhill, Tenn., July 14, 1948; s. Robert Carlton and Ruth Vivian (Becker) W.; m. Evelyn Snyder; 1 child, Sean Phillip. BS in Mech. Engring., U. Tenn., 1971. Designer NASA, Huntsville, Ala., 1968-69, Tenn. Chem. Co., Copperhill, 1969-70; draftsman TVA, Knoxville, 1970-71, design engr., 1971-76, tech. supr., 1976-81, sr. elec. engr., 1981-84, prin. mech. engr., 1984-86, prin. nuclear engr., 1984-86; lead elec. engr. TVA, Chattanooga, 1986-87; asst. chief electrical engr. TVA, Knoxville, 1987-89; instrumentation and controls engrng. mgr. TVA, Knoxville, Tenn., 1989-90, chief elec. engr., 1990-91, chief I&C engr., 1991—. Mem. IEEE, Nat. Mgmt. Assn., Instrument Soc. Am. Republican. Baptist. Avocations: bowling, golf, building model ships. Home: 815 Divot Ct Daisy TN 37379 Office: TVA 1101 Market St Chattanooga TN 37402

WILLIAMS, ROBERT H., environmental scientist. Sr. rsch. scientist ctr. for energy & environ. studies Princeton (N.J.) U. MacArthur fellow John D. and Katherine T. MacArthur Found., 1993; recipient Sadi Carnot award in Energy Conservation, U.S. Dept. Energy, 1991. Office: Princeton U Ctr Energy & Environ Studies P O Box 5263 Princeton NJ 08544*

WILLIAMS, ROBERT HENRY, oil company executive; b. El Paso, Jan. 12, 1946; s. William Frederick and Mary (Page) W.; m. Joanne Marie Mudd, Oct. 22, 1967; children: Lara, Michael, Suzanne, Jennifer. BS in Physics, U. Tex., El Paso, 1968; PhD in Physics, U. Tex., Austin, 1973; MS in Physics, Va. Poly. Inst., 1971. Dir. Gulf Oil R&D, Houston, 1978-81; tech. mgr. Gulf Oil Internat., Houston, 1981-83; exploration mgr. Gulf Oil Co., Houston, 1983-85; mgr. geophys. rsch. Tenneco Oil Co., Houston, 1985-87, mgr., chief geophysicist, 1987-88; founder, mng. dir. Dover Energy, Houston, 1988—; exec. v.p. Tatham Offshore Inc, Houston, 1989—, also bd. dirs.; chmn., chief exec. officer Dover Tech. Inc., Houston, 1989—; cons. Tenneco Inc., Houston, 1989—, Deeptech Internat., 1992—, Ukraine Acad. of Sci., 1993; bd. dirs. Dover Energy, Dove Tech.; bd. dirs., sr. v.p. DeepTech Inc., 1991—; co-founder, chmn., chief exec. officer Castaway Graphite Rods, Inc., 1990—. Contbr. articles to profl. jours. Councilman Boy Scouts Am., Houston, 1989—; patron Mus. Fine Arts, Houston, 1990, 91, 92, 93; leader Girl Scouts U.S.A., Houston, 1989—. Mem. Soc. Exploration Geophysics, Am. Assn. Petroleum Geologists, Am. Geophys. Union. Republican. Avocations: scuba diving, investing, fishing. Office: Dover Tech Tex Commerce Tower 600 Travis Ste # 7500 Houston TX 77002

WILLIAMS, RONALD DAVID, electronics materials executive; b. Marshall, Ark., Mar. 15, 1944; s. Noble Kentucky and Elizabeth (Karns) W.; m. Beth L. Williams, Nov. 1977; children: Stephanie Noble, Keith Michael. BA, Columbia U., 1966, BS, 1967, MBA, 1973. Process engr. DuPont, Deepwater, N.J., 1966; design engr. Combustion Engring. Co., Hartford, 1971; cons. Arthur Andersen & Co., N.Y.C., 1973-76; corp. planner Amax Inc., Greenwich, Conn., 1976-77, group planning adminstr., 1978-80, mgr. corp. planning and analysis, 1980-84, dir. fin. analysis, 1984-86; project mgr. Olin Corp., Stamford, Conn., 1977-78; mgr. ops planning, analysis Savin Corp., Stamford, 1986-88; dir. fin., Bandgap Tech. Corp., Broomfield, Colo., 1988-90; v.p. fin. and adminstrn., 1990-93; dir., 1991—; v.p. and gen. mgr. Bandgap Chem. Corp., 1992—. Served with USN, 1967-70, Vietnam. NASA traineeship, 1971; S.W. Mudd scholar, 1971. Mem. AAAS, Am. Chem. Soc., Am. Mgmt. Assn. Democrat. Club: Appalachian Mountain, Boulder Road Runners. Home: 7361 S Meadow Ct Boulder CO 80301-3951 Office: Bandgap Tech Corp 325 Interlocken Pky Broomfield CO 80021-3437 also: Bandgap Chem Corp 1861 Lefthand Circle Longmont CO 80501

WILLIAMS, RONALD OSCAR, systems engineer; b. Denver, May 10, 1940; s. Oscar H. and Evelyn (Johnson) W. BS in Applied Math., U. Colo. Coll. Engring., 1964, postgrad. U. Colo., U. Denver, George Washington U. Computer programmer Apollo Systems dept., missile and space divsn. Gen. Electric Co., Kennedy Space Ctr., Fla., 1965-67, Manned Spacecraft Ctr., Houston, 1967-68; computer programmer U. Colo., Boulder, 1968-73; computer programmer analyst def. systems divsn. System Devel. Corp. for NORAD, Colorado Springs, 1974-75; engr. def. systems and command-and-info. systems Martin Marietta Aerospace, Denver, 1976-80; systems engr. space and comm. group, def. info. systems divsn. Hughes Aircraft Co., Aurora, Colo., 1980-89. Vol. fireman Clear Lake City (Tex.) Fire Dept., 1968; officer Boulder Emergency Squad, 1969-76, rescue squadman, 1969-76, liaison to cadets, 1971, pers. officer, 1971-76, exec. bd., 1971-76, award of merit, 1971, 72, emergency med. technician 1973—; spl. police officer Boulder Police Dept., 1970-75; spl. dep. sheriff Boulder County Sheriff's Dept., 1970-71; nat. adv. bd. Am. Security Coun., 1979-91, Coalition of Peace through Strength, 1979-91. Served with USMCR, 1958-66. Decorated Organized Res. medal; recipient Cost Improvement Program award Hughes Aircraft Co., 1982, Systems Improvement award, 1982, Top Cost Improvement Program award, 1983. Mem. AAAS, Math. Assn. Am., Am. Math. Soc., Soc. Indsl. and Applied Math., AIAA, Armed Forces Comm. and Electronics Assn., Assn. Old Crows, Am. Def. Preparedness Assn., Marine Corps Assn., Air Force Assn., U.S. Naval Inst., Nat. Geog. Soc., Smithsonian Instn., Met. Opera Guild, Colo. Hist. Soc., Hist. Denver, Inc., Historic Boulder, Inc., Hawaiian Hist. Soc., Denver Art Mus., Denver Botanic Gardens, Denver Mus. Natural History, Denver Zool. Found., Inc., Mensa, Hour of Power Eagles Club. Lutheran.

WILLIAMS, THEODORE P., biophysicist, biology educator; b. Marianna, Pa., May 24, 1933; married; 5 children. BS, Muskingum Coll., 1955; MA, Princeton U., 1957, PhD in Phys. Chemistry, 1959. Rsch. assoc. chemistry Brown U., 1959-61, fellow psychology, 1961-63, asst. prof. biology and med. sci., 1963-66; assoc. prof. biol. sci. Fla. State U., 1966-73, acting chmn. dept., 1970-71, prof. biol. sci., 1973—, co-dir. psychobiol. program, 1971-74, dir. Inst. Moleculat Biophysics, 1985—; prin. investigator numerous grants NIH, NSF, Dept. Energy. Recipient Alexander von Humboldt Sr. Scientist award. Mem. Assn. Rsch. Vision Ophthalmology. Achievements include research in visual processes, sensory mechanisms, fast chemical reactions. Office: Fla State U Inst Molecular Biophysics Tallahase FL 32306*

WILLIAMS, THOMAS EUGENE, pediatric hematologist and oncologist; b. Texarkana, Ark., May 13, 1936; s. Thomas Earle and Frankie Jo (Garner) W.; m. Peggy Jane O'Neill, May 31, 1958; children: Thomas Eugene, Elizabeth Anne, James David. BA, Yale U., 1958; MD, U. Tex. Southwestern Med. Sch., 1962. Diplomate Am. Bd. Pediatrics, Am. Bd. Pediatric Hematology and Oncology. Rotating intern Hermann Hosp., Houston, 1962-63; pediatric resident Children's Med. Ctr., Dallas, 1963-65; fellow pediatric hematology U. Va. Sch. Medicine, Charlottesville, 1967-68; research assoc. Cancer Research Lab., U. Va., Charlottesville, 1968-69; asst. prof. pediatrics and pathology U. Tex. Health Sci. Ctr. at San Antonio, 1969-72, assoc. prof. pediatrics, asst. prof. pathology, 1972-73, assoc. prof. pediatrics and pathology, 1973-79, assoc. prof. pediatrics, 1985—; med. dir. Santa Rosa Children's Hosp. Cancer Research and Treatment Ctr., 1974-79, South Tex. Comprehensive Hemophilia Ctr., 1977-79, dir. pediatric bone marrow transplantation program, 1988—; sr. clin. research scientist Burroughs Wellcome Co., 1979-85; clin. assoc. prof. pediatrics U. N.C. Sch. Medicine, 1979-85; clin. fellow bone marrow transplantation program Johns Hopkins U. Sch. Medicine, Balt., 1985. Contbr. articles to med. jours. Served to lt. comdr. USN, 1965-67. Am. Cancer Soc. advanced clin. fellow, 1968-69, 70-72; Am. Soc. Pharmacology and Exptl. Therapeutics travel awardee, 1968. Mem. Am. Soc. Clin. Oncology, Am. Soc. Hematology, Am. Soc. Pediatric Hematology and Oncology, Am. Acad. Pediatrics, Am. Assn. for Cancer Research, Am. Assn. for Cancer Edn., Am. Fedn. Clin. Research, Pediatric Oncology Group. Episcopalian. Office: 7703 Floyd Curl Dr San Antonio TX 78284-7700

WILLIAMS, THOMAS FRANKLIN, physician, educator; b. Belmont, N.C., Nov. 26, 1921; s. T.F. and Mary L. (Deaton) W.; m. Catharine Carter Catlett, Dec. 15, 1951; children: Mary Wright, Thomas Nelson. BS, U. N.C., 1942; MA, Columbia U., 1943; MD, Harvard U., 1950; DSc (hon.), Med. Coll. Ohio, 1987, U. N.C., 1992. Diplomate Am. Bd. Internal Medicine. Intern Johns Hopkins, Balt., 1950-51; asst. resident physician Johns Hopkins, 1951-53; resident physician Boston VA Hosp., 1953-54; research fellow U. N.C., Chapel Hill, 1954-56; instr. dept. medicine and preventive medicine U. N.C., 1956-57, asst. prof., 1957-61, assoc. prof., 1961-68, prof., 1968; attending physician Strong Meml. Hosp., Rochester, N.Y., 1968—; cons. physician Genesee Hosp., Rochester, N.Y., 1973—, St. Mary's Hosp., Rochester, N.Y., 1974-83, Highland Hosp., Rochester, N.Y., 1973; prof. medicine, preventive medicine and community health U. Rochester, 1968-92, also prof. radiation biology and biophysics, 1968-91, on leave, 1983-91, prof. emeritus, 1992—; mem. adv. bd. U. Rochester (Sch. Medicine and Dentistry), 1968-83; clin. prof. medicine U. Va., 1983-89; lectr. medicine Johns Hopkins U., 1983-89; clin. prof. depts family medicine and medicine Georgetown U., 1983-89; dir. Nat. Inst. on Aging NIH, 1983-91; asst. surgeon gen. USPHS, 1983-91, ret., 1991; med. dir. Monroe Community Hosp., Rochester, 1968-83; mem. rev. coms. Nat. Center for Health Services Research; mem. adv. bd. St. Ann's Home; mem. governing bd. NRC, 1981-83. Contbr. articles on endocrine disorders, diabetes, health care delivery in chronic illness and aging to profl. publs. Served with USNR, 1943-46. USPHS fellow, 1966-67; Markle scholar, 1957-61. Fellow Am. Pub. Health Assn., ACP; mem. AAAS, Inst. Medicine, NAS (coun. 1980-83, governing bd. 1981-83), Assn. Am. Physicians, N.Y. State Med. Soc., Monroe County Med. Soc., Am. Diabetes Assn. (bd. dir. 1974-80), Am. Fedn. Clin. Rsch., Soc. Exptl. Biology and Medicine, Am. Geriatrics Soc., Am. Gerontol. Soc., Rochester Regional Diabetes Assn. (pres. 1977-79), N.C. Coun. for Human Rels. (chmn. 1963-66), Am. Clin. Climatol. Assn. Episcopalian. Home: 287 Dartmouth St Rochester NY 14607-3202 Office: Monroe Community Hosp Office Med Dir Rochester NY 14620

WILLIAMS, THOMAS RHYS, educator, anthropologist; b. Martins Ferry, Ohio, June 13, 1928; s. Harold K. and Dorothy (Lehew) W.; m. Margaret Martin, July 12, 1952; children: Rhys M., Ian T., Tom R. B.A., Miami U., Oxford, Ohio, 1951; M.A., U. Ariz., 1956; Ph.D., Syracuse U., 1956. Asst. prof., asso. prof. anthropology Calif. State U., Sacramento, 1956-65; vis. asso. prof. anthropology U. Calif. Berkeley, 1962; vis. prof. anthropology Stanford U., 1976; prof. anthropology Ohio State U., Columbus, 1965-78; chmn. dept Ohio State U., 1967-71, mem. grad. council, 1969-72, mem. univ. athletic council, 1968-74, chmn. univ. athletic council, 1973-74, exec. com. Coll. Social and Behavior Scis., 1967-71; dean Grad. Sch. George Mason U., Fairfax, Va., 1978-81, prof. anthropology, 1981—; dir. Ctr. for Rsch. and Advanced Studies George Mason U., 1978-81, fed. liaison officer, 1978-81, chmn. faculty adv. bd. grad. degree program in conflict resolution, 1980-86. Author: The Dusun: A North Borneo Society, 1965, Field Methods in the Study of Culture, 1967, A Borneo Childhood: Enculturation in Dusun Society, 1969, Introduction to Socialization: Human Culture Transmitted, 1972; editor, contbr.; Psychological Anthropology, 1975, Socialization and Communication in Primary Groups, 1975, Socialization, 1983, Cultural Anthropology, 1990; contbr. articles to profl. jours. Mem. United Democrats for Humphrey, 1968, Citizens for Humphrey, 1968. Served with USN, 1946-48. Research grantee NSF, 1958, 62, Am. Council Learned Socs.-Social Sci. Research Council, 1959, 63; Ford Found. S.E. Asia, 1974, 76; recipient Disting. Faculty award Calif. State U., Sacramento, 1961, George Mason U., 1983; Disting. Teaching awards Ohio State U., 1968, 76. Fellow Am. Anthrop. Assn., Royal Anthrop. Inst. Gt. Britain; assoc. mem. Current Anthropology; mem. AAAS, Sigma Xi. Office: George Mason U Robinson Hall B-315 4400 University Dr Fairfax VA 22030-4443

WILLIAMS, TREVOR WILSON, aeronautical engineering educator; b. Balt., Sept. 13, 1954; s. George Trevor and Grace Josephine (Saltzer) W.; m. Rosemary Caroline Taylor, June 29, 1985. MA with honors, Oxford (Eng.) U., 1976; MSc with distinction, City U., London, 1978; PhD, Imperial Coll., London, 1981. SERC rsch. assoc. Kingston (Eng.) Poly. Inst., 1981-84, SERC sr. rsch. assoc., 1984-87; NRC sr. rsch. assoc. NASA Langley Rsch. Ctr., Hampton, Va., 1987-89; asst. prof. aero. engring. U. Cin., 1989-92, assoc. prof. aero. engring., 1992—; mem. collateral faculty Ohio Aerospace Inst., Cleve., 1989—; NASA summer faculty fellow Johnson Space Ctr., 1992, USAF, Edwards AFB, 1990. Contbr. articles to profl. publs. Mem. IEEE (sr.; mem. organizing com. conf. on decision and control), AIAA (sr.), Am. Astronautical Soc. (sr.). Achievements include analysis of the zeros of flexible space structures. Office: U Cin Dept Aero Engring Cincinnati OH 45221-0070

WILLIAMS, WINTON HUGH, civil engineer; b. Tampa, Fla., Feb. 14, 1920; s. Herbert DeMain and Alice (Grant) W.; grad. Adj Gens. Sch., Gainesville, Fla., 1943; student U. Tampa, 1948; grad. Transp. Sch., Ft. Eustis, Va., 1949; B.C.E., U. Fla., 1959; grad. Command and Gen. Staff Coll., Ft. Levenworth, Kans., 1964, Engrs. Sch., Ft. Belvoir, 1965, Indsl. Coll. Armed Forces, Washington, 1966, Logistics Mgmt. Center, Ft. Lee, Va., 1972; m. Elizabeth Walser Seelye, Dec. 18, 1949; children—Jan, Dick, Bill, Ann. Constrn. engr. air fields C.E., U.S. Army, McCoy AFB, Fla., 1959-61, Homestead AFB, Miami, Fla., 1961-62; civil engr. C.E., Jacksonville (Fla.) Dist. Office, 1962-64, chief master planning and layout sect., mil. br., engring. div., 1964-70; chief master planning and real estate div. Hdqrs. U.S. Army So. Command, Ft. Amador, C.Z., 1970-75, spl. asst planning and mil. constrn. programming Marine Corps Air Bases Eastern Area, Marine Corps Air Sta., Cherry Point, N.C., 1975-82; cons. engr., Morehead City, N.C., 1982—. Mem. Morehead City Planning Bd., 1982—; mem. Carteret County N.C. Health Bd., 1990—; active Boy Scouts, C.Z.; mem. nat. council U. Tampa. Served with AUS, World War II, Korean War; ETO, Korea; col. Res. Decorated Breast Order of Yun Hi (Republic of China); presdl. citation, Meritorious Service medal (Republic of Korea); eagle scout with gold palm; registered profl. engr., Fla., N.C., C.Z. Fellow ASCE; mem. Res. Officers Assn. (life, v.p. C.Am. and S.Am.), Nat. Soc. Profl. Engrs., Profl. Engrs. N.C., Am. Soc. Photogrammetry, Prestressed Concrete Inst. (profl.), Soc. Am. Mil. Engrs. (engr.), Nat. Eagle Scout Assn., Nat. Rifle Assn. Am., Am. Legion (life), Order Arrow, Theta Chi. Presbyterian. Lion. Clubs: Fort Clayton Riding (pres.), Fort Clayton Golf, Gamboa Golf and Country, Balboa Gun, Am. Bowling Congress. Home and Office: 4322 Coral Pt Morehead City NC 28557-2745

WILLIAMS-ASHMAN, HOWARD GUY, biochemistry educator; b. London, Eng., Sept. 3, 1925; came to U.S., 1950, naturalized, 1962; s. Edward Harold and Violet Rosamund (Sturge) Williams-A.; m. Elisabeth Bächli, Jan. 25, 1959; children—Anne Clare, Christian, Charlotte, Geraldine. B.A., U. Cambridge, 1946; Ph.D., U. London, 1949. From asst. prof. to prof. biochemistry U. Chgo., 1953-64; prof. pharmacology and exptl. therapeutics, also prof. reproductive biology Johns Hopkins Sch. Medicine, 1964-69; prof. biochemistry Ben May Lab. for Cancer Research, U. Chgo., 1969—, Maurice Goldblatt prof., 1973—. Contbr. numerous articles in field to publs. Recipient Research Career award USPHS, 1962-64. Fellow Am. Acad. Arts and Scis. (Amory prize 1975); mem. Am. Soc. Biochemistry and Molecular Biology. Home: 5421 S Cornell Ave Chicago IL 60615-5608 Office: Ben May Inst U Chgo Chicago IL 60637

WILLIAMSON, EDWARD L., retired nuclear engineer, consultant. Various positions Gulf Power Co., Pensacola, Fla., 1949-65; various positions, including sr. v.p. design engring. So. Co. Svcs., Inc., Birmingham, Ala., 1965-88; adj. prof. mech. engring., U. Ala., Birmingham,. Fellow ASME (life, past chmn. N.W. Fla. sect., past chmn. Birmingham sect., bd. nuclear codes and standards, Bernard F. Langer Nuclear Codes and Standards award, 1990, cons.). Home: 1834 Canyon Rd Birmingham AL 35216*

WILLIAMSON, JO ANN, psychologist; b. Wichita, Kans., Feb. 12, 1951; d. Howard T. Murray and Ferryl Arlene (Rumsey) Fleming; m. James Wallace Johnson, Apr. 5, 1984 (div. 1984); m. Michael R. Williamson, Dec. 21, 1990; children: Wesley, Wade. BA, U. Kans., 1973; MA in Psychology, U. Mo., 1974; PhD in Psychology, Auburn U., 1979. Lic. psychologist, Kans. Clin. asst. prof. Ohio State U., Columbus, 1979-80; asst. prof. Chgo. Med. Sch., 1980-81; psychologist U. Mo., Kansas City, 1981-82; psychologist II Rainbow Mental Health Ctr., Kansas City, 1982-83; pvt. practice Wichita, 1983-89; psychologist Iowa Meth. Hosp., Des Moines, 1989-90, Hutchinson (Kans.) Correctional Facility, 1990, Cedarvale, Wichita, 1991, Cowley County Mental Health, Winfield, Kans., 1991; cons. Riverside Hosp., Wichita, 1991-92; mental health psychologist Cowley County, 1991—. Contbr. articles to profl. jours. Mem. APA (div. clin. psychology).

WILLIAMSON, JOHN STEVEN, medicinal chemist; b. Jackson, Miss., Oct. 24, 1958; s. James E. and Jesse R. (Jones) W.; m. Christy M. Wyandt, May 29, 1983; children: Margaret, Elizabeth, Katherine. BS, U. Miss., 1982; PhD, U. Iowa, 1987. Rsch. asst. U.S. Geol. Survey, NSTL Station, Miss., 1977-79; rsch. asst. dept. pharmacognosy U. Miss., University, 1980-82, asst. prof., 1989—; postdoctoral fellow Yale U., New Haven, 1987-89; coord. minority affairs U. Miss., University, 1992—. Named Young Investigator Am. Assn. Colls. and Pharmacies, 1991. Mem. AAAS, Am. Soc. Pharmacognosy (Young Investigator 1990), Am. Chem. Soc. (pres.-elect locl sect. 1992), Am. Soc. Microbiology, Soc. Indsl. Microbiology (book rev. editor 1990—). Episcopalian. Office: U Miss Dept Medicinal Chemistry University MS 38677

WILLIAMSON, KENNETH LEE, chemistry educator; b. Tarentum, Pa., Apr. 13, 1934; s. James D. and Mary June (Becker) W.; m. Mary Louise Hoerner, Sept. 15, 1956; children—Christopher Lee, Tania Louise, Kevin Keith. B.A. cum laude (Nat. scholar), Harvard, 1956; Ph.D. (Allied Chem. and Dye Co. fellow), U. Wis., 1960. Mem. faculty Mt. Holyoke Coll., 1961—, prof. chemistry, 1969—, Mary E. Woolley prof. chemistry, 1984—; mem. Grad. Faculty U. Mass., 1965—; vis. prof. Cornell U., 1966, Dartmouth Coll., 1986-87, Harvard U., 1989-90, U. Trondheim, Norway, 1991, U. Louis Pasteur, Strasbourg, France, 1991, Basel U., Switzerland, 1992, U. Amsterdam, The Netherlands, 1992. Author papers and books in field; patentee in field. Mem. South Hadley Hist. Commn., 1983—. NIH postdoctoral fellow Stanford, 1960-61; NSF sci. faculty fellow U. Liverpool, Eng.; also fellow of univ., 1968-69; Guggenheim fellow, 1975-76; Oxford (Eng.) U. fellow of univ., 1976, 1983; research assoc., Calif. Inst. Tech., 1975, 82. Mem. Am. Chem. Soc., AAAS, Sigma Xi. Home: 43 Woodbridge St South Hadley MA 01075-1138

WILLIAMSON, RONALD EDWIN, waste water plant manager; b. Tarboro, N.C., Mar. 20, 1941; s. Isaac Hubert and Martha (Deal) W.; m. Sharon Lynn Winstead, Nov. 2, 1969; children: Ronda Lynn, Eleanor Renee. Grad. high sch., Tarboro, N.C. Lic. Grade A Water Treatment Operator, Grade IV Wastewater Treatment Operator. Operator water/wastewater Town of Tarboro, N.C., 1960-76; wastewater supt. Town of Chapel Hill, N.C., 1976-77; wastewater supt. Orange Water and Sewer Authority, Carrboro, N.C., 1977-83, water/wastewater plant mgr., 1983—. Mem. Water Environment Fedn., Am. Water Works Assn. Baptist. Achievements include 3 patents for Wastewater Treatment Processes. Home: 515 Arthur Minnis Rd Hillsborough NC 27278 Office: Orange Water and Sewer PO Box 366 400 Jones Ferry Rd Carrboro NC 27278

WILLIAMSON, SAMUEL CHRIS, research ecologist; b. New Braunfels, Tex., Nov. 1, 1946; s. Jens Christian Jr. and Dorothy Marie (Marbach) W.; m. Kathryn Laverne Rutherford, Dec. 28, 1971; children: James Ray, Mark Travis. MS, Tex. Arts and Industries U., 1973; PhD, Colo. State U., 1983. Stats. programmer Tex. Parks and Wildlife Dept., Austin, 1973-77; computer specialist U.S. Fish and Wildlife Svc., Ft. Collins, Colo., 1980-82, rsch. ecologist, 1982; statis. cons. U. Wyo., Baidoa, Somalia, 1988, LGL Alaska Rsch., Inc., Anchorage, 1991; mem. adv. bd. EPA, Las Vegas, Nev., 1989-90; participant confs. and symposia. Contbr. articles to profl. jours., chpt. to manual. With U.S. Army, 1968-70, Vietnam. Decorated Bronze Star medal; scholar Caesar Kleberg Wildlife Found., 1971-73; grantee NSF, 1977-80. Mem. Ecol. Soc. Am., Am. Statis. Assn., Am. Fisheries Soc., Range Soc., Wildlife Soc. Achievements include research, development and improvment of a nationally recognized capability (Instream Flow Incremental Methodology) concentrating on the mathematics and computer modeling of flow-oriented water management related to restoring salmonid population and habitat; development of a conceptual framework and structural design of a cumulative impacts assessment process based on ecological systems analysis and problem solving; research on application of statistics and logic to ecological problems. Office: US Fish and Wildlife Svc 4512 McMurry Ave Fort Collins CO 80525-3400

WILLIAMSON, SAMUEL PERKINS, federal agency administrator, meteorologist; b. Somerville, Tenn., Mar. 5, 1949; s. Julius and Izoula (Smith) W.; m. Brenda Joyce Lee, Sept. 15, 1969 (dec.); children: Keith Ramon, Yulanda Marie. BS, Tenn. State U., 1971, N.C. State U., 1972; MA, Webster U., 1976; postgrad. Engring. Tech. Mgmt., Am. Univ. Weather officer, Ops. Officer, Operational Meteorologist, Unit Commander Air Weather Svc. USAF; phys. scientist Nat. Oceanic Atmospheric Adminstrn. Spl. Project Office; staff meteorologist Office of Meteorology Nat. Weather Svc.; with Next Generation Weather Radar, 1979; dir., dep. dir., chief of Program Planning and Control Joint System Program Office Nat. Oceanic and Atmospheric Admnistrn. U.S. Dept. Commerce; spl. assignment Office of Edn. and Human Resources NSF, 1992—; Lectr. in field. Lt. col. D.C. Air Nat. Guard, commander of Weather Squadron, comptroller; chmn. bd. trustees, mem. bd. fin. Ch. Mem. AAAS, Internat. Electrical and Electronics Engineers Soc., Nat. Tech. Assn., Nat. Guard Assn., Am. Meteorological Soc., Am. Mgmt. Assn., Sr. Exec. Svc. Assn. Achievements include planning, designing, developing and initial deploying of Next Generation Weather Radar's Weather Surveillance Radar (Doppler Weather Radar System), 1979-91. Home: 19121 Barksdale Ct Germantown MD 20874

WILLIAMSON, THOMAS GARNETT, nuclear engineering and engineering physics educator; b. Quincy, Mass., Jan. 27, 1934; s. Robert Burwell and Elizabeth B. (McNeer) W.; m. Kaye Darlan Love, Aug. 16, 1961; children: Allen, Sarah, David. BS, Va. Mil. Inst., 1955; MS, Rensselaer Poly. Inst., 1957; PhD, U. Va., 1960. Asst. prof. nuclear engring. and engring. physics dept. U. Va., Charlottesville, 1960-62, assoc. prof., 1962-69, prof., 1969—, chmn. dept., 1977-90; sr. scientist Westinghouse Savannah River Labs., Aiken, S.C., 1990—; with Gen. Atomic (Calif.), 1965, Combustion Engring., Windsor, Conn., 1970-71, Los Alamos Sci. Lab., 1969, Nat. Bur. Standards, Gaithersburg, Md., 1984-85; cons. Philippine Atomic Energy Commn., 1963, Va. Power Co., 1975-90, Babcock & Wilcox, Lynchburg, Va., 1975-90. Vestryman Ch. of Our Savior, Charlottesville, St. Thaddeus, Aiken, S.C. Fellow Am. Nuclear Soc.; mem. AAAS, Am. Soc. Engring. Edn., Sigma Xi, Tau Beta Pi. Episcopalian. Home: 217 Colleton Ave Aiken SC 29801-7122 Office: Westinghouse Savannah River Labs Aiken SC 22908

WILLIFORD, JOHN FREDERIC, JR., materials research and development scientist; b. Anna, Ill., June 27, 1938; s. John Frederic and Billie Lucille (Willoughby) W.; m. Sylvia Ann Reed, May 1, 1961 (div. 1974); children: Lydia Robin, Ian Edmund; m. Patricia Emadel Buschke, Apr. 5, 1986. AB, Washington U., St. Louis, 1962. Spl. projects technologist Penberthy Electromelt Co., Seattle, 1963-66; sr. rsch. scientist, program mgr. Battelle Meml. Inst., Richland, Wash., 1966-86; v.p. rsch. Tech. Internat. Exch., Inc., Kirkland, Wash., 1987-89; pres. Speed Control Industries, Inc., Kirkland, Richland, 1988-89, JWA div. Emadel Enterprises, Inc., Kirkland, 1989—; cons., 1986—; numerous presentations in field. Contbr. over 100 reports and articles to profl. jours., chpts. to books. Recipient Pierre Jacquet Meml. award Internat. Metallographic Soc., 1971. Achievements include patent for Method for Producing Continuous Lengths of Metal Matrix, Fiber-Reinforced Composites; pioneered use of Nomarski differential interference contrast for metal optic surfaces; devised advanced methods of ultrathinning brittle materials for microstructural evaluation; developed and transferred technologies for ultrathinning of lunar rocks, breccias and soils to the NASA Lunar Receiving Laboratory at Houston; currently conducting research and development in ultracleaning with federal and industrial support, and international technology commercialization work for industrial clients. Home:

7155 NE 126th St Kirkland WA 98034 Office: Emadel Enterprises Inc JWA Div PO Box 2578 Kirkland WA 98083

WILLIFORD, RICHARD ALLEN, oil company executive, flight safety company executive; b. Galveston, Tex., Dec. 24, 1934; s. Walter Hamilton and Marian Lela (Heartfield) W.; m. Mollie Marie Blansett, Feb. 16, 1957; children: Richard Allen Jr., Monica Marie Williford Powell. BS in Petroleum Engring., Tex. A&M U., 1956, BS in Geol. Engring., 1956. Registered profl. engr., Tex. Petroleum/reservoir engr. Gulf Oil Co., La., Tex., 1956-61; mgr. prodn. Tenneco Oil Co., Lafayette, La., and Denver, 1961-73; exec. v.p. Samson Resources Co., Tulsa, 1973-79; chmn., CEO Williford, Inc., Tulsa, 1986—, Safety Tng. Systems, Inc., Tulsa, 1984—, Williford Bldg. Corp., Tulsa, 1990—; chmn. Engineered Equipment Systems, Tulsa, 1989—; past bd. dirs. Samson Resources Co., Tulsa, Tilco, Inc., Tulsa, Intersci. Capital Mgmt. Corp., Tulsa, W-R Leasing Co., Fourth Nat. Bank Tulsa, Sun Belt & Trust Tulsa, Union Nat. Bank Tulsa. Patentee in field. Bd. dirs. Inst. Nautical Archaeology, 12th Man Found., College Station, Tex., 1983—, Tex. A&M Assn. Former Students, College Station, 1985-91, pres., 1989; bd. dirs. Tulsa Opera, Inc., 1985, Thomas Gilcrease Mus. Assn., Tulsa, 1987—, chmn. pers. com., chmn. Rendezvous 1991; chmn. bd. trustees River Parks Authority, Tulsa, 1987—; trustee Tex. A&M Devel. Found., 1991—, Hillcrest Med. Ctr. Found., 1992—, Verde Valley Sch., Sedona, Ariz., 1983-86. Mem. Ind. Petroleum Assn. Am., Soc. Petroleum Engrs., Okla. Ind. Producers Assn., Tex. Ind. Producers & Royalty Owners Assn., Masons, Shriners, Royal Order Jesters, So. Hills Country Club, The Golf Club Okla., Philcrest Tennis Club, The Summit Club, Plaza Club, Tex. A&M Faculty Club, Tau Beta Pi. Republican. Methodist. Avocations: golf, hunting, fishing, flying, scuba diving. Home: 6730 S Evanston Ave Tulsa OK 74136-4509 Office: Williford Cos 6506 S Lewis Ste 220 Tulsa OK 74136

WILLIFORD, ROBERT MARION, civil engineer; b. Murhpysboro, Ill., Nov. 4, 1937; s. Paul Marion and Helen Margaret (Heilig) W.; m. Jewell Eleanor Winkler, Aug. 15, 1959; children: Robert Marion II, Andrew Blaine, Sally Ann. BS, U. Ill., 1960; postgrad., U. Calif., Berkeley, 1971. Registered profl. engr., Calif., Ill., Minn., W.Va., S.C., N.C. Field engr. Bechtel Corp., San Francisco, 1962-66; supt. Bechtel Corp., Weirton, W.Va., 1966-68, Bechtel Pacific Corp., Irian Jaya, Indonesia, 1969-71; tech. mgr. Bechtel Pacific Corp., Queensland, Australia, 1972-73; gen. supt. Bechtel Corp., Virginia, Minn., 1974-77, Bechtel Power Corp., Phoenix, 1977-78; project mgr. BE & K Inc., Birmingham, Ala., 1979-85; pres. CRS Sirrine Svcs. & Power, Greenville, S.C., 1986-88; project dir. Gaylord Container Corp., Antioch, Calif., 1989; mgr. projects BE & K Constrn., Birmingham, Ala., 1990—. Lt. U.S. Army Corps of Engrs., 1960-62. Fellow ASCE; mem. Am. Soc. Mil. Engrs., Am. Soc. Profl. Engrs., Tech. Assn. Pulp & Paper Inst., Constrn. Industry Inst. (tech. com. 1983-85, bd. advisors alternate 1985-87, membership com. 1986-87). Presbyterian. Home: 1674 Wingfield Dr Birmingham AL 35242 Office: BE & K Constrn Co 2000 International Park Dr Birmingham AL 35202

WILLINGHAM, WARREN WILLCOX, psychologist, testing service executive; b. Rome, Ga., Mar. 1, 1930; s. Calder Baynard and Eleanor (Willcox) W.; m. Anna Michal, Mar. 17, 1954; children: Sherry, Judith, Daniel. Student, Ga. Inst. Tech., 1952; PhD, U. Tenn., 1955. Rsch. assoc. World Book Co., N.Y.C., 1959-60; dir. evaluation studies Ga. Inst. Tech., Atlanta, 1960-64; dir. rsch. Coll. Bd., N.Y.C., 1964-68; dir. access rsch. office Coll. Bd., Palo Alto, Calif., 1968-72; asst. v.p., disting. rsch. scientist Ednl. Testing Svc., Princeton, N.J., 1972—; vis. prof. U. Minn., 1988; mem. adv. bd. on ednl. requirements on Sec. Navy, 1968; cons. to numerous schs., colls. U.S. Office Edn. Author: Free Access Higher Education, 1970, Source Book for Higher Education, 1973, College Placement and Exemption, 1974, Assessing Experiential Learning, 1977, Selective Admissions in Higher Education, 1977, Personal Qualities and College Admissions, 1982, Success in College, 1985, Testing Handicapped People, 1988, Predicting College Grades, 1990; editor: Measurement in Education, 1969-72; mem. editorial bds.: Jour. Ednl. Measurement, 1971-75, Alternate Higher Education, 1976-80, Am. Ednl. Research Jour, 1968-71; contbr. articles, tech. reports to profl. jours. Served to lt. USNR, 1955-59. Recipient Ann. award So. Soc. Philosophy and Psychology, 1958. Fellow Am. Psychol. Assn., AAAS; mem. Nat. Council on Measurement in Edn. (dir.), Am. Ednl. Research Assn., Sigma Xi. Office: Ednl Testing Svc Princeton NJ 08540

WILLIS, CLIFFORD LEON, geologist; b. Chanute, Kans., Feb. 20, 1913; s. Arthur Edward and Flossie Duckworth (Fouts) W.; m. Serreta Margaret Thiel, Aug. 21, 1947 (dec.); 1 child, David Gerard. BS in Mining Engring., U. Kans., 1939; PhD, U. Wash., 1950. Geophysicist The Carter Oil Co. (Exxon), Tulsa, 1939-42; instr. U. Wash., Seattle, 1946-50, asst. prof., 1950-54; cons. geologist Harza Engring. Co., Chgo., 1952-54, 80-82, chief geologist, 1954-57, assoc. and chief geologist, 1957-67, v.p., chief geologist, 1967-80; pvt. practice cons. geologist Tucson, Ariz., 1982—; cons. on major dam projects in Iran, Iraq, Pakistan, Greece, Turkey, Ethiopia, Argentina, Venezuela, Colombia, Honduras, El Salvador, Iceland, U.S. Lt. USCG, 1942-46. Recipient Haworth Disting. Alumnus award U. Kans., 1963. Fellow Geol. Soc. Am., Geol. Soc. London; mem. Am. Assn. Petroleum Geologists, Soc. Mining, Metallurgy and Exploration Inc., Assn. Engring. Geologists, Sigma Xi, Tau Beta Pi, Sigma Tau. Republican. Roman Catholic. Avocations: travel, reading. Home: 4795 E Quail Creek Dr Tucson AZ 85718-2630

WILLIS, DAVID PAUL, environmental scientist; b. Hershey, Pa., Dec. 25, 1956; s. Richard Brandt and Lena Marie (Barbini) W.; m. Kathleen Boyd, May 3, 1985; children: Molly, Samuel. BA in Biology, Millersville U., 1979; postgrad., Pa. State U., 1980. Soil conservationist USDA Soil Conservation Svc., Tunkhannock, Pa., 1979-82; environ. scientist Robbins and Assoc., Harrisburg, Pa., 1982-84; project coord. Ohio Wildlife Fedn., Columbus, 1984-86; environ. scientist Gannett Fleming Engrs., Camp Hill., Pa., 1986-88; environ. mgr. Pa. Turnpike Commn., Harrisburg, 1988—. Mem. Pa. Assn. Environ. Profls. (pres. 1991-93, treas. 1989-91), Transp. Rsch. Bd. Com. (chmn. liaison subcom. 1991--). Office: Pa Turnpike Commn PO Box 8531 Harrisburg PA 17105

WILLIS, DONALD J., agricultural research professional; b. Lafayette, Ind., Sept. 30, 1962; s. Larry Gordon and Jo Ann (Slaven) W.; m. Lori Lynn Willis. BS, Purdue U., 1989. Rsch. technician Callahan Enterprises, Inc., Westfield, Ind., 1981-89; rsch. dir. Hytest Seeds, Inc., Buffalo, 1990-92; mgr. rsch. and agronomic svcs. Beachley-Hardy Seeds, Shireminstown, Pa., 1992—; mem. corn & sorghum basic rsch. study and seed improvement com. Am. Seed Trade Assn., 1991-92, soybean div. basic rsch. and seed improvement com., 1991. Mem. Am. Seed Trade Assn. (legis. action com. 1993-94, corn and sorghum basic rsch. com. 1993-94). Republican. Mem. Nazarene Ch.. Home: RD 1 Box 756 Allen Dr Shermans Dale PA 17090 Office: Beachley-Hardy Seeds Co 454 Railroad Ave Shiremanstown PA 17011

WILLIS, GUYE HENRY, JR., soil chemist; b. L.A., July 1, 1937; s. Guye Henry and Esther Mae (Bloomer) W.; m. Phyllis Joy Payne, Dec. 27, 1960; children: Michael Guye, Mark Charles. BS, Okla. State U., 1961; MS, Auburn U., 1963, PhD, 1965. Soil chemist USDA Agrl. Rsch. Svc., Baton Rouge, 1965—. Contbr. over 70 articles to profl. jours. Mem. Am. Soc. Agronomy, Am. Chem. Soc., Am. Soc. Agrl. Engrs. Achievements include discovery that less than 5% of applied pesticides are lost in surface runoff; volatile loss to atmosphere is pathway of greatest loss of surface applied pesticides; rainfall amount more important than rainfall intensity in washing pesticides from plant foliar surfaces. Office: USDA Agrl Rsch Svc 4112 Gourrier Ave Baton Rouge LA 70808

WILLIS, ISAAC, dermatologist, educator; b. Albany, Ga., July 13, 1940; s. R.L. and Susie M. (Miller) W.; m. Alliene Horne, June 12, 1965; children: Isaac Horne, Alliric Isaac. BS, Morehouse Coll., 1961, DSc (hon.), 1989; MD, Howard U., 1965. Diplomate Am. Bd. Dermatology. Intern Phila. Gen. Hosp., 1965-66; fellow Howard U., Washington, 1966-67; resident, fellow U. Pa., Phila., 1967-69, assoc. in dermatology, 1969-70; instr. dept. dermatology U. Calif., San Francisco, 1970-72; asst. prof. Johns Hopkins U. and Johns Hopkins Hosp., Balt., 1972-73, Emory U., Atlanta, 1973-77, assoc. prof., 1977-82; prof. Morehouse Sch. Medicine, Atlanta, 1982—, chief dermatology, 1991—; dep. commdr. of 3297th USA Hosp. (1000B), 1990—; attending staff Phila. Gen. Hosp., 1969-70, Moffit Hosp., U. Calif., 1970-72, Johns Hopkins Hosp., Balt. City Hosp., Good Samaritan Hosp., 1972-74, Crawford W. Long Meml. Hosp., Atlanta, 1974—, West Paces Ferry Hosp., 1974—, others; mem. grants rev. panel EPA, 1986—; mem. gen. medicine group IA sect. NIH, 1985—, mem. nat. adv. bd. Arthritis and Musculoskeletal and Skin Diseases, 1991—; chmn. instl. rev. bd., mem. pharmacy and therapeutic com.; bd. mem. Comml. Bank Gwinnett; West Paces Ferry Hosp.; mem. gov.'s commn. on effectiveness and economy in govt. State of Ga. Human Resources Task Force, 1991—; cons. in field. Bd. dirs. Heritage Bank, Comml. Bank of Ga., chmn. audit rev. com., 1988—. Served to col. USAR, 1983—. EPA grantee, 1985—. Author: Textbook of Dermatology, 1971—; Contbr. articles to profl. jours. Chmn. bd. med. dirs. Lupus Erythematrosus Found., Atlanta, 1975-83; bd. dirs. Jacquelyn McClure Lupus Erythematrosus Clinic, 1982—; bd. med. dirs. Skin Cancer Found., 1980—; trustee Friendship Bapt. Ch., Atlanta, 1980-82 . Nat. Cancer Inst. grantee, 1974-77, 78—; EPA grantee, 1980—. Fellow Am. Acad. Dermatolgy, Am. Dermtol. Assn.; mem. AAAS, Soc. Investigative Dermatology, Am. Fedn. Clin. Research, Am. Soc. Photobiology, Am. Med. Assn., Nat. Med. Assn., Internat. Soc. Tropical Dermatology, Pan Am. Med. Assn., Phi Beta Kappa, Omicron Delta Kappa. Clubs: Frontiers Internat., Sportsman Internat. Subspecialties: dermatology; cancer research (medicine). Home: 1141 Regency Rd NW Atlanta GA 30327-2719 Office: NW Med Ctr 3280 Howell Mill Rd NW Ste 342 Atlanta GA 30327-4109

WILLIS, JEFFREY SCOTT, mechanical engineer, consultant; b. New Orleans, Apr. 18, 1962; s. John Ray and Jane Ann (Broome) W.; m. Linda Faye Sanders, July 30, 1988. BSME, U. Southwest La., 1983-87. Engr. Balco, Boston, 1987-90, R.G. Vanderweil Engrs., Boston, 1990—. Sunday sch. tchr. First Bapt. Ch., Melrose, Mass., 1990-93, bd. Christian edn., 1990-93, property chmn., 1993—. Mem. ASME (assoc. Boston sect., vice chmn. 1991-93, chmn. 1993—), Assn. Energy Engrs. (New England chpt. v.p. 1991, pres. 1992). Home: 30 Lynde Ave Melrose MA 02176 Office: R G Vanderweil Engrs 266 Summer St Boston MA 02210

WILLIS, JUDITH H., zoology educator, researcher; b. Detroit, Jan. 2, 1935; d. J. Shurly and Beatrice (Mintz) Horwitz; m. John S. Willis, Sept. 28, 1958. AB, Cornell U., 1956; PhD, Harvard U., 1961. Instr. U. Ill., Urbana, 1963-64, asst. prof., 1964-68, assoc. prof., 1968-77, prof., 1977-90; head zoology dept. U. Ga., Athens, 1990-93, prof., 1990—; program dir. NSF, Washington, 1983. Contbr. articles to Insect Biochemistry and Molecular Biology, Archives Insect Biochem. Physiology, Devel. Biology. Grantee NIH, 1967-82, 90—, NSF, 1982-90, USDA, 1985-89. Mem. AAAS (chair sect. G 1987-88), Tissue Culture Assn. (exec. bd. 1986-90). Achievements include characterization of diversity of insect cuticular proteins; definition of action of insect juvenile hormone. Office: U Ga Dept Zoology Athens GA 30602

WILLIS, LAWRENCE JACK, chemist; b. Portland, Oreg., Oct. 27, 1948; s. Charles Brown and Edith Adele (Arrington) W.; m. Barbara Ann Zwetschke, June 19, 1971; 1 child, Brian Lawrence. BS in Chemistry, Oreg. State U., 1970; MS in Chemistry, Wright State U., 1977; PhD in Chemistry, Oreg. Grad. Ctr., 1983. Vis. asst. prof. chemistry Willamette U., Salem, Oreg., 1982-83, Lewis and Clark Coll., Portland, 1983-85, U. Portland, 1985-86; lab. dir. Eugene M. Seidel Assocs., Vancouver, Wash., 1986-87; lab. leader Am. Tokyo Kasei (name now TCI Am.), Portland, 1987—. Contbr. articles to profl. publs. 1st lt. USAF, 1970-74. Mem. Am. Chem. Soc. Home: 3510 NW Ashland Pl Aloha OR 97006

WILLIS, RALPH HOUSTON, mathematics educator; b. McMinnville, Tenn., Dec. 26, 1942; s. Carl Houston and Carrie Lee (Hill) W.; m. Gayle Catherine Celestin, June 29, 1973 (div. Apr. 1985); m. Velma Inez Church, Aug. 10, 1985; stepchild, Bobbie Lynn White. BS in Math., Mid. Tenn. State Coll., 1964, MA in Math., 1966. Cert. secondary edn. Instr. math. dept. Western Carolina U., Cullowhee, N.C., 1968-73, asst. prof. math. dept., 1973-83, assoc. prof. math. dept., 1983—. Editor: (newsletters) Abelian Grapevine-Secondary Math, 1970-88, The Child of Mathematics-Elementary-Middle Grade Math., 1972-78; contbr. articles to profl. jours. Dir., coord. Western Carolina U. High Sch. Math. Contest, Cullowhee, 1970—; solicitor-coord. Math. Contest Scholarship Program, Cullowhee, 1971-82; initiator-coord. Math. Dept.'s Vis. Speaker Program, Western Carolina U., Cullowhee, 1974-77; faculty sponsor N.C. Coun. Tchrs. Math. Student Affiliate, Cullowhee, 1988—. 1st lt. U.S. Army, 1966-68. Recipient Paul A. Reid Disting. Svc. award Western Carolina U., 1991, hon. mention N.C. Gov.'s Award for Excellence, 1991. Mem. Nat. Coun. Tchrs. Math., N.C. Coun. Tchrs. Math., Phi Kappa Phi, Kappa Mu Epsilon. Avocations: genealogy, gardening, military history, model building, carpentry. Office: Western Carolina U Math Dept Stillwell Bldg Cullowhee NC 28723

WILLIS, REBECCA LYNN, chemist, educator; b. Fort Baker, Calif., Dec. 15, 1947; d. John Wilfred and Edna Gertrude (Hileman) Irvin; m. James Frederick Willis, Jan. 2, 1977; 1 child, Robert Frederick. BS in Chemistry, Southwestern State Coll., 1970; PhD in Organic Chemistry, Okla. State U., 1974. Asst. prof. chemistry So. Ark. U., Magnolia, 1974-88, assoc. prof., 1988—; vis. prof. U. Ark. for Med. Sci., Little Rock, 1990; grant proposal reviewer NSF, 1992; regional coord. for toxics pub. participation program EPA, Magnolia, 1980-81. Contbr. articles to Jour. Organic Chemistry, Molecular and Cellular Biology. Mem. panel Ethyl Corp. Citizen's Adv. Panel, Magnolia, 1991-92; mem. vestry St. James Episc. Ch., Magnolia, 1990-92. Recipient Rsch. Opportunity awards NSF, 1988, 89, Instrumentation and Lab. Improvement grantee, 1989, 91; named one of Outstanding Young Women Am., 1981. Mem. AAAS, Am. Chem. Soc. (nat. chemistry week coord. 1991, chmn. Ouachita Valley chpt. 1990-91), Coun. on Undergrad. Rsch., Ark. Acad. Sci. Democrat. Office: So Ark U Chemistry Program Magnolia AR 71753

WILLIS, SELENE LOWE, electrical engineer; b. Birmingham, Ala., Mar. 4, 1958; d. Lewis Russell and Bernice (Wilson) Lowe; m. André Maurice Willis, June 12, 1987. BSEE, Tuskegee (Ala.) U., 1980; postgrad. in Computer Programing, UCLA, 1993. Component engr. Hughes Aircraft Corp., El Segundo, Calif., 1980-82; reliability and lead engr. Aero Jet Electro Systems Corp., Azusa, Calif., 1982-84; sr. component engr. Rockwell Internat. Corp., Anaheim, Calif., 1984; design engr. Lockheed Missile & Space Co., Sunnyvale, Calif., 1985-86; property mgr. Penmar Mgmt. Co., L.A., 1987-88; aircraft mechanic McDonnell Douglas Corp., Long Beach, 1989-93; sr. component engr. Gen. Data Communications Corp., Danbury, Conn., 1984-85. Vol. Mercy Hosp. & Children's Hosp., Birmingham, 1972-74; mem. L.A. Gospel Messengers, 1982-84, West Angeles Ch. of God & Christ, L.A., 1990. Bell Labs. scholar, 1976-80; named one of Outstanding Young Women in Am., 1983, 87. Mem. IEEE, ASME, Aerospace and Aircraft Engrs., So. Calif. Profl. Engring. Assn., Tuskegee U. Alumni Assn., Eta Kappa Nu. Mem. Christian Ch. Avocations: piano, computers, softball, real estate.

WILLIS, SHARON WHITE, microbiologist; b. Greensboro, N.C., Mar. 15, 1964; d. Roy Frank and LaVerne (Gilchrest) White. BS, N.C. State U., 1986. Tech. asst. III Ciba-Geigy, Greensboro, N.C., 1986-87; asst. rsch. microbiologist Ciba Vision, Alpharetta, Ga., 1987-92, rsch. microbiologist, 1992—. Mem. Am. Soc. Microbiology. Avocations: reading, billiards, basketball, ceramics. Office: Ciba Vision 5000 McGinnis Ferry Rd Alpharetta GA 30202-3919

WILLIS, SHELBY KENNETH, civil engineer, consultant; b. Alton, Ill., June 9, 1924; s. Shelby Thomas and Mary Sigel (Worsham) W.; m. Ruth Marie Buchanan, Nov. 8, 1945; children: Jan Merry Willis Pocaterra, Jill Marie Willis Gleason, James Shelby (dec.). BSCE, U. Ill., 1945, MS in Engring., 1947. Registered profl. engr., Ala., Ariz., Ark., Colo., Ill., Iowa, Kans., La., Miss., Mo., Nebr., N.Mex., N.D., Ohio, Okla., Pa., Tex., Wash., Wis., Wyo.; registered structural engr., Ill.; registered land surveyor, Kans. Design engr. Western Electric Corp., Duluth, Minn., 1947; design engr., project engr. Howard, Needles, Tammen and Bergendoff, Kansas City, Mo., 1948-57; founding ptnr., rsch. engr., engring. cons. Bucher and Willis, various locations, 1957-83; mng. ptnr. Bucher, Willis and Ratliff, various locations, 1974-88; cons. Bucher, Willis and Ratliff, Engrs., Planners and Architects, Salina, Kans., 1988—. Co-author: Underground Utilization (8 vols.), 1978, Quality in the Constructed Project, 1987. Pes. Salina Optimist Club, 1960-61; trustee Marymount Coll., Salina,1 987-90, Salina Presbyn. Manor Endowment, 1991—; pres. bd. trustees Kans. Coll. Tech. Endowment Assn., Salina, 1987-90. Lt. (j.g.) USN, 1943-46. Named Engr. of Yr. Kans. Engring. Soc., 1981; recipient Disting. Alumnus award U. Ill. Dept. Civil Engring., 1984. Fellow ASCE (life mem.), Am. Cons. Engrs. Coun. (pres. 1983-84, cert. peer reviewer 1984—Engring. Excellence award 1980); mem. NSPE, ASTM, Am. Soc. Photogrammetry and Remote Sensing (life). Republican. Presbyterian. Office: Bucher Willis & Ratliff 609 W North St Salina KS 67401

WILLIS, WILLIAM DARRELL, JR., neurophysiologist, educator; b. Dallas, July 19, 1934; s. William Darrell and Dorcas (Chamberlain) W.; m. Jean Colette Schini, May 28, 1960; 1 child, Thomas Darrell. B.S., B.A., Tex. A&M U., 1956; M.D., U. Tex. Southwestern Med. Sch., 1960; Ph.D., Australian Nat. U., 1963. Postdoctoral research fellow Nat. Inst. Neurol. Diseases and Blindness, Australian Nat. U., 1960-62, Istituto di Fisiologia, U. Pisa, Italy, 1962-63; from asst. prof. to prof. anatomy, chmn. dept. U. Tex. Southwestern Med. Sch., Dallas, 1963-70; chief lab. comparative neurobiology Marine Biomed. Inst., prof. anatomy and physiology U. Tex. Med. Br., Galveston, 1970—; dir. Marine Biomed. Inst. U. Tex. Med. Br., 1978—, chmn. dept. anatomy and neurosci., 1986—, Ashbel Smith prof., 1986—; Mem. neurology B. study sect. NIH, 1968-72, chmn., 1970-72, mem. neurol. disorders Program Project rev. com., 1972-76, Nat. Adv. Neurol. and Communicative Disorders and Stroke Council, 1987-90. Mem. editorial bd. Neurosci. Exptl. Neurology, 1970-90, Archives Italienne Biologie, Neurosci. Letters, 1976-92; chief editor Jour. Neurophysiology, 1978-83, Pain, 1986-89, Jour. Neurosci., 1986-89, editor-in-chief, 1993—; section editor Exptl. Brain Rsch., 1990-92. Mem. AAAS, Am. Assn. Anatomists (exec. com. 1980-86), Am. Pain Soc. (pres. 1982-83), Internat. Assn. Study Pain (coun. 1984-90), Am. Physiol. Soc., Soc. Exptl. Biol. Medicine, Soc. Neurosci. (pres. 1984-85), Internat. Brain Rsch. Orgn., Cajal Club, Sigma Xi, Alpha Omega Alpha. Home: 2925 Beluche Dr Galveston TX 77551-1511 Office: U Tex Med Br Marine Biomed Inst 200 University Blvd Galveston TX 77555-0843

WILLNER, ALAN ELI, electrical engineer, educator; b. Bklyn., Nov. 16, 1962; s. Gerald and Sondra (Bernstein) W.; m. Michelle Frida Green, June 25, 1991. BA, Yeshiva U., 1982; MS, Columbia U., 1984, PhD, 1988. Summer tech. staff David Sarnoff Rsch. Ctr., Princeton, N.J., 1983, 84; grad. rsch. asst. dept. elec. engring. Columbia U., N.Y.C., 1984-88; postdoctoral mem. tech. staff AT&T Bell Labs., Holmdel, N.J., 1988-90; mem. tech. staff Bell Communications Rsch., Red Bank, N.J., 1990-91; asst. prof. U. So. Calif., L.A., 1992—; head del. Harvard Model UN Yeshiva U., 1982; instr. Columbia U., 1987; rev. panel mem. NSF, Washington, 1992, 93. Contbr. articles to IEEE Photonics Tech. Letters, Jour. Optical Engring., Jour. Electrochem. Soc., Electronics Letters, Applied Physics Letters, Applied Optics. Mem. faculty adv. bd. U. So. Calif. Hillel Orgn., 1992. Fellow Semiconductor Rsch. Corp., 1986, NATO/NSF, 1985; Dr. Samuel Belkin scholar Yeshiva U., 1980; recipient Armstrong Found. prize Columbia U., 1985, Young Investigator award NSF, 1992. Mem. SPIE, IEEE Lasers and Electro-Optics Soc. (mem. optical comm. tech. com.), Optical Soc. Am. (vice-chair optical comm. group, symposium organizer ann. meeting 1992), Sigma Xi. Achievements include patents for localized photochemical etching of multilayered semiconductor body, optical star coupler utilizing fiber amplifier technology. Home: Apt 201 1200 S Shenandoah St Los Angeles CA 90035 Office: U So Calif Dept Elec Engring EEB 538 Los Angeles CA 90089-2565

WILLOUGHBY, WILLIAM, II, nuclear engineer; b. Birmingham, Ala., Jan. 14, 1933; s. William and Marion Louise (Hart) W.; m. Doris Jean Lindsey, Oct. 16, 1957; 1 child, William III. BSChemE, MIT, 1954; MS in Nuclear Engring., U. Calif., Berkeley, 1960. Registered profl. engr., S.C. Physicist U. Calif. Lawrence Livermore (Calif.) Lab., 1957-61; tech. support supr. Carolina Va. Nuclear Power Assocs., Parr, S.C., 1962-67; mgr. nuclear engr. S.C. Elec. & Gas, Columbia, 1967-76; sr. project mgr., consulting engr. Stone & Webster Engring., N.Y.C., 1976—; mem. S.C. Nuclear Adv. Coun., Columbia, 1974-76. Contbr. articles to profl. jours. Commodore Columbia Sailing Club, 1974. 1st lt. U.S. Army, 1955-57. Mem. AAAS, Am. Nuclear Soc., Sigma Xi, Tau Beta Pi. Episcopalian. Home: 24 Spruce Dr Wilton CT 06897

WILLOWS, ARTHUR OWEN DENNIS, neurobiologist, zoology educator; b. Winnipeg, Man., Can., Mar. 26, 1941; came to U.S., 1959; s. Danby and Laura Beatrice (Bumstead) W.; m. Karen Dunkelbarger, June 14, 1987; children by previous marriage: Kurt Danby, Keith Stratton, Amy Margaret. B.S., Yale U., 1963; Ph.D., U. Oreg., 1967. Asst. prof. U. Wash., Seattle, 1969-72, assoc. prof., 1972-75, prof. zoology, 1975—, chmn. dept., 1979—; dir. Friday Harbor Labs., Wash., 1973—. Editor: Invertebrate Learning, 2 vols., 1973, Neurobiology and Behavior, 2 vols., 1985, Marine Behavior and Physiology. Guggenheim fellow, 1976. Mem. Soc. for Neurosci., Am. Soc. Zoologists, Western Soc. Naturalists, Am. Soc. Physiologists. Home: Friday Harbor Labs Friday Harbor WA 98250 Office: U Wash Dept Zoology 620 University Rd Friday Harbor WA 98250

WILLSON, WARRACK G., physical chemist. BA in Chemistry & Math., U. No. Colo., 1965; PhD in Phys. Chemistry, U. Wyo., 1970. Rsch. scientist Nat. Resources Rsch. Inst. U. Wyo., Laramie, 1971-73; rsch. engr. div. reactor engring. Engring. Tech. Ctr. E. I. DuPont Nemours Co., Wilmington, Del., 1973-76; group leader coal chem. devel. Occidental Rsch. Corp., La Verne, Calif., 1976-78; mgr. br. gasification and liquefaction Grank Forks (N.D.) Energy Tech. Ctr. Dept. of Energy, 1978-83; dir. Fuels and Process Chemistry Rsch. Inst. and Energy and Environ. Rsch. Ctr. U. N.D., Grand Forks, 1983—; bd. dirs. Internat. Pitts. Coal Conf., 1987-88; tech. dir. Internat. Opportunities in the Synfuels Industry Biennial Symposium, 1988—; vis. prof. coal sci. and tech. Mineral Industry Rsch. Lab. U. Alaska, Fairbanks, 1988-89, adj. prof. coal tech., 1988—; adj. prof. chem. engring. U. N.D., 1980—. Contbr. over 75 articles to profl. jours. Mem. Am. Inst. Chem. Engrs., Am. Chem. Soc., Sigma Xi, Lambda Sigma Tau. Achievements include research and development of fuels production and utilization, of environmental control systems, and of process chemistry. Office: U ND Energy & Environ Rsch Ctr PO Box 8213 Grand Forks ND 58202-8213

WILMAN, DERRY EDWARD VINCENT, medicinal chemist; b. Watford, England, July 29, 1946; s. Herbert Edward and Linda Frances (Pascoe) W.; m. Rosemary Anne Bridgwater, Apr. 4, 1970; children: Paul Darren, Clare Louise. BSc, U. London, Eng., 1968, PhD, 1974. Jr. tech. officer Inst. Cancer Rsch., London, 1968-74, scientist, grade II, 1974-81; scientist, grade I Inst. Cancer Rsch., Sutton, Eng., 1981—; acad. bd. mem. Inst. Cancer Rsch., 1987—, exec. com. British Assn. for Cancer Rsch., 1992—. Editor: Chemistry of Antitumour Agents, 1990; contbr. articles to Jour. chem. Rsch., Jour. Biopharm. Sciss., Current Opinion in Therapeutic Patents 2. Gov. Epsom and Ewell High Sch., Epsom Surrey, Eng., 1989—. Recipient Internat. Cancer Tech. Transfer award, Internat. Union Against Cancer, Pomona Coll., Calif., 1980; grantee NATO (collaborative rsch.), St. Mary's U., Halifax Nova Scotia, Can., 1986-90. Mem. Royal Soc. Chemistry (Career award 1982), Royal Photographic Soc. (lic. 1991). Achievements include major contributions to knowledge of chemistry and anti-cancer activity of aryldialkltriazenes and triazene n-oxides. Office: Inst Cancer Rsch Cotswold Rd, Sutton Surrey SM2 5NG, England

WILMORE, DOUGLAS WAYNE, physician, surgeon; b. Newton, Kans., July 22, 1938; s. Waldo Wayne and Hilda Gard (Adrian) W.; m. Judith Kay Shabert; 1 child, Carol Kristann. BA, Washburn U., 1960; MD, Kans. U., 1964; MS (hon.), Harvard U., 1979. Diplomate Am. Bd. Surgery. Intern Hosp. U. Pa., Phila., 1964-65, resident, fellow, 1965-71; chief clin. rsch. and staff surgeon U.S. Army Inst. Surg. Rsch., Ft. Sam Houston, 1971-79; staff surgeon Brigham and Women's Hosp., Boston, 1979—. Editor: Care of the Surgical Patient, 1988. Lt. Col. U.S. Army, 1971-74. Achievements include development of safe modern techniques for providing parenteral nutrition to critically-ill patients.

WILMUT, CHARLES GORDON, environmental engineer; b. Dallas, Apr. 13, 1947; s. Charlie Ephram and Cathryn (Chapman) W.; m. Janet Gayle Machicek, Jan. 25, 1969; children: Bryan Gregory, Lauren Gabrielle. BS in Aerospace Engring., U. Tex., 1970; MS in Environ. Engring., U. Tex., Arlington, 1977. Registered profl. engr., Tex. Engr. Designers and Planners, Galveston, Tex., 1970-71; project engr. U.S. EPA, Dallas, 1971-73; pres. Gutierrez, Smouse, Wilmut & Assocs., Inc., Dallas, 1973—. Contbr. articles to profl. jours. Active Planning and Zoning Commn., Colleyville, Tex., 1990—. Mem. ASCE, NSPE, Water Pollution Control Fedn., Tex. Soc. Profl. Engrs. (bd. dirs. 1989-92, chmn. environ. coun. 1990-92), Tex.

Water Pollution Control Assn. (sect. rep. 1991-93), Oak Cliff C. of C. (dir. 1989-91). Methodist. Home: 6913 Westcoat Colleyville TX 76034 Office: Gutierrez Smouse Wilmut & Assocs Inc 11117 Shady Trail Dallas TX 75229

WILNER, FREEMAN MARVIN, hematologist, oncologist; b. Detroit, June 14, 1926; s. Jack Burton W. and Belle Gertrude (Goldberg) Weeks; m. Marjorie Louise Tewkesbury, Aug. 29, 1948; children: Jeffrey, Robert, Paul, Laura. BS with honors, Wayne State U., 1950, MD, 1953. Diplomate Am. Bd. Internal Medicine, Am. Bd. Hematology, Am. Bd. Oncology. Intern Detroit Receiving Hosp., Detroit, 1953-54, resident, 1954-57, chief med. resident, 1956; pres. Hematology/Oncology Assocs., Royal Oak, Mich., 1974-86, Hematology/Oncology Cons., Mich., 1986—; med. dir. Rose Cance Inst., chief sect. hematology-oncology Rose Cancer Ctr.-William Beaumont Hosp., Royal Oak, Mich., 1988; presenter, with others, numerous med. seminars; clin. assoc. prof. medicine Wayne State U. Co-author: (with Schneider, John R., Bedell, and Archie) Bleeding and Clotting Disorders, 1981; contbr. articles to profl. jours., case reports and other publs. Bd. dirs. Red Cross Southeastern Mich. Blood Bank, 1991, Mich. Cancer Found., 993. Sgt. USAAF, 1944-47. Recipient Disting. Service Award, School of Medicine , Alumni of Wayne State U., Detroit, 1985; named Tchr. of Yr., House Staff, William Beaumont Hosp., 1969-70, Providence Hosp., 1971. Fellow ACP (Laureate award 1993), Internat. Soc. Hematology, Detroit Acad. of Medicine; mem. AMA, Am. Soc. Hematology, Am. Soc. Clin. Oncology, Mich. State Med. Soc., Mich. Cancer Found. (trustee 1974, bd. dirs.), Wayne County Med. Soc., Oakland County Med. Soc., Leukemia Found. Mich. (adv. bd. 1958-79), World Fedn. Hemophilia. Avocations: photography, bicycling. Office: Hematology Oncology Cons PC 3601 W 13 Mile Rd Royal Oak MI 48073-6769

WILSDON, THOMAS ARTHUR, product development engineer, administrator; b. Waterbury, Conn., Aug. 18, 1942; s. Arthur and Ruth (Wellington) W.; m. Yvonne Jeanne Pettit, June 19, 1964 (div. Apr. 1986); children: Thomas Charles, Beth Jeanne; m. Sharon Diann Culbertson, Feb. 14, 1988; children: Vandee Hyder, Jacklynn Hyder. BSEE, U. Conn., 1964; MBA, SUNY, Buffalo, 1978. Product design engr. Westinghouse Gen. Control Divsn., Buffalo, 1964-78; mgr. product devel. Westinghouse Control Divsn., Asheville, N.C., 1978-87; mgr. Advantage engring. Westinghouse Elec. Components Divsn., Asheville, 1987-93, mgr. control components engring., 1993—. Mem. IEEE, NSPE, Am. Mgmt. Assn. Methodist. Achievements include development of low voltage AC and DC motor starters, ampgart 7200V motor starter components, solid state controlled Advantage motor starters. Home: PO Box 927 Skyland NC 28776-0927 Office: Westinghouse Elec Component Divsn PO Box 5715 Asheville NC 28813

WILSEY, PHILIP ARTHUR, computer scientist; b. Kewanee, Ill., Sept. 24, 1958; s. George A. and Mary Lee (Smith) W.; m. Marilyn L. Hargis, Jan. 2, 1980; children: Patrick A., Zachary E. BS in Math., Ill. State U., 1981; MS in Computer Sci., U. Southwestern La., 1985, PhD in Computer Sci., 1987. Computer programmer Union Ins. Group, Bloomington, Ill., 1980-81, Bob White Computing & Software, Bloomington, 1981-82; rsch. asst. Univ Southwestern La., Lafayette, La., 1983-87; asst. prof. U. Cin., 1987—; cons. MTL, Dayton, 1992—; mem. editorial bd. VLSI Design, 1993-95. Assoc. editor: Potentiais Mag., 1992—; contbr. articles to profl. jours. Mem. AAAS, IEEE, Assn. Computing Machinery, Am. Soc. Elec. Engrs. Home: 4654 Leadwell Ln Cincinnati OH 45242 Office: Computer Architecture Design Lab Dept of ECE ML 0030 Cincinnati OH 45221-0030

WILSHIRE, BRIAN, materials engineering educator, administrator; b. Rhondda, Glamorgan, Wales, May 16, 1937; s. Edmund and Eileen Wilshire. BSc in Metallurgy with 1st class honours, Univ. Coll., Swansea, Wales, 1958, PhD in Metallurgy, 1962, DSc in Metallurgy, 1983. C.Eng., F.I.M. Lectr. Univ. Coll., Swansea, 1960-72, sr. lectr., 1972-78, reader, 1978-82, prof., 1982-85, head of dept., 1985—; bd. dirs. Eidawn Materials Rsch. Ltd., Swansea. Author: Technological and Economic Trends in the Steel Industries, 1983, Creep of Metals & Alloys, 1985, Introduction to Creep, 1993. Acta Metallurgica lectr. Acta Metallurgica, Inc., U.S., 1991. Fellow Royal Acad. Engring., Inst. of Materials. Achievements include Theta projection concept for creep life prediction and remanent life assessment; magnetic flake fingerprint technology. Office: Univ College Swansea, Dept of Materials Engineering, Swansea Wales

WILSON, ALEXANDER THORNTON, journalist; b. Bryn Mawr, Pa., June 20, 1955; s. Conrad and Barbara Allyn (Copp) W.; m. Jerelyn Everett, Nov. 23, 1984; children: Lillian, Frances. BA in Biology, Ithaca Coll., 1977. Assoc. dir. N.Mex. Solar Energy Assn., Santa Fe, 1978-80; exec. dir. N.E. Sustainable Energy Assn., Greenfield, Mass., 1980-85; pres. West River Communications, Inc., Brattlesboro, Vt., 1985—; bd. dirs. N.E. Sustainable Energy Assn., 1987-92. Editor, pub.: (newsletter) Environ. Bldg. News, 1992—; author: Consumer Guide to Home Energy Savings, 1991, 92, Quiet Water Canoe Guide, N.H.-Vt., 1992; contbg. editor: Ind. Energy Mag., 1985-92, Architecture Mag., 1989-92; contbr. articles to Architecture, Jour. Light Constrn., Custom Builder, Ind. Energy, Popular Sci., Consumers Digest, Home Mechanix, Mother Earth News. Vice chmn. Dummerston (Vt.) Planning Commn., 1989-93; active BUHS Sci. Enrichment Com., Brattleboro, 1990-93. Thomas James Meml. scholar Ithaca Coll., 1976. Mem. Soc. Environ. Journalists.

WILSON, BRUCE NORD, agricultural engineer; b. Worthington, Minn., Oct. 4, 1954; s. Arthur Gordon and Delores (Nord) W.; children: Graham Mobeck, Grace Lynnea. MS, U. Minn., 1979; PhD, U. Ky., 1984. Asst. prof. Okla. State U., Stillwater, 1983-87, assoc. prof., 1987-91; asst. prof. U. Minn., St. Paul, 1991—; extension instr. U. Ky., Lexington, 1983-87. Author: (software) SEDIMOT II, 1983; contbr. articles to numerous profl. jours. Mem. Am. Soc. Agrl. Engrs. (Outstanding Tech. Paper, 1989, honorable mention, 1987, 91), Am. Geophys. Union, Soil Conservation Soc. Am., Sigma Xi, Alpha Epsilon. Achievements include demonstration of similarities between drainage networks to small scale erosion processes to that of major river systems; rsch. on hydrology and sedimentology model for surface mining assessments. Office: U Minn 1390 Eckles Ave Saint Paul MN 55108

WILSON, CARL ROBERT, structural engineer; b. Valleyfiedl, Que., Can., June 28, 1940; s. William Robert and Hazel Ann Wilson; m. Verna Lorraine Stewart, Nov. 12, 1965; children: Mark Robert, Tyra Ann. BSCE, U. New Brunswick, Can., 1965; MSCE, U. Alberta, Edmonton, Can., 1967; PhD, Tech. U. Nova Scotia, Halifax, 1971. Registered profl. engr., B.C. Design engr. Regional Municpality, Ottawa, Ont., 1970-72; codes and standards Can. Wood Coun., Ottawa, Ont., 1972-73, dir. tech. svcs., 1973-74; mgr. R & D Coun. Forest Industries, Vancouver, B.C., 1974-80; mgr. rsch. and applied sci. Fletcher Challenge Can., Vancouver, B.C., 1980-90; pres. Carl R. Wilson & Assocs., Ltd., Vancouver, B.C., 1990—; mem. adv. bd. Forintek Can. Corp., Vancouver, 1975-88, Sci. Coun. B.C., Burnabu, 1985-89; chmn. Environ. Can. Pulp and Paper Assn., Montreal, 1985-87. Contbr. articles to Jour. Civil Engring. Chmn. Mt. Seymour Untied Ch., North Vancouver, 1979-82. Fellow Can. Soc. Civil Engrs.; mem. ASTM, Can. Standards Assn., Forest Products Soc., North Shore Winter Club. Achievements include development of preserved wood foundations, first North American probabilistic structural design code for wood, lumber properties program, product acceptance and environmental affairs. Office: Carl R Wilson & Assocs Ltd, 2322 Riverbank Pl, North Vancouver, BC Canada V7H 2L2

WILSON, CHARLES B., neurosurgeon, educator; b. Neosho, Mo., Aug. 31, 1929; married; 3 children. BS, Tulane U., 1951, MD, 1954. Resident pathologist Tulane U., 1955-56, instr. neurosurgery, 1960-61; resident Ochsner Clinic, 1956-60; instr. La. State U., 1961-63; from asst. prof. to prof. U. Ky., 1963-68; prof. neurosurgery U. Calif., San Francisco, 1968—. Mem. Am. Assn. Neurol. Surgery, Am. Assn. Neuropathology, Soc. Neurol. Surgery. Achievements include research in brain tumor chemotherapy. Office: U Calif Med Ctr 505 Parnassus Ave San Francisco CA 94143-0001 also: U Calif Sch Medicine Dept Neurol Surgery San Francisco CA 94143*

WILSON, CHRISTOPHER T., nuclear engineer; b. Reading, Pa., Nov. 25, 1961; s. Harold H. and Lynn (Backus) W.; m. Amy Barrow, Dec. 29, 1984; 1 child, Katharine Marie. BS in Nuclear Engring., U. Mich., 1984; MS in Nuclear Engring., MIT, 1986. Sr. engr. Boston Edison Co., 1986—; Mem. Am. Nuclear Soc., Acad. Model Aeronautics. Home: 18 Nobel Rd Dedham MA 02026 Office: Boston Edison Co 25 Braintree Hill Office Pk Braintree MA 02184

WILSON, DALE WILLIAM, JR., laser technician; b. Madison, Wis., June 12, 1956; s. Dale William Sr. and Elisabeth Meriam (Kolganoff) W.; m. Rhonda McCauley; 1 child, Alexandra Eve. Student, Grafton (W.Va.) High Sch. Rsch. technician EG&G Washington Analytical Svcs. Ctr., Morgantown, W.Va., 1982-91; tech. cons. Spectrum Laser Systems, Masontown, W.Va., 1992—. Mem. Internat. Laser Display Artists Assn., United Chess Fedn. Achievements include patented design for strength particle adhesion strength mechanism, a robotic arm capable of measuring adhesions of molten materials, laser enhanced particle tracking system to measure and track energy plant emission particulates. Office: Spectrum Laser Systems PO Box 727 Masontown WV 26534-0727

WILSON, DAVID CLIFFORD, mathematician, educator; b. Clifton Springs, N.Y., Sept. 23, 1942. BS, U. Wis., 1964; MS, Rutgers U., 1967, PhD, 1969. Mem. Inst. for Advanced Study, Princeton, N.J., 1969-70; asst. prof. math. No. Ill. U., DeKalb, 1970-72; asst. prof. math. U. Fla., Gainesville, 1972-75, assoc. prof. math., 1975-87, prof. math., 1987—. Achievements include patent for automated method for digital echocardiographic imagte quantitation. Office: U Fla Math Dept Gainesville FL 32611

WILSON, DAVID LOUIS, biologist; b. Washington, Jan. 11, 1943; s. Eugene Austin and Helen (Redmond) W.; m. Margaret Sarah Gibbons, July 8, 1967; 1 child, Mariah Elizabeth. BS, U. Md., 1964; PhD, U. Chgo., 1969. Rsch. fellow Calif. Inst. Tech., Pasadena, 1969-72; asst./assoc. prof. U. Miami (Fla.) Sch. Medicine, 1972-81, prof., 1981—, dep. dean acad. affairs, 1982-85; interim dean grad. sch. U. Miami, Coral Gables, Fla., 1983-85, dean arts and scis., 1985-91, prof. biology, 1985—. Contbr. articles to profl. jours. NIH predoctoral fellow U. Chgo., 1964-68, James Franck dissertation fellow, 1968-69, Helen Hay Whitney postdoctoral fellow, 1969-72; NIH rsch. grantee, 1972-86, NSF rsch. grantee, 1982-85. Achievements include development of procedures for gel electrophoresis and analysis of proteins; characterization of biochemical steps occurring during axon regeneration; describing implications of neuroscience for theories of consciousness and epistemology; testing several aging hypotheses. Home: 5125 Riviera Dr Coral Gables FL 33146 Office: U Miami Dept Biology PO Box 249118 Coral Gables FL 33124

WILSON, DEREK EDWARD, chemist; b. Stanmore, Middlesex, Eng., July 7, 1943; s. Walter Edward and Elizabeth Ellen (Smith) W.; m. Kathleen Louise Toms, Sept. 10, 1966; children: Eve, Matthew. BSc with honors, Exeter U., Devon, Eng., 1964, PhD in Phys. Chemistry, 1967. Postdoctoral fellow U. Calgary, Alta., Can., 1967-70; rsch. scientist Harper and Tumstall Ltd., Wellingborough, Eng., 1970-75; group leader Coates Electrographics Ltd., Midsomer Norton, Bath, Eng., 1975-80, tech. mgr., 1980-86, chief scientist, 1986—; presenter at internat. confs. in field. Contbr. articles to profl. publs. Bd. govs. Norton-Radstock Tech. Coll., Near Bath, Eng., 1989—. Mem. Soc. for Imaging Sci. and Tech. (adv. bd., nonimpact printer tech. conf. com. 1992—), Royal Soc. Chemistry, Soc. for Info. Display, Brit. Standards Inst. (graphics improvement panel). Achievements include 6 patents on new advances in ink jet and toner technology. Office: Coates Electrographics Ltd, Norton Hill, Midsomer Norton Bath BA3 4BQ, England

WILSON, DONALD EDWARD, physician, educator; b. Worcester, Mass., Aug. 28, 1936; s. Rivers Rivo and Licine (Bradshaw) W.; m. Patricia C. Littell, Aug. 27, 1977; children: Jeffrey D.E., Sean D., Monique, Sheila L. A.B., Harvard U., 1958; M.D., Tufts U., 1962. Diplomate Am. Bd. Internal Medicine. Intern St. Elizabeth Hosp., Boston, 1962-63; resident in medicine, research fellow in gastroenterology VA Hosp. and Lemuel Shattuck Hosp., Boston, 1963-66; assoc. chief gastroenterology Bklyn. Hosp., 1968-71; instr. medicine SUNY Downstate Med. Center, Bklyn., 1968-71; asst. prof. medicine U. Ill., Chgo., 1971-73; asso. prof. U. Ill., 1973-75, prof., 1975-80, acting head dept. medicine, 1976-77; dir. div. gastroenterology U. Ill. Hosp., Chgo., 1971-80; chief of gastroenterology U. Ill. Hosp., 1973-80, physician-in-chief, 1976-77; prof., chmn. dept. medicine SUNY Downstate Med. Center, Bklyn., 1980-91; physician-in-chief State U. and Kings County Hosp., 1980-91; dean, prof. medicine U. Md.Sch. Medicine, Balt., 1991—; mem. Part II test com. Nat. Bd. Med. Examiners, 1985-88; mem. gen. clin. rsch. ctrs. com. NIH, 1987—; mem. nat. digestive disease adv. bd. NIH, 1985-87, chmn., 1986-87; vis. prof. medicine U. London, Kings Coll. Med. Sch., 1977-78; mem. gastrointestinal drugs adv. bd. FDA, 1985-87, chmn., 1986-87; mem. nat. adv. com. Agy. for Health Care Policy and Rsch., Dept. HHS, 1991—, chmn., 1992—. Contbr. articles to med. jours. Bd. vis. Harvard Sch. of Pub. Health, 1992—. Served to capt. M.C., USAF, 1966-68. Recipient Rsch. award HEW, 1971, 74, Rsch. award John A. Hartford Found., Inc., 1972-79, Rsch. award Distilled Spirits Coun. U.S., 1972-74, Rsch. award VA, 1974. Fellow ACP; mem. NAS, AAAS, Am. Gastroent. Assn., Am. Fedn. Clin. Rsch., Central Soc. Clin. Rsch., Central Rsch. Club, Am. Assn. Study Liver Disease, Chgo. Soc. Gastroenterology (pres. 1978-79), Digestive Disease Found., Midwest Gut Club, Soc. Exptl. Biology and Medicine, N.Y. Acad. Scis., N.Y. Acad. Medicine, N.Y. Soc. Gastroenterology, Chgo. Soc. Gastrointestinal Endoscopy (pres. 1979-80), Assn. Am. Physicians, Assn. for Acad. Minority Physicians (sec./treas. 1986—), Nat. Med. Assn., Assn. Profs. Medicine (sec.-treas. 1990-91), Inst. of Medicine, Med. Club Bklyn., Sigma Pi Phi (grand boule). Club: Harvard (Chgo., N.Y.C.). Home: 2 Whitebridge Ct Baltimore MD 21208 Office: U Md Sch Medicine 655 W Baltimore St Baltimore MD 21201-1559

WILSON, EDWARD OSBORNE, biologist, educator; b. Birmingham, Ala., June 10, 1929; s. Edward Osborne and Inez (Freeman) W.; m. Irene Kelley, Oct. 30, 1955; 1 dau., Catherine Irene. BS, U. Ala., 1949, MS, 1950; PhD, Harvard U., 1955; DS (hon.), Duke U., Grinnell Coll., Lawrence U., U. West Fla.; Oxford U.; DS (hon.), Fitchburg State Coll., Duke U., Grinnell Coll.,, Lawrence U., U. West Fla.,, Fitchburg State Coll.,, Macalester Coll., U. Mass.; DPhil, Uppsala U.; LHD (hon.), U. Ala., Hofstra U.; LLD (hon.), Simon Fraser U. Jr. fellow Soc. Fellows, Harvard U., 1953-56, mem. faculty, 1956—, Baird prof. sci., 1976—, Mellon prof. sci., 1990-93, curator entomology, 1971—; fellow Guggenheim Found., 1978, mem. selection com., 1982-89; bd. dirs. World Wildlife Fund, 1983—, Orgn. Tropical Studies, 1984—, N.Y. Bot. Gardens, 1991—, Am. Mus. Natural History, 1992—, Am. Acad. Liberal Edn., 1993—. Author: The Insect Societies, 1971, Sociobiology: The New Synthesis, 1975, On Human Nature, 1978 (Pulitzer prize for Nonfiction), Promethean Fire, 1983, Biophilia, 1984, (with Bert Holldobler) The Ants, 1990 (Pulitzer prize for Nonfiction), Success and Dominance in Ecosystems, 1990, The Diversity of Life, 1992 (Nat. Wildlife Assn. award). Recipient Cleve.-AAAS rsch. prize, 1967, Nat. Medal Sci., 1976, Leidy medal Acad. Natural Sci., Phila., 1979, Disting. Service award Am. Inst. Biol. Scis., 1976, Mercer award Ecol. Soc. Am., 1971, Founders Meml. award and L.O. Howard award Entomol. Soc. Am., 1972, 85, Archie Carr medal U. Fla., 1978, Disting. Svc. award Am. Humanist Soc., 1982, Tyler ecology prize, 1984, Silver medal Nat. Zool. Park, German Ecol. Inst. prize, 1987, Weaver award scholarly letters Ingersoll Found., 1989, Crafoord prize Royal Swedish Acad. Scis., 1990, Prix d'Inst. de la Vie, Paris, 1990, Revelle medal, 1990, Gold medal Worldwide Fund for Nature, 1990, Achievement award Nat. Wildlife Fedn., 1992, Shaw medal Mo. Botanical Garden, 1993, Internat. prize Biology Govt. of Japan, 1993, others. Fellow Am. Acad. Arts and Scis., Am. Phil. Soc., Deutsche Akad. Naturforsch.; mem. NAS, Am. Genetics Assn. (hon. life), Brit. Ecol. Soc. (hon. life), Entomol. Soc. Am. (hon. life), Zool. Soc. London (hon. life), Am. Humanist Soc. (hon. life), Royal Soc. London, Finnish Acad. Sci. and Letters, Royal Soc. Sci. Uppsala (Sweden). Home: 9 Foster Rd Lexington MA 02173-5505 Office: Harvard U Museum Comparative Zoology Cambridge MA 02138

WILSON, FRANK HENRY, electrical engineer; b. Dinuba, Calif., Dec. 4, 1935; s. Frank Henry and Lurene (Copley) W.; m. Carol B. Greening, Mar. 28, 1964; children: Frank, Scott E. BS, Oreg. State U., 1957. Electronic engr. Varian Assoc., Palo Alto, Calif., 1960-61, Stanford U. Med. Sch., Palo Alto, 1961-68, U. Calif. Med. Sch., Davis, 1968-77, Litronix, Cupertino, Calif., 1978-81, Quantel, Santa Clara, Calif., 1981-87, Heraeus Lasersonics, Milpitas, Calif., 1987-91, Continuum Electro-Optics, Santa Clara, Calif., 1992—. 1st lt. Signal Corps U.S. Army, 1958-60. Mem. IEEE. Home: 3826 Nathan Way Palo Alto CA 94303-4519 Office: Continuum 3150 Central Expwy Santa Clara CA 95051

WILSON, GEORGE PETER, engineer, educator; b. Truro, N.S., Can., Dec. 8, 1939; s. George W. and Madlyn (Merriam) W.; m. Rose Derry Blount, Mar. 17, 1967; children—David, Ian, James, Chris. B.Engring., N.S. Tech. Coll., 1962; diploma engring., Dalhousie U., 1960; M.S., U. Birmingham, Eng., 1964. Registered profl. engr., N.S. Indsl. engr. Can. Nat. Rys., Montreal, Que., 1965-67; dir. Atlantic Indsl. Research Inst., Halifax, N.S., 1972—; prof., head dept. indsl. engring. Tech. U. N.S., Halifax, 1967—; pres. Wilson Fuel Co. Ltd., Truto, N.S., 1976—, Applied Computer Systems Ltd., Truro, 1974—, Wentworth Valley Devels. Ltd, 1987—; bd. dirs. Kerr Controls Ltd., Truro. Contbr. articles on indsl. engring. to profl. jours. Served to lt. Royal Can. Navy Res., 1957-67. Fellow Engring. Inst. Can.; mem. Inst. Indsl. Engrs. (pres. chpt.), Can. Ops. Research Soc., Assn. Profl. Engrs. N.S. Liberal Anglican. Clubs: Wentworth Valley Ski (Halifax) (pres. 1978-80); Shortas Lake Yacht (N.S.) (commodore 1980-82). Office: PO Box 1000, Halifax, NS Canada B3J 2X4

WILSON, HAROLD MARK, power generation developer, marketing specialist; b. Tachikawa AFB, Japan, Feb. 21, 1959; s. Harold Joseph and Cansada Camille (Johnson) W.; m. Wendy Gray, Dec. 30, 1988. BSME, U. Ark., 1982. Prodn. engr. Union Carbide Corp., East Hartford, Conn., 1982-83; energy engr. Conn. Natural Gas Corp., Hartford, Conn., 1983-87; mgr. power generation Am. Gas Assn., Arlington, Va., 1987-91; mktg. mgr. Hydra-Co Enterprises Inc., Syracuse, N.Y., 1991—; advisor Gas. Rsch. Inst., Chgo., 1987-91; co-founder Am. Gas Cooling Ctr., Arlington, 1989-91. Contbr. articles to profl. jours. Vol. Missing and Exploited Children, Arlington, 1990-91. Mem. ASME (assoc.), Assn. Energy Engrs., Assn. Environ. Mgrs. (founding). Achievements include development and founding of Am. Gas Cooling Ctr.

WILSON, JACK, aeronautical engineer; b. Sheffield, Yorkshire, Eng., Jan. 5, 1933; came to U.S., 1956; s. George and Nellie (Place) W.; m. Marjorie Reynolds, June 3, 1961 (div. Jan. 1991); children: Tanya Ruth, Cara. BS in Engring., Imperial Coll., London, 1954; MS in Aero. Engring., Cornell U., 1958, PhD in Aero. Engring., 1962. Sr. scientific officer Royal Aircraft Establishment, Farnborough, Eng., 1962-63; prin. rsch. sci. Avco-Everett Rsch. Lab., Everett, Mass., 1963-72; vis. prof. Inst. de Mecanique des Fluides, Marseille, France, 1972-73; sr. scientist U. Rochester (N.Y.), 1973-80; sr. rsch. assoc. Sohio/BP Am., Cleve., 1980-90; sr. engring. specialist Sverdrup Tech. Inc., Cleve., 1990—. Author: (chpt.) "Gas Lasers" of Applied Optics in Engineering VI, 1980, "Laser Sources" of Techniques in Chemistry XVII, 1982; contbr. articles to profl. jours. Mem. AIAA (sr., mem. tech. com. 1991-92). Achievements include first to demonstrate gas-dynamic laser; patent in application of high speed flow to gas laser media; patent in devel. of antimony dopant sources; measurement of air ionization rate at very high speeds. Office: Sverdrup Tech Inc 2001 Aerospace Pkwy Brookpark OH 44142

WILSON, JAMES LEE, retired geology educator, consultant; b. Waxahachie, Tex., Dec. 1, 1920; s. James Burney and Hallie Christine (Hawkins) W.; m. Della I. Moore, May 8, 1944; children: James Lee Jr., Burney Grant, Dale Ross (dec.). Student, Rice U., 1938-40; BA, U. Tex., 1942, MA, 1944; PhD, Yale U., 1949. Geologist Carter Oil Co., Tulsa, 1943-44; asst. and assoc. prof. U. Tex., Austin, 1949-52; rsch. geologist Shell Devel. Co., Houston, 1952-66; prof. Rice U., Houston, 1966-79, U. Mich., Ann Arbor, 1979-86; geol. cons. New Braunfels, Tex., 1986—; cons. Erico Corp., London, 1985-88, Masera Corp., Tulsa, 1988—, Coyote Geol. Svcs., Boulder, Col., 1990—; adj. prof. Rice U., 1986—,. Author: Carbonate Facies in Geologic History, 1975; contbr. articles to tech. jours.;. With C.E., U.S. Army, 1944-46, Italy. Grantee NSF. Mem. Am. Assn. Petroleum Geologists (hon.), Internat. Sedimentological Soc., Soc. Econ. Paleontology and Minerology (pres. 1972-73, field trip guide books 1989), Paleontological Soc., West Texas Geological Soc., South Tex Geological Soc. Avocations: piano, languages. Home and Office: 1316 Patio Dr New Braunfels TX 78130-8505

WILSON, JAMES RICHARD, mechanical engineer; b. Cin., Apr. 14, 1953; s. James Roland and Barbara June (Bohnekamp) W.; m. Susan Lee Douglas, May 16, 1981. BSME, U. Tenn., 1977. Registered profl. engr., Tenn., Ind. Sales engr. Trane Co., Knoxville, Tenn., 1977-80; mech. engr. Hart, Freeland, Roberts, Inc., Nashville, 1980-84; project engr., assoc. Gresham, Smith & Ptnrs., Nashville, 1984—. Mem. ASHRAE, Am. Soc. Profl. Engrs. Home: 3436 Country Ridge Dr Antioch TN 37013-1037 Office: Gresham Smith & Ptnrs 3310 W End Ave Nashville TN 37203-1058

WILSON, JOHN ANTHONY, physicist; b. Halifax, Yorkshire, Eng., July 16, 1938; s. Harry and Ivy (Shaw) W.; m. Patricia Ann McNulty, Aug. 31, 1963; children: Daniel, Rebecca. BA, U. Cambridge, 1961, MA, 1964; PGCE, U. London, 1962; PhD, U. Cambridge, 1968. Jr. rsch. fellow Cavendish Lab., U. Cambridge, 1969-72; mem. rsch. staff Bell Telephone Labs., Murray Hill, N.J., 1972-79; Royal Soc. sr. rsch. fellow H.H. Wills Physics Lab., U. Bristol, 1979-86, George Wills sr. rsch. fellow, 1986—; mem. rsch. com. Vb Inst. Laue-Langevin, Grenoble, France, 1989—. Asst. editor Jour. Physics, Inst. of Physics, 1982-86; contbr. articles to profl. jours. Mem. Inst. Physics U.K., European Phys. Soc., Am. Chem. Soc., Am. Phys. Soc. Mem. Ch. of Eng. Achievements include research in solid state physics, metal-insular transition, charge density waves, and high temperature superconductivity. Avocations: philately, travel, history. Home: 41 The Dell, Westbury-on-Trym, Bristol BS9 3UF, England Office: U Bristol Dept Physics, Tyndall Ave, Bristol BS8 1TL, England

WILSON, JOHN ERIC, biochemist; b. Champaign, Ill., Dec. 13, 1919; s. William Courtney and Marie Winette (Lytle) W.; m. Marion Ruth Heaton, June 7, 1947; children—Kenneth Heaton, Douglas Courtney, Richard Mosher. S.B., U. Chgo., 1941; M.S., U. Ill., Urbana, 1944; Ph.D., Cornell U., 1948. Research asst. Proxylin Products, Inc., Chgo., summers 1941-42, Gen. Foods Corp., Hoboken, N.J., summer, 1943; asst. in chemistry U. Ill., 1941-44; asst. in biochemistry Cornell U. Med. Coll., N.Y.C., 1944-48, research asso., 1948-50; asst. prof. biochemistry U. N.C., Chapel Hill, 1950-60, asso. prof., 1960-65, prof., 1965-90, prof. emeritus, 1990—, dir. grad. studies, dept. biochemistry, 1965-71, acting dir. neurobiology program, 1968-69, asso. dir., 1969-72, dir., 1972-73; Kenan prof. U. Utrecht, Netherlands, 1978. Mem. editorial bd. Jour. Neurochemistry, 1987—; contbr. numerous articles on biochemistry and neurochemistry to profl. publs. Scoutmaster Occoneechee council Boy Scouts Am., 1959-66; mem. Chapel Hill Twp. Adv. Council, 1978-85, Orange County (N.C.) Planning Bd., 1979-85. Fellow AAAS; mem. Am. Chem. Soc., Am. Soc. for Biochemistry and Molecular Biology, Am. Soc. Neurochemistry, N.C. Acad. Sci., Internat. Soc. for Neurochemistry, Internat. Brain Rsch. Orgn., Soc. Neurosci. (coun. 1969-70, chmn. fin. com. 1973-78, organizer and mem. exec. com. N.C. chpt. 1974-75), Harvey Soc., Sigma Xi, Phi Lambda Upsilon, Alpha Chi Sigma, Beta Theta Pi. Home: 214 Spring Ln Chapel Hill NC 27514-3540 Office: U NC Sch Medicine Dept Biochemistry and Biophysics Chapel Hill NC 27599-7260

WILSON, JOHN KENNETH, optical research professional; b. Wantagh, N.Y., Jan. 8, 1953; s. James L. and Marjorie J. (Riddell) W.; m. Kathleen Ann Ottavi, July 17, 1988; 1 child, Jamie Kathleen. BS, Guilford Coll., 1976. Sales engr. Oriel Corp., Stratford, Conn., 1978-84; regional sales mgr. Ealing Electro Optics, Holliston, Mass., 1984-85, nat. sales mgr., 1985-88; dir. Opt/Optics and Precision Tech. Opt/Optics and Precision Tech., Medway, Mass., 1988-90; pres. T2 Technol. Co., Wilton, Conn., 1990-91, Optical Rsch. Techs., Wilton, 1991—. Bd. dirs. Rep. Town Com., Wilton, 1992—; tenor Wilton Singers, 1992—. Mem. Laser Inst. Am., Optical Soc. Am., Masons. Home: 310 Hurlbutt St Wilton CT 06897

WILSON, KEITH CROOKSTON, chemical engineer; b. Provo, Utah, Dec. 5, 1953; s. M. Lyman Jr. and Dorothy (Crookston) W.; m. Susan Wells, June 24, 1976; children: Mindy, Matthew, Randon, Jonathan. BSChemE, Brigham Young U., 1979. Registered profl. engr., Calif. Engr. Eli Lilly, Indpls., 1979-81; prin. engr. C. F. Braun, Alhambra, Calif., 1981-88; process engr. Arcadian Corp., Augusta, Ga., 1988-91; prodn. mgr. Arcadian Fertilizer L.P., Augusta, Ga., 1991—. Contbr. articles and papers to profl. publs. Office: Arcadian Fertilizer LP PO Box 1483 Augusta GA 30903

WILSON, KENNETH GEDDES, physics research administrator, educator; b. Waltham, Mass., June 8, 1936; s. E. Bright and Emily Fisher (Buckingham) W.; m. Alison Brown, 1982. A.B., Harvard U., 1956, DSc hon., 1981; Ph.D., Calif. Tech. Inst., 1961; Ph.D. (hon.), U. Chgo., 1976. Asst.

prof., then prof. physics Cornell U., Ithaca, N.Y., from 1963, James A. Weeks chmn. in phys. sci., from 1974; Hazel C. Youngberg Trustees Disting prof. The Ohio State U., 1988—. Recipient Nobel prize in physics, 1982, Dannie Heinemann prize, 1973, Boltzmann medal, 1975, Wolf prize, 1980, A.C. Eringen medal, 1984, Franklin medal, 1983, Aneesur Rahman prize, 1993. Mem. NAS, Am. Philos. Soc., Am. Phys. Soc., Am. Acad. Arts and Scis.

WILSON, LEE BRITT, physiology educator; b. Amarillo, Tex., Sept. 10, 1960; s. Robert Warner and Carolyn Marie (Bates) W.; m. Carolyn Martin, July 30, 1983. BS in Biology, West Tex. State U., 1983; PhD in Physiology, La. State U. Med. Ctr., 1988. Pre-doctoral fellow dept. physiology La. State U. Med. Ctr., New Orleans, 1985-88; post-doctoral fellow cardiovascular physiology U. Tex. Southwestern Med. Ctr., Dallas, 1988-91, asst. instr. dept. physiology, 1991-92, asst. prof. dept. physiology, 1992—. Contbr. articles to Jour. Applied Physiology, Brain Rsch., Circulation Rsch., Jour. Physiology. Student Rsch. grantee Am. Heart Assn., 1986-88, Rsch. grantee, 1991—; recipient Nat. Rsch. Svc. Award postdoctoral fellow NIH, 1990-91. Mem. Am. Physiol. Soc., Soc. for Neurosci., N.Y. Acad. Scis., Sigma Xi (assoc.). Office: U Tex Southwestern Med Ctr Dept Physiology 5323 Harry Hines Blvd Dallas TX 75235-9040

WILSON, LEONARD RICHARD, geologist, consultant; b. Superior, Wis., July 23, 1906; s. Ernest and Sarah Jane (Cooke) W.; m. Marian Alice DeWilde, Sept. 1, 1930; children—Richard Graham, Marcia Graham. Ph.B., U. Wis.-Madison, 1930, Ph.M., 1932, Ph.D., 1935. Research assoc. Wis. Geol. and Nat. Hist. Survey, Trout Lake, Wis., 1932-36; instr. to prof. geology Coe Coll., Cedar Rapids, Iowa, 1934-46; head dept. geology and mineralogy U. Mass., Amherst, 1946-56; leader Greenland Ice Cap Am. Geog. Soc., N.Y.C., 1953; prof. geology NYU, N.Y.C., 1956-57; prof. to George L. Cross research prof. geology and geophysics U. Okla., Norman, 1957-77, prof. emeritus, 1977—; geologist Okla. Geol. Survey, Norman, 1957-77, ret. 1977—; cons. in field; research assoc., mem. edn. bd. Mus. Nat. Hist., N.Y.C., 1956-77; curator paleobotany-micropaleontology Okla. Mus. Natural Hist., Norman, 1968—. Contbr. articles to profl. jours. Editor proceedings Iowa Acad. Sci., 1936-46. Melhaup fellow Ohio State U., 1939-40; NSF grantee, 1959-65. Fellow AAAS, Geol. Soc. Am., Palynological Soc. India Coll. of Fellows (Erdtman Internat. medal 1973); mem. Am. Assn. Petroleum Geologists, Am. Assn. Stratigraphic Palynologists (hon.), Nat. Assn. Geology Tchrs. (hon. life), Audubon Soc. (pres. Norman br. 1982-83), Explorers Club, Sigma Xi (pres. Okla. chpt. 1965-66), Phi Beta Kappa (pres. Okla. Alpha chpt. 1978-79), Phi Kappa Phi. Current work: Stratigraphic research in Paleozoic palynology as it relates to hydrocarbon maturation and associated strata. Subspecialties: Geology; Chronobiology.

WILSON, MARK KENT, aerospace engineer; b. Des Moines, Aug. 4, 1947; s. Ronald Tilmer and V. Georgene (Kent) W.; m. Christine Kay Jones, Feb. 10, 1979 (div. May 1982); m. Kathleen Diane Tollefson, Feb. 5, 1983. BS, Purdue U., 1971; MS, U. Dayton, 1987, Stanford U., 1993. Project engr. USAF, Wright-Patterson AFB, Ohio, 1971-81; chief structures br. B2 systems program office USAF, Wright-Patterson AFB, 1981-92, chief flight systems divsn., 1993—. Mem. AIAA (sr.). Achievements include work on B-2 design and acquisition team. Home: 429 Ridge Line Ct Spring Valley OH 45370 Office: USAF ASD/YSOF Wright Patterson AFB OH 45433

WILSON, MARY ELIZABETH, physician; b. Indpls., Nov. 19, 1942; d. Ralph Richard and Catheryn Rebecca (Kurtz) Lausch; m. Harvey Vernon Fineberg, May 16, 1975. AB, Ind. U., 1963; MD, U. Wis., 1971. Diplomate Am. Bd. Internal Medicine, Am. Bd. Infectious Diseases. Tchr. of French and English Marquette Sch., Madison, Wis., 1963-66; intern in medicine Beth Israel Hosp., Boston, 1971-72, resident in medicine, 1972-73, fellow in infectious diseases, 1973-75; physician Albert Schweitzer Hosp., Deschapelles, Haiti, 1974-75, Harvard Health Svcs., Cambridge, Mass., 1974-75; asst. physician Cambridge Hosp., 1975-78; hosp. epidemiologist Mt. Auburn Hosp., Cambridge, 1975-79, chief of infectious diseases, 1978—; adv. com. immunization practices Ctrs. for Disease Control, Atlanta, 1988-92, acad. adv. com. Nat. Inst. Pub. Health, Mex., 1989-91; cons. Ford Found., 1988; instr. in medicine Harvard Med. Sch., Boston, 1975—; lectr. Sultan Qaboos U., Oman, 1991. Author: A World Guide to Infections: Diseases, Distribution, Diagnosis; contbr. articles to profl. jours. Mem. Cambridge Task Force on AIDS, 1987—, Earthwatch, Watertown, Mass., Cultural Survival, Inc., Cambridge; bd. dirs. Horizon Communications, West Cornwall, Conn., 1990. Recipient Lewis E. and Edith Phillips award U. Wis. Med. Sch., 1969, Cora M. and Edward Van Liere award, 1971, Mosby Scholarship Book award, 1971. Fellow ACP, Infectious Diseases Soc. Am., Royal Soc. Tropical Medicine and Hygiene; mem. Am. Soc. Microbiology, N.Y. Acad. Scis., Am. Soc. Tropical Medicine and Hygiene, Mass. Infectious Diseases Soc. (founding mem.), Mass. Med. Soc., Peabody Soc., Internat. Soc. Travel Medicine, Wilderness Med. Soc., Sigma Sigma, Phi Sigma Iota, Alpha Omega Alpha. Avocations: playing the flute, hiking, reading, travel. Office: Mount Auburn Hosp 330 Mt Auburn St Cambridge MA 02138-5502

WILSON, MELVIN NATHANIEL, psychology educator; b. St. Louis, Sept. 27, 1948; s. William F. Wilson and Mary Helen (Warder) Thompson; m. Eunice Clark, June 9, 1969 (div. June 1976); 1 child, Jibri-Akil; m. Angela Maria Davis, July 20, 1980. BA, Millikin U., 1970; MA, U. Ill., 1973, PhD, 1977. Asst. prof. U. Houston, 1976-78; vis. asst. prof. U. Ill., Champaign, 1978-79; asst. prof. U. Va., Charlottesville, 1979-86, assoc. prof., 1986—. Assoc. editor Am. Jour. Community Psychology. Minority Rsch. scholar NSF, 1983-85; fellow Rockefeller Found., 1987, Social Sci. Rsch. Coun., 1991; USPHS grantee, 1970-72. Fellow APA, AM. Psychol. Soc.; mem. Soc. Rsch. on Child Devel., Nat. Coun. Family Rels., Assn. Black Psychologists. Avocation: jogging. Office: U Va Psychology Dept Gilmer Hall Charlottesville VA 22903-2477

WILSON, MYRON ROBERT, JR., former psychiatrist; b. Helena, Mont., Sept. 21, 1932; s. Myron Robert Sr. and Constance Ernestine (Bultman) W. BA, Stanford U., 1954, MD, 1957. Diplomate Am. Bd. Psychiatry and Neurology. Dir. adolescent psychiatry Mayo Clinc, Rochester, Minn., 1965-71; pres. and psychiatrist in chief Wilson Ctr., Faribault, Minn., 1971-86; ret., 1986; chmn. Wilson Ctr., 1986-90; ret., 1990; assoc. clin. prof. psychiatry UCLA, 1985—. Contbr. articles to profl. jours. Chmn., chief exec. officer C.B. Wilson Found., L.A., 1986—; mem. bd. dirs. Pasadena Symphony Orchestra Assn., Calif., 1987. Served to lt. comdr., 1958-60. Fellow Mayo Grad. Sch. Medicine, Rochester, 1960-65. Fellow Am. Psychiat. Assn., Am. Soc. for Adolescent Psychiatry, Internat. Soc. for Adolescent Psychiatry (founder, treas. 1985-88, sec. 1985-88, treas. 1988-92); mem. Soc. Sigma Xi (Mayo Found. chpt.). Episcopalian. Office: Wilson Found 8439 W Sunset Blvd Ste 104 West Hollywood CA 90069-1947

WILSON, RICHARD, physicist, educator; b. London, England, Apr. 29, 1926; came to U.S., 1950; s. Percy and Dorothy (Kingston) W.; m. Andrée Desirée DuMond, Jan. 6, 1952; children: Arthur Christopher, Michael Thomas, Nicholas Graham, Elaine Susan, Annette Adele, Peter James. B.A., Christ Church, Oxford U., 1946, M.A., D.Phil., 1949; M.A., Harvard U., 1957. Research lectr. Christ Ch., Oxford U., 1948-53, research officer Dept. Sci. and Indsl. Research, 1953-55; research assoc. U. Rochester, 1950-51, Stanford U., 1951-52; asst. prof. Harvard U., 1955-57, assoc. prof., 1957-61, prof., 1961—, Mallinkrodt prof. physics, 1983—, chmn. dept. physics, 1982-85, assoc. Adams House, 1971—; cons. NRC; served on numerous govt. adv. coms. Asst. editor 5 books; contbr. over 510 articles to sci. jours. Guggenheim fellow, 1961, 69; Fulbright fellow, 1959, 69. Fellow Am. Phys. Soc. (Forum award 1990); mem. N.Y. Acad. Scis., Am. Acad. Arts and Scis., Soc. Psychical Rsch. (London), Newton Conservators. Home: 15 Bracebridge Rd Newton MA 02159-1787 Office: Harvard U Lyman Lab Cambridge MA 02138

WILSON, ROBERT STOREY, osteopathic physician; b. Vernon, Tex., Aug. 20, 1949; s. Lester Virgil and Kity Virginia (Storey) W.; m. Cynthia Anne Fry, June 16, 1972; children: R. Storey Jr., Jennifer Leigh. BA, So. Meth. U., 1971; DO, Tex. Coll. Osteo. Medicine, 1975. Intern Met. Hosp., Dallas; med. dir. Los Barrios Unitos Clinic (P.H.S.), Dallas, 1976-78; acting dir. student health ctr. North Tex. State U., Denton, 1978; ptnr. Brazos Clinic, Granbury, Tex., 1978—; chief med. staff Hood Gen. Office, Granbury, 1987-89; health officer Hood County, 1993—. Mem. Airport Bd., Granbury, 1987-89. Mem. AAAS, Am. Acad. Family Physicians, N.Y. Acad. Scis., Tex. Acad. Family Physicians, Adults United (founding mem., pres. 1989-91). Republican. Avocations: flying, scuba diving. Home: 221 E Bridge St Granbury TX 76048-2247 Office: Brazos Clinic 305 W Pearl St Granbury TX 76048-2497

WILSON, ROBERT THANIEL, JR., environmental chemist; b. St. Albins, N.Y., June 25, 1957; s. Robert Thaniel and Ida Jean (Gurley) W.; m. Debra Kaye Gilmore, June 25, 1983; 1 child, Helen Rebecca. BS in Chemistry and Biology, Meth. Coll., Fayetteville, N.C., 1979; student, N.C. State U., 1979-80. Cert. EMT, N.C. Grad. teaching asst. N.C. State U., Raleigh, 1979-80; radiation control and test technician II CP&L Harris Energy & Environ. Ctr., Hew Hill, N.C., 1980-82; environ. and chemistry technician II Harris Nuclear Plant, New Hill, N.C., 1982; environ. and chemistry technician I Carolina Power & Light, New Hill, 1982-85, specialsit environ. and chemsitry, 1985-88, sr. specialist environ. and chemistry, 1988—. Co-contbr. articles to profl. jours. Chmn. com. Fuquay-Variner (N.C.) Habitat for Humanity, 1989; adminstrv. bd. Fuquay-Variner United Meth. Ch., 1990-93; lt. Fuquay-Variner Rescue Squad, 1992-93. Named Outstanding Young Men of Am., 1984. Mem. Am. Chem. Soc., Am. Inst. Chemists, Health Physics Soc., Omicron Delta Kappa. Democrat. Methodist. Achievements include development of chemical control program. Office: Carolina Power & Light Co Harris Plant PO Box 165 New Hill NC 27562

WILSON, ROBERT WOODROW, radio astronomer; b. Houston, Tex., Jan. 10, 1936; s. Ralph Woodrow and Fannie May (Willis) W.; m. Elizabeth Rhoads Sawin, Sept. 4, 1958; children—Philip Garrett, Suzanne Katherine, Randal Woodrow. B.A. with honors in Physics, Rice U., 1957; Ph.D., Calif. Inst. Tech., 1962. Research fellow Calif. Inst. Tech., Pasadena, 1962-63; mem. tech. staff AT&T Bell Labs., Holmdel, N.J., 1963-76; head radio physics research dept. AT&T Bell Labs., 1976—. Discoverer 3 deg. k microwave background radiation, 1965; discoverer CO and other molecules in interstellar space using their millimeter wavelength radiation. Recipient Henry Draper medal Royal Astron. Soc., London, 1977, Herschel medal Nat. Acad. Scis., 1977; Nobel prize in physics, 1978; NSF fellow, 1958-61; Cole fellow, 1957-58. Mem. Am. Astron. Soc., Internat. Astron. Union, Am. Phys. Soc., Internat. Sci. Radio Union, Nat. Acad. Scis., Phi Beta Kappa, Sigma Xi. Home: 9 Valley Point Dr Holmdel NJ 07733-1320 Office: AT&T Bell Labs HOH L239 PO Box 400 Holmdel NJ 07733

WILSON, RONALD WAYNE, natural science educator; b. Iowa Falls, Iowa, Aug. 4, 1939; s. Herbert Wayne and Thais Arlene (Lawton) W.; m. Carola Cattani, June 9, 1969 (div. 1978); m. Bonnie J. Fons, Oct. 17, 1987. BS, Iowa State U., 1961; PhD, Mich. State U., 1965. Asst. prof. dept. natural sci. Mich. State U., East Lansing, 1967-70, assoc. prof., 1970-77, prof. dept. natural sci., 1977—. Contbr. articles to profl. jours. Mem. Sigma Xi, Phi Kappa Phi. Office: Mich State U Dept Botany Plant Biol Lab East Lansing MI 48864

WILSON, SCOTT ROLAND, physicist; b. Boulder, Colo., May 15, 1964; s. Raymond William and Dianna Mae (Nevergall) W.; m. Leigh Xu, Aug. 8, 1992. BS in Engring. Physics, U. Colo., 1986; PhD in Physics, U. S.C., 1993. Lab. asst. J&A Assocs., Inc., Golden, Colo., 1984-86; grad. rsch. asst. dept. physics U. S.C., Columbia, 1986—. Mem. Amnesty Internat., Boulder, 1988—. Mem. Am. Phys. Soc., Assn. for Computing Machinery, Soc. Physics Students, Sigma Pi Sigma. Achievements include thesis experiment which set world record limit on existence of 17 kev neutrino. Office: U SC Dept Physics Columbia SC 29208

WILSON, SHERYL A., pharmacist; b. Nashville, Apr. 6, 1957; d. Robert Lewis and Norma Anne (Cox) W. BS in Biology, David Lipscomb U., 1979; BS in Pharmacy, Auburn U., 1985. Lic. pharmacist, Tenn. Student extern/intern East Alabama Med. Ctr., Opelika, Ala., 1982-86; staff pharmacist Metro Nashville Gen. Hosp., 1987—. Flutist Nashville Community Concert Band, 1973—; preschool tchr. Donelson Ch. of Christ, 1988—. Mem. Am. Pharm. Assn., Am. Soc. Hosp. Pharmacists, Tenn. Soc. Hosp. Pharmacists, Mid. Tenn. Soc. Hosp. Pharmacists, Am. Soc. Parenteral and Enteral Nutrition. Democrat. Avocations: art, music, reading, cooking, sewing. Home: 1439 McGavock Pk Nashville TN 37216 Office: Metro Nashville Gen Hosp 72 Hermitage Ave Nashville TN 37210

WILSON, THEODORE HENRY, retired electronics company executive, aerospace engineer; b. Eufaula, Okla., Apr. 23, 1940; s. Theodore V. and Maggie E. (Buie) W.; m. Barbara Ann Tassara, May 16, 1958 (div. 1982); children: Debbie Marie, Nita Leigh, Wilson Axten, Pamela Ann, Brenda Louise, Theodore Henry II, Thomas John; m. Colleen Fagan, Jan. 1, 1983 (div. 1987); m. Karen L. Lerohl, Sept. 26, 1987. BSME, U. Calif., Berkeley, 1962; MSME, U. So. Calif., 1964, MBA, 1970, MSBA, 1971. Sr. rsch. engr. N.Am. Aviation Co. div. Rockwell Internat., Downey, Calif., 1962-65; propulsion analyst, supr. div. applied tech. TRW, Redondo Beach, Calif., 1965-67, mem. devel. staff systems group, 1967-71; sr. fin. analyst worldwide automotive dept. TRW, Cleve., 1971-72; contr. systems and energy group TRW, Redondo Beach, 1972-79; dir. fin. control equipment group TRW, Cleve., 1979-82, v.p. fin. control indsl. and energy group, 1982-85; mem. space and def. group TRW, Redondo Beach, 1985-93, ret., 1993; lectr., mem. com. acctg. curriculum UCLA Extension, 1974-79. Mem. Fin. Execs. Inst. (com. govt. bus.), Machinery and Allied Products Inst. (govt. contracts coun.), Nat. Contract Mgmt. Assn. (bd. advisors), Aerospace Industries Assn. (procurement and fin. coun.), UCLA Chancellors Assocs., Tau Beta Pi, Beta Gamma Sigma, Pi Tau Sigma. Republican. Avocations: golf, bridge. Home: 3617 Via La Selva Palos Verdes Peninsula CA 90274-1115

WILSON, THOMAS LEON, physicist; b. Alpine, Tex., May 21, 1942; s. Homer Marvin and Ogarita Maude (Bailey) W.; m. Joyce Ann Krevosky, May 7, 1978; children—Kenneth Edward Byron, Bailey Elizabeth Victoria. B.A., Rice U., 1964, B.S., 1965, M.A., 1974, Ph.D., 1976. With NASA, Houston, 1965—, astronaut instr., 1965-74, high-energy theoretical physicist, 1969—. Contbr. articles in field to profl. jours. Recipient Hugo Gernsback award IEEE, 1964; NASA fellow, 1969-76. Mem. Am. Phys. Soc., AAAS, N.Y. Acad. Scis., Am. Assn. Physicists in Medicine. Research on grand unified field theory, relativistic quantum field theory, quantum chromodynamics, quantum probability theory, supergravity, quantum cosmology, astrophysics, deep inelastic scattering, neutrino astronomy, authority on neutrino tomography, discoverer classical uncertainty principle. Subspecialty: relativity and gravitation. Patentee in field; contributor to design of NASA's proposed lunar base; originator olive branch as symbol of man's 1st landing on moon (on Susan B. Anthony and Eisenhower dollars). Home: 206 Woodcombe Dr Houston TX 77062-2538 Office: NASA Johnson Space Ctr Houston TX 77058

WILSON, THORNTON ARNOLD, retired aerospace company executive; b. Sikeston, Mo., Feb. 8, 1921; s. Thornton Arnold and Daffodil (Allen) W.; m. Grace Miller, Aug. 5, 1944; children: Thornton Arnold III, Daniel Allen, Sarah Louise Wilson Anderson. Student, Jefferson City (Mo.) Jr. Coll., 1938-40; B.S., Iowa State Coll., 1943; M.S., Calif. Inst. Tech., 1948; M.S. Sloan fellow, MIT, 1952-53. With Boeing Co., Seattle, 1943—, asst. chief tech. staff, project engring. mgr., 1957-58, v.p., mgr. Minuteman br. aerospace div., 1963-64, v.p. ops. and planning, 1964-66, exec. v.p., dir., 1966-68, pres., 1968-72, chief exec. officer, 1969-86, chmn. bd., 1972-88, chmn. emeritus, 1988—; dir. PACCAR, Inc., Weyerhaeuser Co., USX Corp., Hewlett Packard, Inc. Bd. govs. Iowa State U. Found.; trustee Seattle U.; mem. corp. MIT, Nat. Acad. Engring. Named to Nat. Aviation Hall of Fame, 1983, Nat. Bus. Hall of Fame, 1989. Fellow Am. Inst. Aeronautics and Astronautics. Office: The Boeing Co PO Box 3707 Seattle WA 98124-2207

WILSON, WALTER LEROY, architect; b. Pitts., Aug. 2, 1942; s. Walter Clarence and Marie Zella (Wilcox) W.; m. Maxine Davis, June 4, 1968 (div. 1970); m. Lois Mary McCollum, Mar. 24, 1984; 1 child, Hilary Ann. Student, Altus (Okla.) Jr. Coll., 1961, Cen. State U., 1964-67; BArch, B Archtl. Engring., Okla. State U., 1971. Registered architect, Ohio, Tex., Ark., Wis. Architect, intern Trott & Bean Assocs., Columbus, Ohio, 1971-73; mgr. architecture Southwestern Bell, Little Rock, Houston and St. Louis, 1973-80; prin., ptrn. Caradine & Wilson Assocs., North Little Rock, Ark., 1980-82; office mgr., divsn. mgr. Polytech Inc., Milw., 1982-85; prin., owner The Wilson Firm, Milw., 1985-92; pres. Wilson Firm, Ltd., Milw., 1992—; bd. dirs. Milw. Area Tech. Ctr. Archtl. Program; mgr. archtl. and engring. svcs. Housing Authority Dept. City Devel. Mem. Coalition for Econ. Devel. and Justice, Milw., 1986, State Capitol and Exec. Residence Bd., State of Wis., 1986—; del. state Dem. conv., Milw., 1986; apptd. State Com. Ct.-Related Needs of Elderly and Persons with Disabilities. Served with USAF, 1960-64. Mem. AIA, Am. Solar Energy Soc., Wis. Soc. Architects (sec., treas. S.E. chpt. 1990, v.p., pres.-elect 1991, pres. 1992, past pres., co-chair Architecture Awareness Week 1992), Constrn. Specifications Inst. (pub. rels. chmn. 1986, membership com.). Democrat. Presbyterian. Avocations: silk screen printing, basketball, racquetball, computer programming, flying. Home: 7211 W Wabash Ave Milwaukee WI 53223-2608 Office: care of The Wilson Firm 7915 W Appleton Ave Milwaukee WI 53218-4500

WILSON, WILLIAM STANLEY, oceanographer; b. Alexander City, Ala., June 5, 1938; s. Norman W. and Helen C. (Hackemack) W.; m. Anne M. Stout; 1 child, Lauren. BS, William & Mary Coll., 1959, MA, 1965; PhD, Johns Hopkins U., 1972. Marine biol. collector Va. Inst. Marine Sci., Gloucester Point, 1959-62, computer systems analyst, 1964-65; computer systems analyst Chesapeake Bay Inst., Balt., 1965-66; phys. oceanography program mgr. Office of Naval Rsch., Washington, 1972-78; chief oceanic processes program NASA, Washington, 1979-89, program scientist earth observing system, 1989-92; asst. adminstr. for ocean svcs. and coastal zone mgmt. NOAA, Washington, 1992—. Recipient Antarctica Svc. medal NSF, 1961, Superior Civilian Svc. award USN, 1979, Exceptional Sci. Achievement medal NASA, 1981, Disting. Achievement award MTS and Compass Publs., 1989, award Remote Sensing Soc., 1992. Mem. Am. Meteorol. Soc., Am. Geophys. Union (Ocean Scientist award 1984), Oceanography Soc. (com. chmn. 1989-92), Sigma Xi. Avocations: bicycling, hiking, gardening, running. Home: 219 Tunbridge Rd Baltimore MD 21212-3423 Office: N SSMC4 Sta 13632 1305 East West Hwy Silver Spring MD 20910

WILT, JAMES CHRISTOPHER, aerospace engineer; b. Pitts., Oct. 5, 1954; s. James Howard and Isabelle Nicholas (Mouganis) W.; m. Mary Jo Koets, Apr. 24, 1982; children: Elizabeth Ann, Rebecca Lynn, James Christopher Jr. BS in Aerospace Engring., U. Fla., 1976; MSME, MBA, So. Meth. U., 1983. Registered profl. engr., Tex. Assoc. strength engr. McDonnell Aircraft Co., St. Louis, 1976-78; tech. project mgr. Vought Aircraft, Dallas, 1978-85, mgr. program devel., 1987-88, mgr. Dayton (Ohio) office, 1988-90, mgr. C-17 program definition, 1990-92, program mgr. C-17 Nacelle redesign, 1993—. Contbr. to profl. publs. Tenor Dallas Symphony Chorus, 1980—, St. Lukes Episcopal Ch., Dallas, 1985, 87, 90—; vestry mem. St. George Episc. Ch., Dayton, 1989. Mem. NSPE (Congl. fellow 1986), AIAA (assoc. fellow, dir. pub. policy 1987-88, newsletter editor 1975—), Air Force Assn., Airlift Tanker Assn., Tau Beta Pi, Phi Gamma Delta (chpt. historian 1975). Republican. Episcopalian. Home: 1210 N Winnetka Ave Dallas TX 75208 Office: Vought Aircraft Co Mail Stop 6AZ-16 PO Box 655907 Dallas TX 75265-5907

WILT, JOHN ROBERT, chemist; b. Cleve., Mar. 4, 1939; s. Eugene Farwell and Lucille Pearl Williams W.; m. Shirley Jean Harris, Sept. 12, 1959; children: Michael, Timothy, Karl, Nicholas, Paul. BS, Mass. Inst. Tech., 1960; PhD, U. Calif., L.A., 1965. Sr. rsch. chemist Eastman Kodak Rsch. Lab., Rochester, N.Y., 1967-71; mgr. Addressograph-Multigraph Corp., Cleve., 1971-74; asst. to pres. Royal Bus. Machines, Inc., Hartford, Conn., 1974-82; pres. The Thermographic Factory, Inc., Rochester, N.Y., 1982—, Thermopress Mktg., Inc., Rochester, N.Y., 1985—; v.p., pres. Exchange Club of Rochester, N.Y., 1985-88. Capt. U.S. Army, 1965-67. Mem. Greater Wolcott (N.Y.) C. of C., 1992—. Home and Office: 18 Smith St Wolcott NY 14590-1032

WILTSCHKO, DAVID VILANDER, geology educator, tectophysics director; b. Portland, Oreg., Feb. 5, 1949; s. William Wesley and Virginia (Vilander) W.; m. Sherry I. Bame, Oct. 17, 1981; children: Alexander, Elicia. BA, U. Rochester (N.Y.), 1971; MSc, Brown U., 1974, PhD, 1977. Vis. asst. prof. geology U. Mich., Ann Arbor, 1977-79, asst. prof., 1979-84; asst. prof. Tex. A&M U., College Station, 1984-86, assoc. prof., 1986—; assoc. dir. Tex. A&M U. Ctr. for Tectonophysics, College Station, 1987-89, dir., 1989—. Mem. editorial bd. Geology, Boulder, Colo., 1986-89. Office: Tex A&M U Ctr Tectonophysics College Station TX 77843

WIMPRESS, GORDON DUNCAN, JR., corporate consultant, foundation executive; b. Riverside, Calif., Apr. 10, 1922; s. Gordon Duncan and Maude A. (Waldo) W.; m. Jean Margaret Skerry, Nov. 30, 1946; children—Wendy Jo, Victoria Jean, Gordon Duncan III. B.A., U. Oreg., Eugene, 1946, M.A., 1951; Ph.D., U. Denver, 1958; LL.D., Monmouth Coll., Ill., 1970; L.H.D., Tusculum Coll., Greenville, Tenn., 1971. Lic. comml. pilot. Dir. pub. relations, instr. journalism Whittier (Calif.) Coll., 1946-51; asst. to pres. Colo. Sch. Mines, Golden, 1951-59; pres. Monticello Coll., Alton, Ill., 1959-64, Monmouth Coll., Ill., 1964-70, Trinity U., San Antonio, 1970-77; vice chmn. S.W. Found. for Biomed. Rsch., San Antonio, 1977-82, pres., 1982-92, also bd. govs.; pres. Duncan Wimpress & Assocs., Inc., San Antonio, 1992—; commr. Burlington No. R.R. scholarship selection com.; chmn. Valero Energy Corp. scholarship commn.; bd. dirs. Southwest Rsch. Inst. Author: American Journalism Comes of Age, 1950. Bd. dirs. Am. Inst. Character Edn., ARC, Am. Heart Assn., Cancer Therapy and Rsch. Found.; trustee San Antonio Med. Found., Eisenhower Med. Ctr., Rancho Mirage, Calif., Eisenhower Med. Rsch. and Edn. Ctr.; mem. San Antonio Fiesta Commn.; ruling elder United Presbyn. Ch., U.S.A.; mem. adv. bd. Alamo Area chpt. Am. Diabetes Assn. 1st lt. AUS, 1942-45, ETO. Decorated Bronze Star. Mem. Aircraft Owners and Pilots Assn., Am. Acad. Polit. and Social Sci., Am. Assn. Higher Edn., MENSA, Nat. Pilots Assn., Pilots Internat. Assn., Inc., Quiet Birdmen, Greater San Antonio C. of C. (bd. dirs.), North San Antonio C. of C. (bd. dirs.), Assn. Former Intelligence Officers, Confederate Air Force, Pi Gamma Mu, Sigma Delta Chi, Sigma Delta Pi, Sigma Phi Epsilon (trustee found.), Sigma Upsilon, Newcomen Soc. N.Am. Clubs: Argyle, St. Anthony, San Antonio Country, the Dominion (bd. govs.), City , Plaza, San Antonio Golf Assn. Lodge: Rotary (dist. gov. 1983-84, San Antonio). Avocations: golf, skiing, flying. Office: S W Found Biomed Rsch PO Box 78284 San Antonio TX 78278-0818

WINAND, RENÉ FERNAND PAUL, metallurgy educator; b. Ixelles, Belgium, Nov. 15, 1932; s. Fernand and Malvina (Lelievre) W.; m. Christiane Hauer, Mar. 25, 1961; children: Pascaline, Jean-Marc, Henri. BS, Free U. Brussels, 1949, MS in Electromech. Engring., 1954, PhD in Metallurgy, 1960. Engr. Ateliers de Constructions Electriques de Charleroi, Belgium, 1954-55, Brit. Petroleum Belgium, Antwerp, 1956-57; rsch. asst. Free U. Brussels, 1957-61, lectr., 1961-67, prof., 1967—, head dept. metallurgy and electrochemistry, 1968—, chmn. Faculty Applied Scis., 1977-80, coord. Ctr. Indsl. Rsch., 1979—, vice-rector, 1993—; Gollick lectr., U. Mo., Rolla, 1985. Patentee in field; contbr. articles to sci. jours. Chevalier, Order of Couronne, Belgium. Mem. Minerals Metals and Materials Soc. U.S., Electrochem. soc. U.S., Inst. Mining and Metallurgy U.K., Iron and steel Inst. Japan, Can. Inst. Mining and Metallurgy, French Soc. Metallurgy, Robotics Assn. Belgium (hon. chmn. 1988). Avocation: flying. Home: 24 Ave Jean XXIII, B 1330 Rixensart Belgium Office: Free U Brussels, 50 Ave Roosevelt, B 1050 Brussels Belgium

WINCHELL, WILLIAM OLIN, mechanical engineer, educator, lawyer; b. Rochester, N.Y., Dec. 31, 1933; s. Leslie Olin and Hazel Agnes (Apker) W.; m. Doris Jane Martenson, Jan. 19, 1957; children: Jason, Darrell, Kirk. BME, GMI Engring. and Mgmt. Inst., 1956; MSc, Ohio State U., 1970; MBA, U. Detroit, 1976; JD, Detroit Coll. Law, 1980. Bar: Mich. 1981, U.S. Dist. Ct. (ea. dist.) Mich. 1981, U.S. Ct. Appeals (6th cir.) 1982, U.S. Supreme Ct. 1985, N.Y. 1988; registered profl. engr., Mich., N.Y. Cons. Gen. Motors Corp., Detroit and Warren, Mich. and Lockport, N.Y., 1951-87; pvt. practice Royal Oak, Mich., 1981-90; assoc. prof., chmn. dept. indsl. engring. and mech. Alfred (N.Y.) U., 1987-90; assoc. prof. mfg. engring. Ferris State U., Big Rapids, Mich., 1990-91, 93—; vis. assoc. prof., program coord. Purdue U., Warren, Mich., 1991-92. Mem. Royal Oak Long Range Planning Commn., 1980. Served to lt. commdr. USNR, 1956-76. Burton fellow Detroit Coll. Law, 1978. Fellow Am. Soc. Quality Control (v.p. 1985-89); mem. ABA, Mich. Bar Assn., Soc. Mfg. Engrs., Inst. Indsl. Engrs., Am. Soc. Engring. Educators, Tau Beta Pi, Beta Gamma Sigma. Roman Catholic. Club: North Star Sail. Avocations: sailing, woodworking. Office: Purdue U J-108 Macomb 14500 Twelve Mile Rd Warren MI 58093

WINCOR, MICHAEL Z., psychopharmacology educator, clinician, researcher; b. Chgo., Feb. 9, 1946; s. Emanuel and Rose (Kershner) W.; m.

Emily E.M. Smythe; children: Meghan Heather, Katherine Rose. SB in Zoology, U. Chgo., 1966; PharmD, U. So. Calif., 1978. Rsch. project specialist U. Chgo. Sleep Lab., 1968-75; psychiat. pharmacist Brotman Med. Ctr., Culver City, Calif., 1979-83; asst. prof. U. So. Calif., L.A., 1983—; cons. Fed. Bur. Prisons Drug Abuse Program, Terminal Island, Calif., 1978-81, Nat. Inst. Drug Abuse, Bethesda, Md., 1981, The Upjohn Co., Kalamazoo, 1982-87, 91-92, Area XXIV Profl. Standards Rev. Orgn., L.A., 1983, Brotman Med. Ctr., Culver City, Calif., 1983-88, SmithKline Beecham Pharms., Phila., 1990—, Tokyo Coll. of Pharmacy, 1991, G. D. Searle & Co., Chgo., 1992—. Contbr. over 30 articles to profl. jours., chpts. to books, papers presented at nat. and internat. meetings and reviewer. Mem. adv. coun. Franklin Ave. Sch., 1986-89; bd. dirs. K.I. Children's Ctr., 1988-89; trustee the Sequoyah Sch., 1992-93. Recipient Cert. Appreciation, Mayor of L.A., 1981, Bristol Labs Award, 1978; Faculty scholar U. So. Calif. Sch. Pharmacy, 1978. Mem. Am. Coll. Clin. Pharmacy (chmn. constn. and bylaws com. 1983-84, mem. credentials com. 1991-93), Am. Assn. Colls. Pharmacy (focus group on liberalization of the profl. curriculum), Am. Soc. Hosp. Pharmacists (chmn. edn. and tng. adv. working group 1985-88), Am. Pharm. Assn. (del. ann. meeting ho. dels. 1989), Sleep Rsch. Soc., Am. Sleep Disorders Assn., U. So. Calif. Sch. Pharmacy Alumni Assn. (bd. dirs. 1979—), Rho Chi. Avocation: photography. Office: U So Calif 1985 Zonal Ave Los Angeles CA 90033-1086

WINDES, JAMES DUDLEY, psychology educator; b. Phoenix, May 10, 1937; s. William Harvey and Annie Mae (Zumwalt) W.; m. Isabella (Judy) Greenway, Aug. 22, 1958; children: William J., Jill K., John D. BA, Ariz. State U., 1959; MA, U. Ariz., 1963, PhD, 1966. Asst. prof. Psychology No. Ariz. U., Flagstaff, 1966-74, assoc. prof. Psychology, 1975-92; psychol. prof. No. Ariz U., Flagstaff, 1993—. Contbr. articles to profl. jours. Woodrow Wilson fellow Woodrow Wilson Found., 1959, NASA fellow, 1964. Mem. Am. Psychol. Soc., Rocky Mountain Psychol. Assn. Democrat. Home: 1045 W Coy Dr Flagstaff AZ 86001 Office: No Ariz U Dept Psychology Box 15106 Flagstaff AZ 86011

WINDHAGER, ERICH ERNST, physiologist, educator; b. Vienna, Austria, Nov. 11, 1928; came to U.S., 1954; s. Maximilian and Bertha (Feitzinger) W.; m. Helga A. Rapant, June 18, 1956; children: Evelyn Ann, Karen Alice. MD, U. Vienna, 1954. Research fellow in biophysics Harvard Med. Sch., Boston, 1956-58; instr. in physiology Cornell U. Med. Coll., N.Y.C., 1958-61; vis. scientist U. Copenhagen, 1961-63; asst. to prof. physiology Cornell U. Med. Coll., N.Y.C., 1963—, Maxwell M. Upson prof. physiology and biophysics, 1978—, chmn. dept. physiology, 1973—. Recipient Homer W. Smith award N.Y. Heart Assn., 1978. Office: Cornell U Med Coll Dept Physiology 1300 York Ave New York NY 10021-4896

WINDLER, DONALD RICHARD, botanist, educator; b. Centralia, Ill., Feb. 4, 1940; s. Richard Henry William and Minnie Marie (Hanenberger) W.; m. Bonnie Kay Whipkey, June 9, 1963 (div. 1983); 1 child, Erica. MA in Botany, So. Ill. U., 1963-65; PhD in Botany, U. N.C., 1965-70. From asst. prof. to prof. biology Towson (Md.) State U., 1969-84, prof., chmn. dept. biology, 1984-87; acting dean Coll. Natural and Math. Scis. Towson State U., 1987-90; prof., chmn. dept. biology Towson State U., 1990—; bd. dirs. Natural History Soc. Md., Balt., 1985—. Recipient NDEA fellowship, U. N.C., Chapel Hill, 1965-68. Mem. Internat. Assn. for Plant Taxonomists, Am. Soc. of Plant Taxonomy, So. Appalachian Botanical Club (endowment chmn.), Assn. Southeastern Biologists, Torrey Botanical Club, Soc. Econ. Botany. Lutheran. Home: 518 Murdock Rd Towson MD 21212-2020 Office: Towson State U Biology Dept Towson MD 21204-7097

WINDSOR, TED BACON, feed microscopy consultant; b. Montgomery County, Mo., Nov. 10, 1924; s. Clarence Veirs and Floss (McCord) W.; m. Mathilda Marianne Davis, Mar. 7, 1930 (div. Oct. 1968); children: Elizabeth Jo Windsor Stefanon, Rebecca June Windsor Sible, Holly, Paul Davis, Laurel; m. Janet Barr, June 16, 1970 (dec. Oct. 1989); m. Mary Elizabeth Kirtley, Feb. 2, 1993. BS in Chemistry and Biology, St. Louis U., 1950. With Midwest Oil Refining Co., Overland, Mo., 1950-52; chemist, feed microscopist Ralston Purina Co., St. Louis, 1952-55; lab. mgr. Ralston Purina Co., Camp Hill, Pa., 1955-74; chemist Ea. Lab. Svc. Assocs., York, Pa., 1974-75; electrician Harold K. Davis, Grantham, Pa., 1975-77; chemist C.H. Masland & Sons, Carlisle, Pa., 1977-88; feed microscopy cons. Mechanicsburg, Pa., 1974—. Cpl. U.S. Army, 1943-46. Mem. Am. Chem. Soc. (local sect. chmn. 1964-65), Am. Assn. Feed Microscopists (pres. 1962-63), Assn. Analytical Chemists. Achievements include development of unique system for use in microscopic evaluation of animal feedstuffs. Home and Office: 1118 Apple Dr Mechanicsburg PA 17055-3904

WINEGRAD, ALBERT IRVIN, immunologist, educator; b. Phila., 1926. Intern Hosp. U. Pa., Phila., 1952-53, resident in immunology, 1953-55, resident, 1957—; resident in endocrinology Peter Bent Brigham Hosp., Boston, 1955-57; prof. medicine U. Pa. Home: 3400 Spruce St Philadelphia PA 19104-4219 Office: George S Cox Med Rsch Institute 502 Medical Education Bldg 36th & Hamilton Walk Philadelphia PA 19104*

WINELAND, DAVID J., physicist. Recipient Davisson Germer Atomic Surface Physics prize Am. Physical Soc., N.Y., 1990. Office: Natl Inst of Standards & Tech Gaithersburg MD 20899*

WINFIELD, JOHN BUCKNER, rheumatologist, educator; b. Kentfield, Calif., Mar. 19, 1942; s. R. Buckner and Margaret G. (Katterfelt) W.; m. Patricia Nichols (div. 1968); 1 child, Ann Gibson; m. Teresa Lee McGrath, 1969; children: John Buckner III, Virginia Lee. BA, Williams Coll., 1964; MD, Cornell U., 1968. Diplomate Am. Bd. Internal Medicine. Intern in medicine N.Y. Hosp., N.Y.C., 1968-69; staff assoc. LI/Nat. Inst. Allergy and Infectious Diseases NIH, Bethesda, Md., 1969-71; resident in medicine, fellow in rheumatology U. Va. Sch. Medicine, Charlottesville, 1971-73; fellow in immunology Rockefeller U., N.Y.C., 1973-75; asst. prof. medicine U. Va. Sch. Medicine, Charlottesville, 1975-76, assoc. prof. medicine, 1976-78; assoc. prof. medicine U. N.C., Chapel Hill, 1978-81, prof. medicine, 1981—, chief div. rheumatology and immunology, 1978—; dir. Thurston Arthritis Rsch. Ctr. U. N.C. Sch. Medicine, Chapel Hill, 1982—; Smith prof. medicine U. N.C. Sch. Med., Chapel Hill, 1987—; adv. coun. Nat. Inst. Arthritis and Musculoskeletal and Skin Diseases, NIH, 1988-92; chmn. edn. com. Am. Rheumatism Assn., Atlanta, 1980-84; immunol. scis. study sect. NIH, 1979-83, Arthritis Musculoskeletal and Skin study sect., 1992—; vice-chair fellowship com. Arthritis Found., 1982; med. coun. Lupus Found. Am., 1987—. Author more than 100 med. and sci. articles in peer reviewed rheumatology and immunology jours.; mem. editorial bd. Arthritis and Rheumatism, Bull. Rheumatic Diseases, Rheumatology Internat., Clin. Exptl. Rheumatology. Sr. asst. surgeon with USPHS, NIH, Bethesda, Md., 1968-71. Recipient Borden prize Cornell U. Med. Coll., 1964, numerous rsch. grants NIH and Arthritis Found., 1975—, Sr. Investigator award Arthritis Found., 1976-79, Kenan award U. N.C., 1985, NIH merit award, 1992. Fellow ACP; mem. Am. Assn. Immunologists, Am. Coll. Rheumatology, Am. Fedn. Clin. Rsch., Am. Soc. Cin. Investigation, Assn. Am. Physicians, Chapel Hill Country Club. Republican. Episcopalian. Avocations: golf, off-road motorcycling, scuba diving. Home: 801 Kings Mill Rd Chapel Hill NC 27514-4920 Office: U NC Sch Medicine Rheumatology/Immunology Divsn 932 FLOB 231H CB#7280 Chapel Hill NC 27599

WINFREE, ARTHUR TAYLOR, biologist, educator; b. St. Petersburg, Fla., May 15, 1942; s. Charles Van and Dorothy Rose (Scheb) W.; m. Ji-Yun Yang, June 18, 1983; children: Rachael, Erik from previous marriage. B.Engring. in Physics, Cornell U., 1965; Ph.D. in Biology, Princeton U., 1970. Lic. pvt. pilot. Asst. prof. theoretical biology U. Chgo., 1969-72; assoc. prof. biology Purdue U., West Lafayette, Ind., 1972-79; prof. Purdue U., 1979-86; prof. ecology and evolutionary biology U. Ariz., Tucson, 1986-88, Regents' prof., 1989—; pres., dir. research Inst. Natural Philosophy, Inc., 1979-88. Author: The Geometry of Biological Time, 1980, When Time Breaks Down, 1986, The Timing of Biological Clocks, 1987. Recipient Career Devel. award NIH, 1973-78, The Einthoven award Einthoven Found. and Netherlands Royal Acad. Scis., 1989; NSF grantee, 1966—; MacArthur fellow, 1984-89, John Simon Guggenheim Meml. fellow, 1982. Home: 1210 E Placita De Graciela Tucson AZ 85718-2834 Office: Univ Ariz 326 BSW Tucson AZ 85721

WING, THOMAS, micrometrologist, engineer, consultant; b. Shanghai, China, Mar. 12, 1929; came to U.S., 1930, naturalized, 1950; s. Lim and Fong Shee W.; B.S. in Engring. cum laude, Purdue U., 1953, postgrad., 1957-60; postgrad. CCNY, 1953-55; m. Catherine Amajelia Scambia, Nov. 27, 1954; children—Karen Elyse, Thomas Scambia, Robert Frank Joseph, David Anthony. Sr. project mgr. Gulton Industries, Metuchen, N.J., 1960-63; adv. engr. IBM, Lexington, Ky., 1963-65; with guidance and control systems div. Litton Industries, Woodland Hills, Calif., 1966-69, mem. tech. staff, 1976-92; cons. in field, 1992—; staff cons. Devel. Consultants, Cin., 1965-66; image tech. mgr. Fairchild Semiconductor, Mountain View, Calif., 1969-71; gen. mgr. research and devel., v.p., Jade Corp., HLC Mfg. Co., Willow Grove, Pa., 1971-75; pres. Photronic Engring. Labs., Inc., Danbury, Conn., 1975-76; lectr. in micro-photo lithography, Inst. Graphic Communication, Boston; cons. engr. in micro image tech. and metrology; founding mem. Thermo Phys. Properties Research Center, Purdue U.; pres., staff cons. Zantec Inc., North Hollywood, Calif., 1981-86; cons. Kasper Instruments, Sunnyvale, Calif., 1977-79, Quintel Corp., San Jose, Calif., 1979—. Coach, v.p. Roadrunners Hockey Club, 1972-75, founder, 1972, Bristol, Pa.; founder, pres. Eastridge Jr. Hockey Club, San Jose, Calif., 1970-71; coach Belmont (Calif.) Jr. Hockey Club, 1969; v.p. Greater Los Angeles Minor Hockey Assn., 1968-69; v.p., coach West Valley Minor Hockey Club, Tarzana, Calif., 1968-69; coach Topanga Plaza Jr. Hockey Club, 1967. Mem. ASME, ASTM (mem. F-1 com. on microelectronics 1972-75), Soc. Photog. Instrumentation Engrs., Soc. Photog. Scientists and Engrs., Tau Beta Pi, Pi Tau Sigma, Phi Eta Sigma. Organizer first internat. lecture series on micro photo lithography, Boston, 1974; contbr. articles to profl. publs.; patentee currency counter, trimming inductor, damped high frequency accelerometer, three phase dithered pivot, auto focus for step and repeat camera, temperature compensated dither drive for laser gyro, proximity printing mechanism, equilbrator for howitizer. Home and Office: 6261 Jumilla Ave Woodland Hills CA 91367-3822

WINGATE, PHILLIP JEROME, chemist; b. Wingate, Md., Jan. 13, 1913; s. Charles Millard Mead and Cornelia (MacNamara) W.; m. Katharine Smith, July 25, 1942; children: Patricia, Barbara, Douglas. BS in Math., Washington Coll., Chestertown, Md., 1933; MS in Chemistry, U. Md., 1939, PhD in Chemistry, 1942; DSc (hon.), Lehigh U., 1978; LittD (hon.), Washington Coll.; DHL (hon.), Wilmington Coll. Chemist E.I. DuPont de Nemours & Co., Wilmington, Del., 1942-78, ret., 1978. Author: The Colorful DuPont Company, 1982, Before the Bridge, 1985, Bandages of Soft Illusion, 1979, H.L. Mencken's Un-Neglected Anniversary, 1980, The Beasts of Big Business, 1984; contbr. over 250 articles to newspapers and mags. Mem. Am. Chem. Soc., Soc. Chem. Industry. Achievements include 2 patents for dyes. Home: 1 Somero Ln Wilmington DE 19807-1715

WINGO, PAUL GENE, oil company executive; b. Springfield, Mo., Nov. 17, 1945; s. Paul Augustus and Geraldine Georgia (Sloan) W.; m. Sharon Kay Withers, July 24, 1965; children: Lori Rochelle, Shari Lynn. Student, S.W. Mo. State U., 1964. With St. Louis-San Francisco Rwy. Co., Springfield, 1963-74, tax agt., 1974-80; real estate rep. Burlington No. R.R. Co., Springfield, 1980-81; sr. landman Milestone Petroleum, Inc., Billings, Mont., 1981-82, sr. staff landman, asst. sec., 1982-83; land mgr. Meridian Oil Inc., Billings, Mont., 1983-86; dir. land adminstrn. Meridian Oil Inc., Ft. Worth, Tex., 1986—. Mem. Am. Assn. Petroleum Landmen, Ft. Worth Assn. Petroleum Landmen, Desk and Derrick Club of Ft. Worth (advisor 1989-92). Republican. Methodist. Avocations: walking, snow skiing, travel. Office: Meridian Oil Inc 801 Cherry St Fort Worth TX 76102-6803

WINGSTRAND, HANS ANDERS, orthopedic surgeon; b. Göteborg, Sweden, Feb. 20, 1949; s. Karl Anders and Kerstin Elin (Dahl) W.; m. Inga Kristina Bengtsson, June 24, 1974; children: Johan Anders, John Magnus, Anna Maria Kristina. MD, U. Lund, Sweden, 1974, PhD, 1986. Intern Hosp. of Västervik, Sweden, 1974-75, Hosp. of Helsingborg, Sweden, 1975-76; resident Hosp. of Helsingborg, 1976-79; resident in orthopedics Univ. Hosp., Lund, Sweden, 1979-81, 82-87; resident in hand surgery Univ. Hosp., Malmö, Sweden, 1981-82; resident in plastic surgery and neurosurgery Univ. Hosp., Malmö, 1982; rsch. fellow Univ. Hosp., Lund, 1987-91, assoc. prof., 1991—. Co-editor Acta Orthopaedica Scandinavica, 1986—; contbr. articles to profl. jours. Fellow Swedish Orthopedic Soc. (travel grant 1988), Swedish Pediatric Orthopedic Soc., Scandinavian Orthopedic Assn., European Orthopedic Rsch. Soc. Avocations: sailing, hunting. Home: Runstensgrand 8, S-24021 Löddeköpinge Sweden Office: U Hosp, Dept Orthopedics, S-22185 Lund Sweden

WINICK, HERMAN, physicist; b. N.Y.C., June 27, 1932; s. Benjamin and Yetta (Matles) W.; m. Renee Feldman, May 31, 1953; children: Alan Lee, Lisa Frances, Laura Joan. AB, Columbia Coll., 1953; PhD, Columbia U., 1957. Rsch. assoc., instr. U. Rochester, N.Y., 1957-59; from staff physicist to asst. dir. Cambridge (Mass.) Electron Accelerator Harvard U., 1959-73; dep. dir. linear accelerator ctr. Stanford (Calif.) Synchrotron Radiation Lab., 1973—; prof. applied physics Stanford U., 1983—; chairperson tech. rev. com. Synchrotron Radiation Rsch. Ctr., Taiwan, 1984—. Mem. editorial bd.: Nuclear Instruments and Methods, 1982—; co-editor: Synchrotron Radiation Research, 1980; editor: Handbook on Synchrotron Radiation, 1984—. Recipient Humboldt Sr. Scientist award A. von Humboldt Found., 1986, Energy Related Tech. award U.S. Dept. Energy, 1987. Fellow Am. Phys. Soc. (chmn. com. on internat. freedom of scientist 1992); mem. AAAS. Achievements include development of first wiggler and undulator magnets for synchotron radiation research. Home: 853 Tolman Dr Stanford CA 94305 Office: SSRL SLAC Bin 69 PO Box 4349 Stanford CA 94309

WINITZ, HARRIS, pyschology educator; b. White Plains, N.Y., Mar 4, 1933; s. Israel and Ann (Weinshank) W.; m. Shevie Winitz; children: Flora, Simeon, Jennifer. BA, Univ. Vt., 1954; MA, Univ. Iowa, 1956, PhD, 1959. Rsch. assoc. Univ. Kansas, Lawrence, 1959-63; asst prof. Case Western Reserve, Cleve., 1963-65; prof. Univ. Mo., Kansas City, 1965—. Author: Articulatory Acquisition and Behavior, 1959; editor: Comprehension Approach to Foreign Language Instruction, 1981, Human Communication and Its Disorders, 1983; contbr. articles to profl. jours. Bd. dirs. Kansas City Mental Health, 1987-91. Recipient Career Devel. award NIH, 1968-73. Mem. APA, Linguistic Soc. Am., Acoustical Soc. Am., Child Devel. Office: Univ Mo 5100 Rockhill Rd Kansas City MO 64110

WINKLER, GUNTHER, medical executive; b. Laa Thaya, Noe, Austria, Aug. 20, 1957; came to U.S., 1986; s. Kurt and Irmgard (Lahner) W.; m. Maria Toifl, Sept. 11, 1979; children: Claudia Victoria, Marc David. MS in Biochemistry, U. Vienna, Austria, 1983, PhD in Biochemistry, 1986. Rsch. assoc. Inst. Virology U. Vienna, 1982-86; postdoctoral fellow U. Medicine and Dentistry of N.J., Piscataway, 1986-88; rsch. scientist Biogen, Inc., Cambridge, Mass., 1988-91, assoc. dir. med. ops., 1991—. Contbr. articles to profl. jours. Recipient Outstanding Achievement award Austrian Soc. Microbiology, 1986. Mem. AAAS, Am. Assn. Pharm. Scientists, Drug Info. Assn. Achievements include research in virology dealing with structure function relationship of proteins, HIV, flaviviruses; industrial research in CD4, CD4-toxins, complement proteins; 3 patent applications; clinical development of Hirulog and Beta-Interferon from phase I to phase II. Home: 8 Churchill Rd Winchester MA 01890 Office: Biogen Inc 14 Cambridge Ctr Cambridge MA 02142

WINKLER, MARJORIE EVERETT, protein chemist, researcher; b. Suffern, N.Y., July 19, 1954; d. Lucius Theodore and Marian Florence (Greenwood) E.; m. James Roy Winkler, Aug. 23, 1975 (div. 1985); m. Paul Frank Hohenschuh, May 31, 1987; children: William Everett, Charles Theodore. BS, SUNY, Buffalo, 1975, PhD, 1980. Postdoctoral researcher MIT, Cambridge, 1980-82; scientist Genentech Inc., South San Francisco, Calif., 1982-89, sr. scientist, 1989—. Contbr. to profl. publs. NIH fellow, 1980-82. Mem. Am. Chem. Soc., Phi Beta Kappa, Sigma Xi. Achievements include work in protein folding and the recovery of recombinant proteins from E. coli and mammalian cell culture. Home: 2884 Canyon Rd Burlingame CA 94010 Office: Genentech Inc 460 Point San Bruno Blvd South San Francisco CA 94080

WINKLER, MARTIN ALAN, biochemist; b. Ft. Stockton, Tex., June 8, 1950; s. Edward Meyer and Selma Ann (Rolnick) W.; m. Janet Kay Phillips, Aug. 3, 1978. BA, U. Tex., 1974, PhD, 1978. Postdoctoral fellow Rockefeller U., N.Y.C., 1979-81; rsch. asst. St. Jude Children's Rsch. Hosp., Memphis, 1982-84, rsch. assoc., 1985-87; staff scientist Biotherapeutics, Inc., Memphis, 1988, Franklin, Tenn., 1989; sr. rsch. biochemist Abbott Labs., North Chicago, Ill., 1990—; cons. Am. Coll. Testing, Iowa City, 1989-90. Contbr. articles to profl. jours. including Immunochemistry, PNAS, Jour. Biol. Chemistry, Jour. of Chromatography, Cancer Immunology and Immunotherapy. Postdoctoral fellowship Am. Cancer Soc., 1978. Mem. AAAS, Am. Assn. Immunologists, Am. Chem. Soc. Achievements include development of confirmatory immunoassay for antibodies to T.cruzi; rsch. in structure and interaction sites of calmodulin-dep. enzymes, prodn. and analysis of monoclonal antibody heteroconjugates. Office: Abbott Labs 1401 Sheridan Rd North Chicago IL 60064

WINKLER, ROBERT LEWIS, statistics educator, researcher, author, consultant; b. Chgo., Feb. 12, 1943; s. Roy Henry and Catherine Pauline (Fleming) W.; m. Dorothy Marie Hespen, June 13, 1964; children: Kevin Mark, Kristin Lynne. B.S., U. Ill., 1963; Ph.D., U. Chgo., 1966. Asst. prof. Ind. U., Bloomington, 1966-69, assoc. prof., 1969-72, prof., 1972-80, Disting. prof. quantitative bus. analysis, 1980-84; vis. assoc. prof. U. Wash., Seattle, 1970-71; vis. scientist Nat. Ctr. Atmospheric Rsch., Boulder, Colo., 1972; rsch. scientist internat. Inst. Applied Systems Analysis, Laxenburg, Austria, 1973-74; vis. prof. Stanford U., 1974, Institut Européen d'Administration des Affaires, Fontainebleau, France, 1980-81, 90-91; IBM rsch. prof. Duke U., 1984-85, Calvin Bryce Hoover prof., 1985-89, James B. Duke prof., 1989—, sr. assoc. dean for faculty and rsch., 1991—; cons. in field, 1970—. Author: Statistics, 1970, Introduction to Bayesian Inference and Decision, 1972; contbr. numerous articles to profl. jours.; departmental editor Mgmt. Sci., 1981-89; assoc. editor 7 profl. jours., 1970—. Fellow Am. Inst. Decision Scis., Am. Statis. Assn.; mem. Inst. Mgmt. Sci. (v.p. publs. 1982-85), Internat. Inst. Forecasters (pres. 1989-90). Home: 225 Huntington Dr Chapel Hill NC 27514-2419 Office: Duke U Fuqua Sch Bus Durham NC 27706

WINKLER, STEVEN ROBERT, hospital administrator; b. Chattanooga, Tenn., Dec. 4, 1953; s. David Wilfred and Margaret (Tepper) W.; m. Monica Sue Nijoka, July 11, 1987; children: Megan Leigh, Sara Elizabeth. BA, Vanderbilt U., 1976; M in Health Adminstrn., Duke U., 1978. Asst. adminstr. Tepper Hosp. and Clinic, Chattanooga, 1978-79, adminstr., 1979-80; assoc. exec. dir. Humana Hosp.-Brandon (Fla.), 1981-83, Humana Hosp.-Bennett, 1983-84; v.p. ops. Baton Rouge Gen. Med. Ctr., 1984-86, v.p. mktg., 1986-88; v.p. risk mgmt. Gen. Health, Inc., Baton Rouge, 1988—. v.p. Beth Shalom Synagogue, 1990—, pres., 1993-94; active Am. Diabetes Assn. (Baton Rouge, state chpts.), Arthritis Found., United Way of Baton Rouge, 1989. Mem. Am. Coll. Healthcare Execs., Am. Hosp. Assn., Am. Soc. Healthcare Risk Mgmt. (pres midsouth region 1990), Jewish Fedn. of Greater Baton Rouge (v.p. 1990, pres. 1992-93), Tipmasters (pres., Tipmaster of Yr. 1989). Office: Gen Health Inc 5757 Corporate Blvd Ste 200 Baton Rouge LA 70808

WINN, C(OLMAN) BYRON, mechanical engineering educator; b. Canton, Mo., Nov. 21, 1933; s. Colman Kersey and Kiula Elmeda (Ingold) W.; m. Donna Sue Taylor, Aug. 25, 1957; children: Byron, Derek, Julie. BS in Aeronautics, U. Ill., 1958; MS in Aeronautics, Stanford U., 1960, PhD, 1967. Engr. Lockheed Missiles & Space Co., Palo Alto, Calif., 1958-60, sr. engr., 1962-64; rsch. scientist Martin-Marietta, Denver, 1960-62; lectr. Santa Clara (Calif.) U., 1963-65; assoc. prof. Colo. State U., Ft. Collins, 1966-74, prof. mech. engring., 1974—, prof., head dept., 1982—; cons. Space Rsch. Corp., North Troy, Vt., 1969-73. Author: Controls in Solar Energy Systems, 1982, Controls in Solar Energy Systems, 1993. Loaned exec. United Way, Ft. Collins, 1992. With U.S. Army, 1953-55. Named Disting. Alumnus U. Ill., 1984, EPPEC award Platte River Power Authority, 1992. Fellow ASME; mem. AIAA (energy systems award 1992), Internat. Solar Energy Soc. (bd. dirs. 1980-89), Am. Solar Energy Soc. (bd. dirs. 1979-86, solar action com. 1991-93), Tau Beta Pi. Achievements include development of controllers for solar energy systems; design and development of the reconfigurable passive evaluation analysis and test facility; founding of the Energy Analysis and Diagnostic Center, The Waste Minimization Assessment Center, and Manufacturing Excellence Center at Colo. State U. Office: Colo State U Dept Mech Engring Fort Collins CO 80523

WINN, GEORGE MICHAEL, electrical equipment company executive; b. Victoria, Can., 1944. Student U. Wash., 1968. Pres., chief oper. officer, John Fluke Mfg. Co., Everett, Wash., 1988— Office: John Fluke Mfg Co Inc PO Box C9090 6920 Seaway Blvd Everett WA 98206

WINN, WALTER TERRIS, JR., civil/environmental engineer; b. Houston, Aug. 26, 1949; s. Walter Terris and Sue (Colvard) W.; m. Phyllis Ann Hobart, May 22, 1971; children: Holly Kay, Walter Timothy. BS in Civil Engring., Tex. Tech. U., 1972, MS in Civil Engring., 1973. Registered profl. engr., Tex.; diplomate Am. Acad. Environ. Engrs. Design engr. Brown & Root, Inc., Houston, 1973-78; project mgr., v.p. KSA Engrs. Inc., Longview, Tex., 1978—. Youth tchr., leader First United Meth. Ch., Longview, 1986—; pres. Glenwood Water Supply Corp., Gilmer, Tex., 1988—. Named Young Engr. of Yr. Tex. Soc. Profl. Engrs., East Tex. chpt., 1980. Fellow ASCE (br. pres. 1988-89); mem. Water Environ. Fedn. (sect. pres. 1992-93), Am. Water Works Assn. Achievements include projects in fields of civil & environ. engring. in water supply, water distbn., wastewater collection, wastewater treatment, water treatment, solid waste disposal, site remediation, roads, bridges, dams and air pollution control. Office: KSA Engrs Inc PO Box 1552 140 E Tyler #600 Longview TX 75606

WINNETT, A(SA) GEORGE, analytical research chemist, educator; b. N.Y.C., Jan. 4, 1923; s. S.A. and Mary (Klein) W.; m. Elaine Grace Sugerman, Sept. 30, 1948; children: Susan Beth, Steven Mark. BS in Animal Husbandry-Chemistry, Pa. State U., 1947; postgrad., U. Pa., 1947-48; MA in Sci. Edn., NYU, 1951. Chemist Luth. Hosp., N.Y.C., 1948-49, Queens Gen. Hosp., N.Y.C., 1949; high sch. tchr. N.Y.C. Bd. Edn., 1949-52; chemist N.Y.C. Valspar Corp., 1952-55, Reichhold Chem. Corp., 1955-57; asst. prof. dep. agrl. chemistry Rutgers U., New Brunswick, N.J., 1957-62, assoc. prof. dept. environ. sci., 1982-93, prof. emeritus, 1993—; vis. prof. U. Montpellier, France, 1980-81, U. Rennes, France, 1987-88; cons. N.Y. State Civil Svc. Commn., Congressman Richmond, Bklyn,. 1980; cons., residue analyst Smithsonian Inst., 1970-73; reviewer project ARS USDA; mem. environ. law soc. symposium Hofstra U. Sch. Law, 1986; presenter in field. Contbr. articles to profl. jours. including Jour. Econ. Entomology, Bull. Environment and Toxicology, Jour. Agrl. and Food Chemistry, Jour. Chromatographic Sci., Bull. Environ. Contamination and Toxicology, Jour. Chromatography, Sci., Jour. of the Assn. Official Analytical Chemists, Jour. Environ. Sci. Health. With Med. Corps, U.S. Army, 1942-45, ETO. Grantee USDA, 1957-77, 77-88, 89-92, FMC Corp., 1973, Shell Chem. Co., 1969, EPA, 1980, 82, N.J. Dept. Environ. Protection, 1984-85, USDA NAPIAP, 1985-86, 90-91, 91-92, USDA CSRS, 1989-90, 90-91, 92-93; decorated Combat Medics badge, WWII Victory medal. Mem. AAUP (fringe benefit com. chmn. Cook Coll. exec. com., pres. New Brunswick chpt.), Am. Chem. Soc. (pesticide divsn., reviewer jours.), Assn. Official Analytical Chemists, Sigma Xi (Rutgers chpt., sec. 1966-69, v.p. 1969-70, pres. 1970-71). Home: 331 Summit Pl Highland Park NJ 08904 Office: Rutgers U Cook Coll Dept Environ Scis New Brunswick NJ 08903

WINNIE, GLENNA BARBARA, pediatric pulmonologist; b. Lansing, Mich., Oct. 14; d. Robert John and Irene (Fetchik) W.; m. Jeffrey Alan Cooper, Mar. 17, 1990; children: Robert Jefferson Cooper, David Jamison Cooper. BS, Mich. State U., 1973; MD, Vanderbilt U., 1977. Diplomate Am. Bd. Pediatrics, Am. Bd. Pediatric Pumonology. Resident in pediatrics Case Western Res. U./Babies and Childrens Hosp., Cleve., 1977-79, fellow in pediatric pulmonology, 1979-82; asst. prof. pediatrics Albany (N.Y.) Med. Coll., 1982-90, assoc. prof. pediatrcis, 1990—, head pediatric pulmonology sect., 1982—; dir. Albany Pediatric Pulmonary abd Cystic Fibrosis Ctr., 1982—. Contbr. articles to profl. jours. Bd. dirs. Albany Ronald McDonald Ho., 1986-88. Rsch. grantee Nat. Cystic Fibrosis Found., 1984-86, 88-90, NIH, 1987-93. Mem. Am. Acad. Pediatrics, Am. Thoracic Soc. (rsch. fellowship review com. 1989-92), Capital Dist. Pediatric Soc. (treas. 1985-90, pres. 1990-91). Episcopalian. Achievements include description of role of Epstein Barr virus in pulmonary exacerbations in cystic fibrosis. Office: Albany Medical College 47 New Scotland Ave Albany NY 12208

WINOGRAD, NICHOLAS, chemist; b. New London, Conn., Dec. 27, 1945; s. Arthur Selig Winograd and Winifred (Schaefer) Winograd Mayes; m.

Barbara J. Garrison. BS, Rensselaer Poly. Inst., 1967; PhD, Case Western Reserve U., 1970. Asst. prof. chemistry Purdue U., West Lafayette, Ind., 1970-75, assoc. prof. chemistry, 1975-79; prof. chemistry Pa. State U., University Park, 1979-85, Evan Pugh prof. chemistry, 1985—; cons. Shell Devel. Co., Houston, 1975—; mem. chemistry adv. bd. NSF, Washington, 1987-90, analytical chemistry adv. bd., 1986-89. Contbr. articles to profl. jours. A.P. Sloan Found. fellow, 1974; Guggenheim Found. fellow, 1977; recipient Founder's prize Tex. Instruments Found., 1984, Faculty Scholar's Pa. State U., 1985, Bennedetti Pichler award Am. Microchem. Soc., 1991, Outstanding Alumnus award Case Western Res. U., 1991. Fellow AAAS (Sect. award); mem. Am. Chem. Soc. Home: 415 Nimitz Ave State College PA 16801-6412 Office: Pa State U Dept of Chemistry 152 Davey Lab University Park PA 16802-6300

WINOGRAD, SHMUEL, mathematician; b. Tel Aviv, Jan. 4, 1936; came to U.S., 1956, naturalized, 1965; s. Pinchas Mordechai and Rachel Winograd; m. Elaine Ruth Tates, Jan 5, 1958; children: Daniel H., Sharon A. BSEE, MIT, MSEE; PhD in Math., NYU, 1968. Mem. research staff IBM, Yorktown Heights, N.Y., 1961-70, dir. math. sci. dept., 1970-74, 81—; IBM fellow, 1972—; permanent vis. prof. Technion, Israel. Author: (with J.D. Cowan) Reliable Computations in the Presence of Noise; research on complexity of computations and algorithms for signal processing. Fellow IEEE (W. Wallace McDowell award 1974); mem. NAS, Am. Math. Soc., Math. Assn. Am., Am. Philos. Soc., Assn. Computing Machines, Soc. Indsl. and Applied Math., Am. Acad. Arts and Scis. Home: 235 Glendale Rd Scarsdale NY 10583-1533 Office: IBM Research PO Box 218 Yorktown Heights NY 10598-0218

WINOGRAD, TERRY ALLEN, computer science educator; b. Takoma Park, Md., Feb. 24, 1946; m. Carol Hutner; children: Shoshana, Avra. BA in Math., Colo. Coll., 1966, DSc (hon.), 1986; postgrad., U. Coll., London, 1967; PhD in Applied Math., MIT, 1970. Instr. math. MIT, 1970-71, asst. prof. elec. engring, 1971-74; vis. asst. prof. computer sci. and linguistics Stanford (Calif.) U., 1973-74, from asst. prof. to assoc. prof., 1974-79, assoc. prof. computer sci., 1979-89; prof. computer scis., 1989—; cons. Xerox Palo Alto (Calif.) Rsch. Ctr., 1972-85, Hermenet Inc., San Francisco, 1981-83, Action Techs., 1983—; lectr. in field; mem. adv. bd. Ctr. for Teaching and Learning Stanford U., 1979-83, mem. acad. coun. com. on info. tech., 1981-83, mem. undergrad. coun. Sch. Engring., 1985—; adviser State of Calif., 1983. Bd. dirs. Live Oak Inst., Berkeley, Calif., 1980—, pres., 1984—. Grantee Advanced Rsch. Project Agys., 1969-75, NSF, 1975-77, 82-85, Xerox, 1975-80, System Devel. Found., 1982-83; Hon. Woodrow Wilson fellow, 1966, Hon. NSF fellow, 1966, 90-93, Fulbright fellow, 1966-67, Danforth fellow, 1967-70, Mellon Jr. Faculty fellow, 1977. Mem. Computer Profls. for Social Responsibility (mem. nat. exec. com. 1983—, mem. nat. bd. dirs. 1984—, pres. 1987-90), Am. Assn. Artificial Intelligence, Union of Concerned Scientists, Assn. for Computing Machinery. Home: 746 Esplanada Way Palo Alto CA 94305-1013 Office: Stanford U Dept Computer Sci Stanford CA 94305-2140

WINSOR, TRAVIS WALTER, cardiologist, educator; b. San Francisco, Dec. 1, 1914; s. Samuel Wiley and Mabel Edna (Mc Carthy) W.; BA, Stanford U., 1937, MD, 1941; m. Elizabeth Adams, Sept. 1, 1939; children: David Wiley, Susan Elizabeth. Intern, Alameda County Hosp., Oakland, Calif., 1940-41. Diplomate Am. Bd. Internal Medicine, Am. Bd. Cardiovascular Diseases, 1953. asso. fellow and instr. in medicine and cardiology Tulane U. Sch. Medicine, New Orleans, 1941-45; practice medicine specializing in cardiovascular disease; mem. staff L.A. County Hosp., Hosp. of Good Samaritan (hon.), St. Vincent's Med. Ctr., L.A.; clin. instr. in medicine U. So. Calif., L.A., 1945-47, asst. clin. prof. medicine, 1947-61, asso. clin. prof. medicine, 1961-75, clin. prof. medicine, 1975—; dir. Meml. Heart Rsch. Found., Inc., L.A., 1957—. Fellow ACP, Am. Coll. Cardiology, Am. Coll. Chest Physicians, Internat. Coll. Angiology, Am. Coll. Angiology (pres. 1982-83), AMA, Am. Heart Assn.; mem. Calif. Med. Assn., L.A. County Med. Assn., L.A. Soc. Internal Medicine, Am. Coll. Thermmology (pres. 1968-69), Calif. Heart Assn., Internat. Cardiovascular Soc., Sigma Xi. Author: (with George E. Burch) The Plebomanometer, 1943, A Primer of Electrocardiography, 1944, Phlebostatic Axis and Phlebostatic Levels for Venous Pressure Measurements in Man, 1945, Clinical Plethysmography, The Segmental Plethysmograph, 1957; (with A. Sibley and E. Fisher) Electrocardiogram via Telephone, 1960; (with E. Fisher and C. Hyman, and A. Sibley) Influence of Vascular Disease on the Rate of Transmission of Pure Frequency Components of the Arterial Pluse Wave, 1963, Electrocardiogram by Telemetry, 1961; (with C. Hyman) A Primer of Peripheral Vascular Diseases, 1965, An Electric Volume Transducer System for Plethysmographic Recording, 1966, Clinical Techniques for Evaluating drug effects on the Peripheral Circulation, 1967; A Primer of Vectorcardiography, 1972; (with B. Mills, M. Winbury, B. Howe, and H. Berger) Intramyocardial Diversion of Coronary Blood Flow, Effects of Isoproterenol-induced suebendocardial ischemia, 1975, Thermography, Vasodilators: Evaluation and Pharmacology, 1973, Experimental Design for Evaluation of Antianginal Nitrate Drugs, 1973; (with D. Winsor) Physiologic Evaluation of the Cerebral Circulation, 1980; (with A. Sibley and A. Mikail) Computerized Plethysmography and Laser Flowmetry, 1989; (with A. Kappert) Diagnosis of Peripheral Vascular Diseases, 1972, Peripheral Vascular Diseases an Objective Approach, 1958; contbr. articles on cardiovascular diseases to profl. jours. Home: 541 S Lorraine Blvd Los Angeles CA 90020-4817 Office: 4041 Wilshire Blvd Los Angeles CA 90010-3401

WINSTEAD, CAROL JACKSON, mathematics educator; b. Balt., Dec. 23, 1947; d. John Jay and Patricia (Murnaghan) Jackson; m. Thomas Williamson Jr., June 28, 1969; children: Trey, Peter. AA, Pine Manor Coll., Boston, 1967; BA, Goucher Coll., 1969; MS in Edn. with honors, Johns Hopkins U., 1991. Tchr. 3rd grade Bryn Mawr Sch., Balt., 1972-73, tchr. 5th grade, 1973-75, tchr. math., 1975-77; prin. C.J. Enterprises, Balt., 1986-87; tchr. math. St. Pauls Sch. for Girls, Brooklandville, Md., 1988-90; counselor Deer Park Mid. Sch., Randallstown, Md., 1992-93; community counselor, Au Pair Care, 1991. Fundraiser Gilman Sch., Balt., 1990-92; fundraiser Boy's Latin Sch., Balt., 1990-92, mem. sch. bd., 1993. Mem. St. George's Garden Club (treas., admissions com. 1985-91, Horticulture award 1981), Chi Sigma Iota. Avocations: gardening, horticulture, reading, golf. Home and Office: 901 Malvern Ave Baltimore MD 21204-6713

WINTER, DAVID LOUIS, human factors scientist, systems engineer; b. Pitts., July 30, 1930; s. Louis A. and Gladys M. (Quinn) W.; m. Nancy L. Tear, July 1, 1952; children: Leeson, Blaise, Gregory, Lauren. BA, U. Pitts., 1952; MA, Columbia U., 1960; cert. computer sci., Northeastern U., 1971. Assoc. rsch. scientist Am. Insts. Rsch., Washington, 1961-66; sr. rsch. scientist Am. Insts. Rsch., Bedford, Mass., 1966-71, prin. rsch. scientist, 1976—; sr. systems analyst RCA Corp.-Sarnoff Labs., Princeton, N.J., 1971-73; mgr. systems engring. Codon Corp., Bedford, 1973-76; computer systems cons. Mass. Dept. Mental Health, 1971-73. Prs. Mayo Peninsula Civic Assn., Edgewater, Md., 1964-65; v.p. Bedford Human Rels. Coun., 1992—. Capt. USAF, 1952-57. Mem. Am. Acad. Polit. Sci., Human Factors Soc. (dir. New Eng. chpt.), Soc. Ednl. Tech. Democrat. Roman Catholic. Achievements include development of 35 programs physics for computer; design and human factors test for seven USAF electronic, intelligence and backscatter radar systems; design of four computer-assisted training systems (CATS) for E3 AWACS radar, computer displays, communications and navigation subsystems. Home: 27 Gould Rd Bedford MA 01730 Office: Am Insts Rsch 45 North Rd Bedford MA 01730

WINTER, HARLAND STEVEN, pediatric gastroenterologist; b. San Francisco, Sept. 25, 1948; s. Milton and Madeline (Price) W.; m. Susan Weinstein, Oct. 9, 1983. BA, U. Calif., Berkeley, 1970; MD, UCLA, 1974. Intern UCLA, 1974-75, jr. asst. resident pediatrics, 1975-76, sr. asst. resident, 1976-77; clin. fellow gastroenterology Children's Hosp., Boston, 1977-80; rsch. fellow pediatrics Harvard U. Med. Sch., Boston, 1977-79, instr. pediatrics, 1979-82, asst. prof. pediatrics, 1983-91, assoc. prof. pediatrics, 1991—; trustee Crohn's & Colitis Found., N.Y.C., 1991-93, regional med. adv. chmn., 1990—. Editor: (book) Infant Nutrition, the First Three Years, 1984. Named New Investigator, NIH, 1981-84, Prin. Investigator, 1984-87, 87-93.

WINTER, HORST HENNING, chemical engineer, educator; b. Stuttgart, Sept. 9, 1941; s. Simon Wilhelm and Hanna (Schwenn) W.; m. Karin Eckert, Aug. 29, 1969; children: Dirk Christopher, Lisa Susanne, Caroline Elke, Peter Benjamin. D of Engring., U. Stuttgart, 1973. Privatdozent U. Stuttgart, 1976-79; prof. U. Mass., Amherst, 1979—. Editor Rheologica Acta, 1989—; mem. edit. bd. Jour. Rheology, 1989—, Jour. Non-Newtonian Fluid Mechanics, 1989—; contbr. articles to sci. and profl. jours., 1977—. Mem. Am. Soc. Rheology (exec. com. 1990-91), Am. Inst. Chem. Engrs., N.Y. Acad. Sci. Achievements include development of novel rheological techniques, discovery of self-silimar relaxation of polymer critical gels, discovery of universal time spectrum of linear flexible polymers of uniform length. Office: U Mass Dept Chem Engring Amherst MA 01003

WINTER, JIMMY DALE, ecology educator; b. Morris, Minn., Sept. 16, 1946; s. Glenn Arthur and Elva Irene (Sneider) W. BS, U. Minn., 1968, MS, 1970, PhD, 1976. Vis. asst. prof. biology U. Minn., Duluth, 1975-76; postdoctoral assoc. ecology U. Minn., Mpls., 1976-77; asst., assoc. prof., assoc. dir. Environ. Resource Ctr. SUNY, Fredonia, 1978-89, prof., acting dir. Environ. Resources Ctr., 1989-91; assoc. prof. range and wildlife Tex. Tech. U., Lubbock, 1992—. Author: (book chpt.) Fisheries Techniques, 1983; contbr. articles to Am. Fisheries Soc. jour., other profl. publs. Recipient NSF Rsch. Opportunity award, 1984, summer rsch. fellowship, 1970, NIH predoctoral fellowship, 1973-75, postdoctoral fellowship, 1976-77. Mem. Am. Fisheries Soc. (life, pres. N.Y. chpt. 1988-89), Ecol. Soc. Am. (life). Office: Tex Tech U Range & Wildlife Dept Lubbock TX 79409-2125

WINTER, KLAUS H., physicist educator; b. Hamburg, Germany, Dec. 21, 1930; arrived in Switzerland, 1958; s. Franz Hugo August Winter and Luise Frederike Braren-Winter; m. Helga Ertel, Dec. 20, 1957 (dec. 1981); children: Christiane, Sabine; m. Kristina Gmür, Apr. 14, 1984. Diploma in physics, U. Hamburg, 1956; PhD, U. Sorbonne, Paris, 1958. Fellow Coll. de France, Paris, 1955-58; rsch. assoc. CERN, Geneva, 1958-62, sr. physicist, 1962-76, spokesman charm collaboration, 1976-91, spokesman chorus collaboration, 1992—; prof. physics U. Hamburg, 1962—, Humboldt U., Berlin, 1990—. Editor: Neutrino Physics, 1988; editor Physics Letters B, 1971—; contbr. over 200 publs. to refereed sci. jours. Recipient Stern-Gerlach-Preis für Physik, Deutsche Physikalische Gesellschaft, 1993. acm. German Phys. Soc. (Stern-Gerlach medal 1993), European Phys. Soc. Achievements include research in local gauge symmetry of electron and neutrino coupling to Z boson. Office: CERN, CH 1211 Geneva Switzerland

WINTER, MARK LEE, physician; b. Joplin, Mo., Oct. 7, 1950; s. Elza Jr. and Malbryn Lea (Everhard) W.; m. Debbie Wright, Apr. 11, 1987; children: Jason, Tarra, Allison. BA in Biology cum laude, Drury Coll., 1972; MD cum laude, U. Mo., 1976. Resident in otolaryngology U. Tex. Med. Br., 1981; clin. asst. prof. dept. otolaryngology U. Tex. Med. Br., Galveston, 1981—; pvt. practice, 1982—; clin. instr. dept. surgery Tex. Tech U. Health Scis. Ctr., Lubbock, 1984—, clin. asst. prof. dept. med. and surg. neurology, 1987—; adj. prof. Dept. Speech and Hearing Scis. Tex. Tech. U. Helath Scis. Ctr., Lubbock, 1989—. Fellow Am. Acad. Otolaryngology (mem. nat. internat. stds. com. head and neck surg. 1992-94), Am. Acad. Neurotology Soc.; mem. Pan-Am Allergy Soc., Am. Acad. Facial Plastic and Reconstructive Surgery, Lubbock C. of C. (mem. health and medicine com. 1993), Phi Eta Sigma, Phi Mu Alpha. Republican. Avocations: photography, woodworking, antique collecting, draft horse driving. Office: 4002 21st St Lubbock TX 79410

WINTERS, ARTHUR RALPH, JR., chemical and cryogenic engineer, consultant; b. Mercer, Pa., May 17, 1926; s. Arthur Ralph and Rebecca Grace (McLaughry) W.; m. Elizabeth Colt Burgess, Oct. 4, 1952; children: Philip, Andrew, Paul. BS in Chem. Engring. summa cum laude, Lafayette Coll., Easton, Pa., 1948; SM, MIT, 1952. Registered profl. engr., Pa. Devel. engr. Allied Chem. Corp., Morristown, N.J., 1948-50; process engr. Monsanto Co., Boston and St. Louis, 1952-55; specialist Gen. Electric Co., Hudson Falls, N.Y., 1955-62; process mgr. Air Products and Chems. Inc., Trexlertown, Pa., 1963-86; sr. cons. Cryogenic Cons. Inc., Allentown, Pa., 1986—. Contbr. articles to profl. jours. Pres. Saucon Assn. for a Viable Environment, Coopersburg, Pa., 1981-86, other offices, 1976-92. Mem. NSPE, Am. Inst. Chem. Engrs. (chair 1980-81), Am. Chem. Soc., Phi Beta Kappa, Tau Beta Pi. Republican. Presbyterian. Achievements include patents; design and construction of nuclear off-gas clean-up systems for use in some 50 nuclear plants worldwide. Home: 4584 Pleasant View Dr Coopersburg PA 18036 Office: CCI Cryogenics Inc 1176 N Irving St Allentown PA 18103

WINTERS, WENDELL DELOS, microbiology educator, researcher, consultant; b. Herrin, Ill., Sept. 24, 1940; s. Charles D. and Juanita (Burness) W.; m. Bonnie Tomlinson (div. Mar. 1990); children: Timothy, Amma Jean, Craig. BS, U. Ill., 1962; MS, U. Ill., Chgo., 1966, PHD, 1968. Teaching asst. U. Ill. Med. Sch., 1963-68; postdoctoral researcher Mat. Inst. Med. Rsch., London, 1968-70; resarcher Nat. Cancer Inst., NIH, Washington, 1970-71; asst. prof. UCLA Sch. Medicine, 1971-76; assoc. prof. microbiology U. Tex. Health Sci. Ctr., San Antonio, 1977—; cons. Royal Navy, USAF, Nat. Cancer Inst., 1976-82 Am. Med., Kinetic Concepts, Phytotherapeutics, 1983—; chmn. sci. bd., Aloe Rsch. Found., Dallas, Tex., 1989—, pres., 1992—; lectr. Burroughs Wellcome Fund, Cambridge, Eng., 1977, Beckman Rsch., Japan, Australia, 1978, Tomlinson-Burness Fund, Fed. Republic Germany, France, 1980-91, Japan, 1990; rsch. vis. prof. Damon Runyon Cancer Fund, London, 1969-71, NAS, Paris, 1974. Contbr. articles to sci. jours. Mem. West L.A. Housing Bd., 1972-76; mem. scouting com. Boy Scouts Am., San Antonio, 1978-82.. Fellow Am. Acad. Microbiology; mem. Bioelec. Repair and Growth Soc. (program chmn., bd. dirs. 1990-92), Tissue Culture Assn. (chmn. meetings 1972—, v.p. Calif. chpt. 1974-76), Am. Soc. Microbiology, Am. Assn. Cancer Rsch., Am. Assn. Immunologists. Achievements include research in immunotherapy as adjunct in cancer treatment, in vitro assembly of human viruses, bioeffects induced by environmental electromagnetic fields, plant substances as phytotherapeutic agents. Office: U Tex Health Sci Ctr Dept Microbiology 7703 Floyd Curl Dr San Antonio TX 78284-7700

WINTGEN, DIETER, physicist educator. Prof. dept. physics U. Freiburg, Germany. Recipient Gustav Hertz Preis, Deutsche Physikalische Gesellschaft, 1993. Office: U Freiburg-Facultat Physiks, Werthmann Platz, D-7800 Freiburg im Breisgau, Germany*

WINTLE, ROSEMARIE, bio-medical electronic engineer; b. Brigham City, Utah, Sept. 13, 1951; d. DeVere and Kathleen (Layton) W. Student, Weber State Coll., 1972-76, Brigham Young U., 1978-79, U. Utah, 1980-87, ITT Electronic Tech. Inst., 1986-88, Utah State U., 1991-92. Engr. Morton Internat., Brigham City, Utah; computer technician Salt Lake City; engr. Nuclear Med., Mesa, Ariz., 1976-77, U. Utah Hosp. Lab., Salt Lake City, 1980-87; electronic engr. Varian Assocs., Inc., Salt Lake City, 1987-88; electronic bio-med. experiment and rsch. engr. Clin. Rsch. Assocs., Provo, Utah, 1988-89. Contbr. articles to profl. jours. Designer, builder Honeyville (Utah) town playground equipment; designer, mgr. Honeyville town water system. Recipient grant Brigham City. Mem. IEEE (pres.), NSPE, Inst. for Sci. Info., Am. Statis. Assn., Sci. Am. Libr., Computer Club, Amnesty Internat., Libr. of Science, Newbridge Book Club. Mem. LDS Ch.

WIO, HORACIO SERGIO, physicist; b. Buenos Aires, Nov. 14, 1946; s. Welka and Sofia (Beitelmajer) W.; m. Maria Luz Martinez Perez, Feb. 16, 1973; children: Marcelo Gabriel, Mayra Cecilia, Nicolas Bernardo. Tech. Bachellor, Otto Krausse, Buenos Aires, 1965; MSc, Inst. Balseiro, 1972; PhD in Physics, U. Nac. Cuyo, 1977. Rsch. scholar Nat. Rsch. Coun., S.C. Bariloche, 1973-77; asst. researcher Nat. Atomic Energy Commn., S.C. Bariloche, 1977-79; vis. scientist Technischen U., Munich, 1979-80, 87-88; assoc. researcher Nat. Atomic Energy Commn., S.C. Bariloche, 1980-85; invited prof. U. Baleares, Palma de Mallorca, Spain, 1988-89; vis. prof. FAMAF, U.N. Cordoba, Argentina, 1989; assoc. prof. Instituto Balseiro, S.C. Bariloche, 1989—; ind. researcher Centro At. Bariloche, S.C. Bariloche, 1986—; vis. prof. U. Cantabria, Spain, 1990, 92, U. Mar del Plata, Argentina, 1991, 92, 93, U. Freiburg, Germany, 1992; cons. INVAP S.E., S.C. Bariloche, 1977; head stat. physics group Centro Atomico Bariloche, 1986—, head theoretical physics div., 1992—; mem. organizing com. Latin Am. Sch. Physics, La. Plata, 1987, MEDYFINOL Conf., Mar del Plata, 1991. Editor: Connections among..., 1988, Stochastic Processes and Nonequilibrium Statistical Physics Process MEDYFINOL Conf., 1991; author: Introduction to Path Integrals, 1990, An Introduction to Stochastic Processes and Nonequilibrium Statistcal Physics; contbr. over 70 articles to profl. jours. Nat. AEC scholar, 1969-72; Nat. Rsch. Coun. scholar, 1973-77, fellow, 1977—, grantee, 1984, 85, 90, 93; Inst. Coop. Iberoam. Spain joint rsch. grantee, 1990, 92, 93. Fellow Nat. Rsch Coun.; mem. ICTP Italy (assoc. 1992—), Am. Phys. Soc., Argentine Phys. Soc. (treas. 1983-86, mem. com. 1986-88), Club Los Pehuenes. Avocations: chess, swiming, go, judo. Office: Centro Atomico Bariloche, E Bustillo Km 9 1/2, Bariloche Argentina 8400

WIORKOWSKI, JOHN JAMES, mathematics educator; b. Chgo., Sept. 30, 1943; s. John Stanley and Harriet Elizabeth (Bedra) W.; B.S., U. Chgo., 1965, M.S., 1966, Ph.D., 1972; m. Gabrielle K. Hollis, June 4, 1966; 1 child, Fleurette Anne. Research asso. U. Chgo., 1972; asst. prof. Pa. State U., University Park, 1973-74; assoc. prof. U. Tex. at Dallas, Richardson, 1975, assoc. prof. and program head Math Scis. Program, 1979-81, prof., 1981—, asst. to v.p. acad. affairs, 1980-85, asst. v.p. acad. affairs, 1985-91, assoc. v.p. acad. affairs, 1991—; cons. to Fed. Energy Adminstrn., 1975, Tex. Instruments, 1977, Frito-Lay Inc., 1977-78, Republic Nat. Bank, 1979; mem. panel studying 55 mile per hour speed limit Nat. Acad. Sci. Served to capt. U.S. Army, 1968-71. Decorated Army Commendation medal. NSF grantee, 1975—. Am. Council Edn. fellow, 1981-82. Mem. Am. Statis. Assn. (chpt. pres. 1974, v.p. 1977, chpt. pres. 1978), AAAS, Inst. Math. Stats., Biometric Soc., Sigma Xi. Unitarian. Contbr. articles to profl. jours. Home: 428 Bedford Dr Richardson TX 75080-3401 Office: U Tex at Dallas Box 830688 Richardson TX 90161-0688

WIPF, PETER, chemist; b. Aarau, Switzerland, Sept. 5, 1959; came to U.S., 1988; s. Max and Lina (Furter) W.; m. Salla Kaarina Valtanen, July 16, 1987; children: Peter Charles, Heidi Maija-Lina. Diploma in Chemistry, U. Zurich, 1984, PhD, 1987. Rsch. assoc. U. Va., Charlottesville, 1988-90; asst. prof. U. Pitts., 1990—. Contbr. articles to profl. jours.; author book chpts. in field. Recipient Swiss NSF fellowship, 1988-90. Mem. Am. Chem. Soc., Swiss Chem. Soc., AAAS, Pitts. Cancer Inst. Achievements include rsch. in transmetalation reactions and their applications in organic synthesis; devel. of methodology for peptide analogs and total synthesis of biol. activ. products. Office: U Pitts Dept Chemistry Pittsburgh PA 15260

WIRTSCHAFTER, JONATHAN DINE, neuro-ophthalmology educator, scientist; b. Cleve., Apr. 4, 1935; s. Zolton Tilson and Reitza (Dine) W.; m. Carol Lavenstein, Sept. 13, 1959; children: Jacob Daniel, Benjamin Zolton, Joshua Joel, Sara Louise, David Dine, Brooke Ann. Student, UCLA, 1953; BA, Reed Coll., 1956; MD, Harvard U., 1960; MS in Physiology, Linfield Coll., 1963. Diplomate Am. Bd. Ophthalmology (assoc. examiner 1975—), Am. Bd. Neurology. Intern Phila. Gen. Hosp., 1960-61; resident in neurology Good Samaritan Hosp., Portland, Oreg., 1961-63; resident in ophthalmology Johns Hopkins Hosp., Balt., 1963-66; fellow in neurology Columbia-Presbyn. Hosp., N.Y.C., 1966-67; asst. prof. ophthalmology, neurology and neurosurgery U. Ky., Lexington, 1967-70, assoc. prof., 1970-74, prof., chmn. dept., 1974-77, dir. div. ophthalmology, 1967-74; prof. ophthalmology, neurology, neurosurgery U. Minn. Med. Sch., Mpls., 1977—, Frank E. Burch endowed chair in ophthalmology, 1990—; vis. prof. Hadassah-Hebrew U. Med. Ctr., Jerusalem, 1973-74; Earl G. Padfield, Jr., M.D. Meml. lectr. U. Kans., 1986; vis. prof., lectr. numerous other univs.; cons. VA Hosps., Lexington, 1967-77, Mpls., 1977—; spl. cons. Nat. Eye Inst., 1981. Co-author: Ophthalmic Anatomy: A Manual with Some Clinical Applications, 1970, rev. edit., 1981, A Decision-Oriented Manual of Retinoscopy, 1976, Computed Tomography: An Atlas for Ophthalmologists, 1982, Magnetic Resonance Imaging and Computed Tomography: Clinical Neuro-orbital Anatomy, 1992; contbr. numerous articles to profl. jours.; patentee in field. Bd. mem. Temple Israel, Lexington, 1970-73, McPhail Suzuki Music Assn., 1979-81, Mpls. Talmud Torah, 1979-85; founder, bd. mem. Jewish Community Assn. of Lexington, 1969-77; alumni interviewer Reed Coll., 1968—. Grantee Nat. Eye Inst., 1968-71, 78-81, 89—, Fight for Sight, 1974, Benign Essential Belpharospasm Found., 1988. Fellow ACS, N.Am. Neuro-Ophthalmology Soc., Am. Acad. Ophthalmology (course instr. 1981, film rev. com. 1975-78, faculty mem. basic and clin. sci. course 1975-77, mem. subcom. on continuing edn. TV 1972, subcom. on self administered examinations 1985—, Honor award 1980); mem. Am. Acad. Neurology, Am. Ophthal. Soc., AAAS, Assn. for Rsch. in Vision and Ophthalmology, AMA (Hon. Mention award-sci. exhibit 1970), Soc. for Neurosci., Internat. Soc. Neuro-Ophthalmology, Am. Israeli Ophthal. Soc. (bd. mem. 1984—), Am. Assn. for Pediatric Ophthalmology and Strabismus (assoc.), Boylston Med. Soc. Harvard Med. Sch., Alpha Omega Alpha. Democrat. Office: U Minn Hosp 516 Delaware St SE Minneapolis MN 55455-0501

WISCH, DAVID JOHN, structural engineer; b. Jefferson City, Mo., Dec. 6, 1953; s. Theodore A. and Josephine (Lauf) W.; m. Leslie Babin, Oct. 24, 1981; 1 child, Christine. BSCE, U. Mo., Rolla, 1975, MSCE, 1977. Registered profl. engr., La., Calif. Civil engr. Texaco-Ctrl. Offshore Engring., New Orleans, 1977-81, advanced civil engr., 1981-86, sr. project engr., 1986-92; specialist Texaco-Ctrl. Offshore Engring., Bellaire, Tex., 1992—; chmn. fixed systems subcom. Am. Petroleum Inst., Dallas, 1991—, mem. adv. bd. offshore standardization com., 1991—, chmn., 1993—, head U.S. Delegation Internat. Orgn. Standards Tech. Com. 67/Subcom 7, 1993—,mem. Tech. Com. 67/Subcom. 7/AG1, 1993—; mem. structure subcom. Oil Cos. Internat. Exploration and Prodn. Forum, London, 1992—. Author numerous papers/presentations, 1984—. Mem. ASCE, Sigma Xi, Phi Kappa Phi, Tau Beta Pi, Chi Epsilon (chpt. pres. 1975-76). Office: Texaco Ctrl Offshore Engring 4800 Fournace Pl Bellaire TX 77401

WISDOM, GUYRENA KNIGHT, psychologist, educator; b. St. Louis, July 27, 1923; d. Gladys Margaret (Hankins) McCullin. AB, Stowe Tchrs. Coll., 1945; AM, U. Ill., 1951; postgrad. St. Louis U., 1952-53, 58, 62; Washington U., St. Louis, 1959-61; U. Chgo., 1966-67; Drury Coll., 1968; U. Mo., 1971-72; Fontbonne Coll., 1973; Harris-Stowe State Coll., 1974, 81-82. Tchr. elem. sch. St. Louis Pub. Sch. System, 1945-63, psychol. examiner, 1963-68, sch. psychologist, 1968-74, cons. spl. edn., 1974-77, supr. spl. edn. dept., 1977-79, coord. staff devel. div., 1979-81; pvt. tutor, 1971-72; sch. psychologist, 1984-85; pvt. practice psychologist, St. Louis, 1985-88; pvt. practice, 1989—; instr. Harris Tchrs. Coll., St. Louis, 1973-74, Harris-Stowe Coll., 1979. Contbr. articles to profl. jours. Mem. Nat. Assn. Sch. Psychologists, Learning Disabilities Assn., Coun. for Exceptional Children, Assn. Supervision and Curriculum Devel., Pi Lambda Theta, Kappa Delta Pi. Roman Catholic. Home: 5046 Wabada Ave Saint Louis MO 63113-1118

WISE, JAMES ALBERT, psychologist; b. Aberdeen, S.D., Oct. 22, 1944; s. James A. and June Marie (Soike) W.; m. Barbara Antoinette Krysa, Sept. 27, 1974. BS in Psychology, U. Wash., 1965, PhD in Psychology, 1970. Vis. lectr. U. Leiden, Netherlands, 1970; asst. prof. psychology Ohio State U., Columbus, 1970-74; vis. prof. psychology U. Mannheim, Germany, 1974-75; assoc. prof. arch. U. Wash., Seattle, 1975-87; prof. facilities mgmt. Grand Valley State U., Allendale, Mich., 1987-90; staff scientist Battelle Pacific N.W. Lab., Richland, Wash., 1990—; prin. James Wise & Assocs., Seattle, 1978-90; U.S. del. to design methods conf. Polish Acad. Scis., Radjiowicz, 1979; spl. del. Internat. Assn. Space Explorers World Congress, Riyadh, Saudi Arabia, 1989. Co-author: Human Decision Making, 1980, Bank Interiors and Bank Robberies, 1985; contbr. articles to profl. jours.; editorial bd. Jour. Systems Rsch., 1984-92. Liaison officer Greenlake Community Coun., Seattle, 1976-78. With USN, 1962-65. Recipient Design Rsch. award Indsl. Design Mag., 1985, Archtl. Rsch. award Progressive Arch. Mag., 1985. Mem. Human Factors Soc. (chair tech. group on environ. design 1979-84, newsletter editor 1984-90), Systems, Man and Cybernetics Soc. of IEEE (chair internat. conf. 1982), Environ. Design Rsch. Assn., AAAS. Achievements include development of first math model of subjective probability judgments that was non-frequentistic; isovist model a quantitative tool to analyze spatial habitability. Office: Battelle Pacific NW Lab PO Box 999 Richland WA 99352

WISE, JOSEPH FRANCIS, electrical engineer, consultant; b. Randolph, Ohio, May 22, 1931; s. Charles C. and Elva Minnie (Kline) W.; m. Mary Lou McAllister, June 22, 1957; children: Raymond, Cynthia, Robert. BS in Edn., Kent State U., 1953; BSEE, U. Dayton, 1958. High sch. tchr. Brunswick, Ohio, 1953-54; elec. engr. Wright Rsch. & Devel., Wright-Patterson AFB, Ohio, 1958-70, supervisory engr. aerospace propulsion lab., 1970-89; cons. Fairborn, Ohio, 1989—. With U.S. Army, 1954-56. Mem. IEEE (photovoltaic specialist conf. com. 1988). Democrat. Roman Catholic. Home and Office: 845 Warrington Pl Dayton OH 45419

WISE, LESLIE, surgeon, educator; b. Budapest, Hungary, May 3, 1932; came to U.S., 1967; s. Emil and Jolan Weisz; m. Jan Buchanan (div.); children: Nicholas, Julian, Tania. BSc in Medicine, U. Sydney, Australia, 1955; MB, BS, U. Sydney, 1958. Diplomate Am. Bd. Surgery. Intern Royal Newcastle Hosp., New South Wales, Australia, 1958-59, jr. house surgeon, 1959-60, sr. house surgeon, 1960-61; proctor in anatomy Royal Coll. Surgeons, London, 1961-62; surg. resident Hammersmith Hosp., London, 1962-63, St. James Hosp., London, 1963-64; surg. registrar St. Helier Hosp., London, 1964-65; sr. surg. registrar St. Helier Hosp. and St. Mark's Hosp., London, 1965-67; fellow in surgery Johns Hopkins U. Sch. Medicine, Balt., 1967-68; asst. prof. surgery, dir. surg. rsch. labs. Washington U. Sch. Medicine, St. Louis, 1968-69, 1969-71, assoc. prof., 1971-75; attending surgeon Barnes and Allied Hosps., St. Louis, 1969-75, Homer G. Phillips Hosp., John Cochran VA Hosp., St. Louis Children's Hosp., St. Louis County Hosp., 1971-75; prof. surgery SUNY, Stony Brook, 1975-89, Albert Einstein Coll. Medicine, Bronx, N.Y., 1989—; chmn. dept. surgery L.I. Jewish Hosp., New Hyde Park, N.Y., 1975—; surgeon-in-chief Queens Hosp. Ctr., Jamaica, N.Y., 1975-92; program dir. dept. surgery LaGuardia Hosp., Forest Hills, N.Y., 1992—. Editor 3 books; contbr. over 380 articles and abstracts to profl. jours. Mem. senate exec. com. Albert Einstein Coll. Medicine, 1992—. Fellow ACS, Royal Coll. Surgeons (Eng., Hunterian Prof. 1970-71), Royal Soc. Medicine, Collegium Internationale Chirurgiae Digestivae, Internat. Soc. Surgery; mem. AMA, AAUP, Am. Surg. Assn., Am. Soc. Clin. Oncology, Assn. Acad. Surgery, Am. Fedn. Clin. Rsch., Am. Gastroent. Assn., Am. Surg. Assn., Ctrl. Surg. Assn., Soc. Surg. Oncology, Assn. Surgery Alimentary Tract, Am. Gastroent. Assn., Soc. Am. Gastrointestinal Endoscopic Surgeons, Soc. Program Dirs. Surgery, Soc. Laparoendoscopic Surgeons, Soc. Univ. Surgeons, N.Y. Surg. Soc. (pres. elect 1991-92, pres. 1992-93)., N.Y. State Soc. Surgeons, N.Y. State Med. Soc., N.Y. Cancer Soc., N.Y. Met. Breast Cancer Group, Inc., Nassau County Med. Soc. Home: 721 Fifth Ave Apt 38E New York NY 10022 Office: LI Jewish Med Ctr Dept Surgery Lakeville Rd New Hyde Park NY 11042

WISE, ROGER M., chemist; b. Tampa, Fla., Feb. 2, 1946; s. Lawrence W. and Helen M. (Harrington) W.; m. Maureen J. Patrick, Mar. 21, 1970 (div. 1977); m. Lisa R. Fortson, Oct. 29, 1988. BS in Chemistry, Stevens Inst. Tech., 1968; MA in Chemistry, Columbia U., 1969. Chemist sanitary sewers City of Tampa, Fla., 1976—. Mem. Nat. Assn. Environ. Profls., Nat. Assn. Advancement of Sci. Home: 5350 Lake LeClare Rd Lutz FL 33549 Office: City of Tampa Sanitary Sewers 2700 Maritime Blvd Tampa FL 33605

WISH, JAY BARRY, nephrologist, specialist; b. Hartford, Mar. 30, 1950; s. Martin and Evelyn Lillian (Lassman) W.; m. Linda Kristina Hansen, June 29, 1971; (div. 1980); children: Allen Jeremy, Robin Lindsey; m. Diane Elizabeth Perkins, June 5, 1983; children: Jeffrey Bryan, David Phillip. BA, Wesleyan U., 1970; MD, Tufts U., 1974. Diplomate Am. Bd. Internal Medicine, Am. Bd. Nephrology. Resident in medicine New England Med. Ctr., Boston, 1974-79; instr. in medicine Tufts U., Boston, 1978-79; lectr. in health sci. Northeastern U., Boston, 1978-79; asst. prof. of medicine Case Western Res. U., Cleve., 1979-85, assoc. prof. of medicine, 1985—; dir. hemodialysis U. Hosps. of Cleve., 1980—, dir. continuing edn., 1987—; chmn. Med. Adv. Bd. Kidney Found. of Ohio, Cleve., 1985-88. Author: Renal Disease and Hypertension, 1982, Disorders of Potassium, 1984, Metabolic Diseases, 1986, Rheumatic Diseases of the Kidney, 1993, Acid-Base and Electrolyte Disorders in the Critically Ill Patient, 1993; contbr. articles to profl. jours. Chmn. med. rev. bd. End-Stage Renal Disease Network #22, Pitts., 1982-87, End-Stage Renal Disease Network #9, Indpls., 1992—; mem. exec. com. Forum of End-Stage Renal Disease Networks, 1992—; bd. dirs. Renal Phys. Assn., 1993—; mem. Nat. Kidney Found. Fellow Am. Coll. of Physicians; mem. Cleve. Restoration Soc., Cleve. Opera Assocs. (sec.-treas.), Am. Soc. of Nephrology, Internat. Soc. of Nephrology, Alpha Omega Alpha. Democrat. Jewish. Avocation: performing arts. Office: Univ Hosps of Cleve 2074 Abington Rd Cleveland OH 44106-2602

WISKICH, JOSEPH TONY, botany educator, researcher; b. Tully, Queensland, Australia, July 21, 1935; s. Joseph Tony and Vera (Piglic) W.; m. Diane Lesley, Apr. 20, 1968; children: Peter Joseph, Anthony David, Robert Leslie. BSc with honors, U. Sydney (Australia), 1956, PhD, 1960; PhD, U. Adelaide (Australia), 1964. Spl. investigator status. Rsch. assoc. U. Pa., Phila., 1961, UCLA, 1962; rsch. assoc. U. Adelaide, 1963-64, lectr. botany, 1964-71, sr. lectr., 1972-78, assoc. prof., 1979—, mem. univ. coun., 1985—. Mem. editorial bd. Australian Jour. Plant Physiology, 1975-88, Plant Physiology, 1983-92; contbr. over 90 articles to sci. jours. Chmn. Anzaas Inc., Adelaide, 1985-91. Mem. Biochem. Soc. (London), Am. Soc. Plant Physiologists, Australian Biochem. Soc., Scandinavian Soc. Plant Physiologists, Australian Soc. Plant Physiologists, N.Y. Acad. Scis. Achievements include research in bioenergetics and plant metabolic biochemistry. Office: U Adelaide Botany Dept, North TCE, Adelaide 5000, Australia

WISLER, DAVID CHARLES, aerospace engineer, educator; b. Pottstown, Pa., Apr. 21, 1941; s. Lloyd William and Ruth Georgiana (Enos) W.; m. Judith Ann Caleen, Aug. 22, 1964 (dec. Mar. 1979); children: Scott David, Cheryl Lynn; m. Beth Ellen Howard, Jan. 5, 1980; 1 child, Daniel James. BS in Aero Engring., Pa. State U., 1963; MS in Aero. Engring., Cornell U., 1965; PhD in Aero. Engring., U. Colo., 1970. Rsch. engr. GE R & D Ctr., Schenectady, 1965-67; mgr. aerodyn. rsch. lab. GE Aircraft Engines, Evendale, Ohio, 1985—; adj. prof. Iowa State U., Ames, 1987—. Contbr. articles to profl. jours.; patentee in sloped trenches in compressors. Fellow AIAA (assoc.); mem. ASME (chmn. turbomachinery com. 1993—, Melville medal for best tech. paper 1989, Gas Turbine award 1990, 92). Avocation: photography. Home: 1895 Doral Dr Fairfield OH 45011 2625 Office: GE Aircraft Engines 1 Neumann Way MD A323 Cincinnati OH 45215

WISNIEWSKI, HENRYK MIROSLAW, pathology and neuropathology educator, research facility administrator, research scientist; b. Luszkowko, Poland, Feb. 27, 1931; came to U.S., 1966; s. Alexander and Ewa (Korthals) W.; m. Krystyna Wylon, Feb. 14, 1954; children: Alexander (dec.), Thomas. MD, Med. Sch., Gdansk, Poland, 1955; PhD in Exptl. Neuropathology, Med. Sch., Warsaw, Poland, 1960; DSc (hon.), Med. Sch., Gdansk, Poland, 1991, Coll. of Staten Island, 1992. From asst. to assoc. prof., head. lab. exptl. neuropathology, assoc. dir. Inst. Neuropathology Polish Acad. Sci., Warsaw, 1958-66; rsch. assoc., from asst. prof. to prof. Albert Einstein Coll. Medicine, N.Y.C., 1966-75; dir. MRC Demyelinating Diseases Unit, Newcastle upon Tyne, Eng., 1974-76; prof. neuropathology SUNY Health Sci. Ctr., Bklyn., 1976—; dir. N.Y. State Inst. Basic Rsch. in Devel. Disabilities, S.I., 1976—; vis. neuropathologist U. Toronto, Ont., Can., 1961-62; vis. scientist Lab. of Neuropathology Nat. Inst. Neurol. and Communicative Diseases and Stroke, NIH, 1962-63; docent Med. Sch., Warsaw, 1965; past mem. Neuology B study sect. NIH; past mem. mental retardation rsch. com. Nat. Inst. Child Health and Human Devel. Contbr. over 520 articles to profl. jours. and symposia proceedings; mem. editorial bd. Acta Neuropthologica, Neurotoxicology, Jour. Neuropathology and Exptl. Neurology, Devel. Neurosci., Internat. Jour. Geriatric Psychiatry, Alzheimer Disease and Associated Disorders Internat. Jour., Brain Dysfunction, Dementia; mem. editorial adv. bd. Neurobiology of Aging. Recipient N.Y.C. Chpt. award Assn. for Help of Retarded Children, 1984, Welfare League award letchworth Village chpt. Assn. for Help of Retarded Children, 1985, Staten Island Chpt. award Benevolent Soc. for Retarded Children, 1986; named Career Scientist, Health Rsch. Coun. of City of N.Y., Honoris Causa Doctor of Sci., Med. Sch., Gdansk, Poland, 1992, Honoris Causa, Doctor of Sci., Coll. of Staten Island, 1992. Fellow AAAS; mem. Am. Assn. Neuropathologists (pres. 1984, Weil award 1969, Moore award 1972), British Soc. Neuropathology, Can. Assn. Neuropathologists, Am. Assn. Retarded Citizens, Assn. Rsch. in Nervous and Mental Disease, Am. Assn. Mental Deficiency, Internat. Soc. Devel. Neuroscis., Soc. Exptl. Neuropathology, Sigma Xi. Democrat. Roman Catholic. Achievements include research of developmental disabilities, aging, Alzheimer disease; neuronal fibrous protein pathology; demyelinating diseases. Office: NY State Inst Basic Rsch Devel Disabilities 1050 Forest Hill Rd Staten Island NY 10314-6399*

WIŚNIEWSKI, ROLAND, physics educator; b. Warsaw, Poland, Oct. 1, 1929; s. Stephan and Theodora Isabel (Kułakowska) W.; m. Ann Mary Zwolińska, July 5, 1958; 1 child, Darius Thomas. MSc in Mech. Engring., Warsaw U. Tech., 1956, PhD in Electronics, 1964. Asst. lectr. Faculty Electricity, Warsaw U. Tech., 1952-54, lectr., 1954-64, sr. lectr. Inst. Physics, 1964-74, assoc. prof., 1974-84, prof., 1984-90, full prof., 1990—, dir. High Pressure Lab., 1955—, dir. tech. of solid physics dept., 1972—, dep. dir. Inst., 1986-91; cons. Tool Ctr., VIS Co., Warsaw, 1987-92. Co-author: High Pressure: Generation, Measurements, Applications, 1980 (award Ministry Edn. 1981), High Pressure Measurement Techniques, 1983; also over 175 articles; patentee hybridized pressure standard 1GPa, numerous others. Mem. Orgn. for Animals, Warsaw, 1985—. Decorated Order of Polonia Restituta; recipient award for high pressure hydrogen apparatus Ministry Sci. and Higher Edn., 1967, award for sci. achievements, 1972, 73, 74, 83, 86; award for sci. achievements Polish Acad. Sci., 1975, 81; award for investigation of coal under dynamic load, Warsaw U. Tech., 1983, award for sci. achievements, 1987, 90, 93; Commn. Nat. Edn. medal. Mem. European High Pressure Rsch. Group, Polish Mech. Engring. Assn. (sci. adviser), Internat. Assn. for Advancement High Pressure Sci. and Tech. Avocations: sailing, numismatics. Home: Grzybowska 5 - 211, 00-132 Warsaw Poland Office: Warsaw U Tech Inst Physics, Koszykowa 75, 00-662 Warsaw Poland

WISSMAN, MATTHIAS, minister of research and technology; b. Ludwigsburg, Germany, Apr. 15, 1949; s. Paul and Margarete Kalcker. Educator Univs. of Tübingen and Bonn; lawyer Fed. Exec. of CDU, 1975—; mem Bundestag, 1976—; pres. European Union Young Christian Dems., 1976-82; minister of rsch. and tech., 1993—. Author: Zukunftschancen der Jugend, 1979, Einsteigen statt Austeigen, 1983, Marktwirtschahft 2000, 1983. Office: Ministry of Rsrch & Technology, Heinemannstr 2, 5300 Bonn 2, Germany*

WISTRAND, LARS GORAN, chemist; b. Göteborg, Sweden, Aug. 19, 1949; s. Bengt E. and Gunvor (Andersson) W.; m. Ulla Birgitta Jeppsson, Feb. 1, 1980; children: Sara Paulina, Anna Maria, Nilla Cecilia. BS, U. Lund, Sweden, 1972, PhD, 1978; MA, U. Calif., Santa Barbara, 1974. Rsch. assoc. Lund U., 1972-79, 81-84, asst. prof., 1984-90; postdoctoral fellow U. Utah, Salt Lake City, 1979-80; mgr. R&D Nycomed Innovation, Malmö, Sweden, 1990—; cons. AB Astra, Sodertalje, Sweden, 1989-91, Bra Bocker AB, Hoganas, Sweden, 1989—. Referee Acta Chem. Scandinavia, 1981—, Jour. Organic Chem., 1986—; contbr. articles to Jour. Organic Chemistry, Acta Chem. Scandinavia, Organometallics. Grantee Swedish Nat. Sci. Rsch. Coun., 1981-89, Swedish Bd. for Tech. Devel., 1987-91. Mem. Am. Chem. Soc., Swiss Chem. Soc., Swedish Pharm. Soc. Achievements include invention of novel type of contrast agent for MRI. Office: Nycomed Innovation, Ideon Malmo, S-20512 Malmö Sweden

WIT, DAVID EDMUND, software company executive; b. N.Y.C., Feb. 25, 1962; s. Harold Maurice W. and Joan Leta (Rosenthal) Sovern; m. Kathleen Mary Bentley, Sept. 9, 1989. BA summa cum laude, Hamilton Coll., 1985. Rsch. assoc. E.M. Warburg Pincus and Co., N.Y.C., 1985-86; co-chief exec. officer Logicat Inc., N.Y.C., 1986—; bd. dirs. Calif. Energy Co., Omaha, 1987—. Mem. N.Y. Software Industry Assn. (steering com. 1991—), Univ. Club, Phi Beta Kappa. Avocations: running, racquetball, hockey. Home: 736 W End Ave New York NY 10025 Office: Logicat Inc 201 E 16th St New York NY 10003

WITEK, JOHN JAMES, neurologist; b. Chgo., May 19, 1948; s. John James and Betsy Ross (Sisson) W.; m. Margaret Jane Byers, Mar. 27, 1971 (div. 1977); m. Mary Margaret Melbo, June 19, 1981. BA, Northwestern U., 1969; MD, U. Ill., Chgo., 1976. Diplomate Nat. Bd. Med. Examiners, Am. Bd. Psychiatry and Neurology. Intern St. Paul (Minn.) Ramsey Med. Ctr., 1976-77; fellow in behavioral neurology U. Minn./Hennepin County Med. Ctr., Mpls., 1980-82; resident in neurol. U. Minn. Hosp. & Clinics, Mpls., 1977-80; assoc. prof., staff neurologist Mpls. VA Med. Ctr./U. Minn., Mpls., 1982-83; clin. neurologist Mpls. Clinic of Neurology, Golden Valley, Minn., 1983-85, Aspen Med. Group, Mpls./St. Paul, 1985—. Co-author: When You're Sick and Don't Know Why, 1991. Vol. Big Bros./Big Sisters, Mpls., 1989-90. With U.S. Army, 1970-72. Mem. Am. Acad. Neurology, Alpha Omega Alpha. Office: Aspen Med Group 1020 Bandana Blvd W Saint Paul MN 55108

WITENBERG, EARL GEORGE, psychiatrist, educator; b. Middletown, Conn., Aug. 30, 1917; s. Nathan and Goldie (Ruderman) W.; B.A., Wesleyan U., 1937, M.A., 1939; M.D., U. Rochester, 1943; m. Mary Jane Hoffman, Apr. 4, 1948 (dec.); children—William, Susan; m. Carol J. Eagle, June 1, 1986. Intern, Strong Meml. Hosp., Rochester, N.Y., 1943-44; resident in psychiatry Bellevue Hosp., N.Y.C., 1946-48; clin. dir. Postgrad. Center for Psychotherapy, N.Y.C., 1948-49, med. coordinator, 1949-51; chmn. exec. com., fellow W.A. White Inst., 1960-63, dir. inst., 1963-92, E.G. Witenberg chair in psychoanalysis, 1992—; pvt. practice medicine specializing in psychiatry and psychoanalysis, N.Y.C., 1948—; assoc. clin. prof. Albert Einstein Coll. Medicine, 1974-79. Served to lt. M.C., USNR, 1944-46. Fellow Am. Psychiat. Assn.; mem. Am. Acad. Psychoanalysis (pres. 1976-77, William Silverberg award 1977), Am. Coll. Psychoanalysts, AAAS, Sigma Xi, Phi Beta Kappa. Author: How Not to Succeed in Psychotherapy..., 1972; contbr. articles to med. jours., chpts. to books. Home: 215 E 68th St New York NY 10021-5718 Office: 20 W 74th St New York NY 10023

WITHERELL, MICHAEL S., physics educator; b. Toledo, Sept. 22, 1949; s. Thomas W. and Marie (Savage) W.; m. Elizabeth Hall. BS, U. Mich., 1968; MS, U. Wis., 1970, PhD, 1973. Instr. Princeton (N.J.) U., 1973-75, asst. prof., 1975-81; asst. prof. U. Calif., Santa Barbara, 1981-83, assoc. prof., 1983-86, prof., 1986—; mem. high energy physics adv. panel U.S. Dept. Energy, Washington, 1990—; chmn. physics adv. com. Fermi Nat. Accelerator Lab., Batavia, Ill., 1987-89. Guggenheim fellow John S. Guggenheim Found., 1988; recipient W. K. H. Panofsky prize Am. Phys. Soc., 1990. Office: U Calif Dept Physics Santa Barbara CA 93106

WITHEROW, WILLIAM KENNETH, physicist; b. Patuxent, Md., Jan. 12, 1954; s. William W. and Gail E. (Woodle) W.; m. Kathryn C. Price, June 25, 1977; children: Wendy L., William J. BS, U. Tenn., 1977; MS, U. Ala., Huntsville, 1981. Physicist NASA/Marshall Space Flight Ctr., Huntsville, 1973—; mem. SPIE Working Group on Holography. Named Inventor of Yr., NASA, 1990. Mem. Internat. Soc. Optical Engring., Optical Soc. Am., Huntsville Electro-Optical Soc. Achievements include patents on Method of and Apparatus for Double-Exposure Holographic Interferometer, Use of Prism to Compare Two Images for Two Wavelength Holographic System, others. Office: NASA/MSFC ES74 SSL Bldg 4481 Huntsville AL 35812

WITHROW, LUCILLE MONNOT, nursing home administrator; b. Alliance, Ohio, July 28, 1923; d. Charles Edward Monnot and Freda Aldine (Guy) Monnot Cameron; m. Alvin Robert Withrow, June 6, 1945 (dec. 1984); children: Cindi Withrow Johnson, Nancy Withrow Townley, Sharon Withrow Hodgkins, Wendel Alvin. AA in Health Adminstrn., Eastfield Coll., 1976. Lic. nursing home adminstr., Tex.; cert. nursing home ombudsman. Held various clerical positions Dallas, 1950-72; office mgr., asst. adminstr. Christian Care Ctr. Nursing Home, Mesquite, Tex., 1972-76; head adminstr. Christian Care Ctr. Nursing Home and Retirement Complex, Mesquite, 1976-91; nursing home ombudsman Tex. Dept. Aging and Tex. Dept. Health, Dallas, 1991—; mem. com. on geriatric curriculum devel. Eastfield Coll., Mesquite, 1979, 87; mem. ombudsman adv. com. Sr. Citizens Greater Dallas; nursing home cons. Vol. Dallas Arboretum & Botanical Soc., Dallas Summer Musicals Guild; mem. Ombudsman adv. com. Sr. Citizens of Greater Dallas, Health Svcs. Speakers Bur. Recipient Volunteerism award, Tex. Atty. Gen., 1987. Mem. Tex. Assn. Homes for Aging, Am. Assn. Homes for Aging, Health Svcs. Speakers Bur., White Rock Kiwanis. Republican. Mem. Ch. of Christ. Avocations: reading, travel, theater. Home and Office: 11344 Lippitt Ave Dallas TX 75218-1922

WITKIN, MILDRED HOPE FISHER, psychotherapist, educator; b. N.Y.C.; d. Samuel and Sadie (Goldschmidt) Fisher; children: Georgia Hope, Roy Thomas, Laurie Phillips, Kimberly, Nicole, Scott, Joshua, Jennifer; m. Jorge Radovic, Aug. 26, 1983. AB, Hunter Coll., MA, Columbia U., 1968; PhD, NYU, 1973. Diplomate Am. Bd. Sexology, Am. Bd. Sexuality; cert. supr. Head counselor Camp White Lake, Camp Emanuel, Long Beach, N.J.; tchr. econs., polit. sci. Hunter Coll. High Sch.; dir., group leader follow-up program Jewish Vacation Assn., N.Y.C.; investigator N.Y.C. Housing Authority; psychol. counselor Montclair State Coll., Upper Montclair, N.J., 1967-68; mem., lectr. Creative Problem-Solving Inst., U. Buffalo, 1968; psychol. counselor Fairleigh Dickinson U., Teaneck, N.J., 1968, dir. Counseling Center, 1969-74; pvt. practice psychotherapy, N.Y.C., also Westport, Conn.; sr. faculty supr., family therapist and psychotherapist Payne Whitney Psychiat. Clinic, N.Y. Hosp., 1973—; clin. asst. prof. dept. psychiatry Cornell U. Med. Coll., 1974—; assoc. dir. sex therapy and edn. program Cornell-N.Y. Hosp. Med. Ctr., 1974—; sr. cons. Kaplan Inst. for Evaluation and Treatment of Sexual Disorders, 1981—; supr. master's and doctoral candidates, NYU, 1975-82; pvt. practice psychotherapy and sex therapy, N.Y.C., also Westport, Conn.; cons. counselor edn. tng. programs N.Y.C. Bd. Edn., 1971-75; cons. Health Info. Systems, 1972-79; vis. prof. numerous colls. and univs.; chmn. sci. com. 1st Internat. Symposium on Female Sexuality, Buenos Aires, 1984. Exhibited in group shows at Scarsdale (N.Y.) Art Show, 1959, Red Shutter Art Studio, Long Beach, 1968. Edn. legislation chmn. PTA, Yonkers, 1955; publicity chmn. United Jewish Appeal, Scarsdale, 1959-65; Scarsdale chmn. mothers com. Boy Scouts Am., 1961-64; mem. Morrow Assn. on Correction N.J., 1969-91; bd. dirs. Girl Scouts of Am. Recipient Bronze medal for svcs. Hunter Coll.; United Jewish Appeal plaque, 1962; Founders Day award N.Y. U., 1973, citation N.Y. Hosp./Cornell U. Med. Ctr., 1990. Fellow Internat. Coun. Sex Edn. and Parenthood of Am. U., Am. Acad. Clin. Sexologists; mem. AAUW, APA, ACA, Assn. Counseling Supervision, Am. Coll. Personnel Assn., Internat. Assn. Marriage and Family Counselors, Am. Coll. Sexuality (cert.), Women's Med. Assn. N.Y.C., N.Y. Acad. Sci., Am. Coll. Pers. Assn. (nat. mem. commn. II 1973-76), Nat. Assn. Women Deans and Counselors, Am. Assn. Sex Educators, Counselors and Therapists (regional bd., nat. accreditation bd., cert. internat. supr.), Soc. for Sci. Study Sex Therapy and Rsch., Eastern Assn. Sex Therapists, Am. Assn. Marriage and Family Counselors, N.J. Assn. Marriage and Family Counselors, Ackerman Family Inst., Am. Personnel and Guidance Assn., Am., N.Y., N.J. psychol. assns., Creative Edn. Found., Am. Assn. Higher Edn., Assn. Counselor Supervision and Edn., Profl. Women's Caucus, LWV, Am. Assn. counseling and Devel., Am. Women's Med. Assn., Nat. Coun. on Women in Medicine, Argentine Soc. Human Sexuality (hon.), Am. Assn. Sexology (diplomate), Pi Lambda Theta, Kappa Delta Pi, Alpha Chi Alpha. Author: 45-And Single Again, 1985; contbr. articles to profl. jours. and textbooks; lectr. internat. workshops, radio and TV. Home: 9 Sturges Commons Westport CT 06880-2832 Office: 35 Park Ave New York NY 10016-3838

WITKOP, BERNHARD, chemist; b. Freiburg, Baden, Fed. Republic Germany, May 9, 1917; came to U.S., 1947, naturalized, 1953; s. Philipp W. and Hedwig M. (Hirschhorn) W.; m. Marlene Prinz, Aug. 8, 1945; children: Cornelia Johanna, Phyllis, Thomas. Diploma, U. Munich, 1938, PhD, 1940, Golden Dr. Diploma, 1990; ScD, Privat-Dozent, 1947. Matthew T. Mellon research fellow Harvard U., 1947-48, mem. faculty, 1948-50; spl. USPHS fellow Nat. Heart Inst., NIH, 1950-52; vis. scientist Nat. Inst. Arthritis and Metabolic Diseases, 1953, chemist, 1954-55, chief sect. metabolites, 1956—, chief lab. chemistry, 1957-87, scholar, 1987-92, hon. scholar emeritus, 1993; vis. prof. U. Kyoto, Japan, 1961, U. Freiburg, Fed. Repubic Germany, 1962; adj. prof. U. Md. Med. Sch., Balt.; Nobel symposium lectr. Stockholm-Karlskoga, 1981; mem. bd. Internat. Sci. Exchange, 1974; mem. exec. com. NRC, 1975; mem. Com. Internat. Exchange, 1977, Paul Ehrlich Award Com., Frankfurt, 1980. Editor: Fedn. European Biochem. Soc. Letters, 1979-90. Recipient Superior Service award USPHS, 1967; Paul Karrer gold medal U. Zurich, 1971; Kun-ni-to (medal of sci. and culture 2d class) Emperor of Japan, 1975; Alexander von Humboldt award for sr. U.S. scientists, 1978. Mem. NAS, Am. Chem. Soc. (Hillebrand award 1958), Am. Acad. Arts and Sci., Acad. Leopoldina (fgn.), Pharm. Soc. Japan (hon.), Chem. Soc. Japan (hon.), Japanese Biochem. Soc. (hon.). Office: NIH Bethesda MD 20892

WITMER, GARY WILLIAM, research biologist, educator; b. Hamtramck, Mich., Sept. 28, 1951; s. Ralph Robert and Caroline Rose (English) W.; m. Vicki Lynn Ellis, July 6, 1985; children: Brian Timothy, Sarah Kathryn. MS, U. Mich., 1974, Purdue U., 1976; PhD, Oreg. State U., 1981. Area biologist Wash. Dept. Wildlife, Vancouver, 1979-80; rsch. assoc. Oreg. State U., Corvallis, 1980-84; wildlife ecologist Argonne (Ill.) Nat. Lab., 1984-88; asst. prof. biology Pa. State U., DuBois, 1988-91; rsch. wildlife biologist USDA Denver Wildlife Rsch. Ctr., Pullman, Wash., 1991—; mem. deer-orchard task force Wash. Dept. Wildlife, Chelan, 1992—; mem. rodent-agr. task force Agrl. Rsch. Sta., Reno, 1992—. Contbr. chpts. to books, articles to Jour. Wildlife Mgmt., Can. Jour. Zoology, Jour. Pa. Acad. Sci. Vice chair Friends of Tryon Creek State Park, Portland, Oreg., 1987-88; advisor Pa. Fish Commn. Adopt-a-Stream, DuBois, 1990. U. Mich. regents alumni scholar, 1969. Mem. The Wildlife Soc. (Pa. pres. 1991), N.W. Sci. Soc., Soc. for N.W. Vertebrate Biology, Sigma Xi. Achievements include cumulative impact assessment methodologies; method for analysis of mammalian predator food habits; deciphering of bald eagle-salmon relationships. Home: NW 215 Joe St Pullman WA 99163 Office: USDA/APHIS Denver Wildlife Rsch Ctr Natural Resource Sci Wash State U Pullman WA 99164-6410

WITT, AUGUST FERDINAND, aerospace scientist, educator; b. Innsbruck, Austria, 1931; married; three children. PhD in Phys. Chemistry, U. Innsbruck, 1959. Mem. staff MIT, Cambridge, Mass., 1960—, prof. materials sci., 1972-90, TDK prof. materials sci., 1990-93, Ford prof. engring., 1993—; chmn. electonic materials working group NASA, 1982-89; past chmn. Nat. Materials Adv. Bd. Com. on Preparation of Ultra-High Purity, Low Borob Silicon, Gordon Rsch. Conf. on Semiconductor Crystal Growth. Mem. editorial bd. Jour. Crystal Growth, Crystals: Growth, Properties and Applications, Crystal Rsch. and Tech.; co-editor (series) Am. Inst. Physics Series in Growth abd Characterization of Semi-conducting Materials; contbr. numerous articles to profl. and sci. jours. Recipient Outstanding Sci. Achievement award NASA, 1974, Exner Medal for Oustanding Contbns. to Sci. and Tech., Austria, 1976, Space Processing award AIAA, 1992. Mem. Am. Assn. for Crystal Growth (exec. com., pres. 1975-81), Sigma Xi, Tau Beta Pi. Office: MIT 77 Massachusetts Ave Cambridge MA 02139*

WITT, GERHARDT MEYER, hydrogeologist; b. New Haven, Sept. 28, 1953; s. Governor Martin and Florence Elizabeth (Meyer) W.; m. Pamela Sue Copeland, July 30, 1977 (div. June 20, 1990); 1 chld, George Michael. BA in Geology, Furman U., 1976. Registered profl. geologist, Fla., S.C., N.C., Ind., Tenn., Alaska. Hydrogeologist Geraghty & Miller, Inc., West Palm Beach, Fla., 1976-82; sr. hydrogeologist, group mgr. water resource Camp Dresser & McKee Inc., Fort Lauderdale, Fla., 1982-90; mgr. spl. svcs. Versar, Inc., Fort Lauderdale, Fla., 1990; area mgr., supervising hydrogeologist Parsons Brinckerhoff Gore & Storrie, West Palm Beach, 1990-93; pres., sr. hydrogeologist Gerhardt M. Witt and Assocs., Inc., West Palm Beach, 1993—. Contbg. author: Reverse Osmosis: Membrane Technology, Water Chemistry, and Industrial Applications, 1993; contbr. papers to profl. publs. Mem. Dreher Park Zoo, West Palm Beach, 1991, Sci. Mus., West Palm Beach, 1991. Mem. Am. Inst. Profl. Geologists, Geol. Soc. of Am., Nat. Water Well Assn. Presbyterian. Home: 68 Paxford Ln Boynton Beach FL 33462 Office: Gerhardt M Witt and Assocs Inc Ste 205 1870 Forest Hill Blvd West Palm Beach FL 33406

WITTE, JOHN STERLING, solar energy contracting company executive; b. Toledo, Nov. 26, 1953; s. Charles G. and Mary M. (Moulton) W.; m. Lynn M. Vargo, June 13, 1975; children: Evan T., Ryan A. BS in Constrn. Mgmt., Bowling Green State U., 1976; postgrad., U. Toledo, 1990—. Surveyor G.M. Barton Survey Co., Maumee, Ohio, 1976-79; asst. engr. DSET Labs., New River, Ariz., 1979-82; owner, mgr. Sun Shelters, Tempe, Ariz., 1982-86; engring. assoc. Charles Witte and Assocs., Toledo, 1986-88; owner, mgr. CEC Tech., Maumee, 1988—. condr. solar heating design seminar No. Ariz. U., solar design workshops, Ariz. Mem. Am. Solar Energy Soc. Home and Office: 2439 River Rd Maumee OH 43537

WITTE, OWEN NEIL, microbiologist, molecular biologist, educator; b. Bklyn., May 17, 1949. BS, Cornell U., 1971; MD, Stanford U., 1976. Predoctoral fellow Stanford U. Med. Sch., Palo Alto, Calif., 1971-76, MIT Ctr. Cancer Rsch., Cambridge, Mass., 1976-80; asst. prof. UCLA Dept. Microbiology, Molecular Genetics, 1980-82, assoc. prof., 1982-86, prof., 1986—, pres.'s chair in devel. immunology, 1989—; investigator UCLA Howard Hughes Med. Inst., 1986—. Am. Cancer Soc. faculty scholar, 1982-87; recipient Faculty award UCLA, 1990, award in basic cancer rsch. Milken Family Med. Found., 1990, Richard and Hinda Rosenthal Found. award

Am. Assn. Cancer Rsch., 1991; Outstanding Investigator grantee Nat. Cancer Inst.

WITTEN, EDWARD, mathematical physicist; s. Louis W.; m. Chiara Nappi; 3 children. Grad. in history, Brandeis U., 1971; Ph.D., Princeton U., 1976. Prof. physics Princeton Univ., 1980-87; prof. natural scis. Inst. for Advanced Study, Princeton, N. J., 1987—. Contbr. articles to mags. and profl. jours. Recipient Alan Waterman award, NSF, 1986, Fields Medal, 1990; Mac Arthur fellow. Mem. NAS. Office: Inst for Advanced Study Sch Natural Scis Olden Ln Princeton NJ 08540-4920*

WITTEN, LOUIS, physics educator; b. Balt., Apr. 13, 1921; s. Abraham and Bessie (Perman) W.; m. Lorraine Wollach, Mar. 27, 1949 (dec. 1987); children: Edward, Celia, Matthew, Jesse; m. Francis L. White, Jan. 2, 1992. B.E., Johns Hopkins U., 1941, Ph.D., 1951; B.S., NYU, 1944. Research assoc. Princeton U., N.J., 1951-53; research assoc. U. Md., College Park, 1953-54; staff scientist Lincoln Lab., MIT, 1954-55; assoc. dir. Martin Marietta Research Lab., Balt., 1955-68; prof. physics U. Cin., 1968—; trustee Gravity Research Found. Editor: Gravitation: An Introduction to Current Research, 1962, Relativity: Procs. of Relative Conf. in Midwest of 1969, 1970, Symposium on Asymptotic Structure of Space-Time, 1976; patentee in field; contbr. numerous articles to sci. jours. Served to 1st lt. USAF, 1942-46. Fulbright lectr. Weismann Inst. Scis., Rehovot, Israel, 1963-64. Fellow Am. Phys. Soc.; mem. Am. Math. Soc., Am. Astron. Soc., Internat. Astron. Union, AAAS. Office: Univ Cincinnati Dept Physics Cincinnati OH 45221

WITTEN, MARK LEE, lung injury research scientist, educator; b. Amarillo, Tex., June 23, 1953; s. Gerald Lee and Polly Ann (Warren) W.; m. Christine Ann McKee, June 10, 1988; 1 child, Brandon Lee. BS in Phys. Sci., Emporia State U., 1975; PhD, Ind. U., 1983. Postdoctoral fellow U. Ariz., Tucson, 1983-88; instr. in medicine Harvard Med. Sch., Boston, Mass., 1988-90; rsch. asst. prof. U. Ariz., Tucson, 1990—; grant cons. USAF, Washington, 1991—. Contbr. articles to profl. jours. Recipient grant USAF, 1991—, Tng. grant Dept. of Def., 1992—, NIH grant, 1991—, Upjohn Pharm. grant, 1992. Mem. Am. Physiol. Soc., Soc. Critical Care Medicine, Am. Thoracic Soc., N.Y. Acad. Scis. Methodist. Achievements include first animal model of cigarette smoke exposure to show cigarette smoke increases lung permeability; first animal model of passive cigarette smoke. Office: U Ariz Dept Pediatrics AHSC 1501 N Campbell Ave Tucson AZ 85724

WITTEN, THOMAS JEFFERSON, JR., mathematics educator; b. Welch, W.Va., Feb. 10, 1942; s. Thomas Jefferson and Gladys Marium (McMeans) W.; m. Barbara Phyllis Honaker, Feb. 20, 1965; children: Thomas Jefferson III, Rebecca A. Dye, Timothy A., Stephanie L. Dye. BS in Edn., Concord Coll., Athens, W.Va., 1965; MA in Edn., W.Va. U., Morgantown, 1971. Cert. tchr., Va., W.Va. Tchr. math. McDowell County Schs., Gary (W.Va.) High Sch., 1965-71, asst. prin., 1971-73; asst. prin. inst. Tazewell County Schs., Richlands (Va.) High Sch., 1973-87; secondary supr. Jackson County Schs., Ripley, W.Va., 1987-88; asst. prof. math. Southwest Va. Community Coll., Richlands, 1988—; math. cons. S.W. Va. C.C. Computer Math. Grant Project, Richlands, 1988-90; coord. S.W. Va. Tech. Prep Consortium, 1992—. Mem. sch. bd. Tazewell (Va.) County Schs., 1990-91; faculty senate pres. South Va. Community Coll., 1992-94. Recipient K-8 Tchr. Improvement grant, 1992-93. Mem. Va. Community Colls. Assn., Mountain Math Alliance (chmn. 1990—), PTA (life, pres. 1979-81), Richlands Rotary (pres. 1989-91), Masons (jr. deacon 1966-68). Democrat. Methodist. Avocations: painting, reading, computers, old cars, writing. Home: 737 Terry Dr Richlands VA 24641-2616 Office: SW Va C C Box SVCC Richlands VA 24641

WITTENBURG, ROBERT CHARLES, industrial engineer; b. Elgin, Ill.; s. Howard H. and Winifred (McElhose) W.; m. Elana Sue Engleking, June 20, 1970; children: Lynn Ann, Micheal Charles. BS in Indsl. Engring., Northwestern U., 1967, MBA, 1972. Coop. engr. G.D. Searle, Morton Grove, Ill., 1966-67; indsl. engr. Goodyear Tire & Rubber, North Chicago, Ill., 1967-72; project mgr. Sears Roebuck & Co., Chgo., 1972—. Republican. Mem. Evangelical Ch. Home: 1401 Fallcreek Ct Naperville IL 60565 Office: Sears Logistics Svcs 225 Windsor Dr Itasca IL 60143-1223

WITTER, RICHARD LAWRENCE, veterinarian, educator; b. Bangor, Maine, Sept. 10, 1936; s. John Franklin and Verna Harriet (Church) W.; m. Joan Elizabeth Denny, June 30, 1962; children—Jane Katherine, Steven Franklin. B.S., Mich. State U., 1958, D.V.M., 1960; M.S., Cornell U., 1962, Ph.D., 1964. Rsch. veterinarian Agrl. Rsch. Svc., U.S. Dept. Agr., East Lansing, Mich., 1964-75, dir. Avian Disease and Oncology Lab., 1975—; clin. prof. pathology Mich. State U., East Lansing, 1965—. Contbr. articles to profl. jours. Recipient Disting. Alumni award Coll. Vet. Medicine, Mich. State U., 1985, Disting. Service award USDA, 1985. Mem. AVMA, Am. Assn. Avian Pathologists (P.P. Levine award 1967, 81, 88, 92, Upjohn Achievement award 1992), Poultry Sci. Assn. (CPC Internat. award 1976), Mich. Vet. Med. Assn., World Vet. Poultry Assn. (B. Rispens rsch. award 1983). Lodge: Kiwanis. Avocations: piano, hunting, fishing, gardening. Home: 3880 Sheldrake Ave Okemos MI 48864-3646 Office: Avian Disease and Oncology Lab 3606 E Mt Hope Rd East Lansing MI 48823-5338

WITTIG, CURT, chemist, educator. Prof. dept. chemistry U. So. Calif., L.A. Recipient Herbert Broida Atomic Molecular or Chem. Physics prize Am. Physical Soc., 1993. Office: U of Southern California Chemistry Dept University Park Los Angeles CA 90089*

WITTROCK, MERLIN CARL, educational psychologist; b. Twin Falls, Idaho, Jan. 3, 1931; s. Herman C. and Mary Ellen (Baumann) W.; m. Nancy McNulty, Apr. 3, 1953; children: Steven, Catherine, Rebecca. BS in Biology, U. Mo., Columbia, 1953, MS in Ednl. Psychology, 1956; PhD in Ednl. Psychology, U. Ill., Urbana, 1960. Prof. grad. sch. edn. UCLA, 1960—, founder Ctr. Study Evaluation, 1966, chmn. div. ednl. psychology, chmn. faculty, 1991—; fellow Ctr. for Advanced Study in Behavioral Scis., 1967-68; vis. prof. U. Wis., U. Ill., Ind. U., Monash U., Australia; bd. dirs. Far West Labs., San Francisco, 1989—; chmn. com. on evaluation and assessment L.A. Unified Sch. Dist., 1988—; mem. nat. adv. panel for math. scis. NRC of NAS, 1988-89; chmn. nat. bd. Nat. Ctr. for Rsch. in Math. Scis. Edn., 1991—. Author, editor: The Evaluation of Instruction, 1970, Changing Education, 1973, Learning and Instruction, 1977, The Human Brain, 1977, Danish transl., 1980, Spanish transl., 1982, The Brain and Psychology, 1980, Instructional Psychology: Education and Cognitive Processes of the Brain, Neuropsychological and Cognitive Processes of Reading, 1981, Handbook of Research on Teaching, 3d edit., 1986, The Future of Educational Psychology, 1989, Research in Learning and Teaching, 1990, Testing and Cognition, 1991. Capt. USAF. Recipient Thorndike award for outstanding psychol. rsch., 1987, Disting. Tchr. of Univ. award UCLA, 1990; Ford Found. grantee. Fellow AAAS, APA (pres. divsn. ednl. pscyholgy 1984-85, assn. coun. 1988-91, Award for Outstanding Svc. to Ednl. Psychology 1991), Am. Psychol. Soc. (charter), Am. Ednl. Rsch. Assn. (chmn. ann. conv., chmn. publs. 1980-83, assn. coun. 1986-89, bd. dirs. 1987-89, chmn. com. on ednl. TV 1989—, Outstanding Contbns. award 1986, Outstanding Svc. award 1989); mem. Phi Delta Kappa. Office: UCLA 321 Moore Hall Los Angeles CA 90024

WITTRY, DAVID BERYLE, physicist, educator; b. Mason City, Iowa, Feb. 7, 1929; s. Herman Joseph and Edna Pearl (Filbey) W.; m. Mildred Elizabeth DuBois, July 1, 1955; children—James David, Robert Andrew, Kristopher Lee, Diane Marie, Linda Beryle. B.S., U. Wis., 1951; M.S., Calif. Inst. Tech., 1953, Ph.D., 1957. Research fellow Calif. Inst. Tech., Pasadena, 1957-59; asst. prof. U. So. Calif., Los Angeles, 1959-61; assoc. prof. dept. elec. engring. U. So. Calif., 1961-69, prof. dept. materials sci. and elec. engring., 1969—; cons. Hughes Semiconductors, 1958-59, Applied Research Labs., Inc., 1958-83, Exptl. Sta., E.I. du Pont de Nemours & Co., 1962-71, Gen. Telephone and Electronics Research Labs., 1966-72, Autonetics div. N. Am. Aviation, 1961-63, Electronics Research div. Rockwell Internat., 1976-81, Atlantic Richfield Co. Corp. Tech. Lab., 1981-87, Jet Propulsion Lab., 1985-88, Hitachi Instruments, 1989-90. Editor 3 proceedings of cons. Contbr. articles to profl. jours. Patentee in field. Knapp scholar, U. Wis., 1949-51; recipient first award essays on gravity, Gravity Research Found., 1949; Guggenheim fellow, 1967-68; vis. scientist Japan Soc. for Promotion of Sci., U. Osaka Prefecture, 1974. Mem. IEEE, Electron Microscopy Soc. Am. (dir. phys. scis. 1979-81, pres. 1983), Microbeam Analysis Soc. (sec. organizing com. 1966, exec. council 1970-72, pres. 1988, Presdl. award 1980, Birks award 1987, 89, hon. mem.), Am. Phys. Soc., Sigma Xi. Methodist. Office: U So Calif Dept Materials Sci Los Angeles CA 90089-0241

WITWER, RONALD JAMES, mechanical engineer; b. Lancaster, Pa., June 5, 1937; s. Robert James and Anna Margret (Doster) W.; m. Janet Ann Vroom, Jan. 26, 1972; children: Jeanette, Edward, Erik, Julie. AS in Engring. Tech., Wyomissing Poly. Inst., 1958; BSME, Pa. State U., 1963. Registered profl. engr., Del., Pa., W.Va. Apprentice, then designer Textile Machine Works, Wyomissing, Pa., 1955-63; design engr. E.I. DuPont de Nemours, Newark, Del., 1963-68, equipemnt engr., 1968-74, project engr., 1974-79; sr. engr. electronics E.I. DuPont de Nemours, New Cumberland, Pa., 1979-88; sr. engr. polymers E.I. DuPont de Nemours, Parkersburg, W.Va., 1988—. Officer St. Phillip's Luth. Ch., Wilmington, Del., to 1979, St. Paul's Luth. Ch., Hershey, Pa., 1979-88, Mt. Pleasant Meth. Ch., Mineral Wells, W.Va., 1988—. Mem. ASME (sr.), Soc. Plastics Engrs. (sr., bd. dirs. Southeast Ohio chpt.), Soc. Mfg. Engrs., Masons, Shriners. Home: 121 Windsor Estates Mineral Wells WV 26150 Office: EI DuPont de Nemours Polymers Tech PO Box 1217 Parkersburg WV 26102

WITZEL, LOTHAR GUSTAV, physician, gastroenterologist; b. Mannheim, Fed. Republic of Germany, July 27, 1939; s. Gustav and Martha (Pilger) W. MD, U. Freiburg, Fed. Republic of Germany, 1965; PhD, U. Berlin, 1983. Substitute medicine supt. U. Bern (Switzerland) Med. Sch., 1973-77; dir., med. supt. German Red Cross Hosp., Berlin, 1978—. Contbr. articles to profl. jours.; patentee in field. Mem. Indian Soc. Gastroenterology (hon.), European Congress Endoscopy (sec. gen. 1981), Swiss Soc. Gastroenterology (corr. mem. 1989, Award of Gastroenterology 1981). Avocation: jazz music. Office: Koloniestrasse 21, 13359 Berlin Germany also: DRK Kronkenhaus, Drontheimer Str 39, 13359 Berlin Germany

WITZEL, WILLIAM MARTIN, analytical chemist, educator, consultant; b. Amsterdam, N.Y., Mar. 18, 1953; s. Frank and Eunice Aileen (MacLachlan) W.; m. Martha Lou Goetz, May 20, 1978 (div. 1988); children: Adam J., Daniel J., Corinne K. BA in Biology, Hope Coll., 1975, BA in Chemistry, 1978. With BASF, Holland, Mich., 1976—, rsch. chemist, 1981-88, analytical chemist, sect. head, 1988—; mem. analytical steering com. Dry Color Mfg. Assn., Washington, 1988—; instr. Brian Tracy Learning Systems, 1991—. Contbr. articles to profl. publs. Steering com. Coalition for Excellence in Sci. and Math. Edn., Grand Valley State U., 1988—. Grantee NSF, 1990—. Me. Am. Chem. Soc., Beta Beta Beta. Achievements include discovery of major organic by-product in synthesis of alkali blue pigment, development of electrochemical detection method for quantitation of 3,3 dichlonobenzidine by high pressure liquid chromotography. Home: 6544 Bradley Rd Saugatuck MI 49453 Office: BASF 491 Columbia Ave Holland MI 49423

WITZGALL, CHRISTOPH JOHANN, mathematician; b. Hindelang, Bavaria, Germany, Feb. 25, 1929; came to U.S., 1959; s. Otto and Hanna (Schulte-Liese) W.; m. Elizabeth Bingham, Oct. 10, 1964; children: John Chandler, Hanna Elizabeth, George Matheus. PhD in Maths., Universitat Munich, 1958. Rsch. assoc. Princeton (N.J.) U., 1959-61; mathematician Nat. Bur. of Stds., Washington, 1962-66, Boeing Scientific Rsch. Labs., Seattle, 1966-73, Nat. Inst. Stds. and Tech., Gaithersburg, Md., 1973—; acting chief ops. rsch. div. Nat. Inst. Standards and Tech., Gaithersburg, Md., 1979-82; vis. prof. RAND Corp., Santa Monica, Calif., 1961, Argonne Nat. Lab., Joliet, Ill., 1962, U. Tex., Austin, 1971, Universitat Wurzburg, Germany, 1972, U. Md., College Park, 1977, Johns Hopkin's U., Balt., 1985. Co-author: Convexity and Optimization in Finite Dimensions, 1970. Mem. AAAS, Ops. Rsch. Soc. Am., Soc. for Indsl. and Applied Maths., Am. Math. Soc. Office: Nat Inst Stds and Tech Gaithersburg MD 20899

WODARCZYK, FRANCIS JOHN, chemist; b. Chgo., Dec. 11, 1944; s. Sigmund Frank and Josephine Aurelia (Boblak) W. BS, Ill. Inst. Tech., 1966; AM, Harvard U., 1967, PhD, 1971. Postdoctoral U. Calif., Berkeley, 1971-73; rsch. chemist Cambridge Rsch. Labs., Hanscom AFB, Mass., 1973-77; program mgr. Office of Sci. Rsch., Bolling AFB, D.C., 1977-78; mem. tech. staff Rockwell Internat. Sci. Ctr., Thousand Oaks, Calif., 1978-85; program mgr. Office of Sci. Rsch., Bolling AFB, 1985-90; program dir. NSF, Washington, 1990—. Recipient scholarship, George M. Pullman Found., 1962, Ill. Inst. Tech., 1962-66, Ill. State scholarship, 1962-66, NSF fellowship, 1966-71, 71-72. Mem. AAAS, Am. Chem. Soc., Sigma Xi, Alpha Chi Sigma (chpt. v.p. 1965-66, chpt. award 1964). Achievements include development of radio frequency-microwave double resonance spectroscopy; first demonstration cw optically pumped molecular laser; laser-excited electronic to vibrational energy transfer studies. Office: NSF 4201 Wilson Blvd Arlington VA 22230

WOELKERLING, WILLIAM J., botanist educator. Prof. dept. botany La Trobe U., Victoria, Australia. Recipient Gerald W. Prescott award Phycological Soc. Am., 1989. Office: LaTrobe Univ, Dept of Botany, Bundoora Victoria 3063, Australia*

WOERNER, ALFRED IRA, medical device manufacturer, educator; b. Jersey City, N.J., Sept. 21, 1935; s. Theodore and Miriam (Mann) W.; m. Margaret R. Martin, Nov. 27, 1959; children: John, Michael, Judith. DME, Stevens Inst., 1956; MS, Stevens, 1961; MBA, NYU, 1965; LLB, LaSalle U., 1963; PhD, Calif. State U., 1990. Gen. program mgr. Becton Dickenson & Co., Ruthuferd, N.J., 1959-63; group v.p., asst. to pres. Howmet Corp., N.Y.C., 1963-69; gen. mgr., v.p. Wide Range Industries, N.Y.C., 1969-72; pres., owner New World Market Ltd., Westwood, N.J., 1972—, Fairfield Surg. Corp., Stanford, Conn., 1972—; cons. Woerner Assocs., Westwood; prof. Fairleigh Dickinson U., Teaneck, N.J., 1969—. Author: Program Management, 1988. Pres. Bd. Edn., Westwood, 1978-86; adv. Stevens Inst. Tech., Hoboken, N.J., 1972-80. Mem. AMA, ASME, Am. Acad. Cons., Am. Statistical Assn. Achievements include five patents in Medical Industry; development of new process in orthopedic surgery industry. Home: 75 Bergen St Westwood NJ 07675-2332 Office: Fairleigh Dickinson U 1000 River Rd Teaneck NJ 07666-1914

WOERNER, GELDARD HARRY, civil engineer; b. Kansas City, Mo., May 25, 1925; s. Geldard Henry and Vivian Zelda (Harry) W.; m. Olga Ellenora Peterson, Feb. 6, 1947; children: Christine Louise, Geoffry Lee. BSCE, Kans. U., 1945. Registered profl. engr., N.C., Va. Constrn. engr., project engr. Standard Oil Co., Ind., 1947-53; asst. pub. works officer Naval Ammunition Depot, McAlester, Okla., 1957-62; dir. liaison div. Naval Facilities Engring. Command, Alexandria, Va., 1962-71, mil. constrn. advisor, 1971-81; program engr. Beaman Corp., Greensboro, N.C., 1984-85; pvt. practice cons. Sanford, N.C., 1985—; cons. U.S. Congress, Washington, 1982—. Drafted bill and house and senate reports for military construction codification act, public law 97-214. Comdr. USN, 1943-45, 53-57. Recipient Letter of Commendation, Sen. Strom Thurmond, Rear Admiral W. M. Zobel, Congressman Bo Ginn, others. Mem. ASCE, ASTM, Soc. Am. Mil. Engrs., Internat. Conf. Bldg. Ofcls., So. Bldg. Code Congress. Democrat. Lutheran. Home: 317 Lafayette Dr Sanford NC 27330

WOHL, ARMAND JEFFREY, cardiologist; b. Phila., Dec. 11, 1946; s. Herman Lewis and Selma (Paul) W.; m. Marylouise Katherine Giangrossi, Sept. 4, 1977; children: Michael Adam, Todd David. Student, Temple U., 1967; MD, Hahnemann U., 1971. Intern Bexar County Hosp., San Antonio, 1971-72; resident in internal medicine Parkland Hosp., Dallas, 1972-74; fellow in cardiology U. Tex. Southwestern Med. Ctr., Dallas, 1974-76; chief of cardiology USAF Hosp. Elmendorf, Anchorage, 1976-78; chief cardiologist Riverside (Calif.) Med. Clin., 1978-79; cardiologist Grossmont Cardiology Med. Group, La Mesa, Calif., 1980-84; pvt. practice, La Mesa, 1985—; chief of cardiology Grossmont Hosp., La Mesa, 1988-90; asst. clin. prof. Sch. Medicine. U. Calif., San Diego, 1990—. Contbr. articles to profl. jours. Bd. dirs. San Diego County chpt. Am. Heart Assn., 1981-87. Maj. USAF, 1976-78. Fellow Am. Coll. Cardiology (councilor Calif. chpt. 1991—), Am. Coll. Physicians, Coun. on Clin. Cardiology. Avocations: tennis, travel. Office: 5565 Grossmont Center Dr La Mesa CA 91942-3021

WOHLETZ, LEONARD RALPH, soil scientist, consultant; b. Nekoma, N.D., Oct. 22, 1909; s. Frank and Anna (Keifer) W.; m. Jane Geisendorfer, Sept. 1, 1935; children: Mary Jane, Leonard Ralph Jr., Elizabeth Ann, Catherine Ellen, Margaret Lee. BS, U. Calif., Berkeley, 1931, MS, 1933. Jr. soil expert USDA Soil Erosion Svc., Santa Paula, Calif., 1934; asst. regional chief soil surveys USDA Soil Conservation Svc., Santa Paula, 1935; asst. regional chief soil surveys USDA Soil Conservation Svc., Berkeley, 1939-42, soil survey supr., 1942-45, state soil scientist, 1945-68, asst. to state conservationist, 1969-71; cons. soil scientist Berkeley, 1973—. Author: Survey Guide, 1948; contbr. articles to profl. publs. including Know Calif. Land, Soils and Land Use Planning, Planning by Foresight and Hindsight. Mem. Waste Mgmt. Commn., Berkeley, 1981; chmn. com. Rep. for Congress, 8th Dist. Calif., 1980; pres. State and Berkeley Rep. Assembly, 1985—. Recipient Soil Conservationist of Yr., Calif. Wildlife Fedn., 1967. Mem. Soil and Water Conservation Soc. (chmn. organic waste mgmt. com. 1973—, sect. pres., Dist. Svc. award, charter and life mem., Disting. Svc. award 1971, Outstanding Svc. award 1983), Soil Sci. Soc. Am. (emeritus), Internat. Soc. Soil Sci., Profl. Soil Sci. Assn. Calif., Commonwealth Club Calif., San Francisco Farmers Club. Roman Catholic. Achievements include expedition of soil surveys, interpretations, funding and publications in California, improvement of methods of conveying technical soils interpretations, land use, waste management information. Home: 510 Vincente Ave Berkeley CA 94707-1522

WOJTANEK, GUY ANDREW, mechanical engineer; b. Chgo., Dec. 26, 1954; s. Edmund A. and Beatrice Marie (Saugling) W.; m. Jean Marie Sylwestrak, May 19, 1979; 1 child, Devin Charles. BSME, U. Ill., 1977; MBA, Roosevelt U., 1982. Registered profl. engr., Ill. Design/product engr. Littelfuse, Inc., Des Plaines, Ill., 1977-78; project engr. Controls div. Singer Co., Schiller Park, Ill., 1979-84, sr. project engr. Controls div., 1984-87; sr. project engr. Controls div. Eaton Corp., Carol Stream, Ill., 1987—; cons. Non-Destructive Engring. Instruments Inc., Boulder, Colo., 1989—. Patentee in field. Mem. Nat. Soc. Profl. Engrs., Am. Soc. Mech. Engrs. (assoc.). Avocations: riding motorcycles, down-hill skiing. Home: 3N 244 Valewood Dr West Chicago IL 60185 Office: Eaton Controls Div 191 E North Ave Carol Stream IL 60188-2090

WOLF, ALFRED PETER, chemist, educator; b. N.Y.C., Feb. 13, 1923; s. Josef and Margarete (Kunst) W.; m. Elizabeth H. Gross, June 15, 1946; 1 child, Roger O. B.A., Columbia U., 1944, M.A., 1948, Ph.D., 1952; Ph.D. (hon.), U. Uppsala, (Sweden), 1983, U. Rome, 1989. Chemist Brookhaven Nat. Lab., Upton, N.Y., 1951, chemist with tenure, 1957-64, sr. chemist, 1964—, chmn. chemistry dept., 1982-87; adj. prof. chemistry Columbia U., N.Y.C., 1953-83; vis. lectr. U. Calif., Berkeley, 1964; cons. Philip Morris, Inc., Richmond, Va., 1966-91, NIH, Bethesda, Md.; cons., advisor IAEA, Vienna, Austria; advisor Italian NRC, Romse, 1959—, Atomic Rsch. Inst., Julich, Germany, 1981-90; rsch. prof. psychiatry NYU, 1988—. Author: Synthesis of 11C, 18F and 13N Labelled Radiotracers for Biomedical Application, 1982; contbr. numerous articles to profl. jours.; patentee in field; editor: Jour. Labelled Compounds and Radiopharms., Radiochimica Acta. Served with AUS, 1943-46. Recipient JARI award, Pergamon Jours., 1986, The Javits Neurosci. Investigator award, 1986, Georg V. Hevesy Meml. Medal, Georg V. Hevesy Found. of Nuclear Medicine, 1986,. Mem. NAS, Am. Chem. Soc. (Nuclear Applications in Chemistry award 1971, Esselen award 1988), Chem. Soc. (U.K.), Soc. Nuclear Medicine (1982 Paul Aebersold award, pres. radiopharm. council 1980, George Hevesy award 1991), German. Chem. Soc. Home: PO Box 1043 Setauket NY 11733-0803 Office: Brookhaven Nat Lab Dept Chemistry Upton NY 11973

WOLF, JACK KEIL, electrical engineer, educator; b. Newark, Mar. 14, 1935; s. Joseph and Rosaline Miriam (Keil) W.; m. Toby Katz, Sept. 10, 1955; children—Joseph Martin, Jay Steven, Sarah Keil. B.S., U. Pa., 1956; M.S.E., Princeton, 1957, M.A., 1958, Ph.D., 1960. With R.C.A., Princeton, N.J., 1959-60; asso. prof. N.Y. U., 1963-65; from asso. prof. to prof. elec. engring. Poly. Inst. Bklyn., 1965-73; prof. dept. elec. and computer engring. U. Mass., Amherst, 1973-85; chmn. dept. U. Mass., 1973-75; Stephen O. Rice prof. Ctr. Magnetic Rec. Research, dept. elec. engring. and computer sci. U. Calif.-San Diego, La Jolla, 1985—; Mem. tech. staff Bell Telephone Labs., Murray Hill, N.J., 1968-69; engring. assoc. Qualcomm Inc., San Diego, 1985—. Editor for: coding I.E.E.E. Transactions on Information Theory, 1969-72. Served with USAF, 1960-63. NSF sr. postdoctoral fellow, 1971-72; Guggenheim fellow, 1979-80. Fellow IEEE (pres. info. theory group 1974, co-recipient info. theory group prize paper award 1975, co-recipient Comm. Soc. prize paper award 1993), Nat. Acad. Engring.; mem. AAAS, Sigma Xi, Sigma Tau, Eta Kappa Nu, Pi Mu Epsilon, Tau Beta Pi. Achievements include research on information theory, communication theory, computer/communication networks, magnetic recording. Home: 197 Desert Lakes Dr Rancho Mirage CA 92270-4053

WOLF, MARK ALAN, chemical engineer; b. Lincoln, Nebr., July 16, 1958; s. Gary J. and Karen (Hemphill) W.; m. Jane Marie Hojnacki, May 10, 1980; children: Lauren, Brett. BSChemE, Colo. Sch. of Mines, 1980. Registered profl. engr. Sr. process engr. Monsanto Co., St. Louis, 1980-85, Vulcan Chems., Wichita, Kans., 1985-91; prin. process engr. Air Products & Chems., Wichita, Kans., 1991-92, tech. mgr., 1992—. Lay minister Ascension Luth. Ch., Wichita, 1991—. Recipient Engring. Achievement award Monsanto Co., 1983. Mem. AICE. Office: Air Products & Chems Inc 6601 S Ridge Rd Wichita KS 67231-7134

WOLF, MICHAEL ELLIS, clinical psychologist; b. Hutchinson, Kans., Oct. 27, 1950; s. Harold E. and Geraldine F. (Spencer) W.; m. Anne Dullea, July 15, 1978; children: Chris, Brian, Allison. AA, Hutchinson Community Coll., 1970; BA, U. Kans., 1972; MA, Wichita State U., 1976; PhD, Ohio U., 1980. Lic. clin. psychologist, Tex. Psychologist Kaiser Permanente, Dallas, 1980-82; clin. psychologist Dallas Rehab. Inst., 1984-87; clin. psychologist in pvt. practice Dallas, 1982—; cons. inpatient and outpatient chem. dependency and psychiatric treatment programs, Dallas, 1984—. Contbr. articles to profl. jours. Mem., chmn. pre-sch. adv. bd., 1986-88. Franklin Cramm scholar, 1968; Summer Fgn. Lang. Inst. scholar, 1971. Mem. Am. Psychol. Assn., Tex. Psychol. Assn., Dallas Psychol. Assn., Phi Beta Kappa, Phi Theta Kappa. Avocations: sports, camping, reading. Office: 13800 Montfort Dr Ste 200 Dallas TX 75240

WOLF, ROBERT FARKAS, systems and avionics company executive, environmental planning consultant; b. N.Y.C., Feb. 19, 1932; s. Desidar Farkas and Christina (Hodosy) Wolf; m. Victoire M. Cullerot, Oct. 8, 1960. BS in Liberal Studies, SUNY, Albany, 1981; MBA in Econs., Rivier Coll., Nashua, N.H., 1987. Engring. designer Mpls. Honeywell Co., Manchester, N.H., 1956-63; design engr. Sanders Assocs., Nashua, 1963-79; systems analyst Kollsman Instruments Co., Merrimack, N.H., 1979—. Bd. dirs. Webster House Children's Home, Manchester, 1975-80; chmn. Mt Vernon (N.H.) Planning Bd., 1981—, N.H. Regional Planning Commn., Nashua, 1984—; mem. Solid Waste Mgmt. Bd. Mem. Nat. Assn. Regional Couns., N.H. Planners Assn. Republican. Home: 15 S Main St Mont Vernon NH 03057 Office: Kollsman Co 220 Daniel Webster Hwy Merrimack NH 03054-4844

WOLF, ROMAIN MATHIAS, theoretical chemist; b. Wiltz, Luxembourg, Apr. 24, 1954; arrived in Switzerland, 1973; s. Marcel and Irene (Mergen) W.; m. Gabriella Lecis, Mar. 17, 1989. Diploma in chem. engring., ETH, Zürich, Switzerland, 1978, D. in Tech. Sci., 1982. Cert. chem. engring. Postdoctoral fellow MIT, Cambridge, Mass., 1982-83; rsch. scientist Ciba-Geigy, Basel, Switzerland, 1983—. Author: (with others) Solution Behavior of Surfactants, Vol. 2, 1982; contbr. articles to profl. jours including Macromolecules, Jour. Chem. Soc., Perkin Trans. II, Chirality, Jour. Chromatography. Mem. Am. Chem. Soc. (polymer div., computers in chemistry div.), Swiss Polymer Group. Achievements include research in enzymology in hydrocarbon solvents, in separation of enantiomers by chromatography on chiral polymers, in theoretical investigations on the supramolecular structure of polymers. Office: Ciba Geigy, R 1060 7 32, 4002 Basel Switzerland

WOLF, STEWART GEORGE, JR., physician, medical educator; b. Balt., Jan. 12, 1914; s. Stewart George and Angeline (Griffing) W.; m. Virginia Danforth, Aug. 1, 1942; children: Stewart George III, Angeline Griffing, Thomas Danforth. Student, Phillips Acad., 1927-31, Yale U., 1931-33; A.B., Johns Hopkins U., 1934, M.D., 1938; M.D. (hon.), U. Göteborg, Sweden, 1968. Intern N.Y. Hosp., 1938-39, resident medicine, 1939-42, NRC fellow, 1941-42; rsch. fellow Bellevue Hosp., 1939-42, clin. assoc. vis. neuropsychiatrist, 1946-52; rsch. head injury and motion sickness Harvard neurol. unit

Boston City Hosp., 1942-43; asst., then assoc. prof. medicine Cornell U., 1946-52; prof., head dept. medicine U. Okla., 1952-67, Regents prof. medicine, psychiatry and behavioral scis., 1967—, prof. physiology, 1967-69; dir. Marine Biomed. Inst., U. Tex. Med. Br., Galveston, 1969-78; dir. emeritus Marine Biomed. Inst., U. Tex. Med. Br., 1978—, prof. medicine univ., also prof. internal medicine and physiology med. br., 1970-77; prof. medicine Temple U., Phila., 1977—; v.p. med. affairs St. Luke's Hosp., Bethlehem, Pa., 1977-82; dir. Totts Gap Inst., Bangor, Pa., 1958—; supr. clin. activities Okla. Med. Rsch. Found., 1953-55, head psychosomatic and neuromuscular sect., 1952-67, head neurosci. sect., 1967-69; adv. com. Space Medicine and Behavioral Scis., NASA, 1960-61; cons. internal medicine VA Hosp., Oklahoma City, 1952-69; cons. (European Office), Paris, Office Internat. Rsch., NIH, 1963-64; mem. edn. and supply panel Nat. Adv. Commn. on Health Manpower, 1966-67; mem. Nat. Adv. Heart Coun., 1961-65, U.S. Phamacopeia Scope Panel on Gastroenterology, Regent Nat. Libr. Medicine, 1965-69; chmn., 1968-69; mem. Nat. Adv. Environ. Health Scis. Coun., 1978-82; exec. v.p. Frontiers Sci. Found., 1967-69; mem. sci. adv. bd. Muscular Dystrophy Assns. Am., Inc., 1974-91, chmn., 1980-89; mem. gastrointestinal drug adv. com. FDA, 1974-77; bd. Internat. Cardiology Fedn.; mem. bd. visitors dept. biology Boston U., 1978-88; mem. vis. com. Ctr. for Social Rsch., Lehigh U., 1980-90; chmn. adv. com. Wood Inst. on History of Medicine, Coll. Physicians, Phila., 1980-90, mem. program com. Coll. Physicians, 1990-91; dir. Inst. for Advanced Studies in Immunology and Aging, 1988—. Author: Human Gastric Function, 1943, The Stomach, 1965, Social Environment and Health, 1981, others; adv. editor Internat. Dictionary Biology and Medicine, 1978—; editor in chief Integrative Physiol & Behavioral Sci.: The Official Jour. of Pavlovian Soc., 1990—. Pres. Okla. City Symphony Soc., 1956-61; mem. Okla. Sch. of Sci. and Math. Found., 1961—. Recipient Disting. Svc. citation U. Okla., 1968, Dean's award for disting. med. svc., 1992; Horsley Gantt medal Pavlovian Soc., 1987, Hans Selye award Am. Inst. Stress, 1988. Fellow Am. Psychiat. Assn. (disting., trustee 1992—, Hofheimer prize for rsch. 1952); mem. AMA (coun. mental health 1960-64), Am. Soc. Clin. Investigation, Am. Clin. and Climatol. Assn. (pres. 1975-76), Assn. Am. Physicians, Am. Psychosomatic Soc. (pres. 1961-62), Am. Gastroent. Assn. (rsch. award 1943, pres. 1969-70), Am. Heart Assn. (chmn. com. profl. edn., com. internat. program, awards), Coll. Physicians Phila., Collegium Internat. Activitas Nervosae Superioris (exec. com. 1992—, pres. elect 1993), Philos. Soc. Tex., Sigma Xi, Alpha Omega Alpha, Omicron Delta Kappa. Club: Cosmos (Washington). Home: RR 1 Box 1120G Bangor PA 18013-9716 Office: Totts Med Rsch Labs Bangor PA 18013

WOLF, WAYNE HENDRIX, electrical engineering educator; b. Washington, Aug. 12, 1958; s. Jesse David and Carolyn Josephine (Cunningham) W.; m. Nancy Jane Porter, Aug. 12, 1989. BS with distinction, Stanford U., 1980, MS, 1981, PhD, 1984. Lectr. Stanford (Calif.) U., 1984; staff mem. AT&T Bell Labs., Murray Hill, N.J., 1984-89; asst. prof. elec. engring. Princeton (N.J.) U., 1989—. Co-editor: High-Level VLSI Synthesis, 1991; author: (book) Modern VLSI Design, 1983; contbr. Physical Design Automation of VLSI Systems, 1989. Mem. IEEE, Assn. Computing Machinery, Phi Beta Kappa, Tau Beta Pi. Avocations: bicycling, photography, films, flying. Office: Princeton U Dept Elec Engring Princeton NJ 08544

WOLF, WILLIAM MARTIN, computer company executive, consultant; b. Watertown, N.Y., Aug. 29, 1928; s. John and Rose (Emrich) W.; m. Eileen Marie Jolly, Aug. 19, 1952 (div. 1974); children: Rose, Sylvia, William. BS, St. Lawrence U., 1950; MS, U. N.H., 1951; postgrad., U. Pa., 1951-52, MIT, 1952-55. Programmer digital computer lab. MIT, Cambridge, Mass., 1952-54; pres. Wolf R & D Corp., Boston, 1954-69, Wolf Computer Corp., Boston, 1969-76, Planning Systems Internat., Boston, 1976-81, Micro Computer Software Inc., Cambridge, 1981-88, Tech. Acquisition Corp., Boston, 1989-91, Planning Internat., Inc., Boston, 1989—, Wolfsort Corp., Boston, 1989—; dir., exec. v.p.Tech. Capital Network MIT, 1992—; co-founder, pres. Assn. Ind. Software Cos., Washington, 1965-67, Design Sci. Inst., Phila., 1969-73, Nat. Coun. Profl. Svc. Firms, Washington, 1970-75; seminar leader MIT Sloan Sch., Cambridge, 1970; co-founder, bd. dirs. Harbor Nat. Bank, Boston. Author computer program; inventor management system, orbit calculator, sorting method. Co-founder X-10 Orgn., Boston, 1962; trustee Addison Gilbert Hosp., Gloucester, Mass., 1963; v.p. Young Pres. Orgn., Boston, 1970; overseer Mus. Sci., Boston, 1989—; mem. Computer Mus. Named Outstanding Young Man in Boston, Jaycees, 1962; recipient Speaker's award Data Processing Mgmt. Assn., 1966. Mem. World Bus. Coun., MIT Club (Alumni award 1991), Boston Computer Soc., Forty-Niners. Office: Wolfsort Corp 1 Longfellow Pl Apt 3123 Boston MA 02114-2429

WOLFE, ALLAN, physicist; b. Bklyn., Dec. 19, 1942; s. Isidor Irving and Florence (Rosenfeld) W.; m. Marta Elias Boneta, Dec. 30, 1967; 1 child, Daniel Duchaune. BS, Poly. Inst. Bklyn., 1964; MS, U. N.H., 1969, PhD, 1971. Physics chmn. Nasson Coll., Springvale, Maine, 1973-74; prof. physics N.Y.C. Tech. Coll., Bklyn., 1974—; physicist, visitor AT&T Bell Labs., Murray Hill, N.J., 1977—; physics researcher U. L'Aquila, Italy, 1989, Indian Inst. Sci., Bangalore, India, 1990, Japan Soc. Promotion Sci., Tokyo, 1990. Contbr. articles to profl. jours. Avocations: chess, jogging, music, Masonry, Japanese and French langs. Office: NYC Tech Coll 300 Jay St # 812N Brooklyn NY 11201-2902

WOLFE, LOWELL EMERSON, space scientist; b. Springfield, Ohio, Nov. 27, 1953; s. LeRoy and Edna (Christian) W.; m. Joy Wengel, Aug. 21, 1977. BS in Chemistry, Kent State U., 1976. Scientist Life Systems, Inc., Cleve., 1977—. Mem. Am. Chem. Soc. Achievements include design and testing of concepts for improving electrolyzer used on space station Freedom; improving materials, weight reduction and testing of Urine Processor Assembly for use on space station Freedom. Office: Life Systems Inc 24755 Highpoint Rd Cleveland OH 44122

WOLFENSON, AZI U., electrical, mechanical and industrial engineer, consultant; b. Rumania, Aug. 1, 1933; came to Peru, 1937; s. Samuel G. and Polea S. (Ulanowski) W.; m. Rebeca Sterental, Jan. 10, 1983; 1 child, Michael Ben; children by previous marriage: Ida, Jeannette, Ruth, Moises, Alex. Mech., Elec. Engr., Universidad Nacional de Ingenieria, Peru, 1955; MSc in Indsl. Engring., U. Mich., 1966; Indsl. Engr., U. Nacional de Ingenieria, Peru, 1967; PhD in Engring. Mgmt., Pacific Western U., 1983, PhD in Engring. Energy, Century U., 1985, D in Philosophy of Engring. (hon.) World U. Roundtable, Ariz., 1987. Power engr. Peruvian Trading Co., 1956-57; gen. mgr. AMSA Ingenieros S.A., 1957-60; prof. Universidad Nacional de Ingenieria, Peru, 1956-72, dean mech. and elec. engring., 1964-66, dean indsl. engring., 1967-72; dir. SWSA Automotive Parts, Peru, 1954-77; project mgr. Nat. Fin. Corp., Cofide, 1971-73; Peruvian dir. Corporacion Andina de Fomento, CAF, 1971-73; rep. in Peru, CAF, 1973-74; pres. DESPRO cons. firm, 1973-76; exec. pres. Electroperu, 1976-80; cons. engr., 1964—; dir. Tech. Transference Studies, 1971-72. Mem. Superior Coun. Electricity, 1964-66; metal mech. expert for andean group, 1970-71; Nat. coun. Fgn. Investment and Tech. Transfer, 1972-73; councilman at the Concejo Provincial De Lima, 1969-75; mem. Consultive Coun. Ministry Economy and Fin., 1973-74; pres. Peruvian Jewish Community, 1966-70, Peruvian Hebrew Sch., 1976-78; promoter, co-founder, gen. mgr. La Republica Newspaper, Peru, 1981; pres. PROA project promotion AG, Switzerland, 1982—; cofounder El Popular, 1983, El Nacional newspapers, 1985. Recipient awards Order Merit for Disting. Svcs., Peru, 1980, Disting. by City Coun. of Huancayo, 1980, Trujillo, 1978, Huaral, 1979, Piura, 1980, Disting. Contbn. award City of Lima, 1970, 71, Disting. Contbn. to Elec. Devel. in Peru, 1979; others; named 1979 Exec., Gente mag., recognition Israel Govt., 1967, Disting. Comision Integracion Electrica Regional, CIER, medal, 1984. Fellow Inst. Prodn. Engrs., Brit. Inst. Mgmt.; mem. Colegio Ingenieros Peru, Instituto Peruano de Ingenieros Mecanicos (pres. 1965-66, v.p. 1967, dir. 1969, 70, 76), Asociacion Electrotechnica del Peru, ASME, AIIE (sr.), MTM Assn., Am. Soc. Engring. Edn., Am. Inst. Mgmt. Sci., AAAS, Assn. Mgmt. Sci. (dir. 1968), Asociacion Peruana Avance Ciencia, Inst. Administrv. Mgmt., British Inst. Mgmt., Am. Nuclear Soc. (vice chmn. 1988, 90, chmn. Swiss sect. 1991-93), Alumni Assn. of the Mich., Pacific Western and Century U., United Writers Assn., Swiss Soc. Writers, Swiss Sect. PEN Club Internat., others. Author: Work Communications, 1966, Programmed Learning, 1966, Production Planning and Control, 1968, Transfer of Technology, 1971, National Electrical Development, 1977, Energy and Development, 1979, El Gran Desafio, 1981, Hacia una politica economica alternativa, 1982, The Power of Communications: The Media, 1987. Contbr. articles to newspapers and jours. Clubs: Club der 200, FCL, Hebraica. Home: 3781 NE 208 Terr North Miami Beach FL 33180

WOLFENSTEIN, LINCOLN, physicist, educator; b. Cleve., Feb. 10, 1923; s. Leo and Anna (Koppel) W.; m. Wilma Caplin, Feb. 3, 1957; children: Frances, Leonard, Miriam. S.B., U. Chgo., 1943, S.M., 1944, Ph.D., 1949. Physicist Nat. Adv. Com. Aeros., 1944-46; mem. faculty dept. physics Carnegie-Mellon U., Pitts., 1948—; asso. prof. Carnegie-Mellon U., 1957-60, prof., 1960-78, Univ. prof., 1978—. Contbr. articles to profl. jours. Guggenheim fellow, 1973, 83; recipient Sakurai Prize for achievement in particle theory, 1992. Mem. Am. Phys. Soc., Fedn. Am. Scientists, AAAS, Nat. Acad. Sci. Office: Carnegie Mellon U Dept of Physics Pittsburgh PA 15213

WOLFF, RALPH GERALD, civil engineer; b. Dayton, Ky., May 28, 1935; s. John Edward and Charlotte Mabel (Krieg) W.; m. Arline Lucille Dixon, June 11, 1960; children: Cynthia Lynn, David Allen, Kevin Michael. BCE, U. Ky., 1958. Registered profl. engr., Ky.; registered land surveyor, Ky. Asst. resident engr. Ky. Dept. Hwys., Erlanger, 1958-61; resident engr. Ky. Dept. Hwys., Cynthiana, 1962-65; dist. maintenance engr. Ky. Dept. Hwys., Covington, 1965-73; dist. ops. engr. Ky. Dept. Transp., Covington, 1973-86; dist. planning engr. Ky. Transp. Cabinet, Covington, 1986—; state coord. Ky. Engring. Exposure Network, 1990—. Active Kenton County Transp. Task Force, Ft. Mitchell, Ky., 1988—; chmn. OKI (Okla. Ky. Ind.) Hazmat Transp. Com., Cin., 1992; OKI Regional Traffic Mgmt. Tech. Comn., 1993—; No. Ky. Intermodal Task Force, Florence, Ky., 1993—. Mem. NSPE, Ky. Soc. Profl. Engrs., Ky. Assn. Transp. Engrs. Episcopalian. Home: 2936 Campus Dr Crestview Hills KY 41017

WOLFF, ROBERT JOHN, biology educator; b. Marquette, Mich., Jan. 22, 1952; s. Lee Stewart and Mary Joyce (Hamel) W.; m. Marcia Lynn Beugel, May 17, 1974. BA, Hope Coll., 1974, MA, Western Mich. U., 1976; PhD, U. Wis., Milw., 1985. Assoc. prof. biology Trinity Christian Coll., Palos Heights, Ill., 1980—; summer faculty AuSable Inst. Environ. Studies, Mancelona, Mich., 1986-87; faculty rsch. participant Argonne (Ill.) Nat. Lab., 1989—; vis. asst. prof. U. Ill., Chgo., 1990; sci. edn. cons. Sch. Dist. 161, Homewood, Ill., 1990—. Author textbook; contbr. numerous rsch. papers to jours., articles to gen. interest publs. Grantee Smithsonian Instn., Washington, 1985, Morton Arboretum, Lisle, Ill., 1985-87; Field assoc. Field Mus. Natural History, Chgo., 1991. Mem. Am. Arachnological Soc. (chair elections com. 1989, 90), Am. Soc. Zoologists (edn. com. 1990—), Associated Colls. of Chgo. Area (chair biology div. 1983-84, 88-89), Sigma Xi. Mem. Ref. Ch. in Am. Achievements include research in biology and conservation of spiders and other invertebrates. Home: 29 Sorrento Dr Palos Heights IL 60463-1752 Office: Trinity Christian Coll 6601 W College Dr Palos Heights IL 60463-0929

WOLFF, SHELDON, radiobiologist, educator; b. Peabody, Mass., Sept. 22, 1928; s. Henry Herman and Goldie (Lipchitz) W.; m. Frances Faye Farbstein, Oct. 23, 1954; children: Victor Charles, Roger Kenneth, Jessica Raye. B.S. magna cum laude, Tufts U., 1950; M.A., Harvard U., 1951, Ph.D., 1953. Teaching fellow Harvard U., 1951-52; sr. research staff biology div. Oak Ridge Nat. Lab., 1953-66; prof. cytogenetics U. Calif., San Francisco, 1966—, dir. Lab. Radiobiology and Environ. Health, 1983—; vis. prof. radiation biology U. Tenn., 1962, lectr., 1953-65; cons. several fed. sci. agys.; chmn. U.S. Dept. Energy's Health and Environ. Rsch. Adv. Com., 1987—; co-chmn. Joint NIH/Dept. Energy Subcom. on Human Genome, 1989—. Editor: Chromosoma, 1983—; assoc. editor: Cancer Research, 1983—; Editorial bd.: Radiation Research, 1968-72, Photochemistry and Photobiology, 1962-72, Radiation Botany, 1964-86, Mutation Research, 1964—, Caryologia, 1967—, Radiation Effects, 1969-81, Genetics, 1972-85; Contbr. articles to sci. jours. Recipient E.O. Lawrence meml. award U.S. AEC, 1973. Mem. Genetics Soc. Am., Radiation Research Soc. (counselor for biology 1968-72, Failla lectr. 1992, medal 1992), Am. Soc. Naturalists, Am. Soc. Cell Biology, Environmental Mutagen Soc. (council 1972—, pres. 1980-81, award 1982), Internat. Assn. Environ. Mutagen Socs. (treas. 1978-85), Sigma Xi. Democrat. Home: 41 Eugene St Mill Valley CA 94941-1717 Office: U Calif Lab Radiobiology San Francisco CA 94143-0750

WOLFF, SIDNEY CARNE, astronomer, observatory administrator; b. Sioux City, Iowa, June 6, 1941; d. George Albert and Ethel (Smith) Carne; m. Richard J. Wolff, Aug. 29, 1962. BA, Carleton Coll., 1962, DSc (hon.), 1985; PhD, U. Calif., Berkeley, 1966. Postgrad. research fellow Lick Obs, Santa Cruz, Calif., 1969; asst. astronomer U. Hawaii, Honolulu, 1967-71, assoc. astronomer, 1971-76; astronomer, assoc. dir. Inst. Astronomy, Honolulu, 1976-83, acting dir., 1983-84; dir. Kitt Peak Nat. Obs., Tucson, 1984-87, Nat. Optical Astronomy Observatories, 1987—; dir. Gemini Project Gemini 8-Meter Telescopes Project, 1992—. Author: The A-Type Stars--Problems and Perspectives, 1983, (with others) Exploration of the Universe, 1987, Realm of the Universe, 1988, Frontiers of Astronomy, 1990; contbr. articles to profl. jours. Trustee Carleton Coll., 1989—. Research fellow Lick Obs. Santa Cruz, Calif., 1967. Mem. Astron. Soc. Pacific (pres. 1984-86, bd. dirs. 1979-85), Am. Astron. Soc. (coun. 1983-86, pres.-elect 1991, pres. 1992-94). Office: Nat Optical Astronomy Obs PO Box 26732 950 N Cherry Ave Tucson AZ 85726

WOLFFE, ALAN PAUL, molecular embryologist, molecular biologist; b. Burton-on-Trent, Staffordshire, Eng., June 21, 1959; s. Ronald and Mildred (Hasbury) W.; m. Elizabeth Jane Hall, Aug. 21, 1982. BA with honors in Biochemistry, Oxford U., 1981; PhD, MRC London, 1984. Postdoctoral fellow Carnegie Instn., Balt., 1984-86, prin. investigator, 1987; prin. investigator, asst. prof. Nat. Inst. Diabetes and Digestive and Kidney Diseases, Bethesda, Md., 1988-90; assoc. prof. Nat. Inst. Diabetes and Digestive and Kidney Diseases, Bethesda, 1990; chief lab. molecular embryology Nat. Inst. Child Health and Human Devel., Bethesda, 1990—, prof., 1992—; ad hoc reviewer NSF, 1989-92, molecular biology study sect. NIH, 1990; mem. biologyical scis. study sect. NIH, 1991—; mem. biochemistry program panel NSF, 1992-93; chmn. meeting Chromatin Structure and Transcription Fedn. Am. Socs. Exptl. Biology, 1993. Author: Chromatin: Structure and Function, 1992; contbr. articles. Fellow Rsch. Coun. Predoctoral Rsch., 1981-84, European Molecular Biology Orng., 1984-86; grantee Am. Cancer Soc., 1987, Fogarty Internat. Ctr., 1991, NIH, 1992, NSF, 1993. Mem. AAAS, Am. Assn. Cancer Rsch., Am. Soc. Microbiology, Am. Soc. Cell Biology, Brit. Soc. Devel. Biology, Biochemical Soc. (U.K.). Achievements include major contributions to elucidating role of nucleic acid packaging proteins (histones for DNA, Y-box proteins for RNA) in regulating gene expression (transcription of DNA, translation of RNA). Office: Nat Inst Child Health & Human Devel Molecular Embryol Lab Rm B1A13 9000 Rockville Pike Bldg 6 Bethesda MD 20892

WOLFINGER, BERND EMIL, mathematics and computer science educator; b. Pforzheim, Baden, Germany, Feb. 5, 1951; s. Emil Gottlob and Lore Wilhelmine (Waibel) W.; m. Gabriele Margret Maria Bodamer, July 19, 1974; children: Sascha J., Susanne S.A. Student, U. Claude-Bernard, Lyon, France, 1974; diploma in math., U. Karlsruhe, Germany, 1975; PhD, U. Karlsruhe, 1979. Mem. sci. staff Nuclear Rsch. Ctr., Karlsruhe, 1975-80; asst. prof. dept. computer sci. U. Karlsruhe, 1981; prof. U. Hamburg, Germany, 1981—; scientist T.J. Watson Rsch. Ctr., IBM, Yorktown Heights, N.Y., 1985, Internat. Computer Sci. Inst., Berkeley, Calif., 1991; cons. computer and comm. systems; rschr. in field. Contbr. articles to profl. publs.; editor books in field. Mem. IEEE, Assn. Computing Machinery, Gesellschaft fuer Informatik. Avocations: classical music, skiing, hiking, tennis. Home: Ahornweg 98, D-25469 Halstenbek Germany Office: Hamburg U Dept Computer Sci, Vogt-Koelln Str 30, D-22527 Hamburg Germany

WOLFSON, LAWRENCE AARON, hospital administrator; b. Chgo., July 11, 1941; s. Norman William and Doris D. (Brownstein) W.; m. Cheryl Jean Vogel, Feb. 6, 1987; children: Marc David, Sara Elizabeth, Aaron Michael, Ryan Anthony, Ashley Michelle. BA in Biology, Ind. U., South Bend, 1973, MBA, 1980. Sales rep. Gen. Med. Corp., South Bend, 1973-75, Hoechst-Roussel Pharmaceuticals, Somerville, N.J., 1975-79; purchasing agt. Simon Bros., Inc., South Bend, 1979-81; purchasing mgr. Ingalls Meml. Hosp., Harvey, Ill., 1981-83; dir. purchasing Community Hosp., Munster, Ind., 1983-86; corp. purchasing mgr. Columbus-Cuneo-Cabrini Med. Ctr., Chgo., 1986-88; materials mgmt. cons. South Western Med. Ctr., Chgo., 1988-91; asst. dir. materials mgmt. Michael Reese Hosp. and Med. Ctr., Chgo., 1989-91; dir. material mgmt. Regional Med. Ctr. at Memphis, 1991—; mem. editorial bd. Hosp. Material Mgmt. Quarterly, Aspens Pubs. Editorial bd. Hosp. Material Mgmt. Quar. Cubmaster Cub Scouts, South Bend, 1976-78. With USN, 1961-71. Mem. Am. Soc. Hosp. Materials Mgmt., Healthcare Materials Mgmt. Soc. (regional rep. 1984), Nat. Assn. Purchasing Mgmt., Am. Soc. Clin. Pathologists (affiliate), Am. Legion, B'nai Brith (pres. 1980-81). Jewish. Office: Regional Med Ctr at Memphis 877 Jefferson Ave Memphis TN 38103-2897

WOLINSKY, STEVEN MARK, infectious diseases physician, educator; b. New Havewn, Apr. 17, 1953; s. Martin and Anita (Caplan) W.; m. Lorrayne Hinda Stein; children: Samantha Anne, David Ian. BA, Case Western Res. U., 1975; MD, U. Conn., Farmington, 1979. Intern and resident Northwestern U. Med. Sch., Chgo., 1979-82; instr. in medicine U. Rochester, N.Y., 1982-87, Strong Meml. Hosp., Rochester, 1982-87; asst. prof. medicine VA Lakeside Hosp., Chgo., 1987-91, Northwestern Meml. Hosp., Chgo., 1987-93; assoc. prof. medicine and infectious diseases Northwestern U. Med. Sch., Chgo., 1993—. Contbr. articles to profl. publs. Fellow Nat. Found. Infectious Diseases, 1984, Wilmot Cancer Rsch. Found., 1984-87. Mem. AAAS, Am. Soc. Microbiology, N.Y. Acad. Scis., Infectious Disease Soc. Am. Achievements include discovery of polymerase chain reaction for the detection of human papilloma virus, polymerase chain reaction driven in situ hybridization with flow cytometry, perinatal human immunodeficiency virus transmission. Home: 1055 Hohlfelder Glencoe IL 60022 Office: Northwestern U Med Sch Olson 8427 710 N Fairbanks Ct Chicago IL 60611

WOLK, MARTIN, electronic engineer, physicist; b. Long Branch, N.J., Jan. 13, 1930; s. Michael and Tillie (Barron) W.; 1 child, Brett Martin. BS, George Washington U., 1957, MS, 1968; PhD, U. N.Mex., 1973. Physicist Naval Ordnance Lab., White Oak, Md., 1957-59, Nat. Oceanic and Atmospheric Adminstrn., Suitland, Md., 1959-66; solid state physicist Night Vision Lab., Fort Belvoir, Va., 1967-69; rsch. asst. U. N.Mex., Albuquerque, 1969-73; electronics engr. Washington Navy Yard, 1976-83, TRW, Inc., Redondo Beach, Calif., 1983-84; physicist Metrology Engring. Ctr., Pomona, Calif., 1984-85; electronics engr. Naval Aviation Depot North Island, San Diego, 1985—; cons. Marine Corps Logistics Base, Barstow, Calif., 1985—, Naval Weapons Station, Fallbrook, Calif., 1987-89, Naval Weapons Support Ctr., Crane, Ind., 1989—. Contbr. articles to Jour. Quantitative Spectroscopy and Radiative Transfer, Monthly Weather Rev., Proceedings of SPIE. Cpl. U.S. Army, 1946-49, Japan. Mem. IEEE, Soc. Photo-Optical Instrumentation Engring., Sigma Pi Sigma, Sigma Tau. Achievements include development of first Tiros meteorological satellites; research on electron-beam for micro-circuit device fabrication. Home: 740-91 Eastshore Ter Chula Vista CA 91913-2421

WOLYNES, PETER GUY, chemistry researcher, educator; b. Chgo., Apr. 21, 1953; s. Peter and Evelyn Eleanor (Etter) W.; m. Jane Lee Fox, Nov. 26, 1976 (div. 1980); m. Kathleen Cull Bucher, Dec. 22, 1984; children: Margrethe Cull, Eve Cordelia. AB with highest distinction, Ind. U., 1971; AM, Harvard U., 1972, PhD in Chem. Physics, 1976; DSc (hon.), Ind. U., 1988. Research assoc. MIT, Cambridge, 1975-76; asst. prof., assoc. prof. Harvard U., Cambridge, 1976-80; vis. scientist Max Planck Inst. für Biophysikalische Chemie, Gottingen, Fed. Republic Germany, 1977; assoc. prof. chemistry U. Ill., Urbana, 1980-83, prof., 1983—, prof. physics, 1985—; prof. physics and biophysics U. Ill., 1989—; permanent mem. Ctr. for Advanced Study U. Ill., Urbana, 1989—; vis. prof. Inst. for Molecular Sci., Okazaki, Japan, 1982, 87; vis. scientist Inst. for Theoretical Physics, Santa Barbara, Calif., 1987, Ecole normale Supérieure, Paris, 1992, Merski lectr. U. Nebr., 1986, Denkewalter lectr., Loyola U., 1986. Contbr. numerous articles to profl. jours. Mem. Ill. Alliance To Prevent Nuclear War, Champaign, 1981—. Sloan fellow, 1981-83, J.S. Guggenheim fellow, 1986-87; Beckman assoc. Ctr. for Advanced Study, Urbana, 1984-85. Fellow Am. Phys. Soc., Am. Acad. Arts and Scis.; mem. NAS, Am. Chem. Soc. (Pure Chemistry award 1986), AAAS, N.Y. Acad. Scis., Phi Beta Kappa, Sigma Xi, Phi Lambda Upsilon (Fresenius award 1988), Sigma Pi Sigma, Alpha Chi Sigma. Home: 311 W Oregon St Urbana IL 61801-4125 Office: U Ill Sch of Chem Scis 505 S Mathews Ave Urbana IL 61801-3664

WOLYNIC, EDWARD THOMAS, specialty chemicals technology executive; b. Bklyn., May 29, 1948; s. Edward Joseph and Fortunata Wolynic; m. Loraine Cynthia Ciardullo. BS ChemE, Poly. Inst. N.Y., 1969; MS ChemE, Princeton U., 1971, PhD ChemE, 1974. Staff engr. Union Carbide Corp., Bound Brook, N.J., 1974-75; supr. Union Carbide Corp., Tarrytown, N.Y., 1975-77, mgr. mfg. tech., 1977-82, assoc. dir. tech., 1982-85, dir. tech., 1985-88; v.p. rsch. UOP, Tarrytown, 1988-90; v.p. devel. UOP, Des Plaines, Ill., 1990-92; dir. process R&D Internat. Specialty Products Corp., Wayne, N.J., 1993—. Contbr. tech. articles to profl. jours.; patentee in field. Mem. Am. Inst. Chem. Engrs., Indsl. Rsch. Inst., Com. Devel. Assn., N.Am. Catalogue Soc. Avocations: fishing, tennis, skiing. Office: Internat Specialty Products ISP Bldg 1 1361 Alps Rd Wayne NJ 07470

WOMACK, TERRY DEAN, electrical engineer; b. Enid, Okla., Oct. 15, 1953; s. Don Lee and Rosetta Ella (McDowell) W.; divorced; 1 child, Deborah Ann. BSEE, U. Okla., 1976; MBA, Okla. City U., 1985. Registered profl. engr., Okla. Field engr. Schlumberger Well Svcs., Duncan, Okla., 1976-78; engr. Overhead Door S.W. Okla., Lawton, 1978; engr. Halliburton Svcs. Elec. Rsch., Duncan, 1978-80, sr. engr., 1980-82, devel. engr., 1982-83, group supr., 1983-90, sect. supr., 1990—; mem. workstation tech. steering com. Halliburton Co., Dallas, 1990-92, software devel. strategy coun., 1990; chmn. Duncan Telecommunication Commn., 1993—. Chmn. Red River chpt. ARC, Duncan, 1991; chmn. fin. com. 1st United Meth. Ch., Duncan, 1988-90; mem. Jaycees, Duncan, 1991—. Mem. Okla. Soc. Profl. Engrs. (chpt. pres. 1990-91, chpt. Outstanding Engr. 1989), IEEE, Soc. Petroleum Engrs. (econ. and eval. com. 1992—), Internat. Assn. Drilling Contractors (rig floor instrumentation subcom. 1988-92), Elks, AMBUCS, Omicron Delta Kappa. Republican. Achievements include patent in apparatus equipotential housing and plurality of focussed current electrodes for electrically logging a well formation at more than one lateral distance from a borehole. Home: PO Box 750056 Duncan OK 73575-0056 Office: Halliburton Svcs PO Box 1431 Duncan OK 73536-0001

WOMERSLEY, WILLIAM JOHN, physicist; b. Torquay, Devon, Eng., Sept. 22, 1962; came to U.S., 1986; s. Leslie and Patricia (Cox) W. BA, Cambridge (Eng.) U., 1983, MA (hon.), 1987; DPhil, Oxford (Eng.) U., 1986. Rsch. assoc. U. Fla., Gainesville, 1986-89; asst. prof. Fla. State U., Tallahassee, 1989-92; scientist SSC Lab., Dallas, 1993—. Contbr. articles to profl. publs. Mem. Am. Phys. Soc., Brit. Assn. for Advancement of Sci., Fermilab Users' Orgn., SSC Lab. User's Orgn. Achievements include analysis of direct photon production in proton-antiproton collisions using the DØ detector; design of "gem" detector. Office: SSC Lab MS 2005 2550 Beckleymeade Ave Dallas TX 75237

WON, IHN-JAE, geophysicist; b. Yokohama, Japan, Nov. 28, 1943; came to U.S., 1969; s. Y.K. and S.B. W.; m. Susan Mary Thome, Aug. 21, 1971; children: Eugene, Lianne, Henry. BS, Seoul Nat. U., 1967; MS, Columbia U., 1971, PhD, 1973. Registered profl. geologist, N.C. Lt. Korean Army, 1967-69; seismic data interpreter Western Geophys. Co., Houston, 1970; grad. rsch. and teaching asst. Columbia U., N.Y.C., 1969-73; rsch. assoc. Lamont-Doherty Geol. Observatory and Henry Krumb Sch. Mines, N.Y.C., 1973-76; vis. scientist Earthquake Rsch. Inst., Tokyo, Japan, 1981; geophysicist Naval Ocean Rsch. & Devel. Activity Nat. Sci. Lab., Bay St. Louis, Miss., 1984-85; asst. prof. geophysics Dept. Marine, Earth & Atmospheric Sci., N.C. State U., Raleigh, 1976-79, assoc. prof. geophysics, 1979-86, prof. geophysics, 1986-89; pres. Geophex, Ltd., Raleigh, 1983—; cons. in field. Contbr. articles to Jour. Geophys. Rsch., Geophysics, Mining Engring. Jour., Geophys. Rsch. Letter, Marine Geotechnology, IEEE Jour. Ocean Engring. Grantee NSF, NASA and others. Mem. AAAS, Am. Geophys. Union, Soc. Exploration Geophysicists, Carolina Geol. Soc., Internat. Assn. Engring. Geologists, Soc. Mining Engrs. Achievements include patents in Torsional Shearwave Generator; Frequency-domain Airborne Electromagnetic Bathymetry Method. Office: Geophex Ltd 605 Mercury St Raleigh NC 27603-2343

WONG, CHEUK-YIN, physicist; b. Kwangtung, China, Apr. 28, 1941; came to U.S., 1959; s. Hong-Yu and Sau-King (Li) W.; m. Jeanne Pei-Hwa Yang, Feb. 12, 1966; children: Janet, Albert, Lisa. AB, Princeton U., 1961,

PhD, 1966. Rsch. physicist Oak Ridge (Tenn.) Nat. Lab., 1966-68, 70-82, 83-86, sr. physicist, 1986—; rsch. fellow Niels Bohr Inst., Copenhagen, Denmark, 1968-69; vis. scientist MIT, Cambridge, 1982-83; vis. prof. Inst. Nuclear Study, Tokyo, 1988. Contbr. articles to sci. jours. Bd. dirs. E. Tenn. chpt. Orgn. Chinese Ams., Knoxville, 1984-88, pres., 1992. Recipient Publ. award Martin Marietta Energy Systems Inc., 1986. Fellow Am. Phys. Soc.; mem. Princeton Alumni Assn. (chmn. alumni schs. com. East Tenn. 1981-92), Energy Capital Toastmasters Club (adminstrv. v.p. 1991-92), Overseas Chinese Physicists Assn. (life, regional coord. 1992). Achievements include rsch. in theoretical nuclear physics, nuclear shell effects, nuclear fusion cross sections, nuclear shock waves, extension of the time-dependent mean-field approximation, nuclear stopping power at very high energies. Home: 1043 W Outer Dr Oak Ridge TN 37830-8634 Office: Oak Ridge Nat Lab Physics Div Bldg 6003 PO Box 2008 Oak Ridge TN 37831-6373

WONG, CHI-HUEY, chemistry educator; b. Taiwan, Aug. 13, 1948; came to U.S., 1979; m. Yieng-Lii, Mar. 26, 1975; children: Heather, Andrew. BS in Biochemistry, Nat. Taiwan U., 1970, MS in Biochemistry, 1977; PhD in Chemistry, MIT, 1982. Asst. rsch. fellow Inst. Biol. Chemistry Academia Sinica, Taipei, Taiwan, 1974-79; from asst. prof. to prof. chemistry Tex. A&M U., 1983-89; Ernest W. Hahn prof. chemistry Scripps Rsch. Inst., La Jolla, Calif., 1989—; cons. Miles Labs., 1985-88, Dow Chem., 1985-90, G. D. Searle, 1988-90; sci. advisor Anylin, San Diego, 1989—, Cytel, 1990—, Enzymatics, 1990—; head lab. glycosci. frontier rsch. program Riken, Japan, 1991—. Author: Enzymes in Syntheitic Organic Chemistry, 1991; contbr. over 120 articles on enzymatic organic synthesis and bio-organic chemistry; editorial bd. Bicatalysis. Lt. Taiwan Army, 1970-71. Recipient Presdl. Young Investigator in Chemistry award NSF, Washington, 1986; Searle scholar, 1985. Fellow AAAS, Am. Inst. Chemists; mem. Am. Chem. Soc. (Arthur C. Cope scholar award 1993), Am. Soc. Biochemistry and Molecular Biology, N.Y. Acad. Sci. Achievements include 9 patents for enzymatic synthesis; research in bioorganic chemistry, rational design and synthesis of enzyme inhibitors. Home: 13445 Grandvia Pt San Diego CA 92130-1030 Office: Scripps Rsch Inst Dept Chemistry 10666 N Torrey Pines Rd La Jolla CA 92037-1027

WONG, CHING-PING, chemist; b. Canton, China, Mar. 29, 1947; came to U.S., 1966; s. Kwok-Keung and Yun-Kwan (Lo) W.; m. Lorraine Homnack, May 27, 1978; children: Michelle, David. BS in Chemistry, Purdue U., 1969; PhD in Organic and Inorganic Chemistry, Pa. State U., 1975. Postdoctoral scholar Stanford (Calif.) U., 1975-77; mem. rsch. staff AT&T Bell Labs., Princeton, N.J., 1977-82, sr. mem. tech. staff, 1982-87, disting. mem. tech. staff, 1987—; program chmn. 39th Electronic Components Conf., 1989; gen. chmn. 41st Electronic Components and Tech. Conf., 1991; bd. govs. IEEE-Components, Hybrids and Mfg. Tech. Soc., 1987-89, tech. v.p., 1990-91, pres., 1992-93. Author, editor: Polymers for Electronic and Photonic Applications, 1993; contbr. articles to profl. jours. Recipient Outstanding papers and Contbns. award IEEE-Components, Hybrids and Mfg. Tech. Soc., 1991. Recipient Outstanding papers and Contbns. award IEEE-Componenets, Hybrids and Mfg. Tech. Soc., 1991, AT&T Bell Labs Fellow award, 1992. Achievements include over 25 U.S. and numerous internat. patents for integrated device passivation and encepsulation area; pioneer in application of polymers for device reliability without humeticity-a new application on electronic device packaging. Home: 11 Wexford Dr Lawrenceville NJ 08648 Office: AT&T Bell Labs PO Box 900 Princeton NJ 08540

WONG, CHI-SHING, chemical oceanographer; b. Hong Kong, Sept. 1, 1934; s. Shiu-ming and Bing Yu (Kwong) W.; m. Shau-King Emmy Leung, Mar. 5, 1960; children: Adriana Hai-Hua, Calvin Ja-Hua. BSc, U. Hong Kong, 1957, BSc in Chemistry with honours, 1958, MSc in Chemistry, 1961; PhD in Chem. Oceanography, UCSD-Scripps Inst Oceanography, San Diego, 1968. Asst. rsch. officer Fisheries Rsch. Unit, H.K.U., Hong Kong, 1958-60; demonstrator chem. dept. U. Hong Kong, 1960-61; head chem. dept. United Coll., Hong Kong, 1961-62; scientist-in-charge weathership chemistry program Dept. Energy, Mines & Resources, Nanaimo, B.C., Can., 1968-71; head ocean chemistry Dept. Fisheries & Oceans, Sidney, B.C., 1971-89, head Ctr. for Ocean Climate Chemistry, 1989—; chmn. SCOR Working Group on Oceanic CO2, 1982-91; mem. CO2 Panel, 1986—. Chief editor: Trace Metals in Sea Water, 1983, Marine Ecosystem Enclosed Experiments, 1992. Hon. prof. Third Oceanog. Inst., Xiamen, China, 1985. Fellow Royal Soc. Chemistry (U.K.), Chem. Inst. Can.; mem. Can. Meteorol. & Oceanog. Soc., Am. Geophys. Union, Am. Soc. Limnology & Oceanography, Am. Assoc. for the Advancement of Science (AAAS Newcomb Cleveland Prize, 1991-92). Achievements include use of decadal change in C-13 isotope in the ocean to infer that the ocean is a major sink of fossil-fuel CO2; wood burning as a major contribution to atmospheric CO2; notable findings on El Nino effects on air-sea CO2 exchangeand ocean productivity; use of lead isotopes to trace pollution in coastal water, and to trace ocean circulation; findings on marine pollution by oil and metals. Office: Ctr for Ocean Climate Chemistry, Institute of Ocean Sciences, Sidney, BC Canada V8L 4B2

WONG, DAVID CHUNGYAO, scientist; b. Hong Kong, Sept. 26, 1941; came to U.S., 1962; s. Harry J. and Hungar (Chen) W.; m. Kate C. Chen, Sept. 23, 1971; children: Anselm, Harry. B Liberal Arts, Boston U., 1966; MS, U. Mass., 1968, PhD, 1975. Dept. mgr. Arco Solar, Inc., Chatsworth, Calif., 1980-83, sci. advisor, 1983-87; prin. scientist BTU Internat., North Billerica, Mass., 1989—; tech. cons. GPS, Inc., Chatsworth, 1987-88. Recipient Cert. of Recognition for tech. innovation NASA Inventions and Contbns. Bd., 1988. Republican. Achievements include research in liquid dopant diffusion by exlinear laser as drive-in source, dry chem. vapor reposition of titanium dioxide films for solar cell application. Office: BTU Internat 23 Esquire Rd North Billerica MA 01862

WONG, DAVID T., biochemist; b. Hong Kong, Nov. 6, 1935; s. Chi-Keung and Pui-King W.; m. Christina Lee, Dec. 28, 1963; children: Conrad, Melvin, Vincent. Student, Nat. Taiwan U., 1955-56; BS, Seattle Pacific U., 1961; MS, Oreg. State U., 1964; PhD, U. Oreg., 1966. Postdoctoral fellow U. Pa., Phila., 1966-68; sr. biochemist Lilly Rsch. Labs., Indpls., 1968-72, rsch. biochemist, 1973-77, sr. rsch. scientist, 1978-89; rsch. advisor, 1990—; adj. prof. biochemistry and molecular biology Ind. U. Sch. Medicine, 1986—, adj. prof. neurobiology, 1991—. Contbr. numerous articles to sci. jours. Alumnus of Growing Vision Seattle Pacific U., 1989. Recipient Scientist of Yr. award, Pres. award Chinese Neuroscience Soc., 1991, Discoverers award Pharm. Mfr. Assn., 1993. Mem. Am. Soc. Pharmacology and Exptl. Therapeutics, Internat. Soc. Neurochemistry, Am. Soc. Neurochemistry, Soc. Neurosci. (pres. Indpls. chpt. 1987, 88), Soc. Chinese Bioscientists in Am., N.Y. Acad. Scis., Indpls. Assn. Chinese Ams. (pres. 1987), Sigma Xi. Rsch. on biochemistry and pharmacology of neurotransmission; discovery and development of new type of antidepressant drug, Prozac (Fluoxetine) and Dapoxetine, selective inhibitors of serotonin uptake, Tomoxetine, a selective inhibitor of norepinephrine uptake; duloxetine an inhibitor for uptake of serotonin and norepinephrine; studies of potentially useful substances which activate transmission of norepinephrine, dopamine, serotonin, acetylcholine and GABA-neurons; studies of natural products led to the discovery of carboxylic ionophores: Narasin, A28695 and A204, which increase transport of cations across biomembranes. Office: Lilly Rsch Labs Eli Lilly and Co Lilly Corp Ctr Indianapolis IN 46285

WONG, DENNIS KA-CHEONG, physician, physical therapist; b. Hong Kong, Hong Kong, Jan. 23, 1954. BA, Columbia U., 1977, cert. in phys. therapy, 1978, MS in Phys. Therapy, 1982; MD, Am. U. of the Caribbean, Montserrat, Brit. West Indies, 1988. Intern internal medicine dept. SUNY Health Sci. Ctr., Bklyn., 1988-89; Nat. Inst. on Disability and Rehab. Rsch. fellow Harvard-MIT Rehab. Engring. Ctr., Cambridge, Mass., 1989-90; resident rehab. medicine dept Kingsbrook Jewish Med. Ctr., Bklyn., 1990-92, chief resident rehab. medicine dept., 1992—. Mem. Am. Acad. Phys. Medicine and Rehab.

WONG, JAMES BOK, economist, engineer, technologist; b. Canton, China, Dec. 9, 1922; came to U.S., 1938, naturalized, 1962; s. Gen Ham and Chen (Yee) W.; m. Wai Ping Lim, Aug. 3, 1946; children: John, Jane Doris, Julia Ann. BS in Agr., U. Md., 1949, BS in Chem. Engring., 1950; MS, U. Ill., 1951, PhD, 1954. Rsch. asst. U. Ill., Champaign-Urbana, 1950-53; chem. engr. Standard Oil of Ind., Whiting, 1953-55; process design engr., rsch. engr. Shell Devel. Co., Emeryville, Calif., 1955-61; sr. planning engr., prin. planning engr. Chem. Plastics Group, Dart Industries, Inc. (formerly Rexall Drug & Chem. Co.), L.A., 1961-66, supr. planning and econs., 1966-67, mgr. long range planning and econs., 1967, chief economist, 1967-72, dir. econs. and ops. analysis, 1972-78, dir. internat. techs., 1978-81; pres. James B. Wong Assocs., L.A., 1981—; chmn. bd. dirs. United Pacific Bank, 1988—; tech. cons. various corps. Contbr. articles to profl. jours. Bd. dirs., pres. Chinese Am. Citizens Alliance Found.; mem. Asian Am. Edn. Commn., 1971-81. Served with USAAF, 1943-46. Recipient Los Angeles Outstanding Vol. Service award, 1977. Mem. Am. Inst. Chem. Engrs., Am. Chem. Soc., VFW (vice comdr. 1959), Commodores (named to exec. order 1982), Sigma Xi, Tau Beta Pi, Phi Kappa Phi, Pi Mu Epsilon, Phi Lambda Upsilon, Phi Eta Sigma. Home: 2460 Venus Dr Los Angeles CA 90046

WONG, KENNETH LEE, software engineer, consultant; b. L.A., Aug. 15, 1947; s. George Yut and Yue Sam (Lee) W.; m. Betty (Louie) Wong, June 29, 1975; children: Bradford Keith, Karen Beth. BS in Engring., UCLA, 1969, MS in Engring., 1972, postgrad., 1972-73, 76-78. Cert. community coll. instr., Calif. Engring. aide Singer Librascope, Glendale, Calif., 1972-73; computer system design engr. Air Force Avionics Lab., Wright-Patterson AFB, Ohio, 1973-75; mem. tech. staff Hughes Aircraft Co., various cities, Calif., 1976-78, 79-81, TRW Def. and Space Systems Group, Redondo Beach, Calif., 1975-76, 78-79; engring. specialist Northrop Corp., Hawthorne, Calif., 1981-84; mem. tech. staff Jet Propulsion Lab., Pasadena, Calif., 1984-87; software cons. EG&G Spl. Projects, Las Vegas, Nev., 1987, AT&T Bell Labs., Warren, N.J., 1987-88, Westinghouse Electric Corp., Linthicom, Md., 1988, E Systems, Inc., Greenville, Tex., 1988-89; prin. Wong Soft Works, L.A., 1989—. Author tech. reports. Coach, Tigers Youth Club, L.A. 1st lt. USAF, 1973-75. Mem. AIAA, IEEE, Assn. Computing Machinery, Upsilon Pi Epsilon. Republican. Avocations: basketball, photography. Home and Office: Wong Soft Works 3385 Mclaughlin Ave Los Angeles CA 90066-2004

WONG, KWEE CHANG, chemist; s. Liu Hain and Kyin Chee (Hoo) W.; m. Yi Liang, 1981; 1 child, Jane. BSc, U. Rangoon, Burma, 1965. Plating chemist Chamberlain Mfg. Corp., Clinton, Iowa, 1971-75; chief chemist Bonewitz Chem. Svcs., Burlington, Iowa, 1975-79; prin. chemist Dart Industries, Youngstown, Ohio, 1979-83; rsch. chemist Great Lakes Chem. Corp., Costa Mesa, Calif., 1983-87; rsch. assoc. Dexter Corp., Industry, Calif., 1987—. Fellow Am. Inst. Chemists; mem. Am. Chem. Soc., Royal Soc. Chemistry. Achievements include patents in Metal Etching and Dissolution, Wastewater Treatmment for Metal Finishing.

WONG, PATRICK TIN-CHOI, neuroscientist; b. Hong Kong, Mar. 9, 1942; came to U.S., 1960; s. Fook and Chek-Hung (Lee) W.; m. Vivian Wei-Ling Wang, Mar. 25, 1967; children: Patricia, Gerald. BA, U. Calif., Berkeley, 1964; MS, Oreg. State U., 1966, PhD, 1969. Rsch. scientist City of Hope Med. Ctr., Duarte, Calif., 1969-82; sr. rsch. scholar U. So. Calif. Med. Ctr., L.A., 1982-83; scientist, mgr. Newport Corp., Irvine, Calif., 1984—; adv. bd. mem. Devel. Biology Ctr. U. Calif., Irvine, 1987—. Contbr. articles to profl. jours. Mem. Neurosci. Soc. Democrat. Office: Newport Corp 1791 Deere Ave Irvine CA 92714

WONG, RAPHAEL CHAKCHING, biotechnologist, executive; b. Hong Kong, May 3, 1948; came to U.S., 1969; s. Tsai Ko and Man Im Wong; m. Baguio Chien, Aug. 4, 1973; children: Angela, Anthony. BS, U. Wis., 1973, MS, 1976; MBA, U. Calif., Irvine, 1990. Sr. assoc. scientist Miles Labs., Elkhart, Ind., 1976-79; chemist Syva Labs., Palo Alto, Calif., 1979-81; sr. chemist Bio-Rad Labs., Richmond, Calif., 1981, Immutron Inc., Newport Beach, Calif., 1982; v.p. Innometrics, Inc., Santa Ana, Calif., 1982-83; v.p. Bioprobe Internat., Inc., Tustin, Calif., 1983, pres., 1986-88, 90-91, vice-chmn., 1988-90; v.p. UniSyn Techs., Inc., Tustin, 1991; pres. Conrex Pharm. Corp., Malvern, Pa., 1992—. Contbr. articles to Clin. Chemistry, Internat. Jour. Biochemistry. Mem. AAAS, Am. Chem. Soc., Am. Assn. Clin. Chemistry. Achievements include patents for enzyme immunoassay and affinity columns. Office: Conrex Pharm Corp 18 Great Valley Pkwy Ste 120-A Malvern PA 19355

WONG, SUE SIU-WAN, health educator; b. Hong Kong, Apr. 6, 1959; came to U.S., 1966; d. Tin Ho and Yuet Kum (Chan) E. BS, UCLA, 1981; MPH, Loma Linda (Calif.) U., 1990; postgrad., Loma Linda U., 1991—. Cert. health edn. specialist. Asst. to the dir. Project Asia Campus Crusade for Christ, San Bernardino, Calif., 1982-83, Campus Crusade for Christ-Internat. Pers., San Bernardino, 1983-90; health educator San Bernardino County Pub. Health, 1990-92; community lab. instr., rsch. asst. dept. health promotion and edn. Loma Linda (Calif.) U. Sch. Pub. Health, 1992—; rsch. asst., community lab. instr. Sch. Pub. Health Loma Linda U., 1992—. Mem. Minority Health Coalition, San Bernardino, 1990—, Com. for the Culturally Diverse, San Bernardino, 1990—; vol. Am. Cancer Soc.; chair Gt. Am. Smokeout, Inland Empire, 1991. Hulda Crooke scholar Loma Linda U., 1989; named Outstanding Young Woman of Yr., 1983; recipient Am. Cancer Soc. (Calif.) Rose award, 1991, Am. Cancer Soc. (nat.) Gaspar award, 1991. Mem. APHA, Nat. Coun. for Internat. Health, Soc. Pub. Health Edn., Loma Linda U. Alumni Assn. Avocations: travel, reading, volleyball, calligraphy, music.

WONG, THOMAS TANG YUM, engineering educator; b. Hong Kong, July 27, 1952; came to U.S., 1976; s. Kwai Sun and Yee Yuen (Fung) W.; m. Min-i Lee, June 9, 1984; 1 child, Clara Joyce. BSc in Engring., U. Hong Kong, 1975; MS, Northwestern U., Evanston, Ill., 1978, PhD, 1981. Product engr. Motorola Semiconductor, Inc., Hong Kong, 1975-76; teaching asst. Northwestern U., 1976-78, rsch. asst., 1978-80, postdoctoral fellow, 1980-81; asst. prof. Ill. Inst. Tech., Chgo., 1981-86, assoc. prof., 1986—, dir. grad. program dept. elec. engring., 1987—; chmn. Chicagoland Microwave Symposium, 1988; cons. to pvt. industry, 1981—. Author: Fundamentals of Distributed Amplification, 1993; contbr. articles to profl. jours.; book reviewer tech. publs. GE fellow, 1983; rsch. grantee NASA, 1989-91, U.S. Dept. Energy, 1992—, pvt. industry, 1993—. Mem. IEEE (chmn. joint Chgo. chpt. Antenna Propagation and Microwave Theory Techniques Soc. 1987-88, mem. steering com. joint symposium Antennas Propagation Soc./Internat. Union of Radio Sci./Nuclear Electromagnetic Pulse 1992), AAUP, Am. Phys. Soc., Tau Beta Pi, Eta Kappa Nu. Office: Ill Inst Tech Dept Elec/Computer Engring Chicago IL 60616

WONG, TUCK CHUEN, chemist, educator; b. Canton, Guangdong, China, Mar. 28, 1946; came to U.S., 1969; s. Wai King and Yun Tuck (Yuen) W.; m. Kit Lan Tang, June 17, 1972; children: Ellen, Denise, Peter. BSc with honors, Chinese U. Hong Kong, 1969; MS, U. Mich., 1971, PhD, 1974. Asst. prof. Tufts U., Medford, Mass., 1976-81; assoc. prof. U. Mo., Columbia, 1981-89, prof. chemistry, 1989—, dir. Nuclear Magnetic Resonance facility, 1981—; vis. prof. U. Lund, Sweden, 1987-88; cons. Unilever Labs., Edgewater, N.J., 1989—. Contbr. articles to Jour. Am. Chem. Soc., Jour. Phys. Chemistry, Langmuir, others. Grantee Petroleum Rsch. Fund, 1977, 91, NSF, 1981, 82, 89, 92, Rsch. Corp., 1989. Mem. Am. Chem. Soc. (sect. chair U. Mo. sect. 1990-91), Internat. Soc. Magnetic Resonance. Office: Univ Mo Dept Chemistry Columbia MO 65211

WONG, WALLACE, medical supplies company executive, real estate investor; b. Honolulu, July 13, 1941; s. Jack Yung Hung and Theresa (Goo) W.; m. Amy Ju, June 17, 1963; children: Chris, Bradley, Jeffery. Student, UCLA, 1960-63. Chmn., pres. South Bay Coll., Hawthorne, Calif., 1965-86; chmn. Santa Barbara (Calif.) Bus. Coll., 1975—; gen. ptnr. W B Co., Redondo Beach, Calif., 1982—; CEO Cal Am. Med. Supplies, Rancho Santa Margarita, Calif., 1986—, Cal Am. Exports, Inc., Rancho Santa Margarita, 1986—, Pacific Am. Group, Rancho Santa Margarita, 1991—, Cal Am. Technics, Rancho Santa Margarita, 1991—; chmn, CEO Alpine, Inc., Rancho Santa Margarita, 1993—; CEO Alpine, Inc., Rancho Santa Margarita, Calif., 1993—; bd. dirs. Metrobank, L.A., 1981—, Correia Art Glass, Santa Monica, Calif., 1984—. Acting sec. of state State of Calif., Sacramento, 1982; founding mem. Opera Pacific, Orange County, Calif., 1985; mem. Hist. and Cultural Found., Orange County, 1986; v.p. Orange County Chinese Cultural Club, Orange County, 1985. Named for Spirit of Enterprise Resolution, Hist. & Cultural Found., Orange Country, 1987; recipient resolution City of Hawthorne, 1973. Mem. Westren Accred Schs. & Colls. (v.p. 1978-79), Magic Castle (life), Singapore Club. Avocations: traveling, skiing. Office: Alpine Inc 23042 Arroyo Vista Rancho Santa Margarita CA 92688

WONG-CHONG, GEORGE MICHAEL, chemical and environmental engineer; b. Port-of-Spain, Trinidad and Tobago, Mar. 21, 1940; came to U.S., 1969; s. Henry and Tai-You Wong Chong; m. Mari-Lou Santiago, June 21, 1967; children: Micheline, Alexine. BSCE, McGill U., 1967; M of Environ. Engring., U. Western Ont., 1969; PhD, Cornell U., 1974. Asst. prof. U. Hawaii, Honolulu, 1974-75; sr. rsch. fellow Carnegie Mellon Rsch. Inst., Pitts., 1975-78; tech. mgr. ERT Inc. (currently ENSR Cons.), Pitts., 1978-85, Baker TSA Inc. (Baker Environ.), Coraopolis, Pa., 1985-91; mgr. remediation processes ICF Kaiser Engrs., Inc., Pitts., 1991—. Contbr. articles to Water Pollution Control Fedn., Water Rsch. Mem. ASCE (contbr. articles to environ. div.), AICE, Am. Acad. Environ. Engrs. (diplomate), Water Environ. Fedn. (rsch. and indsl. waste com.). Democrat. Roman Catholic. Achievements include patent for phys./chem. treatment of mcpl. indsl. waste waters, direct biol. treatment of high strength ammoniacal waste waters. Office: ICF Kaiser Engrs Inc 4 Gateway Ctr Pittsburgh PA 15222

WONNACOTT, JAMES BRIAN, physician; b. Charlottetown, P.E.I., Can., Feb. 24, 1945; came to U.S., 1978, naturalized, 1984; s. Earl Lepage and Eunice Deborah (Eaton) W.; honors diploma, Prince of Wales Coll., 1964; BSc with honors in Biology, Dalhousie U., 1966, MD, 1972. Diplomate Am. Bd. Family Practice, Coll. Family Physicians Can. Intern, Victoria Gen. Hosp., Halifax, N.S., Can., 1971-72; gen. practice medicine, Summerside, P.E.I., 1975-78; pvt. practice specializing in family practice, Houston; med. dir. alcoholism treatment unit Raleigh Hills Hosp., 1981-83; preceptor teaching staff U. Tex. Med. Sch., Houston, Baylor Coll. of Med., 1984—; exec. med. dir. Oak Forest Med. Ctr., Vis. Nurse Assn., Hospice, Houston, 1991—; mem. med. adv. bd. Med. World News, 1983—. Served as flight surgeon RCAF, 1967-75. Fellow Am. Acad. Family Physicians; mem. AMA, Tex. Med. Assn., Am. Coll. Emergency Physicians, Royal Coll. Medicine, Am. Pub. Health Assn., Tex. Med. Found., Tex. Rsch. Soc. Alcoholism, Soc. USAF Flight Surgeons, Am. Geriatrics Soc., Rotary, Univ. Lodge of Halifax. Methodist. Office: 3515 Oak Forest Dr #3 Houston TX 77018-6121

WOO, BUCK HONG, neuropsychologist; b. Seattle, Jan. 10, 1956; s. Yuen and Jean (Chin) W. BS, U. Wash., 1981; PhD, Calif. Sch. Prof. Psychology, L.A., 1987. Lic. psychologist Mass. Fellow in neuropsychology U. So. Calif., L.A., 1987-89; asst. prof. Dartmouth Med. Sch., Hanover, N.H., 1989-91; dir. brain injury program Whittier Rehab. Hosp., Haverhill, Mass., 1991—. Mem. Internat. Neuropsychol. Soc., Mass. Neuropsychol. Soc., Geriatric Soc. Am. Office: Whittier Rehab Hosp 76 Summer St Haverhill MA 01830

WOO, DAH-CHENG, hydraulic engineer; b. Shanghai, China, Dec. 18, 1921; came to U.S., 1947; s. Pei-Cho and Changtze (Chang) Woo. BS, Hangchow (China) Christ Coll., 1944; MA, U. Mich., 1948, PhD, 1956. Lic. profl. engr. Hydraulic engr. ALN&M Cons. Engrs., Ann Arbor, Mich., 1951-62; sr. hydraulic engr. Fed. Hwy. Adminstrn., Washington, 1962—. Fellow ASCE (urban water resources rsch. coun. 1967—); mem. Am. Geophys. Union, Internat. Water Resources Assn., Internat. Assn. Hydraulic Rsch. Achievements include contbns. to basic knowledge of sheet (or thin) flow over smooth to rough surfaces, in laminar and transition regions, without and with the raindrop impact effect. Home: 2300 Pimmit Dr # 712 Falls Church VA 22043 Office: Fed Hwy Adminstrn 6300 Georgetown Pike Mc Lean VA 22101-2296

WOO, JACKY, immunologist, educator, researcher; b. Hong Kong, June 12, 1965; came to U.S., 1990; s. Chung-San and Lai-Pik (Koo) W. Diploma in Biology with honors, Hong Kong Bapt. Coll., 1987; PhD in Immunology, U. Aberdeen, U.K., 1990. Postdoctoral rsch. assoc. dept. surgery U. Pitts., 1990-92, rsch. asst. prof. immunology, 1992—. Author: The Cytokine Handbook, 1991, The Molecular Biology of Immunosuppression, 1992; contbr. articles to Immunology, Transplantation, Scandinavian Jour. Immunology. Recipient Overseas Rsch. award, U.K., 1988-90; NIH grantee, 1991—. Mem. AAAS, Brit. Soc. Immunologists, Transplant Soc., Am. Assn. Immunologists. Achievements include investigation of the immunosuppressive effect of FK506 in the early development of the drug, leading to the better understanding of its specific activity on T-lymphocytes but not on antigen-presenting cells. Office: U Pitts Dept Surgery Pittsburgh PA 15261

WOO, JONGSIK, mechanical engineer, educator, researcher; b. Taegu, Korea, June 8, 1957; s. Hyungki and Jungae (Bae) W.; m. Kyungja Jeon, Mar. 24, 1985; children: Dohyun, Dowoong. PhD, Ill. Inst. Tech., 1990. Researcher Daewoo Shipbuilding & Heavy Machinery, Ltd., Kyungnam, Korea, 1982-83; researcher Daewoo Shipbuilding & Heavy Machinery, Ltd., Kyungnam, 1984-87, sr. researcher, 1988-91, prin. researcher, 1992—; designer rig platforem Daewoo Shipbuilding & Heavy Machinery, Kyungnam, Korea, 1982-86; lectr. Keoje Coll., Kyungnam, Korea, 1991—. Contbr. AIAA Jour. Mem. AIAA. Office: Daewoo Shipbldg & Heavy Machinery Ltd, 1 Ajoodong, 657-714 Jangseungpo Kyungnam, Republic of Korea

WOO, MING-KO, geographer, educator; b. Hong Kong, Hong Kong, 1941. BA in Geography and Geology, U. Hong Kong, 1964, MA in Geomorphology, 1967; PhD in Hydrology, U. B.C., 1972. Cert. profl. hydrologist Am. Inst. Hydrology. Rsch. programmer Inst. Animal Resource Ecology U. B.C., 1972; asst. prof. dept. geography McMaster U., Hamilton, Ont., 1972-78, assoc. prof., 1978-83, prof., 1983—; particpant various contract projects with Environment Canada, 1973-76, Ont. Hydro, 1974-78, Indian and No. Affairs Can., 1981-87, Internat. Devel. Rsch. Ctr., Can., 1987-89, Atmospheric Environment Svc., 1988-89, 89-91, 92, Wildlife Svc., 1989-91, Ducks Unltd., 1989-91, Can. Electrical Assn., 1992-93; panel chmn. phys. geography panel Ont. Grad. Scholarship, 1984-86; permafrost subcom. Nat. Rsch. Coun. Can., 1985-91; vis. scientist Inst. Glaciology and Geocryology Chinese Acad. Sci., 1986, St. Andrews Univ., Scotland, 1986; mem. Can. nat. com. Internat. Union Geodesy and Geophysics, 1986-88; with study group on regional hydrological responses to global warming Internat. Geographical Union, 1990—; chief del. no. rsch. basins symposium/workshop IHP, 1990—; mem. Can. Nat. Com. World Climate Rsch. Program, 1992—; cons. Ont. Coun. Grad. Studies, 1993; com. mem. hydrology sect. Can. Geophys. Union, 1993—. Contbr. articles to profl. jours. Grantee Natural Sci. and Engring. Rsch. Coun. Can., 1972—, Inland Waters Directorate, 1977-80, Atmospheric Environ. Svc., 1980-82, Social Scis. and Humanities Rsch. Coun. Can., 1983-84, Donner Can. Found., 1982-85, UN Devel. Programme TOTKEN, 1986, 90—, Energy, Mines and Resources, 1989—, Can. Internat. Devel. Agy., 1993—. Mem. Am. Geophysical Union, Am. Meteorology Soc., Can. Assn. Geographers (councillor 1988-91, chmn. internat. biosphere program com. 1989-91, award for Scholarly Distinction, 1992), Geographical Soc. China (corr. mem., com. hydrology). Achievements include research in snow accumulation, snow melt and runoff processes, ice problems in northern Canada, hydrological processes as affected by the presence of permafrost in arctic and western Canada, wetland storage and runoff in termperate, subarctic and Arctic environments, stochastic modelling of droughts, floods and other runoff phenomena, rural water use in the desert fringe of northern Nigeria, tropical soil erosion in southern China, hydrological effects of climatic variability and climate change in Arctic and prairie regions. Office: McMaster U/Dept of Geography, 1280 Main St W, Hamilton, ON Canada L8S 4K1*

WOO, SAVIO LAU-YUEN, bioengineering educator; b. Shanghai, China, June 3, 1942; s. Kwok CHong and Fung Sing (Yu) W.; m. Patricia Tak-kit Cheong, Sept. 6, 1969; children: Kirstin Wei-Chi, Jonathan I-Huei. BSME, Chico State U., 1965; MS, U. Wash., 1966, PhD, 1971. Research assoc. U. Wash., Seattle, 1968-70; asst. research prof. U. Calif.-San Diego, La Jolla, 1970-74, assoc. research prof., 1974-75, assoc. prof., 1975-80, prof. surgery and bioengring., 1980-90; prof. orthopaedic surgery, vice-chmn. orthopaedic surgery U. Pitts., 1990-92, prof. mech. engring., 1990—, Albert B. Ferguson Jr. prof. orthopaedic surgery, 1993—; prin. investigator VA Med. Ctr., San Diego, 1972-90, Pitts., 1990—; cons. bioengr. Childrens Hosp., San Diego, 1973-80; cons med. implant cos., 1978-85; vis. prof. biomechanics Kobe (Japan) U., 1981-82; dir., chief exec. officer M&D Coutts Inst. for Joint Reconstrn. and Rsch., 1984-90; mem. sci. adv. com. Whitaker Found., 1986—, Steadman Sports Medicine Found., 1990—, OsteoArthritis Scis. Inc., 1992—. Assoc. editor Jour. Biochem. Engring., 1979-87, Jour. Biomechanics, 1978—, Jour. Orthopedic Rsch., 1983-92, Materials Sci. Reports, 1990—, Proc. Inst. Mech. Engrs. (Part H), 1990—; mem. editorial adv. bd. Jour. Orthopedic Rsch., 1993—, Materials; contbr. articles to profl. jours. Recipient Elizabeth Winston Lanier Kappa Delta award, 1983, 85, awards for excellence in basic sci. rsch. Am. Orthopaedic Soc. Sports Medicine, 1983, 86, 90, O'Donoghue award, 1990, Wartenweiler Meml. Lectureship Internat. Soc. Biomechs., 1987; citation Am. Coll. Sports Medicine, 1988; Japan Soc. Promotion of Sci. fellow, 1981; Rsch. Career Devel. award NIH, 1977-82. Fellow Am. Inst. Med. and Biol. Engring (founding fellow, chmn. coll. fellows 1992—, bd. dirs. 1992—), ASME (sec., chmn. biomechanics com., chmn. honors com. bioengring. div., mem. exec. com., 1983-88, sec. 1985-86, chmn. 1986-87, H.R. Lissner award 1991); mem. Inst. Medicine NAS, Western Orthopaedic Assn. (hon.), Internat. Soc. Biomed. (mem. coun. 1992—), Biomed. Engring. Soc. (bd. dirs. 1984-86), Am. Acad. Orthopedic Surgeons, Orthopaedic Research Soc. (exec. com. 1983-88, chmn. program com. 1985-86, pres. 1986-87), Am. Soc. Biomechs. (pres. 1985-86, sec. 1977-80, exec. com. 1984-87, Alfonso Borelli award 1993), Internat. Soc. Fractures Repair (bd. dirs. 1984—, v.p. 1987-90, pres. 1990-92), Can. Orthopedic Rsch. Soc. (hon.), European Orthopedic Rsch. Soc. Home: 47 Pleasant View Ln Pittsburgh PA 15238 Office: U Pitts Dept Ortho Surg Liliane Kaufmann Bldg 3471 5th Ave Ste 1010 Pittsburgh PA 15213

WOO, STEVEN EDWARD, aerospace engineer; b. L.A., Dec. 7, 1967. BS in Aerospace Engring., UCLA, 1989. Mission analyst Strategic Def. Ctr. Rockwell Internat. Corp., Seal Beach, Calif., 1989—; mem. tech. staff Space Systems div. Rockwell Internat. Corp., Downey, Calif., 1990. Mem. AIAA, UCLA Alumni Assn. Democrat. Office: Rockwell Internat Corp PO Box 3089 2800 Westminster Blvd Seal Beach CA 90740-2089

WOO, WALTER, computer systems consultant; b. San Antonio, May 12, 1948; s. Foon Foo and Man Yin (Wong) W.; m. Margaret Leong, Aug. 26, 1973; children: Ryan David, Ellery. BA, St. Mary's U., San Antonio, 1971; postgrad., U Houston, 1983. Spl. projects chemist Atlantic Richfield Chem., Channelview, Tex., 1977; programmer/analyst Atlantic Richfield Chem. Co., Houston, 1978-80; systems analyst Aminoil USA, Houston, 1980-81; sr. systems analyst Occidental Petroleum Co., Houston, 1981-82; systems analyst Houston Export Crating Co., 1982-83; systems specialist (mgr.) Ford Aerospace and Comm., Houston, 1983-85; project mgr. Raytheon Corp., Houston, 1985-87; systems cons. Ciber Inc., Houston, 1987-89, Computer Horizons Corp., Houston, 1989-92; ind. cons. Innovative Tech. Info. Systems, Houston, 1992—. Vol. United Way. With Tex. Air N.G. Dow Chem. Co. acad. scholar, 1977. Mem. Data Processing Mgmt. Assn., Golden Key. Republican. Baptist. Avocations: travel, computers, reading, amateur tennis. Home and Office: 2115 Gentryside Dr Houston TX 77077-3601

WOOD, ALBERT E(LMER), biology educator; b. Cape May Ct. House, N.J., Sept. 22, 1910; s. Albert Norton and Edith (Elmer) W.; m. Frances Wright, Jan. 17, 1937 (dec. Jan. 5, 1987); children: Albert Frederick, Roger Conant, Daniel Nixon. BS in Geology, Princeton U., 1930; MA in Zoology, Columbia, 1932, PhD in Geology, 1935; MA, Amherst Coll., 1954. Asst. geologist U.S. Army Engrs., Binghamton, N.Y., 1936-39; assoc. geologist U.S. Army Engrs., Binghamton, 1939-41, geologist, 1941, '46; asst. prof. biology dept. Amherst (Mass.) Coll., 1946-48, assoc. prof., 1948-54, prof., 1954-70. Contbr. over 110 articles to profl. and sci. jours. and sci. symposium proceedings. Lt. col. U.S. Army, 1941-46, ETO. Decorated with Bronze Star, Purple Heart, U.S. Army, ETO, 1941-46; named Cutting Travel fellow, Columbia U., 1934-35, Sr. Postdoctoral fellow NSF, Basle, Switzerland, 1966-67; recipient Disting. Achievement award, Alumni Assn., Poly Prep Country Day Sch., Bklyn., 1961. Fellow AAAS (life), Geol. Soc. Am. (50 yr.); mem. Paleontol. Soc. Am., Soc. Verterbrate Paleontology (past pres., hon.), Am. Soc. Mammalogists. Home: 20 E Mechanic St Cape May Court House NJ 08210

WOOD, FERGUS JAMES, geophysicist, consultant; b. London, Ont., Can., May 13, 1917; came to U.S., 1924, naturalized, 1932; s. Louis Aubrey and Dora Isabel (Elson) W.; student U. Oreg., 1934-36; AB, U. Calif., Berkeley, 1938, postgrad., 1938-39; postgrad. U. Chgo., 1939-40, U. Mich., 1940-42, Calif. Inst. Tech., 1946; m. Doris M. Hack, Sept. 14, 1946; children: Kathryn Celeste Wood Madden, Bonnie Patricia Wood Ward. Teaching asst. U. Mich., 1940-42; instr. in physics and astronomy Pasadena City Coll., 1946-48, John Muir Coll., 1948-49; asst. prof. physics U. Md., 1949-50; assoc. physicist Johns Hopkins U. Applied Physics Lab., 1950-55; sci. editor Ency. Americana, N.Y.C., 1955-60; aero. and space rsch. scientist, sci. asst. to dir. Office Space Flight Programs, Hdqrs., NASA, Washington, 1960-61; program dir. fgn. sci. info. NSF, Washington, 1961-62; phys. scientist, chief sci. and tech. info. staff U.S. Coast and Geodetic Survey, Rockville, Md., 1962-66, phys. scientist Office of Dir., 1967-73, rsch. assoc. Office of Dir., 1973-77, Nat. Ocean Svc.; cons. tidal dynamics, Bonita, Calif., 1978—; mem. Am. Geophys. Union, ICSU-UNESCO Internat. Geol. Correlation Project 274, Working Group #1-Crescendo Events in Coastal Environments, Past and Future (The Millennium Project), 1988—. Capt. USAAF, 1942-46. Recipient Spl. Achievement award Dept. Commerce, NOAA, 1970, 74, 76, 77. Mem. Sigma Pi Sigma, Pi Mu Epsilon, Delta Phi Alpha. Democrat. Presbyterian. Author: The Strategic Role of Perigean Spring Tides in Nautical History and North American Coastal Flooding, 1635-1976, 1978; Tidal Dynamics; Coastal Flooding, and Cycles of Gravitational Force, 1986; contbr. numerous articles to encys., reference sources, profl. jours.; writer, tech. dir. documentary film: Pathfinders from the Stars, 1967; editor-in-chief: The Prince William Sound, Alaska, Earthquake of 1964 and Aftershocks, vols. 1-2A and sci. coordinator vols. 2B, 2C and 3, 1966-69. Home: 3103 Casa Bonita Dr Bonita CA 91902-1735

WOOD, FRANK BRADSHAW, retired astronomy educator; b. Jackson, Tenn., Dec. 21, 1915; s. Thomas Frank and Mary (Bradshaw) W.; m. Elizabeth Hoar Pepper, Oct. 5, 1945; children—Ellen, Eunice, Mary, Stephen. B.S., U. Fla., 1936; postgrad., U. Ariz., 1938-39; A.M., Princeton, 1940, Ph.D., 1941. Research asso. Princeton, 1946; NRC fellow Steward Obs., U. Ariz., also Lick Obs., U. Calif., 1946-47; asst. prof. U. Ariz., 1947-50; asso. prof. U. Pa., 1950, prof. astronomy, 1954-68, chmn. dept., 1954-57, 58-68; exec. dir. Flower and Cook Obs., 1950-54, dir., 1954-68, Flower prof., 1958-68; prof. astronomy, dir. optical astron. observatories U. Fla., Gainesville, 1968-87, assoc. chmn. dept., 1971-77, prof. emeritus, 1989—; established South Pole obs. sta., U. Fla., 1985—. Author: (with J. Sahade) Interacting Binary Stars, 1978; Editor: Astronomical Photoelectric Photometry, 1953, Present and Future of Telescope of Moderate Size, 1958, Photoelectric Astronomy for Amateurs, 1963; Contbr. articles to profl. jours. Col., aide de camp, gov.'s staff, State of Tenn.; served with USNR, 1941-46. Decorated Air medal; Fulbright fellow Australian Nat. U., 1957-58; NATO sr. fellow in sci. U. Canterbury, Christchurch, New Zealand, 1973; Fulbright fellow Instituto de Astronomia y Fisica del Espacio, Buenos Aires, 1977; recipient plaques of appreciation, Govt. South Korea, 1988, Yonsei U., Korean Astron. Soc. Mem. AAAS (sec. sect. D com. on coun. affairs 1970), Am. Astron. Soc. (coun. 1958-61), Royal Astron. Soc. New Zealand (hon.), Internat. Astron. Union (pres. commn. 42 1967-70, v.p. commn. 38 1979-82, pres. 1982-85), Royal Astron. Soc., Fla. Acad. Sci. (chmn. phys. scis. sect. 1974-75, pres. 1983-84, gold medal 1989), Astron. Soc. Can., Astron. Soc. Pacific, Ret. Officers Assn., Internat. Amateur and Profl. Photoelectric Photometry (hon.), Navy League, Explorers Club, Phi Beta Kappa, Sigma Xi. Episcopalian. Clubs: Torch, Kiwanis, Troa. Home: 714 NW 89th St Gainesville FL 32607-1453 Office: Dept Astronomy U Fla Gainesville FL

WOOD, GEORGE MARSHALL, research scientist, program manager; b. Fairfield, Conn., Jan. 20, 1933; s. George Marshall and Claudia Oresta (Ziroli) W.; m. Nancy Ann Litz, May 24, 1952; children: Judith, Nancy, Jennifer. BS in Physics, U. Ga., 1959; MS in Engring. Sci., Rensselaer Poly. Inst., 1968, PhD in Engring. Sci., 1974. Rsch. scientist NASA, Hampton, Va., 1959—, tech. program mgr., 1961—; adj. assoc. prof. Rensselaer Poly. Inst., Troy, N.Y., 1974-83, U. New Orleans, 1982; organizer, chmn. 6 internat. confs. Co-author: Mass Spectrometry: Applications in Science and Engineering, 1986; contbr. numerous articles to profl. jours. Sgt. U.S. Army, 1955-57, Korea. Mem. AIAA, Am. Chem. Soc. (exec. com. Hampton Roads sect., 1983—), Am. Soc. Mass Spectrometry, Sigma Pi Sigma. Achievements include 5 patents in field. Home: 2 Little Bluff Rd Newport News VA 23606 Office: NASA Langley Research Ctr Hampton VA 23665

WOOD, GORDON HARVEY, physicist; b. Trail, B.C., Can.; m. Linda D. McLennan, Dec. 19, 1964; children: Brian, Douglas. M Applied Sci., U. B.C., Vancouver, 1965; PhD, U. British Columbia, Vancouver, 1969. Rsch. officer Nat. Rsch. Coun. Can., Ottawa, Ontario, 1969-80; rsch. coun. officer Nat. Rsch. Coun. Can., Ottawa, 1980—; sec. Can. Nat. Com. for CODATA, Ottawa, 1980-88, chmn., 1988-91; sec.-gen. CODATA, 1990—. Mem. Canadian Assn. Physicists. Mem. Pentecostal Ch. Office: Nat Rsch Coun, Montreal Rd, Ottawa, ON Canada K1A 0S2

WOOD, H(OWARD) JOHN, III, astrophysicist, astronomer; b. Balt., July 19, 1938; s. Howard John Jr. and Cara (Loss) W.; m. Austine Barton Read, June 10, 1961 (div. Jan. 1975); children: Cara Loss, Erika Barton; m. Maria Ilona Kovacs, May 22, 1977; 1 child, Andreas M. BA in Astronomy, Swarthmore Coll., 1960; MA, Ind. U., 1962, PhD, 1965. Lectr., asst. prof. then assoc. prof. U. Va., Charlottesville, 1964-70; staff astronomer European So. Obs., Santiago, Chile, 1970-75; Fulbright Rsch. fellow U. Vienna Obs., 1976-78; rsch. assoc. Ind. U., Bloomington, 1978-81; asst. to the dir. Cerro Tololo Inter-Am. Obs., La Serena, Chile, 1982-83; physicist, astronomer NASA/Goddard Space Flight Ctr., Greenbelt, Md., 1984—; advisor, participant Hubble Space Telescope Allen Comm., NASA, Danbury, Conn., 1990; co-chmn. Hubble Space Telescope Ind. Optical Rev. Panel, Columbia, Md., 1990-91; mem. panel The Townes/SAGE Panel-Jet Propulsion Lab., Pasadena, 1991-92. Co-author: Physics of Ap Stars, 1976; contbr. articles to profl. publs. Recipient 10 rsch. grants NSF, 1965-82. Mem. Am. Astron. Soc. (Rsch. grantee 1978), Internat. Astron. Union (Commn. 29 1962—), Sigma Xi. Achievements include discovery of Balmer-Line variability of Ap stars; discovery of magnetic fields in southern Ap stars; alignment testing and delivery of the DIRBE photometric cryogenic telescope on the COBE spacecraft; alignment and optical prescription for Hubble Space Telescope while in orbit. Office: NASA/Goddard Space Flight Ctr Code 717 Greenbelt MD 20771

WOOD, JAMES DAVID, environmental engineer; b. Kansas City, Mo., Dec. 21, 1953; s. James Maurice Wood and Hazel Fern (Herd) Stidham; m. Lisa Ann Stevens (div. Sept. 1982). BSChemE, U. Mo., Rolla, 1976, MSChemE, 1981; MS in Environ. Engring., Johns Hopkins U., 1982; MS in Engring. Mgmt., Fla. Inst. Tech., 1985. Registered profl. engr., Md.; cert. hazard control mgr.; cert. hazardous material mgr.; cert. hazardous waste specialist; cert. safety and security dir. Sanitary engr. Acad. Health Scis., Ft. Sam Houston, Tex., 1977; sanitary engr. U.S. Army Environ. Hygiene Agy., Aberdeen Proving Ground, Md., 1977-80, section chief air quality mgmt. br., 1993—; sanitary engr. U.S. Army Toxic and Hazardous Materials Agy., Aberdeen Proving Ground, 1980-82, chem. engr., 1982-84, environ. engr., 1984-91, chief air quality mgmt. br., 1991, sr. environ. engr., 1991-93; cons. Project Mgr. PERSHING (Permit Support for INF Treaty between U.S. and former USSR), Redstone Arsenal, Ala., 1988-91; mem. mgmt. steering com. AMCCOM Demilitarization Office, Rock Island (Ill.) Arsenal, 1988—. Mem. Md. Air Control Quality Control Adv. Coun., Balt., 1984-89. Capt. U.S. Army, 1977-82, maj. Res. Mem. AIChE, Md. Soc. Profl. Engrs. (v.p. 1989-90, bd. dirs. 1990—, pres. Susquehann chpt. 1985-86, 87-88, chmn. Engr.-for-a-Day program Susquehann chpt. 1986-91, Young Engr. of Yr. 1988), Soc. Am. Mil. Engrs., Air and Waste Mgmt. Assn., Phi Beta Kappa (Cert. of Excellence 1972). Office: USA Environ Hygiene Agy Air Pollution Engr Divsn Aberdeen Proving Ground MD 21010-5422

WOOD, JEREMY SCOTT, architect, urban designer; b. Glen Ridge, N.J., Oct. 23, 1941; s. William Gamble and Alice-Marguerite (Scott) W.; m. Robin Benensohn-Rosefsky, June 14, 1970; children: Alexis, Jonas, Augusta. AB, Yale U., 1964, M in Architecture, 1970. Registered architect, Ma. Sr. assoc. TAC/The Architects Collaborative, Inc., Cambridge, Mass., 1970—; instr. Boston Archtl. Ctr., 1970-76; head tutor Dept. of Art History and History of Modern Architecture Yale U., 1969-70. Author: (section and chpt. in books) Adaptive reuse: Issues and Case Studies in Building Preservation, 1988, Office Buildings, 1989, Exposed Structure in Building Design, 1993; prin. works include Health Care Internat. Hosp. and Hotel, Clydebank, Glasgow, Scotland, Copley Pl., Boston, The Westin Hotel at Copley Pl., Boston, Liberty Ctr. and Vista Internat. Hotel, Pitts., Wellington Bus. Ctr. Offices, Medford, Mass., Two Portland (Maine) Sq. Office Bldg., One Mifflin Pl., Cambridge, Groton (Mass.) Sch. Dormitories, Coll. Engring. and Applied Sci., Shuwaikh Campus, Kuwait U., Kuwait City, Kuwait; asst. editor Perspecta 11; corr. Architecture and Urbanism, 1976; contbr. articles to profl. jours. Recipient award of Excellence Assn. of Sch. Bus. Offcls., Coun. of Ednl. Facilities Planners, AIA, 1976, Concrete Industry Bd. Spl. Recognition award The Westin Hotel, Boston, 1983, Prestressed Concrete Inst. award, 1983, Honor award Associated Gen. Contractors of Mass., 1985; grantee Urban Land Inst., 1988. Mem. AIA, Boston Soc. Architects, Mass. State Assn. of Architects, Am. Planning Assn., Soc. of Archtl. Hists. (life, New Eng. chpt.), Boston Inst. of Contemporary Art, The Archtl. League of N.Y. Home: 10 Pigeon Hill Rd Weston MA 02193-1620 Office: TAC The Architects Collaborative Inc 46 Brattle St Cambridge MA 02138-3700

WOOD, JOHN MORTIMER, aerospace executive, aeronautical engineer; b. New Orleans, July 7, 1934; s. John Mortimer Sr. and Annie Jeff (Gates) W.; m. Bonnie Ann Blanchette, June 6, 1958 (div. Oct. 1977); m. Barbara Lee Butler, Aug. 12, 1978; 1 child, Mark Douglas. BA in Aero. Engring., U. Tex., 1957. Project engr. Gen. Dynamics/Convair, San Diego, 1957-58, Rocket Power, Inc., Mesa, Ariz., 1961-64; sales mgr. S.E. region Rocket Power, Inc., Huntsville, Ala., 1964-67; dir. mktg. Quantic Industries, San Carlos, Calif., 1967-70; sr. mktg. mgr. Talley Industries of Ariz., Mesa, 1970-77; dir. mktg. Universal Propulsion Co., Inc., Phoenix, 1977-85, v.p. mktg., 1985-91, v.p. contract mgmt., 1992—. 1st lt. USAF, 1958-61. Mem. Am. Def. Preparedness Assn., Assn. for Unmanned Vehicle Sytsems, Tech. Mktg. Soc. of Am., Survival and Flight Equipment Assn. Republican. Home: 111 W Canterbury Ln Phoenix AZ 85023-6252 Office: Universal Propulsion Co Inc 25401 N Central Ave Phoenix AZ 85027-7837

WOOD, JOHN THURSTON, cartographer, jazz musician; b. Chgo., Apr. 29, 1928, s. Clarence Leo and Hilda Bernice (Miller) W.; m. Erma Louise Vogt, July 3, 1957; children: John Thurston Jr., Holly Lynn, Joseph Miller II. BS, Iowa State U., 1952; postgrad., Ohio State U., 1962-64, Indsl. Coll. Armed Forces, Washington, 1974, Fed. Exec. Inst., Charlottesville, Va., 1985. Commd. 2d lt. USAF, 1952, advanced through grades to lt. col., 1968; combat airlift pilot 535th Troop Carrier Squadron, Vung Tau, Vietnam, 1966-67; comdr. USAF Operational Evaluation Detachment, Topeka, 1967-70; aerial and ground survey officer Def. Intelligence Agy., Washington (D.C.), Def. Mapping Agy., Washington (D.C.), Va., 1972-76; ret., 1976; chief user svcs. Nat. Cartographic Info. Ctr. U.S. Geol. Survey, Reston, 1976-83, dep. chief, 1983-85, chief, 1985-89, 1985-89; chief Earth Sci. Info. Office U. S. Geol. Survey, Reston, 1989—. Tuba player Buck Creek Jazz Band, 1977—. Decorated Air medal with four oak leaf clusters. Mem. Am. Soc. for Photogrammetry and Remote Sensing, Am. Congress on Surveying and Mapping, Masons, Shriners. Home: 4007 Terrace Dr Annandale VA 22003-1856 Office: US Geol Survey Nat Ctr Chief ESIO Mail Stop 509 Reston VA 22092

WOOD, LOREN EDWIN, aerospace engineer, consultant; b. Taunton, Mass., Dec. 25, 1927; s. Elmer Roe and Alice Eleanor (Philbrick) W.; m. Ann H. Hamilton, Aug. 6, 1952; children: Joan, Alice, Scott, Carol. BA magna cum laude, Brown U., 1949; AM in Math. Analysis and Applied Math., Cornell U., 1950. Teaching fellow Cornell U., Ithaca, N.Y., 1949-50; mgr. Minuteman silo devel. instrumentation TRW Inc., Redondo beach, Calif., 1958-60; mgr. Apollo engring. and ops. support projects TRW Inc., Houston, 1966-72, mgr. shuttle computer systems and software integration, 1973-74, mgr. shuttle payload flight control studies, 1975-77, mgr. advanced programs and program devel., 1978-87; mgr. space program integration Aerospace Corp., El Segundo, Calif., 1961, mgr. Gemini launch vehicle and target vehicle flight testing, 1962-66; mgr. Apollo mission model devel. and calculations White House Com. to Select Lunar Landing Mission Model, Washington, 1961; project engr. for instrumentation space shuttle program Rockwell Internat., Houston, 1988—; speaker to tech. socs., radio, TV, civic clubs. Author numerous tech. papers and reports. Organizer, adminstr. youth football and baseball leagues, Friendswood, Tex., 1967-69; mem. City Charter Commn., Friendswood, 1970-71, Friendswood City Coun., 1977-83; founding trustee Houston Grad. Sch. Theology, 1983—. Capt. USAF, 1954-56. Named Citizen of Yr., Friendswood C. of C., 1983; recipient Disting. Svc. award U.S. Jaycees, 1983. Fellow AIAA (assoc., chmn. Outstanding Spl. Events 1971-73, chmn. Outstanding sect. 1975-76); mem. Instrumentation Soc. Am. (sr.), Phi Beta Kappa, Sigma Xi. Avocations: family, grandchildren, Christian service, sports. Home: 905 Cowards Creek Dr Friendswood TX 77546-4407

WOOD, ORIN LEW, electronics company executive; b. Hurricane, Utah, Apr. 26, 1936; s. Luwayne and Violet (Spendlove) W.; m. Yvonne E. Silva, Dec. 27, 1957; children: Sharla, Sherri, Brent L., Blaine O. BS, Brigham Young U., 1958; MS, UCLA, 1962; PhD, U. Utah, 1968. Dir. rsch. Biologics, Inc., Salt Lake City, 1967-69; assoc. prof. environ. biophysics Utah State U., Logan, 1969-70; assoc. prof. health occupations, dir. radiol. tech. program Weber State U., Ogden, Utah, 1970-75; owner, cons. O. Lew Wood Assocs., Murray, Utah, 1970—; exec. Quartex, Inc., Salt Lake City, 1978—; chmn. bd., CEO Quartztronics, Inc., Salt Lake City, 1980—, Quartz Kinetics, Inc., Salt Lake City, 1985—, Quartznetics, Inc., Salt Lake City, 1985—. Contbr. numerous articles to profl. jours. including Modern Science and Technology, 1965, Management of Technology II, 1990, Management of Technology III, 1992. Recipient State of Utah Gov.'s medal for Sci. and Tech., 1993; NIH fellow, 1965-68. Mem. IEEE (sr.), Instrument Soc. Am. (sr.), Internat. Assn. Mgmt. of Tech., Licensing Execs. Soc., Tech. Transfer Soc. Achievements include patents for fluidics, biomed. instrumentation, atmospheric particulate sampling and transducers and sensors; development of first high volume cascade impactor for atmospheric sampling; management of group that developed first disposable nebulizer for respiratory therapy; first accurate measurement of whole body rubidium in humans. Home: 811 E Woodshire Circle Murray UT 84107 Office: Quartztronics Inc 1020 Atherton Dr Bldg C Salt Lake City UT 84123

WOOD, SUSANNE GRIFFITHS, analytical environmental chemist; b. Buffalo, N.Y., Dec. 28, 1933; d. John Arnold and Alice Fredericka (Wiede) Griffiths; m. Richard Bruce Wood, Aug. 8, 1970. BA in Biology, SUNY, Buffalo, 1954, MA in Biology, 1957; PhD in Plant Pathology, U. Ill., 1976. Asst. cancer rsch. scientist Roswell Park Meml. Inst., Buffalo, 1957-59, 64-66; clin. microbiologist Deaconess Hosp., Buffalo, 1959-64; teaching and rsch. asst. U. Ill., Urbana-Champaign, 1966-75, rsch. assoc., 1975-80; asst. profl. scientist Ill. Natural History Survey, Champaign, 1980-83, assoc. profl. scientist, 1983-92; lab. supv. Northern Ill. Water Corp., Champaign, Ill., 1993—. Contbr. articles on microbial physiology and environ. chemistry to profl. jours. Assoc. chimesmaster U. Ill., carillonneur U. Luth. Ch., Champaign, 1968—; organist Philo (Ill.) Presbyn. Ch., 1987—; mem. libr. U. Ill. Russian Folk Orch., 1974-92. Mem. AAAS, Am. Chem. Soc., Soc. for Environ. Toxicology and Chemistry, Guild Carillonneurs in N.Am., Am. Acad. Microbiology (reg. med. technologist). Home: PO Box 437 207 S Hayes St Philo IL 61864 Office: Northern Ill Water Corp 206 W White St Champaign IL 61820

WOOD, WARREN WILBUR, hydrologist; b. Pontiac, Mich., Apr. 9, 1937; s. Warren W. and Irma Clair (Moore) W.; m. Anneliese Catherine Funk, June 12, 1961; 1 child, Warren T. BS in Geology, Mich. State U., 1959, PhD in Geology, 1969. Geophysicist Mich. Hwy. Dept., Ann Arbor, 1961-62; hydrologist U.S. Geol. Survey, Lansing, Mich., 1962-70; rsch. hydrologist U.S. Geol. Survey, Lubbock, Tex., 1970-77; supervisory hydrologist U.S. Geol. Survey, Reston, Va., 1977-78; assoc. prof. Tex. Tech U., Lubbock, 1978-81; rsch. hydrologist U.S. Geol. Survey, Reston, 1981—; bd. dirs. Assn. Groundwater Sci. & Engring., Dublin, Ohio, 1982-86, Waterloo Ctr. for Groundwater Rsch., Waterloo, Ontario, Can., 1988-92; adv. coun. Dept. Geol. Scis., East Lansing, Mich., 1989—. Contbr. articles to profl. jours. Troop leader Boy Scouts of Am., Lubbock, Tex., 1973-77. Named Disting. lectr. Assn. Groundwater Sci. & Engring., 1988, Phoebe Hearst lectr. U. Calif. Berkeley, 1989, Birdsall Disting. lectr. Geol. Soc. of Am., 1989; recipient Meritorious Svc. award U.S. Dept. Interior, 1991, Keith Anderson award, 1992. Fellow Geol. Soc. of Am.; mem. Assn. Groundwater Sci. & Engring., Am. Geophys. Union, Geol. Soc. of Washington (membership chmn. 1987-88). Achievements include demonstration that groundwater leakage from evaporating basin controls type and thickness of mineral deposited; provided hypothesis on origin of both large and small lake basins on So. High Plain of Tex. Office: US Geol Survey 431 National Ctr Reston VA 22092

WOOD, WELLINGTON GIBSON, III, biochemistry educator; b. Balt., Dec. 29, 1945; s. Wellington Gibson Jr. and Elsie Bernice (Johnson) W.; m. Beverly Jean Beaver, Feb. 8, 1969; children: Wellington Gibson IV, Katherine Brittingham. BA, Tex. Tech U., 1971, PhD, 1976. Postdoctoral fellow Syracuse (N.Y.) U., 1976-77; staff scientist Bangor (Maine) Mental Health Inst., 1978-80; evaluation coord. VA Med. Ctr., St. Louis, 1980-89; assoc. dir. for edn. and evaluation VA Med. Ctr., Mpls., 1989—; asst. prof. St. Louis U. Sch. Medicine, 1982-87, assoc. prof., 1987-89; assoc. prof. dept. pharmacology U. Minn. Sch. Medicine, Mpls., 1990—; mem. sci. editorial bd. Alcoholism and Drug Rsch. Comm. Ctr., Austin, Tex., 1990—; mem. biochemistry, physiology and medicine study sect. NIH-Nat. Inst. Alcohol Abuse and Alcoholism, 1992-96. Assoc. editor Exptl. Aging Rsch., 1977-82; contbr. numerous articles to profl. jours. Nat. Inst. on Alcohol Abuse and Alcoholism postdoctoral fellow, 1976-77; grantee Nat. Inst. on Alcohol Abuse and Alcoholism, 1987—, Dept. Vets. Affairs, 1981—. Mem. Am. Aging Assn. (bd. dirs. 1984-87), Rsch. Soc. on Alcoholism (chmn. membership com. 1988-91), Internat. Soc. for Biomed. Rsch. on Alcoholism. Achievements include development of a new approach to understanding the molecular actions of alcohol in brain membranes that has shown that alcohol has an asymmetric effect on brain membrane leaflets in vitro and that chronic alcohol consumption impairs the capacity of brain membrane leaflets to maintain structural lipid asymmetry; those effects resulted from the changes in transbilayer distribution of cholesterol domains in the two membrane leaflets. Home: 16091 Huron Path Lakeville MN 55044-8874 Office: VA Med Ctr GRECC 11G Minneapolis MN 55417

WOOD, WILLIAM BARRY, III, biologist, educator; b. Balt., Feb. 19, 1938; s. William Barry, Jr. and Mary Lee (Hutchins) W.; m. Marie-Elisabeth Renate Hartisch, June 30, 1961; children: Oliver Hartisch, Christopher Barry. A.B., Harvard U., 1959; Ph.D., Stanford U., 1963. Asst. prof. biology Calif. Inst. Tech., Pasadena, 1965-68; assoc. prof. Calif. Inst. Tech., 1968-69, prof. biology, 1970-77; prof. molecular, cellular and developmental biology U. Colo., Boulder, 1977—; chmn. dept. U. Colo., 1978-83; mem. panel for developmental biology NSF, 1970-72; physiol. chemistry study sect. NIH, 1974-78; mem. com. on sci. and public policy Nat. Acad. Scis., 1979-80; mem. NIH Cellular and Molecular Basis of Disease Rev. Com., 1984-88. Author: (with J.H. Wilson, R.M. Benbow, L.E. Hood) Biochemistry: A Problems Approach, 2d edit, 1981, (with L.E. Hood and J.H. Wilson) Molecular Biology of Eucaryotic Cells, 1975, (with L.E. Hood and I.L. Weissman) Immunology, 1978, (with L.E. Hood, I.L. Weissman, and J.H. Wilson) Immunology, 2d edit., 1984, (with L.E. Hood and I.L. Weissman) Concepts in Immunology, 1978; editorial rev. bd. Science, 1984—; mem. editorial bd. Cell, 1984-87; contbr. articles to profl. jours. Recipient U.S. Steel Molecular Biology award, 1969; NIH research grantee, 1965—; Guggenheim fellow, 1975-76. Mem. Nat. Acad. Scis., Am. Acad. Arts and Scis., Am. Soc. Biol. Chemists, Genetics Soc. Am., Soc. for Developmental Biology, Soc. Nematology, Am. Soc. Microbiologists, AAAS. Office: Dept MCD Biology Box 347 U Colorado Boulder CO 80309

WOODALL, JAMES BARRY, information systems specialist; b. San Diego, Aug. 15, 1945; s. Jame Franklin and Mildred Lorenia (Nielsen) W.; m. Bonnie Erharda Gloth (div. 1973); m. Kathleen Dee Smith; children: Robert, Mathew, Diane. BA, Randolph-Macon Coll., 1966; MS, Am. U., 1972; MBA, George Washington U., 1978. Cert. prodn. inventory mgr., total quality mgr. Systems engr. RCA, Washington, 1969-72; MIS analyst FNMA, Washington, 1972; spl. project mgr. GTE, Washington and San Carlos, Calif., 1973-83; LOA project mgr. Comsat, Washington, 1978-80; cons. Southcoast Cons., San Diego, 1984-90; mem. sr. tech. staff Litton Co., 1990-91; sr. bus. analyst GEC-Lear Astronics, 1991—; head MIS Royal Saudi Navy and Ministry of Def., 1984-85; dep. program mgr. MIS NASA, Washington, 1986—, Citicorp, H/SOA, GFB, 1988-90. Author: NASA-Strategy on Technology, 1986, Information Analysis & Design, 1989. Precinct capt. Rep. Party; trustee fin. com. United Meth. Ch., Woodland Hills, 1986—, Odyssey Found., L.A., 1988—. Capt. USMCR, 1963-69. Whitehouse fellow, 1972; recipient Letter of Appreciation, White House, 1979, 87. Mem. U.S. Jaycees (chpt. pres. 1975-77, state v.p. community devel. com. 1977-79, nat. v.p. community devel. 1979, 27 awards). Avocation: tennis. Home: 24438 Watt Rd Ramona CA 92065-4157 also: # 110 7109 Farralone Ave Canoga Park CA 91303

WOODALL, LARRY WAYNE, cement company executive; b. Lawrenceburg, Tenn., Aug. 4, 1959; s. C. London and Lula Bell (Gallian) W.; m. Scarlette V., Aug. 4, 1979; children: Poppy, Britni Jade, Noah. Grad., Higher Area Vocat. Sch., 1979. With quality control staff Johnson Controls, Linden, Tenn., 1978-79; ops. mgr. V&W Redi Mix Cement Inc., Hehenwald, Tenn., 1979-90; pres. Mt. Pleasant (Tenn.) Redi Mix. Cement Co., 1990—. Bd. dirs. Rod Bresfield Community Playhouse, 1989-91. Mem. ACI, Jaycees (treas. 1987-88, bd. dirs.). Office: Mt Pleasant Redi Mix Cement 305 Bluegrass Ave Mount Pleasant TN 38474

WOODBURY, DIXON JOHN, physiology educator, researcher; b. Seattle, Dec. 31, 1956; s. John Walter and Betty (Gunderson) W.; m. Susan Diana Harvey, Mar. 20, 1980; children: James Dixon, Thomas Walter, Emily Susan, Kara Leigh. BS in Physics and Chemistry magna cum laude, U. Utah, 1980; PhD in Physiology and Biophysics, U. Calif., Irvine, 1986. Postdoctoral fellow in biochemistry Brandeis U., Waltham, Mass., 1986-89; rsch. assoc. Howard Hughes Med. Inst., Waltham, 1989-90; asst. prof. Wayne State U., Detroit, 1990—. Unit commr. Boy Scouts Am., Orange County, Calif., 1984-86; pres. elder's quorum Ch. of Jesus Christ LDS, Weston Ward, Boston, 1988-90, instr., 1987-88. Fellow U. Calif., 1980, 81, 85, Muscular Dystrophy Assn., 1986-88. Mem. IEEE, Soc. for Neurosci., Biophys. Soc. Office: Wayne State U Sch Medicine Dept Physiology 540 E Canfield St Detroit MI 48201-1998

WOODBURY, FRANKLIN BENNETT WESSLER, metallurgical engineer; b. Joplin, Mo., Dec. 11, 1937; s. Samuel and Pauline Patricia (Bennett) W. AS, Joplin Jr. Coll., 1963; BS in Metall. Engring., U. Mo., Rolla, 1966. Registered profl. engr., Mo., Minn. Assoc. engr. Uranium div. Mallingckrodt Chem., St. Charles, Mo., 1964; rsch. fellow GM Rsch. Lab., Warren, Mich., 1966; asst. instr. metall. engring. U. Mo., Rolla, 1968-71; rsch. metallurgist Twin Cities Rsch. Ctr., Bur. Mines Dept. Interior, Minn., 1971-80; staff engr. office of dir., div. mineral resources tech. Dept. Interior, Washington, 1980-81, participant deptl. exec. managerial devel. program, 1980-81, mgr. substitute materials rsch., 1981-82, mgr. advanced mining tech. div. conservation and devel. mining rsch., 1982-87, sr. staff engr. for minerals and metals, 1987-91, asst. chief div. policy and regulatory analysis, 1991—. Contbr. papers to profl. publs. and confs. Mem. sci. and tech. resource coun. Minn. Legislature, 1977-80. With USAF, 1957-61. Named Engr. of Yr., Minn., 1978; NDEA grad. fellow, 1967-70. Mem. AIME (sec.-treas. Washington sect. 1984-85, 2d v.p. 1986-87, 1st v.p. 1989-90, pres. 1990-91), Nat. Soc. Profl. Engrs. (chmn. nat. task group on engring. mgmt. 1977-80, bd. govs. profl. engrs. in govt. 1976-86, rep. to organizing com. internat. conf. engring. mgmt. 1986—, res. fund com. 1988-89, audit com. 1989-90, budget com. 1989-90, bd. dirs. 1989-90), Nat. Inst. for Engring. Mgt. and Systems (bd. dirs. 1989-93), Minn. Soc. Profl. Engrs. (exec. com. at-large 1978-80, chmn., vice chmn. coms.), Minn. Engring. Socs. Joint Task Com. on Engring. Edn. (chmn. 1977-80), Mo. Soc. Profl. Engrs., Va. Soc. Profl. Engrs. (pres. George Washington chpt. 1983-84, bd. dirs. 1982-86, v.p. for govt. 1984-86, state pres. 1988-89), Washington Soc. Engrs., Scientists and Engrs. Tech. Assessment Coun. Minn. (AIME rep. to bd. dirs. 1976-78, v.p. 1978-80), Am. Soc. Engring. Mgmt. (chmn. Nat. Capitol Area sect. 1985-86, nat. membership chmn. 1985-86 dir.-at.-large, bd. dirs. 1986-88, nat. pres. elect 1988-90, nat. pres. 1990-92, rep. internat. fedn. engring. mgmt. 1992-93, exec. dir. 1993—), Am. Soc. for Metals, KC, Nat. Capital Inter Fraternity Forum (bd. dirs. 1990—), Sigma Xi, Sigma Pi (grand coun. internat., grand herald 1992—, chmn. edn. and scholarship com. 1989-92, internat. chmn. edn. and acad. com., expansion com., province archon Mo., Ark. 1968-70, Va. 1988-92, mid-atlantic region 1992—), Tau Beta Pi, Alpha Sigma Mu. Roman Catholic.

WOODBURY, RICHARD BENJAMIN, anthropologist, educator; b. West Lafayette, Ind., May 16, 1917; s. Charles Goodrich and Marion (Benjamin) W.; m. Nathalie Ferris Sampson, Sept. 18, 1948. Student, Oberlin Coll., 1934-36; BS in Anthropology cum laude, Harvard U., 1939, MA, 1942, PhD, 1949; postgrad., Columbia U., 1939-40. Archeol. research Ariz., 1938, 39, Fla., 1940, Guatemala, 1947-49, El Morro Nat. Monument, N.Mex., 1953-56, Tehuacan, Mex., 1964; archaeologist United Fruit Co. Zaculeu Project, Guatemala, 1947-50; assoc. prof. anthropology U. Ky., 1950-52, Columbia U., 1952-58; rsch. assoc. prof. anthropology interdisciplinary arid lands program U. Ariz., 1959-63; curator archeology and anthropology U.S. Nat. Mus., Smithsonian Instn., Washington, 1963-69, acting. head office anthropology, 1965-66, chmn. office anthropology, 1966-67; prof., chmn. dept. anthropology U. Mass., Amherst, 1969-73; prof. U. Mass., 1973-81, prof. emeritus, 1981—, acting assoc. provost, dean grad. sch., 1973-74; mem. divsn. anthropology and psychology NRC, 1954-57; bd. dirs. Archaeol. Conservancy, 1979-84, Valley Health Plan, Amherst, 1981-84, Mus. of No. Ariz., 1983-90; liason rep. for Smithsonian Instn., Com. for Recovery of Archeol. Remains, 1965-69; assoc. seminar on ecol. systems and cultural evolution Columbia U., 1964-73; mem. exec. com. bd. dirs. Human Relations Area Files, Inc., New Haven, Conn., 1968-70; cons. Conn. Hist. Commn., 1970-72. Author (with A.S. Trik) The Ruins of Zaculeu, Guatemala, 2 vols., 1953, Prehistoric Stone Implements of Northeastern Arizona, 1954, Alfred V. Kidder, 1973, Sixty Years of Southwestern Archaeology, 1993; editor: (with I.A. Sanders) Societies Around the World (2 vols.), 1953, (with others) The Excavation of Hawikuh, 1966, Am. Antiquity, 1954-58, Abstracts of New World Archaeology; editor-in-chief: Am. Anthropologist, 1975-78; mem. editorial bd.: Am. Jour. Archeology, 1957-72. Mem. sch. com., Shutesbury, Mass., 1979-82, chmn., 1980-82, bd. assessors, 1982-85; chmn. finance com. Friends of Amherst Stray Animals, 1983-85. With USAF, 1942-45. Fellow Mus. No. Ariz., 1985. Fellow Am. Anthrop. Assn. (exec. bd. 1963-66, A.V. Kidder award 1989), AAAS (coun. rep. Am. Anthrop. Assn. 1961-63, com. on desert and arid zones rsch. Southwest and Rocky Mountains divsn. 1958-64, vice-chair 1962-64, com. arid. lands 1969-74, sec. 1970-72), Archeol. Inst. Am. (exec. com. 1965-67); mem. Soc. Am. Archeology (treas. 1953-54, pres. 1958-59, chmn. fin. com. 1987-89, Fiftieth Anniversary award 1985, Disting. Svc. award 1988), Ariz. Archeol. and Hist. Soc., Kroeber Anthrop. Soc., Archeol. Conservancy (life), Sigma Xi. Office: U Mass Dept Anthropology Machmer Hall Amherst MA 01003

WOODE, MOSES KWAMENA, medical educator; b. Takoradi, Ghana, June 27, 1947; came to U.S., 1981; s. Emmanuel Kwamena and Georgina Aba (Arthur) W.; m. Eunice Woode, Aug. 24, 1974; children: Linda-Marie, Timothy. BS in Chemistry and Math., U. Ghana, Legon, Accra, 1971, BS in Chemistry with honors, 1972, MS in Chemistry; PhD in Chemistry, Sch. Tech. and Medicine, London, 1978. Assoc. prof. med. edn. and chemistry Sch. Medicine, U. Va., Charlottesville, 1986—, dir. Assisting Students Achieve Med. Degrees, 1987—, asst. dean, acad. support, 1988-91, assoc. dean, acad. support, 1992—, rsch. assoc. prof. chemistry, assoc. prof. ob-gyn., 1986—. Contbr. articles to profl. jours. Grantee NIH, 1987-92, Robert Wood Johnson Found., 1988—, U.S. Dept. HHS, 1990—. Fellow Am. Inst. Chemists; mem. N.Y. Acad. Scis., Va. Acad. Sci., Am. Chem. Soc., Sigma Xi. Achievements include design of academic enrichment programs at high school, college and medical school levels; contribution to understanding of structure-function relationships of biologically important molecules and liquid crystals. Office: U Va Sch Medicine Box 446 Health Scis Ctr Charlottesville VA 22908

WOODHOUSE, DERRICK FERGUS, ophthalmologist; b. Sutton, Surrey, U.K., May 29, 1927; s. Sydney Carver and Erica (Ferguson) W.; m. Jocelyn Laira Perry, Mar. 9, 1957; children: Karen Tace, Iain Kenrick, Gillian Erica. BM, BCh., Oxford U., Eng., 1951; DO, London Coll., 1956. Intern in medicine, surgery, ophthalmology St. Thomas Hosp., Plymouth, Exeter Hosps., London, 1952-53; registrar in ophthalmology Birmingham (Eng.) Eye. Hosp., 1958-60; sr. registrar in ophthalmology Bristol (Eng.) Eye Hosp., 1960-63; cons. eye surgeon Wolverhampton & Midland Counties Eye Infirmary, Eng., 1963-89; staff opthalmologist Liverpool Hosp., NSW, Australia, 1989—. Contbr. articles to profl. jours.; author: Ophthalmic Nursing, 1980. Mem. Wolverhampton Health Authority, 1970-77; treas. Ophthalmic Nursing Bd., 1970-84, chmn., 1984-88. With RAF, 1953-57. Recipient Gold medal, Nepal Med. Assn., 1989. Fellow Royal Coll. Surgeons, Royal Soc. Medicine, U.K. Coll. Ophthalmologists; mem. Irish Coll. Ophthalmologists, Brit. Computer Soc. Mem. Soc. of Friends.

WOODLAND, N. JOSEPH, optical engineer, mechanical engineer; b. Atlantic City, N.J., Sept. 6, 1921. BS in Mech. Engring., Drexel U., 1947; MME, Syracuse U., 1956. Tech. asst. to unit chief, liquid thermal diffusion project for separating uranium isotopes Manhattan Project, Oak Ridge, Tenn., 1943-46; mech. designer Burlington Industries, 1947; lectr. in mech. engring. Drexel U., 1948-49, cons., 1987; cons. in aircraft hydraulics design, 1950; various positions at staff and sr. levels IBM Corp., 1951-87; cons., 1987—. Recipient Nat. Medal Tech. U.S. Dept. Commerce, 1992; named one of Drexel U.'s 100 Most Outstanding Alumni, 1992. Mem. Anthony J. Drexel Soc. Achievements include patent (with Bernard Silver) for Classifying Apparatus and Method, a bar code system, and the creation of the Universal Product Code system. Office: 426 Van Thomas Dr Raleigh NC 27615*

WOODROW-LAFIELD, KAREN ANN, demographer; b. Fairfield, Ill., Oct. 14, 1950; d. Raymond and Margaret Ann (Simpson) Woodrow; m. William Louis Lafield, July 16, 1991. BA, U. Ill., Chgo., 1972; PhD, U. Ill., Urbana, 1984; MA, U. Tenn., 1976. Demographic statistician Population div. U.S. Bur. Census, Washington, 1984-92; rsch. assoc. Ctr. for Social and Demographic Analysis, SUNY, Albany, 1993—. Author: (with others) Undocumented Migration to the United States, 1984—, Immigration, Reform and Undocumented Residents in the United States, 1992; contbr. articles to profl. jours. Mem. AAAS, Population Assn. of Am., Am. Sociol. Assn., Am. Statis. Assn., So. Demographic Assn. Achievements include research on estimates of undocumented aliens living in the U.S., emigration from the U.S., immigration and census coverage issues.

WOODRUFF, NEIL PARKER, agricultural engineer; b. Clyde, Kans., July 25, 1919; s. Charles Scott and Myra (Christian) W.; m. Dorothy Adele Russ, June 15, 1952; children—Timothy C., Thomas S. B.S., Kans. State U., 1949, M.S., 1952; postgrad., Iowa State U., 1959. Agrl. engr. Agrl. Research Service, Dept. Agr., Manhattan, Kans., 1949-63; research leader Agrl. Research Service, Dept. Agr., 1963-75; cons. engr. Manhattan, 1975-77; civil engr. Kans. Dept. Transp., Topeka, 1977-79; prof., mem. grad. faculty Kans. State U., civil engr. facilities planning, 1979-84; mem. sci. exchange team to, Soviet Union, 1974; with W/PT Cons., 1984—. Contbr. articles to tech. jours. and books. Fellow Am. Soc. Agrl. Engrs. (Hancor Soil Water Engring. award 1975); mem. Sigma Xi, Gamma Sigma Delta. Home and Office: 12906 Blue Bonnet Dr Sun City AZ 85375

WOODRUFF, TERESA K., cell biologist; b. Dec. 7, 1963. BA in Zoology and Chemistry summa cum laude, Olivet Nazarene U., 1985; PhD in Molecular Biology, Cellular Biology and Biochemistry, Northwestern U., Evanston, Ill., 1989. Postdoctoral fellow dept. cell structure R&D Genentech, Inc., South San Francisco, 1989-91, scientist dept. cell structure R&D, 1991—; teaching asst. Olivet Nazarene U., 1983-85, Northwestern U., 1987-88; vis. lectr. Nat. Inst. Sci., Beijing, 1991; speaker in field. Author: (with others) Growth Factors and the Ovary, 1989; contbr. articles to profl. jours.; pub. abstracts. Recipient Grad. Fellow award Abbott Labs., 1987, Cornelia Post Channing Meml. award VII Ovarian Workshop, Ares Serono, 1988, NRSA Tng. award NIH, 1988. Mem. AAAS, Am. Chem. Soc., Am. Fertility Soc., Women in Endocrinology, Endocrine Soc., Phi Delta Lamda, Sigma Xi (Grad. Scientist award 1988). Achievements include patent (with others) for Method of Increasing Fertility in Females, Method of Inhibiting Follicular Maturation in Females, TGF-beta Supergene Family of Receptors. Office: Genentech Inc Dept Cell Culture R&D 460 Point San Bruno Blvd South San Francisco CA 94080

WOODS, ARNOLD MARTIN, geologist; b. Richmond, Calif., Mar. 19, 1954; s. Arnold Martin and Jacqueline Adele (Hendricks) W. BA, San Francisco State U., 1976; MA, U. Tex., 1981. Registered geologist. Unit mgr. The Analysts, Inc., Houston, 1976-79; rsch. asst. Geology Dept., U. Tex., Austin, 1980-81; devel. geologist Phillips Petroleum Co., Borger, Tex., 1981-84; cons. geologist Houston, 1984-85; sr. geologist Argyle Energy, Houston, 1985-86; staff geologist Conoco, Inc., Casper, Wyo., 1986—. Coeditor: AAPG Development Geology Manual, 1993; contbr. articles to profl. jours. including AAPG Bull., Jour. Paleontology, WGA Symposium, Geology. Mem. Wyo. Geol. Assn. (sec. 1992, Best Speaker 1991), Am. Assn. Petroleum Geologists (devel. geology com. 1988—, Levorson award 1992), Geol. Soc. of Am. Achievements include rsch. on causes of dinosaur extinction, clastic depositional systems, improved recovery from mature oil and gas fields.

WOODS, JOHN MERLE, mathematics educator, chairman; b. Monroe, Okla., Jan. 10, 1943; s. Harry A. and Vi Ada (Young) W.; m. Thelma Ann Farrar, Jan 24, 1964 (div. Apr. 1976); children: John M. Jr., Roxanne; m. Paula Gail Newman, Jan 23, 1982 (div. Aug. 1991); children: Benjamin, Paul. BS, Okla. State U., 1965; MA in Teaching, Harvard U., 1966; PhD, Fla. State U., 1973. Tchr. math. John Burroughs Sch., Ladue, Mo., 1966-67; prof. math. Okla. Bapt. U., Shawnee, 1967-88, math. series supr. Upward Bound, 1967-70; mathematician, computer analyst USAMERDC, Ft. Belvoir, Va., summer 1968; prof., chmn. math. dept. Southwestern Okla. State U., Weatherford, 1988—, also dir. Title II workshops, 1989—; supr. rsch. grants NSF, Shawnee, 1974-75. Active disaster relief unit Pott-Lincoln Assn., Shawnee, 1984-88; with carpentry missionary So. Bapt. Ch., Billings, Mont., 1984-86; young adult Sunday sch. tchr. Calvary Bapt. Ch., Shawnee, 1982-87; bd. dirs. Okla. Wildlife Fedn., 1992—, also conservation v.p. Danforth Found. fellow, 1970-73, Harvard U. fellow, 1965-66. Mem. AAUP, Okla Coun. Tchrs. of Math. (state v.p., bd. dirs.). Avocations: photography, computers, gardening, hunting, fishing. Home: 517 N 4th St Thomas OK 73669-0381 Office: Southwestern Okla State U 100 Campus Dr Weatherford OK 73096-3098

WOODS, KAREN MARGUERITE, psychologist, human development, institute executive; b. Columbus, Ga., Sept. 17, 1945; d. O. Norman and Nan Catherine (Land) Shands; m. Carl Allen Oberkrom, Sept. 3, 1966 (div. Aug. 1974); children: Kristi Lynn, Jeffrey Michael; m. James Wallace Woods II, Aug. 19, 1978 (div. Feb. 1986); 1 child, Jamie Elizabeth. Student, Mercer U., 1963; BA, William Jewell Coll., 1975; MA, U. Mo., Kansas City, 1976. Lic. psychologist, Mo. Counselor, juvenile officer Platte County Juvenile Ct., Platte City, Mo., 1976-78; sch. psychologist North Kansas City Sch. Dist., Mo., 1978-80; counselor Platte Med. Clinic, Platte City, 1981-84; cofounder/dir. north office Counseling Ctr. for Human Devel., Kansas City, 1984-87; founder, pres. Inst. for Human Devel., Kansas City, 1987—; mem. com. of psychologists State of Mo., 1990—; cons. Platte Valley Spl. Edn. Coop., Smithville, Mo., 1984-88; speaker to area schs. and bus., 1976—, Northland Child Abuse Speakers Bur., 1976-78; dir., founder Northland Women's Resource Ctr., Kansas City, 1989—. Mem. adv. bd. Mo. Div. Family Services, 1981-83; mem. Northland Child Abuse Task Force, Gladstone, Mo., 1981-83. PEO grantee, 1976; named one of Outstanding Young Women in Am., 1978. Mem. NAFE, Am. Psychol. Assn., Mo. Psychol. Assn., Kansas City Mental Health Assn., Zeta Tau Alpha. Democrat. Methodist. Club: Clayview Country. Avocations: traveling, water skiing, writing. Office: Inst for Human Devel 4444 N Belleview Ave Ste 110 Kansas City MO 64116-1507

WOODS, STEPHANIE ELISE, computer information scientist; b. Kans. City, Kans., July 26, 1962; d. Benoyd Myers and Lee Ann (Parks) Ellison; m. Reginald Elbert Woods. BA in Bus. Adminstrn., Wichita State U., 1984. Adv. mktg. support rep. IBM, Houston, 1985—. Mem. NAFE, Alpha Kappa Alpha, Omicron Delta Kappa. Democrat. Methodist. Avocations: sewing, needlework, puzzles, aerobics, weightlifting. Office: IBM 2 Riverway Houston TX 77056

WOODS, THOMAS BRIAN, physician; b. Perth, Australia, Oct. 17, 1938; s. Ivor Eddington and Mary Margaret (Dwyer) W.; m. Clare Taylor Prince, Dec. 23, 1967; children: Stephen Thomas, Daniel Brian, Michael Patrick. B.Sc., U. Western Australia, 1960, MB, BS, 1965; diploma occupational health, U. Sydney, Australia, 1975. Resident med. officer Fremantle Hosp., 1965; sr. med. registrar Repat. Gen. Hosp., Hollywood, Western Australia, 1971-74; cons. physician Osborne Park Hosp. Occupational/Rehab. Medicine, 1976—. Med. officer Royal Australian Air Force, 1966-70. Recipient Res. Force Decoration, 1991. Fellow Royal Australasian Coll. Physicians, Australian Coll. Occupational Medicine; mem. Australian and New Zealand Soc. Occupational Medicine, Aviation Medicine Soc. Australia and New Zealand, Royal Flying Doctor Svc. Australia, Faculty Occupational Medicine (London), Safety Inst. Australia. Roman Catholic. Avocations: literature, art, music. Home: 101 Tyrell St, Nedlands 6009, Australia Office: 106 Outram St, West Perth 6005, Australia

WOODS, THOMAS FABIAN, lawyer; b. St. Paul, June 27, 1956; s. William Fabian and Maxine Elizabeth (Schmit) W.; m. Rona Pilar Quiñanola, Oct. 26, 1985 (div. Dec. 1991); 1 child, Sara Anne; m. Lora Denise Hammers, Aug. 14, 1992. BS in Geophysics, U. Minn., 1979; MS in Geophysics, U. Wyo., 1987; JD, U. Wis., 1992. Bar: Wis. 1992, U.S. Dist. Ct. (we. dist.) Wis. 1992, U.S. Patent Office 1992; registered geophysicist Calif. Jr. wireline engr. Schlumberger Internat., Argentina, 1979-80; gen. velocity engr. Seismograph Svc. Corp., Houston, 1980-81; seismologist, mgr. Seiscom-Delta United Internat., Ltd., Singapore, 1981-83; exploration geophysicist Amoco Corp., Denver, 1986-88; engring. geophysicist Harding-Lawson Assocs., Novato, Calif., 1988; patent law clk. Honeywell, Inc., Mpls., 1990-91; patent law clk. Rayovac Corp., Madison, Wis., 1991-92, patent atty., 1992—; cons. Bison Instruments, Inc., Mpls., 1985, Kans. Oil Co., Manhattan, 1985. Contbr. articles to Oil and Gas Jour. and SEG. Grantee U. Wyo., 1985-86, U. Wis., 1989-90. Mem. ABA, AAAS, AIPLA, WIPLA, Wis. Intellectual Property Law Soc. (pres. 1990-92). Achievements include rsch. in field of seismology; coordination and supervision of 800-level vertical seismic profile experiment in Sunland Park, New Mex. Home: 2658 Cambrian Cir Fitchburg WI 53711 Office: Rayovac Corp 601 Rayovac Ave Madison WI 53711

WOODS, WALTER EARL, biomedical manufacturing executive; b. Phila., Sept. 26, 1944; s. Walter Earl and Janet I. (Ferguson) W.; m. Anna Maria Gianfreda, Dec. 4, 1975; children: Jeffrey, Elaine, Roberto, Carlo. BS in Biology, Del. Valley Coll. Sci. and Agr., 1966. Pilot plant operator Shell Chem., Woodbury, N.J., 1966-67; virologist, tissue culturist 1st U.S. Med. Lab., N.Y.C. and Ft. Meade, Md., 1967-69; virologist Merck, Sharpe & Dohme, West Point, Pa., 1969-70; quality control and assurance supr. Richardson-Merrell Inc., Swiftwater, Pa., 1970-74; cons., dir. influenza vaccine mfg. Richardson-Merrell Inc., Naples, Italy, 1974-75; mgr. biol. prodn. Richardson-Merrell Inc., Swiftwater, 1976-78; mgr. biol. prodn. Connaught Labs., Inc., Swiftwater, 1978-81, dir. vaccine mfg., 1982-84, dir. mfg. resource planning, class A rating, 1984-88, dir. product devel. and mgmt., 1989-91; dir. project mgmt., chmn. bus. groups Pasteur Mérieux Connaught, 1991—; bd. dirs. Connaught-Daiichi Joint Venture, Tokyo; project dir. licensing Acellular Pertussis and Japanese Encepahlitis Vaccines, 1992, H.B. Vaccine, 1993. Bd. dirs. Northeastern Pa. Indsl. Resource Ctr., Wilkes-Barre, Pa., 1988-91. Mem. ASM, Pharm. Mfr.'s. Avocations: soccer, music, gardening, reading. Home: 53 Deerfield Way Scotrun PA 18355 Office: Connaught Labs Inc Box 187 Swiftwater PA 18370

WOODS, WALTER RALPH, animal scientist, research administrator; b. Grant, Va., Dec. 2, 1931; s. John Wythe and Hazel Gladys (Hash) W.; m. Jacqulyn Rose Miller, Sept. 14, 1953; children: Neal Ralph, Diana Lyn. B.S., Murray (Ky.) State U., 1954; M.S., U. Ky., 1955; Ph.D., Okla. State U., 1957. Instr. animal sci. Okla. State U., 1956-57; asst. prof., then assoc. prof. Iowa State U., 1957-62; assoc. prof., then prof. U. Nebr., 1962-71; prof. animal sci., head dept. Purdue U., 1971-85; dean Kans. State U., Manhattan, 1985-92, dir. Agrl. Expt. Sta., 1985-92, dir. Coop. Ext. Svc., 1987-92; asst. adminstr. for regional rsch. CSRS USDA, Washington, 1993—. Author papers, articles in field. Bd. dirs. Ind. 4-H Found., 1979-81, Kans. 4-H Found., 1987-92; mem. leadership coun. Kans. Value Added Ctr., 1990-92; mem. exec. com. Kans. Rural Devel. Coun., 1990-93; chair Coun. Adminstrv. Head Agr., 1992, Great Plains Agr. Coun., 1990-92. Recipient Disting. Agrl. Alumni award Murray State U., 1969, Meritorious Service award Ind. Pork Producers Assn., 1975. Mem. Am. Soc. Animal Sci. (sec.-treas. Midwest sect. 1979-81, pres. Midwest sect. 1983-84), Sigma Xi, Gamma Sigma Delta. Mem. Disciples of Christ Ch. Home: 5901 Mount Eagle Dr # 308 Alexandria VA 22303 Office: USDA Coop States Rsch Svc Aerospace Bldg 901 D St SW Washington DC 20250-2200

WOODSIDE, BERTRAM JOHN, infosystem engineer; b. Danville, Pa., Apr. 20, 1946; s. Cyrus G. and Almerta T. (Kitchen) W.; m. Doreen Knowles; 1 child, Russell. BS, USAF Acad., 1968. Cert. purchasing mgr. Commissioned 2d lt. USAF, 1968, advanced through grades to capt., 1971, resigned, 1976; plant engr. Linde div. Union Carbide Corp., Pitts., 1976, distribution supt., 1977-78; region purchasing mgr. Linde div. Union Carbide Corp., Cleve., 1979-82, region tech. supr., 1983; process analyst Linde div. Union Carbide Corp., Lorain, Ohio, 1984-91; mgr. Process Control Svc. Ctr., Praxair, Inc., Tonawanda, N.Y., 1991—. Decorated DFC with oak leaf cluster, Air medal with seven oak leaf clusters. Mem. Bay Boat Club (sec. 1984-90). Avocations: golf, fishing, sailing. Office: Praxair Inc PO Box 44 Tonawanda NY 14151-0044

WOODSON, HERBERT HORACE, electrical engineering educator; b. Stamford, Tex., Apr. 5, 1925; s. Herbert Viven and Floy (Tunnell) W.; m. Blanche Elizabeth Sears, Aug. 17, 1951; children: William Sears, Robert Sears, Bradford Sears. SB, SM, MIT, 1952, ScD in Elec. Engring., 1956. Registered profl. engr., Tex., Mass. Instr. elec. engring., also project leader magnetics div. Naval Ordnance Lab., 1952-54; mem. faculty M.I.T., 1956-71, prof. elec. engring., 1965-71, Philip Sporn prof. energy processing, 1967-71; prof. elec. engring., chmn. dept. U. Tex., Austin, 1971-81, Alcoa Found. prof., 1972-75, Tex. Atomic Energy Research Found. prof. engring., 1980-82, Ernest H. Cockrell Centennial prof. engring., 1982-93, dir. Center for Energy Studies, 1973-88, assoc. dean devel. and planning Coll. Engring., 1986-87, acting dean, 1987-88, dean' chair for excellence in engring., 1988—; staff engr. elec. engring. div. AEP Service Corp., N.Y.C., 1965-66; cons. to industry, 1956—. Author: (with others) Electromechanical Dynamics, parts I, II, III. Served with USNR, 1943-46. Recipient NSPE Engineer of the Year award, 1990. Fellow IEEE (pres. Power Engring. Soc. 1978-80); mem. Am. Soc. Engring. Edn., Nat. Acad. Engring., AAAS. Achievements include patents in field. Home: 7603 Rustling Rd Austin TX 78731-1333 Office: U Tex Coll Engineering Austin TX 78712

WOODWARD, KENNETH EMERSON, retired mechanical engineer; b. Washington, Oct. 30, 1927; s. George Washington and Mary Josephine (Compton) W.; m. Mary Margaret Eungard, Mar. 29, 1956; children: Stephen Mark, Kristi Lynn. BME, George Washington U., 1949, M Engring. Adminstrn., 1960; MS, U. Md., 1953; PhD, Am. U., 1973. Mech. engr. naval Rsch. Lab., Washington, 1950-54; value engring. program mgr. Harry Diamond Labs., Washington, 1955-74; sci. adviser U.S. Army Med. Bioengring. R & D Lab., Ft. Detrick, Md., 1974-75; mech. engr. Woolcott & Co., Washington, 1975-90; ret., 1990. Author: Solar Energy Applications for the Home, 1978; contbr. to profl. publs. With U.S. Army, 1946-47. Mem. ASME, Am. Soc. for Artificial Intelligence, Am. Soc. Artificial Internal Organs. Republican. Baptist. Achievements include 12 U.S. and 2 foreign patents, development of artifical human heart. Home: 1701 Hunts End Ct Vienna VA 22182

WOODWARD, PAUL RALPH, computational astrophysicist, applied mathematician, educator; b. Rockville Center, N.Y., Aug. 25, 1946; s. William Redin and Edith (Jones) W.; m. Judith Hansburg, Dec. 28, 1972; children: Thomas, Theodore. BA, Cornell U., 1967; PhD, U. Calif., Berkeley, 1973. Physicist Lawrence Livermore Nat. Lab, Livermore, Calif., 1968-71, computational physicist, 1978-85; rsch. assoc. Nat. Radio Astronomy Observatory, Charlottesville, Va., 1974-75; rsch. assoc. observatory Leiden U., The Netherlands, 1975-78; prof. astronomy U. Minn., Mpls., 1985—; dir. graphics and visualization Army High Performance Computing Rsch. Ctr., Mpls., 1990—. National Merit scholar Cornell U., 1963-67; Woodrow Wilson fellow U. Calif., Berkeley, 1967, Whiting fellow, 1968. Fellow Minn. Supercomputer Inst.; mem. Internat. Astronomical Union, Soc. for Indsl. and Apllied Mathematics, N.Y. Acad. Sci., Phi Beta Kappa, Phi Eta Sigma. Research and discoveries (with others) include Simple Line Interface Calculation fluid interface tracking technique for computational fluid dynamics, MUSCL code high-order Godunov method for compressible gas dynamics, Piecewise-Parabolic Method, Fortran-P stylized F-77 for parallel computation high speed computer graphics, animation of computational fluid dynamics data. Office: U Minn APHCRC 1100 Washington Ave S Minneapolis MN 55415

WOODWELL, GEORGE MASTERS, ecologist, educator, author, lecturer; b. Cambridge, Mass., Oct. 23, 1928; s. Philip McIntire and Virginia (Sellers) W.; m. Alice Katharine Rondthaler, June 23, 1955; children: Caroline Alice, Marjorie Virginia, Jane Katharine, John Christopher. AB, Dartmouth Coll., 1950; AM, Duke U., 1956, PhD, 1958; DSc (hon.), Williams Coll., 1977, Miami U., 1984, Carleton Coll., 1988, Muhlenberg Coll., 1990. Mem.

faculty U. Maine, 1957-61, assoc. prof. botany, 1960-61; vis. asst. ecologist, biology dept. Brookhaven Nat. Lab., Upton, N.Y., 1961-62; ecologist Brookhaven Nat. Lab., 1965-67, sr. ecologist, 1967-75; founder, dir. Ecosystems Center, 1975-85; dep. and asst. dir. Marine Biol. Lab., Woods Hole, Mass., 1975-76; founder, pres. and dir. Woods Hole Research Ctr., 1985—; lectr. Yale Sch. Forestry, 1967—; chmn. Conf. on Long Term Biol. Consequences of Nuclear War, 1982-83. Editor: Ecological Effects of Nuclear War, 1965, Diversity and Stability in Ecological Systems, 1969, (with E.V. Pecan) Carbon and the Biosphere, 1973, The Role of Terrestrial Vegetation in the Global Carbon Cycle: Measurement by Remote Sensing, 1984, The Earth in Transition: Patterns and Processes of Biotic Impoverishment, 1990; (with K. Ramakrishna) Forests for the Future, 1993. Founding trustee Environ. Def. Fund, 1967; founding trustee Natural Resources Def. Council, 1970, vice chmn., 1974—; founding trustee World Resources Inst., 1982—; bd. dirs. World Wildlife Fund, 1970-84, chmn., 1980-84; bd. dirs. Conservation Found., 1975-77, Ruth Mott Fund, 1984-91, chmn. 1989-91; bd. dirs. Ctr. for Marine Conservation, 1990—; adv. com. TMI Pub. Health Fund., 1980—. Fellow AAAS, Am. Acad. Arts and Scis.; mem. NAS, Brit. Ecol. Soc., Ecol. Soc. Am. (v.p. 1966-67, pres. 1977-78), Sea Edn. Assn. (bd. dirs. 1980-85), Sigma Xi. Rsch., pub. on structure and function of natural communities, biotic impoverishment, especially ecological effects of ionizing radiation, effects of persistent toxins, world carbon cycle and warming of the earth. Office: Woods Hole Research Ctr Box 296 Fisher House 13 Church St Woods Hole MA 02543

WOOLARD, HENRY WALDO, aerospace engineer; b. Clarksburg, W.Va., June 2, 1917; s. Herbert William and Elsie Marie (Byers) W.; m. Helen Stone Waldron, Aug. 16, 1941; children: Shirley Ann, Robert Waldron. BS in Aero. Engring., U. Mich., 1941; MSME, U. Buffalo, 1954. Aero. engr. NACA, 1941-46; assoc. prof., acting dept. head, aero. engr. W. Va. U., Morgantown, W.Va., 1946-48; rsch. aerodynamicist Cornell Aero. Lab., Buffalo, 1948-57; sr. staff engr. applied physics lab. Johns Hopkins U., Silver Spring, Md., 1957-63; sr. rsch. specialist Lockheed Calif. Co., Burbank, Calif., 1963-67; mem. tech. staff TRW Systems Group, Redondo Beach, Calif., 1967-70; pres. Beta Tech. Co., Palos Verdes, Calif., 1970-71; aero. engr. Air Force Flight Dynamics Lab., Wright-Patterson AFB, Ohio, 1971-85; aero. cons. Dayton, Ohio, 1985-87, Fresno, Calif., 1987—. Contbr. articles to profl. jours. Recipient Scientific Achievement award Air Force Systems Command, 1982. Assoc. fellow AIAA; mem. ASME, Sigma Xi, Sigma Pi Sigma. Office: Consultant Aerospace 1249 W Magill Ave Fresno CA 93711

WOOLDRIDGE, SCOTT ROBERT, semiconductor process engineer; b. Spokane, Wash., Feb. 25, 1960; s. Robert Elmo and Shirley Ann (Schmidt) W.; m. Ginette Marie Gillis, Dec. 29, 1983; children: Robert Scott, James Michael. BS in Chemistry, No. Airz. U., 1982. Plasma etch process engr. Tex. Instruments, Inc., Dallas, 1984—, individual contbr. GaAs plasma etch and deposition process applications. Mem. Am. Chem. Soc., Am. Inst. Chemists, Phi Kappa Phi. Office: Tex Instruments Inc PO Box 226015 M/S 404 Dallas TX 75266

WOOLF, ERIC JOEL, analytical chemist; b. Phila., Nov. 24, 1960; s. Harold Lahn and Judith (Silverman) W. BA, LaSalle Coll., 1982; PhD, Seton Hall U., 1986. Rsch. chemist Berlex Labs., Cedar Knolls, N.J., 1986-88, sr. rsch. chemist, 1988-90; rsch. fellow Merck Rsch. Labs., West Point, Pa., 1990—. Mem. Am. Chem. Soc., Sigma Xi. Office: Merck Rsch Labs WP26-372 West Point PA 19486

WOOLF, NANCY JEAN, neuroscientist, educator; b. Ft. Sill, Okla., July 27, 1954; d. Lee Allen and Rachel Christine (Sedjo) W.; m. Larry Lee Butcher, Dec. 24, 1983; children: Lawson Frederick, Ashley Ellen. BS, UCLA, 1978, PhD, 1983. Grad. researcher UCLA, 1979-83, asst. rsch. neuroscientist, 1984-92, adj. assoc. prof., 1992—, assoc. rsch. neuroscientist, 1992—. Author: (review article) Progress in Neurobiology, 1991; (original rsch.) Proceedings Nat. Acad. Sci., 1993; contbr. 40 articles to various sci. pubs. Recipient Colby prize Sigma Kappa Found., Indpls, 1990; named Woman of Yr. Coll. of the Desert, Palm Desert, Calif., 1976. Mem. AAAS, Assn. Acad. Women (Grad. Woman of Yr. UCLA 1983), Internat. Neural Network Soc., Soc. for Neurosci. Achievements include compilation of complete map of central nervous system neurons that utilize the neurotransmitter acetylcholine; MAP-2 alterations with classical conditioning. Office: U Calif Dept Psychology 405 Hilgard Ave Los Angeles CA 90024-1563

WOOLLAM, JOHN ARTHUR, electrical engineering educator; b. Kalamazoo, Mich., Aug. 10, 1939; s. Arthur Edward and Mildred Edith (Hakes) W.; children: Catherine Jane, Susan June. BA in Physics, Kenyon Coll., 1961; MS in Physics, Mich. State U., 1963, PhD in Solid State Physics, 1967; MSEE, Case-Western Res. U., 1978. Rsch. scientist NASA Lewis Rsch. Ctr., Cleve., 1967-80; prof. U. Nebr., Lincoln, 1979—, dir. Ctr. Microelectronic and Optical Materials Rsch., 1988—; pres. J.A. Woollam co., Inc., Lincoln, 1987—. Editor Jour. Applied Physics Com., 1979—. Grantee NASA, NSF, USAF, Def. Advanced Rsch. Projects Agy. Fellow Am. Phys. Soc.; mem. Am. Vacuum Soc. (chmn. thin film div. 1989-91). Office: U Nebr Dept Elec Engring 209NWSEC Lincoln NE 68588-0511

WOOLLEN, KENNETH LEE, electrical engineer; b. Charlotte, N.C., Aug. 12, 1967; s. Joseph Wesley and Alberta Margaret (Briggs) W.; m. Della Rose Laviner, June 1, 1991. BSEE, N.C. State U., 1990. Co-op Carolina Power & Light, Raleigh, 1987-88; co-op CIA, Washington, 1988-89; mktg. sales acct. Nynex, Raleigh, 1990; dedication engr. MKW Power Systems, Rocky Mount, N.C., 1991-92, project engr., 1992—. Eagle Scout Boy Scouts Am., 1985. Mem. IEEE. Democrat. Presbyterian. Home: 918 Beddingfield Dr Knightdale NC 27545 Office: MKW Power Systems 301 S Church St Rocky Mount NC 27804

WOOLLEY, GAIL SUZANNE, military officer; b. Chgo., Sept. 12, 1965; d. David Alan and Margaret (McWhorter) W. BS in Civil Engring., Rutgers U., 1989. Enlisted U.S. Army, 1989, advanced through grades to lt., 1993. Recipient U.S. Army scholarship, Rutgers U., Piscataway, N.J., 1985, Commendation medal, U.S. Army, Ft. Wood, Mo., 1992, Commendation medal, U.S. Army, Baumholder, Germany, 1991. Mem. Soc. Am. Mil. Engrs. (scholarship recipient 1987), ASCE, U.S. Rowing Assn. Avocations: rowing (crew), running, weight training. Home: 72 Sylvan Dr Morris Plains NJ 07950

WOOLLEY, GEORGE WALTER, biologist, geneticist, educator; b. Osborne, Kans., Nov. 9, 1904; s. George Aitcheson and Nora Belle (Jackson) W.; m. Anne Geneva Collins, Nov. 2, 1936; children: George Aitcheson, Margaret Anne, Lawrence Jackson. B.S., Iowa State U., 1930; M.S., U. Wis., 1931, Ph.D., 1935. Fellow U. Wis., 1935-36; mem. staff Jackson Meml. Lab., Bar Harbor, Maine, 1936-49; bd. dirs. Jackson Meml. Lab., 1937-49, v.p. bd., 1943-47, asst. dir. and sci. adminstr., 1947-49, vis. research assoc., 1949—; mem. chief div. steroid biology Sloan-Kettering Inst., N.Y.C., 1949-58, prof. biology, 1949-58; prof. biology Sloan-Kettering Inst. div. Cornell U. Med. Coll., Ithaca, N.Y., 1951—, chief div. human tumor exptl. chemotherapy, 1958-61, chief div. tumor biology, 1961-66; assoc. scientist Sloan-Kettering Inst. Cancer Research, 1966—; health sci. adminstr., program coordinator, head biol. scis. sect. Nat. Inst. Gen. Medical Scis., NIH, 1966-85; cons. Nat. Edn. Service U.S., Washington, 1961—; spl. cons. to Nat. Cancer Inst., NIH, 1956—; Mem. Expert Panel on Carcinogenicity, unio intern. contra cancerum, 1962—; mem. panel com. on growth NRC, 1945-51; mem. several internat. med. congresses. Author chpts. in med. books; mem. editorial bd. Jour. Nat. Cancer Inst., 1947-50. Trustee Dalton Schs., N.Y.C. Fellow AAAS, N.Y. Acad. Sci.; mem. Am. Mus. Natural History, Nat. Sci. Tchrs. Assn. (cons. 1961—), Am. Assn. Cancer Research (dir. 1951-54), Am. Soc. Human Genetics, Mt. Desert Island Biol. Lab., Soc. Exptl. Biology and Medicine, Am. Inst. Biol. Scis., Am. Assn. Anatomists, Am. Genetic Assn., Wis. Acad. Arts Sci. and Letters, Jackson Lab. Assn., Genetics Soc. Am., Environ. Mutagen Soc., Sigma Xi. Clubs: Bar Harbor, Bar Harbor (Maine) Yacht. Achievements include advanced research, teaching and administration in genetics, from the beginnings of the field to present; contributions to fields of cancer, virology, endocrinology and molecular biology. Home: 5301 Westbard Cir Apt 336 Bethesda MD 20816-1427

WOOLVERTON, CHRISTOPHER JUDE, biopharmaceutical company executive; b. Trenton, N.J., Jan. 9, 1960; s. Paul and Rita (DeBonis) W.; m. Nancy Jo Harshberger, June 29, 1985; children: Lyssa, Samantha, Abbey. BS, Wilkes U., Wilkes-Barre, Pa., 1982; PhD, W.Va. U., 1986. Postdoctoral fellow U. N.C., Chapel Hill, 1986-88; asst. prof. Austin Coll., Sherman, Tex., 1988-92; scientist Scios-Nova Inc., Balt., 1993—; cons. Core Ctr. in Diarrheal Disease, Chapel Hill, 1989. Recipient Phoenix award Am. Heart Assn., Morgantown, W.Va., 1984, Rsch. Grant, U. N.C., Chapel Hill, 1988, Postdoctoral Rsch. award U. N.C., Chapel Hill, 1988; named Disting. Alumnus Wilkes U., Wilkes-Barre, Pa., 1992. Mem. Soc. Mucosal Immunology, Am. Soc. Microbiology (grad. rsch. awards Alleghany br. 1983, 85), Sigma Xi, Psi Chi. Democrat. Methodist. Achievements include research in vet. immunology and immunopathology, agents and actions. Office: Scios-Nova Inc 6200 Freeport Centre Baltimore MD 21224

WOOSLEY, STANFORD EARL, astrophysicist; b. Texarkana, Tex., Dec. 8, 1944; s. Homer Earl and Wanda Faye (Fisher) W.; m. Susan Marie Haas, Aug. 12, 1993. BA in Physics, Rice U., 1966, PhD in Space Sci., 1971. Rsch. assoc. Rice U., Houston, 1971-73; rsch. fellow in physics Kellogg Radiation Lab. Calif. Tech., Pasadena, Calif., 1973-75; asst. prof. U. Calif., Santa Cruz, 1975-78, assoc. prof., 1978-83, prof. astronomy and astrophysics, 1983—, chmn. dept., 1984-87, 89-91; sci. adv. com. Inst. Thoeretical Physics, Santa Barbara, 1991—. Editor: High Energy Transients, 1984, Supernova, 1991, Nuclear Astrophysics, 1993; contbr. over 150 articles to profl. jours. Grantee NASA, NSF, Calif. Space Inst., 1977—. Fellow Am. Phys. Soc.; mem. Am. Astron. Soc. (coun. 1990-93, chair com. to aid astronomy in the FSU). Achievements include research into the lives and deaths of massive stars and supernova, nucleosynthesis and x-ray and gamma ray bursts. Home: 115 Auburn Ave Santa Cruz CA 95060-6231 Office: U Calif Astronomy Dept Santa Cruz CA 95064

WOOTAN, GERALD DON, osteopathic physician, educator; b. Oklahoma City, Nov. 19, 1944; s. Ralph George and Corrinne (Loafman) W. BA, Cen. State U., Edmond, Okla., 1970, BS, 1971; MEd, Cen. State U., 1974; MB, U. Okla., Oklahoma City, 1978; DO, Okla. State U., 1985. Dir. mfg. engring. lab. GE, Oklahoma City, 1965-70; counseling psychologist VA Hosp., Oklahoma City, 1970-76; physician asst. Thomas (Okla.) Med. Clin., 1978-81; pvt. practice, Jenks, Okla., 1986—; intern Tulsa Regional Med. Ctr., 1985-86; assoc. prof. Okla. State U. Coll. Osteo. Medicine, 1986—; sec. Green Country Physicians Group, Inc., Tulsa, 1990-91; chmn. gen. practice quality assurance Tulsa Regional Med. Ctr., 1989-91; v.p. New Horizons Counseling Ctr., Clinton, Okla., 1977-81; sr. aviation med. examiner FAA, Tulsa, 1991—; pres. S.W. Diagnostics, Inc., Tulsa, 1989-91, Okla. Edn. Found. Osteo Medicine, Tulsa, 1988-89; pres., bd. trustees Tulsa Long Term Care Authority. Contbr. articles to profl. jours.; patentee for human restraint. Advancement chmn., chmn. Eagle bd. rev. Boy Scouts Am., Tulsa, 1987-88; trustee Tulsa Long Term Care Authority, 1988-91; trustee Tulsa Community Found. for Indigent Health Care, Inc., 1988-91. With USN, 1962-64. Named Clin. Preceptor of Yr., U. Okla., 1980, Outstanding Alumni award Okla. State U. Coll. Osteo. Medicine, 1990. Mem. Am. Osteo. Assn., Okla. Osteo. Assn., Tulsa Dist. Osteo. Soc. (pres. 1991-92), Am. Acad. Physician Assts., Am. Coll. Gen. Practitioners, Okla. Acad. Gen. Practitioners (v.p.), Am. Coll. Osteo. Family Physicians (bd. cert. 1993, pres. Okla. chpt. 1993-94), Okla. State U. Coll. Osteo. Medicine Alumni Assn. (pres. 1988-89). Avocations: scuba diving, aviation medicine. Home: 4320 E 100th St Tulsa OK 74137 Office: Jenks Health Team 324 W Main St Jenks OK 74037-3774

WOOTEN, FRANK THOMAS, research facility executive; b. Fayetteville, N.C., Sept. 24, 1935; s. Frank Thomas and Katherine (McRae) W.; m. Linda Walker, July 14, 1962; children: Laurin Walker, Patrick Thomas, Ashley Tripp. B.S.E.E., Duke U., 1957, Ph.D., 1964. Engr. Corning Glass Works, Raleigh, N.C., 1964-66; engr. Research Triangle Inst., Research Triangle Park, N.C., 1966-68, mgr. biomed. engring., 1968-75, exec. asst. to pres., 1975-80, v.p., 1980-89; pres. Research Triangle Inst., Research Triangle Park, 1989—; Bd. dirs. N.C. Biotech. Ctr., Microelectronics Ctr. N.C. Contbr. articles on semiconductors and biomed. engring. to profl. publs., 1966-83; patentee semiconductors tech. Bd. dirs. N.C. Biotech. Ctr., 1989—, Microelectronics Ctr., N.C., 1989—; corp. mem. Nat. Inst. Statis. Scis., 1990—. Served to lt. (j.g.) USN, 1957-59. Shell fellow, 1961. Mem. IEEE, Assn. Advancement Med. Instrumentation (chmn. com. on aerospace tech. 1971-77), Strategic Def. Initiative Orgn. (tech. application rev. panel). Baptist. Office: Research Triangle Inst PO Box 12194 3040 W Cornwallis Rd Research Triangle Park NC 27709

WOOTEN, FREDERICK (OLIVER), applied science educator; b. Linwood, Pa., May 16, 1928; s. Frederick Alexander and Martha Emma (Guild) W.; m. Jane Watson MacPherson, Aug. 30, 1952; children: Donald, Bartley. BS in Chemistry, MIT, 1950; PhD in Chemistry, U. Del., 1955. Sr. scientist Lawrence Livermore (Calif.) Lab., 1957-72; prof. applied sci. U. Calif., Davis, 1972—, chmn. dept. applied sci., 1973—; vis. prof. physics Drexel U., Phila, 1964, Chalmers Tekniska Högskola, Göteborg, Sweden, 1967-68, Heriot-Watt U., Edinburgh, Scotland, 1979, Trinity Coll., Dublin, Ireland, 1986; staff physicist All-Am. Engring. Co., Wilmington Del., 1955-57; cons. in field. Author: Optical Properties of Solids, 1972. Mem. AAAS, AM. Phys. Soc., N.Y. Acad. Sci., Materials Rsch. Soc., Sigma Xi. Home: 2328 Alameda Diablo Diablo CA 94528 Office: Univ Calif Dept Applied Sci Davis CA 95616

WOOTTON, JOEL LORIMER, nuclear technologist, health physics consultant; b. Jacksonville, Fla., Jan. 27, 1954; s. James Carter and Jean (Lorimer) W. BA, Fla. Tech. U., 1977; AS, Ctrl. Fla. C.C., Ocala, 1981; postgrad., U. Fla., 1981-86; BS, SUNY, N.Y.C., 1992. Cons., contractor Fla. Power and Light Co., Ft. Pierce, 1981-82; cons., contractor Carolina Power and Light Co., Hartsville, S.C., 1982-84, N.Y. Power Authority, Peekskill, 1984-87, Ariz. Nuclear Power Project, Phoenix, 1987-90, South Tex. Project Houston Light and Power, Palacios, 1990-91, So. Nuclear Operating Co., Dothan, Ala., 1992—; sr. nuclear technologist JLW Nuclear Svcs., Mt. Dora, Fla. Mem. Am. Nuclear Soc., Health Physics Soc. Office: JLW Nuclear Svcs 6363 Dora Dr Mount Dora FL 32757

WOREK, WILLIAM MARTIN, mechanical engineering educator; b. Joliet, Ill., May 7, 1954; s. Joseph and Marjorie Ann (Peterson) W.; m. Mary Jane Hrubos, July 27, 1985; children: Christopher, Michael. BS, Ill. Inst. Tech., 1976, MS, 1977, PhD, 1980. Instr. Ill. Inst. Tech., Chgo., 1977-80, vis. asst. prof., 1980-83, asst. prof., 1983-86; assoc. prof. U. Ill., Chgo., 1986—; cons. Airtite-Airtex Inc., Chgo., 1976—, Atlas Electric Devices Inc., Chgo., 1987—, Eaton Corp., Milw., 1989-91. Contbr. articles to profl. jours. Grantee Kaiser Chems., 1987, LaRoche Chems., 1991, Gas Rsch. Inst., 1988, 92. Mem. ASME (chair 1993-94), ASHRAE (rsch. sub. 1989—), Sigma Xi, Tau Beta Pi. Achievements include development of experimental facilities to test desiccant material samples and desiccant matrix core sections, of experimentally validated numerical simulation routines for open- and closed-cycle adsorption cooling and heat pump systems. Office: U Ill at Chgo Dept Mech Engring M/C 251 Box 4348 Chicago IL 60680-4348

WORGUL, BASIL VLADIMIR, radiation scientist; b. N.Y.C., June 30, 1947; s. John and Stephanie (Litwin) W.; m. Kathleen R. Hennessey, June 14, 1969; children: Ronald Adam, Suzanne Kathleen. BS, U. Miami, Fla., 1969; PhD, U. Vt., 1974. Rsch. assoc. Columbia U., N.Y.C., 1975-78, asst. prof., 1979-84, assoc. prof., 1984-90, dir. Eye Radiation & Environ. Rsch. Lab., 1984—, prof., 1990—; cons. Nat. Coun. on Radiation Protection, Washington, 1988—, Internat. Com. on Radiation Protection, 1988—, NASA, 1990-92; dir. Residents Basic Sci., Dept. Ophthalmology, Columbia U., 1983-92. Co-organizer Citizens to Preserve Warwick, N.Y., 1986. NIH grantee, 1977—, NASA grantee, 1987—, Dept. Energy grantee, 1990—, DNA grantee, 1992—; Robert McCormick Rsch. scholar, 1987-88. Mem. Assn. for Rsch. in Vision and Ophthalmology, Radiation Rsch. Soc., Acad. Scis. Ukraine (fgn.). Achievements include definition of the mechanism of radiation cataract development; discovery of Terminal Body-a cellular organelle; description of the preferred dynamic reorientation of the mitotic spindle in an adult epithelium. Office: Columbia U 630 W 168th St New York NY 10032-3702

WORKMAN, GEORGE HENRY, engineering consultant; b. Muskegon, Mich., Sept. 18, 1939; s. Harvey Merton and Bettie Jane (Meyers) W.; Asso. Sci., Muskegon Community Coll., 1960; B.S.E., U. Mich., 1966, M.S.E., 1966, Ph.D., 1969; m. Vicki Sue Hanish, June 17, 1967; children—Mark, Larry. Prin. engr. Battelle Meml. Inst., Columbus, Ohio, 1969-76; pres. Applied Mechanics Inc., Longboat Key, Fla., 1976—; instr. dept. civil engring. Ohio State U., 1973, 82. Served with USN, 1961-64. Named Outstanding Undergrad. Student, Engring. Mechanics dept. U. Mich., 1965-66, Outstanding Grad. Student, Civil Engring. dept., 1968-69. Registered profl. engr., Ohio. Mem. Am. Acad. of Mechanics, ASME, ASCE, Nat. Soc. Profl. Engrs., Sigma Xi, Chi Epsilon, Phi Kappa Phi, Phi Theta Kappa. Congregationalist. Contbr. tech. papers to nat. and internat. confs. Home and Office: 3431 Bayou Ct Longboat Key FL 34228-3028

WORKMAN, JEROME JAMES, JR., chemist; b. Northfield, Minn., Aug. 6, 1952; s. Jerome James and Louise Mae (Sladek) W.; m. Rebecca Marie Zittel, Aug. 3, 1974; children: Cristina Louise, Stephannie Michelle, Daniel Jerome, Sara Marie, Michael Timothy. BA with honors, St. Mary's Coll., Winona, Minn., 1976, MA, 1980; PhD, Columbia Pacific U., San Rafael, Calif., 1984; postgrad., Columbia U., 1990-91. Pres. Biochem. Cons., Mankato, Minn., 1982-84; sr. chemist Technicon Instruments, Tarrytown, N.Y., 1984-87; supervising scientist Bran & Luebbe/Technicon, Tarrytown, 1987; sr. scientist Hitachi Instruments, Danbury, Conn., 1987-89; mgr. tech. support NIR Systems/Perstorp Analytical, Silver Spring, Md., 1989-90, mgr. mktg., 1990-92, dir. mktg., 1992-93; assoc. advisor Inst. Textile Tech., Charlottesville, Va., 1992—; prin. engr. Real-Time Systems div. Perkin Elmer Corp., Norwalk, Conn., 1993—; instr. Fedn. Analytical Chemistry and Spectroscopy Socs. Author: Near-Infrared Analysis, 1991, Introduction to Near-Infrared Spectroscopy, 1993; co-author: Statistics in Spectroscopy, 1991, (series) Chemometrics in Spectroscopy; co-editor: The Academic Press Handbook of Molecular Spectroscopy, 1993-94; contbg. editor Spectroscopy Mag.; contbr. articles to profl. jours. Am. Heart Assn. H.N. and H.B. Shapira scholar, 1971, 72; grantee NSF, 1977, 78. Fellow Am. Inst. Chemists; mem. ASTM (chmn. quantitative practice working group), Am. Chem. Soc. (instr. course on Practical Near-IR Analysis), Assn. Ofcl. Analytical Chemists Internat., Soc. for Applied Spectroscopy, Coun. Near-Infrared Spectroscopy (sec.), Joint Com. Atomic and Molecular Phys. Data (chmn., exec. coun.), Nat. Honor Soc., Sigma Xi, Delta Epsilon Sigma. Achievements include research in near infrared dispersive instrumentation, statistics and chemometrics. Office: Real-Time Systems Div Perkin Elmer Corp 761 Main Ave Mail Stop 201 Norwalk CT 06859-0001

WORKMAN, JOHN MITCHELL, chemist; b. Uniontown, Pa., Oct. 25, 1949; s. Hugh Lawrence and Mary Louise (Mitchell) W.; m. Gayle Sue Zappin, Nov. 20, 1987. BA in Psychology, Miami U., Oxford, Ohio, 1971; MS in Edn., Kans. State U., 1976; MS in Chemistry, U. Cin., 1985, PhD in Chemistry, 1987. Teaching and rsch. asst. dept. chemistry Wright State U., Dayton, Ohio, 1977-81; grad. teaching asst. U. Cin., 1982-83, grad. rsch. asst., 1983-86; sr. scientist Chemsys Inc., Fairborn, Ohio, 1986-89, dir. elemental analysis, 1989—. Contbr. articles to jours. Analytical Chemistry, Applied Spectroscopy. Vol. VA Med. Ctr., Dayton, Ohio, 1989—. Sgt. U.S. Army, 1972-75. Mem. Am. Chem. Soc., Am. Phys. Soc., Soc. Applied Spectroscopy, Sigma Xi, Sigma Pi Sigma. Roman Catholic. Achievements include research in spectroscopic studies of fundamental plasma characteristics; trace and compositional analysis of aerospace materials including metals and metal alloys by optical emission spectrometry. Home: 1379 Sunset Dr Fairborn OH 45324-5649 Office: Chemsys Inc PO Box 1649 Fairborn OH 45324-7649

WORKS, MADDEN TRAVIS, JR. (PAT WORKS), operations executive, author, skydiving instructor, skydiving publications executive; b. Harris County, Tex., Mar. 17, 1943; s. Madden Travis and Vivian Alle (Browning) W.; m. Janet Elaine Allen, Dec. 19, 1970. Student Tex. A&M U., 1962-64; B.A., U. Houston, 1967, MS C.I.S. Claremont Grad. Sch., 1992. Engr. in tng. Cameron Iron Works, Houston, 1963-65; promotion planner, supr. field advt. Procter & Gamble, Cin., 1967-70; mgr. sta. promotions Union Oil Co. Calif., Chgo., 1970-73; mgr. sales promotion Hunt Wesson Foods, Fullerton, Calif., 1973-76; mgr. promotions Knott's Berry Farm Holdings, 1976-77; owner RWU Parachuting Publs., Fullerton, 1975—; mgr. opers. automation Aerojet-GenCorp, Azusa, 1983—; supr. mfg. engring., chmn. change bd. HITCO, ARMCO, Gardena, Calif., 1981-83; cons. engr. D.A.R. Enterprises; instr. advanced freefall in 9 countries; pub. speaker. Nat. bd. dirs. U.S. Parachute Assn., 1980-86. Served with Army NG, 1962. Recipient medal for merit Australian Parachute Fedn., 1972, Medal for Services French Nat. Team, 1972, certs. appreciation YMCA, 1980, 82, numerous awards and medals related to parachuting. Mem. AIAA (sr.), Computer and Automated Systems Assn. (sr., cert. systems engr., mem. tech. council), Soc. Mfg. Engrs. (sr., cert. mfg. engr., artificial intelligence in mfg. com.), BMW Owners Assn., Am. Motorcycle Assn., U.S. Parachute Assn. (dir. 1980-86), Club: Toastmasters (v.p. 1977). Author: Parachuting, English, German, French and Spanish edits.; Parachuting: United We Fall, 1978; The Art of Freefall, 1979, 2d edit., 1988, CIM Planning Guide, 1986, 2d edit., 1988, Configuration and CIM, 1987; contbr. articles to parachuting publs. Inventor flight suit, parachute line knife; nat. champion Nat. Collegiate Parachuting League, 1967; nat. winner Point of Purchase Profl. Inst., 1974. Home: 1656 E Beechwood Ave Fullerton CA 92635-2149 Office: Aerojet GenCorp 1100 W Hollyvale Azusa CA 91702

WORLEY, DAVID ALLAN, biochemical engineer; b. Alexandria, Va., Jan. 24, 1962; s. Ernest Doyle and Sylvia Joan (Thomas) W. BS, U. Va., 1984; MS, Lehigh U., 1986. Process engr. John Brown E&C Inc., Stamford, Conn., 1986-89; sr. bioprocess engr. Life Scis. Internat., Phila., 1989-90; sr. biochemical engr. Life Scis., Inc., Pleasanton, Calif., 1990-91. Mem. AAAS, Am. Inst. Chem. Engrs., Calif. Acad. Scis., Nat. Wildlife Fedn. Home: 1652 San Luis Rd Walnut Creek CA 94596-3145 Office: Life Scis Internat 7031 Koll Center Pky # 200 Pleasanton CA 94566-3101

WORLEY, JIMMY WELDON, chemist; b. Bowie, Tex., May 2, 1944; s. Elbert Weldon and Della Mae (Winsett) W.; m. Pamela Jean Wood, Aug. 26, 1966; children: Christina, Micah, Amanda. BS, Midwestern U., 1966; PhD, U. Illinois, 1971. Chemist The Agrl. Group of Monsanto, St. Louis, 1971—, rsch. group leader, 1978-82, rsch. mgr., 1982—. Contbr. articles to profl. jours. Mem. Am. Chem. Soc. Achievements include 4 patents. Office: Monsanto Co 800 N Lindbergh Blvd Saint Louis MO 63167

WORLEY, MARVIN GEORGE, JR., architect; b. Oak Park, Ill., Oct. 10, 1934; s. Marvin George and Marie Hyancinth (Donahue) W.; B.Arch., U. Ill., 1958; m. Maryalice Ryan, July 11, 1959; children—Michael Craig, Carrie Ann, Alissa Maria. Project engr. St. Louis area Nike missile bases U.S. Army C.E., Granite City, Ill., 1958-59, architect N.Cen. div. U.S. Army C.E., Chgo., 1960; architect Yerkes & Grunsfeld, architects, Chgo., 1961-65, asso., 1965; asso. Grunsfeld & Assos., architects, Chgo., 1966-85.; prin. Marvin Worley Architects, Oak Park, Ill., 1985—. Dist. architect Oak Park Elementary Schs., Dist. 97, 1973-80. Mem. Oak Park Community Improvement Commn., 1973-75; mem. exec. bd. Oak Park Council PTA, 1970-73, pres., 1971-72. Served with AUS, 1959. Mem. AIA (corporate), Chgo. Assn. Commerce and Industry. Office: 37 South Blvd Oak Park IL 60302-2777

WORMWOOD, RICHARD NAUGHTON, naturalist, theoretical field geologist, field research primatologist; b. Old Forge, N.Y., May 21, 1936; s. F. Earl Hill-Wormwood and Eleanor Bardou-Naughton; m. Michele Gano-Keeney (div.); children: Anneene, Chauncey; m. Donna Rhodes-Harrington, Jan. 30, 1983. Ski dir. Tailored Ski Instruction, Old Forge, 1961-79; interpretive naturalist Adirondack Naturalist Explorations, Warrensburg, N.Y., 1980-91; with Adirondack Park Agency, Newcomb, N.Y., 1991, Omni/Sagamore, Lake George, N.Y., 1992-93; gadfly Pro Ski Instructors of Am., 1965-68. Author: Wilderness Option, 1979, Adirondak Frontier, 1982, Finding Sasquatch, 1982, Adirondack Naturalist, 1989. Civil rights activist, Greenwich Village, Mexico City, 1957; founder Adirondack Naturalist Fellowship. With U.S. Army Mt. Troops, 1958-60. Avocations: wilderness walking, caoneing, wild nature photography, videography. Home: Loon Lake Adirondacks Chestertown NY 12817

WORNE, HOWARD EDWARD, biotechnology company executive; b. Phila., Mar. 1, 1914; s. Edward H. and Lillian G. (Greene) W.; B.S., U. Mex., 1938; M.D., Universidad Libre Mexicana, 1940; Ph.D. in Biochemistry, Instituto Polyté cnico Nacional, 1962; m. Phyllis Dolores Garofalo, Sept. 14, 1962; 1 dau., Elinor D. Pres., Nat. Solvents Corp.,

Elizabeth, N.J., 1942-45; v.p. Synthetic Resins, Inc., Toms River, N.J., 1943-47; v.p. Pentavir div. A.P. DeSanno & Sons, Phoenixville, Pa., 1947-48; sci. dir. Robinson Found., N.Y.C., 1950-55; pres. Pharm. Industries, Inc., Cherry Hill, N.J., 1955-62; pres. Enzymes, Inc., Cherry Hill, 1963-72, dir. Enzymes Japan, Inc., 1969-76; chmn. bd. Inst. Biol. Agr., 1980—; pres., chmn. bd. Worne Biotech., Inc.; chmn. bd. Waste-Energy Corp.; sr. research assoc. Center for Tropical Diseases, U. Lowell. Nat. Grain Yeast Corp fellow, 1935-37. Fellow Am. Inst. Chemists, Am. Phytopath. Soc., Am. Bd. Bioanalysts (diplomate), Am. Coll. Bioanalysts, Nat. Bd. Clin. Bioanalysts; mem. Am. Inst. Scis., Internat. Broncoesophageal Soc., Am. Soc. Microbiology, Soc. Indsl. Microbiology, Am. Chem. Soc., N.J. Acad. Sci., N.Y. Acad. Sci., AAAS, Pan Am. Med. Assn., Soc. Cosmetic Chemists, Soc. Am. Mil. Engrs., Fed. Water Pollution Control Assn. Republican. Presbyterian. Author: Soil Microbiology, 1975; Nonconventional Industrial Use of Microorganisms, 1982; editor in chief Archives of Research, 1953-55, Introduction to Microbial Biotechnology; contbr. sci. papers in field to profl. jours.; patentee in field. Home: 52 Maidstone Pl Vincentown NJ 08088-1251 Office: 1507 Route 206 Mount Holly NJ 08060

WORONOFF, ISRAEL, retired psychology educator; b. Bklyn., Dec. 30, 1926; s. Samuel and Lena (Silverman) W.; m. Fay Goldberg, Feb. 11, 1950; 1 child, Gary. AB in Psychology, U. Mich., 1949, MA in Sociology, 1952, PhD in Edn., 1954. Lic. psychologist, Mich. Instr. Flint (Mich.) Jr. Coll., 1953-54; asst. prof. St. Cloud (Minn.) State Coll., 1954-56; asst. prof. Ea. Mich. U., Ypsilanti, 1956-59, assoc. prof., 1959-62, prof., 1962-92; cons. psychologist Orchard Hills Psychiat. Ctr., Novi, Mich., 1983—, Midwest Mental Health Clinic, Dearborn, Mich., 1978-83. Author: Educator's Guide to Stress Management, 1986. Mem. community rels. com. Jewish Community Assn., Ann Arbor, Mich., 1990—; v.p. edn. Beth Israel Congregation, Ann Arbor, 1985-87. With U.S. Army, 1946-47. Mem. APA, Mich. Psychol. Assn., Am. Ednl. Rsch. Assn. Democrat. Home: 2519 Londonderry Rd Ann Arbor MI 48104-4017

WORRELL, RICHARD VERNON, orthopedic surgeon, college dean; b. Bklyn., June 4, 1931; s. John Elmer and Elaine (Callender) W.; BA, NYU, 1952; MD, Meharry Med. Coll., 1958; m. Audrey Frances Martiny, June 14, 1958; children: Philip Vernon, Amy Elizabeth. Intern Meharry Med. Coll., Nashville, 1958-59; resident gen. surgery Mercy-Douglass Hosp., Phila., 1960-61; resident orthopaedic surgery State U. N.Y. Buffalo Sch. Medicine Affiliated Hosps., 1961-64; resident in orthopaedic pathology Temple U. Med. Ctr., Phila., 1966-67; pvt. practice orthopaedic surgery, Phila., 1964-68; asst. prof. acting head div. orthopaedic surgery U. Conn. Sch. Medicine 1968-70; attending orthopaedic surgeon E.J. Meyer Meml. Hosp., Buffalo, Millard Fillmore Hosp., Buffalo, VA Hosp., Buffalo, Buffalo State Hosp.; clin. instr. orthopaedic surgery SUNY, Buffalo, 1970-74; chief orthopedic surgery VA Hosp., Newington, Conn., 1974-80; asst. prof. surgery (orthopaedics) U. Conn. Sch. Medicine, 1974-77, assoc. prof., 1977-83, asst. dean student affairs, 1980-83; prof. clin. surgery SUNY Downstate Med. Ctr., Bklyn., 1983-86; dir. orthopedic surgery Brookdale Hosp. Med. Ctr., Bklyn., 1983-86; prof. of orthopaedics U. N.Mex. Sch. of Medicine, 1986—; dir. orthopaedic oncology U. N.Mex. Med. Ctr., 1987—; mem. med. staff U. N.Mex. Cancer Ctr., 1987—; chief orthopaedic surgery VA Med. Ctr., Albuquerque, 1987—; cons. in orthopaedic surgery Newington (Conn.) Children's Hosp., 1968-70; mem. sickle cell disease adv. com. NIH, 1982-86. Bd. dirs. Big Bros. Greater Hartford. Served to capt. M.C., U.S. Army Res., 1962-69. Diplomate Am. Bd. Orthopaedic Surgery, Nat. Bd. Med. Examiners. Fellow ACS, Am. Acad. Orthopaedic Surgeons,Royal Soc. Medicine, London; mem. Am. Orthopaedic Assn., Orthopaedic Rsch. Soc., Internat. Soc. Orthopaedic Surgery and Traumatology, AMA, Alpha Omega Alpha.

WORTHINGTON, CHARLES ROY, physics educator; b. Australia, May 17, 1925; came to U.S., 1956; s. Thomas Charles and Mary Ivy (Hodge) W.; m. Alma Rose Burnier, Dec. 12, 1959; children—Laurie Alma, Keith Charles, Ian Roy. B.S., Adelaide U., 1950, Ph.D., 1955. Mem. MRC Biophysics Rsch. unit Kings Coll., London, 1958-61; asst. prof., asso. prof. physics dept. U. Mich., 1961-69; prof. biol. scis. and physics Carnegie-Mellon U., Pitts., 1969—. Contbr. articles to profl. jours. Served with Royal Australian Air Force, 1944-45. Mem. Biophys. Soc. Home: 3024 Sturbridge Ct Allison Park PA 15101-1538 Office: 4400 5th Ave Pittsburgh PA 15213-2683

WOS, CAROL ELAINE, engineer; b. Bremerton, Wash., Apr. 21, 1957; d. Standley Ralph and Janet Estele (Galber) Stocker; m. George Joseph Wos; children: Samuel Harrison, Bridget Monique. BS in Chem., Wash. State U., 1979. Mfg. engr. Internat. Bus. Machines, E. Fishkill, N.Y., 1979-80; process devel. engr. Sperry Corp., Eagan, Minn., 1980-83; sr. process devel. engr. Cray Rsch. Inc., Chippewa Falls, Wis., 1983-90, mem. cleanroom design and constrn. team, 1991-92, bump/tab process engr., 1993—; mem. Cray Wellness Com. Mem. ASTM. Republican. Office: Cray Rsch Inc 900 Lowater Rd Chippewa Falls WI 54729-4401

WOSZCZYK, WIESLAW RICHARD, audio engineering educator, researcher; b. Czestochowa, Poland, Jan. 9, 1951; arrived in Can., 1974; s. Waclaw Konstanty and Krystyna Maria (Malek) W.; m. Trudy Elizabeth Erickson, Dec. 28, 1978; children: Jake, Magda. MA, State Acad. Music, Warsaw, Poland, 1974; PhD, Chopin Acad., Warsaw, 1984. Rsch. asst. McGill U., Montreal, Can., 1974-75; recording engineer Basement Recording Co. Inc., N.Y.C., 1975-76; sound dir. Harry Belafonte Enterprises Inc., N.Y.C., 1977; chief engr. Big Apple Recording Studios, N.Y.C., 1976-78; asst. prof. McGill U., Montreal, 1978-84, chmn. grad. studies in sound recording, 1979—, assoc. prof., 1984-91, full prof., 1991—; tech. dir. McGill Records, Montreal, 1987—; owner, cons. Sonologic Registered, Montreal, 1988—; chmn. internat. conf. TV Sound Today and Tomorrow, 1991. Recording engr. over 50 records and films, 1975-85; prodr. over 30 compact discs, 1985—; mem. rev. bd. Jour. Audio Engring., 1992; contbr. articles to profl. jours. Recipient Grand Prix du Disque award Can. Coun. for Arts, 1978, Bd. Govs. award Audio Engring. Soc., 1991; Major Rsch. SSHRC grantee Rsch. Coun. Can., 1986, 93; Indsl. grantee Sony Classical, 1992, Bruel & Kjaer, 1991. Mem. Audio Engring. Soc. (gov., chmn. com. 1991—), Acoustical Soc. Am., Sigma Xi. Roman Catholic. Achievements include 3 patents on audio transducers; major research on the design and application of transducers for music recording, auditory design in sound recording, mulit-channel sound recording and reproduction. Office: McGill U Faculty Music, 555 Sherbrooke St W, Montreal, PQ Canada H3A 1E3

WOTEKI, CATHERINE ELLEN, nutritionist; b. Fort Leavenworth, Kans., Oct. 7, 1947; d. Joseph Jeremiah and Catherine (Costello) O'Connor; m. Thomas Henry Woteki, June 7, 1969. BS, Mary Washington Coll., 1969; MS, Va. Poly. Inst. and State U., 1971, PhD, 1973. Registered dietitian. Asst. prof. Drexel U., Phila., 1975-77; project dir. Congl. Office of Tech. Assessment, Washington, 1977-80; group leader USDA, Washington, 1980-83; dep. dir. Nat. Ctr. for Health Statis., Washington, 1983-90; dir. Food & Nutrition Bd., Washington, 1990—. Contbr. over 43 articles to profl. jours. Named Outstanding alumna Va. Poly. Inst. and State U., 1987; recipient Elijah White award Nat. Ctr. for Health Statis., 1987, Spl. Recognition award USPHS, 1987, Staff Achievement award Inst. of Medicine, 1991. Mem. Am. Inst. Nutrition, Am. Dietetic Assn. Coun. on Rsch., Inst. Food Technologists, Am. Pub. Health Assn. Office: Food & Nutrition Bd Inst of Medicine 2101 Constitution Ave NW Washington DC 20418

WOUCH, GERALD, materials scientist; b. Phila., Nov. 21, 1939; s. Abraham and Anna (Goloff) W.; m. Lois Lipkin, Dec. 25, 1977; 1 child, Kimberley. BS, Drexel U., 1962, PhD, 1978; MA, Temple U., 1966. Rsch. scientist GE, Phila., 1964-78, Mobil, Princeton, N.J., 1978-81; sr. rsch. scientist R.R. Donnelley & Sons Co., Chgo., 1981—. Contbr. articles to jours., Encyclopedia of Physics, 1991, chpt. to book. Mem. ASTM (printing ink com. 1991—), Am. Chem. Soc., Am. Soc. Materials, Tech. Assn. Pulp and Paper Industry. Achievements include research in containerless processing of materials, crystal growth, thermodynamic properties of metals, materials processing, printing materials, inks, papers, fountain solutions, gravure cylinder engraving, others. Home: 634 Aster Ct Lisle IL 60532 Office: R R Donnelley & Sons Co 750 Warrenville Rd Lisle IL 60532

WOYCZYNSKI, WOJBOR ANDRZEJ, mathematician, educator; b. Czestochowa, Poland, Oct. 24, 1943; came to U.S., 1970; s. Eugeniusz and Otylia Sabina (Borkiewicz) W.; m. Elizabeth W. Holbrook; 1 child, Gregory Holbrook; 1 child by previous marriage, Martin Wojbor. MSEE, Wroclaw (Poland) Poly., 1966; PhD in Math., Wroclaw U., 1968. Asst. prof. Inst. Math. Wroclaw U., 1968-72, assoc. prof., 1972-77; prof. dept. math. Cleve. State U., 1977-82; prof., chmn. dept. math. and stats. Case Western Res. U., Cleve., 1982-91, dir. Ctr. for Stochastic and Chaotic Processes in Sci. and Tech., 1989—; rsch. fellow Inst. Math., Polish Acad. Scis., Warsaw, 1969-76; postdoctoral fellow Carnegie-Mellon U., Pitts., 1970-72; vis. assoc. prof. Northwestern U., Evanston, Ill., 1976-77; vis. prof. Aarhus (Denmark) U., 1972, U. Paris, 1973, U. Wis., Madison, 1976, U. S.C., 1979, U. N.C., Chapel Hill, 1983-84, Göttingen (Germany) U., 1985, 91, U. NSW, Sydney, Australia, 1988, Nagoya (Japan) U., 1992, 93. Dep. editor in chief: Annals of the Polish Math. Soc., 1973-77; assoc. editor: Chemometrics Jour., 1987—, Probability and Mathematical Statistics, 1988—, Annals of Applied Probability, 1989—, Stochastic Processes and their Applications, 1993—; co-editor: Martingale Theory and Harmonic Analysis in Banach Spaces, 1982, Probability Theory and Harmonic Analysis, 1986, Nonlinear Waves and Weak Turbulence, 1993; author: (monograph) Martingales and Geometry in Banach Spaces I, 1975, part II, 1978; co-author: Random Series and Stochastic Integrals: Single and Multiple, 1992. Rsch. grantee NSF, Washington, 1970, 71, 76, 77, 81, 87—, Office of Naval Rsch., Washington, 1985—. Fellow Inst. Math. Stats.; mem. Am. Math. Soc., Am. Statis. Assn., Polish Math. Soc. (Gt. prize 1972), Polish Inst. Arts and Scis., Racquet Club East, Rowfant Club. Roman Catholic. Avocations: tennis, music, skiing, sailing, rare books collecting. Home: 3296 Grenway Rd Cleveland OH 44122-3412 Office: Case Western Res U Dept of Math Cleveland OH 44106

WOYTEK, ANDREW JOSEPH, chemical engineer, researcher; b. Hazleton, Pa., July 12, 1936; s. Steve and Susan (Scavnicky) W.; m. Diane Joan Ritzman, Jan. 21, 1967; children: Judd, Judith. BSChemE, Pa. State U., 1958; MSChemE, Lehigh U., 1974. Chem. engr. Western Electric Co., Reading, Pa., 1958-66, rsch. mgr., 1966-82; dir. rsch. and tech. Air Products and Chem., Allentown, Pa., 1982—; Mem. com. EPA to recommend chem. substitutes for CFCs. Author: (with others) Encyclopedia of Chemical Technology, 3rd and 4th edits., 1980, 93, Fluorine: The First One Hundred Years, 1986. Recipient Kirkpatrick Chem. Engring. Achievement award Chem. Engring. Mag., Chgo., 1977. Mem. AIChE. Roman Catholic. Achievements include patent for process of manufacturing nitrogen trifluoride, number of direct fluorination processes; numerous patents on fluorine process technology. Home: 2893 Meadowbrook Circle S Allentown PA 18103 Office: Air Products and Chem 7201 Hamilton Blvd Allentown PA 18195-1501

WOYTOWITZ, DONALD VINCENT, emergency physician; b. Balt., Sept. 19, 1961; s. Donald Vincent Sr. and Geraldine Marie (Kresslein) W.; m. Karen Faith Burkhart, July 21, 1990; 1 child, Donald Vincent III. BS, U. Md., 1983; MD, U. Md., Balt., 1987. Diplomate Am. Bd. Internal Medicine. Intern U. Miami. (Fla.) Jackson Meml. Hosp., 1987-90; resident in internal medicine U. Miami (Fla.) Jackson Meml. Hosp., 1987-90; attending physician emergency dept. Haileah (Fla.) Hosp., 1990-93; clin. fellow in hematology and oncology U. Va., Charlottesville, 1993—. Mem. ACP, AAAS.

WRANGLÉN, KARL GUSTAF (GÖSTA), applied electrochemistry educator, consultant; b. Örebro, Sweden, Mar. 19, 1923; m. Elvy Brodin, June 27, 1953; 1 son, Hans (dec.). Chem. Engr., Royal Inst. Tech., Stockholm, 1947, D.Engring., 1950, D.Tech., 1955. Guest worker Nat. Bur. Standards, Washington, 1950; research engr. various industries, Sweden, 1951-58; assoc. prof. applied electrochemistry and corrosion sci. Royal Inst. Tech., Stockholm, 1959-63, prof., 1963—; tech. expert Svea Ct. Appeal, 1962-73; cons. Sandvik Steel Works, Sweden, 1963-85. Author: Electrocrystallization of Metals, 1955 (Japanese translation 1956), On the Electrochemistry of World Politics, 1964, On the Corrosion of Contruously Cost Steel, 1967, The Egalitarian Frenzy, 1969, The Rustless Iron Pillar at Delhi, 1970, Corrosion and Protection of Metals, 1972 (translated into Japanese, Polish, Italian, German), 2d edit., 1985, Gold as Incapsulation of Nuclear Waste, 1979, Scientific Societies and Persecuted Scientists, 1980. Mem. Internat. Soc. Electrochemistry (dir. 1969-78, pres. 1975-76). Conservative. Lutheran. Office: Royal Inst Tech, 79 Valhallavägen, Stockholm S-10044, Sweden

WRAPPE, THOMAS KEITH, electrical engineer, product development director; b. Little Rock, June 25, 1956; s. Jarrell Vincent and Susan Carolyn (Keith) W.; m. Melanie Anne Hanssen, Feb. 29, 1992. BSEE, U. Notre Dame, 1978; MBA in Fin., U. Chgo., 1982. Registered profl. engr., Va., Ind. Elec. engr. Westinghouse, Houston, 1978-79; project engr. Westinghouse, Chgo., 1979-83; sr. product mgr. Hughes Network Systems, Germantown, Md., 1983-88; dir. satellite communications Teknekron Communications Systems, Inc., Berkeley, Calif., 1988-91, dir. devel.wireless personal communications products, 1991—. Mem. IEEE, Eta Kappa Nu, Tau Beta Pi. Roman Catholic. Home: 1759 Alabama St San Francisco CA 94110 Office: Teknekron Comm Systems Inc 2121 Allston Way Berkeley CA 94704

WRAY, RICHARD BENGT, systems engineer; b. Stockholm, Feb. 20, 1947; s. Elwood R. Wray and Mildred Diamant Bennett; m. Beryl Marie Kenrick, May 9, 1975; children: Maryann, Matthew, Kristen. BS in Math., Rensselaer Poly. Inst., 1968; MS in Systems Engring., Northwestern U., Evanston, Ill., 1970; MBA in Mgmt., Babson Coll., Wellesley, Mass., 1979; MS in Systems Mgmt., MBA in Govt., Western New Eng. Coll., Springfield, Mass., 1983, 84. Systems analyst Northwestern U., Evanston, 1968-70; hardware systems engr. Analytical Systems Engring. Corp., Burlington, Mass., 1975-77; task leader The Mitre Corp., Bedford, Mass., 1977-81; sr. project mgr. RCA Automated Systems, Burlington, 1981-84; advanced systems engring. specialist Lockheed Corp., Houston, 1984—; cons., Houston, 1990—. Contbr. articles to profl. jours. With USAF, 1970-75, Res., 1975—. Mem. AIAA (sr.), Nat. Coun. on Systems Engring. (v.p. tech. Tex. Gulf Coast chpt. 1992—, chmn. 1991—), Air Force Assn. Achievements include research in space generic avionics architecture, systems engring. and avionics. Office: Lockheed Corp 240 NASA Rd 1 C18 Houston TX 77058

WRBA, HEINRICH, oncologist, research institute administrator; b. Holleischen, Bohemia, Feb. 14, 1922; s. Johannes and Adolfine (Deridiaux) W.; m. Ingeburg Lohndorf, Aug. 31, 1947; children: Petra, Hannes, Sari, Uta. MD, U. Heidelberg, Germany, 1951; D Natural Scis., U. Heidelberg, 1954. Asst. Univ. Clinics Heidelberg, 1951-56; head dept. exptl. pathology Inst. Pathology, U. Munich, 1956-64; dir. Inst. Exptl. Pathology, German Cancer Rsch. Ctr., Heidelberg, 1964-67; head Inst. Cancer Rsch., U. Vienna, Austria, 1967—, dir. Inst. Applied and Exptl. Oncology, 1983—. Contbr. articles to med. jours. With Submarine Svc., German Navy, World War II. Mem. European Orgn. Cancer Rsch. Insts. (founding pres. 1980-83), also others. Office: U Vienna Inst Oncology, Borschkegasse 8 a, A-1090 Vienna Austria

WREBIAK, ANDRZEJ, economic and financial consultant; b. Warsaw, Poland, Oct. 5, 1954; s. Tadeusz and Maria Wrebiak; m. Iwona Ogrodzinska Wrebiak, Dec. 27, 1980; 1 child, Bartosz. MA, Warsaw Sch. Econ., Warsaw, Poland, 1978, PhD, 1987. Fellow, mem. Rsch. Inst. for Developing Countries, Warsaw, Poland, 1978-89; chief expert Polexpert, Warsaw, 1988-89, pres., 1989-91; pres. The Warsaw Cons. Group, 1991—; chmn. supervisory bd. Power Plant, Opole, Poland, 1991—. Fulbright scholar, U. Mich., Ann Arbor, 1987-88. Home: Koszykowa 10/20, 00-456 Warsaw Poland Office: The Warsaw Consulting Group, Gornoslaska 1, 00 443 Warsaw Poland

WREN, ROBERT JAMES, aerospace engineering manager; b. Moline, Ill., May 12, 1935; children: James, Patrick, Kiley. BSCE, U. Tex., 1956; MSCE, So. Meth. U., 1962; doctoral candidate, U. Houston. Registered profl. engr. Tex. Engring. aide Ctrl. Power and Light Co., Corpus Christi, 1954; sta. clk. City of Austin (Tex.) Power Plant, 1954-55; assoc. engr., hydraulic engr. U.S. Bur. of Reclamation, Austin, 1955-57; structural test engr. Gen. Dynamics, Ft. Worth, 1957-62; sr. structural dynamics engr., mgr. vibration and acoustic test facility NASA-Manned Spacecraft Ctr., Houston, 1962-63, 63-66, head exptl. dynamics sect., 1965-70; mgr. Apollo Spacecraft 2TV-1 CSM Test Program, 1966-68, Apollo Lunar Module-2 Drop Test Program, 1968-70; mgr. structural design space sta., space base, lunar base, mars mission NASA-Manned Spacecraft Ctr., Houston, 1970-73; mgr. structural design and devel., space shuttle carrier aircraft-747 NASA Johnson Space Ctr., Houston, 1973-74, mgr. structural div. space shuttle payload systems, 1974-84; mgr. engring. directorate for space shuttle payload safety NASA-Johnson Space Ctr., Houston, 1984—, alternate chmn. space shuttle payload safety review panel, 1990—. Pres. Friendswood Little League Baseball, 1980-83; bd. dirs. Bay Area YMCA, Houston, 1982—, chmn., 1983-84. Recipient Sustained Superior Performance award NASA, Personal Letter of Commendation, George Low NASA Apollo Program, Outstanding Svc. award NASA, Group Achievement awards NASA; Paul Harris fellow Rotary. Mem. Space Ctr. Rotary (dir., treas., sec., v.p. 1979-85, pres. 1985-86), Rotary Dist. 5890 (govt. rep. 1986-87, area coord. 1987-89, zone leader 1988-89, gov.'s adie 1989-90, chmn. dist. assembly 1989-90, 93—, fin. com. 1989-91), Rotary Nat. Award for Space Achievement Found. (co-founder, bd. dirs. 1984—), Rotary World Health Found. Plastic Surgery for Children (co-founder, bd. dirs. 1985—), Rotary Space Meml. Found. (co-founder, bd. dirs. 1986—). Methodist. Avocations: snow and water skiing, running, scuba diving, tennis, sailing. Home: PO Box 1466 Friendswood TX 77546 Office: NASA Johnson Space Ctr Houston TX 77058

WRIGHT, BERRY FRANKLIN, civil engineer; b. Richmond, Va., Dec. 2, 1945; s. Berry Franklin and Nellie (Newton) W.; divorced. BSCE, Va. Mil. Inst., 1968; MSCE in Water Resources, Va. Poly. Inst. and State U., 1980. Registered profl. engr., Va. Engr. trainee Va. Dept. Hwys. and Transp., Richmond, 1968-72; hydraulics engr. design Va. Dept. Hwys., Richmond, 1968-82; chief tech. svcs. Dept. Waste Mgmt., Richmond, 1982—; adj. faculty mem. U. Va. Extension Grad. Sch., Richmond, 1977-92; mem. water quality programming com, Va. Poly. Inst. and State U., Blacksburg, 1989-92. Sponsor youth dept. 1st Bapt. Ch., Ashland, Va., 1985-92; mem. Rep. Nat. Task Force, 1991. Lt. U.S. Army, 1969-71, Vietnam. Decorated Bronze Star; recipient Blood Donor award Friends for Life, 1992. Mem. ASCE (pres. Richmond br. 1980-81), Am. Pub. Works Assn., Richmond Joint Engring. Coun. (treas. 1984-85). Baptist. Achievements include creation of simplified synthetic hydrograph method used to model storms. Home: Rt 1 Box 1362 Ashland VA 23005

WRIGHT, BRADLEY DEAN, chemical engineer, environmental consultant; b. Oakland, Calif., Aug. 4, 1963; s. Alfred Russell and Carol Lee (Peterson) W.; m. Sharon Linn Campbell, June 25, 1988. BS, Kans. State U., 1986. Rsch. asst. Kans. State U., Manhattan, 1986-88; chem. engr. B&V Waste Sci. and Tech. Corp., Kansas City, Mo., 1988-91; environ. cons., 1991—. Mem. Am. Inst. Chem. Engrs., Math. Assn. Am., Hazardous Materials Control Resources Inst. Achievements include computer programming and numerical modeling of treatment processes for hazardous wastes and the transport and fate of contaminants in the environment.

WRIGHT, CHARLES LESLIE, economist; b. Three Rivers, Mich., Nov. 14, 1945; arrived in Brazil, 1968; s. Arden William and Genevieve Elizabeth (Whited) W.; m. Maria da Gloria Miotto, Feb. 11, 1972; children: Marcelo Miotto, Denison Miotto, Elisson M., Alan M. AB, U. Mich., 1968; MS, U. São Paulo (Brazil), 1973; PhD, Ohio State U., 1977. Sr. economist Brazilian Transp. Planning Agy., Brasília, Brazil, 1977-86; assoc. prof. U. Brasília, 1978-91; bilingual tech. editor UNDP report Brazilian Trasnp. Planning Agy.-GEIPOT, Brasília, 1981-82; coord. div. tranps. and communications IPEA Found., Brasília, 1986-89; transport economist Louis Berger Internat. Inc., Manila, Philippines, 1991; assoc. prof. rsch. Nat. Ports and Waterways Inst., La. State U., 1991-93; economist Interamerican Devel. Bank, 1993—; cons. ELCA (UN), Santiago and Brasilia, Govt. of Fed. Dist., Louis Berger Internat., Manila, 1991, Pakistan, P.R. and Nicaragua, 1992; mem. tech. and sci. adv. com. Internat. Fedn. Pedestrians, 1991-92; mem. adv. bd. Inst. for Transp. and Devel. Policy, 1991—. Author: Fast Wheels, Slow Traffic: Urban Transport Choices, 1992, Transporte Rodoviário de Ônibus, 1992, O que é transporte urbano, 1988, Análise Econômica de, 1980; contbr. articles to profl. jours. Vol. Peace Corps, Aquidabã, Brazil, 1968-72; coord. pilot project Nat. Aid Program for Traffic Accident Victims, Curitiba, Brazil, 1987-89; counselor Brazilian Pedestrians Safety Assn., São Paulo, 1990—. Grantee Brazilian Sci. & Tech. Coun., 1989-90. Mem. Am. Econ. Assn., Brazilian Agrl. Econs. Assn., Associacao Nacional Pesquisa Ensino Transportes, Rsch. and Teaching in Transport, Sigma Chi Delta, Phi Kappa Phi, Phi Beta Kappa. Avocations: languages, reading. Home: 1102 Fallsmead Way Rockville MD 20854 Office: Interamerican Devel Bank 1300 N York Ave Washington DC 20577

WRIGHT, CHARLES P., engineering manager; b. Phoenix, Jan. 20, 1945; s. Charles Alan and Marie Jeanine (Allinio) W.; m. Malinda Cowles, Sept. 7, 1985; children: Caroline, Austen. BSME, Ariz. State U., 1966, MS in Measurements, 1968; MS in Systems Mgmt., U. So. Calif., 1979. Measurement engr. TRW Space and Electronics Group, Redland Beach, Calif., 1968-80; mgr. measurments engring. sect. TRW Space & Tech. Group, Redland Beach, Calif., 1980-85, mgr. measurements engring. dept., 1985—; planning com. Aerospace Testing Seminar, L.A., 1988-94; lectr. in field. Author column data acquisition and control, 1991-92; contbg. editor: Personal Engineering and Instrumentation News; contbr. over 30 articles to profl. jours. Mem. Soc. Exptl. Mechanics (chmn. 1980-81, 90-91). Republican. Roman Catholic. Achievements include development of knowledge-based data acquisition systems for mechanical engineering measurements. Home: 2550 Date Circle Torrance CA 90505 Office: TRW Space & Electronics Group 1 Space Pk Redondo Beach CA 90278

WRIGHT, CLARK PHILLIPS, computer systems specialist; b. Orange, Tex., Aug. 30, 1942; s. Madison Brown and Mary Elizabeth (Phillips) W.; m. Stacy Charlotte Klutz, June 5, 1965 (div. Oct. 1979); m. Cora Lou Alexandria Schelling, Oct. 31, 1979; 1 child, Isaac Schelling. BA, U. Tex., 1965. Computer programmer Lockheed Electronics Co., Houston, 1965-67; prin. analyst Control Data Corp., St. Paul, 1967-76; computer scientist DBA Systems, Inc., Lanham, Md., 1976-79; engring. specialist Ford Aerospace Corp., Houston, 1979-90, Loral Aerospace Corp., 1990—. Precinct chmn. Rep. Party of Tex., 1982-86. Mem. IEEE, AIAA, Math Assn. Am., Data Processing Mgmt. Assn., Assn. Computing Machinery, SAR (chartered, sec., treas.), Sons Republic Tex., Info. System Security Assn., Masons, Rotary. Avocations: travel, photography. Home: 5000 Park Ave Dickinson TX 77539-7013 Office: Loral Aerospace Corp PO Box 58487 Houston TX 77258-8487

WRIGHT, DANIEL GODWIN, academic physician; b. St. Louis, July 27, 1945; s. Clayton Whitbeck and Jane Montgomery (Godwin) W.; m. Elizabeth Chalmers, Sept. 4, 1967; 1 child, Christopher Talcot. BA, Yale U., 1967, MD, 1971. Intern Yale-New Haven Hosp., 1971-72, resident in internal medicine, 1972-73; clin. assoc., med. officer Nat. Inst. Allergy and Infectious Diseases NIH, Bethesda, Md., 1973-77; staff investigator Nat. Cancer Inst. NIH, Bethesda, 1978-80; med. fellow Johns Hopkins Hosp., Balt., 1977-78; chief hematology Walter Reed Army Inst. Rsch., Washington, 1979-92; chief hematology, oncology Boston City Hosp./Boston U. Sch. Medicine, 1992—; hematology study sect. mem. NIH, Bethesda, 1984-89. Contbr. articles to profl. jours. Sci. adv. coun. ARC, Rockville, Md., 1989—. Col. U.S. Army M.C., 1979-92. Mem. Am. Soc. Hematology, Am. Soc. for Clin. Investigation, Am. Soc. for Cell Biology, Am. Assn. Immunology. Achievements include research in neutrophil physiology and myelopoiesis. Office: Boston U Sch Medicine 818 Harrison Ave Boston MA 02118

WRIGHT, DONALD LEE, chemist; b. Ruffin, N.C., Mar. 19, 1937; s. John Lawson and Alice S. (McKinney) W.; m. Lyn C. Winebarger, Mar. 15, 1969 (div. 1979); children: Donald L., Graeme C.; m. Marilyn Virginia Hoad, Aug. 11, 1984; children: Heather E., Andrew W. BS in Chemistry, U. N.C., 1958, PhD, 1964. Chemist Rsch. Triangle Inst., Research Triangle Park, N.C., 1963-65; asst. prof. Appalachian State U., Boone, N.C., 1965-69, Tenn. Tech. U., Cookeville, 1969-74; rsch. chemist Dan River, Inc., Danville, Va., 1974-85, group leader, 1985-90; group leader Hickson DanChem. Corp., Danville, Va., 1990—. Mem. Am. Chem. Soc., Phi Beta Kappa. Home: Rt 1 Box 387 Providence NC 27315 Office: Hickson DanChem Corp PO Box 400 Danville VA 24543

WRIGHT, DOUGLAS TYNDALL, former university president, company director, engineering educator; b. Toronto, Ont., Can., Oct. 4, 1927; s. George C. and Etta (Tyndall) W.;. B.A.Sc. with honors in Civil Engring, U. Toronto, 1949; M.S. in Structural Engring, U. Ill., 1952; Ph.D. in Engring, U. Cambridge, 1954; D.Eng. (hon.), Carleton U., 1967; LL.D. (hon.), Brock

U., 1967, Concordia U., 1982; D.Sc. (hon.), Meml. U. Nfld., 1969; DHL (hon.), Northeastern U., 1985; D.Univ. (hon.), Strathclyde U., Glasgow, 1989; D de L'Université, Compiegne U., France, 1991; D Univ. (hon.), Université de Sherbrooke, 1992; DSc, McMaster U., 1993, Queen's U., 1993. Lectr. dept. civil engring. Queen's U., 1954-55, asst. prof., 1955-58, assoc. prof., 1958; prof. civil engring. U. Waterloo, 1958-67, chmn. dept. civil engring., 1958-63, dean engring., 1959-66; chmn. Ont. Com. on Univ. Affairs Govt. of Ont., 1967-72; chmn. Ont. Commn. Post-Secondary Edn., Toronto, 1969-72, dep. provincial sec. for social devel., 1972-79; dep. minister culture and recreation, 1979-80; pres. U. Waterloo, Ont., Ont., 1981-93; prof. engring. U. Waterloo, Ont., 1981—; vis. prof. Universidad Nacional Autónoma de México, 1964, 66, Université de Sherbrooke, 1966-67; cons. engr. Netherlands and Mexican pavilions Expo 1967, Olympic Sports Palace, Mexico City, 1968, Ont. Place Dome and Forum, 1971; tech. adviser Toronto Skydome, 1984-92; dir. ElectroHome Ltd., 1983, Bell Can., 1985—, Westinghouse Can., Can. Venture Founders Ltd., London Life Ins. Co., London Ins. Group, Com. Dev. Ltd., Geometrica Inc., Visible Decision, Meloche, Monnex, Inc., Lac Minerals; mem. Premier's Coun. on Sci. and Tech., Ont., 1985-91, Prime Minister's Nat. Adv. Bd. for Sci. and Tech., 1985-91; Can. rep. Coun. for Internat. Inst. Applied Systems Analysis, Laxenburg, Austria; Prime Min.'s personal rep. to Coun. Mins. of Edn., 1990-91. Contbr. articles to profl. jours. Bd. dirs. African Students Found., Toronto, 1961-66; bd. dirs. Ont. Curriculum Inst., 1964-67; bd. govs. Stratford Shakespearian Festival, 1984-86, mem. senate, 1987. Athlone fellow, 1952-54; named Officer of Order of Can., 1991, Chevalier L'Ordre National du Mérite of France, 1992; recipient Gold medal Ont. Profl. Engrs., 1990, Can. Engrs. Gold Medal award Can. Coun. Profl. Engrs., 1992. Fellow ASCE, Can. Acad. Engring., Engring. Inst. Can. (del. Engrs. Coun. Profl. Devel., N.Y.C. 1961-70); mem. Assn. Profl. Engrs. Province Ont., Internat. Assn. Bridge and Structural Engring., Internat. Assn. Shell Structures, Can. Assn. Latin Am. Studies, Am. Acad. Mechanics, Can. Inst. Internat. Affairs, Can. Inst. Pub. Adminstrn. Clubs: Royal Can. Yacht, Univ. (Toronto). Office: U Waterloo, Waterloo, ON Canada N2L 3G1

WRIGHT, HERBERT E(DGAR), JR., geologist; b. Malden, Mass., Sept. 13, 1917; s. Herbert E. and Annie M. (Richardson) W.; m. Rhea Jane Hahn, June 21, 1943; children—Richard, Jonathan, Stephen, Andrew, Jeffrey. AB, Harvard U., 1939, MA, 1941, PhD, 1943; DSc (hon.), Trinity Coll., Dublin, Ireland, 1966; PhD (hon.), Lund U., Sweden, 1987. Instr. Brown U., 1946-47; asst. prof. geology U. Minn., Mpls., 1947-51, assoc. prof., 1951-59, prof., 1959-74, Regents' prof. geology, ecology and botany, 1974-88; dir. Limnological Research Center, 1963-90. Served to maj. USAAF, 1942-45. Decorated D.F.C., Air medal with 6 oak leaf clusters; recipient Pomerance award Archeol. Inst. Am., 1985, Ann. award Sci. Mus. Minn., 1990; Guggenheim fellow, 1954-55, Wenner-Gren fellow, 1954-55. Fellow AAAS, NAS, Geol. Soc. Am. (Ann. award archeol. divsn. 1989, Disting. Career award geology and geomorphology divsn. 1992), Soc. Am. Archeology (Fryxell award 1993); mem. Ecol. Soc. Am., Am. Soc. Limnology, Oceanography, Am. Quarternary Assn., Arctic Inst., Brit. Ecol. Soc. Research on Quaternary geology, paleoecology, paleolimnology and environ. archaeology in Minn., Wyo., Sweden, Yukon, Labrador, Peru, eastern Mediterranean. Home: 1426 Hythe St Saint Paul MN 55108-1423 Office: U of Pillsbury 221 Pillsbury Hall 310 Pillsbury Dr SE Minneapolis MN 55455

WRIGHT, JOHN CURTIS, chemist, educator; b. Lubbock, Tex., Sept. 17, 1943; s. John Edmund and Jean Irene (Love) W.; m. Carol Louise Swanson, Aug. 17, 1968; children: Dawna Lynn, John David. BS, Union Coll., 1965; PhD, Johns Hopkins U., 1970; postgrad., Purdue U., 1972. From asst. to assoc. prof. U. Wis., Madison, 1972-80, prof., 1980—. Contbr. articles to profl. jours. Recipient William F. Meggers award Applied Spectroscopy Soc. 1981. Mem. Am. Chem. Soc. (Spectrochem. Analysis award 1991), Am. Phys. Soc. Achievements include pioneering work in site selective laser spectroscopy; discovery of mode selective four wave mixing spectroscopy; development of line narrowed nonlinear spectroscopy; patent for Chemical Analysis of Ions Incorporated in Lottices using Coherent Excitation Sources. Office: Univ Wis Dept Chemistry 1101 University Ave Madison WI 53706

WRIGHT, KARA-LYN ANNETTE, software engineer; b. Phila., Feb. 27, 1963; d. Javis Leon and O. Elizabeth (Seals) W. BS in Computer Sci., Drexel U., 1986, MS in Computer Sci., 1992. Intake worker Wheel's Inc., Phila., 1983; staff cons. computer ctr. Drexel U., Phila., 1984, sr. cons., 1984-85; programmer E. I. duPont, Phila., 1985; sr. programmer, computer scientist RMS Techs., Inc., Marlton, N.J., 1986—, tech. mgr. Geophys. Scis. Lab. Author manual: AFCAD Revisited, 1984; co-designer, implementor software. Recipient Letter of Commendation, U.S. Naval Acad., 1990. Mem. IEEE, Assn. for Computing Machinery. Avocations: woodworking, reading, gardening. Home: 42 Medford Ln Willingboro NJ 08046-3121 Office: RMS Tech Inc 5 Eves Dr Marlton NJ 08053-3135

WRIGHT, MAUREEN SMITH, molecular biologist, microbiology educator; b. New Orleans, Dec. 5, 1962; d. Manuel John and Audrey L. (Coustaut) Smith; m. Gregory Gerard Wright, May 31, 1986; 1 child, Jessica Catherine. BS, Xavier U., New Orleans, 1984; PhD, La. State U., 1990. Technician Mycotoxin Lab., So. Regional Rsch. Ctr., Agrl. Rsch. Svc., USDA, New Orleans, 1982-84, molecular biologist, 1990—; adj. prof. microbiology So. U., New Orleans, 1991—. Grad. fellow La. State U. Alumni Fedn., 1986-90. Mem. Am. Soc. for Microbiology, Sigma Xi (chpt. treas. 1992—). Office: USDA ARS SRRC 1100 Robert E Lee Blvd New Orleans LA 70124

WRIGHT, MICHAEL GEORGE, atomic energy company executive; b. Bristol, U.K., Apr. 15, 1939; arrived in Canada, 1962; s. Thomas Greaves and Gladys (Godber) W.; m. Fiona Elliot Crozier, Apr. 18, 1964; children: Craig, Graeme. BS in Metallurgy, U. Wales, U.K., 1960; MSc in Metallurgy, McMaster U., 1964. Metallurgist GE-Simon Carves Atomic Energy Group, Erith, Kent, U.K., 1960-62, Inland Steel Co., East Chicago, Ind., 1964-66, with Atomic Energy of Can., Ltd., Pinawa, Man., 1966—, rsch. scientist, 1966-77, br. mgr., 1977-82, sr. advisor/asst., 1982-84, mgr. engring. div., 1984-86, gen. mgr., 1986—. Office: Atomic Energy Can Ltd/ Whiteshell, Nuclear Rsch Establishment, Pinawa, MB Canada R0E 1L0

WRIGHT, PETER MURRELL, structural engineering educator; b. Toronto, Ont., Can., Sept. 26, 1932; married; 4 children. BS, U. Sask., 1954, MSc, 1961; PhD in Structural Engring., U. Colo., 1968. Engr. Dorman-Long, Ltd., Eng., 1954-57; asst. prof. structural engring. U. Sask., 1957-66; prof. structural engring. U. Toronto, 1968—, assoc. dean engring., 1981-85, acting dean architecture, 1984-88. Recipient Queen's Can. Silver Jubilee medal, 1977, John B. Sterling medal Engring. Inst. Can., 1992. Mem. Can. Soc. Civil Engrs. (pres. 1981-82). Achievements include automatic design of steel building frames with member selection. Office: U Toronto, Dept Civil Engineering, Toronto, ON Canada M5S 1A4*

WRIGHT, PHILIP LINCOLN, zoologist, educator; b. Nashua, N.H., July 9, 1914; s. Clarence H. and Avis (Dary) W.; m. Margaret Ann Halbert, July 21, 1939 (dec. 1988); children: Alden H., Philip L., Ann E. Wright Dwyer; m. Hedwig G. Tourangeau, Dec. 28, 1989. B.S., U. N.H., 1935, M.S., 1937; Ph.D., U. Wis., 1940. Instr. to assoc. prof. zoology U. Mont., 1939-51, prof., 1951-85, emeritus prof., 1985—, chmn. dept. zoology, 1956-69, 70-71; on leave South Africa, 1970; Maytag vis. prof. zoology Ariz. State U., 1980. Author: (with W.H. Nesbitt) Records of North American Big Game, 8th edit., 1981, How to Score Big Game Trophy Heads, 1985; assoc. editor: Jour. Mammalogy, 1956-64; editor gen. notes and revs., 1966-68; contbr. numerous articles on mammalogy and ornithology. Fellow AAAS; mem. Am. Soc. Zoologists, Am. Soc. Mammalogists, Mont. Acad. Scis. (pres. 1957), Soc. for Study Reprodn., Am. Ornithol. Union, Cooper Ornithol. Soc., Wildlife Soc. (editorial bd. jour. Wildlife Mgmt. 1966-68), Explorers Club, Sigma Xi, Phi Sigma, Gamma Alpha. Club: Boone and Crockett (chmn. panel judges 16th award program, hon. mem. 1984).

WRIGHT, RICHARD NEWPORT, III, civil engineer, government official; b. Syracuse, N.Y., May 17, 1932; s. Richard Newport and Carolyn (Baker) W.; m. Teresa Rios, Aug. 23, 1959; children—John Stannard, Carolyn Maria, Elizabeth Rebecca, Edward Newport. B.C.E., Syracuse U., 1953, M.C.E. (Parcel fellow), 1955; Ph.D., U. Ill., 1962. Jr. engr. Pa. R.R., Phila., 1953-55; instr. civil engring. U. Ill., Urbana, 1957-62; asst. prof. U. Ill., 1962-65, assoc. prof., 1965-70, prof., 1970-74, adj. prof., 1974-79; chief structures sect. Bldg. Research div. U.S. Bur. Standards, Washington, 1971-72; pres. Internat. Council for Bldg. Research, Studies and Documentation, 1983-86; dep. dir. tech. Ctr. Bldg. Tech., 1972-73, dir., 1974-91; dir. Bldg. and Fire Rsch. Lab., 1991—. Contbr. articles to profl. jours. Pres. Montgomery Village Found., 1989-90, bd. dirs., 1985—. Served with AUS, 1955-57. Named Fed. Engr. of Yr. Nat. Soc. Profl. Engrs., 1988. Fellow ASCE. Home: 20081 Doolittle St Gaithersburg MD 20879-1354 Office: Nat Inst Standards & Tech Bldg and Fire Rsch Labs Gaithersburg MD 20899

WRIGHT, ROBERT MICHAEL, civil engineer; b. Phila., May 26, 1957; s. Alfred Richard and Patricia Rose (Brinkman) W.; m. Gail Elaine Kellmer, Oct. 2, 1982. BS in Civil Engring., U. Pa., Phila., 1979. Project engr. Phila. Dept. Sts., 1979—; instr. constrn. tech. program Community Coll. Phila., 1988—. Mem. ASCE (Young Govt. Engr. Phila. sect. 1988), Am. Soc. Hwy. Engrs., Am. Pub. Works Assn. Roman Catholic. Home: 730 Livezey Ln Philadelphia PA 19128 Office: Phila Dept Sts 1600 Arch St 10th fl Philadelphia PA 19103

WRIGHT, STEPHEN GAILORD, civil engineering educator, consultant; b. San Diego, Aug. 13, 1943; s. Homer Angelo and Elizabeth Videlle (Ward) W.; m. Ouida Jo Kennedy; children: Michelle, Richard. BSCE, U. Calif., Berkeley, 1966, MSCE, 1967, PhD CE, 1969. Prof. civil engring. U. Tex., Austin, 1969—. Contbr. numerous tech. papers & research reports, 1969—. Mem. ASCE. Republican. Presbyterian. Home: 3406 Shinoak Dr Austin TX 78731-5739 Office: U Tex Dept Civil Engring Austin TX 78712

WRIGHT, STEVEN JAY, environmental engineering educator, consultant; b. Spokane, Wash., June 23, 1949; s. Bennie J. and Margaret Jean (Wiker) W.; m. Dayle Kathleen Wilson, Dec. 26, 1971; children: Glenn, Daniel. BS in Agrl. Engring., Wash. State U., 1971, MS in Hydraulic Engring., 1973; PhD in Civil Engring., Calif. Inst. Tech., 1977. Registered profl. engr., Mich. Rsch. engr. Calif. Inst. Tech., Pasadena, 1977; prof. civil and environ. engring. U. Mich., Ann Arbor, 1977—; mem. exec. com. Huron River Watershed Coun., Ann Arbor, 1979-91; mem. tech. adv. bd. Mich. Great Lakes Protection Fund, Lansing, 1990—; vis. Erskine fellow U. Canterbury, New Zealand, 1992. Author: Essentials of Engineering Fluid Mechanics, 1990; contbr. articles to profl. jours. Leader Boy Scouts Am., Chelsea, Mich., 1985—. Recipient Lorenz Straub award U. Minn., 1978, James R. Rumsey award Mich. Water Pollution Control Fedn., 1983; postdoctoral fellow Swiss Nat. Sci. Found., 1984-85. Mem. ASCE (J.C. Stevens award 1986), Am. Geophys. Union, Internat. Assn. for Hydraulic Rsch., Assn. Ground Water Scientists and Engrs. Achievements include research in mixing of contaminant discharges in rivers and lakes, density intrusions, groundwater contaminant transport, and water infiltration through clay land fill liners. Home: 126 South St Chelsea MI 48118 Office: U Mich 113 Engring 1A Ann Arbor MI 48109-2125

WRIGHT, TERENCE RICHARD, imaging company executive. MA in Chemistry, Oxford U., 1971, DPhil in Chemistry, 1971. Bus. mgr. Europe Kodak Graphics Imaging, 1985-88; dir. rsch. Kodak Ltd., 1989—; dir. rsch. and devel. Kodak in Europe, Harrow, Middlesex, Eng., 1992—. Office: Kodak Ltd, Headstone Dr, HA1 4TY Harrow England

WRIGHT, THEODORE OTIS, forensic engineer; b. Gillette, Wyo., Jan. 17, 1921; s. James Otis and Gladys Mary (Marquiss) W.; m. Phyllis Mae Reeves, June 21, 1942 (div. 1968); children: Mary Suzanne, Theodore Otis Jr., Barbara Joan; m. Edith Marjorie Jewett, May 22, 1968; children: Marjorie Jane, Elizabeth Carter. BSEE, U. Ill., 1951, MS in Engring., 1952; postgrad., Air Command and Staff Coll., 1956-57, UCLA, 1958. Registered profl. engr. Wash. 2d lt. U.S. Air Force, 1942-65, advanced through grades to lt. col., 1957, ret., 1965; dep. for engring. Titon SPO, USAF Sys. Command, L.A., 1957-65; rsch. engr. The Boeing Co., Seattle, 1965-81; pres. The Pretzelwich, Inc., Seattle, 1981—; cons., forensic engr. in pvt. practice Bellevue, Wash., 1988—; adj. prof. U. Wash., Greenriver Jr. Coll., both 1967-68. Contbr. articles to profl. jours. Decorated Purple Heart, Air medal. Mem. U.S. Metric Assn. (life, cert. advanced metrication specialist 1981), Air Force Assn. (life, state pres. 1974-76, 90-91), Jimmy Doolittle fellow 1975), Order Daedalians (life mem.), Nat. Soc. Profl. Engrs. (Western region v.p. 1985-87), Wash. Soc. Profl. Engrs. (state pres. 1981-82, Disting. Svc. award 1980), Sigma Tau, Eta Kappa Nu, Pi Mu Epsilon, Tau Beta Pi. Democrat. Presbyterian. Avocations: flying, photography, classical music, archaeology. Home: 9644 Hilltop Rd Bellevue WA 98004-4006

WRIGHT, THOMAS WILSON, mechanical engineer; b. Fergus Falls, Minn., Oct. 23, 1933; s. Thomas Clarke and Catharine (Wilson) W.; m. Margaret Jean Parkinson, Aug. 27, 1955; children: Andrew Fearon, Charles Townsend, Katharine McClellan. BCE, Cornell U., 1956, MCE, 1957, PhD, 1964. Sr. engr. Aircraft Armaments, Cockeysville, Md., 1959-61; asst. prof. Johns Hopkins U., Balt., 1964-67; mech. engr. Ballistic Rsch. Lab., Aberdeen Proving Ground, Md., 1967-90, sr. rsch. scientist, 1990—; mem. adv. bd. Soc. Engring. Sci., 1988-91, math. scis. inst. Cornell U., Ithaca, N.Y., 1985-86, 88, dept. applied math. and statistics SUNY, Stony Brook, 1992—; part-time lectr., prof. Johns Hopkins U., 1980, 85, 87, 89, 92; referee many jours. of mechanics. Contbr. articles to Jour. Mech. Phys. Solids, Jour. Elasticity, Quar. Jour. Mech. Applied Math.; mem. adv. bd. Internat. Jour. Plasticity, 1992—. 1st lt. U.S. Army, 1957-59. Cooperative fellow NSF, 1963-64, Rsch. fellow Sec. of Army, 1974-75, 90-91; recipient Presdl. Citation, 1977. Fellow ASME; mem. Am. Acad. Mechanics, Soc. for Natural Philosophy. Democrat. Episcopalian. Achievements include discovery of laws for regular shock reflections in solids, evolution of wave fronts in nonlinear elastic solids, research in wave propagation, penetration mechanics, material instability, adiabatic shear bands. Office: Ballistic Rsch Lab Army Rsch Lab Aberdeen Proving Ground MD 21005

WRIGLEY, COLIN WALTER, cereal chemist; b. Sydney, Australia, Dec. 25, 1937; s. Sidney Benjamin and Eva Christina (Macindoe) W.; m. Janice Margaret Saxby, Nov. 5, 1959; children: Christine Ferguson, Jennifer Swanton, Robyn, Margaret. BSc in Biochemistry with honours, U. Sydney, 1959, MSc in Biochemistry, 1961, PhD in Agrl. Chemistry, 1967. Teaching fellow U. Sydney, 1961-63; exptl. scientist CSIRO Wheat Rsch. Unit, Sydney, 1961-63; rsch. rellow U. Sydney, 1964-66; exptl. rsch. scientist CSIRO Wheat Rsch. Unit, Sydney, 1967-69; postdoctoral fellow U. Man., Winnipeg, 1969-70; rsch. scientist CSIRO Wheat Rsch. Unit, Sydney, 1970-83; officer-in-charge CSIRO Grain Quality Rsch. Lab., Sydney, 1983—; bd. dirs. Gradipore Ltd., Sydney; PhD student advisor U. Sydney; mem. editorial bds. 5 internat. jours. Contbr. over 250 publs. to profl. jours.; co-author 7 books on Australian cereal varieties. Bd. dirs. Chs. of Christ Nursing Home (Pendle Hill), Sydney, 1993—. Fellow Royal Australian Chem. Inst. (H.G. Smith medal 1987, F.B. Guthrie medal 1988); mem. Am. Asssn. Cereal Chemists (Thomas Burr Osborne medal 1992), Internat. Electrophoresis Soc. (founding mem.), Australian Biotech. Assn. Achievements include invention of gel isoelectric focusing for fractionation of proteins; 4 patents on the processing and testing of cereal grains. Office: CSIRO Grain Quality Lab, PO Box 7, North Ryde NSW 2113, Australia

WRIST, PETER ELLIS, pulp and paper company executive; b. Mirfield, Eng., Oct. 9, 1927; arrived in Can., 1952; s. Owen and Evelyn (Ellis) W.; m. Mirabelle Harley, Sept. 3, 1955; children: Denise Wrist Parson, Philip, Lydia Wrist Schweizer, Richard; m. Kathryn Idelson. BA, Cambridge (Eng.) U., 1948, MA, 1952; MS, London U., 1952; cert. advanced mgmt. program, Harvard U., 1967; DSc (hon.), U. B.C., 1993. Rsch. physicist Brit. Paper & Bd. Industry Rsch. Assn., Kenley, Eng., 1949-52, Que. North Shore Paper Co., Baie Comeau, Can., 1952-56; rsch. physicist Mead Corp., Dayton, Ohio, 1956-60, assoc. dir. rsch., 1960-61, dir. rsch., 1961-66, mgr. rsch. and engring., 1966-68, v.p. rsch. and engring., 1968-72, v.p. tech., 1972-83; exec. v.p. Pulp and Paper Research Inst. Can., Pointe Claire, Que., 1983-86; pres., chief exec. officer Pulp and Paper Rsch. Inst. Can., Pointe Claire, Que., 1986—; chmn. selection com. Marcus Wallenberg Found.; chmn. Nat. Coun. for Air and Stream Improvement, 1972-75; past mem., chmn. rsch. adv. com. Inst. Paper Chemistry, 1971-83. Contbr. articles to profl. jours. Fellow Tech. Assn. Pulp and Paper Industry (bd. dirs. 1971-77, v.p. 1975-77, pres. 1977-79, Gunnar Nicholson Gold Medal award selection com., Engring. award 1969, Gold medal 1983); mem. Can. Pulp and Paper Assn. (tech. sect., Howard Smith award 1954, Weldon Gold medal 1956), N.Y. Acad. Sci. Clubs: Baie d'Urfé Yacht, Beaconsfield Golf, Forest and Stream. Avocations: skiing, sailing, tennis, gardening. Home: 20722 Gay Cedars, Baie d'Urfe, PQ Canada H9X 2T4 Office: Pulp & Paper Rsch Inst Can, 570 Saint John's Blvd, Pointe Claire, PQ Canada H9R 3J9

WRÓBLEWSKI, RONALD JOHN, physicist; b. N.Y.C., Dec. 24, 1956; s. John Joseph and Florence Yvonne (Davignon) W. BS in Physics, Rensselaer Poly. Inst., 1977; PhD in Physics, Pa. State U., 1984. Staff scientist Arete Assocs., Sherman Oaks, Calif., 1985—. Home: 19610 Sherman Way # 10 Reseda CA 91335 Office: Arete Assocs PO Box 6024 Sherman Oaks CA 91413

WRONA, LEONARD MATHHEW, industrial engineer; b. Buffalo, Feb. 7, 1958; s. Leonard Anthony and Donna Marie (Knowles) W.; m. Kathleen Ann Snyder, Dec. 29, 1979; children: Julia, Elyse, Neil. BS in Indsl. Engring., SUNY, Buffalo, 1980. Registered profl. engr., N.Y., Pa., S.C. Mill engr. Bethlehem Steel Corp., Buffalo, 1980-83; software engr. Ronco Comm. and Electronics, Buffalo, 1983-84; project engr. Ronce Comm. and Electronics, Buffalo, 1984-87, ops. mgr., 1987-88; supervising engr. Hatch Assocs. Cons., Inc., Buffalo, 1988—; installed campus-wide energy mgmt. system Rochester Inst. Tech.; supervised design and constrn. AISI direct steelmaking pilot plant. Mem. NSPE, Inst. of Indsl. Engrs. (sr.), Assn. of Energy Engrs. (sr.), Assn. of Iron and Steel Engrs. Achievements include management of projects for computer automation, process improvements and environmental control for industrial and institutional clients. Office: Hatch Assocs Cons Inc 6215 Sheridan Dr Buffalo NY 14221

WU, ALBERT M., glyco-immunochemistry educator; b. Tainan, Taiwan, Mar. 28, 1940; s. Hong Wu and Miao (Chang) W.; m. June Hsieh, July 1, 1972. BS, Nat. Taiwan U., Taipei, 1965; PhD, N.Y. Med. Coll., N.Y.C., 1975. Staff assoc. Columbia U. Coll. Physicians and Surgeons, N.Y.C., 1976-79, sr. staff assoc., 1979-82; assoc. prof. vet. pathology Tex. A&M U., 1982-89; prof. glyco-immunochemistry Chang-Gung Med. Coll., Tao-Yuan, Taiwan, 1989—. Sr. editor: Molecular Immunology of Complex Carbohydrates, Adv. Exp. Med. Biol., Vol. 228, 1988. Recipient Units award for superior svc. USDA, 1989. Mem. Am. Soc. Biol. Chemists and Molecular Biologists, Soc. for Complex Carbohydrates, N.Am. Taiwanese Prof. Assoc. (senator 1985-87), Sigma Xi. Office: Chang-Gung Med Coll, Glyco-immunochem Lab Area B, Kwei-San Tao-Yuan 33332, Taiwan

WU, CHANG-CHUN, mechanical engineering educator; b. Bengbu, Anhui, China, July 7, 1946; s. Ke-ben and Kuan-ron (Zhang) W.; m. Xue-ying Chen, Aug. 11, 1974; children: Xiao-wen, Dan. Grad., Hefei Poly. U., 1973; MSc, U. Sci. and Tech. of China, Hefei, 1982, PhD, 1987. With Co. of Civil Constrn., Bengbu, 1964-70; teaching asst. Hefei Poly. U., 1974-78; instr. U. Sci. and Tech. of China, 1982-90; Alexander von Humboldt rsch. fellow U. Stuttgart, Fed. Republic of Germany, 1988-89; assoc. prof. U. Sci. and Tech. of China, 1990-92, prof., 1993—; vis. scholar U. Hong Kong, 1989-90, 91-92. Editor Jour. Engring. Mechanics, 1984-91. Recipient Natural Sci. prize Chinese Acad. Sci., 1988, honour medal Nat. Edn. Com. China, 1991, T.H.H. Pian medal ICES'92 Hong Kong, 1992. Fellow East China Network of Solid Mechanics (sec. gen. 1984-88). Avocation: art. Office: U Sci and Tech of China, Dept Modern Mech, Hefei Anhui 230026, China

WU, CHAO-MIN, biomedical engineer; b. Pei-Kang, Taiwan, June 22, 1960; came to U.S., 1986; s. Lung-hwa and Li-Ching (Chen) W.; m. Ling-Chu Lin, June 23, 1986. BS with honors, Chung-Yuan Christian U., Chung-Li, Taiwan, 1982; MS, U. Wyo., 1987. Teaching asst. Chung-Yuan Christian U., Chung-Li, 1984-85; rsch. asst. U. Wyo., Laramie, 1986-87, Pa. State U., State College, 1987-88; from R&D engr. to R&D supr. Suntex Instrumentation Co., Taipei, Taiwan, 1988-90; rsch. assoc. Ohio State U., Columbus, 1990—. Secondary lt. Republic of China Army, 1982-84. Mem. IEEE, Engring. in Medicine and Biology Soc., Acoustic Soc. Am. Office: Ohio State Univ Rm 110 Pressey Hall 1070 Carmack Rd Columbus OH 43210

WU, CHENG YI, physicist; b. Shanghai, People's Republic China, Mar. 18, 1938; arrived in New Zealand, 1984; s. Qin Young and Yung Qi (Lu) Wu; m. Ya Mei Wang, Jan. 1, 1963; children: Da Tong, Da Hao. BSc, Nanking U., Nanking, People's Republic China, 1958; PhD, U. Auckland (New Zealand), 1989. Rsch. asst. Shanghai Acoustics Lab., Academia Sinica, 1959-77, rsch. assoc., 1977-81, rsch. assoc. prof., 1981-84; vis. scientist dept. physics U. Auckland, 1985-89, rsch. fellow dept. physics, 1989—. Contbr. articles to profl. jours. Mem. Acoustics Soc. Am., Physics Soc. Am., Acoustics Soc. China, New Zealand Inst. Physics. Office: U Acukland, Dept of Physics, Auckland New Zealand

WU, DAGUAN, aeroengine design engineer; b. Zhenjiang, Jainsu, Peoples Republic of China, Nov. 13, 1916; s. Jainfei and Jimei (Wang) W.; m. Guo Hua, July 19, 1942; 1 child, Xiaoyun. BS, S.W. United Univ., Kuanming, China, 1942. Pres. and chief designer Rsch. Inst. for Shenyang (China) Aeroengine Design, 1957-78; chief engr. Shenyang Aeroengine Factory, 1973-76; vice chmn. Xian (China) Aeroengine Factory, 1978-82; mem. standing bd. Sci. and Tech. Com. MInistry of Aerospace Industry (MAS), Beijing, 1982—. Editor: Study of Performance of Turbofon Engine and Its Systems, 1986, The Performance Study and Failure Analysis of Gas Turbine Engine in Testing, 1987. Mem. Nat. Com. Chinese Peoples' Polit. Consultative Conf., Beijing, 1978—, standing coun. Chinese Soc. of Engrs. of Thermal Physics, Beijing, 1982—. Named Outstanding Contribution Expert, State Coun., 1990, Ministry of Aerospace Industry, 1992. Mem. AIAA. Achievements include pioneering and organizing design and devel. of first type of aerogas turbine engine in China, 1958—. Home: PO Box 761, Beijing 100012, China Office: Ministry of Aerospace Industry, Sci and Tech Com PO Box 761, Beijing 100012, China

WU, DAN QING, research chemist; b. Fujian, China, July 18, 1960; came to U.S., 1982; m. Juliet Wei, Aug. 7, 1992; 1 child, Michael William. BS, U. Sci. and Tech. China, 1982; PhD, SUNY, Stony Brook, 1990. Rsch. chemist E.I. Du Pont de Nemours & Co., Wilmington, Del., 1989—. Contbr. over 30 articles to sci. jours., including Macromolecules, devel. of scattering method for polymer rsch. Mem. Am. Chem. Soc. (polymer chem. divsn.), Am. Phys. Soc. (high polymer phys. divsn.). Home: 1242 Faun Rd Wilmington DE 19803 Office: EI Du Pont de Nemours & Co Exptl Sta PO Box 80228 Wilmington DE 19880

WU, DAVID LI-MING, cytogeneticist; b. Keelung, Taiwan, Roc, May 14, 1960; came to U.S., 1984; BS, Taipei Medical Coll., Taipei, Taiwan, 1982; MA, Calf. State Univ., 1988. Sr. technologist Cedars-Sinai Medical Ctr., L.A., 1988-92. Mem. Am. Soc. Human Genetics, Assn. Cytogenetic Tech., Chinese Assn. Cytogenetics Southern Calif. (pres.). Office: Cedars-Sinai Med Ctr Med Genetics 8700 Beverly Blvd Los Angeles CA 90047

WU, HSIEN-JUNG, research assistant; b. Taipei, Taiwan, June 13, 1962; s. A-Tien Wu and Tsui-Wan Lin; m. Hsu-Heng Pi, July 19, 1991. MS, U. Mo., 1989; postgrad., Pa. State U., 1990. Prodn. engr. Patent Master Internat. Corp., Taipei, 1987-88. Contbr. articles to profl. jours. Mem. IEEE, Soc. Mfg. Engrs., N.Y. Acad. Scis., Internat. Neural Network Soc., Inst. Indusl. Engrs. Office: Pa State Univ 207 Hammond Bldg University Park PA 16802

WU, JIE, computer science educator; b. Shanghai, China, July 5, 1961; came to the U.S., 1987; s. Zengchang and Yieyi (Shao) W.; m. Ruiguang Zhang, Nov. 16, 1986; 1 child, Stephanie Wu. MS, Shanghai U. Sci. & Engring., 1985; PhD, Fla. Atlantic U., 1989. Lectr. Shanghai U. Sci. & Engring., 1985-87; asst. prof. Fla. Atlantic U., Boca Raton, 1989—; dir. Fla. Atlantic U., 1989—; vis. prof. Ga. Inst. Tech., Atlanta, 1991; cons. SoHar Inc., L.A., 1991. Contbr. articles to profl. jours. Recipient IBM grant, 1990—, Fla. High Tech. grant, 1989-90, Fla. Atlantic U. grant, 1991-92. Mem. IEEE, China Discrete Math. Soc., Can. Soc. Elect., Computer Engring., Upsilon Pi Epsilon. Achievements include discovery of systematic way of determining software fault-tolerant structures, reliable broadcasting schemes on faulty hypercub multicomputers, new computer security models, new interconnection networks for massive parallel computing. Home: 6533 Sweet Maple Ln Boca Raton FL 33433 Office: Dept Computer Sci & Engring Fla Atlantic U Boca Raton FL 33431

WU, JIE-ZHI, fluid mechanist; b. Leshan, Sichuan, China, Jan. 19, 1940; came to U.S., 1989; s. Da-Ren and Shou (Chen) W.; m. Chuan Ji Wu, Apr.

9, 1966; children: Xiao-Hui, Fan. MS, Beijing U. Aeros./Astronautics, China, 1963; PhD, Beijing U. Aeros./Astronautics, 1966. Engr. Chinese Aero. Establishment, Beijing, 1967-79, rsch. engr., 1979-86, assoc. rsch. prof., 1986-87, rsch. prof., 1987—; sr. rsch. scientist, adj. rsch. prof. U. Tenn. Space Inst., Tullahoma, 1991—; hon. fellow U. Minn., Mpls., 1980-82; adj. prof. Jiangsu Inst. Tech., Zhenjiang, China, 1988—; adj. assoc. prof. Chinese U. Sci. and Tech., Hefei, China, 1988—. Author: Introduction to Vorticity and Vortex Dynamics, 1993; contbr. more than 40 articles to profl. jours. Recipient Sci. awards Chinese Aero. Establishment, 1985, 86, 87, Awards of Sci. and Tech. China Ministry of Aero. Industry, 1987, Chinese Govt., 1987, Rsch. Grant NASA Langley Rsch. Ctr., 1987-91. Mem. AIAA, Soc. Indsl. and Applied Math., Chinese Soc. Aerodynamics, Sigma Xi. Achievements include the establishment of a complete theory on the interaction between vorticity field and boundaries (solid boundary, fluid-fluid interface), which has been applied to developing new computational methods and guiding practical Vortex control, and a theoretical foundation for super lift at very high angle of attack and other vortex controls by unsteady excitations based on wave-vortex interacions, receptivity, resonance and streaming. Office: U Tenn Space Inst B H Goethert Pky Tullahoma TN 37388-8897

WU, LIANG TAI, electrical engineer; b. Taipei, Jan. 28, 1948; came to the U.S., 1971; s. Ming-Hsiung and Tso (Kao) W.; m. Joanne Chang, Jan. 16, 1979; children: James, Lisa, Vivian. MA, U. Mich., 1973, MS, 1975, PhD, 1981. Mem. tech. staff Bell Labs., Holmdel, N.J., 1981-83; mem. tech. staff Bell Communications Rsch., Morristown, N.J., 1984-85, dist. mgr., 1985-91, dir., 1991—. Contbr. articles to profl. jours. Asst. scoutmaster Boy Scouts Am., Gladstone, N.J., 1991—. Mem. IEEE. Achievements include 3 U.S. patents, 1 Canadian patent; development of ATM packet networks. Home: 2 Deer Path Gladstone NJ 07934 Office: Bellcore 445 South St Morristown NJ 07960

WU, MEI QIN, mechanical engineer; b. Shanghai, China, Nov. 1, 1953; d. Cheng Qi and Yun Hua (Shen) W.; m. Ping He, Aug. 2, 1983; children: Stewart, Warren. BS, Nanjing U., China, 1977; MS, Tongji U., Shanghai, China, 1981; PhD, Auburn U., 1988. Engr. Shanghai Internal Combustion Engine Rsch. Inst., China, 1977-78, 81-83; instr., rsch. assoc. mech. engring. dept. Auburn (Ala.) U., 1983-88; project cons. Vibron Ltd., Ontario, Can., 1988-92; noise control engr. BVA Systems Ltd., Ontario, Can., 1992-93; sr. devel. engr. Active Noise and Vibration Technologies, Inc., Phoenix, Ariz., 1993—. Contbr. 30 tech. papers to Jour. of Acoustical Soc. of Am., 1986-87, 89-90. 92-93, Noise Control Engring. Jour., 1989, Internat. Noise Control Conf., 1983-93, Internat. Congress on Acoustics, 1983, 86. Mem. ASHRAE, Acoustical Soc. Am., Inst. Noise Control Engring. Achievements include development of the micro-perforation silencer, active concellation of fan noise and the improved cepstrum technique; work experience in machinery noise contro, vibration control and archtl. acoustics. Home: Unit 140 South 2524 W Glenrosa Ave Phoenix AZ 85017 Office: Active Noise Vibration Tech 3811 East Wier Ave Phoenix AZ 85040

WU, ROBERT CHUNG YUNG, space sciences educator; b. Kao-hsiung, Taiwan, Republic of China, Oct. 16, 1943; came to U.S., 1969; s. Liang-Cheng and Grace (Huang) W.; m. Grace S.J. Lee, June 20, 1970; children: David, John, Michael. MS, U. Ill., 1970, PhD, 1973. Postdoctoral fellow U. Iowa, Iowa City, 1976-77; postdoctoral fellow U. So. Calif., L.A., 1973-75, rsch. scientist, 1977-78, asst. rsch. prof., 1978-88, assoc. rsch. prof., 1988—; cons. Rockwell Internat., L.A., 1984, Jet Propulsion Lab., Pasadena, 1985, 89. Contbr. articles to Jour. Chem. Physics, Jour. Geophys. Rsch., Optical Communs., Applied Physics B, Jour. Quantitative Spectroscopy and Radiative Transfer. Mem. Am. Geophys. Union, Optical Soc. Am. Office: U So Calif Space Scis Ctr Los Angeles CA 90089-1341

WU, ROY SHIH SHYONG, biochemist, health scientist administrator; b. Shanghei, China, Nov. 15, 1944; came to U.S. 1956; s. Lung Chung and Chia Ling (Shen) W.; m. Irene Ching Lai Liang, June 8, 1982; 1 child, Michelle. AB, U. Calif., Berkeley, 1967; PhD, Albert Einstein Coll. Med., 1972. Lectr./postdoctoral fellow Dept. Zoology, U. Calif., Berkeley, 1972-75; sr. scientist Biotech Rsch. Labs., Inc., Rockville, Md., 1975-79; sr. staff fellow Div. Cancer Treatment, Nat. Cancer Inst., Bethesda, Md., 1979-82, health scientist adminstr. cancer therapy evaluation, 1986—; cancer expert Lab. of Molecular Pharmacology, Bethesda, Md., 1982-86; mem. PhD theses com. Georgetown U., Washington, 1985. Peer reviewer NSF, 1981, 90, NIH, 1981; contbr. articles to profl. jours. Scoutmaster Boy Scouts Am., Berkeley, 1974-75; com. mem. Orgn. Chinese Ams., Washington, 1977. Formosrn Med. Assn. travel grantee, 1988. Mem. AAAS, Am. Soc. for Biochemistry and Molecular Biology (travel grantee 1985), Am. Assn. Cancer Rsch., Soc. of Chinese Bioscientists in Am. (exec. sec. 1989—), Sigma Chi. Achievements include research in assay for 0 to the 6th degree-alkylguanine-DNA alkyltrasferase using restriction enzyme inhibition. Office: CTEP/DCT/NCI/NIH EPN Rm 734 6130 Executive Blvd Bethesda MD 20892

WU, SHUO-XIAN, acoustics and architecture educator, scientist; b. Quanzhou, Fujian, People's Republic China, May 17, 1947; s. Qiu-Shan Wu and De-Xi Lin; m. Qin-Hui Zhu, Apr. 24, 1977; 1 child, Wu Yan. B.Arch., Qinhua U., Beijing, 1970, M. Engring., 1981, D. Engring., 1984. Technician Xian (People's Republic China) Railroad Bur., 1970-74; asst. engr. Nanchang (People's Republic China) Railroad Bur., 1974-78; lectr. Zhejiang U., Hangzhou, People's Republic China, 1984-86, assoc. prof., 1986-90, prof., 1990—; dir. Archtl. Physics Inst., Hangzhou, 1990—; vis. scholar Sydney (Australia) U., 1987-88, U. Innsbruck, Austria, 1991-92; lectr. internat. confs. Co-author: Design Handbook of Architectural Acoustics, 1989; contbr. articles to profl. jours., including Jour. Acoustical Soc. Am., Jour. Sound and Vibration; finder of image boundary principle in acoustics. Mem. Archtl. Soc. China (acad. com. 1986), Higher Archtl. Edn. China (directive com. 1989), Acoustical Soc. China, Environ. Soc. China, Acoustical Soc. Am. Office: Zhejiang U, Dept Architecture, Hangzhou 310027, China

WU, TSE CHENG, research chemist; b. Hong Kong, Aug. 21, 1923; s. Shau Chuan and Shui (Chan) W.; BS, Yenching U., 1946; MS, U. Ill., 1948; PhD, Iowa State U., 1952; m. Janet Ling, June 14, 1963; children: Alan, Anna, Bernard. Came to U.S., 1947, naturalized, 1962. Prodn. chemist, Yungli Industries, Tangku, China, 1946-47; rsch. asso. Iowa State U., Ames, 1952-53; rsch. chemist duPont Co., Waynesboro, Va., 1953-60, GE, Waterford, N.Y., 1960-71; sr. rsch. chemist Abcor, Inc., Wilmington, Mass., 1971-77; rsch. assoc. Allied-Signal, Inc., Morristown, N.J., 1977-88; cons., 1989—. Mem. Troy Arts Guild, 1968-71, Morris County Art Assn., 1981—. Recipient Gold medallion award for inventions GE, 1967; Allied Corp. patent award, 1983. Eastman Kodak Rsch. fellow, 1951-52. Mem. Am. Chem. Soc., Sigma Xi, Phi Kappa Phi, Phi Lambda Upsilon, Alpha Chi Sigma. Contbr. articles to profl. jours. Patentee in polymer chemistry and organosilicon chemistry. Home: 14E Dorado Dr Morristown NJ 07960-6039

WU, WEN CHUAN, microbial geneticist, educator; b. Taoyuan, Taiwan, Republic of China, Apr. 24, 1931; s. Shiao Tzuon and Mao (Lee) Shieh; m. Yuyen Chen, Nov. 19, 1967; children—Yuan Chenn, Philip Chenn. B.Sc., Nat. Chung Hsing U., 1956; M.Sc., U. Atla. (Can.), 1966; Ph.D., Tokyo U. Agr., 1972. Fgn. research fellow in microbial genetics Nat. Inst. Genetics, Misima, Shizuoka, Japan, 1969-72; postdoctoral fellow in microbial genetics Faculty of Medicine, Meml. U. Nfld. (Can.), St. John's, 1972-75; assoc. prof. microbial genetics, dept. plant pathology Nat. Chung Hsing U., Taichung, Taiwan, Republic of China, 1975-80, prof., 1980—; adj. lectr. biology Toshei Nat. Hosp. Sch. Nursing, Shizuoka, Japan, 1972; vis. prof. Meml. U. Nfld., 1983, U. Guelph (Ont., Can.), 1983-84, U. Toronto, Ont., Can., 1989; vis. scientist U.S. Horticultural Rsch. Lab., Orlando, Fla., 1990. Editor-in-chief: Plant Protection Bull., 1980-82, editorial com., 1978-83, 86-89. Mem. editorial com. Chinese Jour. Microbiology and Immunology, 1981-89, Jour. Genetics and Molecular Biology, 1990—, Plant Pathology Bull., 1992—. Served with Republic of China Marine Corps, 1957-58. Mem. Canadian Soc. Microbiologists, Genetics Soc. Can., Chinese Soc. Genetics, Phytopathological Soc. Republic China, Chinese Soc. Microbiology, Plant Protection Soc. Republic of China, Phytopathol. Soc. Japan, Sigma Xi. Home: 20-3 Minyi St, Taichung 402, Taiwan Office: Nat Chung Hsing U, Grad Inst Dept Plant Pathology, Taichung 402, Taiwan

WU, WEN-TENG, chemical engineer; b. Taiwan, Republic of China, Nov. 18, 1945; m. Alice Huang Wu, Jan. 19, 1974; 1 child, James. Dr. of Engring., Cheng Kung U., 1975. Assoc. prof. Tsing Hua U., Hsin Chu, Taiwan, 1975-81, prof., 1981—, chmn. dept. chem. engring., 1988-91. Contbr. articles to profl. jours. including IEE Procs., Chem. Engring. Comm., Internat. Jour. of Systems Sci., Bio. Process Engring., Biotech. Progress. Recipient Disting. Rsch. award Nat. Sci. Coun., 1991-92. Mem. AICE, Chinese Inst. of Chem. Engring. (Taipei), Sigma Xi. Office: Nat Tsing Hua U, Kun Fu Rd, Hsinchu Taiwan

WU, WILLIAM LUNG-SHEN (YOU-MING WU), aerospace medical engineering design specialist; b. Hangchow, Chekiang Province, China, Sept. 1, 1921; came to U.S., 1941, naturalized, 1955; s. Sing-Chih and Mary (Ju-Mei) Wu. AB in Biochemistry, Stanford U., 1943, MD, 1946; MS in Chemistry and Internal Medicine, Tulane U., 1955; diploma, U.S. Naval Sch. Aviation Medicine, Pensacola, Fla., 1956, USAF Sch. Aviation Medicine, USAF Aerospace Med. Ctr., 1961; cert. of tng. in aviation medicine, UCLA, 1964. Gen. rotating intern U. Iowa Hosps., Iowa City, 1945-46; resident Lincoln (Nebr.) Gen. Hosp., 1946-47, resident in pathology, 1947-48; resident in pathology Bryan Meml. Hosp., Lincoln, 1947-48; fellow, instr. in internal medicine Tulane U., New Orleans, 1948-54; asst. vis. physician Charity Hosp. and Hutchinson Meml. Teaching and Diagnostic and Cancer Detection Clinics, New Orleans, 1948-54; vis. physician Charity Hosp. and Hutchinson Meml. Teaching and Diagnostic Clinics, New Orleans, 1948-54; staff physician Yountville VA Hosp., Napa, Calif., 1958; staff physician Aviation Space and Radiation Med. Group Gen. Dynamics/Convair, San Diego, 1958-61; aerospace med. specialist, med. monitor for Life Sciences Sect. Gen. Dynamics/Astronautics, San Diego, 1961-65; aerospace med. and bioastronautic specialist Lovelace Found. for Med. Edn. and Rsch., Albuquerque, 1965—; staff physician Laguna Honda Hosp., San Francisco, 1968-74; bioastronautics specialist USN, Albuquerque, 1965—; staff physician Kaiser-Permanente Hosp. all-night med. clinic, San Francisco, 1971-73; safety rep. and med. examiner U.S. Civil Aeronaut. Adminstrn., 1959; med. examiner Fed. Aviation Adminstrn., 1961. Author 8 books and 100 tech. papers in field. Comdr., flight surgeon M.C., USNR, 1954-57. Recipient J. Edgar Hoover Gold Disting. Pub. Svc. award Am. Police Hall of Fame, 1991. Fellow San Diego Biomed. Rsch. Inst. (bd. dirs. 1961-65, sec. of fellows 1961-62, chmn. of fellows 1963); mem. AIAA (nom. com. San Diego sect., plant rep. life sci. sect. 1963-65), IEEE (vice chmn. San Diego chpt., profl. tech. group on biomed. electronics 1962-65), N.Y. Acad. Scis., Inst. Environ. Scis., Internat. Univ. Found. (hon. pres.), Internat. Acad. Found. (hon. registrar-sec.), Sigma Xi. Achievements include research of theroetical aspects of cold catalyzed hydrogen fusion nuclear-rocket warm super-conductor hyper-magnetic, hydrogen-fusion space stations. Home: 250 Budd Ave Campbell CA 95008-4063

WU, YING CHU LIN SUSAN, engineering company executive, engineer; b. Beijing, June 23, 1932; came to U.S., 1957; d. Chi-yu and K.C. (Kung) Lin; m. Jain-Ming Wu, June 13, 1959; children: Ernest H., Albert H., Karen H. BSME, Nat. Taiwan U., 1955; MS in Aero. Engring., Ohio State U., 1959; PhD in Aeros., Calif. Inst. Tech., 1963. Sr. engr. Elecro-Optical Systems, Inc., Pasadena, Calif., 1963-65; asst. prof. aero. engring. U. Tenn. Space Inst., Tullahoma, 1965-67, assoc. prof., 1967-73, prof., 1973-88; adminstr. Energy Conversion R&D Programs, Tullahoma, 1981-88; pres., chief exec. officer ERC, Inc., Tullahoma, 1987—. Contbr. over 90 articles to profl. jours. Mem. Better Sch. Task Force, Tullahoma, 1985-86; founding mem. Tullahoma Edn. Found. for Excellence; trustee Rochester Inst. Tech., 1992—. Recipient Chancellor's Rsch. award U. Tenn., 1978. Fellow ASME, AIAA (assoc., chmn. Tenn. sect., H.H. Arnold award 1984); mem. Soc. Women Engrs. (life mem., achievement award 1985), Rotary, Sigma Xi (chmn. U. Tenn. Space Inst. club). Home: 111 Lakewood Dr Tullahoma TN 37388-5227 Office: ERC Inc PO Box 417 Tullahoma TN 37388-0417

WU, ZHENSHAN, economics educator; b. Dalian, Liaoning, China, Mar. 20, 1939; m. Liu Guixiang, Jan. 28, 1967; children: Wu Xiaoqian, Wu Xiaoli. Degree, Dongbei U. Fin. and Econs., Dalian, 1965. Lectr. Dongbei U. Fin. and Econs., 1965-83, assoc. prof., then prof., vice chmn., chmn. fgn. trade dept., 1983—. Author: English Contracts for International Trade, 1986, 2d edit., 1990, International Trade Laws and Practice, 1987, Commercial Communication in English, 1988, International Technology Trade, 1990, Language for International Trade in English, 1991; contbr. articles to profl. publs. Mem. China Internat. Econ. Coop. Soc., Liaoning Internat. Econ. Law Rsch. Assn. (standing councilor 1988—), Liaoning Fgn. Trade Entrepreneurs Assn. (councilor 1991—), China Nat. Coll. and Univ. Br. Learning Group Internat. Econ. Coop. Avocations: music, badminton. Home: Heishijiao Dongbei U Fin, Bldg 8-405, Dalian 116023, China Office: Dongbei U Fin and Econs, Fgn Trade Dept, Dalian 116023, China

WUBAH, DANIEL ASUA, microbiologist; b. Accra, Ghana, Nov. 6, 1960; came to U.S., 1984; s. Daniel Asua and Elizabeth Bruba (Appoe) W. MS, U. Akron, 1988; PhD, U. Ga., 1990. Cert. in hazardous waste site ops. and emergency response health and safety. Postdoctoral fellow U.S. EPA Rsch. Lab., Athens, Ga., 1991-92; asst. prof. Towson (Md.) State U., 1992—. Contbr. articles to profl. publs. Sec. African Students Union, U. Ga., Athens, 1988; v.p. Ghana Soc. of Athens, Ga., 1989. Recipient Paul Acquarone award U. Akron, 1985, Palfrey award U. Ga., 1989, Ruska award Southeastern Electron Microscopy Soc., 1989, Faculty Rsch. grant Towson State U., 1992. Mem. AAAS, Mycological Soc. of Am., Am. Soc. Microbiology, Med. Mycological Soc. of Am., Internat. Soc. for Human and Animal Mycology. Methodist. Achievements include first description of the resting stage of anaerobic zoosporic fungi from rumen; description of a novel morphological development in rumen fungus; demonstration that hitherto fungi belonging to different species were the same. Home: 911 Crosswind Pl Cockeysville MD 21030 Office: Dept Biol Scis Towson State U Towson MD 21204

WUDL, FRED, chemistry educator, consultant; b. Cochabamba, Bolivia, Jan. 8, 1941; came to U.S., 1958; s. Robert and Bertha (Schorr) W.; m. Linda Raimondo, Sept. 2, 1967. BS, UCLA, 1964, PhD, 1967. Postdoctoral rsch. fellow Harvard U., 1967-68; asst. prof. dept. chemistry SUNY, Buffalo, 1968-72; mem. tech. staff AT&T Bell Labs., Murray Hill, N.J., 1972-82; prof. chemistry and physics U. Calif., Santa Barbara, 1982—. Arthur C. Cope scholar award Am. Chem. Soc., 1993. Fellow AAAS. Office: Dept Chemistry U Calif Santa Barbara CA 93106

WUENSCH, BERNHARDT JOHN, ceramic engineering educator; b. Paterson, N.J., Sept. 17, 1933; s. Bernhardt and Ruth Hannah (Slack) W.; m. Mary Jane Harriman, June 4, 1960; children: Stefan Raymond, Katrina Ruth. SB in Physics, MIT, 1955, SM in Physics, 1957, PhD in Crystallography, 1963. Rsch. fellow U. Bern, Switzerland, 1963-64; asst. prof. ceramics MIT, Cambridge, 1964-69, assoc. prof. ceramics, 1969-74, prof., 1974—, TDK chair materials sci. and engring., 1985-90, dir. Ctr. Materials Sci. and Engring., 1988-93, acting dept. head dept. materials sci. and engring., 1980; vis. prof. Crystallographic Inst., U. Saarland, Fed. Republic Germany, 1973; physicist Max Planck Institut für Festkorperforschung, Stuttgart, Fed. Republic Germany, 1981; mem. U.S. nat. com. for crystallography NRC, NAS, 1980-82, 89—; mem. N.E. regional com. for selection of Marshall Scholars, 1970-73, chmn., 1974-80. Co-editor: Modulated Structures, 1979; adv. editor Physics and Chemistry of Minerals, 1976-85; assoc. editor Can. Mineralogist, 1978-80; editor Zeitschrift fuer Kristallographie, 1981-88. Ford Found. postdoctoral fellow, 1964-66. Fellow Am. Ceramic Soc. (Outstanding Educator award 1987), Mineral. Soc. Am.; mem. Am. Crystallographic Assn., Mineral. Assn. Can., Materials Rsch. Soc., Electrochem. Soc. Episcopalian. Home: 190 Southfield Rd Concord MA 01742-3432 Office: MIT Rm 13-4037 Cambridge MA 02139-4307

WUHL, CHARLES MICHAEL, psychiatrist; b. N.Y.C., Sept. 24, 1943; s. Isadore and Sali (Ackner) W.; m. Gail; children—Elise, Amy. M.D., U. Bologna, 1973. Diplomate Am. Bd. Psychiatry and Neurology. Intern, N.Y. Med. Coll., 1975-76, resident in psychiatry, 1976-77; fellow in child psychiatry Columbia Presbyn. Med. Center, 1977-78; practice medicine specializing in psychiatry and child psychiatry, Englewood, N.J., 1978—; attending staff, mem. faculty N.Y. Med. Coll.; psychiatrist NYU, also asst. clin. prof. psychiatry NYU Sch. Medicine. Contbr. to Psychosocial Aspects of Pediatric Care, 1978, World Book Ency., 1980—. Mem. Am. Psychiat Assn., AMA, Am. Acad. Child Psychiatry. Office: 163 Engle St Englewood NJ 07631-2530

WULF, WILLIAM ALLAN, computer information scientist, educator; b. Chgo., Dec. 8, 1939; s. Otto H. and Helen W. (Westermeier) W.; m. Anita K. Jones, July 1, 1977; children: Karin, Ellen. BS, U. Ill., 1961, MSEE, 1963; PhD in Computer Sci., U. Va., 1968. Prof. computer sci. Carnegie-Mellon Univ., Pitts., 1968-81; chmn., chief exec. officer Tartan Labs., Pitts., 1981-87; AT&T prof. computer sci. Univ. Va., Charlottesville, 1988—; asst. dir. Nat. Sci. Found., Washington, 1988-90; chair bd. on computer sci. and telecommunications NRC; bd. dirs. Baker Engrs., Beaver, Pa.; cons. various computer mfrs. Author: Fundamental Structures of Computer Science, 1981. Bd. dirs. Pitts. High Tech. Council,1982-88, Friends of the New Zoo, Pitts. Fellow IEEE, AAAS; mem. ACM (coun.), Nat. Acad. Engring., Computing Rsch. Assn. (bd. dirs.). Avocations: woodworking, photography. Office: U Va Dept Computer Sci Charlottesville VA 22903

WULFERT, ERNST ARNE, research and development director, educator; b. Oslo, Norway, Oct. 24, 1938; s. Karl and Eva (Von Schoultz) W.; m. Danielle Beatrice, July 20, 1964; children: Egon, Alexandra. PhD, U. Oslo, Norway, 1964; DSc, U. Paris, 1968; ISMP, Harvard Bus. Sch., 1985. Rsch. assoc. Pasteur Inst., Paris, France, 1964-68, Med. Rsch. Coun., Ottawa, Canada, 1969-72; dir. R & D Labs. Fournier, Dijon, France, 1972-80, UCB-Pharma, Brussels, Belgium, 1980—; mng. dir. UCB Bioproducts, Brussels, 1985—, v.p., 1990—; vis. prof. Loyola U. Med. Sch., Maywood, Ill., 1987—; bd. dirs. European Fedn. Biotechnology, 1981-87, Sci. Com., Found. for Pharm. Rsch., Paris, 1989—; invited vis. prof. U. Louvain, Brussels, 1991—. Contbr. more than 70 articles to profl. jours. Co-founder European Inst. of Drug Design, Lille, France, 1988. Recipient Rsch. fellowship NATO, 1965, Med. Rsch. Coun., 1969-72, Prix Gallien award, 1978. Mem. Harvard Club of Belgium. Achievements include development of fenofibrate, of ceterizine; research in brain diseases and atherosclerosis. Home: Ave des Aubepines 52, 1180 Brussels Belgium Office: UCB Pharma, Chemin Du Foriest, 1420 Braine L'Alleud Belgium

WULFF, DANIEL LEWIS, molecular biologist, educator; b. Santa Barbara, Calif., Mar. 29, 1937; s. Daniel Reid and Mary (Lewis) W.; m. Bonnie Taylor, Dec. 30, 1957; children: Melissa, Mark, Elise. BS in Chemistry, Calif. Inst. Tech., 1958, PhD in Chemistry, 1962. Postdoctoral fellow Inst. Genetics U. Cologne, Germany, 1962-63; postdoctoral fellow dept. biology Harvard U., Cambridge, Mass., 1963-65; asst. prof. molecular biology U. Calif., Irvine, 1965-68, assoc. prof., 1968-74, prof., 1974-79, assoc. dean biol. scis., 1975-79; dead Coll. Sci. and Math. SUNY, Albany, 1980-93, prof. biol. scis., 1980—. Mem. AAAS, Am. Soc. Microbiology, Genetics Soc. Am., Am. Soc. Biochemistry and Molecular Biology. Office: SUNY Chemistry B27 Albany NY 12222

WULLKOPF, UWE ERICH WALTER, research institute director; b. Hamburg, Germany, Apr. 10, 1940; s. Erich and Christine (Junge) W.; m. Doris Maria Gaide, Aug. 19, 1968; children: Wiebke, Maren. Diploma in econs., U. Hamburg, 1963, D Econs., 1966. Rsch. asst. Grad. Sch. for Econs. and Politics U. Hamburg, 1963-65; town planner Metron, Brugg, Switzerland, 1966-67; vis. scholar Ctr. for Real Estate and Urban Econs. U. Calif., Berkeley, 1967-68; asst. rsch. economist Grad. Sch. Bus. Adminstrn., UCLA, 1968-69; profl. UN Econ. Commn. for Europe, Geneva, 1969-74; dir. Inst. for Housing and Environ., Darmstadt, German, 1974—; mem. com. experts German Social Dem. Party, Bonn, 1979—, Christian Dem. Party, Bonn, 1987—; mem. several profl. groups in housing, urban and regional planning, and environ. Author books on housing, urban and regional devel. and planning; also numerous articles. Avocation: collecting contemporary art. Home: Troyesstrasse 72, 64297 Darmstadt Hessen, Germany Office: Inst Wohnen und Umwelt, Annastrasse 15, 64285 Darmstadt Hessen, Germany

WUNDERLI, WERNER HANS KARL, virologist, researcher; b. Zürich, Switzerland, Apr. 16, 1942; s. Hans Karl and Berta (Funk) Wunderli; m. Heidi Allenspach, July 15, 1971. B of Natural Scis., Fed. Inst. Tech., Zürich, 1970, PhD, 1973. Postdoctoral fellow Inst. Microbiology, Zürich, 1973-74, Ciba Geigy AG, Basel, Switzerland, 1974-75; postdoctoral fellow dept. virology Duke U., Durham, N.C., 1976-78; virologist Nestle, La Tour-de-Peilz, Switzerland, 1978-80; group leader Inst. Med. Microbiology, St. Gallen, Switzerland, 1981-86; dir. diagnostic virology sect. U. Zürich, 1987-92; dir. med. virology Univ. Hosp., Geneva, 1993—; mem. diagnostic commn. Swiss Soc. Microbiology, Berne, 1988—; mem. Swiss workin g group on toxoplasmosis Fed. Office Pub. Health, Berne, 1990—. Mem. European Group for Rapid Viral Diagnosis, Swiss Soc. Microbiology, Union der Schweiz Gesellschaft für Experimental Biology, Am Soc. Microbiology, Aids Soc., Schweiz Verband der Leiter Med. Lab. Reformist. Avocations: lit., music, cross-country skiing. Office: Ctrl Lab Virology, 24 rue Micheli du Crest, 1205 Geneva Switzerland

WUORI, PAUL ADOLF, engineering educator; b. Kauniainen, Finland, Aug. 1, 1933; s. Bruno Adolf and Anna Maria (Nyberg) W.; m. Anne Mari Pihlström, Aug. 17, 1963; children: Eva Maria, Johan Henrik. MS in Engring., Helsinki U. of Tech., Espoo, Finland, 1960, Lic. of Tech., 1969, D in Tech., 1972. Lab engr. Helsinki U. of Tech., Espoo, 1960-67, acting prof., 1967-73, prof. in hydraulic machines, 1973-76, 85—, dean dept. mech. engring., 1976-79, pres., 1979-85; cons. in field. Patentee in field; author: European Journal of Engineering Education, 1987. Served to lt. Finnish Air Force Res., 1959-60. Named Commdr. of the Order of the white rose of Finland, 1983. Mem. European Soc. for Engring. Edn. (mem. adminstrv. council 1979-87, pres. 1985-86), Finnish Acad. Tech. Scis. Lutheran. Club: Helsinki Yacht Soc. Lodges: Ind. Order Odd Fellows, 5 Grankulla (chmn. 1974-75). Avocation: yachting. Home: Tallbackavägen 12, 02700 Grankulla Finland Office: Helsinki Univ of Tech, Otakaari 4 A, 02150 Espoo Finland

WURSTER, DALE ERIC, pharmacy educator; b. Madison, Wis., Jan. 19, 1951; s. Dale Erwin and June M. (Peterson) W.; m. Pamela Ann Marvin, May 31, 1975; children: Elizabeth Ann, Kristin Gail, Dale Edward. BS in Chemistry, U. Wis., 1974; PhD in Phys. Pharmacy, Purdue U., 1979. Asst. prof. Sch. Pharmacy U. N.C., Chapel Hill, N.C., 1979-82; asst. prof. Coll. Pharmacy U. Iowa, Iowa City, 1982-86, assoc. prof. Coll. Pharmacy, 1987—; cons. to pharm. industry; cons. Nat. Assn. Bds. of Pharmacy, Park Ridge, Ill., 1982—. Contbr. articles to profl. jours. Fed. and indsl. grantee. Mem. Am. Assn. Pharm. Scientists, Am. Chem. Soc., Am. Pharm. Assn., Am. Assn. Colls. Pharmacy, Materials Rsch. Soc., Sigma Xi. Achievements include research on surface phenomena; solution and differential scanning calorimetry; chemical kinetics; dissolution kinetics and testing; physics of tablet compression; analytical applications of Fourier transform infrared spectroscopy. Home: 3808 County Down Ln North Liberty IA 52317-9643 Office: U Iowa Coll Pharmacy Iowa City IA 52242

WURSTER, DALE ERWIN, pharmacy educator, university dean emeritus; b. Sparta, Wis., Apr. 10, 1918; s. Edward Emil and Emma Sophia (Steingraeber) W.; m. June Margaret Peterson, June 16, 1944; children: Dale Eric, Susan Gay. BS, U. Wis., 1942, PhD, 1947. With faculty U. Wis. Sch. Pharmacy, 1947-71, prof., 1958-71; prof., dean N.D. State U. Coll. Pharmacy, 1971-72, U. Iowa Coll. Pharmacy, Iowa City, 1972-84, prof., 1972—, interim dean, 1991-92, dean emeritus, 1984—; George B. Kaufman Meml. lectr. Ohio State U., 1968; cons. in field; phys. sci. adminstr. U.S. Navy, 1960-63; sci. adv. Wis. Alumni Rsch. Found., 1968-72; mem. revision com. U.S. Pharmacopoeia, 1961-70, pharmacy rev. com. USPHS, 1966-72. Contbr. articles to profl. jours., chpts. to books; patentee in field. With USNR, 1944-46. Recipient Superior Achievement citation Navy Dept., 1964, merit citation U. Wis., 1976; named Hancher Finkbine Medallion Prof. U. Iowa, 1984; recipient Disting. Alumni award U. Wis. Sch. Pharmacy, 1984. Fellow Am. Assn. Pharm. Scientists (founder, sponsor Dale E. Wurster Rsch. award 1990—, Disting. Pharm. Sci. award 1991); mem. Am. Assn. Colls. Pharmacy (exec. com. 1964-66, chmn. conf. tchrs. 1960-61, vis. scientist 1963-70, recipient Disting. Educator award 1983), Acad. Pharm. Scis. (exec. com. 1967-70, chmn. basic pharmaceutics sect. 1965-67, pres. 1975, Indsl. Pharm. Tech. award 1980), Am. Pharm. Assn. (chmn. sci. sect. 1964-65, Rsch. Achievement award 1965), Wis. (Disting. Service award 1971), Iowa Pharmacists Assn. (Robert G. Gibbs award 1983), Wis. Acad. Scis., Arts and Letters, Soc. Investigative Dermatology, Rumanian Soc. Med. Sci. (hon.), Am. Found. Pharm. Edn. (bd. grants 1987-92), Ea. Va. Med. Sch. Contraceptive Rsch. and Devel. Program (tech. adv. com. 1989—), Am. Assn. Pharm. Scientists (Disting. Scientist award 1991), Sigma Xi, Kappa Psi (past officer), Rho Chi, Phi Lambda Upsilon, Phi Sigma. Home: 16 Brickwood Knoll Iowa City IA 52240

WURTMAN, RICHARD JAY, physician, educator; b. Phila., Mar. 9, 1939; s. Samuel Richard and Hilda (Schreiber) W.; m. Judith Joy Hirschhorn, Nov. 15, 1959; children: Rachael Elisabeth, David Franklin. A.B., U. Pa., 1956; M.D., Harvard U., 1960. Intern Mass. Gen. Hosp., 1960-61, resident, 1961-62, fellow medicine, 1965-66, clin. assoc. in medicine, 1985—; research assoc., med. research officer NIMH, 1962-67; mem. faculty MIT, 1967—, prof. endocrinology and metabolism, 1970-80, prof. neuroendocrine regulation, 1980—, dir. Clin. Research Ctr., 1985—; lectr. medicine Harvard Med. Sch., 1969—; prof. Harvard-MIT Div. Health Scis. and Tech., 1978—; sci. dir. Ctr. for Brain Scis. and Metabolism Charitable Trust, 1981—; invited prof. U. Geneva, 1981; Sterling vis. prof. Boston U., 1981; mem. small grants study sect. NIMH, 1967-69, preclin. psychopharmacolgy study sect., 1971-75; behavioral biology adv. panel NASA, 1969-72; council basic sci. Am. Heart Assn., 1969-74; research adv. bd. Parkinson's Disease Found., 1972-80, Am. Parkinson's Disease Assn., 1978—; com. phototherapy in newborns NRC-Nat. Acad. Scis., 1972-74, com. nutrition, brain devel. and behavior, 1976, mem. space applications bd., 1976-82; mem. task force on drug devel. Muscular Dystrophy Assn., 1980-87; chmn. life scis. adv. com. NASA, 1979-82; mem. adv. bd. Alzheimer's Disease Assn., 1981-84; asso. neuroscis. research program MIT, 1974-82; chmn. life scis. adv. bd. USAF, 1985—; Bennett lectr. Am. Neurol. Assn., 1974; Flexner lectr. U. Pa., 1975; chmn. sci. adv. bd. Interneuron Pharms., 1989—; founder, chmn. sci. adv. bd. Interneuron Pharms. Inc., 1989—; Hans Lindler Meml. lectr. Weizmann Inst., 1993. Author: Catecholamines, 1966, (with others) The Pineal, 1968; editor: (with Judith Wurtman) Nutrition and the Brain, Vols. I and II, 1977, Vols. III, IV, V, 1979, Vol. VI, 1983, Vol. VII, 1986; also articles; mem. editorial bd. Endocrinology, 1967-73, Jour. Pharmacology and Exptl. Therapeutics, 1968-75, Jour. Neural Transmission, 1969-88, Neuroendocrinology, 1969-72, Metabolism, 1970-80, Circulation Research, 1972-77, Jour. Neurochemistry, 1973-82, Life Scis., 1973-81, Brain Research, 1977—. Holder of approximately 40 U.S. patents on new treatments for diseases and conditions. Recipient Alvarenga prize and lectureship Phila. Coll. Physicians, 1970, CIBA-Geigy Drew award in Biomed. Research, 1982, Roger Williams award in Preventative Nutrition, 1987, Roger J. Williams award in Preventive Medicine, 1989, NIMH Merit award, 1989—, Internat. Prize for Modern Nutrition, 1989, Hall of Fame Disting. Alumni award Ctrl. High Sch. Phila., 1992; Disting. lectr. Purdue U., 1984; Rufus Cole lectr. Rockefeller U. 1985; Pfizer lectr. NYU Med. Sch., 1985; Grass Fedn. lectr. U. Ga., 1985, Alan Rothballer Meml. lectr., N.Y. Med. Coll., Valhalla, N.Y., 1989, Gretchen Kerr Green lectr. in the neurosciences, 1989; Chevalier de Malte, Order of the Knights of Malta, 1987; Wellcome Vis. Prof. Washington State U., Pullman, 1989; Julius Axelrod Disting. lectr. in neurosci., CUNY, 1990, Sigma Tau Found. lectr. on aging, Rome, 1990, Disting. lectr. in neurosci., La. State U., 1991, McEwen lectr. Queen's U., Ont., 1991; Plenary lectr. 3d Internat. Symposium on Microdialysis, 1993; Hans Lindner Meml. lectr. Weizmann Inst., 1993. Mem. Am. Soc. Clin. Investigation, Endocrine Soc. (Ernst Oppenheim award 1972), Am. Physiol. Soc., Am. Soc. Biol. Chemists, Am. Soc. Pharmacology and Exptl. Therapeutics (John Jacob Abel award 1968), Am. Soc. Neurochemistry, Soc. Neuroscis., Am. Soc. Clin. Nutrition, Am. Inst. Nutrition (Osborne & Mendel award 1982). Club: Harvard (Boston). Home: 300 Boylston St Boston MA 02116-3923 Office: Mass Inst Tech E25-604 Cambridge MA 02139

WURZBACH, RICHARD NORMAN, utility company engineer; b. Phila., Sept. 10, 1966; s. Carl Norman and Margaret Mary (McDevitt) W. BA in Chemistry, Millersville (Pa.) U., 1988. Predictive maintenance engr. Bechtel Power Corp., Delta, Pa., 1989, Phila. Elec. Co., Delta, 1989—. Contbr. articles to profl. jours. Mem. Internat. Soc. Optical Engring. (steering com. for thermosense 1993—, internat. working group 1991—, co-chmn. predictive maintenance session 1993—), Soc. Tribologists and Lubrication Engrs., U.S. Jaycees (local bd. dirs. 1991—). Roman Catholic. Achievements include development of method for correlating images of emanating thermal energy to metallurgical degredation in lubricated mechanical couplings. Office: Peach Bottom APS A2-2S RD 1 Box 208 Delta PA 17314

WÜTHRICH, KURT, molecular biologist, biophysical chemist; b. Oct. 4, 1938. BS in Chemistry and Physics, U. Bern, Switzerland, 1962; PhD, U. Basel, Switzerland, 1964. Postdoctoral tng. U. Basel, U. Calif., Berkeley, Bell Telephone Labs., Murray Hill, N.J., 1964-69; prof. biophysics Eidgenössische Technische Hochschule Zürich, 1972—; fgn. assoc. NAS. Contbr. 2 monographs and over 400 papers in field. Recipient Friedrich Miescher prize Schweizerische Biochemische Gesellschaft, 1974, shield of faculty of medicine Tokyo U., 1983, P. Bruylants medal Cath. U. Louvain, 1986, Stein and Moore award Protein Soc., 1990, Louisa Gross Horowitz prize Columbia U., 1991, Gilbert N. Lewis medal U. Calif., Berkeley, 1991, Marcel Benoist prize Swiss Confederation, 1992, Disting. Svc. award Miami Winter Symposia, 1993, Prix Louis Jeantet de Médecine, 1993; fgn. fellow Indian Nat. Sci. Acad.; hon. fellow NAS India. Mem. Deutsche Acad. der Naturforscher Leopoldina, EMBO, Academia Europea, Internat. Union Pure and Applied Biophysics (mem. coun. 1975-78, 87-90, sec. gen. 1978-84, v.p. 1984-87), Internat. Coun. Sci. Unions (mem. gen. com. 1980-86, mem. standing com. on free circulation scientists 1982-90), Am. Acad. Arts and Scis. (fgn. hon. mem.). Office: Molecular Biology and Biophysics, ETH Hônggerberg, 8093 Zurich Switzerland

WUTHRICH, PAUL, electrical engineer, researcher, consultant; b. Ziefen, Switzerland, Feb. 25, 1931; came to U.S., 1956; s. Emil and Hulda (Degen) W.; m. Irmgard Ann Garbe, Dec. 3. 1960; children: Christine, Marc. BSEE, U. New Haven, 1970, MSEE, U. Conn., 1973. Prin. engr., mgr. R & D Timex Corp., Waterbury, Conn., 1956-88, staff scientist, 1988—. Mem. Opitcal Soc. Am., Instrument Soc. Am., Phys. Soc. Am. Achievements include 45 patents on watch movement and instrumentation from electric to solid state watches; develop. of first 3D nimslo camera; rsch. on first infusion pump device for med. implant. Home: 760 Hamilton Ave Watertown CT 06795-2311 Office: Timex Corp Waterbury CT 06722

WYANT, JAMES CLAIR, engineering company executive, educator; b. Morenci, Mich., July 31, 1943; s. Clair William and Idah May (Burroughs) W.; m. Louise Doherty, Nov. 20, 1971; 1 child, Clair Frederick. BS, Case Western Reserve, 1965; MS, U. Rochester, 1967, PhD, 1968. Engr. Itek Corp., Lexington, Mass., 1968-74; instr. Lowell (Mass.) Tech. Inst., 1969-74; prof. U. Ariz., Tucson, 1974—; vis. prof. U. Rochester, N.Y., 1983; pres. WYKO Corp., Tucson, 1984—; chmn. Gordon Conf. on Holography Plymouth (N.H.) State Coll., 1984. Editor: Applied Optics and Optical Engineering, vols. VII-X, 1979, 80, 83, 87. Mem. Optical Soc. Am. (bd. dirs. 1979-81, Joseph Fraunhofer award 1992), Soc. Photo-Optical Instrumentation Engring. (pres. 1986). Home: 1881 N King St Tucson AZ 85749-9367 Office: U Ariz Optical Scis Ctr Tucson AZ 85721

WYATT, PHILIP RICHARD, geneticist, physician, researcher; b. St. Louis, Oct. 22, 1951; s. John Poyner and Isabel (Gillespie) W.; m. Sharon Lorraine Parker, June 23, 1978; children: Geoffrey, Kathryn. BS, U. Man., 1972; PhD, U. Man., Winnipeg, Man., Can., 1976; MD, U. Ky., 1980. Lic. Med. Coun. Can. Chief genetics dept. North York Gen. Hosp., Toronto, Ont., Can., 1983—; advisor Ministry Health, Toronto, Ont., Can., 1984-90. Recipient Karger prize Karger Pub., Switzerland, 1976. Fellow Human Biology Coun.; mem. Am. Soc. Human Genetics. Office: Genetics North York, 4001 Leslie St, North York, ON Canada M2K 1T7

WYATT, RICHARD JED, psychiatrist, educator; b. Los Angeles, June 5, 1939; children: Elizabeth, Christopher, Justin. B.A., Johns Hopkins U., 1961, M.D., 1964; M.D. (hon.), Central U. Venezuela, 1977. Intern in pediatrics Western Res. U. Hosp., Cleve., 1964-65; resident in psychiatry Mass. Mental Health Center, Boston, 1965-67; with NIMH, 1967—, asst. dir. intramural research, 1977-87, chief neuropsychiatry br., chief Neurosci. Ctr., 1972—; clin. prof. psychiatry Stanford U. Med. Sch., 1973-74, Duke U. Med. Sch., 1975—, Uniformed Svcs. Sch. Medicine, 1980—, Columbia U., 1987—; practice medicine specializing in psychiatry Washington, 1968—; cons. Psychiatry Shelter Program for the Homeless, Columbia U., 1987—; mem. scil adv. bd. Nat. Alliance for Rsch. Schizophrenia and Depression, 1991. Mem. editorial bd. Advances in Neurosci., 1989, Harvard Rev. Psychiatry, 1992. Recipient Harry Solomon Research award Mass. Mental Health Center, 1968, A.E. Bennett award Soc. Biol. Psychiatry, 1971, Psychopharm. award Am. Psychol. Assn., 1971, Superior Achievement award USPHS, 1980, Dean award Am. Coll. Psychiatrists, McAlpin Research Achievement award Nat. Mental Health Assn., 1986, Arthur P. Noyes award Commonwealth of Pa., 1986, Arieti award Am. Acad. Psychoanalysts, 1989, Robert L. Robinson award for Moods in Music, 1990, Mental Health Bell Media award for Moods in Music Mental Health Assn. Atlanta, 1991, Media award for To Paint the Stars Nat. Mental Health Assn. , 1991. Fellow Am. Psychiat. Assn. (task force on local arrangements 1991—), Am. Coll. Neuropsychopharmacology (nominating com. 1991, Efron award 1983); mem. Washington Psychiat. Assn., Soc. Psychophysiol. Study Sleep, Soc. Biol. Psychology, Am. Assn. Geriatric Psychiatry, Psychiat. Research Soc., Soc. Neuroscis., AMA. Office: NIMH Neurosci Rsch Ctr 2700 Martin Luther King Ave SE Washington DC 20032

WYCOFF, CHARLES COLEMAN, retired anesthesiologist; b. Glazier, Tex., Sept. 2, 1918; s. James Garfield and Ada Sharpe (Braden) W.; m. Gene Marie Henry, May 16, 1942; children: Michelle, Geoffrey, Brian, Roger, Daniel, Norman, Irene, Teresa. AB, U. Calif., Berkeley, 1941; MD, U. Calif., San Francisco, 1943. Diplomate Am. Bd. Anesthesiology. Founder The Wycoff Group of Anesthesiology, San Francisco, 1947-53; chief of anesthesia St. Joseph's Hosp., San Francisco, 1947-52, creator residency tng. programs in anesthesiology, 1950; chief anesthesia San Francisco County Hosp., 1953-54; practice anesthesiology, tchr. Presbyn. Med. Ctr., N.Y.C., 1955-63; asst. prof. anesthesiology Columbia U., N.Y.C., 1955-63; clin. practice anesthesiology St. Francis Meml. Hosp., San Francisco, 1963-84; creator residency tng. programs in anesthesiology San Francisco County Hosp., 1954. Producer, dir. films on regional anesthesia; contbr. articles to sci. jours. Scoutmaster Boy Scouts Am., San Francisco, 1953-55. Capt. M.C., U.S. Army, 1945-47. Mem. Alumni Faculty Assn. Sch. Medicine U. Calif.-San Francisco (councilor-at-large 1979-80). Democrat. Avocations: research in evolution of human behavior, Sierra hiking, gardening. Home: 394 Cross St Napa CA 94559

WYCZALKOWSKI, WOJCIECH ROMAN, mechanical engineer, chemical engineer, researcher; b. Katowice, Poland, Mar. 11, 1946; came to U.S., 1981; s. Pawel Leon and Waclawa (Mossakowska) W.; m. Halina Maria Pankiewicz, Dec. 14, 1969; children: Christopher, Matthew, Alexander. BS in Mech. Engring., Tech. U., Wroclaw, Poland, 1968; MS in Mech. Engring., Tech. U., 1969, PhD in Chem. Engring., 1976. From asst. to assoc. prof. Tech. U., Wroclaw, 1969-81; eng. Starks Assoc., Inc., Buffalo, 1981-82; project mgr. HMW Enterprise, Inc., Etters, Pa., 1982-87; rsch. engr. Kenterprise Rsch., York, Pa., 1987-88; dir. tech. Phila. Mixers, Palmyra, Pa., 1988—; cons. Kenterprise Rsch., York, Pa., 1985-87, SUNY, Buffalo, 1987-90. Contbr. articles to profl. jours. Republican. Roman Catholic. Achievements include 6 Polish patents and 1 U.S. patent; design of industrial computer workstation, many specialized fluid mixer installations; research in foam destruction/industrial applications, application of laser technology in fluid mixing research, chemical engineering. Home: 685 Harvest Dr Harrisburg PA 17111 Office: Phila Mixers Co 1221 E Main St Palmyra PA 17078

WYLIE, CLARENCE RAYMOND, JR., mathematics educator; b. Cin., Sept. 9, 1911; s. Clarence Raymond and Elizabeth M. (Shaw) W.; m. Sikri M. Aho, June 27, 1935 (dec. 1956); children: Chris Raymond (dec.), Charles Victor; m. Ellen F. Rasor, June 25, 1958. B.S., Wayne State U., 1931, A.B., 1931; M.S., Cornell U., 1932, Ph.D., 1934. Instr. Ohio State U., 1934-40, asst. prof., 1940-46; cons. mathematican Propeller Lab., Wright Field, Dayton, Ohio, 1943-46; chmn. dept. math. and acting dean Coll. of Engring., Air Inst. Tech., Wright Field, Dayton, 1946-48; prof. math. U. Utah, Salt Lake City, 1948-69; chmn. dept. U. Utah, 1948-67; prof. math. Furman U., Greenville, S.C., 1969—; William R. Kenan, Jr. prof. math. Furman U., 1971-78, Kenan prof. emeritus, 1978—, chmn. dept. math., 1970-76; part-time cons. math. Gen. Electric Co., Schenectady, Briggs Mfg. Co., Detroit, Aero Products div. Gen. Motors Corp., Dayton; Humble lectr. in sci. Humble Oil Co., 1955; disting. visitor Westminster Coll., Salt Lake City, 1981; cons. mathematician Holloman Air Base, N.Mex., 1955, 56. Author: 101 Puzzles in Thought and Logic, 1958, Advanced Engineering Mathematics, 5th edit, 1982, Introduction to Projective Geometry, 1969, The Wisdom of Eric Lim, 1974, Differential Equations, 1979, Limericks, 1987, The World of Eric Lim, 1991; other books, papers, articles, pub. in math. jours. Strange Havoc; poems, 1956, also miscellaneous poetry. Recipient Disting. Alumni award Wayne State U., 1956, Disting. Engring. Alumni Achievement award, 1985, named to Engring. Coll. Hall of Fame, 1985; Algernon Sydney Sullivan award Furman U., 1982; Bell Tower award Furman U., 1986. Fellow AAAS; mem. Math. Assn. Am. (chmn. Rocky Mountain sect. 1955-56, bd. govs. 1957-60, sect. lectr. Southeastern sect. 1983), Am. Math. Soc., Am. Soc. Engring. Edn. (chmn. math. div. 1949-50, 1957-58), Utah Acad. Sci. Arts and Letters, S.C. Acad. Sci., Phi Beta Kappa, Sigma Xi, Pi Mu Epsilon, Sigma Pi Sigma, Pi Kappa Delta, Delta Sigma Rho, Sigma Delta Psi, Sigma Rho Tau, Phi Kappa Phi, Tau Kappa Alpha. Methodist.

WYNDER, ERNST LUDWIG, science foundation director, epidemiologist; b. Herford, Germany, Apr. 30, 1922; came to U.S., 1938, naturalized 1943; s. Alfred and Therese (Godfrey) W. BA, NYU, 1943; B in Med. Sci., Washington U., 1950, MD, 1950; DSc (hon.), N.Y. Med. Coll., 1992; MD (hon.), Med. Faculty U. Hamburg, 1992. Lic. doctor and surgeon, N.Y. Asst. Sloan-Kettering Inst. for Cancer Rsch., N.Y.C., 1952-54, assoc., 1954-60, assoc. mem., 1960-69, assoc. scientist, 1969-83, adj. mem., 1984—; pres., med. dir. Am. Health Found., N.Y.C., 1969—. Contbr. 600 articles to profl. jours. Decorated officer's cross Order of Merit (Germany), 1991; recipient Bordon Undergrad. Rsch. award Washington U., 1950, N.Y. State Health Edn. and Illness Prevention award N.Y. State, 1981, Max von Pettenkofer medal Max von Pettenkofer Inst., 1988, Alton Ochsner award Alton Ochsner Med. Found., 1988, Lucy Wortham James Clin. Rsch. award Soc. Surg. Oncology, 1989, medal Surgeon Gen. U.S., 1989, Mayor's Ethnic award City of N.Y., 1989, Nathan Pritikin Pioneer award Nathan Pritikin Rsch. Found., 1989, Medal of Honor for clin. rsch. Am. Cancer Soc., 1989, Robert-Koch Gold medal Robert-Koch Stiftung, 1990, Disting. Clinician award Milken Family Med. Found., 1990, CliniOncol Rsch., Julia Hudson Freund lecture award, Wash. U., 1990. Mem. Am. Assn. Cancer Rsch. (hon.), Am. Coll. for Preventive Medicine, Am. Pub. Health Assn., Max Planck Soc. Germany, N.Y. Acad. Scis. (Sarah L. Poiley Meml. award 1979), Am. Soc. Preventive Oncology (Disting. Achievement award 1984). Office: Am Health Found 320 E 43rd St New York NY 10017

WYN-JONES, ALUN (WILLIAM WYN-JONES), software developer, mathematician; b. Tremadoc, Gwynedd, Great Britain, Aug. 15, 1946; came to U.S., 1976; s. Goronwy Wyn and Mai Jones; m. Jocelyn Ripley, July 29, 1977; 1 stepchild, Electra Truman. BSc with honors, U. Manchester, U.K., 1968; MSc, Univ. Coll. London, 1970. Rsch. engr. Marconi-Elliott Computer Labs., Borehamwood, U.K., 1970-71; asst. tutor math. Poly. North London, 1971-72; programmer CRC Info. Systems, Ltd., London, 1972-76; mgr. devel. Warner Computer (now Warner Ins.), N.Y.C., 1976-80; pres., owner, developer Wallsoft Systems, Inc., N.Y.C., 1982-92, Integrity Systems Corp., N.Y.C., 1980—; invited speaker at profl. confs. Author, co-author computer software. Recipient Byte Award Distinction Byte Editors and Columnists, 1988, Readers Choice award Data Based Advisor Readers, 1990, 91. Mem. AAAS, Am. Math. Soc., Math. Assn. Am. Methodist. Achievements include development of template programming in automatic code generation. Home: Apt 14D 609 Columbus Ave New York NY 10024

WYNN, ROBERT RAYMOND, engineer; b. Omaha, Mar. 4, 1929; s. Horace Oscar and Yvonne Cecil (Witters) W.; m. Joann Elizabeth Swicegood, June 28, 1974; children: Kay, William, Frederick, Andrew, Emma, Lawrence, Robert. Diploma in Nuclear Engring., Capitol Radio Engring. Inst., 1964; BSEE, Pacific Internat. Coll. Arts and Scis., 1964; AA in Bus. Adminstrn., Allen Hancock Coll., 1969; MSEE, Pacific Internat. Coll. Arts and Scis., 1971; MSMS, West Coast U., 1975, ASCS, 1985; BSCS, U. State of N.Y., 1985. Registered profl. engr., Calif. Meteorologist United Air Lines, Calif., 1949-53; engring. planner Aircraft Tools Inc., Inglewood, Calif., 1953-55; field service engr. N. Am. Aviation, Inglewood, Calif., 1955-59; R&D engr. Carstedt Research Inc., N. Long Beach, Calif., 1959-60; test engr. Martin Marrietta Corp., Vandenburg AFB, Calif., 1960-64; project engr. Fed. Electric Corp., Vandenburg AFB, Calif., 1965-69; systems engr. Aeronutronic Ford Corp., Pasadena, Calif., 1970-75; MTS Jet Propulsion Lab., Pasadena, Calif., 1975-83; engring. mgr. Space Com., Redondo Beach, Calif., 1983-84; engring. specialist Boeing Service Inc., Pasadena, 1984-86; cons., mem. tech. staff Jet Propulsion Lab., Pasadena, 1986—; instr. computer sci. and CAD, Jet Propulsion Lab., 1980-82. With USAAF, 1946. Mem. Calif. Soc. Profl. Engrs., Exptl. Aircraft Assn. (pres. Lompoc chpt. 1968), W. Coast U. Alumni Assn. Democrat. Avocations: model airplane design and constrn., flying, camping. Home: PO Box 4138 Sunland CA 91041-4138 Office: Jet Propulsion Lab 4800 Oak Grove Dr Pasadena CA 91109-8099

WYRICK, DAVID ALAN, engineering educator, researcher; b. Cheyenne, Wyo., Feb. 11, 1957; s. Carl Howard and Doris Mae (Bruce) W.; m. Joanne Marie Novotny, May 19, 1979; children: Alexandre Patrick, Nathaniel Addison. BSME, U. Wyo., 1979, MSME, 1985; MS in Engring. Mgmt., U. Alaska, Anchorage, 1984; PhD in Engring. Mgmt., U. Mo., Rolla, 1989. Profl. engr., Minn. Ops. engr. ARCO Oil & Gas Co., Houston, 1979-80; facilities engr. ARCO Alaska, Inc., Anchorage, 1980-81; ops./analytical engr. ARCO Exploration Co., Anchorage, 1981-85; sr. ops./analytical engr. ARCO Oil & Gas Co., Midland, Tex., 1985; grad. rsch. asst. U. Wyo., Laramie, 1985-86; grad. teaching asst. U. Mo., Rolla, 1987-89; asst. prof. U. Minn., Duluth, 1989-93; assoc. prof. U. Minn., 1993—. Contbr. articles to profl. jours.; mem. rev. bd. Engring. Mgmt. Jour., 1992—. Head coach Duluth Heights Ice Mite Hockey Team, Duluth, 1992—. Mem. ASME (sec. Alaska sect. 1982-83, chair 1983-84), NSPE, Am. Soc. for Engring. Mgmt. (Merritt Wiliamson award 1990), Am. Soc. for Engring. Edn. Office: UMD Indsl Engring 176 Engr Bldg 10 University Dr Duluth MN 55812-2496

WYRTKI, KLAUS, oceanography educator; b. Tarnowitz, Germany, Feb. 7, 1925; came to U.S., 1961; s. Wilhelm and Margarete (Pacharzina) W.; m. Helga Kocher, June 6, 1954 (div. 1970); children: Undine, Oliver; m. Erika Maassen. PhD magna cum laude, U. Kiel, Germany, 1950. With German Hydrographic Inst., Hamburg, 1950-51; German Rsch. Coun. postdoctoral rsch. fellow U. Kiel, 1951-54; head Inst. Marine Rsch., Djakarta, Indonesia, 1954-57; sr. rsch. officer, then prin. rsch. officer div. fisheries and oceanography Commonwealth Sci. and Indsl. Rsch. Orgn., Sydney, Australia, 1958-61; assoc. rsch. oceanographer, then rsch. oceanographer Scripps Instn. Oceanography, U. Calif., 1961-64; prof. oceanography U. Hawaii, Honolulu, 1964—; chmn. North Pacific Expt., 1974-80, com. on climate changes and ocean Internat. Assn. Phys. Scis. of the Oceans; mem. Spl. Com. on Ocean Rsch. Working Group on Prediction of El Nino, Sci. Working Group on Topography Expt., panel on climate and global change NOAA. Editor: Atlas on Physical Oceanography of International Indian Ocean Expedition; mem. editorial bd. Jour. Phys. Oceanography, 1971-79. Recipient Excellence in Rsch. award U. Hawaii, 1980, Rosenstiel award U. Miami, 1981. Fellow Am. Geophys. Union (Maurice Ewing medal 1989), Am. Meteorol. Soc. (Harald Ulrick Sverdrup Gold medal 1991), Deutsche Meteorologische Gesellschaft (Albert Defant medal 1992). Office: U Hawaii 1000 Pope Rd Honolulu HI 96822-2336

WYSE, DONALD L., agronomist, educator. Prof. dept. agronomy U. Minn., St. Paul. Recipient CIBA-GEIGY/Weed Sci. Soc. Am. award CIBA-GEIGY Corp., 1991. Office: Univ of Minnesota Dept of Agronomy 1991 Buford Circle Saint Paul MN 55108*

WYSLOTSKY, IHOR, engineering company executive; b. Kralovane, Czechoslovakia, Dec. 22, 1930; s. Ivan and Nadia (Alexiew) W.; came to U.S., 1958, naturalized, 1961; M.E., Sch. Aeros., Buenos Aires, Argentina, 1955; m. Marta Farion, 1983; children: Katria, Bohdan, Roman, Alexander. Design engr. Kaiser Industries, Buenos Aires, 1955-58; cons. design engr., Newark, 1959-64; chief engr. Universal Tool Co., Chgo., 1964-69; pres. CBC Devel Co., Inc., Chgo., 1969-74; pres. TEC, Inc., Chgo., 1972-83; pres. REDEX Corp., 1983-89, chmn. 1993—; engring. adviser to bd. Biosystems Insts., Inc., La Jolla, Calif. Co-founder Ukrainian Univ. Studies, U. Ill., Am.-Ukraine Bus. Coun., pres. Mem. Am. Ukraine Bus. Coun. (pres. 1991—), Packaging Inst. U.S.A., Am.-Israeli C. of C. (v.p.), Brit. Engring. Assn. River Plate, Soc. Mfg. Engrs. Patentee in field. Mgmt. adv. bd. Modern Plastics Publs. Home: 6133 N Forest Glen Ave Chicago IL 60646-5015 Office: 5050 Newport Dr Rolling Meadows IL 60008

WYSOCKI, CHARLES JOSEPH, neuroscientist; b. Utica, N.Y., May 4, 1947; s. Charles C. and Helen T. (Szczesna) W.; m. Linda Lorraine Moore, Dec. 21, 1968; children: Tracy Lynn, Theresa Marie, Alexandra Charlene. BA, SUNY, Oswego, 1973; MS, Fla. State U., 1976, PhD, 1978. Grass Found. fellow The Jackson Lab., Bar Harbor, Maine, 1973; postdoctoral tng. Monell Chem. Senses Ctr., Phila., 1978-81, asst. mem., 1980-83, assoc. mem., 1983-90, mem., 1990--; adj. prof. anatomy U. Pa., Phila., 1985--; scientific advisor Fragrance Rsch. Fund, N.Y.C., 1989--; chmn. in tng. Monell Chem. Senses Ctr., Phila., 1989--; project officer, coord. U.S. Russia Scientific Exchange Program in the Chem. Senses, Washington, 1990--. Editor: Chemical Senses Vol. 3: Genetics of Perception and Communication, 1991; contbr. editor: (newsletter) The Armoa-Chology Rev., 1986--; contbr. articles to profl. jours. With U.S Army, 1968-70, Vietnam. NIH grantee 1985--; recipient Kenji Nakanishi Rsch. award Takasago Internat. Corp., 1988. Mem. Assn. for Chemoreception Scis. (membership chmn. 1989-91), N.Y. Acad. Scis., Soc. Neurosci. Internat. Brain Rsch. Orgn., Sigma Xi (Best Student Rsch. award 1976). Office: Monell Chem Senses Ctr 3500 Market St Philadelphia PA 19104-3308

XIE, NAN-ZHU, medical physics educator; b. Guangzhou, Guangdong, People's Republic China, Dec. 2, 1925; s. Jian-Cheng and Rui-Zhen (Tan) X.; m. Pei-Lan Li, Mar. 12, 1952; children: Hong, Dexter Qing, Jimmy Zheng. BS in Physics, Dr. Sun Yat-Sen's U., Guangzhou, 1949. Dir. New Life Film Co., Guangzhou, 1949-52; asst. prof. Guangzhou Normal Coll., 1952-62; assoc. prof. Guangzhou Med. Coll., 1962-80, prof. dept. med. physics, 1980—; med. cons. Welda Med. Apparatus Group Corp., Guangdong, 1990—. Chief editor: Physics for Medical Sciences, 1982, Modern Medical Imaging, 1985, Biomedical Electronics, 1986, Computers in Medicine, 1990. Nat. rep. Chinese Nat. Congress of People's Reps., Beijing, 1983-92. Named Excellent Tchr., Guangdong Govt., 1981. Mem. Chinese Assn. Med. Imaging Tech. (v.p. 1989-92), Chinese Soc. Physics (v.p. Guangdong br. 1984-91), N.Y. Acad. Sci., Am. Assn. Physicists in Medicine (full mem.), Internat. Orgn. for Med. Physics (chmn. devel. countries com. 1989—). Avocations: singing, dancing, swimming, football, tennis. Home and Office: Guangzhou Med Coll, Dong Feng Rd West 195, Guangzhou 510182, China

XIE, WEIPING, mass spectromist; b. Chongqing, Sichuan, People's Republic China, Dec. 22, 1955; came to U.S., 1984; s. Jianyia and Lixiang (Chen) X.; m. Xiaoling Wang, Aug. 1, 1982; 1 child, Xiaoping. MS, U. Minn., 1987, PhD, 1991. Teaching asst. Southwest Agrl. U., Chongqing, 1982-84; rsch. asst. U. Minn., St. Paul, 1985-90, rsch. fellow, 1990-91, rsch. assoc., 1992—. Contbr. articles to Jour. Natural Products, Jour. Agrl. and Food Chemistry, others. Mem. Am. Phytopathological Soc., Am. Soc. Mass Spectrometry, Sigma Xi. Achievements include isolation and structure identification of new derivatives of mycotoxin fasarochromenone from fungal culture. Home: 66 14th Ave SW New Brighton MN 55112 Office: Univ Minn 1991 Buford Cir Saint Paul MN 55108

XIE, YUNBO, oceanographer; b. Guangdong, China, Sept. 28, 1957; arrived in Can., 1986; s. Qianming and Xiouzheng (Zhong) X. BS, Shandong Coll. Oceanography, 1982; MS, Acad. Sinica, Beijing, 1985; PhD, U. B.C., Can., 1991. Rsch. asst. Shanghai Acoustic Lab. Acad. Sinica, 1985-86; postdoctoral fellow U. Victoria, B.C., 1991-92; rsch. assoc. Inst. Ocean Scis., Victoria, 1992—. Contbr. articles to profl. jours. Violinist Hamton Orch., Victoria, 1991-92. Shell scholar, 1987. Mem. Am. Geophys. Union, Can. Geophys. Union, Can. Meteorol. and Oceanographic Soc. (Grad Student prize 1992). Office: Inst of Ocean Scis, 9860 W Saanich Rd, Sidney, BC Canada V8L 4B2

XIE (HSIEH), SHU-SEN, economics and finance educator, consultant; b. Shanghai, China, July 7, 1919; s. Kwei-quan and (Yen) H.; m. Zing-sien Chen, Oct. 10, 1941 (dec. Apr. 1957); children: Dehong, Dehua Xie. BA with honors, St. John's U., Shanghai, People's Republic China, 1941; MA, Columbia U., 1948. Assoc. prof. econs. dept. St. John's U., Shanghai, 1949-52; assoc. prof. Fudan U., Shanghai, 1972-78; assoc. prof. econs. and fin. Shanghai Inst. Fin. and Econs., 1952-72, prof., 1979—; adj. prof. Shanghai Inst. Fgn. Trade, 1983—; vis. prof. De La Salle U., Manila, 1982, 84, 86-87,

Erasmus U., Rotterdam, The Netherlands, Met. State U., St. Paul, 1990; exch. prof. Met. State U., Denver, 1989-90; cons. Shanghai Investment and Trust Corp., 1986—, Bank of Communication, 1988—. Author: Security Markets, 1989, Transnationals and China's Shenzhen, 1989; chief translator: International Economics, 1987; chief editor Survey of International Finance, 1982. Recipient plaque of appreciation De La Salle U., 1986. Mem. China Soc. Fin. (hon. bd. dirs. 1984—), China Soc. Internat. Fin. (bd. dirs. 1984—), Shanghai Soc. Fin. (bd. dirs. 1984—), Shanghai Soc. Internat. Fin. (bd. dirs. 1984—). Office: Shanghai U Fin and Econs, 777 Guoding Rd, Shanghai 200433, China

XIN, LI, physiologist; b. Shenyang, Liaoning, People's Republic of China, June 14, 1955; came to U.S., 1989; s. Zhili and Huiming (Wang) X. MD, China Med. U., Shenyang, Liaoning, People's Republic of China, 1982, MS, 1988. Instr. dept. physiology China Med. U., Shenyang, Liaoning, People's Republic of China, 1983-86, asst. prof. dept. physiology, 1987-89; rsch. asst. prof. dept. physiology U. Tenn., Memphis, 1989-91; rsch. assoc. dept. pharmacology Temple U., Phila., 1992—; coun. standing mem. Youth Scientists Soc., Shenyang, 1988—. Contbr. articles to profl. jours. Recipient Prize for Excellent Paper, Chinese Physiol. Soc., 1985, Visiting fellowship Boehringer Ingelheim Fonds Found. for Basic Rsch. in Medicine, 1989. Mem. AAAS, Am. Physiol. Soc., Soc. for Neurosci., Sigma Xi. Achievements include rsch. on ctrl. control of body temperature regulation, integration by preoptic anterior hypothalamus of brain under both physiol. and pathological conditions as well as actions of pharmacological agts.; mediation of endogenous neuropeptides in analgesic and body temperature responses to opioids. Home: 3900 City Line Ave A-402 Philadelphia PA 19131 Office: Temple U Dept of Pharmacology 3420 N Broad St Philadelphia PA 19140

XU, DAOYI, mathematician, educator; b. Tongnan, Sichuan, Peoples Republic China, Dec. 26, 1947; s. Yuru and Hongfen (Liao) X.; m. Tian Guomin, Jan. 8, 1976; 1 child, Jing. Grad., Nanchong Tchrs. Coll., Sichuan, Peoples Republic China, 1975; postgrad., Huazhong Normal U., Wuhan, Peoples Republic China, 1981-82. Instr. Mianyang Tchrs. Tng. Coll., Sui ming, Peoples Republic China, 1975-78; asst. Mianyang (Peoples Republic China) Tchrs. Coll., 1979-81; lectr. Mianyang Tchrs. Coll., 1982-86, prof., dir. rsch. lab. systems and differential equations, 1987-92; prof. dept. math. Sichuan Normal U., Chengdu, People's Republic of China, 1993—; vis. prof. Germany, Sweden, USSR, 1985, Inst. Math. Academia Sinica, 1989, Japan, 1990; session chmn. 12th IMACS World congress on Sci. Computation, Paris, 1988; reviewer Math. Reviews, U.S. Contbr. tech. papers to field (awards 1986, 1991). Recipient Accomplishment prize Sichuan Govt., 1984, Progress prize, 1988; Chinese Govt. spl. allowance grant, 1991—, Top prize for sci. and tech. progress Mian Yang City Govt., 1991; named Outstanding Sichuan Youth, 1979-89. Mem. Chinese Math. Soc. (coun. mem. Sichuan 1992), Assn. Chinese Automation, Am. Math. Soc. Avocations: music, jogging. Home: Mianyang Tchrs Coll, 16-502 Sichuan Normal U, Chengdu 610066, China Office: Sichuan Normal U, Dept Math, Chengdu 610066, China

XU, DONG, biophysicist, researcher; b. Beijing, Mar. 5, 1965; came to U.S., 1990; s. Hong-qing Xu and Dexia Hu; m. Jingxi Chu, July 27, 1991. BS, Peking U., Beijing, 1987, MS, 1990. Rsch. asst. dept. physics Peking U., 1988-90; teaching asst. dept. physics NYU, 1990; teaching asst. dept. physics U. Ill., Urbana, 1991, rsch. asst. Beckman Inst., 1991—. Contbr. articles to prof. jours. Recipient scholarship Peking U., 1983-87. Mem. Am. Physics Soc. Achievements include development (with Prof. Klaus Schulten) of the spin-boson model for the electron transfer in the photosynthetic reaction center. Office: Univ Ill Beckman Inst Rm 3141 405 N Mathews Urbana IL 61801-3349

XU, GUO-QIN, chemist, educator; b. Changzhou, Peoples Republic of China, Nov. 27, 1961; s. Quanhong and Ahua (Shi) X.; m. Meisheng Zhou, July 7, 1984; 1 child, Ashley. BS, Fudan U., 1982; MA, Princeton U., 1984, PhD, 1987. Rsch. assoc. Brookhaven Nat. Lab., Long Island, 1987-89, U. Toronto, Ontario, Canada, 1989-91; lectr. Nat. U. Singapore, 1991—. Contbr. articles to Jour. Chem. Physics, Jour. Phys. Chemistry. Mem. Singapore Nat. Inst. Chemistry. Achievements include research in gas-surface scattering, photoelectron spectroscopic studies of metal oxidation and surface photochemistry. Office: Nat U Singapore Dept Chemis, 10 Kent Ridge Crescent, Singapore 0511, Singapore

XU, LE, biophysicist, physicist; b. Guilin, Guang Xi, People's Republic of China, Dec. 30, 1941; came to U.S., 1986; s. Xuehan Xu and GuangXi Zhu; m. Qi-Yi Liu, Dec. 16, 1967; 1 child, Song Liu. Student, Nankai U., Tianjin, People's Republic of China, 1958-63, PhD equivalent, 1966. Tchr., lectr., assoc. prof. Nankai U., Tianjin, 1963-86; vis. scholar Mich. State U., East Lansing, 1980-82; vis. asst. prof. U. N.C., Chapel Hill, 1986-88, rsch. scholar, 1988-90, rsch. assoc., 1990—. Contbr. articles to IEEE, Philos. Mag. B, Jour. Electronics, Archives of Biochemistry and Biophysics, Jour. Gen. Physiol. Mem. Am. Phys. Soc., Am. Biophys. Soc. Office: Dept Biochemistry and Biophysics U NC CB # 7260 Chapel Hill NC 27599-7260

XU, SENGEN, biophysicist; b. Zhou Xiang, Zhejiang, China, Jan. 12, 1941; came to U.S. 1988; s. Xiaoming and Caizhu (Gan) X.; m. Zhirong Zhan, May 26, 1968; 1 child, Jiewei Xu. BS, Fudan U., 1964; PhD, Shanghai Inst. Physiology, 1979. Rsch. asst. Shanghai Inst. Physiology, Chinese Acad. Scis., 1964-78, rsch. assoc., 1979-86, rsch. assoc. prof., 1986-89; vis. scholar MRC Lab. of Molecular Biology, Cambridge, Eng., 1982-84; vis. scientist Dept. Physiology, Teikyo U., Tokyo, 1986-87; vis. assoc. Lab. Phys. Biology, NIAMS/NIH, Bethesda, Md., 1988—; acad. adv. Shanghai Inst. Physiology, 1987-89. Contbr. articles to profl. jours. Recipient Award of Rsch. Achievements, Chinese Acad. Scis., 1982, Award for Vis. Scholars, M.F. Perutz Found., Cambridge, Eng., 1984, Award for Fgn. Scientists, Japan Soc. for Promotion of Scis., 1986. Mem. Am. Biophys. Soc. Achievements include discovery that the x-ray diffraction diagram obtained from skinned frog sartorius muscles in the relaxed state at low ionic strength provides evidences for the existence of bound states of myosin crossbridges in which their orientation relative to actin is not sharply defined; discovery that the measurement of radial forces in muscle fibers in presence of various nucleotides provides strong evidences that the radial elasticity of myosin crossbridges is a unique funciton of the state of crossbridges. Home: 4757 Chevy Chase Dr Apt 219 Bethesda MD 20815-6450 Office: Lab of Phys Biology NIAMS NIH Bldg 6 Rm 114 Bethesda MD 20892

XU, ZHAORAN, botanist; b. Guangdong, People's Republic of China, Sept. 3, 1955; came to U.S., 1988; s. LangZhang and ZiLan (Lei) XU; m. Da-an Yu, Dec. 21, 1984; children: Ming, Vangie. BS in Biology, Zhongshan U., Guangzhou, People's Republic of China, 1981, MS in Biology, 1984, PhD in Botany, 1987; student in environ. scis. program, George Mason U., 1990—. Teaching asst. dept. biology Zhongshan U., Guangzhou, 1984-86; lectr. dept. biology, 1987; internat. teaching asst. George Washington U., Washington, summer 1988; postdoctoral rsch. assoc. dept. botany Nat. Mus. Natural History Smithsonian Instn., Washington, 1988-91; mem. teaching team dept. biology George Mason U., 1991; assoc. rsch. biologist dept. biological scis. U. Ala., 1991—; vis. scholar/lectr. instns. including Vienna (Austria) U., Agr. U., Wageningen, The Netherlands, 1987-88, and Royal Botanic Garden, Edinburgh, U.K., 1989. Contbr. articles to profl. jours.; lectr. internat. seminars, symposiums and confs. Grantee Am. Gesneriad & Gloxinia Soc., 1989, WWF Internat., 1990-91, 92-93, Friends of Dyke Marsh, 1991, Washington Biologists' Field Club, 1991, George Washington Meml. Pkwy., Nat. Park Svc., 1991-93. Mem. Am. Gesneriad & Gloxinia Soc., Wash. Botan. Soc., AAAS. Office: U Ala Dept Biological Scis Tuscaloosa AL 35487

XU, ZHIFU, chemist, researcher; b. Qichun, Hubei, China, Nov. 17, 1962; came to U.S., 1988; s. Sixiong Zu and Hanjiao Zhang; m. Suzhen Ruan, Aug. 12, 1987; 1 child, Angela Ruan. BS, Wuhan U., China, 1982; MS, Wuxi Inst. Light Industry, China, 1985; PhD, U. Mich., 1992. Rsch. fellow Zhengzhou Inst. Tech., China, 1985-88; rsch. asst. U. Mich., Ann Arbor, 1988-92, postdoctoral fellow, 1993; rsch. chemist PPG Industries, Inc., Allison Park, PA, 1993—. Co-author: Adv. Dendritic Macromolecules, 1993. Mem. Am. Chem. Soc. Achievements include synthesis and unambiguous characterizations of the world's largest pure hydrocarbon molecule: C1134H1146; efficient synthesis of large-size dendritic molecules up to 12.5 nanometers in diameter. Office: PPG Industries Rsch Ctr 4325 Rosanna Dr Allison Park PA 15101

XUE, MIAO, dentistry educator, biomaterial scientist; b. Shanghai, China, June 24, 1929; s. Kai-Chang and Gui-zhen (Li) X.; m. Zong-Lan Shen, June 15, 1949; children: Jing-Fang, Jie-Fang. DDS, Shangai Med. U. #2, Peoples Republic China, 1953. Asst. dept. prosthetic dentistry sci. Shanghai Med. U. # 2, 1953-60; lectr., asst. prof. dental material rsch. lab. Shanghai Med. U. # 2, 1960-82, vice head, assoc. prof., 1982-87; vice head, assoc. prof. biomaterial rsch. lab. Shanghai Med. U. # 2, 1982-87; head, prof. dept. dental material sci. Shanghai Med. U. # 2, 1987-91, head, prof. dental material and biomaterial rsch. labs., 1988-91; sr. researcher Nat. Testing Ctr. for Med. Polymer, Shangdong, 1988—; dir. Shanghai Biomaterial Rsch. & Testing Ctr., 1989—. Author: (with others) Science of Applied Dental Material, 1963, Material Science of Prosthetic Dentistry, 1987; editor: China Yearbook of Dentistry, 1984—; chief editor Jour. of Dental Materials and Devices, 1991—. Recipient Cert. of Merit, medal, China Ministry of Pub. Health, 1960; Merit cert., Shanghai Sci. & Tech. Commn., 1984, State Edn. Commn. of China, 1988-92, China Ministry of Pub. Health, 1988, Govt. Spl. Subsidy, 1992—. Mem. ASTM, China Dental Material Soc. (bd. dirs.), Shanghai Biomaterial and Product Adminstrn. (bd. dirs.), Shanghai Biomaterial Com. (chief), Shanghai Assn. for Stomatology (com. mem.), Shanghai Med. Engring. for Stomatology (standing com. mem.), China Shape Memory Alloy Com. (com. mem.). Office: Shanghai Biomaterial Rsch Ctr, 716 Xie Tu Rd, Shanghai 200023, China

XUE-YING, DENG, aerodynamicist; b. Jiang Su, Peoples Republic of China, Jan. 4, 1941; s. Yuan-long and Mei-juan (Zhang) Deng; m. Qun Pu, Sept. 28, 1968; children: Wei, Cong. Diploma in aerodynamics, Beijing Inst. Aero. and Astro., 1963. Rsch. assoc. Beijing Inst. Aero. and Astro., 1963-77, lectr., 1978-79, lectr. fluid mechanics inst., 1982-85, prof., dep. dir., 1986-93, v.p. Acad. Affairs, 1993—; vis. fellow dept. mech. and aerospace engring. Princeton (N.J.) U., 1979-82. Author: Strake-Wings and Vortices Separated Flows, 1988, Dynamics of Separated Flows in Engineering, 1991; contbr. articles to AIAA Jour., Jour. Aircraft, Jour. Aeronautics, Jour. Aerodynamics. Recipient Nat. Natural Sci. prize Nat. Sci. and Tech. Com. China, 1987, Award of Sci. and Tech. Progress Nat. Edn. Com. China, 1988, Sci. and Tech. Progress prize Ministry Aeronautics and Astronautics, 1987, 89, 91, Award of Sci. and Tech. Kwang-Hua Sci. and Tech. Fund, 1991. Mem. Chinese Aerodynamics Soc., Chinese Aeronautics Soc. (exec. com. 1988—), Chinese Philately Soc., AIAA. Achievements include research in flows pattern, flow structure, behavior of complicated fluid flows including vortex, separated flows and shock wave/boundary layer interaction. Home: Rm 701 Bldg 916, Zhong Guan Cun Haidian, 100086 Beijing China Office: Beijing U Aero and Astro Fluid Mech Inst, 37 Xueyuan Rd Haidian, 100083 Beijing China

XUN, LUYING, microbiologist, educator; b. Jinan, Peoples Republic China, Mar. 31, 1959; came to U.S., 1984; s. Peixuan Xun and Fangyi Li; m. Peiqin Zhu, Jan. 5, 1987; 1 child, Randy. MSc, U. Man., Winnipeg, Can., 1984; PhD, UCLA, 1989. Postdoctoral fellow U. Idaho, Moscow, 1989-92; asst. prof. microbiology Wash. State U.-Tri Cities, Richland, 1992—. Contbr. articles to Jour. Bacteriology. Mem. AAAS, Am. Soc. Microbiology. Achievements include research on microbial degradation of pentachlorophenol; purified and characterized oxidative dehalogenase and reductive dehalogenase in pentachlorophenol degradation; life cycle of Methanosarcina and related enzymes. Office: Wash State Univ 100 Sprout Rd Richland WA 99352

YABLONSKI, EDWARD ANTHONY, space and terrestrial telecommunications company executive; b. Kulpmont, Pa., June 25, 1936; s. Zigmund and Lottie Yablonski; m. Julie Marie Molesevich, Oct. 25, 1958; children: Mark, Karen. AS, San Antonio Coll., 1963; BSEE (fellow), U. Tex., 1965; MSEE, Poly. Inst. 1967; DEE, Bell Telephone Lab., 1968. Tech. staff Bell Telephone Labs., Holmdel, N.J., 1965-72; asst. engring. mgr. AT&T, 1972-78, mktg. mgr., N.J., 1978-79, bus. mgr. network services, N.J., 1979-80, div. engring. mgr., 1982-83, mgr. ops., 1983-86, dir. strategic planning, internat. telecommunications, 1986-87; regional v.p. Network Systems, 1987-88, dir. internat. engring., 1988-91; v.p. engring. and ops. Am. Mobile Satellite Corp., Washington, 1991—; internat. prof. math. Brookdale Coll., 1970-72, Morris Coll., 1980-81; assoc. prof. math. Monmouth Coll., 1968-70; prof. Somerset Coll., 1981—, chmn. adj. faculty, 1980-82. Sec., Marlboro Twp. Mcpl. Utilities Authority, 1970-74; lt. (j.g.) N.J. Naval Brigade, 1968-75; dept. chmn. United Way, 1978-79. Served with USN, 1953-57. Grad. study program fellow Bell Telephone Labs., 1965-68; registered profl. engr., Tex. Mem. IEEE, Am. Mktg. Assn. (exec.), N.Y. Acad. Sci., VFW, AAAS, Sigma Xi, Tau Beta Pi, Eta Kappa Nu. Republican. Roman Catholic. Club: Tewksbury Optimist (v.p., dir. 1976-79). Author: Logic Reliability Through Redundancy, 1967; researcher time assignment speech interpolation, common channel signalling, computational binary-to-decimal technologies. Home: Woodedge Rd Lebanon NJ 08833-4401 Office: Am Mobile Satellite Corp 1150 Connecticut Ave NW Washington DC 20036-4104

YABLONSKY, HARVEY ALLEN, chemistry educator; b. N.Y.C., Nov. 24, 1933; s. Samuel and Lillian (Pronsky) Y.; married Aug. 23, 1964; children: Leedra Ross, Michael Robert. BS, Bklyn. Coll., 1954, MA, 1958; MS, Stevens Inst. Tech., 1957, PhD, 1964. Teaching asst. Stevens Inst. Tech., Hoboken, N.J., 1956-59, rsch. fellow, 1959-63; lectr. chemistry CUNY, 1960-64; asst. prof. U.S. Merchant Marine Acad., Kings Point, N.Y., 1963-64; lectr. phys. chemistry Rutgers U., Newark, 1967-69; head dept. phys. chemistry Bristol Myers Co., Hillside, N.J., 1964-69; prof. chemistry CUNY, 1969—; prin. H. A. Yablonsky Cons., Cranford, N.J., 1969—. Author: Sorption from Solution on Nylon 66, 1958, Thermal Decomposition of Peroxydisulfate, 1964, Chemistry, 1975, Laboratory Experiments in Chemistry and Physics, 1983, Laboratory Experiments in Earth Sciences, 1986; contbr. articles to profl. jours. Crookes Stanley Rsch. fellow, 1959; named Outstanding Educator Am., Fuller Dees, 1975. Fellow Am. Inst. Chemists (accredited prof.), Chem. Soc.; mem. Am. Chem. Soc., Sigma Xi. Achievements include patent for Thioglycerol-Nitrogen Base Molecular Complex Depilatory Compositions. Office: Kingsborough Coll 2001 Oriental Blvd Brooklyn NY 11235

YADALAM, KASHINATH GANGADHARA, psychiatrist; b. Bangalore, India, Dec. 17, 1954; came to U.S., 1980; s. Gangadhara N. and Ramarathna G. (Daglur) Y.; m. Jyothi Kashinath, Feb. 26, 1981; children: Akhila, Adithya. MD, Kasturba Med. Coll., Manipal, India, 1977. Diplomate, Am. Bd. Psychiatry and Neurology. Resident in psychiatry U. Nebr., Omaha, 1980-83; clin. fellow psychopharmacology Med. Coll. Pa., Phila., 1983-84; instr. Med. Coll. Pa., 1984-85, asst. prof., 1985-89, dir. neuropsychiatry clinic, 1987-91, assoc. prof., 1989-91; med. dir. Diagnostic and Consultation Ctr. Med. Coll. of Pa.; assoc. dir. The Neuropsychiat. Clinic of La., 1991—; med. dir. Schizophrenia Diagnostic and Consultation Ctr., 1990. Author: (with others) Drug Induced Dysfunction in Psychiatry, 1992; contbr. articles to med. jours. Grantee, NIMH, 1987; recipient Young Investigator award, Internat. Congress Schizophrenia Rsch., Balt., 1987, Young Scientist award Winter Workshop on Schizophrenia, Badgastein, Austria, 1990. Mem. Am. Psychiat. Assn., Am. Coll. Clin. Pharmacology, Nat Alliance of the Mentally Ill. Hindu. Avocations: table tennis, chess, squash, fitness. Office: 2829 4th Ave Ste 150 Lake Charles LA 70601-7887

YAGI, FUMIO, mathematician, systems engineer; b. Seattle, July 14, 1917; s. Saihichiro and Kima (Okabe) Y.; m. Shizuko Nakagawa, June 24, 1954. BS, U. Wash., 1938, MS, 1941; PhD, MIT, 1943. Asst. prof. U. Washington, Seattle, 1946-53; app. math. Ballistic Rsch. Lab, Aberdeen Proving Ground, Md., 1953-56; rsch. specialist Jet Prop. Lab, Pasadena, Calif., 1956-63; systems engr., group head Grumman, Bethpage, N.Y., 1963-77; retired, 1977. With U.S. Army, 1944-46, Japan. MIT scholar, 1941-42; Inst. for Advanced Study fellow, 1943, U. Washington fellow, 1940-41. Home: 2914 Sahalee Dr E Redmond WA 98053-6353

YAGI, TAKASHI, physicist, researcher; b. Aomori, Japan, Apr. 20, 1951; s. Hikozo and Fumi Yagi; m. Hiroko Takahashi, Oct. 12, 1983; children: Tanana, Yukon. BS, U. Niigata, Japan, 1974, MS, 1977; PhD, LaTrobe U., Bundoora, Australia, 1983. Physics tchr. Kenyo Sr. High Sch., Kawaguchi, Japan, 1977-78; post-doctoral fellow SUNY, Albany, 1983-84, Geophys. Inst. U., Fairbanks, Alaska, 1984-85; knowledge engr. NTT Software Corp., Yokosuka, Japan, 1986-87; rsch. scientist Inst. Rsch. & Innovation, Kashiwa, Japan, 1987—; involved in nat. project on Excimer Laser, Japan, 1987—. Inventor in field. Mem. Am. Phys. Soc., Phys. Soc. Japan, Japan Soc. Applied Physics, Optical soc. Am. Avocation: fishing. Office: Inst Rsch Innovation Laser, 1201 Takada, Kashiwa, Chiba 277, Japan

YAGNIK, SURESH K., nuclear engineer; b. Banswara, India, Sept. 11, 1954; came to U.S., 1978; s. Harbans Lal and Deveshwari Yagnik; m. Suchitra Yagnik, Dec. 8, 1980; children: Garima, Gunjan. M of Engring., McMaster U., Hamilton, Ont., Can., 1978; MS, U. Calif., Berkeley, 1980, PhD, 1982. Staff scientist Lawrence Berkeley Lab., 1983-87; rsch. engr. Lasercraft, Santa Rosa, Calif., 1987-89; project mgr. Elec. Power Rsch. Inst., Palo Alto, Calif., 1989—; program mgr. Nuclear Fuel Industries Rsch. Group, 1989—. Contbr. articles to profl. jours. Mem. Am. Nuclear Soc., Am. Soc. Testing of Metals. Office: Elec Power Rsch Inst 3412 Hillview Ave Palo Alto CA 94303

YAKOVLEV, VLADISLAV VICTOROVICH, research chemist, consultant; b. Moscow, Mar. 19, 1965; came to U.S., 1991; s. Victor Vasilevich and Valentina Gavrilovna (Tiptzova) Y.; m. Vlada Vladimirovna Simonova, Sept. 26, 1987; 1 child, Victoria. MS in Physics, Moscow State U., 1987, PhD in Physics, 1990. Jr. rschr. Internat. Laser Ctr., Moscow, 1990-91; laser engr., physicist Novatec Laser Systems, San Diego, 1992; rsch. assoc. dept. chemistry U. Calif., San Diego, 1992—; cons. Progress, Moscow, 1988-91, dept. chemistry U. Ariz., Tucson, 1991-92, MXR, Inc., Ann Arbor, Mich., 1992. Contbr. articles to profl. jours. Lenin's fellow Russian Govt., 1985-87. Office: U Calif Dept Chemistry 9500 Gilman Dr La Jolla CA 92093-0339

YAKU, TAKEO TAIRA CHICHIBU KAWAGOE, computer scientist, educator; b. Tokyo, Oct. 21, 1947; s. Masao and Teru (Nagashima) Y. BSc, Jiyu Gakuen Coll., Tokyo, 1970; MSc, Waseda U., Tokyo, 1972, DSc, 1977. Vis. lectr. Tokai U., Hiratsuka, Japan, 1975-76, 85-90, asst. prof., 1976-79, assoc. prof., 1979-85; vis. lectr. Waseda U., 1979-92; assoc. prof. Tokyo Denki U., Hatoyama, Japan, 1985-92; vis. lectr. Toyo U., Kawagoe, Japan, 1991—, Tokyo Denki U., 1992—; prof. Nihon U. Setagaya-Tokyo, 1992—. Author: (with others) Micro Computer Handbook, 1985, (with others) Structured Editors, 1987; editorial bd. mem: Transac. Inst. Electron Information Communications Engrs., 1993—; contbr. articles to profl. jours. Mem. Am. Math. Soc., Assn. Computing Machinery, Japan Soc. Computer Assisted Instrn. (mem. com. 1987—), Internal Resort Svc. Club (Tokyo). Home: 3-10-17 Yagisawa, Hoya-Shi 202, Japan Office: Nihon U, 3-25-40 Sakura-Josui, Setagaya Tokyo 156, Japan

YAKURA, HIDETAKA, immunologist; b. Sapporo, Hokkaido, Japan, Dec. 8, 1947; s. Yasutaroh and Sumiko (Irie) Y. MD, Hokkaido U., 1972, PhD, 1978. Rsch. fellow Farber Cancer Ctr./Harvard Sch. Medicine, Boston, 1976-78; rsch. assoc. Meml. Sloan-Kettering Cancer Ctr., N.Y.C., 1978-83; asst. prof. dept. pathology Asahikawa (Japan) Med. Coll., 1983-85, assoc. prof. dept. pathology, 1985-89; dir., dept. microbiology/immunology Tokyo Met. Inst. for Neurosci., 1989—; vis. investigator Meml. Sloan-Kettering Cancer Ctr., 1988; assoc. scientist Nat. Inst. Neurosci., Nat. Ctr. of Neurology and Psychiatry, Tokyo, 1989—. Mem. Am. Assn. Immunologists. Office: Tokyo Met Inst Neuroscience, 2-6 Musashidai, Fuchu, 183 Tokyo Japan

YALE, JEFFREY FRANKLIN, podiatrist; b. Derby, Conn., Jan. 18, 1943; s. Irving and Bernice (Blume) Y.; m. Lenore Bernsley, Apr. 23, 1987; children: Brian Joseph, Andrew Malcolm. U. Fla., 1960-62; D of Podiatric Medicine, Ill. Coll. Podiatric Medicine, 1966. Diplomate Am. Bd. Podiatric Surgery, Am. Bd. Podiatric Orthopedics, Am. Bd. Med. Quality Assurance and Utilization Rev. Surg. resident Highland Gen. Hosp., Oakland, Calif., 1966-67; capt. U.S. Army Med. Svc., Fort Ord, Calif., 1967-71; instr. masters level Quinnipiac Coll., Hamden, Conn., 1981; cons. surgeon VA Med. Ctr., West Haven, Conn., 1982—; chmn. podiatric surgery Griffin Hosp., Derby, Conn., 1974—; assoc. clin. prof. U. Osteo. Health Scis., Des Moines, 1982—; chmn. Podiatric Medicine Test Com. Nat. Bd. Podiatric Med. Examiners, 1977—; pres. Ct. Examining Bd. in Podiatry, 1979, Am. Acad. Podiatric Sports Medicine, 1986. Author: Firm Footings For the Athlete, 1984, The Arthritic Foot, 1984, Yale's Podiatric Medicine, 3d edit., 1987; contbr. numerous sci. articles to profl. jours. Pres. Yale Podiatry Group, P.C., Ansonia, Conn., 1976—; dir. Podiatry Ins. Co. of Am., Brentwood, Tenn., 1987—; corporator The Savs. Bank of Ansonia, 1982, Griffin Hosp., Derby, Conn., 1982. Capt. U.S. Army, 1967-71. Fellow Am. Acad. Podiatric Sports Medicine, Am. Assn. Hosp. Podiatrists, Am. Coll. Foot Surgeons; mem. New Haven County Podiatric Med. Assn., Conn. Podiatric Med. Assn., Am. Podiatric Med. Assn., Conn. Pub. Health Assn., Am. Pub. HealthAssn., Conn. Examining Bd. in Podiatry. Jewish. Avocations: marble collecting, swimming, gardening. Home: 18 Inwood Rd Woodbridge CT 06525-2558 Office: Yale Podiatry Group PC 364 E Main St Ansonia CT 06401-1995

YALLAMPALLI, CHANDRASEKHAR, medical educator, researcher; b. Diguvamaghan, India, June 12, 1952; came to U.S., 1988; s. Siddaiah Naidu and Munemma (Kuntamukkala) Y.; m. Uma Surapaneni, Apr. 28, 1980; children: Sasidhar, Ragini. B of Vet. Sci., Vet. Sch., Tirupati, India, 1976; PhD, U. Adelaide, Australia, 1986. Rsch. assoc. U. Western Ontario, London, Can., 1986-88; asst. prof. U. Tex. Med. Br., Galveston, 1988—; dir. in vitro fertilization lab. U. Tex. Med. Br., Galveston, 1988—. Contbr. over 50 articles to profl. jours. Recipient Jr. Rsch. scholarship Indian Agrl. Rsch. Coun., New Delhi, 1976-77; scholar U. Adelaide, 1981-85, Med. Rsch. Coun. of Canada fellow U. Western Ontario, 1986-88. Mem. AAAS, Am. Fertility Soc., Soc. Study of Reproduction, Endocrine Soc. Home: 1222 Berkeley Lake Houston TX 77062 Office: Univ Tex Med Br 301 University Dr Galveston TX 77555

YALOW, ROSALYN SUSSMAN, medical physicist; b. N.Y.C., N.Y., July 19, 1921; d. Simon and Clara (Zipper) Sussman; m. A. Aaron Yalow, June 6, 1943; children: Benjamin, Elanna. A.B., Hunter Coll., 1941; M.S., U. Ill., Urbana, 1942, Ph.D., 1945; D.Sc. (hon.), U. Ill., Chgo., 1974, Phila. Coll. Pharmacy and Sci., 1976, N.Y. Med. Coll., 1976, Med. Coll. Wis., Milw., 1977, Yeshiva U., 1977, Southampton (N.Y.) Coll., 1978, Bucknell U., 1978, Princeton U., 1978, Jersey City State Coll., 1979, Med. Coll. Pa., 1979, Manhattan Coll., 1979, U. Vt., 1980, U. Hartford, 1980, Rutgers U., 1980, Rensselaer Poly. Inst., 1980, Colgate U., 1981, U. So. Calif., 1981, Clarkson Coll., 1982, U. Miami, 1983, Washington U., St. Louis, 1983, Adelphi U., 1983, U. Alta. (Can.), 1983, SUNY, 1984, Tel Aviv U., 1985, Claremont (Calif.) U., 1986, Mills Coll., Oakland, Calif., 1986, Cedar Crest Coll., Allentown, Pa., 1988, Drew U., Madison, N.J., 1988, Lehigh U., 1988; L.H.D. (hon.), Hunter Coll., 1978; DSc. (hon.), San Francisco State U., 1989, Technion-Israel Inst. Tech., Haifa, 1989; DSc (hon.), Med. Coll. Ohio Toledo, 1991; L.H.D. (hon.), Sacred Heart U., Conn., 1978, St. Michael's Coll., Winooski Park, Vt., 1979, Johns Hopkins U., 1979, Coll. St. Rose, 1988, Spertus Coll. Judaica, Chgo., 1988; D. honoris causa, U. Rosario, Argentina, 1980, U. Ghent, Belgium, 1984; D. Humanities and Letters (hon.), Columbia U., 1984; D.Phil. honoris causa, Bar-Ilan U., Israel, 1987. Diplomate: Am. Bd. Scis. Lectr., asst. prof. physics Hunter Coll., 1946-50; physicist, asst. chief radioisotope service VA Hosp., Bronx, N.Y., 1950-70, chief nuclear medicine, 1970-80, acting chief radioisotope service, 1968-70; research prof. Mt. Sinai Sch. Medicine, CUNY, 1968-74, Disting. Service prof., 1974-79, Solomon A. Berson Disting. prof.-at-large, 1986—; Disting. prof.-at-large Albert Einstein Coll. Medicine, Yeshiva U., 1979-85, prof. emeritus, 1986—; chmn. dept. clin. scis. Montefiore Med. Ctr., Bronx, 1980-85; cons. Lenox Hill Hosp., N.Y.C., 1956-62, WHO, Bombay, 1978; sec. U.S. Nat. Com. on Med. Physics, 1963-67; mem. nat. com. Radiation Protection, Subcom. 13, 1957; mem. Pres.'s Study Group on Careers for Women, 1966-72; sr. med. investigator VA, 1972-92, sr. med. investigator emeritus, 1992—. Co-editor: Hormone and Metabolic Research, 1973-79; editorial adv. council: Acta Diabetologica Latina, 1975-77, Ency. Universalis, 1978—; editorial bd.: Mt. Sinai Jour. Medicine, 1976-79, Diabetes, 1976, Endocrinology, 1967-72; contbr. numerous articles to profl. jours. Bd. dirs. N.Y. Diabetes Assn., 1974. Recipient VA William S. Middleton Med. Research award, 1960; Eli Lilly award Am. Diabetes Assn., 1961; Van Slyke award N.Y. met. sect. Am. Assn. Clin. Chemists, 1968; award A.C.P., 1971; Dickson prize U. Pitts., 1971; Howard Taylor Ricketts award U. Chgo., 1971; Gairdner Found. Internat. award, 1971; Commemorative medallion Am. Diabetes Assn., 1972;

Bernstein award Med. Soc. State N.Y., 1974; Boehringer-Mannheim Corp. award Am. Assn. Clin. Chemists, 1975; Sci. Achievement award AMA, 1975; Exceptional Service award VA, 1975; A. Cressy Morrison award N.Y. Acad. Scis., 1975; sustaining membership award Assn. Mil. Surgeons, 1975; Distinguished Achievement award Modern Medicine, 1976; Albert Lasker Basic Med. Research award, 1976; La Madonnina Internat. prize Milan, 1977; Golden Plate award Am. Acad. Achievement, 1977; Nobel prize for physiology/medicine, 1977; citation of esteem St. John's U., 1979; G. von Hevesy medal, 1978; Rosalyn S. Yalow Research and Devel. award established Am. Diabetes Assn., 1978; Banting medal, 1978; Torch of Learning award Am. Friends Hebrew U., 1978; Virchow gold medal Virchow-Pirquet Med. Soc., 1978; Gratum Genus Humanum gold medal World Fedn. Nuclear Medicine or Biology, 1978; Jacobi medallion Asso. Alumni Mt. Sinai Sch. Medicine, 1978; Jubilee medal Coll. of New Rochelle, 1978; VA Exceptional Service award, 1978; Fed. Woman's award, 1961; Harvey lectr., 1966; Am. Gastroenterol. Assn. Meml. lectr., 1972; Joslin lectr. New Eng. Diabetes Assn., 1972; Franklin I. Harris Meml. lectr., 1973; 1st Hagedorn Meml. lectr. Acta Endocrinologica Congress, 1973; Sarasota Med. award for achievement and excellence, 1979; gold medal Phi Lambda Kappa, 1980; Achievement in Life award Ency. Brit., 1980; Theobald Smith award, 1982; Pres.'s Cabinet award U. Detroit, 1982; John and Samuel Bard award in medicine and sci. Bard Coll., 1982; Disting. Research award Dallas Assn. Retarded Citizens, 1982, Nat. Medal Sci., 1988; Abram L. Sachar Silver Medallion Brandeis U., Waltham, Mass., 1989, Disting. Scientist of Yr. award ARCS, N.Y.C., 1989, Golden Scroll award The Jewish Advocate, Boston, 1989, spl. award Clin. Ligand Assay Soc., Washington, 1988, numerous others. Fellow N.Y. Acad. Scis. (chmn. biophysics div. 1964-65), Am. Coll. Radiology (asso. in physics), Clin. Soc. N.Y. Diabetes Assn.; mem. Nat. Acad. Scis., Am. Acad. Arts and Scis., Am. Phys. Soc., Radiation Research Soc., Am. Assn. Physicists in Medicine, Biophys. Soc., Soc. Nuclear Medicine, Endocrine Soc. (Koch award 1972, pres. 1978), Am. Physiol. Soc., (hon.) Harvey Soc., (hon.) Med. Assn. Argentina, (hon.) Diabetes Soc. Argentina, (hon.) Am. Coll. Nuclear Physicians, (hon.) The N.Y. Acad. Medicine, (hon.) Am. Gastroent. Assn., (hon.) N.Y. Roentgen Soc., (hon.) Soc. Nuclear Medicine, Phi Beta Kappa, Sigma Xi, Sigma Pi Sigma, Pi Mu Epsilon, Sigma Delta Epsilon, Tau Beta Pi. Office: VA Med Ctr 130 W Kingsbridge Rd Bronx NY 10468-3992

YAMABE, SHIGERU, medical educator; b. Tokyo, July 7, 1923; s. Hiroshi and Jyo (Mihara) Y.; m. Takako Naoi, Apr. 2, 1967; 1 child, Yoko. MS, Osaka (Japan) U., 1946, PhD, 1952. Lectr. Osaka U. Med. Sch., 1953-58; prof. Kobe Coll., Nishinomiya, Japan, 1958-89, hon. prof., 1989—; system dir. Drug Rsch. System Internat., Kobe, 1989—; rsch. exec. mbr. Osaka Seijinbyo Med. Ctr., Higashinariku, Osaka, Japan, 1991—; vis. lectr. Tokyo U., Kyoto (Japan) U., 1966-84; vis. prof. dept. microbiology London Hosp. Med. Coll., 1978—, Case Western Res. U., Cleve., 1982, 84, Harvard Med. Sch., Boston, 1988-89; hon. vis. lectr. London Hosp. Med. Coll., 1990—; invited lectr. U. Paris VI, 1992—; vis. sr. scientist Inst. Pathology, Oxford, 1992—; vis. prof. Grad. Sch. Pub. Health U. Pitts., 1993—; sr. sci. advisor Taiho Pharm. Co., Tokyo, 1991—. Author: Bioenergetics, 1968 (award 1970); internat. jour. editor Antiviral Chemistry and Chemotherapy, 1993—; drug designer, inventor Tazobactam antibiotic, 1991; 40 patents for cancer and AIDS drugs. Fellow Royal Soc. Medicine; mem. Western Pacific Soc. Chemotherapy (organizing com. 1989—). Avocation: poetry. Home: 1-2-7 Kamokogahara, Higashinada, Kobe 658, Japan Office: ESS, Hotel Okura Kobe Chuo-ku, Kobe 650, Japan

YAMADA, KEIICHI, engineering educator, university official; b. Dec. 23, 1931. M in Engring., U. Tokyo, 1956; DSc, U. Göttingen, Fed. Republic of Germany, 1958, U. Freiburg, Fed. Republic of Germany, 1959; D in Engring., U. Tokyo, 1964. Asst. prof. U. Tokyo, 1959-68; assoc. prof. Tokyo Inst. Tech., 1968-75, prof., 1975-77; prof. U. Tsukuba, 1977—; dir. Inst. Socio-Econ. Planning U. Tsukuba, Ibaraki, 1984-86, dir. Rsch. Ctr. for Univ. Studies, 1986-90, provost of colls., 1991-93; vice-dir. Inst. for Policy Scis., Tokyo, 1984—. Author: Life Cycles of Scientific Research, 1986, The World Mountains from the Air, 1987; editor-in-chief Jour. Sci. Policy and Rsch. Mgmt., 1985-88; one-man photog. show Le Montagna del Cielo, Italy, 1989. Mem. Japanese Soc. for Sci. Policy and Rsch. Mgmt. (councilor 1988—), Japan-China Soc. (councilor 1990—). Avocation: alpine aerial photography. Office: U Tsukuba, Tsukuba-shi, Ibaraki 305, Japan

YAMADA, RYOJI, economist; b. Tochigi-Ken, Japan, Sept. 29, 1928; s. Mamoru and Nami Y.; B.A., Aoyama Gakuin U., 1952; M.A., Hitotsubashi U., 1955; Fulbright fellow, Stanford U., 1963-64; postgrad. Brookings Instn., 1964, Internat. Faculty Comparative Econs., Luxemburg, 1964; m. Sadayo Yamada, Aug. 26, 1956; children—Mika, Kazuto. Prof., Aoyama Gakuin U., Tokyo, 1955-69; lectr. Internat. Christian U., Tokyo, 1958-61, 67-70, 80-83, Tokyo Theol. Sem., 1961-63, 67-72, Seikei U., Tokyo, 1971-74; prof. econs. Tokyo Coll. Econs., 1970—, dean, v.p., 1974-76, bd. govs., 1974-76, pres. univ. libr., 1988-91; mem. research com. fin. system in Japan, Fedn. Banker's Assn. in Japan, Tokyo, 1982-84; advanced research works for monetary policy Econ. Planning Bur., 1960-62. Mem. final screening com. grad. applicants Fulbright Com. in Japan, 1965-67, 69; final screening com. grad. applicants East-West Center, U. Hawaii, 1966. Mem. Japan Fin. Assn. (bd. dirs.), Am. Econ. Assn., Can. Econ. Assn. Mem. Nihon Kiristo-Kyodan. Author: An Introduction to Finance, 1957; Financial Structure in Japan, 1967; Lectures on Keynesian Economics, 1970; A Study in International Finance, 1973; An Introduction to Theory of Finance, 1977; Banking in Future, 1983; A Theoretical Study on Financial Structure, 1984, Introduction to Economics, 1988. Home: 7-4 Kiyokawa-cho, Hachioji-shi Tokyo 193, Japan Office: 1-chome 7, Minami-cho, Kokubunji Tokyo 185, Japan

YAMADA, RYUZO, technology company executive; b. Yokohama, Japan, Oct. 20, 1933; s. Yukichi and Shieko (Sakaguchi) Y.; m. Motoko Takeuchi, Oct. 12, 1960; children: Keizo, Kohei, Kuniko. BA, Keio U., Tokyo, 1957; postgrad., Haverford (Pa.) Coll., 1957-59. Mgr. internat. sales Fuji Electric Co. Ltd., Tokyo, 1983, Kyocera Corp., Tokyo, 1983-85; gen. Chiba Sakura Plant Kyocera Corp., 1985—. Mem. Kasumigaseki Country Club, Tokyo Club, Izu Skyline Country Club, Yotsukaido Golf Club. Home: 12-1-2405 Sarugakucho, Shibuya-Ku Tokyo 150, Japan Office: Kyocera Corp, Chiba Sakura Plant, 1-4-3 Ohsaku, Sakura 285, Japan

YAMADA, SHINICHI, mathematician, computer scientist, educator; b. Nagoya, Japan, Jan. 10, 1937; s. Umekichi and Nami (Kawashima) Y.; m. Atsuyo Yamamoto, Oct. 10, 1964; 1 child, Atsushi. BS, U. Tokyo, 1959, DSc, 1988; SM, Harvard U., 1970. System analyst Nippon Univac Co., Tokyo, 1959-70, tech. advisor, 1970-90; lectr. Keio U., Tokyo, 1980-85, Waseda U., Tokyo, 1982—, Chiba U., 1989—, prof. dept. info. sci. Hirosaki U., Japan, 1990-93, prof. dept of information scis., Science U. of Tokyo, 1993—. Author: Sciences of Information Processing, 1984, Micro-Prolog Collection, 1986, Kowalski Logic for Problem Solving, 1987, Encyclopedia of Artificial Intelligence, 1988, Encyclopedia of Computer Systems, 1989. Mem. editorial staff Info. Processing Soc. Japan, 1983-87; contbr. articles to profl. jours. Mem. IEEE, AAAS, Am. Math. Soc., Math. Soc. Japan, Assn. Symbolic Logic, Assn. Computing Machinery, N.Y. Acad. Scis. Clubs: Harvard of Japan. Home: 821-19 Nishi-Fukai, Nagareyama-Shi Chiba 270-01, Japan Office: Sci U Tokyo Dept Info Sci, Yamasaki 2641 Noda-SHI, Chiba 278, Japan

YAMADA, SHOICHIRO, chemistry educator; b. Sakai, Osaka, Japan, Aug. 8, 1922; s. Kyutaro and Kikuno (Yamada) Y. BSc, Osaka (Japan) U., 1944, DSc, 1952. Rsch. assoc. Osaka U., 1946-52, asst. prof., 1952-53, assoc. prof., 1953-64, prof., 1964-86, prof. emeritus, 1986—; vis. prof. U. Florence (Italy), 1964, East China Normal U., Shanghai, People's Republic China, 1984; organizer Internat. Symposium on Coord. Chemistry, Nara, Japan, 1967; hon. rsch. assoc. Univ. Coll., London, 1960-61. Mem. editorial bd. Coordination Chemistry Revs., Amsterdam, 1982-91; assoc. editor Synthesis and Reactivity in Inorganic and Metal-Organic Chemistry, 1967—; mem. adv. bd. Inorganica Chimica Acta, 1967-85; author book on the structure of metal coordination compounds; contbr. articles to profl. jours. Participant internat. confs. on coordination chemistry, Detroit, 1961—. Leverhulme fellow U. Western Australia, Perth, 1972, fgn. fellow Royal Australian Chem. Inst., 1986—. Avocation: stamp collecting. Home: 2-7 Mukonoso 5 chome, Amagasaki 661, Japan

YAMADA, TOSHISHIGE, electrical engineer; b. Miyagi, Japan, Oct. 17, 1958; m. Kunimi; 1 child, Hidenori. BS, Tokyo U., 1981, MS, 1983; PhD, Ariz. State U., 1992. Researcher NEC Microelectronics Rsch. Labs., Kawasaki, Japan, 1983-88; faculty assoc. of elec. engring. Ariz. State U., 1992—. Contbr. articles to profl. jours. Mem. IEEE, Am. Phys. Soc., Phi Kappa Phi, Sigma Xi. Home: # 2126 500 N Metro Blvd Chandler AZ 85226

YAMADORI, TAKASHI, anatomy educator; b. Sendai, Miyagi, Japan, Dec. 4, 1932; s. Toshio Sekiguchi and Kazue Yamadori; m. Fujiko Mohri, Sept. 5, 1965; children: Hideki, Tomoki, Takako. MD, Kobe (Japan) Med. Coll., 1957; PhD, Kobe U., 1965. Intern Yokosuka (Japan) U.S. Naval Hosp., 1957-58; asst. Kobe Med. Coll., 1958-62; Wiss. asst. Freiburg (Fed. Republic of Germany) U., 1962-63; vis. asst. prof. SUNY, Buffalo, 1963-64; asst. prof. Kobe U., 1964-67; assoc. prof. Hirosaki (Japan) U., 1967-70, prof., 1970-80; prof. Kobe U., 1980—; supt. Kobe U. Med. Libr., 1988-92; councilor Kobe U., 1992—; dean Kobe U. Sch. Medicine, 1993—. Author: Learning of Anatomy by Dissection, 1989, (with others) SEM Studies of Brain Ventrical Surfaces, 1978, Human Histology, 1984; author, editor: Manual of Osteology, 1991. Kudo-Fund award Kudo Acad. Found., 1971. Mem. Kobe U. Med. Alumni Assn. (bd. dirs. 1983—), Kobe Cadaver Donating Assn. (bd. dirs. 1980—). Home: Tanaka-cho 4-6-6-705, Higashinda-ku Kobe Japan Office: Kobe U Sch of Medicine, Kusunoki-cho 7-5-1, Chuo-ku Kobe 650, Japan

YAMAGISHI, FREDERICK GEORGE, chemist; b. Reno, Sept. 14, 1943; s. Fred Y. and Grace K. (Watanabe) Y.; m. Joyce S. Ichinotsubo, July 7, 1968; children: Wendy K., Mark K. BS, UCLA, 1965; MS, Calif. State U., L.A., 1967; PhD, UCLA, 1972. Rsch. assoc. U. Pa., Phila., 1972-74; mem. tech. staff Hughes Rsch. Labs., Malibu, Calif., 1974-91, sr. mem. tech. staff, 1991—. Contbr. articles to profl. jours. Mem. Am. Chem. Soc. Achievements include patents in liquid crystals, plasma polymerized thin films. Office: Hughes Rsch Labs 3011 Malibu Canyon Rd Malibu CA 90265

YAMAGUCHI, MASAFUMI, researcher; b. Iwamizawa, Hokkaido, Japan, Feb. 3, 1946; s. Tadashi and Fumi (Aono) Y.; m. Noriko Yamamoto, Nov. 23, 1973; 1 child, Motofumi. BS, Hokkaido Univ., Sapporo, 1968, PhD, 1978. Researcher NTT Electrial Communication Labs., Ibaraki, Japan, 1968-80, supr., 1981-85, section head, 1986—; group leader NTT Opto-electronics Labs., Ibaraki, Japan, 1988—; avd. com. mem. New Energy Devel. Orgn., Tokyo, 1988—; chmn. rsch. com. Photovoltaic Power Generation Assn., Tokyo, 1990—; adv. com. mem. Ministry of Internat. Trade and Industry, Tokyo, 1992—. Author: InP Solar Cells, 1988, ZnSe Blue Emotting Diodes, 1992; contbr. articles to profl. jours. Recipient Paper award Vacuum Soc. Japan, 1986. Mem. IEEE, Materials Rsch. Soc., Japanes Soc. Applied Physics, Inst. Elec. and Communications Engr. of Japan. Achievements include patents in ZnSe Blue Emitting Diodes, InP Solar Cells, GaAs-on-Si Solar Cells photonic functional devices, Tandem Solar Cells; research in superior radiation resistance of InP solar cells. Home: 3-18-17-204 Sugamo, Toshima-Ku 170, Tokyo Japan Office: NTT Opto-electronics Labs, Ibaraki, Tokai-mura 319-11, Japan

YAMAKAWA, HIROMI, polymer chemist, educator; b. Akashi, Hyogo-ken, Japan, Dec. 3, 1931; s. Riichi and Tsugiko (Matsumoto) Y.; m. Emiko Kajiura, Mar. 15, 1964. B.S., Kyoto U., Japan, 1954, M.S., 1956, Ph.D., 1959. Japan Soc. for Promotion of Sci. postdoctoral fellow Kyoto U., Japan, 1959-61; research assoc. James Franck Inst., U. Chgo., 1961-63; instr. dept. polymer chemistry Kyoto U., 1963-64, assoc. prof., 1964-86, prof., 1986—; vis. fellow dept. chemistry Dartmouth Coll., Hanover, N.H., 1971-72. Author: Modern Theory of Polymer Solutions, 1971; mem. editorial adv. bd. Macromolecules, 1987-89; exec. editor Polymer Jour., 1988-89; contbr. articles to profl. jours. Mem. Soc. Polymer Sci. Japan (award in polymer sci. 1969), Phys. Soc. Japan, Chem. Soc. Japan, Am. Phys. Soc., Am. Chem. Soc., N.Y. Acad. Scis. Office: Kyoto U Dept Polymer Chemistry, Kyoto 606-01, Japan

YAMAMOTO, HIRO-AKI, toxicology and pharmacology educator; b. Shimonoseki, Japan, July 9, 1947; s. Hanzō and Haruko (Watanabe) Y.; m. Kyoko Kinoshita, Sept. 24, 1976; children: Takeomi, Mika. Grad. in pharmacy, Fukuoka (Japan) U., 1970; MS, Kyushu U., Fukuoka, 1972, PhD, 1975. Postgrad. researcher in pharmacology U. Calif., San Francisco, 1975-76, vis. scientist, 1979-80; vis. fellow NIH, Bethesda, Md., 1976-77; prof. Fukuyama (Japan) U., 1983-89; asst. prof. toxicology and pharmacology U. Tsukuba, Japan, 1977-83, assoc. prof., 1989—; vis. prof. U. Missouri, 1981-82. Author: Calcium and Biological Systems, 1985, Higienic Chemistry and Public Health, 1990; contbr. articles to sci. jours. Mem. Fukuyama Pollution Com., 1984-89. Mem. N.Y. Acad. Scis. Avocations: tennis, baseball. Home: Matsushiro 1 chome l0l-2, Tsukuba 305, Japan Office: U Tsukuba, Inst Community, Medicine, Tennoudai 1-1-1, Tsukuba 305, Japan

YAMAMOTO, MIKIO, public health and human ecology researcher, educator; b. Shimizu, Japan, May 13, 1913; s. Masaji and Uta (Yamamoto) Y.; m. Masa Azuma, Mar. 13, 1949; children: Masatoshi Yamamoto, Izumi Yamamoto. M.D., Tokyo Imperial U., 1939; D.Med. Sci., Tokyo U., 1951. Prof. Juntendo U., Tokyo, 1956-73; prof. Teikyo U., Tokyo, 1973-84, vis. prof., 1984-89, Boston U., 1991—; pres. Inst. Comprehensive Health Care, Kamakura, Japan, 1984—; v.p. Internat. Union for Health Edn., Paris, 1969-73, 76-79, 91-93; dir. First Asian Regional seminar of Internat. Union for Health Edn., 1973; chmn. Health Edn. com. Japan Med. Assn., 1975-81; pres. First Asian Regional Seminar of Health and Med. Sociology, 1980 (hon. pres. 1986), Soc. Kanto Sociology; bd. dirs. Internat. Fed. Preventive and Social Medicine, Rome, 1984—, founder Japanese Soc. Comprehensive Health Care, Tokyo, 1970—. Author, editor 30 books, 1953-90; regional editor Social Sci. and Medicine, 1966-90; contbr. articles to profl. jours. Mem. populations coun. Japanese Govt., Tokyo, 1965-81; mem. coun. Kanagawa Prefectural Govt., Yokohama, Japan, 1969-90; mem. editorial bd. Shizuoka Shimbun, Shizuoka, Japan, 1984—; mem. nat. com. Diagnosis of Pneumoconiosis, Tokyo, 1952-83; Lt. comdr. Japanese Navy, 1939-46. Recipient Silver medal French Med. Acad., 1973, Blue Ribbon medal Japanese Govt., 1977, Spl. award Ministry of Labour, Tokyo, 1973, Svc. award Kanagawa Prefectural govt., 1983. Fellow Japanese Soc. Pub. Health (hon.), Japan Soc. Industrial Health (hon.), Japanese Med. Soc. Primary Care (hon., conf. chmn. 1983); mem. Internat. Union Health Edn., (v.p. 1969-73, 76-79, 91—), Am. Pub. Health Assn., Japanese Soc. for Hygiene (hon.), Japanese Soc. Public Health (hon.), Japanese Soc. Health and Human Ecology (pres., hon.), Internat. Epidemiological Soc., N.Y. Acad. of Sci. Avocation: painting. Home: Inamuragasaki 1-14-2, Kamakura Kanagawa 248, Japan Office: Inst Comprehensive Healthcare, Inamuragasaki 1-14-2, Kamakura Kanagawa 248, Japan

YAMAMOTO, SHIGERU, educator; b. Kyoto, Japan, July 12, 1929; s. Seiichi and Fusae (Maki) Y.; m. Toshi Kataoka, Mar. 27, 1959. B.A., U. Kyoto, 1955, M.A., 1958. Asst., lit. faculty U. Kyoto, Japan, 1961-62; instr. Kyoto Prefectural U., 1962-64, asst. prof., 1964-81, prof. lit., 1981—, dean, 1986-88. Mem. editorial staff Acta Sumerologica; contbr. articles to profl. jours. Ednl. Ministry of Japan and Kyoto Prefecture grantee, 1974-75. Mem. Soc. for Near Eastern Studies in Japan, Japanese Soc. Western History, Soc. Hist. Research (Kyoto). Home: 15-11 Higashidacho Kamitakano, Sakyoku, Kyoto 606, Japan Office: 1 Hangicho Shimogamo, Sakyoku Kyoto 606, Japan

YAMAMURA, KAZUO, chemist; b. Kyoto, Japan, Dec. 6, 1947. BS, Kyoto U., 1970, MS, 1972, PhD, 1975. Rsch. fellow Chgo. U., 1975-76; asst. prof. Kyushu U., Fukuoka, Japan, 1976-77, Kyoto U., 1977-90; rsch. scientist Dainippon Ink and Chems., Takaishi, Japan, 1990—. Co-author: Facilitated Formation of Tetrahedral Intermediate in Esterase Action, 1971, Water Soluble Cyclophane as Hosts and Catalysts, 1983, Molecular Recognition by Cyclodextrin Hosts, 1990 ; contbr. articles to profl. jours; patentee in field. Grantee Ministry of Edn., 1977, 78, 80, 83, 84, 87, Hattori Houkoukai, 1980, Nippon Kogyo Bank, 1984. Mem. Am. Chem. Soc., Chem. Soc. Japan. Office: Dainippon Ink & Chems, Takasago 1-3, Takaishi Osaka 592, Japan

YAMANA, SHUKICHI, chemistry educator; b. Nichinan-cho, Tottori, Japan, May 28, 1917; s. Katsuji and Tada (Irisawa) Y.; m. Mitsu Fukui, Apr. 13, 1946; children: Hikaru, Manabu, Hajimu. BS, Hiroshima Bunrika U., 1942; DSc in Quantum Chemistry, Kyoto U., 1961. Tchr. Fukushima (Japan) Prefectural Women's Normal Sch., 1942-43; asst. prof. Fukushima Normal Sch., 1943-47, prof., 1947-49; prof. Kyoto (Japan) Normal Sch., 1949; asst. prof. Kyoto Gakugei U., 1949-62, prof., 1962-66; prof. Kyoto U. Edn., 1966-81, prof. emeritus, 1981—; prof. chemistry Kinki U., Higashi Osaka, Japan, 1981-90, lectr., 1990-93; hon. fellow U. Wis., Madison, 1967-68; cons. Colombo Plan, Japan Internat. Coop. Agy., Bangkok, Thailand, 1972. Recipient Rising Sun Third Order of Merit, Japanese Emperor, 1991. Mem. Chem. Soc. Japan (award 1986), Am. Chem. Soc., Japan Sci. Teaching. Avocation: go (Japanese chess), origami. Home: 27-2 Momoyama Tsutsui, Iga Nishi-machi, Fushimi-ku, Kyoto 612, Japan

YAMASHIROYA, HERBERT MITSUGI, microbiologist, educator; b. Honolulu, Hawaii, Sept. 14, 1930; s. Midori and Yasue (Yonemori) Y.; m. Kiyoka Jyone, June 3, 1957; children: Gail, Eliot, Eric, Gary. BA in Bacteriology, U. Hawaii, 1953; MS in Microbiology, U. Ill., Chgo., 1962, PhD in Microbiology, 1965. Supr. clin. labs. Atomic Bomb Casualty Commn. Nat. Acad Scis., Hiroshima, Japan, 1956-58; rsch. and teaching asst. in microbiology U. Ill., Chgo., 1960-64; assoc. to sr. bacteriologist Ill. Inst. Tech. Rsch. Inst., Chgo., 1964-71; assoc. prof., dir. grad. studies dept. pathology U. Ill., Chgo., 1971—; head clin. virology lab. U. Ill. Hosp., Chgo.; lectr. in virology Ill. Coll. Podiatric Medicine, Chgo., 1981-91; lectr., cons. in virology VA West Side Hosp., Chgo., 1982—; mem. Clin. Lab. and Blood Bank adv. bd., State of Ill., 1980-92. Co-author: (textbook) Essentials of Medical Virology, 1975; Author: (with others) Med. Microbiology, 1980. Capt. U.S. Army, 1953-55. Recipient Tanner-Shaughnessey Merit award, Ill. Soc. Microbiology, Chgo. 1982; named Sr. Fellow Inst. for Advanced Biotech., Princeton Junction, N.J., 1991. Mem. AAAS, Am. Soc. Investigative Pathology, Am. Assn. Pathologists, Am. Soc. for Microbiology. Office: U Ill Dept Pathology M/C847 Box 6998 Chicago IL 60680

YAMASHITA, FUMIO, pediatrics educator; b. Oomuta, Fukuoka, Japan, Jan. 27, 1927; s. Kazuo and Hideko (Higashihara) Y.; m. Chieko Yamashita, May 15, 1951; children: Seishiro, Bunshiro, Yushiro. MD, Kyushu U., Fukuoka City, Japan, 1950, PhD, 1960. Cert. physician and pediatrician by Japanese Ministgry Health and Welfare, Japan Pediatric Soc. Intern Kyushu U. Hosp., 1950-51, resident dept. pediatrics, 1951-55, clin. rsch. fellow, 1955-58; rsch. fellow dept. pediatrics Northwestern U., Chgo., 1959-60; chief pediatrician Iizuka Hosp., 1960-62; asst. prof. dept. pediatrics Kyushu U., 1962-67; assoc. prof. dept. pediatrics Kurume (Japan) U., 1967-72, prof., chmn. dept. pediatrics, 1972-92, prof. emeritus, 1992—; vice-dir. Kurume U. Hosp., 1985-87. Mem. Japan Pediatric Soc. (v.p. 1973-75, meeting pres. 1986), Am. Pediatric Soc. (hon.), Internat. Soc. Pediatric Nephrology, Internat. Soc. Nephrology. Home: 2128-3 Mii-Machi, Kurume 830, Japan Office: Kurume Univ, Main Office, 67-Asahi-Machi, Kurume 830, Japan

YAMASHITA, KAZUO, chemistry educator; b. Osaka, Japan, Aug. 17, 1939; s. Senzo and Kazue (Murayama) Y.; m. Misako Sakai, Oct. 7, 1967; children: Masahiro, Yasuyo, Ikuko. BE, Kobe U., Japan, 1962; MS, Kyoto U., Japan, 1964, DSc, 1967. Asst. prof. Hiroshima (Japan) U., 1967-70, assoc. prof., 1970-86; rsch. specialist U. Minn., Mpls., 1973-74; vis. scientist Brookhaven Nat. Lab., N.Y.C., 1978, collaborator, 1981; prof. Hiroshima (Japan) U., 1986—. Author: Organic Photoconductor, 1988, Functionality Materials, 1990; patentee in field. Recipient Awards, Electrochem. Soc. Japan, 1974; grantee Idemitsu Kosan, 1988-90, Mazda Found., 1988-91, Ministry of Edn., 1990-92. Mem. Chem. Soc. Japan, Electrochem. Soc. Japan, Analytical Chem. Soc. Japan, Polarographic Soc. Japan, Am. Chem. Soc., Soc. Japan Applied Physics. Office: Hiroshima U, 1-7-1 Kagamiyama, Higashi Hiroshima 724, Japan

YAMAUCHI, KAZUO, systems engineer, researcher; b. Takatsuki, Osaka, Japan, Nov. 21, 1961; s. Haruo and Ayako (Hiroi) Y.; m. Kumiko Okada, Sept. 27, 1986; 1 child, Eri. AA, Snow Coll., Ephraim, Utah, 1982; BS, Brigham Young U., 1984, MS, 1986. Systems engr. Tex. Instruments Japan Ltd., Hiji-Machi, Oita, Japan, 1986—. Mem. LDS Ch. Avocations: karate, movies, music, driving. Home: 2224-9 Fujiwara Hiji-Machi, Hayami-Gun 879-15, Japan Office: Tex Instruments Japan Ltd, Hiji Plant, 4260 Kawaski, Oita 879-15, Japan

YAMAZAKI, AKIO, biochemistry educator; b. Ashio, Tochigi, Japan, Apr. 10, 1941; came to U.S., 1977; s. Kazuo and Haruko Yamazaki; m. Matsuyo Ito, Apr. 2, 1967; children: Daisuke, Aya. BS, Kyoto (Japan) U., 1969, PhD, 1977. Postdoctoral assoc. Inst. for Biomed. Rsch. U. Tex., Austin, 1977-78; postdoctoral assoc. dept. pathology Med. Sch. Yale U., New Haven, 1978-82; vis. staff mem. Los Alamos Nat. Lab., 1982-85, staff mem., 1985-91; assoc. prof. depts. opthalmology and pharmacology Wayne State U., Detroit, 1991—. Contbr. articles to Jour. Biol. Chemistry, Procs. Nat. Acad. Scis. Grantee NIH; Jules and Doris Stein Professorship. Mem. AAAS, Assn. for Rsch. in Vision and Ophthalmology, Japanese Biochem. Soc., Biophys. Soc., Am. Soc. for Biochemistry and Molecular Biology, Internat. Soc. for Eye Rsch. Achievements include research in cyclic GMP specific binding sites on light activated phosphodiesterase, in activation mechanism of rod GMP phosphodiesterase, in characterization of GTP binding proteins, in interactions between transducing and cyclic GMP phosphodiesterase, and in polymorphism in purified guamylate cyclase from rod photoreceptors. Office: Wayne State U Kresge Eye Inst Detroit MI 48201

YAN, CHEN, electromagnetic physicist; b. Beijing, Oct. 23, 1940; came to U.S., 1990; s. Bin Yan and Fei Zhu; m. Liming Cao, Nov. 11, 1967; 1 child, Ming. M Nuclear Physics, Peking U., Beijing, 1964. Rsch. scientist China Inst. Atomic Energy, Beijing, 1964-86; postdoctoral fellow Nat. Inst. Nuclear Physics, Cantania, Italy, 1986-88; staff scientist physics dept. Tech. U., Munchen, Germany, 1988-90; staff scientist Continuous Electron Beam Accelerator Facility, Newport News, 1990—. Contbr. articles to profl. jours. Mem. Am. Physics Soc., N.Y. Acad. Sci. Achievements include design of magnetic spectrometer, accelerator beam lines. Home: 122 Saddle Dr Newport News VA 23602 Office: CEBAF/SURA 12000 Jefferson Ave Newport News VA 23606

YAN, PEI-YANG, electrical engineer; b. Tianjin, People's Republic of China, July 18, 1957; came to U.S., 1981; d. Zhi-Da and De-Qiu (Yu) Y.; m. Xiao-Chun Mu, June 2, 1984; 1 child, Wendy Mu. MS in Physics, Wayne State U., 1983; PhD in Elec. Engring., Pa. State U., 1988. Sr. process engr. Intel Corp., Santa Clara, Calif., 1988—. Contbr. articles on optical bistability on nonlinear thin film, laser induced nonlocal molecular reorientation, beam amplification via four wave mixings, transmission stability and defect printability of I-line and DUV pellicles, lens aberration effect in both off-axis illumination stepper system and excimer laser stepper system. Mem. Soc. Photo-Optical Instrumentation Engrs. Achievements include theoretical and experimental demonstration of optical transverse bistability in transmission through nonlinear nematic liquid crystal film, thermal grating mediated four wave mixing in nematic liquid crystal film, nonlocal laser included molecular reorientation in nematic liquid crystal film, pulse shortening in nematic liquid crystal film, effect of chromatic aberration in excimer laser lithography, and pellicle defect printability; research in advanced I-line and excimer laser lithography, excimer laser damage properties of different materials used in excimer laser lithography such as chrome, quartz, pellicles, and stepper optics design. Office: Intel Corp PO Box 58119 Santa Clara CA 95052-8119

YANAGISAWA, SHANE HENRY, civil engineer; b. Stamford, Conn., Mar. 24, 1954; s. Samuel Tsuguo and Fern Geraldine (Renar) Y.; m. Cathleen Lynn Farris, Jan. 26, 1988; 1 child, Skye Walker Anfo. BSCE, Rice U., 1979; MS in Engring., U. Tex., 1986. Registered profl. engr., Tex., Calif., Wash. Resident engr. Binkley & Holmes Cons. Engrs., Houston, 1979-81; project engr. Luis Lemus Cons. Engrs., Houston, 1981-85, DMJM, L.A., 1986-88; sr. estimator Losinger USA, Bellevue, Wash., 1988-92; project engr. S.A. Healy, Dallas, 1992—. Mem. ASCE, Am. Concrete Inst., Am. Underground Assn. Achievements include project engineering for first subway tunnel in Dallas; setting of mining record of over 340 foot/day in a 21'-6" tunnel. Office: SA Healy 7708 Chalkstone Dr Dallas TX 75248

YANDERS, ARMON FREDERICK, biological sciences educator, research administrator; b. Lincoln, Nebr., Apr. 12, 1928; s. Fred W. and Beatrice (Pate) Y.; m. Evelyn Louise Gatz, Aug. 1, 1948; children: Mark Frederick,

Kent Michael. A.B., Nebr. State Coll., Peru, 1948; M.S., U. Nebr., 1950, Ph.D., 1953. Research asso. Oak Ridge Nat. Lab. and Northwestern U., 1953-54; biophysicist U.S. Naval Radiol. Def. Lab., San Francisco, 1955-58; asso. geneticist Argonne (Ill.) Nat. Lab., 1958-59; with dept. zoology Mich. State U., 1959-69; prof., asst. dean Mich. State U. (Coll. Natural Sci.), 1963-69; prof. biol. scis. U. Mo., Columbia, 1969—, dean Coll. Arts and Scis., 1969-82, research prof., dir. Environ. Trace Substances Research Ctr., 1983—; research prof., dir. Environ. Trace Substances Research Ctr. and Sinclair Comparative Medicine Research Farm, Columbia, 1984—; Trustee Argonne Univs. Assn., 1965-77, v.p., 1969-73, pres., 1973, 76-77, chmn. bd., 1973-75; bd. dirs. Coun. Colls. Arts and Scis., 1981-82; mem. adv. com. environ. hazards VA, Washington, 1985—, chmn. sci. coun., 1988—. Contbr. articles to profl. jours. Trustee Peru State Coll., 1992. Served from ensign to lt. USNR, 1954-58. Recipient Disting. Svc. award Peru State Coll., 1989. Fellow AAAS; mem. AAUP (Robert W. Martin acad. freedom award 1971), Am. Inst. Biol. Sci., Environ. Mutagen Soc., Genetics Soc. Am., Radiation Research Soc., Soc. Environ. Toxicology and Chemistry. Home: 2405 Ridgefield Rd Columbia MO 65203-1531 Office: U Mo Environ Trace Substances Rsch Ctr/Sinclair Comparative 5450 S Sinclair Rd Columbia MO 65203

YANG, C. C., biochemistry educator, researcher; b. Taipei, Taiwan, Republic of China, July 15, 1927; s. Ching-Chi and Chen Mei Yang; m. Lin Yeh Hsiang, Dec. 18, 1951; five children. MD, Nat. Taiwan U., Taipei, 1950; D Med. Sci., Tokyo Jikei U., 1956. Assoc. prof. Kaohsiung (Taiwan) Med. Coll., 1956-58, prof. biochemistry, 1958-73, pres., 1967-73; concurrent rsch. fellow Inst. Zoology Academia Sinica, Taipei, 1964-87; nat. rsch. chairprof. Nat. Sci. Coun., Republic of China, 1964-67; dir. Inst. Molecular Biology Nat. Tsing Hua U., Hsinchu, Taiwan, 1973-85, prof. Inst. Life Scis., 1985—; rsch. assoc. U. Wis., 1961-62; mem. com. inquiry of div. life sci. Nat. Sci. Coun., 1971—; mem. com. inquiry of arts and sci. Ministry of Edn., Republic of China, 1972—. Mem. editorial coun. Internat. Jour. Toxicon, 1973-90; contbr. numerous articles on snake venom proteins to profl. jours. Recipient Premir's award for the Outstanding Scientists, 1983, Outstanding Rsch. award Nat. Sci. Coun., 1985, 87, 89, 91, Javits Neuroscience Investigator award NIH, 1987-94; named One of Ten Outstanding Young Men, Republic of China, 1965. Mem. AAAS (life), Internat. Soc. Toxinology (coun. mem. 1988-91), Am. Chem. Soc., Protein Soc., N.Y. Acad. Scis., Academia Sinica, Chinese Biochem. Soc. (pres. Taipei chpt. 1979-81), Chinese Chem. Soc. (life), Japanese Biochem. Soc., Formosan Med. Assn. Home: Nat Tsing Hua U, 61-5F West Compound, Hsinchu 30042, Taiwan Office: Nat Tsing Hua U, Inst Life Scis, Hsinchu 30043, Taiwan

YANG, CHEN NING, physicist, educator; b. Hofei, Anhwei, China, Sept. 22, 1922; naturalized, 1964; s. Ke Chuan and Meng Hwa Lo; m. Chih Li Tu, Aug. 26, 1950; children: Franklin, Gilbert, Eulee. BS, Nat. S.W. Assoc. U., China, 1942; PhD, U. Chgo., 1948; DSc (hon.), Princeton U., 1958, Bklyn. Poly. Inst., 1965, U. Wroclaw, Poland, 1974, Gustavus Adolphus Coll., 1975, U. Md., 1979, U. Durham, Eng., 1979, Fudan U., 1984, Eldg. Technische Hochschule, Switzerland, 1987, Moscow State U., 1992. Instr., U. Chgo., 1948-49; mem. Inst. Advanced Study, Princeton U., 1949-55, prof., 1955-66; Albert Einstein prof. SUNY, Stony Brook, 1966—; dir. Inst. Theoretical Physics SUNY, 1966—. Trustee Rockefeller U., 1970-76, Salk Inst., 1978—, Ben Gurion U., 1980—. Recipient Nobel prize for physics, 1957, Rumford prize, 1980, Nat. medal of sci., 1986. Mem. AAAS (bd. dirs. 1975-79), NAS, Am. Phys. Soc., Royal Soc. London (fgn.), Brazilian Acad. Scis., Venezuelan Acad. Scis., Royal Spanish Soc. Scis., Polish Acad. Scis., Am. Philos. Soc., Sigma Xi. Office: SUNY Dept Physics Stony Brook NY 11794

YANG, CHENG-ZHI, petroleum engineer, researcher, educator; b. Henan, People's Republic of China, Aug. 8, 1938; s. Yang Xian-zun and Hou Y.; m. Li Yan-qin, Jan. 8, 1969; children: Yang Song, Yang Lai. BS, Beijing U. Petroleum, 1961. Asst. prof. Beijing U. Petroleum, 1961-75; asst. prof., vice dir. dept. petroleum engring. Sheng-li Coll. Petroleum, 1976-78; engr., sr. researcher, head rsch. Rsch. Inst. Petroleum Exploration and Devel. Beijing, 1979—; head united lab. Acad. Sinica & China Nat. Petroleum Co., 1990—; vis. sr. rsch. engr. Inst. Francais du Petrole, 1979-80, 85-87, 89-90; hon. prof. Da-qing U., 1988—. Author: Petroleum Reservoir Physics, 1975, (with others) World Fine Chemical Engineering Handbook, 1986; contbr. more than 50 articles to profl. jours. Recipient Science-tech. Advance award, 1991; named Internat. Man of Yr., 1991-92. Mem. Soc. Petroleum Engrs., China Petroleum Soc., Beijing Lodge. Achievements include research in enhanced oil recovery, physics-chemistry in oil reservoir, surfactant solution property, absorption of surfactant, colloid and interface chemistry. Home: No 1101, 29 Lodging House, 20 Xue Yuan Rd, Beijing 10083, China Office: Dept EOR RIPED, PO Box 910, Beijing 100083, China

YANG, CHEN-YU, neuroscientist; b. Beijing, Sept. 22, 1934; came to the U.S., 1979; d. Wu-tze and Mo-hwa (Lo) Y.; m. Shih-fang Fan, Apr. 20, 1962; 1 child, Chao-hui. BS, Fu-Tan U., 1958; PhD, SUNY, Stony Brook, 1988. From rsch. asst. to rsch. assoc. Inst. Physiology Chinese Acad. Scis., Shanghai, 1958-79; vis. rsch. scholar dept. anatomy SUNY, Stony Brook, 1979-80, rsch. asst. prof. dept. neurobiology and behavior, 1988—; vis. rsch. scholar dept. ophthalmology NYU, 1980-82. Contbr. articles to Jour. Comparative Neurology, Jour. Gen. Physiology, Neurosci. Letters, Chinese Jour. Physiology, Visual Neurosci., Vision Rsch., others. Mem. Assn. for Rsch. in Vision and Ophthalmology, Sigma Xi. Achievements include research in content of neurotransmitters and their receptors in different types of neurons in vertebrate retina, functional correlations, and anatomical circuitries; blue light sensitive photoreceptors in amphibian retina; scotopic and photopic components of the human electroretinogram (ERG); others. Office: SUNY Dept Neurobiology Behavior Stony Brook NY 11794

YANG, CHIN-PING, chemist, engineering educator; b. Penghu Hsien, Taiwan, Republic of China, Dec. 25, 1931; s. Yee Yang and Chi Chao; m. Ye-ho Hwang, Oct. 10, 1961; children: Chung-Sheng, Chung-Chen, Chung-Cheng. BS in Chemistry, Taiwan Normal U., Taipei, Republic of China, 1956; M. Engring., Tokyo U., 1973; D. Engring., Tokyo Inst. Tech., 1986. Chmn., prof. chem. engring. dept. Tatung Inst. Tech., Taipei, 1970-89, prof. chem. engring. dept., 1990—; chmn. grad. sch. chem. engring. Tatung Inst. Tech., Taipei, 1980-89; cons. Tatung Co., Taipei, 1963—. Contbr. articles to profl. jours.; patentee in field. Mem. Am. Chem. Soc., Japan Chem. Soc., Chinese Chem. Soc., Soc. Polymer Sci. Japan, Soc. Chinese Inst. Chem. Engrs. Home: 126-1 Chien St Shihlin, Taipei Taiwan Office: Tatung Inst of Tech, 40 Chungshan N Rd 3d Sec, Taipei Taiwan

YANG, DI, aerospace engineer, educator; b. Liaoning, Peoples Republic of China, Mar. 7, 1937; s. Weizhang and Yanan (Miao) Y.; m. Zhengshu He, Feb. 1, 1963; children: Hong, Kai, Xu. BS in Engring., Harbin Inst. Tech., 1961. Asst. prof. Harbin (Peoples Republic of China) Inst. Tech., 1961-78, lectr., 1978-82, assoc. prof., 1982-88, prof., 1988—; cons. Ministry of Chemistry, Beijing, Peoples Republic of China, 1973-78. Author: DDZ-III Electronic Regulator Instruments, 1978; contbr. articles to profl. jours. Recipient 2d prize Com. of Sci. and Tech. 1989, 91. Fellow Control Soc. Aerospace and Moving Objects. Office: Harbin Inst Tech, Po Box 137, 150006 Harbin China

YANG, EUGENE L., mechanical engineer; b. Foochow, Fukien, China, Feb. 13, 1935; came to U.S., 1948; s. H.P. and Jean (Hwang) Y.; m. Cynthia Chang, June 12, 1960; children: Keegan, Kyle. BME, Ohio State U., 1957, MSc, 1958, PhD, 1967. registered profl. engr., Ohio. Technical asst. mech. engr. dept Ohio State U., 1955-57, asst. instr., 1959-63, rsch. assoc., 1963-66; staff engr. 1600 projector RCA, Meadowlands, Pa., 1966-70; prin. design engr. Bell & Howell Co., Lincolnwood, Ill., 1971-72; mech. devel. mgr. Addressograph Multigraph Corp., Cleve., 1973-77; sr. tech. specialist 1065 copier and 4850 printer Xerox, Rochester, N.Y., 1977—; sect. leader Xerox, Rochester, 1982-83, team capt., 1988-93. Mem. Midwestern All-Am. Team Collegiate Soccer Conf., 1954; recipient Indsl. Achievement award Indsl. Press, 1957; Benjamin G. Lamme scholar Ohio State U., 1957; Stillman W. Robinson fellow Ohio State U, 1958. Mem. ASME, Soc. Photographic Scientists & Engrs., Audio Engring. Soc. Achievements include quantification of load-bearing capacity of vibrated-in piles; use of center-of-twist to solve projector jitter problem; 5 patents in mechanical-optical systems; electomechanical inventions in means for dual-styli tracking, low friction wrap-post, dual projection paths, diffraction color encoding, paper corrugation, waste combination and software controls; management of synthesis and engineered products in the fields of audio-visual display, photography, lithography, xerographic copiers and color laser printers. Home: 20 Valley Brook Dr Fairport NY 14450-9344 Office: Xerox Corp 1350 Jefferson Rd Henrietta NY 14623-3106

YANG, HENRY TSU YOW, dean, engineering educator; b. Chungking, China, Nov. 29, 1940; s. Chen Pei and Wei Gen Yang; m. Dilling Tsui, Sept. 2, 1966; children: Maria, Martha. BSCE, Nat. Taiwan U., 1962; MSCE, W.Va. U., 1965; PhD, Cornell U., 1968. Rsch. engr. Gilbert Assocs., Reading, Pa., 1968-69; asst. prof. Sch. Aeros. and Astronautics, Purdue U., West Lafayette, Ind., 1969-72, assoc. prof., 1972-76, prof., 1976—, Neil A. Armstrong Disting. prof., 1988—, sch. head, 1979-84; dean engring. Purdue U., 1984—; vis. scientist Flight Dynamics Lab., USAF, 1976; mem. sci. and mfg. bds. Dept. def., 1988-91; mem. sci. adv. bd., USAF, 1985-89; mem. aero. adv. com. NASA, 1985-89; mem. engring. adv. com. NSF, 1988-91; mem. mechs. bd. visitors ONR, 1990-93. Recipient 10 Best Teaching awards Purdue U., 1971-91. Fellow AIAA; mem. NAE, Academia Sinica. Home: 864 Rose St West Lafayette IN 47906-2768 Office: Purdue U Engring Adminstrn Bldg West Lafayette IN 47907

YANG, HONG-QING, mechanical engineer; b. Chang Chun, Jilin, People's Republic China, July 9, 1961; came to U.S., 1983; s. Chang-Ken and Jian-Ming (Pan) Y.; m. Vicky Quian Zhang, June 9, 1992. BS, Shanghai Jiao Tong U., 1982; MS, U. Notre Dame, 1986, PhD, 1987. Rsch. asst. U. Notre Dame, Ind., 1983-87, postdoctoral fellow, 1987-88; project engr. CFD Rsch. Corp., Huntsville, Ala., 1988-90, sr. project engr., 1990—; cons. Miles Lab., Elkhart, Ind., 1983-87. Co-author: Numerical Application in Welding, 1985; also articles. Mem. AIAA, ASME, Am. Phys. Soc. Office: CFD Rsch Corp 3325D Triana Blvd SW Huntsville AL 35805-4643

YANG, HSIN-MING, immunologist; b. Taipei, Taiwan, Dec. 2, 1952; came to U.S., 1980; s. Sze Piao and Yun-Huan (Chang) Y.; m. Yeasing Yeh, June 28, 1980; children: Elaine, Albert. BS, Nat. Taiwan U., 1976, MS, 1983; PhD, U. Wash., 1985. Rsch. assoc. Tri-Svc. Gen. Hosp., Taipei, 1979-80; fellow Scripps Clinic and Rsch. Found., La Jolla, Calif., 1986-88, sr. rsch. assoc., 1988-90; asst. prof. U. Nebr. Med. Ctr., Omaha, 1990-91; sr. rsch. scientist Pacific Biotech, Inc., San Diego, 1991—; lectr. Yun-Pei Coll. Med. Tech., Shinchiu, Taiwan, 1979-80. Contbr. articles to profl. jours. and chpt. to book. Joseph Drown Found. fellow, 1986, Nat. Cancer Ctr. fellow, 1987-88. Mem. Am. Assn. Cancer Rsch., N.Y. Acad. Scis. Avocations: tennis, swimming, table tennis. Office: Pacific Biotech Inc 9050 Camino Santa Fe San Diego CA 92121-3235

YANG, JANE JAN-JAN, engineering executive; b. Taipei, Taiwan, Republic of China, Jan. 26, 1939; came to U.S., 1962; d. Chen-Jeong and En-En (Wong) HSU; m. Tien-Tsai Yang, July 31, 1965; children: John J. L. and David J. S. BS, Nat. Taiwan U., 1961; MS, N.C. State U., 1964; PhD, UCLA, 1971. Teaching asst. Nat. Taiwan U., Taipei, 1961-62, N.C. State U., Raleigh, 1963-64; from teaching asst. to teaching fellow UCLA, 1964-70; tech. staff Rockwell Internat., Anaheim, Calif., 1971-82; sr. scientist group rsch. staff TRW, Inc., Redondo Beach, Calif., 1982-87, dept. mgr. semiconductor device rsch., 1987—; program com., chair Soc. Optical and Quantum Electronics, 1986-90; chair and com. mem. Optical Soc. Am. (chair and com. mem. 1986—), Laser and Electro-optics (chair and com. mem. 1986—); presenter in field. Contbr. articles to profl. jours.; patentee in field. Vol. PTA, L.A., 1964—. Technical fellow TRW, Inc., 1990, Amelia Earhart fellow Zonta Internat., 1966-69. Mem. IEEE (chair and com. mem. 1986—), Optical Soc. Am., SPIE (chair and com. mem. 1986—), UCLA Alumni Assn., Phi Kappa Phi. Avocations: travel, reading, music. Home: 728 S Bristol Ave Los Angeles CA 90049-4902 Office: TRW D/1 1024 One Space Park Redondo Beach CA 90278

YANG, MILDRED SZE-MING, biologist, educator; b. Hong Kong, Jan. 21, 1950; d. Yen-pei and Peggy Teh-ping (Wang) Y.; m. Kam-chuen Lo, Apr. 7, 1988; 1 child, Glenda Wing-yan. BA, U. Calif., San Diego, 1971; PhD, Washington U., St. Louis, 1979. Scientist Sandoz Pharm. Co., Basel, Switzerland, 1979-80; rsch. assoc. Med. Coll. Va., Richmond, 1980-86; lectr. Hong Kong Bapt. Coll., 1987—; bd. dirs. Hong Kong Examination Authority, 1990—. Contbr. articles to Jour. Neurosurgery, Jour. Neurochemistry; contbr. chpt. to book. Mem. Internat. Soc. for Study Xenobiotics, Hong Kong Biochem. Soc., Hong Kong Pharmacology Soc., Orgn. Chinese Ams. (Richmond, Va. chpt., pres. 1985, v.p. 1983-84). Office: Hong Kong Bapt Coll, Dept Biology, 224 Waterloo Rd, Kowloon Hong Kong

YANG, PING-YI, biowaste, wastewater engineering educator; b. Hu-Wei, Taiwan, Jan. 12, 1938; came to U.S., 1967; s. Wen-Fong and Chai-Foung (Lu) Y.; m. Liang-Hua Huang Yang, May 2, 1967; children: Alexander, Anthony. BS, Nat. Taiwan U., Taipei, 1961; MS, Okla. State U., 1969, PhD, 1972. Researcher, sec. chief Wei-Shin Food Co., Taipei, 1962-67; grad. rsch. asst. Okla. State U., Stillwater, 1967-72; rsch. assoc. Cornell U., Ithaca, N.Y., 1972-73; asst. prof. Asian Inst. Tech., Bangkok, Thailand, 1973-75, assoc. prof., 1975-76; assoc. prof. U. Hawaii, Honolulu, 1976-82, prof., 1982-; scientific collaborator Asian Inst. Tech., Bangkok, 1989-91; cons. Nat. Taiwan U., Taipei, 1991-92; mem. Edit. bd. Bioresource Tech., London, 1987-92. Contbr. articles to profl. jours. USEPA Rsch. grantee, 1980-83, USDA Rsch. grantee, 1984--. Mem. Water Environment Fedn., Am. Soc. Agr. Engrs., Am. Soc. Civil Engring., Internat. Assn. on Water Quality. Achievements include the installation and operation, based on rsch. findings of two full scale pig wastewater treatment systems; use of entrapped mixed microbial cell process for removal of organics and nitrogen from contaminated water. Office: U Hawaii Dept Age Engring 3050 Maile Way Honolulu HI 96822

YANG, RUEY-JEN, aerospace engineer; b. Tainan, Taiwan, Jan. 10, 1954; came to U.S., 1977; s. Yin-Wang and Yu-Li (Chen) Y.; m. Stephanie W.L. Wu, Nov. 30, 1985. BS, Nat. Cheng Kung U., Taiwan, 1976; PhD, U. Calif., Berkeley, 1982. Rsch. scientist Scientific Rsch. Assocs., Inc., Glastonbury, Conn., 1982-84; sr. tech. staff Rocketdyne Div. Rockwell Internat. Corp., Canoga Park, Calif., 1984—; referee AIAA, ASME, 1982—. Contbr. articles to Jour. Spacecraft and Rocket, Internat. Jour. for Numerical Methods in Fluids, Jour. of Engring. for Gas Turbines and Power, Jour. Applied Mechanics. Recipient Space Act Tech Brief award NASA, 1991. Mem. AIAA. Mem. Christian Ch. Achievements include contributions in the nat. propulsion tech. devel. programs including Space Shuttle Main Engine, Nat. Launch System and Nat. Aerospace Plane; major contribution in the analysis of flowpath for various propulsion components. Office: Rockwell Internat Corp Rocketdyne Div 6633 Canoga Ave Canoga Park CA 91303

YANG, RUN SHENG, mathematics educator; b. Taixing, Jiang Su, China, June 1, 1942; s. Guang Zhao and Guo Ying (Taing) Y.; m. Shan Hua Zhou, Jan. 23, 1941; 1 child, Ying. Grad., math. diplomate, Nanjing Tchrs. Coll., Peoples Republic China, 1964. Prof. Nanjing Normal U., Peoples Republic China, 1964—; dir. geometry sect dept. math. Nanjing Normal U., 1982—; reviewer Math. Reviews, Providence, U.S., 1986—. Recipient award Nat. Sci Found. Jiangsu Ednl. Com., 1988, 1991. Mem. Math. Soc. Peoples Republic China, Am. Math. Soc. Office: Nianjing Normal U Dept Math, 122 Ning Hai Rd, Nanjing 210024, China

YANG, SAMUEL CHIA-LIN, engineer; b. Taipei, Taiwan, China, Jan. 7, 1969; came to U.S., 1981; s. John Chung and Hannah (Chiu) Y. BSEE, Cornell U., 1990; MSEE, Stanford U., 1991, postgrad., 1994. Systems engr. Hughes Aircraft Co., El Segundo, Calif., 1988—. Hughes doctoral fellow, 1992—, McMullen Dean's scholar Cornell U., 1987-90. Mem. IEEE, AIAA, Tau Beta Pi. Home: 148 Summit Ave Waldwick NJ 07463

YANG, SEN, chemist; b. Kaifeng, Henan, People's Republic of China, Sept. 19, 1960; came to U.S., 1986; s. Xi Wen Yang and Shirun Xu; m. Hong Liao, May 30, 1986. BS, Zhengzhou U., 1982, MS, 1985; PhD, Brown U., 1991. Chemist Arkwright Inst., Fiskeville, R.I., 1990-92, sr. chemist, 1992—. Mem. Am. Chem. Soc. Home: 92 Cindy Ln Warwick RI 02886 Office: Arkwright Inc 538 Main St Fiskeville RI 02823

YANG, SHANG FA, biochemistry educator; b. Tainan, Taiwan, Nov. 10, 1932; came to U.S., 1959, naturalized, 1971; s. Chian-Zuei and En-Liu (Lu) Y.; m. Eleanor Shou-yuan, Sept. 16, 1964; children: Albert, Bryant. BS, Nat. Taiwan U., 1956, MS, 1958; PhD, Utah State U., 1962. Rsch. assoc. U. Calif., Davis, 1962-63, NYU Med. Sch., N.Y.C., 1963-64, U. Calif.-San Diego, La Jolla, 1964-65; asst. biochemist U. Calif.-Davis, 1966-69, assoc. biochemist, 1969-74, prof., biochemist, 1974—, chmn., 1989-90 ; vis. prof. U. Konstanz, Germany, 1974, Nat. Taiwan U., Taipei, 1983, U. Cambridge, U.K., 1983, Nagoya U., Japan, 1988-89. Assoc. editor Jour. Plant Growth Regulation, 1981—. Mem. editorial bd. Plant Physiology, 1974-92, Plant Cell Physiology, 1987-91, Plant Physiology and Biochemistry, 1988—, Acta Phytophysiologica Sinica, 1988—. Recipient Campbell award Am. Inst. Biol. Sci., 1969; Guggenheim fellow, 1982; recipient Internat. Plant growth Substance Assn. Research award, 1985, Wolf Found. prize agriculture, 1991, Outstanding Rsch award Am. Soc. Horticultural Sci., 1992. Mem. NAS, Academia Sinica, Am. Soc. Biochemistry and Molecular Biology, Am. Soc. Plant Physiologists (chmn. Western sect. 1982-84). Home: 1118 Villanova Dr Davis CA 95616-1753 Office: U Calif Dept of Vegetable Corps Davis CA 95616

YANG, VICTOR CHI-MIN, pharmacy educator; b. Shanghai, Kiangsu, China, July 2, 1949; came to U.S., 1975; s. Pei-Nan and Wen-Kou (Pang) Y.; m. Iris Sun Yang, Aug. 25, 1979; children: Joseph, Emily. MS, East Tex. State U., 1977; PhD, Brown U., Providence, 1983. Rsch. asst. East Tex. State U., Commerce, 1975-77; teaching/rsch. asst. Brown U., Providence, 1978-82; co-lectr. MIT, Cambridge, 1984, postdoctoral fellow, 1983-85; asst. prof. U. Mich., Ann Arbor, 1986-90, assoc. prof. Coll. Pharmacy, 1991—; mem. Small Bus. Innovation Rsch. com. NIH, Bethesda, Md., 1988-89; mem. ad. hoc com. Heart, Lung and Blood Inst., NIH, Bethesda, 1988-89. Editor: Cosmetic and Pharmaceutical Application of Polymers, 1992; mem. editorial bd. BioTechniques, 1988—. Pres. Am. Chinese Assn., Brown U., 1979. 2d lt. Taiwan, 1972-74, Taiwan. Recipient NIH FIRST award, Heart, Lung and Blood Inst., 1986, Most Outstanding Manuscript award Am. Assn. Med. Instrumentation, 1989, Biomed. Engring. Grant Whitaker Found., 1987. Mem. AAAS, Am. Chem. Soc. (Arthur Doolittle award 1990), Biomed. Engring. Soc. Achievements include 8 patents in field; devel. of the first protamine-based filter device for extracorporeal blood heparin removal, first electrochem. sensor for heparin, first colorimetric protamine titration method, concept of a bioactive, self-supported artificial lung. Home: 3629 Tanglewood Dr Ann Arbor MI 48105 Office: U Mich Coll Pharmacy 428 Church St Ann Arbor MI 48109-1065

YANG, XIAOWEI, electrical engineer; b. Shanghai, Peoples Republic of China, July 19, 1954; came to the U.S., 1983; s. Yi-Fang and Yi-Min (Wang) Y.; m. Diming Wan, July 22, 1983; children: Yuan, Jessica. BS in Physics, Wuhan U., China, 1978; MS in Electronics, East China Normal U., 1982; MSEE, U. Md., 1988, PhD in Elec. Engring., 1989. Rsch. engr. Wuhan (Peoples Republic of China) U., 1978-79; lectr. East China Normal U., Shanghai, 1982-83; teaching asst. U. Md., College Park, 1984-86, rsch. asst., 1986-89; rsch. assoc. U. Md., College Park, 1989—; sr. systems engr. Cambridge Rsch. Assocs., Inc., McLean, Va., 1993—; adj. asst. prof. U. Md. Univ. Coll., 1993—; mgr. Multichannel Concepts, Inc., Gaithersburg, Md., 1990-93. Contbr. articles to IEEE Transaction on Info. Theory, Biol. Cybernetics, Biophys. Jour., IEEE Transaction on Biomed. Engring., Electronics Tech. SBIR grantee NIH, 1991; Study Abroad fellow China Edn. Ministry, 1983. Mem. IEEE, IEEE Signal Processing Soc., IEEE Communications Soc., IEEE Neural Network Soc. Achievements include patent pending for infrared remote controlled multichannel telemetry system for biophysiological signal recordings. Home: 13904 Grey Colt Dr Gaithersburg MD 20878 Office: Cambridge Rsch Assocs Inc 1430 Spring Hill Rd Ste 200 Mc Lean VA 22102

YANG, ZEREN, obstetrician/gynecologist, researcher; b. Xi Chong, Sichuan, People's Republic of China, July 2, 1958; came to U.S., 1989; s. Di De and Suhua (Zeng) Y.; m. Li Lin Liao, Mar. 30, 1964. MD, West China U. of Med. Scis., Chengdu, Sichuan, 1983, MS, 1986. Intern West China U. of Med. Scis., Chengdu, 1982-83, resident, 1983-88, asst. prof., 1986-88, lectr., 1988—; postdoctoral fellow U. Calif., San Francisco, 1989—. Assoc. editor-in-chief An English-Chinese Dictionary of Ob/Gyn, 1989; contbr. articles to profl. jours. Cheng's scholar U. Calif., 1989; Nat. Com. of Scis. of Beijing, 1987. Mem. Chinese Assn. of Ob./Gyn., Anti-Cancer Assn. of China. Achievements include the discovery of the thrombin receptor; discovery of the regulations of estrogen and progesterone receptors distributions in uterine cervix and cervical cancer; indication that receptors may be useful in estimation of the prognosis of the patient; discovery that the quality of the previous operation on uterus and implantating site of the placenta are very important for the pregnancy with previous uterine scar. Office: Stanford U Sch Medicine CV 194 Falk Cardiovascular Rsch Ctr Stanford CA 94305

YANG KO, LILLIAN YANG, pediatrician; b. An-Ching, An-Hwei, China, Oct. 6, 1944; d. Chao-Fu Yang and Shiu-yun Mo; m. Sai-Cheong Ko, July 1, 1974; children: Mimi Chi-Yang Ko, Yvonne Chi-Yuen Ko, Denise Chi-Heng Ko. M.B.B.S., U. Hong Kong, 1968. Med. and health officer Med. & Health Dept. of Hong Kong, 1970-75, sr. med./health officer, 1976-78, cons. pediatrician, 1979-88; hon. clin. lectr. pediatrics U. Hong Kong, 1981-88, The Chinese U. of Hong Kong, 1985-88; cons. pediatrician in pvt. practice Hong Kong, 1988—; dir., cons. Hong Kong Child Devel. Clinic, Hong Kong, 1988—; cons. in field; lectr. in field. Contbr. articles to profl. jours. Mem. working group on Devel. of Resdl. Child Care Svcs., Hong Kong, 1985-87, Working Group on Child Abuse, 1983-86, others. Paul Harris fellow, Rotary, 1987. Mem. Hong Kong Soc. Child Health and Devel. (founder mem. and vice chmn. 1984—), Hong Kong Coun. for Early Childhood Edn. and Svcs. (chmn. 1988-89, hon. advisor 1989—), Hong Kong Spastic Assn., Hong Kong Med. Assn., Hong Kong Pediatric Soc., Hong Kong Coll. Physicians (founder mem.), Heep Hong Assn. for Presch. Handicapped Children, Hong Kong Coll. Pediatricians (founder). Avocations: reading, music, hiking. Office: Hong Kong Child Devel Clinic, 100 Nathan Rd Rm224 Tung Yin Bldg, Kowloon Hong Kong

YANNARIELLO-BROWN, JUDITH I., biomedical researcher, educator; b. Newark, June 25, 1958; d. Charles Cosmos and Angelina Rose (Migliorelli) Yannariello; m. David Bruce Brown, May 15, 1982; 1 child, Mark David. BS, Rutgers U., Piscataway, N.J., 1980; MS, U. Tex., 1983; PhD, Yale U., 1987. Postdoctoral fellow U. Tex. Med. Br., Galveston, 1987-92, instr., 1992-93, asst. prof. Human Biol. Chemistry and Genetics, 1993—. Co-author: (book chpt.) Endothelial Cell Biology, 1988; contbr. articles to profl. jours. Vice-pres. Grad. Student Assn., U. Tex., Houston, 1981-82; mem. Faculty Women's Caucus, U. Tex. Med. Br., 1992—. Recipient Nat. Rsch. Svc. award Nat. Inst. on Aging, U. Tex. Med. Br., 1988-92, tuition award Charles and June Ross Found., Yale U., 1984-85, travel award J.W. McLaughlin Found., 1988; grantee NSF, 1992-94. Mem. Assn. Women in Sci., Am. Fed. Aging Rsch. (grantee 1991-92), Am. Soc. Cell Biology, Fedn. Am. Soc. for Exptl. Biology, Sigma Xi. Democrat. Mem. Christian Ch. Achievements include research on endothelial cell-extracellular matrix interactions and effects of aging on matrix metabolism and receptor expression. Office: Dept Human Biolog Chemistry and Genetics Univ Tex Med Branch Galveston TX 77555-0647

YANNAS, IOANNIS VASSILIOS, polymer science and engineering educator; b. Athens, Apr. 14, 1935; s. Vassilios Pavlos and Thalia (Sarafoglou) Y.; m. Stamatia Frondistou (div. Oct. 1984); children: Tania, Alexis. AB, Harvard U., 1957; SM, MIT, 1959; MS, Princeton U., 1965, PhD, 1966. Asst. prof. mech. engring. MIT, Cambridge, 1966-68, duPont asst. prof., 1968-69, assoc. prof., 1969-78, prof. polymer sci. and engring. dept. mech. engring., 1978—, prof., dept. materials sci. and engring., 1983—; prof. Harvard-MIT Div. Health Scis. and Tech., Cambridge, 1978—; vis. prof. Royal Inst. Tech., Stockholm, 1974. Mem. editorial bd. Jour. Biomed. Materials Research, 1986—, Jour. Materials Sci. Materials Medicine, 1990—; contbr. over 100 tech. articles; 13 patents in field. Recipient Founders award Soc. for Biomaterials, 1982, Clemson award Soc. for Biomaterials, 1992, Fred O. Conley award Soc. Plastics Engrs., 1982, award in medicine and genetics Sci. Digest/Cutty Sark, 1982, Doolittle award Am. Chem. Soc., 1988; fellow Pub. Health Svc., Princeton U., 1963, Shriners Burns Inst., Mass. Gen. Hosp., Boston, 1980-81. Fellow Am. Inst. Chemists, Am. Inst. of Med. and Biol. Engrs. (founder); mem. Inst. Medicine of Nat. Acad. Scis. Office: MIT Bldg 3-334 77 Massachusetts Ave Cambridge MA 02139-4307

YANO, TADASHI, physics educator; b. Imabari, Ehime Pref, Japan, May 27, 1939; s. Takekiyo and Setsu (Kurokawa) Y.; m. Akemi Otsuki, Oct. 3, 1969; children: Koiti, Tetsuro. BS, Hiroshima U., 1963, MS, 1965, DSc, 1968. Lectr. Rsch. Inst. Fundamental Physics, Kyoto, Japan, 1968; lectr. Ehime U., Matsuyama, Japan, 1968-71, assoc. prof., 1971-83; prof. physics Ehime U., Matsuyama, 1983—. Alexander von Humboldt fellow U. Mainz, Fed. Republic Germany, 1976-77. Mem. Japan Inst. Metals, Phys. Soc. Japan, Physics Edn. Soc. Japan. Avocation: philately. Office: Ehime U Dept Materials Sci, and Engring, Matsuyama 790, Japan

YANOFSKY, CHARLES, biology educator; b. N.Y.C., Apr. 17, 1925; s. Frank and Jennie (Kopatz) Y.; m. Carol Cohen, June 19, 1949, (dec. Dec. 1990); children: Stephen David, Robert Howard, Martin Fred; m. Edna Crawford, Jan. 4, 1992. BS, CCNY, 1948; MS, Yale U., 1950, PhD, 1951, DSc (hon.), 1981; DSc (hon.), U. Chgo., 1980. Rsch. asst. Yale U., 1951-54; asst. prof. microbiology Western Res. U. Med. Sch., 1954-57; mem. faculty Stanford U., 1958—, prof. biology, 1961—, Herzstein prof. biology, 1966—; career investigator Am. Heart Assn., 1969—. Served with AUS, 1944-46. Recipient Lederle Med. Faculty award, 1957; Eli Lilly award bacteriology, 1959; U.S. Steel Co. award molecular biology, 1964; Howard Taylor Ricketts award U. Chgo., 1966; Albert and Mary Lasker award, 1971; Townsend Harris medal Coll. City N.Y., 1973; Louisa Gross Horwitz prize in biology and biochemistry Columbia U., 1976; V.D. Mattia award Roche Inst., 1982; medal Genetics Soc. Am., 1983; Internat. award Gairdner Found., 1985; named Passano Laureate Passano Found., 1992. Mem. NAS (Selman A. Waksman award in microbiology 1972), Am. Acad. Arts and Scis., Genetics Soc. Am. (pres. 1969, Thomas Hunt Morgan medal 1990), Am. Soc. Biol. Chemists (pres. 1984), Royal Soc. (fgn. mem.), Japanese Biochem. Soc. (hon.). Home: 725 Mayfield Ave Stanford CA 94305-1016 Office: Stanford U Dept of Biological Sciences Stanford CA 94305

YANOVSKI, SUSAN ZELITCH, physician, eating disorders specialist; b. Phila., Dec. 18, 1952; d. David Solomon and Lillian (Goldman) Zelitch; m. Jack Adam Yanovski, Dec. 20, 1987; children: Joshua, Rachel. BSW, Widener U., 1978; MD, U. Pa., 1985. Diplomate Am. Bd. Family Practice. Resident in family medicine Thomas Jefferson U., Phila., 1985-88, fellow dept. family medicine, 1988-89; guest researcher clin. neuroendocrinology br. NIMH, Bethesda, Md., 1989-90, rsch. assoc. clin. neuroendocrinology br., 1990-92; spl. expert NIH, Bethesda, 1992—; mem. Collaborative Study Binge Eating Disorder, 1990—; exec. sec. Nat. Task Force on Prevention and Treatment of Obesity, Bethesda, 1992—; clin. asst. prof. family practice Uniformed Svcs. U. Health Scis., Bethesda, 1989—; mem. working group on obesity Nat. Inst. Diabetes and Digestive and Kidney Diseases; cons. work group on eating disorders Am. Psychiat. Assn. Contbr. sci. papers to Am. Jour. Clin. Nutrition, Pediatrics, Am. Family Physician, others; reviewer Archives Family Medicine, Pediatrics, Am. Family Physician; contbr. chpts. on obesity and eating disorders to med. texts. Mem. AAFP (Mead Johnson award for grad. edn. in family practice 1987), N.Am. Assn. Study of Obesity, Am. Inst. Nutrition, Am. Soc. for Clin. Nutrition, Washington Soc. Study of Eating Disorders, Alpha Omega Alpha. Achievements include research on obesity and eating disorders. Office: NIH Bldg 10 3S231 Bethesda MD 20892

YAO, JAMES TSU-PING, civil engineer; b. Shanghai, China, July 7, 1933; came to U.S. 1953; s. C.C. and Mae Jane (Wang) Y.; m. Anna Lee, June 14, 1958; children: Tina Lee, Timothy H.J., Shana Lynn. BSCE, U. Ill., 1957, MSCE, 1958, PhD, 1961. Postdoctoral preceptor Columbia U., N.Y.C., 1964-65; asst. prof. civil engring. U. N.Mex., Albuquerque, 1961-64, assoc. prof., 1965-69, prof., 1969-71; prof. Purdue U., W. Lafayette, Ind., 1971-88, asst. head dept. civil engring., 1983-88, asst. dean grad. sch., 1984-87; prof. Tex. A&M U., College Station, 1988—, head dept. civil engring., 1988-93. Editor Jour. Structural Engring., 1990-92. Max Planck Rsch. award, Alexander Von Humboldt Found., 1990; recipient Civil Engring. Disting. Alumnus award, U. Ill., Urbana, 1991. Fellow ASCE (State-of-the-Art of Civil Engring. award 1973, 83, Alfred M. Freudenthal medal 1990, Richard R. Torrens award 1992). Avocations: volleyball, paperfolding. Office: Tex A&M Univ Dept Civil Engring College Station TX 77843-3136

YAO, XIANG YU, materials scientist; b. Chang Shu, Jiang Su, China, Aug. 16, 1961; came to U.S., 1990; s. Wei Zhi Y. and Zhong Ying Wu; m. Yuwei Li, Dec. 12, 1986; 1 child, Jamie L. MSc, Osaka U., Japan, 1986, PhD, 1989. Teaching asst. Shanghai Jiao-Tong U., 1982-83; rsch. scientist Advanced Materials Processing Inst., Japan, 1989-90; metall. engr. Bender Machine Inc., Vernon, Calif., 1990-91; staff scientist Lawrence Berkeley Lab., Berkeley, Calif., 1992—. Contbr. articles to IEEE Transaction of Plasma Sci. jour., Jour. Applied Physics, Jour. High Temp. Soc. Recipient Acad. award Japan High Temp. Soc., 1988. Achievements include modification of chemical, mech. and optical properties of materials by ion implantation; optical thin film synthesis; devel. of tandem e-beam system applied in materials field; rsch. in observing liquid metal flowusing x-ray. Office: Lawrence Berkeley Lab Bldg 2-308 1 Cyclotron Rd Berkeley CA 94720

YAP, KIE-HAN, engineering executive; b. Yogyakarta, Java, Indonesia, Sept. 16, 1925; s. Hong-Tjoen and Souw-Lien (Tan) Y.; m. Kiauw-Lan The, Mar. 7, 1954; children—Tjay-Hok, Tjay-Yong. Ir, Tech. U. Delft, Netherlands, 1953. Dir., Research Inst. Mgmt. Sci., U. Delft, 1955-61; dir., founder CBO Mgmt. and Tech. Systems Ctr., Rotterdam, Netherlands, 1961—; lectr. various univs., Western Europe and U.S., 1961—. advisor internat. orgns., 1961—. Contbr. articles to profl. jours. Mem. Royal Inst. Engrs. Netherlands. Home: Hoyledesingel 14, 3054 EK Rotterdam The Netherlands

YARAR, BAKI, metallurgical educator; b. Adana, Turkey, Feb. 28, 1941; came to U.S., 1980; s. Salih and Sidika Yarar; m. Ruth G. Yarar; children: Deniz, Defne. BSc in Chemistry, Mid. East Tech. U., Ankara, Turkey, 1961, MSc in Chemistry, 1966; PhD in Surface Chemistry, U. London, 1969; DIC in Mineral Tech., Imperial Coll. London, 1969. Instr. Mid. East Tech. U., 1970-71, asst. prof., 1971-76, assoc. prof., 1976-79; vis. prof. U. B.C., Vancouver, Can., 1979-80; assoc. prof. Colo. Sch. of Mines, Golden, 1980-86, prof., 1986—; pvt. practice cons. in mineral processing, worldwide, 1980—. Author chpts. to books, over 100 papers; editor books. Lt. Turkish Army, 1970-71. Holder numerous awards and certificates of recognition. Mem. Soc. Mining Engrs. (chmn. fundamental com. 1989), Am. Chem. Soc., Materials Rsch. Soc., Sigma Xi (life). Achievements include pioneering work in selective flocculation; invention of the gamma floation process; patent for superconductivity meter device. Home: 13260 Braun Rd Golden CO 80401 Office: Dept Metall Engring Colo Sch Mines Golden CO 80401

YARBROUGH, KAREN MARGUERITE, genetics educator, university official; b. Memphis, Mar. 4, 1938; d. David Williamson and Cleo Marguerite (Hartsfield) Y. BS, Miss. State U., 1961, MS, 1963; PhD, N.C. State U., 1967. Grad. asst. Miss. State U., Starkville, 1961-63; rsch. asst. N.C. State U., Raleigh, 1963-67; asst. prof., assoc. prof. biol. scis. dept. U. So. Miss., Hattiesburg, 1967, prof. biol. scis. dept., 1976, founder, dir. Inst. Genetics, 1971-82, asst. dean, rsch. coord. Coll. Sci. and Tech., 1976-81, acting v.p. acad. affairs, 1981-82, v.p. rsch. and planning, 1982—; bd. dirs. Miss. Rsch. Consortium, 1987—, Miss. U. Rsch. Authority, 1992—, pres., 1992—; bd. dirs. Miss.-Ala. Sea Grant Consortium, 1990—, sec., 1992—; bd. dirs. Exptl. Program to Stimulate Competitive Rsch.,NSF, 1988—, Trustmark Nat. Bank, Hattiesburg; councillor Oak Ridge Associated Univs., 1993—. Contbr. articles to Am. Jour. Mental Deficiency, Clin. Genetics, Am. Jour. Phys. Anthropology, Jour. AVMA, Am. Jour. Human Genetics. Mem. Miss. Econ. Coun., Hattiesburg, 1990—; bd. dirs. Hattiesburg C. of C., 1990-91. Named to Hall of Fame, Miss. Gulf Coast Community Coll. Alumni Assn., 1987, Miss. Outstanding Woman Pres. Commn. on Status Woman, 1993. Mem. Am. Soc. Human Genetics, Soc. Rsch. Adminstrs., Miss. Genetics Assn., Nat. Coun. Univ. Rsch. Adminstrs., Dermatoglyphics Assn. (bd. dirs. 1986-89), Miss. Acad. Scis. (pres. 1983-84), Hattiesburg Area C. of C. (bd. dirs. 1987-91), Sigma Xi, Phi Theta Kappa, Phi Kappa Phi. Office: U So Miss So Sta Box 5116 Hattiesburg MS 39406-5116

YARCHOAN, ROBERT, clinical immunologist, researcher; b. N.Y.C., July 21, 1950; s. Zachary and Anne Mae (Veneroso) Y.; m. Giovana Tosato, Nov. 7; children: Mark, John. BA, Amherst Coll., 1971; MD, U. Pa., 1975. Diplomate Am. Bd. Internal Medicine, Am. Bd. Allergy and Immunology. Resident in medicine U. Minn. Hosps., Mpls., 1975-78; clin. assoc. metabolism br. Nat. Cancer Inst., Bethesda, Md., 1978-80, investigator metabolism br., 1980-83, investigator clin. oncology program, 1983-87, sr. investigator clin. oncology program, 1988-91, head retrovital diseases sect. medicine br., 1991—. Co-author: (chpt.) Cecil Textbook of Medicine, 1992; assoc. editor Jour. Immunology, 1985-89, AIDS Rsch. and Human Retroviruses, 1986—, AIDS, 1990—; sect. editor Thymus, 1992—. Capt. USPHS, 1978—. Recipient Commendation medal USPHS, 1991, Asst. Sec. Health award U.S. Govt. Dept. Health and Human Svcs., 1989, Investors award U.S. Dept. Commerce, 1986, 87. Mem. Am. Assn. Immunologists, Am. Fedn. Clin. Rsch., Clin. Immunology Soc., Am. Soc. for Clin. Investigation, Internat. AIDS Soc. Achievements include development of therapics for AIDS and related disorders including ddC and ddI; research on the interactions between HIV and the immune system. Office: Nat Inst Health Bldg 10 Rm 12N226 9000 Rockville Pike Bethesda MD 20892

YARDLEY, JOHN FINLEY, aerospace engineer; b. St. Louis, Feb. 1, 1925; s. Finley Abna and Johnnie (Patterson) Y.; m. Phyllis Steele, July 25, 1946; children: Kathryn, Robert, Mary, Elizabeth, Susan. B.S., Iowa State Coll., 1944; M.S., Washington U., St. Louis, 1950. Structral and aero. engr. McDonnell Aircraft Corp., St. Louis, 1946-55, chief strength engr., 1956-57; project engr. Mercury spacecraft design, 1958-60; launch ops. mgr. Mercury and Gemini spacecraft, Cape Canaveral, Fla., 1960-64; Gemini tech. dir. Mercury and Gemini spacecraft, 1964-67; v.p., dep. gen. mgr. Eastern div. McDonnell Douglas Astronautics, 1968-72, v.p., gen. mgr., 1973-74, pres., 1981-88, sr. corp. v.p., 1989, ret.; assoc. adminstr. for manned space flight NASA, Washington, 1974-81. Served to ensign USNR, 1943-46. Recipient Achievement award St. Louis sect. Inst. Aerospace Scis., 1961, John J. Montgomery award, 1963, Pub. Service award NASA, 1963, 66 ; profl. achievement citation Iowa State Coll., 1970, Spirit of St. Louis medal, 1973, Alumni citation Washington U., 1975, Disting. Achievement citation Iowa State U., 1976, Presdl. citation as meritorious exec. Sr. Exec. Service, 1980, NASA Disting. Service medal, 1981, Goddard Meml. trophy, 1983, Achievement award Washington U. Engring. Alumni, 1983, Elmer A. Sperry award, 1986; named Engr. of Yr. NASA, 1982; Von Karman Astronautics Lectureship award, 1988. Fellow AIAA (Goddard award 1982), Am. Astronautical Soc. (Space Flight award 1978); mem. Internat. Acad. Astronautics, NASA Alumni League (bd. dirs.), Nat. Acad. Engring., Nat. Space Club (bd. govs.), Tau Beta Pi, Phi Kappa Phi, Phi Eta Sigma, Phi Mu Epsilon. Presbyterian. Home: 14319 Cross Timbers Ct Chesterfield MO 63017-5718

YARGER, JAMES GREGORY, technology company regulatory officer; b. Waverly, Iowa, Sept. 15, 1951; s. Glen Virgil and Lillian Maxine Yarger; m. Jeannie Rae Van Vickle; children: Benjamin, Jason. BA, U. Iowa, 1974; PhD, Brandeis U., 1981. Postdoctoral fellow Harvard U., Cambridge, Mass., 1981-83; sr. rsch. scientist Miles Inc., Elkhart, Ind., 1983-87; staff scientist Amoco Tech. Co., Naperville, Ill., 1987-92, sr. rsch. scientist and regulatory affairs officer, 1993—. Contbr. chpts. to books, articles to profl. jours. including Jour. Biol. Chemistry, Molecular and Cellular Biology, Devels. in Indsl. Microbiology, Jour. Cell. Sci. Participant Gov.'s Voluntary Action Program, Ind., 1987; pres. Hunters Woods Homeowners Assn., St. Charles, Ill., 1991, Charlemagne Homeowners Group, St. Charles, 1992; dep. registrar Kane County, Ill., 1993—. Am. Cancer Soc. fellow, 1981-83. Mem. Am. Soc. Quality Control, Japan Soc. for Biosci., Biotech. and Agr., Sigma Xi. Achievements include patent applications for Transcription effector Sequences, Expression Vector, Phytoene Biosynthesis in Transgenic Hosts, Lycopene Biosynthesis in Genetically Engineered Hosts, Beta-Carotene Biosynthesis in Genetically Engineered Hosts, Zeaxanthin biosynthesis in genetically engineered hosts; discovery of transcription termination activity associated with Saccaromyces cerevisiae URA3 promoter. Home: 713 Indian Way Saint Charles IL 60174-8641 Office: Amoco Tech Co PO Box 3011 Naperville IL 60566-7011

YARIV, AMNON, electrical engineering educator, scientist; b. Tel Aviv, Israel, Apr. 13, 1930; came to U.S., 1951, naturalized, 1964; s. Shraga and Henya (Davidson) Y.; m. Frances Pokras, Apr. 10, 1972; children: Elizabeth, Dana, Gabriela. B.S., U. Calif., Berkeley, 1954, M.S., 1956, Ph.D., 1958. Mem. tech. staff Bell Telephone Labs., 1959-63; dir. laser research Watkins-Johnson Co., 1963-64; mem. faculty Calif. Inst. Tech., 1964—, Thomas G. Myers prof. elec. engring. and applied physics, 1966—; chmn. bd. ORTEL Inc., Accuwave Corp.; cons. in field. Author: Quantum Electronics, 1967, 75, 85, Introduction to Optical Electronics, 1971, 77, 89, Theory and Applications of Quantum Mechanics, Propagation of Light in Crystals. Served with Israeli Army, 1948-50. Recipient Pender award U. Pa., Harvey prize Technion Israel Inst. Tech., 1992. Fellow IEEE (Quantum Electronics award 1980), Am. Optical Soc. (Ives medal 1986), Am. Acad. Arts and Scis.; mem. NAS, NAE, Am. Phys. Soc. Office: 1201 E California Blvd Pasadena CA 91125-0001

YARRIS, STEVE, child psychologist, educator; b. N.Y.C., June 11, 1951; s. John and Rita (Moran) Y. MA in Music Edn., NYU, 1981, MA in Ednl. Psychology, 1983; MA in Sch./Child Psychology, 1988; PsyD in Childhood Psychology, NYU, 1988. Assoc. prof. applied psychology NYU, N.Y.C., 1988—; child psychologist Bd. of Edn., N.Y.C., 1989-92; consulting psychologist State of N.Y. Vocat. & Ednl. Svcs., 1991—. Contbr. articles to profl. jours., chpts. to books. Bd. dirs. Bronx Ind. Living Ctr., 1985, Ctr. for Independence of the Disabled of N.Y., Manhattan, 1986. Mem. APA, N.Y. Soc. Clin. Psychologists, Nat. Register Health Svc. Providers in Psychology, Mensa. Office: 15 Charles St Apt 6-H New York NY 10014

YASSIN, RIHAB R., biomedical researcher, educator; b. Baghdad, Iraq, May 27, 1951; came to U.S., 1979; d. Rafik and Sabiha (Ahmed) Y. BSc in Pharmacy, U. Baghdad, 1976; PhD in Cell Physiology, U. Conn., Farmington, 1985. Lic. pharmacist. Pharmacist Dept. of Health, Iraq, 1976-77; pharmacist, educator U. Baghdad, 1977-79; rsch. fellow Mich. Cancer Found., Detroit, 1985-88; rsch. assoc. Hahnemann U., Phila., 1988-89, rsch. asst. prof., 1989-92, asst. prof., 1992—; lectr., seminars in field. Contbr. articles to Cell Physiology/Biochemistry. Vol. UNICEF, Phila., 1989-90. NIH grantee, 1992—; Hahnemann U. grantee, 1990-91, 92-93. Mem. AAAS, Am. Soc. for Cell Biology, Women in Cancer Rsch., Sigma Xi. Achievements include research on hormonal regulation of cell growth, signal transduction mechanisms mediating cell proliferation. Office: Hahnemann U Broad and Vine Philadelphia PA 19102

YASUDA, MINEO, anatomy educator; b. Kyoto, Japan, July 8, 1937; s. Minoru and Misao (Mochizuki) Y.; m. Iku Akira, Aug. 25, 1968; children: Kumi, Wakana. MD, Kyoto U., 1962, PhD, 1975. Lic. MD, 1963. Instr. anatomy Faculty Medicine Kyoto U., 1963-71, asst. prof. anatomy, 1971-75; head dept. perinatology Inst. Devel. Rsch. Aichi (Japan) Prefectural Colony, 1975-77; prof. anatomy Sch. Medicine Hiroshima (Japan) U., 1977—. Author: Congenital Malformations, 1974, Modern Trends in Human Genetics, 1975, Clinical Human Embryology, 1983; editor Jour. Congenital Anomalies, 1990—. Mem. Teratology Soc., Internat. Soc. Dvel. Biologists, Japanese Teratology Soc. (pres. 1983, bd. dirs. 1977—), Japanese Assn. Anatomists (councilor 1977—, bd. dirs. 1992—), Japanese Soc. Toxicol. Scis. (councilor 1977—), Japan Soc. Human Genetics (councilor 1987—). Avocations: skiing, flute. Home: 13-20 Yahata 2-chome, Fuchu-cho, Aki-gun 735, Japan Office: Hiroshima U Sch Medicine, 2-3 Kasumi 1-chome Minamiku, Hiroshima 734, Japan

YASUE, KUNIO, physicist, applied mathematician; b. Okayama, Japan, Sept. 27, 1951; s. Teruyoshi and Kotoe (Hiramatsu) Y.; m. Saeko Takayanagi, Apr. 18, 1976; children: Ayako, Kanako. BS Tohoku U., Japan, 1974; MS Kyoto U., 1976; PhD Nagoya U., 1979. Researcher, U. Geneva, 1978-82; chief researcher Toshiba Rsch. and Devel. Ctr., Kawasaki, Japan, 1982-84; asst. prof. Notre Dame Seishin U., Okayama, 1984—; prof., dir. Rsch. Inst. for Informatics and Sci.-Notre Dame Seishin U., Okayama. Author: (with Mari Jibu) One Litter of Cosmology, 1991, (with Mari Jibu) One Litter of Quantum Theory, 1991; contbr. articles to profl. jours.; contbr. articles to Discovery of Stochastic Calculus of Variations, (with Mari Jibu) Discovery of Quantum Brain Dynamics. Co-recipient Best Paper award 11th European Meeting on Cybernetics and Systems Rsch., Vienna, 1992; Ministry Edn. Japan grantee, 1985. Mem. Am. Math. Soc. Buddhist. Club: Daito Ryu Aiki Bujutsu (Master Yukiyoshi Sagawa). Home: 1 12 7 Tsugura cho, Okayama 700, Japan Office: Notre Dame Seishin U, 2 16 9 Ifuku cho, Okayama 700, Japan

YASUMOTO, KYODEN, food science educator; b. Kyoto, Japan, Feb. 27, 1934; s. Jukoh and Funsen (Kinoshita) Y.; m. Junko Okamoto, Nov. 5, 1961; children: Takami, Hidemi, Hiroyuki, Tomoharu. B in Agr., Kyoto U., 1956, M in Agr., 1958, D in Agr., 1961. Asst. prof. dept. agrl. chemistry Kyoto U., 1964-68, assoc. prof., 1968-83, prof. Rsch. Inst. Food Sci., 1983—; expert advisor Resources Coun., Sci. & Tech. Agy., Japan, 1985—; mng. dir. interdisciplinary rsch. Inst. Environ. Sci., Kyoto, 1990—; mem. coun. Internat. U. Food & Sci. Tech., 1991—. Author, editor: Encyclopedia Food Agriculture Nutrition, 1981, Food Science, 1987, Nutritional Chemistry, 1987, Food in Living Science, 1990. Mem. Am. Inst. Nutrition, Japan Soc. Food Sci. Nutrition (bd. dirs. Tokyo chpt. 1984—), Inst. Food Technologists (pres. Japan sect. 1989-90). Home: 34-14 Iwakura Nakazaichi-cho, Sakyo-ku, Kyoto 606, Japan Office: Kyoto U Rsch Inst Food Sci, Gokasho, Uji Kyoto 611, Japan

YASUMOTO, TAKESHI, chemistry educator; b. Naha, Okinawa, Japan, Feb. 22, 1935; s. Jitsuga and Kisako (Oshiro) Y.; m. Tomiko Hayashi, May 2, 1963; children: Ken-ichi, Sanehiro, Akiko. BS, Tokyo U., 1957, MS, 1959, PhD, 1966. Asst. prof. Tokyo U., 1960-69; assoc. prof. chemistry Tohoku U., Sendai, Japan, 1969-77, prof., 1977—; cons. on food hygiene Japanese Ministry Health and Welfare, Tokyo, 1985—; cons. on marine biotech. Japanese Ministry Internat. Trade and Industry, Tokyo, 1988—. Author: Marine Natural Products, 1987; editor: Bioactive Marine Products, 1987. Recipient award Symposium on Toxins, Tokyo, 1986, Internat. Conf. on Toxic Marine Phytoplanktons, 1989. Mem. Japanese Soc. Sci. Fisheries (award 1977), Japan Soc. for Biosci., Biotech., and Agrochemistry (award 1992), Internat. Union Pure and Applied Chemistry (chmn. working group). Home: 3-14-13 Nakayama, Aobaku, Sendai 981, Japan Office: Tohoku U Faculty Agr, 1-1 Tsutsumidori-Amamiyamachi, Sendai 981, Japan

YATABE, JON MIKIO, nuclear engineer; b. Berkeley, Calif., Apr. 23, 1937; s. Takeshi and Kuni (Tanaka) Y.; m. Margaret Michiyo Leong, May 14, 1964; children: Eric Clifford, Bradford Roth. BS, U. Calif., Berkeley, 1960; MS, U. Ill., 1962, PhD, 1965. Engring. mgr. Battelle Meml. Inst., Richland, Wash., 1965-70; project mgr. Westinghouse Electric Co., Richland, 1970-82; engring. mgr. Wash. Pub. Power Supply, 1982-83; engring. project mgr. Lawrence Livermore (Calif.) Nat. Lab., 1983—; mem. U.S.-Japan Fuel Fabrication Exch., 1980-82; sr. rep. U.S.-European Fire Studies, France, 1975; sec. meeting on sodium fires Internat. Atomic Energy Agency, Richland, 1976. Contbr. papers to profl. publs; author encyc. article on nuclear fuel fabrication. Mem. San Francisco Opera Assn., 1990, San Francisco Symphony Assn., 1991, Berkeley Repertory Theater Assn., 1991. Mem. Am. Nuclear Soc., Sigma Xi, Tau Beta Pi, Alpha Chi Sigma. Achievements include mgmt. of project to build advanced nuclear fuel fabrication facility for breeder reactor, pioneering large-scale sodium fire safety tests, environ. and engring. documents for laser isotope separation and laser fusion projects. Office: Lawrence Livermore Nat Lab PO Box 808 Livermore CA 99350

YATES, BARRIE JOHN, chemical engineer; b. Crewe, Cheshire, Eng., Aug. 8, 1936; came to U.S., 1961; s. John F. and Annie (Thelwall) Y.; m. Eileen Mary Smith, Sept. 24, 1960; children: David John, Ian Charles. BSc, Birmingham (Eng.) U., 1957; PhD, Manchester (Eng.) U., 1961. Rsch. engr. Dupont Plastics Div., Wilmington, Del., 1961-65; assoc. dir. R&D Cabot Carbon U.K., Stanlow, Cheshire, Eng., 1965-69; R&D group leader Cabot Pampa, Tex., 1969-71; dir. R&D Cabot Europe, Stanlow, 1971-76, Cabot Carbon Black Div., Billerica, Mass., 1976-80; dir. ops. worldwide Cabot Carbon Black Div., Paris, Atlanta, Boston, 1980-90; assoc. gen. mgr. Cabot Chems., Billerica, 1990—. Mem. Instn. Chem. Engrs. Achievements include several patents in the field of carbon black process. Office: Cabot Corp 157 Concord Rd Billerica MA 01821

YATES, KEITH, chemistry educator; b. Preston, Eng., Oct. 22, 1928; came to Can., 1948; s. Harold and Elizabeth Ann (Wilson) Y.; m. Norma June Charter, Aug. 21, 1953; children—Alison, Robyn, Nicola. B.A., U. B.C., Vancouver, 1956, M.Sc., 1957, Ph.D., 1959; D.Phil., Oxford U., England, 1961. Asst. prof. U. Toronto, 1961-64, assoc. prof., 1964-68, prof., 1968-90, prof. emeritus, 1990—, asst. dean, 1967-70, chmn., 1974-85. Author: Huckel Molecular Orbital Theory, 1977; contbr. articles to profl. jours. Served with Royal Navy, 1946-48. Recipient Gen.'s Gold medal Gov. of B.C., 1956; Royal Soc. Can. fellow, 1984. Fellow Chem. Inst. Can. (recipient Syntex award 1984, Palladium medal 1991, Chem. Inst. Can. medal 1991). Avocations: bridge, naval history. Home: 4 Avonwick Gate, Don Mills, ON Canada M3A 2M4 Office: U Toronto, Dept Chemistry, Toronto, ON Canada M5S 1A1

YATES, RENEE HARRIS, economist; b. Oct. 20, 1950; d. Marion and Betty Jane (Edgenton) Harris; m. Earl W. Yates, Sept. 6, 1980 (div. July 1991); 1 child, Clinton Harris Yates. BA, Western Coll. for Women, 1972; MA in Internat. Studies, Johns Hopkins U., 1974. Project devel. officer U.S. Agy. for Internat. Devel., Washington, 1975-81; internat. economist U.S. Treasury Dept./Office Sec. of Internat. Affairs, Washington, 1981-87; pres. World Trade Assocs., Inc., Washington, 1987—; pres. InterFuture, N.Y.C., 1984-87. Mem. TransAfrica, Washington, 1991. Mem. Thursday Luncheon Group. Avocations: sewing, tennis. Office: World Trade Assoc Inc 7320 Carroll Ave Takoma Park MD 20912

YATSU, KIYOSHI, plasma physicist; b. Hachioji, Tokyo, Japan, July 21, 1939; s. Denzaburo and Ei Yatsu; m. Yasuko Takahashi, Mar. 3, 1967; children: Mariko, Takako. B. Engring., U. Electro-Communications, Chofu, Japan, 1963; MS, U. Tokyo, 1965, DSc, 1968. Rsch. assoc. Tokyo U. Edn., 1968-75; asst. prof. U. Tsukuba, 1975-80, assoc. prof., 1980-87, prof., 1987—; chmn. Inst. Physics, 1992—. Mem. Phys. Soc. Japan, Am. Phys. Soc., Atomic Energy Soc. Japan, Japan Soc. Plasma Sci. and Nuclear Fusion Rsch., Physics Soc. Japan. Avocations: airplane, tourism. Home: 1-21-9 Ninomiya, Tsukuba 305, Japan Office: U Tsukuba Inst of Physics, 1-1-1 Tennodai, Tsukuba 305, Japan

YAU, EDWARD TINTAI, toxicologist, pharmacologist; b. Canton, China, Dec. 29, 1944; came to U.S., 1967; s. Wing S. and Fong K. (Wong) Y.; m. Assumpta Koo, July 3, 1979; 1 child, Jonathan C. BS in Biology, Bapt. Coll., Hong Kong, 1967; PhD in Pharmacology, U. Miss., 1974. Diplomate Am. Bd. Toxicology. Postdoctoral fellow, then asst. prof. Purdue U., West Lafayette, Ind., 1974-77; toxicology supr. Wyeth Labs., Great Valley, Pa., 1977-79; sr. toxicologist CIBA-GEIGY Corp., Summit, N.J., 1979-82, mgr., 1982-86, asst. dir., 1986-88, dir., 1988—; adj. prof. U. Miss., Oxford, 1989-92. Contbr. articles to sci. publs. Recipient NSF award, 1970. Mem. Am. Chem. Soc., Am. Coll. Toxicology, Soc. Toxicology, Teratology Soc., Sigma Xi. Republican. Baptist. Home: 67 Grant Ave Clifton NJ 07011-3522 Office: CIBA-GEIGY Corp 556 Morris Ave Summit NJ 07901-1398

YAU, SHING-TUNG, mathematics educator; b. Swatow, China, Apr. 4, 1949; came to U.S., 1969; m. Yu-Yun Kuo; children: Isaac, Michael. PhD, U. Calif., Berkeley, 1971; PhD (hon.), Chinese U. Hong Kong, 1980, Harvard U., 1987. Mem. Inst. Advanced Study, Princeton, N.J., 1971-72; asst. prof. math. SUNY, Stony Brook, 1972-74; prof. math. Stanford (Calif.) U., 1974-79, Inst. Advanced Study, Princeton, 1979-84, U.Calif.-San Diego, La Jolla, 1984-87, Harvard U., Cambridge, Mass., 1987—; vis. prof., chmn. math. dept. U. Tex., Austin, 1986; spl. chair Nat. Tsing Hua U., Hsinchu, Taiwan, 1991—; Wilson T.S. Wang Disting. Vis. Prof. Chinese U. Hong Kong, 1991—. Contbr. numerous articles to profl. publs. Named Honorable prof., Fudan U. China, Academia Sinica China; ScD (hon.), Chinese U. Hong Kong; recipient Sr. Scientist award, Humboldt Found., 1982, Fields medal, 1982, Sci. award, Union Pan Asian Communities, 1985; MacArthur Found. fellow, 1985. Mem. AAAS, NAS, N.Y. Acad. Sci., Acad. Arts and Scis. Boston, Am. Math. Soc., Am. Phys. Soc., Soc. Indsl. Applied Math. Office: Harvard U Dept Math 3d Fl Sci Ctr 1 Oxford St Cambridge MA 02138-2901

YAU, STEPHEN SIK-SANG, electrical engineering educator, computer scientist, researcher; b. Wusei, Kiangsu, China, Aug. 6, 1935; came to U.S., 1958, naturalized, 1968; s. Pen-Chi and Wen-Chum (Shum) Y.; m. Vickie Liu, June 14, 1964; children: Andrew, Philip. BS in Elec. Engring, Nat. Taiwan U., China, 1958; MS in Elec. Engring, U. Ill., Urbana, 1959, PhD, 1961. Rsch. asst. elec. engring. rsch. lab. U. Ill., Urbana, 1959-61; asst. prof. elec. engring. Northwestern U., Evanston, Ill., 1961-64, assoc. prof., 1964-68, prof., 1968-88, prof. computer scis., 1970-88, Walter P. Murphy prof. Elec.

Engring. and Computer Sci., 1986-88, also chmn. dept. computer scis., 1972-77; chmn. dept. elec. engring. and computer sci. Northwestern U., 1977-88; prof. computer and info. sci., chmn. dept. U. Fla., Gainesville, 1988—; conf. chmn. IEEE Computer Conf., Chgo., 1967; symposium chmn. Symposium on feature extraction and selection in pattern recognition Argonne Nat. Lab., 1970; gen. chmn. Nat. Computer Conf., Chgo., 1974, First Internat. Computer Software and Applications Conf., Chgo., 1977; Trustee Nat. Electronics Conf., Inc., 1965-68; chmn. organizing com 11th World Computer Congress, Internat. Fedn. Info. Processing, San Francisco, 1989; gen. co-chmn. Internat. Symposium on Autonomous Decentralized Systems, Japan, 1993. Editor-in-chief Computer mag., 1981-84; editor Journal of Information Sciences, 1983—, IEEE Trans. on Software Engineering, 1988-91; contbr. numerous articles on software engring., distributed and parallel processing systems, computer sci., elec. engring. and related fields to profl. publs.; patentee in field. Recipient Louis E. Levy medal Franklin Inst., 1963, Golden Plate award Am. Acad. of Achievement, 1964, The Silver Core award Internat. Fedn. Info. Processing, 1989, Spl. award, 1989. Fellow IEEE (mem. governing bd. Computer Soc. 1967-76, pres. 1974-75, dir. Inst. 1976-77; Richard E. Merwin award Computer Soc. 1981, Centennial medal 1984, Extraordinary Achievement 1985, Outstanding Contbn. award Computer Sci. Soc. 1985), AAAS, Franklin Inst.; mem. Assn. for Computing Machinery, Am. Fedn. Info.-Processing Socs. (mem. exec. com. 1974-76, 79-82, dir. 1972-82, chmn. awards com. 1979-82, v.p. 1982-84, pres. 1984-86; chmn. Nat. Computer Conf. Bd. 1982-83), Am. Soc. Engring. Edn., Sigma Xi, Tau Beta Pi, Eta Kappa Nu, Pi Mu Epsilon. Office: U Fla Computer & Info Sci Dept Rm 301 Gainesville FL 32611

YAU, TE-LIN, corrosion engineer; b. Ton-Chen, Anhuei, China, Apr. 9, 1945; s. Chiu-Ho and Chih (Yang) Y.; m. Jue-Hua Tsai, Mar. 19, 1979; children: Kai-Huei, Ian-Huei, Jean-Huei. BS in Engring. Sci., Cheng Kung U., Taiwan, 1969; MS in Engring. Sci., Tenn. Technol. U., 1972; PhD in Metall. Engring., Ohio State U., 1979. Mech. engr. Taiwan Shipbuilding Corp., Keelung, Taiwan, 1970-71; corrosion group head Teledyne Wah Chang, Albany, Ore., 1979—; cons. Fontana Corrosion Ctr., Ohio State U., Columbus, 1979. Author: ASTM STP, 1984, 85, 86, 90, 92, ASM Metals Handbook, 1987; contbr. articles to profl. jours. Mem. Nat. Assn. Corrosion Engrs., Electrochem. Soc., TAPPI, Sigma Xi. Achievements include a patent on novel liquid metal seal on zirconium or hafnium reduction apparatus; devel. of new applications for reactive metals, improving the performance of reactive metals by employing metall., electrochem. and mech. techs. Home: 1445 SW Belmont Albany OR 97321 Office: Teledyne Wah Chang Albany 1600 NE Old Salem Albany OR 97321-0460

YAVUZ, TAHIR, engineering educator; b. Tarabzon, Turkey, May 5, 1950; s. Huseyin and Asiye Yayuz; m. Hulya Yayuz, De. 11, 1982; children: H. Gokay, Gizem. BS, Karadenn Tech. U., Trabton, Turkey, 1974; PhD, Leicester U., Eng., 1982. Asst. assoc. prof. Erciyes U., Kayser, Turkey, 1982-85; assoc. prof. Erciyes U., Kayser, 1985-88; assoc. prof. Karadeniz Tech. U., Trabzon, 1988-91, prof., 1991—. Fulbright scholar, 1990. Mem. AIAA. Office: Karadeniz Tech U, Dept Mech Engring, 61080 Trabzon Turkey

YAZDANI, JAMSHED IQBAL, civil engineer; b. Sialkot, Punjab, Pakistan, Sept. 29, 1959; came to U.S., 1987; s. Zafar I. and Anwar (Farzana) Y.; m. Faiqa Sehaam, Dec. 27, 1987; 1 child, Saad I. BS in Civil Engring., U. Engring. and Tech., Lahore, Pakistan, 1983; MS in Civil Engring., Tex. A&M U., 1989. Engr. in tng. Asst. engr. Progressive Cons., Lahore, 1983-84; asst. design engr. Water & Power Devel. Authority, Lahore, 1984-86; grad. rsch. engr. Tex. Transp. Inst., College Station, 1987-89; transp. engr. Calif. Dept. Transp., Marysville, 1989—. Mem. ASCE.

YEAGER, CHARLES ELWOOD (CHUCK YEAGER), retired air force officer; b. Myra, W.Va., Feb. 13, 1923; s. Albert Hal and Susie May (Sizemore) Y.; m. Gennis Faye Dickhouse, Feb. 26, 1945; children: Sharon Yeager Flick, Susan F., Donald C., Michael D. Grad., Air Command and Staff Sch., 1952, Air War Coll., 1961; DSc (hon.), W.Va. U., 1948, Marshall U., Huntington, W.Va., 1969; D in Aero. Sci., Salem Coll., W.Va., 1975. Enlisted in USAAF, 1941; advanced through grades to brig. gen. U.S. Air Force, 1969, fighter pilot, ETO, 1943-46, exptl. flight test pilot, 1945-54; various command assignments U.S. Air Force, U.S., Germany, France and Spain, 1954-62; comdr. 405th Fighter Wing, Seymour Johnson AFB, N.C., 1968-69; vice comdr. 17th Air Force, Ramstein Air Base, Fed. Republic Germany, 1969-71; U.S. def. rep. to Pakistan, 1971-73; spl. asst. to comdr. Air Force Inspection and Safety Ctr., Norton AFB, Calif., 1973, dir. aerospace safety, 1973-75; ret., 1975. Author: (with Leo Janos) Yeager: An Autobiography, 1985, (with Charles Leerhsen) Press On!, 1988. Decorated DSM with oak leaf cluster, Silver Star with oak leaf cluster, Legion of merit with oak leaf cluster, DFC with 2 oak leaf clusters, Bronze star with V device, Air medal with 10 oak leaf clusters, Air Force Commendation medal, Purple Heart; recipient Presdl. medal of Freedom, 1985. First man to fly faster than speed of sound. Home: PO Box 128 Cedar Ridge CA 95924-0128

YEAGER, HAL K., biomedical engineer; b. Camden, N.J., Mar. 16, 1959; s. George W. and Gail (Rubenstein) Y. BS in Natural Sci., Muhlenberg Coll., 1981; MS in Biomed. Engring., Drexel U., 1986. Cardiovascular toxicologist Eli Lilly, Greenfield, Ind., 1986-91; biomed. engring. Eli Lilly, Indpls., 1991-92, sr. project engr., 1992—; software cons. in field. Contbr. articles to profl. jours. Mem. IEEE. Republican. Office: Eli Lilly 8645 Guion Rd Ste F Indianapolis IN 46268

YEAGER, LARRY LEE, civil engineer, consultant; b. Muncie, Ind., Feb. 11, 1948; s. Richard Earl and Elsie Jean (Parks) Y.; m. Carole Elaine Mullineaux, May 27, 1977. BSCE, Purdue U., 1971, MSCE, 1973. Registered profl. engr., Md. Materials engr. Martin Marietta Corp., Rockville, Md., 1973-75; materials rsch. engr. Genstar Stone Products, Inc., Hunt Valley, Md., 1976; dir. engring. Daedalean Assocs., Inc., Woodbine, Md., 1977-81; v.p. engring. Boender Assocs., Inc., Ellicott City, Md., 1982; project engr. Kidde Cons., Inc., Towson, Md., 1983-88, Baltimore County Md., Towson, 1988—; mem. adv. bd. solid waste mgmt. Howard County Pub. Works, Ellicott City, 1976-79; vice chmn. Baltimore County Supervisory Employees Group, Towson, 1988—. Author: Dynamic Modulus of Bituminous Mixtures as Related to Asphalt Type, 1973, A Recommended Procedure for the Determination of the Dynamic Modulus of Asphalt Mixtures, 1975; contbr. articles to profl. jours. Active Wabash Twshp. (Ind.) Vol. Fire Dept. (named Outstanding Vol. 1972), W. Lafayette; coord. Baltimore county March of Dimes-WALKAMERICA, Towson, 1992. Scholar State of Ind., 1966; named Outstanding Profl. Engr., Howard County Coun., 1990. Mem. ASCE (dir. 1990-92, recipient Participation award 1989), NSPE (pres. Howard county chpt. 1989-92, bd. dirs. state chpt. 1990-92). Democrat. Methodist. Achievements include patent in nondestructive test technique utilizing internal friction damping; establishment of dynamic modulus test for bituminous mixtures; development of high early strength cement for highway patches, driveway sealer/leveler for homeowner use, training exercise utilizing computer software for runoff pond design. Home: 2640 Thompson Dr Marriottsville MD 21104 Office: Baltimore County Dept Environ Protection/Res Mgmt 401 Bosley Ave MS 3404 Baltimore MD 21204

YEATTS, DOROTHY ELIZABETH FREEMAN, nurse, retired county official, educator; b. Richmond, Va., Jan. 19, 1925; d. Robert Franklin and Elizabeth Bell (Wiggins) Freeman; m. Roy Earl Yeatts, Nov. 27, 1948; children: Martha Jane Yeatts Couch, Robert Patrick. Diploma in nursing, Stuart Circle Hosp., Richmond, Va., 1947; BS in Nursing, Coll. William and Mary, 1947; cert. pub. health nursing supr., U. N.C., 1974. RN, Va., N.C. Vis. nurse Instructive Vis. Nurses Assn., Richmond, 1947-49; maternity nurse N.C. Bapt. Hosp., Winston-Salem, 1969-71; pub. health nurse I, Forsyth County Health Dept., Winston-Salem, 1971-72, pub. health nurse coord., 1972-74, pub. health nurse supr., 1974-78; instr. ARC, Winston-Salem, 1978-93; vol. Am. Red Cross, 1993—. Pres. Buckingham Park Garden Club, Richmond, 1956-58; elder Trinity Presbyn. Ch., Winston-Salem, 1978-81, Sunday Sch. tchr., 1980-84, circle bible moderator, 1984—; bd. dirs. Forsyth Cancer Soc., Winston-Salem, 1980-86; vol. ARC. Republican. Avocations: arts and crafts, fishing, stamp collecting, woodcarving. Home: 703 Devon Ct Winston Salem NC 27104-1269

YEE, HERMAN TERENCE, pathologist; b. El Paso, Oct. 22, 1959; s. Bun Ling and Kit (Yung) Y. BS, U. So. Calif., 1981, PhD, 1990; MD, McGill U., Montreal, Que., Can., 1990. Resident physician Columbia U., N.Y.C., 1990—. Contbr. articles to profl. jours. Mem. AMA, Am. Chem. Soc., Sierra Club, Wilderness Club, Sigma Xi. Achievements include rsch. in PCR analysis of paraffin embeded DNA, correlation between cytology and ultrasound finding in ovarian cysts. Office: Columbia-Presbyn Med Ctr VC14-215 630 W 168th St New York NY 10032

YEE, LESTER WEY-MING, researcher; b. Malden, Mass., Sept. 23, 1964. BS in Computer Sci., Rensselear Poly. Inst., 1986, MBA, 1987. Programmer MIT, Cambridge, 1984; acct. asst. Lotus Devel. Corp., Cambridge, 1984; programmer IBM, Sterling Forest, N.Y., 1985; cons., programmer Sci. Rsch. Labs., Somerville, Mass., 1986-87; rsch. assoc. Rensselaer Poly. Inst., Troy, N.Y., 1988—. TRW scholar, 1989. Mem. Soc. Mfg. Engr. Office: Rensselaer Poly Inst 9015 CII Troy NY 12180-3590

YEE, PETER BEN-ON, marketing professional; b. Chelsea, Mass., Apr. 13, 1960; s. Walter H.N. and Nancy Young (Seto) Y.; m. Diane Yuu, Aug. 28, 1988. BSEE, Worcester Poly. Inst., 1982; MSEE, Rensselaer Poly. Inst., 1983, MS in Computer and Systems Engring., 1986; MBA, UCLA, 1992. Engr. Hughes Aircraft Co., El Segundo, Calif., TRW, Inc., Redondo Beach, Calif.; product mgr. Tektronix, Inc., Beaverton, Oreg. Mem. Eta Kappa Nu, Tau Beta Pi, Zeta Psi. Home: 1801 Wendy Way Manhattan Beach CA 90266-4140

YEE, RAYMOND KIM, mechanical engineer, researcher; b. Dec. 25, 1956; s. Tung Wing and Yuet Keun (Lee) Y.; m. Wilma Wai Mak, July 8, 1989. BSME with highest honors, Calif. Polytechnic U., 1980; MS in Mech. Engring., U. Calif., Berkley, 1981, PhD in Mech. Engring., 1990. Registered profl. mechanical engr., Calif. Mem. tech. staff AT&T Bell Labs., Naperville, Ill., 1980-84; assoc. instr. U. Calif., Berkley, 1985-87, rsch. asst., 1986-88; rsch. asst. Lawrence Berkley Lab., 1988-90; sr. rsch. and consulting engr. Anamet Labs., Inc., Hayward, Calif., 1990-93; sr. engr., project mgr. APTECH Engring. Svcs., Inc., Sunnyvale, Calif., 1993—. Walter Wells Sr. Meml. scholar, 1978-80; AT&T Bell Labs. Grad. Study fellow, 1981; recipient Math. Achievement award Bank of Am., 1976. Mem. ASME, Tau Beta Pi. Home: 6672 Wooster Ct Castro Valley CA 94552 Office: APTECH Engring Svcs Inc 1282 Reamwood Ave Sunnyvale CA 94089

YEH, HSIEN-YANG, mechanical engineer, educator; b. Herng-Yang, Hounan, China, June 26, 1947; came to U.S., 1970; s. Jih-Min and Huey-Ru (Kao) Y.; m. Ching Hwan, Apr. 30, 1972; children: Shi-Kun Karl, Shi-Pong Norman. BS, Cheng-Kung U, Taiwan, 1969; MS, Brown U., 1971, Columbia U., 1975; PhD, U. So. Calif., 1987. Structural designer Ebasco, Inc., N.Y.C., 1975-76; stress analyst Stone and Webster Engring. Corp., Cherry Hill, N.J., 1976-77; staff engr. Bechtel Power Corp., Norwalk, Calif., 1977-78; mem. tech. staff TRW, Redondo Beach, Calif., 1978-80; sr. staff engr. Hughes Aircraft Co., Torrance, Calif., 1980-86; instr. dept. mech. engring. Calif. State U., Long Beach, 1986-88, assoc. prof., 1988-93; prof., 1993—; Cons. TRW, Hughes Aircraft Co., McDonnell Douglas Aircraft Co. Contbr. tech. papers to profl. jours. including Jour. Reinforced Plastics and Composites, Jour. Electronic Packaging, IEEE Electron Devices. Founder Torrance Chinese Sch., 1989. Grantee NASA-Ames Rsch. Ctr., 1988-89, NSF, 1992-94. Mem. ASME, ASCE, Am. Soc. Engring. Edn., Soc. for Advancement of Material & Process Engring., Sigma Xi. Office: Calif State U Long Beach Dept Mech Engring Long Beach CA 90840

YEH, JAMES TEHCHENG, chemical engineer; b. Fuzhou, Peoples Republic of China, Apr. 19, 1933; came to the U.S., 1958; s. Shusheng and Huihua (Wu) Y.; m. Diana H. Hou, Nov. 18, 1967; 1 child, Theresa Ling. BS, Cheng Kung U., 1956; MS, U. Detroit, 1961; PhD, Stevens Inst. Tech., 1970. Asst. prof. math. and chemistry Philander Smith Coll., Little Rock, Ark., 1963-66; chem. engr. Procedyne Corp., New Brunswick, N.J., 1968-69; sr. chem. engr. Princeton Aqua Sci., New Brunswick, 1970-74; staff system engr. Singer Simulation, Silver Spring, Md., 1974; chem. engr., project leader U.S. Dept. of Energy, Pitts., 1974—; dir. modeling studies Environ. Assessment Coun., Inc., New Brunswick, 1973-74. Chpt. author: Air Pollution Control and Design Handbook, 1974; contbr. articles to profl. jours. Mem. AIChE (session chmn. 1983-92), Am. Chem. Soc., Air and Waste Mgmt. Assn. (lectr. continuing edn. 1983-92), Orgn. Chinese Ams. and Chinese Sch. (bd. dirs., treas. 1983-89). Achievements include patent for NOx control technology. Home: 5705 Glenhill Dr Bethel Park PA 15102 Office: US Dept Energy Cochran Mill Rd Pittsburgh PA 15236

YEH, NOEL K., physicist, educator; b. Malacca, Malaysia, Dec. 15, 1937; m. Yi Yuan; children: Mildred, Andrew S. BA, Williams Coll., 1961; MS, Yale U., 1962, PhD, 1966. Rsch. assoc. Nevis Labs. Columbia U., Irvington, N.Y., 1966-68; instr. physics dept. Columbia U., N.Y.C., 1968-69; from asst. prof. to assoc. prof. physics dept. SUNY, Binghamton, N.Y., 1969-80, prof. physics, 1980—, chmn. physics dept., 1980-86; vis. prof. Max Planck Inst. for Physics and Astrophysics, Munich, 1976-77. Contbr. articles to Phys. Rev., Phys. Rev. Letters, Med. Physics. Mem. Am. Phys. Soc. Office: SUNY Vestal Pkwy E Binghamton NY 13902-6000

YEH, PAUL PAO, electrical and electronics engineer, educator; b. Sung Yang, Chekiang, China, Mar. 25, 1927; came to U.S., 1956; s. Tsung Shan and Shu Huan (Mao) Y.; m. Beverley Pamela Eng, May 15, 1952; children: Judith Elaine, Paul Edmond, Richard Alvin, Ronald Timothy. Student, Nat. Cen. U., Nanking, China, 1946-49; BA Sc in Elec. Engring., U. Toronto (Ont., Can.), 1951; MSEE, U. Pa., 1960, PhD, 1966. Registered profl. engr., Ont. Design engr. Can. Gen. Electric Co., Toronto, 1951-56; asst. prof. SUNY, Binghamton, 1956-57; sr. engr. H.K. Porter, ITE & Kuhlman, Phila. and Detroit, 1957-61; assoc. prof. N.J. Inst. Tech., Newark, 1961-66; supr. rsch. and devel. N.Am. Rockwell, Anaheim, Calif., 1966-70; sr. R&D engr. Lockheed Advanced Devel. Co., Burbank, Calif., 1970-72, 78-89; mem. tech. staff The Aerospace Corp., El Segundo, Calif., 1972-78; chief scientist Advanced Systems Rsch., Pasadena, Calif., 1989—; cons. Consol. Edison Co., N.Y.C., 1963-64; sr. lectr. State U. Calif., Long Beach., 1967-73; vis. prof. Chung Shan Inst. 1989-92; vis. prof. Tsinghua U., 1993; cons. prof. Northwestern Poly. U., 1993—; hon. prof. Beijing U. Aeronautics & Astronautics, 1993—; investigator R & D Stealth tech., electronic warfare, avionics, Aircraft/Missiles, 1966-93. Mem. IEEE (sr.), Nat. Mgmt. Assn., Am. Def. Preparedness Assn., Assn. Old Crows, Chines Am. Engring./Sci. Assn. So. Calif. (pres. 1969-71), Nat. Cen. U. Alumni Assn. (pres. 1977), Nat. Security Indsl. Assn., Beijing Assn. for Sci. & Tech. Exchanges with Fgn. Countries (hon. dir.), Assn. Profl. Engrs. of Ontario Canada.. Republican. Presbyterian. Achievements include patent for Non-Capacitive Transmission Cable. Home: 5555 Via De Campo Yorba Linda CA 92687-4916 Office: Advanced Systems Rsch Inc 33 S Catalina Ave Ste 202 Pasadena CA 91106-2426

YELDANDI, VEERAINDER ANTIAH, engineer, consultant; b. Hyderabad, India, Jan. 31, 1929; came to U.S., 1979; s. Antiah Venkatswamy and Rangamma (Chippa) Y.; m. Suvarna Nagiah Bet, Mar. 15, 1952; children: Vinod, Vivek, Vijay, Vandana. BSCE, Osmania Univ., Hyderabad, India, 1951; MS, Univ. Roorkee, Roorkee, India, 1957. Registered profl. engr., Ill., Wis. Jr. engr. Hyderabad Pub. Works, Hyderabad, 1952-57; lectr. civil engr. Univ. Roorkee, 1957-59; project engr. Bhilai (India) Steel Plant, 1959-79; structure cons. Sargent & Lundy, Chgo., 1980-85, Westinghouse Engring., Chgo., 1986; estimating engr. Chgo. Dept. of Transp., 1987-91; civil engr. Dept. of Water, Chgo., 1991—. Pres. Rotary Club, Bhilai, 1969-70, v.p. Telugu Assn. Chgo., 1986-87. Fellow Am. Soc. Civil Engrs.; mem. Am. Inst. Steel., Am. Water Works Assn. Achievements include preparation of technical report for water supply facilities for Bhilai Steel Plant complex expansion and township, design of the facilities and participated in construction of the project, and design of water mains project for the City of Chicago Dept. Water. Home: 4170 N Marine Dr # 18 K Chicago IL 60613 Office: City of Chgo Dept Water 1000 East Ohio St Chicago IL 60611

YELLIN, HERBERT, science administrator; b. N.Y.C., May 27, 1935; s. Louis and Gussie (Langer) Y.; m. Nancy Ann Nave, Aug. 11, 1963; children: Michael, Marci, Keith, Kim, Craig. BA, CCNY, 1956; PhD, UCLA, 1966. Rsch. physiologist Nat. Inst. Neurol. Diseases and Blindness, NIH, Bethesda, Md., 1965-76; grants assoc. Div. Rsch. Grants, NIH, Bethesda, 1976-77, exec. sec., 1984-86; dir. cataract program Nat. Eye Inst., NIH, Bethesda, 1978-79; exec. sec. Nat. Inst. Neurol. Disorders and Stroke, NIH, Bethesda, 1979-80, staff scientist, 1980-84, sci. rev. adminstr., 1986—; lectr. Schs. Medicine and Dentistry, Georgetown U., Washington, 1967-75; vis. faculty mem. UCLA, summer, 1972; mem. Assembly of Scientists, NIH, Bethesda, 1973-75. Contbr. articles to profl. jours. Mem. Montgomery County (Md.) Parent-Tchr.-Student Assn., 1960's and 70's; pub. sch. trustee, Montgomery County, 1970's; active local civic assns., 1970's. Mem. Soc. for Neurosci., Am. PHysiol. Soc., Am. Assn. Anatomists, N.Y. Acad. Scis. Achievements include descriptions of metabolic heterogeneity of mammalian skeletal muscle fibers, difference in muscle fibers subserving different functions, motoneuron regulation of the metabolism of some skeletal muscle fibers, changes in proprioceptor morphology and function with alterations in muscle status; illustration of survival and function of transplanted neural tissues; developed (with others) small animal models of reproducible spinal cord injury to test therapeutic inventions. Home: 5327 Pooks Hill Rd Bethesda MD 20814

YELLIN, JUDITH, electrologist; b. Balt., Feb. 21; d. Jack and Sarah (Grebow) Levin; m. Sidney Yellin, Jan. 1; children: David, Paul, Tamar. Student U. Md., 1948-50, Catonsville Community Coll., 1969-71. Mgr. credit dept. Lincoln Co., Balt., 1956-59; office mgr. Seaview Constrn. Co., 1960-62; owner, operator Yellin Telephone Soliciting Agy., 1963-65; mgr. Liberty Antique Shop, 1967-69; owner, mgr. Judith Yellin Electrology, 1973—; chief examiner Md. State Bd. Electrology, 1978-81; designer jewelry. Poet: New American Poetry Anthology, 1988, Great Poems of the Western World, Vol. II, 1990. Mem. Am. Electrolysis Assn., Md. Assn. Profl. Electrologists. Avocations: travel, reading, collecting Haitian and art deco, nouveau art and jewelry, poetry, inventing. Home: 6232 Blackstone Ave Baltimore MD 21209-3909 Office: Judith Yellin Electrology 1401 Reisterstown Rd Baltimore MD 21208-3807

YEN, ANTHONY, engineer; b. Shanghai, China, Jan. 22, 1962; came to U.S., 1980; s. Francis Er-Liang and Johanna Hai-Na (Chu) Y. BSEE, Purdue U., 1985; MS, MIT, 1987, EE, 1988, PhD, 1992. Grad. rsch. fellow MIT, Cambridge, 1985-91; mem. tech. staff Tex. Instruments Inc., Dallas, 1991—. Contbr. articles to profl. jours. RCA scholar RCA Corp., 1984-85, Nat. Def. Sci. and Engring. Grad. fellow, 1989-91. Mem. IEEE. Achievements include first to fabricate large area 100nm-period diffraction gratings using achromatic holography. Office: Tex Instruments Inc PO Box 655012 MS 944 Dallas TX 75265

YENICE, MEHMET FIKRI, aeronautics company executive; b. Adana, Turkey, Feb. 2, 1960; s. Necati and Nilufer (Kalak) Y.; m. Esin Zeynep, Dec. 26, 1991; 1 child, Nilufer Ender. BS in Aero. Engring., Embry-Riddle Aero. U., 1985. Structures engr. Monarch Aviation, Inc., Miami, Fla., 1985-86, chief engr., 1986-87; chief engr. Pemco Engrs., Inc., Miami, 1987-89; mng. dir. engring. Pemco Aeroplex, Inc., Birmingham, Ala., 1989-90, v.p. tech. svcs., 1990—; Designated FAA engring. rep. Mem. AIAA. Achievements include design of safer door locking mechanisms on large transport aircraft; rsch. on applicability of propeller theory to very low Reynolds No. propellers. Office: Pemco Aeroplex Inc 1943 N 50th St Birmingham AL 35212

YERBY, JOEL TALBERT, anesthesiologist, pain management consultant; b. Peoria, Ill., Sept. 13, 1957; s. Joel Talbert Sr. and Susan Colleen (Vachon) Y.; m. Karina Schlicht, May 10, 1987. Student, Marquette U., 1975-78; MS in Health Adminstrn., U. Colo., 1991. Diplomate Am. Bd. of Anesthesiology, Am. Acad. of Pain Mgmt. Intern Waltham (Mass.) Hosp., 1982-83; resident in anesthesiology Brigham and Women's Hosp., Boston, 1983-85; fellow in pain mgmt. Columbia Presbyn. Med. Ctr., N.Y.C., 1985-86; anesthesiologist Orlando (Fla.) Regional Med. Ctr., 1986-89, CompHealth, Atlanta, 1989-91; vice-chief anesthesiology Brockton Hosp., Boston, 1991—; dir. adv. bd. Anesthesiology Assocs., Inc., Orlando, 1988-89. Violinist Carnegie Hall, N.Y.C., Symphony Hall, Boston. Violinist Milw. Civic Symphony, 1977-82, Boston Philharmonic Orch., 1983-85. Mem. AAAS, Am. Soc. Anesthesiologists, Internat. Soc. Study of Pain, Am. Coll. Physician Execs., Internat. Soc. Systems Scis., Mensa, Phi Beta Kappa, Alpha Omega Alpha. Home: 341 F Bolivar St Canton MA 02021

YERION, MICHAEL ROSS, civil engineer; b. Jacksonville, Ill., Oct. 18, 1955; s. Billie Ross and Lucia Theresa (Schlëgel) Y.; m. Chalene Helen Hochard, May 2, 1981; children: Gretchen, Erich. BSCE, So. Ill. U., 1979; MS in Engring. Mgmt., U. Mo., 1993. Registered profl. engr., Ill., Mo. Test engr. Panhandle Eastern Pipeline Co., Kansas City, Mo., 1979-80; pipeline engr. Panhandle Eastern Pipeline Co., Springfield, Ill., 1981-83; civil engr. City of St. Peters, Mo., 1983-89, engring. supr., 1989-91, dir. engr., 1991—; chmn. Local Emergency Planning, St. Charles (Mo.) County, 1991—. Dir. O'Fallon Community Band, 1991. Mem. ASCE, Nat. Soc. Profl. Engrs., Am. Pub. Works Assn., Am. Soc. Engr. Mgrs. Roman Catholic. Office: City of Saint Peters 1 City Centre Blvd Saint Peters MO 63376

YESKE, RONALD A., dean; b. Wisconsin Rapids, Wis., Oct. 28, 1946; m. Marilyn Joy (Mantey) Yeske. B in Mech. Engring., Marquette U., 1968. Cert. in engring., Wis., Pa. Asst. prof. U. Ill., Urbana, 1973-78; mgr. corrosion devel. Westinghouse R&D, Pitts., 1978-82; group leader Inst. Paper Chemistry, Appleton, Wis., 1982-86; v.p. Inst. Paper Sci. & Tech., Atlanta, 1986-91; dean engring. U. Wis., Platteville, 1991—. NSF fellow, 1968-73. Office: Coll Engring 575 14th St NW U Wisconsin Platteville WI 53818

YEUNG, EDWARD SZESHING, chemist; b. Hong Kong, Feb. 17, 1948; came to U.S., 1965; s. King Mai Luk and Yu Long Yeung; m. Anna Kunkwok Seto, Sept. 18, 1971; children: Rebecca Tze-Mai, Amanda Tze-Wen. AB magna cum laude, Cornell U., 1968; PhD, U. Calif., Berkeley, 1972. Instr. chemistry Iowa State U., Ames, 1972-74; asst prof. Iowa State U., 1974-77, assoc. prof, 1977-81, prof. chemistry, 1981-89, disting. prof., 1989—. Contbr. articles to profl. jours. Alfred P. Sloan fellow, 1974-76. Fellow AAAS; mem. Soc. Applied Spectrosci. (Lester Strock award 1990), Am. Chem. Soc. (award in Chem. Instrumentation 1987). Home: 1005 Jarrett Cir Ames IA 50014 Office: Iowa State U Gilman Hall Ames IA 50011

YEUNG, TIN-CHUEN, pharmacologist; b. Hong Kong, June 21, 1952; m. Bik Kit Tam, July 21, 1981; children: Louise, Helen. BS, SUNY, Stony Brook, 1975; PhD, SUNY, Bklyn., 1980; M Mgmt., Northwestern U., 1987. Grad. asst. SUNY, Bklyn., 1975-80; rsch. fellow Harvard Med. Sch., Boston, 1980-83; rsch. investigator G.D. Searle & Co., Skokie, Ill., 1983-85; sr. rsch. investigator, rsch. scientist NutraSweet Co., Mount Prospect, Ill., 1986-88; mgr. bus. ventures NutraSweet Co., Deerfield, Ill., 1988-91; mgr. external techs. NutraSweet Co., Mt. Prospect, 1991; mgr. strategic devel. Baxter Healthcare Corp., Round Lake, Ill., 1991—. Contbr. articles to profl. publs. Mem. tech. com., strategic planning com., Niles (Ill.) Sch. Dist., 1992-93. Damon Runyon-Walter Winchell Cancer Found. rsch. fellow, 1981-83; recipient Medallion award Zymark Corp., 1987. Mem. AAAS, Am. Chem. Soc., Assn. Univ. Tech. Mgrs., Licensing Execs. Soc. Achievements include discovery of mechanism of action of the antibacterial and anti-cancer effects of nitroimidazoles. Home: 7831 N Neva Ave Niles IL 60714 Office: Baxter Healthcare Corp Rte 120 & Wilson Rd Round Lake IL 60073

YFF, DAVID ROBERT, chemist; b. Beirut, Lebanon, Mar. 16, 1955; came to U.S., 1974; s. Peter Yff and Ellen (Prodan) Neville; m. Belinda Fay Wilson, Mar. 15, 1986; children: Eric David, Timothy James. AB in Chemistry, Ind. U., 1981. Chemist Am. Synthetic Rubber Corp., Louisville, 1980—. Mem. AIAA, Am. Chem. Soc. Office: Am Synthetic Rubber Corp 4500 Campground Rd Louisville KY 40216

YIN, JOHN, engineering educator; b. Boston, Jan. 6, 1960; s. Richard Y.C. and Helen W.H. (Chang) Y. BA, Columbia Coll., 1982; BS, Columbia U., 1983; PhD, U. Calif., Berkeley, 1988. Rsch. fellow Max-Planck Inst. for Biophys. Chemistry, Goettingen, Germany, 1988-92; asst. prof. Thayer Sch. of Engring., Dartmouth Coll., Hanover, N.H., 1992—. Contbr. articles to Biotech. and Bioengring, Biochem., Biophys. Rsch. Comm., Jour. Bacteriology, Biophys. Jour. Fellowship Max-Planck Soc., 1990, Alexander von Humboldt Found., 1988, E.I. DuPont de Nemours, 1983; recipient Patent Fund award U. Calif., 1987. Mem. AICE, Am. Chem. Soc., Am. Soc. Microbiology. Achievements include patent in cadmium ion-chelating synthetic polymers, research in application of biochemical principles to environmental waste treatment, design, implementation and analysis of reactors for evolving bacteriophages. Office: Thayer Sch Engring Dartmouth Coll Hanover NH 03755

YIN, JUN-JIE, biophysicist; b. Anhui/Anqing, People's Republic of China, Mar. 26, 1944; s. Huan-ran and De-hua (Ding) Y.; m. Fang Bao, July 28, 1972; 1 child, Jia. MS, Chinese Acad. Sci., Beijing, 1981; PhD, Med. Coll. Wis., 1987. Elec. engr. Radio Factory, Huhhot, People's Republic China, 1968-78; rsch. fellow Med. Coll. Wis., Milwaukee, Wis., 1982-86; rsch. scientist Nat. Biomed. ESR Ctr. Med. Coll. Wis., Milw., 1991—; presenter in field. Contbr. articles, abstracts to profl. publs. Mem. AAAS, Biophys. Soc. U.S.A., Nat. ESR Soc. Achievements include development of multifrequency saturation recovery ESR, multifrequency saturation recovery techniques to study the electron spin-lattice relaxation mechanisms, the lipid-lipid, lipid-protein interactions in membranes; research in effect of membrane composition in lateral diffusion and membrane dynamics; measurement of oxygen transport in membranes, using oxygen as a probe and combine saturation recovery ESR technique to study membrane proteins. Home: 313 N 95th St #231 Milwaukee WI 53226 Office: Med Coll Wis/Nat Biomed ESR Ctr Milwaukee WI 53226

YIN, RAYMOND WAH, radiologist; b. Canton, Republic of China, July 2, 1938; came to U.S., 1972; m. Jean Youe Mok, Jan. 29, 1967; children: Linda, Dany, Judy. MD, Sun Yat Sen U., Canton, 1961, Nat. Taiwan U., 1965. Diplomate Am. Bd. Radiology, Am. Bd. Nuclear Medicine. Intern Victoria Gen. Hosp., Dalhausie U., Halifax, Nova Scotia, Can., 1967-68; resident Royal Victoria Hosp., McGill U., Montreal, Que., 1968-72; staff radiologist St. Francis Hosp., Hartford, 1972-75; radiologist St. Joseph Hosp., Bloomington, Ill., 1975-89, Mennonite Brakaw Hosp., Normal, Ill., 1975—; practice medicine specializing in radiology Bloomington, 1975—; asst. prof. radiology U. Ill., Peoria, 1980—, So. Ill. U., 1988—. Fellow Royal Coll. Physicians of Can. Home: 2110 Oakwood Ave Bloomington IL 61704-2412 Office: 200 S Towanda Ave Normal IL 61761-2155

YINGQIAN, QIAN, botanist. Dir. Inst. Botany Acad. Scis., China, 1985—. Office: Academia Sinica, 52 San Li He Rd, Beijing 100864, China*

YIP, CECIL CHEUNG-CHING, biochemist, educator; b. Hong Kong, June 11, 1937; emigrated to Can., 1955; m. Yvette Fung, Oct. 15, 1960; children: Christopher, Adrian. B.Sc., McMaster U., Can., 1959; Ph.D., Rockefeller U., 1963. Research assoc. Rockefeller U., N.Y.C., 1963-64; asst. prof. U. Toronto, Ont., Can., 1964-68, assoc. prof., 1968-74, prof., 1974—, Charles H. Best prof. of med. rsch., 1987-93; chair Banting and Best dept. med. rsch. U. Toronto, Ont., 1990—; vice-dean rsch. U. Toronto, 1992—. Contbr. sci. articles to publs. Recipient Charles H. Best prize Can. Hoechst Diabetes Workshop, 1972. Mem. AAAS, Am. Soc. Biol. Chemists, Am. Chem. Soc., Can. Biochem. Soc. Home: 125 Melrose Ave, Toronto, ON Canada M5M 1Y8 Office: Banting and Best Dept Med Rsch, 112 College St, Toronto, ON Canada M5G 1L6

YLITALO, CAROLINE MELKONIAN, chemical engineer; b. Aleppo, Syria, Jan. 22, 1964; d. Sam and Nadia (Basmaji) Melkonian; m. David Adrian Ylitalo, Dec. 17, 1988. BS, U. Calif., Berkeley, 1987; MS, Stanford U., 1989, PhD, 1991. Post doctoral fellow Lockheed Palo Alto (Calif.) Rsch. Lab., 1991-92; sr. rsch. engr. 3M Co., St. Paul, 1992—. Mem. Soc. Rheology, Phi Beta Kappa, Tau Beta Pi. Home: 930 Norell Ave N Stillwater MN 55082 Office: 3M Co 3M Center Bldg 219-1S-01 Saint Paul MN 55144-1000

YOCHELSON, ELLIS L(EON), paleontologist; b. Washington, Nov. 14, 1928; s. Morris Wolf and Fannie (Botkin) Y.; m. Sally Witt, June 10, 1950; children: Jeffrey, Abby, Charles. B.S., U. Kans., 1949, M.S., 1950; Ph.D., Columbia U., 1955. Paleontologist U.S. Geol. Survey, 1952-85, scientist emeritus, 1991; biostratigrapher, specializing on Paleozoic gastropods and minor classes of extinct mollusks; lectr. night sch. George Washington U., 1962-65; rsch. assoc. dept. paleobiology Smithsonian Instn., Washington, 1965—; lectr. Univ. Coll., U. Md., 1966-74, U. Del., 1981; vis. prof. U. Md., 1986-87; organizer N.Am. Paleontol. Conv., 1969, editor proc., 1970-71; mem. NRC, 1959-68; mem. organizing com. Internat. Congress Systematic and Evolutionary Biology, Boulder, Colo., 1972; research asso. dept. paleobiology U.S. Nat. Mus. Natural History; sec.-gen. IX Internat. Congress Carboniferous Stratigraphy and Geology, 1979; mem. centennial com. U.S. Geol. Survey, 1979. Co-editor: Essays in Paleontology and Stratigraphy, 1967; editor: Scientific Ideas of G.K. Gilbert, 1980; contbr. numerous articles to profl. jours. Fellow AAAS (chmn. sect. E 1971); mem. Soc. Systematic Zoology (sec. 1961-66, councilor 1973), Internat. Palaeontol. Assn. (treas. 1972-76), Paleontol. Soc. (pres. 1976), History of Earth Scis. Soc. (sec.-treas. 1982-85, sec. 1986-87, pres.-elect 1988, pres. 1989), Paleontol. Assn., Sigma Xi. Office: Smithsonian Instn E-304 Mus Natural History Washington DC 20560

YOCOM, JOHN ERWIN, environmental consultant; b. Oberlin, Ohio, Nov. 20, 1922; s. Charles Herbert and Inez (Willis) Y.; m. Elizabeth Clifford, Dec. 21, 1947; 1 child, Judith Anne Yocom Huling. BS in Chem. Engring., MIT, 1947. Registered profl. engr., DEE; cert. indsl. hygienist; diplomate Am. Acad. Environ. Engrs., Am. Bd. Indsl. Hygiene. Project leader Battelle Meml. Inst., Columbus, Ohio, 1947-57; dir. tech. svcs. Bay Area Air Quality Mgmt. Dist., San Francisco, 1957-61; sr. staff mem. Arthur D. Little Inc., San Francisco, 1961-62; sr. project mgr. Kaiser Engrs., Inc., Oakland, Calif., 1962-65; v.p., chief cons. engr. TRC Environ. Co., Windsor, Conn., 1965-90; environ. cons. J.E. Yocom, P.E., CIH, Environ. Cons., West Simsbury, Conn., 1990—. Author: Measuring Indoor Air Quality - A Practical Guide, 1991; contbr. numerous articles to profl. jours., chpts. to books. Mem. Conservation Commn., Simsbury, 1991—. With U.S. Army, 1943-46. Mem. ASHRAE, ASTM, Am. Indsl. Hygiene Assn., Air and Waste Mgmt. Assn. (bd. dirs. 1982-85), Conn. Acad. Sci. and Engring. Home and Office: 12 Fox Den Rd West Simsbury CT 06092

YOCUM, HARRISON GERALD, horticulturist, botanist, educator, researcher; b. Bethlehem, Pa., Apr. 2, 1923; s. Harrison and Bertha May (Meckes) Y. BS, Pa. State U., 1955; MS, Rutgers U., 1961. Horticulture instr. U. Tenn., Martin, 1957-59; biology tchr., libr. asst. high schs., El Paso, Tex., 1959-60; rsch. asst. geochronology lab. U. Ariz., Tucson, 1960-67, rsch. asst. environ. rsch. lab., 1969-76; landscaping supt. Tucson Airport Authority, 1976-82; instr. Pima C.C., Tucson, 1976—. Contbr. articles to profl. jours. Founder Tucson Bot. Gardens, 1964. Mem. Am. Hort. Soc., Men's Garden Club Tucson (pres. 1991), Tucson Cactus & Succulent Soc. (pres. 1991, 92), Internat. Palm Soc. (charter), El Paso Cactus & Rock Club, Tucson Gem & Mineral Soc., Old Pueblo Lapidary Club, Deming Mineral Soc., Nat. Geog. Soc., Ariz.-Sonora Desert Mus., Huachuca Vigilantes, Penn State Alumni Assn., Pa. Club Tucson, Hi-Flyers Toastmasters, Toastmasters, Fraternal Order Police Assocs., Shriners, Masons, Scottish Rite. Lutheran. Home: 1628 N Jefferson Ave Tucson AZ 85712

YODER, BRUCE ALAN, chemist; b. Seward, Nebr., Apr. 29, 1962; s. Elwood John and Elda Raye (Stutzman) Y. BS in Chemistry, Wayne State Coll., 1983. Lab. technician Wayne (Nebr.) State Coll., 1982-83; lab. technician Harris Labs., Lincoln, Nebr., 1984, chemist, 1984; scientist Dorsey Labs., Lincoln, 1984-86, scientist A, 1986-88; product stability analyst Sandoz Pharms., Lincoln, 1988-89, Sandoz Rsch. Inst., Lincoln, 1989-91; mgr. lab. computer ops. Sandoz Pharms., Lincoln, 1991—. Mem. Bellwood Mennonite Ch., Milford, Nebr., 1962—; mem. Lancaster County Young Reps., Lincoln, 1988—, co-chmn., 1990-91, pres. 1991—; mem. exec. com. Lancaster County Rep. Party, 1990—; mem., co-chmn. Def. Adv. Coun. Lancaster County, 1992—; mem. Lincoln Mayors Community Cabinet, 1992—; trustee Wayne State Coll. Found., 1991—. Mem. Am. Inst. Chemists, Am. Chem. Soc., Jaycees. Achievements include design of a sample holder for solid dosage forms when using a hunter color instrument, design of a new computer system for Sandoz Pharmaceuticals laboratory computer operations. Home: 2240 Winding Way Lincoln NE 68506-2846 Office: Sandoz Pharms 10401 Hwy 6 Lincoln NE 68517-9704

YODER, HATTEN SCHUYLER, JR., petrologist; b. Cleve., Mar. 20, 1921; s. Hatten Schuyler and Elizabeth Katherine (Knieling) Y.; m. Elizabeth Marie Bruffey, Aug. 1, 1959; children: Hatten Schuyler III, Karen Marianne. A.A., U. Chgo., 1940, S.B., 1941; student, U. Minn., summer 1941; Ph.D., Mass. Inst. Tech., 1948; Dr. h.c., U. Paris VI, 1981. Petrologist Geophys. Lab., Carnegie Instn., Washington, 1948-71; dir. Geophys. Lab., Carnegie Instn., 1971-86, dir. emeritus, 1986—; cons. Los Alamos Nat. Lab., 1972—. Author: Generation of Basaltic Magma, 1976; editor: The Evolution of the Igneous Rocks: Fiftieth Anniversary Perspectives, 1979; co-editor Jour. of Petrology, 1959-69; assoc. editor Am. Jour. Sci, 1972-90; contbr. articles to sci. jours. Trustee Culver Trust, 1992—. Served to lt. comdr. USNR, 1942-58. Recipient award Mineral. Soc. Am., 1954, Bicentennial medal Columbia, 1954, Arthur L. Day medal Geol. Soc. Am., 1962, A.L. Day prize and lectureship NAS, 1972, A.G. Werner medal German Mineral. Soc., 1972, Golden Plate award Am. Acad. Achievement, 1973, Wollaston medal Geol. Soc. London, 1979; inducted into Lakewood (Ohio) High Sch. Disting. Alumni Hall of Fame, 1990; mineral, yoderite named in his honor. Fellow Geol. Soc. Am. (mem. council 1966-68), Geol. Soc. London (hon.), Geol. Soc. South Africa, Am. Acad. Arts and Scis., Mineral. Soc. Am. (council 1962-64, 69-73, pres. 1971-72, Roebling medal 1992), Am. Geophys. Union (pres. volcanology, geochemistry and petrology sect. 1962-64); mem. Mineral Soc. London (hon. 1983—), Geol. Soc. Edinburgh (corr.), Geol. Soc. Finland, All-Union Mineral. Soc. USSR (hon.), Geochem. Soc. (organizer, founding mem., council 1956-58), Am. Chem. Soc., Mineral. Assn. Can., NAS (chmn. geology sect. 1973-76), Washington Acads. Sci., Geol. Soc. Washington, Chem. Soc. Washington, French Soc. Mineralogy and Crystallography (hon. 1986—), Am. Philos. Soc. (council 1983-85), History Earth Scis. Soc. (pres.-elect, 1993), SAR, Sigma Xi, Phi Delta Theta Golden Legion. Home: 6709 Melody Ln Bethesda MD 20817-3152 Office: Geophys Lab 5251 Broad Branch Rd NW Washington DC 20015-1305

YODER, MARY JANE WARWICK, psychotherapist; b. Corryton, Tenn., Nov. 20, 1933; d. Harry Alonzo and Mary Luzelle (Furches) Warwick; m. Edwin Milton Yoder, Jr., Nov. 1, 1958; children: Anne Daphne, Edwin Warwick. BA, U. N.C., Chapel Hill, 1956; MFA, U. N.C., Greensboro, 1969; MSW, Va. Commonwealth U., 1987; cert. individual psychotherapy, Smith Coll., 1991. Lic. ind. clin. social worker, D.C.; lic. clin. social worker, Va. Editorial asst. Harper & Bros., N.Y.C., 1956-57; flight attendant Pan Am. Airlines, N.Y.C., 1957-59; adj. faculty mem. in ballet Guilford Coll., Greensboro, 1961-64; ballet tchr., adminstr. Jane Yoder Sch. of Ballet, Greensboro, 1964-75; homilitics listener Va. Theol. Sem., Alexandria, 1978-80; social worker, dance therapist Woodbine Nursing Ctr., Alexandria, 1983-87; staff psychotherapist D.C. Inst. Mental Health, 1987-92; pvt. practice Capitol Hill Ctr. Individual and Family Therapy, 1992—. Ballet and book critic Greensboro Daily News, 1961-75. Dancer, choreographer Greensboro Civic Ballet, 1961-75. Mem. Nat. Assn. Social Workers, Greater Washington Soc. for Clin. Social Work, Inc., Washington Sch. Psychiatry, Washington Soc. for Jungian Psychology, Jungian Venture, Army-Navy Country Club. Episcopalian. Avocations: ballet, modern dance, horseback riding, swimming, reading. Office: Capitol Hill Ctr for Individual and Family Therapy 530 Seventh St SE Washington DC 20003

YOERGER, ROGER RAYMOND, agricultural engineer, educator; b. LeMars, Iowa, Feb. 17, 1929; s. Raymond Herman and Crystal Victoria (Ward) Y.; m. Barbara M. Ellison, Feb. 14, 1953; 1 child, Karen Lynne; m. Laura M. Summitt, Dec. 23, 1971; stepchildren—Daniel L. Summitt, Linda Summitt Canull, Anita Summitt Smith. B.S., Iowa State U., 1949, M.S., 1951, Ph.D., 1957. Registered profl. engr., Ill., Pa., Iowa. Instr., asst. prof. agrl. engring. Iowa State U., 1949-56; assoc. prof. agrl. engring. Pa. State U., 1956-58; prof. agrl. engring. U. Ill., Urbana, 1959-85; head agrl. engring. dept. U. Ill., 1978-85, prof. emeritus agrl. engring., 1985—. Contbr. articles to profl. jours. Patentee in field. Mem. Ill. Noise Task Force, 1974-80. Fellow Am. Soc. Agrl. Engrs. (Massey-Ferguson medalist 1989); mem. Am. Soc. Engring. Edn., Phi Kappa Phi (dir. fellowships, dir. 1971-83, pres. elect 1983-86, pres. 1986-89), Rotary, Moose. Roman Catholic. Home: 107 W Holmes St Urbana IL 61801-6614 Office: 1304 W Pennsylvania Ave Urbana IL 61801-4797

YOGANATHAN, AJIT PRITHIVIRAJ, biomedical engineer, educator; b. Colombo, Sri Lanka, Dec. 6, 1951; came to U.S., 1973; s. Ponniah and Mangay (Navaratnam) Y. BSChemE with honors, Univ. Coll., U. London, 1973; PhDChemE, Calif. Tech. U., 1978. Engring. asst. Shell Oil Refinery, Stanlow, Eng., 1972; teaching asst. Calif. Inst. Tech., 1973-74, 1976, rsch. fellow, 1977-79; asst. prof. Ga. Inst. Tech., 1979-83, assoc. prof., 1983-88, chmn. bioengring. com., 1984-88, prof. chem. engring., 1988—, prof. mech. engring., 1989—, co-dir. Emory Biomed. Tech. Ctr., 1992—; adj. assoc. prof. U. Ala., 1985—. Founding fellow Am. Inst. Med. & Biol. Engring., 1992; recipient Edwin Walker prize Brit. Inst. Mech. Engrs., 1988, Humboldt fellowship, 1985, Am. Heart Assn.-Ga. Affiliate Rsch. Investigatorship award, 1980-83, Calif. Inst. Tech. fellowship, 1973-77, Goldsmid Medal and prize Univ. Coll., 1973, 72, Brit. Coun. scholarship, 1971-73. Mem. AICE, ASME, Biomed. Engring. Soc., Am. Soc. Echocardiography (dir. 1987-91). Office: Sch of Chem Engring Ga Tech U Atlanta GA 30332-0100

YOKEL, FELIX YOCHANAN, civil engineer; b. Vienna, July 13, 1922; came to the U.S., 1956; s. Karl and Emma (Hauser) Jokel; m. Susanne Martha Braun, Sept. 20, 1946; children: Uri, Yael, Benjamin Karl. BS, U. Conn., 1959, MS, 1961, PhD, 1963. Registered profl. engr., D.C., Md., N.Y., N.H., Conn., Vt., R.I., Maine. Engr. John Clarkeson, Albany, N.Y., 1961-63; sr. ptnr. Clarkeson, Clough & Yokel, Albany, Boston, 1963-68; prof. SUNY, Binghamton, 1968-70; sr. rsch. engr. NBS/NIST, Gaithersburg, Md., 1970-93, cons. engr., 1993—; adj. prof. George Washington U., Washington, 1984. Contbr. articles to profl. jours. Recipient Silver medal Dept. of Commerce, 1973. Home and Office: Carderock Springs 8208 Fenway Rd Bethesda MD 20817-2731

YOKOZEKI, SHUNSUKE, optical engineering educator; b. Okada, Kagawa, Japan, Feb. 7, 1940; s. Tsutomu and Aki Yokozeki; m. Ikuko Yokozeki, Aug. 10, 1969; children: Satoru, Megumi. BS in Engring., Osaka U., Japan, 1962, MS in Engring., 1967, D of Engring., 1972. Asst. prof. Osaka U., Osaka, Japan, 1967-87, assoc. prof., 1987-90; prof. Kyushu Inst. Tech., Fukuoka, Japan, 1990—. Contbr. articles to prof. jours. Mem. Optical Soc. Am., Internat. Soc. Optical Engring., Japan Soc. Applied Physics, Japan Soc. Precision Engring. Office: Kyushu Inst Tech, Kawazu 680-4, Iizuka Fukuoka 820, Japan

YONDA, ALFRED WILLIAM, mathematician; b. Cambridge, Mass., Aug. 10, 1919; s. Walter and Theophelia (Naruscewicz) Y.; B.S., U. Ala., 1952, M.A. in Math., 1954; m. Mary Jane McManus, Dec. 19, 1949 (dec.); children—Nancy, Kathryn, Elizabeth, John; m. Peggy A. Terrel, June 22, 1975. Mathematician rocket research Redstone Arsenal, Huntsville, Ala., 1953, U.S. Army Ballistic Research Labs., Aberdeen (Md.) Proving Grounds, 1954-56; instr. math. U. Ala., Tuscaloosa, 1954, Temple U., Phila., 1956-57; asso. scientist, research and devel. div. Avco Corp., Wilmington, Mass., 1957-59; sr. mem. tech. staff RCA, Camden, N.J., 1959-66; mgr. computer analysis and programming dept. Raytheon Co. space and information systems div., Sudbury, Mass., 1966-70, mgr. software systems lab., 1969-70, prin. engr. missiles systems div., 1970-73; mgr. systems analysis and programming GTE Govt. Systems Corp., 1973-77, mgr. software engring. Atlantic ops., 1977-82, sr. mem. tech. staff Command Control & Communications Sector, 1983-91; software systems engr. Yonda Software Systems Cons., 1991—. Pres., Milford Area Assn. Retarded Children, 1970-74; vice-chmn. fin. com. Town of Medway, 1973; bd. dirs. Blackstone Valley Mental Health and Retardation Area Bd., 1970-76; trustee Medway Libraries, 1973-82, chmn., 1974-81. Served with USAAF, 1943-46. Hon. fellow Advanced Level Telecommunications Tng. Center, New Delhi, India, 1981. Registered profl. engr. Mem. AAAS, IEEE, Math. Assn. Am., N.Y. Acad. Scis., Sigma Xi, Phi Eta Sigma, Pi Mu Epsilon (pres. Ala. chpt. 1953-54), Sigma Pi Sigma. Contbr. articles to profl. jours. Office: 12 Sunset Dr Medway MA 02053-2008

YONETANI, KAORU, chemist, consultant; b. Kobe, Japan, Mar. 18, 1938; m. Makiko Yonetani, Jan. 3, 1942; children: Goske, Athuko. BS, Kobe Univ., 1960; MS, Northwestern U., 1969, PhD, 1970. Rschr. Matsushita Elect. Industries, Osaka, 1960-61, Sekisiu Chem. Industries, Osaka, 1962-65; rsch. assoc. Northwestern U., Evanston, Ill., 1965-66, post doctoral resident assoc., 1970-71; mgr. Sumitomo Chem. Industries, Osaka, 1971-76; dir. internat. div. Sumika Tech. Consulting Firm, Osaka, 1976—. Home: 6-4-11-2C Sumaura, Suma, Kobe 654, Japan Office: 4-5-33 Kitahama Chuo ward, Osaka 541, Japan

YONEYAMA, HIROSHI, chemistry educator; b. Osaka, Japan, June 21, 1937; s. Katsutoshi and Tsuyako (Hayashi) Y.; m. Takako Doi, Sept. 28, 1965; children: Naoki, Nobuyuki. B in Engring., Osaka U., 1960, Dr. Engring., 1973. Rschr. Matsushita Elec. Indls. Co., Osaka, 1960-67; instr. Osaka U., 1967-73, asst. prof., 1973-76, assoc. prof., 1976-85, prof., 1985—. Editor Denki Kagku, 1991-92, Chemistry Letters, 1993—; contbr. articles to Electrochim Acta, Jour. Electroanal. Chem., Jour. Chem. Soc. Chem. Comm. Achievements include discovery of excellent redox properties of polyaniline; first demonstration of photocleavage of water using p-n photoelectrochemical diode, first electrochemical synthesis of conducting polymer/metal oxide composit films, novel light image formation on polyaniline. Home: 14-213-104 Higashiyamadai 2, Nishinomiya Hyogo 669-11, Japan Office: Osaka U, 2-1 Yamada-oka, Osaka 565, Japan

YOO, KWONG MOW, science educator; b. Tg. Malim, Perak, Malaysia, Nov. 12, 1958; came to U.S., 1983; s. Ah Kwai Yoo and Ng Lang Aik; m. Shu Yun, Oct. 9, 1990; children: Yoo Yen Xin, Yoo Yen Yang. BSc, U. Malaya, 1982; MA in Sci., U. Nebr., 1984; PhD, CUNY, 1990. Asst. prof. CUNY, N.Y.C., 1990-92; lectr. Hong Kong U. Sci. and Tech., 1993—. Mem. Am. Phys. Soc., Optical Soc. Am. Achievements include patents in imaging through random and biomedical media, mirrors, a medical diagnostic tool. Home: 158-15 65 Ave Flushing NY 11365 Office: Hong Kong U Sci and Tech-Physics, Clear Water Bay, Kowloon Hong Kong Hong Kong

YOO, KYUNG HAK, agricultural engineer, educator; b. Seoul, Korea, Dec. 19, 1945; came to U.S., 1972; s. Hyung-Gee and Choon-Wha (Park) Y.; m. Yong-Hye Park, Jan. 9, 1978; 1 child, Sylvia Hye-Jin. BS, Seoul Nat. U., 1971; MS, U. Idaho, 1974, PhD, 1979. Asst. prof. U. Idaho, Moscow, 1980-83; asst. prof. Auburn (Ala.) U., 1983-90, assoc. prof. dept. agrl. engring., 1990—; tech. adviser Internat. Aquaculture Ctr., Auburn; tech. adviser, instr. on-farm irrigation systems mgmt., Sri Lanka, 1982; cons., tech. adviser rainwater harvesting and small-scale irrigation for various pvt. vol. orgns., Somalia, 1986; cons., tech. adviser soil erosion control Cath. Relief Svc., Honduras, 1987, presenter, organizer workshop on water harvesting, watershed mgmt., India, 1988, 89, Morocco, 1989; cons., tech. adviser Save the Children Fedn., Thailand, 1988. Author: Hydrology and Water Supply for Pond Aquaculture, 1993; contbr. to profl. publs. Grantee Ala. Dept. of Econ. and Community Affairs, 1988, USDA, 1990. Mem. Am. Soc. Agrl. Engrs., Soil and Water Conservation Soc. Am., Korean Scientists and Engrs. in Am. Assn., Internat. Com. Irrigation and Drainage, U.S. Com. Irrigation and Drainage, Lions, Sigma Xi, Phi Beta Delta, Gamma Sigma Delta. Achievements include development of a soil erosion prediction simulation model. Office: Auburn Univ Agrl Engring Bldg Rm 208 Auburn AL 36849

YOOD, HAROLD STANLEY, internist; b. Plainfield, N.J., Feb. 23, 1920; s. Raphael and Netta (Newcorn) Y.; m. Helen H. Hull, Nov. 8, 1941; children: Pamela, Patricia Yood Herskovitz, Paula Yood Peterson, Andrew H. BA, U. Va., 1940, MD, 1943. Intern Syracuse (N.Y.) U. Med. Ctr., 1943; pvt. practice Plainfield, N.J., 1946-91; med. dir. Cen. Jersey Individual Physicians Assn.; staff dept. medicine Muhlenberg Hosp., 1946—, pres. staff, 1980-86, cons., 1991—; asst. clin. prof. medicine Robert Wood Johnson Med. Sch. Contbr. articles to Jour. Med. Soc. N.J., Communication for Ciba, others. Bd. govs. Muhlenberg Regional Med. Ctr., 1980-86, exec. com.; trustee, v.p. United Way Plainfield/Fanwood, 1975-81; bd. dirs. United Way Union County, 1978-81; pres. Jewish Community Ctr., Plainfield, 1970-71; v.p. Jewish Fedn. Cen. N.J., 1971-73, Cen. N.J. Jewish Home for Aged, 1973-80. Capt. M.C. AUS, 1944-45, ETO. Decorated Purple Heart, Croix de Guerre (France), Croix de Guerre (Belgium). Fellow Am. Coll. Gastroenterology (sr.), Am. Coll. Angiology (ret.), Internat. Coll. Angiology (ret.); mem. AMA, Med. Soc. N.J. (governing coun. hosp. med. sect. 1983-91, chmn. 1988-90, trustee 1989-90), Union County Med. Soc., Plainfield Area Med. Assn., N.Y. Acad. Sci., Lions (life), N.J. Bridge League. Office: CJIPA 1133 Park Ave Plainfield NJ 07060

YOON, JI-WON, virology, immunology and diabetes educator, research administrator; b. Kang-Jin, Chonnam, Korea, Mar. 28, 1939; came to U.S., 1965; s. Baek-In and Duck-Soon (Lee) Y.; m. Chungja Rhim, Aug. 17, 1968; children: John W., James W. MS, U. Conn., 1971, PhD, 1973. Sr. investigator NIH, Bethesda, Md., 1978-84; prof., chief div. virology U. Calgary, Alta., Can., 1984—, prof., assoc. dir. diabetes rsch. ctr., 1985-90, prof., dir. diabetes rsch. ctr., 1990—; mem. edit. bd. Annual Review Advances Present Rsch. Animal Diabetes, 1990—, Diabetes Rsch. Clin. Practice, 1989—; scientific coord. 10th Internat. Workshop on Immunology Diabetes, Jerusalem, 1989-90; sr. investigator NIH, 1976-84. Contbg. author: Current Topics in Microbiology and Immunology, 1990, Autoimmunity and Pathogenesis of Diabetes, 1990; contbr. articles to New England Jour. Medicine, Jour. Virology, Sci., Nature, The Lancet, Jour. Diabetes. Rsch. fellow Sloan Kettering Cancer Inst., 1973-74, Staff fellow, Sr. Staff fellow NIH, 1974-76, 76-78; recipient NIH Dir. award, 1984, Heritage Med. Scientist award, Alberta Heritage Found. Med. Rsch., 1984, Lectrship. award, 3d Asian Symposium Childhood Diabetes, 1989, 8th Annual Meeting Childhood Diabetes, Osaka, Japan, 1990, 9th Korean/Can. Heritage award, 1989. Mem. Am. Soc. Immunologists, Am. Diabetes Assn., Am. Soc. Microbiology, N.Y. Acad. Sci., Soc. Virology, Internat. Diabetes Fedn. Baptist. Achievements include first isolation of diabetogenic virus from patients with recent onset of IDDM; first demonstration of prevention of virus-induced diabetes by vaccination with nondiabetogenic virus in animals; discovery that autoimmune IDDM can be prevented by depletion of macrophages in autoimmune diabetic NOD mice, certain viral glycoproteins (rubella virus E2 glycoprotein) can induce organ-specific autoimmune disease, research on molecular identification of diabetogenic viral gene in animal models, discovery of a nontoxic organic compound with no side effects that completely prevents type I diabetes in NOD mice, discovery that bacterial superantigens such as staphylococcal enterotozins (SEC1, SEC3) can prevent autoimmune type I diabetes by activation of CD4 suppressor T cells in NOD mice. Home: 206 Edgeview Dr NW, Calgary, AB Canada T3A 4W9 Office: Julia McFarlane Diabetes, Rsch Ctr, 3330 Hospital Dr NW, Calgary, AB Canada T2N 4N1

YOON, YONG-JIN, chemistry educator; b. Chochiwon, Chung-Nam, South Korea, May 10, 1950; s. Chang-Lim and Gui-Rae (Kim) Y.; m. Chong-Hee Kim, May 22, 1977; children: Hyeong-Jae, Hyo-Jae. BS in Chemistry, Sung Kyun Kwan U., 1976, PhD in Organic Chemistry, 1983. Instr. dept. chemistry Gyeongsang Nat. U., Chinju, Korea, 1980-82, asst. prof., 1982-86; assoc. prof. Gyeongsang Nat. U., Chinji, Korea, 1986-91, prof., 1991—; vis. prof. dept. med. chemistry U. Mich., Ann Arbor, 1984-85; vice dean acad. affairs Gyeongsang Nat. U., 1992—. Contbr. articles to profl. jours. Mem. Korean Agrochem. Soc., Korean Chem. Soc. (div. organic chemistry and med. chemistry, Chem. Edn. prize 1987), Am. Chem. Soc. (div. med. chemistry). Office: Dept Chemistry, Gyeongsang Nat U, Chinju 660-701, Republic of Korea

YOREK, MARK ANTHONY, biochemist; b. Mpls., Nov. 2, 1953; s. Al Victor and Arna Victoria (Johnson) Y.; m. Gloria Ruth Hinrichs, Sept. 16, 1954; children: Heather, Matthew. BS, Bemidji State Coll., 1976; PhD, U. N.D., 1981. Postdoctoral fellow in biochemistry U. Iowa, Iowa City, 1981-83, from asst. to assoc. rsch. scientist internal medicine, 1983-87, adj. asst. prof., 1987-91, asst. prof., 1991—; dir. cell and membrane biology U. Iowa Diabetes and Endocrinology Rsch. Ctr., Iowa City, 1986—. Contbr. articles to Diabetes, Jour. Neurochem., Metabolism, Biochem. Biophys. Acta. Asst. Cubmaster Boy Scouts Am., Iowa City, 1988. Rsch. grantee Juvenile Diabetes Found., 1985—, NIH, 1986—, VA, 1987—. Mem. AAAS, Am. Soc. Biol. Chemists, Am. Diabetes Assn., Soc. Neurosci. Roman Catholic. Home: 3533 Maplewood Dr NE Iowa City IA 52240 Office: VA Med Ctr 3E17 Iowa City IA 52246

YORK, HERBERT FRANK, physics educator; b. Rochester, N.Y., Nov. 24, 1921; s. Herbert Frank and Nellie Elizabeth (Lang) Y.; m. Sybil Dunford, Sept. 28, 1947; children: David Winters, Rachel, Cynthia. A.B., U. Rochester, 1942, M.S., 1943; Ph.D., U. Calif.-Berkeley, 1949; D.Sc. (hon.), Case Inst. Tech., 1960; LL.D., U. San Diego, 1964, Claremont Grad. Sch., 1974. Physicst Radiation Lab., U. Calif., Berkeley, 1943-58; assoc. dir. Radiation Lab., U. Calif., 1954-58; asst. prof. physics dept. U. Calif., 1951-54, assoc. prof., 1954-59, prof., 1959-61; dir. Lawrence Radiation Lab., Livermore, 1952-58; chief scientist Advanced Research Project Agy., Dept.

Def., 1958; dir. advanced research projects div. Inst. for Def. Analyses, 1958; dir. def. research and engring. Office Sec. Def., 1958-61; chancellor U. Calif.-San Diego, 1961-64, 70-72, prof. physics, 1964—, chmn. dept. physics, 1968-69, dean grad. studies, 1969-70, dir. program on sci., tech. and pub. affairs, 1972-88; dir. Inst. Global Conflict and Cooperation, 1983-88, dir. emeritus, 1988—; amb. Comprehensive Test Ban Negotiations, 1979-81; trustee Aerospace Corp., Inglewood, Calif., 1961-87; mem. Pres.'s Sci. Adv. Com., 1957-58, 64-68, vice chmn., 1965-67; trustee Inst. Def. Analysis, 1963—; gen. adv. com. ACDA, 1962-69; mem. Def. Sci. Bd., 1977-81; spl. rep. of sec. def. at space arms control talks, 1978-79; cons. Stockholm Internat. Peace Research Inst.; researcher in application atomic energy to nat. def., problems of arms control and disarmament, elementary particles. Author: Race to Oblivion, 1970, Arms Control, 1973, The Advisors, 1976, Making Weapons, Talking Peace, 1987, Does Strategic Defense Breed Offense?, 1987, (with S. Lakoff) A Shield in the Sky, 1989; also numerous articles on arms or disarmament.; bd. dirs. Bull Atomic Scientists. Trustee Bishop's Sch., La Jolla, Calif., 1963-65. Recipient E.O. Lawrence award AEC, 1962; Guggenheim fellow, 1972. Fellow Am. Phys. Soc. (Forum on Physics and Society award 1976), Am. Acad. Arts and Scis.; mem. Internat. Acad. Astronautics, Fedn. Am. Scientists (chmn. 1970-71, mem. exec. com. 1969-76, Pub. Svc. award 1992), Phi Beta Kappa, Sigma Xi. Home: 6110 Camino De La Costa La Jolla CA 92037-6520 Office: U Calif-San Diego Mail Code 0518 La Jolla CA 92093

YOSHIDA, KUNIHISA, chemistry educator, researcher; b. Fukuyama, Hiroshima, Japan, Feb. 22, 1940; s. Hiroshi and Masayo (Iwasa) Y.; m. Akiko Yoshida, Nov. 9, 1969; children: Kumi, Nobuhiro, Kayo. MS, Osaka (Japan) U., 1965, PhD, 1968. Assoc. prof. Osaka U., 1968—. Author: Electooxidaition in Organic Chemistry, 1984; contbr. articles to Jour. Am. Chem. Soc. Home: 4-1-23 Fushiodai, Ikeda Osaka 563, Japan Office: Osaka U, 1-1 Machikaneyama, Toyonoka Osaka 560, Japan

YOSHIDA, TOHRU, science and engineering educator; b. Nagoya, Aichi-Ken, Japan, July 17, 1924; s. Osamu and Sadako (Yaeko) Y.; m. Fusa Ishikawa, Nov. 21, 1949; children: Yoko, Kazuo. Metall. Engring., Nagoya (Japan) Imperial U., 1947, DEng Sci. (hon.), 1961. Metall. engr. Nagoya Mcpl. Indsl. Rsch. Inst., 1951-72, vice-dir., 1973-78; prof. sci. and engring. Meijo U., Tenpaku-Ku Nagoya, Japan, 1979—. Author: Method of Fractography, 1970, Design of Case Hardening, 1971, High Impact Strength Brazed Joint, 1982, Brazing of Graphite, 1991. Recipient Wasserman award Am. Welding. Soc., 1983. Mem. Japan Welding Soc. (spl. mem.), Japan Indsl. Engring. Soc. (chmn. tech. com. on qualifications of welding technique 1978—, chmn. brazing com. 1984-88). Avocations: baseball, golf. Home: Ozone 1 6 13, Kita-ku Nagoya 462, Japan

YOSHIHARA, KEITARO, chemistry educator; b. Nagoya, Aichi, Japan, Nov. 21, 1936; s. Susumu and Shizuko (Veno) Y.; m. Naoko Akashi, May 13, 1967; children: Yutaro, Kenjiro, Shinzaburo. BS, U. of Tokyo, 1960, PhD, 1967. Rsch. assoc. U. of Tokyo, 1965-70; rschr. Inst. Phys. & Chem. Rsch., Wako, Japan, 1970-75; prof. and chmn. dept. electronic structure Inst. Molecular Scis., Okazaki, Japan, 1975—; prof. Grad. U. Advanced Studies, Yokohama, Japan, 1988—. Co-editor Photochemistry and Photobiology, 1984-87; mem. editorial bd. Chemical Physics, 1980—, Laser Chemistry, 1989—, Jour. of Luminescence, 1990—, Chem. Physics Letters, 1992—; co-author: 10 books; contbr. articles to various profl. jours. Recipient Progress award Laser Society, 1983. Achievements include mechanistic studies on photochemical isomerization reaction, ultrafast intermaolecular electron transfer reaction, primary mechanism of photosynthetic reaction of higher plants; development of ultrafast spectroscopy, i.e., femtosecond Raman spectroscopy, widely tunable laser. Home: 14-2 Ishigami, 444 Okazaki Japan Office: Inst Molecular Sci, Myodaiji, 444 Okazaki Japan

YOSHIOKA, MASANORI, pharmaceutical sciences educator; b. Iwagi, Japan, Sept. 4, 1941; s. Takeo and Toshiko (Tasaka) Y.; m. Masae Kurahashi, Mar. 23, 1943; children: Shima, Nami. BA, Showa Coll. Pharm. Scis., Tokyo, 1965; M., U. Tokyo, 1967, PhD, 1970. Cert. pharmacist. Assoc. faculty pharm. sci. U. Tokyo, 1970-73, lectr. faculty pharm. sci., 1977-83, assoc. prof. faculty pharm. sci., 1983; prof. faculty pharm. sci. Setsunan U., Osaka, Japan, 1983—; lectr. faculty medicine Hiroshima U., Hiroshima, Japan, 1985-87, 92; rsch. assoc. sch. medicine Yale U., New Haven, 1970-72; vis. scientist cancer inst. Cairo U., 1980-81; vis. prof. Paris U. XI, Orsay, France, 1986-87, Technion-Israel Inst. Tech., Haifa, 1987. Editor Bunseki, 1981-83, Biogenic Amines, 1983-91, Chem. & Pharm. Bull., 1987-89, 92—, Progress in HPLC, vol. 4, 1989. Mem. com. Japan Soc. for the Promotion Sci., Tokyo, 1980—. Mem. Pharm. Soc. Japan, Japan Soc. for Analytical Chemistry, Japan Soc. Clin. Chemistry, Japan Endocrine Soc., Internat. Soc. Toxinology (Frankfurt am Main, Fed. Republic Germany). Avocations: skiing, tennis. Home: 6-15 Otokoyamayoshii, Yawata Kyoto 614, Japan Office: Setsunan U Faculty Pharm, Sci 52 3 Nagaotogecho, Hirakata 573 01, Japan

YOSHISATO, RANDALL ATSUSHI, chemical engineer; b. Berkeley, Calif., Mar. 7, 1957. BS, U. Calif., Davis, 1979; PhD, U. Iowa, 1985. Instr. Dept. Chem. Engring., U. Iowa, Iowa City, 1983-85, vis. asst. prof., 1985-86; adj. asst. prof. dept. chem. engring. U. Iowa, 1986-89; sr. rsch. engr. Dow Chem. Co., Walnut Creek, Calif., 1989-92; project leader Dow Chem. Co., Pittsburg, Calif., 1992—; clear air task force Santa Clara County Mfg. Group, 1992. Contbr. articles to profl. jours. including Separation Sci. and Tech., Jour. of Chem. Tech. Biotech., Recent Advances in Separation Techniques III., Jour. Membrane Sci., chpt. to book. Active Community Presbyn. Ch., Vallejo, Calif., 1989—; deacon St. Andrew Presbyn. Ch., Iowa City, 1985-88, youth leader, 1981-85. Mem. AICE (Iowa sect. sec. 1986-88), Am. Chem. Soc., Am. Soc. for Engring. Edn., Sigma Xi, Omega Chi Epsilon. Achievements include co-invention of continuous rotating annular electrophoresis column; use of boundary value methods to model membrane separation devices; rsch. on the use of electrokinetic dispersion to improve band sharpening in large scale electrophoresis; development of continuous rotating electrophoresis column design to effect large scale electrophoretic separations on a continuous basis. Office: Dow Chemical Loveridge Rd Pittsburg CA 94565

YOSHIZAKI, SHIRO, medicinal chemist; b. Komatsushima, Tokushima, Japan, Feb. 17, 1944; s. Yoshiyuki and Masu (Matsushima) Y.; m. Ayako Ohno, May 31, 1970; children: Masahiko, Tatsuo. B. Engring., Tokushima (Japan) U., 1966, M. Engring., 1968; PharmD, Osaka (Japan) U., 1981. Cert. cons. engr., Japan. Researcher Otsuka Pharm. Co., Ltd., Tokushima, 1968-81, sr. researcher, 1982-89; sr. researcher NKK Corp., Kawasaki, Japan, 1989—. Contbr. articles to profl. jours.; inventor, patent for procaterol, carteolol, 250 other patents. Fellow Am. Chem. Soc.; mem. AAAS, Pharm. Soc. Japan, Japanese Pharmacol. Soc., Kinki Chem. Soc. (Kagaku Gijyutsu prize 1983), N.Y. Acad. Scis. (Charles Darwin Assocs.),. Buddhist. Avocations: fishing, Go game. Office: NKK Corp, 1-1 Minamiwataridacho, Kawasaki-ku Kawasaki 210, Japan

YOSSIF, GEORGE, psychiatrist; b. Bucharest, Romania, Nov. 18, 1939; came to U.S., 1975; s. Yuan and Eugenia (Paun) Y.; m. Valentina Blanaru, Dec. 20, 1967 (div. 1978); 1 child, Anamaria Verona. MD, Faculty of Medicine, Bucharest, 1962; PhD in Neurophysiology cum laude, Scuola Normale Superiore, Pisa, Italy, 1971. Cert. specialist Endocrinology, Romania; diplomate Am. Bd. Psychiatry and Neurology. Intern in pharmacology Faculty of Medicine of Bucharest, 1962-63, intern in endocrinology, 1963-65; researcher div. neuroendocrinology Inst. Endocrinology, Bucharest, 1965-69; fellow in neurophysiology Scuola Normale Superiore and Istituto di Fisiologia, Pisa, Italy, 1969-71; post doctoral fellow dept. physiology Laval U. Faculty of Medicine, Quebec City, Can., 1971-73, asst. prof. neurophysiology, 1973-74; resident in neurology Ottawa Gen. Hosp., Ont., Can., 1974; research fellow Lipid Research Clinic of St. Michael's Hosp. U. Toronto, Ont., Can., 1974-75; rotating intern Cooper Med. Ctr., Camden, N.J., 1975-76; resident in psychiatry Johns Hopkins Hosp., Balt., 1976-79; instr. psychiatry, dir. Liaison-Cons. Service Dept. Psychiatry Howard U., Washington, 1979-80; staff psychiatrist, chief Intermed. Treatment Unit Bryce State Hosp., Tuscaloosa, Ala., 1980-82; med. dir. Stress Ctr. Lloyd Nolnd Hosp., Fairfield, Ala., 1982-83; practice medicine specializing in psychiatry Birmingham, Ala., 1982-92, Georgetown, Del., 1992—; adj. clin. asst. prof. Coll. Community Health Scis. U. Ala., Tuscaloosa, 1980-82. Contbr. articles to profl. jours. and popular press. Med. Research Council grantee, Laval U., 1973. Fellow Am. Soc. Psychical Rsch.; mem. AAAS, IEEE, Am. Math. Soc., Am. Psychiat. Assn., Assn. Am. Physicians and Surgeons, Internat. Assn. Pvt. and Ind. Doctors, Math. Assn. Am., N.Y. Acad. Scis., Internat. Platform Assn., Am. Textbook Com. (bd. dirs.), Am. Biographical Inst., Rsch. Assn. (bd. govs.). Ea. Orthodox. Office: PO Box 827 PO Box 827 Georgetown DE 19947

YOST, WILLIAM ALBERT, psychology educator, hearing researcher; b. Dallas, Sept. 21, 1944; s. William Jacque and Gladys (Funk) Y.; m. Lee Prater, June 15, 1969; children—Kelley Ann, Alyson Leigh. B.A., Colo. Coll., 1966; Ph.D., Ind. U., 1970. Assoc. prof. psychology U. Fla., Gainesville, 1971-77; dir. sensory physiology and perception program NSF, Washington, 1982-83; prof. psychology Loyola U., Chgo., 1977-89, prof. hearing scis., 1977—, dir. Parmly Hearing Inst., 1977—; adj. prof. otolaryngologist Loyola U., Chgo., 1990—; individual expert bio-acoustics Am. Nat. Standards Inst., 1983—; mem. study sect. Nat. Inst. Deafness and Other Communication Disorders, 1990—; mem. hearing, bioacoustics and biomechanics com. Nat. Rsch. Coun., 1992—. Author: (with others) Fundamentals of Hearing, 1977, 1984; editor (with others) New Directions in Hearing Science, 1985, Directional Hearing, 1987, Auditory Processing of Complex Sounds, 1987, Classification of Complex Sounds, 1989, Psychoacoustics, 1993; ad hoc reviewer NSF, Air Force Office Sci. Rsch., Office Naval Rsch., 1981—; contbr. chpts. to books, articles to profl. jours. Pres. Evanston Tennis Assn., Ill., 1984, 90. Grantee NSF, 1974—, NIH, 1975—, AFOSR, 1983-89. Fellow Acoustical Soc. Am. (assoc. editor jour. 1984-91); mem. NAS (exec. com. on hearing bioacoustics, biomechanics 1981-87), Assn. Rsch. in Otolaryngology (sec.-treas. 1984-87, pres.-elect 1987-88, pres. 1988-89), Acoustics Soc. Am. (chair com. psychol. and physiol. acoustics 1990—), Nat. Inst. Deafness and other Communication Disorders (mem. task force, review panel 1990—), Acoustical Soc. Am. (chair tech. com. 1990—).

YOUD, T. LESLIE, civil engineer; b. Spanish Fork, Utah, Apr. 2, 1938; s. Thomas Leslie and Mary (Evans) Y.; m. Denice Porter, June 26, 1962; children: Verlin, Lance, Melinda, Thomas, Emily. BS, Brigham Young U., 1964; PhD, Iowa State U., 1967. Rsch. civil engr. U.S Geological Survey, Menlo Park, Calif., 1967-84; prof. Brigham Young U., Provo, Utah, 1984--; mem. Utah Earthquake Adv. Bd., Salt Lake City, 1990--, Nat. Res. Coun. Com. on Earthquake Engring., Washington, 1986-92, Nat. Res. Com. on Mitigation Engring., Washington, 1992--. REcipient Master Rsch. award Brigham Young U., 1991. Mem. Am. Soc. Civil Engrs., Earthquake Engring. Rsch. Inst., Am. Soc. for Engring. Edn., Internat. Soc. for Soil Mechanics and Fnd. Engring. Mormon. Achievements include development of techniques for mapping earthquake induced liquefaction hazard and techniques for estimating earthquake induced laterial spread displacements; inventor system for coupling accelarmeters into bore hole casings. Home: 1132 E 1010 North Orem UT 84057 Office: Brigham Young U Dept Civil Engring Provo UT 84602

YOUKER, DAVID EUGENE, biology educator; b. Marshalltown, Iowa, July 4, 1938; s. Elroy J. and Ruby G. (Rank) Y.; m. Judith Ann Reece, July 17, 1959 (div. Nov. 1991); children: Jody Ann Miller, Jayna Lynn Youker; m. Carol Elaine Woessner. AS, Marshalltown C.C., 1958; BA, U. No. Iowa, 1960, MA, 1964. Tchr. (jr. high sch., high sch.) Hubbard (Iowa) Community Sch. Dist., 1960-63; biology, chemistry tchr. (high sch.) Freeport (Ill.) Community Sch. Dist., 1964-66; prof. U. No. Iowa, Cedar Falls, summers 1963-68, No. Ill. U., DeKalb, fall 1972, Sauk Valley C.C., Dixon, Ill., 1966—; instr. NSF Rock River Ecol. Project, No. Ill. U., 1972. Author: (lab. manuals) Principles of Biology, Introduction to Botany, 1978—, (insect identification manual) A Beginner's Guide to the Common Insects, 1967—. Named Acad. Yr., NSF, 1963-64. Mem. NEA, AAAS, Nat. Assn. Biology Tchrs., Am. Inst. of Biol. Scis., Ill. Jr. Coll. Biology Tchrs. Assn., Beta Beta Beta. Home: 1310 Park Ln Dixon IL 61021 Office: Sauk Valley C C 173 Ill Rt 2 Dixon IL 61021

YOUKHARIBACHE, PHILIPPE BIJIN, computational chemist, researcher; b. Paris, Sept. 16, 1955; came to U.S., 1986; s. Amedée Mehdi and Alberte Marie (Baldelli) Y. MS in Phys. Chemistry, U. Paris, Orsay, 1976; PhD in Phys. Sci., U. Paris, 1986. Researcher Ecole Poly., Palaiseau, France, 1978-86; postdoctoral fellow Columbia U., N.Y.C., 1986-87; mgr. Polygen (Europe) Ltd., Paris, 1987-88; dir. product planning Biosym Techs. Inc., San Diego, 1988-90, project dir., 1991—. Mem. AAAS, IEEE, Assn. for Computing Machinery, Biophys. Soc., Protein Soc., Molecular Graphics Soc., Am. Chem. Soc. Achievements include research in vibrational spectroscopy, determination of molecular force fields, protein engineering, membrane protein channels modeling, molecular structure and dynamics analysis, computer assisted drug design, molecular modeling software development. Office: Biosym Techs Inc 9685 Scranton Rd San Diego CA 92121-2777

YOULL, PETER JEROME, environmental chemist; b. Olwein, Iowa, May 13, 1961; s. Jerome V. and Jenny S. (Hertz) Y.; m. Rebecca Ann Price, Dec. 13, 1986; children: Lauren, Paige, Sarah. BS in Chemistry, Ky. Wesleyan Coll., 1984; postgrad., Bowling Green State U., 1984-85. Sr. chemist Aqua Tech. Environ. Labs., Marion, Ohio, 1985—; bd. dirs. Macola Computers, Marion, BMA Tech., Marion. Bd. dirs. Am. Heart Assn. Marion County, 1991—; deacon Emanuel Luth. Ch., Marion, 1991, 92. Office: Aqua Tech Environ Labs 181 S Main St Marion OH 43302

YOUNG, ALAN M., biologist, educator; b. Brockton, Mass., Aug. 28, 1952; s. Merle A. and Elia I. (Alessandri) Y. MS in Marine Sci., U. S.C., 1975, PhD in Marine Sci., 1978. From asst. prof. to assoc. prof. biology Nasson Coll., Springvale, Maine, 1978-83; asst. lab. mgr. Demers Lab., Springvale, 1984-87; vis. assoc. prof. biology Bates Coll., Lewiston, Maine, 1987-88, Hamilton Coll., Clinton, N.Y., 1988-89; vis. asst. prof. marine sci. Stockton State Coll., Pomona, N.J., 1990-91; asst. prof. biology Salem (Mass.) State Coll., 1991—; cons. Wells (Maine) Nat. Estuarine Res.; bd. dirs. Friends of Salem Woods, Inc., 1992—. Contbr. papers to sci. publs. Recipient predoctoral award in marine biology Belle W. Baruch Found., 1977, 78; rsch. award Shell, 1979-80, 82-83; grantee Sigma Xi, 1977; Vollmer fellow Bermuda Biol. Sta., 1977. Mem. Maine Assn. Site Evaluators, Crustacean Soc., New Eng. Estuarine Rsch. Soc., Sigma Xi. Home: 12 Shillaber St Peabody MA 01960 Office: Salem State Coll Dept Biology 352 Lafayette St Salem MA 01970

YOUNG, ALLEN MARCUS, museum curator of zoology, educator, naturalist, consultant, writer; b. Ossining, N.Y., Feb. 23, 1942; s. George Marcus and Margaret Mary (Murphy) Y.; m. Mary Joan Treis, Feb. 17, 1978 (div. Oct. 1990). BA, SUNY, New Paltz, 1964; PhD, U. Chgo., 1968. Postdoctoral fellow Orgn. for Tropical Studies, San Jose, Costa Rica, 1968-70; asst. prof. Lawrence U., Appleton, Wis., 1970-75; curator invertabrate zoology Milw. Pub. Mus., 1975-89, curator zoology, 1989—, v.p. for collections & rsch., 1993—; adj. prof. U. Wis., Milw., 1976—; cons. Chocolate Mfrs., Hersey, Pa., 1980—; freelance writer mags. and newspapers, 1978—. Author: Population Biology of Tropical Insects, 1982, Sarapiqui Chronicle: A Naturalist in Costa Rica, 1991; NSF rain forest gallery exhibitor Milw. Pub. Mus., 1984-89; contbr. over 250 articles to profl. jours. Recipient Rsch. grant NSF, 1972-75, Am. Cocoa Rsch. Inst., 1978—, Am. Assn. Mus. award for outstanding new exhibit gallery, 1988. Mem. AAAS, Assn. for Tropical Biology, Entomol. Soc. Washington, N.Y. Entomol. Soc. Achievements include research in rearing cacao-pollinating midges in higher numbers in Ctrl. Am. cacao plantations, with inexpensive means for small farmers. Office: Milw Pub Mus 800 W Wells St Milwaukee WI 53233

YOUNG, CORNELIUS BRYANT, JR. (C. B. YOUNG), electronics engineer; b. Sardis, Miss., Sept. 2, 1926; s. Cornelius Bryant Sr. and Ethel (Dorr) Y.; m. Marguerite Esther Grosso, May 27, 1950; children: Mark Joseph, Roy Neil, Annette Georgette, Neil Bryant. BEE, Ga. Inst. Tech., 1948; MEE, Bklyn. Poly. Inst., 1954; MS in Computer Sci., Stevens Inst. Tech., 1981. From jr. engr. to dir. applications engring. Western Union, N.Y.C., Upper Saddle River, N.J., 1948-84; sr. engring. specialist ITT Fed. Electric Corp., Paramus, N.J., 1985-89; mgr. commns. and comptrs. The BARC Group, Totawa, N.J., 1990-92; cons. C Squared Systems Engring., Ramsey, N.J., 1990—. Contbr. articles to profl. jours.; inventor microwave lens. Pres. Ramsey (N.J.) Ambulance Corps., 1987-89, 90-91; chmn. Ramapo Valley ARC, Ramsey, 1987-88; chmn. troop 31, Boy Scouts of Am., Ramsey, 1968-76; vol. emergency room Valley Hosp., Ridgewood, N.J., 1980—; mem. Ramsey Bd. Health, 1987—, Soc. of Valley Hosp., 1990—; vol. FISH Network, N.W. Bergen County, 1970—. Lt. (j.g.) USNR, 1944-46, with res. 46-64, ret., 1964. Recipient Citizen of Yr. award Troop 31 Boy Scouts of Am., 1987. Mem. IEEE (life, sr.), Assn. for Computing Machinery, Cons. Networks of N.Y. and N.J., Sigma Xi (assoc.). Republican. Baptist. Avocation: emergency medical services. Home: 68 Deer Trl Ramsey NJ 07446-2110

YOUNG, CRAIG STEVEN, structural engineer; b. Richlands, Va., Apr. 9, 1965; s. Clinton Barnes and Gertrude (Gillman) Y. AS, S.W. Va. C.C., 1987; BSE, Va. Poly. Inst., 1989, MSE, 1990. Inspector, coop. Dept. Transp. State of Va., Bristol, 1988; engr., computer programmer Structural Engrs., Inc., Radford, Va., 1989; rsch. asst. Va. Poly. Inst., Blacksburg, 1989-90; structural engr. Dewberry & Davis, Marion, Va., 1991—. Author: Effects of End Restraint on Steel Deck, 1990; contbr. articles to profl. jours. Recipient John Gundel award for Rsch. Excellence Steel Deck Inst., 1990, Coll. Engring. award Va. Poly. Inst., 1988. Mem. ASCE (assoc.), Am. Concrete Inst., Tau Beta Pi. Achievements include rsch. results which produced a change in the governing specifications/code for analyzing steel deck floor systems, thus reducing initial design cost to steel deck manufacturers. Home: Washington Bldg # 5 101 Magnolia Marion VA 24354 Office: Dewberry & Davis 626 S Main St Marion VA 24354

YOUNG, DANIEL LEE, environmental scientist; b. Cin., Oct. 3, 1962; s. Paul L. and Sophie Wanda (Cole) Y.; m. Brenda Jane Coy, Oct. 2, 1982; children: Daniel L. II, Christina Hope. BS in Earth Sci., Union Inst. U., 1990; MA in Sci. Edn., Antioch U., 1992. Machinist Rheinstahl Tool and Die, Cin., 1978-82; color technologist Sun Chem. Corp., Cin., 1986-90; environ. scientist Westinghouse, Cin., 1990-92; supr. chemistry and indsl. hygiene tng. Flour Daniel, Cin., 1992—; adj. faculty U. Cin., 1991; total quality adv., sci. educator Dept. Energy, Cin., 1990—. With U.S. Army, 1983-86. Recipient George Westinghouse Signature award Westinghouse, 1990. Mem. AAAS, U.S. Cavalry Assn., ASCD, ACGIH, PADI. Republican. Nazarene. Home: 4074 McLean Dr Cincinnati OH 45255 Office: Flour Daniels PO Box 398709 Cincinnati OH 45255

YOUNG, DONALD ROY, pharmacist; b. Belfast, Pa., Oct. 7, 1935; s. Roy Clifford and Gladys Nicholas (Ealer) Y.; m. Joyce Anne Waldridge; children: Donald, Lynda, David. BS in Pharmacy, U. Md., Balt., 1957. Pharmacist Brookside Rhodes Drugs Co., Newark, Del., 1956-57; pharmacist, mgr. Newark Rhodes Drugs Co., 1957-64; pharmacist, owner, mgr. Hudson's Pharmacy, St. Michaels, Md., 1964—; bd. dirs., officer St. Michaels Bank; treas. Calvert Drug Co. Balt., 1970-76. Pres. St. Michaels Improvement Corp., 1966—. Mem. Nat. Assn. Retail Druggists, Ea. Shore Pharm. Assn. (pres. 1967-68, 82-86), Talbot County C. of C. (Outstanding Small Bus. Man of Yr. award 1989), St. Michaels Bus. Assn. (pres.), U. Md. Sch. Pharmacy Alumni Assn. (life), Isaac Walton League, Miles River Yacht Club, Rotary (pres. St. Michaels 1970, Most Outstanding Mem. award 1988), Elks, Masons (32 degree, master 1969-70, apptd. jr. grand deacon of Grand Line 1989-90). Republican. Methodist. Avocations: golf, boating, fishing, travel, hiking. Home: 8118 Tricefields Rd PO Box 130 Saint Michaels MD 21663 Office: Hudson's Pharmacy PO Box 130 Saint Michaels MD 21663

YOUNG, DONALD STIRLING, clinical pathology educator; b. Belfast, N. Ireland, Dec. 17, 1933; s. John Stirling and Ruth Muir (Whipple) Y.; m. Silja Meret; children: Gordon, Robert, Peter. MB, ChB, U. Aberdeen, Scotland, 1957; PhD in Chem. Pathology, U. London, 1962. Terminable lectr. materia medica U. Aberdeen, 1958-59; fellow Postgrad. Med. Sch., U. London, 1959-62, registrar, 1962-64; vis. scientist NIH, Bethesda, Md., 1965-66; chief clin. chemistry service NIH, 1966-77; head clin. chemistry sect. Mayo Clinic, Rochester, Minn., 1977-84; prof. pathology and lab. medicine U. Pa., 1984—; dir. William Pepper Lab. Hosp. of U. Pa., 1984—; past bd. dirs. Nat. Com. Clin. Lab. Standards. Co-editor: Drug Interference and Drug Metabolism in Clinical Chemistry, 1976, Clinician and Chemist, 1979, Chemical Diagnosis of Disease, 1979, Drug Measurement and Drug Effects in Laboratory Health Science, 1980, Interpretation of Clinical Laboratory Tests, 1985, Effects of Drugs on Clinical Laboratory Tests, 1990. Recipient Dir.'s award NIH, 1977, Gerard B. Lambert award, 1974-75, MDS Health Group award Can. Soc. Clin. Chemists, 1978; Roman lectr. Australian Assn. Clin. Biochemists, 1979; Jendrassik award Hungarian Soc. Clin. Pathologists, 1985, ATB award Italian Soc. Clin. Biochemistry, 1987. Mem. Am. Assn. Clin. Chemistry (J.H. Roe award Capital sect. 1973, Bernard Gerulat award N.J. sect. 1977, Ames award 1977, Van Slyke award N.Y. met. sect. 1985, past pres.), Internat. Fedn. Clin. Chemists (past pres.), Acad. Clin. Lab. Physicians and Scientists (past exec. com.), Assn. Clin. Biochemists (Ciba-Corning lectr. 1985). Achievements include research in clinical chemistry; development and application of high resolution analytical techniques in the clinical laboratory; optimized use of the clinical laboratory. Home: 9504 Cable Dr Kensington MD 20895-3622 Office: U Pa William Pepper Lab 3400 Spruce St Philadelphia PA 19104-4220*

YOUNG, EDWIN HAROLD, chemical/metallurgical engineering educator; b. Detroit, Nov. 4, 1918; s. William George and Alice Pearl (Hicks) Y.; m. Ida Signe Soma, June 25, 1944; children—David Harold, Barbara Ellen. B.S. in Chem. Engring, U. Detroit, 1942; M.S. in Chem. Engring, U. Mich., 1949, M.S. in Metall. Engring, 1952. Chem. engr. Wright Air Devel. Center, Dayton, Ohio, 1942-43; instr. U. Mich., Ann Arbor, 1946-52; asst. prof. U. Mich., 1952-56, asso. prof., 1956-59, prof. chem. and metall. engring., 1959-89, prof. emeritus chem. and metall. engring., 1989—; Mem. Mich. Bd. Registration for Profl. Engrs., 1963-78, chmn., 1969-70, 72-73, 75-76; mem. Mich. Bd. Registration for Architects, 1963-78. Author: (with L.E. Brownell) Process Equipment Design, 1959; contbr. articles to profl. jours. Dist. commr. Boy Scouts Am., 1961-64; mem. Wolverine council, 1965-69. Served with USNR, 1943-46; to capt. Res. ret. 1973. Fellow ASME, ASHRAE, Am. Inst. Chemists, Am. Inst. Chem. Engrs. (Donald Q Kern award 1979), Engring. Soc. Detroit; mem. Am. Chem. Soc., Am. Soc. Engring. Edn., Nat. Soc. Profl. Engrs. (pres. 1968-69, award 1977), Mich. Soc. Profl. Engrs. (pres. 1962-63, Engr. of Year award 1976), Mich. Assn. of Professions (pres. 1966, Distinguished award 1970), Nat. Council Engring. Examiners, Naval Res. Assn., Res. Officers Assn., Sigma Xi, Tau Beta Pi, Phi Kappa Phi, Phi Lambda Upsilon, Alpha Chi Sigma. Republican. Baptist. Home: 609 Dartmoor Rd Ann Arbor MI 48103-4513

YOUNG, FREDERIC HISGIN, infosystems executive, data processing consultant; b. Boston, Sept. 7, 1936; s. Ralph Randel Jr. and Wilhelmina Amalia (Imberger) Y.; m. Carol Joan Costello, Sept. 7, 1963 (div. Dec. 1971); children: Tracy Jean, Jodi Ann; m. Kathleen Paula Thorne, Dec. 1, 1984. BBA, U. Mass., 1961; JD, Suffolk U., 1966. Mgr. systems and programs Matrix Corp., Burlington, Mass., 1968-69; sr. cons. Programming Dimensions, Inc., Burlington, 1969-70; bus. mgr. JTB Rehab. Ctr., North Reding, Mass., 1970-75; regional bus. mgr. Mass. Dept. Mental Health, Waltham, 1975-78; dir. personnel mgmt. Mass. Dept. Mental Health, Boston, 1978-81; prin. cons. Lafayette Assocs., Chelsea, Mass., 1980-81; asst. regional dir. Corp. for Applied Systems, Indpls., 1982-84; v.p. cons. svcs. HAS, Inc., Carmel, Ind., 1984-88; v.p. info. sytems Ind. Fed. Credit Union, Anderson, Ind., 1988—. With USN, 1954-56. Republican. Avocation: woodworking. Home: 20447 State Rd 37 N Noblesville IN 46060-9228

YOUNG, GEORGE HANSEN, chemist; b. Columbus, Ohio, Feb. 19, 1962; s. Warren Melvin and Sandra Jean (Shallenberger) Y.; m. Mary Jennifer Justo Torres, June 28, 1986; children: Marisa, Olevia. BS, Youngstown State U., 1983; PhD, Ohio State U., 1989. Grad. teaching asst. Ohio State Univ., Columbus, 1983-89; advanced rsch. and devel. chemist B.F. Goodrich Co., Brecksville, Ohio, 1989—; employment recruiter B.F. Goodrich Co., Brecksville, 1989-92; sr. R&D engr. B.F. Goodrich Co., LaPorte, Tex., 1992—. Named Outstanding Sr. Chemistry Student Am. Inst. Chemists, 1983. Mem. Am. Chem. Soc. (Akron sect.), Youngstown State Univ. Alumni Assn., Ohio State Univ. Alumni Assn. Avocations: basketball, racquetball, canoeing. Home: 124 Harwood Dr League City TX 77573 Office: B F Goodrich Co 2400 Miller Cut Off Rd La Porte TX 77571-9799

YOUNG, HELEN JAMIESON, biologist, educator; b. Denver, Dec. 27, 1955; m. Donald Arthur Stratton; 1 child, Margaret. BA, Washington U., St. Louis, 1977; PhD, SUNY, Stony Brook, 1986. Postdoctoral rsch. asst. U. Calif., Davis, 1986-88; vis. rsch. assoc. Duke U., Durham, N.C., 1988-90; asst. prof. biology Barnard Coll., N.Y.C., 1990—. Contbr. articles to profl. publs. Grantee Jessie Smith Noyes Found., 1984-86, Nat. Geographic Soc.,

1988-90, NSF, 1991-92. Mem. Assn. Tropical Biology, Soc. for Study of Evolution, Bot. Soc. Am., Ecol. Soc. Am. Achievements include research in plant reproduction, male reproductive success in plants, the role of floral odor in pollination. Office: Barnard Coll 3009 Broadway New York NY 10027

YOUNG, JAMES HERBERT, agricultural engineer; b. LaFayette, Ky., Mar. 19, 1941; s. James Monroe and Mary Holmes (Allen) Y.; m. Adrienne Lou Scott, July 20, 1963; children: Leigh Ann Young Sanders, James Scott Young. BS in Agrl. Engring., U. Ky., 1962, MS in Agrl. Engring., 1964; PhD, Okla. State U., 1966. Asst. prof. N.C. State U., Raleigh, 1966-70, assoc. prof., 1970-76, prof., 1976—. Fellow Am. Soc. Agrl. Engrs. (pubs. dir. 1988-90, electric power and processing dir. 1982-84); mem. ASHRAE, Am. Peanut Rsch. and Edn. Soc., Am. Soc. for Engring. Edn., Alpha Epsilon (Outstanding Alumnus 1986). Republican. Baptist. Home: 4104 Pepperton Dr Raleigh NC 27606-1734 Office: NC State U Box 7625 Raleigh NC 27695-7625

YOUNG, JERRY WESLEY, animal nutrition educator; b. Mulberry, Tenn., Aug. 19, 1934; s. Rufus William and Annie Jewell (Sweeney) Y.; m. Charlotte Sullenger, July 8, 1959; children: David, Jeretha. BS, Berry Coll., 1957; MS, N.C. State U., 1959, PhD, 1963. Asst. prof. Iowa State U., Ames, 1965-70, assoc. prof., 1970-74, prof. in animal nutrition, 1974—. Contbr. articles to profl. jours. Postdoctoral fellow NIH, 1963-65. Mem. Am. Dairy Sci. Assn. (Outstanding Dairy Nutrition Rsch. 1987), Am. Inst. Nutrition, Am. Soc. Animal Sci., Sigma Xi, Phi Kappa Phi, Gamma Sigma Delta. Baptist. Office: Iowa State U 313 Kildee Hall Ames IA 50010

YOUNG, JOHN ALAN, electronics company executive; b. Nampa, Idaho, Apr. 24, 1932; s. Lloyd Arthur and Karen Eliza (Miller) Y.; m. Rosemary Murray, Aug. 1, 1954; children: Gregory, Peter, Diana. B.S. in Elec. Engring, Oreg. State U., 1953; M.B.A., Stanford U., 1958. Various mktg. and finance positions Hewlett Packard Co. Inc., Palo Alto, Calif., 1958-63, gen. mgr. microwave div., 1963-68, v.p. electronic products group, 1968-74, exec. v.p., 1974-77, chief oper. officer, 1977-84, pres., 1977-92, chief exec. officer, 1978-92; ret., 1992; bd. dirs. Wells Fargo Bank, Wells Fargo and Co., Chevron Corp. Chmn. ann. fund Stanford U., 1966-73, nat. chmn. corp. gifts, 1973-77, mem. adv. coun. Grad. Sch. Bus., 1967-73, 75-80, univ. trustee, 1977-87; bd. dirs. Mid-Peninsula Urban Coalition, 1971-80, co-chmn., 1975-80; chmn. President's Commn. on Indsl. Competitiveness, 1983-85, Nat. Jr. Achievement, 1983-84; pres. Found. for Malcolm Baldrige Nat. Quality Award; mem. Adv. Com. on Trade Policy and Negotiations. With USAF, 1954-56. Mem. Am. Electronics Assn., Bus. Roundtable (founder, chmn. coun. on competitiveness 1986), Bus. Coun., Pacific Union Club, Palo Alto Club.

YOUNG, JOHN EDWARD, mathematics educator, consultant, writer; b. Wellington, Kans., Feb. 25, 1933; s. Ernest Edgar and Janie Ruth (Vaughn) Y.; m. Rachael Helen Pennington, Aug. 31, 1952; children: Rebecca, Lawrence, Eric, Melinda, Bryan, Kirsten, Sarah, Lucinda. BA, Ottawa (Kans.) U., 1954; MEd, Ohio U., 1957; MA, Emporia (Kans.) State U., 1961; PhD, Kans. State U., 1974. Tchr. high sch. math. and sci. Conway Springs (Kans.) Pub. Schs., 1954-56; tchr. jr. high sch. Kansas City (Kans.) Pub. Schs., 1957-58; tchr. high sch. math. Labette County High Sch., Altamont, Kans., 1958-61; instr. Kent (Ohio) State U., 1962-68; from asst. to full prof. S.E. Mo. State U., Cape Girardeau, 1968—; cons., writer Kindergarten-6th grades sci. and math. grant S.E. Mo. State U., 1985-90; title II cons. in elem. sch. math. numerous sch. dist. in S.E. Mo., 1986-93. Co-author: (textbook) Foundations of Math, 1968, 74, 80, Geometry for Elementary Teachers, 1971, The Mathematics of Business, 1974; co-author: (resource book) Mathematical Activities, 1989, 91. Mem. Gov.'s Adv. Coun. for Mentally Retarded and Devel. Disabled, Jefferson City, Mo., 1984-87; bd. dirs. Cape Girardeau County Assn. for Retarded Citizens, 1979-93. Recipient Oustanding Svc. award Easter Seal Soc., 1985, Jackson County (Mo.) Bd. for Devel. Disabled, 1986, Devel. Disabilities Awareness award Mo. Planning Coun., 1987. Mem. Math. Assn. Am., Nat. Coun. Tchrs. Math., Mo. Coun. Tchrs. Math. (bd. dirs. 1984-86), Mo. Math. Assn. for Advancement Tchr. Tng. (pres. 1983-85). Baptist. Avocations: chess, camping, hiking, gardening. Office: SE Mo State U Math Dept 1 University Plz Cape Girardeau MO 63701-4799

YOUNG, JOHN JACOB, JR., aerospace engineer; b. Moreland, Ga., May 29, 1962; s. John Jacob and Gloria Jewell (Wheelus) Y.; m. Barbara Joan Schleihauf, Oct. 22, 1988; 1 child, Nathan Jacob. BS in Aerospace Engring., Ga. Tech., 1985; MS in Aeronautics and Astronautics, Stanford U., 1987. Engr. in Tng., Ga. Senate intern U.S. Senator Sam Nunn Intern Program, Washington, 1984; engring. coop. Gen. Dynamics Ft. Worth Div., 1980-84; mem. tech. staff The BDM Corp., Huntsville, Ala., 1985-86, Rockwell Internat. Missile Systems Div., Duluth, Ga., 1987-88, Sandia Nat. Labs., Albuquerque, N.Mex., 1988—; AIAA Congl. fellow profl. staff U.S. Senate Appropriations Com., Subcommittee on Def., Washington, 1991—. Organizer, coach BDM Co-ed. Softball team, Huntsville, Ala., 1986; big brother Big Bros., Big Sisters, Huntsville, Ala., 1986; tutor El Dorado High Sch., Albuquerque, N.Mex., 1989-90. Recipient AIAA Gen. Dynamics scholarship, 1984; fellow Coll. Engring, Stanford U., 1985; named Outstanding Young Engr., AIAA Atlanta Section, 1988. Mem. Am. Inst. of Aeronautics and Astronautics, Briaerean Soc., Phi Kappa Phi, Tau Beta Pi, Sigma Gamma Tau, Phi Eta Sigma. Presbyterian. Office: Senate Def Appropriations 119 Dirksen Senate Off Bldg Washington DC 20510

YOUNG, JOHN WATTS, astronaut; b. San Francisco, Sept. 24, 1930; s. William H. Y.; m. Susy Feldman; children by previous marriage: Sandra, John. BS in Aero. Engring, Ga. Inst. Tech., 1952; D Applied Sci. (hon.), Fla. Technol. U., 1970; LLD (hon.), Western State U., 1969; DSc (hon.), U. S.C., 1981, Brown U., 1983. Joined USN, 1952, advanced through grades to capt.; test pilot, program mgr. F4 weapons systems projects, 1959-62; then maintenance officer Fighter Squadron 143, Naval Air Sta., Miramar, Calif., chief astronaut office Flight Ops. Directorate, 1975-87, spl. asst. dir. JSC for engring. ops., safety, 1987—, comdr. 54-hour, 36-orbit 1st flight of Shuttle Space, 1981, and 10 day orbital shuttle 1st flight Space Lab., 1983. Decorated DFC (3), D.S.M. (2); recipient NASA Disting. Svc. medal (3), NASA Exceptional Svc. medal (2), NASA Engring. Achievement medal, 1988, NASA Outstanding Leadership medal 1992; Congl. Space medal of honor, 1981; named Disting. Young Alumnus, Ga. Tech. Inst., 1965, Disting. Alumni Svc. award, 1972; NASA Outstanding Leadership medal, 1992; named to Nat. Aviation Hall of Fame, 1988. Fellow Am. Astronautical Soc. (Flight Achievement award 1972, 81, 83), Soc. Exptl. Test Pilots (Iven Kincheloe award 1972, 81), AIAA (Haley Astronautics award 1973, 82, 84); mem. Sigma Chi. Astronaut NASA, made 1st two-man 3 orbit flight, Gemini 3, Mar. 1965, Gemini 10 3 day flight, 1966, Apollo 10 8-day flight lunar landing dress rehearsal, 1969, Apollo 16 11 day lunar landing and surface exploration, 1972; dir. space shuttle br., astronaut office, 1973-75. Office: NASA Johnson Space Ctr Houston TX 77058

YOUNG, KATHLEEN, clinical psychologist; b. Morristown, N.J., July 11, 1961; d. Lawrence Henry and Sue Ann Y. BA, Conn. Coll., 1983; D of Psychology, Ill. Sch. Profl. Psychology, 1990. Lic. psychologist, Ill., Ind. Head counseling svcs. Greentree Shelter, Bethesda, Md., 1984-85; mental health worker Northwestern Meml. Hosp. Emergency Housing Program, Chgo., 1986-88; extern Northwestern Meml. Hosp. Eating Disorders Program, Chgo., 1987-88; psychology intern U. Ariz. Health Scis. Ctr., Tuscon, 1988-89; therapist Madison Ctr., South Bend, Ind., 1989-90, staff psychologist, 1990-92; psychologist Lake-Cook Psychol. Counseling Assocs., Arlington Hts., Ill., 1992—; clin. psychologist pvt. practice, Chgo., 1992—; affiliate med. staff St. Joseph Hosp., Mishawaka, Ind., 1991-92. Mem. Am. Psychol. Assn., NOW, Internat. Soc. for Study of Multiple Personality and Dissociation, Div. Psychology Women. Office: 2532 N Lincoln Ste 116 Chicago IL 60614

YOUNG, KEITH PRESTON, geologist; b. Buffalo, Wyo., Aug. 18, 1918; married. BA, U. Wyo., 1940, MA, 1942; PhD, U. Wis., 1948. With U. Tex., Austin, 1948—, from asst. prof. to assoc. prof., 1948-58, prof. geology, 1958—, Third Mr. and Mrs. Charles E. Yager prof., 1987—. Mem. Paleontol. Soc., Geol. Soc. Am., Soc. Econ. Paleontology & Minerals, Am. Assn. Petroleum Geology, Am. Inst. Profl. Geologists, Sigma Xi. Office: The Univ of Tex at Austin Dept of Geol Scis Austin TX 78712

YOUNG, LAI-SANG, mathematician educator. Prof. math. UCLA. Recipient Ruth Lyttle Math. prize Am. Math. Soc., 1993. Office: U of California Dept of Mathematics 405 Hilgard Ave Los Angeles CA 90024*

YOUNG, LAWRENCE EUGENE, internist; b. Waterville, Ohio, Mar. 18, 1913; s. William Edward and Ruth Elizabeth (Farnsworth) Y.; m. Martha Annette Briggs, May 25, 1940; children: Carolyn, Beverly, Marjorie, Anderson. BA, Ohio Wesleyan U., 1935, DSc (hon.), 1967; MD, U. Rochester, 1939; DSc (hon.), Med. Coll. Ohio, Toledo, 1977. Diplomate Am. Bd. Internal Medicine. Intern medicine Strong Meml. Hosp., Rochester, N.Y., 1939-40, asst. resident medicine, chief resident, 1940-43, asst. physician, assoc. physician, physician-in-chief, 1943-44, 46-74; asst. bacteriology, asst. medicine Johns Hopkins Sch. Hygiene & Pub. Health Johns Hopkins Hosp., Balt., 1941-42; instr. medicine U. Rochester Sch. Medicine, 1942-44, 46-47, asst. prof., 1948-49, assoc. prof., 1950-57, Dewey prof., chmn. medicine, 1957-74, dir. assoc. hosps. program in internal medicine, 1974-78, prof. medicine emeritus, 1978—; mem. com. on blood NRC, Washington, 1950-53; mem. hematology study sect. NIH, Bethesda, Md., 1953-57; disting. vis. prof. internal medicine U. South Fla. Coll. Medicine, Tampa, 1978—; bd. dirs. Robert Wood Johnson Clin. Scholars Program, Princeton, N.J., 1978-81. Assoc. editor Blood, Jour. Hematology, 1950-56, Am. Jour. Medicine, 1961-74; mem. editorial bd. Jour. Clin. Investigation, 1955-59; contbr. articles to med. jours., chpts. to textbooks. Lt. M.C. USNR, 1944-46. Recipient Alumni Citation, U. Rochester, 1978, Albert David Kaiser award Rochester Acad. Medicine, 1978. Fellow Internat. Soc. Hematology; mem. ACP (life, master, gov. upstate N.Y. 1958-60, regent 1972-78, v.p. 1976-77), Am. Soc. Hematology, Am. Soc. Clin. Investigation, Am. Fedn. Clin. Rsch., Assn. Profs. Medicine (pres. 1966-67, Robert H. Williams Disting. Chmn. Medicine award 1976), Assn. Am. Physicians (pres. 1973-74), Ohio Wesleyan Alumni Assn. (disting. achievement award 1990). Republican. Methodist. Home: 12401 N 22d St A-206 Tampa FL 33612 Office: U South Fla Med Ctr PO Box 19 12901 Bruce B Downs Blvd Tampa FL 33612

YOUNG, MARGARET ELISABETH, physicist; b. London, Dec. 21, 1922; arrived in Can., 1956; d. William and Elizabeth (Hughson) Carr; m. Lawrence Young. BS, U. London, 1943, MS, 1949. Lectr. in physics Royal Free Hosp. Sch. of Medicine/U. London, 1943-49; physicist Med. Rsch. Coun., Harwell, Eng., 1949-51, Ottawa (Can.) Civic Hosp., 1951-52, Charing Cross Hosp., London, 1953-55, British Columbia Cancer Agy., Vancouver, Can., 1956-85; retired. Author: Radiological Physics, 1957, 3d edit. 1983. Mem. Hosp. Physicists Assn., Am. Assn. Physicists in Medicine, Canadian Orgn. Med. Physicists. Home: 3226 W 51st Ave, Vancouver, BC Canada V6N 3V7

YOUNG, MICHAEL WARREN, geneticist, educator; b. Miami, Fla., Mar. 28, 1949; s. Lloyd George and Mildred (Tillery) Y.; m. Laurel Ann Eckhardt, Dec. 27, 1978; children: Natalie, Arissa. BA, U. Tex., 1971, PhD, 1975. NIH postdoctoral fellow Med. Sch., Stanford (Calif.) U., 1975-77; asst. prof. genetics The Rockefeller U., N.Y.C., 1978-83, assoc. prof., 1984-88, prof., 1988—; investigator Howard Hughes Med. Inst., N.Y.C., 1987—; mem. adv. panel on genetic biology NSF, Washington, 1983-87; spl. advisor Am. Cancer Soc., N.Y.C., 1985—; spl. reviewer genetics study sect. NIH, Bethesda, Md., 1990—, mem. cell biology study sect., 1993—. Contbr. articles to profl. jours. Meyer Found. fellow, N.Y.C., 1978-83. Fellow N.Y. Soc. Fellows; mem. AAAS, Genetics Soc. Am., Am. Soc. Microbiologists, N.Y. Acad. Scis., Harvey Soc. Achievements include research on transposable DNA elements, molecular genetics of nerve and muscle development, biological clocks. Home: 16 Pinehill Dr Saddle River NJ 07458-1915 Office: The Rockefeller Univ 1230 York Ave New York NY 10021-6341

YOUNG, MONICA DAWN, electrical engineer; b. Mt. Vernon, Ill., Jan. 5, 1959; d. George Otis and Mary Irene (Minor) Smith; m. Paul Michael Young, Feb. 12, 1977. BS in Elec. Engring., So. Ill. U., 1982, MBA, 1992. Cert. quality auditor; cert. quality engr. Dir., engr. Logistics Enterprises Corp., Mt. Vernon, Ill., 1982-86; instr. Rend Lake Coll., Ina, Ill., 1983—; engr. Gen. Tire, Mt. Vernon, 1986—. Advisor Jr. Achievement, Mt. Vernon, 1987. Mem. IEEE, NAFE, Am. Soc. Quality Control, Assn. for Quailty and Participation. Home: 620 S 34th St Mount Vernon IL 62864 Office: Gen Tire Hwy 142 S Mount Vernon IL 62864

YOUNG, NANCY LEE, medical educator; b. Seattle, July 28, 1963; d. James Clarence and Margaret Rosemary (Brischle) Y.; m. David Christopher Belcher, Dec. 29, 1987; 1 child, David Spencer. BS, Seattle Pacific U., 1986; PhD, Drexel U., 1990. Student tchr. Drexel U., Phila., 1986-89; predoctoral student Fred Hutchinson Cancer Rsch., Seattle, 1989-90, rsch. assoc., 1990—; adj. prof. Seattle Pacific U., 1990—; mem. pre-med. interview bd. Seattle Pacific U., 1993—; career panelist Shoreline Community Coll., Seattle, 1992—; advisor Drexel Alumni Rels., Phila., 1990-91. Contbr. articles to profl. jours. Vol. Phila. Dem. Orgn., 1988. Predoctoral fellow Am. Heart Assn., 1987, 88; named outstanding Young Profl. Seattle Pacific U., 1992. Mem. Fedn. Experimental Biologist, Sigma Xi. Democrat. Home: 5040 42d Ave SW Seattle WA 98136-1227

YOUNG, PAUL RUEL, computer scientist, dean; b. St. Marys, Ohio, Mar. 16, 1936; s. William Raymond and Emma Marie (Steva) Y.; children: Lisa Robin, Neal Eric. BS, Antioch Coll., 1959; PhD, MIT, 1963. Tchr. math. Harley Sch., Rochester, N.Y., 1957; asst. prof. Reed Coll., 1963-66; NSF postdoctoral fellow in math. Stanford U., 1965-66; asst. prof. computer sci. and math. Purdue U., Lafayette, Ind., 1966-67, assoc. prof., 1967-72; chmn., prof. computing and info. scis., prof. math. U. N.Mex., 1978-79; prof. Purdue U., Lafayette, Ind., 1972-83; chmn., prof. computing and info. scis., prof. math. U. N.Mex., 1978-79; chmn., prof. computer sci. U. Wash., 1983-88; assoc dean engring. U. Wash., Seattle, 1991—; vis. prof. elec. engring. and computer sci. U. Calif.-Berkeley, 1972-73, 82-83; Brittingham vis. prof. U. Wis., Madison, 1988-89; mem. adv. com. computer scis. NSF, 1977-80, chmn., 1979-80. Author: (with Michael Machtey) An Introduction to the General Theory of Algorithms, 1978. NSF and Woodrow Wilson fellow, MIT, 1959, 63. Mem. IEEE (chmn. tech. com. on math. founds. of computing 1981-83), Assn. Computing Machinery, Computing Rsch. Assn. (chmn. 1989-91). Office: U Wash Coll Engring 361 Loew Hall FH-10 Seattle WA 98195

YOUNG, RAYMOND H(INCHCLIFFE), JR., chemist; b. Pennsuken, N.J., Nov. 22, 1928; s. Raymond Hinchcliffe and Elisabeth Delores (Hand) Y.; m. Nancy Lee Totten, June 20, 1953 (div. 1975); children: JoAnne, Raymond, Michael, Janet; m. Carol Ann Smith, Dec. 5, 1975; children: Warren, Kimberly. BS, Pa. Mil. Coll., 1953; MS, U. Maine, 1955, PhD, 1961. Specialist Monsanto Co., Springfield, Mass., 1960-70; rsch. chemist Freeport Kaolin Co., Gordon, Ga., 1970-84; sr. devel. assoc. Engelhard Corp., Gordon, 1984—. Contbr. articles to profl. publs. Cpl. USMC, 1946-48. Mem. Am. Chem. Soc. (Arthur K. Doolittle award coatings and plastics div. 1972), Clay Minerals Soc., Soc. for Mining, Metallurgy and Experimentation, Sigma Xi (rsch. award 1955). Achievements include 8 patents for polymeric UV stabilizers, 3 patents for beneficiations process for kaolin clay. Home: 902 River North Blvd Macon GA 31211 Office: Engelhard Corp Briarcliff Rd Gordon GA 31031

YOUNG, ROBERT ALLEN, architectural engineer; b. Portland, Maine, Sept. 18, 1956; s. Raymond William and Marilyn Manola (Stilphen) Y.; m. Deborah Lee Gagnon, July 19, 1980. BS in Civil Engring., U. Maine, 1978; MS in Archtl. Engring., Pa. State U., 1984; MBA, U. Mich., 1991; MS in Historic Preservation, Ea. Mich. U., 1992. Registered profl. engr., Mich. Grad. asst. Pa. State U., University Park, 1979-80; sr. engr. Albert Kahn Assocs., Inc., Detroit, 1980-85, Blount Engrs. Inc., Detroit, 1985-87; project engr. energy cost avoidance project U. Mich., Ann Arbor, 1987—; cons. Workplace Edn. Assocs., Canton, Mich., 1989—; adj. instr. coll. architecture and urban planning U. Mich., 1989-91. Contbr. articles to Nat. Soc. Archtl. Engrs. Times, 1991—. Councilman Christ the King Ch., Livonia, Mich., 1983-86, co-chmn. youth com., 1983-85, chmn. fin. com., 1985-86, renovations/access com., 1989-91). Mem. ASHRAE (Detroit chpt. pres. 1993—, pres.-elect 1992-93, v.p. 1991-92, sec. 1989-91, gov. 1987-89), Nat. Soc. Archtl. Engrs. (founding state dir. 1991—), NSPE (mathcounts com. 1984-91), Assn. Energy Engrs., Nat. Trust for Historic Preservation Forum, Phi Kappa Phi.

YOUNG, ROBERT DONALD, physicist, educator; b. Chgo., Apr. 20, 1940; s. Robert Joseph and Nellie (Krik) Y.; children: Robert Gerald, Jennifer Ann Young Rolinski; m. BJ Marymont, Feb. 14, 1981; 1 child, Emily Marymont. BS in Physics, Ill. Inst. Technology, 1962; MS in Physics, Purdue U., 1965, PhD in Physics, 1967. Devel. engr. Western Elec., Cicero, Ill., 1962; process engr. Nat. Video Corp., Chgo., 1963; asst. prof. physics Ill. State U., Normal, 1967-73, assoc. prof. physics, 1974-78, prof. physics, 1979—; adj. prof. physics U. Ill., Urbana, 1986—. Contbr. articles to Proceedings Nat. Acad. Sci., Ann. Rev. Biophysics, Phys. Rev. Letters, Chemica Scripta, Jour. Phys. Chemistry, Jour. Chem. Physics, Computers in Physics, Physical Rev., Biophysical Jour. Named Researcher of Yr., Ill. State U., 1989. Mem. Am. Phys. Soc., Am. Chem. Soc., Biophys. Soc., Am. Assn. Physics Tchrs. Achievements include research on glassy properties of proteins, usage of computer techniques in physics education, individualized modular approach in physics education. Home: 4 Turner Rd Normal IL 61761-4218 Office: Ill State U Physics Dept Normal IL 61761

YOUNG, ROY ALTON, university administrator, educator; b. McAlister, N.Mex., Mar. 1, 1921; s. John Arthur and Etta Julia (Sprinkle) Y.; m. Marilyn Ruth Sandman, May 22, 1950; children: Janet Elizabeth, Randall Owen. BS, N.Mex. A&M Coll., 1941; MS, Iowa State U., 1942, PhD, 1948; LLD (hon.), N.Mex. State U., 1978. Teaching fellow Iowa State U., 1941-42, instr., 1946-47, Indsl. fellow, 1947-48; asst. prof. Oreg. State U., 1948-50, assoc. prof., 1950-53, prof., 1953—, head dept. botany and plant pathology, 158-66, dean research, 1966-70, acting pres., 1969-70, v.p. for research and grad. studies, 1970-76, dir. Office for Natural Resources Policy, 1986-90; chancellor U. Nebr., Lincoln, 1976-80; mng. dir., pres. Boyce Thompson Inst. Plant Research, Cornell U., Ithaca, N.Y., 1980-86; mem. Commn. on Undergrad. Edn. in Biol. Scis., 1963-68, Gov.'s Sci. Council, 1987-90; cons. State Expt. Stas. div. USDA; chmn. subcom. plant pathogens, agriculture bd. Nat. Acad. Scis.-NRC, 1965-68, mem. exec. com. study on problems of pest control, 1972-75; mem. exec. com. Nat. Govs.' Council on Sci. and Tech., 1970-74; mem. U.S. com. man and biosphere UNESCO, 1973-82; mem. com. to rev. U.S. component Internat. Biol. Program, NAS, 1974-76; mem. adv. panel on post-doctoral fellowships in environ. sci. Rockefeller Found., 1974-78; bd. dirs. Pacific Power & Light Co., 1974-91, PacifiCorp., 1984-91, Boyce Thompson Inst. for Plant Research, 1975—, Boyce Thompson Southwestern Arboretum, 1981-92, Oreg. Grad. Inst., 1987—; mem. adv. com. Directorate for Engring. and Applied Sci., NSF, 1977-81, mem. sea grant adv. panel, 1978-80; mem. policy adv. com. Office of Grants, USDA, 1985-86. Trustee Ithaca Coll., 1982-89. Lt. USNR, 1943-46. Recipient Disting. Svc. award Oreg. State U., 1978. Fellow AAAS (exec. com. Pacific div. 1963-67, pres. div. 1971), Am. Phytopatholgy Soc. (pres. Pacific div. 1957, chmn. spl. com. to develop plans for endowment 1984-86, bd. dirs. 1986-88); mem. Oreg. Acad. Sci., Nat. Assn. State Univs. and Land Grant Colls. (chmn. coun. for rsch. policy and adminstrn. 1970, chmn. standing com. on environment and energy 1974-82, chmn. com. on environment 1984-86), Sigma Xi, Phi Kappa Phi, Phi Sigma, Sigma Alpha Epsilon. Home: 3605 NW Van Buren Ave Corvallis OR 97330-4950

YOUNG, RUSSELL DAWSON, physics consultant; b. Huntington, N.Y., Aug. 17, 1923; s. C. Halsey and Edna (Dawson) Y.; m. Carol Vaughn Jones, Aug. 14, 1954; children: Bessmarie, Gale, Janet, Shari. BS in Physics, Rensselaer Poly. Inst., Troy, N.Y., 1953; PhD in Physics, Pa. State U., 1959. Rsch. assoc. Pa. State U., State College, 1959-61; project leader Nat. Bur. Stds., Gaithersburg, Md., 1961-73, chief optics and micrometrology, 1973-78; chief mech. processing div. Nat. Bur. Stds., Gaithersburg, 1975-80, ind. sys. div. chief, 1980-81, chief mech. prodn. div., 1980-81; pres. R.D. Young Cons., Pasadena, Md., 1981—; mem. tech. bd. Quanscann, Pasadena, 1988—. Contbr. articles to profl. jours.; inventor in field of instrumentation. 1st lt. Signal Corps, U.S. Army, 1943-46. Recipient Edward V. Condon award Dept. Commerce, 1974, Silver medal 1979, Gaede-Langmuir award 1992, Presdl. citation 1986, Wash. Acad. Scis. award 1988. Fellow Internat. Inst. Prodn. Engring. Rsch., Nat. Inst. Standards and Tech. Avocation: boating. Home: 852 Riverside Dr Pasadena MD 21122-1730

YOUNG, STEVE R., chemical engineer; b. Geldrop, The Netherlands, July 19, 1965; s. Robert E. and Mary Ann (Miller) Y.; m. Katherine Ferguson, Sept. 9, 1989; 1 child, Tara Katherine. BSChemE, U. Minn., 1987. Sr. engr. Westinghouse Savannah River Co., Aiken, S.C., 1988—. Mem. Am. Inst. Chem. Engrs. Achievements include research on process development of hazardous waste, mixed-waste treatment technologies, including vitrification, thermal destruction, volume reduction. Office: Westinghouse Savannah River Co Bldg 704-IT Aiken SC 29808

YOUNG, TERI ANN BUTLER, pharmacist; b. Littlefield, Tex., Aug. 22, 1958; d. Doyle Wayne and Bettie May (Lair) Butler; m. James Oren Young, Aug. 1, 1981; children: Andrew Wayne, Aaron Lee. BS in Pharmacy, Southwestern Okla. State U., 1981. Staff pharmacist St. Mary of Plains Hosp., Lubbock, Tex., 1981-84; staff pharmacist West Tex. Hosp., Lubbock, 1984-85, asst. dir. pharmacy, 1985-86; pharmacist cons. for nursing homes Billy D. Davis & Assocs., Lubbock, 1986—; relief pharmacist Prescription Lab., Med. Pharmacy and Foster Infusion Care, Lubbock, 1987-89; staff pharmacist Univ. Med. Ctr., 1990—; relief pharmacist West Tex. Hosp., 1986-91, Highland Hosp., 1990—, Med. Infusion Technology, 1992—. Mem. Lubbock Area Soc. of Hosp. Pharmacists (sec., treas. 1982-83), Lubbock Area Pharm. Assn., West Tex. Pharm. Assn., Am. Soc. Hosp. Pharmacists, Pilot Internat., Lubbock Genealogical Soc. Republican. Baptist. Lodge: Eastern Star. Avocations: needlework, reading, swimming, aerobics. Home: 7410 Toledo Ave Lubbock TX 79424-2214 Office: Univ Med Ctr 602 Indiana Lubbock TX 79415

YOUNG, TERRY WILLIAM, civil engineer; b. Syracuse, N.Y., Sept. 26, 1967; s. Terry H. and Marlene A. (Traub) Y.; m. Beth A. Wheaton, June 9, 1990. AS in Engring. Sci., Onondago C.C., Syracuse, 1987; BS in Civil Engring., U. Buffalo, 1989. Engr. I Blasland & Bouck Engrs. P.C., Syracuse, 1989-92, project engr., 1992—. Mem. ASCE, Am. Inst. Steel Constrn., Am. Concrete Inst., Chi Epsilon. Office: Blasland & Bouck Engrs PC 6723 Towpath Rd Box 66 Syracuse NY 13214

YOUNG, WILLIAM DEAN, psychologist; b. Flushing, N.Y., June 11, 1946; s. Dean and Laura Hervey (Smith) Y.; m. Pamela Sue Williams, Mar. 23, 1968; children: Stephanie D., Matthew W. BA, U. Tulsa, 1968; MEd, U. Md., 1971; EdD, U. Tulsa, 1985. Lic. prof. counselor, Okla.; nat. cert. gerontol. counselor. Mgr. sales adminstrn. Global Internat. Forwarding, Anaheim, Calif., 1972-76; reg. sales mgr. Global Internat. Forwarding, Tulsa, 1976-78; nat. sales mgr. Allied Van Lines, Transvault div., Broadview, Ill., 1978-79; exec. v.p. Hodges Allied Van Lines, Tulsa, 1979-82; gen. mgr. Graebel Okla. Movers, Tulsa, 1982-85; pres. Career Devel. Svcs., Tulsa, 1985—; ptnr. Young, Messer, Koepernik and Assoc., Inc., 1992—; mem. staff, vocat. cons. Laureate Psychiat. Clinic and Hosp., 1990; cons. The Warren Found., Tulsa, 1986, Graebel Van Lines, 1985-87; exec. dir. Okla. Alliance for Mentally Ill, 1992—. Author: Job Search Skills Workbook, 1988; author pamphlet: Moving with Children, 1988. Pres. Mental Health Assn. Tulsa, 1984-85; chmn. bd. dirs. Tulsa County Crisis Ctr., 1986-87, Job Support Ctr., Tulsa, 1987-88; mem. Okla. Mental Health Planning Coun., Oklahoma City, 1988—, chmn., 1990-92; chmn. Tulsa County Planning and Coordinating Bd., 1991-92. Recipient Meritorious Svc. award Mental Health Assn. Tulsa, 1988, Commendation for community svc. Tulsa Psychiat. Ctr., 1987; named Internat. Salesman of the Yr., Allied Van Lines, Chgo., 1980. Mem. AACD, Am. Rehab. Counselors Assn., Assn. for Adult Devel. and Aging (com. chmn. 1986—, mem.-at-large bd. dirs. 1990—), Nat. Rehab. Assn., Nat. Career Assn., Nat. Assn. Forensic Economists. Republican. Episcopalian. Avocations: golf, jogging, squash. Office: Young Messer Koepernik Ste 600 5314 S Yale Ave Tulsa OK 74135

YOUNG, WILLIAM H., federal agency administrator; b. Ilion, N.Y.; m. Betty Young; children: Deborah Young Streeton, William, Elizabeth. BS in Naval Architecture and Marine Engring., Webb Inst. Naval Architecture; MS in Engring., George Washington U. Registered profl. nuclear engr., Calif. Various positions, div. naval reactors AEC, 1962-71, assoc. dir. for submarines, 1968-71, head AEC Office, 1966-68; with Burns & Roe, Inc., 1971-85, v.p. breeder reactor div., 1976-83, also head corp. strategic planning, v.p. project ops., 1984-85; founder, dir. corp. performance and cost evaluations electric utility cos. and nuclear power plants William H. Young & Assocs., Inc., mgmt. cons., from 1985; asst. sec. nuclear energy Dept.

Energy, Washington, 1989-93; past chmn. advanced reactors subcom., nuclear engring. com., ASME. Mem. Am. Nuclear Soc., Am. Mgmt. Assn., Soc. Naval Architects and Marine Engrs. Office: William H Young & Assocs 1442 Sequoia Circle Toms River NJ 08753-2864*

YOUNG, WILLIAM LEWIS, retired mathematics educator; b. Buffalo, July 27, 1929; s. Charles William Young and Ada Laura (Lynch) Stremble. BS, Hartwick Coll., 1951; MA, Pa. State U., 1962; MSEE SUNY-Buffalo, 1978, M.L.S., 1979. Instr. SUNY-Buffalo, 1960-65; asst. prof. State Coll., Fredonia, N.Y., 1965-67; ops. analyst Calspan, Buffalo, 1967-69; prof. math. dept. Erie C.C., Orchard Park, N.Y., 1969-91, coord. math. and computer sci., 1982-84; v.p. faculty fedn., 1974-77. Pres. Aurora Hist. Soc., East Aurora, N.Y., 1975-79, 90—; trustee, 1974-82, 90—; trustee Scheide Mantel bd. Elbert Hubbard Mus., 1990-91, 92—; chmn. Millard Fillmore House Council, East Aurora, 1976-81. Served to 1st lt. USAF, 1951-57. Rsch. fellow On-Line Computer Library Ctr., Columbus, Ohio, 1979. Home: 806 Luther Rd East Aurora NY 14052-9713

YOUNGBERG, GEORGE ANTHONY, pathology educator; b. Chgo., Mar. 14, 1951; s. Gilbert and Barbara (Hoffmann) Y.; m. Rosemary H. Leu, May 12, 1979. BA, Lake Forest Coll., 1973; MD, Northwestern U., 1977. Asst. prof. East Tenn. State U., Johnson City, 1980-85, assoc. prof., 1985-91, prof., 1991—; surg. pathologist U, Physicians Practice Group, Johnson City, 1980—, dir. renal biospy svc., 1988—; rsch. & devel. coord. VA Med. Ctr., Johnson City, 1988—. Contbr. articles to profl. jours. Fellow Coll. Am. Pathologists; mem. Am. Acad. Dermatology (affiliate), U.S. Acad. Pathology, Can. Acad. Pathology, Internat. Soc. Dermatopathology, Phi Beta Kappa. Office: East Tenn State U Dept Pathology Johnson City TN 37614

YOUNGDAHL, PAUL FREDERICK, mechanical engineer; b. Brockway, Pa., Oct. 8, 1921; s. Harry Ludwig and Esther Marie (Carlson) Y.; m. Elinor Louise Jensen, Nov. 27, 1943; children: Mark Erik, Marcia Linnea, Melinda Louise. Student Pa. State U., 1938-40; BS in Engring., U. Mich., 1942, MS in Engring., 1949, PhD, 1962. Indsl. and devel. engr. duPont, Bridgeport, Conn., 1942-43, Carneys Point, N.J., 1946-48; dir. research Mech. Handling Systems, Detroit, 1953-62; prof. U. Mich., Ann Arbor, 1962-74; cons. mech. engr., Palo Alto, Calif., 1974—; dir. Liquid Drive Corp., Holly, Mich. Contbr. articles to profl. jours. Served with USNR, 1943-46. Mem. Mich. Soc. Profl. Engrs., Nat. Soc. Profl. Engrs., ASME, Am. Soc. Engring. Edn., Mich. Assn. Professions, Sigma Xi, Tau Beta Pi, Phi Kappa Phi, Pi Tau Sigma. Methodist. Address: 501 Forest St PH 4 Palo Alto CA 94301

YOUNGELMAN, DAVID ROY, psychologist; b. N.Y.C., June 13, 1959; s. Jerry DeRoma Jr. and Vicki (Marmon) Y.; m. Sandra Lynn Koznetski, May 22, 1991. BA in Psychology and Philosophy, Bucknell U., 1981; MA in Clin. Psychology, Fairleigh Diskinson U., 1983; D of Psychology, Yeshiva U., 1988. Resident in psychhology VA Med. Ctr., Lyons, N.J., 1987-88, staff psychologist, 1988-89, dir. psychology tng., 1990—, asst., chief psychology svc., 1991—; cons. in field; presenter in field. Mem. APA. Office: VA Med Ctr Dept Psychology 116B Lyons NJ 07939

YOUNGS, JACK MARVIN, cost engineer; b. Bklyn., May 2, 1941; s. Jack William and Virginia May (Clark) Y.; BEngring., CCNY, 1964; MBA, San Diego State U., 1973; m. Alexandra Marie Robertson, Oct. 31, 1964; 1 child, Christine Marie. Mass properties engr. Gen. Dynamics Corp., San Diego, 1964-68, rsch. engr., 1968-69, sr. rsch. engr., 1969-80, sr. cost devel. engr., 1980-81, cost devel. engring. specialist, 1981—. Dist. dir. Scripps Ranch Civic Assn., 1976-79; pres. Scripps Ranch Swim Team, 1980-82; dir., 1986-87; judge Greater San Diego Sci. and Engring. Fair, 1981-92. Mem. Princeton U. Parents Assn. Recipient 5th place award World Body Surfing Championships, 1987, 6th place award, 1988. Mem. AIAA, N.Y. Acad. Scis., Alumni Assn, CUNY, Bklyn. Tech. High Sch. Alumni Assn., Inst. Cost Analysis (cert., charter mem., treas. Greater San Diego chpt. 1986-90), Soc. Cost Estimating and Analysis (cert. cost estimator/analyst, pres. San Diego chpt. 1990-91), Internat. Soc. Parametric Analysts (bd. dirs. San Diego chpt. 1987-90), Nat. Mgmt. Assn. (space systems div. charter mem. 1985, award of honor Convair chpt. 1975), Assn. MBA Execs., San Diego State U. Bus. Alumni Assn. (charter mem. 1986), Scripps Ranch Swim and Racquet Club (dir. 1977-80, treas. 1978-79, pres. 1979-80), Beta Gamma Sigma, Chi Epsilon, Sigma Iota Epsilon. Lutheran. Research in life cycle costing and econ. analysis. Home: 11461 Tribuna Ave San Diego CA 92131-1907 Office: PO Box 85990 San Diego CA 92138-5990

YOURISON, KAROLA MARIA, information specialist, librarian; b. Berlin, Germany, June 30, 1937; came to U.S., 1962; m. James E. Yourison, Feb. 29, 1992. BA, U. Pitts., 1974, MLS, 1976. Libr. mgr. Siemens Rsch. & Tech. Lab., Princeton, N.J., 1983-85; mgr. libr. svcs. Software Engring. Inst., Carnegie Mellon U., Pitts., 1986—. Mem. IEEE, Spl. Librs. Assn. (chmn. duplicates exch. com. 1990—, chair-elect sci. and tech. div.), Assn. Computing Machinery. Office: Software Engring Inst 5000 Forbes Ave Pittsburgh PA 15213

YOURTEE, DAVID MERLE, pharmaceutical science educator, molecular toxicologist; b. St. Louis, Oct. 16, 1937; s. Samuel Lambert and Muriel Virginia (Myers) Y.; m. Cynthia Lynn Kirk, Jan. 7, 1981; children: Andrea Nichole, Ryan Chandler, Lara Danielle. BS in Chemistry, U. Mo., 1967; PhD in Pharm. Sci., U. Mo., Kansas City, 1974. Asst. scientist Cancer Rsch. Ctr., Columbia, Mo., 1973-75, assoc. scientist, 1975-77, chmn. dept. biochemistry, 1977-79; prof. pharmacy and medicine U. Mo., Kansas City, 1979—, dir. toxicore, 1983—; mem. Africa adv. panel Coun. for Internat. Exchange Scholars, Washington, 1987-89, chmn. Africa adv. panel, 1988-89; prin. investigator NIH, 1979, 1990, Valles Found. Hodgkins Disease Rsch., 1983. Author: Toxicology of the Aflatoxins, 1993; editor Jour. Toxicology: Toxin Revs., 1989; contbr. articles to profl. jours. Fulbright Sr. Rsch. scholar, 1983, 85. Fellow Mo. Acad. Sci. (pres. 1986-87); mem. Nat. Ctr. Toxicologic Rsch.-Associated Univs. (bd. dirs. 1988-91). Achievements include research in structure mutagenicity relationships and human metabolism of aflatoxins. Office: U Mo Med Sch Bldg 2411 Holmes St # 4co Kansas City MO 64108-2741

YOURZAK, ROBERT JOSEPH, management consultant, engineer, educator; b. Mpls., Aug. 27, 1947; s. Ruth Phyllis Sorenson. BCE, U. Minn., 1969; MSCE, U. Wash., 1971, MBA, 1975. Registered profl. engr., Wash., Minn. Surveyor N.C. Hoium & Assocs., Mpls., 1965-68, Lot Surveys Co., Mpls., 1968-69; site layout engr. Sheehy Constrn. Co., St. Paul, 1968; structural engring. aide Dunham Assocs., Mpls., 1969; aircraft and aerospace structural engr., program rep. Boeing Co., Seattle, 1969-75; engr., estimator Howard S. Wright Constrn. Co., Seattle, 1976-77; dir. project devel. and adminstrn. DeLeuw Cather & Co., Seattle, 1977-78; sr. mgmt. cons. Alexander Grant & Co., Mpls., 1978-79; mgr. project systems dept., project mgr. Henningson, Durham & Richardson, Mpls., 1979-80; dir. project mgmt., regional offices Ellerbe Assocs., Inc., Mpls., 1980-81; pres. Robert Yourzak & Assocs., Inc. Mpls., 1982—; lectr. engring. mgmt. U. Wash., 1977-78; lectr., adj. asst. prof. dept. civil and mineral engring. and mech./indsl. engring. Ctr. For Devel. of Tech. Leadership, Inst. Tech.; mgmt. scis. dept. Sch. Mgmt. U. Minn., 1979-90, bd. adv. inst. tech., 1989—; founding mem., membership com., mem. Univ. of Minn. com. Minn. High Tech. Coun., 1983—; speaker in field. Author: Project Management and Motivating and Managing the Project Team, 1984. Chmn. regional art group experience Seattle Art Mus., 1975-78; mem. Pacific N.W. Arts Council, 1977-78, ex-officio adviser Mus. Week, 1976; bd. dirs. Friends of the Rep. Seattle Repertory Theatre, 1973-77; mem. Symphonics Seattle Symphony Orch., 1975-78. Scholar Boeing Co., 1967-68, Sheehy Constrn. Co., summer 1967. Named An Outstanding Young Man of Am., U.S. Jaycees, 1978. Mem. Am. Soc. Tng. and Devel. (So. Minn. chpt.), Project Mgmt. Inst. (cert. project mgmt. profl., speaker, founding pres. 1985, chmn., adv. com. 1987-89, bd. dirs. 1984-86, program com. chmn. and organizing com. mem. Minn. chpt. 1984, speaker, project mgr. internat. mktg. program 1985-86, chmn. internat. mktg. standing com. 1986, long range and strategic planning com. 1988—, chmn., 1992, v.p. pub. rels. 1987-88, ex-officio dir. 1989, 1992, internat. pres. 1990, chmn. bd. 1991, ex-officio chmn. 1992, internat. bd. dirs., chmn. nominating com. 1992, chmn.), Am. Cons. Engrs. Coun. (peer reviewer 1986-89),Am. Arbitration Assn. (mem. Mpls. panel of constrn. arbitrators), Minn. Surveyors and Engrs. Soc., ASCE (chmn. continuing edn. subcom. Seattle chpt. 1976-79, chmn. program com. 1978, mem. transp. and urban planning tech. group 1978, Edmund Friedman Young Engr. award 1979, chmn. continuing edn. com. 1979-80, chmn. energy com. Minn. chpt. 1980-81, bd. dir. 1981-89, sec. 1981-83, v.p. profl. svcs. info. svcs. 1984-85, pres. 1986-87, past pres. 1987-89, fellow 1988—, speaker), Inst. Indsl. Engrs. (pres. Twin Cities chpt. 1985-86, chmn. program com. 1983-84, bd. dirs. 1985-88, awards com., chmn. 1984-89, speaker), Cons. Engrs. Council Minn. (chmn. pub. rels. com. 1983-85, 88, vice chmn., chmn., 1989, program com. chmn. Midwest engrs. conf. and exposition 1985-90, speaker, Honor award 1992), Mpls. Soc. Fine Arts, Internat. Facility Mgmt. Assn., Am. Soc. Engring. Edn., Rainer Club (co-chmn. Oktoberfest), Sierra Club, Chowder Soc., Mountaineers, North Star Ski Touring, Chi Epsilon (life). Office: 7320 Gallagher Dr Ste 325 Minneapolis MN 55435-4510

YOUTCHEFF, JOHN SHELDON, physicist; b. Newark, Apr. 16, 1925; s. Slav Joseph and Florence Catherine (Davidson) Y.; A.B., Columbia, 1949, B.S., 1950; Ph.D., U. Calif. at Los Angeles, 1953; m. Elsie Marianne, June 17, 1950; children: Karen Janette, John Sheldon, Mark Allen, Heidi Mary Anne, Lisa Ellen. Ops. analyst Gen. Electric Co., Ithaca, N.Y., 1953-56, cons. engr. Missile & Space Div., Phila., 1956-64, mgr. advanced reliability programs, 1964-72; mgr. reliability and maintainability Litton Industries, College Pk., Md., 1972-73; program mgr. U.S. Postal Service Hdqrs., Washington, 1973—; instr. U. Pa., 1965-66, Villanova U., 1957—. Served to lt. USAAF, 1943-46; to comdr. USNR, 1946—. Registered profl. engr., Calif., D.C. Fellow AAAS, British Interplanetary Soc., Am. Inst. Aero. and Astronautics (asso.); mem. IEEE (sr.), Ops. Research Soc., Research Soc. Am., Am. Math. Soc., Am. Physics Soc., Am. Chem. Soc., Am. Astron. Soc., Am. Geol. Soc., Nat. Soc. Profl. Engrs., Engring. and Tech. Socs. Council Del. Valley (speakers bur.), Res. Officers Assn., Am. Legion. Roman Catholic. Clubs: Explorers (N.Y.C.), Optimists Internat. (pres. Valley Forge chpt. 1970-71). Holder 3 U.S. patents; contbr. articles to profl. jours. and proc. Home: Apt B-501 1400 S Joyce St Arlington VA 22202 Office: L'Enfant Plz Washington DC 20260

YU, BRIAN BANGWEI, research physicist, executive; b. Beijing, China, Dec. 2, 1957; came to U.S., 1982; s. Dah-Nien and Min-Xian (Van) Y.; m Jennifer Hou, May 20, 1990; 1 child, Jessica J.Q. AS, AAS, SUNY, 1985; BS, Va. Polytech., 1988, MS, 1990. Rsch. engr. Xaloy Inc., Pulaski, Va., 1988-89; rsch. physicist Am. Rsch. Co., Redford, Va., 1989-90; rsch. engr. Aetna Insulated Wire Co., Virginia Beach, Va., 1990-91, rsch. and devel., quality assurance mgr., 1991—. Author: Electrical Cable Handbook, 1994. Mem. IEEE, Am. Chem. Soc., Internat. Wire Assn. (mgmt. com. 1991—). Home: 5324 Doon St Virginia Beach VA 23464 Office: Aetna Insulated Wire Co 1537 Air Rain Ave Virginia Beach VA 23455

YU, BYUNG PAL, gerontological educator; b. Ham Hung, Korea, June 29, 1931; came to U.S., 1956; s. Hong Soon and Ok Sun (Oum) U.; m. Kyung Hi, June 9, 1959; 1 child, Victor S. BS, Ctrl. Mo. State U., 1960; PhD, U. Ill., 1965. Asst. prof. Med. Coll. of Pa., Phila., 1965-71, assoc. prof., 1971-73; assoc. prof. U. Tex. Health Sci. Ctr., San Antonio, 1973-78, prof., 1978—; pres. Am. Aging Assn., Omaha, 1992—; chair elect biol. scis. Gerontol. Soc. of Am., Washington, 1992; chair sci. adv. bd. Aloe Rsch. Found., Ft. Worth, 1991. Editor: Free Radicals in Aging, 1993, Modulations of Aging Processes by Dietary Restriction, 1994, Aging, Age and Nutrition, AGE, Soc. Biol. Medicine; assoc. editor Jour. Gerontology; contbr. more than 190 publs. to profl. jours. Recipient Nat. Meritorious Medal award, Korea, 1987. Fellow Gerontol. Soc. of Am.; mem. NIH (nutrition study sect.). Home: 9427 Callaghan Rd San Antonio TX 78230 Office: U Tex Health Sci Ctr 7703 Floyd Curl Dr San Antonio TX 78284

YU, CHIEN-CHIH, management information systems educator; b. Keelung, Taiwan, Apr. 2, 1953; s. Shung-Too and Yuan (Kauo) Y.; m. Hsiu-Yuan Hsu, Dec. 25, 1978; children: Anne Ya-Ching, Anthony Che-Young, Angel Dan-Ying. BS, Nat. Cheng Chi U., Taipei, Taiwan, 1975; MS, U. Toledo (Ohio), 1980; MA, U. Tex., 1983, PhD, 1985. Teaching asst. Nat. Cheng Chi U., Taipei, 1977-78, prof., chmn. MIS dept., 1986-90; teaching asst. U. Toledo, 1979-80, U. Tex., Austin, 1980-85; project leader Ministry of Transp. and Comm., Taipei, 1990—, EPA, Taipei, 1993—, Nat. Sci. Coun., Taipei, 1986—; cons. China Info. and Micrographic Inst., Taipei, 1989—, Info. Sci. and Tech. Exhbn. Ctr., Taipei, 1991—, Inst. Info. Industry, 1992—.. Author: An Integrated Expert Decision Support System for Statistics and Simulation, 1990, Integrated Expert Decison Support Systems, Architecture, Development, and Applications; editor MIS Rev., 1991; contbr. articles to profl. jours. 2d lt. Taiwan mil., 1975-77. Grantee Nat. Sci. Coun., 1989. Mem. IEEE (computer soc.), DSI, ACM, Am. Math. Soc., Ops. Rsch. Soc. Am., Chinese Info. Mgmt. Soc. (bd. regents 1990-91), Chinese Indsl. Engring. Soc., Phi Kappa Phi. Avocations: writing poetry, mountain climbing, painting. Office: Nat Cheng Chi U, Dept MIS, Taipei 11623, Taiwan

YU, CONG, biomedical engineer; b. Shanghai, China, July 4, 1962; came to U.S. 1987; d. Hai-Bo and Qi-Xing (Yu) Yang; m. Hong Lou, June 12, 1992. BS, Jiao Tong U., Shanghai, 1984; MS, Jiao Tang U., Shanghai, 1987. Engr. Shanghai Iron and Steel Rsch. Inst., 1987; rsch. asst. U. Iowa, Iowa City, 1987—. Mem. Sigma Xi (assoc.). Home: 1528 Crosby Ln Iowa City IA 52240 Office: Univ of Iowa 1202 Engring Bldg Iowa City IA 52242-1527

YU, HYUK, chemist, educator; b. Kapsan, Korea, Jan. 20, 1933; s. Namjik and Keedong (Shin) Y.; m. Gail Emmens, Jan. 20, 1964; children: Jeffrey, Steven, Douglas. BS in Chem. Engring., Seoul Nat. U., 1955; MS in Organic Chemistry, U. So. Calif., 1958; PhD in Phys. Chemistry, Princeton U., 1962. Rsch. assoc. Dartmouth Coll., Hanover, N.H., 1962-63; rsch. chemist Nat. Inst. Sci. and Tech., Washington, 1963-67; asst. prof. U. Wis., Madison, 1967-69, assoc. prof., 1969-78, prof. chemistry, 1978—, Evan P. Helfaer chair chemistry, 1991—; cons. polymer sci. and standards div. Nat. Inst. Standards and Tech., Gaithersburg, Md., 1967—, Eastman Kodak Co., various locations, 1969—, Procter and Gamble Co., Cin., 1988—, Japan Synthetic Rubber Co., Tsukuba, 1991—; Fulbright-Hays lectr. Inha U., Inchon, Korea, 1972; chmn. Gordon Rsch. Conf. Polymer Physics, 1986, 93. Contbr. articles to refereed publs. John Simon Guggenheim Found. fellow, 1984. Fellow Am. Phys. Soc.; mem. Am. Chem. Soc. (exec. com. polymer chemistry div., editorial adv. bd. 1988-91), N.Y. Acad. Scis., Biophys. Soc., Materials Rsch. Soc., Korean Chem. Soc. (life), Polymer Soc. Korea (life), Sigma Xi (chpt. pres. 1987-88). Office: Univ Wis 1101 University Ave Madison WI 53706

YU, JASON JEEHYON, chemical engineer; b. Taegu, Korea, Apr. 3, 1962; came to U.S., 1980; s. Jaehoon and Pilhee (Chung) Y.; m. Dongyoune Ahn, Feb. 26, 1990; 1 child, James Gibeom. BSChemE, Cooper Union, N.Y.C., 1986; MS, Northwestern U., 1990, PhD, 1992. Cons. G.D. Searle, Skokie, Ill., 1988-1990; sr. engr. Monsanto Co., Pensacola, Fla., 1991—. Contbr. book sect. Catalyst Deactivation, 1991. Mem. AICE. Office: Monsanto Co PO Box 97 Gonzalez FL 32560-0097

YU, KITSON SZEWAI, computer science educator; b. Toishan, Kwangtung, China, Apr. 4, 1950; came to U.S., 1969; s. Ho Yee and Yin Sang (Chan) Y.; m. Mabel Griseldis Wong, July 15, 1972; 1 child, Robin Roberta Emily. BS, Troy State U., 1974, MS, 1977, BS, 1980. Cert. systems profl., data processing educator, Oreg. V.p. Troy (Ala.) Computer Ctr., 1976-81; computer instr. Tory State U., 1980-81, Linn Benton Community Coll., Albany, Oreg., 1981—; dir. real estate program Linn Benton Community Coll., 1985—; mng. broker Kitson Realty, Corvallis, Oreg., 1975—. Vice pres. econ. devel. Daleville C. of C., Ala., 1976; dir. Corvalis Youth Symphony, 1990—. Mem. Data Processing Mgmt. Assn. (bd. dirs. at large 1982—, v.p. 1984-85, pres. 1985-86), Greater Albany Rotary (treas. 1985—), Corvallis Multiple Listing Exch. (bd. dirs. 1990—), Gamma Beta Phi. Home: 2621 NW Lupine Pl Corvallis OR 97330-3537 Office: Linn Benton Community Coll 6500 Pacific Blvd SW Albany OR 97321-3755

YU, MONICA YERK-LIN, medical technologist, consultant; b. Canton, China, July 28, 1958; m. Yun-Tung Lau, Aug. 15, 1990. BS, U. Md., 1981. Med. technologist Georgetown U. Hosp., Washington, 1981-83, Naval Med. Ctr., Bethesda, Md., 1983-86, NIH, Bethesda, 1986—. Mem. Am. Soc. Clin. Pathologists (cert. med. technologist). Office: NIH DTM Bldg 10 RM 1C738 9000 Rockville Pike Bethesda MD 20877

YU, ROGER HONG, physics educator; b. Shanghai, China, Apr. 19, 1960; came to U.S., 1987; s. Rei Qian and Wei-Zen (Zhang) Y.; m. Ting Shi, Sept. 8, 1990; 1 child, William S. BS, Shanghai U. Sci. & Tech., 1982; MS, U. Mo., 1987; PhD, Mont. State U., 1990. Lectr. physics Shanghai U. Sci., 1982-85; teaching asst. U. Mo., Kansas City, 1985-86, rsch. asst., 1986-87; teaching asst. Mont. State U., Bozeman, 1987-88, rsch. asst., 1988-90; prof. physics Ctrl. Wash. U., Ellensburg, 1990—. Contbr. articles to profl. jours.; referee Phys. Rev. B. Mem. Am. Phys. Soc., Coun. Undergrad. Rsch. Office: Ctrl Wash U Dept Physics Ellensburg WA 98926

YU, RU-QIN, chemistry educator; b. Shanghai, China, Nov. 21, 1935; s. Zhen and De-Xun (Yang) Y.; m. Gene-Rong Li, Sept. 2, 1961; children: Zhi-Jing, Zhi-Yuan. BS in Chemistry, Leningrad U., St. Petersburg, Russia, 1959. Rsch. asst. Inst. of Chemistry, Chinese Acad. Scis., Beijing, 1959-61; assoc. prof. Hunan U., Changsha, 1980, prof., 1981—; pres. Hunan U., 1993—; vice dir. of electroanalytical lab. Changchun Inst. of Applied Chemistry, 1989—. Author: Introduction to Chemometrics, 1991, Information-Theoretical Fundamentals of Modern Analytical Chemistry, 1987, Ion-Selective Electrode Analysis Methodology, 1980; regional adv. editor The Analyst, Royal Soc. Chemistry; contbr. numerous articles to profl. jours. Recipient Nat. Natural Sci. award Nat. Sci. and Tech. Com., People's Republic China, 1987, Sci. Progress award nat. Edn. Com., 1986. Mem. Chinese Acad. Scis. (academician 1991—), Chinese Chem. Soc. (bd. dirs. 1986—), Chinese Soc. Analytical Instrumentation (bd. dirs. 1980—), Am. Chem. Soc. Achievements include rsch. on chem. sensors, chemometrics and organic carriers for electrochem. and optical sensors, reported the synthesis of a series of new organic compounds used as carriers for constrn. of pH and other sensors, the use of multivariate calibration methods and neural network in analytical chemistry and chemometrics. Home: Hunan U 12 Fenglincun, Changsha 410012, China Office: Hunan U, Dept Chemistry/Chem Engring, Changsha 410082, China

YUAN, SHAO WEN, aerospace engineer, educator; b. Shanghai, China, Apr. 16, 1914; came to U.S., 1934, naturalized, 1954; s. Ti An and Chieh-huang (Chien) Y.; m. Hui Chih Hu, Nov. 5, 1950. B.S., U. Mich., 1936; M.E., Stanford U., 1939; M.S., Calif. Inst. Tech., 1937, Ph.D., 1941. Rsch. engr. Glenn Martin Co., 1942-43; chief of rsch. Helicopter div. McDonnell Aircraft Corp., 1943-45; instr. Washington U., St. Louis, 1944-45; adj. prof. Poly. Inst. Bklyn., 1946-49, assoc. prof., 1949-54, prof., 1954-57; ptnr. von Kármán, Yuan & Arnold Assocs., 1955-63; prof. aerospace engring. U. Tex., 1958-68; prof., chmn. mech. engring. div. George Washington U., 1968-78, chmn. civil, mech. and environ. dept., 1973-78, 80-81, prof. emeritus, 1984; pres. RISE, Inc., 1977-85; Canadair Chair prof. U. Laval, Can., 1957-58; chmn. adv. com. Joint Inst. for Advancement of Flight Sci., 1970-84; hon. prof. Zhejiang U., 1987—; cons. Edo Aircraft Corp., Aerojet Corp., Cornell Aero. Lab., Dept. of Interior, Oak Ridge Nat. Lab., N.Am., Aviation, Inc., Fairchild-Hiller Corp., McDonnell-Douglas Corp., The World Bank; hon. adviser Nat. Center Research of China, Taiwan, 1958-68; chmn., founder 1st U.S.-China Conf. on Energy, Resources, and Environment, 1982; founder Consortium of Univs. for Promoting Grad. Aerospace Studies, 1984; founder Disting. Lecture Series on Founds. of Aerospace Research and Devel., 1986. Author: Foundations of Fluid Mechanics, 1967; Contbr. to: High Speed Aerodynamics and Jet Propulsion series, 1959, Energy, Resources, and Environment: Procs. at 1st U.S.-China Conf., 1982. Recipient Outstanding Achievements award George Washington U., 1981; named Outstanding Educator of Am., 1970, Outstanding Chinese American, 1983. Fellow AAAS, AIAA; mem. ASME (life), Am. Soc. Engring. Edn., Soc. Engring. Sci. (dir. 1973-78, pres. 1977), Torchbearers of Caltech, Sigma Xi, Phi Kappa Phi, Phi Tau Phi, Sigma Gamma Tau, Pi Tau Sigma, Tau Beta Pi, Tau Xi Sigma. Achievements include patents in field. Home: 1400 Geary Blvd Apt 1505 San Francisco CA 94109-6569

YUDELSON, JERRY MICHAEL, marketing executive; b. Greenwich, Conn., Mar. 9, 1944; s. Coleman E. and Della Mae (Moore) Y.; m. Jessica Stuart, Feb. 16, 1986. BS, Calif. Tech. Inst., 1966; MS, Harvard U., 1968; MBA, U. Oreg., 1993. Registered environ. assessor Calif., Wash., Oreg. Dir. bus. devel. Zond Systems, Tehachapi, Calif., 1985-86; pres. Yudelson Assocs., Costa MEsa, Calif., 1986-87; v.p. sales & mktg. Micro Cogen Systems Inc., Irvine, Calif., 1987-88; regional v.p. Chronar Corp., Santa Ana, Calif., 1988-89; pres. Stuart Assocs. Inc., Garden Grove, Calif., 1989-91; v.p. sales & mktg. Pacific Coast Environ. Inc., Portland, Oreg., 1991-92, Regional Disposal Co., Portland, Oreg., 1992—; dir. Coalition for Clean Air, L.A., 1989-91; mem. State Senate Cost Control Commn., Sacramento, 1989-91, Gov.'s Air Pollution Task Force, Portland, 1992. Candidate U.S. Congress, Orange County, Calif., 1988, Calif. State Assembly, Orange County, 1990. Mem. Air & Waste Mgmt. Assn., Assn. Energy Engrs. (sr. mem.), Portland City Club. Democrat. Office: Regional Disposal Co 317 SW Alder Ste 1185 Portland OR 97204

YUECHIMING, ROGER YUE YUEN SHING, mathematics educator; b. Mauritius, Feb. 25, 1937; s. James and Marie Yuechiming; m. Renée Bethery, Nov. 9, 1963; children: Françoise, Marianne, Isabelle. BSc with 1st class honours, U. Manchester, Eng., 1964, PhD, 1967. Asst. U. Strasbourg, France, 1967-69; lectr. math. U. Paris VII, 1970—; participant math. confs. and seminars in numerous countries; referee various math. jours. Contbr. over 70 articles on ring theory to sci. jours. of numerous countries. Mem. French Math. Soc., Am. Math. Soc., London Math. Soc. Achievements include introduction of concept of p-injective models, new approaches in ring and module theory leading to a better understanding of von Neumann regular rings, V-rings, self-injective rings and generalizations. Home: 38 rue du Surmelin, 75020 Paris France Office: U Paris VII Unite de Recherche Associée 212, Nat Ctr Sci Rsch 2 Pl Jussieu, 75251 Paris France

YUH, CHAO-YI, chemical engineer, electrochemist; b. I-Lan, Taiwan, Feb. 5, 1955; s. Ching-Lung and Zu-Hua (Yang) Y.; m. Yu-Shio Tsai, Feb. 20, 1983; children: Joshua, Tiffany, Bianca. BSChemE, Nat. Tsing-Hua U., Taiwan, 1977; PhD in Chem. Engring., Ill. Inst. Tech., 1985. Grad. assoc. Ill. Inst. Tech., Chgo., 1980-85; mgr. materials devel. Energy Rsch. Corp., Danbury, Conn., 1985—. Contbr. articles to profl. jours. 2d lt. Chinese Air Force, Taiwan, 1977-79. Grad. student grantee Hooker Chems. & Plastics Corp., 1981; SBIR grantee U.S. Dept. Energy, 1989, 93. Mem. AICE, Electrochem. Soc., Nat. Assn. Corrosion Engrs. Achievements include identification of various processes in fuel cell and flow-through electrode via AC-impedance and potential relaxation techniques. Office: Energy Rsch Corp 3 Great Pasture Rd Danbury CT 06813

YUILL, THOMAS MACKAY, academic administrator, microbiology educator; b. Berkeley, Calif., June 14, 1937; s. Joseph Stuart and Louise (Dunlop) Y.; m. Ann Warnes, Aug. 24, 1960; children: Eileen, Gwen. BS, Utah State U., 1959; MS, U. Wis., 1962, PhD, 1964. Lab. officer Walter Reed Army Inst. Rsch., Washington, 1964-66; med. biologist SEATO Med. Research Lab., Bangkok, Thailand, 1966-68; asst. prof. U. Wis.-Madison, 1968-72, assoc. prof., 1972-76, prof., 1976—, dept. chmn., 1979-82, assoc. dean., 1982-93; dir. Inst. Environ. Studies, 1993—; cons. NIH, Bethesda, 1976-86; chmn. Viral Diseases Panel, U.S.-Japan Biomed. Scis. Program, 1979-86, Am. Com. Arbovirology, 1982—; bd. dirs. Cen. Tropical Agrl. Res. Teaching, Turrialba, Costa Rica, 1988—. Contbr. chpts. to books, articles to profl. jours. Served to capt. U.S. Army, 1964-66. Recipient grants state and fed. govts., 1968—. Mem. Orgn. Tropical Studies (pres. 1979-85), Wildlife Disease Assn. (treas. 1980-85, pres. 1985-87), Am. Soc. Tropical Medicine and Hygiene (editorial bd. 1984-92), Royal Soc. Tropical Medicine and Hygiene, Am. Soc. Microbiology, Am. Soc. Virology, Wildlife Soc., Council Agrl. Sci. and Tech., Sigma Xi. Avocations: flying; cross-country skiing; music. Office: U Wis Inst Environ Studies 1007 WARF 610 Walnut St Madison WI 53705

YUKIO, TAKEDA, engineering educator; b. Kobe, Japan, Dec. 2, 1927; m. Misao Takeda, Apr. 24, 1949; 1 child, Hironao. Student, Kobe Tech. Coll., 1948; D. in Engring., Osaka (Japan) U., 1973. Assoc. prof. Marine Tech. Coll. Ministry of Transport (Japan), Kobe, 1948-52; asst. Kobe U. Mercantile Marine, 1952-61, assoc. prof., 1961-75, prof., 1975-91, prof. emeritus, 1991—, curator univ. libr., 1983-85; hon. prof. Shanghai Marine U., 1992. Author: Marine Thyristor Machine, 1972, Study on Marine Electric Power System, 1978, Lightning, Damage and Defence, 1986, Superconducting Engineering, 1991. Hon. mem. Marine Engring. Soc. Japan. Avocations:

painting, table tennis, Go play. Home: 17-3 Higashi Ashiya, Ashiya 659, Japan

YUN, DANIEL DUWHAN, physician, foundation administrator; b. Chinjoo, Korea, Jan. 20, 1933; came to U.S., 1959, naturalized, 1972; s. Kapryong and Woo Im Yun; m. Rebecca Sungja Choi, Apr. 13, 1959; children: Samuel, Lois, Caroline, Judith. BS, Coll. Sci. and Engring., Yonsei U., 1954, MD, 1958; student U. Pa., 1963. Intern, Quincy (Mass.) City Hosp., 1960; resident and fellow Presbyn.-U. Pa. Med. Ctr., Phila., 1961-65; med. dir. Paddon Meml. Hosp., Nfld., Labrador, Can., 1965-66; dir. spl. care unit Rolling Hill Hosp., Elkins Park, Pa., 1967-79; founder, pres. Philip Jaisohn Meml. Found., Inc., Elkins Park, Pa., 1975-85, also med. dir., trustee; clin. prof. medicine U. Xochicalco, 1978. Mem. Bd. Asian Studies Found., U.S. Senatorial Bus. Adv. Bd.; mem. home safety com. Mayor's Commn. on Svcs. to Aging, Phila.; trustee United Way of Southeastern Pa., co-founder Rep. Presdl. Task Force; mem. U.S. Congl. Adv. Bd.; cons. on Korean affairs Phila. City Coun.; hon. mem. adv. coun. Peaceful Unification Policy of Korea; chmn. bd. Korean-Am. Christian Broadcasting of Phila.; mem. Phila. Internat. City Coord. Com.; commr. Pa. Human Rels. Commn., 1991—; founder, pres. Korean Heritage Found., 1991—; amb. City of Phila., 1991. Recipient Phila. award-Human Rights award, 1981, Disting. Community Svc. award Phila. Dist. Atty., 1981, medal of Merit Presdl. Task Force, 1981, Medal of Nat. Order, Republic of Korea, 1984, Nat. Dong Baek medal Republic of Korea, 1987, award City Coun. Phila., 1987, Gov.'s Pa. Heritage awards, 1990, commendation award Pa. Senate, 1991, award Asian Law Ctr., 1991; named to Legion of Honor, The Chapel of Four Chaplains, named Amb. City of Phila., 1991. Mem. AMA, Am. Soc. Internal Medicine, Am. Coll. Cardiology, Am. Heart Assn. (mem. council on clin. cardiology), Pa. Med. Soc., Phila. County Med. Soc., Royal Soc. Health, Am. Coll. Internat. Physicians, World Med. Assn., Fedn. State Med. Bds., Am. Law Enforcement Officers' Assn., Am. Fedn. Police, Internat. Culture Soc. Korea (hon.), Am. Soc. Contemporary Medicine and Surgery. Home: 3903 Somers Dr Huntingdon Valley PA 19006-1913 Office: 60 Township Line Rd Philadelphia PA 19117-2249

YUN-CHOI, HYE SOOK, natural products science educator; b. Seoul, Korea, Sept. 6, 1944; d. Soo Chung and Eae Sook (Lee) Yun; m. Sang Sam Choi, June 21, 1970; children: Yunsun Sumi C., Jay Hoon. MS in Med. Chemistry, U. N.C., 1968, PhD in Med. Chemistry, 1971. Rsch. fellow Ohio State U., Columbus, 1970-71, U. N.C., Chapel Hill, 1972-74, U. Scientifique et Medicale de Grenoble, France, 1975; from instr. to asst. prof. to assoc. prof. Natural Products Rsch. Inst., Seoul Nat. U., 1974—, prof., 1988—. Co-editor: Natural Products Sciences, 1988, New Drug Development from Natural Products, 1989; contbr. articles to profl. jours. including Jour. of Natural Products and Korean Jour. of Pharmacognosy. Grantee Ministry of Health and Social Affairs, 1991-92, Korean Sci. and Engring. Found., 1986, 90, Korea Rsch. Found., 1985. Mem. Korean Soc. Pharmacognosy (sci. sec. 1986-88), Pharm. Soc. of Korea (assoc. sec. 1982), Am. Soc. Pharmacognosy, Am. Chem. Soc. Home: 12-604 Sunkyung, Apt Daichidong, Seoul 135-281, Republic of Korea Office: Nat Products Rsch Inst, Seoul Nat U, Seoul 110-460, Republic of Korea

YURA, JOSEPH ANDREW, civil engineering educator, consulting structural engineer; b. Hazelton, Pa., Apr. 11, 1938; s. Michael and Anna (Sokol) Y.; m. Joan Marie Seman, Aug. 22, 1964; children: Thomas, Christine, Paul, Elizabeth. B.S., Duke U., 1959; M.S., Cornell U., 1961; Ph.D., Lehigh U., 1965. Registered profl. engr., Tex., Pa., Fla. Asst. prof. Lehigh U., Bethlehem, Pa., 1965-66; asst. prof. U. Tex., Austin, 1966-70, assoc. prof., 1970-75, prof., 1975-82, Warren Bellows prof., 1982—. Recipient Gen. dynamics Teaching award U. Tex., 1972; recipient T.R. Higgins Lectureship award Am. Inst. Steel Constrn., 1974, Hussein M. Alharthy Centennial Professorship award U. Tex., Austin, 1987. Mem. ASCE (Raymond C. Reese Rsch. prize 1991), Structural Stability Rsch. Coun., Rsch. Coun. on Structural Connections, Structural Engrs. Assn. Tex., Am. Soc. for Engring Edn., Phi Beta Kappa, Tau Beta Pi. Democrat. Roman Catholic. Home: 5308 Bull Run Austin TX 78727-6608 Office: U Tex Dept Civil Engring 10100 Burnet Rd Austin TX 78758-4497

YURIST, SVETLAN JOSEPH, mechanical engineer; b. Kharkov, USSR, Nov. 20, 1931; came to U.S., 1979, naturalized, 1985; s. Joseph A. and Rosalia S. (Zoilman) Y.; m. Imma Lea Erlikh, Oct. 11, 1960; 1 child, Eugene. M.S. in Mech. Engring. with honors, Poly. Inst., Odessa, USSR, 1954. Engr. designer Welding Equipment Plant, Novaya Utka, USSR, 1954-56; sr. tech. engr. Heavy Duty Automotive Crane Plant, Odessa, 1956-60, asst. chief matallugist, 1971-78; supr. research lab. Inst. Spl. Methods in Foundry Industry, Odessa, 1960-66, project engr. sci. research, 1966-71; engr. designer Teledyne Cast Product, Pomona, Calif., 1979-81; sr. mech. engr. Walt Elliot Disney Enterprises, Glendale, Calif., 1981-83; foundry liaison engr. Pacific Pumps div. Dresser Industries, Inc., Huntington Park, Calif., 1984-86; casting engr. Superior Industries Internat., Inc., Van Nuys, Calif., 1986-89; mech. engr. TAMCO Steel, Rancho Cucamonga, Calif., 1989—. Recipient award for design of automatic lines for casting electric motor parts USSR Ministry Machine Bldg. and Handtools Mfr., 1966, for equipment for permanent mold casting All Union Exhbn. of Nat. Econ. Achievements, 1966-70. Mem. Am. Foundrymen's Soc. Contbr. reports, articles to collections All Union Confs. Spl. Methods in Foundry, USSR; USSR patentee permanent mold casting. Home: 184 W Armstrong Dr Claremont CA 91711-1701 Office: TAMCO Steel 12459 Arrow Hwy Rancho Cucamonga CA 91739-9698

YURKO, JOSEPH ANDREW, chemical engineer; b. Youngstown, Ohio, Mar. 30, 1955; s. Joseph George and Virginia Mary (Cossentino) Y.; m. Valerie Ann Congdon, Sept. 9, 1992; 1 child, Andrew Dale. B in Engring. Sci., Cleve. State U., 1981, B in Chem. Engring., 1981. Lic. profl. engr. Tex. Structural draftsperson HK Ferguson Co., Cleve., 1974-76, architectural draftsperson, 1976-77, process design engr., 1981-84, chem. engr. Chemical Data Systems, Inc., Oxford, Pa., 1984-85; tech. sales engr. Autoclave Data Systems, Inc., Erie, Pa., 1985-86; sr. process design engr. MK Ferguson Co., Cleve., 1987-88, process start up engr., 1988--; cons. EI Dupont de Nemours and Co., Wilmington, Del., 1985-86, Mobil Oil Rsch. Ctr., Princeton, N.J., 1985-86, SmithKline French Labs., Phila., 1985-86, Anheuser-Busch Cos., St. Louis; spkr. in field. Author: (manuals) Aseptic Filtration Technical Operating Procedure, 1990, Waste Stream Evaporator Technical Operating Procedure, 1991, Clean in Place Process Technical Operating Procedure, 1992; editor: (manual) Natural Water Carbonation Technical Operations, 1992. Coach track Cath. Youth Orgn. St. Bridgets Cath. Ch., Parma, Ohio, 1977; campaigner Multiple Sclerosis Soc., Cleve., 1978; counselor Soc. Crippled Children Cuyahoga County, Strongsville, 1979. Mem. Am. Inst. Chem. Engrs. (sec. Cleve. chpt. 1988-89, vice chair Del. Valley chpt. 1986-87), Food Pharm. and Bioengring. Div., Am. Chem. Soc, Cleve. Engring. Soc. Republican. Roman Catholic. Achievements include proposed process conversion of carbon dioxide emissions into oxygen and hydroponically grown products for Anheuser-Busch Inc.; ultraviolet photographic study of particle motion in fluid dynamic study of new Autoclave Engrs. microclave reactor with catalyst basket and internals for heterogeneous catalysis, proposition of ceramically encapsulated metallic beads impregnated with catalyst for novel reactor designed to contain catalyst in magnetic zone. Home: 19099 Hunt Rd Strongsville OH 44136-8415 Office: MK Ferguson Co 1500 W 3rd St Cleveland OH 44113-1406

YUSPA, STUART H., cancer etiologist; b. Balt., July 19, 1941. BS, Johns Hopkins U., 1962; MD, U. Md., 1966. Diplomate Am. Bd. Internal Medicine. Intern Hosp. U. Pa., 1966-67; rsch. assoc. cancer Nat. Cancer Inst., 1967-70; res. internal medicine Hosp. U. Pa., 1970-72; sr. investigator cancer Nat. Cancer Inst., 1972—; chief divsn. cancer etiology Lab. Cellular Carcinogenesis & Tumor Promotion, 1981—; mem. biol. models segment Carcinogenesis Program Nat. Cancer Inst., 1972-78; editor-in-chief Molecular Carcinogenesis. Named Montagna Lectr., 1988; recipient Lila Gruber Cancer Rsch. award, 1989, Elizabeth Miller Meml. Lectr. award, 1990, G.H.A. Clowes Meml. award Am. Assn. fro Cancer Rsch., Phila., 1993. Mem. AAAS, Am. Assn. Cancer Rsch., Am. Soc. Cell Biology, Soc. Investigators Dermatology. Achievements include research in determining mechanisms wherby chemicals initiate or promote malignant transformation of epithelial cells. Office: National Cancer Institute Cancer Etiology 9000 Rockville Pike, Bldg 31 Bethesda MD 20892*

YUSSOUFF, MOHAMMED, physicist, educator; b. Cuttack, India, Aug. 14, 1942; came to U.S., 1991; s. Haji and Nurunnisa Fakhruddin; m. Farhana Begum, Apr. 6, 1969; children: Ashraf, Zeenat, Mustafa. MSc, Delhi U., 1963; PhD, Indian Inst. Tech., Kanpur, 1967. Prof. physics Indian Inst. Tech., Kanpur, 1967-90; vis. prof. physics Mich. State U., East Lansing, 1991—; vis. scientist U. Köln, Germany, 1972-74, U. Western Ont., London, Can., 1990-91; Humboldt scientist Atomic Energy Agy, Jülich, Germany, 1979-81; vis. prof. U. Konstanz, Germany, 1986-89; guest scientist Ford Rsch., Dearborn, Mich., 1991—; mem. com. physics examination Pub. Svc. Commn., Delhi, India, 1976-86, rsch. grants Univ. Grants Commn., Delhi, 1985-90; bd. dirs. Aligarh (India) Muslim U., 1989-90; dir. Internat. Sch. on Band Structure, Indian Inst. Tech., 1986. Editor: Electronic Band Structure and Its Applications, 1987, The Physics of Materials, 1987. Mem. Am. Phys. Soc., Internat. Ctr. Theoretical Physics (assoc.). Islam. Achievements include pioneering work on theory of freezing, theory of disordered systems, exhaust gas sensors and foundations of quantum theory. Home: Apt 45 930 Old Goddard Rd Lincoln Park MI 48146 Office: Mich State Univ Dept Physics East Lansing MI 48824-1116 also: Ford Motor Co Scientific Rsch Lab Dept Physics Dearborn MI 48121-2053

YUSUF, SIAKA OJO, nuclear engineer; b. Okene, Nigeria, Dec. 8, 1960; came to the U.S., 1987; s. Egwo and Salametu (Oniya) Y. BEE, Ahmadu Bello U., Zaria, Nigeria, 1982, MEE, 1986; MS in Nuclear Engring., U. Mich., 1988, PhD in Nuclear Engring., 1993. Vis. scientist Karlsruhe (Germany) Rsch. Ctr., 1982; grad. asst. Ahmadu Bello U., 1983-85, asst. lectr., 1985-86; vis. scientist Internat. Ctr. for Theoretical Physics, Trieste, Italy, 1985; course grader U. Mich., Ann Arbor, 1987-89, teaching asst., 1989-91, Rackham predoctoral fellow, 1991-92, rsch. asst., 1992-93; rsch. fellow GM R & D Ctr., Warren, Mich., 1993—. Contbr. articles and abstracts to Nuclear Sci. and Engring. Am. Nuclear Soc. Transactions. Active Nat. Youth Svc. Corps, Enugu, Nigeria, 1982-83. Scholar Shell Petroleum Co., 1979-82. Mem. Sigma Xi (assoc.), Alpha Nu Sigma. Republican. Achievements include development of dynamic compensator for rhodium self-powered neutron detector, electronic speed transducer for machine control, checking procedure of exam scripts, a shape-independent model of monitor neutron activation analysis. Home: 31379 Mound Rd # E Warren MI 48092 Office: Analytical Chemistry Dept NAO R&D Ctr Warren MI 48090

ZABIN, BURTON ALLEN, chemist; b. Chgo., Mar. 18, 1936. BS, U. Ill., 1957; PhD in Inorganic Chemistry, Purdue U., 1962. Rsch. assoc. chemistry Stanford U., 1962-63; dir. rsch. Bio-Rad Labs., Hercules, Calif., 1963-72, divsn. mgr. chemistry, 1972—. Mem. Am. Chem. Soc. Achievements include research in separations chemistry, including ion exchange resins, gel filtration materials and other column chromatographic materials. Office: Bio-Rad Labs Inc 1000 Alfred nobel Hercules CA 94547*

ZABINSKY, ZELDA BARBARA, operations researcher, industrial engineering educator; b. Tonawanda, N.Y., Oct. 31, 1955; d. Joseph Marvin and Helen Phyllis (Kava) Z.; m. John Clinton Palmer, July 15, 1979; children: Rebecca Ann Zabinsky, Aaron Zeff Palmer. BS, U. Puget Sound, Tacoma, 1977; MS, U. Mich., 1984, PhD, 1985. Tutor math. U. Puget Sound, 1975-77; programmer, analyst Nat. Marine Fisheries, Seattle, 1977, Boeing Computer Svcs., Seattle, 1977-78; sr. systems analyst Vector Rsch. Inc., Ann Arbor, Mich., 1980-84; assoc. prof. indsl. engring. U. Wash., Seattle, 1985-93, 1993—; cons. Boeing Corp., Seattle, 1987, Numerical Methods, Inc., Seattle, 1989-90, METRO, Seattle, 1992. Contbr. articles to tech. jours. Mem. faculty adv. bd. Women in Engring., U. Wash., 1990. Recipient E. Goman Math. award, 1977, Rsch. Initiation award NSF, 1992—; Howarth-Thompson scholar, 1973-77; Benton fellow, 1983-84; rsch. grantee NSF, NASA-Langley, Nat. Forest Svc., NATO, Boeing, 1985—. Mem. Ops. Rsch. Soc. Am., Inst. Indsl. Engrs. (sr.), Math. Programming Soc., Mortar Board, Phi Kappa Phi. Jewish. Avocations: family activities, camping, skiing, windsurfing. Office: U Wash Dept Indsl Engring FU-20 Seattle WA 98195

ZACCARDI, LARRY BRYAN, spectrochemistry scientist, microbiologist; b. Pocatello, Idaho, Nov. 28, 1966; s. Larry Peter Zaccardi and Karen Christine Lee; m. Tamara Page, July 24; children: Jessica Paige, Kara Christine. BS in Microbiology, Idaho State U., 1991, MS, 1993. Rsch. asst. Idaho State U., Pocatello, 1990-92; scientist Westinghouse Idaho Nuclear Co., Idaho Falls, 1992, spectrochemistry scientist, 1992—. Contbr. articles to profl. jours. Recipient Eagle Scout award Boy Scouts Am., 1982; Idaho State U. grantee, 1991. Mem. AAAS, Am. Soc. for Microbiology, Sigma Xi. Republican. Mormon. Home: 409 Kurtwood Dr Pocatello ID 83204 Office: Westinghouse Idaho Nuclear PO Box 4000 MS-5211 Idaho Falls ID 83403

ZACHARIUS, MARTIN PHILIP, mechanical engineer; b. N.Y.C., June 19, 1928; s. Herman Roy and Rebecca (Bengis) Z.; m. Shirley Mofsovitz, Mar. 31, 1948; children: Wendy J. Zacharius Goldstein, Sherrie L. Zacharius. BME, NYU, 1951. Registered profl. engr., N.Y., N.J., Conn., Mass., Fla., Ariz., D.C., also others. Jr. designer Raisler Corp., N.Y.C., 1948; supervising engr. Bechtel Corp., N.Y.C., 1952-53; sr. mech. engr. Krey & Hunt, N.Y.C., 1953-55; prin. M.P. Zacharius & Assocs., N.Y.C., 1955—; pres. Tech. Installations, Inc., N.Y.C., 1963—. Mem. ASHRAE, Am. Arbitration Assn. (life), N.Y. Assn. Cons. Engrs. (Engring Excellence awards in mech. and elec. engring. 1969, 70, 73, 74). Office: 99 Madison Ave New York NY 10016-7419

ZACHARY, LOUIS GEORGE, chemical company consultant; b. Cambridge, Mass., Aug. 14, 1927; s. George E. and Angelike (Hantsis) Zacharakis; AB in Chemistry, Harvard U., 1950; MBA, Columbia U., 1951; m. Lillie Vletas, Apr. 20, 1955; children: Leslie A., Louis George. Prodn. supr. Dewey & Almy Co., Acton, Mass., 1951-52; salesman chem. div. Union Camp Corp., Wayne, N.J., 1952-59, sales mgr. chem. div., 1959-62, gen. mgr. chem. ops., 1962-66, gen. mgr. chem. div., 1970-78, v.p., 1974-78; v.p. Drake Mgmt Co., N.Y.C., 1966-70; sr. v.p. GAF Corp., N.Y.C., 1978-82, mem. office of chmn., 1981-82; cons., 1983-84; chmn., chief exec. officer Universal Die Casting, Inc., Saline, Mich., 1984-90; acting pres. chem. div. Church & Dwight Inc., 1990-91. Mem. vis. com. chem. engring. dept. Johns Hopkins U., Balt., 1981-83. Served with USN, 1945-46. Mem. Chem. Mfrs. Assn. (dir. 1979-83), Synthetic Organic Chem. Mfrs. Assn., Soc. Chem. Industry. Clubs: Baltusrol Golf; Harvard (N.Y.C.). Co-editor: Tall Oil and Its Uses, 1965. Home: 227 Oak Ridge Ave Summit NJ 07901-3258

ZACHERT, VIRGINIA, psychologist, educator; b. Jacksonville, Ala., Mar. 1, 1920; d. R.E. and Cora H. (Massee) Z. Student, Norman Jr. Coll., 1937; A.B., Ga. State Woman's Coll., 1940; M.A., Emory U., 1947; Ph.D., Purdue U., 1949. Diplomate: Am. Bd. Profl. Psychologists. Statistician Davison-Paxon Co., Atlanta, 1941-44; research psychologist Mil. Contracts, Auburn Research Found., Ala. Poly. Inst.; indsl. and research psychologist Sturm & O'Brien (cons. engrs.), 1958-59; research project dir. Western Design, Biloxi, Miss., 1960-61; self-employed cons. psychologist Norman Park, Ga., 1961-71, Good Hope, Ga., 1971—; rsch. assoc. med. edn. Med. Coll. Ga., Augusta, 1963-65, assoc. prof., 1965-70, rsch. prof., 1970-84, rsch. prof. emerita, 1984—, chief learning materials div., 1973-84, mem. faculty senate, 1976-84, mem. acad. coun., 1976-82, pres. acad. coun., 1983, sec., 1978; mem. Ga. Bd. Examiners of Psychologists, 1974-79, v.p., 1977, pres., 1978; adv. bd. Comdr. Gen. ATC USAF, 1967-70; cons. Ga. Silver Haired Legislature, 1980-86, senator, 1987-93, pres. protem, 1987-88, pres. 1989-93, rep. 1993—, speaker protem, 1993—; gov.'s appointee Ga. Coun. on Aging, 1988—; U.S. Senate mem. Fed. Coun. on the Aging, 1990—. Author: (with P.L. Wilds) Essentials of Gynecology-Oncology, 1967, Applications of Gynecology-Oncology, 1967. Del. White House Conf. on Aging, 1981. Served as aerologist USN, 1944-46; aviation psychologist USAF, 1949-54. Fellow AAAS, Am. Psychol. Assn.; mem. AAUP (chpt. pres. 1977-80), Sigma Xi. (chpt. pres. 1980-81). Baptist. Home: 1126 Highland Ave Augusta GA 30904-4628 Office: Med Coll Ga Dept Ob-Gyn Augusta GA 30912

ZAGARA, MAURIZIO, physicist; b. Rome, July 15, 1946; s. Tullio and Liliana (Roberti) Z.; m. Lina Stefani, Dec. 14, 1974; children: Francesco, Annalisa. B.Physics, U. Rome, 1971. TWT designer Elettronica Spa, Rome, 1973-82, I.R. system mgr., 1982-84; rsch. and devel. mgr. E.S.P. Spa., Aprilia, Italy, 1984-86; tech. dir. Serit Spa, Colleferro, Italy, 1986-89; rsch. and devel. mgr. Tecnobiomedica Spa, Pomezia, Italy, 1989—. Patentee in field. Avocations: running, scuba diving, fishing, mountain climbing. Home: Via R Gigliozzi 194, 00128 Rome Italy

ZAGER, RONALD I., chemist, consultant; b. N.Y.C., Dec. 27, 1934; s. Joseph and Theodora (Court) Z; m. Judith Ellen Bilt, Dec. 24, 1961 (div. July 1975); children: Scott Lawrence, Joseph Daniel. BS, Bklyn. Coll., 1955; MS in Chemistry, Stevens Inst. Tech., 1969. Chemist Charles Pfizer & Co., N.Y.C., 1956-58, Halocarbon Products, Hackensack, N.J., 1958-66; devel. chemist Tenneco Chems., Garfield, N.J., 1966-71; sr. chemist Givaudan Corp., Clifton, N.J., 1971-77; tech. dir. Internat. Flavors and Fragrances, Union Beach, N.J., 1977-88; cons. Highlands, N.J., 1988-92, Glen, N.H., 1993—. Mem. Am. Chem. Soc., Assn. Cons. Chemists and Chem. Engrs. (v.p. 1990-92, pres. 1993-94). Achievements include research in aroma chemicals and organic fluorocarbons. Office: Ronald Zager Assocs 85 Glen Ledge Rd Glen NH 03838-1200

ZAGOREN, JOY CARROLL, health facility director, researcher; b. N.Y.C., Oct. 31, 1933; d. Murray Morris and Celia (Donner) Rossman; m. Robert H. Zagoren, June 29, 1958 (div. 1988); children: Glenn, Robin; m. Robert Henry Chester, Apr. 1, 1988; children: Peter, Lisabeth, Melinda, Cecily, Kate. BS, NYU, 1957; MS, Adelphi U., 1969; PhD with distinction, NYU, 1981. Sec. sch. faculty Great Neck (N.Y.) Pub. Schs., 1957-71; rsch. scientist Inst. Psychobiol. Studies, Queens Village, N.Y., 1968-71; rsch. assoc. Albert Einstein Coll. Medicine, Bronx, N.Y., 1971-84; asst. prof. SUNY, Stony Brook, 1984-86; dir. Seriatum, N.Y.C., 1991—; ptnr. Winter Tree Collection. Editor: The Node of Ranvier, 1984; contbr. articles to profl. jours. Chairperson Peace Corps Svc. Coun., Tri-State, 1965-75; pres. Kidney Found., L.I., N.Y., 1965-77; v.p. United Community Fund, L.I., 1970-83; bd. dirs. Jerusalem Mental Health Ctr., N.Y.C., 1986—. Recipient post doctoral fellowship NIH, 1982-84, svc. awards Kidney Found., Kiwanis, and others, 1970-87; named Disting. Alumnus of Yr., Adelphi U., 1986. Mem. Am. Assn. Neuropathologists, Am. Assn. Counseling, Esrath Nashim Hosp. (chairperson 1986—), Kappa Delta Epsilon. Democrat. Jewish. Avocations: art, literature, piano, swimming, gardening. Home: 405 E 82d St New York NY 10028 Office: Seriatum PO Box 371 Livingston Manor NY 12758

ZAH, CHUNG-EN, electrical engineer; b. Taiwan, Republic of China, Jan. 4, 1955; came to U.S., 1981; s. Zung-Kwae and Luan-Chuang (Huang) Z.; m. Li Fung Chang, 1984; children: Angela, Eugenia. BS, Nat. Taiwan U., Taipei, 1977, MS, 1979; MS, Calif. Inst. Tech., 1982, PhD, 1986. Mem. tech. staff Bellcore, Red Bank, N.J., 1985—, disting. mem. tech. staff, 1992—. Fellow Optical Soc. Am.; mem. IEEE (sr.). Achievements include 3 patents on optoelectronic devices. Office: Bellcore NVC3X361 331 Newman Springs Rd Red Bank NJ 07701

ZAHARIEV, GEORGE KOSTADINOV, computer company executive; b. Pazardjik, Bulgaria, Jan. 30, 1941; s. Konstantin Zahariev Nachkov and Ivanka Stojchkova (Yotova) Zaharieva; m. Diana Ivanova Vacheva, Aug. 27, 1966 (div. Sept. 1987); children: Konstantin, Alexander; m. Temenoujka Anguelova Damianova, Dec. 18, 1987; 1 child, Ivan. BS in Mining, Higher Inst. of Mining and, Geology, Sofia, Bulgaria, 1966; MS, Columbia Univ., 1972, D of Engring. Sci., 1975. Chief engr. Bratze Mine/Burgas Copper Mines, M. Tarnoro, Bulgaria, 1966-68; rsch. assoc. Inst. Mining and Geology, Sofia, 1968-70, 75-79, Inst. Cybernetics/Bulgaria Acad. of Sci., Sofia, 1974-80; mktg. mgr. Info. Systems and System Engring. Svcs., Sofia, 1980-82; mng. dir. RDL Programa Bulgaria Acad. of Sci., Sofia, 1982-90; country mgr. Bulgaria Internat. Computers Ltd., U.K., 1990—; cons. UN, N.Y., 1973, 83. Author numerous publs. in field. Mem. Union Bulgarian Engrs., Sofia, 1966, Union Bulgarian Mathematicians, Sofia, 1982. Eastern Orthodox. Avocations: travel, music, books, hiking. Office: ICL-Bulgaria, Acad G Bonchev St, Block 3A, 1113 Sofia Bulgaria

ZAHND, HUGO, retired biochemistry educator; b. Bern, Switzerland, May 16, 1902; came to U.S., 1920, naturalized 1932; m. Rose Theresa Genovese, June 23, 1926 (div. 1968); children: Rosemarie, Richard Hugo; m. Ranka Bekic, Oct. 28, 1968. B.S. in Chemistry, NYU, 1926; A.M. in Chemistry, Columbia U., 1929, Ph.D. in Biochemistry, 1933. Tutor Bklyn. Coll., 1930-33, instr. chemistry, 1933-36, asst. prof., 1936-47, assoc. prof., 1947-55, prof., 1955-72, prof. emeritus, 1972—; cons. in history of chemistry and food and nutrition, 1950—. Contbr. articles to profl. jours. Fellow Am. Inst. Chemists (cert.), AAAS; mem. N.Y. Acad. Sci., Am. Chem. Soc., Isis-History of Sci. Soc., Sigma Xi. Republican. Avocations: keyboard classics pianist, photography. Home: 42 Herbert Ave Port Washington NY 11050-2915

ZAHREDDINE, ZIAD NASSIB, mathematician, educator, researcher; b. Ain-Zhalta, El-Shoof, Lebanon, Aug. 30, 1952; s. Nassib Dawood and Afaf Kattar Zahreddine; m. Reema Mahmood Hassanieh, Jan. 18, 1987; children: Rami, Nizar. BS in Math. with honors, Lebanese U., Hadath, 1975; MS in Math., Sussex (Eng.) U., 1979; PhD in Math., London U., 1983. High sch. tchr. Lebanese Ministry of Edn., Beirut, 1975-78; postdoctoral researcher in math. London U., 1983-84; assoc. prof., researcher math. United Arab Emirates U., Al-Ain, 1984—; reviewer math. revs.; advisor Lebanese Ministry Edn., Beirut, 1975-78; cons. Westfield Coll., London U., 1979-84; mem. various coms. United Arab Emirates U., Al-Ain, 1984—; researcher in stability theory and differential equations. Patentee Boundary Conditions and the Cayley Transform, 1986, An Extension of the Routh Array to the Complex Case, 1990, numerous others. Lebanese U. grantee, 1978. Mem. Am. Math. Soc., London Math. Soc. Avocations: marathons, mountain climbing, squash, skiing, chess. Home: Ain-Zhalta Lebanon Office: United Arab Emirates U, Math Dept, PO Box 17551, Al-Ain United Arab Emirates

ZAIA, JOHN ANTHONY, hematologist; b. Oneida, N.Y., Oct. 28, 1942; married, 1971; children. AB, Holy Cross Coll., 1964; BMSc, Dartmouth Coll., 1966; MD, Harvard U., 1968. Intern St. Louis Child Hosp. and Child Med. Ctr., 1968-71, sr. asst. surgery virology divsn., lab. bur. USPHS, 1971-74; rsch. fellow infectious diseases Sidney Farber Cancer Inst., 1976-77; dir. divsn. pediatrics City of Hope Nat. Med. Ctr., Duarte, Calif., 1980—. Mem. Am. Pediatric Soc., Infectious Disease Soc. Am. Achievements include research on mechanisms of viral pathogenesis; development of anti-viral therapies. Office: City of Hope Nat Med Ctr Dept of Pediatrics 1500 E Duarte Rd Duarte CA 91010*

ZAIDI, IQBAL MEHDI, biochemist, scientist; b. Bijnor, India, June 30, 1957; s. Iqbal Haider and Habib (Zehra) Z.; m. Nuzhat Shikoh, Jan. 2, 1993. BS in Chemistry with honors, Aligarh M. U., 1976, MS in Biochemistry, 1978, PhD in Biochemistry, 1984. Cert. in radiation; cert. in health and happiness. Rsch. fellow Indsl. Toxicology Rsch. Ctr., Lucknow, India, 1979-83; rsch. affiliate N.Y. State Health Dept., Albany, 1984-91; scientist Applied Biosystems div. Perkin Elmer Corp., Foster City, Calif., 1991—. Contbr. articles to profl. jours. Mem. AAAS, Am. Chem. Soc. (biochem. tech. div. 1992—), Shia Assn. Bay Area, N.Y. Acad. Scis. Avocations: photography, swimming, travel, natural history. Office: Applied Biosystems div Perkin Elmer Corp 850 Lincoln Center Dr Foster City CA 94404

ZAK, SHELDON JERRY, mechanical engineer, educator; b. Tczew, Poland, May 28, 1955; came to U.S., 1983; s. Stefan Franciszek and Barbara (Charczuk) Z.; m. Dorothy Zerykier, July 6, 1986; children: Ilana, Helena, Aaron Stefan. BS, Gdansk (Poland) Tech. U., 1978, MSME, 1982. Asst. prof. Gdansk Tech. U., 1980-82, Tuskegee Inst. (Ala.) Sch. Engring., 1983-84; project engr. Syska & Hennessy Engrs., Inc., N.Y.C., 1984-85, Flack & Kurtz Cons. Engrs., N.Y.C., 1986-87, Carlson & Sweat-Monenco, N.Y.C., 1987-88, Edwards & Zuck Cons. Engrs., N.Y.C., 1988-90; prof. Tech. Career Insts., N.Y.C., 1991—; book reviewer Delmar Pub., Inc., Albany, N.Y., 1992—. Author: Heat Transfer, 1980; contbr. articles to profl. jours. Mem. ASME (assoc. mem.), Am. Soc. Heating, Refrigeration and Air-Conditioning Engrs. (assoc. mem.). Republican. Jewish.

ZAKARAUSKAS, PIERRE, physicist; b. Amos, Que., Can., Dec. 25, 1958; s. Joseph and Réjeanne (Latreille) Z.; m. Louise Emily Osborne, May 29, 1982. BSc, U Que., 1980; PhD, U. B.C., Vancouver, 1984. Def. scientist Def. Rsch. Establishment Atlantic, Dartmouth, N.S., 1984-86, Def. Rsch. Establishment Pacific, Victoria, B.C., 1986—; rsch. assoc. dept. psychology U. B.C., 1988-93, asst. prof. dept. ophthalmology, 1993—. Contbr. articles and referee to profl. jours. including Jour. Acoustical Soc. Am., IEEE Transaction on Signal Processing, Neural Network for Ocean Engrng., IEEE Proceedings, Hearing Res., Phys. Rev. D. Mem. Greenpeace, 1990—. Mem. Acoustical Soc. Am., Can. Acoustical Assn. Achievements include research on ice-cracking noise in Arctic, complexity analysis of nearest-neighbor search, applying neural networks to acoustic localization, computational theories of sound localization by mammals, high-energy particle physics. Home: 1252 Woodway Rd, Victoria, BC Canada V9A 6Y6 Office: Def Rsch Establish Pacific, FMO, Victoria, BC Canada V9A 6Y6

ZAKIM, DAVID, biochemist; b. Paterson, N.J., July 10, 1935; s. Sam and Ruth (Surokin) B.; m. Nancy Jane Levine, June 12, 1957 (div. 1976); children: Michael, Eric, Thomas; m. Dagmar Auralia Stanke, July 30, 1978; children: Tamara, Robert. AB in Chemistry, Cornell U., 1956; MD summa cum laude, SUNY, Bklyn., 1961. Diplomate Am. Bd. Internal Medicine. Intern N.Y. Hosp., N.Y.C., 1961-62, asst. resident, 1962-63, fellow, 1963-65; asst. prof. to prof. medicine and pharmacology U. Calif., San Francisco, 1968-83; Vincent Astor Disting. prof. medicine Cornell U. Med. Coll., N.Y.C., 1983—; prof. biochemistry Cornell U. Grad. Sch. Med. Sci., 1983—. Contbr. over 150 articles to profl. jours.; editor: Hepatology: A Textbook of Liver Disease, 1982, 2d edit. 1990, Disorders of Acid Secretion, 1991; editor series: Current Topics in Gastroenterology, 1985—. Capt. U.S. Army, 1965-68. Named Disting. Alumnus, SUNY-Bklyn., 1986. Mem. Am. Assn. Physicians, Am. Soc. Biol. Chemists, Biophysics Soc. Achievements include patent on fusion of proteins with lipid vesicles; demonstration of the lipid-dependence of UDP-glucuronology transferase; elucidation of mechanism for uptake of fatty acids and bilirubin into cells. Home: 15 Cole Dr Armonk NY 10504 Office: Cornell U Med Coll Medicine-Gastroenterology 1300 York Ave New York NY 10021-4805

ZALDASTANI, OTHAR, structural engineer; b. Tbilisi, Georgia, USSR, Aug. 10, 1922; came to U.S., 1946; naturalized, 1956; s. Soliko Nicholas and Mariam Vachnadze (Hirsely) Z.; m. Elizabeth Reily Bailey, June 22, 1963; children: Elizabeth, Anne, Alexander. Diplome D'Ingenieur, Ecole Nationale des Ponts et Chaussees, Paris, 1945; Licencie es Scis., Sorbonne, Paris, 1946; MS in Geotech. Engrng., Harvard U., 1947, DSc in Aerodynamics, 1950. Registered profl. engr., Mass., R.I., Tenn., Mo., N.H. Mem. faculty Harvard U., Cambridge, Mass., 1947-50; ptnr. Nichols, Norton and Zaldastani, Boston, 1952-63; pres. Nichols, Norton and Zaldastani, Inc., Boston, 1964-76, Zaldastani Assocs., Inc., Boston, 1976-88, chmn., 1988—; Gordon McKay vis. lectr. structural mechanics Harvard U., 1961; trustee, 1st v.p. Mass. Constrn. Industry Bd., 1973-76; mem. Mass. Designer Selection Bd., 1976-80. Contbg. author: Advances in Applied Mechanics, vol. 3, 1953. Patentee sound absorbing block, prestressed concrete beam and deck system. Trustee Wheelock Coll., Boston, 1975-81, mem. corp., 1984—; trustee Boston U. Med. Ctr., 1976—; trustee Brooks Sch., North Andover, Mass., 1986—. Recipient awards from various orgns. and agys. including Prestressed Concrete Inst., Cons. Engrs. Coun. New Eng., Am. Inst. Steel Constrn., Concrete Reinforcing Steel Inst., Dept. Transp., Am. Concrete Inst. Fellow ASCE (Ralph W. Horne award), AIAA (assoc.), Am. Concrete Inst.; mem. Georgian Assn. in the U.S. (pres. 1958-65), Sigma Xi, Harvard Club, Harvard Faculty Club (Cambridge), Somerset Club (Boston), Country Club (Brookline, Mass.), Rolling Rock Club (Ligonier, Pa.). Home: 70 Suffolk Rd Chestnut Hill MA 02167-1218 Office: Zaldastani Assocs Inc 7 Water St Boston MA 02109-4511

ZALESKI, MAREK BOHDAN, immunologist; b. Krzemieniec, Poland, Oct. 18, 1936; came to U.S., 1969, naturalized, 1977; s. Stanislaw and Jadwiga (Zienkowicz) Z. M.D., Sch. Medicine, Warsaw, 1960, Dr. Med. Sci., 1963. Instr. dept. histology Sch. Medicine, Warsaw, 1955-60; asst. prof. Sch. Medicine, 1960-69; research asst. prof. (Henry C. and Bertha H. Buswell fellow) dept. microbiology SUNY, Buffalo, 1969-72; assoc. prof. SUNY, 1976-78, prof., 1978—; vis. scientist Inst. Exptl. Biology and Genetics, Czechoslovak Acad. Sci., Prague, 1965; Brit. Council's scholar, research lab. Queen Victoria Hosp., East Grinstead, Eng., 1966-67; asst. prof. dept. anatomy Mich. State U., East Lansing, 1972-75, assoc. prof., 1975-76. Contbg. author: Transplantation and Preservation of Tissues in Human Clinic, 1966, The Man, 1968, Cytophysiology, 1970, Principles of Immunology, 1978, Medical Microbiology, 1982, Molecular Immunology, 1984; co-author: Immunogenetics, 1983, co-editor: Immunobiology of Major Histocompatibility Complex, 1981; co-translator: Spirit of Solidarity, 1984, Marxism and Christianity: The Quarrel and the Dialogue in Poland (J. Tischner), 1987; mem. editorial com. Immunol. Investigations, Immunologia Polska; contbr. articles to med. jours. Former mem. adv. com. for Internat. Rescue Com., Amnesty Internat., Raul Wallenberg Com. USA; bd. dirs., Permanent Chair Polish Culture, Canisius Coll., Bufallo. NIH grantee, 1976-88; NEH grantee, 1985-87. Mem. Polish Anat. Soc., Transplantation Soc., Internat. Soc. Exptl. Hematology, Ernest Witebsky Center Immunology, Am. Assn. Immunologists, Buffalo Collegium of Immunology, N.Y. Acad. Scis., Solidarity and Human Rights Assn. Roman Catholic. Office: SUNY Dept Microbiology Buffalo NY 14214

ZALISKO, EDWARD JOHN, biology educator; b. Peoria, Ill., Jan. 8, 1958; s. Miles Joseph and Elizabeth Rose (Gibala) Z.; m. Amy Walker Brown, June 10, 1982; children: Benjamin Edward, Sarah Elizabeth. BA in Zoology, So. Ill. U., 1980, MA in Zoology, 1982; PhD in Zoology, Wash. State U., 1987. Cert. tchr., Wash. Asst. prof. biology S.E. Mo. State U., Cape Girardeau, 1987-89; prof. biology Blackburn U., Carlinville, Ill., 1989—. Author: Saunders Coll. Pub., Phila., 1993; contbr. articles to profl. jours. Faculty travel grantee Blackburn Coll., Boston, Atlanta, Vancouver, San Antonio, 1989, 90, 91, 92, summer rsch. grantee Blackburn U., 1991, 92, 93, grantee in aid Scanning Microscopy Soc., 1987. Mem. Am. Soc. Zoologists, Am. Soc. Ichthyologists and Herpetologists, Herpetologists League, Soc. for Study of Amphibians and Reptiles. Democrat. Methodist. Office: Blackburn U Dept Biology 700 College Ave Carlinville IL 62626

ZALKOW, LEON H., organic chemistry educator; b. Millen, Ga., Nov. 27, 1929. BCE, Ga. Inst. Tech., 1952, PhD in Chemistry, 1956. Rsch. fellow Wayne State U., 1955-56 57-59; rsch. chemist EI du Pont de Nemours & Co., 1956-57; asst. prof. chem. Okla. State U., 1959-62, assoc. prof., 1962-65; assoc. prof. Ga. Inst. Tech., 1965-69, prof. chem., 1969—; prof. and dept. head U. Negev., 1970-72. Recipient Charles H. Herty medal Am. Chem. Soc., 1993. Mem. AAAS, Am. Chem. Soc., Royal Soc. of Chem. Achievements include research in natural products, conformational analysis, chemistry of bicyclic azides. Office: Georgia Institute of Technology 225 N Ave NW Atlanta GA 30332*

ZALOGA, ROBERT EDWIN, engineering executive; b. Framingham, Mass., June 2, 1962; s. Edwin Anthony and Wanda Stasia (Jakubasz) Z. BS in Marine Engrng., Mass. Maritime Acad., 1984. Prototype support engr. GE, Kapl, Schenectady, N.Y., 1984-85; shift test engr. GE, Kapl, Windsor, Conn., 1985-87; refuel floor system engr. Ga. Power Co. Plant EI Hatch, Baxley, 1987-89; nuclear field engr. GE, Norcorss, Ga., 1989-91, settlement program mgr., 1991—. Lt. USNR, 1984-93. Mem. Am. Nuclear Soc., Soc. Naval Architects (marine engr. 1982—), Mass. Maritime Acad. Alumni, Polo Golf and Country Club, Elfun Soc. Achievements include nuclear industry's first successful remote spent fuel pool leak detection and repair. Office: GE Nuclear Energy Ste 100 22 Technology Park Norcross GA 30092

ZALOOM, VICTOR ANTHONY, industrial engineer, educator; b. Paterson, N.J., June 14, 1944; s. John Anthony and Mary A. (Shammas) Z.; m. Patricia Konvicka, Sept. 5, 1987; children: Sharlene, John, Anna, Michael. BS in Indsl. Engrng., U. Fla., 1966, MS, 1967; PhD, U. Houston, 1970. Registered profl. engr., Tex. Sr. ops. analyst Gen. Dynamics Corp., Fort Worth, 1967-70; assoc. prof. indsl. engrng. Auburn (Ala.) U., 1971-78; prof., chair indsl. engrng. N.C. A&T U., Greensboro, 1978-81, Lamar U., Beaumont, Tex., 1981—; cons. A.F. Logistics Mgmt. Ctr., Montgomery, Ala., 1975-78, L.T.V. Aerospace, Dallas, 1979-82, Dow Chem., Midland, Mich., 1983-92. Author: Statistic Process Quality Control Management, 1987, Statistical Process Quality Operators, 1988; contbr. articles to profl. jours. Mem. Inst. Indsl. Engrs. (sr.), ASME (sr.), Phi Kappa Phi. Office: Lamar U PO Box 10032 Beaumont TX 77710

ZALTA, EDWARD, otorhinolaryngologist, utilization review physician; b. Houston, Mar. 2, 1930; s. Nouri Louis and Marie Zahde (Lizmi) Z.; m. Carolyn Mary Gordon, Oct. 8, 1971; 1 child, Ryan David; children by previous marriage: Nouri Allan, Lori Ann, Barry Thomas, Marci Louise. BS, Tulane U., 1952, MD, 1956. Diplomate Am. Bd. Quality Assurance and Utilization Rev. Physicians. Intern Brooke Army Hosp., San Antonio, 1956-57; resident in otolaryngology U.S. Army Hosp., Ft. Campbell, Ky., 1957-60; practice medicine specializing in otolaryngology Glendora, West Covina and San Dimas, Calif., 1960-82; ENT cons. City of Hope Med. Ctr., 1961-76; mem. staff Foothill Presbyn.; past pres. L.A. Found. Community Svc., L.A. Poison Info. Ctr., So. Calif. Physicians Coun., Inc.; founder, chief exec. officer, chmn. bd. dirs. CAPP CARE, INC.; chmn. bd. MDM; founder Inter-Hosp. Coun. Continuing Med. Edn. Author: (with others) Medicine and Your Money; mem. editorial staff Managed Care Outlook, AAPPO Jour., Med. Interface; contbr. articles to profl. jours. Pres. bd. govs. Glendora Unified Sch. Dist., 1965-71; mem. Calif. Cancer Adv. Council, 1967-71, Commn. of Californias, Los Angeles County Commn. on Economy and Efficiency, U. Calif. Irvine Chief Exec. Roundtable; bd. dirs. U. Calif. Irvine Found. Served to capt. M.C. AUS, 1957-60. Recipient Award of Merit Order St. Lazarus, 1981. Mem. AMA, Calif. Med. Assn., Am. Acad. Otolaryngology, Am. Coun. Otolaryngology, Am. Assn. Preferred Provider Orgns. (past pres.), Am. Coll. Med. Quality, L.A. County Med. Assn. (pres. 1980-81), Kappa Nu, Phi Delta Epsilon, Glendora CountryClub, Centurion Club, Sea Bluff Beach and Racquet Club; Center Club (Costa Mesa, Calif.), Pacific Golf Club (San Juan, Capistrano). Republican. Jewish. Home: 3 Morning Dove Dr Laguna Niguel CA 92677 Office: West Tower 4000 MacArthur Blvd Ste 10000 Newport Beach CA 92660-2526

ZAMAN, MUSHARRAF, civil engineering educator; b. Rajshahi, Bangladesh, Mar. 8, 1952; came to U.S., 1979; s. Md. Gias Uddin and Jahanara Begum; m. Afroza Khanam, Aug. 7, 1981; children: Jessica, Ashiq. MS in Civil Engrng., Carleton U., 1979; PhD in Civil Engrng., U. Ariz., 1982. Registered profl. engr., Okla. Lectr. civil engrng. U. Engrng. and Tech., Dhaka, Bangladesh, 1975-76; rsch. assoc. Asian Inst. Tech., Bangkok, 1976-77; grad. teaching asst. Carleton U., Ottawa, Can., 1977-79; rsch. asst. Va. Tech., Blacksburg, 1979-81; rsch. assoc. U. Ariz., Tucson, 1981-82; asst. prof. U. Okla., Norman, 1982-88, assoc. prof., 1988—; faculty advisor Bangladesh Student Assn. U. Okla., Norman, 1983-91, Muslim Student Assn., 1990-91; summer faculty Argonne (Ill.) Nat. Lab., 1986; v.p. geotech. Consortium Internat., Oklahoma City, 1992—; co-organizer U.S.-Can. Geomechanics Workshop, 1992. Contbr. articles to European Jour. Mechanics, Computer Methods in Applied Mechanics and Engrng., Internat. Jour. Numerical Analysis Met. Geomech., Applied Math. Modelling, ASME Jour. Applied Mechanics. Rsch. grantee NSF, Okla. Dept. Transp., Okla. Advancement of Sci. and Tech., Okla. Dept. Commerce and Energy Resources Inst. Mem. ASCE (mem. deep found. com. 1987—), Internat. Assn. Computer Methods and Advances in Geomechanics (co-chmn. internat. conf. 1992—, co-editor newsletter 1989—). Home: 501 Starbrook Ct Norman OK 73072 Office: U Okla Rm 334 202 W Boyd St Norman OK 73019

ZAMBARDINO, RODOLFO ALFREDO, computing educator; b. Catania, Sicily, Italy, Mar. 17, 1930; arrived in Eng., 1956; s. Eugenio and Antonietta (Defelice) Z.; m. Rosa Greco, Oct. 13, 1954; children: Gabriella, Adrian. Maturita, Liceo Scientifico, Catania, Italy, 1947; D in Indsl. Engrng., Poly., Turin, Italy, 1953. Chartered engr. High voltage engr. Sicilian Electric Authority, Italy, 1953-55; high voltage engr. English Electric Co.-NEL, Stafford, Eng., 1956-62, head math. and computing sect. transformer div., 1962-69; prin. lectr. computing dept. Staffordshire Poly., Stafford, 1969—; external examiner, adviser of several polytechs. and univs., Eng., 1972—. Contbr. articles to profl. jours. Mem. 3 subject bds. Coun. Acadmic Awards, London, 1978-86. Fellow Instn. Elec. Engrs., British Computer Soc. (examiner, moderator 1972—, mem. accreditation com. 1975-89), Inst. Maths. and Applications; mem. Assn. Computing Machinery, Royal Soc. Arts. Avocations: Asian art and architecture, travel, rambling. Home: 10 Greenfield Rd, Stafford ST17 OPU, England Office: Staffordshire Univ Computing Sch, Beaconside Ln, Stafford ST18 OAD, England

ZAMECNIK, PAUL CHARLES, oncologist, medical research scientist; b. Cleve., Ohio, Nov. 22, 1912; m.; 3 children. AB, Dartmouth Coll., 1933; MD, Harvard U., 1936; DSc (hon.), U. Utrecht, Utrecht, 1966; hon. DSc., Columbia U., 1971; Harvard U., 1982, Roger Williams Coll., 1983, Dartmouth Coll., 1988. Resident C.P. Huntington Meml. Hos., Boston, MA, 1936-37; intern. U. Hosps., Cleve., Ohio, 1938-39; traveling fellow from Harvard U. at Carlsberg Lab., Copenhagen, 1939-40; Finney-Howell fellow Rockefeller Inst., 1941-42; instr., assoc. prof. Harvard U., 1942-56, Collis P. Huntington prof., 1956-79; dir. J.C. Warren Labs., 1956-79; emeritus prof. oncological medicine Sch. Medicine, 1979—; prin. sci. Worcester Found. Experimental Biology, 1979—; physician Mass. Ge. Hosp., 1956-79; hon. physician Mass. Gen. Hosp., 1979—. Recipient Warren Triennial prize, 1946, 50, Ewing award, 1962, Borden award, 1965, Am. Cancer Soc. Nat. award, 1968, Passano award, 1970, Nat. Medal of Sci. NSF, 1991, Hudson Hoagland award, 1992. mem. Am. Acad. Arts. & Sci., NAS, Am. Soc. Biol. Chemists, Am. Assn. Cancer Rsch. (pres., 1964-65). Office: Worcester Found Exptl Biology 222 Maple Ave Shrewsbury MA 01545-2732

ZAMRINI, EDWARD YOUSSEF, behavioral neurologist; b. Cairo, Apr. 27, 1958; s. Joseph Boutros and Therese Edouard (Artin) Z. BSc, Am. U. Beirut, 1980, MD, 1984. Dir. geriatric neuropsychiatry assessment unit VA Med. Ctr., Augusta, Ga., 1990—. Contbr. articles to profl. jours. Mem. AMA, ACP, N.Y. Acad. Sci., Am. Acad. Neurology, Alzheimer's Disease and Related Disorders Assn. (bd. dirs. 1990—, chair edn. com. 1990-92, columnist 1989—, pres. Augusta chpt. 1993—), Amnesty Internat. (chpt. rep. 1990). Office: VA Med Ctr 1 Freedom Way Augusta GA 30904-6285

ZANA, DONALD DOMINICK, information resource manager; b. Pitts., Oct. 9, 1942; s. Dominick Jr. and Kathryn (Turino) Z.; m. Janet Ann Kundrak, May 30, 1964; children: Christine Ann, Michael Steven, Anthony Phillip. BS in Edn., Indiana U. of Pa., 1964; M of Commerce in Mgmt., U. Richmond, Va., 1972; grad. with honors, Army Logistics Mgmt. Ctr., 1973, Army Command and Gen. Staff Coll., 1978. Commd. 2nd lt. U.S. Army, 1964, advanced through grades to lt. col., resigned, 1984; prin. systems analyst Teledyne Brown Engrng., Huntsville, Ala., 1984-85, sr. systems analyst, 1985-86, br. mgr. software acquisition and evaluation, 1986-90, mgr. automation integration, 1990-93, dir. computing resources and tech., 1993—. Mem. IEEE, Assn. for Computing Machinery, Assn. of U.S. Army, Ret. Officers Assn. Roman Catholic. Home: 7807 Springbrook Dr SE Huntsville AL 35802-3321 Office: Teledyne Brown Engring 300 Sparkman Dr NW # 17 Huntsville AL 35805-7007

ZANBAK, CANER, mining engineer, geologist; b. Iskenderun, Hatay, Turkey, Sept. 15, 1949; came to U.S., 1980; s. Kemal and Lamia (Kiziler) Z.; m. Figen Demirel, Dec. 3, 1979; children: Cengiz Kemal, Pinar Türkan. BS in Mining Engrng., Istanbul (Turkey) Tech. U., 1970, MS in Mining Engring., 1971; PhD in Geology, U. Ill., 1978. Cert. profl. geologist; registered profl. geologist, S.C. Lectr., rschr. Istanbul Tech. U., 1971-74, asst. prof., 1978-80; rschr. U. Ill., Champaign, 1974-78, trustee Geothrust com., 1990—; asst. prof. geology Kent (Ohio) State U., 1980-82; assoc. prof. geology S.D. Sch. Mines, Rapid City, 1982-84; prin. geologist Terraform Engrs., Naperville, Ill., 1984-85; assoc., ptnr. Woodward-Clyde Cons., Chgo., 1985-93; pres. EMCZ, Inc. Environ. and Mining cons., Streamwood, Ill., 1993—; adj. prof. Ill. Inst. Tech., Chgo., 1988—. Contbr. articles to Internat. Jour. Rock Mechanics and Mining Scis. Fulbright scholar, 1974. Mem. Am. Inst. Profl. Geologists, Ill. Assn. Environ. Profls., Assn. Engring. Geologists (ASTM liason 1991—), Soc. Mining Engrs. Achievements include rsch. in rock mechs, engrng. geology, hazardous waste mgmt., environ. concerns, surface mining and chem. industries. Home and Office: EMCZ Inc 551 Ascot Ln Ste 100 Streamwood IL 60107

ZANDE, MICHAEL DOMINIC, clinical psychologist, consultant; b. Columbus, Ohio, Feb. 11, 1960; s. Richard Dominic and Marilyn (Love) Z.; m. Laura Marie Goettl, July 27, 1985. BA in Psychology, Bowling Green State U., 1982; MA in Clin. Psychology, Nova U., 1984, PhD in Clin. Psychology, 1988. Lic. psychologist, N.C. Clin. psychology intern Durham (N.C.) VA Med. Ctr., 1987-88, Duke U. Med. Ctr., Durham, 1987-88; clin. psychologist Raleigh (N.C.) Community Hosp., 1987—; clin. dir. Raleigh Psychology, 1989-91; founder, owner The Zande Psychology Group, Raleigh, N.C., 1991—; dir. psychol. svcs. Charter Northridge Hosp., Raleigh, 1989—; consulting clin. supr. employee assistance program Rex Hosp. Author: (tng. program) Life Skills Tng., 1989. Bd. dirs., mem. community outreach com. Am. Cancer Soc., Wake County, N.C., 1990-91; area coord. Pajcic for Gov., Lauderhill, Fla., 1986. Mem APA, N.C. Psychol. Assn. (bd. dirs. 1990—, chair pub. advocacy com. 1989—). Democrat. Avocations: gardening, cooking, music, dog tng. Home: 2401 Mt Vernon Church Rd Raleigh NC 27614 Office: The Zande Psychology Group 3722 Benson Dr Ste 100 Raleigh NC 27609-7321

ZANDER, JOSEF, gynecologist; b. Juelich, Fed. Republic Germany, June 19, 1918; s. Karl and Gertrud (Mueller) Z.; M.D., U. Tuebingen, 1946, postgrad. dept. of pathology, 1946-47; postgrad., Kaiser Wilhelm Inst. Biochemistry, 1947-49; M.D. (hon.) U. Innsbrück, Austria, 1986; children: Karl, Gabriele, Ferdinand, Jan, Susanne, Katharina. Tng. in ob-gyn, univs. Marburg and Cologne med. schs., 1949-56; rsch. assoc. biochemistry U. Utah, 1956-57; mem. faculty U. Cologne Med. Sch., 1958-64; prof. ob-gyn, chmn. dept. U. Heidelberg Med. Sch., 1964-69; prof., chmn. 1st dept. ob-gyn, U. Munich Med. Sch., 1970-87. Fellow ACS, Am. Coll. Ob-gyn. (hon.), Am. Gynecol. Soc. (hon.), Felix Rutledge Soc.; mem. German Soc. Endocrinology, Endocrine Soc. U.S., Am. Soc. Pelvic Surgeons, German Acad. Natural Scientists Leopoldina, Bavarian Acad. Scis., German Ob-Gyn Soc. (hon), Hungarian Ob-Gyn Soc. (hon.), Bavarian Ob-Gyn Soc. (hon.), Italian Ob-Gyn Soc. (hon.), Austrian Ob-Gyn Soc. (hon.), Munich Med. Soc. (hon.), Rotary. Roman Catholic. Author, editor books; contbr. articles in field to profl. jours. Office: 11 Maistrasse, 80337 Munich 2, Germany

ZANGENEH, FEREYDOUN, pediatrics educator, pediatric endocrinologist; b. Tehran, Iran, July 11, 1937; came to U.S., 1962, 86; married. MD, U. Tehran, Iran, 1961. Diplomate Am. Bd. Pediatrics and Pediatric Endocrinology. Intern Washington Hosp. Ctr., 1962; resident in pediatrics U. Chgo. Hosps. & Clinics, 1963-65; fellow in pediatric endocrinology Children's Memorial Hosp., Chgo., 1965-66, U. Wash., Seattle, 1966-68; asst. prof. to prof. U. Tehran, Iran, 1968-86; assoc. prof. pediatrics Marshall U. Sch. of Medicine, Huntington, W.Va., 1986-89; assoc. prof. pediatrics W.Va. U., Charleston, 1989—, dir. pediatric endocrinology, 1989—; instr. pediatrics U. Wash., Seattle, 1967-68; dir. dept. endocrinology Children's Hosp. Med. Ctr., Tehran, Iran, 1968-86; dir. pediatric endocrinology Marshall U. Sch. Medicine, Huntington, 1986-89; mem. W.Va. Adv. Com. on Newborn Metabolic Screening, W.Va. Adv. Com. on Diabetes Mellitus. Author: (with others) Pediatric Endocrinology and Metabolism; contbr. articles and abstracts to profl. jours. Bd. dirs., rsch. com., camp com. Am. Diabetes Assn., W.Va. affiliate, Charleston, active in Huntington W. Va. affiliate, 1986-89. Capt. Iranian Army Med. Corps, 1969-71. Recipient Gharib Rsch. award, Iranian Pediatric Soc., Tehran, 1976. Fellow Am. Acad. Pediatrics; mem. Internat. Pediatric Assn. (adv. expert panel 1977-83), Union Mid. Ea. and Mediterranean Pediatric Socs. (exec. coun. 1979-85), Lawson Wilkins Pediatric Endocrine Soc., Endocrine Soc., Am. Assn. Clin. Endocrinologists. Office: WVa U Dept Pediatrics 830 Pennsylvania Ave Ste 104 Charleston WV 25302-3389

ZANONI, MICHAEL MCNEAL, forensic scientist; b. San Francisco, Jan. 6, 1948; s. Mario and Ethel (Corbitt) Z.; m. Sheron Lee Chic. BA in Police Sci., San Jose State U., 1970, MS in Criminal Justice, 1979; PhD in Criminology, Columbia-Pacific U., 1982. Police officer City of San Jose, Calif., 1969-75; pvt. investigator Palo Alto, Calif., 1975-83; forensic sci. cons. San Jose, 1983—; document examiner in field. Contbr. articles to profl. jours. Bd. dirs. Calif. chpt. ARC, Palo Alto, 1976-83; bd. dirs. Palo Alto Humane Soc., 1991—; mem. Bay Area Mountain Rescue Unit, San Mateo, Calif., 1991—. Fellow Am. Acad. Forensic Scis.; mem. Can. Soc. Forensic Scis., Internat. Assn. for Identification. Buddhist. Achievements include research demonstrating relationship of semiotics and linguistics to unconscious processes occurring when document or handwriting examiner formulates opinions. Office: Box 369 San Carlos CA 94070-0369

ZAPPOLI, BERNAND, aerospace engineer; b. Batna, Algeria, French, Apr. 2, 1951; s. Orfeo and Simone (Barbarin) Z.; m. Denise Valentini, July 30, 1973; children: Pierre, Anne-Claire. Thése d'Etat, U. Marseilles, France, 1979. Habilitation a diriger des recherches. Engr. Soc. Europenne Propulsion, Vernon, 1975-81, CNES, Toulouse, France, 1981-89; program mgr. CNES, Toulouse, 1989–; prof. mechanics Paul Sabatier U., Toulouse, 1992–; chmn. Migro-G Com. Internat. Astro. Fed., Paris, 1992. Office: CNES, 18 Ave E Belin, F-31500 Toulouse France

ZARAGOZA, FEDERICO MAYOR, protective agency administrator, biochemist; b. Barcelona, Spain, Jan. 27, 1934. Grad. Pharm., U. Madrid, 1956, PhD in Pharm., 1958. Prof. biochemistry, faculty pharmacy U. Granada, Spain, 1963-73, dir. interfaculty dept. pharmacy, 1967-68, rector, 1968-72, hon. rector, 1972—; prof. faculty sci. U. Madrid, 1973—, dir. ctr. molecular biology Severo Ochoa, higher coun. sci. rsch., 1973-78, sci. chmn., 1983—; undersec. Ministry Edn. and Sci., Spain, 1974-76; mem. Parliament, Spain, 1977-78; dep. dir.-gen. UNESCO, 1978-81; dir. gen. UNESCO, Paris, 1984—; min. Edn. and Sci. Spain, 1981-82; mem. adv. com. European Ctr. Higher Edn. UNESCO, Bucharest, 1974-78, mem. adv. com. sci. rsch. and human needs, 1974—, mem. various groups experts, 1981—; chmn. adv. com. sci. and tech. rsch. Office of Prime Min., Spain, 1974-78; chmn. commn. edn. and sci. Spanish Congress, 1977-78; mem. various groups experts Internat. Fedn. Insts. Advanced Study, 1981—; dir. Inst. Scis. Man, 1983—; Spanish mem. European Parliament. Author: Mañana siempre es tarde, 1987; editor: La lucha contra la enfermedad, 1986; co-editor: (with Severo Ochoa and M. Barbacid) Oncogenes y Patología Molecular, 1987; contbr. articles to profl. jours. Decorated Grand Cross, Alfonso X el Sabio, Grand Cross, Orden Civil de Sanidad, Gran Cruz de Carlos III (Spain); Gran Cruz de Caro y Cuervo (Colombia); Encomienda con Placa del Libertador (Venezuela); Grand Officer, Nat. Order of Merit of Republic (France). Mem. AAAS, Spanish Soc. Biochemistry (pres. 1970-74), Spanish Royal Acad. Pharmacy, Société Française de Chimie Biologique, Biochemical Soc., Inst. Scis. Man (dir. 1983—), Acad. Européenne des Arts, des Scis. et des Lettres (founding), World Acad. Arts and Scis., Acad. Française de Pharmacie (corr.), Policy Interaction Coun., Forum Issyk-Kul (founding), Club of Rome. Office: UNESCO, 7 place de Fontenoy, 75700 Paris France*

ZARE, RICHARD NEIL, chemistry educator; b. Cleve., Nov. 19, 1939; s. Milton and Dorothy (Amdur) Z.; m. Susan Leigh Shively, Apr. 20, 1963; children: Bethany Jean, Bonnie Sue, Rachel Amdur. BA, Harvard, 1961; postgrad., U. Calif. at Berkeley, 1961-63; PhD (NSF predoctoral fellow), Harvard, 1964; DS (hon.), U. Ariz., 1990, Northwestern U., 1993. Postdoctoral fellow Harvard, 1964; postdoctoral research asso. Joint Inst. for Lab. Astrophysics, 1964-65; asst. prof. chemistry Mass. Inst. Tech., 1965-66; asst. prof. dept. physics and astrophysics U. Colo., 1966-68, assoc. prof. physics and astrophysics, asso. prof. chemistry, 1968-69; prof. chemistry Columbia, 1969-77, Higgins prof. natural sci., 1975-77; prof. Stanford U., 1977—, Shell Disting. prof. chemistry, 1980-85, Marguerite Blake Wilbur prof. chemistry, 1987—, prof. physics, 1992—; cons. Aeronomy Lab., NOAA, 1966-77, radio standards physics div. Nat. Bur. Standards, 1968-77, Lawrence Livermore Lab., U. Calif., 1974—, SRI, Internat., 1974—, Los Alamos Sci. Lab., U. Calif., 1975—; fellow adjoint, Joint Inst. Lab. Astrophysics, U. Colo.; mem. IBM Sci. Adv. Com., 1977-92; chmn. commn. on physical scis., mathematics, and applications Nat. Rsch. Coun., 1992—; researcher and author publs. on laser chemistry and chem. physics. editor Chem. Physics Letters, 1982-85. Recipient Fresenius award Phi Lambda Upsilon, 1974, Michael Polanyi medal, 1979, Nat. Medal Sci., 1983, Soroptimist Internat. Santa Cruz award Spectroscopy Soc. Pitts., 1983, Michelson-Morley award Case Inst. Tech. Case We. Res. U., 1986, ISCO award for Significant Contbns. to Instrumentation for Biochem. Separations, 1990, The Harvey prize, 1993; nonresident fellow Joint Inst. for Lab. Astrophysics, 1970—, Alfred P. Sloan fellow, 1967-69, Christensen fellow St. Catherine's Coll., Oxford U., 1982, Stanford U. fellow, 1984-86. Fellow AAAS, Calif. Acad. Scis. (hon.); mem. Nat. Acad. Sci. (Chem. Scis. award 1991), Nat. Scis. Bd., Am. Acad. Arts and Scis., Am. Phys. Soc. (Earle K. Plyler prize 1981, Irving Langmuir prize 1985), Am. Chem. Soc. (Harrison Howe award Rochester chpt. 1985, Remsen award Md. chpt. 1985, Kirkwood award, Yale U. New Haven chpt. 1986, Willard Gibbs medal Chgo. chpt. 1990, Peter Debye award in phys. chemistry 1991, Linus Pauling Medal, 1993), Am. Philos. Soc., Chem. Soc. London, Phi Beta Kappa. Office: Stanford U Dept Chemistry Stanford CA 94305-5080

ZARET, EFREM HERBERT, chemist; b. Chgo., Nov. 23, 1941; s. Abraham and Pearl (Pollack) Z.; m. Norrie Silverstein, Aug. 1, 1964; children: Deborah, David. BS in Chemistry, Ill. Inst. Tech., 1964; PhD, U. Wis., Milw., 1976. V.p. Hartz Mountain Corp., Harrison, N.J., 1972-87, Boyle Midway Household Products, N.Y.C., 1987-90; pres. EZ Assocs., Inc., Berkeley Heights, N.J., 1990—. Contbr. articles to Organic Chemistry. Mem., chmn. Berkeley Heights Planning Bd., 1984-91; active Berkeley Heights Bd. Edn., 1985-91. Mem. Assn. Cons. Chemists and Chem. Engrs., Soc. Packaging Profls., Am. Entomological Soc., Am. Chem. Soc. Office: EZ Assocs Inc 135 Briarwood Dr E Berkeley Heights NJ 07922

ZARLENGA, DANTE SAM, JR., molecular biologist; b. Youngstown, Ohio, Oct. 15, 1953; s. Dante Sam Sr. and Joann (Dilallo) Z.; m. Gloria Ann Glass, July 18, 1981. AB in Chemistry, Youngstown State U., 1976; MS in Biochemistry, U. Cin., 1979, PhD in Biochemistry, 1982. NIH postdoctoral fellow Johns Hopkins U., Balt., 1982-84; rsch. assoc. USDA, Beltsville, Md., 1984-87, sr. scientist, 1987—; cons. RM Nardone & Assocs., Silver Spring, Md., 1983—; instr., course dir. Ctr. for Advanced Tng. in Cell and Molecular Biology, Cath. U., Washington, 1983—; sci. reviewer AID, Washington, 1985—. Contbr. articles to profl. jours. Named Early Career Scientist of Yr., Beltsville region ARS, USDA, 1991; NIH postdoctoral fellow Johns Hopkins U., 1982-84. Mem. AAAS, Helminthological Soc. Washington (exec. com. 1989-91), Am. Assn. Vet. Parasitologists, Am. Soc. Parasitologists. Democrat. Roman Catholic. Achievements include research and patents in Trichinella Spiralis Antigens for use as immunodiagnostic reagents or vaccines, DNA Sequences Encoding Diagnostic Antigens for Cysticercosis. Home: 9555 Michaels Way Ellicott City MD 21042-2463 Office: USDA ARS LPSI BPL Bldg 1180 BARC East Beltsville MD 20705

ZARLING, JOHN PAUL, mechanical engineering educator; b. Elmhurst, Ill., Mar. 15, 1942; s. Earnie W. and Lorraine M. (Feustel) Z.; m. Fran E. Carlson, Nov. 27, 1965; children: Matthew, John, Katie. BSMechE, Mich. Tech. U., 1964, MSEM, 1966, PhD in EM, 1970. Instr. U. Wis., Madison, 1966-68; asst. prof. U. Wis., Kenosha, 1971-73, assoc. prof., 1973-76; assoc. prof. U. Alaska, Fairbanks, 1976-79, prof., 1979—, assoc. dean, 1986—; asst. vice chancellor U. Wis., 1974-76; prin. Zarling Engring., Fairbanks, 1978—. Editor conf. procs.; contbr. articles, monographs, procs. to profl. publs. Active Fairbanks Youth Hockey, 1976-83. Mem. ASHRAE, ASME (chmn. Milw. sect. 1975), Am. Soc. Engring. Edn. (bd. dirs. 1986-88). Lutheran. Achievements include patents for Apparatus for Containing Toxic Spills Employing Hybrid Thermosyphon, Thermosyphon Condensate Return Device, Passive-Active Hybrid Thermosyphon. Office: U Alaska Fairbanks Inst Northern Engring Fairbanks AK 99775

ZARTMAN, DAVID LESTER, dairy science educator, researcher; b. Albuquerque, July 6, 1940; s. Lester Grant and Mary Elizabeth (Kitchel) Z.; m. Micheal Aline Plemmons, July 6, 1963; children: Kami Renee, Dalan Lee. BS, N.Mex. State U., 1962; MS, Ohio State U., 1966, PhD, 1968. Jr. ptnr. Marlea Guernsey Farm, Albuquerque, 1962-64; grad. rsch. assoc. Ohio State U., Columbus, 1964-68; asst. prof. dairy sci. N.Mex. State U., Las Cruces, 1968-71, assoc. prof., 1971-79, prof., 1979-84; prof., chmn. dept. Ohio State U., Columbus, 1984—; pres. Mary K. Zartman, Inc., Albuquerque, 1976-84; cons. Bio-Med. Electronics, Inc., San Diego, 1984-89, Zartemp, Inc., Northbrook, Ill., 1990, Recom Applied Solutions, 1993—. Contbr. articles to profl. jours.; patentee in field. Recipient State Regional Outstanding Young Farmer award Jaycees, 1963, Disting. Rsch. award N.Mex. State U. Coll. Agr. and Home Econs., 1983; named one of Top 100 Agr. Alumni, N.Mex. State U. Centennial, 1987; spl. postdoctoral fellow NIH, New Zealand, 1973; Fulbright-Hayes lectr., Malaysia, 1976. Fellow AAAS; mem. Am. Dairy Sci. Assn., Am. Soc. Animal Sci., Dairy Shrine Club, Ohio Extension Profs. Assn., Gamma Sigma Delta, Alpha Gamma Rho (1st Outstanding Alumnus N.Mex. chpt. 1985), Sigma Xi, Alpha Zeta, Phi Kappa Phi. Home: 7671 Deer Creek Dr Columbus OH 43085-1551 Office: Ohio State U 2027 Coffey Rd Columbus OH 43210-1094

ZASK, ARIE, chemist, researcher; b. Tel Aviv, Israel, Dec. 20, 1956. BS, SUNY, Stony Brook, 1977; PhD, Princeton U., 1982. Postdoctoral fellow Columbia U., N.Y.C., 1982-84; sr. scientist Wyeth-Ayerst Rsch., Princeton, N.J., 1984-88, rsch. scientist, 1988-91, prin. scientist, 1991—; referee Jour. Organic Chemistry, 1989—. Contbr. articles to profl. jours. Mem. AAAS, Am. Chem. Soc. Achievements include patents for diabetes and osteoporosis therapies; development of organic synthesis methodology based on classical and organometallic chemistry. Office: Wyeth-Ayerst Rsch CN 8000 Princeton NJ 08543-8000

ZASLAVSKY, GEORGE MOISEEVICH, theoretical physicist, researcher; b. Odessa, USSR, May 31, 1935; came to U.S., 1991; s. Moisey Davidovich and Polina (Begagoen) Z.; m. Anna Benjaminovna Herman, Jan. 21, 1985; 1 child, Mickael G. D of Theoretical Physics, Krasnoyarsk (USSR) Inst. Physics, 1974. Sr. rschr. Inst. Nuclear Physics, Novosibirsk, USSR, 1964-71; head dept. Inst. Physics, Krasnoyarsk, USSR, 1971-84; head theoretical dept. Space Rsch. Inst., Moscow, 1984-92; prof. physics and math. Courant Inst. Math., NYU, N.Y.C., 1992—. Author: Chaos in Dynamic Systems, 1985, (with others) Nonlinear Physics, 1989, Weak Chaos and Quasiregular Patterns, 1991; editor Jour. CHAOS, 1991. Office: Courant Inst Math NYU Apt 13R 1 Washington Sq Village New York NY 10012

ZAU, GAVIN CHWAN HAW, chemical engineer; b. London, Oct. 10, 1963; came to the U.S., 1987; s. Ronny and Alice (Chan) Z. BSChE, Princeton U., 1985; MS in Mgmt., MIT, 1991, postgrad., 1987—. Rsch. asst. Siemens Rsch. and Tech. Lab., Princeton, N.J., 1985; rsch. engr. New Japan Radio Co. Ltd., Kamifukuoka, Japan, 1986-87; rsch. asst. MIT, Cambridge, 1987—. Contbr. articles to Jour. Electrochem. Soc. Arthur D. Little fellow MIT, 1987. Mem. Am. Prodn. and Inventory Control Soc., Sigma Xi. Home: 16 Lilac Ct Cambridge MA 02141-1912 Office: MIT 66-225 25 Ames St Cambridge MA 02139

ZAVON, PETER LAWRENCE, industrial hygienist; b. Johnson City, N.Y., Sept. 18, 1949; s. Mitchell R. and Faith R. (Shottenfeld) Z. BA in Astronomy, Boston U., 1976; MS, U. Cin., 1984. Cert. indsl. hygienist. Summer intern Shell Chem. Co., Deer Park, Tex., 1976; asst. indsl. hygienist Princeton (N.J.) U., 1977-84; sr. cons. Occusafe Inc., Wheeling, Ill., 1984-87; sr. indsl. hygienist Xerox Corp., Webster, N.Y., 1987—; adj. asst. prof. U. Rochester Dept. Environ. Medicine, Rochester, 1992—. Mem. Am. Indsl. Hygiene Assn., Am. Conf. of Govt. Indsl. Hygienists, British Occupational Hygiene Soc., Am. Acad. Indsl. Hygiene, Western N.Y. Sect. Am. Indsl. Hygiene Assn. (pres. 1989-90, N.J. sect. dir. 1980-81). Home: 30 Woodline Dr Penfield NY 14526 Office: Xerox Corp Ops Safety 800 Salt Rd 843-16S Webster NY 14580

ZAVRTANIK, DANILO, physicist, researcher; b. Nova Gorica, Slovenia, Aug. 15, 1953; s. Avgust and Alojzija Zavrtanik; m. Pia Bratina, Sept. 16, 1972; 1 child, Marko. BS in Physics, U. Ljubljana, Slovenia, 1979, MS in Physics, 1984, PhD in Physics, 1987. Sci. assoc. European Lab. for Particle Physics CERN, Geneva, Switzerland, 1990-92; postgrad. researcher Inst. Jožef Stefan, Ljubljana, 1979-87, researcher, 1987-90, 92—, dir. gen., 1992—; prof. physics U. Ljubljana, 1992—; mem. sci. com. Ctrs. of Excellence of Ctrl. European Initiative, 1992. Contbr. 50 articles to sci. jours. Mem. Slovene Biomed. Soc., Slovene Math., Phys. and Astron. Soc., European Phys. Soc., Slovene C. of C. (mem. governing bd. 1992). Office: J Stefan Inst, Jamova 39 pp box 100, Ljubljana SI-61111, Slovenia

ZAWADA, EDWARD THADDEUS, JR., physician, educator; b. Chgo., Oct. 3, 1947; s. Edward Thaddeus and Evelyn Mary (Kovarek) Z.; m. Nancy Ann Stephen, Mar. 26, 1977; children: Elizabeth, Nicholas, Victoria, Alexandra. BS summa cum laude, Loyola U., Chgo., 1969; MD summa cum laude, Loyola-Stritch Sch. Medicine, 1973. Diplomate Am. Bd. Internal Medicine, Am. Bd. Nephrology, Am. Bd. Nutrition, Am. Bd. Critical Care, Am. Bd. Geriatrics, Am. Bd. Clin. Pharm. Intern UCLA Hosp., 1973, resident, 1974-76; asst. prof. medicine UCLA, 1978-79, U. Utah, Salt Lake City, 1979-81; assoc. prof. medicine Med. Coll. Va., Richmond, 1981-83; assoc. prof. medicine, physiology & pharmacology U. S.D. Sch. Medicine, Sioux Falls, 1983-86, Freeman prof., chmn. dept. Internal Medicine, 1987—, chief div. nephrology and hypertension, 1983-88, pres. univ. physician's practice plan, 1992—; chief renal sect. Salt Lake VA Med. Ctr., 1980-81; asst. chief med. service McGuire VA Med. Ctr., Richmond, 1981-83. Editor: Geriatric Nephrology and Urology, 1984; contbr. articles to profl. publs. Pres. Minnehaha div. Am. Heart Assn., 1984-87, pres. Dakota affiliate Am. Heart Assn., 1989-91. VA Hosp. System grantee, 1981-85, 85-88. Fellow ACP, Am. Coll. Chest Physicians, Am. Coll. Nutrition, Am. Coll. Clin. Pharmacology, Internat. Coll. Angiology, Am. Coll. Angiology, Am. Coll. Clin. Pharmacology, Royal Soc. Medicine; mem. Internat. Soc. Nephrology, Am. Soc. Nephrology, Am. Soc. Pharmacology and Exptl. Therapeutics, Am. Physiol. Soc., Am. Inst. Nutrition, Am. Soc. Clin. Nutrition, Am. Geriatric Soc., Westward Ho Country Club. Democrat. Roman Catholic. Avocations: golf, tennis, skiing, cinema, music. Home: 2908 Duchess Sioux Falls SD 57103-4551 Office: U SD Sch Medicine 2501 W 22nd St Sioux Falls SD 57105

ZAWADSKY, JOSEPH PETER, physician; b. South River, N.J., Jan. 16, 1930; m. Marilyn Mark; 6 children. BA, Princeton U., 1951; MD, Columbia U., 1955. Diplomate Am. Bd. Orthopaedic Surgery. Intern Columbia-Presbyn. Med. Ctr., N.Y.C., 1955-56; gen. practice South River, 1958-61; asst. resident in othropaedic surgery N.Y. Orthopaedic Hosp./Columbia-Presbyn. Med. Ctr., N.Y.C., 1961-63, jr. Annie C. Kane fellow, resident, 1963-64; instr. orthopaedic surgery Columbia U., N.Y.C., 1964-72; mem. med. staff Middlesex Rehab. Hosp., North Brunswick, N.J., 1964-74, pres. staff, 1968-69; prof. surgery, program dir. orthopaedic residency program UMDNJ-Robert Wood Johnson Med. Sch., New Brunswick, N.J., also numerous coms.; chief orthopaedic surgery Robert Wood Johnson Univ. Hosp., New Brunswick; sr. attending orthopaedic surgeon St. Peter's Med. Ctr., New Brunswick; mem. courtesy staff Med. Ctr. at Princeton, N.J.; chief divsn. orthopaedics dept. surgery Raritan Valley Hosp., Green Brook, N.J., 1970-81; asst. attending orthopaedic surgeon Vanderbilt Clinic Presbyn. Hosp./Columbia-Presbyn. Med. Ctr., N.Y.C., 1964-72; cons. orthopaedic surgeon McCosh Infirmary Princeton U./Rutgers U.; orthopaedic surgeon Princeton & Rutgers football teams. Co-author numerous abstracts; contbr. articles to profl. jours.; mem. editorial adv. bd. Surg. Rounds for Orthopaedics. Site visitor, mem. examiners com. Am. Bd. Orthopaedic Surgery, 1989—; mem. State of N.J. Athletic Tng. Bd. Trustees, 1986—; mem. med. bd. trustees, rsch. evaluation com. and grant planning & priorities com. Musculoskeletal Transplant Found.; mem. State of N.J. Athletic Tng. Adv. Commn.; course dir. Forum Medicus Postgrad. Advance in Sports Medicine, 1985-88. Capt. surg. svc. USAF. Recipient Disting. Am. award Delaware Valley chpt. Nat. Football Found. and Hall of Fame, 1974; named to Wall of Fame South River Bd. Edn., 1988; grantee Orthopaedic Rsch. & Edn. Found., 1989-92 (4 grants), Musculoskeletal Transplant Found., 1990, 91, 92, Whitaker Found., 1991-93, Found. of UMDNJ, 1992, N.J. Ctr. for Biomaterials and Med. Devices, 1992. Fellow ACS, Am. Acad. Orthopaedic Surgeons; mem. AMA, Am. Orthopaedic Assn. (del.-at-large exec. com.), Am. Coll. Sports Medicine, Ea. Orthopaedic Assn., Am. Orthopaedic Soc. for Sports Medicine, Internat. Coll. Surgeons, Orthopaedic Rsch. Soc., Acad. Orthopaedic Soc., Assn. Ivy League Team Physicians, So. Orthopaedic Assn., N.J. Orthopaedic Soc., Acad. Medicine N.J., Assn. for Arthritic Hip and Knee Surgery, Stinchfield Soc., Middlesex County Med. Soc. (v.p. 1983-84, pres. 1985-86). Home: 161 Hodge Rd Princeton NJ 08540 Office: 215 Easton Ave New Brunswick NJ 08901

ZEALBERG, JOSEPH JAMES, psychiatry and behavioral sciences educator. BS in Organic Chemistry, Temple U., 1973; MD, Med. Coll. of Pa., 1985. Diplomate Am. Bd. Psychiatry and Neurology. Resident Dept. Behavioral Medicine and Psychiatry U. Va. Med. Ctr., Charlottesville, 1981-84, chief resident, 1984-85; asst. prof. Dept. Behavioral Medicine and Psychiatry U. Va. Med. Ctr., Marion, Va., 1985-87; staff psychiatrist admissions unit Southwestern State Hosp., Marion, Va., 1985-87; assoc. prof. psychiatry and behavioral scis. Med. U. S.C., Charleston, 1987—, staff psychiatrist, 1987—; staff psychiatrist Charleston Meml. Hosp., 1987—, Ralph Johnson VA Med. Ctr., 1987—; cons. Overlook Mental Health Ctr., Knoxville, 1991, Crisis Intervention Ctr., Inc., Nashville, 1991, Memphis Mental Health Ctr., 1991, Wake county-Mental Health Div., Raleigh, N.C., 1990, Dixon Com., 1990; lectr. in field. Author: (with others) The Role of Mobile Crisis Teams in Community Disasters, in Responding to Disaster: A Guide for Mental Health Professionals, 1992, Providing Psychiatric Emergency Care During Disasters: Hurricane Hugo in Charleston, South Carolina, 1990; contbr. articles to profl. jours. including Focus on Emergency Psychiatry, Comtemporary Psychiatry, Am. Assn. for Cancer Rsch., 1977, many others. Cons. to all area local police depts., Charleston County; cons. Charleston's Interfaith Crisis Ministry; presenter Nat. Depression Screening Day, WCIV-TV; presenter various civic ctrs. Recipient numerous grants. Mem. AAAS, Am. Psychiat. Assn., Va. NeuroOpsychiat. Soc., Southwestern Va. Med. Soc., Am. Assn. for Emergency Psychiatry, S.C. Psychiat. Assn. Home: 1007 Cummings Circle Mount Pleasant SC 29464 Office: Dept Psychiatry Med U of SC 171 Ashley Ave Charleston SC 29425

ZEAMER, RICHARD JERE, engineer, executive; b. Orange, N.J., May 13, 1921; s. Jay and Margery Lilly (Herman) Z.; m. Jean Catherine Hellens, July 8, 1944 (div. 1966); children: Audrie Dagna, Richard Warwick, Geoffrey Hellens; m. Theresa Elizabeth Taborsky, Mar. 27, 1969; children: Emily Elizabeth, Charlotte Anne. BSME, MIT, 1943, MSCE, 1948; PhD in Mech. Engring., U. Utah, 1975. Registered profl. engr., Utah. Civil engr. Morton C. Tuttle, Boston, 1949-53; process design engr. Nekoosa Edwards Paper Co., Port Edwards, Wis., 1953-55; process engr. W.Va. Pulp and Paper Co., Luke, Md., 1955-60; rocket engr., supr. Allegany Ballistics Lab., Rocket Ctr., W.Va., 1960-65; engring. supr. Hercules Powder Co., Magna, Utah, 1965-69; engr. structures, heat flow, combustion & failure analysis Hercules Rocket Plant, Magna, 1969-83; project engring. mgr. Hercules Aerospace Div., Magna, 1983-89; pres., mgr. Applied Sci. Assocs., Salt Lake City, 1989—; chmn. policy studies UN Assn. Utah, 1990—; project leader world problem analyses, 1990—. Contbr. papers, articles, reports to profl. publs. Judge sci. fair, Salt Lake County, Utah, 1985—; chmn. citizens policy panel Utah chpt. UN Assn., U.S.A., N.Y.C., 1990—; mem. Utah State Hist. Soc., Salt Lake City, 1968-91, Mil. History Soc. Utah, Salt Lake City, 1990—. 1st lt. U.S. Army, 1943-46. Recipient commendation for presentation on world population problem Utah's Forum on Global Environ., 1992. Fellow AIAA (astronautics assoc.); mem. Cons. Engrs. Coun. Utah (article award 1992), League Utah Writers, Wasatch Mountain Club (hike leader 1987—). Achievements include patent for improved artillery ammunition, successful application of engineering approach to analysis of history. Home and Office: Applied Sci Assocs 843 13th Ave Salt Lake City UT 84103

ZEBOUNI, NADEEM GABRIEL, structural engineer; b. Baghdad, Iraq, Mar. 29, 1948; came to U.S., 1969; s. Gabriel Antoon and Cecile A.R. (Alaka) Z.; m. G. Terry Bryant, July 17, 1988; children: Matthew, Paul, Nadia. BScCE, Al-Hickma U., Baghdad, 1969; M. Engring. in Structural Engring., U. Fla., 1971. Registered profl. engr., Fla. Project engr. Rowe Holmes & Assocs., Tampa, Fla., 1971-75; pres. Zebouni & Assocs., Jacksonville, Fla., 1976-77; project engr. Reynolds, Smith & Hills, Jacksonville, 1978-79; cons. Zebouni Engring., Ponte Vedra Beach, Fla., 1979—; pub. speaker on Mid. East, engring. and constrn.; restaurateur Cheelo's Cafe, 1983-85; seminar speaker advanced engring. tech. Iraqi Embassy, U.S.A., 1989. Mem. Bd. Appeals-Code Enforcement, St. Johns County, 1992-96. Named for Outstanding Svc., Am. Concrete Inst., 1986-88. Mem. ASCE, Fla. Engring. Soc. (chair community action com., Outstanding Svc. award 1983), Kiwanis (Jacksonville Beach, Fla. chpt., sr. mem., Disting. Svc. 1988). Achievements include invention of plastic design of irregular multistory steel frames design method, of design of Canadian precast floor system for manufacturing in Florida. Home: 2579 S Ponte Vedra Blvd Ponte Vedra Beach FL 32082 Office: Zebouni Engring 200 Executive Way Ste 216 Ponte Vedra Beach FL 32082

ZEBROWITZ, LESLIE ANN, psychology educator; b. Detroit, Nov. 8, 1944; d. Aaron Harry and Esther (Milgrom) Z.; m. A. Verne McArthur (div. July 1988); children: Caleb Jonathan McArthur, Loren Zachary McArthur. BA, U. Wis., 1966; MS, Yale U., 1968, PhD, 1970. Asst. prof. psychology Brandeis U., Waltham, Mass., 1970-76, assoc. prof., 1976-82, prof., 1982—, chmn. dept., 1986-91; Manuel Yellen prof. social rels. Brandeis U., Waltham, 1989—; vis. scholar Henry Murray Rsch. Ctr. Radcliffe Coll., Waltham, 1991-92. Author: Social Perception, 1991; contbr. numerous articles to sci. jours. Ford. Found. faculty fellow, 1973-74; NIMH rsch. grantee, 1975-81, 87—. Fellow Am. Psychol. Assn.; mem. Am. Psychol. Soc. (charter), Soc. for Exptl. Social Psychology, Ea. Psychol. Assn., Phi Beta Kappa. Office: Brandeis U Dept of Psychology Waltham MA 02254

ZECCHINI, EDWARD JOHN, chemical engineer, researcher; b. Shelbyville, Ind., Apr. 15, 1964; s. Charles Robert and Jannette (Furman) Z.; m. Cathy Hopkins, Dec. 28, 1985; 1 child, Matthew Ryan. BS, Rensselaer Poly. Inst., Troy, N.Y., 1985; MS, Okla. State U., 1986, PhD, 1990. Instr. Okla. State Univ., Stillwater, 1989; vis. asst. prof. Okla. State Univ., 1990-91; staff rsch. engr. Phillips Petroleum Co., Bartlesville, Okla., 1991—; cons. Phillips Petroleum Co., Bartlesville, 1990-91, Kennicott, Great Britain, 1990-91. Contbr. articles to profl. jours. Mem. AICE, NSPE, Sigma Xi, Omega Chi Epsilon (Mu chpt. treas. 1986-87). Home: 501 Chisholm Cir Bartlesville OK 74006 Office: Phillips Petroleum Co 342A PL PRC Bartlesville OK 74004

ZEE, DAVID SAMUEL, neurologist; b. Chgo., Aug. 14, 1944; s. Harry and Pearl (Taube) Z.; m. Paulette Victoria Thompson, July 1, 1983; children: Nathaniel, William. BA, Northwestern U., 1965; MD, Johns Hopkins U., 1969. Intern N.Y.C. Hosp., Bethesda, Md., 1969-70; resident Johns Hopkins Hosp., Balt., 1970-73; clin. assoc. USPHS/NIH, Bethesda, Md., 1973-75; asst. prof. Johns Hopkins Sch. Medicine, Balt., 1975-78; assoc. prof. Johns Hopkins Sch. Medicine, 1978-85, prof., 1985—; mem. study sect. NIH, 1989-92, chair, 1992—. Co-author: Neurology of Eye Movement, 1983, 2d edit., 1991; co-editor 140 jour. articles, 3 textbooks. Lt. comdr. USPHS, 1973-75. Named Morris Bender lectr. Mt. Sinai Med. Sch., 1992, Silversides lectr. U. Toronto, 1985, Jerome Merlis lectr. U. Md., 1992. Mem. Am. Neurol. Assn., Barany Soc. Achievements include discoveries related to normal control of eye movement and vestibular function. Office: Johns Hopkins Med Sch 600 N Wolfe St Baltimore MD 21287

ZEFFREN, EUGENE, toiletries company executive; b. St. Louis, Nov. 21, 1941; s. Harry Morris and Bess (Dennis) Z.; m. Steccia Leigh Stern, Feb. 2, 1964; children: Maryl Renee, Bradley Cruvant. AB, Washington U., 1963; MS, U. Chgo., 1965, PhD, 1967. Research chemist Procter & Gamble Co., Cin., 1967-75, sect. head, 1975-77, assoc. dir., 1977-79; v.p. research and devel. Helene Curtis, Inc., Chgo., 1979—. Co-author: The Study of Enzyme Mechanisms, 1973; contbr. articles to profl. jours.; patentee in field of enzymes and hair care. Mem. AAAS, Am. Chem. Soc., Soc. Cosmetic Chemists, Cosmetic Toiletry and Fragrance Assn. (mem. sci. adv. com. 1979—, vice chmn. sci. adv. com. 1984-88, chmn. 1988-90), Omicron Delta Kappa. Republican. Jewish. Avocations: tennis, swimming, skiing, reading adventure and espionage novels. Office: Helene Curtis Inc 4401 W North Ave Chicago IL 60639-4706

ZEH, HEINZ-DIETER, retired theoretical physics educator; b. Braunschweig, Fed. Republic Germany, May 8, 1932; s. Otto and Irmgard (Stöckner) Z.; m. Sigrid Besch, July 5, 1941. Diploma, U. Heidelberg, Fed. Republic Germany, 1960, D. in Natural Scis., 1962, D. Habilitation, 1966. Rsch. assoc. Calif. Inst. Tech., Pasadena, 1964-65, U. Calif. San Diego, La Jolla, 1965-66; from lectr. to prof. U. Heidelberg, 1966-89. Author: Physik der Zeitrichtung, 1984, Physical Basis of the Direction of Time, 1989, 92. Home: Gaiberger Strasse 38, D-69151 Waldhilsbach Germany

ZEHE, ALFRED FRITZ KARL, physics educator; b. Farnstaedt, Halle, Germany, May 23, 1939; came to Mexico, 1990; s. Alfred and Lina Olga (Kuhnt) Z.; m. Ruth Schorrig, July 30, 1960; children: Axel, Peter. MSc, U. Leipzig, Germany, 1964, D Natural Scis., 1969, Habilitation, 1974; D honoris causa, U. Puebla, Mexico, 1980. Researcher U. Leipzig, 1964-68, asst. prof., 1968-71, assoc. prof., 1971-75; prof. Tech. U. Zwickau, Germany, 1975-80, Tech. U. Dresden, Germany, 1980-91; prof. physics U. Puebla, 1991—; dir. dept. math. and scis. Tech. U. Zwickau, Germany, 1975-76; dir. dept. exptl. solid-state physics U. Puebla, 1977-80; dir. vacuum-physics chair Tech. U. Dresden, 1980-90. Author: Zehe-Roepke Rule for Electromigration, 1985; editor: Fisica y Tecnologia de Semiconductores, 1982; contbr. articles to profl. jours.; patentee in field. Recipient Cincuentenario medal of honor U. PUebla, 1989, Disting. Prof. award Inst. Superior de Informatica, Santo Domingo, 1991; named Disting. Citizen, City Coun., Puebla, 1990. Mem. Phys. Soc. Germany, Phys. Soc. Mex., Math. Soc. Republic Dominicana, Acad. Materials Scis. Mex., Acad. Scis. Republic Dominicana, N.Y. Acad. Scis. Avocations: music, travel, science. Home: Schillerstr 19, 01326 Dresden Germany Office: U Autonoma de Puebla, Apartado Postal # 1505, 72000 Puebla Mexico

ZEHL, OTIS GEORGE, optical physicist; b. Elizabeth, N.J., July 9, 1946; s. Otis George and Elizabeth Theresa (Blehart) Z.; m. Ellen Mayo Hummel, May 18, 1985; 1 child, Elizabeth. BA in Physics with high honors, Rutgers U., 1968; PhD in Physics, U. Md., 1978. Teaching asst. grad. physics lab. U. Md., College Park, Md., 1971-73; doctoral rsch. asst. solid state physics U. Md., 5, Md., 1973-78; sr. scientist Amecom Litton Systems, 5, Md., 1979-91; sr. optical physicist Sachs Freeman Assn., Landover, Md., 1991—; dir. Planning Design Implementation, Inc., Vienna, Va., 1987—. Contbr. articles to profl. jours. Dir. Rosslyn Children's Ctr., Arlington, Va., 1991—. With U.S. Army, 1969-71. Supporter Friends of the Libraries-Rutgers, 1991—. Achievements include co-invention and five patents in areas of acousto-optics and electro-optics; tech. mgmt. and contbr. to state of art Bragg cell fabrication and test; rsch. and devel. of optical components to def. systems. Home: 10113 Windy Knoll Ln Vienna VA 22182 Office: Sachs Freeman Assocs 1401 McCormick Dr Landover MD 20785

ZEHNER, LEE RANDALL, environmental service executive, research director; b. Darby, Pa., Mar. 15, 1947; s. Warren L. and Alycia G. (Van Riper) Z.; m. Susan D. Hovland, June 23, 1973; children: Adam, Erica. BS in Chemistry, U. Pa., 1968; PhD in Organic Chemistry, U. Minn., 1973. Sr. rsch. chemist Arco Chem. Co., Glenolden, Pa., 1973-78; rsch. group leader Ashland Chem. Co., Dublin, Ohio, 1978-82; mgr. organic rsch. W.R. Grace & Co., Clarksville, Md., 1982-85; dir. biotech. programs Biospherics Inc., Rockville, Md., 1985—, v.p. sci. svcs., 1991—. Author various tech. publs.; patentee in field. Mem. AICE, Am. Chem. Soc., Inst. Food Technologists, Am. Mgmt. Assn. Avocations: swimming, gardening, chess. Home: 131 Brinkwood Rd Brookeville MD 20833 Office: Biospherics Inc 12051 Indian Creek Ct Beltsville MD 20705

ZEHNWIRTH, BEN ZION, mathematician consultant; b. Petah Tiqua, Israel, Dec. 23, 1947; s. Zalman and Genia (Lipschitz) Z.; m. Faye Sarah Waysman, Nov. 27, 1973; children: Zeta Bella, Sol Joseph. BS with honors, Melbourne U., Australia, 1969, MS, 1971, PhD, 1973. Lectr. Royal Melbourne Inst. Tech., Victoria, Australia, 1974, Macquarie U., Sydney, NSW, Australia, 1975-80; lektor U. Copenhagen, Denmark, 1980; sr. lectr. Macquarie U., Sydney, NSW, 1981-86, assoc. prof., 1987-90; cons. Insureware Pty. Ltd., Melbourne, Victoria, 1990—; vis. prof. U. B.C., Vancouver, 1980, U. Waterloo, Ontario, Can., 1986, 88; cons. Mercer, Sydney, 1986-87. Co-author: Introductory Statistics with Applications in General Insurance, 1983; author: Computer Software Program, 1985, 90—; contbr. over 30 articles to profl. jours. Recipient Clarence Arthur Culp Meml. award Am. Risk Ins. Assn., 1985—, Dwight prize for Math. Statistics, 1969, Wheat Industry Rsch. Coun. grant, 1979-80, 80-81. Mem. Statis. Soc. Australia, Inst. Math. Statistics, Inst. Actuaries, Royal Statis. Soc. London, Internat. Assn. Actuaries, Casualty Actuarial Soc. Avocations: chess, table tennis, bridge. Home: 15 Sidwell Ave, East Saint Kilda 3183, Australia Office: Insureware P/L, 15 Sidwell Ave, East Saint Kilda 3183, Australia

ZEI, ROBERT WILLIAM, geology educator, consultant; b. Chgo., May 4, 1954; s. Bruno Jr. and Mabel Elaine (Duntze) Z.; m. Annette L. Mario, May 30, 1987. BA in Biology, Northwestern U., 1976; PhD in Geology, U. Pitts., 1991. Asst. prof. phys. sci. Kutztown (Pa.) U., 1985-87, 92—; instr. Johns Hopkins U., Balt., 1986—, West Chester (Pa.) U. Geology Dept., 1989-92. Mem. Geol. Soc. Am., Nat. Assn. Geology Tchrs., Pa. Acad. Sci., Paleontolog. Soc., Soc. Sedimentary Geology, Sigma Xi. Office: Phys Sci Dept Kutztown U Kutztown PA 19350

ZEICHICK, ALAN L., computer journalist, computer scientist; b. Framingham, Mass.; s. Herbert H. and Gloria (Raymond) Z.; m. Carole A. Sacharin. BS, U. Maine, 1984. Systems analyst U. Maine, Orono, 1982-84; editor Camden (Maine) Communications, 1984-87, Internat. Data Group,

Peterborough, N.H., 1987-90; editorial dir. Miller Freeman, Inc., San Francisco, 1990—. Mem. editorial bd. Internat. Assn. Knowledge Engrs., 1990—; editorial dir. Computer Security Jour., Mathematica Jour.; editor-in-chief Cadence Mag., 05/2 Computing; editor AI Expert Jour.; contbr. articles to profl. jours. Mem. IEEE, ACM, Internat. Assn. Knowledge Engrs., Am. Assn. Artificial Intelligence, Internat. Neural Network Soc. Office: Miller Freeman Inc 600 Harrison St San Francisco CA 94107

ZEILINGER, PHILIP THOMAS, aeronautical engineer; b. David City, Nebr., Feb. 13, 1940; s. Thomas Leroy and Sylvia Dorothy Zeilinger; m. Elna Rae Simpson, June 13, 1970; children: Shari, Chris. AS, Wentworth Mil. Acad., Lexington, Mo., 1959; BSME, Kans. U., 1962. Estimator, engr. Reynolds Electronics and Engring. Co., El Paso, Tex., 1966-68; accessories coord. ITI Garrrett, Phoenix, 1974-79, cntrl. access engr., 1968—, controls coord. ITEC, 1983-84, integrated support specialist ITEC, 1984-86, mgr. systems software light helo turbine engring. co. div., 1986-91, FAA designated engr. rep. engine div., 1991—; chmn. Light Helicopter Turbine Engine Company Computer Aided Acquistion and Logistics Working Group. V.p. Indsl. Devel. Authority, Tempe, Ariz., 1979-84; pres. Univ. Royal Garden Homes Assn., Tempe, 1984-90. 1st lt. U.S. Army, 1962-66. Recipient Vol. Svc. award City of Tempe, 1984, Grand Cross of Color, Internat. Order of Rainbow Girls, 1978. Mem. AIAA, Aircraft Owners and Pilots Assn., Explt. Aircraft Assn. (v.p. chpt. 228 1974-79), Masons (master 1990-92, chmn. statewide picnic 1992, Mason of the Yr. 1992). Democrat. Unitarian. Achievements include patent for Airesearch/Garrett. Home: 760 N Sycamore Pl Chandler AZ 85224 Office: 11 S 34th St Phoenix AZ 85010

ZEISSIG, HILMAR RICHARD, oil and gas executive; b. Berlin, Germany, June 29, 1938; came to U.S., 1978; s. Richard W. and Sigrid (Von Knauer) Z.; m. Heidi Harlandt, Sept. 6, 1967 (div. Oct. 1982); m. Ria M. Lobert, Oct. 4, 1983; children: Philipp, Richard, Isabel, Alexander. Student, U. Kiel, Fed. Republic Germany, 1959-60, U. Lausaunne, Switzerland, 1961; JD, U. Bonn, Fed. Republic Germany, 1964; PhD, U. Cologne, Fed. Republic Germany, 1969. Rsch. assoc. UN Econ. Commn. for Latin Am. and the Carribean, Santiago, Chile, 1965-66; legal advisor Export Industry Assn., Hamburg, Fed. Republic Germany, 1967-69; corp. counsel Gelsenberg A.G., Essen, Fed. Republic Germany, 1970; chief negotiator Deminex Oil Co., Dusseldorf, Fed. Republic Germany, 1971-73; gen. mgr. Deminex Peru, Lima, 1973-75, Deminex Egypt, Cairo, 1976-77; pres., chief exec. officer Lingen Oil & Gas Inc., Houston, 1978-84; chmn., pres. Houston Internat. Bus. Corp., 1982—; exec. v.p., dir. Natural Res. Group, Inc., Houston, 1985—; officer, dir. Control Ambietal Integral S.A. de D.V., Mexico City; U.S. rep. Prakla-Seismos AG Germany, Houston, 1986-92; energy cons. Econ. Commn. for Latin Am. and Caribbaen, UN, Mex. and Ctrl. Am., 1986—; environ. cons. Germany and Mexican Govts., 1989—. Author annual study on Ctrl. Am. energy problems, 1986-93. Bd. dirs. German-Peruvian C. of C., Lima, 1973-75; active Econ. Devel. Com., Sugar Land, Tex., 1990—. Mem. Ibero Am. Verein Hamburg, Houston C. of C. (chmn. various coms. 1980-86), Houston World Trade Assn. (bd. dirs. 1989—), German Am. C. of C. (bd. dirs. 1979—), Rotary of Sugar Land, Rotary Internat. (chmn. group study exch. dist. 589 1986-92). Avocations: sailing, reading political and historical publications. Home: 823 Alhambra Ct Sugar Land TX 77478-4003 Office: Houston Internat Bus Corp 5 Post Oak Park Ste 330 Houston TX 77027-3413

ZEITLIN, BRUCE ALLEN, superconducting material technology executive; b. N.Y.C., July 31, 1943; s. Lester and Rae (Benson) Z.; m. Amy Joy Kozan, Aug. 29, 1965; children: Laurence, Jessica, Andrea. BS, Rensselaer Poly. Tech., 1965; MS, Stevens Inst. Tech., 1968. Scientist Airco Cen. Rsch. Lab., Murray Hill, N.J., 1965-70; tech. dir. Magnetic Corp. of Am., Waltham, Mass., 1970-72; v.p. IGC/Advanced Superconductors, Waterbury, Conn., 1986—, Intermagnetics Gen. Corp., Guilderland, N.Y., 1986—; bd. dirs. Alsthom Intermagnets SA, Paris. Patentee in field. Mem. Am. Phys. Soc. Avocation: astronomy. Office: IGC/Advanced Superconductor 1875 Thomaston Ave Waterbury CT 06704-1034

ZELBY, LEON WOLF, electrical engineering educator, consulting engineer; b. Sosnowiec, Poland, Mar. 26, 1925; came to U.S., 1946, naturalized, 1951; s. Herszel and Helen (Wajnryb) Zylberberg; m. Rachel Kupfermintz, Dec. 28, 1954; children: Laurie Susan, Andrew Stephen. BSEE, Moore Sch. Elec. Engring., 1956; MS, Calif. Inst. Tech., 1957; PhD, U. Pa., 1961. Registered profl. engr., Pa., Okla. Mem. staff RCA, Hughes R & D Labs., Lincoln Lab., MIT, Sandia Corp., Argonne (Ill.) Nat. Labs., Inst. for Energy Analysis; mem. faculty U. Pa., 1959-67, assoc. prof., 1964-67; assoc. dir. plasma engring. Inst. Direct Energy Conversion, 1962-67; prof. U. Okla., Norman, 1967—, dir. Sch. Elec. Engring., 1967-71; cons. RCA, 1961-67, Moore Sch. Elec. Engring., 1967-68, also pvt. firms. Editor Tech. and Soc. mag., 1990-93; contbr. articles on energy-associated problems and issues to profl. jours. With AUS, 1946-47. Cons. Electrodynamic Corp. fellow Calif. Inst. Tech., 1957, Mpls.-Honeywell fellow U. Pa., 1957-58, Harrison fellow, 1958. Mem. IEEE, Franklin Inst., Sigma Xi, Tau Beta Pi, Eta Kappa Nu, Pi Mu Epsilon, Sigma Tau, Phi Kappa Phi. Home: 1009 Whispering Pines Dr Norman OK 73072-6912 Office: U Okla Norman OK 73019

ZELDES, ILYA MICHAEL, forensic scientist; b. Baku, Azerbaidjan, USSR, Mar. 15, 1933; came to U.S., 1976; s. Michael B. and Pauline L. (Ainbinder) Z.; m. Emma S. Kryss, Nov. 5, 1957; 1 child, Irina Zeldes Rieser. JD, U. Azerbaidjan, Baku, 1955; PhD in Forensic Scis., U. Moscow, 1969. Expert-criminalist Med. Examiner's Bur., Baku, 1954-57; rsch. assoc. Criminalistics Lab., Moscow, 1958-62; sr. rsch. assoc. All-Union Sci. Rsch. Inst. Forensic Expertise, Moscow, 1962-75; chief forensic scientist S.D. Forensic Lab., Pierre, 1977-93; owner Forensic Scientist's Svcs., Pierre, 1977-93. Author: Physical-Technical Examination, 1968, Complex Examination, 1971, The Problems of Crime, 1981; contbr. numerous articles to profl. publs. in Australia, Austria, Bulgaria, Can., Eng., Germany, Holland, India, Ireland, Israel, Rep. of China, Taiwan, U.S. and USSR. Mem. Internat. Assn. Identification (rep. S.D. chpt. 1979-93, chmn. forensic lab. analysis subcom. 1991-92), Assn. Firearm and Tool Mark Examiners (disting.). Avocation: travel. Home: # 1 5735 Foxlake Dr North Fort Myers FL 33917

ZELENAKAS, KEN WALTER, clinical scientist; b. Yonkers, N.Y., Sept. 9, 1950; s. Joseph Albert and Irene (Panek) Z.; m. Wilhelmina Nowak, Nov. 24, 1973; children: Andrea, Karl. BS in Biology, Iona Coll., 1976; MS in Biology, NYU, 1979. Sci. unit mgr. Am. Health Found., Valhalla, N.Y., 1976-81; toxicology Ciba-Ceigy, Summit, N.Y., 1981-85, clin. scientist, 1985—. Contbr. articles to profl. jours. Mem. ch. coun. Zion Luth. Ch., Rahway, N.J., 1988; coach Pop Warner Football, Woodbridge, N.J., 1992. Mem. AAAS, Assn. Clin. Pharmacology. Democrat. Lutheran. Achievements include research in chemical carcinogenesis, DNA interaction, DNA repair. Home: 638 Ridgedale Ave Woodbridge NJ 07095 Office: Ciba-Geigy Corp 556 Morris Ave Summit NJ 07901

ZELEZNY, WILLIAM FRANCIS, retired physical chemist; b. Rollins, Mont., Sept. 5, 1918; s. Joseph Matthew and Birdie Estelle (Loder) Z.; m. Virginia Lee Scarcliff, Sept. 14, 1949. BS in Chemistry, Mont. State Coll., 1940; MS in Metallurgy, Mont. Sch. Mines, 1941; PhD in Phys. Chemistry, State U. Iowa, 1951. Scientist NACA, Cleve., 1951-54; metallurgist div. indsl. research Wash. State Coll., Pullman, 1954-57; scientist atomic energy div. Phillips Petroleum Co., Idaho Falls, Id., 1957-66, Idaho Nuclear Corp., Idaho Falls, 1966-70; mem. staff Los Alamos (N.Mex.) Sci. Lab., 1970-80; instr. metallurgy State U. Iowa, Iowa City, 1948-49; asst. prof. metallurgy Wash. State Coll., 1956-57; instr. U. Idaho, Idaho Falls, 1960-68. Contbr. articles to profl. jours.; patentee in field. Served with AUS, 1944-46. Mem. Am. Chem. Soc. (sec. N.Mex. sect. 1978-79), Microbeam Analysis Soc., Am. Soc. Metals, The Minerals, Metals & Materials Soc., Sigma Xi, Alpha Chi Sigma. Democrat. Methodist. Avocation: gardening. Home: PO Box 37 Rollins MT 59931-0037

ZELINSKI, JOSEPH JOHN, engineering educator, consultant; b. Glen Lyon, Pa., Dec. 30, 1922; s. John Joseph and Lottie Mary (Oshinski) Z.; m. Mildred G. Sirois, July 22, 1946; children: Douglas John, Peter David. BS, Pa. State U., 1944, PhD, 1950. Grad. fellow Pa. State U., University Park, 1946-50; project supr. applied physics lab. Johns Hopkins U., Silver Spring, Md., 1950-58; staff scientist Space Tech. Labs. (now TRW, Inc.), Redondo Beach, Calif., 1958-60; head chem. tech. div. Ops. Evaluation Group MIT, Cambridge, 1960-62; prin. rsch. scientist Avco Everett (Mass.) Rsch. Lab., 1962-64; prof. mech. engring. Northeastern U., Boston, 1964-85, prof. emeritus, 1985—; pres. World Edn. Resources, Ltd., Tampa, Fla., 1984—; cons. Avco Everett Rsch. Lab., 1964-71, Pratt & Whitney Aircraft, East Hartford, Conn., 1966-70, Modern Electric Products and Phys. Scis. Co., Inc., Boston, 1980-82, Morrison, Mahoney and Miller, Boston, 1984; vice-chmn., chmn. exec. com. Univ. Grad. Coun., Northeastern U., Boston, 1980-84, dir. mech. engring. grad. program, 1982-85; del. 4th World Conf. Continuing Engring. Edn., Beijing China People to People, Spokane, Wash., 1989. Contbr. articles to profl. jours. Prin. Confraternity Christian Doctrine, Andover, Mass., 1961-64; pres. Andover Edn. Coun., 1962-64; vice chmn. Dem. Town Com., Boxford, Mass., 1980-84. Lt. (j.g.) USNR, 1943-46, PTO. Mem. AAAS, ASME, Am. Chem. Soc., N.Y. Acad. Scis., Combustion Inst. Democrat. Roman Catholic. Achievements include U.S. and foreign patents for coal combustion system for magnetohydrodynamic power generation, for fuel-cooled combustion systems for jet engines flying at high Mach numbers; prediction of optical observables of re-entry vehicles from analysis of decomposition mechanisms of heat-shield materials; invention of high-temperature furnace for production of crystalline graphite; development and verification of a design method for ramjet combustors. Home: 9207 Jubilee Ct Hunters Green Tampa FL 33647

ZELLER, CLAUDE, physicist, researcher; b. Aulnay, France, Dec. 11, 1940; came to U.S., 1976; m. Elisabeth Kreib, 1962 (div. 1967); 1 child, Frederic; m. Florence Labour, Oct. 14, 1967; children: Caroline, Elisabeth. PhD, Univ. Nancy, France, 1968. Rsch. physicist Univ. Nancy, France, 1968-76; visiting rsch. faculty Univ. Pa., Phila., 1976-79; sr. physicist Pitney Bowes R&D, Norwalk, Conn., 1979-84, mgr. applied physics, 1984-91; fellow Pitney Bowes, Shelton, Conn., 1992—; adv. bd. CNRS, Paris, 1969-71; sec. scientific bd. Univ. Nancy, 1971-76; cons. Bruker-Spectrospin, Wissembourg, France, 1970-75. Contbr. articles to profl. jours. Recipient sr. fellowship, NATO, 1976. Mem. IEEE, N.Y. Acad. Scis., Am. Physical Soc., Appalachian Mountain Club. Roman Catholic. Achievements include 3 patents and 74 published articles in area of research in U.S. and European journals. Notable findings include synthetic conductors whose electrical conductivity is comparable to copper, work cited in Physics Today. Home: 97 Fan Hill Rd Monroe CT 06468-1831 Office: Pitney Bowes Inc 35 Waterview Dr Shelton CT 06484-3000

ZELLER, MICHAEL JAMES, psychologist, educator; b. Des Moines, Dec. 3, 1939; s. George and Lila (Fitch) Z. BS, Iowa State U., 1962, MS, 1967. Instr. psychology Mankato (Minn.) State U., 1967-73, asst. prof. psychology, 1974-89, assoc. prof. psychology, 1990—; mem. social sci. edn. coun. Mankato State U., 1976—; ednl. cons. Random House, Scott Foresman, West Pub. Editor: Test Item File to Accompany Introduction to Psychology, 6th edit., 1992; co-author: Test Item File to Accompany Introduction to Psychology, 5th edit., 1989, Test File for Psychology, 3d edit., 1988, Unit Mastery Workbook, 1st, edit., 1974, 2d edit., 1976, Test Item File to Accompany Psychology, 1st edit., 1974, 2d edit., 1976; editor: Psychology: A Personal Approach, 1st edit., 1982, 2d edit., 1984. With USAR, 1964-70. Mem. APA, Minn. Psychol. Assn., Inter-Faculty Orgn., Psi Chi (award 1988). Achievements include development and research on educational materials, methods of instruction and career opportunities for psychology majors. Home: PO Box 1958 Mankato MN 56002 Office: Mankato State Univ MSU 35 PO Box 8400 Mankato MN 56002

ZELLERS, ROBERT CHARLES, construction materials engineer, consultant; b. Youngstown, Ohio, June 13, 1943; s. Charles Newton and Beatrice Eleanor (Snavely) Z.; m. Patricia Ann Ockerman, Nov. 27, 1965; children: Derek, Shannon, Robyn. BEng in Civil Engring., Youngstown State U., 1967. Registered profl. engr., Pa., Ohio; registered profl. land surveyor, Pa. Materials engr. Standard Slag Co., Youngstown, 1966-72; asst. chief engr. The Duquesne Slag Co., Pitts., 1972-81; exec. sec. Pa. Slag Assn., Pitts., 1972-81; v.p. engring. Forta Corp., Grove City, Pa., 1978-82, exec. v.p., 1983—; owner, mgr. Zellers Design Group, Mercer, Pa., 1982—; owner Zellers Galleries, Mercer, 1979—; prin. IMTEK, Grove City, Pa., 1988-91, ICEMS, Toledo, 1992—; presenter in field at confs., seminars and profl. orgn. meetings. Contbr. articles to profl. jours. Mem. NSPE, ASTM (various coms.), Pa. Soc. Profl. Engrs., Engring. Soc. Western Pa., Am. Concrete Inst. (various coms.), Am. Mgmt. Assn., Constrn. Specifications Inst., Internat. Congress Bldg. Ofcls., So. Bldg. Code Congress Internat., Nat. Ready Mixed Concrete Assn., Synthetic Fiber Assn. (co-founder, pres.), Corvette Owners Assn., Grove City Country Club, Lions. Office: Forta Corp 100 Forta Dr Grove City PA 16127-9099

ZEMANSKY, GILBERT MAREK, hydrogeologist, water quality engineer; b. L.A., Feb. 15, 1944; s. Stanley Donald and Anne Alice (Person) Z.; m. Ellen Ruth Kroeker, Apr. 19, 1980; children: Rebekah, Peter. BS, U.S. Naval Acad., 1965; MS, U. Colo., 1973; PhD, U. Wash., 1983. Registered ground water profl., Iowa. Commd. lt. USN, 1965, served in various locations including Vietnam, 1965-71, resigned, 1971; asst. engring. specialist State Water Resources Control Bd., Sacramento, 1973-74; pipeline monitor Alaska Dept. Environ. Cons., Fairbanks, 1974-75; rsch. environ. engr., instr. U. Alaska, Fairbanks, 1975-76; ind. cons. Seattle and Corvallis, Oreg., 1976-85; chief sci. and tech. sect. Ill. Pollution Control Bd., Chgo., 1985-87; dir. water pollution control Hall-Kimbrell Environ. Svcs., Lawrence, Kans., 1988-90; prin. hydrogeologist Terracon Environ., Inc., Kansas City, Mo., 1990—; tech. com. Metro, Seattle, 1978-80; mem. placer mining adv. group Alaska Dept. Environ. Cons., Juneau, 1983-84; chmn. water quality task force Seattle Water Dept., 1983-84; mem. water resources tech. adv. com. Northeast Ill. Planning Commn., Chgo., 1985-87; tech. com. Hazardous Waste Adv. Com., Washington, 1991—. Contbr. articles to profl. jours. Mem. Nat. Ground Water Assn. (jour. reviewer), Am. Water Resources Assn. (jour. reviewer), Am. Inst. Hydrology (cert. profl. ground water hydrologist, ASTM (mem. subcom. D-18.21 on ground water and vadose zoning investigations). Achievements include focus of regulatory attention on mining-related water pollution in Alaska leading to the development of national effluent guidelines for placer mining, assisting to control pesticide impacts on water quality in Pacific Northwest and in assessment and remediation of ground water and soil quality at various sites. Home: 3000 W 19th Ct Lawrence KS 66047 Office: Terracon Environ Inc 7810 NW 100th St Kansas City MO 64153

ZEMTSOV, ALEXANDER, dermatology and biochemistry educator; b. Baku, USSR, Nov. 9, 1959; came to U.S., 1977; s. Ilya and Maryn (Dubinsky) Z.; m. Tali Giveon, Oct. 17, 1987; children: Raquel Karen, Gregory Ethan. BA magna cum laude, Temple U., 1981; MSc, U. Pa., 1982; MD with honors, NYU, 1986. Diplomate Am. Bd. Dermatology. Intern, then resident Cleve. Clinic Hosp. Found., 1989-90; asst. prof. dermatology and biochemistry Tex. Tech. Sch. Medicine, Lubbock, 1990—. Contbr. articles to profl. jours. and books. Recipient Am. Soc. Dermatol. Surgery award, 1989; Cert. Appreciation, Ohio Dermatol. Soc., 1990. Fellow Am. Acad. Dermatology, Am. Contact Dermatitis Soc.; mem. Soc. Magnetic Resonance, Internat. Soc. for Digital Imaging of Skin (pres.), Kiwanis, Lubbock Club. Jewish. Avocations: stamp collecting, hiking, swimming. Office: Tex Tech Sch Medicine 3601 4th St Lubbock TX 79430-0001

ZEN, E-AN, research geologist; b. Peking, China, May 31, 1928; came to U.S., 1946, naturalized, 1963; s. Hung-chun and Heng-chi'h (Chen) Z. AB, Cornell U., 1951; MA, Harvard U., 1952, PhD, 1955. Research fellow Woods Hole Oceanographic Inst., 1955-56, research asso., 1956-58; asst. prof. U. N.C., 1958-59; geologist U.S. Geol. Survey, 1959-80, rsch. geologist, 1981-89; adj. rsch. prof. geology U. Md., 1990—; vis. assoc. prof. Calif. Inst. Tech., 1962; Crosby vis. prof. MIT, 1973; Harry H. Hess sr. vis. fellow Princeton U., 1981; counselor 28th Internat. Geol. Congress. 1986-89. Contbr. articles to profl. jours. Fellow AAAS, Am. Acad. Arts and Scis., Geol. Soc. Am. (councillor 1985-88, v.p. 1991, pres. 1992), Mineral. Soc. Am. (coun. 1975-77, pres. 1975-76, Roebling medal 1991); mem. NAS, Geol. Soc. Washington (pres. 1973), Mineral. Assn. Can. Office: U Md Dept Geology College Park MD 20742

ZENDLE, HOWARD MARK, software development researcher; b. Binghamton, N.Y., June 8, 1949; s. Abraham and Evelyn (Hershowitz) Z. BA in Physics summa cum laude, SUNY, Binghamton, 1972, MA in Physics, 1976; MSEE, Syracuse U., 1987. With IBM, Owego, N.Y., 1974—, staff programmer, 1978-83, mgr. microprocessor applications software, 1979-81, mgr. tactical avionics software, 1981-82, adv. programmer, 1983-86, sr. programmer, 1986—; mem. Fed. Sector div. Mktg. Conf. IBM, 1991. Sec. Men's Club Beth David Synagogue, Binghamton, 1984-85, v.p., 1986-88; bd. dirs. Jewish Community Ctr., Binghamton, 1983-86. Recipient Informal awards IBM, 1975, 78, 81, 83, 91, 92. Mem. IEEE, Assn. for Computing Machinery, Cen. Electric Railfan's Assn., Masons, Phi Beta Kappa, Sigma Pi Sigma. Republican. Avocations: railfanning, research into history of industrial development in America.

ZENIERIS, PETROS EFSTRATIOS, civil engineer; b. Athens, Feb. 20, 1955; came to the U.S., 1980; s. Efstratios P. and Persefoni (Stefanidis) Z.; m. Jer D. Billimoria, Sept. 21, 1990; 1 child, Alexander. BCE, Poly. Inst., Greece, 1979; MS in Geotech. and Environ. Engring., U. Okla., 1988. Registered profl. engr., Okla., Greece. Researcher U. Okla., Norman, 1983-86; field and lab engr. Standard Testing and Engring. Co., Oklahoma City, 1986-87, lead geotech. engr., 1987-88, mgr. geotech. engring., 1988—. Author: The Feasibility of Using Fly Ash as a Binder in Fine and Coarse Aggregate Bases, 1988; contbr. articles to profl. jours. Mem. ASCE (bd. registration for profl. engrs. 1991—), Okla. Water Resources Bd. (lic.), Okla. Soc. Environ. Profls., Greek Soc. Profl. Engrs. and Land Surveyors. Achievements include research in fly ash and its chemical properties. Office: Standard Testing Engring Co 4300 N Lincoln Blvd Oklahoma City OK 73105

ZENNER, HANS PETER, otolaryngologist; b. Essen, Germany, Nov. 13, 1947; s. Hans and Eleonore (Lang) Z.; m. Birgit Zenner, 1977; 3 children. MD, U. Mainz, Germany, 1972, PhD, 1974; Dr.habil., U. Wuerzburg, Germany, 1981. Wiss. asst. U. Wuerzburg, 1974-81, dozent, 1981-86, prof., 1986-88; prof. orolaryngology, chmn. dept. U. Tübingen, Germany, 1988—; vis. scientist U. Mich., Ann Arbor, 1985, Washington U., St. Louis, 1987. Author: Allergologie, 1987, Therapie HNO, 1993, Physiologie; editor: All. Atemwegserkrank., 1988. Pres. Inst. Sonderhoerhilfe, Munich, 1990—; advisor Govt. of Germany, 1992—. Recipient Troeltsch award German Acad. Otolayngology, 1982, Sandor-Cseresmes medal Hungarian Triological Soc., 1985, Four Centennial prize, U. Würzburg, 1985, Leibniz award German Rsch. Coun., 1986, Haymann prize German Triological Soc., 1988. Mem. Rotary. Home: Silcherstr 5, W-7400 Tübingen Germany Office: U Tübingen, Dept Otolaryngology, W-7400 Tübingen Germany

ZENONE, JOHN MARK, system programmer, technical analyst; b. Phila., June 20, 1958. Student, U. Ark., Little Rock; grad., Pulaski Tech. Coll., 1980. With First Nat. Bank, Little Rock, 1979-81; objective text coord./cons., lead computer operator U. Ark. for Med. Scis., Little Rock, 1981-87, applications programmer 1, 1981-87; tech. analyst/ programmer Alltel, Systematics Info. Svcs., Inc., Little Rock, 1987—.

ZENTMYER, GEORGE AUBREY, plant pathology educator; b. North Platte, Nebr., Aug. 9, 1913; s. George Aubrey and Mary Elizabeth (Strahorn) Z.; m. Dorothy Anne Dudley, May 24, 1941; children: Elizabeth Zentmyer Dossa, Jane Zentmyer Fernald, Susan Dudley. A.B., UCLA, 1935; M.S., U. Calif., 1936, Ph.D., 1938. Asst. forest pathologist U.S. Dept. Agr., San Francisco, 1937-40; asst. pathologist Conn. Agrl. Expt. Sta., New Haven, 1940-44; asst. plant pathologist to plant pathologist U. Calif., Riverside, 1944-62, prof. plant pathology, 1962—, prof. emeritus, 1981—, faculty rsch. lectr., 1964, chmn. dept., 1968-73; cons. NSF, Trust Ty. of Pacific Islands, 1964, 66, Commonwealth of Australia Forest and Timber Bur., 1968, AID, Ghana and Nigeria, 1969, Govt. of South Africa, 1980, Govt. of Israel, 1983, Govt. of Western Australia, 1983, Ministry Agriculture and U. Cordoba, Spain, 1989; mem. NRC panels, 1968-73. Author: Plant Disease Development and Control, 1968, Recent Advances in Pest Control, 1957, Plant Pathology, an Advanced Treatise, 1977, The Soil-Root Interface, 1979, Phytophthora: Its Biology, Taxonomy, Ecology and Pathology, 1983, Ecology and Management of Soilborne Plant Pathogens, 1985; assoc. editor: Ann. Rev. of Phytopathology, 1971—, Jour. Phytopathology, 1951-54; contbr. articles to profl. jours. Bd. dirs. Riverside YMCA, 1949-58, Friends of Mission Inn, 1981—, pres., 1991-93, Calif. Mus. Photography, 1988—; pres. Town and Gown Orgn., Riverside, 1962; bd. dirs. Riverside Hospice, 1982-85, pres., 1984-85; bd. dirs. Friends U. Calif. Riverside Botanic Garden, 1985-89, 91—, pres., 1987-89. Recipient award of honor Calif. Avocado Soc., 1954, spl. award of honor, 1981; recipient Emeritus Faculty award U. Calif., Riverside, 1991; Guggenheim fellow, Australia, 1964-65, NATO sr. sci. fellow, Eng., 1971; NSF rsch. grantee, 1963, 68, 71, 74, 78; Bellagio scholar Rockefeller Found., 1985. Fellow AAAS (pres. Pacific div. 1974-75), Am. Phytopath. Soc. (pres. 1966, pres. Pacific div. 1955, found. bd. dirs. 1987—, v.p. 1991, Award of Distinction 1983, award of Merit Caribbean div. 1972, Lifetime Achievement award Pacific div. 1991), Explorers Club; mem. NAS, Mycol. Soc. Am., Am. Inst. Biol. Scis., Bot. Soc. Am., Brit. Mycol. Soc., Australasian Plant Pathology Soc., Philippine Phytopath. Soc., Indian Phytopath. Soc., Assn. Tropical Biology, Internat. Soc. Plant Pathology (councilor 1973-78), Pacific Assn. Tropical Phytopathology, Sigma Xi, Gamma Sigma Alpha. Home: 708 Via La Paloma Riverside CA 92507-6465

ZEPF, THOMAS HERMAN, physics educator, researcher; b. Cin., Feb. 13, 1935; s. Paul A. and Agnes J. (Schulz) Z. BS summa cum laude, Xavier U., 1957; MS, St. Louis U., 1960, PhD, 1963. Asst. prof. physics Creighton U., Omaha, 1962-67, assoc. prof., 1967-75, prof., 1975—, acting chmn. dept. physics, 1963-66, chmn., 1966-73, 81-93, coord. allied health programs, 1975-76, coord. pre-health scis. advising, 1976-81; cons. physicist VA Hosp., Omaha, 1966-71; vis. prof. physics St. Louis U., 1973-74; program evaluator Am. Coun. on Edn., 1988—. Contbr. articles and abstracts to Surface Sci., Bull. Am. Phys. Soc., Proceedings Nebr. Acad. Sci., The Physics Tchr. jour., others. Chmn. physics judging com. Greater Nebr. Sci. and Engring. Fair, 1973-85. Recipient Cert. Recognition award Phi Beta Kappa U. Cin. chpt., 1953, Disting. Faculty Svc. award Creighton U., 1987. Mem. AAAS, Am. Phys. Soc., Am. Assn. Physics Tchrs. (pres. Nebr. sect. 1978), Nebr. Acad. Sci. (life, chmn. physics sect. 1985—), Internat. Brotherhood Magicians, Soc. Am. Magicians (pres. assembly #7, 1964-65), KC, Sigma Xi (Achievement award for rsch. St. Louis chpt. 1963, pres. Omaha chpt. 1993-94), Sigma Pi Sigma. Roman Catholic. Office: Creighton U Dept Physics Omaha NE 68178

ZEPP, LAWRENCE PETER, mechanical engineer; b. N.Y.C., June 26, 1952; s. Clarence Peter and Muriel (Elwin) Z.; m. Pamela Renée, May 21, 1977; children: Jason Douglas, Deanne Elizabeth. BA in Physics, SUNY, Geneseo, 1974. Tchr. physics Napoleon (Ohio) High Sch., 1974-75; project engr. Prestolite Co., Toledo, 1975-81, resident engr., 1981-82; mgr. pump engring. Internat. Hydraulic Systems, Southgate, Mich., 1982-84; mgr. power unit Dura Corp., Toledo, 1984-86; mgr. pump group Xolox Corp., Ft. Wayne, Ind., 1986-91; mgr. fluid products Xolox Corp., Ft. Wayne, 1991—. Den leader Boy Scouts Am., Ft. Wayne, 1988—, troop leader, 1991-92; v.p. Glenwood Community Assn., Ft. Wayne, 1988—. Recipient inventor award Allied Chem. Corp., 1982. Mem. Soc. Automotive Engrs., Soc. Plastics Engrs., Exptl. Aircraft Assn. (chpt. pres. 1977, Outstanding Mem. award 1977). Lutheran. Achievements include patent on hydraulic valve, hydraulic pressure amplifier, linear potientiometer, boat trim tab systems, quick connect hydraulic fitting, ratio mixing device, pump for viscous fluids, also patents pending. Office: Xolox Corp 6932 Gettysburg Pike Fort Wayne IN 46804

ZEROUG, SMAINE, electrophysicist; b. Ouled Djellal, Biskra, Algeria, Dec. 12, 1962; came to U.S., 1985; s. Ahmed and Aicha (Zaddam) Z. BS in Physics, U. Algiers, Bab-Ezzouar, 1985; MS in Electrophysics, Poly. U., 1989, PhD in Electrophysics, 1993. Rsch. fellow Poly. U., Farmingdale, N.Y., 1986-90, grad. asst., 1990-92; rsch. scientist Schlumberger-Doll Rsch., Ridgefield, Conn., 1992—. Contbr. articles to profl. publs. Mem. IEEE, Acoustical Soc. of Am., Sigma Xi. Achievements include contribution to analytical modeling of ultrasonic beam propagation and scattering in elastic structures. Home: Imm Patrimoine Algerois, Kouba 16050, Algeria Office: Schlumberger-Doll Rsch Old Quarry Rd Ridgefield CT 06877

ZERVAS, NICHOLAS THEMISTOCLES, neurosurgeon; b. Lynn, Mass., Mar. 9, 1929; s. Themistocles and Demetra P. (Stasinopoulos) Z.; m. Thalia Poleway, Feb. 15, 1959; children—T. Nicholas, Christopher Louis, Rhea. A.B., Harvard U., 1950; M.D., U. Chgo., 1954. Intern N.Y. Hosp., 1955; resident in neurology Montreal Neurol. Inst., 1956; resident in neurosurgery Mass. Gen. Hosp., Boston, 1958-62; fellow in stereotaxic cerebral surgery U. Paris, 1960-61; asst. attending surgeon, asso. neurosurgery Jefferson Med. Coll., Phila., 1962-67; asso. prof. surgery Harvard U., 1971-77; also chief neurosurg. service Beth Israel Hosp., Boston, 1967-77; prof. surgery Harvard U., 1977—; also chief neurosurg. service Mass. Gen. Hosp., 1977—; Higgins prof. neurosurgery Harvard U., 1987—. Contbr. numerous articles to sci. jours. Chmn. Mass. Coun. Arts and Humanities, 1983-91; trustee Boston Symphony Orch., 1990—, vice chmn., 1993—. Capt. M.C. AUS, 1956-58. Mem. Am. Acad. Neurol. Surgery (pres. 1990-91), Am. Assn. Neurol. Surgeons, Soc. Neurol. Surgeons, Am. Neurol. Assn., Am. Bd. Neurol. Surgery (chmn. 1990-91), Inst. Medicine Nat. Acad. Scis., Sigma Xi. Home: 100 Canton Ave Milton MA 02186-3507 Office: Mass Gen Hosp 32 Fruit St Boston MA 02114-2698

ZEUNER, RAYMOND ALFRED, manufacturing engineer; b. Buffalo, N.Y., May 2, 1937; s. Alfred R. and Erika I. (Vogel) Z.; m. Mary Ellen Zeppenfeld, June 15, 1963; children: Susan Ann, Sandra Lee, Jeffrey Raymond, Raymond Joseph. BSEE, SUNY, Buffalo, 1965. Registered profl. engr., N.Y., Ohio. Jr. engr. Republic Steel Corp., Buffalo, 1963-67, project engr., 1967-72, asst. chief engr., 1973-81; chief engr. Republic Steel Corp., Warren, Ohio, 1981-84; works engr. LTV Steel Corp., Warren, Ohio, 1984-88; chief engr. WCI Steel Corp., Warren, Ohio, 1988—. With USN, 1957-61. Mem. Assn. Iron and Steel Engrs. (chmn. Niagra chpt. 1979-80, Youngstown chpt. 1989, 93, bd. dirs. local chpt. 1981, 88, 93), Cortland Conservation Club (pres. 1984). Republican. Presbyterian. Achievements include research in hot mill delay table heat retention panels, steel making precipitator rehabilitation. Office: WCI Steel Co 1040 Pine Ave SE Warren OH 44483-6528

ZEWAIL, AHMED HASSAN, chemistry and chemical engineering educator, editor, consultant; b. Damanhour, Egypt, Feb. 26, 1946; came to U.S., 1969, naturalized, 1982; s. Hassan A. Zewail and Rawhia Dar; m. Dema Zewail; children: Maha, Amani, Nabeel. B.S., Alexandria U., Egypt, 1967, M.S., 1969; Ph.D., U. Pa., 1974; MA (hon.), Oxford U., 1991; DS (hon.), Am. U., Cairo, 1993. Teaching asst. U. Pa., Phila., 1969-70; IBM fellow U. Calif., Berkeley, 1974-76; asst. prof. chemistry and chem. engring. Calif. Inst. Tech., Pasadena, 1976-78, assoc. prof., 1978-82, prof., 1982-89, Linus Pauling prof. chem. physics, 1990—; cons. Xerox Corp., Webster, N.Y., 1977-80, ARCO Solar, Inc., Calif., 1978-81. Editor Laser Chemistry jour., 1981-85, Photochemistry and Photobiology, cols. I and II, Advances in Laser Chemistry, Advances in Laser Spectroscopy, Ultrafast Phenomena VII and VIII, The Chemical Bond: Structure and Dynamics; contbr. more than 200 articles in rsch. and devel. in lasers and applications to sci. jour.; patentee solar energy field. Recipient Tchr.-scholar award Dreyfus Found., 1979-85, Alexander von Humboldt Sr. U.S. Scientist award, 1983. John Simon Guggenheim Meml. Found. award, 1987, King Faisal Internat. prize in sci., 1989, Wolf Found. prize in chemistry, 1993; Sloan Found. rsch. fellow, 1978-82, Egyptian- Am. Person of Yr., NASA award, Faraday Pub. Discourse, Sir Cyril Hinshelwood chair Lectureship, 1st AMM Achievement award, Nobel Laureate Signature award, Carl Zeiss award, Medal and Shield of Honor. Fellow Am. Phys. Soc.; mem. NAS, Am. Acad. Arts and Scis. (medal of the Royal Netherlands Acad. Arts and Scis.), Third World Acad. Scis., Am. Chem. Soc. (Harrison Howe award 1989, Hochest prize 1990), Earle K. Plyler prize, Am. Physical Soc., 1993, Wolf prize, Wolf Foundation, 1993, Sigma Xi. Office: Calif Inst Tech Div Chemistry and Chem Engring Mail Code 127 72 Pasadena CA 91125

ZEYEN, RICHARD LEO, III, analytical chemist, educator; b. Fostoria, Ohio, Apr. 6, 1957; s. Richard Leo Jr. and Elizabeth Ann (Park) Z.; m. Diane Cory Saenz, Dec. 22, 1978; children: Gabriel, Amanda. BS, Bowling Green State U., 1985. Analytical chemist Cooper Tire & Rubber Co., Findlay, Ohio, 1985—; adj. prof. chemistry U. Findlay, 1988—. Contbr. articles to profl. jours. With USN, 1976-80. Mem. Am. Chem. Soc., Toledo Zool. Soc. Achievements include research on practical applications for chemical reconstruction of rubber compounds, correlation of specific heat values of rubber compounds predicted by methods of mixture model to specific heat values determined by direct measurement. Office: Cooper Tire & Rubber Co Lima and Western Aves Findlay OH 45840

ZGONC, JANICE ANN, computer specialist; b. Greensburg, Pa., Apr. 8, 1956; d. Joseph Paul and Jennie (Yaniszeski) Z. BA, Edinboro U. of Pa., 1979. Computer specialist Dept. Def./U.S. Army, Washington, 1981—. With U.S. Army, 1974-76. Mem. NAFE, Capital PC Users Group, Assn. Old Crows, Data Processing Mgmt. Assn., Armed Forces Communication and Electronics Assn. Home: 1304 S Thomas St Arlington VA 22204

ZHANG, BINGLIN, physics educator; b. Hebei, People's Republic China, Nov. 7, 1937; s. Dongchang Zhang and Jizhen Hu; m. Genping Zhu, July 26, 1965; children: Mei, Lan. Grad., Zhengzhou U., Zhengzhou, People's Republic China, 1961. Asst. prof. Zhengzhou U., 1961-78, lectr. physics, 1979-83; vis. scientist U. Mo., Columbia, 1983-85, U. Auckland, 1986, U. Waikato, New Zealand, 1986; assoc. prof. Zhengzhou U., 1986-90; vis. prof. Iowa State U., Ames, 1991; prof. physics Zhengzhou U., 1992—; Assoc. dir. Henan Fundamental and Applied Sci. Rsch. Inst., Henan Province, People's Republic of China, 1989—; participant 1st Internat. Conf. on Micro System Technology, Berlin, 1990. Inventor BeO Argon laser, fluorescence enhancement and frequency shift of Rh6G; contbr. articles to profl. jours. Mem. Optical Soc. Henan (v.p. 1983—), Phys. Soc. China (light scattering com. 1981—), Am. Phys. Soc., Optical Soc. Am., N.Y. Acad. Scis. Office: Zhengzhou U Physics Dept, Zhengzhou 450052, China

ZHANG, CHENG-YUE, astrophysicist; b. Zhenjiang, Jiangsu, People's Republic of China, Mar. 26, 1945; arrived in Can., 1988; s. En-Shou and Ming-Tang (Fu) Z., m. June Xue, Aug. 20, 1969, 1 child, Ke-Qing. MSc, U. Nanjing, People's Republic of China, 1981; PhD, U. Groningen, The Netherlands, 1988. Rsch. assoc. Purple Mountain Observatory, Academia Sinica, Nanjing, 1982-90; vis. scholar Kapteyn Astron. Inst. and Lab. for Space Rsch., Groningen, 1985-88; postdoctoral fellow dept. physics and astronomy U. Calgary, Alta., Can., 1988-91; postdoctoral rsch. assoc. U. Tex., Austin, 1991-93; sci. programmer, asst. scientist Space Telescope Sci. Inst., Balt., 1993—. Contbr. articles to Astrophys. Jour., Bull. of Am. Astron. Soc., Astronomy and Astrophysics, Astron. Jour., Astrophysical Jour. Mem. Can. Astron. Soc., Internat. Astron. Union, Am. Astron. Soc., Astron. Soc. of the Pacific. Achievements include discovery that the core mass and age of central stars of planetary nebulae can be determined according to their positions on a model diagram of the radio-surface-brightness-temperature versus the effective temperature without having to rely on distance assumptions, that model fitting to spectral energy distribution of 66 planetary nebulae from 0.1 to 100 mu-m leads to a detailed breakdown of the three components of the system, of the co-existence of C- and O- rich dust in some young planetary nebulae, of extended far-infrared emission of the star-forming regions in Serpens and Lambda Orionis, of HI gas towards Lambda Orionis showing a deficiency in HI equal to the mass of ionized gas in the HII region S264, of combined acceleration mechanism of charged particles by a DC electric field and turbulence in the neutral sheet of the solar active regions. Home: 103 W 39th Apt 1212 Baltimore MD 21210 Office: Space Telescope Sci Inst 3700 San Martin Dr Baltimore MD 21218

ZHANG, CHEN-ZHI, project engineer; b. Chengdu, Sichuan, China, May 20, 1957; came to U.S., 1987; s. Shi-Min and Zhong-Hua (Zhou) Z.; m. Xiaoxia Tang, Nov. 8, 1984; 1 child, Susan Y. BS, U. Elec. Sci. and Tech., Chengdu, MS; PhD, U. Tex. Device engr. Inst. Microelectronics, Chengdu, 1984-87; project engr. Aurora Assocs., Santa Clara, Calif., 1991—. Contbr. articles to Applied Optics, Optical Engring., other profl. jours. Achievements include analysis of interaction of Nd:YAG laser with optical materials, sensors and imaging array; design of high power MOS devices; design and testing various Acousto-Optic and Electro-Optic devices and relevant signal processing systems. Home: 2911 Creek Point Dr San Jose CA 95133-2924 Office: Aurora Assocs Bldg 30 3350 Scott Blvd Santa Clara CA 95054

ZHANG, FU-XUE, scientist; b. Yunnan, People's Republic China, Jan. 13, 1939; s. Zhang Wei-Qi and Duo-Ding (Liu); m. Xiu-Luan Wang, Oct. 1, 1967; children: Wei, Lei. BS in Physics, Yunnan U., 1961. Project vice-dir. 11th Lab. in the 10th Rsch. Inst. Sci. and Technol. Com. on Nat. Def., Beijing, 1961-66; project mgr. 10th Lab. in the 14th Rsch. Inst. Sci. and Technol. Com. on Nat. Def., Guangzhou, People's Republic China, 1966-74; vice-dir., sr. engr. 26th lab. 14th rsch. inst. Sci. and Technol. Com. on Nat. Def., Sichuan, People's Republic China, 1974-85; dir., prof. sensor electronics sect. Beijing Info. Tech. Inst., 1985—; cons. Beijing Info. Tech. Inst., 1985—; cons. Beijing Mcpl. Govt. and Govt. of Sichuan Province, Sichuan; hon. prof. Nanjing (People's Republic China) Aeronautical Engring. Inst.; hon. dir. Bejing Zhonghui Sensing Tech. Applications Inst., Beijing Pingu Jinghai High-Tech Application Inst. Author: Piezoelectricity (books I and II), 1988 (Excellent Sci. and Tech. Books award 1988); contbr. 227 papers to profl. jours.; author 15 books in field. Candidate exec. All-china Fedn. of Trade Unions, Beijing, 1979-84, exec., 1984-89. Named Advanced Individual in Guangdong, Govt. of Guangdong Province, 1973, Advanced Individual in Sichuan, Govt. of Sichuan Province, 1978, Nat. Advanced Sci. and Tech. Worker, China Sci. Com., 1978, Nat. Model Worker, China State Coun. 1979, State Expert Making Great Contbn., State Ministry of Labor and Personal Affairs, 1984; recipient 5 Nat. Inventive and Progressive awards, 20 ministerial progressive awards including Nat. Sci. Conf., 1978, Nat. Inventive award, 1983, 84, Silver award Nat. High Quality Product, Spl. Allowance cert. Chinese State Coun., 1991. Mem. China Assn. Inertial Tech. (councillor), Sensor Assn. (permanent councillor), China Assn. of Electronics (sr. mem., v.p. electronic sensitive tech. br.), IEEE (sr.), China Assn. of Electronics Quality Mgmt. (councillor). Achievements include development of the piezo-crystal rate gyro, piezoelectric fluidic rate sensor, and other devices applied to navigations, weapons, and robotics fields, theory that the human body consists of electric dipoles which, under static electric fields, turn to the field direction and move along the field direction; invention of gas pendulum inclination sensor, gas acceletometer, electric field therapeutical device which can cut short the healing time of born wounded by 2-3 times and has remarkable curative effects on long-time bone fractures, unhealing born, soft-tissue injury, disease in cervical vertebra, inflamation in shoulder periphery and arthritis; 10 pantents in U.S., UK, and China. Office: Beijing Info Tech Inst, Dewai, Beijing 100101, China

ZHANG, GUO HE, physiologist; b. Yuanping, Shanxi, China, Aug. 27, 1953; came to U.S., 1988; s. Runtang Zhang and Runyu Jia; m. Ping Dang, Jan. 10, 1981; 1 child, Yanyan. MSc, Shanxi Med. Coll., 1981; PhD, Beijing (China) Med. U., 1986. Rsch. assoc. Shanxi Med. Coll., Taiyuan, 1977-78, Beijing Inst. Nutritional Sources, 1982-83; lectr. Beijing Med. U., 1987; postdoctoral rsch. assoc. U. Tex., Houston, 1988-89; postdoctoral rsch. assoc. U. Rochester, N.Y., 1990-91, scientist, 1992—. Contbr. articles to profl. jours. Office: U Rochester Box 611 601 Elmwood Ave Rochester NY 14642

ZHANG, HONG TU, physics educator; b. Jinxi, Liaoning, China, June 6, 1932; m. Lu de Han, Mar. 8, 1963; 2 children. Diploma, Jilin U., China, 1954; MA in Physics, Beijing U., China, 1957. Lectr. assoc. prof., dept. physics Lanzhou U., China, 1957-83, prof., chmn. dept. materials sci., 1984-88; dir., divsn. materials rsch. Qingdao U., China; scientific guest Max Planck Inst. Metals Rsch., Germany, 1985-86. Contbr. to profl. jours. Recipient Sci. and Tech. prize Gansu Province, 1985, Sci. and Tech. prize Nat. Edn. Com., 1986. Mem. Chinese Soc. Physics, Metals, and Mech. Engring., Internat. Ctr. Materials Physics (mem. academic com.), Academia Sinica. Achievements include work on the inclusion theory and applications in study of fractures and applications to composite and ceramic materials. Office: Academia Sinica, 52 San Li He Rd, Beijing 100864, China*

ZHANG, JIANHONG, research engineer; b. Wanzai, China, May 17, 1958; d. Taolin and Kangxuan (Wang) Z.; m. Jianhua Zhou, May 17, 1985; 1 child, Sophia Zhou. MS, Nanjing Aero. Inst., 1985; PhD, U. Mich., 1990. Project engr. Tianjing (China) Inst. Chemistry, 1982; rsch. asst. Nanjing (China) Aero. Inst., 1982-85; teaching asst. U. Mich., Ann Arbor, 1985-87, rsch. asst., 1989-90; tech. cons. Ford Motor Co., Dearborn, Mich., 1987-88, rsch. engr. sr., 1990-. Contbr. articles to profl. jours. Mem. AIAA, NAFE, Soc. Automotive Engring., Sigma Xi, Tau Beta Pi.

ZHANG, JIPING, engineering analyst; b. Changsha, Hunan, China, Apr. 29, 1954; came to U.S., 1983; s. Zhenhui Zhang and Qiaoling Yang; m. Ping-Ping Zhou, May 29, 1981; children: David, Thomas. BS in Mech. Engring., Tsinghua U., Beijing, China, 1980; MS in Materials Engring., U. Wis., Milw., 1986, PhD in Engring., 1991. Designer Hunan Machine Tool Works, Changsha, China, 1976-77; asst. instr. Xiangtan (China) U., 1980-83; vis. scholar U. Wis., Milw., 1983-85, rsch. and teaching asst., 1986-91; metallurgist Charter Wire Mfg., Milw., 1986; engring. analyst Siemens Power Corp., West Allis, Wis., 1991—. Contbr. articles to profl. jours. Grad. sch. fellow U. Wis., Milw., 1987, dissertation fellow, 1988. Mem. ASME, Am. Soc. Metals, Am. Acad. Mechanics. Achievements include research on crack tip displacement factors in both LEFM and EPFM, fatigue analysis, finite element method, boundary element method, sliding wear, turbine/generator components, slip line field analysis, anisotropic plastic yielding, warm forming of metals. Home: W133S6824 Bristlecone Ct Muskego WI 53150 Office: Siemens Power Corp 1040 S 70th St West Allis WI 53214

ZHANG, JUN YI, chemist, educator; b. Fenyang, Shanxi, Peoples Republic of China, Oct. 29, 1929; s. Zhang Yuankai and Wang Ailan; m. Xin Min Ren; children: Ling Ren, Ying Ren. BS, Fu Ren U., Peking, 1950. Instr. Shanxi U., Taiyuan, Peoples Republic of China, 1950-52; instr. dept. pharmacology Beijing Med. Coll., 1952-55; asst. prof. Inst. Chemistry Academia Sinica, Beijing, 1955-78; assoc. prof. Inst. Photographic Chemistry Academia Sinica, Beijing, 1978-88, prof., 1989—. Translator: Modern Molecular Photochemistry, 1987; contbr. articles to Jour. Chemistry, Chem. Bull., Applied Chemistry, Chinese Physics Letters. Mem. Beijing Chem. Soc. (bd. dirs. 1991—), Am. Chem. Soc., Chinese Soc. for Photographic Sci. and Engring. Achievements include research in new additives in photographic films, new color-couplers. Home: 919-408, Haidian Zhongguan Cun, Beijing 100080, China Office: Inst Photographic Chemistry, Dewai Bei Sha Tan, Beijing 100101, China

ZHANG, LI-XING, physician, medical facility executive; b. Tianjing, China, May 6, 1934; s. Kui-Dong and Jing-Yu (Liu) Z.; m. Xiu-Ping Wang, Apr. 30, 1972; 1 child, Man. MD, Beijing Med. Coll., 1956. Diplomate in medicine. Tng. Tb Surveillance Rsch. Unit, The Hague, The Netherlands, 1985-86; tng. Tb bacteriology course, Ottawa, Ont., Can., 1987; chief epidemiology Beijing Tb Ctr., 1956-62, chief Tb control, 1963-81, vice dir., 1981-84, dir., 1987—; advisor on Tb, Chinese Ministry Health, Beijing, 1987—; cons. Mcpl. Bur. Pub. Health, Beijing, 1989—. Contbr. articles to med. jours. Named Prominent Scientist, Beijing City Govt., 1990. Mem. Internat. Union Against Tb and Lung Diseases (sci. com. on Tb control 1988—, award ea. region 1990), Chinese Anti-Tb Assn. (v.p. 1986—), Japan Anti-Tb Assn. (hon.). Office: Beijing Rsch Inst for, Tuberculosis Control Xin Jie Kou, Beijing 100035, China

ZHANG, NAIQIAN, agricultural engineer, educator; b. Jiangling, Hubei, Peoples Republic China, Oct. 19, 1946; came to U.S., 1981; s. Shuzu and Fulin (Yang) Z.; m. Yabao Zhang; 1 child, Jian. BS, Beijing U. Agrl. Engring., 1970; MS in Agrl. Engring., Purdue U., 1984; PhD, Va. Poly. Inst. and State U., 1987. Asst. engr. Agr. Machinery Repair Factory, Huailai, Hebei, China, 1971-74, Internal Combustion Engine Factory, Baoding, Hebei, 1974-78, Agr. Machinery Rsch. Inst., Tianjin, China, 1978-81; postdoctoral researcher U. Calif., Davis, 1987; vis. asst. prof. Va. Poly. Inst. and State U., Blacksburg, 1987-90; asst. prof. dept. agrl. engring. Kans. State U., Manhattan, 1990—; guest prof. Bonn (Fed. Republic Germany) U., 1988-89. Contbr. articles to profl. publs. Grantee USDA, 1992, Kans. Dept. Transp., 1992. Mem. Am. Soc. Agrl. Engrs., Soc. Automotive Engrs., Am. Soc. Engring. Edn., Sigma Xi. Achievements include development of control systems on tractors, algorithms used in geographic information systems, research on site-specific chemical application, optical sensors and signal processing. Home: 716 Humboldt St Manhattan KS 66502 Office: Kans State U Dept Agrl Engring Manhattan KS 66506

ZHANG, PING, aerospace engineering educator; b. Wuxi, Jiangsu, China, Sept. 24, 1937; s. Di-Su Zhang and Jing-Xin Zhu; m. Ke-Xian Hu, Jan. 26, 1962; children: Zhong, Bei. Undergrad. student mech. engring., Beijing Inst. Tech., 1954-59, grad. student mech. engring., 1960-62. Instr. dept. flight vehicle engring. Beijing Inst. Tech., 1963-77, asst. prof. dept. flight vehicle engring., 1978-79; vis. scholar Purdue U. Sch. Mech./Aero. and Astron. Engring., West Lafayette, Ind., 1979-81; assoc. head dept. flight vehicle engring. Beijing Inst. Tech., 1982-86, assoc. prof. dept. flight vehicle engring., 1986-90, prof., dir. propulsion divsn. dept. flight vehicle engring., 1990—; acad. committeeman Beijing Inst. Tech., 1988—. Editorial committeeman Jour. Propulsion Tech., Ministry of Aerospace, Beijing, 1990—; author: Combustion Diagnostics, 1988, Solid Rocket Propulsion, 1992; contbr. articles to profl. jours. including Jour. Propulsion Tech., Jour. Aerospace Power. Recipient 3d award in progress of sci. and tech. Ministry of Machinery and Electronic Industries, People's Republic China, 1986, 91. Fellow Soc. Beijing Engring. Thermodynamics; mem. China Aero. and Astronautical Soc. (Outstanding Article 1988), Am. Inst. Aeronautics. Achievements include patents in field. Home: Unit 45 No 10 PO Box 327, Beijing 100081, China Office: Beijing Inst Tech, Flight Veh Engr PO Box 327, Beijing 100081, China

ZHANG, TAO, physics educator, director; b. Laiyang, Shandong, People's Rep. of China, Sept. 9, 1942; s. Xueji and Fengying (Jiang) Z.; m. Xu Jian, Apr. 2, 1971; children: Rui, Xu. BS in Engring., Tianjin (Peoples Rep. China) U, 1968; MSc, Henan Norman U., Xinxiang, People's Rep. of China, 1981. Engr. Yongxing Radio Equipment Factory, Sichuan, People's Rep. of China, 1968-78; instr. Henan Normal U., 1981-85, prof., 1986—, dir. condensed matter physics lab., 1986—; vis. asst. prof. U. Waterloo, Ont., Can., 1985-86; concurrent prof., dir. condensed matter physics lab. Zhengzhou (People's Rep. of China) U., Henan, 1989—; vis. prof. Internat. Ctr. for Theoretical Physics, Trieste, Italy, 1988, Il Universita Degli Studi di Roma, 1988, U. Waterloo, 1991-92; participant 5th Internat. Conf. on Solid Films and Surfaces, Brown U./U. R.I., 1990; v.p. Henan Sci. and Tech. Commn., 1991—. Contbr. numerous articles to profl. jours. Mem. standing com. Chinese People's Polit. Consultative Conf. of Henan Province, Zhengzhou, 1988. Recipient Nat. Outstanding Achievement award, 1991. Mem. Am. Phys. soc. (life, participant Mar. meeting Las Vegas, Nev. 1986, Asia-Pacific symposium on surface physics Shanghai 1987), Henan Assn. Translators (pres. 1991—), Henan Assn. for Sci. and Tech. (chmn. 1992—), Assn. Sci. and Technol. Info. Henan (hon. chmn. 1991—), China Assn. for Sci. and Tech. (com. 1993—), Nat. Com. Chinese People's Polit. Consultative Conf. Avocation: classical music. Home and Office: Henan Assn for Sci and Tech, Bldg 4 Zheng Yi St, 450003 Zhengzhou 450003, China

ZHANG, THEODORE TIAN-ZE, oncologist, health association administrator; b. Shenyang, Liaoning, People's Republic of China, Apr. 2, 1920. MBChB, Christie Meml. Med. Coll., Shen-yang, People's Republic of China, 1943. Intern Christie Meml. Med. Coll. Hosp., 1944, resident, 1945-48; chief resident surg. dept. Cen. Hosp., Lanzhou, People's Republic of China, 1949-50; vis. surgeon Mackenzie Meml. Hosp., Tianjin, People's Republic of China, 1951-52; dep. chief surg. oncology People's Hosp., Tianjin, People's Republic of China, 1953-75; chief surg. oncology Cancer Hosp., Tianjin, 1975-85; dir. Tianjin Cancer Inst. and Hosp., 1983-92; sec. gen. China Anti-Cancer Assn., 1985-87; pres. China Anti-Cancer Assn., Tianjin, 1988—, Asian Pacific Fedn. Orgns. for Cancer Rsch. and Control, 1989-91. Contbg. author: Oncology, Chinese Medical Encyclopedia, 1983, Gastric Cancer, 1987, Recent Advances in Cancer Chemotherapy, 1987, Researches on Breast Cancer, vol. 1, 1987, vol. 2, 1989; contbr. more than 120 articles to profl. jours. Mem. Bridge Assn. (v.p. Tianjin chpt.). Avocation: bridge. Office: Tianjin Cancer Inst and Hosp, Huan-Hu-Xi Rd, Ti-Yuan-Bei, Tianjin 300060, China

ZHANG, TIANYOU, biomedical engineer; b. Yichang, Hubei, China, Aug. 7, 1938; s. Zhenhan and Jiaoxian (Deng) Z.; m. Guirong Zhang, Apr. 12, 1980; 1 child, Wenjin. MS, Beijing Inst. Tech., 1959, PhD, 1979. Engr. Acad. Sinica, Beijing, 1959-77; chief engr., rsch. prof. Inst. New Tech. Application, Beijing, 1977—; dir. lab. biochem. separation and detection Bejing Acad. Sci., 1988—; guest researcher Nat. Heart, Lung, Blood Inst., NIH, Bethesda, Md., 1987-88; councillor Analytical Instrument Soc. China, Beijing, 1981—. Author: Countercurrent Chromatography, 1991, High-Speed C.C.C., 1993; contbr. articles on bioengring. to numerous publs. Recipient prizes Municipality of Beijing, 1980-87, 92, Chinese Mil., 1990. Achievements include patents on non-synchronous coil planet centrifuge, multi-layer high-speed C.C.C., U.S. patent on cross-=axis CPC for large scale preparative countercurrent chromatography. Office: Beijing Inst New Tech, Applications, 16 Xizhmen Nandajei, Beijing 100035, China

ZHANG, YONG-HANG, research electrical engineer; b. Nanjing, Jiangsu, China, May 8, 1959; came to U.S., 1991; s. Ru-Ying and Aili (Wang) Z.; m. Zhiruo Shuai, May 2, 1985; 1 child, Elisa Tiannuo. MSc, Inst. Semicondrs., Beijing, 1987; D rer. nat., Stuttgart (Germany) U., 1991. Teaching asst. Nanjing Normal U., 1982-84; rscher. Max-Planck-Inst. for Solid States, Stuttgart, 1988-91; asst. rsch. engr. U. Calif., Santa Barbara, 1991—; referee Phys. Rev. B. Contbr. articles to Phys. Rev. B., Applied Physics, Applied Physics Letters. Mem. IEEE, Am. Phys. Soc. Office: Phelps Hall 1413 U Calif Santa Barbara CA 93106

ZHAO, JIAN HUI, electrical and computer engineering educator; b. Anxi, Fujian, China, Aug. 2, 1959; came to U.S., 1983; s. Yumao Zhao and Su Qing Chen. BS in Physics, Amoy U., China, 1982; MS in Physics, U. Toledo, 1985; PhD in E.E., Carnegie Mellon U., 1988. Rsch. asst. Nonlinear Optics Lab. U. Toledo, Ohio, 1983-85; rsch. asst. Solid State Elec. Lab. Carnegie Mellon U., Pitts., 1985-88; asst. prof. electric and computer engring. dept. Rutgers U., New Brunswick, N.J., 1988-93, assoc. prof. electric and computer engring. dept., 1993—; cons. Army Rsch. Lab., Ft. Monmouth, 1991—, NZ Applied Techs., Boston. Contbr. articles to profl. jours. Henry Rutgers fellow Rutgers U., New Brunswick, N.J., 1989, 90; recipient Initiation award NSF, 1990. Mem. IEEE, Am. Phys. Soc., Materials Rsch. Soc., N.Y. Acad. Sci. Achievements include patents for InP/InGaAsP optoelectronic high speed thyristor, an AlGaAs/GaAs-based optothyristor for ultra-high power switching, field effect real space transfer transistor, electrical tunable superlattice detector for wavelength division demultiplexing applications. Office: Rutgers U Elec and Computer Engring Dept Brett & Bowser Rds Box 909 Piscataway NJ 08855

ZHAO, LUE PING, biostatistician, genetic epidemiologist; b. Shanghai, People's Republic of China, Jan. 22, 1961; s. Yu-Ming and Mu-zhong (Pan) Z.; m. Yi Su Huang, Aug. 30, 1990; 1 child, Derek. MS, U. Wash., 1987, PhD, 1989. Staff scientist Fred Hutchinson Cancer Rsch. Ctr., Seattle, 1989-90; fellow Harvard Sch. Pub. Health, Boston, 1990-91; asst. researcher Cancer Rsch. Ctr. Hawaii, U. Honolulu, 1991-93, Fred Hutchinson Cancer Rsch. Ctr., Seattle, 1993—. Contbr. articles to profl. jours. Recipient First award Nat. Cancer Inst., 1992. Home: 2640 Dole St # E206 Honolulu HI 96822 Office: Fred Hutchinson Cancer Rsch Ctr 1124 Columbia St Seattle WA 98104

ZHAO, MEISHAN, chemical physics educator, researcher; b. Shanxian, Shandong, People's Republic of China, Nov. 5, 1958; came to U.S., 1984; s. Zhong Chen Zhao and Ming Rong Zhang; m. Linlin Cai, Sept. 2, 1983; children: Fang, Yuan. MS in Physics, U. Minn., 1986, PhD in Chem. Physics, 1989. Lectr. physics S.E. U. China, Nanjing, 1982-84; teaching asst., rsch. asst. U. Minn., Mpls., 1984-89; rschr. James Franck Inst. U. Chgo., 1990—. Contbr. articles to profl. jours. Mem. Am. Phys. Soc. (internat. editorial bd. Internat. Physics Edn., Chinese ed., 1991-92). Home: 5642 S Drexel Ave Chicago IL 60637 Office: Univ Chgo James Franck Inst 5640 Ellis Ave Chicago IL 60637

ZHAO, RU HE, mechanical engineer; b. Canton, Guangdong, China, Oct. 1, 1956; arrived in Can., 1989; s. Xian Fu and Cai Juan (Li) Z.; m. Yu Ying Yang, Oct. 25, 1985; 1 child, Ting. MSc, South China U. Tech., Canton, 1985; PhD, South China U. Tech., 1987, Oxford (Eng.) U., 1987. Teaching asst. South China U. Tech., 1982-85, lectr., 1987-88; vis. scientist UMIST, Manchester, Eng., 1988-89; postdoctoral fellow PAPRICAN, Vancouver, B.C., Can., 1989-91; rsch. engr. U. B.C., Vancouver, 1992—. Contbr. articles to profl. publs. Mem. Can. Pulp and Paper Assn., Rheology Soc. Achievements include patent pending in field. Office: U B C Pulp & Paper Ctr, 2385 East Mall, Vancouver, BC Canada V6T 1Z4

ZHENG, BAOHUA, electronics engineer, consultant; b. Yang Xian, Shanxi, China, Nov. 26, 1959; came to U.S., 1982; s. Shu Zhen Zheng and Shu E Zhang; m. Tung Li, May 17, 1985; children: Albert, Andrew. BS, Shanxi Mech. Engring. Inst., Xian, 1982; MSc, Washington U., St. Louis, 1983, DSc, 1989. Rsch. engr. Washington U. Ctrl. Inst., 1987-89; systems engr. Perma-Graphics, Inc., St. Louis, 1989-92; sr. devel. engr. TALX Corp., St. Louis, 1992—; cons. TALX Corp., 1990-92, Perma-Graphics, Inc., 1992—, DSP Applications, St. Louis, 1990—. Mem. IEEE (co-author transcript

1988), St. Louis Engrs. Club. Achievements include patent (with other) for digital hearing aid techniques; patent pending for multi-frequency signal receivers; design of practical inspection machine for the credit card industry. Home: 809 Westwood Dr # 1S Saint Louis MO 63105 Office: TALX Corp 1850 Borman Ct Saint Louis MO 63146

ZHENG, MAGGIE (XIAOCI), materials scientist; b. Shanghai, China, Apr. 21, 1949; came to U.S., 1986; d. George and Helen (Chou) Cheng; divorced; 1 child, Dee. BS in Physics, Qutu Normal U., Shangdong, China, 1981; MSEE, U. Sci. and Tech. China, Beijing, 1984; MS in Materials Sci., U. Wis., 1988, PhD in Materials Sci., 1991. Lectr. Tsinghua U., Beijing, 1984-86; assoc. scientist United Techs., East Hartford, Conn., 1991-92; staff scientist Engineered Coatings, Inc., Rocky Hill, Conn., 1992-93; materials and coating process engr. Chromalloy Turbine Techs., Middletown, N.Y., 1993—; rsch. asst. U. Wis., Madison, 1986-91. Contbr. articles in profl. publs.; patentee in field. Mem. NAFE, Am. Metal Soc., Minerals, Metals & Materials Soc. Office: Chromalloy Turbine Techs 105 Tower Dr Middletown NY 10940

ZHENG, YOULU, computer scientist, educator; b. Hangzhou, Zhejiang, People's Republic of China, Nov. 3, 1942; came to U.S., 1984; s. Fat-Lai and Yi (Tong) Cheng;m. Li Sun, June 20, 1970; 1 child, Nan. BS, Sichuan (People's Republic of China) U., 1967; MS, Zhejiang U., 1981; PhD, Wash. State U., 1987. Engr. Chongqing (People's Republic of China) Automobile Co., 1967-75, Yunnan (People's Republic of China) Electronic Equipment Co., 1975-79, ISC Systems Co., Spokane, Wash., 1986; lectr. Chengdu (People's Republic of China) U. Sci. and Tech., 1981-84; computer scientist, sr. engr. EXP Group, Inc., Fremont, Calif., 1990—; prof. U. Montana, Missoula, 1987—; cons. Meswell Tech., San Leandro, Calif., 1990—; advisor SunLabs, Missoula, Mont., 1991—; dir. computer graphics and visualization lab. U. Mont., 1992—; lectr. info. tech. summer sch. Chinese Acad. Scis., Beijing, 1993; adj. prof. Shantou UU., People's Republic of China, 1993—. Contbr. articles on computer sci. to acad. jours., 1981-93. NSF grantee and prin. investigator. Mem. Math. Assn. Am., Assn. Computer Machinery. Home: 110 Ben Hogan Dr Missoula MT 59803 Office: U Mont Missoula MT 59812

ZHENG, ZHONG-ZHI See CHENG, J. S.

ZHONG, JIANHUI, physicist; b. Guiyang, Guizhou, China, Mar. 9, 1956; s. Luonghua and Wenqing (Wang) Z.; m. Tong Wang, Mar. 22, 1986; children: R. Ming, E. Wen. BS, Nanjing U., 1982; PhD, Brown U., 1988. Teaching asst. Brown U., Providence, 1982-83, rsch. asst., 1983-87; postdoctoral fellow Yale U., New Haven, Conn., 1988-89, rsch. scientist, 1990, asst. prof., 1990—. Author: (with others) Diffusion and Perfusion: Magnetic Resonance Imaging, 1992; peer reviewer magnetic Resonance Imaging, 1988—, Jour. Magnetic Resonance Imaging; contbr. over 20 articles to profl. jours. Grantee Whitaker Found., 1991-94, Yale Med. Sch., 1991-92. mem. Am. Assn. Physicists in Medicine, Soc. Magnetic Resonance Imaging (basic sci. com. on relaxometry and biophysics 1992—), Soc. Magnetic Resonance in Medicine. Achievements include research in water relaxation and diffusion in protein solutions and tissues with NMR spectroscopy and imaging experiments, and elucidated in particular the important roles of water-macromolecular interaction for tissue relaxation and diffusion processes. Office: Yale U Sch Medicine Dept Diagnostic Radiology 333 Cedar St Fitkin B New Haven CT 06510

ZHONG, WILLIAM JIANG SHENG, ceramic engineer; b. Fujian, China, Sept. 3, 1934; Came to U.S. 1980; s. Lu Zai and Bertha (Fang) Djung; m. Baoru Liu Zhong, Sept. 14, 1964; children: Charles H., Joan H. BChemE, S. China Poly. Coll., Guang Zhou, 1953; MS in Mineralogy, Geology, Miami U., Oxford, Ohio, 1982. Asst. rsch. scientist Rsch. Inst. Optics and Fine Mechanics, Changchun, China, 1953-62; lectr. Coll. of Optics and Fine Mechanics, Changchun, China, 1959-61; assoc. rsch. scientist Rsch. Inst. Optics and Fine Mechanics, Changchun, 1962-70, Shanghai Light Ind. Bur., China, 1970-72; dep. chief engr. Xin Hu Glass Wks., Shanghai, 1972-80; tech. dir. Kigre, Inc., Toledo, Ohio, 1982-86; prin. glass scientist Circon ACMI, Stamford, Conn., 1986—. Co-author: Optical Glass (in Chinese), 1964; contbr. articles to profl. jours. Recipient Sci. & Tech. Achievement in high transmittance lead glasses, Shanghai Sci. and Tech. Com., 1978, in 2.3 meter telescope mirror disk, 1979. Mem. Am. Ceramic Soc. Achievements include joint establishment of optical glass production technologies in China; zero expansion glass ceramic large astronomic telescope mirror blank manufacture technology in China; patents in laser glass, microchannel plate glasses. Office: Circon ACMI 300 Stillwater Ave Stamford CT 06902-3695

ZHOU, CHIPING, mathematician, educator; b. Shanghai, People's Republic of China, Jan. 21, 1957; s. Xingui Zhou and Qi Zhu; m. Xiaoyu He, June 22, 1986; children: Kevin K., Brandon K. BS, Fudan U., Shanghai, 1983, MS, 1986; PhD, U. Hawaii, 1990. Asst. prof. Fudan U., Shanghai, 1986—; lectr. Chaminade U., Honolulu, 1990; instr. U. Hawaii, Honolulu, 1990–. Author: Some Problems for Elliptic and Hyperbolic Equations, 1986, Maximum Principles and Liouville Theorems for Elliptic Partial Differential Equations, 1991; contbr. articles to profl. jours. Recipient rsch. fellowship Rsch. Corp. of U. Hawaii, 1989. Mem. Am. Math. Soc., Math. Assn. Am. Achievements include research in partial differential equations and their applications, Clifford algebras in analysis; discovery of generalized maximum principles for elliptic and parabolic systems. Office: Univ of Hawaii - HCC Math Dept 874 Dillingham Honolulu HI 96817

ZHOU, JING-RONG, electrical engineer, researcher; b. Qian Wei, Sichuan, People's Republic of China, Jan. 30, 1955; came to U.S. 1986; s. Kai-Xue and Ming-Xiu (Ren) Z.; m. Yu-Shu Wu, Oct. 1, 1982; 1 child, Michael Shujie. BS, S.W. Jiaotong U.; MS, Chengdu Inst. Radio Engring., China, 1985; PhD, Ariz. State U., 1991. Asst. lectr. Chendu Inst. Radio Engring., 1985-86; vis. faculty Ariz. State U., Tempe, 1986-87, faculty rsch. assoc., 1991—. Contbr. articles to profl. jours. Mem. IEEE, Sigma Xi. Achievements include research on simulated quantum effects in ultrasmall MESFET and HEMT devices using quantum hydrodynamic equations the first time. Office: Ariz State Univ Dept Elec Engring Tempe AZ 85287

ZHOU, MING DE, aeronautical scientist, educator; b. Zhejiang, Peoples Republic of China, June 26, 1937; s. Pin Xiang and Ang Din (Xia) Z.; m. Zhuang Yuhua, Aug. 12, 1936; children: Zhengyu, Yan Zhuang. BS, Beijing Aero. Inst., 1962; MS, Northwestern U. Tech., 1967; PhD, Internat. Edn. Rsch. Found., 1992. Tchr. Harbin (China) U. Tech., 1962-64, 67-73; from lectr. to prof. Nanjing (China) Aero. Inst., 1973-86, 86—, dean bd. postgrad. studies, 1985-89; nationally qualified PhD advisor China, 1989—; rsch. scientist U. Ariz., Tucson, 1991-93, rsch. prof., 1993—; vis. scholar Cambridge (England) U., 1980-82; guest scientist Inst. Exptl. Fluid Mechanics, Göttingen, Germany, 1983-84, 85, 87; sr. vis. scientist Tech. U. Berlin, 1988, 90; rsch. assoc. U. So. Calif., L.A., 1989-90. Mem. editorial com. Chinese Jour. Exptl. Mechanics, 1986-89; author: (with others) Viscous Flows and Their Measurements, 1988, (with others) Introduction to Vorticity and Vortex Dynamics, 1992; contbr. articles to Aero. Jour. U.K., Experiments in Fluids, AIAA Jour., Chinese Jour. Aeronautics. Co-recipient Nat. award Progress in Sci. and Tech. first class, Peoples Republic of China, 1985. Mem. AIAA (sr.), Am. Phys. Soc., Chinese Soc. Aeronautics, Chinese Soc. Mechanics (mem. acad. group exptl. fluid mechanics 1986-89), Chinese Soc. Aerodynamic Rsch. (acad. group unsteady flow and vortex control 1985-89). Achievements include patent for techniques and device of artificial boundary layer transition.

ZHOU, SHAO-MIN, chemist, educator; b. Jin-Jiang, Fujian, Peoples Republic of China, Nov. 18, 1921; m. Fu Su-wen, Jan. 1, 1948; children: Luwen, Hai-wen, Yue. BS, Amoy U., 1945; MS, Mendeleev Coll., USSR, 1957. Asst. dept. chemistry Amoy U., Amoy, Peoples Republic of China, 1946-50, lectr., 1950-53, assoc. prof., 1957-78; prof. Xiamen (Amoy) U., Peoples Republic of China, 1978—; researcher Moscow D.I. Mendeleev Coll. Chem. Tech., 1954-57. Author: Electrodeposition of Metal-Principle and Experimental Methods, 1987; editor: Advances in Electrochemical Methods, 1988; contbr. over 150 articles to profl. jours. Recipient 2d Rank Prize of Progress in Sci. and Tech. Nat. Edn. Com., 1988. Achievements include patents in iridescent chromium plating, plating additives concentration analyzer. Office: Xiamen U, Dept Chemistry, 361005 Xiamen China

ZHOU, SIMON ZHENGZHUO, laser scientist; b. Shanghai, China, Jan. 31, 1942; came to U.S., 1987; s. Ming-Qing and Yue-Chun (Hu) Z.; m. Peggy B. Chen, May 16, 1985; children: Shiyun, Stanley. BS, U. Sci. and Tech., Shanghai, 1965, MS, 1967. Engr. Shanghai Metall. Factory, 1968-73; scientist Shanghai Inst. Laser Tech., 1974-87; sr. scientist Florod Corp., Gardena, Calif., 1988—. Contbr. articles to profl. jours. Recipient Invention prize Sci. Com. China, 1991, Sci. Progress prize Sci. Com. Shanghai, China, 1989, Honored Thesis award, 1985, Small Bus. Innovation Rsch. Phase I fund NSF, 1992. Mem. Internat. Soc. Optical Engring., Optical Soc. China, Laser Soc. Shanghai. Achievements include patent for first longitudinal discharged XeCl excimer laser with new preionization device and longest one-gas-filling life time of more than 1 year; discovery of titanium laser and 33 laser lines; first deposition of several new metal films by LCVD. Office: Florod Corp 17360 Gramercy Pl Gardena CA 90247-5263

ZHOU, THEODORE XI, physicist; b. Shanghai, China, Oct. 4, 1949; came to U.S., 1980; m. Denise Ding Liu, Sept. 30, 1988. AB, Princeton U., 1983; PhD, Brown U., 1989. Postdoctoral fellow Inst. Energy Conversion, U. Del., Newark, 1988-91, Microelectronics Rsch. Ctr., Iowa State U., Ames, 1991; staff scientist Solar Cells, Inc., Toledo, 1991—. Office: Solar Cells Inc 2650 N Reynolds Rd Toledo OH 43615

ZHOU, ZHEN-HONG, electrical engineer; b. Ji-An, Peoples Republic of China, Oct. 16, 1968; came to the U.S., 1983; s. Li Jiang and Chang Li (Hu) Z. BEE, Poly. U., 1989; MEE, MIT, 1991, PhD, 1993. Rsch. asst. Polytech. U., Bklyn., 1986-87; researcher IBM Rsch., Yorktown Heights, N.Y., 1987-89; rsch. fellow MIT Microsystems Labs., Cambridge, 1989-91; mem. tech. staff AT&T Bell Labs., Murray Hill, N.J., ;, 1991-92; rsch. fellow MIT Tech. Rsch. Lab., Cambridge, 1992—; pres. ACT Rsch. Corp., Cambridge, 1991-93. Contbr. articles to Jour. Applied Physics, Applied Physics Letters, Jour. Vacuum Sci. and Tech., Jour. Electrochem. Soc., Jour. Electronic Materials. Recipient Achievement award NASA, 1986. Mem. IEEE, Sigma Xi, Eta Kappa Nu, Tau Beta Pi. Home: 3102 Sands Pl Bronx NY 10461

ZHOU, ZONG-YUAN, physics educator; b. Ningbo, China, Mar. 22, 1942; came to U.S., 1981; s. Chuen-Yuen Chow and Shu-Zhen Chen; m. Bi-He Wu, Sept. 30, 1967; children: Jing-Song, Ning. Diploma, Nanjing U., 1964; PhD equivlant cert. in Physics, Utah State U., 1992. Asst. prof. Nanjing (China) U., 1964-80, lectr., 1981-84, assoc. prof. physics, 1985-91; vis. scholar Lawrence Berkeley (Calif.) Lab., 1981-84; adj. prof. Inst. Modern Physics of Academia Sinica, Lanzhou, China, 1986-91; vis. prof. Los Alamos (N.Mex.) Nat. Lab., 1987-91; rsch. scientist Utah State U., Logan, 1991—; vis. assoc. prof. physics Idaho State U., 1993—; mem. internat. adv. com. 5th Internat. Conf. on Nuclei Far From Stability, Rosseau Lake, Ont., Can., 1987. Co-author: Nuclear Shapes and Nuclear Structure at Low Excitation Energies, 1992; contbr. articles to profl. jours. Recipient Silver award China Invention Soc., 1986. Mem. Chinese Physics Soc., Am. Physics Soc. Achievements include Chinese patent for probe of the moisture-density combined gauge using neutron and gamma ray sources. Home: 2012-D 23d St Los Alamos NM 87544 Office: Los Alamos Nat Lab Los Alamos NM 87545

ZHU, JIANCHAO, computer and control scientist, engineer, educator; b. Handan, Hebei, People's Republic of China, July 12, 1952; came to U.S., 1982; s. Jiezi and Jiyun (Lu) Z.; m. Yan Ma, Sept. 16, 1986. BSEE, Beijing Polytech. U., 1976; MSEE, U. Ala., 1984, MA in Math., 1986, PhD, 1989. Tchr., mem. rsch. staff Beijing Polytech. U., 1977-82; rsch. assoc. U. Ala., Huntsville, 1989-90; asst. prof. elec. computer engring. La. State U., Baton Rouge, 1990—; rsch. faculty mem. Remote Sensing and Image Processing Lab., La. State U., Baton Rouge, 1990—; prin., co-prin. investigator on various fed. and state govt. and indsl. rsch. contracts and grants. Contbr. unified eigenvalue theory for time-varying linear dynamical systems and articles to profl. jours. Recipient Rsch. Initiation award NSF, 1991. Mem. IEEE (mem. steering com. southeastern symposium on systems therapy 1990—, exec. com. mem. Baton Rouge sect. 1991-92, vice-chmn. Baton Rouge sect. 1992-93), Soc. Indsl. and Applied Math. Office: La State U Elec Computer Engring Dept Elec Computer Engring Baton Rouge LA 70803

ZHU, JIANPING, mathematics educator; b. Beijing, People's Republic of China, May 16, 1958; came to U.S., 1986; s. Shunqian and Yuhua (Li) Z.; m. Yan Wang, May 22, 1986; 1 child, Lily Ann. BS, Zhejiang U., Hangzhou, China, 1982; MS, Dalian Inst. Tech., China, 1984; PhD, SUNY, Stony Brook, 1990. Lectr. math. Shanghai Jiaotong U., Shanghai, China, 1984-86; research asst. math. SUNY, Stony Brook, 1986-90; asst. prof. math. Miss. State U., 1990-93, assoc. prof., 1993—; cons. Shanghai Mcpl. Constrn. Corp., 1984-86. Contbr. articles to profl. jours. SUNY-Stony Brook fellow, 1986-90, rsch. fellow Intel U. Ptnrs., 1992; recipient 2d prize IBM Supercomputing Competition, 1990. Mem. Am. Math. Soc., Soc. of Indsl. and Applied Math. Office: Miss State U Dept Math Mississippi State MS 39762

ZHU, QING, software engineer; b. NingBo, People's Republic of China, Oct. 30, 1962; came to U.S., 1988; s. Guo Fu and Zhi Lian (Xia) Z.; m. Yiming Yu, Aug. 19, 1988. BS, Shanghai Jiao Tong U., People's Republic of China, 1983, MS, 1986; PhD, Kans. State U., 1992. Rsch. asst. Kans. State U., Manhattan, 1988-92; software engr. Polychrome Systems Inc., Lenexa, Kans., 1992—. Contbr. chpts. to books and articles to profl. jours. Mem. Ops. Rsch. Soc. Am., Sigma Xi, Alpha Pi Mu. Achievements include rsch. on uncertainty reasoning methodology for both two-valued logic systems and multivalued logic systems based on Dempster and Shafer's evidence theory. Office: Polychrome Systems Inc 10930 Lackman Rd Lenexa KS 66219

ZHU, XINMING, optical engineer; b. Jiaxing, Zhejiang, China, July 20, 1947; came to U.S. 1987; BS in Physics, U. Sci. & Tech. of China, Beijing, 1968; MS in Optics, Shanghai Inst. Optics and Fine Mechanics, 1981. Tchr. U. Sci. & Tech. of China, Hefei, 1973-78; rsch. assoc. Shanghai Inst. Optics & Fine Mechanics, 1982-86; rsch. assoc. dept. chemistry Dept. Chemistry, Northwestern U., Evanston, Ill., 1987-90; rsch. fellow in optics dept. physics Dept. Physics, Howard U., Washington, 1991—. Contbr. articles to profl. jours. Mem. Internat. Soc. Optical Engring. Achievements include research on new design of opto-electronic switch in semiconductors; turnable infrared picosecond laser system; laser induced fluorescence of free radicals; picosecond laser spectroscopy; others. Office: Howard Univ Dept Physics 2355 6th St NW Washington DC 20059

ZHU, YUNPING, clinical physicist; b. Nantong, Jiangsu, China, Nov. 9, 1959; came to Can., 1990; s. Xue and Xinfen (Huang) Z.; m. Liqun Zhang, July 24, 1985. MSc, Rice U., 1985, PhD, 1987. Diplomate Am. Bd. Radiology. Project investigator M.D. Anderson Cancer Ctr., Houston, 1987-89, asst. physicist, 1989-90; clin. physicist Ont. Cancer Inst., Toronto, 1990—; asst. prof. U. Toronto, Toronto, Can., 1991—. Reviewer Internat. Jour. Radiation Oncology, 1989; referee Med. Physics, 1991—; contbr. articles to profl. jours. Predoctoral fellow Robert A. Welch Found., Rice U., 1983-87; rsch. fellow M.D. Anderson Cancer Ctr., 1987-89; major rsch. grantee Med. Rsch. Coun. Can., 1992—. Mem. Am. Soc. for Therapeutic Radiology, Am. Assn. of Physicists in Medicine, Optical Soc. Am. Achievements include development of fast fourier transform techniques which can rapidly model fully three-dimensional x-ray dose distributions while accounting for inhomogeneities of the human body to an acceptable accuracy; research on determination of multileaf collimator position. Office: Ontario Cancer Inst, 500 Sherbourne St, Toronto, ON Canada M4X 1K9

ZIABICKI, ANDRZEJ JOZEF, chemist; b. Gdynia, Poland, Sept. 9, 1933; s. Modest and Alicja (Szenberg) Z.; m. Leslawa Kedzierska, Aug. 4, 1960; children: Jacek, Joanna. BS, Poly. U., Wroclaw, Poland, 1953, MS, 1956; PhD, Poly. U., Lodz, Poland, 1960; DSc, Poly. U., Warsaw, 1965. Rsch. fellow Inst. Synthetic Fibers, Jelenia Gora, Poland, 1956-59; engr. nylon spinning factory Poland; sr. rsch. fellow, head lab. polymer physics Rsch. Inst. Gen. Chemistry, Warsaw, 1959-67; prof. inst. fundamental tech. rsch. Polish Acad. Scis., Warsaw, 1967—; vis. prof. dept. polymer chemistry Kyoto U., 1981-82; cons. in field, 1970-91; mem. state com. on sci. and tech., 1989-91, govtl. adv. com. on legis. in sci. and rsch., 1992—. Author: Poliamidy, 1964, Fizyka Procesow Formowania Wlokien, 1970, Fundamentals of Fiber Formation, 1976, Russian edit., 1979, Chinese edit., 1983, High-Speed Fiber Spinning, 1986, Russian edit., 1988, Chinese edit., 1992; contbr.: Man-Made Fibers, Science and Technology, 1968, Applied Fiber Science, 1979; contbr. articles to Jour. Chem. Physics, Colloid and Polymer Sci., Sen-I Gakkaishi, Jour. Non-Newtonian Fluid Mechanics, others. Active Solidarity Movement; bd. dirs. Stefan Batory Found., 1989—. Recipient scientific awards Polish Acad. Scis., 1978, 81, 86, S.G. Smith Meml. medal Textile Inst., 1987. Mem. Soc. for Advancement Scis. and Arts (bd. dirs. 1980—). Achievements include research in theories of spinnability of fluids, diffusional theory of crystal nucleation. Office: Polish Acad Scis, 21 Swietokrzyska St, 00-049 Warsaw Poland

ZIEGELMAN, ROBERT LEE, architect; b. Detroit, Apr. 20, 1936; s. Samuel Ziegelman and Annabelle (Davis) Benton; m. Esther C. Adelson, June 20, 1957 (div.); children: Shelly Dawn, John David, Adam. M. BArch, U. Mich., 1958; MArch, MIT, 1959. Registered architect, Mich., Pa., Okla., D.C., N.Y., Ga., Tex., Md., Mo., Ark., Toronto, Ont. Project designer Eero Saarinen and Assocs., Birmingham, Mich., 1959-62; project designer Minoru Yamasaki and Assocs., Birmingham, 1962-63; prin. Ziegelman and Ziegelman, Architects, Birmingham, 1963-72, Robert L. Ziegelman, Architect, Birmingham, 1972-80; chmn. bd. Luckenbach/Ziegelman and Ptnrs. Inc., Birmingham, 1980—; adj. prof. U. Mich., 1976-85; design lectr. MIT, 1980; chmn. bd. Insta-Bldgs. Inc., Birmingham, 1966-80. Prin. works include Guatemala Hosp. and Surg. Ctr. (award of excellence, 1974, Am. Inst. Steel Constr.), Westinghouse Electric Pace Control Ctrs (Gov.'s award 1980), William Bell Townhouses (Record Apts. of Yr. award Archtl. Record, 1982, award of merit, AIA, Housing Mag., 1982, Homes for Better Living award Housing Mag. 1982, Handleman Distbn. Ctrs. (award of honor Mich. Soc. Architects 1983), Medallion Office Bldg. (award of honor Mich. Soc. Architects 1983), North Valley Office Pk. (honor award Detroit chpt. AIA 1986), Handleman Corp. Offices (honor award Detroit chpt. AIA 1985), Prefabricated Bank Bldg. (honor award Detroit chpt. AIA 1970), 260 Brown (M award Masonry Inst. Am., Mich. Soc. Architects 1987), Thal residence (M award Mason Inst. Am., Mich. Soc. Architects 1985), U. Mich. (M award Masonry Inst. Am., Mich. Soc. Architects 1985, AIA Honor award 1990); contbr. articles to profl. jours.; patentee in field. Bd. dirs. Boys Club Am., Detroit, 1980-82, Birmingham, Bloomfield Art Assn., 1990; chmn. Housing Bd. Appeals, Birmingham, 1989; mem. Civic Design Com., Birmingham, 1964, Dean Search Com., U. Mich., 1986. Fellow AIA (pres. Detroit chpt. 1992); mem. Nat. Coun. Archtl. Registrations Bds. Office: Luckenbach/Ziegelman 115 W Brown St Birmingham MI 48009-6019

ZIEGLER, ALBERT H., mechanical engineer; b. Staten Island, N.Y., May 24, 1957; s. Warren H. and Susanna (Heiligenthaler) Z.; m. Janet Lynn Mook, June 3, 1990. B. Engring., CCNY, 1980; M. Engring., Widener U., 1984. Registered profl. engr., Pa., N.J., Del. Project engr. Gulf Oil Co. Inc., Phila., 1980-84; mech. engr. Allstates Design, Longhorne, Pa., 1984-86; engring. mgr. Synerfac Inc., Longhorne, 1987-88; engring. dir. Racs Assocs., Inc., Trevose, Pa., 1989—; bd. dirs. Pennington (N.J.) Tech. Inc.; cons. Air Products and Chems., Inc., Paulsboro, N.J., 1992—, Miller-Remick Corp., Haddon Heights, N.J., 1987-91. Mem. ASME, ASHRAE, NSPE, Assn. Energy Engrs. (sr.), Tau Beta Pi. Republican. Lutheran. Achievements include research in the application of heat transfer and fluid mechanics for the process and air conditioning industries; development of air conditioning systems for critical applications. Office: RACS Assocs 3610 Old Lincoln Hwy Trevose PA 19053

ZIEGLER, CHRISTINE BERNADETTE, psychology educator, consultant; b. Syracuse, N.Y., Mar. 22, 1951; d. Salvatore and Beverlie (Hopkins) Capozzi; m. Steven Jon Ziegler, Jan. 7,1979;1 child. Justin. Bs, SUNY, Brockport, 1978; MS, Syracuse U., 1980, PhD, 1982. Adj. asst. prof. SUNY, Cortland, N.Y., 1983-86, Syracuse (N.Y.) U., 1984-86, LeMoyne Coll., Syracuse, 1986-87; rsch. cons. Syracuse (N.Y.) U., 1982-86; assoc. prof. Kennesaw (Ga.) State Coll., 1987—; parent facilitator Ga. Coun. Child Abuse, Atlanta, 1990—; cons. Dissertation Rsch. UGA, Atlanta; cons. aggression reduction tng. Northwest Regional Hosp., Rome, Ga.; presenter numerous profl. confs. in field. contbr. articles to profl. jours. Mem. Juvenile Ct. Panel, Health Children's Initiative; mem. sch. adv. com., Marietta, Ga., 1989, mem. sch. bond com., 1989. Recipient fellowship Syracuse U., 1982. Mem. APA, AAAS, NAS, Ga. Psychol. Assn., Southeastern Psychol. Assn., Soc. Soc. Philosophy and Psychology, Soc. for Rsch. in Child Devel. Avocations: camping, gardening, environmental protection, conservation. Home: 1408 Dewberry Trl Marietta GA 30062-4013 Office: Kennesaw State College 3455 Steve Frey Rd Kennesaw GA 30144

ZIEGLER, JOHN BENJAMIN, chemist, lepidopterist; b. Rochester, N.Y., Jan. 2, 1917; s. John Benjamin Sr. and Sarah Jeanette (Murrell) Z.; m. Dorothy Mary Zucker, June 29, 1946 (dec. July 1985); children: Katherine Lois, Jeffrey Benjamin, Conrad Lawrence. BS in Chemistry, U. Rochester, 1939; MS in Chemistry, U. Ill., 1940, PhD in Organic Chemistry, 1946. Cert. chemist. Jr. chemist Merck & Co., Inc., Rahway, N.J., 1940-43; spl. rsch. asst. U. Ill., Urbana, 1943-46; chemist J.T. Baker Chem. Co., Phillipsburg, N.J., 1946-48; assoc. chemist CIBA Pharm. Co., Summit, N.J., 1948-51, sr. chemist, 1951-53, supr. labs., 1953-64, mgr. process rsch. lab., 1964-69; dir. chem. devel. CIBA-GEIGY Pharm. Co., Summit, 1969-75, dir. process rsch., 1975-76, sr. staff scientist, 1976-80; ret.; bus. adminstr. sci. dept. Seton Hall U., South Orange, N.J., 1980-82. Contbr. articles to profl. jours.; patentee in field. Chmn. com. Union county Tech. inst., 1963; past mem. N.J. Coun. R & D. Mem. Am. Chem. Soc., Lepidoptera Rsch. Found., Lepidopterists' Soc. (exec. coun., past treas., other offices), Assn. for Tropical Lepidoptera, Alpha Chi Sigma (pres. Zeta chpt. 1945-46), Phi Lambda Upsilon, Sigma Xi. Republican. Avocations: lepidoptera rsch., tennis, hiking, reading, current events. Home: 64 Canoe Brook Pky Summit NJ 07901-1434

ZIELIŃSKI, JERZY STANISŁAW, scientist, electrical engineering educator; b. Łódź, Poland, Oct. 27, 1933; s. Jakub and Janina (Bocheńska) Z.; m. Jadwiga Wesołowska), Sept. 1, 1961; 1 child, Wojciech. MSc CEng, Tech. U., Łódź, 1956, PhD, 1964, DSc, 1969. Asst. Tech. U., Łódź, 1956-60, asst. lectr., 1960-64, lectr., 1964-70, asst. prof. engring., 1970-80; asst. prof. engring. Tech. U., Lublin, Poland, 1976-82; assoc. prof. Tech. U., Lublin, 1982-86; assoc. prof. U. Łódź, 1982-90, prof., head dept. informatics, 1990—; cons. Power Inst., Warsaw, 1973, Rsch. Ctr. Automatic, Łódź, 1987-91, Power Co., Zamość; rsch. dir. Rsch. Ctr. Automatic, Łódź, 1991—. Author: Trans. Analysis in Electrical Power Systems with Application of the Method of Characteristics, 1975, Overvoltage in Electrical Power Systems Computation with Application of Analog and Digital Computers, 1985; (textbook) Analog and Digital Modeling, 1980, System Engineering, 1984; author, co-author more than 200 papers and reports. Recipient Sci award Polish Acad. Sci. 1966, Sci. award Minister Higher Edn. 1965, 76, Sci. award Polish Soc. Theoretical and Applied Electrotechnics 1963-68, Sci. award Rectors of the Univs: Tech. U. Łódź 1969, 70, 73, Tech. U. Lublin 1981-84, U. Łódź 1985-88. Mem. Assn. Polish Electricians (expert, Disting. Silver and Gold decorations)., Polish Soc. Informatics., Polish Soc. Theoretical and Applied Electrotechnics. Roman Catholic. Avocations: music, cycling, walking, stamp-collecting. Home: 99/101m 63 Narutowicza, 90-145 Łódź Poland Office: U Łódź Informatics, 39 Rewolucji 1905r, 90-214 Lodz Poland

ZIELINSKI, PAUL BERNARD, grant program administrator, civil engineer; b. West Allis, Wis., Sept. 9, 1932; s. Stanley Charles and Lottie Charlotte (Pliskiewicz) Z.; m. Monica Theresa Beres, July 13, 1957; children: Daniel Paul, Gregory John, Robert Mathias, Sarah Anne. BSCE, Marquette U., 1956; MS, U. Wis., 1961, PhD, 1965. Registered profl. engr., Wis., S.C. Asst. instr. engring. mechanics Marquette U., Milw., 1956-59, asst. prof., 1964-67; instr. civil engring. U. Wis., Madison, 1959-64; from asst. prof. to prof. Clemson (S.C.) U., 1967-78, prof. environ. and systems engring., 1982-90, prof. civil engring., 1982-90, prof. emeritus, 1991—; dir. S.C. Water Resources Rsch. Inst., Clemson, 1978-90; assoc. dir. associateship grant program Nat. Rsch. Coun., Washington, 1990—; cons. Am. Pub. Works Assn., Chgo., 1973-76, Nat. Coun. Examiners of Engring. and Surveying, Clemson, 1973—. Chmn. Clemson City Planning Commn., 1971-74; ex-officio mem. S.C. Water Resources Commn., Columbia, 1978-90. Mem. ASCE, Am. Soc. for Engring. Edn., Sigma Xi. Roman Catholic. Home: 2111 Wisconsin Ave NW # 717 Washington DC 20007 Office: Nat Rsch Coun 2201 Constitution Ave NW Washington DC 20418

ZIEMER, RODGER EDMUND, electrical engineering educator, consultant; b. Sargeant, Minn., Aug. 22, 1937; s. Arnold Edmund and Ruth Ann (Rush) Z.; m. Sandra Lorann Person, June 23, 1960; children: Mark Edmund, Amy Lorann, Norma Jean, Sandra Lynn. B.S., U. Minn., 1960, M.S., 1962, Ph.D., 1965. Registered profl. engr., Mo. Research asst. U. Minn., Mpls., 1960-62; research assoc. U. Minn., 1962; prof. elec. engring. U. Mo., Rolla, 1968-83; prof. elec. engring. U. Colo., Colorado Springs, 1984—, chmn. dept. elec. engring., 1984—; cons. Emerson Electric Co., St. Louis, 1972—, Mid-Am. Regional Council, Kansas City, Mo., 1974, Motorola, Inc., Scottsdale, Ariz., 1980-84, Martin Marietta, Orlando, 1980-81, TRW, Colorado Springs, summer 1985, Sperry, Phoenix, 1986; sabbatical at Motorola, Scottsdale, Ariz., 1991. Author: Principles of Communications, 1976, 2d edit., 1985, 3d edit., 1990, Signals and Systems, 1983, 2d edit., 1989, 3rd edit., 1993, Digital Communications and Spread Spectrum Systems, 1985, Introduction to Digital Communication, 1992; editor: IEEE Jour. on Selected Areas in Communications, 1989, 92, IEEE Communications Mag., 1991. Served to capt. USAF, 1965-68. Scholar Western Electric, 1957-59; trainee NASA, 1962-65. Fellow IEEE; mem. Am. Soc. Engring. Edn., Armed Forces Communications and Electronics Assn., Sigma Xi, Tau Beta Pi, Eta Kappa Nu. Lutheran. Home: 8315 Pilot Ct Colorado Springs CO 80920-4412 Office: Univ Colo PO Box 7150 Colorado Springs CO 80933-7150

ZIER, J(OHN) ALBERT, chemical engineer; b. Vermilion, Ohio, Nov. 21, 1919; s. George Benjamin and Ola Louise (Leimbach) Z.; m. Catherine Marie Pollak, July 3, 1948 (dec. Feb. 1966); children: Mark A., Christine A.; m. Martha Antoinette Akers, June 14, 1970; stepchildren: James, Thomas, Wendy. B. CH E, Ohio State U., 1941. Registered profl. engr., Ohio. Head control lab. Glidden Co., Cleve., 1941-44, chemist, chem. engr., 1946-58, head solvent resin lab., 1958-62, pilot plant mgr., 1962-70, mgr. resin mfg., 1970-84, cons., 1985, 89—. Author: (with others) Ohio Society, Sons of the American Revolution, Centennial Register, 1989. Pastor Wesley Meth. Ch., Chgo., 1952-53. With U.S. Army, 1944-46, ETO. Fellow Am. Inst. Chemists; mem. Am. Inst. Chem. Engrs., Soc. for Paint Tech., Am. Chem. Soc., SAR (pres. Western Res. Soc. 1991). Methodist. Achievements include 4 patents. Home: 13733 Newton Rd Cleveland OH 44130-2737

ZIERDT, CHARLES HENRY, microbiologist; b. Pitts., Apr. 24, 1922; s. Conrad Henry and Nancy Leora (Harshberger) Z.; m. Margaret May Wise, June 1, 1942 (div. 1962); children—Charles Henry, Jr., Carolyn, Douglas, Richard; m. Willadene Smith, Sept. 30, 1967. B.S., Pa. State U., 1943; M.S., U. Mich., 1945; Ph.D., George Washington U., 1967. Rsch. assoc. Parke-Davis & Co., Detroit, 1945-48; microbiologist Henry Ford Hosp., Detroit, 1948-53, USPHS, Detroit, 1953-56; rsch. microbiologist NIH, Bethesda, Md., 1956—. Scientist sponsor U. Md., 1975—; instr. Found. Advanced Edn. Scis., Bethesda, 1978—. Author: Glucose Nonfermenting Gram Negative Bacteria in Clinical Microbiology, 1978; Non-fermentative Gram Negative Rods: Laboratory Identification and Clinical Aspects, 1985; McGraw-Hill Yearbook of Science and Technology, 1986; Diagnostic Procedures for Bacterial Infections, 1987; contbr. over 100 articles to profl. jours. Patentee in field. Active PTA. Fellow Am. Acad. Microbiology; mem. Am. Soc. Microbiology (chpt. pres. 1976), U.S. Fedn. Culture Collections (membership chmn. 1985), Avanti Owners Assn. Internat., Mensa, Model A Ford Club of Am. (Fairfax, Va. chpt. pres. 1985), Model T Ford Club Internat., Sigma Xi. Republican. Achievements include the classification and pathogenesis of Blastocystis Hominis. Avocations: gardening; antique car restoration. Home: 4100 Norbeck Rd Rockville MD 20853-1869 Office: NIH Bethesda MD 20892

ZIETLOW, DANIEL H., electrical engineer; b. Milw., Mar. 18, 1963. BSEE, U. Wis., Platteville, 1988. Elec. systems engr. nuclear power dept. Wis. Electric Power Co., Milw., 1989-91, elec. engr. start-up inspection div. engring. & constrn., 1991—. Mem. IEEE, Am. Nuclear Soc. Office: Wis Electric Power Co PO Box 2046 231 W Michigan St Milwaukee WI 53201

ZIFERSTEIN, ISIDORE, psychoanalyst, educator, consultant; b. Klinkowitz, Bessarabia, Russia, Aug. 10, 1909; came to U.S., 1920; s. Samuel David and Anna (Russler) Z.; m. Barbara Shapiro, June 21, 1935; children: D. Gail, J. Dan. BA, Columbia U., 1931, MD, 1935; PhD, So. Calif. Psychoanalytic Inst, 1977. Intern Jewish Hosp. of Bklyn., N.Y., 1935-37; staff psychiatrist Mt. Pleasant (Iowa) State Hosp., 1937-41; chief resident psychiatrist Psychiat. Inst. of Grasslands Hosp. of Westchester County, Valhalla, N.Y., 1941-44; pvt. practice, psychoanalysis and psychiatry N.Y.C., 1944-47, L.A., 1947—; mem. faculty So. Calif. Psychoanalytic Inst., L.A., 1951-70, mem. bd. trustees, 1953-57, mem. edn. com., 1953-57; cons. L.A. (Calif.) Psychiat. Svcs., 1954-63; researcher The Psychiat. & Psychosomatic Rsch. Inst., Mt. Sinai Hosp., L.A., 1955-65; assoc. clin. prof. of psychiatry Univ. So. Calif., L.A., 1960-64, Univ. Calif., L.A., 1970-77; rsch. cons. Postgrad. Ctr. for Mental Health, N.Y.C., 1962-75; attending staff dept. psychiatry Cedars-Sinai Med. Ctr., L.A., 1975—; psychiat. cons. Iowa State Penitentiary, Ft. Madison; lectr. on transcultural psychiatry, group psychotherapy and group dynamics UCLA, U. Calif., Berkeley, USC, U. Wash., Willamette Coll., Eugene, Oreg., U. Oreg., U. B.C., U. Md., U. Wis., U. Judaism, U. Mex., Wayne State U., U. Pitts., Chgo. Med. Coll., Ctr. for Study of Democratic Instns., U. Leningrad, Bekhterev Psychoneurol. Rsch. Inst., Leningrad, BBC, San Francisco State Coll., others. Contbr. over 65 articles to Am. Jour. Psychiatry, Am. Jour. Orthopsychiatry, Internat. Jour. Group Psychotherapy, Praxis Der Kinderpsychologie Und Kinderpsychiatrie, and others. Bd. dirs. Viewer-Sponsored TV Found., L.A., 1960, Nat. Assn. for Better Broadcasting, L.A., 1962-75, ACLU So. Calif. chpt., L.A., 1962-77; pres. Peace Edn. Coun., Pasadena, Calif., 1960; del. to state conv. Calif. Dem. Coun., Sacramento, 1960; mem. del. to Soviet Union, Promoting Enduring Peace, New Haven, 1959; participant Conf. of Scientists for Peace, Oslo, Norway, 1962; del. to "Pacem in Terris" Convocation, SANE, N.Y.C., 1963, mem. nat. bd., 1970-74, and many others. Recipient Pulitzer scholarship award Pulitzer Found., N.Y.C., 1927, Green Prize for Outstanding scholarship Columbia Coll., N.Y.C., 1930, Peace award Women for Legis. Action, L.A., 1962, grant for rsch. in transcultural psychiatry Founds. Fund for Rsch. in Psychiatry, New Haven, 1963, grant for continuing rsch. in transcultural psychiatry NIMH, Bethesda, Md., 1969, Pawlowski Peace Prize, Pawlowski Peace Found., Inc., Wakefield, Mass., 1974. Fellow Am. Psychiat. Assn. (life, fellowship medal 1970), AAAS (life); mem. AMA (life), Am. Psychoanalytic Assn. (life, cert. in psychoanalysis by bd. profl. standards), Internat. Psychoanalytical Assn., World Fedn. for Mental Health, Westside Jewish Culture Club (lectr.), Physicians for Social Responsibility, Sierra Club, Nat. Wildlife Fedn., Environ. Def. Fund, Common Cause, MADD, Phi Beta Kappa. Democrat. Jewish. Achievements include research on transcultural psychiatry, group psychotherapy and group dynamics. Office: 1819 N Curson Ave West Hollywood CA 90046-2205

ZIGAMENT, JOHN CHARLES, mechanical engineer; b. Evanston, Ill., June 29, 1965; s. Don J. and Ruth (Mann) Z. BA cum laude, Augustana Coll., Rock Island, Ill., 1987; BSME, Purdue U., 1989. Mech. engr. Bechtel Corp., Naperville, Ill., 1990—. Precinct committeeman City of Rock Island, 1984-86; pres. Ranchview Condominium Assn., Naperville, 1992—; election judge DuPage County Bd. Elections, Naperville, 1992—. Mem. ASME. Home: 1813 Appaloosa Dr Naperville IL 60565-1794 Office: Bechtel Corp 1240 E Diehl Naperville IL 60565

ZIGER, DAVID HOWARD, chemical engineer; b. Milw., July 28, 1956; s. Bernard and Betty Joyce (Blankstein) Z.; m. Ellen Cohen, June 17, 1979; children: Brian, Rebecca, Daniel. BSChemE, U. Wis., 1978; MS, U. Ill., Champaign, PhD in Chem. Engring. Mem. tech. staff AT&T Bell Labs., Princeton, N.J., 1983-90, Allentown, Pa., 1990—; AT&T assignee to Sematech Consortium, Austin, Tex., 1989-91; mem. adv. bd. KTI Microelectronics Symposium, San Diego, 1989; mentor lithography programs, U. Calif., Stanford, 1989-91. Author articles on correlation of solubilities in various fluids, thermodynamic properties of supercritical fluid mix, extraction of thin films with supercritical fluids, control of labile thermal processes, and generalized approach to lithography chemistry; referee tech. publs. on thermodynamics and semicondr. processing. Mem. Am. Phys. Soc. Jewish. Achievements include patents in semiconductor processing ; measurement of partial molar volume of mixtures at infinite dilution under supercritical fluid conditions, this parameter is key to understanding the physical chemistry of supercritical fluid mixtures; generalization of approach to modeling lithographic performance in the manufacture of semiconductors; research in generalized correlation of solubility in supercritical fluid mixtures. Office: AT&T Bell Labs 9333 John Young Pkwy Orlando FL 32819

ZIGLER, EDWARD FRANK, educator, psychologist; b. Kansas City, Mo., Mar. 1, 1930; s. Louis and Gertrude (Gleitman) Z.; m. Bernice Gorelick, Aug. 28, 1955; 1 child, Scott. BA, U. Mo.-Kansas City, 1954; PhD, U. Tex., 1958; MA (hon.), Yale, 1967; DSc (hon.), Boston Coll., 1985; LHD (hon.), Bank St. Coll. Edn., 1989, U. New Haven, 1991, St. Joseph Coll., 1991; PhD (hon.), U. Mo., 1993. Psychol. intern Worcester (Mass.) State Hosp., 1957-59; asst. prof. psychology U. Mo., 1958-59; mem. faculty Yale U., 1959—, prof. psychology and child study center, 1967—, Sterling prof., 1976—, dir. child devel. program, 1961-76, chmn. dept. psychology, 1973-74; head psychology sect. Yale Child Study Center, 1967—; dir. Bush Center in Child Devel. and Social Policy, 1977—; dir. Office Child Devel.; chief Children's Bur., HEW, Washington, 1970-72; cons. in field, 1962—; Mem. nat. steering com. Project Head Start, 1965-70, chmn. 15th anniversary Head Start com., 1980; mem. adv. com. Head Start quality and expansion U.S. Dept. Health and Human Svcs., 1993—, nat. adv. com. Nat. Lab. Early Childhood Edn., 1967-70; nat. research adv. bd. Nat. Assn. Retarded Children, 1968-73; nat. research council Project Follow-Through, 1968-70; chmn. Vietnamese Children's Resettlement Adv. Com., 1975; mem. President's Com. on Mental Retardation, 1980; joint appointee Yale U. Sch. Medicine, 1982—, chmn. Yale Infant Care Leave Commn., 1983-85, New Parents as Tchrs., 1986—. Author, co-author, editor books and monographs; contbr. articles to profl. jours. Served with AUS, 1951-53. Recipient Gunnar Dybwad Disting. scholar in behavioral and social sci. award Nat. Assn. Retarded Children, 1964, 69, Social Sci. Aux. award, 1962, Alumni Achievement award U. Mo., 1965, Alumnus of Yr. award, 1972, C. Anderson Aldrich award Am. Acad. Pediatrics, 1985, Nat. Achievement award Assn. for Advancement Psychology, 1985, Dorthea Lynde Dix Humanitarian award for svc. to handicapped Elwyn Inst., 1987, Sci. Leadership award Joseph P. Kennedy Jr. Found., 1990, Mensa Edn. and Rsch. Found. award for excellence, 1990, Nat. head Start Assn. award, 1990, Bldg. dedication Edward Zigler Head Start Ctr., 1990, As They Grow award in edn. Parents mag., 1990, Excellence in Edn. award Pi Lambda Theta, 1991, Friend of Edn. award Conn. Edn. Assn., 1991, Loyola-Mellon Social Sci. award 1991, President's award Conn. Assn. Human Svcs., 1991, Harold W. McGraw, Jr. prize in Edn., 1992, Disting. Achievement in Rsch. award Internat. Assn. Scientific Study of Mental Deficiency, 1992; named Hon. Commr. Internat. Yr. of Child, 1979. Fellow Am. Orthopsychiat. Assn. (Blanche F. Ittleson award 1989), Am. Psychol. Assn. (pres. div. 7 1974-75, G. Stanley Hall award 1979, award for disting. contbns. to psychology in pub. interest 1982, Nicholas Hobbs award 1985, award for disting. profl. contbns. to knowledge 1986, Edgar A. Doll award 1986, award for disting. contbn. to community psychology and community mental health 1989); mem. Inst. Medicine of NAS, AAAS, Am. Acad. Mental Retardation (career rsch. award 1982), Sigma Xi. Home: 177 Ridgewood Ave North Haven CT 06473-4442 Office: Yale U Dept Psychology Hill Louse Ave Box 11A New Haven CT 06515-2251

ZIL, JOHN STEPHEN, psychiatrist, physiologist; b. Chgo., Oct. 8, 1947; s. Stephen Vincent and Marilyn Charlotte (Jackson) Zilius; 1 child, Charlene-Elena. BS magna cum laude, U. Redlands, 1969; MD, U. Calif., San Diego, 1973; MPH, Yale U., 1977; JD with honors, Jefferson Coll., 1985. Intern, resident in psychiatry and neurology U. Ariz., 1973-75; fellow in psychiatry, advanced fellow in social and community psychiatry, Yale community cons. to Conn. State Dept. Corrections, Yale U., 1975-77, instr. psychiatry and physiology, 1976-77; instr. physiology U. Mass., 1976-77; acting unit chief Inpatient and Day Hosp. Conn. Mental Health Ctr., Yale-New Haven Hosp. Inc., 1975-76, unit chief, 1976-77; asst. prof. psychiatry U. Calif., San Francisco, 1977-82, assoc. prof. psychiatry and medicine, 1982-86, vice-chmn. dept. psychiatry, 1983-86; adj. prof. Calif. State U., 1985-87; assoc. prof. bioengring. U. Calif., Berkely and San Francisco, 1982—, clin. faculty, Davis, 1991—; chief psychiatry and neurology VA Med. Ctr., Fresno, Calif., 1977-86, prin. investigator Sleep Rsch. & Physiology Lab., 1980-86; dir. dept. psychiatry and neurology U. Calif.-San Francisco, Fresno-Cen. San Joaquin Valley Med. Edn. Program and Affiliated Hosps. and Clinics, 1983-86; chief psychiatrist State of Calif. Dept. Corrections cen. office, 1986—; chmn. State of Calif. Inter-Agy. Tech. Adv. com. on Mentally Ill Inmates & Parolees, 1986—; mem. med. adv. com. Calif. State Personnel Bd., 1986—; appointed councillor Calif. State Mental Health Plan, 1988—; cons. Nat. Inst. Corrections, 1992—; invited faculty contbr. and editor Am. Coll. Psychiatrist's Resident in Tng. Exam., 1981—. Author: The Case of the Sleepwalking Rapist, 1992, Mentally Disordered Criminal Offenders, 5 vols., 1989, reprinted, 1991; contbg. author: The Measurement Mandate: On the Road to Performance Improvement in Health Care, 1993; assoc. editor Corrective and Social Psychiatry Jour., 1978—, referee, 1980—, reviewer, 1981—; contbr. articles in field to profl. jours. Nat. Merit scholar, 1965; recipient Nat. Recognition award Bank of Am., 1965, Julian Lee Roberts award U. Redlands, 1969, Kendall award Internat. Symposium in Biochemistry Research, 1970, Campus-Wide Profl. Achievement award U. Calif., 1992. Fellow Royal Soc. Health, Am. Assn. Social Psychiatry; mem. Am. Assn. Mental Health Profls. in Corrections (nat. pres. 1978—), Calif. Scholarship Fedn. (past pres.), AAUP, Am. Psychiat. Assn., Nat. Council on Crime and Delinquency, Am. Pub. Health Assn., Delta Alpha, Alpha Epislon Delta. Office: PO Box 163359 Sacramento CA 95816-9359

ZIMET, CARL NORMAN, psychologist, educator; b. Vienna, Austria, June 3, 1925; came to U.S., 1943, naturalized, 1945; s. Leon and Gisela (Kosser) Z.; m. Sara F. Goodman, June 4, 1950; children: Andrew, Gregory. BA, Cornell U., 1949; PhD, Syracuse U., 1953; postdoctoral fellow, Standard U., 1953-55. Diplomate in clin. psychology Am. Bd. Profl. Psychology (trustee 1966-74). Instr., then asst. prof. psychology and psychiatry Yale U., 1955-63; mem. faculty U. Colo. Med. Center, 1963—, prof. clin. psychology, 1965—, head div., 1963—; Mem. Colo. Bd. Psychol. Examiners, 1966-72, Colo. Mental Health Planning Commn., 1964-66; mem. acad. adv. com. John F. Kennedy Child Devel. Center, U. Colo., 1966—; chmn. Council for Nat. Register of Health Service Providers in Psychology, 1975-85, pres., mem. exec. bd. div. psychotherapy, 1970—; chair exec. com. Assn. Psychol. Internship Ctrs., 1988-91. Bd. editors: Jour. Clin. Psychology, 1962—, Jour. Clin. and Cons. Psychology, 1964-73, Psychotherapy, 1967—, Profl. Psychology, 1969—. With USNR, 1943-46. Recipient Disting. Service award Colo. Psychol. Assn., 1976. Fellow Am. Psychol. Assn. (council reps. 1969-72, 73—, bd. dirs. 1985-88, Disting. award for profl. contbn. 1987, div. psychotherapy and div. clin. psychology), Soc. Personality Assessment (pres. 1975-76, bd. dirs., chair gen. psychol. services 1987—); mem. Am. Acad. Clin. Psychology (pres. 1993—), Denver Psychoanalytic (trustee 1968-71), Am. Group Psychotherapy Assn., Med. Sch. Profs. Psychology (pres. 1992-94). Home: 4325 E 6th Ave Denver CO 80220-4939

ZIMM, BRUNO HASBROUCK, physical chemistry educator; b. Woodstock, N.Y., Oct. 31, 1920; s. Bruno L. and Louise S. (Hasbrouck) Z.; m. Georgianna S. Grevatt, June 17, 1944; children: Louis H., Carl B. Grad., Kent (Conn.) Sch., 1938; A.B., Columbia U., 1941, M.S., 1943, Ph.D., 1944. Research assoc. Columbia U., 1944; research assoc., instr. Polytech. Inst. Bklyn., 1944-46; instr. chemistry U. Calif. at Berkeley, 1946-47, asst. prof., 1947-50, assoc. prof., 1950-51; vis. lectr. Harvard U., 1950-51; research assoc. research lab. Gen. Electric Co., 1951-60; prof. chemistry U. Calif., La Jolla, 1960-91, prof. emeritus, 1991—. Assoc. editor: Jour. Chem. Physics, 1947-49; adv. bd.: Jour. Polymer Sci, 1953-62, Jour. Bio-Rheology, 1962-73, Jour. Biopolymers, 1963—, Jour. Phys. Chemistry, 1963-68, Jour. Biophys. Chemistry, 1973—. Recipient Bingham Medal Soc. Rheology, 1960, High Polymer Physics prize Am. Phys. Soc., 1963; Kirkwood medal Yale U., 1982. Mem. Biophys. Soc., Am. Soc. Biol. Chemists, Am. Chem. Soc. (Baekeland award 1957), Nat. Acad. Scis. (award in Chem. Scis. 1981), Am. Acad. Arts and Scis., Am. Phys. Soc.

ZIMMER, JAMES PETER, laboratory executive, consultant; b. St. Louis, Nov. 25, 1952; s. Arthur James and Valerie M. Zimmer; m. Patricia Ann Reed, Jan. 25, 1977; children: Nicholas J., Julie M. Electronic technician Flight Safety Inst., Flushing, N.Y., 1975-77; sr. field svc. engr. Technicon Inst., Tarrytown, N.Y., 1977-81; field svc. mgr. V.T.I., Fountain Valley, Calif., 1981-83, Telsar Labs., Godfrey, Ill., 1985-88; pres., CEO, Telsar Labs., Alton, Ill., 1988—; mem. adv. bd. Jefferson C.C., Hillsboro, Mo., 1988—. Mem. Mo. Athletic Club, Harbor Point Yacht Club, Alton Motorboat Club, Rotary. Home: 4809 Kaskaskia Trail Godfrey IL 62025 Office: Telsar Labs 319-321 Ridge St Alton IL 62002

ZIMMERMAN, CHARLES LEONARD, systems engineer; b. Waterloo, Iowa, Oct. 14, 1964; s. Carl Browning and Lou Anne (Leonard) Z.; m. Melissa Lynn Tiemann, Dec. 19, 1992. BS in Indsl. and Mgmt. Engring., U. Iowa, 1987, MS in Indsl. and Mgmt. Engring., 1989. Systems engr. Caterpillar, Inc., Aurora, Ill., 1989—. Mem. NSPE, Alpha Phi Omega (Disting. Svc. Key 1986, regional dir. 1990—, nat. bd. dirs. 1990—). Republican. Presbyterian. Home: 225 N Oakhurst Dr # 35 Aurora IL 60504 Office: Caterpillar Inc PO Box 348 Aurora IL 60507

ZIMMERMAN, DANIEL D., chemical engineer; b. Oct. 1, 1935; came to U.S., 1954; s. Zigmund and Sophie (Gottesfeld) Z.; m. Bella Neuman, June 25, 1961; children: Jean Marc, Philippe. BS in Chem. Engring., CCNY, 1960; MS in Chem. Engring., NYU, N.Y.C., 1962. Registered profl. engr., N.J., N.Y. Process engr. Maxwell House, Hoboken, N.J., 1963-67; engr. R&D Celanese Corp., Clark, N.J., 1967-70; sr. engr. R&D Celanese Corp., Summit, N.J., 1970-73, group leader R&D tech. svc., 1973-82; mgr. new product devel. Cyro Industries, Orange, Conn., 1983—. Contbr. articles to profl. jours. Fellow Am. Chem. Soc., Soc. Plastics Engrs., Polymer Processing Soc. Achievements include 11 patents for reinforced microporous films, prodn. of novel celled open microporous films, polyalkylene terephthalate resins, high melt index microporous films, process for preparing a thermoplastic film involving a cold stretching step and multiple hot stretching, corona treated microporous films, simultaneous stretching of multiple plies of polymeric films, isocyanate-coupled reinforced oxymethylene polymers, blends of polymers particularly polycarbonates/acrylate-modified rubber compositions, and blends of acrylate and modified rubber grafts. Home: 614 A Cherokee Ln Stratford CT 06497

ZIMMERMAN, DELANO ELMER, physician; b. Fond du Lac, Wis., Mar. 21, 1933; s. Elmer Herbert and Agatha Angeline (Freund) Z.; m. Nancy Margaret Garry, Aug. 13, 1966; children: Kate Zimmerman Lennard, Joseph, Nick. BS, U. Wis., 1961, MD, 1965. Diplomate Am. Soc. Profl. Disability Cons. Intern, Hennepin County Hosp., Mpls., 1965,; physician, surgeon Winnebago (Wis.) State Hosp., 1966-67; gen. practice medicine, Neenah, Wis., 1967-73; emergency room physician Community Emergency Svcs. , Appleton, 1973-77, Meml. Med. Center, Springfield, Ill., 1977-92; faculty So. Ill. U. Sch. Medicine, Springfield, 1977-93; mem., bd. dirs. nominating com. Sangamon Valley chpt. ARC. With USN, 1951-56. Mem. Am. Coll. Emergency Physicians, Ill. Coll. Emergency Physicians (bd. dirs., awards com., fin. com., mem.-at-large, govt. affairs com.), Soc. for Acad. Emergency Medicine. Roman Catholic. Home: The Cottage 1467 Cowling Bay Rd Neenah WI 54956-9720

ZIMMERMAN, DORIS LUCILE, chemist; b. L.A., July 30, 1942; d. Walter Merritt and Letta Minnie (Reese) Briggs; m. Christopher Scott Zimmerman, June 5, 1964; children: Susan Christina, David Scott, Brian Allan. BS in Chemistry, Carnegie Mellon U., 1964; MS in Chemistry, Youngstown State U., 1989, MS in Materials Engring., 1992; postgrad., Kent (Ohio) State U. High sch. tchr. Ohio County Schs., Vienna and Campbell, 1983-87; sr. chemist Konwal, Warren, Ohio, 1988-91; limited faculty mem. Kent (Ohio) State U., 1991—; substitute tchr. County Schs. of Ohio, Warren, 1972-82; tutor, 1965—. Active ARC, WSI, Lifeguarding, Warren, 1965—; com. chmn. AAUW, Warren, 1982-87; chmn. Trumbull Mobile Meals, Warren, 1977—, Pink Thumb Garden Club, Warren, 1965, 88. Recipient Svc. award ARD, 1981, Trumbull Mobile Meals, 1985. Mem. Am. Chem. Soc. (sec. 1985-90, chmn. elect 1990, chmn. 1991, alternate councilor 1992—, Commendation award 1990), Carnegie Mellon Alumni Assn. (admissions councilor, Svc. award 1981), Phi Lambda Upsilon, Phi Kappa Phi, Sigma Xi. Republican. Methodist. Avocations: masters' swimming, sailboat racing, tennis, bridge. Home: 2944 Coit Dr NW Warren OH 44485-1444

ZIMMERMAN, EARL ABRAM, physician, scientist, educator, neuroendocrinology researcher; b. Harrisburg, Pa., May 5, 1937; s. Earl Beckley and Hazel Marie (Myers) Z.; m. Diane Leenheer, Sept. 14, 1960 (div. Aug. 1982); m. Poppy Ann Warren, Sept. 5, 1982. B.S. in Chemistry, Franklin and Marshall Coll., 1959; M.D., U. Pa., 1963. Diplomate Am. Bd. Psychiatry and Neurology, Am. Bd. Internal Medicine. Intern Presbyn. Hosp., N.Y.C., 1963-64, resident, 1964-65; resident Neurol. Inst. CPMC, N.Y.C., 1965-68, research fellow endocrinology, 1970-72; asst. prof. to prof. neurology Columbia U., N.Y.C., 1972-85; prof., chmn. dept. neurology Oreg. Health Sci. U., Portland, 1985—; dir. neurology Helen Hayes Hosp., Haverstraw, N.Y., 1982-83. Mem. editorial bd.: Jour. Histochem. Cytochemistry, 1980-85, 87, Neuroendocrinology, 1985-88, Annals of Neurology, 1985; contbr. numerous articles to profl. jours. Maj. USAF, 1968-70. Rsch. grantee NIH, 1977—. Mem. Am. Neurol. Assn. (program chmn. 1980-82), Am. Acad. Neurology (Wartenber lectr. 1985), Endocrine Soc. Democrat. Mem. United Ch. of Christ. Avocations: woodworking; gardening; theatre; music; art; skiing; tennis. Home: Apt 17B 3181 SW Sam Jackson Park Rd # 226L Portland OR 97201-3011 Office: Oreg Health Sci U Dept Neurology 3181 SW Sam Jackson Park Rd Portland OR 97201-3011

ZIMMERMAN, HOWARD ELLIOT, chemist, educator; b. N.Y.C., July 5, 1926; s. Charles and May (Cohen) Z.; m. Jane Kirschenheiter, June 3, 1950 (dec. Jan. 1975); children: Robert, Steven, James; m. Martha L. Bailey Kaufman, Nov. 7, 1975 (div. Oct. 1990); m. Margaret J. Vick, Oct. 1991; stepchildren: Peter and Tanya Kaufman. B.S., Yale U., 1950, Ph.D., 1953. NRC fellow Harvard U., 1953-54; faculty Northwestern U., 1954-60, asst. prof., 1955-60; assoc. prof. U. Wis., Madison, 1960-61, prof. chemistry U. Wis., 1961—, Arthur C. Cope and Hilldale prof. chemistry, 1975—; Chmn. 4th Internat. Union Pure and Applied Chemistry Symposium on Photochemistry, 1972. Author: Quantum Mechanics for Organic Chemists, 1975; mem. editorial bd.: Jour. Organic Chemistry, 1967-71, Molecular Photochemistry, 1969-73, Jour. Am. Chem. Soc., 1982-85, Revs. Reactive Intermediates, 1984-89; contbr. articles to profl. jours. Recipient Halpern award for photochemistry N.Y. Acad. Scis., 1979, Chem. Pioneer award Am. Inst. Chemists, 1986, Sr. Alexander von Humboldt award, 1988, Hilldale award U. Wis., 1988-89, 90. Mem. NAS, Am. Chem. Soc. (James Flack Norris award 1976, Arthur C. Cope Scholar award 1991), Chem. Soc. London, German Chem. Soc., Inter-Am. Photochemistry Assn. (co-chmn. organic div. 1977-79, exec. com. 1979-86), Phi Beta Kappa, Sigma Xi. Home: 1 Oconto Ct Madison WI 53705-4925 Office: U Wis Chemistry Dept 1101 University Ave Madison WI 53706-1396

ZIMMERMAN, JEANNINE, crime laboratory specialist, researcher; b. St. Joseph, Mo., Oct. 29, 1930; d. Victor Hugo and Audrey Louise (Abrams) Drake; (div.); children: Matthew, Mark, Brian, Jeremy. BA in Psychology summa cum laude, Colo. Women's Coll., 1975; MA in Criminal Justice, U. Colo., Denver, 1982. Cert. law enforcement tchr. Questioned document examiner Aurora, Colo., 1972-77; crime lab. specialist Aurora Police Dept., 1977—. Contbr. articles to profl. jours. Recipient Excellence award City of Aurora, 1986. Fellow Am. Acad. Forensic Scientists; mem. Southwestern Assn. Forensic Document Examiners (Colo. rep. 1988-89). Republican. Office: Aurora Police Dept 15001 E Alameda Dr Aurora CO 80012

ZIMMERMAN, JOHN WAYNE, wildlife ecologist; b. Elkhart, Ind., Dec. 26, 1953; s. Cecil Paul and Helen Louise (Probus) Z.; m. Margaret Kaye Dunmire, Dec. 30, 1972; 1 child, Scot Mikeljon. BS in Wildlife Ecology, Okla. State U., 1979, MS, 1982; PhD in Zoology, N.C. State U., 1992. Cert. assoc. wildlife biologist. Extension instr. Wildlands Rsch. Inst., Santa Cruz, Calif., 1983-84; rsch. asst. dept. zoology N.C. State U., Raleigh, 1982-84, teaching technician dept. biol. scis., 1986-91, vis. asst. prof. dept. biol. scis., 1992—, summer instr. dept. zoology, 1987-93. Contbr. articles to profl. jours.; co-author software. With U.S. Army, 1972-76, Germany. N.C. State U. grantee, 1991-92, N.C. Biotech. Ctr. grantee, 1992. Mem. Am. Soc. Mammalogists, The Wildlife Soc., Sigma Xi. Office: NC State U 2717 Bostian Raleigh NC 27695-7611

ZIMMERMAN, MICHAEL RAYMOND, pathology educator, anthropologist; b. Newark, Dec. 26, 1937; s. Edward Louis and Bessie (Herman) Z.; m. Barbara Elaine Hoffman, Dec. 25, 1960; children: Jill Robin, Wendy Vida. BA, Washington and Jefferson Coll., 1959; MD, NYU, 1963; PhD, U. Pa., 1976. Diplomate Am. Bd. Pathology. Intern in surgery Kings County

Hosp., Bklyn., 1963-64; resident-fellow in pathology NYU- Bellevue Med. Ctr., N.Y.C., 1964-68; pathologist Lankenau Hosp., Phila., 1970-72; asst. prof. pathology U. Pa., Phila., 1972-77; assoc. prof. pathology and anthropology U. Mich., Ann Arbor, 1977-80; assoc. prof. pathology Hahnemann U., Phila., 1980-82; pathologist Jeanes Hosp., Phila., 1982-85; dir. pathology and lab. svcs. Coney Island Hosp., Bklyn., 1985-87; prof. anthropology U. Pa., Phila., 1985—; assoc. prof. pathology Hahnemann U., Phila., 1980-87, prof. pathology and lab. medicine, 1987—, chmn. exec. faculty, 1991—; cons. pathologist U. Scranton, Pa., 1985—; rsch. assoc. Univ. Mus., Phila., 1982—; mem. faculty paleopathology course Armed Forces Inst. Pathology, Washington, 1992. Author: Foundations of Medical Anthropology, 1980; co-author: Atlas of Human Paleopathology, 1982; co-editor, author: Dating and Age Determination of Biological Materials, 1986; mem. editorial adv. bd. Advance for Adminstrn. Labs., 1992—; also articles in Arctic Anthropology, Annals Clin. Lab. Sci. Trustee Martin's Run Life Care Retirement Community, Broomall, Pa., 1988—. Maj. USAR, 1968-70. Rackham faculty rsch. grantee U. Mich., 1978-80. Mem. Am. Soc. Clin. Pathologists, Coll. Am. Pathologists (insp. 1988—), Am. Assn. Phys. Anthropologists, Paleopathology Assn. (charter). Achievements include research in paleopathology; field work in Egypt and Alaska. Office: Hahnemann U MS 435 Broad and Vine Philadelphia PA 19102

ZIMMERMAN, RICHARD ORIN, safety engineer; b. Newberg, Oreg., Mar. 2, 1956; s. Orin Frank and Eva Josephine (Schumann) Z.; m. Susan Scharn, Aug. 27, 1977; children: Andrea J., Chad O. BS, Oreg. State U., 1977, MS, 1978; EdD, Wash. State U., 1988. Cert. safety profl., hazards control mgr., accident investigator. Grad. teaching asst. Oreg. State U., Corvallis, 1977-78; safety coord. Iowa Beef Processors, Inc., Pasco, Wash., 1978-80; fire and safety engr. Westinghouse Hanford Co., Richland, Wash., 1980-88; mgr. emergency preparedness Westinghouse Savannah River Co., Aiken, S.C., 1989-91; mgr. indsl. hygiene Westinghouse Hanford Co., Richland, Wash., 1991—; mem. steering com. trade emergency preparedness spl. interet group U.S. Dept. Energy, Washington, 1988-91; speaker nat. confs., 1981—. Instr. ARC, Kennewick, Wash., 1978-85, Am. Heart Assn., Richland, 1978-85; mem. Tri-City Tech. Coun., Richland, 1983-85; trustee Mid-Columbia Libr., Pasco, 1988. Recipient outstanding svc. award Oreg. Vocat. Indsl. Clubs Am., 1977, outstanding vol. award ARC, 1983. Mem. Am. Soc. Safety Engrs. (prof., sect. chmn. 1983-85, chpt. pres. 1989, regional oper. com. 1989-91). Republican. Roman Catholic. Avocations: woodworking, aikido, volleyball. Home: 220 Orchard Way Richland WA 99352-9659 Office: Westinghouse Hanford Co PO Box 1970 Richland WA 99352-0539

ZIMMERMAN, S(AMUEL) MORT(ON), electrical and electronics engineering executive; b. Paterson, N.J., Mar. 18, 1927; s. Solomon Zimmerman and Miriam (Feder) Glatzer; m. Marion Patricia Bogue, Sept. 15, 1951; children: Judy, Suzy, Sharon, Dan. Student, Ga. Inst. Tech., 1942-44, 46-48, Oglethorpe U., 1948-51; BSEE, Pacific Internat. U., L.A., 1958. Pres. Comml. Electronics Corp., Dallas, 1954-56, Electron Corp. subs. LTV Corp., 1956-65; chmn. bd., pres. Capital Bancshares, Inc., 1965-66; chmn. bd. Capital Nat. Bank Tampa (formerly Springs Nat. Bank), Fla., 1965, Capital Nat. Bank Miami (name now Peoples Downtown Bank), Fla., 1966, Merc. Nat. Bank Miami Beach (name now Barnett Bank), Fla., 1967, Underwriters Bank & Trust Co. N.Y. (name now Banco Cen.), 1968; chmn. bd., pres. Capital Gen. Corp., 1967, Comml. Tech., Inc., 1977—, Petro Imperial Corp. and subs. DOL Resources and Tech.-Star, Dallas, 1983—; chmn. bd., pres. Trans Exchange Corp., 1965—, Electric & Gas Tech., Inc., 1985—, also chmn. 3 subs. cos.; chmn. bd. Video Sci Tech., Inc., 1981-92, Interfederal Capital Inc., 1990—. Patentee TV camera video amplifier and blanking circuits, electronic thermometer, video x-ray image system and methods, video system and method for presentation and reproduction x-ray film images, electromagnetic radio frequency lighting system, laser display of electronically generated image signal. Petty oficer USN, 1942-45. Recipient Interfaith award City of N.Y. Mem. IEEE, Brookhaven Country Club. Republican. Jewish. Home: 5629 Meaders Ln Dallas TX 75229-6655 Office: Electric & Gas Tech Inc 13636 Neutron Rd Dallas TX 75244-4410

ZIMMERMAN, STANLEY WILLIAM, retired electrical engineering educator; b. Detroit, July 30, 1907; s. William Richard and Martha (Gebhardt) Z.; m. Evelyn Raney, Oct. 1, 1932 (dec. 1972); children: Dorothy Zimmerman Bynack, Jo Anne Zimmerman Busch, William S., Richard L. BS, MS, U. Mich., 1930. Registered profl. engr., Mass., N.Y. Elec. engr. Detroit Edison, 1929-30, GE, Schenectady, Pittsfield, N.Y., Mass, 1930-45; prof elec. engring. Cornell U., Ithaca, N.Y., 1945-72; cons. to numerous orgns. including Argone Nat. Labs., Lawrence Radiation Labs., Westinghouse, Victor Insulator and many others, 1945-72. Vol. for Boy Scouts, Libr., other civic projects, Ithaca, 1945-72. Achievements include development of measurement techniques microseconds and natural lighting; high voltage system protection; design of HV generation and transmission system systems nuclear rsch. Fellow IEEE; mem. N.Y. Soc. Profl. Engrs. Achievements include development of techniques to measure microseconds and natural ligtning; to reduce risks with high voltage; design of HV transformer and others that are confidential or proprietary. Home: 102 Valley Rd Ithaca NY 14850

ZIMMERMAN, STEVEN CHARLES, chemistry educator; b. Chgo., Oct. 8, 1957; s. Howard Elliot and Jane (Kirschenheiter) Z.; m. Sharon Shavitt, Aug. 5, 1990. BS, U. Wis., 1979; MA, Columbia U., MPhil, PhD, 1983. Asst. prof. chemistry U. Ill., Urbana, 1985-91, assoc. prof. chemistry, 1991—. Contbr. articles to profl. publs. Office: U Ill 1209 W California St Urbana IL 61801

ZIMMERMAN, STUART ODELL, biomathematician, educator; b. Chgo., July 27, 1935; s. O.C. and Matilda (Kribs) Z.; m. Mary Joan Spiegel, Aug. 8, 1959; 1 child, Kurt Zimmerman. BA, U. Chgo., 1954, PhD, 1964. From instr. to asst. prof. U. Chgo., 1964-66; assoc. prof. M.D. Anderson Tumor Inst. U. Tex., Houston, 1967-72, prof., chmn. M.D. Anderson Cancer Ctr., 1968—, Mattie A. Fair Rsch. chmn. Anderson Cancer Ctr., 1988—; mem. chmn. NIH, Biomed. Rsch. Tech. Program Study Sect., Washington, 1987-91. Author: (with others) Textbook of Dental Biochemistry, 1968, 2d edit., 1976; assoc. editor Jour. Computers in Biology and Medicine, 1970—; mem. pub. bd. Bull. Math. Biology, 1976-82; contbr. articles to profl. jours. Zoller fellow, 1960-64. Mem. AAAS, Am. Statistical Assn., Soc. Math. Biology, Assn. for Computing Machinery (treas. 1983-85, bd. dirs.), Forum Club Houston. Roman Catholic. Achievements include research in non-linear dynamics of populations; development of model describing how cyclins and their complexes regulate S&M phases of cell cycles, methods for matching human chromosome homologues and for characterization of isolated metaphase and prophase chromosome bands. Office: U Tex M D Anderson Cancer Ctr 1515 Holcombe Box 237 Houston TX 77030

ZIMMERMAN, WILLIAM JAMES, biologist, educator; b. Dallas, Tex., Nov. 10, 1952; s. Charles David and Kazuko (Takabayashi) Z. MS in Botany, Wash. State U., 1979; PhD in Biol. Scis., U. Mo., 1984. Adj. faculty Washington U., St. Louis, 1984-85; postdoctoral fellow J. Blaustein Desert Inst. Ben Gurion U., Sede Boqer, Israel, 1985-86; postdoctoral rsch. assoc. Wash. State U., Pullman, 1987-90; asst. prof. biology U. Mich., Dearborn, 1990—. Contbr. chpts. to books, articles to Plant and Soil, New Phytologist, Microbial Ecology. Mem. Amnesty Internat. Fulbright grantee, 1979, USAID rsch. grantee, 1989, H. Rackham faculty grantee U. Mich., 1991. Mem. Am. Soc. for Microbiology, Bot. Soc. Am., Phycological Soc. Am., Sigma Xi. Democrat. Baptist. Achievements include elucidation of possible synonymy among new world species of the aquatic fern Azolla Lam; accessional fingerprinting of genetic variation in nitrogen biofertilizers for use in rice fields, and in GSM or MIB-producing CBA. Home: 524 Tobin Dr Apt 312 Inkster MI 48141 Office: U Mich Dearborn Dept Natural Scis 4901 Evergreen Rd Dearborn MI 48128

ZIMMERMANN, (ARTHUR) GERHARD, chemist, educator, researcher; b. Markranstädt, Fed. Republic Germany, Aug. 7, 1930; s. Oswald Arthur and Clara Marie (Teichmann) Z.; m. Jütta Maria Haûsmann, June 7, 1957 (dec. June 1989); children: Evelyn Spindler, Elke Ebert. Diploma in chemistry, U. Leipzig, Fed. Republic Germany, 1954, D in Organic Chemistry, 1957. Diplomate chemistry. Researcher Leûna (Fed. Republic Germany)-Werke, 1954-58, mgr., 1959-66, head dept. rsch. and devel., 1967-73; prof. chemistry Tech. U. Mersebûrg, Fed. Republic Germany, 1969-88; head. dept. rsch. Acad. Scis. German Dem. Republic, Leipzig, 1974-91; lectr. chemistry U. Leipzig, 1963-69; cons. various indsl. and sci. orgns., Fed. Republic Germany, 1964—. Co-author: (handbook) Physical Methods of Organic Chemistry, 1963; (textbook) Industrial Chemistry, 1974; (monograph) Industrial Organic Chemistry, 1991; mem. editorial bd. several sci. jours.; contbr. articles to profl. jours.; patentee in field. Avocations: gardening, painting, wood carving. Home: Naunhofer Str 14, O 7027 Leipzig Germany Office: U Leipzig Inst Chem Tech, Permoserstr 15, O 7050 Leipzig Germany

ZIMMERMANN, HORST ERNST FRIEDRICH, economics educator; b. Krefeld, Germany, Mar. 11, 1934; s. Arthur and Elfriede Z.; m. Amrei Moehl, June 24, 1967; children: Roland, Anne, Lisa. Dr. rer. pol., U. Cologne, 1963; postgrad. student, U. Munich, 1957-58, Northwestern U., 1958-59; postdoctoral scholar, U. Pa., 1965-66. Asst. dept. pub. fin. U. Cologne, Fed. Republic Germany, 1960-65; rsch. scholar U. Cologne, 1965-68; prof. econs. Phillips U., Marburg, Fed. Republic of Germany, 1969—; counsel Fed. Agy. Regional Rsch. and Area Planning, 1977-82, Coun. Experts on Environ. Problems, 1981-90, Adv. Coun. to Fed. Fin. Ministry, 1986—; prof. econs. Philipps U., Marburg, Fed. Republic Germany, 1969—; vis. scholar Brookings Instn., Washington, 1972-73, Urban Inst., Washington, 1977—, Cambridge (Eng.) U., 1986; mem. Fed. Govt. Global Change Coun., 1992—. Author: Public Aid to Developing Countries, 1963, Public Expenditure and Regional Economic Development, 1970, Regional Preferences, 1973, Public Finance, 6th edit., 1990, Regional Incidence of Fiscal Flows, 1981, Studies in Comparative Federalism, 1981, Future of Government Finances, 1988. Recipient August Loesch prize for regional rsch., 1976. Mem. Acad. Regional Rsch. and Planning, Am. Econ. Assn., Soc. Econ. and Social Scis. Office: Am Plan 2, D-3550 Marburg 1, Germany

ZIMMERMANN, MICHAEL LOUIS, chemist; b. Arlington, Va., June 2, 1954; s. Emile L. and Jeannine (Simpson) Z.; m. Tammy Gray, Dec. 25, 1990; 1 child, Anya Katrin. BS in Biology and Chemistry, U. Va., 1975; MS in Biology, Va. Commonwealth U., 1983, PhD in Organic Chemistry, 1993. Lab. specialist II Med. Coll. Va., Richmond, 1976-80; assoc. chemist Philip Morris USA, Richmond, 1980-82, scientist, 1982-85, project leader, 1985—, rsch. scientist, 1985—. Mem. Maymont Found. Guild, Richmond, 1985—. K.C. scholar, 1972. Mem. Am. Chem. Soc., Am. Soc. Pharmacognosy, Am. Soc. Mass Spectroscopy, Guild Am. Luthiers, Am. Violin Soc. Home: 4041 Dorset Rd Richmond VA 23234 Office: Philip Morris PO Box 26583 Richmond VA 23261

ZIMMERMANN, WOLFGANG KARL, biologist, researcher; b. Heidelberg, Baden, Germany, Nov. 7, 1953; s. Erwin Karl and Augusta Liselotte (Kriegers) Z. Abitur, Bunsen Gymnasium, Heidelberg, Germany, 1974; Diplom, U. Heidelberg, Heidelberg, Germany, 1981, Dr. rer. nat., 1984. Researcher Fed. Biology Rsch. Inst. Agrl. and Forestry, Heidelberg, 1981-84; postdoctoral fellow U. Manchester, Inst. Sci. and Tech., Eng., 1985-87; rsch. group leader Inst. of Biotechnology, ETH-Zürich, Switzerland, 1988-93; prof. biotechnology Aalborg Univ., Denmark, 1993—; researcher Wash. State U., Pullman, 1982; lectr. Swiss Fed. Inst. Tech., Zürich, Switzerland, 1988-93; del. European Community, 1988-92. Contbr. articles to profl. jours.; patentee in field. German Acad. Exch. Orgn. grantee, 1982. Mem. Swiss Soc. Microbiology, Vereinigung für Allgemeine and Angewandte Mikrobiologie. Office: Aalborg University, Dept Civil Engineering, 9000 Aalborg Denmark

ZIMMIE, THOMAS FRANK, civil engineer, educator; b. Scranton, Pa., Jan. 24, 1939; s. Thomas and Stella Josephine (Price) Z.; m. Patricia Joyce Kelly, June 8, 1962 (div. 1979); 1 child, David Thomas; m. Judith Anne Braden, July 13, 1989. BSCE, Worcester Poly. Inst., 1960; MSCE, U. Conn., 1962, PhD in Geotech. Engring., 1972. Registered profl. engr., N.Y., Conn. Staff engr. Union Carbide Corp. (Linde div.), Buffalo, 1964-68; profl. engr. Town of Mansfield, Conn., 1968-72; ptnr. Wang and Zimmie Cons., Troy, N.Y., 1973-80; v.p. Arch Engring. Cons., Troy, 1984-88; program dir. NSF, Washington, 1988-90; pres., CEO Civrotech Cons. Engrs., Inc., Troy, 1993—; mem. faculty dept. civil engring Rensselaer Poly. Inst., Troy, 1973—; postdoctoral researcher Norwegian Geotech. Inst., Oslo, 1972-73; geotech. engr. N.Y. Dept. Environ. Conservation, Albany, 1983-85; town engr. Town of North Greenbush, N.Y., 1985-88. Editor: Permeability and Groundwater Contamination, 1981. 1st lt. U.S. Army, 1962-64. Mem. ASCE (Outstanding Svc. award 1986, 87), ASTM (Spl. Svc. award 1980, Charles Dudley award 1984), Transp. Rsch. Bd., Am. Rd. and Transp. Builders Assn. Achievements include research in environmental geotechnology. Home: 39 Zelenke Dr Wynantskill NY 12198 Office: Civil/Environ Engring Dept Soil Mechanics Lab Rensselaer Poly Inst Troy NY 12180

ZIMNISKI, STEPHEN JOSEPH, biologist, educator; b. Biddeford, Maine, Oct. 9, 1948; s. Victor Joseph and Mary (Consalvo) Z.; 1 child, Kristen Anne. BS in Biology, U. Maine, 1970; MS in Zoology, U. Mo., 1973; PhD in Physiology, Boston U., 1981. Instr. U. Maine, Portland, 1975-78, Northeastern U., Boston, 1979-81; postdoctoral fellow Vanderbilt U., Nashville, Tenn., 1981-83; from asst. to assoc. prof. biochemistry U. Miami, Fla., 1983-90; assoc. prof. ob.-gyn. U. Kans., Wichita, 1990—. Contbr. articles to profl. jours. Mem. Am. Assn. Cancer Rsch., Am. Soc. for Biol. Chemistry and Molecular Biology, Endocrine Soc., Soc. for Study Reproduction, AAAS. Achievements include research in induction of hormone independent breast tumors by antiestrogen tamoxifen, dissociation of estrogen production from substrate or enzyme concentrations. Office: Womens Rsch Inst 2903 E Central Wichita KS 67214

ZIMOLONG, BERNHARD MICHAEL, psychologist, educator; b. Breslau, Germany, Apr. 26, 1944; s. Hans Joachim and Hiltraud (John) Z.; m. Ursula Eva-Maria Herbst, Aug. 5, 1966; 1 child, Andreas. Diploma, U. Munster, Germany, 1970; PhD, U. Braunschweig, Germany, 1974, habilitation, 1981. Asst. prof. U. Braunschweig, 1972-82; prof. U. Bochum, Germany, 1984—; speaker Spl. Rsch. Ctr. 187, 1992—; vis. prof. Purdue U., West-Lafayette, Ind., 1983-84, Decision Rsch., Eugene, Oreg., 1988-89. Author books on engring. psychology, human reliability, safety mgmt. and occupational safety and prevention; contbr. articles to profl. jours. Lt. German Army, 1964-66. Grantee Heisenberg Deutsche Forschungsgemeinschaft, 1982-84. Mem. Human Factors Soc., Berufsverband Deutscher Psychologen, Deutsche Gesellschaft Psychologie. Office: Ruhr U Bochum / Psychology, 4630 Bochum Germany

ZINBERG, STANLEY, physician, educator; b. N.Y.C., Aug. 18, 1934; s. Phillip M. and Etta (Beck) Z.; m. Margaret T. McNally; children: Lloyd M., Randi Ellen, Gregory A. BA, Columbia Coll., 1955; MD, SUNY, 1959; MS, NYU, 1990. Diplomate Am. Bd. Obstetrics and Gynecology. Intern Cornell Med. div. Bellevue Hosp., N.Y.C., 1959-60; resident in ob-gyn. NYU Bellevue Med. Ctr., N.Y.C., 1960-64; assoc. prof. ob-gyn. NYU Sch. Medicine, N.Y.C., 1966—; chief gynecology Bellevue Hosp., N.Y.C., 1975-81; chief ob-gyn. N.Y. Downtown Hosp., N.Y.C., 1981—; mem. staff NYU Hosp.; examiner Am. Bd. Ob-gyn., 1976—; mem. Residency Rev. Com. for Ob-gyn., 1987-92; chmn. faculty coun. NYU Sch. Medicine, N.Y.C., 1978-79; pres. med. staff N.Y. Downtown Hosp., 1991-92. Contbr. articles to profl. jours. Capt. U.S. Army, 1964-66. Fellow Am. Coll. Obstetricians and Gynecologists (Manhattan sect. chmn. 1979-82), N.Y. Obstet. Soc. (pres. 1989-90), N.Y. Acad. Medicine (chmn. sect. on ob-gyn. 1985-86), N.Y. Gynecol. Soc.; mem. Assn. Profs. of Gynecology and Obstetrics (affiliate), Soc. Alumni Bellevue Hosp., Bellevue Obstet. and Gynecol. Soc. (pres. 1988-92), N.Y. State Bd. for Profl. Med. Conduct, 1993. Avocations: painting, photography, tennis. Home: 1365 York Ave 33K New York NY 10021 Office: NY Downtown Hosp 170 William St New York NY 10038

ZINCK, BARBARA BAREIS, chemist; b. Bryn Mawr, Pa., Feb. 10, 1955; d. Herman Frederick and Barbara Jean (Brenner) Bareis; m. Gerald Thomas Zinck, Jan. 19, 1980; children: Jennifer Jean, Christian Frederick. BS in Chemistry, Muhlenberg Coll., 1977. Quality control chemist Wyeth Labs., West Chester, Pa., 1978-80; analytical rsch. chemist Wyeth Labs., Radnor, Pa., 1980-85; quality control supr. Noramco (Johnson & Johnson), Wilmington, Del., 1985-86; stability supr. ICI Pharm. Group, Newark, Del., 1986-89, IMS supr., 1989-91; sr. analytical chemist ICI Pharm. Group, Wilmington, 1991-92; quality assurance/compliance mgr. Zeneca Pharm. Group, Wilmington, 1992—. Sec. Chesapeake Isle Civic Assn., North East, Md., 1987-88, v.p., 1988-89, pres., 1989-90; chair worship com. North East United Meth. Ch., 1992—. Recipient Outstanding Performance award Dale Carnegie, 1990. Mem. Am. Chem. Soc., Am. Soc. for Quality Control, Chromatography Forum of Delaware Valley, Am. Assn. Pharm. Scientists. Home: Chesapeake Isle 144 Rolling Ave North East MD 21901 Office: Zeneca Pham Group 1800 Concord Pk Wilmington DE 19897

ZINGARELLI, ROBERT ALAN, research physicist; b. Apalachicola, Fla., Oct. 25, 1962; s. Genaro Angelo and Allie (Weinel) Z.; m. Jennifer George Zingarelli, Feb. 14, 1988; children: Maxwell Duncan, Veronica Beryl; 1 stepchild, Alex Bliss. BS, Carnegie-Mellon U., 1984; PhD, Fla. State U., 1990. Rsch. physicist Naval Rsch. Lab., Bay St. Louis, Miss., 1990—; interviewer Carnegie-Mellon Admissions Coun., Pitts., 1991—. Contbr. articles to profl. jours. Mem. Acoustic Soc. Am., Young Sci. Network. Republican. Office: Naval Rsch Lab Code 7181 Bay Saint Louis MS 39529

ZINKAND, WILLIAM COLLIER, neuropharmacologist; b. Hagerstown, Md., Mar. 11, 1959; s. Marshall William and Dorothy Mabel (Newey) Z.; m. Dorothy Elizabeth Crabb, Aug. 14, 1982; children: Andrew William (dec.), Benjamin Jack. BS, Va. Poly. Inst. and State U., 1981. Rsch. assoc. dept. psychology U. Md., Catonsville, 1981-84; rsch. assoc. dept. physiology U. Md., Balt., 1984-88; jr. pharmacologist ICI Americas, Wilmington, Del., 1988-90, pharmacologist II, 1990-92, sr. pharmacologist, 1992—. Contbr. articles to profl. jours. Pres. Cecil County Men's Shelter, Inc., Elkton, Md., 1992; trustee West Nottingham Presbyn. Ch., Colora, Md., 1990—. Mem. AAAS, Soc. Neurosci. (pres. Del. area chpt. 1992—, sec.-treas. 1993), Mid Atlantic Pharmacol. Soc., Sigma Xi. Republican. Home: 281 Connelly Rd Rising Sun MD 21911 Office: ICI Pharm Group LW-211 Wilmington DE 19897

ZINKERNAGEL, ROLF MARTIN, immunologist educator; b. Basle, Switzerland, Jan. 6, 1944; s. Robert W. and Suzanne (Staehlin) Z.; m. Kathrin G. Lüdin, Mar. 11, 1968; children: Christine, Annelies, Martin. MD, U. Basel, 1968. Intern in surgery Claraspital, Basel, 1968-69; postdoctoral Inst. Biochemistry, Lausanne, 1970-72, Dept. Microbiology, ANU, Canberra, Australia, 1973-75; asst. prof. Dept. Immunopathology, Scripps U., La Jolla, Calif., 1975-80, mem., 1978-79; assoc. prof. Dept. Pathology, Div. Exptl. Pathology, U. Zurich, 1979-92; full prof. Dept. Pathology, Inst. Exptl. Immunology, U. Zurich, 1992—. Editorial bd. Exptl. Cell Biology, 1976-88, Immunogenetics, 1977—, Parasite Immunology, 1978-84, Jour. of Immunology, 1978-80, Thymus, 1979-89, Antiviral Rsch., 1980-88, Jour. of Exptl. Medicine, 1981-84, Cellular Immunology, 1983—, European Jour. of Immunology, 1981—, Jour. of Environ. Pathology Toxicology and Oncology, 1981—, Internat. Jour. of Microbiology, 1983—, and others. Mem. Swiss Soc. of Allergy and Immunology, Australian Soc. for Immunology, Am. Assn. of Immunoloists, Am. Assn. of Pathologists, Scandinavian Soc. of Immunology (hon.), Soc. Fracaise d'Immunolgie (hon.), Swiss Soc. of Pathology, Swiss Soc. of Microbiology, Swiss Soc. of Cell and Molecular Biology, Acadmia Euopea, Internat. Soc. for Antiviral Rsch., ENI European Network of Immunol. Instns., Deutsche Gesellschaft fur Immunologie, Deutsche Gesellschaft fur Virologie, others. Achievements include co-discovery of MHC-restricted T cell recognition; discovery of the tymus role in determining MHC-restricted T-cell specficity, NK-cell activity in virus infections, T-cell epitope escape virus mutants, tolerances to viruses; rsch. on role of virus-specific T-cells in causing immunopathology. Office: Dept Pathology Inst, Exptl Immunology, U Hosp, 8091 Zurich Switzerland

ZINKLE, STEVEN JOHN, engineer, researcher; b. Prairie du Chien, Wis., Nov. 5, 1958; s. Aloysius Peter and Katherine Edith (Brownlee) Z.; m. Teresa Allen Medford, May 26, 1990; 1 child, Austin Chase. BS, U. Wis., 1980, PhD, 1985. Rsch. staff Oak Ridge (Tenn.) Nat. Lab., 1985—; vis. scientist Forschungszentrum Jülich, Germany, 1991-92, Risø Nat. Lab., Roskilde, 1991-92. Recipient Rsch. Publ. award Martin Marietta Energy Systems, Oak Ridge, 1991, David Rose Excellence in Fusion Engring. award Fusion Power Assocs., Gaithersburg, Md., 1992. Mem. Am. Soc. Metals Internat., Am. Ceramic Soc., Sigma Xi, Phi Kappa Phi. Office: Oak Ridge Nat Lab PO Box 2008 Oak Ridge TN 37831-6376

ZINN, BEN T., engineer, educator, consultant; b. Tel Aviv, Apr. 21, 1937; came to U.S., 1957; s. Samuel and Fridah (Gelbfish) Cynowicz; children: Edward R., Leslie H. B.S. in Mech. Engring. cum laude, NYU, 1961; M.S. in Mech. Engring., Stanford U., 1962; M.S. in Aerospace Engring., Princeton U., 1963, Ph.D. in Aerospace Engring. and Mech. Scis., 1965. Asst. research Princeton U., 1964-65; asst. prof. Ga. Inst. Tech., Atlanta, 1965-67, assoc. prof., 1967-70, prof., 1970-73, Regents prof., 1973—, Disting. prof., 1990, Davis S. Lewis, Jr. chair Sch. Aerospace Engring., 1992—; research scientist research div. Am. Standard Co., New Brunswick, N.J., summer 1976; cons. Brasilian Space Research Inst., Sao Jose dos Campos, Brazil, Aetna Casualty & Sr. Co., Atlanta. Recipient David Orr Mech. Engring. prize NYU, 1961; recipient Founder's Day NYU, 1961, Cert. of Recognition NASA, 1974; Ford fellow, 1962-63. Fellow AIAA (assoc. editor jour. 1982—), Combustion Inst. (past. pres. Eastern sect.), Nat. Fire Acad. (bd. visitors 1979-82), Am. Tech. Soc. (v.p. Atlanta chpt. 1980-84, pres. 1984-86), Sigma Xi (rsch. award 1969, sustained rsch. award 1976), Tau Beta Pi, Pi Tau Sigma. Office: Ga Inst Tech Aerospace Engring Atlanta GA 30332

ZINN, MICHAEL WALLACE, aerospace engineer; b. Washington, Dec. 30, 1962; s. Wallace Bernard and Frances E. AA, Charles County C.C., La Plata, Md., 1983; BS, Tri-State U., Angola, Ind., 1986. Coop student Naval Ordnance Sta., Indian Head, Md., 1980-86; mine decoy engr. Naval Ordnance Sta., Indian Head, 1986-87, airbreathing propulsion engr., 1987-92; airbreathing propulsion engr. Naval Surface Warfare Ctr., Indian Head, 1992—; mem. Joint Army-Navy-NASA-Air Force airbreathing com., expendable engine subcom., Laurel, Md., 1987—. Author several tech. papers for AIAA and JANNAF. Pres. Port Tobacco Players, Inc., La Plata, Md., 1992-93. Mem. AIAA, Am. Def. Preparedness Assn. Achievements include work on aging surveillance programs for expendable gas turbine engines, on aging properties of expendable engines and solid propellant gas generators. Home: 4 Somerset St La Plata MD 20646

ZINNER, LÉA BARBIERI, inorganic chemistry educator, researcher; b. São Paulo, Brazil, Apr. 2, 1945; d. Carlo and Maria Eunice (Lascala) Barbieri; m. Klaus Zinner, Sept. 5, 1968; children: Claudia Renate, Martina. Grad., U. São Paulo, 1967, PhD, 1970. Asst. U. São Paulo, 1968-70, pvt. docent, 1978-84, assoc. prof., 1984-91, full prof., 1991—; vis. prof. U. Puerto Rico, 1978; vis. researcher U. Würzburg, Fed. Republic of Germany, 1991. Editor book, 1990; contbr. articles to profl. jours. Postdoctoral fellow Wayne State U., Detroit, 1976. Mem. Am. Chem. Soc., N.Y. Acad. Sci., Acad. Sci., Sigma Xi. Achievements include Brazilian patent; research in lanthanide chemistry, including new compounds and applications. Office: Univ of São Paulo, Inst of Chem CP 20780, 01498-970 São Paulo Brazil

ZIPF, WILLIAM BYRON, pediatric endocrinologist, educator; b. Dayton, Ohio, Mar. 20, 1946; s. Robert Eugene and Merium (Murr) Z.; m. Joanne Fisher, Sept. 20, 1969; children: William Byron Jr., Thanda Lynn, Robert E. II. BA, Denison U., 1968; MD, Ohio State U., 1972. Diplomate Nat. Bd. Med. Examiners, Am. Bd. Pediatrics, Am. Bd. Pediatric Endocrinology. Intern in pediatrics Mott Children's Hosp./U. Mich., Ann Arbor, 1972-73, resident in pediatrics, 1973-75, clin. fellow in pediatric endocrinology, 1975-76, rsch. fellow, 1976-78; asst. prof. dept. pediatrics and physiology Ohio State U., Columbus, 1978-83, assoc. prof., 1983-89, prof., 1989—; dir. clin. study ctr. Children's Hosp./Ohio State U., Columbus, 1982—, vice-chmn. dept. pediatrics, 1989—, dir. pediatric endocrinology, 1990—. Contbr. chpts. on endocrine diseases of children to books, articles to profl. jours. Grantee NIH, 1980-84, Cystic Fibrosis Found., 1987-92. Fellow Am. Acad. Pediatrics, Nat. Med. Bd.; mem. Soc. Pediatric Rsch., Endocrine Soc., Lawson Wilkins Soc. Pediatric Endocrinolgoy. Achievements include discovery of endocrine abnormalities associated with conditions of altered nutrition in children; definition of role of pancreatic peptide abnormalities in specific appetite disorders. Office: Childrens Hosp 700 Childrens Dr Columbus OH 43205-2696

ZIRIN, HAROLD, astronomer, educator; b. Boston, Oct. 7, 1929; s. Jack and Anna (Buchwalter) Z.; m. Mary Noble Fleming, Apr. 20, 1957; children: Daniel Meyer, Dana Mary. A.B., Harvard U., 1950, A.M., 1951, Ph.D., 1952. Asst. phys. scientist RAND Corp., 1952-53; lectr. Harvard, 1953-55; research staff High Altitude Obs., Boulder, Colo., 1955-64; prof. astrophysics

Calif. Inst. Tech., 1964—; staff mem. Hale Obs., 1964-80; chief astronomer Big Bear Solar Obs., 1969-80, dir., 1980—; U.S.- USSR exchange scientist, 1960-61; vis. prof. Coll. de France, 1986, Japan Soc. P. Sci., 1992. Author: The Solar Atmosphere, 1966, Astrophysics of the Sun, 1987; co-translator: Five Billion Vodka Bottles to the Moon, 1991; adv. editor: Soviet Astronomy, 1965-69; editor Magnetic and Velocity Fields of Solar Active Regions. Trustee Polique Canyon Assn., 1977-90. Agassiz fellow, 1951-52; Sloan fellow, 1958-60; Guggenheim fellow, 1960-61. Mem. Am. Astron. Soc., Internat. Astron. Union, AURA (dir. 1977-83). Home: 1178 Sonoma Dr Altadena CA 91001-3150 Office: Calif Inst Tech Big Bear Solar Observatory 40386 N Shore Ln Big Bear City CA 92314

ZIRKIND, RALPH, physicist; b. N.Y.C., Oct. 20, 1918; s. Isaac and Zicel (Lifshitz) Z.; m. Ann Goldman, Nov. 22, 1940; children: Sheila Zirkind Knopf, Elaine Zirkind Gorman, Edward I. B.S., CCNY, 1940; M.S., Ill. Inst. Tech., 1945; postgrad., George Washington U., 1946-47; Ph.D., U. Md., 1950; D.Sc., U. R.I., 1968. Physicist Navy Dept., 1945-50, chief physicist, 1951-60; physicist Oak Ridge Nat. Lab., 1950-51, Advanced Research Project Agy., Washington, 1960-63; prof. Poly. Inst. Bklyn., 1963-70, U. R.I., Kingston, 1970-72; adj. prof. U. R.I., 1972—; physicist Advanced Research Projects Agy., Arlington, Va., 1972-74; cons. Advanced Rsch. Projects Agy., and industry, Arlington, 1974—; lectr. U. Md., 1948-50, George Washington U., 1952-53, U. Mich., 1966; cons. ACDA, Jet Propulsion Lab., Calif. Inst. Tech. Contbg. author: Jet Propulsion Series, 1952, FAR Infrared Properties of Materials, 1968; editor: Electromagnetic Sensing of Earth, 1967; mem. editorial bd.: Infrared Physics, 1963—; contbr. articles profl. jours. Recipient Meritorious Civilian Service award Navy Dept., 1957, Meritorious Civilian Service award Dept. Def., 1970. Mem. Am. Phys. Soc., N.Y. Acad. Scis., Sigma Xi, Sigma Pi Sigma, Eta Kappa Nu. Home: 820 Hillsboro Dr Silver Spring MD 20902-3202 Office: 4001 Fairfax Dr Ste 700 Arlington VA 22203-1618

ZIRPS, FOTENA ANATOLIA, psychologist, researcher; b. Pitts., Mar. 27, 1958; d. George T. and Barbara F. (Skinner) Z. BA, U. Akron, 1983, MA, 1987; PhD, Fla. State U., 1990. Sch. psychologist Canton (Ohio) City Schs., 1985-86; sch. psychologist Leon County Schs., Tallahassee, 1986-88, program evaluator, 1988-90; cons. Evaluation Systems Design, Inc., Tallahassee, 1990-91; pres. Zirps, Vella and Assocs., Inc., Tallahassee, 1991—; dir. program evaluation Families First, Atlanta, 1991—; tchr. Fla. State, Tallahassee, summers 1988, 89, 90, 91, grant coord., 1989-90. Author: Sun and Moon, 1991, (with others) Computer Models of Reading, 1989; author, cartoonist: (slide show/audio tape) Human Rights, 1986. Chmn. grad. student adv. com. Fla. State, 1986-88. Mem. Am. Psychol. Assn., Am. Evaluation Assn., Am. Ednl. Rsch. Assn. (Disting. Presenter 1991), Nat. Coun. Rsch. in Child Welfare, Fla. Ednl. Rsch. Assn. (Disting. Author 1990). Quaker. Avocations: running, racquetball, tennis, reading. Office: Families First 1105 W Peachtree St NW Atlanta GA 30309-3695

ZIRVI, KARIMULLAH ABD, experimental oncologist, educator; b. Lahore, Punjab, Pakistan, May 20, 1940; s. Khudabakhash Abd and Amatul Karim Zirvi; m. Amatul Latif Shaukat, Sept. 26, 1966; children: Nasir, Sumra, Monib, Khalid. BSc, Punjab U., Lahore, 1959, BEd, 1961; MSc, Karachi (Pakistan) U., 1963; PhD, U. Louisville, 1968. Sr. rsch. officer Pakistan Coun. Sci. and Indsl. Rsch. Labs., Peshawar, Pakistan, 1969-72; assoc. prof. Pahlavi U., Shiraz, Iran, 1972-78; vis. scientist U. Louisville, 1978-79; rsch. assoc. U. Calif., San Diego, 1979-81, Vanderbilt U., Nashville, 1981-82; asst. prof. UMDNJ-N.J. Med. Sch., Newark, 1982-89; assoc. prof. U. Medicine Dentistry N.J.-N.J. Med. Sch., Newark, 1989—; rsch. tng. fellow Internat. Cancer Rsch. Union, Heidelberg, Fed. Republic Germany, 1978. Contbr. over 65 rsch. articles to profl. publs. Fulbright-Hays fellow, 1964; grantee NIH, 1984, N.J. Cancer Commn., 1988, Hazardous Substance Mgmt. Rsch., 1989. Mem. AAAS, Am. Assn. for Cancer Rsch., Metastasis Rsch. Soc., Ahmadiyya Muslim Med. Assn. (sec. 1986—). Achievements include patent for N-substituted cyclobutane-carboxamides, as muscle relaxant with lessened side effects. Home: 14-21 Saddle River Rd Fair Lawn NJ 07410 Office: UMDNJ-NJ Dental Sch 110 Bergen St Newark NJ 07103

ZISA, DAVID ANTHONY, pharmaceutical researcher; b. N.Y.C., July 1, 1959; s. Robert John and Marilyn Eugenia (Pappert) Z.; m. Diane M. LoTempio, Sept. 21, 1981; children; Sabrina M., David C., Christopher A., Stephanie M. BA in Chemistry and Biology, SUNY, Buffalo, 1981; MS in Pharmaceutics, Bklyn. Coll. Pharmacy, 1992. Analyt. chemist Lederle Labs., Pearl River, N.Y., 1981-83, devel. chemist, 1983-85, sr. rsch. pharmacist, 1985-90; sr. registration assoc. Am. Cyanamid Med. Rsch., Pearl River, N.Y., 1990—; faculty lectr. Land of Lakes conf. U. Wis., Madison, 1992—. Mem. Am. Assn. Pharm. Scientist, Parenteral Drug Assn. Republican. Roman Catholic. Achievements include development and characterization of an model vacuum fluid bed processor, analysis of biologics via plasma emission spectroscopy. Office: Am Cyanamid Med Rsch Div Bldg 140 Rm 327 N Middletown Rd Pearl River NY 10965

ZISKIN, JAY HERSELL, psychologist; b. Mpls., Apr. 27, 1920; s. Cyrus E. and Sally (Sugarman) Z.; m. Mae Lee Billet, July 31, 1955; children: Kenneth, Laura, Nina, Randolph. LLB, U. So. Calif., L.A., 1946, PhD, 1962. Pvt. practice psychology L.A., 1968—. Co-author: Coping with Psychiatric and sychological Testimony, 4th edit. 1988, Brain Damage Claims, 1992. Fellow APA, Am Psychol. Soc.; mem. Am. Psychology-Law Soc. (1st pres. 1970-71). Office: PO Box 24219 Los Angeles CA 90024

ZISLIS, PAUL MARTIN, software engineering executive; b. Chgo., Feb. 8, 1948; s. Harold Solomon and Beatrice (Bossen) Z.; m. Sharon Margo Kaufmann, June 8, 1969; children: Daniel, Benjamin, Rachel. BS in Computer Sci., U. Ill., 1969; SM in Info. Sci., U. Chgo., 1971; PhD in Computer Sci., Purdue U., 1974. Mem. tech. staff AT&T Bell Labs., Naperville, Ill., 1969-72, 74-77; supr. data network devel. AT&T Bell Labs., Holmdel, N.J., 1977-81, dept. head data network architecture, 1981-82; dept. head advanced software tech. AT&T Bell Labs., Naperville, 1982-90; dir. software engring. Raynet Corp., Menlo Park, Calif., 1990-92, dir. product validation, 1993—. Contbr. articles to profl. jours. Grad. fellow IBM, 1971-72. Mem. IEEE Computer Soc., Assn. for Computing Machinery, Phi Beta Kappa. Achievements include two patents for telecommunications software. Office: Raynet Corp 155 Constitution Dr Menlo Park CA 94025-1106

ZLATKIS, ALBERT, chemistry educator; b. Pomorzany, Poland, Mar. 27, 1924; came to Can. 1927, U.S., 1949, naturalized, 1959; s. Louis and Zisel (Nable) Z.; m. Esther Shessel, June 15, 1947; children: Lori, Robert, Debra. B.A.Sc., U. Toronto, 1947, M.A.Sc., 1948; Ph.D., Wayne State U., 1952. Rsch. chemist Shell Oil Co., Houston, 1953-55; asst. prof. chemistry U. Houston, 1955-58, assoc. prof., 1958-63, chmn. chemistry dept., 1958-62, prof. chemistry, 1963—; adj. prof. chemistry Baylor Coll. Medicine, 1975—; tour speaker Am. Chem. Soc., 1961, 63, 66, South Africa Chem. Inst., 1971, USSR Acad. Scis., 1973, Polish Acad. Scis., 1977; chmn. Internat. Symposium on Advances in Chromatography, 1963—. Author: Practice of Gas Chromatography, 1967, Preparative Gas Chromatography, 1971, Advances in Chromatography, 25 vols., 1963—, A Concise Introduction to Organic Chemistry, 1973, High Performance Thin-Layer Chromatography, 1977, 75 Years of Chromatography—A Historical Dialogue, 1979, Instrumental HPTLC, 1980, Electron Capture—Theory and Practice in Chromatography, 1981; contbr. articles to profl. jours.; mem. editorial bd. Jour. High Resolution Chromatography, Jour. Chromatograpic Science, Chromatographia, Jour. Chromatography. Recipient Analyst of Yr. award Dallas Soc. Analytical Chemistry, 1977, Tswett Meml. medal USSR Acad. Scis., 1980, Tswett Chromatography medal, 1983, NASA Tech. award, 1975, 80, NASA Patent award, 1978, Disting. Tex. Scientist award Tex. Acad. Sci., 1985, James L. Waters Pioneers in Devel. Analytical Instrumentation award, 1990; grantee USPHS, 1966, NASA, 1964-76, Welch Found., 1969, AEC, 1967, NSF, 1978, EPA, 1979, U.S. Army Rsch. Office, 1982, Gulf Coast Hazardous Substance Rsch. Ctr., 1989, Tex. Advanced Rsch. Program, 1991. Mem. Am. Chem. Soc. (chmn. S.E. Tex. sect. 1964, award in chromatography 1973, S.W. regional award 1988), Groupement pour l'Avancement des Méthodes Spectrographiques (France), Sigma Xi, Phi Lambda Upsilon. Home: 22 Sandalwood Dr Houston TX 77024-7122

ZLOTNIK, ALBERT, immunologist, researcher; b. Mexico City, Nov. 29, 1954; came to U.S., 1977; s. Abraham Y. and Aurora (Espinosa) Z.; m. Linda M. Hollerung, Feb. 20, 1982; children: Michael, Alexander, Sara. MS, U. Colo., 1980, PhD in Immunology, 1981. Staff scientist DNAX Rsch. Inst., Palo Alto, Calif., 1984-89, sr. staff scientist, 1989—. Assoc. editor Jour. Immunology, 1987-91, Lymphokine and Cytokine Rsch., 1992—. Mem. AAAS, Am. Assn. Immunologists, Soc. Leukocyte Biology. Office: DNAX Rsch Inst 901 California Ave Palo Alto CA 94304

ZMEU, BOGDAN, structural engineer, consultant; b. Bucharest, Romania, Dec. 6, 1956; came to U.S., 1981; s. Ioan and Irina (Papadisea) Z.; m. Cristina Virginia Orasanu, Aug. 20, 1983. B of Engring., Cooper Union, 1986, M of Engring., 1987. Registered profl. engr., N.Y. Structural engr. Frederic R. Harris, Inc., N.Y.C., 1987—. Mem. Chi Epsilon, Tau Beta Pi. Office: Frederic R Harris Inc 300 E 42d St New York NY 10017

ZMUDZKA, BARBARA ZOFIA, researcher; b. Cracow, Poland, Apr. 1, 1934; came to U.S., 1981; d. Roman Lugowski and Wanda (Tyszko) Lugowska; m. Andrzej Zmudzki, Apr. 17, 1962 (div. 1992); children: Marcin, Anna Maria. MSc, U. Warsaw, Poland, 1956; PhD, State Inst. Hygiene, Warsaw, 1965; DSc, Polish Acad. Sci., Warsaw, 1974. Asst. State Inst. Hygiene, 1956-65; assoc. prof. Inst. Oncology, Warsaw, 1965-81; vis. scientist Uniformed Svcs. U. Health Sci., Bethesda, Md., 1981-83, Nat. Cancer Inst., NIH, Bethesda, 1983-90; rsch. scientist ORAU program FDA, Rockville, Md., 1990—. Contbr. over 50 articles to profl. publs. WHO grantee, 1967; recipient Sci. Achievement award Polish Acad. Sci., 1972, 81, Outstanding Basic Rsch. award Polish Ministry of Health, 1973. Achievements include studies on polynucleotides, on activation of HIV by radiation, cloning and molecular studies on mammalian DNA polymerase beta. Home: 12112 Portree Dr Rockville MD 20852 Office: ORAU program CDRH FDA 12709 Twinbrook Pkwy Rockville MD 20852

ZOBERI, NADIM BIN-ASAD, management consultant; b. Karachi, Pakistan, July 20, 1951; came to U.S., 1973; s. Asad Ahmad and Nawab Bano Zoberi; m. Samira Khalid, Mar. 24, 1989; 1 child, Noor Jehan. BS in Math., Physics and Chemistry, U. Karachi, 1971; B in Computer Sci., U. Wis., River Falls, 1981, BBA, 1980. Indsl., project engr. ADC Telecommunications, Mpls., 1979-84, supr. prodn. and inventory control, 1984-87; cons. Coopers & Lybrand, Mpls., 1988-89; dir. mfg. Daig Corp., Minnetonka, Minn., 1989-90; dir. quality internat. op. N.W. Airlines, St. Paul, 1990-92; mgmt. cons. KPMG Peat Marwick, Mpls., 1992—. Exec. advisor Jr. Achievement, Mpls., 1985-87. Mem. Assn. Mfg. Excellence, Inst. of Indsl. Engrs. Muslim. Avocations: tennis, travel, reading, music. Home: 4126 Meadowlark Way Saint Paul MN 55122-1779 Office: KPMG Peat Marwick 4200 Norwest Ctr 90 S 7th St Minneapolis MN 55402

ZOGHBI, SAMI SPIRIDON, radiochemist; b. Beirut, Lebanon, May 29, 1948; came to U.S., 1969; s. Spiridon Simon and Tamam (Abou-Rjeily) Z.; m. Lena Fouad Chaghoury, Aug. 24, 1971; children: Jennifer, Benjamin. BA, Anderson U., 1971; MS, Ball State U., 1974; PhD, Purdue U., 1978. Rsch. assoc. Yale U. Sch. Medicine, New Haven, 1978-84, asst. prof., 1984-90, rsch. scientist in nuclear medicine, 1990—, dir. radiochem. rsch. lab., 1981—, mem. radiation safety com., 1988—; mem. radioactive drug rsch. com. Yale New Haven Hosp., 1986—, mem. med. staff, 1982—. Mem. Sigma Xi (life). Home: 195 Colony Rd New Haven CT 06511 Office: Yale U Sch Medicine 333 Cedar St New Haven CT 06510

ZOHOURIAN MASHMOUL, MOHAMMAD JALAL-OD-DIN, chemistry educator; b. Mashhad, Khorassan, Iran, Jan. 26, 1965; s. Gholamreza Zohourian Mashmoul and Bibi Marzieh Khadembashi. BS in Teaching Chemistry, Tchrs. Tng. U., Tehran, Iran, 1987; MS in Chemistry, Sharif U. Tech., Tehran, 1990. Instr. chemistry Mil. Acad., Tehran; master rschr. in RIPI, Nat. Iranian Oil Co., Tehran, 1992—. Translator: Preparative Methods of Polymer Chemistry, 1992. 1st lt. Iran Inf., 1990—. Mem. Am. Chem. Soc., Iranian Chem. and Engring. Soc. Achievements include research in linear aliphatic unsaturated polyesters, drug reducer agents for the crude oil pipelines, engineering polyacetals. Office: Nat Irainian Oil Co Rsch Inst Petroleum Industry, Polymer Dept PO Box 1836, Tehran 11365-95, Iran

ZOIDIS, ANN MARGARET, biologist, researcher; b. Bronx., N.Y., Feb. 19, 1962; d. John and Margaret (Barbarto) Z. BA, Smith Coll., 1983; MS, San Francisco State U., 1989. Rschr. dept. geology Smith Coll., Northampton, Mass., 1980-83; rschr. Quest N.W., Seattle, 1981-83, San Francisco Zoo, 1986-89; teaching asst. San Francisco State U., 1986-89; fundraiser/devel. Allied Whale, Bar Harbor, Maine, 1992, rsch. biologist, 1992—. Recipient Rsch. scholarship Scholarships Found. Inc., 1989, Biol. scholarship San Francisco State U., 1988, Youth Activity Fund grant Explorers Club, 1982. Mem. AAAS, Am. Cetacean Soc., Sigma Xi (rsch. grant 1988). Office: Allied Whale 105 Eden St Bar Harbor ME 04607

ZOIS, CONSTANTINE NICHOLAS ATHANASIOS, meteorology educator; b. Newark, Feb. 21, 1938; s. Athanasios Konstantinos and Asimina (Speros-Blekas) Z.; m. Elyse Stein, Dec. 26, 1971; children: Jennifer, Jonathan. BA, Rutgers U., 1961; MS, Fla. State U., 1965; PhD, Rutgers U., 1980. Draftsman Babcock and Wilcox Corp., Newark, 1956; designer Foster Wheeler Corp., Carteret, N.J., 1956; instr. Rutgers U., New Brunswick, N.J., 1961-62; grad. asst. Fla. State U., Tallahassee, 1962-65; rsch. meteorologist Nat. Weather Svc., Garden City, L.I., N.Y., 1965-67; prof. Kean Coll. N.J., Union, 1967—; founder meteorology program Kean Coll. N.J.; cons. Connell, Foley and Geiser, Roseland, N.J., 1986-88. Author and Editor: Papers in Marine Science, 1971; author: Dynamical and Physical Oceanography, 1988, Atmospheric Dynamics: Exercises and Problems, 1988, Climatology Workbook, 1988. Weather Map Folio, 1988. Mem. AAAS, Nat. Weather Assn., Am. Meteorol. Soc. (pres. N.J. chpt. 1980-81), N.Y. Acad. Scis. (chmn. atmospheric scis. sect. 1987-88), N.J. Marine Scis. Consortium, Phi Beta Kappa. Republican. Greek Orthodox. Home: 2798 Carol Rd Union NJ 07083-4831 Office: Kean Coll of NJ Morris Ave Union NJ 07083-7117

ZOLLER, MICHAEL, otolaryngologist; b. New Orleans, July 21, 1947; s. Harry and Mildred (Daitch) Z.; m. Linda Kramer, Dec. 21, 1974; children: Rebecca, Jonathan. BS, U. New Orleans, 1968; MD, Tulane U., 1972. Resident in gen. surgery Jewish Hosp., St. Louis, 1972-74, Washington U. Sch. Medicine; resident in otolaryngology Mass. Eye and Ear Infirmary, Harvard U., Boston, 1974-77; pres. Ear, Nose and Throat Assocs., Savannah, Ga., 1977—; asst. clin. prof. surgery Med. Coll. Ga., Augusta, 1982-93. Chmn. med. divsn. United Way, Savannah, 1990, chmn. profl. divsn., 1991; mem. edn. com. Am. Cancer Assn., Savannah, 1991, mem. bd. dirs., 1993—; mem. bd. dirs. Savannah Country Day Sch., 1993—; pres. Savannah Jewish Fedn., 1991-93. Recipient Young Leadership award Savannah Jewish Fedn., 1985, Boss of Yr. award Savannah Jaycees, 1993; Harvard U. Med. Sch. fellow, 1976-77. Fellow ACS; mem. AMA, Am. Acad. Head and Neck Soc., Am. Head and Neck Soc., Am. Neurotology Soc., Ga. Med. Soc. (sec. endowment fund 1992, pres. 1992), 1st Dist. Med. Assn. (pres. 1987-88), Med. Assn. Ga. (alt. bd. dirs., mem. ho. dels., Ga. Cup award 1993), So. Med. Assn., Savannah C. of C., Leadership Savannah. Office: Ear Nose and Throat Assocs 5201 Frederick St Savannah GA 31405-4596

ZOLOTOV, YURII ALEXANDER, chemist; b. Vysokovskoe, Klin Dist., Russia, Oct. 4, 1932; s. Alexander Georgievich Zolotov and Alexandra Mikhailovna Zolotova; m. Galina Alexeevna Ivanova, Apr. 19, 1959; children: Michael Yu, Maria Yu. Diploma Chemist, Moscow U., 1955; Candidate of Sci., Vernadskii Inst., Moscow, 1959, DSc, 1966. Cert. chemistry rschr. Jr. scientist Vernadskii Inst. Geochemical and Analytical Chemistry, Moscow, 1958-62, sr. scientist, 1962-68, dep. dir., 1968-79, head lab. solvent extraction, 1972-89; dir. Kurnakov Inst. Gen. and Inorganic Chemistry, Moscow, 1989—; prof. chem. dept. Moscow U., 1978-89, head analytical chemistry divsn., 1989—. Author: Extraction of Chelate Compounds, 1968 (Mendeleev Chem. Soc. award 1969), Ion Chromatography in Water, 1988, Preconcentration Analysis of Trace Elements, 1990, Analytical Chemistry: Problems and Achievements, 1992, others; contbr. over 500 sci. papers to profl. jours. Recipient USSR State prize, 1972, F. Emich Gold medal Austrian Soc. Analytical Chemistry, 1990, Russian State prize, 1991. Mem. Russian Acad. Scis. (pres. sci. coun. analytical chemistry 1988—), Mendeleev Russian Chem. Soc. (pres. 1991—, Mendeleev Gold medal 1993), Japan Soc. Analytical Chemistry (hon.). Achievements include 50 patents in field. Office: Kurnakov Inst, 31 Leninskii Prospect, 117907 Moscow Russia

ZOLTICK, BRAD J., physicist, mathematics educator; b. Trenton, N.J., Jan. 10, 1957. BA in Physics, Math., Hamilton Coll., Clinton, N.Y., 1978; MA in Physics, Washington U., St. Louis, 1983; MSEE, Case Western Res. U., 1992. Cert. physics tchr., N.J. Sr. rsch. technician Princeton (N.J.) Applied Rsch. EG & G, 1981; teaching asst. dept. physics Washington U., St. Louis, 1981-84; engring. tech. writer Technicomp, Inc., Cleve., 1985; teaching and rsch. asst. elec. engring. applied physics dept. Case Western Res. U., Cleve., 1985-87; computer programmer divsn. neurosci. Walter Reed Army Inst. Rsch., Washington, 1987; math. instr. Found. Advanced Edn. in Scis., NIH, Bethesda, Md., 1988—; physicist, computer programmer Nat. Eye Inst., Bethesda, 1988—. Fellow Weizmann Inst. Sci., 1978-80; Benjamin Walworth Arnold scholar Hamilton Coll., 1975, Fayerweather scholar, 1975, Acad. scholar, 1974-78. Mem. Sigma Pi Sigma. Home: 4000 Tunlaw Rd NW Apt 912A Washington DC 20007 Office: Nat Eye Inst Lab Sensorimotor Rsch Bldg 10 Room 10C101 Bethesda MD 20892

ZOOK, MARTHA FRANCES HARRIS, retired nursing administrator; b. Topeka, Nov. 15, 1921; d. Dwight Thacher and Helen Muriel (Houston) Harris; m. Paul Warren Zook, July 2, 1948; children: Mark Warren (dec.), Mary Elizabeth Zook Hughey. RN, Meriden (Conn.) Hosp. Sch. Nursing, 1947; student U. Kans., 1948-49, Kans. State U., 1960-61, Barton County Community Coll., 1970-73; BA, Stephens Coll., 1977; postgrad. Ft. Hays State U., 1978-79. Staff nurse Stormont Hosp., Topeka, 1947-48; staff nurse Watkins Meml. Hosp., Lawrence, Kans., 1948-49; nursing supr. Larned State Hosp., 1949-53, sect. supr., 1956-57, dir. nursing, 1958-61, 83-86; sect. nurse Sedgewick Sect., 1961-76, clin. instr. nursing edn., 1976-77, dir. nursing edn., 1977-83; clinic nurse for podiatrist; sect. supr. Dillon Bldg., Larned, 1957-58; Vol. Am. Cancer Soc., ARC, Welcome Inn, Sr. Citizens' Ctr. Pawnee County, Larned grade sch. children's drug info. program. Mem. AAUW, DAR, Sacred Heart Altar Soc. Democrat. Roman Catholic. Home: 1109 Johnson Ave Larned KS 67550-2232

ZOOK, MERLIN WAYNE, meteorologist; b. Connellsville, Pa., July 2, 1937; s. Ellrose Durr and Frances Adeline (Loucks) Z.; m. Maxine Beatrice Hartzler, May 1, 1965; children: Kevin Ray, Kathleen Joy. BA, Goshen (Ind.) Coll., 1959; MS, Pa. State U., 1961. Cert. consulting meteorologist, environ. profl. Rsch. assoc. U. Mich., Ann Arbor, 1958; grad. asst. Pa. State U., University Park, 1960-61; audio-visual asst., staff meteorologist Mennonite Cen. Com., Akron, Pa., 1961-63; air quality program specialist Pa. Dept. Environ. Resources, Harrisburg, 1963—; book reviewer Sci. Edn. Dept. Boston U., 1990-92, Nat. Weather Assn., Temple Hills, Md., 1983-88, book rev. editor, 1988-92; scientist, participant AAAS-Bell Atlantic Found., Washington, 1989-90. Author, contbr.: (chpt.) Behind the Dim Unknown, 1966. Guest lectr. Millersville (Pa.) State U., 1988, 90, Boy Scouts Am., Camp Hill, Pa., 1990, Pa. State U., Middletown, 1990, 91—, Cub Scouts Am., Camp Hill, 1991—. Mem. Am. Meteorol. Soc., Nat. Assn. Environ. Profls., Union of Concerned Scientists, Pa. State U. Alumni Assn., Sigma Xi. Achievements include development of models for the daily prediction of the Pollutant Standards Index in Pa., of collection of cloud type photographs with classifications for study of cloud characteristics/physics; research in meso-scale meteorology and localized forecasting, on the relationship between solar radiation and formation of ozone in urban areas in Pa., on migratory patterns of local birds influenced by meteorological conditions. Home: 105 June Dr Camp Hill PA 17011-5069 Office: Pa Bur Air Quality Dept Environ Resources 400 Market St Harrisburg PA 17105

ZOON, KATHRYN EGLOFF, biochemist; b. Yonkers, N.Y., Nov. 6, 1948; d. August R. and Violet T. (Pollock) Egloff; B.S. (N.Y. State Regents fellow), Rensselaer Poly. Inst., 1970; Ph.D. (fellow), Johns Hopkins U., 1975; m. Robert A. Zoon, Aug. 22, 1970; children—Christine K, Jennifer R. Interferon research fellow NIH, Bethesda, Md., 1975-77, staff fellow, 1977-79, sr. staff fellow, 1979-80; sr. staff fellow div. biochem. biophysics Bur. Biologics, FDA, Bethesda, 1980-83; rsch. chemist divsn. biochem. biophysics, 1983-84, rsch. chemist divsn. virology, 1984-88, rsch. chemist div. cytokine biology, Ctr. for Biologics Evaluation and Rsch., FDA, 1988—, div. dir., 1989-92; dir. Ctr. for Biologics Evaluation and Rsch., 1992. Mem. Am. Soc. Biochemistry and Molecular Biology, Internat. Soc. Interferon Research. Roman Catholic. Contbr. numerous articles on research in biol. chemistry to sci. jours.; editor Interferon Research, 1980—. Office: CBER 1401 Rockville Pike Rockville MD 20852

ZOROWSKI, CARL FRANK, engineering educator, university administrator; b. Pitts., July 14, 1930; s. Stanley and Mary Josephine (Kozuch) Z.; m. Sarah Jane Crossley, Aug. 7, 1954 (dec. 1983); children: Kathleen Ann, Karl Alan, Kristine Alaine; m. Louise Parrish Lockwood, Apr. 13, 1985. BSME, Carnegie Inst. Tech., 1952, MSME, 1953, PhD, 1956. Instr. Carnegie Inst. Tech., Pitts., 1952-56, asst. prof., 1956-61, assoc. prof., 1961-62; prof. mech. and aero. engring. dept. N.C. State U., Raleigh, 1964-66, R.J. Reynolds Industries prof., 1966—, assoc. dept. head, 1964-72, dept. head, 1972-79, assoc. dean acad. affairs Sch. Engring., 1979-85, dir. Integrated Mfg. Systems Inst., 1986-92, dept. head, 1992-93; dir. Succeed/NSF Coalition, 1993—. Served to 2d lt. USAR, 1952-58. Recipient research award Sigma Xi, 1967. Fellow ASME (Richards Meml. award 1975); mem. Am. Soc. Engring. Edn. (Western Electric award 1968), Fiber Soc. (Achievement award 1970). Contbr. pubs. to profl. lit.; patentee in field. Home: 103 Windyrush Ln Cary NC 27511 Office: NC State U Box 7901 Raleigh NC 27695

ZOU, YUN, mathematician; b. Xian, Shaanxi, People's Republic of China, Sept. 28, 1962; s. Jiunru and Shunhua (Xu) Z.; m XiaoLing Tian, June 25, 1987. BS, North Western U., Xian, 1983; M.Tech., East China Inst. Tech., Nanjing, 1987, PhD, 1990. Asst. engr. 7th Designing Inst. of Chinese Nuclear Ministry, Taiyuan, 1983-84; lectr. math. East China Inst. Tech., Nanjing, 1990-92; assoc. prof. math. Nanjing U. Sci. & Tech., 1992-93, prof. applied math., 1993—. Reviewer Math. Revs., 1990—; contbr. articles to profl. jours. Nat. Natural Sci. Fund of Peoples Republic of China grantee, 1991. Mem. Am. Math. Soc., Chinese Automation Assn., Acad. Activity Facilitating Group of E. China Inst. Tech. Avocations: volleyball, writing short stories and poetry. Home: Xiao Lingwei 200, Bldg 515-11, Nanjing Jiangsu, China Office: Nanjing U Sci & Tech, 810 Rsch Group, Nanjing China 210014

ZSIGO, JOZSEF MIHALY, chemist; b. Kiskunfelegyhaza, Hungary, Sept. 29, 1951; came to U.S., 1989; s. Jozsef and Erzsebet (Messzi) Z.; m. Annamaria Ban, June 28, 1977. MS, Szeged U., Szeged, Hungary, 1976, PhD, 1981; postdoctoral, Max-Planck Inst., Frankfurt, Germany, 1987-88. Rsch. co-worker Hungarian Acad. of Sci., Biol. Rsch. Ctr., Szeged, Hungary, 1976-81; asst. A. Szent-Gyorgyi U., Szeged, Hungary, 1981-87; postdoctoral fellow Max-Plank Inst., Frankfurt, Germany, 1987-88; asst. prof. Szeged U., 1988-89, 92—, Tulane U., New Orleans, 1989-92. Contbr. articles to profl. jours. Mem. Am. Chem. Soc. Hungarian Chem. Soc., Am. Peptide Soc. Achievements include patents on superactive GRF antagonists, long acting GRF antagonist peptides with non-natural residues. Home: 2000 N Court St # 9H Fairfield IA 52556

ZUBKOWSKI, JEFFREY DENNIS, chemistry educator; b. Ellwood City, Pa., July 23, 1957; s. Walter and JoAnne (Miller) Z.; m. Dell Westerbeck, Aug. 12, 1989. BS, U. Pitts., 1979; PhD, Ind. U., 1983. Postdoctoral fellow U. Toronto, Ond., Can., 1983-85; asst. prof. Jackson (Miss.) State U., 1985-91, assoc. prof., 1991—; advanced tutor Office of Minority Affairs, U. Miss. Med. Ctr., Jackson, 1985—; vis. prof. Hinds Community Coll. Raymond, miss., 1992—. Co-author: Experiments in General Chemistry, 1990, Introduction to Chemistry Laboratory, 1992, Review Manual for the DAT Exam, 1992. Mem. Am. Chem. Soc., Am. Crystallography Soc., Miss. Acad. of Sci. (chem. div. chmn. 1991-92). Achievements include rsch. interests in molecular recognition in chival resolving systems, use of arene ligands with transition metals as new electronic materials. Office: Jackson State Univ PO Box 17190 1400 Lynch St Jackson MS 39217

ZUBROFF, LEONARD SAUL, surgeon; b. Minersville, Pa., Mar. 27, 1925; s. Abe and Fannie (Freedline) Z.; BA, Wayne State U., 1945, MD, 1949. Diplomate Am. Bd. Surgery. Intern Garfield Hosp., Washington, 1949-50, resident in surgery, 1951-55, chief resident surgery, 1954-55; pvt. practice medicine specializing in surgery, 1958-76; med. dir. Chevrolet Gear and Axle Plant, Chevrolet Forge Plant, GM, Detroit, 1977-78, divisional med. dir. Detroit Diesel Allison div., 1978-87, regional med. dir. GM, 1987-89; ret.,

1989; bd. trustees LeVine Found.; mem. staff Hutzel Hosp., Detroit Meml. Hosp.; chief of surgery, chief profl. svcs. N.E. Air Command, Pepperell AFB, Newfoundland. With USAF, 1956-58. Fellow ACS; mem. Mich. State, Wayne County med. socs., Acad. Surgery Detroit, Am. Coll. Occupational Medicine, Mich. Occupational Med. Assn. (pres. 1990-91), Detroit Occupational Physicians Assn. (former pres.), Masons (33 degree), Phi Lambda Kappa. Home and Office: 16233 W 9 Mile Rd Apt 201 Southfield MI 48075-5927

ZUCKER, ARNOLD HARRIS, psychiatrist; b. Bklyn., July 29, 1930; s. Charles Israel and Bertha (Leff) Z.; m. Marilyn Pistreich, June 10, 1962; children: Harvey, Deborah, Shoshanna, David. BA, Bklyn. Coll., 1950; MD, SUNY, Bklyn., 1954. Diplomate, Am. Bd. Psychiatry and Neurology. Intern USPHS, Staten Island, N.Y., 1954-55; resident Kings County Hosp., Bklyn., 1955-56, Southwestern Med. Sch., Dallas, 1958-59, Albert Einstein Coll. Medicine, Bronx, N.Y., 1959-60; asst. clin. prof. psychiatry Albert Einstein Coll. Medicine, 1960-72; pvt. practice Mt. Vernon, N.Y., 1960—; assoc. attending psychiatrist, Mt. Vernon Hosp.; assoc. prof. pastoral counseling, Iona Coll., New Rochelle, N.Y., 1968—. Contbr. articles to profl. jours. Surgeon, USPHS, 1956-58. Fellow Am. Psychiat. Assn., Am. Acad. Psychoanalysis; mem. Am. Psychoanalytic Assn., Assn. Psychoanalytic Medicine, AMA, Westchester Psychoanalytic Soc., Phi Beta Kappa. Democrat. Jewish. Avocation: religious studies. Office: 120 E Prospect Ave Mount Vernon NY 10550-2205

ZUCKER, HOWARD ALAN, pediatric cardiologist, intensivist, anesthesiologist; b. N.Y.C., Sept. 6, 1959; s. Saul and Phyllis (Goldblatt) Z.. BS, McGill U., Montreal, Quebec, Can., 1979; MD, George Washington U., 1982. Pediatric intern Johns Hopkins Hosp., Balt., 1982-83, pediatric resident, 1983-85; anesthesiology resident Hosp. of U. Pa., Phila., 1985-87; pediatric critical care fellow Children's Hosp. of Phila., 1987-88; asst. prof. anesthesiology and pediatrics Yale U. Sch. Medicine, New Haven, Conn., 1988-90; pediatric cardiology fellow Children's Hosp., Harvard Med. Sch., Boston, 1990-92; asst. prof. pediatrics and anesthesiology Columbia U. Coll. Physicians and Surgeons, N.Y.C., 1992—; pediatric dir. ICU Columbia Presbyn. Med. Ctr. Babies Hosp., N.Y.C., 1992—; involved with crew tng. of NASA Space Shuttle STS-1 Mission, 1978-80. Fellow Am. Acad. Pediatrics; mem. AMA, Am. Soc. Anesthesiologists, Am. Heart Assn., Soc. Critical Care Medicine. Jewish. Achievements include research in adaptation to zero gravity, cardiac critical care. Home: 1500 Palisade Ave Fort Lee NJ 07024 Office: Columbia Presbyn Med Ctr Babies Hosp 3959 Broadway New York NY 10032

ZUCKERKANDL, EMILE, molecular evolutionary biologist, scientific institute executive; b. Vienna, Austria, July 4, 1922; came to U.S., 1975; s. Frederic and Gertrude (Stekel) Z.; m. Jane Gammon Metz, June 2, 1950. M.S., U. Ill., 1947; Ph.D., Sorbonne, Paris, 1959. Postdoctoral research fellow Calif. Inst. Tech., Pasadena, 1959-64; research dir. CNRS, Montpellier, France, 1967-80, dir. Ctr. Macromolecular Biochemistry, 1965-75; pres. Linus Pauling Inst., Palo Alto, Calif., 1980-92, Inst. Molecular Med. Scis., Palo Alto, Calif., 1992—; cons. in genetics Stanford U., 1963, vis. prof., 1964; vis. prof. U. Del., 1976. Contbg. author: Evolving Genes and Proteins, 1965; co-author: Genetique des Populations, 1976; editor Jour. Molecular Evolution, 1971—. Decorated Order of Merit (France). Fellow AAAS; mem. Societe de Chimie Biologique, N.Y. Acad. Scis., Internat. Soc. Study Origin of Life. Home: 565 Arastradero Rd Palo Alto CA 94306-4339 Office: Inst Molecular Med Scis 460 Page Mill Rd Palo Alto CA 94306-2025

ZUHDI, NAZIH, surgeon; b. Beirut, May 19, 1925; came to U.S., 1950; s. Omar and Lutfiye (Atef) Z.; children by previous marriage—Omar, Nabil; m. 2d, Annette McMichael; children—Adam, Leyla, Zachariah. BA, Am. U., Beirut, 1946, MD, 1950. Diplomate Am. Bd. Surgery, Am. Bd. Thoracic Surgery. Intern St. Vincent's Hosp., S.I., N.Y., 1950-51, Presbyn.-Columbia Med. Ctr., N.Y.C., 1951-52; resident Kings County SUNY Med. Ctr., N.Y.C., 1952-56; fellow SUNY Downstate Med. Ctr., Bklyn., 1953-54; resident Univ. Hosp., Mpls., 1956; resident Univ. Hosp., Oklahoma City, 1957-58, practice surgery specializing in cardiovascular and thoracic, 1958-87, practice in heart transplantation, lung transplantation and heart-lung transplantation, 1987—; founder, dir. Transplantation Inst. Bapt. Med. Ctr., 1984—, transplantation surgeon in chief Bapt. Hosp., Oklahoma City; founder, chmn. Okla. Cardiovascular Inst., Oklahoma City, 1983-84, Okla. Heart Ctr., Oklahoma City, 1984-85. Contbg. author Cardiac Surgery, 1967, 2d edit., 1972; contbr. articles to profl. jours.; developer numerous med. devices, techniques, rsch. and publs. on cardiopulmonary bypass, internal hypothermia, assisted circulation, heart surgery and transplantation of thoracic organs; developer heart-lung machines; designer, use of exptl. plastic bypass hearts; originator use of banked citrated blood for cardiopulmonary bypass for open heart surgery, of clin. non-hemic primes of heart-lung machines producing intentional hemodilution, at present, the universally accepted principle of cardiopulmonary bypass for partial and total body perfusion; researcher in cardiovascular studies. Muslim scholar, lectr.; named Hon. Citizen Brazil. Fellow ACS; mem. AMA, Am. Thoracic Soc., Okla. Thoracic Soc., So. Med. Assn., Okla. Med. Assn., Internat., Coll. Angiology, Am. Coll. Chest Physicians, Oklahoma City C. of C., Oklahoma County Med. Soc., Oklahoma City Clin. Soc., Okla. Surg. Assn., Oklahoma City Surg. Soc., Southwestern Surg. Congress, Am. Coll. Cardiology, Am. Soc. Artificial Internal Organs, Soc. Thoracic Surgeons (founder mem.), Am. Assn. for Thoracic Surgery, Internat. Cardiovascular Soc., Okla. State Heart Assn., Osler Soc., So. Thoracic Surg. Assn., Lillehei Surg. Soc., Internat. Soc. Heart Transplantation, Dwight Harken's Founder's Group Cardiac Surgery, Internat. Soc. Cardiothoracic Surgery (Japan)(founder), Am. Soc. Transplant Surgeons, Milestones of Cardiology of Am. Coll. Cardiology, Okla. City Golf and Country Club. Club: Oklahoma City Golf and Country. Achievements include work on first banked citrated blood for cardiopulmonary bypass for open heart surgery; invention of clin. non-hemic primes of heart-lung machines producing intentional hemodilution. Home: 7305 Lancet Ct Oklahoma City OK 73120-1430

ZUKAS, JONAS ALGIMANTAS, physicist; b. Kaunas, Lithuania, Nov. 12, 1939; came to U.S., 1949; s. Jonas and Elena (Efimavicius) Z.; m. Ann Russell, Jan. 29, 1972; children: John Vincent, Victoria Elena. BS in Physics, Poly. Inst. Bklyn., 1962; MS in Applied Sci., U. Del., 1969; PhD in Engring. Mechanics, U. Ariz., 1973. Rsch. physicist U.S. Army Ballistic Rsch. Lab., Aberdeen Proving Ground, Md., 1966-93; pres. Computational Mechanics Cons., Balt., Md., 1987—; adv. bd. Internat. Jour. Mech. Scis., Manchester, Eng., 1986-92, Internat. Jour. Impact Engring., Liverpool, Eng., 1986—. Co-author: Impact Dynamics, 1982, Fundamentals of Shaped Charges, 1989; editor: High Velocity Impact Dynamics, 1990; contbr. chpts. to books, articles to profl. jours. 1st lt. U.S. Army, 1964-66. Mem. ASME, AIAA. Achievements include development of analytical models for laminated anisotropic structures; also for the dynamic response of fabric body armors; directing development of 2D and 3D finite element software on personal computers and workstations for wave propagation and impact studies; organizing and teaching seminars on material response to ultra-high loading. Office: Computational Mechanics Consultants Inc PO Box 11314 Baltimore MD 21239-0314

ZUKIN, STEPHEN RANDOLPH, psychiatrist; b. Phila., Aug. 15, 1948; s. Solomon G. and Rose (Katz) Z.; children: Valerie Anne, Heather Nicole. BA, Haverford Coll., 1970; MD, Johns Hopkins U., 1974. Diplomate Am. Bd. Psychiatry and Neurology. Intern and resident in psychiatry Mt. Zion Med. Ctr., San Francisco, 1974-77; asst. prof. of psychiatry SUNY - Downstate, Bklyn., 1977-79, Mt. Sinai Sch. Medicine, N.Y.C., 1979-82; assoc. prof. psychiatry Albert Einstein Coll. Medicine, Bronx, N.Y., 1982-87, assoc. prof. neuroscience, 1984-87, prof. psychiatry and neuroscience, 1987—; dir. rsch. Bronx Psychiat. Ctr., 1983—; attending psychiatrist Montefiore Med. Ctr., Bronx, 1982—. Author more than 75 scientific publs., 1974—. Rsch. grantee Nat. Inst. on Drug Abuse, 1980—; recipient Kempf Fund award for R&D in Psychol. Psychiatry, 1992. Mem. Am. Psychiatric Assn., Soc. for Neuroscience, Am. Coll. Neuropsychopharmacology, Am. Soc. Pharmacology and Exptl. Therapeutics, N.Y. Acad. Scis. (conf. com. 1992—). Achievements include co-discovery of phencyclidine/NMDA receptor of brain; research in phencyclidine/NMDA hypothesis of schizophrenia.

ZUKOWSKI, CHARLES ALBERT, electrical engineering educator; b. Buffalo, Aug. 17, 1959; s. Stanley Paul and Ruth Elizabeth (Hudson) Z.; m. Deborra Jean Hills, June 10, 1983; children: Claire, Sara. BS, MS, MIT, 1982, PhD, 1985. Rsch. assoc. MIT, Cambridge, Mass., 1985; asst. prof. Columbia U., N.Y.C., 1985-90, assoc. prof., 1991—; cons. IBM Corp., Yorktown Heights, N.Y., 1988—. Author: The Bounding Approach to VLSI Circuit Simulation, 1986. Mem. Conservation Adv. Bd., Yorktown, 1988—. Named Presdl. Young Investigator, NSF, 1987. Mem. IEEE, Sigma Xi, Eta Kappa Nu. Achievements include one patent in field. Office: Columbia U 500 W 120th St #1312 New York NY 10027-6699

ZUMINO, BRUNO, physics educator, researcher; b. Rome, Apr. 28, 1923; came to U.S., 1951; naturalized, 1962; divorced. D in Math. Scis., U. Rome, 1945. From asst. prof. to prof. NYU, 1953-69; staff mem. European Orgn. for Nuclear Rsch. (CERN), Geneva, 1969-81; prof. physics U. Calif., Berkeley, 1981—, Alfred C. and Mary Sprague Miller rsch. prof., 1989; Loeb lectr. Harvard U., Cambridge, Mass., spring 1966; vis. prof. Columbia U., N.Y.C., fall 1978; disting. vis. prof. Enrico Fermi Inst., U. Chgo., spring 1983. Recipient Max Planck medal German Phys. Soc., 1989, Wigner medal Found. for Group Theory and Fundamental Physics, Am. Nuclear Soc., 1992; co-recipient Dirac medal Internat. Ctr. for Theoretical Physics, Trieste, Italy, 1987; Guggenheim Found. fellow, 1968-69, 87-88. Fellow Am. Phys. Soc. (co-recipient Heineman prize 1988), Am. Acad. Arts and Scis.; mem. NAS, Italian Phys. Soc. Office: U Calif Dept Physics Berkeley CA 94720

ZUMPE, DORIS, psychiatry researcher, educator; b. Berlin, May 18, 1940; came to U.S., 1972; d. Herman Frank and Eva (Wagner) Z. BSc, U. London, 1961, PhD, 1970. Asst. to K.Z. Lorenz, Max-Planck-Inst. für Verhaltensphysiologie, Seewiesen, Fed. Republic Germany, 1961-64; rsch. asst. and assoc., lectr. Inst. Psychiatry, U. London, 1965-72; rsch. assoc. Emory U. Sch. Medicine, Atlanta, 1972-74, asst. prof. psychiatry (ethology), 1974-77, assoc. prof., 1977-87, prof., 1987—; reviewer NSF, 5 sci. jours. Contbr. over 120 articles to profl. jours. NIMH grantee, 1971—. Mem. AAAS, Internat. Soc. Psychoneuroendocrinology, Internat. Primatological Soc., Internat. Soc. for Human Ethology, Soc. for Study of Reprodn., Am. Soc. Primatologists, N.Y. Acad. Scis., Earl Music Am., Viola da Gamba Soc. Am. Avocation: semi-professional musician. Office: Emory U Sch Medicine Dept Psychiatry Atlanta GA 30322

ZUNDE, PRANAS, information science educator, researcher; b. Kaunas, Lithuania, Nov. 26, 1923; came to U.S., 1960, naturalized, 1964; s. Pranas and Elzbieta (Lisajevic) Z.; m. Alge R. Bizauskas, May 29, 1945; children: Alge R., Audronis K., Aurelia R., Aidis L., Gytis J. Dipl. Ing., Hannover Inst. Tech., 1947; MS, George Washington U., 1965; PhD, Ga. Inst. Tech., 1968. Dir. project Documentation Inc., Bethesda, Md., 1961-64; mgr. mgmt. info. system Documentation Inc., Bethesda, 1964-65; sr. research scientist Ga. Inst. Tech., Atlanta, 1965-68, assoc. prof., 1968-72, prof. dept. computer sci., 1973-91, prof. emeritus, 1991—; cons. UNESCO, Caracas, Venezuela, 1970-72, Esquela Polit. Nacional, Quito, Ecuador, 1974-75, State of Ga., Atlanta, 1976-78, Royal Sch. Librarianship, Copenhagen, 1985-87, Clemson (N.C.) U., 1987—, N.Y. State Dept. Edn., 1993; vis. prof. Simon Bolivar U., Caracas, 1976, J. Kepler U., Austria, 1981; vis. scientist Riso Nat. Lab., Roskilde, Denmark, 1983, Lithuanian Acad. Scis., Vilnius, 1988-89. Author: Agriculture in Soviet Lithuania, 1962, National Science Information Systems in Eastern Europe, 1972; editor: Procs. Info. Utilities, 1974, Procs. Founds. of Info. and Software Sci., 1983-90; contbr. articles to tech. and sci. jours. Mem. senate Vitoldus Magnus U., Kaunas, Lithuania, 1990—. NSF grantee; Fulbright prof. NAS, 1975. Mem. Am. Soc. Info. Sci., Semiotic Soc. Am., Soc. Sigma Xi. Roman Catholic. Office: Ga Inst Tech Coll Computing North Ave Atlanta GA 30332

ZUNICH, KATHRYN MARGARET, physician; b. Bklyn., Feb. 12, 1953; d. Peter and Margaret (Pezzarossi) Z. BA, Fordham U., 1975; MD, SUNY, Syracuse, 1978. Diplomate Am. Bd. Internal Medicine, Am. Bd. Allergy & Immunology. Intern St. Vincent's Hosp. & Med. Ctr., N.Y.C., 1978-79, resident in internal medicine, 1980-83; fellow in allergy and immunology U. Colo. Health Sci. Ctr., Denver, 1985-88, Nat. Jewish Ctr. for Immunology and Respiratory Medicine, Denver, 1985-88; med. officer lab. of immunoregulation Nat. Inst. Allergy & Infectious Diseases, Bethesda, Md., 1988-90, chief transplantation sect., 1990-91, chief asthma sect., 1991-92; assoc. med. dir. Inst. Clin. Immunology & Infectious Diseases, Syntex Rsch., Palo Alto, Calif., 1992—. Contbr. articles to profl. jours. V.p. D.C. chpt. Fieri, 1990-91. Mem. Am. Coll. Physicians, Am. Coll. Allergy & Immunology, Am. Acad. Allergy & Immunology. Democrat. Roman Catholic. Office: Syntex Rsch 3401 Hillview Ave A4-11D Palo Alto CA 94303

ZUSPAN, FREDERICK PAUL, obstetrician/gynecologist, educator; b. Richwood, Ohio, Jan. 20, 1922; s. Irl Goff and Kathryn (Speyer) Z.; m. Mary Jane Cox, Nov. 23, 1943; children: Mark Frederick, Kathryn Jane, Bethany Anne. B.A., Ohio State U., 1947, M.D., 1951. Intern Univ. Hosps., Columbus, Ohio, 1951-52; resident Univ. Hosps., 1952-54, Western Res. U., Cleve., 1954-56; Oblebay fellow Western Res. U., 1958-60, asst. prof., 1958-60; chmn. dept. ob-gyn McDowell (Ky.) Meml. Hosp., 1956-58, chief clin. services, 1957-58; prof., chmn. dept. ob-gyn Med. Coll. Ga., Augusta, 1960-66; Joseph Boliver DeLee prof. ob-gyn, chmn. dept. U. Chgo., 1966-75; obstetrician, gynecologist in chief Chgo. Lying-In Hosp., 1966-75; prof., chmn. dept. ob-gyn Ohio State U., Columbus, 1975-87; R.L. Meiling prof. in ob-gyn Ohio State U. Sch. Medicine, Columbus, 1984-90; prof. emeritus Ohio State U., Columbus, 1991—. Founding editor: Lying In, Jour. Reproductive Medicine; editor-in-chief: Am. Jour. Ob-Gyn and Ob-Gyn. Reports, (with Lindheimer and Katz) Hypertension in Pregnancy, 1976; Current Developments in Perinatology, 1977, (with Quilligan) Operative Obstetrics, 1981, 89, Manual of Practical Obstetrics, 1981, 90, Am. Jour. Ob-Gyn., Clin. and Exptl. Hypertension in Pregnancy, 1979-86, (with Rayburn) Drug Therapy in Ob-Gyn., 1981, 2d edit. 1986; editor: (with Christian) Controversies in Obstetrics and Gynecology; editor in chief Ob-Gyn. Reports, 1988-90; editor in chief Am. Jour. Ob-Gyn.; contbr. articles to med. jours., chpts. to books. Pres. Barren Found., 1974-76. Served with USNR, 1942-43; to 1st lt. USMCR, 1943-45. Decorated DFC, Air medal wth 10 oak leaf clusters. Mem. Soc. Gynecol. Investigation, Chgo. Gynecol. Soc., Am. Assn. Ob-Gyn, Columbus Ob-gyn Soc. (pres. 1984-85), Am. Acad. Reproductive Medicine (pres.), Am. Coll. Obstetricians and Gynecologists, Assn. Profs. Gynecology and Obstetrics (pres. 1972), South Atlantic Assn. Obstetricians and Gynecologists (Found. prize for research 1962), Central Assn. Ob-Gyn (cert. of merit, research prize 1970), Am. Soc. Clin. Exptl. Hypnosis (exec. sec. 1968, v.p. 1970), So. Gynecol. Investigation, Internat. Soc. Study Hypertension in Pregnancy (pres. 1981-83), Am. Gynecology and Obstetrics Soc. (pres. 1986-87), Soc. Perinatal Obstetrics, Perinatal Research Soc., Sigma Xi, Alpha Omega Alpha, Alpha Kappa Kappa. Home: Upper Arlington 2400 Coventry Rd Columbus OH 43211

ZUXUN, SUN, academic administrator, physicist; b. Wuhan, Hubei, China, Nov. 15, 1937; s. Zhichang Sun and Manyun Yang; m. Daohong Zhu, Oct. 5, 1967 (dec. Mar. 1992); 1 child, Qian; m. Jianting Lu, Mar. 30, 1993. Bachelor degree, Tsing Hua U., Beijing, China, 1961, PhD, 1965. Rsch. asst., group leader China Inst. of Atomic Energy, Beijing, 1965-78, assoc. prof., sect. leader, 1978-82, prof., 1986, pres., 1985—; vis. scientist dept. physics Munich U., 1978-80, Los Alamos Nat. Lab., 1982-83. Editor: Fast Neutron Physics, 1992; contbr. articles on few body reactions, low energy polarization reactions and high energy polarization reactions to jours. including Chinese Jour. of Nuclear Physics, Phys. Rev. Letters, Phys. Rev. Mem. Chinese Nuclear Soc. (exec. councillor 1990—), Chinese Phys. Soc. (exec. councillor 1986—), Chinese Nuclear Physics Soc. (v.p., pres. 1987—). Achievements include development of first on-line particle identification and multi-parameter data acquisition system in China; measurement of light nuclei reactions data for 20 years; measurement of low-energy and immediate energy polarization nuclear reactions data; discovery of low-energy quasi-free scattering; leader of construction of the first tandem accelerator nuclear physics laboratory in China; measurement of the longitudinal components of spin-rotation parameters of p-cl for the first time. Office: China Inst Atomic Energy, PO Box 275, Beijing 102413, China

ZUZOLO, RALPH CARMINE, biologist, educator, researcher, consultant; b. Dente Cane, Avellino, Italy, Sept. 5, 1929; came to U.S., 1930; s. Antonio and Assunta (Nardone) Z.; m. Betty Ann Fong, July 22, 1972. PhD in Cell/ Cellular Physiology and Micro-Surgery, NYU, 1960. Asst. prof. CCNY of CUNY, 1965-75, assoc. prof., 1975-93, prof., 1993—; co-dir. Robert Chambers Lab. for Cell Micro-Manipulation and microinjection CCNY, 1978—, supr. dept. biology Sch. Gen. Studies, 1968—; adj. rsch. scientist NYU, N.Y.C., 1965-78; dir. course and standing CCNY, 1988—; cons. Robert Chambers Lab. CCNY, 1978—. Author: (manuals) Cellular Micromanipulation, 1992, Manual of Ruby Laser Microsurgery, 1993. Cpl. USMC, 1951-52, Korea. Recipient Rsch. fellowship U. Tex., 1966-68, grant Olympus Corp. of Am., 1981—. Mem. AAUP, Nat. Tissue Culture Assn., Applied Spectroscopy, N.Y. Microscopical Soc., Sigma Xi. Achievements include development of instrumentation and methods for microinjection of macromolecules into single plant and animal cells and embryos. Office: CCNY Convent Ave at 138th St New York NY 10031

ZVETINA, JAMES RAYMOND, pulmonary physician; b. Chgo., Oct. 14, 1913; s. John and Jennie (Albrecht) Z.; m. Florence Courtney, Feb. 4, 1944. BS, Loyola U., 1940; MD, U. Ill., 1943. Intern West Suburban Hosp., Oak Park, Ill., 1944, resident physician, 1944-45; asst. ward med. officer USNH, NOB, Norfolk, Va., 1945; staff physician Pulmonary TB Svc. VA Med. Hosp., Hines, Ill., 1946-54; asst. chief Pulmonary Svc. VA Med. Hosp., Hines, Ill., 1954-68, sect. chief, 1968-88, attending physician, 1988-91, cons., 1992—; clin. prof. medicine Coll. Medicine, U. Ill., Chgo., 1978—; mem. adv. bd. Coll. Medicine, U. Ill., 1985—; rep. Rsch. Conf. in Pulmonary Disease, VA Armed Forces, 1946-74. Contbr. articles to profl. jours. V.p. Chgo. Cath. Physicians, 1979, pres., 1978. Comdr. USNR, 1945-46, med. officer USNR, ret. Recipient Svc. award 40 Yrs. VA Adminstrn., 1985, Svc. award 30 Yrs. U. Ill. Med. Sch., 1978. Fellow Am. Coll. Chest Physicians; mem. AMA, Ill. State Med. Soc., Chgo. Med. Soc. Roman Catholic. Achievements include research in area of pulmonary infections. Home: 96 Forest Ave Riverside IL 60546-1977 Office: VA Hines Hines IL 60141

ZWANZIG, ROBERT WALTER, chemist, physical science educator; b. Bklyn., Apr. 9, 1928; s. Walter and Bertha (Weil) Z.; m. H. Frances Ryder, June 6, 1953; children: Elizabeth Ann, Carl Philip. B.S., Poly. Inst. Bklyn., 1948; M.S., U. So. Calif., 1950; Ph.D., Calif. Inst. Tech., 1952. Rsch. assoc. Yale, 1951-54; asst. prof. Johns Hopkins, 1954-58; phys. chemist Nat. Bur. Standards, 1958-66; rsch. prof. U. Md., College Park, 1966—; disting. prof. phys. sci. U. Md., 1980-88, disting. prof. emeritus, 1988—; rsch. chemist NIH, Bethesda, Md., 1988—. Asso. editor: Jour. of Chem. Physics, 1965-67, Jour. of Math. Physics, 1968-70, Transport Theory and Statistical Physics, 1970, Chem. Physics, 1973—; Contbr. articles profl. jours. Dept. Commerce Silver medal, 1965. Fellow Am. Phys. Soc., Am. Acad. Arts and Scis., AAAS; mem. Nat. Acad. Scis., Am. Chem. Soc. (Peter Debye award in phys. chemistry 1976; Irving Langmuir award 1984). Home: 5314 Sangamore Rd Bethesda MD 20816-2355 Office: NIH Nat Inst Digestive and Diabetes and Kidney Diseases Bethesda MD 20892

ZWASS, VLADIMIR, computer scientist, educator; b. Lvov, USSR, Feb. 3, 1946; came to U.S., 1970, naturalized, 1979; s. Adam and Friderike (Getzler) Z.; m. Alicia Kogut, Apr. 24, 1977; 1 child, Joshua Jonathan. M.S., Moscow Inst. Energetics, 1969; M.Ph., Columbia U., 1974, Ph.D., 1975. Mem. profl. staff IAEA, Vienna, Austria, 1970; asst. prof. computer sci. Fairleigh Dickinson U., 1975-79, assoc. prof., 1979-84, prof., 1984—, prof. computer sci. and mgmt. info. systems, 1990—, chmn. com. computer sci., 1976—; cons. U.S. Govt., Met. Life Ins. Co., Citibank, Diebold Group; seminar assoc. Columbia U., 1986—; speaker nat. and internat. meetings. Author: Introduction to Computer Science, 1981, Programming in Fortran, 1981, Programming in Pascal, 1985, Programming in Basic, 1986, Management Information Systems, 1992; editor-in-chief Jour. Mgmt. Info. Systems, 1983—; contbr. articles to profl. jours. and publs., Ency. Britannica, N.Y. Times, chpts. to books. Columbia U. fellow, 1970-71; Helena Rubinstein Found. scholar, 1971-75; grantee USN, other agys. Mem. Assn. Computer Machinery, IEEE, Sigma Xi, Eta Kappa Nu. Home: 538 Churchill Rd Teaneck NJ 07666-2900 Office: Fairleigh Dickinson U Teaneck NJ 07666

ZWEIDLER, ALFRED, medical scientist, educator; b. Dübendorf, Switzerland, Feb. 1, 1937; s. Hermann and Mina (Nyffeler) Z.; m. Irene Erika Bärtschi, Apr. 6, 1963; children: Regula, Patrick, Stephan. Bus. Diploma, Kaufmännischer Verein, ZUrich, 1955; Mat. Dip. Type B, Inst. Juventus, ZUrich, 1960; Dr. phil. II, U. Zurich, 1966. Rsch. assoc. Experimental Cancer Res. U., Zurich, 1965-66; postdoctoral fellow Cell Rsch. Inst. U. Tex., Austin, 1966-69; vis. scientist Inst. Cancer Rsch., Phila., 1969-70; sr. rsch. assoc. Dept. Botany U. Tex, Austin, 1970-71; asst. mem. Inst. Cancer Rsch., Phila., 1972-76; assoc. mem. Fox Chase Cancer Ctr., Phila., 1976-78, mem., 1980—; adj. assoc. prof. U. Pa. Med. Sch., Phila., 1984—. Co-author: Histone Genes and Histone Gene Expression, 1984; editor: Histones and Cancer, 1979; contbr. articles to profl. jours. Pres. New Helvetic Soc., Phila., 1978-84; dir. M.Rohrbach Cultural Fund, Phila., 1984—. Recipient Outstanding Student Rsch. award U. Zurich, 1964; U. Tex. fellow, 1966-69; Nat. Cancer Inst. grantee, 1972-91. Mem. Am. Assn. for the Advancement Sci., Am. Soc. for Devel. and Cell Biology, The Protein Soc., Swiss Genetic Soc., Sigma Xi. Achievements include patents for the process for semiautomatic separation of organs from fly larvae; discovered most discriminating method for protein separation on the basis of mass, charge and hydrophobicity called AUT-PAGE; discovered nonallelic histone variants in vertebrates; developed methods for studying protein-protein interaction called protein footprinting. Home: 304 Ashbourne Rd Elkins Park PA 19117 Office: Fox Chase Cancer Ctr 7701 Burholme Ave Philadelphia PA 19111

ZWEIFEL, TERRY L., aeronautical engineer, researcher; b. Phoenix, Oct. 24, 1942; s. Robert Rudy and Ruby Mae (Toliver) Z.; m. Carol Jean Vogt, June 2, 1965. BS in Aero. Engring., U. Ariz., 1965; student, L.A. Valley Coll., 1971-75. Cert. controls engr., Calif. Tech. mgr. Lockheed Aero. Systems Co., Burbank, Calif., 1965-79; chief engr. Simmonds Precision Products Inc., Vergennes, Vt., 1979-81; sr. fellow Honeywell Inc., Phoenix, 1981—; lectr. to profl. groups. Contbr. numerous articles to profl. jours.; patentee windshear detection, flight guidance in windshear, cruise airspeed control, numerous other patents in field. Mem. AIAA (Atmospheric Environment award 1989). Avocations: computer software, chess, travel. Home: 7250 N 30th Dr Phoenix AZ 85051-7513 Office: Honeywell Inc 21111 N 19th Ave Phoenix AZ 85027-2700

ZWEIMAN, BURTON, physician, scientist, educator; b. N.Y.C., June 7, 1931; s. Charles and Gertrude (Levine) Z.; m. Claire Traig, Dec. 30, 1962; children: Amy Beth, Diane Susan. A.B., U. Pa., 1952, M.D., 1956. Intern Mt. Sinai Hosp., N.Y.C.; resident in medicine Hosp. U. Pa., Bellevue Hosp. Center, 1957-60; fellow NYU Sch. Medicine, 1960-61; mem. faculty dept. medicine U. Pa. Sch. Medicine, Phila., 1963—; prof. medicine, chief allergy and immunology sect. U. Pa. Sch. Medicine, 1975—; cons. U.S. Army, NIH; co-chmn. Am. Bd. Allergy and Immunology, 1979-81. Editor Jour. Allergy Clin. Immunology, 1988-93; contbr. articles to med. jours. Served with M.C., USNR, 1961-63. Allergy Found. Am. fellow, 1959-61. Fellow ACP, Am. Acad. Allergy (pres.-elect exec. com.); mem. Am. Assn. Immunologists, Am. Fedn. Clin. Rsch., Phi Beta Kappa, Alpha Omega Alpha. Office: U Pa Sch Medicine 512 Johnson Pavilion 36th and Hamilton Walk Philadelphia PA 19104

ZWEIZIG, J(OHN) RODERICK, biomedical engineer; b. Zion, Ill., Aug. 28, 1923; s. John Harold and Ethel Maud (Nicholson) Z.; m. Alene Margery Gustavson, June 9, 1951; children: Margery Anne Zweizig Atkinson, Peter Zweizig Atkinson, John George, Susan Lee Zweizig Bowers. BSEE, Northwestern U., 1948; M Engring., UCLA, 1975; postgrad., U. Calif., Santa Barbara, 1975-82. Mechanician dept. psychology Northwestern U., Evanston, Ill., 1948-51; rsch. engr. jet propulsion lab. Caltech, Pasadena, Calif., 1951-55; asst. engr. dept. engring. UCLA, Westwood, 1955-60, asst. engr. Brain Rsch. Inst., 1960-75, historian Brain Rsch. Inst., 1975—; participant internat. meetings on math. computer sci., neural computing, biomed. computing. Contbr. to profl. publs. With USN, 1944-46, PTO. Mem. IEEE, AFCEA, Am. Legion, Sigma Xi, Eta Kappa Nu. Unitarian.

ZWERNEMAN, FARREL JON, engineering educator; b. Giddings, Tex., Nov. 20, 1954; s. Leroy Quintus and Lavern Lila (Wiederhold) Z.; m. Marjorie Anne Reilly, Sept. 15, 1979; children: Ryan, Craig, John. MS, U. Tex., 1983, PhD, 1985. Registered profl. engr., Okla. Design engr. Petro-Marine Engring., Houston, 1978-80; sr. design engr. Omega Marine Svcs., Houston, 1980-81; asst. prof. Okla. State U., Stillwater, 1985-89, assoc. prof.,

1989—; scorer Nat. Coun. Engring. Examiners, Clemson, S.C., 1987-88; lectr. short course on fracture mechanics and fatigue Union Coll., Schenectady, N.Y., 1990. Author: Fracture Mechanics: Perspectives and Directions, 1989, Structures to Repeated Loading, 1991; contbr. articles to Jour. Structural Engring., Engring. Fracture Mechanics. Recipient 2d place engring. design competition Lincoln Arc Welding Found., 1983. Mem. ASCE, NSPE, Am. Inst. Steel Constrn. (fellow 1982). Office: Okla State U Civil Engring 207 Eng South Stillwater OK 74078

ZWICKNAGL, GERTRUD, physicist. Researcher Inst. Fur Festkorper, Stuttgart, Germany. Recipient Walter-Schottky-Preis für Festkorperforschung, Deutsche Physikalische Gesellschaft, 1993. Office: Inst fur Festkorper-Forschung, Heisenbergstr 1 Postfach 800665, D-7000 Stuttgart 80, Germany*

ZWIENER, JAMES MILTON, physicist; b. Kansas City, Mo., Sept. 9, 1942; s. Royal Dean Zwiener and Betty Sue Karr/Roney Ford; m. Donna Kay Roberts, Aug. 28, 1967; children: Mark, Albert, Gwendolyn. BS in Physics, U. Mo., Rolla, 1967. Physicist George C. Marshall Space Flight Ctr., Huntsville, Ala., 1963-88, team leader contamination control, 1988-90, br. chief space environ. effects br., 1990—. Contbr. articles to profl. jours. Achievements include patents in real time reflectometer. Office: EH15/NASA/MSFC Marshall Space Flight Ctr Huntsville AL 35812

ZWOYER, EUGENE MILTON, consulting engineering executive; b. Plainfield, N.J., Sept. 8, 1926; s. Paul Ellsworth and Marie Susan (Britt) Z.; m. Dorothy Lucille Seward, Feb. 23, 1946; children: Gregory, Jeffrey, Douglas. Student, U. Notre Dame, 1944, Mo. Valley Coll., 1944-45; BS, U. N.Mex., 1947; MS, Ill. Inst. Tech., 1949; PhD, U. Ill., 1953. Mem. faculty U. N.Mex., Albuquerque, 1948-71, prof. civil engring., dir. Eric Wang Civil Engring. Rsch. Facility, 1961-70; rsch. assoc. U. Ill., Urbana, 1951-53; owner, cons. engr. Eugene Zwoyer & Assocs., Albuquerque, 1954-72; exec. dir., sec. ASCE, N.Y.C., 1972-82; pres. Am. Assn. Engring. Socs., N.Y.C., 1982-84; exec. v.p. T.Y. Lin Internat., San Francisco, 1984-86, pres., 1986-89; owner Eugene Zwoyer Cons. Engr., 1989—; chief oper. officer, treas. Polar Molecular Corp., Saginaw, Mich., 1990, exec. v.p., 1991-92. Trustee Small Bus. Research Corp., 1976-80; trustee Engring. Info., Inc., 1981-84; internat. trustee People-to-People Internat. 1974-86; v.p. World Fedn. Engring. Orgns., 1982-85. Served to lt. (j.g.) USN, 1944-46. Named Outstanding Engr. of Yr. Albuquerque chpt. N.Mex Soc. Profl. Engrs., 1969, One Who Served the Best Interests of the Constrn. Industry, Engring. News Record, 1980; recipient Disting. Alumnus award the Civil Engring. Alumni Assn. at U. Ill., 1979, Disting. Alumnus award Engring. Coll. Alumni Assn., U. N.Mex., 1982, Can.-Am. Civil Engring. Amity award Am. Soc. Civil Engrs., 1988, Award for Outstanding Profl. Contbns. and Leadership Coll. Engring. U. N.Mex., 1989. Mem. AAAS, ASCE (dist. bd. dirs. 1968-71), NSPE, Am. Soc. Engring. Edn., Am. Concrete Inst., Nat. Acad. Code Adminstrn. (trustee, mem. exec. com. 1973-79), Engrs. Joint Coun. (bd. dirs. 1978-79), Engring. Soc. Commn. on Energy (bd. dirs. 1977-82), Sigma Xi, Sigma Tau, Chi Epsilon. Home: 6363 Christie Ave Apt 1326 Emeryville CA 94608-1940 Office: Ste 200C 1172 San Pablo Ave Berkeley CA 94706-2245

ZYGMUNT, ZIELIŃSKI, civil engineer, educator; b. Żurawniki, Poland, Nov. 6, 1939; s. Feliks and Helena (Biernacka) Z.; m. Janina Niezgocka, Sept. 1, 1965; children: Monika, Izabela. Degree in civil engring., Wrocław (Poland) Tech. U., 1961, D in Tech. Sci., 1970, PhD in Tech. Sci., 1978. Asst. adj. prof. Wrocław Tech. U., 1962-74; docent Szczecin (Poland) Tech. U., 1974-91, prof., 1991—, head rsch. ctr., 1974—, dep. dir. civil engring. inst., 1976-78, 84-87, prodean bldg. and archtl. faculty, 1981-84; bldg. inspector, Wrocław, 1967-70; verifier Marine Bldg. Design Office, Poland, 1988—. Author: Projektowanie Dróg, 1980, 2d edit., 1989, Roboty Ziemne, 1977; contbr. articles to profl. publs. Recipient Gold Cross of Merit, Govt. of Poland, 1982, award Polish Minister of Sci., Higher Edn. and Technics, 1978, 82, Knigh's Cross, Polonia Restituta Order, Pres. Polen, 1991. Mem. Polish Union Bldg. Engrs. and Technicians (expert 1988—), Assn. Communication and Technicians (vice chmn. tech. sect. 1971-74). Avocations: stamps, travel, motoring. Home: ul Zawadzkiego 164 m 4, 71-246 Szczecin Poland

ZYPMAN-NIECHONSKI, FREDY RUBEN, physicist; b. Montevideo, Uruguay, May 25, 1960; came to U.S., 1985; s. Michael and Julia (Niechonski) Z. BS in Engring, U. de la Repùblica, Montevideo, Uruguay, 1982; MS, Case Western U., 1987, PhD, 1988. Lic. physicist U. Simon Bolivar, Caracas, Venezuela. Rsch. asst. Case Western Res. U., Cleve., 1985-88; faculty assoc. U. N.C., Charlotte, 1988-91; asst. prof. U. P.R., Humacao, 1991—; cons. Technicom, Humacao, P.R., 1991-92, Picker Internat., Cleve., 1988; reviewer Founds. of Physics, Humacao, 1992. Contbr. articles to profl. jours. Recipient fellowship CONICIT, 1983, Case Western Res. U., 1985-88. Mem. IEEE, Am. Phys. Soc., N.Y. Acad. Scis., Am. Inst. Physics. Achievements include discovery of supercoherent states; standars in magnetic resonance imaging; dissipation in quantum wells; scanning tunneling spectroscopy. Office: Dept of Physics Univ of Puerto Rico Humacao PR 00791

ZYROFF, ELLEN SLOTOROFF, information scientist, classicist; b. Atlantic City, N.J., Aug. 1, 1946; d. Joseph George and Sylvia Beverly (Roth) Slotoroff; m. Jack Zyroff, June 21, 1970; children: Dena Rachel, David Aaron. AB, Barnard Coll., 1968; MA, The Johns Hopkins U., 1969, PhD, 1971; MS, Columbia U., 1973. Instr. The Johns Hopkins U., Balt., 1970-71, Yeshiva U., N.Y.C., 1971-72, Bklyn Coll., 1971-72; libr., instr. U. Calif., 1979, 81, 91, San Diego State U., 1981-85; prof. San Diego Mesa Coll., 1981—; dir. The Reference Desk Rsch. Svcs., La Jolla, Calif., 1983—; prin. libr. San Diego County Libr., 1985—; v.p. Archeol. Soc. Am., Balt., 1970-71; chairperson div. coms., Am. Library Assn., 1981--. Author: The Author's Apostrophe in Epic from Homer Through Lucan, 1971, Cooperative Library Instruction for Maximum Benefit, 1989. Pres. Women's Am. ORT, San Diego, 1979-81. Mem. ALA, Am. Philol. Assn., Calif. Libr. Assn., Am. Soc. Info. Sci., Am. Classical League, Toastmasters, Beta Phi Mu. Office: PO Box 12122 La Jolla CA 92039

ZYZNEWSKY, WLADIMIR A., neurologist; b. Venezuela, Feb. 21, 1948; came to U.S., 1972; s. Julian and Alexandra (Schabakewich) Z.; m. Rosa Rivera, Aug. 6, 1981; children: Oksana, Bohdan, Alexandra. MD, U. de Carabobo, Venezuela, 1971. Diplomate Am. Bd. Psychiatry and Neurology. Intern Brooklyn Jewish Hosp., N.Y., 1972-73; resident in phychiatry Beth Israel Hosp., N.Y.C., 1973-76; resident in neurology N.Y. U. Med. Ctr., N.Y.C., 1976-77; mem. staff U. Md., Baltimore, 1977-79; pvt. practice U. Pitts. Hosp., Wheeling, W.Va., 1981—. Office: 2115 Chapline St Ste 101 Wheeling WV 26003

Geographic Index

UNITED STATES

ALABAMA

Alexander City
Powers, Runas, Jr. rheumatologist

Auburn
Barrett, Ronald Martin aerospace engineer
Clark, Terrence Patrick veterinarian, researcher
Crocker, Malcolm John mechanical engineer, noise control engineer, educator
Daron, Harlow H. biochemist
Dobson, F. Stephen ecologist
Gandhi, Shailesh Ramesh biotechnologist
Gill, William Robert soil scientist
Herring, Bruce E. engineering educator
Jaeger, Richard Charles electrical engineer, educator, science center director
Lemke, Paul Arenz botany educator
Mehta, Jagjivan Ram research scientist
Molz, Fred John, III hydrologist, educator
Parsons, Daniel Lankester pharmaceutics educator
Rostohar, Raymond chemist, chromatographer
Schafer, Robert Louis agricultural engineer, researcher
Seidman, Stephen Benjamin computer scientist
Siginer, Dennis A. mechanical engineering educator, researcher
Tippur, Hareesh V. mechanical engineer
Turnquist, Paul Kenneth agricultural engineer, educator
Vodyanoy, Vitaly Jacob biophysicist, educator
Yoo, Kyung Hak agricultural engineer, educator

Birmingham
Adams, Alfred Bernard, Jr. environmental engineer
Allen, James Madison family practice physician, consultant
Anderson, Peter Glennie research pathologist, educator
Balschi, James Alvin medical educator
Barton, James Clyde, Jr. hematologist, medical oncologist
Benos, Dale John physiology educator
Bhugra, Bindu microbiologist, researcher
Birkedal-Hansen, Henning dentist, educator
Briles, David E(lwood) microbiology educator
Brown, Ronnie Jeffrey biologist, researcher
Bueschen, Anton Joslyn physician, educator
Bugg, Charles Edward biochemistry educator, scientist
Bunt, Randolph Cedric mechanical engineer
Cannon, Richard Alan chemical process engineer
Cooper, Max Dale physician, medical educator, researcher
Dahl, Hilbert Douglas mining company executive
Dahlin, Robert Steven chemical engineer
Damato, David Joseph electrical engineer
Durant, John Ridgeway physician
Elgavish, Ada biochemist
Elgavish, Gabriel Andreas physical biochemistry educator
Elkourie, Paul telecommunications company manager
Finley, Wayne House medical educator
Francis, Kennon Thompson physiologist
Friedlander, Michael J. neuroscientist, animal physiologist, medical educator
Giddens, John Madison, Jr. nuclear engineer
Hammond, C(larke) Randolph healthcare executive
Hirschowitz, Basil Isaac physician
Jannett, Thomas Cottongim electrical engineering educator
Johnson, James Hodge engineering company executive
Kahlon, Jasbir Brar viral epidemiologist, researcher
Kelly, David Reid pathologist
Koopman, William James medical educator, internist, immunologist
Kumar, Manish engineer
Lammertsma, Koop chemist, consultant
McCarl, Henry N. economics and geology educator
Miller, George McCord electrical engineer
Moore, William Gower Innes biochemist
Mowry, Robert Wilbur pathologist, educator
Nuckols, Frank Joseph psychiatrist
Palmisano, Paul Anthony pediatrician, educator
Parks, Rodney Keith mechanical engineer, consultant
Pitt, Robert Ervin environmental engineer, educator
Pittman, Constance Shen physician, educator
Pittman, James Allen, Jr. endocrinologist, dean, educator
Pohost, Gerald M. cardiologist, medical educator
Prejean, J. David toxicologist
Pritchard, David Graham research scientist, educator
Pulliam, Terry Lester chemical engineer
Quinn, Charles Layton, Jr. electrical engineer
Ramey, Craig T. pschology educator
Richards, J. Scott rehabilitation medicine professional
Rigney, E. Douglas biomedical engineer
Rose, Lucy McCombs chemist
Rouse, John Wilson, Jr. research institute administrator
Sanders, Michael Kevin hypertension researcher
Scott, Owen Myers, Jr. nuclear engineer
Seaman, Duncan Campbell civil engineer
Sekar, M. Chandra pharmacology researcher, educator
Shanks, Stephen Ray safety engineering consultant
Shaw, Johnny Harvey civil engineer
Shealy, Y. Fulmer biochemist
Singh, Raj Kumar biochemist, researcher
Skalka, Harold Walter ophthalmologist, educator
Smith, John Stephen educational administrator
Sobol, Wlad Theodore physicist
Spears, Randall Lynn civil engineer
Stephens, Deborah Lynn health facility executive
Stover, Samuel Landis physiatrist
Strickler, Howard Martin physician
Warnock, David Gene nephrologist, pharmacology educator
Wideman, Gilder LeVaugh obstetrician/gynecologist
Williamson, Edward L. retired nuclear engineer, consultant
Williford, Robert Marion civil engineer
Yenice, Mehmet Fikri aeronautics company executive

Dauphin Island
Plakas, Steven Michael biological research scientist

Dozier
Grantham, Charles Edward broadcast engineer

Florence
Burford, Alexander Mitchell, Jr. physician, pathologist

Fort McClellan
Tran, Toan Vu electronics engineer

Gadsden
Hanson, Ronald Windell cardiologist

Gurley
Souder, Edith Irene information scientist

Harvest
Colbert, Robert Floyd aerospace engineer

Huntsville
Ballard, Richard Owen rocket propulsion engineer
Blackmon, James Bertram aerospace engineer
Boody, Frederick Parker, Jr. nuclear engineer, optical engineer
Bramon, Christopher John aerospace engineer
Buckelew, Robin Browne aerospace engineer, manager
Buddington, Patricia Arrington engineer
Burns, Pat Ackerman Gonia infosytems specialist, software engineer
Bushman, David Mark aerospace engineer
Campbell, Jonathan Wesley astrophysicist, aerospace engineer
Carter, Regina Roberts physicist
Chappell, Charles Richard space scientist
Chasman, Daniel Benzion aerospace engineer
Chassay, Roger Paul, Jr. engineering executive, project manager
Cowan, Penelope Sims materials engineer
Craig, Thomas Franklin electrical engineer
Curreri, Peter Angelo materials scientist
Dannenberg, Konrad K. aeronautical engineer
Daussman, Grover Frederick industrial engineer, consultant
Davis, Stephan Rowan aerospace engineer
Dayton, Deane Kraybill computer company executive
Decher, Rudolf physicist
Derrickson, James Harrison astrophysicist
Douillard, Paul Arthur engineering and financial executive, consultant
Duncan, Lisa Sandra engineer
Dunn, Karl Lindemann electrical engineer
Eagles, David M. physicist
Emerson, William Kary engineering company executive
Farley, Wayne Curtis mechanical engineer
Gore, James William civil engineer
Griner, Donald Burke engineer
Hadaway, James Benjamin physicist
Haeussermann, Walter systems engineer
Hung, Ru J. engineering educator
Imtiaz, Kauser Syed aerospace engineer
Jaenisch, Holger Marcel physicist
Jones, Jack Allen aerospace engineer
Kestle, Wendell Russell cost and economic analyst, consultant
Kim, Young Kil aerospace engineer
Krishnan, Anantha mechanical engineer
Lee, Thomas J. aerospace scientist
LeMaster, Robert Allen mechanical engineer
Leonard, Kathleen Mary environmental engineering educator
Liggett, Mark William mechanical engineer
Lundquist, Charles Arthur university official
Marx, Richard Brian forensic scientist
Mauldin, Charles Robert aerospace engineer
McAuley, Van Alfon aerospace mathematician
McCollough, Michael Leon astronomer
Modlin, James Michael mechanical engineer, army officer
Montgomery, Willard Wayne physicist
Musielak, Zdzislaw Edward physicist, educator
Perkins, James Francis physicist
Polites, Michael Edward aerospace engineer
Powers, Clifford Blake, Jr. communications researcher
Preston, R. Kevin software engineer
Pruitt, Alice Fay mathematician, engineer
Reiss, Donald Andrew physicist
Roberts, Thomas George retired physicist
Rubin, Bradley Craig astrophysicist
Savage, Julian Michele civil engineer
Scarl, Ethan Adam computer scientist
Schroer, Bernard Jon industrial engineering educator
Sharpe, Mitchell Raymond science writer
Stevens, Dale Marlin civil engineer
Stone, Richard John physicist
Tietke, Wilhelm gastroenterologist
Tiwari, Subhash Ramadhar chemical engineer
Torres, Antonio civil engineer
Vaughan, Otha H., Jr. aerospace engineer, research scientist
Vinz, Frank Louis electrical engineer
Wang, Zhi Jian aerospace engineer
Watson, Raymond Coke, Jr. college president, engineering educator, mathematics educator
Weeks, David Jamison electrical engineer
Witherow, William Kenneth physicist
Yang, Hong-Qing mechanical engineer
Zana, Donald Dominick information resource manager
Zwiener, James Milton physicist

Madison
Frakes, Lawrence Wright program analyst, logistics engineer
Hawk, Clark Wiliams mechanical engineering educator
Rosenberger, Franz Ernst physics educator
Sarphie, David Francis biomedical engineer

Maxwell AFB
Frank, Paul Sardo, Jr. forester, air force officer

McIntosh
Von Tersch, Frances Knight analytical chemist

Mobile
Beeman, Curt Pletcher research chemist
Boulian, Charles Joseph civil-structural engineer
Campbell, Naomi Flowers biochemist
Chandler, Thomas Eugene industrial engineer
Esham, Richard Henry internal medicine and geriatrics educator
Harpen, Michael Dennis physicist, educator
Lambert, James LeBeau chemistry educator
Madura, Jeffry David chemistry educator
Nettles, Joseph Lee dentist
Omar, Husam Anwar civil engineering educator
Parmley, Loren Francis, Jr. medical educator
Raider, Louis physician, radiologist
Taylor, Washington Theophilus mathematics educator
Vitulli, William Francis psychology educator
Webber, David Michael civil engineer
Wilkerson, William Edward, Jr. civil engineer
Wilkinson, Ronald Eugene engineer

Montevallo
Braid, Malcolm Ross biology educator

Montgomery
Adams, James Henry chemical engineer
Bezoari, Massimo Daniel chemistry educator, writer
Cain, Steven Lyle aerospace engineer
Fulmer, Kevin Michael environmental engineer
Greene, Ernest Rinaldo, Jr. anesthesiologist, chemical engineer
Harrell, Barbara Williams public health administrator
Hornsby, Andrew Preston, Jr. human services administrator
Longuet, Gregory Arthur automation engineer, consultant
Parmer, Dan Gerald veterinarian
Tan, Boen Hie biochemist

Moss Point
Harris, Kenneth Kelley chemical engineer

Normal
Caulfield, Henry John physics educator
Coleman, Tommy Lee soil science educator, researcher, laboratory director
Mays, David Arthur agronomy educator, small business owner
Pacumbaba, R.P. plant pathologist, educator
Sabota, Catherine Marie horticulturist, educator
Wang, Jai-Ching physics educator

Pelham
Turner, Malcolm Elijah biomathematician, educator

Pennington
Broussard, Peter Allen energy engineer

Redstone Arsenal
Cotney, Carol Ann research aerospace engineer
Glenn, Mark William operations research analyst
Hollowell, Monte J. engineer, operations research analyst
Jaklitsch, Donald John materials engineer, consultant
Miller, Walter Edward physical scientist, researcher
Pittman, William Claude electrical engineer
Smith, Troy Alvin aerospace research engineer

Selma
Collins, Eugene Boyd chemist, molecular pathologist, consultant

Theodore
Smith, David Floyd chemical engineer

Trussville
Davey, James Joseph process control engineer

Tuscaloosa
Carter, Olice Cleveland, Jr. civil engineer
Churchill, Sharon Anne-Kernicky research engineer, consultant
Clavelli, Louis John physicist
Cooper, Eugene Bruce speech-language pathologist, educator
Curtner, Mary Elizabeth psychologist
Darden, William Howard, Jr. biology educator
Flinn, David R. federal agency research director
Jones, Jerry Edward family physician, educator
LaMoreaux, Philip Elmer geologist, hydrogeologist, consultant
Mancini, Ernest Anthony geologist, educator, researcher
Martin, James Arthur aerospace educator
Moody, Maxwell, Jr. retired physician
Stitt, Kathleen Roberta nutrition educator
Todd, Beth Ann mechanical engineer, educator
Weaver, Jerry Reece management scientist, educator
Williams, Maurice clinical psychologist
Xu, ZhaoRan botanist

Tuskegee
Adeyeye, Samuel Oyewole nutritional biochemist
Almazan, Aurea Malabag biochemist
Datiri, Benjamin Chumang soil scientist
Koons, Lawrence Franklin chemistry educator

ALASKA

Anchorage
Armstrong, Richard Scott mechanical engineer
Condy, Sylvia Robbins psychologist
Crawford, E(dwin) Ben psychologist
Gier, Karan Hancock counseling psychologist
Guinn, Janet Martin psychologist, consultant
Kinney, Donald Gregory civil engineer
Mann, Lester Perry mathematics educator
Watts, Michael Arthur materials engineer

Fairbanks
Duffy, Lawrence Kevin biochemist, educator
Fathauer, Theodore Frederick meteorologist
Guthrie, Russell Dale vertebrate paleontologist
Hoskins, L. Claron chemistry educator
Jewett, Stephen Carl marine biologist, researcher, consultant
Kessel, Brina educator, ornithologist
Nelson, Michael Gordon mining engineer, educator, consultant
Osborne, Daniel Lloyd electronics engineer
Sengupta, Mritunjoy mining engineer, educator
Stragier, Cynthia Andreas pharmacist
Zarling, John Paul mechanical engineering educator

Juneau
Angelo, Michael Arnold information scientist
Mala, Theodore Anthony physician, state official
Ritter, Grant L. water operations supervisor

Kenai
Morse, Aaron Holt chemist, consultant

Tuntutuliak
Bond, Ward C. mathematics and computer educator

ARIZONA

Amado
Weekes, Trevor C. astrophysicist

Buckeye
Burton, Edward Lewis industrial procedures and training consultant

Carefree
Johnson, Charles Foreman architect, architectural photographer, planning architecture and system engineering consultant
Robbins, Conrad W. naval architect

Casa Grande
Krauss, Sue Elizabeth radiological medical management technologist
McGillicuddy, Joan Marie psychotherapist, consultant

Cave Creek
MacKay, John mechanical engineer

Chandler
Basilier, Erik Nils software scientist and engineer
Hatch, Jeffrey Scott chemist
Newman, Brett aerospace engineer
Vondra, Lawrence Steven aerospace engineer
Yamada, Toshishige electrical engineer

Claypool
Halstead, John Irvin, II analytical chemist
Lowe, Douglas George analytical chemist

Coolidge
Hiller, William Clark physics educator, engineering educator, consultant

Cortaro
Lindsey, Douglas trauma surgeon

Flagstaff
Avery, Charles Carrington forestry educator, researcher
Colbert, Edwin Harris paleontologist, museum curator
Hammond, Howard David retired editor
Irving, Douglas Dorset behavioral scientist, consultant
Millis, Robert Lowell astronomer
Phillips, Arthur Morton, III botanist, consultant
Shoemaker, Eugene Merle geologist
Venedam, Richard Joseph chemist, educator
Windes, James Dudley psychology educator

Florence
Girtman, Gregory Iverson psychologist

Fort Huachuca
Clark, Brian Thomas mathematical statistician, operations research analyst
Riordon, John Arthur electrical engineer

Gilbert
Lamb, Edward Allen, Jr. business owner

Goodyear
Cabaret, Joseph Ronald electronics company executive

Green Valley
Fischer, Harry William radiologist, educator

Litchfield Park
Kramer, Rex W., Jr. former naval officer, business executive
Miller, Kenneth Edward mechanical engineer, consultant
Reid, Ralph Ralston, Jr. electronics executive, engineer

Maricopa
Ellsworth, Peter Campbell entomologist

Mesa
Boisjoly, Roger Mark structural engineer
Boren, Kenneth Ray endocrinologist
Bunchman, Herbert Harry, II plastic surgeon
Fogle, Homer William, Jr. electrical engineer, inventor
Horne, Jeremy fiber optics, computer research executive
Joardar, Kuntal electrical engineer
Kunz, Donald Lee aeronautical engineer
Scofield, Larry Allan civil engineer
Stemple, Alan Douglas aerospace engineer

Miami
McWaters, Thomas David mining engineer, consultant

Oracle
Augustine, Margret L. architect, designer

Page
Leus McFarlen, Patricia Cheryl water chemist

Paradise Valley
Butler, Byron Clinton physician, cosmologist, gemologist, scientist
Gookin, William Scudder hydrologist, consultant

Peoria
McKee, Margaret Crile pulmonary medicine and critical care physician

Phoenix
Adler, Eugene Victor forensic toxicologist, consultant
Allen, John Rybolt L. chemist, biochemist
Archer, Gregory Alan clinical psychologist, writer
Atutis, Bernard P. manufacturing company executive
Bachus, Benson Floyd mechanical engineer, consultant
Bouwer, Herman laboratory executive
Brittingham, James Calvin nuclear engineer
Castaneda, Mario chemical engineer, educator
Chan, Michael Chiu-Hon chiropractor
Charlton, John Kipp pediatrician
Chisholm, Tom Shepherd environmental engineer
Christensen, Karl Reed aerospace engineer
Clay, Ambrose Whitlock Winston telecommunications company executive
Dittemore, David H. aerospace engineer, engineering executive
Doyle, Michael Phillip civil engineer
Drea, Edward Joseph pharmacist
Elien, Mona Marie air transportation professional
Faul, Gary Lyle electrical engineering supervisor
Flitman, Stephen Samuel neurologist
Freyermuth, Clifford L. structural engineering consultant
Garcia, Richard Louis plant physiologist/micrometeorologist, researcher
Hancock, Thomas Emerson educational psychologist, educator
Helms, Mary Ann critical care nurse, consultant
Kleyman, Henry Semyon electrical engineer
Landrum, Larry James computer engineer
Lorenzen, Robert Frederick ophthalmologist
Massey, L. Edward chemical marketing executive
Meinhart, Robert David analytical toxicologist
Myers, Gregory Edwin aerospace engineer
Peterson, Stephen Cary mechanical engineer
Roop, Mark Edward process control applications engineer
Sauer, Barry W. medical research center administrator, bioengineering educator
Schiffner, Charles Robert architect
Schmitt, Louis Alfred engineer
Siegel, Richard Steven water resource specialist
Smarandache, Florentin mathematics researcher, writer
Struble, Donald Edward mechanical engineer
Thomas, Harold William avionics systems engineer, flight instructor
Williams, Damon S. civil engineer
Wood, John Mortimer aerospace executive, aeronautical engineer
Wu, Mei Qin mechanical engineer
Zeilinger, Philip Thomas aeronautical engineer
Zweifel, Terry L. aeronautical engineer, researcher

Prescott
Davidson-Moore, Kathy Louise psychologist
Hasbrook, A. Howard aviation safety engineer, consultant
Longfellow, Layne psychologist
Moses, Elbert Raymond, Jr. speech and dramatic arts educator

Prescott Valley
Beck, John Roland environmental consultant

Riviera
Jones, Vernon Quentin surveyor

Scottsdale
Aybar, Charles Anton aviation executive
Barbee, Joe Ed lawyer
DeHaven, Kenneth Le Moyne retired physician
Friedman, Shelly Arnold cosmetic surgeon
Gall, Donald Alan data processing executive
Gilson, Arnold Leslie engineering executive
Kline, Arthur Jonathan electronics engineer
Leeland, Steven Brian electronics engineer
Leeser, David O. materials engineer, metallurgist
Millett, Merlin Lyle aerospace consultant, educator
Newman, Marc Alan electrical engineer
Pomeroy, Kent Lytle physical medicine and rehabilitation physician
Teletzke, Gerald Howard environmental engineer

Sedona
D'Javid, Ismail Faridoon surgeon
Dorrell, Vernon Andrew engineering executive
Eggert, Robert John, Sr. economist
Silvern, Leonard Charles engineering executive
Stamm, Robert Franz research physicist

Sells
Rubendall, Richard Arthur civil engineer

Sierra Vista
Fletcher, Craig Steven electronic technician
Ricco, Raymond Joseph, Jr. computer systems engineer

Sun City
Cannady, Edward Wyatt, Jr. retired physician
Dale, Martha Ericson clinical psychologist, educator
Dapples, Edward Charles geologist, educator
Sabanas-Wells, Alvina Olga orthopedic surgeon
Vander Molen, Jack Jacobus engineering executive, consultant
Woodruff, Neil Parker agricultural engineer

Sun City West
Calderwood, William Arthur physician

Tempe
Ahmed, Kazem Uddin technical staff member, consultant
Anand, Suresh Chandra physician
Backus, Charles Edward engineering educator, researcher
Berman, Neil Sheldon chemical engineering educator
Bilimoria, Karl Dhunjishaw aerospace engineer
Blankenship, Robert Eugene chemistry educator
Brazel, Anthony James geographer, climatologist
Bristol, Stanley David mathematics educator
Carpenter, Ray Warren materials scientist and engineer, educator
Chaffee, Frederic H., Jr. astronomer
Chattopadhyay, Aditi aerospace engineering educator
Coleman, Edwin DeWitt, III electronics engineer
Cowley, John Maxwell physics educator
Downs, Floyd L. mathematics educator
Fernando, Harindra Joseph engineering educator
Ferreira, Jay Michael mechanical engineer
Garcia, Antonio Agustin chemical engineer, bioengineer, educator
Greenfield, James Dwight technical writer
Guilbeau, Eric J. biomedical engineer, electrical engineer, educator
Hageman, Brian Charles researcher
Hashemi-Yeganeh, Shahrokh electrical engineering educator, researcher
Hickson, Robin Julian mining company executive
Ihrig, Edwin Charles, Jr. mathematics educator
Johnson, Ross Jeffrey statistician
Laananen, David Horton mechanical engineer, educator
Marcus, David Alan physicist
Marusiak, Ronald John quality engineer, electronics executive
McCarthy, David Edward energy conservation engineer
Metcalf, Virgil Alonzo economics educator
Metzger, Darryl Eugene mechanical and aerospace engineering educator
Mock, Peter Allen hydrogeologist
Moore, Carleton Bryant geochemistry educator
Morgan, J. Ronald aerospace company executive
Nigam, Bishan Perkash physics educator
Noce, Robert Henry neuropsychiatrist, educator
Opie, Jane Maria audiologist
Oscherwitz, Steven Lee internist, infectious disease physician
Parkhurst, Charles Lloyd electronics company executive
Pettit, George Robert chemistry educator, cancer researcher
Péwé, Troy Lewis geologist, educator
Roberts, Peter Christopher Tudor engineering executive
Robertson, Samuel Harry, III transportation safety research engineer, educator
Rodriguez, Armando Antonio electrical engineering educator
Salmirs, Seymour aeronautical engineer, educator
Schneller, Eugene S. sociology educator
Shaw, Milton Clayton mechanical engineering educator
Singhal, Avinash Chandra engineering educator
Skibitzke, Herbert Ernst, Jr. hydrologist
Swink, Laurence Nim chemist, consultant
Thums, Charles William designer, consultant
Tillery, Bill W. physics educator
Vandenberg, Edwin James chemist, educator
Zhou, Jing-Rong electrical engineer, researcher

Tucson
Angel, James Roger Prior astronomer
Anthes, Clifford Charles retired mechanical engineer, consultant
Arabyan, Ara mechanical engineer
Brusseau, Mark Lewis environmental educator, researcher
Burgess, Kathryn Hoy biologist
Burrows, Adam Seth physicist
Burrows, Benjamin physician, educator
Carruthers, Peter Ambler physicist, educator
Carter, L. Philip neurosurgeon, consultant
Chalfoun, Nader Victor architect, educator
Chau, Lai-Kwan research chemist
Cheddar, Donville Glen chemist, educator
Cogut, Theodore Louis environmental specialist, meteorologist
Cortner, Hanna Joan science administrator, research scientist, educator
Craghead, James Douglas civil engineer
De Young, David Spencer astrophysicist
Enriquez, Francisco Javier immunologist
Feeney, Craig Michael chemist
Fink, James Brewster geophysicist, consultant
Foster, Kennith Earl life sciences educator
Fritts, Harold Clark dendrochronology educator, researcher
Gall, Eric Papineau physician educator
Gerba, Charles Peter microbiologist, educator
Glick, Michael Andrew aerospace engineer
Glisky, Elizabeth Louise psychology educator
Goodall, Jane ethnologist
Hagedorn, Henry Howard entomology educator
Hatfield, (David) Brooke analytical chemist, spectroscopist
Hetrick, David LeRoy nuclear engineering educator
Hill, Henry Allen physicist, educator
Hofstadter, Daniel Samuel aerospace engineer
Homburg, Jeffrey Allan geoarchaeologist
Howard, Robert Franklin observatory administrator, astronomer
Hunten, Donald Mount planetary scientist, educator
Isenhower, William Martin civil engineering educator
Jackson, Kenneth Arthur physicist, researcher
Jefferies, John Trevor astronomer, astrophysicist, observatory administrator
Jeter, Wayburn Stewart retired microbiology educator, microbiologist
Jones, Roger Clyde electrical engineer, educator
Kaltenbach, Carl Colin agriculturist, educator
Kamilli, Robert Joseph geologist
Karson, Catherine June computer programmer, consultant
Katakkar, Suresh Balaji hematologist, oncologist
Kececioglu, Dimitri Basil mechanical engineering educator
Kessler, John Otto physicist, educator
Kiersch, George Alfred geological consultant, educator emeritus
Kilkson, Rein physics educator
Kingery, William David ceramics and anthropology educator
Kischer, Clayton Ward embryologist, educator
Krider, E. Philip atmospheric scientist, educator
Lamb, Lowell David physicist
Lamb, Willis Eugene, Jr. physicist, educator
Lane, Leonard J. hydrologist
Leaman, Gordon James, Jr. chemical engineer
Lebl, Michal peptide chemist
Leibacher, John William astronomer
Levy, Eugene Howard planetary sciences educator, researcher
Long, Austin geosciences educator
Manciet, Lorraine Hanna physiologist, educator
Marchalonis, John Jacob immunologist, educator
Massey, Lawrence Jeremiah broadband engineer
McCluskey, Kevin fungal molecular geneticist
Merilan, Jean Elizabeth statistics educator
Milberg, Morton Edwin chemist
Mollova, Nevena Nikolova chemist
Monteiro, Renato Duarte Carneiro industrial engineer, educator, researcher
More, Syver Wakeman geologist
Netting, Robert M. anthropology educator
Parks, Robert Edson optical engineer
Payne, Claire Margaret molecular and cellular biologist
Peyghambarian, Nasser optical science educator
Potts, Albert M. ophthalmologist
Ran, Chongwei geotechnical engineer
Regan, John Ward molecular pharmacologist
Ricke, P. Scott obstetrician/gynecologist
Roemer, Elizabeth astronomer, educator
Rose, Hugh management consultant
Rosenzweig, Michael Leo ecology educator
Sears, William Rees engineering educator
Shannon, Robert Rennie optical sciences center administrator, educator
Shropshire, Donald Gray hospital executive
Slack, Marion Kimball pharmacy educator
Smerdon, Ernest Thomas academic administrator
Soren, David archaeology educator, administrator
Speas, Robert Dixon aeronautical engineer, aviation company executive
Strittmatter, Peter Albert astronomer, educator
Thomson, Cynthia Ann clinical nutrition research specialist
Tifft, William Grant astronomer
Trueblood, Mark computer programmer, author
Verdery, Roy Burton, III gerontologist, consultant
Wait, James Richard electrical engineering educator, scientist
Wallace, Terry Charles, Jr. geophysicist, educator
Willis, Clifford Leon geologist
Winfree, Arthur Taylor biologist, educator
Witten, Mark Lee lung injury research scientist, educator
Wolff, Sidney Carne astronomer, observatory administrator
Wyant, James Clair engineering company executive, educator
Yocum, Harrison Gerald horticulturist, botanist, educator, researcher

ARKANSAS

Bauxite
Watson, David Raymond chemist

Calico Rock
Grasse, John M., Jr. physician, missionary

Conway
Hamblin, Daniel Morgan economist
Holt, Frank Ross retired aerospace engineer
McCarron, Robert Frederick, II orthopedic surgeon
Spatz, Kenneth Chris statistics educator

De Valls Bluff
Jones, Robert Eugene physician

El Dorado
Dlabach, Gregory Wayne mathematics educator
Reames, Thomas Eugene chemical engineer

Fayetteville
Beyrouty, Craig A. agronomist, educator
Bhunia, Arun Kumar microbiologist, immunologist, researcher
Combs, Linda Jones business administration educator, researcher
Etges, William James biologist
Gaddy, James Leoma chemical engineer, educator
Green, Otis Michael electrical engineer, consultant
Jones, Fay architect
Lacy, Claud H. Sandberg astronomer
LeFevre, Elbert Walter, Jr. civil engineering educator
Morris, Justin Roy food scientist, enologist, consultant
Revels, Mia Renea science educator
Schmitt, Neil Martin biomedical engineer, electrical engineering educator
Selvam, Rathinam Panneer civil engineering educator
Taylor, Gaylon Don industrial engineering educator
Templeton, George Earl, II plant pathologist

Fort Smith
Ashley, Ella Jane (Ella Jane Rader) medical technologist
Coleman, Michael Dortch nephrologist
Snider, James Rhodes radiologist
Still, Eugene Fontaine, II plastic surgeon, educator

Huntsville
Carr, Gerald Paul former astronaut, business executive, former marine officer

Jefferson
Heflich, Robert Henry microbiologist
Sutherland, John Bruce, IV microbiologist

Jonesboro
Stalcup, Thomas Eugene, Sr. civil engineer

Little Rock
Bannon, Gary Anthony molecular geneticist
Benson, Barton Kenneth civil engineer
Chang, Louis Wai-wah medical educator, researcher
Compadre, Cesar Manuel chemistry educator
Darsey, Jerome Anthony chemistry educator, researcher
Dykman, Roscoe Arnold psychologist, educator
Fribourgh, James Henry university administrator
Garcia-Rill, Edgar Enrique neuroscientist
Hardin, James Webb molecular biologist, medical center administrator
Hauser, Simon Petrus hematologist
Hinson, Jack Allsbrook research toxicologist, educator
Hocott, Joe Bill chemical engineer
Liu, Shi Jesse physiologist, researcher
Lyublinskaya, Irina E. physicist, researcher
Maloney, Francis Patrick physiatrist
McAllister, Russell Benton biologist, researcher
McCoy, John Greene computer applications consultant
Sotomora-von Ahn, Ricardo Federico pediatrician, educator
Talburt, John Randolph computer science educator
Townsend, James Willis computer scientist
Wessinger, William David pharmacologist

Magnolia
Willis, Rebecca Lynn chemist, educator

Marianna
Carroll, Stephen Douglas chemist, research specialist

Monticello
Cain, Michael Dean research forester

Morrilton
Adams, Earle Myles technical representative

North Little Rock
Amick, S. Eugene military engineer
Biondo, Raymond Vitus dermatologist
Lawson, William Bradford psychiatrist

Paragould
Pickney, Charles Edward environmental engineer

Rogers
Parris, Luther Allen gemologist, goldsmith

Russellville
Albert, Milton John retired chemist, microbiologist
Allen, Rhouis Eric electrical engineer
Cook, Stephen Patterson physical sciences and astronomy educator, researcher
Cooper, Robert Michael nuclear energy industry specialist
Krenke, Frederick William electrical engineer
Meeks, Lisa Kaye hydrogeologist, researcher
Proffitt, Alan Wayne engineering educator

Springdale
Pogue, William Reid former astronaut, foundation executive, business and aerospace consultant

State University
Bednarz, James C. wildlife ecologist educator

Stuttgart
Miller, Thomas Nathan chemical engineer

Texarkana
Harrison, James Wilburn gynecologist

CALIFORNIA

Agoura Hills
Bleiberg, Leon William surgical podiatrist
Chang, Chong Eun chemical engineer
Meeks, Crawford Russell, Jr. mechanical engineer

Alameda
Billings, Thomas Neal computer and publishing executive, management consultant
Brown, Stephen Lawrence environmental consultant
Mandel, Ronald James neuroscientist
Paustenbach, Dennis James environmental toxicologist
Schoeler, George Bernard medical entomologist

Albany
Chook, Edward Kongyen disaster medicine educator
Sullivan, Robert Scott architect, graphic designer

Alhambra
Anderson, Gordon MacKenzie petroleum service contractors executive
Kister, Henry Z. chemical engineer
Luk, King Sing engineering company executive, educator
Moeller, Ronald Scott mechanical engineer

Sepulveda, Eduardo Solideo *chemical engineer*

Alta Loma
Myers, Edward E. *retired biology and anthropology educator*

Altadena
Fairbanks, Mary Kathleen *data analyst, researcher*
Norton, Harry Neugebauer *aerospace engineer*

Anaheim
Aronowitz, Frederick *physicist*
Bailey, Don Matthew *aerospace company executive*
Glad, Dain Sturgis *aerospace engineer*
Hubbard, Charles Ronald *engineering executive*
Liem, Annie *pediatrician*
McLuckey, John Alexander, Jr. *electronics company executive*
Prince, Warren Victor *mechanical engineer*
Rivera, Armando Remonte *utilities engineer*
Schlose, William Timothy *health care executive*
Shah, Kirti Jayantilal *electrical engineer*

Arcadia
Anderson, Holly Geis *medical clinic executive, radio personality*
Broderick, Donald Leland *electronics engineer*
Proctor, Richard J. *geologist, consultant*
Sloane, Beverly LeBov *educational writer, consultant*

Arcata
Chaney, Ronald Claire *environmental engineering educator, consultant*
Hu, Senqi *psychologist, educator*
Russell, John Blair *chemistry educator*

Artesia
Kurtz, Russell Marc *laser/optics engineer*

Auburn
DeCosta, William Joseph *manufacturing engineer*

Avila Beach
Riches, Kenneth William *nuclear regulatory engineer*

Azusa
Works, Madden Travis, Jr. (Pat Works) *operations executive, author, skydiving instructor, skydiving publications executive*

Bakersfield
Gillie, Michelle Francoise *industrial hygienist*
Starzer, Michael Ray *petroleum engineer*

Barstow
Sutterby, Larry Quentin *internist*

Bellflower
Richardson, Bryan Kevin *mathematics educator, naval aviator*

Belmont
Jaffe, David Henry *computer systems engineer*
Schreiber, Andrew *psychotherapist*

Belvedere Tiburon
Hinshaw, Horton Corwin *physician*

Berkeley
Alonso, Jose Ramon *physicist*
Alpen, Edward Lewis *biophysicist, educator*
Ames, Bruce N(athan) *biochemistry and molecular biology educator*
Attwood, David Thomas *physicist, educator*
Baker, Herbert George *botany educator*
Bergman, Robert George *chemist, educator*
Berry, William Benjamin Newell *geologist, educator, former museum administrator*
Bickel, Peter John *statistician, educator*
Block, Jack *psychology educator*
Bolt, Bruce Alan *seismologist, educator*
Budinger, Thomas Francis *radiologist, educator*
Calvin, Melvin *chemist, educator*
Canfield, Judy Ohlbaum *psychologist*
Carmichael, Ian Stuart Edward *geologist, educator*
Casson, Andrew J. *mathematician educator*
Castro, Joseph Ronald *physician, oncology researcher, educator*
Chamberlain, Owen *nuclear physicist*
Chern, Shiing-Shen *mathematics educator*
Cline, Thomas Warren *molecular biologist, educator*
Cozzarelli, Nicholas R. *molecular biologist, educator*
Daftari, Inder Krishan *physicist, researcher*
Dake, Karl Manning *research psychologist*
Dauben, William Garfield *chemist, educator*
Day, Lucille Elizabeth *laboratory administrator, educator, author*
Denn, Morton Mace *chemical engineering educator*
DePaolo, Donald James *earth science educator*
Desoer, Charles Auguste *electrical engineer*
Diamond, Richard Martin *nuclear chemist*
Dornfeld, David A. *engineering educator*
Fisher, Robert Amos *physical chemist*
Fowler, Thomas Kenneth *physicist*
Fox, Joan Phyllis *environmental engineer*
Fuhs, G(eorg) Wolfgang *medical research scientist*
Gaillard, Mary Katharine *physics educator*
Gardner, Wilford Robert *physicist, educator*
Glaser, Donald A(rthur) *physicist*
Goldhaber, Gerson *physicist, educator*
Goodman, Corey Scott *neurobiology educator, researcher*
Gough, Harrison Gould *psychologist, educator*
Graham, Susan Lois *computer science educator, consultant*
Grube, Joel William *psychologist, researcher*
Gurgin, Vonnie Ann *social scientist*
Hahn, Erwin Louis *physicist, educator*
Hall, Bronwyn Hughes *economics educator*
Haller, Eugene Ernest *materials scientist, educator*
Harris, Guy Hendrickson *chemical research engineer*
Head-Gordon, Teresa Lyn *chemist*
Heathcock, Clayton Howell *chemistry educator, researcher*
Helmholz, August Carl *physicist, educator emeritus*
Hoffman, Darleane Christian *chemistry educator*
Holder, Harold D. *public health administrator, communications specialist, educator*
Hsu, Chieh Su *applied mechanics engineering educator, researcher*
Huff, Welcome Rex Anthony *chemical researcher*
Jeanloz, Raymond *geophysicist, educator*
Jewell, William Sylvester *engineering educator*
Johanson, Donald Carl *physical anthropologist*
Johnson, Ned Keith *ornithologist, educator*
Johnston, Harold S(ledge) *chemistry educator*
Kim, Sung-Hou *chemistry educator, biophysical and biological chemist*
King, Mary-Claire *epidemiologist, educator, geneticist*
Komvopoulos, Kyriakos *mechanical engineer, educator*
Koshland, Daniel Edward, Jr. *biochemist, educator*
Koshland, Marian Elliott *immunologist, educator*
Kubler-Ross, Elisabeth *physician*
Laitone, Edmund Victor *mechanical engineer*
Lee, Yuan T(seh) *chemistry educator*
Leemans, Wim Pieter *physicist*
Lemmon, Hal Elmont *computer scientist*
Leopold, Luna Bergere *geology educator*
Lidicker, William Zander, Jr. *zoologist, educator*
Lucas, Donald *physical chemist*
Macher, Janet Marie *industrial hygienist*
Madabhushi, Govindachari Venkata *civil engineer*
Mandelstam, Stanley *physicist*
Marsden, Jerrold Eldon *mathematician, educator*
McKinney, Wayne Richard *physicist*
Mikesell, Walter R., Jr. *mechanical engineer, engineering executive*
Miller, William Hughes *theoretical chemist, educator*
Mitchell, James Kenneth *civil engineer, educator*
Monismith, Carl Leroy *civil engineering educator*
Mote, Clayton Daniel, Jr. *mechanical engineer, educator*
Muller, Richard Stephen *electrical engineer, educator*
Ott, David Michael *engineering company executive*
Park, Roderic Bruce *university official*
Patterson, David Andrew *computer scientist, educator, consultant*
Penry, Deborah L. *biologist, educator*
Pfeiffer, Juergen Wolfgang *biologist*
Pines, Alexander *chemistry educator, researcher*
Pitelka, Frank Alois *zoologist, educator*
Pitzer, Kenneth Sanborn *chemist, educator*
Poinar, George Orlo, Jr. *insect pathologist and paleontologist, educator*
Popov, Egor Paul *engineering educator*
Profet, Margie *biomedical researcher*
Quesada, Mark Alejandro *physical chemist*
Ratner, Marina *mathematician, educator, researcher*
Rice, John Andrew *statistician*
Ritchie, Robert Oliver *materials science educator*
Rock, Irvin *research scientist*
Rosenblatt, Gerd Matthew *chemist*
Rosenzweig, Mark Richard *psychology educator*
Rubin, Gerald Mayer *molecular biologist, biochemistry educator*
Sadoulet, Bernard *astrophysicist, educator*
Saykally, Richard James *chemistry educator*
Schultz, Peter G. *chemistry educator*
Seaborg, Glenn Theodore *chemistry educator*
Shank, Charles Vernon *science administrator, educator*
Sheen, Portia Yunn-ling *retired physician*
Shen, Hsieh Wen *civil engineer, consultant, educator*
Shen, Yuen-Ron *physics educator*
Silva, Paul Claude *botanist*
Siri, William E. *physicist*
Smelser, Neil Joseph *sociologist*
Somorjai, Gabor Arpad *chemist, educator*
Spinrad, Hyron *astronomer*
Still, Gerald G. *plant physiologist, research director*
Streitwieser, Andrew, Jr. *chemistry educator*
Suzuki, Yuichiro Justin *biomedical scientist*
Symons, Timothy James McNeil *physicist*
Teece, David John *economics and management educator*
Thomas, Gareth *metallurgy educator*
Tien, Chang-Lin *chancellor*
Tinoco, Ignacio, Jr. *chemist, educator*
Tomizuka, Masayoshi *mechanical engineering educator, researcher*
Townes, Charles Hard *physics educator*
Vaux, Henry James *forest economist, educator*
Vojta, Paul Alan *mathematics educator*
Wake, David Burton *biology educator, researcher*
Walsh, Jane Ellen McCann *health care executive*
Weber, Eicke Richard *physicist*
Welch, William John *astronomer, educator*
Westphal, Andrew Jonathan *physicist, researcher*
Whinnery, John Roy *electrical engineering educator*
Wohletz, Leonard Ralph *soil scientist, consultant*
Wrappe, Thomas Keith *electrical engineer, product development director*
Yao, Xiang Yu *materials scientist*
Zumino, Bruno *physics educator, researcher*
Zwoyer, Eugene Milton *consulting engineering executive*

Bermuda Dunes
Smith, Walter J. *engineering consultant*

Beverly Hills
Alhanati, Shelley *clinical psychologist*
Evans, Louise *psychologist*
Fein, William *ophthalmologist*
Fox, David Peter *psychologist, educator*
Hyatt, Walter Jones *aerospace engineer*
Klein, Arnold William *dermatologist*
Lee, Nancy Francine *psychologist*

Big Bear City
Zirin, Harold *astronomer, educator*

Big Bear Lake
Mac Iver, Douglas Yaney *geologist*

Bishop
Groeneveld, David Paul *plant ecologist*

Bodega Bay
Hand, Cadet Hammond, Jr. *marine biologist, educator*

Bolinas
Wayburn, Laurie Andrea *environmental and wildlife foundation administrator, conservationist*

Bonita
Wood, Fergus James *geophysicist, consultant*

Borrego Springs
Scannell, William Edward *aerospace company executive, consultant, educator*

Boulder Creek
Ruch, Wayne Eugene *microlithography engineer*

Brawley
Jaquith, George Oakes *ophthalmologist*

Brea
Shen, Gene Giin-Yuan *organic chemist*
Wheeler, Annemarie Ruth *biochemist*

Buena Park
Parker, Larry Lee *electronics company executive, consultant*
Wiersema, Harold LeRoy *aerospace engineer*

Burbank
Mullin, Sherman N. *engrineering executive*

Burlingame
Bender, Michael David *gastroenterologist*
Hotz, Henry Palmer *physicist*

Byron
Greengrass, Roy Myron *plant engineer*

Calabasas
Drezner, Stephen M. *policy analyst, engineer*
Goodman, William Alfred *materials engineer*
Tellep, Daniel Michael *aerospace executive, mechanical engineer*

Calimesa
Kaplan, Ozer Benjamin *environmental health specialist, consultant*

Camarillo
Alexander, John Charles *editor, writer*
MacDonald, Norval (Woodrow) *safety engineer*
McConnel, Richard Appleton *aerospace company official*
Parker, Theodore Clifford *electronics engineer*

Cambria
DuFresne, Armand Frederick *management and engineering consultant*

Cameron Park
House, Lon William *engineer, energy economist*

Camp Connell
Soule, Thayer *documentary film maker*

Campbell
Wu, William Lung-Shen (You-Ming Wu) *aerospace medical engineering design specialist*

Canoga Park
Carman, Robert Lincoln, Jr. *physicist*
Hanneman, Timothy John *aerospace engineer*
Haymaker, Carlton Luther, Jr. (Bud Haymaker) *metrology engineer*
Holmes, Richard Brooks *mathematical physicist*
Lu, Cheng-yi *mechanical engineer*
Muniz, Benigno, Jr. *aerospace engineer, space advocate*
Norman, Arnold McCallum, Jr. *project engineer*
Peinemann, Manfred K.A. *marketing executive*
Yang, Ruey-Jen *aerospace engineer*

Canyon Lake
Schilling, Frederick Augustus, Jr. *geologist, consultant*

Capo Beach
Roemer, Edward Pier *neurologist*

Cardiff by the Sea
Elsner, Frederick Harvey *chemical scientist*

Carlsbad
Fountain, William David *laser/optics engineer*
Graham, Robert Klark *lens manufacturer*
Halberg, Charles John August, Jr. *mathematics educator*
Pantos, William Pantazes *mechanical engineer, consultant*
Raphael, George Farid *chemical engineer*

Carmel
Brahtz, John Frederick Peel *civil engineering educator*
Doe, Richard Philip *physician, educator*
Smith, Jeffry Alan *public health administrator, physician, consultant*

Carmel Valley
Lorenzen, Coby *emeritus engineering educator*

Carmichael
Wagner, Carruth John *physician*

Carson
Gittleman, Morris *consultant, metallurgist*

Castro Valley
Charney, Philip *dermatologist*

Cedar Ridge
Yeager, Charles Elwood (Chuck Yeager) *retired air force officer*

Cerritos
Acosta, Nelson John *civil engineer*

Chatsworth
Bearnson, William R. *systems engineer*
Levine, Arnold Milton *retired electrical engineer, documentary filmmaker*
Lin, Ching-Fang *engineering executive*
Rizvi, Javed *facilities engineer*

Chico
Kistner, David Harold *biology educator*
Welter, Lee Orrin *anesthesiologist, biomedical engineer*

China Lake
Meyer, Steven John *electrical engineer*

Chino
Ellington, James Willard *mechanical design engineer*

Chula Vista
Allen, Henry Wesley *biomedical researcher*
Rusconi, Louis Joseph *marine engineer*
Smith, Peggy O'Doniel *physicist, educator*
Sullivan, Lawrence Jerome *aerospace engineer*
Wolk, Martin *electronic engineer, physicist*

Claremont
Brown, Robert James Sidford *physicist*
Cha, Philip Dao *engineering educator*
Chang, Y(uan)-F(eng) *mathematics and computer science educator*
Eriksen, Clyde Hedman *ecology educator*
Gable, Robert S. *psychology educator*
Grieman, Fred Joseph *chemist, educator*
Johnson, James Lawrence *clinical psychologist, writer*
Johnson, Jerome Linné *cardiologist*
Leeb, Charles Samuel *clinical psychologist*
Long, Franklin Asbury *chemistry educator*
Musson, Donald George *chemist*
Wents, Doris Roberta *psychologist*

Clovis
Ensminger, Marion Eugene *animal science educator, author*

Coloma
Wall, Sonja Eloise *nurse, administrator*

Colton
Halstead, Bruce Walter *biotoxicologist*

Commerce
Olbrantz, Don Lee *chemist*

Concord
Cole, Timothy David *chemist*
Edgerton, Robert Frank *optical scientist*
Ivy, Edward Everett *entomologist, consultant*
Middleton, Michael John *civil engineer*

Corona
Gjerde, Andrea Jo *entomologist*

Corona Del Mar
Brokaw, Charles Jacob *educator, cellular biologist*
Tether, Anthony John *aerospace executive*

Coronado
Crilly, Eugene Richard *engineering consultant*

Corte Madera
Epstein, William Louis *dermatologist, educator*

Costa Mesa
Bono, Philip *aerospace consultant*
Carpenter, Frank Charles, Jr. *retired electronics engineer*
Cross, Glenn Laban *engineering executive, development planner*
Ferguson, Bruce R. *information systems consultant*
Lorance, Elmer Donald *organic chemistry educator*
Richmond, Ronald LeRoy *aerospace engineer*

Covina
Jameson, Julianne *clinical psychologist*
Schneider, Calvin *physician*
Takei, Toshihisa *otolaryngologist*

Crescent City
Ghosh, Sid *telecommunications engineer*

Culver City
Kamins, David Stone *programmer, analyst*
Sussman, Deborah Evelyn *designer, company executive*
Wiegand, Stanley Byron *engineer*

Cupertino
Iodice, Elaine *software engineer*
Lee, Yuen Fung *mechanical engineer*
Schmid, Anthony Peter *materials scientist, engineer*
Starkweather, Gary Keith *optical engineer, computer company executive*
Treybig, James G. *computer company executive*

Danville
Belding, William Anson *ceramic engineer*
Peterson, Jack Milton *retired physicist*

Davis
Addicott, Fredrick Taylor *retired botany educator*
Akesson, Norman Berndt *agricultural engineer, emeritus educator*
Aldredge, Ralph Curtis, III *mechanical engineer, educator*
Allard, Robert Wayne *geneticist, educator*
Armstrong, Peter Brownell *biologist*
Baldwin, Ransom Leland *animal science educator*
Bernoco, Domenico *immunogeneticist, educator*
Cahill, Thomas Andrew *physicist, educator*
Chancellor, William Joseph *agricultural engineering educator*
Chetelat, Roger Topping *geneticist*
Enders, Allen Coffin *anatomy educator*
Feeney, Robert Earl *research biochemist*
Flint, Mary Louise *entomologist*
Fowler, William Mayo, Jr. *rehabilitation medicine physician*
Fridley, Robert Bruce *agricultural engineering educator, academic administrator*
Gardner, William Allen *electrical engineering educator*
Gates, Bruce Clark *chemical engineer, educator*
Ghausi, Mohammed Shuaib *electrical engineering educator, university dean*
Gifford, Ernest Milton *biologist, educator*
Greenwood, M. R. C. *college dean, biologist, nutrition educator*
Grey, Robert Dean *biology educator*
Hendrickx, Andrew George *anatomy educator*
Hess, Charles Edward *environmental horticulture educator*
Hess, Ronald Andrew *aerospace engineer, educator*
Hill, Fredric William *nutritionist, poultry scientist*
Hollinger, Mannfred Alan *pharmacologist, educator, toxicologist*
Horowitz, Isaac M. *control research consultant, writer*
Hunt, Charles Edward *electrical engineer, educator*

Kepner, Robert Allen *agricultural engineering researcher, educator*
Kester, Dale Emmert *pomologist, educator*
Klasing, Susan Allen *environmental toxicologist, consultant*
Knox, William Jordan *physicist*
Laben, Dorothy Lobb *volunteer nutrition educator, consultant*
Laidlaw, Harry Hyde, Jr. *entomology educator*
Lotter, Donald Willard *environmental educator*
Marois, Jim *plant pathologist, educator*
Miller, Milton David *agronomist, educator*
Moores, Eldridge Morton *geology educator*
Mulase, Motohico *mathematics educator*
Natsoulas, Thomas *psychology educator*
O'Donnell, Sean *zoologist*
Olsson, Ronald Arthur *computer science educator*
Overstreet, James Wilkins *obstetrics and gynecology educator, administrator*
Piedrahita, Raul Humberto *aquacultural engineer*
Puckett, Elbridge Gerry *mathematician, educator*
Rappaport, Lawrence *plant physiology and horticulture educator*
Rick, Charles Madeira, Jr. *geneticist, educator*
Ryu, Dewey Doo Young *biochemical engineering educator*
Schenker, Marc Benet *medical educator*
Skinner, G(eorge) William *anthropologist, educator*
Stebbins, George Ledyard *research botanist, retired educator*
Stemler, Alan James *plant biology educator*
Stewart, James Ian *agrometeorologist*
Stumpf, Paul Karl *former biochemistry educator*
Syvanen, Michael *geneticist, educator*
Utts, Jessica Marie *statistics educator*
Uyemoto, Jerry Kazumitsu *plant pathologist, educator*
Van Bruggen, Ariena Hendrika Cornelia *plant pathologist*
Waller, Niels Gordon *psychologist*
Wooten, Frederick (Oliver) *applied science educator*
Yang, Shang Fa *biochemistry educator, plant physiologist*

Del Mar
Reid, Joseph Lee *physical oceanographer, educator*

Downey
Adegbola, Sikiru Kolawole *aerospace engineer, educator*
Austin, James Albert *healthcare executive, obstetrician-gynecologist*
Baumann, Theodore Robert *aerospace engineer, consultant, army officer*
Bienhoff, Dallas Gene *aerospace engineer*
Demarchi, Ernest Nicholas *aerospace company executive*
Flagg, Robert Finch *research aerospace engineer*
Frassinelli, Guido Joseph *aerospace engineer*
Redeker, Allan Grant *physician, medical educator*
Weinberger, Frank *information systems advisor*

Duarte
DuBridge, Lee Alvin *physicist*
Shapero, Sanford Marvin *hospital executive, rabbi*
Smith, Steven Sidney *molecular biologist*
Zaia, John Anthony *hematologist*

Edwards
Deets, Dwain Aaron *aeronautical research engineer*
Henry, Gary Norman *astronautical engineer, educator*
Thompson, Milton Orville *aeronautical engineer*

Edwards AFB
Bachus, Blaine Louis *pilot, aircraft mechanical engineer*
Hodges, Vernon Wray *mechanical engineer*
Mead, Franklin Braidwood, Jr. *aerospace engineer*
Morehart, James Henry *mechanical engineer*

El Cajon
Haubert, Roy A. *former engineer, educator*
Lanflisi, Raymond Robert *aeronautical engineer*
Rose, Raymond Allen *computer scientist*
Symes, Clifford E. *telecommunications engineer*

El Centro
Goldsberry, Richard Eugene *mobile intensive care paramedic, registered nurse*

El Cerrito
Gwinn, William Dulaney *physical chemist, educator, consultant*

El Dorado Hills
Todd, Terry Ray *physicist*

El Segundo
Ajmera, Kishore Tarachand *civil engineer*
Aldridge, Edward C., Jr. *aerospace transportation executive*
Dement, Franklin Leroy, Jr. *aerospace engineer*
Gaylord, Robert Stephen *aerospace engineering manager*
Jacobson, Alexander Donald *technological company executive*
Kashar, Lawrence Joseph *metallurgical engineer, consultant*
Kramer, Gordon *mechanical engineer*
Lantz, Norman Foster *electrical engineer*
Liaw, Haw-Ming (Charles) *physicist*
Lomheim, Terrence Scott *physicist*
Lotrick, Joseph *aeronautical engineer*
Mehlman, Lon Douglas *information systems specialist*
Mitchell, John Noyes, Jr. *electrical engineer*
Mo, Roger Shih-Yah *electronics engineering manager*
Olsen, Donald Paul *communications system engineer*
Plummer, James Walter *engineering company executive*
Radys, Raymond George *laser scientist*
Ryniker, Bruce Walter Durland *industrial designer, manufacturing executive*
Tamrat, Befecadu *aeronautical engineer*

Emeryville
Masri, Merle Sid *biochemist, consultant*
McKereghan, Peter Fleming *hydrogeologist, consultant*

Encinitas
Manganiello, Eugene Joseph *retired aerospace engineer*
Rummerfield, Philip Sheridan *medical physicist*

Encino
Hawthorne, Marion Frederick *chemistry educator*

Escondido
Damsbo, Ann Marie *psychologist*
Grew, Raymond Edward *mechanical engineer*

Eureka
Conversano, Guy John *civil engineer*
Roberts, Robert Chadwick *ecologist, environmental scientist, consultant*

Fairfield
Edwards, Richard Charles *oral and maxillofacial surgeon*
Martin, Clyde Verne *psychiatrist*

Fallbrook
Tess, Roy William Henry *chemist*

Felton
Denton, Dorothea Mary *electronics company manager*

Fontana
Poulsen, Dennis Robert *environmentalist*

Foothill Ranch
Pierson, Steve Douglas *aerospace engineer*

Foster City
Andresen, Mark Nils *electrical engineer*
Baselt, Randall Clint *toxicologist*
Graf, Edward Dutton *grouting consultant*
Iovannisci, David Mark *biochemist researcher*
Ting, Chen-Hanson *software engineer*
Wiktorowicz, John Edward *research biochemist*
Zaidi, Iqbal Mehdi *biochemist, scientist*

Fountain Valley
Nakagawa, John Edward *aeronautical engineer*
Turkel, Solomon Henry *physicist*

Fremont
Berry, Michael James *chemist*
Gupta, Rajesh *industrial engineer, quality assurance specialist*
Khan, Mahbub R. *physicist*
Ritter, Terry Lee *electrical engineer, educator*
Wang, Ying Zhe *mechanical and optical engineer*

Fresno
Archer, Thomas John *aeronautical engineer*
Botwin, Michael David *psychologist, researcher*
Chandler, Bruce Frederick *internist*
Gray, Jennifer Emily *biology educator*
Hanna, George Parker *environmental engineer*
Harvey, John Marshall *agricultural scientist*
Hickey, Rosemary Becker *retired podiatrist, lecturer, writer*
Leigh, Hoyle *psychiatrist, educator*
Mallory, Mary Edith *psychology educator*
O'Brien, John Conway *economist, educator, writer*
Rolfs, Kirk Alan *agronomist*
Stiffler, Kevin Lee *engineering technologist*
Swanson, Steven Clifford *clinical psychologist*
Templer, Donald Irvin *psychologist, educator*
Woolard, Henry Waldo *aerospace engineer*

Fullerton
Murray, Steven Nelsen *marine biologist, educator*
Silverman, Paul Hyman *parasitologist, former university official*
Solomon, Allen Louis *chemist*

Garden Grove
Vodonick, Emil J. *engineer*

Gardena
Hu, Steve Seng-Chiu *scientific research company executive, academic administrator*
Stuart, Jay William *engineer*
Zhou, Simon Zhengzhuo *laser scientist*

Gilroy
Barham, Warren S. *horticulturist*

Glendale
Clemens, Roger Allyn *medical and scientific affairs manager*
Dent, Ernest DuBose, Jr. *pathologist*
De Santis, James Joseph *clinical psychologist*
Farmer, Crofton Bernard *atmospheric physicist*
Jernazian, Levon Noubar *psychologist*
Knoop, Vern Thomas *civil engineer, consultant*
Kohler, Dylan Whitaker *software engineer, animator*
Leonard, Constance Joanne *civil, environmental engineer*
Oppenheimer, Preston Carl *psychotherapist, counseling agency administrator, psychodiagnostician*

Glendora
Ahern, John Edward *mechanical engineer, consultant*
Haile, Benjamin Carroll, Jr. *chemical engineer, mechanical engineer*

Gold River
Forbes, Kenneth Albert Faucher *urological surgeon*

Goleta
Jones, Colin Elliott *physicist, educator*
Mangaser, Amante Apostol *computer engineer*

Granada Hills
Brown, Alan Charlton *retired aeronautical engineer*
Shoemaker, Harold Lloyd *infosystem specialist*

Grand Terrace
Addington, William Hampton *civil engineer*

Half Moon Bay
Sedlak, Bonnie Joy *university official*

Harbor City
Bornino-Glusac, Anna Maria *mathematics educator*

Hawthorne
Bullard, Roger Dale *aerospace engineer*
Mockaitis, Joseph Peter *logistics engineer*
Weiss, Max Tibor *aerospace company executive*

Hayward
Baalman, Robert Joseph *biology educator*
Duncan, Doris Gottschalk *information systems educator*
Hunnicutt, Richard Pearce *metallurgical engineer*
Kirkland, Shari Lynn *clinical psychologist*
Pearce-Percy, Henry Thomas *physicist*
Whalen, Thomas Earl *psychology educator*

Hemet
Violet, Woodrow Wilson, Jr. *retired chiropractor*

Hercules
Emmanuel, Jorge Agustin *chemical engineer, environmental consultant*
Heffelfinger, David Mark *optical engineer*
Zabin, Burton Allen *chemist*

Hermosa Beach
Wickwire, Patricia Joanne Nellor *psychologist, educator*

Hillsborough
Blume, John August *consulting civil engineer*

Hollywood
Bratcher, Twila Langdon *conchologist, malacologist*

Huntington Beach
Adam, Steven Jeffrey *chemical engineer*
Anderson, Raymond Hartwell, Jr. *metallurgical engineer*
Bozanic, Jeffrey Evan *marine sciences research center director*
Falcon, Joseph A. *mechanical engineering consultant*
Goodrich, Craig Robert *business analyst*
Gustavino, Stephen Ray *aerospace engineer*
Owen, Thomas J. *environmental engineer*
Smith, Joseph Frank *spacecraft electronics and systems engineer*

Imperial
Angelo, Gayle-Jean *mathematics and physical sciences educator*

Inglewood
Hankins, Hesterly G., III *computer systems analyst*
Lewis, Roy Roosevelt *space physicist*
Sukov, Richard Joel *radiologist*

Irvine
Abajian, Henry Krikor *mechanical engineer*
Atwong, Matthew Kok Lun *chemical engineer*
Beckman, Arnold Orville *analytical instrument manufacturing company executive*
Bennett, Albert Farrell *biology educator*
Berns, Michael W. *cell biologist, educator*
Bhalla, Deepak Kumar *cell biologist, toxicologist, educator*
Bradic, Zdravko *chemist*
Bryant, Peter James *biologist, educator*
Caffey, Benjamin Franklin *civil engineer*
Carpenter, F. Lynn *biology educator*
Charles, M. Arthur *endocrinologist, educator*
Chicz-DeMet, Aleksandra *science educator, consultant*
Chrysikopoulos, Constantinos Vassilios *environmental engineering educator*
Colmenares, Jorge Eliecer *mechanical engineer*
Connolly, John Earle *surgeon, educator*
Demetrescu, Mihai Constantin *computer company executive, scientist*
Dunlap, Dale Richard *civil engineer*
Edwards, Donald Kenneth *mechanical engineer, educator*
Gess, Albin Horst *lawyer*
Graves, Joseph Lewis, Jr. *evolutionary biologist, educator*
Gupta, Sudhir *immunologist, educator*
Hess, Cecil F. *engineering executive*
Hoffman, Donald David *cognitive and computer science educator*
Howland, Paul *industrial engineer*
Juberg, Richard Kent *mathematician, educator*
Juhasz, Tibor *physicist, researcher*
Kar Roy, Arjun *electrical engineer, educator*
Kinsman, Robert Preston *biomedical plastics engineer*
Kontny, Vincent *engineering and construction company executive*
Lavernia, Enrique Jose *materials science and engineering educator*
Mc Gaugh, James Lafayette *psychobiologist*
Moe, Michael K. *physicist*
Phalan, Robert F. *environmental scientist*
Rauscher, Frances Helen *psychologist*
Real, Thomas Michael *civil engineer*
Reines, Frederick *physicist, educator*
Rentzepis, Peter M. *chemistry educator*
Rowland, Frank Sherwood *chemistry educator*
Russell, Roger Wolcott *psychobiologist, educator, researcher*
Sheldon, Mark Scott *research engineer*
Sirignano, William Alfonso *aerospace and mechanical engineer, educator*
Sperling, George *cognitive scientist, educator*
Spiberg, Philippe Frederic *research engineer*
Sprimont, Thomas Eugene *computing executive*
Stubberud, Allen Roger *electrical engineering educator*
Swanberg, Christopher Gerard *environmental engineer*
Villaverde, Roberto *civil engineer*
Wang, Ran-Hong Raymond *optical engineer, scientist*
Werner, Roy Anthony *aerospace executive*
Wong, Patrick Tin-Choi *neuroscientist*

Irwindale
Saless, Fathieh Molaparast *biochemist*

Kensington
Connick, Robert Elwell *chemistry educator*

Kingsburg
Blanton, Roy Edgar *sanitation agency executive*

La Canada Flintridge
Byrne, George Melvin *physician*

La Habra
Roberts, Liona Russell, Jr. *electronics engineer, executive*

La Jolla
Abarbanel, Henry Don Isaac *physicist, academic director*
Alvarez, Pablo *neuroscientist*
Anderson, Victor Charles *applied physics educator*
Anel, Alberto *biochemist*
Barnett, Tim P. *meteorologist*
Bartsch, Dirk-Uwe Guenther *research bioengineer, consultant*
Benson, Andrew Alm *biochemistry educator*
Boger, Dale L. *chemistry educator*
Braun, Hans-Benjamin *physicist*
Buckingham, Michael John *oceanography educator*
Burbidge, Geoffrey *astrophysicist, educator*
Campbell, Iain Leslie *biomedical scientist*
Chien, Shu *physiology and bioengineering educator*
Covington, Stephanie Stewart *psychotherapist, author*
Cox, Charles Shipley *oceanography researcher, educator*
Deane, Grant Biden *physicist*
Driscoll, Charles F. *research physicist*
Dulbecco, Renato *biologist, educator*
Edelman, Gerald Maurice *biochemist, educator*
Feher, George *physics and biophysics scientist, educator*
Fernandez, Fernando Lawrence *research company executive, aeronautical engineer*
Fischer, Peter *research immunologist*
Freedman, Michael Hartley *mathematician, educator*
Fung, Yuan-Cheng Bertram *bioengineering educator, author*
Gilbert, James Freeman *geophysics educator*
Gittes, Ruben Foster *urological surgeon*
Goldbaum, Michael Henry *ophthalmologist*
Goodkind, John Morton *physics educator*
Guillemin, Roger C. L. *physiologist*
Hostetler, Karl Yoder *internist, endocrinologist, educator*
Huntley, Mark Edward *biological oceanographer*
Ishizaka, Kimishige *immunologist, educator*
Kearns, David Richard *chemistry educator*
Kerr, Donald MacLean, Jr. *physicist*
Kitada, Shinichi *biochemist*
Knox, Robert Arthur *oceanographer, academic director*
Lerner, Richard Alan *chemistry educator, scientist*
Morikis, Dimitrios *physicist*
Morrow, William John Woodroofe *immunologist*
Mullis, Kary Banks *biochemist*
Munk, Walter Heinrich *geophysics educator*
Nicolaou, Kyriacos Costa *chemistry educator*
Nyhan, William Leo *pediatrician, educator*
O'Neil, Thomas Michael *physicist, educator*
Papadimitriou, Christos *computer science educator*
Penhune, John Paul *science company executive, electrical engineer*
Polich, John Michael *experimental psychologist*
Reissner, Eric (Max Erich Reissner) *applied mechanics educator*
Ride, Sally Kristen *physics educator, scientist, former astronaut*
Riedinger, Alan Blair *chemical engineer, consultant*
Rosenblatt, Murray *mathematics educator*
Rotenberg, Manuel *physics educator*
Schmid-Schoenbein, Geert Wilfried *biomedical engineer, educator*
Sclater, John George *geophysics educator*
Seidman, Stephanie Lenore *lawyer*
Sham, Lu Jeu *physics educator*
Sharpless, K. Barry *chemist*
Skalak, Richard *engineering mechanics educator, researcher*
Somerville, Richard Chapin James *science educator*
Spiro, Melford Elliot *anthropology educator*
Stern, Martin O(scar) *physicist, consultant*
Sung, Kuo-Li Paul *bioengineering educator*
Tara, (Tara Singh) *research chemist*
Terry, Robert Davis *neuropathologist, educator*
Tio, Kek-Kiong *mechanical engineering researcher*
Tsien, Roger Yonchien *chemist and cell biologist*
Tu, Charles Wuching *electrical and computer engineering educator*
Walker, Sydney, III *pharmacologist, psychiatric administrator*
Wall, Frederick Theodore *chemistry educator*
Williams, Forman Arthur *combustion theorist, engineering science educator*
Wong, Chi-Huey *chemistry educator*
Yakovlev, Vladislav Victorovich *research chemist, consultant*
York, Herbert Frank *physics educator*
Zyroff, Ellen Slotoroff *information scientist, classicist*

La Mesa
Bourke, Lyle James *electronics company executive, small business owner*
Butler, Paul Clyde *psychologist*
Kropotoff, George Alex *civil engineer*
Schelar, Virginia Mae *chemistry educator, consultant*
Wohl, Armand Jeffrey *cardiologist*

La Palma
Haldane, George French *mechanical engineer, consultant*

La Puente
Goldberg, David Bryan *biomedical researcher*

La Verne
Good, Harvey Frederick *biologist, educator*
Hwang, Cordelia Jong *chemist*
Sasaki, Shusuke *electromechanical engineer*

Laguna Hills
Batdorf, Samuel B(urbridge) *physicist*
Daily, Augustus Dee, Jr. *psychologist, retired*

Laguna Niguel
Nelson, Alfred John *retired pharmaceutical company executive*

Lake Arrowhead
Beckman, James Wallace Bim *economist, marketing executive*

Lakewood
Bogdan, James Thomas *secondary education educator, electronics researcher and developer*

Lancaster
Harwood, Kirk Edward *aeronautical engineer*

Lindsay
Webster, John Robert *chemical engineer*

Livermore
Alder, Berni Julian *physicist*
Amemiya, Chris Tsuyoshi *geneticist, biomedical scientist*
Baldwin, David E. *physicist*
Batzel, Roger Elwood *chemist*
Batzer, Mark Andrew *molecular geneticist*
Blattner, Meera McCuaig *computer science educator*
Brereton, Sandra Joy *engineer*
Chapline, George Frederick, Jr. *theoretical physicist*
Christensen, Richard Monson *mechanical engineer, materials engineer*
Correll, Donald Lee, Jr. *physicist*
Crow, Neil Byrne *geologist*
Drake, Richard Paul *physicist, educator*
Eby, Frank Shilling *research scientist*
Ellsaesser, Hugh Walter *retired atmospheric scientist*
Erskine, David John *physicist*
Fortner, Richard J. *physical scientist*
Hall, Howard Lewis *nuclear chemist*
Hammer, James Henry *physicist*
Hauber, Janet Elaine *mechanical engineer*
Kidder, Ray Edward *physicist, consultant*
Kulander, Kenneth Charles *physicist*
Lewis, Daniel Lee *mechanical engineer*
Lindner, Duane Lee *materials science management professional*
McMillan, Charles Frederick *physicist*
Milanovich, Fred Paul *physicist*
Nuckolls, John Hopkins *physicist, researcher*
Olsen, Clifford Wayne *physical chemist*
Orel, Ann Elizabeth *applied science educator, researcher*
Ornellas, Donald Louis *chemist researcher*
Picraux, Samuel Thomas *physics researcher*
Price, Clifford Warren *metallurgist, researcher*
Reitze, David Howard *physicist*
Sheem, Sang Keun *optical engineering researcher*
Shotts, Wayne J. *nuclear scientist, federal agency administrator*
Struble, Gordon Lee *physicist, researcher*
Swiger, Roy Raymond *biomedical scientist*
Van Devender, J. Pace *physical scientist*
Verry, William Robert *mathematics researcher*
Warshaw, Stephen Isaac *physicist*
Yatabe, Jon Mikio *nuclear engineer*

Loma Linda
Bailey, Leonard Lee *surgeon*
Betancourt, Hector Mainhard *psychology scientist, educator*
Castelaz, Patrick Frank *computer scientist*
Hill, Kelvin Arthur Willoughby *biochemistry educator*
Longo, Lawrence Daniel *physiologist, gynecologist*
Mace, John Weldon *pediatrician*
Slater, James Munro *radiation oncologist*
Taylor, Barry Llewellyn *microbiologist, educator*
Tosk, Jeffrey Morton *biophysicist, educator*

Lompoc
Peltekof, Stephan *systems engineer*

Long Beach
Blanche, Joe Advincula *aerospace engineer*
Bos, John Arthur *aircraft manufacturing executive*
Calkins, Robert Bruce *aerospace engineer*
Cynar, Sandra Jean *electrical engineering educator*
de Soto, Simon *mechanical engineer*
Dillon, Michael Earl *engineering executive*
Fung, Henry Chong *microbiologist, educator, administrator*
Gehring, George Joseph, Jr. *dentist*
Heavin, Myron Gene *aeronautical engineer*
Hildebrant, Andy McClellan *electrical engineer*
Kumar, Rajendra *electrical engineering educator*
McCauley, Hugh Wayne *human factors engineer, industrial designer*
McIntosh, Gregory Cecil *manufacturing engineer*
Rozenbergs, John *electronics engineer*
Seymour, Janet Martha *psychologist*
Simpson, Myles Alan *acoustics researcher*
Smith, William Ray *biophysicist, engineer*
Tabrisky, Phyllis Page *physiatrist, educator*
Wayne, Lawrence Gershon *microbiologist, researcher*
Yeh, Hsien-Yang *mechanical engineer, educator*

Los Alamitos
Iravanchy, Shawn *aerospace engineer, astrophysicist, consultant*
Karkia, Mohammad Reza *energy engineer, educator*

Los Altos
Bell, Chester Gordon *computer engineering company executive*
Bergrun, Norman Riley *aerospace executive*
Carr, Jacquelyn B. *psychologist, educator*
Elkind, Jerome Isaac *computer scientist, company executive*
Fondahl, John Walker *civil engineering educator*
Fraknoi, Andrew *astronomy educator, astronomical society executive*
Jones, Robert Thomas *aerospace scientist*
Menke, James Michael *chiropractor*
Saraf, Dilip Govind *electronics executive*

Los Altos Hills
van Tamelen, Eugene Earle *chemist, educator*

Los Angeles
Abrahamson, James Alan *retired air force officer*
Allen, Frederick Graham *consulting engineer*
Allerton, Samuel Ellsworth *biochemist*
Alwan, Abeer Abdul-Hussain *electrical engineering educator*
Arnold, Arthur Palmer *neurobiologist*
Ashley, Sharon Anita *pediatric anesthesiologist*
Bao, Joseph Yue-Se *orthopaedist, microsurgeon, educator*
Baumhefner, Robert Walter *neurologist*
Bayes, Kyle David *chemistry educator*
Beam, William Washington, III *data coordinator*
Beamer, Lesa Jean *molecular biologist*
Benson, Sidney William *chemistry researcher*
Bernstein, Sol *cardiologist, medical services administrator*
Biles, John Alexander *pharmaceutical chemistry educator*
Bittenbender, Brad James *environmental safety and industrial hygiene manager*
Black, Craig Call *museum administrator*
Boado, Ruben Jose *biochemist*
Boehm, Barry William *computer science educator*
Bono, Anthony Salvatore Emanuel, II *data processing executive*
Bottjer, David John *geological sciences educator*
Braginsky, Stanislav Iosifovich *physicist, geophysicist, researcher*
Brett, Cliff *software developer*
Brinegar, Claude Stout *oil company executive*
Buffington, Gary Lee Roy *safety standards engineer, construction executive*
Burns, Marcelline *psychologist, researcher*
Burns, Marshall *physicist*
Byers, Nina *physics educator*
Caldwell, Allan Blair *emergency health services company executive*
Campos, Joaquin Paul, III *chemical physicist, regulatory affairs specialist*
Chacko, George Kuttickal *systems science educator, consultant*
Chapman, Orville Lamar *chemist, educator*
Chen, Francis F. *physics and engineering educator*
Chen, Irvin Shao Yu *microbiologist, educator*
Chilingarian, George Varos *petroleum and civil engineering educator*
Chobotov, Vladimir Alexander *mechanical engineer*
Christie, Steven Lee *aerospace engineering researcher*
Clemente, Carmine Domenic *anatomist, educator*
Cline, David Bruce *physicist, educator*
Cohen, Randy Wade *neuroscientist*
Coleman, Paul Jerome, Jr. *physicist, educator*
Conn, Robert William *applied physics educator*
Cornwall, John Michael *physics educator, consultant, researcher*
Cosner, Christopher Mark *engineer*
Couldwell, William Tupper *neurosurgeon*
Cram, Donald James *chemistry educator*
Dawson, John Myrick *plasma physics educator*
Decyk, Viktor Konstantyn *research physicist, consultant*
Detels, Roger *epidemiologist, physician, former university dean*
Dhir, Vijay K. *mechanical engineering educator*
Dodds, Dale Irvin *chemicals executive*
Domaradzki, Julian Andrzej *physics educator*
Doo, Yi-Chung *aerospace engineer*
Dougherty, Elmer Lloyd, Jr. *chemical engineering educator, consultant*
Dunn, Bruce Sidney *materials science educator*
Edwards, Kenneth Neil *chemist, consultant*
Edwards, Ward Dennis *psychology and industrial engineering educator*
Egbuonu, Zephyrinus Chiedu *civil engineer, electrical engineer*
Einav, Shmuel *biomedical engineering educator*
El-Sayed, Mostafa Amr *chemistry educator*
Engel, Jerome, Jr. *neurologist, neuroscientist, educator*
Esfandiari-Fard, Omid David *microbiologist*
Fahey, John Leslie *immunologist*
Felker, Peter *chemistry educator*
Feyl, Susan *safety engineer, educator*
Finegold, Sydney Martin *microbiology and immunology educator*
Fischer, Alfred George *geology educator*
Friedhoff, Richard Mark *computer scientist, entrepreneur*
Garmire, Elsa Meints *electrical engineering educator, consultant*
Garratty, George *immunohematologist*
Geoffrion, Arthur Minot *management scientist*
Goulding, Merrill Keith *engineer, consultant*
Green, Richard *psychiatrist, lawyer, educator*
Greenfield, Moses A. *medical physicist, educator*
Griffith, Carl David *civil engineer*
Grinnell, Alan Dale *neurobiologist, educator, researcher*
Gross, Sharon Ruth *psychology educator, researcher*
Gruntman, Michael A. *physicist, researcher, educator*
Gundersen, Martin A. *electrical engineering and physics educator*
Hahn, Hong Thomas *mechanical engineering educator*
Haskell, Charles Mortimer *medical oncologist, educator*
Hawkins, Shane V. *electronics and computer engineer*
Henderson, Brian Edmond *physician, educator*
Hesse, Christian August *mining industry consultant*
Hobel, Calvin John *obstetrician/gynecologist*
Horowitz, Ben *medical center executive*
Horwitz, David A. *medicine and microbiology educator*
House, John W. *otologist*
Huang, Sung-cheng *electrical engineering educator*
Ilgen, Marc Robert *aerospace engineer*
Jelliffe, Roger Woodham *cardiologist, clinical pharmacologist*
Jen, Tien-Chien *mechanical engineer*
Jenden, Donald James *pharmacologist, educator*
Jiang, Hongwen *physicist*
Johnson, Cage Saul *hematologist, educator*
Johnson, Charles Erik *physics researcher*
Jordan, Gregory Wayne *aeronautical engineer*
Ju, Jiann-Wen *mechanics educator, researcher*
Kaplan, Samuel *pediatric cardiologist*
Kardously, George J. *chemical engineer*
Katchur, Marlene Martha *nursing administrator*
Kay, Alan *computer scientist*
Kennel, Charles Frederick *physicist, educator*
Kivelson, Margaret Galland *physicist*
Klevatt, Steve *production software developer*
Knize, Randall James *physics educator*
Krupp, Edwin Charles *astronomer*
Kumar, Anil *nuclear engineer*
Kunc, Joseph Anthony *physics and engineering educator, consultant*
Laaly, Heshmat Ollah *research chemist, roofing consultant, author*
Langer, Glenn Arthur *cellular physiologist, educator*
Leal, George D. *engineering company executive*
Lee, David Woon *chemist, lawyer*
Levine, Raphael David *chemistry educator*
Lewin, Klaus J. *pathologist, educator*
Liberman, Robert Paul *psychiatry educator, researcher, writer*
Loeblich, Helen Nina Tappan *paleontologist, educator*
Lunt, Owen Raynal *biologist, educator*
Madlang, Rodolfo Mojica *urologic surgeon*
Maki, Kazumi *physicist, educator*
Maksymowicz, John *electrical engineer*
Maronde, Robert Francis *internist, clinical pharmacologist, educator*
Martinez, Miguel Acevedo *urologist, consultant, lecturer*
Mathias, Mildred Esther *botany educator*
Maxworthy, Tony *mechanical and aerospace engineering educator*
McClure, William Owen *biologist*
McLaughlin, John *production company technical director*
McLean, Ian Small *astronomer, physics educator*
Merchant, Roland Samuel, Sr. *hospital administrator, educator*
Mettler, Ruben Frederick *former electronics and engineering company executive*
Miller, Wendell Smith *chemist, consultant*
Mohr, John Luther *biologist, environmental consultant*
Monkewitz, Peter Alexis *mechanical and aerospace engineer, educator*
Morales, Cynthia Torres *clinical psychologist, consultant*
Mortensen, Richard Edgar *engineering educator*
Moss, Charles Norman *physician*
Moy, Ronald Leonard *dermatologist, surgeon*
Muntz, Eric Phillip *aerospace engineering and radiology educator, consultant*
Muntz, Richard Robert *computer scientist, educator*
Nathwani, Bharat Narottam *pathologist, consultant*
Okrent, David *engineering educator*
Olah, George Andrew *chemist, educator*
O'Neil, Harold Francis, Jr. *psychologist, educator*
O'Neill, Russell Richard *engineering educator*
Partow-Navid, Parviz *information systems educator*
Patel, Chandra Kumar Naranbhai *communications company executive, researcher*
Peplinski, Daniel Raymond *project engineer*
Perlmutter, Milton Manuel *chemist, accountant, financial consultant, property manager, real estate appraiser, home building inspector*
Portenier, Walter James *aerospace engineer*
Price, Zane Herbert *cell biologist, research microscopist*
Pugay, Jeffrey Ibanez *mechanical engineer*
Putt, John Ward *propulsion engineer, retired, consultant*
Ramo, Simon *engineering executive*
Rauch, Lawrence Lee *aerospace and electrical engineer, educator*
Raven, Bertram H(erbert) *psychology educator*
Raymond, Arthur Emmons *aerospace engineer*
Razouk, Rashad Elias *retired chemistry educator*
Rechtin, Eberhardt *aerospace educator*
Redheffer, Raymond Moos *mathematician, educator*
Reiss, Howard *chemistry educator*
Remillard, Richard Louis *automotive engineer*
Re Velle, Jack B(oyer) *consulting statistician*
Rimoin, David Lawrence *physician, geneticist*
Rogoway, Lawrence Paul *civil engineer, consultant*
Ropchan, Jim R. *research chemist, administrator*
Rosenblatt, Joseph David *hematologist, oncologist*
Rosenblum, Robert *computer graphics software developer*
Rosenthal, John Thomas *surgeon, transplantation surgeon*
Ryan, Stephen Joseph, Jr. *ophthalmology educator, university dean*
Samudlo, Jeffrey Bryan *architect, educator, planner, business owner*
Saravanja-Fabris, Neda *mechanical engineering educator*
Sasao, Toshiaki *psychologist, researcher*
Scheibel, Arnold Bernard *psychiatrist, educator, researcher*
Schep, Raymond Albert *chemist*
Schneider, Edward Lewis *academic administrator, research administrator*
Schubert, Gerald *planetary and geophysics educator*
Schwinger, Julian *physicist, educator*
Sears, David O'Keefe *psychology educator*
Secor, Stephen Molyneux *physiological ecologist*
Senitzky, Israel Ralph *physicist*
Senkan, Selim M. *chemical engineering educator*
Sepmeyer, Ludwig William *systems engineer, consultant*
Shacks, Samuel James *physician*
Shahab, Salman *systems engineer*
Shapiro, Isadore *materials scientist, consultant*
Shlian, Deborah Matchar *physician*
Siegel, Michael Elliot *nuclear medicine physician, educator*
Signoretti, Rudolph George *propulsion engineer, consultant*
Sinatra, Frank Raymond *pediatric gastroenterologist*
Slaughter, John Brooks *university president*
Slavkin, Harold C. *biologist*
Solomon, David Harris *physician, educator*
Sprague, Norman Frederick, Jr. *surgeon, educator*
Steckel, Richard J. *radiologist, academic administrator*
Stewart, Brent Kevin *radiological science educator*
Straatsma, Bradley Ralph *ophthalmologist, educator*
Sung, Yun-Chen *computer graphics software developer*
Sve, Charles *mechanical engineer*
Thomas, Duncan Campbell *biostatistics educator*
Tran, Johan-Chanh Minh *research scientist*
Treiman, David Murray *neurology educator*
Tuckson, Reed V. *university president*
Tyndall, Terry Scott *electrical engineer, biomedical engineer*
Udwadia, Firdaus Erach *engineering educator, consultant*
Ufimtsev, Pyotr Yakovlevich *radio engineer, educator*
Uijtdehaage, Sebastian Hendricus J. *psychophysiology researcher*
Urena-Alexiades, Jose Luis *electrical engineer*
Urist, Marshall Raymond *orthopedic surgeon, researcher*
Van Der Meulen, Joseph Pierre *neurologist*
Vasquez, Rodolfo Anthony *protein chemist, researcher*
Villani, Daniel Dexter *aerospace engineer*
Walker, Raymond John *physicist*
Walsh, John Harley *medical educator*
Watanabe, Richard Megumi *medical research assistant*
Webster, Jeffery Norman *technology policy analyst*
West, Charles David *chemistry educator*
Wilcox, Rano Roger *psychology educator*
Williams, Bruce Warren *psychologist*
Willner, Alan Eli *electrical engineer, educator*
Wincor, Michael Z. *psychopharmacology educator, clinician, researcher*
Winsor, Travis Walter *cardiologist, educator*
Wittig, Curt *chemist, educator*
Wittrock, Merlin Carl *educational psychologist*
Wittry, David Beryle *physicist, educator*
Wong, James Bok *economist, engineer, technologist*
Wong, Kenneth Lee *software engineer, consultant*
Woolf, Nancy Jean *neuroscientist, educator*
Wu, David Li-Ming *cytogeneticist*
Wu, Robert Chung Yung *space sciences educator*
Young, Lai-Sang *mathematician educator*
Ziskin, Jay Hersell *psychologist*

Los Angeles AFB
DiDomenico, Paul B. *military officer*
Faudree, Edward Franklin, Jr. *military officer*
Moe, Osborne Kenneth *physicist*
Shaver, Marc Steven *aerospace engineer, air force executive officer*

Los Gatos
Burnett, James Ray *physicist, consultant*
Cusick, Joseph David *science administrator, retired*
Dumas, Jeffrey Mack *lawyer*
Fujitani, Martin Tomio *software quality engineer*
Leung, Charles Cheung-Wan *technological company executive*
Nitz, Frederic William *electronics company executive*

Magalia
Kincheloe, William Ladd *mechanical engineer*

Malibu
Chester, Arthur Noble *physicist*
Forer, Bertram Robin *psychologist, researcher*
Hasenberg, Thomas Charles *physicist*
Yamagishi, Frederick George *chemist*

Manhattan Beach
Blanton, John Arthur *architect*
Davis, Dean Earl *aerospace engineer*
Pringle, Ronald Sandy Alexander *seismic inspector*
Yee, Peter Ben-On *marketing professional*

Manteca
Rainey, Barbara Ann *sensory evaluation consultant*

Martinez
Efron, Robert *neurology educator, research institute administrator*
Geokas, Michael C. *gastroenterologist*
Miller, Nicole Gabrielle *clinical psychologist*
Nielsen-Bohlman, Lynn Tracy *neuroscientist*

McClellan AFB
Frank, Christopher Lynd *mechanical engineer*

Menlo Park
Alexander, Theron *psychologist, writer*
Bukry, John David *geologist*
Carr, Michael *secondary education educator*
Carr, Michael Harold *geologist*
Dalrymple, Gary Brent *research geologist*
Darrah, James Gore *physicist, financial executive, real estate developer*
Dolberg, David Spencer *business executive, lawyer, scientist*
Fergason, James L. *optical company executive*
Glushko, Victor *medical products executive, biochemist*
Hem, John David *research chemist*
Hodges, James Clark *engineer*
Hogan, Clarence Lester *retired electronics executive*
Johnston, Brian Howard *molecular biologist*
Kandt, Ronald Kirk *computer software company executive, consultant*
Kohne, Richard Edward *retired engineering executive*
Lindh, Allan Goddard *seismologist*
MacGregor, James Thomas *toxicologist*
Meyers, Paul Allan *chemist*
Morimoto, Roderick Blaine *research engineer*
Nell, Janine Marie *metallurgical and materials engineer*
Neumann, Peter Gabriel *computer scientist*
Newcomb, Robert Whitney *biotechnologist, neuroscience researcher*
Nichols, Frederic Hone *oceanographer*
Niksa, Stephen Joseph *research chemical engineer, consultant*
Saldich, Robert Joseph *electronics company executive*
Thiers, Eugene Andres *economist*
Tietjen, James *research institute administrator*
Tokheim, Robert Edward *physicist*
Wallace, Robert Earl *geologist*
Zislis, Paul Martin *software engineering executive*

Merced
McCullough, John James, III *civil engineer*
Olsen, David Magnor *science educator*

Mill Valley
Wallerstein, Robert Solomon *psychiatrist*

Millbrae
Gamlen, James Eli, Jr. *corrosion engineer*

Milpitas
Fonck, Eugene Jason *electronics engineer, physicist*
Lee, Kenneth *physicist*

Mission Hills
Cramer, Frank Brown *engineering executive, combustion engineer, systems consultant*

Modesto
Lipomi, Michael Joseph *health facility administrator*
Steffan, Wallace Allan *entomologist, educator*

Moffett Field
Albers, James Arthur *engineering executive*
Baldwin, Betty Jo *computer specialist*
Carlson, Richard Merrill *aeronautical engineer, research executive*
Cohen, Malcolm Martin *psychologist, researcher*
Compton, Dale Leonard *space agency administrator*
Cook, Anthony Malcolm *aerospace engineer*
Ellis, Stephen Roger *research scientist*
Greenleaf, John Edward *research physiologist*
Korsmeyer, David Jerome *aerospace engineer*
Mihalov, John Donald *aerospace research scientist*
O'Handley, Douglas Alexander *astronomer*

Peterson, Victor Lowell *aerospace engineer, research center administrator*
Ross, Muriel Dorothy *research scientist*
Seiff, Alvin *planetary scientist, atmosphere physics and aerodynamics consultant*
Statler, Irving Carl *aerospace engineer*
Strawa, Anthony Walter *researcher*
Wignarajah, Kanapathipillai *plant physiologist, researcher, educator*

Mojave
Rutan, Elbert L. (Burt Rutan) *aircraft designer*

Monrovia
Pray, Ralph Emerson *metallurgical engineer*
Rudnyk, Marian E. *planetary photogeologist*

Monte Sereno
Dalton, Peter John *electronics executive*

Montecito
Wheelon, Albert Dewell *physicist*

Monterey
Atchley, Anthony Armstrong *physicist, educator*
Boger, Dan Calvin *economics educator, statistical and economic consultant*
Burl, Jeffrey Brian *electrical engineer, educator*
Faulkner, Frank David *mathematics educator*
Gordis, Joshua Haim *mechanical engineer, researcher*
Newberry, Conrad Floyde *aerospace engineering educator*
Rockower, Edward B. *physicist, operations researcher, consultant*
Sarpkaya, Turgut *mechanical engineering educator*
Shull, Harrison *chemist, educator*

Moorpark
Hovanec, Timothy Arthur *aquatic researcher*
Monteiro, Sergio Lara *physics educator*

Moreno Valley
Hamill, Carol *biologist, writing instructor*
Trainor, Paul Vincent *weapon systems engineer, genealogy researcher*

Morgan Hill
Gehman, Bruce Lawrence *materials scientist*
Mancini, Robert Karl *computer analyst, consultant*

Mountain View
Ashford, Robert Louis *computer professional*
Cooper, David Robert *geotechnical testing professional*
Couloures, Kevin Gottlieb *process development researcher*
Cusumano, James Anthony *chemical company executive, former recording artist*
Donoho, Laurel Roberta *industry analyst*
Grimes, Craig Alan *research scientist*
Hagmann, Robert Brian *computer scientist*
Harber, M(ichael) Eric *industrial engineer*
Johnson, Noel Lars *biomedical engineer*
Lu, Wuan-Tsun *microbiologist, immunologist*
Mutch, James Donald *pharmaceutical executive*
Pendleton, Joan Marie *microprocessor designer*
Perrella, Anthony Joseph *electronics engineer, consultant*
Saifer, Mark Gary Pierce *pharmaceutical executive*
Warren, Richard Wayne *obstetrician, gynecologist*
Watson, David Colquitt *electrical engineer, educator*

Napa
Ianziti, Adelbert John *industrial designer*
Wycoff, Charles Coleman *retired anesthesiologist*

National City
Morgan, Jacob Richard *cardiologist*

Nevada City
Vaughan-Kroeker, Nadine *psychologist*

Newark
Nystrom, Gustav Adolph *mechanical engineer*

Newbury Park
Calderone, Marlene Elizabeth *toxicology technician*

Newport Beach
Jones, Roger Wayne *electronics executive*
Maddock, Thomas Smothers *engineering company executive, civil engineer*
Sharbaugh, W(illiam) James *plastics engineer, consultant*
Whittemore, Paul Baxter *psychologist*
Zalta, Edward *otorhinolaryngologist, utilization review physician*

Norco
McClanahan, Michael Nelson *digital systems analyst*

North Hollywood
Kaplan, Martin Nathan *electrical and electronic engineer*
McGee, Sam *laser scientist*
Predescu, Viorel N. *electrical engineer*
Thomson, John Ansel Armstrong *biochemist*

Northridge
Cleary, James W. *retired university administrator*
Court, Arnold *climatologist*
Fidell, Linda Selzer *psychology educator, consultant*
Golerkansky, Peter Joseph *electrical engineer*
Jakobsen, Jakob Knudsen *mechanical engineer*
Rengarajan, Sembiam Rajagopal *electrical engineering educator, researcher, consultant*
Stout, Thomas Melville *control system engineer*
Taylor, Kent Douglas *molecular biologist*

Novato
Reed, Dwayne Milton *medical epidemiologist, educator*
Sacks, Colin Hamilton *psychologist educator*
Swanson, Lee Richard *computer security executive*

Oakland
Aubert, Allan Charles *aerospace engineer, consultant*
Cumming, Janice Dorothy *clinical psychologist*
Ellison, Carol Rinkleib *psychologist, educator*
Friedman, Gary David *epidemiologist, health facility administrator*
Jukes, Thomas Hughes *biological chemist, educator*
Kakade, Ashok Madhav *civil engineer, consultant*
Killebrew, Ellen Jane (Mrs. Edward S. Graves) *cardiologist*
Mikalow, Alfred Alexander, II *deep sea diver, marine surveyor, marine diving consultant*
Morier, Dean Michael *psychology educator*
Musihin, Konstantin K. *electrical engineer*
Nebelkopf, Ethan *psychologist*
Reitz, Richard Elmer *physician*
Schell, Farrel Loy *transportation engineer*
Solomon, Daniel *psychologist*
Tsztoo, David Fong *civil engineer*
Whitsel, Richard Harry *biologist*

Oceanside
Hofmann, Frieder Karl *biotechnologist, consultant*

Ontario
Johnson, Maurice Verner, Jr. *agricultural research and development executive*
Sainath, Ramaiyer *electrical engineer*

Orange
Appelbaum, Bruce David *physician*
Becker, Juliette *psychologist, marriage and family therapist*
DiSaia, Philip John *gynecologist, obstetrician, radiology educator*
Furnas, David William *plastic surgeon*
Hodges, Robert Edgar *physician, educator*
Knoth, Russell Laine *psychologist, educator*
Letzring, Tracy John *civil engineer*
Lott, Ira Totz *pediatric neurologist*
Nguyen, Quan A. *medical physicist*
Silverman, Benjamin K. *pediatrician, educator*
Talbott, George Robert *physicist, mathematician, educator*
Thompson, William Benbow, Jr. *obstetrician/gynecologist, educator*
Toeppe, William Joseph, Jr. *retired aerospace engineer*
Varsanyi-Nagy, Maria *biochemist*
Vice, Charles Loren *electromechanical engineer*

Orangevale
Gibson, Gordon Ronald *chemist*

Orinda
Heftmann, Erich *biochemist*

Oroville
Shelton, Joel Edward *clinical psychologist*

Pacific Grove
Beidleman, Richard Gooch *biologist, educator*
Lindstrom, Kris Peter *environmental consultant*

Pacific Palisades
Claes, Daniel John *physician*
Csendes, Ernest *chemist, corporate and financial executive*
Gregor, Eduard *laser physicist*
Jennings, Marcella Grady *rancher, investor*
Shapiro, Nathan *acoustical engineer, retired*

Pacifica
Hitz, C. Breck *optics and laser scientist*
Reich, Randi Ruth Novak *software engineer*

Palm Desert
Sausman, Karen *zoological park administrator*

Palm Springs
Krick, Irving Parkhurst *meteorologist*

Palo Alto
Aberth, William Henry *physicist*
Andersen, Torben Brender *optical researcher, astronomer*
Balzhiser, Richard Earl *research and development company executive*
Beretta, Giordano Bruno *computer scientist, researcher*
Bienenstock, Arthur Irwin *physicist, educator*
Briggs, Winslow Russell *plant biologist, educator*
Cooper, Allen David *research scientist, educator*
Culler, Floyd LeRoy, Jr. *chemical engineer*
Cutler, Leonard Samuel *physicist*
Datlowe, Dayton Wood *space scientist, physicist*
Drell, Sidney David *physicist, educator*
Eckroad, Steven Wallace *electrical engineer*
Farber, Eugene Mark *psoriasis research institute administrator*
Feigenbaum, Edward Albert *computer science educator*
Fisher, Thornton Roberts *physicist*
Fried, John H. *chemist*
Garland, Harry Thomas *research administrator*
Gilbert, Keith Duncan *electronics executive*
Hartman, Keith Walter *physicist*
Henry, Kent Douglas *analytical chemist, hardware engineer*
Hermsen, Robert William *engineering consultant*
Hewlett, William (Redington) *manufacturing company executive, electrical engineer*
Hodge, Philip Gibson, Jr. *mechanical and aerospace engineering educator*
Holman, Halsted Reid *medical educator*
Holmes, Thomas Joseph *aerospace engineering consultant*
Ingham, David R. *physicist*
Itnyre, Jacqueline Harriet *programmer*
Kazan, Benjamin *research engineer*
Kelley, Robert Franklin *systems analyst, consultant*
Lane, William Kenneth *physician*
MacDonald, Alexander Daniel *physics consultant*
Mattern, Douglas James *electronics reliability engineer*
McCloskey, Thomas Henry *mechanical engineer, consultant*
McKeever, Sheila A. *utilities executive*
Menon, Padmanabhan *aerospace engineer*
Merz, Antony Willits *aerospace engineer*
Moll, John Lewis *electronics engineer*
Neil, Gary Lawrence *pharmaceutical company research executive, biochemical pharmacologist*
Neou, In-Meei Ching-yuan *mechanical engineering educator, consultant*
Oliver, Bernard More *electrical engineer, technical consultant*
Packard, David *manufacturing company executive, electrical engineer*
Panofsky, Wolfgang Kurt Hermann *physicist, educator*
Pauling, Linus Carl *chemistry educator*
Quate, Calvin Forrest *engineering educator*
Richter, Burton *physicist, educator*
Riley, Paul Eugene *electronics engineer*
Roberts, Charles S. *software engineer*
Sandmeier, Ruedi Beat *agricultural research executive*
Schneider, Thomas R(ichard) *physicist*
Schreiber, Everett Charles, Jr. *chemist, educator*
Shumway, Norman Edward *surgeon, educator*
Spinrad, Robert Joseph *computer scientist*
Stringer, John *materials scientist*
Taimuty, Samuel Isaac *physicist*
Taylor, John Joseph *nuclear engineer*
Taylor, Richard Edward *physicist, educator*
Theeuwes, Felix *physical chemist*
Ullman, Edwin Fisher *research chemist*
Van de Walle, Chris Gilbert *physicist*
Varney, Robert Nathan *retired physicist, researcher*
Verdonk, Edward Dennis *physicist*
Warne, William Elmo *irrigationist*
Yagnik, Suresh K. *nuclear engineer*
Youngdahl, Paul Frederick *mechanical engineer*
Zlotnik, Albert *immunologist, researcher*
Zuckerkandl, Emile *molecular evolutionary biologist, scientific institute executive*
Zunich, Kathryn Margaret *physician*

Palos Verdes Estates
Aro, Glenn Scott *environmental and safety executive*
Basnight, Arvin Odell *public administrator, aviation consultant*
Joshi, Satish Devdas *organic chemist*

Palos Verdes Peninsula
Pfund, Edward Theodore, Jr. *electronics company executive*
Waaland, Irving Theodore *retired aerospace design executive*
Weiss, Herbert Klemm *aeronautical engineer*
Wilson, Theodore Henry *retired electronics company executive, aerospace engineer*

Pasadena
Albee, Arden Leroy *geologist, educator*
Anderson, Don Lynn *geophysicist, educator*
Anson, Fred Colvig *chemistry educator*
Atkinson, David John *computer scientist*
Babcock, Horace W. *astronomer*
Baldeschwieler, John Dickson *chemist, educator*
Barnard, William Marion *psychiatrist*
Beauchamp, Jesse Lee (Jack Beauchamp) *chemistry educator*
Bercaw, John Edward *chemistry educator, consultant*
Bjorck, Jeffrey Paul *psychology educator, clinical psychologist*
Boulos, Paul Fares *civil and environmental engineer*
Brennen, Christopher E. *fluid mechanics educator*
Cagin, Tahir *physicist*
Caine, Stephen Howard *data processing executive*
Casani, John Richard *electrical engineer*
Celniker, Susan Elizabeth *geneticist, molecular biologist*
Chahine, Moustafa Toufic *atmospheric scientist*
Chang, Daniel Hsing-Nan *aerospace engineer*
Chiang, Richard Yi-Ning *aerospace engineer*
Court, John Hugh *psychology educator*
Crisp, Joy Anne *research scientist*
Cuk, Slobodan *engineering educator*
Culick, Fred Ellsworth Clow *physics and engineering educator*
Damji, Karim Sadrudin *environmental toxicologist, consultant*
Dervan, Peter Brendan *chemistry educator*
Doupnik, Craig Allen *physiologist*
Everhart, Thomas Eugene *university president, engineering educator*
Flagan, Richard Charles *chemical engineering educator*
Fowler, William Alfred *retired physics educator*
Frautschi, Steven Clark *physicist, educator*
Gaskell, Robert Weyand *physicist*
Geller, Gary Neil *systems engineer*
Gell-Mann, Murray *theoretical physicist, educator*
Goddard, Ralph Edward *electrical engineer*
Goddard, William Andrew, III *chemist, applied physicist, educator*
Goei, Bernard Thwan-Poo (Bert Goei) *architectural firm executive*
Goldreich, Peter Martin *astrophysics and planetary physics educator*
Goodwin, David George *mechanical engineering educator*
Gorsuch, Richard Lee *psychologist, educator*
Gray, Harry Barkus *chemistry educator*
Grubbs, Robert H. *chemistry educator*
Hatheway, Alson Earle *mechanical engineer*
Henninger, Polly *neuropsychologist, researcher*
Hess, Ann Marie *systems engineer, electronic data processing specialist*
Hitlin, David George *physicist, educator*
Hopfield, John Joseph *biophysicist, educator*
Horvath, Joan Catherine *aeronautical engineer*
Housner, George William *civil engineering educator, consultant*
Isenberg, John Frederick *optical engineer*
Jastrow, Robert *physicist*
Joffe, Benjamin *mechanical engineer*
Kanamori, Hiroo *physics and astronomy educator*
Kannan, Rangaramanujam *polymer physicist, chemical engineer*
Kirby, Shaun Keven *physicist*
Kulkarni, Shrinivas R. *astronomy educator*
Leonard, Nelson Jordan *chemistry educator*
Lesh, James Richard *engineering manager*
Lewis, Edward B. *biology educator*
Libbrecht, Kenneth *astronomy educator*
Liepmann, Hans Wolfgang *physicist, educator*
Lo, Shui-yin *physicist*
Lugg, Marlene Martha *health information systems specialist*
Man, Kin Fung *physicist, researcher*
Marcus, Rudolph Arthur *chemist, educator*
Mathias, Alice Irene *health plan company executive*
McGill, Thomas Conley *physics educator*
Middlebrook, Robert David *electronics educator*
Moore, William Vincent *electronics and communications systems engineer*
Morari, Manfred *chemical engineer, educator*
Myers, Andrew Gordon *chemistry educator*
Najm, Issam Nasri *environmental engineer*
Neugebauer, Gerry *astrophysicist, educator*
Nghiem, Son Van *electrical engineer*
Nguyen, Tien Manh *communications systems engineer*
Niebur, Ernst Dietrich *computational neuroscientist*
Otoshi, Tom Yasuo *electrical engineer*
Owen, Ray David *biology educator*
Perez, Reinaldo Joseph *electrical engineer*
Pottorff, Beau Backus *astronautical engineer*
Prasad, K. Venkatesh *electrical and computer engineer*
Reid, Macgregor Stewart *aeronautics and microwave engineer*
Roberts, John D. *chemist, educator*
Rothenberg, Ellen *biologist*
Sahu, Ranajit *engineer*
Sandage, Allan Rex *astronomer*
Sargent, Wallace Leslie William *astronomer, educator*
Schmidt, Maarten *astronomy educator*
Schober, Robert Charles *electrical engineer*
Schwarz, John Henry *theoretical physicist, educator*
Searle, Leonard *astronomer, researcher*
Seinfeld, John Hersh *chemical engineering educator*
Shalack, Joan Helen *psychiatrist*
Sharp, Robert Phillip *geology educator, researcher*
Sinha, Mahadeva Prasad *chemist, consultant*
Smith, Edward John *geophysicist, physicist*
Smith, Louis *maintenance engineer*
Sperry, Roger Wolcott *neurobiologist, educator*
Sternberg, Paul Warren *biologist, educator*
Stone, Edward Carroll *physicist, educator*
Story, Randall Mark *biochemist*
Swain, Robert Victor *environmental engineer*
Sweetser, Theodore Higgins *mathematician*
Taylor, Hugh Pettingill, Jr. *geologist, educator*
Tekippe, Rudy Joseph *civil engineer*
Terhune, Robert William *optics scientist*
Thompson, Robert James *physicist*
Thorne, Kip Stephen *physicist, educator*
Varshavsky, Alexander Jacob *molecular biologist*
Vining, Cronin Beals *physicist*
Vogt, Rochus Eugen *physicist, educator*
Wang, Joseph Jiong *astrophysicist*
Wasserburg, Gerald Joseph *geology and geophysics educator*
Weisbin, Charles Richard *nuclear engineer*
Weymann, Ray J. *astronomy educator*
Wynn, Robert Raymond *engineer*
Yariv, Amnon *electrical engineering educator, scientist*
Yeh, Paul Pao *electrical and electronics engineer, educator*
Zewail, Ahmed Hassan *chemistry and chemical engineering educator, editor, consultant*

Paso Robles
Schaberg, Burl Rowland, Jr. *engineering company executive*
Still, Harold Henry, Jr. *engineering company executive*

Pebble Beach
Keene, Clifford Henry *medical administrator*

Penn Valley
Klohs, Murle William *chemist, consultant*

Penryn
Bryson, Vern Elrick *nuclear engineer*

Pescadero
Nelson, Kay Yarborough *author, columnist*

Petaluma
Belmares, Hector *chemist*

Pico Rivera
Banuk, Ronald Edward *mechanical engineer*
Bunch, Jeffrey Omer *mechanical engineer*
Gardner, Stanley *forensic engineer, expert witness*
Ling, Rung Tai *physicist*

Piedmont
Collen, Morris Frank *physician*
Elgin, Gita *psychologist*

Pine Valley
Liddiard, Glen Edwin *physicist*

Pismo Beach
Saveker, David Richard *naval and marine architectural engineering executive*

Pittsburg
Weed, Ronald De Vern *engineering consulting company executive*
Yoshisato, Randall Atsushi *chemical engineer*

Placerville
Craib, Kenneth Bryden *resource development executive, physicist, economist*

Playa Del Rey
Kaelin, Barney James *technological artist*
Tai, Frank *aerospace engineer*

Pleasant Hill
Hopkins, Robert Arthur *retired industrial engineer*

Pleasanton
Choy, Clement Kin-Man *research scientist*
Llenado, Ramon *chemist, chemicals executive*
Longmuir, Alan Gordon *manufacturing executive*
Shen, Mason Ming-Sun *pain and stress management center administrator*
Worley, David Allan *biochemical engineer*

Pomona
Aurilia, Antonio *physicist*
Eagleton, Robert Don *physics educator*
Freeland, John Chester, III *neuropsychologist*
Kauser, Fazal Bakhsh *aerospace engineer, educator*

Port Hueneme
Matigan, Robert *electrical engineer*

Poway
Aschenbrenner, Frank Aloysious *former diversified manufacturing company executive*
Shelby, William Ray Murray *logistics engineer*

Ramona
Woodall, James Barry *information systems specialist*

Rancho Cucamonga
Wells, Jon Barrett engineer
Yurist, Svetlan Joseph mechanical engineer

Rancho Mirage
Bolton, Merle Ray, Jr. cardiologist
Deiter, Newton Elliott clinical psychologist
McCrea, Russell James mechanical engineer, consultant
Savin, Ronald Richard chemical company executive, inventor
Wolf, Jack Keil electrical engineer, educator

Rancho Palos Verdes
Morizumi, Shigenori James applied mathematician
Park, Kwan Il energy engineer

Rancho Santa Fe
Schirra, Walter Marty, Jr. business consultant, former astronaut

Rancho Santa Margarita
Sharp, William Charles systems engineer
Wong, Wallace medical supplies company executive, real estate investor

Redondo Beach
Ball, William Paul physicist, engineer
Burris, Harrison Robert computer and software developer
Chazen, Melvin Leonard chemical engineer
Everett, Robert William environmental engineer
Fesq, Lorraine Mae aerospace and computer engineer
Foster, John Stuart, Jr. physicist, former defense industry executive
Gehrlein, Michael Timothy air force officer
Hughes, James Arthur electrical engineer
Roth, Thomas J. physicist
Sackheim, Robert Lewis aerospace engineer, educator
Schwarzbek, Stephen Mark physicist
Wright, Charles P. engineering manager
Yang, Jane Jan-Jan engineering executive

Redwood City
Bluher, Grigory computer scientist, mathematician
Hirschhorn, Elizabeth Ann environmental engineer, consultant
Jobs, Steven Paul computer corporation executive
Murthy, Sudha Akula physicist, consultant
Nacht, Sergio biochemist
Rohde, James Vincent software systems company executive
Sharma, Kuldeepak Bhardwaj pharmaceutical scientist
Taenzer, Jon Charles scientist, electronics engineering consultant

Reseda
Ross, Amy Ann experimental pathologist

Rialto
Merrill, Steven William research and development executive

Richmond
Balakrishnan, Krishna (Balki Balakrishnan) biotechnologist, corporate executive
Hobbs, Kevin David software engineer
Meyer, Jarold Alan oil company research executive
Moehle, Jack P. civil engineer, engineering executive

Ridgecrest
Pearson, John mechanical engineer
St. Amand, Pierre geophysicist

Riverside
Beni, Gerardo electrical and computer engineering educator, robotics scientist
Chang, Andrew C. engineer, educator
Chang, Sylvia Tan health facility administrator, educator
Dammann, David Patrick electronics engineer
Embleton, Tom William horticultural science educator
Fung, Sun-Yiu Samuel physics educator
Ghosh, Subir statistician
Hackwood, Susan electrical and computer engineering educator
Hinkle, Nancy C. veterinary entomologist
Karlow, Edwin Anthony physicist, educator
Mathaudhu, Sukhdev Singh mechanical engineer
Miller, Thomas Albert entomology educator
Mudd, John Brian biochemist
Nothnagel, Eugene Alfred plant cell biology educator
Parke, Ross Duke psychology educator
Taylor, Randy Steven hospital administrator
Van Gundy, Seymour Dean nematologist, plant pathologist, educator
White, Robert Stephen physics educator
Zentmyer, George Aubrey plant pathology educator

Rocklin
Vande Streek, Penny Robillard nuclear medicine physician, researcher, educator

Rolling Hills Estates
Bellis, Carroll Joseph surgeon
Lee, James King technology corporation executive

Rosamond
Stephens, John Walter facility engineer

Rosemead
Heath, Ted Harris environmental engineer

Sacramento
Bailey, Thomas Everett engineering company executive
Bogren, Hugo Gunnar radiology educator
Cavigli, Henry James petroleum engineer
Collins, William Leroy telecommunications engineer
Cummings, Belinda construction engineer
Dedrick, Kent Gentry retired physicist, researcher
Devontine, Julie E(lizabeth) J(acqueline) (The Marchioness de Roe Devon) systems analyst, consultant
Fairall, Richard Snowden mechanical engineer
Futernick, Kenneth David education educator, software developer
Gilmore, Allan Emory forensic consultant
Guyer, J. Paul civil engineer, architect, consultant
Hartman, Howard Levi mining engineering educator, consultant
Helander, Clifford John state agency administrator
Hwang, Jenn-Shin civil engineer
Jakovac, John Paul construction executive
Lathi, Bhagawandas Pannalal electrical engineering educator
Little, Robert John, Jr. botanist, environmental consultant
Mallarapu, Rupa Latha transportation engineer, consultant
Maurath, Garry Caldwell hydrogeologist
Meral, Gerald Harvey biologist
Moore, David Sumner forensic document examiner
Morrow, Joseph Eugene psychology educator
Nussenbaum, Siegfried Fred chemistry educator
Panuschka, Gerhard civil and sanitary engineer, inventor
Pryor, Douglas Keith clinical psychologist, consultant
Reardon, Frederick Henry mechanical engineer, educator
Rosenberg, Dan Yale retired plant pathologist
Rubin, Andrew Lawrence toxicologist, cell biologist
Sharma, Arjun Dutta cardiologist
Udvardy, Miklos Dezso Ferenc biology educator
Zil, John Stephen psychiatrist, physiologist

Saint Helena
Amerine, Maynard Andrew enologist, educator

Salinas
Eifler, Carl Frederick retired psychologist
Finnberg, Elaine Agnes psychologist, editor
Lee, Gilbert Brooks retired ophthalmology engineer
Reichenbach, Thomas veterinarian
Stubblefield, James Irvin physician, surgeon, army officer

San Anselmo
Bodington, Charles E. chemical engineer

San Bernardino
Bauer, Steven Michael cost containment engineer
Holtz, Tobenette aerospace engineer
Russo, Alvin Leon obstetrician/gynecologist
Skoog, William Arthur oncologist

San Bruno
Bradley, Charles William podiatrist

San Carlos
Hoffman, Paul Jerome psychologist
Zanoni, Michael McNeal forensic scientist

San Clemente
Arora, Sam Sunder mechanical engineer
Crider, Hoyt health care executive
Ditty, Marilyn Louise gerontologist, educator
Tran, Dean opto-electronic devices engineer

San Diego
Abalos, Ted Quinto electronics engineer
Archibald, James David biology educator, paleontologist
Beyster, John Robert engineering company executive
Blelloch, Paul Andrew mechanical engineer
Brog, Tov Binyamin electrical engineer
Brusca, Richard C. zoologist, museum curator, researcher, educator
Butler, Geoffrey Scott systems engineer, educator, consultant
Cameron, Charles Bruce naval officer, electrical engineer
Caulder, Jerry Dale weed scientist
Chambers, Henry George surgeon
Chouinard, Warren E. engineer
Coleman, Dale Lynn electronics engineer, aviator, educator
Crick, Francis Harry Compton biologist, educator
Dayson, Rodney Andrew chemical engineer
Dean, Richard Anthony mechanical engineer
Deuble, John L., Jr. environmental science and engineering services consultant
Eckhart, Walter molecular biologist, educator
Edwards, Darrel psychologist
Edwards, Jack Elmer personnel researcher
Ennis, Joel Brian physicist
Ewig, Carl Stephen chemist
Fraitag, Leonard Alan mechanical and design engineer
Galbraith, Nanette Elaine Gerks forensic and management sciences company executive
Gigli, Irma physician, educator, academic administrator
Goldstein, Mark Kingston Levin high technology company executive, researcher
Greene, John M. physicist
Grossberg, John Morris psychologist, educator
Guy, Teresa Ann aerospace engineer
Hannaman, George William, Jr. nuclear/electrical engineer
Hansbrough, John Fenwick surgery educator
Harriett, Judy Anne medical equipment company executive
Harwood, Ivan Richmond pediatric pulmonologist
Helinski, Donald Raymond biologist, educator
Inoue, Michael Shigeru industrial engineer, electrical engineer
Jamieson, Stuart William cardiologist, educator
Karin, Sidney research and development executive
Kent, Theodore Charles psychologist
Klimowicz, Thomas F. metallurgist
Klump, Wolfgang Manfred biochemist
Kobus, David Allan research psychologist, consultant
Kosmatka, John Benedict aerospace engineer, educator
Landersman, Stuart David engineer
Langlais, Philip Joseph psychologist, educator
Lerner, Sheldon plastic surgeon
Lewis, Gregory Williams neuroscientist
Linnville, Steven Emory psychologist, navy officer
Lockwood Hourani, Laurel Lee epidemiologist
Loper, Warren Edward computer scientist
Lovelace, Alan Mathieson aerospace company executive
Lowrey, D('Orvey) Preston, III mechanical engineer
Luce, Timothy Charles physicist
Mestril, Ruben biochemist, researcher
Meyer, Thomas Robert television product executive
Mooradian, Gregory Charles physicist
Moser, Kenneth Miles physician
Northup, T. Eugene nuclear engineer
Obenour, Jerry Lee scientific company executive
Oliphant, Charles Romig physician
O'Rourke, Ronald Eugene computer and electronics industry executive
Overskei, David physicist
Parrish, Clyde Robin, III electronics engineer
Paz, Nils physicist
Peters, Raymond Eugene computer systems company executive
Prescott, Lawrence Malcolm medical and health science writer
Razzaghi, Mahmoud mechanical engineer
Reed, Sharon Lee infectious disease consultant
Rosen, Peter health facility administrator, emergency physician, educator
Ross, John, Jr. physician, educator
Royston, Ivor scientific director
Ryder, Oliver Allison geneticist, conservation biologist, educator
Salk, Jonas Edward physician, scientist
Schmidt, Joseph David urologist
Schmidtmann, Victor Henry instrumentation engineer
Schorr, Martin Mark psychologist, educator, writer
Schryver, Bruce John safety engineer
Schwertly, Harvey Kenneth, Jr. computer electronics educator
Sejnowski, Terrence Joseph science educator
Sell, Robert Emerson electrical engineer
Shah, Nipulkumar principal design engineer
Shoemaker, Patrick Allen engineer
Simpson, Henry Kertan, Jr. psychologist, educator
Smith, Raymond Edward health care administrator
Smith, Richard Alan neurologist, medical association administrator
Smith, Steven Cole computer scientist
Stowell, Larry Joseph agricultural consultant
Thomas, Charles Allen, Jr. molecular biologist, educator
Trejo, Leonard Joseph psychologist
Tretter, James Ray pharmaceutical company executive
Trout, Monroe Eugene hospital systems executive
Tuccio, Sam Anthony aerospace executive, physicist
Viterbi, Andrew James electrical engineering and computer science educator, business executive
Waltz, Ronald Edward physicist
Wang, Kuo Chang aerospace engineering educator
Weinrich, James Donald psychobiologist, educator
Yang, Hsin-Ming immunologist
Youkharibache, Philippe Bijin computational chemist, researcher
Youngs, Jack Marvin cost engineer

San Dimas
Smith, Michael Steven data processing executive
Thompson, Allan Robert sales engineer

San Francisco
Aird, Robert Burns neurologist, educator
Arieff, Allen Ives physician
Backus, John computer scientist
Barondes, Samuel Herbert psychiatrist, educator
Baxter, John Darling physician, educator, health facility administrator
Bechtel, Stephen Davison, Jr. engineering company executive
Benet, Leslie Zachary pharmacokineticist
Bernstein, Harold Seth pediatric cardiologist, molecular geneticist
Betlach, Mary Carolyn biochemist, molecular biology researcher
Bigler, William Norman chemistry educator, biochemist
Bishop, John Michael biomedical research scientist, educator
Blackburn, Elizabeth Helen molecular biologist
Blaevoet, Jeffrey Paul mechanical engineer
Boisvert, William Andrew nutritional biochemist, researcher
Borson, Daniel Benjamin physiology educator, inventor, researcher
Boyer, Herbert Wayne biochemist
Brandin, Alf Elvin retired mining and shipping company executive
Brovold, Frederick Norman geotechnical and civil engineer
Brown, H. William urban economist, private banker
Brown, Walter Creighton biologist
Burlingame, Alma Lyman chemist, educator
Calvin, Allen David psychologist, educator
Cape, Ronald Elliot biotechnology company executive
Cheng, Kwong Man structural/bridge engineer
Christensen, David William mathematician, engineer
Clark, Stuart Alan mechanical engineer
Cline, Carolyn Joan plastic and reconstructive surgeon
David, George psychiatrist, economic theory lecturer
Dawdy, David Russell hydrologist
Dawson, Chandler R. opththalmologist, educator
Debas, Haile T. gastrointestinal surgeon, physiologist, educator
Denning, Michael Marion computer company executive
Dill, Kenneth Austin pharmaceutical chemistry educator
Dolby, Ray Milton engineering company executive, electrical engineer
Düzgünes, Nejat A. biophysicist
Eastwood, Susan medical scientific editor
Engleman, Ephraim Philip physician
Farrell, Edward Joseph mathematics educator
Fishman, Robert Allen educator, neurologist
Forsen, Harold Kay engineering executive
Frane, James Thomas mechanical engineer, author
Goldie, Peter electronic publisher, computer consultant
Gortner, Susan Reichert nursing educator
Gottfried, Eugene Leslie physician, educator
Harary, Keith psychologist
Havel, Richard Joseph physician, educator
Heyman, Melvin Bernard pediatric gastroenterologist
Iacono, James Michael research center administrator
Jaffe, Robert Benton obstetrician-gynecologist, reproductive endocrinologist
Jampolsky, Arthur ophthalmologist
Jeffrey, Robert Asahel, Jr. pathologist
Johnson, Herman Leonall research nutritionist
Kan, Yuet Wai physician, investigator
Katzung, Bertram George pharmacologist
Khosla, Ved Mitter oral and maxillofacial surgeon, educator
Kirschner, Marc Wallace biochemist, cell biologist
Kreitzberg, Fred Charles construction management company executive
Kuck, Marie Elizabeth Bukovsky retired pharmacist
Kuo, Chang Keng electrical engineer
Lee, Philip Randolph medical educator
Lin, Tung Yen civil engineer, educator
Mannino, J. Davis psychotherapist
Mason, Dean Towle cardiologist
McKnight, Steven Lanier molecular biologist
Morrin, Thomas Harvey engineering research company executive
Neelakantan, Ganapathy Subramanian geotechnical engineer
Nguyen, Ann Cac Khue pharmaceutical and medicinal chemist
Nicholson, William Joseph forest products company executive
Papalexopoulos, Alex Democrates electrical engineer
Peirce, Pamela Kay horticulturist, writer
Peterlin, Boris Matija physician
Randall, Janet Ann biology educator, researcher
Rutter, William J. biochemist, educator
Schmidt, Robert Milton physician, scientist, educator
Schweickart, Russell L. communications executive, astronaut
Shumate, Charles Albert retired dermatologist
Shushkewich, Kenneth Wayne structural engineer
Sowder, Robert Robertson architect
Vreeland, Robert Wilder electronics engineer
Warner, Rollin Miles, Jr. economics educator, financial planner, real estate broker
Westerdahl, John Brian nutritionist, health educator
Wilson, Charles B. neurosurgeon, educator
Wolff, Sheldon radiobiologist, educator
Yuan, Shao Wen aerospace engineer, educator
Zeichick, Alan L. computer journalist, computer scientist

San Jose
Adamovics, Andris toxicologist, consultant
Adams, William John, Jr. mechanical engineer
Aylesworth, John Richard software professional
Baker, James Reginald aerospace company executive, engineer
Blanch, Roy Lavern electrical engineer
Bryhan, Anthony James metallurgist
Chen, Wen H. laboratory executive, educator
Coleman, Brian Fitzgerald electrical engineer
Conrad, Jeffrey Philip mechanical engineer, manager
Eigler, Donald Mark physicist
Evans, Donald Foster electronics technician, computer consultant
Forster, Michael Lockwood (Mike Forster) software engineer
Glasgow, J. C. (John Carl Glasgow) systems engineering executive
Gorum, Victoria computer engineer
Gruber, John Balsbaugh physics educator, university administrator
Gunter, William Dayle, Jr. physicist
Hill, Robert James mechanical engineer
Jutamulia, Suganda electro-optic scientist
Karis, Thomas Edward chemical engineer
Kasai, Paul Haruo chemist
Lakkaraju, Harinarayana Sarma physics educator, consultant
Lin, Tao electronics engineering manager
Lippe, Philipp Maria neurosurgeon, educator
McCarthy, Mary Ann Bartley electrical engineer
McDowell, Jennifer sociologist, composer, playwright, publisher
Mizrah, Len Leonid physicist
Nimmagadda, Rao Rajagopala materials scientist, researcher
Oak, Ronald Stuart health and safety administrator
Parker, Michael Andrew materials scientist
Pellegrini, Robert J. psychology educator
Peltzer, Douglas Lea semiconductor device manufacturing company executive
Rabolt, John Francis optics scientist
Rao, Atambir Singh nuclear engineer
Rudge, William Edwin, IV computational physicist
Schweikert, Daniel George electrical engineer, administrator
Scifres, Donald Ray physics and engineering administrator
Shatney, Clayton Henry surgeon
Smith, Harry James mathematician
Stahl, Desmond Eugene systems engineer
Stein, Arthur Oscar pediatrician
Stockton, Anderson Berrian electronics company executive, consultant, genealogist
Wilkins, Daniel R. nuclear engineer, nuclear energy industry executive

San Juan Capistrano
Burrows, Barbara Ann veterinarian
Fisher, Delbert Arthur physician, educator
Huta, Henry Nicholaus manufacturing and service company executive
Testa, Stephen Michael geologist, consultant

San Lorenzo
Lantz, Charles Alan chiropractor, researcher
Thompson, Lyle Eugene electrical engineer

San Luis Obispo
Moazzami, Sara civil engineering educator
Race, Lisa Anne environmental chemist
Schumann, Thomas Gerald physics educator

San Marino
Benzer, Seymour neurosciences educator
Lashley, Virginia Stephenson Hughes retired computer science educator

San Mateo
Burke, Richard James optical physicist, consultant
Horwitz, David Larry pharmaceuticals company executive, researcher, educator
Pappas, Costas Ernest aeronautical engineer, consultant

San Pedro
Ellis, George Edwin, Jr. chemical engineer
Lowi, Alvin, Jr. mechanical engineer, consultant

San Rafael
Badgley, John Roy architect
Danse, Ilene Homnick Raisfeld physician, educator, toxicologist
Hudy, John Joseph psychologist
Jindrich, Ervin James municipal government official
Marmann, Sigrid software development company executive
Starks, Michael Richard stereographer, consultant
Taylor, Irving mechanical engineer, consultant

San Ramon
Brennan, Sean Michael *cancer research scientist, educator*
Bugno, Walter Thomas *civil engineer*
Montgomery, Elizabeth Ann *clinical research consultant*
Morris, Jay Kevin *chemical engineer*
Ratiu, Mircea Dimitrie *mechanical engineer*

San Ysidro
Ito, Shigemasa *electronics executive*

Santa Ana
Amies, Alex Phillip *civil engineer*
Bennett, Brian O'Leary *utilities executive*
Bentley, William Arthur *engineer, electro-optical consultant*
Cagle, Thomas Marquis *electronics engineer*
Daniel, Ramon, Jr. *psychologist, consultant, bilingual educator*
Do, Tai Huu *mechanical engineer*
Groner, Paul S(tephen) *electrical engineer*
Gross, Herbert Gerald *space physicist*
Hilmy, Said Ibrahim *structural engineer, consultant*
Labbe, Armand Joseph *museum curator, anthropologist*
London, Ray William *clinical and forensic psychologist*
Nelson, Richard David *electro-optics professional*
Thompson, Malcolm Francis *electrical engineer*

Santa Ana Heights
George, Kattunilathu Oommen *homoeopathic physician, educator*
Jungren, Jon Erik *civil engineer*

Santa Barbara
Anderson, Stephen Thomas *computer design engineer*
Atwater, Tanya Maria *marine geophysicist, educator*
Awschalom, David Daniel *physicist*
Beutler, Larry Edward *psychology educator*
Blaskó, Andrei *chemist, researcher*
Butler, Alison *chemist, educator*
Casey, Steven Michael *ergonomist, human factors engineer*
Duarte, Ramon Gonzalez *nurse, educator, researcher*
Dudziak, Walter Francis *physicist*
Fisher, Steven Kay *neurobiology eductor*
Ford, Peter C. *chemistry educator*
Goodchild, Michael *geographer, educator*
Grunke, Andrew Frederick *astronomy educator, instrument designer*
Gutsche, Steven Lyle *physicist*
Hatherill, John Robert *toxicologist, educator*
Heeger, Alan Jay *physicist*
Israelachvili, Jacob Nissim *chemical engineer*
Jennings, David Thomas, III *electronics executive, consultant*
Jiang, Wenbin *electrical engineer*
Kendler, Tracy Seedman *psychology educator*
Kennett, James Peter *geology and zoology educator*
Kohn, Walter *educator, physicist*
Lawrance, Charles H. *civil and sanitary engineer*
Lee, Hua *electrical engineering educator*
Lennox Buchthal, Margaret Agnes *neurophysiologist*
Lick, Wilbert James *mechanical engineering educator*
Luyendyk, Bruce Peter *geophysicist, educator, institution administrator*
Majumdar, Arunava *mechanical engineer, educator*
Mathews, Barbara Edith *gynecologist*
Meinel, Aden Baker *scientist*
Merz, James Logan *electrical engineering and materials educator, researcher*
Metiu, Horia Ion *chemistry educator*
Mitra, Sanjit Kumar *electrical and computer engineering educator*
Morrison, Rollin John *physicist, educator*
Mueller, George E. *corporation executive*
Peale, Stanton Jerrold *physics educator*
Peterson, Charles Marquis *medical educator*
Pincus, Philip A. *chemical engineering educator*
Russell, Charles Roberts *chemical engineer*
Schofield, Keith *research chemist*
Sommers, Adele Ann *engineering specialist, technical trainer*
Swalley, Robert Farrell *structural engineer, consultant*
Witherell, Michael S. *physics educator*
Wudl, Fred *chemistry educator, consultant*
Zhang, Yong-Hang *research electrical engineer*

Santa Clara
Edwards, George Henry *technical writer*
Falgiano, Victor Joseph *electrical engineer, consultant*
Field, Alexander James *economics educator, dean*
Flinn, Paul Anthony *materials scientist*
Gozani, Tsahi *nuclear physicist*
Greene, Frank Sullivan, Jr. *business executive*
Gupta, Anand *mechanical engineer*
Klein, Harold Paul *microbiologist*
Mand, Ranjit Singh *device physicist, educator*
McDonald, Mark Douglas *electrical engineer*
Mohr, Siegfried Heinrich *mechanical and optical engineer*
Morgan, James C. *electronics executive*
Nowak, Romuald *physicist*
Oliver, David Jarrell *electronic and systems engineer*
Shoup, Terry Emerson *university dean, engineering educator*
Wilson, Frank Henry *electrical engineer*
Yan, Pei-Yang *electrical engineer*
Zhang, Chen-Zhi *project engineer*

Santa Clarita
Abbott, John Rodger *electrical engineer*
Buck, Douglas Earl *chemist*
Granlund, Thomas Arthur *engineering executive, consultant*
Mahler, David *chemical company executive*

Santa Cruz
Bunnett, Joseph Frederick *chemist, educator*
Chance, Hugh Nicholas *mechanical engineer*
Drake, Frank Donald *astronomy educator*
Dunn, Robert Leland *energy engineer*
Faber, Sandra Moore *astronomer, educator*
Goldbeck, Robert Arthur, Jr. *physical chemist*
Griggs, Gary Bruce *earth sciences educator, oceanographer, geologist, consultant*
Kraft, Robert Paul *astronomer, educator*
Lay, Thorne *geosciences educator*
Lilly, Les J. *micronautics engineer*
Machotka, Pavel *psychology and art educator*
Noller, Harry Francis, Jr. *biochemist, educator*
Osterbrock, Donald E(dward) *astronomy educator*
Talamantes, Frank J. *biology educator*
Williams, Quentin Christopher *geophysicist*
Woosley, Stanford Earl *astrophysicist*

Santa Fe Springs
Hovanec, B. Michael *chemist, researcher*
McCarty, William Britt *natural resource company executive, educator*

Santa Maria
Geise, Harry Fremont *retired meteorologist*
Raich, Abraham Leonard *rabbi, quality control professional*

Santa Monica
Coleman, William Eliah *psychologist*
Crain, Cullen Malone *electrical engineer*
Gupta, Rishab Kumar *medical association administrator, educator, researcher*
Intriligator, Devrie Shapiro *physicist*
Shipbaugh, Calvin LeRoy *physicist*
Smith, Roberts Angus *biochemist, educator*
Spivack, Herman M. *fluid dynamicist, aerospace engineer*
Veneklasen, Paul Schuelke *physicist, acoustics consultant*

Santa Rosa
Anderson, Terry Marlene *civil engineer*
Canfield, Philip Charles *electrical engineer*
de Wys, Egbert Christiaan *geochemist*
Dwight, Herbert M., Jr. *optical engineer, manufacturing executive*
Mackay, Kenneth Donald *environmental services company executive*
Rancourt, James Daniel *optical engineer*
Sibley, Charles Gald *biologist, educator*

Saratoga
Mihnea, Tatiana *mathematics educator*

Sausalito
Elion, Herbert A. *optoelectronics and bioengineering executive, physicist*

Scotts Valley
Snyder, Charles Theodore *geologist*

Seal Beach
Gironda, A. John, III *engineer*
Uda, Robert Takeo *aerospace engineer*
Woo, Steven Edward *aerospace engineer*

Shaver Lake
Warren, Barbara Kathleen (Sue Warren) *wildlife biologist*

Sherman Oaks
Catran, Jack *aerospace and physiology scientist, writer*
Pothitt, Kathleen Marie *physical oceanography researcher*
Rosen, Alexander Carl *psychologist, consultant*
Wróblewski, Ronald John *physicist*

Simi Valley
Ahsanullah, Omar Faruk *computer engineer, consultant*

Solana Beach
Agnew, Harold Melvin *physicist*

South San Francisco
Gaertner, Alfred Ludwig *biochemist*
Gonda, Igor *pharmaceutical scientist*
Levinson, Arthur David *molecular biologist*
Masover, Gerald Kenneth *microbiologist*
Nuwaysir, Lydia Marie *analytical chemist*
Pennica, Diane *molecular biologist*
Pitcher, Wayne Harold, Jr. *biotechnology company executive*
Winkler, Marjorie Everett *protein chemist, researcher*
Woodruff, Teresa K. *cell biologist*

Springville
Pugh, Paul Franklin *engineer consultant*

Stanford
Andersen, Hans Christian *chemistry educator*
Anderson, Martin Carl *economist*
Angell, James Browne *electrical engineering educator*
Arrow, Kenneth Joseph *economist, educator*
Aziz, Khalid *petroleum engineering educator*
Bai, Taeil Albert *research physicist*
Baker, Bruce S. *molecular biologist*
Bashaw, Matthew Charles *physicist*
Bauer, Eugene Andrew *dermatologist, educator*
Baylor, Denis Aristide *neurobiology educator*
Beasley, Malcolm Roy *physics educator*
Berg, Paul *biochemist, educator*
Botstein, David *geneticist, educator*
Boudart, Michel *chemist, chemical engineer, educator*
Bracewell, Ronald Newbold *electrical engineering and computer science educator*
Bradshaw, Peter *engineering educator*
Brauman, John I. *chemist, educator*
Brown, J. Martin *oncologist, educator*
Chang-Hasnain, Constance Jui-Hua *educator*
Chu, Steven *physics educator*
Cleary, Michael *pathologist, educator*
Coleman, Robert Griffin *geology educator*
Contag, Christopher Heinz *biomedical researcher*
Contag, Pamela Reilly *biologist, researcher*
Cutler, Cassius Chapin *physicist, educator*
Dantzig, George Bernard *applied mathematics educator*
Davis, Mark M. *microbiologist, educator*
Djerassi, Carl *chemist, educator, writer*
Efron, Bradley *mathematics educator*
Ehrlich, Paul Ralph *biology educator*
Ernst, Wallace Gary *geology educator, dean*
Eshleman, Von Russel *electrical engineering educator*
Franklin, Gene Farthing *electrical engineering educator, consultant*
Gast, Alice P. *chemical engineering educator*
Glazer, Gary Mark *radiology educator*
Gliner, Erast Boris *theoretical physicist*
Golub, Gene Howard *computer science educator, researcher*
Goodman, Joseph Wilfred *electrical engineering educator*
Greenberg, Joseph H. *anthropologist*
Hagstrom, Stig Bernt *materials science and engineering educator*
Harrison, Walter Ashley *physicist, educator*
Herrmann, George *mechanical engineering educator*
Herrmannsfeldt, William Bernard *physicist*
Honig, Lawrence Sterling *neuroscientist, neurologist*
Hughes, Thomas Joseph *mechanical engineering educator, consultant*
Huntington, Hillard Griswold *economist*
Jardetzky, Oleg *medical educator, scientist*
Johnson, William Summer *chemistry educator*
Kaiser, Armin Dale *biochemist, educator*
Karlin, Samuel *mathematics educator, researcher*
Keller, Joseph Bishop *mathematician, educator*
Knuth, Donald Ervin *computer sciences educator*
Kornberg, Arthur *biochemist*
Korner, Anneliese F. *psychology research scientist*
Kruger, Paul *nuclear civil engineering educator*
Lathrop, Kaye Don *nuclear scientist, educator*
Levine, Seymour *psychology educator, researcher*
Levy, Ronald *medical educator, researcher*
Lieberman, Gerald J. *statistics educator*
Liu, Guosong *neurobiologist*
Long, Sharon Rugel *molecular biologist, plant biology educator*
Macovski, Albert *electrical engineering educator*
Matin, Abdul *microbiology educator, consultant*
McCarthy, John *computer scientist, educator*
McCluskey, Edward Joseph *engineering educator*
McConnell, Harden Marsden *biophysical chemistry researcher, chemistry educator*
McDevitt, Hugh O'Neill *immunology educator, physician*
Melmon, Kenneth Lloyd *physician, biologist, pharmacologist, consultant*
Mosher, Harry Stone *chemistry educator*
Nix, William Dale *materials scientist, educator*
Olson, Darin S. *research scientist*
Ott, Wayne Robert *environmental engineer*
Pearson, Scott Roberts *economics educator*
Piquette, Gary Norman *reproductive endocrinologist*
Plummer, James D. *electrical engineering educator*
Pratto, Felicia *psychologist*
Ramey, Henry Jackson, Jr. *petroleum engineering educator*
Reynolds, William Craig *mechanical engineer, educator*
Ricardo-Campbell, Rita *economist, educator*
Roberts, Eric Stenius *computer science educator*
Rosenberg, Saul Allen *oncologist, educator*
Ross, John *physical chemist, educator*
Rott, Nicholas *fluid mechanics educator*
Sa, Luiz Augusto Discher *physicist*
Schawlow, Arthur Leonard *physicist, educator*
Scott, Matthew Peter *biology educator*
Siegman, Anthony Edward *electrical engineer, educator*
Small, Peter McMichael *physician, researcher*
Springer, George Stephen *mechanical engineering educator*
Steele, Claude Mason *psychology educator*
Stevens, Mary Elizabeth *biomedical engineer, consultant*
Stryer, Lubert *biochemist, educator*
Sturrock, Peter Andrew *space science and astrophysics educator*
Taube, Henry *chemistry educator*
Teicholz, Paul M. *civil engineering educator, administrator*
Teller, Edward *physicist*
Thompson, David Alfred *industrial engineer*
Thompson, George Albert *geophysics educator*
Trost, Barry Martin *chemist, educator*
Wagoner, Robert Vernon *astrophysicist, educator*
Weyant, John Peter *engineering economic systems educator*
Winick, Herman *physicist*
Winograd, Terry Allen *computer science educator*
Yang, Zeren *obstetrician/gynecologist, researcher*
Yanofsky, Charles *biology educator*
Zare, Richard Neil *chemistry educator*

Stockton
Biddle, Donald Ray *aerospace company executive*
Fletcher, David Quentin *civil engineering educator*
Gross, Paul Hans *chemistry educator*
Renson, Jean Felix *psychiatry educator*

Sunland
Karney, James Lynn *physicist, optical engineering consultant*

Sunnyvale
Altamura, Michael Victor *physician*
Breiner, Sheldon *geophysics educator, business executive*
Bullis, W(illiam) Murray *physicist, consultant*
Devgan, Onkar Dave N. *technologist, consultant*
Floersheim, Robert Bruce *military engineer*
Herman, Michael Harry *physicist, researcher*
Holbrook, Anthony *manufacturing company executive*
Ma, Fengchow Clarence *agricultural engineering consultant*
Martinez-Galarce, Dennis Stanley *physicist*
Miller, Joseph Arthur *manufacturing engineer, educator, consultant*
Pugmire, Gregg Thomas *optical engineer*
Rugge, Henry Ferdinand *medical products executive*
Sankar, Subramanian Vaidya *aerospace engineer*
Saxena, Amol *podiatrist, consultant*
Schlobohm, John Carl *electrical engineer*
Silvestri, Antonio Michael (Tony) *electrical engineer*
Stevens, John Lawrence *quality assurance professional*
Yee, Raymond Kim *mechanical engineer, researcher*

Sylmar
Levine, Paul Allan *cardiologist*
Sholder, Jason Allen *medical products company executive*

Tarzana
Macmillan, Robert Smith *electronics engineer*

Temecula
Feltz, Charles Henderson *former mechanical engineer, consultant*
Roemmele, Brian Karl *electronics, publishing, financial and real estate executive*

Templeton
Gandsey, Louis John *petroleum and environmental consultant*

Thermal
Johnson, Charles Wayne *mining engineer, mining executive*

Thousand Oaks
Andrews, Angus Percy *mathematician, writer*
Bahn, Gilbert Schuyler *retired mechanical engineer, researcher*
Chen, Shih-Hsiung *aerospace engineer*
Conant, David Arthur *architectural acoustician, educator, consultant*
Gentry, James Frederick *chemical engineer, consultant*
Longo, Joseph Thomas *electronics executive*
Malmuth, Norman David *program manager*
Newman, Paul Richard *physicist*
Rohrbach, Jay William *biotechnology company supervisor*
Sachdev, Raj Kumar *biochemical engineer*
Shankar, Vijaya V. *aeronautical engineer*
Wang, I-Tung *atmospheric scientist*

Tiburon
Heacox, Russel Louis *mechanical engineer*
Pratt, Jeremy *human ecologist, researcher*

Topanga
Anderson, Donald Norton, Jr. *retired electrical engineer*

Torrance
Black, Suzanne Alexandra *clinical psychologist, clinical neuropsychologist, researcher*
Chao, Conrad Russell *obstetrician*
French, William J. *cardiologist, educator*
Howard, Donivan R(ichard) *engineering executive*
Krueger, Kurt Edward *environmental management company official*
Leake, Donald Lewis *oral and maxillofacial surgeon, oboist*
Malhotra, Vijay Kumar *mathematics educator*
Mann, Michael Martin *electronics company executive*
Rizkin, Alexander *photonics researcher, educator*
Schipper, Leon *mechanical engineer*
Shirk, Kevin William *fiber optic product manager*
Sorstokke, Susan Eileen *systems engineer*
Sun, Zongjian *physicist, researcher*
Wen, Cheng Paul *electrical engineer*

Tracy
Christon, Mark Allen *mechanical engineer*

Travis AFB
Mamula, Mark *aerospace engineer*

Trona
Laire, Howard George *chemist*

Tustin
Charley, Philip James *testing laboratory executive*
Cruzen, Matt Earl *research biochemist*
Sinnette, John Townsend, Jr. *research scientist, consultant*

Ukiah
Pinoli, Burt Arthur *airline executive*

Upland
Robinson, Hurley *surgeon*

Vacaville
Coulson, Kinsell Leroy *meteorologist*

Valencia
Davison, Arthur Lee *scientific instrument manufacturing company executive, engineer*
Golijanin, Danilo M. *materials scientist, physics educator*
Leung, Kam H. *medical company executive*

Valley Ford
Clowes, Garth Anthony *electronics executive, consultant*

Van Nuys
Ahuja, Anil *energy engineer*
Altshiller, Arthur Leonard *physics educator*
Chusid, Michael Thomas *architect*
Cooper, Leroy Gordon, Jr. *former astronaut, business consultant*
Schmetzer, William Montgomery *aeronautical engineer*

Ventura
Cowen, Stanton Jonathan *air pollution control engineer*
Gaynor, Joseph *chemical engineering consultant*
Hendel, Frank J(oseph) *chemical and aerospace engineer, educator, technical consultant*
Naurath, David Allison *engineering psychologist, researcher*
Wheeler, Harold Alden *retired radio engineer*

Victorville
Syed, Moinuddin *electrical engineer*

Villa Grande
Shirilau, Jeffery Micheal *engineering executive*
Shirilau, Mark Steven *utilities executive*

Visalia
Riegel, Byron William *ophthalmologist*

Vista
Maggay, Isidore, III *engineering executive, food processing engineer*

Walnut
Lee, Peter Y. *electrical engineer, consultant*

Walnut Creek
Burgarino, Anthony Emanuel *environmental engineer, consultant*
Farr, Lee Edward *physician*
Garrett, Suzanne Thornton *management educator*

Loyonnet, Georges-Claude *engineer*
McKnight, Lenore Ravin *child psychiatrist*
Minshall, Greg *computer programmer*
Oakeshott, Gordon B(laisdell) *geologist*
Sanborn, Charles Evan *retired chemical engineer*
VanDenburgh, Chris Allan *electrical engineer*
Van Maerssen, Otto L. *aerospace engineer, consulting firm executive*

Weed
Kyle, Chester Richard *mechanical engineer*

West Hollywood
Wilson, Myron Robert, Jr. *former psychiatrist*
Ziferstein, Isidore *psychoanalyst, educator, consultant*

West Los Angeles
Van Zak, David Bruce *psychologist*

Westlake Village
Hokana, Gregory Howard *engineering executive*
Huestis, Marilyn Ann *toxicologist, clinical chemist*

Westminster
Allen, Merrill James *marine biologist*
Armstrong, Gene Lee *retired aerospace company executive*

Westwood
Brydon, Harold Wesley *entomologist, writer*

Whittier
Prickett, David Clinton *physician*

Wilton
Shapero, Harris Joel *pediatrician*

Woodland
Phan, Chuong Van *biotechnologist*

Woodland Hills
Blanchard, William Henry *psychologist*
Darnell, Alfred J(erome) *chemical engineer, consultant*
Fox, Stuart Ira *physiologist*
Higginbotham, Lloyd William *mechanical engineer*
Holguin, Librado Malacara (Lee Holguin) *civil engineer*
Merin, Robert Lynn *periodontist*
Piersol, Allan Gerald *engineer*
Sharma, Brahama Datta *chemistry educator*
Wing, Thomas *micrometrologist, consultant*

Woodside
Ashley, Holt *aerospace scientist, educator*

COLORADO

Akron
Halvorson, Ardell David *research leader, soil scientist*

Aurora
Bilett, Jane Louise *clinical psychologist*
Bingham, Paris Edward, Jr. *electrical engineer, computer consultant*
Lewey, Scot Michael *gastroenterologist*
McClendon, Irvin Lee, Sr. *technical writer, editor*
Schwartz, Lawrence *aeronautical engineer*
Zimmerman, Jeannine *crime laboratory specialist, researcher*

Boulder
Albritton, Daniel L. *aeronomist*
Archambeau, Charles Bruce *physics educator, geophysics research scientist*
Avery, Susan Kathryn *electrical engineering educator, researcher*
Barnes, Frank Stephenson *electrical engineer, educator*
Baylor, Jill S(tein) *electrical engineer*
Begelman, Mitchell C. *astrophysicist, educator*
Benz, Samuel Paul *physicist*
Birks, John William *chemistry educator*
Born, George H. *aerospace engineer, educator*
Brues, Alice Mossie *physical anthropologist, educator*
Burleski, Joseph Anthony, Jr. *information services professional*
Butler, James Hall *oceanographer, atmospheric chemist*
Byerly, Radford, Jr. *science policy official*
Cantrell, Christopher Allen *atmospheric chemist*
Carlson, Lawrence Evan *mechanical engineering educator*
Cathey, Wade Thomas *electrical engineering educator*
Cech, Thomas Robert *chemistry and biochemistry educator*
Chung, Young Chu *chemist*
Clifford, Steven Francis *science research director*
Cornman, Larry Bruce *physicist*
Cowing, Thomas William *architectural, civil engineer*
Cristol, Stanley Jerome *chemistry educator*
Cuntz, Manfred Adolf *astrophysicist, researcher*
Daughenbaugh, Randall Jay *chemical company executive*
De Fries, John Clarence *behavioral genetics educator, institute administrator*
Dunn, Gordon Harold *physicist*
Fleener, Terry Noel *marketing professional*
Forest, Carl Anthony *lawyer*
Forssander, Paul Richard *inventor, artist, entrepreneur*
Frehlich, Rodney George *engineer, researcher*
Gebbie, Katharine Blodgett *astrophysicist*
Glover, Fred William *artificial intelligence and optimization research director, educator*
Goldfarb, Ronald B. *research physicist*
Gu, Youfan *cryogenic researcher*
Gupta, Kuldip Chand *electrical and computer engineering educator, researcher*
Haenggi, Dieter Christoph *psychologist, researcher*
Hall, John Lewis *physicist, researcher*
Hansen, Ross N. *electrical engineer*
Hildner, Ernest Gotthold, III *solar physicist, science administrator*
Holdsworth, Janet Nott *women's health nurse*
Jerritts, Stephen G. *computer company executive*
Kamper, Robert Andrew *physicist*
Kauffman, Erle Galen *geologist, paleontologist*
Kellogg, William Welch *meteorologist*
Kisslinger, Carl *geophysicist, educator*
Lieberman, James Lance *chemistry educator*
Lineberger, William Carl *chemistry educator*
Little, Charles Gordon *geophysicist*
Low, Boon Chye *physicist*
Maclay, Timothy Dean *aerospace engineer*
Masterson, Linda Histen *medical company executive*
McCray, Richard Alan *astrophysicist, educator*
McFadden, Pamela Ann *architect*
Meehl, Gerald Allen *research climatologist*
Meier, Mark F. *research scientist, glaciologist, educator*
Miller, Harold William *nuclear geochemist*
Mohl, James Brian *aerospace engineer*
Morris, Alvin Lee *meteorologist, retired consulting corporation executive*
Norcross, David Warren *physicist, researcher*
Phelps, Arthur Van Rensselaer *physicist, consultant*
Prodan, Richard Stephen *electrical engineer*
Pryor, Wayne Robert *astronomer*
Rotunno, Richard *meteorologist*
Sani, Robert LeRoy *chemical engineering educator*
Schneider, Stephen Henry *climatologist, researcher*
Serafin, Robert Joseph *science center administrator, electrical engineer*
Smith, Ernest Ketcham *electrical engineer*
Smythe, William Rodman *physicist, educator*
Snow, Theodore Peck, Jr. *astrophysicist, author*
Sparks, Larry Leon *physicist*
Tary, John Joseph *engineer*
Usman, Nassim *research chemist and biochemist*
Walton, Harold Frederic *retired chemistry educator*
Wieman, Carl E. *physics educator*
Wood, William Barry, III *biologist, educator*

Brighton
Kohlmeier, Sharon Louise *medical laboratory administrator, medical technologist*

Broomfield
Simon, Wayne Eugene *engineer, mathematician*
Williams, Ronald David *electronics materials executive*

Canon City
Mc Bride, John Alexander *retired chemical engineer*

Carbondale
Cowgill, Ursula Moser *biologist, educator, environmental consultant*

Cheyenne Wells
Palmer, Rayetta J. *computer educator*

Colorado Springs
Allen, James L. *electrical engineer*
Burciaga, Juan Ramon *physics educator*
Carroll, David Todd *computer engineer*
Caughlin, Donald Joseph, Jr. *engineering educator, research scientist, experimental test pilot*
Cole, Julian Wayne (Perry Cole) *computer educator, consultant, programmer, analyst*
Dougherty, Harry Melville, III *reliability, maintainability engineer*
Edmonds, Richard Lee *air force officer*
Eller, Thomas Julian *aerospace company executive, astronautical engineer, computer scientist*
Freeman, Eugene Edward *electronics company executive*
Ginnett, Robert Charles *organizational psychologist*
Henrickson, Eiler Leonard *geologist, educator*
Herzog, Catherine Anita *process development engineer*
Jacobsmeyer, Jay Michael *electrical engineer*
Jones, Richard Eric, Jr. *aerospace engineer*
Kelly, Emery Leonard *bioenvironmental engineer*
Kubida, William Joseph *lawyer*
Lee, Kotik Kai *physicist*
MacLeod, Richard Patrick *foundation administrator*
Reynolds, Michael Floyd *civil engineer, educator*
Rydbeck, Bruce Vernon *civil engineer*
Stienmier, Saundra Kay Young *aviation educator*
Tanzer, Andrew Ethan *mechanical engineer*
Tenney, Leon Walter *civil engineer*
Tutt, Charles Leaming, Jr. *educational administrator, former mechanical engineering educator*
Vorndam, Paul Eric *analytical and organic chemist*
Wainionpaa, John William *systems engineer*
Ziemer, Rodger Edmund *electrical engineering educator, consultant*

Colorado State University
Cordain, Loren *physical education educator, graduate program director, researcher*
Moore, Janice Kay *biology educator*
Whicker, Floyd Ward *biology educator, ecologist*

Crested Butte
Green, Walter Verney *materials scientist*

Crestone
Temple, Lee Brett *architect*

Denver
Aikawa, Jerry Kazuo *physician, educator*
Anderson, Robert *environmental specialist, physician*
Barth, David Victor *computer systems designer*
Barz, Richard L. *microbiologist*
Berens, Randolph Lee *microbiologist, educator*
Bradford, Phillips Verner *engineering research executive*
Bradley, James Alexander *software engineer, researcher*
Brown, Mark Steven *medical physicist*
Burgess, Larry Lee *aerospace executive*
Callejo, Gerald Rodriguez *aerospace engineer*
Cevenini, Roberto Mauro *gas and oil industry executive, entrepreneur*
Cobban, William Aubrey *paleontologist*
Colvis, John Paris *aerospace engineer, mathematician, scientist*
Creech-Eakman, Michelle Jeanne *physicist, educator*
Cullum, Colin Munro *psychiatry and neurology educator*
Devitt, John Lawrence *consulting engineer*
Drăgoi, Dănut *physicist*
Dunn, Andrea Lee *biomedical researcher*
East, Donald Robert *civil engineer*
Eaton, Gareth Richard *chemistry educator, university dean*
Ferguson, Lloyd Elbert *manufacturing engineer*
Ferrell, Rebecca V. *biology educator*
Frevert, Donald Kent *hydraulic engineer*
Fryt, Monte Stanislaus *petroleum company executive, speaker, advisor*
Garske, Jay Toring *geologist, oil and minerals consultant*
Goldfield, Joseph *environmental engineer*
Gordon, Joseph Wallace *nuclear engineer*
Handelsman, Mitchell M. *psychologist, educator*
Henson, Peter Mitchell *physician, immunology and respiratory medicine executive*
Jacobson, Olof Hildebrand *forensic engineer*
Johnson, Walter Earl *geophysicist*
Kassan, Stuart S. *rheumatologist*
Keating, Larry Grant *electrical engineer, educator*
Kellogg, Karl Stuart *geologist*
Komdat, John Raymond *data processing consultant*
Kopke, Monte Ford *engineering executive*
Krikos, George Alexander *pathologist, educator*
Landon, Susan Melinda *petroleum geologist*
Liu, Chaoqun *staff scientist*
Lubeck, Marvin Jay *ophthalmologist*
Marshall, Charles Francis *aerospace engineer*
Marts, Kenneth Patrick *materials engineer*
McCammon, Donald Lee *civil engineer*
Mc Candless, Bruce, II *engineer, former astronaut*
McShane, Eugene Mac *psychologist*
Meldrum, Daniel Richard *general surgeon, physician*
Mendez, Celestino Galo *mathematics educator*
Mooney, Dennis John *forensic document examiner*
Murcray, Frank James *physicist*
Nelson, Nancy Eleanor *pediatrician, educator*
Olsen, Harold William *research civil engineer*
Peregrine, David Seymour *astronomer, consultant*
Perez, Jean-Yves *engineering company executive*
Perreten, Frank Arnold *surgeon, ophthalmologist*
Pfenninger, Karl H. *cell biology and neuroscience educator*
Poirot, James Wesley *engineering company executive*
Puck, Theodore Thomas *geneticist, biophysicist, educator*
Rendu, Jean-Michel Marie *mining executive*
Repine, John E. *pediatrician, educator*
Rhodes, Chuck William *electrical engineer*
Riese, Arthur Carl *environmental engineering company executive, consultant*
Rogers, Dale Arthur *electrical engineer*
Rubin, Betsy Claire *clinical psychologist, researcher*
Rumack, Barry H. *physician, toxicologist, pediatrician*
Schreiber, Edward *computer scientist*
Seidel, Frank Arthur *chemical engineer*
Sherman, David Michael *geologist*
Stephens, Larry Dean *engineer*
Sussman, Karl Edgar *physician*
Tabakoff, Boris *pharmacologist educator*
Talmage, David Wilson *physician, microbiology and medical educator, former university administrator*
Todd, Donald Frederick *geologist*
Wamboldt, Marianne Zdeblick *psychiatrist*
Washington, Reginald Louis *pediatric cardiologist*
Weiner, Norman *pharmacology educator*
Zimet, Carl Norman *psychologist, educator*

Durango
Jones, Janet Lee *psychology educator, cognitive scientist*
Lauth, Robert Edward *geologist*
Osterhoudt, Walter Jabez *geophysical and geological exploration consultant*
Spencer, Donald Clayton *mathematician*
Thurston, William Richardson *oil and gas industry executive, geologist*

Englewood
Breitenbach, Allan Joseph *geotechnical engineer*
Cowen, Donald Eugene *physician*
Lazarus, Steven S. *management consultant, marketing consultant*
Le, Khanh Tuong *utility executive*
McLellon, Richard Steven *aerospace engineer, consultant*
Prichard, Robert Alexander, Jr. *telecommunications engineer*
Schirmer, Howard August, Jr. *civil engineer*
Scott, Eugene Ray *electrical engineer, consultant*
Sexton, Amy Manerbino *computer analyst*
Ward, Milton Hawkins *mining company executive*

Estes Park
Blumrich, Josef Franz *aerospace engineer*
Friedman, Sander Berl *engineering educator*
Moore, Omar Khayyam *experimental sociologist*

Evergreen
Heyl, Allen Van, Jr. *geologist*
Phillips, Adran Abner (Abe Phillips) *geologist, oil and gas exploration consultant*
Pullen, Margaret I. *genetic physicist*

Falcon AFB
Chan, John Doddson *aerospace engineer*
Palermo, Christopher John *aerospace engineer*

Fort Carson
Tygret, James William *civil engineer*

Fort Collins
Bamburg, James Robert *biochemistry educator*
Benjamin, Stephen Alfred *veterinary medicine educator, environmental pathologist, researcher*
Bernstein, Elliot Roy *chemistry educator*
Bigiani, Albertino Roberto *electrophysiologist, researcher*
Black, William Cormack, IV *insect geneticist, statistician*
Booth, Karla Ann Smith *biochemist*
Boyd, Landis Lee *agricultural engineer, educator*
Coffin, Debra Peters *ecologist, researcher*
Connell, James Roger *atmospheric turbulence researcher, educator*
Crist, Thomas Owen *ecologist*
Criswell, Marvin Eugene *civil engineering educator, consultant*
Eberhart, Steve A. *federal agency administrator, research geneticist*
Elkind, Mortimer Murray *biophysicist, educator*
Emslie, William Arthur *electrical engineer*
Erslev, Eric Allan *geologist, educator*
Fixman, Marshall *chemist, educator*
Garvey, Daniel Cyril *mechanical engineer*
Gubler, Duane J. *research scientist, administrator*
Harper, Judson Morse *university administrator, consultant, educator*
Hecker, Richard Jacob *research geneticist*
Ishimaru, Carol Anne *plant pathologist*
Jayasumana, Anura Padmananda *electrical engineering educator*
Johnson, Robert Britten *geology educator*
Lameiro, Gerard Francis *research institute director*
Mader, Douglas Paul *quality educator*
Maga, Joseph Andrew *food science educator*
McHugh, Helen Frances *university dean, home economist*
Mesloh, Warren Henry *civil, environmental engineer*
Mielke, Paul William, Jr. *statistician*
Milhous, Robert Thurlow *hydraulic engineer*
Niswender, Gordon Dean *physiologist, educator*
Ogg, James Elvis *microbiologist, educator*
Richardson, Everett Vern *hydraulic engineer, educator, administrator*
Rolston, Holmes, III *theologian, educator, philosopher*
Smith, Gary Chester *meat scientist, researcher*
Smith, Ralph Earl *virologist*
Thurman, Pamela Jumper *research scientist*
Turzillo, Adele Marie *reproductive physiologist, researcher*
Ward, James Vernon *biologist, educator*
Wilbur, Paul James *mechanical engineering educator*
Williamson, Samuel Chris *research ecologist*
Winn, C(olman) Byron *mechanical engineering educator*

Frisco
Power, Walter Robert *geologist*

Glenwood Springs
Violette, Glenn Phillip *construction engineer*

Golden
Ansell, George Stephen *metallurgical engineering educator, academic administrator*
Bertness, Kristine Ann *physicist*
Brennan, Ann Herlevich *communications professional*
Collins, Heather Lynne *government official*
Coors, William K. *brewery executive*
Cowley, Scott West *chemist, educator*
Cummins, Nancyellen Heckeroth *electronics engineer*
Edwards, Kenneth Ward *chemistry educator*
Halstead, Philip Hubert *geologist, consultant*
Hubbard, Harold Mead *research institute executive*
Jafari, Bahram Amir *petroleum engineer, consultant*
Jones, Leonard Dale *facilities engineer*
Lachel, Dennis John *geologist, engineer, consultant*
Mangone, Ralph Joseph *metallurgical engineer*
McNeill, William *environmental scientist*
Mishra, Brajendra *metallurgical engineering educator, researcher*
Morgan, Gary Patrick *energy engineer*
Morrison, Roger Barron *geologist, executive*
Muljadi, Eduard Benedictus *electrical engineer*
Myers, Daryl Ronald *metrology engineer*
Packey, Daniel J. *economist, researcher*
Patino, Hugo *food science research engineer*
Poettmann, Frederick Heinz *retired petroleum engineering educator*
Salamon, Miklos Dezso Gyorgy *mining educator*
Sloan, Earle Dendy, Jr. *chemical engineering educator*
Sunderman, Duane Neuman *chemist, research institute executive*
Tangler, James Louis *aeronautical engineer*
Tankelevich, Roman Lvovich *computer scientist*
Tegtmeier, Ronald Eugene *physician, surgeon*
Togerson, John Dennis *computer software company executive*
Yarar, Baki *metallurgical educator*

Granby
McGrath, Richard William *osteopathic physician*

Grand Junction
Agapito, J. F. T. *mining engineer, mineralogist*
Misra, Prasanta Kumar *physics educator*
Skogen, Haven Sherman *oil company executive*

Greeley
Dingeman, Thomas Edward *wastewater treatment plant administrator*
Searls, Donald Turner *statistician, educator*

Gunnison
Drake, Roger Allan *psychology educator*

Hayden
Dunn, Steven Allen *chemist*

Lakewood
Civish, Gayle Ann *psychologist*
Hill, Walter Edward, Jr. *geochemist, extractive metallurgist*
Klima, Jon Edward *civil engineering educator, consultant*
Lu, Paul Haihsing *mining engineer, geotechnical consultant*
McElwee, Dennis John *pharmaceutical company executive, attorney*
Mueller, Raymond Jay *software development executive*
Pitts, John Roland *physicist*

LaSalle
Stevenson, James Ralph *school psychologist, author*

Littleton
Anderson, Bruce Morgan *computer scientist*
Brown, Elizabeth Eleanor *retired librarian*
Brychel, Rudolph Myron *engineer*
Chamberlin, Paul Davis *metallurgical engineer*
Gilman, James Russell *petroleum engineer*
Jargon, Jerry Robert *chemical engineer*
Nefzger, Charles LeRoy *astronautics company executive*
Ross, Reuben James, Jr. *paleontologist*
Scruggs, David Wayne *aerospace engineer*
Ulrich, John Ross Gerald *aerospace engineer*

Longmont
Jones, Beverly Ann Miller *nursing administrator, patient services administrator*
Ulrich, John August *microbiology educator*

Louisville
Matthews, Shaw Hall, III *reliability engineer*
Pesacreta, George Joseph *optical metrology engineer*
Schemmel, Terence Dean *physicist*
Williams, Marsha Kay *data processing executive*

Loveland
Rosander, Arlyn Custer *mathematical statistician, management consultant*

Parker
Smith, Keith Scott *electrical engineer*

Peterson AFB
Mercier, Daniel Edmond *military officer, astronautical engineer*
Rainey, Larry Bruce *systems engineer*
Taylor, Mark Jesse *military engineer*

Pueblo
Deming, Wendy Anne *mental health professional*

Rangely
DeWitt, James Howard *water treatment technician*

Silverthorne
Dillon, Ray William *engineering technician*
Robertson, James Mueller *civil engineer, educator*

Snowmass Village
Diamond, Edward *gynecologist, infertility specialist, clinician*

U S A F Academy
Bossert, David Edward *electrical engineer*
Haupt, Randy Larry *electrical engineering educator*
Quan, Ralph W. *engineer, researcher*
Webb, Steven Garnett *engineering educator*

University Of Colorado
Beylkin, Gregory *mathematician*
Greene, Chris H. *physicist, educator*
Ranker, Tom A. *botanist, educator*

Ward
Benedict, Audrey DeLella *biologist, educator*

Westminster
Banerjee, Bejoy Kumar *mechanical engineer*
Dotson, Gerald Richard *biology educator*

Wheat Ridge
Henehan, Joan *chemical engineer, consultant*
Scherich, Erwin Thomas *civil engineer, consultant*

CONNECTICUT

Ansonia
Yale, Jeffrey Franklin *podiatrist*

Avon
Goodson, Richard Carle, Jr. *chemist, hazardous waste management consultant*
Mc Ilveen, Walter *mechanical engineer*
Weiss, Robert Michael *dentist*

Bethany
Bergen, Robert Ludlum, Jr. *materials scientist*

Bloomfield
Brooks, Douglas Lee *retired oceanographic and atmospheric policy analyst, environmental policy consultant*
Cohen, Patricia Ann *biochemist*
Dobay, Donald G. *chemical engineer*
Fine, Morton Samuel *civil engineer*
Johnson, Linda Thelma *computer consultant*
Oszurek, Paul John *industrial chemist*

Branford
Sinton, Christopher Michael *neurophysiologist*

Bridgeport
Bernstein, Larry Howard *clinical pathologist*
Brunale, Vito John *aerospace engineer*
Chih, Chung-Ying *physicist, consultant*
Dolan, John Patrick *psychiatrist*
Hmurcik, Lawrence Vincent *electrical engineering educator*
Reed, Charles Eli *retired chemist, chemical engineer*
Skowron, Tadeusz Adam *physician*
Tucci, James Vincent *physicist*
van der Kroef, Justus Maria *political science educator*

Danbury
Folchetti, J. Robert *water, wastewater engineer*
Gagnon, John Harvey *psychotherapist, educator*
Goldstein, Joel *management science educator, researcher*
Joyce, William H. *chemist*
Kuryla, William Collier *chemist, consultant*
Malwitz, Nelson Edward *chemical engineer*
Pastor, Stephen Daniel *chemistry educator, researecher*
Rivera, Luis Ruben *electrical engineer*
Schweitzer, Mark Andrew *civil engineer*
Tittman, Jay *physicist, consultant*
Vossler, John Albert *civil engineer*
Yuh, Chao-Yi *chemical engineer, electrochemist*

Danielson
Sondak, Steven David *electrical engineer*

Darien
Glenn, Roland Douglas *chemical engineer*
Weiss, Robert Franklin *podiatrist*

Dayville
Violette, Carol Ann *chemist, environmental compliance consultant*

East Berlin
Holdsworth, Robert Leo, Jr. *emergency medical services consultant*

East Granby
Kimberley, John A. *mechanical engineer, consultant*

East Hampton
Goncarovs, Gunti *radio chemist*

East Hartford
Cassidy, John Francis, Jr. *research center executive*
Davis, Roger L. *aeronautical engineer*
Hobbs, David E. *mechanical engineer*
Johnson, Bruce Virgil *mechanical engineer, physicist, researcher*
Meinzer, Richard A. *research scientist*
Nelson, Chad Matthew *chemical engineer, researcher*
Sears, Sandra Lee *computer consultant*
Tanaka, Richard I. *computer products company executive*

East Haven
Wasserman, Harry Hershal *chemistry educator*

Enfield
Oliver, Bruce Lawrence *information systems specialist, educator*

Fairfield
Burd, Robert Meyer *hematologist, educator*
Dillingham, Catherine Knight *environmental consultant*
Goodrich, David Charles *management psychologist*

Farmington
Coykendall, Alan Littlefield *dentist, educator*
Donaldson, James Oswell, III *neurology educator*
Jones, Thomas Gordon *internist*
Kegeles, S. Stephen *behavioral science educator*
Loew, Leslie Max *biophysicist*
Maranzano, Miguel Franscisco *engineer*
Mulhearn, Cynthia Ann *industrial/organizational psychology researcher*
Olson, David P. *physicist*
Rothfield, Naomi Fox *physician*
Sunderman, F(rederick) William, Jr. *toxicologist, educator, pathologist*
Viner, Mark William *psychiatrist*
Waknine, Samuel *dental materials scientist, researcher, educator*

Georgetown
Duvivier, Jean Fernand *management consultant*

Glastonbury
Bates, Stephen Cuyler *research engineering executive*
Grubin, Harold Lewis *physicist, researcher*
Hujar, Randal Joseph *software company executive, consultant*
Magnavita, Jeffrey Joseph *psychologist*
Roy, Kenneth Russell *school system administrator, educator*

Greenwich
Burnham, Virginia Schroeder *medical writer*
Dahl, Andrew Wilbur *health services executive*
Foraste, Roland *psychiatrist*
Gargiulo, Gerald John *psychoanalyst, writer*
Langley, Patricia Coffroth *psychiatric social worker*
Mock, Robert Claude *architect*
Nadel, Norman Allen *civil engineer*

Groton
Bernatowicz, Felix Jan Brzozowski *mechanical engineer, consultant*
Cooper, Richard Arthur *oceanographer*
Fitzgerald, William F. *chemical oceanographer, educator*
Helm, John Leslie *mechanical engineer, company executive*
Katz, Lori Susan *toxicologist*
Kelly, Sarah Elizabeth *chemist*
Lincoln, Walter Butler, Jr. *marine engineer, educator*
Milburn, Darrell Edward *ocean engineer, coast guard officer*
Vinick, Fredric James *chemist*

Guilford
Chatt, Allen Barrett *psychologist, neuroscientist*
Duffield, Albert J. *mechanical engineer*

Hamden
Dubois, Normand Rene *microbiologist, researcher*
Leckman, James Frederick *psychiatry and pediatrics educator*
Roche, (Eamonn) Kevin *architect*

Hartford
Braddock, John William *safety engineer*
Brauer, Rima Lois *psychiatrist*
Bronzino, Joseph Daniel *electrical engineer*
Casale, Joseph Wilbert *environmental organic chemist, researcher*
Cheng, Wing-Tai Savio *medical technology consultant*
Child, Frank Malcolm *biologist educator*
Dembek, Zygmunt Francis *epidemiologist*
Hajek, Thomas J. *aerospace engineer*
Kang, Juliana Haeng-Cha *anesthesiologist*
Knibbs, David Ralph *electron microscopist*
Lindsay, Robert *physicist, educator*
Margulies, Robert Allan *physician, educator, emergency medical service director*
Moyer, Ralph Owen, Jr. *chemist, educator*
Pastuszak, William Theodore *hematopathologist*
Roberts, Melville Parker, Jr. *neurosurgeon, educator*
Smith, Donald Arthur *mechanical engineer, researcher*
Szymonik, Peter Ted *computer systems coordinator*
Wagner, Joel H. *aerospace engineer*
Weingold, Harris D. *aerospace engineer*

Higganum
Marcus, Jules Alexander *physicist, educator*

Lebanon
Feldman, Kathleen Ann *microbiologist, researcher*

Madison
Lehberger, Charles Wayne *mechanical engineer, engineering executive*

Manchester
Milewski, Stanislaw Antoni *ophthalmologist, educator*

Mansfield
Shaw, Montgomery Throop *chemical engineering educator*

Meriden
McLeod, John Arthur Sr. *mechanical engineer*
Merwin, June Rae *research scientist, cell biologist*

Middlebury
Davis, Robert Glenn *research scientist*
Jancis, Elmar Harry *chemist*

Middletown
Horne, Gregory Stuart *geologist, educator*
Wilkes-Gibbs, Deanna Lynn *psychologist, educator*

Milford
Dmytruk, Maksym, Jr. *mechanical engineer*
Koch, Evamaria Wysk *oceanographer, educator*
Robohm, Richard Arthur *microbiologist, researcher*

Monroe
Turko, Alexander Anthony *biology educator*

Mystic
Chiang, Albert Chinfa *polymer chemist*
Parberry, Edward Allen *mathematician, consultant*
Thompson, Robert Allan *aerospace engineer*

New Britain
Baskerville, Charles Alexander *geologist, educator*
Charkiewicz, Mitchell Michael, Jr. *economics and finance educator*
Cotten-Huston, Annie Laura *psychologist, educator*
Czajkowski, Eva Anna *aerospace engineer, educator*
Margiotta, Mary-Lou Ann *software engineer*

New Canaan
Dean, Robert Bruce *architect*
Gottlieb, Arnold *dentist*

New Fairfield
Daukshus, A. Joseph *systems engineer*

New Haven
Adair, Robert Kemp *physicist, educator*
Altman, Sidney *biology educator*
Baldwin, John Charles *surgeon, researcher*
Bauman, Natan *audiologist, acoustical engineer*
Berson, Jerome Abraham *chemistry educator*
Bromley, David Allan *physicist, educator*
Brünger, Axel Thomas *biophysicist, researcher, educator*
Bunney, Benjamin Stephenson *psychiatrist*
Buss, Leo William *biologist, educator*
Chubukov, Andrey Vadim *physicist*
Clark, Sydney Procter *geophysics educator*
Coca-Prados, Miguel *molecular biologist*
Cofrancesco, Donald George *health facility administrator*
Cohen, Donald Jay *pediatrics, psychiatry and psychology educator, administrator*
Comer, James Pierpont *psychiatrist*
Cometto-Muñiz, Jorge Enrique *biochemist*
Condon, Thomas Brian (Brian Condon) *hospital executive*
Conklin, Harold Colyer *anthropologist, educator*
Crothers, Donald Morris *biochemist, educator*
De Rose, Sandra Michele *psychotherapist, educator, supervisor, administrator*
DeVita, Vincent Theodore, Jr. *oncologist*
Dingman, Douglas Wayne *microbiologist*
Dolan, Thomas F., Jr. *pediatrician, educator*
DuBois, Arthur Brooks *physiologist, educator*
Elias, Jack Angel *physician, educator*
Feinstein, Alvan Richard *physician*
Friedlaender, Gary Elliott *orthopedist, educator*
Genel, Myron *pediatrician, educator*
Goldsmith, Mary Helen M. *biology educator*
Heninger, George R. *psychiatry educator, researcher*
Hoffleit, Ellen Dorrit *astronomer*
Horváth, Csaba *chemical engineering educator, researcher*
Iachello, Francesco *physicist educator*
Ibbott, Geoffrey Stephen *physicist*
Jacoby, Robert Ottinger *comparative medicine educator*
Klein, Martin Jesse *physicist, educator, historian of science*
Konigsberg, William Henry *molecular biophysics and biochemistry educator, administrator*
Lentz, Thomas Lawrence *biomedical educator, dean*
Letsou, George Vasilios *cardiothoracic surgeon*
Li, Jianming *molecular and cellular biologist*
Lonergan, Brian Joseph *economist, planner*
Manuelidis, Laura *pathologist, neuropathologist, experimentalist*
Marmor, Theodore Richard *political science and public management educator*
Miller, Joan G. *psychology educator*
Mjolsness, Eric Daniel *computer science educator*
Mostow, George Daniel *mathematics educator*
Narendra, Kumpati Subrahmanya *electrical engineer, educator*
Nath, Ravinder *physicist*
Oemler, Augustus, Jr. *astronomy educator*
Parker, Peter D.M. *physicist, educator, researcher*
Piatetski-Shapiro, Ilya *mathematics educator*
Pober, Jordan S. *pathologist, educator*
Pospisil, Leopold Jaroslav *anthropology educator*
Price, Lawrence H(oward) *psychiatrist, researcher, educator*
Pruett, Kyle Dean *psychiatrist, writer, educator*
Ransom, Bruce Robert *neurologist, neurophysiologist, educator*
Ravikumar, Thanjavur Subramaniam *surgical oncologist*
Reyes, Marcia Stygles *medical technologist*
Rose, Marian Henrietta *physics educator*
Rouse, Irving *anthropologist, emeritus educator*
Ryder, Robert Winsor *medical epidemiologist*
Salovey, Peter *psychology educator*
Saltzman, Barry *meteorologist, educator*
Sartorelli, Alan Clayton *pharmacology educator*
Seigel, Arthur Michael *neurologist, educator*
Seilacher, Adolf *biologist, educator*
Shope, Robert Ellis *epidemiology educator*
Sigler, Paul Benjamin *molecular biology educator, protein crystallographer*
Silverstone, David Edward *ophthalmologist*
Singer, Burton Herbert *statistics educator*
Skinner, Helen Catherine Wild *biomineralogist*
Smith, Brian Richard *hematologist, oncologist, immunologist*
Smooke, Mitchell David *mechanical engineering educator, consultant*
Snyder, Michael *biology educator*
Solnit, Albert Jay *commissioner, physician, educator*
Sontheimer, Harald Wolfgang *scientist, cell biology researcher*
Squinto, Stephen Paul *molecular biologist, biochemist*
Sreenivasan, Katepalli Raju *mechanical engineering educator*
Steitz, Joan Argetsinger *biochemistry educator*
Weiss, Robert M. *urologist, educator*
Wiberg, Kenneth Berle *chemist, educator*
Zhong, Jianhui *physicist*
Zigler, Edward Frank *educator, psychologist*
Zoghbi, Sami Spiridon *radiochemist*

New London
Browning, David Gunter *physicist*
Christian, Carl Franz *electromechanical engineer*
Haas, Thomas Joseph *coast guard officer*
Horgan, Peter James *energy engineer*
Moffett, Mark Beyer *physicist*
Visich, Karen Michelle *mechanical engineer*
Von Winkle, William Anton *electrical engineer, educator*

New Milford
Lee, Eldon Chen-Hsiung *chemist*

New Preston
Duffis, Allen Jacobus *polymer chemistry extrusion specialist*

Newington
Gates, Marvin *construction engineer*

Newtown
Cullen, Ernest André *chemist, researcher*

Niantic
Konrad, William Lawrence *electrical engineer*

Norfolk
Egler, Frank Edwin *ecologist, administrator*

North Haven
Pfeifer, Howard Melford *mechanical engineer*

North Stonington
Mollegen, Albert Theodore, Jr. *engineering company executive*

Norwalk
Coates, John Peter *marketing professional*
Greenberg, Sheldon Burt *plastic and reconstructive surgeon*
Hannah, Robert Wesley *infrared spectroscopist*
Kelley, Gaynor Nathaniel *instrumentation manufacturing company executive*
Needham, Charles William *neurosurgeon*
Reiss, Betti *biological and medical researcher, medical writer*
Rose, Gilbert Jacob *psychiatrist, writer, psychoanalyst*
Twist-Rudolph, Donna Joy *neurophysiology and psychology researcher*
Workman, Jerome James, Jr. *chemist*

Old Lyme
Anderson, Theodore Robert *physicist*

Orange
Douskey, Theresa Kathryn *health facility administrator*
Lobay, Ivan *mechanical engineering educator*

Plainfield
Singh, Reepu Daman *civil engineer*

Putnam
Epstein, Sandra Gail *psychologist*

Quaker Hill
Conover, Lloyd Hillyard *retired pharmaceutical research scientist and executive*

Ridgefield
Beck, Vernon David *physicist, consultant*
Byrne, Daniel William *computer specialist, medical researcher, biostatistician, educator*
Letts, Lindsay Gordon *pharmacologist, educator*
Preeg, William Edward *oil company executive*
Sadow, Harvey S. *health care company executive*
Sen, Pabitra N. *physicist, researcher*
Sobol, Bruce J. *internist, educator*
Wahl, Martha Stoessel *mathematics educator*
Zeroug, Smaine *electrophysicist*

Rocky Hill
Chuang, Frank Shiunn-Jea *engineering executive, consultant*
Griesé, John William, III *astronomer*

Sandy Hook
Cristino, Joseph Anthony *electrical engineer*
Lukeris, Spiro *engineer, consultant*

Shelton
Adank, James P. *physicist, administrator*
Fedor, George Matthew, III *industrial engineer*
Novick, Ronald Padrov *electrical engineer*
Zeller, Claude *physicist, researcher*

Southbury
Wescott, Roger Williams *anthropology educator*

Southington
Barry, Richard William *chemist, consultant*
Scherziger, Keith Joseph *civil engineer*

Southport
Hill, David Lawrence *research corporation executive*

Stamford
Allaire, Paul Arthur *office equipment company executive*
Asmar, Charles Edmond *structural engineer, consultant*
Cottle, Robert Duquemin *plastic surgeon, otolaryngologist*
Deneberg, Jeffrey N. *engineering executive*
Goodhue, Peter Ames *obstetrician/gynecologist, educator*

Gupta, Dharam V. *chemical engineer*
Johnstone, John William, Jr. *chemical company executive*
Miller, Wilbur Hobart *business diversification consultant*
Ostroff, Norman *chemical engineer*
Panzer, Hans Peter *chemist*
Robert, Debra Ann *chemical engineer*
Rodriguez, J. Louis *civil engineer, land surveyor*
Rosenberg, Charles Harvey *otorhinolaryngologist*
Schroeder, Richard Philip *quality control executive, consultant*
Sheard, Charles, III *dermatologist*
Walsh, Thomas Joseph *neuro-ophthalmologist*
Zhong, William Jiang Sheng *ceramic engineer*

Storrs
Daniels, Stephen Bushnell *geneticist*
Devereux, Owen Francis *metallurgy educator*
Johnston, E. Russell, Jr. *civil engineer, educator*
Stephens, Jack Edward *civil engineer, educator*

Storrs Mansfield
Ammar, Reda Anwar *computer science engineering educator*
Brown, Scott Wilson *educational psychology educator*
Cetegen, Baki M. *mechanical engineering educator*
DiBenedetto, Anthony Thomas *engineering educator*
Hiestand, Nancy Laura *biology researcher*
Hinckley, Lynn Schellig *microbiologist*
Howes, Trevor Denis *metallurgical engineering educator, researcher*
Kelleher, William Joseph *pharmaceutical consultant*
Malla, Ramesh Babu *structural and mechanical engineering educator*
McEachern, William Archibald *economics educator*
Tang, Qing *optical engineer*

Stratford
Beres, Milan *surgeon*
Casper, Patricia A. *human factors research engineer*
Kmetz, Christopher Paul *lubrication systems development engineer*
Lain, Allen Warren *reliability and maintainability engineer*
Sena, Kanaga Nitchinga *neurologist*
Tarulli, Frank James *aerodynamicist*
Zimmerman, Daniel D. *chemical engineer*

Suffield
Chung, Douglas Chu *pharmacist, consultant*

Tolland
Butterworth, Michael *computer programmer*
Roberge, Lawrence Francis *neuroscientist, biotechnology consultant*

Torrington
Artushenia, Marilyn Joanne *internist, educator*

Trumbull
Gladki, Hanna Zofia *civil engineer, hydraulic mixer specialist*
Kontos, Emmanuel George *polymer chemist*
Schmitt, William Howard *cosmetics company executive*
Vaidya, Kirit Rameshchandra *anesthesiologist, physician*

Uncasville
Javor, Edward Richard *electrical engineer*

Wallingford
Dunkle, Lisa Marie *pediatrics educator*

Waterbury
Fica, Juan *endocrinologist*
Garsten, Joel Jay *gastroenterologist*
Phillips, Walter Mills, III *psychologist, educator*
Williams, Oliver Alan, Jr. *computer scientist*
Wuthrich, Paul *electrical engineer, researcher, consultant*
Zeitlin, Bruce Allen *superconducting material technology executive*

Waterford
Hinkle, Muriel Ruth Nelson *naval warfare analysis company executive*
Johnson, Gary William *environmental scientist, consultant*
Kandetzki, Carl Arthur *engineer*

Watertown
Jacovich, Stephen William *electronics engineer, consultant*

West Hartford
Cornell, Robert Witherspoon *engineering consultant*
Giannaros, Demetrios Spiros *economist, educator*
Goldstein, Rubin *consulting physicist*
Markham, Sister M(aria) Clare *chemistry educator, college administrator*
Rimiller, Ronald Wayne *podiatrist*

West Haven
Cooney, Ned Lyhne *psychologist*
Debeyssey, Mark Sammer *molecular and cellular biologist*
Huguenel, Edward David *microbiologist*
Pickle, Willian Neel, II *molecular biologist, researcher*
Sarris, John *mechanical engineering educator*
Scheyer, Richard David *neurologist, researcher*
Wheeler, George Lawrence *chemist, educator*

West Redding
Foster, Edward John *engineering physicist*
Goldman, Ernest Harold *computer engineering educator*
Paltauf, Rudolf Charles *aerospace engineer*

West Simsbury
Yocom, John Erwin *environmental consultant*

Westport
Cassetta, Sebastian Ernest *industry executive*
Lopker, Anita Mae *psychiatrist*

Wilton
Juran, Joseph Moses *engineer*
Willoughby, William, II *nuclear engineer*
Wilson, John Kenneth *optical research professional*

Winchester
Carter, John Angus *geologist, geochemist, environmental engineer*

Windsor
Cowen, Bruce David *environmental service company executive*
Garde, Anand Madhav *materials scientist*
Gianakos, Nicholas *engineering executive, consultant*
McDonald, Michael Shawn *nuclear engineer, consultant*
Scherer, A. Edward *nuclear engineering executive*

Wolcott
Hillsman, Regina Onie *orthopedic surgeon*

Woodbury
Roberts, Robert Clark *sales engineer*

DELAWARE

Bethany Beach
Wanzer, Mary Kathryn *computer company executive, consultant*

Claymont
Sommermann, Jeffrey Herbert *mechanical engineer*

Dover
Johnson, Denise Doreen *electrical engineer*
Ofosu, Mildred Dean *immunogeneticist*
Porterfield, Craig Allen *psychologist, consultant*
Vawter, William Snyder *computer software consultant*

Fenwick Island
Montone, Liber Joseph *engineering consultant*

Georgetown
Yossif, George *psychiatrist*

Greenville
Schroeder, Herman Elbert *scientific consultant*

Hockessin
Bischoff, Joyce Arlene *information systems consultant, lecturer*
Bischoff, Kenneth Bruce *chemical engineer, educator*

Lewes
Curtis, Lawrence Andrew *biologist, educator*
Gaffney, Patrick Michael *marine biologist, aquacultural geneticist*
Saliba, Anis Khalil *surgeon*

Milford
Quinn, Edward Francis, III *orthopedic surgeon*

New Castle
Empson, Cheryl Diane *validation engineer*
Khatib, Nazih Mahmoud *mechanical engineer*
Searle, Roger Blaine *manufacturing computer system consultant*

Newark
Adams, Joseph Brian *engineer, mathematics educator*
Barton, Lyndon O'Dowd *mechanical engineer*
Bilinsky, Yaroslav *political scientist*
Böer, Karl Wolfgang *physicist, educator*
Bragagnolo, Julio Alfredo *physicist*
Campbell, Linzy Leon *microbiologist, educator*
Cheung, Wilson D. *electrical engineer*
Cooper, Stuart Leonard *chemical engineering educator, researcher, consultant*
Crowley, Michael Joseph *mechanical engineer*
Daniels, William Burton *physicist, educator*
DeCherney, George Stephen *research scientist, research facility administrator*
Dentel, Steven Keith *environmental engineer, educator*
Evans, Dennis Hyde *chemist, educator*
Eyler, Edward Eugene *physicist, educator*
Faghri, Ardeshir *civil engineering educator*
Gangopadhyay, Sunita Bhardwaj *physicist, researcher*
Gorski, Robert Alexander *chemist, consultant*
Henry, Norman Whitfield, III *research chemist*
Hopkins, Mark Willard *mechanical engineer*
Hutton, David Glenn *environmental consultant, chemical engineer*
Jordan, Robert R. *geologist, educator*
Kikuchi, Shinya *transportation engineer*
Kirby, James Thornton, Jr. *civil engineering educator*
Klein, Michael Tully *chemical engineering educator, consultant*
Kobayashi, Nobuhisa *civil and coastal engineer, educator*
McCullough, Roy Lynn *chemical engineering educator*
McIntosh, Donald Harry *electrical engineering consultant*
Meakin, John David *university research executive, educator*
Mendelsohn, Ray Langer *chemical engineering consultant*
Morgan, John Davis *physicist*
Ness, Norman Frederick *astrophysicist, educator, administrator*
Olagunju, David Olarewaju *mathematician, educator*
Olson, Carl Marcus *chemist retired*
Plata-Salamán, Carlos Ramon *neuroscientist*
Ravi, Vilupanur Alwar *materials scientist*
Reinhardt, Charles Francis *toxicologist*
Russell, Thomas William Fraser *chemical engineering educator*
Schultz, Jerold Marvin *materials scientist, educator*
Smith, Charles William *physicist*
Sten, Johannes Walter *controls system engineer, consultant*
Uricheck-Holzapfel, Maryanne *pharmacist*
Urquhart, Andrew Willard *engineering executive, metallurgist*
Valdata, Patricia *English language educator, aviation writer*
Wagner, Norman Joseph, III *chemical engineering educator*
Weinacht, Richard Jay *mathematician*
Wetlaufer, Donald Burton *biochemist, educator*

Rockland
Myoda, Toshio Timothy *microbiologist, consultant, educator*

Smyrna
Hutchison, James Arthur, Jr. *engineering company executive*

Wilmington
Banville, Debra Lee *research chemist*
Bhagat, Ashok Kumar *chemical engineer*
Birewar, Deepak Baburao *chemical engineer, researcher*
Blanchet-Fincher, Graciela Beatriz *physicist*
Busche, Robert Marion *chemical engineer, consultant*
Coulston, George William *chemical engineer, researcher*
Dax, Scott Louis *chemist, researcher*
DeBlieu, Ivan Knowlton *plastic pipe company executive, consultant*
Fullerton, Jesse Wilson *applications specialist*
Galli, Paolo *chemical company executive*
Gossage, Thomas Layton *chemical company executive*
Hartzell, Charles R. *research administrator, biochemist, cell biologist*
Hesp, B. *pharmaceutical executive*
Hoegger, Erhard Fritz *chemist, consultant*
Hollingsworth, David Southerland *chemical company executive*
Howe, Burton Brower *pharmacologist*
Irr, Joseph David *geneticist*
Irrthum, Henri Emile *chemicals executive, engineer*
Kaye, Neil S. *psychiatry educator*
Kennedy, William James *pharmaceutical company executive*
Kinney, Anthony John *biochemist, researcher*
Kissa, Erik *chemist*
Kleinpeter, Joseph Andrew *computer operations specialist*
Kobsa, Henry *research physicist*
Krape, Philip Joseph *chemist*
Kuebeler, Glenn Charles *engineering executive*
Levitt, George *retired chemist*
Levy, Stanley Burton *mechanical engineer, marketing executive*
MacLachlan, Alexander *chemical company executive*
Marcali, Jean Gregory *chemist*
Marsh, Robert Harry *chemical company executive*
Moore, Carl Gordon *chemist, educator*
Nair, K. Manikantan *materials scientist*
Obrinski, Virginia Wallin *retired school psychologist*
O'Bryan, Saundra M. *clinical chemist, biosafety officer*
Pan, Henry Yue-Ming *clinical pharmacologist*
Parshall, George William *research chemist*
Pih, Norman *chemical engineer*
Rodgers, Rhonda Lee *health facility administrator*
Rudolph, Jeffrey Stewart *pharmacist, chemist*
Sands, Howard *pharmacologist, biochemist, research scientist*
Sciance, Carroll Thomas *chemical engineer*
Simmons, Howard Ensign, Jr. *chemist, research administrator*
Smith, Marvin Schade *civil engineer, land surveyor*
Wang, Wen-Ying Lee *systems specialist*
Watts, Malcolm L. *chemist*
Webster, Owen Wright *chemist*
Wingate, Phillip Jerome *chemist*
Wu, Dan Qing *research chemist*
Zinck, Barbara Bareis *chemist*
Zinkand, William Collier *neuropharmacologist*

DISTRICT OF COLUMBIA

Washington
Abelson, Philip Hauge *physicist*
Abraham, George *research physicist, engineer*
Abrahamson, George R. *civilian military scientist*
Adams, Robert Edward *journalist*
Alberts, Bruce Michael *foundation administrator*
Alic, John Anthony *engineer, policy analyst*
Alleva, John James *research biologist*
Amey, Earle Bartley *materials engineer*
Anderson, Donald Morgan *entomologist*
Andrews, Wallace Henry *microbiologist*
Angus, Sara Goodwin *environmental engineer*
Arkilic, Galip Mehmet *mechanical engineer, educator*
Arling, Bryan Jeremy *internist*
Arrowsmith-Lowe, Thomas *federal agency administrator, medical educator*
Asker, James Robert *journalist*
Bachkosky, John M. (Jack Bachkosky) *federal agency administrator*
Bailey, Norman Alishan *economist*
Bainum, Peter Montgomery *aerospace engineer, consultant*
Bakalian, Alexander Edward *public health engineer*
Baker, D. James *federal agency administrator*
Baker, Donald James *federal official, oceanographer, administrator*
Baker, Louis Coombs Weller *chemistry educator, researcher*
Bamford, Donald Lawrence (Larry Bamford) *computer scientist*
Basinger, William Daniel *computer programmer*
Beall, James Howard *physicist, educator*
Beckler, David Zander *government official, science administrator*
Beehler, Bruce McPherson *research zoologist, ornithologist*
Begle, Douglas Pierce *ichthyologist*
Belk, Keith E. *international marketing specialist, researcher*
Bellanti, Joseph A. *microbiologist, educator*
Benz, Robert Daniel *toxicologist*
Bernthal, Frederick Michael *federal agency administrator*
Billings, Linda *writer, analyst*
Bishop, Walton Burrell *science writer, consultant*
Bjoro, Edwin Francis, Jr. *nuclear engineer*
Blair, Patricia Wohlgemuth *economics writer*
Blanc, Theodore Von Sickle *physicist*
Bohi, Douglas Ray *economist*
Bonsignore, Joseph John *publishing company executive*
Bosnak, Robert J. *mechanical engineer, federal agency administrator*
Boyce, Peter Bradford *astronomer, professional association executive*
Brayton, Peter Russell *program director, scientist*
Britt, Harold C. *physicist*
Brodrick, James Ray *mechanical engineer*
Brown, George Edward, Jr. *congressman*
Brown, Harold *corporate director, consultant, former secretary of defense*
Brown, James Harvey *neuroscientist, government research administrator*
Brown, Louis *physicist, researcher*
Browner, Carol *federal official*
Burcham, Jeffrey Anthony *nuclear engineer*
Burka, Maria Karpati *chemical engineer*
Burke, Jerry Alan *retired chemist*
Burnfield, Daniel Lee *engineering program manager*
Burris, James Frederick *academic dean, educator*
Burton, William Joseph *engineering executive*
Callender, Clive Orville *surgeon*
Cambel, Ali Bulent *engineering educator*
Cameron, Maryellen *science association administrator, geologist, educator*
Carlo, George Louis *epidemiologist*
Carrico, John Paul *physicist*
Carroll, J. Raymond *engineering educator*
Carson, Regina Edwards *pharmacy administrator, educator*
Cassidy, James Edward *chemistry consultant*
Casterline, James Larkin, Jr. *research biochemist*
Catoe, Bette Lorrina *physician, health educator*
Caywood, James Alexander, III *transportation engineering company executive, civil engineer*
Challinor, David *scientific institute administrator*
Chalmers, Franklin Stevens, Jr. *engineering consultant*
Chan, Wai-Yee *molecular geneticist*
Chang, Ker-Chi *environmental engineer, civil engineer*
Charlton, Kevin Michael *aerospace engineer, support contractor*
Chen, Ho-Hong H. H. *industrial engineering executive, educator*
Cherniavsky, John Charles *computer scientist*
Childs, Sadie L. *mathematician, chemist, patent agent*
Choi, Junho *electronic engineer*
Chow, Gan-Moog *physicist*
Chubb, Scott Robinson *research physicist*
Clarkson, John J. *dentist, dental association administrator*
Cleave, Mary L. *environmental engineer, former astronaut*
Clutter, Mary Elizabeth *federal official*
Coffey, Timothy *physicist*
Colberg-Poley, Anamaris Martha *virologist*
Cole, Dean Allen *biomedical researcher*
Comstock, M(ary) Joan *publishing executive, chemist*
Cooper, Benita Ann *federal agency administrator*
Cooper, Chester Lawrence *research administrator*
Cooper, Khershed Pessie *metallurgist*
Corell, Robert Walden *science administration educator*
Cowdry, Rex William *physician, researcher*
Cramer, James Perry *association executive, publisher, educator, architectural historian*
Crandall, David Hugh *physicist*
Crannell, Hall Leinster *physics educator*
Crawford, Lester Mills, Jr. *veterinarian*
Crum, John Kistler *chemical society director*
Culbertson, James Clifford *physicist*
Cunningham, LeMoine Julius *physicist*
Curfman, David Ralph *neurological surgeon, musician*
Curry, Robert Michael *broadcast engineer*
Daay, Badie Peter *electrical engineer*
Dailey, Victoria Ann *economist, policy analyst*
Dale, Charles Jeffrey *operations research analyst*
Day, Agnes Adeline *microbiology educator*
Deason, Jonathan P. *environmental engineer, federal agency administrator*
De Fabo, Edward Charles *photobiology and photoimmunology, research scientist, educator*
DeLuca, Mary *telecommunications engineer*
DeMars, Bruce *naval administrator*
Deming, W(illiam) Edwards *statistics educator, consultant*
Denny, Robert William, Jr. *electrical and electronics engineer*
de Queiroz, Kevin *zoologist*
Deutch, John Mark *federal official, chemist, academic administrator*
DeVilbiss, Jonathan Frederick *product marketing administrator*
Devine, Katherine *publisher, environmental consultant*
DiBona, Charles Joseph *association executive*
Dickman, Robert Laurence *physicist, researcher*
Dinberg, Michael David *industrial engineer*
Doman, Elvira *science administrator*
Downey, Richard Morgan *lawyer*
Dressick, Walter J. *chemist*
Durocher, Cort Louis *aerospace engineer, association executive*
Dusold, Laurence Richard *chemist, computer specialist*
Dutta, Subijoy *environmental engineer*
Edelstein, Alan Shane *physicist*
Egan, John Thomas *computer scientist*
Eghbal, Morad *geologist, lawyer*
Eisner, Howard *engineering educator, engineering executive*
Elias, Thomas Sam *botanist, author*
Emerson, Susan *oil company executive*
Endahl, Lowell Jerome *retired electrical cooperative executive*
Eom, Kie-Bum *computer engineering educator*
Epstein, Gerald Lewis *technology policy analyst*
Erb, Karl Albert *physicist, government official*
Faeth, Lisa Ellen *chemical engineer*
Fainberg, Anthony *physicist*
Falci, Kenneth J. *food and nutrition scientist*
Feldman, Bruce Allen *otolaryngologist*
Feldman, Charles *physicist*
Fell, James Carlton *scientific and technical affairs executive, consultant*
Feller, William Frank *surgery educator*
Felsher, Murray *geologist, publisher*
Feuillet, Christian Patrice *botanist*
Feulner, Edwin John, Jr. *research foundation executive*
Filbin, Gerald Joseph *ecologist*
Finerty, Martin Joseph, Jr. *military officer, researcher*
Finger, Larry Wayne *crystallographer, mineralogist*
Finney, Essex Eugene, Jr. *science executive*
Firestone, David *chemist*
Fischl, Myron Arthur *psychologist*
Fisher, Farley *chemist, federal agency administrator*
Fisk, Lennard Ayres *physicist, educator*
Fiske, Richard Sewell *geologist*
Fogleman, Guy Carroll *physicist, mathematician, educator*
Fontana, David A. *chemical engineer, military officer*
Fortenberry, Jeffrey Kenton *nuclear engineer*

Frank, Richard Stephen chemist
Freitag, Robert Frederick government official
Friedman, Herbert physicist
Friedman, Moshe research physicist
Fuller, Kathryn Scott environmental association executive, lawyer
Furiga, Richard Daniel government official
Gardner, Bruce Lynn agricultural economist
Gay-Bryant, Claudine Moss physician
Gebbie, Kristine Moore health official
Geller, Harold Arthur earth and space sciences executive
Genega, Stanley G. career officer, federal agency administrator
Gergely, Tomas astronomer
Giallorenzi, Thomas Gaetano optical engineer
Gibbons, John Howard (Jack Gibbons) physicist, government official
Gilbreath, William Pollock federal agency administrator
Glenn, John Herschel, Jr. senator
Godwin, Stephen Rountree not-for-profit organization administrator
Goffi, Richard James nuclear engineer
Goldberg, Kirsten Boyd science journalist
Goldin, Daniel S. government agency administrator
Goldson, Alfred Lloyd oncologist educator
Goldstein, Allan Leonard biochemist, educator
Good, Mary Lowe (Mrs. Billy Jewel Good) government official
Goodwin, Irwin magazine editor
Gottlieb, H. David podiatrist
Grafton, Robert Bruce science foundation official
Grant, Wendell Carver, II civil engineer
Gray, Mary Wheat statistician, lawyer
Green, Edward Crocker health consulting firm executive
Greenberg, Richard Alan psychiatrist, educator
Greenewalt, David geophysicist
Gregory, Robert Scott pharmacist
Griffith, Jerry Dice government official, nuclear engineer
Grimm, Curt David anthropologist
Gronbeck, Christopher Elliott energy engineer
Gross, Thomas Paul lawyer
Grossman, Nathan physician
Grosvenor, Gilbert Melville journalist, educator, business executive
Grua, Charles government official
Grundy, Richard David engineer
Gueller, Samuel civil and environmental engineer
Guimond, Richard Joseph federal agency executive, environmental scientist
Gwal, Ajit Kumar electrical engineer, federal government official
Haber, Paul Adrian Life geriatrician
Haggerty, James Joseph writer
Hair, Jay Dee association executive
Hallgren, Richard Edwin meteorologist
Haq, Bilal Ul national science foundation program director, researcher
Harriman, Philip Darling geneticist, science foundation executive
Harris, Leonard Andrew civil engineer
Harris, Wesley L. federal agency administrator
Harrison, Edward Thomas, Jr. chemist
Harshbarger, John Carl pathobiologist
Harwit, Martin Otto astrophysicist, educator, museum director
Haskins, Caryl Parker scientist, author
Hassan, Aftab Syed scientific research director
Hazelrigg, George Arthur, Jr. engineer
Heacock, E(arl) Larry electrical engineer
Heineken, Frederick G. biochemical engineer
Henkin, Robert Irwin neurobiologist, internal medicine, nutrition and neurology educator, scientific products company executive
Herman, Barbara Helen pediatric psychiatrist, educator
Herring, Kenneth Lee editor scientific society publications
Hershey, Robert Lewis mechanical engineer, management consultant
Hess, LaVerne Derryl research laboratory scientist
Hess, Wilmot Norton science administrator
Hicks, Jocelyn Muriel laboratory medicine specialist
Higgins, Peter Thomas government information management executive
Hillery, Robert Charles naval engineer, management consultant
Hodes, Richard Michael internist, educator
Hoffman, Robert S. federal agency administrator
Holloway, Harry aerospace medical doctor
Holmes, George Edward molecular biologist, researcher, educator
Homziak, Jurij environment, aquaculture and marine resources specialist, consultant
Hope, Ammie Deloris computer programmer, systems analyst
Howell, Richard Paul, Sr. transportation engineer
Huey, Beverly Messick psychologist
Huntress, Wesley Theodore, Jr. government official
Ifft, Edward Milton government official
Imam, M. Ashraf materials scientist, educator
Inyang, Hilary Inyang geoenvironmental engineer, researcher
Jaron, Dov biomedical engineer, educator
Jernigan, Robert Wayne statistics educator
Jewett, David Stuart federal agency administrator
Johnson, David Simonds meteorologist
Johnson, George Patrick science policy analyst
Johnson, Omotunde Evan George economist
Johnston, Gordon Innes aerospace program manager
Jones, Howard St. Claire, Jr. electronics engineering executive
Jordan, Arthur Kent electronics engineer
Jordan, George Eugene air force officer
Karle, Isabella chemist
Karle, Jean Marianne chemist
Karle, Jerome research physicist
Kearns, David Todd federal agency administrator
Kellogg, Charles Gary civil engineer
Kelly, Douglas Elliott biomedical researcher, association administrator
King, John LaVerne, III facilities engineer, energy management specialist
Kinzer, Robert Lee astrophysicist
Kirchhoff, William Hayes chemical physicist
Kirkbride, Chalmer Gatlin chemical engineer
Kirwan, Gayle M. physics educator
Kleiman, Devra Gail zoologist, zoological park administrator
Klein, Philipp Hillel physical chemistry consultant
Kolb, Charles Chester humanities administrator
Koomanoff, Frederick Alan systems management engineer, researcher
Kooyoomjian, K. Jack environmental engineer
Kopecko, Dennis Jon microbiologist, researcher
Kostka, Madonna Lou (Donna Kostka) naturalist, environmental scientist, ecologist
Kramer, Jay Harlan physiologist, biochemist, researcher, educator
Krauss, Robert Wallfar botanist, university dean
Krug, Edward Charles environmental scientist
Krugman, Stanley Liebert science administrator, geneticist
Lambert, James Morrison physics educator
Langley, Rolland Ament, Jr. construction and engineering company executive
Larew, Hiram Gordon, III research coordinator
LaRosa, John Charles internist, educator, researcher
Lauber, John K. research psychologist
Laukaran, Virginia Hight epidemiologist
Laureno, Robert neurologist
Ledley, Robert Steven biophysicist
Leggon, Cheryl Bernadette sociologist, staff officer
Lenoir, William Benjamin aeronautical scientist-astronaut
Leonard, Joel I. biomedical engineer
Level, Allison Vickers science librarian
Lewis, Forbes Downer computer science educator, researcher
Lewis, Gwendolyn L. sociologist, policy analyst
Liebenson, Gloria Krasnow interior design executive
Liebowitz, Harold aeronautical engineering educator, university dean
Lindsey, Alfred Walter federal agency official, environmental engineer
Lipkin, Richard Martin journalist, science writer
Lippman, Marc Estes pharmacology educator
Lizotte, Michael Peter aquatic ecologist, researcher
Lorber, Mortimer physiology educator, researcher
Lord, Norman William physicist, consultant
Lovejoy, Thomas Eugene tropical and conservation biologist, association executive
Lozansky, Edward Dmitry physicist, consultant
Lynch, Charles Theodore, Sr. materials science engineering researcher, administrator, educator
Ma, David I electronics engineer
MacDonald, Mhairi Graham neonatologist
Maciorowski, Anthony Francis ecological toxicologist
Maddock, Jerome Torrence information services specialist
Mahan, Clarence federal agency administrator, writer
Mandel, Elliott David structural engineer, consultant
Mann, Oscar physician, internist, educator
Mao, Ho-kwang geophysicist, educator
Mariotte, Michael Lee environmental activist, environmental publication director
Marlay, Robert Charles physicist, engineer
Mattingly, Thomas K. astronaut
Maudlin, Robert V. economics and government affairs consultant
McAlexander, Thomas Victor government executive
McDonald, Bernard Robert federal agency administrator
McGrath, Kenneth James chemist
McGrory, Joseph Bennett physicist
McPherson, Ronald P. federal agency administrator
Meikle, Philip G. government agency executive
Meyers, Wayne Marvin microbiologist
Miller, Allen Richard mathematician
Mirick, Robert Allen military officer
Misner, Robert David electronic warfare and magnetic recording consultant, electro-mechanical company executive
Mock, John Edwin science administrator, nuclear engineer
Moraff, Howard science foundation director
Moran, Ricardo Julio economist
Morehouse, David Frank geologist
Mowbray, Robert Norman forest ecologist, government agricultural and natural resource development officer
Munasinghe, Mohan development economist
Murphy, Robert Earl scientist, government agency administrator
Myers, Dale DeHaven government, industry, aeronautics and space agency administrator
Nabholz, Joseph Vincent biologist, ecologist
Natsukari, Naoki psychiatrist, neurochemist
Nelson, Carl Michael construction executive
Nelson, David Brian physicist
Newhouse, Alan Russell federal government executive
Newhouse, Quentin, Jr. social psychologist, educator, researcher
Nguyen, Hao Marc aerospace engineer
Norcross, Marvin Augustus veterinarian, government agency official
Notario, Vicente molecular biology educator, researcher
Novello, Antonia Coello U.S. surgeon general
Obenauer, John Charles physicist
Ochs, Walter J. civil engineer, drainage adviser
O'Connell, Daniel Craig psychology educator
O'Donnell, Brendan James naval officer
Oertel, Goetz K. H. physicist, professional association administrator
Okay, John Louis information scientist
Oler, Wesley Marion, III physician, educator
Olsavsky, John George aerospace engineer
Olsen, Kathie Lynn neuroscientist, administrator
Over, Jana Thais program analyst
Parker, Robert Allan Ridley astronaut
Parks, Vincent Joseph civil engineering educator
Patrick, Janet Cline medical society administrator
Pearson, Jeremiah W., III military career officer, federal agency official
Pecora, Louis Michael physicist
Perich, Michael Joseph medical entomologist consultant
Perry, Daniel Patrick science association administrator
Perry, Seymour Monroe physician
Peters, Frank Albert chemical engineer
Petersdorf, Robert George medical educator, association executive
Petersen, Richard Herman government executive, aeronautical engineer
Peterson, Paul Michael agrostologist
Peterson, William Frank physician
Pettis, Francis Joseph, Jr. electrical engineer
Phillips, Gary W. psychometrician
Pine, David Jonah aerospace engineer
Pinstrup-Andersen, Per educational administrator
Pittinger, Charles Bernard, Jr. civil engineer
Pitts, Nathaniel Gilbert science foundation director
Plowman, R. Dean federal agriculture agency administrator
Plucknett, Donald Lovelle scientific advisor
Podolny, Walter, Jr. structural engineer
Pollack, Murray Michael physician
Powell, Margaret Ann Simmons computer scientist
Prakash, Ravi scientific counselor, biomedical engineering educator
Prasad, Surya Sattiraju environmental systems engineer
Press, Frank geophysicist, educator
Prewitt, Charles Thompson geochemist
Price, Jonathan G. geologist
Prothro, Edwin Terry psychologist educator
Pyke, Thomas Nicholas, Jr. government science and engineering administrator
Pyle, Thomas Edward oceanographer, academic director
Qadri, Syed Burhanullah physicist
Quinn, Jarus William physicist, association executive
Raab, Harry Frederick, Jr. physicist
Rabson, Robert plant physiologist, administrator
Raine, William Alexis physicist
Randall, Robert L(ee) industrial economist
Rao, Venigalla Basaveswara biology educator
Reaman, Gregory Harold pediatric hematologist, oncologist
Reck, Francis James electronics engineer
Redman, Robert Shelton pathologist, dentist
Reis, Victor H. mechanical engineer, government official
Reynick, Robert J. materials scientist
Reynolds, Robert Joel economist, consultant
Rife, Jack Clark physicist
Ritter, Donald Lawrence congressman, scientist
Roberts, Howard Richard food scientist, association administrator
Robinowitz, Carolyn Bauer psychiatrist, educator
Robinson, Emily Worth computer systems analyst, mathematician
Rockefeller, John Davison, IV (Jay Rockefeller) senator, former governor
Roco, Mihail Constantin mechanical engineer, educator
Rodenhuis, David Roy meteorologist, educator
Rodgers, James Earl physicist, educator
Rollwagen, John A. federal official
Romanowski, Thomas Andrew physics educator
Rosenberg, Joel Barry government economist
Ross, Bruce Mitchell psychology educator
Rourk, Christopher John electrical engineer
Roy, Robin K. government official
Rucker, Joshua E. environmental engineer
Russell, Ted McKinnies electronics technician
Sabshin, Melvin psychiatrist, educator, medical association administrator
Salmon, William Cooper mechanical engineer, engineering academy executive
Sampson, Robert Neil association executive
San Martin, Robert L. federal official
Saunders, Jimmy Dale aerospace engineer, physicist, naval officer
Scarr, Harry Alan federal agency administrator
Schad, Theodore MacNeeve science research administrator, consultant
Schafrik, Robert Edward materials engineer, information technologist
Schulze, Norman Ronnie aerospace engineer
Schwartz, Joel epidemiologist
Scott, Roland Boyd pediatrician
Sexton, Ken environmental health scientist
Shamaiengar, Muthu chemist
Shank, Fred Ross federal agency administrator
Shay, Edward Griffin chemical engineer
Sheehan, Jerrard Robert technology policy analyst, electrical engineer
Shen, Yuan-Yuan mathematics educator
Shetler, Stanwyn Gerald botanist, museum official
Shine, Kenneth I. cardiologist, educator
Short, Elizabeth M. physician, educator, federal agency administrator
Siegel, Jack S. federal official
Silverman, Lester Paul economist, energy industry consultant
Simon, Gary Leonard internist, educator
Simpers, Glen Richard aerospace engineer
Singer, Maxine Frank biochemist
Sivasubramanian, Kolinjavadi Nagarajan neonatologist, educator
Skeen, David Ray computer systems administrator
Skinner, Maurice Edward, IV information security system specialist
Small, Albert Harrison engineering executive
Smith, Janet Sue systems specialist
Smith, Philip Meek research organization executive
Smith, Richard Melvyn government official
Snyder, Jed Cobb foreign affairs specialist
Soderberg, David Lawrence chemist
Solimando, Dominic Anthony, Jr. pharmacist
Solomon, Elinor Harris economics educator
Sorrows, Howard Earle executive, physicist
Sparrowe, Rollin D. wildlife biologist
Spilhaus, Athelstan Frederick, Jr. oceanographer, association executive
Stanford, Dennis Joe archaeologist, museum curator
Stanley, Thomas P. chief engineer
Starrs, James Edward law and forensics educator, consultant
Stauffer, Ronald Jay project engineer, aerospace engineer
Steinberg, Marcia Irene science foundation program director
Stever, Horton Guyford aerospace scientist and engineer, educator, consultant
Stewart, Frank Maurice federal agency administrator
Stine, Jeffrey K. science historian, curator
Strand, Kaj Aage astronomer
Stransky, Robert Joseph, Jr. nuclear engineer
Streb, Alan Joseph government official, engineer
Suarez Quian, Carlos Andrés biology educator
Sunderlin, Charles Eugene consultant
Taggart, G. Bruce professional society administrator
Talbot, Frank Hamilton museum director, marine researcher
Thomas, Richard Dean toxicologist, pathologist
Thompson, H. Brian telecommunications executive
Thompson, James Robert, Jr. federal space center executive
Thonnard, Ernst internist, researcher
Timm, Gary Everett science administrator, chemist
Todd, Richard Henry retired physican, investor
Toll, John Sampson association administrator, former university administrator, physics educator
Toma, Joseph S. defense analyst, retired military officer
Tortora, Robert D. mathematician
Tousey, Richard physicist
Treichler, Ray agricultural chemist
Truly, Richard H. federal agency administrator
Tsao, John Chur materials engineer, government regulator
Turner, John Andrew economist
Tyner, C. Fred federal agency administrator
Tyrrell, Albert Ray government liaison for industry
Udeinya, Iroka Joseph pharmacologist, researcher
Umminger, Bruce Lynn government official, scientist, educator
Umpleby, Stuart Anspach management science educator
Vandiver, Pamela Bowren research scientist
Veatch, Robert Marlin medical ethics researcher, philosophy educator
Vernikos, Joan science association director
Vidic, Branislav cell biologist
Viscuso, Susan Rice psychologist, researcher
Vladavsky, Lyubov computer scientist, educator
Waggoner, Lee Reynolds federal agency administrator
Wagner, Glenn Norman pathologist
Wakelyn, Phillip Jeffrey chemist, consultant
Walden, Omi Gail public affairs and government relations executive, energy resources specialist
Wallace, Jane House geologist
Wallace, Joan S. psychologist
Wallis, W(ilson) Allen economist, educator, statistician
Walters, John Linton electronics engineer, consultant
Wang, Franklin Fu Yen materials scientist, educator
Ward, Wanda Elaine psychologist
Warnick, Walter Lee mechanical engineer
Watkins, James David government official, naval officer
Watters, Thomas Robert geologist, museum administrator
Weaver, Carolyn Leslie economist, public policy researcher
Weiler, Kurt Walter radio astronomer
Weiss, Richard Gerald chemist educator
Weiss, Stanley Alan mining, chemicals and refractory company executive
Weisz, Adrian chemist
Weitzman, Stanley Howard ichthyologist
Wepman, Barry Jay psychologist
Werbos, Paul John neural research director
Wermiel, Jared Sam mechanical engineer
Wessel, James Kenneth engineering executive
West, Robert MacLellan science education consultant
Wetherill, George West geophysicist, planetary scientist
Whitcomb, James Hall geophysicist, foundation administrator
White, Robert Marshall physicist, government official, educator
Wickland, Diane Elizbeth ecologist
Wilcox, Richard Hoag information scientist
Wilkinson, Ronald Sterne science administrator and historian
Wilkniss, Peter E. foundation administrator, researcher
Williams, Arthur E. federal agency administrator
Williams, Earl George acoustics scientist
Williams, Frederick Wallace research chemist
Williams, Luther Steward biologist, federal agency administrator
Williams, Peter Maclellan nuclear engineer
Woods, Walter Ralph animal scientist, research administrator
Woteki, Catherine Ellen nutritionist
Wright, Charles Leslie economist
Wyatt, Richard Jed psychiatrist, educator
Yablonski, Edward Anthony space and terrestrial telecommunications company executive, educator
Yochelson, Ellis L(eon) paleontologist
Yoder, Hatten Schuyler, Jr. petrologist
Yoder, Mary Jane Warwick psychotherapist
Young, John Jacob, Jr. aerospace engineer
Youtcheff, John Sheldon physicist
Zhu, Xinming optical engineer
Zielinski, Paul Bernard grant program administrator, civil engineer

FLORIDA

Alachua
Thornton, J. Ronald technologist

Arcadia
Durkin, John Charles agriculturist

Atlantic Beach
Engelmann, Rudolph Herman electronics consultant
Wheeler, William Crawford agricultural engineer, educator

Atlantis
Newmark, Emanuel ophthalmologist

Auburndale
Ellis, Robert Jeffry health facility executive

Bal Harbour
Radford, Linda Robertson psychologist

Bartow
McFarlin, Richard Francis industrial chemist, researcher

Bay Pines
Keskiner, Ali psychiatrist
Laven, David Lawrence nuclear and radiologic pharmacist, consultant

Boca Raton
Boer, F. Peter chemical company executive
Ding, Mingzhou physicist
Finkl, Charles William, II geologist, educator
Gagliardi, Raymond Alfred physician
Grant, John Alexander, Jr. engineering consultant
Han, Chingping Jim industrial engineer, educator
Lin, Y. K. engineer, educator
Luca-Moretti, Maurizio nutrition scientist, researcher
More, Kane Jean science educator
Mussenden, Georg Antonio electronics engineer
Qiu, Shen Li physicist
Richardson, Deborah Ruth psychology educator
Schroeck, Franklin Emmett, Jr. mathematical physicist
Spurlock, Paul Andrew nuclear engineer
Su, Tsung-Chow Joe engineering educator
Tsai, Chi-Tay mechanical engineering educator
Vijayabhaskar, Rajagopal Coimbatore chemist, educator
Wu, Jie computer science educator

Bonita Springs
Holler, John Raymond electronics engineer

Boynton Beach
Mittel, John J. *economist, corporate executive*
Turner, William Benjamin *electrical engineer*

Bradenton
Ambrusko, John Stephen *retired surgeon, county official*
Balsley, Howard Lloyd *economist*
Waters, Will Estel *horticulturist, researcher, educator*

Brandon
Mussenden, Gerald *psychologist*

Brooksville
Farrell, Mark David *environmental engineer*

Cape Canaveral
Clark, John F. *aerospace research and engineering educator*
Hurt, George Richard *electrical engineer*
McCluney, (William) Ross *physics researcher, technical consultant*
Preece, Raymond George *electrical engineer*

Cape Coral
Franklin, Keith Jerome *electrical engineer*

Cassadaga
Boland, Lois Walker *retired mathematician and computer systems analyst*

Chuluota
Hatton, Thurman Timbrook, Jr. *retired horticulturist, consultant*

Clearwater
Baker, James Burnell *systems engineer*
Davidson, Joan Gather *psychologist*
Fenderson, Caroline Houston *psychotherapist*
Fromhagen, Carl, Jr. *obstetrician/gynecologist*
Horton, Donna Alberg *technical writer*
King, Charles Herbert, Jr. *civil engineer*
Klingbiel, Paul Herman *information science consultant*
Loewenstein, George Wolfgang *retired physician, UN consultant*
Rinde, John Jacques *internist*
Tanner, Craig Richard *fire and explosion engineer*
Wheat, Myron William, Jr. *cardiothoracic surgeon*

Cocoa
Kendrick, Pamela Ann *mathematics educator*

Cocoa Beach
Gurr, Clifton Lee *engineering executive*
Strickland, Christopher Alan *statistician, systems analyst*
Williams, Judith Renee *architect*

Coral Gables
Fung, Kee-Ying *engineer, educator, researcher*
Howard, Bernard Eufinger *mathematics and computer science educator*
Jacobson, Leonard I. *psychologist, educator*
Kaifer, Angel Emilio *chemistry educator, researcher*
Namini, Ahmad Hossein *structural engineer, educator*
Saffir, Herbert Seymour *structural engineer, consultant*
Stewart, Harris Bates, Jr. *oceanographer*
Walsh-McGehee, Martha Bosse *conservationist*
Warburton, Ralph Joseph *architect, engineer, planner, educator*
Wilson, David Louis *biologist*

Coral Springs
Sanders, Marc Andrew *computer technical consultant*

Crystal River
Skramstad, Phillip James *chemist*

Dade City
McBath, Donald Linus *osteopathic physician*

Dania
Apter, Nathaniel Stanley *psychiatrist, educator, retired, researcher*
Messing, Charles Garrett *zoologist*

Daytona Beach
Helfrick, Albert Darlington *electronics engineering educator, consultant*
Page, Armand Ernest *data processing director, consultant*
Rubin, Mark Stephen *ophthalmic surgeon*
Simatos, Nicholas Jerry *aerospace sciences educator, consultant*
Wehner, Henry Otto, III *pharmacist, consultant*

De Land
Liberman, Michael *metallurgist, researcher*

Deerfield Beach
Areskog, Donald Clinton *chiropractor*

Deland
Pershe, Edward Richard *civil engineer*

Delray Beach
Charyk, Joseph Vincent *retired satellite telecommunications executive*
Rowland, Robert Charles *clinical psychotherapist, writer, researcher*

Dunedin
Iaquinto, Joseph Francis *electrical engineer*

Dunnellon
Lapp, Roger James *consulting pharmacist*

Eglin A F B
Ewing, Craig Michael *aerospace engineer*
Guida, James John *manufacturing engineer, consultant*
Hopmeier, Michael Jonathon *mechanical engineer, troubleshooter*
Porter, John Louis *aerospace engineer*
Prewitt, Nathan Coleman *aerospace engineer*
Swift, Gerald Allan *aeronautical engineer*

Englewood
Suiter, John William *industrial engineer, consultant*

Fern Park
Arnowitz, Leonard *engineering executive*

Fort Lauderdale
Bas Csatary, Laszlo Kalman *anesthesiologist, cancer researcher*
Berwig, Newton Urbano *aerospace executive*
Clark, Jeffrey Alan *mechanical engineer*
Fishe, Gerald Raymond Aylmer *engineering executive*
Gomez, Jaime G. *neurosurgeon*
Laursen, Brett Paul *psychology educator*
Lodwick, Gwilym Savage *radiologist, educator*
McGreevy, Mary *retired psychology educator*
Morris, James Bruce *internist*
Munzer, Martha Eiseman *writer*
Ondrejack, John Joseph *mechanical engineer*
Parsons, Bruce Andrew *biomedical engineer, scientist*
Price, Alexander *retired osteopathic physician*
Sklar, Alexander *electrical company executive*
Strom-Paikin, Joyce Elizabeth *nursing administrator*
Swiller, Randolph Jacob *internist*

Fort Myers
Hoppmeyer, Calvin Carl, Jr. *civil engineer*
Kastenbaum, Marvin Aaron *statistician*
March, Jacqueline Front *chemist*
Moeschl, Stanley Francis *electrical engineer, management consultant*

Fort Myers Beach
Arneson, Harold Elias Grant *manufacturing engineer, consultant*

Fort Pierce
Calvert, David Victor *soil science educator*
Killday, K. Brian *organic chemist*
Mooney, John Bradford, Jr. *oceanographer, engineer, consultant*
Rice, Mary Esther *biologist*

Gainesville
Abbaschian, Reza *materials science and engineering educator*
Agrios, George Nicholas *botanist*
Anderson, Richard McLemore *internist*
Anghaie, Samim *nuclear engineer, educator*
Bennett, Thomas Peter *museum director, educator, biologist*
Bertholf, Roger Lloyd *chemist, toxicologist*
Bodor, Nicholas Stephen *medicinal chemistry researcher, educator, consultant*
Boehm, Eric Walter Albert *molecular mycology researcher*
Brown, William Samuel, Jr. *communication processes and disorders educator*
Brungard, Martin Alan *civil engineer*
Buchholz, Richard *ethologist*
Burridge, Michael John *veterinarian, educator, academic administrator*
Cantliffe, Daniel James *horticulture educator*
Chen, Wayne H. *electrical engineer, educator*
Childers, Norman Franklin *horticulture educator*
Christie, Richard G. *plant pathologist*
Clark, William Burton, IV *dentist, educator*
Cluff, Leighton Eggertsen *physician*
Colgate, Samuel Oran *chemistry educator*
Das, Utpal *electrical engineer, researcher*
Delfino, Joseph John *environmental engineering sciences educator*
Derrickson, William Borden *business executive*
Dewar, Michael James Steuart *chemistry educator*
Diaz, Nils Juan *nuclear engineering educator*
Drucker, Daniel Charles *engineer, educator*
Drury, Kenneth Clayton *biological scientist*
Edwardson, John Richard *agronomist*
Ernsberger, Fred Martin *former materials scientist*
Evans, Dwight Landis *psychiatrist, educator*
Ewel, Katherine Carter *ecologist, educator*
Focks, Dana Alan *medical entomologist, epidemiologist*
Fournier, Donald Joseph, Jr. *mechanical engineer*
Frost, Susan Cooke *biochemistry educator*
Garcelon, John Herrick *structural engineer, researcher*
Green, David Marvin *psychology educator, researcher, consultant*
Halperin, Joseph *nuclear chemist*
Hansen, Peter James *reproductive physiologist, researcher*
Hanson, Harold Palmer *physicist, government official, editor, academic administrator*
Harrison, Willard W. *chemist, educator*
Hiebert, Ernest *plant pathologist, educator*
Hoit, Marc I. *civil engineer, educator*
Holloway, Paul Howard *materials science educator*
Hope, George Marion *vision scientist*
Hoy, Marjorie Ann *entomology educator, researcher*
Iwata, Brian Anthony *psychologist*
Jaeger, Marc Julius *physiology educator, researcher*
Jones, David Alwyn *geneticist, botany educator*
Khan, Saeed Rehman *cell biologist, researcher*
Kiltie, Richard Alan *zoology educator*
Kurzweg, Ulrich Hermann *engineering science educator*
Lawless, John Joseph *anatomy educator*
Maples, William Ross *anthropology educator, consultant*
Mead, Frank Waldreth *taxonomic entomologist*
Meredith, Julia Alice *nematologist, biologist, researcher*
Micha, David Allan *chemistry and physics educator*
Mishra, Vishnu S. *research scientist*
Moore, G(eorge) Paul *speech pathologist, educator*
Morse, Martin A. *surgeon*
Mussenden, Maria Elisabeth *psychologist, substance abuse counselor*
Ohrn, Nils Yngve *chemistry and physics educator*
Olila, Oscar Gesta *soil and water scientist*
Orazem, Mark Edward *chemical engineering educator*
Palovcik, Reinhard Anton *research neurophysiologist*
Pop, Emil *research chemist*
Price, Donald Ray *agricultural engineer, university administrator*
Primack, Alice Lefler *librarian*
Proctor, Charles Lafayette, II *mechanical engineer, educator, consultant*
Purcifull, Dan Elwood *plant virologist, educator*
Putnam, Hugh Dyer *environmental engineer, educator, consultant*
Roan, Vernon Parker, Jr. *mechanical engineer, educator*
Rubin, Melvin Lynne *ophthalmologist, educator*
Schelske, Claire L. *limnologist*
Seale, James Lawrence, Jr. *agricultural economics educator, international trade researcher*
Seiberling, Lucy Elizabeth *physicist*
Semple-Rowland, Susan Lynn *neuroscientist*
Shaw, Lawrance Neil *agricultural engineer, educator*
Sherif, S. A. *mechanical engineering educator*
Shortelle, Kevin James *systems engineer, consultant*
Shyy, Wei *aerospace, mechanical engineering educator*
Smith, Alexander Goudy *physics and astronomy educator*
Smith, John James, Jr. *environmental engineering laboratory director*
Sorkin, Robert Daniel *psychologist, educator*
Strates, Basil Stavros *biomedical educator, researcher*
Sullivan, Neil Samuel *physicist, researcher, educator*
Tanner, David Burnham *physics educator*
Tavano, Frank *physicist, electrical engineer*
Tou, Julius T. *electrical and computer engineering educator*
Trickey, Samuel Baldwin *physics educator, researcher*
Viessman, Warren, Jr. *academic dean, civil engineering educator, researcher*
Vitali, Juan A. *nuclear engineer, consultant*
White, Susie Mae *school psychologist*
Widmer, Charles Glenn *dentist, researcher*
Wilcox, Charles Julian *geneticist, educator*
Williams, Norris Hagan, Jr. *biologist, educator, curator*
Wilson, David Clifford *mathematician, educator*
Wood, Frank Bradshaw *retired astronomy educator*
Yau, Stephen Sik-sang *electrical engineering educator, computer scientist, researcher*

Gonzalez
Plischke, Le Moyne Wilfred *research chemist*
Yu, Jason Jeehyon *chemical engineer*

Hallandale
Haspel, Arthur Carl *podiatrist, surgeon*

Hialeah
Fisher, Barbara Turk *school psychologist*

Hollywood
Connolly, Connie Christine *nurse anesthetist*
Kuzmanović, Bogdan Ognjan *structural engineer*

Homestead
Byrd, Mary Laager *life science researcher*
Dudik, Rollie M. *healthcare executive*
Ramos-Ledon, Leandro Juan *biologist*
Schaffer, Bruce Alan *plant physiologist, educator*

Indialantic
Carroll, Charles Lemuel, Jr. *mathematician*
Preece, Betty P. *engineer, educator*

Indian Harbour Beach
Denaburg, Charles Robert *metallurgical engineer, retired government official*

Inverness
Esquibel, Edward Valdez *psychiatrist, clinical medical program developer*
Mavros, George S. *clinical laboratory director*

Jacksonville
Bodkin, Lawrence Edward *inventor, research development company executive, gemologist*
Brashier, Edward Martin *environmental consultant*
Bremer, Alfonso M. *neurosurgeon, neuro-oncologist*
Colon-Otero, Gerardo *hematologist, oncologist*
Givens, Stephen Bruce *medical physicist*
Goldberg, William K. *chemist*
Halley, James Alfred *physician*
Hawkins, James Douglas, Jr. *structural engineer, architect*
Hudson, Thomas *robotics company executive*
Hutchins, Paul Francis, Jr. *energy engineer*
Jernigan, John Milton *chemist, chemical engineer*
Joyce, Edward Rowen *chemical engineer, educator*
Keifer, Orion Paul *naval officer, mechanical engineer*
O'Neal, Kathleen Len *financial administrator*
Paryani, Shyam Bhojraj *radiologist*
Prempree, Thongbliew *oncology radiologist*
Reynolds, Ellis W. *chemist*
Robinson, Christine Marie *mathematics educator*
Russell, David Emerson *mechanical engineer, consultant*
Shoup, James Raymond *computer systems consultant*
Tardona, Daniel Richard *marine naturalist, park ranger, educator*
Ulrich, Alfred Daniel, III *chemist*
Vallort, Ronald Peter *mechanical engineer*
Welch, Philip Burland *electronics company executive*

Jay
Brecke, Barry John *weed scientist, researcher*
Peacock, Hugh Anthony *agricultural research director, educator*

Jensen Beach
Martin, Roy Erik *marine biologist, consultant*

Juno Beach
Vaughn, Eddie Michael *mechanical engineer*

Jupiter
Nessmith, H(erbert) Alva *dentist*

Kennedy Space Center
Austin, Lisa Susan Coleman *mechanical engineer*
Baylis, William Thomas *systems logistics engineer*
Crippen, Robert Laurel *naval officer, former astronaut*
Evenson, Michael Donald *software/hardware engineer*
Head, Julie Etta *computer engineer*
Hosken, Richard Bruce *systems engineer*
Peck, Joan Kay *systems engineer*
Russell, Ray Lamar *project manager*
Shaw, Brewster Hopkinson, Jr. *astronaut*

Key West
Benavides, Jaime Miguel *orthopedist*
Devereaux, Christian Windsor, III *computer scientist*

Lake Placid
Rew, William Edmund *civil engineer*

Lake Worth
Kline, Gordon Mabey *chemist, editor*

Lakeland
Campbell, Doris Klein *retired psychology educator*
Higby, Edward Julian *safety engineer*
Walters, Marten Doig *chemical engineer*

Largo
Brown, Warren Joseph *physician*
Caffee, Marcus Pat *computer consulting executive*
Cook, Robert Eugene, Sr. *engineer*
Dolan, John E. *consultant, retired utility executive*

Lecanto
McLean-Wainwright, Pamela Lynne *educational consultant, college educator, counselor, program developer*

Lehigh Acres
Smiley, Joseph Elbert, Jr. *evaluation engineer, librarian*
Whelahan, Yvette Ann *nursing administrator, consultant*

Lighthouse Point
Farho, James Henry, Jr. *mechanical engineer, consultant*

Live Oak
Exley, Sheck *geologist*

Longboat Key
Brown, Henry *chemist*
Workman, George Henry *engineer, consultant*

Longwood
Crawford, Ian Drummond *business executive, electronic engineer*
McAvoy, Gilbert Paul *retired engineer*

Lynn Haven
Coggeshall, Norman David *former oil company executive, research director, investment executive*

Maitland
Fincher, Daryl Wayne *fire protection engineer*
O'Connor, Kevin Patrick *civil engineer*
Von Hilsheimer, George Edwin, III *neuropsychologist*

Mango
Spencer, Francis Montgomery James *pharmacist*

Marco Island
Hyde, James Franklin *industrial chemical consultant*
Jacobson, Linda S(ue) *astronomy educator*

Melbourne
Arnold, Toni Lavalle *designer*
Baney, Richard Neil *physician, internist*
Broussard, William Joseph *physician, cattleman, environmentalist*
Collins, Emmanuel Gye *aerospace engineer*
Davidson, Keith Dewayne *corrosion engineer*
Doi, Shinobu *research electrical engineer, consultant*
Droll, Raymond John *engineer*
Favero, Kenneth Edward *medical administrator*
Harrell, Thomas Hicks *psychologist*
Iacoponi, Michael Joseph *systems engineer*
Minor, Mark William *allergist*
Nelson, Gordon Leigh *chemist, educator*
Shaikh, Muzaffar Abid *management science educator*
Subramanian, Chelakara Suryanarayanan *mechanical and aerospace engineering educator*
Swalm, Thomas Sterling *aerospace executive, retired military officer*

Melbourne Beach
Belefant, Arthur *engineer*

Miami
Arango, Richard Steven *architect, graphic and industrial designer*
Argibay, Jorge Luis *information systems firm executive and founder*
Ashkenazi, Josef *physicist*
Bachmeyer, Thomas John *fundraising executive*
Beck, Morris *allergist*
Bellero, Chiaffredo John *civil engineer*
Bezdek, Hugo Frank *scientific laboratory administrator*
Blanco, Luciano-Nilo *physicist*
Bunge, Richard Paul *cell biologist, educator*
Burkett, Marjorie Theresa *nursing educator, gerontology nurse*
Byrnes, John Joseph *medical educator*
Chang, Francis *medical physicist*
Chang, Hsuan Hung *dosimetrist*
Chen, Chun-fan *biology educator*
Cherepanov, Genady Petrovich *mathematician, mechanical engineer*
Clarkson, Kenneth Wright *economics educator*
Cooper, William James *chemist*
Dewanjee, Mrinal Kanti *radiopharmaceutical chemist*
Dorst, Neal Martin *meteorologist, computer programmer*
Farcus, Joseph Jay *architect, interior designer*
Fernandez Stigliano, Ariel *chemistry educator*
Freshwater, Michael Felix *hand surgeon*
Ganz, William I. *radiology educator*
Getz, Morton Ernest *internist, gastroenterologist*
Gibby, Mabel Enid Kunce *psychologist*
Ginsberg, Myron David *neurologist*
Goodnick, Paul Joel *psychiatrist*
Hawkins, Pamela Leigh Huffman *biochemist*
Haynes, Duncan Harold *pharmacology educator*
Herman, Larry Marvin *psychotherapist*
Hornicek, Francis John, Jr. *cell biophysicist, orthopedic surgeon*
Howell, Ralph Rodney *pediatrician, educator*
Humphries, Joan Ropes *psychologist, educator*
Huysman, Arlene Weiss *psychologist, educator*
Itzhak, Yossef *neuropharmacologist*
Iver, Robert Drew *dentist*
Jones, William Kinzy *materials engineering educator*
Lapidus, Morris *retired architect, interior designer*

Levin, Robert Bruce *industrial engineer, executive, consultant*
Mandri, Daniel Francisco *psychiatrist*
Marcus, Joy John *pharmacist, educator, consultant*
Mertz, Patricia Mann *dermatology educator*
Mettinger, Karl Lennart *neurologist*
Mintz, Daniel Harvey *endocrinologist, educator, academic administrator*
Monsalve, Martha Eugenia *pharmacist*
Mooers, Christopher Northrup Kennard *physical oceanographer, educator*
Nagel, Joachim Hans *biomedical engineer, educator*
Nicholson, William Mac *naval architect, marine engineer, consultant*
Parra-Diaz, Dennisse *biophysical chemist*
Pasquel, Joaquin *chemical engineer*
Peck, Michael Dickens *burn surgeon*
Prager, Michael Haskell *fishery population dynamicist*
Reark, John Benson *consulting ecologist, landscape architect*
Reik, Rita Ann Fitzpatrick *pathologist*
Ricordi, Camillo *transplant surgeon, diabetes researcher, educator*
Rosendahl, Bruce Ray *dean, geophysicist, educator*
Russell, Elbert Winslow *neuropsychologist*
Ryan, James Walter *physician, medical researcher*
Salazar-Carrillo, Jorge *economics educator*
Sanchez, Javier Alberto *industrial engineer*
Sánchez-Ramos, Juan Ramon *neurologist, researcher*
Sapan, Christine Vogel *protein biochemist, researcher*
Sheets, Robert Chester *meteorologist*
Shenouda, George Samaan *engineering executive, consultant*
Sichewski, Vernon Roger *physician*
Slonim, Ralph Joseph, Jr. *cardiologist*
Suarez, George Michael *urologist*
Tansel, Ibrahim Nur *mechanical engineer, educator*
Tappert, Frederick Drach *physicist*
Thornton, Thomas Elton *psychologist, consultant*
Ugwu, Martin Cornelius *pharmacist*
Van Dyk, Michael Anthony *software designer*
Varley, Reed Brian *fire protection engineer, consultant*
Venet, Claude Henry *architect, acoustics consultant*
Veziroglu, Turhan Nejat *mechanical engineering educator, energy researcher*
Vought, Franklin Kipling *pharmaceuticals chemist, researcher*
Weiner, William Jerrold *neurologist, educator*
Westman, Wesley Charles *psychologist*
Wheeler, Donald Keith *community and economic development specialist*

Miami Beach
Lehrman, David *orthopedic surgeon*

Miami Shores
Cherry, Andrew L., Jr. *social work educator, researcher*

Mount Dora
Staats, Dee Ann *toxicologist*
Wootton, Joel Lorimer *nuclear technologist, health physics consultant*

Mulberry
Sabatino, David Matthew *chemical engineer*

Naples
Archibald, Frederick Ratcliffe *chemical engineer*
Benson, Ronald Edward, Jr. *environmental engineer*
Bush, John William *business executive, federal official*
Gahagan, Thomas Gail *obstetrician/gynecologist*
Halvorson, William Arthur *economic research consultant*
Leverenz, Humboldt Walter *retired chemical research engineer*
Lister, Charles Allan *electrical engineer, consulting engineer*
Long, James Alvin *exploration geophysicist*
McCollom, Jean Margaret *ecologist*
Rehak, James Richard *orthodontist*

New Port Richey
Eldred, Nelson Richards *chemist, consultant*
Hauber, Frederick August *ophthalmologist*

Niceville
Soben, Robert Sidney *systems scientist*

North Fort Myers
Underwood, George Alfred *mechanical engineer*
Zeldes, Ilya Michael *forensic scientist*

North Miami Beach
Bernstein, Sheldon *biochemist*
Wolfenson, Azi U. *electrical engineer*

North Palm Beach
Mathavan, Sudershan Kumar *nuclear power engineer*
Rothstein, Arnold Joel *marine engineer, mechanical engineer*

Ocala
Corwin, William *psychiatrist*
Johnson, Winston Conrad *mathematics educator*
Killian, Ruth Selvey *home economist*

Odessa
Lambos, William Andrew *computer consultant, social science educator*

Ona
Pate, Findlay Moye *agriculture educator, university center director*

Orange Park
Stroud, Debra Sue *medical technologist*
Walsh, Gregory Sheehan *optical systems professional*

Orlando
Agid, Steven Jay *aerospace engineer*
Blue, Joseph Edward *physicist*
Breakfield, Paul Thomas, III *physicist*
Bunting, Gary Glenn *operations research analyst, educator*
Carpenter, Robert Van, Jr. *laser and engineering technician*
Carter, Thomas Allen *engineering executive, consultant*
Casanova-Lucena, Maria Antonia *computer engineer*
Cates, Harold Thomas *aircraft and electronics company executive*
Deaton, John Earl *aerospace experimental psychologist*
Debnath, Lokenath *mathematician, educator*
Dewey, Cameron Boss *civil engineer*
Eichner, Gregory Thomas *computer engineer, consultant*
Eligon, Ann Marie Paula *physicist*
Huber, Gary Arthur *aerospace engineer*
Johnson, Tesla Francis *data processing executive, educator*
Kalphat-Lopez, Henriet Michelle *electric and electronics engineer*
Kennedy, Robert Samuel *experimental psychologist, consultant*
Kersten, Robert Donavon *engineering educator, consultant*
Kira, Gerald Glenn *engineering executive*
Koch, Kevin Robert *metallurgical engineer*
Li-Kam-Wa, Patrick *research optics scientist, consultant*
Madhanagopal, Thiruvengadathan *environmental engineer*
Mangold, Vernon Lee *physicist*
Mayer, Richard Thomas *laboratory director, entomologist*
Mengel, Lynn Irene Sheets *health science research coordinator*
Metevier, Christopher John *electrical engineer*
Miller, Harvey Alfred *botanist, educator*
Mirmiran, Amir *civil engineer*
Mortazawi, Amir *electrical engineerring educator*
Nayfeh, Jamal Faris *mechanincal and aerospace engineering educator*
Norris, Franklin Gray *thoracic and cardiovascular surgeon*
Piquette, Jean Conrad *physicist*
Rafetto, John *podiatrist*
Sasseen, George Thiery *aerospace engineering executive*
Schroeter, Dirk Joachim *mechanical engineer*
Smetheram, Herbert Edwin *company executive*
Thomas, Garland Leon *aerospace engineer*
Ting, Robert Yen-ying *physicist*
Touffaire, Pierre Julien *physician*
Trytek, Linda Faye *microbiologist*
Verwey, Timothy Andrew *structural engineer, consultant*
Ziger, David Howard *chemical engineer*

Palatka
Jenab, S. Abe *civil and water resources engineer*

Palm Bay
Bellstedt, Olaf *software engineer*
Cox, David Leon *telecommunications company executive*
Walden, W. Thomas *lead software engineer, consultant*

Palm Beach
Epley, Marion Jay *oil company executive*
Hopper, Arthur Frederick *biological science educator*
Symons, George Edgar *environmental engineering editor*

Palm Beach Gardens
Jacobson, Alan Leonard *otolaryngologist*
Silver, Gordon Hoffman *metallurgical engineering consultant*

Palm Coast
Van Dusen, James *cardiologist*

Palm Harbor
Morris, David Brian *chemical engineer*
Schafer, Edward Albert, Jr. *data processing executive*

Panama City
Cilek, James Edwin *medical and veterinary entomologist*
Leitheiser, James Victor *environmental specialist*
Sowell, James Adolf *quality assurance professional*
Walker, Richard, Jr. *nephrologist, internist*

Pembroke Pines
Schwartzberg, Leo Mark *civil engineer, consultant*

Pensacola
Clare, George *safety engineer, systems safety consultant*
Davis, Leslie Shannon *research chemist, chemistry educator*
Gudry, Frederick E., Jr. *aerospace medical researcher*
Jones, Walter Harrison *chemist*
McSwain, Richard Horace *materials engineer, consultant*
Mohrherr, Carl Joseph *biologist*
Noland, Robert Edgar *dentist*
Olsen, Richard Galen *biomedical engineer, researcher*
Platz, Terrance Oscar *utilities company executive*
Prochaska, Otto *engineer*

Pinellas Park
Tower, Alton G., Jr. *pharmacist*

Plantation
Charles, Joel *audio and video tape analyst, voice identification consultant*
Hollifield, Christopher Stanford *engineer, consultant*
Shim, Jack V. *civil engineer, consultant*

Pompano Beach
Francis, William Kevin *civil engineer*
Kavasoglu, Abdulkadir Yekta *civil engineer*

Ponte Vedra Beach
ReMine, William Hervey, Jr. *surgeon*
Zebouni, Nadeem Gabriel *structural engineer*

Port Charlotte
Munger, Elmer Lewis *civil engineer, educator*
Norris, Dolores June *computer specialist*
Parvin, Philip E. *retired agricultural researcher and educator*

Port Richey
Essex, Douglas Michael *optical engineer*

Port Saint Lucie
Wertheimer, David Eliot *medical facility administrator, cardiologist*

Punta Gorda
Beever, James William, III *biologist*
Beever, Lisa Britt-Dodd *environmental planner*
Dewey, Alan H. *electronic engineer*

Riviera Beach
Dominick, Paul Scott *chemist, researcher*
Kazimir, Donald Joseph *industrial engineer*

Saint Augustine
Gerling, Gerard Michael *neurologist*
Koger, Mildred Emmelene Nichols *educational psychologist*

Saint Marks
Faintich, Stephen Robert *chemical engineer*

Saint Petersburg
Bradley, William Guy *molecular virologist*
Castle, Raymond Nielson *chemist, educator*
Clarke, Kit Hansen *radiologist*
Eastridge, Michael Dwayne *clinical psychologist*
Eldridge, Peter John *fishery biologist*
Ferguson, John Carruthers *biologist*
Fishman, Mark Brian *computer scientist, educator*
Good, Robert Alan *physician, educator*
Kannan, Ramanuja Chari *civil engineer*
Kazor, Walter Robert *statistical process control and quality assurance consultant*
Kormilev, Nicholas Alexander *retired entomologist*
Rester, Alfred Carl, Jr. *physicist*
Serrie, Hendrick *anthropology and international business educator*
Waddell, David Garrett *electrical engineer, consultant*
Willey, Edward Norburn *physician*

Sarasota
Angelotti, Richard H. *science administrator, banker*
Greenstein, Jerome *mechanical engineer*
Grodberg, Marcus Gordon *drug research consultant*
Hendon, Marvin Keith *psychologist*
Klutzow, Friedrich Wilhelm *neuropathologist*
Lewis, Brian Kreglow *computer consultant*
Liu, Suyi *biophysicist*
Minette, Dennis Jerome *financial computing consultant*
Myerson, Albert Leon *physical chemist*
Petrie, George Whitefield, III *retired mathematics educator*

Satellite Beach
Nunnally, Stephens Watson *civil engineer*
Speaker, Edwin Ellis *retired aerospace engineer*

Seminole
Schwartzberg, Roger Kerry *osteopath, internist*

South Miami
Remmer, Harry Thomas, Jr. *obstetrician/gynecologist*

Stuart
Hahn, Walter George *naval architect*
Jaller, Michael M. *retired orthopaedic surgeon*

Sun City Center
Edwards, Paul Beverly *retired science and engineering educator*

Sunrise
Tay, Roger Yew-Siow *electromagnetic engineer*
Tielens, Steven Robert *information specialist*

Surfside
Polley, Richard Donald *microbiologist, polymer chemist*

Tallahassee
Anderson, John Roy *grouting engineer*
Arce, Pedro Edgardo *chemical engineering educator*
Ashler, Philip Frederic *international trade and development advisor*
Berg, Bernd Albert *physics educator*
Brennan, Leonard Alfred *research scientist, administrator*
Brigham, John Carl *psychology educator*
Choppin, Gregory Robert *chemistry educator*
Collins, Angelo *science educator*
Coloney, Wayne Herndon *civil engineer*
De Forest, Sherwood Searle *agricultural engineer, agribusiness services executive*
Figley, Charles Ray *psychology educator*
Friedmann, E(merich) Imre *biologist, educator*
Galbraith, Lissa Ruth *industrial engineer*
Gilmer, Robert *mathematics educator*
Glenn, Rogers *psychologist, student advisor, consultant*
Hall, Houghton Alexander *engineering professional*
Hartsfield, Brent David *environmental engineer*
Henderson, Kaye Neil *civil engineer, business executive*
Herndon, Roy Clifford *physicist*
James, Francis Crews *zoology educator*
Jones, Gladys Hurt *retired mathematics educator*
Kemper, Kirby Wayne *physics educator*
Leavell, Michael Ray *computer programmer and analyst*
Maguire, Charlotte Edwards *retired physician*
Mandelkern, Leo *biophysics and chemistry educator*
Means, Donald Bruce *environmental educator, research ecologist*
Miller, John Richard *scientist*
Moilanen, Michael David *civil engineer*
Nichols, Eugene Douglas *mathematics educator*
O'Brien, James Joseph *meteorology and oceanography educator*
Onokpise, Oghenekome Ukrakpo *agronomist, educator, forest geneticist, agroforester*
Owens, Joseph Francis, III *physics educator*
Pfeffer, Richard Lawrence *geophysics educator*
Rhodes, Roberta Ann *dietitian*
Schrieffer, John Robert *physics educator, science administrator*
Stivers, Marshall Lee *civil engineer*
Subbuswamy, Muthuswamy *environmental engineer, researcher, consultant*
Taylor, J(ames) Herbert *cell biology educator*
Walton, Jeffrey Howard *physicist*
Whitney, Glayde Dennis *psychologist, educator, geneticist*
Williams, Theodore P. *biophysicist, biology educator*

Tampa
Barness, Lewis Abraham *physician*
Behnke, Roy Herbert *physician, educator*
Bussone, David Eben *hospital administrator*
Calderon, Eduardo *general practice physician, researcher*
Cho, Jai Hang *internist, hematologist, educator*
Clark, Michael Earl *psychologist*
Couret, Rafael Manuel *electrical engineering executive*
Crane, Roger Alan *engineering educator*
del Regato, Juan Angel *radio-therapeutist and oncologist, educator*
Dunigan, David Deeds *biochemist, virologist, educator*
Gergess, Antoine Nicolas *civil engineer*
Hickman, Hugh Vernon *physics educator*
Jacobson, Howard Newman *obstetrics/gynecology educator, researcher*
Johnson, Anthony O'Leary (Andy Johnson) *meteorologist, consultant*
Kalmaz, Ekrem Errol *environmental scientist*
Liller, Karen DeSafey *health education educator*
Masters, Eugene Richard *environmental engineer*
Meddin, Jeffrey Dean *safety executive*
Miller, Ronald Lewis *research hydrologist, chemist*
Mines, Richard Oliver, Jr. *civil and environmental engineer*
Mullins, Michael Drew *environmental educator*
Muroff, Lawrence Ross *nuclear medicine physician*
Olanow, C(harles) Warren *neurologist, educator*
Pasetti, Louis Oscar *dentist*
Perret, Gerard Anthony, Jr. *orthodontist*
Phares, Vicky *psychology educator*
Price, Douglas Armstrong *chiropractor*
Richards, Ira Steven *toxicology educator*
Rodts, Gerald Edward *computer scientist*
Saff, Edward Barry *mathematics educator*
Santti, Gary Allen *hazardous waste engineer*
Schuh, Sandra Anderson *ethics educator*
Shah, Harish Hiralal *mechanical engineer*
Shenefelt, Philip David *dermatologist*
Shephard, Bruce Dennis *obstetrician, medical writer*
Spellacy, William Nelson *obstetrician-gynecologist, educator*
Swan, Charles Wesley *psychoneuroimmunologist*
Trunnell, Thomas Newton *dermatologist*
Varlotta, David *anesthesiologist*
Wade, Thomas Edward *university research administrator, electrical engineering educator*
Werner, Mark Henry *neurologist, researcher*
Wienker, Curtis Wakefield *physical anthropologist, educator*
Wiest, John Andrew *dentist*
Wise, Roger M. *chemist*
Young, Lawrence Eugene *internist*
Zelinski, Joseph John *engineering educator, consultant*

Tavernier
Grove, Jack Stein *naturalist, marine biologist*

Temple Terrace
DeReus, Harry Bruce *mechanical engineer, consultant*
Dobrowolski, Kathleen *data processing executive*
Gagliardo, Victor Arthur *environmental engineer*

Titusville
Claridge, Richard Allen *structural engineer*
Luecke, Conrad John *aerospace educator*
Sipos, Charles Andrew *manufacturing executive*

Tyndall AFB
Herrlinger, Stephen Paul *flight test engineer, air force officer, educator*

Venice
Hardenburg, Robert Earle *horticulturist*
Hays, Herschel Martin *electrical engineer*
Shaw, Bryce Robert *author*

Vero Beach
Calmes, John Wintle *architect*
Cooke, Robert Edmond *physician, educator, former college president*
Hribar, Lawrence Joseph *entomologist, researcher*
Hungerford, Herbert Eugene *nuclear engineering educator*
Small, Wilfred Thomas *surgeon, educator*

West Palm Beach
Balaguer, John P. *aircraft manufacturing executive*
Bower, Ruth Lawther *mathematics educator*
Freudenthal, Ralph Ira *toxicology consultant*
Giacco, Alexander Fortunatus *chemical industry executive*
Gillette, Frank C., Jr. *aeronautical engineer*
Koff, Bernard L. *engineering executive*
Laura, Robert Anthony *coastal engineer, consultant*
Olsak, Ivan Karel *civil engineer*
Roberts, Hyman Jacob *internist, researcher, author, publisher*
Shipman, James Melton *propulsion engineer*
Still, Mary Jane (M. J. Still) *mathematics educator*
Whitfield, Graham Frank *orthopedic surgeon*
Witt, Gerhardt Meyer *hydrogeologist*

Winter Haven
Grierson, William *retired agriculture educator, consultant*
Mandal, Krishna Pada *radiation physicist*

Winter Park
Granberry, Edwin Phillips, Jr. *safety engineer, consultant*
Hartmann, Rudolf *electro-optical engineer*
Kerr, James Wilson *engineer*
Mackey, Robert Eugene *environmental engineer*
McAlpine, Kenneth Donald *systems engineer, researcher*
Pollack, Robert William *psychiatrist*
Sayed, Sayed M. *engineering executive*
Thampi, Mohan Varghese *environmental health and civil engineer*

GEORGIA

Albany
McManus, James William *chemist, researcher*

Alpharetta
Clay, Stanton Tower *technical consultant*
Hung, William Mo-Wei *chemist*
Miller, Robert Allen *software engineering educator*
Rettig, Terry *veterinarian, wildlife consultant*
Willis, Sharon White *microbiologist*

Americus
Barnes, David Benton *school psychologist*

Athens
Allinger, Norman Louis *chemistry educator*
Allsbrook, Ogden Olmstead, Jr. *economics educator*
Annis, Patricia Anne *textile scientist, educator, researcher*
Boerma, Henry Roger *agronomist, educator*
Boudinot, Frank Douglas *pharmaceutics educator*
Bounous, Denise Ida *veterinary pathology educator*
Bowen, John Metcalf *pharmacologist, toxicologist, educator*
Compton, Mark Melville *poultry science educator*
Darvill, Alan G. *biochemist, botanist, educator*
Eriksson, Karl-Erik Lennart *biochemist, educator*
Field, Dorothy *gerontologist, educator*
Grossman, Gary David *ecological educator*
Jackson, Robert Benton, IV *environmental engineer*
Jones, Samuel B., Jr. *botany educator*
Kraszewski, Andrzej Wojciech *electrical engineer, researcher*
Lewis, David Lamar *research microbiologist*
Martin, Leonard Louis *social psychologist, educator*
McCutcheon, Steven Clifton *environmental engineer*
Plummer, Gayther L(ynn) *climatologist, ecologist, researcher*
Pulliam, Howard Ronald *ecology educator*
Straw, William Russell *environmental scientist*
Sumner, Malcom Edward *agronomist, educator*
Tollner, Ernest William *agricultural engineering educator, agricultural radiology consultant*
Torrance, Ellis Paul *psychologist, educator*
Willis, Judith H. *zoology educator, researcher*

Atlanta
Aidun, Cyrus Khodarahm *mechanical engineering educator*
Anderson, Alan Julian *database consultant*
Anderson, Gloria Long *chemistry educator*
Antolovich, Stephen Dale *engineering educator*
Aral, Mustafa Mehmet *civil engineer*
Armanios, Erian Abdelmessih *aerospace engineer, educator*
Ashby, Eugene Christopher *chemistry educator, consultant*
Atluri, Satya N(adham) *aerospace engineering educator*
Bakay, Roy Arpad Earle *neurosurgeon, educator*
Barker, Michael Dean *nuclear engineer*
Barnett, Crawford Fannin, Jr. *internist, educator, cardiologist*
Barnett, Robert Neal *physicist, researcher*
Beasley, Ernest William, Jr. *endocrinologist*
Berbenich, William Alfred *electronics engineer*
Berg, George G. *toxicologist*
Bonnet, Henri *engineering manager*
Boothe, Ronald George *psychology and ophthalmology educator*
Bowman, Albert W. *structural engineer*
Boylan, Glenn Gerard *environmental company official, consultant*
Bragdon, Clifford Richardson *city planner, educator*
Breck, Katherine Anne *technical writer*
Bridgeman, David Marsh *chemical company executive*
Brighton, John Austin *mechanical engineer, educator*
Bryan, Spencer Maurice *cost engineer*
Calise, Anthony John *aerospace engineering educator*
Carey, Gerald John, Jr. *former air force officer, research institute director*
Cavallaro, Joseph John *microbiologist*
Chaifetz, L. John *industrial engineer*
Chambers, James Patrick *mechanical engineer*
Chisholm, Tommy *lawyer, utility company executive*
Chou, Mie-Yin *physicist*
Clark, Benjamin Cates, Jr. *flavor chemist, organic research chemist*
Clay, Kelli Suzanne *mechanical engineer*
Clements, James David *psychiatry educator, physician*
Clerke, William Henry III *forester*
Clifton, David Samuel, Jr. *research executive, economist*
Coluccio, Lynne M. *biochemistry educator*
Connor, Charles William *airline pilot*
Cooper, Gerald Rice *clinical pathologist*
Cramer, Howard Ross *geologist, environmental consultant*
Crowe, Devon George *physicist, engineering consultant*
Cunefare, Kenneth Arthur *mechanical engineer, educator*
Damken, John August *computer systems engineer*
Davis, Elizabeth Emily Louise Thorpe *vision psychophysicist, psychologist and computer scientist*
Davis, Lawrence William *radiation oncologist*
Dennison, Daniel B. *chemist*
Dobes, William Lamar, Jr. *dermatologist*
Doetsch, Paul William *biochemist, educator*
Dowdle, Walter Reid *microbiologist, medical center administrator*
Driscoll, Terence Patrick *environmental engineer*
Duffell, James Michael *computer systems analyst*
Dysart, Benjamin Clay, III *environmental management consultant, conservationist, engineer*
Ebneter, Stewart Dwight *engineer*
Edelhauser, Henry F. *physiologist, ophthalmologist, medical educator*
Ehrlich, Margaret Isabella Gorley *systems engineer, mathematics educator, consultant*
Eichholz, Geoffrey G(unther) *physics educator*
Erbil, Ahmet *physics educator*
Evans, Edwin Curtis *internist, educator, geriatrician*
Fash, William Leonard *architecture educator, retired college dean*
Ferguson, Erik Tillman *transportation planning educator*
Fitzgerald, John Edmund *civil engineering educator, dean*
Fleming, Julian Denver, Jr. *lawyer*
Foley, James David *computer science educator, consultant*
Fontaine, Arnold Anthony *mechanical engineer*
Foster, Roger Sherman, Jr. *surgeon, educator, health facility administrator*
Fox, Ronald Forrest *physics educator*
Freedman, Louis Martin *dentist*
Garcia, Ernest Victor *medical physicist*
Gatley, Donald Perkins *mechanical engineer, building scientist*
Gentry, David Raymond *engineer*
Gersch, Harold Arthur *physics educator*
Gerst, Steven Richard *healthcare director, physician*
Glasman, Michael Morris *metallurgical engineer*
Greenberg, Raymond Seth *dean, educator*
Griesser, James Albert (Jamie Griesser) *computer scientist*
Grissom, Raymond Earl, Jr. *toxicologist*
Gu, Yuanchao *biomedical engineer, geneticist*
Hanna, M(axcy) G(rover), Jr. *civil engineer*
Hartzell, Harrison Criss, Jr. *biomedical research educator*
Haulbrook, Robert Wayne *computer systems engineer*
Higgins, Richard J. *microelectrical engineer*
Hill, Craig Livingston *chemistry educator, consultant*
Hollinger, Charlotte Elizabeth *medical technologist, tree farmer*
Huber, Douglas Crawford *pathologist*
Hutcheson, Philip Charles *computer programmer, analyst*
Iacobucci, Guillermo Arturo *chemist*
Ibegbu, Chris Chidozie *immunologist, virologist*
Johnson, Barry Lee *public health research administrator*
Jones, Mark Mitchell *plastic surgeon*
Kardomateas, George Alexander *aerospace engineering educator*
Khudenko, Boris Mikhail *environmental engineer*
Killorin, Edward Wylly *lawyer, tree farmer*
King, Frederick Alexander *neuroscientist, educator*
King, K(imberly) N(elson) *computer science educator*
Kiser, Don Curtis *immunologist*
Kobelski, Robert John *analytical chemist*
La Farge, Timothy *plant geneticist*
Levy, Daniel *economics educator*
Lichtenwalner, Owen C. *engineer, telecommunications executive*
Liew, Chong Shing *structural engineer*
Lin, Jung-Chung *microbiologist, researcher*
Lollar, John Sherman, III (Pete Lollar) *hematologist*
Lowry, Robin Pearce *nephrologist, educator*
Lucchesi, John C. *biology educator*
Lumpkin, Alva Moore, III *electrical engineer, marketing professional*
Macari, Emir José *civil engineering educator*
Manley, Lance Filson *data processing consultant*
Martin, Kevin Paul *physicist*
Matula, Richard A(llan) *academic administrator*
McBroom, Thomas William *mechanical engineer*
McClure, Harold Monroe *veterinary pathologist*
McMillan, Robert Walker *physicist, consultant*
Mumtaz, Mohammad Moizuddin *toxicologist, researcher*
Nash, William Wray, Jr. *retired city planning educator*
Nerem, Robert Michael *engineering educator, consultant*
Neylan, John Francis, III *nephrologist, educator*
O'Driscoll, Jeremiah Joseph, Jr. *safety engineer, consultant*
Olkowski, Zbigniew Lech *physician, educator*
Parsonson, Peter Sterling *civil engineer, educator, consultant*
Pitts, Marcellus Theadore *civil engineer, consultant*
Plotsky, Paul Mitchell *neuroscientist, educator*
Preble, Darrell W. *systems analyst*
Pruett, Clayton Dunklin *biotechnical company executive*
Queen, Brian Charles *chemist*
Rashford, David R. *physicist, educator*
Ratliff, Hugh Donald *industrial engineering educator*
Reed, Diane Gray *business information service company executive*
Rink, Christopher Lee *information technology consultant, photographer*
Robinson, Jeffery Herbert *modular building company executive*
Rumbaugh, Duane M. *psychology educator*
Schafer, Ronald William *electrical engineering educator*
Schultz, Linda Jane *epidemiologist*
Schwartz, David Alan *infectious disease and placental pathologist*
Seifert, Alvin Ronald *psychologist*
Sezgin, Mesut *environmental engineer*
Shen, Kangkang *research physicist*
Siegler, Melody Victoria Stephanie *neuroscientist*
Simmons, Samuel William *retired public health official*
Smith, David Carr *organic chemist*
Smith, David Doyle *engineer, consultant*
Smith, Marc Kevin *mechanical engineering educator*
Snarey, John Robert *psychologist, researcher, educator*
Spurlin, Lisa Turner *biologist*
Stafford, Patrick Morgan *biophysicist*
Stelson, Arthur Wesley *research chemical engineer, atmospheric scientist*
Strahle, Warren Charles *aerospace engineer, educator*
Su, Kendall Ling-Chiao *engineering educator*
Sullivan, James Hargrove, Jr. *mechanical engineer, utility engineer*
Sullivan, Louis Wade *former secretary health and human services, physician*
Talkington, Deborah Frances *microbiologist*
Tamin, Azaibi *molecular virologist, researcher*
Taylor, Laura Marie *interior designer*
Thakker, Ashok *aerospace engineering company executive*
Toler, James C. *electrical engineer*
Tucker, Robert Arnold *electrical engineer*
Turner, John Sidney, Jr. *otolaryngologist, educator*
Tutumluer, Erol *civil engineer*
Tyler, Carl Walter, Jr. *physician, health research administrator*
Underwood, Arthur Louis, Jr. *chemistry educator, researcher*
Waller, John Louis *anesthesiology educator*
Walters, Lori Antionette *electrical engineer, consultant*
White, John Austin, Jr. *engineering educator, dean, consultant*
Williams, Charles Murray *computer information systems educator, consultant*
Willis, Isaac *dermatologist, educator*
Yoganathan, Ajit Prithiviraj *biomedical engineer, educator*
Zalkow, Leon H. *organic chemistry educator*
Zinn, Ben T. *engineer, educator, consultant*
Zirps, Fotena Anatolia *psychologist, researcher*
Zumpe, Doris *psychiatry researcher, educator*
Zunde, Pranas *information science educator, researcher*

Attapulgus
Phan, Justin Triquang *electrical project engineer*

Augusta
Allen, Marshall Bonner, Jr. *neurosurgeon, consultant*
Colborn, Gene Louis *anatomy educator, researcher*
Daniell, Laura Christine *pharmacology and toxicology educator*
Devoe, Lawrence Daniel *obstetrician/gynecologist educator*
Feldman, Elaine Bossak *medical nutritionist, educator*
Gambrell, Richard Donald, Jr. *endocrinologist, educator*
Huang, Denis K. *chemical engineer, consultant*
Huisman, Titus Hendrik Jan *biochemist, educator*
Karp, Warren Bill *medical and dental educator, researcher*
Katz, David Yale *electronics engineering technology educator*
Krauss, Jonathan Seth *pathologist*
Loomis, Earl Alfred, Jr. *psychiatrist*
Loring, David William *neuropsychologist, researcher*
Mahesh, Virendra Bhushan *endocrinologist*
Merin, Robert Gillespie *anesthesiology educator*
Nosek, Thomas Michael *physiologist, educator*
Oleskowicz, Jeanette *physician*
Pallas, Christopher William *cardiologist*
Plouffe, Leo, Jr. *endocrinologist*
Rausch, Jeffrey Lynn *psychiatrist, psychopharmacologist*
Rubin, Joseph William *surgeon, educator*
Smith, Randolph Relihan *plastic surgeon*
Traven, Kevin Charles *structural engineer*
Wilson, Keith Crookston *chemical engineer*
Zachert, Virginia *psychologist, educator*
Zamrini, Edward Youssef *behavioral neurologist*

Austell
Prather, Brenda Joyce *librarian*

Baxley
Belcher, Ronald Anthony *nuclear energy educator*

Bogart
Butts, David Phillip *science educator*

Brunswick
Crowe, Hal Scott *chiropractor*
Iannicelli, Joseph *chemical company executive, consultant*
McDonald, Julian LeRoy, Jr. *manufacturing director*

Byron
Nyczepir, Andrew Peter *nematologist*

Carrollton
Gustin, Ann Winifred *psychologist*

Clarkston
Grimes, Richard Allen *economics educator*
Saint-Come, Claude Marc *science educator, consultant*

Columbus
Braswell, J(ames) Randall *retired quality control professional*
Massey, Carolyn Shavers *mathematics educator*
Moussa, Khalil Mahmoud *polymer photochemist, researcher*
Riggsby, Ernest Duward *science educator*

Conyers
Kilkelly, Brian Holten *lighting company executive, real estate partner*

Cornelia
Jeffers, Dale Welborn *computer specialist, systems analyst*

Decatur
Alderete, Joseph Frank *psychiatrist, medical service adminstrator*
Crews, John Eric *rehabilitation administrator*
Kann, Herbert Ellis, Jr. *hematologist, oncologist*
Mills, Terry, III *forensic chemist*
Pittman, Victor Fred *computer systems engineer*
Sadun, Alberto Carlo *astrophysicist, physics educator*
Stivers, Theodore Edward *food products executive, consultant*

Demorest
Wainberg, Robert Howard *biology educator*

Duluth
Engleman, Dennis Eugene *electrical engineer*
Hessman, Frederick William *aeronautical engineer*
Mowrey-McKee, Mary Flowers *biochemist*
Savatsky, Bruce Jon *chemical engineer*
Stewart, Alan Frederick *electrical engineer*

Dunwoody
Bartolo, Donna M. *hospital administrator, nurse*

Fort Gordon
Craft, David Walton *clinical microbiologist, army officer*
Latiff-Bolet, Ligia *psychologist*

Fort Stewart
Houston, Thomas Dewey *environmental engineer*

Fort Valley
Lewis, Larry *mathematics educator*

Gordon
Young, Raymond H(inchcliffe), Jr. *chemist*

Griffin
Arkin, Gerald Franklin *agricultural research administrator, educator*
Braman, S. Kristine *entomologist*
Carter, Edward Fenton, III *pathologist, medical examiner*
Doyle, Michael Patrick *food microbiologist, educator, researcher*
Gillaspie, Athey Graves, Jr. *pathologist, researcher*
Trimmer, Brenda Kay *pharmacist*

Jersey
Batchelor, Joseph Brooklyn, Jr. *electronics engineer, consultant*

Kennesaw
Ghosh, Deepak Ranjan *chemical/environmental engineer*
McCoy, R. Wesley *biology educator*
Polmann, Donald Jeffrey *water resources engineer, environmental consultant*
Ziegler, Christine Bernadette *psychology educator, consultant*

Kingsland
Shockley, Carol Frances *psychologist, psychotherapist*

La Grange
Toth, Danny Andrew *pharmacist*
West, John Thomas *surgeon*

Lawrenceville
Adair, Lila McGahee *physics educator*
Carter, Dale William *psychologist*
Kretsch, Michael Gene *civil engineer*

Lilburn
Neumann, Thomas William *archaeologist*

Lithia Springs
Hebel, Gail Suzette *obstetrician/gynecologist*

Locust Grove
Smith, Al Jackson, Jr. *environmental engineer, lawyer*

Macon
Hails, Robert Emmet *aerospace consultant, business executive, former air force officer*
Harrison, James Ostelle *ecologist*
Volpe, Erminio Peter *biologist, educator*

Marietta
Breese, John Allen *chemical industry executive*
Engelstad, Stephen Phillip *mechanical engineer*
Fox, Marian Cavender *mathematics educator*
Garrett, Joseph Edward *aerospace engineer*
Gerhardt, Rosario Alejandrina *materials scientist*
Hamilton, Stephen Stewart *mechanical engineer, consultant*
Harbort, Robert Adolph, Jr. *computer science educator*
Hayes, Robert Deming *electrical engineer, consultant*
Holtz, Noel *neurologist*
Houser, Vincent Paul *medical research director*
Johnson, Samuel Walter, II *mechanical engineer*
Jordan, George Washington, Jr. *engineering company executive*
Miles, Thomas Caswell *engineer*
Oakes, Thomas Wyatt *environmental engineer*
Peake, Thaddeus Andrew, III *environmental engineer*
Pearson, Hugh Stephen *mechanical and metallurgical engineer*
Rowe, Alvin George *structural engineer, consultant*
Sloan, Mary Jean *media specialist*
Swift, John Lionel *civil engineer*
Voots, Terry Lynne *technical sales engineer*
Wilks, Ronald *engineer*

Martinez
Butts, William Randolph *nuclear medicine technologist*

Milledgeville
Jones, Harold Charles *retired biologist*

Monroe
Gammans, James Patrick *ceramic engineer*

Newnan
Jensen, Bruce Alan *control engineer*

Norcross
Adams, Dee Briane *hydrologist, civil engineer*
Buchanan, Diane Kay *pharmacist*
Chan, David Chuk *software engineer*
Crymes, Ronald Jack *structural steel detailer, draftsman*
Dibb, David Walter *research association administrator*
Duncan, Phillip Charles *research scientist*
Greene, Don Howard *product designer*
Javaheri, Kathleen Dakin *construction and environmental professional*
Johnson, Roger Warren *chemical engineer*
Moore, Christopher Barry *industrial engineer*
Nelson, Larry Keith *questioned document examiner*
Newbern, Laura Lynn *forestry association executive, editor*
Pippin, John Eldon *electronics engineer, electronics company executive*
Stockwell, Albert H. *procurement professional*
Zaloga, Robert Edwin *engineering executive*

Norristown
McCord, James Richard, III *chemical engineer, mathematician*

Robins AFB
Christian, Thomas Franklin, Jr. *aerospace engineer, educator*
Washington, David Earl *computer systems professional*

Roswell
Arnold, Jack Waldo *physicist*
Bristow, Preston Abner, Jr. *civil engineer, environmental engineer*
Jensen, Mogens Reimer *psychologist*
Sanner, George Elwood *electrical engineer*
Walczak, Zbigniew Kazimierz *polymer science and engineering researcher*

Sandersville
Malla, Prakash Babu *research materials chemist*

Sapelo Island
Alberts, James Joseph *scientist, researcher*

Savannah
Brown, Thomas Edward *chemist*
Dixon, Sandra Wise *aerospace engineer*
Hsu, Ming-Yu *engineer, educator*
Menzel, David Washington *oceanographer*
Nawrocki, H(enry) Franz *propulsion technology scientist*
Simonaitis, Richard Ambrose *chemist*
Throne, James Edward *entomologist*
Zoller, Michael *otolaryngologist*

Smyrna
Stevenson, Earl, Jr. *civil engineer*

Snellville
Hudgens, Kimberlyn Nan *industrial engineer*

Statesboro
Dean, Cleon Eugene *physicist*
Hanson, Roland Stuart *industrial engineer, educator*
Hurst, Michael Owen *biochemistry educator*
Lefcort, Hugh George *zoologist*
Mobley, Cleon Marion, Jr. (Chip Mobley) *physics educator, real estate executive*
Parrish, John Wesley, Jr. *physiologist, biology educator*
Vives, Stephen Paul *biology educator, fish biologist*

Stone Mountain
Quang, Eiping *nuclear engineer*
Rogers, James Virgil, Jr. *radiologist, educator*

Swainsboro
Watt, (Arthur) Dwight, Jr. *computer programming and microcomputer specialist, educator*

Thomasville
Buckner, James Lee *forester, biologist*
Haynes, Harold Eugene, Jr. *medical physicist*

Tifton
Butler, James Lee *agricultural engineer, researcher*
Rogers, Charlie Ellic *entomologist*
Thomas, Adrian Wesley *laboratory director*

Tucker
O'Neil, Daniel Joseph *academic administrator, research executive*
Traina, Paul Joseph *environmental engineer*

Valdosta
Hoff, Edwin Frank, Jr. *research chemist*
Lankford, Mary Angeline Gruver *pharmacist*
Mares, Joseph Thomas *entomologist*

Waycross
Nienow, James Anthony *biologist, educator*

Waynesboro
Legrand, Ronald Lyn *nuclear facility executive*

Winterville
Anderson, David Prewitt *university dean*

HAWAII

Aiea
Brassfield, Patricia Ann *psychologist*
Heinz, Don J. *agronomist*
Ma, Jeanetta Ping Chan *elementary education educator*

Camp Smith
Surface, Stephen Walter *water treatment chemist, environmental protection specialist*

Ewa Beach
Dizon, Jose Solomon *planning engineer*

Hilo
Taniguchi, Tokuso *surgeon*

Honolulu
Abbott, Isabella Aiona *biology educator*
Abdul, Corinna Gay *software engineer, consultant*
Alicata, Joseph Everett *microbiology researcher, parasitologist*
Brock, James Melmuth *engineer, venture capitalist*
Cattell, Heather Birkett *psychologist*
Chambers, Kenneth Carter *astronomer*
Ching, Chauncey Tai Kin *agricultural economics educator*
Ching, Daniel Gerald *civil engineer*
Chiu, Arthur Nang Lick *engineering educator*
Chock, Clifford Yet-Chong *family practice physician*
Chun, Lowell Koon Wa *architect*
Duckworth, Walter Donald *museum executive, entomologist*
Duncan, John Wiley *mathematics and computer educator, retired air force officer*
Edwards, John Wesley, Jr. *urologist*
Edwards, Margo H. *marine geophysicist, researcher*
Flannelly, Kevin J. *psychologist, research analyst*
Flannelly, Laura T. *mental health nurse, nursing educator, researcher*
Foster, Stephen Roch *civil engineer*
Fuchs, Roland John *geography educator, university administrator*
Fujioka, Roger Sadao *research microbiology educator*
Fukumoto, Neal Susumu *civil engineer*
Gilbert, Fred Ivan, Jr. *physician, researcher*
Grace, George William *linguistics educator*
Hall, Donald Norman Blake *astronomer*
Hays, Ronald Jackson *naval officer*
Helsley, Charles Everett *geologist, geophysicist*
Ishikawa-Fullmer, Janet Satomi *psychologist, educator*
Jongeward, George Ronald *systems analyst*
Kamemoto, Haruyuki *horticulture educator*
Kay, Elizabeth Alison *zoology educator*
Keil, Klaus *geology educator, consultant*
Khan, Mohammad Asad *geophysicist, educator, former energy minister and senator of Pakistan*
Klink, Paul L. *computer company executive*
Kop, Tim M. *psychologist*
Krock, Hans-Jurgen *civil engineer*
Kwong, James Kin-Ping *geological engineer*
Laney, Leroy Olan *economist, banker*
Lashlee, JoLynne Van Marsdon *army officer, nursing administrator*
Mandel, Morton *molecular biologist*
McCarthy, Laurence James *physician, pathologist*
Meech, Karen Jean *astronomer*
Mercier, John René *nuclear engineer, health physicist*
Mizokami, Iris Chieko *mechanical engineer*
Nelson, Jeanne Francess *mathematics educator*
Olipares, Hubert Barut *biological safety officer*
Pang, Herbert George *ophthalmologist*
Rezachek, David Allen *energy and environmental engineer*
Scheuer, Paul Josef *chemistry educator*
Seifert, Josef *chemist, educator*
Shefchick, Thomas Peter *forensic electrical engineer*
Smith, Albert Charles *biologist, educator*
Smith-Kayode, Timi *food technologist*
Staff, Robert James, Jr. *international economist, consultant*
Stevens, Stephen Edward *psychiatrist*
Sugiki, Shigemi *ophthalmologist, educator*
Swanson, Richard William *statistician*
Tamaye, Elaine E. *coastal/ocean engineer*
Toyomura, Dennis Takeshi *architect*
Uyehara, Catherine Fay Takako (Yamauchi) *physiologist, educator, pharmacologist*
Vargas, Roger Irvin *entomologist, ecologist*
Wessel, Paul *geology and geophysics educator*
Wheeler, Carl *mathematics educator*
Wyrtki, Klaus *oceanography educator*
Yang, Ping-Yi *biowaste, wastewater engineering educator*
Zhou, Chiping *mathematician, educator*

Kahului
Hughes, Arleigh Bruce *microbiologist, educator*

Kamuela
Hamilton, John Carl *astronomer, telescope operator*
Stillings, Dennis Otto *research director*

Kaneohe
Ahmed, Iqbal *psychiatrist, consultant*
Hanson, Richard Edwin *civil engineer*
May, Richard Paul *data processing professional*

Koloa
Donohugh, Donald Lee *physician*

Lihue
Tabata, Lyle Mikio *mechanical engineer*

Wahiawa
Imperial, John Vince *systems engineer, consultant*

Wailuku
Savona, Michael Richard *physician*

Waimanalo
Divakaran, Subramaniam *biochemist*

Waipahu
Caldwell, Peter Derek *pediatrician, pediatric cardiologist*

IDAHO

Blackfoot
Goodyear, Jack Dale *electronic educator*

Boise
Burton, Lawrence DeVere *agriculturist, educator*
Corder, Loren David (Zeke Corder) *quality assurance engineer*
Del Carmen, Rene Jover *process and environmental engineer*
Habben, David Marshall *state official*
Marks, Ernest E. *maintenance engineer*
Olson, Richard Dean *researcher, pharmacology educator*
Pon-Brown, Kay Migyoku *information systems specialist*

Bonners Ferry
McClintock, William Thomas *health care administrator*

Harrison
Skidmore, Eric Arthur *industrial engineer*

Hayden
Hundhausen, Robert John *mining engineer*

Hayden Lake
Lehrer, William Peter, Jr. *animal scientist*

Idaho Falls
Buden, David *aerospace power engineer, nuclear power researcher*
Cott, Donald Wing *aerospace engineer*
Crawford, Thomas Mark *laser/optics physicist, business owner, consultant*
Elias, Thomas Ittan *mechanical engineer*
Epstein, Jonathan Stone *engineering executive*
Harris, Robert James *engineer*
Hicks, Michael David *nuclear engineer*
Holcombe, Homer Wayne *nuclear quality assurance professional*
Ischay, Christopher Patrick *engineer*
Kerr, Thomas Andrew *senior program engineer*
Kuan, Pui *nuclear engineer*
Long, John Kelley *nuclear reactor physicist, consultant*
McCarthy, Jeremiah Justin *environmental engineer*
Motloch, Chester George *nuclear engineer*
Newman, Stanley Ray *oil refining company executive*
Ramshaw, John David *chemical physicist, mechanical engineer*
Tamashiro, Thomas Koyei *retired electrical engineer*
Zaccardi, Larry Bryan *spectrochemistry scientist, microbiologist*

Kimberly
Carter, David LaVere *soil scientist, researcher, consultant*
Trout, Thomas James *agricultural engineer*

Lewiston
Bjerke, Robert Keith *chemist*
Heidorn, Douglas Bruce *medical physicist*

Moscow
Bartlett, Robert Watkins *academic dean, metallurgist*
DeShazer, James Arthur *agricultural engineer, educator, administrator*
Hendee, John Clare *college dean, natural resources educator*
Jacobsen, Richard T. *mechanical engineering educator*
Marshall, John David *forest biologist*
McGeehan, Steven Lewis *soil scientist*
Miller, Maynard Malcolm *geologist, educator, research foundation director, explorer, state legislator*
Peterson, Charles Loren *agricultural engineer, educator*
Roberts, Lorin Watson *botanist, educator*
Scott, J(ames) Michael *research biologist*
Stauffer, Larry Allen *mechanical engineer, educator, consultant*
Stumpf, Bernhard Josef *physicist*

Mountain Home
Meyr, Shari Louise *information consultant*

Pocatello
Crawford, Kevan Charles *nuclear engineer, educator*
Dykes, Fred William *retired nuclear scientist*
Huck, Matthew L. *process development engineer*
Lee, Richard A. *mechanical engineer, consultant*
Moore, Kevin L. *electrical engineering educator*
Smith, John Julian *agronomist*
Stuffle, Linda Robertson *electrical engineer*

Post Falls
Brede, Andrew Douglas *research director, plant breeder*

Priest River
Freibott, George August *physician, chemist, priest*

Shelley
Kimmel, Richard John *engineer*

Tuttle
Ravenscroft, Bryan Dale *alternate energy research company executive*

ILLINOIS

Abbott Park
Allen, Steven Paul *microbiologist*
Boyd, Steven Armen *medicinal chemist*
Jeng, Tzyy-Wen *biochemist*
Peterson, Bryan Charles *biochemist*
Shipkowitz, Nathan L. *microbiologist*
Swift, Kerry Michael *physical chemist*
Trivedi, Jay Sanjay *chemist*
Wideburg, Norman Earl *biochemist*

Addison
Findling, David Martin *application engineer*

Algonquin
Stanek, Donald George, Jr. *computer analyst*

Alton
Zimmer, James Peter *laboratory executive, consultant*

Argo
Totten, Venita Laverne *chemist*

Argonne
Alexander, Dale Edward *materials scientist*
Barrett, Gregory Lawrence *environmental scientist*
Bauer, Theodore Henry *nuclear engineer*
Bhatti, Neeloo *environmental scientist*
Blair, Robert Eugene *physicist, researcher*
Braun, Joseph Carl *nuclear engineer, scientist*
Coffey, Howard Thomas *physicist*
Demirgian, Jack Charles *analytical chemist*
Depiante, Eduardo Victor *nuclear engineer*
Doss, Ezzat Danial *mechanical engineer, researcher*
Drucker, Harvey *biologist*
Dunford, Robert Walter *physicist*
Erdemir, Ali *materials scientist*
Garner, Patrick Lynn *nuclear engineer*
Heine, James Arthur *utilities plant manager*
Holtzman, Richard Beves *health physicist, chemist*
Jorgensen, James Douglas *research physicist*
Kini, Aravinda Mattar *materials chemist*
Lawson, Robert Davis *theoretical nuclear physicist*
Miller, Shelby Alexander *chemical engineer, educator*
Myles, Kevin Michael *metallurgical engineer*
Regalbuto, Monica Cristina *chemical engineer, research scientist*
Reifman, Jaques *nuclear engineer, researcher*
Routbort, Jules Lazar *physicist, editor*
Schriesheim, Alan *research administrator*
Seefeldt, Waldemar Bernhard *chemical engineer*
Tang, Yu *structural engineer*
Tatar, John Joseph *computer scientist*
Thorn, Robert Jerome *chemist*
Toppel, Bert Jack *reactor physicist*

Arlington Heights
Enright, John Carl *occupational health engineer*
Jenny, Daniel P. *retired engineer*
McGuire, Mark William *electrical engineer*
Monti, Laura Anne *psychology researcher, educator*
Shetty, Mulki Radhakrishna *oncologist, consultant*

Ashland
Benz, Donald Ray *nuclear safety engineer, researcher*

Aurora
Ball, William James *pediatrician*
Bleck, Phyllis Claire *surgeon, musician*
Freyberg, Dale Wayne *technical trainer*
Ika, Prasad Venkata *chemist*
Quinn, Richard Kendall *environmental engineer*
Zimmerman, Charles Leonard *systems engineer*

Barrington
Groesch, John William, Jr. *marketing research consultant*
Perry, I. Chet *petroleum company executive*
Sandu, Constantine *development engineer*
Vandeberg, John Thomas *chemical company executive*

Bartlett
FitzSimons, Christopher *design engineer*

Batavia
Cooper, Peter Semler *physicist*
Rapidis, Petros A. *research physicist*
Tollestrup, Alvin Virgil *physicist*
Tweedy, Robert Hugh *equipment company executive*

Belleville
Steffen, Alan Leslie *entomologist*

Bellwood
Gregory, Vance Peter, Jr. *chemist*

Berwyn
De Lerno, Manuel Joseph *electrical engineer*
Misurec, Rudolf *physician, surgeon*

Bloomington
Jaggi, Narendra K. *physics educator, researcher*
Switzer, Jon Rex *architect*
Weber, David Frederick *genetics educator*

Bradley
O'Flaherty, Gerald Nolan *secondary school educator, consultant*

Breese
Anderson, Donald Thomas, Jr. *environmental consultant*

Brookfield
Rabb, George Bernard *zoologist*

Buffalo Grove
Ahmed, Osman *mechanical engineer*
Kalvin, Douglas Mark *research chemist*
O'Sullivan, Michael Anthony *civil engineer*

Burr Ridge
Bathina, Harinath Babu *chemist, researcher*
Mockaitis, Algis Peter *mechanical engineer*

Cahokia
Herrick, Paul E. *aerospace technology educator, researcher*
Redmount, Ian H. *physicist*

Calumet City
Kovach, Joseph William *management consultant, psychologist, educator*

Carbondale
Ali, Naushad *physicist*
Bates, Sharon Ann *plant and soil scientist, educator*
Bozzola, John Joseph *botany educator, researcher*
Burr, Brooks Milo *zoology educator*
Chandrashekar, Varadaraj *endocrinologist*
Chen, Tian-Jie *physicist, educator*
Chugh, Yoginder Paul *mining engineering educator*
DiLalla, Lisabeth Anne *developmental psychology researcher, educator*
Gumerman, George John *archaeologist*
Hatziadoniu, Constantine Ioannis *electrical engineer, educator*
Mohlenbrock, Robert Herman, Jr. *botanist, educator*
Tao, Rongjia *physicist, educator*

Carlinville
Zalisko, Edward John *biology educator*

Carol Stream
Wojtanek, Guy Andrew *mechanical engineer*

Carterville
Feldmann, Herman Fred *chemical engineer*
Honea, Franklin Ivan *chemical engineer*

Cary
White, William *research physicist*

Centralia
Davidson, Karen Sue *computer software designer*

Champaign
Boddu, Veera Mallu *chemical engineering researcher*
Cartwright, Keros *hydrogeologist, researcher*
Donchin, Emanuel *psychologist, educator*
Eriksen, Charles Walter *psychologist, educator*
Helm, Charles George *entomologist, researcher*
Hirsch, Jerry *psychology and biology educator*
Hodge, Winifred *environmental scientist, researcher*
Hsu, Chien-Yeh *electrical engineer, speech and hearing scientist*
Joncich, David Michael *energy engineer*
Khan, Latif Akbar *mineral engineer*
Komorita, Samuel Shozo *psychology educator*
Kruger, William Arnold *consulting civil engineer*
Kuck, David Jerome *computer system researcher, administrator*
Laughlin, Patrick Ray *psychologist*
Page, Lawrence Merle *ichthyologist, educator*
Portnoy, Stephen Lane *statistician*
Puckett, Hoyle Brooks *agricultural engineer, research scientist, consultant*
Sanderson, Glen Charles *science director*
Schwartz, Mark William *ecologist*
Semonin, Richard Gerard *state official*
Shahin, M. Y. *engineering*
Slichter, Charles Pence *physicist, educator*
Smarr, Larry Lee *science administrator, educator, astrophysicist*
Wasserman, Stanley *statistician, educator*
Wood, Susanne Griffiths *analytical environmental chemist*

Chicago
Acs, Joseph Steven *transportation engineering consultant*

Al-Salqan, Yahya Yousef computer science researcher
Altmann, Jeanne zoologist, educator
Arnason, Barry Gilbert Wyatt neurologist, educator
Arnsdorf, Morton Frank cardiologist, educator
Arzbaecher, Robert C(harles) research institute executive, electrical engineer, researcher
Bachicha, Joseph Alfred obstetrician/gynecologist, educator
Backman, Ari Ismo electronics company executive
Bailey, Mark William electrical engineer
Baker, M(ervin) Duane healthcare, biotechnology marketing and business consultant
Banerjee, Prashant industrial engineering educator, researcher
Barter, James T. psychiatrist, educator
Basch, Michael Franz psychiatrist, psychoanalyst, educator
Batlivala, Robert Bomi D. oil company executive, economist educator
Batlle, Daniel Campi nephrologist
Baumgartner, Donald Lawrence entomologist, educator
Beach, David Duncan naval architect
Beck, Robert N. nuclear medicine educator
Becker, Gary Stanley economist, educator
Beitler, Stephen Seth cosmetics company executive
Bieler, Rudiger biologist, curator
Biljetina, Richard chemical engineering researcher
Boggs, Joseph Dodridge pediatric pathologist, educator
Bookstein, Abraham information science educator
Braker, William Paul aquarium executive, ichthyologist
Brask, Gerald Irving sanitary engineer
Bryant, Fred Boyd psychology educator
Budenholzer, Roland Anthony mechanical engineering educator
Burt, Robert Norcross diversified manufacturing company executive
Calderon, Alberto P. mathematician, educator
Camras, Marvin electrical and computer engineering educator, inventor
Carney, Jean Kathryn psychologist
Cassel, Christine Karen physician
Castignetti, Domenic microbiologist, biology educator
Chaiyabhat, Win engineering manager
Chakravarthy, Sreenathan Ramanna physicist
Chambers, Donald Arthur biochemistry educator
Chandler, John W. biochemistry educator, ophthalmology educator
Chandrasekhar, Subrahmanyan theoretical astrophysicist, educator
Chang, Huai Ted environmental engineer, educator, researcher
Chang, Wei medical physicist
Chen, Gui-Qiang mathematician, educator, researcher
Chernick, Cedric L. research grants administrator
Chiou, Win Loung pharmacokinetics educator, director
Chmell, Samuel Jay orthopedic surgeon
Chu, James Chien-Hua medical physicist, educator
Chung, Paul Myungha mechanical engineer, educator
Clark, Raymond John mechanical engineer
Clevidence, Derek Edward biologist
Cohen, Melvin R. physician, educator
Coopersmith, Bernard Ira obstetrician/gynecologist, educator
Corbett, Jules John microbiology educator
Crawford, Isiaah clinical psychologist
Cronin, James Watson physicist, educator
Croon, Gregory Steven nuclear engineer
Datta, Syamal Kumar medical educator, researcher
Deorio, Anthony Joseph surgeon
DeSantiago, Michael Francis mechanical engineer
Diamond, Seymour physician
Driskell, Claude Evans dentist
Dumbacher, John Philip evolutionary biologist
Dwyer, Dennis D. information technology executive
Ernest, J. Terry ocular physiologist, educator
Feingold, Daniel Leon physician, consultant
Fowler, George Selton, Jr. architect
Freed, Karl Frederick chemistry educator
Fried, Josef chemist, educator
Fritzsche, Hellmut physics educator
Fujita, Tetsuya Theodore educator, meteorologist
Gasparotto, Renso civil engineer
Gelston, John Herbert electrical engineer
Gerstner, Robert William structural engineering educator, consultant
Giannopoulos, Joanne pharmacist, consultant
Giovacchini, Peter Louis psychoanalyst
Gofron, Kazimierz Jan physicist
Goldberg, Arnold Irving psychoanalyst, educator
Gomer, Robert chemistry educator
Goodman, Harold pediatrician, consultant
Gottlander, Robert Jan Lars dental company executive
Greaves, William Webster chemist, patent analyst
Griffith, B(ezaleel) Herold physician, educator, plastic surgeon
Gruba-McCallister, Frank Peter psychologist
Gupta, Hem Chander mechanical and electrical engineer
Guralnick, Sidney Aaron civil engineering educator
Halpern, Jack chemist, educator
Hamby, Peter Norman nuclear engineer
Handwerker, A. M. retired transportation executive
Hanson, Floyd Bliss applied mathematician, computational scientist, mathematical biologist
Harris, Chauncy Dennison geographer, educator
Harris, Jules Eli medical educator, physician, clinical scientist, administrator
Hartwell, Robert Carl chemical engineer
Harwood, John Simon chemist, university official, consultant
Healy, Kevin E. biomedical engineering educator
Hendrix, Ronald Wayne physician, radiologist
Hildebranski, Robert Joseph civil engineer
Hinojosa, Raul physician, ear pathology researcher
Hobbs, Marvin engineering executive
Hoff, Gerald Charles transportation engineer, planner
Hoppert, Gloria Jean food products executive
Hsiao, Ming-Yuan nuclear engineer, researcher
Hudik, Martin Francis hospital administrator, educator
Huggins, Charles Brenton surgical educator
Hughes, John Russell physician, educator
Iqbal, Zafar Mohd cancer researcher, biochemist, pharmacologist, toxicologist, consultant
Jadvar, Hossein biomedical engineer, oncologist
Jeyendran, Rajasingam Sivaperagasam andrologist, researcher
Jia, Hong programmer, analyst
Jilhewar, Ashok gastroenterologist
Kadanoff, Leo Philip physicist
Karri, Surya B. Reddy chemical engineer
Kass, Leon Richard life sciences educator
Kelly, Gerald Wayne chemical coatings company executive
Kiefer, John Harold chemical engineering educator
Kiel, William Frederick architectural specifications consultant
Knospe, William Herbert medical educator
Koch, Alisa Erika biomedical researcher, rheumatologist
Koncel, James E. electrical engineer
Kornel, Ludwig medical educator, physician, scientist
Koster van Groos, August Ferdinand geology educator
Krawetz, Arthur Altshuler chemist, science administrator
Krishnamachari, Sadagopa Iyengar mechanical engineer, consultant
Lammers, Thomas Gerard museum curator
Landsberg, Lewis endocrinologist, medical researcher
Langdon, Paul Eugene, Jr. consulting civil engineer
Larsen, Janine Louise electrical engineering educator
Larson, Nancy Celeste computer systems analyst, music educator
Lederman, Leon Max physicist, educator
Lee, Bernard Shing-Shu research company executive
Levine, Michael William vision researcher, educator
Levy, Donald Harris chemistry educator
Linden, Henry Robert chemical engineering research executive
Lombardo, Janice Ellen microbiologist
Long, Kevin Jay medicolegal consultant
Lopez, Carolyn Catherine physician
Lovell, James A., Jr. business executive, former astronaut
Lu, Catherine Mean Hoa electrical engineer
Lundberg, George David, II medical editor, pathologist
Machinis, Peter Alexander civil engineer
Marotta, Sabath Fred physiology educator
Mazenko, Gene Francis physics educator
McCrone, Walter Cox research institute executive
McCullagh, Peter statistician, mathematician
McCullough, Henry G(lenn) L(uther) nuclear engineer
McHenry, Keith Welles, Jr. oil company executive
McKee, Keith Earl manufacturing technology executive
McLawhon, Ronald William pathology educator, biochemist
Megaridis, Constantine Michael mechanical engineering educator, researcher
Mehlenbacher, Dohn Harlow civil engineer, consultant
Meier, Paul statistician, mathematics educator
Meyer, Paul Reims, Jr. orthopaedic surgeon
Meyer, Peter physicist, educator
Miller, Michael Carl chemist, researcher
Miller, Stephen Douglas immunologist, educator
Minkowycz, W. J. engineering educator
Moawad, Atef obstetrician-gynecologist, educator
Monyak, Wendell Peter pharmacist
Moskal, Joseph Russell biochemist
Muller, Dietrich Alfred Helmut physicist, educator
Mullins, Obera microbiologist
Mundlak, Yair agriculture and economics educator
Murphy, Thomas Joseph chemistry educator
Nair, Sudhakar Edayillam mechanical and aerospace engineering educator
Narahashi, Toshio pharmacology educator
Neimark, Vassa interior architect
Nicholas, Ralph Wallace anthropologist, educator
Nizamuddin, Khawaja structural engineer
Norris, James Rufus, Jr. chemist, educator, consultant
Norton, Steven David immunologist, researcher
Nowak, Chester John, Jr. infosystems specialist
O'Connor, James John utility company executive
Oshiyoye, Adekunle Emmanuel physician, realtor
Osiyoye, Adekunle obstetrician/gynecologist, educator
Osowiec, Darlene Ann psychologist, consultant, educator
Ostrow, Jay Donald gastroenterology educator, researcher
Papatheofanis, Frank John biochemist
Parker, Eugene Newman physicist, educator
Patterson, Roy physician, educator
Pavelka, Elaine Blanche mathematics educator
Peebles, William Reginald, Jr. mechanical engineer
Pierce, Shelby Crawford oil company executive
Pluth, Joseph John chemist, consultant
Polydoris, Nicholas George electronics executive
Pope, Richard M. rheumatologist
Porter, Robert William mechanical engineering educator
Ramsey, Glenn Eugene pathologist, blood bank physician
Raney, Leland Wayne nuclear engineer
Rapoza, Norbert Pacheco medical association administrator, virologist
Ray, Richard Eugene osteopath, psychiatrist
Rikoski, Richard Anthony engineering executive, electrical engineer
Rizzo, Thomas Anthony psychologist
Robbins, Kenneth Carl biochemist
Rodgers, James Foster association executive, economist
Rokach, Abraham Jacob structural engineering and computer software consultant
Rosen, Steven Terry oncologist, hematologist
Rosenberg, Sheldon psychologist, educator
Rubenstein, Arthur Harold physician, educator
Russo, Gilberto engineering educator
Ryan, Hugh William engineer, consultant
Rymer, William Zev research scientist, administrator
Safar, Michal information scientist
Saweikis, Matthew A. information scientist
Schramm, David Norman astrophysicist, educator
Schutt, Jeffry Allen physicist
Schwartzberg, Joanne Gilbert physician
Sebastian, Scott Joseph civil engineer
Seifert, Timothy Michael infosystems specialist
Sereno, Paul C. paleontologist, educator
Shafer, Kevin Lee water resources engineer
Shapiro, James A. bacterial geneticist, educator
Shawhan, Peter Sven physicist
Shulman, Stanford Taylor pediatrics educator, infectious disease researcher
Siegler, Mark internist, educator
Simpson, John Alexander physicist
Singh, Manmohan orthopedic surgeon, educator
Skiba, Aurelia Ellen parochial school educator
Skimina, Timothy Anthony electrical engineer, computer consultant
Skolnick, Andrew Abraham science and medical journalist, photographer
Sresty, Guggilam Chalamaiah environmental engineer
Stafford, Fred Ezra science adminstrator
Stamler, Jeremiah physician, educator
Stark, Benjamin Chapman biology educator, molecular biologist
Stecich, John Patrick structural engineer
Steiner, Donald Frederick biochemist, physician, educator
Stigler, Stephen Mack statistician, educator
Surgi, Elizabeth Benson veterinarian
Swerdlow, Martin Abraham physician, pathologist
Taira, Frances Snow nurse, educator
Tao, Mariano biochemistry educator
Thomas, Leona Marlene health information educator
Thompson, Robert W. theoretical physicist
Tijunelis, Donatas engineering executive
Tipp, Karen Lynn Wagner school psychologist
Tobin, Calvin Jay architect
Todd, James S. surgeon, educator, medical association administrator
Trapani, Catherine special education and educational psychology educator
Trenary, Michael chemistry educator
Triplett, Kelly B. chemist
Turkevich, Anthony Leonid chemist, educator
Vasa, Rohitkumar Bhupatrai pediatrician, neonatologist
Visotsky, Harold Meryle psychiatrist, educator
Wadden, Richard Albert environmental science educator, consultant, researcher
Wade, Michael John ecology and evolution educator, researcher
Wall, Jacqueline Remondet industrial psychologist, rehabilitation counselor
Walshe, Brian Francis management consultant
Weichselbaum, Ralph R. oncologist chairman
Wied, George Ludwig physician
Williams, Jack Raymond civil engineer
Williams, Philip Copelain obstetrician, gynecologist
Williams-Ashman, Howard Guy biochemistry educator
Wolinsky, Steven Mark infectious diseases physician, educator
Wong, Thomas Tang Yum engineering educator
Worek, William Martin mechanical engineering educator
Yamashiroya, Herbert Mitsugi microbiologist, educator
Yeldandi, Veerainder Antiah engineer, consultant
Young, Kathleen clinical psychologist
Zeffren, Eugene toiletries company executive
Zhao, Meishan chemical physics educator, researcher

Chicago Heights
Carpenter, Kenneth Russell international trading executive

Clinton
Ramanuja, Teralandur Krishnaswamy structural engineer

Collinsville
Modeer, Victor Albert, Jr. civil engineer

Crete
Crowley, Michael Summers ceramic engineer

Crystal Lake
Dabkowski, John electrical engineer, consultant, researcher

Danville
Boone, Thomas John metallurgical engineer
Craig, Hurshel Eugene agronomist
Elghammer, Richard William psychologist

De Kalb
Bow, Sing Tze engineer, educator
Field, Robert Eugene mechanical engineer, educator, consulant, researcher
Kerr, Sandra Lee psychology educator
Robison, Norman Glenn tropical research director
Rosenmann, Daniel physicist, educator
Rossing, Thomas D. physics educator
Stoia, Dennis Vasile industrial engineering educator
Troyer, Alvah Forrest seed corn company executive, plant breeder
Williams, Eddie R. university administrator

Decatur
Graf, Karl Rockwell nuclear engineer
Houran, James Patrick counselor, psychiatric technician
Lambirth, Thomas Thurman clinical psychologist
Schlueter, Gerald Francis mechanical engineering educator
Weatherbee, Carl retired chemistry educator, genealogist

Deerfield
Loucks, Vernon R., Jr. health care products and services company executive
Runkle, Robert Scott environmental company executive
Schechter, Paul J. pharmacologist

Des Plaines
Arena, Blaise Joseph research chemist
Cloutier, Stephen Edward chemical engineer
Flakne, Dawn Gayle electronics engineer
Laban, Saad Lotfy medicinal chemist
Price, Richard George electrical engineer
Surgi, Marion Rene chemist
Vogl, Anna Katharina chemical engineer
Wilks, Alan Delbert chemical research and technology executive, researcher

Dixon
Youker, David Eugene biology educator

Downers Grove
Boese, Mark Alan forensic scientist, chemist, educator
Boese, Robert Alan forensic chemist
Currie, Bruce LaMonte pharmaceutical sciences educator, medicinal chemistry researcher
Klassen, Jane Frances electrical engineer
LaRocca, Patricia Darlene McAleer mathematics educator
Mayer, Alejandro Miguel biologist, educator
Shah, Bankim software consultant

Dundee
Burger, George Vanderkarr wildlife ecologist, researcher

Dunlap
Reinsma, Harold Lawrence design consultant, engineer

East Moline
Bosworth, Douglas LeRoy farm implement company executive

Edwardsville
Skitka, Linda Jean psychology educator
Teralandur, Parthasarathy Krishnaswamy audiologist

Elgin
Alvarez, Guillermo A. research scientist
Arndt, Bruce Allen mechanical engineer

Elk Grove Village
DeAngelis, Catherine D. pediatrics educator

Elmhurst
Balcerzak, Marion John mechanical engineer
Betinis, Emanuel James physics and mathematics educator
Fornatto, Elio Joseph otolaryngologist, educator
Muellner, William Charles computer scientist, physicist
Simpson, Gene Milton, III mechanical engineer, manufacturing engineer

Evanston
Achenbach, Jan Drewes engineer
Basolo, Fred chemistry educator
Bowman, Monroe Bengt architect
Braeutigam, Ronald Ray economics educator
Brown, James Allison anthropology educator
Chang, R. P. H. materials science educator
Cheng, Herbert Su-Yuen mechanical engineering educator
Cohen, Jerome Bernard materials science educator
Colton, Frank Benjamin retired chemist
Dau, Peter Caine neurologist, immunologist, educator
Dutta, Pulak physicist, educator
Galbreath, Gary John biology educator, researcher
Garg, Anupam K. physicist
Goldberg, Erwin biochemistry educator
Howard, Kenneth Irwin psychology educator
Ibers, James Arthur chemist, educator
Ionescu Tulcea, Cassius research mathematician, educator
Keer, Leon Morris engineering educator
Klotz, Irving Myron chemist, educator
Lambert, Joseph Buckley chemistry educator
Letsinger, Robert Lewis chemistry educator
Lueptow, Richard Michael mechanical engineer, educator
Marks, Tobin Jay chemistry educator
Matalon, Moshe applied mathematician, educator
Matkowsky, Bernard Judah applied mathematician, educator
Meshii, Masahiro materials science educator
Offner, Franklin Faller biophysics educator
Olmstead, William Edward mathematics educator
Petrich, Mark Anton chemical engineer, educator
Pople, John Anthony chemistry educator
Quintanilla, Antonio Paulet physician, educator
Rath, Gustave J(oseph) industrial engineering educator, psychologist
Reiter, Stanley economist, educator
Revelle, William Roger psychology educator
Rutledge, Janet Caroline electrical engineer, educator
Sachtler, Wolfgang Max Hugo chemistry educator
Saxena, Renu medical physicist
Schneider-Criezis, Susan Marie architect
Schonfeld, Eugene Paul investment company executive, software designer
Shah, Surendra Poonamchand engineering educator, researcher
Taflove, Allen electrical engineer, educator, researcher, consultant
Teng, Mao-Hua materials scientist, researcher
Udler, Dmitry physicist, educator
Vaynman, Semyon materials scientist
Weertman, Johannes materials science educator

Flora
Gill, Henry Leonard civil engineer, consultant

Forest Park
Johnson, Calvin Keith research executive, chemist
Orland, Frank Jay oral microbiologist, educator

Fox Lake
Diprizio, Rosario Peter mathematics educator

Freeport
Fitzsimmons, John Patrick electrical engineer
Lockwood, John Patterson electrical engineer
Trickel, Neal Edward chemist, chemical researcher

Galesburg
Brodl, Mark Raymond biology educator
Kooser, Robert Galen chemical educator

Glen Ellyn
Agruss, Neil Stuart cardiologist
Kelada, Nabih Philobbos chemist, consultant
Shah, Nirodh nuclear engineer

Glencoe
Taylor, Roy Lewis botanist

Glenview
Bell, Charles Eugene, Jr. industrial engineer
Hutter, Gary Michael environmental engineer
Logani, Kulbhushan Lal civil and structural engineer
Savic, Stanley Dimitrius physicist
Slayton, Ransom Dunn engineer, consultant
Sowa, Paul Edward research engineer

Godfrey
McDaniels, John Louis mathematics educator

Grayslake
Dulmes, Steven Lee computer science educator
Rieken, Danny Michael naval officer

Gurnee
Krueger, Darrell George *nuclear power industry consultant*
Vandevender, Robert Lee, II *nuclear engineering consultant*

Harvey
Heilicser, Bernard Jay *emergency physician*
Liem, Khian Kioe *medical entomologist*

Hazel Crest
Prentice, Robert Craig *cardiologist*

Highland Park
Dobkin, Irving Bern *entomologist, sculptor*

Hines
Kanofsky, Jeffrey Ronald *physician, educator*
Palmer, Martha Jane *computer specialist*
Trimble, John Leonard *sensor psychophysicist, biomedical engineer*
Zvetina, James Raymond *pulmonary physician*

Hinsdale
Karplus, Henry Berthold *physicist, research engineer*
Kazan, Robert Peter *neurosurgeon*
Martin, Jeffrey Alan *chemical company executive*
Morello, Josephine A. *microbiology educator, pathology educator*
Robertson, Abel L., Jr. *pathologist*

Homewood
Grunwald, Arnold Paul *communications executive, engineer*

Island Lake
O'Day, Kathleen Louise *food products executive*

Itasca
Wittenburg, Robert Charles *industrial engineer*

Jacksonville
Hainline, Adrian, Jr. *biochemist*

Joliet
Nazos, Demetri Eleftherios *obstetrician, gynecologist, medical facility executive*

Kankakee
Armstrong, Douglas *organic chemist, educator*
Schroeder, David Harold *health care facility executive*
Smith, Charles Hayden *utilities executive*

La Grange
Cooke, Steven John *scientist, chemical engineer, consultant*

La Grange Park
Webster, Lois Shand *association executive*

Lake Bluff
Fortuna, William Frank *architect, architectural engineer*
Kelly, Daniel John *physician*

Lake Forest
Davidson, Richard Alan *data communications company executive*
Lambert, John Boyd *chemical engineer, consultant*

Lake Zurich
Grychowski, Jerry Richard *mechanical engineer*

Lanark
Gray, Gary Gene *ecologist, educator*

Lansing
Mandich, Nenad Vojinov *chemical industry executive*

Lemont
Haupt, H. James *mechanical design engineer*
Katz, Joseph Jacob *chemist, educator*
Melnikov, Paul *analytical chemist, instrumentation engineer*
Williams, Jack Marvin *chemist*

Lena
Vickery, Eugene Livingstone *retired physician, writer*

Libertyville
Burrows, Brian William *research and development manufacturing executive*
Mishra, Ajay Kumar *software engineer*
Munson, Norma Frances *biologist, ecologist, nutritionist, educator*
Nichols, Thomas Robert *biostatistician, consultant*

Lincolnshire
West, Dennis Paul *pharmacologist, pharmacist, educator*

Lisle
Wouch, Gerald *materials scientist*

Lombard
Papakyriakou, Michael John *biomechanical engineer*
Velardo, Joseph Thomas *molecular biology and endocrinology educator*

Long Grove
Dajani, Esam Zapher *pharmacologist*
Davé, Vipul Bhupendra *polymer engineer*

Loves Park
Lyon, Roger Wayne *information scientist*
Pearson, Roger Alan *chemical engineer*

Macomb
Anderson, Richard Vernon *ecology educator, researcher*
Harris, Karen L. *psychologist*
Rao, Vaman *economics educator*
Sather, J. Henry *biologist*
Stidd, Benton Maurice *biologist, educator*

Marengo
Jones, Jack Hugh *applications engineer, educator*

Maywood
Albala, David Mois *urologist, educator*
Amero, Sally Ann *molecular biologist, researcher*
Gamelli, Richard L. *surgeon, educator*
Haschke, Paul Charles *analytical chemist*
Kovacs, Elizabeth J. *medical educator*
McNulty, John Alexander *anatomy educator*
Relwani, Nirmalkumar Murlidhar (Nick Relwani) *mechanical engineer*
Schultz, Richard Michael *biochemistry educator*
Van De Kar, Louis David *pharmacologist, educator*

McHenry
Sturm, Richard E. *occupational physician*

Moline
Milas, Robert Wayne *neurosurgeon*
Stowe, David Henry, Jr. *agricultural and industrial equipment company executive*

Morris
Mirabella, Francis Michael, Jr. *polymer scientist*

Mount Prospect
Basar, Ronald John *research engineer, engineering executive*
Breitsameter, Frank John *safety engineer*
Rasmussen, James Michael *research mechanical engineer, inventor*
Scott, Norman L. *engineering consultant*

Mount Vernon
Young, Monica Dawn *electrical engineer*

Murphysboro
McCormack, Robert Paul *environmental engineer*

Naperville
Carrera, Martin Enrique *research scientist*
Craigo, Gordon Earl *engineer*
Dieterle, Robert *chemist*
Fields, Ellis Kirby *research chemist*
Furchtgott, David Grover *computer engineer*
Geisel, Charles Edward *industrial engineer, consultant*
Hacker, David Solomon *chemical engineer, researcher*
Harms, David Jacob *agricultural consultant*
Hauptmann, Randal Mark *molecular biologist*
Koeppe, Eugene Charles, Jr. *electrical engineer*
Kopala, Peter Steven *mechanical engineer*
Kurth, Paul DuWayne *biotechnical services executive*
Lin, Chi-Hung *chemical engineer*
Meyer, Delbert Henry *organic chemist, researcher*
Narutis, Vytas *chemist, researcher*
Niebuhr, Christopher *chemical engineer*
Pector, Scott Walter *telecommunications engineer*
Sellers, Lucia Sunhee *systems engineer*
Shannon, James Edward *water chemist, consultant*
Sherren, Anne Terry *chemistry educator*
Tucker, Beverly Sowers *information specialist*
Vora, Manu Kishandas *chemical engineer, quality consultant*
Weinstein, David Ira *industrial chemist*
Wiatr, Christopher L. *microbiologist*
Yarger, James Gregory *technology company regulatory officer*
Zigament, John Charles *mechanical engineer*

Niles
Koci, Henry James *manufacturing company executive*
Parikh, Dilip *quality assurance engineer*
Rasouli, Firooz *chemical engineer, researcher*

Normal
Anderson, Roger Clark *biology educator*
Morse, Philip Dexter, II *chemist, educator*
Preston, Robert Leslie *cell physiologist, educator*
Yin, Raymond Wah *radiologist*
Young, Robert Donald *physicist, educator*

North Chicago
Burnham, Duane Lee *pharmaceutical company executive*
Bush, Eugene Nyle *pharmacologist, research scientist*
Carney, Ronald Eugene *chemist*
Chu, Alexander Hang-Torng *chemical engineer*
Hindo, Walid Afram *radiology educator, researcher*
Kim, Yoon Berm *immunologist, educator*
McCandless, David Wayne *neuroscientist, anatomy educator*
Nair, Velayudhan *pharmacologist, medical educator*
Sapienza, Anthony Rosario *physician, educator, dean ambulatory facilities*
Schwartz, Robert David *fermentation microbiologist, bioengineer*
Thompson, Richard Edward *biochemist*
Walters, D. Eric *biochemistry educator*
Weil, Max Harry *physician, medical educator, medical scientist*
Winkler, Martin Alan *biochemist*

Northbrook
Dout, Anne Jacqueline *manufacturing company executive*
Hamilton, Robert Burns *civil engineer*
Polsky, Michael Peter *mechanical engineer*
Williams, David Allan *dentist, educator*

Northfield
Gunderson, Edward Lynn *environmental engineer*
Rockwell, Ned M. *chemical engineer*

O'Fallon
Jenner, William Alexander *meteorologist, educator*

Oak Brook
Armbruster, Walter Joseph *foundation administrator*
Haupt, Carl P. *retail drugs executive*
Maides-Keane, Shirley Allen *psychologist*
Merola, Raymond Anthony *engineering executive*

Oak Forest
Kogut, Kenneth Joseph *consulting engineer*
Lekberg, Robert David *chemist*

Oak Lawn
Byrnes, Michael Francis *podiatrist*
Karr, Joseph Peter *podiatrist*
Rathi, Manohar *neonatologist, pediatrician*

Oak Park
Brackett, Edward Boone, III *orthopedic surgeon*
Golden, Leslie Morris *software development company executive*
Jabagi, Habib Daoud *consulting civil engineer*
Schoen, Robert Dennis *civil engineer*
Vandervoort, Kurt George *physicist*
Worley, Marvin George, Jr. *architect*

Olympia Fields
Kasimos, John Nicholas *pathologist*

Oneida
Lawson, Larry Dale *environmental consultant*

Orland Park
English, Floyd Leroy *telecommunications company executive*
Germino, Felix Joseph *chemist, research-development company executive*

Palatine
Novak, Robert Louis *civil engineer, pavement management consultant*
Rhoades, Douglas Duane *chemical engineer*

Palos Heights
Wolff, Robert John *biology educator*

Palos Hills
Maciulis, Linda S. *computer coordinator, consultant*

Park Forest
Orr, Marcia *child development researcher, child care consultant*

Park Ridge
Darling, Cheryl MacLeod *health facility administrator, researcher*
Kleckner, Dean Ralph *trade association executive*
Manzi, Joseph Edward *construction executive*
Tomaszkiewicz, Francis Xavier *imaging technology educator*

Pekin
Petricola, Anthony John *chemical engineer*

Peoria
Chamberlain, Joseph Miles *astronomer, educator*
Cunningham, Raymond Leo *research chemist*
Ehmke, Dale William *agriculturist*
Fites, Donald Vester *tractor company executive*
Herrmann, Judith Ann *microbiologist*
King, Jerry Wayne *research chemist*
Kurtzman, Cletus Paul *microbiologist*
Lin, Shundar *sanitary engineer*
Okamura, Kiyohisa *mechanical engineer, educator*
Shareef, Iqbal *mechanical engineer, educator*
Watkins, George M. *surgeon, educator*

Plainfield
Martens, Frederick Hilbert *nuclear engineer*

Prophetstown
Sanders, Gary Glenn *electronics engineer, consultant*

Quincy
Del Castillo, Julio Cesar *neurosurgeon*

Rantoul
Valencia, Rogelio Pasco *electronics engineer*

Ringwood
Stresen-Reuter, Frederick Arthur, II *chemical company communications executive*

Robinson
Garrett-Perry, Nanette Dawn *chemical engineer*

Rock Island
De Vos, Alois J. *civil engineer*
Forlini, Frank John, Jr. *cardiologist*
Hohl, Martin D. *electrical engineer*
Johnson, George Edwin *hydraulic engineer*
Poppen, Andrew Gerard *environmental engineer*

Rockford
Bixby, Mark Ellis *city official*
Bradley, Charles MacArthur *architect*
Casagranda, Robert Charles *industrial engineer*
Gaylord, Edson I. *manufacturing company executive*
Ostrom, Charles Curtis *financial consultant, former military officer*
Tebrinke, Kevin Richard *manufacturing engineer*

Rolling Meadows
Wyslotsky, Ihor *engineering company executive*

Roscoe
Jacobs, Richard Dearborn *consulting engineering firm executive*

Rosemont
Cisko, George Joseph, Jr. *applied mechanics engineer, research lab manager*
Martin, Scott Lawrence *psychologist*

Round Lake
Yeung, Tin-Chuen *pharmacologist*

Saint Charles
Haugen, Robert Kenneth *product developer*

Sandwich
Petridis, Petros Antonios *electrical engineer*

Sauget
Baltz, Richard Arthur *chemical engineer*

Savoy
Ridgway, Marcella Davies *veterinarian*

Schaumburg
Galvin, Robert W. *electronics executive*
Kenig, Noe *electronics company executive*
Kennedy, Patrick Michael *fire analyst*
Loucks, Eric David *water resources engineer, consultant*

Scott AFB
Gibson, David Allen *civil engineer, career officer*

Skokie
Chang, Shi-Kuo *electrical engineering and computer science educator, novelist*
Corley, William Gene *engineering research executive*
Gutterman, Milton M. *operations research analyst*
Heuer, Margaret B. *data processing coordinator*
Lavenda, Nathan *physiology educator*
Nisperos, Arturo Galvez *engineering geologist, petrographer*
Russell, Henry George *structural engineer*
Siegal, Burton Lee *product designer, consultant*
Stittsworth, James Dale *neuroscientist*

South Holland
Poprick, Mary Ann *psychologist*

Spring Grove
Durrett, Andrew Manning *industrial designer*

Springfield
Albright, Deborah Elaine *emergency physician*
Amador, Armando Gerardo *medical educator*
Aylward, Glen Philip *psychologist*
Ballenger, Hurley René *electrical engineer*
Campbell, Kathleen Charlotte Murphey *audiology educator*
Feldman, Bruce Alan *psychiatrist*
Fields, Joseph Newton, III *oncologist*
Gallina, Charles Onofrio *nuclear regulatory official*
Hahin, Christopher *metallurgical engineer, corrosion engineer*
Henebry, Michael Stevens *toxicologist*
Henry, Theodore Lynn *nuclear scientist, educator*
Khardori, Nancy *infectious disease specialist*
Lyons, John Rolland *civil engineer*
Munyer, Edward A. *zoologist, museum adminstrator*
Reed, John Charles *chemical engineer*
Somani, Satu Motilal *pharmacologist, toxicologist, educator*
Stonecipher, Larry Dale *mathematics educator*

Streamwood
Zanbak, Caner *mining engineer, geologist*

Urbana
Anastasio, Thomas Joseph *neuroscientist, educator, researcher*
Aref, Hassan *fluid mechanics educator*
Assanis, Dennis N. (Dionissios Assanis) *mechanical engineering educator*
Axford, Roy Arthur *nuclear engineering educator*
Azzi, Daniel W. *mathematician*
Bahr, Janice Mary *reproductive physiologist*
Basar, Tangul Ünerdem *electrical engineering educator, researcher*
Bayne, James Wilmer *mechanical engineering educator*
Beak, Peter Andrew *chemistry educator*
Bergman, Lawrence Alan *engineering educator*
Briskin, Donald Phillip *biochemist*
Brown, Theodore Lawrence *chemistry educator*
Buckius, Richard O. *mechanical and industrial engineering educator*
Burkholder, Donald Lyman *mathematician, educator*
Burton, Rodney Lane *engineering educator, researcher*
Chang, Ruey-Jang *life science researcher*
Chao, Bei Tse *mechanical engineering educator*
Chato, John Clark *mechanical engineering educator*
Chiang, Tai-Chang *physics educator*
Choe, Won-Ho (Wayne Choe) *plasma physicist*
Crofts, Antony Richard *biophysics educator*
Curtin, David Yarrow *chemist, educator*
Cusano, Cristino *mechanical engineer, educator*
Dantzig, Jonathan A. *mechanical engineer, educator*
Devadoss, Chelladurai *physical chemist*
DeVor, Richard Earl *mechanical and industrial engineering educator*
Dovring, Folke *land economics educator, consultant*
Drickamer, Harry George *retired chemistry educator*
Ducoff, Howard S. *radiation biologist*
Dutton, J. Craig *mechanical engineering educator*
Edelsbrunner, Herbert *computer scientist, mathematician*
Eden, James Gary *electrical engineering and physics educator, researcher*
Engelbrecht, Richard Stevens *environmental engineering educator*
Fossum, Robert Merle *mathematician, educator*
Gaskins, H. Rex *animal sciences educator*
Ginsberg, Donald Maurice *educator, physicist*
Goldbart, Paul Mark *theoretical physicist, educator*
Govindjee *biophysics and biology educator*
Greene, Laura Helen *physicist*
Gruebele, Martin *chemistry educator*
Gutowsky, H. S. *chemistry educator*
Hagberg, Daniel Scott *ceramic engineer*
Hager, Lowell Paul *biochemistry educator*
Hein, Ilmar Arthur *electrical engineer*
Henson, C. Ward *mathematician, educator*
Holonyak, Nick, Jr. *electrical engineering educator*
Holt, Donald A. *university administrator, agronomist, consultant, researcher*
Horwitz, Alan Fredrick *cell and molecular biology educator*
Huang, Zhi-Yong *honey bee biologist*
Hunt, Donnell Ray *agricultural engineering educator*
Iben, Icko, Jr. *astrophysicist, educator*
Isaacson, Richard Evan *microbiologist*
Jakobsson, Eric Gunnar *biophysicist, educator*
Jenkins, William Kenneth *electrical engineering educator*
Jonas, Jiri *chemistry educator*
Jones, Benjamin Angus, Jr. *retired agricultural engineering educator, administrator*
Jones, Robert Lewis *soil mineralogy and ecology educator*
Katzenellenbogen, John Albert *chemistry educator*
Klein, Miles Vincent *physics educator*
Klemperer, Walter George *chemistry educator, researcher*
Korban, Schuyler Safi *plant geneticist*
Larson, Carl Shipley *engineering educator, consultant*
Larson, Reed William *psychologist, educator*
Lauterbur, Paul C(hristian) *chemistry educator*
Lee, Ki Dong *aeronautical engineer, educator*
Linowes, David Francis *political economist, educator*
Liu, Zi-Chao *aerospace engineering educator*

Lo, Kwok-Yung *astronomer*
Lyding, Joseph William *electrical and computer engineer*
Mazumder, Jyotirmoy *mechanical and industrial engineering educator*
Miley, George Hunter *nuclear engineering educator*
Minear, Roger Allan *chemist, educator*
Morkoc, Hadis *electrical engineer, educator*
Nayfeh, Munir Hasan *physicist*
Ngai, Ka-Leung *biochemist, researcher*
Peters, James Empson *mechanical and industrial engineering educator*
Rebeiz, Constantin Anis *plant physiology educator*
Rich, Robert F. *political sciences educator, science administrator*
Rockett, Angus Alexander *materials science educator*
Schmidt, Stephen Christopher *agricultural economist, educator*
Schweizer, Kenneth Steven *physics educator*
Shurtleff, Malcolm C. *plant pathologist, consultant, educator, extension specialist*
Splittstoesser, Walter Emil *plant physiologist*
Stork, Wilmer Dean, II *physical chemist, researcher*
Stout, Glenn Emanuel *water resources center administrator*
Struble, Leslie Jeanne *civil engineer, educator*
Taylor, Henry L. *aerospace psychologist, educator*
Trigger, Kenneth James *manufacturing engineering educator*
Vakakis, Alexander F. *mechanical engineering educator*
Veeraraghavan, Dharmaraj Tharuvai *materials scientist, researcher*
Visek, Willard James *nutritionist, animal scientist, physician, educator*
Walker, John Scott *mechanical engineering educator*
Wert, Charles Allen *metallurgical and mining engineering educator*
White, Scott Ray *engineering educator*
White, W(illiam) Arthur *geologist*
Wolynes, Peter Guy *chemistry researcher, educator*
Xu, Dong *biophysicist, researcher*
Yoerger, Roger Raymond *agricultural engineer, educator*
Zimmerman, Steven Charles *chemistry educator*

Vernon Hills
Raisman, Allan Leslie *food products executive*
Wikarski, Nancy Susan *information technology executive*

Villa Park
O'Leary, Dennis Sophian *medical organization executive*
Wiede, William, Jr. *chemical engineer, consultant*

Washington
Hallinan, John Cornelius *mechanical engineering consultant*

Waukegan
Curtis, Clark Britten *software engineer*
Dayal, Sandeep *marketing professional*
Weisz, Reuben R. *neurology educator*

West Chicago
Jeppesen, C. Larry *lighting company executive*

Westchester
Hernandez, Medardo Concepcion *chemist*

Western Springs
Swiatek, Kenneth Robert *neuroscientist*

Westmont
Forbes, Bo Crosby *clinical psychologist*
Jones, Dale Leslie *nuclear engineer*
Kudrna, Frank Louis, Jr. *civil engineer, consultant*
McConnell, Patricia Ann *health facility administrator*
McIntosh, Don Leslie *electrical engineer*
Mendelsohn, Avrum Joseph *psychologist*

Wheaton
Bogdonoff, Maurice Lambert *physician*
El-Moursi, Houssam Hafez *civil engineer*
Sloan, Michael Lee *physics and computer science educator, author*

Willowbrook
Rothman, Alan Bernard *consultant, materials and components technologist*

Wilmette
Hamilton, Wallis Sylvester *hydraulic engineer, consultant*
Muhlenbruch, Carl W. *civil engineer*
Veneziano, Philip Paul *biologist, educator*

Wilmington
Jackson, Bennie, Jr. *nuclear engineer*

Winnetka
Seymour, Frederick Prescott, Jr. *industrial engineer, consultant*
Weber, John Bertram *architect*

Wood River
Stevens, Robert Edward *engineering company executive*

Woodstock
Leibhardt, Edward *optics industry professional*
McKittrick, Philip Thomas, Jr. *analytical chemist*

Zion
Scharping, Brian Wayne *mechanical engineer*
Solomon, Patrick Michael *mechanical engineer*

INDIANA

Anderson
Panchanathan, Viswanathan *product development specialist*

Angola
Lin, Ping-Wha *educator*

Avilla
Sneary, Max Eugene *physician*

Bloomington
Anderson, Brenda Jean *biological psychologist*
Caldwell, Lynton Keith *social scientist, educator*
Chisholm, Malcolm Harold *chemistry educator*
Conneally, P. Michael *medical educator*
Davidson, Ernest Roy *chemist, educator*
Gest, Howard *microbiologist, educator*
Hammel, Harold Theodore *physiology and biophysics educator, researcher*
Hieftje, Gary Martin *analytical chemist, educator*
Hofstadter, Douglas Richard *cognitive, computer scientist, educator*
Johnson, Sidney Malcolm *foreign language educator*
Kohr, Roland Ellsworth *hospital administrator*
Magnus, Philip Douglas *chemistry educator*
Mead, Sean Michael *anthropological researcher, consultant*
Novotny, Milos V. *chemistry educator*
Parmenter, Charles Stedman *chemistry educator*
Pollock, Robert Elwood *nuclear physicist*
Purdom, Paul Walton, Jr. *computer scientist*
Schwartz, Drew *geneticist, educator*
Szymanski, John James *physicist, educator*
Townsend, James Tarlton *psychologist*

Butler
Ford, Lee Ellen *scientist, educator, retired lawyer*

Carbon
Robinson, Glenn Hugh *soil scientist*

Carmel
Haslanger, Martin Frederick *pharmaceutical industry professional, researcher*
Roche, James Richard *pediatric dentist, university dean*

Chesterton
Wiemann, Marion Russell, Jr. *retired executive, biologist, microscopist*

Clarksville
Mu, Eduardo *electrical engineer*

Columbus
Bedapudi, Prakash *aerospace engineer*
Hartley, James Michaelis *aerospace systems, printing and hardwood products manufacturing executive*
Kamo, Roy *engineering company executive*
Kubo, Isoroku *mechanical engineer*
Lucke, John Edward *mechanical engineer*
Totten, Gary Allen *spectroscopist*

Crane
Thomas, Greg Hamilton *electronics engineer*
Waggoner, Susan Marie *electronics engineer*

Crown Point
Lee, Robert Jeffrey *municipal utility professional*

Danville
Shartle, Stanley Musgrave *consulting engineer, land surveyor*

Decatur
Coalson, James A. *grain company executive, researcher*

East Chicago
Hughes, Ian Frank *steel company executive*

Elkhart
Arlook, Theodore David *dermatologist*
Atchison, Arthur Mark *industrial, research and development engineer*
Byrd, William Garlen *clinical pharmacist, medical researcher*
Chism, James Arthur *information systems executive*
Drzewiecki, David Samuel *mechanical engineer*
Free, Alfred Henry *clinical chemist, consultant*
Free, Helen M. *chemist, consultant*
Rogers, Robert Wayne *electronics engineer*

Evansville
Hartsaw, William O. *mechanical engineering educator*
Knott, John Robert *mathematics educator*
Moody, Brian Wayne *chemist*

Fort Wayne
Beineke, Lowell Wayne *mathematics educator*
Ferguson, Susan Katharine Stover *nurse, psychotherapist, consultant*
Frantz, Dean Leslie *psychotherapist*
Gillespie, Robert Bruce *biology educator*
Lyons, Jerry Lee *mechanical engineer*
Mills, Rodney Daniel *engineering company executive*
Richardson, Joseph Hill *physician, educator*
Szuhaj, Bernard Francis *food research director*
Taylor, Mike Allen *computer systems analyst*
Weatherford, George Edward *civil engineer*
Weinswig, Shepard Arnold *optical engineer*
Zepp, Lawrence Peter *mechanical engineer*

Franklin
Launey, George Volney, III *economics educator*

Garrett
Baker, Suzon Lynne *mathematics educator*

Gary
Echtenkamp, Stephen Frederick *biomedical researcher*
Meyerson, Seymour *retired chemist*
Stephens, Paul Alfred *dentist*

Goshen
Heap, James Clarence *retired mechanical engineer*

Granger
Chmiel, Chester T. *adhesive chemist, consultant*

Greenfield
Kuyatt, Brian Lee *toxicologist, scientific systems analyst*

Hammond
Neff, Gregory Pall *manufacturing engineering educator, consultant*
Vojcak, Edward Daniel *metallurgist*

Hanover
Krantz, John Howell *psychology educator*

Hobart
Seeley, Mark *agronomist*

Huntingburg
Rossmann, Charles Boris *obstetrician/gynecologist*
Von Taaffe-Rossmann, Cosima T. *physician, writer, inventor*

Indianapolis
Alford, Joseph Savage, Jr. *chemical engineer, research scientist*
Allen, Stephen D(ean) *pathologist, microbiologist*
Amundson, Merle Edward *pharmaceuticals executive*
Arps, David Foster *electronics engineer*
Ashmore, Robert Winston *computational pharmacologist*
Bannister, Lance Terry *applications engineer*
Belagaje, Rama M. *biotechnology scientist*
Besch, Henry Roland, Jr. *pharmacologist, educator*
Bonate, Peter Lawrence *pharmacologist*
Brady, Mary Sue *pediatric dietitian, educator*
Broxmeyer, Hal Edward *medical educator*
Campbell, Judith Lowe *child psychiatrist*
Caraher, Michael Edward *systems analyst*
Carr, Floyd Eugene *sales engineer*
Chern, Jiun-Der *medical physicist*
Chernish, Stanley Michael *physician*
Childers, Richard Herbert, Jr. *chemist*
Christian, Joe Clark *medical genetics researcher, educator*
Cleary, Robert Emmet *gynecologist, infertility specialist*
Cliff, Johnnie Marie *mathematics and chemistry educator*
Cones, Van Buren *electronics engineer, consultant*
Cooley, Rick Eugene *chemist*
Daily, William Allen *retired microbiologist*
Davis, Robert Drummond, Sr. *chemist, researcher*
DeLong, Allyn Frank *biochemist*
Dere, Willard Honglen *internist, educator*
Desai, Mukund Ramanlal *research and development chemist*
Dillon, Howard Burton *civil engineer*
Ehringer, William Dennis *membrane biophysicist*
Eigen, Howard *pediatrician, educator*
Escobar, Luis Fernando *pediatrician, geneticist*
Evans, Richard James *mechanical engineer*
Faulk, Ward Page *immunologist*
Fer, Ahmet F. *electrical engineer, educator*
Fife, Wilmer Krafft *chemistry educator*
Gable, Robert William, Jr. *aerospace engineer*
Gehlert, Donald Richard *pharmacologist*
Gregory, Richard Lee *immunologist*
Hathaway, David Roger *physician, medical educator*
Henry, Matthew James *biochemist, plant pathologist*
Hildebrand, William Clayton *chemist*
Hurley, Thomas Daniel *biochemist*
Jones, James Lamar *psychologist*
Jones, Katharine Jean *research physicist*
Kauffman, Raymond Francis *biochemical pharmacologist*
Kwon, Byoung Se *geneticist, educator*
Lahiri, Debomoy Kumar *molecular neurobiologist, educator*
Lin, Zhen-Biao *acoustical and electrical engineer*
Lindseth, Richard Emil *orthopaedic surgeon*
Liu, Pingyu *physicist, educator*
Long, Eric Charles *biochemist*
Mair, Bruce Logan *interior designer, company executive*
Manders, Karl Lee *neurosurgeon*
Marlin, Donnell Charles *dental educator*
Marshall, Frederick Joseph *retired research chemist*
McBride, Angela Barron *nursing educator*
McKain, Theodore F. *mechanical engineer*
Merritt, Doris Honig *pediatrics educator*
Miyamoto, Richard Takashi *otolaryngologist*
Mundell, John Anthony *environmental engineer, consultant*
Norins, Arthur Leonard *physician, educator*
Nurnberger, John I., Jr. *psychiatrist, educator*
Oakeson, David Oscar *aerospace engineer*
Reilly, Jeanette P. *clinical psychologist*
Reilly, Peter C. *chemical company executive*
Roberts, Wilbur Eugene *dental educator, research scientist*
Rohn, Robert Jones *internist, educator*
Ross, Edward *cardiologist*
Schaible, Robert Hilton *biologist*
Scopatz, Stephen David *engineering executive, educator*
Scott, William Leonard *research chemist, educator*
Sowers, Edward Eugene *lawyer*
Stephens, Thomas Wesley *biochemist*
Stookey, George Kenneth *research institute administrator, dental educator*
Sullivan, John Lawrence, III *psychiatrist*
Svoboda, Gordon Howard *pharmacognosist, consultant*
Thomas, Jerry Arthur *soil scientist*
Todd, Zane Grey *utility executive*
Torres-Olivencia, Noel R. *aeronautical engineer*
Traicoff, Ellen Braden *psychologist*
Vlach, Jeffrey Allen *environmental engineer*
Wallace, F. Blake *aerospace executive, mechanical engineer*
Wallace, Robert Eugene, II *rail transit systems executive*
Walther, Joseph Edward *health facility administrator, retired physician*
Watt, Jeffrey Xavier *mathematics sciences educator, researcher*
Weaver, Michael Anthony *mining engineer, consultant*
Weber, George *oncology and pharmacology researcher, educator*
Weinberger, Myron Hilmar *medical educator*
Wong, David T. *biochemist*
Yeager, Hal K. *biomedical engineer*

Jasper
Lents, Thomas Alan *waste water treatment company executive*
Wendholt, Norman William *civil engineer*

Kendallville
Caldwell, Andrew Brian *quality control engineer*
Shelby, Beverly Jean *quality assurance professional*

Kokomo
Almquist, Donald John *retired electronics company executive*
Kramer, Geoffrey Philip *psychology educator*
Lai, George Ying-Dean *metallurgist*
Miller, Robert Frank *retired electronics engineer, educator*
Shimanek, Ronald Wenzel *engineering executive*
Sissom, John Douglas *systems engineer*

Lafayette
Berman, Michael Allan *biomedical clinical engineer*
Bowers, Conrad Paul *chemist*
Chandrasekaran, Rengaswami *biochemist, educator*
Hamlin, Kurt Wesley *manufacturing engineer*
Mc Laughlin, John Francis *civil engineer, educator*
Mertz, Edwin Theodore *biochemist, emeritus educator*
Ott, Karl Otto *nuclear engineering educator, consultant*
Pyer, John Clayton *analytical chemist*
Speir, Jeffrey Alan *biophysicist*
Weaver, Michael John *chemist, educator*

Lawrenceburg
Wakeman, Thomas George *mechanical engineer*

Marion
Green, Robert Frederick *physician, photographer*
Hall, Charles Adams *infosystems specialist*

Merrillville
Blaschke, Lawrence Raymond *utility company professional*
Chang, Kai Siung *medical physicist*

Michigan City
Mothkur, Sridhar Rao *radiologist*

Morgantown
Jones, Barbara Ewer *school psychologist*

Mount Vernon
Bader, Keith Bryan *chemical engineer*
Sommerfield, Thomas A. *process engineer*

Muncie
Costill, David Lee *physiologist, educator*
Harris, Joseph McAllister *chemist*
Lang, Patricia Louise *chemistry educator, vibrational spectroscopist*
Mertens, Thomas Robert *biology educator*

New Albany
Baxter, Joseph Diedrich *dentist*

Noblesville
Young, Frederic Hisgin *infosystems executive, data processing consultant*

North Vernon
Karkut, Richard Theodore *clinical psychologist*
Siener, Joseph Frank *utilities supervisor*

Notre Dame
Bender, Harvey A. *biology educator*
Berry, William Bernard *engineering educator, researcher*
Buyer, Linda Susan *psychologist, educator*
Carberry, James John *chemical engineer, educator*
Cheng, Minquan *chemical engineer*
Craig, George Brownlee, Jr. *entomologist*
Fehlner, Thomas Patrick *chemistry educator*
Fraser, Malcolm James, Jr. *biological sciences educator*
Gad-el-Hak, Mohamed *aerospace and mechanical engineering educator, scientist*
Gutschick, Raymond Charles *geology educator, researcher, micropaleontologist*
Hayes, Robert Green *chemical educator, researcher*
Huber, Paul William *biochemistry educator, researcher*
Manier, August Edward *philosophy of biology educator*
McLinden, James Hugh *molecular biologist*
O'Meara, Onorato Timothy *university administrator, mathematician*
Sain, Michael Kent *electrical engineering educator*
Schuler, Robert Hugo *chemist, educator*
Skaar, Steven Baard *engineering educator*
Su, Yali *chemist*
Thomas, John Kerry *chemistry educator*
Trozzolo, Anthony Marion *chemistry educator*
Varma, Arvind *chemical engineering educator, researcher*

Pendleton
Leonard, Elizabeth Ann *veterinarian*

Plainfield
Meyer, Duane Russell *civil and cost engineer, consultant*

Princeton
Mullins, Richard Austin *chemical engineer*

Richmond
Sabine, Neil B. *ecology educator*

Rochester
Grooms, John Merril *research and development engineer*

Rockville
Swaim, John Franklin *physician, health care executive*

Schererville
Ontto, Donald Edward *environmental analytical chemist, wastewater treatment consultant*

Scottsburg
Kho, Eusebio *surgeon*

Shelbyville
Deaton, Timothy Lee *computer systems integrator*

South Bend
Apostolides, Anthony Demetrios *economist, educator*

Cottle, Eugene Thomas aerospace engineer
Farkas, Thomas secondary education educator

Terre Haute
Duong, Taihung anatomist
English, Robert Eugene manufacturing engineering educator
Grant, Michael Joseph electrical engineering educator, consultant
Guthrie, Frank Albert chemistry educator
LaMagna, John Thomas chemical engineer
Malooley, David Joseph electronics and computer technology educator
Stoffer, Barbara Jean research laboratory technician
Western, Arthur Boyd, Jr. physics educator

Upland
Whipple, Andrew Powell biology educator

Valparaiso
Cook, Addison Gilbert chemistry educator
Shipley-Phillips, Jeanette Kay aquatic biologist
Tarhini, Kassim Mohamad civil engineering educator

Walton
Chu, Johnson Chin Sheng physician

Warsaw
Bradt, Rexford Hale chemical engineer
Cupp, Jon Michael environmental scientist

West Lafayette
Baird, William McKenzie chemical carcinogenesis researcher, biochemistry educator
Baumgardt, Billy Ray university official, agriculturist
Bement, Arden Lee, Jr. engineering educator
Brown, Herbert Charles chemistry educator
Chao, Kwang-Chu chemical engineer, educator
Cochrane, Thomas Thurston tropical soil scientist, agronomist
Cohen, Raymond mechanical engineer, educator
Connor, John Murray agricultural economics educator
Davidson, Terry Lee experimental psychology educator
Delp, Edward John, III electrical engineer, educator
Drnevich, Vincent Paul civil engineering educator
Edwards, Charles Richard entomology and pest management educator
Erickson, Homer Theodore horticulture educator
Farris, Thomas N. engineering educator, researcher
Ferreira, Paulo Alexandre molecular biologist
Gorenstein, David G. chemistry educator
Hall, Stephen Grow research neuroscientist
Hanks, Alan R. chemistry educator
Hayes, John Marion civil engineer
Heister, Stephen Douglas aerospace propulsion educator, researcher
Jagacinski, Carolyn Mary psychology educator
Janick, Jules horticultural scientist, educator
Johannsen, Chris Jakob agronomist, educator, administrator
Johnston, Clifford Thomas soil and environmental chemistry educator
Kampen, Emerson chemical company executive
Kessler, Wayne Vincent health sciences educator, researcher, consultant
Kokini, Klod mechanical engineer
Krockover, Gerald Howard science educator
Larkins, Brian Allen botany educator
Leap, Darrell Ivan hydrogeologist
Madanat, Samer Michel civil engineer, educator
Margerum, Dale William chemistry educator
McClelland, Thomas Melville meteorologist, researcher
McLaughlin, Gerald Lee parasitology educator
Monke, Edwin John agricultural engineering educator
Morrison, Harry chemistry educator, university dean
Nelson, Philip Edwin food scientist, educator
Overhauser, Albert Warner physicist
Pritsker, A. Alan B. engineering executive, educator
Ramadhyani, Satish mechanical engineering educator
Richey, Clarence Bentley agricultural engineering educator
Sadeghi, Farshid engineering educator
St. John, Charles Virgil retired pharmaceutical company executive
Sato, Hiroshi materials science educator
Schneegurt, Mark Allen biochemist, researcher
Schwartz, Richard John electrical engineering educator, researcher
Schweickert, Richard Justus psychologist, educator
Sherman, Louis Allen biologist, researcher
Short, Dennis Ray engineering educator
Singh, Rakesh Kumar process engineer, educator
Sivathanu, Yudaya Raju aerospace engineer
Solberg, James Joseph industrial engineering educator
Tacker, Willis Arnold, Jr academic administrator, medical educator, researcher
Tenorio, Manoel Fernando da Mota computer engineering educator
Thomas, Marlin Uluess industrial engineering educator
Thompson, Howard Doyle mechanical engineer
Tiffany, Stephen Thomas psychologist
Viskanta, Raymond mechanical engineering educator
Wasserman, Gerald Steward psychobiology educator
Weiner, Andrew Marc electrical engineering educator, laser researcher
Yang, Henry Tsu Yow dean, engineering educator

Westville
Das, Purna Chandra physics and mathematics educator
Spores, John Michael psychologist

Winchester
Anderson, Gary Alan waste water plant executive

Zionsville
Heck, David Alan orthopaedic surgery educator, mechanical engineering educator

IOWA

Ames
Anderson, Lloyd Lee animal science educator
Bastiaans, Glenn John analytical chemist, researcher
Boylan, David Ray retired chemical engineer, educator
Bremner, John McColl agronomy and biochemistry educator
Brown, Robert Grover engineering educator
Buchele, Wesley Fisher agricultural engineering educator, consultant
Bullen, Daniel Bernard nuclear engineering educator
Cao, Zhijun physicist
Cate, Rodney Michael university dean
Cink, James Henry chemical safety consultant, educator
Clem, John Richard physicist, educator
Colvin, Thomas Stuart agricultural engineer, farmer
Curry, Norval Herbert retired agricultural engineer
Dahiya, Rajbir Singh mathematics educator, researcher
Dayal, Vinay aerospace engineer, educator
DePristo, Andrew E. chemist, educator
DeYong, Gregory Donald chemist
Fox, Karl August economist, eco-behavioral scientist
Franzen, Hugo Friedrich chemistry educator, researcher
Freeman, Albert E. agricultural science educator
Gaertner, Richard Francis manufacturing research center executive
Greve, John Henry veterinary parasitologist, educator
Hallauer, Arnel Roy geneticist
Hammond, Earl Gullette food science educator
Han, Sang Hyun metallurgist
Hanisch, Kathy Ann psychologist
Hill, John Christian physics educator
Honavar, Vasant Gajanan computer scientist, educator
Houk, Robert Samuel chemistry educator
Huston, Jeffrey Charles mechanical engineer, educator
Isely, Duane biology and botany educator
Iversen, James Delano aerospace engineering educator, consultant
Johnson, Howard Paul agricultural engineering educator
Johnson, Lawrence Alan cereal technologist, educator, researcher, administrator
Johnson, Stanley R. economist, educator
Johnson, Willie Roy industrial psychology educator
Keeney, Dennis Raymond soil science educator
Kelly, William Harold physicist, physics educator
Knox, Ralph David physicist
Lane, Orris John, Jr. engineer
Luban, Marshall physicist
Mischke, Charles Russell mechanical engineering educator
Moon, Harley William veterinarian
Moyer, James Wallace biophysicist
Ostenson, Jerome Edward physicist
Owen, Michael agronomist, educator
Porter, Max L. engineering educator
Rieger, Phillip Warren aquatic ecology educator, researcher
Riley, William Franklin mechanical engineering educator
Ruedenberg, Klaus theoretical chemist, educator
Seaton, Vaughn Allen veterinary pathology educator
Stahr, Henry Michael analytical toxicology
Stevens, Mark Gregory immunologist
Sturges, Leroy D. engineering educator
Tabatabai, M. Ali agronomist
Topel, David Glen college dean, animal science educator
Vermeer, Mark Ellis project engineer
Wesley, Irene Varelas research microbiologist
Wilhelm, Harley A. mechanical engineer
Yeung, Edward Szeshing chemist
Young, Jerry Wesley animal nutrition educator

Ankeny
Irwin, Donald Berl psychology educator
Weigel, Ollie J. dentist, mayor

Bettendorf
Heyderman, Arthur Jerome engineer, civilian military employee

Birmingham
Goudy, James Joseph Ralph electronics executive, educator

Cedar Falls
Wiens, Darrell John biologist, educator

Cedar Rapids
Ashbacher, Charles David computer programmer, educator
Bechler, Ronald Jerry structural engineer
Dvorak, Clarence Allen microbiologist
McCall, Daryl Lynn avionics engineer
Rydell, Earl Everett electrical engineer

Center Junction
Antons, Pauline Marie mathematics educator

Clinton
Martin, Dennis Charles dentist

Clive
Iruvanti, Pran Rao endocrine biochemist, researcher

Davenport
Bartlett, Peter Greenough engineering company executive
Bhatti, Iftikhar Hamid chiropractic educator
Lemke, Cindy Ann support center founder and administrator
Sandry, Karla Kay Foreman industrial engineering educator

Decorah
Dengler, Madison Luther psychologist, educator
Voltmer, Michael Dale electric company executive

Des Moines
Canby, Craig Allen anatomy educator
Meetz, Gerald David anatomist
Pandeya, Nirmalendu Kumar plastic and flight surgeon, military officer
Rotert, Kelly Eugene engineer, consultant
Seifert, Robert P. agricultural products company executive
Smith, David Welton mechanical engineer, director research

Dubuque
Kane, Kevin Thomas editor

Schaefer, Joseph Albert physics and engineering educator, consultant

Fairfield
Hagelin, John Samuel theoretical physicist
Perez, Jose Luis computer scientist, engineer, philosopher
Zsigo, Jozsef Mihaly chemist

Grinnell
Campbell, David George ecologist

Hills
Tomlinson, G. Richard industrial engineer

Iowa City
Abboud, Francois Mitry physician, educator
Alipour-Haghighi, Fariborz mechanical engineer
Andreasen, Nancy Coover psychiatrist, educator
Bar, Robert S. endocrinologist
Berg, Mary Jaylene pharmacy educator, researcher
Bertolatus, John Andrew physician, educator
Block, Robert I. psychologist, researcher, educator
Cooper, Reginald Rudyard orthopaedic surgeon, educator
Corson, John Duncan vascular surgeon
Dexter, Franklin anesthesiologist
Eckstein, John William physician, educator
Fellows, Robert Ellis medical educator, medical scientist
Folk, George Edgar, Jr. environmental physiology educator
Folkins, John William speech scientist, educator
Forsythe, Robert Elliott economics educator
Fulton, Alice Bordwell biochemist, educator
Goel, Vijay Kumar biomedical engineer
Goff, Harold Milton chemistry educator
Grassian, Vicki Helene chemistry educator
Hammond, Harold Logan pathology educator, oral pathologist
Haug, Edward Joseph, Jr. mechanical engineering educator, simulation research engineer
Husted, Russell Forest research scientist
Johnson, Eugene Walter mathematician
Koch, Donald LeRoy geologist, state agency administrator
Lee, Shyan Jer physical chemist
Lim, Ramon (Khe-Siong) neuroscience educator
Marshall, Jeffrey Scott mechanical engineer, educator
Milkman, Roger Dawson genetics educator, molecular evolution researcher
Miller, Richard Keith engineering educator
Pennington, Edward Charles medical physicist
Pickett, Jolene Sue aerospace research scientist
Rajagopal, Rangaswamy geography and engineering educator
Reckase, Mark Daniel psychometrician
Routh, Joseph Isaac biochemist
Sharp, Charles Paul electronics technician
Shibata, Erwin Fumio cardiovascular physiologist
Solursh, Michael biology educator, researcher
Stegink, Lewis Dale biochemist, educator
Tye-Murray, Nancy research scientist
Van Allen, James Alfred physicist, educator
Voo, Liming M. biomedical engineer, researcher
Wang, Semyung mechanical engineer, researcher
Weiner, George Jay internist
Wurster, Dale Eric pharmacy educator
Wurster, Dale Erwin pharmacy educator, university dean emeritus
Yorek, Mark Anthony biochemist
Yu, Cong biomedical engineer

Larchwood
Onet, Virginia veterinary parasitologist, researcher, educator

Lidderdale
Hagemann, Dolores Ann water company official

Marion
Stover, Donald Rae software engineering executive, retired

Marshalltown
McCann, Michael John industrial engineer, educator, consultant
Packer, Karen Gilliland cancer patient educator, researcher
Sheeler, John Briggs chemical engineer

Mason City
Chanco, Amado Garcia surgeon
Hughes, Mark Lee pharmacist
Rosenberg, Dale Norman psychology educator

Mount Pleasant
Sandy, Edward Allen obstetrician/gynecologist

Muscatine
Johnson, Donald Lee agricultural materials processing company executive
Stanley, Richard Holt consulting environmental engineer

Newton
Durant, Gerald Wayne materials engineer

North Liberty
Glenister, Brian Frederick geologist, educator

Palo
Martin, Robert Anthony mechanical engineer

Parkersburg
Boukerrou, Lakhdar agricultural researcher

Shenandoah
Elliott, John Earl water plant administrator

Sioux City
Petersen, Perry Marvin agronomist
Spellman, George Geneser, Sr. internist
Uphoff, John Vincent mechanical engineer
Vaught, Richard Loren urologist
Walker, Jimmie Kent mechanical engineer

Tabor
Reese, William Albert, III psychologist

West Des Moines
Brunk, Samuel Frederick oncologist

Wilton
Cronbaugh, Kurt Allen industrial electrician

KANSAS

Atchinson
Chinnaswamy, Rangan cereal chemist

Bonner Springs
Elliott-Watson, Doris Jean psychiatric, mental health and gerontological nurse educator

Burlington
Dingler, Maurice Eugene civil engineer

Emporia
Boor, Myron Vernon psychologist, educator
Schrock, John Richard biology educator

Garden City
Dick, Gary Lowell agricultural consultant, educator

Hamilton
Lockard, Walter Junior petroleum company executive

Hays
Coyne, Patrick Ivan physiological ecologist

Hutchinson
Haag, Joel Edward architect

Independence
Barbi, Josef Walter engineering, manufacturing and export companies executive

Industrial Airport
Hiner, Thomas Joseph tractor manufacturing administrator

Kansas City
Andrews, Glen K. biochemist
Arakawa, Kasumi physician, educator
Cuppage, Francis Edward physician, educator
Dunn, Marvin Irvin physician
Grantham, Jared James nephrologist, educator
Gray, Donald Lee chemist
Grisolia, Santiago biochemistry educator
Hung, Kuen-Shan anatomy educator
Lee, Kyo Rak radiologist
Mathewson, Hugh Spalding anesthesiologist, educator
Moore, Wayne V. pediatrician, educator, endocrinologist
Pierce, John Thomas industrial hygienist, toxicologist
Samson, Frederick Eugene, Jr. neuroscientist, educator
Smith, Donald Dean biologist, educator
Smith, Peter Guy neuroscience educator, researcher
Taylor, Brenda Carol computer specialist

Larned
Zook, Martha Frances Harris retired nursing administrator

Lawrence
Alexander, Byron Allen insect systematist
Armitage, Kenneth Barclay biology educator, ecologist
Bennett, Stephen Christopher biology educator
Binns, William Arthur clinical psychologist
Brehm, Jack Williams social psychologist, educator
Crandall, Christian Stuart social psychology educator
Enos, Paul geologist, educator
Farokhi, Saeed aerospace engineering educator, consultant
Gerhard, Lee Clarence geologist, educator
Green, Don Wesley chemical and petroleum engineering educator
Harmony, Marlin Dale chemistry educator, researcher
Haufler, Christopher Hardin botany educator
Haugh, Dan Anthony mechanical engineer
Hersh, Robert Tweed biology educator
Kwak, Nowhan physics educator
Lane, Meredith Anne botany educator, museum director
McCabe, John Lee engineer, educator, writer
McCabe, Steven Lee structural engineer
Mc Kinney, Ross Erwin civil engineering educator
Merriam, Daniel F(rancis) geologist
Michener, Charles Duncan entomologist, biologist, educator
Moore, Richard Kerr electrical engineering educator
Olea, Ricardo Antonio geological researcher, engineer
Ralston, John Peter theoretical physicist, educator
Roddis, Winifred Mary Kim structural engineering educator
Sanders, Robert B. biochemistry educator
Stella, Valentino John pharmaceutical chemistry educator
Vossoughi, Shapour chemical and petroleum engineering educator

Leawood
Mosher, Alan Dale chemical engineer

Lenexa
Johannes, Richard Dale civil engineering executive
Zhu, Qing software engineer

Manhattan
Acevedo, Edmund Osvaldo physical education educator
Babcock, Michael Ward economics educator
Barkley, Theodore Mitchell biology educator
Bechtel, Donald Bruce biologist, educator, research chemist
Cogley, Allen C. mechanical engineering educator, administrator
Faw, Richard Earl nuclear engineering educator
Fitch, Gregory Kent biologist
Hagen, Lawrence Jacob agricultural engineer
Hahn, Richard Ray academic administrator
Ham, George Eldon soil microbiologist, educator

Hoseney, Russell Carl *chemistry educator*
Hosni, Mohammad Hosein *mechanical engineering educator*
Jardine, Douglas Joseph *plant pathologist, educator*
Johnson, Terry Charles *biologist, researcher*
Johnson, William Howard *agricultural engineer, educator*
Kaufman, Donald Wayne *research ecologist*
Knupfer, Nancy Nelson *computer science educator*
Krishnamoorthi, Ramaswamy *biochemistry educator*
Kyle, Benjamin Gayle *chemical engineer, educator*
Lee, E(ugene) Stanley *industrial engineer, mathematician, educator*
Oehme, Frederick Wolfgang *medical researcher and educator*
Posler, Gerry Lynn *agronomist, educator*
Roberts, Thomas Carrol *management consultant*
Samelson, Franz *psychologist*
Shanteau, James *psychology educator, researcher*
Simons, Gale Gene *nuclear engineering educator, university administrator*
Stalheim-Smith, Ann *biology educator*
Stiefel, Vernon Leo *entomology educator*
Streeter, John Willis *information systems manager*
Troyer, Deryl Lee *life sciences educator*
Vorhies, Mahlon Wesley *veterinary pathologist, educator*
Wang, Xuemin *biochemistry educator*
Zhang, Naiqian *agricultural engineer, educator*

Mission
Mortensen, Kenneth Peter *materials engineer, water treatment engineer*
Van Veldhuizen, Peter Jay *oncologist, hematologist*

North Newton
Lehman, Thomas Alan *chemistry educator*

Olathe
Early, Marvin Milford, Jr. *computer scientist*
Picou, Gary Lee *technical publication editor*
Poston, Ann Genevieve *psychotherapist, nurse*
Ruf, Jacob Frederick *chemical engineer*

Overland Park
Baker, Charles H. *engineering company executive*
Burger, Henry G. *anthropologist, vocabulary scientist, publisher*
Cook, Ronald Lee *petroleum engineer*
Georgiana, John Thomas *electrical engineer*
Hansen, Christian Gregory *architectural engineer*
Hoffman, Jerry Carl *civil engineer*
Hsu, Peter Cheazone *chemical engineer*
Levere, James Gordon *electrical engineer*

Paola
Banks, Robert Earl *family practice physician*

Parsons
Lomas, Lyle Wayne *agricultural research administrator, educator*
Spradlin, Joseph E. *embryologist, psychologist, medical educator*

Pittsburg
Foresman, James Buckey *geologist, geochemist, industrial hygienist*

Pratt
Rice, Ramona Gail *physiologist, psychologist, educator, consultant*

Saint John
Robinson, Alexander Jacob *clinical psychologist*

Salina
Piper, Jon Kingsbury *ecologist, researcher*
Reh, John W. *engineer, consultant*
Selm, Robert Prickett *consulting engineer*
Willis, Shelby Kenneth *civil engineer, consultant*

Shawnee Mission
Boone, David Daniel *electrical engineer*
Dockhorn, Robert John *physician*
Jones, George Humphrey *retired healthcare executive, hospital facilities and communications consultant*
Sternberg, David Edward *psychiatrist*

Stilwell
Fought, Lorianne *chemist*

Topeka
Blue, Jeffrey Kenneth *neuromuscular and skeletal researcher*
Gerard, Mark Edward *mechanical engineer*
Gonzalez, William Joseph *sales executive, geosynthetic engineer*
Hammerschmidt, Ronald Francis *environmental director*
Menninger, William Walter *psychiatrist*
Slemmons, Robert Sheldon *architect*
Spohn, Herbert Emil *psychologist*

University Of Kansas
Humphrey, Philip Strong *university museum director*

Wichita
Alagić, Suad *computer science educator, researcher, consultant*
Arnold, David Dean *electrical engineer*
Bagwell, Kathleen Kay *infosystems specialist*
Cho, Sechin *medical geneticist*
Clark, Susan Matthews *psychologist*
Davis, John Stewart *medical educator, endocrinology researcher*
Dyck, George *psychiatry educator*
Egbert, Robert Iman *electrical engineering educator, academic administrator*
Ellis, David R. *aeronautical researcher*
Green, James Anton *systems engineer*
Greywall, Mahesh Inder-Singh *mechanical engineer*
Gumnick, James Louis *research institute executive*
Haynes, Deborah Gene *physician*
Hoffman, Bernard Douglas, Jr. *petroleum engineer, company executive*
Kice, John Edward *engineering executive*
Mc Kee, George Moffitt, Jr. *civil engineer, consultant*
McVicar, Robert William, Jr. *industrial engineer*
Paarmann, Larry Dean *electrical engineering educator*
Schalon, Charles Lawrence *psychologist*
Wattson, Robert Kean *aeronautical engineer*
Wolf, Mark Alan *chemical engineer*
Zimniski, Stephen Joseph *biologist, educator*

Winfield
Docherty, Robert Kelliehan, III *quality assurance engineer, computer instructor*

KENTUCKY

Ashland
Alexander, William Robert *chemist*
Johnson, Bobby Joe *electronic instrumentation technician*
Kovar, Dan Rada *pharmacist*

Bowling Green
Bryson, Keith *obstetrician/gynecologist*
Hartman, David Robert *chemistry educator*
Marshall, Willis Henry *psychiatrist*

Calvert City
Gray, Carol Hickson *chemical engineer*
Upshaw, Timothy Alan *chemical engineer*

Carrollton
Tatera, James Frank *chemist, process analysis specialist*

Catlettsburg
Fischer, Robert Lee *electrical engineering executive, educator*

Corbin
Steenbergen, Gary Lewis *computer aided design educator*

Crestview Hills
Boehm, Lawrence Edward *psychologist, educator*
Wolff, Ralph Gerald *civil engineer*

Edgewood
Mick, Elizabeth Ellen *medical technologist*

Elkton
Manthey, Frank Anthony *physician, director*

Erlanger
Richmond, Raymond Dean *aerospace engineer*

Frankfort
Franke, George Edward *highway engineer*
Grubbs, Jeffrey Thomas *environmental biologist*

Georgetown
Tieman, Robert Scott *chemist*

Hopkinsville
Spencer, Lewis VanClief *retired physicist*

LaGrange
Leslie, Robert Fremont *mobile testing executive, non-destructive testing inspector*

Lexington
Adams, Larry Dell *mining engineering executive*
Anderson, Timothy Lee *electrical engineer*
Bachas, Leonidas Gregory *chemistry educator*
Butterfield, D. Allan *chemistry educator, university administrator*
DeLuca, Patrick Phillip *pharmaceutical scientist, educator, administrator*
Diana, John Nicholas *physiologist*
Drake, David Lee *engineer*
Dutt, Ray Horn *reproduction physiologist, educator*
Gilliam, M(elvin) Randolph *urologist, educator*
Hauser, Kurt Francis *anatomy and neurobiology educator*
Hodes, Jonathan Ezra *neurosurgeon, educator*
Kang, Bann C. *immunologist*
Knapp, Frederick Whiton *entomologist, educator*
Kraman, Steve Seth *physician, educator*
Lineberry, Gene Thomas *mining engineering educator*
Liu, Yong-Biao *entomologist*
Man, Chi-Sing *mathematician, educator*
Markesbery, William R. *neurology and pathology educator, physician*
Maruyama, Yosh *physician, educator*
Ng, Kwok-Wai *physics educator*
Partington, John Edwin *retired psychologist*
Porter, William Hudson *clinical chemist*
Rangnekar, Vivek Mangesh *molecular biologist, researcher*
Reed, Michael Robert *agricultural economist*
Robertson, Larry Wayne *toxicologist, educator*
Rodriguez, Lorraine Ditzler *biologist, consultant*
Sadler, John Peter *mechanical engineering educator*
Schmitt, Frederick Adrian *gerontologist, neuropsychologist*
Sih, Andrew *biologist, educator*
Spearman, Terence Neil *biochemist*
Standler, Ronald B. *physicist*
Thio, Alan PooAn *chemist*
Walker, John Neal *agricultural engineering educator*

London
Geiger, Daniel Jay *mining engineer*

Louisville
Al-ibrahim, Hussam Sadiq *electrical engineer*
Aronhime, Peter Byron *electrical engineering educator, researcher*
Aronoff, George Rodger *medicine and pharmacology educator*
Atlas, Ronald M. *microbiologist educator, ecologist*
Bhattacherjee, Parimal *pharmacologist*
Bloch, James Phillips *clinical psychologist*
Blumenreich, Martin Sigvart *medical educator*
Colley, Thomas Elbert, Jr. *psychologist*
Cornelius, Wayne Anderson *engineering technology educator, consultant*
Cox, Hollace Lawton *electrical engineering educator*
Danzl, Daniel Frank *emergency physician*
Deshpande, Pradeep Bapusaheb *chemical engineer, educator, consultant*
Downard, Daniel Patrick *electrical engineer, consultant*
Effinger, Charles Edward, Jr. *environmental/energy engineer*
Fleischman, Marvin *chemical and environmental engineering educator*
Frye, Raymond Eugene *geotechnical engineer*
Garcia, Rafael Jorge *chemical engineer*
Garretson, Henry David *neurosurgeon*
Greaver, Joanne Hutchins *mathematics educator, author*
Harris, Patrick Donald *physiology educator*
Hasselbacher, Peter *rheumatologist, educator*
Hawkins, Darroll Lee *civil engineer*
Hoye, Robert Earl *urban policy educator, health care management consultant*
Hurst, Harrell Emerson *pharmacology and toxicology educator*
Hutchison, George Barkley *epidemiologist, educator*
Jacob, Robert Allen *surgeon*
Kibiloski, Floyd Terry *business and computer consultant, editor*
Kohnhorst, Earl Eugene *tobacco company executive*
Kotwal, Girish Jayant *biochemist, educator*
Krainer, Edward Frank *engineering executive*
La Rocca, Renato V. *oncologist, researcher*
Livingood, Marvin Duane *chemical engineer*
Marshall, Gary Scott *pediatrician, educator*
Mather, Elizabeth Vivian *health care executive*
Mueller, Robert William *process engineer*
Olson, William Henry *neurology educator, administrator*
Park, Chang Hwan *physicist, educator*
Petry, Heywood Megson *psychology educator*
Ralston, Patricia A. Stark *engineering and computer science educator*
Schneider, Arthur *computer graphics specialist*
Smith, Robert F., Jr. *civil engineer*
Syed, Ibrahim Bijli *medical educator, theologist*
Thind, Gurdarshan S. *medical educator*
Thorp, James Harrison, III *aquatic ecologist*
Tran, Long Trieu *industrial engineer*
Uhde, George Irvin *physician*
Wesley, Stephen Burton *energy services executive*
Wheeler, Thomas Jay *biochemist*
Yff, David Robert *chemist*

Morehead
Mann, James Darwin *mathematics educator*
Rajaravivarma, Rathika *electrical engineer*

Murray
McCreary, Terry Wade *chemist, educator*

Owensboro
Watts, Kenneth Michael *computer scientist*

Paducah
Bassi, Joseph Arthur *physician*
Calhoun, Jesse, Jr. *electrical engineer*
Walden, Robert Thomas *physicist, consultant*

Pikeville
Christensen, Ralph J. *chemistry educator*

Prestonsburg
Mc Aninch, Robert Danford *philosophy and government affairs educator*

Richmond
Calie, Patrick Joseph *molecular biologist*
Howard, Aughtum Luciel Smith *retired mathematics educator*
Schnare, Paul Stewart *computer scientist, mathematician*

Scottsville
Claiborne, Jimmy David *electrical engineer*

Whitesville
Robertson, Clifford Houston *physician*

Wickliffe
Ford, Victor Lavann *research scientist, forest biologist*

Wilmore
Nonneman, Arthur J. *psychology educator*

Winchester
Studebaker, John Milton *utilities engineer, consultant, educator*

LOUISIANA

Alexandria
Moore, James Neal *design engineer*
Rogers, James Edwin *geology and hydrology consultant*
Slipman, (Samuel) Ronald *hospital administrator*

Baker
Moody, Lamon Lamar, Jr. *civil engineer*

Baton Rouge
Acar, Yalcin Bekir *civil engineer, soil remediation technology executive, educator*
Ajmera, Pratul Kumar *electrical engineer*
Beckman, Joseph Alfred *research and development administrator*
Besch, Everett Dickman *veterinarian, university dean emeritus*
Besing, Joan Marie *audiology and hearing science educator*
Brown, Leo Dale *inorganic chemist*
Callaway, Joseph *physics educator*
Chapman, Russell Leonard *botany educator*
Chen, Peter Pin-Shan *electrical engineering and computer science educator, data processing executive*
Christian, Frederick Ade *entomologist, physiologist, biology educator*
Constant, William David *chemical engineer, educator*
Copes, John Carson, III *mechanical engineer*
Corripio, Armando Benito *chemical engineering educator*
Costner, Charles Lynn *civil engineer*
Craft, Benjamin Cole, III *accelerator physicist*
Cullen, John Knox, Jr. *hearing science educator*
Dartez, Charles Bennett *environmental consultant*
Desbrandes, Robert *petroleum engineering educator, consultant*
Eaton, Harvill Carlton *university administrator*
Fan, J.D. (Jiangdi) *physics educator*
Finney, Clifton Donald *inventor, manufacturing executive*
Gooden, Robert *chemist*
Greaney, John Patrick *civil engineer*
Hajare, Raju Padmakar *chemical engineer*
Hansel, William *biology educator*
Head, Jonathan Frederick *cell biologist*
Imhoff, Donald Wilbur *chemist*
Iyer, Poorni Ramchandran *toxicologist*
Kazmann, Raphael Gabriel *civil and hydrologic engineer*
Killebrew, Charles Joseph *biologist*
Lambremont, Edward Nelson, Jr. *nuclear science educator*
Luckett, John Paul, Jr. *medical radiation physicist*
Lynam, Donald Ray *environmental health scientist*
Mailander, Michael Peter *biological and agricultural engineering educator*
Martin, Freddie Anthony *agronomist, educator*
Marvel, John Thomas *chemical company executive*
Marx, Brian *statistician*
McLaury, Ralph Leon *physician*
Moody, Gene Byron *engineering executive, small business owner*
Osborne, William Galloway, Jr. *chemical engineer*
Osterberger, James Sheldon, Jr. *internist, educator*
Parent, Richard Alfred *toxicologist, consultant*
Parish, Richard Lee *engineer, consultant*
Pentas, Herodotos Antreas *civil engineer*
Pezeshki, Sadrodin Reza *plant ecophysiology educator*
Rabideau, Peter Wayne *university dean, chemistry educator*
Remsen, James Vanderbeek, Jr. *biologist, museum curator*
Romero, Jorge Antonio *neurologist, educator*
Rupnik, Kresimir *physicist, researcher*
Russell, Craig John *management educator*
Sajo, Erno *nuclear engineer, educator, physicist*
Shaheen, Eli Esber *environmental engineer*
Steiner, Dirk Douglas *psychology educator*
Stopher, Peter Robert *transportation executive, consultant*
Tipton, Kenneth Warren *agricultural administrator, researcher*
Tucker, Shirley Lois Cotter *botany educator, researcher*
Tumay, Mehmet Taner *geotechnical consultant, educator, research administrator*
Valsaraj, Kalliat Thazhathuveetil *chemical engineering educator*
Van Lopik, Jack Richard *geologist, educator*
Voyiadjis, George Zino *civil engineer, educator*
Waters, William Frederick *psychology educator*
West, Philip William *chemistry educator*
Wiley, Samuel Kay *chemical engineer*
Willis, Guye Henry, Jr. *soil chemist*
Winkler, Steven Robert *hospital administrator*
Zhu, Jianchao *computer and control scientist, engineer, educator*

Belle Chasse
Arimura, Akira *biomedical research laboratory administrator, educator*

Burnside
Drescher, Edwin Anthony *chemical engineer*

Calhoun
Robbins, Marion Le Ron *agricultural research executive*

Covington
Cowden, Ronald Reed *biomedical educator, cell biologist*

Dixie
Smoak, Karl Randal *civil engineer*

Donaldsonville
Carville, Thomas Edward *environmental engineer*

Eunice
Landreneau, Rodney Edmund, Jr. *physician*

Fort Polk
Riley, Francena *nurse, military officer*

Hammond
Hejtmancik, Milton Rudolph *medical educator*
Holt-Ochsner, Liana Kay *psychology educator, researcher*

Harahan
Mahl, George John, III *electrical engineering consultant*

Harvey
Broussard, Malcolm Joseph *pharmacist, consultant*

Haughton
De Ment, James Alderson *soil scientist*

Houma
Conrad, Harold Theodore *psychiatrist*

Independence
Ryerson, Sunny Ann *entomologist*

Kenner
deMonsabert, Winston Russel *chemist*
Pratt, Mark Ernest *mechanical engineer*

Lacombe
Mangus, Carl William *technical safety and standards consultant, engineer*

Lafayette
Billeaud, Manning Francis *civil engineer*
Chieri, Pericle Adriano Carlo *educator, consulting mechanical and aeronautical engineer, naval architect*
Deaton, Lewis Edward *biology educator*
Donovan, Brian Joseph *maritime industry executive*
Jeansonne, Gloria Janelle *laboratory administrator, medical technologist*
Kavanaugh, Howard Van Zant *medical physicist*
Richard, Sylvan Joseph *electrical engineer*
Salters, Richard Stewart *engineering company executive*

Lake Charles
Inman, James Carlton, Jr. *psychological counselor*
Levingston, Ernest Lee *engineering executive*
Nam, Tin (Tonny Nam) *chemical engineer*
Roy, Francis Charles *electrical engineer*
Yadalam, Kashinath Gangadhara *psychiatrist*

LaPlace
Brodt, Burton Pardee *chemical engineer, researcher*

Mandeville
Knoepfler, Nestor Beyer *chemical engineer*
Pollock, Jack Paden *biology and dental educator, consultant, free-lance writer, retired army officer*

Metairie
Flettrich, Carl Flaspoller *structural engineer*
Harell, George S. *radiologist*
Hartman, James Austin *geologist*
Horkowitz, Sylvester Peter *chemist*
Huber, John Henry, III *economic scientist, researcher*
Munchmeyer, Frederick Clarke *naval architect, marine engineer*
Nicoladis, Michael Frank *engineering company executive*
N'Vietson, Tung Thanh *civil engineer*
Sibley, Deborah Ellen Thurston *immunochemist*

Monroe
Baum, Lawrence Stephen *biologist, educator*
Coon, Fred Albert, III *mechanical engineer*
Fouts, James Fremont *mining company executive*
Glawe, Lloyd Neil *geology educator*
Pope, Carey Nat *neurotoxicologist*

New Orleans
Alexander, Beverly Moore *mechanical engineer*
Andrews, Bethlehem Kottes *research chemist*
Angelides, Demosthenes Constantinos *civil engineer*
Barbee, Robert Wayne *cardiovascular physiologist*
Bautista, Abraham Parana *immunologist*
Beckerman, Robert Cy *pediatrician, educator*
Berenson, Gerald Sanders *physician*
Berlin, Charles I. *otolaryngologist, educator*
Bertrand, William Ellis *public health educator, international health center administrator*
Bundy, Kirk Jon *biomaterials educator, researcher, consultant*
Cairo, Jimmy Michael *physiologist*
Carter, Rebecca Davilene *surgical oncology educator*
Chattree, Mayank *mechanical engineering educator, researcher*
Collins, Harry David *forensic engineering specialist, mechanical and nuclear engineer, retired army officer*
Cook, Julia Lea *geneticist*
Corrigan, James John, Jr. *pediatrician*
Dalferes, Edward Roosevelt, Jr. *biochemical researcher*
Dimitrios, Don Fedon *civil engineer, land surveyor*
Flettrich, Alvin Schaaf, Jr. *civil engineer*
French, Alfred Dexter *chemist*
Gerber, Michael Albert *pathologist, researcher*
Gottlieb, A(braham) Arthur *medical educator, biotechnology corporate executive*
Granger, Wesley Miles *medical educator*
Hallila, Bruce Allan *welding engineer*
Harper, Robert John, Jr. *chemist, researcher*
Harwood, Robin Louise *psychologist*
Hebert, Leonard Bernard, Jr. *contractor*
Howard, Richard Ralston, II *medical health advisor, researcher, financier*
Huot, Rachel Irene *cell biologist*
Hyman, Albert Lewis *cardiologist*
Incaprera, Frank Philip *internist*
Kline, David Gellinger *neurosurgery educator*
Kreisman, Norman Richard *physiologist*
Latorre, Robert George *naval architecture and engineering educator*
Low, Frank Norman *anatomist, educator*
Martin, Louis Frank *surgery and physiology educator*
McGuire, James Horton *physics educator*
McManis, Kenneth Louis *civil engineer, educator*
Millikan, Larry Edward *dermatologist*
Nakamoto, Tetsuo *nutritional physiology educator*
Nichols, Ronald Lee *surgeon, educator*
Ochsner, Seymour Fiske *radiologist, editor*
O'Connor, Kim Claire *chemical engineering and biotechnology educator*
Phillips, John Benton *chemical engineer*
Pittman, Jacquelyn *mental health nurse, nursing educator*
Prasad, Chandan *neuroscientist*
Re, Richard N. *endocrinologist*
Riddick, Frank Adams, Jr. *physician, health care facility administrator*
Roheim, Paul Samuel *physiology educator*
Rosensteel, George T. *physics educator, nuclear physicist*
Salvaggio, John Edmond *physician, educator*
Schally, Andrew Victor *biochemist, researcher*
Sizemore, Robert Carlen *immunologist*
Smolek, Michael Kevin *optics scientist*
Striegel, Andre Michael *chemist*
Sumrell, Gene *research chemist*
Svenson, Ernest Olander *psychiatrist, psychoanalyst*
Tang, Jinke *physicist*
Veith, Robert Woody *hematologist*
Weill, Hans *physician, educator*
Wiewiorowski, Tadeusz Karol *research chemist, consultant*
Wijesundera, Vishaka *civil engineer, scientist*
Wright, Maureen Smith *molecular biologist, microbiology educator*

Pineville
Swearingen, David Clarke *general practice physician*

Plaquemine
Whigham, Mark Anthony *computer scientist*

Princeton
Tollefsen, Gerald Elmer *chemical engineer*

River Ridge
Chatry, Frederic Metzinger *civil engineer*

Ruston
Abdelhamied, Kadry A. *biomedical engineer*
Dorsett, Charles Irvin *mathematics educator*
Hale, Paul Nolen, Jr. *engineering administrator, educator*
Livingston, Mary M. *psychology educator*
McCall, Richard Powell *physics educator*
Reneau, Daniel D. *university administrator*
Robbins, Jackie Wayne Darmon *agricultural and irrigation engineer*
Walker, Harrell Lynn *plant pathologist, botany educator, researcher*
Warrington, Robert O'Neil, Jr. *mechanical engineering educator and administrator, researcher*

Saint Gabriel
Bayer, Arthur Craig *organic chemist*
Das, Dilip Kumar *chemical engineer*

Shreveport
Griffith, Robert Charles *allergist, educator*
Mancini, Mary Catherine *cardiothoracic surgeon, researcher*
Norwood, Keith Edward *civil engineer*
Paull, William Bernard *standards engineer, biologist*
Shelby, James Stanford *cardiovascular surgeon*
Sloan, Wayne Francis *mechanical engineer*

Slaughter
Gremillion, Curtis Lionel, Jr. *psychologist, hospital administrator, musician*

Slidell
Levenson, Maria Nijole *medical technologist*
Muller, Robert Joseph *gynecologist*
Sanders, Georgia Elizabeth *science and mathematics educator*
Stiffey, Arthur Van Buren *microbiologist*

Thibodaux
Pope, William David, III *pipeline engineer*

MAINE

Acton
Lotz, William Allen *consulting engineer*

Alfred
Pepin, John Nelson *materials research and design engineer*

Augusta
Gensheimer, Kathleen Friend *epidemiologist*

Bangor
Beaupain, Elaine Shapiro *psychiatric social worker*
Fiori, Michael J. *pharmacist*
Sawyer, James Lawrence *architect*

Bar Harbor
Davisson, Muriel Trask *geneticist*
Fox, Richard Romaine *geneticist, consultant*
Guidi, John Neil *scientific software engineer*
Johnson, Eric Walter *neuroscientist*
Snell, George Davis *geneticist*
Zoidis, Ann Margaret *biologist, researcher*

Bath
Dreher, Lawrence John *mechanical engineer*
Watts, Helen Caswell *civil engineer*

Bristol
Hochgraf, Norman Nicolai *retired chemical company executive*

Bucks Harbor
Bandurski, Bruce Lord *ecological and environmental scientist*

Camden
Spock, Benjamin McLane *physician, educator*

Cape Neddick
Ulan, Martin Sylvester *retired hospital administrator, health services consultant*

Castone
Otto, Fred Bishop *physics educator*

Chebeague Island
Allen, Clayton Hamilton *physicist, acoustician*

Damariscotta
Emerson, William Stevenson *retired chemist, consultant, writer*

East Boothbay
Eldred, Kenneth McKechnie *acoustical consultant*

East Winthrop
Wiesendanger, J(ohn) Ulrich *consulting engineer*

Freeport
Hebson, Charles Stephan *hydrogeologist, civil engineer*

Glen Cove
Shepardson, Jed Phillip *computer consultant*

Greenville
Sommerman, Kathryn Martha *retired entomologist*

Hollis Center
Page, Edward Crozer, Jr. *chemical engineer, consultant*

Lewiston
Gleason, Thomas Clifford *university official*
Semon, Mark David *physicist, educator*
Tighe, Thomas James Gasson, Jr. *healthcare executive*

Loring AFB
Mitchell, Philip Michael *aerospace engineer, air force officer*

Orono
Boyle, Kevin John *economics educator, consultant*
Bushway, Rodney John *food science educator*
Clapham, William Montgomery *plant physiologist*
Farthing, G. William *psychology educator*
Hill, Richard Conrad *engineering educator, energy consultant*
Kiran, Erdogan *chemical engineering educator*
Nazmy, Aly Sadek *structural engineering educator, consultant*
O'Connor, Raymond Joseph *wildlife educator*
Tarr, Charles Edwin *physicist, educator*
Tyler, Seth *zoology educator*

Phillips
Appell, George Nathan *social anthropologist*

Portland
Becker, Seymour *hazardous materials and wastes specialist*
Dubé, Richard Lawrence *landscape specialist, consultant*
Ikalainen, Allen James *environmental engineer*

Presque Isle
Reeves, Alvin Frederick, II *genetics educator*

Scarborough
Barrantes, Denny Manny *biochemist*

Skowhegan
Hornstein, Louis Sidney *retired emergency room physician*

South Portland
Iennaco, John Joseph *civil engineer, consultant*
Tewhey, John David *geochemist, environmental consultant*

South Windham
DuBourdieu, Daniel John *biotechnologist, researcher*

Surry
Pickett, James McPherson *speech scientist*

Togus
Sensenig, David Martin *surgeon*

Union
Buchan, Ronald Forbes *preventive medicine physician*

West Boothbay Harbor
Field, John Douglas *marine biologist*
Sieracki, Michael Edward *biological oceanographer*

West Enfield
Ramsdell, Richard Adoniram *marine engineer*

Winthrop
O'Brien, Ellen K. *hydrologist*

York Harbor
Curtis, Edward Joseph, Jr. *gas industry executive, management consultant*
McIntosh, Donald Waldron *retired mechanical engineer*

MARYLAND

Aberdeen
Anderson, Alfred Oliver *mathematician, consultant*
Fought, Sheryl Kristine *environmental scientist, engineer*
Levy, Ronald Barnett *aeronautical engineer*
Walters, William Place *aerospace engineer*

Aberdeen Proving Ground
Anderson, William Robert *physicist*
Armstrong, Robert Don *toxicologist*
Arroyo, Carmen Milagros *research chemist*
Chien, Lung-Siaen (Larry Chien) *mechanical engineer*
D'Amico, William Peter, Jr. *mechanical engineer*
Evans, Edward Spencer, Jr. *entomologist*
Grayson, Richard Andrew *aerospace engineer*
Mackay, Raymond Arthur *chemist*
Monty, Richard Arthur *experimental psychologist*
Porter, Dale Wayne *biochemist*
Steger, Ralph James *chemist*
Stopa, Peter Joseph *biochemist, microbiologist*
Stuempfle, Arthur Karl *physical science manager*
Wood, James David *environmental engineer*
Wright, Thomas Wilson *mechanical engineer*

Adamstown
Simmins, John James *ceramic engineer, consultant*

Adelphi
Miller, Raymond Jarvis *agronomist, college dean, university official*
Torrieri, Don Joseph *electronics engineer, mathematician, researcher*

Annapolis
Anderson, William Carl *association executive, environmental engineer, consultant*
Aprigliano, Louis Francis *metallurgist, researcher*
Cane, Guy *engineering test pilot, aviation consultant*
Clampitt, Otis Clinton, Jr. *health agency executive*
Crawford, Claude Cecil, III *biomedical consultant*
DiAiso, Robert Joseph *civil engineer*
Dickey, Joseph Waldo *physicist*
Gillmer, Thomas Charles *naval architect*
Greene, Eric *naval architect, marine engineer*
Heiner, Lee Francis *physicist*
Heller, Austin Norman *chemical and environmental engineer*
Henderson, William Boyd *engineering consulting company executive*
Hendrick, John Morton *science and engineering editor*
Ho, Louis Ting *mechanical engineer*
Hospodor, Andrew Thomas *electronics executive*
Jansson, John Phillip *architect, consultant*
Lindler, Keith William *marine engineering educator, consultant*
Linker, Lewis Craig *environmental engineer*
Miller, Richards Thorn *naval architect, engineer*
Papet, Louis M. *federal official, civil engineer*
Payne, Winfield Scott *national security policy research executive*
Sheppard, John Wilbur *computer research scientist*
Siatkowski, Ronald E. *chemistry educator*

Annapolis Junction
McClure, Richard Bruce *satellite communications engineer*

Baltimore
Adjei, Alex Asiedu *internist, pharmacologist*
Aisner, Joseph *oncologist, physician*
Alexander, Melvin Taylor *quality assurance engineer, statistician*
Anfinsen, Christian Boehmer *biochemist*
Arsham, Hossein *information scientist educator*
Auchincloss, Peter Eric *water quality improvement executive, consultant*
Baltazar, Romulo Flores *cardiologist*
Baum, Stefi Alison *astronomer*
Bausell, R. Barker, Jr. *research methodology educator*
Becker, Lewis Charles *cardiology educator*
Berg, Jeremy M. *chemistry educator*
Berger, Bruce Warren *physician, urologist*
Bochner, Bruce Scott *immunologist, educator*
Bosma, James Frederick *pediatrician*
Brock, Mary Anne *research biologist, consultant*
Brown, Donald David *biology educator*
Bylsma, Frederick Wilburn *neuropsychologist*
Caplan, Yale Howard *toxicologist, consultant*
Chapanis, Alphonse *human factors engineer, ergonomist*
Clark, Patricia *molecular biologist*
Coleman, Richard Walter *biology educator*
Collins, Oliver Michael *electrical engineer, educator, consultant, researcher*
Cone, Richard Allen *biophysics educator*
Corotis, Ross Barry *civil engineering educator, academic administrator*
Crain, Barbara Jean *pathologist, educator*
De Hoff, John Burling *physician, consultant*
Deoul, Neal *electronics company executive*
Diehl, Myron Herbert, Jr. *engineer*
Djordjevic, Borislav Boro *materials scientist, researcher*
Donahoo, Melvin Lawrence *aerospace management consultant, industrial engineer*
Dong, Dennis Long-Yu *biochemist*
Donnay, Albert Hamburger *environmental health engineer*
Donohue, Marc David *chemical engineering educator*
Drachman, Daniel Bruce *neurologist*
Feldstein, Stanley *psychologist*
Fenselau, Catherine Clarke *chemistry educator*
Fishman, Jacob Robert *psychiatrist, educator, corporate executive, investor*
Flagle, Charles Lawrence *pharmaceutical industry software firm executive*
Friedman, Marion *internist, family physician, medical administrator*
Friedrich, Christopher Andrew *internist, geneticist*
Funk Orsini, Paula Ann *pharmaceutical administration educator*
Gaber, Robert *psychologist*
Giambra, Leonard Michael *psychologist*
Giddens, Don Peyton *engineering educator, researcher*
Goldberg, Morton Falk *ophthalmologist, educator*
Gottfredson, Gary Don *psychologist*
Graves, John Fred, III *microelectronics engineer*
Green, Robert Edward, Jr. *physicist, educator*
Grossman, Lawrence *biochemist, educator*
Gruninger, Robert Martin *civil engineer*
Habermann, Helen Margaret *plant physiologist, educator*
Harrington, William Fields *biochemist, educator*
Hart, Helen Mavis *planetary astronomer*
Hartman, Philip Emil *biology educator*
Henry, Howell George *electrical engineer*
Heselton, Kenneth Emery *energy engineer*
Holland, John Lewis *psychologist*
Horn, Janet *physician*
Horowitz, Emanuel *materials science and engineering consultant*
Hutton, Larrie Van *cognitive scientist, neural networker*
Inglehart, Lorretta Jeannette *physicist*
Jani, Sushma Niranjan *child and adolescent psychiatrist*
Jastreboff, Pawel Jerzy *neuroscientist, educator*
Jensen, Arthur Seigfried *consulting engineering physicist*
Jensen, Soren Stistrup *mathematics educator*
Johns, Michael Marieb Edward *otolaryngologist, university dean*
Johnson, Kenneth Peter *neurologist, medical researcher*
Kaplan, Alexander Efimovich *physics educator, engineering educator*
Khazan, Naim *pharmacology educator*
Kimmel, Melvin Joel *psychologist*
Knox, David LaLonde *ophthalmologist*
Koliatsos, Vassilis Eleftherios *neurobiologist*
Kowal, Charles Thomas *astronomer*
Kruger, Jerome *materials science educator, consultant*
Krushat, William Mark *mathematical statistician*
Kwiterovich, Peter Oscar, Jr. *medical science educator, researcher, physician*
Lamberg, Lynne Friedman *medical journalist*
Lee, Carlton K. K. *clinical pharmacist, consultant, educator*
Lee, Yung-Keun *physicist, educator*
Leshko, Brian Joseph *civil engineer*
Lesser, Ronald Peter *neurologist*
Littlefield, John Walley *physiology educator, geneticist, cell biologist, pediatrician*
Lozovatsky, Mikhail *civil engineer, consultant*
Mahesh, Mahadevappa Mysore *medical physicist, researcher*
Marcellas, Thomas Wilson *electronics company executive*
Markowska, Alicja Lidia *neuroscientist, researcher*
Martin, George Reilly *federal agency administrator*
Maumenee, Irene Hussels *opthalmology educator*
McCarty, Richard Earl *biochemist, biochemistry educator*
McGowan, George Vincent *public utility executive*
Mc Hugh, Paul R. *psychiatrist, neurologist, educator*
Moses, Hamilton, III *neurology educator, hospital executive*
Mouton, Peter Randolph *neuroscientist, biologist*
Murphy, Douglas Blakeney *cell biology educator*
Myslinski, Norbert Raymond *medical educator*
Nagey, David Augustus *physician, researcher*
Natarajan, Thyagarajan *civil engineer*
Nathans, Daniel *biologist*
Noar, Mark David *internist, gastroenterologist, therapeutic endoscopist, consultant, inventor*
Orlitzky, Robert *engineer*
Owens, Albert Henry, Jr. *oncologist, educator*

Palmer, C(harles) Harvey electrical engineer, educator
Patel, Manu Ambalal environmental engineer
Poehler, Theodore Otto university provost, engineer, researcher
Pollard, Thomas Dean biologist, educator
Posner, Gary Herbert chemist, educator
Prendergast, Robert A. pathologist educator
Price, Thomas Ransone neurologist, educator
Raben, Daniel Max biochemist
Racusen, Lorraine Claire pathologist, researcher
Ramesh, K(aliat) T(hazhathveetil) mechanical engineer, educator
Rao, Gopala Utukuru radiological physicist, educator
Rayson, Glendon Ennes internist, preventive medicine specialist, writer
Reddi, A. Hari orthopaedics and biological chemistry educator
Roberts, Randolph Wilson health science educator
Rose, Noel Richard immunologist, microbiologist, educator
Saffer, Linda Diane biology researcher
Salisbury, John William research geophysicist, consultant
Schnaar, Ronald Lee biomedical researcher, educator
Schuman, William John, Jr. mechanical engineer
Schuster, Frank Feist neurologist, social services administrator
Schwandt, Leslie Marion psychologist, researcher
Shah, Shirish Kalyanbhai computer science, chemistry and environmental science educator
Sharpe, William Norman, Jr. mechanical engineer, educator
Sherman, Alan Theodore computer science educator
Sherwin, Roger William medical educator
Sibeck, David G. geophysicist
Silbergeld, Ellen Kovner environmental epidemiologist and toxicologist
Smith, Hamilton Othanel molecular biologist, educator
Snyder, Solomon Halbert psychiatrist, pharmacologist
Soeken, Karen Lynne research methods educator, researcher
Sommer, Alfred public health professional, epidemiologist
Stacy, Pheriba archaeologist
Steiner, Robert Frank biochemist
Steinwachs, Donald Michael public health educator
Stolberg, Ernest Milton environmental engineer, consultant
Suskind, Sigmund Richard microbiology educator
Swift, David Leslie environmental health educator, consultant
Tewari, Kirti Prakash biophysicist
Thakor, Nitish Vyomesh biomedical engineering educator
Trpis, Milan entomologist, scientist, educator
Turney, Stephen Zachary cardiothoracic surgeon, educator
Vogelstein, Bert oncology educator
Wagner, Henry Nicholas, Jr. physician
Walch, Marianne microbiologist
Walton, Kimberly Ann medical laboratory technician
Weaver, Kenneth Newcomer geologist, state official
Weber, David Joseph biochemist, educator
Westerhout, Gart retired astronomer
Whalen, Barbara Rhoads mechanical engineer, consultant
White, Emil Henry chemistry educator
Wierman, John Charles mathematician, educator
Wilber, John Franklin endocrinologist, educator
Wilson, Donald Edward physician, educator
Winstead, Carol Jackson mathematics educator
Woolverton, Christopher Jude biopharmaceutical company executive
yeager, larry lee civil engineer, consultant
Yellin, Judith electrologist
Zee, David Samuel neurologist
Zhang, Cheng-Yue astrophysicist
Zukas, Jonas Algimantas physicist

Bel Air
Cash, (Cynthia) LaVerne physicist
Powers, Doris Hurt engineering company executive

Beltsville
Endo, Burton Yoshiaki research plant pathologist
Hackett, Kevin James insect pathologist
Hoberg, Eric Paul parasitologist
Howe, Juliette Coupain chemist
Kearney, Philip Charles biochemist
Klun, Jerome Anthony entomologist, researcher
Lincicome, David Richard biomedical and animal scientist
Norris, Karl Howard optics scientist, agricultural engineer
Quirk, Frank Joseph management consulting company executive
Schmidt, Walter Friedrich research chemist
Soper, Thomas Sherwood biochemist
Starr, James LeRoy soil scientist
Stern, Michael chemist
van Schilfgaarde, Jan agricultural engineer, government agricultural service administrator
Velasco, James chemist, consultant
Zarlenga, Dante Sam, Jr. molecular biologist
Zehner, Lee Randall environmental service executive, research director

Benedict
Sanders, James Grady biogeochemist

Berlin
Brodsky, Allen radiological and health physicist, consultant
Passwater, Richard Albert biochemist, writer

Bethesda
Abrams, Scott Irwin immunologist
Adamson, Richard Henry biochemist
Aldroubi, Akram mathematician, researcher
Alexander, Duane Frederick pediatrician, research administrator
Alexander, Nancy Jeanne health science facility administrator, researcher
Amouzadeh, Hamid R. toxicologist
Augustine, Norman Ralph industrial executive
Axelrod, Julius biochemist, pharmacologist
Backus, Robert Coburn biophysical chemist
Bak, Martin Joseph biomedical engineer
Baldwin, Wendy Harmer social demographer
Barber, Ann McDonald internist
Bartky, Ian Robertson physical chemist
Beall, Robert Joseph foundation executive
Bender, Erwin Rader, Jr. air force officer
Bernstein, Bernard engineering executive
Bernstein, Lori Robin biochemist, molecular biologist, educator
Bick, Katherine Livingstone scientist, international liaison consultant
Blush, Steven Michael nuclear scientist, safety consultant
Brady, Roscoe Owen neurogeneticist, educator
Brinley, F(loyd) J(ohn), Jr. health science institution executive, physician
Broder, Christopher Charles microbiologist
Broder, Samuel federal agency administrator
Bunger, Rolf physiology educator
Burg, Maurice Benjamin renal physiologist, physician
Burns, Brian Patrick electrical engineer
Bynum, Barbara Stewart federal health institute administrator
Camp, Frances Spencer retired nurse
Cantoni, Giulio Leonardo biochemist, government official
Cath, Stanley Howard psychiatrist, psychoanalyst
Cheever, Allen Williams pathologist
Coelho, Anthony Mendes, Jr. health science administrator
Cohen, Lois Ruth Kushner government research institute official
Collins, Francis S. medical research scientist
Contois, David Francis biochemist
Creveling, Cyrus Robbins chemist, neuroscientist
Dai, Charles Mun-Hong mechanical engineer, consultant
Danforth, David Newton, Jr. physician, scientist
Datiles, Manuel Bernaldes, III ophthalmologist
Dawid, Igor Bert biologist
DeFilippes, Mary Wolpert pharmacologist
Dickler, Howard Byron biomedical administrator, research physician
Didsbury, Howard Francis futurist educator, lecturer, consultant
Di Paolo, Joseph Amadeo geneticist
Donahue, Robert Edward veterinarian, researcher
Duncan, Constance Catharine psychologist
Durant, Frederick Clark, III aerospace history and space art consultant
Durek, Thomas Andrew computer company executive
Ehrenstein, Gerald biophysicist
Elin, Ronald John pathologist
Elson, Hannah Friedman research biologist
Everstine, Gordon Carl mathematician, educator
Fales, Henry Marshall chemist
Fauci, Anthony Stephen health facility administrator, physician
Fisher, Suzanne Eileen health scientist administrator
Fordyce, Samuel Wesley electrical engineer, communications company executive, consultant
Fraumeni, Joseph F., Jr. scientific researcher, medical educator, physician, military officer
Freedman, Laurence Stuart statistician
Friedman, Robert Morris pathologist, molecular biologist
Furey, Deborah Ann aerospace engineer, naval architect
Gallin, John Isaac health science association administrator
Gallo, Robert Charles research scientist
Gelboin, Harry Victor biochemistry educator, researcher
Geller, Ronald Gene health administrator
Gibson, Sam Thompson internist, educator
Gohagan, John Kenneth medical institute administrator, educator
Gold, Philip William neurobiologist
Goldstein, Murray health organization official
Gottesman, Michael Marc biomedical researcher
Grant, Angus John immunologist
Greenwald, Peter physician, government medical research director
Gruber, Jack medical virologist, biomedical research administrator
Gschwindt de Gyor, Peter George economist
Hagen, Stephen James physicist
Hallett, Mark physician, neurologist, health research institute administrator
Halula, Madelon Clair microbiologist
Haseltine, Florence Pat research administrator, obstetrician, gynecologist
Hauck, Frederick Hamilton retired naval officer, astronaut, business executive
Hausman, Steven Jack health science administrator
Hennet, Remy Jean-Claude geochemist, environmental consultant
Hicks, Thomas Erasmo nuclear engineer
Hoak, John Charles physician, educator
Holmes, Margaret F. health science administrator, researcher
Hook, Vivian Yuan-Wen Ho biochemist, neuroscientist
Hurd, Suzanne Sheldon federal agency health science director
Jacobson, Arthur Eli research chemist
Jamieson, John Anthony engineering consulting company executive
Johnson, Joyce Marie psychiatrist, epidemiologist
Kaliner, Michael Aron immunologist, allergist, medical association administrator
Kandasamy, Sathasiva Balakrishna pharmacologist
Kaufman, Seymour biochemist
Kaufmann, Peter G. research biomedical scientist, psychologist, educator
Kawakami, Yutaka biomedical researcher, hematologist
Keith, Jerry M. molecular biologist
King, Ron Glen neuroscientist
Kirschstein, Ruth Lillian physician
Korn, Edward David biochemist
Kozak, Robert William immunologist
Krause, Richard Michael medical scientist, government official, educator
Kresina, Thomas Francis immunologist, educator
Kupfer, Carl ophthalmologist, science administrator
Lange, Nicholas Theodore statistician
Lee, Fang-Jen Scott biochemist, molecular biologist
Leikin, Sergey L. biophysicist, researcher
Lenfant, Claude Jean-Marie physician
Leventhal, Carl M. neurologist
Levine, Arthur Samuel physician, scientist
LeVine, Donald Jay electrical engineer
Lindberg, Donald Allan Bror library administrator, pathologist, educator
Liotta, Lance A. pathologist
Lipkin, Bernice Sacks computer science educator
Lipton, James Abbott epidemiologist, researcher
Lisack, John, Jr. cartographer, executive
Lockshin, Michael Dan rheumatologist
Löe, Harald dentist, educator, researcher
Maher, John Francis physician, educator
Malech, Harry Lewis immunologist, researcher
Malone, Winfred Francis toxicologist
Mason, James Ostermann public health administrator
McCarthy, Justin Huntly naval architect
McIlvain, Jess Hall architect, consultant
McKinney, Cynthia Eileen molecular biologist
McLellan, Katharine Esther health physicist, consultant
McMurphy, Michael Allen energy company executive, lawyer
Metzger, Henry federal research institution administrator
Miller, Alexandra Cecile radiation biologist, researcher, educator
Miller, Barry Alan epidemiologist, cancer researcher
Mishkin, Mortimer neuropsychologist
Mojcik, Christopher Francis internist
Moss, Bernard virologist, researcher
Moss, Joel medical researcher
Murayama, Makio biochemist
Myers, Lawrence Stanley, Jr. radiation biologist
Nash, Howard Allen biochemist, researcher
Nelson, Lawrence Merle reproductive endocrinologist
Nelson, Thomas John research biochemist
Neva, Franklin Allen physician, educator
Nirenberg, Marshall Warren biochemist
Noguchi, Constance Tom molecular biologist, researcher
Notkins, Abner Louis physician, researcher
Obrams, Gunta Iris health facility administrator
Odeyale, Charles Olajide biomedical engineer
Ognibene, Frederick Peter internist
Okunieff, Paul radiation oncologist, physician
Onufrock, Richard Shade pharmacist, researcher
Osterman, Lisa Ellen geologist, researcher
Papaioannou, Evangelia-Lilly psychologist, researcher
Pastan, Ira Harry biomedical science researcher
Pastor, Richard Walter research chemist
Paul, William Erwin immunologist, researcher
Plaut, Marshall physician, medical administrator, educator
Poston, John Michael biochemist
Potter, Michael genetics researcher, medical researcher
Purcell, Robert Harry virologist
Quinnan, Gerald Vincent, Jr. medical educator
Rabson, Alan Saul physician, educator
Rapoport, Judith psychiatrist
Reese, Thomas Sargent neurobiology educator and researcher
Reeves, Richard Allen government aerospace program executive, lawyer
Regdos, Shane Lawrence nuclear chemist
Reid, Janet Warner biologist consultant
Rich, Harry Louis physicist, marine engineering consultant
Robbins, Keith Cranston research scientist
Robinson, David Mason physiologist
Robison, Wilbur Gerald, Jr. research biologist
Rodbard, David endocrinologist, biophysicist
Rosenberg, Jacob Joseph orthodontist
Rosenberg, Steven Aaron surgeon, medical researcher
Saffiotti, Umberto pathologist
Salisbury, Tamara Paula foundation executive
Sausville, Edward Anthony medical oncologist
Saville, Michael Wayne physician, scientist
Schlom, Jeffrey Bert research scientist
Sheridan, Philip Henry pediatrician, neurologist
Shulman, Lawrence Edward biomedical research administrator, rheumatologist
Silberberg, Rein nuclear astrophysicist, researcher
Sinclair, Warren Keith radiation biophysicist, organization executive, consultant
Sirén, Anna-Leena Kaarina neuroscientist
Smith, Thomas Graves, Jr. image processing scientist, neurophysiologist
Snitch, Thomas Harold science educator, consultant
Snow, James Byron, Jr. physician, research administrator
Sobering, Geoffrey Simon biomedical scientist
Sokoloff, Louis physiologist, neurochemist
Spangler, Miller Brant science and technology analyst, planner, consultant
Sporn, Michael Benjamin cancer etiologist
Stadtman, Earl Reece biochemist
Stadtman, Thressa Campbell biochemist
Stone, Philip M. physicist, nuclear engineer
Tabor, Edward physician, researcher
Talbot, Bernard government medical research facility official, physician
Templeton, Nancy Valentine Smyth research scientist
Theuer, Charles Philip surgeon
Thom, Arleen Kaye surgeon
Thorgeirsson, Snorri Sveinn physician, pharmacologist
Tseng, Christopher Kuo-Hou health scientist, administrator, chemist
Vaitukaitis, Judith Louise medical research administrator
Van Cott, Harold Porter human factors professional
Varmus, Harold Eliot microbiologist, educator
Vaughan, Martha biochemist
Vickers, James Hudson veterinarian, research pathologist
Waldmann, Thomas Alexander medical research scientist, physician
Webster, Henry deForest experimental neuropathologist
Weiss, George Herbert mathematician, consultant
Witkop, Bernhard chemist
Wolffe, Alan Paul molecular embryologist, molecular biologist
Woolley, George Walter biologist, geneticist, educator
Wu, Roy Shih Shyong biochemist, health scientist administrator
Xu, Sengen biophysicist
Yanovski, Susan Zelitch physician, eating disorders specialist
Yarchoan, Robert clinical immunologist, researcher
Yellin, Herbert science administrator
Yokel, Felix Yochanan civil engineer
Yu, Monica Yerk-lin medical technologist, consultant
Yuspa, Stuart H. cancer etiologist
Zierdt, Charles Henry microbiologist
Zoltick, Brad J. physicist, mathematics educator
Zwanzig, Robert Walter chemist, physical science educator

Bowie
Reichmann, Péter Iván mathematician

Brooklandville
Azola, Martin P. civil engineer, construction manager

Burtonsville
Lo, Chun-Lau John aerospace engineer, consultant

Cambridge
Shipe, Gary Thomas ceramic engineer

Chevy Chase
Choppin, Purnell Whittington research administrator, virology researcher, educator
Cowan, William Maxwell neurobiologist
Harter, Donald Harry research administrator, medical educator
Raifman, Irving clinical psychologist
Wallerstein, Leibert Benet economist

Claiborne
Guinness, Kenelm L. civil engineer

Clarksburg
Arnold, Jay engineering executive
Evans, John Vaughan satellite laboratory executive, physicist
Townsend, John William, Jr. physicist, retired federal aerospace agency executive

Clinton
Sizemore, Carolyn Lee nuclear medicine technologist

Cockeysville Hunt Valley
Bilodeau, John Lowell engineer
Cohen, Kenneth Allan chemist
Evans, George Leonard microbiologist
Gift, James J. toxicologist
Hirsch, Richard Arthur mechanical engineer
Marsik, Frederic John microbiologist
Morhardt, Josef Emil, IV environmentalist, engineering company executive
Morhardt, Sia S. environmental scientist
Orlidge, Leslie Arthur electrical engineer

Colesville
Aidar, Nelson civil engineer, consultant

College Park
Aloimonos, Yiannis John computer sciences educator
Andrews, Kevin Paul physical chemist
Arsenault, Richard Joseph materials science and engineering educator
Ayyub, Bilal Mohammed civil engineering educator
Busby, Kenneth Owen applied physicist
Cohen, Michael Lee statistics educator
Colwell, Rita Rossi microbiologist, molecular biologist
De Souza-Machado, Sergio Guilherme physics research assistant
Dieter, George E., Jr. university dean
Dragt, Alexander James physicist
Embody, Daniel Robert biometrician
Fetter, Steve physicist
Fisher, Michael Ellis mathematical physicist, chemist
Fox, Christopher educational administrator
Gomez, Romel Del Rosario physicist
Granatstein, Victor Lawrence electrical engineer, educator
Greene, Charles Arthur materials engineer
Griem, Hans Rudolf physicist, educator
Han, Ruijing chemical engineer
Howard, Russell Duane aerospace engineer
Imberski, Richard Bernard zoology educator
Jacobs, Dan biometrician, ecologist
Kaylor, Jefferson Daniel, Jr. electronics executive
Kerr, Frank John astronomer, educator
Kissell, Kenneth Eugene astronomer
Kopp, Debra Lynn manufacturing engineer, consultant
Kuenzel, Wayne John avian physiologist, neuroscientist
Kung, Shain-dow molecular biologist, educator
Lau, Yun-Tung physicist, consultant
Loss, John C. architect, educator
Marcus, Steven Irl electrical engineering educator
Moore, John Hays chemistry educator
Ohadi, Michael M. mechanical engineering educator
Peaslee, David Chase physics educator
Ponnamperuma, Cyril Andrew chemist
Prentice, Ann Ethelynd academic adminstrator
Rasmusson, Eugene Martin meteorology researcher
Redish, Edward Frederick physicist, educator
Reggia, James Allen computer scientist, educator
Rosenfeld, Azriel computer science educator, consultant
Sachs, Harvey M. policy analyst
Sagdeev, Roald Zinnurovi physicist educator
Sengers, Jan Vincent physicist
Skuja, Andris physics educator
Teglasi, Hedwig psychologist, educator
Wang, Jian Guang engineering physicist
Wang, Liqin materials scientist
Weeks, John David chemistry and physical science educator
Wilkinson, Gerald Stewart zoology educator
Zen, E-an research geologist

Columbia
Arndt, Julie Anne Preuss aeronautical engineer
Barton, William Elliott biologist, administrator
Beach, Robert Mark biologist
Bremerman, Michael Vance reliability engineer
Bustard, Thomas Stratton engineering company executive
Clark, Billy Pat physicist
Dale, Charlene Boothe international health administrator
Falk, Steven Mitchell biomedical engineer, consultant
Fan, Xiyun polymer engineer
Fryberger, Theodore Kevin mechanical and ocean engineer
Fu, Gang nuclear engineer
Grabowsky, Craig aerospace engineer
Hegedus, L. Louis chemical engineer, research and development executive
Hurwitch, Jonathan William energy consultant
Imre, Paul David mental health administrator
Kalavapudi, Murali civil engineer
Khare, Mohan chemist
Krausman, John Anthony mechanical engineer
Lain, David Cornelius health scientist, researcher
Miller, Hartman Cyril, Jr. chemical hazardous material training specialist

Raksis, Joseph W. *chemist*
Rapkin, Jerome *defense industry executive*
Rezaiyan, A. John *chemical engineer, consultant*
Rice, Roy Warren *ceramic engineer*
Roby, Richard Joseph *research engineering executive, educator*
Roomsburg, Judy Dennis *industrial/organizational psychologist*
Shiue, Gong-Huey *chemist*
Taylor, Scott Thomas *microwave engineer*
Vanderlinde, William Edward *materials engineer*
Vu, Cung *chemical engineer*
Walter, James Frederic *biochemical engineer*

Crofton
Watson, Robert Tanner *physical scientist*

Cumberland
Goss, Patricia Lynn *information specialist*

Darnestown
Gottlieb, Julius Judah *podiatrist*

Easton
Hurley, Richard Keith *chemist*
Thompson, Robert Campbell *orthopaedic surgeon*

Edgewood
Collins, William Henry *environmental scientist*
Russell, William Alexander, Jr. *environmental scientist*

Elkton
Harrell, Douglas Gaines *chemical engineer*

Ellicott City
Davis, Patricia Mahoney *software engineer*
Elgort, Andrew Charles *school psychologist*

Fort Detrick
Nelson, James Harold *health sciences administrator*
Schaad, Norman Werth *plant pathologist*
Shaffer, Stephen M. *computer system analyst*

Fort Meade
Ewell, Allen Elmer, Jr. *naval officer*
Madden, Robert William *physicist*

Fort Washington
Godette, Stanley Rickford *microbiologist*

Frederick
Bolon, Brad Newland *veterinary pathologist*
Boyd, V(irginia) Ann Lewis *biology educator*
Bryan, John Leland *retired engineering educator*
Cragg, Gordon Mitchell *government chemist*
Creasia, Donald Anthony *toxicologist, researcher*
Henchal, Erik Alexander *microbiologist*
Kappe, David Syme *environmental chemist*
Knisely, Ralph Franklin *retired microbiologist*
Moore, Sharon Pauline *biologist*
Ochoa, Augusto Carlos *immunologist, researcher*
Olsen, Daren Wayne *electrical engineer*
Waalkes, Michael Phillip *toxicologist*
Wellner, Robert Brian *physiologist*

Gaithersburg
Ballard, Lowell Douglas *mechanical engineer*
Bartholomew, Richard William *mechanical engineer*
Beltracchi, Leo *engineer*
Bonnell, David William *research chemist*
Brinckman, Frederick Edward, Jr. *research chemist, consultant*
Bruening, Robert John *physicist, researcher*
Burrows, James H. *computer scientist*
Cahn, John Werner *metallurgist, educator*
Caplin, Jerrold Leon *health physicist*
Carasso, Alfred Sam *mathematician*
Casella, Russell Carl *physicist*
Coleman, Mark David *mechanical engineer*
Epler, Katherine Susan *chemist*
Evenson, Kenneth M. *physicist*
Feric, Gordan *mechanical engineer*
Frahm, Veryl Harvey, Jr. *laboratory manager*
French, Judson Cull *government official*
Hall, Dale Edward *electrochemist*
Hamer, Walter Jay *chemical consultant, science writer*
Hellwig, Helmut Wilhelm *air force research director*
Hoffer, James Brian *physicist, consultant*
Hoppes, Harrison Neil *corporate executive, chemical engineer*
Johnson, Donald Rex *research institute administrator*
Juliano, John Joseph *energy systems engineer*
Kolstad, George Andrew *physicist, geoscientist*
Kramer, Thomas Rollin *automation researcher*
Lide, David Reynolds *science editor*
Lowke, George E. *biotechnology executive*
Lu, Kwang-Tzu *physicist*
Marshall, Richard Dale *structural engineer*
Masters, Larry William *physical science administrator*
McClelland, Alan *molecular biologist, laboratory administrator*
Nakatani, Alan Isamu *physicist*
Nguyen, Tinh *materials scientist*
Oran, Gary Carl *computer scientist*
Phelan, Frederick Rossiter, Jr. *chemical engineer*
Prabhakar, Arati *federal administration research director, electrical engineer*
Rabinow, Jacob *electrical engineer, consultant*
Reader, Joseph *physicist*
Ricker, Richard Edmond *metallurgical scientist*
Rollence, Michele Lynette *molecular biologist*
Rosenblatt, Joan Raup *mathematical statistician*
Sayer, John Samuel *information systems consultant*
Schwartz, Lyle H. *materials scientist, government official*
Seiler, David George *physicist*
Semerjian, Hratch Gregory *research and development executive*
Simons, David Stuart *physicist*
Simpson, Stephen Lee *consulting civil engineer*
Sobers, David George *environmentalist*
Teague, Edgar Clayton *physicist*
Top, Franklin Henry, Jr. *physician, researcher*
Vorburger, Theodore Vincent *physicist, metrologist*
Weber, Alfons *physicist*
Wineland, David J. *physicist*
Witzgall, Christoph Johann *mathematician*
Wright, Richard Newport, III *civil engineer, government official*

Germantown
Charlton, Gordon Randolph *physicist*
Dorsey, William Walter *aerospace engineer, engineering executive*
Matthews, Dale Samuel *information scientist*
Olson, James Hilding *software engineer*
Van Houten, Robert *nuclear engineer, consultant*
Williamson, Samuel Perkins *federal agency administrator, meteorologist*

Gibson Island
Kiddoo, Richard Clyde *retired oil company executive*

Glenelg
Williams, Donald John *research physicist*

Grasonville
Prout, George Russell, Jr. *medical educator, urologist*

Greenbelt
Alexander, Joseph Kunkle, Jr. *physicist*
Blackburn, James Kent *astrophysicist*
Chan, Clara Suet-Phang *physician*
Comiso, Josefino Cacas *physical scientist*
Danks, Anthony Cyril *detector scientist, research astronomer*
Day, John H. *physicist*
Degnan, John James, III *physicist*
England, Martin Nicholas *astro/geophysicist*
Fitzmaurice, Michael William *electrical and mechanical engineer*
Haxton, Donovan Merle, Jr. *astronomer*
Kalshoven, James Edward, Jr. *electronics engineer*
Ku, Jentung *mechanical and aerospace engineer*
Liu, Han-Shou *space scientist, researcher*
Lynch, John Patrick *aerospace engineer*
Maran, Stephen Paul *astronomer*
Mather, John Cromwell *astrophysicist*
McGee, Thomas Joseph *atmospheric scientist*
Mumma, Michael Jon *physicist*
Ozernoy, Leonid Moissey *astrophysicist*
Palmer, David Michael Oliver *astronomer*
Parkinson, Claire Lucille *climatologist*
Petuchowski, Samuel Judah *physicist*
Schatten, Kenneth Howard *physicist*
Simpson, Joanne Malkus *meteorologist*
Steiner, Mark David *engineering manager*
Tilton, James Charles *computer engineer*
Whitlock, Laura Alice *research scientist*
Wood, H(oward) John, III *astrophysicist, astronomer*

Hanover
Kistler, Alan Lewis *engineering executive*

Havre De Grace
Wiegand, Gordon William *flow cytometrist, electro-optic consultant*

Highland
MacFadden, Kenneth Orville *chemist*

Hyattsville
Guadagno, Mary Ann Noecker *social scientist, consultant*
Kohn, Barbara Ann *veterinarian*
McLin, William Merriman *foundation administrator*
Piccinino, Linda Jeanne *statistician*
Pickle, Linda Williams *biostatistician*
Pittarelli, George William *agronomist, researcher*
Sindoris, Arthur Richard *electronics engineer, government official*

Jessup
Bazzazieh, Nader *environmental engineer*

Kensington
Aronson, Casper Jacob *physicist*

La Plata
Zinn, Michael Wallace *aerospace engineer*

Landover
Fabunmi, James Ayinde *aeronautical engineer*
Ho, Jin-Meng *acoustician*
Tripp, Herbert Alan *systems engineer*
Zehl, Otis George *optical physicist*

Lanham
Black, Clinton James *engineering company executive*
Dittberner, Gerald John *engineer, meteorologist, space scientist*
Hirschhorn, Joel Stephen *engineer*
Nithianandam, Jeyasingh *physicist*

Lanham Seabrook
Lesikar, James Daniel, II *physicist*

Laurel
Babin, Steven Michael *atmospheric scientist, researcher*
Basappa, Shivanand *computer engineer*
Benson, Richard Charles *chemist*
Billig, Frederick Stucky *mechanical engineer*
Blum, Norman Allen *physicist*
Bostrom, Carl Otto *physicist, laboratory director*
Budman, Charles Avrom *aerospace engineer*
Eaton, Alvin Ralph *research and development administrator, aeronautical and systems engineer*
Ellis, David H. *biologist, research behaviorist*
Farrell, William James, Jr. *mathematician*
Ford, Byron Milton *computer consultant*
Halushynsky, George Dobroslav *systems engineer*
Hamill, Bruce W. *psychologist*
Henderson, Daniel Gardner *electrical engineer*
Hoffman, David John *physiologist*
Krimigis, Stamatios Mike *physicist, researcher, space science/engineering manager, consultant*
Linevsky, Milton Joshua *physical chemist*
Maurer, Richard Hornsby *physicist*
Meyer, James Henry *meteorologist*
Moorjani, Kishin *physicist, researcher*
Ramos, Angel Salvador *veterinarian*
Sundaresan, P. Ramnathan *research chemist, consultant*
Thomas, Michael Eugene *electrical engineer, researcher, educator*
Voss, Paul Joseph *physicist*
White, Michael Elias *aerospace engineer*

Lexington Park
Nannery, Michael Alan *civil engineer*

Linthicum
Massimini, Joseph Nicholas *software engineer*

Lusby
Roxey, Timothy Errol *nuclear engineer, biomedical consultant*

Lutherville
Barton, Meta Packard *business executive, medical science research executive*

Monkton
Doepkens, Frederick Henry *agriscience educator*

Monrovia
Atanasoff, John Vincent *physicist*

Mount Savage
Warren, D. Elayne *environmental sanitarian*

North Potomac
Agarwal, Duli Chand *mechanical engineer*
Finger, Stanley Melvin *chemical engineer, consultant*

Oakland
Farrar, Richard Bartlett, Jr. *secondary education educator, wildlife biology consultant*

Odenton
Mucha, John Frank *data processing professional*

Olney
Baker, Carl Gwin *science educator*

Owings Mills
Elkins, Robert N. *association executive*
Knisbacher, Jeffrey Mark *computer scientist*

Pasadena
Young, Russell Dawson *physics consultant*

Patuxent River
Eastburg, Steven Roger *naval officer, aeronautical engineer*
Tipton, Thomas Wesley *flight engineer*
Ullom, Lawrence Charles, Jr. *engineer*

Poolesville
Chang, Zhao Hua *biomedical engineer*
Newman, John Dennis *neuroethologist, biomedical researcher*

Potomac
Lawrence, Robert Edward *electrical engineer*
Mc Bryde, Felix Webster *geographer, ecologist, consultant*
Peters, Carol Beattie Taylor (Mrs. Frank Albert Peters) *mathematician*

Princess Anne
Adams, James Alfred *natural science educator*
Joshi, Jagmohan *agronomist, consultant*

Reisterstown
Bond, Nelson Leighton, Jr. *health care executive*

Riverdale
Guetzkow, Daniel Steere *computer company executive*

Rockville
Abron, Lilia A. *engineer*
Banfield, William Gethin *physician*
Barnes, John Maurice *plant pathologist*
Beattie, Donald A. *energy scientist, consultant*
Beckjord, Eric Stephen *energy researcher, nuclear engineering educator*
Bridge, T(homas) Peter *psychiatrist, researcher*
Brumback, Gary Bruce *industrial and organizational psychologist*
Burdick, William MacDonald *biomedical engineer*
Cannon, Grace Bert *immunologist*
Chakravarty, Dipto *computer-performance engineer*
Clark, William Anthony *neuropharmacologist*
Corley, Daniel Martin *physicist*
Cunningham, Keith Allen, II *computer services company executive*
Donahue, Mary Rosenberg *psychologist*
Fenton, Wayne S. *psychiatrist*
Gabelnick, Henry Lewis *medical research director*
Goodwin, Frederick King *psychiatrist*
Gougé, Susan Cornelia Jones *microbiologist*
Grady, Lee Timothy *pharmaceutical chemist*
Grist, Clarence Richard *chemist, precious metals investor*
Gruber, Kenneth Allen *health scientist, administrator*
Guest, Gerald Bentley *veterinarian*
Hackett, Joseph Leo *microbiologist, clinical pathologist*
Harkonen, Wesley Scott *physician*
Jacobson-Kram, David *toxicologist*
Ji, Xinhua *physical chemist, educator*
Josephs, Melvin Jay *professional society administrator*
Kerwin, Courtney Michael *public health scientist, administrator*
Kessler, David A. *health services commissioner*
Kindt, Thomas James *chemist*
Koplaski, John *structural engineer*
Lewis, Benjamin Pershing, Jr. *pharmacist, public health service officer*
Lewis, Paul Martin *engineering psychologist*
Lightfoote, Marilyn Madry *molecular immunologist*
Liu, Charles Chung-Cha *transportation engineer, consultant*
Long, Cedric William *health research facility executive*
McCormick, Kathleen Ann Krym *geriatrics nurse, federal agency administrator*
Monaghan, W(illiam) Patrick *immunohematologist, retired naval officer, health educator, consultant*
Munzner, Robert Frederick *biomedical engineer*
Murray, Peter *metallurgist, manufacturing company executive*
Nora, James Jackson *physician, author, educator*
Norbedo, Anthony Julius *engineering executive*
O'Rangers, John Joseph *biochemist*
Paul, Steven M. *psychiatrist*
Phillips, Mark Douglas *aerospace company executive*
Rheinstein, Peter Howard *government official, physician, lawyer*
Rinkenberger, Richard Krug *physical scientist*
Rosenstein, Marvin *public health association administrator*
Ruggera, Paul Stephen *biomedical engineer*
Scearce, P. Jennings, Jr. *engineering executive*
Schmuff, Norman Robert *organic chemist*
Seagle, Edgar Franklin *environmental engineer, consultant*
Seltser, Raymond *epidemiologist, educator*
Sorensen, John Noble *mechanical and nuclear engineer*
Tabibi, S. Esmail *pharmaceutical researcher, educator*
Teske, Richard Henry *veterinarian*
Van Arsdel, William Campbell, III *pharmacologist*
Weaver, Christopher S(cot) *scientist*
Whitney, Robert A., Jr. *veterinarian, government public health executive*
Zmudzka, Barbara Zofia *researcher*
Zoon, Kathryn Egloff *biochemist*

Ruxton
Duer, Ellen Ann Dagon *anesthesiologist*

Saint Michaels
Young, Donald Roy *pharmacist*

Salisbury
Flory, Charles David *retired psychologist, consultant*
May, Everette Lee, Jr. *pharmacologist, educator*

Sandy Spring
Kanarowski, Stanley Martin *chemist, chemical engineer, government official*

Seabrook
Browning, Ronald Kenneth *aerospace engineer*

Severn
Fowler, Floyd Earl *national security consultant*

Severna Park
Davis, John Adams, Jr. *electrical engineer, roboticist, executive*
Mallory, Charles William *consulting engineer, marketing professional*
Retterer, Bernard Lee *electronic engineering consultant*

Silver Spring
Attaway, David Henry *federal research administrator, oceanographer*
Baer, Ledolph *oceanographer, meteorologist*
Baker, Scott Ralph *toxicologist*
Briscoe, Melbourne G. *oceanographer, administrator*
Burke, Margaret Ann *computer and communications company specialist*
Ceasor, Augusta Casey *medical technologist, microbiologist*
Cramer, Mercade Adonis, Jr. *computer company executive*
Dalton, Robert Edgar *mathematician, computer scientist*
Forbes, Jerry Wayne *research physicist*
Foster, Nancy Marie *environmental analyst, government official*
Friday, Elbert Walter, Jr. *federal agency administrator, meteorologist*
Gaunaurd, Guillermo C. *physicist, engineer, researcher*
Green, Robert Lamar *consulting agricultural engineer*
Grossberg, David Burton *cardiologist*
Hattis, David Ben-Ami *architect*
Hoch, Peggy Marie *computer scientist*
Hua, Lulin *technological company executive, research scientist*
Kidwell, Michael Eades *engineering executive*
Levy, William Joel *endocrinologist*
Lippman, Muriel Marianne *biomedical scientist*
Lunin, Jesse *retired soil scientist*
Lynch, Sonia *data processing consultant*
Lynt, Richard King *microbiologist*
Mansky, Arthur William *computer engineer*
Mathur, Veerendra Kumar *physicist, researcher, project manager*
Mok, Carson Kwok-Chi *structural engineer*
Ostenso, Ned Allen *oceanographer, government official*
Resch, Lawrence Randel *mechanical engineer*
Restorff, James Brian *physicist*
Riel, Gordon Kienzie *research physicist*
Rodgers, Imogene Sevin *toxicologist*
Romberger, John Albert *scientist*
Schonholtz, George Jerome *orthopaedic surgeon*
Scipio, L(ouis) Albert, II *aerospace science engineering educator, architect, military historian*
Sharon, Michael *endocrinologist*
Shelton, William Chastain *retired statistician*
Short, Steve Eugene *engineer*
Telesetsky, Walter *government official*
Thorpe, Jack Victor *energy consultant*
Tokar, John Michael *oceanographer, ocean engineer*
Whitten, Charles Arthur *geodetic consultant*
Wilson, William Stanley *oceanographer*

Solomons
Baranowski, Paul Joseph *nuclear instrumentation technician*

Sparks
Nelson, John Howard *food company research executive*

Stevensville
Trescott, Sara Lou *water resources engineer*

Suitland
Guss, Paul Phillip *physicist*
Speier, Peter Michael *mathematics educator*
Vojtech, Richard Joseph *nuclear physicist*

Takoma Park
Yates, Renee Harris *economist*

Tantallon
Dickens, Doris Lee *psychiatrist*

Temple Hills
Day, Mary Jane Thomas cartographer
Strauss, Simon Wolf technical consultant

Timonium
Gupta, Ramesh Chandra geotechnical engineer, consultant

Towson
Huang, Joseph Chen-Huan civil engineer
Oppenheimer, Joel K. civil engineer
Redd, Rudolph James chemical engineer, consultant
Weaver, Kerry Alan construction engineer
Windler, Donald Richard botanist, educator
Wubah, Daniel Asua microbiologist

Upper Marlboro
Lisle, Martha Oglesby mathematics educator

Waldorf
Cochran, Ada data specialist, writer
Salvador, Mark Z. system safety engineer, aerospace engineer

Wheaton
Fawcett, Howard Hoy chemical health and safety consultant
Ghosh, Arun Kumar economics, social sciences and accounting educator

Woodbine
Daniels, Frederick Thomas reactor engineer
Herrick, Elbert Charles chemist, consultant

Woodstock
Ballweber, Hettie Lou archaeologist

MASSACHUSETTS

Acton
Egan, Bruce A. engineering consultant
Flood, Harold William chemical engineer, educator
Golden, John Joseph, Jr. manufacturing company executive
Richter, Edwin William physicist, editor, consultant

Amherst
Barde, Digambar Krushnaji manufacturing executive
Breger, Dwayne Steven solar energy research engineer
Bromery, Randolph Wilson geologist, educator
Cardé, Ring Richard Tomlinson entomologist, educator
DeVries, Geert Jan science educator
Godfrey, Paul Joseph science foundation director
Goldsby, Richard Allen biochemistry educator
Haensel, Vladimir chemical engineering educator
Hillel, Daniel soil physics and hydrology educator, researcher, consultant
MacKnight, William John chemist, educator
Maroney, Michael James chemistry educator
McBride, Thomas Craig physician, medical director
Morbey, Graham Kenneth management educator
O'Connor, Kevin Neal psychologist
Popplestone, Robin John computer scientist, educator
Popstefanija, Ivan electrical engineer
Ralph, James R. physician
Romer, Robert Horton physicist, educator
Sclove, Richard Evan technology educator, writer, consultant
Scott, David Knight physicist, university administrator
Sokolik, Igor physicist
Strom, Stephen Eric astronomer
Voigtman, Edward George, Jr. chemist educator
Walker, Robert Wyman environmental sciences educator
Winter, Horst Henning chemical engineer, educator
Woodbury, Richard Benjamin anthropologist, educator

Andover
Burke, Shawn Edmund mechanical engineer
Mac Neish, Richard Stockton archaeologist, educator
McCauley, Charles Irvin manufacturing engineer
Peck, A. William aerospace engineer
Rhoads, Kevin George consulting engineer
Tangarone, Bruce Steven biochemist

Arlington
Leonardos, Gregory chemist, odor consultant

Attleboro
Griffin, Edwin H., Jr. (Hank Griffin) chemist

Bedford
Alarcon, Rogelio Alfonso physician, researcher
Bounar, Khaled Hosie electrical engineer, consultant
Eldering, Herman George physicist
Frederickson, Arthur Robb physicist
Goel, Aditya Prasad engineering educator, consultant
Greenwald, Anton Carl physicist, researcher
Hicks, Walter Joseph electrical engineer, consultant
Hogan, Stephen John electrical engineer
Horowitz, Barry Martin systems research and engineering company executive
Jelalian, Albert V. electrical engineer
Johansen, Jack T. engineering company executive
Kim, Kwang Ho electrical engineer
Lackoff, Martin Robert engineer, physical scientist, researcher
Latimer, James Hearn systems engineer
Lopatin, George research chemist
Maravelias, Peter systems engineer
Nickerson, Raymond Stephen psychologist
Ren, Chung-Li engineer
Roche, Kerry Lee microbiologist
Siwek, Donald Fancher neuroanatomist
Smith, James Bigelow, Sr. electrical engineer
Winter, David Louis human factors scientist, systems engineer

Belmont
Chiu, Tak-Ming magnetic resonance physicist
Glines, Stephen Ramey software industry executive
Hauser, George biochemist, educator
Lex, Barbara Wendy medical anthropologist
McHenry, Douglas Bruce naturalist
Merrill, Edward Wilson chemical engineering educator
Montealegre, José Ramiro information systems consultant
Pope, Harrison Graham, Jr. psychiatrist, educator
Popper, Charles William child and adolescent psychiatrist
Washburn, Barbara cartography researcher

Beverly
Harris, Miles Fitzgerald meteorologist
Keating, Carole Joanna biotechnologist

Billerica
Bathey, Balakrishnan R. materials scientist
Brebbia, Carlos Alberto computational mechanic, engineering consultant
Carolan, Donald Bartley Abraham, Jr. electrical engineer
Conoby, Joseph Francis chemist
Holmes, Carl Kenneth mechanical engineer
KuLesza, Frank William chemical engineer
Miller, Dawn Marie meteorologist, product marketing specialist
Schmidt, James Robert facilities engineer
Yates, Barrie John chemical engineer

Boston
Abou-Samra, Abdul Badi biomedical researcher, physician
Adelstein, S(tanley) James physician, educator
Aldoori, Walid Hamid researcher
Alexander-Bridges, Maria Carmalita medical researcher
Allen, Ryne Cunliffe electrical engineer
Andrews, Sally May healthcare administrator
Anthony, Ethan architect
Avery, Mary Ellen pediatrician, educator
Baleja, James Donald biochemist, educator
Banerjee, Ajoy Kumar engineer, constructor, consultant
Baron, David Hume science journalist
Belkind-Gerson, Jaime gastroenterologist, nutritionist, researcher
Benacerraf, Baruj pathologist, educator
Bern, Murray Morris hematologist, oncologist
Berson, Eliot Lawrence ophthalmologist, medical educator
Bettinger, Jeffrie Allen chemical engineer
Bieber, Frederick Robert medical educator
Black, Paul Henry medical educator, researcher
Blout, Elkan Rogers biological chemistry educator, university dean
Bougas, James Andrew physician, surgeon
Brecher, Kenneth astrophysicist
Brenner, Barry Morton physician
Brenner, Brian Raymond structural engineer
Browning, William Elgar, Jr. chemistry consultant
Bucher, Gail Phillips cosmetic chemist
Bullock, Daniel Hugh computational neuroscience educator, psychologist
Burke, John Francis surgeon, educator, researcher
Burns, Padraic physician, psychiatrist, psychoanalyst, educator
Bush, John Burchard, Jr. consumer products company executive
Call, Katherine Mary biologist
Canalas, Robert Anthony nuclear engineer
Caplan, David Norman neurology educator
Carr, Daniel Barry anesthesiologist, endocrinologist, medical researcher
Carradini, Lawrence comparative biologist, science administrator
Carr-Locke, David Leslie gastroenterologist
Cave, David Ralph gastroenterologist, educator
Chishti, Athar H. biochemist
Chobanian, Aram Van physician
Cohen, Alan Seymour internist
Cohen, Robert Sonné physicist, philosopher, educator
Cristiani, Vincent Anthony counseling psychology educator
Crocker, Allen Carrol pediatrician
Crowley, William Francis, Jr. medical educator
Cushing, Steven educator, researcher, consultant
D'Agostino, Ralph Benedict mathematician, statistician, educator, consultant
De Cherney, Alan Hersh obstetrics and gynecology educator
Deligianis, Anthony chemical engineer
De Meyere, Robert Emmet civil engineer
Deresiewicz, Robert Leslie molecular biologist, physician, educator
Deutsch, Thomas Frederick physicist
Dilts, David Alan mechanical design engineer
Doherty, Robert Francis, Jr. aerospace industry professional
Dolnikowski, Gregory Gordon scientist
Donaldson, Robert Louis computer systems professional
Doody, Daniel Patrick pediatric surgeon
Doyle, Patrick Francis utility company executive
Druss, David Lloyd geotechnical engineer
Egdahl, Richard Harrison surgeon, medical educator, health science administrator
Ehrmann, Robert Lincoln pathologist
El-Baz, Farouk program director, educator
Elkowitz, Allan Barry information systems manager
Espy-Wilson, Carol Yvonne electrical engineer, educator
Fields, Bernard Nathan microbiologist, physician
Fitzpatrick, Thomas Bernard dermatologist, educator
Flansburgh, Earl Robert architect
Folkman, Moses Judah surgeon
Foote, Warren Edgar neuroscientist, psychologist, educator
Friedman, Lawrence Samuel gastroenterologist, educator
Gaintner, J(ohn) Richard health facility executive, medical educator
Gergely, John biochemistry educator
Gilmore, Maurice Eugene mathematics educator
Gilmore, Thomas David biologist
Glimcher, Melvin Jacob orthopedic surgeon
Goldin, Barry Ralph biochemistry researcher, educator
Gottlieb, Leonard Solomon pathology educator
Grenzebach, William Southwood nuclear engineer
Grous, John Joseph hematologist, oncologist, educator
Hagar, William Gardner, III photobiology educator
Hanley, Michelle Boucher civil engineer
Harrison, Robert Hunter psychology educator
Hay, Elizabeth Dexter embryology researcher, educator
Hayes, A(ndrew) Wallace toxicologist
Hein, John William dentist, educator
Hilkert, Robert Joseph cardiologist
Hoop, Bernard physicist, researcher, educator
Hornig, Donald Frederick scientist
Hubel, David Hunter physiologist, educator
Hutchinson, Martha LuClare pathologist
Ingber, Donald Elliot pathology and cell biology educator
Isberg, Ralph molecular biologist, educator
Isselbacher, Kurt Julius physician, educator
Jackson, Earl, Jr. medical technologist
Jacobson, Gary Ronald biology educator, researcher
Joyce-Brady, Martin Francis medical educator, physician, researcher
Karnovsky, Morris John pathologist, biologist
Kaye, Kenneth Marc physician, educator
Kazemi, Homayoun physician, medical educator
Kennedy, Eugene Patrick biochemist, educator
Kim, Samuel Homer pediatric surgeon
Kimura, Robert Shigetsugu otologic researcher
Kunkel, Louis Martens research scientist, educator
Lam, Bing Kit pharmacologist
Langdon, Geoffrey Moore architect, educator
Lanner, Michael research administrator, consultant
Lavine, Leroy Stanley orthopedist, surgeon, consultant
Leach, Robert Ellis physician, educator
Leaf, Alexander physician, educator
Leeman, Susan Epstein neuroscientist
LeMay, Marjorie Jeannette neuroradiologist
Levinsky, Norman George physician, educator
Levy, Boris chemistry educator
Linden, Lynette Lois bioelectrical engineer
Little, John Bertram physician, radiobiology educator, researcher
Loughlin, Kevin Raymond urologic surgeon
Lown, Bernard cardiologist, educator
Lu, Hsiao-ming physicist
Malenka, Bertram Julian physicist, educator
Mandell, Robert Lindsay periodontist, researcher
Mannick, John Anthony surgeon
Marcum, James Arthur physiology educator
Mekalanos, John J. microbiology educator
Michaud, Richard Omer financial economist, researcher
Moellering, Robert Charles, Jr. internist, educator
Moore, Francis Daniels, Jr. surgeon
Moore, Richard Lawrence structural engineer, consultant
Morgentaler, Abraham urologist, researcher
Murray, Mary Katherine reproductive biologist, educator
Nadim, Ali aerospace and mechanical engineering educator
Naeser, Margaret Ann linguist, medical researcher
Naimi, Shapur cardiologist, educator
Nathan, David Gordon physician, educator
Nathanson, James A neurologist
Newhouse, Joseph Paul economics educator
Nugent, Matthew Alfred biochemist
Nyberg, Stanley Eric cognitive scientist
Oas, John Gilbert neurologist, researcher
Pandelidis, Ioannis O. engineering manager
Pardee, Arthur Beck biochemist, educator
Parrish, John Albert dermatologist, research administrator
Potts, John Thomas, Jr. physician, educator
Prabhudas, Mercy Ratnavathy microbiologist, immunologist
Prescott, John Hernage aquarium executive
Puliafito, Carmen Anthony ophthalmologist, laser researcher
Rashkovetsky, Leonid biochemist
Raviola, Elio anatomist, neurobiologist
Reznicek, Bernard William power company executive
Rice, Peter Alan physician, scientist
Robertson, Erle Shervinton virologist, molecular biologist
Rodriguez, Agustin Antonio surgeon
Rosenberg, Irwin Harold physician, educator
Rühlmann, Andreas Carl-Erich Conrad biochemist
Ruvkun, Gary B. molecular geneticist
Ryan, Kenneth John physician, educator
Sage, James Timothy physicist, educator
Sager, Ruth geneticist
Savage, Deborah Ellen chemical engineer, environmental policy researcher
Schaaf, John Urban communication management specialist
Schaffer, Priscilla Ann virologist
Schaller, Jane Green pediatrician
Scheuzger, Thomas Peter audio engineer
Schlossman, Stuart Franklin physician, educator, researcher
Schneeberger, Eveline Elsa pathologist, cell biologist, educator
Scuderi, Louis Anthony climatology educator
Selker, Harry Paul medical educator
Sessoms, Allen Lee academic administrator, former diplomat, physicist
Shields, Lawrence Thornton orthopedic surgeon, educator
Siegel, Richard Allen economist
Sienkiewicz, Frank Frederick observatory curator
Sinai, Allen Leo economist, educator
Sledge, Clement Blount orthopedic surgeon, educator
Smith, Wendy Anne biologist, educator
Spengler, Kenneth C. meteorologist, professional society administrator
Stachel, John Jay physicist, educator
Stanley, H(arry) Eugene physicist, educator
Steele, Glenn Daniel, Jr. surgical oncologist
Strominger, Jack Leonard biochemist
Struhl, Kevin molecular biologist, educator
Stull, Donald LeRoy architect
Stygar, Michael David civil engineer
Surman, Owen Stanley psychiatrist
Swartz, Morton Norman medical educator
Thomas, Peter biochemistry educator
Thorburn, John Thomas, III marine engineer, consultant
Timmerman, Robert Wilson engineering executive
Tucker, Katherine Louise nutritional epidemiologist, educator
Turner, Raymond Edward chemistry educator, researcher
Vandam, Leroy David anesthesiologist, educator
Velkov, Simeon Hristov civil engineer
Vickers, Amy engineer
Von Fischer, George Herman social psychologist, unified social systems scientist, management consultant, data processing executive
Von Goeler, Eberhard physics educator
Walsh, Christopher Thomas biochemist, department chairman
Warth, James Arthur physician, researcher
Washburn, H. Bradford, Jr. museum administrator, cartographer, photographer
Wayland, Sarah Catherine cognitive psychologist
Weber, Georg Franz immunologist
Webster, Edward William medical physicist
White, George Edward pedodontist, educator
Willis, Jeffrey Scott mechanical engineer, consultant
Wolf, William Martin computer company executive, consultant
Wright, Daniel Godwin academic physician
Zaldastani, Othar structural engineer
Zervas, Nicholas Themistocles neurosurgeon

Boxford
Hill, Joseph Caldwell microwave engineer, consultant
Laderoute, Charles David engineer, economist, consultant

Braintree
Wilson, Christopher T. nuclear engineer

Brewster
Gumpright, Herbert Lawrence, Jr. dentist

Bridgewater
Daley, Henry Owen, Jr. chemist, educator

Brighton
Cohen, Jonathan orthopedic surgery educator, researcher
Stanton, Joseph Robert physician

Brockton
D'Agostino, Paul Anthony electrical engineer
Hodge-Spencer, Cheryl Ann orthodontist
O'Farrell, Timothy James psychologist
Park, Byiung Jun textile engineer

Brookline
Gutoff, Edgar Benjamin chemical engineer
Jacobson, Murray M. chemical engineer
Jakab, Irene psychiatrist
Katz, Israel engineering educator, retired
Kraut, Joel Arthur ophthalmologist
Perry, Frederick Sayward, Jr. corporate executive
Ross, Robert Nathan medical writer, consultant
Sand, Michael industrial designer
Strauss, Bruce Paul engineering physicist, consultant

Burlington
Anaebonam, Aloysius Onyeabo pharmacist
Olsen, Allen Neil civil engineer
Rabinowitz, Arthur Philip hematologist
Shaikh, Naimuddin medical physicist

Cambridge
Alberty, Robert Arnold chemistry educator
Anderson, James Gilbert chemistry educator
Ashford, Nicholas Askounes technology and policy educator
Augerson, William Sinclair internist
Ayyadurai, V.A. Shiva software engineer
Baker, James Gilbert optics scientist
Baldwin, Daniel Flanagan mechanical engineer
Baltimore, David microbiologist, educator
Barger, James Edwin physicist
Barry, Brenda Elizabeth respiratory biologist
Baum, Peter Samuel research director
Bazzaz, Fakhri A. plant biology educator, administrator
Benedek, George Bernard physicist, educator
Beranek, Leo Leroy scientific foundation executive, engineering consultant
Blair, Kim Billy mechanical engineer, researcher
Bloch, Konrad Emil biochemist
Bloembergen, Nicolaas physicist, educator
Bogorad, Lawrence biologist
Bott, Raoul mathematician, educator
Braida, Louis Benjamin Daniel electrical engineering educator
Branscomb, Lewis McAdory physicist
Brower, Michael Chadbourne research administrator
Brown, Robert Arthur chemical engineering educator
Brusch, John Lynch physician
Buchi, George Hermann chemistry educator
Burchfiel, Burrell Clark geology educator
Burke, Bernard Flood physicist, educator
Burns, Roger George mineralogist, educator
Canizares, Claude Roger astrophysicist, educator
Carrier, George Francis applied mathematics educator
Cheatham, Thomas Edward, Jr. computer scientist, educator
Chen, Peter chemistry educator
Chin, Aland Kwang-Yu physicist
Ciappenelli, Donald John chemist, academic administrator
Coate, David Edward acoustician consultant
Coleman, Sidney Richard physicist, educator
Colton, Clark Kenneth chemical engineering educator
Cook, Andrew Robert experimental chemical physicist
Cooper, Mary Campbell information services executive
Corbato, Fernando Jose electrical engineer and computer science educator
Corey, Elias James chemistry educator
Covert, Eugene Edzards aerophysics educator
Crandall, Stephen Harry engineering educator
Cumings, Edwin Harlan biology educator
Dame, Thomas Michael radio astronomer
Davidson, Frank Paul macro-engineer, lawyer
Davis, Edgar Glenn science and health policy executive
Demain, Arnold Lester microbiologist, educator
Diaconis, Persi W. mathematical statistician, educator
DiBerardinis, Louis Joseph industrial hygiene engineer, consultant, educator
Dickman, Steven Gary science writer
Dietz, Albert George Henry engineering educator
Doering, William von Eggers organic chemist, educator
Downes, Gregory architectural organization executive
Dresselhaus, Mildred Spiewak physics and engineering educator
Eagar, Thomas Waddy metallurgist, educator
Elkies, Noam D. mathematics educator
Emanuel, Kerry Andrew earth sciences educator
Erikson, Raymond Leo biology educator
Evans, David A(lbert) chemistry educator
Evans, Robley Dunglison physicist
Fang, Yue statistician
Farber, Neal Mark biotechnologist, molecular biologist

Fell, Barry (Howard Barraclough Fell) *marine biologist, educator*
Feshbach, Herman *physicist, educator*
Field, George Brooks *theoretical astrophysicist*
Field, Robert Warren *chemistry educator*
Flatté, Michael Edward *physicist, researcher*
Flemings, Merton Corson *engineer, materials scientist, educator*
Fox, Maurice Sanford *molecular biologist, educator*
French, Anthony Philip *physicist, educator*
Friedman, Jerome Isaac *physics educator, researcher*
Friend, Cynthia M. *chemist, educator*
Frosch, Robert Alan *retired automobile manufacturing executive, physicist*
Gagliardi, Ugo Oscar *systems software architect, educator*
Gallager, Robert Gray *electrical engineering educator*
Gardner, Howard Earl *psychologist, author*
Gay, David Holden *project technician*
Geller, Margaret Joan *astrophysicist, educator*
Gilbert, Walter *molecular biologist, educator*
Goldberg, Ray Allan *agribusiness educator*
Goldman, Peter *health science, chemistry, molecular pharmacology educator*
Goldstone, Jeffrey *physicist*
Golovchenko, Jene Andrew *physics and applied physics educator*
Gordon, George Stanwood, Jr. *physicist*
Gordon, Roy Gerald *chemistry educator*
Gordon, Steven Jeffrey *mechanical engineer*
Greenberg, Paul Ernest *economics consultant*
Gregory, Bruce Nicholas *astrophysicist, educator*
Greitzer, Edward Marc *aeronautical engineering educator, consultant*
Grindlay, Jonathan Ellis *astrophysics educator*
Grosz, Barbara Jean *computer science educator*
Grover, Mark Donald *computer scientist*
Guth, Alan Harvey *physicist, educator*
Hansen, Robert Joseph *civil engineer*
Harling, Otto Karl *nuclear engineering educator, researcher*
Hartl, Daniel Lee *genetics educator*
Hebert, Benjamin Francis *mechanical engineer*
Helgason, Sigurdur *mathematician, educator*
Herschbach, Dudley Robert *chemistry educator*
Hillis, William Daniel *scientist, engineer*
Holm, Richard Hadley *chemist, educator*
Holmberg, Eva Birgitta *research scientist*
Holmes, Donna Jean *biologist*
Holzman, Philip Seidman *psychologist, educator*
Howitt, Andrew Wilson *scientist*
Hsiao, William C. *economist, actuary educator*
Hwa, Terence Tai-Li *physicist*
Hynes, Richard Olding *biology educator*
Jacobsen, Edward Hastings *physicist*
Jacobson, Ralph Henry *laboratory executive, former air force officer*
Jaffe, Arthur Michael *physicist, mathematician, educator*
Johnson, Mark Allan *electrical engineer*
Jones, Richard Victor *physics and engineering educator*
Joshi, Amol Prabhatchandra *chemical engineer*
Joskow, Paul Lewis *economist, educator*
Kapor, Mitchell David *foundation executive*
Karplus, Martin *chemistry educator*
Keck, James Collyer *physicist, educator*
Kelley, Albert Joseph *management educator, executive consultant*
Kendall, Henry Way *physicist*
Kennedy, Stephen Dandridge *economist, researcher*
Keramas, James George *engineering educator*
Kerman, Arthur Kent *physicist, educator*
Khorana, Har Gobind *chemist, educator*
Kim, Peter Sung-bai *biochemistry educator*
King, Jonathan Alan *molecular biology educator*
King, Ronold Wyeth Percival *physics educator*
Kirshner, Robert P. *astrophysicist, educator*
Klaubert, Earl Christian *chemical and general engineering administrator, consultant*
Kleppner, Daniel *physicist, educator*
Klibanov, Alexander Maxim *chemistry educator*
Knowles, Jeremy Randall *chemist, educator*
Kobus, Richard Lawrence *architectural company executive*
Kung, H. T. *computer science and engineering educator, consultant*
Lada, Elizabeth A. *astronomer*
Ladd, Charles Cushing, III *civil engineering educator*
Ladino, Cynthia Anne *cell biologist*
La Mantia, Charles Robert *management consulting company executive*
Lamberg-Karlovsky, Clifford Charles *anthropologist, archaeologist*
Lampson, Butler Wright *computer scientist*
Langer, Robert Samuel *chemical, biochemical engineering educator*
Latanision, Ronald Michael *materials science and engineering educator, consultant*
Layzer, David *astrophysicist, educator*
Lee, Patrick A. *physics educator*
Lee, Thomas Henry *electrical engineer, educator*
LeMessurier, William James *structural engineer*
Li, Yonghong *physics researcher*
Lindzen, Richard Siegmund *meteorologist, educator*
Lippard, Stephen James *chemist, educator*
Lipscomb, William Nunn, Jr. *retired physical chemistry educator*
Litster, James David *physics educator, dean*
Liu, Jianguo *ecologist*
Lomet, David Bruce *computer scientist*
Lorenz, Edward Norton *meteorologist, educator*
Luu, Jane *astronomer*
Lynch, Harry James *biologist*
Lynch, Nancy Ann *computer scientist, educator*
Lyon, Richard Harold *educator, physicist*
Mackey, George Whitelaw *educator, mathematician*
MacPherson, Robert Duncan *mathematician, educator*
Magasanik, Boris *microbiology educator*
Magnanti, Thomas L. *business management educator*
Mann, Robert Wellesley *biomedical engineer, educator*
Manoharan, Ramasamy *biomedical scientist*
Marks, David Hunter *civil engineering educator*
Marsden, Brian Geoffrey *astronomer*
Martin, Paul Cecil *physicist, educator*
Martinez, Luis Enrique, Jr. *chemist researcher*
Mayr, Ernst *emeritus zoology educator, author*
McCarthy, James Joseph *oceanography educator*
Mc Cune, William James, Jr. *manufacturing company executive*
McElroy, Michael *physicist, researcher*
Meador, Charles Lawrence *management and systems consultant, educator*
Mehrabi, M. Reza *chemical engineer*
Mendelsohn, Everett Irwin *history of science educator*
Meselson, Matthew Stanley *biochemist, educator*
Miao, Shili *plant biology educator, plant ecology researcher*
Minsky, Marvin Lee *mathematician, educator*
Mitter, Sanjoy K. *electrical engineering educator*
Molina, Mario Jose *physical chemist, educator*
Moniz, Ernest Jeffrey *physics educator*
Moore, Sally Falk *anthropology educator*
Morel, François M.M. *civil and environmental engineering educator*
Morgenthaler, Ann Welke *electrical engineer*
Mosteller, Frederick *mathematical statistician, educator*
Mulligan, Richard C. *molecular biology educator*
Mumford, David Bryant *mathematics educator*
Narayan, Ramesh *astronomy educator*
Negele, John William *physics educator, consultant*
Oppenheim, Irwin *chemical physicist, educator*
Orlen, Joel *association executive*
Papert, Seymour Aubrey *mathematician, educator, writer*
Paul, William *physicist, educator*
Pestana-Nascimento, Juan M. *civil, geotechnical and geoenvironmental engineer, consultant*
Pfister, Donald Henry *biology educator*
Pian, Theodore Hsueh-Huang *engineering educator, consultant*
Piattelli-Palmarini, Massimo *cognitive scientist*
Pierce, Naomi Ellen *biology educator, researcher*
Pilbeam, David Roger *paleoanthropology educator*
Pineau, Daniel Robert *structural engineer*
Pinker, Steven A. *cognitive science educator*
Poggio, Tomaso Armando *physicist, educator, computer scientist, researcher*
Poon, Chi-Sang *biomedical engineering researcher*
Porkolab, Miklos *physics educator, researcher*
Postol, Theodore A. *physicist, educator*
Press, William Henry *astrophysicist, computer scientist*
Pritchard, David Edward *physics educator*
Purcell, Edward Mills *physics educator*
Rabin, Michael O. *computer scientist, mathematician*
Rahman, Anwarur *physicist*
Ramsey, Norman F. *physicist, educator*
Rasmussen, Norman Carl *nuclear engineer*
Raymond, John Charles *physicist*
Revol, Jean-Pierre Charles *physicist*
Rha, ChoKyun *biomaterials scientist and engineer, researcher, educator, inventor*
Rhodes, William George, III *physicist*
Robbins, Phillips Wesley *biology educator*
Roberts, Edward Baer *technology management educator*
Rockart, John Fralick *information systems reseacher*
Rosenof, Howard Paul *electrical engineer*
Rosenthal, Robert *psychology educator*
Rowe, Stephen Cooper *venture capitalist, entrepreneur*
Runge, Erich Karl Rainer *physicist*
Russell, Kenneth Calvin *metallurgical engineer, educator*
Samanta Roy, Robie Isaac *aerospace engineer*
Sanders, John Lyell, Jr. *educator, researcher*
Saponaro, Joseph A. *company executive*
Schimmel, Paul Reinhard *biochemist, biophysicist, educator*
Schreiber, Stuart L. *chemist, educator*
Schultes, Richard Evans *ethnobotanist, museum executive, educator, conservationist*
Schwedock, Julie *molecular biologist*
Segal, Irving Ezra *mathematics educator*
Shapiro, Irwin Ira *physicist, educator*
Sharp, Phillip Allen *academic administrator, biologist, educator*
Siever, Raymond *geology educator*
Singer, Isadore Manuel *mathematician, educator*
Smith, Peter Lloyd *physicist*
Snipes, Joseph Allan *research scientist, physicist*
Soydemir, Cetin *geotechnical engineer, earthquake engineer*
Spaepen, Frans August *applied physics researcher, educator*
Steinfeld, Jeffrey Irwin *chemistry educator, consultant, author*
Stephanopoulos, Gregory *chemical engineering educator, consultant, researcher*
Stevens, Kenneth Noble *electrical engineering educator*
Stewart, Sue Ellen *molecular biologist*
Storch, Joel Abraham *mathematician*
Stout, Philip John *chemist*
Stubbe, JoAnne *chemistry educator*
Sussman, Gerald Jay *electrical engineering educator*
Sweeney, Bryan Philip *structural and geotechnical engineer*
Swets, John Arthur *psychologist, scientist*
Szekely, Julian *materials engineering educator*
Taubes, Clifford H. *mathematician educator*
Thomas, Edwin L. *materials engineering educator*
Thompson, Benjamin *architect*
Ting, Samuel Chao Chung *physicist, educator*
Tonegawa, Susumu *biology educator*
Toomre, Alar *mathematics educator, theoretical astronomer*
Trilling, Leon *aeronautical engineering educator*
Turkle, Sherry *sociologist, psychologist, educator*
Valiant, Leslie Gabriel *computer scientist*
Vander Velde, Wallace Earl *aeronautical and astronautical educator*
vanRiper, William John *computer graphics scientist, consultant*
Venkatesh, Yeldur Padmanabha *biochemist, researcher*
Vercelli, Donata *immunologist, educator*
Vest, Charles Marstiller *university administrator*
Vincent, James Louis *biotechnology company executive*
Virgil, Scott Christopher *chemistry educator*
Wacker, Warren Ernest Clyde *physician, educator*
Wald, George *biochemist, educator*
Walker, Graham Charles *biology educator*
Wang, James Chuo *biochemistry and molecular biology educator*
Wang, Jian-Sheng *materials scientist*
Watson, Joyce Margaret *observatory librarian*
Waugh, John Stewart *chemist, educator*
Weiler, Paul Cronin *law educator*
Weimar, Robert Alden *environmental engineering executive*
Weinberg, Robert Allan *biochemist, educator*
Westheimer, Frank Henry *chemist, educator*
Whitesides, George McClelland *chemistry educator*
Wiegand, Thomas Edward von *psychophysicist, consultant*
Wiesner, Jerome Bert *engineering educator, researcher*
Wiley, Don Craig *biochemistry and biophysics educator*
Wilson, Edward Osborne *biologist, educator*
Wilson, Mary Elizabeth *physician*
Wilson, Richard *physicist, educator*
Winkler, Gunther *medical executive*
Witt, August Ferdinand *aerospace scientist, educator*
Wood, Jeremy Scott *architect, urban designer*
Wuensch, Bernhardt John *ceramic engineering educator*
Wurtman, Richard Jay *physician, educator*
Yannas, Ioannis Vassilios *polymer science and engineering educator*
Yau, Shing-Tung *mathematics educator*
Zau, Gavin Chwan Haw *chemical engineer*

Canton
Yerby, Joel Talbert *anesthesiologist, pain management consultant*

Carlisle
Guarnaccia, David Guy *mechanical engineer*

Centerville
Rocher, Edouard Yves *computer engineer*

Charlestown
Chung-Welch, Nancy Yuen Ming *biologist*
Lizak, Martin James *physicist*

Chelmsford
Lerner, James Peter *software engineer*
Sepucha, Robert C. *chemical physicist, optics scientist*

Chestnut Hill
Bakshi, Pradip M. *physicist*
Bennett, Ovell Francis *chemistry educator, researcher, consultant*
Gonsalves, Robert Arthur *electrical engineering educator, consultant*
Hoffman, Charles Stuart *molecular geneticist*
Kosasky, Harold Jack *gynecologist*
Mohanty, Udayan *chemical physicist, theoretical chemist*
Stanbury, John Bruton *physician, educator*

Concord
Figwer, Jozef Jacek *acoustics consultant*
Hogan, Daniel Bolten *management consultant*
Palay, Sanford Louis *retired scientist, educator*
Plauger, P.J. *science writer*

Conway
Powsner, Gary *computer consultant*

Danvers
Dutta, Arunava *chemical engineer*
Hathaway, Alden Moinet, II *electrical engineer*
Newell, Philip Bruce *physicist, lighting engineer*
Tiernan, Robert Joseph *research and development engineer*

Dighton
Chu, David Yuk *chemical engineer*

Dover
Bonis, Laszlo Joseph *business executive, scientist*

Duxbury
Hillman, Robert Edward *biologist*

East Falmouth
Dixon, Brian Gilbert *chemist*

Fall River
Weltman, Joel Kenneth *immunologist*

Falmouth
Allen, Duff Shederic, Jr. *retired chemist*
Hollister, Charles Davis *oceanographer*
Sato, Kazuyoshi *pathologist*

Feeding Hills
Norris, Pamela *school psychologist*

Fitchburg
Jackson, Jimmy Lynn *engineer, consulting spectroscopist*
Nomishan, Daniel Apesuur *science and mathematics educator*
Wickstrom, Carl Webster *analytical chemist*

Foxboro
Ghosh, Asish *control engineer, consultant*
Pierce, Francis Casimir *civil engineer*
Ryskamp, Carroll Joseph *chemical engineer*

Framingham
Bose, Amar Gopal *electrical engineering educator*
Castelli, William *cardiovascular epidemiologist, educator*
Cornell, Charles Alfred *engineering executive*
Eastman, Robert Eugene *electronic engineer, consultant*
Fedorowicz, Jane *information systems educator*
Roe, Georgeanne Thomas *information brokerage executive*

Gardner
Wagenknecht, Walter Chappell *radiologist*

Georgetown
O'Brien, John Steininger *clinical psychologist*

Granby
Shaheen, William A. *civil engineering educator, consultant*

Groton
Kirshen, Paul Howard *water resources consultant*

Hanscom AFB
Ahmadjian, Mark *chemist*
Coté, Marc George *electrical engineer*
Kirkwood, Robert Keith *applied physicist*
Lai, Shu Tim *physicist*
Paulson, John Frederick *research chemist*
Sargent, Douglas Robert *air force officer, engineer*
Shepherd, Freeman Daniel *physicist*
Van Tassel, Roger Alan *infrared phenomenologist, geophysicist*

Harwich
Volpicelli, Frederick Gabriel *computer scientist*

Hatfield
Borgatti, Douglas Richard *environmental engineer*

Haverhill
DeSchuytner, Edward Alphonse *biochemist, educator*
Niccolini, Drew George *gastroenterologist*
Woo, Buck Hong *neuropsychologist*

Holbrook
Noyes, Walter Omar *tree surgeon*

Holden
Gordon, Steven B. *chemist*

Holliston
Epstein, Scott Mitchell *engineering executive*

Hopkinton
Leamon, Tom B. *industrial engineer, educator*
Liou, Jenn-Chorng *electronics engineer*
MacLean, Mary Elise *chemist*
Novich, Bruce Eric *materials engineer*
Svrluga, Richard Charles *science and technology executive*

Hudson
Raina, Rajesh *computer engineer*

Hyannis
Dubuque, Gregory Lee *medical physicist*
Williams, Ann Meagher *hospital administrator*

Ipswich
Jennings, Frederic Beach, Jr. *economist, consultant*

Jamaica Plain
Arbeit, Robert David *physician*
Cook, Robert Edward *educator, plant ecology researcher*
Pierce, Chester Middlebrook *psychiatrist, educator*

Kingston
Slot, Larry Lee *molecular biologist*

Lenox
Krofta, Milos *engineer*
Shammas, Nazih Kheirallah *environmental engineering educator*

Lexington
Bartlett, Paul Doughty *chemist, educator*
Berstein, Irving Aaron *biotechnology and medical technology executive*
Blanchard, Robert Lorne *engineering physicist*
Bowden, Denise Lynn *civil engineer*
Brown, Elliott Rowe *physicist*
Buchanan, John Machlin *biochemistry educator*
Callerame, Joseph *physicist*
Cathou, Renata Egone *chemist, consultant*
Cooper, William Eugene *consulting engineering executive*
Dionne, Gerald Francis *research physicist*
Ehrlich, Daniel Jacob *optical engineer, optical scientist*
Eng, Richard Shen *electrical engineer*
Gibbs, Martin *biologist, educator*
Guivens, Norman Roy, Jr. *mathematician, engineer*
Halberstam, Isidore Meir *meteorologist*
Haldeman, Charles Waldo, III *aeronautical engineer*
Hardy, John W. *optics scientist*
Heath, Gregory Ernest *electrical engineer*
Jensen, Mona Dickson *chemist researcher*
Kautz, Frederick Alton, II *aerospace engineer*
Keicher, William Eugene *electrical engineer*
Kestigian, Michael *scientist*
McClure, David Woodard *electrical engineer*
Morrow, Walter Edwin, Jr. *electrical engineer, university laboratory administrator*
Nuthmann, Conrad Christopher *civil engineer, consultant*
Samour, Carlos M. *chemist*
Schloemann, Ernst Fritz (Rudolf August) *physicist, engineer*
Schonhorn, Harold *chemist, researcher*
Silva-Tulla, Francisco *civil engineer*
Stiglitz, Martin Richard *electrical engineer, writer*
Tustison, Randal Wayne *materials scientist*
Weiss, Scott Jeffrey *civil engineer*

Lincoln
Dunwiddie, Peter William *plant ecologist*
LeGates, John Crews Boulton *information scientist*

Longmeadow
Ferris, Theodore Vincent *chemical engineer, consulting technologist*

Lowell
Awerbuch, Shimon *research consultant*
Chandra, Kavitha *electrical engineer, educator*
Colling, David Allen *industrial engineer, educator*
Hojnacki, Jerome Louis *university dean*
Jourjine, Alexander N. *theoretical physicist*
Kyriacou, Demetrios *electro-organic chemist, researcher, consultant*
Martín, José Ginoris *nuclear engineer*
McCabe, Douglas Raymond *analytical chemist*
O'Connor, Denis *mechanical engineer*
Reinisch, Bodo Walter *electrical engineering educator*
Sebastian, Kunnat Joseph *physics educator*
Shina, Sammy Gourgy *engineering educator, consultant*

Ludlow
Budnick, Thomas Peter *social worker*

Lunenburg
Rhodin, Anders G.J. *orthopaedic surgeon, chelonian researcher and herpetologist*

Lynn
Brogunier, Claude Rowland *environmental engineer*

Ehrich, Fredric F. *aeronautical engineer*
Hatch, Mark Bruce *software engineer*

Lynnfield
Paradis, Richard Robert *energy conservation engineer*

Malden
Rollins, Scott Franklin *chemist*

Manchester
Gaythwaite, John William *civil engineer*

Manomet
Castro, Gonzalo *ecologist*

Marblehead
Krebs, James Norton *retired electric power industry executive*

Marlborough
Pittack, Uwe Jens *engineer, physicist*
Szmanda, Charles Raymond *chemist*
Tramontozzi, Louis Robert *electrical engineer*
Walton, Thomas Cody *chemical plastics engineer*

Medfield
Gradijan, Jack Robertson *software engineer*
Slovacek, Rudolf Edward *biochemist*

Medford
Balabanian, Norman *electrical engineering educator*
Blanco, Frances Rebecca Briones *chemist, medical technologist*
Cormack, Allan MacLeod *physicist, educator*
Garrelick, Joel Marc *acoustical scientist, consultant*
Hankour, Rachid *civil engineer*
Junger, Miguel Chapero *acoustics researcher*
Kafka, Tomas *physicist*
Kinsbourne, Marcel *neurologist, behavioral neuroscientist*
Liszczak, Theodore Michael *university administrator*
Nelson, Frederick Carl *mechanical engineering educator, academic administrator*
Trefethen, Lloyd MacGregor *engineering educator*

Medway
Yonda, Alfred William *mathematician*

Milford
Nicoson, Steven Wayne *geotechnical engineer*

Millbury
Pan, Coda H. T. *mechanical engineering educator, consultant, researcher*

Milton
Hanley, Joseph Andrew *civil engineer*

Nahant
Lavalli, Kari Lee *marine biologist*

Natick
Akkara, Joseph Augustine *biochemist, educator*
Andreotti, Raymond Edward *chemical engineering technologist*
DeCosta, Peter F. *chemical engineer*
Gionfriddo, Maurice Paul *research and development manager, aeronautical engineer*
Jones, Bruce Hovey *physician, researcher*
Lee, Calvin K. *aerospace engineer*
Neumeyer, John Leopold *research company administrator, chemistry educator*
Spada, Marianne Rina *medicinal chemist*
Stillman, Michael James *neuroscientist*
Wang, Chia Ping *physicist, educator*

Needham
Weller, Thomas Huckle *physician, emeritus educator*

New Bedford
Merolla, Michele Edward *chiropractor*

Newton
Adams, Onie H. Powers (Onie H. Powers) *retired biochemist*
Barclay, Stanton Dewitt *engineering executive, consultant*
Cavicchi, Leslie Scott *health facility administrator*
Heyn, Arno Harry Albert *retired chemistry educator*
Quinlan, Kenneth Paul *chemist*
Thompson, Stephen Arthur *publishing executive*
Vasilakis, Andrew D. *mechanical engineer*
Wadzinski, Henry Teofil *physicist*
Weisskopf, Victor Frederick *physicist*

Newton Center
Weiner, Melvin Milton *electrical engineer*

Newton Upper Falls
Tucker, Richard Douglas *computer engineer, consultant*

North Andover
Gladstein, Martin Keith *electrical engineer*
Siller, Curtis Albert *electrical engineer*

North Attleboro
Corridori, Anthony Joseph *engineering executive*

North Billerica
Wong, David Chungyao *scientist*

North Chatham
Hiscock, Richard Carson *marine safety investigator*

North Dartmouth
Caverly, Robert H. *electrical engineer, educator*
Cory, Lester Warren *electrical engineering educator*
Lemay, Gerald J. *electrical engineer, educator*
Read, Dorothy Louise *biology educator*
Sauro, Joseph Pio *physics educator*

North Grafton
Loew, Franklin Martin *veterinary medical and biological scientist, university dean*
Rowan, Andrew Nicholas *biologist, educator*

Northampton
Carfora, John Michael *economics and political science educator*
Decowski, Piotr *physicist*
Hayssen, Virginia *mammalogist, educator*
Mintzer, Paul *ophthalmologist, educator*
Sommer, Alfred Hermann *retired physical chemist*
Tallent, Norman *psychologist*
White, Virginia *chemistry educator*

Northborough
Jeas, William C. *electronics and aerospace engineering executive*

Norwood
Imbault, James Joseph *electromechanical engineering executive*
Seder, Richard Henry *physician*

Oxford
Schur, Walter Robert *physician*

Peabody
Lipman, Richard Paul *pediatrician*
Meader, John Leon *environmental engineer, consultant*
Peters, Leo Francis *environmental engineer*
Terlizzi, James Vincent, Jr. *secondary education educator*

Pembroke
Heinmiller, Robert H., Jr. *communications company executive*

Pittsfield
Stall, Trisha Marie *mechanical engineer*
Wheelock, Kenneth Steven *chemist*

Quincy
Bennett, Bruce Anthony *civil engineer*
Bernardone, Jeffrey John *podiatrist*
Giberson, Karl Willard *physics and philosophy educator*
Mancini, Rocco Anthony *civil engineer*
Shalit, Bernard Lawrence *dentist*

Randolph
Jelley, Scott Allen *microbiologist*

Reading
Carlotto, Mark Joseph *electrical and computer engineer, researcher*
Gelb, Arthur *business executive, electrical and systems engineer*
Melconian, Jerry Ohanes *engineering executive*
Tuttle, David Bauman *data processing executive*

Rockland
Doucette, Paul Stanislaus *environmental scientist*
Karabots, Joseph William *environmental engineering company executive*
Manns, Roy Lokumal *polymer technologist, researcher, executive*

Rockport
Gavelis, Jonas Rimvydas *dentist, educator*
Hull, Gordon Ferrie *physicist*

Roxbury
Berman, Marlene Oscar *neuropsychologist, educator*
Peters, Alan *anatomy educator*

Salem
Brown, Walter Redvers John *physicist*
Kaiser, Kurt Boye *physicist*
Khattak, Chandra Prakash *materials scientist*
Phillips, Bessie Gertrude Wright *school system administrator, museum trustee*
Vaughan, Margaret Evelyn *psychologist, consultant*
Young, Alan M. *biologist, educator*

Sandwich
Chambers, Edward Allen *aerophysical systems designer, consultant engineer*
Mattson, Clarence Russell *safety engineer*

Sharon
Cahn, Glenn Evan *psychologist*
Honikman, Larry Howard *pediatrician*

Shrewsbury
Aghajanian, John Gregory *electron microscopist, cell biologist*
Garabedian, Charles, Jr. *mathematics educator*
Paredes, Eduardo *medical physicist*
Pederson, Thoru Judd *biologist, research institute director*
Sassen, Georgia *psychologist*
Zamecnik, Paul Charles *oncologist, medical research scientist*

Shutesbury
Smulski, Stephen John *wood scientist, consultant*

Somerville
Cordingley, James John *systems engineer*
Kaltsos, Angelo John *electronics executive, educator, photographer*
Leventis, Nicholas *chemist, consultant*
Resmini, Richard *electrical engineer*
Stahovich, Thomas Frank *mechanical engineer, researcher*
Stoddard, Forrest Shaffer *aerospace engineer, educator*

South Dennis
Svikla, Alius Julius *pharmacist*

South Hadley
Harrison, Anna Jane *chemist, educator*
Williamson, Kenneth Lee *chemistry educator*

South Walpole
Leung, Woon-Fong *mechanical engineer, research scientist*

Southborough
Madras, Bertha Kalifon *neuroscientist, consultant*

Springfield
Andrzejewski, Chester, Jr. *immunologist, research scientist*
Cohen, Saul Mark *chemist, retired*
Farkas, Paul Stephen *gastroenterologist*
Frankel, Kenneth Mark *thoracic surgeon*
Friedmann, Paul *surgeon, educator*
Harris, Roger Scott *energy consultant*
Kottamasu, Mohan Rao *physician*
Masi, James Vincent *electrical engineering educator*
Oestreicher, Michael Christopher *architect*
Rahnamai, Kourosh Jonathan *electrical engineering educator*
Reed, William Piper, Jr. *surgeon, educator*

Stockbridge
Rothenberg, Albert *psychiatrist, educator*

Stoughton
Lamarque, Maurice Patrick Jean *health products executive*

Stow
Kinsella, Daniel John *electrical engineer*

Sturbridge
Feller, Winthrop Bruce *physicist*

Sudbury
Blackey, Edwin Arthur, Jr. *geologist*
Burgarella, John Paul *electronics engineer, consultant*
McManus, John Gerard *software engineer*
Wallace, James Jr. *engineering executive, researcher*
White, Bertram Milton *chemicals executive*

Swampscott
Neumann, Gerhard *mechanical engineer*

Taunton
Barbour, Robert Charles *technology executive*
Donovan, Stephen James *mechanical engineer*
Shastry, Shambhu Kadhambiny *scientist, engineering executive*

Tewksbury
Dumont, Michael Gerard *electro-optical engineer*

Topsfield
Isler, Norman John *aircraft engine company administrator, consultant*
Viviani, Gary Lee *electrical engineer*

Townsend
Greene, Roland *chiropractor*

Vineyard Haven
Billingham, Rupert Everett *zoologist, educator*
Knowles, Christopher Allan *healthcare executive*

Waban
Aisner, Mark *internist*

Wakefield
By, Andre Bernard *engineering executive, research scientist*
Goldberg, Harold Seymour *electrical engineer, academic administrator*
Magaw, Jeffrey Donald *civil engineer*
Robinson, Harold Wendell, Jr. *systems engineer*
Scarpa, Robert Louis *civil engineer*

Waltham
Caspar, Donald Louis Dvorak *physics and structural biology educator*
Chanenchuk, Claire Ann *chemical engineer, market developer*
Deser, Stanley *educator, physicist*
Fasman, Gerald David *biochemistry educator*
Ganong, William Francis, III *speech sciences research executive*
Hitchcock, Christopher Brian *computer and environmental scientist, policy services executive*
Huxley, Hugh Esmor *molecular biologist, educator*
Jencks, William Platt *biochemist, educator*
Kustin, Kenneth *chemist*
Lackner, James Robert *aerospace medicine educator*
McBrearty, Sally Ann *archaeologist*
Mitchell, Janet Brew *health services researcher*
Nandy, Subas *chemical engineer, consultant*
O'Donnell, Teresa Hohol *software development engineer, antennas engineer*
Perez-Ramirez, Bernardo *biochemist, researcher, educator*
Petit, William *chemist*
Roth, Peter Hans *chemical engineer*
Timasheff, Serge Nicholas *chemist, educator*
Tobet, Stuart Allen *neurobiologist*
Zebrowitz, Leslie Ann *psychology educator*

Waquoit
Saunders, John Warren, Jr. *biology educator, consultant*

Watertown
Kim, Byung Kyu *hematologist, consultant*
Lin, Alice Lee Lan *physicist, researcher, educator*
Moskowitz, Richard *physician*
Oakes, Carlton Elsworth *physicist*
Sahatjian, Ronald Alexander *science foundation executive*

Wayland
Clayton, John *engineering executive*
Drevinsky, David Matthew *civil engineer*
Kant, Arthur *scientist emeritus, physical chemist*
Puzella, Angelo *microwave engineer*

Wellesley
Murray, Joseph Edward *plastic surgeon*
Picardi, Anthony Charles *systems engineer, market research consultant*
Reif, Arnold E. *medical educator*

Wellfleet
Jentz, John Macdonald *engineer, travel executive*

West Bridgewater
Kirby, Kevin Andrew *utilities company executive*

West Roxbury
Charness, Michael Edward *neurologist, neuroscientist*

West Springfield
McKenzie, Rita Lynn *psychologist*

Westborough
Cardoza, James Ernest *wildlife biologist*
Lindquist, Dana Rae *mechanical engineer*
Oliver, David Edwin *physicist*
Robinson, Donald Edward *electrical power industry executive*
Tessier, Mark A. *electrical engineer*
Tse, Po Yin *electrical engineer*

Westfield
Buckmore, Alvah Clarence, Jr. *computer scientist, ballistician*
Taylor, James Kenneth *biology educator, science teaching consultant*

Westford
Haramundanis, Katherine Leonora *computer and data processing company executive*
Sobie, Walter Richard *semiconductor engineer*

Weston
Resden, Ronald Everette *medical devices product development engineer*
Wells, Lionelle Dudley *psychiatrist*

Westport
Stevens, Herbert Howe *mechanical engineer, consultant*

Westwood
Thomas, Abdelnour Simon *software company executive*

Weymouth
Johnson, Mark David *anesthesiologist, educator*

Whitman
Thompson, Andrew Ernest *mathematics educator*

Wilbraham
Dailey, Franklyn Edward, Jr. *electronic image technology company executive, consultant*
Lovell, Walter Carl *engineer, inventor*

Williamstown
Strait, Jefferson *physicist, educator*
Williams, Heather *neuroethologist, educator*

Wilmington
Boxleitner, Warren James *electrical engineer, researcher*
Colby, George Vincent, Jr. *electrical engineer*
Faccini, Ernest Carlo *mechanical engineer*
McCard, Harold Kenneth *aerospace company executive*
Reeves, Barry Lucas *research engineer*

Winchester
Agrawal, Vimal Kumar *structural engineer*
Jabre, Eddy-Marco *architect*
Joseph, Paul Gerard *civil engineer, consultant*

Woburn
Cox, Terrence Guy *manufacturing automation executive*
Driedger, Paul Edwin *pharmaceutical researcher*
Mehra, Raman Kumar *data processing executive, automation and control engineering researcher*
Steinbrecher, Donald Harley *electrical engineer*

Woods Hole
Anderson, Donald Mark *biological oceanographer*
Ballard, Robert Duane *marine scientist*
Berggren, William Alfred *geologist, research micropaleontologist, educator*
Broadus, James Matthew *research center administrator*
Burris, John Edward *biologist*
Caswell, Hal *mathematical ecologist*
Dorman, Craig Emery *oceanographer, academic administrator*
Ebert, James David *research biologist, educator*
Hart, Stanley Robert *geochemist, educator*
Inoué, Shinya *microscopy and cell biology scientist, educator*
Milliman, John D. *oceanographer, geologist*
Stanton, Timothy Kevin *physicist*
Steele, John Hyslop *marine scientist, oceanographic institute administrator*
Stewart, William Kenneth, Jr. *ocean engineer*
Stone, Thomas Alan *geologist*
Uchupi, Elazar *geologist, researcher*
Woodwell, George Masters *ecologist, educator, author, lecturer*

Worcester
Andis, Michael D. *urban entomologist*
Apelian, Diran *materials scientist, provost*
Bell, Peter Mayo *geophysicist*
Berrios, Javier *electrical engineer, consultant*
Boss, Michael Alan *biologist*
Escott, Shoolah Hope *microbiologist*
Latham, Eleanor Ruth Earthrowl *neuropsychology therapist*
Levin, Peter Lawrence *electrical engineering educator*
Michelson, Alan David *pediatric hematologist*
Norton, Robert Leo, Sr. *mechanical engineering educator, researcher*
Och, Mohamad Rachid *psychiatrist, consultant*
Reilly, Judith Gladding *physics educator*
Rotithor, Hemant Govind *electrical engineering educator*
Shalhoub, Victoria Aman *molecular biochemist, researcher*
Stavnezer, Janet Marie *immunologist*
Temsamani, Jamal *molecular biologist, researcher*
Theroux, Steven Joseph *biologist, educator*
Volkert, Michael Rudolf *molecular geneticist*

MICHIGAN

Ada
Calvert, George David consumer products company executive

Adrian
Cox, Chad William medical technologist
Haddad, Inad physician
Scioly, Anthony Joseph chemistry and physics educator

Allen Park
Kaldor, George pathologist, educator

Alma
Reed, Bruce Cameron physics educator, astronomy researcher

Ann Arbor
Agranoff, Bernard William biochemist, educator
Anderson, William R. biologist, educator, curator, director
Arts, Henry Alexander otolaryngologist
Ash, Major McKinley, Jr. dentist, educator
Atreya, Arvind mechanical engineering educator
Aupperle, Eric Max data network center administrator, research scientist, engineering educator
Banks, Peter Morgan electrical engineering educator
Beeton, Alfred Merle laboratory director, limnologist, educator
Bhattacharya, Pallab Kumar electrical engineering educator, researcher
Bilello, John Charles materials science and engineering educator
Brakel, Linda A. Wimer psychoanalyst, researcher
Breck, James Edward fisheries research biologist
Brereton, Giles John mechanical engineering researcher
Brock, Thomas Gregory plant physiologist
Brooks, Sharon Lynn dentist, educator
Bucksbaum, Philip Howard physicist
Burke, Robert Harry surgeon, educator
Cantrall, Irving J(ames) entomologist, educator
Carnahan, Brice chemical engineer, educator
Chaffin, Donald B. industrial engineer, researcher
Chen, Michael Ming mechanical engineering educator
Chupp, Timothy E. physicist, educator, nuclear scientist, academic administrator
Clarke, Roy physicist, educator
Conway, Lynn Ann computer scientist, educator
Coran, Arnold Gerald pediatric surgeon, educator
Craig, Robert George dental science educator
Crane, Horace Richard educator, physicist
Curl, Rane Locke chemical engineering educator, consultant
Dawson, William Ryan zoology educator
DeWitt, Sheila Hobbs research chemist
DeYoung, Raymond conservation behavior educator
Donahue, Thomas Michael physics educator
Dougherty, Richard Martin library and information science educator
Dow, William Gould electrical engineer, educator
Duderstadt, James Johnson university president
Easter, Stephen Sherman, Jr. biology educator
Eby, David W. research psychologist
England, Anthony Wayne electrical engineering and computer science educator, astronaut, geophysicist
Faeth, Gerard Michael aerospace engineering educator, researcher
Feng, Hsien Wen biochemistry educator, researcher
Filisko, Frank Edward physicist, educator
Fink, William Lee ichthyologist, systematist
Flannery, Kent V. anthropologist, educator
Fox, James Carroll aerospace engineer, program manager
Gamota, Daniel Roman materials engineer
Gaston, Hugh Philip marriage counselor, educator
Gelehrter, Thomas David medical and genetics educator, physician
Gillespie, Thomas David mechanical engineer, researcher
Gingerich, Philip Derstine paleontologist, evolutionary biologist, educator
Glazko, Anthony J(oachim) pharmaceutical consultant
Goldstein, Irwin Joseph medical research executive
Goldstein, Steven Alan medical educator, engineering educator
Grant, Michael Peter electrical engineer
Greden, John Francis psychiatrist, educator
Greene, Douglas A. internist, educator
Gruppen, Larry Dale psychologist, educational researcher
Gurnis, Michael geological sciences educator
Haddad, George Ilyas engineering educator, research scientist
Haddock, Fred T. astronomer, educator
Haddox, Mark electronic engineer
Hansen, Will civil engineer, educator, consultant
Hawkins, Joseph Elmer, Jr. acoustic physiologist
Hays, Paul B. science educator, researcher
Hess, Ida Irene statistician
Hill, Bruce Marvin statistician, scientist, educator
Hochster, Melvin mathematician, educator
Hoff, Julian Theodore physician, educator
Hryciw, Roman D. civil engineering educator
Huelke, Donald Fred anatomy and cell biology educator, research scientist
Humes, H(arvey) David nephrologist, educator
Islam, Mohammed N. optics scientist
Jones, James Ray architectural engineering research associate
Joscelyn, Kent Buckley criminologist, research scientist, lawyer
Jove, Richard molecular biologist
Julius, Stevo physician, educator, physiologist
Kauffman, Charles William aerospace engineer
Kaviany, Massoud mechanical engineer educator
Kelly, William Crowley geological sciences educator
Kluger, Matthew Jay physiologist, educator
Knox, Eric botanist, educator
Kuhl, David Edmund physician, radiology educator
Kurnit, David Martin pediatrician, educator
La Du, Bert Nichols, Jr. physician, educator
LaJeunesse, Robert Paul design engineer
Lee, Stephen chemist, educator
Lichter, Paul Richard ophthalmology educator
Long, Michael William cell/molecular biologist, educator
Losada, Marcial Francisco research scientist, psychologist
Lowe, John Burton molecular biology educator, pathologist
Markel, Dorene Samuels geneticist
Martin, David Charles materials science engineering educator
Mavrikakis, Manos chemical engineer
McBee, Gary L. chemist
McNally, Patrick Joseph aerospace engineer, technical program manager
Midgley, A(lvin) Rees, Jr. reproductive endocrinology educator, researcher
Miller, Josef M. otolaryngologist, educator
Monteith, David Keith Brisson toxicologist, researcher
Moore, Thomas Edwin museum director, biology educator
Morrel-Samuels, Palmer experimental social psychologist
Morris, Michael David chemistry educator
Ning, Xue-Han (Hsueh-Han Ning) physiologist, researcher
Nisbett, Richard Eugene psychology educator
O'Donnell, Matthew electrical engineering, computer science educator
Oncley, John Lawrence biophysics educator, consultant
Payne, Anita Hart reproductive endocrinologist, researcher
Pender, Nola J. community health nursing educator, researcher
Peterson, Lauren Michael physicist, educator
Pitt, Bertram cardiologist, consultant
Pollack, Henry Nathan geophysics educator
Rand, Stephen Colby physicist
Rauchmiller, Robert Frank, Jr. physicist
Reznicek, Anton Albert plant systematist
Ringler, Daniel Howard lab animal medicine educator
Roe, Byron Paul physics educator
Schottenfeld, David epidemiologist, educator
Schultz, Albert Barry engineering educator
Schwank, Johannes Walter chemical engineering educator
Scott, Norman Ross electrical engineering educator
Shappirio, David Gordon biologist, educator
Smith, Donald Norbert engineering executive
Soslowsky, Louis Jeffrey bioengineering educator, researcher
Stevenson, Harold William psychology educator
Stickels, Charles Arthur metallurgical engineer
Sudijono, John Leonard physicist
Tandon, Rajiv psychiatrist, educator
Taylor, Andrew Christopher electronics and electrical engineer
Thall, Aron David cell biologist
Trujillo, Keith Arnold psychopharmacologist
Ulaby, Fawwaz Tayssir electrical engineering and computer science educator, research center administrator
Vakalo, Emmanuel-George architecture and planning educator, researcher
Van Vlack, Lawrence Hall engineering educator
Veltman, Martinus J. physics educator
Wagner, Warren Herbert, Jr. educator, botanist
Wang, Kevin Ka-Wang pharmaceutical biochemist
Weg, John Gerard physician
Wicha, Max S. oncologist, educator
Wiggins, Roger C. internist, educator
Woronoff, Israel retired psychology educator
Wright, Steven Jay environmental engineering educator, consultant
Yang, Victor Chi-Min pharmacy educator
Young, Edwin Harold chemical/metallurgical engineering educator

Auburn Hills
Laperriere, Francis William electrical engineer, audio recording engineer
Nusholtz, Guy Samuel research engineer
Verma, Dhirendra civil engineer

Bad Axe
Rosenfeld, Joel ophthalmologist

Battle Creek
O'Brien, John F(rancis) civil engineer
Robie, Donna Jean chemist
Waite, Lawrence Wesley osteopathic physician

Bay City
Pearsall, Harry James dentist

Benton Harbor
De Long, Dale Ray chemicals executive

Berrien Springs
Stokes, Charles Junius economist, educator

Big Rapids
Murnik, Mary Rengo biology educator
Thapa, Khagendra survey engineering educator
Weinlander, Max Martin retired psychologist

Birmingham
Miley, Hugh Howard retired physician
Williams, Charles Wilmot engineer, consultant
Ziegelman, Robert Lee architect

Bloomfield Hills
Cohen, Alberto cardiologist
Heinen, Charles M. retired chemical and materials engineer
Jacobowitz, Ellen Sue museum administrator

Brighton
Gillespie, Shane Patrick chassis engineer
Miller, Hugh Thomas computer consultant

Canton
Olsen, Gary Alvin design engineer

Cass City
Reeder, Mike Fredrick materials engineer, consultant

Chelsea
Scott, James Noel quality assurance professional

Clare
Samorek, Alexander Henry electrical engineer, mathematics and technology educator

Copemish
Wells, Herschel James physician, former hospital administrator

Dearborn
Boulos, Edward Nashed transportation specialist
Carter, Roscoe Owen, III chemist
Chou, Clifford Chi Fong research engineering executive
Coburn, Ronald Murray ophthalmic surgeon, researcher
Duffy, James Joseph engineer
Faunce, Mark David product design engineer
Ginder, John Matthew physicist
Jaworski, David Joseph project engineer
Kelly, Carol Johnson chemist
Morris, John Michael energy technology educator
Patil, Prabhakar Bapusaheb electrical and electronic research manager
Plee, Steven Leonard mechanical engineer
Reeve, Lorraine Ellen biochemist, researcher
Rokosz, Susan Marie environmental engineer
Sengupta, Subrata dean, engineering educator
Sone, Masazumi electrical engineer
Suchy, Susanne N. nursing educator
Zimmerman, William James biologist, educator

Detroit
Amladi, Prasad Ganesh management consulting executive, health care consultant, researcher
Baker, Laurence Howard oncology educator
Bassett, Leland Kinsey communications company executive
Batcha, George mechanical and nuclear engineer
Belfield, Kevin Donald chemistry educator
Benjamins, Joyce Ann neurology educator
Bowlby, Richard Eric computer systems analyst
Cantoni, Louis Joseph psychologist, poet, sculptor
Catey, Laurie Lynn mechanical engineer
Cerny, Joseph Charles urologist, educator
Diaz, Fernando Gustavo neurosurgeon
Fitzgerald, Robert Hannon, Jr. orthopedic surgeon
Gill, Mohammad Akram civil engineer
Gleichman, John Alan safety and loss control executive
Gupta, Suraj Narayan physicist, educator
Hsu, Julie Man-ching pediatric pulmonologist
Hunter, Robert Fabio civil engineer
Ingrody, Pamela Theresa mechanical engineer
Iverson, Robert Louis, Jr. internist, physician, intensive care administrator, medical educator
Jampel, Robert Steven ophthalmologist, educator
Johnson, Carl Randolph chemist, educator
Kaplan, Joseph pediatrician
King, Albert I. bioengineering educator
Kopp, Monica biologist, educator
Krawetz, Stephen Andrew molecular biology and genetics educator
Krull, Edward Alexander dermatologist
Lai, Ming-Chia mechanical engineering educator, researcher
Leyh, George Francis association executive
Lupulescu, Aurel Peter medical educator, researcher, physician
Mahmud, Syed Masud engineering educator, researcher
Mayes, Maureen Davidica physician, educator
McCarroll, Kathleen Ann radiologist, educator
Meilgaard, Morten Christian food products executive, international consultant
Moore, W. James civil engineer, engineering executive
Novak, Raymond Francis research institute director, pharmacology educator
Person, Victoria Bernadett civil engineer, consultant
Phillis, John Whitfield physiologist, educator
Porter, Arthur T. oncologist, educator
Pratt, Robert George electrical engineer
Putatunda, Susil Kumar metallurgy educator
Putchakayala, Hari Babu chemical engineer
Rajlich, Vaclav Thomas computer scientist, researcher, educator
Rickel, Annette Urso psychology educator
Riser, Bruce L. research pathologist
Romano, Louis James chemist, educator
Roth, Thomas psychiatry educator
Ryntz, Rose Ann chemist
Saravolatz, Louis Donald epidemiologist, educator
Schaffler, Mitchell Barry research scientist, anatomist, educator
Schmidt, Robert mechanical and civil engineering educator
Shantz, Carolyn Uhlinger psychology educator
Shih, Jing-Luen Allen medical physicist
Sloane, Bonnie Fiedorek pharmacology and cancer biology educator, researcher
Sun, Jing electrical engineering educator
Swanborg, Robert Harry immunology educator
Tamimi, Nasser Taher educator, medical physicist
Tewari, Kewal Krishan chemist, researcher
Thomas, Robert Leighton physicist, researcher
Tunac, Josefino Ballesteros biotechnology administrator
Wierzbicki, Jacek Gabriel physicist, researcher
Woodbury, Dixon John physiology educator, researcher
Yamazaki, Akio biochemistry educator

Dexter
Roberts, Walter Arthur computer systems scientist

East Lansing
Andersland, Orlando Baldwin civil engineering educator
Anderson, James Henry university dean
Antaya, Timothy Allen physicist
Asmussen, Jes, Jr. electrical engineer
Atchison, William David pharmacology educator
Austin, Sam M. physics educator
Beckmeyer, Henry Ernest anesthesiologist, medical educator
Blosser, Henry Gabriel physicist
Bowerman, William Wesley, IV biologist, researcher
Case, Eldon Darrel materials science educator
Dewhurst, Charles Kurt museum director, curator, folklorist, English educator
Dilley, David Ross plant physiologist, researcher
Dye, James Louis chemistry educator
Fischer, Lawrence Joseph toxicologist, educator
Fulbright, Dennis Wayne plant pathologist, educator
Gonzalez, Michael John nutrition educator, nutriologist
Goodman, Erik David engineering educator
Grant, Rhoda biomedical researcher, educator, medical physiologist
Haack, Robert Allen forest entomologist
Hackel, Emanuel science educator
Hoganson, Curtis Wendell physical chemist
Hoppensteadt, Frank Charles mathematician, university dean
Hull, Jerome, Jr. horticultural extension specialist
Ilgen, Daniel Richard psychology educator
Jasiuk, Iwona Maria engineering educator
Johnson, John Irwin, Jr. neuroscientist
Kamdem, Donatien Pascal chemistry educator
Kerr, Norbert Lee experimental social psychologist, educator
Kevern, Niles Russell aquatic ecologist, educator
Leigh, Linda Diane psychologist, clinical neuropsychologist
Lindell, Michael Keith psychology educator
McIntyre, John Philip, Jr. physics educator
Ohlrogge, John B. botany and plant pathology educator
Pence, Thomas James mechanical engineer, educator, consultant
Petrides, George Athan ecologist, educator
Pollack, Gerald Leslie physicist, educator
Preiss, Jack biochemistry educator
Reddy, Chilecampalli Adinarayana microbiology educator
Reinhart, Mary Ann medical association administrator
Sauer, Harold John physician, educator
Saul, William Edward academic administrator, civil engineering educator
Schemmel, Rachel Anne food science and human nutrition educator, researcher
Szerszen, Jedrzej Bogumil (Andrew Szerszen) plant pathologist, educator
Tasker, John Baker veterinary medical educator, college dean
Tiedje, James Michael microbiology educator, ecologist
von Bernuth, Robert Dean agricultural engineering educator, consultant
Walker, Bruce Edward anatomy educator
Wilson, Ronald Wayne natural science educator
Witter, Richard Lawrence veterinarian, educator
Yussouff, Mohammed physicist, educator

Escanaba
Cooper, Janelle Lunette neurologist

Farmington
Hanisko, John-Cyril Patrick electronics engineer, physicist
Koshy, Vettithara Cherian chemistry educator, technical director/formulator
Salabounis, Manuel computer information scientist, mathematician, scientist
Theodore, Ares Nicholas research scientist, educator, entrepreneur

Farmington Hills
Dragun, James soil chemist
Rope, Barry Stuart packaging engineer, consultant

Ferndale
Hyder, Ghulam Muhammad Ali physicist

Flint
Adams, Paul Allison biologist, educator
Gratch, Serge mechanical engineering educator
Himes, George Elliott pathologist
Vaishnava, Prem Prakash engineering educator
Villaire, William Louis chemical engineer

Franklin
Sax, Mary Randolph speech pathologist

Fruitport
Anderson, Frances Swem nuclear medical technologist

Grand Blanc
Wasfie, Tarik Jawad surgeon, educator

Grand Haven
Parmelee, Walker Michael psychologist

Grand Rapids
Bartek, Gordon Luke radiologist
MacDonald, David Richard industrial psychologist
Scott, Richard Lynn data processing executive
Tomlinson, Gary Earl museum curator
Van Zytveld, John Bos physicist, educator

Grosse Ile
Frisch, Kurt Charles educator, administrator

Grosse Pointe
Beierwaltes, William Henry physician, educator
Wayland, Marilyn Ticknor medical researcher, evaluator, educator

Grosse Pointe Woods
Barth, Carolyn Lou hospital administrator, microbiologist

Hemlock
Wheelock, Scott A. physicist

Hickory Corners
Fitzstephens, Donna Marie biologist

Highland Park
Crittenden, Mary Lynne science educator

Holland
Benko, James John chemist
Hamstra, Stephen Arthur mechanical engineer
McConnell, Anthony polymer scientist, consultant
Witzel, William Martin analytical chemist, educator, consultant

Houghton
Caneba, Gerard Tablada chemical engineering educator
Huang, Eugene Yuching civil engineer, educator
Jaszczak, John Anthony physicist
Lumsdaine, Edward mechanical engineering educator, university dean
Pandit, Sudhakar Madhavrao engineering educator
Podila, Gopi Krishna biochemistry and molecular biology educator
Reynolds, Terry Scott social science educator
Thangaraj, Ayyakannu Raj mechanical engineering educator

Howell
Dombkowski, Joseph John water treatment specialist

Jackson
Kendall, Kay Lynn *interior designer*

Kalamazoo
Arwashan, Naji *structural engineer*
Askew, Thomas Rendall *physics educator, researcher, consultant*
Athappilly, Kuriakose Kunjuvarkey *computers and quantitative methods educator*
Baker, Carolyn Ann *research biochemist*
Carmichael, Charles Wesley *industrial engineer*
Conder, George Anthony *parasitologist*
Davis, Charles Alexander *electrical engineer, educator*
Dietz, Alma *microbiologist*
Elliott, Marc Eldon *civil engineer*
Elrod, David Wayne *computational chemist, information scientist*
Engelmann, Paul Victor *plastics engineering educator*
Finzel, Barry Craig *research scientist*
Greenfield, John Charles *bio-organic chemist*
Klein, Ronald Don *molecular biologist*
Marotti, Keith Richard *molecular biologist, researcher*
Mizsak, Stephen Andrew *chemist*
Reilly, Michael Thomas *biochemical engineer*
Robertson, John Harvey *microbiologist*
Shebuski, Ronald John *pharmaceutical company executive*
Stapleton, Susan Rebecca *biochemistry educator*
Stiver, James Frederick *pharmacist, health physicist, administrator, scientist*
Sundick, Robert Ira *anthropologist, educator*
Tsai, Ti-Dao *electrophysiologist*

Lake Linden
Campbell, Wilbur Harold *research plant biochemist, educator*

Lakeside
Price, Robert A. *mechanical engineer*

Lansing
Chazell, Russell Earl *environmental chemist*
Harvey, John Ashmore *civil engineer*
Ip, John H. *cardiologist*
Kozlowski, Steve W.J. *organizational psychologist*
Kumar, Sanjay *systems engineer*
Vincent, Frederick Michael *neurologist, electromyographer, educational administrator*

Livermore
Johnson, Roy Ragnar *electrical engineer*

Livonia
Anas, Julianne Kay *administrative laboratory director*
Gordon, Craig Jeffrey *oncologist*
Ladouceur, Harold Abel *mechanical engineer, consultant*
Rahman, Ahmed Assem *project engineer, stress analyst*

Madison Heights
Chapman, Gilbert Bryant *physicist*
Shah, Jayprakash Balvantrai *civil engineer*

Mancelona
Whelan, Joseph L. *neurologist*

Marquette
Twohey, Michael Brian *fishery biologist*

Mason
Fisher, Marye Jill *physical therapist, educator*
McDonald, James Harold *chemical engineer, energy specialist*
Toekes, Barna *chemical engineer, polymers consultant*

Metamora
Blass, Gerhard Alois *physics educator*

Midland
Bernius, Mark T. *research physicist*
Carson, Gordon Bloom *engineering executive*
Cobel, George Bassett *chemical engineer*
Collard, Randle Scott *chemist*
Cuthbert, Robert Lowell *product specialist*
Davidson, John Hunter *agriculturist*
Doan, Herbert Dow *technical business consultant*
Dreyfuss, Patricia *chemist, researcher*
Habermann, David Andrew *chemical engineer*
Heiny, Richard Lloyd *chemical engineer*
Hilty, Terrence Keith *analytical sciences manager*
Leng, Marguerite Lambert *regulatory consultant, biochemist*
Lipowitz, Jonathan *materials scientist*
McDade, Joseph John *microbiologist*
Meister, Bernard John *chemical engineer*
Morgan, Roger John *research scientist*
Rudolf, Philip Reinhold *chemist, crystallographer*
Schultz, Dale Herbert *chemical process control engineer*
Serrano, Myrna *materials scientist, chemical engineer*
Shastri, Ranganath Krishna *materials scientist*
Snow, Steven Ashley *chemist*
Speier, John Leo, Jr. *chemist*
Stull, Daniel Richard *research thermochemist, educator, consultant*
Swinehart, Frederic Melvin *chemical engineer*
Tabor, Theodore Emmett *chemical company research executive*
Wells, James Douglas *chemical engineer*
Weyenberg, Donald Richard *chemist*

Milford
Prostak, Arnold S. *chemical instrumentation engineer*
Shedlowsky, James Paul *engineer*

Mount Clemens
McGregor, Theodore Anthony *chemical company executive*

Mount Pleasant
Colarelli, Stephen Michael *psychology educator, organizational psychologist*
Dunbar, Gary Leo *psychology educator*
Logomarsino, Jack *nutrition educator*
Novitski, Charles Edward *biology educator*
Rubin, Stuart Harvey *computer science educator, researcher*
VanHouten, Jacob Wesley *environmental project manager, consultant*

Muskegon
Heyen, Beatrice J. *psychotherapist*
Kuhn, Robert Herman *public works and utilities executive, engineer*
Meilinger, Peter Martin *quality assurance manager, analytical chemist*

Newport
Johnson, Rodney William *utility executive*
Kirkland, Matthew Carl *nuclear engineer*

Northville
Abbasi, Tariq Afzal *psychiatrist, educator*

Novi
Coloske, Steven Robert *mechanical engineer*
Lewes, Kenneth Allen *clinical psychologist*
Parks, Steven James *aerospace engineer*
Singh, Jaswant *environmental company executive*

Oak Park
Borovoy, Marc Allen *podiatrist*

Okemos
Gillespie, Gary Don *physician*
Loconto, Paul Ralph *chemist, consultant*

Olivet
Seabrook, Barry Steven *environmental scientist, consultant*

Plainwell
Hultmark, Gordon Alan *civil engineer*
Kleckner, Marlin Dallas *veterinarian*

Plymouth
Clark, Kenneth William *mechanical engineer*
Grannan, William Stephen *safety engineer, consultant*
Harless, James Malcolm *corporate executive, environmental consultant*

Pontiac
Van Den Boom, Wayne Jerome *industrial engineer*

Portage
Riesenberger, John Richard *strategic marketing company executive*

Rapid River
Olson, Marian Edna *nurse, social psychologist*

Rochester
Braunstein, Daniel Norman *management psychologist, educator, consultant*
Chaudhry, G. Rasul *molecular biologist, educator*
Eliezer, Isaac *chemistry educator*
Nag, Asish Chandra *cell biology educator*
Reddy, Venkat Narsimha *ophthalmalogist, researcher*
Riley, Douglas Scott *quality assurance specialist, biochemist*
Walia, Satish Kumar *microbiologist, educator*

Rochester Hills
Hicks, George William *automotive and mechanical engineer*
Wertenberger, Steven Bruce *laser applications engineer*

Roseville
Carr, Doleen Pellett *computer service and environmental specialist, consultant*

Royal Oak
Klosinski, Deanna Dupree *medical laboratory sciences educator*
Langer, Steve Gerhardt *biomedical physicist, consultant*
Meyer, Gregory Joseph *power company executive*
Robbins, Thomas Owen *pathologist, educator*
Wilner, Freeman Marvin *hematologist, oncologist*

Saginaw
Faubel, Gerald Lee *agronomist, golf course superintendent*
Namboodiri, Krishnan *chemist*

Saint Clair Shores
Johns, Gary Christopher *electronic technologies educator*
Petz, Thomas Joseph *internist*
Rownd, Robert Harvey *biochemistry and molecular biology educator*
Walker, Frank Banghart *pathologist*

Saint Joseph
Butt, Jimmy Lee *retired association executive*
Castenson, Roger R. *agricultural engineer, association executive*
Maley, Wayne Allen *engineering consultant*

Saline
Lamson, Evonne Viola *computer software company executive, computer consultant, pastor, Christian education administrator*

Sandusky
Keeler, Lynne Livingston Mills *psychologist, educator, consultant*

Smiths Creek
Snyder, George Robert *engineer*

South Lyon
Guthrie, Michael Steele *magnetic circuit design engineer*

Southfield
Akinmusuru, Joseph Olugbenga *civil engineer*
Arlinghaus, William Charles *mathematics educator*
Hassan, Mohammad Hassan *electrical engineering educator*
Mackey, Robert Joseph *video publisher*
Morales, Raul Hector *physician*
Prasuhn, Alan Lee *civil engineer, educator*
Ramesh, Swaminathan *chemist*
Rosenzweig, Norman *psychiatry educator*
Sadasivan, Mahavijayan *chemical engineering manager*
Zubroff, Leonard Saul *surgeon*

Sterling Heights
Burke, Thomas Joseph *civil engineer*
Scott, David Lawrence *mechanical engineer*

Stevensville
Martin, Larry J. *health educator, counselor, naturalist*

Sturgis
Mackay, Edward *engineer*

Tecumseh
Chapman, Darik Ray *chemical process engineer*
Herrick, Todd W. *manufacturing company executive*

Temperance
Piniewski, Robert James *geologist*

Three Rivers
Boyer, Nicodemus Elijah *organic-polymer chemist, consultant*
Johnson, William Herbert *emergency medicine physician, aerospace physician, retired air force officer*

Troy
Ahmed, Imthyas Abdul *computer scientist*
Chrisman, Vince Darrell *information scientist*
Gardon, John Leslie *paint company research executive*
Gopalan, Muhundan *software engineer*
Helmle, Ralph Peter *computer systems developer, manager*
Jiao, Jianzhong, Sr. *development engineer, educator*
Nagata, Isao *chemist*
Purcell, Jerry *chemist*
Ross, Eric Alan *civil engineer*
Slocum, Lester Edwin *environmental engineer*

University Center
Pelzer, Charles Francis *molecular geneticist, biology educator, cancer researcher*

Warren
Abdul, Abdul Shaheed *hydrogeologist*
Brayer, Robert Marvin *program manager, engineer*
Buck, M. Scott *electrical engineer*
Deak, Charles Karol *chemist*
Druschitz, Alan Peter *research engineer*
Franetovic, Vjekoslav *physicist*
Furey, Robert L. *research chemist*
Gallopoulos, Nicholas Efstratios *chemical engineer*
Ginsberg, Myron *research scientist*
Goenka, Pawan Kumar *mechanical engineer*
Heremans, Joseph Pierre *physicist*
Horton, William David, Jr. *army officer*
Kia, Sheila Farrokhalaee *chemical engineer, researcher*
Kuhns, James Howard *communications engineer*
Mance, Andrew Mark *chemist*
Meng, Wen Jin *materials scientist*
Schreck, Richard Michael *biomedical engineer, consultant*
Sell, Jeffrey Alan *physicist*
Smith, John Robert *materials scientist*
Vaz, Nuno Artur *physicist*
Winchell, William Olin *mechanical engineer, educator, lawyer*
Yusuf, Siaka Ojo *nuclear engineer*

West Bloomfield
Sarwer-Foner, Gerald Jacob *physician, educator*

Whitmore Lake
Stanny, Gary *infosystems specialist, rocket scientist*

Ypsilanti
Bonem, Elliott Jeffrey *psychology educator*
Kennelly, William James *chemist*
Lou, Zhong (David) *mechanical engineer, biomedical engineer*
Randolph, Linda Jane *mathematics educator*

Zeeland
Hollingsworth, Cornelia Ann *food scientist*

MINNESOTA

Arden Hills
Spinelli, Julio Cesar *biomedical engineer*

Austin
Schmid, Harald Heinrich Otto *biochemistry educator, academic director*

Biddeford
Carter, Herbert Jacque *biologist, educator*

Bloomington
Bristow, Julian Paul Gregory *electrical engineer*
Drewek, Gerard Alan *mechanical engineer*
Seashore, Stanley E(manuel) *social and organizational psychology researcher*

Brooklyn Park
Peterson, Donn Neal *forensic engineer*

Burnsville
Geminn, Walter Lawrence, Jr. *computer programmer, analyst*

Duluth
Gallian, Joseph Anthony *mathematics educator*
Haller, Edwin Wolfgang *physiologist, educator*
Hoffman, Richard George *psychologist*
Johnston, Carol Arlene *ecological researcher*
Kubista, Theodore Paul *surgeon*
Lindquist, Edward Lee *biological scientist, ecologist*
Sebastian, James Albert *obstetrician/gynecologist, educator*
Thomborson, Clark David (Clark David Thompson) *computer scientist, educator*
Wyrick, David Alan *engineering educator, researcher*

Eagan
Berg, Dean Michael *applications engineer*
Ernst, Gregory Alan *energy consultant*

Eden Prairie
Gertis, Neill Allan *writer*
Hawley, Sandra Sue *electrical engineer*
Penn, Sherry Eve *communication psychologist, educator*

Elgin
Meyer, Robert Verner *farmer*

Faribault
Powers, Kim Dean *optical engineer*

Grand Rapids
Garshelis, David Lance *wildlife biologist*

Hopkins
Tempero, Kenneth Floyd *pharmaceutical company executive, physician, clinical pharmacologist*

Jordan
Lark, Ronald Edwin *logistics support engineer*

Lake Elmo
Vivona, Daniel Nicholas *chemist*

Mankato
Khaliq, Muhammad Abdul *electrical engineer*
Kvamme, John Peder *electric technology company executive*
Sachau, Daniel Arthur *psychology educator*
Zeller, Michael James *psychologist, educator*

Maple Grove
Daniel, John Mahendra Kumar *biomedical engineer, researcher*
Griffith, Patrick Theodore *systems engineer*

Maple Plain
Erdmann, John Baird *environmental engineer*

Marshall
Carberry, Edward Andrew *chemistry educator*

Mendota Heights
Schoon, David Jacob *electronic engineer*

Minneapolis
Abrahamson, Scott David *mechanical engineer, educator*
Alving, Amy Elsa *aerospace engineering educator*
Arndt, Roger Edward Anthony *hydraulic engineer, educator*
Atluru, Durgaprasadarao *veterinarian, educator*
Baisch, Steven Dale *pediatrician*
Baker, John Stevenson (Michael Dyregrov) *writer*
Barden, Robert Christopher *psychologist, educator, lawyer*
Behnke, James Ralph *food company executive*
Berg, Stanton Oneal *firearms and ballistics consultant*
Born, David Omar *psychologist, educator*
Boudreau, Robert James *nuclear medicine physician, researcher*
Bukonda, Ngoyi K. Zacharie *public health administrator*
Caldwell, Michael DeFoix *surgeon, educator*
Carlson, Marvin *analytical chemist*
Carlson, Richard Raymond *statistician, consultant*
Chilton, William David *architect*
Chipman, John Somerset *economist, educator*
Chornenky, Victor Ivan *laser scientist, researcher*
Clayton, Paula Jean *psychiatry educator*
Cline, James Michael *physicist*
Cornelissen-Guillaume, Germaine Gabrielle *chronobiologist, physicist*
Craig, James Lynn *physician, consumer products company executive*
Crouch, Steven L. *mining engineer*
Culter, John Dougherty *chemical engineer*
Cunningham, Thomas B. *aerospace engineer*
Dahl, Gerald LuVern *psychotherapist, educator, consultant, writer*
Decoursey, William Leslie *engineer*
Du, Ding Zhu *mathematician, educator*
Dworkin, Martin *microbiologist, educator*
Eberly, Raina Elaine *psychologist, educator*
Engdahl, Brian Edward *psychologist*
Exe, David Allen *electrical engineer*
Fiedler, Robert Max *management consultant*
Friedman, Avner *mathematician, educator*
Fryd, David Steven *biostatistician, consultant*
Galambos, Theodore Victor *civil engineer, educator*
Gale, Stephan Marc *civil engineer*
Galeazza, Marc Thomas *neuroscientist*
Garry, Vincent Ferrer *environmental toxicology researcher, educator*
Geise, Richard Allen *medical physicist*
Giese, Clayton Frederick *physics educator, researcher*
Goldman, Allen Marshall *physics educator*
Goldstein, Richard Jay *mechanical engineer, educator*
Gorham, Eville *scientist, educator*
Greaves, Ian Alexander *occupational physician*
Gudmundson, Barbara Rohrke *ecologist*
Gusek, Todd Walter *food scientist*
Haase, Ashley Thomson *microbiology educator, scientist*
Hamermesh, Morton *physicist, educator*
Healton, Bruce Carney *data processing executive*
Heller, Kenneth Jeffrey *physicist*
Helmes, Leslie Scott *architect*
Hobbie, Russell Klyver *physicist*
Hurwicz, Leonid *economist, educator*
Jackson, Robert Loring *science and mathematics educator, academic administrator*
James, Walter *retired computer information specialist*
Jesness, Bradley L. *psychology educator, testing and professional selection consultant*
Johnson, David Wolcott *psychologist, educator*
Johnson, Robert Glenn *physics educator*
Joseph, Daniel Donald *aeronautical engineer, educator*
Karato, Shun-ichiro *geophysicist*

Keane, William Francis nephrology educator, research foundation executive
Keefe, William Robert mechanical engineer
Kralewski, John Edward health service administration educator
Kroll, Mark William electrical engineer
Kuhi, Leonard Vello astronomer, university administrator
Kvalseth, Tarald Oddvar mechanical engineer, educator
Lentz, Richard David psychiatrist
Luepker, Russell Vincent epidemiology educator
Luiso, Anthony international food company executive
Malchow, Douglas Byron engineering executive
Marshak, Marvin Lloyd physicist, educator
Maxton, Robert Connell metallurgical engineer
McAloon, Todd Richard food microbiologist
Michael, Alfred Frederick, Jr. physician, medical educator
Miller, Jeffrey Steven hematologist, researcher
Miller, Robert Francis physiologist, educator
Mitchell, Eugene Alexander safety consultant
Moscowitz, Albert Joseph chemist, educator
Mulich, Steve Francis safety engineer
Najarian, John Sarkis surgeon, educator
Nier, Alfred Otto Carl physicist
Nitsche, Johannes Carl Christian mathematics educator
O'Malley, William David architect
Patankar, Suhas V. engineering educator
Patel, Jayant Ramanlal mechanical engineer administrator
Patterson, Richard George retired aerospace engineer
Perkuhn, Gaylen Lee civil/structural engineer
Peterson, Douglas Arthur physician
Porter, William L. electrical engineer
Rahman, Yueh-Erh biologist
Rauch, Andrew Martin civil/structural engineer
Santi, Peter Alan neuroanatomist, educator
Serstock, Doris Shay retired microbiologist, educator, civic worker
Sheikh, Suneel Ismail aerospace engineer, researcher
Shumway, Sara J. cardiothoracic surgeon
Smyrl, William H. chemistry educator
Sonsteby, Kristi Lee healthcare consultant
Stenwick, Michael William internist, geriatric medicine consultant
Swaiman, Kenneth Fred pediatric neurologist, educator
Truhlar, Donald Gene chemist, educator
Warwick, Warren J. pediatrics educator
Weber, Lowell Wyckoff internist
Weir, Edward Kenneth cardiologist
Wilgen, Francis Joseph mechanical engineer
Will, Thomas Eric psychologist
Wirtschafter, Jonathan Dine neuro-ophthalmology educator, scientist
Wood, Wellington Gibson, III biochemistry educator
Woodward, Paul Ralph computational astrophysicist, applied mathematician, educator
Wright, Herbert E(dgar), Jr. geologist
Yourzak, Robert Joseph management consultant, engineer, educator
Zoberi, Nadim Bin-Asad management consultant

Minnetonka
Andrus, W(infield) Scott scientist, consultant
Grivna, Edward Lewis electronics engineer, consultant
Johnson, Lennart Ingemar materials engineering consultant

Monticello
Beres, Joel Edward electrical engineer, consultant

Moorhead
Emmel, Bruce Henry mathematics educator
Larson, Betty Jean dietitian, educator
Nezhad, Hameed Gholam energy management educator
Samaraweera, Upasiri research chemist
Sun, Li-Teh economics educator

Northfield
Hardgrove, George Lind, Jr. chemistry educator
Huff, Charles William psychologist, educator

Osseo
Haun, James William chemical engineer, retired food company executive, consultant

Park Rapids
Tonn, Robert James entomologist

Plymouth
Ding, Ni chemist
Levine, Leon chemical engineer

Red Wing
Lee, Gordon Melvin electrical engineering consultant

Richfield
Feldsien, Lawrence Frank civil engineer

Robbinsdale
Dannewitz, Stephen Richard emergency physician, consultant, toxicologist

Rochester
Beahrs, Oliver Howard surgeon
Betts, Douglas Norman civil engineer
Bodine, Peter Van Nest biochemist
Carmichael, Stephen Webb anatomist, educator
Engel, Andrew George neurologist
Gehling, Michael Paul engineering executive
Gleich, Gerald Joseph immunologist, medical scientist
Huse, Diane Marie dietitian
Kyle, Robert Arthur medical educator, oncologist
LaRusso, Nicholas F. gastroenterologist, educator
Lockhart, John Campbell bioengineer, physiologist
Mayberry, William Eugene retired physician
Moyer, Thomas Phillip biochemist
Naessens, James Michael biostatistician
Pittelkow, Mark Robert dermatology educator, researcher
Randall, Barbara Ann computer design engineer
Robb, Richard Arlin biophysics educator, scientist
Silva, Norberto DeJesus biophysicist
Steiner, Jeffery Allen project engineer, executive
Whelen, Andrew Christian microbiologist

Saint Cloud
Ellis, Bruce W. electrical engineering educator
Kirick, Daniel John agronomist

Saint Louis Park
Speirs, Robert Frank logistics engineer

Saint Paul
Alm, Roger Russell chemist
Caldwell, Elwood Fleming food science educator, researcher, editor
Cheng, H(wei) H(sien) agriculture and environmental science educator
Christensen, Clyde Martin plant pathology educator
Davis, Mark Avery ecologist, educator
De Simone, Livio Diego diversified manufacturing company executive
Downing, Michael William pharmaceutical company executive
Emeagwali, Dale Brown molecular biologist
Erickson, Brice Carl chemist
Evans, Roger Lynwood scientist, patent liaison
Fingerson, Leroy Malvin corporate executive, engineer
Fisch, Richard S. physicist, psychophysicist
Goodell, John Dewitte electromechanical engineer
Goodman, Lawrence Eugene structural analyst, educator
Graf, Timothy L. mechanical engineer
Halvorsen, Thomas Glen mechanical engineer
Heuer, Marvin Arthur physician, science foundation executive
Hicks, Dale R. agronomist, educator
Hults, Scott Samuel engineer
Hunter, Alan Graham reproductive physiologist, educator
Johnson, Brian Dennis engineer
Jones, Charles Weldon biologist, educator, researcher
Karl, Daniel William biochemist, researcher, consultant
Kommedahl, Thor plant pathology educator
Lampert, Leonard Franklin mechanical engineer
Lee, Charles C. physicist
Leonard, Kurt John plant pathologist, university program director
Ling, Joseph Tso-Ti manufacturing company executive, environmental engineer
Marcellus, Manley Clark, Jr. chemical engineer
McKinnell, Robert Gilmore zoology, genetics and cell biology educator
Mech, Lucyan David research biologist, conservationist
Mitchell, William Cobbey physicist
Newman, Raymond Melvin biologist, educator
Ng, Lewis Yok-Hoi civil engineer
Padmanabhan, Mahesh food and chemical engineer, researcher
Perry, James Alfred natural resources director, researcher, educator, consultant
Rasmusson, Donald C. agronomist, educator
Ritschel, James Allan computer research specialist
Rzepecki, Edward Louis packaging management educator
Schmitt, Michael A. agronomist, educator
Schumer, Douglas Brian physicist
Sheehan, Richard Laurence, Jr. packaging engineer
Stebbings, William Lee chemist, researcher
Swain, Edward Balcom environmental research scientist
Tate, Jeffrey L. biology institute administrator
Thenen, Shirley Warnock nutritional biochemistry educator
Uban, Stephen Alan mechanical engineer
Voss, Steven Ronald environmental engineer
Watson, James Edwin physicist
Wendt, Hans Werner life scientist, educator
White, Michael Ernest animal scientist
Wilson, Bruce Nord agricultural engineer
Witek, John James neurologist
Wyse, Donald L. agronomist, educator
Xie, Weiping mass spectromist
Ylitalo, Caroline Melkonian chemical engineer

Saint Peter
Fuller, Richard Milton physics educator
Thompson, H. Bradford chemist, educator

Spring Lake Park
Powell, Christopher Robert systems analyst

Stillwater
Anderson, Geraldine Louise laboratory scientist

Vadrais Heights
Weyler, Walter Eugen, Jr. process engineer

Warroad
Gouin, Warner Peter electrical engineer

Welch
Hellen, Paul Eric electrical engineer

White Bear Lake
Holmen, Reynold Emanuel chemist

Worthington
Carlson, Rolf Stanley psychologist

MISSISSIPPI

Bay Saint Louis
Quinn, John Michael physicist, geophysicist
Skramstad, Robert Allen oceanographer
Zingarelli, Robert Alan research physicist

Biloxi
Deegen, Uwe Frederick marine biologist
Ransom, Perry Sylvester civil engineer
Wasserman, Karen Boling clinical psychologist, nursing consultant

Brandon
Cooley, Sheila Leanne psychologist, consultant
King, Kenneth Vernon, Jr. pharmacist
Mitchell, Roy Devoy industrial engineer

Cleveland
Strahan, Jimmie Rose mathematics educator

Columbus
Gray, Stanley Randolph, Jr. systems engineer
Lockhart, Frank David healthcare company executive

Escatawpa
Chapel, Theron Theodore quality assurance engineer

Gulfport
Doudrick, Robert Lawrence research plant pathologist

Hattiesburg
Biesiot, Patricia Marie biology educator, researcher
Miller, James Edward computer scientist, educator
Nicholson, James Allen orthodontist, inventor
Noblin, Charles Donald clinical psychologist, educator
Santangelo, George Michael molecular geneticist
West, Michael Howard dentist
Yarbrough, Karen Marguerite genetics educator, university official

Hazlehurst
Lowenkamp, William Charles, Jr. medical device engineer, researcher, consultant

Horn Lake
Schadrack, William Charles, III design engineer

Jackson
Adams-Hilliard, Beverly Lynn chemist
Burns, Robert, Jr. architect, freelance writer, artist
Cai, Zhengwei biomedical researcher
Currier, Robert David neurologist
Das, Suman Kumar plastic surgeon, researcher
Didlake, Ralph Hunter, Jr. surgeon
Dunsford, Harold Atkinson pathologist, researcher
Finley, Richard Wade internist, educator
Forks, Thomas Paul osteopath
Goddard, Jerome medical entomologist
Hutchins, James Blair neuroscientist
Kliesch, William Frank physician
Lee, Daniel Kuhn economist
Russell, Robert Pritchard ophthalmologist
Sinning, Allan Ray anatomy educator, researcher
Skelton, Gordon William data processing executive, educator
Snodgrass, Samuel Robert neurologist
Sullivan, John Fallon, Jr. government official
Zubkowski, Jeffrey Dennis chemistry educator

Laurel
Lindstrom, Eric Everett ophthalmologist

Lorman
Hylander, Walter Raymond, Jr. retired civil engineer
Williams, Richard, Jr. animal scientist

Meridian
Mutziger, John Charles physician

Mississippi State
Agba, Emmanuel Ikechukwu mechanical engineering educator
Chakroun, Walid mechanical engineering educator
King, Roger Lee electrical engineer, consultant
McGilberry, Joe H. food service executive
McMillen, David L. psychology educator
Monts, David Lee scientist, educator
Powe, Ralph Elward university administrator
Rais-Rohani, Masoud aerospace and engineering mechanics educator
Sawyer, David Neal petroleum industry executive
Thompson, Joe Floyd aerospace engineer, researcher
Thompson, Warren S. dean, academic administrator
Truax, Dennis Dale civil engineer, educator, consultant
Usher, John Mark industrial engineering educator
Zhu, Jianping mathematics educator

Ocean Springs
Gunter, Gordon zoologist
McNulty, Matthew Francis, Jr. health care administration educator, university administrator, consultant, horse and cattle breeder

Oxford
Breazeale, Mack Alfred physics educator
Meyer, L. Donald agricultural engineer, researcher, educator
Mutchler, Calvin Kendal hydraulic research engineer

Pascagoula
Corben, Herbert Charles physicist, educator
Skipper, Adrian chemical engineer
Swint, Joseph Ellis computer systems analyst

Picayune
Lowrie, Allen geologist, oceanographer

Pickens
Patton, John Anthony chemical engineer

Poplarville
Edwards, Ned Carmack, Jr. agronomist, university program director

Ridgeland
Evans, Wayne Edward environmental microbiologist, researcher

Seminary
Frazier, Ronald Gerald, Jr. mechanical engineer

Starkville
Friend, Alexander Lloyd forester educator
Priest, Melville Stanton consulting hydraulic engineer

Stennis Space Center
Hurlburt, Harley Ernest oceanographer
Lewando, Alfred Gerard, Jr. oceanographer
Nunez, Stephen Christopher aerospace technologist
Nunn, James Ross engineering executive
Smith, Mary Kay Wilhelm safety engineer
Sprague, Vance Glover, Jr. oceanography executive, naval reserve officer
Teng, Chung-Chu ocean engineer, researcher

Stoneville
Duke, Stephen Oscar physiologist, researcher
Ranney, Carleton David plant pathology researcher, administrator
Rutger, J. Neil agronomy research administrator

University
Aughenbaugh, Nolan Blaine engineering educator, consultant
Benson, William Hazlehurst environmental toxicologist
Elsherbeni, Atef Zakaria electrical engineering educator
Hackett, Robert Moore engineering educator
Jung, Mankil organic synthetic chemist, educator
Kushlan, James A. biology educator
McChesney, James Dewey pharmaceutical scientist
Parsons, Glenn Ray biology educator
Sufka, Kenneth Joseph psychology educator
Sukanek, Peter Charles chemical engineering educator, researcher
Uddin, Waheed civil engineer
Williamson, John Steven medicinal chemist

Vicksburg
Ahlvin, Richard Glen civil engineer
Albritton, Gayle Edward structural engineer
Bryant, Larry Michael structural engineer, researcher
Demirbilek, Zeki research hydraulic engineer
Ethridge, Loyde Timothy hydraulic engineer, consultant
Fehl, Barry Dean civil engineer
Lashlee, Jon David physical scientist
McRae, John Leonidas civil engineer
Middlebusher, Mark Alan computer scientist
Olsen, Richard Scott geotechnical engineer
Scott, Angela Freeman civil engineer

MISSOURI

Annapolis
Goehman, M. Conway mechanical engineer

Ava
Murray, Delbert Milton engineer

Ballwin
López-Candales, Angel cardiologist, researcher

Bridgeton
January, Daniel Bruce electronics engineer

Camdenton
Krehbiel, Darren David civil engineer

Cape Girardeau
Close, Edward Roy hydrogeologist, environmental engineer, physicist
Dahiya, Jai Narain physics educator, researcher
Hathaway, Ruth Ann chemist
Young, John Edward mathematics educator, consultant, writer

Chesterfield
Biggerstaff, Randy Lee sports medicine consultant
Bockserman, Robert Julian chemist
Franz, John E. bio-organic chemist, researcher
Lang, James Douglas aerospace engineer, educator
Preissner, Edgar Daryl engineering executive
Schuermann, Mark Harry chemist, educator
Smith, Lawrence Abner aeronautical engineer
Yardley, John Finley aerospace engineer

Clayton
Osterloh, Everett William county official

Columbia
Adams, John Ewart chemistry educator
Allen, William Cecil physician, educator
Altomari, Mark G. clinical psychologist
Backus, Elaine Athene entomologist, educator
Barbero, Giulio John physician, educator
Beckwith, Catherine S. veterinarian
Beem, John Kelly mathematician, educator
Bevins, Robert Jackson agricultural economics educator
Boley, Mark S. physicist, mathematician
Brown, Olen Ray medical microbiology research educator
Carrel, James Elliott biologist
Chang, Jian Cherng research analyst
Coe, Edward Harold, Jr. agronomist, educator, geneticist
Corcoran, Michael John orthopedic surgeon
Cowan, Nelson cognitive psychologist, researcher
Day, Cecil LeRoy agricultural engineering educator
Douty, Richard Thomas structural engineer
Eisenstark, Abraham research director, microbiologist
El-Bayya, Majed Mohammed civil engineer
Fahim, Mostafa Safwat reproductive biologist, consultant
Finkelstein, Richard Alan microbiologist
Heldman, Dennis Ray engineering educator
Hess, Leonard Wayne obstetrician/gynecologist, perinatologist
Khojasteh, Ali medical oncologist, hematologist
Kimel, William Robert engineering educator, university dean
Merilan, Charles Preston dairy husbandry scientist
Misfeldt, Michael Lee immunologist, educator
Mitchell, Roger Lowry agronomy educator
Morehouse, Lawrence Glen veterinarian, educational administrator
Niblack, Terry L. nematologist, plant pathology educator
Poehlmann, Carl John agronomist, researcher
Popovici, Galina physicist
Shelton, Kevin L. geology educator
Stalling, David Laurence research chemist
Stonnington, Henry Herbert physician, medical executive, educator
Viswanath, Dabir Srikantiah chemical engineer
Wagner, Joseph Edward veterinarian, educator
Waidelich, Donald Long electrical engineer, consultant
Walkenbach, Ronald Joseph foundation executive, pharmacology educator
Wong, Tuck Chuen chemist, educator
Yanders, Armon Frederick biological sciences educator, research administrator

Earth City
Puetz, William Charles *engineering company executive*

Eureka
Coles, Richard W(arren) *biology educator, research administrator*

Fenton
Bubash, James Edward *engineering executive, entrepreneur, inventor*
Richardson, Thomas Hampton *design consulting engineer*

Ferguson
Fieldhammer, Eugene Louis *civil engineer*

Florissant
Burns, Donald Raymond *structural and mechanical design engineer*
Cook, Alfred Alden *chemical engineer*
Schwarze, Robert Francis *osteopath, dermatologist*
Tomazi, George Donald *electrical engineer*

Fort Leonard Wood
Porter, Bruce Jackman *military engineer, computer software engineer*

Fortuna
Ramer, James LeRoy *civil engineer*

Fulton
Chapman, Garry Von *mechanical engineer*

Gladstone
Moffitt, Christopher Edward *physicist*

Grandin
Wallace, Louise Margaret *critical care nurse*

Hallsville
McFate, Kenneth Leverne *association administrator*

Imperial
Auld, Robert Henry, Jr. *biomedical engineer, educator, consultant, author*

Independence
Cady, Elwyn Loomis, Jr. *medicolegal consultant, educator*
Johnson, Cecil Kirk, Jr. *cement chemist*
Murray, Thomas Reed *aerospace engineer*
Sturges, Sidney James *pharmacist, educator, investment and development company executive*

Jefferson City
Neumann, Donald Lee *civil engineer, environmental specialist*
Nordstrom, James William *nutritionist, educator*
Weithman, Allan Stephen *fisheries biologist*

Joplin
Kerr, Frank Floyd *retired electrical engineer*
Whittle, Philip Rodger *chemistry educator, crime laboratory director*

Kansas City
Bergman, Carla Elaine *hydrologist, consultant*
Boyd, John Addison, Jr. *civil engineer*
Bretzke, Virginia Louise *civil engineer, environmental engineer*
Changho, Casto Ong *power plant construction executive*
Ching, Wai Yim *physics educator, researcher*
De Blas, Angel Luis *biologist, educator*
Dileepan, Kottarappat Narayanan *biochemist, researcher, educator*
Donahey, Rex Craig *structural engineer*
Eddy, Charles Alan *chiropractor*
Eidemiller, Donald Roy *mechanical engineer*
Fry-Wendt, Sherri Diane *psychologist*
Gale, George Daniel, Jr. *philosophy of science educator, researcher*
Gier, Audra May Calhoon *environmental chemist*
Grosskreutz, Joseph Charles *physicist, engineering researcher, educator*
Haar, Andrew John *environmental engineer*
Hagsten, Ib *animal scientist, educator*
Heath, Timothy Gordon *bioanalytical chemist*
Heausler, Thomas Folse *structural engineer*
Herrick, Thomas Edward *aeronautical/astronautical engineer*
Hilpman, Paul Lorenz *geology educator*
Hobson, Keith Lee *civil engineer, consultant*
Johnson, Richard Dean *pharmaceutical consultant, educator*
Keith, Dale Martin *utilities consultant*
Kinsey, John Scott *environmental scientist*
Kuhn, Leigh Ann *air pollution control engineer*
Mc Kelvey, John Clifford *research institute executive*
McNulty, Richard Paul *meteorologist*
Murphy, Richard David *physics educator*
Ostby, Frederick Paul, Jr. *meteorologist, government official*
Rosner, Ronald Alan *mechanical engineer*
Santoro, Alex *infosystems specialist*
Sauer, Gordon Chenoweth *physician, educator*
Schoolman, Arnold *neurological surgeon*
Schuler, Martin Luke *geotechnical engineer*
Shi, Zheng *radiological physicist*
Signorelli, Joseph *control systems engineer*
Simmons, Robert Marvin *environmental scientist, consultant*
Smoot, John Eldon *mechanical engineer*
Stannard, William George *civil engineer*
Stern, Thomas Lee *physician, educator, medical association administrator*
Van Booven, Judy Lee *data processing manager*
von Kehl, Inge *toxicologist-pharmacologist*
Wade, Robert Glenn *engineering executive*
Waterborg, Jakob Harm *biochemistry educator*
Wegst, Audrey V. *physicist, consulting firm executive*
Welling, Larry Wayne *pathologist, educator, physiologist*
Wickham-St. Germain, Margaret Edna *mass spectrometrist*
Wieliczka, David Michael *physics educator, researcher*
Winitz, Harris *pyschology educator*
Woods, Karen Marguerite *psychologist, human development, institute executive*
Yourtee, David Merle *pharmaceutical science educator, molecular toxicologist*
Zemansky, Gilbert Marek *hydrogeologist, water quality engineer*

Kirksville
Allen, Stephen Louis *electrical engineer, educator*
Hilgartner, C(harles) Andrew *theorist*
Martin, John Richard *pharmacology educator, researcher*
Rearick, James Isaac *biochemist, educator*
Twining, Linda Carol *biologist, educator*

Lake Saint Louis
Samuelson, Robert Donald *retired combat aircraft executive*

Lake Sherwood
Struckhoff, Ronald Robert *manufacturing engineer*

Lee's Summit
LaCombe, Ronald Dean *mechanical engineer*

Liberty
Philpot, John Lee *physics educator*

Manchester
Purdy, Donald Gilbert, Jr. *soil scientist*

Marshfield
Hopper, Kevin Andrew *electrical engineer, educator*

Maryland Heights
Chinn, Rex Arlyn *chemist*
Dahiya, Jai Bhagwan *chemist*
Goldfarb, Marvin Al *civil engineer*
Scarponcini, Paul *computer scientist*

Mexico
Kessler, Donna Kay Ens *mathematics educator*

Mountain View
Olszewski, Lee Michael *instrument company executive*

Neosho
Jefferson, Michael L *environmental educator*

Nevada
Kuchta, Steven Jerry *psychologist*

New Bloomfield
Quay, Wilbur Brooks *biologist*

Point Lookout
Allen, Jerry Pat *aviation science educator*

Rolla
Belarbi, Abdeldjelil *civil engineering educator, researcher*
Bolon, Albert Eugene *nuclear engineer, educator*
Cox, Norman Roy *engineer, educator, consultant*
Crosbie, Alfred Linden *mechanical engineering educator*
Dare, Charles Ernest *civil engineer, educator*
Datz, Israel Mortimer *information systems specialist*
Day, Delbert Edwin *ceramic engineering educator*
Dharani, Lokeswarappa Rudrappa *engineering educator*
Finaish, Fathi Ali *aeronautical engineering educator*
Haemmerlie, Frances Montgomery *psychology educator, consultant*
Hatheway, Allen Wayne *geological engineer, educator*
Hunt, Theodore William *civil engineer*
Ingram, William Thomas, III *mathematics educator*
Look, Dwight Chester, Jr. *mechanical engineering educator, researcher*
Munger, Paul R. *civil engineering educator*
Numbere, Daopu Thompson *petroleum engineer, educator*
Peaslee, Kent Dean *metallurgical engineer*
Rao, Vittal Srirangam *electrical engineering educator*
Sarchet, Bernard Reginald *retired chemical engineering educator*
Sauer, Harry John, Jr. *mechanical engineering educator, university administrator*
Schulz, Michael *physicist*
Shrestha, Bijaya *nuclear scientist*
Sparlin, Don Merle *physicist*

Saint Charles
Brahmbhatt, Sudhirkumar *chemical company executive*
Kim, Kyong-Min *semiconductor materials scientist*
Lapinski, John Ralph, Jr. *aerospace engineer*
Ruwwe, William Otto *automotive engineer*

Saint James
Spurgeon, Earl E. *systems engineer*

Saint Joseph
Brooks, Steven Doyle *environmental engineer*

Saint Louis
Antonacci, Anthony Eugene *food corporation engineer*
Armbruster, Barbara Louise *botanist*
Bachman, Clifford Albert *engineering specialist, technical consultant*
Baile, Clifton A. *biologist, researcher*
Balestra, Chester Lee *electrical engineer, educator*
Baum, Janet Suzanne *architect*
Beck, Lois Grant *anthropologist, educator*
Bell, Laura Jeane *retired nurse*
Benson, Robert John *computer science educator, administrator*
Binns, Walter Robert *physics researcher*
Bird, Harrie Waldo, Jr. *psychiatrist, educator*
Birman, Victor Mark *mechanical and aerospace engineering educator*
Bohne, Jeanette Kathryn *mathematics and science educator*
Bourne, Carol Elizabeth Mulligan *biology educator, phycologist*
Briles, John Christopher *civil engineer*
Brown, Jay Wright *food manufacturing company executive*
Brumbaugh, Philip S. *minerals manager, quality control consultant*
Bruns, Billy Lee *consulting electrical engineer*
Brunstrom, Gerald Ray *engineering executive, consultant*
Byrnes, Christopher Ian *academic dean, researcher*
Carlsson, Anders Einar *physicist*
Cawns, Albert Edward *computer systems consultant*
Churchill, Ralph John *environmental chemist*
Clark, Carl Arthur *retired psychology educator, researcher*
Conradi, Mark Stephen *physicist, educator*
Cooling, Thomas Lee *civil engineer, consultant*
Cooper, Stephen Randolph *chemistry educator and industrial research manager*
Cottler, Linda Bauer *epidemiologist*
Cowsik, Ramanath *physics educator*
Crosby, Marshall Robert *botanist, educator*
Du Bois, Philip Hunter *psychologist, educator*
Emerick, Josephine L. *engineer*
Erickson, Robert Anders *optical engineer, physicist*
Evans, David Myrddin *aerospace engineer*
Farrior, Gilbert Mitchell *metallurgical engineer*
Fernandez-Pol, Jose Alberto *physician, radiology and nuclear medicine educator*
Fleming, Timothy Peter *molecular biologist*
Folkerts, Dennis Michael *telecommunications specialist*
Francis, Faith Ellen *biochemist*
Frazier, Kimberlee Gonterman *veterinarian*
Gardner, Gregory Allen *industrial engineer*
Gfeller, Donna Kvinge *clinical psychologist*
Gould, Phillip L. *civil engineering educator, consultant*
Graham, Donald James *food technologist*
Green, Maurice *molecular biologist, virologist, educator*
Green, Samuel Isaac *optoelectronic engineer*
Gregory, Patricia Jeanne *corporate relations director*
Haimo, Deborah Tepper *mathematics educator*
Hakkinen, Raimo Jaakko *aeronautical engineer, scientist*
Hamper, Bruce Cameron *organic chemist, agricultural chemist*
Harris, James C II *aerospace engineer*
Hellmuth, George Francis *architect*
Hemming, Bruce Clark *plant pathologist*
Henis, Jay Myls Stuart *research scientist, research director*
Hess, Linda Candace *process control engineer*
Hile, Matthew George *psychologist, researcher*
Hirsch, Ira J. *otolaryngologist, educator*
Houston, Devin Burl *biomedical scientist, educator*
Howard, Susan Carol Pearcy *biochemist*
Hsu, Chung Yi *neurologist*
Huddleston, Philip Lee *physicist*
Husar, Rudolf Bertalan *mechanical engineering educator*
Jakschik, Barbara A. *science educator, researcher*
Jenks, Gerald Erwin *aerospace company executive*
Jobe, Muriel Ida *medical technologist*
Johnston, Gerald Andrew *aerospace company executive*
Kagan, Sioma *economics educator*
Kaufmann, Gary Bryan *chemist*
Kelton, Kenneth Franklin *physicist, educator*
Khomami, Bamin *chemical engineer, educator*
Khuhro, Shafiq Ahmed *biomedical engineer*
Kiser, Karen Maureen *medical technologist, educator*
Kornfeld, Stuart A. *hematology educator*
Kuhlman, Robert E. *orthopedic surgeon*
Kurtz, Michael E. *medical educator*
Landau, William Milton *neurologist*
Larsen, Alvin Henry *chemical engineer*
Long, Christopher *toxicologist*
Lucchesi, Lionel Louis *lawyer*
Lucking, Peter Stephen *industrial engineering consultant*
Luecke, Kenn Robert *software engineer*
Lupo, Michael Vincent *aviation engineer*
Macon, Irene Elizabeth *interior designer, consultant*
Mattison, Richard *psychiatry educator*
McFadden, James Frederick, Jr. *surgeon*
McIntosh, Helen Horton *research scientist*
Montague, Michael James *plant physiologist*
Moran, Sharon Joyce *chemist*
Murray, Patrick Robert *microbiologist, educator*
Murray, Robert Wallace *chemistry educator*
Norberg, Richard Edwin *physicist, educator*
Norris, John Robert *design engineer*
Nowak, Felicia Veronika *endocrinologist, molecular biologist, educator*
O'Neill, James William *structural engineer*
Orton, George Frederick *aerospace engineer*
Partridge, Nicola Chennell *physiology educator*
Patterson, Miles Lawrence *psychology educator*
Perlmutter, Lawrence David *aerospace engineer*
Pestronk, Alan *neurologist*
Prickett, Gordon Odin *mining, mineral and energy engineer*
Puri, Pushpinder Singh *chemical engineer*
Quicksall, Carl Owen *chemist, researcher*
Ray, James Allen *aerospace engineer*
Redmore, Derek *chemist*
Reh, Thomas Edward *radiologist, educator*
Rice, Treva Kay *genetic epidemiologist*
Richards, Diana Lyn *psychologist*
Ripp, Bryan Jerome *geological engineer*
Robins, Eli *psychiatrist, biochemist, educator*
Rogers, John Russell *engineer*
Sanes, Joshua Richard *biologist, researcher*
Santiago, Julio V. *medical educator, medical association administrator*
Scarborough, George Edward *aerospace engineer, researcher*
Schoenhard, William Charles, Jr. *health care executive*
Schonfeld, Gustav *medical educator*
Schreiber, Robert John *environmental engineer*
Schulz, Raymond Charles *electrical engineer*
Sharp, Dexter Brian *organic chemist, consultant*
Shelke, Kantha *cereal chemist*
Shrage, Sydney *aeronautical engineering consultant*
Silverman, David Charles *materials engineer*
Slatopolsky, Eduardo *nephrologist, educator*
Sly, William S. *biochemist, educator*
Sontag, Glennon Christy *electrical engineering consultant, travel industry executive*
Sorrell, Wilfred Henry *astrophysics educator*
Spray, Thomas L. *surgeon, educator*
Sullentrup, Michael Gerard *structural engineer, consultant*
Suydam, Peter R. *clinical engineer, consultant*
Symington, Janey Studt *cell and molecular biologist*
Takes, Peter Arthur *immunologist*
Tarasidis, Jamie Burnette *aerospace engineer*
Taylor, Morris Anthony *chemistry educator*
Ter-Pogossian, Michel Mathew *radiation sciences educator*
Thach, Robert Edwards *biology educator*
Thach, William Thomas, Jr. *neurobiology and neurology educator*
Thalden, Barry R. *architect*
Thomas, Lewis Jones, Jr. *anesthesiology educator, biomedical researcher*
Tolan, Robert Warren *pediatric infectious disease specialist*
Torno, Laurent Jean, Jr. *architect*
Varner, Joseph Elmer *biology educator, researcher*
Visser, Matthew Joseph *physicist*
Walker, Earl E. *manufacturing executive*
Walker, Robert Mowbray *physicist, educator*
Weaver, Charles Lyndell, Jr. *architect*
Wisdom, Guyrena Knight *psychologist, educator*
Worley, Jimmy Weldon *chemist*
Zheng, Baohua *electronics engineer, consultant*

Saint Peters
Warren, Joan Leigh *pediatrician*
Yerion, Michael Ross *civil engineer*

Springfield
Criswell, Charles Harrison *analytical chemist, environmental and forensic consultant and executive*
Hackett, Earl Randolph *neurologist*
Havel, John Edward *biology educator*
Liu, Yuan Hsiung *drafting and design educator*
Miller, James Frederick *geologist, educator*
Nuccitelli, Saul Arnold *civil engineer, consultant*
Rogers, Roddy *geotechnical engineer*

University City
Glaenzer, Richard Howard *electrical engineer*

Warrensburg
Belshe, John Francis *zoology and ecology educator*
Voorhees, Frank Ray *biology educator*

MONTANA

Bigfork
Thomas, Robert Glenn *biophysicist*

Billings
Bütz, Michael Ray *psychologist*
Darrow, George F. *natural resources company owner, consultant*
Gerlach, Thurlo Thompson *electrical engineer*
Gumper, Lindell Lewis *psychologist*
Mueller, Kenneth Howard *pathologist*

Bozeman
Block, Richard Atten *psychology educator*
Characklis, William Gregory *research center director, engineering educator*
Gray, Philip Howard *psychologist, educator*
Horner, John Robert *paleontologist, researcher*
Maxwell, Bruce Dale *plant ecologist, educator*
Onsager, Jerome Andrew *research entomologist*
Remington, Scott Alan *laser engineer*
Suit, D. James *civil engineer*

Butte
Murray, Joseph *chemistry educator*
Nelson, Gordon Leon, Jr. *aeronautical engineer*
Ruppel, Edward Thompson *geologist*

Columbia Falls
Spade, George Lawrence *scientist*

East Helena
Blossom, Neal William *chemical engineer*

Florence
Campbell, Charles *geologist*

Great Falls
Tufte, Erling Arden *civil engineer*

Hamilton
Garon, Claude Francis *laboratory administrator, researcher*
Rudbach, Jon Anthony *biotechnical company executive*

Helena
Johnson, David Sellie *civil engineer*
Johnson, Qulan Adrian *software engineer*
Warren, Christopher Charles *electronics executive*

Miles City
Birk-Updyke, Dawn Marie *psychologist*
Heitschmidt, Rodney Keith *rangeland ecologist*

Missoula
Craighead, John Johnson *wildlife biologist*
Gritzner, Jeffrey Allman *geographer, educator*
Newman, Jan Bristow *surgeon*
Peterson, James Algert *geologist, educator*
Rice, Steven Dale *electronics educator*
Strobel, David Allen *psychology educator*
Turman, George *former lieutenant governor*
Weisel, George Ferdinand *retired zoology educator*
Zheng, Youlu *computer scientist, educator*

Polson
Stanford, Jack Arthur *biological station administrator*

Rollins
Zelezny, William Francis *retired physical chemist*

Troy
Sherman, Signe Lidfeldt *securities analyst, former research chemist*

Whitehall
Clark, Steven Joseph *energy engineer, business owner, inventor, author*
Gallagher, Neil Paul *metallurgical engineer*

Wolf Point
Listerud, Mark Boyd *surgeon*

NEBRASKA

Brownville
Dingman, Norman Ray *engineering executive*

Chadron
Hardy, Joyce Margaret Phillips *plant physiologist, educator*

Clay Center
Hahn, George LeRoy *agricultural engineer, biometeorologist*
Roeth, Frederick Warren *agronomy educator*

Columbus
Selig, Phyllis Sims *architect*

Elkhorn
Graves, Harris Breiner *physician, hospital administrator*

Fairbury
Greenwood, Richard P. *wastewater superintendent*

Grafton
Benorden, Robert Roy *utility executive*

Grand Island
Bosley, Warren Guy *pediatrician*

Hastings
Wilhelm, Dallas Eugene, Jr. *biology educator*

Kearney
Goddard, David Benjamin *physician assistant, clinical perfusionist*
Miller, Richard Lee *psychology educator*
Tillotson, Dwight Keith *biologist*

Las Vegas
Haas, Robert John *aerospace engineer*

Lincoln
Arumuganathan, Kathiravepillai *plant flow cytometrist, cell and molecular biologist*
Crews, Patricia Cox *textile scientist, educator*
Elias, Samy E. G. *engineering executive*
Francis, Charles Andrew *agronomy educator, consultant*
Genoways, Hugh Howard *museum director*
Grew, Priscilla Croswell *academic administrator, geology educator*
Gross, Michael Lawrence *chemistry educator*
Hanna, Milford A. *agricultural engineering educator*
Hirai, Denitsu *surgeon*
Hubbard, Kenneth Gene *climatologist*
Koszewski, Bohdan Julius *internist, medical educator*
Krause, Joseph Lee, Jr. *electrical engineer*
Lawson, Merlin Paul *dean, climatologist*
Maher, Robert Crawford *electrical engineer, educator*
Nelson, Darrell Wayne *university administrator*
Nelson, Don Jerome *electrical engineering and computer science educator*
Newhouse, Norman Lynn *mechanical engineer*
Paek, James Joon-Hong *civil engineer, construction management educator*
Sellmyer, David Julian *physicist, educator*
Sohaili, Aspi Isfandiar *physiologist, educator*
Spanier, Graham Basil *university administrator, family sociologist*
Splinter, William Eldon *agricultural engineering educator*
Steinkamp, Keith Kendall *electrical engineer*
Summers, James Donald *agricultural engineering consultant*
Taylor, Steve Lloyd *food scientist, educator, consultant*
Weaver, Arthur Lawrence *physician*
Woollam, John Arthur *electrical engineering educator*
Yoder, Bruce Alan *chemist*

Nebraska City
Hammerschmidt, Ben L. *environmental scientist*

North Platte
Northup, Brian Keith *ecologist*
Seymour, Ronald Clement *entomologist, researcher*

Offutt AFB
Feingold, Mark Lawrence *electronic warfare officer*

Omaha
Alsharif, Naser Zaki *toxicologist, educator, researcher*
Badeer, Henry Sarkis *physiology educator*
Ben-Yaacov, Gideon *computer system designer*
Bierman, Philip Jay *physician, reseracher, educator*
Bouda, David William *insurance medical officer*
Casey, Murray Joseph *obstetrician/gynecologist, educator*
Chan, Wing-Chung *pathologist, educator*
Clements, Luther Davis, Jr. *chemical engineer, educator*
Coy, William Raymond *civil engineer*
Dalrymple, Glenn Vogt *radiologist*
DeJonge, Christopher John *obstetrics/gynecology educator*
Ehrhardt, Anton F. *medical microbiology educator*
Eilts, Susanne Elizabeth *physician*
Ertl, Ronald Frank *research coordinator*
Frisse, Ronald Joseph *telecommunications engineer*
Futrell, Nancy Nielson *neurologist*
Gendelman, Howard Eliot *biomedical researcher, physician*
Godfrey, Maurice *biomedical scientist*
Harter, David John *radiation oncologist*
Hasterlo, John S. *civil engineer, researcher*
Klassen, Lynell W. *rheumatologist, transplant immunologist*
Korbitz, Bernard Carl *oncologist, hematologist, educator, consultant*
Lenz, Charles Eldon *electrical engineering consultant*
Mardis, Hal Kennedy *urological surgeon, educator, researcher*
Maystrick, David Paul *engineering executive*
McIntire, Matilda Stewart *pediatrician, educator, retired*
Meyer, Karl V. *civil engineer, consultant*
Nair, Chandra Kunju Pillai *internist, educator*
Newton, Sean Curry *cell biologist*
Rossbach, Philip Edward *civil engineer*
St. John, Margaret Kay *research coordinator*
Schalles, John Frederick *biology educator*
Schwartz, Rodney Jay *civil, mechanical and electrical engineer*
Silberberg, Steven Richard *meteorology educator*
Simmons, Lee Guyton, Jr. *zoological park director*
Sketch, Michael Hugh *cardiologist, educator*
Thompson, William Charles *civil engineer*
Thorson, James Alden *gerontologist, author, consultant*
Tunnicliff, David George *civil engineer*
Vasiliades, John *chemist*
Zepf, Thomas Herman *physics educator, researcher*

Valley
Chapman, John Arthur *agricultural engineering executive*

Wayne
Johar, Jogindar Singh *chemistry educator*

NEVADA

Boulder City
Tryon, John Griggs *electrical engineer educator*

Carson City
Colombo, Michael Patrick *mechanical engineer*
Fischer, Michael John *ophthalmologist, physician*
Hayes, Gordon Glenn *civil engineer*
Wadman, William Wood, III *health physicist, consulting company executive*

Elko
Moren, Leslie Arthur *physician*

Ely
Alderman, Minnis Amelia *psychologist, educator, small business owner*

Fallon
Bolen, Terry Lee *optometrist*
Terry, Charles James *metallurgical engineer*

Gardnerville
Burns, James Kent (Jasper) *science illustrator*

Incline Village
Strack, Harold Arthur *retired electronics company executive, retired air force officer, planner, analyst, musician*

Las Vegas
Barth, Delbert Sylvester *environmental studies educator*
Beggs, James Harry *electrical engineer*
Broca, Laurent Antoine *aerospace scientist*
Buzard, Kurt Andre *ophthalmologist*
Carper, Stephen William *biochemist, researcher*
Chiang, Tom Chuan-Hsien *biochemist*
Faley, Robert Lawrence *instruments company executive*
Francis, Timothy Duane *chiropractor*
Freiberg, Jeffrey Joseph *civil engineer*
Fyfe, Richard Warren *electro-optics executive*
Harpster, Robert Eugene *engineering geologist*
Hattem, Albert Worth *physician*
Kurlinski, John Parker *physician*
Lanni, Joseph Anthony *military officer*
LaRubio, Daniel Paul, Jr. *civil engineer*
Le Fave, Gene Marion *polymer amd chemical company executive*
Levich, Robert Alan *geologist*
McWhirter, Joan Brighton *psychologist*
Messenger, George Clement *engineering executive, consultant*
Owens, Mark Jeffrey *geotechnical engineer*
Peck, Gaillard Ray, Jr. *aerospace business and healthcare consultant*
Pridham, Thomas Grenville *research microbiologist*
Rask, Michael Raymond *orthopaedist*
Seggev, Joram Simon *allergist, clinical immumologist*
Singh, Sahjendra Narain *electrical engineering educator, researcher*
Snaper, Alvin Allyn *engineer*
Trabia, Mohamed Bahaa Eldeen *mechanical engineer*
Wade, Stacy Lynn *computer specialist*
Walker, Lawrence Reddeford *ecologist, educator*

Minden
Bently, Donald Emery *electrical engineer*

Reno
Bautista, Renato Go *chemical engineer, educator*
Bryan, Raymond Guy *mechanical engineer, consultant*
Cummings, Nicholas Andrew *psychologist*
Fox, Carl Alan *research institute executive*
Gifford, Gerald Frederic *environmental program director*
Hoelzer, Guy Andrew *biologist*
MacKintosh, Frederick Roy *oncologist*
Meeker, Lawrence Edwin *civil engineer*
Nassiriharand, Amir *systems engineer*
Pierson, William R. *chemist*
Reddy, Rajasekara L. *mechanical engineer*
Ritter, Dale F. *geologist, research association administrator*
Smith, Aaron *research director, clinical psychologist*
Snyder, Martin Bradford *mechanical engineering educator*
Taranik, James Vladimir *geologist, educator*

Tonopah
Lathrop, Lester Wayne *mechanical engineer*

NEW HAMPSHIRE

Amherst
Pihl, Lawrence Edward *electrical engineer*

Bedford
Hall, Pamela S. *environmental consulting firm executive*

Berlin
Cabaup, Joseph John *geology educator*
Davis, Alvin Robert, Jr. *structural engineer*

Center Sandwich
Shoup, Carl Sumner *retired economist*

Concord
Rines, Robert Harvey *lawyer, inventor, law center executive, educator*
Smart, Melissa Bedor *environmental consulting company executive*

Durham
Bowden, William Breckenridge *natural resources educator*
Burdick, David Maaloe *marine ecological researcher*
Distelbrink, Jan Hendrik *physicist, researcher*
Pilgrim, Sidney Alfred Leslie *soil science educator*
Seitz, William Rudolf *chemistry educator*

Exeter
Gray, Christopher Donald *software researcher, author, consultant*

Freedom
Lamb, Henry Grodon *safety engineer*

Glen
Zager, Ronald I. *chemist, consultant*

Grantham
Amick, Charles L. *electrical engineer, consultant*

Hampstead
Hunt, Philip George *computer consultant*

Hanover
Baumgartner, James Earl *mathematics educator*
Boyce, Richard Lee *forest ecologist, researcher*
Brar, Gurdarshan Singh *soil scientist, researcher*
Bzik, David John *parasitologist, researcher*
Gilbert, John Jouett *aquatic ecologist, educator*
Gosselin, Robert Edmond *pharmacologist, educator*
Jacobs, Nicholas Joseph *microbiology educator*
Kantrowitz, Arthur *physicist, educator*
Kleck, Robert Eldon *psychology educator*
Koop, Charles Everett *surgeon, government official*
Kovacs, Austin *research engineer*
Long, Carl Ferdinand *engineering educator*
Naumann, Robert Bruno Alexander *chemist, physicist, educator*
Oreskes, Naomi *earth sciences educator, historian*
Queneau, Paul Etienne *metallurgical engineer, educator*
Stearns, Stephen Russell *civil engineer, forensic engineer, educator*
Stockmayer, Walter H(ugo) *chemistry educator*
Stukel, Therese Anne *biostatistician, educator*
Sturge, Michael Dudley *physicist*
Wegner, Gary Alan *astronomer*
Yin, John *engineering educator*

Hillsborough
Pearson, William Rowland *retired nuclear engineer*

Hollis
Palance, David M(ichael) *electrical engineer*
Riccobono, Juanita Rae *solar energy engineer*

Jaffrey
Robinson, Lawrence Wiswall *project engineer*
Walling, Cheves Thomson *chemistry educator*

Keene
Koontz, James L. *manufacturing executive*

Lebanon
Emery, Virginia Olga Beattie *psychologist, researcher*
Kelley, Maurice Leslie, Jr. *gastroenterologist, educator*
Rous, Stephen Norman *urologist, educator, editor*
Sox, Harold Carleton, Jr. *physician, educator*
Spencer-Green, George Thomas *medical educator*

Lee
Latham, Paul Walker, II *electrical engineer*

Lyme
Kelemen, Denis George *physical chemist, consultant*

Manchester
Blake, Jeannette Belisle *psychotherapist*
DesRochers, Gerard Camille *surgeon*
Dokla, Carl Phillip John *psychobiologist, educator*
Emery, Paul Emile *psychiatrist*
Khazei, Amir Mohsen *surgeon, oncologist*
Poklemba, Ronald Steven *engineer*
Prew, Diane Schmidt *information systems executive*
Seidman, Robert Howard *computer information specialist, educator*
Warfel, Christopher George *mechanical engineer, design consultant*

Merrimack
Anthony, Gregory Milton *infosystems executive*
Hower, Philip Leland *semiconductor device engineer*
Malley, James Henry Michael *industrial engineer*
Wolf, Robert Farkas *systems and avionics company executive, environmental planning consultant*

Nashua
Blatt, Stephen Robert *systems engineer*
Dick, Ellen A. *computer engineer*
Fallet, George *civil engineer*
Klarman, Karl Joseph *electromechanical engineer*
Moskowitz, Ronald *electronics executive*

New London
Morrey, Walter Thomas *electronics engineer*

Newport
Gibbs, Elizabeth Dorothea *developmental psychologist*

North Sutton
Springsteen, Arthur William *organic chemist*

Plymouth
Chabot, Christopher Cleaves *biology educator*

Portsmouth
Bavicchi, Robert Ferris *construction materials technician, concrete batchplant operator*
Faughnan, William Anthony, Jr. *electrical/ electronics engineer*
Fernald, James Michael *engineer*
Fields, William Alexander *naval officer, mechanical engineer*
Klotz, Louis Herman *structural engineer, educator, consultant*
Perkins, Robert Bennett *mechanical engineer*
Powers, Eva Agoston *clinical psychologist*
Rodriguez, Fabio Enrique *electrical engineer*

Rye
Heald, Paul Francis *mechanical engineer, consultant*

Stratham
Bjorkman, Gordon Stuart, Jr. *structural engineer, consultant*

Weare
Dombrowski, Frank Paul, Jr. *pharmacist*
Pierce, John Alvin *physics researcher*

Windham
Hurst, Michael William *psychologist*

NEW JERSEY

Allendale
Macaya, Roman Federico *biochemist*

Allenhurst
Calabro, Joseph John, III *physician*

Alpine
Sommers, Sheldon Charles *pathologist*

Annandale
Flannery, Brian Paul *physicist, educator*
Gorbaty, Martin Leo *chemist, researcher*
Sinfelt, John Henry *chemist*
Varadaraj, Ramesh *research chemist*

Atlantic City
Flournoy, Thomas Henry *mechanical engineer*
Vansuetendael, Nancy Jean *physicist*

Avenel
Mazurkiewicz, John Anthony *plant process engineer*

Barnegat
Lowe, Angela Maria *civil engineer*

Basking Ridge
McCall, David W. *chemist, administrator, materials consultant*
Milcarek, William Francis *marketing professional*

Bayonne
Olszewski, Jerzy Adam *electrical engineer*
Steele Clapp, Jonathan Charles *chemical engineer*

Bayville
Scancella, Robert J. *civil engineer*

Bedminster
Beiman, Elliott *research pharmaceutical chemist*
Bovey, Frank Alden *research chemist*
Collins, George Joseph *chemist*
Hart, Terry Jonathan *communications executive*

Belle Mead
Singley, Mark Eldridge *agricultural engineering educator*

Belleville
Caputo, Wayne James *surgeon, podiatrist*
Czirbik, Rudolf Joseph *cell biologist*
Parikh, Manor Madanmohan *mechanical engineer*
Torre, Bennett Patrick *computer scientist, consultant*

Berkeley Heights
Hirsch, Mark J. *computer engineer*
Townsend, Palmer Wilson *chemical engineer, consultant*
Zaret, Efrem Herbert *chemist*

Birmingham
McGarvey, Francis Xavier *chemical engineer*

Blairstown
Martin, James Walter *chemist, technology executive*

Bloomfield
Hutcheon, Cifford Robert *engineer*
Kwon, Joon Taek *chemistry researcher*
Meenakshi Sundaram, Kandasamy *chemical engineer*
Monahan, Edward James *geotechnical engineer*

Bound Brook
Gould, Donald Everett *chemical company executive*

Brick
Kowalski, Lynn Mary *podiatrist*

Bridgewater
Albrethsen, Adrian Edysel *metallurgist, consultant*
Iovine, Carmine P. *chemicals executive*
Ramharack, Roopram *research scientist*
Roehrenbeck, Paul William *marketing professional*
Twardowski, Thomas Edward, Jr. *development chemist*
Weingast, Marvin *laboratory director*
Wilkinson, Clifford Steven *civil engineer*

Brielle
Kirby, Gary Neil *metallurgical and materials engineer, consultant*

Browns Mills
Gu, Jiang *biomedical scientist*

Budd Lake
Krause, Lois Ruth Breur *chemistry educator, engineer*

Caldwell
Kapusinski, Albert Thomas *economist, educator*

Camden
Ances, I. G(eorge) *obstetrician/gynecologist, educator*
Beck, David Paul *biochemist*
Foglesong, Paul David *microbiology educator*
Johnson, David Willis *food products executive*
Simon, Frederick Edward *chemist*

Cape May Court House
Wood, Albert E(lmer) *biology educator*

Chatham
Hinderliter, Richard Glenn *electrical engineer*
KixMiller, Richard Wood *chemist, corporate executive*

Cherry Hill
Davis, Kevin Jon *chemical engineer, consulting engineer*
Fuentevilla, Manuel Edward *chemical engineer*
Hayasi, Nisiki *physicist*
McCabe, Robert James *engineering executive*
Olearchyk, Andrew S. *cardiothoracic surgeon, educator*
Parker, Jack Royal *engineering executive*

Cinnaminson
Evans, David A. *plant geneticist*

Clark
Apelian, Virginia Matosian *psychologist, assertiveness training instructor, lecturer, consultant*
Walsh, Daniel Stephen *systems engineering consultant*

Clifton
Fowler, Cecile Ann *nurse, professional soloist*
Held, George Anthony *architect*
Srinivasachari, Samavedam *chemical engineer*
Swystun-Rives, Bohdana Alexandra *dentist*

Clinton
Bosch, Robert John, Jr. *mechanical engineer*

Closter
Hillman, Leon *electrical engineer*

Collingswood
Carty, John Brooks *civil engineer*

Cranbury
Emont, George Daniel *healthcare executive*
Larson, Eric Heath *chemical engineer*
Rosenberg, Alberto *pharmaceutical company executive, physician*
Sofia, R. D. *pharmacologist*
Wang, Chih Chun *chemist, scientific administrator*

Cranford
Dashevsky, Sidney George *clinical and industrial psychologist*

Deepwater
Millican, David Wayne *analytical chemist*

Denville
Brooks, Marc Benjamin *civil engineer*
Price, Robert Edmunds *civil engineer*

Dover
Betta, Mary Beth *engineer*
Capellos, Chris Spiridon *chemist*
Dombroski, Lee Anne Zarger *medical physicist*
Tatyrek, Alfred Frank *materials engineer, research chemist*

East Brunswick
Chang, Stephen S. *food scientist, educator, researcher, inventor*
Irgon, Joseph *physical chemist*
Lund, Daryl B. *college dean*
Steiger, Fred Harold *chemist*

East Hanover
Klemann, Lawrence Paul *chemical scientist*

East Millstone
Smith, Jacqueline Hagan *toxicologist*

East Orange
Astor, Peter H. *environmental consultant, mathematician*
Brundage, Gertrude Barnes *pediatrician*
Sibilski, Peter John *chemical engineer*

Edgewater
Ahtchi-Ali, Badreddine *chemical engineer*
Schiltz, John Raymond *biochemist, researcher*
Spaltro, Suree Methmanus *chemist, researcher*

Edison
Farazdel, Abbas *chemist*
Lin, Yi-Hua *environmental chemist*
Mosher, Frederick Kenneth *engineering consultant*
Pagdon, William Harry *optical systems technician*
Parsons, William Hugh *chemist, researcher*
Patel, Bhavin R. *chemical engineer*
Sapoff, Meyer *electronics component manufacturer*
Shattuck, Douglas E. *mechanical engineer, consultant*

Elizabeth
Buonanni, Brian Francis *health care facility administrator, consultant*
Hynes, Nancy Ellen *reading educator*
Marticorena, Ernesto Jesus *mechanical engineer, educator*
Poch, Herbert Edward *pediatrician, educator*
Sananman, Michael Lawrence *neurologist*
Trawick, Lafayette James, Jr. *research scientist*

Elmer
Robinson, John Abbott *mechanical engineer*

Elmwood Park
Ogden, Michael Richard *civil engineer, consultant*
Semeraro, Michael Archangel, Jr. *civil engineer*

Englewood
Grier, Nathaniel *chemist*
Rabin, Aaron *neurologist*
Wuhl, Charles Michael *psychiatrist*

Englewood Cliffs
Dobrowolski, Francis Joseph *environmental engineer, consultant*
DoVale, Fern Louise *civil engineer*
Farrell, Patricia Ann *psychologist, educator*

Fair Lawn
Gal, Aaron *electronics engineer*
Golaski, Nicholas John *information scientist*
Humiec, Frank Stanley *chemist*
Kannankeril, Charles Paul *chemical engineer*
Mills, Lester Stephen *chemist*
Motin, Revell Judith *data processing executive*
Nizin, Joel Scott *surgeon*

Fairfield
Regino, Thomas Charles *technology assessment, R&D benchmarking specialist*

Fanwood
McNally, Harry John, Jr. *engineer, construction and real estate executive, consultant, researcher, accountant*

Far Hills
Barnum, Mary Ann Mook *information management manager*

Flemington
Easter, Charles Henry *technical writer*
Mitry, Mark *chemist*
Scasta, David Lynn *psychiatrist*

Florham Park
Amabile, Michael John *mechanical engineer, consultant*
Bhagat, Phiroz Maneck *mechanical engineer*
Jameson, J(ames) Larry *cable company executive*
Matsen, John Morris *engineer*
Naimark, George Modell *marketing and management consultant*
Nash, Lillian Dorothy *gynecologist, reproductive endocrinologist*
Perham, Roy Gates, III *industrial psychologist*
Richter, Jeffrey Alan *chemical engineer*
Sprow, Frank Barker *oil company executive*

Fords
Chryss, George *chemical company executive, consultant*
Lynch, Charles Andrew *chemical company executive*

Fort Hancock
Klein, George D. *geologist, business executive*

Fort Lee
Screpetis, Dennis *nuclear engineer, consultant*

Fort Monmouth
Kronenberg, Stanley *research physicist*
McBride, George Gustave *electronics engineer*

Franklin Lakes
Hetzel, Donald Stanford *chemist*
Lovell, Theodore *electrical engineer, consultant*

Freehold
Stirrat, William Albert *electronics engineer*

Gillette
Nathanson, Linda Sue *technical writer, software training specialist*

Glassboro
Kaufman, Denise Norma *psychologist, addictions counselor, educator*
Otooni, Monde A. *materials scientist, physicist*
Schultz, Charles William *chemistry educator*

Glasser
Suplicki, John Clarke *chemical and metallurgical engineer, administrator*

Green Brook
Elias, Donald Francis *environmental consultant*

Green Village
Castenschiold, Rene *engineering company executive, author, consultant*

Hackensack
Case, Gerard Ramon *drafting technician*
Gross, Peter Alan *epidemiologist, researcher*
Margron, Frederick Joseph *environmental engineer*
Mavrovic, Ivo *chemical engineer*
Skovronek, Herbert Samuel *environmental scientist*
Ten Kate, Peter Cornelius *civil engineer*

Hackettstown
Elejalde, Cesar Carlos *food engineer*

Haddon Heights
Narin, Francis *research company executive*
Oliver, Daniel Anthony *electrical engineer, company executive*

Haddonfield
Reisner, John Henry, Jr. *electronic research physicist*

Hammonton
Stephanick, Carol Ann *dentist, consultant*
Weigel, Henry Donald *civil engineer*

Hancocks Bridge
Bond, Randall Clay *quality assurance engineer*
Trencher, Gary Joseph *nuclear engineer*

Hazlet
Chudzik, Douglas Walter *internist*

Hewitt
Selwyn, Donald *engineering administrator, researcher, inventor, educator*

Hillside
Patell, Mahesh *pharmacist, researcher*
Tencza, Thomas Michael *chemist*

Hoboken
Avgousti, Marios *chemical engineer*
Bernstein, Jeremy *physicist, educator*
Bose, Ajay Kumar *chemistry educator*
Bruno, Michael Stephen *ocean engineering educator, researcher*
Brzosko, Jan Stefan *physicist*
Chirlian, Paul Michael *electrical engineering educator*
Cogan, Kenneth George *quality assurance engineer*
Donskoy, Dimitri Michailovitch *physicist, researcher*
Gastwirt, Lawrence E. *chemical engineer*
Kovenklioglu, Suphan Remzi *chemical engineering educator*
Kunhardt, Erich Enrique *physicist, educator*
Prasad, Marehalli Gopalan *mechanical engineering educator*
Raveché, Harold Joseph *university administrator, physical chemist*
Swern, Frederic Lee *engineering educator*
Talimcioglu, Nazmi Mete *civil and environmental engineer, educator*
Tsenoglou, Christos *chemical engineer*

Holmdel
Abate, John E. *electrical/electronic engineer, communications consultant*
Colmenares, Narses Jose *electrical engineer*
Denker, John Stewart *physicist*
Francis, Eulalie Marie *psychologist*
Frenkiel, Richard Henry *electronics company research and development executive*
Haskell, Barry Geoffry *communications company research administrator*
Kogelnik, Herwig Werner *electronics company executive*
Lawser, John Jutten *electrical engineer*
Linker, Kerrie Lynn *systems engineer*
McCallum, Charles John, Jr. *communications company executive*
Merrill, Aubrey James *systems and electrical engineer*
Miller, Robert Alan *physicist*
Murthy, Srinivasa K. *engineering corporation executive*
Stolen, Rogers Hall *optics scientist*
Tkach, Robert William *physicist*
West, Earle Huddleston *communications company professional*
Wilson, Robert Woodrow *radio astronomer*

Hopatcong
Ferderber-Hersonski, Boris Constantin *process engineer*

Hopewell
Basavanhally, Nagesh Ramamoorthy *opto-electronics packaging engineer*

Iselin
Garfinkel, Harmon Mark *specialty chemicals company executive*
Gondek, John Richard *quality assurance professional*
Smith, Orin Robert *chemical company executive*

Jersey City
Alfano, Michael Charles *pharmaceutical company executive*
Chatterjee, Amit *structural engineer*
Giuliani, Eleanor Regina *biology educator*
Gudema, Norman H. *civil engineer*
Hernon, Richard Francis *engineer*
Mastro, Victor John *mathematician, educator*
Nakhla, Atif Mounir *biochemist*

Keansburg
Margolis, James Mark *international industrial developer*

Kenilworth
Colucci, Robert Dominick *cardiovascular pharmacologist*
DiGiacomo, Ruth Ann *research scientist*
Evans, Charlie Anderson *chemist*
Filippone Steinbrick, Gay *pharmacist, educator*
Ganguly, Ashit Kumar *organic chemist*
Pachter, Jonathan Alan *biochemist, researcher*
Pellerito-Bessette, Frances *research biologist*
Wagman, Gerald Howard *library administrator*

Kinnelon
Preston, Andrew Joseph *pharmacist, drug company executive*

Lambertville
Cohen, Edward *private consultant*

Lawrenceville
Campbell, David Bruce *biology educator, researcher*
Enegess, David Norman *chemical engineer*
Oana, Harry Jerome *optical engineer*
Romero, Ralph *physicist*

Leonia
Armstrong, Edward Bradford, Jr. *oral and maxillofacial surgeon, educator*
Husar, Emile *civil engineer, consultant*

Linden
Bedrick, Bernice *retired science educator, consultant*
Dietz, Thomas Gordon *chemist*
Duvdevani, Ilan *chemical engineer, researcher*
Musillo, Joseph H. *electrical engineer*
Pink, Harry Stuard *petroleum chemist*

Linwood
Godfrey, George Cheeseman, II *surgeon*

Little Falls
Dohr, Donald R. *metallurgical engineer, researcher*
Stiles, John Callender *physicist*

Livingston
Caballes, Romeo Lopez *pathologist, bone tumor researcher*
Gupta, Gopal Das *mechanical engineer, researcher*
Heilmeier, George Harry *electrical engineer, researcher*
Koester, J. Anthony *science publication editor*
Peterson, Glenn Stephen *chemist*

Lodi
Budwani, Ramesh Nebhandas *consultant*
Melignano, Carmine (Emanuel Melignano) *video engineer*

Long Branch
Makhija, Mohan *nuclear medicine physician*

Lyndhurst
Moese, Mark Douglas *environmental consultant*

Lyons
Youngelman, David Roy *psychologist*

Mahwah
Kierstead, James Allan *computer scientist*
Landeck, Robin Elaine *environmental engineer*
Sulla, Nancy *computer coordinator, consultant, business owner*

Manville
Koscelnick, Jeanne *computer scientist*

Maplewood
Leal, Joseph Rogers *chemist*
Lev, Alexander Shulim *mechanical engineer*
Marlow, Joseph *automation engineer*
Shuttleworth, Anne Margaret *psychiatrist*
Stillman, Gerald Israel *electrical engineer, science administrator*

Marlton
Wright, Kara-Lyn Annette *software engineer*

Maywood
Kennedy, Joanne Patricia *technical service representative*

Mendham
Desjardins, Raoul *medical association administrator, financial consultant*

Metuchen
Gerardi, Roy G., Jr. *mechanical engineer*
Stapley, Edward Olley *retired microbiologist, research administrator*

Middletown
Anania, William Christian *podiatrist*
Broer, Matthijs Meno *physicist*
Erfani, Shervin *electrical engineer*
Swett, Susan *chemical engineer*

Millburn
Heistein, Robert Kenneth *obstetrician/gynecologist*
Nicol, Marjorie Carmichael *research psychologist*

Milltown
Sacharow, Stanley *chemist, consultant*

Monmouth Junction
Drzewinski, Michael Anthony *materials scientist*
Ellerbusch, Fred *environmental engineer*
Molnar-Kimber, Katherine Lu *immunologist, molecular biologist*
Mylonakis, Stamatios Gregory *chemist*

Montclair
Hutchins, Carleen Maley *acoustical engineer, violin maker, consultant*

Montvale
Bowman, Patricia Imig *microbiologist*

Moorestown
Druckman, William Frank *mechanical engineer*
Lipscomb, Thomas Heber, Jr. *civil engineer, consultant*
Nespor, Jerald Daniel *electrical engineer, educator*

Morris Plains
Degady, Marc *chemical engineer*
Fielding, Stuart *psychopharmacologist*
Goodes, Melvin Russell *manufacturing company executive*
Kumar, Surinder *food company executive*
Montana, Anthony James *analytical chemist*
Moroni, Antonio *chemist, consultant, international coordinator*
Otani, Mike *optical company executive*
Patel, Kanti Shamjibhai *civil engineer*
Woolley, Gail Suzanne *military officer*

Morristown
Baron, Hazen Jay *dental scientist*
Barpal, Isaac Ruben *technology and operations executive*
Bockian, James Bernard *computer systems executive*
Chen, Walter Yi-Chen *electrical engineer*
de Mauret, Kevin John *geologist*
Derelanko, Michael Joseph *toxicologist*
Dillon, Joseph Francis, Jr. *physicist, consultant*
Epstein, Samuel Seth *computer scientist, researcher*
Golecki, Ilan *physicist, researcher, educator*
Goodman, Matthew Samuel *physicist*
Iqbal, Zafar *physicist*
Kavesh, Sheldon *chemical engineer*
Lemberg, Howard Lee *telecommunications network research executive*
Nielsen, Jakob *user interface engineer*
Pepper, David J. *electrical engineer*
Reimschuessel, Herbert Kurt *chemist, consultant*
Stephens, William Edward *electronics engineer*
Tenopyr, Mary Louise Welsh (Mrs. Joseph Tenopyr) *psychologist*
Traeger, Donna Jean *health facility manager*
Wei, Victor Keh *electrical engineer*
Wu, Liang Tai *electrical engineer*
Wu, Tse Cheng *research chemist*

Mount Holly
Stabenau, Walter Frank *systems engineer*
Worne, Howard Edward *biotechnology company executive*

Mount Laurel
Barba, Evans Michael *civil engineer*
Case, Mark Edward *geotechnical engineer, consultant*
Cazes, Jack *chemist, marketing consultant, editor*
Melick, George Fleury *mechanical engineer, educator*
Passarella, Louis Anthony *systems engineer*

Mountainside
Buccini, Frank John *molecular geneticist*
Ricciardi, Antonio *prosthodontist, educator*

Murray Hill
Atal, Bishnu Saroop *speech research executive*
Baker, William Oliver *research chemist, educator*
Batlogg, Bertram *physicist*
Cho, Alfred Yi *electrical engineer*
D'Asaro, Lucian Arthur *physicist*
Geusic, Joseph Edward *physicist*
Gordon, James Power *optics scientist*
Graham, Ronald Lewis *mathematician*
Johnson, Bertrand H. *optical engineer*
Johnson, David W., Jr. *ceramic scientist, researcher*
Krishnamurthy, Ramachandran (Krish) *chemical engineer, researcher*
Larson, Ronald Gary *chemical engineer*
Lemaire, Paul Joseph *scientist*
Litman, Diane Judith *computer scientist*
Mayo, John Sullivan *telecommunications company executive*
Pinczuk, Aron *physicist*
Quan, Xina *polymer engineer*
Sali, Vlad Naim *chemical engineer*
Sheu, Lien-Lung *chemical engineer*
Simpkins, Peter G. *applied physicist, researcher*
Sullivan, Paul Andrew *research electrical engineer*
Tseng, Chia-Jeng *computer engineer*
Tully, John Charles *research chemical physicist*

Neptune
Brown, Paul Leighton *electrical engineer*

Neshanic Station
Castellon, Christine New *information systems manager*

New Brunswick
Borah, Gregory Louis *plastic and reconstructive surgeon*
Cizewski, Jolie Antonia *physics educator, researcher*
Contrada, Richard J(ude) *psychologist*
Daubechies, Ingrid *mathematics educator*
Day, Peter Rodney *geneticist, educator*
Dill, Ellis Harold *university dean*
Duvoisin, Roger Clair *physician, medical educator*
Ehrenfeld, David William *biology educator, author*
Gaylor, James Leroy *biomedical research director*
Glass, Arnold Lewis *psychology educator*
Greco, Ralph Steven *surgeon, researcher, medical educator*
Gussin, Robert Zalmon *health care company executive*
Hadani, Itzhak *experimental psychologist, human factors engineer*
Hayakawa, Kan-Ichi *food science educator*
Ho, Chi-Tang *food chemistry educator*
Kruskal, Martin David *mathematical physicist, educator*
Kulikowski, Casimir Alexander *computer science educator, research program director*
Lebowitz, Joel Louis *physicist, educator*
Lerner, Henry Hyam *chemist*
Martin, Frank Scott *chemist*
McGuire, John Lawrence *pharmaceuticals research executive*
Montville, Thomas Joseph *food microbiologist, educator*
Nawy, Edward George *civil engineer, educator*
Norris, Peter Edward *semiconductor physicist*
Pandey, Ramesh Chandra *chemist*
Psuty, Norbert Phillip *marine sciences educator*
Rosen, Joseph David *chemist, educator*
Rosen, Robert Thomas *analytical and food chemist*
Sachs, Clifford Jay *research scientist*
Solberg, Myron *food scientist, educator*
Strauss, Ulrich Paul *educator, chemist*
Strawderman, William E. *statistics educator*
Torrey, Henry Cutler *physicist*
Treves, Jean-François *mathematician educator*
Vrijenhoek, Robert Charles *biologist*
Wainright, Sam Chapman *marine scientist, educator*
Walsh, Thomas Joseph *neuroscientist, educator*
Winnett, A(sa) George *analytical research chemist, educator*
Zawadsky, Joseph Peter *physician*

New Providence
Bishop, David John *physicist*
Chandross, Edwin A. *chemist, polymer researcher*
Koszi, Louis A. *electronics engineer*
Kotynek, George Roy *mechanical engineer, educator, marketing executive*
Li, Hong *physicist*
MacChesney, John Burnette *materials scientist, researcher*
Passner, Albert *physicist, researcher*
Pearton, Stephen John *physicist*
Phillips, Julia Mae *physicist*
Sharp, Louis *scientist*
Shepp, Lawrence Alan *mathematician, educator*
Smith, Neville Vincent *physicist*
Stillinger, Frank Henry *chemist, educator*
Thompson, Larry Flack *chemical company executive*

Newark
Armenante, Piero M. *chemical engineering educator*
Bar-Ness, Yeheskel *electrical engineer, educator*
Beyer-Mears, Annette *physiologist*
Bigley, William Joseph, Jr. *control engineer*
Blount, Alice McDaniel *museum curator*
Cheremisinoff, Paul Nicholas *environmental engineer, educator*
Fadem, Barbara H. *psychobiologist, psychiatry educator*
Friedland, Bernard *engineer, educator*
Fu, Shou-Cheng Joseph *biomedicine educator*
Garfinkle, Devra *mathematician, educator*
Geskin, Ernest S(amuel) *science administrator, consultant*
Goldenberg, David Milton *experimental pathologist, oncologist*
Greene, Clifford *psychologist*
Hochberg, Mark S. *cardiac surgeon*
Hrycak, Peter *mechanical engineer, educator*
Hsu, Cheng-Tzu Thomas *civil engineering educator*
Iffy, Leslie *medical educator*
Jandinski, John Joseph *dentist, immunologist*
Kantor, Mel Lewis *dental educator, researcher*
Karvelas, Dennis E. *computers and information science educator*
Kazem, Ismail *radiation oncologist, educator, health science facility administrator*
Klein, Marshall S. *health facility administrator*
Kunicki, Jan Ireneusz *physicist*
Ledeen, Robert Wagner *neurochemist, educator*
Marcus, Robert Boris *physicist*
Materna, Thomas Walter *ophthalmologist*
Mistry, Kishorkumar Purushottamdas *biomedical scientist, educator*
Park, Min-Yong *human factors and safety engineer, educator*
Pignataro, Louis James *engineering educator*
Porter, Michael Blair *applied mathematics educator*
Rosenblatt, Jay Seth *psychobiologist, educator*
Ryzlak, Maria Teresa *biochemist, educator*
Scherzer, Norman Alan *medical educator*
Shaw, Henry *chemical engineering educator*
Skurnick, Joan Hardy *biostatistician, educator*
Steinberg, Reuben Benjamin *utility management engineer*
Suchow, Lawrence *chemistry educator, researcher, consultant*
Sun, Benedict Ching-San *engineering educator, consultant*
Suresh, Bangalore Ananthaswami *information systems educator*
Tallal, Paula *psychologist*
Waelde, Lawrence Richard *chemist*
Zirvi, Karimullah Abd *experimental oncologist, educator*

North Branch
Kravec, Cynthia Vallen *microbiologist*

North Brunswick
Awan, Ahmad Noor *civil engineer*
Schnitzler, Paul *electronics engineering manager*

North Plainfield
Stoner, Henry Raymond *retired chemical engineer*

Nutley
Abbondanzo, Susan Jane *research geneticist*
Amornmarn, Lina *chemist*
Bautz, Gordon Thomas *information scientist, researcher*
Behl, Charanjit R. *pharmaceutical scientist*
Connor, John Arthur *neuroscientist*
Curran, Thomas *molecular biologist, educator*
Dennin, Robert Aloysius, Jr. *pharmaceutical research scientist*
Douvan-Kulesha, Irina *chemist*
Drews, Jürgen *pharmaceutical researcher*
Enthoven, Dirk *clinical pharmacologist, researcher*
Gately, Maurice Kent *research immunologist*
Hall, Clifford Charles *pharmaceutical biochemist*
Liu, Yu-Jih *engineer*
Magee, Richard Stephen *mechanical engineering educator*
Mostillo, Ralph *medical association executive*
Pruess, David Louis *biochemist*

Oakhurst
Rossignol, Roger John *coatings company executive*

Oakland
Bacaloglu, Radu *chemical engineer*
Pepe, Teri-Anne *development chemist*
Spataro, Vincent John *electrical engineer*

Ocean City
Reiter, William Martin *chemical engineer*

Old Bridge
Brennan, George Gerard *pediatrician*
Pflug, Leo Joseph, Jr. *civil engineer*

Oradell
Tong, Mary Powderly *mathematician educator, retired*

Paramus
Aronson, Miriam Klausner *gerontologist, consultant, researcher, educator*
Bard, Jonathan Adam *molecular biologist*
Greenberg, William Michael *psychiatrist*
Klarreich, Susan Rae *chemistry educator*
Sergi, Anthony Robert *physician, surgeon*

Parsippany
Boulos, Atef Zekry *chemist*
Clark, Philip Raymond *nuclear utility executive, engineer*
Gilby, Steve *metallurgical engineering researcher*
Hasegawa, Ryusuke *materials scientist*
Labriola, Joseph Arthur *environmental biologist*
Liebermann, Howard Horst *metallurgical engineering executive*
Marscher, William Donnelly *engineering company executive*
Wadey, Brian Leu *polymer engineer*

Passaic
Baum, Howard Barry *physician*
Lindholm, Clifford Falstrom, II *engineering executive, mayor*

Paterson
DeBari, Vincent Anthony *medical researcher, educator*
Williams, Dennis Thomas *civil engineer*

Paulsboro
Domingue, Raymond Pierre *chemist, consultant, educator*
Dwyer, Francis Gerard *chemical engineer, researcher*
Garwood, William Everett *chemist researcher*

Pennington
Krupowicz, John Joseph *metallurgical engineer*
Marder, William Zev *technological consultant*

Pennsauken
Alday, Paul Stackhouse, Jr. *mechanical engineer*

Pennsville
Ibrahim, Fayez Barsoum *chemist*
Ryan, Timothy William *analytical chemist*

Pequannock
MacMurren, Harold Henry, Jr. *psychologist, lawyer*

Perth Amboy
Hall, Pamela Elizabeth *psychologist*

Phillipsburg
Guinzburg, Adiel *mechanical engineer*
Kim, Ih Chin *pediatrician*

Piscataway
Alekman, Stanley Lawrence *chemical company executive*
Boucher, Thomas Owen *engineering educator, researcher*
Briggs, David Griffith *mechanical engineer, educator*
Chien, Yie Wen *pharmaceutics educator*
Dayton, John Thomas, Jr. *computer researcher*
De Chino, Karen Linnia *engineering association administrator*
Denhardt, David Tilton *molecular and cell biology educator*
Donnelly, Kim Frances *computer scientist*
Dornburg, Ralph Christoph *biology educator*
Dunn, Stanley Martin *biomedical engineering educator*
Flanagan, James Loton *electrical engineer, educator*
Freidkin, Evgenii S. *physicist*
Gaeta, Vincent Ettore *laboratory technologist*
Gaffar, Abdul *research scientist, administrator*
Hara, Masanori *chemist*
Hellenbrecht, Edward Paul *mechanical engineer*
Hung, George Kit *biomedical engineering educator*
Julesz, Bela *experimental psychologist, educator, electrical engineer*
Kear, Bernard Henry *materials scientist*
Keller, Mark *medical educator*
Mammone, Richard James *engineering educator*
Mehlman, Myron A. *environmental and occupational medicine educator, environmental toxicologist*
Messing, Joachim Wilhelm *molecular biology educator*
Neishlos, Arye Leon *chemist*
Nilsson, Bo Ingvar *hematologist*
Pandina, Robert John *neuropsychologist*
Paraskevopoulos, Nicholas George *electrical engineering educator*
Polinsky, Joseph Thomas *purchasing manager*
Robbins, Allen Bishop *physics educator*
Ruh, Edwin *ceramic engineer, consultant, researcher*
Salkind, Alvin J. *electrochemical engineer, educator*
Semmlow, John Leonard *biomedical engineer, research scientist*
Shatkin, Aaron Jeffrey *biochemistry educator*
Sinko, Patrick J. *pharmacist, educator*
Snitzer, Elias *physicist*
Sohal, Iqbal Singh *structural engineer, educator*
Spence, Donald Pond *psychologist, psychoanalyst*
Sugerman, Abraham Arthur *psychiatrist*
Szewczyk, Martin Joseph *chemical engineer*
Thomas, Thresia K. *biomedical researcher, biochemistry educator*
Urban, Cathleen Andrea *client technical support consultant*
Walker, James Calvin *computer software systems developer, politician*
Williams, James Richard *human factors engineering psychologist*
Zhao, Jian Hui *electrical and computer engineering educator*

Plainfield
Yood, Harold Stanley *internist*

Plainsboro
Pinninti, Krishna Rao *economist*
Royds, Robert Bruce *physician*

Port Newark
McKenna, James Emmet *chemist*

Pottersville
Konecky, Milton Stuart *chemist*

Princeton
Adler, Stephen Louis *physicist*
Aizenman, Michael *mathematics and physics educator*
Anderson, Philip Warren *physicist*
Bahcall, Neta Assaf *astrophysicist*
Bair, William Alois *engineer*
Balakrishna, Subash *chemical engineer*
Bentz, Bryan Lloyd *chemist*
Birnbaum, Jerome *pharmaceutical company executive*
Blair, David William *mechanical engineer*
Bogdonoff, Seymour Moses *aeronautical engineer*
Borel, Armand *mathematics educator*
Bosacchi, Bruno *physicist*
Breland, Hunter Mansfield *psychologist*
Caffarelli, Luis Angel *mathematician, educator*
Christodoulou, Demetrios *mathematics educator*
Cohen, Isaac Louis (Ike Cohen) *data processing executive*
Cryer, Dennis Robert *pharmaceutical company executive, researcher*
Davidson, Ronald Crosby *physicist, educator*
Debenedetti, Pablo Gaston *chemical engineer*
Deligné, Pierre R. *mathematician*
Dinkevich, Solomon *structural engineer*
Dyson, Freeman John *physicist*
Etz, Lois Kapelsohn *architectural company executive*
Fefferman, Charles Louis *mathematics educator*
Fernandes, Prabhavathi Bhat *molecular biologist*
File, Joseph *research physics engineer*
Fisch, Nathaniel Joseph *physicist*
Fitch, Val Logsdon *physics educator*
Fitton, Gary Michael *electronics engineer*
Freeman, Marjorie Kler *interior designer*
Furth, Harold Paul *physicist, educator*
Ghosh, Chuni Lal *physicist*
Ghosh, Malathi *physicist*
Glassman, Irvin *mechanical and aeronautical engineering educator, consultant*
Gogulski, Paul *construction consultant*
Gott, J. Richard, III *astrophysicist*
Grant, Peter Raymond *biologist, researcher, educator*
Green, Joseph *chemist*
Griffiths, Phillip A. *mathematician, academic administrator*
Groves, John Taylor, III *chemist, educator*
Haldane, Frederick Duncan Michael *physics educator*
Happer, William, Jr. *physicist, educator*
Harnad, Stevan Robert *cognitive scientist*
Henderson, John Goodchilde Norie *electrical engineer*
Hill, Russell Gibson *chemical engineer, consultant*
Hillier, James *communications executive, researcher*
Hoebel, Bartley Gore *psychology educator*
Hulse, Russell Alan *astrophysicist, plasma physicist*
Hut, Piet *astrophysics educator*
Jagerman, David Lewis *mathematician*
Jardin, Stephen Charles *plasma physicist*
Jones, Maitland, Jr. *chemistry educator*
Kane, Michael Joel *physician*
Karol, Reuben Hirsh *civil engineer, sculptor*
Katz, Irvin Ronald *research scientist*
Kauzmann, Walter Joseph *chemistry educator*
Kevrekidis, Yannis George *chemical engineer*
Kilian, Robert Joseph *chemist*
Klein, Leonard *chemist*
Kramer, Richard Harry, Jr. *biomedical engineer*
Kulkarni, Sanjeev Ramesh *electrical engineering educator*
Kurtz, Andrew Dallas *chemical engineer*
Kyin, Saw William *chemist, consultant*
Langlands, Robert Phelan *mathematician*
Law, Chung King *aerospace engineering educator, researcher*
Lawrence, Robert Michael *research chemist*
Lemelson, Jerome H. *inventor*
Lemonick, Aaron *physicist, educator*
Levine, Arnold Jay *molecular biology educator, researcher*
Lewis, Alan James *pharmaceutical executive, pharmacologist*
Liao, Hsiang Peng *chemist*
Libchaber, Albert Joseph *physics educator*
Lieb, Elliott Hershel *physicist, mathematician, educator*
Lipscombe, Trevor Charles Edmund *physical science editor, researcher*
Little, Dorothy Marion Sheila *chemist*
Liu, Edward Chang-Kai *biochemist*
Los, Marinus *agrochemical researcher*
Mahlman, Jerry David *research meteorologist*
Majda, Andrew J. *mathematician, educator*
Manabe, Syukuro *climatologist*
Marchese, Anthony John *research engineer*
Marshall, Carol Joyce *clinical research data coordinator*
Mc Clure, Donald Stuart *physical chemist, educator*
McKearn, Thomas Joseph *immunology and pathology educator, scientist*
Melamed, Benjamin *computer scientist*
Miller, George Armitage *psychologist, educator*
Miyamoto, Maylene Hu *computer science educator*
Molloy, Christopher John *molecular cell biologist, pharmacist*
Mueller, Dennis Warren *physicist*
Mueller, Peter Sterling *psychiatrist, educator*
Navrotsky, Alexandra *geophysicist*
Novotny, Jiri *biophysicist*
Ondetti, Miguel Angel *chemist, consultant*
Oort, Abraham Hans *meteorologist, researcher, educator*
Orszag, Steven Alan *applied mathematician, educator*
Ostriker, Jeremiah Paul *astrophysicist, educator*
Paczynski, Bohdan *astrophysicist, educator*
Patel, Mukund Ranchhodlal *electrical engineer, researcher*
Platt, Judith Roberta *electrical engineer*
Plummer, Ernest Lockhart *industrial research chemist*
Pollock, Adrian Anthony *physicist*
Pothos, Emmanuel *neuroscientist*
Purohit, Milind Vasant *physicist*
Ramaprasad, Kackadasam Raghavachar *physical chemist*
Rebenfeld, Ludwig *chemist*
Register, Richard Alan *chemical engineering educator*
Reinhardt, Uwe Ernst *economist, educator*
Rhimes, Richard David *civil engineer*
Rolle, F. Robert *health care consultant*
Rothwell, Timothy Gordon *pharmaceutical company executive*
Scairpon, Sharon Cecilia *information scientist*
Schoen, Alvin E., Jr. *environmental engineer*
Schreiner, Ceinwen Ann *mammalian and genetic toxicologist*
Schroeder, Alfred Christian *electronics research engineer*
Scoles, Giacinto *chemistry educator*
Seizinger, Bernd Robert *molecular geneticist, physician, researcher*
Shafir, Eldar *psychology educator*
Shenk, Thomas Eugene *molecular biology educator*
Sherr, Rubby *nuclear physicist, educator*
Shiber, Mary Claire *biochemist*
Silver, Lee Merrill *science educator*
Sinai, Yakov G. *theoretical mathematician, educator*
Skalecki, Lisa Marie *aerospace engineer*
Smagorinsky, Joseph *meteorologist*
Smith, Arthur John Stewart *physicist, educator*
Socolow, Robert Harry *mechanical and aerospace engineering educator, scientist*
Song, Limin *acoustical engineer*
Spencer, Thomas C. *mathmatician*
Spiro, Thomas George *chemistry educator*
Stabenau, M. Catherine *engineering executive*
Steinberg, Malcolm Saul *biologist, educator*
Stengel, Robert Frank *mechanical and aerospace engineering educator*
Stratton, Brentley Clarke *research physicist*
Strike, Donald Peter *pharmaceutical research director, research chemist*
Tang, Chao *physicist*
Taylor, Edward Curtis *chemistry educator*
Taylor, Joseph Hooton, Jr. *radio astronomer, physicist*
Tilghman, Shirley Marie *biology educator*
Treiman, Sam Bard *physics educator*
Triscari, Joseph *clinical research director*
Umscheid, Ludwig Joseph *government computer specialist*
Vahaviolos, Sotirios John *electrical engineer, scientist, corporate executive*
Vann, Joseph Mc Alpin *nuclear engineer*
Von Hippel, Frank Niels *public and international affairs educator*

Wei, James *chemical engineering educator, academic dean*
Weigmann, Hans-Dietrich H. *chemist*
Weimer, Paul K(essler) *electrical engineer*
Weiner, Myra Lee *toxicologist*
Wickramanayake, Godage Bandula *environmental engineer*
Wightman, Arthur Strong *physicist, educator*
Wigner, Eugene Paul *physicist, educator*
Wilczek, Frank Anthony *physics educator*
Wiles, Andrew J. *mathematican, educator*
Williams, Robert H. *environmental scientist*
Willingham, Warren Willcox *psychologist, testing service executive*
Witten, Edward *mathematical physicist*
Wolf, Wayne Hendrix *electrical engineering educator*
Wong, Ching-Ping *chemist*
Zask, Arie *chemist, researcher*

Princeton Junction
Haddad, James Henry *chemical engineering consultant*
Lee, Lihsyng Stanford *medical researcher, biotechnical consultant*
Lull, William Paul *engineering consultant*
Marchisotto, Robert *pharmacologist*

Rahway
Beckhusen, Eric Herman *chemical engineer, consultant*
Monaghan, Richard Leo *microbiologist*
Monek, Donna Marie *pharmacist*
Patchett, Arthur Allan *medicinal chemist, pharmaceutical executive*
Scolnick, Edward Mark *science administrator*
Shapiro, Bennett Michaels *biochemist, educator*
Tobert, Jonathan Andrew *clinical pharmacologist*
Tung, Hsien-Hsin *chemical engineer*
Van der Ploeg, Leonardus Harke Theresia *molecular* biologist

Ramsey
Young, Cornelius Bryant, Jr. (C. B. Young) *electronics engineer*

Randolph
Lem, Kwok Wai *research scientist*
Lichtig, Leo Kenneth *health economist*
Price, Elizabeth Cain *environmental chemist*

Raritan
Bottenus, Ralph Edward *biochemist*
Haller, William Paul *analytical chemist, robotics specialist*
Hill, Paul W. *biochemical engineer*
Johnson, Dana Lee *biophysical chemist*
Stuting, Hans Helmuth *chemist, researcher, consultant*
Weinstein, Norman Jacob *chemical engineer, consultant*

Red Bank
Chang, Chuan Chung *physicist*
Chynoweth, Alan Gerald *telecommunications research executive*
Cox, Donald Clyde *electrical engineer*
Hertz, Daniel Leroy, Jr. *entrepreneur*
Link, Frederick Charles *systems engineer*
Schwarz, Steven Allan *electrical engineer*
Sniffen, Paul Harvey *electronic engineer*
Zah, Chung-En *electrical engineer*

Ridgefield
Aybar, Romeo *architect*
Goldman, Arnold Ira *biophysicist, statistical analyst*

Ridgefield Park
Eul, Wilfried Ludwig *chemist*

Ridgewood
Hartmann, Gregor Louis *technical translator, writer*
Hassialis, Menelaos Dimitiou *mineral engineer*
Tsapatsaris, Nicholas *civil engineer, consultant*

Ringoes
Overton, Santford Vance *applications chemist*

Rochelle Park
Grossman, Martin Bernard *data systems consultant*
Long, Timothy Scott *chemist, consultant*
Schapiro, Jerome Bentley *chemical company executive*
Vernick, Arnold Sander *environmental engineer*

Rockaway
DeBlasi, Robert Vincent *school system administrator*

Rocky Hill
Ahmed, S. Basheer *research company executive, educator*

Roseland
Patterson, Scott Paul *civil engineer, consultant*

Roselle Park
Wilchins, Sidney A. *gynecologist*

Rumson
Pflum, William John *physician*
Rowe, Harrison Edward *electrical engineer*
Soneira, Raymond M. *computer company executive, scientist*

Rutherford
Brooks, Richard M. *mechanical engineer*
Fallon, David Michael *mechanical engineer*
Rubano, Richard Frank *civil engineer*
Szukics, James Charles *mechanical engineer*

Saddle River
Cappitella, Mauro John *architect*

Salem
Crymble, John Frederick *chemical engineer, consultant*

Scotch Plains
Sweeney, Lucy Graham *psychologist*

Sea Bright
Plummer, Dirk Arnold *electrical engineer*

Sea Girt
Moore, Penelope Ann *forensic scientist*

Secaucus
Brown, Ira Bernard *data processing executive*
Johansen, Robert John *electrical engineer*
Newton, V. Miller *medical psychotherapist, writer*

Sewell
Detofsky, Louis Bennett *secondary education educator*

Short Hills
Gerstein, Mark Bender *biophysicist*

Sicklerville
Waldow, Stephen Michael *radiation biologist, educational consultant*

Skillman
Ma, Tony Yong *chemist, international business consultant*
Shah, Hash N. *plastics technologist, researcher*

Somerset
Avolio, John *chemist, inventor*
Heller, Donald Franklin *chemical and laser physicist*
Long, Gilbert Morris *chemical engineer*
Oh, Nam Hwan *food scientist*
Rosenstraus, Maurice Jay *biologist*
Singh, Krishna Deo *toxicologist*
Tompa, Gary Steven *systems technology director, material scientist*

Somerville
McGregor, Walter *medical products company designer, inventor, consultant, educator*
O'Neill, Patrick J. *pharmacologist*
Paulson, John Daniel *toxicologist*
Pennington, Rodney Lee *engineer*
Shive, Richard Byron *architect*

South Orange
Joel, Amos Edward, Jr. *telecommunications consultant*

South Plainfield
Borah, Kripanath *pharmacist*
Friedman, Mark *physician, consultant*
Neumann, Robert William *engineer*
Schlossman, Mitchell Lloyd *cosmetics and chemical specialties executive*

South River
Geczik, Ronald Joseph *pharmaceutical researcher*

Sparta
Mott, Peter Andrew *biologist*
Truran, William R. *electrical engineer*

Spring Lake
Burke, J(ohn) Michael *environmental company executive*

Springfield
Adams, James Mills *chemicals executive*
Goldstein, Irving Robert *mechanical and industrial engineer, educator, consultant*
Kim, Myunghee *psychiatrist, child psychiatrist, psychoanalyst*
Lofredo, Antony *chemical engineer*
Perilstein, Fred Michael *consulting electrical engineer*

Stewartsville
Eckert, John Andrew *chemist, technical consultant*

Summit
Amemiya, Kenjie *toxicologist*
Bernhard, Michael Ian *pharmaceutical company executive*
Burbank, Robinson Derry *crystallographer*
Charbonneau, Larry Francis *research chemist*
Fukui, Hatsuaki *electrical engineer, art historian*
Kitsopoulos, Sotirios C. *electrical engineer, consultant*
Makhija, Subhash *research chemical and polymer engineer*
Mathis, James Forrest *retired petroleum company executive*
Phillips, James Charles *physicist, educator*
Sawyer, Linda Claire *materials scientist*
Van Cleef, Jabez Lindsay *marketing professional*
van Roosbroeck, Willy Werner *physicist*
Yau, Edward Tintai *toxicologist, pharmacologist*
Zachary, Louis George *chemical company consultant*
Zelenakas, Ken Walter *clinical scientist*
Ziegler, John Benjamin *chemist, lepidopterist*

Surf City
Aurner, Robert Ray, II *oil company, auto diagnostic and restaurant franchise and company development executive*

Teaneck
Churg, Jacob *pathologist*
Friedhoff, Lawrence Tim *internist, bio-medical researcher*
Rhodes, Rondell Horace *biology educator*
Shein, Samuel T. *clinical psychologist*
Woerner, Alfred Ira *medical device manufacturer, educator*
Zwass, Vladimir *computer scientist, educator*

Tenafly
Cosgriff, Stuart Worcester *internist, consultant*

Teterboro
Ficalora, Joseph Paul *optical engineer*
O'Brien, Joseph Edward *laboratory director*

Tinton Falls
Orlando, Carl *medical research and development executive*
Trigg, Clifton Thomas *information systems consultant*

Toms River
Curreri, John Robert *mechanical engineer, consultant*
Fanuele, Michael Anthony *electronics engineer, research engineer*
Young, William H. *federal agency administrator*

Totowa
Kennedy, John William *manufacturing company executive*

Towaco
Pasquale, Frank Anthony *engineering executive*

Trenton
Barclay, Robert, Jr. *chemist*
Colley, Roger J. *environmental biotechnology research company executive*
DeMarco, Peter Vincent *standards assurance executive*
Ellis, Clifford Aubrey *civil engineer*
Evers, Martin Louis *internist*
Kaus, Edward Guy *occupational health consultant*
Oliu, Ramon *chemical engineer*
Parsa, Bahman *nuclear chemist*
Rau, Eric *chemist*
Shumila, Michael John *electrical engineer*
Simpson, Raymond William *electronics engineer*
Singh, Allan *psychology educator*
Troy, Marleen Abbie *environmental engineer*
Tucker, Robert Keith *environmental scientist, research administrator*

Union
Davison, Glenn Alan *podiatric surgeon*
Lewandowski, Andrew Anthony *utilities executive, consultant*
Zois, Constantine Nicholas Athanasios *meteorology educator*

Union Beach
Schreck, Lisa Tanzman *sensory analyst*
Schulman, Marvin *chemical engineer*

Upper Montclair
Aronson, David *chemical and mechanical engineer*
Townsend, David John *psychologist*

Upper Saddle River
Cooper, Douglas Winslow *research physicist*
Paley, Steven Jann *engineering executive*
Pavlik, Thomas Joseph *quality control engineer, statistician*

Verona
France, Samuel Ewing Hill *mechanical engineer*

Vineland
Abendroth, Reinhard Paul *materials engineer*
Clinton, Lawrence Paul *psychiatrist*
Steward, Mollie Aileen *mathematics and computer science educator*

Voorhees
Fischette, Michael Thomas *engineering executive*

Waldwick
Yang, Samuel Chia-Lin *engineer*

Wall
Montgomery, John Harold *environmentalist*

Wallington
Melzacki, Krzysztof *physicist*

Warren
Barer, Sol Joseph *biotechnology company executive*
Ropp, Richard Claude *chemist*

Washington
Schlink, Frederick John *retired mechanical engineer*

Wayne
Anandan, Munisamy *physicist*
Arturi, Anthony Joseph *engineering executive, consultant*
Burlant, William Jack *chemical company executive*
Drossman, Jay Lewis *aerospace executive*
Kressel, Neil Jeffrey *psychologist*
Perry, Robert Ryan *manufacturing engineer*
Predpall, Daniel Francis *environmental engineering executive, consultant*
Shedden, James Reid *avionics engineer, consultant*
Sidman, Kenneth Robert *chemical engineer*
Trice, William Henry *paper company executive*
White, Doris Gnauck *science educator, biochemical and biophysics researcher*
Wolynic, Edward Thomas *specialty chemicals technology executive*

West Caldwell
DeFilippis, Carl William *engineer, meteorologist*
Nittoli, Thomas *chemist, consultant*

West Keansburg
Smith, Keith Brunton *civil engineer*

West Long Branch
Aguiar, Adam Martin *chemist, educator*
Dworkin, Larry Udell *electrical engineer, research director, consultant*

West Milford
Parent, Edward George *biochemist*

West Orange
Brodkin, Roger Harrison *dermatologist, educator*
Erdreich, John *acoustician, consultant*
Feigenbaum, Abraham Samuel *nutritional biochemist*
Ghali, Anwar Youssef *psychiatrist, educator*
Schoen, Howard Franklin *computer programmer, analyst*
Wiedl, Sheila Colleen *biologist*

West Trenton
Roman, Cecelia Florence *cardiologist*
Skrobacz, Edwin Stanley *structural engineer, educator*

Westfield
Feret, Adam Edward, Jr. *dentist*
Schrager, Gloria Ogur *pediatrician*
Specht, Gordon Dean *retired petroleum executive*
Stoudt, Thomas Henry *research microbiologist*

Westwood
Fabrikant, Craig Steven *psychologist*

Whippany
Andry, Steven Craig *electrical engineer, educator*
Pickering, Barbara Ann *pharmaceutical sales representative, nurse*
Schmidtmann, Lucie Ann *engineer*

Whitehouse Station
Nusim, Stanley Herbert *chemical engineer, executive director*

Willingboro
Ingerman, Peter Zilahy *infosystems consultant*
Schnapf, Abraham *aerospace engineer, consultant*
Shelton, Thomas McKinley *electronics engineer*

Woodbridge
Adams, James Derek *construction engineer*
Fee, Geraldine Julia *psychophysiologist*

Woodbury
Stambaugh, John Edgar *oncologist, hematologist, pharmacologist, educator*
Stanzione, Kaydon Al *aerospace industry executive, advisor*

Wyckoff
Van Dyk, Robert *health care center executive*

Yardville
Grace, Michael Judd *immunologist*

NEW MEXICO

Alamogordo
Stapp, John Paul *surgeon, former air force officer*

Albuquerque
Abrums, John Denise *internist*
Arvizu, Dan Eliab *mechanical engineer*
Asbury, Charles Theodore, Jr. *civil engineering consultant*
Ashbaugh, Scott Gregory *mechanical engineer, consultant*
Baer, Stephen Cooper *manufacturing executive*
Banning, Ronald Ray *systems engineer*
Bolie, Victor Wayne *electrical and computer engineering educator*
Cabaniss, Gerry Henderson *geophysicist, analyst*
Cahill, Paul Augustine *chemistry researcher*
Campbell, James Edward *physicist, consultant*
Carter, Charleata A. *cancer researcher, developmental biologist, cell biologist, toxicologist*
Chen, Zhen *research engineer*
Chronister, Richard Davis *physicist*
Clark, Teresa Watkins *psychotherapist, mental health counselor*
Cummings, John Chester, Jr. *research professional*
Davis, Robin L. *pharmacy educator*
Dencoff, John Edgar *research technologist*
Doerr, Stephen Eugene *research engineer*
Dorko, Ernest Alexander *chemist, researcher*
Evans, Pauline D. *physicist, educator*
Ewing, Rodney Charles *mineralogist, geology educator, materials scientist*
Findley, James Smith *biology and zoology educator, museum director*
Fleury, Paul Aimé *physicist*
Fuchs, Beth Ann *research technician*
Gartling, David Keith *mechanical engineer*
Gourley, Paul Lee *physicist*
Gruchalla, Michael Emeric *electronics engineer*
Guardia, David King *obstetrics/gynecology educator*
Haddad, Edward Raouf *civil engineer, consultant*
Hancock, Don Ray *researcher*
Hardy, Gregg Edmund *aerospace engineer*
Harrison, Charles Wagner, Jr. *applied physicist*
Hobbs, Michael Lane *chemical engineer, researcher*
Hudson, Jack Alan *electrical engineer*
Hylko, James Mark *health physicist*
Ingber, Marc Stuart *applied mechanics educator*
Johnson, Ralph Theodore, Jr. *physicist*
Johnson, Stewart Willard *civil engineer*
Kerley, Gerald Irwin *physicist*
Lawrence, Raymond Jeffery *physicist*
Leigh, Gerald Garrett *research engineer*
Maji, Arup Kanti *civil engineering educator*
Majumder, Sabir Ahmed *biophysicist*
Mauzy, Michael Philip *environmental consultant, chemical engineer*
McGowan, John Joseph *energy manager*
Morris, Don Melvin *surgical oncologist*
Narath, Albert *national laboratory director*
Omer, George Elbert, Jr. *orthopaedic surgeon, educator*
Ortiz, Joseph Vincent *chemistry educator*
Ottensmeyer, David Joseph *neurosurgeon, health care executive*
Owen, Thomas Edward *environmental engineer*
Palmer, Miles R *engineering scientist, consultant*
Papike, James Joseph *geology educator, science institute director*
Pepper, William Burton, Jr. *aeronautical engineer, consultant*
Prinja, Anil Kant *nuclear engineering educator*
Reed, Ray Paul *engineering mechanics measurement consultant*
Regener, Victor H. *physicist*
Rhodes, Doug *geologist*
Robb, Jeffery Michael *forensic toxicologist*
Romig, Alton Dale, Jr. *metallurgist, educator*
Sanford, Thomas William Louis *physicist*
Santillanes, Simon Paul *analytical chemist, biotoxicologist*
Schmitt, Harrison Hagan *former senator, geologist, astronaut, consultant*
Schuster, Lawrence Joseph *mechanical engineer*
Smith, Jean *interior design firm executive*
Spake, Robert Wright *safety engineer, consultant*
Tucker, Roy Anthony *electro-optical instrumentation engineer, consultant*
Vianco, Paul Thomas *metallurgist*
Wengerd, Sherman Alexander *geologist, educator*

Artesia
Horner, Elaine Evelyn *elementary education educator*

Belen
Toliver, Lee *mechanical engineer*

Carlsbad
Houghton, Woods Edward *agricultural science educator*
Watts, Marvin Lee *minerals company executive, chemist, educator*

Farmington
MacCallum, (Edythe) Lorene *pharmacist*
Norvelle, Norman Reese *environmental and industrial chemist*

Hobbs
Garey, Donald Lee *pipeline and oil company executive*

Kirtland AFB
Anderson, Christine Marlene *software engineer*
Baum, Carl Edward *electromagnetic theorist*
Godfrey, Brendan Berry *physicist*
Koslover, Robert Avner *physicist*
Lederer, John Martin *aeronautical engineer*
Miller, Leonard Doy *army officer*
Sneegas, Stanley Alan *air force officer, aerospace engineer*

Las Cruces
Boyle, Francis William, Jr. *computer company executive, chemistry educator*
Cho, Michael Yongkook *biochemical engineer, educator, consultant*
Colbaugh, Richard Donald *mechanical engineer, educator, researcher*
Foster, Robert Edwin *energy efficiency engineer, consultant*
Gregory, W. Larry *social psychology educator*
Harary, Frank *mathematician, computer science educator*
Hunt, Darwin Paul *psychology educator*
Kemp, John Daniel *biochemist, educator*
Kilmer, Neal Harold *physical scientist*
Matthews, Larryl Kent *mechanical engineering educator*
McCaslin, Bobby D. *soil scientist, educator*
Morales, Richard *structural engineer*
Naser, Najih A. *chemistry educator, researcher*
Reeves, Billy Dean *obstetrics/gynecology educator emeritus*
Shouman, Ahmad Raafat *mechanical engineering educator*
Stout, Larry John *civil engineer, consultant*
Vick, Austin Lafayette *civil engineer*
Weigle, Robert Edward *mechanical engineer, research director*
Wilkinson, William Kunkel *psychologist, educator*

Las Vegas
Riley, Carroll Lavern *anthropology educator*
Shaw, Mary Elizabeth *plant pathologist*

Los Alamos
Ahluwalia, Dharam Vir *physicist*
Balatsky, Alexander Vasilievitch *physicist*
Barfield, Walter David *physicist*
Bish, David Lee *mineralogist*
Bradbury, Norris Edwin *physicist*
Briesmeister, Richard Arthur *chemist*
Castain, Ralph Henri *physicist*
Colgate, Stirling Auchincloss *physicist*
Connellee-Clay, Barbara *quality assurance auditor, laboratory administrator*
Donohoe, Robert James *spectroscopist*
Durkee, Joe Worthington, Jr. *nuclear engineer*
Ecke, Robert Everett *physicist*
Fishbine, Brian Howard *physicist*
Fisk, Zachary *physical scientist*
Frauenfelder, Hans *physicist, educator*
Funsten, Herbert Oliver, III *physicist*
Fusco, Penny Plummer *mechanical engineer*
Goldstone, Philip David *physicist*
Guell, David Charles *chemical engineer*
Hakkila, Eero Arnold *nuclear safeguards technology chemist*
Hall, Michael L. *nuclear engineer, computational physicist*
Hanson, Kenneth Merrill *physicist*
Hecker, Siegfried Stephen *metallurgist*
Hill, Mary Ann *metallurgical engineer*
Hill, Roger Eugene *physicist*
Johnson, James Norman *physicist*
Jolly, Edward Lee *physicist, pulsed power engineer*
Keepin, George Robert, Jr. *physicist*
Kellner, Richard George *mathematician, computer scientist*
Kubas, Gregory Joseph *research chemist*
Kung, Pang-Jen *materials scientist, electrical engineer*
Linford, Rulon Kesler *physicist, program director*
Maraman, William Joseph *nuclear engineering company executive*
Mark, Kathleen Abbott *writer*
McDonald, Thomas Edwin, Jr. *electrical engineer*
McNaughton, Michael Walford *physicist, educator*
Miller, Warner Allen *physicist*
Mischke, Richard Evans *physicist*
Moore, Gregory James *physicist*
Mordechai, Shaul *physicist*
Mullen, Ken Ian *chemist*
Nunz, Gregory Joseph *aerospace engineer, mathematics educator, technical manager*
Oakley, Marta Tlapova *psychologist*
Olinger, Chad Tracy *physicist*
Onstott, Edward Irvin *research chemist*
Peratt, Anthony Lee *electrical engineer, physicist*
Preston, Dean Laverne *physicist*
Pynn, Roger *physicist*
Robinson, Robert Alan *physicist*
Rosen, Louis *physicist*
Rosocha, Louis Andrew *physicist*
Sappey, Andrew David *chemical physicist*
Tait, Carleton Drew *geochemist*
Takeda, Harunori *physicist*
Thompson, Lois Jean Heidke Ore *industrial psychologist*
Trocki, Linda Katherine *geoscientist, natural resource economist*
Turner, Leaf *physics researcher*
Wade, Rodger Grant *financial systems analyst*
Wallace, Terry Charles, Sr. *technical administrator, researcher, consultant*
Zhou, Zong-Yuan *physics educator*

Mesilla
Cryder, Cathy M. *plant geneticist*

Mesilla Park
Tombaugh, Clyde William *astronomer, educator*

Peralta
Diebold, Charles Harbou *seed company executive*

Placitas
Pirkl, James Joseph *industrial designer*
Silk, Marshall Bruce *emergency physician*

Portales
Lyon, Betty Clayton *mathematics educator*

Questa
Lamb, Margaret Weldon *lawyer*

Rodeo
Scholes, Robert Thornton *allergist, research administrator*

Santa Fe
Albach, Carl Rudolph *consulting electrical engineer*
Buck, Christian Brevoort Zabriskie *independent oil operator*
Campbell, Ashley Sawyer *retired mechanical engineering educator*
Conron, John Phelan *architect*
Cowan, George Arthur *chemist, bank executive, director*
Davidson, James Madison, III *engineer, technical manager*
Fisher, Robert Alan *laser physicist*
King, Lionel Detlev Percival *retired nuclear physicist*
Knapp, Edward Alan *scientist, government administrator*
Marcy, Willard *chemist, chemical engineer, retired*
Mendez, C. Beatriz *obstetrician/gynecologist*
Moellenbeck, Albert John, Jr. *engineering executive*
Pickrell, Thomas Richard *retired oil company executive*
Schwartz, George R. *physician*
Sheehan, William Francis *chemist educator*
Stevenson, Robert Edwin *microbiologist, culture collection executive*

Socorro
Arterburn, David Roe *mathematics educator*
Gullapalli, Pratap *chemist, researcher, educator*
Heller, John Phillip *petroleum scientist, educator*
Kieft, Thomas Lamar *biology educator*
Kottlowski, Frank Edward *geologist*
Lancaster, John Howard *civil engineer*
Popp, Carl J. *university administrator, chemistry educator*

Taos
Reynolds, Michael Everett *architect*
Viceps, Karlis David *solar residential designer*

Tijeras
Saiz, Bernadette Louise *morphology technician*
Sholtis, Joseph Arnold, Jr. *nuclear engineer, engineering executive, retired military officer*
Vizcaino, Henry P. *mining engineer, consultant*

Univ Of New Mexico
Velk, Robert James *psychologist*

White Sands Missile Range
Clelland, Michael Darr *electrical engineer*
Norman, James Harold *mathematician, researcher*
Rosen, David Lawrence *physicist, researcher*

NEW YORK

Afton
Church, Richard Dwight *electrical engineer*

Albany
Apostle, Christos Nicholas *social psychologist*
Azam, Farooq *chemist, researcher*
Beik, Mostafa Ali-Akbar *laser, electro-optical and fiber optics research engineer*
Bosart, Lance F. *meteorology educator*
Csiza, Charles Karoly *veterinarian, microbiologist*
Davies, Kelvin James Anthony *research scientist, medical educator, consultant*
Demerjian, Kenneth L. *atmospheric science educator, research center director*
DeNuzzo, Rinaldo Vincent *pharmacy educator*
Dougherty, James *orthopedic surgeon, educator*
Falk, Dean *anthropology educator*
Frisch, Harry Lloyd *chemist, educator*
Geoffroy, Donald Noel *civil engineer, consultant*
Grasso, Patricia Gaetana *biochemist, educator*
Hinge, Adam William *energy efficiency engineer*
Hoffmeister, Jana Marie *cardiologist*
Israel, Allen Charles *psychology educator*
Kim, Jai Soo *physics educator*
Levine, Louis David *museum director, archaeologist*
Mascarenhas, Joseph Peter *biologist, educator, researcher, consultant*
McKay, Donald Arthur *mechanical engineer*
Paravati, Michael Peter *medico-legal investigator, forensic sciences program administrator, correction specialist*
Posner, Norman Ames *medical educator*
Reichert, Leo Edmund, Jr. *biochemist, endocrinologist*
Robinson, David Ashley *family physician*
Sanchez de la Peña, Salvador Alfonso *biomedical chronobiologist*
Sandu, Bogdan Mihai *aeronautical engineer*
Schmidt, John Thomas *neurobiologist*
Stellrecht Burns, Kathleen Anne *virologist*
Stewart, Margaret McBride *biology educator, researcher*
Tedeschi, Henry *bioscience educator*
Toombs, Russ William *laboratory director*
Winnie, Glenna Barbara *pediatric pulmonologist*
Wulff, Daniel Lewis *molecular biologist, educator*

Albertson
Campbell, Henry J., Jr. *mechanical engineering consultant*

Alfred
Bayya, Shyam Sundar *ceramic engineer*
Frechette, Van Derck *ceramic engineer*
Fukuda, Steven Ken *materials science educator*
Lovelace, Eugene Arthur *psychology educator*
McCauley, James Weymann *ceramics engineer, educator*
Shelby, James Elbert *materials scientist, educator*
Snyder, Robert Lyman *ceramic scientist, educator*
Spriggs, Richard Moore *ceramic engineer, research center administrator*

Amherst
Halvorsen, Stanley Warren *neuropharmacologist, researcher*
Lee, George C. *civil engineer, university administrator*
Vaughan, John Thomas *biology educator*
Wakoff, Robert *electrical engineer, consultant*

Amityville
Bogorad, Barbara Ellen *psychologist*
Liang, Vera Beh-Yuin Tsai *psychiatrist, educator*
Sodaro, Edward Richard *psychiatrist*
Upadhyay, Yogendra Nath *physician, educator*

Annandale
Ferguson, John Barclay *biology educator*

Ardsley
Gayle, Joseph Central, Jr. *computer information professional*
Kochak, Gregory Michael *biophysical pharmacy/pharmacokinetics researcher*
Redalieu, Elliot *pharmacokinetics executive*

Armonk
Kuehler, Jack Dwyer *computer company executive*
Mc Groddy, James Cleary *computer company executive*

Astoria
Atkinson, Holly Gail *physician, journalist, author, lecturer*

Auburn
Daggett, Wesley John *electrical engineer*

Babylon
Lopez, Joseph Jack *oil company executive, consultant*

Baldwin
Lister, Bruce Alcott *food scientist, consultant*

Bath
Brabham, Dale Edwin *product engineer*
Huang, Edwin I-Chuen *physician, environmental researcher*
Sandt, John Joseph *psychiatrist*

Bay Shore
Pinsker, Walter *allergist, immunologist*

Bayport
Courant, Ernest David *physicist*

Bayside
Pshtissky, Yacov *electrical engineer*

Beacon
Robison, Peter Donald *biochemist*

Bearsville
Mullette, Julienne Patricia *television personality and producer, astrologer, author, health center administrator*

Beechhurst Flushing
Biegen, Elaine Ruth *psychologist*

Bellmore
Harris, Ira Stephen *medical and health sciences educator*

Bellport
Barton, Mark Quayle *physicist*

Bethpage
Baglio, Vincent Paul *aeronautical engineer*
Brown, James Kenneth *computer engineer*
Caporali, Renso L. *aerospace executive*
Cheung, Lim Hung *physicist*
Litke, John David *computer scientist*
Marrone, Daniel Scott *business, production and quality management educator*
Nordin, Paul *physicist, system engineer*
Rockensies, John William *mechanical engineer*
Scheuing, Richard Albert *aerospace corporate executive*
Tindell, Runyon Howard *aerospace engineer*

Binghamton
Bethje, Robert *general surgeon, retired*
Buckley, John Leo *retired environmental biologist*
Burright, Richard George *psychology educator*
Chatterjee, Monish Ranjan *electrical engineering educator*
Doetschman, David Charles *chemistry educator*
Eisch, John Joseph *chemist, educator*
Greiner, Thomas Moseley *physical anthropologist, archaeologist, consultant*
Hogan, Joseph Thomas *podiatrist*
Hopkins, Douglas Charles *electrical engineer*
Huie, Carmen Wah-Kit *chemistry educator*
Levis, Donald James *psychologist, educator*
Ligi, Barbara Jean *architectural and interior designer*
Prime, Roger Carl *marketing professional*
Sackman, George Lawrence *educator*
Taylor, Kenneth Douglas *finance and computer consultant, educator*
Whittingham, M(ichael) Stanley *chemist*
Yeh, Noel K. *physicist, educator*

Blauvelt
Prysch, Peter *electrical engineer*

Bohemia
Kern, Harry *developmental engineer*
Krichever, Mark *optical engineer*

Brentwood
Spinillo, Peter Arsenio *energy engineer*

Brewster
Vigdor, Martin George *psychologist*

Briarcliff Manor
Bingham, J. Peter *electronics executive*
Frair, Wayne Franklin *biologist, educator*
Mehrotra, Vivek *materials scientist*
Price, Ronald Franklin *air transportation executive*

Brockport
Wallnau, Larry Brownstein *psychologist educator*

Bronx
Abbey, Leland Russell *internist*
Alexander, Christina Lillian *pharmacist*
Bassett, C(harles) Andrew L(oockerman) *orthopaedic surgeon, educator*
Berger, Frederick Jerome *electrical engineer, educator*
Bloom, Alan Arthur *gastroenterologist, educator*
Burk, Robert David *physician, medical educator*
Charry, Jonathan M. *psychologist*
Chibbaro, Anthony Joseph *environmental and occupational health and safety professional*
Davidson, Michael *psychiatrist, neuroscientist*
Dutcher, Janice Jean Phillips *medical oncologist*
Farley, Rosemary Carroll *mathematics and computer science educator*
Fernandez-Pol, Blanca Dora *psychiatrist, researcher*
Fleischer, Norman *director of endocrinology, medical educator*
Foldes, Francis Ferenc *anesthesiologist*
Frenz, Dorothy Ann *cell and developmental biologist*
Galterio, Louis *healthcare information executive*
Goldberg, Myron Allen *physician, psychiatrist*
Gottfried, David Scott *chemist researcher*
Hart, John Amasa *wildlife biologist, conservationist*
Herbert, Victor Daniel *medical educator*
Hom, Wayne Chiu *civil engineer*
Horwitz, Susan Band *molecular pharmacologist*
Hovnanian, H. Philip *biomedical engineer*
Kanofsky, Jacob Daniel *psychiatrist, educator*
Linden, Barnard Jay *electrical engineer*
Lopez, Rafael *pediatrician*
Lubkin, Virginia Leila *ophthalmologist*
Lyles, Anna Marie *zoo curator, ornithologist*
Michelsen, W(olfgang) Jost *neurosurgeon, educator*
Murthy, Vadiraja Venkatesa *biochemist, researcher, educator*
Neuspiel, Daniel Robert *pediatrician, epidemiologist*
Peck, Fred Neil *economist, educator*
Perez, Luz Lillian *psychologist*
Peterson, Eric Scott *physical chemist*
Philipp, Manfred Hans Wilhelm *biochemist, educator*
Purpura, Dominick P. *neuroscientist, university dean*
Rogler, Charles Edward *medical educator*
Rothstein, Howard *biology educator*
Ruben, Robert Joel *physician, educator*
Rubinstein, Arye *pediatrician, microbiology and immunology educator*
Schaller, George Beals *zoologist*
Scharff, Matthew Daniel *immunologist, cell biologist, educator*
Scharrer, Berta Vogel *anatomy and neuroscience educator*
Shafritz, David Andrew *physician, research scientist*
Sircar, Ratna *neurobiology educator, researcher*
Stein, Ruth Elizabeth Klein *physician*
Sun, Emily M. *economics educator*
Van De Water, Thomas Roger *neuroscientist, educator*
Visco, Ferdinand Joseph *cardiologist, educator*
Waelsch, Salome Gluecksohn *geneticist, educator*
Waltz, Joseph McKendree *neurosurgeon, educator*
Weiner, Richard Lenard *hospital administrator, educator, pediatrician*
Wharton, Danny Carroll *zoo biologist*
Wiernik, Peter Harris *oncologist, educator*
Yalow, Rosalyn Sussman *medical physicist*
Zhou, Zhen-Hong *electrical engineer*

Bronxville
Noble, James Kendrick, Jr. *media industry consultant*

Brooklyn
Altura, Bella T. *physiologist, educator*
Altura, Burton Myron *physiologist, educator*
Astwood, William Peter *psychotherapist*
Bae, Ben Hee Chan *microbiologist*
Baird, Rosemarie Annette *pharmacist*
Banks, Ephraim *chemistry educator, consultant*
Beach, Arthur Thomas *mechanical engineer*
Begleiter, Henri *psychiatry educator*
Benes, Solomon *biomedical scientist, physician*
Benton, Peter Montgomery *business development and information science consultant*
Bolonkin, Alexander Alexandrovich *mathematician*
Bradford, Susan Kay *nutritionist*
Choudhury, Deo Chand *physicist, educator*
Crum, Albert Byrd *psychiatrist, consultant*
Dobrin, Raymond Allen *psychometrician*
Edemeka, Udo Edemeka *surgeon*
El Kodsi, Baroukh *gastroenterologist, educator*
Ettrick, Marco Antonio *theoretical physicist*
Feldman, Felix *pediatric hematologist, oncologist*
Feldmar, Gabriel Gabor *psychologist, educator*
Folan, Lorcan Michael *physicist*
Friedman, Gerald Manfred *geologist, educator*
Friedman, Howard Samuel *cardiologist, educator*
Frisch, Ivan Thomas *computer and communications company executive*
Garibaldi, Louis *aquarium administrator*
Gintautas, Jonas *physician, scientist, administrator*
Gotta, Alexander Walter *anesthesiologist, educator*
Greenberg, Bernard *pediatrician*
Guy, Matthew Joel *gastroenterologist*
Hakola, John Wayne *mechanical engineer, consultant, small business owner*
Hardy, Major Preston, Jr. *analytical chemist*
Heppa, Douglas Van *computer specialist*
Hilsenrath, Joel Alan *computer scientist, consultant*
Hirsch, Warren Mitchell *chemistry educator*
Janah, Arjun *physicist, educator*
Kaplan, Mitchell Alan *sociologist, researcher*
Karamouz, Mohammad *engineering educator*
Kravath, Richard Elliot *pediatrician, educator*

Langer, Arthur Mark *mineralogist*
Lee, Brendan *geneticist*
Levinton, Michael Jay *electrical engineer, energy consultant*
Levy, Melvin *tunnel engineer, union executive*
Levy, Sidney *psychologist*
Lichstein, Edgar *cardiologist*
Lohmann, George Young, Jr. *neurosurgeon, hospital executive*
LoPinto, Robert Anthony *environmental engineer*
Lubowsky, Jack *academic director*
Luke, Sunny *medical geneticist, researcher*
Lyubavina, Olga Samuilovna *economist*
Ma, Tsu Sheng *chemist, educator, consultant*
Macroe-Wiegand, Viola Lucille (Countess Des Escherolles) *psychiatrist, psychoanalyst*
Marazzo, Joseph John *civil engineer*
Margolin, Harold *metallurgical educator*
McLean, William Ronald *electrical engineer, consultant*
Mendelson, Sol *physical science educator, consultant*
Mendez, Hermann Armando *pediatrician, educator*
Mesiha, Mounir Sobhy *industrial pharmacy educator, consultant*
Milbury, Thomas Giberson *engineering executive*
Milhorat, Thomas Herrick *neurosurgeon*
Myerson, Allan Stuart *chemical engineering educator, university dean*
Namba, Tatsuji *physician, researcher*
Nandivada, Nagendra Nath *biochemist, researcher*
Nolan, Robert Patrick *chemistry educator*
Norstrand, Iris Fletcher *psychiatrist, neurologist, educator*
Nurhussein, Mohammed Alamin *internist, geriatrician, educator*
Ong, Say Kee *environmental engineering educator*
Ortiz, Mary Theresa *biomedical engineer, educator*
Othmer, Donald Frederick *chemical engineer, educator*
Pagala, Murali Krishna *physiologist*
Pan, Huo-Hsi *mechanical engineer, educator*
Pascali, Raresh *mechanical engineering educator*
Payyapilli, John *engineer*
Pearce, Eli M. *chemistry educator, administrator*
Pennisten, John William *computer scientist, linguist, actuary*
Peters, Mercedes *psychoanalyst*
Rabinowitz, Simon S. *physician, scientist, pediatric gastroenterologist*
Raffanielo, Robert Donald *research scientist, educator*
Ravitz, Leonard J., Jr. *physician, scientist, consultant*
Ronn, Avigdor Meir *chemical physics educator, consultant, researcher*
Rumore, Martha Mary *pharmacist, educator*
Saphire, Gary Steven *podiatrist*
Schwarz, Richard Howard *obstetrician, gynecologist, educator*
Schweikert, Edgar Oskar *dentist*
Shalita, Alan Remi *dermatologist*
Sultzer, Barnet Martin *microbiology and immunology educator*
Tamir, Theodor *electrophysics researcher, educator*
Tanacredi, John T(homas) *ecotoxicologist*
Teraoka, Iwao *polymer scientist*
Verma, Ram Sagar *geneticist, educator, author, administrator*
Viswanathan, Ramaswamy *physician, educator*
Vogl, Otto *polymer science and engineering educator*
Vroman, Georgine Marie *medical anthropologist*
Wang, Yao *engineering educator*
Weinstein, Marie Pastore *psychologist*
Whang, Sung H. *metallurgical science educator*
Wolfe, Allan *physicist*
Yablonsky, Harvey Allen *chemistry educator*

Brookville
Heimer, Walter Irwin *psychologist, educator*

Buchanan
Hayes, Charles Victor *nuclear engineer*

Buffalo
Abate, Ralph Francis *structural engineer*
Adams, Donald E. *mechanical engineer*
Anbar, Michael *biophysics educator*
Anderson, Wayne Arthur *electrical engineering educator*
Atwood, Theodore *marine engineer, researcher*
Azizkhan, Richard George *pediatric surgeon*
Batty, J. Michael *geographer, educator*
Beasley, Charles Alfred *mining engineer, educator*
Belgrader, Phillip *molecular biologist*
Borst, Lyle Benjamin *physicist, educator*
Brooks, John Samuel Joseph *pathologist, researcher*
Calkins, Evan *physician, educator*
Dandona, Paresh *endocrinologist*
Derechin, Moises *biochemistry educator*
Duax, William Leo *biological researcher*
Enhorning, Goran *obstetrician/gynecologist, educator*
Frandina, Philip Frank *civil engineer, consultant*
Freudenheim, Jo L. *social and preventive medicine educator*
Garvey, James Francis *physical chemist*
Gogan, Catherine Mary *dental educator*
Guo, Dongyao *crystallographer, researcher*
Hare, Daphne Kean *medical association director, educator*
Hauptman, Herbert Aaron *mathematician, educator, researcher*
Hida, George Tiberiu *ceramic engineer*
Horoszewicz, Juliusz Stanislaw *oncologist, cancer researcher, laboratory administrator*
Jerge, Dale Robert *loss control specialist, industrial hygienist*
Kiser, Kenneth M(aynard) *academic dean, chemical engineering educator*
Kostyniak, Paul John *toxicology educator*
Kurlan, Marvin Zeft *surgeon*
Leland, Harold Robert *research and development corporation executive, electronics engineer*
Levy, Harold James *physician, psychiatrist*
Lobo, Roy Francis *structural engineer*
Loomis, Ronald Earl *biophysicist*
Maloney, Milford Charles *internal medicine educator*
Masling, Joseph Melvin *psychology educator*
Mayhew, Eric George *cancer researcher, educator*
McHale, Magda Cordell *academic administrator, trend analyst*
McWilliams, C. Paul, Jr. *engineering executive*
Mihich, Enrico *medical researcher*
Milgrom, Felix *immunologist, educator*
Moran, Joseph John *psychology educator*
Nair, Madhavan Puthiya Veethil *immunologist, consultant*
Naughton, John Patrick *cardiologist, medical school administrator*
Ohrt, Jean Marie *chemist*
Okhi, Shinpei *biophysicist*
Patel, Suresh *chemist, researcher*
Perl, Andras *immunologist, educator*
Piver, M. Steven *gynecologic oncologist*
Pollock, Donald Kerr *anthropologist*
Ruckenstein, Eli *chemical engineering educator*
Sarjeant, Walter James *electrical and computer engineering educator*
Shanahan, Thomas Cornelius *immunologist*
Sharma, Minoti *chemist, researcher*
Shedd, Donald Pomroy *surgeon*
Shui, Xiaoping *materials scientist*
Steward, A(lma) Ruth *chemistry educator, researcher*
Suess, James Francis *clinical psychologist*
Surianello, Frank Domenic *civil engineer*
Tedlock, Dennis *anthropology and literature educator*
Thompson, John James *chemist, researcher, consultant*
Tomasi, Thomas B. *cell biologist, administrator*
Trevisan, Maurizio *epidemiologist, researcher*
Tsekanovskii, Eduard Ruvimovich *mathematician, educator*
Weber, Thomas William *chemical engineering educator*
Weller, Sol William *chemical engineering educator*
White, Thomas David *biology educator*
Wiesenberg, Russel John *statistician*
Williams, James Samuel, Jr. *chemist*
Wrona, Leonard Matthew *industrial engineer*
Zaleski, Marek Bohdan *immunologist*

Calverton
Racaniello, Lori Kuck *cellular and molecular biologist*
Sun, Homer Ko *electrical engineer*

Canaan
Bell, James Milton *psychiatrist*

Canton
Romey, William Dowden *geologist, educator*

Carmel
Bardell, Paul Harold, Jr. *electrical engineer*
Strojny, Norman *analytical chemist*

Castle Point
Mehta, Rakesh Kumar *physician, consultant*

Cazenovia
Madhoun, Fadi Salah *civil engineer*

Centereach
Daley, John Patrick *physicist*

Chappaqua
Demuth, Joseph E. *physicist, research administrator*

Chenango Bridge
Fisher, Dale Dunbar *animal scientist, dairy nutritionist*

Chestertown
Wormwood, Richard Naughton *naturalist, theoretical field geologist, field research primatologist*

Chestnut Ridge
Day, Stacey Biswas *physician, educator*
Huntoon, Robert Brian *chemist, food industry consultant*

City Island
Ward, Joan Gaye *psychologist*

Clarence
Greatbatch, Wilson *biomedical engineer*

Clifton Park
Benner, Ronald Allen, Jr. *mechanical engineer*
Scher, Robert Sander *instrument design company executive*
Schmitt, Roland Walter *retired academic administrator*

Clifton Springs
Bramlet, Roland Charles *radiation physicist*

Clinton
McKee, Francis John *medical association executive, lawyer*

Cobleskill
Ingels, Jack Edward *horticulture educator*

Cold Spring Harbor
Chmelev, Sandra D'Arcangelo *laboratory administrator*
Roberts, Richard John *molecular biologist*
Watson, James Dewey *molecular biologist, educator*
Wigler, Michael H. *molecular biologist*

College Point
Judge, Joseph B. *clinical psychologist*

Commack
Frankenberger, Glenn F(rances) *lawyer*

Cooperstown
Harman, Willard Nelson *malacologist, educator*
Pearson, Thomas Arthur *epidemiologist, educator*
Peters, Theodore, Jr. *research biochemist, consultant*

Corning
Beall, George Halsey *ceramic engineer*
Gulati, Suresh Thakurdas *glass scientist, researcher*
Havewala, Noshir Behram *chemical engineer*
Houghton, James Richardson *glass manufacturing company executive*
Johnson, Janet LeAnn Moe *statistician, project manager*
Josbeno, Larry Joseph *physics educator*
Keck, Donald Bruce *physicist*
Miller, Roger Allen *physicist*
Ryszytiwskyj, William Paul *mechanical engineer*
Whitney, William Percy, II *materials scientist*
Williams, Jimmie Lewis *research chemist*

Cortland
Klotz, Richard Lawrence *biology educator*

Cragsmoor
Foster, Lanny Gordon *writer, publisher*

Deer Park
Rosenblum, Judith Barbara *psychologist*
Wernick, Justin *podiatrist, educator*

Dobbs Ferry
Hoey, Michael Dennis *organic chemist*

Dundee
Vandyne, Bruce Dewitt *quality control executive*

East Aurora
Croston, Arthur Michael *pollution control engineer*
Young, William Lewis *retired mathematics educator*

East Fishkill
Poschmann, Andrew William *information systems and management consultant*

East Greenbush
Davenport, Francis Leo *chemical engineer*

East Hampton
Garrett, Charles Geoffrey Blythe *engineering consultant*
Hellman, Harriet Louise *pediatric nurse practitioner*

East Islip
Dinstber, George Charles *construction design engineer, consultant*

East Meadow
Albert, Gerald *clinical psychologist*
Kurian, Pius *physician*
Malamud, Herbert *physicist*
Rachlin, Stephen Leonard *psychiatrist*

East Patchogue
Metz, Donald Joseph *scientist*

East Setauket
Duff, Ronald G. *research scientist*

East Syracuse
Harju, Wayne *electrical engineer*
Landsberg, Dennis Robert *engineering executive, consultant*
Santanam, Suresh *chemical engineer, environmental consultant*

Edmeston
Price, James Melford *physician*

Elmhurst
Barron, Charles Thomas *psychiatrist*
Jethwani, Mohan *civil engineer*
Kekatos, Deppie-Tinny Z. *microbiologist, researcher, lab technologist*

Elmira
Orsillo, James Edward *computer systems engineer, company executive*

Elmont
Highland, Harold Joseph *computer scientist*

Elmsford
Sklarew, Robert Jay *biomedical research educator, consultant*

Endicott
Creasy, William Russel *chemist, writer*

Fairport
DeStefano, Paul Richard *optical engineer*
Oldshue, James Y. *chemical engineering consultant*
Phillips, Anthony *optical engineer*
Stohr, Erich Charles *biomedical engineer*

Far Rockaway
Neches, Richard Brooks *cardiologist, educator*

Farmingdale
Fuchs, Sheldon James *plant engineer*
Nolan, Peter John *physics educator*
Oyibo, Gabriel A. *aeronautics and astronautics educator*
Purandare, Yeshwant K. *chemistry educator, consultant*

Fishkill
Brocks, Eric Randy *ophthalmologist, surgeon*

Floral Park
Dalto, Michael *medical microbiologist*
Schwartz, Teri J(ean) *clinical psychologist*

Flushing
Artzt, Alice Feldman *mathematics educator*
Commoner, Barry *biologist, educator*
Finks, Robert Melvin *paleontologist, educator*
Furst, George *forensic dentist*
Gezelter, Robert L. *computer systems consultant*
Kurtz, Max *civil engineer, consultant*
Lifschitz, Karl *sales executive*
Lorber, Daniel Louis *endocrinologist, educator*
Roze, Uldis *biologist, author*
Schnall, Edith Lea (Mrs. Herbert Schnall) *microbiologist, educator*
Shteinfeld, Joseph *electrical engineer*
Weintraub, Joseph *computer company executive*
Weiss, George Arthur *orthodontist*
Weiss, Joseph *physician*

Forest Hills
Hasselriis, Floyd Norbert *mechanical engineer*
Vulis, Dimitri Lvovich *computer consultancy executive*

Fort Drum
Whiteman, Wayne Edward *army officer*

Fredonia
Tomlinson, Bruce Lloyd *biology educator, researcher*

Freeport
Burstein, Stephen David *neurosurgeon*

Garden City
Eickelberg, W. Warren Barbour *academic administrator*
Goodman, Daniel *electrical engineer, mathematical physicist*
Kirsch, Robert *director analytical research and development*
Singer, Jeffrey Michael *organic analytical chemist*
Williams, Irving Laurence *physics educator*

Geneseo
Olczak, Paul Vincent *psychologist*

Geneva
Dickson, James Edwin, II *obstetrician/gynecologist*
Gardner, Audrey V. *chemist, researcher*
Roelofs, Wendell Lee *biochemistry educator, consultant*
Seem, Robert Charles *plant pathologist*
Siebert, Karl Joseph *food science educator, consultant*
Wilcox, Wayne F. *plant pathologist, educator, researcher*

Glen Cove
Casem, Conrado Sibayan *civil/structural engineer*
Grant, Anthony Victor *biology educator*
Hardy, Maurice G. *medical and industrial equipment manufacturing company executive*
Mainhardt, Douglas Robert *chemical engineer*
Weitzmann, Carl Joseph *biochemist*

Glen Oaks
Snyder, Peter Rubin *neuropsychologist*

Glens Falls
Clements, Brian Matthew *computer consultant, educator*
Elton, Richard Kenneth *polymer chemist*
Rist, Harold Ernest *consulting engineer*

Glenville
Anderson, Roy Everett *electrical engineering consultant*

Grand Island
Epstein, David Aaron *biochemist*
Faracca, Michael Patrick *engineer*
Rader, Charles George *chemical company executive*

Great Neck
Elkowitz, Lloyd Kent *dental anesthesiologist, dentist, pharmacist*
Gillin, John F. *quality/test engineer*
Ratner, Harold *pediatrician, educator*
Ratner, Lillian Gross *psychiatrist*
Shaw, Martin Andrew *clinical and research psychologist*
Wachsman, Harvey Frederick *lawyer, neurosurgeon*
Wang, Jian-Ming *research optics scientist*

Greenlawn
Stevens, John Richard *architectural historian*

Greenvale
Pall, David B. *manufacturing company executive, chemist*

Greenwich
Briggs, George Madison *civil engineer*

Hamilton
Dovidio, John Francis *psychology educator*
Edmonston, William Edward, Jr. *publisher*
Haines, Michael Robert *economist, educator*
Kessler, Dietrich *biology educator*
Tierney, Ann Jane *neuroscientist*

Hampton Bays
Hoberman, Shirley E. *speech pathologist, audiologist*

Harriman
Parikh, Hemant Bhupendra *chemical engineer*

Harrison
Schulz, Helmut Wilhelm *chemical engineer, environmental executive*

Hastings On Hudson
Weil, Edward David *chemist*

Hauppauge
Bhide, Dan Bhagwatprasad Dinkar *chemical engineer*
Cohen, Martin Gilbert *physicist*
de Lanerolle, Nimal Gerard *process engineer*
Fine, Stanley Sidney *pharmaceuticals and chemicals executive*
Kelly, Alexander Joseph *electrical engineer*
Minasy, Arthur John *aerospace and electronic detection systems executive*

Hawthorne
McConnell, John Edward *electrical engineer, company executive*
Mitchell, Joan LaVerne *research scientist*

Hempstead
Hastings, Harold Morris *mathematics educator, researcher, author*
Krauze, Tadeusz Karol *sociologist, educator*
Laano, Archie Bienvenido Maaño *cardiologist*
Maier, Henry B. *environmental engineer*

Henrietta
Yang, Eugene L. *mechanical engineer*

Hewlett
Lowenstein, Alfred Samuel *cardiologist*

Hicksville
Gross, A. Christopher *environmental manager*
Hallak, Joseph *optometrist*

Hart, Dean Evan *research optometrist*

Highland
Rosenberger, David A. *research scientist, cooperative extension specialist*

Honeoye Falls
Cuffney, Robert Howard *electro-optical engineer*

Hopewell Junction
Beyene, Wendemagegnehu Tsegaye *electrical engineer*
LaPlante, Mark Joseph *laser and electro-optics engineer*
Marcotte, Vincent Charles *metallurgist*
Mohammad, Shaikh Noor *electronics engineer, educator*
Puttlitz, Karl Joseph, Sr. *metallurgical engineer, consultant*

Howard Beach
Istrico, Richard Arthur *physician*

Hudson
DeCrosta, Edward Francis, Jr. *former paper products company executive, consultant*

Huntington
Grossfeld, Michael L. *hospital administrator*
Schwarz, Robert Charles *aerospace engineer, consultant*
Trager, Gary Alan *endocrinologist, diabetologist*

Huntington Station
Lanzano, Ralph Eugene *civil engineer*

Hurley
Soltanoff, Jack *nutritionist, chiropractor*

Hyde Park
Moore, Roger Stephenson *chemical and energy engineer*

Inwood
Fine, Sidney Gilbert *chemist*

Irvington
Elbaum, Marek *electro-optical sciences executive, researcher*
Ray, James Henry *nuclear engineer*
Rembar, James Carlson *psychologist*

Islip
Takach, Peter Edward *process engineer*

Ithaca
Alexander, Martin *educator, researcher*
Altman, David Wayne *geneticist*
Ast, Dieter Gerhard *materials science educator*
Ballantyne, Joseph Merrill *science educator, program administrator, researcher*
Barber, Edmund Amaral, Jr. *retired mechanical engineer*
Barker, Robert *biochemistry educator*
Bates, David Martin *botanist, educator*
Batterman, Boris William *physicist, educator, academic director*
Bauman, Dale Elton *nutritional biochemistry educator*
Beins, Bernard Charles *psychology educator*
Berkelman, Karl *physics educator*
Bethe, Hans Albrecht *physicist, educator*
Billera, Louis J(oseph) *mathematics educator*
Bowman, Dwight Douglas *parasitologist*
Bramble, James Henry *mathematician, educator*
Callister, John Richard *mechanical engineer*
Carlin, Herbert J. *electrical engineering educator, researcher*
Craighead, Harold G. *physics educator*
Crepet, William Louis *botanist, educator*
Darlington, Richard Benjamin *psychology educator*
Davies, Peter John *plant physiology educator, researcher*
De Boer, Pieter Cornelis Tobias *mechanical and aerospace engineering educator*
Di Salvo, Francis Joseph *chemistry educator*
Dynkin, Eugene B. *mathematics educator*
Eisner, Thomas *biologist, educator*
Evans, Gary William *human ecology educator*
Fay, Robert Clinton *chemist, educator*
Gergely, Peter *structural engineering educator*
Gilbert, Robert Owen *veterinary educator, researcher*
Gottfried, Kurt *physicist, educator*
Gubbins, Keith Edmund *chemical engineering educator*
Guckenheimer, John *mathematician*
Hajek, Ann Elizabeth *insect pathologist*
Halpern, Bruce Peter *physiologist, consultant*
Hardy, Ralph W. F. *biochemist, biotechnology executive*
Harrison, Aidan Timothy *chemist*
Hartquist, E(dwin) Eugene *electrical engineer*
Isaacson, Michael Saul *physics educator, researcher*
Jagendorf, Andre Tridon *plant physiologist*
Jenkins, James Thomas *mechanical engineering researcher*
Jones, Barclay Gibbs *regional economics researcher*
Jones, Edward David *plant pathologist*
Kingsbury, John Merriam *botanist, educator*
Kinoshita, Toichiro *physicist*
Koh, Carolyn Ann *chemical engineer*
Kroha, Johann *physicist*
Krumhansl, Carol Lynne *psychologist, educator*
Last, Robert Louis *plant geneticist*
Ledford, Richard Allison *food science educator, food microbiologist*
Leibovich, Sidney *engineering educator*
Lopez, Jorge Washington *veterinary virologist*
Lumley, John Leask *physicist, educator*
Marko, John Frederick *theoretical physicist*
Maxwell, William Laughlin *industrial engineering educator*
McLafferty, Fred Warren *chemist, educator*
Meinwald, Jerrold *chemist, educator*
Mermin, N. David *physicist, educator, essayist*
Mirabito, Michael Mark *communications educator*
Morrison, George Harold *chemist, educator*
Mortlock, Robert Paul *microbiologist, educator*
Nathanielsz, Peter William *physiologist*
Nation, John Arthur *electrical engineering educator, researcher*
Nerode, Anil *mathematician, educator*
O'Rourke, Thomas Denis *civil engineer, educator*
Pohl, Robert Otto *physics educator*
Quimby, Fred William *pathology educator, veterinarian*
Rehkugler, Gerald Edwin *agricultural engineering educator, consultant*
Richardson, Robert Coleman *physics educator, researcher*
Rodríguez, Ferdinand *chemical engineer, educator*
Sagan, Carl Edward *astronomer, educator, author*
Salpeter, Edwin Ernest *physical sciences educator*
Scheraga, Harold Abraham *physical chemistry educator*
Seeley, John George *horticulture educator*
Smith, Charles Robert *conservationist, naturalist, ornithologist, educator*
Smith, Julian Cleveland, Jr. *chemical engineering educator*
Stedinger, Jery Russell *civil and environmental engineer, researcher*
Steinkraus, Keith Hartley *microbiology educator*
Sudan, Ravindra Nath *electrical engineer, physicist, educator*
Thiel, Daniel Joseph *physicist, researcher*
Thompson, Larry Joseph *veterinary toxicologist, consultant*
Tigner, Maurice *physicist, educator*
Tomek, William Goodrich *agricultural economist*
Uhl, Charles Harrison *botanist, plant cytologist*
Van Campen, Darrell Robert *chemist*
Von Berg, Robert Lee *chemical engineer educator, nuclear engineer*
Wang, Kuo-King *manufacturing engineer, educator*
Wasserman, Robert Harold *biology educator*
Webb, Watt Wetmore *physicist, educator*
Weinstein, Leonard Harlan *institute program director*
White, Kenneth William *electrical engineering consultant*
Widom, Benjamin *chemistry educator*
Williams, David Vandergrift *organizational psychologist*
Zimmerman, Stanley William *retired electrical engineering educator*

Jackson Heights
Michaelson, Herbert Bernard *technical communications consultant*
Olmsted, Robert Amson *civil engineer*

Jamaica
Ali, Khwaja Mohammed *aeronautical engineer*
Bradlow, Herbert Leon *endocrinologist, educator*
Dubroff, Jerome M. *cardiologist*
Finkel, Robert Warren *physicist*
Ford, Sue Marie *toxicology educator, researcher*
Kocivar, Ben *aviation specialist, journalist*
Lyons, Patrick Joseph *management educator*
Muresanu, Violeta Ana *civil, structural engineer*
Srinivasan, Mandayam Paramekanthi *software services executive*
Sun, Siao Fang *chemistry educator*
Testa, Anthony Carmine *chemistry educator*

Katonah
Mooney, Robert Michael *ophthalmologist*

Kingston
Dennison, Robert Abel, III *civil engineer*
Martin, Joanne Lea *computer scientist, researcher*
Mohanty, Ajaya K. *physicist*

Lake Placid
Fearn, Jeffrey Charles *biochemist*
Sato, Gordon Hisashi *retired biologist, researcher*

Lake Success
Bladykas, Michael P. *mechanical engineering consultant*

Lancaster
Neumaier, Gerhard John *environment consulting company executive*

Latham
Sciabica, Vincent Samuel *chemist, researcher*

Little Neck
Pearse, James Newburg *electrical engineer*

Liverpool
DiFino, Santo Michael *hematologist*
Neidhardt, Jean Sliva *systems integration professional, sales executive*

Livingston Manor
Zagoren, Joy Carroll *health facility director, researcher*

Lockport
Schultz, Gerald Alfred *chemical company executive*

Locust Valley
Coleman, John Howard *physicist*
Schor, Joseph Martin *pharmaceutical executive, biochemist*

Long Beach
Pellegrino, Charles Robert *author*

Long Island City
Hancock, William Marvin *engineering executive*
Jablowsky, Albert Isaac *civil engineer*
McCotter, Michael Wayne *civil engineer*
Stevenson, John O'Farrell, Jr. *dean*

Malverne
Ryan, Suzanne Irene *nursing educator*

Mamaroneck
Ekizian, Harry *civil engineering consultant*
Mazzola, Claude Joseph *physicist, small business owner*

Manhasset
Cerami, Anthony *biochemistry educator*
Chirmule, Narendra Bhalchandra *immunologist, educator*
Gal, David *gynecologic oncologist, obstetrician, gynecologist*
Halperin, John Jacob *neurology educator, researcher*
Hashimoto, Shiori *biomedical scientist*
Hinds, Glester Samuel *program specialist, tax consultant*
McCloskey, Thomas Warren *flow cytometrist, immunologist*
Mullin, Gerard Emmanuel *physician, educator, researcher*
Nelson, Roy Leslie *cardiac surgeon, researcher, educator*
Powell, Saul Reuben *research scientist, surgical educator*
Scherr, Lawrence *physician, educator*
Spater-Zimmerman, Susan *psychiatrist, educator*

Maspeth
Merjan, Stanley *civil engineer, inventor*

Massapequa
Margulies, Andrew Michael *chiropractor*

Melville
Chan, Jack-Kang *anti-submarine warfare engineer, mathematician*
Kaneko, Hisashi *business executive, electrical engineer*
Lofaso, Anthony Julius *mechanical engineer*
Marchesano, John Edward *electro-optical engineer*
Pelle, Edward Gerard *biochemist*
Ray, Gordon Thompson *communications executive*
Ullmann, Edward Hans *systems analyst*

Middle Village
Miele, Joel Arthur, Sr. *civil engineer*

Middletown
Anderman, Irving Ingersoll *dentist*
Edelhertz, Helaine Wolfson *mathematics educator*
Herries, Edward Matthew *civil engineer*
Zheng, Maggie (Xiaoci) *materials scientist*

Millbrook
Cole, Jonathan Jay *aquatic scientist, researcher*

Miller Place
Gresser, Mark Geoffrey *podiatrist*

Mineola
Pascucci, John Joseph *environmental engineer*

Mohegan Lake
Paik, John Kee *structural engineer*

Monroe
Werzberger, Alan *pediatrician*

Monticello
Lauterstein, Joseph *cardiologist*

Montrose
Reber, Raymond Andrew *chemical engineer*

Morrisville
Rouse, Robert Moorefield *mathematician, educator*

Mount Kisco
Gutstein, Sidney *gastroenterologist*
Laster, Richard *biotechnology executive*
Schneider, Robert Jay *oncologist*

Mount Morris
Sala, Martin Andrew *biophysicist*

Mount Vernon
Brenner, Amy Rebecca *podiatrist*
Zucker, Arnold Harris *psychiatrist*

Nanuet
Savitz, Martin Harold *neurosurgeon*

Nesconset
Feldman, Gary Marc *nutritionist, consultant*

New Hartford
Maurer, Gernant Elmer *metallurgical executive, consultant*

New Hyde Park
Biddle, David *neurologist*
Epstein, Joseph Allen *neurosurgeon*
Lee, Won Jay *radiologist*
Prisco, Douglas Louis *physician*
Stein, Theodore Anthony *biochemist, educator*
Wise, Leslie *surgeon, educator*

New Paltz
Huth, Paul Curtis *ecosystem scientist, botanist*
Schnell, George Adam *geographer, educator*
Sperber, Irwin *sociologist*

New Rochelle
Parmer, Edgar Alan *radiologist, musician*
Resnick, Henry Roy *pharmacist*
Rivlin, Richard Saul *physician, educator*

New York
Abdelnoor, Alexander Michael *immunologist, educator*
Acrivos, Andreas *chemical engineering educator*
Adamson, John William *hematologist*
Ahmad, Jameel *civil engineer, researcher, educator*
Alexander, Harold *bioengineer, educator*
Allison, David Bradley *psychologist*
Alpert, Warren *oil company executive, philanthropist*
Anderson, Paul *product management executive*
Anderson, Samuel Wentworth *research scientist*
Arceo, Thelma Llave *energy engineer*
Asanuma, Hiroshi *physician, educator*
Ashen, Philip *chemist*
Audin, Lindsay Peter *energy conservation professional*
Aviv, David Gordon *electronics engineering executive*
Aviv, Jonathan Enoch *otolaryngologist, educator*
Baden, Michael M. *pathologist, educator*
Baer, Rudolf Lewis *dermatologist, educator*
Banks, Russell *chemical company executive*
Bardin, Clyde Wayne *biomedical researcher and developer of contraceptives*
Barron, Susan *clinical psychologist*
Bartlett, Elizabeth Easton *interior designer*
Bartlett, Elsa Jaffe *neuropsychologist, educator*
Bass, Mikhail *electrical engineer*
Baum, Richard Theodore *engineering executive*
Becker, Herbert P. *mechanical engineer*
Bedrij, Orest J. *industrialist, scientist*
Bekesi, Julis George *medical researcher*
Belden, David Leigh *professional association executive, engineering educator*
Bendixen, Henrik Holt *physician, educator, dean*
Berle, Peter Adolf Augustus *lawyer, association executive*
Berns, Kenneth Ira *physician*
Bertino, Joseph Rocco *physician, educator*
Betti, Raimondo *civil engineering educator*
Beube, Frank Edward *periodontist, educator*
Biedler, June L. *oncologist*
Billig, Robert Emmanuel *psychiatric social worker*
Birbari, Adil Elias *physician, educator*
Bird, Mary Lynne Miller *association executive*
Birman, Joseph Leon *physics educator*
Blewett, John Paul *physicist*
Blitzer, Andrew *otolaryngologist, educator*
Boksay, Istvan Janos Endre *geriatric psychiatrist*
Boley, Bruno Adrian *engineering educator*
Born, Allen *mining executive*
Boshkov, Stefan Hristov *mining engineer, educator*
Brandt, Kathleen Weil-Garris *art history educator*
Breslow, Jan Leslie *scientist, educator, physician*
Breslow, Ronald Charles *chemist, educator*
Brodsky, Stanley Martin *engineering technology educator, researcher*
Broecker, Wallace S. *geophysics educator*
Brookler, Kenneth Haskell *otolaryngologist, educator*
Brown, Arthur Edward *physician*
Brown, Edward James *utility executive*
Brown, Jason Walter *neurologist, educator, researcher*
Brunetto, Frank *electrical engineer*
Brust, John Calvin Morrison *neurology educator*
Bryt, Albert *psychiatrist*
Buhks, Ephraim *electronics educator, researcher, consultant*
Buttke, Thomas Frederick *mathematics educator*
Callender, Robert Howard *biophysics educator, research scientist*
Cammarata, Angelo *surgical oncologist*
Cancro, Robert *psychiatrist*
Candia, Oscar A. *ophthalmologist, physiology educator*
Cannon, Paul Jude *physician, educator*
Cantilli, Edmund Joseph *safety engineer educator, author*
Carmiciano, Mario *electrical engineer*
Caroline, Leona Ruth *retired microbiologist*
Carter, David Martin *dermatologist*
Case, Hadley *oil company executive*
Cassedy, Edward Spencer, Jr. *electrical engineering educator*
Castro-Blanco, David Raphael *clinical psychologist, researcher*
Chan, Eric Ping-Pang *industrial designer*
Chandra, Ramesh *medical physics educator*
Chargaff, Erwin *biochemistry educator emeritus, writer*
Chase, Merrill Wallace *immunologist, educator*
Cheng, Chuen Yan *biochemist*
Cherepakhov, Galina *metallurgical and chemical engineer*
Christman, Edward Arthur *physicist*
Clancy, Thomas Joseph *civil engineer*
Clark, Lynne Wilson *speech and language pathology educator*
Cochrane, James Louis *economist*
Cohen, Edward *civil engineer*
Cohen, Noel Lee *otolaryngologist, educator*
Cohen, Richard Lawrence *electrical engineer, mechanical engineer*
Coleman, Lester Laudy *otolaryngologist*
Conley, John Joseph *otolaryngologist*
Cookson, Albert Ernest *telephone and telegraph company executive*
Corbett, Gerard Francis *electronics executive*
Cornell, James S. *critical care and pulmonary physician*
Cowin, Stephen Corteen *biomedical engineering educator, consultant*
Cranefield, Paul Frederic *pharmacology educator, physician, scientist*
Cross, George Alan Martin *biochemistry educator, researcher*
Crosson, Joseph Patrick *metallurgical engineer, consultant*
Cunningham, Dorothy Jane *physiology educator*
Daken, Richard Joseph, Jr. *electrical engineer*
Danaher, Frank Erwin *transportation technologist*
Daniele, Joan O'Donnell *clinical psychologist*
Dantuono, Louise Mildred *obstetrician/gynecologist*
Darnell, James Edwin, Jr. *molecular biologist, educator*
Davis, Kenneth Leon *psychiatrist, pharmacologist, medical educator*
Dayanim, Farangis *occupational and environmental medicine consultant*
Deane, John Herbert *technologist*
DeAngelis, Lisa Marie *neurologist, educator*
de Duve, Christian René *chemist, biologist, educator*
DeFlorio, Mary Lucy *physician, psychiatrist*
Dell, Ralph Bishop *pediatrician, researcher*
Deresiewicz, Herbert *mechanical engineering educator*
Dickey, David G. *electrical engineer*
Dieterich, Douglas Thomas *gastroenterologist, researcher*
Ding, Aihao *medical educator, researcher*
Dolan-Baldwin, Colleen Anne *global technology executive*
Dole, Vincent Paul *medical research executive, educator*
Doorish, John Francis *physicist, mathematician*
Druss, Richard George *psychiatrist, educator*
Eaton, Richard Gillette *surgeon, educator*
Edmunds, Robert Thomas *retired surgeon*
Ego-Aguirre, Ernesto *surgeon*
Eichen, Marc Alan *computer support*
Eisele, Carolyn *mathematician*
Eliasson, Kerstin Elisabeth *science and education policy advisor*
Ellis, John Taylor *pathologist, educator*
Emmert, Richard Eugene *professional association executive*
Englot, Joseph Michael *structural engineer*
Epstein, Seth Paul *immunologist, researcher*
Ergas, Enrique *orthopedic surgeon*
Fanning, William Henry, Jr. *computer scientist*
Feigenbaum, Mitchell Jay *physics educator*
Ferin, Michel Jacques *reproductive endocrinologist, educator*
Ferrier, Joseph John *atmospheric physicist*
Field, Michael Stanley *information services company executive*

Field, Steven Philip medical educator
Fins, Joseph Jack internist, medical ethicist
Fitzsimmons, Sophie Sonia interior designer
Fodstad, Harald neurosurgeon
Fogel, Irving Martin consulting engineering
Foley, Kathleen M. neurologist, educator, researcher
Fox, Daniel Michael foundation administrator, author
Freedman, Alfred Mordecai pscyhiatrist, educator
Freidenbergs, Ingrid psychologist
Freudenstein, Ferdinand mechanical engineering educator
Fried, Robert psychology educator
Friedhoff, Arnold J. psychiatrist, medical scientist
Friedlander, Ralph thoracic and vascular surgeon
Friedman, Robert Jay physician
Furneaux, Henry Morrice biochemist, educator
Gaerlan, Pureza Flor Monzon pediatrician
Galanopoulos, Kelly biomedical engineer
Gannon, Jane Frances information scientist, researcher
Garvey, Michael Steven veterinarian, educator
Gelb, Richard Lee pharmaceutical corporation executive
Gelbart, Abe mathematician, educator
Gelernt, Irwin M. surgeon, educator
Gellman, Isaiah association executive
Genaro, Donald Michael industrial designer
Gershengorn, Marvin Carl physician, educator
Gertler, Menard M. physician, educator
Giancotti, Francesca Romana immunologist, cancer research scientist
Gilbert, Richard Michael physician, educator
Ginsberg-Fellner, Fredda pediatric endocrinologist, researcher
Gitlow, Stanley Edward internist, educator
Glanzer, Murray psychology educator
Godbold, James Homer, Jr. biostatistician, educator
Goff, Kenneth Alan engineering executive
Goldberg, Harold Howard psychologist, educator
Goldfarb, Richard Charles radiologist
Goldreich, Joseph Daniel consulting structural engineer
Goldsmith, Michael Allen medical oncologist, educator
Golomb, Frederick Martin surgeon, educator
Gomory, Ralph Edward mathematician, manufacturing company executive, foundation executive
Goodrich, James Tait neuroscientist, pediatric neurosurgeon
Gordon, Ronnie Roslyn pediatrics educator, consultant
Gotschlich, Emil Claus physician, educator
Goulianos, Konstantin physics educator
Graf, Jeffrey Howard cardiologist
Green, Maurice Richard neuropsychiatrist
Greenfield, Bruce Paul investment analyst, biology researcher
Greengard, Paul neuroscientist
Greenleaf, Marcia Diane psychologist, writer, educator
Greer, R. Douglas behaviorology and education educator
Griff, Irene Carol cell biology researcher
Griffis, Fletcher Hughes civil engineering educator, engineering executive
Grindea, Daniel international economist
Grossman, Jacob S. structural engineer
Grumet, Martin biomedical researcher
Grunes, Robert Lewis engineering consulting firm executive
Guillen, Michael Arthur mathematical physicist, educator, writer, television journalist
Gumbs, Godfrey Anthony physicist, educator, researcher
Guo, Chu chemistry educator
Haddad, Heskel Marshall ophthalmologist
Haddock, Robert Lynn information services entrepreneur, writer
Hajjar, David P. biochemist, educator
Hajjar, Katherine Amberson physician, pediatrician
Hamburg, David A. psychiatrist, foundation executive
Hamilton, Linda Helen clinical psychologist
Hanafusa, Hidesaburo virologist
Handel, Yitzchak S. psychologist, educator
Hansen, James E. physicist, meteorologist, federal agencey administrator
Harlap, Susan epidemiologist, educator
Hashim, George A. immunologist, biomedical researcher, educator
Haywood, H(erbert) Carl(ton) psychologist
Heal, Geoffrey Martin economics educator
Heckscher, August journalist, author, foundation executive
Hegde, Ashok Narayan neuroscientist
Heisel, Ralph Arthur architect
Hendrickson, Wayne A(rthur) biochemist, educator
Hennessy, John Francis, III engineering executive, mechanical engineer
Heyerdahl, Thor anthropologist, explorer, author
Hillery, Mark Stephen physicist
Hirano, Arlene Akiko neurobiologist, research scientist
Hirsch, Joseph Allen psychology and pharmacology educator
Hirsch, Jules physician, biochemistry educator
Hirschhorn, Kurt pediatrics educator
Hirschhorn, Rochelle medical educator
Hirschy, James Conrad radiologist
Hochberg, Irving audiologist, educator
Hof, Patrick Raymond neurobiologist
Hoffberg, Steven Mark lawyer
Hogan, Charles Carlton psychiatrist
Hommes, Frits Aukustinus biology educator
Horvath, Diana Meredith plant molecular biologist
Hoskins, William John obstetrician/gynecologist, educator
Huettner, Richard Alfred lawyer
Hyman, Bruce Malcolm ophthalmologist
Hyman, Leonard Stephen finanical executive, economist, author
Ivanov, Lyuben Dimitrov naval architecture researcher, educator
Ivanovitch, Michael Stevo economist
Jackson, Sherry Diane internist
Jackson, Terrance Sheldon editor
Jacobs, Jonathan Lewis physician
Jacobson, Willard James science educator
Jaffe, William J(ulian) industrial engineer, educator
Jagiello, Georgiana M. geneticist, educator
James, Gary Douglas biological anthropologist, educator, researcher
Jewelewicz, Raphael obstetrician/gynecologist, educator
Jonas, Ruth Haber psychologist
Jonas, Saran neurologist, educator
Joskow, Renee W. dentist, educator
Kabat, Elvin Abraham immunochemist, biochemist, educator
Kaku, Michio theoretical nuclear physicist
Kalamotousakis, George John economist
Kao, Richard Juichang biostatistician
Kappas, Attallah physician, medical scientist
Karouna, Kir George chemist, consultant
Katz, Jose cardiologist, theoretical physicist
Katz, Steven Edward psychiatrist, state health official
Kaunitz, Hans physician, pathologist
Kauth, Benjamin podiatric consultant
Keefe, Deborah Lynn cardiologist, educator
Keller, Stephen chemist educator
Kelman, Charles D. ophthalmologist, educator
Khuri, Nicola Najib physicist, educator
Kidd, Julie Johnson museum director
Kilburn, Penelope White data processing executive
Kirshenbaum, Richard Irving public health physician
Kleiman, Norman Jay molecular biologist, biochemist
Klein, Morton industrial engineer, educator
Klempner, Larry Brian network engineer, income tax preparer
Kliment, Robert Michael architect
Kline, Milton Vance psychologist, educator
Knowles, Richard James Robert medical physicist, educator, consultant
Kocherlakota, Sreedhar manufacturing engineer
Koestler, Robert John conservation research scientist, biologist
Komisar, Arnold otolaryngologist, educator
Kopelman, Arthur Harold biology educator, population ecologist
Kothera, Lynne M. clinical neuropsychologist
Krasnow, Maurice psychoanalyst, educator
Kreek, Mary Jeanne physician
Kromidas, Lambros cell biologist, physical scientist
Kucic, Joseph management consultant, industrial engineer
Kulkarni, Ravi Shripad mathematics educator, researcher
Kushner, Brian Harris pediatric oncologist
Kvint, Vladimir Lev mining engineer, economist, educator
Lamirande, Arthur Gordon editor
Lane, Joseph M. orthopaedic surgeon, oncologist
Langdon, George Dorland, Jr. museum administrator
LaQuaglia, Michael Patrick pediatric surgeon, neuroblastoma researcher
Laragh, John Henry physician, scientist, educator
Lattis, Richard Lynn zoo director
Lau, Harry Hung-Kwan acoustical and interior designer, consultant
Lax, Peter David mathematician
Lazar, Judith Tockman pharmaceutical company researcher
Lederberg, Joshua geneticist, educator
Lee, Martin Yongho mechanical engineer
Lee, Mathew Hung Mun physiatrist
Lee, Tsung-Dao physicist, educator
Leshnower, Alan Lee podiatrist
Levy, Miguel physicist
Lewis, Jonathan Joseph surgeon, molecular biologist
Lewis, Stuart Weslie surgeon
Lewis, William Scheer electrical engineer
Li, Yao science educator
Lichter, Robert Louis science foundation administrator, chemist
Lieberman, Harvey Michael hepatologist, gastroenterologist, educator
Lieberstein, Melvin administrative engineer
Lipkin, George dermatologist, researcher
Lipkin, Martin physician, scientist
Lipton, Lester ophthalmologist, entrepreneur
Liu, Si-kwang veterinary pathologist
Lu, Ruth structural engineer
Lucchesi, Arsete Joseph mathematician, university dean
Lymberis, Costas Triantafillos environmental engineer
Lynch, John Joseph city and regional planner, consultant
Macken, Daniel Loos cardiologist, educator
Mader, Bryn John vertebrate paleontologist
Mahadeva, Wijeyaraj Anandakumar information company executive
Maiese, Kenneth neurologist
Manger, William Muir internist
Manly, Carol Ann speech pathologist
Marcus, Eric Robert psychiatrist
Markiewicz, Leszek research biochemist
Marks, Andrew Robert molecular biologist
Marks, Paul Alan oncologist, cell biologist
Marlin, John Tepper economist, writer, consultant
Marshall, John association administrator
Martin, Lenore Marie bioorganic researcher, educator
Matthaei, Gay Humphrey interior designer
Matzat, Gregory Mark naval architect
Mayo, Joan Bradley microbiologist, epidemiologist
McCabe, John Cordell surgeon
McCarthy, Joseph Gerald plastic surgeon, educator
McCarty, Maclyn medical scientist
McClelland, Shearwood Junior orthopaedic surgeon
McCormick, Douglas Kramer editor
McGovern, John Hugh urologist, educator
McKean, Henry P. mathematics institute administrator
Mc Kenna, Malcolm Carnegie vertebrate paleontologist, curator, educator
Mc Murtry, James Gilmer, III neurosurgeon
Megherbi, Dalila electrical and computer engineer, researcher
Melamid, Alexander economics educator, consultant
Mendelsohn, John oncologist, hematologist, educator
Merrifield, Robert Bruce biochemist, educator
Mezic, Richard Joseph engineer, consultant
Michnovicz, Jon Joseph physician, research endocrinologist
Middleton, David physicist, applied mathematician, educator
Miller, Alan software executive, management specialist
Miller, Cate psychologist, educator
Millman, Robert Barnet psychiatry and public health educator
Mininberg, David T. pediatric urology surgeon, educator
Mischel, Harriet Nerlove psychologist, educator
Mishkin, Frederic Stanley economics educator
Miskovitz, Paul Frederick gastroenterologist
Mittleman, Marvin Harold physicist, educator
Mizrahi, Abraham Mordechay cosmetics and health care company executive, physician
Momtaheni, Mohsen oral and maxillofacial surgeon, academician, clinician
Morawetz, Cathleen Synge mathematician
Morfopoulos, V. metallurgical engineer, materials engineer
Morishima, Akira physician, director, educator, consultant
Morse, Stephen Scott virologist, immunologist
Moses, Johnnie, Jr. microbiologist
Mow, Van C. engineering educator, researcher
Muchnick, Richard Stuart ophthalmologist, educator
Mukherjee, Trishit reproductive infertility specialist
Murray, William food products executive
Nagamiya, Shoji physicist, educator
Nakanishi, Koji chemistry educator, research institute administrator
Nauert, Roger Charles health care executive
Navratil, Gerald Anton physicist, educator
Nelkin, Dorothy sociology and science policy educator, researcher
Newbold, Herbert Leon, Jr. psychiatrist, writer
Nirenberg, Louis mathematician, educator
Norell, Mark Allen paleontology educator
Norwick, Braham textile specialist, consultant, columnist
Noz, Marilyn Eileen radiology educator
Ordorica, Steven Anthony obstetrician/gynecologist, educator
Osborne, Michael Piers surgeon, health facility administrator
Osbourn, Gordon Cecil materials scientist
Osgood, Richard M., Jr. applied physics and electrical engineering educator, research administrator
Overweg, Norbert Ido Albert physician
Owen, Thomas Joseph acoustical engineer
Paaswell, Robert Emil civil engineer, educator
Pais, Abraham physicist, educator
Panoutsopoulos, Basile electrical engineer
Pappas, John clinical psychologist
Pardes, Herbert psychiatrist, educator
Parkin, Gerard Francis Ralph chemistry educator, researcher
Pavis, Jesse Andrew sociology educator
Pedley, Timothy Asbury, IV neurologist, educator, researcher
Penraat, Jaap architect
Percus, Jerome Kenneth physicist, educator
Pestano, Gary Anthony biologist
Pfrang, Edward Oscar association executive
Philipson, Lennart Carl microbiologist, science administrator
Poliacof, Michael Mircea electrical engineer
Pollack, Robert Elliot biological sciences educator, writer, scientist
Posamentier, Alfred Steven mathematics educator, university administrator
Poshni, Iqbal Ahmed microbiologist
Posner, Jerome Beebe neurologist, educator
Principe, Joseph Vincent, Jr. environmental engineer
Psaltakis, Emanuel P. environmental engineer
Pye, Lenwood David materials science educator, researcher, consultant
Queen, Daniel acoustical engineer, consultant
Qureshi, Sajjad Aslam biologist, researcher
Racaniello, Vincent Raimondi microbiologist, medical educator
Rahimian, Ahmad structural engineer
Rampino, Michael R. earth scientist
Rankin, James Edwin computer systems consultant
Rebelsky, Leonid physicist
Reeke, George Norman, Jr. neuroscientist, crystallographer, educator
Reis, Donald Jeffery neurologist, neurobiologist, educator
Reisner, Milton psychiatrist
Ristich, Miodrag psychiatrist
Roberts, 2ames Lewis medical sciences educator
Robinson, James LeRoy architect, educator, developer
Roepe, Paul David biophysical chemist
Roman, Stanford Augustus, Jr. medical educator, dean
Rosendorff, Clive cardiologist
Rosenfeld, Louis biochemist
Rosenfield, Richard Ernest emeritus medical educator
Ross, Donald Edward engineering company executive
Rosskothen, Heinz Dieter engineer
Rothenberg, Robert Edward physician, surgeon, author
Rothman, David J. history and medical educator
Rothman, James Edward cell biologist, educator
Rothschild, Nan Askin archaeologist
Rubin, Gustav orthopedic surgeon, consultant, researcher
Sachar, David Bernard gastroenterologist, medical educator
Sachdev, Ved Parkash neurosurgeon
Sacks, Henry S. medical researcher, infectious disease physician
Safier, Lenore Beryl research chemist
Saha, Dhanonjoy Chandra biomedical research scientist
Saito, Mitsuru civil engineer, educator
Salvadori, Mario mathematical engineer
Sanchez, Miguel Ramon dermatologist, educator
Sapse, Anne-Marie chemistry educator
Schaefer, Steven David head and neck surgeon, physiologist
Scheinberg, Labe Charles physician, educator
Schiavi, Raul Constante psychiatrist, educator, researcher
Schlessinger, Joseph pharmacology educator
Schmeltz, Edward James engineering executive
Schneck, Jerome M. psychiatrist, medical historian, educator
Schoenfeld, Robert Louis biomedical engineer
Schwimmer, David physician, educator
Sciacca, Kathleen psychologist
Scopp, Irwin Walter periodontist, educator
Seadler, Stephen Edward business and computer consultant, social scientist
Seitz, Frederick university president emeritus
Sendax, Victor Irven dentist, educator, dental implant researcher
Shapiro, Michael Harold health care executive, consultant, publisher
Shapiro, Murray structural engineer
Sharp, Victoria Lee medical director
Shelanski, Michael L. cell biologist, educator
Shinnar, Reuel chemical engineering educator, industrial consultant
Shrout, Patrick Elliot psychometrician, educator
Sobel, Kenneth Mark electrical engineer, educator
Somasundaran, Ponisseril mineral engineering and applied science educator, consultant, researcher
Sonnabend, Joseph Adolph microbiologist
Sorota, Steve biomedical researcher
Sorrel, William Edwin psychiatrist, educator, psychoanalyst
Speth, James Gustave United Nations executive, lawyer
Spiegel, Herbert psychiatrist, educator
Spielberger, Lawrence physician, educator
Spielman, Andrew Ian biochemist
Spingarn, Clifford Leroy internist, educator
Spruch, Larry physicist, educator
Stahl, Frank Ludwig civil engineer
Stasior, William F. engineering company executive
Steinglass, Peter Joseph psychiatrist, educator
Stellman, Steven Dale epidemiologist
Stephens, Olin James, II naval architect, yacht designer
Sternlieb, Cheryl Marcia internist
Stoopler, Mark Benjamin physician
Stork, Gilbert (Josse) chemistry educator, investigator
Stotzky, Guenther microbiologist, educator
Straus, David Jeremy hematologist, educator
Stroke, Hinko Henry physicist, educator
Stutman, Leonard Jay research scientist, cardiologist
Sugiyama, Kazunori music producer
Sun, Tung-Tien medical science educator
Sylla, Richard Eugene economics educator
Tabler, William Benjamin architect
Tallent, Marc Andrew clinical psychologist
Tamm, Igor biomedical scientist, educator
Taylor, Patricia Elsie epidemiologist
Teich, Malvin Carl electrical engineering educator
Teplitzky, Philip Herman computer consultant
Terwilliger, Joseph Douglas statistical geneticist
Tester, Leonard Wayne psychology educator
Tichenor, Wellington Shelton physician
Tilley, Shermaine Ann molecular immunologist, educator
Tilson, Dorothy Ruth word processing executive
Tolete-Velcek, Francisca Agatep pediatric surgeon, surgery educator
Torigian, Puzant Crossley clinical research pharmacist, pharmaceutical company executive
Tsuji, Moriya immunologist
Turro, Nicholas John chemistry educator
Tyrl, Paul mathematics educator, researcher, consultant
Upton, Arthur Canfield retired experimental pathologist
Urso, Charles Joseph physician
Valentini, James Joseph chemistry educator
Van Hemmen, Hendrik Fokko vehicle engineer
Verdesca, Arthur Salvatore internist, corporate medical director
Vilcek, Jan Tomas medical educator
Vivera, Arsenio Bondoc allergist
Voorsanger, Bartholomew architect
Watanabe, Kyoichi A(loysius) chemist, researcher, pharmacology educator
Watkins, Charles Booker, Jr. mechanical engineering educator
Wazneh, Leila Hussein organic chemist
Weinberg, Samuel pediatric dermatologist
Weinsaft, Paul Phineas retired physician, administrator
Weinstein, Herbert chemical engineer, educator
Weinstein, I. Bernard physician
Weiss, Samuel Abraham psychologist, psychoanalyst
Weksler, Babette Barbash hematologist
Werner, Andrew Joseph physician, endocrinologist, musicologist
Wharton, Ralph Nathaniel psychiatrist, educator
Wheeler, Wesley Dreer marine engineer, naval architect, consultant
Whelan, Elizabeth Ann Murphy epidemiologist
Wiemer-Sumner, Anne-Marie psychotherapist, educational administrator
Wiesel, Torsten Nils neurobiologist, educator
Windhager, Erich Ernst physiologist, educator
Wit, David Edmund software company executive
Witenberg, Earl George psychiatrist, educator
Witkin, Mildred Hope Fisher psychotherapist, educator
Worgul, Basil Vladimir radiation scientist
Wynder, Ernst Ludwig science foundation director, epidemiologist
Wyn-Jones, Alun (William Wyn-Jones) software developer, mathematician
Yarris, Steve child psychologist, educator
Yee, Herman Terence pathologist
Young, Helen Jamieson biologist, educator
Young, Michael Warren geneticist, educator
Zacharius, Martin Philip mechanical engineer
Zakim, David biochemist
Zaslavsky, George Moiseevich theoretical physicist, researcher
Zinberg, Stanley physician, educator
Zmeu, Bogdan structural engineer, consultant
Zucker, Howard Alan pediatric cardiologist, intensivist, anesthesiologist
Zukowski, Charles Albert electrical engineering educator
Zuzolo, Ralph Carmine biologist, educator, researcher, consultant

Newton Falls
Hunter, William Schmidt engineering executive, environmental engineer

Niagara Falls
Butry, Paul John engineer
Dojka, Edwin Sigmund civil engineer

Niagara University
Osberg, Timothy M. psychologist, educator, researcher, clinician

North Babylon
Tipirneni, Tirumala Rao metallurgical engineer

North Tarrytown
Schippa, Joseph Thomas, Jr. school psychologist, educational consultant, hypnotherapist

Northport
Gebhard, David Fairchild aeronautical engineer, consultant
Gray, James Lee systems analyst

Norwich
Brooks, Robert Raymond pharmacologist
Garzione, John Edward physical therapist
King, Alison Beth pharmaceutical company executive
Sietsema, William Kendall biochemist

Nyack
Esser, Aristide Henri *psychiatrist*
Lee, Lillian Vanessa *microbiologist*
Ryan, William B. F. *geologist*

Oakdale
Panzarella, John Edward *water quality chemist*

Old Westbury
Andrews, Mark Anthony William *physiologist, educator*
Colef, Michael *engineering educator*
Koenig, Gottlieb *mechanical engineering educator*
Ozelli, Tunch *economics educator, consultant*

Olean
Catalano, Robert Anthony *ophthalmologist, physician, hospital administrator, writer*
Datta, Prasanta *mechanical engineer*

Oneida
Muschenheim, Frederick *pathologist*

Oneonta
Helser, Terry Lee *biochemistry educator*
Hickey, Francis Roger *physicist, educator*
Merilan, Michael Preston *astrophysicist, educator*
Trotti, Lisa Onorato *psychology educator*

Orangeburg
Lajtha, Abel *biochemist*
Reilly, Margaret Anne *pharmacologist, educator*
Squires, Richard Felt *research scientist*

Oriskany Falls
Michels, Richard Steven *microbiologist*

Ossining
Baylor, Sandra Johnson *electrical engineer*
Campbell, Harry L. *systems analyst*
Miron, Amihai *electronic systems executive, electrical engineer*
Phelps, Glenn Howard *mechanical carbon process engineer*
Troubetzkoy, Eugene Serge *physicist*

Oswego
Gooding, Charles Thomas *psychology educator, college dean*
Seago, James Lynn *biologist, educator*
Sudhakaran, Gubbi Ramarao *physicist, educator*
Tulve, Nicolle Suzanne *researcher*

Owego
Genova, Vincent Joseph *physicist*
Kemp, Eugene Thomas *veterinarian*

Oyster Bay
Hatch, Mary Wendell Vander Poel *laboratory executive, interior decorator*

Ozone Park
Catalfo, Betty Marie *health service executive, nutritionist*

Painted Post
Benjamin, Keith Edward *mechanical engineer*
Hammond, George Simms *chemist*
Stookey, Stanley Donald *chemist*

Palisades
Brusa, Douglas Peter *purchasing executive*
Cane, Mark Alan *oceanography and climate researcher*
Eaton, Gordon Pryor *geologist, research director*
Kent, Dennis Vladimir *geophysicist, researcher*

Pearl River
Barrett, James Edward *biology educator, research*
Citardi, Mattio H. *chemist, researcher, system manager*
Dimmig, Bruce David *architect*
Galante, Joseph Anthony, Jr. *computer programmer*
Kolor, Michael Garrett *research chemist*
Trust, Ronald Irving *organic chemist*
Zisa, David Anthony *pharmaceutical researcher*

Peekskill
Harding, Charlton Matthew *electrical engineer*

Penfield
Baxter, Meriwether Lewis, Jr. *gear engineer, consultant*
Hutteman, Robert William *civil engineer*

Pittsford
Marshall, Joseph Frank *electronic engineer*
Saini, Vasant Durgadas *computer software company executive*

Plainview
Gresser, Herbert David *physicist*

Plattsburgh
Graziadei, William Daniel, III *biology educator, researcher*
Heintz, Roger Lewis *biochemist, educator, researcher*
Helinger, Michael Green *mathematics educator*
Smith, Noel Wilson *psychology educator*

Pleasantville
Nabirahni, David M.A. *chemist, educator*
Pike, John Nazarian *optical engineering consultant*

Pomona
Ciaccio, Leonard Louis *chemist researcher, science administrator*
Glassman, Lawrence S. *plastic surgeon*
Gordon, Edmund Wyatt *psychologist, educator*
Jaffrey, Ira *oncologist, educator*
Masters, Robert Edward Lee *neural re-education researcher, psychotherapist, human potential educator*

Port Chester
Ailloni-Charas, Miriam Clara *interior designer, consultant*

Port Jefferson
Elman, Howard Lawrence *aeronautical engineer*
Hirschl, Simon *pathologist*
Paskin, Arthur *physicist, consultant*

Port Washington
Gaddis, M. Francis *mechanical and marine engineer, environmental scientist*
Hayslett, Paul Joseph *computer programmer, consultant*
Zahnd, Hugo *emeritus biochemistry educator*

Potsdam
Chin, Der-Tau *chemical engineer, educator*
Gallagher, Richard Hugo *university official, engineer*
Glasser, Joshua David *computer scientist*
Glauser, Mark Nelson *mechanical and aeronautical engineering educator*
Herman, William Elsworth *psychology educator*
Kane, James Harry *mechanical engineer, educator, researcher*
Matijevic, Egon *chemistry educator, researcher, consultant*
Mochel, Myron George *mechanical engineer, educator*
Ortmeyer, Thomas Howard *electrical engineering educator*
Phillips, William Robert *fluid dynamics engineering educator*
Wells, David John *university official, mechanical engineer*

Poughkeepsie
De Cusatis, Casimer Maurice *fiber optics engineer*
Flanagan, Frederick James *water systems engineer*
Golden, Reynold Stephen *family practice physician, educator*
Henry, Charles Jay *library director*
Maling, George Croswell, Jr. *physicist*
McEnroe, Caroline Ann *legal assistant*
Milano, Charles Thomas *obstetrician, legal medicine consultant*
Puretz, Donald Harris *educator*
Rossi, Miriam *chemistry educator, researcher*
Vakirtzis, Adamantios Montos *systems analyst*
Vita, James Paul *software engineer*

Pound Ridge
Landis, Pamela Ann Youngman *tribologist*

Purchase
Daniel, Charles Timothy *transportation engineer, consultant*
Lucas, Billy Joe *philosophy educator*
Santucci, Anthony Charles *neuroscientist, educator*

Queens
Farkas, Edward Barrister *airport administrator, electrical engineer*

Queensbury
Perry, Leland Charles *quality engineer, quality assurance analyst*

Ray Brook
Roy, Karen Mary *limnologist, state government regulator*

Rensselaer
Krasney, Ethel Levin *research chemist*
LaBrie, Teresa Kathleen *research scientist*

Rhinebeck
Hellerman, Leo *retired computer scientist and mathematician*

Rhinecliff
Conklin, John Roger *electronics company executive*

Ridge
Begley, Anthony Martin *research physicist*

Riverdale
Friedman, Ronald Marvin *cellular biologist*
Jha, Nand Kishore *engineering educator, researcher*

Riverhead
Senesac, Andrew Frederick *weed scientist*

Rochester
Abramowicz, Jacques Sylvain *obstetrician, perinatologist*
Arden, Bruce Wesley *computer science and electrical engineering educator*
Bambury, Ronald Edward *polymer chemist*
Barton, Russell William *psychiatrist, author*
Begenisich, Ted Bert *physiology educator*
Blumberg, Benjamin Mautner *virologist*
Borch, Richard Frederic *pharmacology and chemistry educator*
Bowen, William Henry *dental researcher, dental educator*
Brzustowicz, Richard John *neurosurgeon, educator*
Burns, Peter David *imaging scientist*
Cain, B(urton) Edward *chemistry educator*
Castle, William Eugene *academic administrator*
Chan, Donald Pin-Kwan *orthopaedic surgeon, educator*
Cline, Douglas *physicist, educator*
Coburn, Theodore James *retired physicist*
Colby, Ralph Hayes *chemical engineer*
Crino, Marjanne Helen *anesthesiologist*
Dieck, William Wallace Sandford *chemical engineer*
Dumont, Mark Eliot *biochemist, educator*
Ewing, James Francis *biochemist, researcher*
Factor, Ronda Ellen *research chemist*
Finnigan, James Francis *civil engineer*
Fisher, George Myles Cordell *electronics equipment company executive, mathematician, engineer*
Forbes, Gilbert Burnett *physician, educator*
Ford, Loretta C. *retired university dean, nurse, educator*
Ford, Mary Elizabeth (Libby Ford) *environmental health engineer*
Gates, Marshall DeMotte, Jr. *chemistry educator*
Goldsmith, Lowell Alan *medical educator*
Goldstein, David Arthur *biophysicist, educator*
Greener, Jehuda *polymer scientist*
Hailstone, Richard Kenneth *chemist, educator*
Harris, Eric William *neuroscientist, pharmaceutical executive*
Harris, Howard Alan *laboratory director*
Hayes, Charles Franklin, III *museum research director*
Hejazi, Shahram *biomedical engineer*
Hioe, Foek Tjiong *physics educator, researcher*
Holzbach, James Francis *civil engineer*
Hopkins, Thomas Duvall *economics educator*
Horbatuck, Suzanne Marie *optical engineer*
Huizenga, John Robert *nuclear chemist, educator*
Iglewski, Barbara Hotham *microbiologist, educator*
Jansen, Kathryn Lynn *chemist*
Jenekhe, Samson Ally *chemical engineering educator, polymer scientist*
Johnson, Jean Elaine *nursing educator*
Johnston, Frank C. *psychologist*
Jones, Charles Lee *industrial designer*
Kadin, Alan Mitchell *physicist*
Kingslake, Rudolf *retired optical designer*
Krogh-Jespersen, Mary-Beth *chemist, educator*
La Celle, Paul Louis *biophysics educator*
Langworthy, Harold Frederick *manufacturing company executive*
Lanzafame, Raymond Joseph *surgeon, researcher*
Levy, Harold David *psycholinguist*
Li, James Chen Min *materials science educator*
Lubinsky, Anthony Richard *physicist*
Luckey, George William *research chemist*
Margevich, Douglas Edward *spectroscopist*
Martic, Peter Ante *research chemist*
McCormack, Grace *retired microbiology educator*
McCrory, Robert Lee *physicist, mechanical engineering educator*
McHugh, William Dennis *dental educator, researcher*
Melissinos, Adrian Constantin *physicist, educator*
Miles, William Robert *structural engineer*
Moore, Duncan Thomas *optics educator*
Ogut, Ali *mechanical engineering educator*
Papadakos, Peter John *critical care physician*
Paradowski, Robert John *history of science educator*
Parker, Kevin James *electrical engineer educator*
Parker, Thomas Sherman *civil engineer*
Paterson, Eileen *radiation oncologist, educator*
Paz-Pujalt, Gustavo Roberto *physical chemist*
Pettee, Daniel Starr *neurologist*
Pfendt, Henry George *retired information systems executive, management consultant*
Plosser, Charles Irving *economics educator*
Rao, Joseph Michael *chemist*
Rubin, Bruce Joel *chemical engineer, electrical engineer*
Saunders, William Hundley, Jr. *chemist, educator*
Setchell, John Stanford, Jr. *color systems engineer*
Simon, Albert *physicist, engineer, educator*
Snell, Karen Black *audiologist, educator*
Soures, John M. *physicist, researcher*
Su, Qichang *physicist*
Thomas, Leo J. *manufacturing company executive*
Thomas, Telfer Lawson *chemist, researcher*
Vernarelli, Michael Joseph *economics educator, consultant*
Wey, Jong Shinn *research laboratory executive*
Whitten, David George *chemistry educator, researcher*
Wiley, Jason LaRue, Jr. *neurosurgeon*
Willey, James Campbell *internist, educator*
Williams, Thomas Franklin *physician, educator*
Zhang, Guo He *physiologist*

Rockville Centre
Epel, Lidia Marmurek *dentist*
Silecchia, Jerome A. *mechanical engineer*
Weber, Arthur Phineas *chemical engineer*

Ronkonkoma
Castrogiovanni, Anthony G. *aerospace engineer, researcher*
Phillips, Kevin John *consulting engineer*
Ranpuria, Kishor Prajaram *wire and cable engineer*

Roosevelt Island
Montemayor, Jesus Samson *physician*

Roslyn
Silverstein, Seth *physician*

Rouses Point
Al-Hakim, Ali Hussein *chemist*

Rush
Eastman, Carolyn Ann *microbiology company executive*

Rye
Reader, George Gordon *physician, educator*

Rye Brook
Aquino, Joseph Mario *clinical psychologist*

Saint James
Bigeleisen, Jacob *chemist, educator*
Irvine, Thomas Francis, Jr. *mechanical engineering educator*

Salt Point
Lackey, Mary Michele *physician assistant*

Sanborn
Mowrey, Timothy James *management and financial consultant*

Sands Point
Lear, Erwin *anesthesiologist, educator*

Saranac Lake
North, Robert John *biologist*

Saratoga Springs
Dorsey, James Baker *surgeon, lawyer*
Houghton, Raymond Carl, Jr. *computer science educator*
Walter, Paul Hermann Lawrence *chemistry educator*

Sayville
Pagano, Alphonse Frederick *obstetrician/gynecologist*

Scarsdale
Cohen, Irwin *economist*
Cox, Robert Hames *chemist, scientific consultant*

Schenectady
Alpher, Ralph Asher *physicist*
Anthony, Thomas Richard *research physicist*
Billmeyer, Fred Wallace, Jr. *chemist, educator*
Bolebruch, Jeffrey John *sales executive*
Boyer, John Frederick *biology educator*
Bucinell, Ronald Blaise *mechanical engineer*
Chestnut, Harold *foundation executive, electrical engineer*
DeBono, Kenneth George *psychology educator*
Dumoulin, Charles Lucian *physicist*
Fagg, William Harrison *retired infosystems specialist*
Felak, Richard Peter *electric power industry consultant*
German, Marjorie DaCosta *mechanical engineer*
Huening, Walter Carl, Jr. *retired consulting application engineer*
Kolb, Mark Andrew *aerospace engineer*
Mafi, Mohammad *civil engineer, educator*
Marchione, Sharyn Lee *computer scientist*
Mead, Kathryn Nadia *astrophysicist, educator*
Meiklejohn, William Henry *physicist*
Pasamanick, Benjamin *psychiatrist, educator*
Peak, David *physicist, educator, researcher*
Philip, A. G. Davis *astronomer, editor, educator*
Robb, Walter Lee *retired electric company executive, management company executive*
Roloff, Thomas Paul *combustion engineer*
Sohie, Guy Rose Louis *electrical engineer, researcher*
Stone, William C. *mathematics educator*
Thomann, Gary Calvin *electrical engineer*
Tolpadi, Anil Kumar *mechanical engineer, researcher*
Trabold, Thomas Aquinas *chemical engineer*
Wheeler, George Charles *consulting company executive*
Wicks, Frank Eugene *mechanical engineer, educator*

Sea Cliff
Rich, Charles Anthony *geologist*

Setauket
Gelinas, Paul Joseph *psychologist, author*

Somers
van Ryn, Ted Mattheus *electrical engineer*

South Huntington
De Lucia-Weinberg, Diane Marie *systems analyst*

Southampton
Melter, Robert Alan *mathematics educator, researcher*
Shumway, Sandra Elisabeth *shellfish biologist*

Southold
Knight, Harold Edwin Holm, Jr. *utility company executive*

Sparkill
Rosko, John James *biology educator*

Spencerport
Astill, Bernard Douglas *environmental health and safety consultant*

Spring Valley
McCormick, Jerry Robert Daniel *chemistry consultant*

Staten Island
Berger, Herbert *retired internist, educator*
Butler, Kevin Cornell *physicist*
Chauhan, Ved Pal Singh *biochemist, researcher*
Gokarn, Vijay Murlidhar *pharmacist, consultant*
Greenfield, Val Shea *ophthalmologist*
Meltzer, Yale Leon *economist, educator*
Rajakaruna, Lalith Asoka *civil engineer*
Sabido, Almeda Alice *mental health facility administrator*
Singer, Edward Nathan *engineer, consultant*
Tsui, Chia-Chi *electrical engineer, educator*
Wisniewski, Henryk Miroslaw *pathology and neuropathology educator, research facility administrator, research scientist*

Stony Brook
Baldo, George Jesse *biophysicist, physiology educator*
Bao, Zhenlei *physicist*
Brown, Gerald Edward *physicist, educator*
Carlson, Elof Axel *genetics educator*
Dhadwal, Harbans Singh *electrical engineer, consultant*
Franklin, Nancy Jo *psychology educator*
Futuyma, Douglas Joel *ecology educator*
Glimm, James Gilbert *mathematician*
Guilak, Farshid *biomedical engineering researcher, educator*
Henn, Fritz Albert *psychiatrist*
Huang, Peisen Simon *mechanical engineer*
Hyman, Leslie Gaye *epidemiologist*
Kaplan, Allen P. *physician, educator, academic administrator*
Kuchner, Eugene Frederick *neurosurgeon, educator*
Levine, Marvin *psychologist, author*
Lissauer, Jack Jonathan *astronomy educator*
Magda, Margareta Tatiana *physicist*
McLennan, Scott Mellin *geochemist, educator*
Michelsohn, Marie-Louise *mathematician, educator*
Morgan, Steven Gaines *marine ecology researcher, educator*
Poppers, Paul Jules *anesthesiologist, educator*
Pritchard, Donald William *oceanographer*
Rachlin, Howard *psychologist, educator*
Rohlf, F. James *biometrist, educator*
Scarlata, Suzanne Frances *biophysical chemist*
Schoenfeld, Elinor Randi *epidemiologist*
Schoonen, Martin Adrianus Arnoldus *geology educator*
Schubel, Jerry Robert *marine science educator, scientist, university dean and official*
Scranton, Mary Isabelle *oceanographer*
Shuryak, Edward Vladimirovich *physicist*
Smaldone, Gerald Christopher *physiologist*
Steigbigel, Roy Theodore *infectious disease physician and scientist, educator*
Strecker, Robert Edwin *neuroscientist, educator*
Tewarson, Reginald Prabhakar *mathematics educator, consultant*
Truxal, John Groff *electrical engineering educator*
Tycko, Daniel H. *physicist*
Volkman, David J. *immunology educator*
Weidner, Donald J. *geophysicist educator*
Williams, George Christopher *biologist, ecology and evolution educator*
Yang, Chen Ning *physicist, educator*
Yang, Chen-yu *neuroscientist*

Stormville
McGee, James Patrick *information industry executive*

Stuyvesant
Lincoln, Janet Elizabeth *experimental psychologist, consultant*

Suffern
Eoga, Michael Gerard *network systems programmer*
Sutherland, George Leslie *retired chemical company executive*

Syosset
Hershey, Alfred Day *geneticist*
Vida, Stephen Robert *environmental engineer*
Wachspress, Melvin Harold *electrical engineer, consultant*

Syracuse
Baker, Bruce Edward *orthopedic surgeon, consultant*
Ballingall, James McLean, III *physicist*
Bellanger, Barbara Doris Hoysak *biomedical research technologist*
Birge, Robert Richards *chemistry educator*
Boghosian, Paula der *educator, consultant*
Boston, Louis Russell *packaging engineer*
Burgess, Robert Lewis *educator, ecologist*
Cabasso, Israel *polymer science educator*
Cheng, David Keun *educator*
Clausen, Jerry Lee *psychiatrist*
Cooper, John Edwin *fisheries biologist*
Csermely, Thomas John *computer engineer, physicist*
Elligott, Linda A. *environmental scientist*
Evans, Judy Anne *health center administrator*
Fiske, Sandra Rappaport *psychologist, educator*
Gass, Tyler Evan *hydrogeologist*
Goel, Amrit Lal *engineering educator*
Gold, Joseph *medical researcher*
Graver, Jack Edward *mathematics educator*
Haas, David Colton *neurologist, educator*
Hamlett, James Gordon *electronics engineer, management consultant and educator*
Hansen, Per Brinch *computer scientist*
Hyla, James Franklin *rheumatologist, educator*
Kodali, Hari Prasad *electrical engineer*
LaLonde, Robert Thomas *chemistry educator*
Lane, Michael Joseph *microbiology educator*
LePage, Wilbur Reed *electrical engineering educator*
Lui, Eric Mun *civil engineering educator, practitioner*
Marshall, David *orthodontist*
Mi, Yongli *polymer engineer, researcher*
Miller, Douglas Alan *civil engineer*
Miron, Murray Samuel *psychologist, educator*
Pearse, George Ancell, Jr. *chemistry educator, researcher*
Pelczer, István *spectroscopist*
Pennock, Donald William *retired mechanical engineer*
Phillips, Arthur William, Jr. *biology educator*
Poiesz, Bernard Joseph *oncologist*
Russell-Hunter, W(illiam) D(evigne) *zoology educator, research biologist, writer*
Schenerman, Mark Allen *research chemist, educator*
Schiff, Eric Allan *physics educator*
Sleezer, Paul David *organic chemist*
Slepecky, Norma B. *cell biologist*
Smardon, Richard Clay *landscape architecture/environmental studies educator*
Sprafkin, Robert Peter *psychologist, educator*
Strait, Bradley Justus *electrical engineering educator*
Szasz, Thomas Stephen *psychiatrist, educator, writer*
Tabor, John Malcolm *genetic engineer*
Todd, Roger Harold *metallurgical engineer, failure analyst*
Verrillo, Ronald Thomas *neuroscientist*
Wadden, Thomas Antony *psychology educator*
Ware, Benjamin Ray *university administrator*
Waterman, Daniel *mathematician, educator*
Young, Terry William *civil engineer*

Tarrytown
Bartoo, Richard Kieth *chemical engineer, consultant*
Belanger, Ronald Louis *chemist*
Cedarbaum, Jesse Michael *neurologist*
Davis-Bruno, Karen L. *pharmacologist, researcher*
Farrell, Gregory Alan *biomedical engineer*
Flanigen, Edith Marie *materials scientist*
Kinigakis, Panagiotis *research scientist, engineer, author*
Klein, Marcia Schneiderman *analyst, programmer, consultant*
Lauterbach, Hans *pharmaceutical company executive*
Margoshes, Marvin *chemist, consultant*
Salop, Arnold *internist*

Thornwood
Douglas, Patricia Jeanne *systems designer*

Tonawanda
Bernard, John P. *chemical engineer*
Drozdziel, Marion John *aeronautical engineer*
Rovison, John Michael, Jr. *chemical engineer*
Woodside, Bertram John *infosystem engineer*

Troy
Anderson, John Bailey *electrical engineering educator*
Bergles, Arthur Edward *mechanical engineering educator*
Blanchet, Thierry Alain *mechanical engineering educator*
Brunelle, Eugene John, Jr. *mechanical engineering educator*
Cole, Julian D. *mathematician, educator*
Cook, Raymond Douglas *biomedical engineer*
Doremus, Robert Heward *glass and ceramics processing educator*
Dvorak, George J. *materials engineering educator*
Fleischer, Robert Louis *physics educator*
Fradkov, Valery Eugene *materials scientist*
Ghandhi, Sorab Khushro *electrical engineering educator*
Giaever, Ivar *physicist*
Gill, William Nelson *chemical engineering educator*
Glicksman, Martin Eden *materials engineering educator*
Graves, Robert John *industrial engineering educator*
Hayes, David Kirk *nuclear engineer*
Ho, Yifong *environmental engineer*
Hsu, Cheng *decision sciences and engineering systems educator*
Jordan, Mark Henry *consulting civil engineer*
Kalsher, Michael John *psychology educator*
Krause, Sonja *chemistry educator, researcher*
Lane, Adelaide Irene *computer systems specialist, researcher*
Loewy, Robert Gustav *engineering educator, aeronautical engineering executive*
McDonald, John Francis Patrick *electrical engineering educator*
Miller, Donald Spencer *geologist, educator*
Modestino, James William *electrical engineering educator*
Ostrogorsky, Aleksandar George *mechanical engineer, educator*
Pegna, Joseph *engineer, educator*
Przybycien, Todd Michael *chemical engineering educator*
Roy, Rob J. *biomedical engineer, anesthesiologist*
Roysam, Badrinath *computer scientist, educator*
Saridis, George Nicholas *electrical engineer*
Savic, Michael I. *engineering educator, signal and speech researcher*
Saxena, Arjun Nath *physicist*
Sopko, Stephen Joseph *structural engineer*
Soracco, Reginald John *microbiological biochemist*
Sperber, Daniel *physicist*
Tartaglia, James Joseph *molecular virologist*
Tham, Fook Siong *crystallographer*
Umstadter, Karl Robert *engineering physicist*
Ushe, Zwelonke Ian *development engineer*
Yee, Lester Wey-Ming *researcher*
Zimmie, Thomas Frank *civil engineer, educator*

Tuckahoe
Reyes-Guerra, Antonio *dental surgeon*

Tuxedo Park
Flescher, Eliezer *immunologist*
Hall, Frederick Keith *chemist*
Noel, Dale Leon *chemist*
Vahey, David William *measurement scientist, physicist*

Upton
Curtiss, Joseph August *electrical engineer, consultant*
Czajkowski, Carl Joseph *metallurgist, engineer*
Foerster, Conrad Louis *project engineer*
Friedlander, Gerhart *nuclear chemist*
Fthenakis, Vasilis *chemical engineer, consultant, educator*
Goldhaber, Gertrude Scharff *physicist*
Goldhaber, Maurice *physicist*
Gunther, William Edward *electrical engineer, researcher*
Hamilton, Leonard Derwent *physician, molecular biologist*
Hendrie, Joseph Mallam *physicist, nuclear engineer, government official*
Holden, Norman Edward *nuclear physicist, philosopher, historian*
Kato, Walter Yoneo *physicist*
Kim, Inn Seock *nuclear engineer*
Laster, Brenda Hope *radiobiologist*
Lewin, Keith Frederic *environmental science research associate*
Lindenbaum, S(eymour) J(oseph) *physicist*
Lowenstein, Derek Irving *physicist*
McWhan, Denis Bayman *physicist*
Montag, Mordechai *mechanical engineer*
Morris, Samuel Cary *environmental scientist, consultant, educator*
Mubayi, Vinod *physicist*
Nawrocky, Roman Jaroslaw *research electrical engineer*
Ozaki, Satoshi *physicist*
Pogorelsky, Igor Vladislav *laser physicist*
Rau, Ralph Ronald *physicist*
Ruckman, Mark Warren *physicist*
Samios, Nicholas Peter *physicist*
Schwartz, Melvin *physics educator, laboratory administrator*
Setlow, Richard Burton *biophysicist*
Souw, Bernard Eng-Kie *physicist*
Steinberg, Meyer *chemical engineer*
Studier, Frederick William *biophysicist*
Sutherland, John Clark *physicist, researcher*
Sutin, Norman *chemistry educator, scientist*
Tang, Ignatius Ning-Bang *chemical engineer*
Weber, David Alexander *physicist*
Weeks, John Randel, IV *nuclear engineer, metallurgist*
Wolf, Alfred Peter *chemist, educator*

Utica
Antzelevitch, Charles *research center executive*
Grisewood, Norman Curtis *chemical engineer*
Harney, Patrick Joseph Dwyer *meteorologist, consultant*
Labuz, Ronald Matthew *design educator*
Pulliam, Curtis Richard *chemistry educator*
Ragusa, Paul Carmen *mechanical and structural engineer*

Vails Gate
Timm, Walter Clement *chemical engineer, researcher*

Valhalla
Brook, Judith Suzanne *psychiatry educator*
Chung, Fung-Lung *cancer research scientist*
Ehrenkrantz, David *medical researcher*
Gross, Stanislaw *environmental sciences educator, activist*
Levere, Richard David *physician, academic administrator, educator*
Levy, Norman B. *psychiatrist, educator*
McClung, John Arthur *cardiologist*
Rosenfeld, Louis *biochemist, educator*
Williams, Gary Murray *medical researcher, pathology educator*

Valley Stream
Sapadin, Linda Alice *psychologist, writer*
Tartell, Robert Morris *dentist*

Vernon
Glover, Anthony Richard *engineer*

Vestal
Lazaroff, Norman *microbiologist, researcher*

Waddington
Olszewski, James Frederick *metallurgical engineer*

Wallkill
Bittner, Ronald Joseph *computer systems analyst*

Wantagh
Malden, Joan Williams *physical therapist*

Wappingers Falls
Engelman, Melvin Alkon *retired dentist, business executive, scientist*
McCamy, Calvin Samuel *optics scientist*
Ninnie, Eugene Dante *civil engineer*

Watertown
Dimmick, Kris Douglas *civil engineer*
Miller, David William *geologist*

Webster
Bluhm, Terry Lee *materials scientist, chemist*
Drumheller, Jerry Paul *laser scientist, development engineer*
Duke, Charles Bryan *research and development manufacturing executive, physics educator*
Hwang, Shyshung *mechanical engineer*
Machell, Arthur R. *retired mechanical engineer, association board member*
O'Neill, James Francis *chemist*
Pan, David Han-Kuang *polymer scientist*
Zavon, Peter Lawrence *industrial hygienist*

Wellsville
Hoover, Herbert William, Sr. *foundry engineer*

West Haverstraw
Cochran, George Van Brunt *physician, surgery educator, researcher*

West Nyack
Hornik, Joseph William *civil engineer*

West Point
Arkin, Joseph *mathematician, lecturer*
Oldaker, Bruce Gordon *physicist, military officer*
Samples, Jerry Wayne *military officer*
Shoop, Barry LeRoy *electrical engineer*

West Seneca
McMahon, James Francis *quality control engineer*

West Valley
Palmer, Ronald Alan *materials scientist, consultant*

Westbury
Munsinger, Roger Alan *marketing executive*
O'Neill, Peter Thadeus *software engineer*
Stalzer, John Francis *environmental biologist*

White Plains
Blass, John Paul *medical educator, physician*
Bosco, Paul D. *computer scientist*
Cheng, Alexander Lihdar *computer scientist, researcher*
Cho, Nam Min *physicist*
Dalessio, Anthony Thomas *industrial and organization psychologist*
Glassman, George Morton *dermatologist*
Greenberg, Sharon Gail *neuroscientist, researcher*
Machover, Carl *computer graphics consultant*
Marshall, L(uther) Gerald *acoustic consultant*
Maslansky, Carol Jeanne *toxicologist*
Mclean, John *architect, industrial designer*
Meehan, Robert Henry *utilities executive, human resources executive, business educator*
Meyers-Jouan, Michael Stuart *computer engineer*
Papp, Laszlo George *architect*
Parker, Robert Andrew *psychologist, computer consultant*
Pfeffer, Cynthia Roberta *psychiatrist, educator*
Riha, William Edwin *beverage company executive*
Rozenfeld, Gregory *petroleum engineer*
Samii, Abdol Hossein *physician, educator*
Seelye, Edward Eggleston *psychiatrist*
Shapiro, Robert M. *electronics company executive*
Theisz, Erwin Jan *scientist*
Valentino, Joseph Vincent *engineering consultant*

Whitesboro
Bulman, William Patrick *data processing executive*
Serling, Joel Martin *educational psychologist*

Whitestone
Juszczak, Nicholas Mauro *psychology educator*
Panides, Elias *mechanical engineer*

Williamsville
Hertzog, Robert William *pathologist, consultant, educator*
May, Kenneth Myron *information services professional, consultant*
Paladino, Joseph Anthony *clinical pharmacist*

Williston Park
Drago, Robert John *psychologist*

Wolcott
Wilt, John Robert *chemist*

Woodbury
Bleicher, Sheldon Joseph *endocrinologist, medical educator*
Gibbons, Edward Francis *psychobiologist*
Herczynski, Andrzej *physicist, educator*
Kitzis, Gary David *periodontics educator, periodontist*
Memoli, Michael Anthony *environmental engineer*
Schmid, Charles Ernest *acoustician, administrator*

Woodmere
Erdberg, Mindel Ruth *psychiatrist*

Woodside
Bashias, Norman Jack *software engineer, computer consultant*
Gurian, Martin Edward *textile engineer*

Woodstock
Villchur, Edgar *audiology research scientist*

Yonkers
Anderson, Bror Ernest *chemist*
Arnone, Mary Grace *radiology technologist*
D'Angelo, Andrew William *civil engineer*
LoPinto, Charles Adam *chemical engineer*
Lupiani, Donald Anthony *psychologist*
Varma, Baidya Nath *sociologist, broadcaster*

Yorktown Heights
Avouris, Phaedon *chemical physicist*
Bhaskar, Ramamoorthi *artifical intelligence scientist, researcher*
Buchwalter, Stephen L. *chemist, researcher*
Cifuentes, Arturo Ovalle *civil and mechanical engineer*
Clevenger, Larry Alfred *electronics company researcher*
Cocke, John *computer scientist*
Colgan, Evan George *materials scientist*
Cuomo, Jerome John *materials scientist*
Dennard, Robert Heath *engineering executive, scientist*
Doganata, Yurdaer Nezihi *research scientist*
Green, Paul Eliot, Jr. *communications scientist*
Gutzwiller, Martin Charles *theoretical physicist, research scientist*
Haller, Ivan *chemist*
Hamaguchi, Satoshi *mathematician*
Hartstein, Allan Mark *physicist*
Heinz, Tony Frederick *physicist*
Henle, Robert Athanasius *engineer*
Hoffman, Alan Jerome *mathematician, educator*
Holtzberg, Frederic *chemist, solid state researcher*
Kanicki, Jerzy *chemist, researcher*
Kash, Jeffrey Alan *physicist, researcher*
Landauer, Rolf William *physicist*
Mandelbrot, Benoit B. *mathematician, scientist, educator*
Nguyen, Philong *electrical engineer*
Ning, Tak Hung *physicist, microelectric technologist*
Romankiw, Lubomyr Taras *materials engineer*
Sachs, Martin William *computer scientist*
Tsang, James Chen-Hsiang *physicist*
Tu, King-Ning *materials science and engineering educator*
Webb, Richard A. *physicist*
Winograd, Shmuel *mathematician*

NORTH CAROLINA

Archdale
Riddick, Douglas Smith *horticultural industrialist, industrial designer*

Arden
Price, Jeffrey Brian *quality engineer*

Asheville
Haggard, William Henry *meteorologist*
Honeycutt, Michael Allen *computer consultant*
Squibb, Samuel Dexter *chemistry educator*
Stevens, John Gehret *chemistry educator*
Van Engelen, Debra Lynn *chemistry educator*
Von Viczay, Marika (Ilona) *naturopathic medical doctor, physician*
Wilsdon, Thomas Arthur *product development engineer, administrator*

Beaufort
Bonaventura, Joseph *biochemist, educator, research center director*

Biscoe
Kelly, Terry Lee *computer systems analyst*

Boone
Shull, Julian Kenneth, Jr. *biology educator*

Brevard
Valentini, Jose Esteban *scientist, educator, chemical engineering director*

Buies Creek
Nemecz, George *biochemist*

Burlington
Flagg, Raymond Osbourn *biology executive*
Golden, Carole Ann *immunologist, microbiologist*
Mason, James Michael *biomedical laboratories executive*

Carrboro
Williamson, Ronald Edwin *waste water plant manager*

Cary
Amtoft-Nielsen, Joan Theresa *physician, educator, researcher*
Bursiek, Ralph David *information systems company executive*
Chignell, Colin Francis *pharmacologist*
Harvey, Louis James *architect*
Janezic, Darrell John *environmental engineer*
Krasner, Paul R. *psychologist*
Myers, Donald Richard *biomedical engineer*
Nyce, David Scott *electronics company executive*
Rauf, Robert Charles, Sr. *systems specialist*
Robertson, William Bell, Jr. *environmental scientist*

Chapel Hill
Augustine, Brian Howard *electronics researcher*
Baerg, Richard Henry *podiatrist, educator, consultant*
Bartlett, Jeffrey Stanton *molecular biologist*
Behrman, Jack Newton *economist*
Bowman, Frederick Oscar, Jr. *cardiothoracic surgeon, retired*
Brinkhous, Kenneth Merle *pathologist, educator*
Chang, Jerjang *veterinarian, educator*
Chi, Vernon L. *computer science educator, administrator*
Colvin, John Trevor *physicist*
Coulter, Elizabeth Jackson *biostatistician, educator*
Diliberto, Pamela Allen *pharmacologist, researcher*
Douglas, Michael Gilbert *cell biologist, educator*
Driscol, Jeffrey William *chemist, researcher*
Droessler, Earl G. *geophysicist educator*
Eliel, Ernest Ludwig *chemist, educator*
Finkler, Kaja *anthropologist, educator*
Frankenberg, Dirk *marine scientist*
Goldman, Leonard Manuel *physicist, engineering educator*
Greeley, Richard Folsom *retired sanitary engineer*
Grisham, Joe Wheeler *pathologist, educator*
Hackney, Anthony C. *nutrition and physiology educator, researcher*
Hairston, Nelson George *animal ecologist*

Handler, Enid Irene *health care administrator, consultant*
Hawkins, David Rollo, Sr. *psychiatrist, educator*
Henson, Anna Miriam *otolaryngology researcher, medical educator*
Hernandez, John Peter *physicist, educator*
Howell, William Everett *computer professional, consultant*
Huang, Eng-Shang *virology educator, biomedical engineer*
Hutchison, Clyde Allen, III *microbiology educator*
Jorgenson, James Wallace *chromatographer, educator*
Kallianpur, Gopinath *statistician*
Kier, William McKee *biologist, educator*
Krasny, Harvey Charles *research biochemist, medical research scientist*
Kuenzler, Edward Julian *ecologist and environmental biologist*
Lockett, Stephen John *medical biophysicist*
Macdonald, James Ross *physicist, educator*
Maroni, Gustavo Primo *geneticist, educator*
McBay, Arthur John *toxicologist, consultant*
Merzbacher, Eugen *physicist, educator*
Miller, Daniel Newton, Jr. *geologist, consultant*
Murray, Royce Wilton *chemistry educator*
Neumann, Andrew Conrad *geological oceanography educator*
Okun, Daniel Alexander *environmental engineering educator, consulting engineer*
Pagano, Joseph Stephen *physician, researcher, educator*
Parr, Robert Ghormley *chemistry educator*
Peters, Robert William *speech and hearing sciences educator*
Peterson, Gary *child psychiatrist*
Pfouts, Ralph William *economist, consultant*
Popkin, Carol Lederhaus *epidemiologist*
Powell, Judith Carol *clinical psychologist*
Roberts, Harold Ross *medical educator, hematologist*
Roberts, Louis Douglas *physics educator, researcher*
Rogers, John James William *geology educator*
Rosenman, Julian Gary *radiation oncologist*
Skatrud, David Dale *physicist*
Smithies, Oliver *pathologist, educator*
Stephens, Laurence David, Jr. *linguist, consultant*
Suzuki, Kunihiko *biomedical educator, researcher*
Topal, Michael David *biochemistry educator*
Ulshen, Martin Howard *pediatric gastroenterologist, researcher*
Warren, Donald William *physiology educator, dentistry educator*
Williams, Mark Edward *geriatrician*
Wilson, John Eric *biochemist*
Winfield, John Buckner *rheumatologist, educator*
Xu, Le *biophysicist, physicist*

Charlotte
Babić, Davorin *electrical engineer, researcher*
Brown, James Eugene, III *business executive*
Browning, Burt Oliver *mechanical engineer, consultant*
Cornell, James Fraser, Jr. *biologist, educator*
Duffy, Sally M. *psychologist*
Foss, Ralph Scot *mechanical engineer*
Gothard, Michael Eugene *industrial engineer*
Hartley, Joe David *communications specialist*
Iverson, Francis Kenneth *metals company executive*
Jones, Daniel Silas *chemistry educator*
Kuykendall, Terry Allen *environmental and chemical engineer*
Lee, William States *utility executive*
Monroe, Frederick Leroy *chemist*
Nemzek, Thomas Alexander *nuclear engineer*
Owen, Kenneth Dale *orthodontist*
Priestley, G. T. Eric *manufacturing company executive*
Ruckterstuhl, Russell M(ilton) *mechanical engineer, consulting engineer*
Sellers, Macklyn Rhett, Jr. *architect*
Van Wallendael, Lori Robinson *psychology educator*

Clayton
Anderson, Pamela Boyette *quality assurance professional*

Cullowhee
Willis, Ralph Houston *mathematics educator*

Davidson
Alday, John Hane *mechanical engineer*
Barton, Cole *psychologist, educator*
Ramirez, Julio Jesus *neuroscientist*
Schuh, Merlyn Duane *chemist, educator*

Duck
Majewski, Theodore Eugene *chemist*

Durham
Abdel-Rahman, Mohamed *plant pathologist, physiologist*
Adams, Dolph O. *pathologist, educator*
Amos, Dennis B. *immunologist*
Auciello, Orlando Hector *physicist*
Baker, Lenox Dial *orthopaedist, genealogist*
Bast, Robert Clinton, Jr. *research scientist, medical educator*
Bejan, Adrian *mechanical engineering educator*
Bennett, Peter Brian *researcher, anesthesiology educator*
Billings, William Dwight *ecology educator*
Bilpuch, Edward George *nuclear physicist, educator*
Brandon, Robert Norton *zoology and philosophy educator*
Bryant, Robert Leamon *mathematics educator*
Buckley, Rebecca Hatcher *physician*
Bunch, Michael Brannen *psychologist, educator*
Bush, Mark Bennett *ecologist, educator*
Casseday, John Herbert *neurobiologist*
Chilton, Mary-Dell Matchett *chemical company executive*
Cohen, Harvey Jay *physician, educator*
Coury, Louis Albert, Jr. *chemistry educator, researcher*
Cruze, Alvin M. *research institute executive*
Culberson, William Louis *botany educator*
Daniels-Race, Theda Marcelle *electrical engineering educator*
Davis, James Evans *general and thoracic surgeon, parliamentarian, author*
Dawson, Jeffrey Robert *immunology educator*
DeBusk, George Henry, Jr. *paleoecologist*
Diamond, Irving T. *physiology educator*
Dowell, Earl Hugh *university dean, aerospace and mechanical engineering educator*
Falletta, John Matthew *pediatrician*
Flurchick, Kenneth Michael *computational physicist*
Fouts, James Ralph *pharmacologist, educator, clergyman*
Fraser-Reid, Bertram Oliver *chemistry educator*
Garg, Devendra Prakash *mechanical engineer, educator*
Goodwin, Frank Erik *materials engineer*
Gross, Samson Richard *geneticist, biochemist, educator*
Hamaker, Richard Franklin *engineer*
Hammes, Gordon G. *chemistry educator*
Hammond, Charles Bessellieu *obstetrician-gynecologist, educator*
Hayes, Brian Paul *editor, writer*
Herbel, LeRoy Alec, Jr. *telecommunications engineer*
Hill, Robert Lee *biochemistry educator, administrator*
James, Lawrence Hoy *nuclear physicist*
Jaszczak, Ronald Jack *physicist, researcher, consultant*
Jennings, Robert Burgess *experimental pathologist, medical educator*
Johnson, Edward A. *physician, educator*
Judd, William Reid *computer engineer, graphic artist*
Keefe, Francis Joseph *psychology educator*
Krystal, Andrew Darrell *psychiatrist, biomedical engineer*
Kurlander, Roger Jay *medical educator, researcher*
Lambert, Dennis Michael *virologist, researcher*
Littman, Susan Joy *physician*
Livingstone, Daniel Archibald *zoology educator*
Lockhead, Gregory Roger *psychology educator*
Massaro, Edward Joseph *cell biology, biochemistry research scientist, experimental pathology educator*
Massoud, Hisham Zakaria *electrical engineering educator*
McClellan, Roger Orville *toxicologist*
McKinney, Collin Jo *electrical engineer*
Merchenthaler, Istvan Jozsef *anatomist*
Meyer, Horst *physics educator*
Miller, Charles Gregory *biomedical researcher*
Miller, David Edmond *physician*
Murray, William James *anesthesiology educator, clinical pharmacologist*
Pilkington, Theo Clyde *biomedical and electrical engineering educator*
Pirrung, Michael Craig *chemistry educator, consultant*
Qualls, Robert Gerald *ecologist*
Reller, L. Barth *microbiologist, educator*
Sabiston, David Coston, Jr. *educator, surgeon*
Schmidt-Nielson, Knut *physiologist, educator*
Simons, Elwyn LaVerne *physical anthropologist, primatologist, paleontologist, educator*
Snyderman, Ralph *medical educator, physician*
Staddon, John Eric Rayner *psychology, zoology, neurobiology educator*
Straub, Karl David *biochemist researcher*
Tien, Robert Deryang *radiologist, educator*
Wainwright, Stephen A. *zoology educator, design consultant*
Wilbur, Karl Milton *zoologist, educator*
Wilbur, Robert Lynch *botanist, educator*
Williams, Redford Brown *medical educator*
Winkler, Robert Lewis *statistics educator, researcher, author, consultant*

Elizabeth City
Oriaku, Ebere Agwu *economics educator*

Elm City
Parker, Josephus Derward *corporation executive*

Emerald Isle
Hardy, Sally Maria *retired biological sciences educator*

Enka
Barbour, Eric S. *electrical engineer*
Stewart, Ronald *chemical engineer*

Fayetteville
Baldwin, George Michael *industrial marketing professional*
Sloggy, John Edward *engineering executive*

Flat Rock
Matteson, Thomas Dickens *aeronautical engineer, consultant*

Fort Bragg
Moss, Kenneth Wayne *neurologist*
Sears, Catherine Marie *osteopath, radiologist*

Fuquay-Varina
Jarman, Scott Allen *plastics engineer*

Gastonia
Kiser, Clyde Vernon *retired demographer*
Prince, George Edward *pediatrician*

Goldsboro
Gravely, Jane Candace *computer company executive*

Gorner
Jones, Stephen Yates *pharmacist*

Greensboro
Adams, Chester Z. *sales engineer*
Adelberger, Rexford Earle *physics educator*
Bailey, William Nathan *systems analyst*
Baird, Haynes Wallace *pathologist*
Banegas, Estevan Brown *agricultural biotechnology executive*
Berggren, Thage *automotive executive*
Blanchet-Sadri, Francine *mathematician*
Cotter, John Burley *ophthalmologist, corneal specialist*
Garibay, Joseph Michael *mechanical engineer*
Heck, Jonathan Daniel *toxicologist*
Ilias, Shamsuddin *chemical engineer, educator*
Johnson, Jeffrey Allan *industrial engineer*
Kurepa, Alexandra *mathematician*
Lange, Garrett Warren *psychology educator*
Lund, Harold Howard *ceramic engineer, civil engineer, consultant*
Mosier, Stephen Russell *college program director, physicist*
O'Hara, Robert James *evolutionary biologist*
Reid, Jack Richard *research executive*
Robinson, Edward Norwood, Jr. *physician, educator*
Stein, Richard Martin *mechanical engineer*
Tate, John Edward *chemical engineer*
Truesdale, Gerald Lynn *plastic and reconstructive surgeon*

Greenville
Blackmon, Margaret Lee *pharmaceutical chemist*
Dar, Mohammad Saeed *pharmacologist, educator*
Deal, Jo Anne McCoy *quality control professional*
Jain, Sunil *pharmaceutical scientist*
Johnson, Ronald Sanders *physical biochemist*
Lee, Kenneth Stuart *neurosurgeon*
Lieberman, Edward Marvin *biomedical scientist, educator*
Lust, Robert Maurice, Jr. *physiologist, educator, researcher*
Marks, Richard Henry Lee *biochemist, educator*
Metzger, Walter James, Jr. *physician, educator*
Odabasi, Halis *physics educator*
Volkman, Alvin *pathologist, educator*
Waugh, William Howard *biomedical educator*
Webster, Raymond Earl *psychology educator, psychotherapist*
Wiley, John Edwin *cytogeneticist*

Hampstead
Solomon, Robert Douglas *pathology educator*

Hendersonvlle
Kehr, August Ernest *geneticist, researcher*

Hickory
Kuehnert, Deborah Anne *medical center administrator*
Seaman, William Daniel *biology educator, clergyman*
Sears, Frederick Mark *research manager, mechanical engineer*
Steelman, Sanford Lewis *research scientist, biochemist*

High Point
Huston, Fred John *automotive engineer*
Schwarz, Saul Samuel *neurosurgeon*

Highland
Sandor, George Nason *mechanical engineer, educator*

Hillsborough
Cooley, Philip Chester *computer modeller*

Hope Mills
Baylor, John Patrick *nurse*

Huntersville
Acheson, Scott Allen *research biochemist*

Kernersville
Farrer-Meschan, Rachel (Mrs. Isadore Meschan) *obstetrics/gynecology educator*
Meschan, Isadore *radiologist, educator*

Kinston
Arcino, Manuel Dagan *microbiologist, consultant*

Laurinburg
Hardin, William Beamon, Jr. *electrical engineer*

Lewisville
Bolz, Roger William *mechanical engineer, consultant*

Linwood
Barnes, Melver Raymond *chemist*

Mocksville
Townsend, Arthur Simeon *manufacturing engineer*

Monroe
Smith, Jeffrey Alan *occupational medicine physician, toxicologist*

Morehead City
Williams, Winton Hugh *civil engineer*

Morganton
Faunce, William Dale *clinical psychologist, researcher*

Morrisville
Odum, Jeffery Neal *mechanical engineer*
Voisin-Lestringant, Emmanuelle Marie *pharmacologist, consultant*

Mount Airy
Ratliff, Robert Barns, Jr. *mechanical engineer*

Mount Holly
Roberts, Warren Hoyle, Jr. *chemist*

Murfreesboro
McLawhorn, Rebecca Lawrence *mathematics educator*

Nags Head
Rogallo, Francis Melvin *mechanical, aeronautical engineer*

New Bern
Whitehurst, Brooks Morris *chemical engineer*

New Hill
Weber, Michael Howard *nuclear control operator*
Wilson, Robert Thaniel, Jr. *environmental chemist*

Oriental
Rowell, John Thomas *psychologist, consultant*

Oteen
Chapman, William Edward *pathologist*

Oxford
Jackson, D. Michael *research entomologist, educator*

Pine Knoll Shores
Baker, Edward George *retired mechanical engineer*

Raleigh
Alman, Ted Irwin *engineer, consultant*
Arya, Satya Pal Singh *meteorology educator*
Baliga, Bantval Jayant *electrical engineering educator, consultant*
Barnhardt, Robert Alexander *college dean*
Benson, D(avid) Michael *plant pathologist*
Berry, Joni Ingram *hospice pharmacist, educator*
Bilbro, Griff Luhrs *electronics engineering educator*
Boblett, Mark Anthony *civil engineering technician*
Borden, Roy Herbert, Jr. *civil engineer, geotechnical consultant*
Bourham, Mohamed Abdelhay *nuclear engineer, educator*
Brown, John David *physicist*
Campbell, Charles Lee *plant pathologist*
Carbonell, Ruben Guillermo *chemical engineering educator*
Church, Kern Everidge *engineer, consultant*
Ciraulo, Stephen Joseph *nurse, anesthetist*
Clifton, Marcella Dawn *dentist*
Daub, Margaret E. *plant pathologist, educator*
Davey, Charles Bingham *soil science educator*
De Hertogh, August Albert *horticulture educator, researcher*
Dudziak, Donald John *nuclear engineer, educator*
Dunphy, Edward James *crop science extension specialist*
Eberhardt, Allen Craig *biomedical engineer, mechanical engineer*
Fang, Shu-Cherng *industrial engineering and operations research educator*
Gardner, Robin Pierce *engineering educator*
Gillani, Noor Velshi *mechanical engineer, researcher, educator*
Gilligan, John Gerard *nuclear engineer, educator*
Godwin, Lars Duvall *civil engineer*
Grossfeld, Robert Michael *physiologist, zoologist, educator, neurobiologist*
Hanson, John M. *civil engineering and construction educator*
Hardin, James W. *botanist, herbarium curator, educator*
Hassan, Hosni Moustafa *biochemistry, toxicology and microbiology educator, biologist*
Havner, Kerry Shuford *civil engineering and materials science educator*
Hubbard, Bessie Renee *mechanical engineer, mathematician*
Janowitz, Gerald Saul *geophysicist, educator*
Kelly, Richard Lee Woods *civil engineer*
Kolbas, Robert Michael *electrical engineering educator*
Kriz, George James *agricultural research administrator, educator*
Kuhr, Ronald John *entomology and toxicology educator*
MacConnell, Gary Scott *environmental engineer*
Mayo, Robert Michael *nuclear engineering educator, physicist*
McAllister, David Franklin *computer science educator*
Meier, Wilbur Leroy, Jr. *industrial engineer, educator, former university*
Mock, Gary Norman *textile engineering educator*
Monteiro-Riviere, Nancy Ann *biologist, educator*
Moody, Roger Wayne *civil engineer*
Narayan, Jagdish *materials science educator*
Olson, Neil Chester *physiologist, educator*
Osburn, Carlton Morris *electrical and computer engineering educator*
Overcash, Michael Ray *chemical engineering educator*
Owens, Tyler Benjamin *chemist*
Petitte, James Nicholas *poultry science educator*
Pollock, Kenneth Hugh *statistics educator*
Poole, Marion Ronald *civil engineering executive*
Poran, Chaim Jehuda *civil and environmental engineer, educator*
Rohrbach, Roger Phillip *agricultural engineer, educator*
Rose, Thoma Hadley *environmental consultant*
Scandalios, John George *geneticist, educator*
Sederoff, Ronald Ross *geneticist*
Skaggs, Richard Wayne *agricultural engineering educator*
Starkey, Russell Bruce, Jr. *utilities executive*
Stuber, Charles William *genetics educator, researcher*
Tove, Samuel B. *biochemistry educator*
Tyczkowska, Krystyna Liszewska *chemist*
Wetsch, John Robert *computer systems administrator*
Whitten, Jerry Lynn *chemistry educator*
Williams, Hugh Alexander, Jr. *retired mechanical engineer, consultant*
Won, Ihn-Jae *geophysicist*
Woodland, N. Joseph *optical engineer, mechanical engineer*
Young, James Herbert *agricultural engineer*
Zande, Michael Dominic *clinical psychologist, consultant*
Zimmerman, John Wayne *wildlife ecologist*
Zorowski, Carl Frank *engineering educator, university administrator*

Research Triangle Park
Agarwal, Sanjay Krishna *chemical engineer*
Barry, David Walter *infectious diseases physician, researcher*
Chadwick, Robert William *toxicologist*
Chao, James Lee *chemist*
DeVeaugh-Geiss, Joseph *psychiatrist*
Dibner, Mark Douglas *research executive, industry analyst*
Dobbin, Ronald Denny *federal agency administrator, occupational hygienist, researcher*
Drake, John Walter *geneticist*
Elion, Gertrude Belle *research scientist, pharmacology educator*
Everitt, Henry Olin, III *physicist*
Franklin, David Lee *economist*
Friedland, Beth Rena *ophthalmologist*
Gallagher, Edward Joseph *scientific marketing executive*
Garner, Jasper Henry Barkdoll *ecologist*
Gerald, Nash Ogden *environmental engineer*
Graves, Joan Page *biologist*
Hajian, Gerald *biostatistician, engineer*
Hitchings, George Herbert *retired pharmaceutical company executive, educator*
Holland, Charles Edward *corporate executive*
Huang, Jim Jay *chemist*
Huber, Brian Edward *molecular biologist*
Hyatt, David Eric *environmental scientist*
Judd, Burke Haycock *geneticist*
King, Theodore M. *obstetrician, gynecologist, educator*
Kohn, Michael Charles *theoretical biochemistry professional*
Kramer, David Alan *biomathematician*
Kuhn, Matthew *engineering company executive*

Lange, Robert William *immunotoxicologist*
Larsen, Ralph Irving *environmental engineer*
Lee, Paul Huk-Kai *biomedical research scientist*
Lewin, Anita Hana *research chemist*
Mann, David Mark *researcher*
Martin, William Royall, Jr. *association executive*
May, Michael Lee *magazine editor*
Mayes, Mark Edward *molecular toxicologist*
Metz, Alan *psychiatrist*
Miller, Wayne Howard *biochemist*
Odman, Mehmet Talat *mechanical engineer, researcher*
Olden, Kenneth *public health service administrator, researcher*
Profeta, Salvatore, Jr. *chemist*
Richmond, James Arthur *entomologist*
Smith, Luther A. *statistician*
Spiegel, Ronald John *electrical engineer*
Stelling, John Henry Edward *chemical engineer*
Sykes, Richard Brook *microbiologist*
Tucker, William Gene *environmental engineer*
Van Winkle, Jon *chemical engineer*
Venkatasubramanian, Rama *electrical engineer*
Wooten, Frank Thomas *research facility executive*

Rocky Mount
Matthews, Drexel Gene *quality control executive*
Woollen, Kenneth Lee *electrical engineer*

Salemburg
Baugh, Charles Milton *biochemistry educator, college dean*

Salisbury
Kiser, Glenn Augustus *pediatrician*

Sanford
Woerner, Geldard Harry *civil engineer*

Skyland
Connolly, William Michael *civil and structural engineer*

Spring Hope
Lavatelli, Leo Silvio *retired physicist, educator*

Swannanoa
Stuck, Roger Dean *electrical engineering educator*

Warrenton
Padgett, Bobby Lee, II *chemist*

Washington
Thomson, Stuart McGuire, Jr. *science educator*

Wilmington
Cannon, Albert Earl, Jr. *electrical engineer*
Gillen, Howard William *neurologist, medical historian*
Lloyd, Donald Grey, Jr. *civil engineer*
McFall, Gregory Brennon *marine biologist*
Merritt, James Francis *biological sciences educator*
White, Larry Keith *electrical engineer*

Wilson
Kushner, Michael James *neurologist, consultant*

Winston Salem
Allen, Nina Strömgren *biology educator*
Blatchley, Brett Lance *computer engineer*
Bowman, Marjorie Ann *physician, academic administrator*
Clarkson, Thomas Boston *comparative medicine educator*
Dobbins, James Talmage, Jr. *analytical chemist, researcher*
Ehmann, Carl William *consumer products executive, researcher*
Esch, Gerald Wisler *biology educator*
Everhart, Francis Grover, Jr. *manufacturing company executive*
Flory, Walter S., Jr. *geneticist, botanist, educator*
Henderson, Richard Martin *chemical engineer*
Howell, Charles Maitland *dermatologist*
Hutcherson, Karen Fulghum *nursing administrator*
Jerome, Walter Gray *cell biologist*
Kaufman, William *internist*
Li, Linxi *biomedical scientist*
Penry, James Kiffin *physician, neurology educator*
Simonsen, John Charles *exercise physiologist*
Sorci-Thomas, Mary Gay *biomedical researcher, educator*
Toole, James Francis *medical educator*
Yeatts, Dorothy Elizabeth Freeman *nurse, retired county official, educator*

NORTH DAKOTA

Bismarck
Carmichael, Virgil Wesly *mining, civil and geological engineer, former coal company executive*
Ogaard, Louis Adolph *environmental administrator, computer consultant*
Spilman, Timothy Frank *utilities engineer*

Fargo
Colby, S(tanley) Brent *medical physicist*
Cross, Harold Zane *agronomist, educator*
Fatland, Charlotte Lee *chemist*
Hadley, Mary *nutritionist*
Hahn, Benjamin Daniel *research executive*
Li, Kam Wu *mechanical engineer, educator*
Rogers, David Anthony *electrical engineer, educator, researcher*
Stanislao, Joseph *engineering educator, academic administrator, industrial consultant*
Urban, Marek Wojciech *chemist educator, consultant*

Grand Forks
Bolonchuk, William Walter *physical educator*
De Remer, Edgar Dale *aviation educator*
Gjovig, Bruce Quentin *manufacturing consultant*
Honts, Charles Robert *psychology educator*
Hunt, Janet Ross *research nutritionist*
Ljubicic, Blazo *mechanical engineering researcher*
Long, William McMurray *physiology educator*
Messiha, Fathy S *pharmacologist, toxicologist, educator*
Nielsen, Forrest Harold *research nutritionist*
Nordlie, Robert Conrad *biochemistry educator*
Willson, Warrack G. *physical chemist*

Mandan
Halvorson, Gary Alfred *soil scientist*
Reichman, George Albert *soil scientist, educator, consultant*

Minot
Morgan, Rose Marie *biology educator, researcher*
Royer, Ronald Alan *entomologist, educator*

OHIO

Ada
Elliott, Robert Betzel *physician*
Ward, Robert Lee *civil engineering educator*

Akron
Calderon, Nissim *tire and rubber company executive*
Cheng, Stephen Zheng Di *chemistry educator, polymeric material researcher*
Czachura, Kimberly Ann Napua *electrical engineer*
Dwenger, Thomas Andrew *engineer*
Elliott, Jarrell Richard, Jr. *chemical engineer, educator*
Franck, Ardath Amond *psychologist*
Galiatsatos, Vassilios *chemist, educator*
Garczewski, Ronald James *transportation engineer*
Hollis, William Frederick *information scientist*
Holtman, Mark Steven *chemist*
Huckstep, April Yvette *chemist*
Kelley, Frank Nicholas *university dean*
Livigni, Russell A. *polymer chemist*
MacCracken, Mary Jo *physical education educator*
Rukovena, Frank, Jr. *chemical engineer*
Schooley, Arthur Thomas *chemical engineer*
Scott, Mary Ellen Ann *chemist*
Staines, Michael Laurence *oil and gas production executive*
Thayer, Ronda Renee Bayer *chemical engineer*
Uscheek, David Petrovich *chemist*
Verstraete, Mary Clare *biomedical engineering educator*

Alexandria
Palmer, Melville Louis *retired agricultural engineering educator*

Alledonia
Bartsch, David Leo *environmental engineer*

Alliance
Clark, Gregory Alton *research chemist*
Gleixner, Richard Anthony *materials engineer*
Kitto, John Buck, Jr. *mechanical engineer*
Walters, Tracy Wayne *mechanical engineer*

Amelia
Scott, Steven Mike *mechanical engineer*

Ashtabula
Bonner, David Calhoun *chemical company executive*

Athens
Beale, William Taylor *engineering company executive*
Gallaway, Lowell Eugene *economist, educator*
Hedges, Richard H. *epidemiologist*
Irwin, Richard Dennis *electrical engineering educator*
Kordesch, Martin Eric *physicist, educator*
Nance, Richard Damian *geologist*
Nurre, Joseph Henry *engineering educator*
Palmer, Brent David *microanatomy educator, reproductive biologist*
Reed, Michael Alan *scientist*
Skinner, Ray, Jr. *earth science educator, consultant*

Avon Lake
Farkas, Julius *chemist*
Maresca, Louis M. *chemicals executive*
Martin, Christine Kaler *chemist*
Tseng, Hsiung Scott *engineer*

Bay Village
Ferraro, Vincent *mechanical engineer, civil engineer*

Beachwood
Malik, Tariq Mahmood *scientist, chemical engineer*

Beavercreek
Chang, Won Soon *research scientist, mechanical engineer*
Havasy, Charles Kukenis *microelectronics engineer*
Leonard, Thomas Allen *physicist*

Bellbrook
Maneggio, Lizette *civil engineer*

Bellefontaine
Myers, Daniel Lee *manufacturing engineer*
Reames, Spencer Eugene *science educator*

Berea
Little, Richard Allen *mathematics and computer science educator*
Miller, Dennis Dixon *economics educator*

Boardman
Price, William Anthony *psychiatrist*

Bowling Green
Church, Robert Max, Jr. *sales executive*
Guion, Robert Morgan *psychologist, educator*
Hann, William Donald *microbiologist, biology educator*
Midden, William Robert *chemist*
Newman, Elsie Louise *mathematics educator*
Perry, Robert Lee *sociologist, ethnologist*
Smith, Stan Lee *biology educator*
Walker, Daniel Jay *biologist*

Brecksville
Crozier, David Wayne *mechanical engineer*

Brookpark
Gooch, Lawrence Lee *astronautical engineering executive*
Wilson, Jack *aeronautical engineer*

Canal Winchester
Burrier, Gail Warren *physician*

Canton
Arora, Sardari Lal *chemistry educator*
Janson, Richard Wilford *manufacturing company executive*
Maioriello, Richard Patrick *otolaryngologist*
Nisly, Kenneth Eugene *chemical engineer, environmental engineer*

Cederville
Kennel, Elliot Byron *nuclear engineer*

Celina
Fanning, Ronald Heath *architect, engineer*

Centerville
Keating, Tristan Jack *retired aeronautical engineer*

Chagrin Falls
Grasselli, Jeanette Gecsy *university official*
Keyes, Marion Alvah, IV *manufacturing company executive*
Swaney, Cynthia Ann *medical computer service sales executive, business consultant*

Chillicothe
Chen, Wen Fu *otolaryngologist*
Rutledge, Wyman Cy *research physicist*

Cincinnati
Adams, Donald Scott *research scientist, pharmacist*
Albert, Roy Ernest *environmental health educator, researcher*
Albrecht, Helmut Heinrich *medical director*
Anderson, Jerry William, Jr. *technical and business consulting executive, educator*
Arantes, José Carlos *industrial engineer, educator*
Ashley, Kevin Edward *research chemist*
Ball, William James, Jr. *biochemistry educator*
Behnke, Erica Jean *physiologist*
Bosley, David Calvin *design engineering executive*
Braman, Heather Ruth *technical writer, editor, consultant, antiques dealer*
Brand, Larry Milton *biochemist*
Brankamp, Robert George *research biochemist*
Bry, Pierre François *engineering manager*
Buncher, Charles Ralph *epidemiologist, educator*
Burrows, Richard Steven *chemist*
Cabe, Jerry Lynn *mechanical engineer, researcher*
Cahay, Marc Michel *electrical engineer, educator*
Carr, Albert Anthony *organic chemist*
Chin, NeeOo Wong *reproductive endocrinologist*
Cole, Theodore John *osteopathic physician*
Derstadt, Ronald Theodore *health care administrator*
Devitt, John William *physicist*
Dodd, Steven Louis *systems engineer*
Enrico, David Russell *mechanical engineer*
Fairobent, Douglas Kevin *computer programmer*
Fanger, Bradford Otto *biochemist*
Fenoglio-Preiser, Cecilia Mettler *pathologist, educator*
Fody, Edward Paul *pathologist*
Fried, Joel Robert *chemical engineering educator*
Frinak, Sheila Jo *engineer*
Gallagher, Joan Shodder *research immunologist*
Gerner, Frank Matthew *mechanical engineering educator*
Gimpel, Rodney Frederick *chemical engineer*
Graen, George B. *psychologist, researcher*
Günther, Marian W(aclaw) J(an) *theoretical physicist*
Hamilton, Michael Bruce *engineer*
Heimlich, Henry Jay *physician, surgeon*
Heineman, William Richard *chemistry educator*
Hensgen, Herbert Thomas *medical technologist*
Hochstrasser, John Michael *environmental engineer, industrial hygienist*
Hubbard, Arthur Thornton *chemistry educator, electro-surface chemist*
Hurst, Christon James *microbiologist*
Hutchison, Robert B. *chemist*
Kardes, Frank Robert *marketing educator*
Kawahara, Fred Katsumi *research chemist*
Kielb, Robert Evans *propulsion engineer*
Kiser, Thelma Kay *analytical chemist*
Kitzmiller, Karl William *dermatologist*
Kordenbrock, Douglas William *biomedical electronics technician*
Kuchibhotla, Sudhakar *environmental engineer, consultant*
Kupper, Philip Lloyd *chemist*
LaBath, Octave Aaron *mechanical engineer*
Lacy, Mark Edward *computational biologist, systems scientist*
Lakes, Stephen Charles *research chemist, educator*
Leusch, Mark Steven *microbiologist*
Leylek, James H. *aerospace engineer*
Liang, Nong *chemist, researcher*
Luckner, Herman Richard, III *interior designer*
Luffy, Ronald Jon *mechanical engineer*
Lyman, Howard B(urbeck) *psychologist*
Madson, Philip Ward *engineering executive*
Magarill, Simon *optical engineer*
Maxon, Harry Russell, III *nuclear medicine physician*
McCarthy, James Ray *organic chemist*
Meal, Larie *chemistry educator, consultant*
Meese, Ernest Harold *thoracic and cardiovascular surgeon*
Murphy, Eugene F. *aerospace, communications and electronics executive*
Nelson, Sandra Lynn *biochemist*
Newman, Simon Louis *immunologist, educator*
Niemoller, Arthur B. *electrical engineer*
Orkwis, Paul David *aerospace engineering educator*
Pancheri, Eugene Joseph *chemical engineer*
Papadakis, Constantine N. *engineering educator, dean*
Patel, Kishor Manubhai *nuclear medicine physicist*
Pavelic, Zlatko P. *physician, pathologist*
Qian, Yongjia *physicist*
Quo, Phillip C. *mechanical engineering educator*
Reasoner, Donald J. *microbiologist*
Redlinger, Samuel Edward *chemical engineering consultant*
Rickabaugh, Janet Fraley *environmental chemistry educator*
Rosenthal, Susan Leslie *psychologist*
Rouan, Gregory W. *internal medicine physician, educator*
Rubinstein, Jack Herbert *health center administrator, pediatrics educator*
Schiff, Gilbert Martin *virologist, microbiologist, medical educator*
Schubert, William Kuenneth *hospital medical center executive*
Shroff, Ramesh Naginlal *research scientist, physicist*
Smith, Philip Luther *research biochemist*
Solé, Pedro *chemical engineer*
Sperelakis, Nicholas *physiology and biophysics educator, researcher*
Stanifer, Robert Dale *system analyst, manager*
Swaine, Robert Leslie, Jr. *chemist*
Tricoli, James Vincent *cancer genetics educator*
Trimpe, Michael Anthony *forensic scientist*
Ungers, Leslie Joseph *engineering executive*
Warden, Glenn Donald *burn surgeon*
Weisman, Joel *nuclear engineering educator, engineering consultant*
Williams, Trevor Wilson *aeronautical engineering educator*
Wilsey, Philip Arthur *computer scientist*
Wisler, David Charles *aerospace engineer, educator*
Witten, Louis *physics educator*
Young, Daniel Lee *environmental scientist*

Circleville
Cooper, John Edgar, Sr. *research technician*

Cleveland
Abraham, Tonson *chemist, researcher*
Banerjee, Amiya Kumar *biochemist*
Banks, Bruce A. *engineer, physicist, researcher*
Bansal, Narottam Prasad *ceramic research engineer*
Bate, Brian R. *psychologist*
Berridge, Marc Sheldon *chemist, educator*
Berzins, Erna Marija *physician*
Blackwell, John *polymer scientist, educator*
Bloch, Edward Henry *scientist, retired anatomy educator*
Brosilow, Rosalie *engineering editor*
Canterbury, Ronald A. *biologist*
Caplan, Arnold I. *biology educator*
Carek, Gerald Allen *mechanical engineer*
Chamis, Christos Constantinos *aerospace scientist, educator*
Collura, Thomas Francis *biomedical engineer*
Cook, William R., Jr. *chemist*
Coroneos, Rula Mavrakis *mathematician*
Corrado, David Joseph *systems engineer*
Corrigan, Victor Gerard *automotive technology executive*
Dahm, Arnold Jay *physicist, educator*
Davis, Pamela Bowes *pediatric pulmonologist*
Decker, Arthur John *optical physicist, researcher*
DeGuire, Mark Robert *materials scientist, educator*
Dendy, Roger Paul *communications engineer*
Denko, Joanne D. *psychiatrist, writer*
Difiore, Julliann Marie *biomedical engineer*
Duckworth, Donald Reid *oil company executive*
Duffy, Stephen Francis *civil engineer, educator*
Dunbar, Robert Copeland *chemist, educator*
Eagleton, Robert Lee *civil engineer*
Elliott, W(illiam) Crawford *geology researcher*
Ellis, Brenda Lee *mathematician, computer scientist, consultant, educator*
Ernsberger, Paul Roos *research biologist, neuropharmacologist*
Ferguson, Sheila Alease *psychologist, consultant, researcher*
Fernandez, René *aerospace engineer*
Fordyce, James Stuart *federal agency administrator*
Goffman, William *mathematician, educator*
Goldstein, Marvin Emanuel *aerospace scientist, research center administrator*
Goodman, Donald Joseph *dentist*
Haas, Jeffrey Edward *aerospace engineer*
Halford, Gary Ross *aerospace engineer, researcher*
Hancock, James Beaty *interior designer*
Hardy, Richard Allen *mechanical engineer, diesel fuel engine specialist*
Healy, Bernadine P. *physician, educator, federal agency administrator*
Herrington, Daniel Robert *chemist*
Herrup, Karl *neurobiologist*
Hoag, David H. *steel company executive*
Holzbach, Raymond Thomas *gastroenterologist, author, educator*
Hwang, Danny Pang *aerospace engineer*
Ibrahim, Mounir Boshra *mechanical engineering educator*
Jain, Raj Kumar *electrical engineering researcher*
Jarroll, Edward Lee *biology educator*
Kenat, Thomas Arthur *chemical engineer, consultant*
Klineberg, John Michael *federal agency administrator, aerospace researcher*
Kocka, Thomas John *mechanical engineer*
Koenig, Jack L. *chemist, educator*
Krieger, Irvin Mitchell *chemistry educator, consultant*
Lakshmanan, Mark Chandrakant *physiologist, physician*
Madden, James Desmond *forensic engineer*
Martinek, Frank Joseph *chemical company executive*
Maximovich, Michael Joseph *chemist, consultant*
Mazza, A. J. *electrical engineer, consultant*
Mc Henry, Martin Christopher *physician*
Meaney, Thomas Francis *radiologist*
Moore, John James Cunningham *neonatologist*
Mortimer, J. Thomas *biomedical engineering educator*
Myers, Ronald Eugene *research chemist*
Neuger, Sanford *orthodontics educator*
Neuman, Michael Robert *biomedical engineer*
Noebe, Ronald Dean *materials research engineer*
Noneman, Edward E. *engineering executive*
Ostrach, Simon *engineering educator*
Peckham, P. Hunter *biomedical engineer, educator*
Phillips, Stephen Marshall *electrical engineering educator*
Ramsey, Leland Jay *clinical electrical engineer*
Reshotko, Eli *aerospace engineer, educator*
Resnick, Phillip Jacob *psychiatrist*
Robbins, Frederick Chapman *physician, medical school dean emeritus*
Ross, Lawrence John *federal agency administrator*
Ruff, Robert Louis *neurologist, physiology researcher*
Rutledge, Sharon Kay *research engineer*
Savinell, Robert F. *engineering educator*
Scarpa, Antonio *medicine educator, biomedical scientist*
Schloemer, Paul George *diversified manufacturing company executive*
Schlosser, Herbert *theoretical physicist*
Singer, Kenneth David *physicist, educator*
Steinke, Ronald Joseph *aerospace engineer, consultant*
Tarnowski, Kenneth J. *psychologist*

Tinker, H(arold) Burnham *chemical company executive*
Tracht, Allen Eric *electronics executive*
Trefts, Albert S. *mechanical engineer, consultant*
Varley, John Owen *engineer*
Wang, Shi-Qing *physicist*
Webb, John Allen, Jr. *engineering executive*
Westlock, Jeannine Marie *health care consultant*
Wish, Jay Barry *nephrologist, specialist*
Wolfe, Lowell Emerson *space scientist*
Woyczynski, Wojbor Andrzej *mathematician, educator*
Yurko, Joseph Andrew *chemical engineer*
Zier, J(ohn) Albert *chemical engineer*

Columbiana
Richman, John Emmett *architect*

Columbus
Adams, David Parrish *historian, educator*
Akbar, Sheikh Ali *materials science and engineering educator*
Aldemir, Tunc *nuclear engineering educator*
Alexander, Carl Albert *ceramic engineer*
Alouani, Mebarek *physicist, research specialist*
Amato, Vincent Edward *geotechnical engineer, consultant*
Apel, John Paul *civil engineer*
Araki, Takaharu *editor*
Bai, Sungchul Charles *nutritionist*
Baker, Gregory Richard *mathematician*
Balcerzak, Stanley Paul *physician, educator*
Bechtel, Stephen E. *mechanical engineer, educator*
Bedford, Keith Wilson *civil engineering and atmospheric science educator*
Behrman, Edward Joseph *biochemistry educator*
Bernays, Peter Michael *retired chemical editor*
Berntson, Gary Glen *psychiatry, psychology and pediatrics educator*
Bhushan, Bharat *mechanical engineer*
Bianchine, Joseph Raymond *pharmacologist*
Bondurant, Byron Lee *agricultural engineering educator*
Buhac, H(rvoje) Joseph *geotechnical engineer*
Busick, Robert James *computer scientist*
Camboni, Silvana Maria *environmental sociologist*
Capen, Charles Chabert *veterinary pathology educator, researcher*
Carlton, Robert L. *clinical psychologist*
Chen, Roger Ko-chung *electronics executive*
Chloupek, Frank Ray *physicist*
Chovan, John David *biomedical engineer*
Christoforidis, A. John *radiologist, educator*
Clymer, Bradley Dean *electrical engineering educator*
Cosens, Kenneth Wayne *sanitary engineer, educator*
Covault, Lloyd R., Jr. *hospital administrator, psychiatrist*
Crenshaw, Michael Douglas *chemist, researcher*
Cruz, Jose Bejar, Jr. *electrical engineering educator, dean*
Culver, David Alan *aquatic ecology educator*
Daehn, Glenn Steven *materials scientist*
D'Amato, Anthony Salvatore *food products company executive*
Davis, June Leah *psychologist*
Davis, Keith Robert *plant biology educator*
Ensminger, Dale *mechanical engineer, electrical engineer*
Evans, Michael Leigh *physiologist*
Fass, Robert J. *epidemiologist, academic administrator*
Feller, Dennis Rudolph *pharmacology educator*
Fentiman, Audeen Walters *nuclear engineer, educator*
Fraser, Jane Marian *operations research specialist, educator*
Fryczkowski, Andrzej Witold *ophthalmologist, educator, business executive*
Furste, Wesley Leonard, II *surgeon, educator*
Gatewood, Buford Echols *retired educator, aeronautical and astronautical engineer*
Gehring, Richard Webster *structural engineer*
Golightly, Danold Wayne *chemist*
Goodman, Hubert Thorman *psychiatrist, consultant*
Greenlee, Kenneth William *chemical consultant*
Guezennec, Yann Guillaume *mechanical engineering educator*
Gupta, Prabhat Kumar *materials scientist*
Hayton, William Leroy *pharmacology educator*
Hill, Jack Douglas *electrical engineer*
Hilliard, Kirk Loveland, Jr. *osteopathic physician, educator*
Hom, Theresa Maria *osteopathic physician*
Houser, Donald Russell *mechanical engineering educator, consultant*
Howden, David Gordon *engineering educator*
Huang, Jason Jianzhong *ceramic engineer*
Jackson, Curtis Maitland *metallurgical engineer*
Jackson, Michel Tah-Tung *phonetician, linguist*
Janus, Mark David *priest, psychologist, researcher, consulant*
Jezek, Kenneth Charles *geophysicist, educator, researcher*
Johnson, Lee Frederick *molecular geneticist*
Jolly, Daniel Ehs *dental educator*
Kannel, Jerrold Williams *mechanical engineer*
Kapadia, Mehernosh Minocheher *engineering executive*
Kaplan, Paul Elias *physiatrist, educator*
Kapral, Frank Albert *medical microbiology and immunology educator*
Kipp, Carl Robert *mechanical engineer*
Kolattukudy, Pappachan Ettoop *biochemist, educator*
Langley, Teddy Lee *engineering executive*
Leissa, Arthur William *mechanical engineering educator*
Levine, Zachary Howard *physicist*
Love, Daniel Michael *manufacturing engineer*
Lytle, John Arden *chemical engineer*
Marshall, Alan George *chemistry and biochemistry educator*
McNulty, Frank John *laboratory coordinator*
McSweeny, Paul Edward *research technologist*
Meites, Samuel *clinical chemist, educator*
Merritt, Joy Ellen *chemist, editor*
Miller, Don Wilson *nuclear engineering educator*
Miller, Terry Alan *chemistry educator*
Mills, Robert Laurence *physicist, educator*
Morrow, Grant, III *geneticist*
Nahar, Sultana Nurun *research physicist*
Namboodiri, Krishnan *sociology educator*
Nelson, Gordon Leon *agricultural engineering educator*
Olesen, Douglas Eugene *research institute executive*
Olsen, Richard George *microbiology educator*
Orin, David Edward *electrical engineer*
Ozkan, Umit Sivrioglu *chemical engineering educator*
Parsons, Donald Oscar *economics educator*
Pathak, Dev S. *pharmaceutical administrator, marketing educator*
Peterle, Tony John *zoologist, educator*
Peters, Leon, Jr. *electrical engineering educator, research administ*
Petty, Richard Edward *psychologist, educator, researcher*
Pfau, Richard Olin *forensic chemist, forensic science educator*
Platika, Doros *neurologist*
Rapp, Robert Anthony *metallurgical engineering educator, consultant*
Redmond, Robert Francis *nuclear engineering educator*
Reece, Robert William *zoological park administrator*
Relle, Ferenc Matyas *chemist*
Richards, Ernest William *clinical nutritionist*
Rokhlin, Stanislav Iosef *engineering educator*
Ruan, Ju-Ai *physics and tribology researcher*
Rubin, Karl Cooper *mathematician*
Ruggles, Harvey Richard *civil engineer*
St. Pierre, George Roland, Jr. *materials science and engineering administrator, educator*
Schmalbrock, Petra *medical physicist, researcher, educator*
Schwab, Glenn Orville *retired agricultural engineering educator, consultant*
Serraglio, Mario *architect*
Shannon, Larry James *civil engineer*
Sherman, William Michael *physiology educator, researcher*
Shupe, Lloyd Merle *chemist, consultant*
Singer, Sherwin Jeffrey *theoretical chemist, chemistry educator*
Singh, Rajendra *mechanical engineering educator*
Skidmore, Duane Richard *chemical engineering educator, researcher*
Skiest, Eugene Norman *food company executive*
Slonim, Arnold Robert *biochemist, physiologist*
Stage, Richard Lee *utilities executive*
Stephens, Sheryl Lynne *family practice physician*
Stuckert, Gregory Kent *mechanical engineer, researcher*
Sweeney, Thomas Leonard *chemical engineering educator, researcher*
Taiganides, E. Paul *agricultural-environmental engineer, consultant*
Taylor, Celianna I. *information systems specialist*
Toth, Karoly Charles *electrical engineer, researcher*
Turchi, Peter John *aerospace and electrical engineer, educator, scientist*
Tzagournis, Manuel *physician, educator, university dean and official*
Velagaleti, Ranga Rao *agronomist, environmental scientist*
Waldron, Kenneth John *mechanical engineering educator, researcher*
Wali, Mohan Kishen *environmental science and natural resources educator*
Weeks, Thomas J. *chemist*
Whitford, Philip Clason *biology educator*
Wiechel, John F. *mechanical engineer*
Wu, Chao-Min *biomedical engineer*
Zartman, David Lester *dairy science educator, researcher*
Zipf, William Byron *pediatric endocrinologist, educator*
Zuspan, Frederick Paul *obstetrician/gynecologist, educator*

Dayton
Addison, Wallace Lloyd *aerospace engineer*
Arn, Kenneth Dale *physician, city official*
Ballal, Dilip Ramchandra *mechanical engineering educator*
Beam, Jerry Edward *electrical engineer*
Blommel, Scot Anthony *environmental engineer*
Brand, Vance Devoe *astronaut, government official*
Byczkowski, Janusz Zbigniew *toxicologist*
Carson, Richard McKee *chemical engineer*
Cartmell, James V. *research and development executive*
Chan, Yupo *engineer*
Chang, Jae Chan *internist, educator*
Christensen, Julien Martin *psychologist, educator*
Chuck, Leon *materials scientist*
Donaldson, Steven Lee *materials research engineer*
Dugan, John Patrick *optical engineer*
Emrick, Donald Day *chemist, consultant*
Fulton, Darrell Nelson *infosystems specialist*
Gray, James Randolph, III *air force officer*
Gregor, Clunie Bryan *geology educator*
Heneghan, Shawn Patrick *research scientist, educator*
Henley, Terry Lew *computer company executive*
Jain, Surinder Mohan *electronics engineering educator*
Jones, Hobert W *health physicist*
Khalimsky, Efim *mathematics and computer science educator*
Krug, Maurice F. *engineering company executive*
Kumar, Binod *materials engineer, educator*
Loughran, Gerard Andrew *chemistry consultant, polymer scientist*
Manasreh, Omar M. *electronics engineer, physicist*
Mathews, David *foundation executive*
Mehta, Rajendra *chemist, researcher*
Muir, Herman Stanley, III *lawyer*
Murphy, Martin Joseph, Jr. *cancer research center executive*
Nixon, Charles William *bioacoustician*
Premus, Robert *economics educator*
Ramalingam, Mysore Loganathan *mechanical engineer*
Reichel, Lee Elmer *mechanical engineer*
Repperger, Daniel William *electrical engineer*
Rowe, Joseph Everett *electrical engineering educator, administrator*
Saliba, Joseph Elias *civil engineering educator*
Saxer, Richard Karl *metallurgical engineer, retired air force officer*
Shaffer, Jill *clinical psychologist*
Spicer, John Austin *physicist*
Standley, Paul Melvin *chemist*
Takahashi, Fumiaki *research mechanical engineer*
Tighe-Moore, Barbara Jeanne *electronics executive*
Tsonis, Panagiotis Antonios *biologist, researcher*
Vander Wiel, Kenneth Carlton *computer services company executive*
von Ohain, Hans Joachim P. *aerospace scientist*
Warden, Gary George *computer engineer*
West, Johnny Carl *aeronautical engineer*
Wise, Joseph Francis *electrical engineer, consultant*

Delaware
Burtt, Edward Howland, Jr. *ornithologist, natural history educator*

Dublin
Blakley, Brent Alan *polymer chemist, computer programmer*
Lamp, Benson J. *tractor company executive*
Mueller, Donald Scott *chemist*
Notowidigdo, Musinggih Hartoko *information systems executive*
Spies, Phyllis Bova *information services company executive*
Viezer, Timothy Wayne *economist*

Elmore
Kaczynski, Don *metallurgical engineer*

Elyria
Mahjoub, Elisabeth Mueller *health facility administrator*

Enon
Cyphers, Daniel Clarence *aerospace engineer*

Fairborn
Conklin, Robert Eugene *electronics engineer*
Moore, Edmund Harvey *materials science and engineering engineer*
Workman, John Mitchell *chemist*

Findlay
Zeyen, Richard Leo, III *analytical chemist, educator*

Fostoria
Curlis, David Alan *civil engineer*

Fremont
Laurer, Timothy James *electrical engineer, consultant*
Sattler, Nancy Joan *math and physical science educator*

Galion
Reisner, Andrew Douglas *psychologist, chief clinical officer*

Girard
German, Norton Isaiah *pathologist, educator*

Granville
Carr, Thomas Michael *analytical chemist*
Choudhary, Manoj Kumar *chemical engineer, researcher*
Grant, Roderick McLellan, Jr. *physicist, educator*
Hare, Keith William *computer consultant*
Ploetz, Lawrence Jeffrey *ceramic engineer*

Greenfield
Jenkins, James William *osteopath*

Grove City
Blazic, Martin Louis *manufacturing engineer*
Leybovich, Alexander Yevgeny *electrophysicist, engineer*

Hamilton
Fulero, Solomon M. *psychologist, educator, lawyer*
Hergert, David Joseph *engineering educator*
Johnson, Pauline Benge *nurse, anesthetist*

Hartville
Miday, Stephen Paul *quality assurance engineer*

Highland Heights
Pavlovich, Donald *technical writer*

Hinckley
Sarbach, Donald Victor *retired chemist*

Huber Heights
Strong, Richard Allen *environmental safety engineer*

Hudson
Carman, Charles Jerry *chemical company executive*
Kempe, Robert Aron *venture management executive*
Kirchner, James William *electrical engineer*

Independence
Emling, William Harold *metallurgical engineer*
Evans, George Frederick *consulting engineer*
Purcell, Thomas Owen, Jr. *chemical company executive*
Toth, James Michael *electronics engineer*

Ironton
Mitchell, Maurice McClellan, Jr. *chemist*

Kent
Cooke, G. Dennis *biological science educator*
Cooperrider, Tom Smith *botanist*
Kahana, David Ewan *physicist*
Li, Zili *research physicist*
McKee, David Lannen *economics educator*
Palffy-Muhoray, Peter *physicist, educator*
Stevenson, J. Ross *biological sciences educator, researcher*

Kettering
Haas, Trice Walter *researcher*
Mantil, Joseph Chacko *nuclear medicine physician, researcher*

Kinsman
Alfonsi, William E. *interior designer, funeral industry consultant*

Lima
Johnson, Patricia Lyn *mathematics educator*

Logan
Frost, Jack Martin *civil engineer*

Loveland
Anderson, Roy Alan *chemical engineer*
Masters, Ron Anthony *research chemist*

Madison
Spiesman, Benjamin Lewis *mechanical design engineer*

Mansfield
Gregory, Thomas Bradford *mathematics educator*

Marietta
Chase, Robert William *petroleum engineering educator, consultant*
Goyle, Rajinder Kumar *civil engineer, consultant*
Kirkland, Ned Matthews *chemical engineering manager*
Putnam, Robert Ervin *chemistry educator*
Tipton, Jon Paul *allergist*

Marion
Lim, Shun Ping *cardiologist*
Youll, Peter Jerome *environmental chemist*

Massillon
Hatzilabrou, Labros *mechanical engineer, metallurgical engineer*

Materials Park
Wasch, Allan Denzel *technical publication editor*

Maumee
Mohler, Terence John *psychologist*
Witte, John Sterling *solar energy contracting company executive*

Mayfield Heights
Fatemi, Seyyed Hossein *cell biologist, physician*

Melmore
Cox, James Grady *chemist*

Miamisburg
Attalla, Albert *chemist*
Hughes, Thomas William *technical company executive*
Nease, Allan Bruce *research chemist*
Peterson, George P. *mechanical engineer, research and development firm executive*
Terezakis, Terry Nicholas *power/utility engineer*

Middletown
Easley, Michael Wayne *public health professional*
Sabata, Ashok *materials engineer*

Milan
Henry, Joseph Patrick *chemical company executive*

Milford
Green, David Richard *development chemist*
Kinstle, James Francis *polymer scientist*
Klosterman, Albert Leonard *technical development business executive, mechanical engineer*

Mount Vernon
Schaub, Fred S. *mechanical engineer*

Newark
Green, John David *engineering executive*
Swope, Robert J. *physical science laboratory administrator*
Tiburcio, Astrophel Castillo *polymer chemist*

Newbury
Kirman, Lyle Edward *chemist, engineer, consultant*

North Canton
George, Donald James *architecture educator, administrator*

North Olmsted
Galysh, Robert Alan *information systems analyst*
Puhk, Heino *chemist, researcher*

North Royalton
Mohler, Stanley Ross, Jr. *aeronautical research engineer*

Norwalk
Gutowicz, Matthew Francis, Jr. *radiologist*

Norwich
Ely, Wayne Harrison *broadcast engineer*

Oberlin
Ricker, Alison Scott *science librarian*
Schmucker, Bruce Owen *geotechnical engineer*

Ostrander
Smith, Rick A. *mechanical engineer, consultant*

Oxford
Danielson, Neil David *chemistry educator*
Eshbaugh, W(illiam) Hardy *botanist, educator*
Haley-Oliphant, Ann Elizabeth *science educator*
Harris, Yvette Renee *psychologist, educator*
Pfohl, Dawn Gertrude *laboratory executive*
Powell, Martha Jane *botany educator*
Willeke, Gene E. *environmental engineer, educator*

Painesville
Scozzie, James Anthony *chemist*

Parma
Laughlin, Ethelreda Ross *chemistry educator*

Paulding
Riggenbach, Duane Lee *maintenance engineer*

Pickerington
Blackman, Edwin Jackson *software engineer*

Piketon
Patton, Finis S., Jr. *nuclear chemist*

Piqua
Anderson, Christine Lee *analytical chemist*
Nagarajan, Sundaram *metallurgical engineer*

Portsmouth
Hamilton, Virginia Mae *mathematics educator, consultant*
Hubler, H. Clark *writer, retired educator*
Rudolph, Brian Albert *computer science educator*

Powell
Adeli, Hojjat *civil engineer, computer scientist, educator*
Manchester, Carol Ann Freshwater *psychologist*

Ripley
McMillan, James Albert *electronics engineer, educator*

Rootstown
Chopko, Bohdan Wolodymyr *neuroscientist*
Maron, Michael Brent *physiologist*
Westerman, Philip William *biomedical researcher, medical educator*

Ross
Nelson, Dennis George Anthony *dental researcher, life scientist*

Shaker Heights
Blue, Reginald C. *psychologist, educator*

Sidney
Schumann, Stanley Paul *mechanical engineer*

Solon
Ambrose, Robert Micheal *application engineer, consultant*
Williams, Peter Charles *engineer*

Springboro
Dawson, Brian Robert *chemist, environmentalist*

Springfield
Hobbs, Horton Holcombe, III *biology educator*
Orhon, Necdet Kadri *physician*
Ryu, Kyoo-Hai Lee *physiologist*

Strongsville
Opplt, Jan Jiri *pathologist, educator*
Schroeder, Stanley Brian *chemist, coating application engineer*

Sylvania
Kurek, Dolores Bodnar *physical science and mathematics educator*

Tiffin
Baker, David B. *environmental scientist*
Einsel, David William, Jr. *consultant, retired army officer*

Tipp City
Glassmeyer, James Milton *aerospace and electronics engineer*
Klimkowski, Robert John *photo reproduction process technical executive*

Toledo
Bagley, Brian G. *materials science educator, researcher*
Barrett, Michael John *anesthesiologist*
Benham, Linda Sue *civil engineer*
Chrysochoos, John *chemistry educator*
Como, Francis W. *plastics manufacturing executive*
Duran, Emilio *molecular biologist*
Edwards, Jimmie Garvin *chemistry educator, consultant*
Gerhardinger, Peter F. *engineer*
Gibson, Raymond Novarro *computer programmer, analyst*
Guo, Hua *chemist, educator*
Harris, James Herman *pathologist*
Hood, Douglas Crary *electronics educator*
Jankun, Jerzy Witold *biochemist*
Keith, Theo Gordon, Jr. *mechanical engineering educator*
Lohmann, John J. *quality assurance engineer*
McSweeney, Austin John *psychology educator, researcher*
Munger, Harold Charles *architect*
Pansky, Ben *anatomy educator, science researcher*
Randolph, Brian Walter *civil engineer, educator*
Reese, Herschel Henry *mechanical engineer*
Riseley, Martha Suzannah Heater (Mrs. Charles Riseley) *psychologist, educator*
Rubin, Allan Maier *physician, surgeon*
Skrzypczak-Jankun, Ewa *crystallographer*
Stoner, Gary David *pathology educator*
Taylor, Douglas Floyd *mechanical engineer*
Waldfogel, LaRue Verl *electrical engineering executive*
Zhou, Theodore Xi *physicist*

University Heights
Skolnik, Leonard *chemical engineer*

Wakeman
Arhar, Joseph Ronald *chemist*

Warren
Alli, Richard James, Sr. *electronics executive, service executive*
Zeuner, Raymond Alfred *manufacturing engineer*
Zimmerman, Doris Lucile *chemist*

Westchester
Kosti, Carl Michael *dentist, researcher*

Westerville
Bilisoly, Roger Sessa *statistician*

Westlake
Huff, Ronald Garland *mechanical engineer*
Kroll, Casimer V. *engineering executive*
Sminchak, David William *civil engineer, consultant*

Wickliffe
Bares, William G. *chemical company executive*
Dunn, Horton, Jr. *organic chemist*
Kornbrekke, Ralph Erik *colloid chemist*

Willoughby
Hassell, Peter Albert *electrical and metallurgical engineer*
Sieglaff, Charles Lewis *chemist*
Stevens, William Frederick, III *software engineer*

Wooster
Ferree, David Curtis *horticultural researcher*
Geho, Walter Blair *biomedical research executive*
Madden, Laurence Vincent *plant pathology educator*
Shafer, Berman Joseph *oil company executive*

Worthington
Elgin, Charles Robert *chief technology officer*

Wright Patterson AFB
Blake, William Bruce *aerospace engineer*
Bowman, William Jerry *air force officer*
Eisenhauer, William Joseph, Jr. *aerospace engineer*
Goesch, William Holbrook *aeronautical engineer*
Henry, Leanne Joan *physicist*
Knoedler, Andrew James *aerospace engineer*
Mehalic, Mark Andrew *electrical engineer, air force officer*
Moore, Thomas Joseph *research psychologist*
Myers, Jerry Alan *computer engineer*
Ruscello, Anthony *aerospace engineer*
Schell, Allan Carter *electrical engineer*
Sellin, M. Derek *aerospace engineer*
Wallace, Robert Luther, II *engineer*
Wilson, Mark Kent *aerospace engineer*

Wyoming
Jarzembski, William Bernard *biomedical engineer*

Xenia
Fussichen, Kenneth *computer scientist*

Yellow Springs
Schmidt, William Joseph *engineering executive*
Spokane, Robert Bruce *biophysical chemist*

Youngstown
Bearer, Cynthia Frances *neonatologist*
Mastriana, Robert Alan *architect*
McLennan, Donald Elmore *physicist*
Stahl, Joel Sol *plastic-chemical engineer*
Suchora, Daniel Henry *mechanical engineering educator, consultant*
Varma, Raj Narayan *nutrition educator, researcher*

Zanesville
Ray, John Walker *otolaryngologist, educator*
Wilcox, Roger Clark *psychologist, researcher*

OKLAHOMA

Ada
Stephens, David Basil *civil engineer*
Thompson, Rahmona Ann *plant taxonomist*
Van Burkleo, Bill Ben *osteopath, emergency physician*

Ardmore
Olsen, Thomas William *geologist*
Patterson, Manford K(enneth), Jr. *foundation administrator, researcher, scientist*

Bartlesville
Beever, William Herbert *chemist*
Brannan, Michael Steven *civil, environmental, automotive engineer*
Byers, Jim Don *research chemist*
Clay, Harris Aubrey *chemical engineer*
Gao, Hong Wen *chemical engineer*
Gill, Anna Margherita Anya *application specialist*
Guillory, Jack Paul *chemist, researcher*
Hankinson, Risdon William *chemical engineer*
Hunt, Harold Ray *chemical engineer*
Lee, Fu-Ming *chemical engineer*
Lew, Lawrence Edward *chemical engineer*
Mihm, John Clifford *chemical engineer*
Moczygemba, George Anthony *research chemist*
Risley, Allyn W(ayne) *petroleum engineer, manager*
Rutledge, Kathleen Pillsbury *sensory scientist, researcher*
Zecchini, Edward John *chemical engineer, researcher*

Bethany
Arnold, Donald Smith *chemical engineer, consultant*
Reinschmiedt, Anne Tierney *nurse, health care executive*

Chickasha
Rienne, Dozie Ignatius *technologist*

Claremore
Kelly, Vincent Michael, Jr. *orthodontist*

Duncan
King, Randall Kent *mechanical engineer*
Womack, Terry Dean *electrical engineer*

Durant
Craige, Danny Dwaine *dentist*

Edmond
Bass, Thomas David *biology educator*
Hanson-Painton, Olivia Lou *biochemist, educator*
Radke, William John *biology educator*

Enid
Ward, Llewellyn O(rcutt), III *oil producer*

Fort Sill
Evans, Paul *osteopath*

Fort Towson
Pike, Thomas Harrison *plant chemist*

Guthrie
Bell, Thomas Eugene *psychologist, educational administrator*

Healdton
Eck, Kenneth Frank *pharmacist*

Jenks
Leming, W(illiam) Vaughn *electronics engineer*
Wootan, Gerald Don *osteopathic physician, educator*

Keota
Davis, Thomas Pinkney *mathematics educator*

Lane
Edelson, Jonathan Victor *entomologist*

Lawton
Webb, O(rville) Lynn *physician, pharmacologist, educator*

Mcalester
Alles, Rodney Neal, Sr. *information management executive*
McSherry, Frank D(avid), Jr. *writer, editor*

McLoud
Whinery, Michael Albert *physician*

Moore
Moore, Dalton, Jr. *petroleum engineer*

Muskogee
Dandridge, William Shelton *orthopedic surgeon*
Hatley, Larry J. *plant manager*
Kent, Bartis Milton *physician*
Washington, Allen Reed *chemist*

Norman
Allen, Jonathan Dean *chemist, researcher*
Applegarth, Ronald Wilbert *engineer*
Bemben, Michael George *exercise physiologist*
Bert, Charles Wesley *mechanical and aerospace engineer, educator*
Ciereszko, Leon Stanley *chemistry educator*
Cronenwett, William Treadwell *electrical engineering educator, consultant*
Dille, John Robert *physician*
Egle, Davis Max *mechanical engineering educator*
Eilts, Michael Dean *research meteorologist, manager*
Everett, Jess Walter *environmental engineering educator, researcher*
Gal-Chen, Tzvi *geophysicist*
Gronlund, Scott Douglas *psychology educator, researcher*
Hutchison, Victor Hobbs *biologist, educator*
Lamb, Peter James *meteorology educator, researcher, consultant*
Lyberopoulos, Athanasios Nikolaos *mathematician, engineer*
Maddox, Robert Alan *atmospheric scientist*
Mankin, Charles John *geology educator*
McClanahan, Walter Val *psychologist*
Menzie, Donald E. *petroleum engineer, educator*
Nicewander, Walter Alan *psychology educator*
O'Rear, Edgar Allen, III *chemical engineering educator*
Ortiz-Leduc, William *plant biologist, educator, researcher*
Stoltenberg, Cal Dale *psychology educator*
Zaman, Musharraf *civil engineering educator*
Zelby, Leon Wolf *electrical engineering educator, consulting engineer*

Nowata
Osborn, Ann George *retired chemist*

Oklahoma City
Abernathy, Jack Harvey *petroleum, utility company and banking executive*
Alaupovic, Petar *biochemist, educator*
Allen, James Harmon, Jr. *civil engineer*
Anderson, Kenneth Edwin *technical writer*
Barber, Susan Carrol *biology educator*
Blackwell, John Adrian, Jr. *computer company executive*
Bourdeau, James Edward *nephrologist, researcher*
Bradford, Reagan Howard, Jr. *ophthalmology educator*
Branch, John Curtis *biology educator, lawyer*
Chen, Wei R. *physics educator*
Dones, Maria Margarita *anatomist, educator*
D'Souza, Maximian Felix *medical physicist*
England, Gary Alan *television meteorologist*
Everett, Royice Bert *obstetrician/gynecologist*
Gavaler, Judith Ann Stohr Van Thiel *epidemiologist*
Hamm, Robert MacGowan *psychologist*
Hampton, James Wilburn *hematologist, medical oncologist*
Hough, Jack Van Doren *otologist*
Jones, Jeffery Lynn *software engineer*
Jones, Renee Kauerauf *health care administrator*
Kimerer, Neil Banard, Sr. *psychiatrist, educator*
McClintic, George Vance, III *petroleum engineer, real estate broker*
Miller, Herbert Dell *petroleum engineer*
Morgan, Robert Steve *mechanical engineer*
Muchmore, John Stephen *endocrinologist*
Oehlert, William Herbert, Jr. *cardiologist, educator*
O'Keeffe, Hugh Williams *oil industry executive*
Parke, David W., II *ophthalmologist, educator, healthcare executive*
Parker, John R. *physician, radiologist*
Prasad, B.H. *utility company executive*
Robison, Clarence, Jr. *surgeon*
Rohrig, Timothy Patrick *toxicologist, educator*
Rossavik, Ivar Kristian *obstetrician/gynecologist*
Sanders, Gilbert Otis *educational and research psychologist, addictions treatment therapist, consultant, educator*
Snow, Clyde Collins *anthropologist*
Thompson, Ann Marie *neuroscientist, researcher, educator*
Thompson, Guy Thomas *safety engineer*
Thurman, William Gentry *medical research foundation executive, pediatric hematology and oncology physician, educator*
Wallis, Robert Joe *pharmacist, retail executive*
Watson, Steven Edward *family physician*
Zenieris, Petros Efstratios *civil engineer*
Zuhdi, Nazih *surgeon*

Oologah
Knight, Gary Charles *mechanical engineer*

Ponca City
Cole, Jack Howard *mechanical engineer*
Lewald, Peter Andrew *process engineer*
Oster, Pamela Ann *radiologic technologist*
Wann, Laymond Doyle *retired petroleum research scientist*

Sand Springs
Atiyeh, Elia Mtanos *civil engineer*

Sapulpa
Powers, Eldon Nathaniel *data processing executive*
Welcher, Ronnie Dean *waste management and environmental services executive*

Stillwater
Browning, Charles Benton *university dean, agricultural educator*
Brusewitz, Gerald Henry *agricultural engineering educator, researcher*
Buck, Richard Forde *physicist*
Campbell, John Roy *animal scientist educator, academic administrator*
Davis, Gordon Dale, II *biochemist*
Faulkner, Lloyd C. *veterinary medicine educator*
Foldvari, Istvan *physicist*
Foutch, Gary Lynn *chemical engineering educator*
Fox, Joseph Carl *veterinary medicine educator, researcher, parasitologist*
Grischkowsky, Daniel Richard *research scientist*
Haan, Charles Thomas *agricultural engineering educator*
Holt, Elizabeth Manners *chemistry educator*
Joshi, R. Malatesha *reading education educator*
Kennedy, Douglas Wayne *physicist*
Kocan, Katherine Mautz *veterinary educator, researcher*
Lucca, Don Anthony *mechanical engineer*
Martin, Joel Jerome *physics educator*
Minahen, Timothy Malcolm *engineering educator*
Misawa, Eduardo Akira *mechanical engineer, educator*
Mize, Joe Henry *industrial engineer, educator*
Noyes, Ronald T. *agricultural engineering educator*
Pennington, Rodney Edward *molecular biologist*
Poole, Richard William *economics educator*
Qualls, Charles Wayne, Jr. *veterinary pathology educator*
Spivey, Howard Olin *biochemistry and physical chemistry educator*
Thompson, David Russell *agricultural engineering educator, academic dean*
Tree, David Alan *chemical engineering educator*
Whaley, Max Weldon *chemist*
Zwerneman, Farrel Jon *engineering educator*

Tahlequah
Wickham, M(arvin) Gary *optometry educator*

Tecumseh
Lowe, Ronald Dean *aerospace engineer, consultant*

Tinker AFB
Pray, Donald George *aerospace engineer*

Tulsa
Achterberg, Ernest Reginald *mining engineer*
Alexander, John Robert *hospital administrator, internist*
Banik, Niranjan Chandra-Dutta *physicist, researcher*
Banks, James Daniel *chemical engineer*
Bennett, Curtis Owen *research engineer*
Bennison, Allan Parnell *geological consultant*
Blais, Roger Nathaniel *physics educator*
Blanton, Roger Edmund *mechanical engineer*
Brenner, George Marvin *pharmacologist*
Chin, Alexander Foster *electronics educator*
Cobbs, James Harold *engineer, consultant*
Dix, Fred Andrew, Jr. *professional society executive*
Duncan, Lewis Mannan, III *physicist, education administrator*
Earlougher, Robert Charles, Sr. *petroleum engineer*
Greer, Clayton Andrew *electrical engineer*
Hall, George Joseph, Jr. *geologist, educator, consultant, geotechnical engineer*
Härtel, Charmine Emma Jean *industrial/ organizational psychology educator, consultant*
Hensley, Jarvis Alan *petroleum engineer*
Johnson, Gerald, III *cardiovascular physiologist, researcher*
Kalbfleisch, John McDowell *cardiologist, educator*
Kemp, Sarah (Sally Leech) *neurodevelopment specialist*
King, Joseph Willet *child psychiatrist*
Leder, Frederic *chemical engineer*
Lewis, Ceylon Smith, Jr. *physician*
Mabry, Samuel Stewart *petroleum engineer, consultant*
McCaw, Valerie Sue *civil engineer*
McCullough, Robert Dale, II *osteopath*
Mersch, Carol Linda *information systems specialist*
Meyerhoff, Arthur Augustus *geologist, consultant*
Mitchell, Donald Hearin *computer scientist*
Nebergall, Robert William *orthopedic surgeon, educator*
Nutter, Dale E. *mechanical engineer*
Okada, Robert Dean *cardiologist*
Quillin, Patrick *nutritionist, writer*
Robinson, Enders Anthony *geophysics educator, writer*
Rotenberg, Don Harris *chemist*
Schmidt, Gregory Martin *osteopathic physician, independent oil producer*
Smith, Vernon Soruix *neonatologist, pediatrician, educator*
Strafuss, David Louis *mechanical engineer*
Thompson, Carla Jo Horn *mathematics educator*
Tubbs, David Eugene *mechanical engineer, marketing professional*
Warren, Tommy Melvin *petroleum engineer*
Williford, Richard Allen *oil company executive, flight safety company executive*
Young, William Dean *psychologist*

Vinita
Beavers, Roy L. *utility executive*
Neer, Charles Sumner, II *orthopaedic surgeon, educator*

Waynoka
Olson, Rex Melton *oil and gas company executive*

Weatherford
Grant, Peter Michael *biologist, educator*
Nail, Paul Reid *psychology educator*
Woods, John Merle *mathematics educator, chairman*

Woodward
Boucher, Darrell A., Jr. *vocational education educator*

Yukon
Huynh, Nam Hoang *physics educator*

OREGON

Albany
Norman, E. Gladys *business computer educator, consultant*
Yau, Te-Lin *corrosion engineer*
Yu, Kitson Szewai *computer science educator*

Aloha
Rojhantalab, Hossein Mohammad *chemical engineer, researcher*
Willis, Lawrence Jack *chemist*

Ashland
Ferrero, Thomas Paul *engineering geologist*
Goddard, Kenneth William *forensic scientist*
Grover, James Robb *retired chemist, editor*

Bandon
Lindquist, Louis William *artist, writer*

Beaverton
Allen, Paul C. *physicist*
Cipale, Joseph Michael *software engineer*
Cole, Samuel Joseph *computational chemist*
Critchlow, B. Vaughn *research facility administrator, researcher*
Gerlach, Robert Louis *research and development executive, physicist*
Haluska, George Joseph *biomedical scientist*
Hughes, Laurel Ellen *psychologist, educator, writer*
Montagna, William *scientist*
Murdock, Bruce *physicist*
Purvis, George Dewey, III *computational chemist*

Bend
Riegel, Gregg Mason *ecologist, researcher*

Charleston
Shapiro, Lynda P. *biology educator, director*

Corvallis
Arp, Daniel J. *biochemistry educator*
Bernieri, Frank John *social psychology educator*
Byrne, John Vincent *academic administrator*
Caldwell, Douglas Ray *oceanographer, educator*
Chambers, Kenton Lee *botany educator*
Chen, Chaur-Fong *civil engineer, educator*
Dixon, Robert Keith *plant physiologist, researcher*
Farkas, Daniel Frederick *food science and technology educator*
Fontana, Peter Robert *physics educator*
Ford, Mary Spencer (Jesse) *ecologist, writer*
Fuchigami, Leslie Hirao *horticulturist, researcher*
Gillis, John Simon *psychologist, educator*
Goodnick, Stephen Marshall *electrical engineer, educator*
Hashimoto, Andrew Ginji *bioresource engineer*
Ingham, Elaine Ruth *ecology educator*
Johnston, Larea Dennis *taxonomist*
Krantz, Gerald William *entomology educator*
Laver, Murray Lane *chemist, educator*
Liegel, Leon Herman *soil scientist, forester*
Liston, Aaron Irving *botanist*
Lubchenco, Jane *marine biologist, educator*
Mason, Robert Thomas *zoologist*
Miner, John Ronald *agricultural engineer*
Muir, Patricia Susan *biology educator, researcher*
Neilson, Ronald Price *ecology educator*
Oldfield, James Edmund *nutrition educator*
Parks, Harold Raymond *mathematician, educator*
Phillips, Donald Lundahl *research ecologist*
Rapier, Pascal Moran *chemical engineer, physicist*
Reed, Donald James *biochemistry educator*
Rieth, Peter Allan *business executive*
Shoemaker, Clara Brink *retired chemistry educator*
Skelton, John Edward *computer science educator, consultant*
Sollitt, Charles Kevin *ocean engineering educator, laboratory director*
Welty, James R. *mechanical engineer, educator*
Young, Roy Alton *university administrator, educator*

Eugene
Boekelheide, Virgil Carl *chemistry educator*
Goldberg, Lewis Robert *psychology educator, researcher*
Kovtynovich, Dan *civil engineer*
Noyes, Richard Macy *physical chemist, educator*
Pearson, John Mark *civil engineer*
Reed, Diane Marie *psychologist*
Schellman, John A. *chemistry educator*
Slovic, Stewart Paul *psychologist*
Watson, Mary Ellen *ophthalmic technologist*
Wessells, Norman Keith *biologist, educator, university administrator*
Williams, Francis Leon *engineering executive*

Florence
Ericksen, Jerald Laverne *educator, engineering scientist*

Grants Pass
Selinger, Rosemary Celeste Lee *medical psychotherapist*

Hillsboro
Bhagwan, Sudhir *computer industry and research executive, consultant*
Chen, James Jen-Chuan *electrical engineer*
Krampits, Mark William *power management software specialist*
Stewart, Kenneth Ray *food scientist*

Independence
Bhat, Bal Krishen *geneticist, plant breeder*

La Grande
Betts, Burr Joseph *biology educator*
Mason, Richard Randolph *entomologist*
Thomas, Jack Ward *wildlife biologist*
Tiedemann, Arthur Ralph *ecologist, researcher*

Mcminnville
Blodgett, Forrest Clinton *economics educator*
Deer, James William *physicist*

Medford
Kunkle, Donald Edward *physicist*
Linn, Carole Anne *dietitian*

Milwaukie
Schafer, Walter Warren *dentist*

Newport
Weber, Lavern John *marine life administrator, educator*

Oregon City
Wall, Brian Raymond *forest economist, consultant*

Pendleton
Klepper, Elizabeth Lee *physiologist*
Smiley, Richard Wayne *research center administrator, researcher*

Portland
Antoch, Zdenek Vincent *electrical engineering educator*
Bacon, Vicky Lee *lighting services executive*
Baker, Timothy Alan *healthcare administrator, educator, consultant*
Barmack, Neal Herbert *neuroscientist*
Bortner, James Bradley *wildlife biologist*
Durland, Sven O. *research and development engineer*
Grimsbo, Raymond Allen *forensic scientist*
Hagenstein, William David *consulting forester*
Hall, Howard Pickering *mathematics educator*
Hazel, Joanie Beverly *elementary educator*
Jacob, Stanley Wallace *surgeon, educator*
Janssen, Andrew Gerard *civil engineer*
Jarrell, Wesley Michael *soil and ecosystem science educator, researcher, consultant*
Lall, B. Kent *civil engineering educator*
Lambert, Richard William *mathematics educator*
Lewitt, Miles Martin *computer engineering company executive*
Lezak, Muriel Deutsch *psychology, neurology and psychiatry educator*
Loehr, Thomas Michael *chemist, educator*
Lum, Paul *writer*
McCartney, Bruce Lloyd *hydraulic engineer, consultant*
Mozena, John Daniel *podiatrist*
Newell, Nanette *biotechnologist*
North, Richard Alan *neuropharmacologist*
Peyton, David Harold *biochemistry educator*
Pillers, De-Ann Margaret *neonatologist*
Prendergast, William John *ophthalmologist*
Raniere, Lawrence Charles *environmental scientist*
Rhodes, Robert LeRoy *systems engineer*
Rosenbaum, James Todd *rheumatologst, educator*
Rosenfeld, Ron Gershon *pediatrics educator*
Ruben, Laurens Norman *biology educator*
Sangrey, Dwight A. *civil engineer, educator*
Sklovsky, Robert Joel *pharmacology educator*
Spencer, Peter Simner *neurotoxicologist*
Stalnaker, John Hulbert *physician*
Swan, Kenneth Carl *physician, surgeon*
Swank, Roy Laver *physician, educator, inventor*
Vernon, Jack Allen *otolaryngology educator, laboratory administrator*
von Linsowe, Marina Dorothy *information systems consultant*
Watkins, Charles Reynolds *medical equipment company executive*
Yudelson, Jerry Michael *marketing executive*
Zimmerman, Earl Abram *physician, scientist, educator, neuroendocrinology researcher*

Roseburg
Pendleton, Verne H., Jr. *geologist*

Salem
Butts, Edward P. *civil engineer, consultant*
Dixon, Robert Gene *engineering company executive, engineering educator*
Edge, James Edward *health care administrator*
Gillette, (Philip) Roger *physicist, systems engineer*
Lambert, Ralph William *civil engineer*
Pugh, Tim Francis, II *software engineer, aeronautical engineer*

Seaside
See, Paul DeWitt *geology educator*

Selma
Roy, Harold Edward *research chemist*

Springfield
Detlefsen, William David, Jr. *chemist, administrator*

Sunriver
Clough, Ray William, Jr. *civil engineering educator*

Tualatin
Webster, Merlyn Hugh, Jr. *manufacturing engineer, information systems consultant*

Wilsonville
Isberg, Reuben Albert *radio communications engineer*
Poindexter, Kim M. *electrical engineer, consultant*
Willett, Kenneth G. *software engineer*

PENNSYLVANIA

Abington
Ayoub, Ayoub Barsoum *mathematician, educator*

Alcoa Center
Auses, John Paul (Jay) *technical specialist*
Azarkhin, Alexander *mechanical engineer, mathematician*
Bonewitz, Robert Allen *chemist, manufacturing executive*
Dobbs, Charles Luther *analytical chemist*
Fussell, Paul Stephen *mechanical engineer*
Kinosz, Donald Lee *quality manager*
Nordmark, Glenn Everett *civil engineer*
Pien, Shyh-Jye John *mechanical engineer*
Warchol, Mark Francis Andrew *design engineer*

Allentown
Armor, John N. *chemical company research manager*
Bannon, George *economics educator, department chairman*
Daniels, Charles Joseph, III *electrical engineer*
Foster, Edward Paul (Ted Foster) *process industries executive*
Gaylor, Donald Hughes *surgeon, educator*
Graham, Kenneth Robert *psychologist, educator*
Hansel, James Gordon *chemical engineer*
Kratzer, Guy Livingston *surgeon*
Levin, Ken *radiologist*
Morris, Stanley M. *research and engineering executive*
Orphanides, Gus George *chemical company official*
Schweighardt, Frank Kenneth *chemist*
Sheesley, John Henry *statistician*
Szklenski, Theodore Paul *electrical engineer*
Webley, Paul Anthony *chemical engineer*
Winters, Arthur Ralph, Jr. *chemical and cryogenic engineer, consultant*
Woytek, Andrew Joseph *chemical engineer, researcher*

Allison Park
Sartori, David Ezio *statistician, consultant*
Xu, Zhifu *chemist, researcher*

Altoona
Arbitell, Michelle Reneé *clinical psychologist*
Gannon, Michael Robert *biology educator*
Hoppel, Thomas O'Marah *electrical engineer*

Ambler
Sorrentino, John Anthony *environmental economics educator, consultant*
Vanyo, Edward Alan *environmental engineer*

Annville
Cullari, Salvatore Santino *clinical psychologist, educator*

Aston
Kingrea, James Irvin, Jr. *electromechanical engineer*

Avis
Hillyard, Richie Doak *wastewater treatment plant operator*

Bala Cynwyd
Burtle, James Lindley *economist, educator*
Katz, Julian *gastroenterologist, educator*
Kirschner, Ronald Allen *osteopathic plastic surgeon, otolaryngologist, educator*
Robertson, Kenneth McLeod *mechanical engineer*

Bangor
Wolf, Stewart George, Jr. *physician, medical educator*

Barnesboro
Moore, David Austin *pharmaceutical company executive, consultant*

Bath
Ayers, Joseph Williams *chemical company executive*

Beaver
Hodge, Philip Tully *structural engineer*

Bellefonte
Grindall, Emerson Jon *civil engineer*

Bensalem
Bontempo, Daniel *civil engineer*

Berwyn
Burch, John Walter *mining equipment company executive*
Dickson, Brian *physician, researcher, educator*
Lund, George Edward *retired electrical engineer*
Salvatore, Scott Richard *ecologist*
Triegel, Elly Kirsten *geologist*

Bethel Park
Korchynsky, Michael *metallurgical engineer*

Bethlehem
Advani, Sunder *engineering educator, university dean*
Alhadeff, Jack Abraham *biochemist, educator*
Fisher, John William *civil engineering educator*
Frankel, Barbara Brown *cultural anthropologist*
Friedman, Sharon Mae *science journalism educator*
Georgakis, Christos *chemical engineer educator, consultant, researcher*
Horvath, Vincent Victor *electrical engineer*
Jain, Himanshu *materials science engineering educator*
Kugelman, Irwin Jay *civil engineering educator*
Lennon, Gerard Patrick *civil engineering educator, researcher*
Levy, Edward K. *mechanical engineering educator*
Licini, Jerome Carl *physicist, educator*
Lyman, Charles Edson *materials scientist, educator*
Odrey, Nicholas Gerald *industrial engineer, educator*
Ostapenko, Alexis *civil engineer, educator*
Pense, Alan Wiggins *metallurgical engineer, academic administrator*
Roberts, Malcolm John *steel company executive*
Roberts, Richard *mechanical engineering educator*
Snyder, John Mendenhall *medical administrator, retired thoracic surgeon*
Stephenson, Edward Thomas *consulting metallurgist*
Voutsas, Apostolos Theoharis *chemical engineer*
Wazontek, Stella Catherine *computer programmer, analyst, software engineer*
Wei, Robert Peh-Ying *mechanics educator*
Williams, David Bernard *metallurgical engineer*

Biglerville
Saliu, Ion *software developer, computer programmer*

Birdsboro
Moyer, David Lee *veterinarian*

Blakeslee
Hayes, Alberta Phyllis Wildrick *retired health service executive*

Bloomsburg
Waggoner, John Edward *psychology educator*

Boalsburg
Gettig, Martin Winthrop *retired mechanical engineer*

Boothwyn
McLaughlin, Edward David *surgeon, educator*

Brackenridge
Houze, Gerald Lucian, Jr. *metallurgist*

Bradford Woods
Allardice, John McCarrell *coatings manufacturing company executive*

Breinigsville
Reynolds, C(laude) Lewis, Jr. *materials scientist, researcher*

Bridgeville
Glidden, Bruce *structural engineer, construction consultant*
Pearlman, Seth Leonard *civil engineer*
Pierce, Jeffrey Leo *power systems engineer, consultant*

Bristol
Raughley, Dean Aaron *process engineer*

Broomall
Emplit, Raymond Henry *electrical engineer*
Schonbach, Bernard Harvey *engineering executive*

Bryn Mawr
Crawford, Maria Luisa Buse *geology educator*
Friedman, Arnold Carl *diagnostic radiologist*
Pettit, Horace *allergist, consultant*

Butler
Conforti, Ronald Anthony, Jr. *communications and electronics consultant*

California
Newhouse, Joseph Robert *plant pathologist, mycologist*

Carlisle
Egolf, Kenneth Lee *chemistry educator*
Jones, Randall Marvin *chemist*
Lewis, Gregory Lee *gastroenterologist*

Carnegie
Reinhart, Robert Karl *control engineer, consultant*

Chadds Ford
DeVries, Frederick William *chemical engineer*

Chalfont
Ballod, Martin Charles *civil engineer, mechanical engineer*
Mendlowski, Bronislaw *retired pathologist*

Charleroi
Teaford, Norman Baker *engineering manager*

Chester
Brown, Vicki Lee *civil engineering educator*

Chester Springs
Humphrey, Leonard Claude *electrical engineer*

Clairton
Maddalena, Frederick Louis *chemical engineer*

Clarks Summit
Miniutti, Robert Leonard *engineering company executive*

Coatesville
Gehring, David Austin *physician, adminstrator, cardiologist*
McNally, Mark Matthew *control systems engineer*
Nocks, James Jay *psychiatrist*

Collegeville
Behrens, Rudolph *mechanical engineer*
Cordes, Eugene Harold *biochemist*
Farmar, Robert Melville *medical scientist, educator*

Colmar
Alexander, Wayne Andrew *product engineer*

Conshohocken
Cohen, Alan *civil engineer*
Rutolo, James Daniel *mechanical engineer*
Shanbaky, Ivna Oliveira *physicist*

Cooksburg
Meley, Robert Wayne *structural engineer*

Coopersburg
Peserik, James E. *electrical engineer*
Siess, Alfred Albert, Jr. *engineering consultant, management executive*

Coraopolis
Kimes, Mark Edward *civil engineer*
Manuel, Phillip Earnest *meteorologist*
Rabosky, Joseph George *engineering consulting company executive*
Skovira, Robert Joseph *information scientist, educator*
Stewart, Richard Allan *civil engineer*

Coudersport
Sproull, Wayne Treber *consultant*

Danville
Kleponis, Jerome Albert *dentist*
Murphy, Stephan David *electrical engineer*
Schuller, Diane Ethel *allergist, immunologist, educator*

Delta
Wurzbach, Richard Norman *utility company engineer*

Downingtown
Tarpley, William Beverly, Jr. physics and chemistry consultant

Doylestown
Dimond, Roberta Ralston psychology and sociology educator
Rapaport, Martin Baruch safety engineer
Samuelson, Andrew Lief civil engineer
Tinnerino, Natale Francis electronics engineer
Weber, Charles Walton chemistry educator

Drexel Hill
Amoroso, Marie Dorothy retired medical technologist

East Greenville
Grubb, David Conway mechanical engineer

East Stroudsburg
Bergo, Conrad Hunter chemistry educator
Rymon, Larry Maring environmental scientist, educator

Easton
Dennis, Ronald Marvin chemical engineering educator, consultant
Lee, Terry James environmental engineer, consultant

Ebensburg
Anderson, Mark Edward mining engineer, consultant

Edinboro
Snyder, Donald Benjamin biology educator

Erie
Clark, Gordon Hostetter, Jr. physician
Crankshaw, John Hamilton mechanical engineer
Czarnecki, Gregory James ecologist
Gray, Robert Beckwith engineer
Hsu, Bertrand Dahung mechanical engineer
Jones, William V(incent) health center administrator
Kalkhof, Thomas Corrigan physician
Kohli, Tejbans Singh chemical engineer
Macosko, Paul John, II psychotherapist
Makarowski, William Stephen rheumatologist
Michaelides, Doros Nikita internist, educator

Export
Fisher, David Marc software engineer

Exton
Piatkowski, Steven Mark environmental engineer
Scott, Alan electrical engineer, consultant

Fairless Hills
Campbell, Sharon Lynn communications company executive
Rosella, John Daniel psychologist

Fairviewvill
Kisielowski, Eugene engineering executive

Fleetwood
Pangrazi, Ronald Joseph chemist

Flourtown
Di Maria, Charles Walter mechanical and automation engineer, consultant
Kissileff, Alfred veterinarian, researcher

Forest City
Fleming, Lawrence Thomas engineering executive

Fort Washington
Buescher, Adolph Ernst (Dolph Buescher) aerospace company executive
Newton, Crystal Hoffman mechanical engineer

Gaines
Beller, Martin Leonard retired orthopaedic surgeon

Gettysburg
Darnell, Lonnie Lee electronic engineer
Holland, Koren Alayne chemistry educator

Gibsonia
Cauna, Nikolajs physician, medical educator
Shoub, Earle Phelps chemical engineer, educator

Gilbertsville
Ladisch, Thomas Peter chemical engineer

Gladwyne
Stick, Alyce Cushing information systems consultant

Greensburg
Austin, George Lynn surgeon
Catalano, Louis William, Jr. neurologist
Coffield, Ronald Dale mechanical engineer
Kriss, Joseph James data processing executive

Greenville
Kelly, Richard Michael metallurgical engineer

Grove City
Zellers, Robert Charles construction materials engineer, consultant

Hanover
Davis, Ruth C. pharmacy educator

Harrisburg
Gallaher, William Marshall dental laboratory technician
Grandon, Raymond Charles physician, educator
Kimmel, Robert Irving corporate communication design consultant, former state government official
Rosen, Herman civil engineer
Shatto, Ellen Latherow mathematics educator
Stroh, Oscar Henry agricultural, civil and industrial engineer, consultant
Wei, I-Yuan research and development manager
Willis, David Paul environmental scientist
Zook, Merlin Wayne meteorologist

Hatboro
Bickel, John Frederick consulting engineer

Hatfield
Kish, Michael Stephen safety professional

Haverford
Blumberg, Herbert Haskell psychology educator
Goppelt, John Walter physician, psychiatrist
Keith, Frederick W., Jr. retired chemical engineer
Lazar, Anna chemist

Havertown
Tassone, Bruce Anthony chemical company executive

Hawley
Bruck, James Alvin telecommunications professional

Hazleton
Smith, David chemistry educator

Hershey
Benowitz, Steven Ira science-medical writer
Davis, Dwight cardiologist, educator
Demers, Laurence Maurice science educator
Duncan, Charles Lee food products company executive
King, Steven Harold health physicist
Marshall, Wayne Keith anesthesiology educator
Norgren, Ralph neuroscientist
Pierce, William Schuler cardiac surgeon, educator
Rapp, Fred virologist
St. Hilaire, Catherine Lillian food company executive
Stanley, Bruce Alan molecular biologist
Stumpf, David Michael food scientist
Tan, Tjiauw-Ling psychiatrist, educator

Holtwood
Liebman, Shirley Anne analytical research scientist

Horsham
Derby, Christopher William civil engineer
Johnston, James Bennett biochemist
Logue, John Joseph psychologist
Roth, Stephen biochemist, educator

Houston
Ali, Ashraf civil engineer
Rajakumar, Charles mechanical engineer

Huntingdon
Schock, William Wallace pediatrician

Huntingdon Valley
Blume, Horst Karl manufacturing engineer
Godfrey, John Carl medicinal chemist

Indiana
Cale, William Graham, Jr. environmental sciences educator, university administrator, researcher
Fapohunda, Babatunde Olusegun energy specialist
Jones, Shelley Pryce chemical company executive
Lord, Thomas Reeves biology educator
Purdy, David Lawrence biotechnical company executive

Jenkintown
Mifsud, Lewis electrical engineer, fire origin investigator, physicist
Robinson, Mark Louis metallurgical engineer

Johnstown
Brezovec, Paul John environmental specialist
Dombrowski, John Micheal architectural consulting engineer
Kintner, Elisabeth Turner chemistry educator
Kosker, Leon Kevin electrical engineer
McNair, Dennis Michael biology educator
Pandiarajan, Vijayakumar industrial engineer
Salem, Kenneth George theoretical physicist

King Of Prussia
Bilofsky, Howard Steven scientist
Black, Jonathan bioengineering educator
Bramson, Robert Sherman lawyer
Del Tito, Benjamin John, Jr. pharmaceuticals researcher
Dolce, Kathleen Ann health physicist, inspector
Eggleston, Drake Stephen pharmaceutical researcher
Gitlitz, Melvin Hyman chemist, researcher
Hawes, Nancy Elizabeth secondary educator
Peerce-Landers, Pamela Jane chemical researcher

Kingston
Fierman, Gerald Shea electrical distribution company executive

Kutztown
Dougherty, Percy H. geographer, educator
Janke, Rhonda Rae agronomist, educator
Zei, Robert William geology educator, consultant

Lancaster
Dunlap, Lawrence Hallowell museum curator
Eastman, G. Yale aeronautical engineer
Falk, Robert Barclay, Jr. anesthesiologist, educator
Hess, Earl Hollinger laboratory executive, chemist
Murthy, Kamalakara Akula chemical engineer, researcher
Schwartz, John Howard poultry science and extension educator
Spalding, George Robert acoustician, consultant
Stewart, Arlene Jean Golden designer, stylist

Landenberg
Aldrich, Nancy Armstrong psychotherapist, clinical social worker

Langhorne
Shah, Mubarik Ahmad surgeon
Venable, Robert Ellis crop scientist

Lansdale
Schnable, George Luther chemist

Latrobe
Conley, Edward Vincent, Jr. research engineer

Lehigh Valley
Golden, Timothy Christopher chemist, researcher

Lehman
Alston, Steven Gail physicist, educator
Nouri-Moghadam, Mohamad Reza mathematics and physics educator

Lester
Cahill, Lawrence Bernard environmental engineer

Lewisburg
Gidh, Kedar Keshav chemical engineer
Kozick, Richard James electrical engineering educator
Orbison, James Graham civil engineer, educator
Veening, Hans chemistry educator

Ligonier
Mattern, Gerry A. engineering consultant

Linwood
Balale, Emanuel Michael chemical engineer

Lock Haven
Smith, Augustine Joseph physicist, educator
Sweeny, Charles David chemist

Lower Burrell
Englehart, Edwin Thomas metals company executive

Malvern
Fisher, Sallie Ann chemist
Holveck, David P. pharmaceutical company executive
Langrall, Harrison Morton, Jr. internist
Nardone, Robert Carmen cell biologist
Wong, Raphael Chakching biotechnologist, executive

Mars
Pantelis, John Andrew, Jr. civil engineer

McMurray
Mortimer, James Winslow analytical chemist
Richards, John Lewis civil engineer, educator

Meadville
Cable, Charles Allen mathematician
Wharton, William Polk psychologist, consultant

Mechanicsburg
Beard, Robert Douglas geologist
Hunter, Larry Lee electrical engineer
Windsor, Ted Bacon feed microscopy consultant

Media
Ataiifar, Ali Akbar economist, educator
Behbehanian, Mahin Fazeli surgeon
Harnish, Richard John psychologist
McClintic, Fred Frazier simulation engineer
Voltz, Sterling Ernest physical chemist, researcher

Mehoopany
Miller, Cliff engineer

Mendenhall
Mehne, Paul Herbert chemical engineer, engineering consultant

Merion Station
Amado, Ralph David physics educator

Middletown
Fox, Barry Howard nuclear engineer
Foxx, Richard Michael psychology educator
Hartman, Charles Eugene nuclear engineer
McClintock, Samuel Alan engineering educator

Millersville
Greco, Thomas G. chemist, educator

Monessen
Lementowski, Michal surgeon

Monroeville
Campbell, Donald Acheson nuclear engineer, consultant
Cremeens, David Lynn soil scientist, consultant
Fedak, Mitchel George chemistry educator
Lin, Ming Shek allergist, immunologist
Liu, Yinshi metallurgical engineer
Mandel, Herbert Maurice civil engineer
McDermott, John Francis, IV civil engineer, consultant
Nemeth, Edward Joseph process research specialist
Parker, James Roger chemist
Penman, Paul Duane nuclear power laboratory executive

Morrisville
Jarvi, George Albert chemical engineer

Moylan
Eberl, James Joseph physical chemist, consultant

Muncy
Lowe, John Raymond, Jr. mechanical engineer

Murrysville
Linn, Paul Anthony nuclear engineer

Nazareth
Bader, William Alan computer engineer

New Castle
Flannery, Wilbur Eugene health science association administrator, internist

New Galilee
Randza, Jason Michael aerospace engineer, consultant

New Holland
Crane, Jack Wilbur agricultural and mechnical engineer
Ekis, Imants materials engineer, consultant

New Hope
Bertele, William environmental engineer

New Kensington
Bridenbaugh, Peter Reese industrial research executive
Dando, Neal Richard chemist
Evancho, Joseph William engineering executive, metallurgical engineer
Hunsicker, Harold Yundt metallurgical engineer
Oder, Robin Roy physicist

Newtown
Hammonds, Elizabeth Ann environmental engineer
Long, Harry (On-Yuen Eng) chemist, rubber science and technology consultant
Marton, Joseph paper chemistry consultant, educator

Norristown
Tornetta, Frank Joseph anesthesiologist, educator, consultant
Weiner, Harold M. radiologist

North Huntingdon
Metala, Michael Joseph mechanical engineer

North Wales
Gray, David Marshall chemical engineer

Oakdale
Dean, Frank Warren, Jr. chemist, pet food company executive
Sieger, Edward Regis software engineer

Palmyra
Moyer, John Henry, III physician, educator
Singer, William Harry computer/software engineer, expert systems designer, consultant, entrepreneur
Wyczalkowski, Wojciech Roman mechanical engineer, chemical engineer, researcher

Philadelphia
Adom, Edwin Nii Amalai psychiatrist
Altamirano, Anibal Alberto physiologist
Altshuler, David Thomas computer scientist
Amsterdam, Jay D. psychiatry educator, researcher
Andes, Charles Lovett museum executive, technology association executive
Anni, Eleni (Helen Anni) biochemistry and biophysics educator
Anyanwu, Chukwukre alcohol and drug abuse facility administrator
Aravas, Nikolaos mechanical engineering educator
Asbury, Arthur Knight neurologist, educator
Austrian, Robert physician, educator
Axel, Leon radiologist, educator
Barchi, Robert Lawrence neuroscience educator, clinical neurologist, neuroscientist
Barker, Clyde Frederick surgeon, educator
Batterman, Steven Charles mechanical engineering, bioengineering educator
Beacham, Dorothy Ann medical research scientist
Beauchamp, Gary Keith physiologist
Beck, Aaron Temkin psychiatrist
Berg, Ivar Elis, Jr. social science educator
Black, Perry neurological surgeon, educator
Blumberg, Baruch Samuel academic administrator
Bolt, Eugene Albert, Jr. historic natural science museum curator, historian
Bozenhardt, Herman Fredrick control technologist, automation system designer
Brayman, Kenneth Lewis surgeon
Brecher, Ephraim Fred structural engineering consultant
Breitman, Joseph B. prosthodonist, dental educator
Bryan, Henry Collier private school mathematics educator
Buerk, Donald Gene medical educator
Calabi, Eugenio mathematician, educator
Chance, Britton biophysics and physical chemistry educator emeritus
Cho, Young Il mechanical engineering educator
Chu, Mon-Li Hsiung molecular biologist
Clifford, Maurice Cecil physician, former college president, foundation executive
Cohen, Ira Myron aeronautical and mechanical engineering educator
Cohn, Mildred biochemist, educator
Colman, Robert Wolf physician, medical educator
Comer, Nathan Lawrence psychiatrist, educator
Comis, Robert Leo oncologist, educator
Composto, Russell John materials science engineering educator, researcher
Conn, Rex Boland, Jr. physician, educator
Cox, Robert Harold physiology educator
Craig, James Clifford, Jr. chemical engineer
Cutler, Neal Evan gerontologist, educator
Dai, Hai-Lung physical chemist, researcher
Davis, Alan Robert molecular biologist
Davis, Raymond, Jr. chemist, researcher, educator
DeHoratius, Raphael Joseph rheumatologist
De La Cadena, Raul Alvarez physician, pathology and thrombosis educator
DellaVecchia, Michael Anthony ophthalmologist
Dementis, Katharine Hopkins interior designer
Doman, Janet Joy association executive
Doty, Richard Leroy medical researcher
Dumsha, David Allen chemical company executive
Dymicky, Michael retired chemist
Eisenstein, Bruce Allan electrical engineering educator
Elesh, David Bert sociology educator
Engheta, Nader electrical engineering educator, researcher
Fisher, Aron Baer physiology and medicine educator
Fletcher, Jeffrey Edward biochemist, researcher
Forsyth, Keith William optical engineer
Frankel, Sherman physicist
Frankl, William Stewart cardiologist, educator
Franz, Craig Joseph biology educator
Freed, Edmond Lee podiatrist
Fromm, Eli engineering educator
Frucht, Harold physician
Genis, Vladimir I. electrical engineer
Gibson, William Charles neuropsychologist
Glick, John H. oncologist, medical educator
Goldman, Yale E. physiologist, educator
Gozum, Marvin Enriquez internist
Grebner, Eugene Ernest biochemist, educator
Greene, Joyce Marie biology educator
Gross, Kaufman Kennard applications engineer, consultant
Hameka, Hendrik Frederik chemist, educator
Hand, Peter James neurobiologist, educator
Heyman, Julius Scott neuroscientist
Hoenigswald, Henry Max linguist, educator
Holmes, Edward W. physician, educator
Hopson, Kevin Mathew chemist

Hung, Paul Porwen *biotechnologist, educator, consultant*
Iskandrian, Ami Simon (Edward) *cardiologist, educator*
Jackson, Fred *oil executive*
Jaggard, Dwight L(incoln) *electrical engineering educator*
Jarosz, Boleslaw Francis *mechanical engineer*
Johnson, Bonnie Jean *neuroscientist*
Johnson, Elmer Marshall *toxicologist, teratologist*
Johnson, Mark Dee *pharmacologist, researcher*
Joseph, Peter Maron *physics educator*
Kadison, Richard Vincent *mathematician, educator*
Kaji, Akira *microbiology scientist, educator*
Kaplan, Martin Nathan *electrical engineer*
Kauffman, Joel Mervin *chemistry educator, researcher, consultant*
Kelley, William Nimmons *physician, educator*
Kirsch, Ted Michael *pathobiologist*
Kissick, William Lee *physician, educator*
Klaver, Martin Arnold, Jr. *business administrator*
Kleban, Morton Harold *psychologist*
Knudson, Alfred George, Jr. *medical geneticist*
Kollros, Peter Richard *child neurology educator, researcher*
Koprowski, Hilary *microbiology educator, medical scientist*
Kresh, J. Yasha *cardiovascular researcher, educator*
Kritchevsky, David *biochemist, educator*
Ku, Y. H. *engineering educator*
Kundel, Harold Louis *radiologist, educator*
Kurtz, Alfred Bernard *radiologist*
Lantos, Peter R(ichard) *industrial consultant, chemical engineer*
Lefer, Allan Mark *physiologist*
Lehmann, John Charles *neuropharmacologist*
Leodore, Richard Anthony *electronic manufacturing company executive*
Levitt, Israel Monroe *astronomer*
Levy, Robert Isaac *physician, educator, research director*
Lewis, Paul Le Roy *pathology educator*
Liebman, Paul Arno *biophysicist, educator*
Lief, Harold Isaiah *psychiatrist*
London, William Thomas *internist*
Lorenz, Beth June *mechanical engineer*
MacDiarmid, Alan Graham *metallurgist, educator*
Madaio, Michael Peter *medical educator*
Malis, Bernard Jay *pharmacologist*
Mancall, Elliott Lee *neurologist, educator*
Mancini, Nicholas Angelo *psychologist*
Mansour, Farid Fam *civil engineer*
Marino, Paul Lawrence *physician, researcher*
Mastroianni, Luigi, Jr. *physician, educator*
Mayock, Robert Lee *internist*
McEachron, Donald Lynn *biology educator, researcher*
Ming, Si-Chun *pathologist, educator*
Molino-Bonagura, Lory Jean *neurobiologist*
Morrison, Adrian Russell *veterinarian educator*
Munson, Janis Elizabeth Tremblay *engineering company executive*
Nezu, Christine Maguth *clinical psychologist, educator*
Niewiarowski, Stefan *physiology educator, biomedical research scientist*
Nimoityn, Philip *cardiologist*
Norwood, Carol Ruth *research laboratory administrator*
Onaral, Banu Kum *electrical/biomedical engineering educator*
Ortiz-Arduan, Alberto *nephrologist*
Otvos, Laszlo Istvan, Jr. *organic chemist*
Owens, Gary Mitchell *physician, educator*
Patrick, Ruth (Mrs. Charles Hodge) *limnologist, diatom taxonomist, educator*
Peck, Robert McCracken *naturalist, science historian, writer*
Pfeffer, Philip Elliot *biophysicist*
Pickands, James, III *mathematical statistician, educator*
Pollack, Howard Martin *radiologist, teacher, author, researcher*
Price, Karen Overstreet *pharmacist, medical editor*
Prockop, Darwin Johnson *biochemist, physician*
Quinn, John Albert *chemical engineering educator*
Raftery, M. Daniel *chemistry researcher*
Rawson, Nancy Ellen *neurobiology researcher*
Reid, John Mitchell *biomedical engineer*
Rescorla, Robert Arthur *psychology educator*
Rhoads, Jonathan Evans *surgeon*
Ridenour, Marcella V. *motor development educator*
Roder, Heinrich *biophysicist, educator*
Rohrlich, George Friedrich *social economist*
Rosan, Burton *microbiology educator*
Rosenberg, Robert Allen *psychologist, educator, optometrist*
Rothwarf, Allen *electrical engineering educator*
Rubik, Beverly Anne *university administrator*
Rubin, Stephen Curtis *gynecologic oncologist, educator*
Sague, John E(lmer) *mechanical engineer*
Sanders, Robert Walter *ecologist, researcher*
Sato, Takami *pediatrician, medical oncologist*
Scedrov, Andre *mathematics and computer science researcher, educator*
Schidlow, Daniel *pediatrician, medical association administrator*
Schoen, Allen Harry *aerospace engineering executive*
Schumacher, H(arry) Ralph *internist, researcher, medical educator*
Schwan, Herman Paul *electrical engineering and physical science educator, research scientist*
Schwartz, Elias *pediatrician*
Schwartz, Gordon Francis *surgeon, educator*
Seidmon, E. James *urologist*
Selzer, Michael Edgar *neurologist*
Shapiro, Sandor Solomon *hematologist*
Sheffield, Joel Bensen *biology educator*
Shen, Benjamin Shih-Ping *scientist, engineer, educator*
Shipkin, Paul M. *neurologist*
Shockman, Gerald David *microbiologist, educator*
Silage, Dennis Alex *electrical engineering educator*
Silberberg, Donald H. *neurologist*
Skalka, Anna Marie *molecular biologist, virologist*
Smith, Amos Brittain, III *chemist, educator*
Snyder, Donald Carl, Jr. *physics educator*
Sorensen, Henrik Vittrup *electrical engineering educator*
Soslow, Arnold *quality consultant*
Spamer, Earle Edward *museum executive*
Spector, Harvey M. *osteopathic physician*
Sridhara, Channarayapatna Ramakrishna Setty *health facility administrator, consultant*
Stunkard, Albert James *physician, educator*
Suntharalingam, Nagalingam *radiation therapy educator*
Sutnick, Alton Ivan *medical school dean, educator, researcher, physician*
Taichman, Norton Stanley *pathology educator*
Tantala, Albert Martin *civil engineer*
Terzian, Karnig Yervant *civil engineer*
Thorn, James Douglas *safety engineer*
Tran, John Kim-Son Tan *chemical senses executive, research administrator*
Tudor, John Julian *microbiologist*
Tyagi, Som Dev *physics educator*
Wagner, Mary Emma *geologist educator, researcher*
Walinsky, Paul *cardiology educator*
Wein, Alan Jerome *urologist, educator, researcher*
Weisz, Paul B(urg) *physicist, chemical engineer*
Wiesner, Robert *aeronautical engineer*
Winegrad, Albert Irvin *immunologist, educator*
Wright, Robert Michael *civil engineer*
Wysocki, Charles Joseph *neuroscientist*
Xin, Li *physiologist*
Yassin, Rihab R. *biomedical researcher, educator*
Young, Donald Stirling *clinical pathology educator*
Yun, Daniel Duwhan *physician, foundation administrator*
Zimmerman, Michael Raymond *pathology educator, anthropologist*
Zweidler, Alfred *medical scientist, educator*
Zweiman, Burton *physician, scientist, educator*

Pittsburgh
Abdelhak, Sherif Samy *health science executive*
Akay, Adnan *mechanical engineer educator*
Allen, David Woodroffe *computer scientist*
Allen, Thomas E. *obstetrician/gynecologist*
Allstot, David James *electrical and computer engineering educator*
Amon, Cristina Hortensia *mechanical engineer, educator*
Andersen, Theodore Selmer *engineering manager*
Anderson, Russell Karl, Jr. *physicist, horse breeder*
Becherer, Richard John *architectural educator*
Behrend, William Louis *electrical engineer*
Berliner, Hans Jack *computer scientist*
Blackburn, Ruth Elizabeth *biomedical research scientist*
Bloom, William Millard *industrial design engineer*
Boczkaj, Bohdan Karol *structural engineer*
Boward, Joseph Frank *civil engineer*
Brungraber, Louis Edward *retired mechanical engineer*
Burbea, Jacob N. *mathematics educator*
Cagan, Jonathan *mechanical engineering educator*
Cao, You Sheng *materials science and engineering researcher*
Cavalet, James Roger *engineering executive, consultant*
Claycamp, Henry Gregg *radiobiologist educator*
Cobb, James Temple, Jr. *chemical engineer, engineering educator*
Collins, Charles Curtis *pharmacist, educator*
Connolly, Patricia Ann Stacy *physical education educator*
Constantino-Bana, Rose Eva *nursing educator, researcher*
Curtis, Bill *software engineering researcher*
Cutkosky, Richard Edwin *physicist, educator*
deGroat, William Chesney *pharmacology educator*
Director, Stephen William *electrical engineering educator, researcher*
Donovan, John Edward *psychologist*
Duong, Victor (Viet) Hong *nuclear engineer*
Dutt, David Alan *physicist*
Falk, Joel *electrical engineering educator*
Ferguson, Jackson Robert, Jr. *astronautical engineer*
Fichman, Mark *industrial/organizational psychologist*
Fischhoff, Baruch *psychologist, educator*
Fisher, Bernard *surgeon, researcher, educator*
Frederick, Edward Russell *chemical engineering consultant*
Friedberg, Simeon Adlow *physicist, educator*
Frieze, Alan Michael *mathematician, educator*
Froehlich, Fritz Edgar *telecommunications educator and scientist*
Garrett, James Henry, Jr. *civil engineering educator*
Geiwitz, (Peter) James *psychologist, writer, researcher*
George, John Michael *chemical engineer*
Gilbert, Ralph Whitmel, Jr. *engineering company executive*
Glencer, Suzanne Thomson *science educator*
Green, Mayer Albert *physician*
Griffin, Donald S. *nuclear engineer, consultant*
Griffiths, Robert Budington *physics educator*
Grossmann, Ignacio Emilio *chemical engineering educator*
Hall, Charles Allan *numerical analyst, educator*
Hammack, William S. *chemical engineering educator*
Held, William James *civil engineer*
Henry, Keith Edward *chemical engineer*
Hercules, David Michael *chemistry educator, consultant*
Herndon, James Henry *orthopedic surgeon, educator*
Hillebrand, Julie Ann *biotechnology executive*
Ho, Chuen-hwei Nelson *environmental engineer*
Hoburg, James Frederick *electrical engineering educator*
Hollister, Floyd Hill *electrical engineer, engineering executive*
Huang, Mei Qing *physics educator, researcher*
Jobes, Christopher Charles *mechanical engineer*
Johnson, Mark Henry *psychology educator*
Joyner, Claude Reuben, Jr. *physician, medical educator*
Kanal, Emanuel *radiologist*
Kannan, Ravi *mathematician educator*
Kappmeyer, Keith K. *manufacturing company executive*
Kaufman, William Morris *research institute director, engineer*
Kenkel, James Lawrence *economics educator*
Khonsari, Michael M. *engineering educator*
Kotovsky, Kenneth *psychology educator*
Kreithen, Melvin Louis *biologist, educator*
Krutz, Ronald L. *computer engineer*
Kumar, Bhagavatula Vijaya *electrical and computer engineering educator*
Laughlin, David Eugene *materials science educator, metallurgical consultant*
Lawrence, Margery H(ulings) *utilities executive*
Leney, George Willard *environmental administrator, consulting geologist*
Lengyel, Joseph William *engineering manager*
Liberman, Irving *optical engineer*
Luthy, Richard Godfrey *environmental engineering educator*
MacDonald, Hubert Clarence *analytical chemist*
Marino, Ignazio Roberto *transplant surgeon, researcher*
Markussen, Joanne Marie *chemical engineer*
Matway, Roy Joseph *material scientist*
McKnight, Jennifer Lee Cowles *molecular biologist*
McNamee, William Lawrence *industrial engineer*
Mehrabian, Robert *academic administrator*
Morgan, Daniel Carl *civil engineer*
Murphy, Robert Francis *biology educator*
Neufeld, Ronald David *environmental engineer*
Neuman, Charles P. *electrical and computer engineering educator, consultant*
O'Donnell, William James *engineering executive*
Oh, Kook Sang *diagnostic radiologist, pediatric radiologist*
Ohlsson, Stellan *scientist, researcher*
Omiros, George James *medical foundation executive*
Orbison, David Vaillant *clinical psychologist, consultant*
Osterle, John Fletcher *mechanical engineering educator*
Owens, Gregory Randolph *physician, medical educator*
Page, Lorne Albert *physicist, educator*
Palmer, Alan Michael *neuroscientist*
Perloff, Robert *psychologist, educator*
Perper, Joshua Arte *forensic pathologist*
Phillips, James Macilduff *material handling company executive, engineering and manufacturing executive*
Pohland, Frederick George *environmental engineering educator, researcher*
Putman, Thomas Harold *electrical engineer, consultant*
Pyeritz, Reed Edwin *internist*
Ramezan, Massood *mechanical engineer*
Reder, Lynne Marie *cognitive science educator*
Redgate, Edward Stewart *physiologist, educator*
Reichek, Nathaniel *cardiologist*
Reid, Robert H. *engineering consultant*
Rettura, Guy *civil engineer*
Rigatti, Brian Walter *psychiatric researcher*
Rohrer, Ronald Alan *electrical and computer engineering educator, consultant*
Russell, Alan James *chemical engineering and biotechnology educator*
Sabloff, Jeremy Arac *archaeologist*
Safar, Peter *emergency health care facility administrator, educator*
Sax, Martin *crystallographer*
Schlabach, Leland A. *electrical engineer*
Schonhardt, Carl Mario *analytical chemist*
Schultz, Hyman *analytical chemist*
Schwass, Gary L. *utilities executive*
Schwemmer, David Eugene *structural engineer*
Schwendeman, Louis Paul *civil engineer, retired*
Silverstein, Alan Jay *physician*
Simaan, Marwan A. *electrical engineering educator*
Sinharoy, Samar *physicist, researcher*
Smith, Paul Christian *mechanical engineer*
Sniderman, Marvin *dentist*
Sorensen, Raymond Andrew *physics educator*
Stancil, Daniel Dean *electrical engineering educator*
Stanko, Ronald Thomas *physician, medical educator, clinical nutritionist*
Suzuki, Jon Byron *periodontist, educator*
Taylor, D. Lansing *cell biology educator*
Taylor, Lyle H. *physicist*
Tilton, Robert Daymond *chemical engineer*
Touhill, C. Joseph *environmental engineer*
Tran, Phuoc Xuan *mechanical engineer*
Treadwell, Kenneth Myron *mechanical engineer, consultant*
Turbeville, Robert Morris *engineering executive*
Turnbull, Gordon Keith *metal company executive, metallurgical engineer*
Valoski, Michael Peter *industrial hygienist*
Vidovich, Danko Victor *neurosurgeon, researcher*
Voas, Sharon Joyce *environment and science reporter*
Vogeley, Clyde Eicher, Jr. *engineering educator, consultant*
Wang, Allan Zuwu *cell biologist*
Warner, Richard David *research foundation executive*
Wehmeier, Helge H. *chemical company executive*
Whiteside, Theresa Listowski *immunologist, educator*
Williams, Max Lea, Jr. *engineer, educator*
Wipf, Peter *chemist*
Wolfenstein, Lincoln *physicist, educator*
Wong-Chong, George Michael *chemical and environmental engineer*
Woo, Jacky *immunologist, educator, researcher*
Woo, Savio Lau-Yuen *bioengineering educator*
Worthington, Charles Roy *physics educator*
Yeh, James Tehcheng *chemical engineer*
Yourison, Karola Maria *information specialist, librarian*

Plumsteadville
Vaughan, Stephen Owens *project chemist*

Plymouth Meeting
Friedman, Philip Harvey *psychologist*
Nobel, Joel J. *physician*
Parker, Jon Irving *ecologist*

Point Pleasant
Moss, Herbert Irwin *chemist*

Polk
Hall, Richard Clayton *psychologist, consultant, researcher*

Pottstown
Diana, Guy Dominic *chemist*
Messics, Mark Craig *civil engineer*
Weathington, Billy Christopher *analytical chemist*

Quarryville
Schuck, Terry Karl *chemist, environmental consultant*

Radnor
Marland, Alkis Joseph *leasing company executive, computer science educator, financial planner*
Mizutani, Satoshi *research administrator*
Ohnishi, Stanley Tsuyoshi *biomedical director, biophysicist*

Reading
Backenstoss, Henry Brightbill *electrical engineer, consultant*
Bell, Frances Louise *medical technologist*
Broome, Kenneth Reginald *civil engineer, consultant*
Dulski, Thomas R. *chemist, writer*
Feeman, James Frederic *chemist, consultant*
Lusch, Charles Jack *physician*
Richart, Douglas Stephen *chemist*
Rowe, Jay E., Jr. *research and development director*

Rheems
Adams, James Lee *poultry scientist*

Rockledge
Polakoff, Pedro Paul, II *neurosurgeon*

Saint Marys
Kasaback, Ronald Lawrence *mechanical engineer*
Paxton, R(alph) Robert *chemical engineer*

Sayre
Beezhold, Donald Harry *immunologist*
Thomas, John Melvin *surgeon*

Schwenksville
Frech, Bruce *mathematician*

Scranton
Fahey, Paul Farrell *college administrator*
McLean, William George *engineering education consultant*
Rhiew, Francis Changnam *radiologist*

Sharon
Anttila, Samuel David *environmental scientist*

Shippensburg
Clark, Mary Diane *psychologist, educator*

Shippingport
Bly, Charles Albert *nuclear engineer, research scientist*

Shiremanstown
Willis, Donald J. *agricultural research professional*

Slippery Rock
Fallon, L(ouis) Fleming, Jr. *public health educator, researcher*
Hart, Robert Gerald *physiology educator*

Somerset
Critchfield, Harold Samuel *retired civil engineer*
Nair, Velupillai Krishnan *cardiologist*
Weaver, Craig Lee *civil engineer*

Spring Grove
Carter, Cecil Neal *environmental engineer*
Gleim, Jeffrey Eugene *research chemist*

Spring House
Andrade-Gordon, Patricia *biological scientist*
Caldwell, Gary Wayne *chemist, researcher*
De Jong, Gary Joel *chemist*
Emmons, William David *chemist*
Frederick, Clay Bruce *toxicologist, researcher*
Greer, Edward Cooper *chemist*
Klotz, Wendy Lynnett *analytical chemist*
Payn, Clyde Francis *technology company executive, consultant*

Springfield
Mirman, Merrill Jay *physician, surgeon*

State College
Baker, Dale Eugene *soil chemist*
Caldwell, Curtis Irvin *acoustical engineer*
Ginoza, William *retired biophysics educator*
Hall, David Lee *engineering executive*
Hansen, Robert J. *mechanical engineer*
Hettche, L. Raymond *research director*
Lannin, Jeffrey S. *physicist, educator*
Lauchle, Gerald Clyde *acoustics educator*
Mao, Xiaoping *fiber optics engineer, researcher*
Marktukanitz, Richard Peter *metallurgical engineer*
Moon, Marla Lynn *optometrist*
Ruud, Clayton Olaf *engineering educator*
Wang, Qunzhen *mechanical engineer*

Swarthmore
Hiltz, Arnold Aubrey *former chemist*
Pasternack, Robert Francis *chemistry educator*
Voet, Judith Greenwald *chemistry educator*

Swiftwater
Anthony, Damon Sherman *biologist, researcher*
Lee, Chung Keel *biologist*
Woods, Walter Earl *biomedical manufacturing executive*

Tamaqua
Cusatis, John Anthony *chemist*

Thorndale
Hodess, Arthur Bart *cardiologist*

Tobyhanna
Weinstein, William Steven *technical engineer*

Trafford
Hufton, Jeffrey Raymond *chemical engineer*

Trainer
Martin, Richard Douglas *nuclear engineer, consultant*

Trevose
Ziegler, Albert H. *mechanical engineer*

Trout Run
Michaels, Gordon Joseph *metals company executive*

Uniontown
Dimitric, Ivko Milan *mathematician, educator*

University Park
Achterberg, Cheryl Lynn *nutrition educator*
Amateau, Maurice Francis *materials scientist, educator*
Aplan, Frank Fulton *metallurgical engineering educator*
Badzian, Andrzej Ryszard *physicist*
Barnes, Hubert Lloyd *geochemistry educator*
Baum, Paul Frank *mathematics educator*
Berg, Clyde Clarence *research agronomist*
Biederman, Edwin Williams, Jr. *petroleum geologist*

Bieniawski, Zdzislaw Tadeusz mineral engineer, educator, consultant
Brenchley, Jean Elnora microbiologist, researcher
Buffington, Dennis Elvin agricultural engineering educator
Buskirk, Elsworth Robert physiologist, educator
Buss, Edward George geneticist
Carim, Altaf Hyder materials science and engineering educator
Córdova, France Anne-Dominic astrophysics educator
Cosgrove, Daniel Joseph biology educator
Cuffey, Roger J. paleontology educator
Dong, Cheng bioengineering educator
Dunson, William Albert biology educator
Edwards, Robert Mitchell nuclear engineering educator
Fisher, Charles Raymond, Jr. marine biologist
Fonash, Stephen Joseph engineering educator
German, Randall Michael materials science educator, consultant
Geselowitz, David Beryl bioengineering educator
Gu, Claire Xiang-Guang physicist
Hagen, Daniel Russell physiologist
Haghighat, Alireza nuclear engineering educator
Ham, Inyong industrial engineering educator
Hardy, Henry Reginald, Jr. geophysicist, educator
Hegarty, William Patrick chemical engineering educator
Hood, Lamartine Frain college dean
Jurs, Peter Christian chemistry educator
Koopmann, Gary Hugo educational center administrator, mechanical engineering educator
Kroger, Manfred food science educator
Kurtz, Stewart Kendall physics educator, researcher
Lakshminarayana, Budugur aerospace engineering educator
Leath, Kenneth Thomas research plant pathologist, educator
Macdonald, Digby Donald scientist, science administrator
Malone, Charles Trescott photovoltaic specialist
Manbeck, Harvey B. agriculturist, educator
Mathews, John David electrical engineering educator, consultant
McCormick, Barnes Warnock aerospace engineering educator
McDonnell, Archie Joseph environmental engineer
McKenna, Mark Joseph physicist, educator, researcher
Newnham, Robert Everest materials scientist, department chairman
Olivero, John Joseph, Jr. physics educator
Ovaert, Timothy Christopher mechanical engineering educator
Pal, Sibtosh mechanical engineer, researcher
Pennypacker, Barbara White plant pathologist
Proctor, Robert Neel biologist, historian, educator
Ramani, Raja Venkat mining engineering educator
Rashid, Kamal A. university international programs director, educator
Reischman, Michael Mack university official
Rolls, Barbara Jean biobehavioral health educator, laboratory director
Roy, Rustum interdisciplinary materials researcher, educator
Scanlon, Andrew structural engineering educator
Settles, Gary Stuart fluid dynamics educator
Skinner, Mark Andrew astrophysicist
Smith, Deane Kingsley, Jr. mineralogy educator
Sommer, Henry Joseph, III mechanical engineering educator
Stern, Robert Morris psychology educator and psychophysiology researcher
Stoneking, Mark Allen anthropologist
Tammen, James F. plant pathologist, educator
Taylor, Alan Henry geography educator, ecological consultant
Thomas, Joab Langston academic administrator, biology educator
Thomas King, McCubbin, Jr. physicist
Usher, Peter Denis astronomy educator
Vannice, M. Albert chemical engineering educator, researcher
Villafranca, Joseph J. biochemistry educator
Vrentas, Christine Mary chemical engineer, researcher
White, William Blaine geochemist, educator
Winograd, Nicholas chemist
Wu, Hsien-Jung research assistant

Upland
Green, Lawrence neurologist, educator

Valley Forge
Erb, Doretta Louise Barker polymer applications scientist

Vandergrift
Kulick, Richard John computer scientist, researcher

Villanova
Bush, David Frederic psychologist, educator
Edwards, John Ralph chemist, educator
McLaughlin, Philip VanDoren, Jr. mechanical engineering educator, researcher, consultant
Melby, Edward Carlos, Jr. veterinarian
Nataraj, Chandrasekhar mechanical engineering educator, researcher

Warminster
Leiby, Craig Duane control systems engineer, electrical design engineer
Mohl, David Bruce electronics engineer
Slawiatynsky, Marion Michael biomedical electronics engineer, software consultant
Tatnall, George Jacob aeronautical engineer
Vanderbeek, Frederick Hallet, Jr. chemical engineer

Warren
Sternberg, Mark Edward chemical engineer

Warrendale
Karrs, Stanley Richard chemical engineer
Scott, Alexander Robinson engineering association executive

Waverly
Matthews, Richard J. pharmaceutical research company executive

Wayne
Martino, Peter Dominic software company executive, military officer
McArdle, Joan Terruso parochial school mathematics and science educator
Murray, Pamela Alison quality data processing executive
Swank, Annette Marie computer software designer
West, Mark David software engineer

Waynesboro
Cryer, Theodore Hudson ophthalmologist, educator

Waynesburg
Maguire, Mildred May chemistry educator, magnetic resonance researcher

Wellsboro
Rottiers, Donald Victor physiologist

Wellsville
Shott, Edward Earl engineer, researcher

West Chester
Dionne, Ovila Joseph physicist
Siery, Raymond Alexander laboratory administrator
Skeath, Ann Regina mathematics educator
Vinokur, Roman Yudkovich physicist, engineer

West Chester University
Renner, Michael John psychologist, biologist, educator

West Mifflin
Ardash, Garin mechanical engineer
Smith, Stewart Edward physical chemist
Starmack, John Robert mathematics educator

West Point
Callahan Graham, Pia Laaster medical researcher, virology researcher
Donnelly, John James, III immunologist, blood banker
Gross, Dennis Michael pharmacologist
Hilleman, Maurice Ralph virus research scientist
Johnson, Robert Gahagen, Jr. medical researcher, educator
Kalejta, Paul Edward chemist
Katrinak, Thomas Paul analytical chemist
Maglaty, Joseph Louis data processor, consultant
Shafer, Jules Alan pharmaceutical company executive
Tomassini, Joanne Elizabeth virologist, researcher
Woolf, Eric Joel analytical chemist

Wheatland
Bolt, Michael Gerald metallurgist
Gruber, Jack Alan chemist

Wilkes Barre
Brooks, Charles Irving psychology educator
Bush, Kirk Bowen electrical engineering educator, consultant
Hayes, Wilbur Frank biology educator
Hernandez, Wilbert Eduardo physician
Kerns, James Albert structural engineer
Shukla, Kapil P. medical physicist

Williamsport
Fisher, David George physics educator
Leisey, April Louise Snyder chemist
Neff, Harold Parker environmental engineer, consultant
Owens, Edwin Geynet mathematics educator
Shuch, H. Paul engineering educator
Sutliff, Kimberly Ann psychologist

Willow Grove
Chatterjee, Hem Chandra electrical engineer
Schiffman, Louis F. management consultant
Spikes, John Jefferson, Sr. forensic toxicologist, pharmacologist

Willow Street
Ebling, Glenn Russell energy conservation executive

Wilmerding
Cunkelman, Brian Lee mechanical engineer

Wynnewood
Bordogna, Joseph educator, engineer
Harkins, Herbert Perrin otolaryngologist, educator
Schmaus, Siegfried H. A. engineering executive
Shurina, Robert David educator

Yardley
Tams, Thomas Walter civil and structural engineer, land surveyor

York
Casteel, Mark Allen psychology educator
Jacobs, Donald Warren dentist
Korenberg, Jacob mechanical engineer
Miller, Donald Kenneth engineering consultant
Monson, Raymond Edwin welding engineer
Moore, Walter Calvin chemical engineer
Pease, Howard Franklin structural engineer

RHODE ISLAND

Bristol
Von Riesen, Daniel D. chemistry educator

Cranston
Basu, Prithwish mechanical engineering researcher
Botelho, Robert Gilbert energy engineer
Fang, Pen Jeng engineering executive and consultant
Mruk, Charles Karzimer agronomist
Watt, Norman Ramsay chemistry educator, researcher

East Greenwich
O'Neill, Shawn Thomas waste water treatment executive

Esmund
Caduto, Ralph nuclear biochemist

Fiskeville
Yang, Sen chemist

Foster
Schuyler, Peter R. biomedical electrical engineer, educator, researcher

Kingston
Boothroyd, Geoffrey industrial and manufacturing engineering educator
Dewhurst, Peter industrial engineering educator
Driver, Rodney David mathematics educator, state legislator
Freeman, David Laurence chemist, educator
Goos, Roger Delmon mycologist
Laux, David Charles microbiologist, educator
McCorkle, Richard Anthony physicist
Palm, William John mechanical engineering educator
Patton, Alexander James engineer, consultant
Rossi, Joseph Stephen research psychologist
Sigurdsson, Haraldur oceanography educator, researcher
White, Frank M. mechanical engineer, educator

Lincoln
Chakoian, George aerospace engineer

Narragansett
Arimoto, Richard atmospheric chemist
Jacob, Ninni Sarah health physicist
Leinen, Margaret Sandra oceanographic researcher
Pilson, Michael Edward Quinton oceanography educator

Newport
Connery, Steven Charles computer scientist
Koch, Robert Michael research scientist, consultant
Mellberg, Leonard Evert physicist
Swamy, Deepak Nanjunda electrical engineer
Viana, Thomas Arnold computer scientist

North Providence
Stankiewicz, Andrzej Jerzy physician, biochemistry educator

North Scituate
Dupree, Thomas Andrew forester, state official

Providence
Amaral, Joseph Ferreira surgeon
Avissar, Yael Julia molecular biology educator
Biron, Christine Anne medical science educator, researcher
Capone, Antonio psychiatrist
Cooper, Leon N. physicist, educator
Dafermos, Constantine Michael applied mathematics educator
Dawicki, Doloretta Diane research biochemist, educator
Dempsey, Raymond Leo, Jr. radio and television producer, moderator, writer
DePetrillo, Paolo Bartolomeo medical educator
DiCamillo, Peter John software engineer
Dowben, Robert Morris physician, scientist
Elbaum, Charles physicist, educator, researcher
Erikson, George Emil (Erik Erikson) anatomist, archivist, historian, educator, information specialist
Freiberger, Walter Frederick mathematics educator, actuarial science consultant, educator
Freund, Lambert Ben engineering educator, researcher, consultant
Galletti, Pierre Marie artificial organ scientist, medical science educator
Gottschalk, Walter Helbig mathematician, educator
Hazeltine, Barrett electrical engineer, educator
Hitti, Youssef Samir biologist
Jaco, William H. mathematical association executive
Kates, Robert William geographer, educator
Knopf, Paul Mark immunoparasitologist
Lonks, John Richard physician
Martin, Horace Feleciano pathologist, law educator
Mehlman, Edwin Stephen endodontist
Merlino, Anthony Frank orthopedic surgeon
Mustard, John Fraser geologist, educator
North, David Lee medical physicist
Ocrant, Ian pediatric endocrinologist
O'Keeffe, Mary Kathleen psychology educator
Parks, Robert Emmett, Jr. medical science educator
Petrucci, Jane Margaret medical technologist, laboratory director
Rieger, Phillip Henri chemistry educator, researcher
Silverman, Harvey Fox engineering educator, dean
Stopa, Edward Gregory neuropathologist
Tabenkin, Alexander Nathan metrologist
Tauc, Jan physics educator
Vezeridis, Michael Panagiotis surgeon, researcher, educator
Walecki, Wojciech Jan physicist, engineer

Saunderstown
Knauss, John Atkinson federal agency administrator, oceanographer, educator, former university dean

Smithfield
Morahan-Martin, Janet May psychologist, educator

Wakefield
Fair, Charles Maitland neuroscientist, author
Mason, Scott MacGregor entrepreneur, inventor, consultant
Tarzwell, Clarence Matthew aquatic biologist

Warwick
Dickinson, Katherine Diana microbiologist
Dubois, Janice Ann primatologist, educator
Hunt, James H., Jr. safety and environmental executive
Jacoby, Margaret Mary astronomer, educator
Solomon, Richard Strean psychologist, educator

West Greenwich
Breakstone, Robert Albert consumer products company executive

Westerly
Varnhagen, Melvin Jay mechanical engineer

SOUTH CAROLINA

Abbeville
Cellura, Angele Raymond psychologist

Aiken
Bradley, Robert Foster chemical engineer
Coleman, Jerry Todd chemist
Danko, Edward Thomas mechanical and industrial engineer
Gleichauf, John George ophthalmologist
Groce, William Henry, III environmental engineer, consultant
Hernady, Bertalan Fred thermonuclear engineer
Hernandez-Martich, Jose David population and conservation biologist, educator
Hootman, Harry Edward nuclear engineer, consultant
Hyder, Monte Lee chemist
Miller, Phillip Edward environmental scientist
Murphy, Edward Thomas engineering executive
Novak, Charles R. computer scientist
Rood, Robert Eugene construction engineering executive, consultant
Smith, Michael Howard ecologist
Stewart, Michael Kenneth quality assurance professional
Voss, Terence J. human factors scientist, educator
Wilder, James Robbins mechanical engineer
Williamson, Thomas Garnett nuclear engineering and engineering physics educator
Young, Steve R. chemical engineer

Anderson
Goodner, Homer Wade safety risk analysis specialist, industrial process system failure risk consultant
Pflieger, Kenneth John architect

Arcadia
Hubbell, Douglas Osborne chemical engineering consultant

Charleston
Adinoff, Bryon Harlen psychiatrist
Apple, David Joseph ophthalmology educator
Arrigo, Salvatore Joseph biologist
Beidel, Deborah Casamassa clinical psychologist, researcher
Bennett, Jay Brett medical equipment company executive
Bissada, Nabil Kaddis urologist, educator, researcher, author
Bolin, Edmund Mike electrical engineer, franchise engineering consultant
Burrell, Victor Gregory, Jr. marine scientist
Cheng, Thomas Clement parasitologist, immunologist, educator, author
Delli Colli, Humbert Thomas chemist, product development specialist
Favaro, Mary Kaye Asperheim (Mrs. Biagino Philip Favaro) pediatrician
Fenn, Jimmy O'Neil physicist
Greene, George Chester, III chemical engineer
Hoel, David Gerhard state administrator, statistician, scientist
Johnson, Dewey E(dward) dentist
LeRoy, Edward Carwile rheumatologist
Margolius, Harry Stephen pharmacologist, physician
Ohning, Bryan Lawrence neonatologist, educator
Parker, Charles Dean, Jr. engineering company executive
Pharr, Pamela Northington physiology educator, researcher
Schuman, Stanley epidemiologist, educator
Scoggin, James Franklin, Jr. electrical engineering educator
Sens, Mary Ann pathology educator
Stroud, Sally Dawley nursing educator, researcher
Zealberg, Joseph James psychiatry and behavioral sciences educator

Cheraw
Williams, Morgan Lewis pharmacist

Clemson
Bunn, Joe Millard agricultural engineering educator
Clayton, Donald Delbert astrophysicist, nuclear physicist, educator
DeVault, William Leonard orthopedic surgeon
Fadel, Georges Michel mechanical engineering educator
Ferrell, William Garland, Jr. industrial engineering educator
Fescemyer, Howard William entomology educator, insect physiology reseacher
Figliola, Richard Stephen aerospace, mechanical engineer
Haile, James Mitchell engineering educator
Kapat, Jayanta Sankar mechanical engineer
Kimbler, Delbert Lee, Jr. industrial engineering educator
Lee, Daniel Dixon, Jr. nutritionist, educator
McNulty, Peter J. physics educator
Nevitt, Michael Vogt materials scientist, educator
Paul, Frank Waters mechanical engineer, educator, consultant
Taylor, Joseph Christopher audio systems engineer

Clinton
Buggie, Stephen Edward psychology educator

Columbia
Ahmed, Hassan Juma nuclear engineer
Bates, William Lawrence civil engineer
Belka, Kevin Lee electrical engineer
Brooker, Jeff Zeigler cardiologist
Cannon, Randy Ray civil engineer
Coleman, Robert Samuel chemistry researcher, educator
Cooper, William Allen, Jr. audiologist
Coull, Bruce Charles marine ecology educator
Datta, Timir physicist, solid state/materials consultant
Dawson, Wallace Douglas, Jr. geneticist
Dilworth, Stephen James mathematics educator
Dorgay, Charles Kenneth chemical engineer
Edge, Ronald Dovaston physics educator
Ernst, Edward Willis electrical engineering educator
Gandy, James Thomas meteorologist
Kanes, William Henry geology educator, research center administrator
Lee, Alexandra Saimovici civil engineer
Lin, Tu endocrinologist, educator, researcher, academic administrator
Linyard, Samuel Edward Goldsmith civil engineer
Lovell, Charles Rickey biologist, educator
Madden, Arthur Allen nuclear pharmacist
Newton, Rhonwen Leonard microcomputer consultant
Page, Tonya Fair cogeneration facility coordinator

Preedom, Barry Mason *physicist, educator*
Pumariega, JoAnne Buttacavoli *mathematics educator*
Ramsey, Bonnie Jeanne *mental health facility administrator, psychiatrist*
Rathbun, Ted Allan *anthropologist, educator*
Ritter, James Anthony *chemical engineer, educator, consultant*
Schuette, Oswald Francis *physics educator*
Schwarz, Ferdinand (Fred Schwarz) *ophthalmologist, ophthalmic plastic surgeon*
Shafer, John Milton *hydrologist, consultant, software developer*
Shea, Mary Elizabeth Craig *psychologist, educator*
Vernberg, Frank John *marine and biological sciences educator*
Whitlock, Edward Madison, Jr. *civil engineer*
Wilson, Scott Roland *physicist*

Conway
Skinner, Samuel Ballou, III *physics educator, researcher*

Duncan
Swafford, Stephen Scott *electrical engineer*

Eastover
Williams, Anthony M. *chemist*

Florence
Hunter, Nancy Quintero *obstetrician/gynecologist*
Kittrell, Benjamin Upchurch *agronomist*

Fort Mill
Hodge, Bobby Lynn *mechanical engineering director*
Montgomery, Terry Gray *textiles researcher*

Gaffney
Jones, Nancy Gale *retired biology educator*

Goose Creek
Reckamp, Douglas E. *military officer*
Sullivan, James *consultant*

Greenville
Akpan, Edward *metallurgical engineer*
Carlson, William Scott *structural engineer*
Chandler, Henry William *environmental engineer*
Cronemeyer, Donald Charles *physicist*
Csernak, Stephen Francis *structural engineer*
Cureton, Claudette Hazel Chapman *biology educator*
Gamble, Francoise Yoko *structural engineer*
Henning, Kathleen Ann *manufacturing systems engineer*
Holland, David Lee *environmental scientist, consultant*
Kilgore, Donald Gibson, Jr. *pathologist*
King, David Steven *quality control executive*
Mitchell, William Avery, Jr. *orthodontist*
Plumstead, William Charles *testing engineer, consultant*
Schneider, George William *retired aircraft design engineer*
Simrall, Dorothy Van Winkle *psychologist*
Stutes, Anthony Wayne *mechanical engineer*
Toth, James Joseph *chemical engineer*

Greer
Baker, Stephen Holbrooke *quality engineering executive*
Hawkins, Janet Lynn *school psychologist*
Roldan, Luis Gonzalez *materials scientist*

Hampton
Platts, Francis Holbrook *plastics engineer*

Hartsville
Edson, Herbert Robbins *hospital executive*
Menius, Espie Flynn, Jr. *electrical engineer*
Terry, Stuart L(ee) *plastics engineer*

Hilton Head Island
Hamlin, Scott Jeffrey *physicist*
Hogan, Joseph Charles *university dean*
Huckins, Harold Aaron *chemical engineer*
Stockard, Joe Lee *public health service officer, consultant*

Hollywood
Hull, Edward Whaley Seabrook *freelance writer, consultant*

Hopkins
Clarkson, Jocelyn Adrene *medical technologist*

Jenkinsville
Loignon, Gerald Arthur, Jr. *nuclear engineer*

Lake City
TruLuck, James Paul, Jr. *dentist, vintner*

Mount Pleasant
Carnes, Robert Mann *aerospace company executive*
Lambert, Richard Dale, Jr. *structural engineer, consultant*
Taylor, Jon Guerry *civil engineer*

Mullins
Rogers, Colonel Hoyt *agricultural consultant*

Myrtle Beach
Madory, James Richard *hospital administrator, former air force officer*

North Augusta
Dhom, Robert Charles *chemical engineer*
Ferguson, Kenneth Lee *nuclear engineer, engineering manager*
McRee, John Browning, Jr. *physician*
Rajendran, Narasimhan *civil engineer*

North Myrtle Beach
Atkinson, Harold Witherspoon *utilities consultant, real estate broker*

Orangeburg
Graule, Raymond S(iegfried) *metallurgical engineer*
Isa, Saliman Alhaji *electrical engineering educator*

Pawleys Island
Cepluch, Robert J. *retired mechanical engineer*

Rock Hill
Evans, Wallace Rockwell, Jr. *mechanical engineer*
Mattison, George Chester, Jr. *chemical company executive, consultant*
Mitchell, Paula Levin *biology educator, editor*
Prus, Joseph Stanley *psychology educator, consultant*
Waked, Robert Jean *chemical engineering executive*

Spartanburg
Armstrong, Joanne Marie *clinical psychologist, family mediator*
Fundenberg, Herman Hugh *research scientist*
Guthrie, John Robert *physician, health science facility administrator*
Hilton, Theodore Craig *computer scientist, computer executive*
Lee, Gary L. *engineering executive*
Stroup, David Richard *architect*
Vassy, David Leon, Jr. *radiological physicist*

Summerville
Orvin, George Henry *psychiatrist*
Singleton, Joy Ann *quality systems professional*

Sumter
Bryant, Wendy Sims *medical physicist*

West Columbia
Faust, John William, Jr. *electrical engineer, educator*
Salthouse, Thomas Newton *cell biologist, biomaterial researcher*

West Union
Klutz, Anthony Aloysius, Jr. *safety and environmental manager*

Winnsboro
Williams, Charles Oliver, Jr. *dentist*

York
Fritz, Edward William *mechanical engineer*

SOUTH DAKOTA

Aberdeen
Markanda, Raj Kumar *mathematics educator*

Brookings
Anderson, Gary Arlen *agricultural engineering educator, consultant*
Bretsch, Carey Lane *civil/architectural engineering administrator*
Duffey, George Henry *physics educator*
Hecht, Harry George *chemistry educator*
Morgan, Walter *retired poultry science educator*
O'Brien, John Joseph *physicist*
Swiden, Ladell Ray *research center director*

Mitchell
Tatina, Robert Edward *biology educator*

Rapid City
Gowen, Richard Joseph *electrical engineering educator, college president*
Islam, M. Rafiqul *petroleum engineering educator*
Johnson, L. Ronald *geophysicist*
Puszynski, Jan Alojzy *chemical engineer, educator*
Ramakrishnan, Venkataswamy *civil engineer, educator*
Smith, Paul Letton, Jr. *research scientist*

Sioux Falls
Sadler, James Bertram *psychologist, clergyman*
Van Demark, Robert Eugene, Sr. *orthopedic surgeon*
Zawada, Edward Thaddeus, Jr. *physician, educator*

Spearfish
Cox, Thomas Patrick *biological psychology educator*

Vermillion
Chadima, Sarah Anne *geologist*
Hammond, Richard Horace *geologist*

TENNESSEE

Alcoa
Cormia, Frank Howard *industrial engineering administrator*

Arlington
Van Horn, Mary Reneé *interior designer*

Arnold AFB
Daniel, Donald Clifton *aerospace engineer*

Bartlett
Orsak, Joseph Cyril *civil engineer*

Brentwood
Alley, E. Roberts *environmental engineer*
McCormick, Robert Jefferson *chemical engineer*

Bristol
Dirlam, David Kirk *psychology educator*
Hyer, Charles Terry *civil and mining engineer*

Brownsville
Kalin, Robert *retired mathematics educator*

Brunswick
Panicker, Mathew Mathai *nuclear engineer, educator*

Chapel Hill
Christman, Luther Parmalee *university dean emeritus, consultant*

Chattanooga
Clark, Jeff Ray *economist*
Durham, Lawrence Bradley *nuclear engineer, consultant, mediator*
Gibson, Kathy Halvey *nuclear reactor technology educator*
Johnson, Joseph Erle *mathematician*
Manaker, Arnold Martin *mechanical engineer, consultant*
Mason, Michael Edward *mechanical engineer*
Matthews, Michael Roland *environmental engineer*
Powell, Patricia Ann *mathematics and business educator*
Rogers, Ross Frederick, III *nuclear engineer*
Scrudder, Eugene Owen *chemist, environmental specialist*
Stevens, Donna Jo *nuclear power plant administrator*
Warren, Amye Richelle *psychologist, educator*
Williams, Robert Carlton *electrical engineer*

Clarksville
Franklin, Keith Barry *entrepreneur, technical consultant, former military officer*
Frayser, Michael Keith *electrical engineer*

Columbia
Cowan, Michael John *civil engineer*

Cookeville
Boswell, Fred C. *retired soil science educator, researcher*
Chamkha, Ali Jawad *research engineer*
Chowbey, Sanjay Kumar *mechanical engineer*
Chowdhuri, Pritindra *electrical engineer, educator*
Darvennes, Corinne Marcelle *mechanical engineering educator, researcher*
Kumar, Krishna *physics educator*
Lerner, Joseph *dean*
Mason, John Thomas, III *chemical engineering educator, consultant*
Russell, Josette Renee *industrial engineer*
Salmon, Fay Tian *mechanical engineering educator*
Tsatsaronis, George *mechanical engineering educator, researcher*
Ventrice, Marie Busck *engineering educator*

Dresden
Betz, Norman L. *science educator, consultant*

Dyersburg
Baker, Kerry Allen *household products company executive*
Bell, Helen Cherry *chemistry educator*

Erwin
Thompson, Mark Alan *environmental engineer*

Harrison
Fisher, Paul Douglas *psychologist, program director*

Hixson
Duckworth, Jerrell James *electrical engineer*

Jackson
Bearb, Michael Edwin *anesthesiologist*
Harwood, Thomas Riegel *pathologist*

Johnson City
Ferslew, Kenneth Emil *pharmacology and toxicology educator*
Isaac, Walter Lon *psychology educator*
Johnson, Dan Myron *biology educator*
Robbins, Charles Michael *teratologist, research developmental biologist*
Youngberg, George Anthony *pathology educator*

Jonesborough
Weaver, Kenneth *gynecologist, researcher*

Kingsport
Bagrodia, Shriram *chemical engineer*
Denton, David Lee *laboratory manager, chemical engineer*
Embree, Norris Dean *chemist, consultant*
Gray, T(heodore) Flint, Jr. *chemist*
Head, William Iverson, Sr. *retired chemical company executive*
Hendrix, Kenneth Allen *systems analyst*
Kashdan, David Stuart *chemist*
Lee, Kwan Rim *statistician*
Maggard, Bill Neal *mechanical engineer*
Mahaffey, Richard Roberts *information analyst*
Papas, Andreas Michael *nutritional biochemist*
Pecorini, Thomas Joseph *materials engineer*
Ryans, James Lee *chemical engineer*
Scott, H(erbert) Andrew *retired chemical engineer*
Sharma, Mahendra Kumar *chemist*
Siirola, Jeffrey John *chemical engineer*
Tant, Martin Ray *chemical engineer*
Watkins, William H(enry) *electrical engineer*

Kingston
Goranson, Harvey Edward *fire protection engineer*
Manly, William Donald *metallurgist*
Shacter, John *manager, technology/strategic planning consultant*

Knoxville
Bell, Corinne Reed *psychologist*
Borie, Bernard Simon, Jr. *physicist, educator*
Bressler, Marcus N. *science administrator*
Busmann, Thomas Gary *chemical engineer*
Campbell, William Buford, Jr. *materials engineer, chemist, forensic consultant*
Carroll, Roger Clinton *medical biology educator*
Chadha, Navneet *chemical engineer, environmental consultant*
Chen, Fang Chu *mechanical engineer*
Chen, James Pai-fun *biology educator, researcher*
Clark, Joseph Daniel *ecologist*
Cliff, Steven Burris *engineer*
Copper, Christine Leigh *chemist*
Danko, Joseph Christopher *metals engineer, university official*
Dean, John Aurie *chemist, author, chemistry educator emeritus*
Deeds, William Edward *physicist, educator*
Eryurek, Evren *nuclear engineer, researcher*
Gresshoff, Peter Michael *molecular geneticist, educator*
Hammer, Donald Arthur *ecologist*
Iwanski, Myron Leonard *environmental engineer*
Kidd, Janice Lee *nutritionist, consultant*
Kliefoth, A(rthur) Bernhard, III *neurosurgeon*
Kovac, Jeffrey Dean *chemistry educator*
Laroussi, Mounir *electrical engineer*
Martin, H(arry) Lee *mechanical engineer, robotics company executive*
Mashburn, John Walter *quality control engineer*
McGuire, John Albert *dentist*
Mc Hargue, Carl Jack *research laboratory administator*
Morton, Terry Wayne *architect*
Panjehpour, Masoud *research scientist, educator*
Parks, James Edgar *physicist, consultant*
Phelps, James Edward *electrical engineer, consultant*
Pitman, Frank Albert *aerospace engineer*
Prados, John William *educational administrator*
Richardson, Don Orland *agricultural educator*
Schuler, Theodore Anthony *civil engineer, city official*
Schweitzer, George Keene *chemistry educator*
Sherrill, Ronald Nolan *pharmacist, consultant*
Swingle, Homer Dale *horticulturist, educator*
Trigiano, Robert Nicholas *biotechnologist, educator*
White, David Cleaveland *microbial ecologist, environmental toxicologist*

Manchester
Howard, Robert P. *propulsion engineering scientist*

Maryville
Crisp, Polly Lenore *psychologist*
Lucas, Melinda Ann *health facility director*
Sievert, Lynnette Carlson *biologist, educator*

Memphis
Anthony, Frank Andrew *biochemist*
Buster, John Edmond *gynecologist, medical researcher*
Chung, King-Thom *microbiologist, educator*
Fain, John Nicholas *biochemistry educator*
Fang, Chunchang *physical chemist, chemical engineer*
Farrington, Joseph Kirby *microbiologist*
Gibson, Clifford William *military officer*
Godsey, William Cole *physician*
Guenter, Thomas Edward *retired chemical engineer*
Hawk, Charles Silas *computer programmer, analyst*
Heitmeyer, Mickey E. *conservationist*
Howe, Daniel Bo *aviation executive, retired naval officer*
Howe, Martha Morgan *microbiologist, educator*
Huggins, James Anthony *biology educator*
Jahan, Muhammad Shah *physicist*
Johnson, Johnny *research psychologist, consultant*
Johnston, Archibald Currie *geophysics educator, research director*
Kossmann, Charles Edward *cardiologist*
Kress, Albert Otto, Jr. *polymer chemist*
Kreuz, Roger James *psychology educator*
Lasslo, Andrew *medicinal chemist, educator*
Lazar, Rande Harris *otolaryngologist*
Lyman, Beverly Ann *biochemical toxicologist*
Mauer, Alvin Marx *physician, medical educator*
Mendel, Maurice *audiologist, educator*
Monroe, Daniel Milton, Jr. *biologist, chemist*
Nienhuis, Arthur Wesley *physician, researcher*
Ohman, Marianne *medical technologist*
Olcott, Richard Jay *data processing professional*
Pabst, Michael John *immunologist, dental researcher*
Patel, Tarun R. *pharmaceutical scientist*
Ramey, Harmon Hobson, Jr. *materials scientist*
Randle, Ronald Eugene *structural engineer*
Shanklin, Douglas Radford *physician*
Sullivan, Jay Michael *medical educator*
Watson, Donald Charles *cardiothoracic surgeon, educator*
Williams, Edward F(oster), III *environmental engineer*
Wolfson, Lawrence Aaron *hospital administrator*

Millington
Melcher, Jerry William Cooper *clinical psychologist, army officer*

Morristown
Culvern, Julian Brewer *chemist, educator, writer-naturalist*
Johnson, John Robert *petroleum company executive*

Mount Juliet
Auld, Bernie Dyson *civil engineer, consultant*

Mount Pleasant
Woodall, Larry Wayne *cement company executive*

Murfreesboro
Mitchell, Jerry Calvin *environmental company executive*
Schmidt, Constance Rojko *psychology educator, researcher*

Nashville
Abernethy, Virginia Deane *population and environment educator*
Ballou, Christopher Aaron *civil engineer*
Basu, Prodyot Kumar *civil engineer, educator*
Brase, David Arthur *neuropharmacologist*
Byron, Joseph Winston *pharmacologist*
Casillas, Robert Patrick *research toxicologist*
Charlton, Clivel George *neuroscientist, educator*
Cohen, Stanley *biochemistry educator*
Crofford, Oscar Bledsoe, Jr. *internist, medical educator*
Cullen, Marion Permilla *nutritionist*
Das, Salil Kumar *biochemist*
Eads, Billy Gene *electrical engineer*
Giorgio, Todd Donald *chemical engineering educator*
Grugel, Richard Nelson *materials scientist*
Guengerich, Frederick Peter *biochemistry educator, toxicologist, researcher*
Harlan, James Phillip *environmental engineer*
Hollemweguer, Enoc Juan *immunologist*
Hondeghem, Luc M. *cardiovascular and pharmacology educator*
Inagami, Tadashi *biochemist, educator*
Jones, Regi Wilson *data processing executive*
Klein, Christopher Carnahan *economist*
Martin, James Cullen *chemistry educator*
McClanahan, Larry Duncan *civil engineer, consultant*
Mellor, Arthur M(Cleod) *mechanical engineering educator*
Newman, John Hughes *medical educator*
Pan, Zhengda *physicist, researcher*
Parrish, Edward Alton, Jr. *electrical engineering educator, researcher*
Pearsall, Sam Haff *engineer, consultant*
Phillips, John A(tlas), III *geneticist, educator*
Pincus, Theodore *microbiologist, educator*

Pribor, Hugo Casimer physician
Quinn, Robert William physician, educator
Reinisch, Lou medical physics researcher, educator
Richmond, Ann White cell and molecular biologist, educator
Robertson, David clinical pharmacologist, physician, educator
Ross, Joseph Comer physician, educator, academic administrator
Russell, William Evans pediatric endocrinologist
Sephel, Gregory Charles biochemist, educator
Sloan, Paula Rackoff mathematics educator
Stewart, William Timothy environmental engineer
Stone, Michael Paul biophysical chemist, researcher
Tarbell, Dean Stanley chemistry educator
Taylor, Robert Bonds instructional designer
Thompson, Travis psychology educator, administrator, researcher
Wang, Taylor Gunjin science administrator, astronaut, educator
Westbrook, Fred Emerson agronomist, educator
Williams, Marsha Rhea computer scientist, educator, researcher, consultant
Wilson, James Richard mechanical engineer
Wilson, Sheryl A. pharmacist

Oak Ridge
Auerbach, Stanley Irving ecologist, environmental scientist, educator
Barkley, Linda Kay chemical analyst, spectroscopist
Barnes, Paul Randall electrical engineer
Basaran, Osman A. chemical engineer
Beecher, Stephen Clinton physicist
Bicehouse, Henry James health physicist
Blank, Merle Leonard biochemist, researcher
Burtis, Carl A., Jr. chemist
Carlsmith, Roger Snedden chemistry and energy conservation researcher
Cawley, Charles Nash enviromental scientist
Chen, Chung-Hsuan research physicist
Chen, Gwo-Liang physicist, researcher
Clayton, Dwight Alan electrical engineer
Cochran, Henry Douglas chemical engineer
Das, Sujit policy analyst
Davenport, Clyde McCall development engineer
Fontana, Mario H. nuclear engineer
Greene, David Lloyd transportation energy researcher
Grimes, James Gordon geologist
Gude, William D. retired biologist
Hightower, Jesse Robert chemical engineer, researcher
Holcombe, Cressie Earl, Jr. ceramic engineer
Kasten, Paul Rudolph nuclear engineer, educator
Kerchner, Harold Richard physicist, researcher
Korsah, Kofi nuclear engineer
Larson, Bennett Charles solid state physicist, researcher
Marshall, William Leitch chemist
Marz, Loren Carl environmental engineer and scientist, chemist
Mehrotra, Subhash Chandra hydrologist, consultant
Miller, John Cameron research chemist
Moore, Marcus Lamar mechanical engineer, consultant
Morrow, Roy Wayne chemist
Postma, Herman physicist, consultant
Rasor, Elizabeth Ann hydrogeologist, environmental scientist
Reed, Robert Marshall ecologist
Reichle, David Edward ecologist, biophysicist
Renshaw, Amanda Frances nuclear engineer
Roop, Robert Dickinson biologist
Sauers, Isidor physicist, researcher
Shapiro, Theodore chemical engineer
Snyder, Fred Leonard health sciences administrator
Specht, Eliot David physicist
Spray, Paul surgeon
Taylor, Paul Allen chemical engineer
Tonn, Bruce Edward social scientist, researcher
Toth, Louis McKenna chemist
Trivelpiece, Alvin William physicist, corporate executive
Veigel, Jon Michael corporate professional
Waters, Dean A. engineering executive
Weinberg, Alvin Martin physicist
Wesolowski, David Jude geochemist
Whealton, John H. physicist
Wong, Cheuk-Yin physicist
Zinkle, Steven John engineer, researcher

Ripley
Nunn, Jenny Wren pharmacist

Sewanee
Palisano, John Raymond biologist, educator

Soddy Daisy
Smith, Eual Randall mechanical engineer

South Pittsburg
Cordell, Francis Merritt instrument engineer, consultant

Trenton
McCullough, Kathryn T. Baker utilities executive

Tullahoma
Dahotre, Narendra Bapurao materials scientist, researcher, educator
Li, Liqiang physicist
Pate, Samuel Ralph engineering corporation executive
Sheth, Atul Chandravadan chemical engineering educator, researcher
Wu, Jie-Zhi fluid mechanist
Wu, Ying Chu Lin Susan engineering company executive, engineer

TEXAS

Abilene
Hennig, Charles William psychology educator
Pickens, Jimmy Burton earth and life science educator, military officer

Amarillo
Ayad, Joseph Magdy psychologist
Keaton, Lawrence Cluer engineer, consultant
Smith, Clinton W. civil engineer, consultant
Von Eschen, Robert Leroy electrical engineer, consultant

Arlington
Argento, Vittorio Karl environmental engineer
Caffey, James Enoch civil engineer
Cox, Verne Caperton psychology educator
Damuth, John Erwin marine geologist
Denny, Thomas Albert product development specialist, researcher
Ellwood, Brooks Beresford geophysicist, educator
Ickes, William psychologist, educator
Keller, Ben Robert, Jr. gynecologist
Kendall, Jillian D. information systems specialist, program developer, educator
Rayburn, Ray Arthur audio engineer
Rollins, Albert Williamson civil engineer, consultant
Smith, Russell Lamar biochemistry educator
Stanovsky, Joseph Jerry aerospace engineering educator
Tokerud, Robert Eugene electrical engineer
Wiig, Karl Martin management scientist, systems engineer

Austin
Abraham, Jacob A. computer engineering educator, consultant
Aggarwal, Jagdishkumar Keshoram electrical and computer engineering educator, research director
Albin, Leslie Owens biology educator
Arroyo-Vazquez, Bryan wildlife biologist
Baghai, Nina Lucille geology researcher
Baker, Lee Edward biomedical engineering educator
Bard, Allen Joseph chemist, educator
Bard, Jonathan F. mechanical engineering and operations research educator
Barlow, Joel William chemical engineering educator
Barrow, Arthur Ray chemical engineer
Bash, Frank Ness astronomer, educator
Biesele, John Julius biologist, educator
Bilby, Curt computer systems executive
Bishop, Robert Harold aerospace engineering educator
Bledsoe, Woodrow Wilson mathematics and computer sciences educator
Bohm, Arno Rudolf physicist
Boyer, Robert Ernst geologist, educator
Breen, John Edward civil engineer, educator
Britt, Chester Olen electrical engineer, consultant
Bronaugh, Edwin Lee research center administrator
Brown, Dennis Taylor molecular biology educator
Brown, Richard Malcolm, Jr. botany educator
Browne, James Clayton computer science educator
Burns, Ned Hamilton civil engineering educator
Burns, Robert Wayne computer scientist
Butzer, Karl W. archaeology and geography educator
Carrasquillo, Ramon Luis civil engineering educator, consultant
Charbeneau, Randall J. environmental and civil engineer
Chotiros, Nicholas Pornchai research engineer
Cobb, John Winston plasma physicist
Collier, Steven Edward utilities executive, consultant
Curle, Robin Lea computer software industry executive
Curran, Dian Beard physicist, consultant
Daniel, Mark Paul fiber optics network technician
Diehl, Randy Lee psychology educator, researcher
Dijkstra, Edsger Wybe computer science educator, mathematician
Downer, Michael C. physicist
Drake, Stephen Douglas clinical psychologist, health facility administrator
Drummond, William Eckel physics educator
Duncombe, Raynor Lockwood astronomer
Durkin, Anthony Joseph biomedical engineer, researcher
Eldredge-Thompson, Linda Gaile psychologist
Elequin, Cleto, Jr. retired physician
Estes, L(ola) Caroline aquarium store owner, operator
Fahrenthold, Eric Paul engineering educator
Fair, James Rutherford, Jr. chemical engineering educator, consultant
Fisher, William Lawrence geologist, educator
Flynn, William Thomas civil engineer
Folkers, Karl August chemistry educator
Fonken, Gerhard Joseph chemistry educator, university administrator
Fowler, David Wayne architectural engineering educator
Frank, Karl H. civil engineer, educator
Freid, James Martin mechanical engineer
Garner, Douglas Russell science writer
Gavande, Sampat Anand agricultural engineer, soil scientist
Gentle, Kenneth William physicist
Gillman, Leonard mathematician, educator
Gloyna, Earnest Frederick environmental engineer, educator
Gramann, Richard Anthony aerospace engineer
Guzik, Michael Anthony electrical engineer, consultant
Hammond, Charles Earl chemist, researcher
Hart, John Fincher construction management company executive
Hazeltine, Richard Deimel physics educator, university institute director
Hendricks, Terry Joseph mechanical engineer
Herb, Cynthia Johnson programmer
Herman, Robert physics educator
Hilburn, John Charles geologist, geophysicist
Himmelblau, David Mautner chemical engineer
Holley, Edward R. civil engineering educator
Holtzman, Wayne Harold psychologist, educator
Howell, John Reid mechanical engineer, educator
Hubbs, Clark zoologist, researcher
Hudson, William Ronald transportation engineering educator
Hull, David George aerospace engineering educator, researcher
Hurley, Laurence Harold medicinal chemistry educator
Jensen, Paul Allen electrical engineer
Jirsa, James Otis civil engineering educator
Kalthoff, Klaus Otto zoology educator
Keith, Roger Horn chemical engineer
Klein, Dale Edward nuclear engineering educator
Koros, William John chemical engineering educator
Kozmetsky, George computer science educator
Lake, Larry Wayne petroleum engineer
Lam, Simon Shin-Sing computer science educator
Lambert, David L. astronomy educator
Langerman, Scott Miles geotechnical engineer
Langston, Wann, Jr. paleontologist, educator, researcher emeritus
Larky, Steven Philip electrical engineer
Lewis, Richard Van chemist
Liljestrand, Howard Michael environmental engineering educator
Lobb, Michael Louis psychologist
Lopreato, Joseph sociology educator, author
Lundelius, Ernest Luther, Jr. vertebrate paleontologist, educator
Marcus, Harris Leon mechanical engineering and materials science educator
Mark, Hans Michael aerospace engineering educator, physicist
Maxwell, Arthur Eugene oceanographer, marine geophysicist, educator
McCullough, Benjamin Franklin transportation researcher, educator
McKenna, Thomas Edward hydrogeologist
Mc Ketta, John J., Jr. chemical engineering educator
McKinney, Daene Claude environmental engineering educator
Misra, Jayadev computer science educator
Moulthrop, James Sylvester research engineer, consultant
Murthy, Vanukuri Radha Krishna civil engineer
Newberger, Barry Stephen physicist, research scientist
Nguyen, Truc Chinh analytical chemist
Nicholas, Nickie Lee industrial hygienist
Nichols, Steven Parks mechanical engineer, academic administrator
Northington, David K. research center director, botanist, educator
Norton, Jerry Don biology educator
Oakes, Melvin Ervin Louis physics educator
Oden, John Tinsley engineering mechanics educator, consultant
Olson, Roy Edwin civil engineering educator
Park, Thomas Joseph biology researcher, educator
Paul, Donald Ross chemical engineer, educator
Pickhardt, Carl Emile, III psychologist
Popovich, Robert P. biochemical engineer, educator
Posey, Daniel Earl analytical chemist
Powell, Brian Hill software engineer
Pradzynski, Andrzej Henryk chemist
Prigogine, Vicomte Ilya physics educator
Psaris, Amy Celia manufacturing engineer
Pulich, Warren Mark, Jr. plant ecologist, coastal biologist
Ray, Robert Landon nuclear physicist
Reed, Lester James biochemist, educator
Reese, Lymon Clifton civil engineering educator
Richards-Kortum, Rebecca Rae biomedical engineering educator
Rylander, Henry Grady, Jr. mechanical engineering educator
Schade, Mark Lynn psychologist
Schechter, Robert Samuel chemical engineer, educator
Schless, James Murray internist
Sharp, John Malcolm, Jr. geology educator
Simpson, Beryl B. botany educator
Smith, Daniel Montague engineer
Sprinkle, James Thomas paleontologist, educator
Starr, Richard Cawthon botany educator
Stearns, Fred LeRoy controls engineer
Steinfink, Hugo chemical engineering educator
Stoffa, Paul L. geophysicist, educator
Straiton, Archie Waugh electrical engineering educator
Streetman, Ben Garland electrical engineering educator
Studer, James Edward geological engineer
Sturdevant, Wayne Alan engineering manager
Sutton, Harry Eldon geneticist, educator
Szebehely, Victor G. aeronautical engineer
Tapley, Byron Dean aerospace engineer, educator
Tesar, Delbert machine systems and robotics educator, researcher, manufacturing consultant
Thompson, Lawrence Franklin, Jr. computer corporation executive
Thornton, Joseph Scott research institute executive, materials scientist
Tucker, Richard Lee civil engineer, educator
Turner, Billie Lee botanist, educator
Uhlenbeck, Karen Keskulla mathematician, educator
Vishniac, Ethan Tecumseh astronomy educator
Wagner, William Michael data processing department administrator
Walton, Charles Michael civil engineering educator
Wehring, Bernard William nuclear engineering educator
Welch, Ashley James engineering educator
Weldon, William Forrest electrical and mechanical engineer, educator
Wenner, Edward James, III microelectronics executive, facility/safety engineer
Wheeler, John Craig astrophysicist, writer
Woodson, Herbert Horace electrical engineering educator
Wright, Stephen Gailord civil engineering educator, consultant
Young, Keith Preston geologist
Yura, Joseph Andrew civil engineering educator, consulting structural engineer

Barker
Hranitzky, E. Burnell medical physicist

Bastrop
O'Connell, Walter Edward psychologist

Baytown
Davis, Phillip Eugene oil company executive, chemical engineer
Johnson, Malcolm Pratt marketing professional
Lander, Deborah Rosemary chemist

Beaumont
Bianchi, Thomas Stephen biology and oceanography educator
Bollich, Charles N. agronomist
Hopper, Jack Rudd chemical engineering educator
Krehbiel, David Kent chemical engineer
Long, Alfred B. retired oil company executive, consultant
Pizzo, Joe physics educator
Robertson, Ivan Denzil, III chemical engineer
Roller, Richard Allen marine invertebrate physiologist
Schenck, Jack Lee electric utility executive
Streeper, Robert William environmental chemist
Walker, John Michael consulting environmental hydrogeologist
Williams, Curt Alan mechanical engineer
Zaloom, Victor Anthony industrial engineer, educator

Bellaire
Guest, Weldon S. biomedical products and services executive
Mayo, Clyde Calvin organizational psychologist
McKinzie, Howard Lee petroleum engineer
Pokorny, Alex Daniel psychiatrist
Wisch, David John structural engineer

Big Sandy
Mitchell, Deborah Jane educator

Big Spring
Fryrear, Donald William agricultural engineer

Boerne
Mitchelhill, James Moffat civil engineer

Bowie
Reynolds, Don William geologist

Brooks AFB
Donovan, Lawrence physicist
Goettl, Barry Patrick personnel research psychologist
Goodman, Howard Alan human factors engineer, air force officer
Nikolai, Timothy John aerospace engineer
Spector, Jonathan Michael research psychologist, cognitive scientist
Stamper, David Andrew psychologist
Tirre, William Charles research psychologist

Brownsville
Farst, Don David zoo director, veterinarian
Pena, Eleuterio utility executive

Bryan
Kellett, William Hiram, Jr. retired architect, engineer, educator
Van Arsdel, Eugene Parr tree pathologist, consultant meteorologist

Bushland
Howell, Terry Allen agricultural engineer

Camp Wood
Triplett, William Carryl physician, researcher

Carrollton
Graham, Richard Douglas computer company executive, consultant
Laurent, Duane Giles software engineer
Wang, Peter Zhenming physicist

Cleveland
Dolney, Tabatha Ann physics educator

College Station
Anderson, Duwayne Marlo earth and polar scientist, university administrator
Archer, Steven Ronald ecology educator
Barton, Derek Harold Richard chemist
Baskharone, Erian Aziz mechanical and aerospace engineering educator
Bass, George Fletcher archaeology educator
Batchelor, Bill civil engineering educator
Berg, Robert Raymond geologist, educator
Bradley, Walter Lee mechanical engineer, educator, researcher, consultant
Carter, Craig Nash veterinary epidemiologist, educator, researcher, software developer
Chiou, George Chung-Yih pharmacologist, educator
Christiansen, Dennis Lee transportation engineer
Coté, Gerard Laurence biomedical engineering educator
Cotton, Frank Albert chemist, educator
Duggal, Arun Sanjay ocean engineer
Ehsani, Mehrdad electrical engineering educator, consultant
Fletcher, Leroy Stevenson mechanical engineer, educator
Fliss, Albert Edward, Jr. chemical engineer
Frenz, Bertram Anton crystallographer
Gallaway, Bob M. consulting engineer
Gaston, Jerry Collins sociology educator
Goodman, David Wayne research chemist
Granger, Harris Joseph physiologist, educator
Hall, Timothy C. biology educator, consultant
Han, Je-Chin mechanical engineering educator
Hiler, Edward Allan academic administrator, agricultural engineering educator
Holtzapple, Mark Thomas biochemical engineer, educator
Jack, Steven Bruce forest ecologist, educator
Jaric, Marko Vukobrat physicist, educator, researcher
Kihm, Kyung D. (Ken Kihm) mechanical engineering educator
Kirk, Wiley Price, Jr. physics and electrical engineering educator
Kohel, Russell James geneticist
Koppa, Rodger Joseph industrial engineering educator
Kotzabassis, Constantinos agricultural engineering educator
Kunze, Otto Robert retired agricultural engineering educator
Latimer, George Webster, Jr. chemist
Lee, William John petroleum engineering educator, consultant
Lin, Guang Hai research scientist, consultant
Lowery, Lee Leon, Jr. civil engineer
Lusas, Edmund William food processing research executive
Lytton, Robert Leonard civil engineer, educator
Magnuson, Charles Emil physicist
Mathewson, Christopher Colville engineering geologist, educator
McIntyre, Peter Mastin physicist, educator
Morrison, Gerald Lee mechanical engineering educator
Mouchaty, Georges physicist
Murphy, Kathleen Jane psychologist
Natowitz, Joseph B. chemistry educator, administrator
Neff, Ray Quinn electric power consultant
O'Neal, Dennis Lee mechanical engineering educator
Pandey, Raghvendra Kumar physicist, educator
Parlos, Alexander George systems and control engineer
Peddicord, Kenneth Lee academic administrator
Reddell, Donald Lee agricultural engineer
Rezak, Richard geology and oceanography educator
Riggs, Penny Kaye cytogeneticist

Rohack, John James *cardiologist*
Rowe, Gilbert Thomas *oceanography educator*
Stanker, Larry Henry *biochemist*
Steffy, John Richard *nautical archaeologist, educator*
Stermer, Raymond Andrew *agricultural engineer*
Stipanovic, Robert Douglas *chemist, researcher*
Summers, Max (Duanne) *entomologist, scientist, educator*
Wagner, John Philip *safety engineering educator, science researcher*
Walker, Duncan Moore Henry *electrical engineer*
Willard, Scott Thomas *reproductive physiologist*
Williams, Karl Morgan *instrumentation engineer, researcher*
Wiltschko, David Vilander *geology educator, tectophysics director*
Yao, James Tsu-Ping *civil engineer*

Colleyville
Newton, Richard Wayne *food products executive*

Commerce
Betts, James Gordon *biology educator*

Conroe
Robbins, Jessie Earl *metallurgist*

Corinth
O'Keefe, Joseph Kirk *systems engineer*

Corpus Christi
Azopardi, Korita Marie *mathematics educator*
Cox, William Andrew *cardiovascular thoracic surgeon*
Cutlip, Randall Brower *retired psychologist, former college president*
Hammer, John Morgan *surgeon*
Lim, Alexander Rufasta *neurologist*
Plunkett, Roy J. *retired chemical engineer*
Smith, Jeremy Owen *chemist*

Cypress
Attles, LeRoy *aerospace engineer*
Day, Robert Michael *oil company executive*
Taylor, Paul Duane, Jr. *chemical engineer*

Dallas
Alberthal, Lester M., Jr. *information processing services executive*
Al-Hashimi, Ibtisam *oral scientist, educator*
Anderson, Douglas Warren *optics scientist*
Anderson, Jack Roy *health care company executive*
Austin, Robert Brendon *civil engineer*
Barnett, Peter Ralph *health science facility administrator, dentist*
Berbary, Maurice Shehadeh *physician, military officer, hospital administrator, educator*
Blackburn, Charles Lee *oil company executive*
Blattner, Wolfram Georg Michael *meteorologist*
Blessing, Edward Warfield *petroleum company executive*
Bowcock, Anne Mary *medical educator*
Breslau, Neil Art *endocrinologist, researcher, educator*
Brooks, James Elwood *geologist, educator*
Brown, Michael Stuart *geneticist*
Browne, Richard Harold *statistician, consultant*
Butler, Donald Philip *electrical engineer, educator*
Chang, Cheng-Hui (Karen) *medical physicist*
Chapman, Sandra Bond *neurolinguist, researcher*
Chatterjee, Amitava *electronics engineer*
Chen, Zhan *physicist*
Chu, Ting Li *electrical engineering educator, consultant*
Dance, William Elijah *industrial neutron radiologist, researcher*
Dean, Gary Neal *architect*
Dees, Tom Moore *internist*
Dudley, George William *behavioral scientist, writer*
Eidels, Leon *biochemistry educator*
Ellison, Luther Frederick *oil company executive*
Esin, Joseph Okon *computer information systems educator*
Farmer, Jim L(ee) *civil engineer, consultant*
Fischer, John Gregory *mechanical engineer*
Fischer Lindahl, Kirsten *biologist, educator*
Fordtran, John Satterfield *physician*
Frank, Steven Neil *chemist*
Free, Mary Moore *anthropologist*
Frenkel, Eugene Phillip *physician*
Garbers, David Lorn *biochemist*
Gerken, George Manz *neuroscientist, educator*
Gibbs, James Alanson *geologist*
Giesen, Herman Mills *engineering executive, consultant, mechanical forensic engineer*
Gilbert, Paul H. *engineer, consultant*
Gilman, Alfred Goodman *pharmacologist, educator*
Giniecki, Kathleen Anne *environmental engineer*
Goldstein, Joseph Leonard *physician, medical educator, molecular genetics scientist*
Gross, William Spargo *civil engineer*
Gyongyossy, Leslie Laszlo *mechanical engineer, consultant*
Hamzehee, Hossein Gholi *nuclear engineer*
Hanratty, Carin Gale *pediatric nurse practitioner*
Harbaugh, Lois Jensen *secondary science educator*
Hosmane, Narayan Sadashiv *chemistry educator*
Hudson, Courtney Morley *interior landscape designer*
Johnson, Gifford Kenneth *testing laboratory executive*
Junkins, Jerry R. *electronics company executive*
Karlson, Kevin Wade *trial, forensic, and clinical psychologist, consultant*
Kaufman, Tina Marie *physician recruiter, research consultant, presentation graphics consultant*
Kilby, Jack St. Clair *electrical engineer*
Klein, Edward Robert *pharmacist*
Kokkinakis, Demetrius Michael *biochemist, researcher*
Kress, Gerard Clayton, Jr. *psychologist, educator*
Kurt, Thomas Lee *medical toxicologist*
Lackey, James Franklin, Jr. *civil engineer*
Land, Geoffrey Allison *science administrator*
Langton, Michael John *mechanical engineer, consultant*
Lutz, Robert Brady, Jr. *engineering executive, consultant*
Maasoumi, Esfandiar *economics educator*
Margolin, Solomon Begelfor *pharmacologist*
Marks, James Frederic *pediatric endocrinologist, educator*
Marshall, John Harris, Jr. *geologist, oil company executive*
McCormick, J. Philip *natural gas company executive*
McCracken, Alexander Walker *pathologist*
McCracken, George H. *microbiologist*
McWhorter, Kathleen *orthodontist*
Menter, M(artin) Alan *dermatologist*
Metzler, Jerry Don *nursing administrator*
Miller, William *science administrator*
Mitchell, Jere Holloway *physiologist, researcher, medical educator*
Mize, Charles Edward *academic pediatrician*
Montgomery, Philip O'Bryan, Jr. *pathologist*
Mooz, Elizabeth Dodd *biochemist*
Murphy, John Carter *economics educator*
Norgard, Michael Vincent *microbiology educator, researcher*
O'Neill, Mark Joseph *solar energy engineer*
Oppenheim, Victor Eduard *consulting geologist*
Pakes, Steven P. *medical school administrator*
Parkey, Robert Wayne *radiology and nuclear medicine educator, research radiologist*
Peek, Leon Ashley *psychologist*
Poon, William Wai-Lik *program analyst*
Prihoda, James Sheldon *endocrinologist*
Qian, Dahong *laser development engineer*
Ramsbacher, Scott Blane *civil engineer*
Ravichandran, Kurumbail Gopalakrishnan *chemist*
Read, Leslie Webster *electrical engineer*
Reinert, James A. *entomologist, educator*
Rice, Charles Lane *surgical educator*
Ries, Edward Richard *petroleum geologist, consultant*
Roeser, Ross Joseph *audiologist, educator*
Rohrich, Rodney James *plastic surgeon, educator*
Sampath, Krishnaswamy *reservoir engineer*
Sanchez, Dorothea Yialamas *neuroscientist*
Schulze, Richard Hans *engineering executive, environmental engineer*
Schwitters, Roy Frederick *physicist, educator*
Scrivner, James Daniel *electrical engineer, consultant*
Senkayi, Abu Lwanga *environmental soil scientist*
Shaw, Dean Alvin *architect*
Shepard, Mark Louis *animal scientist*
Simon, Solomon Henry *artificial intelligence manager*
Simon, Theodore Ronald *physician, medical educator*
Smagula, Cynthia Scott *molecular biologist*
Srere, Paul A. *biochemist, educator*
Stacy, Dennis William *architect*
Stone, Marvin Jules *immunologist, educator*
Suppes, Trisha *neuroscientist*
Tew, E. James, Jr. *electronics company executive*
Tiernan, J(anice) Carter Matheney *mathematics specialist, consultant*
Toohig, Timothy E. *physicist*
Trahern, Charles Garrett *physicist*
Urquhart, Sally Ann *environmental scientist, chemist*
Vestal, Tommy Ray *lawyer*
Vitetta, Ellen S. *microbiologist educator, immunologist*
Ward, Derek William *lead systems analyst*
Wheeler, Clarence Joseph, Jr. *physician*
Wilmut, Charles Gordon *environmental engineer*
Wilson, Lee Britt *physiology educator*
Wilt, James Christopher *aerospace engineer*
Withrow, Lucille Monnot *nursing home administrator*
Wolf, Michael Ellis *clinical psychologist*
Womersley, William John *physicist*
Wooldridge, Scott Robert *semiconductor process engineer*
Yanagisawa, Shane Henry *civil engineer*
Yen, Anthony *engineer*
Zimmerman, S(amuel) Mort(on) *electrical and electronics engineering executive*

Dayton
Baysinger, Stephen Michael *air force officer*

De Soto
Marsh, Herbert Rhea, Jr. *dentist*

Deer Park
Brigman, James Gemeny *chemical engineer*
Taggart, Austin Dale, II *chemist*

Denton
Braterman, Paul Sydney *chemistry educator*
Das, Sajal Kumar *computer science educator, researcher*
Elder, Mark Lee *university research administrator, writer*
Goggin, Noreen Louise *kinesiology educator*
Head, Gregory Alan *mechanical engineer, consultant*
Hurdis, Everett Cushing *chemistry educator*
Kennelly, Kevin Joseph *psychology educator*
Kim, Yong-Dal *physicist*
Lana, Philip K. *emergency response educator, consultant*
Lancaster, Francine Elaine *neurobiologist*
Renka, Robert Joseph *computer science educator, consultant*
Schafer, Rollie Randolph, Jr. *neuroscientist*
Schwalm, Fritz Ekkehardt *biology educator*
Wagers, William Delbert, Jr. *theoretical geneticist*

El Campo
Tomlinson, Thomas King *petroleum engineer*

El Paso
Baptist, James Noel *biochemist*
Coleman, Howard S. *engineer, physicist*
Cuevas, David *psychologist*
Dalton, Oren Navarro *mathematician*
Harris, Arthur Horne *biology educator*
Johnson, Jerry Douglas *biology educator, researcher*
Malpass, Roy Southwell *psychology educator*
McClure, John Casper *materials scientist, educator*
Mitchell, Paula Rae *nursing educator*
Pearl, William Richard Emden *pediatric cardiologist*
Seaman, Edwin Dwight *physician, laboratory director*
Shadaram, Mehdi *electrical engineering educator*
Torres, Israel *oral and maxillofacial surgeon*
Wang, Paul Weily *physics educator*

Euless
Tunnell, Clida Diane *air transportation specialist*

Flower Mound
Baker, Scott Preston *computer engineer*

Fort Worth
Agarwal, Neeraj *anatomy and biology educator*
Andrews, Harvey Wellington *medical company executive*
Bakintas, Konstantine *civil, environmental design engineer*
Bhatia, Deepak Hazarilal *engineering executive*
Bonakdar, Mojtaba *chemistry educator*
Bradley, Richard Gordon, Jr. *aerospace engineer, engineering director*
Brooks, Lloyd William, Jr. *osteopath, interventional cardiologist, educator*
Buckner, John Kendrick *aerospace engineer*
Caldwell, Billy Ray *geologist*
Cunningham, Atlee Marion, Jr. *aeronautical engineer*
Demaree, Robert Glenn *psychologist, educator*
Dickman, Dean Anthony *mechanical engineer*
Doran, Robert Stuart *mathematics educator*
Estes, Jacob Thomas, Jr. *pharmacist, consultant*
Ewell, Wallace Edmund *transportation engineer*
Fischer, Marsha Leigh *civil engineer*
Gutsche, Carl David *chemistry educator*
Hines, Dwight Allen, II *family practice physician*
Hughes, Michael Wayne *systems engineer*
Jensen, Harlan Ellsworth *veterinarian, educator*
Kent, D. Randall, Jr. *engineerng company executive*
Knight, Kevin Kyle *research psychologist*
Madan, Sudhir Yashpal *chemical engineer*
Miller, Bruce Neil *physicist*
Mills, John James *research director, mechanical engineering educator*
Minter, David Edward *chemistry educator*
Moon, Billy G. *electrical engineer*
Narramore, Jimmy Charles *aerospace engineer*
Nation, Laura Crockett *electrical engineer*
Oakford, Lawrence Xavier *electron microscopist, laboratory administrator*
Papini, Mauricio Roberto *psychologist*
Parker, Robert Hallett *ecologist*
Raval, Dilip N. *physical biochemist*
Romine, Thomas Beeson, Jr. *consulting engineering executive*
Simpson, Dennis Dwayne *psychologist, educator*
Simrin, Harry S. *aerospace operations specialist*
Smith, Thomas Hunter *ophthalmologist, ophthalmic plastic and orbital surgeon*
Suba, Steven Antonio *obstetrician/gynecologist*
Szal, Grace Rowan *research scientist*
Vick, John *engineering executive*
Whiteley, James Morris *retired aerospace engineer*
Wingo, Paul Gene *oil company executive*

Freeport
Mercer, William Edward, II *chemical research technician*

Friendswood
Welch, George Osman *electrical engineer*
Wood, Loren Edwin *aerospace engineer, consultant*

Fulshear
Lurix, Paul Leslie, Jr. *chemist*

Galveston
Anderson, Karl Elmo *educator*
Bhat, Hari Krishen *biochemical toxicologist*
Budelmann, Bernd Ulrich *zoologist, educator*
Caillouet, Charles Wax, Jr. *fisheries scientist*
Dawson, Earl Bliss *obstetrics and gynecology educator*
Desai, Manubhai Haribhai *surgeon*
Felthous, Alan Robert *psychiatrist*
Fons, Michael Patrick *virologist*
Freeman, Daniel Herbert, Jr. *biostatistician*
Giam, Choo-Seng *marine science educator*
Gold, Daniel Howard *ophthalmologist, educator*
Head, Elizabeth Spoor *retired medical technologist*
Heggers, John Paul *surgery and microbiology educator, microbiologist, retired army officer*
Herndon, David N. *surgeon*
James, Thomas Naum *cardiologist, educator*
Koeppe, Patsy Poduska *internist, educator*
Liehr, Joachim Georg *pharmacology educator, cancer researcher*
McCombs, Jerome Lester *clinical cytogenetics laboratory director*
McKendall, Robert Roland *neurologist, virologist, educator*
McTigue, Teresa Ann *biologist, researcher, educator*
Miller, Todd Q. *social psychology educator*
Otis, John James *civil engineer*
Prakash, Louise *biophysics educator*
Puzdrowski, Richard Leo *neuroscientist*
Richardson, Carol Joan *pediatrician*
Santschi, Peter Hans *marine sciences educator*
Schmidly, David J. *dean*
Sheppard, Louis Clarke *biomedical engineer, educator*
Suzuki, Fujio *immunologist, educator, researcher*
Terrebonne, Annie Marie *medical technologist, educator, clinical laboratory scientist*
Weigel, Paul Henry *biochemistry educator, consultant*
Willis, William Darrell, Jr. *neurophysiologist, educator*
Yallampalli, Chandrasekhar *medical educator, researcher*
Yannariello-Brown, Judith I. *biomedical researcher, educator*

Garland
Jackson, Edwin L. *electrical engineer*

Georgetown
Camp, Thomas Harley *economist*
Gerding, Thomas Graham *medical products company executive*

Glen Rose
Ragan, James Otis *engineer, consultant*

Granbury
Crittenden, Calvin Clyde *retired engineering executive*
Wilson, Robert Storey *osteopathic physician*

Grand Prairie
Griffith, Carl Dean *electronics engineer*
Payne, Anthony Glen *clinical nutritionist, naturopathic physician*

Grapevine
Baker, Scott Michael *industrial engineer*
McIlheran, Mark Lane *mechanical engineer*

Greenville
Daniel, Eddy Wayne *mechanical engineer*
Hanson, David Alan *software engineer*
Johnston, John Thomas *engineering executive*

Haltom City
Kundu, Debabrata *mechanical engineer*

Harlingen
Ryall, A(lbert) Lloyd *horticulturist, refrigeration engineer*

Hereford
Carlile, Lynne *private school educator*

Horseshoe Bay
Ramey, James Melton *chemist*

Houston
Ahmad, Salahuddin *nuclear scientist*
Allman, Mark C. *physicist*
Appel, Stanley Hersh *neurologist*
Armeniades, Constantine D. *chemical engineer, educator*
Ashley, William Hilton, Jr. *software engineer*
Askew, William Earl *chemist, educator*
Aslam, Muhammed Javed *physician*
Bains, Elizabeth Miller *aerospace engineer*
Bally, Albert W. *geology educator*
Barlow, Nadine Gail *planetary geoscientist*
Barnes, Ronald Francis *mathematics educator*
Barrett, Bernard Morris, Jr. *plastic and reconstructive surgeon*
Barrow, Thomas Davies *oil and mining company executive*
Beauchamp, Jeffery Oliver *mechanical engineer*
Becker, Frederick Fenimore *cancer center administrator, pathologist*
Bell, Jerome Albert *aerospace engineer*
Billingsley, David Stuart *researcher*
Bishop, David Nolan *electrical engineer*
Bishop, Thomas Ray *retired mechanical engineer*
Black, David Charles *astrophysicist*
Boles, Jeffrey Oakley *biochemist, researcher*
Bonchev, Danail Georgiev *chemist, educator*
Boyce, Meherwan Phiroz *engineering executive, consultant*
Brandenstein, Daniel Charles *astronaut, naval officer*
Brandt, I. Marvin *chemical engineer*
Brody, Baruch Alter *medical educator, academic center administrator*
Brooks, Philip Russell *chemistry educator, researcher*
Brown, Jack Harold Upton *physiology educator, university official, biomedical engineer*
Bruce, Robert Douglas *acoustics, noise and vibration control consultant*
Bruner, Janet M. *neuropathologist*
Burns, Sally Ann *medical association administrator*
Cadwalder, Hugh Maurice *psychology educator*
Cameron, William Duncan *plastic company executive*
Cantwell, Thomas *geophysicist, electrical engineer*
Caprioli, Richard Michael *biochemist, educator*
Carter, James Sumter *oil company executive, tree farmer*
Chaku, Pran Nath *metallurgist*
Chamberlain, Joseph Wyan *astronomer, educator*
Chandra, Ajey *chemical engineer*
Chen, Guanrong (Ron Chen) *electrical engineer, educator, applied mathematician*
Chen, Min-Chu *mechanical engineer*
Cizek, John Gary *safety and fire engineer*
Clark, Carolyn Archer *technologist, scientist*
Clunie, Thomas John *chemical engineer*
Cohen, Aaron *aerospace engineer*
Coleman, Samuel Ebow *chemist, engineer*
Collipp, Bruce Garfield *ocean engineer, consultant*
Cooley, Denton Arthur *surgeon, educator*
Couch, Robert Barnard *physician, educator*
Cowen, Joseph Eugene *mechanical engineer*
Cox, Frank D. (Buddy Cox) *oil company executive, exploration consultant*
Crook, Troy Norman *geophysicist, consultant*
Cunningham, R. Walter *venture capitalist*
Curl, Robert Floyd, Jr. *chemistry educator*
Daily, Louis *ophthalmologist*
Danburg, Jerome Samuel *oil company executive*
Daniels, Cindy Lou *space agency executive*
Davids, Robert Norman *petroleum exploration geologist*
Davis, Barry Robert *biometry educator, physician*
Davis, Brian Richard *chemical engineer*
Dawn, Frederic Samuel *chemical, textile engineer*
Dawood, Mohamed Yusoff *obstetrician/gynecologist*
DeBakey, Michael Ellis *cardiovascular surgeon, educator*
Dennis, John Emory, Jr. *mathematics educator*
Dessler, Alexander Jack *space physics and astronomy educator, scientist*
Dhanabalan, Parthiban *electrical engineer*
Dice, Bruce Burton *exploration company executive*
Dickens, Thomas Allen *physicist*
Dilly, Ronald Lee *civil engineer, construction technology educator*
Dilsaver, Steven Charles *psychiatry educator*
Dougan, Deborah Rae *neuropsychology professional*
Downs, Hartley H., III *chemist*
Dreizen, Samuel *oncologist*
Dudrick, Stanley John *surgeon, educator*
Duke, Michael B. *aerospace scientist*
Dukler, Abraham Emanuel *chemical engineer*
Dunbar, Bonnie J. *engineer, astronaut*
Dwyer, John James *mechanical engineer*
Edwards, Victor Henry *chemical engineer*
Eichberger, LeRoy Carl *stress analyst, mechanical engineering consultant*
Elizardo, Kelly Patricia *chemical engineer*
Elliot, Douglas Gene *chemical engineer, engineering company executive, consultant*
Engel, James Harry *computer company executive*
Epright, Charles John *engineer*
Fabricant, Jill Diane *medical technology company executive*
Farach-Carson, Mary Cynthia *biochemist, educator*
Farley, Martin Birtell *geologist*
Farmer, Joe Sam *petroleum company executive*
Ferguson, Richard Peter *project engineer*
Fisher, Anna Lee *physician, astronaut*
Frenger, Paul Fred *medical computer consultant, physician*
Frost, John Elliott *minerals company executive*
Fukuyama, Tohru *organic chemistry educator*
Garnes, Delbert Franklin *clinical and consulting psychologist, educator*
Gest, Robbie Dale *aerospace engineer*
Gibson, Kathleen Rita *anatomy and anthropology educator*
Gibson, Michael Addison *chemical engineering company executive*
Gibson, Robert Lee *astronaut*
Gilbert, David Wallace *aerospace engineer*

Goerss, James Malcolm *statistician, electrical engineer*
Golbraykh, Isaak German *geologist*
Golubitsky, Martin Aaron *mathematician, educator*
Gonzalez, David Alfonso *environmental engineer*
Goodman, Herbert Irwin *petroleum company executive*
Grau, Raphael Anthony *reliability engineer*
Greenberg, Frank *clinical geneticist, educator, academic administrator*
Guinn, David Crittenden *petroleum engineer, drilling and exploration company executive*
Hackemesser, Larry Gene *mechanical engineer*
Hackerman, Norman *retired university president, chemist*
Haddox, Mari Kristine *biomedical scientist*
Hardt, Robert Miller *mathematics educator*
Hartsfield, Henry Warren, Jr. *astronaut*
Havrilek, Christopher Moore *technical specialist*
Heird, William Carroll *pediatrician, educator*
Heit, Raymond Anthony *civil engineer*
Henize, Karl Gordon *astronaut, astronomy educator*
Herz, Josef Edward *chemist*
Hollyfield, Joe G. *ophthalmology educator*
Holmquest, Donald Lee *physician, astronaut, lawyer*
Hong, Waun Ki *medical oncologist, clinical investigator*
Horton, William Arnold *medical geneticist, educator*
Housholder, Glenn Tholen *pharmacology educator*
Hurwitz, Charles Edwin *oil company executive*
Ivins, Marsha S. *aerospace engineer, astronaut*
Jacks, Jean-Pierre Georges Yves *chemical engineering sales executive*
Jackson, Douglas Webster *environmental scientist, consultant*
Jankovic, Joseph *neurologist, educator, scientist*
Jonke, Erica Elizabeth *aerospace engineer*
Jordan, Neal Francis *geophysicist, researcher*
Jorden, James Roy *oil company engineering executive*
Justice, (David) Blair *psychology educator, author*
Kachel, Wayne M. *environmental engineer*
Karacan, Ismet *psychiatrist, educator*
Karim, Amin H. *cardiologist*
Kellaway, Peter *neurophysiologist, researcher*
Kennedy, Ken *computer science educator*
Kershaw, Carol Jean *psychologist*
Kerwin, Joseph Peter *physician, former astronaut*
Kimmel, Marek *biomathematician, educator*
Kinsey, James Lloyd *chemist, educator*
Klausmeyer, David Michael *scientific instruments manufacturing company executive*
Kochi, Jay Kazuo *chemist, educator*
Krause, Kurt Lamont *biochemistry educator*
Kuhlke, William Charles *plastics engineer*
Kuntz, Hal Goggan *petroleum exploration company executive*
Kuo, Peter Te *cardiologist*
Kwok, Lance Stephen *optometry educator, vision researcher*
Lam, Daniel Haw *chemical engineer*
Landrum, Hugh Linson, Jr. *engineering executive*
Lane, Neal Francis *university provost, physics researcher*
Lawrence, Carolyn Marie *engineer*
Ledley, Tamara Shapiro *earth system scientist, climatologist*
Leonard, Gilbert Stanley *oil company executive*
Libshitz, Herman I. *radiologist, educator*
Little, Jack Edward *oil company executive*
Lopez-Nakazono, Benito *chemical engineer*
Lotze, Evie Daniel *psychodramatist*
Lotzová, Eva *immunologist, researcher, educator*
Lucid, Shannon W. *biochemist, astronaut*
Maligas, Manuel Nick *metallurgical engineer*
Man, Xiuting Cheng *acoustic and ultrasonic engineer, consultant*
Marcus, Donald Martin *internist, educator*
Margrave, John Lee *chemist, educator, university administrator*
Mariotto, Marco Jerome *psychology educator, researcher*
Mateker, Emil Joseph, Jr. *geophysicist*
Matney, William Brooks, VII *electrical engineer, marine engineer*
Matthews, Charles Sedwick *petroleum engineering consultant, research advisor*
McBride, Jon Andrew *astronaut, aerospace engineer*
Mc Bride, Raymond Andrew *pathologist, physician, educator*
McCants, Malcolm Thomas *chemical engineer*
McClay, Harvey Curtis *data processing executive*
McCown, Shaun Michael Patrick *chemist, consultant*
McKay, Colin Bernard *biomedical engineer*
McKechnie, John Charles *gastroenterologist, educator*
McLaughlin, Thomas Daniel *aerospace engineer*
Mehra, Jagdish *physicist*
Meyer, John Stirling *neurologist, educator*
Mian, Farouk Aslam *chemical engineer, educator*
Middleditch, Brian Stanley *biochemistry educator*
Miele, Angelo *engineering educator, researcher, consultant, author*
Mifflin, Richard Thomas *applied mathematician*
Milam, John Daniel *pathologist, educator*
Miller, Janel Howell *psychologist*
Mohrmann, Leonard Edward, Jr. *chemical engineer*
Montijo, Ralph Elias, Jr. *engineering executive*
Moore, Fay Linda *systems analyst*
Moore, Walter Parker, Jr. *civil engineering company executive*
Mosher, Donald Raymond *chemical engineer, consultant*
Moss, Simon Charles *physics educator*
Mossavar-Rahmani, Bijan *oil and gas company executive*
Mountain, Clifton Fletcher *surgeon, educator*
Naqvi, Sarwar *aerospace engineer*
Neidell, Norman Samson *oil and gas exploration consultant*
Nelson, David Loren *geneticist, educator*
Newman, Paul Wayne *dentist, cattle rancher*
Nicastro, David Harlan *forensic engineer, consultant, author*
Nichols, Buford Lee, Jr. *physiologist*
Nordgren, Ronald Paul *civil engineering educator, research engineer*
Nordlander, Jan Peter Arne *physicist, educator*
NosÉ, Yukihiko *surgeon, educator*
Novy, Diane Marie *psychologist*
Okafor, Michael Chukwuemeka *chemical engineer*
O'Malley, Bert William *cell biologist, educator, physician*
O'Neill, Michael James *medical physicist*
O'Neill, Michael Wayne *civil engineer, educator*
Ordonez, Nelson Gonzalo *pathologist*
Overmyer, Robert Franklyn *astronaut, marine corps officer*
Oxer, John Paul Daniell *civil engineer*
Page, Valda Denise *epidemologist, researcher, nutritionist*
Parker, Norman Neil, Jr. *software engineer, mathematics educator*
Patten, Bernard Michael *neurologist, educator*
Patterson, Donald Eugene *research scientist*
Peng, Liang-Chuan *mechanical engineer*
Pharr, George Mathews *materials science and engineering educator*
Phinney, William Charles *geologist*
Pierce, George Foster, Jr. *architect*
Piper, Lloyd Llewellyn, II *engineer, service industry executive*
Pollock, Raphael Etomar *surgeon, educator*
Pomerantz, James Robert *psychology educator, academic admininstrator*
Popek, Edwina Jane *pathologist*
Porcher, Frank Bryan, II *systems engineer*
Powell, Alan *mechanical engineer, scientist, educator*
Prather, Rita Catherine *psychology educator*
Pratt, David Lee *oil company executive*
Price, Frederick Clinton *chemical engineer, editorial consultant*
Puddy, Donald Ray *mechanical engineer*
Rabson, Thomas Avelyn *electrical engineering educator, researcher*
Rajapaksa, Yatendra Ramyakanthi Perera *structural engineer*
Raskin, Steven Allen *academic program director*
Rensimer, Edward R. *internist, educator*
Rochelle, William Curson *aerospace engineer*
Roney, Lynn Karol *psychologist*
Rosenthal, Alan Irwin *geophysicist*
Ross, Patti Jayne *obstetrics and gynecology educator*
Rubio, Pedro A. *cardiovascular surgeon*
Rudolph, Andrew Henry *dermatologist, educator*
Rudolph, Frederick Byron *biochemistry educator*
Rypien, David Vincent *materials research engineer*
Salerno, Philip Adams *infosystems specialist*
Sandstrum, Steve D. *engineer, marketing manager*
Saunders, Grady F. *biochemistry and biology educator, researcher*
Saunders, Richard Wayne *pipeline engineer*
Schachtel, Barbara Harriet Levin *epidemiologist, educator*
Schroepfer, George John, Jr. *biochemistry educator*
Schull, William J. *genetecist, educator*
Schultz, Philip Stephen *engineering executive*
Schultz, Stanley George *physiologist, educator*
Scott, Carl Douglas *aerospace engineer*
Sell, Stewart *pathologist, immunologist, educator*
Shakes, Diane Carol *developmental biologist, educator*
Sharp, David Paul *computer technologist, researcher*
Shearer, William T. *pediatrician, educator*
Shen, Liang Chi *electrical engineer, educator, researcher*
Sinclair, A(lbert) Richard *petroleum engineer*
Sloan, Harold David *chemical engineering consultant*
Smalley, Arthur Louis, Jr. *engineering and construction company executive*
Smalley, Richard Errett *chemistry and physics educator, researcher*
Smith, Lloyd Hilton *independent oil and gas producer*
Smith, Thomas Bradford *engineering administrator*
Soileau, Kerry Michael *aerospace technologist, researcher*
Soltau, Ronald Charles *chemical engineer*
Spanos, Pol Dimitrios *engineering educator*
Stark, George Edward *software reliability engineer, statistician*
Stehlin, John Sebastian, Jr. *surgeon*
Stewart, Robert Jackson *software development engineer, researcher*
Stimson, Paul Gary *pathologist*
Sullivan, Kathryn D. *geologist, astronaut*
Sun, Yanyi *research scientist*
Sung, Ming *chemical engineer*
Talwani, Manik *geophysicist, educator*
Tellez, George Henry *safety professional, consultant*
Thagard, Norman E. *astronaut*
Thomas, David Wayne *engineer*
Thornton, Kathryn C. *physicist, astronaut*
Ting, Paul Cheng Tung *research scientist, inventor*
Tiras, Herbert Gerald *engineering executive*
Tomasovic, Stephen Peter *radiobiologist, educator*
Trkula, David *biophysics educator*
Tucker, Randolph Wadsworth *engineering executive*
Umrigar, Dara Nariman *civil engineer*
Vail, Peter Robbins *geologist*
Valencia, Jaime Alfonso *chemical engineer*
Valrand, Carlos Bruno *aerospace engineer*
Van Sickels, Martin John *chemical engineer*
Vassilopoulou-Sellin, Rena *medical educator*
Vipulanandan, Cumaraswamy *civil engineer, educator*
Wade, James William *aerospace engineer*
Walker, Esper Lafayette, Jr. *civil engineer*
Wang, Chao-Cheng *mathematician, engineer*
Weinstein, Roy *physics educator, researcher*
Weinstock, George Matthew *biology educator, researcher*
Wesselski, Clarence J. *aerospace engineer*
Westby, Timothy Scott *oil company research engineer*
Wilcox, Richard Cecil *information systems executive*
Will, Edward Edmund *oil company executive*
Williams, M. Wright *psychologist, educator*
Williams, Robert Henry *oil company executive*
Wilson, Thomas Leon *physicist*
Wonnacott, James Brian *physician*
Woo, Walter *computer systems consultant*
Woods, Stephanie Elise *computer information scientist*
Wray, Richard Bengt *systems engineer*
Wren, Robert James *aerospace engineering manager*
Wright, Clark Phillips *computer systems specialist*
Young, John Watts *astronaut*
Zeissig, Hilmar Richard *oil and gas executive*
Zimmerman, Stuart ODell *biomathematician, educator*
Zlatkis, Albert *chemistry educator*

Humble
Brown, Samuel Joseph, Jr. *mechanical engineer*
Hahne, C. E. (Gene Hahne) *computer services executive*
Stevens, Elizabeth *psychotherapist*
Trowbridge, John Parks *physician, nutritional medicine specialist, joint treatment specialist*

Huntsville
Vick, Marie *retired health science educator*

Hurst
Bishara, Amin Tawadros *mechanical engineer, technical services executive*

Ingelside
Naismith, James Pomeroy *civil engineer*

Irving
Donnelly, Barbara Schettler *medical technologist*
Fukui, George Masaaki *microbiology consultant*
Garcia, Raymond Lloyd *dermatologist*
Halter, Edmund John *mechanical engineer*
Holdar, Robert Martin *chemist*
McBrayer, H. Eugene *retired petroleum industry executive*
McCormack, Grace Lynette *engineering technician*
McVay, Barbara Chaves *mathematics educator*
Rees, Frank William, Jr. *architect*
Spies, Jacob John *health care executive*
Taylor, Lawrence Stanton *software engineer*

Katy
Tal, Jacob *obstetrician/gynecologist*

Kerrville
Drummond, Roger Otto *livestock entomologist*
Kunz, Sidney *entomologist*
Taylor, William Logan *electrical engineer*

Kilgore
Springer, Andrea Paulette Ryan *physical therapist, biology educator*

Killeen
Roberts, Mary Lou *school psychologist*
Wells, James David, Jr. *military officer*

Kingsville
Diersing, Robert Joseph *computer science educator*
Kruse, Olan E. *physics educator*
Morey, Philip Stockton, Jr. *mathematics educator*
Perez, John Carlos *biology educator*
Suson, Daniel Jeffrey *physicist*

Kingwood
Bowman, Stephen Wayne *quality assurance engineer, consultant*
Burghduff, John Brian *mathematics educator*
Heldenbrand, David William *civil engineer*
Rademacher, John Martin *sanitary engineer*

La Porte
Wienert, James Wilbur *chemical engineer*
Young, George Hansen *chemist*

Lackland AFB
Charlesworth, Ernest Neal *immunologist, educator*
Cody, John Thomas *forensic toxicologist, biological chemist researcher*
Dixon, Patricia Sue *medical biotechnologist, researcher*
Ward, William Wade *clinical immunologist*

LaGrange
Riehs, John Daryl *state agency administrator*

League City
Kanuth, James Gordan *chemical engineer*
Lamar, James Lewis, Jr. *chemical engineer*

Lewisville
Ross, Lesa Moore *quality assurance engineer*

Lindale
Bockhop, Clarence William *retired agricultural engineer*

Longview
Canfield, Glenn, Jr. *metallurgical engineer*
Robinson, Alfred G. *petroleum chemist*
Turner, Carl Jeane *international business consultant, electronics engineer*
Winn, Walter Terris, Jr. *civil/environmental engineer*

Lubbock
Amir-Moez, Ali Reza *mathematician, educator*
Curl, Samuel Everett *university dean, agricultural scientist*
Dasgupta, Purnendu Kumar *chemist, educator*
Doris, Peter A. *biomedical scientist*
Elder, Bessie Ruth *pharmacist*
Engel, Thomas Gregory *electrical engineer, educator*
Green, Joseph Barnet *neurologist*
Hennessey, Audrey Kathleen *information systems educator*
Hentges, David John *microbiology educator*
Huynh, Alex Vu *electrical engineer*
Illner-Canizaro, Hana *physician, oral surgeon, researcher*
Ishihara, Osamu *electrical engineer, physicist, educator*
Jackson, Raymond Carl *cytogeneticist*
Jorgensen, Eric Edward *wildlife ecologist, researcher*
Kiesling, Ernst Willie *civil engineering educator*
Koh, Pun Kien *retired educator, metallurgist, consultant*
Laing, Malcolm Brian *geologist, consultant*
Lundberg, Alan Dale *electrical engineer*
Mathews, Nancy Ellen *wildlife ecologist, educator*
McGlone, John James *biologist*
Rummel, Don *agronomist*
Shires, George Thomas *surgeon, physician, educator*
Winter, Jimmy Dale *ecology educator*
Winter, Mark Lee *physician*
Young, Teri Ann Butler *pharmacist*
Zemtsov, Alexander *dermatology and biochemistry educator*

Lufkin
Perry, Lewis Charles *emergency medicine physician, osteopath*

Marshall
Ford, Clyde Gilpin *chemistry educator*

Mcallen
Farias, Fred, III *optometrist*
Julian, Elmo Clayton *analytical chemist*

McGregor
Johnson, Gary Wayne *aerospace engineer*

McKinney
Emge, Thomas Michael *electrical engineer*

Meadows
Jeffrey, Marcus Fannin *control systems engineer*

Mesquite
Jacobs, Mark Elliott *electronics engineer*

Midland
Groce, James Freelan *petroleum engineer*
Grover, Rosalind Redfern *oil and gas company executive*
Riebe, Susan Jane *environmental engineer*

Missouri City
Edwards-Hollaway, Sheri Ann *civil engineer*

Mont Belvieu
Raczkowski, Cynthia Lea *chemist*

Montgomery
Falkingham, Donald Herbert *oil company executive*

Nacogdoches
Brennan, Thomas George, Jr. *audiologist, speech-language pathologist*
Clagett, Arthur F(rank), Jr. *psychologist, sociologist, qualitative research writer, retired sociology educator*
Ludorf, Mark Robert *cognitive psychologist, educator*
Speer, James Ramsey *developmental psychologist*

Nassau Bay
Kimzey, John Howard *chemical and aerospace engineer*

New Braunfels
Brown, Marvin Lee *retired air force officer*
Wilson, James Lee *retired geology educator, consultant*

Odessa
Armstrong, Gerald Carver *mining engineer*
Kurtz, Edwin Bernard *biology educator, researcher*
McCullough, Gary William *psychology educator, researcher*
Reeves, Robert Grier LeFevre *geology educator, scientist*

Orange
Cardner, David Victor *chemical engineer*
Oyekan, Soni Olufemi *chemical engineer*
Russell, Kenneth William *chemical engineer*

Pasadena
Brown, Robert Griffith *chemist, chemical engineer, chemicals executive*
Cheng, Chung P. *chemical engineer*
Griffin, John Joseph, Jr. *chemist, video producer*
Keyworth, Donald Arthur *technical development executive*
Meyer, Kathleen Anne *school psychologist*
Stephens, Sidney Dee *chemical manufacturing company executive*
Tagoe, Christopher Cecil *chemical engineer*

Pearland
Oman, Paul Richard *entrepreneur*

Pflugerville
Johnson, David Lee *mechanical building systems engineer*

Pittsburg
Vierra, Frank Huey *environmental engineer, consultant*

Plano
Andrews, Judy Coker *electronics company executive*
Broyles, Michael Lee *geophysics and physics educator*
Franklin, Thomas Doyal, Jr. *medical research administrator*
Good, David Michael *engineer*
Hinton, Norman Wayne *information services executive*
Kean, James Allen *petroleum engineer*
Montgomery, John Henry *electronics company executive*
Reid, William Michael *mechanical engineer*
Tolle, Glen Conrad *mechanical engineer*
Westfall, Linda Louise *software engineer, metrics analyst*

Point Comfort
Martinez, Mario Antonio *chemical engineer*

Port Aransas
Koepsel, Wellington Wesley *electrical engineering educator*

Port Lavaca
Anfosso, Christian Lorenz *analytical chemist*

Port Neches
Ruth, Stan M. *chemical engineer*

Pottsboro
Hanning, Gary William *utility executive, consultant*

Prairie View
Braithwaite, Cleantis Esewanu *molecular biologist*

Randolph AFB
Ling, Edward Hugo *mechanical engineer*

Raymondville
Montgomery-Davis, Joseph *osteopathic physician*

Richardson
Adamson, Dan Klinglesmith *association executive*
Baker, Thaddeous Joseph *electronics engineer*
Balsara, Poras Tehmurasp *electrical engineering educator*
Gauthier, Jon Lawrence *telecommunications engineer*
Gerhardt, Douglas L. *computer scientist*

Hanson, William Bert physics educator, science administrator
Hope, James Dennis, Sr. components engineer
Kinsman, Frank Ellwood engineering executive
O'Neal, Stephen Michael information systems consultant
Ramasamy, Ravichandran chemistry educator
Ward-McLemore, Ethel research geophysicist, mathematician
Wiorkowski, John James mathematics educator

Richmond
Hay, Richard Carman anesthesiologist

Roanoke
Dodson, George W. computer company executive, consultant
Raz, Tzvi industrial engineer

Rockport
Jones, Lawrence Ryman retired research scientist

Rockwall
Rickards, Michael Anthony aerospace engineer
Sparks, Sherman Paul osteopathic physician

San Angelo
Cline, David Christopher geologist
Dunham, Gregory Mark obstetrician/gynecologist
Harrington, William Palmer retired civil engineer
Menzies, Carl Stephen agricultural research administrator, ruminant nutritionist

San Antonio
Abramson, Hyman Norman engineering and science research executive
Agrawal, Chandra Mauli biomaterials engineer, educator
Aldridge, Daniel aerospace engineer
Bach, Stephan Bruno Heinrich chemistry educator
Baker, Helen Marie health services executive
Barnes, Betty Rae counselor
Baughman, John Thomas data base designer, educator
Blaylock, Neil Wingfield, Jr. applied statistician, educator
Bloom, Wallace psychologist
Bose, Animesh materials scientist, engineer
Budalur, Thyagarajan Subbanarayan chemistry educator
Burno, John Gordon, Jr. microbiologist
Burnside, Otis Halbert mechanical engineer
Butler, Clark Michael aerospace engineer
Cardenas, John I. electrical engineer, consultant
Chan, Kwai Shing materials engineer, researcher
Cook, Harold Rodney military officer, medical facility administrator
Deviney, Marvin Lee, Jr. research institute scientist, program manager
Di Maio, Vincent Joseph Martin forensic pathologist
Doane, Thomas Roy environmental toxicologist
Donaldson, Willis Lyle research institute administrator
Edlund, Carl E. physicist
Esparza, Edward Duran mechanical engineer
Fodor, George Emeric chemist
Goland, Martin research institute executive
Gomes, Norman Vincent retired industrial engineer
Grubb, Robert Lynn computer system designer
Hall, Brad Bailey orthopaedic surgeon, educator
Harms, John Martin electrical engineer
Hausheer, Frederick Herman internist, cancer researcher, pharmaceutical company officer
Henderson, Arvis Burl data processing executive, biochemist
Horton, Granville Eugene nuclear medicine physician, retired air force officer
Jensen, Andrew Oden obstetrician/gynecologist
Jones, James Ogden geologist, educator
Jorgensen, James H. pathologist, educator, microbiologist
Juhasz, Stephen editor, consultant
Keeler, Jill Rolf pharmacologist, army officer, consultant
Lyle, Robert Edward chemist
Masters, Bettie Sue Siler biochemist, educator
Miles, Paula Effette mechanical engineer
Murthy, Krishna Kesava infectious desease and immunology scientist
Mylar, J(ames) Lewis psychologist, consultant
Ostmo, David Charles engineering executive
Owen, Thomas Edwin electrical engineer
Page, Richard Allen materials scientist
Panda, Markandeswar chemistry researcher
Patterson, Jan Evans epidemiologist, educator
Petty, Olive Scott geophysical engineer
Phillips, Thomas Dean mechanical engineer
Phillips, William Thomas nuclear medicine physician, educator
Poarch, Mary Hope Edmondson science educator
Quinn, Mary Ellen science educator
Renthal, Robert David biochemist, educator
Ribble, Ronald George psychologist, educator, writer
Romero, Emilio Felipe psychiatry educator, psychotherapist, hospital administrator
Sablik, Martin John research physicist
Schnitzler, Robert Neil cardiologist
Sherman, James Owen psychologist
Skelton, John Goss, Jr. psychologist
Sloan, Tod Burns anesthesiologist, researcher
Smith, John Marvin, III surgeon, educator
Smith, Reginald Brian Furness anesthesiologist, educator
Stebbins, Richard Henderson electronics engineer, peace officer, security consultant
Thacker, Ben Howard civil engineer
Tokoly, Mary Andree microbiologist
Townsend, Frank Marion pathology educator
Trench, William Frederick mathematics educator
Tsin, Andrew Tsang Cheung biochemistry educator
Vafeades, Peter mechanical engineering educator
Walker, Timothy John aerospace engineer
Weinbrenner, George Ryan aeronautical engineer
Wieland, David Henry aerospace structural engineer
Williams, Thomas Eugene pediatric hematologist and oncologist
Wimpress, Gordon Duncan, Jr. corporate consultant, foundation executive
Winters, Wendell Delos microbiology educator, researcher, consultant
Yu, Byung Pal gerontological educator

San Marcos
Fletcher, John Lynn psychology educator
Longley, Glenn biology educator, research director

Sanger
Gervasi, Anne language professional, English language educator

Sheppard AFB
Fittante, Philip Russell air force officer, pilot

Smithville
Rundhaug, Joyce Elizabeth biochemist

South Padre Island
Judd, Frank Wayne population ecologist, physiological ecologist

Spring
Forester, David Roger research scientist
Green, Sharon Jordan interior decorator
Joy, David Anthony computer consultant
Kust, Roger Nayland chemist
Wentzler, Thomas H. chemical company executive

Springtown
Brister, Donald Wayne mechanical engineer

Stafford
Sparks, William Sidney biologist

Sugar Land
Dye, Robert Fulton chemical engineer
Mata, Zoila chemist
Phares, Lindsey James consultant, retired physicist and engineer
Solomon, David sales representative
Verret, Douglas Peter semiconductor engineer

Temple
Brasher, George Walter physician
Gaa, Peter Charles organic chemist, researcher
Malone, Stephen Robert plant physiologist
Palmer, William Alan entomologist

Texas City
Chen, Yuan James chemical company executive
Fuchs, Owen George chemist
Tandon, Don Ashoka electrical engineer

The Woodlands
Maxwell, Steve A. molecular biologist, researcher
Ojwang, Joshua Odoyo molecular biologist
Savir, Etan mathematics educator
Stanford, Michael Francis scientist, engineer

Tyler
Cohen, Allen Barry health science administrator, biochemist
D'Andrea, Mark radiation oncologist
Dodson, Ronald Franklin electron microscopist, administrator
James, Harold Lee biochemist, researcher
Loughmiller, Grover Campbell psychologist, consultant
Mattern, James Michael physicist
Rudd, Leo Slaton psychology educator, minister
Smith, James Edward petroleum engineer, consultant
Walsh, Kenneth Albert chemist

Univ Of Texas At Arlington
Wang, Shu-Shaw (Peter Wang) educator

Valley View
Wallace, Donald John, III rancher, former pest control company executive

Victoria
Wauer, Roland Horst biologist

Waco
Foster, Ottis Charles civil engineer
Held, Colbert Colgate retired diplomat
Hynan, Linda Susan psychology educator
Pierce, Benjamin Allen biologist, educator
Rolf, Howard Leroy mathematician, educator
Sivam, Thangavel Parama aerospace engineer
Stallings, Frank, Jr. industrial engineer
Wilcox, Walter Mark elementary particle physicist, educator

Wadsworth
Haralson, John Olen utility company executive

Webster
Rappaport, Martin Paul internist, nephrologist, educator
Stephens, Douglas Kimble chemical engineer

Weslaco
Amador, Jose Manuel plant pathologist, research center administrator
Collins, Anita Marguerite research geneticist

Wichita Falls
Hinton, Troy Dean civil engineer
Peterson, Holger Martin electrical engineer

Willis
Gilbert, Nathan mechanical engineer

Woodsboro
Rooke, Allen Driscoll, Jr. civil engineer

Wylie
Rigali, Joseph Leo quality assurance professional

UTAH

Brigham City
George, Russell Joseph design engineer
Mathias, Edward Charles aerospace engineer

Clearfield
Ashmead, Harve DeWayne nutritionist, executive, educator

Kaysville
Ashmead, Allez Morrill speech-hearing-language pathologist, orofacial myologist, consultant

Logan
Anderson, Jay LaMar horticulture educator, researcher, consultant
Dorst, Howard Earl entomologist
Emert, George Henry biochemist, academic administrator
Folkman, Steven Lee engineering educator
Ford, Robert Elden natural resources educator, geographer, consultant
Scouten, William Henry chemistry educator, academic administrator
Sidle, Roy C. research hydrologist
Sigler, William Franklin environmental consultant
Tariq, Athar Mohammad soil microbiologist
Vandenberg, John Donald entomologist
Van Dusen, Lani Marie psychologist

Midvale
Morris, Stephen Blaine clinical psychologist

Ogden
Davidson, Thomas Ferguson chemical engineer
Earley, Charles Willard biologist
Strand, Tena Joy civil engineer

Orem
Ghent, Robert Maynard, Jr. clinical audiologist, engineer, consultant

Provo
Barker, Dee H. chemical engineering educator
Benson, Alvin K. geophysicist, consultant, educator
Eatough, Craig Norman mechanical engineer
Lang, William Edward mathematics educator
McArthur, Eldon Durant geneticist, researcher
Riley, James Alvin civil engineer
Shiozawa, Dennis Kenji zoology educator
Smith, Howard Duane zoology educator
Smoot, Leon Douglas university dean, chemical engineering educator
Thorup, Richard Maxwell soil scientist
Youd, T. Leslie civil engineer

Salt Lake City
Abildskov, J. A. cardiologist, educator
Adler, Frederick Russell mathematical ecologist
Allison, Merle Lee geologist
Anderson, Charles Ross civil engineer
Anderson, Jeffrey Lance cardiologist, educator
Bauer, A(ugust) Robert, Jr. surgeon, educator
Belliston, Edward Glen medical facility administrator, consultant
Bennion, John Stradling nuclear engineer, consultant
Bhayani, Kiran Lilachand environmental engineer, programs manager
Bjorkman, David Jess gastroenterologist, educator
Boyd, Richard Hays chemistry educator
Capecchi, Mario Renato genetecist, educator
Coffill, Charles Frederick, Jr. aerospace engineer, educator
Conceicao, Josie chemistry researcher
Coon, Hilary Huntington psychiatric genetics educator, researcher
Corley, Jean Arnette Leister infosystems executive
Daynes, Raymond Austin immunology educator
De Vries, Kenneth Lawrence mechanical engineer, educator
Dworzanski, Jacek Pawel analytical biochemist, researcher
Epperson, Vaughn Elmo civil engineer
Eyring, Edward Marcus chemical educator
Feucht, Donald Lee research institute executive
Gandhi, Om Parkash electrical engineer
Giddings, J. Calvin chemistry educator
Gill, Ajit Singh civil engineer
Gregersen, Max A. structural, earthquake and civil engineer
Grissom, Charles Buell chemistry educator
Hansen, Dale J. science administrator, plant biochemist
Hogan, Mervin Booth mechanical engineer, educator
Huber, Robert John electrical engineering educator
Hunt, Charles Butler geologist
Jacobsen, Stephen Charles biomedical engineer, educator
Jensen, Gordon Fred university administrator
Kim, Sung Wan pharmacology educator
Major, Thomas D. academic program director
Matsen, John Martin pathology educator, microbiologist
McArthur, Gregory Robert biomedical engineer
McCusker, Charles Frederick psychologist, consultant
Meyer, Frank Henry physicist
Middleton, Anthony Wayne, Jr. urologist, educator
Miller, Jan Dean metallurgy educator
Miller, Joel Steven solid state scientist
Moser, Royce, Jr. physician, medical educator
Musci, Teresa Stella developmental biologist, researcher
Myers, Marcus Norville research educator
Newport, Brian John physicist
Oakeson, Ralph Willard pharmaceutical and materials scientist
Odell, William Douglas physician, scientist, educator
Olsen, Donald Bert biomedical engineer, experimental surgeon, research facility director
Olson, Ferron Allred metallurgist, educator
Parry, Robert Walter chemistry educator
Rogers, Vern Child engineering company executive
St. Dennis, Bruce John computer engineer, software researcher
Salmon, David Charles mechanical engineer
Sandquist, Gary Marlin engineering educator
Saunders, Barry Collins civil engineer
Schoeff, Larry educator
Silver, Barnard Stewart mechanical engineer, consultant
Sohn, Hong Yong metallurgical and chemical engineering educator
Speck, Kenneth Richard materials scientist
Stang, Peter John organic chemist
Straight, Richard Coleman photobiologist
Strickley, Robert Gordon pharmaceutical chemist
Swerdlow, Harold biomedical engineer
Thompson, Elbert Orson retired dentist, consultant
Volberg, Herman William electronic engineer, consultant
Wood, Orin Lew electronics company executive
Zeamer, Richard Jere engineer, executive

Sandy
Jarvis, Kent Graham electrical engineer
Jorgensen, Leland Howard aerospace research engineer
Lambert, James Michael chemist, researcher

South Jordan
Brinkerhoff, Lorin C. nuclear engineer, management and safety consultant

Thiokol
Carlisle, Alan Robert aeronautical design engineer

Vernal
Folks, F(rancis) Neil biologist, researcher
Johnson, Marlin Deon research facility administrator
Remington, Delwin Woolley soil conservationist

VERMONT

Barre
Delano, A(rthur) Brookins, Jr. civil engineer, consultant

Brattleboro
Gorman, Robert Saul architect
Kotkov, Benjamin clinical psychologist

Burlington
Albertini, Richard Joseph molecular geneticist, educator
Allard, Judith Louise secondary education educator
Andosca, Robert George engineering educator
Banschbach, Valerie Suzanne biologist
Beliveau, Jean-Guy Lionel civil engineering educator
Bickel, Warren Kurt psychiatry and psychology educator
Francklyn, Christopher Steward molecular biologist
Held, Paul G. molecular biologist
Henson, Earl Bennette biologist
Houston, John F. chemist
Low, Robert B. physiology educator
Mann, Kenneth Gerard biochemist, educator
Morrow, Paul Lowell forensic pathologist
Sachs, Thomas Dudley biomedical engineering scientist
Sentell, Karen Belinda chemist
Tracy, Russell Peter biomedical researcher, educator
Wilder, David Gould orthopaedic biomechanics researcher

Essex Junction
Albaugh, Kevin Bruce chemical engineer
Bernard, Ronald Allan computer performance analyst
Linde, Harold George chemist

Johnson
Genter, Robert Brian biology educator

Middlebury
Dunham, Jeffrey Solon physicist, educator
Gibson, Eleanor Jack (Mrs. James J. Gibson) psychology educator

Morrisville
Lechevalier, Mary Pfeil retired microbiologist, educator

Northfield
Barnard, William Howard, Jr. biologist, educator

Norwich
Japikse, David mechanical engineer, manufacturing executive
Snapper, Ernst mathematics educator

Pittsford
Betts, Alan Keith atmospheric scientist

Pownal
Dequasie, Andrew Eugene chemical engineer

Saint Johnsbury
Taylor, Leland Alan chemical and materials scientist, researcher

South Burlington
Kenney, Timothy P. computer scientist

Stowe
Eisenberg, Howard Edward physician, psychotherapist, educator, consultant
Springer-Miller, John Holt computer systems developer

Vergennes
Harris, Martin Sebastian, Jr. architect

White River Junction
Colice, Gene Leslie physician

Williston
Dakin, Robert Edwin electrical engineer
Lambert, Michael Irving systems engineer

VIRGINIA

Abingdon
Blake, John Charles, Jr. environmental engineer

Alexandria
Ackerman, Roy Alan research and development executive
Baker, George Harold, III physicist
Bockwoldt, Todd Shane nuclear engineer
Brandell, Sol Richard electrical power and control system engineer, research mathematician
Brickell, Charles Hennessey, Jr. marine engineer, retired military officer
Briggs, Jeffrey Lawrence ecologist
Bui, James defense industry researcher
Chatelier, Paul Richard aviation psychologist, training company officer
Corson, Walter Harris sociologist
Del Fosse, Claude Marie aerospace software executive
Dighton, Robert Duane military operations analyst
Doeppner, Thomas Walter electrical engineer, educator, consultant
Doerry, Norbert Henry naval engineer
Eckhart, Myron, Jr. marine engineer
Ellison, Thorleif consulting engineer

Fabian, John M. *former astronaut, air force officer*
Fore, Claude Harvel, III *scientist, information analyst*
Graves, Michael Leon, II *aeronautical engineer*
Grogan, Edward Joseph *clinical engineer*
Howerton, Robert Melton *telecommunications engineer*
Huffman, D. C., Jr *pharmacology educator, health science association administrator*
Judkins, William Sutton *environmental engineer*
Lasser, Howard Gilbert *chemical engineer, consultant*
Lau, K(wan) P(ang) *independent power developer*
Litzsinger, Orville Jack *aerospace executive*
McDaniel, William Howard Taft, Jr. *computer information systems educator*
Murray, Russell, II *aeronautical engineer, defense analyst, consultant*
Ostrowski, Peter Phillip *aerospace engineer, researcher*
Perchik, Benjamin Ivan *operations research analyst*
Perry, John Stephen *meteorologist*
Rafey, Larry Dean *microbiologist*
Rall, Lloyd Louis *civil engineer*
Reeker, Larry H. *computer scientist, educator*
Rimson, Ira Jay *forensic engineer*
Robson, Anthony Emerson *plasma physicist*
Rogers, McKinley Bradford *mechanical engineer*
Shapiro, Maurice Mandel *astrophysicist*
Stafford, Thomas Patten *retired military officer, former astronaut*
Straus, Leon Stephan *physicist*
Taylor, William Brockenbrough *engineer, consultant, management consultant*
Thiel, Thomas Joseph *information scientist, consultant*
Thompson, LeRoy, Jr. *radio engineer, army reserve officer*

Annandale
Abdellah, Faye Glenn *retired public health service executive*
Berg, Lillian Douglas *chemistry educator*
Cravens, Gerald McAdoo *telecommunications engineer*
Crawford, Mark Harwood *technology and energy journalist*
Dugan, John Vincent, Jr. *research and development manager, scientist*
Faraday, Bruce John *scientific research company executive, physicist*
Matuszko, Anthony Joseph *research chemist, administrator*
Neubauer, Werner George *physicist*
Schlegelmilch, Reuben Orville *electrical engineer, consultant*
Simonian, Simon John *surgeon, scientist, educator*

Arlington
Adams, Donald Edward *biology researcher*
Ary, T. S. *federal official*
Bandopadhyaya, Amitava (Amit Bando) *economist, consultant, educator*
Barbato, Joseph Allen *writer*
Bartlett, Randolph W. *engineer*
Barton, Henry David *operations research analyst*
Batchelor, Barry Lee *software engineer*
Baughman, George Washington *aeronautical operations research scientist*
Benefield, Jeniefer Len *software and systems engineer*
Bikales, Norbert M. *chemist, science administrator*
Bogue, Sean Kenneth *civil engineer*
Chase, Ramon L. *aerospace engineer*
Chawla, Manmohan Singh (Monte Chawla) *armor, weapon system and arms control technology specialist*
Cobb, Richard E. *mechanical engineer*
Cottrell, Daniel Edward *systems engineer*
Evans, Roger Michael *acoustical engineer*
Flowers, Harold Lee *consulting aerospace engineer*
Flynn, Thomas R. *engineering executive*
Gentner, Paul LeFoe *architect, consultant*
Gilbert, Arthur Charles *aerospace engineer, consulting engineer*
Hall, Carl William *agricultural and mechanical engineer*
Jutras, Larry Mark *engineer*
Keeler, Roger Norris *physicist*
Kessenich, Karl Otto *aerospace engineer*
Knipling, Edward Fred *retired research entomologist, agricultural administrator*
Lambert, Richard Bowles, Jr *science foundation program director, oceanographer*
Lazich, Daniel *aerospace engineer*
Leibowitz, Alan Jay *environmental scientist*
Lesko, John Nicholas, Jr. *research scientist*
London, J. Phillip *professional services company executive*
McCarthy, Wilbert Alan *mechanical engineer*
McDivitt, James Alton *defense and aerospace executive, astronaut*
Melichar, John Ancell *aerospace engineer*
Mott, Charles Davis *civil engineer*
Murray, Jeanne Morris *computer scientist, educator, consultant*
Nelson, Gary Rohde *computer systems executive*
Nguyen-Dinh, Thanh *internist, geriatrician*
Perkins, Richard Burle, II *environmental engineer, international consultant*
Pollock, Neal Jay *electronics executive*
Rahman, Muhammad Abdur *mechanical engineer*
Ratchford, Joseph Thomas *technical policy educator, consultant*
Reierson, James Dutton *systems engineer, physicist*
Riegel, Kurt Wetherhold *environmental protection, occupational safety and health*
Rossotti, Charles Ossola *computer software company executive*
Roy, Gabriel Delvis *scientific researcher, research manager*
Seymour, Charles Wilfred *electronics engineer*
Sheridan, Andrew James, III *ophthalmologist*
Staffa, Judy Anne *epidemiologist*
Stevens, Donald King *aeronautical engineer*
Stollnitz, Fred *comparative psychologist, government program administrator*
Stone, Jacqueline Marie *biotechnology patent examiner*
Stuart, Charles Edward *electrical engineer, oceanographer*
Sutton, George W. *aerospace company executive*
Terry, Becky Faye *electrical engineer, executive*
Tilton, James Joseph *research librarian*
Unver, Erdal Ali *research mechanical engineer*
Vasudévan, Asuri Krishnaswami *materials scientist*
Vu, Bien Quang *mechanical engineer*
Waring, John Alfred *retired research writer, lecturer, consultant*
Watt, William Stewart *physical chemist*
Wayland, Russell Gibson, Jr. *geology consultant, retired government official*
Wodarczyk, Francis John *chemist*
Zgonc, Janice Ann *computer specialist*
Zirkind, Ralph *physicist*

Ashland
Falls, Elsa Queen *biologist, educator*
Wright, Berry Franklin *civil engineer*

Berryville
White, Eugene Vaden *pharmacist*

Big Stone Gap
Ogbonnaya, Chuks Alfred *entomologist, agronomist, educator*

Blacksburg
Alley, Marcus M. *agronomy educator*
Ash, Philip *psychologist*
Boardman, Gregory Dale *environmental engineer, educator*
Bocvarov, Spiro *aerospace and electrical engineer*
Brown, Eugene Francis *mechanical engineer, educator*
Bryant, Clifton Dow *sociologist, educator*
Bursal, Faruk Halil *mechanical engineer, educator*
Cairns, John, Jr. *environmental science educator, researcher*
Collins, George Briggs *retired physicist*
Cropper, André Dominic *electrical engineering educator*
De Datta, Surajit Kumar *soil scientist, agronomist, educator*
Fabrycky, Wolter Joseph *engineering educator, author, industrial and systems engineer*
Giurgiutiu, Victor *aeronautical engineer, educator*
Glasser, Wolfgang Gerhard *forest products and chemical engineering researcher, educator*
Gray, Festus Gail *electrical engineer, educator, researcher*
Graybeal, Jack Daniel *chemist, educator*
Ha, Sam Bong *molecular geneticist*
Harvey, Aubrey Eaton, III *industrial engineer*
Haugh, Clarence Gene *agricultural engineering educator*
Hinkelmann, Klaus Heinrich *statistician, educator*
Inman, Daniel John *mechanical engineer, educator*
Jones, James Beverly *mechanical engineering educator*
Jones, Robert Millard *engineering educator*
Kornhauser, Alan Abram *mechanical engineer*
McGee, Henry Alexander, Jr. *chemical engineering educator*
Morgan, George Emir, III *financial economics educator*
Nichols, James Robbs *university dean*
Perumpral, John Verghese *agricultural engineer, administrator, educator*
Phadke, Arun G. *electrical engineering educator*
Price, Dennis Lee *industrial engineer, educator*
Rappaport, Theodore Scott *electrical engineering educator*
Reddy, J. Narasimh *mechanical engineering educator*
Rogers, Craig Alan *mechanical engineering educator, university program director*
Ross, Mary Harvey *entomology educator, researcher*
Safaai-Jazi, Ahmad *electrical engineering educator, researcher*
Salom, Scott Michael *forest entomologist*
Samaranayake, Gamini Saratchandra *chemist, researcher*
Smith, Charles William *mechanical engineer, educator*
Song, Ohseop *research engineer*
Stern, E. George *architect, educator*
Wilkins, Tracy Dale *microbiologist, educator*

Bluefield
Mullennex, Ronald Hale *geologist, consultant*

Burke
Bursik, David James *software engineer*
Uwujaren, Gilbert Patrick *economist, consultant*

Centreville
Hanson, Lowell Knute *seminar developer and leader, information systems consultant*

Chantilly
Lobel, Steven Alan *immunologist*
Macionski, Lawrence Edward *electronics engineer*
McCormick, Thomas Jay *infosystems engineer*
Neitz, David Allan *information scientist*
Srivastava, Kailash Chandra *microbiologist*

Charlottesville
Abse, David Wilfred *psychiatrist, psychologist*
Averill, Bruce Alan *chemistry educator*
Baragiola, Raul Antonio *physicist*
Barrett, Eugene J. *researcher, medical educator, physician*
Bleck, Thomas Pritchett *neurologist, neuroscientist, educator*
Carey, Robert Munson *university dean, physician*
Carta, Giorgio *chemical engineering educator*
Chapman, Martin Dudley *immunochemist*
Chevalier, Roger Alan *astronomy educator, consultant*
Chialvo, Ariel Augusto *chemical engineering research scientist*
Davis, John Staige, IV *physician*
Dorning, John Joseph *nuclear engineering, engineering physics and applied mathematics educator*
Edlich, Richard French *biomedical engineering educator*
Gaskin, Felicia *biochemist, educator*
Gottesman, Irving Isadore *psychiatric genetics educator, consultant*
Greer, David Llewellyn *electrical engineer*
Gwaltney, Jack Merrit, Jr. *physician, educator, scientist*
Hanson, Brett Allen *nuclear engineer*
Hartman, Mark Leopold *internist, endocrinologist*
Herr, John Christian *cell biologist, educator*
Hutchinson, Thomas Eugene *biomedical engineering educator*
Hymes, Dell Hathaway *anthropologist*
Jacobson, Ira David *aerospace engineer, educator, researcher*
Jenkins, Alan Deloss *urologic surgeon, educator*
Johnson, Robert Alan *materials science educator*
Kellermann, Kenneth Irwin *astronomer*
Koshiya, Naihiro *neuroscientist*
MacDonald, Timothy Lee *chemistry educator*
Mansfield, Lois Edna *mathematics educator, researcher*
Martin, Robert Bruce *chemistry educator*
Mattauch, Robert Joseph *electrical engineering educator*
McCarty, Richard Charles *psychology educator*
Morton, Harold S(ylvanus), Jr. *retired mechanical and aerospace engineering educator*
Murphy, Richard Alan *physiology educator*
Murray, Joseph James, Jr. *biology educator, zoologist*
Nolan, Stanton Peelle *surgeon, educator*
Owens, Justine Elizabeth *cognitive psychology educator*
Parshall, Karen Virginia Hunger *mathematics and science historian*
Phillips, Larry H., II *neurologist*
Reppucci, Nicholas Dickon *psychologist, educator*
Rosenblum, Marvin *mathematics educator*
Scarr, Sandra Wood *psychology educator, researcher*
Scheld, William Michael *internist, educator*
Scott, Leonard Lewy, Jr. *mathematician, educator*
Shen, Tsung Ying *medicinal chemistry educator*
Shugart, Herman Henry *environmental sciences educator, researcher*
Snavely, William Pennington *economics educator*
Song, Xiaotong *physicist*
Taylor, Samuel James *mathematics educator*
Thornton, Earl Arthur *mechanical and aerospace engineering educator*
Townsend, Miles Averill *aerospace and mechanical engineering educator*
Vanden Bout, Paul Adrian *astronomer, physicist, educator*
Wadlington, Walter James *legal educator*
Waxman, Ronald *computer engineer*
Wilson, Melvin Nathaniel *psychology educator*
Woode, Moses Kwamena *medical educator*
Wulf, William Allan *computer information scientist, educator*

Chesapeake
Crane, Richard Turner *otolaryngologist*
Jaques, James Alfred, III *communications engineer*
Little, W. Ken, Jr. *structural engineer*
Perry, Harry Montford *computer engineering executive, retired officer*
White, Eugene Thomas, III *nuclear project engineer*

Clifton
Gawrylowicz, Henry Thaddeus *aerospace engineer*
Hoffman, Karla Leigh *mathematician*

Covington
Foster, James Joseph *chemical engineer*

Crystal City
Lintz, Paul Rodgers *physicist, patent examiner*

Dahlgren
Boyer, Charles Thomas, Jr. *mechanical engineer*
Delurey, Michael William *mechanical engineer*
Evans, Alan George *electrical engineer*
Holt, William Henry *physicist, researcher*
Knowles, Stephen Howard *space scientist*
Westbrook, Susan Elizabeth *horticulturist*

Danville
Hoffman, Allan Augustus *retired urologist, consultant*
Wright, Donald Lee *chemist*

Dayton
Couch, James Vance *psychology educator*

Dillwyn
Moseley, John Marshall *nurseryman*

Dulles
Lindberg, Robert E., Jr. *space systems engineer, controls researcher*
Thompson, David Walker *astronautics company executive*

Dyke
Hart, Jean Hardy *information systems specialist, consultant*

Elkton
Ghorpade, Ajit Kisanrao *chemical engineer*

Fairfax
Adamson, Sandra Lee *anthropologist*
Beale, Guy Otis *engineering educator, consultant*
Boone, James Virgil *engineering executive*
Bosco, Ronald F. *engineering company executive*
Dam, A. Scott *nuclear/mechanical engineer*
Davis, Walter Barry *quality control professional*
Dietz, Thomas Michael *human ecology educator and researcher*
Elias, Antonio L. *aeronautical engineer, aerospace executive*
Hampton, Matthew Joseph *engineering technician*
Harte, Rebecca Elizabeth *computer scientist, consultant*
Lovell, Robert R(oland) *engineering executive*
Morowitz, Harold Joseph *biophysicist, educator*
Ney, Ronald Ellroy, Jr. *chemist*
Palmer, James Daniel *information technology educator*
Payne, Fred J. *physician, educator*
Peters, Esther Caroline *aquatic toxicologist, pathobiologist, consultant*
Pixley, John Sherman, Sr. *research company executive*
Reed, Christopher Robert *civil engineer*
Sage, Andrew Patrick, Jr. *systems information and software engineering educator*
Sowder, Donald Dillard *chemicals executive*
Soyfer, Valery Nikolayevich *biology educator*
Ward, Charles Raymond *systems engineer*
Ware, Lawrence Leslie, Jr. *microbiologist*
Williams, Thomas Rhys *educator, anthropologist*

Fairfax Station
Coaker, James Whitfield *mechanical engineer*

Falls Church
Adams, Nancy R. *nurse, military officer*
Benedict, Jeffrey Dean *financial and technical consultant, mathematics educator*
Evans, Peter Yoshio *ophthalmologist, educator*
Feldmann, Edward George *pharmaceutical chemist*
Fink, Charles Augustin *behavioral systems scientist*
Frankel, Irwin *chemical engineer*
Ho, Hien Van *pediatrician*
Hodge, Donald Ray *systems engineer*
LaNoue, Alcide Moodie *medical corps officer, health care administrator*
McEwan, Robert Neal *molecular biologist*
Nashman, Alvin Eli *computer company executive*
Nickle, Dennis Edwin *electronics engineer, church deacon*
Robertson, Laurie Luissa *computer scientist*
Rosen, Coleman William *radiological physicist*
Schilling, William Richard *aerospace engineer, research and development company executive*
Shriner, Robert Dale *economist, management consultant*
Simpson, John Arol *retired government executive, physicist*
Spindel, William *chemistry educator, scientist, educational administrator*
Traceski, Frank Theodore *materials engineer*
Villarreal, Carlos Castaneda *engineering executive*
Weiss, Armand Berl *economist, association management executive*

Farmville
Breil, David Allen *botany educator*

Forest
Durand, James Howard *mechanical engineer*

Fort Belvoir
Austin, Charles Louis *software engineer*
Blackwell, Neal Elwood *agricultural and mechanical engineer, researcher*
Diercks, Frederick Otto *government official*
Heberlein, David Craig *physicist*

Fort Monroe
Weiland, Peter Lawrence *military officer*

Franklin
Raymond, Dale Rodney *chemical engineer*

Fredericksburg
Moorman, William Jacob *agronomist, consultant*

Gainesville
Derstine, Mark Stephen *mechanical engineer*
Steger, Edward Herman *chemist*

Gasburg
Morrow, Bruce W. *business executive, consultant*

Glen Allen
Seymour, Harlan Francis *computer services company executive*
Smith, Noval Albert, Jr. *systems engineer*

Gloucester Point
Penhale, Polly Ann *marine biologist, educator*
Schaffner, Linda Carol *biological oceanography educator*

Grafton
Dischinger, Hugh Charles *civil engineer*

Great Falls
Dillon, Kathleen Gereaux *dentist*
Haines, Andrew Lyon *engineering company executive*

Greenwood
Casero, Joseph Manuel *civil engineer*
Findlay, John Wilson *retired physicist*

Halifax
Martin, Wesley Davis, Jr. *entomologist, research biologist*

Hampden Sydney
Joyner, Weyland Thomas *physicist, educator, business consultant*
Porterfield, William Wendell *chemist, educator*
Shear, William Albert *biology educator*

Hampton
Bayer, Janice Ilene *mechanical engineer*
Burner, Alpheus Wilson, Jr. *optical engineer*
Bushnell, Dennis Meyer *mechanical engineer, researcher*
Christhilf, David Michael *aerospace engineer*
Deepak, Adarsh *meteorologist, atmospheric optician*
Duberg, John Edward *aeronautical engineer, educator*
Farrukh, Usamah Omar *electrical engineering educator, researcher*
Ganoe, George Grant *electrical engineer*
Glass, David Eugene *mechanical engineer, educator*
He, Chengjian *aerospace engineer*
Hefner, Jerry Ned *aerospace engineer*
Holloway, Paul Fayette *aerospace executive*
Hunt, Patricia Jacqueline *mathematician, system manager, graphics programmer*
Jegley, Dawn Catherine *aerospace engineer*
Kelly, Jeffrey Jennings *mechanical engineer*
Liceaga, Carlos Arturo *research computer engineer*
Mackley, Ernest Albert *aeropropulsion engineer, consultant*
Martin, Roderick Harry *aeronautics research scientist*
Noblitt, Nancy Anne *aerospace engineer*
Pruett, Charles David *aerospace research scientist, educator*
Quinto, P. Frank *aerospace engineer*
Rivers, Horace Kevin *aerospace engineer, researcher, thermal-structural analyst*
Rivers, Robert Allen *research pilot, aerospace engineer*
Schutte, Paul Cameron *research scientist*
Shenhar, Joram *mechanical engineer*
Smith, Mary-Ann Hrivnak *meteorologist, geophysicist*
Sun, Keun Jenn *physicist*
Voigt, Robert G. *numerical analyst*
Wang, Liang-guo *research scientist*
Weinstein, Leonard Murrey *aerospace engineer, consultant, researcher*

Wood, George Marshall *research scientist, program manager*

Harrisonburg
Giovanetti, Kevin Louis *physicist, educator*

Herndon
Alton, Cecil Claude *computer scientist*
Day, Melvin Sherman *information company executive*
Foster, Linda Ann *biomaterials research scientist*
Kagan, Marvin Bernard *architect*
Marshall, Larry Ronald *laser physicist*
Mehring, James Warren *electrical engineer*
Peck, Dallas Lynn *geologist*
Ruddell, James Thomas *civil engineer, consultant*
Stirewalt, Edward Neale *chemist, scientific analyst*

Kilmarnock
Gilruth, Robert Rowe *aerospace consultant*

Kinsale
Gould, Gordon *physicist, retired optical communications executive*

Langley AFB
Osborn, James Henshaw *operations research analyst*

Leesburg
McDow, Russell Edward, Jr. *surgeon*
Mokhtarzadeh, Ahmad Agha *agronomist, consultant*

Lexington
Arthur, James Howard *mechanical engineering educator*
Spencer, Edgar Winston *geology educator*
Tierney, Michael John *mathematics and computer science educator*
Tyree, Lewis, Jr. *retired compressed gas company executive, inventor, technical consultant*

Locust Grove
Stein, Richard Louis *chemist, educator*

Lorton
Koschny, Theresa Mary *environmental biologist*

Luray
Tessler, Steven *ecologist*

Lynchburg
Cresson, David Homer, Jr. *pathologist*
Fath, George R. *electrical engineer, communications executive*
Kovach, James Michael *engineering executive*
Latimer, Paul Jerry *non-destructive testing engineer*
Looney, Thomas Albert *psychologist, educator*
Morgan, Evan *chemist*
Peters, Ralph Irwin, Jr. *biology educator, researcher*

Manakin Sabot
Brickey, James Allan *owner environmental testing laboratory, consultant*

Manassas
Carvalho, Julie Ann *psychologist*

Marion
Young, Craig Steven *structural engineer*

Mc Lean
Airst, Malcolm Jeffrey *electronics engineer*
Amr, Asad Tamer *environmental engineer, consultant*
Arnold, Kent Lowry *electrical engineer*
Beene, Kirk D. *systems engineer*
Bizzigotti, George Ora *environmental chemist, consultant*
Bluitt, Karen *computer program manager*
Dedrick, Robert Lyle *toxicologist, biomedical engineer*
Dickerson, Michael Joe *telecommunications engineer*
Field, Francis Edward *electrical engineer, educator*
Fowler, Thomas Benton, Jr. *electrical engineering education, consultant*
Frankum, Ronald Bruce *communications executive, entrepreneur*
Gerasch, Thomas Ernest *computer scientist*
Harkins, William Douglas *mechanical engineer*
Kratz, Ruediger *neurologist, researcher*
Lane, Susan Nancy *structural engineer*
Ligon, Daisy Matutina *systems engineer*
Marinenko, George *chemist*
McCambridge, John James *civil engineer*
McConathy, Donald Reed, Jr. *meteorologist, remote sensing program manager, systems engineer*
Myers, Kenneth Alan *air force officer, aerospace engineer*
Parrish, Thomas Dennison *computer systems engineer*
Paul, Anton Dilo *chemical engineering consultant, researcher*
Perry, Dennis Gordon *computer scientist*
Schmeidler, Neal Francis *engineering executive*
Schneck, Paul Bennett *computer scientist*
Sinha, Agam Nath *engineering management executive*
Starr, Stuart Howard *systems engineer, long range planner*
Stendahl, Steven James *mathematician, system engineer*
Torres, Rigo Romualdo *electrical engineer*
Waesche, R(ichard) H(enley) Woodward *combustion research scientist*
Watts, Helena Roselle *military analyst*
Woo, Dah-Cheng *hydraulic engineer*
Yang, Xiaowei *electrical engineer*

Meadowview
Hebard, Frederick V. *plant pathologist*

Mechanicsville
McCahill, Thomas Day *physician*

Middletown
Kisak, Paul Francis *engineering company executive*

Midlothian
Hauxwell, Gerald Dean *chemical engineer*

Newington
Foster, Eugene Lewis *engineering executive*

Newport News
Deleo, Richard *engineering executive*
Earnhardt, Daniel Edwin *automotive engineer*
Giles, Glenn Ernest, Jr. *nuclear engineer*
Hartline, Beverly Karplus *physicist, science educator*
Hempfling, Gregory Jay *mechanical engineer*
Montane, Jean Joseph *mechanical engineer*
Neil, George Randall *physicist*
Pohl, John Joseph, Jr. *retired mechanical engineer*
Ranellone, Richard Francis *shipbuilding company executive*
Schatzel, Robert Mathew *logistics engineer*
Yan, Chen *electromagnetic physicist*

Norfolk
Anderson, Freedolph Deryl *gynecologist*
Bellenkes, Andrew Hilary *aerospace experimental psychologist*
Brown, Kenneth Gerald *chemistry educator*
Burtoft, John Nelson, Jr. *cardiovascular physician assistant*
Carpenter, Allan Lee *civil engineer*
Corl, William Edward *environmental chemist*
Csanady, Gabriel Tibor *oceanographer, meteorologist, environmental engineer*
Hou, Jiashi *mathematician, educator*
Kandil, Osama Abd El Mohsin *mechanical and aerospace engineering educator*
Lakdawala, Vishnu Keshavlal *electrical and computer engineering educator*
McLaren, John Paterson, Jr. *civil engineer*
Oelberg, David George *neonatologist, biomedical researcher*
Overby, Veriti Page *chemist, environmental protection specialist*
Pariser, Robert Jay *dermatologist*
Pittenger, Gary Lynn *biomedical educator*
Redondo, Diego Ramon *health, physical education and recreation educator*
Stokes, Thomas Lane, Jr. *biologist, consultant*
Vušković, Leposava *physicist, educator*

Palmyra
Mulckhuyse, Jacob John *energy conservation and environmental consultant*
Weiss-Wunder, Linda Teresa *neuroscience research consultant*

Pearisburg
Morse, F. D., Jr. *dentist*
Stafford, Steven Ward *civil engineer*

Petersburg
Chandler, Paul Anderson *physicist, researcher*
Correale, Steven Thomas *materials scientist*

Portsmouth
Cox, William Walter *dentist*
Snyder, Peter James *nuclear engineer*

Radford
Cabbage, William Austin *chemical engineer*
Hudspeth, William Jean *neuroscientist*
McGraw, Katherine Annette *fisheries biologist, environmental consultant*
Pribram, Karl Harry *psychology educator, researcher*

Reston
Cohen, Philip *hydrogeologist*
Ethridge, Max Michael *civil engineer*
Grass, Judith Ellen *computer scientist*
Guptill, Stephen Charles *physical scientist*
Hanna, William Francis *geophysicist*
Hartong, Mark Worthington *military officer, engineer*
Hu, Tsay-Hsin Gilbert *aerospace engineer*
Huebner, John Stephen *geologist*
Jaffe, Russell Merritt *pathologist, research director*
Kahn, Robert E. *electrical engineer*
Miller, Lynne Marie *environmental company executive*
Mumzhiu, Alexander *machine vision systems engineer*
Riehle, James Ronald *volcanologist*
Sato, Motoaki *geologist, researcher*
Subasic, Christine Ann *architectural engineer*
Wetzel, John Paul *structural dynamics engineer*
Wood, John Thurston *cartographer, jazz musician*
Wood, Warren Wilbur *hydrologist*

Richlands
Witten, Thomas Jefferson, Jr. *mathematics educator*

Richmond
Bradley, Sterling Gaylen *microbiology and pharmacology educator*
Buck, Gregory Allen *molecular biology educator*
Chakravorty, Krishna Pada *chemist, spectroscopist*
Christie, Laurence Glenn, Jr. *surgeon*
Deevi, Seetharama C. *materials scientist*
Dimitriou, Michael Anthony *biochemist, sales and marketing manager*
Duong, Minh Truc *project engineer*
Elmore, Stanley McDowell *orthopaedic surgeon*
Fleisher, Paul *elementary education educator*
Freund, Emma Frances *medical technologist*
Gandy, Gerald Larmon *rehabilitation counseling educator*
Goldman, Israel David *hematologist, oncologist*
Hadfield, M. Gary *neuropathologist, educator*
Haines, Michael James *asphalt company official*
Hanneman, Rodney Elton *metallurgical engineer*
Hardage, Page Taylor *health care administrator*
Harmon, Charles Winston *energy engineer*
Harris, Louis Selig *pharmacologist, researcher*
Harris, Robert Bernard *biochemist*
Hayden, W(alter) John *botanist, educator*
Kalen, Joseph David *physicist, researcher*
Kendig, Edwin Lawrence, Jr. *physician, educator*
Kinsley, Craig Howard *neuroscientist*
Kinsley, Homan Benjamin, Jr. *chemist, chemical engineer*
Kornstein, Michael Jeffrey *pathologist*
Marsee, Dewey Robert *chemical engineer*
McLelland, Slaten Anthony *electrical engineer*
Murdoch-Kitt, Norma Hood *psychologist*
Neale, Michael Churton *behavior geneticist*
Neifeld, James Paul *surgical oncologist*
Owen, Duncan Shaw, Jr. *physician, medical educator*
Palik, Robert Richard *mechanical engineer*
Pandurangi, Ananda Krishna *psychiatrist*
Perdue, Pamela Price *computer engineer*
Prasad, Ravi *chemical engineer*
Reynolds, Bradford Charles *industrial engineer, management consultant*
Reynolds, Thomas Robert *scientist, biotechnology company executive*
Robinson, Susan Estes *pharmacology educator*
Roth, Karl Sebastian *pediatrician*
Rutan, Sarah Cooper *chemistry educator*
Safo, Martin Kwasi *chemist*
Singh, Nirbhay Nand *psychology educator, researcher*
Sirica, Alphonse Eugene *pathology educator*
Sprinkle, William Melvin *engineering administrator, audio-acoustical engineer*
Thomas, Charles Edwin *chemist, researcher*
Totten, Arthur Irving, Jr. *retired metals company executive, consultant*
Walsh, Scott Wesley *reproductive physiologist, researcher*
Ward, John Wesley *pharmacologist*
Zimmermann, Michael Louis *chemist*

Riner
Foster, Joy Via *library media specialist*

Roanoke
Al-Zubaidi, Amer Aziz *physicist, educator*
Enright, Michael Joseph *radiologist*
McKenna, John Dennis *environmental testing engineer*
Nicholas, James Thomas *electronics engineer*
Phillips, Earle Norman *electro-optical engineer*
Smith, Gary Lee *chemical engineer*
Uhm, Dan *process engineer*

Rosslyn
Fisher, Daniel Robert *consultant*

Salem
Fisher, Charles Harold *chemistry educator, researcher*

Seven Corners
Goncz, Douglas Dana *information broker*

Springfield
Adams, William B. *consultant*
Bryan, Hayes Richard *aerospace engineer*
Bush, Norman *research and development executive*
Duff, William Grierson *electrical engineer*
Elbarbary, Ibrahim Abdel Tawab *chemist*
McMillan, Ronald Therow *optician*
Reed, Charles Kenneth *physicist*
Welty, Kenneth Harry *civil engineer*

Sterling
Bergeman, George William *mathematics educator, software author*
Edwards, Stephen Glenn *air force officer, astronautical engineer*
Fothergill, John Wesley, Jr. *systems engineering and design company executive*
Ghazarian, Rouben *structural engineer*
Hansen, Alan Lee *architect*
O'Rourke, Thomas Joseph *aerospace engineer, consultant*

Suffolk
Tinto, Joseph Vincent *electrical engineer*

Surry
Enroughty, Christopher James *nuclear chemistry technician*
Johnson, Keith Edward *chemist*

Sweet Briar
Hyman, Scott David *physicist, educator*
Wassell, Stephen Robert *mathematics educator, researcher*

Vienna
Austin, Frank Hutches, Jr. *aerospace physician, educator*
Edwards, John William *physicist*
Koutrouvelis, Panos George *radiologist*
Murray, Arthur Joseph *engineering consultant, researcher*
Nicklas, John G. *systems analyst*
Ross, John R., III *aerospace engineer*
Roth, James *engineering company executive*
Smith, Esther Thomas *editor*
Woodward, Kenneth Emerson *retired mechanical engineer*

Virginia Beach
Barranco, Sam Christopher *biologist, researcher*
Kornylak, Harold John *osteopathic physician*
Lichtenberg, Byron K. *futurist, manufacturing executive, space flight consultant*
Lowe, Cameron Anderson *dentist, endodontist, educator*
Morgan, Michael Joseph *advanced information technology executive*
Oswaks, Roy Michael *surgeon, educator*
Stephan, Charles Robert *retired ocean engineering educator, consultant*
Sweet, Rita Genevieve *civil engineer*
Switzer, Terence Lee *civil engineer*
Van de Riet, James Lee *environmental engineer*
Yu, Brian Bangwei *research physicist, executive*

Washington
Ayers, Jack Duane *metallurgist*

Weyers Cave
Levin, Bernard H. *psychologist educator*

Williamsburg
Dunn, Ronald Holland *civil engineer, management executive, railway consultant*
Hinders, Mark Karl *mechanical engineer*
Krakauer, Henry *physics educator*
Muller, Julius Frederick *chemist, business administrator*
Refinetti, Roberto *physiological psychologist*
Rogers, Verna Aileen *mechanical and biomedical engineer, researcher*
Starnes, William Herbert, Jr. *chemist, educator*
Vossel, Richard Alan *systems engineer*

Winchester
Cleland, Ned Murray *civil engineer*
Gordon, Richard Warner *naval propulsion engineer*
Horsburgh, Robert Laurie *entomologist*
Ludwig, George Harry *physicist*
Murtagh, John Edward *chemist, consultant*
Teal, Gilbert Earle *industrial engineer*
Turner, William Richard *retired aeronautical engineer, consultant*

Wise
Frank, Mary Lou Bryant *psychologist, educator*
Low, Emmet Francis, Jr. *mathematics educator*

Woodbridge
Campbell, Robert P. *information scientist*
Peck, Dianne Kawecki *architect*
Reha, William Christopher *urologic surgeon*
Schaefer, Carl George, Jr. *aerospace engineer*

Woodstock
Vachher, Prehlad Singh *psychiatrist*

Wytheville
McConnell, James Joseph *internist*

Yorktown
Ambur, Damodar Reddy *aerospace engineer*

WASHINGTON

Anacortes
Sulkin, Stephen David *marine biology educator*

Bainbridge Is
Hanson, Robert James *electrical test engineer, consultant*
Whitener, Philip Charles *aeronautical engineer, consultant*

Bellevue
Boike, Shawn Paul *aerospace project/design engineer*
Carlson, Curtis Eugene *orthodontist, periodontist*
Chen, Ching-Hong *medical biochemist, researcher*
Genskow, John Robert *civil engineer, consultant*
Hackett, Carol Ann Hedden *physician*
Killgore, Mark William *civil engineer*
Liang, Jeffrey Der-Shing *retired electrical engineer, civil worker*
Lipkin, Mary Castleman Davis (Mrs. Arthur Bennett Lipkin) *retired psychiatric social worker*
Randish, Joan Marie *dentist*
Roselle, Richard Donaldson *industrial, marine and interior designer*
Sharp, Kevan Denton *civil engineer*
Wright, Theodore Otis *forensic engineer*

Bellingham
Albrecht, Albert Pearson *electronics engineer, consultant*
Landis, Wayne G. *environmental toxicologist*
Lippman, Louis Grombacher *psychology educator*
Ross, Charles Alexander *geologist*
Thorndike, Robert Mann *psychology educator*

Bothell
Daigle, Ronald Elvin *medical imaging scientist, researcher*
Hadjicostis, Andreas Nicholas *physicist*
Kosterman, Richard Jay *political psychologist, political consultant*
Pihl, James Melvin *electrical engineer*
Taylor, Dean Perron *biotechnologist, researcher*

Bremerton
Thovson, Brett Lorin *physicist*

Cathlamet
Torget, Arne O. *electrical engineer*

Colville
Culton, Sarah Alexander *psychologist, writer*

Des Moines
Sinon, John Adelbert, Jr. *electrical engineer*

Eastsound
Anders, William Alison *aerospace and diversified manufacturing company executive, former astronaut, former ambassador*

Ellensburg
Mitchell, Robert Curtis *physicist, educator*
Rosell, Sharon Lynn *physics and chemistry educator, researcher*
Yu, Roger Hong *physics educator*

Ephrata
Levey, Sandra Collins *civil engineer*

Everett
Deboo, Behram Savakshaw *microbiologist*
Hitomi, Georgia Kay *mechanical engineer*
Winn, George Michael *electrical equipment company executive*

Fairchild AFB
McDonnell, John Patrick *military officer*

Federal Way
Cunningham, John Randolph *systems analyst*
Hansen, Michael Roy *chemist*
King, Don E. *air transportation executive*

Friday Harbor
Willows, Arthur Owen Dennis *neurobiologist, zoology educator*

Hansville
Griffin, DeWitt James *architect, real estate developer*
Strahilevitz, Meir *inventor, researcher, psychiatry educator*

Issaquah
Ford, Dennis Harcourt *laser systems consultant*

Kalama
Liang, Jason Chia *research chemist*

Kennewick
Carroll, John Moore *retired electrical engineer*
Cobb, William Thompson *environmental consultant*
Williams, Brian James *geotechnical engineer*

Kent
McClure, Allan Howard *materials engineer, space contamination specialist, space materials consultant*
Raymond, Eugene Thomas *technical writer, retired aircraft engineer*

Kingston
Pichal, Henri Thomas *electronics engineer, physicist, consultant*

Kirkland
Branca, Andrew Angelo *biochemist, educator*
Evans, Robert Vincent *engineering executive*
Mitchell, Joseph Patrick *architect*
Pearson, Mark Landell *molecular biologist*
Steinmann, John Colburn *architect*
Williford, John Frederic, Jr. *materials research and development scientist*

Longview
Sandstrom, Robert Edward *physician, pathologist*

Lynnwood
Olsen, Kenneth Harold *geophysicist, astrophysicist*

Manchester
Fearon, Lee Charles *chemist*

Mercer Island
Bridgforth, Robert Moore, Jr. *aerospace engineer*

Moses Lake
Leadbetter, Mark Renton, Jr. *orthopaedic surgeon*

Mukilteo
Bohn, Dennis Allen *electrical engineer, consultant, writer*

Oak Harbor
Meaux, Alan Douglas *facilities technician, sculptor*

Ocean Shores
Morgan, Audrey *architect*

Olympia
Boose, Richard Bradshaw *biochemist*
Flemming, Stanley Lalit Kumar *family practice physician, state legislator*
Hirsch, Gary Mark *energy policy specialist, cogeneration manager*
Jamison, David W. *marine scientist*
Penney, Robert Allan *engineering executive*
Saari, Albin Toivo *electronics engineer*
Snow, Blaine Arlie *language educator*
Steiger, Gretchen Helene *marine mammalogist, research biologist*
Tudor, Gregory Scott *land information system analyst*

Port Orchard
Kammin, William Robert *environmental laboratory director*

Port Townsend
Barnard, Maxwell Kay *inventor, computer systems executive*
Speser, Philip Lester *social scientist, consultant*

Poulsbo
Kolb, James A. *science association director, writer*

Prosser
Proebsting, Edward Louis, Jr. *research horticulturist*

Pullman
Banas, Emil Mike *physicist, educator*
Barnes, Charles D. *neuroscientist, educator*
Carper, Kenneth Lynn *architect, educator*
Davis, William Charles *veterinary microbiology educator*
Dunlap, Riley E. *sociologist*
Gurusiddaiah, Sarangamat *plant scientist*
Klavano, Paul Arthur *veterinary pharmacologist, educator, anesthesiologist*
Lewis, Norman G. *academic administrator, researcher, consultant*
Pandey, Lakshmi Narayan *physicist, researcher*
Pellerin, Roy Francis *civil engineer, educator*
Ryan, Clarence Augustine, Jr. *biochemistry educator*
Thompson, John N. *ecology, evolutionary biology educator, researcher*
Troutt, Timothy Ray *mechanical engineer, educator*
Witmer, Gary William *research biologist, educator*

Puyallup
Stark, John David *pesticide toxicologist*
Walize, Reuben Thompson, III *health research administrator*

Redmond
Gates, William Henry, III *software company executive*
Kelman, Bruce Jerry *toxicologist, consultant*
Ransdell, Tod Elliot *pharmaceutical, in vitro diagnostics biotechnologist*
Wang, Lin *physicist, computer science educator, computer software consultant*
Yagi, Fumio *mathematician, systems engineer*

Renton
LaDue, Robin Annette *psychologist, educator*
Lockridge, Alice Ann *secondary education educator*

Richland
Austin, Inez J. *nuclear scientist*
Bair, William J. *radiation biologist*
Brenden, Byron Byrne *physicist, optical engineer*
Compton, James Allan *chemical engineer*
Fisher, Darrell Reed *medical physicist, researcher*
Gates, Thomas Edward *civil engineer, waste management administrator*
Henager, Charles Henry *civil engineer*
Huckaby, James L. *chemical engineer*
Jacobsen, Gerald Bernhardt *biochemist*
Keska, Jerry Kazimierz *mechanical engineering educator*
Konkel, R(ichard) Steven *environmental and social science consultant*
Lemon, Douglas Karl *physicist*
Link, Steven Otto *environmental scientist, statistician*
Onishi, Yasuo *environmental researcher*
Ramesh, Kalahasti Subrahmanyam *materials scientist*
Rebagay, Teofila Velasco *chemist, chemical engineer*
Ritter, Gerald Lee *chemical engineer*
Roop, Joseph McLeod *economist*
Sen, Jyotirmoy (Joe Sen) *control engineer*
Smith, Richard Dale *chemist, researcher*
Wiley, William Rodney *microbiologist, administrator*
Wise, James Albert *psychologist*
Xun, Luying *microbiologist, educator*
Zimmerman, Richard Orin *safety engineer*

Seattle
Barrett, Michael Wayne *product development company executive*
Bateman, Robert Edwin *aeronautical engineer*
Becker, Virginia Grafton *psychologist*
Bernard, Eddie Nolan *oceanographer*
Bowen, Jewell Ray *academic dean, chemical engineering educator*
Brotherton, Vince Morgan *research chemist*
Brown, Craig William *chemist*
Butler, John Ben, III *physician, computer specialist*
Calvin, William Howard *neurophysiologist*
Cardenas, Diana Delia *physician, educator*
Cassady, Philip Earl *mechanical engineer*
Castanes, James Christopher *architect*
Clarren, Sterling Keith *pediatrician*
Cosgrove, Benjamin A. *retired aerospace company executive*
Cramer, John Gleason, Jr. *physics educator, experimental physicist*
Creager, Joe Scott *geology and oceanography educator*
Criminale, William Oliver, Jr. *applied mathematics educator*
Dale, Philip Scott *psychologist educator*
Davis, Terry Lee *communications systems engineer*
Dawson, Geraldine *psychologist, educator*
Day, Robert Winsor *research administrator*
Dehmelt, Hans Georg *experimental physicist*
Deyrup-Olsen, Ingrith Johnson *physiologist, educator*
Diessner, Daniel Joseph *fiber optics engineer*
Dorwart, Brian Curtis *geotechnical engineer, consultant*
Eberhard, Marc Olivier *civil engineering educator*
Edmondson, W(allace) Thomas *limnologist, educator*
Fadden, Delmar McLean *electrical engineer*
Fischer, Edmond Henri *biochemistry educator*
Fischer, Gregory Robert *geotechnical engineer*
Foe, Elizabeth *biologist, educator*
Galster, Richard W. *civil engineer*
Gartz, Paul Ebner *systems engineer*
Gelb, Michael H. *chemistry educator*
Gillis, Steven *biotechnology company executive*
Goodall, Frances Louise *nurse, production company assistant*
Greenwald, Anthony Galt *psychology educator*
Gruger, Edward Hart, Jr. *retired chemist*
Gunderson, Donald Raymond *fisheries educator and researcher*
Haglund, Michael Martin *neurosurgeon*
Harder, Ruel Tan *mechanical engineer*
Hartwell, Leland Harrison *geneticist, educator*
Henley, Ernest Mark *physics educator, university dean emeritus*
Hertzberg, Abraham *aeronautical engineering educator, university research scientist*
Hol, Wim Gerardus Jozef *biophysical chemist*
Hood, Leroy Edward *biologist*
Hornbein, Thomas Frederic *anesthesiologist*
Hutchinson, William Burke *surgeon, research center director*
Ii, Jack Morito *aerospace engineer*
Ishimaru, Akira *electrical engineering educator*
Johnston, Allan Hugh *physicist*
Kantowitz, Barry Howard *ergonomist, researcher*
Karr, James Richard *ecologist, researcher, educator*
Kim, Yongmin *electrical engineering educator*
Klebanoff, Seymour Joseph *medical educator*
Klee, Victor La Rue *mathematician, educator*
Klepper, John Richard *physiologist, educator*
Koff, Andrew *microbiologist, biomedical researcher*
Krebs, Edwin Gerhard *biochemistry educator*
Krohn, Kenneth Albert *radiology educator*
Kukull, Walter Anthony *epidemiologist, educator*
Kunze, Eric *physical oceanographer, educator*
Mao, Kent Keqiang *consulting civil engineer*
Martin, George M. *pathologist, gerontologist*
McCormick, Norman Joseph *mechanical engineer, nuclear engineer, educator*
Merrill, Ronald Thomas *geophysicist, educator*
Miles, Edward Lancelot *marine studies educator, consultant, director*
Mills, David Michael *physicist*
Morgan, Jeff *research engineer*
Nelson, George Driver *astronomy and education educator, former astronaut*
Nelson, James Alonzo *radiologist, educator*
Nelson, Walter William *computer programmer, consultant*
Neppe, Vernon Michael *neuropsychiatrist, author, educator*
Nester, Eugene William *microbiology educator, immunology educator*
Oman, Henry *retired electrical engineer, engineering executive*
Osborne, Mason Scott *mathematician, educator*
Owens, Edward Henry *geologist, consultant*
Page, Roy Christopher *periodontist, educator*
Plisetskaya, Erika Michael *biology and physiology educator*
Pocker, Yeshayau *chemistry, biochemistry educator*
Porte, Daniel, Jr. *physician, educator, health facility administrator*
Porter, Robert Philip *electrical engineering educator*
Quay, Paul Douglas *oceanography educator*
Quimby, George Irving *anthropologist, former museum director*
Raisbeck, James David *aircraft design executive*
Ratner, Buddy Dennis *bioengineer, educator*
Ray, Sankar *opto-electroic device scientist*
Rimbey, Peter Raymond *physicist, mathematical engineer*
Robertson, William Osborne *physician*
Robinson, Gary Dale *aerospace company executive*
Roe, Michael Dean *psychologist*
Ross, Russell *pathologist, educator*
Roubal, William Theodore *biophysicist, educator*
Rubbert, Paul Edward *engineering executive*
Rubel, Edwin W *neurobiologist*
Schellenberg, Gerard David *geneticist, researcher*
Schiffrin, Milton Julius *physiologist*
Schomaker, Verner *chemist, educator*
Somani, Arun Kumar *electrical engineer, educator*
Spindel, Robert Charles *electrical engineering educator*
Stanback, Mark Thomas *behavioral ecologist, researcher*
Sutter, Joseph F. *aeronautical engineer, consultant, retired airline company executive*
Taylor, Ronald Russell *engineer*
Terrell, W(illiam) Glenn *university president emeritus*
Thiry, Paul *architect*
Thomas, E(dward) Donnall *physician, researcher*
Thompson, Leigh Lassiter *psychologist, educator*
Thouless, David James *physicist, educator*
Todaro, George Joseph *pathologist*
Tolliver, Gerald Arthur *research psychologist, behavioral counselor*
Tripp, Leonard Lee *software engineer*
Tsai, Mavis *clinical psychologist*
Tukey, Harold Bradford, Jr. *horticulture educator*
Van Hoosier, Gerald Leonard *veterinary science educator*
Viggers, Robert Frederick *mechanical engineering educator*
Wallace, John Michael *meteorology educator*
Wallem, Daniel Ray *metallurgical engineer*
Wang, Ji-Ping *pharmacist, researcher*
Weintraub, Harold M. *geneticist*
Whitman, Alexander H., Jr. *civil engineer, consultant*
Wiker, Steven Forrester *industrial engineering educator*
Wiktor, Peter Jan *engineer*
Wilcox, Paul Horne *academic administrator, researcher*
Wilets, Lawrence *physics educator*
Wilson, Thornton Arnold *retired aerospace company executive*
Young, Nancy Lee *medical educator*
Young, Paul Ruel *computer scientist, academic administrator*
Zabinsky, Zelda Barbara *operations researcher, industrial engineering educator*
Zhao, Lue Ping *biostatistician, genetic epidemiologist*

South Bend
Heinz, Roney Allen *civil engineering consultant*

Spokane
Benson, Allen B. *chemist, educator, consultant*
Budinger, Frederick Charles *geotechnical engineer*
Cameron, Robert H. *water association administrator*
Delistraty, Damon Andrew *toxicologist*
Dunau, Andrew T. *mathematics, science and technology consultant*
Esvelt, Larry Allen *environmental engineer*
Mackay, Alexander Russell *surgeon*
McClellan, David Lawrence *physician, medical facility administrator*
McGrew, Stephen Paul *physicist, entrepreneur*
Nandagopal, Mallur R. *engineer*
Steadman, Robert Kempton *oral and maxillofacial surgeon*
Stone, Michael David *landscape architect*
Vandervert, Larry Raymond *psychologist, educator, writer*
Waller, James Edward, Jr. *experimental psychologist, educator*

Stanwood
Lipscomb, David Milton *audiologist, consultant*

Steilacoom
Morgenthaler, John Herbert *chemical engineer*

Tacoma
Champ, Stanley Gordon *scientific company executive*
Grenley, Philip *urologist*
Hamner, Homer Howell *economist, educator*
Harding, Karen Elaine *chemistry educator and department chair*
Hoffmann, Gunther F. *forest products executive*
Ketchersid, Wayne Lester, Jr. *medical technologist*
MacGinitie, Laura Anne *electrical engineer*
Runde, Douglas Edward *wildlife biologist*
Schauss, Alexander George *psychologist, researcher*
Thomas, Steven Joseph *structural engineer*
Tison, David Lawrence *microbiologist*

Tumwater
Miles, Donald Geoffrey *economist*

Vancouver
Dietze, Gerald Roger *chemist*
Hartmann, David Peter *electrical engineer, educator, consultant*
Iverson, Richard Matthew *earth scientist*
Kelley, Charles Ray *psychologist*
Mairose, Paul Timothy *mechanical engineer, consultant*
Redman, Steven Phillip *engineer*

Walla Walla
DeBroeck, Dennis Alan *design engineer*
Gary, Walter J(oseph) *entomologist, educator*
Schrader, Ernest Karl *engineer*

Wallula
Cox, Dennis Joseph *engineer*

Washougal
Vogel, Ronald Bruce *food products executive*

Wenatchee
Hoyt, Stanley Charles *retired research administrator, entomologist*

West Richland
Cole, Charles R. *hydrologist, researcher*

Woodinville
Kaufman, William Carl *biophysicist*
McGavin, Jock Campbell *airframe design engineer*

Woodland
Brown, Alan Johnson *chemicals executive*

Yakima
Anderson, Gregory Martin *medical company representative*
Brown, Randy Lee *systems engineer*
LaFontaine, Thomas E. *chemical engineer*
Sutphen, Robert Ray *mechanical engineer, manager*

WEST VIRGINIA

Apple Grove
Burkett, Eugene John *chemical engineer*

Bethany
Cooey, William Randolph *economics educator*

Bluefield
Blaydes, James Elliott *ophthalmologist*

Charleston
Stone, Jennings Edward *applications engineer*
Televantos, John Yiannakis *engineering executive*
Zangeneh, Fereydoun *pediatrics educator, pediatric endocrinologist*

Culloden
Dixon, Victor Lee *wire company executive*

Elkview
Hudson, Kenneth Shane *electrical engineer*

Fairmont
Fulda, Michael *political science educator, space policy researcher*

Falling Waters
Schellhaas, Linda Jean *scientist, consultant*
Schellhaas, Robert Wesley *data processing executive, composer, musician*

Green Bank
Parker, David Hiram *physicist, electrical engineer*

Huntington
Cocke, William Marvin, Jr. *plastic surgeon, educator*
Driscoll, Henry Keane *endocrinologist, researcher*
Foley, Arvil Eugene *mechanical engineer*
Green, Todd Lachlan *physiologist, educator*
Harrison, Marcia Ann *biology educator*
Leppla, David Charles *pathology educator*
Mallory, David Stanton *physiology educator*
Mufson, Maurice Albert *physician, educator*
Simmons, Harry Dady *pathologist*

Lewisburg
Cardis, Thomas Michael *chemist*
Moore, Joan L. *radiology educator, physician*

Man
Bofill, Rano Solidum *physician*

Masontown
Wilson, Dale William, Jr. *laser technician*

Mineral Wells
Prather, Denzil Lewis *petroleum engineer*

Montgomery
Thornton, Stafford Earl *civil engineering educator*

Morgantown
Brown, Paul B. *physiologist educator*
Butcher, Fred R. *biochemistry educator, university administrator*
Chase, Philip Noyes *psychologist*
Christiansen, Tim Alan *ecologist*
Das, Kamalendu *chemist*
Eck, Ronald Warren *civil engineer, educator*
Einzig, Stanley *pediatric cardiologist, researcher*
Gasper-Galvin, Lee DeLong *chemical engineer*
Gladfelter, Wilbert Eugene *physiology educator*
Gopalakrishnan, Bhaskaran *manufacturing engineer*
Gottschalk, Kurt William *research forester*
Guthrie, Hugh Delmar *chemical engineer*
Hill, Ronald Charles *surgeon, educator*
Lamm, Donald Lee *urologist*
Lattal, Kennon Andy *psychology educator*
Maxwell, Robert Haworth *college dean*
Ponte, Charles Dennis *pharmacist, educator*
Prucz, Jacky Carol *mechanical and aerospace engineer, educator*
Schroder, John L., Jr. *retired mining engineer*
Shaeiwitz, Joseph Alan *chemical engineering educator*
Shoemaker, Harold Dee *mechanical engineer*
Tavaglione, David *computer scientist, consultant*
Weinstein, George William *ophthalmology educator*

Parkersburg
Lindstrom, Timothy Rhea *chemist*
Witwer, Ronald James *mechanical engineer*

Rainelle
Scott, Pamela Moyers *physician assistant*

Rocket Center
Hartman, Kenneth Owen *chemist*

Ronceverte
Johnson, Bret Gossard *dietitian, consultant*

Scott Depot
Butch, James Nicholas *electronics company executive*

South Charleston
Britton, Laurence George *research scientist*
Nielsen, Kenneth Andrew *chemical engineer*

Triadelphia
Mc Cullough, John Phillip *management consultant, educator*

Wheeling
Coast, Morgan K. *civil engineer, consultant*
Rivers, Lee Walter *chemical engineer*
Rogo, Kathleen *safety engineer*
Zyznewsky, Wladimir A. *neurologist*

WISCONSIN

Appleton
Aziz, Salman *chemical engineer, company executive*
De Stasio, Elizabeth Ann *biology educator*
Van den Akker, Johannes Archibald *physicist*

Bayfield
Gallinat, Michael Paul *fisheries biologist*

Beldenville
Mullenax, Charles Howard *veterinarian, researcher*

Brookfield
Curfman, Floyd Edwin *engineering educator*
Diesem, John Lawrence *information systems specialist*

Chippewa
Conger, Jeffrey Scott *electrical engineer*

Chippewa Falls
Wos, Carol Elaine *engineer*

DeForest
Miller, Paul Dean *breeding company executive, geneticist*

Delafield
Mudek, Arthur Peter *automation engineer, consultant*

Eau Claire
St. Louis, Robert Vincent *chemist, educator*

Fond Du Lac
Fife, William J., Jr. *metal products executive*
Seichter, Daniel John *electrical engineer*

Fort McCoy
Truthan, Charles Edwin *physician*

Grafton
Eber, Lorenz *civil engineer*

Green Bay
Davis, Gregory John *mathematician, educator*
Hudson, Halbert Austin, Jr. *retired manufacturing engineer, consultant*
Swetlik, William Philip *orthodontist*
von Heimburg, Roger Lyle *surgeon*
Wiley, Dale Stephen *mechanical engineer*

Hartford
Janzen, Norine Madelyn Quinlan *medical technologist*

Hartland
Vitek, Richard Kenneth *scientific instrument company executive*

Hudson
Fahning, Melvyn Luverne *veterinary educator*

Janesville
Hornby, Robert Ray *mechanical engineer*
Morgan, Donna Jean *psychotherapist*

Juneau
Sindelar, Robert Albert *civil engineer*

Kaukauna
Janssen, Gail Edwin *banking executive*

Kenosha
Greenebaum, Ben *physicist*
Harris, Benjamin *psychologist*
Potente, Eugene, Jr. *interior designer*

Kimberly
Bressers, Daniel Joseph *utility executive*

La Crosse
Abts, Daniel Carl *computer scientist*
Burmaster, Mark Joseph *software engineer*
Costakos, Dennis Theodore *neonatologist, researcher*
Davy, Michael Francis *civil engineer, consultant*
Lindesmith, Larry Alan *physician, administrator*
Meinertz, Jeffery Robert *physiologist*
Monfre, Joseph Paul *mechanical engineer, consultant*
Silva, Paul Douglas *reproductive endocrinologist*
Smith, Martin Jay *physician, biomedical research scientist*

Lake Geneva
Craft, Timothy George *utility company executive*

Madison
Amundson, Clyde Howard *engineering educator, researcher*
Anderson, Frederic Simon B. *physicist*
Askey, Richard Allen *mathematician*
Bagchi, Amalendu *environmental engineer*
Barlow, Kent Michael *mechanical engineer*
Barnes, Robert F. *agronomist*
Berthouex, Paul Mac *civil and environmental engineer, educator*
Berven, Norman Lee *counselor, psychologist, educator*
Bollinger, John Gustave *engineering educator, college dean*
Borisy, Gary G. *molecular biology educator*
Briggs, Rodney Arthur *agronomist, consultant*
Bruhn, Hjalmar Diehl *retired agricultural engineer, educator*
Burkholder, Wendell Eugene *entomologist*
Carbon, Max William *nuclear engineering educator*
Carbone, Paul Peter *oncologist, educator, administrator*
Carter, Paul R. *agronomist consultant*
Churchwell, Edward Bruce *astronomer, educator*
Clark, David Leigh *marine geologist, educator*
Cleland, W(illiam) Wallace *biochemistry educator*
Dahl, Lawrence Frederick *chemistry educator, researcher*
Daie, Jaleh *science educator, researcher, administrator*
Dasgupta, Ranjit Kumar *virologist*
de Boor, Carl *mathematician*
Dentine, Margaret Raab *animal geneticist, educator*
Desautels, Edouard Joseph *computer science educator*
DeVries, Marvin Frank *mechanical engineering educator*
Dott, Robert Henry, Jr. *geologist, educator*
Duffie, John Atwater *chemical engineer, educator*
Duffie, Neil Arthur *mechanical engineering educator, researcher*
Easterday, Bernard Carlyle *veterinary medicine educator*
Ediger, Mark D. *chemistry educator*
Ellis, Arthur Baron *chemist, educator*
Eloranta, Edwin Walter *meteorologist, researcher*
Erickson, John Ronald *research administrator*
Evert, Ray Franklin *botany educator*
Farrar, Thomas C. *chemist, educator*
Fennema, Owen Richard *food chemistry educator*
Ferry, John Douglass *chemist*
Fettiplace, Robert *neurophysiologist*
Fowler, John Francis *radiobiologist*
Freudenburg, William R. *sociology educator*
Gordon, Mark Elliott *environmental engineer*
Gorski, Jack *biochemistry educator*
Graham, James Miller *physiology researcher*
Gustafson, David Harold *industrial engineering and preventive medicine educator*
Hailman, Jack Parker *zoology educator*
Haller, Archibald Orben *sociologist, educator*
Helgeson, John Paul *physiologist, researcher*
Hesse, Thurman Dale *welding metallurgy educator, consultant*
Hokin, Lowell Edward *biochemist, educator*
Hong, Richard *pediatric immunologist, educator*
Hopen, Herbert John *horticulture educator*
Iltis, Hugh Hellmut *plant taxonomist and evolutionist, educator*
Jackson, Carl Robert *obstetrician/gynecologist*
Jeanne, Robert Lawrence *entomologist, educator, researcher*
Jefferson, James Walter *psychiatry educator*
Johnson, Richard Warren *chemist*
Kaul, David Glenn *civil engineer*
Kemnitz, Joseph William *physiologist, researcher*
Kim, Sangtae *chemical engineering educator*
Kirk, Thomas Kent *research scientist*
Kleene, Stephen Cole *retired mathematician, educator*
Knutson, Lynn Douglas *physics educator*
Kuzmic, Petr *chemist*
Laessig, Ronald Harold *pathology educator, state official*
Lagally, Max Gunter *physics educator*
Lanher, Bertrand Simon *biological spectroscopist*
Lawler, James Edward *physics educator*
Loper, Carl Richard, Jr. *metallurgical engineer, educator*
Maher, Louis James, Jr. *geologist, educator*
Maki, Dennis G. *medical educator, researcher, clinician*
Miller, Michael Beach *schizophrenia researcher*
Monroe, John Robert *electrical engineer*
Morton, Stephen Dana *chemist*
Moses, Gregory Allen *engineering educator*
Mozdziak, Paul Edward *growth biologist*
Muller, Daniel *biomedical scientist*
Nordby, Eugene Jorgen *orthopedic surgeon*
Olson, Hector Monroy *research support engineer*
Olson, Norman Fredrick *food science educator*
Pampel, Roland D. *computer company executive*
Pariza, Michael Willard *research institute executive, microbiology and toxicology educator*
Perkowski, Casimir Anthony *biopharmaceutical executive, consultant*
Peterson, David Maurice *plant physiologist, researcher*
Pitot, Henry Clement, III *physician, educator*
Pray, Lloyd Charles *geologist, educator*
Radwin, Robert Gerry *science educator, researcher, consultant*
Ramesh, Krishnan *mechanical engineer*
Rasmussen, Dennis Robert *behavioral ecologist*
Rich, Daniel Hulbert *chemist*
Richards, Hugh Taylor *physics educator*
Richardson, Kevin William *civil and environmental engineer*
Roberts, Leigh Milton *psychiatrist*
Russell, Jeffrey Scott *civil engineering educator*
Schatten, Gerald Phillip *cell biologist, educator*
Schuler, Ronald Theodore *agricultural engineering educator*
Sedgwick, Julie Beth *immunologist*
Sheffield, Lewis Glosson *physiologist*
Shohet, Juda Leon *electrical and computer engineering educator, researcher, high technology company executive*
Sih, Charles John *pharmaceutical chemistry educator*
Skinner, James Lauriston *chemist, educator*
Skoog, Folke Karl *botany educator*
Smith, Matthew Jay *chemist*
Smith, William Leo *meteorologist, researcher, educator*
Southard, James Hewitt *biochemist, researcher*
Sufit, Robert Louis *neurologist, educator*
Temin, Howard Martin *scientist, educator*
Tran, Tri Duc *chemical engineer*
Vailas, Arthur C. *biomechanics educator*
Webster, John Goodwin *biomedical engineering educator, researcher*
West, Robert Culbertson *chemistry educator*
Westphal, Klaus Wilhelm *university museum director*
Whitlon, Donna Sue *neuroscientist, researcher*
Whitmyer, Robert Wayne *soil scientist, consultant, researcher*
Woods, Thomas Fabian *lawyer*
Wright, John Curtis *chemist, educator*
Yu, Hyuk *chemist, educator*
Yuill, Thomas MacKay *university administrator, microbiology educator*
Zimmerman, Howard Elliot *chemist, educator*

Marshfield
Kelman, Donald Brian *neurosurgeon*
Kitchell, Shawn Ray *plant engineer, educator*
Marx, James John *immunologist*
Stueland, Dean Theodore *emergency physician*

Menasha
Mahnke, Kurt Luther *psychotherapist, clergyman*
Sutter, Carl Clifford *civil engineer*

Menomonee Falls
Markowski, Roberta Jean *electrical engineer*
Moberg, Clifford Allen *mold products company executive*
Schommer, Gerard Edward *mechanical engineer*

Menomonie
Seaborn, Carol Dean *nutrition researcher*
Swanson, Helen Anne *psychology educator*

Middleton
Adney, James Richard *physicist*
Haynes, Joel Robert *molecular biologist*
Herb, Raymond G. *physicist, manufacturing company executive*

Milwaukee
Allgaier, Glen Robert *electronics engineer, researcher*
Anderson, Rebecca Cogwell *psychologist*
Arkadan, Abdul-Rahman Ahmad *electrical engineer, educator*
Atlee, John Light, III *physician*
Babcock, Janice Beatrice *health care coordinator*
Bacon, John Stuart *biochemical engineer*
Baker, John E. *cardiac biochemist, educator*
Bashford, James Adney, Jr. *experimental psychology researcher, educator*
Blomquist, Michael Allen *civil engineer*
Brauer, John Robert *electrical engineer*
Bub, Alexander David *acoustical engineer*
Carroll, Edward William *anatomist, educator*
Chow, John Lap Hong *biomedical engineer*
Cronin, Vincent Sean *geologist*
Davis, Thomas William *college administrator, electrical engineering educator*
Dorff, Gerald J. *physician*
Doumas, Basil Thomas *chemist, researcher, educator*
Effros, Richard Matthew *medical educator, physician*
Frantzides, Constantine Themis *general surgeon*
Funahashi, Akira *physician, educator*
Garimella, Suresh Venkata *mechanical engineering educator*
Gonnering, Russell Stephen *ophthalmic plastic surgeon*
Gopal, Raj *energy systems engineer*
Gorelick, Jeffrey Bruce *physician, educator*
Greenler, Robert George *physics educator, researcher*
Hanson, Curtis Jay *structural project engineer*
Heinen, James Albin *electrical engineering educator*
Humber, Wilbur James *psychologist*
Hutz, Reinhold Josef *physiologist*
Johnson, F. Michael *electrical and automation systems professional*
Karkheck, John Peter *physics educator, researcher*
Kloehn, Ralph Anthony *plastic surgeon*
Knasel, Thomas Lowell *information systems consultant*
Kraut, Joanne Lenora *computer programmer, analyst*
Kusumi, Akihiro *scientist, educator*
Landis, Fred *mechanical engineering educator*
Laubenheimer, Jeffrey John *civil engineer*
Levy, Moises *physics educator*
Libnoch, Joseph Anthony *physician, educator*
Lindner, Albert Michael *structural engineer*
Lord, Guy Russell, Jr. *psychiatry educator*
McKinney, Bryan Lee *chemist*
Modlinski, Neal David *computer systems analyst*
Montgomery, Robert Renwick *medical association administrator, educator*
Moore, Gary Thomas *aerospace architect*
Morgan, Donald George *magnetic separation engineer*
Morris, Robert DuBois *epidemiologist*
Nagarkatti, Jai Prakash *chemical company executive*
Namdari, Bahram *surgeon*
Newman, Robert Wyckoff *research engineer*
Perez, Ronald A. *mechanical engineering educator*
Petering, David Harold *chemistry educator*
Petersen, Ralph Allen *chemist*
Puta, Diane Fay *medical staff services director*
Reid, Robert Lelon *college dean, mechanical engineer*
Remsen, Charles Cornell, III *microbiologist, research administrator, educator*
Renken, Kevin James *mechanical engineering educator*
Rosen, Barry Howard *museum director, history educator*
Silverberg, James Mark *anthropology educator, researcher*
Simms, John Carson *logic, mathematics and computer science educator*
Smith, Michael Lawrence *knowledge engineer, researcher*
Story, Michael Thomas *biomedical researcher, educator*
Strickler, John Rudi *biological oceanographer*
Tekkanat, Bora *materials scientist*
Tisser, Clifford Roy *electrical engineer*
Vetro, James Paul *electrical project engineer*
Wagner, Marvin *general and vascular surgeon, educator*
Warren, Richard M. *experimental psychologist, educator*
Wilson, Walter Leroy *architect*
Yin, Jun-jie *biophysicist*
Young, Allen Marcus *museum curator of zoology, educator, naturalist, consultant, writer*
Zietlow, Daniel H. *electrical engineer*

Muskego
Brown, Serena Marie *mathematics and home economics educator*

Neenah
Polley, William Alphonse *power systems engineer*
Underhill, Robert Alan *consumer products company executive*
Zimmerman, Delano Elmer *physician*

New Berlin
DeBaker, Brian Glenn *mechanical engineer*
Risberg, Robert Lawrence, Sr. *electronics executive, consultant, engineer*

North Freedom
Fausett, Robert Julian *engineering geologist, consultant*

Oconomowoc
Luedke, Patricia Georgianne *microbiologist*
Raether, Scott Edward *mechanical engineer*

Onalaska
Dukerschein, Jeanne Therese *aquatic biologist, educator, researcher*
Soballe, David Michael *limnologist*

Oshkosh
Rouf, Mohammed Abdur *microbiology educator*

Pewaukee
Kelly, A(llan) James *environmental engineer*

Platteville
Balachandran, Swaminathan *industrial engineering educator*
Yeske, Ronald A. *dean*

Racine
Guntly, Leon Arnold *mechanical engineer*
Kim, Zaezeung *allergist, immunologist, educator*
Neumiller, Phillip Joseph, III *research scientist*
Rooney, John Connell *consulting civil engineer*

Rothschild
Drew, Richard Allen *electrical and instrument engineer*

Salem
Lambert, James Allen *industrial electrician*

Schofield
Adams, James William *retired chemist*

Sheboygan
Bockius, Thomas John *mechanical engineer*
Golubski, Joseph Frank *pathologist, physician*
Marr, Kathleen Mary *biologist, educator*

South Milwaukee
Van Dusen, Harold Alan, Jr. *electrical engineer*

Stoughton
Huber, David Lawrence *physicist, educator*

Sussex
Dewey, Craig Douglas *engineering executive*

Waukesha
Otu, Joseph Obi *mathematical physicist*
Svetic, Ralph E. *electrical engineer, mathematician*

West Allis
Zhang, Jiping *engineering analyst*

Whitewater
Newman, Lisa Ann *speech pathologist, educator*
Stekel, Frank Donald *physics educator*

Williams Bay
Harper, Doyal Alexander, Jr. *astronomer, educator*
Hobbs, Lewis Mankin *astronomer*
Kron, Richard G. *astrophysicist, educator*

WYOMING

Buffalo
Velasquez, Pablo *mining executive*

Casper
Cole, Malvin *neurologist, educator*
Rasmussen, Niels Lee *geologist, environmental chemist*

Cheyenne
Beaven, Thornton Ray *physical scientist*
Feusner, LeRoy Carroll *chemical engineer*
Flick, William Fredrick *surgeon*
Laycock, Anita Simon *psychotherapist*
Rust, Lynn Eugene *geologist*

Gillette
Frederick, James Paul *chemical engineer*

Jackson
Davis, Randy L. *soil scientist*
Werner, Frank David *aeronautical engineer*

Laramie
Chai, Winberg *political science educator, foundation chair*
Cronkleton, Thomas Eugene *physician*
Forster, Bruce Alexander *economics educator*
Grandy, Walter Thomas, Jr. *physicist*
Hinds, Frank Crossman *animal science educator*
Laman, Jerry Thomas *mining company executive*
Meyer, Edmond Gerald *energy and natural resources educator, resources scientist, entrepreneur, former chemistry educator*
Speight, James Glassford *research company executive*

Moose
Craighead, Frank Cooper, Jr. *ecologist*

Rock Springs
Mitchell, Sandra Louise *biology educator, researcher*

Torrington
Jensen, Christopher Douglas *civil engineer*

TERRITORIES OF THE UNITED STATES

GUAM

Tamuning
Mayer, Peter Conrad *economics educator*

PUERTO RICO

Bay
Arce-Cacho, Eric Amaury *solar energy engineer, consultant*

Bayamon
El-Khatib, Shukri Muhammed *biochemist*
Juarbe, Charles *otolaryngologist, neck surgeon*

Caguas
Tulenko, Maria Josefina pharmacist

Corozal
Rodriguez Garcia, Jose A. agronomist, investigator

Gurabo
Curet-Ramos, José Antonio internist

Hato Rey
Edwards-Vidal, Dimas Francisco mechanical engineer, consultant
Rosario-Guardiola, Reinaldo dermatologist

Humacao
Esteban, Ernesto Pedro physicist
Love, James Brewster pharmaceutical engineer
Zypman-Niechonski, Fredy Ruben physicist

Luquillo
Arnizaut de Mattos, Ana Beatriz veterinarian

Manati
Garcia, Pedro Ivan psychologist
Silva-Ruíz, Sergio Andrés biochemist

Mayaguez
Collins, Dennis Glenn mathematics educator
Mandavilli, Satya Narayana chemical engineering educator, researcher
Ramirez Cancel, Carlos Manuel psychologist, educator
Rodríguez-Arias, Jorge H. retired agricultural engineering educator
Souto Bachiller, Fernando Alberto chemistry educator
Suarez, Luis Edgardo mechanical engineering educator

Ponce
Torres-Aybar, Francisco Gualberto medical educator

Rio Piedras
Perez, Victor medical technologist, laboratory director
Pinilla, Ana Rita neuropsychologist, researcher
Sardina, Rafael Herminio nuclear engineer
Toranzos, Gary Antonio microbiology educator

San German
Quintero, Héctor Enrique science educator

San Juan
Benton, Stephen Richard civil and mechanical engineer
De Jesús, Nydia Rosa physician, anesthesiologist
Fernandez-Repollet, Emma D. pharmacology educator
Jiménez, Braulio Dueño toxicologist
Lugo, Ariel E. botanist, federal agency administrator
Opava-Stitzer, Susan Catherine physiologist, researcher
Prevor, Ruth Claire psychologist
Quiñones, Jose Antonio structural engineer, consultant
Ramírez-Ronda, Carlos Héctor physician
Rodriguez Arroyo, Jesus gynecologic oncologist
Rodriguez-del Valle, Nuri microbiology educator
Sahai, Hardeo medical educator

Santurce
Fernandez-Martinez, Jose physician

VIRGIN ISLANDS

Kingshill
Crossman, Stafford Mac Arthur agronomist, researcher

Saint Croix
Pierce, Lambert Reid architect

MILITARY ADDRESSES OF THE UNITED STATES

ATLANTIC

APO
De Roux, Tomas E. electrical engineer
Knowlton, Nancy biologist
Powers, Nelson Roger entomologist
Rubinoff, Ira biologist, research administrator, conservationist

FPO
English, Gary Emery military officer

EUROPE

APO
Carioti, Bruno Mario civil engineer
Foster, Kirk Anthony emergency medical service administrator, educator, consultant
Freeman, Brian S. electrical engineer
Goodwin, Richard Clarke military analyst
Hatton, Daniel Kelly computer scientist
Huffman, Kenneth Alan operations researcher, mathematician
Scheltema, Robert William military officer

FPO
Lahr, Brian Scott pilot

PACIFIC

APO
Heppner, Donald Gray, Jr. immunology research physician

FPO
Carlisle, Mark Ross naval aviator

CANADA

ALBERTA

Beaverlodge
McElgunn, James Douglas agriculturist, researcher

Calgary
Codding, Penelope Wixson chemistry educator
Cooke, David Lawrence chemical engineer
de Krasinski, Joseph Stanislas mechanical engineering educator
Dixon, Gordon Henry biochemist
Farries, John Keith petroleum engineering company executive
Ghali, Amin civil engineering educator
Goren, Howard Joseph biochemistry educator
Jones, Geoffrey Melvill physiology research educator
Kentfield, John Alan mechanical engineering educator
Kimberley, Barry Paull ear surgeon
Kwok, Sun astronomer
MacCulloch, Patrick C. oil industry executive
Maier, Gerald James natural gas transmission and marketing company executive
Merta De Velehrad, Jan diving and safety engineer, scientist, psychologist, inventor, educator, civil servant
Milone, Eugene Frank astronomer, educator
Mungan, Necmettin petroleum consultant
Nghiem, Long Xuan computer company executive
Nigg, Benno M. biomechanics educator
Nowlan, Godfrey S. geologist
Sreenivasan, Sreenivasa Ranga physicist, educator
Stell, William Kenyon neuroscientist, educator
Venkatesan, Doraswamy astrophysicist, physics educator
Wallace, John Lawrence immunophysiologist, educator
Yoon, Ji-Won virology, immunology and diabetes educator, research administrator

Edmonton
Bach, Lars wood products engineer, researcher
Bartlett, Fred Michael Pearce structural engineer
Cossins, Edwin Albert biology educator, academic administrator
Craggs, Anthony mechanical engineer
de Guzman, Roman de Lara chemical engineer, consultant
Dewhurst, William George physician, psychiatrist, educator, researcher
Freeman, Gordon Russel chemistry educator
Hiruki, Chuji plant virologist, science educator
Hrudey, Steve E. civil engineer, educator
Israel, Werner physics educator
Kalra, Yash Pal soil chemist
Kanasewich, Ernest Roman physics educator
Kay, Cyril Max biochemist
Kennedy, D. J. Laurie civil engineering educator
Khan, Abdul Quasim chemistry researcher
Khanna, Faqir Chand physics educator
Kitching, Peter physics educator
Krotki, Karol Jozef sociology educator, demographer
Lemieux, Raymond Urgel chemistry educator
Morgenstern, Norbert Rubin civil engineering educator
Page, Don Nelson theoretical gravitational physics educator
Rajotte, Ray V. biomedical engineer, researcher
Rao, Ming chemical engineering and computer science educator
Rostoker, Gordon physicist, educator
Stanley, S. J. civil engineering educator
Stollery, Robert construction company executive

Lethbridge
Johnson, Daniel Lloyd biogeographer
Simmons, Robert Arthur engineer, consultant
Sonntag, Bernard H. agrologist, research executive

BRITISH COLUMBIA

Agassiz
Molnar, Joseph Michael plant physiologist, research director

Bamfield
Druehl, Louis Dix biology educator

Burnaby
Baille, David L. biologist, educator
Brandhorst, Bruce Peter biology educator
Copes, Parzival economist, researcher
Einstein, Frederick William Boldt chemistry educator
Forgacs, Otto Lionel forest products company executive
Han, Jiawei computer scientist, educator
Roitberg, Bernard David biology educator
Runka, Gary G. agricultural company executive
Saif, Mehrdad electrical engineering educator

Kaleden
Swales, John E. (Ted) retired horticulturalist

Nanaimo
Ricker, William Edwin biologist

North Vancouver
Buckland, Peter Graham structural engineer
Morgenstern, Brian D. civil engineer
Wilson, Carl Robert structural engineer

Sidney
Best, Melvyn Edward geophysicist, researcher
Lanterman, William Stanley, III plant pathologist, researcher, administrator
Scrimger, Joseph Arnold research company executive
van den Bergh, Sidney astronomer
Weichert, Dieter Horst seismologist, researcher
Wong, Chi-Shing chemical oceanographer
Xie, Yunbo oceanographer

Summerland
Dueck, John agricultural researcher, plant pathologist
Looney, Norman Earl pomologist, plant physiologist

Surrey
Stiemer, Siegfried F. civil engineer

Vancouver
Affleck, Ian Keith physics educator
Burhenne, Hans Joachim physician, radiology educator
Chase, Richard Lionel St. Lucian geology and oceanography educator
Chow, Anthony Wei-Chik physician
Chu, Allen Yum-Ching automation company executive, systems consultant
Clement, Douglas Bruce medical educator
Drance, Stephen Michael ophthalmologist, educator
Eaves, Allen Charles Edward hematologist, medical agency administrator
Fryzuk, Michael Daniel chemistry educator
Hallbauer, Robert Edward mining company executive
Hardy, Walter Newbold physics educator, researcher
Hatzikiriakos, Savvas Georgios chemical engineer
Healey, Michael Charles fishery ecologist, educator
Hochachka, Peter William biology educator
Isaacson, Michael civil engineering educator
Isman, Murray entomology educator
Jones, David Robert zoology educator
Keevil, Norman Bell mining executive
Kieffer, Susan Werner geology educator
Kiefl, Robert Frances physicist, educator
Larkin, Peter Anthony zoology educator, university dean and official
Lavkulich, Leslie Michael soil science educator
Mathews, William Henry geologist, educator
Mavinic, Donald Stephen civil engineering educator
Nemetz, Peter Newman policy analysis educator, economics researcher
Owen, Bruce Douglas animal physiologist
Randall, David John physiologist, zoologist, educator
Roy, Chunilal psychiatrist
Schultz, Kirk R. pediatric hematology-oncology educator
Seymour, Brian Richard mathematician
Silver, Hulbert K.B. physician, educator
Smith, Michael biochemistry educator
Soregaroli, A(rthur) E(arl) mining company executive, geologist
Vogt, Erich Wolfgang physicist, university administrator
Wedepohl, Leonhard M. electrical engineering educator
Young, Margaret Elisabeth physicist
Zhao, Ru He mechanical engineer

Victoria
Hesser, James Edward astronomy researcher
Hoffman, Paul Felix geologist, educator
Loring, Thomas Joseph forest ecologist
Wilkinson, R. L. agriculturalist
Zakarauskas, Pierre physicist

MANITOBA

Anola
de Nevers, Roy Olaf retired aerospace company executive

Pinawa
Wright, Michael George atomic energy company executive

Winnipeg
Angel, Aubie physician, academic administrator
Anthonisen, Nicholas R. respiratory physiologist
Bushuk, Walter agricultural studies educator
Cohen, Harley civil engineer, science educator
Fielding, Ronald Roy aeronautical engineer
Greenberg, Arnold H. pediatrics educator, cell biologist
Hamerton, John Laurence geneticist, educator
Lang, Otto E. business executive, former Canadian cabinet minister
Mauro, Arthur financial executive, university chancellor
McKee, James Stanley Colton physics educator
Petersmeyer, John Clinton architect
Secco, Anthony Silvio chemistry educator
Smith, Ian Cormack Palmer biophysicist
Steele, John Wiseman pharmacy educator
Storgaard, Anna K. agriculturalist

NEW BRUNSWICK

Fredericton
Armstrong, Robin Louis university official, physicist
Boorman, Roy Slater science administrator, geologist
Douglas, Robert Andrew civil engineer, educator
Faig, Wolfgang survey engineer, engineering educator
Grotterod, Knut retired paper company executive
Neill, Robert D. engineering executive
Unger, Israel dean, chemistry educator

Moncton
Hanson, John Mark ecologist, researcher

Sackville
Dekster, Boris Veniamin mathematician, educator

Saint John
Thomas, Martin Lewis H. marine ecologist, educator

NEWFOUNDLAND

Saint John's
Clark, Jack I. civil engineer, researcher
Datta, Indranath naval architect, educator
Sheath, Robert Gordon botanist

NOVA SCOTIA

Halifax
Andrew, John Wallace medical physicist
Borgese, Elisabeth Mann political science educator, author
Huggard, Richard James federal agency administrator
Mufti, Aftab A. civil engineering educator
O'Dor, Ron physiologist, marine biologist
Scaratt, David J. marine biologist
Wilson, George Peter industrial engineer

Wolfville
Ogilvie, Kelvin Kenneth chemistry educator

ONTARIO

Almonte
Morrison, Angus Curran aviation executive

Bramalea
Hornby-Anderson, Sara Ann metallurgical engineer, marketing professional

Burlington
Krishnappan, Bommanna Gounder fluid mechanics engineer

Chalk River
Dolling, Gerald physicist, research executive

Chatham
Shakhmundes, Lev mathematician

Cold Lake
Clarke, Thomas Edward research and development management educator

Cornwall
McIntee, Gilbert George materials testing engineer

Delhi
Court, William Arthur chemist, researcher
Whitfield, Gary Hugh research director

Downsview
Kim, Tae-Chul foundation engineer
Tennyson, Roderick C. aerospace scientist

Etobicoke
Bahadur, Birendra display specialist, liquid crystal researcher
Ibrahim, Mohammad Fathy Kahlil aerospace engineer
Mufti, Navaid Ahmed network design engineer
Stojanowski, Wiktor J. mechanical engineer

Gloucester
Mykytiuk, Alex P. chemist

Guelph
Burnside, Edward Blair geneticist, educator, administrator
Karl, Gabriel physics educator
Lougheed, Everett Charles retired horticulture educator, researcher
Miller, Murray Henry soil science educator
Stevens, (Ernest) Donald zoology educator

Hamilton
Basmajian, John Varoujan medical scientist, educator, physician
Bienenstock, John physician, educator
Collins, Malcolm Frank physicist, educator
Kenney-Wallace, Geraldine chemistry and physics educator
Kenny-Wallace, G. A. chemical engineer
Kingwood, Alfred E. geologist, educator
McNutt, R. H. geologist, geochemist, educator
Purdey, Gary Rush materials science and engineering educator, dean
Steiner, George information systems and management science educator, researcher
Taylor, James Hutchings chancellor
Thode, Henry George chemistry educator
Vlachopoulos, John chemical engineering educator
Walker, Roger Geoffrey geology educator, consultant
Woo, Ming-Ko geographer, educator

Kingston
Benidickson, Agnes university chancellor
McGeer, James Peter research executive, consultant
Meisel, John political scientist
Smith, Roy Edward mechanical engineer, rail vehicle consultant
Smol, John Paul limnologist, educator
Stewart, Alec Thompson physicist

Kitchener
Mundy, Phillip Carl engineer

London
Bancroft, George Michael chemical physicist, educator
Barnett, Henry Joseph Macaulay neurologist
Battista, Jerry Joseph medical physicist
Bauer, Michael Anthony computer scientist, educator
Broadwell, Charles E. retired agricultural products company executive
Brown, James Douglas materials engineer, researcher
Choy, Patrick C. biochemistry educator
Fyfe, William Sefton geochemist, educator
Mathur, Radhey Mohan electrical engineering educator, dean
Valberg, Leslie Stephen medical educator, physician, researcher

Manotick
Hobson, George Donald retired geophysicist

Mississauga
Crewe, Katherine engineer
Elfstrom, Gary Macdonald aerospace engineer, consultant
Errampalli, Deena molecular plant pathologist
Evans, Essi H. research scientist
Evans, John Robert former university president, physician
Lawford, G. Ross research and development company executive
Milligan, Victor civil engineer, consultant

North York
Buick, Fred J.R. physiologist, researcher
Godson, Warren Lehman meteorologist
Nicholls, Ralph William physicist, educator

Vafopoulou, Xanthe *biologist*
Wyatt, Philip Richard *geneticist, physician, researcher*

Orleans
Vanier, Jacques *physicist*

Ottawa
Alper, Anne Elizabeth *professional association executive*
Altman, Samuel Pinover *mechanical engineer, research consultant*
Armstrong, David William *biotechnologist, microbiologist*
Atif, Morad Rachid *architect*
Babcock, Elkanah Andrew *geologist*
Baltacioglu, Mehmet Necip *civil engineer*
Baum, Bernard Rene *biosystematist*
Bharghava, Vijay *engineer*
Bozozuk, Michael *civil engineer*
Brasier, Steven Paul *publishing professional*
Carey, Paul Richard *biophysicist, scientific administrator*
Coleman, John Morley *transportation research director*
Connelly, Alan B. *career officer, engineer*
Dagum, Camilo *economist, educator*
de Bold, Adolfo J. *pathology and physiology educator, research scientist*
Dence, Michael Robert *research director*
Devereaux, William A. *engineer*
Dlab, Vlastimil *mathematics educator, researcher*
Dubey, Ram Janam *inorganic and environmental chemist*
Emery, Alan Roy *museum director*
Friesen, Henry George *endocrinologist, educator*
Georganas, Nicolas D. *electrical engineering educator*
Gold, Lorne W. *Canadian government official*
Goldmann, Nahum *product development executive*
Halstead, Ronald Lawrence *soil scientist*
Herzberg, Gerhard *physicist*
Ingold, Keith Usherwood *chemist, educator*
Keith, Stephen Ernest *acoustical researcher*
Keon, Wilbert Joseph *cardiologist, surgeon, educator*
Lees, Ron Milne *physicist, educator*
Lister, E. Edward *animal science consultant*
Lockwood, David John *physicist, researcher*
MacLeod, John Munroe *radio astronomer, academic administrator*
McFarlane, James Ross *mechanical engineer, educator*
McLaren, Digby Johns *geologist, educator*
Mirza, Shaukat *engineering educator, researcher, consultant*
Morand, Peter *research agency executive*
Morton, Donald Charles *astronomer*
Perron, Pierre O. *science administrator*
Ramsay, Donald Allan *physical chemist*
Rummery, Terrance Edward *nuclear engineering executive, researcher*
Seaden, George *civil engineer*
Siebrand, Willem *theoretical chemist, science editor*
Stinson, Michael Roy *physicist*
Tomlinson, Roger W. *geographer*
Veizer, Ján *geology educator*
Vézina, Monique *Canadian government official*
Wood, Gordon Harvey *physicist*

Ottawa-Hull
Valcourt, Bernard *Canadian government official, lawyer*

Petawawa
Elchuk, Steve *chemist*

Saint Catharines
Jolly, Wayne Travis *geologist, educator*
Picken, Harry Belfrage *aerospace engineer*

Sault Sainte Marie
Banerjee, Samarendranath *orthopaedic surgeon*

Scarborough
Teitsma, Albert *physicist*

Simcoe
Collver, Keith Russell *agricultural products exective*

Toronto
Broder, Irvin *pathologist, educator*
Brumer, Paul William *chemical physicist, educator*
Chasin, Marshall Lewis *audiologist, educator*
De Nil, Luc Frans *speech-language pathologist*
Ham, James Milton *engineering educator*
Harrison, Robert Victor *auditory physiologist*
Heinke, Gerhard William (Gary Heinke) *environmental engineering educator*
Kaiser, Nicholas *physicist, educator*
Kerbel, Robert Stephen *cell biologist, cancer researcher*
Kerr, Peter Donald *geography educator emeritus*
Khouw, Boen Tie *biochemist*
Kunov, Hans *biomedical engineering educator, electrical engineering educator*
Langton, Maurice C. *marketing and business services entrepreneur, consultant*
Lehma, Alfred Baker *mathematician, educator*
Ling, Victor *oncologist, educator*
MacLennan, David Herman *scientist, educator*
Mak, Tak Wah *biochemist*
Masui, Yoshio *zoology educator*
Mc Culloch, Ernest Armstrong *physician, educator*
Mustard, James Fraser *research institute executive*
Nesbitt, Lloyd Ivan *podiatrist*
Orlowski, Stanislaw Tadeusz *architect*
Ostry, Sylvia *Canadian public servant, economist*
Paul, Leendert Cornelis *medical educator*
Polanyi, John Charles *chemist, educator*
Powis, Alfred *natural resources company executive*
Ramakrishnan, Ramani *acoustician*
Riordan, John Richard *chemist*
Rynard, Hugh C. *engineer, engineering executive*
San, Nguyen Duy *psychiatrist*
Scott, Steven Donald *geology educator, researcher*
Seaquist, Ernest Raymond *astronomy educator*
Seeman, Philip *pharmacology educator, neurochemistry researcher*
Siminovitch, Louis *biophysics educator, scientist*
Simmons, James *geography educator*
Smith, Kenneth Carless *electrical engineering educator*
Springfield, J. *civil engineer*
Tang, You-Zhi *chemist, researcher*
Templeton, John Marks, Jr. *pediatric surgeon, financial service executive*
Till, James Edgar *scientist*
Tobe, Stephen Solomon *zoology educator*
Tremaine, Scott Duncan *astrophysicist*
Venetsanopoulos, Anastasios Nicolaos *electrical engineer, educator*
Venter, Ronald Daniel *mechanical engineering educator, researcher, administrator*
Volpé, Robert *endocrinologist*
Wiggins, Glenn B. *entomologist*
Wright, Peter Murrell *structural engineering educator*
Yates, Keith *chemistry educator*
Yip, Cecil Cheung-Ching *biochemist, educator*
Zhu, Yunping *clinical physicist*

Toronto-Etobicoke
Kurys, Jurij-Georgius *environmental engineer, scientist, consultant*

Unionville
Godfrey, David Wilfred Holland *aeronautical engineer, communication educator*

Waterloo
Downer, Roger George H. *biologist*
Hewitt, Kenneth *geography educator*
Kalbfleisch, John David *statistics educator, dean*
Morgan, Alan Vivian *geologist, educator*
Roulston, David John *engineering educator*
Rudin, Alfred *chemistry educator emeritus*
Thompson, John Eveleigh *horticulturist, educator*
Thomson, N. R. *civil engineering educator*
Warner, Barry Gregory *geographer*
Wright, Douglas Tyndall *former university president, company director, engineering educator*

Willowdale
Rhodes, Wayne Robert *ergonomist, consultant*

Windsor
Barron, Ronald Michael *applied mathematician, educator, researcher*
Courtenay, Irene Doris *nursing consultant*
Hageniers, Omer Leon *mechanical engineer*
Monforton, Gerard Roland *civil engineer, educator*
Sale, Peter Francis *biology educator, marine ecologist*

PRINCE EDWARD ISLAND

Charlottetown
Gupta, Umesh Chandra *agriculturist, soil scientist*
McCreath, Peter S. *Canadian government official, civil engineer*

North Rustico
MacDonald, Jerome Edward *consultant, school psychologist*

QUEBEC

Boucherville
Marcotte, Michel Claude *geotechnical engineer, consultant*
Utracki, L. Adam *polymer engineer*
Venne, Louise Marguerite *librarian*

Dorval
El-Duweini, Aadel Khalaf *clinical pharmacologist, information scientist*

Lac Beauport
Lane, Peter *ornithologist*

Laval
Bellini, Francesco *chemist*
Frisque, Gilles *forestry engineer*
Kluepfel, Dieter *microbiologist*
Pichette, Claude *former banking executive, university rector, research executive*
Trudel, Michel *virologist*

Lennoxville
Deschenes, Jean-Marie *agriculturist, researcher*

Montreal
Bailar, John Christian, III *public health educator, physician, statistician*
Barrette, Jean *physicist, researcher*
Belanger, Pierre Rolland *university dean, electrical engineering educator*
Carbonneau, Come *mining company executive*
Carignan, Claude *astronomer, educator*
Carriere, Serge *physiologist, physician, educator*
Chan, Tak Hang *chemist, educator*
Chang, Thomas Ming Swi *biotechnologist, medical scientist*
Chretien, Michel *physician, educator, administrator*
Cohen, Montague *medical physics educator*
Conrad, Bruce R. *earth scientist*
Corinthios, Michael Jean George *electrical engineering educator*
Cruess, Richard Leigh *surgeon, university dean*
Cuello, Augusto Claudio Guillermo *medical research scientist, author*
Dubé, Ghyslain *earth scientist*
Eisenberg, Adi *chemist*
Filiatrault, Andre *civil engineering educator*
Fortin, Joseph André *forestry educator, researcher*
Fyffe, Les *earth scientist*
Gallagher, Tanya Marie *speech pathologist, educator*
Gjedde, Albert Hellmut *neuroscientist, neurology educator*
Gold, Phil *physician, educator*
Gulrajani, Ramesh Mulchand *biomedical engineer, educator*
Healey, Chris M. *earth scientist*
Hoffman, Kevin William *aerospace engineer*
Imorde, Henry K. *earth scientist*
Kalff, Jacob *biology educator*
Kalman, Calvin Shea *physicist*
Kamal, Musa Rasim *chemical engineer, consultant*
Karpati, George *neurologist*
Kerby, R. C. *earth scientist*
Lamarre, Bernard *engineering, contracting and manufacturing advisor*
Lee, Robert Gum Hong *chemical company executive*
Leggett, William C. *biology educator, educational administrator*
Leitch, Craig H. B. *earth scientist*
Leroy, Claude *physics educator, researcher*
Levine, Martin David *computer science and electrical engineering educator*
MacFarlane, David B. *physicist, educator*
Matziorinis, Kenneth N. *economist*
Michaud, Georges Joseph *astrophysics educator*
Milic-Emili, Joseph *physician, educator*
Mirza, M. Saeed *civil engineering educator*
Morrissette, Jean Fernand *electronics company executive*
Pellemans, Nicolas *patent agent, mechanical engineer*
Pelletier, Claude Henri *biomedical engineer*
Pinsky, Leonard *geneticist*
Plaa, Gabriel Leon *toxicologist, educator*
Prichard, Roger Kingsley *university administrator, dean, educator*
Roberge, Fernand Adrien *biomedical researcher*
Rowland, Helen *geographer*
Savard, G. S. *earth scientist*
Selvadurai, Antony Patrick Sinnappa *civil engineering educator, applied mathematician, consultant*
Shea, William Rene *historian, philosopher of science, educator*
Skup, Daniel *molecular biologist, educator, researcher*
Solomon, Samuel *biochemistry educator*
Somerville, Margaret Anne Ganley *law educator*
Sourkes, Theodore Lionel *biochemistry educator*
Stanners, Clifford Paul *molecular biologist, cell biologist, biochemistry educator*
Stern, Eric Petru *chemist*
Stewart, Jane *psychology educator*
Sutherland, C. A. *metallurgical engineer*
Swinden, H. Scott *earth scientist*
Townshend, Brent Scott *electrical engineer*
Weir, D. Robert *metallurgical engineer, engineering executive*
Wesemael, François *physics educator*
Woszczyk, Wieslaw Richard *audio engineering educator, researcher*

Otterburn Park
Roth, Annemarie *conservationalist*

Outremont
Levesque, Rene Jules Albert *former physicist*

Pointe Claire
Brodniewicz, Teresa Maria *biochemist, researcher*
De Brouwer, Nathalie *librarian*
Kubanek, George R. *chemical engineer*
Mitchell, Denis *civil engineer*
Sherry, Cameron William *occupational hygienist, consultant*
Wrist, Peter Ellis *pulp and paper company executive*

Quebec
Cheng, Li *mechanical engineer, educator*
Engel, Charles Robert *chemist, educator*
Lessard, Roger A. *physicist, educator*

Rimouski
Jean, Roger V. *mathematician, educator*
Walton, Alan *oceanographer*

Saint-Hubert
Doré, Roland *dean, science association director*

Saint Jean-sur-Richelieu
Côté, Jean-Charles *research molecular biologist*

Saint Jerome
Joly, Jean-Gil *medical biochemist, internist, administrator, researcher, educator*

Sainte Anne de Bellevue
Broughton, Robert Stephen *irrigation and drainage engineering educator, consultant*
Buckland, Roger Basil *university dean, educator, vice principal*
Davies, Roger *geoscience educator*

Sainte-Croix
Grenier, Fernand *geographer, consultant*

Sainte-Foy
Beaulieu, Jacques Alexandre *physicist*
Denis, Paul-Yves *geography educator*
LeDuy, Anh *engineering educator*
St-Yves, Angèle *agricultural engineer*

Sherbrooke
Bonn, Ferdinand J. *geography educator, environmental scientist*
Lecomte, Roger *physicist*
Paultre, Patrick *civil engineering educator*

Sillery
Tassé, Yvon Roma *engineer*

Trois Rivières
Lavallee, H.-Claude *chemical engineer, researcher*
Leblanc, Roger Maurice *chemistry educator*

Varennes
Vijh, Ashok Kumar *chemistry educator, researcher*

Vaudreuil
Webb, Paul *physicist*

Verdun
Bielby, Gregory John *electrical engineer*

Ville de Laval
Siemiatycki, Jack *epidemiologist, biostatistician, educator*
Tijssen, Peter H. T. *molecular virology educator, researcher*

Westmount
Dunbar, Maxwell John *oceanographer, educator*

SASKATCHEWAN

Regina
Kybett, Brian David *chemist*
Mollard, John Douglas *engineering and geology executive*
Seshadri, Rangaswamy *engineering dean*
Sharp, James J. *civil engineer*
Symes, Lawrence Richard *university dean, computer science educator*
Webster, Alexander James *agrologist*

Saskatoon
Babiuk, Lorne Alan *virologist, immunologist, research administrator*
Beaton, James Duncan *soil scientist*
Braidek, John George *agriculturist*
Harvey, Bryan Laurence *crop science educator*
Huang, Pan Ming *soil science educator*
Kells, J. A. *civil engineer*
Khachatourians, George Gharadaghi *microbiology educator*
Nikiforuk, Peter Nick *university dean*
Oelck, Michael M. *plant geneticist, researcher*
Patience, John Francis *nutritionist*
Quail, John Wilson *chemist, educator*
Smith, C. D. *civil engineering educator*
Storey, Gary Garfield *agricultural studies educator*
Tao, Yong-Xin *mechanical engineer, researcher*

MEXICO

Aguascalientes
Santana-Garcia, Mario A. *plant physiologist*

Aristoteles
Akel, Ollie James *oil company executive*

Juarez
Torres Medina, Emilio *oncologist, consultant*

La Paz
Ortega-Rubio, Alfredo *ecologist, researcher*

Leon
Aboites, Vicente *physicist*

Mexicali
Diaz Vela, Luis H(umberto) *computer company executive*

Mexico City
Arizpe, Lourdes *anthropologist, researcher*
Asomoza, Rene *physicist*
Ceballos, Gerardo *biology educator, researcher*
Figueroa, Juan Manuel *physicist*
González Flores, Agustín Eduardo *physicist*
Morales-Acevedo, Arturo *electrical engineering researcher, educator*
Palacios, Joaquin Alquisira *polymer scientist*
Rosenblueth, Emilio *structural engineer*
Sánchez, Luis Ruben *environmental engineer*
Tamariz, Joaquin *chemist, educator*

Monterrey
Muci Küchler, Karim Heinz *mechanical engineering educator*

Monterrey Nuevo Leon
Garcia Martinez, Ricardo Javier *architect*

Pabellon-Arteaga
De Alba-Avila, Abraham *plant ecologist*

Puebla
Chatterjee, Tapan Kumar *astrophysics researcher*
Zehe, Alfred Fritz Karl *physics educator*

Tehuacán
Romero, Miguel A. *animal nutrition director*

ALBANIA

Tiranë
Buda, Aleks *science administrator, history researcher, educator*

ANDORRA

Andorra
Bastida, Daniel *data processing facility administrator, educator*

Andorra la Vella
Mestre, S(olana) Daniel *economics consultant*

ARGENTINA

Bahía Blanca
Barrantes, Francisco Jose *biochemist, educator*
Cardozo, Miguel Angel *telecommunications engineering educator*
Panzone, Rafael *mathematics educator*

Bariloche
Barbero, José Alfredo *physics researcher*
Wio, Horacio Sergio *physicist*

Buenos Aires
Balve, Beba Carmen *research center administrator*
Buscaglia, Adolfo Edgardo *economist, educator*
Cernuschi-Frias, Bruno *electrical engineer*
De Leon, Pablo Gabriel *aerospace engineer*
Diaz, Alberto *biotechnologist*
Di Russo, Erasmo Victor *aeronautical engineer, educator, consultant*
Florin-Christensen, Jorge *biologist*
Gaggioli, Nestor Gustavo *physicist, researcher, educator*
Macon, Jorge *fiscal economist*
Martin, Osvaldo Jose *investment consultant, entrepreneur*
Milano, Antonio *engineering executive*

Nitka, Hermann Guillermo *hospital administrator*
Paneth, Thomas *retired physicist*
Piccione, Nicolas Antonio *economist, educator*
Saracco, Guillermo Jorge *optical and medical products executive*
Stoppani, Andres Oscar Manuel *director research center, educator*
Wais de Badgen, Irene Rut *limnologist*

La Plata
Scalise, Osvaldo Hector *physics researcher*
Vilche, Jorge Roberto *physical chemistry educator*

Mendoza
Branham, Richard Lacy, Jr. *astronomer*

Rosario
Rotolo, Vilma Stolfi *immunology researcher*

Salta
Barbarán, Francisco Ramón *educator, researcher*
Gonzo, Elio Emilio *chemical engineer, educator*

Santa Fe
Deiber, Julio Alcides *chemical engineering educator*
Idelsohn, Sergio Rodolfo *mechanical engineering educator*

Tucuman
Valentinuzzi, Max Eugene *bioengineering and physiology educator, researcher*

AUSTRALIA

Adelaide
Possingham, Hugh Philip *mathematical ecologist*
Shearwin, Keith Edward *biochemist*
Twidale, C(harles) R(owland) *geomorphologist, educator*
Wiskich, Joseph Tony *botany educator, researcher*

Ascot Vale
Bish, Robert Leonard *applied mathematician, metallurgist, researcher*

Brisbane
English, Francis Peter *ophthalmologist, educator*
Knight, Alan Edward Whitmarsh *physical chemistry educator*
Wentrup, Curt *chemist, educator*

Bundoora
James, Bruce David *chemistry educator*
Woelkerling, William J. *botanist educator*

Canberra
Bennett, Martin Arthur *chemist, educator*
Craig, David Parker *science academy executive, emeritus educator*
Evans, Denis James *research scientist*
Free, Ross Vincent *federal official*
Godara, Lal Chand *electrical engineering educator*
Neumann, Bernhard Hermann *mathematician*
Wagner, M(ax) Michael *computer scientist, educator*

Caulfield East
Hart, Barry Thomas *environmental chemist*

Caulfield South
Macesic, Nedeljko *electrical engineer*

City Beach
Pelczar, Otto *electrical engineer*

Clayton
Bray, Andrew Malcolm *chemist*
Evans, Richard Alexander *research chemist*
Hearn, Milton Thomas *biomedical scientist*
Meijs, Gordon Francis *chemist, research scientist*

East Saint Kilda
Zehnwirth, Ben Zion *mathematician consultant*

Fishermen's Bend
D'Cruz, Jonathan *aeronautical engineer*
Scott, Murray Leslie *aerospace engineer*

Fremantle
Flacks, Louis Michael *consulting physician*

Greenwith
Raymont, Warwick Deane *chemical engineer, environmental engineer*

Hobart
Murfet, Ian Campbell *botany educator*
Tilbrook, Bronte David *research scientist*

Hughes
Bolonkin, Kirill Andrew *aeronautical engineer*

Kensington
Bradley, Susan M. *chemistry educator*

Kirribilli
Phillips, Shelley *psychologist, writer*

Kwinana
Grocott, Stephen Charles *industrial research chemist*

Lane Cove
Jackson, Peter Edward *chemist*

Lindfield
Chiang, Kin Seng *optical physicist, engineer*

Lucas Heights
Collins, Richard Edward *physicist*
Nowotny, Janusz *materials scientist*

Melbourne
Allardice, David John *fuel technologist*
Banwell, Martin Gerhardt *chemistry researcher*
Kemp, Bruce E. *protein chemist*
Prince, Stephen *software developer, researcher*
Przelozny, Zbigniew *physicist*
Rich, Thomas Hewitt *curator*
Story, David Frederick *pharmacologist, educator*
Taylor, Hugh Ringland *ophthalmologist, educator*
Tran-Cong, Ton *applied mathematician, researcher*

Milton
Brown, Trevor Ernest *environment risk consultant*

Mulgrave
Porter, Colin Andrew *optics scientist*

Murdoch
Bauchspiess, Karl Rudolf *physicist*

Nedlands
Cottingham, Marion Scott *computer scientist, educator*
Oxnard, Charles Ernest *anatomist, anthropologist, human biologist, educator*

North Melbourne
Roberts, Godwin *medical products manager, consultant*

North Ryde
Wrigley, Colin Walter *cereal chemist*

Parkville
Foote, Simon James *molecular biologist*
Lim, Kieran Fergus *chemistry educator*
Shann, Frank Athol *paediatrician*
Tucker, Rodney Stuart *electrical and electronic engineering educator, consultant*

Perth
Kieronska, Dorota Helena *computer science educator*

Port Melbourne
Schofield, William Hunter *research aerodynamicist*

Prospect
Mayo, Oliver *biology researcher*

Pyrmont
Lawrence, Martin William *physicist*

Queensland
Connell, Desley William *chemist, educator, administrator*
Cooper, William Thomas *natural history artist*
Lawn, Ian David *marine biologist*
Pickles, James Oliver *physiologist*

Randwick
Lance, James Waldo *neurologist*

Rockhampton
Lynch, Thomas Brendan *pathologist*
Warner, Lesley Rae *biology educator*

Ryde
Caffin, Roger Neil *research scientist, consultant*

Saint Albans
Orbell, John Donald *chemist, educator*

Saint Lucia
Page, Arthur Anthony *astronomer*

Salisbury
Bedford, Anthony John *defense science executive*
Hermann, John Arthur *physicist*

Somers
Gifkins, Robert Cecil *materials engineer*

Sydney
Anderson, Donald Thomas *zoologist, educator*
Barnard, Peter Deane *dentist*
Brinck, Keith *computer scientist*
Carter, John Phillip *civil engineering educator, consultant, researcher*
Cowan, Henry Jacob *architectural engineer, educator*
Cram, Lawrence Edward *astrophysicist*
Ehrlich, Frederick *surgery consultant, orthopedist, rehabilitation specialist*
Hinde, John Gordon *lawyer, solicitor*
Lay, Peter Andrew *chemistry educator*
Mackie, John Charles *chemistry educator*
McBride, William Griffith *research gynecologist*
Okabe, Mitsuaki *economist*
Shaw, Keith Moffatt *mechanical engineer*

Tamworth
Cook, John Bell *chemist*

Townsville
Lucas, John Stewart *marine biologist*

Wembley
Graham, James *mineralogist*

West Perth
Brine, John Alfred Seymour *physician, consultant*
Woods, Thomas Brian *physician*

Westmead
Touyz, Stephen William *clinical psychologist, educator*

Woden
Nikolic, George *cardiologist, consultant*
Sinnett, Peter Frank *physician, geriatrics educator*

Wollongong
Kohoutek, Richard *civil engineer*

AUSTRIA

Baden
Lukas, Elsa Victoria *radiobiologist, radiobiochemist*

Graz
List, Hans C. *manufacturing engineer*

Innsbruck
Kräutler, Bernhard *chemistry educator*

Kapfenberg
Mitter, Werner Sepp *physicist, researcher, educator*

Klagenfurt
Melezinek, Adolf *engineering educator*

Krems
Hirner, Johann Josef *mechanical engineer*

Laxenburg
Nakicenovic, Nebojsa *economist, interdisciplinary researcher*

Lenzing
Gamerith, Gernot *chemist, researcher*

Leoben
Fettweis, Günter Bernhard Leo *mining engineering educator*
Lederer, Klaus *macromolecular chemistry educator*
Schmidt, Walter J. *exploration and mineral economist*

Leonding
Wechsberg, Manfred Ingo *chemical researcher*

Linz
Lell, Eberhard *retired inorganic chemist*
Pilz, Günter Franz *mathematics educator*

Mondsee
Dokulil, Marin *limnologist*

Salzburg
Rasssem, Mohammed Hassan *sociology and cultural science educator*

Vienna
Aulitzky, Herbert *retired erosion and avalanche control educator*
Blix, Hans Martin *international atomic energy official*
Ceska, Miroslav *biochemist, researcher*
Czernilofsky, Armin Peter *biochemist*
Karigl, Günther *mathematician*
Koss, Peter *research administrator*
Leibetseder, Josef Leopold *nutritionist, educator*
Lim, Youngil *economist*
Margarétha, Herbert Moriz Paul Maria *chemical consultant*
Moreno-Lopez, Jorge *virologist, educator*
Mosser, Hans Matthias *radiologist*
Niederreiter, Harald Guenther *mathematician, researcher*
Pichler, J(ohann) Hanns *economics educator*
Purgathofer, Werner *computer science educator*
Resch, Helmuth *environmental science educator*
Sammad, Mohamed Abdel *cardiovascular surgeon, consultant*
Sekyra, Hugo Michael *industrial executive*
Selberherr, Siegfried *university dean, educator, researcher, consultant*
Skritek, Paul *electrical engineering educator, consultant*
Stadlbauer, Harald Stefan *engineer*
Tappeiner, Gerhard *dermatologist, educator*
Varga, Thomas *mechanical engineering educator*
Welzig, Werner *philologist*
Wrba, Heinrich *oncologist, research institute administrator*

BAHRAIN

Isa Town
Ali Mohamed, Ahmed Yusuf *chemistry educator, researcher*

Manama
Hassan, Jawad Ebrahim *power engineer, consultant*

BANGLADESH

Chittagong
Islam, Jamal Nazrul *mathematics and physics educator, director*

Dhaka
Abeyesundere, Nihal Anton Aelian *health organization representative*
Ahmed, Abu *economics educator*

BARBADOS

Bridgetown
Headley, Oliver St. Clair *chemistry educator*

Christchurch
Ramsahoye, Lyttleton Estil *geophysicist, consultant, educator*

BELGIUM

Aalst
De Loof, Jef Emiel Elodie *general physician*

Antwerp
De Wandeleer, Patrick Jules *electronics executive*
Kuyk, Willem *mathematics educator*
Metdepenninghen, Carlos Maurits W. *radiologist*
Uyttenbroeck, Frans Joseph *gynecologic oncologist*

Asse
Lorijn, Johannes Albertus *economist*

Beerse
Janssen, Paul Adriaan Jan *pharmaceutical company executive*

Braine L'Alleud
Wulfert, Ernst Arne *research and development director, educator*

Brussels
Auerbacher, Peter *cancer research organization administrator*
Baron, Gino Victor *chemical engineering educator*
Bourdeau, Philippe *environmental scientist*
Corsi, Patrick *computer scientist*
Degrave, Alex G. *computer engineer, consultant*
Dehousse, Jean-Maurice *federal official*
Fasella, Paolo Maria *general science researcher, development facility director*
Godfraind, Theophile Joseph *pharmacologist educator*
Goffinet, Serge *neuropsychiatrist, researcher*
Ledic, Michèle *economist*
Luxen, Andre Jules Marie *chemist*
Mosselmans, Jean-Marc *physician*
Nicolis, Gregoire *science educator*
Nisolle, Etienne *industrial engineer*
Pouleur, Hubert Gustave *cardiologist*
Scheelen, André Joannes *chemical researcher*
Vestmar, Brigel Johannes Ahlmann *information systems agency adviser, scientist*
Vissol, Thierry-Louis *senior economist, researcher*
Winand, René Fernand Paul *metallurgy educator*

Ghent
Colardyn, Francis Achille *physician*
de Leenheer, Andreas Prudent *medical biochemistry and toxicology educator*
de Thibault de Boesinghe, Léopold Baron *physician*
Goethals, Eric Jozef *chemistry educator*
Ringoir, Severin Maria Ghislenus *medical educator, physician*
Walschot, Leopold Gustave *conservator*

Ghislenghien
Englebienne, Patrick P. *biochemist, consultant*

Haren
De Ceuster, Luc Frans *avionics educator*

Heverlee
L'abbe, Gerrit Karel *chemist*
Mewis, Joannes J(oanna) *chemical engineering educator*
Van Assche, Frans Jan Maurits *economics educator*

Jemeppe-Sur-Sambre
Carlier, Jean Joachim *cardiologist, educator, administrator*

La Hulpe
Eber, Michel *information technology company executive*

Leuven
De Ranter, Camiel Joseph *chemist, educator*
Moldenaers, Paula Fernande *chemical engineer, educator*
Peeters, Theo Louis *biochemical engineering educator*
Van Geyt, Henri Louis *architect*

Leuven-Heverlee
Belmans, Ronnie Jozef Maria *foundation adminstrator, researcher*

Liège
Alexandre, Gilbert Fernand A.E. *surgeon*
Battisti, Oreste Guerino *pediatrician*
Calvaer, Andre J. *electrical science educator, consultant*
Perdang, Jean Marcel *astrophysicist*

Louvain
Van Rompuy, Paul Frans *economics educator*

Louvain-la-Neuve
Sintzoff, Michel *computer scientist, educator*

Melsbroek
Becuwe, Ivan Gerard *aeronautical engineer*

Namur
Cornelis, Eric Rene *mathematician*

Rhode-Saint-Genese
Sarma, Gabbita Sundara Rama *aerospace scientist*

Rixensart
Thrower, Keith James *chemist*

Turnhout
Caruso, Nancy Jean *chemist*

Wilrijk
Van Ooteghem, Marc Michel Martin *pharmacology educator*

BENIN

Cotonou
Ezin, Jean-Pierre Onvêhoun *mathematician*

BOLIVIA

La Paz
Hartmann, Luis Felipe *health science association administrator, endocrinologist*

Sucre
Larrazábal Antezana, Erik *economics educator, organization executive*

BRAZIL

Barueri
Ghizoni, César Celeste *electrical engineer*

Belo Horizonte
Pena, Sergio Danilo Junho *physician*
Piló-Veloso, Dorila *chemistry educator, researcher*

Campinas
Brandalise, Silvia Regina *pediatrician*
Brito Cruz, Carlos Henrique *physicist, science administrator*
Gonçalves da Silva, Cylon Eudóxio Tricot *physics educator*

Cotia
Araujo, Marcio Santos Silva *chemical engineer*

Curitiba
Berman, Marcelo Samuel *mathematics and physics educator, cosmology researcher*

Itaguai
Braz-Filho, Raimundo *organic chemistry educator*

Minas Gerais
Cimbleris, Borisas *engineering educator, writer*

Passo Fundo
Baier, Augusto Carlos *plant researcher*

Petrópolis
Gomide, Fernando de Mello *physics educator, researcher*

Recife
Coelho, Hélio Teixeira *physics researcher, consultant*
Pragana, Rildo José Da Costa *information processing company executive*

Rio Claro
Potter, Paul Edwin *geologist, educator, consultant*

Rio de Janeiro
Barreiro, Eliezer Jesus *medicinal chemistry educator*
Costa Neto, Adelina *chemistry educator, consultant*
de Biasi, Ronaldo Sergio *materials science educator*
Döbereiner, Johanna *soil biology scientist*
Espinola, Aïda *chemical engineer, educator*
Felcman, Judith *chemistry educator*
Leite, Carlos Alberto *physician*
Matos, Jose Gilvomar Rocha *civil engineer*
Souza, Marco Antonio *civil engineer, educator*

São Paulo
Akkermann, Schaia *mechanical, electrical and civil engineer*
Arruda-Neto, João Dias de Toledo *nuclear physicist*
Barbuy, Beatriz *astronomy educator*
Bergamini, Eduardo Whitaker *electrical engineer*
da Costa, Newton Carneiro Affonso *mathematics educator*
Fernicola, Nilda Alicia Gallego Gándara de *pharmacist, biochemist*
Gabbai, Alberto Alain *neurology educator, researcher*
Goldemberg, Jose *educator*
Hersztajn Moldau, Juan *economist*
Korolkovas, Andrejus *pharmaceutical chemistry educator*
Mariotto, Paulo Antonio *electronics educator*
Paris, Tania de Faria Gellert *information technology service executive*
Sadi, Marcus Vinicius *urologist*
Zinner, Léa Barbieri *inorganic chemistry educator, researcher*

Salvador
Andrade, Francisco Alvaro Conceicao *chemistry educator*
Silva, Benedicto Alves de Castro *surgeon, educator*

Sao Carlos
Motheo, Artur de Jesus *chemist, educator*

Sao Jose dos Campos
Bismarck-Nasr, Maher Nasr *aeronautical engineering educator*
Lutterbach, Rogerio Alves *aeronautical engineer, consultant, researcher*
Souza, Marcelo Lopes *aerospace engineer, researcher, consultant*

Uberlandia
Prado, Neilton Gonçalves *urologist*

BRUNEI

Bandai Seri Begawan
Roy, Bimalendu Narayan *ceramic engineering educator*

BULGARIA

Sofia
Hadjikov, Lyuben Manolov *civil engineer*
Khristov, Khristo Yankov *physicist*
Radev, Ivan Stefanov *electronics engineer*
Zahariev, George Kostadinov *computer company executive*

CAMEROON

Douala
Obenson, Philip *computer science educator*

CAYMAN ISLANDS

Georgetown
Husemann, Anthony James *science educator*

CHILE

La Serena
Eggen, Olin Jeuck *astrophysicist, administrator*

Santiago
Allende, Jorge Eduardo *biochemist, molecular biologist*
Arratia-Perez, Ramiro *chemistry educator*
Cabrera, Alejandro Leopoldo *physics researcher and educator*
Jara Diaz, Sergio R. *transport economics educator*
Loeb, Barbara L. *chemistry educator*
Parada, Jaime Alfonso *mechanical engineer, consultant*
Seelenberger, Sergio Hernan *chemical company executive*

Temuco
Vogel, Eugenio Emilio *physics educator*

Valparaiso
Hernandez-Sanchez, Juan Longino *electrical engineering educator*

CHINA

Beijing
Chen, Gong Ning *mathematics educator*
Chen, Naixing *thermal science educator, researcher*
Fan, Jiaxiang *economics educator, researcher*
Guangzhao, Zhou *theoretical physicist*
Henggao, Ding *federal official*
Hua, Tong-Wen *chemistry educator, researcher*
Jian, Song *government official, science administrator*
Leng, Xin-Fu *chemist, educator*
Liang, Junxiang *aeronautics and astronautics engineer, educator*
Lin, Justin Yifu *economist, educator*
Liu, Bai-Xin *materials scientist, educator*
Liu, Yuanfang *chemistry educator*
Mao, Yu-shi *economist, engineer, educator*
Shao, Wenjie *librarian*
Shu, Sun *geologist*
Situ, Ming *aerospace engineering researcher*
Su, Dongzhuang *computer science educator, university official*
Wang, Yibing *physicist*
Wei, Gaoyuan *chemistry educator*
Wu, Daguan *aeroengine design engineer*
Xue-ying, Deng *aerodynamicist*
Yang, Cheng-Zhi *petroleum engineer, researcher, educator*
Yingqian, Qian *botanist*
Zhang, Fu-Xue *scientist*
Zhang, Hong Tu *physics educator*
Zhang, Jun Yi *chemist, educator*
Zhang, Li-Xing *physician, medical facility executive*
Zhang, Ping *aerospace engineering educator*
Zhang, Tianyou *biomedical engineer*
Zuxun, Sun *academic administrator, physicist*

Changsha
Yu, Ru-Qin *chemistry educator*

Chengdu
Xu, Daoyi *mathematician, educator*

Dalian
Wu, Zhenshan *economics educator*

Guangzhou
Hou, Guang Kun *computer science educator*
Xie, Nan-Zhu *medical physics educator*

Hangzhou
Wu, Shuo-Xian *acoustics and architecture educator, scientist*

Harbin
Ma, Zuguang *optical scientist, quantum electronics educator*
Yang, Di *aerospace engineer, educator*

Hefei
Ge, Li-Feng *engineer*
Wu, Chang-chun *mechanical engineering educator*

Huhot
Chen, Jian Ning *mathematics educator*

Lanzhou
Gao, Yi-Tian *physicist, educator, astronomer*

Linfen
Hou, Jin Chuan *mathematics educator*

Nanjing
Du, Gonghuan *acoustics educator*
He, Duo-Min *physics educator*
Lin, You Ju *physicist*
Ton, Dao-Rong *mathematics educator*
Yang, Run Sheng *mathematics educator*
Zou, Yun *mathematician*

Nanning
Guo, Xin Kang *mathematics educator*

Shanghai
Cao, Jia Ding *mathematics educator*
Cheng, Ansheng *mathematics educator, researcher*
Cheng, J. S. (Zhong-Zhi Zheng) *nuclear astrophysical chemist, educator*
Ding, Hai *mathematician*
Ge, Guang Ping *mathematics educator, statistician*
Gong, Yitai *information specialist*
Wang, Ji-Qing (Chi-Ching Wong) *acoustician, educator*
Wei, Musheng *mathematics educator*
Xie (Hsieh), Shu-Sen *economics and finance educator, consultant*
Xue, Miao *dentistry educator, biomaterial scientist*

Tianjin
Zhang, Theodore Tian-ze *oncologist, health association administrator*

Wuhan
Cha, Chuan-sin *chemistry educator*

Wuhu
Mo, Jiaqi *mathematics educator*

Xi'an
Chen, Shilu *flight mechanics educator*
Fan, Changxin *electrical engineering educator*
Su, Yaoxi *fluid mechanics educator*

Xiamen
Zhou, Shao-min *chemist, educator*

Zhengzhou
Zhang, Binglin *physics educator*
Zhang, Tao *physics educator, director*

COLOMBIA

Armenia
Pareja-Heredia, Diego *mathematics educator, bookseller consultant*

Bogota
Garcia Martinez, Hernando *agrochemical company executive, consultant*
Leon Dub, Marcelo *chemicals executive, entrepreneur*
Pulido, Carlos Orlando *electronics engineer, economist, consultant*

Cali
Cock, James Heywood *agricultural scientist*
Voysest, Oswaldo *agronomist, researcher*

COSTA RICA

San Jose
Bergold, Orm *medical educator*
Gongora-Trejos, Enrique *mathematician, educator*
Sauter, Franz Fabian *structural engineer*

Turrialba
Lastra, Jose Ramon *plant virologist*

CROATIA

Dubrovnik
Vukovic, Drago Vuko *electronics engineer*

Split
Boschi, Srdjan *radiologist, nephrologist*

Zagreb
Bozicevic, Juraj *educator*
Catic, Igor Julio *mechanical engineering educator*
Čosović, M. Božena *chemist, researcher*
Janković, Slobodan *aerodynamics educator, researcher*
Kojić-Prodić, Biserka *chemist*
Kučan, Željko *biochemistry educator*
Novaković, Branko Mane *engineering educator*
Ursic, Srebrenka *information technology researcher, manager, consultant*

CUBA

Havana
Fernández Miranda, Jorge *physico-mathematician, researcher*
Kouri, Gustavo Pedro *virologist*

CYPRUS

Nicosia
Aristodemou, Loucas Elias *industrial engineer, manufacturing company executive*
Chacholiades, Miltiades *economics educator*
Vrachas, Constantinos Aghisilaou *aeronautical engineer*

CZECH REPUBLIC

Brno
Vesely, Karel *chemist, educator*

Ceske Budejovice
Dolejs, Petr *aquatic scientist, consultant*
Sláma, Karel *biologist, zoologist*

Neratovice
Hlubuček, Vratislav *chemical engineer*

Olomouc
Šimánek, Vilím *chemist, educator*

Plzen
Wagner, Karel, Jr. *nuclear engineer, educator*

Prague
Brdička, Miroslav *retired physicist*
Červeny, Libor *chemistry educator*
Kočka, Jan Vilém *physicist*
Kodym, Miloslav *psychologist, researcher*
Macek, Karel *analytical chemistry educator*
Merhaut, Josef *electroacoustics educator*
Procházka, Karel *chemistry educator*
Říman, Josef *biology educator*
Šesták, Jiří Vladimír *mechanical engineering educator*

DENMARK

Aalborg
Bagchi, Kallol Kumar *computer science and engineering educator, researcher*
Zimmermann, Wolfgang Karl *biologist, researcher*

Aarhus
Bundgaard, Nils *sound and acoustics professional*
Pedersen, Bent Carl Christian *cytogeneticist*
Straede, Christen Andersen *research center administrator*

Beder
Fjerdingstad, Ejnar Jules *retired biological scientist and educator*

Bronshoj
Skylv, Grethe Krogh *rheumatologist, anthropologist*

Charlottenlund
Langer, Jerk Wang *physician, medical journalist*

Copenhagen
Andersen, Leif Percival *physician*
Bohr, Aage Niels *physicist*
Christensen, Søren Brøgger *medicinal chemist*
Dal, Erik *former editorial association administrator*
Hansen, Finn *electrical engineer*
Jensen, Bjarne Sloth *economist*
Langer, Seppo Wang *physician, registrar, medical journalist*
Larsen, Ib Hyldstrup *engineer*
Mottelson, Ben R. *physicist*
Nielsen, Peter Eigil *biochemist, educator*
Olesen, Mogens Norgaard *mathematics educator*
Rose-Hansen, John *geologist*
Wilkins, Cornelius Kendall *chemist, researcher*

Fredericia
Jensen, Uffe Steiner *nuclear engineer*

Frederiksberg
Bjørn-Andersen, Niels *information systems researcher*
Skibsted, Leif Horsfelt *food chemistry researcher*

Gentofte
Jaroszewski, Jerzy W. *chemist*
Larsen, Jesper Kampmann *applied mathematician, educator*

Hellerup
Jacobsen, Grete Krag *pathologist*

Hoersholm
Jensen, Ole *energy researcher*

Holte
Niordson, Frithiof Igor *mechanical engineering educator*

Kalundborg
Bitsch-Larsen, Lars Kristian *anaesthesia and intensive care specialist*

Lyngby
Alting, Leo Larsen *manufacturing engineer, educator*
Bay, Niels *manufacturing engineering educator*
Bjørnø, Leif *industrial acoustics educator*
Kozhevnikova, Irina N. *physicist*
Tøndering, Claus *software engineer*

Nakskov
Christensen, Ole *general practice physician*

Odense
Kemp, Ejvind *internist*
Oddershede, Jens Norgaard *chemistry educator*

Roskilde
Engvild, Kjeld Christensen *plant physiologist*
Heydorn, Kaj *science laboratory administrator*

Skodsborg
Rasmussen, Gunnar *engineer*

DOMINICA

Portsmouth
Cooles, Philip Edward *physician*

ECUADOR

Quito
Bedoya, Michael Julian *veterinarian*
Casals, Juan Federico *economist, consultant*
Kozioł, Michael John *biochemist, researcher*
Torres, Guido Adolfo *water treatment company executive*

EGYPT

Cairo
El-Gammal, Abdel-Aziz Mohamed *aero-mechanical engineer*
Elnomrossy, Mokhtar Malek *aeronautical engineer*
El-Sayed, Karimat Mahmoud *physics and crystallography educator*
Korayem, Essam Ali *computer company executive*
Moftah, Mounir Amin *engineering executive*
Rafea, Ahmed Abdelwahed *computer science educator*

Dokki
Nasser, Essam *electrical engineer, physicist*

Giza
Kassas, Mohamed *desert ecologist, environmental consultant*
Salem, Ibrahim Ahmed *electrical engineer, consultant, educator*

Maadi-Cairo
Amer, Magid Hashim *oncologist, physician*

Monsura
El Nahass, Mohammed Refat Ahmed *physiology educator, researcher*

Nasr City
El-Hamalaway, Mohamed-Younis Abd-El-Samie *computer engineering educator*

Sharkia
El-Sayed, Ahmed Fayez *aeronautical engineering educator*

ENGLAND

Abingdon
Denne-Hinnov, Gerd Boël *physicist*

Amersham
Finlan, Martin Francis *physicist, consultant*

Barnwood
Collier, John Gordon *nuclear scientist*

Basingstoke
Butcher, Brian Ronald *software company executive*

Bath
Buchanan, John Grant *chemistry educator*
Wilson, Derek Edward *chemist*

Bedford
Hadfield, James Irvine Havelock *surgeon*

Birmingham
Campbell, John *engineering educator*

Blackpool
Avasthi, Ram Bandhu *physician, general practitioner*

Bocking
Axon, Kenneth Stuart *chemist*

Bradford
Gardner, Michael Leopold George *biochemistry and physiology researcher, educator*

Brighton
Cornforth, Sir John Warcup *chemist*
Darwin, Christopher John *experimental psychologist*
Kroto, Harold Walter *science educator*
Luo, Hong Yue *biomedical engineering researcher*
Spratt, Brian Geoffrey *microbiologist, educator, researcher*

Bristol
Albone, Eric Stephen *chemist, educator*
Round, Frank E. *botanist, educator*
Wilson, John Anthony *physicist*

Burton-on-Trent
Page, Dennis *coal scientist, consultant*

Byfleet
Mervyn, Leonard *biochemist*

Camberley
Al-Zubaidi, Ali Abdul Jabbar *engineer*
Ayres, Raymond Maurice *mechanical engineer*

Cambridge
Ashby, Michael Farries *engineering educator*
Barlow, Horace Basil *physiologist*
Brenner, Sydney *biologist, educator*
Cottrell, Alan *materials scientist*
Dowling, Ann Patricia *mechanical engineering educator, researcher*
Hall, Christopher *materials scientist, researcher*
Hawking, Stephen W. *astrophysicist, mathematician*
Hawthorne, Sir William (Rede) *aerospace and mechanical engineer, educator*
Hodgkin, Sir Alan Lloyd *biophysicist*
Huxley, Sir Andrew (Fielding) *physiologist, educator*
Johnson, Kenneth Langstreth *engineer, educator*
Josephson, Brian David *physicist*
Klug, Aaron *molecular biologist*
Kornberg, Sir Hans Leo *biochemist*
Lynden-Bell, Donald *astronomer*
Mair, William Austyn *aeronautical engineer*
Martin, Archer John Porter *retired chemistry educator*
Milstein, César *molecular biologist*
Needham, Joseph *biochemist, historian of science, Orientalist*
Perutz, Max Ferdinand *molecular biologist*
Rees, David Charles *chemist, researcher*
Rees, Martin John *astronomy educator*
Sanger, Frederick *retired molecular biologist*
Tabor, David *physics educator*
Todd, Alexander Robertus (Baron Todd of Trumpington) *chemistry educator*

Cambridgeshire
Ford, Brian J. *research biologist, author, broadcaster*

Canterbury
Eager, Robert Donald *computer science educator*

Cathay Park
Temperley, H. N. V. *mathematician educator*

Chelmsford
Cross, Trevor Arthur *physicist*

Colchester
Allen, Leslie *physics educator*
Boyd, Thomas James Morrow *physicist*
Loudon, Rodney *optics scientist, educator*

Coventry
Qin, Ning *engineering educator*
Rowlands, George *physics educator*

Cranfield
Turner, Anthony Peter Francis *biotechnologist, educator*

Crawley
Van Schuyver, Connie Jo *geophysicist*

Dewsbury
Mishra, Arun Kumar *ophthalmologist*

Durham
Feast, William James *polymer chemist*
Parker, David *chemistry educator*
Richards, Randal William *chemist, educator*
Robson, Geoffrey Robert *geologist, seismologist, consultant*

Essex
Kneen, Geoffrey *chemist*
Thompson, Haydn Ashley *electronics engineer*

Falmouth
Batchelor, Anthony Stephen *geotechnical engineer, consultant*

Farnborough
Benson, Alan James *avaiation medical doctor*
Treble, Frederick Christopher *electrical engineer, consultant*

Filton
Herring, Paul George Colin *aeronautical engineer*
Joyce, Terence Thomas *aerospace engineer*

Guildford
Spier, Raymond Eric *microbial engineer*

Hampshire
Perera, Uduwanage Dayaratna *plant reliability specialist, engineer*
Suhrbier, Klaus Rudolf *hydrodynamicist, naval architect*

Harrow
Wright, Terence Richard *imaging company executive*

Hayes
Hounsfield, Godfrey Newbold *scientist*

Herefordshire
Sandhu, Bachittar Singh *ophthalmologist*

Highfield
Hanna, David *optics scientist, educator*

Huddersfield
Jewsbury, Roger Alan *chemistry educator*

Hull
Shorter, John *chemistry lecturer*

Kingston-Upon-Thames
Jatala, Shahid Majid *electronics engineer*

Leeds
Hambling, Milton Herbert *medical virologist consultant, lecturer*
Slechta, Jiri *theoretical physicist*

Leicester
Light, William Allan *mathematics educator*
Riach, Peter Andrew *economist*

Linton
Kendrew, John Cowdery *molecular biologist, former college president*

Liverpool
Bulman Page, Philip Charles *chemistry researcher*
Twin, Peter John *physics educator*

Livingston
Sime, James Thomson *consultant psychologist*

London
Arnott, Eric John *ophthalmologist*
Asfoury, Zakaria Mohammed *physician*
Bodmer, Walter Fred *cancer research administrator*
Choyce, David Peter *ophthalmologist*
Drake, Simon Robert *inorganic chemistry*
Eddo, James Ekundayo *industrial chemist, chemicals executive*
Gahan, Peter Brian *cell biologist, researcher*
Hall, Trevor James *physicist, educator, consultant*
Irvine, John Henry *science analyst*
Kakati, Dinesh Chandra *physician*
Kassam, Amirali Hassanali *agricultural scientist*
Kathirgamanathan, Poopathy *chemist*
Katz, Sir Bernard *physiologist*
Kolb, Noel Joseph *research scientist*
Mason, Stephen Finney *chemistry educator*
McLean, Malcolm *materials scientist, educator, researcher*
Miller, Paul Andrew *editor*
Pavlovic, Milija N. *structural engineer, educator*
Pecorino, Lauren Teresa *biologist*
Popper, Karl (Raimund) *author*
Porter, Lord George *chemist, educator*
Reese, Colin Bernard *chemistry educator*
Retsas, Spyros *oncologist*
Ross, Euan Macdonald *pediatrician, educator*
Rubens, Robert David *physician, educator*
Sandergaard, Theodore Jorgensen *information technology director*
Sapsford, Ralph Neville *cardiothoracic surgeon*
Searl, John Roy Robert *engineering executive*
Seaton, Michael John *physicist*
Stelle, Kellogg Sheffield *physicist*
Thomas, John Meurig *chemist, scientist*
van den Muyzenberg, Laurens *engineer*
Vane, John Robert *pharmacologist*
Wallace, Bonnie Ann *biophysics educator, researcher*
White, Norman Arthur *engineer, corporate executive, educator*
Wilkinson, Geoffrey *chemist, educator*

Loughborough
Arvanitoyannis, Ioannis *polymer chemistry researcher*

Macclesfield
Cohen, Raymond James *radio astronomy educator*
Graham, Dorothy Ruth *software engineering consultant*

Manchester
Banks, Ronald Eric *chemistry educator*
Poll, David Ian Alistair *aerospace engineering educator, consultant*
Wilkinson, Peter Maurice *physician consultant, pharmacologist*

Middlesex
Bulkin, Bernard Joseph *oil industry executive*

Newcastle upon Tyne
Gibson, Gerald John *physician*

North Wales
Barlow, Robert Dudley *medical biochemist*

North Yorkshire
Martin, Carl Nigel *biotechnologist*

Northwood
Smith, Paul John *plastic and reconstructive surgeon, consultant*

Norwich
Hopwood, David Alan *biotechnologist, geneticist, educator*
Synge, Richard Laurence Millington *biochemist*

Nottingham
Choi, Kwing-So *physics and engineering educator*
Logan, Norman *chemistry researcher, educator*
Soldatos, Kostas P. *mathematician*

Olney
Edmonds, Andrew Nicola *software engineer*
Keith-Lucas, David *retired aeronautical engineer*

Ormskirk
Machlachlan, Julia Bronwyn *materials scientist, researcher*

Oxford
Atkins, Peter William *chemistry educator*
Barnes, Peter John *food scientist*
Binney, James Jeffrey *physicist*
Carlow, John Sydney *research physicist, consultant*
Dewey, John F. *geologist, educator*
Hamilton, William Donald *zoologist, educator*
Hirsch, Peter Bernhard *metallurgist*
Hodgkin, Dorothy Crowfoot *chemist*
Penrose, Roger *mathematics educator*
Reading, Harold G. *geology educator*

Pontefract
Robinson, Melvyn Roland *urologist*

Portsmouth
Croset, Michel Roger *electronics engineer*
Hinds, James William *aeronautical engineer*

Preston
Gaze, Nigel Raymond *plastic surgeon*
O'Shea, John Anthony *consulting electrical engineer*

Reading
Almond, Matthew John *chemistry educator, researcher, writer*
Webb, Gregory Frank *retired aeronautical engineer*

Saint Annes-On-Sea
Teakle, Neil William *engineering executive*

Salford
Stone, Robert John *research psychologist*

Sevenoaks
Luttman, Horace Charles *retired aeronautical engineer*

Sheffield
Henshilwood, James Andrew *chemist, researcher*
Jones, Frank Ralph *materials scientist engineer*
Willett, Peter *information science researcher*

Sittingbourne
Thomas, David Burton *chemical company executive*

Southampton
Gambling, William Alexander *optoelectronics research center director*
Markvart, Tomas *physicist*

Southhampton
Fleischmann, Martin *chemistry educator*
Payne, David N. *optics scientist*

Stafford
Zambardino, Rodolfo Alfredo *computing educator*

Sunbury on Thames
Armstrong, Keith Bernard *engineering consultant*
Packer, Kenneth John *spectroscopist*

Sussex
Bunyard, Alan Donald *designer, inventor*

Sutton
Wilman, Derry Edward Vincent *medicinal chemist*

Swindon
Blundell, Thomas Leon *scientist, science administrator*

Talybont-on-Usk
Crabtree, Lewis Frederick *aeronautical engineer, educator*

Teddington
McCartney, Louis Neil *research scientist*

Wallington
Green, Maurice Berkeley *agrochemical research consultant*

Warrington
Dunbavin, Philip Richard *acoustical consultant*

Welwyn
Eckers, Christine *mass spectrometry scientist*

West Clandon
Lumsden, James Gerard *computer scientist*

Yeovil
Kokkalis, Anastasios *aeronautical engineer, consultant*

York
Waddington, David James *chemistry educator*

ETHIOPIA

Addis Ababa
Haile Giorgis, Workneh *civil engineer*

FINLAND

Espoo
Heikkinen, Raimo Allan *electronics executive*
Korhonen, Antti Samuli *metallurgist, educator*
Wuori, Paul Adolf *engineering educator*

Hämeenlinna
Kivinen, Seppo Tapio *obstetrician/gynecologist, hospital administrator*

Hautjärvi
Alanko, Matti Lauri Juhani *animal reproduction educator*

Helsinki
Allardt, Erik Anders *academic administrator, educator*
Illman, Sören Arnold *mathematician, educator*
Kalimo, Esko Antero *research institute administrator, educator*
Mustakallio, Kimmo Kalervo *dermatologist*
Partinen, Markku Mikael *neurologist*
Perheentupa, Jaakko Pentti *pediatrician*
Pulliainen, Erkki Ossi Olavi *zoology educator, legislator*
Routti, Jorma Tapio *federal agency administrator, engineering educator*
Siltanen, Pentti Kustaa Pietari *cardiologist*
Sinclair, John David *psychologist*
Sipinen, Seppo Antero *obstetrician/gynecologist*
Tervo, Timo Martti *ophthalmologist*
Varmavuori, Anneli *chemical society administrator, chemist*
von Haartman, Harry Ulf *international transportation company executive*

Kotka
Cranston, Wilber Charles *engineering company executive*

Kuopio
Hakola, Hannu Panu Aukusti *psychiatry educator*
Maenpaa, Pekka Heikki *biochemistry educator*
Matilainen, Riitta Marja *pediatrician, pediatric neurologist*

Nastola
Jarvenkyla, Jyri Jaakko *chemical engineer*

Oulu
Arjas, Elja *statistician*
Koskinen, Ari Mauri Petri *organic chemistry educator*

Tampere
Merikoski, Jorma Kaarlo *mathematics educator*
Nikkari, Tapio Urho *medical biochemistry educator*

Turku
Collan, Yrjö Urho *pathologist, medical educator, physician, toxicopathology consultant*
Eerola, Osmo Tapio *research electrical engineer*
Suoninen, Eero Juhani *materials science educator*
Toivanen, Paavo Uuras *immunologist, microbiologist, educator*
Toivonen, Hannu Tapio *control engineering educator*

FRANCE

Aix Les Bains
Tabau, Robert Louis *rheumatologist, researcher*

Allauch
Lai, Richard Jean *research administrator*

Amiens
Chacron, Joseph *mathematics educator*

Angers
Monroche, André Victor Jacques *physician*

Antony
Jolles, Georges Edgar René *scientist*
Mazzo, David Joseph *analytical chemist*
Seiersen, Nicholas Steen *data processing executive*

Aulnay-sous-Bois
Shahin, Majdi Musa *biologist*

Bagneux
Jegham, Samir *chemist*
Wang, Zhao Zhong *physics researcher*

Ballan Mire
Delbarre, Bernard *pharmacologist, consultant*

Beaujon
Pariente, René Guillaume *physician, educator*

Besançon
Boillat, Guy Maurice Georges *mathematical physicist*
Olsem, Jean-Pierre *economics educator*
Pelissier, Edouard-Pierre *surgeon*

Bicêtre
Baulieu, Etienne-Emile *endocrinologist*

Bondy
Beaugrand, Michel *physician, educator*

Boulogne
Dubourg, Olivier Jean *cardiology educator, researcher*
Reboul, Jaques Regis *computer company executive*

Bourgoin
Le, Quang Nam *engineering researcher*

Brest
Floch, Herve Alexander *medical biologist*

Brignoles
Thomas, Raymond Jean *computer consultant*

Bures-sur-Yvette
Ruelle, David Pierre *physicist*

Caen
Calvino, Philippe Andre Marie *engineer, manufacturing executive*

Castillon-du-Gard
Jerne, Niels Kaj *scientist*

Chartres
Benoit, Jean-Pierre Robert *pneumologist, consultant*

Chatenay-Malabry
Fabre, Raoul François *electronics company executive*
Perrault, Georges Gabriel *chemical engineer*

Corseul
Guerin, Patrick Gerard *veterinary surgeon, consultant*

Creteil
Coscas, Gabriel Josue *ophthalmologist, educator*
Robert, Leslie Ladislas *research center administrator, consultant*

Crolles
Reader, Alec Harold *physicist, researcher*

Dijon
Chanussot, Guy *physics educator*

Duvy
Brevignon, Jean-Pierre *physicist*

Eure et Loir
Adjamah, Kokouvi Michel *physician*

Fontainebleau
Saillon, Alfred *psychiatrist*

Garches
Durigon, Michel Louis *pathologist, forensic medicine educator*

Gargilesse
Barda, Jean Francis *electronics engineer, corporate executive*

Gif sur Yvette
Cotton, Jean-Pierre Aimé *physicist*
Daoud, Mohamed *physicist*
Fournet, Gerard Lucien *physics educator*
Jouzel, Jean *researcher*
Pierre, Françoise *physics educator*
Radvanyi, Pierre Charles *physicist*

La Montagne
Rocaboy, Françoise Marie Jeanne *acoustical engineer*

LaGaude
Pignal, Pierre Ivan *computer scientist, educator*

Landivisiau
Floch-Baillet, Daniele Luce *ophthalmologist*

Laval
Sauvé, Georges *surgeon*

Lavera
Boniface, Christian Pierre *chemical engineer, oil industry executive*

Laxou
Anghileri, Leopoldo José *researcher*

Le Cannet des Maures
Gudin, Serge *plant breeder*

Le Mans
Castagnede, Bernard Roger *physicist, educator*

Le Pecq
Mallevialle, Joël Christian *marine engineer*

Le Vesinet Yvelines
Hillion, Pierre Théodore Marie *mathematical physicist*

Lille
Gillet, Roland *financial economist*
Santoro, Ferrucio Fontes *microbiologist*

Limoges
Moreau, Hugues Andre *physician*

Longjumeau
Kapandji, Adalbert Ibrahim *orthopedic surgeon*

Lyons
Meunier, Pierre Jean *medical educator*

Maisons-Lafitte
Obadia, Andre Isaac *surgeon*

Marcq en Baroeul
Choain, Jean Georges *physician, acupuncturist*

Marseilles
Azzopardi, Marc Antoine *astrophysicist, scientist*
Dumitrescu, Lucien Z. *aerospace researcher*
Favre, Alexandre Jean *physics educator*
Poitout, Dominique Gilbert M. *orthopedic surgeon, educator*
Vague, Jean Marie *endocrinologist*

Metz
Tran-Viet, Tu *cardiovascular and thoracic surgeon*

Meudon
Mamon, Gary Allan *astrophysicist*
Sculfort, Jean-Lou *chemistry and physics educator*

Meudon-Bellevue
Neel, Louis Eugene Felix *physicist*

Mont-Saint-Aignan
Gouesbet, Gerard *systems and process engineering educator*

Montpellier
Dauchez, Pierre Guislain *roboticist*
Michel, Henri Marie *medical educator*
Pasteur, Nicole *population geneticist*

Mulhouse
Decker, Christian Lucien *research chemist*
Fouassier, Jean-Pierre *chemist, educator*
Riess, Henri Gerard *chemist, educator*

Nanterre
Berquez, Gérard Paul *psychiatrist, psychoanalyst*
Buffet, Pierre *computer scientist*
Chambolle, Thierry Jean-Francois *environmental scientist*
Morin-Postel, Christine *international operations executive*
Nguyen-Trong, Hoang *physician, consultant*

Nantes
Peerhossaini, Mohammad Hassan *engineering and physics educator*
Tardy, Daniel Louis *mechanical engineer, education consultant*

Neuilly sur Seine
Wennerstrom, Arthur John *aeronautical engineer*

Nice
Auwerx, Johan Henri *molecular biologist, medical educator*
Nicolay, Jean Honoré *cardiologist*
Sornette, Didier Paul Charles Robert *physicist*

Nord
Lablanche, Jean-Marc Andre *cardiologist, educator*

Nouzilly
Plommet, Michel Georges *microbiologist, researcher*

Nyons
Bottero, Philippe Bernard *general practitioner*

Orsay
Deutsch, Claude David *physicist, educator*
Fiszer-Szafarz, Berta (Berta Safars) *research scientist*
Guillaumont, Robert *chemist, educator*
Liepkalns, Vis Argots *biochemist*
Reich, Robert Claude *metallurgist, physicist*
Seyden-Penne, Jacqueline *research chemist*

Oullins
Louisot, Pierre Auguste Alphonse *biochemist*

Paris
Aaron, Jean-Jacques *chemist, educator*
Atlan, Paul *gynecologist*
Bathias, Claude *materials science educator, consultant*
Ben Amor, Ismäil *obstetrician/gynecologist*
Bénassy, Jean-Pascal *economist, researcher, educator*
Chahid-Nourai, Behrouz J.P. *economist, corporate executive*
Chany, Charles *microbiology educator*
Cohen-Tannoudji, Claude Nessim *physics educator*
Courtois, Yves *biologist*
Cousteau, Jacques-Yves *marine explorer, film producer, writer*
Curien, Hubert *mineralogy educator*
Daligand, Daniel *engineer, association executive, expert witness*
Dausset, Jean *immunologist*
de Gennes, Pierre Gilles *physicist, educator*
Delacour, Yves Jean Claude Marie *technology information executive*
Delahay, Paul *chemistry educator*
De Vitry D'Avaucourt, Arnaud *engineer*
Dewar, James McEwen *agricultural executive, consultant*
DuBois, Jean Gabriel *pharmaceutical executive, pharmacist*
Ferrando, Raymond *animal nutrition scientist, educator*
Fournier, Jean Pierre *architect, real estate developer*
Friedman, Kenneth Michael *energy policy analyst*
Gauvenet, André Jean *engineering educator*
Genet, Jean Pierre *chemistry educator, researcher*
Giraudet, Michele *mathematics educator, researcher*
Gontier, Jean Roger *internist, physiology educator, consultant*
Goupy, Jacques Louis *chemiometrics engineer*
Haroche, Serge *optics scientist*
Jacob, François *biologist*
Jolles, Pierre *biochemist*
Lefebure, Alain Paul *family physician*
LeGoffic, Francois *biotechnology educator*
Lehn, Jean-Marie Pierre *chemistry educator*
Lehner, Gerhard Hans *engineer, consultant*
Levy, Etienne Paul Louis *surgical department administrator*
Lions, Jacques Louis *physicist*
Luton, Jean Marie *space agency administrator*
Lwoff, André Michel *retired microbiologist, virologist*
Malherbe, Bernard *surgeon*
Mareschi, Jean Pierre *biochemical engineer*
McAdams, Stephen Edward *experimental psychologist, educator*
Meyer, Jean-Pierre *psychiatrist*
Mitz, Vladimir *plastic surgeon*
Nême, Jacques *economist*
Nestvold, Elwood Olaf *oil service company executive*
Ohayon, Roger Jean *aerospace engineering educator, scientific deputy*
Ojcius, David Marcelo *biochemist, researcher*
Pfau, Michel Alexandre *chemist, researcher, consultant*
Rips, Richard Maurice *chemist, pharmacologist*
Rosenschein, Guy Raoul *pediatric and visceral surgeon, airline pilot*
Roudybush, Franklin *diplomat, educator*
Rouvillois, Philippe André Marie *science administrator*
Sirat, Gabriel Yeshoua *physics educator*
Spitz, Erich *electronics industry executive*
Tarantola, Albert *geophysicist, educator*
Vieillard-Baron, Bertrand Louis *engineering executive*
Wautier, Jean Luc *hematologist*
Yuechiming, Roger Yue Yuen Shing *mathematics educator*
Zaragoza, Federico Mayor *protective agency administrator, biochemist*

Paris La Defense
Farge, Yves Marie *physicist*
Faure, François Michel *metallurgical engineer*

Périgueux
Delluc, Gilles *physician, researcher*

Port-Marly
Dégeilh, Robert *retired chemist*

Puteaux
Bally, Laurent Marie Joseph *software engineering company executive*

Rantigny
Langlais, Catherine Renee *science administrator*

Reims
Mathlouthi, Mohamed *chemistry educator*

Rennes
Genetet, Bernard *hematologist, immunologist, educator*
Lerman, Israël César *data classification and processing researcher*

Rognac
Castel, Gérard Joseph *physician*

Rueil Malmaison
Rondeau, Jacques Antoine *marketing specialist, chemical engineer*

Saclay
Soukiassian, Patrick Gilles *physics educator, physicist*
Teixier, Annie Mireille J. *research scientist*

Saint-Cloud
Mallet, Michel Francois *numerical analyst*
Perrier, Piérre Claude *aeronautical engineer, researcher*

Saint Etienne
Soustelle, Michel Marcel Philippe *chemistry educator, researcher*

Saint Symphorien Ozo
Bernasconi, Christian *chemical engineer*

Strasbourg
Aunis, Dominique *neurobiologist*
Broll, Norbert *physicist, consultant*
Danzin, Charles Marie *enzymologist*
Dirheimer, Guy *biochemist, educator*
Drain, Charles Michael *biophysical chemistry educator*
Elkomoss, Sabry Gobran *physicist*
Heck, André *astronomer*
Meyer, Richard *psychiatrist*
Muller, Jean-Claude *nuclear engineer*
Petrovic, Alexandre Gabriel *physician, physiology educator, medical research director*
Schlegel, Justin J. *psychological consultant*
Thierry, Robert Charles *microbiologist*

Suresnes
Donguy, Paul Joseph *aerospace engineer*
Pouliquen, Marcel François *space propulsion engineer, educator*

Toulouse
Conte, Jean Jacques *medical educator, nephrologist*
Diaz, A. Michel *computer and control science researcher, administrator*
Fonta, Caroline *biologist*
Lattes, Armand *educator, researcher*
Portal, Jean-Claude *physicist*
Zappoli, Bernand *aerospace engineer*

Tours
Coursaget, Pierre Louis *virologist*
Renoux, Gerard Eugene *immunologist*

Velizy-Villacoublay
Musikas, Claude *chemical researcher*

Vernon
Lehé, Jean Robert *mechanical engineer*

Villejuif
Hatzfeld, Jacques Alexandre *biologist, researcher*

Villeurbanne
Campigotto, Corrado Marco *physicist*
Godet, Maurice *mechanical engineer*
Roche, Alain Andre *research chemist*

Villiers Le Bel
Verillon, Francis Charles *chemist*

Wissembourg
Grisar, Johann Martin *research chemist*

GABON

Libreville
Berre, Andre Dieudonne *federal agency executive, oil company executive*

GERMANY

Aachen
Benetschik, Hannes *mechanical engineer*
Jakobs, Kai *computer scientist*
Meixner, Josef *emeritus physics educator*
Nastase, Adriana *aerospace scientist, educator, researcher*

Allgau
Lichtenheld, Frank Robert *physician, plastic and reconstructive surgeon, urologist*

Bad Nauheim
Hueting, Juergen *internist*

Baden
Musshoff, Karl Albert *radiation oncologist*

Bamberg
Krauss, Hans Lŭdwig *emeritus educator*

Bavaria
Ehler, Herbert *computer scientist*

Bayern
Boldt, Heinz *aerospace engineer*

Bayreuth
Esquinazi, Pablo David *physicist*

Berlin
Albach, Horst *economist*
Becker, Uwe Eugen *physicist*
Bernard, Herbert Fritz *aerospace engineer*
Brätter, Peter *chemist, educator*
Ehrig, Hartmut *computer science educator, mathematician*
Eichstädt, Hermann Werner *cardiology educator*
Fobbe, Franz Caspar *radiologist*
Franzke, Hans-Hermann *engineering scientist, educator*
Gobrecht, Heinrich Friedrich *physicist, educator*
Henglein, Friedrich Arnim *chemistry educator*
Klapoetke, Thomas Matthias *chemist*
Klinkmuller, Erich *economist*
Krauter, Stefan Christof Werner *electrical engineer*
Lemke, Heinz Ulrich *computer science educator*
Mauritz, Karl Heinz *neurology educator*
Möller, Detlev *atmospheric chemist*
Poehlmann, Gerhard Manfred *cartography educator*
Priebe, Stefan *psychiatrist*
Rauls, Walter Matthias *engineering executive*
Saenger, Wolfram Heinrich Edmund *crystallography educator*
Schulze, Matthias Michael *chemist*
Selle, Burkhardt Herbert Richard *physicist*
Spur, Günter *manufacturing engineering educator*
Tiedemann, Heinz *biochemist*
Vorbrueggen, Helmut Ferdinand *organic chemist*
Wallner, Franz *engineer, educator*
Witzel, Lothar Gustav *physician, gastroenterologist*

Bielefeld
Lauven, Peter Michael *anesthesiologist*
Muller, Achim *chemistry educator*

Blumberg
Wesley, James Paul *theoretical physicist, lecturer, consultant*

Böbliagen
Fischer, Bernhard Franz *physicist*
Mühe, Erich *surgical educator*

Bochum
Blauert, Jens Peter *acoustician, educator*
Kaernbach, Christian *psychophysicist*
Kneller, Eckart Friedrich *materials scientist, electrical engineer, educator, researcher*
Kunze, Hans-Joachim Dieter *physics educator*
Mergner, Hans Konrad *zoology educator*
Ponosov, Arcady Vladimirovitch *mathematician*
Wieghardt, Karl Ernst *chemistry educator*
Zimolong, Bernhard Michael *psychologist, educator*

Bonn
Cloes, Roger Arthur Josef *economist, consultant*
Eckmiller, Rolf Eberhard *neuroscientist, educator*
Giannis, Athanassios *chemist, physician*
Henkel, Christian Johann *astrophysicist*
Klingen, Leo H. *computer science educator*
Korte, Bernhard Hermann *mathematician, educator*
Paul, Wolfgang *physics educator*
Riesenhuber, Heinz Friedrich *German minister for research and technology*
Schwerdtfeger, Walter Kurt *public health official, researcher*
Shaw, John Andrew *information systems executive*
Wehrberger, Klaus Herbert *physics educator, research manager*
Weiling, Franz Joseph Bernard *retired botany and biometry educator*
Wissman, Matthias *minister of research and technology*

Borstel
Loppnow, Harald *biologist*

Braunschweig
Blazek, Jiri *aerospace engineer, researcher*
Boldt, Peter *chemistry educator, researcher*
Collins, John *molecular genetics educator, researcher*
Hahn, Klaus-Uwe *aerospace engineer*
Hollmann, Rudolf Werner *geologist, palaeontology researcher*
Radespiel, Rolf Ernst *aerospace engineer*
Schomburg, Dietmar *chemist, researcher*
Voss, Werner Konrad Karl *architect, engineer*

Bremen
Baykut, Mehmet Gökhan *chemist*
Hansohm, Dirk Christian *economics educator, editor*
Heinz, Walter Richard *sociology educator*
Iglseder, Heinrich Rudolf *space engineer, researcher*
Mertens, Josef Wilhelm *engineer*
Theile, Burkhard *physicist*
Valentine-Thon, Elizabeth Anne *biologist*

Bremerhaven
Crawford, Richard M. *botanist*

Clausthal
Bauer, Ernst Georg *physicist, educator*

Cologne
Merten, Utz Peter *physician*
Stuhl, Oskar Paul *organic chemist*
Teodorescu, George *industrial designer, educator, consultant*

Darmstadt
Clausen, Thomas Hans Wilhelm *chemist*
Fetting, Fritz *chemistry educator*
Kozhuharov, Christophor *physicist*
Lichtenthaler, Frieder Wilhelm *chemist, educator*
Theobald, Jürgen Peter *physics educator*
Wullkopf, Uwe Erich Walter *research institute director*

Dessaŭ
Bach, Günther *organic chemist, researcher*

Dortmund
Freund, Eckhard *electrical engineering educator*
Neumann, Wilhelm Paul *chemistry educator*

Dresden
Pfitzmann, Andreas *computer security educator*
Reinschke, Kurt Johannes *mathematician, educator*

Düsseldorf
Nickel, Horst Wilhelm *psychology educator*
Schadewaldt, Hans *medical educator*
Strehblow, Hans-Henning Steffen *chemistry educator*

Duisburg
Franke, Hilmar *physics educator*

Emden
Gombler, Willy Hans *chemistry educator*

Erlangen
Brajder, Antonio *electrical engineer, engineering executive*
Hartmann, Werner *physicist*

Essen
Erbel, Raimund *physician, educator*
Hoffmann, Günter Georg *chemist*
Schilling, Hartmut *engineer*
Stein, Gerald *metallurgical engineer*
van Wissen, Gerardus Wilhelmus Johannes Maria *consulting engineering company executive*

Fellbach
Puettmer, Marcus Armin *metallurgical engineer*

Frankfurt
Duus, Peter *neurology educator*
Engel, Juergen Kurt *chemist, researcher*
Entian, Karl-Dieter *microbiology educator*
Greiner, Walter Albin Erhard *physicist*
Hilgenfeld, Rolf *chemist*
Ilten, David Frederick *chemist*
Klöpffer, Walter *chemist, educator*
Lüthi, Bruno *physicist, educator*
Sewell, Adrian Clive *clinical biochemist*
Tholey, Paul Nikolaus *psychology educator, physical education educator*

Frankfurt am Main
Baur, Werner Heinz *mineralogist, educator*
Lorenz, Ruediger *neurosurgeon*
Michel, Hartmut *biochemist*
Vogel, Gerhard, Hans *pharmacologist, toxicologist*

Freiburg
Gao, Hong-Bo *physics educator, researcher*
Häussinger, Dieter Lothar *medical educator*
Schaefer, Hans-Eckart *pathologist*
Wagner, Bernhard Rupert *computer scientist*
Wintgen, Dieter *physicist educator*

Friedrichshafen
Wilczek, Elmar Ulrich *aviation professional*

Fulda
Stegmann, Thomas Joseph *physician*

Garching
Grieger, Günter *physicist*
Mossbauer, Rudolf Ludwig *physicist, educator*
Schlag, Edward William *chemistry educator*
Scott, Bruce Douglas *physicist*
Walther, Herbert *physicist, educator*

Germersheim
Pollmer, Jost Udo *food chemist*

Giessen
Hoppe, Rudolf Reinhold Otto *chemist, educator*
Sell, Friedrich Leopold *economics educator, researcher*

Gilching
Jung, Reinhard Paul *computer system company executive*

Gladbeck
Geisler, Linus Sebastian *physician, educator*

Göttingen
Eigen, Manfred *physicist*
Faubel, Manfred *physicist*
Gandhi, Suketu Ramesh *chemist*
Lorenz-Meyer, Wolfgang *aeronautical engineer*
Neher, Erwin *biophysicist*
Nitzsche, Fred *aeronautical engineer*
Oellerich, Michael *chemistry educator, chemical pathologist*
Roesky, Herbert Walter *chemistry educator*
Sheldrick, George Michael *chemistry educator, crystallographer*
Tietze, Lutz Friedjan *chemist, educator*
Wedemeyer, Erich Hans *physicist*

Gommern
Voigt, Hans-Dieter *oil company executive, researcher, educator*

Greifswald
Gaab, Michael Robert *neurosurgery educator, consultant*
Teuscher, Eberhard *pharmacist*

Hagen
Struecker, Gerhard *analytical chemist*

Haina
Müller-Isberner, Joachim Rüdiger *psychiatrist, educator*

Halle
Benker, Hans Otto *mathematics educator*
Goldschmidt, Bernd *mathematics educator*

Hamburg
Bruns, Michael Willi Erich *virologist, veterinarian*
Comberg, Hans-Ulrich *physician*
Jensen, Elwood Vernon *biochemist*
Maurer, Hans Hilarius *pharmacology educator, researcher*
Scheurle, Jurgen Karl *mathematician, educator*
Sprecher, Gustav Ewald *pharmacy educator*
Trinks, Hauke Gerhard *physicist, researcher*
Tscheuschner, Ralf Dietrich *theoretical physicist*
Voss, Jürgen *chemistry educator*
Wiik, Björn H. *physicist researcher, director*
Wolfinger, Bernd Emil *mathematics and computer science educator*

Hanau
Vasak, David Jiři Jan *physicist, consultant*

Hannover
Döhler, Klaus Dieter *pharmaceutical and development executive*
Habermehl, Gerhard Georg *chemist, educator*
Kallfelz, Hans Carlo *cardiologist, pediatrics and pediatric cardiology educator*
Liebler, Elisabeth M. *veterinary pathologist*
Phan-Tan, Tai *scientist, researcher, educator*
Von Zur Mühlen, Alexander Meinhard *physician, internal medicine educator*

Heidelberg
Abel, Ulrich Rainer *biometrician, researcher, financial consultant*
Bode, Christoph Albert-Maria *cardiology educator, researcher*
Dihle, Albrecht Gottfried Ferdinand *professional society administrator, classics educator*
Duschl, Wolfgang Josef *astrophysicist*
Oberdorfer, Franz *chemist*
Papavassiliou, Athanasios George *molecular biologist, researcher*
Sakmann, Bert *physician, cell physiologist*
Serrano, Luis Felipe *biochemist, researcher*
Staab, Heinz A. *chemist*

Heilbronn
Pschunder, Willi *semiconductor engineer*

Hemsbach
Froessl, Horst Waldemar *business executive, data processing developer*

Hennef
Wette, Eduard Wilhelm *mathematician*

Heppenheim
Singer, Peter *physician, researcher, consultant*

Hirschberg
Kreuter, Konrad Franz *software engineer*

Iserlohn
Paradies, Hasko Henrich *chemistry educator*

Jena
Krätzel, Ekkehard *mathematics educator*

Jülich
Feinendegen, Ludwig Emil *retired hospital and research institute director*
Krasser, Hans Wolfgang *physicist*
Mann, Ulrich *petroleum geologist*
Qaim, Syed Muhammad *nuclear chemist, researcher, educator*
Rogister, Andre Lambert *physicist*
Stengel, Eberhard Friedrich Otto *botanist*

Kaiserslautern
Fan, Tian-You *applied mathematician, educator*

Karlsruhe
Hohmeyer, Olav Hans *economist, researcher*
Schulz, Paul *physicist*

Kassel
Bosbach, Bruno *mathematics educator*

Katlenburg-Lindau
Hagfors, Tor *national astronomy center director*

Kiel
Brockhoff, Klaus K.L. *marketing and management educator*
Meissner, Rudolf Otto *geophysicist, educator*

Kleinseebach
Kalender, Willi Alfred *medical physicist*

Köln
Funken, Karl-Heinz *chemist*

Königswinter
Ternyik, Stephen *polytechnology researcher*

Koblenz
Kuntze, Herbert Kurt Erwin *aeronautical engineer*

Konstanz
Jäger-Waldau, Arnulf Albert *physicist*

Kruft
Lekim, Dac *chemist*

Ladenburg
Traub, Peter *biochemist*

Landsberg
Panizza, Michael *civil engineer*

Leipzig
Mössner, Joachim *internist, gastroenterologist*
Mueller, Rudhard Klaus *toxicologist*
Zimmermann, (Arthur) Gerhard *chemist, educator, researcher*

Leonberg
Syblik, Detlev Adolf *computer engineer*

Lohhof
Arndt, Norbert Karl Erhard *mechanical engineer*

Lüneburg
Linde, Robert Hermann *economics educator*

Ludwigsburg
Queeney, David *computer engineer*

Ludwigshafen
Hibst, Hartmut *scientist, chemistry educator*

Ludwigshafen am Rhein
Laun, Hans Martin *rheology researcher*

Luebeck
Gober, Hans Joachim *physics educator*

Magdeburg
Sabel, Bernhard August Maria *research neuroscientist, psychologist*

Mainz
Egel, Christoph *computer scientist*
Gütlich, Philipp *chemistry educator*
Meisel, Werner Paul Ernst *physicist*
Rabe, Jürgen P. *chemical physicist*
Schirmacher, Peter *molecular pathologist, educator*
Schmitt, Heinz-Josef *physician*
Schönberger, Winfried Josef *pediatrics educator*

Mannheim
Nürnberger, Günther *mathematician*
Steffens, Franz Eugen Aloys *computer science educator*

Marburg
Brandt, Reinhard *chemist*
Gemsa, Diethard *immunologist*
Mannheim, Walter *medical microbiologist*
Patzelt, Paul *nuclear chemist*
Rienhoff, Otto *physician, medical informatics educator*
Wesemann, Wolfgang *biochemistry educator*
Zimmermann, Horst Ernst Friedrich *economics educator*

Martinsried
Huber, Robert *biochemist*

Mittelberg
Nanz, Claus Ernest *economist, consultant*

Mûlhein
Wilke, Gunther *chemistry educator*

Münster
Bonus, Holger *economics educator*
Hubert, Walter *psychologist*
Jeitschko, Wolfgang Karl *chemistry educator*
Kordes, Hagen *education researcher, author*
Rüter, Ingo *research chemist, toxicology consultant*
Thurm, Ulrich *zoology educator*
Voss, Werner *dermatologist*

Munich
Berg, Jan Mikael *science educator*
Bergsteiner, Harald *architect*
Binnig, Gerd Karl *physicist*
Bodlaj, Viktor *electrical engineer*
Bohn, Horst-Ulrich *physicist*
Born, Gunthard Karl *aerospace executive*
Cleve, Hartwig Karl *medical educator*
Dühmke, Eckhart *radiation oncology educator*
Fastl, Hugo Michael *acoustical engineer, educator*
Fischer, Ernst Otto *chemist, educator*
Freymann, Raymond Florent *aeronautical engineer*
Fuhrmann, Horst *science administrator*
Giacconi, Riccardo *astrophysicist, educator*
Gottzein, Eveline *aerospace engineer*
Grünewald, Michael *physics researcher, research program coordinator*
Hein, Fritz Eugen *engineer, consultant, architect*
Liepsch, Dieter Walter *engineering educator*
Pääbo, Svante *molecular biologist, biochemist*
Paumgartner, Gustav *hepatologist, educator*
Pfleiderer, Hans Markus *physicist*
Porkert, Manfred (Bruno) *medical sciences educator, author*
Reviczky, Janos *mathematician, researcher*
Schlee, Walter *mathematician*
Schmidt, Stefan *mechanical engineer, economist*
Sebestyén, István *computer scientist, educator, consultant*
Stiegler, Karl Drago *mathematician*
Stroke, George Wilhelm *physicist, educator*
Toft, Jürgen Herbert *orthopedic surgeon*
Wagner, Thomas Alfred *chemical engineer*
Weis, Serge *neuropathologist*
Weisweiler, Peter *physician, educator*
Wess, Julius *nuclear scientist*
Wessjohann, Ludger Aloisius *chemistry educator, consultant*
Zander, Josef *gynecologist*

Neckargemuend
Kirchmayer-Hilprecht, Martin *geologist*

Neubiberg
Triftshäuser, Werner *physics educator*

Nordrhein
Grauer, Manfred *computer engineering educator, researcher*

Paderborn
Belli, Fevzi *computing science educator, consultant*
Frank, Helmar Gunter *educational cyberneticist*
Krohn, Karsten *chemistry educator*

Putzbrunn
Gregoriou, Gregor Georg *aeronautical engineer*

Regensburg
Steinborn, E(rnst) Otto H. *physicist, educator*

Reutlingen
Hartmann, Jürgen Heinrich *physicist*

Rostock
Loehr, Johannes-Matthias *physician*

Saarbrücken
König, Heinz Johannes Erdmann *mathematics educator*
Kornadt, Hans-Joachim Kurt *psychologist, researcher*
Siekmann, Jörg Hans *computer science educator*
Veith, Michael *chemist*

Schwaebisch
Khan, Hamid Raza *physicist*

Steinfôrde
Kolditz, Lothar *chemistry educator*

Steinfurt
Niederdrenk, Klaus *mathematician, educational administrator, educator*

Stuttgart
Dittrich, Herbert *mineralogist, researcher*
Klitzing, Klaus von *institute administrator, physicist*
Konuma, Mitsuharu *materials scientist*
Kramer, Horst Emil Adolf *physical chemist*
Krueger, Ronald *aerospace engineer*
Luckenbach, Alexander Heinrich *dentist*
Ning, Cun-Zheng *physicist*
Pfisterer, Fritz *physicist*
Pollmer, Wolfgang Gerhard *agronomy educator*
Schiehlen, Werner Otto *mechanical engineer*
Wagner, Henri Paul *aerospace engineer*
Zwicknagl, Gertrud *physicist*

Sulzbach
Deutsch, Hans-Peter Walter *physicist*
Mester, Ulrich *ophthalmologist*

Triefenstein
Makowski, Gerd *aerospace engineer*

Tübingen
Frommhold, Walter *radiologist*
Hirnle, Peter *gynecologist, obstetrician, radiation oncologist, cancer researcher*
Nüsslein-Volhard, Christiane *medical researcher*
Pawelec, Graham Peter *immunologist*
Schaumburg-Lever, Gundula Maria *dermatologist*
Zenner, Hans Peter *otolaryngologist*

Ulm
Presting, Hartmut *physicist*
Reineker, Peter *physics educator*

Volklingen
Reinhardt, Kurt *retired radiological and nuclear physician*

Waldbrunn
Goedde, Josef *physiologist*

Waldhilsbach
Zeh, Heinz-Dieter *retired theoretical physics educator*

Wiesbaden
Adeniji-Fashola, Adeyemo Ayodele *mechanical engineer, consultant*
Braun, Michael Walter *technology consultant*
Elben, Ulrich *chemist*
Sachse, Guenther *health facility administrator, medical educator*

Wildberg
Wild, Hans Jochen *systems engineering executive*

Witten
an der Heiden, Wulf-Uwe *mathematics educator*

Wülfrath
Mullen, Alexander *information scientist, chemist*

Wuerzburg
Buntrock, Gerhard Friedrich Richard *mathematician*
Reiners, Karlheinz *neurologist, educator*
Schaefer, Roland Michael *nephrologist, consultant*

Wuppertal
Schubert, Guenther Erich *pathologist*
Vienken, Joerg Hans *chemical engineer*
Von Weizsäcker, Ernst Ulrich *environmental scientist*

GHANA

Kumasi
Owusu-Ansah, Twum *mathematics educator*

GREECE

Athens
Androutsellis-Theotokis, Paul *civil engineer*
Antoniou, Panayotis A. *surgeon*
Balaras, Constantinos Agelou *mechanical engineer*
Barbatis, Calypso *histopathologist*
Deliyannis, Constantine Christos *economist, mathematician, educator*
Demetriou, Ioannes Constantine *mathematics educator, researcher*
Dritsas, George Vassilios *electronics engineer*

Frangakis, Gerassimos P. *electronic engineer*
Gotzoyannis, Stavros Eleutherios *cardiologist*
Halkias, Christos Constantine *electronics educator, consultant*
Hatzakis, Michael *electrical engineer, research executive*
Jeftic, Ljubomir Mile *marine scientist*
Kalathas, John (Ioannis) *airline pilot, physicist, oceanographer*
Katsikadelis, John *civil engineering educator*
Koupparis, Michael Andreas *analytical chemistry educator*
Koutras, Demetrios A. *physician, endocrinology investigator, educator*
Mantas, John *health informatics educator*
Markatos, Nicolas-Chris Gregory *chemical engineering educator*
Paleos, Constantinos Marcos *chemist*
Panaretos, John *mathematics and statistics educator*
Papandreou, Constantine *tele-informatics scientist, educator*
Papathanasiou, Athanasios George *mechanical engineer*
Petropoulos, Labros S. *physicist, researcher*
Rapidis, Alexander Demetrius *maxillofacial surgeon*
Rekkas, Christos Michail *mechanical engineer*
Sarantopoulos, Theodore *physician, cardiologist, pathologist*
Screttas, Constantinos George *chemistry educator*
Sekeris, Constantine Evangelos *biochemistry educator*
Singhellakis, Panagiotis Nicolaos *endocrinologist, educator*
Stergiou, Konstantinos *biological oceanographer, researcher*
Tsakas, Spyros Christos *genetics educator*

Crete
Economou, Eleftherios Nickolas *physics educator, researcher*

Ekali
Sotirakos, Iannis *civil engineer*

Elefsing
El-Husban, Tayseer Khalaf *internist, consultant*

Ioannina
Albanis, Triadafillos Athanasios *chemistry educator, researcher*
Alexandropoulos, Nikolaos *physics educator*
Mikropoulos, Anastassios (Tassos Mikropoulos) *physicist, researcher*

Kalamaki
Cacouris, Elias Michael *economist, consultant*

Kallithea
Thomadakis, Panagiotis Evangelos *computer company executive*

Karlovassi
Katsikas, Sokratis Konstantine *electrical engineer, educator*

Koridallos
Cambalouris, Michael Dimitrios *aircraft engineer*

Patras
Anagnostopoulos, Stavros Aristidou *earthquake engineer, educator*
Dassios, George Theodore *mathematician, educator, researcher*
Makios, Vasilios *electronics educator*

Thessaloniki
Agorastos, Theodoros *obstetrics and gynecology educator*
Glavopoulos, Christos Dimitrios *mechanical engineering consultant*
Panagiotopoulos, Panagiotis Dionysios *mechanical engineering educator*
Panayiotou, Constantinos *chemical engineering educator*
Papanastasiou, Tasos Charilaou *chemical engineering educator*
Sakellaropoulos, George Panayotis *chemical engineering educator*
Smokovitis, Athanassios A. *physiologist, educator*

GRENADA

Saint George's
Brunson, Joel Garrett *pathologist, educator*

GUATEMALA

Mixco
Silva, Fernando Arturo *civil engineer*

HONG KONG

Clear Water Bay
Che, Chun-Tao *chemistry educator*

Hong Kong
Chan, Raymond Honfu *mathematics educator*
Chung, Cho Man *psychiatrist*
Fong, Wang-Fun *biochemist*
Ge, Weikun *physicist, educator*
Kwok, Raymond Hung Fai *economics educator*
Kwok, Russell Chi-Yan *retail company executive*
Lee, Edward King Pang *dermatologist, public health service officer*
Lochovsky, Frederick Horst *computer science educator*
Lui, Ming Wah *electronics executive*
Tam, Alfred Yat-Cheung *pediatrician, consultant*
Tang, Pui Fun Louisa *fragrance research administrator*
Yoo, Kwong Mow *science educator*

Kowloon
Chang, Donald Choy *biophysicist*
Chang, Leroy L. *physicist*
Chin, James Kee-Hong *surgeon*
Chow, Stephen Heung Wing *physician*
Chu, Wayne Shu-Wing *food industry entrepreneur, researcher*
Demokan, Muhtesem Süleyman *electrical engineering educator*
Huang, Ju-Chang *civil engineering educator*
Lui, Ng Ying Bik *engineering educator, consultant*
Tin, Kam Chung *industrial chemist, educator*
Yang, Mildred Sze-ming *biologist, educator*
Yang Ko, Lillian Yang *pediatrician*

Kowloon Bay
Chan, Andrew Mancheong *engineering executive*

Kwun Tong
Chou, Chung Lim *electrical engineer*

Quarry Bay
Bast, Albert John, III *civil engineer*

Sha Tin
Kao, Charles Kuen *electrical engineer, educator*
King, Walter Wing-Keung *surgeon, head and neck surgery consultant*
Lu, Wudu *mathematics educator*

Sha Tin NT
Cheung, Kwok-wai *optical network researcher*
Cheung, Wilkin Wai-Kuen *entomologist, educator*

Tai Wai
Barbalas, Lorina Cheng *chemist*

Wanchai
Chen, Concordia Chao *mathematician*
van Hoften, James Dougal Adrianus *business executive, former astronaut*

HUNGARY

Budapest
Ábrahám, György *mechanical engineering educator*
Braun, Tibor *chemist*
Deák, Peter *physicist, educator*
Deak, Tibor *microbiologist*
Hajos, Zoltan George *chemist*
Kis-Tamás, Attila *chemist*
Kosáry, Domokos *historian*
Lovro, Istvań *aircraft engineer*
Markó, László *chemist*
Mészáros, Ernö *meteorologist, researcher, science administrator*
Pungor, Erno *chemist, educator*
Starosolszky, Ődön *civil engineer*
Stefanovits, Pál *agriculturalist*
Szasz, Andras István *physicist, educator, researcher*
Vajda, György *electrical engineer*
Vámos, George A. *mechanical engineer, consultant*

Debrecen
Sztrik, János *mathematics educator, researcher*

Szeged
Dékány, Imre Lajos *chemistry educator*

Tiszavasvari
Timar, Tibor *chemist*

ICELAND

Reykjavik
Gudjonsson, Birgir *physician*
Kvaran, Agust *chemistry educator, research scientist*
Pind, Jörgen Leonhard *computational linguist, psycholinguist*
Sigurdsson, Thordur Baldur *data processing executive*
Thorarensen, Oddur C.S. *pharmacist*
Thorgeirsson, Gudmundur *physician, cardiologist*

INDIA

Bangalore
Chakrabartty, Sunil Kumar *mathematician*
Kantheti, Badari Narayana *aeronautical engineer*
Narasimha, Roddam *laboratory director, educator*
Ramasubbu, Sunder *aeronautical engineer, researcher*
Rao, K. Prabhakara *aerospace engineering educator*
Reddy, Narayana Muniswamy *aerospace educator*

Barabanki
Suryavanshi, O. P. S. *corporate executive*

Bathinda
Bansal, Satish Kumar *civil engineer, educator*

Bhavnagar
Natarajan, Paramasivam *chemistry educator*

Bombay
Holla, Kadambar Seetharam *chemist, educator*
Khopkar, Shripad Moreshwer *chemistry educator*
Krishnamurthy, Suresh Kumar *chemist, researcher*
Paul, Biraja Bilash *engineering consulting company executive*
Shah, Natverlal Jagjivandas *cardiologist*
Shikarkhane, Naren Shriram *laser scientist*
Tarafdar, Shankar Prosad *astrophysicist, educator*

Chandigarh
Chawla, Amrik Singh *chemist, educator*
Singh, Harkishan *chemist, educator*

Gorakhpur
Das, Ishwar *chemistry educator*

Hosur
Ramani, Narayan *mechanical engineer*

Hyderabad
Vathsal, Srinivasan *electrical engineer, scientist*

Kanpur
Tewari, Ashish *aerospace scientist, consultant*

Madras
Chandra Sekharan, Pakkirisamy *forensic scientist*
Pethachi, Muthiah Chidambaram *textile industry executive*

Madurai
Raju, Perumal Reddy *chemist*

Nagpur
Sehgal, Jawaharlal *environmental company executive*

New Delhi
Adholeya, Alok *biotechnologist, researcher*
Hillary, Edmund Percival *diplomat, explorer, bee farmer*
Menon, Mambillikalathil Govind Kumar *physicist*
Roy, Tuhin Kumar *engineering company executive*

Patiala
Bajpai, Pramod Kumar *chemical and biochemical engineer*

Pune
Gogate, Kamalakar Chintaman *metallurgist*
Purohit, Sharad Chandra *mathematician*

Rasayani
Modak, Chintamani Krishna *information and documentation officer*

Tiruchirapalli
Jeyaraman, Ramasubbu *chemist, educator*

Varanasi
Srivastava, Om Prakash *soil science and agricultural chemistry educator*

Vellore
Balasubramanian, Aiylam Subramaniaier *biochemistry educator*

INDONESIA

Bandung
Norup, Kim Stefan *civil engineer*
Wijaya, Andi *clinical laboratory executive, clinical chemistry educator*

Jakarta
Cheron, James Clinton *mechanical engineer*
Lembong, Johannes Tarcicius *cardiologist*

Pangkalanbun
Galdikas, Birute *primatologist*

Serpong
Akil, Husein Avicenna *acoustical engineer*

IRAN

Babolsar
Mohanazadeh, Farajollah Bakhtiari *chemist, educator*

Tehran
Ahmadinejad, Behrouz *chemist, educator*
Eslami, Mohammad Reza *mechanical engineering educator*
Kaghazchi, Tahereh *chemical engineering educator*
Mirdamadi, Hamid Reza *structural engineering educator, researcher*
Mozaffari, Mojtaba *computer science educator*
Sharifi, Iraj Alagha *organic chemistry educator*
Sohrabi, Morteza *chemical engineering educator*
Vahabzadeh, Farzaneh *food scientist, educator, researcher*
Zohourian Mashmoul, Mohammad Jalal-od-din *chemistry educator*

IRELAND

Carlow
Kavanagh, Yvonne Marie *physicist*

Cork
Burke, Laurence Declan *chemistry educator*
Hill, Martin Jude *engineer, consultant*

Dublin
Balfe, Alan *biochemist*
Bonnar, John *obstetrics/gynecology educator, consultant*
Bunni, Nael Georges *engineering consultant, international arbitrator, conciliator*
Mac An Airchinnigh, Mícheál *computer science educator*
Norton, Desmond Anthony *economics educator*
Walton, Ernest Thomas Sinton *physicist*

Galway
Lavelle, Seän Marius *clinical informatics educator*

Limerick
Eaton, Malachy Michael *computer systems educator*

Maynooth
Whittaker, Peter Anthony *biology educator*

Shannon
Prendergast, Walter Gerard *computer engineer*

Wicklow
Wayman, Patrick Arthur *astronomer*

ISLE OF MAN

Douglas
Lamming, Robert Love *retired surgeon*

ISRAEL

Beer Sheva
Gottlieb, Moshe *chemical engineer, educator*

Bet Dagan
Halperin, Joseph *entomologist*

Binyamina
Ran, Josef *electronics executive, engineer*

Haifa
Apeloig, Yitzhak *chemistry educator, researcher*
Idan, Moshe *aerospace engineer*
Natan, Benveniste *aeronautical engineer, educator, researcher*
Nissim, Eliahu *aerospace engineer, researcher*
Shpilrain, Vladimir Evald *mathematician, educator*
Singer, Josef *aerospace engineer, educator*
Weihs, Daniel *engineering educator*

Jerusalem
Berns, Donald Sheldon *research scientist*
Isaacs, Philip Klein *retired chemist*
Lipman, Daniel Gordon *neuropsychiatrist*
Marcus, Yizhak *chemistry educator*
Metzger, Gershon *chemist, patent attorney*
Shaik, Sason Sabakh *chemistry educator*

Migdal
Selivansky, Dror *chemistry educator, resarcher*

Nes Ziona
Schmell, Eli David *biotechnologist*

Ramallar
Laila, Abduhameed Abdelrahman *chemistry educator*

Ramat Gan
Hassner, Alfred *chemistry educator*
Steinberger, Yosef *ecologist, biologist*

Ramat-Hasharon
Ben-Asher, Joseph Zalman *aeronautical engineer*

Rehovot
Cahen, David *materials chemist*
Halevy, Abraham Hayim *horticulturist, plant physiologist*

Technion City
Halevi, Emil Amitai *chemistry educator*

Tel Aviv
Jortner, Joshua *physical chemistry scientist, educator*
Kaldor, Uzi *chemistry educator*
Seifert, Avi *aerodynamic engineering educator and researcher*

Tel Hashomer
Belkin, Michael *ophthalmologist, educator, researcher*

Yavne
Caner, Marc *physicist*
Fish, Falk *microbiologist, immunologist, researcher, inventor*

ITALY

Ancona
Marchesi, Gian Franco *psychiatry educator*
Milani-Comparetti, Marco Severo *geneticist, bioethicist*

Assago
Affaticati, Giuseppe Eugenio *telecommunications, instrumentation engineer*

Bari
Nicastro, Francesco Vito Mario *agricultural science educator*
Pizzoli, Elsa Maria *chemistry educator*
Recchi, Vincenzo *manufacturing executive*
Schena, Francesco Paolo *nephrology educator*

Bologna
Chierici, Gian Luigi *petroleum engineering executive, educator*
Foraboschi, Franco Paolo *chemical engineering educator*
Susi, Enrichetta *chemist, researcher*

Brescia
Bosio, Angelo *pharmacologist, psychiatrist*
Villa, Roberto Riccardo *chemist*

Cagliari
Arca, Giuseppe *mathematician, educator*

Camerino
Miyake, Akio *biologist, educator*

Catania
Montaudo, Giorgio *chemistry educator, researcher*

Ferrara
Manfredini, Stefano *medicinal chemistry educator*
Pandini, Davide *electrical engineer*

Florence
Bennici, Andrea *botany educator*
Brandi, Maria Luisa *endocrinologist, educator*
Giolitti, Alessandro *chemist*
Verga Sheggi, Annamaria *physicist*

Frascati
Paparazzo, Ernesto *chemist*

Gorizia
Valdemarin, Livio *computer peripherals company executive*

Ispra
Blaesser, Gerd *theoretical physics researcher*

Crutzen, Yves Robert *engineer, scientist, educator*
Rickerby, David George *physicist, materials scientist*

Macerata
Fruzzetti, Oreste Giorgio *geologist*

Messina
Nigro, Aldo *physiology and psychology educator*

Milan
Barassi, Dario *management consultant*
Bellobono, Ignazio Renato *chemist, educator*
Benfenati, Emilio *chemist, researcher*
Boehm, Günther *pediatrician*
Boeri, Renato Raimondo *neurologist*
Bonadonna, Gianni *oncologist*
Bondi, Enrico *engineer*
Chan, Ah Wing Edith *chemist*
Colombo, Antonio *cardiologist*
D'Amico, Giuseppe *nephrologist*
Gambarini, Grazia Lavinia *engineering educator*
Gavezzotti, Angelo *chemistry educator*
Lugiato, Luigi Alberto *physics educator*
Mandorini, Vittorio *research physicist*
Mandrioli, Dino Giusto *computer scientist, educator*
Martegani, Enzo *molecular biology educator*
Montesano, Aldo Maria *economics educator*
Pelosi, Giancarlo *cardiologist*
Poluzzi, Amleto *chemical company consultant*
Pontiroli, Antonio Ettore *medical educator*
Resnati, Giuseppe Paolo *chemistry researcher, chemistry educator*
Rodda, Luca *computer company executive, researcher*
Silani, Vincenzo *neurology and neuroscience educator*
Sindoni, Elio *physics educator*
Soria, Marco Raffaello *molecular and cellular biologist*
Trasatti, Sergio *chemistry educator*
Valcavi, Umberto *chemistry educator*

Moncalieri
La Rocca, Aldo Vittorio *mechanical engineer*

Muggia
Parmesani, Rolando Romano *engineer*

Naples
Carlomagno, Giovanni Maria *engineering educator*
Coiro, Domenico Pietro *research scientist, educator*
Covello, Aldo *physics educator*
Mirra, Carlo *aerospace engineer, researcher*
Tarro, Giulio *virologist*

Novara
Cerofolini, Gianfranco *laboratory administrator*
Pernicone, Nicola *catalyst consultant*

Padua
Mammi, Mario *chemist, educator*
Mirandola, Alberto *mechanical engineering educator*
Rosati, Mario *mathematician, educator*
Schrefler, Bernhard Aribo *civil engineering educator*

Palermo
San Biagio, Pier Luigi *physicist*

Parma
Bertolini, Fernando *mathematics educator, Italian embassy cultural administrator*
Cita, Maria *geology educator*
Rizzolatti, Giacomo *neuroscientist*

Pavia
di Jeso, Fernando *biochemistry educator*

Perugia
Laganà, Antonio *chemical kinetics educator*

Pianezza
Badetti, Rolando Emilio *health science facility administrator*

Pisa
Cimatti, Giovanni Ermanno *rational mechanics educator*
De Giorgi, Ennio *mathematics educator*
Simonetti, Ignazio *cardiologist*

Potenza
Battaglia, Franco *chemistry educator*
Korchmaros, Gabor Gabriele *mathematics educator*

Povo-Trento
Pietra, Francesco *chemist*

Rome
Aiello, Pietro *cardiologist*
Colombo, Umberto Paolo *Italian government official*
Cosmi, Ermelando Vinicio *obstetrics/gynecology educator, consultant*
Depasquale, Francesco *aerospace engineer*
Ferracuti, Stefano Eugenio *forensic psychiatrist*
Frati, Luigi *oncologist, pathologist*
Frova, Andrea Fausto *physicist, author*
Giomini, Marcello *chemistry educator*
Lavenda, Bernard Howard *chemical physics educator, scientist*
Levi-Montalcini, Rita *neurobiologist, researcher*
Lopez-Portillo, José Ramon *economist, international government representative*
Manfredi, Mario Erminio *neurologist educator*
Marini Bettolo, Giovanni Battista *chemistry educator, researcher*
Marinuzzi, Francesco *computer science educator*
Mercurio, Antonino Marco *anthropologist, artist*
Mignani, Roberto *physics educator, researcher*
Parisi, Giorgio *physicist, educator*
Perrotta, Giorgio *engineering executive*
Rossmiller, George Eddie *agricultural economist*
Rotondo, Gaetano Mario *aerospace medicine physician, retired military officer*
Salvini, Giorgio *physicist, educator*
Tao, Kar-Ling James *physiologist, researcher*
Vulpetti, Giovanni *physicist*
Zagara, Maurizio *physicist*

San Donato
Bellussi, Giuseppe Carlo *chemical research manager*
Donati, Gianni *chemical engineer, administrator*
Roggero, Arnaldo *polymer chemistry executive*

Siena
Ghiara, Paolo *immunopharmacologist*

Torino
Comoglio, Paolo Maria *cell and molecular biologist, educator*
Elia, Michele *mathematics educator*
Gasco, Alberto *medicinal chemistry educator*
Grillo, Maria Angelica *biochemist*
Rossi, Guido A(ntonio) *applied mathematics educator, researcher*

Treviso
Scardellato, Adriano *software production company executive*

Trieste
Nistri, Andrea *pharmacology educator*
Panza, Giuliano Francesco *seismologist, educator*
Salam, Abdus *physicist, educator*

Udine
Dikranjan, Dikran Nishan *mathematics educator*
Sorrentino, Dario Rosario *medical educator*

Varese
Mezzanotte, Paolo Alessandro *aeronautical engineer*

Verona
Frossi, Paolo *engineer, consultant*
Tridente, Giuseppe *immunoligist educator*

Villanterio
Mendoliera, Salvatore *electrical engineer*

Vittorio Veneto
Albrizio, Francesco *chemical consultant, chemistry educator*

JAMAICA

Kingston
Persaud, Bishnodat *sustainable development educator*
Szentpály, László Von *chemistry educator*

JAPAN

Aichi
Okada, Akane *chemist*
Sakai, Toshihiko *engineer*
Sekimura, Toshio *theoretical biophysics educator*
Takizawa, Akira *engineering educator*
Toyoda, Tadashi *physicist*
Tsuda, Takao *chemistry educator*

Amagasaki
Yamada, Shoichiro *chemistry educator*

Ashiya
Maeda, Yukio *engineering educator*
Yukio, Takeda *engineering educator*

Atsugi
Ikuma, Yasuro *ceramics educator, researcher*

Bunkyo
Kaneko, Masao *chemist*
Sasaki, Taizo *physicist*

Chiba
Fujii, Akira *pharmacology educator*
Ihara, Hirokzu *systems engineer*
Kitamura, Toshinori *psychiatrist*
Kondo, Jun *physicist*
Kuroda, Rokuro *chemist, educator*
Maruyama, Koshi *pathologist, educator*
Nishimura, Chiaki *molecular virologist educator*
Noguchi, Hiroshi *structural engineering educator*
Yagi, Takashi *physicist, researcher*
Yamada, Shinichi *mathematician, computer scientist, educator*

Chikusa-ku
Iizuka, Jugoro *physics educator, researcher*

Chiyoda-ku
Satsumabayashi, Sadayoshi *dental educator*
Tanigawa, Kanzo *Japanese minister of science and technology*
Uchida, Hideo *engineer*

Chofu
Nozaki, Shinji *engineering educator*
Tomita, Etsuji *computer science educator*

Chuo-ku
Yamadori, Takashi *anatomy educator*

Daito
Sakai, Shogo *theoretical chemist*

Ebetsu
Kanoh, Minami *neuropsychologist, clinical psychologist*

Fuchu
Funabiki, Ryuhei *nutritional biochemist*
Takaki, Ryuji *physics educator*

Fukuoka
Aizawa, Keio *biology educator*
El-Agraa, Ali M. *economics educator*
Goto, Norihiro *aeronautical engineering educator*
Hattori, Akira *economics educator*
Hirokawa, Shoji *chemistry educator*
Omura, Tsuneo *medical educator*
Shirai, Takeshi *physician*
Takahashi, Iichiro *economics educator*
Yokozeki, Shunsuke *optical engineering educator*

Funabashi
Katayama, Tetsuya *geologist, materials research petrographer*
Sasaki, Wataru *physics educator*

Gamagouri-shi
Kojima, Ryuichi O. *computer educator*

Gifu
Hatada, Kazuyuki *mathematician, educator*

Gotsu
Hirayama, Chisato *healthcare facility administrator, physician, educator*

Hachioji
Shimoji, Sadao *applied mathematics educator, engineer*

Hakodate
Suzuki, Tetsuya *biochemistry educator*

Hamakita
Tsuchiya, Yutaka *photonics engineer, researcher*

Higashi
Katsuki, Hirohiko *biochemistry educator*

Hirakata
Nakanishi, Tsutomu *pharmaceutical science educator*
Yoshioka, Masanori *pharmaceutical sciences educator*

Hiratsuka
Asada, Toshi *seismologist, educator*
Manabe, Shunji *control engineering educator*

Hiroshima
Ban, Sadayuki *radiation geneticist*
Ebara, Ryuichiro *metallurgist, researcher*
Hiroyasu, Hiroyuki *mechanical engineering educator*
Otsuka, Hideaki *chemistry educator*
Shohoji, Takao *statistician*
Tahara, Eiichi *pathologist, educator*
Yamashita, Kazuo *chemistry educator*
Yasuda, Mineo *anatomy educator*

Hitachi
Hashimoto, Tsuneyuki *materials scientist*
Oku, Tatsuo *mechanical engineering educator*

Hokkaido
Saito, Shuzo *electrical engineering educator*
Suzuki, Akira *chemistry educator*

Hyogo
Terasawa, Mititaka *physics educator*

Ibaragi
Usuki, Satoshi *physician, educator*

Ibaraki
Futai, Masamitsu *biochemistry researcher, educator*
Kishimoto, Kazuo *mathematical engineering educator*
Kurobane, Itsuo *chemical company executive*
Murao, Kenji *chemist*
Numai, Takahiro *scientist, electrical engineer*
Takahashi, Tsutomu *chemist*
Yamada, Keiichi *engineering educator, university official*

Ichihara
Niu, Keishiro *science educator*

Iizuka
Fujii, Masayuki *chemistry educator*

Iizuka City
Nishimura, Manabu *nephrologist*

Ise
Hasegawa, Akinori *chemistry educator*
Hayashi, Takemi *physics educator*

Isehara
Fujita, Tsuneo *systems analysis educator*
Morita, Kazutoshi *psychology educator, consultant*
Oka, Tetsuo *medical educator*

Ishikawa
Fukuda, Ichiro *astronomer, researcher*
Konishi, Kenji *geology educator*
Nasu, Shoichi *electrical engineering educator*

Ishinomaki
Hiwatashi, Koichi *biologist, educator*
Mizutani, Hiroshi *biogeochemist, researcher*

Itogun
Sakurai, Takeo *surgery educator*

Iwate
Kawauchi, Hiroshi *hormone science educator*

Iyomishima
Honda, Toshio *cardiologist*

Kagawa
Kobayashi, Yoshinari *polymer chemist*

Kagoshima
Itahara, Toshio *chemistry educator*
Tanaka, Toshijiro *physicist*

Kamakura
Ishida, Osami *microwave engineer*
Yamamoto, Mikio *public health and human ecology researcher, educator*

Kamigori
Terabe, Shigeru *chemistry educator*

Kanagawa
Hankins, Raleigh Walter *microbiologist*
Kamijo, Fumihiko *computer science and information processing educator*
Karatsu, Osamu *telephone company research and development executive*
Kinoe, Yosuke *computer company researcher, engineer*
Maeda, Toshihide Munenobu *spacecraft system engineer*
Matsubara, Tomoo *software scientist*
Nakajima, Amane *computer engineer, researcher*
Saitoh, Tamotsu *pharmacology educator*
Takeoka, Tsuneyuki *neurology educator*
Wada, Akiyoshi *physicist*

Kanazawa
Kawamura, Mitsunori *material scientist, civil engineering educator*

Kawagoe-Shi
Endo, Hajime *research chemist*

Kawaguchi
Matsumoto, Hiroshi *mechanical engineer*

Kawasaki
Asajima, Shoichi *economics educator*
Taniuchi, Kiyoshi *mechanical engineering educator*
Terada, Yoshinaga *economist*
Yoshizaki, Shiro *medicinal chemist*

Kishiwada
Tateyama, Ichiro *gynecologist*

Kita-ku
Ohnami, Masateru *mechanical engineering educator*

Kitakyushu
Kuryu, Masao *economics educator*
Nishi, Michihiro *mechanical engineering educator*
Tachibana, Takeshi *mechanical engineer, educator*
Takeda, Yoshiyuki *chemical company executive*

Kobe
Aonuma, Tatsuo *management sciences educator*
Baba, Yoshinobu *biotechnologist*
Funaba, Masatomi *political economy educator*
Hirooka, Masaaki *economics educator*
Homma, Morio *microbiology educator*
Masai, Mitsuo *chemical engineer, educator*
Takao, Hama *physiological chemistry educator*
Tani, Shohei *pharmacy educator*
Yamabe, Shigeru *medical educator*

Kochi
Hojo, Masashi *chemistry educator*
Kotsuki, Hiyoshizo *chemist, educator*

Koganei
Akiyama, Masayasu *chemistry educator*
Hasegawa, Tadashi *chemical educator*

Kokubunji
Shimahara, Kenzo *applied microbiology educator*

Komagane-shi
Endoh, Ryohei *cardiologist*

Koriyama
Ohama, Yoshihiko *architectural engineer, educator*

Kumamoto
Kida, Sigeo *chemistry educator*
Maeda, Hiroshi *medical educator*
Mitarai, Osamu *physics educator*
Tsusue, Akio *geology educator*

Kurashiki
Itami, Jinroh *physician*
Masamoto, Junzo *chemist, researcher*

Kurume
Yamashita, Fumio *pediatrics educator*

Kyoto
Araki, Takeo *chemistry educator*
Einaga, Yoshiyuki *chemist*
Fujita, Eiichi *educator emeritus*
Fukui, Kenichi *chemist*
Hanai, Toshihiko *chemist*
Hayashi, Takao *mathematics educator, historian, Indologist*
Iwashimizu, Yukio *mechanical engineering educator*
Kawabata, Nariyoshi *chemistry educator*
Kitagawa, Toshikazu *chemistry educator*
Kobayashi, Naomasa *biology educator*
Kunugi, Shigeru *scientist, researcher, educator*
Makigami, Yasuji *transportation engineering educator*
Mori, Shigejumi *mathematician educator*
Murazawa, Tadashi *mathematics educator*
Nagao, Makoto *electrical engineering educator*
Oku, Akira *chemical engineer, educator*
Saegusa, Takeo *polymer scientist*
Saito, Isao *chemist*
Sakurai, Akira *nuclear engineer*
Seki, Hiroharu *political science educator, researcher*
Shibayama, Mitsuhiro *materials science educator*
Tachibana, Akitomo *chemistry educator*
Tachiwaki, Tokumatsu *chemistry educator*
Tamura, Imao *retired engineering educator*
Watanabe, Yoshihito *chemistry educator*
Yamakawa, Hiromi *polymer chemist, educator*
Yamana, Shukichi *chemistry educator*

Matsue
Kamiya, Noriaki *research mathematician*

Matsuyama
Fukui, Yasuyuki *psychology educator*
Yano, Tadashi *physics educator*

Meguro-ku
Nakayama, Wataru *engineering educator*
Sakamoto, Munenori *engineer educator, researcher, chemist*

Miki-Iyo
Fujime, Yukihiro *horticultural science educator, researcher*

Minamata-shi
Inoue, Takeshi *psychiatrist*

Minato-ku
Sawada, Hideo *polymer chemistry consultant, chemist*

Sekimoto, Tadahiro *electronics company executive*

Mishima
Nakatsuji, Norio *biologist*

Mitaka
Fukuzumi, Naoyoshi (Hai-chin Chen) *pathology educator*
Kaifu, Norio *astronomer*

Mito
Misawa, Susumu *physicist, educator*

Miyagi
Miyajima, Hiroshi *aerospace engineer*
Nakazawa, Mitsuru *ophthalmologist, educator*

Miyakonojo
Eto, Morifusa *chemistry educator*

Muroran
Fu, Yuan Chin *chemical engineering educator*

Nagano
Ito, Kentaro *electrical engineering educator*
Matsuda, Yasuhiro *computer scientist*

Nagaoka
Matsuno, Koichiro *biophysics educator*

Nagasaki
Ariyoshi, Toshihiko *toxicologist, educator*
Hamada, Keinosuke *chemistry educator emeritus*

Nagoya
Abe, Yoshihiro *ceramic engineering educator*
Ando, Shigeru *biology educator*
Aoki, Keizo *chemistry educator*
Esaki, Toshiyuki *pharmaceutical chemist*
Fukui, Yasuo *astronomer*
Hida, Takeyuki *mathematics researcher, educator*
Kato, Michinobu *chemistry educator*
Matsuzaki, Yuji *aerospace and biomechanics educator, researcher*
Mori, Shigeya *economist, educator*
Murakami, Edahiko *chemistry educator*
Nitta, Kyoko *aerospace engineering educator*
Ohta, Hirobumi *aeronautical engineer, educator*
Okamoto, Yoshio *chemistry educator, researcher*
Shioiri, Takayuki *pharmaceutical science educator*
Shoji, Eguchi *organic chemistry educator*
Takagi, Shigeru *chemistry educator*
Tsuge, Shin *chemistry educator*
Yoshida, Tohru *science and engineering educator*

Nankoku
Takeshima, Yoichi *psychology educator*

Nara
Hayashi, Tadao *engineering educator*
Shigesada, Nanako *mathematical biology educator, researcher*

Narashino
Inazumi, Hikoji *chemical engineering educator*

Neyagawa
Motoba, Toshio *physics educator*

Niigata
Asakura, Hitoshi *internal medicine educator*
Satsumabayashi, Koko *chemistry educator*

Niimi-shi
Tanabe, Yo *chemistry researcher*

Nishihara
Tawata, Shinkichi *agricultural engineering educator*

Nishiku
Mataga, Noboru *scientist*

Nishinomiya
Imamura, Tsutomu *physicist, educator*
Saito, Yoshitaka *biochemistry educator*
Takeda, Yasuhiro *chemistry educator*

Noda
Kwon, Glenn S. *pharmaceutical scientist*
Suzuki, Taira *physics educator*
Takenaka, Tadashi *electrical engineer, educator*

Obihiro
Goto, Ken *chronobiologist, educator*

Ohbu
Higashida, Yoshisuke *mechanical engineer*

Oita
Honda, Natsuo *anesthesiologist, educator*
Kimura, Haruo *aeronautical engineering educator*
Yamauchi, Kazuo *systems engineer, researcher*

Okayama
Hamada, Hiroki *science educator*
Nagata, Hiroshi *psycholinguistics educator*
Oda, Takuzo *biochemistry educator*
Okada, Shigeru *pathology educator*
Torii, Sigeru *chemistry educator*
Ubuka, Toshihiko *biochemistry educator*
Yasue, Kunio *physicist, applied mathematician*

Okazaki
Mitsuke, Koichiro *chemistry educator*
Yoshihara, Keitaro *chemistry educator*

Okinawa
Higa, Tatsuo *marine science educator*
Noda, Yutaka *physician, otolaryngologist*
Sakanashi, Matao *pharmacology educator*
Tako, Masakuni *chemistry researcher and educator*

Okubo
Nishimura, Susumu *biologist*

Osaka
Aoki, Ichiro *theoretical biophysics educator, researcher*
Fueno, Takayuki *chemistry educator*
Horiuchi, Atsushi *physician, educator*
Ikeda, Kazuyosi *physicist, poet*
Iwata, Kazuaki *manufacturing engineer educator*
Iwata, Kazuo *business executive*
Iyoda, Mitsuhiko *economics educator*
Kinoshita, Shigeru *ophthalmologist*
Kitano, Kazuaki *microbiologist, researcher*
Kobayashi, Mitsue *chemistry educator*
Kuwata, Kazuhiro *chemist*
Masuo, Ryuichi *mechanical engineering educator*
Matsumura, Fumitake *economics educator*
Murai, Shinji *chemistry educator*
Nakagawa, Yuzo *laboratory administrator*
Nakai, Hiroshi *civil engineering educator*
Nakazato, Hiroshi *molecular biologist*
Nishiyama, Toshiyuki *physics educator*
Nishizawa, Teiji *computer engineer*
Oka, Kunio *chemistry educator*
Okawa, Yoshikuni *computer science educator*
Osumi, Masato *utility company executive*
Sakaguchi, Genji *food microbiologist, educator*
Sakamoto, Ichitaro *oceanologist, consultant*
Shimbo, Masaki *technology company executive*
Sonogashira, Kenkichi *chemistry educator*
Suga, Hiroshi *chemistry educator*
Sugawara, Tamio *chemist*
Susumu, Kamata *organic and medicinal chemist*
Takagi, Shinji *economist, educator*
Takino, Masuichi *physician*
Ueda, Einosuke *physician*
Umeki, Shigenobu *physician, researcher*
Watanabe, Toshiharu *ecologist, educator*
Yonetani, Kaoru *chemist, consultant*
Yoneyama, Hiroshi *chemistry educator*
Yoshida, Kunihisa *chemistry educator, researcher*

Ota-ku
Sano, Keiji *neurosurgeon*

Otsu
Ando, Takashi *chemistry educator*
Matsushita, Keiichiro *sociology educator*
Matsuura, Teruo *chemistry educator*
Ohara, Akito *biology educator*
Takemoto, Kiichi *chemistry educator*
Wada, Eitaro *biogeochemist*

Sagamihara
Fujii, Kozo *engineering educator*
Okui, Kazumitsu *biology educator*
Sakai, Kunikazu *chemistry researcher*

Saitama
Sugawara, Isamu *molecular pathology educator*

Sakai-Gun
Ise, Norio *chemistry educator*

Sakura
Yamada, Ryuzo *technology company executive*

Sakyoku
Yamamoto, Shigeru *educator*

Sambu-Gun
Morishige, Fukumi *surgeon*

Sapporo
Fukuda, Morimichi *medical educator*
Giga, Yoshikazu *mathematician*
Hasegawa, Hideki *electrical engineering educator*
Iwamoto, Masakazu *chemistry educator*
Maeno, Norikazu *geophysics educator*
Mizutani, Junya *chemist, educator*
Okada, Fumihiko *psychiatrist*

Sendai
Abiko, Takashi *peptide chemist*
Hiroyuki, Hashimoto *engineering educator*
Niitu, Yasutaka *pediatrician, educator*
Oikawa, Atsushi *pharmacology educator*
Oikawa, Hiroshi *materials science educator*
Okuyama, Shinichi *physician*
Shoji, Sadao *soil scientist*
Sone, Toshio *acoustical engineering educator*
Takahashi, Kazuko *organic chemistry researcher, educator*
Yasumoto, Takeshi *chemistry educator*

Setagaya-ku
Iijima, Shigetaka *astronomy educator*

Seto
Rin, Zengi *economic history educator*

Shibuya-ku
Torii, Shuko *psychology educator*

Shimizu
Isono, Kiyoshi *biochemistry educator*

Shimotsuga
Ichimura, Tohju *pediatrician, educator*

Shinagawa-ku
Kittaka, Atsushi *chemist*
Shiozaki, Masao *synthetic and organic chemist*

Shinjyuku-ku
Honami, Shinji *mechanical engineer educator*
Takeoka, Shinji *chemist*

Shiso-Gun
Domae, Takashi *cereal chemist*

Shizuoka
Isemura, Mamoru *biochemist, educator*
Shirai, Yasuto *information science educator, researcher*
Tabata, Yukio *engineering researcher*

Showa
Einaga, Hisahiko *chemistry educator*

Suita
Amino, Nobuyuki *endocrinologist, educator*
Hayaishi, Osamu *director science institute*
Iwatani, Yoshinori *physician, educator, researcher*
Masuhara, Hiroshi *chemist, educator*
Murata, Yasuo *economist, educator*
Shimazaki, Yasuhisa *cardiac surgeon*
Shioya, Suteaki *biochemical engineering educator*

Takahashi
Matsui, Eiichi *economics and sociology educator*

Takaishi
Yamamura, Kazuo *chemist*

Takarazuka
Imado, Fumiaki *control engineer*
Miyakado, Masakazu *biochemist, chemist*

Tama-shi
Kawano, Hiroshi *computer science and aesthetics educator*

Tenri
Koizumi, Shunzo *surgeon*

Tochigi
Iida, Shuichi *educator, physicist*
Ishizaki, Tatsushi *physician, parasitologist*
Takasaki, Etsuji *urology educator*

Togane
Uchiyama, Shoichi *mechanical engineer*

Tokai-mura
Yamaguchi, Masafumi *researcher*

Tokonaka
Tashiro, Kohji *macromolecular scientist*

Tokorozawa
Hashimoto, Nobuyuki *electro-optical engineer*
Nakamura, Hiroshi *urology educator*

Tokushima
Kaneshina, Shoji *biophysical chemistry educator*
Nishimoto, Nobushige *pharmacognosy educator*
Oshiro, Yasuo *medicinal chemist*
Tori, Motoo *chemist, educator*

Tokyo
Aihara, Kazuyuki *mathematical engineering educator*
Aoki, Junjiro *chemical engineer*
Aoki, Masamitsu *chemical company executive*
Aoyama, Hiroyuki *structural engineering educator*
Arima, Akito *academic administrator*
Azuma, Takamitsu *architect, educator*
Chang, Chia-Wun *chemist, researcher*
Chiba, Kiyoshi *chemist*
Crist, B. Vincent *chemist*
Emori, Richard Ichiro *engineering executive*
Fujii, Hironori Aliga *aerospace engineer, educator*
Fukai, Yuh *physics educator*
Fukuyama, Yukio *child neurologist, pediatrics educator*
Furuya, Tsutomu *plant chemist and biochemist, educator*
Hamada, Nobuhiro A. *electrical engineer, researcher*
Harada, Yoshiya *chemistry educator*
Hashimoto, Kunio *architect, educator*
Hatanaka, Hiroshi *neurosurgeon*
Hayashi, George Yoichi *airline executive*
Hayashi, Taizo *hydraulics researcher, educator*
Hayashi, Yoshihiro *architect*
Hideaki, Okada *information systems specialist*
Higashiguchi, Minoru *aerospace electronic engineering educator*
Hirota, Jitsuya *reactor physicist*
Honda, Hiroshi *engineer, energy economist*
Hori, Yukio *scientific association administrator, emeritus engineering educator*
Ibayashi, Tsuguio *legal educator*
Iida, Yukisato *lawyer*
Imai, Noriyoshi *chemist, researcher*
Inada, Tadahico *mechanical engineer*
Iri, Masao *mathematical engineering educator*
Ishii, Akira *medical parasitologist, malariologist, allergist*
Ishii, Yoshinori *geophysics educator*
Ishiwa, Sadao Chigusa *geneticist, educator*
Iwakura, Yoshio *chemistry educator*
Iwasaki, Toshio *hydraulics engineer, investigator*
Iwashita, Takeki *physics educator, researcher*
Joh, Yasushi *science administrator, chemist*
Kamiya, Yoshio *research chemist, educator*
Kawahara, Mutsuto *civil engineer, educator*
Kawamata, Motoo *chemical company executive*
Kigoshi, Kunihiko *geochemistry educator*
Kinoshita, Tomio *economics educator*
Kitahara, Shizuo *allergist*
Kitani, Osamu *agriculture educator*
Kitazawa, Koichi *materials science educator*
Kobayashi, Susumu *data processing executive, super computer consultant*
Koshi, Masaki *engineering educator*
Kozai, Yoshihide *astronomer*
Kurita, Chushiro *writer, engineering educator, researcher*
Kurokawa, Kisho *architect*
Kurusu, Yasuhiko *chemistry educator*
Maekawa, Mamoru *computer science educator*
Maki, Atsushi *economics educator*
Masuda, Gohta *physician, educator*
Matsui, Iwao *chemical engineer*
Matsumoto, Kazuko *chemistry educator*
Miura, Tanetoshi *acoustics educator*
Miyake, Akira *physics educator*
Morishita, Etsuo *aeronautical, mechanical engineering educator*
Muramatsu, Ichiro *chemistry educator*
Musha, Toshimitsu *physicist, educator*
Nagai, Tsuneji *pharmaceutics educator*
Nagasaka, Kyosuke *chemical engineer*
Nagasawa, Yuko *aerospace psychiatrist*
Nagata, Minoru *electronics engineer*
Nakagaki, Masayuki *chemist*
Nakahara, Masayoshi *chemist*
Nishi, Kenji *electrical engineer*
Noguchi, Shun *chemistry educator*
Nomura, Shigeaki *aerospace engineer*
Nozoe, Tetsuo *organic chemist, research consultant*
Odawara, Ken'ichi *economist, educator*
Ohe, Shuzo *chemical engineer, educator*
Ohta, Hiroshi *economics educator*
Okitani, Akihiro *food chemistry educator*
Onishi, Akira *economics educator*
Onishi, Ryoichi *aeronautical engineer, designer*
Ori, Kan *political science educator*
Otsuka, Kanji *information science educator*
Ozawa, Keiya *hematologist, researcher*
Ranney, Maurice William *chemical company executive*
Saima, Atsushi *science and engineering educator*
Saitoh, Tadashi *electrical engineering educator*
Sakai, Katsuo *electrophotographic engineer*
Sakata, Kimio *aerospace engineer, researcher*
Sakurada, Yutaka *chemist*
Sakurai, Kiyoshi *economics educator*
Sakuta, Manabu *neurologist, educator*
Sato, Noriaki *engineering educator*
Shibata, Akikazu *semiconductor scientist*
Shikata, Jun-ichi *surgery educator*
Shima, Mikiko *polymer scientist*
Shimada, Shinji *dermatology educator, researcher*
Suami, Tetsuo *chemistry educator*
Takahashi, Keiichi *zoology educator*
Takasaki, Yoshitaka *telecommunications scientist, electrical engineer*
Takashio, Masachika *biochemist*
Terao, Toshio *physician, educator*
Torii, Tetsuya *retired science educator*
Tsuchida, Eishun *chemistry educator*
Tsuda, Kyosuke *organic chemist, science association administrator*
Wakimura, Yoshitaro *economics educator*
Watanabe, Kouichi *pharmacologist, educator*
Yaku, Takeo Taira Chichibu Kawagoe *computer scientist, educator*
Yakura, Hidetaka *immunologist*
Yamada, Ryoji *economist*

Tondabayashi
Nozato, Ryoichi *metallurgy educator, researcher*

Toshima-Ku
Furuichi, Susumu *physics researcher*

Tottori
Morishima, Isao *biochemistry educator*

Toyama
Hamada, Jin *biologist*
Hayashi, Mitsuhiko *physics educator*
Matsugo, Seiichi *chemistry educator, chemistry researcher*
Tsuji, Haruo *orthopaedics educator*

Toyohashi
Kobayashi, Toshiro *materials science educator*
Sasaki, Shin-Ichi *university president, microbiologist, educator*

Toyonaka
Ikeuchi, Satoru *astrophysicist, educator*
Kishimoto, Uichiro *biophysicist*
Miyamoto, Sigenori *astronomy educator, researcher*

Toyota
Miyachi, Iwao *electrical engineering educator*

Tsuchiura
Ishikawa, Yuichi *mechanical engineer, researcher*

Tsukuba
Araki, Hiroshi *chemist*
Esaki, Leo *physicist*
Hara, Yasuo *physics educator and researcher*
Hirano, Ken-ichi *metallurgist, educator*
Iguchi, Ienari *physicist, educator*
Imamura, Toru *molecular cell biologist*
Ishimaru, Hajime *physics and engineering educator*
Kato, Daisuke *electronics researcher*
Koga, Tatsuzo *aerospace engineer*
Nannichi, Yasuo *engineering educator*
Onuki, Hideo *physicist*
Takagi, Hideaki *computer scientist, mathematician*
Takahashi, Masayuki *aquatic ecologist*
Yamamoto, Hiro-Aki *toxicology and pharmacology educator*
Yatsu, Kiyoshi *plasma physicist*

Ube
Kageyama, Yoshiro *retired mechanical engineering educator*
Sekitani, Toru *otolaryngologist, educator*

Uji
Hata, Koichi *heat transfer researcher*
Shiotsu, Masahiro *engineering educator*
Yasumoto, Kyoden *food science educator*

Urawa
Hiyama, Tetsuo *biochemistry educator*
Narasaki, Hisatake *analytical chemist*
Rhodes, James Richard *economics educator*

Utsunomiya
Gunzo, Izawa *chemistry educator*

Wako
Ogawa, Tomoya *chemist, veterinary medical science educator*

Yamaguchi
Chiba, Yoshihiko *biology educator*
Suzuki, Nobutaka *chemistry educator*

Yamakita
Mita, Itura *polymer chemist*

Yamanashi-ken
Onaya, Toshimasa *internal medicine educator*

Yamazaki
Shiraiwa, Kenichi *mathematician*

Yokohama
Aika, Ken-ichi *chemistry educator*
Aizawa, Masuo *bioengineering educator*
Asawa, Tatsurō *chemistry researcher*
Hirota, Minoru *chemistry educator*
Ibata, Koichi *chemistry research manager*
Inoue, Yoshio *engineering educator*
Kaneko, Yoshihiro *cardiologist, researcher*
Kunieda, Hironobu *physical chemistry educator*
Matsuzaki, Takao *chemist*
Momoki, Kozo *analytical chemist, educator*
Niki, Katsumi *chemist, educator*
Ogawa, Seiichiro *chemistry educator*
Oi, Ryu *chemist*

Soetanto, Kawan *biomedical-electrical engineering educator*
Tanaka, Nobuyoshi *consulting engineering*
Toyama, Takahisa *oil company executive, engineer*
Tsuruta, Yutaka *building materials researcher*

Yokosuka
Nakagawa, Kiyoshi *communications engineer*
Sasaki, Masafumi *electrical engineering educator*
Sekioka, Mitsuru *geoscience educator*

Yoshii
Hayashi, Shizuo *physics educator*

JORDAN

Amman
Majdalawi, Fouad Farouk *aeronautical engineer*
Philippi, Edmond Jean *airline executive*

Irbid
Bashir, Nabil Ahmad *biochemist, educator*
Mansour, Awad Rasheed *chemical engineering educator*

KENYA

Eldoret
Patel, Kiritkumar Natwerbhai *researcher, chemist*

Mbita
Khan, Zeyaur Rahman *entomologist*

Nairobi
Leakey, Mary Douglas *archaeologist, anthropologist*
Onyonka, Zachary *federal agency administrator*
Sanchez, Pedro Antonio, Jr. *soil scientist, administrator*
Swallow, Brent Murray *agricultural economist, researcher*

KIRIBATI

Tarawa
Kaitaake, Anterea *minister of education science and technology*

LATVIA

Riga
Silins, Andrejs Roberts *physics educator*

LEBANON

Beirut
Rouayheb, George Michael *scientific research council advisor*

LIBYA

Tripoli
Abuzakouk, Marai Mohammad *civil aviation engineer*

LIECHTENSTEIN

Schaan
Uebleis, Andreas Michael *engineer*

MACAU

Macau
Leong, Mang Su *economist*

Taipa
Li, Yiping (Y.P.) *applied mathematics educator*

MALAYSIA

Johor Bahru
Tan, Yoke San *mechanical engineer*

Kuala Lumpur
Chan, Wan Choon *mining company executive*
Chue, Seck Hong *mechanical engineer*
Egbogah, Emmanuel Onu *petroleum engineer, geologist*
Oguntoye, Ferdinand Abayomi *economist, statistician, computer consultant*
Rajadurai, Pathmanathan *pathologist, educator, consultant*

Penang
Das, Kumudeswar *food and biochemical engineering educator*
Ibrahim, Kamarulazizi *physics educator, researcher*

Petaling
Ong, Boon Kheng *electronics executive*

Selangor
Tan, Chai Tiam *civil engineer, precast concrete company executive*

MARTINIQUE

Fort-de-France
Bucher, Bernard Jean-Marie *immunopathologist, researcher, consultant*

MONACO

Monaco
Kalaidjian, Berj Boghos *civil engineer*

MONGOLIAN PEOPLE'S REPUBLIC

Ulan Bator
Batbayar, Bat-Erdeniin *microbiologist*

NAMIBIA

Kalkfeld
Oelofse, Jan Harm *game rancher, wildlife management consultant*

NAURU

Nauru
Tun, Maung Myint Thein *analytical chemist*

NEPAL

Kathmandu
Poudyal, Sri Ram *economics educator, consultant*

THE NETHERLANDS

Amsterdam
Drenth, Pieter Johan Diederk *psychology educator, consultant*
Dreschler, Wouter Albert *audiologist, researcher*
Fortuin, Johannes Martinus H. *chemical engineer*
Geerlings, Peter Johannes *psychiatrist, psychoanalyst*
Hazewinkel, Michiel *mathematician, educator*
Koetsier, Johan Carel *clinical neurology educator*
Kuyper, Paul *physicist*
Lodder, Adrianus *physics educator*
Newling, Donald William *urological surgeon*
Seal, Michael *physicist, industrial diamond consultant*
Tanenbaum, Andrew Stuart *computer scientist, educator, author*
Van Os, Nico Maria *research chemist*
Wiegman, Lenore Ho *chemist*

Apeldoorn
Cramer, Jacqueline Marian *environmental scientist, researcher*

Arnhem
Sikkema, Doetze Jakob *chemist*

Bunnik
van Dyke, Jacob *civil engineer*

Delft
Citroen, Charles Louis *information scientist, consultant*
Katgerman, Laurens *materials scientist*
Koekoek, Roelof *mathematics educator*
Moulijn, Jacob A. *chemical technology educator*
Peters, Charles Martin *research and development scientist, consultant*
Prasad, Ramjee *telecommunications scientist, electrical engineer*

Den Helder
Van Der Meij, Govert Pieter *physics educator*

Eindhoven
Brussaard, Gerrit *telecommunications engineering educator*
Butterweck, Hans Juergen *electrical engineering educator*
Cowern, Nicholas Edward Benedict *physical scientist*
De Bokx, Pieter Klaas *research chemist*
Eykhoff, Pieter *electrical engineering educator*
Kroesen, Gerrit Maria Wilhelmus *physicist, educator*
Van Santen, Rutger Anthony *catalysis educator*

Enschede
Seshan, Kulathu Iyer *chemistry educator, researcher*

Goor
Bonting, Sjoerd Lieuwe *biochemist, priest*

Groningen
van der Meer, Jan *hematologist*

Haren
Den Otter, Cornelis Johannes *animal physiologist, educator*

Hengelo
Penninger, Johannes Mathieu L. *chemical engineer*

Hilversum
Eppink, Andreas *psychologist*

Huizen
le Comte, Corstiaan *radar system engineer*

Leiden
Bedeaux, Dick *chemist*
den Breejen, Jan-Dirk *computer integrated manufacturing educator*
Lugtenburg, Johan *chemistry educator*
Maassen, Johannes Antonie *chemistry educator, consultant*
Miley, George Kildare *astronomy educator*
Shane, William Whitney *astronomer*
Spaan, Willy Josephus *molecular virologist*

Maastricht
De Baets, Marc Hubert *immunologist, internist*
Janevski, Blagoja Kame *radiologist, educator*
Van Den Herik, Hendrik Jacob *computer science and artificial intelligence educator*
Verheyen, Marcel Mathieu *homoeopathist, consultant*

Muiderberg
Van Waning, Willem Ernst *computer scientist*

Nijmegen
Hermans, Hubert John *psychologist, researcher*

Noordwijk
Caswell, Robert Douglas *aerospace engineer*
Kletzkine, Philippe *aerospace engineer*
Novara, Mauro *space system engineer*
Panin, Fabio Massimo *mechanical engineer*
Racca, Giuseppe Domenico *aerospace engineer*

Petten
Veltman, Arie Taeke *electronic engineer*

Ra
Van Dishoect, Edwine *physicist educator*

Rotterdam
Yap, Kie-Han *engineering executive*

Schiedam
Bray, William Harold *naval architect*

Sittard
Schreuder, Hein *chemical company executive, business administration educator*

Soesterberg
Smoorenburg, Guido Franciscus *biophysicist, educator*

Te Utrecht
Vliegenthart, Johannes Frederik G. *bio-organic chemistry educator*

The Hague
Irvin, George William *economics educator*

Tilburg
Maat, Benjamin *radiation oncologist*

Utrecht
Gooskens, Robert Henricus Johannus *pediatric neurologist*
Inoué, Takao *logician, philosopher*
Muradin-Szweykowska, Maria *physicist*
Van Zutphen, Lambertus F.M. (Bert Van Zutphen) *geneticist, educator*

Wageningen
Coolman, Fiepko *retired agricultural engineer, researcher, consultant*

Zoetermeer
Ritzen, Jozef Maria Mathias *economist*

NEW ZEALAND

Auckland
Marshall, Arthur Harold *architectural engineer*
Schwerdtfeger, Peter Adolf *research chemist*
Wu, Cheng Yi *physicist*

Canterbury
Kulasiri, Gamalathge Don *computer educator, mechanical and agricultural engineer*

Dunedin
Adams, Duncan Dartrey *medical researcher, physician*
Bairam, Erkin Ibrahim *economics educator*
Dodd, John Newton *retired physics educator*

Manurewa
McCreadie, Allan Robert *chemical engineer, consultant*

Palmerston North
Keen, Alan Robert *flavor scientist*
Parton, Kathleen *veterinarian*

Upper Hutt
Atkinson, Paul Henry *cell biologist*

Wellington
Blakeley, Roger William George *engineering executive*

NIGERIA

Ikoyi
Madakson, Peter Bitrus *physicist, engineer*

Ilorin
Adediran, Suara Adedeji *chemistry educator*
Singh, Sardul *physicist*

Kano
Otokpa, Augustine Emmanuel Ogaba, Jr. *research scientist, consultant*

NORTHERN IRELAND

Belfast
Raghunathan, Raghu Srinivasan *aeronautical engineer, educator, researcher*

Dungannon
Peyton, James William Rodney *consultant surgeon*

Lurgan
Johnston, T. Miles G. *broadcast engineer*

NORWAY

As
Thue-Hansen, Vidar *physics educator*

Bergen
Kløve, Torleiv *informatics educator*
Nerdal, Willy *research chemist*
Ramirez-Garcia, Eduardo Agustin *biomedical engineer*
Stamnes, Jakob Johan *physicist educator*
Svebak, Sven Egil *psychology educator*

Grimstad
Conway, John Thomas *computational fluid dynamacist, educator*

Hjelset
Hartmann-Johnsen, Olaf Johan *internist*

Honefoss
Pettersen, Bjørn Ragnvald *astronomer, researcher*

Hvasser
Reinert, Erik Steenfeldt *economist, researcher, administrator*

Kjeller
Bjorvatten, Tor Anderson *chemical engineer*

Kristiansand
Csángó, Péter András *microbiologist*

Nesbru
Viktil, Martin *electrical engineer*

Oslo
Barton, Nick *rock mechanics engineer*
Blomhoff, Rune *biochemist educator, researcher*
Brady, M(elvin) Michael *engineer, writer*
Ekeland, Arne Erling *surgeon, educator*
Fagerhol, Magne Kristoffer *immunologist*
Fenstad, Jens Erik *mathematics educator*
Flottorp, Gordon *audiophysicist*
Guttormsen, Magne Sveen *nuclear physicist*
Haug, Roar Brandt *architect*
Hernes, Gudmund *federal official*
Knudsen, Knud-Endre *civil engineer, consultant*
Østrem, Gunnar Muldrup *glaciologist*
Retterstol, Nils *psychiatrist*
Singh, Devendra Pal *polymer scientist*
Skjaerstad, Ragnar *electronics company executive*
Waaler, Bjarne Arentz *physician, educator*

Sandvika
Jacob, George (Prasad) *electronics and telecommunications engineer*

Ski
Omland, Tov *physician, medical microbiologist*

Stavanger
Papatzacos, Paul George *mathematical physicist, educator*
Ruoff, Heinz Peter *chemistry educator*

Tromsø
El-Gewely, M. Raafat *biology educator*

Trondheim
Chao, Koung-An *physics educator*
Forssell, Börje Andreas *electronics engineer, educator, consultant*
Roaldset, Elen *geologist*
Rokstad, Odd Arne *chemical engineer*
Saether, Ola Magne *geochemist*
Sundnes, Gunnar *retired marine biology educator*
Svaasand, Lars Othar *electronics researcher*

OMAN

Muscat
Khair, Abul *chemistry educator*

PAKISTAN

Badamibagh
Murtaza, Ghulam *chemist, consultant*

Faisalabad
Irfan, Muhammad *pathology educator*

Gulberg
Haq, Iftikhar Ul *mechanical engineer, consultant*

Hyderabad
Ali, Syed Wajahat *physical chemist, researcher*

Islamabad
Afzal, Mohammad *biologist*
Awan, Ghulam Mustafa *economist, political scientist, educator*
Bahadur, Khawaja Ali *mechanical engineer, consultant*
Qureshi, Iqbal Hussain *nuclear chemist*
Soomro, Ellahi Bukhsh *Pakistani federal minister*

Lahore
Azhar, Barkat Ali *economic adviser, researcher*
Chawla, Lal Muhammad *mathematics educator*
Ghani, Ashraf Muhammad *mechanical engineer, business and engineering consultant*
Majeed, Abdul *mathematics educator*

Larkana
Soomro, Akbar Haider *ophthalmologist, educator*

Multan
Siddiqui, Maqbool Ahmad *engineering consultant and executive*

Peshawar
Qazilbash, Imtiaz Ali *engineering executive, consultant*

Rawalpindi
Malik, Abdul Hamid *engineering executive*

Sheikhupura
Ahmad, Khalil *medical practitioner*

PANAMA

Balboa
Smith, Alan Paul *plant ecologist and physiologist*

Panama
Rognoni, Paulina Amelia *cardiologist*
Tarte, Rodrigo *agriculture educator, researcher, consultant*

PAPUA NEW GUINEA

Port Moresby
Koyyalamudi, Sundarrao *chemistry educator*

PARAGUAY

Asunción
Ferreira Falcon, Magno *economist*
Vera Garcia, Rafael *food and nutrition biochemist*

PERU

Lima
French, Edward Ronald *plant pathologist*
Moesgen, Karl John *electronics engineer, science educator*

THE PHILIPPINES

Batangas
Malabanan, Ernesto Herella *internist*

Binondo
Gan, Felisa So *physician*

Cebu City
Mendoza, Genaro Tumamak *mechanical engineering consultant*

Diffun
Temanel, Billy Estoque *agronomy research director, educator, consultant*

Iloilo
Penecilla, Gerard Ledesma *botany educator*
Tuburan, Isidra Bombeo *aquaculturist*

Makati
Juliano, Petronilo Ochoa *chemist*

Malabon Manila
Pizarro, Antonio Crisostomo *agricultural educator, researcher*

Manila
Benitez, Isidro Basa *obstetrician/gynecologist, oncologist*
Khush, Gurdev Singh *geneticist*
Lim, Joseph Dy *oral surgeon*
Maher, Francis Randolph *engineer, consultant*
Tan, John K. *chemical company executive, educator*
Tangco, Ambrosio Flores *health administrator, surgeon, orthopedist*

Pangasinah
Posadas, Martin Posadas *physician, educator, businessman*

Quezon City
Alarcon, Minella Clutario *physics educator, researcher*
Bisnar, Miguel Chiong *water pollution engineer*
David, Lourdes Tenmatay *librarian*
Javier, Aileen Riego *pathologist*
Solidum, Emilio Solidum *electronics and communications engineer*

POLAND

Bielsko-Biala
Lazar, Maciej Alan *electronics executive, engineer*

Bydgoszcz
Paczkowski, Jerzy *chemistry educator*

Gdańsk
Jagoda, Jerzy Antoni *marine engineer*
Kubale, Marek Edward *computer scientist, educator*

Gliwice
Szewczyk, Pawel *research institute administrator, educator*

Katowice
Kokot, Franciszek Józef *physician*

Kielce
Ripinsky-Naxon, Michael *archaeologist, art historian, ethnologist*

Kraków
Broda, Rafal Jan *physicist*
Jaskula, Marian Józef *chemist*
Kowalczyk, Maciej Stanislaw *obstetrician/gynecologist*
Noga, Marian *mechanical engineer*
Pytko, Stanislaw Jerzy *mechanical engineering educator*

Lodz
Guzek, Jan Wojciech *physiology educator*
Kmieć, Bogumil Leon *embryologist, educator, histologist*
Penczek, Stanislaw *chemistry educator*
Zieliński, Jerzy Stanisław *scientist, electrical engineering educator*

Lublin
Karczmarz, Kazimierz *botany educator*
Kubacki, Krzysztof Stefan *mathematics educator*
Rogalski, Jerzy Marian *biochemist, educator*

Olsztyn
Gazinski, Benon *agricultural economics educator*

Poznan
Dabrowski, Adam Miroslaw *digital and analog signal processing scientist*
Golab, Wlodzimierz Andrzej *biologist, geographer, librarian*
Knapowski, Jan Boleslaw *pathophysiology educator, physician, researcher*
Szulc, Roman Władysław *physician*

Szczecin
Flejterski, Stanislaw *economist, educator, consultant*
Zygmunt, Zieliński *civil engineer, educator*

Warsaw
Koscielak, Jerzy *scientist, science administrator*
Romaniuk, Ryszard Stanislaw *electrical engineering educator, consultant*
Semadeni, Zbigniew Wladyslaw *mathematician, educator*
Tarnecki, Remigiusz Leszek *neurophysiology educator, laboratory director*
Wilczynski, Ryszard Leslaw *economist, educator*
Wiśniewski, Roland *physics educator*
Wrebiak, Andrzej *economic and financial consultant*
Ziabicki, Andrzej Jozef *chemist*

Wroclaw
Kabacik, Pawel *research electrical engineer*
Lugowski, Andrzej Mieczyslaw *electrical engineer*
Staniszewski, Andrzej Marek *surgeon, educator*
Wilgocki, Michal *electrochemist, chemistry educator*

Zabrze
Gwóźdź, Bolesław Michael *physiologist*

Zielona Gora
Gil, Janusz Andrzej *physics educator*

PORTUGAL

Aveiro
Borrego, Carlos Soares *environmental engineering educator*

Braga
Cruz-Pinto, Jose Joaquim C. *chemical and materials engineering educator, researcher*

Coimbra
Antunes, Carlos Lemos *electrical engineering educator*
Cunha-Vaz, Jose Guilherme Fernandes *ophthalmologist*
Varandas, Antonio Joaquim De Campos *chemistry educator*

Lisbon
Campos, Luís Manuel Braga da Costa *mathematics, physics, acoustics and aeronautics educator*
Costa, Luís Chaves da *engineering executive*
De Aguiar, Ricardo Jorge Frutuoso *research geophysicist*
De Almeida, Antonio Castro Mendes *surgery educator*
Martins, Ana Paula *economics educator*
Nogueira e Silva, Jose Afonso *engineering executive*
Pimenta, Gervásio Manuel *chemist, researcher*
Portela, Antonio Gouvea *retired mechanical engineering researcher*
Reis, João Carlos Ribeiro *chemistry educator*
Salvador, Armindo Jose Alves Silva *biochemist*
Santo, Harold Paul *engineer, educator, researcher, consultant, designer*
Soares, Eusebio Lopes *anesthesiologist*
Teixeira da Cruz, Antonio *pharmaceutical company executive*
Villax, Ivan Emeric *chemical engineer, researcher*

Oeiras
Gomes, João Fernando Pereira *chemical engineer*

Porto
Guedes-Silva, António Alberto Matos *economist*
Machado, Adelio Alcino Sampaio Castro *chemistry educator*

QATAR

Doha
Bu-Hulaiga, Mohammed-Ihsan Ali *information systems educator*

REPUBLIC OF KOREA

Andong
Park, Byung-Soo *chemistry educator, dean*

Buk-gu
Han, Oksoo *biochemistry educator*

Cheongju
Kang, Sang Joon *ecologist, educator*

Chinju
Nam, Jung Wan *mathematics educator*
Yoon, Yong-Jin *chemistry educator*

Chonju
Kang, Sung Kyew *medical educator*
Park, Byeong-Jeon *engineering educator*

Daejeon
Kim, Wan Joo *medicinal chemist*
Lee, Sung Taick *aerospace engineer*
Son, Ki Sub *health facility administrator, surgeon*
Whang, Kyu-Young *computer science educator, director*

Daejon
Cho, Chang Gi *polymer scientist*
Kim, Heesook Park *chemist*
Lee, Elhang Howard *physicist, researcher, educator*

Inchon
Kim, Jong Soo *polymer scientist*
Lee, Usik *mechanical engineer, educator*

Iri
Kil, Bong-Seop *biology educator*

Jangseungpo
Woo, Jongsik *mechanical engineer, educator, researcher*

Jochiwon
Lee, Chi-Woo *chemistry educator*

Kyunggi-Do
Minn, Young Key *astronomer*

Pohang
Chung, Sung-Kee *chemistry educator*
Hwang, Woonbong *mechanical engineering educator, consultant*
Lee, Sun Bok *biochemical engineering educator*

Pusan
Sung, Dae Dong *chemistry educator*

Seoul
Baek, Se-Min *plastic surgeon*
Cha, Dong Se *economist, research institute administrator*
Choi, Dae Hyun *precision company executive*
Chung, Hwan Yung *neurosurgeon*
Ha, Sung Kyu *mechanical engineer, educator*
Hong, Yong Shik *aerospace engineer*
Jeung, In-Seuck *aerospace engineering educator*
Kang, Bin Goo *biologist*
Kang, Shin Il *economist*
Kim, Moon-Il *metallurgical engineering educator*
Kim, Myung Soo *chemist, educator*
Kim, Youdan *aerospace engineer*
Ko, Myoung-Sam *control engineering educator*
Lee, Sang-Gak *astronomy educator*
Lee, Young Ki *economist, researcher*
Lee, Young Moo *polymer chemist*
Paik, Young-Ki *biochemist, nuclear biologist*
Park, Sang-Chul *molecular biologist, educator*
Park, Won-Hoon *chemical engineer*
Rhee, Hyun-Ku *chemical engineering educator*
Surh, Dae Suk *engineering consulting company executive*
Yun-Choi, Hye Sook *natural products science educator*

Suwon
Kim, Byung-Dong *molecular biology educator*

Taegu
Lee, Zuk-Nae *psychiatry educator, psychotherapist*
Park, Yong-Tae *chemistry educator*

Taejeon
Bok, Song Hae *biotechnologist, researcher*

Taejon
Kim, Jae Nyoung *chemist*
Kim, Sung Chul *polymer engineering educator*
Lee, Choochon *physics educator, researcher*
Park, O Ok *chemical engineering educator*

Yongsan-Gu Seoul
Kang, Minho *engineering executive*

Yusung
Kim, Jin-Keun *engineering educator*

ROMANIA

Bucharest
Cristescu, Romulus *mathematician, educator, science administrator*
Drăgănescu, Mihai *electronic engineering educator*
Mihaileanu, Andrei Calin *energy researcher*
Milu, Constantin Gheorghe *physicist*
Popa, Petru *federal agency administrator, engineering executive, educator*
Spacu, Petru George *chemistry educator*

Cluj-Napoca
Farkas, György-Miklós *chemist, researcher*

RUSSIA

Barnaul
Karakozov, Sergei Dmitrievich *mathematics educator*

Kaliningrad
Anfimov, Nikolai *aerospace researcher*

Moscow
Anodina, Tatyana Grigoryevna *aviation expert*
Basov, Nikolai Gennadievich *physicist*
Bochkarev, Nikolai Gennadievich *astrophysics researcher*
Bogdanov, Nikita Alexeevich *geology educator*
Boyarchuk, Alexander *astronomer*
Bulgak, Vladimir Borisovich *telecommunications engineer*
Cherenkov, Pavel Alexeyevich *physicist*
Chernyak, Boris Victor *biochemist, researcher*
Feschenko, Alexander *nuclear scientist*
Frank, Ilya Mikhailovich *physicist*
Frolov, Konstantin Vasilievitch *mechanical engineer, science administrator*
Fursikov, Andrei Vladimirovich *mathematics educator*
Galeyev, Albert Abubakirovich *physicist*
Goldanskii, Vitalii Iosifovich *chemist, physicist*
Golitsyn, Georgiy *research institute director*
Gribov, Vladimir N. *physicist*
Hohlov, Yuri Eugenievich *mathematician*
Itskevich, Efim Solomonovich *physicist*
Kefeli, Valentin Ilich *biologist*
Keldysh, Leonid Veniaminovich *physics educator*
Khalatnikov, Isaac Markovich *theoretical physicist, educator*
Kliouev, Vladimir Vladimirovitch *control systems scientist*
Knipper, Andrei Lvovich *geologist, administrator, researcher*
Kotlyakov, Vladimir Michailovich *geographer, glaciologist researcher*
Laverov, Nikolai Pavlovitch *science foundation executive*
Liakishev, Nikolai Pavlovich *metallurgist, materials scientist*
Litvinchev, Igor Semionovich *mathematician, educator*
Lobashev, Vladimir Mikhailovich *physicist*
Maslov, Viktor Pavlovich *mathematician, educator*
Mirzabekov, Andrey Daryevich *molecular biologist*
Mokhov, Oleg Ivanovich *mathematician*
Olevskii, Victor Marcovich *chemical engineer, educator*
Osipov, Yurii *mathematician, mechanical scientist, educator*
Osipyan, Yuri Andreyevich *physicist*
Ossipyan Yuriy, Andrew *physicist, metallurgist, educator*
Ostrovsky, Alexey *geophysicist, researcher*
Platé, Nicolai A. *polymer chemist*
Poglazov, Boris Fedorovich *biochemist, researcher, administrator*
Prokhorov, Aleksandr Mikhailovich *radiophysicist*
Razborov, Alexander A. *mathematician*
Saltykov, Boris Georgievich *economist*
Shestakov, Sergey Vasiliyevich *geneticist, biotechnologist*
Strakhov, Vladimir Nikolayevich *geophysics educator*
Syromiatnikov, Vladimir Sergeevich *aerospace engineer*
Tartakovskiy, Vladimir Alexandrovich *chemist, researcher*
Tatarinov, Leonid Petrovich *science administrator, paleontologist*
Valiev, Kamil Akhmetovich *engineering educator*
Vladimirov, Vasiliy Sergeyevich *mathematician*
Zolotov, Yurii Alexander *chemist*

Nizhniy Novgorod
Gaponov-Grekhov, Andrey Viktorovich *physicist*

Novosibirsk
Aleksandrov, Leonid Naumovitsh *physicist, educator, researcher*
Koshelev, Yuriy Grigoryevich *mathematics educator*

Podolsk
Nikishenko, Semion Boris *chemical research engineer*

Saint Petersburg
Abalakin, Viktor Kuz'mich *astronomer*
Amusia, Miron Ya *physics educator*
Andreev, Vacheslav Mikchaylovitch *physicist*
Faddeev, Ludwig D. *theoretical mathematician*
Faddeyev, Ludvig Dmitriyevich *mathematician, educator*
Finkelshtein, Andrey Michailovich *astronomy educator*
Golant, Victor Evgen'evich *physicist, researcher, educator*
Shevchenko, Sergey Markovich *organic chemist*
Sokolsky, Andrej Georgiyevich *mathematician, academic administrator*
Svidersky, Vladimir Leonidovich *neurophysiologist*
Takhtadzhyan, Armen Leonovich *botanist*

Sverdlovsk
Oshtrakh, Michael Iosifovich *physicist, biophysicist*

Taganrog
Markov, Vladimir Vasilievich *radio engineer*

Troitzk
Letokhov, Vladilen Stepanovich *physicist, educator*

Voronezh
Kostin, Vladimir Alexeevich *mathematics educator*

RWANDA

Kigali
Fox, Emile *physician*

SAUDI ARABIA

Alkhobar
Balasubramanian, Krishna *chemistry educator, consultant*

Dhahran
Al-Afaleq, Eljazi *organic chemistry educator, administrator*
Ali, Mohammad Farhat *chemistry educator*
Espy, James William *chemicals executive*
Fayad, Nabil Mohamed *chemist, researcher*
Hamid, Syed Halim *chemical engineer*
Harruff, Lewis Gregory *industrial chemist*
Milad, Moheb Fawzy *consultant, urologist*
Raju, Krishnam *chemist*
Warne, Ronson Joseph *mathematics educator*

Jeddah
Peeran, Syed Muneer *airline executive*

Jubail
Gray, Kenneth Wayne *chemist, researcher*

Makkah
Muathen, Hussni Ahmad *chemistry educator*

Riyadh
Al-Mohawes, Nasser Abdullah *electrical engineering educator*
Al-Ohali, Khalid Suliman *aeronautical engineer, researcher, industrial development specialist*
Al-Sari, Ahmad Mohammad *data processing executive*
Battistelli, Joseph John *electronics executive*
Chaudhary, Shaukat Ali *plant taxonomist, ecologist*
Deeik, Khalil George *economist, financing company executive*

Ghadia, Suresh Kantilal *chemical engineer*
Isbister, William Hugh *surgeon*
Lewis, Don Alan *civil, sanitary engineer, consultant*
Mousa, Alyaa Mohammed Ali *neurobiologist, researcher*
Moutaery, Khalaf Reden *neurosurgeon*
Tabba, Mohammad Myassar *civil engineer, manager, educator*
Vora, Manhar Morarji *radiopharmaceutical scientist*
Weheba, Abdulsalam Mohamad *immunologist, allergist, consultant*

Sharjah
Ali, Syed Ibrahim *civil engineer*

Unizah
Konsowa, Mokhtar Hassan *mathematics educator*

Yanbu Al Sin
Choudary, Shaukat Hussain *science administrator, chemist*

SCOTLAND

Aberdeen
Cresser, Malcolm Stewart *soil scientist, educator*

Carnoustie
Harris, William Joseph *genetics educator*

Dundee
Black, Sir James (Whyte) *pharmacologist*

East Kilbride
Hann, James *science administrator*

Edinburgh
Boulter, Patrick Stewart *surgery educator*
Chandley, Ann Chester *research scientist, cytogeneticist*
Simone, James Nicholas *physicist*

Glasgow
Cohen, Stephen Douglas *mathematician, educator*
Courtney, James McNiven *chemist*
Sandham, William Allan *electronics and electrical engineer, educator*
Vickerman, Keith *biologist*

Peebles
Hooper, John Edward *retired physicist, researcher*

Saint Andrews
Tanner, Joanne Elizabeth *psychologist, researcher*

Stirling
Cosgrove, Raymond Francis *science company executive*
Kleinpoppen, Hans Johann Willi *physics educator, researcher*

SEYCHELLES

Victoria
Shamlaye, Conrad Francois *health facility administrator, epidemiologist*

SIERRA LEONE

Freetown
Tahilramani, Sham Atmaram *mechanical engineer*

SINGAPORE

Kent Ridge
Wijeysundera, Nihal Ekanayake *mechanical engineering educator, industrial consultant*

Singapore
Adaikan, Ganesan Periannan *pharmacologist, medical scientist*
Chapman, Mark St. John *chemist*
Gan, Leong-Huat *chemist, educator*
Kan, Bill Yuet Him *aeronautical engineer*
Leng, Gerard Siew-Bing *mechanical and production engineering educator*
Liu, Chang Yu *engineering educator*
Mak, Koon Hou *cardiologist, researcher*
Ng, Kheng Siang *cardiologist, educator, researcher*
Ohta, Hideaki *economics and finance researcher*
Sebastian, Marcus *software developer*
Sim, Ah Tee *engineer*
Sundararajan, Narasimhan *electrical engineering educator*
Tan, Kim Leong *pediatrician, neonatologist, medical educator*
Tan, Kok Tin *aerospace engineer*
Taylor, Leland Harris, Jr. *civil engineer*
Tong, Yit Chow *research scientist*
Xu, Guo-Qin *chemist, educator*

SLOVAK REPUBLIC

Bratislava
Daněk, Vladimir *chemistry scientist*
Ertl, Peter *chemist, researcher*

SLOVENIA

Ljubljana
Paulin, Andrej *metallurgical engineer, educator*
Ribaric, Marijan *physicist, mathematician*
Sicherl, Pavle *economics educator, consultant*
Stusek, Anton *mechanical engineer, researcher*
Tancig, Peter *federal official*
Zavrtanik, Danilo *physicist, researcher*

Novo Mesto
Sokolić, Milenko *biotechnologist, researcher*

Škofja Loka
Roblek, Branko *physics educator*

SOUTH AFRICA

Bellville
Gihwala, Dherendra Isver *chemist*

Brooklyn
Smith, Edwin David *electrical engineer*

Cape Town
Berman, Mervyn Clive *biochemist*
Ellis, George Francis Rayner *astronomy educator*

Halfway House
Grieve, Graham Robert *civil engineering executive*

Johannesburg
Kirsten, Hendrik Albertus *geotechnical engineer*
Mendelsohn, Dennis *chemical pathology educator, consultant*
Preston, Charles Brian *orthodontist, school administrator*
Staskun, Benjamin *retired chemistry educator*

Northlands
Stacey, Thomas Richard *geotechnical engineer*

Pretoria
Stevenson, William John *mechanical engineer*
Stumpf, Waldo Edmund *nuclear energy industry executive*

Randburg
Dean, John Francis *astronomer, researcher*

Umtata
Rutherford, John Stewart *chemistry educator*

SPAIN

Alcala de Henares
Saiz, Enrique *chemist, educator*

Arrecife
Jerndal, Jens *holistic medicine educator, health center promoter*

Badajoz
Olivares Del Valle, Francisco Javier *physical chemistry educator*

Barcelona
Balcázar, José Luis *computer scientist, educator*
Barbat, Alex Horia *civil engineer*
Casañ-Pastor, Nieves *chemist, researcher*
Collins, Paul Andrew *industrial designer*
Gatell, Jose Maria *physician*
Geijo, Fernando Antonio *chemist*
Mira Galiana, Jaime Jose Juan *economist, consultant*
Sigalés, Bartomeu *chemical engineering educator*
Subirana, Juan Antonio *polymer chemist, educator*
Wagensberg, Jorge *physicist*

Bellaterra
Terradas, Jaume *ecologist*

Bilbao
Arias, Pedro Luis *chemical engineering educator, researcher*

Blanes
Lopez de Mantaras, Ramon *computer scientist, researcher*

Burjasot
Garcia, Domingo *mathematics educator*
Miracle, Maria Rosa *ecology educator*

Cordoba
Caridad, Jose Maria *computer scientist, educator*

Galdácano
Collazos Gonzalez, Julio *internist, researcher*

Gijon
Jimenez, Juan Ignacio *electrical engineer, consultant*

Girona
Carbo, Ramon *chemistry educator, researcher*

La Coruna
Nunez-Centella, Ramon Antonio *chemist, science educator, journalist*

Las Arenas
Erce, Ignacio, III *aerospace engineer*

Las Palmas
Saavedra, Juan Ortega *chemical engineering educator*

Lleida
Montoliu, Jesus *nephrologist, educator*

Madrid
Aguilera, Jorge *electrical and solar energy engineer*
Alfonseca, Manuel *computer scientist, educator*
Algora del Valle, Carlos *physics educator, researcher*
Ayala, Juan Alfonso *molecular biology research, biochemistry educator*
Bonilla, Luis Lopez *physics and mathematics educator*
Campo, J. M. *mechanical engineer*
Cárabe López, Julio *physicist*
Carbo-Fite, Rafael *physicist, researcher*
Coma-Canella, Isabel *cardiologist*
Consejo, Eduardo *aeronautics engineer*
Delgado-Barrio, Gerardo *physics educator, researcher*
Dominguez Ortega, Luis *medical educator, health facility administrator*
Galindo, Miguel Angel *economics educator*
Gallego-Juarez, Juan Antonio *ultrasonics research scientist*
Garrido Perez, Mercedes *chemist*
Gil-Loyzaga, Pablo Enrique *neurobiologist, researcher, educator*
González Ureña, Angel *chemistry researcher and educator*
Leon-Sanz, Miguel *physician, endocrinologist, nutritionist*
Martin de Agar, Pilar Maria *ecology educator*
Martinez Sanz, Antonio F. *chemist*
Milla Gravalos, Emilio *industrial engineer*
Moya de Guerra, Elvira *physics educator*
Najera, Rafael *virologist*
Pedregal, Pablo *mathematician, educator*
Rasines, Isidoro *chemist*
Sanchez Sudon, Fernando *renewable energy company executive*
Saura-Calixto, Fulgencio *food chemist, educator*
Velasco Negueruela, Arturo *biology educator*

Manises
Sosa-Riera, Raul *air transportation executive*

Manresa
Castella, Xavier *critical care physician, researcher*

Oviedo
Garcia-Moran, Manuel *surgeon*

Palma Mallo
Petitpierre, Eduard *genetics educator*

Salamanca
Cervero, Jose Maria *physics educator*
Jaraiz, Eladio Maldonado *chemical engineering educator*

San Sebastian
Echenique, Pedro Miguel *physicist, educator*

Santiago
Alonso-Fernandez, José Ramón *chemist*
Díez, José Alberto *business educator*
Nieto-Roig, Juan Jose *mathematician, educator, researcher*

Seville
Ventosa, Antonio *microbiologist, educator*

Tacoronte
Kardas, Sigmund Joseph, Jr. *secondary education educator*

Valencia
Bonet, Jose *mathematics educator, researcher*
Corma, Avelino *chemist, researcher*
Peiro, Jose Maria *psychologist, educator*
Sentandreu, Rafael *microbiologist*

Vitoria
Gracia, Juan Miguel *mathematics educator*
López De Viñaspre Urquiola, Teodoro *engineering company executive*

SRI LANKA

Colombo
Arambewela, Lakshmi Sriyani Rajapakse *chemist*
Fernando, Cecil T. *engineering executive*
Madanayake, Lalith Prasanna *marine engineer*

Hambantota
Abeysundara, Urugamuwe Gamacharige Yasantha *civil engineer*

Nugegoda
Somadasa, Hettiwatte *mathematician, educator*

SWEDEN

Älvsjö
Hausman, Bogumil *computer scientist*
Jauhiainen, Pertti Juhani *software engineer*

Bålsta
Bergström, Sten Rudolf *psychology educator*

Danderyd
Hast, Robert *hematologist*

Fagersta
Vuorinen Ruppi, Sakari Antero *materials scientist*

Göteborg
Andersson, Stig Ingvar *physicist*
Malmcrona-Friberg, Karin Elisabet *microbiologist*
Wallin, B(engt) Gunnar *neurophysiology educator*

Gothenburg
Bona, Christian Maximilian *dentist, psychotherapist*
Hansson, Gunnar Claes *medical biochemist*
Nordén, Bengt Johan Fredrik *chemist*
Wikman, Georg Karl *institution administrator*

Kista
Jacobson, Ivar Hjalmar *data processing executive*

Knivsta
Sigfried, Stefan Bertil *software development company executive, consultant*

Linköping
af Ekenstam, Adolf W. *mathematics educator*
Andersson, Tommy Evert *researcher*
Borch, Kurt Esben *surgeon*
Grubbström, Karl Robert William *economist*
Steinvall, Kurt Ove *physicist*
Willander, Lars Magnus *physics educator*

Lulea
Shahnavaz, Houshang *ergonomics educator*

Lund
Abdulla, Mohamed *physician, educator*
Akselsson, Kjell Roland *environment technology educator, researcher*
Grimmeiss, Hermann Georg *physics educator, researcher*
Johannesson, Thomas Rolf *engineering educator*
Larsson, Anders Lars *anesthesiologist*
Malmberg, Torsten *human ecologist*
Nilsson, Kurt Gösta Ingemar *biotechnology and enzyme technology scientist*
Shinoda, Kōzō *chemistry educator, researcher*
Sundler, Frank Eskil Georg *histology educator, biomedical scientist*
Wadsö, B. Ingemar *science educator*
Wingstrand, Hans Anders *orthopedic surgeon*

Malmö
Andersson, Ulf Gòran Christer *biochemist*
Cronberg, Stig *infectious diseases educator*
Manthorpe, Rolf *medical educator*
Wistrand, Lars Goran *chemist*

Nässjö
Schultz, Per-Olov *occupational health physician, consultant*

Nykoping
Grounes, Mikael *metallurgist*
Kivikas, Töivelemb *physicist, executive*

Östersund
Jonsson, Lars Olov *hospital administrator*

Östhammar
Karlsson, Ingemar Harry *engineer*

Saltsjöbaden
Elvius, Aina Margareta *retired astronomer*

Södertälje
Kock, Lars Anders Wolfram *physician, educator*

Spånga
Abramczuk, Tomasz *computer image processing specialist*

Stockholm
Alfvén, Hannes Olof Gosta *physicist*
Arbin, Astrid Valborg *chemist*
Backenroth-Ohsako, Gunnel Anne Maj *psychologist, educator, researcher*
Bergström, K. Sune D. *biochemist*
Carlson, Per J. *physics educator, academic administrator*
Collins, Vincent Peter *pathologist*
Cronhjort, Bjorn Torvald *systems analyst*
Daneholt, Per Bertil Edvard *molecular geneticist*
Enflo, Anita Margarita *physicist, chemist*
Fernandez-Moran, Humberto *biophysicist*
Fertner, Antoni *electrical engineer*
Frankenhaeuser, Marianne *psychology educator*
Fransson, Torsten Henry *mechanical engineering educator, researcher*
Fuchs, Laszlo Jehoshua *mathematician*
Garegg, Per Johan *chemist, educator*
Glader, Mats Lennart *economics educator, researcher*
Håfors, Aina Birgitta *wood conservation chemist*
Hagson, Carl Allan *utilities executive*
Henriksson, Jan Hugo Lennart *architect, educator*
Jörsäter, Steven Bertil *astronomer, educator*
Lilliehöök, J(ohan) Björn O(lof) *health science association administrator*
Lindqvist, Jens Harry *physician*
Lundquist, Per Birger *physician, medical illustrator, artist, surgeon*
Magneli, Arne *chemist*
McNown, John Stephenson *hydraulic engineer, educator*
Mellstedt, Håkan Sören Thure *oncologist, medical facility administrator*
Möller, Göran *immunology educator*
Neilson, Alasdair Hewitt *microbiologist*
Öfverholm, Stefan *electrical engineer*
Olsson, Nils Urban *research manager*
Palmer, Edward Emery *economics educator*
Petermann, Gotz Eike *mathematics educator, researcher*
Reizenstein, Peter Georg *hematologist*
Samuelsson, Bengt Ingemar *medical chemist*
Sjoberg, Berndt Olof Harald *chemist, research administrator*
Söderström, Hans Tson *economist*
Sundberg, Johan Emil Fredrik *communications educator*
Unckel, Per *minister of education*
Wranglén, Karl Gustaf (Gösta) *chemist*

Strömstad
Vigmo, Josef *retired geriatrician*

Sundsvall
Sarafoglou, Nikias *economist, educator*

Ulricehamn
Alfredsson, Mats Lennart *physics researcher, administrator*

Umeå
Nilsson, Dan Christer *plant ecology educator, researcher*
Sandström, Gunnar Emanuel *microbiologist*

Uppsala
Bäckvall, Jan-Erling *chemist*
Dubiel, Thomas Wieslaw *cardiothoracic surgeon, educator*
Ekelöf, Tord Johan Carl *elementary particle physicist*
Gustavson, Karl-Henrik *physician*
Laurent, Torvard Claude *biochemist, educator*
Öberg, Kjell Erik *physician, educator, researcher*
Taube, Adam A.S. *biometrician, educator*
Vedung, Evert Oskar *political science educator*
Wiklund, K. Lars C. *anesthesiologist, scientist, educator*

Vällingby
Hallberg, Bengt Olof *fiber optic company executive and specialist*

Västerås
Saha, Murari Mohan *electrical engineer*

SWITZERLAND

Aubonne
Rodriguez, Moises-Enrique *industrial engineer*

Baar
Maurer, René *engineer, economist*

Baden
Arnal, Michel Philippe *mechanical engineer, consultant*
Pozzi, Angelo *executive, civil engineer*

Basel
Arber, Werner *microbiologist*
Fattinger, Christof Peter *physicist*
Le-Van, Ngo *research chemist*
Odavic, Ranko *physician, pharmaceutical company executive*
Reichstein, Tadeus *botanist, scientist, educator*
Töglhofer, Wolfgang *medical director, researcher*
von Sprecher, Andreas *chemist*
Wolf, Romain Mathias *theoretical chemist*

Bern
Burri, Peter Hermann *anatomy, histology and embryology educator*
Fleisch, Herbert André *pathophysiologist*
Gabutti, Alberto *physicist*
Kislovski, Andre Serge *electronics engineer*
Reichen, Jürg *pharmacology educator*
Shiner, John Stewart *biophysicist educator*

Buchs
Jütz, Jakob Johann *applied optics engineering educator*

Burgdorf
Haeberlin, Heinrich Rudolf *electrical engineering educator*

Carouge Geneva
Giordan, Andre Jean Pierre Henri *biologist researcher*

Contra
Eccles, Sir John Carew *physiologist*

Dubendorf
Hug, Rudolf Peter *chemist*

Fribourg
Bydzovsky, Viktor *surgeon*
Hatschek, Rudolf Alexander *electronics company executive*

Gais
Langenegger, Otto *hydrogeologist*

Geneva
Bellaiche, Charles Roger *computer company executive*
Berclaz, Theodore M. *chemistry educator*
Bloch, Antoine *cardiologist*
Charpak, Georges *physicist*
Harigel, Gert Günter *physicist*
Helland, Douglas Rolf *intergovernmental organization computer executive*
Héritier, Charles André *physicist, computer systems consultant, educator*
Hofmann, Albert Josef *physicist*
Jowett, John Martin *physicist*
Kündig, Ernst Peter *chemist, researcher, educator*
Lawrence, Roderick John *architect, social science educator, researcher, consultant*
Mueller, Paul *chemist, educator*
Muller, Paul-Emile *electrical engineer*
Parthé, Erwin *crystallographer, educator*
Perrenoud, Jean Jacques *cardiologist, educator*
Piot, Peter *medical microbiologist, epidemiologist*
Polunin, Nicholas *environmentalist, author, editor*
Rabinowicz, Théodore *neuropathology educator*
Rochaix, Jean-David *molecular biologist educator*
Rubbia, Carlo *physicist*
Sethuraman, Salem Venkataraman *economist*
Stahel, Walter Rudolf *industrial analyst, consultant*
Steinberger, Jack *physicist, educator*
Telegdi, Valentine Louis *physicist*
Winter, Klaus H. *physicist educator*
Wunderli, Werner Hans Karl *virologist, researcher*

Grandevent
Karpinski, Jacek *computer company executive*

Grenchen
Siraut, Philippe C. *watch and electronics company executive*

Hinterkappelen
Keller, Laurent *biologist, researcher*

La Rippe
Johnsen, Kjell *accelerator physicist, educator*
Stevens, Prescott Allen *environmental health engineer, consultant*

Lausanne
Borel, Georges Antoine *gastroenterologist, consultant*
Delaloye, Bernard *nuclear medicine physician*
Guillemin, Michel Pierre *occupational hygienist*
Howling, Alan Arthur *physicist, researcher*
Schneider, Wolf-Dieter *physicist, educator*
Thalmann, Daniel *computer science educator*

Le Mont
Bachmann, Fedor Wolfgang *hematology educator, laboratory director*

Locarno
Moresi, Remo P. *mathematician*

Neuchâtel
Neier, Reinhard Werner *chemistry educator*
Schmid, Hans Dieter *civil engineering consultant*

Nyon
Jung, André *internist*

Porrentruy
Chevalier, Jean *physics educator*

Romanel
Ellis, Brian Norman *engineering executive*

Rueschlikon
Müller, Karl Alexander *physicist, researcher*
Rohrer, Heinrich *physicist*

Ruschlikon Zurich
Bednorz, J. Georg *crystallographer*

Schweizerhalle
Hadjistamov, Dimiter *chemical engineer*

Uetikon
Pfenninger, Armin *chemist, researcher*

Villigen
Reddy, Guvvala Nagabhushana *neurochemist*
Rehwald, Walther R. *physicist, researcher*

Wallisellen
Kolbe, Hellmuth Walter *acoustical engineer, sound recording engineer*

Winterthur
Bührer, Heiner Georg *chemist, educator*

Zollikofen
Muntwyler, Urs Walter *electronics engineer*

Zug
Hannema, Dirk *information technology executive*

Zurich
Baladi, Viviane *mathematician*
Dechmann, Manfred *psychotherapist*
Dunitz, Jack David *retired chemistry educator, researcher*
Ernst, Richard Robert *chemist, educator*
Fröhlich, Jürg Martin *physicist, educator*
Geier, Gerhard *chemistry educator*
Giger, Peter *engineer*
Groh, Gabor Gyula *mathematician, educator*
Gruen, Armin *photogrammetry educator*
Hastings, S. Robert *architect, building researcher*
Hauser, Helmut Otmar *biochemistry educator*
Hepguler, Yasar Metin *architectural engineering educator, consultant*
Huston, Rima *chemist*
Kalman, Rudolf Emil *research mathematician, systems scientist*
Klatte, Diethard W. *mathematician*
Laube, Thomas *chemist, researcher*
Mueller, Stephan *geophysicist, educator*
Prelog, Vladimir *chemist*
Quack, Martin *physical chemistry educator*
von Schuller-Goetzburg, Viktorin Wolfgang *economist, consultant*
von Segesser, Ludwig Karl *cardiovascular surgeon*
Weber, Donald Charles *entomologist*
Wüthrich, Kurt *molecular biologist, biophysical chemist*
Zinkernagel, Rolf Martin *immunologist educator*

TAIWAN

Chang-Hwa
Tai, Dar Fu *chemist*

Chung-Li
Chen, Wen-Yih *chemical engineer*
Hong, Zuu-Chang *engineering educator*
Tseng, Tien-Jiunn *physics educator*

Hsinchu
Chen, Chu-Chin *food chemist*
Chen, Lih-Juann *materials science educator*
Chou, Lih-Hsin *materials science and engineering educator*
Huang, Yang-Tung *electronics educator, consultant*
Hwu, Reuben Jih-Ru *chemistry educator*
Lee, Sanboh *materials scientist*
Liu, Ti Lang *physics educator*
Lu, Tian-Huey *physics educator*
Wu, Wen-Teng *chemical engineer*
Yang, C. C. *biochemistry educator, researcher*

Kaohsiung
Chen, Chen-Tung Arthur *chemistry and oceanography educator*
Hsu, Zuey-Shin *physiology educator*
Lin, Li-Min *dentistry educator*

Keelung
Chien, Yew-Hu *aquaculture educator*

Kwei-San
Wu, Albert M. *glyco-immunochemistry educator*

Lungtan
Cheng, Sheng-San *chemistry research scientist, consultant*
Chern, Jeng-Shing *aerospace engineer*

Nankang
Lin, Fei-Jann *zoology educator, researcher*

Pingtung
Wang, Bor-Tsuen *mechanical engineering educator*

Taichung
Chen, Chih-Ying *mathematics and computer science educator, researcher*
Chi, Hsin *entomology and ecology educator*
Kuo, Tung-Yao *industrial engineering educator*
Lee, Yung-Ming *educator*
Wu, Wen Chuan *microbial geneticist, educator*

Tainan
Chang, Keh-Chin *aerospace engineer, educator*
Chao, Yei-chin *aerospace engineering educator*
Choi, Siu-Tong *aeronautical engineering educator*
Hsiao, Fei-bin *aerospace engineering educator*
Jing, Hung-Sying *aerospace engineering educator*
Kung, Ling-Yang *electronics engineer, educator*
Low, Chow-Eng *chemistry educator*

Taipei
Chang, Chun-hsing *psychologist, educator*
Chen, Yang-Fang *physicist, educator*
Chern, Jenn-Chuan *civil engineering educator*
Chern, Ji-Wang *pharmacy educator*
Chiang, Cheng-Wen *internal medicine educator, physician*
Chou, Tein-chen *economics educator*
Dai, Peter Kuang-Hsun *government official, aerospace executive*
Hwang, Woei-Yann Pauchy *physics educator*
Lee, Tsu Tian *electrical engineering educator*
Liao, Chung Min *agricultural engineer, educator*
Liao, Kevin Chii Wen *data processing executive*
Lin, Yeou-Lin *systems engineer, consultant*
Lin, Ying-Chih *chemistry educator*
Ma, Chueng-Shyang (Robert Ma) *reproductive physiology educator, geneticist*
Pan, Tzu-Ming *biotechnologist*
Su, Yuh-Long *chemistry educator, researcher*
Wang, Dahong *architect, consultant*
Wang, Ling Danny *research scientist*
Wei, Yau-Huei *biomedical research scientist*
Yang, Chin-Ping *chemist, engineering educator*
Yu, Chien-Chih *management information systems educator*

TANZANIA

Dar es Salaam
Paalman, Maria Elisabeth Monica *public health executive*

THAILAND

Bangkok
Himathongkam, Thep *endocrinologist*
Huynh Ngoc, Phien *computer science educator, consultant*
Issaragrisil, Surapol *hematologist*
Petchclai, Bencha *physician, researcher, inventor*
Pongsiri, Nutavoot *chemical engineer*
Punnapayak, Hunsa *microbiologist, educator*
San Juan, German Moral *structural engineer, international consultant*
Singh, Gajendra *agricultural engineering educator*

Khon Kaen
Lumbiganon, Pisake *obstetrician/gynecologist*

TRINIDAD AND TOBAGO

Port of Spain
Williams, Allan Nathaniel *agricultural economist*

Saint Augustine
Osborne, Robin William *civil engineer, educator*

San Fernando
Constance, Mervyn *utility executive*
Sawh, Lall Ramnath *urologist*

Trinidad
Parris, C(harles) Deighton *television engineer, consultant*

TUNISIA

Mahrajane
Ghazali, Salem *linguist, educator*

TURKEY

Ankara
Ertem, Özcan *aeronautical engineer*
Tekelioglu, Meral *physician, educator*

Cankaya
Özmen, Atilla *federal agency administrator, physics educator*

Izmir
Figen, I. Sevki *computer company executive*

Maslak-Istanbul
Bekâroğlu, özer *chemist, educator*

Pendik-Istanbul
Rosenberg, Sherman *program director*

Trabzon
Yavuz, Tahir *engineering educator*

UGANDA

Kampala
Kaddu, John Baptist *parasitologist, consultant*

UKRAINE

Dniepropetrovsk
Kukushkin, Vladimir Ivanovich *aviation engineer, educator*

Kiev
Gamarnik, Moisey Yankelevich *solid state physicist*
Gotovchits, Georgy Olexandrovich *electrical engineer, educator*
Kostyuk, Platon Grigorevich *physiologist*
Martynyuk, Anatoly Andreevich *mathematician*

UNITED ARAB EMIRATES

Al-Ain
Kiwan, Abdul Mageed Metwally *chemistry educator*
Zahreddine, Ziad Nassib *mathematician, educator, researcher*

Dubai
Beg, Mirza Umair *toxicologist*

URUGUAY

Montevideo
Ventura, Oscar Nestor *chemistry educator, researcher*

UZBEKISTAN

Toshkent
Salakhitdinov, Makhmud *mathematics educator*

VENEZUELA

Caracas
Berti, Arturo Luis *sanitary engineer*
Cerrolaza, Miguel Enrique *civil engineer, educator*
Chang-Mota, Roberto *electrical engineer*
Matienzo, Rafael Antonio *computer engineer*
Nakano, Tatsuhiko *chemist, researcher, educator*
Rangel-Aldao, Rafael *biochemist*
Sáez, Alberto M. *physics educator*
Vanegas, Horacio *neurobiology educator, director*

Caripe
Pereira, Jose Francisco *plant physiologist*

Cumana
Velasquez Perez, Jose R. *cardiologist*

WALES

Aberystwyth Dyfe
Roberts, William James Cynfab *physician*

Cardiff
Morris, William Allan *engineer*
Vingoe, Francis James *clinical psychologist*

Clwyd
Nichol, Douglas *geologist*

Swansea
Morgan, Kenneth *civil engineering educator, researcher*
Weatherill, Nigel Peter *mathematician, researcher*
Wilshire, Brian *materials engineering educator, administrator*

YUGOSLAVIA

Belgrade
Kanazir, Dušan *molecular biologist, biochemist, educator*
Miljevic, Vujo I(lija) *physicist, researcher*
Mladenovic, Nikola Sreten *mechanical engineer*
Ocvirk, Andrej *Slovene federal official*
Solaja, Bogdan Aleksandar *chemist, educator*

Novi Sad
Dokic, Petar *chemist, educator*

ADDRESS UNPUBLISHED

Abanero, Jose Nelito Talavera *aerospace engineer*
Abbott, Regina A. *neurodiagnostic technologist, consultant, business owner*
Abela, George Samih *medical educator, internist, cardiologist*
Abella, Isaac David *physicist, educator*
Adelman, Richard Charles *gerontology educator, researcher*
Afsarmanesh, Hamideh *computer science educator, research scientist*
Agar, John Russell, Jr. *school district supervisor*
Ahearne, John Francis *scientific research society director, researcher*
Akasofu, Syun-Ichi *geophysicist*
Albagli, Louise Martha *psychologist*
Alberts, Allison Christine *biologist*
Albrecht, Allan James *computer scientist*
Alcazar, Antonio *electrical engineer*
Alden, Ingemar Bengt *pharmaceuticals executive*
Aldrin, Buzz *former astronaut, science consultant*
Alexander, Jonathan *cardiologist, consultant*
Allen, Burkley *mechanical engineer*
Allen, Charles Eugene *college administrator, agriculturist*
Allen, Leatrice Delorice *psychologist*
Allison, John McComb *aeronautical engineer, retired*
Alpher, Victor Seth *clinical psychologist, consultant*
Al-Qadi, Imad Lutfi *civil engineering educator, researcher*
Altan, Taylan *engineering educator, mechanical engineer, consultant*
Alter, Blanche Pearl *physician, educator*
Alzofon, Julia *laboratory administrator*
Amiel, David *orthopaedic surgery educator*
Ancheta, Caesar Paul *software engineer*
Andrea, Mario Iacobucci *engineer, scientist, gemologist, appraiser*
Andrews, David Charles *energy and power consultant*
Anton, Walter Foster *civil engineer*
Archibald, David William *virologist, dentist*
Arisman, Ruth Kathleen *environmental manager*
Armstrong, Andrew Thurman *chemist*
Armstrong, Elmer Franklin *information systems and data processing educator*
Armstrong, Neil A. *former astronaut*
Arnett, Edward McCollin *chemistry educator, researcher*
Arnold Hubert, Nancy Kay *writer*
Asch, David Kent *medical educator*
Austin, Ralph Leroy *chemicals executive*
Avendano, Tania *software engineer*
Azaryan, Anahit Vazgenovna *biochemist, researcher*

Bacani, Nicanor-Guglielmo Vila civil and structural engineer, consultant, real estate investor
Badinger, Michael Albert computer applications designer, inventor, consultant
Baggerly, John Lynwood nuclear engineer
Bailey, Arthur Emery natural products chemist
Balaban, Edward Paul physician
Balettie, Roger Eugene aerospace engineer
Ball, Edna Marion cardiac rehabilitation nurse educator
Bamberger, Joseph Alexander mechanical engineer, educator
Bangs, Allan Philip computer scientist, software engineer, consultant
Bartholomew, Donald Dekle engineering executive, inventor
Bartley, Thomas Lee electronics engineer
Bascom, Willard Newell research engineer, scientist
Basham-Tooker, Janet Brooks geropsychologist, educator
Basu, Janet Else journalist
Bateson, Mary Catherine anthropology educator
Batson, Susan Catherine chemistry educator
Bauer, Victor John pharmaceutical company executive
Bauman, Jan Georgius Josef cell biologist, histochemist
Baxter, Loran Richard civil engineer
Baym, Gordon Alan physicist, educator
Becker, Brooks management consultant, chemist
Becker, Bruce Carl, II physician, educator
Beckman, Donald A. engineer
Behm, Dennis Arthur engineering specialist
Belgau, Robert Joseph aeronautical engineer
Belman, Anita Leggold pediatric neurologist
Benmark, Leslie Ann chemical company executive
Bercel, Nicholas Anthony neurologist, neurophysiologist
Bergfield, Gene Raymond engineering educator
Berry, Chester Ridlon physicist
Berry, Kim Lamar energy engineer
Berry, Richard Stephen chemist
Bertolett, Craig Randolph mechanical engineer consultant
Binder, Kurt physicist, educator
Bini, Dante Natale architect, industrial designer
Bjorndahl, David Lee electrical engineer
Bjornsson, Johannes pathologist
Black, Kristine Mary physicist
Black, Patricia Jean medical technologist
Blandino, Ramon Arturo psychologist, consultant, researcher
Bloch, Erich electrical engineer, former science foundation administrator
Block, Barbara Ann biology educator
Blum, Gregory Lee civil engineer
Boardman, Rosanne Virginia logistics consultant
Bois, Pierre former medical research organization executive
Bolton, Julia Gooden hospital administrator
Borg, Stefan Lennart psychiatrist, educator
Bose, Anjan electrical engineering educator, researcher, consultant
Bova, Michael Anthony physicist
Boxenhorn, Burton aerospace engineer
Boyd, Edward Lee physicist, computer scientist
Bratton, William Edward electronics executive, management consultant
Brent, Robert Leonard physician, educator
Brimacombe, James Keith metallurgical engineering educator, researcher, consultant
Broadwell, Milton Edward nurse, anesthetist, educator
Brooke, Ralph Ian dental educator, vice provost, university dean
Browder, Felix Earl mathematician, educator
Brown, Barbara S. environmental scientist
Brown, Charles Durward aerospace engineer
Brown, David Edwin retired metallurgist, consultant
Brown, Elizabeth Ruth neonatologist
Brown, Frances Susan scientific publication editor
Brown, Geraldine nurse, freelance writer
Brown, Jerry Milford medical company executive
Brown, Linda Joan psychotherapist, psychoanalyst
Brown, Rhonda Rochelle chemist, health facility administrator
Bruneau, Angus A. engineer
Bryant, Donald Ashley molecular biologist
Buckner, Harry Benjamin civil engineer
Budinski, Kenneth Gerard metallurgist
Bullock, Theodore Holmes biologist, educator
Burdett, Barbra Elaine biology educator
Burhans, Frank Malcolm mechanical engineer
Burns, William Edgar chemical engineer
Butenandt, Adolf Friedrich Johann physiological chemist
Butler, Robert Russell systems engineer
Bynes, Frank Howard, Jr. physician
Cahn, Robert Nathan physicist
Cairns, Theodore LeSueur chemist
Calvert, William Preston radiologist
Camper, Jeffrey Douglas biology educator
Carey, Martin Conrad gastroenterologist, molecular biophysicist, educator
Carpenter, Malcolm Scott astronaut, oceanographer
Carreker, John Russell retired agricultural engineer
Carrera, Rodolfo nuclear engineer
Carta, Franklin Oliver retired aeronautical engineer
Carver, Ron G(eorge) chemical engineer
Casella, Peter F(iore) patent and licensing executive
Castracane, James physicist
Chaikof, Elliot Lorne vascular surgeon
Champlin, William Glen clinical microbiologist-immunologist
Chang, Charles Shing biology educator, researcher
Chao, James Min-Tzu architect
Chastain, Denise Jean process engineer, consultant
Chegini, Nasser cell biology educator, endocrinologist
Chen, Chaozong chemistry educator
Chen, Chung Long software engineer
Chen, Henry systems analyst
Chin, Hong Woo oncologist, educator, researcher
Chin, Robert Allen engineering graphics educator
Chorpenning, Frank Winslow immunology educator, researcher
Chou, David Chih-Kuang analytical chemist
Chu, Benjamin Thomas Peng-Nien chemistry educator
Chu, Shirley Shan-Chi retired educator
Chyu, Jih-Jiang structural engineer, consultant
Claessens, Pierre chemist, researcher
Clapp, Beverly Booker university administrator, accountant
Clarke, Nicholas Charles metallurgical executive, mineral technologist
Clemendor, Anthony Arnold obstetrician, gynecologist, educator
Clements, Michael Craig health services consulting executive, retired renal dialysis technician
Clogg, Clifford Collier statistician, educator
Coffey, Kevin Robert materials scientist, researcher
Cohen, Robert Fadian research engineer
Collins, Eileen Marie astronaut
Collins, Erik psychologist, researcher
Colodny, Edwin Irving lawyer, retired airline executive
Commito, Richard William podiatrist
Compton, W. Dale physicist
Constant, Clinton chemical engineer, consultant
Coohill, Thomas Patrick biophysicist, photobiologist
Cork, Randall Charles anesthesiology educator
Costin, Daniel Patrick mechanical engineer
Cowan, Mary Elizabeth medical technologist
Cox, James Carl, Jr. chemist, researcher, consultant
Crabtree, Robert Howard chemistry educator, consultant
Crain, Danny B. microbiologist
Creelman, Marjorie Broer psychologist
Critchfield, Jeffrey Moore immunology researcher
Critoph, Eugene retired physicist, nuclear research company executive
Crossley, Francis Rendel Erskine engineering educator
Crowley, Joseph Michael electrical engineer, educator
Crowther, Richard Layton architect, consultant, researcher, author, lecturer
Crumpton, Evelyn psychologist, educator
Császár, Ákos mathematician
Cuthbert, Versie chemical engineer
Dahlstrom, Norman Herbert engineering executive
Dalderup, Louise Maria medical and chemical toxicologist, nutritionist
Daley, George Quentin internist, biomedical research scientist
Dalton, Robert Lowry, Jr. petroleum engineer
Dandashi, Fayad Alexander applied scientist
Dao, Larry computer software engineer
Darlington, Julian Trueheart biology educator retired
Davis, Kenneth Earl, Sr. environmentalist
Davoodi, Hamid mechanical engineering educator
Dawson, Gerald Lee engineering company executive
DeAngelis, Thomas P. chemist
Deisenhofer, Johann biochemistry educator, researcher
Delap, Bill Jay engineer, consultant
Del Castillo, Jaime economic planning consultant, educator
DeLuca, John neuropsychologist
Denton, M. Bonner research chemistry educator
DePalma, Ralph George surgeon, educator
De Salva, Salvatore Joseph pharmacologist, toxicologist
de Séguin des Hons, Luc Donald physician, medical biologist
De Sofi, Oliver Julius data processing executive
de Wolff, Frederik Albert toxicology educator
Dicke, Robert Henry educator, physicist
Dickinson, William Richard geologist, educator
Dickson, Paul Wesley, Jr. physicist
Dionigi, Christopher Paul plant physiologist
Dishmon, Samuel Quinton, Jr. computer scientist
Doane, Woolson Whitney internist
Dobbel, Rodger Francis interior designer
Dodson, James Noland geologist
Dombrowski, Anne Wesseling microbiologist, researcher
Dong, Linda Yanling optical engineer
Dorros, Irwin retired telecommunications executive
Doubledee, Deanna Gail software engineer, consultant
Drebus, Richard William pharmaceutical company executive
Drozd, Joseph Duane aerospace engineer
Dryden, Richie Sloan physician
Duarte, Cristobal G. nephrologist, educator
Dudley, Lonnie LeRoy information scientist
Duffy, Martin Edward management consultant, economist
Dugan, Patrick Raymond microbiologist, university dean
Dumarot, Dan Peter electrical engineer
Dunski, Jonathan Frank biologist, educator
Dutson, Thayne R. college dean
Dwyer, Gerald Paul, Jr. economics educator, consultant
Dybvig, Douglas Howard manufacturing executive, researcher
Dziewanowska, Zofia Elizabeth neuropsychiatrist, researcher, physician
Ebeling-Koning, Derek Bram nuclear engineer
Edgar, Thomas Flynn chemical engineering educator
Edwards, Harold Hugh, Jr. aerospace engineer, management consultant
Egan, William Thomas information systems consultant
Eiden, Michael Josef aerospace engineer
Ein, Daniel allergist
Ellis, Michael David aerospace engineer
Ellis, Wayne Enoch nurse, anesthetist, air force officer
Ellner, Paul D. clinical microbiologist
El-Saiedi, Ali Fahmy science foundation administrator
Emery, Mark Lewis civil engineer
Engel, Joanne Netter internist, educator
English, Bruce Vaughan museum director and executive, environmental consultant
Englund, John Arthur research company executive
Enns, Kevin Scott architect, artist
Erteza, Ireena Ahmed electrical engineer
Esterman, Benjamin ophthalmologist
Ewankowich, Stephen Frank, Jr. mechanical engineer
Ewen, H.I. physicist
Fajardo, Julius Escalante plant pathologist
Fang, Joseph Pe Yong chemistry educator
Fanjul, Rafael James, Jr. engineer, mathematician
Fargnoli, Gregory E. safety engineer
Farnsworth, Michael Edward mechanical engineer
Feldman, Jacob Alex electrical engineer
Fellner-Feldegg, Hugo Robert scientific consultant
Ferguson, James Clarke mathematician, algorithmist
Fish, John Perry oceanographic company executive, historian
Fisher, Michael Alan air force officer, mechanical engineer
Fitchett, Vernon Harold retired physician, surgeon, educator
Flor, Loy Lorenz chemist, corrosion engineer, consultant
Flores-Lopez, Auremir microbiologist
Fluss, Harold Shrage engineer
Flynn, Patrick Joesph aerospace engineer
Flynt, Clifton William computer programmer, software designer
Folkens, Alan Theodore microbiologist
Fontana, Alessandro Italian minister of universities and research
Foster, Barbara Melanie microscopist, consultant
Foster, Edward Joseph structural engineer
Foster, Lynn Irma physicist
Foster, Norman Holland geologist
Foulkes, William David psychologist, educator
Fowler, Nancy Crowley government economist
Fox, Michael Wilson veterinarian, animal behaviorist
Frank, Michael Victor risk assessment engineer
Frank, Sanders Thalheimer physician, educator
Frankel, Michael S. telecommunication and automation sciences executive
Franklin, Howard David chemical engineer
Franks, Paul Todd laboratory manager
Frederick, Norman L., Jr. electrical engineer
Frederick, Ronald David aerospace engineer
Fredine, C(larence) Gordon biologist, former government agency official
Freitag, Peter Roy transportation specialist
Frere, Maurice Herbert soil scientist
Fritcher, Earl Edwin civil engineer, consultant
Fromlet, K. Hubert banking economist
Fu, Albert Joseph astrophysicist
Gaitonde, Sunil Sharadchandra computer engineer
Gajdusek, Daniel Carleton pediatrician, research virologist
Galas, David John molecular biology educator, researcher
Gardner, Clyde Edward health care executive, consultant, educator
Garvey, Justine Spring immunochemistry educator, biology educator
Gass, Arthur Edward chemist
Gatlin, Larry Alan pharmaceutical professional
Gault, Donald Eiker planetary geologist
Geiger, Louis Charles electrical engineer, retired
Gerber, Arthur Mitchell chemist
Gerhardt, Jon Stuart mechanical engineer, engineering educator
Giannini, A. James psychiatrist, educator, researcher
Gilbert, Charles D. neurobiologist
Gilbert, Jo psychologist
Gillett, John Bledsoe civil engineer
Gilmartin, Malvern oceanographer
Girouard, Kenneth civil engineer
Glashow, Sheldon Lee physicist, educator
Glass, Thomas Graham, Jr. retired general surgeon, educator
Glassman, Armand Barry physician, pathologist, scientist, educator, administrator
Glassock, Richard James nephrologist
Glover, Everett William, Jr. environmental engineer
Godo, Einar computer engineer
Goffman, Thomas Edward radiobiologist, researcher
Goldberger, Arthur Earl, Jr. industrial engineer, consultant
Goldman, Charles Remington environmental scientist, educator
Goldstein, Walter Elliott biotechnology executive
Goodman, Alan Noel physician
Gopal, Pradip Goolab engineer, environmental consultant
Gordon, Helmut Albert biomedical researcher, pharmacology educator
Gordon, Jacqueline Regina pharmacist
Gordon, Richard Edwards psychiatrist
Gorenberg, Norman Bernard aeronautical engineer, consultant
Govier, George Wheeler petroleum engineer
Graham, Kirsten R. information service executive
Gramines, Versa Rice telecommunications professional, retired
Grandi, Attilio engineering consultant
Grebenc, Joseph D. mechanical engineer, consultant
Green, Donald Ross research psychologist
Greene, Laurence Whitridge, Jr. surgical educator
Greenwood, Frank information scientist
Gretzinger, James engineering consultant
Griffith, Charles Russell nuclear operations consultant
Griffith, Steven Lee research physiologist
Griffitts, James John physician
Grigger, David John chemical engineer
Gross, Seymour Paul retired civil engineer
Grotzinger, Timothy Lee electronics engineer
Groves, Sheridon Hale orthopedic surgeon
Grow, Robert Theodore economist, association executive
Grusky, David Bryan sociology educator
Guéritée, Nicolas endocrinologist
Gusdon, John Paul, Jr. obstetrics/gynecology educator, physician
Guthrie, Robert Val psychologist
Haber, Paul health psychologist, educator
Haberman, Charles Morris mechanical engineer, educator
Hadley, Glen L. electrical engineer
Hakim-Elahi, Enayat obstetrician/gynecologist, educator
Hall, Grace Rosalie physicist, educator, literary scholar
Halliday, William Ross retired physician, speleologist, writer
Halpin, Daniel William civil engineering educator, consultant
Halvorson, Harlyn Odell marine life administrator, biological sciences educator
Hamel, David Charles health and safety engineer
Hanamirian, Varujan mechanical engineer, educator, journalist, publisher
Hancock, John C. pharmacologist
Hansen, Shirley Jean energy consulting executive, professional association administrator
Hanson, Hugh ecologist, educator
Hanton, Scott Douglas physical chemist
Hardin, Clifford Morris retired executive
Harijan, Ram technology transfer researcher
Harris, Martin Stephen aerospace engineering executive
Harris, Shari Lea mathematics educator
Harrison, Barry economics educator, researcher, author
Hasha, Dennis Lloyd research chemist, spectroscopist
Haskin, Larry Allen earth and planetary scientist, educator
Hatano, Sadashi molecular biology educator
Hauben, Manfred physician
Hauser, Julius retired drug regulatory official
He, Xing-Fei physicist
Heckadon, Robert Gordon plastic surgeon
Hedge, Jeanne Colleen health physicist
Heffron, W(alter) Gordon computer consultant
Heinemeyer, Paul Hugh quality assurance specialist
Hemann, Raymond Glenn aerospace research company executive
Hendricks, Leonard D. emergency medicine physician, consultant
Hendrickson, William George business executive
Hermanson, Theodore Harry software engineer
Herrmann, Walter retired laboratory administrator
Hertz, Patricia Ritenour analytical chemist
Herz, George Peter industrial consultant, chemical engineer
Herzfeld, Charles Maria physicist
Hess, Ulrich Edward electrical engineer
Hessen, Margaret Trexler internist, educator
Hewitt, John Stringer nuclear engineer
Hickox, Gary Randolph pulp and paper engineer
Higdon, Charles Anthony elementary education educator
Higginson, Roy Patrick psychologist
Hill, Patrick Ray power quality consultant
Hinton, David Owen electrical engineer
Hinton, Wilburt Hartley, II construction engineering executive
Hirning, Harvey James agricultural engineer, educator
Hirose, Teruo Terry surgeon
Hirsch, Walter economist, researcher
Hirsh, Norman Barry management consultant
Ho, John Wing-Shing chemistry educator, researcher
Hochenegg, Leonhard physician
Hodgson, Kenneth P. mining executive, real estate investor
Hoey, David Joseph aerospace design engineer
Hoffmann, Roald chemist, educator
Holden, Eric George microbiologist
Ho-Le, Ken Khoa software engineer
Holland, Michael James computer services administrator
Holldobler, Berthold Karl zoologist, educator
Holmes, Robert Wayne service executive, consultant, biological historian
Holub, Robert Frantisek nuclear chemist, physicist
Holzman, David Carl journalist
Hopfield, Jessica F. neuroscientist, researcher, administrator
Hopson, Janet Louise writer, educator
Horowitz, Joseph civil engineer
Horton, Wilfred Henry mathematics educator
Horton, William Alan structural designer
Hotes, Robert William cognitive scientist
Howe, John Prentice, III health science center executive, physician
Hu, Ximing engineering educator
Huffman, Henry Samuel building mechanical engineer
Hughson, Mary Helen electrical engineer
Hunt, Ronald Duncan veterinarian, educator, pathologist
Hunter, Richard Grant, Jr. neurologist, executive
Hutchinson, John Woodside applied mechanics educator, consultant
Huvos, Piroska Eva molecular biology researcher, chemistry educator
Huzel, Dieter Karl retired aerospace engineer
Ing, Dean Charles novelist, consultant
Ingle, James Chesney, Jr. geology educator
Inman, Cullen Langdon telecommunications scientist
Inouye, David William zoology educator
Ishida, Andrew neurobiologist
Jacobs, Ralph Raymond physicist
Jahn, Billie Jane nursing educator, consultant
James, Earl Eugene, Jr. aerospace engineering executive
Jayne, Cynthia Elizabeth psychologist
Jeck, Richard Kahr research meteorologist
Jee, Melvin Wie On mechanical engineer, civilian military engineer
Jeffries, Robert Alan physicist
Jennings, Jerry L. psychologist
Jensen, Marvin Eli retired agricultural engineer
Jewell, Thomas Keith civil engineering educator
Jia, Quanxi electrical and material scientist, researcher
Jim, Kam Fook medical writer
John, Leonard Keith aerospace and mechanical engineer
Johnson, Deborah Crosland Wright mathematics educator
Johnson, Robert Andrew ecologist
Johnson, Warren Eliot wildlife ecologist
Johnston, Ralph Kennedy, Sr. aerospace engineer
Jokl, Alois Louis electrical engineer
Jones, David Allan electronics engineer
Jones, George Richard physicist
Jones, Robert William physicist
Jones, Walton Linton internist, former government official
Joseph, Lura Ellen geologist
Jurczyk, Joanne Monica technical analyst
Kaczanowski, Carl Henry podiatrist, educator
Kaliski, Mary psychologist
Kandula, Max mechanical engineer, aerospace engineer
Kane, John Vincent, Jr. nuclear physicist, researcher
Karapostoles, Demetrios Aristides civil engineer
Kasper, Horst Manfred lawyer
Kathman, R. Deedee biologist, educator
Kaufmann, Walter Erwin neurologist, neuropathologist, educator
Kays, Stanley J. horticulturist, educator
Kebblish, John Basil retired coal company executive, consultant
Keigler, John E. aerospace engineer
Keller, George Henry research administrator, consulting biochemist
Kemnitz, Josef Blazek pathologist, scientist
Kenny, Douglas Timothy psychology educator, former university president
Kent, Howard Lees obstetrician/gynecologist
Kent, Jan Georg computer consultant
Kerns, Allen Dennis energy and environmental manager
Kersey, Terry L(ee) astronautical engineer
Kieselmann, Gerhard Maria data processing executive
King, Patricia Ann law educator
Kinzie, Daniel Joseph superconducting technology and mechanical engineer
Klein, Martin ocean engineering consultant
Klement, Haim electronics engineer
Knapp, Mark Israel industrial and management engineer
Knapp, William Bernard cardiologist
Knoll, Florence Schust architect, designer
Knottenbelt, Hans Jorgen economist
Knowles, David Eugene chemist
Knowles, Richard Norris chemist
Kocaoglu, Dundar F. engineering executive, industrial and civil engineering educator
Koelmel, Lorna Lee data processing executive
Köhler, Georges J. F. scientist, immunologist

Kokopeli, Peter Heine space designer
Koltai, Stephen Miklos mechanical engineer, consultant, economist
Koppany, Charles Robert chemical engineer
Kornfeld, Peter internist
Kotha, Subbaramaiah biochemist
Kowlessar, Muriel retired pediatric educator
Krappinger, Herbert Ernst economist
Kravitz, Rubin chemist
Krimm, Martin Christian electrical engineer, educator
Krùus, Harri Kullervo physician
Ku, Thomas Hsiu-Heng biochemical and specialty chemical company executive
Kuesel, Thomas Robert civil engineer
Kuiperi, Hans Cornelis chemical trading company executive
Kuper, George Henry research and development institute executive
Kwan, Henry King-Hong pharmaceutical chemist
Lacerna, Leocadio Valderrama research physician
Lagerlof, Ronald Stephen sound recording engineer
Lambert, Jon Kelly engineer
LaMunyon, Craig Willis biology researcher
Lande, Alexander physicist, educator
Lang, Derek Edward aerospace engineer
Langmore, John Preston biophysicist, educator
Laor, Herzel physicist
Largman, Kenneth strategic analyst, strategic defense analysis company executive
La Rivière, Jan Willem Maurits environmental biology educator
Lasry, Jean-Michel mathematics educator
Latz, John Paul aerospace engineer
Lawrence, Jordan psychologist
Leakey, Richard Erskine paleoanthropologist, museum director
Leder, Philip geneticist, educator
Lee, Lawrence Cho commodities advisor
Lee, Sung Jai medicinal chemist
Leigh, Michael Charles electrical engineer
Leighton, Charles Raymond construction inspector
LeMarbe, Edward Stanley engineering manager, engineer
Lerman, Gerald Steven software company executive
Lerner, Armand acoustical consultant
Letowski, Tomasz Rajmund acoustical engineer, educator
Lewalski, Elzbieta Ewa mechanical engineer, consultant
Lewis, Graham Thomas analytical inorganic chemist
Lichstein, Herman Carlton microbiology educator emeritus
Liekhus, Kevin James chemical engineer
Ligotti, Eugene Ferdinand retired dentist
Lin, Otto Chui Chau materials scientist, educator
Lindquist, Anders Gunnar applied mathematician, educator
Linton, William Sidney marketing research professional
Linz, Anthony James osteopathic physician, consultant, educator
Liskov, Barbara Huberman software engineering educator
Liu, Alan Fong-Ching mechanical engineer
Liu, Young King biomedical engineering educator
Livengood, Timothy Austin astronomer
Livingston, Robert Burr neuroscientist, educator
Lloyd, Joseph Wesley physicist, researcher
Loftin, Karin Christiane biomedical specialist, researcher
Logan, Bruce David physician
Lolis, Elias biomedical researcher
Lonergan, Thomas Francis, III criminal justice consultant
Lorenzino, Gerardo Augusto linguist
Lowy, Israel internist, educator
Loy, Richard Franklin civil engineer
Lu, Yingzhong nuclear engineer, educator, researcher
Ludden, John Franklin financial economist
MacDonald, Stewart Dixon ornithologist, ecologist, biologist
MacDuffee, Robert Colton family physician, pathologist
MacQueen, Robert Moffat solar physicist
Madueme, Godswill C. nuclear scientist, international safeguards agency administrator
Main, Myrna Joan mathematics educator
Maki, Fumihiko architect, educator
Makins, James Edward retired dentist, dental educator, educational administrator
Malluche, Hartmut Horst nephrologist, medical educator
Mangels, Ann Reed nutrition educator, researcher
Mann, Laura Susan aerospace engineer
Marchessault, Robert H. chemical engineer
Marinas, Manuel Guillermo, Jr. psychiatrist
Markle, Douglas Frank ichthyologist, educator
Markoe, Arnold Michael radiation oncologist
Markovich-Treece, Patricia economist, art consultant
Maropis, Nicholas engineering executive
Martin, Charles Raymond chemist, educator
Martin, JoAnne Diodato consulting engineer
Martin, John Brand engineering educator, researcher
Martin, Michael Ray transportation engineer
Martinez, Salvador electronic technology educator
Marwill, Robert Douglas aircraft design engineer
Maskell, Donald Andrew contracts administrator
Materson, Richard Stephen physician, educator
Matheny, Adam Pence, Jr. child psychologist, educator, consultant, researcher
Matossian, Jesse Nerses physicist
Matthes, Howard Kurt computer consultant and researcher
Mattoussi, Hedi Mohamed physicist
Maurer, Robert (Stanley) osteopathic physician
Mazarakis, Michael Gerassimos physicist, researcher
McCutchan, Marcus Gene water utility executive
McDermott, Kevin J. engineering educator, consultant
McDonough, John Glennon project engineer
McEnnan, James Judd physicist
McEntire, B. Joseph mechanical engineer
McGervey, Teresa Ann cartographer
McHugh, Earl Stephen dentist
McKinley, John McKeen retired physics educator
McLean, Ryan John technical service professional
McMartin, Kenneth Esler toxicology educator
McQuarrie, Terry Scott technical executive
McQueen, Rebecca Hodges health care executive, consultant
McQueney, Patricia Ann biologist, researcher
Medzihradsky, Fedor biochemist, educator
Mehta Malani, Hina biostatistician, educator
Melia, Angela Therese biologist, pharmacokineticist
Merenbloom, Robert Barry hospital and medical school administrator
Mershimer, Robert John chemical engineer
Mertz, Walter retired government research executive
Metlay, Michael Peter nuclear physicist
Metsger, Robert William geologist
Meyer, Harold Louis mechanical engineer
Mian, Guo electrical engineer
Michaelis, Elias K. neurochemist
Mickelson, Elliot Spencer quality assurance professional
Middleton, Gerard Viner geology educator
Mignella, Amy Tighe environmental engineering researcher
Miller, Herman Lunden retired physicist
Miller, Merle Leroy retired manufacturing company executive
Milo, Frank Anthony manufacturing company executive
Mil'shtein, Samson semiconductor physicist
Mitchell, Neil Charles geophysicist
Mittal, Manoj aerospace engineer
Moffatt, Hugh McCulloch, Jr. hospital administrator, physical therapist
Moliere, Jeffrey Michael cardio-pulmonary administrator
Moll, David Carter civil engineer
Molloy, William Earl, Jr. systems engineer
Monahan, Edward Charles academic administrator, marine science educator
Monninger, Robert Harold George ophthalmologist, educator
Mooneyhan, Esther Louise nurse, educator
Moore, Charles Willard architect, educator
Moore, James Allan agricultural engineering educator
Moore, Sandra architect, environmental designer, educator
Moores, Anita Jean Young computer consultant
Morgan, Linda Claire industrial engineer
Morishita, Teresa Yukiko veterinarian, consultant, researcher
Morrison, Robert Thomas engineering consultant
Mosjidis, Cecilia O'Hara botanist, researcher
Mott, Sir Nevill (Francis Mott) physicist, educator, author
Mudar, M(arian) J(ean) biologist, environmental scientist
Muranaka, Ken-ichiro biophysicist
Murrell, Kenneth Darwin research administrator, microbiologist
Musmanni, Sergio chemist, researcher
Myers, David Francis chemical engineer
Myers, Eric Arthur physicist
Nagel, Max Richard retired, applied optics physicist
Nagys, Elizabeth Ann environmental issues educator
Namboodri, Chettoor Govindan mechanical engineer
Narayan, K(avassey) Sureswaran physicist
Nason, Dolores Irene computer company executive, counselor, eucharistic minister
Neal, Stephen Wayne electrical engineer
Necula, Nicholas electrical engineering educator, researcher
Nedoluha, Alfred Karl Franz physicist
Neher, Leslie Irwin engineer, former air force officer
Nelson, Peter Edward energy engineer
Neufeld, Murray Jerome aerospace scientist/engineer, consultant
Newton, Jeffrey F. project engineer
Nicholls, Richard Aurelius obstetrician, gynecologist
Nichols, C(laude) Alan mechanical engineer
Nix, Martin Eugene engineer
Noci, Giovanni Enrico electronics engineer
Nocks, Randall Ian systems engineer
Nolte, Marty Dee nuclear power plant training manager
Nyman, Bruce Mitchell electrical engineer
Oakes, Ellen Ruth psychotherapist, health institute administrator
Ochoa, Severo biochemist
O'Connor, Diane Geralyn Ott engineer
Odink, Debra Alida chemist, researcher
Okuda, Kunio emeritus medical educator
Oldham, Timothy Richard physicist
Olstowski, Franciszek chemical engineer, consultant
Oppenheimer, Larry Eric physical chemist
Opperman, Danny Gene packaging professional, consultant
Oprandy, John Jay research scientist, molecular biologist
O'Reilly, John Joseph engineer
Ormasa, John utility executive
Ormiston, Timothy Shawn mechanical engineer
Ortolano, Ralph J. engineering consultant
Oster, Ludwig Friedrich physicist
Page, Philip Ronald chemist
Palade, George Emil cell biologist, educator
Palmer, Ashley Joanne aerospace engineer
Paproski, Ronald James structural engineer
Paris, David Andrew dentist
Park, Jon Keith dentist, educator
Parkinson, William Quillian paleontologist
Parsegian, V. Adrian biophysicist
Partheniades, Emmanuel civil engineer, educator
Pathak, Vibhav Gautam electrical engineer
Patterson, Patricia Lynn applied mathematician, geophysicist, engineer
Paul, Vera Maxine mathematics educator
Paxton, John Wesley electronics company executive
Pearlmutter, Florence Nichols psychologist, therapist
Pearson, Ralph Gottfrid chemistry educator
Peled, Israel electrical engineer, consultant
Penner, Karen Marie civil engineer
Penzias, Arno Allan astrophysicist, research scientist, information systems specialist
Pepper, Dorothy Mae nurse
Perry, George Wilson oil and gas company executive
Peters, Randy Alan scientist
Peterson, Dwight Malcolm chemist
Peterson, Eric Follett engineering executive
Pfalser, Ivan Lewis civil engineer
Pfister, Daniel F. electrical engineer
Phaup, Arthelius Augustus, III structural engineer
Phelps, James Solomon, III astrodynamic engineer
Pick, James Block management and sociology educator
Pierce, Charles Earl software engineer
Pierce, Robert Raymond materials engineer, consultant
Pigott, Melissa Ann social psychologist
Pinkert, Dorothy Minna chemist
Pniakowski, Andrew Frank structural engineer
Polasek, Edward John electrical engineer, consultant
Polkosnik, Walter physicist
Popp, Dale D. orthopedic surgeon
Porter, Irene Rae civil engineer
Post, Laura Cynthia biologist
Pound, Robert Vivian physics educator
Prasad, Satish C(handra) physicist
Prell, Martin mechanical engineer
Presley, Alice Ruth Weiss physicist, researcher
Priester, Gayle Boller engineer, consultant
Pritzker, Andreas Eugen Max physicist, administrator, author
Prusiner, Stanley Ben neurology and biochemistry educator, researcher
Pullin, Jorge Alfredo physics researcher
Pütsep, Peeter Ervin electronics executive
Pytlinski, Jerzy Teodor physicist, research administrator, educator
Quilès, Paul French federal official
Rabó, Jule Anthony chemical research administrator, consultant
Rabon, William James, Jr. architect
Rahman, Khandaker Mohammad Abdur engineering educator
Rahman, Sami Ur environmental engineer
Rakutis, Ruta chemical economist
Ralston, Roy B. petroleum consultant
Ramachandran, Narayanan aerospace scientist
Ramsey, William Dale, Jr. petroleum company executive
Rapin, Charles René Jules computer science educator
Redlich, Robert Walter physicist, electrical engineer, consultant
Redmond, Gail Elizabeth chemical company consultant
Reifsnider, Kenneth Leonard metallurgist, educator
Rendina, George chemistry educator
Rice, Eric Edward technologies executive
Rice, Stuart Alan chemist, educator
Rich, Raphael Z. mechanical engineer, researcher
Richardson, Donald Charles engineer, consultant
Richardson, Jasper Edgar nuclear physicist
Richmond, Julius Benjamin retired physician, health policy educator emeritus
Riedner, Werner Ludwig Fritz retired chemicals executive, industrial consultant
Rihani, Sarmad Albert (Sam Rihani) civil engineer
Riordan, William John manufacturing process designer, consultant
Roberts, Earl John carbohydrate chemist
Roberts, Marie Dyer computer systems specialist
Robertson, Jerry Lewis chemical engineer
Robinson, Bruce Butler physicist
Robinson, David Allen computer engineer
Robinson, Rudyard Livingstone economist, financial analyst
Roby, Christina Yen data processing specialist, instructor
Rockwell, Benjamin Allen physicist
Rogers, Kate Ellen interior design educator
Rohrer, Richard Joseph nuclear engineer
Root, M. Belinda chemist
Rosario, Myra Odette molecular biologist, pharmacist, educator
Rose, James Turner aerospace consultant
Rose, William Cudebec electrophysiologist
Roseig, Esther Marian veterinary researcher
Rosemberg, Eugenia physician, scientist, educator, medical research administrator
Rosenkilde, Carl Edward physicist
Rosenkoetter, Gerald Edwin engineering and construction company executive
Rothfeld, Leonard Benjamin chemical/environmental engineer
Roychoudhuri, Chandrasekhar physicist
Rubin, Vera Cooper research astronomer
Rudert, Cynthia Sue gastroenterologist
Rudzki, Eugeniusz Maciej chemical engineer, consultant
Ruegg, Stephen Lawrence quality engineer, chemist
Rüetschi, Paul electrochemist
Rugge, Hugo Robert physicist
Rutstrom, Dante Joseph chemist
Sadusky, Maria Christine environmental scientist
Saines, Marvin hydrogeologist
St. Cyr, John Albert, II cardiovascular and thoracic surgeon
Samson, John Roscoe, Jr. electrical engineer
Sancaktar, Erol engineering educator
Sanchez Muñoz, Carlos Eduardo physicist
Saneto, Russell Patrick neurobiologist
Sastry, Padma Krishnamurthy electrical engineer, consultant
Sauvage, Lester Rosaire health facility administrator, cardiovascular surgeon
Sawlivich, Wayne Bradstreet biochemist
Sayigh, Laela Suad biologist
Scardera, Michael Paul air force officer
Schaller, John Walter electrical engineer
Scheirman, William Lynn chemical engineer
Schenkel, Susan psychologist, educator, author
Schiess, Klaus Joachim mechanical engineer
Schneider, Eleonora Frey physician
Schreiner, Christina Maria emergency physician
Schwarzschild, Martin astronomer, educator
Schwinn, Donald Edwin environmental engineer
Scott, Amy Annette Holloway nursing educator
Scott, Ian Laurence biomedical engineer
Scott, Larry Marcus aerospace engineer, mathematician
Sebek, Kenneth David aerospace engineer, air force officer
Segre, Diego veterinary pathology educator, retired
Sella, George John, Jr. chemical company executive
Sestini, Virgil Andrew secondary education educator
Shalabi, Mazen Ahmad chemical engineer, educator
Shariff, Asghar J. geologist
Sharp, William Wheeler geologist
Shaw, Mary Ann psychologist
Shepard, Alan Bartlett, Jr. astronaut, real estate developer
Sheppard, William Vernon transportation engineer
Shuart, Mark James aerospace engineer
Shukla, Mahesh structural engineer
Shumick, Diana Lynn computer executive
Shupler, Ronald Steven environmental engineer
Siddayao, Corazon Morales economist, educator
Siegbahn, Kai Manne Börje physicist, educator
Sieloff, Christina Lyne nurse
Sills, Richard Reynolds scientist, educator
Simeón Negrín, Rosa Elena veterinary educator
Simmons, Shalon Girlee electrical engineer
Simon, Melvin I. molecular biologist, educator
Simon, Michele Johanna computer systems specialist
Siyan, Karanjit Saint Germain Singh software engineer
Sjostrand, Fritiof Stig biologist, educator
Skala, Gary Dennis electric and gas utilities executive management consultant
Skinner, James Stanford physiologist, educator
Slaughter, Freeman Cluff retired dentist
Sliger, Rebecca North nuclear engineer
Smelser, Ronald Eugene mechanical engineer
Smith, Bodrell Joer'dan architect, city planner
Smith, Christie Parker operations researcher
Smith, David Mitchell systems and software researcher, consultant
Smith, James Lanning engineering executive
Smith, Joe Mauk chemical engineer, educator
Smith, Mildred Cassandra systems engineer
Smith, Susan Finnegan computer management coordinator
Smoot, George Fitzgerald, III astrophysicist
Snow, W. Sterling biology and chemistry educator
Snyder, Melissa Rosemary biochemist
Solomon, Julius Oscar Lee pharmacist, hypnotherapist
Solomon, Susan chemist, scientist
Somes, Grant William statistician, biomedical researcher
Soroush, Masoud chemical engineer
Southwick, Charles Henry zoologist, educator
Souza Mendes, Paulo Roberto de mechanical engineering educator
Sovde-Pennell, Barbara Ann sonographer
Spencer, William Stewart radiologic technologist
Sponable, Jess M. astronautical engineer, physicist
Sprengnether, Ronald John civil engineer
Spurr, Paul Raymond organic chemist
Stanevich, Kenneth William mechanical design engineer
Starek, Rodger William chemist
Stavely, Homer Eaton, Jr. psychologist educator
Steinberg, Fred Lyle radiologist
Steinert, Leon Albert mathematical physicist
Stelzer, John Friedrich nuclear engineer, researcher
Stephanedes, Yorgos J. transportation educator
Stern, Robin Lauri medical physicist
Stokes, Charles Anderson chemical engineer, consultant
Stone, James Robert surgeon
Strangway, David William university president
Strier, Murray Paul chemist, consultant
Stroud, John Franklin engineering educator, scientist
Stuart, James Davies analytical chemist, educator
Stuart, Sandra Joyce computer information scientist
Studness, Charles Michael economist
Suppes, Patrick statistics, education, philosophy and psychology educator
Swartzlander, Earl Eugene, Jr. engineering educator, former electronics company executive
Synek, M. physics educator, researcher
Tamor, Stephen physicist
Tang, Dah-Lain Almon automatic control engineer, researcher
Tappan, Clay McConnell environmental engineer, consultant
Tate, Manford Ben guided missile scientist, investor
Taylor, Lesli Ann pediatric surgery educator
Teal, Edwin Earl engineering physicist, consultant
Temam, Roger M. mathematician
Temes, Clifford Lawrence electrical engineer
Tenney, Stephen Marsh physiologist, educator
Terris, Susan physician, cardiologist
Thews, Gerhard physiology educator
Thomson, Grace Marie nurse, minister
Thornton, Peter Brittin mechanical engineer
Thuillier, Richard Howard meteorologist
Thuning-Robinson, Claire oncologist
Tice, David Charles aerodynamics engineer
Timmer, David Hart civil engineer
Tirkel, Anatol Zygmunt physicist
Tisdale, Patrick David retired pediatrician
Tokue, Ikuo chemist, researcher
Toledo-Pereyra, Luis Horacio transplant surgeon, researcher, educator
Tolonen, Risto Markus systems software engineer
Tompkins, Laurie biologist, educator
Toner, Walter Joseph, Jr. transportation engineer, financial consultant
Torres, Manuel aerospace engineer
Tourtillott, Eleanor Alice nurse, educational consultant
Tran, Nang Tri electrical engineer, physicist
Tribble, Alan Charles physicist
Trigoboff, Daniel Howard psychologist
Trott, Keith Dennis electrical engineer, researcher
Tsai, Tom Chunghu chemical engineer
Tsai, Wen-Ying sculptor, painter, engineer
Turner, John Freeland foundation administrator, former federal agency administrator, former state senator
Turner, William Oliver, III civil engineer
Tweed, Paul Basset chemical engineer, explosives and suicidology consultant
Tyler, Ewen William John retired mining company executive, consulting geologist
Vanderford, Frank Josire physicist, computer scientist consultant
van der Meer, Simon accelerator physicist
Vanderwalker, Diane Mary materials scientist
Van Winkle, Edgar Walling electrical engineer, computer consultant
Varon, Dan electrical engineer
Varvak, Mark mathematician, researcher
Vary, James Patrick physics educator
Vernon, Sidney physician
Victor, Andrew Crost physicist, consultant, small business owner
Vigfusson, Johannes Orn scientific officer
Vigler, Mildred Sceiford retired chemist
Villarrubia, John Steven physicist
Vogel, H. Victoria psychotherapist, educator
Voldman, Steven Howard electrical engineer
von Kutzleben, Siegfried Edwin engineering consultant
Vook, Frederick Werner electrical engineer
Wald, Francine Joy Weintraub (Mrs. Bernard J. Wald) physicist, academic administrator
Walters, Kenn David scientist, company executive
Wang, Bor-Jenq mechanical engineer
Wang, Ruqing research physicist
Wang, Shih-Liang mechanical engineering educator, researcher
Wardwell, James Charles computer and management consultant
Warfel, John Hiatt medical educator, retired
Warner, Janet Claire software design engineer
Waters, Robert George laser engineer
Watson, Robert Barden physicist
Watson, Stuart Lansing chemist
Weinberg, Steven physics educator
Weingarten, Joseph Leonard aerospace engineer
Weinrich, Stanley David chemical engineer
Weiss, Michael James chemistry educator
Weissmann, Heidi Seitelblum radiologist, educator
Welber, Irwin research laboratory executive
Welsh, Elizabeth Ann immunologist, research scientist
Werbach, Melvyn Roy physician, writer
Wessel, Morris Arthur pediatrician
West, Jack Henry petroleum geologist
West, William Ward aerospace engineer

Professional Index Listings

Professional Index

ARCHITECTURE & DESIGN

UNITED STATES

Philadelphia
Dementis, Katharine Hopkins *interior designer*

Pittsburgh
Becherer, Richard John *architectural educator*

SOUTH CAROLINA

Anderson
Pflieger, Kenneth John *architect*

Spartanburg
Stroup, David Richard *architect*

TENNESSEE

Arlington
Van Horn, Mary Reneé *interior designer*

Knoxville
Morton, Terry Wayne *architect*

TEXAS

Bryan
Kellett, William Hiram, Jr. *retired architect, engineer, educator*

Dallas
Dean, Gary Neal *architect*
Hudson, Courtney Morley *interior landscape designer*
Shaw, Dean Alvin *architect*
Stacy, Dennis William *architect*

Houston
Pierce, George Foster, Jr. *architect*

Irving
Rees, Frank William, Jr. *architect*

VERMONT

Brattleboro
Gorman, Robert Saul *architect*

Vergennes
Harris, Martin Sebastian, Jr. *architect*

VIRGINIA

Arlington
Gentner, Paul LeFoe *architect, consultant*

Blacksburg
Stern, E. George *architect, educator*

Herndon
Kagan, Marvin Bernard *architect*

Sterling
Hansen, Alan Lee *architect*

Woodbridge
Peck, Dianne Kawecki *architect*

WASHINGTON

Bellevue
Roselle, Richard Donaldson *industrial, marine and interior designer*

Hansville
Griffin, DeWitt James *architect, real estate developer*

Kirkland
Mitchell, Joseph Patrick *architect*
Steinmann, John Colburn *architect*

Ocean Shores
Morgan, Audrey *architect*

Pullman
Carper, Kenneth Lynn *architect, educator*

Seattle
Castanes, James Christopher *architect*
Thiry, Paul *architect*

Spokane
Stone, Michael David *landscape architect*

WISCONSIN

Kenosha
Potente, Eugene, Jr. *interior designer*

Milwaukee
Moore, Gary Thomas *aerospace architect*
Wilson, Walter Leroy *architect*

TERRITORIES OF THE UNITED STATES

VIRGIN ISLANDS

Saint Croix
Pierce, Lambert Reid *architect*

CANADA

MANITOBA

Winnipeg
Petersmeyer, John Clinton *architect*

NEWFOUNDLAND

Saint John's
Datta, Indranath *naval architect, educator*

ONTARIO

Ottawa
Atif, Morad Rachid *architect*

Toronto
Orlowski, Stanislaw Tadeusz *architect*

MEXICO

Monterrey Nuevo Leon
Garcia Martinez, Ricardo Javier *architect*

BELGIUM

Leuven
Van Geyt, Henri Louis *architect*

CHINA

Hangzhou
Wu, Shuo-Xian *acoustics and architecture educator, scientist*

ENGLAND

Sussex
Bunyard, Alan Donald *designer, inventor*

FRANCE

Paris
Fournier, Jean Pierre *architect, real estate developer*

GERMANY

Braunschweig
Voss, Werner Konrad Karl *architect, engineer*

Cologne
Teodorescu, George *industrial designer, educator, consultant*

Munich
Bergsteiner, Harald *architect*

JAPAN

Tokyo
Azuma, Takamitsu *architect, educator*
Hashimoto, Kunio *architect, educator*
Hayashi, Yoshihiro *architect*
Kurokawa, Kisho *architect*

THE NETHERLANDS

Schiedam
Bray, William Harold *naval architect*

NORWAY

Oslo
Haug, Roar Brandt *architect*

SPAIN

Barcelona
Collins, Paul Andrew *industrial designer*

SWEDEN

Stockholm
Henriksson, Jan Hugo Lennart *architect, educator*

SWITZERLAND

Geneva
Lawrence, Roderick John *architect, social science educator, researcher, consultant*

Zurich
Hastings, S. Robert *architect, building researcher*

TAIWAN

Taipei
Wang, Dahong *architect, consultant*

ADDRESS UNPUBLISHED

Bini, Dante Natale *architect, industrial designer*
Chao, James Min-Tzu *architect*
Crowther, Richard Layton *architect, consultant, researcher, author, lecturer*
Dobbel, Rodger Francis *interior designer*
Enns, Kevin Scott *architect, artist*
Horton, William Alan *structural designer*
Knoll, Florence Schust *architect, designer*
Kokopeli, Peter Heine *space designer*
Maki, Fumihiko *architect, educator*
Moore, Charles Willard *architect, educator*
Moore, Sandra *architect, environmental designer, educator*
Rabon, William James, Jr. *architect*
Rogers, Kate Ellen *interior design educator*
Smith, Bodrell Joer'dan *architect, city planner*
Williams, John Howard *architect, retired*

COMMUNICATIONS: SCIENCE MEDIA

UNITED STATES

ALABAMA

Huntsville
Powers, Clifford Blake, Jr. *communications researcher*
Sharpe, Mitchell Raymond *science writer*

ARIZONA

Flagstaff
Hammond, Howard David *retired editor*

Tempe
Greenfield, James Dwight *technical writer*

Tucson
Kingery, William David *ceramics and anthropology educator*

CALIFORNIA

Arcadia
Sloane, Beverly LeBov *educational writer, consultant*

Camarillo
Alexander, John Charles *editor, writer*

Camp Connell
Soule, Thayer *documentary film maker*

Los Angeles
McLaughlin, John *production company technical director*

Pescadero
Nelson, Kay Yarborough *author, columnist*

Playa Del Rey
Kaelin, Barney James *technological artist*

San Diego
Prescott, Lawrence Malcolm *medical and health science writer*

San Francisco
Eastwood, Susan *medical scientific editor*
Schweickart, Russell L. *communications executive, astronaut*

Santa Clara
Edwards, George Henry *technical writer*

COLORADO

Aurora
McClendon, Irvin Lee, Sr. *technical writer, editor*

CONNECTICUT

Greenwich
Burnham, Virginia Schroeder *medical writer*

DISTRICT OF COLUMBIA

Washington
Adams, Robert Edward *journalist*
Asker, James Robert *journalist*
Billings, Linda *writer, analyst*
Bishop, Walton Burrell *science writer, consultant*
Blair, Patricia Wohlgemuth *economics writer*
Goldberg, Kirsten Boyd *science journalist*
Goodwin, Irwin *magazine editor*
Grosvenor, Gilbert Melville *journalist, educator, business executive*
Haggerty, James Joseph *writer*
Hassan, Aftab Syed *scientific research director*
Lipkin, Richard Martin *journalist, science writer*
Mariotte, Michael Lee *environmental activist, environmental publication director*

FLORIDA

Clearwater
Horton, Donna Alberg *technical writer*

Fort Lauderdale
Munzer, Martha Eiseman *writer*

Palm Beach
Symons, George Edgar *environmental engineering editor*

Venice
Shaw, Bryce Robert *author*

GEORGIA

Atlanta
Breck, Katherine Anne *technical writer*

ILLINOIS

Chicago
Lundberg, George David, II *medical editor, pathologist*
Skolnick, Andrew Abraham *science and medical journalist, photographer*

Homewood
Grunwald, Arnold Paul *communications executive, engineer*

IOWA

Dubuque
Kane, Kevin Thomas *editor*

KANSAS

Olathe
Picou, Gary Lee *technical publication editor*

MARYLAND

Annapolis
Hendrick, John Morton *science and engineering editor*

Baltimore
Lamberg, Lynne Friedman *medical journalist*

Gaithersburg
Lide, David Reynolds *science editor*

MASSACHUSETTS

Boston
Baron, David Hume *science journalist*

Brookline
Ross, Robert Nathan *medical writer, consultant*

Cambridge
Dickman, Steven Gary *science writer*

Concord
Plauger, P.J. *science writer*

MICHIGAN

Southfield
Mackey, Robert Joseph *video publisher*

MINNESOTA

Eden Prairie
Gertis, Neill Allan *writer*

Minneapolis
Baker, John Stevenson (Michael Dyregrov) *writer*

NEVADA

Gardnerville
Burns, James Kent (Jasper) *science illustrator*

NEW JERSEY

Flemington
Easter, Charles Henry *technical writer*

Livingston
Koester, J. Anthony *science publication editor*

Princeton
Lipscombe, Trevor Charles Edmund *physical science editor, researcher*

NEW MEXICO

Los Alamos
Mark, Kathleen Abbott *writer*

NEW YORK

Bearsville
Mullette, Julienne Patricia *television personality and producer, astrologer, author, health center administrator*

Bronxville
Noble, James Kendrick, Jr. *media industry consultant*

Cragsmoor
Foster, Lanny Gordon *writer, publisher*

Hamilton
Edmonston, William Edward, Jr. *publisher*

Odessa
Lambos, William Andrew *computer consultant, social science educator*

Palm Bay
Bellstedt, Olaf *software engineer*

Saint Petersburg
Fishman, Mark Brian *computer scientist, educator*

Sunrise
Tielens, Steven Robert *information specialist*

Tallahassee
Leavell, Michael Ray *computer programmer and analyst*

Tampa
Rodts, Gerald Edward *computer scientist*

GEORGIA

Atlanta
Duffell, James Michael *computer systems analyst*
Foley, James David *computer science educator, consultant*
Griesser, James Albert (Jamie) *computer scientist*
Hutcheson, Philip Charles *computer programmer, analyst*
King, K(imberly) N(elson) *computer science educator*
Preble, Darrell W. *systems analyst*
Williams, Charles Murray *computer information systems educator, consultant*

Cornelia
Jeffers, Dale Welborn *computer specialist, systems analyst*

Marietta
Harbort, Robert Adolph, Jr. *computer science educator*

Norcross
Chan, David Chuk *software engineer*
Duncan, Phillip Charles *research scientist*

Robins AFB
Washington, David Earl *computer systems professional*

Swainsboro
Watt, (Arthur) Dwight, Jr. *computer programming and microcomputer specialist, educator*

HAWAII

Honolulu
Abdul, Corinna Gay *software engineer, consultant*

IDAHO

Mountain Home
Meyr, Shari Louise *information consultant*

ILLINOIS

Addison
Findling, David Martin *application engineer*

Algonquin
Stanek, Donald George, Jr. *computer analyst*

Argonne
Tatar, John Joseph *computer scientist*

Centralia
Davidson, Karen Sue *computer software designer*

Champaign
Kuck, David Jerome *computer system researcher, administrator*

Chicago
Al-Salqan, Yahya Yousef *computer science researcher*
Jia, Hong *programmer, analyst*
Larson, Nancy Celeste *computer systems analyst, music educator*
Safar, Michal *information scientist*
Saweikis, Matthew A. *information scientist*

Downers Grove
Shah, Bankim *software consultant*

Elmhurst
Muellner, William Charles *computer scientist, physicist*

Evanston
Schonfeld, Eugene Paul *investment company executive, software designer*

Grayslake
Dulmes, Steven Lee *computer science educator*

Hines
Palmer, Martha Jane *computer specialist*

Libertyville
Mishra, Ajay Kumar *software engineer*

Loves Park
Lyon, Roger Wayne *information scientist*

Palos Hills
Maciulis, Linda S. *computer coordinator, consultant*

Skokie
Gutterman, Milton M. *operations research analyst*

Urbana
Edelsbrunner, Herbert *computer scientist, mathematician*

INDIANA

Bloomington
Purdom, Paul Walton, Jr. *computer scientist*

Fort Wayne
Taylor, Mike Allen *computer systems analyst*

Indianapolis
Caraher, Michael Edward *systems analyst*

Shelbyville
Deaton, Timothy Lee *computer systems integrator*

IOWA

Ames
Honavar, Vasant Gajanan *computer scientist, educator*

Cedar Rapids
Ashbacher, Charles David *computer programmer, educator*

Fairfield
Perez, Jose Luis *computer scientist, engineer, philosopher*

Marion
Stover, Donald Rae *software engineering executive, retired*

KANSAS

Kansas City
Taylor, Brenda Carol *computer specialist*

Lenexa
Zhu, Qing *software engineer*

Manhattan
Knupfer, Nancy Nelson *computer science educator*

Olathe
Early, Marvin Milford, Jr. *computer scientist*

Wichita
Alagić, Suad *computer science educator, researcher, consultant*

KENTUCKY

Corbin
Steenbergen, Gary Lewis *computer aided design educator*

Louisville
Kibiloski, Floyd Terry *business and computer consultant, editor*

Owensboro
Watts, Kenneth Michael *computer scientist*

Richmond
Schnare, Paul Stewart *computer scientist, mathematician*

LOUISIANA

Plaquemine
Whigham, Mark Anthony *computer scientist*

MAINE

Bar Harbor
Guidi, John Neil *scientific software engineer*

Glen Cove
Shepardson, Jed Phillip *computer consultant*

MARYLAND

Annapolis
Sheppard, John Wilbur *computer research scientist*

Baltimore
Arsham, Hossein *information scientist educator*
Sherman, Alan Theodore *computer science educator*

Bethesda
Lipkin, Bernice Sacks *computer science educator*

College Park
Aloimonos, Yiannis John *computer sciences educator*
Reggia, James Allen *computer scientist, educator*
Rosenfeld, Azriel *computer science educator, consultant*
Sachs, Harvey M. *policy analyst*

Cumberland
Goss, Patricia Lynn *information specialist*

Ellicott City
Davis, Patricia Mahoney *software engineer*

Fort Detrick
Shaffer, Stephen M. *computer system analyst*

Gaithersburg
Burrows, James H. *computer scientist*
Oran, Gary Carl *computer scientist*

Germantown
Matthews, Dale Samuel *information scientist*

Laurel
Ford, Byron Milton *computer consultant*

Linthicum
Massimini, Joseph Nicholas *software engineer*

Owings Mills
Knisbacher, Jeffrey Mark *computer scientist*

Silver Spring
Hoch, Peggy Marie *computer scientist*

MASSACHUSETTS

Amherst
Popplestone, Robin John *computer scientist, educator*
Sclove, Richard Evan *technology educator, writer, consultant*

Boston
Cushing, Steven *educator, researcher, consultant*
Donaldson, Robert Louis *computer systems professional*
Nyberg, Stanley Eric *cognitive scientist*

Cambridge
Baum, Peter Samuel *research director*
Cheatham, Thomas Edward, Jr. *computer scientist, educator*
Corbato, Fernando Jose *electrical engineer and computer science educator*
Gagliardi, Ugo Oscar *systems software architect, educator*
Grosz, Barbara Jean *computer science educator*
Grover, Mark Donald *computer scientist*
Hillis, William Daniel *scientist, engineer*
Kung, H. T. *computer science and engineering educator, consultant*
Lampson, Butler Wright *computer scientist*
Lomet, David Bruce *computer scientist*
Lynch, Nancy Ann *computer scientist, educator*
Roberts, Edward Baer *technology management educator*
Rockart, John Fralick *information systems reseacher*
Valiant, Leslie Gabriel *computer scientist*
vanRiper, William John *computer graphics scientist, consultant*

Chelmsford
Lerner, James Peter *software engineer*

Conway
Powsner, Gary *computer consultant*

Framingham
Fedorowicz, Jane *information systems educator*

Harwich
Volpicelli, Frederick Gabriel *computer scientist*

Lincoln
LeGates, John Crews Boulton *information scientist*

Lynn
Hatch, Mark Bruce *software engineer*

Medfield
Gradijan, Jack Robertson *software engineer*

Sudbury
McManus, John Gerard *software engineer*

Waltham
Hitchcock, Christopher Brian *computer and environmental scientist, policy services executive*
O'Donnell, Teresa Hohol *software development engineer, antennas engineer*

Westfield
Buckmore, Alvah Clarence, Jr. *computer scientist, ballistician*

MICHIGAN

Ann Arbor
Aupperle, Eric Max *data network center administrator, research scientist, engineering educator*
Conway, Lynn Ann *computer scientist, educator*

Detroit
Rajlich, Vaclav Thomas *computer scientist, researcher, educator*

Dexter
Roberts, Walter Arthur *computer systems scientist*

Farmington
Salabounis, Manuel *computer information scientist, mathematician, scientist*

Kalamazoo
Athappilly, Kuriakose Kunjuvarkey *computers and quantitative methods educator*

Midland
Hilty, Terrence Keith *analytical sciences manager*

Mount Pleasant
Rubin, Stuart Harvey *computer science educator, researcher*

Roseville
Carr, Doleen Pellett *computer service and environmental specialist, consultant*

Troy
Ahmed, Imthyas Abdul *computer scientist*
Chrisman, Vince Darrell *information scientist*
Gopalan, Muhundan *software engineer*

Warren
Ginsberg, Myron *research scientist*

MINNESOTA

Burnsville
Geminn, Walter Lawrence, Jr. *computer programmer, analyst*

Duluth
Thomborson, Clark David (Clark David Thompson) *computer scientist, educator*

Minneapolis
James, Walter *retired computer information specialist*

Minnetonka
Andrus, W(infield) Scott *scientist, consultant*

Saint Paul
Evans, Roger Lynwood *scientist, patent liaison*
Ritschel, James Allan *computer research specialist*

Spring Lake Park
Powell, Christopher Robert *systems analyst*

MISSISSIPPI

Hattiesburg
Miller, James Edward *computer scientist, educator*

Pascagoula
Swint, Joseph Ellis *computer systems analyst*

Vicksburg
Middlebusher, Mark Alan *computer scientist*

MISSOURI

Maryland Heights
Scarponcini, Paul *computer scientist*

Saint Louis
Benson, Robert John *computer science educator, administrator*
Cawns, Albert Edward *computer systems consultant*

MONTANA

Missoula
Zheng, Youlu *computer scientist, educator*

NEVADA

Las Vegas
Wade, Stacy Lynn *computer specialist*

NEW HAMPSHIRE

Hampstead
Hunt, Philip George *computer consultant*

Manchester
Seidman, Robert Howard *computer information specialist, educator*

NEW JERSEY

Belleville
Torre, Bennett Patrick *computer scientist, consultant*

Fair Lawn
Golaski, Nicholas John *information scientist*

Fairfield
Regino, Thomas Charles *technology assessment, R&D benchmarking specialist*

Mahwah
Kierstead, James Allan *computer scientist*
Sulla, Nancy *computer coordinator, consultant, business owner*

Manville
Koscelnick, Jeanne *computer scientist*

Marlton
Wright, Kara-Lyn Annette *software engineer*

Morristown
Epstein, Samuel Seth *computer scientist, researcher*

Murray Hill
Litman, Diane Judith *computer scientist*

New Brunswick
Kulikowski, Casimir A. *computer science educator, research program director*

Newark
Karvelas, Dennis E. *computers and information science educator*

Nutley
Bautz, Gordon Thomas *information scientist, researcher*

Piscataway
Dayton, John Thomas, Jr. *computer researcher*
Donnelly, Kim Frances *computer scientist*
Urban, Cathleen Andrea *client technical support consultant*
Walker, James Calvin *computer software systems developer, politician*

Princeton
Katz, Irvin Ronald *research scientist*
Melamed, Benjamin *computer scientist*
Miyamoto, Maylene Hu *computer science educator*
Scairpon, Sharon Cecilia *information scientist*
Umscheid, Ludwig Joseph *government computer specialist*

Rochelle Park
Grossman, Martin Bernard *data systems consultant*

Teaneck
Zwass, Vladimir *computer scientist, educator*

Tinton Falls
Trigg, Clifton Thomas *information systems consultant*

West Orange
Schoen, Howard Franklin *computer programmer, analyst*

Willingboro
Ingerman, Peter Zilahy *infosystems consultant*

NEW MEXICO

Los Alamos
Wade, Rodger Grant *financial systems analyst*

NEW YORK

Ardsley
Gayle, Joseph Central, Jr. *computer information professional*

Bethpage
Litke, John David *computer scientist*

Binghamton
Taylor, Kenneth Douglas *finance and computer consultant, educator*

Brooklyn
Heppa, Douglas Van *computer specialist*
Hilsenrath, Joel Alan *computer scientist, consultant*
Pennisten, John William *computer scientist, linguist, actuary*

Elmont
Highland, Harold Joseph *computer scientist*

Flushing
Gezelter, Robert L. *computer systems consultant*

Forest Hills
Vulis, Dimitri Lvovich *computer consultancy executive*

Glens Falls
Clements, Brian Matthew *computer consultant, educator*

Kingston
Martin, Joanne Lea *computer scientist, researcher*

Liverpool
Neidhardt, Jean Sliva *systems integration professional, sales executive*

Manhasset
Hinds, Glester Samuel *program specialist, tax consultant*

Melville
Ullmann, Edward Hans *systems analyst*

New York
Deane, John Herbert *technologist*
Eichen, Marc Alan *computer support*
Fanning, William Henry, Jr. *computer scientist*
Gannon, Jane Frances *information scientist, researcher*
Rankin, James Edwin *computer systems consultant*
Teplitzky, Philip Herman *computer consultant*
Wyn-Jones, Alun (William Wyn-Jones) *software developer, mathematician*

Northport
Gray, James Lee *systems analyst*

Ossining
Campbell, Harry L. *systems analyst*

Pearl River
Galante, Joseph Anthony, Jr. *computer programmer*

Port Washington
Hayslett, Paul Joseph *computer programmer, consultant*

Potsdam
Glasser, Joshua David *computer scientist*

Poughkeepsie
Vakirtzis, Adamantios Montos *systems analyst*
Vita, James Paul *software engineer*

Rhinebeck
Hellerman, Leo *retired computer scientist and mathematician*

Rochester
Arden, Bruce Wesley *computer science and electrical engineering educator*

Saratoga Springs
Houghton, Raymond Carl, Jr. *computer science educator*

Schenectady
Fagg, William Harrison *retired infosystems specialist*
Marchione, Sharyn Lee *computer scientist*

Suffern
Eoga, Michael Gerard *network systems programmer*

Syracuse
Hansen, Per Brinch *computer scientist*

Tarrytown
Klein, Marcia Schneiderman *analyst, programmer, consultant*

Troy
Roysam, Badrinath *computer scientist, educator*
Yee, Lester Wey-Ming *researcher*

Wallkill
Bittner, Ronald Joseph *computer systems analyst*

Westbury
O'Neill, Peter Thadeus *software engineer*

White Plains
Bosco, Paul D. *computer scientist*
Cheng, Alexander Lihdar *computer scientist, researcher*

Woodside
Bashias, Norman Jack *software engineer, computer consultant*

Yorktown Heights
Bhaskar, Ramamoorthi *artifical intelligence scientist, researcher*
Cocke, John *computer scientist*
Sachs, Martin William *computer scientist*

NORTH CAROLINA

Asheville
Honeycutt, Michael Allen *computer consultant*

Biscoe
Kelly, Terry Lee *computer systems analyst*

Cary
Rauf, Robert Charles, Sr. *systems specialist*

Chapel Hill
Chi, Vernon L. *computer science educator, administrator*
Howell, William Everett *computer professional, consultant*

Greensboro
Bailey, William Nathan *systems analyst*

Hillsborough
Cooley, Philip Chester *computer modeller*

Raleigh
McAllister, David Franklin *computer science educator*
Wetsch, John Robert *computer systems administrator*

OHIO

Akron
Hollis, William Frederick *information scientist*

Cincinnati
Fairobent, Douglas Kevin *computer programmer*
Stanifer, Robert Dale *system analyst, manager*
Wilsey, Philip Arthur *computer scientist*

Columbus
Busick, Robert James *computer scientist*
Fraser, Jane Marian *operations research specialist, educator*

Pickerington
Blackman, Edwin Jackson *software engineer*

Portsmouth
Rudolph, Brian Albert *computer science educator*

Toledo
Gibson, Raymond Novarro *computer programmer, analyst*

Willoughby
Stevens, William Frederick, III *software engineer*

Xenia
Fussichen, Kenneth *computer scientist*

OKLAHOMA

Oklahoma City
Jones, Jeffery Lynn *software engineer*

Tulsa
Mitchell, Donald Hearin *computer scientist*

OREGON

Albany
Yu, Kitson Szewai *computer science educator*

Beaverton
Cipale, Joseph Michael *software engineer*

Corvallis
Skelton, John Edward *computer science educator, consultant*

Hillsboro
Krampits, Mark William *power management software specialist*

Salem
Pugh, Tim Francis, II *software engineer, aeronautical engineer*

Wilsonville
Willett, Kenneth G. *software engineer*

PENNSYLVANIA

Bethlehem
Wazontek, Stella Catherine *computer programmer, analyst, software engineer*

Biglerville
Saliu, Ion *software developer, computer programmer*

Coraopolis
Skovira, Robert Joseph *information scientist, educator*

Monroeville
Nemeth, Edward Joseph *process research specialist*

Oakdale
Sieger, Edward Regis *software engineer*

Philadelphia
Altshuler, David Thomas *computer scientist*
Klaver, Martin Arnold, Jr. *business administrator*

Pittsburgh
Allen, David Woodroffe *computer scientist*
Berliner, Hans Jack *computer scientist*
Curtis, Bill *software engineering researcher*
Froehlich, Fritz Edgar *telecommunications educator and scientist*

Vandergrift
Kulick, Richard John *computer scientist, researcher*

Wayne
Swank, Annette Marie *computer software designer*
West, Mark David *software engineer*

West Point
Maglaty, Joseph Louis *data processor, consultant*

RHODE ISLAND

Newport
Connery, Steven Charles *computer scientist*
Viana, Thomas Arnold *computer scientist*

SOUTH CAROLINA

Aiken
Novak, Charles R. *computer scientist*

Spartanburg
Hilton, Theodore Craig *computer scientist, computer executive*

TENNESSEE

Kingsport
Hendrix, Kenneth Allen *systems analyst*
Mahaffey, Richard Roberts *information analyst*

Memphis
Hawk, Charles Silas *computer programmer, analyst*

Nashville
Williams, Marsha Rhea *computer scientist, educator, researcher, consultant*

TEXAS

Arlington
Kendall, Jillian D. *information systems specialist, program developer, educator*
Wiig, Karl Martin *management scientist, systems engineer*

Austin
Browne, James Clayton *computer science educator*
Burns, Robert Wayne *computer scientist*
Dijkstra, Edsger Wybe *computer science educator, mathematician*
Herb, Cynthia Johnson *programmer*
Kozmetsky, George *computer science educator*
Lam, Simon Shin-Sing *computer science educator*
Misra, Jayadev *computer science educator*

Dallas
Esin, Joseph Okon *computer information systems educator*
Poon, William Wai-Lik *program analyst*
Qian, Dahong *laser development engineer*
Simon, Solomon Henry *artificial intelligence manager*

Denton
Das, Sajal Kumar *computer science educator, researcher*
Renka, Robert Joseph *computer science educator, consultant*

Greenville
Hanson, David Alan *software engineer*

Houston
Ashley, William Hilton, Jr. *software engineer*
Billingsley, David Stuart *researcher*
Kennedy, Ken *computer science educator*
Moore, Fay Linda *systems analyst*
Parker, Norman Neil, Jr. *software engineer, mathematics educator*
Sharp, David Paul *computer technologist, researcher*
Stark, George Edward *software reliability engineer, statistician*

Stewart, Robert Jackson *software development engineer, researcher*
Woo, Walter *computer systems consultant*
Woods, Stephanie Elise *computer information scientist*
Wright, Clark Phillips *computer systems specialist*

Irving
Taylor, Lawrence Stanton *software engineer*

Kingsville
Diersing, Robert Joseph *computer science educator*

Lubbock
Hennessey, Audrey Kathleen *information systems educator*

Plano
Westfall, Linda Louise *software engineer, metrics analyst*

Richardson
Gerhardt, Douglas L. *computer scientist*

San Antonio
Baughman, John Thomas *data base designer, educator*
Grubb, Robert Lynn *computer system designer*

Spring
Joy, David Anthony *computer consultant*

UTAH

Salt Lake City
St. Dennis, Bruce John *computer engineer, software researcher*

VERMONT

Essex Junction
Bernard, Ronald Allan *computer performance analyst*

South Burlington
Kenney, Timothy P. *computer scientist*

VIRGINIA

Alexandria
Fore, Claude Harvel, III *scientist, information analyst*
McDaniel, William Howard Taft, Jr. *computer information systems educator*
Perchik, Benjamin Ivan *operations research analyst*
Reeker, Larry H. *computer scientist, educator*
Thiel, Thomas Joseph *information scientist, consultant*

Arlington
Barton, Henry David *operations research analyst*
Batchelor, Barry Lee *software engineer*
Murray, Jeanne Morris *computer scientist, educator, consultant*
Zgonc, Janice Ann *computer specialist*

Burke
Bursik, David James *software engineer*

Chantilly
Neitz, David Allan *information scientist*

Charlottesville
Wulf, William Allan *computer information scientist, educator*

Fairfax
Harte, Rebecca Elizabeth *computer scientist, consultant*
Sage, Andrew Patrick, Jr. *systems information and software engineering educator*

Falls Church
Robertson, Laurie Luissa *computer scientist*

Fort Belvoir
Austin, Charles Louis *software engineer*

Herndon
Alton, Cecil Claude *computer scientist*

Langley AFB
Osborn, James Henshaw *operations research analyst*

Mc Lean
Gerasch, Thomas Ernest *computer scientist*
Perry, Dennis Gordon *computer scientist*
Schneck, Paul Bennett *computer scientist*

Reston
Grass, Judith Ellen *computer scientist*

Seven Corners
Goncz, Douglas Dana *information broker*

Vienna
Nicklas, John G. *systems analyst*

WASHINGTON

Seattle
Nelson, Walter William *computer programmer, consultant*
Tripp, Leonard Lee *software engineer*
Young, Paul Ruel *computer scientist, dean*
Zabinsky, Zelda Barbara *operations researcher, industrial engineering educator*

Spokane
Dunau, Andrew T. *mathematics, science and technology consultant*

WEST VIRGINIA

Falling Waters
Schellhaas, Linda Jean *scientist, consultant*

Morgantown
Tavaglione, David *computer scientist, consultant*

WISCONSIN

Brookfield
Diesem, John Lawrence *information systems specialist*

La Crosse
Abts, Daniel Carl *computer scientist*

Madison
Desautels, Edouard Joseph *computer science educator*

Milwaukee
Knasel, Thomas Lowell *information systems consultant*
Kraut, Joanne Lenora *computer programmer, analyst*
Smith, Michael Lawrence *knowledge engineer, researcher*

MILITARY ADDRESSES OF THE UNITED STATES

EUROPE

APO
Hatton, Daniel Kelly *computer scientist*

CANADA

BRITISH COLUMBIA

Burnaby
Han, Jiawei *computer scientist, educator*

ONTARIO

Etobicoke
Bahadur, Birendra *display specialist, liquid crystal researcher*

Hamilton
Steiner, George *information systems and management science educator, researcher*

London
Bauer, Michael Anthony *computer scientist, educator*

QUEBEC

Montreal
Cohen, Montague *medical physics educator*
Levine, Martin David *computer science and electrical engineering educator*

ANDORRA

Andorra
Bastida, Daniel *data processing facility administrator, educator*

AUSTRALIA

Canberra
Wagner, M(ax) Michael *computer scientist, educator*

Melbourne
Prince, Stephen *software developer, researcher*

Nedlands
Cottingham, Marion Scott *computer scientist, educator*

Perth
Kieronska, Dorota Helena *computer science educator*

Sydney
Brinck, Keith *computer scientist*

AUSTRIA

Vienna
Purgathofer, Werner *computer science educator*

BELGIUM

Brussels
Corsi, Patrick *computer scientist*

Louvain-la-Neuve
Sintzoff, Michel *computer scientist, educator*

CAMEROON

Douala
Obenson, Philip *computer science educator*

CHINA

Beijing
Su, Dongzhuang *computer science educator, university official*

Guangzhou
Hou, Guang Kun *computer science educator*

Shanghai
Gong, Yitai *information specialist*

DENMARK

Aalborg
Bagchi, Kallol Kumar *computer science and engineering educator, researcher*

Frederiksberg
Bjørn-Andersen, Niels *information systems researcher*

Lyngby
Tøndering, Claus *software engineer*

EGYPT

Cairo
Rafea, Ahmed Abdelwahed *computer science educator*

Nasr City
El-Hamalaway, Mohamed-Younis Abd-El-Samie *computer engineering educator*

ENGLAND

Canterbury
Eager, Robert Donald *computer science educator*

London
Sandergaard, Theodore Jorgensen *information technology director*

Macclesfield
Graham, Dorothy Ruth *software engineering consultant*

Olney
Edmonds, Andrew Nicola *software engineer*

Sheffield
Willett, Peter *information science researcher*

Stafford
Zambardino, Rodolfo Alfredo *computing educator*

West Clandon
Lumsden, James Gerard *computer scientist*

FRANCE

Brignoles
Thomas, Raymond Jean *computer consultant*

LaGaude
Pignal, Pierre Ivan *computer scientist, educator*

Nanterre
Buffet, Pierre *computer scientist*

Paris
Delacour, Yves Jean Claude Marie *technology information executive*

Rennes
Lerman, Israël César *data classification and processing researcher*

Toulouse
Diaz, A. Michel *computer and control science researcher, administrator*

GERMANY

Aachen
Jakobs, Kai *computer scientist*

Bavaria
Ehler, Herbert *computer scientist*

Berlin
Ehrig, Hartmut *computer science educator, mathematician*
Lemke, Heinz Ulrich *computer science educator*

Bonn
Klingen, Leo H. *computer science educator*

Dresden
Pfitzmann, Andreas *computer security educator*

Freiburg
Wagner, Bernhard Rupert *computer scientist*

Hirschberg
Kreuter, Konrad Franz *software engineer*

Mainz
Egel, Christoph *computer scientist*

Mannheim
Steffens, Franz Eugen Aloys *computer science educator*

Munich
Sebestyén, István *computer scientist, educator, consultant*

Paderborn
Belli, Fevzi *computing science educator, consultant*

Saarbrücken
Siekmann, Jörg Hans *computer science educator*

Wülfrath
Mullen, Alexander *information scientist, chemist*

HONG KONG

Hong Kong
Lochovsky, Frederick Horst *computer science educator*

IRAN

Tehran
Mozaffari, Mojtaba *computer science educator*

IRELAND

Dublin
Mac An Airchinnigh, Mícheál *computer science educator*

Limerick
Eaton, Malachy Michael *computer systems educator*

ITALY

Rome
Marinuzzi, Francesco *computer science educator*

JAPAN

Chofu
Tomita, Etsuji *computer science educator*

Gamagouri-shi
Kojima, Ryuichi O. *computer educator*

Kanagawa
Kamijo, Fumihiko *computer science and information processing educator*
Matsubara, Tomoo *software scientist*

Kobe
Aonuma, Tatsuo *management sciences educator*

Nagano
Matsuda, Yasuhiro *computer scientist*

Osaka
Okawa, Yoshikuni *computer science educator*

Shizuoka
Shirai, Yasuto *information science educator, researcher*

Tama-shi
Kawano, Hiroshi *computer science and aesthetics educator*

Tokyo
Maekawa, Mamoru *computer science educator*
Otsuka, Kanji *information science educator*
Shibata, Akikazu *semiconductor scientist*
Yaku, Takeo Taira Chichibu Kawagoe *computer scientist, educator*

Tsukuba
Kato, Daisuke *electronics researcher*
Takagi, Hideaki *computer scientist, mathematician*

THE NETHERLANDS

Amsterdam
Tanenbaum, Andrew Stuart *computer scientist, educator, author*

Delft
Citroen, Charles Louis *information scientist, consultant*

Maastricht
Van Den Herik, Hendrik Jacob *computer science and artificial intelligence educator*

Muiderberg
Van Waning, Willem Ernst *computer scientist*

NEW ZEALAND

Canterbury
Kulasiri, Gamalathge Don *computer educator, mechanical and agricultural engineer*

NORWAY

Bergen
Kløve, Torleiv *informatics educator*

POLAND

Gdańsk
Kubale, Marek Edward *computer scientist, educator*

Poznan
Dabrowski, Adam Miroslaw *digital and analog signal processing scientist*

QATAR

Doha
Bu-Hulaiga, Mohammed-Ihsan Ali *information systems educator*

REPUBLIC OF KOREA

Daejeon
Whang, Kyu-Young *computer science educator, director*

SPAIN

Barcelona
Balcázar, José Luis *computer scientist, educator*

Blanes
Lopez de Mantaras, Ramon *computer scientist, researcher*

Cordoba
Caridad, Jose Maria *computer scientist, educator*

Madrid
Alfonseca, Manuel *computer scientist, educator*

SWEDEN

Älvsjö
Hausman, Bogumil *computer scientist*
Jauhiainen, Pertti Juhani *software engineer*

SWITZERLAND

Lausanne
Thalmann, Daniel *computer science educator*

THAILAND

Bangkok
Huynh Ngoc, Phien *computer science educator, consultant*

ADDRESS UNPUBLISHED

Afsarmanesh, Hamideh *computer science educator, research scientist*
Albrecht, Allan James *computer scientist*
Ancheta, Caesar Paul *software engineer*
Armstrong, Elmer Franklin *information systems and data processing educator*
Avendano, Tania *software engineer*
Badinger, Michael Albert *computer applications designer, inventor, consultant*
Bangs, Allan Philip *computer scientist, software engineer, consultant*
Boardman, Rosanne Virginia *logistics consultant*
Chen, Chung Long *software engineer*
Chen, Henry *systems analyst*
Dishmon, Samuel Quinton, Jr. *computer scientist*
Doubledee, Deanna Gail *software engineer, consultant*
Dudley, Lonnie LeRoy *information scientist*
Egar, William Thomas *information systems consultant*
Flynt, Clifton William *computer programmer, software designer*
Greenwood, Frank *information scientist*
Harijan, Ram *technology transfer researcher*
Heffron, W(alter) Gordon *computer consultant*
Hermanson, Theodore Harry *software engineer*
Ho-Le, Ken Khoa *software engineer*
Holland, Michael James *computer services administrator*
Liskov, Barbara Huberman *software engineering educator*
Martinez, Salvador *electronic technology educator*
Matthes, Howard Kurt *computer consultant and researcher*
Moores, Anita Jean Young *computer consultant*
Pierce, Charles Earl *software engineer*
Rapin, Charles René Jules *computer science educator*
Roberts, Marie Dyer *computer systems specialist*
Smith, David Mitchell *systems and software researcher, consultant*
Stuart, Sandra Joyce *computer information scientist*
Tolonen, Risto Markus *systems software engineer*
Walters, Kenn David *scientist, company executive*
Wardwell, James Charles *computer and management consultant*
Warner, Janet Claire *software design engineer*
Whitner, Jane Marvin *scientific applications programmer*
Williams, Kourt Durell *operations analyst*
Zendle, Howard Mark *software development researcher*
Zenone, John Mark *system programmer, technical analyst*

EDUCATION

UNITED STATES

ALABAMA

Birmingham
Smith, John Stephen *educational administrator*

Huntsville
Lundquist, Charles Arthur *university official*
Watson, Raymond Coke, Jr. *college president, engineering educator, mathematics educator*

ARIZONA

Tempe
Bristol, Stanley David *mathematics educator*

Tucson
Smerdon, Ernest Thomas *academic administrator*

ARKANSAS

Little Rock
Fribourgh, James Henry *university administrator*

CALIFORNIA

Berkeley
Park, Roderic Bruce *university official*
Tien, Chang-Lin *chancellor*

Davis
Greenwood, M. R. C. *college dean, biologist, nutrition educator*

Half Moon Bay
Sedlak, Bonnie Joy *university official*

Lakewood
Bogdan, James Thomas *secondary education educator, electronics researcher and developer*

Los Angeles
Schneider, Edward Lewis *academic administrator, research administrator*
Slaughter, John Brooks *university president*
Tuckson, Reed V. *university president*

Menlo Park
Carr, Michael *secondary education educator*

Northridge
Cleary, James W. *retired university administrator*

Pasadena
Everhart, Thomas Eugene *university president, engineering educator*

Sacramento
Futernick, Kenneth David *education educator, software developer*

Van Nuys
Altshiller, Arthur Leonard *physics educator*

COLORADO

Cheyenne Wells
Palmer, Rayetta J. *computer educator*

Colorado State University
Cordain, Loren *physical education educator, graduate program director, researcher*

Fort Collins
Harper, Judson Morse *university administrator, consultant, educator*
McHugh, Helen Frances *university dean, home economist*

CONNECTICUT

Glastonbury
Roy, Kenneth Russell *school system administrator, educator*

DISTRICT OF COLUMBIA

Washington
Burris, James Frederick *academic dean, educator*

FLORIDA

Alachua
Thornton, J. Ronald *technologist*

Boca Raton
More, Kane Jean *science educator*

Gainesville
Viessman, Warren, Jr. *academic dean, civil engineering educator, researcher*

Jay
Peacock, Hugh Anthony *agricultural research director, educator*

Miami
Rosendahl, Bruce Ray *dean, geophysicist, educator*

Port Charlotte
Norris, Dolores June *computer specialist*

Tampa
Mullins, Michael Drew *environmental educator*
Wade, Thomas Edward *university research administrator, electrical engineering educator*

GEORGIA

Atlanta
Greenberg, Raymond Seth *dean, educator*
Matula, Richard A(llan) *academic administrator*

Columbus
Massey, Carolyn Shavers *mathematics educator*

Lawrenceville
Adair, Lila McGahee *physics educator*

Tucker
O'Neil, Daniel Joseph *academic administrator, research executive*

Winterville
Anderson, David Prewitt *university dean*

HAWAII

Aiea
Ma, Jeanetta Ping Chan *elementary education educator*

Honolulu
Nelson, Jeanne Francess *mathematics educator*

IDAHO

Moscow
Bartlett, Robert Watkins *academic dean, metallurgist*
Hendee, John Clare *college dean, natural resources educator*

ILLINOIS

Bradley
O'Flaherty, Gerald Nolan *secondary school educator, consultant*

Chicago
Chernick, Cedric L. *research grants administrator*
Skiba, Aurelia Ellen *parochial school educator*

De Kalb
Williams, Eddie R. *university administrator*

Urbana
Holt, Donald A. *university administrator, agronomist, consultant, researcher*

INDIANA

Angola
Lin, Ping-Wha *educator*

Garrett
Baker, Suzon Lynne *mathematics educator*

Notre Dame
O'Meara, Onorato Timothy *university administrator, mathematician*

South Bend
Farkas, Thomas *secondary education educator*

West Lafayette
Baumgardt, Billy Ray *university official, agriculturist*
Tacker, Willis Arnold, Jr *academic administrator, medical educator, researcher*
Yang, Henry Tsu Yow *dean, engineering educator*

IOWA

Ames
Cate, Rodney Michael *university dean*
Topel, David Glen *college dean, animal science educator*

KANSAS

Manhattan
Acevedo, Edmund Osvaldo *physical education educator*
Hahn, Richard Ray *academic administrator*

KENTUCKY

Louisville
Hoye, Robert Earl *urban policy educator, health care management consultant*

LOUISIANA

Baton Rouge
Eaton, Harvill Carlton *university administrator*
Rabideau, Peter Wayne *university dean, chemistry educator*

Ruston
Reneau, Daniel D. *university administrator*

Slidell
Sanders, Georgia Elizabeth *science and mathematics educator*

MAINE

Lewiston
Gleason, Thomas Clifford *university official*

MARYLAND

Baltimore
Bausell, R. Barker, Jr. *research methodology educator*
Poehler, Theodore Otto *university provost, engineer, researcher*

Bethesda
Cohen, Lois Ruth Kushner *government research institute official*

College Park
Dieter, George E., Jr. *university dean*
Fox, Christopher *educational administrator*
Prentice, Ann Ethelynd *academic adminstrator*

Oakland
Farrar, Richard Bartlett, Jr. *secondary education educator, wildlife biology consultant*

Suitland
Speier, Peter Michael *mathematics educator*

MASSACHUSETTS

Boston
El-Baz, Farouk *program director, educator*
Sessoms, Allen Lee *academic administrator, former diplomat, physicist*

Cambridge
Sharp, Phillip Allen *academic administrator, biologist, educator*
Vest, Charles Marstiller *university administrator*

Lowell
Hojnacki, Jerome Louis *university dean*

Medford
Liszczak, Theodore Michael *university administrator*

Peabody
Terlizzi, James Vincent, Jr. *secondary education educator*

Salem
Phillips, Bessie Gertrude Wright *school system administrator, museum trustee*

Shrewsbury
Garabedian, Charles, Jr. *mathematics educator*

MICHIGAN

Ann Arbor
Duderstadt, James Johnson *university president*

Dearborn
Sengupta, Subrata *dean, engineering educator*

East Lansing
Anderson, James Henry *university dean*
Saul, William Edward *academic administrator, civil engineering educator*

Saint Clair Shores
Johns, Gary Christopher *electronic technologies educator*

Stevensville
Martin, Larry J. *health educator, counselor, naturalist*

MINNESOTA

Moorhead
Emmel, Bruce Henry *mathematics educator*

Saint Paul
Perry, James Alfred *natural resources director, researcher, educator, consultant*

MISSISSIPPI

Mississippi State
Powe, Ralph Elward *university administrator*
Thompson, Warren S. *dean, academic administrator*

MISSOURI

Saint Louis
Bohne, Jeanette Kathryn *mathematics and science educator*
Byrnes, Christopher Ian *academic dean, researcher*
Gregory, Patricia Jeanne *corporate relations director*

MONTANA

Bozeman
Characklis, William Gregory *research center director, engineering educator*

NEBRASKA

Lincoln
Grew, Priscilla Croswell *academic administrator, geology educator*
Lawson, Merlin Paul *dean, climatologist*
Nelson, Darrell Wayne *university administrator*
Spanier, Graham Basil *university administrator, family sociologist*

NEW JERSEY

East Brunswick
Lund, Daryl B. *college dean*

Elizabeth
Hynes, Nancy Ellen *reading educator*

Hoboken
Raveché, Harold Joseph *university administrator, physical chemist*

Linden
Bedrick, Bernice *retired science educator, consultant*

New Brunswick
Dill, Ellis Harold *university dean*

Rockaway
DeBlasi, Robert Vincent *school system administrator*

Sewell
Detofsky, Louis Bennett *secondary education educator*

NEW MEXICO

Artesia
Horner, Elaine Evelyn *elementary education educator*

Socorro
Popp, Carl J. *university administrator, chemistry educator*

NEW YORK

Bronx
Fleischer, Norman *director of endocrinology, medical educator*

Brooklyn
Lubowsky, Jack *academic director*

Buffalo
Kiser, Kenneth M(aynard) *academic dean, chemical engineering educator*

Clifton Park
Schmitt, Roland Walter *retired academic administrator*

Garden City
Eickelberg, W. Warren Barbour *academic administrator*

Ithaca
Weinstein, Leonard Harlan *institute program director*

Long Island City
Stevenson, John O'Farrell, Jr. *dean*

New York
Audin, Lindsay Peter *energy conservation professional*
Eliasson, Kerstin Elisabeth *science and education policy advisor*
Greer, R. Douglas *behaviorology and education educator*
Seitz, Frederick *university president emeritus*

Potsdam
Gallagher, Richard Hugo *university official, engineer*
Wells, David John *university official, mechanical engineer*

Poughkeepsie
Purota, Donald Harris *educator*

Rochester
Castle, William Eugene *academic administrator*
Ford, Loretta C. *retired university dean, nurse educator*

Syracuse
Ware, Benjamin Ray *university administrator*

NORTH CAROLINA

Brevard
Valentini, Jose Esteban *scientist, educator, chemical engineering director*

Durham
Dowell, Earl Hugh *university dean, aerospace and mechanical engineering educator*

Greensboro
Mosier, Stephen Russell *college program director, physicist*

Raleigh
Barnhardt, Robert Alexander *college dean*

OHIO

Akron
Kelley, Frank Nicholas *university dean*
MacCracken, Mary Jo *physical education educator*

Chagrin Falls
Grasselli, Jeanette Gecsy *university official*

Dayton
Rowe, Joseph Everett *electrical engineering educator, administrator*

Fremont
Sattler, Nancy Joan *math and physical science educator*

OKLAHOMA

Keota
Davis, Thomas Pinkney *mathematics educator*

Stillwater
Browning, Charles Benton *university dean, agricultural educator*
Joshi, R. Malatesha *reading education educator*

Woodward
Boucher, Darrell A., Jr. *vocational education educator*

OREGON

Corvallis
Byrne, John Vincent *academic administrator*
Young, Roy Alton *university administrator, educator*

Portland
Hazel, Joanie Beverly *elementary educator*

PENNSYLVANIA

King Of Prussia
Hawes, Nancy Elizabeth *secondary educator*

Philadelphia
Blumberg, Baruch Samuel *academic administrator*
Bryan, Henry Collier *private school mathematics educator*
Rubik, Beverly Anne *university administrator*
Snyder, Donald Carl, Jr. *physics educator*
Sutnick, Alton Ivan *medical school dean, educator, researcher, physician*

Pittsburgh
Connolly, Patricia Ann Stacy *physical education educator*
Mehrabian, Robert *academic administrator*

Scranton
Fahey, Paul Farrell *college administrator*

University Park
Hood, Lamartine Frain *college dean*
Koopmann, Gary Hugo *educational center administrator, mechanical engineering educator*
Rashid, Kamal A. *university international programs director, educator*
Reischman, Michael Mack *university official*
Thomas, Joab Langston *academic administrator, biology educator*

Wayne
McArdle, Joan Terruso *parochial school mathematics and science educator*

SOUTH CAROLINA

Hilton Head Island
Hogan, Joseph Charles *university dean*

TENNESSEE

Chapel Hill
Christman, Luther Parmalee *university dean emeritus, consultant*

Cookeville
Lerner, Joseph *dean*

Knoxville
Prados, John William *educational administrator*

TEXAS

Big Sandy
Mitchell, Deborah Jane *educator*

College Station
Hiler, Edward Allan *academic administrator, agricultural engineering educator*
Peddicord, Kenneth Lee *academic administrator*

Corpus Christi
Azopardi, Korita Marie *mathematics educator*

Denton
Elder, Mark Lee *university research administrator, writer*
Lana, Philip K. *emergency response educator, consultant*

Fort Worth
Mills, John James *research director, mechanical engineering educator*

Galveston
Schmidly, David J. *dean*

Hereford
Carlile, Lynne *private school educator*

Houston
Hackerman, Norman *retired university president, chemist*
Lane, Neal Francis *university provost, physics researcher*
Raskin, Steven Allen *academic program director*

Irving
McVay, Barbara Chaves *mathematics educator*

Lubbock
Curl, Samuel Everett *university dean, agricultural scientist*

San Antonio
Poarch, Mary Hope Edmondson *science educator*

UTAH

Provo
Smoot, Leon Douglas *university dean, chemical engineering educator*

Salt Lake City
Jensen, Gordon Fred *university administrator*
Major, Thomas D. *academic program director*

VERMONT

Burlington
Allard, Judith Louise *secondary education educator*

VIRGINIA

Blacksburg
Nichols, James Robbs *university dean*

Charlottesville
Carey, Robert Munson *university dean, physician*

Norfolk
Redondo, Diego Ramon *health, physical education and recreation educator*

Richmond
Fleisher, Paul *elementary education educator*

WASHINGTON

Pullman
Lewis, Norman G. *academic administrator, researcher, consultant*

Renton
Lockridge, Alice Ann *secondary education educator*

Seattle
Bowen, Jewell Ray *academic dean, chemical engineering educator*
Terrell, W(illiam) Glenn *university president emeritus*
Wilcox, Paul Horne *academic administrator, researcher*

WEST VIRGINIA

Morgantown
Maxwell, Robert Haworth *college dean*

WISCONSIN

Madison
Yuill, Thomas MacKay *university administrator, microbiology educator*

Milwaukee
Davis, Thomas William *college administrator, electrical engineering educator*
Reid, Robert Lelon *college dean, mechanical engineer*

Platteville
Yeske, Ronald A. *dean*

CANADA

NEW BRUNSWICK

Fredericton
Armstrong, Robin Louis *university official, physicist*
Unger, Israel *dean, chemistry educator*

ONTARIO

Cold Lake
Clarke, Thomas Edward *research and development management educator*

Hamilton
Taylor, James Hutchings *chancellor*

Kingston
Benidickson, Agnes *university chancellor*

Mississauga
Evans, John Robert *former university president, physician*

Waterloo
Wright, Douglas Tyndall *former university president, company director, engineering educator*

QUEBEC

Montreal
Belanger, Pierre Rolland *university dean, electrical engineering educator*
Prichard, Roger Kingsley *university administrator, dean, educator*

Saint-Hubert
Doré, Roland *dean, science association director*

Sainte Anne de Bellevue
Buckland, Roger Basil *university dean, educator, vice principal*

SASKATCHEWAN

Regina
Symes, Lawrence Richard *university dean, computer science educator*

Saskatoon
Nikiforuk, Peter Nick *university dean*

ARGENTINA

Buenos Aires
Stoppani, Andres Oscar Manuel *director research center, educator*

AUSTRIA

Vienna
Resch, Helmuth *environmental science educator*
Selberherr, Siegfried *university dean, educator, researcher, consultant*

BRAZIL

São Paulo
Goldemberg, Jose *educator*

CHINA

Beijing
Zuxun, Sun *academic administrator, physicist*

FINLAND

Helsinki
Allardt, Erik Anders *academic administrator, educator*

GERMANY

Münster
Kordes, Hagen *education researcher, author*

JAPAN

Kyoto
Fujita, Eiichi *educator emeritus*

Sakyoku
Yamamoto, Shigeru *educator*

Tokyo
Arima, Akito *academic administrator*

Toyohashi
Sasaki, Shin-Ichi *university president, microbiologist, educator*

RUSSIA

Moscow
Osipov, Yurii *mathematician, mechanical scientist, educator*

SPAIN

Tacoronte
Kardas, Sigmund Joseph, Jr. *secondary education educator*

TURKEY

Pendik-Istanbul
Rosenberg, Sherman *program director*

ADDRESS UNPUBLISHED

Agar, John Russell, Jr. *school district supervisor*
Allen, Charles Eugene *college administrator, agriculturist*
Chu, Shirley Shan-Chi *retired educator*
Clapp, Beverly Booker *university administrator, accountant*
Dutson, Thayne R. *college dean*
Higdon, Charles Anthony *elementary education educator*
Monahan, Edward Charles *academic administrator, marine science educator*
Sestini, Virgil Andrew *secondary education educator*
Snow, W. Sterling *biology and chemistry educator*
Strangway, David William *university president*

ENGINEERING: ACOUSTICAL

UNITED STATES

CALIFORNIA

Long Beach
Simpson, Myles Alan *acoustics researcher*

Los Angeles
Ufimtsev, Pyotr Yakovlevich *radio engineer, educator*

Pacific Palisades
Shapiro, Nathan *acoustical engineer, retired*

MARYLAND

Landover
Ho, Jin-Meng *acoustician*

MASSACHUSETTS

Boston
Scheuzger, Thomas Peter *audio engineer*

Concord
Figwer, Jozef Jacek *acoustics consultant*

Medford
Garrelick, Joel Marc *acoustical scientist, consultant*

NEW JERSEY

Montclair
Hutchins, Carleen Maley *acoustical engineer, violin maker, consultant*

Princeton
Song, Limin *acoustical engineer*

West Orange
Erdreich, John *acoustician, consultant*

NEW YORK

New York
Owen, Thomas Joseph *acoustical engineer*
Queen, Daniel *acoustical engineer, consultant*

Woodbury
Schmid, Charles Ernest *acoustician, administrator*

PENNSYLVANIA

Lancaster
Spalding, George Robert *acoustician, consultant*

State College
Caldwell, Curtis Irvin *acoustical engineer*
Lauchle, Gerald Clyde *acoustics educator*

SOUTH CAROLINA

Clemson
Taylor, Joseph Christopher *audio systems engineer*

TEXAS

Arlington
Rayburn, Ray Arthur *audio engineer*

Houston
Bruce, Robert Douglas *acoustics, noise and vibration control consultant*
Man, Xiuting Cheng *acoustic and ultrasonic engineer, consultant*

VIRGINIA

Arlington
Evans, Roger Michael *acoustical engineer*

WISCONSIN

Milwaukee
Bub, Alexander David *acoustical engineer*

CANADA

BRITISH COLUMBIA

Sidney
Scrimger, Joseph Arnold *research company executive*

ONTARIO

Ottawa
Keith, Stephen Ernest *acoustical researcher*

QUEBEC

Montreal
Woszczyk, Wieslaw Richard *audio engineering educator, researcher*

CZECH REPUBLIC

Prague
Merhaut, Josef *electoacoustics educator*

DENMARK

Aarhus
Bundgaard, Nils *sound and acoustics professional*

ENGLAND

Warrington
Dunbavin, Philip Richard *acoustical consultant*

FRANCE

La Montagne
Rocaboy, Françoise Marie Jeanne *acoustical engineer*

GERMANY

Munich
Fastl, Hugo Michael *acoustical engineer, educator*

INDONESIA

Serpong
Akil, Husein Avicenna *acoustical engineer*

JAPAN

Sendai
Sone, Toshio *acoustical engineering educator*

Tokyo
Miura, Tanetoshi *acoustics educator*

SWITZERLAND

Wallisellen
Kolbe, Hellmuth Walter *acoustical engineer, sound recording engineer*

ADDRESS UNPUBLISHED

Lagerlof, Ronald Stephen *sound recording engineer*
Lerner, Armand *acoustical consultant*
Letowski, Tomasz Rajmund *acoustical engineer, educator*

ENGINEERING: AEROSPACE & AERONAUTICAL

UNITED STATES

ALABAMA

Auburn
Barrett, Ronald Martin *aerospace engineer*

Harvest
Colbert, Robert Floyd *aerospace engineer*

Huntsville
Ballard, Richard Owen *rocket propulsion engineer*
Blackmon, James Bertram *aerospace engineer*
Bramon, Christopher John *aerospace engineer*
Buckelew, Robin Browne *aerospace engineer, manager*
Bushman, David Mark *aerospace engineer*
Chasman, Daniel Benzion *aerospace engineer*
Dannenberg, Konrad K. *aeronautical engineer*
Davis, Stephan Rowan *aerospace engineer*
Imtiaz, Kauser Syed *aerospace engineer*
Jones, Jack Allen *aerospace engineer*
Kim, Young Kil *aerospace engineer*
Mauldin, Charles Robert *aerospace engineer*
Polites, Michael Edward *aerospace engineer*
Vaughan, Otha H., Jr. *aerospace engineer, research scientist*
Wang, Zhi Jian *aerospace engineer*

Montgomery
Cain, Steven Lyle *aerospace engineer*

Redstone Arsenal
Cotney, Carol Ann *research aerospace engineer*
Smith, Troy Alvin *aerospace research engineer*

Tuscaloosa
Martin, James Arthur *aerospace educator*

ARIZONA

Chandler
Newman, Brett *aerospace engineer*
Vondra, Lawrence Steven *aerospace engineer*

Mesa
Kunz, Donald Lee *aeronautical engineer*
Stemple, Alan Douglas *aerospace engineer*

Phoenix
Christensen, Karl Reed *aerospace engineer*
Dittemore, David H. *aerospace engineer, engineering executive*
Myers, Gregory Edwin *aerospace engineer*
Thomas, Harold William *avionics systems engineer, flight instructor*
Zeilinger, Philip Thomas *aeronautical engineer*
Zweifel, Terry L. *aeronautical engineer, researcher*

Scottsdale
Millett, Merlin Lyle *aerospace consultant, educator*

Tempe
Bilimoria, Karl Dhunjishaw *aerospace engineer*
Chattopadhyay, Aditi *aerospace engineering educator*
Salmirs, Seymour *aeronautical engineer, educator*

Tucson
Glick, Michael Andrew *aerospace engineer*
Hofstadter, Daniel Samuel *aerospace engineer*
Speas, Robert Dixon *aeronautical engineer, aviation company executive*

ARKANSAS

Conway
Holt, Frank Ross *retired aerospace engineer*

CALIFORNIA

Altadena
Norton, Harry Neugebauer *aerospace engineer*

Anaheim
Glad, Dain Sturgis *aerospace engineer*

Beverly Hills
Hyatt, Walter Jones *aerospace engineer*

Buena Park
Wiersema, Harold LeRoy *aerospace engineer*

Camarillo
McConnel, Richard Appleton *aerospace company official*

Canoga Park
Hanneman, Timothy John *aerospace engineer*
Muniz, Benigno, Jr. *aerospace engineer, space advocate*
Norman, Arnold McCallum, Jr. *project engineer*
Yang, Ruey-Jen *aerospace engineer*

Chula Vista
Sullivan, Lawrence Jerome *aerospace engineer*

Costa Mesa
Bono, Philip *aerospace consultant*
Richmond, Ronald LeRoy *aerospace engineer*

Davis
Hess, Ronald Andrew *aerospace engineer, educator*

Downey
Adegbola, Sikiru Kolawole *aerospace engineer, educator*
Baumann, Theodore Robert *aerospace engineer, consultant, army officer*
Bienhoff, Dallas Gene *aerospace engineer*
Demarchi, Ernest Nicholas *aerospace company executive*
Flagg, Robert Finch *research aerospace engineer*
Frassinelli, Guido Joseph *aerospace engineer*

Edwards
Deets, Dwain Aaron *aeronautical research engineer*
Henry, Gary Norman *astronautical engineer, educator*
Thompson, Milton Orville *aeronautical engineer*

Edwards AFB
Mead, Franklin Braidwood, Jr. *aerospace engineer*

El Cajon
Lanflisi, Raymond Robert *aeronautical engineer*

El Segundo
Dement, Franklin Leroy, Jr. *aerospace engineer*
Gaylord, Robert Stephen *aerospace engineering manager*
Lotrick, Joseph *aeronautical engineer*
Tamrat, Befecadu *aeronautical engineer*

Encinitas
Manganiello, Eugene Joseph *retired aerospace engineer*

Foothill Ranch
Pierson, Steve Douglas *aerospace engineer*

Fountain Valley
Nakagawa, John Edward *aeronautical engineer*

Fresno
Archer, Thomas John *aeronautical engineer*
Woolard, Henry Waldo *aerospace engineer*

Garden Grove
Vodonick, Emil J. *engineer*

Granada Hills
Brown, Alan Charlton *retired aeronautical engineer*

Hawthorne
Bullard, Roger Dale *aerospace engineer*

Huntington Beach
Gustavino, Stephen Ray *aerospace engineer*
Smith, Joseph Frank *spacecraft electronics and systems engineer*

Irvine
Sirignano, William Alfonso *aerospace and mechanical engineer, educator*
Werner, Roy Anthony *aerospace executive*

La Jolla
Fernandez, Fernando Lawrence *research company executive, aeronautical engineer*

Lancaster
Harwood, Kirk Edward *aeronautical engineer*

Long Beach
Blanche, Joe Advincula *aerospace engineer*
Calkins, Robert Bruce *aerospace engineer*
Heavin, Myron Gene *aeronautical engineer*

Los Alamitos
Iravanchy, Shawn *aerospace engineer, astrophysicist, consultant*

Los Altos
Jones, Robert Thomas *aerospace scientist*

Los Angeles
Christie, Steven Lee *aerospace engineering researcher*
Doo, Yi-Chung *aerospace engineer*
Ilgen, Marc Robert *aerospace engineer*
Jordan, Gregory Wayne *aeronautical engineer*
Monkewitz, Peter Alexis *mechanical and aerospace engineer, educator*
Muntz, Eric Phillip *aerospace engineering and radiology educator, consultant*
Peplinski, Daniel Raymond *project engineer*
Portenier, Walter James *aerospace engineer*
Rauch, Lawrence Lee *aerospace and electrical engineer, educator*
Raymond, Arthur Emmons *aerospace engineer*
Rechtin, Eberhardt *aerospace educator*
Schubert, Gerald *planetary and geophysics educator*
Villani, Daniel Dexter *aerospace engineer*

Los Angeles AFB
Shaver, Marc Steven *aerospace engineer, air force executive officer*

Manhattan Beach
Davis, Dean Earl *aerospace engineer*

Moffett Field
Carlson, Richard Merrill *aeronautical engineer, research executive*
Cook, Anthony Malcolm *aerospace engineer*
Korsmeyer, David Jerome *aerospace engineer*
Mihalov, John Donald *aerospace research scientist*
Peterson, Victor Lowell *aerospace engineer, research center administrator*
Statler, Irving Carl *aerospace engineer*

Monterey
Newberry, Conrad Floyde *aerospace engineering educator*

Oakland
Aubert, Allan Charles *aerospace engineer, consultant*

Orange
Toeppe, William Joseph, Jr. *retired aerospace engineer*

Palo Alto
Holmes, Thomas Joseph *aerospace engineering consultant*
Menon, Padmanabhan *aerospace engineer*
Merz, Antony Willits *aerospace engineer*

Palos Verdes Peninsula
Weiss, Herbert Klemm *aeronautical engineer*

Pasadena
Chang, Daniel Hsing-Nan *aerospace engineer*
Chiang, Richard Yi-Ning *aerospace engineer*
Horvath, Joan Catherine *aeronautical engineer*
Pottorff, Beau Backus *astronautical engineer*
Reid, Macgregor Stewart *aeronautics and microwave engineer*

Playa Del Rey
Tai, Frank *aerospace engineer*

Pomona
Kauser, Fazal Bakhsh *aerospace engineer, educator*

Redondo Beach
Fesq, Lorraine Mae *aerospace and computer engineer*
Sackheim, Robert Lewis *aerospace engineer, educator*

San Bernardino
Holtz, Tobenette *aerospace engineer*

San Diego
Guy, Teresa Ann *aerospace engineer*
Kosmatka, John Benedict *aerospace engineer, educator*
Wang, Kuo Chang *aerospace engineering educator*

San Francisco
Yuan, Shao Wen *aerospace engineer, educator*

San Mateo
Pappas, Costas Ernest *aeronautical engineer, consultant*

Seal Beach
Uda, Robert Takeo *aerospace engineer*
Woo, Steven Edward *aerospace engineer*

Sherman Oaks
Catran, Jack *aerospace and physiology scientist, writer*

Sunnyvale
Sankar, Subramanian Vaidya *aerospace engineer*

Thousand Oaks
Chen, Shih-Hsiung *aerospace engineer*
Shankar, Vijaya V. *aeronautical engineer*

Travis AFB
Mamula, Mark *aerospace engineer*

Van Nuys
Schmetzer, William Montgomery *aeronautical engineer*

Walnut Creek
Van Maerssen, Otto L. *aerospace engineer, consulting firm executive*

COLORADO

Aurora
Schwartz, Lawrence *aeronautical engineer*

Boulder
Born, George H. *aerospace engineer, educator*
Maclay, Timothy Dean *aerospace engineer*
Mohl, James Brian *aerospace engineer*

Colorado Springs
Caughlin, Donald Joseph, Jr. *engineering educator, research scientist, experimental test pilot*
Jones, Richard Eric, Jr. *aerospace engineer*

Denver
Callejo, Gerald Rodriguez *aerospace engineer*
Colvis, John Paris *aerospace engineer, mathematician, scientist*
Marshall, Charles Francis *aerospace engineer*

Englewood
McLellon, Richard Steven *aerospace engineer, consultant*

Estes Park
Blumrich, Josef Franz *aerospace engineer*

Falcon AFB
Chan, John Doddson *aerospace engineer*
Palermo, Christopher John *aerospace engineer*

Golden
Tangler, James Louis *aeronautical engineer*

Littleton
Scruggs, David Wayne *aerospace engineer*
Ulrich, John Ross Gerald *aerospace engineer*

CONNECTICUT

Bridgeport
Brunale, Vito John *aerospace engineer*

East Hartford
Davis, Roger L. *aeronautical engineer*

Hartford
Hajek, Thomas J. *aerospace engineer*
Wagner, Joel H. *aerospace engineer*
Weingold, Harris D. *aerospace engineer*

Mystic
Thompson, Robert Allan *aerospace engineer*

New Britain
Czajkowski, Eva Anna *aerospace engineer, educator*

Stratford
Kmetz, Christopher Paul *lubrication systems development engineer*
Tarulli, Frank James *aerodynamicist*

West Redding
Paltauf, Rudolf Charles *aerospace engineer*

DISTRICT OF COLUMBIA

Washington
Barnum, Peter Montgomery *aerospace engineer, consultant*
Charlton, Kevin Michael *aerospace engineer, support contractor*
Durocher, Cort Louis *aerospace engineer, association executive*
Lenoir, William Benjamin *aeronautical scientist-astronaut*
Liebowitz, Harold *aeronautical engineering educator, university dean*
Nguyen, Hao Marc *aerospace engineer*
Olsavsky, John George *aerospace engineer*
Pine, David Jonah *aerospace engineer*
Saunders, Jimmy Dale *aerospace engineer, physicist, naval officer*
Schulze, Norman Ronnie *aerospace engineer*
Simpers, Glen Richard *aerospace engineer*
Stauffer, Ronald Jay *project engineer, aerospace engineer*
Stever, Horton Guyford *aerospace scientist and engineer, educator, consultant*
Young, John Jacob, Jr. *aerospace engineer*

FLORIDA

Cape Canaveral
Clark, John F. *aerospace research and engineering educator*

Eglin A F B
Ewing, Craig Michael *aerospace engineer*
Porter, John Louis *aerospace engineer*
Prewitt, Nathan Coleman *aerospace engineer*
Swift, Gerald Allan *aeronautical engineer*

Gainesville
Shyy, Wei *aerospace, mechanical engineering educator*

Kennedy Space Center
Russell, Ray Lamar *project manager*

Melbourne
Collins, Emmanuel Gye *aerospace engineer*
Swalm, Thomas Sterling *aerospace executive, retired military officer*

Orlando
Agid, Steven Jay *aerospace engineer*
Huber, Gary Arthur *aerospace engineer*
Thomas, Garland Leon *aerospace engineer*

Satellite Beach
Speaker, Edwin Ellis *retired aerospace engineer*

Titusville
Luecke, Conrad John *aerospace educator*

Tyndall AFB
Herrlinger, Stephen Paul *flight test engineer, air force officer, educator*

West Palm Beach
Gillette, Frank C., Jr. *aeronautical engineer*
Shipman, James Melton *propulsion engineer*

GEORGIA

Atlanta
Armanios, Erian Abdelmessih *aerospace engineer, educator*
Atluri, Satya N(adham) *aerospace engineering educator*
Calise, Anthony John *aerospace engineering educator*
Kardomateas, George Alexander *aerospace engineering educator*
Strahle, Warren Charles *aerospace engineer, educator*

Duluth
Hessman, Frederick William *aeronautical engineer*

Marietta
Garrett, Joseph Edward *aerospace engineer*

Robins AFB
Christian, Thomas Franklin, Jr. *aerospace engineer, educator*

Savannah
Dixon, Sandra Wise *aerospace engineer*
Nawrocki, H(enry) Franz *propulsion technology scientist*

IDAHO

Idaho Falls
Buden, David *aerospace power engineer, nuclear power researcher*
Cott, Donald Wing *aerospace engineer*

ILLINOIS

Bartlett
FitzSimons, Christopher *design engineer*

Cahokia
Herrick, Paul E. *aerospace technology educator, researcher*

Urbana
Burton, Rodney Lane *engineering educator, researcher*
Lee, Ki Dong *aeronautical engineer, educator*
Liu, Zi-Chao *aerospace engineering educator*

INDIANA

Columbus
Bedapudi, Prakash *aerospace engineer*

Indianapolis
Gable, Robert William, Jr. *aerospace engineer*
Oakeson, David Oscar *aerospace engineer*
Torres-Olivencia, Noel R. *aeronautical engineer*

South Bend
Cottle, Eugene Thomas *aerospace engineer*

West Lafayette
Heister, Stephen Douglas *aerospace propulsion educator, researcher*
Sivathanu, Yudaya Raju *aerospace engineer*

IOWA

Ames
Dayal, Vinay *aerospace engineer, educator*
Iversen, James Delano *aerospace engineering educator, consultant*

Iowa City
Pickett, Jolene Sue *aerospace research scientist*

KANSAS

Lawrence
Farokhi, Saeed *aerospace engineering educator, consultant*

Wichita
Ellis, David R. *aeronautical researcher*
Wattson, Robert Kean *aeronautical engineer*

KENTUCKY

Erlanger
Richmond, Raymond Dean *aerospace engineer*

MAINE

Loring AFB
Mitchell, Philip Michael *aerospace engineer, air force officer*

MARYLAND

Aberdeen
Levy, Ronald Barnett *aeronautical engineer*
Walters, William Place *aerospace engineer*

Aberdeen Proving Ground
Grayson, Richard Andrew *aerospace engineer*

Baltimore
Donahoo, Melvin Lawrence *aerospace management consultant, industrial engineer*
Giddens, Don Peyton *engineering educator, researcher*

Bethesda
Durant, Frederick Clark, III *aerospace history and space art consultant*
Furey, Deborah Ann *aerospace engineer, naval architect*

Burtonsville
Lo, Chun-Lau John *aerospace engineer, consultant*

College Park
Howard, Russell Duane *aerospace engineer*

Columbia
Arndt, Julie Anne Preuss *aeronautical engineer*
Grabowsky, Craig *aerospace engineer*

Germantown
Dorsey, William Walter *aerospace engineer, engineering executive*

Greenbelt
Ku, Jentung *mechanical and aerospace engineer*
Lynch, John Patrick *aerospace engineer*

La Plata
Zinn, Michael Wallace *aerospace engineer*

Landover
Fabunmi, James Ayinde *aeronautical engineer*

Laurel
Budman, Charles Avrom *aerospace engineer*
White, Michael Elias *aerospace engineer*

Patuxent River
Tipton, Thomas Wesley *flight engineer*

Seabrook
Browning, Ronald Kenneth *aerospace engineer*

Silver Spring
Scipio, L(ouis) Albert, II *aerospace science engineering educator, architect, military historian*

MASSACHUSETTS

Andover
Peck, A. William *aerospace engineer*

Boston
Nadim, Ali *aerospace and mechanical engineering educator*

Cambridge
Greitzer, Edward Marc *aeronautical engineering educator, consultant*
Samanta Roy, Robie Isaac *aerospace engineer*
Trilling, Leon *aeronautical engineering educator*
Vander Velde, Wallace Earl *aeronautical and astronautical educator*

Lexington
Haldeman, Charles Waldo, III *aeronautical engineer*
Kautz, Frederick Alton, II *aerospace engineer*

Lynn
Ehrich, Fredric F. *aeronautical engineer*

Natick
Gionfriddo, Maurice Paul *research and development manager, aeronautical engineer*
Lee, Calvin K. *aerospace engineer*

Northborough
Jeas, William C. *electronics and aerospace engineering executive*

Sandwich
Chambers, Edward Allen *aerophysical systems designer, consultant engineer*

Somerville
Stoddard, Forrest Shaffer *aerospace engineer, educator*

Waltham
Lackner, James Robert *aerospace medicine educator*

MICHIGAN

Ann Arbor
Faeth, Gerard Michael *aerospace engineering educator, researcher*
Fox, James Carroll *aerospace engineer, program manager*
Kauffman, Charles William *aerospace engineer*
McNally, Patrick Joseph *aerospace engineer, technical program manager*

Novi
Parks, Steven James *aerospace engineer*

MINNESOTA

Minneapolis
Alving, Amy Elsa *aerospace engineering educator*
Cunningham, Thomas B. *aerospace engineer*
Joseph, Daniel Donald *aeronautical engineer, educator*
Patterson, Richard George *retired aerospace engineer*
Sheikh, Suneel Ismail *aerospace engineer, researcher*

MISSISSIPPI

Mississippi State
Rais-Rohani, Masoud *aerospace and engineering mechanics educator*
Thompson, Joe Floyd *aerospace engineer, researcher*

Stennis Space Center
Nunez, Stephen Christopher *aerospace technologist*

MISSOURI

Chesterfield
Lang, James Douglas *aerospace engineer, educator*
Smith, Lawrence Abner *aeronautical engineer*
Yardley, John Finley *aerospace engineer*

Independence
Murray, Thomas Reed *aerospace engineer*

Kansas City
Herrick, Thomas Edward *aeronautical/astronautical engineer*

Rolla
Finaish, Fathi Ali *aeronautical engineering educator*

Saint Charles
Lapinski, John Ralph, Jr. *aerospace engineer*

Saint Louis
Evans, David Myrddin *aerospace engineer*
Hakkinen, Raimo Jaakko *aeronautical engineer, scientist*
Harris, James C II *aerospace engineer*
Lupo, Michael Vincent *aviation engineer*
Orton, George Frederick *aerospace engineer*
Perlmutter, Lawrence David *aerospace engineer*
Ray, James Allen *aerospace engineer*
Scarborough, George Edward *aerospace engineer, researcher*
Shrage, Sydney *aeronautical engineering consultant*
Tarasidis, Jamie Burnette *aerospace engineer*

MONTANA

Butte
Nelson, Gordon Leon, Jr. *aeronautical engineer*

NEBRASKA

Las Vegas
Haas, Robert John *aerospace engineer*

NEVADA

Las Vegas
Broca, Laurent Antoine *aerospace scientist*

NEW JERSEY

Princeton
Bogdonoff, Seymour Moses *aeronautical engineer*
Law, Chung King *aerospace engineering educator, researcher*
Skalecki, Lisa Marie *aerospace engineer*

Waldwick
Yang, Samuel Chia-Lin *engineer*

Wayne
Shedden, James Reid *avionics engineer, consultant*

Willingboro
Schnapf, Abraham *aerospace engineer, consultant*

NEW MEXICO

Albuquerque
Hardy, Gregg Edmund *aerospace engineer*
Pepper, William Burton, Jr. *aeronautical engineer, consultant*

Kirtland AFB
Lederer, John Martin *aeronautical engineer*
Sneegas, Stanley Alan *air force officer, aerospace engineer*

Los Alamos
Nunz, Gregory Joseph *aerospace engineer, mathematics educator, technical manager*

NEW YORK

Albany
Sandu, Bogdan Mihai *aeronautical engineer*

Bethpage
Baglio, Vincent Paul *aeronautical engineer*
Scheuing, Richard Albert *aerospace corporate executive*
Tindell, Runyon Howard *aerospace engineer*

Farmingdale
Oyibo, Gabriel A. *aeronautics and astronautics educator*

Huntington
Schwarz, Robert Charles *aerospace engineer, consultant*

Jamaica
Ali, Khwaja Mohammed *aeronautical engineer*

Northport
Gebhard, David Fairchild *aeronautical engineer, consultant*

Port Jefferson
Elman, Howard Lawrence *aeronautical engineer*

Ronkonkoma
Castrogiovanni, Anthony G. *aerospace engineer, researcher*

Schenectady
Kolb, Mark Andrew *aerospace engineer*

Tonawanda
Drozdziel, Marion John *aeronautical engineer*

Troy
Loewy, Robert Gustav *engineering educator, aeronautical engineering executive*

NORTH CAROLINA

Flat Rock
Matteson, Thomas Dickens *aeronautical engineer, consultant*

OHIO

Brookpark
Wilson, Jack *aeronautical engineer*

Centerville
Keating, Tristan Jack *retired aeronautical engineer*

Cincinnati
Kielb, Robert Evans *propulsion engineer*
Leylek, James H. *aerospace engineer*
Orkwis, Paul David *aerospace engineering educator*
Williams, Trevor Wilson *aeronautical engineering educator*
Wisler, David Charles *aerospace engineer, educator*

Cleveland
Chamis, Christos Constantinos *aerospace scientist, educator*
Fernandez, René *aerospace engineer*
Goldstein, Marvin Emanuel *aerospace scientist, research center administrator*
Haas, Jeffrey Edward *aerospace engineer*
Halford, Gary Ross *aerospace engineer, researcher*
Hwang, Danny Pang *aerospace engineer*
Reshotko, Eli *aerospace engineer, educator*
Steinke, Ronald Joseph *aerospace engineer, consultant*

Columbus
Gatewood, Buford Echols *retired educator, aeronautical and astronautical engineer*
Turchi, Peter John *aerospace and electrical engineer, educator, scientist*

Dayton
Addison, Wallace Lloyd *aerospace engineer*
Chan, Yupo *engineer*
West, Johnny Carl *aeronautical engineer*

Enon
Cyphers, Daniel Clarence *aerospace engineer*

North Royalton
Mohler, Stanley Ross, Jr. *aeronautical research engineer*

Tipp City
Glassmeyer, James Milton *aerospace and electronics engineer*

Wright Patterson AFB
Blake, William Bruce *aerospace engineer*
Eisenhauer, William Joseph, Jr. *aerospace engineer*
Goesch, William Holbrook *aeronautical engineer*
Knoedler, Andrew James *aerospace engineer*
Ruscello, Anthony *aerospace engineer*
Sellin, M. Derek *aerospace engineer*
Wilson, Mark Kent *aerospace engineer*

OKLAHOMA

Tecumseh
Lowe, Ronald Dean *aerospace engineer, consultant*

Tinker AFB
Pray, Donald George *aerospace engineer*

PENNSYLVANIA

Fort Washington
Buescher, Adolph Ernst (Dolph Buescher) *aerospace company executive*

Lancaster
Eastman, G. Yale *aeronautical engineer*

New Galilee
Randza, Jason Michael *aerospace engineer, consultant*

Philadelphia
Cohen, Ira Myron *aeronautical and mechanical engineering educator*
Wiesner, Robert *aeronautical engineer*

Pittsburgh
Ferguson, Jackson Robert, Jr. *astronautical engineer*

University Park
Lakshminarayana, Budugur *aerospace engineering educator*
McCormick, Barnes Warnock *aerospace engineering educator*

Warminster
Tatnall, George Jacob *aeronautical engineer*

RHODE ISLAND

Lincoln
Chakoian, George *aerospace engineer*

SOUTH CAROLINA

Clemson
Figliola, Richard Stephen *aerospace, mechanical engineer*

Greenville
Schneider, George William *retired aircraft design engineer*

TENNESSEE

Arnold AFB
Daniel, Donald Clifton *aerospace engineer*

Knoxville
Pitman, Frank Albert *aerospace engineer*

TEXAS

Arlington
Stanovsky, Joseph Jerry *aerospace engineering educator*

Austin
Bishop, Robert Harold *aerospace engineering educator*
Gramann, Richard Anthony *aerospace engineer*
Hull, David George *aerospace engineering educator, researcher*
Mark, Hans Michael *aerospace engineering educator, physicist*
Szebehely, Victor G. *aeronautical engineer*
Tapley, Byron Dean *aerospace engineer, educator*

Brooks AFB
Nikolai, Timothy John *aerospace engineer*

Cypress
Attles, LeRoy *aerospace engineer*

Dallas
Wilt, James Christopher *aerospace engineer*

Fort Worth
Bradley, Richard Gordon, Jr. *aerospace engineer, engineering director*
Buckner, John Kendrick *aerospace engineer*
Cunningham, Atlee Marion, Jr. *aeronautical engineer*
Narramore, Jimmy Charles *aerospace engineer*
Whiteley, James Morris *retired aerospace engineer*

Friendswood
Wood, Loren Edwin *aerospace engineer, consultant*

Houston
Bains, Elizabeth Miller *aerospace engineer*
Bell, Jerome Albert *aerospace engineer*
Cohen, Aaron *aerospace engineer*
Duke, Michael B. *aerospace scientist*
Dunbar, Bonnie J. *engineer, astronaut*
Epright, Charles John *engineer*
Gest, Robbie Dale *aerospace engineer*
Gilbert, David Wallace *aerospace engineer*
Ivins, Marsha S. *aerospace engineer, astronaut*
Jonke, Erica Elizabeth *aerospace engineer*
McLaughlin, Thomas Daniel *aerospace engineer*
Naqvi, Sarwar *aerospace engineer*
Rochelle, William Curson *aerospace engineer*
Scott, Carl Douglas *aerospace engineer*
Valrand, Carlos Bruno *aerospace engineer*
Wade, James William *aerospace engineer*
Wesselski, Clarence J. *aerospace engineer*
Wren, Robert James *aerospace engineering manager*

McGregor
Johnson, Gary Wayne *aerospace engineer*

Rockwall
Rickards, Michael Anthony *aerospace engineer*

San Antonio
Aldridge, Daniel *aerospace engineer*
Butler, Clark Michael *aerospace engineer*
Walker, Timothy John *aerospace engineer*
Weinbrenner, George Ryan *aeronautical engineer*
Wieland, David Henry *aerospace structural engineer*

Waco
Sivam, Thangavel Parama *aerospace engineer*

UTAH

Brigham City
Mathias, Edward Charles *aerospace engineer*

Salt Lake City
Coffill, Charles Frederick, Jr. *aerospace engineer, educator*

Sandy
Jorgensen, Leland Howard *aerospace research engineer*

Thiokol
Carlisle, Alan Robert *aeronautical design engineer*

VIRGINIA

Alexandria
Graves, Michael Leon, II *aeronautical engineer*
Murray, Russell, II *aeronautical engineer, defense analyst, consultant*
Ostrowski, Peter Phillip *aerospace engineer, researcher*

Arlington
Baughman, George Washington *aeronautical operations research scientist*
Chase, Ramon L. *aerospace engineer*
Gilbert, Arthur Charles *aerospace engineer, consulting engineer*
Kessenich, Karl Otto *aerospace engineer*
Lazich, Daniel *aerospace engineer*
Melichar, John Ancell *aerospace engineer*
Stevens, Donald King *aeronautical engineer*

Blacksburg
Bocvarov, Spiro *aerospace and electrical engineer*
Giurgiutiu, Victor *aeronautical engineer, educator*

Charlottesville
Jacobson, Ira David *aerospace engineer, educator, researcher*
Townsend, Miles Averill *aerospace and mechanical engineering educator*

Clifton
Gawrylowicz, Henry Thaddeus *aerospace engineer*

Dulles
Lindberg, Robert E., Jr. *space systems engineer, controls researcher*

Fairfax
Elias, Antonio L. *aeronautical engineer, aerospace executive*

Falls Church
Schilling, William Richard *aerospace engineer, research and development company executive*

Hampton
Christhilf, David Michael *aerospace engineer*
Duberg, John Edward *aeronautical engineer, educator*
He, Chengjian *aerospace engineer*
Hefner, Jerry Ned *aerospace engineer*
Jegley, Dawn Catherine *aerospace engineer*
Mackley, Ernest Albert *aeropropulsion engineer, consultant*
Martin, Roderick Harry *aeronautics research scientist*
Noblitt, Nancy Anne *aerospace engineer*
Pruett, Charles David *aerospace research scientist, educator*
Quinto, P. Frank *aerospace engineer*
Rivers, Horace Kevin *aerospace engineer, researcher, thermal-structural analyst*
Rivers, Robert Allen *research pilot, aerospace engineer*
Weinstein, Leonard Murrey *aerospace engineer, consultant, researcher*
Wood, George Marshall *research scientist, program manager*

Kilmarnock
Gilruth, Robert Rowe *aerospace consultant*

Mc Lean
Waesche, R(ichard) H(enley) Woodward *combustion research scientist*

Reston
Hu, Tsay-Hsin Gilbert *aerospace engineer*

Springfield
Bryan, Hayes Richard *aerospace engineer*

Sterling
O'Rourke, Thomas Joseph *aerospace engineer, consultant*

Vienna
Ross, John R., III *aerospace engineer*

Winchester
Turner, William Richard *retired aeronautical engineer, consultant*

Woodbridge
Schaefer, Carl George, Jr. *aerospace engineer*

Yorktown
Ambur, Damodar Reddy *aerospace engineer*

WASHINGTON

Bainbridge Is
Whitener, Philip Charles *aeronautical engineer, consultant*

Bellevue
Boike, Shawn Paul *aerospace project/design engineer*

Kent
Raymond, Eugene Thomas *technical writer, retired aircraft engineer*

Mercer Island
Bridgforth, Robert Moore, Jr. *aerospace engineer*

Seattle
Bateman, Robert Edwin *aeronautical engineer*
Hertzberg, Abraham *aeronautical engineering educator, university research scientist*
Ii, Jack Morito *aerospace engineer*
Sutter, Joseph F. *aeronautical engineer, consultant, retired airline company executive*
Wiktor, Peter Jan *engineer*

WYOMING

Jackson
Werner, Frank David *aeronautical engineer*

CANADA

MANITOBA

Anola
de Nevers, Roy Olaf *retired aerospace company executive*

Winnipeg
Fielding, Ronald Roy *aeronautical engineer*

ONTARIO

Etobicoke
Ibrahim, Mohammad Fathy Kahlil *aerospace engineer*

Mississauga
Elfstrom, Gary Macdonald *aerospace engineer, consultant*

Saint Catharines
Picken, Harry Belfrage *aerospace engineer*

Unionville
Godfrey, David Wilfred Holland *aeronautical engineer, communication educator*

QUEBEC

Montreal
Hoffman, Kevin William *aerospace engineer*

ARGENTINA

Buenos Aires
De Leon, Pablo Gabriel *aerospace engineer*
Di Russo, Erasmo Victor *aeronautical engineer, educator, consultant*

AUSTRALIA

Fishermen's Bend
D'Cruz, Jonathan *aeronautical engineer*
Scott, Murray Leslie *aerospace engineer*

Hughes
Bolonkin, Kirill Andrew *aeronautical engineer*

Port Melbourne
Schofield, William Hunter *research aerodynamicist*

BELGIUM

Haren
De Ceuster, Luc Frans *avionics educator*

Melsbroek
Becuwe, Ivan Gerard *aeronautical engineer*

Rhode-Saint-Genese
Sarma, Gabbita Sundara Rama *aerospace scientist*

BRAZIL

Sao Jose dos Campos
Bismarck-Nasr, Maher Nasr *aeronautical engineering educator*
Lutterbach, Rogerio Alves *aeronautical engineer, consultant, researcher*
Souza, Marcelo Lopes *aerospace engineer, researcher, consultant*

CHINA

Beijing
Liang, Junxiang *aeronautics and astronautics engineer, educator*
Situ, Ming *aerospace engineering researcher*
Wu, Daguan *aeroengine design engineer*
Xue-ying, Deng *aerodynamicist*
Zhang, Ping *aerospace engineering educator*

Harbin
Yang, Di *aerospace engineer, educator*

Xi'an
Chen, Shilu *flight mechanics educator*

CROATIA

Zagreb
Janković, Slobodan *aerodynamics educator, researcher*

CYPRUS

Nicosia
Vrachas, Constantinos Aghisilaou *aeronautical engineer*

EGYPT

Cairo
Elnomrossy, Mokhtar Malek *aeronautical engineer*

Sharkia
El-Sayed, Ahmed Fayez *aeronautical engineering educator*

ENGLAND

Cambridge
Mair, William Austyn *aeronautical engineer*

Filton
Herring, Paul George Colin *aeronautical engineer*
Joyce, Terence Thomas *aerospace engineer*

Manchester
Poll, David Ian Alistair *aerospace engineering educator, consultant*

Olney
Keith-Lucas, David *retired aeronautical engineer*

Portsmouth
Hinds, James William *aeronautical engineer*

Reading
Webb, Gregory Frank *retired aeronautical engineer*

Sevenoaks
Luttman, Horace Charles *retired aeronautical engineer*

Talybont-on-Usk
Crabtree, Lewis Frederick *aeronautical engineer, educator*

Yeovil
Kokkalis, Anastasios *aeronautical engineer, consultant*

FRANCE

Marseilles
Dumitrescu, Lucien Z. *aerospace researcher*

Neuilly sur Seine
Wennerstrom, Arthur John *aeronautical engineer*

Paris
Ohayon, Roger Jean *aerospace engineering educator, scientific deputy*

Saint-Cloud
Perrier, Piérre Claude *aeronautical engineer, researcher*

Suresnes
Donguy, Paul Joseph *aerospace engineer*
Pouliquen, Marcel François *space propulsion engineer, educator*

Toulouse
Zappoli, Bernand *aerospace engineer*

GERMANY

Bayern
Boldt, Heinz *aerospace engineer*

Berlin
Bernard, Herbert Fritz *aerospace engineer*

Braunschweig
Blazek, Jiri *aerospace engineer, researcher*
Hahn, Klaus-Uwe *aerospace engineer*
Radespiel, Rolf Ernst *aerospace engineer*

Bremen
Iglseder, Heinrich Rudolf *space engineer, researcher*
Mertens, Josef Wilhelm *engineer*

Göttingen
Lorenz-Meyer, Wolfgang *aeronautical engineer*
Nitzsche, Fred *aeronautical engineer*

Koblenz
Kuntze, Herbert Kurt Erwin *aeronautical engineer*

Munich
Freymann, Raymond Florent *aeronautical engineer*
Gottzein, Eveline *aerospace engineer*

Putzbrunn
Gregoriou, Gregor Georg *aeronautical engineer*

Stuttgart
Krueger, Ronald *aerospace engineer*
Wagner, Henri Paul *aerospace engineer*

Triefenstein
Makowski, Gerd *aerospace engineer*

GREECE

Koridallos
Cambalouris, Michael Dimitrios *aircraft engineer*

HUNGARY

Budapest
Lovro, István *aircraft engineer*

INDIA

Bangalore
Kantheti, Badari Narayana *aeronautical engineer*
Narasimha, Roddam *laboratory director, educator*
Ramasubbu, Sunder *aeronautical engineer, researcher*
Rao, K. Prabhakara *aerospace engineering educator*
Reddy, Narayana Muniswamy *aerospace educator*

Kanpur
Tewari, Ashish *aerospace scientist, consultant*

ISRAEL

Haifa
Idan, Moshe *aerospace engineer*
Natan, Benveniste *aeronautical engineer, educator, researcher*
Nissim, Eliahu *aerospace engineer, researcher*
Singer, Josef *aerospace engineer, educator*

Ramat-Hasharon
Ben-Asher, Joseph Zalman *aeronautical engineer*

Tel Aviv
Seifert, Avi *aerodynamic engineering educator and researcher*

ITALY

Naples
Coiro, Domenico Pietro *research scientist, educator*
Mirra, Carlo *aerospace engineer, researcher*

Rome
Depasquale, Francesco *aerospace engineer*

Varese
Mezzanotte, Paolo Alessandro *aeronautical engineer*

JAPAN

Fukuoka
Goto, Norihiro *aeronautical engineering educator*

Kanagawa
Maeda, Toshihide Munenobu *spacecraft system engineer*

Miyagi
Miyajima, Hiroshi *aerospace engineer*

Nagoya
Matsuzaki, Yuji *aerospace and biomechanics educator, researcher*
Nitta, Kyoko *aerospace engineering educator*
Ohta, Hirobumi *aeronautical engineer, educator*

Oita
Kimura, Haruo *aeronautical engineering educator*

Tokyo
Fujii, Hironori Aliga *aerospace engineer, educator*
Higashiguchi, Minoru *aerospace electronic engineering educator*
Morishita, Etsuo *aeronautical, mechanical engineering educator*
Nomura, Shigeaki *aerospace engineer*
Onishi, Ryoichi *aeronautical engineer, designer*
Sakata, Kimio *aerospace engineer, researcher*

Tsukuba
Koga, Tatsuzo *aerospace engineer*

JORDAN

Amman
Majdalawi, Fouad Farouk *aeronautical engineer*

THE NETHERLANDS

Noordwijk
Caswell, Robert Douglas *aerospace engineer*
Kletzkine, Philippe *aerospace engineer*
Novara, Mauro *space system engineer*
Racca, Giuseppe Domenico *aerospace engineer*

NORTHERN IRELAND

Belfast
Raghunathan, Raghu Srinivasan *aeronautical engineer, educator, researcher*

REPUBLIC OF KOREA

Daejeon
Lee, Sung Taick *aerospace engineer*

Seoul
Hong, Yong Shik *aerospace engineer*
Jeung, In-Seuck *aerospace engineering educator*
Kim, Youdan *aerospace engineer*

RUSSIA

Kaliningrad
Anfimov, Nikolai *aerospace researcher*

Moscow
Syromiatnikov, Vladimir Sergeevich *aerospace engineer*

SAUDI ARABIA

Riyadh
Al-Ohali, Khalid Suliman *aeronautical engineer, researcher, industrial development specialist*

SINGAPORE

Singapore
Kan, Bill Yuet Him *aeronautical engineer*
Tan, Kok Tin *aerospace engineer*

SPAIN

Las Arenas
Erce, Ignacio, III *aerospace engineer*

Madrid
Consejo, Eduardo *aeronautics engineer*

TAIWAN

Lungtan
Chern, Jeng-Shing *aerospace engineer*

Tainan
Chang, Keh-Chin *aerospace engineer, educator*
Chao, Yei-chin *aerospace engineering educator*
Choi, Siu-Tong *aeronautical engineering educator*
Hsiao, Fei-bin *aerospace engineering educator*
Jing, Hung-Sying *aerospace engineering educator*

TURKEY

Ankara
Ertem, Özcan *aeronautical engineer*

ADDRESS UNPUBLISHED

Abanero, Jose Nelito Talavera *aerospace engineer*
Allison, John McComb *aeronautical engineer, retired*
Balettie, Roger Eugene *aerospace engineer*
Behm, Dennis Arthur *engineering specialist*
Belgau, Robert Joseph *aeronautical engineer*
Boxenhorn, Burton *aerospace engineer*
Brown, Charles Durward *aerospace engineer*
Carta, Franklin Oliver *retired aeronautical engineer*
Drozd, Joseph Duane *aerospace engineer*
Edwards, Harold Hugh, Jr. *aerospace engineer, management consultant*
Eiden, Michael Josef *aerospace engineer*
Ellis, Michael David *aerospace engineer*
Flynn, Patrick Joesph *aerospace engineer*
Frederick, Ronald David *aerospace engineer*
Gorenberg, Norman Bernard *aeronautical engineer, consultant*
Harris, Martin Stephen *aerospace engineering executive*
Hoey, David Joseph *aerospace design engineer*
Huzel, Dieter Karl *retired aerospace engineer*
James, Earl Eugene, Jr. *aerospace engineering executive*
John, Leonard Keith *aerospace and mechanical engineer*
Johnston, Ralph Kennedy, Sr. *aerospace engineer*
Keigler, John E. *aerospace engineer*
Kersey, Terry L(ee) *astronautical engineer*
Lang, Derek Edward *aerospace engineer*
Latz, John Paul *aerospace engineer*
Mann, Laura Susan *aerospace engineer*
Mittal, Manoj *aerospace engineer*
Neufeld, Murray Jerome *aerospace scientist/engineer, consultant*
Newton, Jeffrey F. *project engineer*
Nix, Martin Eugene *engineer*
Palmer, Ashley Joanne *aerospace engineer*
Scott, Larry Marcus *aerospace engineer, mathematician*
Sebek, Kenneth David *aerospace engineer, air force officer*
Shuart, Mark James *aerospace engineer*
Sponable, Jess M. *astronautical engineer, physicist*
Tice, David Charles *aerodynamics engineer*
Torres, Manuel *aerospace engineer*
Weingarten, Joseph Leonard *aerospace engineer*
West, William Ward *aerospace engineer*
Whittle, Frank *aeronautical engineer*
Wilkinson, Todd Thomas *project engineer*

ENGINEERING: AGRICULTURAL

UNITED STATES

ALABAMA

Auburn
Schafer, Robert Louis *agricultural engineer, researcher*
Turnquist, Paul Kenneth *agricultural engineer, educator*
Yoo, Kyung Hak *agricultural engineer, educator*

ARIZONA

Sun City
Woodruff, Neil Parker *agricultural engineer*

CALIFORNIA

Carmel Valley
Lorenzen, Coby *emeritus engineering educator*

Davis
Akesson, Norman Berndt *agricultural engineer, emeritus educator*
Chancellor, William Joseph *agricultural engineering educator*
Fridley, Robert Bruce *agricultural engineering educator, academic administrator*
Kepner, Robert Allen *agricultural engineering researcher, educator*

Ontario
Johnson, Maurice Verner, Jr. *agricultural research and development executive*

Sunnyvale
Ma, Fengchow Clarence *agricultural engineering consultant*

COLORADO

Fort Collins
Boyd, Landis Lee *agricultural engineer, educator*

FLORIDA

Atlantic Beach
Wheeler, William Crawford *agricultural engineer, educator*

Gainesville
Price, Donald Ray *agricultural engineer, university administrator*
Shaw, Lawrance Neil *agricultural engineer, educator*

Tallahassee
De Forest, Sherwood Searle *agricultural engineer, agribusiness services executive*

GEORGIA

Athens
Tollner, Ernest William *agricultural engineering educator, agricultural radiology consultant*

Tifton
Butler, James Lee *agricultural engineer, researcher*

IDAHO

Kimberly
Trout, Thomas James *agricultural engineer*

Moscow
DeShazer, James Arthur *agricultural engineer, educator, administrator*
Peterson, Charles Loren *agricultural engineer, educator*

ILLINOIS

Champaign
Puckett, Hoyle Brooks *agricultural engineer, research scientist, consultant*

Urbana
Hunt, Donnell Ray *agricultural engineering educator*
Jones, Benjamin Angus, Jr. *retired agricultural engineering educator, administrator*
Yoerger, Roger Raymond *agricultural engineer, educator*

INDIANA

West Lafayette
Monke, Edwin John *agricultural engineering educator*
Richey, Clarence Bentley *agricultural engineering educator*

IOWA

Ames
Buchele, Wesley Fisher *agricultural engineering educator, consultant*
Colvin, Thomas Stuart *agricultural engineer, farmer*
Curry, Norval Herbert *retired agricultural engineer*
Johnson, Howard Paul *agricultural engineering educator*

KANSAS

Manhattan
Hagen, Lawrence Jacob *agricultural engineer*
Johnson, William Howard *agricultural engineer, educator*
Zhang, Naiqian *agricultural engineer, educator*

KENTUCKY

Lexington
Walker, John Neal *agricultural engineering educator*

LOUISIANA

Baton Rouge
Tipton, Kenneth Warren *agricultural administrator, researcher*

Ruston
Robbins, Jackie Wayne Darmon *agricultural and irrigation engineer*

MARYLAND

Silver Spring
Green, Robert Lamar *consulting agricultural engineer*

MICHIGAN

East Lansing
von Bernuth, Robert Dean *agricultural engineering educator, consultant*

Saint Joseph
Castenson, Roger R. *agricultural engineer, association executive*

MINNESOTA

Saint Paul
Wilson, Bruce Nord *agricultural engineer*

MISSISSIPPI

Oxford
Meyer, L. Donald *agricultural engineer, researcher, educator*
Mutchler, Calvin Kendal *hydraulic research engineer*

MISSOURI

Columbia
Day, Cecil LeRoy *agricultural engineering educator*

NEBRASKA

Clay Center
Hahn, George LeRoy *agricultural engineer, biometeorologist*

Lincoln
Hanna, Milford A. *agricultural engineering educator*
Splinter, William Eldon *agricultural engineering educator*
Summers, James Donald *agricultural engineering consultant*

Valley
Chapman, John Arthur *agricultural engineering executive*

NEW JERSEY

Belle Mead
Singley, Mark Eldridge *agricultural engineering educator*

NEW YORK

Ithaca
Rehkugler, Gerald Edwin *agricultural engineering educator, consultant*

NORTH CAROLINA

Raleigh
Kriz, George James *agricultural research administrator, educator*
Rohrbach, Roger Phillip *agricultural engineer, educator*
Skaggs, Richard Wayne *agricultural engineering educator*
Young, James Herbert *agricultural engineer*

OHIO

Alexandria
Palmer, Melville Louis *retired agricultural engineering educator*

Columbus
Bondurant, Byron Lee *agricultural engineering educator*
Nelson, Gordon Leon *agricultural engineering educator*
Schwab, Glenn Orville *retired agricultural engineering educator, consultant*
Taiganides, E. Paul *agricultural-environmental engineer, consultant*

OKLAHOMA

Stillwater
Brusewitz, Gerald Henry *agricultural engineering educator, researcher*
Haan, Charles Thomas *agricultural engineering educator*
Noyes, Ronald T. *agricultural engineering educator*
Thompson, David Russell *agricultural engineering educator, academic dean*

OREGON

Corvallis
Hashimoto, Andrew Ginji *bioresource engineer*
Miner, John Ronald *agricultural engineer*

PENNSYLVANIA

Harrisburg
Stroh, Oscar Henry *agricultural, civil and industrial engineer, consultant*

University Park
Buffington, Dennis Elvin *agricultural engineering educator*

SOUTH CAROLINA

Clemson
Bunn, Joe Millard *agricultural engineering educator*

SOUTH DAKOTA

Brookings
Anderson, Gary Arlen *agricultural engineering educator, consultant*

TENNESSEE

Knoxville
Richardson, Don Orland *agricultural educator*

TEXAS

Austin
Gavande, Sampat Anand *agricultural engineer, soil scientist*

Big Spring
Fryrear, Donald William *agricultural engineer*

Bushland
Howell, Terry Allen *agricultural engineer*

College Station
Kotzabassis, Constantinos *agricultural engineering educator*
Kunze, Otto Robert *retired agricultural engineering educator*
Reddell, Donald Lee *agricultural engineer*
Stermer, Raymond Andrew *agricultural engineer*

Lindale
Bockhop, Clarence William *retired agricultural engineer*

VIRGINIA

Arlington
Hall, Carl William *agricultural and mechanical engineer*

Blacksburg
Haugh, Clarence Gene *agricultural engineering educator*
Perumpral, John Verghese *agricultural engineer, administrator, educator*

Fort Belvoir
Blackwell, Neal Elwood *agricultural and mechanical engineer, researcher*

WISCONSIN

Madison
Bruhn, Hjalmar Diehl *retired agricultural engineer, educator*
Schuler, Ronald Theodore *agricultural engineering educator*

TERRITORIES OF THE UNITED STATES

PUERTO RICO

Mayaguez
Rodríguez-Arias, Jorge H. *retired agricultural engineering educator*

CANADA

ONTARIO

Delhi
Whitfield, Gary Hugh *research director*

QUEBEC

Sainte-Foy
St-Yves, Angèle *agricultural engineer*

JAPAN

Nishihara
Tawata, Shinkichi *agricultural engineering educator*

KENYA

Nairobi
Swallow, Brent Murray *agricultural economist, researcher*

THE NETHERLANDS

Wageningen
Coolman, Fiepko *retired agricultural engineer, researcher, consultant*

TAIWAN

Taipei
Liao, Chung Min *agricultural engineer, educator*

THAILAND

Bangkok
Singh, Gajendra *agricultural engineering educator*

ADDRESS UNPUBLISHED

Carreker, John Russell *retired agricultural engineer*
Hirning, Harvey James *agricultural engineer, educator*
Jensen, Marvin Eli *retired agricultural engineer*
Moore, James Allan *agricultural engineering educator*

ENGINEERING: BIOMEDICAL

UNITED STATES

ALABAMA

Birmingham
Rigney, E. Douglas *biomedical engineer*

Madison
Sarphie, David Francis *biomedical engineer*

ARIZONA

Tempe
Guilbeau, Eric J. *biomedical engineer, electrical engineer, educator*

ARKANSAS

Fayetteville
Schmitt, Neil Martin *biomedical engineer, electrical engineering educator*

CALIFORNIA

Irvine
Kinsman, Robert Preston *biomedical plastics engineer*

La Jolla
Bartsch, Dirk-Uwe Guenther *research bioengineer, consultant*
Chien, Shu *physiology and bioengineering educator*
Fung, Yuan-Cheng Bertram *bioengineering educator, author*
Schmid-Schoenbein, Geert Wilfried *biomedical engineer, educator*

Los Angeles
Einav, Shmuel *biomedical engineering educator*

Mountain View
Johnson, Noel Lars *biomedical engineer*

Stanford
Contag, Christopher Heinz *biomedical researcher*
Stevens, Mary Elizabeth *biomedical engineer, consultant*

DISTRICT OF COLUMBIA

Washington
Jaron, Dov *biomedical engineer, educator*
Leonard, Joel I. *biomedical engineer*

FLORIDA

Fort Lauderdale
Parsons, Bruce Andrew *biomedical engineer, scientist*

Miami
Nagel, Joachim Hans *biomedical engineer, educator*

Pensacola
Olsen, Richard Galen *biomedical engineer, researcher*

GEORGIA

Atlanta
Gu, Yuanchao *biomedical engineer, geneticist*
Hartzell, Harrison Criss, Jr. *biomedical research educator*
Yoganathan, Ajit Prithiviraj *biomedical engineer, educator*

ILLINOIS

Chicago
Healy, Kevin E. *biomedical engineering educator*
Jadvar, Hossein *biomedical engineer, oncologist*

INDIANA

Indianapolis
Yeager, Hal K. *biomedical engineer*

Lafayette
Berman, Michael Allan *biomedical clinical engineer*

IOWA

Iowa City
Goel, Vijay Kumar *biomedical engineer*
Voo, Liming M. *biomedical engineer, researcher*
Yu, Cong *biomedical engineer*

LOUISIANA

Ruston
Abdelhamied, Kadry A. *biomedical engineer*

MARYLAND

Annapolis
Crawford, Claude Cecil, III *biomedical consultant*

Baltimore
Thakor, Nitish Vyomesh *biomedical engineering educator*

Bethesda
Bak, Martin Joseph *biomedical engineer*
Kawakami, Yutaka *biomedical researcher, hematologist*
Odeyale, Charles Olajide *biomedical engineer*
Sobering, Geoffrey Simon *biomedical scientist*

Columbia
Falk, Steven Mitchell *biomedical engineer, consultant*

Poolesville
Chang, Zhao Hua *biomedical engineer*

Rockville
Burdick, William MacDonald *biomedical engineer*
Munzner, Robert Frederick *biomedical engineer*
Ruggera, Paul Stephen *biomedical engineer*

MASSACHUSETTS

Cambridge
Mann, Robert Wellesley *biomedical engineer, educator*
Poon, Chi-Sang *biomedical engineering researcher*

Weston
Resden, Ronald Everette *medical devices product development engineer*

MICHIGAN

Ann Arbor
Soslowsky, Louis Jeffrey *bioengineering educator, researcher*

Warren
Schreck, Richard Michael *biomedical engineer, consultant*

MINNESOTA

Arden Hills
Spinelli, Julio Cesar *biomedical engineer*

Maple Grove
Daniel, John Mahendra Kumar *biomedical engineer, researcher*

MISSOURI

Imperial
Auld, Robert Henry, Jr. *biomedical engineer, educator, consultant, author*

Saint Louis
Khuhro, Shafiq Ahmed *biomedical engineer*
Suydam, Peter R. *clinical engineer, consultant*

NEBRASKA

Omaha
Gendelman, Howard Eliot *biomedical researcher, physician*

NEW JERSEY

Piscataway
Dunn, Stanley Martin *biomedical engineering educator*
Hung, George Kit *biomedical engineering educator*
Semmlow, John Leonard *biomedical engineer, research scientist*

Princeton
Kramer, Richard Harry, Jr. *biomedical engineer*

NEW YORK

Bronx
Hovnanian, H. Philip *biomedical engineer*

Brooklyn
Ortiz, Mary Theresa *biomedical engineer, educator*

Clarence
Greatbatch, Wilson *biomedical engineer*

Elmsford
Sklarew, Robert Jay *biomedical research educator, consultant*

Fairport
Stohr, Erich Charles *biomedical engineer*

New York
Alexander, Harold *bioengineer, educator*
Bardin, Clyde Wayne *biomedical researcher and developer of contraceptives*
Cowin, Stephen Corteen *biomedical engineering educator, consultant*
Galanopoulos, Kelly *biomedical engineer*
Rosskothen, Heinz Dieter *engineer*
Schoenfeld, Robert Louis *biomedical engineer*

Rochester
Hejazi, Shahram *biomedical engineer*

Stony Brook
Guilak, Farshid *biomedical engineering researcher, educator*

Syracuse
Tabor, John Malcolm *genetic engineer*

Tarrytown
Farrell, Gregory Alan *biomedical engineer*

Troy
Cook, Raymond Douglas *biomedical engineer*
Roy, Rob J. *biomedical engineer, anesthesiologist*

NORTH CAROLINA

Cary
Myers, Donald Richard *biomedical engineer*

Durham
Miller, Charles Gregory *biomedical researcher*
Pilkington, Theo Clyde *biomedical and electrical engineering educator*

Raleigh
Eberhardt, Allen Craig *biomedical engineer, mechanical engineer*

OHIO

Akron
Verstraete, Mary Clare *biomedical engineering educator*

Cleveland
Collura, Thomas Francis *biomedical engineer*
Difiore, Juliann Marie *biomedical engineer*
Mortimer, J. Thomas *biomedical engineering educator*
Neuman, Michael Robert *biomedical engineer*
Peckham, P. Hunter *biomedical engineer, educator*

Columbus
Chovan, John David *biomedical engineer*
Wu, Chao-Min *biomedical engineer*

Wyoming
Jarzembski, William Bernard *biomedical engineer*

PENNSYLVANIA

Philadelphia
Reid, John Mitchell *biomedical engineer*

Pittsburgh
Woo, Savio Lau-Yuen *bioengineering educator*

University Park
Geselowitz, David Beryl *bioengineering educator*

Warminster
Slawiatynsky, Marion Michael *biomedical electronics engineer, software consultant*

RHODE ISLAND

Foster
Schuyler, Peter R. *biomedical electrical engineer, educator, researcher*

TEXAS

Austin
Baker, Lee Edward *biomedical engineering educator*
Durkin, Anthony Joseph *biomedical engineer, researcher*
Richards-Kortum, Rebecca Rae *biomedical engineering educator*

College Station
Coté, Gerard Laurence *biomedical engineering educator*

Galveston
Sheppard, Louis Clarke *biomedical engineer, educator*

Houston
McKay, Colin Bernard *biomedical engineer*

San Antonio
Agrawal, Chandra Mauli *biomaterials engineer, educator*

UTAH

Salt Lake City
Jacobsen, Stephen Charles *biomedical engineer, educator*
McArthur, Gregory Robert *biomedical engineer*
Olsen, Donald Bert *biomedical engineer, experimental surgeon, research facility director*
Swerdlow, Harold *biomedical engineer*

VERMONT

Burlington
Sachs, Thomas Dudley *biomedical engineering scientist*

VIRGINIA

Alexandria
Grogan, Edward Joseph *clinical engineer*

Charlottesville
Edlich, Richard French *biomedical engineering educator*
Hutchinson, Thomas Eugene *biomedical engineering educator*

Norfolk
Pittenger, Gary Lynn *biomedical educator*

WISCONSIN

Madison
Webster, John Goodwin *biomedical engineering educator, researcher*

Milwaukee
Chow, John Lap Hong *biomedical engineer*
Newman, Robert Wyckoff *research engineer*

CANADA

ALBERTA

Edmonton
Rajotte, Ray V. *biomedical engineer, researcher*

ONTARIO

Toronto
Kunov, Hans *biomedical engineering educator, electrical engineering educator*

QUEBEC

Montreal
Gulrajani, Ramesh Mulchand *biomedical engineer, educator*
Pelletier, Claude Henri *biomedical engineer*

CHINA

Beijing
Zhang, Tianyou *biomedical engineer*

ENGLAND

Brighton
Luo, Hong Yue *biomedical engineering researcher*

JAPAN

Yokohama
Aizawa, Masuo *bioengineering educator*
Soetanto, Kawan *biomedical-electrical engineering educator*

NORWAY

Bergen
Ramirez-Garcia, Eduardo Agustin *biomedical engineer*

ADDRESS UNPUBLISHED

Liu, Young King *biomedical engineering educator*
Scott, Ian Laurence *biomedical engineer*
Zweizig, J(ohn) Roderick *biomedical engineer*

ENGINEERING: CHEMICAL

UNITED STATES

ALABAMA

Birmingham
Cannon, Richard Alan *chemical process engineer*
Dahlin, Robert Steven *chemical engineer*
Pulliam, Terry Lester *chemical engineer*

Huntsville
Tiwari, Subhash Ramadhar *chemical engineer*

Montgomery
Adams, James Henry *chemical engineer*

Moss Point
Harris, Kenneth Kelley *chemical engineer*

Theodore
Smith, David Floyd *chemical engineer*

ARIZONA

Phoenix
Castaneda, Mario *chemical engineer, educator*

Tempe
Berman, Neil Sheldon *chemical engineering educator*
Garcia, Antonio Agustin *chemical engineer, bioengineer, educator*

Tucson
Leaman, Gordon James, Jr. *chemical engineer*

ARKANSAS

El Dorado
Reames, Thomas Eugene *chemical engineer*

Fayetteville
Gaddy, James Leoma *chemical engineer, educator*

Little Rock
Hocott, Joe Bill *chemical engineer*

Stuttgart
Miller, Thomas Nathan *chemical engineer*

CALIFORNIA

Agoura Hills
Chang, Chong Eun *chemical engineer*

Alhambra
Kister, Henry Z. *chemical engineer*
Sepulveda, Eduardo Solideo *chemical engineer*

Berkeley
Denn, Morton Mace *chemical engineering educator*
Harris, Guy Hendrickson *chemical research engineer*

Carlsbad
Raphael, George Farid *chemical engineer*

Davis
Gates, Bruce Clark *chemical engineer, educator*

Glendora
Haile, Benjamin Carroll, Jr. *chemical engineer, mechanical engineer*

Hercules
Emmanuel, Jorge Agustin *chemical engineer, environmental consultant*

Huntington Beach
Adam, Steven Jeffrey *chemical engineer*

Irvine
Atwong, Matthew Kok Lun *chemical engineer*

La Jolla
Riedinger, Alan Blair *chemical engineer, consultant*

Lindsay
Webster, John Robert *chemical engineer*

Los Angeles
Dougherty, Elmer Lloyd, Jr. *chemical engineering educator, consultant*
Kardously, George J. *chemical engineer*
Senkan, Selim M. *chemical engineering educator*

Menlo Park
Niksa, Stephen Joseph *research chemical engineer, consultant*

Palo Alto
Culler, Floyd LeRoy, Jr. *chemical engineer*

Pasadena
Flagan, Richard Charles *chemical engineering educator*
Morari, Manfred *chemical engineer, educator*
Seinfeld, John Hersh *chemical engineering educator*

Pittsburg
Yoshisato, Randall Atsushi *chemical engineer*

Pleasanton
Worley, David Allan *biochemical engineer*

Redondo Beach
Chazen, Melvin Leonard *chemical engineer*

San Anselmo
Bodington, Charles E. *chemical engineer*

San Diego
Dayson, Rodney Andrew *chemical engineer*

San Jose
Karis, Thomas Edward *chemical engineer*

San Pedro
Ellis, George Edwin, Jr. *chemical engineer*

San Ramon
Morris, Jay Kevin *chemical engineer*

Santa Barbara
Israelachvili, Jacob Nissim *chemical engineer*
Pincus, Philip A. *chemical engineering educator*
Russell, Charles Roberts *chemical engineer*

Stanford
Boudart, Michel *chemist, chemical engineer, educator*
Gast, Alice P. *chemical engineering educator*

Thousand Oaks
Gentry, James Frederick *chemical engineer, consultant*
Sachdev, Raj Kumar *biochemical engineer*

Ventura
Gaynor, Joseph *chemical engineering consultant*
Hendel, Frank J(oseph) *chemical and aerospace engineer, educator, technical consultant*

Walnut Creek
Sanborn, Charles Evan *retired chemical engineer*

Woodland Hills
Darnell, Alfred J(erome) *chemical engineer, consultant*

COLORADO

Boulder
Sani, Robert LeRoy *chemical engineering educator*

Canon City
Mc Bride, John Alexander *retired chemical engineer*

Denver
Perez, Jean-Yves *engineering company executive*
Seidel, Frank Arthur *chemical engineer*

Golden
Sloan, Earle Dendy, Jr. *chemical engineering educator*

Littleton
Jargon, Jerry Robert *chemical engineer*

Wheat Ridge
Henehan, Joan *chemical engineer, consultant*

CONNECTICUT

Bloomfield
Dobay, Donald G. *chemical engineer*

Danbury
Malwitz, Nelson Edward *chemical engineer*
Yuh, Chao-Yi *chemical engineer, electrochemist*

Darien
Glenn, Roland Douglas *chemical engineer*

East Hartford
Nelson, Chad Matthew *chemical engineer, researcher*

Mansfield
Shaw, Montgomery Throop *chemical engineering educator*

New Haven
Horváth, Csaba *chemical engineering educator, researcher*

Stamford
Gupta, Dharam V. *chemical engineer*
Ostroff, Norman *chemical engineer*
Robert, Debra Ann *chemical engineer*

Stratford
Zimmerman, Daniel D. *chemical engineer*

West Hartford
Markham, Sister M(aria) Clare *chemistry educator, college administrator*

DELAWARE

Hockessin
Bischoff, Kenneth Bruce *chemical engineer, educator*

Newark
Cooper, Stuart Leonard *chemical engineering educator, researcher, consultant*
Klein, Michael Tully *chemical engineering educator, consultant*
McCullough, Roy Lynn *chemical engineering educator*
Mendelsohn, Ray Langer *chemical engineering consultant*
Russell, Thomas William Fraser *chemical engineering educator*
Wagner, Norman Joseph, III *chemical engineering educator*

Wilmington
Bhagat, Ashok Kumar *chemical engineer*
Birewar, Deepak Baburao *chemical engineer, researcher*
Busche, Robert Marion *chemical engineer, consultant*
Coulston, George William *chemical engineer, researcher*
Pih, Norman *chemical engineer*
Sciance, Carroll Thomas *chemical engineer*

DISTRICT OF COLUMBIA

Washington
Burka, Maria Karpati *chemical engineer*
Faeth, Lisa Ellen *chemical engineer*
Fontana, David A. *chemical engineer, military officer*
Kirkbride, Chalmer Gatlin *chemical engineer*
Peters, Frank Albert *chemical engineer*
Shay, Edward Griffin *chemical engineer*

FLORIDA

Gainesville
Orazem, Mark Edward *chemical engineering educator*

Gonzalez
Yu, Jason Jeehyon *chemical engineer*

Jacksonville
Jernigan, John Milton *chemist, chemical engineer*
Joyce, Edward Rowen *chemical engineer, educator*

Lakeland
Walters, Marten Doig *chemical engineer*

Miami
Pasquel, Joaquin *chemical engineer*

Mulberry
Sabatino, David Matthew *chemical engineer*

Naples
Archibald, Frederick Ratcliffe *chemical engineer*
Leverenz, Humboldt Walter *retired chemical research engineer*

Orlando
Ziger, David Howard *chemical engineer*

Palm Harbor
Morris, David Brian *chemical engineer*

Saint Marks
Faintich, Stephen Robert *chemical engineer*

Tallahassee
Arce, Pedro Edgardo *chemical engineering educator*

GEORGIA

Atlanta
Stelson, Arthur Wesley *research chemical engineer, atmospheric scientist*

Augusta
Huang, Denis K. *chemical engineer, consultant*
Wilson, Keith Crookston *chemical engineer*

Duluth
Savatsky, Bruce Jon *chemical engineer*

Kennesaw
Ghosh, Deepak Ranjan *chemical/environmental engineer*

Norcross
Johnson, Roger Warren *chemical engineer*

Norristown
McCord, James Richard, III *chemical engineer, mathematician*

ILLINOIS

Argonne
Miller, Shelby Alexander *chemical engineer, educator*
Regalbuto, Monica Cristina *chemical engineer, research scientist*
Seefeldt, Waldemar Bernhard *chemical engineer*

Carterville
Feldmann, Herman Fred *chemical engineer*
Honea, Franklin Ivan *chemical engineer*

Champaign
Boddu, Veera Mallu *chemical engineering researcher*

Chicago
Biljetina, Richard *chemical engineering researcher*
Hartwell, Robert Carl *chemical engineer*
Karri, Surya B. Reddy *chemical engineer*
Kiefer, John Harold *chemical engineering educator*
Linden, Henry Robert *chemical engineering research executive*

Des Plaines
Cloutier, Stephen Edward *chemical engineer*
Vogl, Anna Katharina *chemical engineer*

Evanston
Petrich, Mark Anton *chemical engineer, educator*

La Grange
Cooke, Steven John *scientist, chemical engineer, consultant*

Lake Forest
Lambert, John Boyd *chemical engineer, consultant*

Loves Park
Pearson, Roger Alan *chemical engineer*

Naperville
Hacker, David Solomon *chemical engineer, researcher*
Lin, Chi-Hung *chemical engineer*
Niebuhr, Christopher *chemical engineer*
Vora, Manu Kishandas *chemical engineer, quality consultant*

Niles
Rasouli, Firooz *chemical engineer, researcher*

North Chicago
Chu, Alexander Hang-Torng *chemical engineer*

Northfield
Rockwell, Ned M. *chemical engineer*

Palatine
Rhoades, Douglas Duane *chemical engineer*

Pekin
Petricola, Anthony John *chemical engineer*

Robinson
Garrett-Perry, Nanette Dawn *chemical engineer*

Sauget
Baltz, Richard Arthur *chemical engineer*

Springfield
Reed, John Charles *chemical engineer*

Villa Park
Wiede, William, Jr. *chemical engineer, consultant*

INDIANA

Indianapolis
Alford, Joseph Savage, Jr. *chemical engineer, research scientist*

Mount Vernon
Bader, Keith Bryan *chemical engineer*

Notre Dame
Carberry, James John *chemical engineer, educator*
Cheng, Minquan *chemical engineer*
Varma, Arvind *chemical engineering educator, researcher*

Princeton
Mullins, Richard Austin *chemical engineer*

Terre Haute
LaMagna, John Thomas *chemical engineer*

Warsaw
Bradt, Rexford Hale *chemical engineer*

West Lafayette
Chao, Kwang-Chu *chemical engineer, educator*

IOWA

Ames
Boylan, David Ray *retired chemical engineer, educator*

Marshalltown
Sheeler, John Briggs *chemical engineer*

KANSAS

Lawrence
Green, Don Wesley *chemical and petroleum engineering educator*
Vossoughi, Shapour *chemical and petroleum engineering educator*

Leawood
Mosher, Alan Dale *chemical engineer*

Manhattan
Kyle, Benjamin Gayle *chemical engineer, educator*

Olathe
Ruf, Jacob Frederick *chemical engineer*

Overland Park
Hsu, Peter Cheazone *chemical engineer*

Wichita
Wolf, Mark Alan *chemical engineer*

KENTUCKY

Calvert City
Gray, Carol Hickson *chemical engineer*
Upshaw, Timothy Alan *chemical engineer*

Louisville
Deshpande, Pradeep Bapusaheb *chemical engineer, educator, consultant*
Fleischman, Marvin *chemical and environmental engineering educator*
Garcia, Rafael Jorge *chemical engineer*
Livingood, Marvin Duane *chemical engineer*

LOUISIANA

Baton Rouge
Constant, William David *chemical engineer, educator*
Corripio, Armando Benito *chemical engineering educator*
Hajare, Raju Padmakar *chemical engineer*
Osborne, William Galloway, Jr. *chemical engineer*
Valsaraj, Kalliat Thazhathuveetil *chemical engineering educator*
Wiley, Samuel Kay *chemical engineer*

Burnside
Drescher, Edwin Anthony *chemical engineer*

Lake Charles
Nam, Tin (Tonny Nam) *chemical engineer*

LaPlace
Brodt, Burton Pardee *chemical engineer, researcher*

Mandeville
Knoepfler, Nestor Beyer *chemical engineer*

New Orleans
O'Connor, Kim Claire *chemical engineering and biotechnology educator*
Phillips, John Benton *chemical engineer*

Princeton
Tollefsen, Gerald Elmer *chemical engineer*

Saint Gabriel
Das, Dilip Kumar *chemical engineer*

MAINE

Hollis Center
Page, Edward Crozer, Jr. *chemical engineer, consultant*

Orono
Kiran, Erdogan *chemical engineering educator*

MARYLAND

Annapolis
Heller, Austin Norman *chemical and environmental engineer*

Baltimore
Donohue, Marc David *chemical engineering educator*

College Park
Han, Ruijing *chemical engineer*

Columbia
Hegedus, L. Louis *chemical engineer, research and development executive*
Rezaiyan, A. John *chemical engineer, consultant*
Vu, Cung *chemical engineer*

Elkton
Harrell, Douglas Gaines *chemical engineer*

Gaithersburg
Hoppes, Harrison Neil *corporate executive, chemical engineer*
Phelan, Frederick Rossiter, Jr. *chemical engineer*

North Potomac
Finger, Stanley Melvin *chemical engineer, consultant*

Rockville
Abron, Lilia A. *engineer*

Towson
Redd, Rudolph James *chemical engineer, consultant*

MASSACHUSETTS

Acton
Flood, Harold William *chemical engineer, educator*

Amherst
Haensel, Vladimir *chemical engineering educator*
Winter, Horst Henning *chemical engineer, educator*

Belmont
Merrill, Edward Wilson *chemical engineering educator*

Billerica
KuLesza, Frank William *chemical engineer*
Yates, Barrie John *chemical engineer*

Boston
Bettinger, Jeffrie Allen *chemical engineer*
Deligianis, Anthony *chemical engineer*
Savage, Deborah Ellen *chemical engineer, environmental policy researcher*

Brookline
Gutoff, Edgar Benjamin *chemical engineer*
Jacobson, Murray M. *chemical engineer*

Cambridge
Brown, Robert Arthur *chemical engineering educator*
Colton, Clark Kenneth *chemical engineering educator*
Joshi, Amol Prabhatchandra *chemical engineer*
Klaubert, Earl Christian *chemical and general engineering administrator, consultant*
Langer, Robert Samuel *chemical, biochemical engineering educator*
Mehrabi, M. Reza *chemical engineer*
Stephanopoulos, Gregory *chemical engineering educator, consultant, researcher*
Zau, Gavin Chwan Haw *chemical engineer*

Danvers
Dutta, Arunava *chemical engineer*

Dighton
Chu, David Yuk *chemical engineer*

Foxboro
Ryskamp, Carroll Joseph *chemical engineer*

Longmeadow
Ferris, Theodore Vincent *chemical engineer, consulting technologist*

Marlborough
Walton, Thomas Cody *chemical plastics engineer*

Natick
Andreotti, Raymond Edward *chemical engineering technologist*
DeCosta, Peter F. *chemical engineer*

Waltham
Chanenchuk, Claire Ann *chemical engineer, market developer*
Nandy, Subas *chemical engineer, consultant*
Roth, Peter Hans *chemical engineer*

MICHIGAN

Ann Arbor
Carnahan, Brice *chemical engineer, educator*
Curl, Rane Locke *chemical engineering educator, consultant*
Mavrikakis, Manos *chemical engineer*
Schwank, Johannes Walter *chemical engineering educator*
Young, Edwin Harold *chemical/metallurgical engineering educator*

Bloomfield Hills
Heinen, Charles M. *retired chemical and materials engineer*

Detroit
Putchakayala, Hari Babu *chemical engineer*

Flint
Villaire, William Louis *chemical engineer*

Houghton
Caneba, Gerard Tablada *chemical engineering educator*

Kalamazoo
Reilly, Michael Thomas *biochemical engineer*

Mason
McDonald, James Harold *chemical engineer, energy specialist*
Toekes, Barna *chemical engineer, polymers consultant*

Midland
Cobel, George Bassett *chemical engineer*
Habermann, David Andrew *chemical engineer*
Heiny, Richard Lloyd *chemical engineer*
Meister, Bernard John *chemical engineer*
Schultz, Dale Herbert *chemical process control engineer*
Swinehart, Frederic Melvin *chemical engineer*
Wells, James Douglas *chemical engineer*

Milford
Prostak, Arnold S. *chemical instrumentation engineer*

Southfield
Sadasivan, Mahavijayan *chemical engineering manager*

Tecumseh
Chapman, Darik Ray *chemical process engineer*

Warren
Gallopoulos, Nicholas Efstratios *chemical engineer*
Kia, Sheila Farrokhalaee *chemical engineer, researcher*

MINNESOTA

Minneapolis
Culter, John Dougherty *chemical engineer*

Osseo
Haun, James William *chemical engineer, retired food company executive, consultant*

Plymouth
Levine, Leon *chemical engineer*

Saint Paul
Marcellus, Manley Clark, Jr. *chemical engineer*
Ylitalo, Caroline Melkonian *chemical engineer*

MISSISSIPPI

Pascagoula
Skipper, Adrian *chemical engineer*

Pickens
Patton, John Anthony *chemical engineer*

University
Sukanek, Peter Charles *chemical engineering educator, researcher*

MISSOURI

Columbia
Viswanath, Dabir Srikantiah *chemical engineer*

Florissant
Cook, Alfred Alden *chemical engineer*

Rolla
Sarchet, Bernard Reginald *retired chemical engineering educator*

Saint Louis
Khomami, Bamin *chemical engineer, educator*
Larsen, Alvin Henry *chemical engineer*
Puri, Pushpinder Singh *chemical engineer*

MONTANA

East Helena
Blossom, Neal William *chemical engineer*

NEBRASKA

Omaha
Clements, Luther Davis, Jr. *chemical engineer, educator*

NEVADA

Reno
Bautista, Renato Go *chemical engineer, educator*

NEW JERSEY

Bayonne
Steele Clapp, Jonathan Charles *chemical engineer*

Berkeley Heights
Townsend, Palmer Wilson *chemical engineer, consultant*

Birmingham
McGarvey, Francis Xavier *chemical engineer*

Bloomfield
Meenakshi Sundaram, Kandasamy *chemical engineer*

Cherry Hill
Davis, Kevin Jon *chemical engineer, consulting engineer*
Fuentevilla, Manuel Edward *chemical engineer*

Clifton
Srinivasachari, Samavedam *chemical engineer*

Cranbury
Larson, Eric Heath *chemical engineer*

East Orange
Sibilski, Peter John *chemical engineer*

Edgewater
Ahtchi-Ali, Badreddine *chemical engineer*

Edison
Patel, Bhavin R. *chemical engineer*

Fair Lawn
Kannankeril, Charles Paul *chemical engineer*

Florham Park
Richter, Jeffrey Alan *chemical engineer*

Glasser
Suplicki, John Clarke *chemical and metallurgical engineer, administrator*

Hackensack
Mavrovic, Ivo *chemical engineer*

Hoboken
Avgousti, Marios *chemical engineer*
Gastwirt, Lawrence E. *chemical engineer*
Kovenklioglu, Suphan Remzi *chemical engineering educator*
Tsenoglou, Christos *chemical engineer*

Lawrenceville
Enegess, David Norman *chemical engineer*

Linden
Duvdevani, Ilan *chemical engineer, researcher*

Middletown
Swett, Susan *chemical engineer*

Morris Plains
Degady, Marc *chemical engineer*

Morristown
Kavesh, Sheldon *chemical engineer*

Murray Hill
Krishnamurthy, Ramachandran (Krish) *chemical engineer, researcher*
Larson, Ronald Gary *chemical engineer*
Sali, Vlad Naim *chemical engineer*
Sheu, Lien-Lung *chemical engineer*

Newark
Armenante, Piero M. *chemical engineering educator*
Shaw, Henry *chemical engineering educator*

North Plainfield
Stoner, Henry Raymond *retired chemical engineer*

Oakland
Bacaloglu, Radu *chemical engineer*

Ocean City
Reiter, William Martin *chemical engineer*

Paulsboro
Dwyer, Francis Gerard *chemical engineer, researcher*

Piscataway
Salkind, Alvin J. *electrochemical engineer, educator*
Szewczyk, Martin Joseph *chemical engineer*

Princeton
Balakrishna, Subash *chemical engineer*
Debenedetti, Pablo Gaston *chemical engineer*
Hill, Russell Gibson *chemical engineer, consultant*
Kevrekidis, Yannis George *chemical engineer*
Kurtz, Andrew Dallas *chemical engineer*
Register, Richard Alan *chemical engineering educator*
Wei, James *chemical engineering educator, academic dean*

Princeton Junction
Haddad, James Henry *chemical engineering consultant*

Rahway
Beckhusen, Eric Herman *chemical engineer, consultant*
Tung, Hsien-Hsin *chemical engineer*

Raritan
Hill, Paul W. *biochemical engineer*
Weinstein, Norman Jacob *chemical engineer, consultant*

Salem
Crymble, John Frederick *chemical engineer, consultant*

Somerset
Long, Gilbert Morris *chemical engineer*

Springfield
Lofredo, Antony *chemical engineer*

Summit
Makhija, Subhash *research chemical and polymer engineer*

Trenton
Oliu, Ramon *chemical engineer*

Union Beach
Schulman, Marvin *chemical engineer*

Upper Montclair
Aronson, David *chemical and mechanical engineer*

Wayne
Sidman, Kenneth Robert *chemical engineer*

Whitehouse Station
Nusim, Stanley Herbert *chemical engineer, executive director*

NEW MEXICO

Albuquerque
Hobbs, Michael Lane *chemical engineer, researcher*

Los Alamos
Guell, David Charles *chemical engineer*

NEW YORK

Brooklyn
Myerson, Allan Stuart *chemical engineering educator, university dean*
Othmer, Donald Frederick *chemical engineer, educator*

Buffalo
Ruckenstein, Eli *chemical engineering educator*
Weber, Thomas William *chemical engineering educator*
Weller, Sol William *chemical engineering educator*

Corning
Havewala, Noshir Behram *chemical engineer*

East Greenbush
Davenport, Francis Leo *chemical engineer*

East Syracuse
Santanam, Suresh *chemical engineer, environmental consultant*

Fairport
Oldshue, James Y. *chemical engineering consultant*

Glen Cove
Mainhardt, Douglas Robert *chemical engineer*

Harriman
Parikh, Hemant Bhupendra *chemical engineer*

Harrison
Schulz, Helmut Wilhelm *chemical engineer, environmental executive*

Hauppauge
Bhide, Dan Bhagwatprasad Dinkar *chemical engineer*

Hyde Park
Moore, Roger Stephenson *chemical and energy engineer*

Ithaca
Gubbins, Keith Edmund *chemical engineering educator*
Koh, Carolyn Ann *chemical engineer*
Rodríguez, Ferdinand *chemical engineer, educator*
Smith, Julian Cleveland, Jr. *chemical engineering educator*
Von Berg, Robert Lee *chemical engineer educator, nuclear engineer*

Montrose
Reber, Raymond Andrew *chemical engineer*

New York
Acrivos, Andreas *chemical engineering educator*
Shinnar, Reuel *chemical engineering educator, industrial consultant*
Weinstein, Herbert *chemical engineer, educator*

Potsdam
Chin, Der-Tau *chemical engineer, educator*

Rochester
Colby, Ralph Hayes *chemical engineer*
Dieck, William Wallace Sandford *chemical engineer*
Jenekhe, Samson Ally *chemical engineering educator, polymer scientist*
Rubin, Bruce Joel *chemical engineer, electrical engineer*

Rockville Centre
Weber, Arthur Phineas *chemical engineer*

Schenectady
Trabold, Thomas Aquinas *chemical engineer*

Tarrytown
Bartoo, Richard Keith *chemical engineer, consultant*

Tonawanda
Bernard, John P. *chemical engineer*
Rovison, John Michael, Jr. *chemical engineer*

Troy
Gill, William Nelson *chemical engineering educator*
Przybycien, Todd Michael *chemical engineering educator*
Ushe, Zwelonke Ian *development engineer*

Upton
Fthenakis, Vasilis *chemical engineer, consultant, educator*
Steinberg, Meyer *chemical engineer*
Tang, Ignatius Ning-Bang *chemical engineer*

Utica
Grisewood, Norman Curtis *chemical engineer*

Vails Gate
Timm, Walter Clement *chemical engineer, researcher*

Yonkers
LoPinto, Charles Adam *chemical engineer*

NORTH CAROLINA

Enka
Stewart, Ronald *chemical engineer*

Greensboro
Ilias, Shamsuddin *chemical engineer, educator*
Tate, John Edward *chemical engineer*

New Bern
Whitehurst, Brooks Morris *chemical engineer*

Raleigh
Carbonell, Ruben Guillermo *chemical engineering educator*
Overcash, Michael Ray *chemical engineering educator*

Research Triangle Park
Agarwal, Sanjay Krishna *chemical engineer*
Stelling, John Henry Edward *chemical engineer*
Van Winkle, Jon *chemical engineer*

Winston Salem
Henderson, Richard Martin *chemical engineer*

OHIO

Akron
Elliott, Jarrell Richard, Jr. *chemical engineer, educator*
Rukovena, Frank, Jr. *chemical engineer*
Schooley, Arthur Thomas *chemical engineer*
Thayer, Ronda Renee Bayer *chemical engineer*

Beachwood
Malik, Tariq Mahmood *scientist, chemical engineer*

Canton
Nisly, Kenneth Eugene *chemical engineer, environmental engineer*

Cincinnati
Fried, Joel Robert *chemical engineering educator*
Gimpel, Rodney Frederick *chemical engineer*
Pancheri, Eugene Joseph *chemical engineer*
Redlinger, Samuel Edward *chemical engineering consultant*
Solé, Pedro *chemical engineer*

Cleveland
Kenat, Thomas Arthur *chemical engineer, consultant*
Yurko, Joseph Andrew *chemical engineer*
Zier, J(ohn) Albert *chemical engineer*

Columbus
Lytle, John Arden *chemical engineer*
Ozkan, Umit Sivrioglu *chemical engineering educator*
Skidmore, Duane Richard *chemical engineering educator, researcher*
Sweeney, Thomas Leonard *chemical engineering educator, researcher*

Dayton
Carson, Richard McKee *chemical engineer*

Granville
Choudhary, Manoj Kumar *chemical engineer, researcher*

Loveland
Anderson, Roy Alan *chemical engineer*

Marietta
Kirkland, Ned Matthews *chemical engineering manager*

University Heights
Skolnik, Leonard *chemical engineer*

Youngstown
Stahl, Joel Sol *plastic-chemical engineer*

OKLAHOMA

Bartlesville
Clay, Harris Aubrey *chemical engineer*
Gao, Hong Wen *chemical engineer*
Hankinson, Risdon William *chemical engineer*
Hunt, Harold Ray *chemical engineer*
Lee, Fu-Ming *chemical engineer*
Lew, Lawrence Edward *chemical engineer*
Mihm, John Clifford *chemical engineer*
Zecchini, Edward John *chemical engineer, researcher*

Bethany
Arnold, Donald Smith *chemical engineer, consultant*

Norman
O'Rear, Edgar Allen, III *chemical engineering educator*

Stillwater
Foutch, Gary Lynn *chemical engineering educator*
Tree, David Alan *chemical engineering educator*

Tulsa
Banks, James Daniel *chemical engineer*
Leder, Frederic *chemical engineer*

OREGON

Aloha
Rojhantalab, Hossein Mohammad *chemical engineer, researcher*

Corvallis
Rapier, Pascal Moran *chemical engineer, physicist*

PENNSYLVANIA

Allentown
Hansel, James Gordon *chemical engineer*
Webley, Paul Anthony *chemical engineer*
Winters, Arthur Ralph, Jr. *chemical and cryogenic engineer, consultant*
Woytek, Andrew Joseph *chemical engineer, researcher*

Bethlehem
Georgakis, Christos *chemical engineer educator, consultant, researcher*
Voutsas, Apostolos Theoharis *chemical engineer*

Chadds Ford
DeVries, Frederick William *chemical engineer*

Clairton
Maddalena, Frederick Louis *chemical engineer*

Easton
Dennis, Ronald Marvin *chemical engineering educator, consultant*

Erie
Kohli, Tejbans Singh *chemical engineer*

Gibsonia
Shoub, Earle Phelps *chemical engineer, educator*

Gilbertsville
Ladisch, Thomas Peter *chemical engineer*

Haverford
Keith, Frederick W., Jr. *retired chemical engineer*

Lancaster
Murthy, Kamalakara Akula *chemical engineer, researcher*

Lewisburg
Gidh, Kedar Keshav *chemical engineer*

Linwood
Balale, Emanuel Michael *chemical engineer*

Mendenhall
Mehne, Paul Herbert *chemical engineer, engineering consultant*

Morrisville
Jarvi, George Albert *chemical engineer*

North Wales
Gray, David Marshall *chemical engineer*

Philadelphia
Craig, James Clifford, Jr. *chemical engineer*
Lantos, Peter R(ichard) *industrial consultant, chemical engineer*
Quinn, John Albert *chemical engineering educator*

Pittsburgh
Cobb, James Temple, Jr. *chemical engineer, engineering educator*
Frederick, Edward Russell *chemical engineering consultant*
George, John Michael *chemical engineer*
Grossmann, Ignacio Emilio *chemical engineering educator*
Henry, Keith Edward *chemical engineer*
Markussen, Joanne Marie *chemical engineer*
Russell, Alan James *chemical engineering and biotechnology educator*
Tilton, Robert Daymond *chemical engineer*
Yeh, James Tehcheng *chemical engineer*

Saint Marys
Paxton, R(alph) Robert *chemical engineer*

Trafford
Hufton, Jeffrey Raymond *chemical engineer*

University Park
Hegarty, William Patrick *chemical engineering educator*
Vannice, M. Albert *chemical engineering educator, researcher*
Vrentas, Christine Mary *chemical engineer, researcher*

Warminster
Vanderbeek, Frederick Hallet, Jr. *chemical engineer*

Warren
Sternberg, Mark Edward *chemical engineer*

Warrendale
Karrs, Stanley Richard *chemical engineer*

York
Moore, Walter Calvin *chemical engineer*

SOUTH CAROLINA

Aiken
Bradley, Robert Foster *chemical engineer*
Young, Steve R. *chemical engineer*

Arcadia
Hubbell, Douglas Osborne *chemical engineering consultant*

Charleston
Greene, George Chester, III *chemical engineer*

Clemson
Haile, James Mitchell *engineering educator*

Columbia
Dorgay, Charles Kenneth *chemical engineer*
Ritter, James Anthony *chemical engineer, educator, consultant*

Greenville
Toth, James Joseph *chemical engineer*

Hilton Head Island
Huckins, Harold Aaron *chemical engineer*

North Augusta
Dhom, Robert Charles *chemical engineer*

SOUTH DAKOTA

Rapid City
Puszynski, Jan Alojzy *chemical engineer, educator*

TENNESSEE

Brentwood
McCormick, Robert Jefferson *chemical engineer*

Cookeville
Mason, John Thomas, III *chemical engineering educator, consultant*

Kingsport
Bagrodia, Shriram *chemical engineer*
Ryans, James Lee *chemical engineer*
Scott, H(erbert) Andrew *retired chemical engineer*
Siirola, Jeffrey John *chemical engineer*
Tant, Martin Ray *chemical engineer*

Knoxville
Busmann, Thomas Gary *chemical engineer*
Chadha, Navneet *chemical engineer, environmental consultant*

Memphis
Guenter, Thomas Edward *retired chemical engineer*

Nashville
Giorgio, Todd Donald *chemical engineering educator*

Oak Ridge
Basaran, Osman A. *chemical engineer*
Cochran, Henry Douglas *chemical engineer*
Hightower, Jesse Robert *chemical engineer, researcher*
Shapiro, Theodore *chemical engineer*
Taylor, Paul Allen *chemical engineer*

Tullahoma
Sheth, Atul Chandravadan *chemical engineering educator, researcher*

TEXAS

Austin
Barlow, Joel William *chemical engineering educator*
Barrow, Arthur Ray *chemical engineer*
Fair, James Rutherford, Jr. *chemical engineering educator, consultant*
Himmelblau, David Mautner *chemical engineer*
Keith, Roger Horn *chemical engineer*
Koros, William John *chemical engineering educator*
Mc Ketta, John J., Jr. *chemical engineering educator*
Paul, Donald Ross *chemical engineer, educator*
Popovich, Robert P. *biochemical engineer, educator*
Schechter, Robert Samuel *chemical engineer, educator*
Steinfink, Hugo *chemical engineering educator*

Beaumont
Hopper, Jack Rudd *chemical engineering educator*
Krehbiel, David Kent *chemical engineer*
Robertson, Ivan Denzil, III *chemical engineer*

College Station
Fliss, Albert Edward, Jr. *chemical engineer*

Corpus Christi
Plunkett, Roy J. *retired chemical engineer*

Cypress
Taylor, Paul Duane, Jr. *chemical engineer*

Deer Park
Brigman, James Gemeny *chemical engineer*

Fort Worth
Madan, Sudhir Yashpal *chemical engineer*

Houston
Armeniades, Constantine D. *chemical engineer, educator*
Brandt, I. Marvin *chemical engineer*
Chandra, Ajey *chemical engineer*
Clunie, Thomas John *chemical engineer*
Davis, Brian Richard *chemical engineer*
Dawn, Frederic Samuel *chemical, textile engineer*
Dukler, Abraham Emanuel *chemical engineer*
Edwards, Victor Henry *chemical engineer*
Elizardo, Kelly Patricia *chemical engineer*

Elliot, Douglas Gene *chemical engineer, engineering company executive, consultant*
Jacks, Jean-Pierre Georges Yves *chemical engineering sales executive*
Lam, Daniel Haw *chemical engineer*
Lopez-Nakazono, Benito *chemical engineer*
McCants, Malcolm Thomas *chemical engineer*
Mian, Farouk Aslam *chemical engineer, educator*
Mohrmann, Leonard Edward, Jr. *chemical engineer*
Mosher, Donald Raymond *chemical engineer, consultant*
Okafor, Michael Chukwuemeka *chemical engineer*
Price, Frederick Clinton *chemical engineer, editorial consultant*
Sloan, Harold David *chemical engineering consultant*
Soltau, Ronald Charles *chemical engineer*
Sung, Ming *chemical engineer*
Valencia, Jaime Alfonso *chemical engineer*
Van Sickels, Martin John *chemical engineer*

La Porte
Wienert, James Wilbur *chemical engineer*

League City
Kanuth, James Gordan *chemical engineer*
Lamar, James Lewis, Jr. *chemical engineer*

Nassau Bay
Kimzey, John Howard *chemical and aerospace engineer*

Orange
Cardner, David Victor *chemical engineer*
Oyekan, Soni Olufemi *chemical engineer*
Russell, Kenneth William *chemical engineer*

Pasadena
Cheng, Chung P. *chemical engineer*
Tagoe, Christopher Cecil *chemical engineer*

Point Comfort
Martinez, Mario Antonio *chemical engineer*

Port Neches
Ruth, Stan M. *chemical engineer*

Sugar Land
Dye, Robert Fulton *chemical engineer*

Webster
Stephens, Douglas Kimble *chemical engineer*

UTAH

Ogden
Davidson, Thomas Ferguson *chemical engineer*

Provo
Barker, Dee H. *chemical engineering educator*

VERMONT

Essex Junction
Albaugh, Kevin Bruce *chemical engineer*

Pownal
Dequasie, Andrew Eugene *chemical engineer*

VIRGINIA

Alexandria
Lasser, Howard Gilbert *chemical engineer, consultant*

Blacksburg
Glasser, Wolfgang Gerhard *forest products and chemical engineering researcher, educator*
McGee, Henry Alexander, Jr. *chemical engineering educator*

Charlottesville
Carta, Giorgio *chemical engineering educator*
Chialvo, Ariel Augusto *chemical engineering research scientist*

Covington
Foster, James Joseph *chemical engineer*

Elkton
Ghorpade, Ajit Kisanrao *chemical engineer*

Falls Church
Frankel, Irwin *chemical engineer*

Franklin
Raymond, Dale Rodney *chemical engineer*

Mc Lean
Paul, Anton Dilo *chemical engineering consultant, researcher*

Midlothian
Hauxwell, Gerald Dean *chemical engineer*

Radford
Cabbage, William Austin *chemical engineer*

Richmond
Marsee, Dewey Robert *chemical engineer*
Prasad, Ravi *chemical engineer*

Roanoke
Smith, Gary Lee *chemical engineer*

WASHINGTON

Richland
Compton, James Allan *chemical engineer*
Huckaby, James L. *chemical engineer*
Ritter, Gerald Lee *chemical engineer*

Steilacoom
Morgenthaler, John Herbert *chemical engineer*

Yakima
LaFontaine, Thomas E. *chemical engineer*

WEST VIRGINIA

Apple Grove
Burkett, Eugene John *chemical engineer*

Morgantown
Gasper-Galvin, Lee DeLong *chemical engineer*
Guthrie, Hugh Delmar *chemical engineer*
Shaeiwitz, Joseph Alan *chemical engineering educator*

South Charleston
Nielsen, Kenneth Andrew *chemical engineer*

Wheeling
Rivers, Lee Walter *chemical engineer*

WISCONSIN

Appleton
Aziz, Salman *chemical engineer, company executive*

Madison
Duffie, John Atwater *chemical engineer, educator*
Kim, Sangtae *chemical engineering educator*
Tran, Tri Duc *chemical engineer*

WYOMING

Cheyenne
Feusner, LeRoy Carroll *chemical engineer*

Gillette
Frederick, James Paul *chemical engineer*

TERRITORIES OF THE UNITED STATES

PUERTO RICO

Mayaguez
Mandavilli, Satya Narayana *chemical engineering educator, researcher*

CANADA

ALBERTA

Calgary
Cooke, David Lawrence *chemical engineer*

Edmonton
de Guzman, Roman de Lara *chemical engineer, consultant*
Rao, Ming *chemical engineering and computer science educator*

BRITISH COLUMBIA

Vancouver
Hatzikiriakos, Savvas Georgios *chemical engineer*

ONTARIO

Hamilton
Kenny-Wallace, G. A. *chemical engineer*
Vlachopoulos, John *chemical engineering educator*

QUEBEC

Montreal
Kamal, Musa Rasim *chemical engineer, consultant*

Pointe Claire
Kubanek, George R. *chemical engineer*

Trois Rivières
Lavallee, H.-Claude *chemical engineer, researcher*

ARGENTINA

Salta
Gonzo, Elio Emilio *chemical engineer, educator*

Santa Fe
Deiber, Julio Alcides *chemical engineering educator*

AUSTRALIA

Greenwith
Raymont, Warwick Deane *chemical engineer, environmental engineer*

BELGIUM

Brussels
Baron, Gino Victor *chemical engineering educator*

Heverlee
Mewis, Joannes J(oanna) *chemical engineering educator*

Leuven
Moldenaers, Paula Fernande *chemical engineer, educator*

BRAZIL

Cotia
Araujo, Marcio Santos Silva *chemical engineer*

Rio de Janeiro
Espinola, Aïda *chemical engineer, educator*

CZECH REPUBLIC

Neratovice
Hlubuček, Vratislav *chemical engineer*

Prague
Červeny, Libor *chemistry educator*

FINLAND

Nastola
Jarvenkyla, Jyri Jaakko *chemical engineer*

FRANCE

Chatenay-Malabry
Perrault, Georges Gabriel *chemical engineer*

Lavera
Boniface, Christian Pierre *chemical engineer, oil industry executive*

Paris
Mareschi, Jean Pierre *biochemical engineer*

Saint Symphorien Ozo
Bernasconi, Christian *chemical engineer*

Velizy-Villacoublay
Musikas, Claude *chemical researcher*

GERMANY

Munich
Wagner, Thomas Alfred *chemical engineer*

Wuppertal
Vienken, Joerg Hans *chemical engineer*

GREECE

Athens
Markatos, Nicolas-Chris Gregory *chemical engineering educator*

Thessaloniki
Panayiotou, Constantinos *chemical engineering educator*
Papanastasiou, Tasos Charilaou *chemical engineering educator*
Sakellaropoulos, George Panayotis *chemical engineering educator*

INDIA

Patiala
Bajpai, Pramod Kumar *chemical and biochemical engineer*

IRAN

Tehran
Kaghazchi, Tahereh *chemical engineering educator*
Sohrabi, Morteza *chemical engineering educator*

ISRAEL

Beer Sheva
Gottlieb, Moshe *chemical engineer, educator*

ITALY

Bologna
Foraboschi, Franco Paolo *chemical engineering educator*

San Donato
Donati, Gianni *chemical engineer, administrator*

JAPAN

Kobe
Masai, Mitsuo *chemical engineer, educator*

Kyoto
Oku, Akira *chemical engineer, educator*

Muroran
Fu, Yuan Chin *chemical engineering educator*

Narashino
Inazumi, Hikoji *chemical engineering educator*

Suita
Shioya, Suteaki *biochemical engineering educator*

Tokyo
Aoki, Junjiro *chemical engineer*
Matsui, Iwao *chemical engineer*
Nagasaka, Kyosuke *chemical engineer*
Ohe, Shuzo *chemical engineer, educator*

JORDAN

Irbid
Mansour, Awad Rasheed *chemical engineering educator*

THE NETHERLANDS

Amsterdam
Fortuin, Johannes Martinus H. *chemical engineer*

Hengelo
Penninger, Johannes Mathieu L. *chemical engineer*

NEW ZEALAND

Manurewa
McCreadie, Allan Robert *chemical engineer, consultant*

NORWAY

Kjeller
Bjorvatten, Tor Anderson *chemical engineer*

Trondheim
Rokstad, Odd Arne *chemical engineer*

PORTUGAL

Braga
Cruz-Pinto, Jose Joaquim C. *chemical and materials engineering educator, researcher*

Lisbon
Villax, Ivan Emeric *chemical engineer, researcher*

Oeiras
Gomes, João Fernando Pereira *chemical engineer*

REPUBLIC OF KOREA

Pohang
Lee, Sun Bok *biochemical engineering educator*

Seoul
Park, Won-Hoon *chemical engineer*
Rhee, Hyun-Ku *chemical engineering educator*

Taejon
Park, O Ok *chemical engineering educator*

RUSSIA

Moscow
Olevskii, Victor Marcovich *chemical engineer, educator*

Podolsk
Nikishenko, Semion Boris *chemical research engineer*

SAUDI ARABIA

Dhahran
Hamid, Syed Halim *chemical engineer*

Riyadh
Ghadia, Suresh Kantilal *chemical engineer*

SPAIN

Barcelona
Sigalés, Bartomeu *chemical engineering educator*

Bilbao
Arias, Pedro Luis *chemical engineering educator, researcher*

Las Palmas
Saavedra, Juan Ortega *chemical engineering educator*

Salamanca
Jaraiz, Eladio Maldonado *chemical engineering educator*

SWITZERLAND

Schweizerhalle
Hadjistamov, Dimiter *chemical engineer*

TAIWAN

Chung-Li
Chen, Wen-Yih *chemical engineer*

Hsinchu
Wu, Wen-Teng *chemical engineer*

THAILAND

Bangkok
Pongsiri, Nutavoot *chemical engineer*

ADDRESS UNPUBLISHED

Burns, William Edgar *chemical engineer*
Carver, Ron G(eorge) *chemical engineer*
Constant, Clinton *chemical engineer, consultant*
Cuthbert, Versie *chemical engineer*
Edgar, Thomas Flynn *chemical engineering educator*
Franklin, Howard David *chemical engineer*
Grigger, David John *chemical engineer*
Herz, George Peter *industrial consultant, chemical engineer*
Koppany, Charles Robert *chemical engineer*
Liekhus, Kevin James *chemical engineer*
Marchessault, Robert H. *chemical engineer*
Mershimer, Robert John *chemical engineer*
Myers, David Francis *chemical engineer*
Olstowski, Franciszek *chemical engineer, consultant*
Rabó, Jule Anthony *chemical research administrator, consultant*
Robertson, Jerry Lewis *chemical engineer*
Rothfeld, Leonard Benjamin *chemical/environmental engineer*
Rudzki, Eugeniusz Maciej *chemical engineer, consultant*
Scheirman, William Lynn *chemical engineer*
Shalabi, Mazen Ahmad *chemical engineer, educator*
Smith, Joe Mauk *chemical engineer, educator*
Soroush, Masoud *chemical engineer*
Stokes, Charles Anderson *chemical engineer, consultant*
Tsai, Tom Chunghu *chemical engineer*
Tweed, Paul Basset *chemical engineer, explosives and suicidology consultant*
Weinrich, Stanley David *chemical engineer*
Wicks, Moye, III *chemical engineer, researcher*
Wright, Bradley Dean *chemical engineer, environmental consultant*

ENGINEERING: CIVIL & STRUCTURAL

UNITED STATES

ALABAMA

Birmingham
Seaman, Duncan Campbell *civil engineer*
Shaw, Johnny Harvey *civil engineer*
Spears, Randall Lynn *civil engineer*
Williford, Robert Marion *civil engineer*

Huntsville
Gore, James William *civil engineer*
Savage, Julian Michele *civil engineer*
Stevens, Dale Marlin *civil engineer*
Torres, Antonio *civil engineer*

Mobile
Boulian, Charles Joseph *civil-structural engineer*
Omar, Husam Anwar *civil engineering educator*
Webber, David Michael *civil engineer*
Wilkerson, William Edward, Jr. *civil engineer*

Tuscaloosa
Carter, Olice Cleveland, Jr. *civil engineer*

ALASKA

Anchorage
Kinney, Donald Gregory *civil engineer*

ARIZONA

Mesa
Boisjoly, Roger Mark *structural engineer*
Scofield, Larry Allan *civil engineer*

Paradise Valley
Gookin, William Scudder *hydrologist, consultant*

Phoenix
Doyle, Michael Phillip *civil engineer*
Freyermuth, Clifford L. *structural engineering consultant*
Williams, Damon S. *civil engineer*

Sells
Rubendall, Richard Arthur *civil engineer*

Tucson
Craghead, James Douglas *civil engineer*
Isenhower, William Martin *civil engineering educator*
Ran, Chongwei *geotechnical engineer*

ARKANSAS

Fayetteville
LeFevre, Elbert Walter, Jr. *civil engineering educator*
Selvam, Rathinam Panneer *civil engineering educator*

Jonesboro
Stalcup, Thomas Eugene, Sr. *civil engineer*

Little Rock
Benson, Barton Kenneth *civil engineer*

CALIFORNIA

Berkeley
Madabhushi, Govindachari Venkata *civil engineer*
Mitchell, James Kenneth *civil engineer, educator*
Monismith, Carl Leroy *civil engineering educator*
Shen, Hsieh Wen *civil engineer, consultant, educator*

Carmel
Brahtz, John Frederick Peel *civil engineering educator*

Cerritos
Acosta, Nelson John *civil engineer*

Concord
Middleton, Michael John *civil engineer*

El Segundo
Ajmera, Kishore Tarachand *civil engineer*

Eureka
Conversano, Guy John *civil engineer*

Foster City
Graf, Edward Dutton *grouting consultant*

Glendale
Knoop, Vern Thomas *civil engineer, consultant*
Leonard, Constance Joanne *civil, environmental engineer*

Grand Terrace
Addington, William Hampton *civil engineer*

Hillsborough
Blume, John August *consulting civil engineer*

Irvine
Caffey, Benjamin Franklin *civil engineer*
Dunlap, Dale Richard *civil engineer*
Real, Thomas Michael *civil engineer*
Villaverde, Roberto *civil engineer*

La Mesa
Kropotoff, George Alex *civil engineer*

Los Altos
Fondahl, John Walker *civil engineering educator*

Los Angeles
Egbuonu, Zephyrinus Chiedu *civil engineer, electrical engineer*
Griffith, Carl David *civil engineer*
Rogoway, Lawrence Paul *civil engineer, consultant*

Merced
McCullough, John James, III *civil engineer*

Oakland
Kakade, Ashok Madhav *civil engineer, consultant*
Tsztoo, David Fong *civil engineer*

Orange
Letzring, Tracy John *civil engineer*

Pasadena
Boulos, Paul Fares *civil and environmental engineer*
Housner, George William *civil engineering educator, consultant*
Tekippe, Rudy Joseph *civil engineer*

Pismo Beach
Saveker, David Richard *naval and marine architectural engineering executive*

Richmond
Moehle, Jack P. *civil engineer, engineering executive*

Sacramento
Guyer, J. Paul *civil engineer, architect, consultant*
Hwang, Jenn-Shin *civil engineer*
Panuschka, Gerhard *civil and sanitary engineer, inventor*

San Francisco
Brovold, Frederick Norman *geotechnical and civil engineer*
Cheng, Kwong Man *structural/bridge engineer*
Lin, Tung Yen *civil engineer, educator*
Neelakantan, Ganapathy Subramanian *geotechnical engineer*
Shushkewich, Kenneth Wayne *structural engineer*

San Luis Obispo
Moazzami, Sara *civil engineering educator*

San Ramon
Bugno, Walter Thomas *civil engineer*

Santa Ana
Amies, Alex Phillip *civil engineer*
Hilmy, Said Ibrahim *structural engineer, consultant*

Santa Ana Heights
Jungren, Jon Erik *civil engineer*

Santa Barbara
Lawrance, Charles H. *civil and sanitary engineer*
Swalley, Robert Farrell *structural engineer, consultant*

Santa Rosa
Anderson, Terry Marlene *civil engineer*

Stanford
Kruger, Paul *nuclear civil engineering educator*
Teicholz, Paul M. *civil engineering educator, administrator*

Stockton
Fletcher, David Quentin *civil engineering educator*

Woodland Hills
Holguin, Librado Malacara (Lee Holguin) *civil engineer*

COLORADO

Boulder
Cowing, Thomas William *architectural, civil engineer*

Colorado Springs
Reynolds, Michael Floyd *civil engineer, educator*
Rydbeck, Bruce Vernon *civil engineer*
Tenney, Leon Walter *civil engineer*

Denver
East, Donald Robert *civil engineer*
Frevert, Donald Kent *hydraulic engineer*
McCammon, Donald Lee *civil engineer*
Olsen, Harold William *research civil engineer*

Englewood
Schirmer, Howard August, Jr. *civil engineer*

Fort Carson
Tygret, James William *civil engineer*

Fort Collins
Criswell, Marvin Eugene *civil engineering educator, consultant*
Mesloh, Warren Henry *civil, environmental engineer*
Richardson, Everett Vern *hydraulic engineer, educator, administrator*

Glenwood Springs
Violette, Glenn Phillip *construction engineer*

Lakewood
Klima, Jon Edward *civil engineering educator, consultant*

Silverthorne
Robertson, James Mueller *civil engineer, educator*

Wheat Ridge
Scherich, Erwin Thomas *civil engineer, consultant*

CONNECTICUT

Bloomfield
Fine, Morton Samuel *civil engineer*

Danbury
Schweitzer, Mark Andrew *civil engineer*
Vossler, John Albert *civil engineer*

Greenwich
Nadel, Norman Allen *civil engineer*

Newington
Gates, Marvin *construction engineer*

Plainfield
Singh, Reepu Daman *civil engineer*

Southington
Scherziger, Keith Joseph *civil engineer*

Stamford
Asmar, Charles Edmond *structural engineer, consultant*
Rodriguez, J. Louis *civil engineer, land surveyor*

Storrs
Johnston, E. Russell, Jr. *civil engineer, educator*
Stephens, Jack Edward *civil engineer, educator*

Storrs Mansfield
Malla, Ramesh Babu *structural and mechanical engineering educator*

Trumbull
Gladki, Hanna Zofia *civil engineer, hydraulic mixer specialist*

DELAWARE

Newark
Faghri, Ardeshir *civil engineering educator*
Kirby, James Thornton, Jr. *civil engineering educator*
Kobayashi, Nobuhisa *civil and coastal engineer, educator*

Wilmington
Smith, Marvin Schade *civil engineer, land surveyor*

DISTRICT OF COLUMBIA

Washington
Caywood, James Alexander, III *transportation engineering company executive, civil engineer*
Grant, Wendell Carver, II *civil engineer*
Gueller, Samuel *civil and environmental engineer*
Harris, Leonard Andrew *civil engineer*
Kellogg, Charles Gary *civil engineer*
Mandel, Elliott David *structural engineer, consultant*
Ochs, Walter J. *civil engineer, drainage adviser*
Parks, Vincent Joseph *civil engineering educator*
Pittinger, Charles Bernard, Jr. *civil engineer*
Podolny, Walter, Jr. *structural engineer*
Zielinski, Paul Bernard *grant program administrator, civil engineer*

FLORIDA

Clearwater
King, Charles Herbert, Jr. *civil engineer*

Coral Gables
Namini, Ahmad Hossein *structural engineer, educator*
Saffir, Herbert Seymour *structural engineer, consultant*

Deland
Pershe, Edward Richard *civil engineer*

Fort Myers
Hoppmeyer, Calvin Carl, Jr. *civil engineer*

Gainesville
Brungard, Martin Alan *civil engineer*
Garcelon, John Herrick *structural engineer, researcher*
Hoit, Marc I. *civil engineer, educator*

Hollywood
Kuzmanović, Bogdan Ognjan *structural engineer*

Jacksonville
Hawkins, James Douglas, Jr. *structural engineer, architect*

Lake Placid
Rew, William Edmund *civil engineer*

Maitland
O'Connor, Kevin Patrick *civil engineer*

Miami
Bellero, Chiaffredo John *civil engineer*

Orlando
Dewey, Cameron Boss *civil engineer*
Mirmiran, Amir *civil engineer*
Verwey, Timothy Andrew *structural engineer, consultant*

Palatka
Jenab, S. Abe *civil and water resources engineer*

Pembroke Pines
Schwartzberg, Leo Mark *civil engineer, consultant*

Plantation
Shim, Jack V. *civil engineer, consultant*

Pompano Beach
Francis, William Kevin *civil engineer*
Kavasoglu, Abdulkadir Yekta *civil engineer*

Ponte Vedra Beach
Zebouni, Nadeem Gabriel *structural engineer*

Port Charlotte
Munger, Elmer Lewis *civil engineer, educator*

Saint Petersburg
Kannan, Ramanuja Chari *civil engineer*

Satellite Beach
Nunnally, Stephens Watson *civil engineer*

Tallahassee
Anderson, John Roy *grouting engineer*
Coloney, Wayne Herndon *civil engineer*
Henderson, Kaye Neil *civil engineer, business executive*
Moilanen, Michael David *civil engineer*
Stivers, Marshall Lee *civil engineer*

Tampa
Gergess, Antoine Nicolas *civil engineer*
Mines, Richard Oliver, Jr. *civil and environmental engineer*

Titusville
Claridge, Richard Allen *structural engineer*

West Palm Beach
Olsak, Ivan Karel *civil engineer*

GEORGIA

Atlanta
Aral, Mustafa Mehmet *civil engineer*
Bowman, Albert W. *structural engineer*
Fitzgerald, John Edmund *civil engineering educator, dean*
Hanna, M(axcy) G(rover), Jr. *civil engineer*
Liew, Chong Shing *structural engineer*
Macari, Emir José *civil engineering educator*
Parsonson, Peter Sterling *civil engineer, educator, consultant*
Pitts, Marcellus Theadore *civil engineer, consultant*
Tutumluer, Erol *civil engineer*

Augusta
Traven, Kevin Charles *structural engineer*

Kennesaw
Polmann, Donald Jeffrey *water resources engineer, environmental consultant*

Lawrenceville
Kretsch, Michael Gene *civil engineer*

Marietta
Rowe, Alvin George *structural engineer, consultant*
Swift, John Lionel *civil engineer*

Norcross
Adams, Dee Briane *hydrologist, civil engineer*
Javaheri, Kathleen Dakin *construction and environmental professional*

Roswell
Bristow, Preston Abner, Jr. *civil engineer, environmental engineer*

Smyrna
Stevenson, Earl, Jr. *civil engineer*

HAWAII

Honolulu
Ching, Daniel Gerald *civil engineer*
Chiu, Arthur Nang Lick *engineering educator*

Foster, Stephen Roch *civil engineer*
Fukumoto, Neal Susumu *civil engineer*
Krock, Hans-Jurgen *civil engineer*

Kaneohe
Hanson, Richard Edwin *civil engineer*

ILLINOIS

Argonne
Tang, Yu *structural engineer*

Buffalo Grove
O'Sullivan, Michael Anthony *civil engineer*

Champaign
Kruger, William Arnold *consulting civil engineer*

Chicago
Gasparotto, Renso *civil engineer*
Gerstner, Robert William *structural engineering educator, consultant*
Guralnick, Sidney Aaron *civil engineering educator*
Hildebranski, Robert Joseph *civil engineer*
Langdon, Paul Eugene, Jr. *consulting civil engineer*
Machinis, Peter Alexander *civil engineer*
Mehlenbacher, Dohn Harlow *civil engineer, consultant*
Nizamuddin, Khawaja *structural engineer*
Rokach, Abraham Jacob *structural engineering and computer software consultant*
Sebastian, Scott Joseph *civil engineer*
Stecich, John Patrick *structural engineer*
Williams, Jack Raymond *civil engineer*

Clinton
Ramanuja, Teralandur Krishnaswamy *structural engineer*

Collinsville
Modeer, Victor Albert, Jr. *civil engineer*

Dunlap
Reinsma, Harold Lawrence *design consultant, engineer*

Evanston
Achenbach, Jan Drewes *engineer*

Flora
Gill, Henry Leonard *civil engineer, consultant*

Glenview
Logani, Kulbhushan Lal *civil and structural engineer*

Lake Bluff
Fortuna, William Frank *architect, architectural engineer*

Northbrook
Hamilton, Robert Burns *civil engineer*

Oak Park
Jabagi, Habib Daoud *consulting civil engineer*
Schoen, Robert Dennis *civil engineer*

Palatine
Novak, Robert Louis *civil engineer, pavement management consultant*

Rock Island
De Vos, Alois J. *civil engineer*
Johnson, George Edwin *hydraulic engineer*

Scott AFB
Gibson, David Allen *civil engineer, career officer*

Skokie
Russell, Henry George *structural engineer*

Springfield
Lyons, John Rolland *civil engineer*

Urbana
Struble, Leslie Jeanne *civil engineer, educator*

Westmont
Kudrna, Frank Louis, Jr. *civil engineer, consultant*

Wheaton
El-Moursi, Houssam Hafez *civil engineer*

Wilmette
Muhlenbruch, Carl W. *civil engineer*

INDIANA

Danville
Shartle, Stanley Musgrave *consulting engineer, land surveyor*

Fort Wayne
Weatherford, George Edward *civil engineer*

Indianapolis
Dillon, Howard Burton *civil engineer*

Jasper
Wendholt, Norman William *civil engineer*

Lafayette
Mc Laughlin, John Francis *civil engineer, educator*

Plainfield
Meyer, Duane Russell *civil and cost engineer, consultant*

Valparaiso
Tarhini, Kassim Mohamad *civil engineering educator*

West Lafayette
Drnevich, Vincent Paul *civil engineering educator*
Hayes, John Marion *civil engineer*
Madanat, Samer Michel *civil engineer, educator*

IOWA

Ames
Lane, Orris John, Jr. *engineer*
Vermeer, Mark Ellis *project engineer*

Cedar Rapids
Bechler, Ronald Jerry *structural engineer*

Des Moines
Rotert, Kelly Eugene *engineer, consultant*

KANSAS

Burlington
Dingler, Maurice Eugene *civil engineer*

Lawrence
McCabe, Steven Lee *structural engineer*
Mc Kinney, Ross Erwin *civil engineering educator*
Roddis, Winifred Mary Kim *structural engineering educator*

Lenexa
Johannes, Richard Dale *civil engineering executive*

Overland Park
Hansen, Christian Gregory *architectural engineer*
Hoffman, Jerry Carl *civil engineer*

Salina
Willis, Shelby Kenneth *civil engineer, consultant*

Wichita
Mc Kee, George Moffitt, Jr. *civil engineer, consultant*

KENTUCKY

Crestview Hills
Wolff, Ralph Gerald *civil engineer*

Louisville
Hawkins, Darroll Lee *civil engineer*
Smith, Robert F., Jr. *civil engineer*

LOUISIANA

Baker
Moody, Lamon Lamar, Jr. *civil engineer*

Baton Rouge
Acar, Yalcin Bekir *civil engineer, soil remediation technology executive, educator*
Costner, Charles Lynn *civil engineer*
Greaney, John Patrick *civil engineer*
Kazmann, Raphael Gabriel *civil and hydrologic engineer*
Pentas, Herodotos Antreas *civil engineer*
Tumay, Mehmet Taner *geotechnical consultant, educator, research administrator*
Voyiadjis, George Zino *civil engineer, educator*

Dixie
Smoak, Karl Randal *civil engineer*

Lafayette
Billeaud, Manning Francis *civil engineer*

Metairie
Flettrich, Carl Flaspoller *structural engineer*
N'Vietson, Tung Thanh *civil engineer*

New Orleans
Angelides, Demosthenes Constantinos *civil engineer*
Dimitrios, Don Pedon *civil engineer, land surveyor*
Flettrich, Alvin Schaaf, Jr. *civil engineer*
McManis, Kenneth Louis *civil engineer, educator*
Wijesundera, Vishaka *civil engineer, scientist*

River Ridge
Chatry, Frederic Metzinger *civil engineer*

Shreveport
Norwood, Keith Edward *civil engineer*

MAINE

Bath
Watts, Helen Caswell *civil engineer*

East Winthrop
Wiesendanger, J(ohn) Ulrich *consulting engineer*

Orono
Nazmy, Aly Sadek *structural engineering educator, consultant*

South Portland
Iennaco, John Joseph *civil engineer, consultant*

MARYLAND

Annapolis
DiAiso, Robert Joseph *civil engineer*

Baltimore
Corotis, Ross Barry *civil engineering educator, academic administrator*
Gruninger, Robert Martin *civil engineer*
Leshko, Brian Joseph *civil engineer*
Lozovatsky, Mikhail *civil engineer, consultant*
Natarajan, Thyagarajan *civil engineer*
yeager, larry lee *civil engineer, consultant*

Bethesda
Yokel, Felix Yochanan *civil engineer*

Brooklandville
Azola, Martin P. *civil engineer, construction manager*

Claiborne
Guinness, Kenelm L. *civil engineer*

Colesville
Aidar, Nelson *civil engineer, consultant*

College Park
Ayyub, Bilal Mohammed *civil engineering educator*

Columbia
Kalavapudi, Murali *civil engineer*

Gaithersburg
Marshall, Richard Dale *structural engineer*
Simpson, Stephen Lee *consulting civil engineer*
Wright, Richard Newport, III *civil engineer, government official*

Lexington Park
Nannery, Michael Alan *civil engineer*

Rockville
Koplaski, John *structural engineer*

Silver Spring
Mok, Carson Kwok-Chi *structural engineer*

Towson
Huang, Joseph Chen-Huan *civil engineer*
Oppenheimer, Joel K. *civil engineer*

MASSACHUSETTS

Billerica
Brebbia, Carlos Alberto *computational mechanic, engineering consultant*

Boston
Brenner, Brian Raymond *structural engineer*
De Meyere, Robert Emmet *civil engineer*
Druss, David Lloyd *geotechnical engineer*
Hanley, Michelle Boucher *civil engineer*
Moore, Richard Lawrence *structural engineer, consultant*
Stygar, Michael David *civil engineer*
Velkov, Simeon Hristov *civil engineer*
Zaldastani, Othar *structural engineer*

Burlington
Olsen, Allen Neil *civil engineer*

Cambridge
Hansen, Robert Joseph *civil engineer*
Ladd, Charles Cushing, III *civil engineering educator*
LeMessurier, William James *structural engineer*
Marks, David Hunter *civil engineering educator*
Morel, François M.M. *civil and environmental engineering educator*
Pestana-Nascimento, Juan M. *civil, geotechnical and geoenvironmental engineer, consultant*
Pineau, Daniel Robert *structural engineer*
Soydemir, Cetin *geotechnical engineer, earthquake engineer*
Sweeney, Bryan Philip *structural and geotechnical engineer*

Foxboro
Pierce, Francis Casimir *civil engineer*

Granby
Shaheen, William A. *civil engineering educator, consultant*

Lexington
Bowden, Denise Lynn *civil engineer*
Nuthmann, Conrad Christopher *civil engineer, consultant*
Silva-Tulla, Francisco *civil engineer*
Weiss, Scott Jeffrey *civil engineer*

Manchester
Gaythwaite, John William *civil engineer*

Medford
Hankour, Rachid *civil engineer*

Milford
Nicoson, Steven Wayne *geotechnical engineer*

Milton
Hanley, Joseph Andrew *civil engineer*

Quincy
Bennett, Bruce Anthony *civil engineer*
Mancini, Rocco Anthony *civil engineer*

Wakefield
Magaw, Jeffrey Donald *civil engineer*
Scarpa, Robert Louis *civil engineer*

Wayland
Drevinsky, David Matthew *civil engineer*

Winchester
Agrawal, Vimal Kumar *structural engineer*
Joseph, Paul Gerard *civil engineer, consultant*

MICHIGAN

Ann Arbor
Hansen, Will *civil engineer, educator, consultant*
Hryciw, Roman D. *civil engineering educator*
Jones, James Ray *architectural engineering research associate*

Auburn Hills
Verma, Dhirendra *civil engineer*

Battle Creek
O'Brien, John F(rancis) *civil engineer*

Detroit
Gill, Mohammad Akram *civil engineer*
Hunter, Robert Fabio *civil engineer*
Moore, W. James *civil engineer, engineering executive*
Person, Victoria Bernadett *civil engineer, consultant*

East Lansing
Andersland, Orlando Baldwin *civil engineering educator*

Houghton
Huang, Eugene Yuching *civil engineer, educator*

Kalamazoo
Arwashan, Naji *structural engineer*
Elliott, Marc Eldon *civil engineer*

Lansing
Harvey, John Ashmore *civil engineer*

Madison Heights
Shah, Jayprakash Balvantrai *civil engineer*

Plainwell
Hultmark, Gordon Alan *civil engineer*

Southfield
Akinmusuru, Joseph Olugbenga *civil engineer*
Prasuhn, Alan Lee *civil engineer, educator*

Sterling Heights
Burke, Thomas Joseph *civil engineer*

Troy
Ross, Eric Alan *civil engineer*

MINNESOTA

Minneapolis
Arndt, Roger Edward Anthony *hydraulic engineer, educator*
Galambos, Theodore Victor *civil engineer, educator*
Gale, Stephan Marc *civil engineer*
Perkuhn, Gaylen Lee *civil/structural engineer*
Rauch, Andrew Martin *civil/structural engineer*

Richfield
Feldsien, Lawrence Frank *civil engineer*

Rochester
Betts, Douglas Norman *civil engineer*
Steiner, Jeffery Allen *project engineer, executive*

Saint Paul
Goodman, Lawrence Eugene *structural analyst, educator*
Ng, Lewis Yok-Hoi *civil engineer*

MISSISSIPPI

Biloxi
Ransom, Perry Sylvester *civil engineer*

Lorman
Hylander, Walter Raymond, Jr. *retired civil engineer*

Mississippi State
Truax, Dennis Dale *civil engineer, educator, consultant*

Starkville
Priest, Melville Stanton *consulting hydraulic engineer*

University
Hackett, Robert Moore *engineering educator*
Uddin, Waheed *civil engineer*

Vicksburg
Ahlvin, Richard Glen *civil engineer*
Albritton, Gayle Edward *structural engineer*
Bryant, Larry Michael *structural engineer, researcher*
Fehl, Barry Dean *civil engineer*
McRae, John Leonidas *civil engineer*
Olsen, Richard Scott *geotechnical engineer*
Scott, Angela Freeman *civil engineer*

MISSOURI

Camdenton
Krehbiel, Darren David *civil engineer*

Columbia
Douty, Richard Thomas *structural engineer*
El-Bayya, Majed Mohammed *civil engineer*

Ferguson
Fieldhammer, Eugene Louis *civil engineer*

Fortuna
Ramer, James LeRoy *civil engineer*

Jefferson City
Neumann, Donald Lee *civil engineer, environmental specialist*

Kansas City
Boyd, John Addison, Jr. *civil engineer*
Bretzke, Virginia Louise *civil engineer, environmental engineer*
Donahey, Rex Craig *structural engineer*
Heausler, Thomas Folse *structural engineer*
Hobson, Keith Lee *civil engineer, consultant*
Schuler, Martin Luke *geotechnical engineer*
Stannard, William George *civil engineer*

Maryland Heights
Goldfarb, Marvin Al *civil engineer*

Rolla
Belarbi, Abdeldjelil *civil engineering educator, researcher*
Dare, Charles Ernest *civil engineer, educator*
Hunt, Theodore William *civil engineer*
Munger, Paul R. *civil engineering educator*

Saint Louis
Briles, John Christopher *civil engineer*
Cooling, Thomas Lee *civil engineer, consultant*
Gould, Phillip L. *civil engineering educator, consultant*
O'Neill, James William *structural engineer*
Sullentrup, Michael Gerard *structural engineer, consultant*

Saint Peters
Yerion, Michael Ross *civil engineer*

Springfield
Nuccitelli, Saul Arnold *civil engineer, consultant*

MONTANA

Bozeman
Suit, D. James *civil engineer*

Great Falls
Tufte, Erling Arden *civil engineer*

Helena
Johnson, David Sellie *civil engineer*

NEBRASKA

Lincoln
Paek, James Joon-Hong *civil engineer, construction management educator*

Omaha
Coy, William Raymond *civil engineer*
Hasterlo, John S. *civil engineer, researcher*
Meyer, Karl V. *civil engineer, consultant*
Rossbach, Philip Edward *civil engineer*
Schwartz, Rodney Jay *civil, mechanical and electrical engineer*
Thompson, William Charles *civil engineer*
Tunnicliff, David George *civil engineer*

NEVADA

Carson City
Hayes, Gordon Glenn *civil engineer*

Las Vegas
Freiberg, Jeffrey Joseph *civil engineer*
LaRubio, Daniel Paul, Jr. *civil engineer*
Owens, Mark Jeffrey *geotechnical engineer*

Reno
Meeker, Lawrence Edwin *civil engineer*

NEW HAMPSHIRE

Berlin
Davis, Alvin Robert, Jr. *structural engineer*

Hanover
Kovacs, Austin *research engineer*
Stearns, Stephen Russell *civil engineer, forensic engineer, educator*

Nashua
Fallet, George *civil engineer*

Portsmouth
Klotz, Louis Herman *structural engineer, educator, consultant*

Stratham
Bjorkman, Gordon Stuart, Jr. *structural engineer, consultant*

NEW JERSEY

Barnegat
Lowe, Angela Maria *civil engineer*

Bayville
Scancella, Robert J. *civil engineer*

Bloomfield
Monahan, Edward James *geotechnical engineer*

Bridgewater
Wilkinson, Clifford Steven *civil engineer*

Collingswood
Carty, John Brooks *civil engineer*

Denville
Brooks, Marc Benjamin *civil engineer*
Price, Robert Edmunds *civil engineer*

Elmwood Park
Ogden, Michael Richard *civil engineer, consultant*
Semeraro, Michael Archangel, Jr. *civil engineer*

Englewood Cliffs
DoVale, Fern Louise *civil engineer*

Hackensack
Ten Kate, Peter Cornelius *civil engineer*

Hammonton
Weigel, Henry Donald *civil engineer*

Hoboken
Talimcioglu, Nazmi Mete *civil and environmental engineer, educator*

Jersey City
Chatterjee, Amit *structural engineer*
Gudema, Norman H. *civil engineer*

Leonia
Husar, Emile *civil engineer, consultant*

Moorestown
Lipscomb, Thomas Heber, Jr. *civil engineer, consultant*

Morris Plains
Patel, Kanti Shamjibhai *civil engineer*

Mount Laurel
Barba, Evans Michael *civil engineer*
Case, Mark Edward *geotechnical engineer, consultant*

New Brunswick
Nawy, Edward George *civil engineer, educator*

Newark
Hsu, Cheng-Tzu Thomas *civil engineering educator*

North Brunswick
Awan, Ahmad Noor *civil engineer*

Old Bridge
Pflug, Leo Joseph, Jr. *civil engineer*

Paterson
Williams, Dennis Thomas *civil engineer*

Piscataway
Sohal, Iqbal Singh *structural engineer, educator*

Princeton
Dinkevich, Solomon *structural engineer*
Karol, Reuben Hirsh *civil engineer, sculptor*
Rhimes, Richard David *civil engineer*

Ridgewood
Tsapatsaris, Nicholas *civil engineer, consultant*

Roseland
Patterson, Scott Paul *civil engineer, consultant*

Rutherford
Rubano, Richard Frank *civil engineer*

Trenton
Ellis, Clifford Aubrey *civil engineer*

West Keansburg
Smith, Keith Brunton *civil engineer*

West Trenton
Skrobacz, Edwin Stanley *structural engineer, educator*

Woodbridge
Adams, James Derek *construction engineer*

NEW MEXICO

Albuquerque
Asbury, Charles Theodore, Jr. *civil engineering consultant*
Haddad, Edward Raouf *civil engineer, consultant*
Johnson, Stewart Willard *civil engineer*
Maji, Arup Kanti *civil engineering educator*

Las Cruces
Morales, Richard *structural engineer*
Stout, Larry John *civil engineer, consultant*
Vick, Austin Lafayette *civil engineer*

Socorro
Lancaster, John Howard *civil engineer*

NEW YORK

Albany
Geoffroy, Donald Noel *civil engineer, consultant*

Amherst
Lee, George C. *civil engineer, university administrator*

Bronx
Hom, Wayne Chiu *civil engineer*

Brooklyn
Karamouz, Mohammad *engineering educator*
Levy, Melvin *tunnel engineer, union executive*
Marazzo, Joseph John *civil engineer*

Buffalo
Abate, Ralph Francis *structural engineer*
Frandina, Philip Frank *civil engineer, consultant*
Lobo, Roy Francis *structural engineer*
Surianello, Frank Domenic *civil engineer*

Cazenovia
Madhoun, Fadi Salah *civil engineer*

Elmhurst
Jethwani, Mohan *civil engineer*

Flushing
Kurtz, Max *civil engineer, consultant*

Glen Cove
Casem, Conrado Sibayan *civil/structural engineer*

Greenwich
Briggs, George Madison *civil engineer*

Huntington Station
Lanzano, Ralph Eugene *civil engineer*

Ithaca
Gergely, Peter *structural engineering educator*
O'Rourke, Thomas Denis *civil engineer, educator*
Stedinger, Jery Russell *civil and environmental engineer, researcher*

Jackson Heights
Olmsted, Robert Amson *civil engineer*

Jamaica
Muresanu, Violeta Ana *civil, structural engineer*

Kingston
Dennison, Robert Abel, III *civil engineer*

Long Island City
Jablowsky, Albert Isaac *civil engineer*
McCotter, Michael Wayne *civil engineer*

Mamaroneck
Ekizian, Harry *civil engineering consultant*

Maspeth
Merjan, Stanley *civil engineer, inventor*

Middle Village
Miele, Joel Arthur, Sr. *civil engineer*

Middletown
Herries, Edward Matthew *civil engineer*

Mohegan Lake
Paik, John Kee *structural engineer*

New York
Ahmad, Jameel *civil engineer, researcher, educator*
Betti, Raimondo *civil engineering educator*
Clancy, Thomas Joseph *civil engineer*
Cohen, Edward *civil engineer*
Englot, Joseph Michael *structural engineer*
Goff, Kenneth Alan *engineering executive*
Goldreich, Joseph Daniel *consulting structural engineer*
Griffis, Fletcher Hughes *civil engineering educator, engineering executive*
Grossman, Jacob S. *structural engineer*
Lu, Ruth *structural engineer*
Paaswell, Robert Emil *civil engineer, educator*
Rahimian, Ahmad *structural engineer*
Saito, Mitsuru *civil engineer, educator*
Shapiro, Murray *structural engineer*
Stahl, Frank Ludwig *civil engineer*
Zmeu, Bogdan *structural engineer, consultant*

Niagara Falls
Dojka, Edwin Sigmund *civil engineer*

Penfield
Hutteman, Robert William *civil engineer*

Rochester
Finnigan, James Francis *civil engineer*
Holzbach, James Francis *civil engineer*
Miles, William Robert *structural engineer*
Parker, Thomas Sherman *civil engineer*

Schenectady
Mafi, Mohammad *civil engineer, educator*

Staten Island
Rajakaruna, Lalith Asoka *civil engineer*

Syracuse
Lui, Eric Mun *civil engineering educator, practitioner*
Miller, Douglas Alan *civil engineer*
Young, Terry William *civil engineer*

Troy
Jordan, Mark Henry *consulting civil engineer*
Sopko, Stephen Joseph *structural engineer*
Zimmie, Thomas Frank *civil engineer, educator*

Wappingers Falls
Ninnie, Eugene Dante *civil engineer*

Watertown
Dimmick, Kris Douglas *civil engineer*

West Nyack
Hornik, Joseph William *civil engineer*

Yonkers
D'Angelo, Andrew William *civil engineer*

Yorktown Heights
Cifuentes, Arturo Ovalle *civil and mechanical engineer*

NORTH CAROLINA

Morehead City
Williams, Winton Hugh *civil engineer*

Raleigh
Boblett, Mark Anthony *civil engineering technician*
Borden, Roy Herbert, Jr. *civil engineer, geotechnical consultant*
Godwin, Lars Duvall *civil engineer*
Hanson, John M. *civil engineering and construction educator*
Havner, Kerry Shuford *civil engineering and materials science educator*
Kelly, Richard Lee Woods *civil engineer*
Moody, Roger Wayne *civil engineer*
Poole, Marion Ronald *civil engineering executive*
Poran, Chaim Jehuda *civil and environmental engineer, educator*

Sanford
Woerner, Geldard Harry *civil engineer*

Skyland
Connolly, William Michael *civil and structural engineer*

Wilmington
Lloyd, Donald Grey, Jr. *civil engineer*

OHIO

Ada
Ward, Robert Lee *civil engineering educator*

Bellbrook
Maneggio, Lizette *civil engineer*

Cleveland
Duffy, Stephen Francis *civil engineer, educator*
Eagleton, Robert Lee *civil engineer*

Columbus
Amato, Vincent Edward *geotechnical engineer, consultant*
Apel, John Paul *civil engineer*
Bedford, Keith Wilson *civil engineering and atmospheric science educator*
Buhac, H(rvoje) Joseph *geotechnical engineer*
Gehring, Richard Webster *structural engineer*
Ruggles, Harvey Richard *civil engineer*
Shannon, Larry James *civil engineer*

Dayton
Saliba, Joseph Elias *civil engineering educator*

Fostoria
Curlis, David Alan *civil engineer*

Logan
Frost, Jack Martin *civil engineer*

Marietta
Goyle, Rajinder Kumar *civil engineer, consultant*

Oberlin
Schmucker, Bruce Owen *geotechnical engineer*

Powell
Adeli, Hojjat *civil engineer, computer scientist, educator*

Toledo
Benham, Linda Sue *civil engineer*
Randolph, Brian Walter *civil engineer, educator*

Westlake
Sminchak, David William *civil engineer, consultant*

OKLAHOMA

Ada
Stephens, David Basil *civil engineer*

Bartlesville
Brannan, Michael Steven *civil, environmental, automotive engineer*

Norman
Zaman, Musharraf *civil engineering educator*

Oklahoma City
Allen, James Harmon, Jr. *civil engineer*
Zenieris, Petros Efstratios *civil engineer*

Sand Springs
Atiyeh, Elia Mtanos *civil engineer*

Tulsa
McCaw, Valerie Sue *civil engineer*

OREGON

Corvallis
Chen, Chaur-Fong *civil engineer, educator*

Eugene
Kovtynovich, Dan *civil engineer*
Pearson, John Mark *civil engineer*

Portland
Janssen, Andrew Gerard *civil engineer*
Lall, B. Kent *civil engineering educator*
Sangrey, Dwight A. *civil engineer, educator*

Salem
Butts, Edward P. *civil engineer, consultant*
Lambert, Ralph William *civil engineer*

Sunriver
Clough, Ray William, Jr. *civil engineering educator*

PENNSYLVANIA

Alcoa Center
Nordmark, Glenn Everett *civil engineer*

Beaver
Hodge, Philip Tully *structural engineer*

Bellefonte
Grindall, Emerson Jon *civil engineer*

Bensalem
Bontempo, Daniel *civil engineer*

Bethlehem
Fisher, John William *civil engineering educator*
Kugelman, Irwin Jay *civil engineering educator*
Lennon, Gerard Patrick *civil engineering educator, researcher*
Ostapenko, Alexis *civil engineer, educator*

Bridgeville
Glidden, Bruce *structural engineer, construction consultant*
Pearlman, Seth Leonard *civil engineer*

Chalfont
Ballod, Martin Charles *civil engineer, mechanical engineer*

Chester
Brown, Vicki Lee *civil engineering educator*

Conshohocken
Cohen, Alan *civil engineer*

Cooksburg
Meley, Robert Wayne *structural engineer*

Coraopolis
Kimes, Mark Edward *civil engineer*
Stewart, Richard Allan *civil engineer*

Doylestown
Samuelson, Andrew Lief *civil engineer*

Harrisburg
Rosen, Herman *civil engineer*

Horsham
Derby, Christopher William *civil engineer*

Houston
Ali, Ashraf *civil engineer*

Johnstown
Dombrowski, John Micheal *architectural consulting engineer*

Lewisburg
Orbison, James Graham *civil engineer, educator*

Mars
Pantelis, John Andrew, Jr. *civil engineer*

McMurray
Richards, John Lewis *civil engineer, educator*

Monroeville
Mandel, Herbert Maurice *civil engineer*
McDermott, John Francis, IV *civil engineer, consultant*

Philadelphia
Brecher, Ephraim Fred *structural engineering consultant*
Mansour, Farid Fam *civil engineer*
Tantala, Albert Martin *civil engineer*
Terzian, Karnig Yervant *civil engineer*
Wright, Robert Michael *civil engineer*

Pittsburgh
Boczkaj, Bohdan Karol *structural engineer*
Boward, Joseph Frank *civil engineer*
Garrett, James Henry, Jr. *civil engineering educator*
Held, William James *civil engineer*
Morgan, Daniel Carl *civil engineer*
Rettura, Guy *civil engineer*
Schwemmer, David Eugene *structural engineer*
Schwendeman, Louis Paul *civil engineer, retired*

Pottstown
Messics, Mark Craig *civil engineer*

Reading
Broome, Kenneth Reginald *civil engineer, consultant*

Somerset
Critchfield, Harold Samuel *retired civil engineer*
Weaver, Craig Lee *civil engineer*

University Park
Scanlon, Andrew *structural engineering educator*

Wilkes Barre
Kerns, James Albert *structural engineer*

Yardley
Tams, Thomas Walter *civil and structural engineer, land surveyor*

York
Pease, Howard Franklin *structural engineer*

SOUTH CAROLINA

Columbia
Bates, William Lawrence *civil engineer*
Cannon, Randy Ray *civil engineer*
Lee, Alexandra Saimovici *civil engineer*
Linyard, Samuel Edward Goldsmith *civil engineer*
Whitlock, Edward Madison, Jr. *civil engineer*

Greenville
Carlson, William Scott *structural engineer*
Csernak, Stephen Francis *structural engineer*
Gamble, Francoise Yoko *structural engineer*

Mount Pleasant
Lambert, Richard Dale, Jr. *structural engineer, consultant*
Taylor, Jon Guerry *civil engineer*

North Augusta
Rajendran, Narasimhan *civil engineer*

SOUTH DAKOTA

Brookings
Bretsch, Carey Lane *civil/architectural engineering administrator*

Rapid City
Ramakrishnan, Venkataswamy *civil engineer, educator*

TENNESSEE

Bartlett
Orsak, Joseph Cyril *civil engineer*

Bristol
Hyer, Charles Terry *civil and mining engineer*

Columbia
Cowan, Michael John *civil engineer*

Knoxville
Schuler, Theodore Anthony *civil engineer, city official*

Memphis
Randle, Ronald Eugene *structural engineer*

Mount Juliet
Auld, Bernie Dyson *civil engineer, consultant*

Nashville
Ballou, Christopher Aaron *civil engineer*
Basu, Prodyot Kumar *civil engineer, educator*
McClanahan, Larry Duncan *civil engineer, consultant*

TEXAS

Amarillo
Smith, Clinton W. *civil engineer, consultant*

Arlington
Caffey, James Enoch *civil engineer*
Rollins, Albert Williamson *civil engineer, consultant*

Austin
Breen, John Edward *civil engineer, educator*
Burns, Ned Hamilton *civil engineering educator*
Carrasquillo, Ramon Luis *civil engineering educator, consultant*
Flynn, William Thomas *civil engineer*
Fowler, David Wayne *architectural engineering educator*
Frank, Karl H. *civil engineer, educator*
Holley, Edward R. *civil engineering educator*
Jensen, Paul Allen *electrical engineer*
Jirsa, James Otis *civil engineering educator*
Murthy, Vanukuri Radha Krishna *civil engineer*
Olson, Roy Edwin *civil engineering educator*
Reese, Lymon Clifton *civil engineering educator*
Studer, James Edward *geological engineer*
Tucker, Richard Lee *civil engineer, educator*
Walton, Charles Michael *civil engineering educator*
Wright, Stephen Gailord *civil engineering educator, consultant*
Yura, Joseph Andrew *civil engineering educator, consulting structural engineer*

Bellaire
Wisch, David John *structural engineer*

Boerne
Mitchelhill, James Moffat *civil engineer*

College Station
Batchelor, Bill *civil engineering educator*
Gallaway, Bob M. *consulting engineer*
Lowery, Lee Leon, Jr. *civil engineer*
Lytton, Robert Leonard *civil engineer, educator*
Yao, James Tsu-Ping *civil engineer*

Dallas
Austin, Robert Brendon *civil engineer*
Farmer, Jim L(ee) *civil engineer, consultant*
Gross, William Spargo *civil engineer*
Lackey, James Franklin, Jr. *civil engineer*
Ramsbacher, Scott Blane *civil engineer*
Yanagisawa, Shane Henry *civil engineer*

Fort Worth
Bakintas, Konstantine *civil, environmental design engineer*
Fischer, Marsha Leigh *civil engineer*

Galveston
Otis, John James *civil engineer*

Houston
Dilly, Ronald Lee *civil engineer, construction technology educator*
Ferguson, Richard Peter *project engineer*
Heit, Raymond Anthony *civil engineer*
Moore, Walter Parker, Jr. *civil engineering company executive*
Nordgren, Ronald Paul *civil engineering educator, research engineer*
O'Neill, Michael Wayne *civil engineer, educator*
Oxer, John Paul Daniell *civil engineer*
Rajapaksa, Yatendra Ramyakanthi Perera *structural engineer*
Saunders, Richard Wayne *pipeline engineer*
Umrigar, Dara Nariman *civil engineer*
Vipulanandan, Cumaraswamy *civil engineer, educator*
Walker, Esper Lafayette, Jr. *civil engineer*

Ingelside
Naismith, James Pomeroy *civil engineer*

Irving
McCormack, Grace Lynette *engineering technician*

Kingwood
Heldenbrand, David William *civil engineer*

Longview
Winn, Walter Terris, Jr. *civil/environmental engineer*

Lubbock
Kiesling, Ernst Willie *civil engineering educator*

Missouri City
Edwards-Hollaway, Sheri Ann *civil engineer*

San Angelo
Harrington, William Palmer *retired civil engineer*

San Antonio
Thacker, Ben Howard *civil engineer*

Waco
Foster, Ottis Charles *civil engineer*

Wichita Falls
Hinton, Troy Dean *civil engineer*

Woodsboro
Rooke, Allen Driscoll, Jr. *civil engineer*

UTAH

Ogden
Strand, Tena Joy *civil engineer*

Provo
Riley, James Alvin *civil engineer*
Youd, T. Leslie *civil engineer*

Salt Lake City
Anderson, Charles Ross *civil engineer*
Epperson, Vaughn Elmo *civil engineer*
Gill, Ajit Singh *civil engineer*
Gregersen, Max A. *structural, earthquake and civil engineer*
Saunders, Barry Collins *civil engineer*

VERMONT

Barre
Delano, A(rthur) Brookins, Jr. *civil engineer, consultant*

Burlington
Beliveau, Jean-Guy Lionel *civil engineering educator*

VIRGINIA

Alexandria
Rall, Lloyd Louis *civil engineer*

Arlington
Bogue, Sean Kenneth *civil engineer*
Mott, Charles Davis *civil engineer*

Ashland
Wright, Berry Franklin *civil engineer*

Chesapeake
Little, W. Ken, Jr. *structural engineer*

Fairfax
Reed, Christopher Robert *civil engineer*

Grafton
Dischinger, Hugh Charles *civil engineer*

Greenwood
Casero, Joseph Manuel *civil engineer*

Herndon
Ruddell, James Thomas *civil engineer, consultant*

Marion
Young, Craig Steven *structural engineer*

Mc Lean
Lane, Susan Nancy *structural engineer*
McCambridge, John James *civil engineer*
Woo, Dah-Cheng *hydraulic engineer*

Norfolk
Carpenter, Allan Lee *civil engineer*
McLaren, John Paterson, Jr. *civil engineer*

Pearlsburg
Stafford, Steven Ward *civil engineer*

Reston
Ethridge, Max Michael *civil engineer*
Subasic, Christine Ann *architectural engineer*

Springfield
Welty, Kenneth Harry *civil engineer*

Sterling
Ghazarian, Rouben *structural engineer*

Virginia Beach
Sweet, Rita Genevieve *civil engineer*
Switzer, Terence Lee *civil engineer*

Williamsburg
Dunn, Ronald Holland *civil engineer, management executive, railway consultant*

Winchester
Cleland, Ned Murray *civil engineer*

WASHINGTON

Bellevue
Genskow, John Robert *civil engineer, consultant*
Killgore, Mark William *civil engineer*
Sharp, Kevan Denton *civil engineer*

Ephrata
Levey, Sandra Collins *civil engineer*

Kennewick
Williams, Brian James *geotechnical engineer*

Olympia
Tudor, Gregory Scott *land information system analyst*

Pullman
Pellerin, Roy Francis *civil engineer, educator*

Richland
Gates, Thomas Edward *civil engineer, waste management administrator*
Henager, Charles Henry *civil engineer*

Seattle
Dorwart, Brian Curtis *geotechnical engineer, consultant*
Eberhard, Marc Olivier *civil engineering educator*
Galster, Richard W. *civil engineer*
Mao, Kent Keqiang *consulting civil engineer*
Whitman, Alexander H., Jr. *civil engineer, consultant*

South Bend
Heinz, Roney Allen *civil engineering consultant*

Tacoma
Thomas, Steven Joseph *structural engineer*

Walla Walla
Schrader, Ernest Karl *engineer*

WEST VIRGINIA

Montgomery
Thornton, Stafford Earl *civil engineering educator*

Morgantown
Eck, Ronald Warren *civil engineer, educator*

Wheeling
Coast, Morgan K. *civil engineer, consultant*

WISCONSIN

Brookfield
Curfman, Floyd Edwin *engineering educator*

Grafton
Eber, Lorenz *civil engineer*

Juneau
Sindelar, Robert Albert *civil engineer*

La Crosse
Davy, Michael Francis *civil engineer, consultant*

Madison
Kaul, David Glenn *civil engineer*
Richardson, Kevin William *civil and environmental engineer*
Russell, Jeffrey Scott *civil engineering educator*

Menasha
Sutter, Carl Clifford *civil engineer*

Milwaukee
Blomquist, Michael Allen *civil engineer*
Hanson, Curtis Jay *structural project engineer*
Laubenheimer, Jeffrey John *civil engineer*
Lindner, Albert Michael *structural engineer*

Racine
Rooney, John Connell *consulting civil engineer*

WYOMING

Torrington
Jensen, Christopher Douglas *civil engineer*

TERRITORIES OF THE UNITED STATES

PUERTO RICO

San Juan
Benton, Stephen Richard *civil and mechanical engineer*
Quiñones, Jose Antonio *structural engineer, consultant*

MILITARY ADDRESSES OF THE UNITED STATES

EUROPE

APO
Carioti, Bruno Mario *civil engineer*

CANADA

ALBERTA

Calgary
Ghali, Amin *civil engineering educator*

Edmonton
Bartlett, Fred Michael Pearce *structural engineer*
Hrudey, Steve E. *civil engineer, educator*
Kennedy, D. J. Laurie *civil engineering educator*
Morgenstern, Norbert Rubin *civil engineering educator*
Stanley, S. J. *civil engineering educator*

BRITISH COLUMBIA

North Vancouver
Buckland, Peter Graham *structural engineer*
Morgenstern, Brian D. *civil engineer*
Wilson, Carl Robert *structural engineer*

Surrey
Stiemer, Siegfried F. *civil engineer*

Vancouver
Isaacson, Michael *civil engineering educator*
Mavinic, Donald Stephen *civil engineering educator*

MANITOBA

Winnipeg
Cohen, Harley *civil engineer, science educator*

NEW BRUNSWICK

Fredericton
Douglas, Robert Andrew *civil engineer, educator*

NEWFOUNDLAND

Saint John's
Clark, Jack I. *civil engineer, researcher*

NOVA SCOTIA

Halifax
Mufti, Aftab A. *civil engineering educator*

ONTARIO

Mississauga
Milligan, Victor *civil engineer, consultant*

Ottawa
Baltacioglu, Mehmet Necip *civil engineer*
Bozozuk, Michael *civil engineer*
Seaden, George *civil engineer*

Toronto
Springfield, J. *civil engineer*
Wright, Peter Murrell *structural engineering educator*

Waterloo
Thomson, N. R. *civil engineering educator*

Windsor
Monforton, Gerard Roland *civil engineer, educator*

PRINCE EDWARD ISLAND

Charlottetown
McCreath, Peter S. *Canadian government official, civil engineer*

QUEBEC

Montreal
Filiatrault, Andre *civil engineering educator*
Mirza, M. Saeed *civil engineering educator*
Selvadurai, Antony Patrick Sinnappa *civil engineering educator, applied mathematician, consultant*

Pointe Claire
Mitchell, Denis *civil engineer*

Sainte Anne de Bellevue
Broughton, Robert Stephen *irrigation and drainage engineering educator, consultant*

Sherbrooke
Paultre, Patrick *civil engineering educator*

SASKATCHEWAN

Regina
Sharp, James J. *civil engineer*

Saskatoon
Kells, J. A. *civil engineer*
Smith, C. D. *civil engineering educator*

MEXICO

Mexico City
Rosenblueth, Emilio *structural engineer*

AUSTRALIA

Sydney
Carter, John Phillip *civil engineering educator, consultant, researcher*
Cowan, Henry Jacob *architectural engineer, educator*

Wollongong
Kohoutek, Richard *civil engineer*

BRAZIL

Rio de Janeiro
Matos, Jose Gilvomar Rocha *civil engineer*
Souza, Marco Antonio *civil engineer, educator*

BULGARIA

Sofia
Hadjikov, Lyuben Manolov *civil engineer*

COSTA RICA

San Jose
Sauter, Franz Fabian *structural engineer*

ENGLAND

Falmouth
Batchelor, Anthony Stephen *geotechnical engineer, consultant*

London
Pavlovic, Milija N. *structural engineer, educator*

ETHIOPIA

Addis Ababa
Haile Giorgis, Workneh *civil engineer*

GERMANY

Landsberg
Panizza, Michael *civil engineer*

GREECE

Athens
Androutsellis-Theotokis, Paul *civil engineer*
Katsikadelis, John *civil engineering educator*

Ekali
Sotirakos, Iannis *civil engineer*

GUATEMALA

Mixco
Silva, Fernando Arturo *civil engineer*

HONG KONG

Kowloon
Huang, Ju-Chang *civil engineering educator*

Quarry Bay
Bast, Albert John, III *civil engineer*

HUNGARY

Budapest
Starosolszky, Ödön *civil engineer*

INDIA

Bathinda
Bansal, Satish Kumar *civil engineer, educator*

INDONESIA

Bandung
Norup, Kim Stefan *civil engineer*

IRAN

Tehran
Mirdamadi, Hamid Reza *structural engineering educator, researcher*

ITALY

Padua
Schrefler, Bernhard Aribo *civil engineering educator*

JAPAN

Chiba
Noguchi, Hiroshi *structural engineering educator*

Koriyama
Ohama, Yoshihiko *architectural engineer, educator*

Osaka
Nakai, Hiroshi *civil engineering educator*

Tokyo
Aoyama, Hiroyuki *structural engineering educator*
Iwasaki, Toshio *hydraulics engineer, investigator*
Kawahara, Mutsuto *civil engineer, educator*

LIBYA

Tripoli
Abuzakouk, Marai Mohammad *civil aviation engineer*

MALAYSIA

Selangor
Tan, Chai Tiam *civil engineer, precast concrete company executive*

MONACO

Monaco
Kalaidjian, Berj Boghos *civil engineer*

THE NETHERLANDS

Bunnik
van Dyke, Jacob *civil engineer*

NEW ZEALAND

Auckland
Marshall, Arthur Harold *architectural engineer*

NORWAY

Oslo
Barton, Nick *rock mechanics engineer*
Knudsen, Knud-Endre *civil engineer, consultant*

POLAND

Szczecin
Zygmunt, Zieliński *civil engineer, educator*

SAUDI ARABIA

Riyadh
Lewis, Don Alan *civil, sanitary engineer, consultant*
Tabba, Mohammad Myassar *civil engineer, manager, educator*

Sharjah
Ali, Syed Ibrahim *civil engineer*

SINGAPORE

Singapore
Taylor, Leland Harris, Jr. *civil engineer*

SOUTH AFRICA

Halfway House
Grieve, Graham Robert *civil engineering executive*

SPAIN

Barcelona
Barbat, Alex Horia *civil engineer*

SRI LANKA

Hambantota
Abeysundara, Urugamuwe Gamacharige Yasantha *civil engineer*

SWEDEN

Stockholm
McNown, John Stephenson *hydraulic engineer, educator*

SWITZERLAND

Baden
Pozzi, Angelo *executive, civil engineer*

Neuchâtel
Schmid, Hans Dieter *civil engineering consultant*

Zurich
Hepguler, Yasar Metin *architectural engineering educator, consultant*

TAIWAN

Taipei
Chern, Jenn-Chuan *civil engineering educator*

THAILAND

Bangkok
San Juan, German Moral *structural engineer, international consultant*

TRINIDAD AND TOBAGO

Saint Augustine
Osborne, Robin William *civil engineer, educator*

VENEZUELA

Caracas
Cerrolaza, Miguel Enrique *civil engineer, educator*

WALES

Swansea
Morgan, Kenneth *civil engineering educator, researcher*

ADDRESS UNPUBLISHED

Al-Qadi, Imad Lutfi *civil engineering educator, researcher*
Anton, Walter Foster *civil engineer*
Bacani, Nicanor-Guglielmo Vila *civil and structural engineer, consultant, real estate investor*
Baxter, Loran Richard *civil engineer*
Blum, Gregory Lee *civil engineer*
Buckner, Harry Benjamin *civil engineer*
Chyu, Jih-Jiang *structural engineer, consultant*
Emery, Mark Lewis *civil engineer*
Foster, Edward Joseph *structural engineer*
Fritcher, Earl Edwin *civil engineer, consultant*
Gillett, John Bledsoe *civil engineer*
Girouard, Kenneth *civil engineer*
Gross, Seymour Paul *retired civil engineer*
Halpin, Daniel William *civil engineering educator, consultant*
Horowitz, Joseph *civil engineer*
Jewell, Thomas Keith *civil engineering educator*
Karapostoles, Demetrios Aristides *civil engineer*
Kuesel, Thomas Robert *civil engineer*
Loy, Richard Franklin *civil engineer*
Moll, David Carter *civil engineer*
Paproski, Ronald James *structural engineer*
Partheniades, Emmanuel *civil engineer, educator*
Penner, Karen Marie *civil engineer*
Pfalser, Ivan Lewis *civil engineer*
Phaup, Arthelius Augustus, III *structural engineer*
Pniakowski, Andrew Frank *structural engineer*
Porter, Irene Rae *civil engineer*
Rihani, Sarmad Albert (Sam Rihani) *civil engineer*
Shukla, Mahesh *structural engineer*
Sprengnether, Ronald John *civil engineer*
Timmer, David Hart *civil engineer*
Turner, William Oliver, III *civil engineer*
Yazdani, Jamshed Iqbal *civil engineer*
Young, Robert Allen *architectural engineer*

ENGINEERING: COMPUTER

UNITED STATES

ARIZONA

Chandler
Basilier, Erik Nils *software scientist and engineer*

Phoenix
Landrum, Larry James *computer engineer*

Sierra Vista
Ricco, Raymond Joseph, Jr. *computer systems engineer*

CALIFORNIA

Belmont
Jaffe, David Henry *computer systems engineer*

Foster City
Ting, Chen-Hanson *software engineer*

Goleta
Mangaser, Amante Apostol *computer engineer*

Los Angeles
Wong, Kenneth Lee *software engineer, consultant*

Los Gatos
Fujitani, Martin Tomio *software quality engineer*

San Jose
Gorum, Victoria *computer engineer*

Santa Barbara
Anderson, Stephen Thomas *computer design engineer*

Simi Valley
Ahsanullah, Omar Faruk *computer engineer, consultant*

COLORADO

Colorado Springs
Carroll, David Todd *computer engineer*

CONNECTICUT

Storrs Mansfield
Ammar, Reda Anwar *computer science engineering educator*

West Redding
Goldman, Ernest Harold *computer engineering educator*

DISTRICT OF COLUMBIA

Washington
Eom, Kie-Bum *computer engineering educator*

FLORIDA

Kennedy Space Center
Head, Julie Etta *computer engineer*

Orlando
Casanova-Lucena, Maria Antonia *computer engineer*
Eichner, Gregory Thomas *computer engineer, consultant*

Palm Bay
Walden, W. Thomas *lead software engineer, consultant*

GEORGIA

Alpharetta
Miller, Robert Allen *software engineering educator*

Atlanta
Anderson, Alan Julian *database consultant*
Damken, John August *computer systems engineer*
Haulbrook, Robert Wayne *computer systems engineer*

Decatur
Pittman, Victor Fred *computer systems engineer*

ILLINOIS

Naperville
Furchtgott, David Grover *computer engineer*

Waukegan
Curtis, Clark Britten *software engineer*

INDIANA

West Lafayette
Tenorio, Manoel Fernando da Mota *computer engineering educator*

LOUISIANA

Baton Rouge
Zhu, Jianchao *computer and control scientist, engineer, educator*

MARYLAND

Germantown
Olson, James Hilding *software engineer*

Greenbelt
Tilton, James Charles *computer engineer*

Laurel
Basappa, Shivanand *computer engineer*

Rockville
Chakravarty, Dipto *computer-performance engineer*

Silver Spring
Mansky, Arthur William *computer engineer*

MASSACHUSETTS

Cambridge
Ayyadurai, V.A. Shiva *software engineer*

Centerville
Rocher, Edouard Yves *computer engineer*

Hudson
Raina, Rajesh *computer engineer*

Newton Upper Falls
Tucker, Richard Douglas *computer engineer, consultant*

Westford
Sobie, Walter Richard *semiconductor engineer*

MICHIGAN

Troy
Helmle, Ralph Peter *computer systems developer, manager*

MINNESOTA

Eagan
Berg, Dean Michael *applications engineer*

Rochester
Randall, Barbara Ann *computer design engineer*

MISSOURI

Saint Louis
Luecke, Kenn Robert *software engineer*

MONTANA

Helena
Johnson, Qulan Adrian *software engineer*

NEBRASKA

Omaha
Ben-Yaacov, Gideon *computer system designer*

NEW HAMPSHIRE

Nashua
Dick, Ellen A. *computer engineer*

NEW JERSEY

Berkeley Heights
Hirsch, Mark J. *computer engineer*

Haddon Heights
Narin, Francis *research company executive*

Morristown
Nielsen, Jakob *user interface engineer*

Murray Hill
Tseng, Chia-Jeng *computer engineer*

NEW MEXICO

Kirtland AFB
Anderson, Christine Marlene *software engineer*

NEW YORK

Bethpage
Brown, James Kenneth *computer engineer*

Elmira
Orsillo, James Edward *computer systems engineer, company executive*

Syracuse
Csermely, Thomas John *computer engineer, physicist*

Tonawanda
Woodside, Bertram John *infosystem engineer*

White Plains
Meyers-Jouan, Michael Stuart *computer engineer*

NORTH CAROLINA

Durham
Hamaker, Richard Franklin *engineer*
Judd, William Reid *computer engineer, graphic artist*

Winston Salem
Blatchley, Brett Lance *computer engineer*

OHIO

Dayton
Warden, Gary George *computer engineer*

Granville
Hare, Keith William *computer consultant*

Wright Patterson AFB
Myers, Jerry Alan *computer engineer*

PENNSYLVANIA

Export
Fisher, David Marc *software engineer*

Nazareth
Bader, William Alan *computer engineer*

Palmyra
Singer, William Harry *computer/software engineer, expert systems designer, consultant, entrepreneur*

Pittsburgh
Kritz, Ronald L. *computer engineer*

RHODE ISLAND

Providence
DiCamillo, Peter John *software engineer*

TEXAS

Austin
Abraham, Jacob A. *computer engineering educator, consultant*
Powell, Brian Hill *software engineer*
Tesar, Delbert *machine systems and robotics educator, researcher, manufacturing consultant*

Carrollton
Laurent, Duane Giles *software engineer*

Flower Mound
Baker, Scott Preston *computer engineer*

VERMONT

Stowe
Springer-Miller, John Holt *computer systems developer*

VIRGINIA

Chantilly
McCormick, Thomas Jay *infosystems engineer*

Charlottesville
Waxman, Ronald *computer engineer*

Hampton
Liceaga, Carlos Arturo *research computer engineer*

Mc Lean
Parrish, Thomas Dennison *computer systems engineer*

Richmond
Perdue, Pamela Price *computer engineer*

WISCONSIN

La Crosse
Burmaster, Mark Joseph *software engineer*

Milwaukee
Modlinski, Neal David *computer systems analyst*

BELGIUM

Brussels
Degrave, Alex G. *computer engineer, consultant*

GERMANY

Leonberg
Syblik, Detlev Adolf *computer engineer*

Ludwigsburg
Queeney, David *computer engineer*

Nordrhein
Grauer, Manfred *computer engineering educator, researcher*

IRELAND

Shannon
Prendergast, Walter Gerard *computer engineer*

ITALY

Milan
Mandrioli, Dino Giusto *computer scientist, educator*

JAPAN

Kanagawa
Kinoe, Yosuke *computer company researcher, engineer*
Nakajima, Amane *computer engineer, researcher*

Osaka
Nishizawa, Teiji *computer engineer*

VENEZUELA

Caracas
Matienzo, Rafael Antonio *computer engineer*

ADDRESS UNPUBLISHED

Dao, Larry *computer software engineer*
Gaitonde, Sunil Sharadchandra *computer engineer*
Godo, Einar *computer engineer*
Molloy, William Earl, Jr. *systems engineer*
Robinson, David Allen *computer engineer*
Siyan, Karanjit Saint Germain Singh *software engineer*

ENGINEERING: ELECTRICAL

UNITED STATES

ALABAMA

Auburn
Jaeger, Richard Charles *electrical engineer, educator, science center director*

Birmingham
Damato, David Joseph *electrical engineer*
Jannett, Thomas Cottongim *electrical engineering educator*
Miller, George McCord *electrical engineer*
Quinn, Charles Layton, Jr. *electrical engineer*

Fort McClellan
Tran, Toan Vu *electronics engineer*

Huntsville
Craig, Thomas Franklin *electrical engineer*
Dunn, Karl Lindemann *electrical engineer*
Vinz, Frank Louis *electrical engineer*
Weeks, David Jamison *electrical engineer*

Redstone Arsenal
Pittman, William Claude *electrical engineer*

ALASKA

Fairbanks
Osborne, Daniel Lloyd *electronics engineer*

ARIZONA

Chandler
Yamada, Toshishige *electrical engineer*

Fort Huachuca
Riordon, John Arthur *electrical engineer*

Mesa
Fogle, Homer William, Jr. *electrical engineer, inventor*
Horne, Jeremy *fiber optics, computer research executive*
Joardar, Kuntal *electrical engineer*

Phoenix
Faul, Gary Lyle *electrical engineering supervisor*
Kleyman, Henry Semyon *electrical engineer*

Scottsdale
Kline, Arthur Jonathan *electronics engineer*
Leeland, Steven Brian *electronics engineer*
Newman, Marc Alan *electrical engineer*

Tempe
Backus, Charles Edward *engineering educator, researcher*
Coleman, Edwin DeWitt, III *electronics engineer*
Hashemi-Yeganeh, Shahrokh *electrical engineering educator, researcher*
Rodriguez, Armando Antonio *electrical engineering educator*
Zhou, Jing-Rong *electrical engineer, researcher*

Tucson
Jones, Roger Clyde *electrical engineer, educator*
Wait, James Richard *electrical engineering educator, scientist*

ARKANSAS

Fayetteville
Green, Otis Michael *electrical engineer, consultant*

Russellville
Allen, Rhouis Eric *electrical engineer*
Krenke, Frederick William *electrical engineer*

CALIFORNIA

Anaheim
Shah, Kirti Jayantilal *electrical engineer*

Arcadia
Broderick, Donald Leland *electronics engineer*

Berkeley
Desoer, Charles Auguste *electrical engineer*
Muller, Richard Stephen *electrical engineer, educator*
Whinnery, John Roy *electrical engineering educator*
Wrappe, Thomas Keith *electrical engineer, product development director*

Bermuda Dunes
Smith, Walter J. *engineering consultant*

Camarillo
Parker, Theodore Clifford *electronics engineer*

Chatsworth
Levine, Arnold Milton *retired electrical engineer, documentary filmmaker*

China Lake
Meyer, Steven John *electrical engineer*

Chula Vista
Wolk, Martin *electronic engineer, physicist*

Costa Mesa
Carpenter, Frank Charles, Jr. *retired electronics engineer*

Davis
Gardner, William Allen *electrical engineering educator*
Ghausi, Mohammed Shuaib *electrical engineering educator, university dean*
Hunt, Charles Edward *electrical engineer, educator*

El Segundo
Lantz, Norman Foster *electrical engineer*
Mitchell, John Noyes, Jr. *electrical engineer*
Mo, Roger Shih-Yah *electronics engineering manager*

Foster City
Andresen, Mark Nils *electrical engineer*

Fremont
Ritter, Terry Lee *electrical engineer, educator*

Irvine
Kar Roy, Arjun *electrical engineer, educator*
Stubberud, Allen Roger *electrical engineering educator*

La Jolla
Tu, Charles Wuching *electrical and computer engineering educator*

La Verne
Sasaki, Shusuke *electromechanical engineer*

Long Beach
Cynar, Sandra Jean *electrical engineering educator*
Hildebrant, Andy McClellan *electrical engineer*
Kumar, Rajendra *electrical engineering educator*
Rozenbergs, John *electronics engineer*

Los Angeles
Alwan, Abeer Abdul-Hussain *electrical engineering educator*
Garmire, Elsa Meints *electrical engineering educator, consultant*

Gundersen, Martin A. *electrical engineering and physics educator*
Hawkins, Shane V. *electronics and computer engineer*
Huang, Sung-cheng *electrical engineering educator*
Maksymowicz, John *electrical engineer*
Tyndall, Terry Scott *electrical engineer, biomedical engineer*
Urena-Alexiades, Jose Luis *electrical engineer*
Willner, Alan Eli *electrical engineer, educator*

Milpitas
Fonck, Eugene Jason *electronics engineer, physicist*

Monterey
Burl, Jeffrey Brian *electrical engineer, educator*

Mountain View
Perrella, Anthony Joseph *electronics engineer, consultant*
Watson, David Colquitt *electrical engineer, educator*

North Hollywood
Kaplan, Martin Nathan *electrical and electronic engineer*
Predescu, Viorel N. *electrical engineer*

Northridge
Golerkansky, Peter Joseph *electrical engineer*
Rengarajan, Sembiam Rajagopal *electrical engineering educator, researcher, consultant*

Oakland
Musihin, Konstantin K. *electrical engineer*

Ontario
Sainath, Ramaiyer *electrical engineer*

Palo Alto
Eckroad, Steven Wallace *electrical engineer*
Kazan, Benjamin *research engineer*
Mattern, Douglas James *electronics reliability engineer*
Moll, John Lewis *electronics engineer*
Oliver, Bernard More *electrical engineer, technical consultant*
Riley, Paul Eugene *electronics engineer*

Pasadena
Casani, John Richard *electrical engineer*
Goddard, Ralph Edward *electrical engineer*
Middlebrook, Robert David *electronics educator*
Moore, William Vincent *electronics and communications systems engineer*
Nghiem, Son Van *electrical engineer*
Otoshi, Tom Yasuo *electrical engineer*
Perez, Reinaldo Joseph *electrical engineer*
Prasad, K. Venkatesh *electrical and computer engineer*
Schober, Robert Charles *electrical engineer*
Yariv, Amnon *electrical engineering educator, scientist*
Yeh, Paul Pao *electrical and electronics engineer, educator*

Port Hueneme
Matigan, Robert *electrical engineer*

Rancho Cucamonga
Wells, Jon Barrett *engineer*

Rancho Mirage
Wolf, Jack Keil *electrical engineer, educator*

Redondo Beach
Hughes, James Arthur *electrical engineer*

Redwood City
Taenzer, Jon Charles *scientist, electronics engineering consultant*

Riverside
Beni, Gerardo *electrical and computer engineering educator, robotics scientist*
Dammann, David Patrick *electronics engineer*
Hackwood, Susan *electrical and computer engineering educator*

Sacramento
Lathi, Bhagawandas Pannalal *electrical engineering educator*

San Clemente
Tran, Dean *opto-electronic devices engineer*

San Diego
Abalos, Ted Quinto *electronics engineer*
Brog, Tov Binyamin *electrical engineer*
Coleman, Dale Lynn *electronics engineer, aviator, educator*
Parrish, Clyde Robin, III *electronics engineer*
Sell, Robert Emerson *electrical engineer*
Viterbi, Andrew James *electrical engineering and computer science educator, business executive*

San Dimas
Thompson, Allan Robert *sales engineer*

San Francisco
Kuo, Chang Keng *electrical engineer*
Papalexopoulos, Alex Democrates *electrical engineer*
Vreeland, Robert Wilder *electronics engineer*

San Jose
Blanch, Roy Lavern *electrical engineer*
Chen, Wen H. *laboratory executive, educator*
Coleman, Brian Fitzgerald *electrical engineer*
Lin, Tao *electronics engineering manager*
McCarthy, Mary Ann Bartley *electrical engineer*
Schweikert, Daniel George *electrical engineer, administrator*

San Lorenzo
Thompson, Lyle Eugene *electrical engineer*

Santa Ana
Cagle, Thomas Marquis *electronics engineer*
Groner, Paul S(tephen) *electrical engineer*
Thompson, Malcolm Francis *electrical engineer*

Santa Barbara
Jiang, Wenbin *electrical engineer*
Lee, Hua *electrical engineering educator*
Merz, James Logan *electrical engineering and materials educator, researcher*
Mitra, Sanjit Kumar *electrical and computer engineering educator*
Zhang, Yong-Hang *research electrical engineer*

Santa Clara
Falgiano, Victor Joseph *electrical engineer, consultant*
McDonald, Mark Douglas *electrical engineer*
Wilson, Frank Henry *electrical engineer*
Yan, Pei-Yang *electrical engineer*
Zhang, Chen-Zhi *project engineer*

Santa Clarita
Abbott, John Rodger *electrical engineer*

Santa Monica
Crain, Cullen Malone *electrical engineer*

Santa Rosa
Canfield, Philip Charles *electrical engineer*

Stanford
Angell, James Browne *electrical engineering educator*
Bracewell, Ronald Newbold *electrical engineering and computer science educator*
Eshleman, Von Russel *electrical engineering educator*
Franklin, Gene Farthing *electrical engineering educator, consultant*
Goodman, Joseph Wilfred *electrical engineering educator*
Macovski, Albert *electrical engineering educator*
Plummer, James D. *electrical engineering educator*
Siegman, Anthony Edward *electrical engineer, educator*

Sunnyvale
Schlobohm, John Carl *electrical engineer*
Silvestri, Antonio Michael (Tony) *electrical engineer*

Tarzana
Macmillan, Robert Smith *electronics engineer*

Topanga
Anderson, Donald Norton, Jr. *retired electrical engineer*

Torrance
Rizkin, Alexander *photonics researcher, educator*
Wen, Cheng Paul *electrical engineer*

Victorville
Syed, Moinuddin *electrical engineer*

Walnut
Lee, Peter Y. *electrical engineer, consultant*

Walnut Creek
VanDenburgh, Chris Allan *electrical engineer*

COLORADO

Aurora
Bingham, Paris Edward, Jr. *electrical engineer, computer consultant*

Boulder
Avery, Susan Kathryn *electrical engineering educator, researcher*
Barnes, Frank Stephenson *electrical engineer, educator*
Baylor, Jill S(tein) *electrical engineer*
Cathey, Wade Thomas *electrical engineering educator*
Gupta, Kuldip Chand *electrical and computer engineering educator, researcher*
Hansen, Ross N. *electrical engineer*
Prodan, Richard Stephen *electrical engineer*
Smith, Ernest Ketcham *electrical engineer*

Colorado Springs
Allen, James L. *electrical engineer*
Freeman, Eugene Edward *electronics company executive*
Jacobsmeyer, Jay Michael *electrical engineer*
Ziemer, Rodger Edmund *electrical engineering educator, consultant*

Denver
Keating, Larry Grant *electrical engineer, educator*
Rhodes, Chuck William *electrical engineer*
Rogers, Dale Arthur *electrical engineer*

Englewood
Scott, Eugene Ray *electrical engineer, consultant*

Fort Collins
Emslie, William Arthur *electrical engineer*
Jayasumana, Anura Padmananda *electrical engineering educator*

Golden
Cummins, Nancyellen Heckeroth *electronics engineer*
Muljadi, Eduard Benedictus *electrical engineer*

Parker
Smith, Keith Scott *electrical engineer*

U S A F Academy
Bossert, David Edward *electrical engineer*
Haupt, Randy Larry *electrical engineering educator*

CONNECTICUT

Bridgeport
Hmurcik, Lawrence Vincent *electrical engineering educator*

Danbury
Rivera, Luis Ruben *electrical engineer*

Danielson
Sondak, Steven David *electrical engineer*

Hartford
Bronzino, Joseph Daniel *electrical engineer*

New Haven
Narendra, Kumpati Subrahmanya *electrical engineer, educator*

New London
Christian, Carl Franz *electromechanical engineer*
Von Winkle, William Anton *electrical engineer, educator*

Niantic
Konrad, William Lawrence *electrical engineer*

Sandy Hook
Cristino, Joseph Anthony *electrical engineer*

Shelton
Novick, Ronald Padrov *electrical engineer*

Uncasville
Javor, Edward Richard *electrical engineer*

Waterbury
Wuthrich, Paul *electrical engineer, researcher, consultant*

Watertown
Jacovich, Stephen William *electronics engineer, consultant*

DELAWARE

Dover
Johnson, Denise Doreen *electrical engineer*

Newark
Cheung, Wilson D. *electrical engineer*
McIntosh, Donald Harry *electrical engineering consultant*

DISTRICT OF COLUMBIA

Washington
Choi, Junho *electronic engineer*
Daay, Badie Peter *electrical engineer*
Denny, Robert William, Jr. *electrical and electronics engineer*
Gwal, Ajit Kumar *electrical engineer, federal government official*
Heacock, E(arl) Larry *electrical engineer*
Jones, Howard St. Claire, Jr. *electronics engineering executive*
Jordan, Arthur Kent *electronics engineer*
Ma, David I *electronics engineer*
Misner, Robert David *electronic warfare and magnetic recording consultant, electro-mechanical company executive*
Pettis, Francis Joseph, Jr. *electrical engineer*
Reck, Francis James *electronics engineer*
Rourk, Christopher John *electrical engineer*
Russell, Ted McKinnies *electronics technician*
Walters, John Linton *electronics engineer, consultant*

FLORIDA

Atlantic Beach
Engelmann, Rudolph Herman *electronics consultant*

Boca Raton
Mussenden, Georg Antonio *electronics engineer*

Bonita Springs
Holler, John Raymond *electronics engineer*

Boynton Beach
Turner, William Benjamin *electrical engineer*

Cape Canaveral
Hurt, George Richard *electrical engineer*
Preece, Raymond George *electrical engineer*

Cape Coral
Franklin, Keith Jerome *electrical engineer*

Daytona Beach
Helfrick, Albert Darlington *electronics engineering educator, consultant*

Dunedin
Iaquinto, Joseph Francis *electrical engineer*

Fort Myers
Moeschl, Stanley Francis *electrical engineer, management consultant*

Gainesville
Chen, Wayne H. *electrical engineer, educator*
Das, Utpal *electrical engineer, researcher*
Yau, Stephen Sik-sang *electrical engineering educator, computer scientist, researcher*

Indialantic
Preece, Betty P. *engineer, educator*

Longwood
Crawford, Ian Drummond *business executive, electronic engineer*

Melbourne
Doi, Shinobu *research electrical engineer, consultant*

Naples
Lister, Charles Allan *electrical engineer, consulting engineer*

North Miami Beach
Wolfenson, Azi U. *electrical engineer*

Orlando
Kalphat-Lopez, Henriet Michelle *electric and electronics engineer*
Metevier, Christopher John *electrical engineer*
Mortazawi, Amir *electrical engineerring educator*

Punta Gorda
Dewey, Alan H. *electronic engineer*

Saint Petersburg
Waddell, David Garrett *electrical engineer, consultant*

Sunrise
Tay, Roger Yew-Siow *electromagnetic engineer*

Venice
Hays, Herschel Martin *electrical engineer*

GEORGIA

Athens
Kraszewski, Andrzej Wojciech *electrical engineer, researcher*

Atlanta
Bryan, Spencer Maurice *cost engineer*
Higgins, Richard J. *microelectrical engineer*
Lumpkin, Alva Moore, III *electrical engineer, marketing professional*
Schafer, Ronald William *electrical engineering educator*
Toler, James C. *electrical engineer*
Tucker, Robert Arnold *electrical engineer*
Walters, Lori Antionette *electrical engineer, consultant*

Attapulgus
Phan, Justin Triquang *electrical project engineer*

Augusta
Katz, David Yale *electronics engineering technology educator*

Duluth
Engleman, Dennis Eugene *electrical engineer*
Stewart, Alan Frederick *electrical engineer*

Jersey
Batchelor, Joseph Brooklyn, Jr. *electronics engineer, consultant*

Marietta
Hayes, Robert Deming *electrical engineer, consultant*

Roswell
Sanner, George Elwood *electrical engineer*

HAWAII

Honolulu
Shefchick, Thomas Peter *forensic electrical engineer*

IDAHO

Blackfoot
Goodyear, Jack Dale *electronic educator*

Idaho Falls
Tamashiro, Thomas Koyei *retired electrical engineer*

Pocatello
Moore, Kevin L. *electrical engineering educator*
Stuffle, Linda Robertson *electrical engineer*

ILLINOIS

Arlington Heights
McGuire, Mark William *electrical engineer*

Berwyn
De Lerno, Manuel Joseph *electrical engineer*

Carbondale
Hatziadoniu, Constantine Ioannis *electrical engineer, educator*

Champaign
Hsu, Chien-Yeh *electrical engineer, speech and hearing scientist*

Chicago
Bailey, Mark William *electrical engineer*
Camras, Marvin *electrical and computer engineering educator, inventor*
Gelston, John Herbert *electrical engineer*
Gupta, Hem Chander *mechanical and electrical engineer*
Koncel, James E. *electrical engineer*
Larsen, Janine Louise *electrical engineering educator*
Lu, Catherine Mean Hoa *electrical engineer*
Skimina, Timothy Anthony *electrical engineer, computer consultant*

Crystal Lake
Dabkowski, John *electrical engineer, consultant, researcher*

Des Plaines
Flakne, Dawn Gayle *electronics engineer*
Price, Richard George *electrical engineer*

Downers Grove
Klassen, Jane Frances *electrical engineer*

Evanston
Rutledge, Janet Caroline *electrical engineer, educator*
Taflove, Allen *electrical engineer, educator, researcher, consultant*

Freeport
Fitzsimmons, John Patrick *electrical engineer*
Lockwood, John Patterson *electrical engineer*

Glenview
Slayton, Ransom Dunn *engineer, consultant*

Marengo
Jones, Jack Hugh *applications engineer, educator*

Mount Vernon
Young, Monica Dawn *electrical engineer*

Naperville
Koeppe, Eugene Charles, Jr. *electrical engineer*
Pector, Scott Walter *telecommunications engineer*

Prophetstown
Sanders, Gary Glenn *electronics engineer, consultant*

Rantoul
Valencia, Rogelio Pasco *electronics engineer*

Rock Island
Hohl, Martin D. *electrical engineer*

Sandwich
Petridis, Petros Antonios *electrical engineer*

Skokie
Chang, Shi-Kuo *electrical engineering and computer science educator, novelist*

Springfield
Ballenger, Hurley René *electrical engineer*

Urbana
Basar, Tangul Ünerdem *electrical engineering educator, researcher*
Eden, James Gary *electrical engineering and physics educator, researcher*
Hein, Ilmar Arthur *electrical engineer*
Holonyak, Nick, Jr. *electrical engineering educator*
Jenkins, William Kenneth *electrical engineering educator*
Lyding, Joseph William *electrical and computer engineer*
Morkoc, Hadis *electrical engineer, educator*

Westmont
McIntosh, Don Leslie *electrical engineer*

INDIANA

Clarksville
Mu, Eduardo *electrical engineer*

Crane
Thomas, Greg Hamilton *electronics engineer*
Waggoner, Susan Marie *electronics engineer*

Elkhart
Rogers, Robert Wayne *electronics engineer*

Indianapolis
Arps, David Foster *electronics engineer*
Cones, Van Buren *electronics engineer, consultant*
Fer, Ahmet F. *electrical engineer, educator*
Lin, Zhen-Biao *acoustical and electrical engineer*

Kokomo
Miller, Robert Frank *retired electronics engineer, educator*

Notre Dame
Berry, William Bernard *engineering educator, researcher*
Sain, Michael Kent *electrical engineering educator*

Terre Haute
Grant, Michael Joseph *electrical engineering educator, consultant*
Malooley, David Joseph *electronics and computer technology educator*

West Lafayette
Delp, Edward John, III *electrical engineer, educator*
Schwartz, Richard John *electrical engineering educator, researcher*
Weiner, Andrew Marc *electrical engineering educator, laser researcher*

IOWA

Cedar Rapids
Rydell, Earl Everett *electrical engineer*

Iowa City
Sharp, Charles Paul *electronics technician*

Wilton
Cronbaugh, Kurt Allen *industrial electrician*

KANSAS

Lawrence
Moore, Richard Kerr *electrical engineering educator*

Overland Park
Georgiana, John Thomas *electrical engineer*
Levere, James Gordon *electrical engineer*

Shawnee Mission
Boone, David Daniel *electrical engineer*

Wichita
Arnold, David Dean *electrical engineer*
Egbert, Robert Iman *electrical engineering educator, academic administrator*
Paarmann, Larry Dean *electrical engineering educator*

KENTUCKY

Catlettsburg
Fischer, Robert Lee *electrical engineering executive, educator*

Lexington
Anderson, Timothy Lee *electrical engineer*
Drake, David Lee *engineer*

Louisville
Al-ibrahim, Hussam Sadiq *electrical engineer*
Aronhime, Peter Byron *electrical engineering educator, researcher*
Cox, Hollace Lawton *electrical engineering educator*
Downard, Daniel Patrick *electrical engineer, consultant*

Morehead
Rajaravivarma, Rathika *electrical engineer*

Paducah
Calhoun, Jesse, Jr. *electrical engineer*

Scottsville
Claiborne, Jimmy David *electrical engineer*

LOUISIANA

Baton Rouge
Ajmera, Pratul Kumar *electrical engineer*
Chen, Peter Pin-Shan *electrical engineering and computer science educator, data processing executive*

Harahan
Mahl, George John, III *electrical engineering consultant*

Lafayette
Richard, Sylvan Joseph *electrical engineer*

Lake Charles
Roy, Francis Charles *electrical engineer*

MARYLAND

Adelphi
Torrieri, Don Joseph *electronics engineer, mathematician, researcher*

Baltimore
Collins, Oliver Michael *electrical engineer, educator, consultant, researcher*
Graves, John Fred, III *microelectronics engineer*
Henry, Howell George *electrical engineer*
Palmer, C(harles) Harvey *electrical engineer, educator*

Bethesda
Burns, Brian Patrick *electrical engineer*
Fordyce, Samuel Wesley *electrical engineer, communications company executive, consultant*
LeVine, Donald Jay *electrical engineer*

Cockeysville Hunt Valley
Orlidge, Leslie Arthur *electrical engineer*

College Park
Granatstein, Victor Lawrence *electrical engineer, educator*
Marcus, Steven Irl *electrical engineering educator*

Frederick
Olsen, Daren Wayne *electrical engineer*

Gaithersburg
Rabinow, Jacob *electrical engineer, consultant*

Greenbelt
Fitzmaurice, Michael William *electrical and mechanical engineer*
Kalshoven, James Edward, Jr. *electronics engineer*

Hyattsville
Sindoris, Arthur Richard *electronics engineer, government official*

Lanham
Dittberner, Gerald John *engineer, meteorologist, space scientist*

Laurel
Henderson, Daniel Gardner *electrical engineer*
Thomas, Michael Eugene *electrical engineer, researcher, educator*

Potomac
Lawrence, Robert Edward *electrical engineer*

Rockville
Weaver, Christopher S(cot) *scientist*

Severna Park
Davis, John Adams, Jr. *electrical engineer, roboticist, executive*
Retterer, Bernard Lee *electronic engineering consultant*

MASSACHUSETTS

Amherst
Popstefanija, Ivan *electrical engineer*

Bedford
Bounar, Khaled Hosie *electrical engineer, consultant*
Goel, Aditya Prasad *engineering educator, consultant*
Hicks, Walter Joseph *electrical engineer, consultant*
Hogan, Stephen John *electrical engineer*
Jelalian, Albert V. *electrical engineer*
Kim, Kwang Ho *electrical engineer*
Lackoff, Martin Robert *engineer, physical scientist, researcher*
Smith, James Bigelow, Sr. *electrical engineer*

Billerica
Carolan, Donald Bartley Abraham, Jr. *electrical engineer*

Boston
Allen, Ryne Cunliffe *electrical engineer*
Espy-Wilson, Carol Yvonne *electrical engineer, educator*
Linden, Lynette Lois *bioelectrical engineer*

Boxford
Hill, Joseph Caldwell *microwave engineer, consultant*

Brockton
D'Agostino, Paul Anthony *electrical engineer*

Cambridge
Braida, Louis Benjamin Daniel *electrical engineering educator*
Gallager, Robert Gray *electrical engineering educator*
Johnson, Mark Allan *electrical engineer*
Lee, Thomas Henry *electrical engineer, educator*
Mitter, Sanjoy K. *electrical engineering educator*
Morgenthaler, Ann Welke *electrical engineer*
Rosenof, Howard Paul *electrical engineer*
Stevens, Kenneth Noble *electrical engineering educator*
Sussman, Gerald Jay *electrical engineering educator*

Chestnut Hill
Gonsalves, Robert Arthur *electrical engineering educator, consultant*

Danvers
Hathaway, Alden Moinet, II *electrical engineer*

Framingham
Bose, Amar Gopal *electrical engineering educator*
Eastman, Robert Eugene *electronic engineer, consultant*

Hanscom AFB
Coté, Marc George *electrical engineer*

Hopkinton
Liou, Jenn-Chorng *electronics engineer*

Lexington
Eng, Richard Shen *electrical engineer*
Heath, Gregory Ernest *electrical engineer*
Keicher, William Eugene *electrical engineer*
McClure, David Woodard *electrical engineer*
Morrow, Walter Edwin, Jr. *electrical engineer, university laboratory administrator*
Stiglitz, Martin Richard *electrical engineer, writer*

Lowell
Chandra, Kavitha *electrical engineer, educator*
Reinisch, Bodo Walter *electrical engineering educator*

Marlborough
Tramontozzi, Louis Robert *electrical engineer*

Medford
Balabanian, Norman *electrical engineering educator*

Newton Center
Weiner, Melvin Milton *electrical engineer*

North Andover
Gladstein, Martin Keith *electrical engineer*
Siller, Curtis Albert *electrical engineer*

North Dartmouth
Caverly, Robert H. *electrical engineer, educator*
Cory, Lester Warren *electrical engineering educator*
Lemay, Gerald J. *electrical engineer, educator*

Reading
Carlotto, Mark Joseph *electrical and computer engineer, researcher*

Somerville
Resmini, Richard *electrical engineer*

Springfield
Masi, James Vincent *electrical engineering educator*
Rahnamai, Kourosh Jonathan *electrical engineering educator*

Stow
Kinsella, Daniel John *electrical engineer*

Sudbury
Burgarella, John Paul *electronics engineer, consultant*

Topsfield
Viviani, Gary Lee *electrical engineer*

Wakefield
Goldberg, Harold Seymour *electrical engineer, academic administrator*

Westborough
Robinson, Donald Edward *electrical power industry executive*
Tessier, Mark A. *electrical engineer*
Tse, Po Yin *electrical engineer*

Wilbraham
Dailey, Franklyn Edward, Jr. *electronic image technology company executive, consultant*

Wilmington
Boxleitner, Warren James *electrical engineer, researcher*
Colby, George Vincent, Jr. *electrical engineer*

Woburn
Steinbrecher, Donald Harley *electrical engineer*

Worcester
Berrios, Javier *electrical engineer, consultant*
Levin, Peter Lawrence *electrical engineering educator*
Rotithor, Hemant Govind *electrical engineering educator*

MICHIGAN

Ann Arbor
Banks, Peter Morgan *electrical engineering educator*
Bhattacharya, Pallab Kumar *electrical engineering educator, researcher*
Dow, William Gould *electrical engineer, educator*
England, Anthony Wayne *electrical engineering and computer science educator, astronaut, geophysicist*
Grant, Michael Peter *electrical engineer*
Haddox, Mark *electronic engineer*
LaJeunesse, Robert Paul *design engineer*
O'Donnell, Matthew *electrical engineering, computer science educator*
Scott, Norman Ross *electrical engineering educator*
Taylor, Andrew Christopher *electronics and electrical engineer*
Ulaby, Fawwaz Tayssir *electrical engineering and computer science educator, research center administrator*

Auburn Hills
Laperriere, Francis William *electrical engineer, audio recording engineer*

Clare
Samorek, Alexander Henry *electrical engineer, mathematics and technology educator*

Dearborn
Jaworski, David Joseph *project engineer*
Patil, Prabhakar Bapusaheb *electrical and electronic research manager*
Sone, Masazumi *electrical engineer*

Detroit
Mahmud, Syed Masud *engineering educator, researcher*
Pratt, Robert George *electrical engineer*
Sun, Jing *electrical engineering educator*

East Lansing
Asmussen, Jes, Jr. *electrical engineer*

Farmington
Hanisko, John-Cyril Patrick *electronics engineer, physicist*

Kalamazoo
Davis, Charles Alexander *electrical engineer, educator*

Livermore
Johnson, Roy Ragnar *electrical engineer*

South Lyon
Guthrie, Michael Steele *magnetic circuit design engineer*

Southfield
Hassan, Mohammad Hassan *electrical engineering educator*

Warren
Buck, M. Scott *electrical engineer*

MINNESOTA

Bloomington
Bristow, Julian Paul Gregory *electrical engineer*

Eden Prairie
Hawley, Sandra Sue *electrical engineer*

Mankato
Khaliq, Muhammad Abdul *electrical engineer*

Mendota Heights
Schoon, David Jacob *electronic engineer*

Minneapolis
Exe, David Allen *electrical engineer*
Kroll, Mark William *electrical engineer*
Porter, William L. *electrical engineer*

Minnetonka
Grivna, Edward Lewis *electronics engineer, consultant*

Monticello
Beres, Joel Edward *electrical engineer, consultant*

Red Wing
Lee, Gordon Melvin *electrical engineering consultant*

Saint Cloud
Ellis, Bruce W. *electrical engineering educator*

Saint Paul
Goodell, John Dewitte *electromechanical engineer*

Warroad
Gouin, Warner Peter *electrical engineer*

Welch
Hellen, Paul Eric *electrical engineer*

MISSISSIPPI

Mississippi State
King, Roger Lee *electrical engineer, consultant*

University
Elsherbeni, Atef Zakaria *electrical engineering educator*

MISSOURI

Bridgeton
January, Daniel Bruce *electronics engineer*

Columbia
Waidelich, Donald Long *electrical engineer, consultant*

Florissant
Tomazi, George Donald *electrical engineer*

Joplin
Kerr, Frank Floyd *retired electrical engineer*

Kirksville
Allen, Stephen Louis *electrical engineer, educator*

Marshfield
Hopper, Kevin Andrew *electrical engineer, educator*

Rolla
Cox, Norman Roy *engineer, educator, consultant*
Rao, Vittal Srirangam *electrical engineering educator*

Saint Charles
Kim, Kyong-Min *semiconductor materials scientist*

Saint Louis
Balestra, Chester Lee *electrical engineer, educator*
Bruns, Billy Lee *consulting electrical engineer*
Schulz, Raymond Charles *electrical engineer*
Sontag, Glennon Christy *electrical engineering consultant, travel industry executive*
Zheng, Baohua *electronics engineer, consultant*

University City
Glaenzer, Richard Howard *electrical engineer*

MONTANA

Billings
Gerlach, Thurlo Thompson *electrical engineer*

Missoula
Rice, Steven Dale *electronics educator*

NEBRASKA

Lincoln
Krause, Joseph Lee, Jr. *electrical engineer*
Maher, Robert Crawford *electrical engineer, educator*
Nelson, Don Jerome *electrical engineering and computer science educator*
Steinkamp, Keith Kendall *electrical engineer*
Woollam, John Arthur *electrical engineering educator*

Omaha
Lenz, Charles Eldon *electrical engineering consultant*

NEVADA

Boulder City
Tryon, John Griggs *electrical engineer educator*

Las Vegas
Beggs, James Harry *electrical engineer*
Singh, Sahjendra Narain *electrical engineering educator, researcher*

Minden
Bently, Donald Emery *electrical engineer*

NEW HAMPSHIRE

Amherst
Pihl, Lawrence Edward *electrical engineer*

Grantham
Amick, Charles L. *electrical engineer, consultant*

Hollis
Palance, David M(ichael) *electrical engineer*

Lee
Latham, Paul Walker, II *electrical engineer*

New London
Morrey, Walter Thomas *electronics engineer*

Portsmouth
Faughnan, William Anthony, Jr. *electrical/electronics engineer*
Rodriguez, Fabio Enrique *electrical engineer*

NEW JERSEY

Bayonne
Olszewski, Jerzy Adam *electrical engineer*

Chatham
Hinderliter, Richard Glenn *electrical engineer*

Closter
Hillman, Leon *electrical engineer*

Fair Lawn
Gal, Aaron *electronics engineer*

Fort Monmouth
McBride, George Gustave *electronics engineer*

Franklin Lakes
Lovell, Theodore *electrical engineer, consultant*

Freehold
Stirrat, William Albert *electronics engineer*

Haddon Heights
Oliver, Daniel Anthony *electrical engineer, company executive*

Hoboken
Chirlian, Paul Michael *electrical engineering educator*

Holmdel
Abate, John E. *electrical/electronic engineer, communications consultant*
Colmenares, Narses Jose *electrical engineer*
Lawser, John Jutten *electrical engineer*
Murthy, Srinivasa K. *engineering corporation executive*

Linden
Musillo, Joseph H. *electrical engineer*

Livingston
Heilmeier, George Harry *electrical engineer, researcher*

Maplewood
Stillman, Gerald Israel *electrical engineer, science administrator*

Middletown
Erfani, Shervin *electrical engineer*

Moorestown
Nespor, Jerald Daniel *electrical engineer, educator*

Morristown
Chen, Walter Yi-Chen *electrical engineer*
Pepper, David J. *electrical engineer*
Stephens, William Edward *electronics engineer*
Wei, Victor Keh *electrical engineer*
Wu, Liang Tai *electrical engineer*

Murray Hill
Cho, Alfred Yi *electrical engineer*
Sullivan, Paul Andrew *research electrical engineer*

Neptune
Brown, Paul Leighton *electrical engineer*

New Providence
Koszi, Louis A. *electronics engineer*

Newark
Bar-Ness, Yeheskel *electrical engineer, educator*
Bigley, William Joseph, Jr. *control engineer*

North Brunswick
Schnitzler, Paul *electronics engineering manager*

Oakland
Spataro, Vincent John *electrical engineer*

Piscataway
Flanagan, James Loton *electrical engineer, educator*
Paraskevopoulos, Nicholas George *electrical engineering educator*
Zhao, Jian Hui *electrical and computer engineering educator*

Princeton
Fitton, Gary Michael *electronics engineer*
Henderson, John Goodchilde Norie *electrical engineer*
Kulkarni, Sanjeev Ramesh *electrical engineering educator*
Patel, Mukund Ranchhodlal *electrical engineer, researcher*
Platt, Judith Roberta *electrical engineer*
Schroeder, Alfred Christian *electronics research engineer*
Vahaviolos, Sotirios John *electrical engineer, scientist, corporate executive*
Weimer, Paul K(essler) *electrical engineer*
Wolf, Wayne Hendrix *electrical engineering educator*

Ramsey
Young, Cornelius Bryant, Jr. (C. B. Young) *electronics engineer*

Red Bank
Cox, Donald Clyde *electrical engineer*
Schwarz, Steven Allan *electrical engineer*
Sniffen, Paul Harvey *electronic engineer*
Zah, Chung-En *electrical engineer*

Rumson
Rowe, Harrison Edward *electrical engineer*

Sea Bright
Plummer, Dirk Arnold *electrical engineer*

Secaucus
Johansen, Robert John *electrical engineer*

Sparta
Truran, William R. *electrical engineer*

Springfield
Perilstein, Fred Michael *consulting electrical engineer*

Summit
Fukui, Hatsuaki *electrical engineer, art historian*
Kitsopoulos, Sotirios C. *electrical engineer, consultant*

Toms River
Fanuele, Michael Anthony *electronics engineer, research engineer*

Trenton
Shumila, Michael John *electrical engineer*
Simpson, Raymond William *electronics engineer*

West Long Branch
Dworkin, Larry Udell *electrical engineer, research director, consultant*

Whippany
Andry, Steven Craig *electrical engineer, educator*

Willingboro
Shelton, Thomas McKinley *electronics engineer*

NEW MEXICO

Albuquerque
Bolie, Victor Wayne *electrical and computer engineering educator*
Gruchalla, Michael Emeric *electronics engineer*
Hudson, Jack Alan *electrical engineer*
Tucker, Roy Anthony *electro-optical instrumentation engineer, consultant*

Kirtland AFB
Baum, Carl Edward *electromagnetic theorist*

Los Alamos
McDonald, Thomas Edwin, Jr. *electrical engineer*
Peratt, Anthony Lee *electrical engineer, physicist*

Santa Fe
Albach, Carl Rudolph *consulting electrical engineer*

White Sands Missile Range
Clelland, Michael Darr *electrical engineer*

NEW YORK

Afton
Church, Richard Dwight *electrical engineer*

Amherst
Wakoff, Robert *electrical engineer, consultant*

Auburn
Daggett, Wesley John *electrical engineer*

Bayside
Pshtissky, Yacov *electrical engineer*

Binghamton
Chatterjee, Monish Ranjan *electrical engineering educator*
Hopkins, Douglas Charles *electrical engineer*
Sackman, George Lawrence *educator*

Blauvelt
Prysch, Peter *electrical engineer*

Bronx
Berger, Frederick Jerome *electrical engineer, educator*
Linden, Barnard Jay *electrical engineer*
Zhou, Zhen-Hong *electrical engineer*

Brooklyn
Levinton, Michael Jay *electrical engineer, energy consultant*
McLean, William Ronald *electrical engineer, consultant*

Buffalo
Anderson, Wayne Arthur *electrical engineering educator*
Sarjeant, Walter James *electrical and computer engineering educator*

Calverton
Sun, Homer Ko *electrical engineer*

Carmel
Bardell, Paul Harold, Jr. *electrical engineer*

East Syracuse
Harju, Wayne *electrical engineer*

Flushing
Shteinfeld, Joseph *electrical engineer*

Garden City
Goodman, Daniel *electrical engineer, mathematical physicist*

Glenville
Anderson, Roy Everett *electrical engineering consultant*

Hauppauge
Kelly, Alexander Joseph *electrical engineer*

Hawthorne
McConnell, John Edward *electrical engineer, company executive*

Hopewell Junction
Beyene, Wendemagegnehu Tsegaye *electrical engineer*
Mohammad, Shaikh Noor *electronics engineer, educator*

Ithaca
Carlin, Herbert J. *electrical engineering educator, researcher*
Hartquist, E(dwin) Eugene *electrical engineer*
Nation, John Arthur *electrical engineering educator, researcher*
Sudan, Ravindra Nath *electrical engineer, physicist, educator*
White, Kenneth William *electrical engineering consultant*
Zimmerman, Stanley William *retired electrical engineering educator*

Little Neck
Pearse, James Newburg *electrical engineer*

New York
Bass, Mikhail *electrical engineer*
Brunetto, Frank *electrical engineer*
Carmiciano, Mario *electrical engineer*
Cassedy, Edward Spencer, Jr. *electrical engineering educator*
Cohen, Richard Lawrence *electrical engineer, mechanical engineer*
Daken, Richard Joseph, Jr. *electrical engineer*
Dickey, David G. *electrical engineer*
Klempner, Larry Brian *network engineer, income tax preparer*
Lewis, William Scheer *electrical engineer*
Megherbi, Dalila *electrical and computer engineer, researcher*
Panoutsopoulos, Basile *electrical engineer*
Poliacof, Michael Mircea *electrical engineer*
Sobel, Kenneth Mark *electrical engineer, educator*
Teich, Malvin Carl *electrical engineering educator*
Zukowski, Charles Albert *electrical engineering educator*

Ossining
Baylor, Sandra Johnson *electrical engineer*
Miron, Amihai *electronic systems executive, electrical engineer*

Peekskill
Harding, Charlton Matthew *electrical engineer*

Pittsford
Marshall, Joseph Frank *electronic engineer*

Potsdam
Ortmeyer, Thomas Howard *electrical engineering educator*

Rochester
Parker, Kevin James *electrical engineer educator*

Schenectady
Chestnut, Harold *foundation executive, electrical engineer*
Sohie, Guy Rose Louis *electrical engineer, researcher*
Thomann, Gary Calvin *electrical engineer*

Somers
van Ryn, Ted Mattheus *electrical engineer*

Staten Island
Singer, Edward Nathan *engineer, consultant*
Tsui, Chia-Chi *electrical engineer, educator*

Stony Brook
Dhadwal, Harbans Singh *electrical engineer, consultant*
Truxal, John Groff *electrical engineering educator*

Syosset
Wachspress, Melvin Harold *electrical engineer, consultant*

Syracuse
Cheng, David Keun *educator*
Goel, Amrit Lal *engineering educator*
Hamlett, James Gordon *electronics engineer, management consultant and educator*
Kodali, Hari Prasad *electrical engineer*
LePage, Wilbur Reed *electrical engineering educator*
Strait, Bradley Justus *electrical engineering educator*

Troy
Anderson, John Bailey *electrical engineering educator*
Ghandhi, Sorab Khushro *electrical engineering educator*
McDonald, John Francis Patrick *electrical engineering educator*
Modestino, James William *electrical engineering educator*
Saridis, George Nicholas *electrical engineer*

Upton
Curtiss, Joseph August *electrical engineer, consultant*
Gunther, William Edward *electrical engineer, researcher*
Nawrocky, Roman Jaroslaw *research electrical engineer*

West Point
Shoop, Barry LeRoy *electrical engineer*

Yorktown Heights
Henle, Robert Athanasius *engineer*
Nguyen, Philong *electrical engineer*

NORTH CAROLINA

Asheville
Wilsdon, Thomas Arthur *product development engineer, administrator*

Charlotte
Babić, Davorin *electrical engineer, researcher*

Durham
Daniels-Race, Theda Marcelle *electrical engineering educator*
Massoud, Hisham Zakaria *electrical engineering educator*
McKinney, Collin Jo *electrical engineer*

Enka
Barbour, Eric S. *electrical engineer*

Laurinburg
Hardin, William Beamon, Jr. *electrical engineer*

Raleigh
Baliga, Bantval Jayant *electrical engineering educator, consultant*
Bilbro, Griff Luhrs *electronics engineering educator*
Kolbas, Robert Michael *electrical engineering educator*
Osburn, Carlton Morris *electrical and computer engineering educator*

Research Triangle Park
Spiegel, Ronald John *electrical engineer*
Venkatasubramanian, Rama *electrical engineer*

Rocky Mount
Woollen, Kenneth Lee *electrical engineer*

Lynchburg
Fath, George R. *electrical engineer, communications executive*

Mc Lean
Airst, Malcolm Jeffrey *electronics engineer*
Arnold, Kent Lowry *electrical engineer*
Field, Francis Edward *electrical engineer, educator*
Fowler, Thomas Benton, Jr. *electrical engineering education, consultant*
Torres, Rigo Romualdo *electrical engineer*
Yang, Xiaowei *electrical engineer*

Norfolk
Lakdawala, Vishnu Keshavlal *electrical and computer engineering educator*

Reston
Kahn, Robert E. *electrical engineer*

Richmond
McLelland, Slaten Anthony *electrical engineer*

Roanoke
Nicholas, James Thomas *electronics engineer*
Phillips, Earle Norman *electro-optical engineer*

Springfield
Duff, William Grierson *electrical engineer*

Suffolk
Tinto, Joseph Vincent *electrical engineer*

Vienna
Murray, Arthur Joseph *engineering consultant, researcher*

WASHINGTON

Bainbridge Is
Hanson, Robert James *electrical test engineer, consultant*

Bellevue
Liang, Jeffrey Der-Shing *retired electrical engineer, civil worker*

Bellingham
Albrecht, Albert Pearson *electronics engineer, consultant*

Bothell
Pihl, James Melvin *electrical engineer*

Cathlamet
Torget, Arne O. *electrical engineer*

Des Moines
Sinon, John Adelbert, Jr. *electrical engineer*

Kennewick
Carroll, John Moore *retired electrical engineer*

Kingston
Pichal, Henri Thomas *electronics engineer, physicist, consultant*

Mukilteo
Bohn, Dennis Allen *electrical engineer, consultant, writer*

Olympia
Saari, Albin Toivo *electronics engineer*

Seattle
Barrett, Michael Wayne *product development company executive*
Fadden, Delmar McLean *electrical engineer*
Ishimaru, Akira *electrical engineering educator*
Kim, Yongmin *electrical engineering educator*
Oman, Henry *retired electrical engineer, engineering executive*
Porter, Robert Philip *electrical engineering educator*
Somani, Arun Kumar *electrical engineer, educator*
Spindel, Robert Charles *electrical engineering educator*

Tacoma
MacGinitie, Laura Anne *electrical engineer*

Vancouver
Hartmann, David Peter *electrical engineer, educator, consultant*

WEST VIRGINIA

Elkview
Hudson, Kenneth Shane *electrical engineer*

WISCONSIN

Chippewa
Conger, Jeffrey Scott *electrical engineer*

Delafield
Mudek, Arthur Peter *automation engineer, consultant*

Fond Du Lac
Seichter, Daniel John *electrical engineer*

Madison
Monroe, John Robert *electrical engineer*
Shohet, Juda Leon *electrical and computer engineering educator, researcher, high technology company executive*

Menomonee Falls
Markowski, Roberta Jean *electrical engineer*

Milwaukee
Allgaier, Glen Robert *electronics engineer, researcher*
Arkadan, Abdul-Rahman Ahmad *electrical engineer, educator*
Brauer, John Robert *electrical engineer*
Heinen, James Albin *electrical engineering educator*
Johnson, F. Michael *electrical and automation systems professional*
Tisser, Clifford Roy *electrical engineer*
Vetro, James Paul *electrical project engineer*
Zietlow, Daniel H. *electrical engineer*

Rothschild
Drew, Richard Allen *electrical and instrument engineer*

South Milwaukee
Van Dusen, Harold Alan, Jr. *electrical engineer*

Waukesha
Svetic, Ralph E. *electrical engineer, mathematician*

MILITARY ADDRESSES OF THE UNITED STATES

ATLANTIC

APO
De Roux, Tomas E. *electrical engineer*

EUROPE

APO
Freeman, Brian S. *electrical engineer*

CANADA

BRITISH COLUMBIA

Burnaby
Saif, Mehrdad *electrical engineering educator*

Vancouver
Wedepohl, Leonhard M. *electrical engineering educator*

ONTARIO

Etobicoke
Mufti, Navaid Ahmed *network design engineer*

London
Mathur, Radhey Mohan *electrical engineering educator, dean*

Ottawa
Georganas, Nicolas D. *electrical engineering educator*

Toronto
Smith, Kenneth Carless *electrical engineering educator*
Venetsanopoulos, Anastasios Nicolaos *electrical engineer, educator*

QUEBEC

Montreal
Corinthios, Michael Jean George *electrical engineering educator*
Townshend, Brent Scott *electrical engineer*

Verdun
Bielby, Gregory John *electrical engineer*

MEXICO

Mexico City
Morales-Acevedo, Arturo *electrical engineering researcher, educator*

ARGENTINA

Buenos Aires
Cernuschi-Frias, Bruno *electrical engineer*

AUSTRALIA

Canberra
Godara, Lal Chand *electrical engineering educator*

Caulfield South
Macesic, Nedeljko *electrical engineer*

City Beach
Pelczar, Otto *electrical engineer*

Parkville
Tucker, Rodney Stuart *electrical and electronic engineering educator, consultant*

AUSTRIA

Vienna
Skritek, Paul *electrical engineering educator, consultant*

BELGIUM

Liège
Calvaer, Andre J. *electrical science educator, consultant*

BRAZIL

Barueri
Ghizoni, César Celeste *electrical engineer*

São Paulo
Bergamini, Eduardo Whitaker *electrical engineer*
Mariotto, Paulo Antonio *electronics educator*

CHILE

Valparaiso
Hernandez-Sanchez, Juan Longino *electrical engineering educator*

CHINA

Xi'an
Fan, Changxin *electrical engineering educator*

COLOMBIA

Bogota
Pulido, Carlos Orlando *electronics engineer, economist, consultant*

CROATIA

Dubrovnik
Vukovic, Drago Vuko *electronics engineer*

Zagreb
Bozicevic, Juraj *educator*
Ursic, Srebrenka *information technology researcher, manager, consultant*

DENMARK

Copenhagen
Hansen, Finn *electrical engineer*

EGYPT

Dokki
Nasser, Essam *electrical engineer, physicist*

Giza
Salem, Ibrahim Ahmed *electrical engineer, consultant, educator*

ENGLAND

Essex
Thompson, Haydn Ashley *electronics engineer*

Farnborough
Treble, Frederick Christopher *electrical engineer, consultant*

Kingston-Upon-Thames
Jatala, Shahid Majid *electronics engineer*

Portsmouth
Croset, Michel Roger *electronics engineer*

Preston
O'Shea, John Anthony *consulting electrical engineer*

Southampton
Gambling, William Alexander *optoelectronics research center director*

FINLAND

Turku
Eerola, Osmo Tapio *research electrical engineer*

FRANCE

Caen
Calvino, Philippe Andre Marie *engineer, manufacturing executive*

Gargilesse
Barda, Jean Francis *electronics engineer, corporate executive*

Montpellier
Dauchez, Pierre Guislain *roboticist*

Paris
Lehner, Gerhard Hans *engineer, consultant*

GERMANY

Berlin
Krauter, Stefan Christof Werner *electrical engineer*

Dortmund
Freund, Eckhard *electrical engineering educator*

Erlangen
Brajder, Antonio *electrical engineer, engineering executive*

Heilbronn
Pschunder, Willi *semiconductor engineer*

Munich
Bodlaj, Viktor *electrical engineer*

GREECE

Athens
Dritsas, George Vassilios *electronics engineer*
Frangakis, Gerassimos P. *electronic engineer*
Halkias, Christos Constantine *electronics educator, consultant*
Hatzakis, Michael *electrical engineer, research executive*

Karlovassi
Katsikas, Sokratis Konstantine *electrical engineer, educator*

Patras
Makios, Vasilios *electronics educator*

HONG KONG

Kowloon
Demokan, Muhtesem Süleyman *electrical engineering educator*

Kwun Tong
Chou, Chung Lim *electrical engineer*

Sha Tin
Kao, Charles Kuen *electrical engineer, educator*

HUNGARY

Budapest
Vajda, György *electrical engineer*

INDIA

Hyderabad
Vathsal, Srinivasan *electrical engineer, scientist*

ITALY

Ferrara
Pandini, Davide *electrical engineer*

Villanterio
Mendoliera, Salvatore *electrical engineer*

JAPAN

Hokkaido
Saito, Shuzo *electrical engineering educator*

Ibaraki
Numai, Takahiro *scientist, electrical engineer*

Ishikawa
Nasu, Shoichi *electrical engineering educator*

Kyoto
Nagao, Makoto *electrical engineering educator*

Nagano
Ito, Kentaro *electrical engineering educator*

Noda
Takenaka, Tadashi *electrical engineer, educator*

Sapporo
Hasegawa, Hideki *electrical engineering educator*

Tokyo
Hamada, Nobuhiro A. *electrical engineer, researcher*
Nagata, Minoru *electronics engineer*
Nishi, Kenji *electrical engineer*
Saitoh, Tadashi *electrical engineering educator*

Toyota
Miyachi, Iwao *electrical engineering educator*

Yokosuka
Sasaki, Masafumi *electrical engineering educator*

THE NETHERLANDS

Eindhoven
Butterweck, Hans Juergen *electrical engineering educator*
Eykhoff, Pieter *electrical engineering educator*

Petten
Veltman, Arie Taeke *electronic engineer*

NORWAY

Nesbru
Viktil, Martin *electrical engineer*

Sandvika
Jacob, George (Prasad) *electronics and telecommunications engineer*

Trondheim
Forssell, Börje Andreas *electronics engineer, educator, consultant*
Svaasand, Lars Othar *electronics researcher*

PERU

Lima
Moesgen, Karl John *electronics engineer, science educator*

THE PHILIPPINES

Quezon City
Solidum, Emilio Solidum *electronics and communications engineer*

POLAND

Lodz
Zieliński, Jerzy Stanisław *scientist, electrical engineering educator*

Warsaw
Romaniuk, Ryszard Stanislaw *electrical engineering educator, consultant*

Wroclaw
Kabacik, Pawel *research electrical engineer*
Lugowski, Andrzej Mieczyslaw *electrical engineer*

PORTUGAL

Coimbra
Antunes, Carlos Lemos *electrical engineering educator*

REPUBLIC OF KOREA

Seoul
Ko, Myoung-Sam *control engineering educator*

ROMANIA

Bucharest
Drăgănescu, Mihai *electronic engineering educator*

RUSSIA

Taganrog
Markov, Vladimir Vasilievich *radio engineer*

SAUDI ARABIA

Riyadh
Al-Mohawes, Nasser Abdullah *electrical engineering educator*

SCOTLAND

Glasgow
Sandham, William Allan *electronics and electrical engineer, educator*

SINGAPORE

Singapore
Sundararajan, Narasimhan *electrical engineering educator*

SOUTH AFRICA

Brooklyn
Smith, Edwin David *electrical engineer*

SPAIN

Gijon
Jimenez, Juan Ignacio *electrical engineer, consultant*

Madrid
Aguilera, Jorge *electrical and solar energy engineer*

SWEDEN

Stockholm
Fertner, Antoni *electrical engineer*
Öfverholm, Stefan *electrical engineer*

Västerås
Saha, Murari Mohan *electrical engineer*

SWITZERLAND

Bern
Kislovski, Andre Serge *electronics engineer*

Burgdorf
Haeberlin, Heinrich Rudolf *electrical engineering educator*

Geneva
Muller, Paul-Emile *electrical engineer*

Zollikofen
Muntwyler, Urs Walter *electronics engineer*

TAIWAN

Tainan
Kung, Ling-Yang *electronics engineer, educator*

Taipei
Lee, Tsu Tian *electrical engineering educator*

TRINIDAD AND TOBAGO

Trinidad
Parris, C(harles) Deighton *television engineer, consultant*

UKRAINE

Kiev
Gotovchits, Georgy Olexandrovich *electrical engineer, educator*

VENEZUELA

Caracas
Chang-Mota, Roberto *electrical engineer*

ADDRESS UNPUBLISHED

Alcazar, Antonio *electrical engineer*
Bartley, Thomas Lee *electronics engineer*
Bjorndahl, David Lee *electrical engineer*
Bloch, Erich *electrical engineer, former science foundation administrator*
Bose, Anjan *electrical engineering educator, researcher, consultant*
Crowley, Joseph Michael *electrical engineer, educator*
Dumarot, Dan Peter *electrical engineer*
Erteza, Ireena Ahmed *electrical engineer*
Fanjul, Rafael James, Jr. *engineer, mathematician*
Feldman, Jacob Alex *electrical engineer*
Frederick, Norman L., Jr. *electrical engineer*
Geiger, Louis Charles *electrical engineer, retired*
Grotzinger, Timothy Lee *electronics engineer*
Hadley, Glen L. *electrical engineer*
Hess, Ulrich Edward *electrical engineer*
Hinton, David Owen *electrical engineer*
Hughson, Mary Helen *electrical engineer*
Jokl, Alois Louis *electrical engineer*
Jones, David Allan *electronics engineer*
Klement, Haim *electronics engineer*
Krimm, Martin Christian *electrical engineer, educator*
Leigh, Michael Charles *electrical engineer*
Martin, John Brand *engineering educator, researcher*
Mian, Guo *electrical engineer*
Neal, Stephen Wayne *electrical engineer*
Necula, Nicholas *electrical engineering educator, researcher*
Noci, Giovanni Enrico *electronics engineer*
Nyman, Bruce Mitchell *electrical engineer*
Pathak, Vibhav Gautam *electrical engineer*
Peled, Israel *electrical engineer, consultant*
Pfister, Daniel F. *electrical engineer*
Polasek, Edward John *electrical engineer, consultant*
Rahman, Khandaker Mohammad Abdur *engineering educator*
Samson, John Roscoe, Jr. *electrical engineer*
Sastry, Padma Krishnamurthy *electrical engineer, consultant*
Schaller, John Walter *electrical engineer*
Simmons, Shalon Girlee *electrical engineer*
Temes, Clifford Lawrence *electrical engineer*
Tran, Nang Tri *electrical engineer, physicist*
Trott, Keith Dennis *electrical engineer, researcher*
Van Winkle, Edgar Walling *electrical engineer, computer consultant*
Varon, Dan *electrical engineer*
Voldman, Steven Howard *electrical engineer*
Vook, Frederick Werner *electrical engineer*
Willis, Selene Lowe *electrical engineer*
Wintle, Rosemarie *bio-medical electronic engineer*

ENGINEERING: ENVIRONMENTAL & SANITARY

UNITED STATES

ALABAMA

Birmingham
Adams, Alfred Bernard, Jr. *environmental engineer*
Pitt, Robert Ervin *environmental engineer, educator*

Huntsville
Leonard, Kathleen Mary *environmental engineering educator*

Montgomery
Fulmer, Kevin Michael *environmental engineer*

Pennington
Broussard, Peter Allen *energy engineer*

ARIZONA

Phoenix
Chisholm, Tom Shepherd *environmental engineer*

Prescott Valley
Beck, John Roland *environmental consultant*

Scottsdale
Teletzke, Gerald Howard *environmental engineer*

ARKANSAS

Paragould
Pickney, Charles Edward *environmental engineer*

CALIFORNIA

Alameda
Brown, Stephen Lawrence *environmental consultant*

Arcata
Chaney, Ronald Claire *environmental engineering educator, consultant*

Bakersfield
Gillie, Michelle Francoise *industrial hygienist*

Berkeley
Fox, Joan Phyllis *environmental engineer*

Calimesa
Kaplan, Ozer Benjamin *environmental health specialist, consultant*

Davis
Lotter, Donald Willard *environmental educator*

Fresno
Hanna, George Parker *environmental engineer*
Stiffler, Kevin Lee *engineering technologist*

Huntington Beach
Owen, Thomas J. *environmental engineer*

Irvine
Chrysikopoulos, Constantinos Vassilios *environmental engineering educator*
Swanberg, Christopher Gerard *environmental engineer*

Los Angeles
Bittenbender, Brad James *environmental safety and industrial hygiene manager*

Palos Verdes Estates
Aro, Glenn Scott *environmental and safety executive*

Pasadena
Najm, Issam Nasri *environmental engineer*
Swain, Robert Victor *environmental engineer*

Redondo Beach
Everett, Robert William *environmental engineer*

Redwood City
Hirschhorn, Elizabeth Ann *environmental engineer, consultant*

Rosemead
Heath, Ted Harris *environmental engineer*

Stanford
Ott, Wayne Robert *environmental engineer*

Ventura
Cowen, Stanton Jonathan *air pollution control engineer*

Walnut Creek
Burgarino, Anthony Emanuel *environmental engineer, consultant*

COLORADO

Colorado Springs
Kelly, Emery Leonard *bioenvironmental engineer*

Denver
Goldfield, Joseph *environmental engineer*
Riese, Arthur Carl *environmental engineering company executive, consultant*

CONNECTICUT

West Simsbury
Yocom, John Erwin *environmental consultant*

DELAWARE

Newark
Dentel, Steven Keith *environmental engineer, educator*

DISTRICT OF COLUMBIA

Washington
Angus, Sara Goodwin *environmental engineer*
Chang, Ker-Chi *environmental engineer, civil engineer*
Cleave, Mary L. *environmental engineer, former astronaut*
Deason, Jonathan P. *environmental engineer, federal agency administrator*
Dutta, Subijoy *environmental engineer*
Inyang, Hilary Inyang *geoenvironmental engineer, researcher*
Kooyoomjian, K. Jack *environmental engineer*
Prasad, Surya Sattiraju *environmental systems engineer*
Rucker, Joshua E. *environmental engineer*

FLORIDA

Brooksville
Farrell, Mark David *environmental engineer*

Gainesville
Putnam, Hugh Dyer *environmental engineer, educator, consultant*
Smith, John James, Jr. *environmental engineering laboratory director*

Jacksonville
Brashier, Edward Martin *environmental consultant*

Naples
Benson, Ronald Edward, Jr. *environmental engineer*

Orlando
Madhanagopal, Thiruvengadathan *environmental engineer*

Tallahassee
Hartsfield, Brent David *environmental engineer*
Subbuswamy, Muthuswamy *environmental engineer, researcher, consultant*

Tampa
Masters, Eugene Richard *environmental engineer*
Santti, Gary Allen *hazardous waste engineer*

Temple Terrace
Gagliardo, Victor Arthur *environmental engineer*

Winter Park
Mackey, Robert Eugene *environmental engineer*
Thampi, Mohan Varghese *environmental health and civil engineer*

GEORGIA

Athens
Jackson, Robert Benton, IV *environmental engineer*
McCutcheon, Steven Clifton *environmental engineer*

Atlanta
Driscoll, Terence Patrick *environmental engineer*
Khudenko, Boris Mikhail *environmental engineer*
Sezgin, Mesut *environmental engineer*

Fort Stewart
Houston, Thomas Dewey *environmental engineer*

Locust Grove
Smith, Al Jackson, Jr. *environmental engineer, lawyer*

Marietta
Oakes, Thomas Wyatt *environmental engineer*
Peake, Thaddeus Andrew, III *environmental engineer*

Tucker
Traina, Paul Joseph *environmental engineer*

HAWAII

Honolulu
Rezachek, David Allen *energy and environmental engineer*
Yang, Ping-Yi *biowaste, wastewater engineering educator*

IDAHO

Boise
Del Carmen, Rene Jover *process and environmental engineer*

Idaho Falls
McCarthy, Jeremiah Justin *environmental engineer*

ILLINOIS

Aurora
Quinn, Richard Kendall *environmental engineer*

Chicago
Brask, Gerald Irving *sanitary engineer*
Chang, Huai Ted *environmental engineer, educator, researcher*
Sresty, Guggilam Chalamaiah *environmental engineer*

Glenview
Hutter, Gary Michael *environmental engineer*

Murphysboro
McCormack, Robert Paul *environmental engineer*

Northfield
Gunderson, Edward Lynn *environmental engineer*

Oneida
Lawson, Larry Dale *environmental consultant*

Peoria
Lin, Shundar *sanitary engineer*

Rock Island
Poppen, Andrew Gerard *environmental engineer*

Urbana
Engelbrecht, Richard Stevens *environmental engineering educator*

INDIANA

Indianapolis
Mundell, John Anthony *environmental engineer, consultant*
Vlach, Jeffrey Allen *environmental engineer*

IOWA

Muscatine
Stanley, Richard Holt *consulting environmental engineer*

KENTUCKY

Louisville
Effinger, Charles Edward, Jr. *environmental/energy engineer*

LOUISIANA

Baton Rouge
Shaheen, Eli Esber *environmental engineer*

Donaldsonville
Carville, Thomas Edward *environmental engineer*

MAINE

Portland
Ikalainen, Allen James *environmental engineer*

MARYLAND

Aberdeen Proving Ground
Wood, James David *environmental engineer*

Annapolis
Linker, Lewis Craig *environmental engineer*

Baltimore
Patel, Manu Ambalal *environmental engineer*
Stolberg, Ernest Milton *environmental engineer, consultant*

Edgewood
Collins, William Henry *environmental scientist*

Jessup
Bazzazieh, Nader *environmental engineer*

Rockville
Seagle, Edgar Franklin *environmental engineer, consultant*

Stevensville
Trescott, Sara Lou *water resources engineer*

MASSACHUSETTS

Hatfield
Borgatti, Douglas Richard *environmental engineer*

Lenox
Shammas, Nazih Kheirallah *environmental engineering educator*

Lynn
Brogunier, Claude Rowland *environmental engineer*

Peabody
Meader, John Leon *environmental engineer, consultant*
Peters, Leo Francis *environmental engineer*

Rockland
Karabots, Joseph William *environmental engineering company executive*

MICHIGAN

Ann Arbor
Wright, Steven Jay *environmental engineering educator, consultant*

Dearborn
Rokosz, Susan Marie *environmental engineer*

Novi
Singh, Jaswant *environmental company executive*

Troy
Slocum, Lester Edwin *environmental engineer*

MINNESOTA

Maple Plain
Erdmann, John Baird *environmental engineer*

Saint Paul
Voss, Steven Ronald *environmental engineer*

MISSOURI

Kansas City
Haar, Andrew John *environmental engineer*
Kuhn, Leigh Ann *air pollution control engineer*

Saint Joseph
Brooks, Steven Doyle *environmental engineer*

Saint Louis
Schreiber, Robert John *environmental engineer*

NEBRASKA

Nebraska City
Hammerschmidt, Ben L. *environmental scientist*

NEW HAMPSHIRE

Concord
Smart, Melissa Bedor *environmental consulting company executive*

NEW JERSEY

Englewood Cliffs
Dobrowolski, Francis Joseph *environmental engineer, consultant*

Green Brook
Elias, Donald Francis *environmental consultant*

Hackensack
Margron, Frederick Joseph *environmental engineer*

Mahwah
Landeck, Robin Elaine *environmental engineer*

Monmouth Junction
Ellerbusch, Fred *environmental engineer*

Newark
Cheremisinoff, Paul Nicholas *environmental engineer, educator*

Princeton
Schoen, Alvin E., Jr. *environmental engineer*
Wickramanayake, Godage Bandula *environmental engineer*

Rochelle Park
Vernick, Arnold Sander *environmental engineer*

Trenton
Troy, Marleen Abbie *environmental engineer*

Wayne
Predpall, Daniel Francis *environmental engineering executive, consultant*

NEW MEXICO

Albuquerque
Owen, Thomas Edward *environmental engineer*

NEW YORK

Brooklyn
LoPinto, Robert Anthony *environmental engineer*
Ong, Say Kee *environmental engineering educator*

East Aurora
Croston, Arthur Michael *pollution control engineer*

Hempstead
Maier, Henry B. *environmental engineer*

Mineola
Pascucci, John Joseph *environmental engineer*

New York
Lymberis, Costas Triantafillos *environmental engineer*
Principe, Joseph Vincent, Jr. *environmental engineer*
Psaltakis, Emanuel P. *environmental engineer*

Poughkeepsie
Flanagan, Frederick James *water systems engineer*

Rochester
Ford, Mary Elizabeth (Libby Ford) *environmental health engineer*

Syosset
Vida, Stephen Robert *environmental engineer*

Troy
Ho, Yifong *environmental engineer*

Webster
Zavon, Peter Lawrence *industrial hygienist*

Woodbury
Memoli, Michael Anthony *environmental engineer*

NORTH CAROLINA

Cary
Janezic, Darrell John *environmental engineer*

Chapel Hill
Greeley, Richard Folsom *retired sanitary engineer*
Okun, Daniel Alexander *environmental engineering educator, consulting engineer*

Charlotte
Kuykendall, Terry Allen *environmental and chemical engineer*

Raleigh
MacConnell, Gary Scott *environmental engineer*

Research Triangle Park
Gerald, Nash Ogden *environmental engineer*
Larsen, Ralph Irving *environmental engineer*
Tucker, William Gene *environmental engineer*

OHIO

Alledonia
Bartsch, David Leo *environmental engineer*

Cincinnati
Hochstrasser, John Michael *environmental engineer, industrial hygienist*
Kuchibhotla, Sudhakar *environmental engineer, consultant*

Columbus
Cosens, Kenneth Wayne *sanitary engineer, educator*

Dayton
Blommel, Scot Anthony *environmental engineer*

Huber Heights
Strong, Richard Allen *environmental safety engineer*

Oxford
Willeke, Gene E. *environmental engineer, educator*

OKLAHOMA

Norman
Everett, Jess Walter *environmental engineering educator, researcher*

PENNSYLVANIA

Ambler
Vanyo, Edward Alan *environmental engineer*

Easton
Lee, Terry James *environmental engineer, consultant*

Exton
Piatkowski, Steven Mark *environmental engineer*

Lester
Cahill, Lawrence Bernard *environmental engineer*

New Hope
Bertele, William *environmental engineer*

Newtown
Hammonds, Elizabeth Ann *environmental engineer*

Pittsburgh
Ho, Chuen-hwei Nelson *environmental engineer*
Luthy, Richard Godfrey *environmental engineering educator*
Neufeld, Ronald David *environmental engineer*
Pohland, Frederick George *environmental engineering educator, researcher*
Touhill, C. Joseph *environmental engineer*
Wong-Chong, George Michael *chemical and environmental engineer*

Spring Grove
Carter, Cecil Neal *environmental engineer*

University Park
McDonnell, Archie Joseph *environmental engineer*

Williamsport
Neff, Harold Parker *environmental engineer, consultant*

SOUTH CAROLINA

Aiken
Groce, William Henry, III *environmental engineer, consultant*

Greenville
Chandler, Henry William *environmental engineer*

TENNESSEE

Brentwood
Alley, E. Roberts *environmental engineer*

Chattanooga
Matthews, Michael Roland *environmental engineer*

Erwin
Thompson, Mark Alan *environmental engineer*

Knoxville
Iwanski, Myron Leonard *environmental engineer*

Memphis
Williams, Edward F(oster), III *environmental engineer*

Nashville
Harlan, James Phillip *environmental engineer*
Stewart, William Timothy *environmental engineer*

Oak Ridge
Marz, Loren Carl *environmental engineer and scientist, chemist*

TEXAS

Arlington
Argento, Vittorio Karl *environmental engineer*

Austin
Charbeneau, Randall J. *environmental and civil engineer*
Gloyna, Earnest Frederick *environmental engineer, educator*
Liljestrand, Howard Michael *environmental engineering educator*
McKinney, Daene Claude *environmental engineering educator*

Dallas
Giniecki, Kathleen Anne *environmental engineer*
Sampath, Krishnaswamy *reservoir engineer*
Wilmut, Charles Gordon *environmental engineer*

Houston
Gonzalez, David Alfonso *environmental engineer*
Kachel, Wayne M. *environmental engineer*

Kingwood
Rademacher, John Martin *sanitary engineer*

Midland
Riebe, Susan Jane *environmental engineer*

Pittsburg
Vierra, Frank Huey *environmental engineer, consultant*

UTAH

Salt Lake City
Bhayani, Kiran Lilachand *environmental engineer, programs manager*

VIRGINIA

Abingdon
Blake, John Charles, Jr. *environmental engineer*

Alexandria
Judkins, William Sutton *environmental engineer*

Arlington
Perkins, Richard Burle, II *environmental engineer, international consultant*

Blacksburg
Boardman, Gregory Dale *environmental engineer, educator*

Mc Lean
Amr, Asad Tamer *environmental engineer, consultant*

Roanoke
McKenna, John Dennis *environmental testing engineer*

Virginia Beach
Van de Riet, James Lee *environmental engineer*

WASHINGTON

Kennewick
Cobb, William Thompson *environmental consultant*

Spokane
Esvelt, Larry Allen *environmental engineer*

WISCONSIN

Madison
Bagchi, Amalendu *environmental engineer*
Berthouex, Paul Mac *civil and environmental engineer, educator*
Gordon, Mark Elliott *environmental engineer*

Pewaukee
Kelly, A(llan) James *environmental engineer*

CANADA

ONTARIO

Toronto
Heinke, Gerhard William (Gary Heinke) *environmental engineering educator*

Toronto-Etobicoke
Kurys, Jurij-Georgius *environmental engineer, scientist, consultant*

MEXICO

Mexico City
Sánchez, Luis Ruben *environmental engineer*

ECUADOR

Quito
Torres, Guido Adolfo *water treatment company executive*

ENGLAND

Camberley
Al-Zubaidi, Ali Abdul Jabbar *engineer*

IRELAND

Cork
Hill, Martin Jude *engineer, consultant*

THE PHILIPPINES

Quezon City
Bisnar, Miguel Chiong *water pollution engineer*

PORTUGAL

Aveiro
Borrego, Carlos Soares *environmental engineering educator*

SWITZERLAND

La Rippe
Stevens, Prescott Allen *environmental health engineer, consultant*

VENEZUELA

Caracas
Berti, Arturo Luis *sanitary engineer*

ADDRESS UNPUBLISHED

Andrews, David Charles *energy and power consultant*
Glover, Everett William, Jr. *environmental engineer*
Gopal, Pradip Goolab *engineer, environmental consultant*
Mignella, Amy Tighe *environmental engineering researcher*
Rahman, Sami Ur *environmental engineer*
Schwinn, Donald Edwin *environmental engineer*
Shupler, Ronald Steven *environmental engineer*
Tappan, Clay McConnell *environmental engineer, consultant*

ENGINEERING: INDUSTRIAL

UNITED STATES

ALABAMA

Huntsville
Daussman, Grover Frederick *industrial engineer, consultant*
Schroer, Bernard Jon *industrial engineering educator*

Mobile
Chandler, Thomas Eugene *industrial engineer*

ARIZONA

Tucson
Monteiro, Renato Duarte Carneiro *industrial engineer, educator, researcher*

ARKANSAS

Fayetteville
Taylor, Gaylon Don *industrial engineering educator*

CALIFORNIA

Fremont
Gupta, Rajesh *industrial engineer, quality assurance specialist*

Irvine
Howland, Paul *industrial engineer*

Mountain View
Harber, M(ichael) Eric *industrial engineer*

Pleasant Hill
Hopkins, Robert Arthur *retired industrial engineer*

San Diego
Inoue, Michael Shigeru *industrial engineer, electrical engineer*

Stanford
Thompson, David Alfred *industrial engineer*

CONNECTICUT

Shelton
Fedor, George Matthew, III *industrial engineer*

DISTRICT OF COLUMBIA

Washington
Dinberg, Michael David *industrial engineer*

FLORIDA

Boca Raton
Han, Chingping Jim *industrial engineer, educator*

Englewood
Suiter, John William *industrial engineer, consultant*

Miami
Levin, Robert Bruce *industrial engineer, executive, consultant*
Sanchez, Javier Alberto *industrial engineer*

Riviera Beach
Kazimir, Donald Joseph *industrial engineer*

Tallahassee
Galbraith, Lissa Ruth *industrial engineer*

GEORGIA

Atlanta
Chaifetz, L. John *industrial engineer*
Ratliff, Hugh Donald *industrial engineering educator*

Norcross
Moore, Christopher Barry *industrial engineer*

Snellville
Hudgens, Kimberlyn Nan *industrial engineer*

Statesboro
Hanson, Roland Stuart *industrial engineer, educator*

IDAHO

Harrison
Skidmore, Eric Arthur *industrial engineer*

ILLINOIS

Chicago
Banerjee, Prashant *industrial engineering educator, researcher*

De Kalb
Stoia, Dennis Vasile *industrial engineering educator*

Evanston
Rath, Gustave Joseph *industrial engineering educator*

Glenview
Bell, Charles Eugene, Jr. *industrial engineer*

Itasca
Wittenburg, Robert Charles *industrial engineer*

Naperville
Geisel, Charles Edward *industrial engineer, consultant*

Rockford
Casagranda, Robert Charles *industrial engineer*

Winnetka
Seymour, Frederick Prescott, Jr. *industrial engineer, consultant*

INDIANA

Elkhart
Atchison, Arthur Mark *industrial, research and development engineer*

Kendallville
Shelby, Beverly Jean *quality assurance professional*

West Lafayette
Solberg, James Joseph *industrial engineering educator*
Thomas, Marlin Uluess *industrial engineering educator*

IOWA

Davenport
Sandry, Karla Kay Foreman *industrial engineering educator*

Hills
Tomlinson, G. Richard *industrial engineer*

Marshalltown
McCann, Michael John *industrial engineer, educator, consultant*

KANSAS

Manhattan
Lee, E(ugene) Stanley *industrial engineer, mathematician, educator*

Wichita
McVicar, Robert William, Jr. *industrial engineer*

KENTUCKY

Louisville
Tran, Long Trieu *industrial engineer*

LOUISIANA

Ruston
Hale, Paul Nolen, Jr. *engineering administrator, educator*

MARYLAND

Silver Spring
Thorpe, Jack Victor *energy consultant*

MASSACHUSETTS

Cambridge
DiBerardinis, Louis Joseph *industrial hygiene engineer, consultant, educator*

Hopkinton
Leamon, Tom B. *industrial engineer, educator*

Lowell
Colling, David Allen *industrial engineer, educator*

MICHIGAN

Ann Arbor
Chaffin, Donald B. *industrial engineer, researcher*

Kalamazoo
Carmichael, Charles Wesley *industrial engineer*

Pontiac
Van Den Boom, Wayne Jerome *industrial engineer*

Smiths Creek
Snyder, George Robert *engineer*

MISSISSIPPI

Brandon
Mitchell, Roy Devoy *industrial engineer*

Mississippi State
Usher, John Mark *industrial engineering educator*

MISSOURI

Saint Louis
Gardner, Gregory Allen *industrial engineer*
Lucking, Peter Stephen *industrial engineering consultant*

NEW HAMPSHIRE

Jaffrey
Robinson, Lawrence Wiswall *project engineer*

Merrimack
Malley, James Henry Michael *industrial engineer*

NEW JERSEY

Avenel
Mazurkiewicz, John Anthony *plant process engineer*

Hoboken
Cogan, Kenneth George *quality assurance engineer*

NEW YORK

Buffalo
Wrona, Leonard Mathhew *industrial engineer*

Ithaca
Maxwell, William Laughlin *industrial engineering educator*

New York
Bedrij, Orest J. *industrialist, scientist*
Jaffe, William J(ulian) *industrial engineer, educator*
Klein, Morton *industrial engineer, educator*

Troy
Graves, Robert John *industrial engineering educator*

NORTH CAROLINA

Charlotte
Gothard, Michael Eugene *industrial engineer*

Greensboro
Johnson, Jeffrey Allan *industrial engineer*

Raleigh
Fang, Shu-Cherng *industrial engineering and operations research educator*
Meier, Wilbur Leroy, Jr. *industrial engineer, educator, former university chancellor*

OHIO

Cincinnati
Arantes, José Carlos *industrial engineer, educator*

OKLAHOMA

Stillwater
Mize, Joe Henry *industrial engineer, educator*

PENNSYLVANIA

Bethlehem
Odrey, Nicholas Gerald *industrial engineer, educator*

Johnstown
Pandiarajan, Vijayakumar *industrial engineer*

Media
McClintic, Fred Frazier *simulation engineer*

Pittsburgh
Bloom, William Millard *industrial design engineer*
McNamee, William Lawrence *industrial engineer*

University Park
Ham, Inyong *industrial engineering educator*

RHODE ISLAND

Kingston
Boothroyd, Geoffrey *industrial and manufacturing engineering educator*
Dewhurst, Peter *industrial engineering educator*

SOUTH CAROLINA

Clemson
Ferrell, William Garland, Jr. *industrial engineering educator*
Kimbler, Delbert Lee, Jr. *industrial engineering educator*

TENNESSEE

Alcoa
Cormia, Frank Howard *industrial engineering administrator*

Cookeville
Russell, Josette Renee *industrial engineer*

TEXAS

Beaumont
Zaloom, Victor Anthony *industrial engineer, educator*

College Station
Koppa, Rodger Joseph *industrial engineering educator*

Grapevine
Baker, Scott Michael *industrial engineer*

Houston
Piper, Lloyd Llewellyn, II *engineer, service industry executive*

Roanoke
Raz, Tzvi *industrial engineer*

San Antonio
Gomes, Norman Vincent *retired industrial engineer*

Waco
Stallings, Frank, Jr. *industrial engineer*

VIRGINIA

Blacksburg
Harvey, Aubrey Eaton, III *industrial engineer*
Price, Dennis Lee *industrial engineer, educator*

Richmond
Reynolds, Bradford Charles *industrial engineer, management consultant*

Winchester
Teal, Gilbert Earle *industrial engineer*

WASHINGTON

Seattle
Wiker, Steven Forrester *industrial engineering educator*

WISCONSIN

Madison
Gustafson, David Harold *industrial engineering and preventive medicine educator*

Platteville
Balachandran, Swaminathan *industrial engineering educator*

CANADA

NOVA SCOTIA

Halifax
Wilson, George Peter *industrial engineer*

BELGIUM

Brussels
Nisolle, Etienne *industrial engineer*

CYPRUS

Nicosia
Aristodemou, Loucas Elias *industrial engineer, manufacturing company executive*

REPUBLIC OF KOREA

Seoul
Surh, Dae Suk *engineering consulting company executive*

SPAIN

Madrid
Milla Gravalos, Emilio *industrial engineer*

SWITZERLAND

Aubonne
Rodriguez, Moises-Enrique *industrial engineer*

TAIWAN

Taichung
Kuo, Tung-Yao *industrial engineering educator*

ADDRESS UNPUBLISHED

Goldberger, Arthur Earl, Jr. *industrial engineer, consultant*
Knapp, Mark Israel *industrial and management engineer*
Morgan, Linda Claire *industrial engineer*

ENGINEERING: MANUFACTURING

UNITED STATES

CALIFORNIA

Auburn
DeCosta, William Joseph *manufacturing engineer*

Long Beach
McIntosh, Gregory Cecil *manufacturing engineer*

Sunnyvale
Miller, Joseph Arthur *manufacturing engineer, educator, consultant*

COLORADO

Denver
Ferguson, Lloyd Elbert *manufacturing engineer*

FLORIDA

Eglin A F B
Guida, James John *manufacturing engineer, consultant*

Fort Myers Beach
Arneson, Harold Elias Grant *manufacturing engineer, consultant*

IDAHO

Boise
Marks, Ernest E. *maintenance engineer*

ILLINOIS

Rockford
Tebrinke, Kevin Richard *manufacturing engineer*

Urbana
Trigger, Kenneth James *manufacturing engineering educator*

INDIANA

Hammond
Neff, Gregory Pall *manufacturing engineering educator, consultant*

Lafayette
Hamlin, Kurt Wesley *manufacturing engineer*

Terre Haute
English, Robert Eugene *manufacturing engineering educator*

MARYLAND

College Park
Kopp, Debra Lynn *manufacturing engineer, consultant*

MASSACHUSETTS

Andover
McCauley, Charles Irvin *manufacturing engineer*

MISSOURI

Ava
Murray, Delbert Milton *engineer*

Lake Sherwood
Struckhoff, Ronald Robert *manufacturing engineer*

Saint Louis
Rogers, John Russell *engineer*

NEW JERSEY

Wayne
Perry, Robert Ryan *manufacturing engineer*

NEW YORK

Ithaca
Wang, Kuo-King *manufacturing engineer, educator*

New York
Kocherlakota, Sreedhar *manufacturing engineer*

NORTH CAROLINA

Mocksville
Townsend, Arthur Simeon *manufacturing engineer*

OHIO

Bellefontaine
Myers, Daniel Lee *manufacturing engineer*

Columbus
Love, Daniel Michael *manufacturing engineer*

Grove City
Blazic, Martin Louis *manufacturing engineer*

Warren
Zeuner, Raymond Alfred *manufacturing engineer*

OREGON

Tualatin
Webster, Merlyn Hugh, Jr. *manufacturing engineer, information systems consultant*

PENNSYLVANIA

Huntingdon Valley
Blume, Horst Karl *manufacturing engineer*

SOUTH CAROLINA

Greenville
Henning, Kathleen Ann *manufacturing systems engineer*

TEXAS

Austin
Psaris, Amy Celia *manufacturing engineer*

WEST VIRGINIA

Morgantown
Gopalakrishnan, Bhaskaran *manufacturing engineer*

WISCONSIN

Green Bay
Hudson, Halbert Austin, Jr. *retired manufacturing engineer, consultant*

AUSTRIA

Graz
List, Hans C. *manufacturing engineer*

DENMARK

Lyngby
Alting, Leo Larsen *manufacturing engineer, educator*

GERMANY

Berlin
Spur, Günter *manufacturing engineering educator*

JAPAN

Osaka
Iwata, Kazuaki *manufacturing engineer educator*

ENGINEERING: MARINE

UNITED STATES

CALIFORNIA

Chula Vista
Rusconi, Louis Joseph *marine engineer*

CONNECTICUT

Groton
Lincoln, Walter Butler, Jr. *marine engineer, educator*
Milburn, Darrell Edward *ocean engineer, coast guard officer*

DISTRICT OF COLUMBIA

Washington
Hillery, Robert Charles *naval engineer, management consultant*

FLORIDA

North Palm Beach
Rothstein, Arnold Joel *marine engineer, mechanical engineer*

West Palm Beach
Laura, Robert Anthony *coastal engineer, consultant*

HAWAII

Honolulu
Tamaye, Elaine E. *coastal/ocean engineer*

LOUISIANA

Metairie
Munchmeyer, Frederick Clarke *naval architect, marine engineer*

New Orleans
Latorre, Robert George *naval architecture and engineering educator*

MAINE

West Enfield
Ramsdell, Richard Adoniram *marine engineer*

MARYLAND

Annapolis
Greene, Eric *naval architect, marine engineer*
Lindler, Keith William *marine engineering educator, consultant*

Columbia
Rapkin, Jerome *defense industry executive*

MASSACHUSETTS

Boston
Thorburn, John Thomas, III *marine engineer, consultant*

MISSISSIPPI

Stennis Space Center
Teng, Chung-Chu *ocean engineer, researcher*

NEW JERSEY

Hoboken
Bruno, Michael Stephen *ocean engineering educator, researcher*

NEW YORK

Buffalo
Atwood, Theodore *marine engineer, researcher*

New York
Wheeler, Wesley Dreer *marine engineer, naval architect, consultant*

OREGON

Corvallis
Sollitt, Charles Kevin *ocean engineering educator, laboratory director*

TEXAS

College Station
Duggal, Arun Sanjay *ocean engineer*

VIRGINIA

Alexandria
Brickell, Charles Hennessey, Jr. *marine engineer, retired military officer*
Doerry, Norbert Henry *naval engineer*
Eckhart, Myron, Jr. *marine engineer*

Virginia Beach
Stephan, Charles Robert *retired ocean engineering educator, consultant*

Winchester
Gordon, Richard Warner *naval propulsion engineer*

ENGLAND

Hampshire
Suhrbier, Klaus Rudolf *hydrodynamicist, naval architect*

FRANCE

Le Pecq
Mallevialle, Joël Christian *marine engineer*

POLAND

Gdańsk
Jagoda, Jerzy Antoni *marine engineer*

SRI LANKA

Colombo
Madanayake, Lalith Prasanna *marine engineer*

ADDRESS UNPUBLISHED

Klein, Martin *ocean engineering consultant*

ENGINEERING: MECHANICAL

UNITED STATES

ALABAMA

Auburn
Crocker, Malcolm John *mechanical engineer, noise control engineer, educator*
Herring, Bruce E. *engineering educator*
Siginer, Dennis A. *mechanical engineering educator, researcher*
Tippur, Hareesh V. *mechanical engineer*

Birmingham
Bunt, Randolph Cedric *mechanical engineer*
Parks, Rodney Keith *mechanical engineer, consultant*

Huntsville
Farley, Wayne Curtis *mechanical engineer*
Krishnan, Anantha *mechanical engineer*
LeMaster, Robert Allen *mechanical engineer*
Liggett, Mark William *mechanical engineer*
Modlin, James Michael *mechanical engineer, army officer*
Yang, Hong-Qing *mechanical engineer*

Madison
Hawk, Clark Wiliams *mechanical engineering educator*

Montgomery
Longuet, Gregory Arthur *automation engineer, consultant*

Tuscaloosa
Todd, Beth Ann *mechanical engineer, educator*

ALASKA

Anchorage
Armstrong, Richard Scott *mechanical engineer*

Fairbanks
Zarling, John Paul *mechanical engineering educator*

ARIZONA

Cave Creek
MacKay, John *mechanical engineer*

Litchfield Park
Miller, Kenneth Edward *mechanical engineer, consultant*

Phoenix
Bachus, Benson Floyd *mechanical engineer, consultant*
Peterson, Stephen Cary *mechanical engineer*
Struble, Donald Edward *mechanical engineer*
Wu, Mei Qin *mechanical engineer*

Tempe
Ferreira, Jay Michael *mechanical engineer*
Laananen, David Horton *mechanical engineer, educator*
Metzger, Darryl Eugene *mechanical and aerospace engineering educator*
Shaw, Milton Clayton *mechanical engineering educator*

Tucson
Anthes, Clifford Charles *retired mechanical engineer, consultant*
Arabyan, Ara *mechanical engineer*
Kececioglu, Dimitri Basil *mechanical engineering educator*

CALIFORNIA

Agoura Hills
Meeks, Crawford Russell, Jr. *mechanical engineer*

Alhambra
Moeller, Ronald Scott *mechanical engineer*

Anaheim
Prince, Warren Victor *mechanical engineer*

Berkeley
Hsu, Chieh Su *applied mechanics engineering educator, researcher*
Komvopoulos, Kyriakos *mechanical engineer, educator*
Laitone, Edmund Victor *mechanical engineer*
Mikesell, Walter R., Jr. *mechanical engineer, engineering executive*
Mote, Clayton Daniel, Jr. *mechanical engineer, educator*

Tomizuka, Masayoshi *mechanical engineering educator, researcher*

Canoga Park
Lu, Cheng-yi *mechanical engineer*

Carlsbad
Pantos, William Pantazes *mechanical engineer, consultant*

Chino
Ellington, James Willard *mechanical design engineer*

Cupertino
Lee, Yuen Fung *mechanical engineer*

Davis
Aldredge, Ralph Curtis, III *mechanical engineer, educator*

Edwards AFB
Hodges, Vernon Wray *mechanical engineer*
Morehart, James Henry *mechanical engineer*

El Segundo
Kramer, Gordon *mechanical engineer*

Escondido
Grew, Raymond Edward *mechanical engineer*

Fremont
Wang, Ying Zhe *mechanical and optical engineer*

Glendora
Ahern, John Edward *mechanical engineer, consultant*

Huntington Beach
Falcon, Joseph A. *mechanical engineering consultant*

Irvine
Abajian, Henry Krikor *mechanical engineer*
Colmenares, Jorge Eliecer *mechanical engineer*
Edwards, Donald Kenneth *mechanical engineer, educator*
Sheldon, Mark Scott *research engineer*

La Jolla
Tio, Kek-Kiong *mechanical engineering researcher*

La Palma
Haldane, George French *mechanical engineer, consultant*

Livermore
Christensen, Richard Monson *mechanical engineer, materials engineer*
Hauber, Janet Elaine *mechanical engineer*
Lewis, Daniel Lee *mechanical engineer*

Long Beach
de Soto, Simon *mechanical engineer*
Yeh, Hsien-Yang *mechanical engineer, educator*

Los Angeles
Chobotov, Vladimir Alexander *mechanical engineer*
Dhir, Vijay K. *mechanical engineering educator*
Hahn, Hong Thomas *mechanical engineering educator*
Jen, Tien-Chien *mechanical engineer*
Ju, Jiann-Wen *mechanics educator, researcher*
Maxworthy, Tony *mechanical and aerospace engineering educator*
Pugay, Jeffrey Ibanez *mechanical engineer*
Saravanja-Fabris, Neda *mechanical engineering educator*
Sve, Charles *mechanical engineer*

Magalia
Kincheloe, William Ladd *mechanical engineer*

McClellan AFB
Frank, Christopher Lynd *mechanical engineer*

Monterey
Gordis, Joshua Haim *mechanical engineer, researcher*
Sarpkaya, Turgut *mechanical engineering educator*

Newark
Nystrom, Gustav Adolph *mechanical engineer*

Northridge
Jakobsen, Jakob Knudsen *mechanical engineer*

Orange
Vice, Charles Loren *electromechanical engineer*

Palo Alto
Hodge, Philip Gibson, Jr. *mechanical and aerospace engineering educator*
McCloskey, Thomas Henry *mechanical engineer, consultant*
Neou, In-Meei Ching-yuan *mechanical engineering educator, consultant*
Youngdahl, Paul Frederick *mechanical engineer*

Pasadena
Brennen, Christopher E. *fluid mechanics educator*
Goodwin, David George *mechanical engineering educator*
Hatheway, Alson Earle *mechanical engineer*
Joffe, Benjamin *mechanical engineer*

Pico Rivera
Banuk, Ronald Edward *mechanical engineer*
Bunch, Jeffrey Omer *mechanical engineer*

Rancho Cucamonga
Yurist, Svetlan Joseph *mechanical engineer*

Rancho Mirage
McCrea, Russell James *mechanical engineer, consultant*

Ridgecrest
Pearson, John *mechanical engineer*

Riverside
Mathaudhu, Sukhdev Singh *mechanical engineer*

Sacramento
Fairall, Richard Snowden *mechanical engineer*
Reardon, Frederick Henry *mechanical engineer, educator*

San Clemente
Arora, Sam Sunder *mechanical engineer*

San Diego
Blelloch, Paul Andrew *mechanical engineer*
Dean, Richard Anthony *mechanical engineer*
Fraitag, Leonard Alan *mechanical and design engineer*
Lowrey, D('Orvey) Preston, III *mechanical engineer*
Razzaghi, Mahmoud *mechanical engineer*

San Francisco
Blaevoet, Jeffrey Paul *mechanical engineer*
Clark, Stuart Alan *mechanical engineer*
Frane, James Thomas *mechanical engineer, author*

San Jose
Adams, William John, Jr. *mechanical engineer*
Conrad, Jeffrey Philip *mechanical engineer, manager*
Hill, Robert James *mechanical engineer*

San Pedro
Lowi, Alvin, Jr. *mechanical engineer, consultant*

San Rafael
Taylor, Irving *mechanical engineer, consultant*

San Ramon
Ratiu, Mircea Dimitrie *mechanical engineer*

Santa Ana
Do, Tai Huu *mechanical engineer*

Santa Barbara
Lick, Wilbert James *mechanical engineering educator*
Majumdar, Arunava *mechanical engineer, educator*

Santa Clara
Gupta, Anand *mechanical engineer*
Shoup, Terry Emerson *university dean, engineering educator*

Santa Cruz
Chance, Hugh Nicholas *mechanical engineer*

Santa Monica
Spivack, Herman M. *fluid dynamicist, aerospace engineer*

Stanford
Herrmann, George *mechanical engineering educator*
Hughes, Thomas Joseph *mechanical engineering educator, consultant*
Reynolds, William Craig *mechanical engineer, educator*
Rott, Nicholas *fluid mechanics educator*
Springer, George Stephen *mechanical engineering educator*

Sunnyvale
Yee, Raymond Kim *mechanical engineer, researcher*

Temecula
Feltz, Charles Henderson *former mechanical engineer, consultant*

Thousand Oaks
Bahn, Gilbert Schuyler *retired mechanical engineer, researcher*

Tiburon
Heacox, Russel Louis *mechanical engineer*

Torrance
Schipper, Leon *mechanical engineer*

Tracy
Christon, Mark Allen *mechanical engineer*

Weed
Kyle, Chester Richard *mechanical engineer*

Woodland Hills
Higginbotham, Lloyd William *mechanical engineer*

COLORADO

Boulder
Carlson, Lawrence Evan *mechanical engineering educator*

Colorado Springs
Tanzer, Andrew Ethan *mechanical engineer*

Fort Collins
Garvey, Daniel Cyril *mechanical engineer*
Wilbur, Paul James *mechanical engineering educator*
Winn, C(olman) Byron *mechanical engineering educator*

Westminster
Banerjee, Bejoy Kumar *mechanical engineer*

CONNECTICUT

Avon
Mc Ilveen, Walter *mechanical engineer*

East Granby
Kimberley, John A. *mechanical engineer, consultant*

East Hartford
Hobbs, David E. *mechanical engineer*
Johnson, Bruce Virgil *mechanical engineer, physicist, researcher*

Groton
Bernatowicz, Felix Jan Brzozowski *mechanical engineer, consultant*
Helm, John Leslie *mechanical engineer, company executive*

Guilford
Duffield, Albert J. *mechanical engineer*

Hartford
Smith, Donald Arthur *mechanical engineer, researcher*

Madison
Lehberger, Charles Wayne *mechanical engineer, engineering executive*

Meriden
McLeod, John Arthur Sr. *mechanical engineer*

Milford
Dmytruk, Maksym, Jr. *mechanical engineer*

New Haven
Smooke, Mitchell David *mechanical engineering educator, consultant*
Sreenivasan, Katepalli Raju *mechanical engineering educator*

New London
Visich, Karen Michelle *mechanical engineer*

North Haven
Pfeifer, Howard Melford *mechanical engineer*

Orange
Lobay, Ivan *mechanical engineering educator*

Storrs Mansfield
Cetegen, Baki M. *mechanical engineering educator*

West Haven
Sarris, John *mechanical engineering educator*

DELAWARE

Claymont
Sommermann, Jeffrey Herbert *mechanical engineer*

New Castle
Khatib, Nazih Mahmoud *mechanical engineer*

Newark
Adams, Joseph Brian *engineer, mathematics educator*
Barton, Lyndon O'Dowd *mechanical engineer*
Crowley, Michael Joseph *mechanical engineer*
Hopkins, Mark Willard *mechanical engineer*

Wilmington
Levy, Stanley Burton *mechanical engineer, marketing executive*

DISTRICT OF COLUMBIA

Washington
Arkilic, Galip Mehmet *mechanical engineer, educator*
Bosnak, Robert J. *mechanical engineer, federal agency administrator*
Brodrick, James Ray *mechanical engineer*
Hershey, Robert Lewis *mechanical engineer, management consultant*
Reis, Victor H. *mechanical engineer, government official*
Roco, Mihail Constantin *mechanical engineer, educator*
Salmon, William Cooper *mechanical engineer, engineering academy executive*
Warnick, Walter Lee *mechanical engineer*
Wermiel, Jared Sam *mechanical engineer*

FLORIDA

Boca Raton
Tsai, Chi-Tay *mechanical engineering educator*

Eglin A F B
Hopmeier, Michael Jonathon *mechanical engineer, troubleshooter*

Fort Lauderdale
Clark, Jeffrey Alan *mechanical engineer*
Ondrejack, John Joseph *mechanical engineer*

Gainesville
Fournier, Donald Joseph, Jr. *mechanical engineer*
Proctor, Charles Lafayette, II *mechanical engineer, educator, consultant*
Roan, Vernon Parker, Jr. *mechanical engineer, educator*
Sherif, S. A. *mechanical engineering educator*

Jacksonville
Keifer, Orion Paul *naval officer, mechanical engineer*
Russell, David Emerson *mechanical engineer, consultant*
Vallort, Ronald Peter *mechanical engineer*

Juno Beach
Vaughn, Eddie Michael *mechanical engineer*

Kennedy Space Center
Austin, Lisa Susan Coleman *mechanical engineer*

Lighthouse Point
Farho, James Henry, Jr. *mechanical engineer, consultant*

Melbourne
Subramanian, Chelakara Suryanarayanan *mechanical and aerospace engineering educator*

Miami
Tansel, Ibrahim Nur *mechanical engineer, educator*
Veziroglu, Turhan Nejat *mechanical engineering educator, energy researcher*

North Fort Myers
Underwood, George Alfred *mechanical engineer*

Orlando
Nayfeh, Jamal Faris *mechanincal and aerospace engineering educator*
Schroeter, Dirk Joachim *mechanical engineer*

Sarasota
Greenstein, Jerome *mechanical engineer*

Tampa
Shah, Harish Hiralal *mechanical engineer*

Temple Terrace
DeReus, Harry Bruce *mechanical engineer, consultant*

GEORGIA

Atlanta
Aidun, Cyrus Khodarahm *mechanical engineering educator*
Brighton, John Austin *mechanical engineer, educator*
Chambers, James Patrick *mechanical engineer*
Clay, Kelli Suzanne *mechanical engineer*
Cunefare, Kenneth Arthur *mechanical engineer, educator*
Fontaine, Arnold Anthony *mechanical engineer*
Gatley, Donald Perkins *mechanical engineer, building scientist*
Lichtenwalner, Owen C. *engineer, telecommunications executive*
McBroom, Thomas William *mechanical engineer*
Smith, Marc Kevin *mechanical engineering educator*
Sullivan, James Hargrove, Jr. *mechanical engineer, utility engineer*
Zinn, Ben T. *engineer, educator, consultant*

Marietta
Engelstad, Stephen Phillip *mechanical engineer*
Hamilton, Stephen Stewart *mechanical engineer, consultant*
Johnson, Samuel Walter, II *mechanical engineer*
Miles, Thomas Caswell *engineer*
Pearson, Hugh Stephen *mechanical and metallurgical engineer*

HAWAII

Honolulu
Mizokami, Iris Chieko *mechanical engineer*

Lihue
Tabata, Lyle Mikio *mechanical engineer*

IDAHO

Idaho Falls
Elias, Thomas Ittan *mechanical engineer*

Moscow
Jacobsen, Richard T. *mechanical engineering educator*
Stauffer, Larry Allen *mechanical engineer, educator, consultant*

Pocatello
Lee, Richard A. *mechanical engineer, consultant*

ILLINOIS

Argonne
Doss, Ezzat Danial *mechanical engineer, researcher*

Buffalo Grove
Ahmed, Osman *mechanical engineer*

Burr Ridge
Mockaitis, Algis Peter *mechanical engineer*

Carol Stream
Wojtanek, Guy Andrew *mechanical engineer*

Chicago
Budenholzer, Roland Anthony *mechanical engineering educator*
Chung, Paul Myungha *mechanical engineer, educator*
Clark, Raymond John *mechanical engineer*
DeSantiago, Michael Francis *mechanical engineer*
Krishnamachari, Sadagopa Iyengar *mechanical engineer, consultant*
Megaridis, Constantine Michael *mechanical engineering educator, researcher*
Nair, Sudhakar Edayillam *mechanical and aerospace engineering educator*
Peebles, William Reginald, Jr. *mechanical engineer*
Porter, Robert William *mechanical engineering educator*
Worek, William Martin *mechanical engineering educator*

De Kalb
Field, Robert Eugene *mechanical engineer, educator, consulant, researcher*

Decatur
Schlueter, Gerald Francis *mechanical engineering educator*

Elgin
Arndt, Bruce Allen *mechanical engineer*

Elmhurst
Balcerzak, Marion John *mechanical engineer*
Simpson, Gene Milton, III *mechanical engineer, manufacturing engineer*

Evanston
Cheng, Herbert Su-Yuen *mechanical engineering educator*
Lueptow, Richard Michael *mechanical engineer, educator*

Lake Zurich
Grychowski, Jerry Richard *mechanical engineer*

Lemont
Haupt, H. James *mechanical design engineer*

Lombard
Papakyriakou, Michael John *biomechanical engineer*

Maywood
Relwani, Nirmalkumar Murlidhar (Nick Relwani) *mechanical engineer*

Mount Prospect
Rasmussen, James Michael *research mechanical engineer, inventor*

Naperville
Kopala, Peter Steven *mechanical engineer*
Zigament, John Charles *mechanical engineer*

Northbrook
Polsky, Michael Peter *mechanical engineer*

Peoria
Okamura, Kiyohisa *mechanical engineer, educator*
Shareef, Iqbal *mechanical engineer, educator*

Rosemont
Cisko, George Joseph, Jr. *applied mechanics engineer, research lab manager*

Urbana
Assanis, Dennis N. (Dionissios Assanis) *mechanical engineering educator*
Bayne, James Wilmer *mechanical engineering educator*
Buckius, Richard O. *mechanical and industrial engineering educator*
Chao, Bei Tse *mechanical engineering educator*
Chato, John Clark *mechanical engineering educator*
Cusano, Cristino *mechanical engineer, educator*
Dantzig, Jonathan A. *mechanical engineer, educator*
DeVor, Richard Earl *mechanical and industrial engineering educator*
Dutton, J. Craig *mechanical engineering educator*
Mazumder, Jyotirmoy *mechanical and industrial engineering educator*
Peters, James Empson *mechanical and industrial engineering educator*
Vakakis, Alexander F. *mechanical engineering educator*
Walker, John Scott *mechanical engineering educator*

Washington
Hallinan, John Cornelius *mechanical engineering consultant*

Zion
Scharping, Brian Wayne *mechanical engineer*
Solomon, Patrick Michael *mechanical engineer*

INDIANA

Columbus
Kubo, Isoroku *mechanical engineer*
Lucke, John Edward *mechanical engineer*

Elkhart
Drzewiecki, David Samuel *mechanical engineer*

Evansville
Hartsaw, William O. *mechanical engineering educator*

Fort Wayne
Lyons, Jerry Lee *mechanical engineer*
Zepp, Lawrence Peter *mechanical engineer*

Goshen
Heap, James Clarence *retired mechanical engineer*

Indianapolis
Evans, Richard James *mechanical engineer*
McKain, Theodore F. *mechanical engineer*

Lawrenceburg
Wakeman, Thomas George *mechanical engineer*

Notre Dame
Gad-el-Hak, Mohamed *aerospace and mechanical engineering educator, scientist*

West Lafayette
Cohen, Raymond *mechanical engineer, educator*
Kokini, Klod *mechanical engineer*
Ramadhyani, Satish *mechanical engineering educator*
Thompson, Howard Doyle *mechanical engineer*
Viskanta, Raymond *mechanical engineering educator*

IOWA

Ames
Huston, Jeffrey Charles *mechanical engineer, educator*
Mischke, Charles Russell *mechanical engineering educator*
Riley, William Franklin *mechanical engineering educator*
Wilhelm, Harley A. *mechanical engineer*

Des Moines
Smith, David Welton *mechanical engineer, director research*

Iowa City
Alipour-Haghighi, Fariborz *mechanical engineer*
Haug, Edward Joseph, Jr. *mechanical engineering educator, simulation research engineer*

Marshall, Jeffrey Scott *mechanical engineer, educator*
Wang, Semyung *mechanical engineer, researcher*

Palo
Martin, Robert Anthony *mechanical engineer*

Sioux City
Uphoff, John Vincent *mechanical engineer*
Walker, Jimmie Kent *mechanical engineer*

KANSAS

Lawrence
Haugh, Dan Anthony *mechanical engineer*

Manhattan
Cogley, Allen C. *mechanical engineering educator, administrator*
Hosni, Mohammad Hosein *mechanical engineering educator*

Topeka
Gerard, Mark Edward *mechanical engineer*

Wichita
Greywall, Mahesh Inder-Singh *mechanical engineer*

KENTUCKY

Lexington
Sadler, John Peter *mechanical engineering educator*

LOUISIANA

Baton Rouge
Copes, John Carson, III *mechanical engineer*

Kenner
Pratt, Mark Ernest *mechanical engineer*

Lafayette
Chieri, Pericle Adriano Carlo *educator, consulting mechanical and aeronautical engineer, naval architect*

Monroe
Coon, Fred Albert, III *mechanical engineer*

New Orleans
Alexander, Beverly Moore *mechanical engineer*
Chattree, Mayank *mechanical engineering educator, researcher*

Ruston
Warrington, Robert O'Neil, Jr. *mechanical engineering educator and administrator, researcher*

Shreveport
Sloan, Wayne Francis *mechanical engineer*

MAINE

Acton
Lotz, William Allen *consulting engineer*

Bath
Dreher, Lawrence John *mechanical engineer*

York Harbor
McIntosh, Donald Waldron *retired mechanical engineer*

MARYLAND

Aberdeen Proving Ground
Chien, Lung-Siaen (Larry Chien) *mechanical engineer*
D'Amico, William Peter, Jr. *mechanical engineer*
Wright, Thomas Wilson *mechanical engineer*

Annapolis
Ho, Louis Ting *mechanical engineer*

Baltimore
Diehl, Myron Herbert, Jr. *engineer*
Ramesh, K(aliat) T(hazhathveetil) *mechanical engineer, educator*
Schuman, William John, Jr. *mechanical engineer*
Sharpe, William Norman, Jr. *mechanical engineer, educator*
Whalen, Barbara Rhoads *mechanical engineer, consultant*

Bethesda
Dai, Charles Mun-Hong *mechanical engineer, consultant*

Cockeysville Hunt Valley
Hirsch, Richard Arthur *mechanical engineer*

College Park
Ohadi, Michael M. *mechanical engineering educator*

Columbia
Fryberger, Theodore Kevin *mechanical and ocean engineer*
Krausman, John Anthony *mechanical engineer*

Gaithersburg
Ballard, Lowell Douglas *mechanical engineer*
Bartholomew, Richard William *mechanical engineer*
Coleman, Mark David *mechanical engineer*
Feric, Gordan *mechanical engineer*

Laurel
Billig, Frederick Stucky *mechanical engineer*

North Potomac
Agarwal, Duli Chand *mechanical engineer*

Rockville
Sorensen, John Noble *mechanical and nuclear engineer*

Silver Spring
Resch, Lawrence Randel *mechanical engineer*

MASSACHUSETTS

Andover
Burke, Shawn Edmund *mechanical engineer*

Billerica
Holmes, Carl Kenneth *mechanical engineer*

Boston
Dilts, David Alan *mechanical design engineer*
Willis, Jeffrey Scott *mechanical engineer, consultant*

Cambridge
Baldwin, Daniel Flanagan *mechanical engineer*
Blair, Kim Billy *mechanical engineer, researcher*
Gordon, Steven Jeffrey *mechanical engineer*
Hebert, Benjamin Francis *mechanical engineer*

Carlisle
Guarnaccia, David Guy *mechanical engineer*

Lowell
O'Connor, Denis *mechanical engineer*

Medford
Nelson, Frederick Carl *mechanical engineering educator, academic administrator*
Trefethen, Lloyd MacGregor *engineering educator*

Millbury
Pan, Coda H. T. *mechanical engineering educator, consultant, researcher*

Newton
Vasilakis, Andrew D. *mechanical engineer*

Norwood
Imbault, James Joseph *electromechanical engineering executive*

Pittsfield
Stall, Trisha Marie *mechanical engineer*

Somerville
Stahovich, Thomas Frank *mechanical engineer, researcher*

South Walpole
Leung, Woon-Fong *mechanical engineer, research scientist*

Swampscott
Neumann, Gerhard *mechanical engineer*

Taunton
Donovan, Stephen James *mechanical engineer*

Westborough
Lindquist, Dana Rae *mechanical engineer*

Westport
Stevens, Herbert Howe *mechanical engineer, consultant*

Wilmington
Faccini, Ernest Carlo *mechanical engineer*

Worcester
Norton, Robert Leo, Sr. *mechanical engineering educator, researcher*

MICHIGAN

Ann Arbor
Atreya, Arvind *mechanical engineering educator*
Brereton, Giles John *mechanical engineering researcher*
Chen, Michael Ming *mechanical engineering educator*
Gillespie, Thomas David *mechanical engineer, researcher*
Kaviany, Massoud *mechanical engineer educator*

Dearborn
Plee, Steven Leonard *mechanical engineer*

Detroit
Batcha, George *mechanical and nuclear engineer*
Catey, Laurie Lynn *mechanical engineer*
Ingrody, Pamela Theresa *mechanical engineer*
Lai, Ming-Chia *mechanical engineering educator, researcher*
Schmidt, Robert *mechanical and civil engineering educator*

East Lansing
Pence, Thomas James *mechanical engineer, educator, consultant*

Flint
Gratch, Serge *mechanical engineering educator*

Holland
Hamstra, Stephen Arthur *mechanical engineer*

Houghton
Lumsdaine, Edward *mechanical engineering educator, university dean*
Thangaraj, Ayyakannu Raj *mechanical engineering educator*

Lakeside
Price, Robert A. *mechanical engineer*

Livonia
Ladouceur, Harold Abel *mechanical engineer, consultant*

Rahman, Ahmed Assem *project engineer, stress analyst*

Novi
Coloske, Steven Robert *mechanical engineer*

Plymouth
Clark, Kenneth William *mechanical engineer*

Sterling Heights
Scott, David Lawrence *mechanical engineer*

Warren
Goenka, Pawan Kumar *mechanical engineer*
Winchell, William Olin *mechanical engineer, educator, lawyer*

Ypsilanti
Lou, Zheng (David) *mechanical engineer, biomedical engineer*

MINNESOTA

Bloomington
Drewek, Gerard Alan *mechanical engineer*

Minneapolis
Abrahamson, Scott David *mechanical engineer, educator*
Goldstein, Richard Jay *mechanical engineer, educator*
Keefe, William Robert *mechanical engineer*
Kvalseth, Tarald Oddvar *mechanical engineer, educator*
Patel, Jayant Ramanlal *mechanical engineer administrator*
Wilgen, Francis Joseph *mechanical engineer*

Saint Paul
Graf, Timothy L. *mechanical engineer*
Halvorsen, Thomas Glen *mechanical engineer*
Lampert, Leonard Franklin *mechanical engineer*
Uban, Stephen Alan *mechanical engineer*

MISSISSIPPI

Mississippi State
Agba, Emmanuel Ikechukwu *mechanical engineering educator*
Chakroun, Walid *mechanical engineering educator*

Seminary
Frazier, Ronald Gerald, Jr. *mechanical engineer*

MISSOURI

Annapolis
Goehman, M. Conway *mechanical engineer*

Florissant
Burns, Donald Raymond *structural and mechanical design engineer*

Fulton
Chapman, Garry Von *mechanical engineer*

Kansas City
Eidemiller, Donald Roy *mechanical engineer*
Rosner, Ronald Alan *mechanical engineer*
Smoot, John Eldon *mechanical engineer*

Lee's Summit
LaCombe, Ronald Dean *mechanical engineer*

Rolla
Crosbie, Alfred Linden *mechanical engineering educator*
Dharani, Lokeswarappa Rudrappa *engineering educator*
Look, Dwight Chester, Jr. *mechanical engineering educator, researcher*
Sauer, Harry John, Jr. *mechanical engineering educator, university administrator*

Saint Louis
Birman, Victor Mark *mechanical and aerospace engineering educator*
Husar, Rudolf Bertalan *mechanical engineering educator*
Norris, John Robert *design engineer*

NEBRASKA

Lincoln
Newhouse, Norman Lynn *mechanical engineer*

NEVADA

Carson City
Colombo, Michael Patrick *mechanical engineer*

Las Vegas
Trabia, Mohamed Bahaa Eldeen *mechanical engineer*

Reno
Bryan, Raymond Guy *mechanical engineer, consultant*
Reddy, Rajasekara L. *mechanical engineer*
Snyder, Martin Bradford *mechanical engineering educator*

Tonopah
Lathrop, Lester Wayne *mechanical engineer*

NEW HAMPSHIRE

Manchester
Warfel, Christopher George *mechanical engineer, design consultant*

Nashua
Klarman, Karl Joseph *electromechanical engineer*

Portsmouth
Perkins, Robert Bennett *mechanical engineer*

Rye
Heald, Paul Francis *mechanical engineer, consultant*

NEW JERSEY

Atlantic City
Flournoy, Thomas Henry *mechanical engineer*

Belleville
Parikh, Manor Madanmohan *mechanical engineer*

Clinton
Bosch, Robert John, Jr. *mechanical engineer*

Edison
Shattuck, Douglas E. *mechanical engineer, consultant*

Elizabeth
Marticorena, Ernesto Jesus *mechanical engineer, educator*

Elmer
Robinson, John Abbott *mechanical engineer*

Florham Park
Amabile, Michael John *mechanical engineer, consultant*
Bhagat, Phiroz Maneck *mechanical engineer*

Hoboken
Prasad, Marehalli Gopalan *mechanical engineering educator*

Livingston
Gupta, Gopal Das *mechanical engineer, researcher*

Maplewood
Lev, Alexander Shulim *mechanical engineer*
Marlow, Joseph *automation engineer*

Metuchen
Gerardi, Roy G., Jr. *mechanical engineer*

Moorestown
Druckman, William Frank *mechanical engineer*

Mount Laurel
Melick, George Fleury *mechanical engineer, educator*

New Providence
Kotynek, George Roy *mechanical engineer, educator, marketing executive*

Newark
Hrycak, Peter *mechanical engineer, educator*

Nutley
Magee, Richard Stephen *mechanical engineering educator*

Pennington
Marder, William Zev *technological consultant*

Pennsauken
Alday, Paul Stackhouse, Jr. *mechanical engineer*

Phillipsburg
Guinzburg, Adiel *mechanical engineer*

Piscataway
Briggs, David Griffith *mechanical engineer, educator*
Hellenbrecht, Edward Paul *mechanical engineer*

Princeton
Blair, David William *mechanical engineer*
Marchese, Anthony John *research engineer*
Socolow, Robert Harry *mechanical and aerospace engineering educator, scientist*
Stengel, Robert Frank *mechanical and aerospace engineering educator*

Rutherford
Brooks, Richard M. *mechanical engineer*
Fallon, David Michael *mechanical engineer*
Szukics, James Charles *mechanical engineer*

Springfield
Goldstein, Irving Robert *mechanical and industrial engineer, educator, consultant*

Toms River
Curreri, John Robert *mechanical engineer, consultant*

Verona
France, Samuel Ewing Hill *mechanical engineer*

Washington
Schlink, Frederick John *retired mechanical engineer*

NEW MEXICO

Albuquerque
Arvizu, Dan Eliab *mechanical engineer*
Ashbaugh, Scott Gregory *mechanical engineer, consultant*
Gartling, David Keith *mechanical engineer*
Ingber, Marc Stuart *applied mechanics educator*
Reed, Ray Paul *engineering mechanics measurement consultant*
Schuster, Lawrence Joseph *mechanical engineer*

Belen
Toliver, Lee *mechanical engineer*

Las Cruces
Colbaugh, Richard Donald *mechanical engineer, educator, researcher*
Matthews, Larryl Kent *mechanical engineering educator*
Shouman, Ahmad Raafat *mechanical engineering educator*
Weigle, Robert Edward *mechanical engineer, research director*

Los Alamos
Fusco, Penny Plummer *mechanical engineer*

Santa Fe
Campbell, Ashley Sawyer *retired mechanical engineering educator*

NEW YORK

Albany
McKay, Donald Arthur *mechanical engineer*

Albertson
Campbell, Henry J., Jr. *mechanical engineering consultant*

Bethpage
Rockensies, John William *mechanical engineer*

Brooklyn
Beach, Arthur Thomas *mechanical engineer*
Hakola, John Wayne *mechanical engineer, consultant, small business owner*
Pan, Huo-Hsi *mechanical engineer, educator*
Pascali, Raresh *mechanical engineering educator*

Buffalo
Adams, Donald E. *mechanical engineer*

Clifton Park
Benner, Ronald Allen, Jr. *mechanical engineer*

Corning
Ryszytiwskyj, William Paul *mechanical engineer*

Forest Hills
Hasselriis, Floyd Norbert *mechanical engineer*

Henrietta
Yang, Eugene L. *mechanical engineer*

Ithaca
Barber, Edmund Amaral, Jr. *retired mechanical engineer*
Callister, John Richard *mechanical engineer*
De Boer, Pieter Cornelis Tobias *mechanical and aerospace engineering educator*
Jenkins, James Thomas *mechanical engineering researcher*

Lake Success
Bladykas, Michael P. *mechanical engineering consultant*

Melville
Lofaso, Anthony Julius *mechanical engineer*

New York
Becker, Herbert P. *mechanical engineer*
Deresiewicz, Herbert *mechanical engineering educator*
Freudenstein, Ferdinand *mechanical engineering educator*
Lee, Martin Yongho *mechanical engineer*
Mezic, Richard Joseph *engineer, consultant*
Watkins, Charles Booker, Jr. *mechanical engineering educator*
Zacharius, Martin Philip *mechanical engineer*

Old Westbury
Koenig, Gottlieb *mechanical engineering educator*

Olean
Datta, Prasanta *mechanical engineer*

Ossining
Phelps, Glenn Howard *mechanical carbon process engineer*

Painted Post
Benjamin, Keith Edward *mechanical engineer*

Penfield
Baxter, Meriwether Lewis, Jr. *gear engineer, consultant*

Port Washington
Gaddis, M. Francis *mechanical and marine engineer, environmental scientist*

Potsdam
Glauser, Mark Nelson *mechanical and aeronautical engineering educator*
Kane, James Harry *mechanical engineer, educator, researcher*
Mochel, Myron George *mechanical engineer, educator*
Phillips, William Robert *fluid dynamics engineering educator*

Rochester
Ogut, Ali *mechanical engineering educator*

Rockville Centre
Silecchia, Jerome A. *mechanical engineer*

Saint James
Irvine, Thomas Francis, Jr. *mechanical engineering educator*

Schenectady
Bucinell, Ronald Blaise *mechanical engineer*
German, Marjorie DaCosta *mechanical engineer*
Roloff, Thomas Paul *combustion engineer*
Tolpadi, Anil Kumar *mechanical engineer, researcher*
Wicks, Frank Eugene *mechanical engineer, educator*

Stony Brook
Huang, Peisen Simon *mechanical engineer*

Syracuse
Pennock, Donald William *retired mechanical engineer*

Troy
Bergles, Arthur Edward *mechanical engineering educator*
Blanchet, Thierry Alain *mechanical engineering educator*
Brunelle, Eugene John, Jr. *mechanical engineering educator*
Ostrogorsky, Aleksandar George *mechanical engineer, educator*

Upton
Montag, Mordechai *mechanical engineer*

Utica
Ragusa, Paul Carmen *mechanical and structural engineer*

Webster
Hwang, Shyshung *mechanical engineer*
Machell, Arthur R. *retired mechanical engineer, association board member*

Whitestone
Panides, Elias *mechanical engineer*

NORTH CAROLINA

Charlotte
Browning, Burt Oliver *mechanical engineer, consultant*
Foss, Ralph Scot *mechanical engineer*
Ruckterstuhl, Russell M(ilton) *mechanical engineer, consulting engineer*

Davidson
Alday, John Hane *mechanical engineer*

Durham
Bejan, Adrian *mechanical engineering educator*
Garg, Devendra Prakash *mechanical engineer, educator*

Greensboro
Garibay, Joseph Michael *mechanical engineer*
Stein, Richard Martin *mechanical engineer*

Highland
Sandor, George Nason *mechanical engineer, educator*

Lewisville
Bolz, Roger William *mechanical engineer, consultant*

Morrisville
Odum, Jeffery Neal *mechanical engineer*

Mount Airy
Ratliff, Robert Barns, Jr. *mechanical engineer*

Nags Head
Rogallo, Francis Melvin *mechanical, aeronautical engineer*

Pine Knoll Shores
Baker, Edward George *retired mechanical engineer*

Raleigh
Gillani, Noor Velshi *mechanical engineer, researcher, educator*
Hubbard, Bessie Renee *mechanical engineer, mathematician*
Williams, Hugh Alexander, Jr. *retired mechanical engineer, consultant*

Research Triangle Park
Odman, Mehmet Talat *mechanical engineer, researcher*

NORTH DAKOTA

Fargo
Li, Kam Wu *mechanical engineer, educator*

Grand Forks
Ljubicic, Blazo *mechanical engineering researcher*

OHIO

Alliance
Kitto, John Buck, Jr. *mechanical engineer*
Walters, Tracy Wayne *mechanical engineer*

Amelia
Scott, Steven Mike *mechanical engineer*

Bay Village
Ferraro, Vincent *mechanical engineer, civil engineer*

Beavercreek
Chang, Won Soon *research scientist, mechanical engineer*

Brecksville
Crozier, David Wayne *mechanical engineer*

Cincinnati
Cabe, Jerry Lynn *mechanical engineer, researcher*
Enrico, David Russell *mechanical engineer*
Gerner, Frank Matthew *mechanical engineering educator*
Hamilton, Michael Bruce *engineer*
LaBath, Octave Aaron *mechanical engineer*
Luffy, Ronald Jon *mechanical engineer*
Quo, Phillip C. *mechanical engineering educator*

Cleveland
Carek, Gerald Allen *mechanical engineer*
Hardy, Richard Allen *mechanical engineer, diesel fuel engine specialist*
Ibrahim, Mounir Boshra *mechanical engineering educator*
Kocka, Thomas John *mechanical engineer*
Trefts, Albert S. *mechanical engineer, consultant*

Columbus
Bechtel, Stephen E. *mechanical engineer, educator*
Bhushan, Bharat *mechanical engineer*
Ensminger, Dale *mechanical engineer, electrical engineer*
Guezennec, Yann Guillaume *mechanical engineering educator*
Houser, Donald Russell *mechanical engineering educator, consultant*
Kannel, Jerrold Williams *mechanical engineer*
Kipp, Carl Robert *mechanical engineer*
Leissa, Arthur William *mechanical engineering educator*
Singh, Rajendra *mechanical engineering educator*
Stuckert, Gregory Kent *mechanical engineer, researcher*
Waldron, Kenneth John *mechanical engineering educator, researcher*
Wiechel, John F. *mechanical engineer*

Dayton
Ballal, Dilip Ramchandra *mechanical engineering educator*
Ramalingam, Mysore Loganathan *mechanical engineer*
Reichel, Lee Elmer *mechanical engineer*
Takahashi, Fumiaki *research mechanical engineer*

Madison
Spiesman, Benjamin Lewis *mechanical design engineer*

Massillon
Hatzilabrou, Labros *mechanical engineer, metallurgical engineer*

Miamisburg
Peterson, George P. *mechanical engineer, research and development firm executive*

Mount Vernon
Schaub, Fred S. *mechanical engineer*

Ostrander
Smith, Rick A. *mechanical engineer, consultant*

Sidney
Schumann, Stanley Paul *mechanical engineer*

Solon
Williams, Peter Charles *engineer*

Toledo
Keith, Theo Gordon, Jr. *mechanical engineering educator*
Reese, Herschel Henry *mechanical engineer*
Taylor, Douglas Floyd *mechanical engineer*

Westlake
Huff, Ronald Garland *mechanical engineer*

Youngstown
Suchora, Daniel Henry *mechanical engineering educator, consultant*

OKLAHOMA

Duncan
King, Randall Kent *mechanical engineer*

Norman
Bert, Charles Wesley *mechanical and aerospace engineer, educator*
Egle, Davis Max *mechanical engineering educator*

Oklahoma City
Morgan, Robert Steve *mechanical engineer*

Oologah
Knight, Gary Charles *mechanical engineer*

Ponca City
Cole, Jack Howard *mechanical engineer*

Stillwater
Lucca, Don Anthony *mechanical engineer*
Misawa, Eduardo Akira *mechanical engineer, educator*

Tulsa
Blanton, Roger Edmund *mechanical engineer*
Nutter, Dale E. *mechanical engineer*
Strafuss, David Louis *mechanical engineer*
Tubbs, David Eugene *mechanical engineer, marketing professional*

OREGON

Corvallis
Welty, James R. *mechanical engineer, educator*

PENNSYLVANIA

Alcoa Center
Azarkhin, Alexander *mechanical engineer, mathematician*
Fussell, Paul Stephen *mechanical engineer*
Pien, Shyh-Jye John *mechanical engineer*

Aston
Kingrea, James Irvin, Jr. *electromechanical engineer*

Bala Cynwyd
Robertson, Kenneth McLeod *mechanical engineer*

Bethlehem
Levy, Edward K. *mechanical engineering educator*

Roberts, Richard *mechanical engineering educator*
Wei, Robert Peh-Ying *mechanics educator*

Boalsburg
Gettig, Martin Winthrop *retired mechanical engineer*

Collegeville
Behrens, Rudolph *mechanical engineer*

Conshohocken
Rutolo, James Daniel *mechanical engineer*

East Greenville
Grubb, David Conway *mechanical engineer*

Erie
Crankshaw, John Hamilton *mechanical engineer*
Hsu, Bertrand Dahung *mechanical engineer*

Flourtown
Di Maria, Charles Walter *mechanical and automation engineer, consultant*

Fort Washington
Newton, Crystal Hoffman *mechanical engineer*

Greensburg
Coffield, Ronald Dale *mechanical engineer*

Houston
Rajakumar, Charles *mechanical engineer*

Muncy
Lowe, John Raymond, Jr. *mechanical engineer*

New Holland
Crane, Jack Wilbur *agricultural and mechnical engineer*

North Huntingdon
Metala, Michael Joseph *mechanical engineer*

Palmyra
Wyczalkowski, Wojciech Roman *mechanical engineer, chemical engineer, researcher*

Philadelphia
Aravas, Nikolaos *mechanical engineering educator*
Batterman, Steven Charles *mechanical engineering, bioengineering educator*
Cho, Young Il *mechanical engineering educator*
Jarosz, Boleslaw Francis *mechanical engineer*
Lorenz, Beth June *mechanical engineer*
Sague, John E(lmer) *mechanical engineer*

Pittsburgh
Akay, Adnan *mechanical engineer educator*
Amon, Cristina Hortensia *mechanical engineer, educator*
Brungraber, Louis Edward *retired mechanical engineer*
Cagan, Jonathan *mechanical engineering educator*
Jobes, Christopher Charles *mechanical engineer*
Osterle, John Fletcher *mechanical engineering educator*
Ramezan, Massood *mechanical engineer*
Smith, Paul Christian *mechanical engineer*
Tran, Phuoc Xuan *mechanical engineer*
Treadwell, Kenneth Myron *mechanical engineer, consultant*

Saint Marys
Kasaback, Ronald Lawrence *mechanical engineer*

State College
Hansen, Robert J. *mechanical engineer*
Wang, Qunzhen *mechanical engineer*

Trevose
Ziegler, Albert H. *mechanical engineer*

University Park
Ovaert, Timothy Christopher *mechanical engineering educator*
Pal, Sibtosh *mechanical engineer, researcher*
Settles, Gary Stuart *fluid dynamics educator*
Sommer, Henry Joseph, III *mechanical engineering educator*

Villanova
McLaughlin, Philip VanDoren, Jr. *mechanical engineering educator, researcher, consultant*
Nataraj, Chandrasekhar *mechanical engineering educator, researcher*

West Mifflin
Ardash, Garin *mechanical engineer*

Wilmerding
Cunkelman, Brian Lee *mechanical engineer*

York
Korenberg, Jacob *mechanical engineer*

RHODE ISLAND

Cranston
Basu, Prithwish *mechanical engineering researcher*

Kingston
Palm, William John *mechanical engineering educator*
White, Frank M. *mechanical engineer, educator*

Westerly
Varnhagen, Melvin Jay *mechanical engineer*

SOUTH CAROLINA

Aiken
Danko, Edward Thomas *mechanical and industrial engineer*
Wilder, James Robbins *mechanical engineer*

Clemson
Fadel, Georges Michel *mechanical engineering educator*
Kapat, Jayanta Sankar *mechanical engineer*
Paul, Frank Waters *mechanical engineer, educator, consultant*

Fort Mill
Hodge, Bobby Lynn *mechanical engineering director*

Greenville
Stutes, Anthony Wayne *mechanical engineer*

Pawleys Island
Cepluch, Robert J. *retired mechanical engineer*

Rock Hill
Evans, Wallace Rockwell, Jr. *mechanical engineer*

York
Fritz, Edward William *mechanical engineer*

TENNESSEE

Chattanooga
Manaker, Arnold Martin *mechanical engineer, consultant*
Mason, Michael Edward *mechanical engineer*

Cookeville
Chowbey, Sanjay Kumar *mechanical engineer*
Darvennes, Corinne Marcelle *mechanical engineering educator, researcher*
Salmon, Fay Tian *mechanical engineering educator*
Tsatsaronis, George *mechanical engineering educator, researcher*

Kingsport
Maggard, Bill Neal *mechanical engineer*

Knoxville
Chen, Fang Chu *mechanical engineer*
Martin, H(arry) Lee *mechanical engineer, robotics company executive*

Nashville
Mellor, Arthur M(Cleod) *mechanical engineering educator*
Wilson, James Richard *mechanical engineer*

Oak Ridge
Moore, Marcus Lamar *mechanical engineer, consultant*

Soddy Daisy
Smith, Eual Randall *mechanical engineer*

TEXAS

Austin
Bard, Jonathan F. *mechanical engineering and operations research educator*
Freid, James Martin *mechanical engineer*
Hendricks, Terry Joseph *mechanical engineer*
Howell, John Reid *mechanical engineer, educator*
Nichols, Steven Parks *mechanical engineer, academic administrator*
Oden, John Tinsley *engineering mechanics educator, consultant*
Rylander, Henry Grady, Jr. *mechanical engineering educator*

Beaumont
Williams, Curt Alan *mechanical engineer*

College Station
Baskharone, Erian Aziz *mechanical and aerospace engineering educator*
Bradley, Walter Lee *mechanical engineer, educator, researcher, consultant*
Fletcher, Leroy Stevenson *mechanical engineer, educator*
Han, Je-Chin *mechanical engineering educator*
Kihm, Kyung D. (Ken Kihm) *mechanical engineering educator*
Morrison, Gerald Lee *mechanical engineering educator*
O'Neal, Dennis Lee *mechanical engineering educator*

Dallas
Fischer, John Gregory *mechanical engineer*
Gyongyossy, Leslie Laszlo *mechanical engineer, consultant*
Langton, Michael John *mechanical engineer, consultant*

Denton
Head, Gregory Alan *mechanical engineer, consultant*

Fort Worth
Dickman, Dean Anthony *mechanical engineer*

Grapevine
McIlheran, Mark Lane *mechanical engineer*

Greenville
Daniel, Eddy Wayne *mechanical engineer*

Haltom City
Kundu, Debabrata *mechanical engineer*

Houston
Beauchamp, Jeffery Oliver *mechanical engineer*
Bishop, Thomas Ray *retired mechanical engineer*
Chen, Min-Chu *mechanical engineer*
Cowen, Joseph Eugene *mechanical engineer*
Dwyer, John James *mechanical engineer*
Eichberger, LeRoy Carl *stress analyst, mechanical engineering consultant*
Hackemesser, Larry Gene *mechanical engineer*
Peng, Liang-Chuan *mechanical engineer*
Powell, Alan *mechanical engineer, scientist, educator*
Puddy, Donald Ray *mechanical engineer*
Smith, Thomas Bradford *engineering administrator*

Humble
Brown, Samuel Joseph, Jr. *mechanical engineer*

Hurst
Bishara, Amin Tawadros *mechanical engineer, technical services executive*

Irving
Halter, Edmund John *mechanical engineer*

Lubbock
Koh, Pun Kien *retired educator, metallurgist, consultant*

Pflugerville
Johnson, David Lee *mechanical building systems engineer*

Plano
Reid, William Michael *mechanical engineer*
Tolle, Glen Conrad *mechanical engineer*

Randolph AFB
Ling, Edward Hugo *mechanical engineer*

San Antonio
Burnside, Otis Halbert *mechanical engineer*
Esparza, Edward Duran *mechanical engineer*
Miles, Paula Effette *mechanical engineer*
Phillips, Thomas Dean *mechanical engineer*
Vafeades, Peter *mechanical engineering educator*

Springtown
Brister, Donald Wayne *mechanical engineer*

Willis
Gilbert, Nathan *mechanical engineer*

UTAH

Provo
Eatough, Craig Norman *mechanical engineer*

Salt Lake City
De Vries, Kenneth Lawrence *mechanical engineer, educator*
Hogan, Mervin Booth *mechanical engineer, educator*
Salmon, David Charles *mechanical engineer*
Silver, Barnard Stewart *mechanical engineer, consultant*

VERMONT

Norwich
Japikse, David *mechanical engineer, manufacturing executive*

VIRGINIA

Alexandria
Rogers, McKinley Bradford *mechanical engineer*

Arlington
Cobb, Richard E. *mechanical engineer*
McCarthy, Wilbert Alan *mechanical engineer*
Rahman, Muhammad Abdur *mechanical engineer*
Unver, Erdal Ali *research mechanical engineer*
Vu, Bien Quang *mechanical engineer*

Blacksburg
Brown, Eugene Francis *mechanical engineer, educator*
Bursal, Faruk Halil *mechanical engineer, educator*
Inman, Daniel John *mechanical engineer, educator*
Jones, James Beverly *mechanical engineering educator*
Kornhauser, Alan Abram *mechanical engineer*
Reddy, J. Narasimh *mechanical engineering educator*
Rogers, Craig Alan *mechanical engineering educator, university program director*
Smith, Charles William *mechanical engineer, educator*

Charlottesville
Morton, Harold S(ylvanus), Jr. *retired mechanical and aerospace engineering educator*
Thornton, Earl Arthur *mechanical and aerospace engineering educator*

Dahlgren
Boyer, Charles Thomas, Jr. *mechanical engineer*
Delurey, Michael William *mechanical engineer*

Fairfax Station
Coaker, James Whitfield *mechanical engineer*

Forest
Durand, James Howard *mechanical engineer*

Gainesville
Derstine, Mark Stephen *mechanical engineer*

Hampton
Bayer, Janice Ilene *mechanical engineer*
Bushnell, Dennis Meyer *mechanical engineer, researcher*
Glass, David Eugene *mechanical engineer, educator*
Kelly, Jeffrey Jennings *mechanical engineer*
Shenhar, Joram *mechanical engineer*

Lexington
Arthur, James Howard *mechanical engineering educator*

Mc Lean
Harkins, William Douglas *mechanical engineer*

Newport News
Hempfling, Gregory Jay *mechanical engineer*
Montane, Jean Joseph *mechanical engineer*
Pohl, John Joseph, Jr. *retired mechanical engineer*

Norfolk
Kandil, Osama Abd El Mohsin *mechanical and aerospace engineering educator*

Reston
Wetzel, John Paul *structural dynamics engineer*

Richmond
Duong, Minh Truc *project engineer*
Palik, Robert Richard *mechanical engineer*

Vienna
Woodward, Kenneth Emerson *retired mechanical engineer*

Williamsburg
Hinders, Mark Karl *mechanical engineer*
Rogers, Verna Aileen *mechanical and biomedical engineer, researcher*

WASHINGTON

Everett
Hitomi, Georgia Kay *mechanical engineer*

Pullman
Troutt, Timothy Ray *mechanical engineer, educator*

Richland
Keska, Jerry Kazimierz *mechanical engineering educator*

Seattle
Cassady, Philip Earl *mechanical engineer*
Harder, Ruel Tan *mechanical engineer*
McCormick, Norman Joseph *mechanical engineer, nuclear engineer, educator*
Viggers, Robert Frederick *mechanical engineering educator*

Vancouver
Mairose, Paul Timothy *mechanical engineer, consultant*

Yakima
Sutphen, Robert Ray *mechanical engineer, manager*

WEST VIRGINIA

Huntington
Foley, Arvil Eugene *mechanical engineer*

Morgantown
Prucz, Jacky Carol *mechanical and aerospace engineer, educator*
Shoemaker, Harold Dee *mechanical engineer*

Parkersburg
Witwer, Ronald James *mechanical engineer*

WISCONSIN

Green Bay
Wiley, Dale Stephen *mechanical engineer*

Janesville
Hornby, Robert Ray *mechanical engineer*

La Crosse
Monfre, Joseph Paul *mechanical engineer, consultant*

Madison
Barlow, Kent Michael *mechanical engineer*
Bollinger, John Gustave *engineering educator, college dean*
DeVries, Marvin Frank *mechanical engineering educator*
Duffie, Neil Arthur *mechanical engineering educator, researcher*
Ramesh, Krishnan *mechanical engineer*

Marshfield
Kitchell, Shawn Ray *plant engineer, educator*

Menomonee Falls
Schommer, Gerard Edward *mechanical engineer*

Milwaukee
Garimella, Suresh Venkata *mechanical engineering educator*
Landis, Fred *mechanical engineering educator*
Perez, Ronald A. *mechanical engineering educator*
Renken, Kevin James *mechanical engineering educator*

New Berlin
DeBaker, Brian Glenn *mechanical engineer*

Oconomowoc
Raether, Scott Edward *mechanical engineer*

Racine
Guntly, Leon Arnold *mechanical engineer*

Sheboygan
Bockius, Thomas John *mechanical engineer*

West Allis
Zhang, Jiping *engineering analyst*

TERRITORIES OF THE UNITED STATES

PUERTO RICO

Hato Rey
Edwards-Vidal, Dimas Francisco *mechanical engineer, consultant*

Mayaguez
Suarez, Luis Edgardo *mechanical engineering educator*

CANADA

ALBERTA

Calgary
de Krasinski, Joseph Stanislas *mechanical engineering educator*
Kentfield, John Alan *mechanical engineering educator*

Edmonton
Craggs, Anthony *mechanical engineer*

BRITISH COLUMBIA

Vancouver
Zhao, Ru He *mechanical engineer*

ONTARIO

Burlington
Krishnappan, Bommanna Gounder *fluid mechanics engineer*

Etobicoke
Stojanowski, Wiktor J. *mechanical engineer*

Kingston
Smith, Roy Edward *mechanical engineer, rail vehicle consultant*

Ottawa
Altman, Samuel Pinover *mechanical engineer, research consultant*
McFarlane, James Ross *mechanical engineer, educator*

Toronto
Venter, Ronald Daniel *mechanical engineering educator, researcher, administrator*

Windsor
Hageniers, Omer Leon *mechanical engineer*

QUEBEC

Montreal
Pellemans, Nicolas *patent agent, mechanical engineer*

Quebec
Cheng, Li *mechanical engineer, educator*

SASKATCHEWAN

Saskatoon
Tao, Yong-Xin *mechanical engineer, researcher*

MEXICO

Monterrey
Muci Küchler, Karim Heinz *mechanical engineering educator*

ARGENTINA

Santa Fe
Idelsohn, Sergio Rodolfo *mechanical engineering educator*

AUSTRALIA

Sydney
Shaw, Keith Moffatt *mechanical engineer*

AUSTRIA

Krems
Hirner, Johann Josef *mechanical engineer*

Vienna
Varga, Thomas *mechanical engineering educator*

BRAZIL

São Paulo
Akkermann, Schaia *mechanical, electrical and civil engineer*

CHILE

Santiago
Parada, Jaime Alfonso *mechanical engineer, consultant*

CHINA

Hefei
Wu, Chang-chun *mechanical engineering educator*

Xi'an
Su, Yaoxi *fluid mechanics educator*

CROATIA

Zagreb
Čatic, Igor Julio *mechanical engineering educator*

CZECH REPUBLIC

Prague
Šesták, Jiří Vladimír *mechanical engineering educator*

DENMARK

Holte
Niordson, Frithiof Igor *mechanical engineering educator*

EGYPT

Cairo
El-Gammal, Abdel-Aziz Mohamed *aero-mechanical engineer*

ENGLAND

Camberley
Ayres, Raymond Maurice *mechanical engineer*

Cambridge
Dowling, Ann Patricia *mechanical engineering educator, researcher*
Hawthorne, Sir William (Rede) *aerospace and mechanical engineer, educator*

FRANCE

Nantes
Tardy, Daniel Louis *mechanical engineer, education consultant*

Vernon
Lehé, Jean Robert *mechanical engineer*

Villeurbanne
Godet, Maurice *mechanical engineer*

GERMANY

Aachen
Benetschik, Hannes *mechanical engineer*

Lohhof
Arndt, Norbert Karl Erhard *mechanical engineer*

Munich
Schmidt, Stefan *mechanical engineer, economist*

Stuttgart
Schiehlen, Werner Otto *mechanical engineer*

Wiesbaden
Adeniji-Fashola, Adeyemo Ayodele *mechanical engineer, consultant*

GREECE

Athens
Balaras, Constantinos Agelou *mechanical engineer*
Papathanasiou, Athanasios George *mechanical engineer*
Rekkas, Christos Michail *mechanical engineer*

Thessaloniki
Glavopoulos, Christos Dimitrios *mechanical engineering consultant*
Panagiotopoulos, Panagiotis Dionysios *mechanical engineering educator*

HUNGARY

Budapest
Àbrahám, György *mechanical engineering educator*
Vámos, George A. *mechanical engineer, consultant*

INDIA

Hosur
Ramani, Narayan *mechanical engineer*

INDONESIA

Jakarta
Cheron, James Clinton *mechanical engineer*

IRAN

Tehran
Eslami, Mohammad Reza *mechanical engineering educator*

ITALY

Moncalieri
La Rocca, Aldo Vittorio *mechanical engineer*

Padua
Mirandola, Alberto *mechanical engineering educator*

JAPAN

Hiroshima
Hiroyasu, Hiroyuki *mechanical engineering educator*

Hitachi
Oku, Tatsuo *mechanical engineering educator*

Kawaguchi
Matsumoto, Hiroshi *mechanical engineer*

Kawasaki
Taniuchi, Kiyoshi *mechanical engineering educator*

Kita-ku
Ohnami, Masateru *mechanical engineering educator*

Kitakyushu
Nishi, Michihiro *mechanical engineering educator*
Tachibana, Takeshi *mechanical engineer, educator*

Kyoto
Iwashimizu, Yukio *mechanical engineering educator*

Ohbu
Higashida, Yoshisuke *mechanical engineer*

Osaka
Masuo, Ryuichi *mechanical engineering educator*

Shinjyuku-ku
Honami, Shinji *mechanical engineer educator*

Togane
Uchiyama, Shoichi *mechanical engineer*

Tokyo
Inada, Tadahico *mechanical engineer*

Tsuchiura
Ishikawa, Yuichi *mechanical engineer, researcher*

Ube
Kageyama, Yoshiro *retired mechanical engineering educator*

MALAYSIA

Johor Bahru
Tan, Yoke San *mechanical engineer*

Kuala Lumpur
Chue, Seck Hong *mechanical engineer*

THE NETHERLANDS

Noordwijk
Panin, Fabio Massimo *mechanical engineer*

PAKISTAN

Gulberg
Haq, Iftikhar Ul *mechanical engineer, consultant*

Islamabad
Bahadur, Khawaja Ali *mechanical engineer, consultant*

Lahore
Ghani, Ashraf Muhammad *mechanical engineer, business and engineering consultant*

THE PHILIPPINES

Cebu City
Mendoza, Genaro Tumamak *mechanical engineering consultant*

POLAND

Kraków
Noga, Marian *mechanical engineer*
Pytko, Stanislaw Jerzy *mechanical engineering educator*

PORTUGAL

Lisbon
Portela, Antonio Gouvea *retired mechanical engineering researcher*

REPUBLIC OF KOREA

Inchon
Lee, Usik *mechanical engineer, educator*

Jangseungpo
Woo, Jongsik *mechanical engineer, educator, researcher*

Pohang
Hwang, Woonbong *mechanical engineering educator, consultant*

Seoul
Ha, Sung Kyu *mechanical engineer, educator*

RUSSIA

Moscow
Frolov, Konstantin Vasilievitch *mechanical engineer, science administrator*

SIERRA LEONE

Freetown
Tahilramani, Sham Atmaram *mechanical engineer*

SINGAPORE

Kent Ridge
Wijeysundera, Nihal Ekanayake *mechanical engineering educator, industrial consultant*

Singapore
Leng, Gerard Siew-Bing *mechanical and production engineering educator*

SLOVENIA

Ljubljana
Stusek, Anton *mechanical engineer, researcher*

SOUTH AFRICA

Pretoria
Stevenson, William John *mechanical engineer*

SPAIN

Madrid
Campo, J. M. *mechanical engineer*

SWEDEN

Stockholm
Fransson, Torsten Henry *mechanical engineering educator, researcher*

SWITZERLAND

Baden
Arnal, Michel Philippe *mechanical engineer, consultant*

TAIWAN

Pingtung
Wang, Bor-Tsuen *mechanical engineering educator*

YUGOSLAVIA

Belgrade
Mladenovic, Nikola Sreten *mechanical engineer*

ADDRESS UNPUBLISHED

Allen, Burkley *mechanical engineer*
Altan, Taylan *engineering educator, mechanical engineer, consultant*
Bamberger, Joseph Alexander *mechanical engineer, educator*
Bertolett, Craig Randolph *mechanical engineer consultant*
Burhans, Frank Malcolm *mechanical engineer*
Costin, Daniel Patrick *mechanical engineer*
Davoodi, Hamid *mechanical engineering educator*
Ewankowich, Stephen Frank, Jr. *mechanical engineer*
Farnsworth, Michael Edward *mechanical engineer*
Gerhardt, Jon Stuart *mechanical engineer, engineering educator*
Grebenc, Joseph D. *mechanical engineer, consultant*
Haberman, Charles Morris *mechanical engineer, educator*
Hanamirian, Varujan *mechanical engineer, educator, journalist, publisher*
Huffman, Henry Samuel *building mechanical engineer*
Hutchinson, John Woodside *applied mechanics educator, consultant*
Jee, Melvin Wie On *mechanical engineer, civilian military engineer*
Kandula, Max *mechanical engineer, aerospace engineer*
Kinzie, Daniel Joseph *superconducting technology and mechanical engineer*
Koltai, Stephen Miklos *mechanical engineer, consultant, economist*
Lambert, Jon Kelly *engineer*
Lewalski, Elzbieta Ewa *mechanical engineer, consultant*
Liu, Alan Fong-Ching *mechanical engineer*
Marwill, Robert Douglas *aircraft design engineer*
McEntire, B. Joseph *mechanical engineer*
Meyer, Harold Louis *mechanical engineer*
Namboodri, Chettoor Govindan *mechanical engineer*
Nichols, C(laude) Alan *mechanical engineer*
O'Reilly, John Joseph *engineer*
Ormiston, Timothy Shawn *mechanical engineer*
Prell, Martin *mechanical engineer*
Rich, Raphael Z. *mechanical engineer, researcher*
Sancaktar, Erol *engineering educator*
Schiess, Klaus Joachim *mechanical engineer*
Smelser, Ronald Eugene *mechanical engineer*
Souza Mendes, Paulo Roberto de *mechanical engineering educator*
Stanevich, Kenneth William *mechanical design engineer*
Thornton, Peter Brittin *mechanical engineer*
Wang, Bor-Jenq *mechanical engineer*
Wang, Shih-Liang *mechanical engineering educator, researcher*
Zak, Sheldon Jerry *mechanical engineer, educator*

ENGINEERING: METALLURGICAL & MATERIALS

UNITED STATES

ALABAMA

Huntsville
Cowan, Penelope Sims *materials engineer*
Curreri, Peter Angelo *materials scientist*

Redstone Arsenal
Jaklitsch, Donald John *materials engineer, consultant*

ALASKA

Anchorage
Watts, Michael Arthur *materials engineer*

ARIZONA

Scottsdale
Leeser, David O. *materials engineer, metallurgist*

Tempe
Carpenter, Ray Warren *materials scientist and engineer, educator*

CALIFORNIA

Calabasas
Goodman, William Alfred *materials engineer*

Danville
Belding, William Anson *ceramic engineer*

El Segundo
Kashar, Lawrence Joseph *metallurgical engineer, consultant*

Hayward
Hunnicutt, Richard Pearce *metallurgical engineer*

Huntington Beach
Anderson, Raymond Hartwell, Jr. *metallurgical engineer*

Irvine
Lavernia, Enrique Jose *materials science and engineering educator*

Menlo Park
Nell, Janine Marie *metallurgical and materials engineer*

Monrovia
Pray, Ralph Emerson *metallurgical engineer*

Newport Beach
Sharbaugh, W(illiam) James *plastics engineer, consultant*

COLORADO

Denver
Marts, Kenneth Patrick *materials engineer*

Golden
Ansell, George Stephen *metallurgical engineering educator, academic administrator*
Mangone, Ralph Joseph *metallurgical engineer*
Mishra, Brajendra *metallurgical engineering educator, researcher*

Littleton
Chamberlin, Paul Davis *metallurgical engineer*

CONNECTICUT

Stamford
Zhong, William Jiang Sheng *ceramic engineer*

Storrs Mansfield
Howes, Trevor Denis *metallurgical engineering educator, researcher*

DISTRICT OF COLUMBIA

Washington
Amey, Earle Bartley *materials engineer*
Lynch, Charles Theodore, Sr. *materials science engineering researcher, administrator, educator*
Reynick, Robert J. *materials scientist*
Schafrik, Robert Edward *materials engineer, information technologist*
Tsao, John Chur *materials engineer, government regulator*

FLORIDA

Indian Harbour Beach
Denaburg, Charles Robert *metallurgical engineer, retired government official*

Miami
Jones, William Kinzy *materials engineering educator*

Orlando
Koch, Kevin Robert *metallurgical engineer*

Palm Beach Gardens
Silver, Gordon Hoffman *metallurgical engineering consultant*

Pensacola
McSwain, Richard Horace *materials engineer, consultant*

GEORGIA

Atlanta
Glasman, Michael Morris *metallurgical engineer*

Monroe
Gammans, James Patrick *ceramic engineer*

Roswell
Walczak, Zbigniew Kazimierz *polymer science and engineering researcher*

ILLINOIS

Argonne
Myles, Kevin Michael *metallurgical engineer*

Crete
Crowley, Michael Summers *ceramic engineer*

Danville
Boone, Thomas John *metallurgical engineer*

Long Grove
Davé, Vipul Bhupendra *polymer engineer*

Springfield
Hahin, Christopher *metallurgical engineer, corrosion engineer*

Urbana
Hagberg, Daniel Scott *ceramic engineer*
Wert, Charles Allen *metallurgical and mining engineering educator*

INDIANA

Kokomo
Lai, George Ying-Dean *metallurgist*

West Lafayette
Bement, Arden Lee, Jr. *engineering educator*

IOWA

Newton
Durant, Gerald Wayne *materials engineer*

KANSAS

Mission
Mortensen, Kenneth Peter *materials engineer, water treatment engineer*

MAINE

Alfred
Pepin, John Nelson *materials research and design engineer*

MARYLAND

Adamstown
Simmins, John James *ceramic engineer, consultant*

Annapolis
Aprigliano, Louis Francis *metallurgist, researcher*

Baltimore
Horowitz, Emanuel *materials science and engineering consultant*

Cambridge
Shipe, Gary Thomas *ceramic engineer*

College Park
Arsenault, Richard Joseph *materials science and engineering educator*
Greene, Charles Arthur *materials engineer*

Columbia
Fan, Xiyun *polymer engineer*
Rice, Roy Warren *ceramic engineer*
Vanderlinde, William Edward *materials engineer*

Gaithersburg
Ricker, Richard Edmond *metallurgical scientist*

MASSACHUSETTS

Brockton
Park, Byiung Jun *textile engineer*

Cambridge
Latanision, Ronald Michael *materials science and engineering educator, consultant*
Rha, ChoKyun *biomaterials scientist and engineer, researcher, educator, inventor*
Russell, Kenneth Calvin *metallurgical engineer, educator*
Szekely, Julian *materials engineering educator*
Thomas, Edwin L. *materials engineering educator*
Wuensch, Bernhardt John *ceramic engineering educator*
Yannas, Ioannis Vassilios *polymer science and engineering educator*

Hopkinton
Novich, Bruce Eric *materials engineer*

Rockland
Manns, Roy Lokumal *polymer technologist, researcher, executive*

MICHIGAN

Ann Arbor
Gamota, Daniel Roman *materials engineer*
Martin, David Charles *materials science engineering educator*
Stickels, Charles Arthur *metallurgical engineer*
Van Vlack, Lawrence Hall *engineering educator*

Cass City
Reeder, Mike Fredrick *materials engineer, consultant*

Detroit
Putatunda, Susil Kumar *metallurgy educator*

MINNESOTA

Minneapolis
Maxton, Robert Connell *metallurgical engineer*

Minnetonka
Johnson, Lennart Ingemar *materials engineering consultant*

MISSOURI

Rolla
Day, Delbert Edwin *ceramic engineering educator*
Peaslee, Kent Dean *metallurgical engineer*

Saint Louis
Farrior, Gilbert Mitchell *metallurgical engineer*
Silverman, David Charles *materials engineer*

MONTANA

Whitehall
Gallagher, Neil Paul *metallurgical engineer*

NEBRASKA

Lincoln
Crews, Patricia Cox *textile scientist, educator*

NEVADA

Fallon
Terry, Charles James *metallurgical engineer*

NEW HAMPSHIRE

Hanover
Queneau, Paul Etienne *metallurgical engineer, educator*

NEW JERSEY

Brielle
Kirby, Gary Neil *metallurgical and materials engineer, consultant*

Dover
Tatyrek, Alfred Frank *materials engineer, research chemist*

Little Falls
Dohr, Donald R. *metallurgical engineer, researcher*

Murray Hill
Quan, Xina *polymer engineer*

Parsippany
Gilby, Steve *metallurgical engineering researcher*
Liebermann, Howard Horst *metallurgical engineering executive*
Wadey, Brian Leu *polymer engineer*

Pennington
Krupowicz, John Joseph *metallurgical engineer*

Piscataway
Ruh, Edwin *ceramic engineer, consultant, researcher*

Vineland
Abendroth, Reinhard Paul *materials engineer*

NEW MEXICO

Los Alamos
Hill, Mary Ann *metallurgical engineer*

NEW YORK

Alfred
Bayya, Shyam Sundar *ceramic engineer*
Frechette, Van Derck *ceramic engineer*
McCauley, James Weymann *ceramics engineer, educator*
Spriggs, Richard Moore *ceramic engineer, research center administrator*

Brooklyn
Margolin, Harold *metallurgical educator*

Buffalo
Hida, George Tiberiu *ceramic engineer*

Corning
Beall, George Halsey *ceramic engineer*

Hopewell Junction
Puttlitz, Karl Joseph, Sr. *metallurgical engineer, consultant*

New Hartford
Maurer, Gernant Elmer *metallurgical executive, consultant*

New York
Cherepakhov, Galina *metallurgical and chemical engineer*
Crosson, Joseph Patrick *metallurgical engineer, consultant*
Morfopoulos, V. *metallurgical engineer, materials engineer*

North Babylon
Tipirneni, Tirumala Rao *metallurgical engineer*

Syracuse
Cabasso, Israel *polymer science educator*
Mi, Yongli *polymer engineer, researcher*
Todd, Roger Harold *metallurgical engineer, failure analyst*

Troy
Dvorak, George J. *materials engineering educator*
Glicksman, Martin Eden *materials engineering educator*

Upton
Czajkowski, Carl Joseph *metallurgist, engineer*
Weeks, John Randel, IV *nuclear engineer, metallurgist*

Waddington
Olszewski, James Frederick *metallurgical engineer*

Wellsville
Hoover, Herbert William, Sr. *foundry engineer*

Woodside
Gurian, Martin Edward *textile engineer*

Yorktown Heights
Romankiw, Lubomyr Taras *materials engineer*
Tu, King-Ning *materials science and engineering educator*

NORTH CAROLINA

Durham
Goodwin, Frank Erik *materials engineer*

Greensboro
Lund, Harold Howard *ceramic engineer, civil engineer, consultant*

Raleigh
Mock, Gary Norman *textile engineering educator*

OHIO

Alliance
Gleixner, Richard Anthony *materials engineer*

Cleveland
Bansal, Narottam Prasad *ceramic research engineer*
Noebe, Ronald Dean *materials research engineer*
Rutledge, Sharon Kay *research engineer*

Columbus
Akbar, Sheikh Ali *materials science and engineering educator*
Alexander, Carl Albert *ceramic engineer*
Huang, Jason Jianzhong *ceramic engineer*
Jackson, Curtis Maitland *metallurgical engineer*
Rapp, Robert Anthony *metallurgical engineering educator, consultant*
St. Pierre, George Roland, Jr. *materials science and engineering administrator, educator*

Dayton
Donaldson, Steven Lee *materials research engineer*
Kumar, Binod *materials engineer, educator*
Saxer, Richard Karl *metallurgical engineer, retired air force officer*

Elmore
Kaczynski, Don *metallurgical engineer*

Fairborn
Moore, Edmund Harvey *materials science and engineering engineer*

Granville
Ploetz, Lawrence Jeffrey *ceramic engineer*

Independence
Emling, William Harold *metallurgical engineer*

Middletown
Sabata, Ashok *materials engineer*

Piqua
Nagarajan, Sundaram *metallurgical engineer*

PENNSYLVANIA

Bethel Park
Korchynsky, Michael *metallurgical engineer*

Bethlehem
Jain, Himanshu *materials science engineering educator*
Pense, Alan Wiggins *metallurgical engineer, academic administrator*

Williams, David Bernard *metallurgical engineer*

Greenville
Kelly, Richard Michael *metallurgical engineer*

Grove City
Zellers, Robert Charles *construction materials engineer, consultant*

Jenkintown
Robinson, Mark Louis *metallurgical engineer*

Latrobe
Conley, Edward Vincent, Jr. *research engineer*

Monroeville
Liu, Yinshi *metallurgical engineer*

New Holland
Ekis, Imants *materials engineer, consultant*

New Kensington
Hunsicker, Harold Yundt *metallurgical engineer*

Philadelphia
Composto, Russell John *materials science engineering educator, researcher*

State College
Marktukanitz, Richard Peter *metallurgical engineer*

University Park
Aplan, Frank Fulton *metallurgical engineering educator*
Carim, Altaf Hyder *materials science and engineering educator*

SOUTH CAROLINA

Greenville
Akpan, Edward *metallurgical engineer*

Hampton
Platts, Francis Holbrook *plastics engineer*

Orangeburg
Graule, Raymond S(iegfried) *metallurgical engineer*

TENNESSEE

Kingsport
Pecorini, Thomas Joseph *materials engineer*

Knoxville
Campbell, William Buford, Jr. *materials engineer, chemist, forensic consultant*
Danko, Joseph Christopher *metals engineer, university official*

Oak Ridge
Holcombe, Cressie Earl, Jr. *ceramic engineer*

TEXAS

Austin
Marcus, Harris Leon *mechanical engineering and materials science educator*
Moulthrop, James Sylvester *research engineer, consultant*

Houston
Kuhlke, William Charles *plastics engineer*
Maligas, Manuel Nick *metallurgical engineer*
Pharr, George Mathews *materials science and engineering educator*
Rypien, David Vincent *materials research engineer*
Sandstrum, Steve D. *engineer, marketing manager*

Longview
Canfield, Glenn, Jr. *metallurgical engineer*

San Antonio
Bose, Animesh *materials scientist, engineer*
Chan, Kwai Shing *materials engineer, researcher*

UTAH

Salt Lake City
Sohn, Hong Yong *metallurgical and chemical engineering educator*

VIRGINIA

Falls Church
Traceski, Frank Theodore *materials engineer*

Richmond
Hanneman, Rodney Elton *metallurgical engineer*

WASHINGTON

Kent
McClure, Allan Howard *materials engineer, space contamination specialist, space materials consultant*

Seattle
Wallem, Daniel Ray *metallurgical engineer*

WISCONSIN

Madison
Loper, Carl Richard, Jr. *metallurgical engineer, educator*

CANADA

ONTARIO

Bramalea
Hornby-Anderson, Sara Ann *metallurgical engineer, marketing professional*

Cornwall
McIntee, Gilbert George *materials testing engineer*

London
Brown, James Douglas *materials engineer, researcher*

QUEBEC

Boucherville
Utracki, L. Adam *polymer engineer*

Montreal
Sutherland, C. A. *metallurgical engineer*
Weir, D. Robert *metallurgical engineer, engineering executive*

MEXICO

Mexico City
Palacios, Joaquin Alquisira *polymer scientist*

AUSTRALIA

Somers
Gifkins, Robert Cecil *materials engineer*

BELGIUM

Brussels
Winand, René Fernand Paul *metallurgy educator*

BRUNEI

Bandai Seri Begawan
Roy, Bimalendu Narayan *ceramic engineering educator*

ENGLAND

Sheffield
Jones, Frank Ralph *materials scientist engineer*

FRANCE

Paris La Defense
Faure, François Michel *metallurgical engineer*

GERMANY

Essen
Stein, Gerald *metallurgical engineer*

Fellbach
Puettmer, Marcus Armin *metallurgical engineer*

INDIA

Pune
Gogate, Kamalakar Chintaman *metallurgist*

JAPAN

Atsugi
Ikuma, Yasuro *ceramics educator, researcher*

Kanazawa
Kawamura, Mitsunori *material scientist, civil engineering educator*

Kyoto
Saegusa, Takeo *polymer scientist*

Nagoya
Abe, Yoshihiro *ceramic engineering educator*

NORWAY

Oslo
Singh, Devendra Pal *polymer scientist*

REPUBLIC OF KOREA

Seoul
Kim, Moon-Il *metallurgical engineering educator*

Taejon
Kim, Sung Chul *polymer engineering educator*

SLOVENIA

Ljubljana
Paulin, Andrej *metallurgical engineer, educator*

SOUTH AFRICA

Northlands
Stacey, Thomas Richard *geotechnical engineer*

TAIWAN

Hsinchu
Chou, Lih-Hsin *materials science and engineering educator*

WALES

Swansea
Wilshire, Brian *materials engineering educator, administrator*

ADDRESS UNPUBLISHED

Brimacombe, James Keith *metallurgical engineering educator, researcher, consultant*
Clarke, Nicholas Charles *metallurgical executive, mineral technologist*
O'Connor, Diane Geralyn Ott *engineer*
Pierce, Robert Raymond *materials engineer, consultant*
Reifsnider, Kenneth Leonard *metallurgist, educator*
Woodbury, Franklin Bennett Wessler *metallurgical engineer*

ENGINEERING: MINING

UNITED STATES

ALASKA

Fairbanks
Nelson, Michael Gordon *mining engineer, educator, consultant*
Sengupta, Mritunjoy *mining engineer, educator*

ARIZONA

Miami
McWaters, Thomas David *mining engineer, consultant*

CALIFORNIA

Sacramento
Hartman, Howard Levi *mining engineering educator, consultant*

Thermal
Johnson, Charles Wayne *mining engineer, mining executive*

COLORADO

Golden
Salamon, Miklos Dezso Gyorgy *mining educator*

Grand Junction
Agapito, J. F. T. *mining engineer, mineralogist*

Lakewood
Lu, Paul Haihsing *mining engineer, geotechnical consultant*

IDAHO

Hayden
Hundhausen, Robert John *mining engineer*

ILLINOIS

Carbondale
Chugh, Yoginder Paul *mining engineering educator*

Champaign
Khan, Latif Akbar *mineral engineer*

Streamwood
Zanbak, Caner *mining engineer, geologist*

INDIANA

Indianapolis
Weaver, Michael Anthony *mining engineer, consultant*

KENTUCKY

Lexington
Adams, Larry Dell *mining engineering executive*
Lineberry, Gene Thomas *mining engineering educator*

London
Geiger, Daniel Jay *mining engineer*

MINNESOTA

Minneapolis
Crouch, Steven L. *mining engineer*

MISSOURI

Saint Louis
Prickett, Gordon Odin *mining, mineral and energy engineer*
Ripp, Bryan Jerome *geological engineer*

NEW MEXICO

Tijeras
Vizcaino, Henry P. *mining engineer, consultant*

NEW YORK

Buffalo
Beasley, Charles Alfred *mining engineer, educator*

New York
Boshkov, Stefan Hristov *mining engineer, educator*
Kvint, Vladimir Lev *mining engineer, economist, educator*
Somasundaran, Ponisseril *mineral engineering and applied science educator, consultant, researcher*

NORTH DAKOTA

Bismarck
Carmichael, Virgil Wesly *mining, civil and geological engineer, former coal company executive*

OKLAHOMA

Tulsa
Achterberg, Ernest Reginald *mining engineer*

PENNSYLVANIA

Ebensburg
Anderson, Mark Edward *mining engineer, consultant*

University Park
Bieniawski, Zdzislaw Tadeusz *mineral engineer, educator, consultant*
Ramani, Raja Venkat *mining engineering educator*

TEXAS

Odessa
Armstrong, Gerald Carver *mining engineer*

WEST VIRGINIA

Morgantown
Schroder, John L., Jr. *retired mining engineer*

AUSTRIA

Leoben
Fettweis, Günter Bernhard Leo *mining engineering educator*

ENGINEERING: NUCLEAR

UNITED STATES

ALABAMA

Birmingham
Giddens, John Madison, Jr. *nuclear engineer*
Scott, Owen Myers, Jr. *nuclear engineer*
Williamson, Edward L. *retired nuclear engineer, consultant*

Huntsville
Boody, Frederick Parker, Jr. *nuclear engineer, optical engineer*

ARIZONA

Phoenix
Brittingham, James Calvin *nuclear engineer*

Tucson
Hetrick, David LeRoy *nuclear engineering educator*

CALIFORNIA

Avila Beach
Riches, Kenneth William *nuclear regulatory engineer*

Livermore
Yatabe, Jon Mikio *nuclear engineer*

Los Angeles
Kumar, Anil *nuclear engineer*

Palo Alto
Taylor, John Joseph *nuclear engineer*
Yagnik, Suresh K. *nuclear engineer*

Pasadena
Weisbin, Charles Richard *nuclear engineer*

Penryn
Bryson, Vern Elrick *nuclear engineer*

San Bernardino
Bauer, Steven Michael *cost containment engineer*

San Diego
Hannaman, George William, Jr. *nuclear/electrical engineer*
Northup, T. Eugene *nuclear engineer*

San Jose
Rao, Atambir Singh *nuclear engineer*
Wilkins, Daniel R. *nuclear engineer, nuclear energy industry executive*

COLORADO

Denver
Gordon, Joseph Wallace *nuclear engineer*

CONNECTICUT

Wilton
Willoughby, William, II *nuclear engineer*

Windsor
McDonald, Michael Shawn *nuclear engineer, consultant*
Scherer, A. Edward *nuclear engineering executive*

DISTRICT OF COLUMBIA

Washington
Bjoro, Edwin Francis, Jr. *nuclear engineer*
Burcham, Jeffrey Anthony *nuclear engineer*
Burnfield, Daniel Lee *engineering program manager*
Fortenberry, Jeffrey Kenton *nuclear engineer*
Goffi, Richard James *nuclear engineer*
Stransky, Robert Joseph, Jr. *nuclear engineer*
Williams, Peter Maclellan *nuclear engineer*

FLORIDA

Boca Raton
Spurlock, Paul Andrew *nuclear engineer*

Gainesville
Anghaie, Samim *nuclear engineer, educator*
Diaz, Nils Juan *nuclear engineering educator*
Vitali, Juan A. *nuclear engineer, consultant*

North Palm Beach
Mathavan, Sudershan Kumar *nuclear power engineer*

Vero Beach
Hungerford, Herbert Eugene *nuclear engineering educator*

GEORGIA

Atlanta
Barker, Michael Dean *nuclear engineer*

Stone Mountain
Quang, Eiping *nuclear engineer*

HAWAII

Honolulu
Mercier, John René *nuclear engineer, health physicist*

IDAHO

Idaho Falls
Hicks, Michael David *nuclear engineer*
Holcombe, Homer Wayne *nuclear quality assurance professional*
Kuan, Pui *nuclear engineer*
Motloch, Chester George *nuclear engineer*

Pocatello
Crawford, Kevan Charles *nuclear engineer, educator*

ILLINOIS

Argonne
Bauer, Theodore Henry *nuclear engineer*
Braun, Joseph Carl *nuclear engineer, scientist*
Depiante, Eduardo Victor *nuclear engineer*
Garner, Patrick Lynn *nuclear engineer*
Reifman, Jaques *nuclear engineer, researcher*

Ashland
Benz, Donald Ray *nuclear safety engineer, researcher*

Chicago
Croon, Gregory Steven *nuclear engineer*
Hamby, Peter Norman *nuclear engineer*
Hsiao, Ming-Yuan *nuclear engineer, researcher*
McCullough, Henry G(lenn) L(uther) *nuclear engineer*
Raney, Leland Wayne *nuclear engineer*

Decatur
Graf, Karl Rockwell *nuclear engineer*

Glen Ellyn
Shah, Nirodh *nuclear engineer*

Gurnee
Vandevender, Robert Lee, II *nuclear engineering consultant*

Plainfield
Martens, Frederick Hilbert *nuclear engineer*

Urbana
Axford, Roy Arthur *nuclear engineering educator*
Miley, George Hunter *nuclear engineering educator*

Westmont
Jones, Dale Leslie *nuclear engineer*

Wilmington
Jackson, Bennie, Jr. *nuclear engineer*

INDIANA

Lafayette
Ott, Karl Otto *nuclear engineering educator, consultant*

IOWA

Ames
Bullen, Daniel Bernard *nuclear engineering educator*

KANSAS

Manhattan
Faw, Richard Earl *nuclear engineering educator*
Simons, Gale Gene *nuclear engineering educator, university administrator*

LOUISIANA

Baton Rouge
Sajo, Erno *nuclear engineer, educator, physicist*

MARYLAND

Bethesda
Hicks, Thomas Erasmo *nuclear engineer*

Columbia
Fu, Gang *nuclear engineer*

Gaithersburg
Juliano, John Joseph *energy systems engineer*

Germantown
Van Houten, Robert *nuclear engineer, consultant*

Lusby
Roxey, Timothy Errol *nuclear engineer, biomedical consultant*

Rockville
Beckjord, Eric Stephen *energy researcher, nuclear engineering educator*

Woodbine
Daniels, Frederick Thomas *reactor engineer*

MASSACHUSETTS

Boston
Canalas, Robert Anthony *nuclear engineer*
Grenzebach, William Southwood *nuclear engineer*

Braintree
Wilson, Christopher T. *nuclear engineer*

Cambridge
Harling, Otto Karl *nuclear engineering educator, researcher*
Rasmussen, Norman Carl *nuclear engineer*

Lowell
Martín, José Ginoris *nuclear engineer*

MICHIGAN

Newport
Kirkland, Matthew Carl *nuclear engineer*

Warren
Yusuf, Siaka Ojo *nuclear engineer*

MISSOURI

Columbia
Kimel, William Robert *engineering educator, university dean*

Rolla
Bolon, Albert Eugene *nuclear engineer, educator*

NEW HAMPSHIRE

Hillsborough
Pearson, William Rowland *retired nuclear engineer*

NEW JERSEY

Fort Lee
Screpetis, Dennis *nuclear engineer, consultant*

Hancocks Bridge
Trencher, Gary Joseph *nuclear engineer*

Princeton
Vann, Joseph Mc Alpin *nuclear engineer*

NEW MEXICO

Albuquerque
Prinja, Anil Kant *nuclear engineering educator*

Los Alamos
Durkee, Joe Worthington, Jr. *nuclear engineer*
Hall, Michael L. *nuclear engineer, computational physicist*
Maraman, William Joseph *nuclear engineering company executive*

Tijeras
Sholtis, Joseph Arnold, Jr. *nuclear engineer, engineering executive, retired military officer*

NEW YORK

Buchanan
Hayes, Charles Victor *nuclear engineer*

Irvington
Ray, James Henry *nuclear engineer*

Troy
Hayes, David Kirk *nuclear engineer*

Upton
Kim, Inn Seock *nuclear engineer*

NORTH CAROLINA

Charlotte
Nemzek, Thomas Alexander *nuclear engineer*

New Hill
Weber, Michael Howard *nuclear control operator*

Raleigh
Bourham, Mohamed Abdelhay *nuclear engineer, educator*
Dudziak, Donald John *nuclear engineer, educator*
Gilligan, John Gerard *nuclear engineer, educator*
Mayo, Robert Michael *nuclear engineering educator, physicist*

OHIO

Cederville
Kennel, Elliot Byron *nuclear engineer*

Cincinnati
Weisman, Joel *nuclear engineering educator, engineering consultant*

Columbus
Aldemir, Tunc *nuclear engineering educator*
Fentiman, Audeen Walters *nuclear engineer, educator*
Miller, Don Wilson *nuclear engineering educator*
Redmond, Robert Francis *nuclear engineering educator*

PENNSYLVANIA

Middletown
Fox, Barry Howard *nuclear engineer*
Hartman, Charles Eugene *nuclear engineer*

Monroeville
Campbell, Donald Acheson *nuclear engineer, consultant*
Penman, Paul Duane *nuclear power laboratory executive*

Murrysville
Linn, Paul Anthony *nuclear engineer*

Pittsburgh
Duong, Victor (Viet) Hong *nuclear engineer*
Griffin, Donald S. *nuclear engineer, consultant*

Shippingport
Bly, Charles Albert *nuclear engineer, research scientist*

Trainer
Martin, Richard Douglas *nuclear engineer, consultant*

University Park
Edwards, Robert Mitchell *nuclear engineering educator*
Haghighat, Alireza *nuclear engineering educator*

SOUTH CAROLINA

Aiken
Hernady, Bertalan Fred *thermonuclear engineer*
Hootman, Harry Edward *nuclear engineer, consultant*
Williamson, Thomas Garnett *nuclear engineering and engineering physics educator*

Columbia
Ahmed, Hassan Juma *nuclear engineer*

Jenkinsville
Loignon, Gerald Arthur, Jr. *nuclear engineer*

North Augusta
Ferguson, Kenneth Lee *nuclear engineer, engineering manager*

TENNESSEE

Brunswick
Panicker, Mathew Mathai *nuclear engineer, educator*

Chattanooga
Durham, Lawrence Bradley *nuclear engineer, consultant, mediator*
Gibson, Kathy Halvey *nuclear reactor technology educator*
Rogers, Ross Frederick, III *nuclear engineer*

Knoxville
Eryurek, Evren *nuclear engineer, researcher*

Oak Ridge
Fontana, Mario H. *nuclear engineer*
Kasten, Paul Rudolph *nuclear engineer, educator*
Korsah, Kofi *nuclear engineer*
Renshaw, Amanda Frances *nuclear engineer*

TEXAS

Austin
Klein, Dale Edward *nuclear engineering educator*
Wehring, Bernard William *nuclear engineering educator*

Dallas
Hamzehee, Hossein Gholi *nuclear engineer*

Houston
Ahmad, Salahuddin *nuclear scientist*

UTAH

Salt Lake City
Bennion, John Stradling *nuclear engineer, consultant*

South Jordan
Brinkerhoff, Lorin C. *nuclear engineer, management and safety consultant*

VIRGINIA

Alexandria
Bockwoldt, Todd Shane *nuclear engineer*

Charlottesville
Dorning, John Joseph *nuclear engineering, engineering physics and applied mathematics educator*
Hanson, Brett Allen *nuclear engineer*

Chesapeake
White, Eugene Thomas, III *nuclear project engineer*

Fairfax
Dam, A. Scott *nuclear/mechanical engineer*

Newport News
Giles, Glenn Ernest, Jr. *nuclear engineer*

Portsmouth
Snyder, Peter James *nuclear engineer*

WISCONSIN

Madison
Carbon, Max William *nuclear engineering educator*

TERRITORIES OF THE UNITED STATES

PUERTO RICO

Rio Piedras
Sardina, Rafael Herminio *nuclear engineer*

CANADA

ONTARIO

Ottawa
Rummery, Terrance Edward *nuclear engineering executive, researcher*

CZECH REPUBLIC

Plzen
Wagner, Karel, Jr. *nuclear engineer, educator*

DENMARK

Fredericia
Jensen, Uffe Steiner *nuclear engineer*

FRANCE

Strasbourg
Muller, Jean-Claude *nuclear engineer*

JAPAN

Kyoto
Sakurai, Akira *nuclear engineer*

ADDRESS UNPUBLISHED

Baggerly, John Lynwood *nuclear engineer*
Carrera, Rodolfo *nuclear engineer*
Ebeling-Koning, Derek Bram *nuclear engineer*

Griffith, Charles Russell *nuclear operations consultant*
Hewitt, John Stringer *nuclear engineer*
Lu, Yingzhong *nuclear engineer, educator, researcher*
Martin, JoAnne Diodato *consulting engineer*
Rohrer, Richard Joseph *nuclear engineer*
Sliger, Rebecca North *nuclear engineer*
Stelzer, John Friedrich *nuclear engineer, researcher*

ENGINEERING: OPTICAL

UNITED STATES

ARIZONA

Tucson
Parks, Robert Edson *optical engineer*

CALIFORNIA

Artesia
Kurtz, Russell Marc *laser/optics engineer*

Carlsbad
Fountain, William David *laser/optics engineer*

Cupertino
Starkweather, Gary Keith *optical engineer, computer company executive*

Hercules
Heffelfinger, David Mark *optical engineer*

Irvine
Wang, Ran-Hong Raymond *optical engineer, scientist*

Livermore
Sheem, Sang Keun *optical engineering researcher*

Pasadena
Isenberg, John Frederick *optical engineer*

Santa Ana
Bentley, William Arthur *engineer, electro-optical consultant*

Santa Clara
Mohr, Siegfried Heinrich *mechanical and optical engineer*

Santa Rosa
Dwight, Herbert M., Jr. *optical engineer, manufacturing executive*
Rancourt, James Daniel *optical engineer*

Sunnyvale
Pugmire, Gregg Thomas *optical engineer*

COLORADO

Louisville
Pesacreta, George Joseph *optical metrology engineer*

CONNECTICUT

Storrs Mansfield
Tang, Qing *optical engineer*

Wilton
Wilson, John Kenneth *optical research professional*

DISTRICT OF COLUMBIA

Washington
Giallorenzi, Thomas Gaetano *optical engineer*
Zhu, Xinming *optical engineer*

FLORIDA

Orange Park
Walsh, Gregory Sheehan *optical systems professional*

Port Richey
Essex, Douglas Michael *optical engineer*

Winter Park
Hartmann, Rudolf *electro-optical engineer*

IDAHO

Idaho Falls
Crawford, Thomas Mark *laser/optics physicist, business owner, consultant*

ILLINOIS

Woodstock
Leibhardt, Edward *optics industry professional*

INDIANA

Fort Wayne
Weinswig, Shepard Arnold *optical engineer*

MARYLAND

Greenbelt
Danks, Anthony Cyril *detector scientist, research astronomer*

MASSACHUSETTS

Lexington
Ehrlich, Daniel Jacob *optical engineer, optical scientist*

MINNESOTA

Faribault
Powers, Kim Dean *optical engineer*

MISSOURI

Saint Louis
Erickson, Robert Anders *optical engineer, physicist*

NEW JERSEY

Lawrenceville
Oana, Harry Jerome *optical engineer*

Murray Hill
Johnson, Bertrand H. *optical engineer*

Teterboro
Ficalora, Joseph Paul *optical engineer*

NEW YORK

Albany
Beik, Mostafa Ali-Akbar *laser, electro-optical and fiber optics research engineer*

Bohemia
Krichever, Mark *optical engineer*

Fairport
DeStefano, Paul Richard *optical engineer*
Phillips, Anthony *optical engineer*

Honeoye Falls
Cuffney, Robert Howard *electro-optical engineer*

Melville
Marchesano, John Edward *electro-optical engineer*

Pleasantville
Pike, John Nazarian *optical engineering consultant*

Poughkeepsie
De Cusatis, Casimer Maurice *fiber optics engineer*

Rochester
Horbatuck, Suzanne Marie *optical engineer*

NORTH CAROLINA

Raleigh
Woodland, N. Joseph *optical engineer, mechanical engineer*

OHIO

Cincinnati
Magarill, Simon *optical engineer*

Dayton
Dugan, John Patrick *optical engineer*

PENNSYLVANIA

Philadelphia
Forsyth, Keith William *optical engineer*

Pittsburgh
Liberman, Irving *optical engineer*

State College
Mao, Xiaoping *fiber optics engineer, researcher*

University Park
Malone, Charles Trescott *photovoltaic specialist*

Wellsville
Shott, Edward Earl *engineer, researcher*

VIRGINIA

Hampton
Burner, Alpheus Wilson, Jr. *optical engineer*

WASHINGTON

Seattle
Diessner, Daniel Joseph *fiber optics engineer*

CHINA

Harbin
Ma, Zuguang *optical scientist, quantum electronics educator*

JAPAN

Tokorozawa
Hashimoto, Nobuyuki *electro-optical engineer*

Tokyo
Sakai, Katsuo *electrophotographic engineer*

SWITZERLAND

Buchs
Jütz, Jakob Johann *applied optics engineering educator*

ADDRESS UNPUBLISHED

Dong, Linda Yanling *optical engineer*

ENGINEERING: OTHER SPECIALTIES

UNITED STATES

ALABAMA

Birmingham
Kumar, Manish *engineer*

Dozier
Grantham, Charles Edward *broadcast engineer*

Huntsville
Buddington, Patricia Arrington *engineer*
Griner, Donald Burke *engineer*
Hung, Ru J. *engineering educator*

Mobile
Wilkinson, Ronald Eugene *engineer*

Redstone Arsenal
Hollowell, Monte J. *engineer, operations research analyst*

Trussville
Davey, James Joseph *process control engineer*

Tuscaloosa
Churchill, Sharon Anne-Kernicky *research engineer, consultant*

ARIZONA

Phoenix
Roop, Mark Edward *process control applications engineer*
Schmitt, Louis Alfred *engineer*

Tempe
Fernando, Harindra Joseph *engineering educator*
McCarthy, David Edward *energy conservation engineer*
Singhal, Avinash Chandra *engineering educator*

Tucson
Massey, Lawrence Jeremiah *broadband engineer*
Sears, William Rees *engineering educator*

ARKANSAS

Russellville
Proffitt, Alan Wayne *engineering educator*

CALIFORNIA

Anaheim
Rivera, Armando Remonte *utilities engineer*

Berkeley
Dornfeld, David A. *engineering educator*
Jewell, William Sylvester *engineering educator*
Popov, Egor Paul *engineering educator*

Boulder Creek
Ruch, Wayne Eugene *microlithography engineer*

Byron
Greengrass, Roy Myron *plant engineer*

Cameron Park
House, Lon William *engineer, energy economist*

Canoga Park
Haymaker, Carlton Luther, Jr. (Bud Haymaker) *metrology engineer*

Chatsworth
Rizvi, Javed *facilities engineer*

Claremont
Cha, Philip Dao *engineering educator*

Coronado
Crilly, Eugene Richard *engineering consultant*

Crescent City
Ghosh, Sid *telecommunications engineer*

Culver City
Wiegand, Stanley Byron *engineer*

Davis
Piedrahita, Raul Humberto *aquacultural engineer*
Ryu, Dewey Doo Young *biochemical engineering educator*

El Cajon
Haubert, Roy A. *former engineer, educator*
Symes, Clifford E. *telecommunications engineer*

El Segundo
Olsen, Donald Paul *communications system engineer*

Gardena
Stuart, Jay William *engineer*

Hawthorne
Mockaitis, Joseph Peter *logistics engineer*

Huntington Beach
Goodrich, Craig Robert *business analyst*

Irvine
Spiberg, Philippe Frederic *research engineer*

La Jolla
Skalak, Richard *engineering mechanics educator, researcher*
Sung, Kuo-Li Paul *bioengineering educator*
Williams, Forman Arthur *combustion theorist, engineering science educator*

Livermore
Brereton, Sandra Joy *engineer*

Long Beach
McCauley, Hugh Wayne *human factors engineer, industrial designer*

Los Alamitos
Karkia, Mohammad Reza *energy engineer, educator*

Los Angeles
Allen, Frederick Graham *consulting engineer*
Cosner, Christopher Mark *engineer*
Goulding, Merrill Keith *engineer, consultant*
Mortensen, Richard Edgar *engineering educator*
Okrent, David *engineering educator*
O'Neill, Russell Richard *engineering educator*
Putt, John Ward *propulsion engineer, retired, consultant*
Remillard, Richard Louis *automotive engineer*
Signoretti, Rudolph George *propulsion engineer, consultant*
Udwadia, Firdaus Erach *engineering educator, consultant*

Menlo Park
Hodges, James Clark *engineer*
Morimoto, Roderick Blaine *research engineer*

Millbrae
Gamlen, James Eli, Jr. *corrosion engineer*

Moreno Valley
Trainor, Paul Vincent *weapon systems engineer, genealogy researcher*

Mountain View
Cooper, David Robert *geotechnical testing professional*

Oakland
Schell, Farrel Loy *transportation engineer*

Palo Alto
Hermsen, Robert William *engineering consultant*
Quate, Calvin Forrest *engineering educator*

Pasadena
Cuk, Slobodan *engineering educator*
Nguyen, Tien Manh *communications systems engineer*
Sahu, Ranajit *engineer*
Smith, Louis *maintenance engineer*
Wynn, Robert Raymond *engineer*

Pico Rivera
Gardner, Stanley *forensic engineer, expert witness*

Poway
Shelby, William Ray Murray *logistics engineer*

Rancho Palos Verdes
Park, Kwan Il *energy engineer*

Riverside
Chang, Andrew C. *engineer, educator*

Rosamond
Stephens, John Walter *facility engineer*

Sacramento
Collins, William Leroy *telecommunications engineer*
Cummings, Belinda *construction engineer*
Mallarapu, Rupa Latha *transportation engineer, consultant*

San Diego
Chouinard, Warren E. *engineer*
Landersman, Stuart David *engineer*
Schmidtmann, Victor Henry *instrumentation engineer*
Shah, Nipulkumar *principal design engineer*
Shoemaker, Patrick Allen *engineer*
Youngs, Jack Marvin *cost engineer*

Santa Cruz
Dunn, Robert Leland *energy engineer*
Lilly, Les J. *micronautics engineer*

Seal Beach
Gironda, A. John, III *engineer*

Springville
Pugh, Paul Franklin *engineer consultant*

Stanford
Bradshaw, Peter *engineering educator*
McCluskey, Edward Joseph *engineering educator*

Weyant, John Peter *engineering economic systems educator*

Sunnyvale
Floersheim, Robert Bruce *military engineer*

Van Nuys
Ahuja, Anil *energy engineer*

Ventura
Wheeler, Harold Alden *retired radio engineer*

Walnut Creek
Loyonnet, Georges-Claude *engineer*

Woodland Hills
Piersol, Allan Gerald *engineer*

COLORADO

Boulder
Frehlich, Rodney George *engineer, researcher*
Gu, Youfan *cryogenic researcher*
Tary, John Joseph *engineer*

Broomfield
Simon, Wayne Eugene *engineer, mathematician*

Colorado Springs
Dougherty, Harry Melville, III *reliability, maintainability engineer*
Herzog, Catherine Anita *process development engineer*

Denver
Devitt, John Lawrence *consulting engineer*
Jacobson, Olof Hildebrand *forensic engineer*
Mc Candless, Bruce, II *engineer, former astronaut*
Stephens, Larry Dean *engineer*

Englewood
Breitenbach, Allan Joseph *geotechnical engineer*
Prichard, Robert Alexander, Jr. *telecommunications engineer*

Estes Park
Friedman, Sander Berl *engineering educator*

Fort Collins
Mader, Douglas Paul *quality educator*
Milhous, Robert Thurlow *hydraulic engineer*

Golden
Jones, Leonard Dale *facilities engineer*
Morgan, Gary Patrick *energy engineer*
Myers, Daryl Ronald *metrology engineer*
Patino, Hugo *food science research engineer*

Littleton
Brychel, Rudolph Myron *engineer*

Louisville
Matthews, Shaw Hall, III *reliability engineer*

Silverthorne
Dillon, Ray William *engineering technician*

U S A F Academy
Quan, Ralph W. *engineer, researcher*
Webb, Steven Garnett *engineering educator*

CONNECTICUT

Danbury
Folchetti, J. Robert *water, wastewater engineer*

Farmington
Maranzano, Miguel Franscisco *engineer*

Hartford
Knibbs, David Ralph *electron microscopist*

New London
Horgan, Peter James *energy engineer*

Sandy Hook
Lukeris, Spiro *engineer, consultant*

Storrs Mansfield
DiBenedetto, Anthony Thomas *engineering educator*

Stratford
Casper, Patricia A. *human factors research engineer*
Lain, Allen Warren *reliability and maintainability engineer*

Waterford
Kandetzki, Carl Arthur *engineer*

West Hartford
Cornell, Robert Witherspoon *engineering consultant*

Wilton
Juran, Joseph Moses *engineer*

DELAWARE

Fenwick Island
Montone, Liber Joseph *engineering consultant*

New Castle
Empson, Cheryl Diane *validation engineer*

Newark
Kikuchi, Shinya *transportation engineer*

Wilmington
Fullerton, Jesse Wilson *applications specialist*

DISTRICT OF COLUMBIA

Washington
Alic, John Anthony *engineer, policy analyst*
Bakalian, Alexander Edward *public health engineer*
Cambel, Ali Bulent *engineering educator*
Carroll, J. Raymond *engineering educator*
Chalmers, Franklin Stevens, Jr. *engineering consultant*
Curry, Robert Michael *broadcast engineer*
DeLuca, Mary *telecommunications engineer*
Eisner, Howard *engineering educator, engineering executive*
Gronbeck, Christopher Elliott *energy engineer*
Grundy, Richard David *engineer*
Hazelrigg, George Arthur, Jr. *engineer*
Howell, Richard Paul, Sr. *transportation engineer*
King, John LaVerne, III *facilities engineer, energy management specialist*
Stanley, Thomas P. *chief engineer*

FLORIDA

Boca Raton
Grant, John Alexander, Jr. *engineering consultant*
Lin, Y. K. *engineer, educator*
Su, Tsung-Chow Joe *engineering educator*

Clearwater
Tanner, Craig Richard *fire and explosion engineer*

Coral Gables
Fung, Kee-Ying *engineer, educator, researcher*

Gainesville
Delfino, Joseph John *environmental engineering sciences educator*
Drucker, Daniel Charles *engineer, educator*
Kurzweg, Ulrich Hermann *engineering science educator*

Jacksonville
Hutchins, Paul Francis, Jr. *energy engineer*

Largo
Cook, Robert Eugene, Sr. *engineer*

Lehigh Acres
Smiley, Joseph Elbert, Jr. *evaluation engineer, librarian*

Longboat Key
Workman, George Henry *engineer, consultant*

Longwood
McAvoy, Gilbert Paul *retired engineer*

Maitland
Fincher, Daryl Wayne *fire protection engineer*

Melbourne
Davidson, Keith Dewayne *corrosion engineer*
Droll, Raymond John *engineer*

Melbourne Beach
Belefant, Arthur *engineer*

Miami
Varley, Reed Brian *fire protection engineer, consultant*

Orlando
Carpenter, Robert Van, Jr. *laser and engineering technician*
Kersten, Robert Donavon *engineering educator, consultant*

Pensacola
Prochaska, Otto *engineer*

Plantation
Hollifield, Christopher Stanford *engineer, consultant*

Sun City Center
Edwards, Paul Beverly *retired science and engineering educator*

Tallahassee
Hall, Houghton Alexander *engineering professional*

Tampa
Crane, Roger Alan *engineering educator*
Zelinski, Joseph John *engineering educator, consultant*

Winter Park
Kerr, James Wilson *engineer*

GEORGIA

Atlanta
Antolovich, Stephen Dale *engineering educator*
Berbenich, William Alfred *electronics engineer*
Gentry, David Raymond *engineer*
Nerem, Robert Michael *engineering educator, consultant*
Smith, David Doyle *engineer, consultant*
Su, Kendall Ling-Chiao *engineering educator*
White, John Austin, Jr. *engineering educator, dean, consultant*

Marietta
Voots, Terry Lynne *technical sales engineer*
Wilks, Ronald *engineer*

Newnan
Jensen, Bruce Alan *control engineer*

Savannah
Hsu, Ming-Yu *engineer, educator*

HAWAII

Ewa Beach
Dizon, Jose Solomon *planning engineer*

Honolulu
Brock, James Melmuth *engineer, venture capitalist*

IDAHO

Idaho Falls
Harris, Robert James *engineer*
Ischay, Christopher Patrick *engineer*
Kerr, Thomas Andrew *senior program engineer*

Pocatello
Huck, Matthew L. *process development engineer*

Shelley
Kimmel, Richard John *engineer*

ILLINOIS

Arlington Heights
Jenny, Daniel P. *retired engineer*

Barrington
Sandu, Constantine *development engineer*

Champaign
Joncich, David Michael *energy engineer*
Shahin, M. Y. *engineering*

Chicago
Acs, Joseph Steven *transportation engineering consultant*
Hoff, Gerald Charles *transportation engineer, planner*
Minkowycz, W. J. *engineering educator*
Russo, Gilberto *engineering educator*
Ryan, Hugh William *engineer, consultant*
Shafer, Kevin Lee *water resources engineer*
Wong, Thomas Tang Yum *engineering educator*
Yeldandi, Veeraínder Antiah *engineer, consultant*

De Kalb
Bow, Sing Tze *engineer, educator*

Evanston
Keer, Leon Morris *engineering educator*
Shah, Surendra Poonamchand *engineering educator, researcher*

Glenview
Sowa, Paul Edward *research engineer*

Mount Prospect
Basar, Ronald John *research engineer, engineering executive*
Scott, Norman L. *engineering consultant*

Naperville
Craigo, Gordon Earl *engineer*

Oak Forest
Kogut, Kenneth Joseph *consulting engineer*

Schaumburg
Loucks, Eric David *water resources engineer, consultant*

Urbana
Bergman, Lawrence Alan *engineering educator*
Larson, Carl Shipley *engineering educator, consultant*
White, Scott Ray *engineering educator*

Wilmette
Hamilton, Wallis Sylvester *hydraulic engineer, consultant*

INDIANA

Anderson
Panchanathan, Viswanathan *product development specialist*

Indianapolis
Bannister, Lance Terry *applications engineer*
Carr, Floyd Eugene *sales engineer*

Kendallville
Caldwell, Andrew Brian *quality control engineer*

Mount Vernon
Sommerfield, Thomas A. *process engineer*

Notre Dame
Skaar, Steven Baard *engineering educator*

Rochester
Grooms, John Merril *research and development engineer*

West Lafayette
Farris, Thomas N. *engineering educator, researcher*
Sadeghi, Farshid *engineering educator*
Short, Dennis Ray *engineering educator*
Singh, Rakesh Kumar *process engineer, educator*

IOWA

Ames
Brown, Robert Grover *engineering educator*
Porter, Max L. *engineering educator*
Sturges, Leroy D. *engineering educator*

Bettendorf
Heyderman, Arthur Jerome *engineer, civilian military employee*

Cedar Rapids
McCall, Daryl Lynn *avionics engineer*

Iowa City
Miller, Richard Keith *engineering educator*

KANSAS

Lawrence
McCabe, John Lee *engineer, educator, writer*

Salina
Reh, John W. *engineer, consultant*
Selm, Robert Prickett *consulting engineer*

Winfield
Docherty, Robert Kelliehan, III *quality assurance engineer, computer instructor*

KENTUCKY

Frankfort
Franke, George Edward *highway engineer*

Louisville
Cornelius, Wayne Anderson *engineering technology educator, consultant*
Frye, Raymond Eugene *geotechnical engineer*
Mueller, Robert William *process engineer*
Ralston, Patricia A. Stark *engineering and computer science educator*

Winchester
Studebaker, John Milton *utilities engineer, consultant, educator*

LOUISIANA

Alexandria
Moore, James Neal *design engineer*

Baton Rouge
Mailander, Michael Peter *biological and agricultural engineering educator*
Parish, Richard Lee *engineer, consultant*

New Orleans
Collins, Harry David *forensic engineering specialist, mechanical and nuclear engineer, retired army officer*
Hallila, Bruce Allan *welding engineer*

Shreveport
Paull, William Bernard *standards engineer, biologist*

MAINE

Orono
Hill, Richard Conrad *engineering educator, energy consultant*

MARYLAND

Annapolis Junction
McClure, Richard Bruce *satellite communications engineer*

Baltimore
Heselton, Kenneth Emery *energy engineer*
Orlitzky, Robert *engineer*
Shah, Shirish Kalyanbhai *computer science, chemistry and environmental science educator*

Cockeysville Hunt Valley
Bilodeau, John Lowell *engineer*

Columbia
Taylor, Scott Thomas *microwave engineer*

Frederick
Bryan, John Leland *retired engineering educator*

Gaithersburg
Beltracchi, Leo *engineer*

Lanham
Hirschhorn, Joel Stephen *engineer*

Patuxent River
Ullom, Lawrence Charles, Jr. *engineer*

Rockville
Liu, Charles Chung-Cha *transportation engineer, consultant*

Severna Park
Mallory, Charles William *consulting engineer, marketing professional*

Silver Spring
Short, Steve Eugene *engineer*

Timonium
Gupta, Ramesh Chandra *geotechnical engineer, consultant*

MASSACHUSETTS

Acton
Egan, Bruce A. *engineering consultant*

Amherst
Breger, Dwayne Steven *solar energy research engineer*

Andover
Rhoads, Kevin George *consulting engineer*

Bedford
Ren, Chung-Li *engineer*

Billerica
Schmidt, James Robert *facilities engineer*

Boston
Banerjee, Ajoy Kumar *engineer, constructor, consultant*
Vickers, Amy *engineer*

Brookline
Katz, Israel *engineering educator, retired*

Cambridge
Crandall, Stephen Harry *engineering educator*
Davidson, Frank Paul *macro-engineer, lawyer*
Dietz, Albert George Henry *engineering educator*
Flemings, Merton Corson *engineer, materials scientist, educator*
Keramas, James George *engineering educator*
Pian, Theodore Hsueh-Huang *engineering educator, consultant*
Wiesner, Jerome Bert *engineering educator, researcher*

Danvers
Tiernan, Robert Joseph *research and development engineer*

Fitchburg
Jackson, Jimmy Lynn *engineer, consulting spectroscopist*

Lenox
Krofta, Milos *engineer*

Lowell
Shina, Sammy Gourgy *engineering educator, consultant*

Lynnfield
Paradis, Richard Robert *energy conservation engineer*

Tewksbury
Dumont, Michael Gerard *electro-optical engineer*

Wayland
Puzella, Angelo *microwave engineer*

Wellfleet
Jentz, John Macdonald *engineer, travel executive*

Wilbraham
Lovell, Walter Carl *engineer, inventor*

Wilmington
Reeves, Barry Lucas *research engineer*

MICHIGAN

Ann Arbor
Bilello, John Charles *materials science and engineering educator*
Haddad, George Ilyas *engineering educator, research scientist*
Schultz, Albert Barry *engineering educator*

Auburn Hills
Nusholtz, Guy Samuel *research engineer*

Big Rapids
Thapa, Khagendra *survey engineering educator*

Birmingham
Williams, Charles Wilmot *engineer, consultant*

Brighton
Gillespie, Shane Patrick *chassis engineer*

Canton
Olsen, Gary Alvin *design engineer*

Dearborn
Duffy, James Joseph *engineer*
Faunce, Mark David *product design engineer*

Detroit
King, Albert I. *bioengineering educator*

East Lansing
Goodman, Erik David *engineering educator*
Jasiuk, Iwona Maria *engineering educator*

Farmington Hills
Rope, Barry Stuart *packaging engineer, consultant*

Flint
Vaishnava, Prem Prakash *engineering educator*

Houghton
Pandit, Sudhakar Madhavrao *engineering educator*

Kalamazoo
Engelmann, Paul Victor *plastics engineering educator*

Milford
Shedlowsky, James Paul *engineer*

Rochester Hills
Hicks, George William *automotive and mechanical engineer*
Wertenberger, Steven Bruce *laser applications engineer*

Saint Joseph
Maley, Wayne Allen *engineering consultant*

Sturgis
Mackay, Edward *engineer*

Troy
Jiao, Jianzhong, Sr. *development engineer, educator*

Warren
Brayer, Robert Marvin *program manager, engineer*
Druschitz, Alan Peter *research engineer*
Kuhns, James Howard *communications engineer*

MINNESOTA

Brooklyn Park
Peterson, Donn Neal *forensic engineer*

Duluth
Wyrick, David Alan *engineering educator, researcher*

Jordan
Lark, Ronald Edwin *logistics support engineer*

Minneapolis
Decoursey, William Leslie *engineer*
Patankar, Suhas V. *engineering educator*

Moorhead
Nezhad, Hameed Gholam *energy management educator*

Saint Louis Park
Speirs, Robert Frank *logistics engineer*

Saint Paul
Hults, Scott Samuel *engineer*
Johnson, Brian Dennis *engineer*
Sheehan, Richard Laurence, Jr. *packaging engineer*

Vadrais Heights
Weyler, Walter Eugen, Jr. *process engineer*

MISSISSIPPI

Escatawpa
Chapel, Theron Theodore *quality assurance engineer*

Hazlehurst
Lowenkamp, William Charles, Jr. *medical device engineer, researcher, consultant*

Horn Lake
Schadrack, William Charles, III *design engineer*

University
Aughenbaugh, Nolan Blaine *engineering educator, consultant*

Vicksburg
Demirbilek, Zeki *research hydraulic engineer*
Ethridge, Loyde Timothy *hydraulic engineer, consultant*

MISSOURI

Columbia
Heldman, Dennis Ray *engineering educator*

Fenton
Richardson, Thomas Hampton *design consulting engineer*

Fort Leonard Wood
Porter, Bruce Jackman *military engineer, computer software engineer*

Kansas City
Signorelli, Joseph *control systems engineer*
Zemansky, Gilbert Marek *hydrogeologist, water quality engineer*

Point Lookout
Allen, Jerry Pat *aviation science educator*

Saint Charles
Ruwwe, William Otto *automotive engineer*

Saint Louis
Antonacci, Anthony Eugene *food corporation engineer*
Bachman, Clifford Albert *engineering specialist, technical consultant*
Emerick, Josephine L. *engineer*
Green, Samuel Isaac *optoelectronic engineer*
Hess, Linda Candace *process control engineer*

Springfield
Rogers, Roddy *geotechnical engineer*

MONTANA

Bozeman
Remington, Scott Alan *laser engineer*

Whitehall
Clark, Steven Joseph *energy engineer, business owner, inventor, author*

NEBRASKA

Omaha
Frisse, Ronald Joseph *telecommunications engineer*

NEVADA

Las Vegas
Snaper, Alvin Allyn *engineer*

NEW HAMPSHIRE

Hanover
Long, Carl Ferdinand *engineering educator*
Yin, John *engineering educator*

Hollis
Riccobono, Juanita Rae *solar energy engineer*

Manchester
Poklemba, Ronald Steven *engineer*

Merrimack
Hower, Philip Leland *semiconductor device engineer*

Portsmouth
Fernald, James Michael *engineer*

NEW JERSEY

Bloomfield
Hutcheon, Clifford Robert *engineer*

Dover
Betta, Mary Beth *engineer*

Edison
Mosher, Frederick Kenneth *engineering consultant*

Fanwood
McNally, Harry John, Jr. *engineer, construction and real estate executive, consultant, researcher, accountant*

Florham Park
Matsen, John Morris *engineer*

Hackensack
Case, Gerard Ramon *drafting technician*

Hoboken
Swern, Frederic Lee *engineering educator*

Hopatcong
Ferderber-Hersonski, Boris Constantin *process engineer*

Hopewell
Basavanhally, Nagesh Ramamoorthy *opto-electronics packaging engineer*

Jersey City
Hernon, Richard Francis *engineer*

Lodi
Melignano, Carmine (Emanuel Melignano) *video engineer*

Newark
Friedland, Bernard *engineer, educator*
Pignataro, Louis James *engineering educator*
Sun, Benedict Ching-San *engineering educator, consultant*

Nutley
Liu, Yu-Jih *engineer*

Piscataway
Boucher, Thomas Owen *engineering educator, researcher*
Mammone, Richard James *engineering educator*

Princeton
Bair, William Alois *engineer*
File, Joseph *research physics engineer*
Glassman, Irvin *mechanical and aeronautical engineering educator, consultant*

Princeton Junction
Lull, William Paul *engineering consultant*

Randolph
Lem, Kwok Wai *research scientist*

Ridgewood
Hassialis, Menelaos Dimitiou *mineral engineer*

Skillman
Shah, Hash N. *plastics technologist, researcher*

Somerville
Pennington, Rodney Lee *engineer*

South Plainfield
Neumann, Robert William *engineer*

West Caldwell
DeFilippis, Carl William *engineer, meteorologist*

Whippany
Schmidtmann, Lucie Ann *engineer*

NEW MEXICO

Albuquerque
Chen, Zhen *research engineer*
Doerr, Stephen Eugene *research engineer*
Fuchs, Beth Ann *research technician*
Leigh, Gerald Garrett *research engineer*
McGowan, John Joseph *energy manager*
Palmer, Miles R *engineering scientist, consultant*

Las Cruces
Foster, Robert Edwin *energy efficiency engineer, consultant*

Santa Fe
Davidson, James Madison, III *engineer, technical manager*

NEW YORK

Albany
Hinge, Adam William *energy efficiency engineer*

Bath
Brabham, Dale Edwin *product engineer*

Bohemia
Kern, Harry *developmental engineer*

Brentwood
Spinillo, Peter Arsenio *energy engineer*

Brooklyn
Payyapilli, John *engineer*
Wang, Yao *engineering educator*

East Hampton
Garrett, Charles Geoffrey Blythe *engineering consultant*

Farmingdale
Fuchs, Sheldon James *plant engineer*

Glens Falls
Rist, Harold Ernest *consulting engineer*

Hauppauge
de Lanerolle, Nimal Gerard *process engineer*

Hopewell Junction
LaPlante, Mark Joseph *laser and electro-optics engineer*

Islip
Takach, Peter Edward *process engineer*

Ithaca
Leibovich, Sidney *engineering educator*

Melville
Chan, Jack-Kang *anti-submarine warfare engineer, mathematician*

New York
Arceo, Thelma Llave *energy engineer*
Boley, Bruno Adrian *engineering educator*
Brodsky, Stanley Martin *engineering technology educator, researcher*
Li, Yao *science educator*
Lieberstein, Melvin *administrative engineer*
Mow, Van C. *engineering educator, researcher*
Salvadori, Mario *mathematical engineer*
Van Hemmen, Hendrik Fokko *vehicle engineer*

Niagara Falls
Butry, Paul John *engineer*

Old Westbury
Colef, Michael *engineering educator*

Pound Ridge
Landis, Pamela Ann Youngman *tribologist*

Purchase
Daniel, Charles Timothy *transportation engineer, consultant*

Riverdale
Jha, Nand Kishore *engineering educator, researcher*

Rochester
Setchell, John Stanford, Jr. *color systems engineer*

Ronkonkoma
Phillips, Kevin John *consulting engineer*
Ranpuria, Kishor Prajaram *wire and cable engineer*

Schenectady
Huening, Walter Carl, Jr. *retired consulting application engineer*

Syracuse
Boston, Louis Russell *packaging engineer*

Troy
Hsu, Cheng *decision sciences and engineering systems educator*
Pegna, Joseph *engineer, educator*
Savic, Michael I. *engineering educator, signal and speech researcher*

Upton
Foerster, Conrad Louis *project engineer*

Vernon
Glover, Anthony Richard *engineer*

White Plains
Valentino, Joseph Vincent *engineering consultant*

NORTH CAROLINA

Durham
Herbel, LeRoy Alec, Jr. *telecommunications engineer*

Fuquay-Varina
Jarman, Scott Allen *plastics engineer*

Greensboro
Adams, Chester Z. *sales engineer*

High Point
Huston, Fred John *automotive engineer*

Raleigh
Gardner, Robin Pierce *engineering educator*
Zorowski, Carl Frank *engineering educator, university administrator*

NORTH DAKOTA

Fargo
Stanislao, Joseph *engineering educator, academic administrator, industrial consultant*

OHIO

Akron
Dwenger, Thomas Andrew *engineer*
Garczewski, Ronald James *transportation engineer*

Athens
Nurre, Joseph Henry *engineering educator*

Avon Lake
Tseng, Hsiung Scott *engineer*

Cincinnati
Frinak, Sheila Jo *engineer*
Papadakis, Constantine N. *engineering educator, dean*

Cleveland
Dendy, Roger Paul *communications engineer*
Madden, James Desmond *forensic engineer*
Ostrach, Simon *engineering educator*
Savinell, Robert F. *engineering educator*
Varley, John Owen *engineer*

Columbus
Howden, David Gordon *engineering educator*
Rokhlin, Stanislav Iosef *engineering educator*

Hamilton
Hergert, David Joseph *engineering educator*

Independence
Evans, George Frederick *consulting engineer*

Miamisburg
Terezakis, Terry Nicholas *power/utility engineer*

Norwich
Ely, Wayne Harrison *broadcast engineer*

Paulding
Riggenbach, Duane Lee *maintenance engineer*

Solon
Ambrose, Robert Micheal *application engineer, consultant*

Toledo
Gerhardinger, Peter F. *engineer*

Worthington
Elgin, Charles Robert *chief technology officer*

Wright Patterson AFB
Wallace, Robert Luther, II *engineer*

OKLAHOMA

Bartlesville
Gill, Anna Margherita Anya *application specialist*

Muskogee
Hatley, Larry J. *plant manager*

Norman
Applegarth, Ronald Wilbert *engineer*

Ponca City
Lewald, Peter Andrew *process engineer*

Stillwater
Minahen, Timothy Malcolm *engineering educator*
Zwerneman, Farrel Jon *engineering educator*

OREGON

Albany
Yau, Te-Lin *corrosion engineer*

Florence
Ericksen, Jerald Laverne *educator, engineering scientist*

Portland
Durland, Sven O. *research and development engineer*
McCartney, Bruce Lloyd *hydraulic engineer, consultant*

PENNSYLVANIA

Alcoa Center
Warchol, Mark Francis Andrew *design engineer*

Bethlehem
Advani, Sunder *engineering educator, university dean*

Bridgeville
Pierce, Jeffrey Leo *power systems engineer, consultant*

Bristol
Raughley, Dean Aaron *process engineer*

Carnegie
Reinhart, Robert Karl *control engineer, consultant*

Colmar
Alexander, Wayne Andrew *product engineer*

Coopersburg
Siess, Alfred Albert, Jr. *engineering consultant, management executive*

Hatboro
Bickel, John Frederick *consulting engineer*

Indiana
Fapohunda, Babatunde Olusegun *energy specialist*

King Of Prussia
Black, Jonathan *bioengineering educator*

Ligonier
Mattern, Gerry A. *engineering consultant*

Mehoopany
Miller, Cliff *engineer*

Middletown
McClintock, Samuel Alan *engineering educator*

Philadelphia
Bozenhardt, Herman Fredrick *control technologist, automation system designer*
Fromm, Eli *engineering educator*
Gross, Kaufman Kennard *applications engineer, consultant*
Ku, Y. H. *engineering educator*

Pittsburgh
Hammack, William S. *chemical engineering educator*
Khonsari, Michael M. *engineering educator*
Reid, Robert H. *engineering consultant*
Vogeley, Clyde Eicher, Jr. *engineering educator, consultant*
Williams, Max Lea, Jr. *engineer, educator*

Scranton
McLean, William George *engineering education consultant*

State College
Ruud, Clayton Olaf *engineering educator*

Tobyhanna
Weinstein, William Steven *technical engineer*

University Park
Dong, Cheng *bioengineering educator*
Fonash, Stephen Joseph *engineering educator*
Wu, Hsien-Jung *research assistant*

Williamsport
Shuch, H. Paul *engineering educator*

Wynnewood
Bordogna, Joseph *educator, engineer*

York
Miller, Donald Kenneth *engineering consultant*
Monson, Raymond Edwin *welding engineer*

RHODE ISLAND

Cranston
Botelho, Robert Gilbert *energy engineer*

Kingston
Patton, Alexander James *engineer, consultant*

Providence
Freund, Lambert Ben *engineering educator, researcher, consultant*
Silverman, Harvey Fox *engineering educator, dean*

SOUTH CAROLINA

Goose Creek
Sullivan, James *consultant*

Greenville
Plumstead, William Charles *testing engineer, consultant*

Hartsville
Terry, Stuart L(ee) *plastics engineer*

TENNESSEE

Cookeville
Chamkha, Ali Jawad *research engineer*
Ventrice, Marie Busck *engineering educator*

Knoxville
Cliff, Steven Burris *engineer*
Mashburn, John Walter *quality control engineer*

Manchester
Howard, Robert P. *propulsion engineering scientist*

Oak Ridge
Davenport, Clyde McCall *development engineer*
Greene, David Lloyd *transportation energy researcher*
Zinkle, Steven John *engineer, researcher*

South Pittsburg
Cordell, Francis Merritt *instrument engineer, consultant*

TEXAS

Austin
Chotiros, Nicholas Pornchai *research engineer*
Fahrenthold, Eric Paul *engineering educator*
Langerman, Scott Miles *geotechnical engineer*
Smith, Daniel Montague *engineer*
Stearns, Fred LeRoy *controls engineer*
Welch, Ashley James *engineering educator*

Brooks AFB
Goodman, Howard Alan *human factors engineer, air force officer*

College Station
Christiansen, Dennis Lee *transportation engineer*
Mathewson, Christopher Colville *engineering geologist, educator*
Williams, Karl Morgan *instrumentation engineer, researcher*

Dallas
Gilbert, Paul H. *engineer, consultant*
O'Neill, Mark Joseph *solar energy engineer*
Wooldridge, Scott Robert *semiconductor process engineer*

El Paso
Coleman, Howard S. *engineer, physicist*

Fort Worth
Ewell, Wallace Edmund *transportation engineer*

Glen Rose
Ragan, James Otis *engineer, consultant*

Houston
Collipp, Bruce Garfield *ocean engineer, consultant*
Grau, Raphael Anthony *reliability engineer*
Lawrence, Carolyn Marie *engineer*
Miele, Angelo *engineering educator, researcher, consultant, author*
Nicastro, David Harlan *forensic engineer, consultant, author*
Smalley, Arthur Louis, Jr. *engineering and construction company executive*
Spanos, Pol Dimitrios *engineering educator*
Thomas, David Wayne *engineer*
Ting, Paul Cheng Tung *research scientist, inventor*

Plano
Good, David Michael *engineer*

Richardson
Gauthier, Jon Lawrence *telecommunications engineer*

San Antonio
Petty, Olive Scott *geophysical engineer*

Sugar Land
Verret, Douglas Peter *semiconductor engineer*

UTAH

Brigham City
George, Russell Joseph *design engineer*

Logan
Folkman, Steven Lee *engineering educator*

Orem
Ghent, Robert Maynard, Jr. *clinical audiologist, engineer, consultant*

Salt Lake City
Huber, Robert John *electrical engineering educator*
Sandquist, Gary Marlin *engineering educator*

VERMONT

Burlington
Andosca, Robert George *engineering educator*

VIRGINIA

Alexandria
Ellison, Thorleif *consulting engineer*
Rimson, Ira Jay *forensic engineer*
Taylor, William Brockenbrough *engineer, consultant, management consultant*

Annandale
Cravens, Gerald McAdoo *telecommunications engineer*

Arlington
Bartlett, Randolph W. *engineer*
Jutras, Larry Mark *engineer*

Blacksburg
Fabrycky, Wolter Joseph *engineering educator, author, industrial and systems engineer*
Jones, Robert Millard *engineering educator*
Song, Ohseop *research engineer*

Chesapeake
Jaques, James Alfred, III *communications engineer*

Fairfax
Bosco, Ronald F. *engineering company executive*
Hampton, Matthew Joseph *engineering technician*

Lynchburg
Latimer, Paul Jerry *non-destructive testing engineer*

Mc Lean
Dickerson, Michael Joe *telecommunications engineer*

Newport News
Earnhardt, Daniel Edwin *automotive engineer*
Schatzel, Robert Mathew *logistics engineer*

Reston
Mumzhiu, Alexander *machine vision systems engineer*

Richmond
Harmon, Charles Winston *energy engineer*

Roanoke
Uhm, Dan *process engineer*

WASHINGTON

Bellevue
Wright, Theodore Otis *forensic engineer*

Richland
Sen, Jyotirmoy (Joe Sen) *control engineer*

Seattle
Davis, Terry Lee *communications systems engineer*
Fischer, Gregory Robert *geotechnical engineer*
Morgan, Jeff *research engineer*
Taylor, Ronald Russell *engineer*

Spokane
Budinger, Frederick Charles *geotechnical engineer*
Nandagopal, Mallur R. *engineer*

Vancouver
Redman, Steven Phillip *engineer*

Walla Walla
DeBroeck, Dennis Alan *design engineer*

Wallula
Cox, Dennis Joseph *engineer*

Woodinville
McGavin, Jock Campbell *airframe design engineer*

WEST VIRGINIA

Charleston
Stone, Jennings Edward *applications engineer*

WISCONSIN

Chippewa Falls
Wos, Carol Elaine *engineer*

Madison
Amundson, Clyde Howard *engineering educator, researcher*
Moses, Gregory Allen *engineering educator*
Olson, Hector Monroy *research support engineer*

Milwaukee
Gopal, Raj *energy systems engineer*
Morgan, Donald George *magnetic separation engineer*

Neenah
Polley, William Alphonse *power systems engineer*

North Freedom
Fausett, Robert Julian *engineering geologist, consultant*

TERRITORIES OF THE UNITED STATES

PUERTO RICO

Bay
Arce-Cacho, Eric Amaury *solar energy engineer, consultant*

Humacao
Love, James Brewster *pharmaceutical engineer*

CANADA

ALBERTA

Edmonton
Bach, Lars *wood products engineer, researcher*

Lethbridge
Simmons, Robert Arthur *engineer, consultant*

NEW BRUNSWICK

Fredericton
Faig, Wolfgang *survey engineer, engineering educator*

ONTARIO

Downsview
Kim, Tae-Chul *foundation engineer*

Kitchener
Mundy, Phillip Carl *engineer*

Mississauga
Crewe, Katherine *engineer*

Ottawa
Bharghava, Vijay *engineer*
Devereaux, William A. *engineer*
Mirza, Shaukat *engineering educator, researcher, consultant*

Toronto
Ham, James Milton *engineering educator*
Rynard, Hugh C. *engineer, engineering executive*

Waterloo
Roulston, David John *engineering educator*

QUEBEC

Boucherville
Marcotte, Michel Claude *geotechnical engineer, consultant*

Sainte-Foy
LeDuy, Anh *engineering educator*

Sillery
Tassé, Yvon Roma *engineer*

SASKATCHEWAN

Regina
Seshadri, Rangaswamy *engineering dean*

ARGENTINA

Bahía Blanca
Cardozo, Miguel Angel *telecommunications engineering educator*

AUSTRIA

Klagenfurt
Melezinek, Adolf *engineering educator*

Vienna
Stadlbauer, Harald Stefan *engineer*

BAHRAIN

Manama
Hassan, Jawad Ebrahim *power engineer, consultant*

BRAZIL

Minas Gerais
Cimbleris, Borisas *engineering educator, writer*

BULGARIA

Sofia
Radev, Ivan Stefanov *electronics engineer*

CHINA

Hefei
Ge, Li-Feng *engineer*

CROATIA

Zagreb
Novaković, Branko Mane *engineering educator*

DENMARK

Copenhagen
Larsen, Ib Hyldstrup *engineer*

Lyngby
Bay, Niels *manufacturing engineering educator*

Skodsborg
Rasmussen, Gunnar *engineer*

ENGLAND

Birmingham
Campbell, John *engineering educator*

Cambridge
Ashby, Michael Farries *engineering educator*
Johnson, Kenneth Langstreth *engineer, educator*

Coventry
Qin, Ning *engineering educator*

Guildford
Spier, Raymond Eric *microbial engineer*

Hampshire
Perera, Uduwanage Dayaratna *plant reliability specialist, engineer*

London
van den Muyzenberg, Laurens *engineer*

Sunbury on Thames
Armstrong, Keith Bernard *engineering consultant*

FINLAND

Espoo
Wuori, Paul Adolf *engineering educator*

Turku
Toivonen, Hannu Tapio *control engineering educator*

FRANCE

Bourgoin
Le, Quang Nam *engineering researcher*

Nantes
Peerhossaini, Mohammad Hassan *engineering and physics educator*

Paris
Daligand, Daniel *engineer, association executive, expert witness*
De Vitry D'Avaucourt, Arnaud *engineer*
Gauvenet, André Jean *engineering educator*
Goupy, Jacques Louis *chemiometrics engineer*

GERMANY

Berlin
Franzke, Hans-Hermann *engineering scientist, educator*
Wallner, Franz *engineer, educator*

Essen
Schilling, Hartmut *engineer*

Munich
Hein, Fritz Eugen *engineer, consultant, architect*
Liepsch, Dieter Walter *engineering educator*

HONG KONG

Kowloon
Lui, Ng Ying Bik *engineering educator, consultant*

IRELAND

Dublin
Bunni, Nael Georges *engineering consultant, international arbitrator, conciliator*

ISRAEL

Haifa
Weihs, Daniel *engineering educator*

ITALY

Assago
Affaticati, Giuseppe Eugenio *telecommunications, instrumentation engineer*

Ispra
Crutzen, Yves Robert *engineer, scientist, educator*

Milan
Bondi, Enrico *engineer*
Gambarini, Grazia Lavinia *engineering educator*

Muggia
Parmesani, Rolando Romano *engineer*

Naples
Carlomagno, Giovanni Maria *engineering educator*

Verona
Frossi, Paolo *engineer, consultant*

JAPAN

Aichi
Sakai, Toshihiko *engineer*
Takizawa, Akira *engineering educator*

Ashiya
Maeda, Yukio *engineering educator*
Yukio, Takeda *engineering educator*

Chiyoda-ku
Uchida, Hideo *engineer*

Chofu
Nozaki, Shinji *engineering educator*

Fukuoka
Yokozeki, Shunsuke *optical engineering educator*

Hamakita
Tsuchiya, Yutaka *photonics engineer, researcher*

Hiratsuka
Manabe, Shunji *control engineering educator*

Ibaraki
Yamada, Keiichi *engineering educator, university official*

Kamakura
Ishida, Osami *microwave engineer*

Kyoto
Makigami, Yasuji *transportation engineering educator*
Tamura, Imao *retired engineering educator*

Meguro-ku
Nakayama, Wataru *engineering educator*
Sakamoto, Munenori *engineer educator, researcher, chemist*

Nagoya
Yoshida, Tohru *science and engineering educator*

Nara
Hayashi, Tadao *engineering educator*

Sagamihara
Fujii, Kozo *engineering educator*

Sendai
Hiroyuki, Hashimoto *engineering educator*

Shizuoka
Tabata, Yukio *engineering researcher*

Takarazuka
Imado, Fumiaki *control engineer*

Tokai-mura
Yamaguchi, Masafumi *researcher*

Tokyo
Aihara, Kazuyuki *mathematical engineering educator*
Hayashi, Taizo *hydraulics researcher, educator*
Iri, Masao *mathematical engineering educator*
Koshi, Masaki *engineering educator*
Saima, Atsushi *science and engineering educator*
Sato, Noriaki *engineering educator*

Tsukuba
Nannichi, Yasuo *engineering educator*

Uji
Shiotsu, Masahiro *engineering educator*

Yokohama
Inoue, Yoshio *engineering educator*
Tanaka, Nobuyoshi *consulting engineering*

Yokosuka
Nakagawa, Kiyoshi *communications engineer*

LIECHTENSTEIN

Schaan
Uebleis, Andreas Michael *engineer*

MALAYSIA

Penang
Das, Kumudeswar *food and biochemical engineering educator*

THE NETHERLANDS

Delft
Prasad, Ramjee *telecommunications scientist, electrical engineer*

Eindhoven
Brussaard, Gerrit *telecommunications engineering educator*

Huizen
le Comte, Corstiaan *radar system engineer*

NORTHERN IRELAND

Lurgan
Johnston, T. Miles G. *broadcast engineer*

NORWAY

Oslo
Brady, M(elvin) Michael *engineer, writer*

THE PHILIPPINES

Manila
Maher, Francis Randolph *engineer, consultant*

PORTUGAL

Lisbon
Santo, Harold Paul *engineer, educator, researcher, consultant, designer*

REPUBLIC OF KOREA

Chonju
Park, Byeong-Jeon *engineering educator*

Yusung
Kim, Jin-Keun *engineering educator*

RUSSIA

Moscow
Bulgak, Vladimir Borisovich *telecommunications engineer*
Valiev, Kamil Akhmetovich *engineering educator*

SINGAPORE

Singapore
Liu, Chang Yu *engineering educator*
Sim, Ah Tee *engineer*

SOUTH AFRICA

Johannesburg
Kirsten, Hendrik Albertus *geotechnical engineer*

SWEDEN

Lulea
Shahnavaz, Houshang *ergonomics educator*

Lund
Johannesson, Thomas Rolf *engineering educator*

Östhammar
Karlsson, Ingemar Harry *engineer*

SWITZERLAND

Baar
Maurer, René *engineer, economist*

Zurich
Giger, Peter *engineer*

TAIWAN

Chung-Li
Hong, Zuu-Chang *engineering educator*

Taichung
Lee, Yung-Ming *educator*

TURKEY

Trabzon
Yavuz, Tahir *engineering educator*

UKRAINE

Dniepropetrovsk
Kukushkin, Vladimir Ivanovich *aviation engineer, educator*

WALES

Cardiff
Morris, William Allan *engineer*

ADDRESS UNPUBLISHED

Andrea, Mario Iacobucci *engineer, scientist, gemologist, appraiser*
Bascom, Willard Newell *research engineer, scientist*
Beckman, Donald A. *engineer*
Bergfield, Gene Raymond *engineering educator*
Berry, Kim Lamar *energy engineer*
Bruneau, Angus A. *engineer*
Chastain, Denise Jean *process engineer, consultant*
Chin, Robert Allen *engineering graphics educator*
Cohen, Robert Fadian *research engineer*
Crossley, Francis Rendel Erskine *engineering educator*
Delap, Bill Jay *engineer, consultant*
Fluss, Harold Shrage *engineer*
Grandi, Attilio *engineering consultant*
Gretzinger, James *engineering consultant*
Hickox, Gary Randolph *pulp and paper engineer*
Hu, Ximing *engineering educator*
Jurczyk, Joanne Monica *technical analyst*
LeMarbe, Edward Stanley *engineering manager, engineer*
Martin, Michael Ray *transportation engineer*
McDermott, Kevin J. *engineering educator, consultant*
McDonough, John Glennon *project engineer*
Morrison, Robert Thomas *engineering consultant*
Neher, Leslie Irwin *engineer, former air force officer*
Nelson, Peter Edward *energy engineer*
Phelps, James Solomon, III *astrodynamic engineer*
Priester, Gayle Boller *engineer, consultant*
Richardson, Donald Charles *engineer, consultant*
Sheppard, William Vernon *transportation engineer*
Stroud, John Franklin *engineering educator, scientist*
Swartzlander, Earl Eugene, Jr. *engineering educator, former electronics company executive*
Tang, Dah-Lain Almon *automatic control engineer, researcher*
Toner, Walter Joseph, Jr. *transportation engineer, financial consultant*
Tsai, Wen-Ying *sculptor, painter, engineer*
von Kutzleben, Siegfried Edwin *engineering consultant*
Waters, Robert George *laser engineer*
Wilson, Harold Mark *power generation developer, marketing specialist*
Zhang, Jianhong *research engineer*

ENGINEERING: PETROLEUM

UNITED STATES

CALIFORNIA

Bakersfield
Starzer, Michael Ray *petroleum engineer*

Los Angeles
Chilingarian, George Varos *petroleum and civil engineering educator*

Sacramento
Cavigli, Henry James *petroleum engineer*

Stanford
Aziz, Khalid *petroleum engineering educator*
Ramey, Henry Jackson, Jr. *petroleum engineering educator*

COLORADO

Golden
Jafari, Bahram Amir *petroleum engineer, consultant*
Poettmann, Frederick Heinz *retired petroleum engineering educator*

Littleton
Gilman, James Russell *petroleum engineer*

KANSAS

Overland Park
Cook, Ronald Lee *petroleum engineer*

Wichita
Hoffman, Bernard Douglas, Jr. *petroleum engineer, company executive*

LOUISIANA

Baton Rouge
Desbrandes, Robert *petroleum engineering educator, consultant*

Thibodaux
Pope, William David, III *pipeline engineer*

MISSISSIPPI

Mississippi State
Sawyer, David Neal *petroleum industry executive*

MISSOURI

Rolla
Numbere, Daopu Thompson *petroleum engineer, educator*

NEW MEXICO

Socorro
Heller, John Phillip *petroleum scientist, educator*

NEW YORK

White Plains
Rozenfeld, Gregory *petroleum engineer*

OHIO

Marietta
Chase, Robert William *petroleum engineering educator, consultant*

OKLAHOMA

Bartlesville
Risley, Allyn W(ayne) *petroleum engineer, manager*

Moore
Moore, Dalton, Jr. *petroleum engineer*

Norman
Menzie, Donald E. *petroleum engineer, educator*

Oklahoma City
McClintic, George Vance, III *petroleum engineer, real estate broker*
Miller, Herbert Dell *petroleum engineer*

Tulsa
Bennett, Curtis Owen *research engineer*
Cobbs, James Harold *engineer, consultant*
Earlougher, Robert Charles, Sr. *petroleum engineer*
Hensley, Jarvis Alan *petroleum engineer*
Mabry, Samuel Stewart *petroleum engineer, consultant*
Warren, Tommy Melvin *petroleum engineer*

SOUTH DAKOTA

Rapid City
Islam, M. Rafiqul *petroleum engineering educator*

TEXAS

Austin
Lake, Larry Wayne *petroleum engineer*

Bellaire
McKinzie, Howard Lee *petroleum engineer*

College Station
Lee, William John *petroleum engineering educator, consultant*

El Campo
Tomlinson, Thomas King *petroleum engineer*

Houston
Matthews, Charles Sedwick *petroleum engineering consultant, research advisor*
Sinclair, A(lbert) Richard *petroleum engineer*
Westby, Timothy Scott *oil company research engineer*

Midland
Groce, James Freelan *petroleum engineer*

Plano
Kean, James Allen *petroleum engineer*

Tyler
Smith, James Edward *petroleum engineer, consultant*

WEST VIRGINIA

Mineral Wells
Prather, Denzil Lewis *petroleum engineer*

CHINA

Beijing
Yang, Cheng-Zhi *petroleum engineer, researcher, educator*

MALAYSIA

Kuala Lumpur
Egbogah, Emmanuel Onu *petroleum engineer, geologist*

ADDRESS UNPUBLISHED

Dalton, Robert Lowry, Jr. *petroleum engineer*
Govier, George Wheeler *petroleum engineer*

ENGINEERING: SAFETY & QUALITY CONTROL

UNITED STATES

ALABAMA

Birmingham
Shanks, Stephen Ray *safety engineering consultant*

ARIZONA

Prescott
Hasbrook, A. Howard *aviation safety engineer, consultant*

CALIFORNIA

Camarillo
MacDonald, Norval (Woodrow) *safety engineer*

Los Angeles
Buffington, Gary Lee Roy *safety standards engineer, construction executive*
Feyl, Susan *safety engineer, educator*

San Diego
Schryver, Bruce John *safety engineer*

CONNECTICUT

Hartford
Braddock, John William *safety engineer*

FLORIDA

Lakeland
Higby, Edward Julian *safety engineer*

Pensacola
Clare, George *safety engineer, systems safety consultant*

Tampa
Meddin, Jeffrey Dean *safety executive*

Winter Park
Granberry, Edwin Phillips, Jr. *safety engineer, consultant*

GEORGIA

Atlanta
O'Driscoll, Jeremiah Joseph, Jr. *safety engineer, consultant*

IDAHO

Boise
Corder, Loren David (Zeke Corder) *quality assurance engineer*

ILLINOIS

Mount Prospect
Breitsameter, Frank John *safety engineer*

Niles
Parikh, Dilip *quality assurance engineer*

LOUISIANA

Lacombe
Mangus, Carl William *technical safety and standards consultant, engineer*

MARYLAND

Baltimore
Alexander, Melvin Taylor *quality assurance engineer, statistician*

Columbia
Bremerman, Michael Vance *reliability engineer*
Miller, Hartman Cyril, Jr. *chemical hazardous material training specialist*

Waldorf
Salvador, Mark Z. *system safety engineer, aerospace engineer*

MASSACHUSETTS

North Chatham
Hiscock, Richard Carson *marine safety investigator*

Sandwich
Mattson, Clarence Russell *safety engineer*

MICHIGAN

Detroit
Gleichman, John Alan *safety and loss control executive*

Plymouth
Grannan, William Stephen *safety engineer, consultant*

MINNESOTA

Minneapolis
Mitchell, Eugene Alexander *safety consultant*
Mulich, Steve Francis *safety engineer*

MISSISSIPPI

Stennis Space Center
Smith, Mary Kay Wilhelm *safety engineer*

MISSOURI

Saint Louis
Brumbaugh, Philip S. *minerals manager, quality control consultant*

NEW HAMPSHIRE

Freedom
Lamb, Henry Grodon *safety engineer*

NEW JERSEY

Hancocks Bridge
Bond, Randall Clay *quality assurance engineer*

Newark
Park, Min-Yong *human factors and safety engineer, educator*

Upper Saddle River
Pavlik, Thomas Joseph *quality control engineer, statistician*

NEW MEXICO

Albuquerque
Spake, Robert Wright *safety engineer, consultant*

NEW YORK

Bronx
Chibbaro, Anthony Joseph *environmental and occupational health and safety professional*

Buffalo
Jerge, Dale Robert *loss control specialist, industrial hygienist*

Great Neck
Gillin, John F. *quality/test engineer*

New York
Cantilli, Edmund Joseph *safety engineer educator, author*

Queensbury
Perry, Leland Charles *quality engineer, quality assurance analyst*

West Seneca
McMahon, James Francis *quality control engineer*

NORTH CAROLINA

Arden
Price, Jeffrey Brian *quality engineer*

Raleigh
Church, Kern Everidge *engineer, consultant*

Rocky Mount
Matthews, Drexel Gene *quality control executive*

OHIO

Hartville
Miday, Stephen Paul *quality assurance engineer*

Toledo
Lohmann, John J. *quality assurance engineer*

OKLAHOMA

Oklahoma City
Thompson, Guy Thomas *safety engineer*

PENNSYLVANIA

Doylestown
Rapaport, Martin Baruch *safety engineer*

Hatfield
Kish, Michael Stephen *safety professional*

Philadelphia
Thorn, James Douglas *safety engineer*

RHODE ISLAND

Warwick
Hunt, James H., Jr. *safety and environmental executive*

SOUTH CAROLINA

Aiken
Voss, Terence J. *human factors scientist, educator*

Anderson
Goodner, Homer Wade *safety risk analysis specialis industrial process system failure risk consultant*

Summerville
Singleton, Joy Ann *quality systems professional*

West Union
Klutz, Anthony Aloysius, Jr. *safety and environmental manager*

TENNESSEE

Kingston
Goranson, Harvey Edward *fire protection engineer*

TEXAS

Amarillo
Keaton, Lawrence Cluer *engineer, consultant*

College Station
Wagner, John Philip *safety engineering educator, science researcher*

Houston
Cizek, John Gary *safety and fire engineer*
Tellez, George Henry *safety professional, consultant*

Kingwood
Bowman, Stephen Wayne *quality assurance engineer, consultant*

Lewisville
Ross, Lesa Moore *quality assurance engineer*

WASHINGTON

Richland
Zimmerman, Richard Orin *safety engineer*

WEST VIRGINIA

Wheeling
Rogo, Kathleen *safety engineer*

CANADA

ALBERTA

Calgary
Merta De Velehrad, Jan *diving and safety engineer, scientist, psychologist, inventor, educator, civil servant*

ADDRESS UNPUBLISHED

Fargnoli, Gregory E. *safety engineer*
Frank, Michael Victor *risk assessment engineer*
Hamel, David Charles *health and safety engineer*
Ruegg, Stephen Lawrence *quality engineer, chemist*

ENGINEERING: SYSTEMS

UNITED STATES

ALABAMA

Huntsville
Duncan, Lisa Sandra *engineer*
Haeussermann, Walter *systems engineer*

CALIFORNIA

Chatsworth
Bearnson, William R. *systems engineer*

Lompoc
Peltekof, Stephan *systems engineer*

Los Angeles
Sepmeyer, Ludwig William *systems engineer, consultant*
Shahab, Salman *systems engineer*

Northridge
Stout, Thomas Melville *control system engineer*

Pasadena
Hess, Ann Marie *systems engineer, electronic data processing specialist*

Rancho Santa Margarita
Sharp, William Charles *systems engineer*

San Diego
Butler, Geoffrey Scott *systems engineer, educator, consultant*

San Jose
Glasgow, J. C. (John Carl Glasgow) *systems engineering executive*
Stahl, Desmond Eugene *systems engineer*

Santa Clara
Oliver, David Jarrell *electronic and systems engineer*

Torrance
Sorstokke, Susan Eileen *systems engineer*

COLORADO

Colorado Springs
Wainionpaa, John William *systems engineer*

Peterson AFB
Rainey, Larry Bruce *systems engineer*

CONNECTICUT

New Fairfield
Daukshus, A. Joseph *systems engineer*

DELAWARE

Newark
Sten, Johannes Walter *controls system engineer, consultant*

DISTRICT OF COLUMBIA

Washington
Koomanoff, Frederick Alan *systems management engineer, researcher*

FLORIDA

Clearwater
Baker, James Burnell *systems engineer*

Gainesville
Shortelle, Kevin James *systems engineer, consultant*

Kennedy Space Center
Baylis, William Thomas *systems logistics engineer*
Hosken, Richard Bruce *systems engineer*
Peck, Joan Kay *systems engineer*

Melbourne
Iacoponi, Michael Joseph *systems engineer*

Winter Park
McAlpine, Kenneth Donald *systems engineer, researcher*

HAWAII

Wahiawa
Imperial, John Vince *systems engineer, consultant*

ILLINOIS

Aurora
Zimmerman, Charles Leonard *systems engineer*

Naperville
Sellers, Lucia Sunhee *systems engineer*

INDIANA

Kokomo
Sissom, John Douglas *systems engineer*

KANSAS

Wichita
Green, James Anton *systems engineer*

MARYLAND

Landover
Tripp, Herbert Alan *systems engineer*

Laurel
Halushynsky, George Dobroslav *systems engineer*

MASSACHUSETTS

Bedford
Latimer, James Hearn *systems engineer*
Maravelias, Peter *systems engineer*
Winter, David Louis *human factors scientist, systems engineer*

Foxboro
Ghosh, Asish *control engineer, consultant*

Somerville
Cordingley, James John *systems engineer*

Wakefield
Robinson, Harold Wendell, Jr. *systems engineer*

Wellesley
Picardi, Anthony Charles *systems engineer, market research consultant*

MICHIGAN

Lansing
Kumar, Sanjay *systems engineer*

MINNESOTA

Maple Grove
Griffith, Patrick Theodore *systems engineer*

MISSISSIPPI

Columbus
Gray, Stanley Randolph, Jr. *systems engineer*

MISSOURI

Saint James
Spurgeon, Earl E. *systems engineer*

NEVADA

Reno
Nassirharand, Amir *systems engineer*

NEW HAMPSHIRE

Nashua
Blatt, Stephen Robert *systems engineer*

NEW JERSEY

Clark
Walsh, Daniel Stephen *systems engineering consultant*

Holmdel
Linker, Kerrie Lynn *systems engineer*
Merrill, Aubrey James *systems and electrical engineer*

Mount Holly
Stabenau, Walter Frank *systems engineer*

Mount Laurel
Passarella, Louis Anthony *systems engineer*

Red Bank
Link, Frederick Charles *systems engineer*

Somerset
Tompa, Gary Steven *systems technology director, material scientist*

NEW MEXICO

Albuquerque
Banning, Ronald Ray *systems engineer*

NEW YORK

Grand Island
Faracca, Michael Patrick *engineer*

OHIO

Cincinnati
Dodd, Steven Louis *systems engineer*

Cleveland
Corrado, David Joseph *systems engineer*

OREGON

Portland
Rhodes, Robert LeRoy *systems engineer*

PENNSYLVANIA

Coatesville
McNally, Mark Matthew *control systems engineer*

Warminster
Leiby, Craig Duane *control systems engineer, electrical design engineer*

TEXAS

College Station
Parlos, Alexander George *systems and control engineer*

Corinth
O'Keefe, Joseph Kirk *systems engineer*

Dallas
Ward, Derek William *lead systems analyst*

Fort Worth
Hughes, Michael Wayne *systems engineer*

Houston
Porcher, Frank Bryan, II *systems engineer*
Wray, Richard Bengt *systems engineer*

Meadows
Jeffrey, Marcus Fannin *control systems engineer*

VERMONT

Williston
Lambert, Michael Irving *systems engineer*

VIRGINIA

Arlington
Benefield, Jeniefer Len *software and systems engineer*
Cottrell, Daniel Edward *systems engineer*
Reierson, James Dutton *systems engineer, physicist*

Fairfax
Ward, Charles Raymond *systems engineer*

Falls Church
Hodge, Donald Ray *systems engineer*

Glen Allen
Smith, Noval Albert, Jr. *systems engineer*

Mc Lean
Beene, Kirk D. *systems engineer*
Ligon, Daisy Matutina *systems engineer*
Starr, Stuart Howard *systems engineer, long range planner*

Williamsburg
Vossel, Richard Alan *systems engineer*

WASHINGTON

Federal Way
Cunningham, John Randolph *systems analyst*

Seattle
Gartz, Paul Ebner *systems engineer*

Yakima
Brown, Randy Lee *systems engineer*

FRANCE

Mont-Saint-Aignan
Gouesbet, Gerard *systems and process engineering educator*

JAPAN

Chiba
Ihara, Hirokzu *systems engineer*

Oita
Yamauchi, Kazuo *systems engineer, researcher*

TAIWAN

Taipei
Lin, Yeou-Lin *systems engineer, consultant*

ADDRESS UNPUBLISHED

Butler, Robert Russell *systems engineer*
Nocks, Randall Ian *systems engineer*
Smith, Mildred Cassandra *systems engineer*
Williams, Ronald Oscar *systems engineer*

EXECUTIVES & SPECIALISTS: CHEMICALS

UNITED STATES

CALIFORNIA

Los Angeles
Dodds, Dale Irvin *chemicals executive*

Mountain View
Cusumano, James Anthony *chemical company executive, former recording artist*

Rancho Mirage
Savin, Ronald Richard *chemical company executive, inventor*

Santa Clarita
Mahler, David *chemical company executive*

COLORADO

Boulder
Daughenbaugh, Randall Jay *chemical company executive*

CONNECTICUT

Stamford
Johnstone, John William, Jr. *chemical company executive*

DELAWARE

Wilmington
DeBlieu, Ivan Knowlton *plastic pipe company executive, consultant*
Galli, Paolo *chemical company executive*
Gossage, Thomas Layton *chemical company executive*
Hollingsworth, David Southerland *chemical company executive*
Irrthum, Henri Emile *chemicals executive, engineer*
MacLachlan, Alexander *chemical company executive*
Marsh, Robert Harry *chemical company executive*

DISTRICT OF COLUMBIA

Washington
Baker, Louis Coombs Weller *chemistry educator, researcher*

FLORIDA

Boca Raton
Boer, F. Peter *chemical company executive*

Sarasota
Grodberg, Marcus Gordon *drug research consultant*

West Palm Beach
Giacco, Alexander Fortunatus *chemical industry executive*

GEORGIA

Atlanta
Bridgeman, David Marsh *chemical company executive*

Brunswick
Iannicelli, Joseph *chemical company executive, consultant*

Marietta
Breese, John Allen *chemical industry executive*

ILLINOIS

Barrington
Vandeberg, John Thomas *chemical company executive*

Chicago
Kelly, Gerald Wayne *chemical coatings company executive*

Des Plaines
Wilks, Alan Delbert *chemical research and technology executive, researcher*

Hinsdale
Martin, Jeffrey Alan *chemical company executive*

Lansing
Mandich, Nenad Vojinov *chemical industry executive*

Northbrook
Dout, Anne Jacqueline *manufacturing company executive*

Ringwood
Stresen-Reuter, Frederick Arthur, II *chemical company communications executive*

INDIANA

Indianapolis
Reilly, Peter C. *chemical company executive*

West Lafayette
Kampen, Emerson *chemical company executive*

LOUISIANA

Baton Rouge
Marvel, John Thomas *chemical company executive*

MAINE

Bristol
Hochgraf, Norman Nicolai *retired chemical company executive*

MASSACHUSETTS

Sudbury
White, Bertram Milton *chemicals executive*

MICHIGAN

Benton Harbor
De Long, Dale Ray *chemicals executive*

Grosse Ile
Frisch, Kurt Charles *educator, administrator*

Midland
Tabor, Theodore Emmett *chemical company research executive*

Mount Clemens
McGregor, Theodore Anthony *chemical company executive*

MISSOURI

Saint Charles
Brahmbhatt, Sudhirkumar *chemical company executive*

NEVADA

Las Vegas
Le Fave, Gene Marion *polymer amd chemical company executive*

NEW JERSEY

Bound Brook
Gould, Donald Everett *chemical company executive*

Bridgewater
Iovine, Carmine P. *chemicals executive*
Weingast, Marvin *laboratory director*

Fords
Chryss, George *chemical company executive, consultant*
Lynch, Charles Andrew *chemical company executive*

Iselin
Garfinkel, Harmon Mark *specialty chemicals company executive*
Smith, Orin Robert *chemical company executive*

New Providence
Thompson, Larry Flack *chemical company executive*

Oakhurst
Rossignol, Roger John *coatings company executive*

Piscataway
Alekman, Stanley Lawrence *chemical company executive*

Rochelle Park
Schapiro, Jerome Bentley *chemical company executive*

Springfield
Adams, James Mills *chemicals executive*

Warren
Barer, Sol Joseph *biotechnology company executive*

Wayne
Burlant, William Jack *chemical company executive*
Wolynic, Edward Thomas *specialty chemicals technology executive*

NEW YORK

Grand Island
Rader, Charles George *chemical company executive*

Hauppauge
Fine, Stanley Sidney *pharmaceuticals and chemicals executive*

Lockport
Schultz, Gerald Alfred *chemical company executive*

New York
Banks, Russell *chemical company executive*

Rush
Eastman, Carolyn Ann *microbiology company executive*

Suffern
Sutherland, George Leslie *retired chemical company executive*

NORTH CAROLINA

Durham
Chilton, Mary-Dell Matchett *chemical company executive*

OHIO

Ashtabula
Bonner, David Calhoun *chemical company executive*

Avon Lake
Maresca, Louis M. *chemicals executive*

Cleveland
Martinek, Frank Joseph *chemical company executive*
Tinker, H(arold) Burnham *chemical company executive*

Hudson
Carman, Charles Jerry *chemical company executive*

Independence
Purcell, Thomas Owen, Jr. *chemical company executive*

Milan
Henry, Joseph Patrick *chemical company executive*

Wickliffe
Bares, William G. *chemical company executive*

PENNSYLVANIA

Allentown
Armor, John N. *chemical company research manager*
Foster, Edward Paul (Ted Foster) *process industries executive*

Bath
Ayers, Joseph Williams *chemical company executive*

Bradford Woods
Allardice, John McCarrell *coatings manufacturing company executive*

Havertown
Tassone, Bruce Anthony *chemical company executive*

Indiana
Jones, Shelley Pryce *chemical company executive*

Malvern
Wong, Raphael Chakching *biotechnologist, executive*

Philadelphia
Dumsha, David Allen *chemical company executive*
Tran, John Kim-Son Tan *chemical senses executive, research administrator*

Pittsburgh
Wehmeier, Helge H. *chemical company executive*

SOUTH CAROLINA

Rock Hill
Mattison, George Chester, Jr. *chemical company executive, consultant*

TENNESSEE

Kingsport
Head, William Iverson, Sr. *retired chemical company executive*

TEXAS

Houston
Cameron, William Duncan *plastic company executive*
Gibson, Michael Addison *chemical engineering company executive*

Pasadena
Brown, Robert Griffith *chemist, chemical engineer, chemicals executive*
Keyworth, Donald Arthur *technical development executive*
Stephens, Sidney Dee *chemical manufacturing company executive*

Spring
Wentzler, Thomas H. *chemical company executive*

Texas City
Chen, Yuan James *chemical company executive*

VIRGINIA

Fairfax
Sowder, Donald Dillard *chemicals executive*

WASHINGTON

Woodland
Brown, Alan Johnson *chemicals executive*

WISCONSIN

Milwaukee
Nagarkatti, Jai Prakash *chemical company executive*

CANADA

QUEBEC

Montreal
Lee, Robert Gum Hong *chemical company executive*

CHILE

Santiago
Seelenberger, Sergio Hernan *chemical company executive*

COLOMBIA

Bogota
Garcia Martinez, Hernando *agrochemical company executive, consultant*
Leon Dub, Marcelo *chemicals executive, entrepreneur*

ENGLAND

Sittingbourne
Thomas, David Burton *chemical company executive*

ITALY

Milan
Poluzzi, Amleto *chemical company consultant*

JAPAN

Ibaraki
Kurobane, Itsuo *chemical company executive*

Kitakyushu
Takeda, Yoshiyuki *chemical company executive*

Minato-ku
Sawada, Hideo *polymer chemistry consultant, chemist*

Tokyo
Aoki, Masamitsu *chemical company executive*
Kawamata, Motoo *chemical company executive*
Ranney, Maurice William *chemical company executive*

THE NETHERLANDS

Sittard
Schreuder, Hein *chemical company executive, business administration educator*

THE PHILIPPINES

Manila
Tan, John K. *chemical company executive, educator*

SAUDI ARABIA

Dhahran
Espy, James William *chemicals executive*

ADDRESS UNPUBLISHED

Austin, Ralph Leroy *chemicals executive*
Benmark, Leslie Ann *chemical company executive*
Ku, Thomas Hsiu-Heng *biochemical and specialty chemical company executive*
Kuiperi, Hans Cornelis *chemical trading company executive*
Riedner, Werner Ludwig Fritz *retired chemicals executive, industrial consultant*
Sella, George John, Jr. *chemical company executive*

EXECUTIVES & SPECIALISTS: COMMUNICATIONS

UNITED STATES

ALABAMA

Birmingham
Elkourie, Paul *telecommunications company manager*

ARIZONA

Phoenix
Clay, Ambrose Whitlock Winston *telecommunications company executive*

CALIFORNIA

Los Angeles
Patel, Chandra Kumar Naranbhai *communications company executive, researcher*

COLORADO

Golden
Brennan, Ann Herlevich *communications professional*

DISTRICT OF COLUMBIA

Washington
Bonsignore, Joseph John *publishing company executive*
Comstock, M(ary) Joan *publishing executive, chemist*
Devine, Katherine *publisher, environmental consultant*
Thompson, H. Brian *telecommunications executive*
Yablonski, Edward Anthony *space and terrestrial telecommunications company executive, educator*

FLORIDA

Delray Beach
Charyk, Joseph Vincent *retired satellite telecommunications executive*

Palm Bay
Cox, David Leon *telecommunications company executive*

ILLINOIS

Orland Park
English, Floyd Leroy *telecommunications company executive*

MASSACHUSETTS

Newton
Thompson, Stephen Arthur *publishing executive*

Pembroke
Heinmiller, Robert H., Jr. *communications company executive*

MICHIGAN

Detroit
Bassett, Leland Kinsey *communications company executive*

MISSOURI

Saint Louis
Folkerts, Dennis Michael *telecommunications specialist*

NEW JERSEY

Bedminster
Hart, Terry Jonathan *communications executive*

Holmdel
Haskell, Barry Geoffry *communications company research administrator*
McCallum, Charles John, Jr. *communications company executive*
West, Earle Huddleston *communications company professional*

Morristown
Lemberg, Howard Lee *telecommunications network research executive*

Murray Hill
Mayo, John Sullivan *telecommunications company executive*

Princeton
Hillier, James *communications executive, researcher*

Red Bank
Chynoweth, Alan Gerald *telecommunications research executive*

South Orange
Joel, Amos Edward, Jr. *telecommunications consultant*

NEW YORK

Brooklyn
Frisch, Ivan Thomas *computer and communications company executive*

Jackson Heights
Michaelson, Herbert Bernard *technical communications consultant*

Melville
Ray, Gordon Thompson *communications executive*

New York
Cookson, Albert Ernest *telephone and telegraph company executive*

NORTH CAROLINA

Charlotte
Hartley, Joe David *communications specialist*

PENNSYLVANIA

Fairless Hills
Campbell, Sharon Lynn *communications company executive*

Hawley
Bruck, James Alvin *telecommunications professional*

TEXAS

Austin
Daniel, Mark Paul *fiber optics network technician*

VIRGINIA

Mc Lean
Frankum, Ronald Bruce *communications executive, entrepreneur*

Springfield
Adams, William B. *consultant*

JAPAN

Kanagawa
Karatsu, Osamu *telephone company research and development executive*

ADDRESS UNPUBLISHED

Dorros, Irwin *retired telecommunications executive*
Frankel, Michael S. *telecommunication and automation sciences executive*
Gramines, Versa Rice *telecommunications professional, retired*
Inman, Cullen Langdon *telecommunications scientist*

EXECUTIVES & SPECIALISTS: COMPUTERS

UNITED STATES

ALABAMA

Huntsville
Dayton, Deane Kraybill *computer company executive*

ARIZONA

Scottsdale
Gall, Donald Alan *data processing executive*

CALIFORNIA

Alameda
Billings, Thomas Neal *computer and publishing executive, management consultant*

Cupertino
Treybig, James G. *computer company executive*

El Segundo
Jacobson, Alexander Donald *technological company executive*

Irvine
Sprimont, Thomas Eugene *computing executive*

Los Altos
Bell, Chester Gordon *computer engineering company executive*

Los Angeles
Bono, Anthony Salvatore Emanuel, II *data processing executive*

Menlo Park
Kandt, Ronald Kirk *computer software company executive, consultant*

Moffett Field
Baldwin, Betty Jo *computer specialist*

Mountain View
Ashford, Robert Louis *computer professional*
Pendleton, Joan Marie *microprocessor designer*

Pasadena
Caine, Stephen Howard *data processing executive*

Redwood City
Jobs, Steven Paul *computer corporation executive*
Rohde, James Vincent *software systems company executive*

San Diego
O'Rourke, Ronald Eugene *computer and electronics industry executive*

San Dimas
Smith, Michael Steven *data processing executive*

San Rafael
Marmann, Sigrid *software development company executive*

COLORADO

Boulder
Jerritts, Stephen G. *computer company executive*

Colorado Springs
Cole, Julian Wayne (Perry Cole) *computer educator, consultant, programmer, analyst*

Golden
Togerson, John Dennis *computer software company executive*

Louisville
Williams, Marsha Kay *data processing executive*

CONNECTICUT

Bloomfield
Johnson, Linda Thelma *computer consultant*

East Hartford
Tanaka, Richard I. *computer products company executive*

Glastonbury
Hujar, Randal Joseph *software company executive, consultant*

DELAWARE

Bethany Beach
Wanzer, Mary Kathryn *computer company executive, consultant*

FLORIDA

Daytona Beach
Page, Armand Ernest *data processing director, consultant*

Jacksonville
Shoup, James Raymond *computer systems consultant*

Orlando
Bunting, Gary Glenn *operations research analyst, educator*
Johnson, Tesla Francis *data processing executive, educator*

Palm Harbor
Schafer, Edward Albert, Jr. *data processing executive*

Sarasota
Lewis, Brian Kreglow *computer consultant*

Temple Terrace
Dobrowolski, Kathleen *data processing executive*

GEORGIA

Atlanta
Reed, Diane Gray *business information service company executive*

HAWAII

Honolulu
Klink, Paul L. *computer company executive*

Kaneohe
May, Richard Paul *data processing professional*

ILLINOIS

Lake Forest
Davidson, Richard Alan *data communications company executive*

Oak Park
Golden, Leslie Morris *software development company executive*

Skokie
Heuer, Margaret B. *data processing coordinator*

MARYLAND

Baltimore
Flagle, Charles Lawrence *pharmaceutical industry software firm executive*

Bethesda
Durek, Thomas Andrew *computer company executive*

Odenton
Mucha, John Frank *data processing professional*

Riverdale
Guetzkow, Daniel Steere *computer company executive*

Rockville
Cunningham, Keith Allen, II *computer services company executive*

Silver Spring
Cramer, Mercade Adonis, Jr. *computer company executive*
Hua, Lulin *technological company executive, research scientist*
Lynch, Sonia *data processing consultant*

Waldorf
Cochran, Ada *data specialist, writer*

MASSACHUSETTS

Bedford
Horowitz, Barry Martin *systems research and engineering company executive*

Belmont
Glines, Stephen Ramey *software industry executive*

Boston
Wolf, William Martin *computer company executive, consultant*

Cambridge
Gay, David Holden *project technician*

Reading
Tuttle, David Bauman *data processing executive*

Westford
Haramundanis, Katherine Leonora *computer and data processing company executive*

Westwood
Thomas, Abdelnour Simon *software company executive*

Woburn
Mehra, Raman Kumar *data processing executive, automation and control engineering researcher*

MICHIGAN

Brighton
Miller, Hugh Thomas *computer consultant*

Grand Rapids
Scott, Richard Lynn *data processing executive*

Saline
Lamson, Evonne Viola *computer software company executive, computer consultant, pastor, Christian education administrator*

MINNESOTA

Minneapolis
Healton, Bruce Carney *data processing executive*

MISSISSIPPI

Jackson
Skelton, Gordon William *data processing executive, educator*

MISSOURI

Kansas City
Van Booven, Judy Lee *data processing manager*

NEW HAMPSHIRE

Exeter
Gray, Christopher Donald *software researcher, author, consultant*

Manchester
Prew, Diane Schmidt *information systems executive*

NEW JERSEY

Fair Lawn
Motin, Revell Judith *data processing executive*

Far Hills
Barnum, Mary Ann Mook *information management manager*

Morristown
Bockian, James Bernard *computer systems executive*

Piscataway
Polinsky, Joseph Thomas *purchasing manager*

Princeton
Cohen, Isaac Louis (Ike Cohen) *data processing executive*

Rumson
Soneira, Raymond M. *computer company executive, scientist*

Secaucus
Brown, Ira Bernard *data processing executive*

NEW MEXICO

Las Cruces
Boyle, Francis William, Jr. *computer company executive, chemistry educator*

NEW YORK

Armonk
Kuehler, Jack Dwyer *computer company executive*
Mc Groddy, James Cleary *computer company executive*

Bethpage
Marrone, Daniel Scott *business, production and quality management educator*

Bronx
Galterio, Louis *healthcare information executive*

Flushing
Weintraub, Joseph *computer company executive*

Jamaica
Srinivasan, Mandayam Paramekanthi *software services executive*

New York
Kilburn, Penelope White *data processing executive*
Miller, Alan *software executive, management specialist*
Tilson, Dorothy Ruth *word processing executive*
Wit, David Edmund *software company executive*

Pittsford
Saini, Vasant Durgadas *computer software company executive*

Stormville
McGee, James Patrick *information industry executive*

Syracuse
Boghosian, Paula der *educator, consultant*

White Plains
Machover, Carl *computer graphics consultant*

Whitesboro
Bulman, William Patrick *data processing executive*

NORTH CAROLINA

Goldsboro
Gravely, Jane Candace *computer company executive*

OHIO

Chagrin Falls
Swaney, Cynthia Ann *medical computer service sales executive, business consultant*

Dayton
Henley, Terry Lew *computer company executive*
Vander Wiel, Kenneth Carlton *computer services company executive*

OKLAHOMA

Oklahoma City
Blackwell, John Adrian, Jr. *computer company executive*

Sapulpa
Powers, Eldon Nathaniel *data processing executive*

OREGON

Albany
Norman, E. Gladys *business computer educator, consultant*

Portland
Lewitt, Miles Martin *computer engineering company executive*

PENNSYLVANIA

Greensburg
Kriss, Joseph James *data processing executive*

Wayne
Martino, Peter Dominic *software company executive, military officer*
Murray, Pamela Alison *quality data processing executive*

RHODE ISLAND

West Greenwich
Breakstone, Robert Albert *consumer products company executive*

SOUTH CAROLINA

Columbia
Newton, Rhonwen Leonard *microcomputer consultant*

TENNESSEE

Memphis
Olcott, Richard Jay *data processing professional*

Nashville
Jones, Regi Wilson *data processing executive*

TEXAS

Austin
Bilby, Curt *computer systems executive*
Curle, Robin Lea *computer software industry executive*
Wagner, William Michael *data processing department administrator*

Carrollton
Graham, Richard Douglas *computer company executive, consultant*

Houston
Engel, James Harry *computer company executive*
McClay, Harvey Curtis *data processing executive*

Humble
Hahne, C. E. (Gene Hahne) *computer services executive*

Richardson
O'Neal, Stephen Michael *information systems consultant*

Roanoke
Dodson, George W. *computer company executive, consultant*

VIRGINIA

Arlington
Nelson, Gary Rohde *computer systems executive*
Rossotti, Charles Ossola *computer software company executive*

Falls Church
Nashman, Alvin Eli *computer company executive*

Gasburg
Morrow, Bruce W. *business executive, consultant*

Glen Allen
Seymour, Harlan Francis *computer services company executive*

Mc Lean
Bluitt, Karen *computer program manager*

Sterling
Fothergill, John Wesley, Jr. *systems engineering and design company executive*

WASHINGTON

Port Townsend
Barnard, Maxwell Kay *inventor, computer systems executive*

Redmond
Gates, William Henry, III *software company executive*

WEST VIRGINIA

Falling Waters
Schellhaas, Robert Wesley *data processing executive, composer, musician*

WISCONSIN

Madison
Pampel, Roland D. *computer company executive*

CANADA

ALBERTA

Calgary
Nghiem, Long Xuan *computer company executive*

BRITISH COLUMBIA

Vancouver
Chu, Allen Yum-Ching *automation company executive, systems consultant*

MEXICO

Mexicali
Diaz Vela, Luis H(umberto) *computer company executive*

BRAZIL

São Paulo
Paris, Tania de Faria Gellert *information technology service executive*

BULGARIA

Sofia
Zahariev, George Kostadinov *computer company executive*

ENGLAND

Basingstoke
Butcher, Brian Ronald *software company executive*

FRANCE

Antony
Seiersen, Nicholas Steen *data processing executive*

Boulogne
Reboul, Jaques Regis *computer company executive*

Puteaux
Bally, Laurent Marie Joseph *software engineering company executive*

GERMANY

Gilching
Jung, Reinhard Paul *computer system company executive*

Hemsbach
Froessl, Horst Waldemar *business executive, data processing developer*

ICELAND

Reykjavik
Sigurdsson, Thordur Baldur *data processing executive*

ITALY

Gorizia
Valdemarin, Livio *computer peripherals company executive*

Treviso
Scardellato, Adriano *software production company executive*

JAPAN

Tokyo
Kobayashi, Susumu *data processing executive, super computer consultant*

SAUDI ARABIA

Riyadh
Al-Sari, Ahmad Mohammad *data processing executive*

SINGAPORE

Singapore
Sebastian, Marcus *software developer*

SWEDEN

Kista
Jacobson, Ivar Hjalmar *data processing executive*

Knivsta
Sigfried, Stefan Bertil *software development company executive, consultant*

SWITZERLAND

Geneva
Bellaiche, Charles Roger *computer company executive*
Helland, Douglas Rolf *intergovernmental organization computer executive*

TAIWAN

Taipei
Liao, Kevin Chii Wen *data processing executive*

TURKEY

Izmir
Figen, I. Sevki *computer company executive*

ADDRESS UNPUBLISHED

De Sofi, Oliver Julius *data processing executive*
Graham, Kirsten R. *information service executive*
Hill, Patrick Ray *power quality consultant*
Kent, Jan Georg *computer consultant*
Kieselmann, Gerhard Maria *data processing executive*
Koelmel, Lorna Lee *data processing executive*
Lerman, Gerald Steven *software company executive*
Nason, Dolores Irene *computer company executive, counselor, eucharistic minister*
Roby, Christina Yen *data processing specialist, educator*
Shumick, Diana Lynn *computer executive*
Smith, Susan Finnegan *computer management coordinator*

EXECUTIVES & SPECIALISTS: ELECTRONICS

UNITED STATES

ARIZONA

Goodyear
Cabaret, Joseph Ronald *electronics company executive*

Litchfield Park
Reid, Ralph Ralston, Jr. *electronics executive, engineer*

Sierra Vista
Fletcher, Craig Steven *electronic technician*

Tempe
Marusiak, Ronald John *quality engineer, electronics executive*
Parkhurst, Charles Lloyd *electronics company executive*

CALIFORNIA

Anaheim
McLuckey, John Alexander, Jr. *electronics company executive*

Buena Park
Parker, Larry Lee *electronics company executive, consultant*

Felton
Denton, Dorothea Mary *electronics company manager*

La Habra
Roberts, Liona Russell, Jr. *electronics engineer, executive*

La Jolla
Penhune, John Paul *science company executive, electrical engineer*

La Mesa
Bourke, Lyle James *electronics company executive, small business owner*

Los Altos
Saraf, Dilip Govind *electronics executive*

Los Angeles
Mettler, Ruben Frederick *former electronics and engineering company executive*

Los Gatos
Nitz, Frederic William *electronics company executive*

Menlo Park
Hogan, Clarence Lester *retired electronics executive*
Saldich, Robert Joseph *electronics company executive*

Monte Sereno
Dalton, Peter John *electronics executive*

Newport Beach
Jones, Roger Wayne *electronics executive*

Novato
Swanson, Lee Richard *computer security executive*

Palo Alto
Gilbert, Keith Duncan *electronics executive*

Palos Verdes Peninsula
Pfund, Edward Theodore, Jr. *electronics company executive*
Wilson, Theodore Henry *retired electronics company executive, aerospace engineer*

Rolling Hills Estates
Lee, James King *technology corporation executive*

San Diego
Meyer, Thomas Robert *television product executive*

San Francisco
Denning, Michael Marion *computer company executive*

San Jose
Peltzer, Douglas Lea *semiconductor device manufacturing company executive*
Stockton, Anderson Berrian *electronics company executive, consultant, genealogist*

San Ysidro
Ito, Shigemasa *electronics executive*

Santa Barbara
Jennings, David Thomas, III *electronics executive, consultant*

Santa Clara
Morgan, James C. *electronics executive*

Sausalito
Elion, Herbert A. *optoelectronics and bioengineering executive, physicist*

Temecula
Roemmele, Brian Karl *electronics, publishing, financial and real estate executive*

Thousand Oaks
Longo, Joseph Thomas *electronics executive*

Torrance
Mann, Michael Martin *electronics company executive*

Valley Ford
Clowes, Garth Anthony *electronics executive, consultant*

COLORADO

Broomfield
Williams, Ronald David *electronics materials executive*

CONNECTICUT

Stamford
Schroeder, Richard Philip *quality control executive, consultant*

Waterford
Hinkle, Muriel Ruth Nelson *naval warfare analysis company executive*

FLORIDA

Fort Lauderdale
Sklar, Alexander *electrical company executive*

Jacksonville
Hudson, Thomas *robotics company executive*
Welch, Philip Burland *electronics company executive*

Orlando
Cates, Harold Thomas *aircraft and electronics company executive*

GEORGIA

Conyers
Kilkelly, Brian Holten *lighting company executive, real estate partner*

Norcross
Pippin, John Eldon *electronics engineer, electronics company executive*

ILLINOIS

Chicago
Backman, Ari Ismo *electronics company executive*
Polydoris, Nicholas George *electronics executive*

Schaumburg
Galvin, Robert W. *electronics executive*
Kenig, Noe *electronics company executive*

INDIANA

Kokomo
Almquist, Donald John *retired electronics company executive*

IOWA

Birmingham
Goudy, James Joseph Ralph *electronics executive, educator*

Decorah
Voltmer, Michael Dale *electric company executive*

MARYLAND

Annapolis
Hospodor, Andrew Thomas *electronics executive*

Baltimore
Deoul, Neal *electronics company executive*
Marcellas, Thomas Wilson *electronics company executive*

Clarksburg
Evans, John Vaughan *satellite laboratory executive, physicist*

College Park
Kaylor, Jefferson Daniel, Jr. *electronics executive*

Laurel
Eaton, Alvin Ralph *research and development administrator, aeronautical and systems engineer*

MASSACHUSETTS

Somerville
Kaltsos, Angelo John *electronics executive, educator, photographer*

MONTANA

Helena
Warren, Christopher Charles *electronics executive*

NEVADA

Incline Village
Strack, Harold Arthur *retired electronics company executive, retired air force officer, planner, analyst, musician*

Las Vegas
Fyfe, Richard Warren *electro-optics executive*

NEW HAMPSHIRE

Nashua
Moskowitz, Ronald *electronics executive*

NEW JERSEY

Edison
Sapoff, Meyer *electronics component manufacturer*

Holmdel
Frenkiel, Richard Henry *electronics company research and development executive*
Kogelnik, Herwig Werner *electronics company executive*

Morristown
Barpal, Isaac Ruben *technology and operations executive*

NEW YORK

Briarcliff Manor
Bingham, J. Peter *electronics executive*

Buffalo
Leland, Harold Robert *research and development corporation executive, electronics engineer*

Hauppauge
Minasy, Arthur John *aerospace and electronic detection systems executive*

New York
Buhks, Ephraim *electronics educator, researcher, consultant*

Rhinecliff
Conklin, John Roger *electronics company executive*

Thornwood
Douglas, Patricia Jeanne *systems designer*

Webster
Duke, Charles Bryan *research and development manufacturing executive, physics educator*

White Plains
Shapiro, Robert M. *electronics company executive*

Yorktown Heights
Clevenger, Larry Alfred *electronics company researcher*

NORTH CAROLINA

Cary
Nyce, David Scott *electronics company executive*

Chapel Hill
Augustine, Brian Howard *electronics researcher*

OHIO

Cincinnati
Anderson, Jerry William, Jr. *technical and business consulting executive, educator*

Cleveland
Tracht, Allen Eric *electronics executive*

Columbus
Chen, Roger Ko-chung *electronics executive*

Dayton
Tighe-Moore, Barbara Jeanne *electronics executive*

Warren
Alli, Richard James, Sr. *electronics executive, service executive*

OREGON

Beaverton
Gerlach, Robert Louis *research and development executive, physicist*

PENNSYLVANIA

Butler
Conforti, Ronald Anthony, Jr. *communications and electronics consultant*

Kingston
Fierman, Gerald Shea *electrical distribution company executive*

Philadelphia
Leodore, Richard Anthony *electronic manufacturing company executive*

Pittsburgh
Andersen, Theodore Selmer *engineering manager*

TEXAS

Austin
Thompson, Lawrence Franklin, Jr. *computer corporation executive*

Dallas
Junkins, Jerry R. *electronics company executive*
Tew, E. James, Jr. *electronics company executive*
Zimmerman, S(amuel) Mort(on) *electrical and electronics engineering executive*

Houston
Klausmeyer, David Michael *scientific instruments manufacturing company executive*

Plano
Andrews, Judy Coker *electronics company executive*
Montgomery, John Henry *electronics company executive*

VIRGINIA

Arlington
McDivitt, James Alton *defense and aerospace executive, astronaut*
Pollock, Neal Jay *electronics executive*

Chesapeake
Perry, Harry Montford *computer engineering executive, retired officer*

WEST VIRGINIA

Scott Depot
Butch, James Nicholas *electronics company executive*

WISCONSIN

New Berlin
Risberg, Robert Lawrence, Sr. *electronics executive, consultant, engineer*

Salem
Lambert, James Allen *industrial electrician*

CANADA

ONTARIO

Ottawa
Goldmann, Nahum *product development executive*

QUEBEC

Montreal
Morrissette, Jean Fernand *electronics company executive*

BELGIUM

Antwerp
De Wandeleer, Patrick Jules *electronics executive*

FINLAND

Espoo
Heikkinen, Raimo Allan *electronics executive*

FRANCE

Chatenay-Malabry
Fabre, Raoul François *electronics company executive*

Paris
Spitz, Erich *electronics industry executive*

HONG KONG

Hong Kong
Lui, Ming Wah *electronics executive*

ISRAEL

Binyamina
Ran, Josef *electronics executive, engineer*

ITALY

Milan
Rodda, Luca *computer company executive, researcher*

JAPAN

Minato-ku
Sekimoto, Tadahiro *electronics company executive*

MALAYSIA

Petaling
Ong, Boon Kheng *electronics executive*

THE NETHERLANDS

Leiden
den Breejen, Jan-Dirk *computer integrated manufacturing educator*

NORWAY

Oslo
Skjaerstad, Ragnar *electronics company executive*

POLAND

Bielsko-Biala
Lazar, Maciej Alan *electronics executive, engineer*

SWEDEN

Vällingby
Hallberg, Bengt Olof *fiber optic company executive and specialist*

SWITZERLAND

Fribourg
Hatschek, Rudolf Alexander *electronics company executive*

Grandevent
Karpinski, Jacek *computer company executive*

Grenchen
Siraut, Philippe C. *watch and electronics company executive*

TAIWAN

Hsinchu
Huang, Yang-Tung *electronics educator, consultant*

ADDRESS UNPUBLISHED

Bratton, William Edward *electronics executive, management consultant*
Paxton, John Wesley *electronics company executive*
Pütsep, Peeter Ervin *electronics executive*
Rice, Eric Edward *technologies executive*
Young, John Alan *electronics company executive*

EXECUTIVES & SPECIALISTS: ENGINEERING

UNITED STATES

ALABAMA

Birmingham
Johnson, James Hodge *engineering company executive*

Huntsville
Chassay, Roger Paul, Jr. *engineering executive, project manager*
Douillard, Paul Arthur *engineering and financial executive, consultant*
Emerson, William Kary *engineering company executive*

ARIZONA

Scottsdale
Gilson, Arnold Leslie *engineering executive*

Sedona
Dorrell, Vernon Andrew *engineering executive*
Silvern, Leonard Charles *engineering executive*

Sun City
Vander Molen, Jack Jacobus *engineering executive, consultant*

Tempe
Roberts, Peter Christopher Tudor *engineering executive*

Tucson
Wyant, James Clair *engineering company executive, educator*

CALIFORNIA

Alhambra
Luk, King Sing *engineering company executive, educator*

Anaheim
Hubbard, Charles Ronald *engineering executive*

Berkeley
Ott, David Michael *engineering company executive*
Zwoyer, Eugene Milton *consulting engineering executive*

Burbank
Mullin, Sherman N. *engineering executive*

Chatsworth
Lin, Ching-Fang *engineering executive*

Costa Mesa
Cross, Glenn Laban *engineering executive, development planner*

El Segundo
Plummer, James Walter *engineering company executive*

Irvine
Hess, Cecil F. *engineering executive*
Kontny, Vincent *engineering and construction company executive*

Long Beach
Dillon, Michael Earl *engineering executive*

Los Altos
Bergrun, Norman Riley *aerospace executive*

Los Angeles
Leal, George D. *engineering company executive*
Ramo, Simon *engineering executive*

Menlo Park
Kohne, Richard Edward *retired engineering executive*

Mission Hills
Cramer, Frank Brown *engineering executive, combustion engineer, systems consultant*

Moffett Field
Albers, James Arthur *engineering executive*

Newport Beach
Maddock, Thomas Smothers *engineering company executive, civil engineer*

Pasadena
Lesh, James Richard *engineering manager*

Paso Robles
Schaberg, Burl Rowland, Jr. *engineering company executive*
Still, Harold Henry, Jr. *engineering company executive*

Pittsburg
Weed, Ronald De Vern *engineering consulting company executive*

Redondo Beach
Wright, Charles P. *engineering manager*
Yang, Jane Jan-Jan *engineering executive*

Sacramento
Bailey, Thomas Everett *engineering company executive*

San Diego
Beyster, John Robert *engineering company executive*

San Francisco
Bechtel, Stephen Davison, Jr. *engineering company executive*
Dolby, Ray Milton *engineering company executive, electrical engineer*
Forsen, Harold Kay *engineering executive*
Morrin, Thomas Harvey *engineering research company executive*

Santa Clarita
Granlund, Thomas Arthur *engineering executive, consultant*

Torrance
Howard, Donivan R(ichard) *engineering executive*

Villa Grande
Shirilau, Jeffery Micheal *engineering executive*

Vista
Maggay, Isidore, III *engineering executive, food processing engineer*

Westlake Village
Hokana, Gregory Howard *engineering executive*

COLORADO

Denver
Bradford, Phillips Verner *engineering research executive*
Kopke, Monte Ford *engineering executive*
Poirot, James Wesley *engineering company executive*

CONNECTICUT

East Hartford
Cassidy, John Francis, Jr. *research center executive*

Glastonbury
Bates, Stephen Cuyler *research engineering executive*

North Stonington
Mollegen, Albert Theodore, Jr. *engineering company executive*

Rocky Hill
Chuang, Frank Shiunn-Jea *engineering executive, consultant*

Stamford
Deneberg, Jeffrey N. *engineering executive*

Windsor
Gianakos, Nicholas *engineering executive, consultant*

DELAWARE

Newark
Urquhart, Andrew Willard *engineering executive, metallurgist*

Smyrna
Hutchison, James Arthur, Jr. *engineering company executive*

Wilmington
Kuebeler, Glenn Charles *engineering executive*

DISTRICT OF COLUMBIA

Washington
Burton, William Joseph *engineering executive*
Chen, Ho-Hong H. H. *industrial engineering executive, educator*
Small, Albert Harrison *engineering executive*
Wessel, James Kenneth *engineering executive*

FLORIDA

Cocoa Beach
Gurr, Clifton Lee *engineering executive*

Fern Park
Arnowitz, Leonard *engineering executive*

Fort Lauderdale
Fishe, Gerald Raymond Aylmer *engineering executive*

Miami
Shenouda, George Samaan *engineering executive, consultant*

Orlando
Carter, Thomas Allen *engineering executive, consultant*
Kira, Gerald Glenn *engineering executive*

Tampa
Couret, Rafael Manuel *electrical engineering executive*

West Palm Beach
Koff, Bernard L. *engineering executive*

Winter Park
Sayed, Sayed M. *engineering executive*

GEORGIA

Atlanta
Bonnet, Henri *engineering manager*

Marietta
Jordan, George Washington, Jr. *engineering company executive*

Norcross
Zaloga, Robert Edwin *engineering executive*

IDAHO

Idaho Falls
Epstein, Jonathan Stone *engineering executive*

ILLINOIS

Chicago
Chaiyabhat, Win *engineering manager*
Hobbs, Marvin *engineering executive*
Rikoski, Richard Anthony *engineering executive, electrical engineer*
Tijunelis, Donatas *engineering executive*

Oak Brook
Merola, Raymond Anthony *engineering executive*

Rolling Meadows
Wyslotsky, Ihor *engineering company executive*

Roscoe
Jacobs, Richard Dearborn *consulting engineering firm executive*

Skokie
Corley, William Gene *engineering research executive*

Wood River
Stevens, Robert Edward *engineering company executive*

INDIANA

Columbus
Kamo, Roy *engineering company executive*

Fort Wayne
Mills, Rodney Daniel *engineering company executive*

Indianapolis
Scopatz, Stephen David *engineering executive, educator*

Kokomo
Shimanek, Ronald Wenzel *engineering executive*

West Lafayette
Pritsker, A. Alan B. *engineering executive, educator*

IOWA

Davenport
Bartlett, Peter Greenough *engineering company executive*

KANSAS

Independence
Barbi, Josef Walter *engineering, manufacturing and export companies executive*

Overland Park
Baker, Charles H. *engineering company executive*

Wichita
Kice, John Edward *engineering executive*

KENTUCKY

Louisville
Krainer, Edward Frank *engineering executive*

LOUISIANA

Baton Rouge
Moody, Gene Byron *engineering executive, small business owner*

Lafayette
Salters, Richard Stewart *engineering company executive*

Lake Charles
Levingston, Ernest Lee *engineering executive*

Metairie
Nicoladis, Michael Frank *engineering company executive*

MARYLAND

Annapolis
Henderson, William Boyd *engineering consulting company executive*

Bel Air
Powers, Doris Hurt *engineering company executive*

Bethesda
Bernstein, Bernard *engineering executive*
Jamieson, John Anthony *engineering consulting company executive*

Clarksburg
Arnold, Jay *engineering executive*

Columbia
Bustard, Thomas Stratton *engineering company executive*
Roby, Richard Joseph *research engineering executive, educator*

Greenbelt
Steiner, Mark David *engineering manager*

Hanover
Kistler, Alan Lewis *engineering executive*

Lanham
Black, Clinton James *engineering company executive*

Rockville
Norbedo, Anthony Julius *engineering executive*
Scearce, P. Jennings, Jr. *engineering executive*

Silver Spring
Kidwell, Michael Eades *engineering executive*

MASSACHUSETTS

Bedford
Johansen, Jack T. *engineering company executive*

Boston
Pandelidis, Ioannis O. *engineering manager*
Timmerman, Robert Wilson *engineering executive*

Cambridge
Weimar, Robert Alden *environmental engineering executive*

Framingham
Cornell, Charles Alfred *engineering executive*

Holliston
Epstein, Scott Mitchell *engineering executive*

Lexington
Cooper, William Eugene *consulting engineering executive*

Newton
Barclay, Stanton Dewitt *engineering executive, consultant*

North Attleboro
Corridori, Anthony Joseph *engineering executive*

Reading
Melconian, Jerry Ohanes *engineering executive*

Sudbury
Wallace, James Jr. *engineering executive, researcher*

Wakefield
By, Andre Bernard *engineering executive, research scientist*

Wayland
Clayton, John *engineering executive*

MICHIGAN

Ann Arbor
Smith, Donald Norbert *engineering executive*

Dearborn
Chou, Clifford Chi Fong *research engineering executive*

Midland
Carson, Gordon Bloom *engineering executive*

MINNESOTA

Minneapolis
Malchow, Douglas Byron *engineering executive*

Rochester
Gehling, Michael Paul *engineering executive*

MISSISSIPPI

Stennis Space Center
Nunn, James Ross *engineering executive*

MISSOURI

Chesterfield
Preissner, Edgar Daryl *engineering executive*

Earth City
Puetz, William Charles *engineering company executive*

Fenton
Bubash, James Edward *engineering executive, entrepreneur, inventor*

Kansas City
Wade, Robert Glenn *engineering executive*

Saint Louis
Brunstrom, Gerald Ray *engineering executive, consultant*

NEBRASKA

Brownville
Dingman, Norman Ray *engineering executive*

Lincoln
Elias, Samy E. G. *engineering executive*

Omaha
Maystrick, David Paul *engineering executive*

NEVADA

Las Vegas
Messenger, George Clement *engineering executive, consultant*

NEW JERSEY

Cherry Hill
McCabe, Robert James *engineering executive*
Parker, Jack Royal *engineering executive*

Green Village
Castenschiold, Rene *engineering company executive, author, consultant*

Hewitt
Selwyn, Donald *engineering administrator, researcher, inventor, educator*

Parsippany
Marscher, William Donnelly *engineering company executive*

Passaic
Lindholm, Clifford Falstrom, II *engineering executive, mayor*

Princeton
Stabenau, M. Catherine *engineering executive*

Towaco
Pasquale, Frank Anthony *engineering executive*

Upper Saddle River
Paley, Steven Jann *engineering executive*

Voorhees
Fischette, Michael Thomas *engineering executive*

Wayne
Arturi, Anthony Joseph *engineering executive, consultant*

NEW MEXICO

Santa Fe
Moellenbeck, Albert John, Jr. *engineering executive*

NEW YORK

Brooklyn
Milbury, Thomas Giberson *engineering executive*

Buffalo
McWilliams, C. Paul, Jr. *engineering executive*

East Syracuse
Landsberg, Dennis Robert *engineering executive, consultant*

Long Island City
Hancock, William Marvin *engineering executive*

New York
Aviv, David Gordon *electronics engineering executive*
Baum, Richard Theodore *engineering executive*
Fogel, Irving Martin *consulting engineering*
Grunes, Robert Lewis *engineering consulting firm executive*
Hennessy, John Francis, III *engineering executive, mechanical engineer*
Ross, Donald Edward *engineering company executive*
Schmeltz, Edward James *engineering executive*
Stasior, William F. *engineering company executive*

Newton Falls
Hunter, William Schmidt *engineering executive, environmental engineer*

Schenectady
Wheeler, George Charles *consulting company executive*

Yorktown Heights
Dennard, Robert Heath *engineering executive, scientist*

NORTH CAROLINA

Fayetteville
Sloggy, John Edward *engineering executive*

Research Triangle Park
Kuhn, Matthew *engineering company executive*

OHIO

Athens
Beale, William Taylor *engineering company executive*

Brookpark
Gooch, Lawrence Lee *astronautical engineering executive*

Cincinnati
Bosley, David Calvin *design engineering executive*
Bry, Pierre François *engineering manager*
Madson, Philip Ward *engineering executive*
Ungers, Leslie Joseph *engineering executive*

Cleveland
Noneman, Edward E. *engineering executive*
Webb, John Allen, Jr. *engineering executive*

Columbus
Kapadia, Mehernosh Minocheher *engineering executive*

Dayton
Krug, Maurice F. *engineering company executive*

Newark
Green, John David *engineering executive*

Toledo
Waldfogel, LaRue Verl *electrical engineering executive*

Westlake
Kroll, Casimer V. *engineering executive*

Yellow Springs
Schmidt, William Joseph *engineering executive*

OREGON

Eugene
Williams, Francis Leon *engineering executive*

Salem
Dixon, Robert Gene *engineering company executive, engineering educator*

PENNSYLVANIA

Broomall
Schonbach, Bernard Harvey *engineering executive*

Charleroi
Teaford, Norman Baker *engineering manager*

Clarks Summit
Miniutti, Robert Leonard *engineering company executive*

Coraopolis
Rabosky, Joseph George *engineering consulting company executive*

Fairviewvill
Kisielowski, Eugene *engineering executive*

Forest City
Fleming, Lawrence Thomas *engineering executive*

New Kensington
Evancho, Joseph William *engineering executive, metallurgical engineer*

Philadelphia
Munson, Janis Elizabeth Tremblay *engineering company executive*

Pittsburgh
Cavalet, James Roger *engineering executive, consultant*
Gilbert, Ralph Whitmel, Jr. *engineering company executive*
Lengyel, Joseph William *engineering manager*
O'Donnell, William James *engineering executive*
Turbeville, Robert Morris *engineering executive*

State College
Hall, David Lee *engineering executive*

Willow Street
Ebling, Glenn Russell *energy conservation executive*

Wynnewood
Schmaus, Siegfried H. A. *engineering executive*

RHODE ISLAND

Cranston
Fang, Pen Jeng *engineering executive and consultant*

SOUTH CAROLINA

Aiken
Murphy, Edward Thomas *engineering executive*
Rood, Robert Eugene *construction engineering executive, consultant*

Charleston
Parker, Charles Dean, Jr. *engineering company executive*

Greer
Baker, Stephen Holbrooke *quality engineering executive*

Rock Hill
Waked, Robert Jean *chemical engineering executive*

Spartanburg
Lee, Gary L. *engineering executive*

TENNESSEE

Oak Ridge
Waters, Dean A. *engineering executive*

Tullahoma
Pate, Samuel Ralph *engineering corporation executive*
Wu, Ying Chu Lin Susan *engineering company executive, engineer*

TEXAS

Austin
Sturdevant, Wayne Alan *engineering manager*

Dallas
Giesen, Herman Mills *engineering executive, consultant, mechanical forensic engineer*
Lutz, Robert Brady, Jr. *engineering executive, consultant*
Schulze, Richard Hans *engineering executive, environmental engineer*

Fort Worth
Bhatia, Deepak Hazarilal *engineering executive*
Kent, D. Randall, Jr. *engineerng company executive*
Romine, Thomas Beeson, Jr. *consulting engineering executive*
Vick, John *engineering executive*

Granbury
Crittenden, Calvin Clyde *retired engineering executive*

Greenville
Johnston, John Thomas *engineering executive*

Houston
Boyce, Meherwan Phiroz *engineering executive, consultant*
Landrum, Hugh Linson, Jr. *engineering executive*
Montijo, Ralph Elias, Jr. *engineering executive*
Schultz, Philip Stephen *engineering executive*
Tiras, Herbert Gerald *engineering executive*
Tucker, Randolph Wadsworth *engineering executive*

Richardson
Kinsman, Frank Ellwood *engineering executive*

San Antonio
Abramson, Hyman Norman *engineering and science research executive*
Ostmo, David Charles *engineering executive*

UTAH

Salt Lake City
Zeamer, Richard Jere *engineer, executive*

VIRGINIA

Alexandria
Ackerman, Roy Alan *research and development executive*

Arlington
Flynn, Thomas R. *engineering executive*

Fairfax
Boone, James Virgil *engineering executive*
Lovell, Robert R(oland) *engineering executive*

Falls Church
Villarreal, Carlos Castaneda *engineering executive*

Great Falls
Haines, Andrew Lyon *engineering company executive*

Lynchburg
Kovach, James Michael *engineering executive*

Mc Lean
Schmeidler, Neal Francis *engineering executive*
Sinha, Agam Nath *engineering management executive*

Middletown
Kisak, Paul Francis *engineering company executive*

Newington
Foster, Eugene Lewis *engineering executive*

Newport News
Deleo, Richard *engineering executive*

Richmond
Sprinkle, William Melvin *engineering administrator, audio-acoustical engineer*

Vienna
Roth, James *engineering company executive*

WASHINGTON

Kirkland
Evans, Robert Vincent *engineering executive*

Olympia
Penney, Robert Allan *engineering executive*

Seattle
Rubbert, Paul Edward *engineering executive*

WEST VIRGINIA

Charleston
Televantos, John Yiannakis *engineering executive*

WISCONSIN

Sussex
Dewey, Craig Douglas *engineering executive*

CANADA

NEW BRUNSWICK

Fredericton
Neill, Robert D. *engineering executive*

QUEBEC

Montreal
Lamarre, Bernard *engineering, contracting and manufacturing advisor*

SASKATCHEWAN

Regina
Mollard, John Douglas *engineering and geology executive*

ARGENTINA

Buenos Aires
Milano, Antonio *engineering executive*

EGYPT

Cairo
Moftah, Mounir Amin *engineering executive*

ENGLAND

London
Searl, John Roy Robert *engineering executive*

Saint Annes-On-Sea
Teakle, Neil William *engineering executive*

FINLAND

Kotka
Cranston, Wilber Charles *engineering company executive*

FRANCE

Paris
Vieillard-Baron, Bertrand Louis *engineering executive*

GERMANY

Berlin
Rauls, Walter Matthias *engineering executive*

Essen
van Wissen, Gerardus Wilhelmus Johannes Maria *consulting engineering company executive*

Wildberg
Wild, Hans Jochen *systems engineering executive*

HONG KONG

Kowloon Bay
Chan, Andrew Mancheong *engineering executive*

INDIA

Bombay
Paul, Biraja Bilash *engineering consulting company executive*

New Delhi
Roy, Tuhin Kumar *engineering company executive*

ITALY

Rome
Perrotta, Giorgio *engineering executive*

JAPAN

Tokyo
Emori, Richard Ichiro *engineering executive*

THE NETHERLANDS

Rotterdam
Yap, Kie-Han *engineering executive*

NEW ZEALAND

Wellington
Blakeley, Roger William George *engineering executive*

PAKISTAN

Multan
Siddiqui, Maqbool Ahmad *engineering consultant and executive*

Peshawar
Qazilbash, Imtiaz Ali *engineering executive, consultant*

Rawalpindi
Malik, Abdul Hamid *engineering executive*

PORTUGAL

Lisbon
Costa, Luís Chaves da *engineering executive*
Nogueira e Silva, Jose Afonso *engineering executive*

REPUBLIC OF KOREA

Seoul
Choi, Dae Hyun *precision company executive*

Yongsan-Gu Seoul
Kang, Minho *engineering executive*

SAUDI ARABIA

Riyadh
Battistelli, Joseph John *electronics executive*

SPAIN

Vitoria
López De Viñaspre Urquiola, Teodoro *engineering company executive*

SRI LANKA

Colombo
Fernando, Cecil T. *engineering executive*

SWITZERLAND

Romanel
Ellis, Brian Norman *engineering executive*

ADDRESS UNPUBLISHED

Bartholomew, Donald Dekle *engineering executive, inventor*
Dahlstrom, Norman Herbert *engineering executive*
Dandashi, Fayad Alexander *applied scientist*
Dawson, Gerald Lee *engineering company executive*
Hinton, Wilburt Hartley, II *construction engineering executive*
Kocaoglu, Dundar F. *engineering executive, industrial and civil engineering educator*
Maropis, Nicholas *engineering executive*
Ortolano, Ralph J. *engineering consultant*
Peterson, Eric Follett *engineering executive*
Rosenkoetter, Gerald Edwin *engineering and construction company executive*
Smith, James Lanning *engineering executive*

EXECUTIVES & SPECIALISTS: ENVIRONMENT

UNITED STATES

ARIZONA

Phoenix
Siegel, Richard Steven *water resource specialist*

CALIFORNIA

Kingsburg
Blanton, Roy Edgar *sanitation agency executive*

San Diego
Deuble, John L., Jr. *environmental science and engineering services consultant*

Santa Rosa
Mackay, Kenneth Donald *environmental services company executive*

COLORADO

Denver
Anderson, Robert *environmental specialist, physician*

Rangely
DeWitt, James Howard *water treatment technician*

CONNECTICUT

Fairfield
Dillingham, Catherine Knight *environmental consultant*

Windsor
Cowen, Bruce David *environmental service company executive*

GEORGIA

Atlanta
Boylan, Glenn Gerard *environmental company official, consultant*
Dysart, Benjamin Clay, III *environmental management consultant, conservationist, engineer*

INDIANA

Jasper
Lents, Thomas Alan *waste water treatment company executive*

Winchester
Anderson, Gary Alan *waste water plant executive*

MAINE

Portland
Becker, Seymour *hazardous materials and wastes specialist*

MARYLAND

Baltimore
Auchincloss, Peter Eric *water quality improvement executive, consultant*

Beltsville
Zehner, Lee Randall *environmental service executive, research director*

NEBRASKA

Fairbury
Greenwood, Richard P. *wastewater superintendent*

NEW HAMPSHIRE

Bedford
Hall, Pamela S. *environmental consulting firm executive*

NEW YORK

Hicksville
Gross, A. Christopher *environmental manager*

Lancaster
Neumaier, Gerhard John *environment consulting company executive*

NORTH CAROLINA

Carrboro
Williamson, Ronald Edwin *waste water plant manager*

OKLAHOMA

Sapulpa
Welcher, Ronnie Dean *waste management and environmental services executive*

PENNSYLVANIA

Avis
Hillyard, Richie Doak *wastewater treatment plant operator*

RHODE ISLAND

East Greenwich
O'Neill, Shawn Thomas *waste water treatment executive*

VIRGINIA

Arlington
Riegel, Kurt Wetherhold *environmental protection, occupational safety and health*

INDIA

Nagpur
Sehgal, Jawaharlal *environmental company executive*

ADDRESS UNPUBLISHED

Kerns, Allen Dennis *energy and environmental manager*

EXECUTIVES & SPECIALISTS: FOOD PRODUCTS

UNITED STATES

COLORADO

Golden
Coors, William K. *brewery executive*

GEORGIA

Decatur
Stivers, Theodore Edward *food products executive, consultant*

ILLINOIS

Chicago
Hoppert, Gloria Jean *food products executive*

Island Lake
O'Day, Kathleen Louise *food products executive*

Vernon Hills
Raisman, Allan Leslie *food products executive*

MARYLAND

Sparks
Nelson, John Howard *food company research executive*

MICHIGAN

Detroit
Meilgaard, Morten Christian *food products executive, international consultant*

MINNESOTA

Minneapolis
Behnke, James Ralph *food company executive*
Gusek, Todd Walter *food scientist*
Luiso, Anthony *international food company executive*

MISSOURI

Saint Louis
Brown, Jay Wright *food manufacturing company executive*

NEW JERSEY

Camden
Johnson, David Willis *food products executive*

Hackettstown
Elejalde, Cesar Carlos *food engineer*

Morris Plains
Kumar, Surinder *food company executive*

NEW YORK

Dundee
Vandyne, Bruce Dewitt *quality control executive*

New York
Murray, William *food products executive*

Tarrytown
Kinigakis, Panagiotis *research scientist, engineer, author*

White Plains
Riha, William Edwin *beverage company executive*

OHIO

Columbus
D'Amato, Anthony Salvatore *food products company executive*
Skiest, Eugene Norman *food company executive*

PENNSYLVANIA

Hershey
Duncan, Charles Lee *food products company executive*
St. Hilaire, Catherine Lillian *food company executive*

TEXAS

College Station
Lusas, Edmund William *food processing research executive*

Colleyville
Newton, Richard Wayne *food products executive*

WASHINGTON

Washougal
Vogel, Ronald Bruce *food products executive*

HONG KONG

Kowloon
Chu, Wayne Shu-Wing *food industry entrepreneur, researcher*

EXECUTIVES & SPECIALISTS: HEALTH PRODUCTS

UNITED STATES

ALABAMA

Birmingham
Hammond, C(larke) Randolph *healthcare executive*

CALIFORNIA

Anaheim
Schlose, William Timothy *health care executive*

Menlo Park
Fergason, James L. *optical company executive*
Glushko, Victor *medical products executive, biochemist*

Pasadena
Mathias, Alice Irene *health plan company executive*

Rancho Santa Margarita
Wong, Wallace *medical supplies company executive, real estate investor*

San Clemente
Crider, Hoyt *health care executive*

San Diego
Harriett, Judy Anne *medical equipment company executive*

Sunnyvale
Rugge, Henry Ferdinand *medical products executive*

Sylmar
Sholder, Jason Allen *medical products company executive*

Thousand Oaks
Rohrbach, Jay William *biotechnology company supervisor*

Valencia
Leung, Kam H. *medical company executive*

COLORADO

Boulder
Masterson, Linda Histen *medical company executive*

CONNECTICUT

Ridgefield
Sadow, Harvey S. *health care company executive*

DISTRICT OF COLUMBIA

Washington
Green, Edward Crocker *health consulting firm executive*

FLORIDA

Homestead
Dudik, Rollie M. *healthcare executive*

GEORGIA

Atlanta
Pruett, Clayton Dunklin *biotechnical company executive*

ILLINOIS

Chicago
Gottlander, Robert Jan Lars *dental company executive*

Chicago Heights
Carpenter, Kenneth Russell *international trading executive*

Deerfield
Loucks, Vernon R., Jr. *health care products and services company executive*

Naperville
Kurth, Paul DuWayne *biotechnical services executive*

Oak Brook
Haupt, Carl P. *retail drugs executive*

KENTUCKY

Louisville
Mather, Elizabeth Vivian *health care executive*

MAINE

Lewiston
Tighe, Thomas James Gasson, Jr. *healthcare executive*

MARYLAND

Reisterstown
Bond, Nelson Leighton, Jr. *health care executive*

MASSACHUSETTS

Cambridge
Vincent, James Louis *biotechnology company executive*
Winkler, Gunther *medical executive*

Lexington
Berstein, Irving Aaron *biotechnology and medical technology executive*

Stoughton
Lamarque, Maurice Patrick Jean *health products executive*

MISSISSIPPI

Columbus
Lockhart, Frank David *healthcare company executive*

NEW JERSEY

Cranbury
Emont, George Daniel *healthcare executive*

Morris Plains
Otani, Mike *optical company executive*

New Brunswick
Gussin, Robert Zalmon *health care company executive*

Somerville
McGregor, Walter *medical products company designer, inventor, consultant, educator*

Teaneck
Woerner, Alfred Ira *medical device manufacturer, educator*

Tinton Falls
Orlando, Carl *medical research and development executive*

Whippany
Pickering, Barbara Ann *pharmaceutical sales representative, nurse*

NEW YORK

Glen Cove
Hardy, Maurice G. *medical and industrial equipment manufacturing company executive*

Mount Kisco
Laster, Richard *biotechnology executive*

New York
Mizrahi, Abraham Mordechay *cosmetics and health care company executive, physician*
Shapiro, Michael Harold *health care executive, consultant, publisher*

NORTH CAROLINA

Burlington
Flagg, Raymond Osbourn *biology executive*

OHIO

Cincinnati
Derstadt, Ronald Theodore *health care administrator*

OREGON

Portland
Watkins, Charles Reynolds *medical equipment company executive*

PENNSYLVANIA

Swiftwater
Woods, Walter Earl *biomedical manufacturing executive*

SOUTH CAROLINA

Charleston
Bennett, Jay Brett *medical equipment company executive*

TEXAS

Bellaire
Guest, Weldon S. *biomedical products and services executive*

Dallas
Anderson, Jack Roy *health care company executive*

Fort Worth
Andrews, Harvey Wellington *medical company executive*

Georgetown
Gerding, Thomas Graham *medical products company executive*

Houston
Fabricant, Jill Diane *medical technology company executive*

Irving
Spies, Jacob John *health care executive*

San Antonio
Baker, Helen Marie *health services executive*

WASHINGTON

Seattle
Gillis, Steven *biotechnology company executive*

Yakima
Anderson, Gregory Martin *medical company representative*

ARGENTINA

Buenos Aires
Saracco, Guillermo Jorge *optical and medical products executive*

AUSTRALIA

North Melbourne
Roberts, Godwin *medical products manager, consultant*

SWITZERLAND

Basel
Töglhofer, Wolfgang *medical director, researcher*

ADDRESS UNPUBLISHED

Brown, Jerry Milford *medical company executive*
Gardner, Clyde Edward *health care executive, consultant, educator*
McQueen, Rebecca Hodges *health care executive, consultant*

EXECUTIVES & SPECIALISTS: INFOSYSTEMS

UNITED STATES

ALABAMA

Huntsville
Burns, Pat Ackerman Gonia *infosytems specialist, software engineer*
Zana, Donald Dominick *information resource manager*

CALIFORNIA

Altadena
Fairbanks, Mary Kathleen *data analyst, researcher*

El Segundo
Mehlman, Lon Douglas *information systems specialist*

Granada Hills
Shoemaker, Harold Lloyd *infosystem specialist*

Hayward
Duncan, Doris Gottschalk *information systems educator*

Inglewood
Hankins, Hesterly G., III *computer systems analyst*

Pasadena
Lugg, Marlene Martha *health information systems specialist*

Ramona
Woodall, James Barry *information systems specialist*

San Diego
Peters, Raymond Eugene *computer systems company executive*

San Marino
Lashley, Virginia Stephenson Hughes *retired computer science educator*

COLORADO

Lakewood
Mueller, Raymond Jay *software development executive*

CONNECTICUT

Enfield
Oliver, Bruce Lawrence *information systems specialist, educator*

DELAWARE

Hockessin
Bischoff, Joyce Arlene *information systems consultant, lecturer*

DISTRICT OF COLUMBIA

Washington
Higgins, Peter Thomas *government information management executive*
Maddock, Jerome Torrence *information services specialist*
Over, Jana Thais *program analyst*
Skeen, David Ray *computer systems administrator*
Skinner, Maurice Edward, IV *information security system specialist*
Smith, Janet Sue *systems specialist*

FLORIDA

Jacksonville
O'Neal, Kathleen Len *financial administrator*

Largo
Caffee, Marcus Pat *computer consulting executive*

Miami
Argibay, Jorge Luis *information systems firm executive and founder*

Sarasota
Minette, Dennis Jerome *financial computing consultant*

GEORGIA

Atlanta
Manley, Lance Filson *data processing consultant*
Rink, Christopher Lee *information technology consultant, photographer*
Zunde, Pranas *information science educator, researcher*

HAWAII

Honolulu
Jongeward, George Ronald *systems analyst*

IDAHO

Boise
Pon-Brown, Kay Migyoku *information systems specialist*

ILLINOIS

Chicago
Bookstein, Abraham *information science educator*
Dwyer, Dennis D. *information technology executive*
Nowak, Chester John, Jr. *infosystems specialist*
Seifert, Timothy Michael *infosystems specialist*

Vernon Hills
Wikarski, Nancy Susan *information technology executive*

INDIANA

Elkhart
Chism, James Arthur *information systems executive*

Marion
Hall, Charles Adams *infosystems specialist*

Noblesville
Young, Frederic Hisgin *infosystems executive, data processing consultant*

KANSAS

Manhattan
Streeter, John Willis *information systems manager*

Wichita
Bagwell, Kathleen Kay *infosystems specialist*

KENTUCKY

Louisville
Schneider, Arthur *computer graphics specialist*

MARYLAND

Gaithersburg
Sayer, John Samuel *information systems consultant*

Silver Spring
Burke, Margaret Ann *computer and communications company specialist*

MASSACHUSETTS

Acton
Golden, John Joseph, Jr. *manufacturing company executive*

Belmont
Montealegre, José Ramiro *information systems consultant*

Boston
Elkowitz, Allan Barry *information systems manager*
Schaaf, John Urban *communication management specialist*

Cambridge
Cooper, Mary Campbell *information services executive*
Meador, Charles Lawrence *management and systems consultant, educator*

Framingham
Roe, Georgeanne Thomas *information brokerage executive*

MICHIGAN

Detroit
Bowlby, Richard Eric *computer systems analyst*

Whitmore Lake
Stanny, Gary *infosystems specialist, rocket scientist*

MISSOURI

Kansas City
Santoro, Alex *infosystems specialist*

Rolla
Datz, Israel Mortimer *information systems specialist*

NEW HAMPSHIRE

Merrimack
Anthony, Gregory Milton *infosystems executive*
Wolf, Robert Farkas *systems and avionics company executive, environmental planning consultant*

NEW JERSEY

Gillette
Nathanson, Linda Sue *technical writer, software training specialist*

Neshanic Station
Castellon, Christine New *information systems manager*

Newark
Suresh, Bangalore Ananthaswami *information systems educator*

NEW YORK

New York
Field, Michael Stanley *information services company executive*
Haddock, Robert Lynn *information services entrepreneur, writer*
Mahadeva, Wijeyaraj Anandakumar *information company executive*

Rochester
Pfendt, Henry George *retired information systems executive, management consultant*

South Huntington
De Lucia-Weinberg, Diane Marie *systems analyst*

Troy
Lane, Adelaide Irene *computer systems specialist, researcher*

Williamsville
May, Kenneth Myron *information services professional, consultant*

Yorktown Heights
Green, Paul Eliot, Jr. *communications scientist*

NORTH CAROLINA

Cary
Bursiek, Ralph David *information systems company executive*

OHIO

Columbus
Taylor, Celianna I. *information systems specialist*

Dayton
Fulton, Darrell Nelson *infosystems specialist*

Dublin
Notowidigdo, Musinggih Hartoko *information systems executive*
Spies, Phyllis Bova *information services company executive*

North Olmsted
Galysh, Robert Alan *information systems analyst*

OKLAHOMA

Mcalester
Alles, Rodney Neal, Sr. *information management executive*

Tulsa
Mersch, Carol Linda *information systems specialist*

OREGON

Portland
von Linsowe, Marina Dorothy *information systems consultant*

PENNSYLVANIA

Gladwyne
Stick, Alyce Cushing *information systems consultant*

Harrisburg
Kimmel, Robert Irving *corporate communication design consultant, former state government official*

TENNESSEE

Nashville
Taylor, Robert Bonds *instructional designer*

TEXAS

Dallas
Alberthal, Lester M., Jr. *information processing services executive*

Houston
Salerno, Philip Adams *infosystems specialist*
Wilcox, Richard Cecil *information systems executive*

Plano
Hinton, Norman Wayne *information services executive*

UTAH

Salt Lake City
Corley, Jean Arnette Leister *infosystems executive*

VIRGINIA

Centreville
Hanson, Lowell Knute *seminar developer and leader, information systems consultant*

Dyke
Hart, Jean Hardy *information systems specialist, consultant*

Fairfax
Palmer, James Daniel *information technology educator*

Herndon
Day, Melvin Sherman *information company executive*

Virginia Beach
Morgan, Michael Joseph *advanced information technology executive*

BELGIUM

Brussels
Vestmar, Brigel Johannes Ahlmann *information systems agency adviser, scientist*

La Hulpe
Eber, Michel *information technology company executive*

BRAZIL

Recife
Pragana, Rildo José Da Costa *information processing company executive*

EGYPT

Cairo
Korayem, Essam Ali *computer company executive*

GERMANY

Bonn
Shaw, John Andrew *information systems executive*

GREECE

Athens
Papandreou, Constantine *tele-informatics scientist, educator*

Kallithea
Thomadakis, Panagiotis Evangelos *computer company executive*

JAPAN

Isehara
Fujita, Tsuneo *systems analysis educator*

Tokyo
Hideaki, Okada *information systems specialist*

SWEDEN

Spänga
Abramczuk, Tomasz *computer image processing specialist*

Stockholm
Cronhjort, Bjorn Torvald *systems analyst*

SWITZERLAND

Zug
Hannema, Dirk *information technology executive*

TAIWAN

Taipei
Yu, Chien-Chih *management information systems educator*

ADDRESS UNPUBLISHED

Simon, Michele Johanna *computer systems specialist*

EXECUTIVES & SPECIALISTS: MANUFACTURING

UNITED STATES

ARIZONA

Phoenix
Atutis, Bernard P. *manufacturing company executive*

CALIFORNIA

Carlsbad
Graham, Robert Klark *lens manufacturer*

Hawthorne
Weiss, Max Tibor *aerospace company executive*

Irvine
Beckman, Arnold Orville *analytical instrument manufacturing company executive*

Long Beach
Bos, John Arthur *aircraft manufacturing executive*

Menlo Park
Dolberg, David Spencer *business executive, lawyer, scientist*

Palo Alto
Hewlett, William (Redington) *manufacturing company executive, electrical engineer*
Packard, David *manufacturing company executive, electrical engineer*

Pleasanton
Longmuir, Alan Gordon *manufacturing executive*

Poway
Aschenbrenner, Frank Aloysious *former diversified manufacturing company executive*

San Diego
Lovelace, Alan Mathieson *aerospace company executive*
Obenour, Jerry Lee *scientific company executive*

San Francisco
Nicholson, William Joseph *forest products company executive*

San Juan Capistrano
Huta, Henry Nicholaus *manufacturing and service company executive*

Santa Barbara
Mueller, George E. *corporation executive*

Sunnyvale
Holbrook, Anthony *manufacturing company executive*

Valencia
Davison, Arthur Lee *scientific instrument manufacturing company executive, engineer*

CONNECTICUT

Norwalk
Kelley, Gaynor Nathaniel *instrumentation manufacturing company executive*

Stamford
Allaire, Paul Arthur *office equipment company executive*

Westport
Cassetta, Sebastian Ernest *industry executive*

FLORIDA

Gainesville
Derrickson, William Borden *business executive*

Titusville
Sipos, Charles Andrew *manufacturing executive*

West Palm Beach
Balaguer, John P. *aircraft manufacturing executive*

GEORGIA

Brunswick
McDonald, Julian LeRoy, Jr. *manufacturing director*

ILLINOIS

Batavia
Tweedy, Robert Hugh *equipment company executive*

Chicago
Burt, Robert Norcross *diversified manufacturing company executive*
Lovell, James A., Jr. *business executive, former astronaut*
McKee, Keith Earl *manufacturing technology executive*

East Moline
Bosworth, Douglas LeRoy *farm implement company executive*

Libertyville
Burrows, Brian William *research and development manufacturing executive*

Moline
Stowe, David Henry, Jr. *agricultural and industrial equipment company executive*

Niles
Koci, Henry James *manufacturing company executive*

Peoria
Fites, Donald Vester *tractor company executive*

Rockford
Gaylord, Edson I. *manufacturing company executive*

Saint Charles
Haugen, Robert Kenneth *product developer*

Schaumburg
Fisher, George Myles Cordell *electronics equipment company executive, mathematician, engineer*

West Chicago
Jeppesen, C. Larry *lighting company executive*

IOWA

Ames
Gaertner, Richard Francis *manufacturing research center executive*

KANSAS

Industrial Airport
Hiner, Thomas Joseph *tractor manufacturing administrator*

LOUISIANA

Baton Rouge
Finney, Clifton Donald *inventor, manufacturing executive*

MARYLAND

Bethesda
Augustine, Norman Ralph *industrial executive*

Gaithersburg
Lowke, George E. *biotechnology executive*

MASSACHUSETTS

Amherst
Barde, Digambar Krushnaji *manufacturing executive*

Brookline
Perry, Frederick Sayward, Jr. *corporate executive*

Cambridge
Mc Cune, William James, Jr. *manufacturing company executive*
Saponaro, Joseph A. *company executive*

Dover
Bonis, Laszlo Joseph *business executive, scientist*

Topsfield
Isler, Norman John *aircraft engine company administrator, consultant*

Wilmington
McCard, Harold Kenneth *aerospace company executive*

Woburn
Cox, Terrence Guy *manufacturing automation executive*

MICHIGAN

Midland
Cuthbert, Robert Lowell *product specialist*

Tecumseh
Herrick, Todd W. *manufacturing company executive*

Troy
Gardon, John Leslie *paint company research executive*

MINNESOTA

Minneapolis
Berg, Stanton Oneal *firearms and ballistics consultant*

Saint Paul
De Simone, Livio Diego *diversified manufacturing company executive*
Fingerson, Leroy Malvin *corporate executive, engineer*
Ling, Joseph Tso-Ti *manufacturing company executive, environmental engineer*

MISSOURI

Mountain View
Olszewski, Lee Michael *instrument company executive*

Saint Louis
Walker, Earl E. *manufacturing executive*

NEW HAMPSHIRE

Keene
Koontz, James L. *manufacturing executive*

NEW JERSEY

Florham Park
Jameson, J(ames) Larry *cable company executive*

Keansburg
Margolis, James Mark *international industrial developer*

Morris Plains
Goodes, Melvin Russell *manufacturing company executive*

Red Bank
Hertz, Daniel Leroy, Jr. *entrepreneur*

Totowa
Kennedy, John William *manufacturing company executive*

Trenton
DeMarco, Peter Vincent *standards assurance executive*

Wayne
Trice, William Henry *paper company executive*

NEW MEXICO

Albuquerque
Baer, Stephen Cooper *manufacturing executive*

NEW YORK

Clifton Park
Scher, Robert Sander *instrument design company executive*

Corning
Houghton, James Richardson *glass manufacturing company executive*

Greenvale
Pall, David B. *manufacturing company executive, chemist*

Hudson
DeCrosta, Edward Francis, Jr. *former paper products company executive, consultant*

Melville
Kaneko, Hisashi *business executive, electrical engineer*

Rochester
Langworthy, Harold Frederick *manufacturing company executive*
Thomas, Leo J. *manufacturing company executive*
Wey, Jong Shinn *research laboratory executive*

Troy
Doremus, Robert Heward *glass and ceramics processing educator*

NORTH CAROLINA

Charlotte
Brown, James Eugene, III *business executive*
Priestley, G. T. Eric *manufacturing company executive*

Research Triangle Park
Holland, Charles Edward *corporate executive*

Winston Salem
Everhart, Francis Grover, Jr. *manufacturing company executive*

NORTH DAKOTA

Grand Forks
Gjovig, Bruce Quentin *manufacturing consultant*

OHIO

Akron
Calderon, Nissim *tire and rubber company executive*

Canton
Janson, Richard Wilford *manufacturing company executive*

Chagrin Falls
Keyes, Marion Alvah, IV *manufacturing company executive*

Cleveland
Corrigan, Victor Gerard *automotive technology executive*
Schloemer, Paul George *diversified manufacturing company executive*

Dublin
Lamp, Benson J. *tractor company executive*

Hudson
Kempe, Robert Aron *venture management executive*

Milford
Klosterman, Albert Leonard *technical development business executive, mechanical engineer*

Toledo
Como, Francis W. *plastics manufacturing executive*

PENNSYLVANIA

Pittsburgh
Kappmeyer, Keith K. *manufacturing company executive*
Phillips, James Macilduff *material handling company executive, engineering and manufacturing executive*

Spring House
Payn, Clyde Francis *technology company executive, consultant*

VIRGINIA

Hampton
Holloway, Paul Fayette *aerospace executive*

Newport News
Ranellone, Richard Francis *shipbuilding company executive*

Richmond
Haines, Michael James *asphalt company official*

Springfield
Bush, Norman *research and development executive*

WASHINGTON

Eastsound
Anders, William Alison *aerospace and diversified manufacturing company executive, former astronaut, former ambassador*

Everett
Winn, George Michael *electrical equipment company executive*

Seattle
Wilson, Thornton Arnold *retired aerospace company executive*

Tacoma
Hoffmann, Gunther F. *forest products executive*

WEST VIRGINIA

Culloden
Dixon, Victor Lee *wire company executive*

WISCONSIN

Hartland
Vitek, Richard Kenneth *scientific instrument company executive*

Menomonee Falls
Moberg, Clifford Allen *mold products company executive*

CANADA

BRITISH COLUMBIA

Burnaby
Forgacs, Otto Lionel *forest products company executive*

NEW BRUNSWICK

Fredericton
Grotterod, Knut *retired paper company executive*

QUEBEC

Pointe Claire
Wrist, Peter Ellis *pulp and paper company executive*

AUSTRIA

Vienna
Sekyra, Hugo Michael *industrial executive*

ENGLAND

Harrow
Wright, Terence Richard *imaging company executive*

HONG KONG

Wanchai
van Hoften, James Dougal Adrianus *business executive, former astronaut*

ITALY

Bari
Recchi, Vincenzo *manufacturing executive*

JAPAN

Osaka
Iwata, Kazuo *business executive*

Sakura
Yamada, Ryuzo *technology company executive*

ADDRESS UNPUBLISHED

Dybvig, Douglas Howard *manufacturing executive, researcher*
Hardin, Clifford Morris *retired executive*
Hemann, Raymond Glenn *aerospace research company executive*
Maskell, Donald Andrew *contracts administrator*
Miller, Merle Leroy *retired manufacturing company executive*
Milo, Frank Anthony *manufacturing company executive*
Riordan, William John *manufacturing process designer, consultant*

EXECUTIVES & SPECIALISTS: MINING

UNITED STATES

ALABAMA

Birmingham
Dahl, Hilbert Douglas *mining company executive*

ARIZONA

Tempe
Hickson, Robin Julian *mining company executive*

CALIFORNIA

Los Angeles
Hesse, Christian August *mining industry consultant*

COLORADO

Denver
Rendu, Jean-Michel Marie *mining executive*

Englewood
Ward, Milton Hawkins *mining company executive*

DISTRICT OF COLUMBIA

Washington
Weiss, Stanley Alan *mining, chemicals and refractory company executive*

LOUISIANA

Monroe
Fouts, James Fremont *mining company executive*

NEW YORK

New York
Born, Allen *mining executive*

PENNSYLVANIA

Pittsburgh
Valoski, Michael Peter *industrial hygienist*

WYOMING

Buffalo
Velasquez, Pablo *mining executive*

Laramie
Laman, Jerry Thomas *mining company executive*

CANADA

BRITISH COLUMBIA

Vancouver
Hallbauer, Robert Edward *mining company executive*
Keevil, Norman Bell *mining executive*
Soregaroli, A(rthur) E(arl) *mining company executive, geologist*

QUEBEC

Montreal
Carbonneau, Come *mining company executive*

MALAYSIA

Kuala Lumpur
Chan, Wan Choon *mining company executive*

ADDRESS UNPUBLISHED

Hodgson, Kenneth P. *mining executive, real estate investor*
Tyler, Ewen William John *retired mining company executive, consulting geologist*

EXECUTIVES & SPECIALISTS: OTHER SPECIALTIES

UNITED STATES

ALABAMA

Tuscaloosa
Weaver, Jerry Reece *management scientist, educator*

ARIZONA

Gilbert
Lamb, Edward Allen, Jr. *business owner*

Phoenix
Massey, L. Edward *chemical marketing executive*

Tucson
Rose, Hugh *management consultant*

ARKANSAS

Fayetteville
Combs, Linda Jones *business administration educator, researcher*

Morrilton
Adams, Earle Myles *technical representative*

CALIFORNIA

Cambria
DuFresne, Armand Frederick *management and engineering consultant*

Canoga Park
Peinemann, Manfred K.A. *marketing executive*

Los Angeles
Geoffrion, Arthur Minot *management scientist*
Webster, Jeffery Norman *technology policy analyst*

Los Gatos
Leung, Charles Cheung-Wan *technological company executive*

Manhattan Beach
Yee, Peter Ben-On *marketing professional*

Moorpark
Hovanec, Timothy Arthur *aquatic researcher*

Palm Desert
Sausman, Karen *zoological park administrator*

Rancho Santa Fe
Schirra, Walter Marty, Jr. *business consultant, former astronaut*

Rialto
Merrill, Steven William *research and development executive*

Sacramento
Jakovac, John Paul *construction executive*

San Francisco
Kreitzberg, Fred Charles *construction management company executive*

Santa Clara
Greene, Frank Sullivan, Jr. *business executive*

Santa Maria
Raich, Abraham Leonard *rabbi, quality control professional*

Sunnyvale
Stevens, John Lawrence *quality assurance professional*

Walnut Creek
Garrett, Suzanne Thornton *management educator*

COLORADO

Boulder
Fleener, Terry Noel *marketing professional*

Englewood
Lazarus, Steven S. *management consultant, marketing consultant*

CONNECTICUT

Danbury
Goldstein, Joel *management science educator, researcher*

Georgetown
Duvivier, Jean Fernand *management consultant*

Norwalk
Coates, John Peter *marketing professional*

Stamford
Miller, Wilbur Hobart *business diversification consultant*

Trumbull
Schmitt, William Howard *cosmetics company executive*

Waterbury
Zeitlin, Bruce Allen *superconducting material technology executive*

Woodbury
Roberts, Robert Clark *sales engineer*

DISTRICT OF COLUMBIA

Washington
Belk, Keith E. *international marketing specialist, researcher*
DeVilbiss, Jonathan Frederick *product marketing administrator*
Langley, Rolland Ament, Jr. *construction and engineering company executive*
Nelson, Carl Michael *construction executive*
Sunderlin, Charles Eugene *consultant*
Tyrrell, Albert Ray *government liaison for industry*
Umpleby, Stuart Anspach *management science educator*

FLORIDA

Melbourne
Shaikh, Muzaffar Abid *management science educator*

Miami
Prager, Michael Haskell *fishery population dynamicist*

Panama City
Sowell, James Adolf *quality assurance professional*

Plantation
Charles, Joel *audio and video tape analyst, voice identification consultant*

Sarasota
Angelotti, Richard H. *science administrator, banker*

GEORGIA

Atlanta
Robinson, Jeffery Herbert *modular building company executive*

Columbus
Braswell, J(ames) Randall *retired quality control professional*

Norcross
Stockwell, Albert H. *procurement professional*

ILLINOIS

Aurora
Freyberg, Dale Wayne *technical trainer*

Calumet City
Kovach, Joseph William *management consultant, psychologist, educator*

Chicago
Beitler, Stephen Seth *cosmetics company executive*
Braker, William Paul *aquarium executive, ichthyologist*
Walshe, Brian Francis *management consultant*
Zeffren, Eugene *toiletries company executive*

De Kalb
Troyer, Alvah Forrest *seed corn company executive, plant breeder*

Park Forest
Orr, Marcia *child development researcher, child care consultant*

Park Ridge
Manzi, Joseph Edward *construction executive*

Rockford
Ostrom, Charles Curtis *financial consultant, former military officer*

Waukegan
Dayal, Sandeep *marketing professional*

INDIANA

Decatur
Coalson, James A. *grain company executive, researcher*

East Chicago
Hughes, Ian Frank *steel company executive*

IOWA

Des Moines
Seifert, Robert P. *agricultural products company executive*

Muscatine
Johnson, Donald Lee *agricultural materials processing company executive*

KANSAS

Manhattan
Roberts, Thomas Carrol *management consultant*

Topeka
Gonzalez, William Joseph *sales executive, geosynthetic engineer*

KENTUCKY

LaGrange
Leslie, Robert Fremont *mobile testing executive, non-destructive testing inspector*

Louisville
Kohnhorst, Earl Eugene *tobacco company executive*

LOUISIANA

Baton Rouge
Russell, Craig John *management educator*

Calhoun
Robbins, Marion Le Ron *agricultural research executive*

New Orleans
Hebert, Leonard Bernard, Jr. *contractor*

MARYLAND

Baltimore
Yellin, Judith *electrologist*

Beltsville
Quirk, Frank Joseph *management consulting company executive*

Bethesda
Van Cott, Harold Porter *human factors professional*

Severn
Fowler, Floyd Earl *national security consultant*

Towson
Weaver, Kerry Alan *construction engineer*

MASSACHUSETTS

Amherst
Morbey, Graham Kenneth *management educator*

Boston
Bush, John Burchard, Jr. *consumer products company executive*
Prescott, John Hernage *aquarium executive*
Von Fischer, George Herman *social psychologist, unified social systems scientist, management consultant, data processing executive*

Cambridge
Frosch, Robert Alan *retired automobile manufacturing executive, physicist*
Kelley, Albert Joseph *management educator, executive consultant*
La Mantia, Charles Robert *management consulting company executive*
Magnanti, Thomas L. *business management educator*
Rowe, Stephen Cooper *venture capitalist, entrepreneur*

Concord
Hogan, Daniel Bolten *management consultant*

Hopkinton
Svrluga, Richard Charles *science and technology executive*

Jamaica Plain
Cook, Robert Edward *educator, plant ecology researcher*

Taunton
Barbour, Robert Charles *technology executive*

MICHIGAN

Ada
Calvert, George David *consumer products company executive*

Chelsea
Scott, James Noel *quality assurance professional*

Detroit
Amladi, Prasad Ganesh *management consulting executive, health care consultant, researcher*

Midland
Doan, Herbert Dow *technical business consultant*

Muskegon
Meilinger, Peter Martin *quality assurance manager, analytical chemist*

Portage
Riesenberger, John Richard *strategic marketing company executive*

Rochester
Riley, Douglas Scott *quality assurance specialist, biochemist*

MINNESOTA

Minneapolis
Fiedler, Robert Max *management consultant*
Yourzak, Robert Joseph *management consultant, engineer, educator*
Zoberi, Nadim Bin-Asad *management consultant*

MISSISSIPPI

Mississippi State
McGilberry, Joe H. *food service executive*

NEBRASKA

Omaha
Simmons, Lee Guyton, Jr. *zoological park director*

NEVADA

Las Vegas
Faley, Robert Lawrence *instruments company executive*

NEW HAMPSHIRE

Portsmouth
Bavicchi, Robert Ferris *construction materials technician, concrete batchplant operator*

NEW JERSEY

Basking Ridge
Milcarek, William Francis *marketing professional*

Bridgewater
Roehrenbeck, Paul William *marketing professional*

Florham Park
Naimark, George Modell *marketing and management consultant*

Iselin
Gondek, John Richard *quality assurance professional*

Lambertville
Cohen, Edward *private consultant*

Lodi
Budwani, Ramesh Nebhandas *consultant*

Princeton
Gogulski, Paul *construction consultant*

South Plainfield
Schlossman, Mitchell Lloyd *cosmetics and chemical specialties executive*

Summit
Van Cleef, Jabez Lindsay *marketing professional*

NEW MEXICO

Albuquerque
Hancock, Don Ray *researcher*

Carlsbad
Watts, Marvin Lee *minerals company executive, chemist, educator*

Los Alamos
Connellee-Clay, Barbara *quality assurance auditor, laboratory administrator*

NEW YORK

Binghamton
Prime, Roger Carl *marketing professional*

Bronx
Lyles, Anna Marie *zoo curator, ornithologist*

Brooklyn
Garibaldi, Louis *aquarium administrator*

Flushing
Lifschitz, Karl *sales executive*

Jamaica
Lyons, Patrick Joseph *management educator*

New York
Anderson, Paul *product management executive*
Kucic, Joseph *management consultant, industrial engineer*
Lattis, Richard Lynn *zoo director*
Norwick, Braham *textile specialist, consultant, columnist*
Seadler, Stephen Edward *business and computer consultant, social scientist*

Palisades
Brusa, Douglas Peter *purchasing executive*

Sanborn
Mowrey, Timothy James *management and financial consultant*

Schenectady
Bolebruch, Jeffrey John *sales executive*

Westbury
Munsinger, Roger Alan *marketing executive*

NORTH CAROLINA

Charlotte
Iverson, Francis Kenneth *metals company executive*

Clayton
Anderson, Pamela Boyette *quality assurance professional*

Greensboro
Banegas, Estevan Brown *agricultural biotechnology executive*
Berggren, Thage *automotive executive*
Reid, Jack Richard *research executive*

Greenville
Deal, Jo Anne McCoy *quality control professional*

Research Triangle Park
Gallagher, Edward Joseph *scientific marketing executive*

Winston Salem
Ehmann, Carl William *consumer products executive, researcher*

OHIO

Bowling Green
Church, Robert Max, Jr. *sales executive*

Cincinnati
Kardes, Frank Robert *marketing educator*

Circleville
Cooper, John Edgar, Sr. *research technician*

Cleveland
Hoag, David H. *steel company executive*
Westlock, Jeannine Marie *health care consultant*

Columbus
Pathak, Dev S. *pharmaceutical administrator, marketing educator*
Reece, Robert William *zoological park administrator*

Maumee
Witte, John Sterling *solar energy contracting company executive*

Miamisburg
Hughes, Thomas William *technical company executive*

Tiffin
Einsel, David William, Jr. *consultant, retired army officer*

OKLAHOMA

Chickasha
Rienne, Dozie Ignatius *technologist*

OREGON

Corvallis
Rieth, Peter Allan *business executive*

Newport
Weber, Lavern John *marine life administrator, educator*

Portland
Yudelson, Jerry Michael *marketing executive*

PENNSYLVANIA

Alcoa Center
Kinosz, Donald Lee *quality manager*

Bethlehem
Roberts, Malcolm John *steel company executive*

Lower Burrell
Englehart, Edwin Thomas *metals company executive*

New Kensington
Bridenbaugh, Peter Reese *industrial research executive*

Philadelphia
Soslow, Arnold *quality consultant*

Pittsburgh
Turnbull, Gordon Keith *metal company executive, metallurgical engineer*

Radnor
Marland, Alkis Joseph *leasing company executive, computer science educator, financial planner*

Trout Run
Michaels, Gordon Joseph *metals company executive*

Willow Grove
Schiffman, Louis F. *management consultant*

RHODE ISLAND

Wakefield
Mason, Scott MacGregor *entrepreneur, inventor, consultant*

SOUTH CAROLINA

Aiken
Stewart, Michael Kenneth *quality assurance professional*

Fort Mill
Montgomery, Terry Gray *textiles researcher*

Greenville
King, David Steven *quality control executive*

TENNESSEE

Clarksville
Franklin, Keith Barry *entrepreneur, technical consultant, former military officer*

Dyersburg
Baker, Kerry Allen *household products company executive*

Kingston
Shacter, John *manager, technology/strategic planning consultant*

Mount Pleasant
Woodall, Larry Wayne *cement company executive*

TEXAS

Austin
Estes, L(ola) Caroline *aquarium store owner, operator*
Hart, John Fincher *construction management company executive*

Baytown
Johnson, Malcolm Pratt *marketing professional*

Brownsville
Farst, Don David *zoo director, veterinarian*

Houston
Cunningham, R. Walter *venture capitalist*

Longview
Turner, Carl Jeane *international business consultant, electronics engineer*

Pearland
Oman, Paul Richard *entrepreneur*

San Antonio
Wimpress, Gordon Duncan, Jr. *corporate consultant, foundation executive*

Spring
Green, Sharon Jordan *interior decorator*

Sugar Land
Solomon, David *sales representative*

Wylie
Rigali, Joseph Leo *quality assurance professional*

VIRGINIA

Arlington
London, J. Phillip *professional services company executive*

Fairfax
Davis, Walter Barry *quality control professional*

Falls Church
Benedict, Jeffrey Dean *financial and technical consultant, mathematics educator*

Richmond
Totten, Arthur Irving, Jr. *retired metals company executive, consultant*

Rosslyn
Fisher, Daniel Robert *consultant*

WEST VIRGINIA

Triadelphia
Mc Cullough, John Phillip *management consultant, educator*

WISCONSIN

Fond Du Lac
Fife, William J., Jr. *metal products executive*

Kaukauna
Janssen, Gail Edwin *banking executive*

Neenah
Underhill, Robert Alan *consumer products company executive*

CANADA

ALBERTA

Edmonton
Stollery, Robert *construction company executive*

MANITOBA

Winnipeg
Mauro, Arthur *financial executive, university chancellor*

ONTARIO

London
Broadwell, Charles E. *retired agricultural products company executive*

Simcoe
Collver, Keith Russell *agricultural products exective*

Toronto
Langton, Maurice C. *marketing and business services entrepreneur, consultant*

ARGENTINA

Buenos Aires
Martin, Osvaldo Jose *investment consultant, entrepreneur*

FRANCE

Paris
Dewar, James McEwen *agricultural executive, consultant*

Rueil Malmaison
Rondeau, Jacques Antoine *marketing specialist, chemical engineer*

GERMANY

Kiel
Brockhoff, Klaus K.L. *marketing and management educator*

HONG KONG

Hong Kong
Kwok, Russell Chi-Yan *retail company executive*
Tang, Pui Fun Louisa *fragrance research administrator*

INDIA

Barabanki
Suryavanshi, O. P. S. *corporate executive*

Madras
Pethachi, Muthiah Chidambaram *textile industry executive*

ITALY

Milan
Barassi, Dario *management consultant*

SCOTLAND

Stirling
Cosgrove, Raymond Francis *science company executive*

ADDRESS UNPUBLISHED

Becker, Brooks *management consultant, chemist*
Halvorson, Harlyn Odell *marine life administrator, biological sciences educator*
Heinemeyer, Paul Hugh *quality assurance specialist*
Hendrickson, William George *business executive*
Hirsh, Norman Barry *management consultant*
Holmes, Robert Wayne *service executive, consultant, biological historian*
Lee, Lawrence Cho *commodities advisor*
Leighton, Charles Raymond *construction inspector*
Linton, William Sidney *marketing research professional*
McLean, Ryan John *technical service professional*
McQuarrie, Terry Scott *technical executive*
Mickelson, Elliot Spencer *quality assurance professional*
Opperman, Danny Gene *packaging professional, consultant*
Pick, James Block *management and sociology educator*
Redmond, Gail Elizabeth *chemical company consultant*
Wierbowski, Cynthia Ann *quality systems manager*

EXECUTIVES & SPECIALISTS: PHARMACEUTICALS

UNITED STATES

CALIFORNIA

Laguna Niguel
Nelson, Alfred John *retired pharmaceutical company executive*

Mountain View
Mutch, James Donald *pharmaceutical executive*
Saifer, Mark Gary Pierce *pharmaceutical executive*

Palo Alto
Neil, Gary Lawrence *pharmaceutical company research executive, biochemical pharmacologist*

San Diego
Tretter, James Ray *pharmaceutical company executive*

San Mateo
Horwitz, David Larry *pharmaceuticals company executive, researcher, educator*

COLORADO

Lakewood
McElwee, Dennis John *pharmaceutical company executive, attorney*

DELAWARE

Wilmington
Hesp, B. *pharmaceutical executive*
Kennedy, William James *pharmaceutical company executive*

ILLINOIS

North Chicago
Burnham, Duane Lee *pharmaceutical company executive*

INDIANA

Carmel
Haslanger, Martin Frederick *pharmaceutical industry professional, researcher*

Indianapolis
Amundson, Merle Edward *pharmaceuticals executive*

West Lafayette
St. John, Charles Virgil *retired pharmaceutical company executive*

MARYLAND

Baltimore
Woolverton, Christopher Jude *biopharmaceutical company executive*

MASSACHUSETTS

Woburn
Driedger, Paul Edwin *pharmaceutical researcher*

MICHIGAN

Ann Arbor
Glazko, Anthony J(oachim) *pharmaceutical consultant*

Kalamazoo
Shebuski, Ronald John *pharmaceutical company executive*

MINNESOTA

Hopkins
Tempero, Kenneth Floyd *pharmaceutical company executive, physician, clinical pharmacologist*

Saint Paul
Downing, Michael William *pharmaceutical company executive*

MISSOURI

Kansas City
Johnson, Richard Dean *pharmaceutical consultant, educator*

NEW JERSEY

Cranbury
Rosenberg, Alberto *pharmaceutical company executive, physician*

Jersey City
Alfano, Michael Charles *pharmaceutical company executive*

New Brunswick
McGuire, John Lawrence *pharmaceuticals research executive*

Nutley
Behl, Charanjit R. *pharmaceutical scientist*

Princeton
Birnbaum, Jerome *pharmaceutical company executive*
Cryer, Dennis Robert *pharmaceutical company executive, researcher*
Lewis, Alan James *pharmaceutical executive, pharmacologist*
Rothwell, Timothy Gordon *pharmaceutical company executive*
Triscari, Joseph *clinical research director*

Summit
Bernhard, Michael Ian *pharmaceutical company executive*

NEW YORK

Ardsley
Redalieu, Elliot *pharmacokinetics executive*

Locust Valley
Schor, Joseph Martin *pharmaceutical executive, biochemist*

New York
Gelb, Richard Lee *pharmaceutical corporation executive*
Lazar, Judith Tockman *pharmaceutical company researcher*

Norwich
King, Alison Beth *pharmaceutical company executive*

Tarrytown
Lauterbach, Hans *pharmaceutical company executive*

NORTH CAROLINA

Research Triangle Park
Hitchings, George Herbert *retired pharmaceutical company executive, educator*

PENNSYLVANIA

Barnesboro
Moore, David Austin *pharmaceutical company executive, consultant*

King Of Prussia
Eggleston, Drake Stephen *pharmaceutical researcher*

Malvern
Holveck, David P. *pharmaceutical company executive*

Waverly
Matthews, Richard J. *pharmaceutical research company executive*

West Point
Shafer, Jules Alan *pharmaceutical company executive*

VIRGINIA

Charlottesville
Shen, Tsung Ying *medicinal chemistry educator*

WISCONSIN

Madison
Perkowski, Casimir Anthony *biopharmaceutical executive, consultant*

BELGIUM

Beerse
Janssen, Paul Adriaan Jan *pharmaceutical company executive*

FRANCE

Paris
DuBois, Jean Gabriel *pharmaceutical executive, pharmacist*

PORTUGAL

Lisbon
Teixeira da Cruz, Antonio *pharmaceutical company executive*

ADDRESS UNPUBLISHED

Alden, Ingemar Bengt *pharmaceuticals executive*
Bauer, Victor John *pharmaceutical company executive*
Drebus, Richard William *pharmaceutical company executive*
Gatlin, Larry Alan *pharmaceutical professional*

EXECUTIVES & SPECIALISTS: RESEARCH

UNITED STATES

ALABAMA

Birmingham
Rouse, John Wilson, Jr. *research institute administrator*

ARIZONA

Phoenix
Bouwer, Herman *laboratory executive*

Tempe
Hageman, Brian Charles *researcher*

Tucson
Cortner, Hanna Joan *science administrator, research scientist, educator*
Shannon, Robert Rennie *optical sciences center administrator, educator*

CALIFORNIA

Berkeley
Shank, Charles Vernon *science administrator, educator*
Still, Gerald G. *plant physiologist, research director*

Gardena
Hu, Steve Seng-Chiu *scientific research company executive, academic administrator*

Huntington Beach
Bozanic, Jeffrey Evan *marine sciences research center director*

Livermore
Eby, Frank Shilling *research scientist*

Los Gatos
Cusick, Joseph David *science administrator, retired*

Menlo Park
Tietjen, James *research institute administrator*

Moffett Field
Ross, Muriel Dorothy *research scientist*

Mountain View
Grimes, Craig Alan *research scientist*

Palo Alto
Balzhiser, Richard Earl *research and development company executive*
Garland, Harry Thomas *research administrator*
Zuckerkandl, Emile *molecular evolutionary biologist, scientific institute executive*

San Francisco
Iacono, James Michael *research center administrator*

Thousand Oaks
Malmuth, Norman David *program manager*

Tustin
Charley, Philip James *testing laboratory executive*

COLORADO

Boulder
Byerly, Radford, Jr. *science policy official*
Clifford, Steven Francis *science research director*
Glover, Fred William *artificial intelligence and optimization research director, educator*
Serafin, Robert Joseph *science center administrator, electrical engineer*

Fort Collins
Lameiro, Gerard Francis *research institute director*

Golden
Hubbard, Harold Mead *research institute executive*

CONNECTICUT

Southport
Hill, David Lawrence *research corporation executive*

DELAWARE

Newark
Meakin, John David *university research executive, educator*

Wilmington
Hartzell, Charles R. *research administrator, biochemist, cell biologist*

DISTRICT OF COLUMBIA

Washington
Cameron, Maryellen *science association administrator, geologist, educator*
Challinor, David *scientific institute administrator*
Cooper, Chester Lawrence *research administrator*
Corell, Robert Walden *science administration educator*
Crum, John Kistler *chemical society director*
Dale, Charles Jeffrey *operations research analyst*
Feulner, Edwin John, Jr. *research foundation executive*
Geller, Harold Arthur *earth and space sciences executive*
Grafton, Robert Bruce *science foundation official*
Haq, Bilal Ul *national science foundation program director, researcher*
Hess, LaVerne Derryl *research laboratory scientist*
Hess, Wilmot Norton *science administrator*
Johnson, George Patrick *science policy analyst*
Krugman, Stanley Liebert *science administrator, geneticist*
Mock, John Edwin *science administrator, nuclear engineer*
Moraff, Howard *science foundation director*
Murphy, Robert Earl *scientist, government agency administrator*
Perry, Daniel Patrick *science association administrator*
Pitts, Nathaniel Gilbert *science foundation director*
Pyke, Thomas Nicholas, Jr. *government science and engineering administrator*
Schad, Theodore MacNeeve *science research administrator, consultant*
Smith, Philip Meek *research organization executive*
Steinberg, Marcia Irene *science foundation program director*
Timm, Gary Everett *science administrator, chemist*
Vernikos, Joan *science association director*

FLORIDA

Jacksonville
Bodkin, Lawrence Edward *inventor, research development company executive, gemologist*

Miami
Bezdek, Hugo Frank *scientific laboratory administrator*

Niceville
Soben, Robert Sidney *systems scientist*

Tallahassee
Brennan, Leonard Alfred *research scientist, administrator*

GEORGIA

Atlanta
Clifton, David Samuel, Jr. *research executive, economist*
Johnson, Barry Lee *public health research administrator*

Norcross
Dibb, David Walter *research association administrator*
Newbern, Laura Lynn *forestry association executive, editor*

IDAHO

Post Falls
Brede, Andrew Douglas *research director, plant breeder*

ILLINOIS

Alton
Zimmer, James Peter *laboratory executive, consultant*

Argonne
Schriesheim, Alan *research administrator*

Champaign
Sanderson, Glen Charles *science director*
Smarr, Larry Lee *science administrator, educator, astrophysicist*

Chicago
Arzbaecher, Robert C(harles) *research institute executive, electrical engineer, researcher*
Lee, Bernard Shing-Shu *research company executive*
McCrone, Walter Cox *research institute executive*
Rymer, William Zev *research scientist, administrator*
Stafford, Fred Ezra *science adminstrator*

Urbana
Stout, Glenn Emanuel *water resources center administrator*

KANSAS

Wichita
Gumnick, James Louis *research institute executive*

LOUISIANA

Baton Rouge
Beckman, Joseph Alfred *research and development administrator*

MARYLAND

Annapolis
Anderson, William Carl *association executive, environmental engineer, consultant*

Bethesda
Bick, Katherine Livingstone *scientist, international liaison consultant*
Brinley, F(loyd) J(ohn), Jr. *health science institution executive, physician*
Goldstein, Murray *health organization official*
Hausman, Steven Jack *health science administrator*
Shulman, Lawrence Edward *biomedical research administrator, rheumatologist*
Spangler, Miller Brant *science and technology analyst, planner, consultant*
Tseng, Christopher Kuo-Hou *health scientist, administrator, chemist*
Yellin, Herbert *science administrator*

Chevy Chase
Choppin, Purnell Whittington *research administrator, virology researcher, educator*
Harter, Donald Harry *research administrator, medical educator*

Columbia
Lain, David Cornelius *health scientist, researcher*

Gaithersburg
Kramer, Thomas Rollin *automation researcher*
Semerjian, Hratch Gregory *research and development executive*

Lutherville
Barton, Meta Packard *business executive, medical science research executive*

MASSACHUSETTS

Amherst
Godfrey, Paul Joseph *science foundation director*

Boston
Hornig, Donald Frederick *scientist*
Lanner, Michael *research administrator, consultant*

Cambridge
Beranek, Leo Leroy *scientific foundation executive, engineering consultant*
Jacobson, Ralph Henry *laboratory executive, former air force officer*

Lowell
Awerbuch, Shimon *research consultant*

Natick
Neumeyer, John Leopold *research company administrator, chemistry educator*

Reading
Gelb, Arthur *business executive, electrical and systems engineer*

Waltham
Ganong, William Francis, III *speech sciences research executive*

Watertown
Sahatjian, Ronald Alexander *science foundation executive*

Woods Hole
Broadus, James Matthew *research center administrator*

MICHIGAN

Ann Arbor
Beeton, Alfred Merle *laboratory director, limnologist, educator*

Detroit
Novak, Raymond Francis *research institute director, pharmacology educator*
Tunac, Josefino Ballesteros *biotechnology administrator*

Plymouth
Harless, James Malcolm *corporate executive, environmental consultant*

MISSOURI

Kansas City
Mc Kelvey, John Clifford *research institute executive*

MONTANA

Hamilton
Rudbach, Jon Anthony *biotechnical company executive*

Polson
Stanford, Jack Arthur *biological station administrator*

NEVADA

Reno
Fox, Carl Alan *research institute executive*

NEW JERSEY

New Brunswick
Gaylor, James Leroy *biomedical research director*

Newark
Geskin, Ernest S(amuel) *science administrator, consultant*

Princeton
Klein, Leonard *chemist*
Marshall, Carol Joyce *clinical research data coordinator*

Rahway
Scolnick, Edward Mark *science administrator*

Rocky Hill
Ahmed, S. Basheer *research company executive, educator*

Trenton
Colley, Roger J. *environmental biotechnology research company executive*

NEW MEXICO

Albuquerque
Cummings, John Chester, Jr. *research professional*
Narath, Albert *national laboratory director*

Los Alamos
Wallace, Terry Charles, Sr. *technical administrator, researcher, consultant*

NEW YORK

Albany
Toombs, Russ William *laboratory director*

Buffalo
McHale, Magda Cordell *academic administrator, trend analyst*

New York
Gellman, Isaiah *association executive*
Lichter, Robert Louis *science foundation administrator, chemist*
Speth, James Gustave *United Nations executive, lawyer*
Wynder, Ernst Ludwig *science foundation director, epidemiologist*

Staten Island
Wisniewski, Henryk Miroslaw *pathology and neuropathology educator, research facility administrator, research scientist*

Utica
Antzelevitch, Charles *research center executive*

NORTH CAROLINA

Durham
Cruze, Alvin M. *research institute executive*

Hickory
Sears, Frederick Mark *research manager, mechanical engineer*

Research Triangle Park
Wooten, Frank Thomas *research facility executive*

NORTH DAKOTA

Bismarck
Ogaard, Louis Adolph *environmental administrator, computer consultant*

Fargo
Hahn, Benjamin Daniel *research executive*

OHIO

Columbus
McSweeny, Paul Edward *research technologist*
Olesen, Douglas Eugene *research institute executive*

Dayton
Cartmell, James V. *research and development executive*

OKLAHOMA

Ardmore
Patterson, Manford K(enneth), Jr. *foundation administrator, researcher, scientist*

Stillwater
Grischkowsky, Daniel Richard *research scientist*

OREGON

Beaverton
Critchlow, B. Vaughn *research facility administrator, researcher*
Montagna, William *scientist*

Hillsboro
Bhagwan, Sudhir *computer industry and research executive, consultant*

Pendleton
Smiley, Richard Wayne *research center administrator, researcher*

PENNSYLVANIA

Allentown
Morris, Stanley M. *research and engineering executive*

Pittsburgh
Kaufman, William Morris *research institute director, engineer*
Warner, Richard David *research foundation executive*

Radnor
Mizutani, Satoshi *research administrator*

State College
Hettche, L. Raymond *research director*

University Park
Macdonald, Digby Donald *scientist, science administrator*

West Chester
Siery, Raymond Alexander *laboratory administrator*

SOUTH CAROLINA

Columbia
Page, Tonya Fair *cogeneration facility coordinator*

SOUTH DAKOTA

Brookings
Swiden, Ladell Ray *research center director*

TENNESSEE

Knoxville
Bressler, Marcus N. *science administrator*
Mc Hargue, Carl Jack *research laboratory administator*

Nashville
Wang, Taylor Gunjin *science administrator, astronaut, educator*

Oak Ridge
Veigel, Jon Michael *corporate professional*

TEXAS

Austin
Bronaugh, Edwin Lee *research center administrator*
Northington, David K. *research center director, botanist, educator*
Thornton, Joseph Scott *research institute executive, materials scientist*

Dallas
Johnson, Gifford Kenneth *testing laboratory executive*
Land, Geoffrey Allison *science administrator*
Miller, William *science administrator*
Pakes, Steven P. *medical school administrator*

San Antonio
Deviney, Marvin Lee, Jr. *research institute scientist, program manager*
Donaldson, Willis Lyle *research institute administrator*
Goland, Martin *research institute executive*
Henderson, Arvis Burl *data processing executive, biochemist*

Tyler
Cohen, Allen Barry *health science administrator, biochemist*

UTAH

Salt Lake City
Feucht, Donald Lee *research institute executive*
Hansen, Dale J. *science administrator, plant biochemist*

Vernal
Johnson, Marlin Deon *research facility administrator*

VIRGINIA

Annandale
Dugan, John Vincent, Jr. *research and development manager, scientist*
Faraday, Bruce John *scientific research company executive, physicist*

Arlington
Lambert, Richard Bowles, Jr *science foundation program director, oceanographer*
Lesko, John Nicholas, Jr. *research scientist*

Fairfax
Pixley, John Sherman, Sr. *research company executive*

Falls Church
Simpson, John Arol *retired government executive, physicist*

WASHINGTON

Poulsbo
Kolb, James A. *science association director, writer*

Spokane
McClellan, David Lawrence *physician, medical facility administrator*

Tacoma
Champ, Stanley Gordon *scientific company executive*

WISCONSIN

Madison
Erickson, John Ronald *research administrator*
Kirk, Thomas Kent *research scientist*

WYOMING

Laramie
Speight, James Glassford *research company executive*

CANADA

NEW BRUNSWICK

Fredericton
Boorman, Roy Slater *science administrator, geologist*

ONTARIO

Mississauga
Lawford, G. Ross *research and development company executive*

Ottawa
Dence, Michael Robert *research director*
Morand, Peter *research agency executive*
Perron, Pierre O. *science administrator*

Toronto
Mustard, James Fraser *research institute executive*

ALBANIA

Tiranë
Buda, Aleks *science administrator, history researcher, educator*

ARGENTINA

Buenos Aires
Balve, Beba Carmen *research center administrator*

AUSTRALIA

Salisbury
Bedford, Anthony John *defense science executive*

AUSTRIA

Vienna
Koss, Peter *research administrator*

BELGIUM

Brussels
Auerbacher, Peter *cancer research organization administrator*
Fasella, Paolo Maria *general science researcher, development facility director*

CHINA

Beijing
Zhang, Fu-Xue *scientist*

DENMARK

Aarhus
Straede, Christen Andersen *research center administrator*

Roskilde
Heydorn, Kaj *science laboratory administrator*

ENGLAND

London
Irvine, John Henry *science analyst*

Swindon
Blundell, Thomas Leon *scientist, science administrator*

Teddington
McCartney, Louis Neil *research scientist*

FINLAND

Helsinki
Kalimo, Esko Antero *research institute administrator, educator*

FRANCE

Allauch
Lai, Richard Jean *research administrator*

Creteil
Robert, Leslie Ladislas *research center administrator, consultant*

Nanterre
Morin-Postel, Christine *international operations executive*

Orsay
Fiszer-Szafarz, Berta (Berta Safars) *research scientist*

Paris
Luton, Jean Marie *space agency administrator*
Rouvillois, Philippe André Marie *science administrator*

Rantigny
Langlais, Catherine Renee *science administrator*

GERMANY

Darmstadt
Wullkopf, Uwe Erich Walter *research institute director*

Hannover
Döhler, Klaus Dieter *pharmaceutical and development executive*

Munich
Fuhrmann, Horst *science administrator*

JAPAN

Osaka
Shimbo, Masaki *technology company executive*

Suita
Hayaishi, Osamu *director science institute*

Tokyo
Hori, Yukio *scientific association administrator, emeritus engineering educator*
Joh, Yasushi *science administrator, chemist*
Takasaki, Yoshitaka *telecommunications scientist, electrical engineer*
Tsuda, Kyosuke *organic chemist, science association administrator*

LEBANON

Beirut
Rouayheb, George Michael *scientific research council advisor*

NORWAY

Ski
Omland, Tov *physician, medical microbiologist*

POLAND

Gliwice
Szewczyk, Pawel *research institute administrator, educator*

Warsaw
Koscielak, Jerzy *scientist, science administrator*

RUSSIA

Moscow
Golitsyn, Georgiy *research institute director*
Laverov, Nikolai Pavlovitch *science foundation executive*
Tatarinov, Leonid Petrovich *science administrator, paleontologist*

SAUDI ARABIA

Yanbu Al Sin
Choudary, Shaukat Hussain *science administrator, chemist*

SCOTLAND

East Kilbride
Hann, James *science administrator*

ADDRESS UNPUBLISHED

Ahearne, John Francis *scientific research society director, researcher*
El-Saiedi, Ali Fahmy *science foundation administrator*
Englund, John Arthur *research company executive*
Goldstein, Walter Elliott *biotechnology executive*
Kuper, George Henry *research and development institute executive*
Mertz, Walter *retired government research executive*
Murrell, Kenneth Darwin *research administrator, microbiologist*
Sills, Richard Reynolds *scientist, educator*
Vigfusson, Johannes Orn *scientific officer*
Welber, Irwin *research laboratory executive*

EXECUTIVES & SPECIALISTS: TRANSPORTATION

UNITED STATES

ALABAMA

Birmingham
Yenice, Mehmet Fikri *aeronautics company executive*

ALASKA

Juneau
Ritter, Grant L. *water operations supervisor*

ARIZONA

Phoenix
Elien, Mona Marie *air transportation professional*
Wood, John Mortimer *aerospace executive, aeronautical engineer*

Scottsdale
Aybar, Charles Anton *aviation executive*

Tempe
Morgan, J. Ronald *aerospace company executive*
Robertson, Samuel Harry, III *transportation safety research engineer, educator*

ARKANSAS

Huntsville
Carr, Gerald Paul *former astronaut, business executive, former marine officer*

Springdale
Pogue, William Reid *former astronaut, foundation executive, business and aerospace consultant*

CALIFORNIA

Anaheim
Bailey, Don Matthew *aerospace company executive*

Borrego Springs
Scannell, William Edward *aerospace company executive, consultant, educator*

Calabasas
Tellep, Daniel Michael *aerospace executive, mechanical engineer*

Corona Del Mar
Tether, Anthony John *aerospace executive*

Edwards AFB
Bachus, Blaine Louis *pilot, aircraft mechanical engineer*

El Segundo
Aldridge, Edward C., Jr. *aerospace transportation executive*

Palos Verdes Peninsula
Waaland, Irving Theodore *retired aerospace design executive*

San Diego
Tuccio, Sam Anthony *aerospace executive, physicist*

San Jose
Baker, James Reginald *aerospace company executive, engineer*

Stockton
Biddle, Donald Ray *aerospace company executive*

Ukiah
Pinoli, Burt Arthur *airline executive*

Van Nuys
Cooper, Leroy Gordon, Jr. *former astronaut, business consultant*

Westminster
Armstrong, Gene Lee *retired aerospace company executive*

COLORADO

Colorado Springs
Eller, Thomas Julian *aerospace company executive, astronautical engineer, computer scientist*
Stienmier, Saundra Kay Young *aviation educator*

Denver
Burgess, Larry Lee *aerospace executive*

Littleton
Nefzger, Charles LeRoy *astronautics company executive*

DISTRICT OF COLUMBIA

Washington
Mattingly, Thomas K. *astronaut*
Parker, Robert Allan Ridley *astronaut*

FLORIDA

Fort Lauderdale
Berwig, Newton Urbano *aerospace executive*

Kennedy Space Center
Shaw, Brewster Hopkinson, Jr. *astronaut*

Orlando
Sasseen, George Thiery *aerospace engineering executive*
Smetheram, Herbert Edwin *company executive*

GEORGIA

Atlanta
Connor, Charles William *airline pilot*
Ferguson, Erik Tillman *transportation planning educator*
Thakker, Ashok *aerospace engineering company executive*

Macon
Hails, Robert Emmet *aerospace consultant, business executive, former air force officer*

ILLINOIS

Chicago
Handwerker, A. M. *retired transportation executive*

INDIANA

Columbus
Hartley, James Michaelis *aerospace systems, printing and hardwood products manufacturing executive*

Indianapolis
Wallace, F. Blake *aerospace executive, mechanical engineer*
Wallace, Robert Eugene, II *rail transit systems executive*

LOUISIANA

Baton Rouge
Stopher, Peter Robert *transportation executive, consultant*

MARYLAND

Annapolis
Cane, Guy *engineering test pilot, aviation consultant*

Rockville
Phillips, Mark Douglas *aerospace company executive*

MASSACHUSETTS

Boston
Doherty, Robert Francis, Jr. *aerospace industry professional*

MICHIGAN

Dearborn
Boulos, Edward Nashed *transportation specialist*

MISSOURI

Lake Saint Louis
Samuelson, Robert Donald *retired combat aircraft executive*

Saint Louis
Jenks, Gerald Erwin *aerospace company executive*
Johnston, Gerald Andrew *aerospace company executive*

NEW JERSEY

Wayne
Drossman, Jay Lewis *aerospace executive*

Woodbury
Stanzione, Kaydon Al *aerospace industry executive, advisor*

NEW YORK

Bethpage
Caporali, Renso L. *aerospace executive*

Briarcliff Manor
Price, Ronald Franklin *air transportation executive*

Jamaica
Kocivar, Ben *aviation specialist, journalist*

New York
Danaher, Frank Erwin *transportation technologist*

Queens
Farkas, Edward Barrister *airport administrator, electrical engineer*

NORTH CAROLINA

Raleigh
Alman, Ted Irwin *engineer, consultant*

NORTH DAKOTA

Grand Forks
De Remer, Edgar Dale *aviation educator*

OHIO

Cincinnati
Murphy, Eugene F. *aerospace, communications and electronics executive*

Dayton
Brand, Vance Devoe *astronaut, government official*

PENNSYLVANIA

Philadelphia
Schoen, Allen Harry *aerospace engineering executive*

SOUTH CAROLINA

Mount Pleasant
Carnes, Robert Mann *aerospace company executive*

TENNESSEE

Memphis
Howe, Daniel Bo *aviation executive, retired naval officer*

TEXAS

Austin
Hudson, William Ronald *transportation engineering educator*
McCullough, Benjamin Franklin *transportation researcher, educator*

Euless
Tunnell, Clida Diane *air transportation specialist*

Fort Worth
Simrin, Harry S. *aerospace operations specialist*

Houston
Brandenstein, Daniel Charles *astronaut, naval officer*
Hartsfield, Henry Warren, Jr. *astronaut*
Henize, Karl Gordon *astronaut, astronomy educator*
McBride, Jon Andrew *astronaut, aerospace engineer*
Thagard, Norman E. *astronaut*
Young, John Watts *astronaut*

VIRGINIA

Alexandria
Del Fosse, Claude Marie *aerospace software executive*
Fabian, John M. *former astronaut, air force officer*
Litzsinger, Orville Jack *aerospace executive*

Arlington
Sutton, George W. *aerospace company executive*

Dulles
Thompson, David Walker *astronautics company executive*

WASHINGTON

Federal Way
King, Don E. *air transportation executive*

Seattle
Cosgrove, Benjamin A. *retired aerospace company executive*
Raisbeck, James David *aircraft design executive*
Robinson, Gary Dale *aerospace company executive*

MILITARY ADDRESSES OF THE UNITED STATES

EUROPE

FPO
Lahr, Brian Scott *pilot*

CANADA

ONTARIO

Almonte
Morrison, Angus Curran *aviation executive*

Ottawa
Coleman, John Morley *transportation research director*

FINLAND

Helsinki
von Haartman, Harry Ulf *international transportation company executive*

GERMANY

Friedrichshafen
Wilczek, Elmar Ulrich *aviation professional*

Munich
Born, Gunthard Karl *aerospace executive*

GREECE

Athens
Kalathas, John (Ioannis) *airline pilot, physicist, oceanographer*

JAPAN

Tokyo
Hayashi, George Yoichi *airline executive*

JORDAN

Amman
Philippi, Edmond Jean *airline executive*

SAUDI ARABIA

Jeddah
Peeran, Syed Muneer *airline executive*

SPAIN

Manises
Sosa-Riera, Raul *air transportation executive*

ADDRESS UNPUBLISHED

Aldrin, Buzz *former astronaut, science consultant*
Armstrong, Neil A. *former astronaut*
Carpenter, Malcolm Scott *astronaut, oceanographer*
Collins, Eileen Marie *astronaut*
Freitag, Peter Roy *transportation specialist*
Rose, James Turner *aerospace consultant*
Shepard, Alan Bartlett, Jr. *astronaut, real estate developer*
Stephanedes, Yorgos J. *transportation educator*

EXECUTIVES & SPECIALISTS: UTILITIES

UNITED STATES

ARIZONA

Buckeye
Burton, Edward Lewis *industrial procedures and training consultant*

ARKANSAS

Russellville
Cooper, Robert Michael *nuclear energy industry specialist*

CALIFORNIA

Alhambra
Anderson, Gordon MacKenzie *petroleum service contractors executive*

Los Angeles
Brinegar, Claude Stout *oil company executive*

Palo Alto
McKeever, Sheila A. *utilities executive*

Richmond
Meyer, Jarold Alan *oil company research executive*

San Francisco
Brandin, Alf Elvin *retired mining and shipping company executive*

Santa Ana
Bennett, Brian O'Leary *utilities executive*

Santa Fe Springs
McCarty, William Britt *natural resource company executive, educator*

Templeton
Gandsey, Louis John *petroleum and environmental consultant*

Villa Grande
Shirilau, Mark Steven *utilities executive*

COLORADO

Denver
Cevenini, Roberto Mauro *gas and oil industry executive, entrepreneur*
Fryt, Monte Stanislaus *petroleum company executive, speaker, advisor*

Durango
Thurston, William Richardson *oil and gas industry executive, geologist*

Englewood
Le, Khanh Tuong *utility executive*

Grand Junction
Skogen, Haven Sherman *oil company executive*

CONNECTICUT

Ridgefield
Preeg, William Edward *oil company executive*

DISTRICT OF COLUMBIA

Washington
Emerson, Susan *oil company executive*
Endahl, Lowell Jerome *retired electrical cooperative executive*
Walden, Omi Gail *public affairs and government relations executive, energy resources specialist*

FLORIDA

Largo
Dolan, John E. *consultant, retired utility executive*

Lynn Haven
Coggeshall, Norman David *former oil company executive, research director, investment executive*

Mount Dora
Wootton, Joel Lorimer *nuclear technologist, health physics consultant*

Naples
Bush, John William *business executive, federal official*

Palm Beach
Epley, Marion Jay *oil company executive*

Pensacola
Platz, Terrance Oscar *utilities company executive*

GEORGIA

Baxley
Belcher, Ronald Anthony *nuclear energy educator*

Waynesboro
Legrand, Ronald Lyn *nuclear facility executive*

IDAHO

Idaho Falls
Newman, Stanley Ray *oil refining company executive*

Tuttle
Ravenscroft, Bryan Dale *alternate energy research company executive*

ILLINOIS

Argonne
Heine, James Arthur *utilities plant manager*

Barrington
Groesch, John William, Jr. *marketing research consultant*
Perry, I. Chet *petroleum company executive*

Chicago
Batlivala, Robert Bomi D. *oil company executive, economist educator*
McHenry, Keith Welles, Jr. *oil company executive*
O'Connor, James John *utility company executive*
Pierce, Shelby Crawford *oil company executive*

Gurnee
Krueger, Darrell George *nuclear power industry consultant*

Kankakee
Smith, Charles Hayden *utilities executive*

Springfield
Gallina, Charles Onofrio *nuclear regulatory official*

INDIANA

Indianapolis
Todd, Zane Grey *utility executive*

Merrillville
Blaschke, Lawrence Raymond *utility company professional*

North Vernon
Siener, Joseph Frank *utilities supervisor*

IOWA

Lidderdale
Hagemann, Dolores Ann *water company official*

Shenandoah
Elliott, John Earl *water plant administrator*

KANSAS

Hamilton
Lockard, Walter Junior *petroleum company executive*

KENTUCKY

Ashland
Johnson, Bobby Joe *electronic instrumentation technician*

Louisville
Wesley, Stephen Burton *energy services executive*

LOUISIANA

Lafayette
Donovan, Brian Joseph *maritime industry executive*

MAINE

York Harbor
Curtis, Edward Joseph, Jr. *gas industry executive, management consultant*

MARYLAND

Baltimore
McGowan, George Vincent *public utility executive*

Bethesda
McMurphy, Michael Allen *energy company executive, lawyer*

Columbia
Hurwitch, Jonathan William *energy consultant*

Gibson Island
Kiddoo, Richard Clyde *retired oil company executive*

MASSACHUSETTS

Boston
Doyle, Patrick Francis *utility company executive*
Reznicek, Bernard William *power company executive*

Marblehead
Krebs, James Norton *retired electric power industry executive*

Springfield
Harris, Roger Scott *energy consultant*

West Bridgewater
Kirby, Kevin Andrew *utilities company executive*

MICHIGAN

Muskegon
Kuhn, Robert Herman *public works and utilities executive, engineer*

Newport
Johnson, Rodney William *utility executive*

Royal Oak
Meyer, Gregory Joseph *power company executive*

MINNESOTA

Eagan
Ernst, Gregory Alan *energy consultant*

Mankato
Kvamme, John Peder *electric technology company executive*

MISSOURI

Kansas City
Changho, Casto Ong *power plant construction executive*
Keith, Dale Martin *utilities consultant*

NEBRASKA

Grafton
Benorden, Robert Roy *utility executive*

NEW JERSEY

Florham Park
Sprow, Frank Barker *oil company executive*

Newark
Steinberg, Reuben Benjamin *utility management engineer*

Parsippany
Clark, Philip Raymond *nuclear utility executive, engineer*

Summit
Mathis, James Forrest *retired petroleum company executive*

Surf City
Aurner, Robert Ray, II *oil company, auto diagnostic and restaurant franchise and company development executive*

Union
Lewandowski, Andrew Anthony *utilities executive, consultant*

Westfield
Specht, Gordon Dean *retired petroleum executive*

NEW MEXICO

Hobbs
Garey, Donald Lee *pipeline and oil company executive*

Santa Fe
Buck, Christian Brevoort Zabriskie *independent oil operator*
Pickrell, Thomas Richard *retired oil company executive*

NEW YORK

Babylon
Lopez, Joseph Jack *oil company executive, consultant*

East Islip
Dinstber, George Charles *construction design engineer, consultant*

New York
Alpert, Warren *oil company executive, philanthropist*
Brown, Edward James *utility executive*
Case, Hadley *oil company executive*
Dolan-Baldwin, Colleen Anne *global technology executive*

Schenectady
Felak, Richard Peter *electric power industry consultant*
Robb, Walter Lee *retired electric company executive, management company executive*

Southold
Knight, Harold Edwin Holm, Jr. *utility company executive*

White Plains
Meehan, Robert Henry *utilities executive, human resources executive, business educator*

NORTH CAROLINA

Charlotte
Lee, William States *utility executive*

Fayetteville
Baldwin, George Michael *industrial marketing professional*

Raleigh
Starkey, Russell Bruce, Jr. *utilities executive*

NORTH DAKOTA

Bismarck
Spilman, Timothy Frank *utilities engineer*

OHIO

Akron
Staines, Michael Laurence *oil and gas production executive*

Cleveland
Duckworth, Donald Reid *oil company executive*

Columbus
Stage, Richard Lee *utilities executive*

Wooster
Shafer, Berman Joseph *oil company executive*

OKLAHOMA

Enid
Ward, Llewellyn O(rcutt), III *oil producer*

Oklahoma City
Abernathy, Jack Harvey *petroleum, utility company and banking executive*
O'Keeffe, Hugh Williams *oil industry executive*
Prasad, B.H. *utility company executive*

Tulsa
Williford, Richard Allen *oil company executive, flight safety company executive*

Vinita
Beavers, Roy L. *utility executive*

Waynoka
Olson, Rex Melton *oil and gas company executive*

OREGON

Portland
Bacon, Vicky Lee *lighting services executive*

PENNSYLVANIA

Berwyn
Burch, John Walter *mining equipment company executive*

Coudersport
Sproull, Wayne Treber *consultant*

Delta
Wurzbach, Richard Norman *utility company engineer*

Philadelphia
Jackson, Fred *oil executive*

Pittsburgh
Lawrence, Margery H(ulings) *utilities executive*
Schwass, Gary L. *utilities executive*

SOUTH CAROLINA

North Myrtle Beach
Atkinson, Harold Witherspoon *utilities consultant, real estate broker*

TENNESSEE

Chattanooga
Stevens, Donna Jo *nuclear power plant administrator*

Morristown
Johnson, John Robert *petroleum company executive*

Trenton
McCullough, Kathryn T. Baker *utilities executive*

TEXAS

Austin
Collier, Steven Edward *utilities executive, consultant*

Baytown
Davis, Phillip Eugene *oil company executive, chemical engineer*

Beaumont
Long, Alfred B. *retired oil company executive, consultant*
Schenck, Jack Lee *electric utility executive*

Brownsville
Pena, Eleuterio *utility executive*

College Station
Neff, Ray Quinn *electric power consultant*

Cypress
Day, Robert Michael *oil company executive*

Dallas
Blackburn, Charles Lee *oil company executive*
Blessing, Edward Warfield *petroleum company executive*
Ellison, Luther Frederick *oil company executive*
McCormick, J. Philip *natural gas company executive*

Fort Worth
Wingo, Paul Gene *oil company executive*

Houston
Barrow, Thomas Davies *oil and mining company executive*
Carter, James Sumter *oil company executive, tree farmer*
Cox, Frank D. (Buddy Cox) *oil company executive, exploration consultant*
Danburg, Jerome Samuel *oil company executive*
Dice, Bruce Burton *exploration company executive*
Farmer, Joe Sam *petroleum company executive*
Frost, John Elliott *minerals company executive*
Goodman, Herbert Irwin *petroleum company executive*
Guinn, David Crittenden *petroleum engineer, drilling and exploration company executive*
Hurwitz, Charles Edwin *oil company executive*
Jorden, James Roy *oil company engineering executive*
Kuntz, Hal Goggan *petroleum exploration company executive*
Leonard, Gilbert Stanley *oil company executive*
Little, Jack Edward *oil company executive*
Mossavar-Rahmani, Bijan *oil and gas company executive*
Neidell, Norman Samson *oil and gas exploration consultant*
Pratt, David Lee *oil company executive*
Smith, Lloyd Hilton *independent oil and gas producer*
Will, Edward Edmund *oil company executive*
Williams, Robert Henry *oil company executive*
Zeissig, Hilmar Richard *oil and gas executive*

Irving
McBrayer, H. Eugene *retired petroleum industry executive*

Midland
Grover, Rosalind Redfern *oil and gas company executive*

Montgomery
Falkingham, Donald Herbert *oil company executive*

Pottsboro
Hanning, Gary William *utility executive, consultant*

Wadsworth
Haralson, John Olen *utility company executive*

VIRGINIA

Alexandria
Lau, K(wan) P(ang) *independent power developer*

Lexington
Tyree, Lewis, Jr. *retired compressed gas company executive, inventor, technical consultant*

WASHINGTON

Spokane
Cameron, Robert H. *water association administrator*

WISCONSIN

Kimberly
Bressers, Daniel Joseph *utility executive*

Lake Geneva
Craft, Timothy George *utility company executive*

CANADA

ALBERTA

Calgary
Farries, John Keith *petroleum engineering company executive*
MacCulloch, Patrick C. *oil industry executive*
Maier, Gerald James *natural gas transmission and marketing company executive*
Mungan, Necmettin *petroleum consultant*

MANITOBA

Pinawa
Wright, Michael George *atomic energy company executive*

Winnipeg
Lang, Otto E. *business executive, former Canadian cabinet minister*

ONTARIO

Toronto
Powis, Alfred *natural resources company executive*

MEXICO

Aristoteles
Akel, Ollie James *oil company executive*

ENGLAND

London
White, Norman Arthur *engineer, corporate executive, educator*

Middlesex
Bulkin, Bernard Joseph *oil industry executive*

FRANCE

Paris
Nestvold, Elwood Olaf *oil service company executive*

GERMANY

Gommern
Voigt, Hans-Dieter *oil company executive, researcher, educator*

ITALY

Bologna
Chierici, Gian Luigi *petroleum engineering executive, educator*

JAPAN

Osaka
Osumi, Masato *utility company executive*

Yokohama
Toyama, Takahisa *oil company executive, engineer*

SOUTH AFRICA

Pretoria
Stumpf, Waldo Edmund *nuclear energy industry executive*

SPAIN

Madrid
Sanchez Sudon, Fernando *renewable energy company executive*

SWEDEN

Stockholm
Hagson, Carl Allan *utilities executive*

TRINIDAD AND TOBAGO

San Fernando
Constance, Mervyn *utility executive*

ADDRESS UNPUBLISHED

Hansen, Shirley Jean *energy consulting executive, professional association administrator*
Kebblish, John Basil *retired coal company executive, consultant*
McCutchan, Marcus Gene *water utility executive*
Nolte, Marty Dee *nuclear power plant training manager*
Ormasa, John *utility executive*
Perry, George Wilson *oil and gas company executive*
Ralston, Roy B. *petroleum consultant*
Ramsey, William Dale, Jr. *petroleum company executive*
Skala, Gary Dennis *electric and gas utilities executive management consultant*

FOUNDATIONS & ASSOCIATIONS

UNITED STATES

CALIFORNIA

Bolinas
Wayburn, Laurie Andrea *environmental and wildlife foundation administrator, conservationist*

Calabasas
Drezner, Stephen M. *policy analyst, engineer*

COLORADO

Colorado Springs
MacLeod, Richard Patrick *foundation administrator*
Tutt, Charles Leaming, Jr. *educational administrator, former mechanical engineering educator*

DISTRICT OF COLUMBIA

Washington
Alberts, Bruce Michael *foundation administrator*
Cramer, James Perry *association executive, publisher, educator, architectural historian*
DiBona, Charles Joseph *association executive*
Doman, Elvira *science administrator*
Fuller, Kathryn Scott *environmental association executive, lawyer*
Godwin, Stephen Rountree *not-for-profit organization administrator*
Hair, Jay Dee *association executive*
Herring, Kenneth Lee *editor scientific society publications*
Kolb, Charles Chester *humanities administrator*
Pinstrup-Andersen, Per *educational administrator*
Sampson, Robert Neil *association executive*
Taggart, G. Bruce *professional society administrator*
Toll, John Sampson *association administrator, former university administrator, physics educator*
Wilkniss, Peter E. *foundation administrator, researcher*

ILLINOIS

Chicago
Rodgers, James Foster *association executive, economist*

La Grange Park
Webster, Lois Shand *association executive*

Oak Brook
Armbruster, Walter Joseph *foundation administrator*

Park Ridge
Kleckner, Dean Ralph *trade association executive*

Villa Park
O'Leary, Dennis Sophian *medical organization executive*

MARYLAND

Annapolis
Payne, Winfield Scott *national security policy research executive*

Bethesda
Beall, Robert Joseph *foundation executive*
Salisbury, Tamara Paula *foundation executive*

Gaithersburg
Johnson, Donald Rex *research institute administrator*

Hyattsville
McLin, William Merriman *foundation administrator*

Rockville
Josephs, Melvin Jay *professional society administrator*

MASSACHUSETTS

Cambridge
Brower, Michael Chadbourne *research administrator*
Kapor, Mitchell David *foundation executive*
Orlen, Joel *association executive*

MICHIGAN

Detroit
Leyh, George Francis *association executive*

Saint Joseph
Butt, Jimmy Lee *retired association executive*

MISSOURI

Columbia
Walkenbach, Ronald Joseph *foundation executive, pharmacology educator*

Hallsville
McFate, Kenneth Leverne *association administrator*

Kirksville
Hilgartner, C(harles) Andrew *theorist*

NEW JERSEY

Piscataway
De Chino, Karen Linnia *engineering association administrator*

NEW YORK

New York
Belden, David Leigh *professional association executive, engineering educator*
Bird, Mary Lynne Miller *association executive*
Emmert, Richard Eugene *professional association executive*
Fox, Daniel Michael *foundation administrator, author*
Marshall, John *association administrator*
Pfrang, Edward Oscar *association executive*

Oyster Bay
Hatch, Mary Wendell Vander Poel *laboratory executive, interior decorator*

NORTH CAROLINA

Research Triangle Park
Martin, William Royall, Jr. *association executive*

OHIO

Dayton
Mathews, David *foundation executive*

OKLAHOMA

Tulsa
Dix, Fred Andrew, Jr. *professional society executive*

PENNSYLVANIA

Philadelphia
Doman, Janet Joy *association executive*

Warrendale
Scott, Alexander Robinson *engineering association executive*

RHODE ISLAND

Providence
Jaco, William H. *mathematical association executive*

TEXAS

Richardson
Adamson, Dan Klinglesmith *association executive*

CANADA

ONTARIO

Ottawa
Alper, Anne Elizabeth *professional association executive*

QUEBEC

Laval
Pichette, Claude *former banking executive, university rector, research executive*

AUSTRALIA

Canberra
Craig, David Parker *science academy executive, emeritus educator*

BELGIUM

Leuven-Heverlee
Belmans, Ronnie Jozef Maria *foundation adminstrator, researcher*

DENMARK

Copenhagen
Dal, Erik *former editorial association administrator*

FINLAND

Helsinki
Varmavuori, Anneli *chemical society administrator, chemist*

ADDRESS UNPUBLISHED

Largman, Kenneth *strategic analyst, strategic defense analysis company executive*
Turner, John Freeland *foundation administrator, former federal agency administrator, former state senator*

GOVERNMENT & MILITARY

UNITED STATES

ALABAMA

Tuscaloosa
Flinn, David R. *federal agency research director*

ARIZONA

Litchfield Park
Kramer, Rex W., Jr. *former naval officer, business executive*

ARKANSAS

North Little Rock
Amick, S. Eugene *military engineer*

CALIFORNIA

Cedar Ridge
Yeager, Charles Elwood (Chuck Yeager) *retired air force officer*

Los Angeles
Abrahamson, James Alan *retired air force officer*

Los Angeles AFB
DiDomenico, Paul B. *military officer*
Faudree, Edward Franklin, Jr. *military officer*

Moffett Field
Compton, Dale Leonard *space agency administrator*

Palos Verdes Estates
Basnight, Arvin Odell *public administrator, aviation consultant*

Redondo Beach
Gehrlein, Michael Timothy *air force officer*

Sacramento
Helander, Clifford John *state agency administrator*

San Diego
Cameron, Charles Bruce *naval officer, electrical engineer*

San Rafael
Jindrich, Ervin James *municipal government official*

COLORADO

Colorado Springs
Edmonds, Richard Lee *air force officer*

Fort Collins
Eberhart, Steve A. *federal agency administrator, research geneticist*

Golden
Collins, Heather Lynne *government official*

Peterson AFB
Mercier, Daniel Edmond *military officer, astronautical engineer*
Taylor, Mark Jesse *military engineer*

CONNECTICUT

New London
Haas, Thomas Joseph *coast guard officer*

DISTRICT OF COLUMBIA

Washington
Abrahamson, George R. *civilian military scientist*
Arrowsmith-Lowe, Thomas *federal agency administrator, medical educator*
Bachkosky, John M. (Jack Bachkosky) *federal agency administrator*
Baker, D. James *federal agency administrator*
Baker, Donald James *federal official, oceanographer, administrator*
Beckler, David Zander *government official, science administrator*
Bernthal, Frederick Michael *federal agency administrator*
Brayton, Peter Russell *program director, scientist*
Brown, George Edward, Jr. *congressman*
Brown, Harold *corporate director, consultant, former secretary of defense*
Browner, Carol *federal official*
Clutter, Mary Elizabeth *federal official*
Cooper, Benita Ann *federal agency administrator*
DeMars, Bruce *naval administrator*

Deutch, John Mark *federal official, chemist, academic administrator*
Fell, James Carlton *scientific and technical affairs executive, consultant*
Finerty, Martin Joseph, Jr. *military officer, researcher*
Finney, Essex Eugene, Jr. *science executive*
Fisher, Farley *chemist, federal agency administrator*
Freitag, Robert Frederick *government official*
Furiga, Richard Daniel *government official*
Gebbie, Kristine Moore *health official*
Genega, Stanley G. *career officer, federal agency administrator*
Gilbreath, William Pollock *federal agency administrator*
Glenn, John Herschel, Jr. *senator*
Goldin, Daniel S. *government agency administrator*
Good, Mary Lowe (Mrs. Billy Jewel Good) *government official*
Griffith, Jerry Dice *government official, nuclear engineer*
Grua, Charles *government official*
Guimond, Richard Joseph *federal agency executive, environmental scientist*
Harris, Wesley L. *federal agency administrator*
Hoffman, Robert S. *federal agency administrator*
Huntress, Wesley Theodore, Jr. *government official*
Ifft, Edward Milton *government official*
Jewett, David Stuart *federal agency administrator*
Jordan, George Eugene *air force officer*
Kearns, David Todd *federal agency administrator*
Larew, Hiram Gordon, III *research coordinator*
Lindsey, Alfred Walter *federal agency official, environmental engineer*
Mahan, Clarence *federal agency administrator, writer*
McAlexander, Thomas Victor *government executive*
McDonald, Bernard Robert *federal agency administrator*
McPherson, Ronald P. *federal agency administrator*
Meikle, Philip G. *government agency executive*
Mirick, Robert Allen *military officer*
Myers, Dale DeHaven *government, industry, aeronautics and space agency administrator*
Newhouse, Alan Russell *federal government executive*
Novello, Antonia Coello *U.S. surgeon general*
O'Donnell, Brendan James *naval officer*
Pearson, Jeremiah W., III *military career officer, federal agency official*
Petersen, Richard Herman *government executive, aeronautical engineer*
Plowman, R. Dean *federal agriculture agency administrator*
Prakash, Ravi *scientific counselor, biomedical engineering educator*
Ritter, Donald Lawrence *congressman, scientist*
Rockefeller, John Davison, IV (Jay Rockefeller) *senator, former governor*
Rollwagen, John A. *federal official*
Roy, Robin K. *government official*
San Martin, Robert L. *federal official*
Scarr, Harry Alan *federal agency administrator*
Shank, Fred Ross *federal agency administrator*
Sheehan, Jerrard Robert *technology policy analyst, electrical engineer*
Siegel, Jack S. *federal official*
Smith, Richard Melvyn *government official*
Stewart, Frank Maurice *federal agency administrator*
Streb, Alan Joseph *government official, engineer*
Thompson, James Robert, Jr. *federal space center executive*
Toma, Joseph S. *defense analyst, retired military officer*
Truly, Richard H. *federal agency administrator*
Tyner, C. Fred *federal agency administrator*
Umminger, Bruce Lynn *government official, scientist, educator*
Waggoner, Lee Reynolds *federal agency administrator*
Watkins, James David *government official, naval officer*
Williams, Arthur E. *federal agency administrator*

FLORIDA

Kennedy Space Center
Crippen, Robert Laurel *naval officer, former astronaut*

GEORGIA

Atlanta
Carey, Gerald John, Jr. *former air force officer, research institute director*
Ebneter, Stewart Dwight *engineer*
Sullivan, Louis Wade *former secretary health and human services, physician*

HAWAII

Honolulu
Hays, Ronald Jackson *naval officer*
Lashlee, JoLynne Van Marsdon *army officer, nursing administrator*

IDAHO

Boise
Habben, David Marshall *state official*

ILLINOIS

Champaign
Semonin, Richard Gerard *state official*

Grayslake
Rieken, Danny Michael *naval officer*

Rockford
Bixby, Mark Ellis *city official*

INDIANA

Crown Point
Lee, Robert Jeffrey *municipal utility professional*

KANSAS

Topeka
Hammerschmidt, Ronald Francis *environmental director*

MARYLAND

Annapolis
Papet, Louis M. *federal official, civil engineer*

Baltimore
Martin, George Reilly *federal agency administrator*

Beltsville
van Schilfgaarde, Jan *agricultural engineer, government agricultural service administrator*

Bethesda
Bender, Erwin Rader, Jr. *air force officer*
Broder, Samuel *federal agency administrator*
Bynum, Barbara Stewart *federal health institute administrator*
Hauck, Frederick Hamilton *retired naval officer, astronaut, business executive*
Reeves, Richard Allen *government aerospace program executive, lawyer*

Fort Meade
Ewell, Allen Elmer, Jr. *naval officer*

Gaithersburg
French, Judson Cull *government official*
Hellwig, Helmut Wilhelm *air force research director*
Prabhakar, Arati *federal administration research director, electrical engineer*

Germantown
Williamson, Samuel Perkins *federal agency administrator, meteorologist*

Patuxent River
Eastburg, Steven Roger *naval officer, aeronautical engineer*

Rockville
Kessler, David A. *health services commissioner*
Rheinstein, Peter Howard *government official, physician, lawyer*

Silver Spring
Attaway, David Henry *federal research administrator, oceanographer*
Foster, Nancy Marie *environmental analyst, government official*
Friday, Elbert Walter, Jr. *federal agency administrator, meteorologist*
Telesetsky, Walter *government official*

MASSACHUSETTS

Hanscom AFB
Sargent, Douglas Robert *air force officer, engineer*

MICHIGAN

Warren
Horton, William David, Jr. *army officer*

MISSISSIPPI

Jackson
Sullivan, John Fallon, Jr. *government official*

MISSOURI

Clayton
Osterloh, Everett William *county official*

MONTANA

Missoula
Turman, George *former lieutenant governor*

NEBRASKA

Offutt AFB
Feingold, Mark Lawrence *electronic warfare officer*

NEVADA

Las Vegas
Lanni, Joseph Anthony *military officer*

NEW HAMPSHIRE

Portsmouth
Fields, William Alexander *naval officer, mechanical engineer*

NEW JERSEY

Morris Plains
Woolley, Gail Suzanne *military officer*

Toms River
Young, William H. *federal agency administrator*

NEW MEXICO

Albuquerque
Schmitt, Harrison Hagan *former senator, geologist, astronaut, consultant*

Kirtland AFB
Miller, Leonard Doy *army officer*

Santa Fe
Knapp, Edward Alan *scientist, government administrator*

NEW YORK

Fort Drum
Whiteman, Wayne Edward *army officer*

West Point
Samples, Jerry Wayne *military officer*

NORTH CAROLINA

Research Triangle Park
Dobbin, Ronald Denny *federal agency administrator, occupational hygienist, researcher*
Mann, David Mark *researcher*

OHIO

Cleveland
Fordyce, James Stuart *federal agency administrator*
Klineberg, John Michael *federal agency administrator, aerospace researcher*
Ross, Lawrence John *federal agency administrator*

Dayton
Gray, James Randolph, III *air force officer*

Wright Patterson AFB
Bowman, William Jerry *air force officer*

RHODE ISLAND

Saunderstown
Knauss, John Atkinson *federal agency administrator, oceanographer, educator, former university dean*

SOUTH CAROLINA

Goose Creek
Reckamp, Douglas E. *military officer*

TENNESSEE

Memphis
Gibson, Clifford William *military officer*

TEXAS

Dayton
Baysinger, Stephen Michael *air force officer*

Houston
Daniels, Cindy Lou *space agency executive*
Overmyer, Robert Franklyn *astronaut, marine corps officer*

Killeen
Wells, James David, Jr. *military officer*

LaGrange
Riehs, John Daryl *state agency administrator*

New Braunfels
Brown, Marvin Lee *retired air force officer*

Sheppard AFB
Fittante, Philip Russell *air force officer, pilot*

Waco
Held, Colbert Colgate *retired diplomat*

VIRGINIA

Alexandria
Dighton, Robert Duane *military operations analyst*
Stafford, Thomas Patten *retired military officer, former astronaut*

Arlington
Ary, T. S. *federal official*
Chawla, Manmohan Singh (Monte Chawla) *armor, weapon system and arms control technology specialist*

Falls Church
LaNoue, Alcide Moodie *medical corps officer, health care administrator*

Fort Belvoir
Diercks, Frederick Otto *government official*

Fort Monroe
Weiland, Peter Lawrence *military officer*

Mc Lean
Myers, Kenneth Alan *air force officer, aerospace engineer*

Watts, Helena Roselle *military analyst*

Reston
Hartong, Mark Worthington *military officer, engineer*

Sterling
Edwards, Stephen Glenn *air force officer, astronautical engineer*

WASHINGTON

Fairchild AFB
McDonnell, John Patrick *military officer*

Oak Harbor
Meaux, Alan Douglas *facilities technician, sculptor*

Olympia
Hirsch, Gary Mark *energy policy specialist, cogeneration manager*

MILITARY ADDRESSES OF THE UNITED STATES

ATLANTIC

FPO
English, Gary Emery *military officer*

EUROPE

APO
Goodwin, Richard Clarke *military analyst*
Scheltema, Robert William *military officer*

PACIFIC

FPO
Carlisle, Mark Ross *naval aviator*

CANADA

NOVA SCOTIA

Halifax
Huggard, Richard James *federal agency administrator*

ONTARIO

Ottawa
Connelly, Alan B. *career officer, engineer*
Gold, Lorne W. *Canadian government official*
Vézina, Monique *Canadian government official*

Ottawa-Hull
Valcourt, Bernard *Canadian government official, lawyer*

Toronto
Ostry, Sylvia *Canadian public servant, economist*

AUSTRALIA

Canberra
Free, Ross Vincent *federal official*

AUSTRIA

Vienna
Blix, Hans Martin *international atomic energy official*

BELGIUM

Brussels
Dehousse, Jean-Maurice *federal official*

CHINA

Beijing
Henggao, Ding *federal official*
Jian, Song *government official, science administrator*

FINLAND

Helsinki
Routti, Jorma Tapio *federal agency administrator, engineering educator*

FRANCE

Paris
Friedman, Kenneth Michael *energy policy analyst*
Roudybush, Franklin *diplomat, educator*
Zaragoza, Federico Mayor *protective agency administrator, biochemist*

GABON

Libreville
Berre, Andre Dieudonne *federal agency executive, oil company executive*

GERMANY

Bonn
Riesenhuber, Heinz Friedrich *German minister for research and technology*
Wissman, Matthias *minister of research and technology*

INDIA

New Delhi
Hillary, Edmund Percival *diplomat, explorer, bee farmer*

ITALY

Rome
Colombo, Umberto Paolo *Italian government official*

JAPAN

Chiyoda-ku
Tanigawa, Kanzo *Japanese minister of science and technology*

KENYA

Nairobi
Onyonka, Zachary *federal agency administrator*

KIRIBATI

Tarawa
Kaitaake, Anterea *minister of education science and technology*

NORWAY

Oslo
Hernes, Gudmund *federal official*

PAKISTAN

Islamabad
Soomro, Ellahi Bukhsh *Pakistani federal minister*

ROMANIA

Bucharest
Popa, Petru *federal agency administrator, engineering executive, educator*

SLOVENIA

Ljubljana
Tancig, Peter *federal official*

SWEDEN

Stockholm
Unckel, Per *minister of education*

TAIWAN

Taipei
Dai, Peter Kuang-Hsun *government official, aerospace executive*

TURKEY

Çankaya
Özmen, Atilla *federal agency administrator, physics educator*

YUGOSLAVIA

Belgrade
Ocvirk, Andrej *Slovene federal official*

ADDRESS UNPUBLISHED

Fisher, Michael Alan *air force officer, mechanical engineer*
Fontana, Alessandro *Italian minister of universities and research*
Hauser, Julius *retired drug regulatory official*
Quilès, Paul *French federal official*
Scardera, Michael Paul *air force officer*
Smith, Christie Parker *operations researcher*
Whitehead, Nelson Peter *foreign service officer*

HEALTHCARE SERVICES

UNITED STATES

ALABAMA

Birmingham
Richards, J. Scott *rehabilitation medicine professional*
Stephens, Deborah Lynn *health facility executive*

Montgomery
Harrell, Barbara Williams *public health administrator*
Hornsby, Andrew Preston, Jr. *human services administrator*

Tuscaloosa
Cooper, Eugene Bruce *speech-language pathologist, educator*
Stitt, Kathleen Roberta *nutrition educator*

ARIZONA

Casa Grande
Krauss, Sue Elizabeth *radiological medical management technologist*
McGillicuddy, Joan Marie *psychotherapist, consultant*

Phoenix
Chan, Michael Chiu-Hon *chiropractor*
Helms, Mary Ann *critical care nurse, consultant*
Sauer, Barry W. *medical research center administrator, bioengineering educator*

Tempe
Opie, Jane Maria *audiologist*

Tucson
Shropshire, Donald Gray *hospital executive*
Thomson, Cynthia Ann *clinical nutrition research specialist*

ARKANSAS

Fort Smith
Ashley, Ella Jane (Ella Jane Rader) *medical technologist*

CALIFORNIA

Albany
Chook, Edward Kongyen *disaster medicine educator*

Arcadia
Anderson, Holly Geis *medical clinic executive, radio personality*

Belmont
Schreiber, Andrew *psychotherapist*

Berkeley
Day, Lucille Elizabeth *laboratory administrator, educator, author*
Holder, Harold D. *public health administrator, communications specialist, educator*
Macher, Janet Marie *industrial hygienist*
Walsh, Jane Ellen McCann *health care executive*

Carmel
Smith, Jeffry Alan *public health administrator, physician, consultant*

Coloma
Wall, Sonja Eloise *nurse, administrator*

Danville
Shen, Mason Ming-Sun *pain and stress management center administrator*

Davis
Hill, Fredric William *nutritionist, poultry scientist*
Laben, Dorothy Lobb *volunteer nutrition educator, consultant*

Downey
Austin, James Albert *healthcare executive, obstetrician-gynecologist*

Duarte
Shapero, Sanford Marvin *hospital executive, rabbi*

El Centro
Goldeberry, Richard Eugene *mobile intensive care paramedic, registered nurse*

Glendale
Clemens, Roger Allyn *medical and scientific affairs manager*
Oppenheimer, Preston Carl *psychotherapist, counseling agency administrator, psychodiagnostician*

Hemet
Violet, Woodrow Wilson, Jr. *retired chiropractor*

La Jolla
Covington, Stephanie Stewart *psychotherapist, author*

Los Altos
Menke, James Michael *chiropractor*

Los Angeles
Horowitz, Ben *medical center executive*
Katchur, Marlene Martha *nursing administrator*
Merchant, Roland Samuel, Sr. *hospital administrator, educator*

Modesto
Lipomi, Michael Joseph *health facility administrator*

Newbury Park
Calderone, Marlene Elizabeth *toxicology technician*

Palo Alto
Andersen, Torben Brender *optical researcher, astronomer*

Pebble Beach
Keene, Clifford Henry *medical administrator*

Riverside
Chang, Sylvia Tan *health facility administrator, educator*
Taylor, Randy Steven *hospital administrator*

San Diego
Galbraith, Nanette Elaine Gerks *forensic and management sciences company executive*
Rosen, Peter *health facility administrator, emergency physician, educator*
Royston, Ivor *scientific director*
Smith, Raymond Edward *health care administrator*
Trout, Monroe Eugene *hospital systems executive*

San Francisco
Gortner, Susan Reichert *nursing educator*
Johnson, Herman Leonall *research nutritionist*
Mannino, J. Davis *psychotherapist*
Westerdahl, John Brian *nutritionist, health educator*

San Jose
Oak, Ronald Stuart *health and safety administrator*

San Lorenzo
Lantz, Charles Alan *chiropractor, researcher*

Santa Barbara
Duarte, Ramon Gonzalez *nurse, educator, researcher*

Santa Monica
Gupta, Rishab Kumar *medical association administrator, educator, researcher*

Thousand Oaks
Conant, David Arthur *architectural acoustician, educator, consultant*

West Hollywood
Ziferstein, Isidore *psychoanalyst, educator, consultant*

COLORADO

Boulder
Holdsworth, Janet Nott *women's health nurse*

Brighton
Kohlmeier, Sharon Louise *medical laboratory administrator, medical technologist*

Fort Collins
Gubler, Duane J. *research scientist, administrator*

Longmont
Jones, Beverly Ann Miller *nursing administrator, patient services administrator*

Pueblo
Deming, Wendy Anne *mental health professional*

CONNECTICUT

Danbury
Gagnon, John Harvey *psychotherapist, educator*

East Berlin
Holdsworth, Robert Leo, Jr. *emergency medical services consultant*

Greenwich
Dahl, Andrew Wilbur *health services executive*
Langley, Patricia Coffroth *psychiatric social worker*

New Haven
Bauman, Natan *audiologist, acoustical engineer*
Cofrancesco, Donald George *health facility administrator*
Condon, Thomas Brian (Brian Condon) *hospital executive*
De Rose, Sandra Michele *psychotherapist, educator, supervisor, administrator*
Reyes, Marcia Stygles *medical technologist*

Orange
Douskey, Theresa Kathryn *health facility administrator*

DELAWARE

Wilmington
Rodgers, Rhonda Lee *health facility administrator*

DISTRICT OF COLUMBIA

Washington
Patrick, Janet Cline *medical society administrator*
Woteki, Catherine Ellen *nutritionist*
Yoder, Mary Jane Warwick *psychotherapist*

FLORIDA

Auburndale
Ellis, Robert Jeffry *health facility executive*

Boca Raton
Luca-Moretti, Maurizio *nutrition scientist, researcher*

Clearwater
Fenderson, Caroline Houston *psychotherapist*

Deerfield Beach
Areskog, Donald Clinton *chiropractor*

Delray Beach
Rowland, Robert Charles *clinical psychotherapist, writer, researcher*

Fort Lauderdale
Strom-Paikin, Joyce Elizabeth *nursing administrator*

Gainesville
Brown, William Samuel, Jr. *communication processes and disorders educator*
Moore, G(eorge) Paul *speech pathologist, educator*

Hollywood
Connolly, Connie Christine *nurse anesthetist*

Inverness
Mavros, George S. *clinical laboratory director*

Lehigh Acres
Whelahan, Yvette Ann *nursing administrator, consultant*

Melbourne
Favero, Kenneth Edward *medical administrator*

Miami
Bachmeyer, Thomas John *fundraising executive*
Burkett, Marjorie Theresa *nursing educator, gerontology nurse*
Herman, Larry Marvin *psychotherapist*

Miami Shores
Cherry, Andrew L., Jr. *social work educator, researcher*

North Fort Myers
Zeldes, Ilya Michael *forensic scientist*

Orange Park
Stroud, Debra Sue *medical technologist*

Orlando
Mayer, Richard Thomas *laboratory director, entomologist*
Mengel, Lynn Irene Sheets *health science research coordinator*

Port Saint Lucie
Wertheimer, David Eliot *medical facility administrator, cardiologist*

Tallahassee
Rhodes, Roberta Ann *dietitian*

Tampa
Bussone, David Eben *hospital administrator*
Liller, Karen DeSafey *health education educator*
Price, Douglas Armstrong *chiropractor*

GEORGIA

Atlanta
Gerst, Steven Richard *healthcare director, physician*
Hollinger, Charlotte Elizabeth *medical technologist, tree farmer*
Simmons, Samuel William *retired public health official*

Augusta
Feldman, Elaine Bossak *medical nutritionist, educator*

Brunswick
Crowe, Hal Scott *chiropractor*

Decatur
Crews, John Eric *rehabilitation administrator*

Dunwoody
Bartolo, Donna M. *hospital administrator, nurse*

Martinez
Butts, William Randolph *nuclear medicine technologist*

Tifton
Thomas, Adrian Wesley *laboratory director*

HAWAII

Honolulu
Flannelly, Laura T. *mental health nurse, nursing educator, researcher*
Olipares, Hubert Barut *biological safety officer*

IDAHO

Bonners Ferry
McClintock, William Thomas *health care administrator*

ILLINOIS

Arlington Heights
Enright, John Carl *occupational health engineer*

Chicago
Hudik, Martin Francis *hospital administrator, educator*
Rapoza, Norbert Pacheco *medical association administrator, virologist*
Taira, Frances Snow *nurse, educator*
Thomas, Leona Marlene *health information educator*

Decatur
Houran, James Patrick *counselor, psychiatric technician*

Edwardsville
Teralandur, Parthasarathy Krishnaswamy *audiologist*

Kankakee
Schroeder, David Harold *health care facility executive*

Park Ridge
Darling, Cheryl MacLeod *health facility administrator, researcher*

Springfield
Campbell, Kathleen Charlotte Murphey *audiology educator*
Khardori, Nancy *infectious disease specialist*

Urbana
Visek, Willard James *nutritionist, animal scientist, physician, educator*

Westmont
McConnell, Patricia Ann *health facility administrator*

INDIANA

Bloomington
Kohr, Roland Ellsworth *hospital administrator*

Fort Wayne
Ferguson, Susan Katharine Stover *nurse, psychotherapist, consultant*
Frantz, Dean Leslie *psychotherapist*

Indianapolis
McBride, Angela Barron *nursing educator*
Stookey, George Kenneth *research institute administrator, dental educator*
Walther, Joseph Edward *health facility administrator, retired physician*

Terre Haute
Stoffer, Barbara Jean *research laboratory technician*

IOWA

Davenport
Bhatti, Iftikhar Hamid *chiropractic educator*
Lemke, Cindy Ann *support center founder and administrator*

Iowa City
Folkins, John William *speech scientist, educator*

Marshalltown
Packer, Karen Gilliland *cancer patient educator, researcher*

KANSAS

Bonner Springs
Elliott-Watson, Doris Jean *psychiatric, mental health and gerontological nurse educator*

Larned
Zook, Martha Frances Harris *retired nursing administrator*

Olathe
Poston, Ann Genevieve *psychotherapist, nurse*

Shawnee Mission
Jones, George Humphrey *retired healthcare executive, hospital facilities and communications consultant*

Topeka
Blue, Jeffrey Kenneth *neuromuscular and skeletal researcher*

KENTUCKY

Edgewood
Mick, Elizabeth Ellen *medical technologist*

LOUISIANA

Alexandria
Slipman, (Samuel) Ronald *hospital administrator*

Baton Rouge
Besing, Joan Marie *audiology and hearing science educator*
Cullen, John Knox, Jr. *hearing science educator*
Winkler, Steven Robert *hospital administrator*

Belle Chasse
Arimura, Akira *biomedical research laboratory administrator, educator*

Fort Polk
Riley, Francena *nurse, military officer*

Lafayette
Jeansonne, Gloria Janelle *laboratory administrator, medical technologist*

New Orleans
Howard, Richard Ralston, II *medical health advisor, researcher, financier*
Nakamoto, Tetsuo *nutritional physiology educator*
Pittman, Jacquelyn *mental health nurse, nursing educator*

Slidell
Levenson, Maria Nijole *medical technologist*

MAINE

Bangor
Beaupain, Elaine Shapiro *psychiatric social worker*

Cape Neddick
Ulan, Martin Sylvester *retired hospital administrator, health services consultant*

East Boothbay
Eldred, Kenneth McKechnie *acoustical consultant*

MARYLAND

Annapolis
Clampitt, Otis Clinton, Jr. *health agency executive*

Baltimore
Donnay, Albert Hamburger *environmental health engineer*
Sommer, Alfred *public health professional, epidemiologist*
Steinwachs, Donald Michael *public health educator*
Walton, Kimberly Ann *medical laboratory technician*

Bethesda
Alexander, Nancy Jeanne *health science facility administrator, researcher*
Camp, Frances Spencer *retired nurse*
Coelho, Anthony Mendes, Jr. *health science administrator*
Dickler, Howard Byron *biomedical administrator, research physician*
Fauci, Anthony Stephen *health facility administrator, physician*
Fisher, Suzanne Eileen *health scientist administrator*
Gallin, John Isaac *health science association administrator*
Geller, Ronald Gene *health administrator*
Gohagan, John Kenneth *medical institute administrator, educator*
Holmes, Margaret E. *health science administrator, researcher*
Hurd, Suzanne Sheldon *federal agency health science director*
Mason, James Ostermann *public health administrator*
Metzger, Henry *federal research institution administrator*
Notkins, Abner Louis *physician, researcher*
Obrams, Gunta Iris *health facility administrator*
Quinnan, Gerald Vincent, Jr. *medical educator*
Talbot, Bernard *government medical research facility official, physician*
Vaitukaitis, Judith Louise *medical research administrator*
Yu, Monica Yerk-lin *medical technologist, consultant*

Clinton
Sizemore, Carolyn Lee *nuclear medicine technologist*

Columbia
Dale, Charlene Boothe *international health administrator*
Imre, Paul David *mental health administrator*

Fort Detrick
Nelson, James Harold *health sciences administrator*

Owings Mills
Elkins, Robert N. *association executive*

Rockville
Gabelnick, Henry Lewis *medical research director*
Gruber, Kenneth Allen *health scientist, administrator*
Kerwin, Courtney Michael *public health scientist, administrator*
Long, Cedric William *health research facility executive*
McCormick, Kathleen Ann Krym *geriatrics nurse, federal agency administrator*
Rosenstein, Marvin *public health association administrator*

Silver Spring
Ceasor, Augusta Casey *medical technologist, microbiologist*

Temple Hills
Strauss, Simon Wolf *technical consultant*

Wheaton
Fawcett, Howard Hoy *chemical health and safety consultant*

MASSACHUSETTS

Amherst
McBride, Thomas Craig *physician, medical director*

Boston
Andrews, Sally May *healthcare administrator*
Gaintner, J(ohn) Richard *health facility executive, medical educator*
Jackson, Earl, Jr. *medical technologist*

Cambridge
Coate, David Edward *acoustician consultant*
Davis, Edgar Glenn *science and health policy executive*
Goldman, Peter *health science, chemistry, molecular pharmacology educator*

Hyannis
Williams, Ann Meagher *hospital administrator*

Ludlow
Budnick, Thomas Peter *social worker*

Medford
Junger, Miguel Chapero *acoustics researcher*

New Bedford
Merolla, Michele Edward *chiropractor*

Newton
Cavicchi, Leslie Scott *health facility administrator*

Townsend
Greene, Roland *chiropractor*

Vineyard Haven
Knowles, Christopher Allan *healthcare executive*

Waltham
Mitchell, Janet Brew *health services researcher*

Worcester
Latham, Eleanor Ruth Earthrowl *neuropsychology therapist*

MICHIGAN

Adrian
Cox, Chad William *medical technologist*

Ann Arbor
Brakel, Linda A. Wimer *psychoanalyst, researcher*
Gaston, Hugh Philip *marriage counselor, educator*
Pender, Nola J. *community health nursing educator, researcher*

Dearborn
Suchy, Susanne N. *nursing educator*

East Lansing
Gonzalez, Michael John *nutrition educator, nutriologist*
Reinhart, Mary Ann *medical association administrator*
Schemmel, Rachel Anne *food science and human nutrition educator, researcher*

Franklin
Sax, Mary Randolph *speech pathologist*

Fruitport
Anderson, Frances Swem *nuclear medical technologist*

Grosse Pointe Woods
Barth, Carolyn Lou *hospital administrator, microbiologist*

Livonia
Anas, Julianne Kay *administrative laboratory director*

Mason
Fisher, Marye Jill *physical therapist, educator*

Mount Pleasant
Logomarsino, Jack *nutrition educator*

Muskegon
Heyen, Beatrice J. *psychotherapist*

Rapid River
Olson, Marian Edna *nurse, social psychologist*

Royal Oak
Klosinski, Deanna Dupree *medical laboratory sciences educator*

MINNESOTA

Minneapolis
Bukonda, Ngoyi K. Zacharie *public health administrator*
Dahl, Gerald LuVern *psychotherapist, educator, consultant, writer*
Kralewski, John Edward *health service administration educator*
Sonsteby, Kristi Lee *healthcare consultant*

Moorhead
Larson, Betty Jean *dietitian, educator*

Rochester
Huse, Diane Marie *dietitian*

Stillwater
Anderson, Geraldine Louise *laboratory scientist*

MISSISSIPPI

Ocean Springs
McNulty, Matthew Francis, Jr. *health care administration educator, university administrator, consultant, horse and cattle breeder*

MISSOURI

Chesterfield
Biggerstaff, Randy Lee *sports medicine consultant*

Grandin
Wallace, Louise Margaret *critical care nurse*

Jefferson City
Nordstrom, James William *nutritionist, educator*

Kansas City
Eddy, Charles Alan *chiropractor*
Stern, Thomas Lee *physician, educator, medical association administrator*

Saint Louis
Bell, Laura Jeane *retired nurse*
Jobe, Muriel Ida *medical technologist*
Kiser, Karen Maureen *medical technologist, educator*
Schoenhard, William Charles, Jr. *health care executive*

NEBRASKA

Kearney
Goddard, David Benjamin *physician assistant, clinical perfusionist*

NEVADA

Fallon
Bolen, Terry Lee *optometrist*

Las Vegas
Francis, Timothy Duane *chiropractor*

NEW HAMPSHIRE

Manchester
Blake, Jeannette Belisle *psychotherapist*

NEW JERSEY

Clifton
Fowler, Cecile Ann *nurse, professional soloist*

Elizabeth
Buonanni, Brian Francis *health care facility administrator, consultant*

Mendham
Desjardins, Raoul *medical association administrator, financial consultant*

Morristown
Traeger, Donna Jean *health facility manager*

Murray Hill
Atal, Bishnu Saroop *speech research executive*

Newark
Klein, Marshall S. *health facility administrator*

Nutley
Mostillo, Ralph *medical association executive*

Piscataway
Gaeta, Vincent Ettore *laboratory technologist*

Princeton
Rolle, F. Robert *health care consultant*

Secaucus
Newton, V. Miller *medical psychotherapist, writer*

Teterboro
O'Brien, Joseph Edward *laboratory director*

Trenton
Kaus, Edward Guy *occupational health consultant*

Wyckoff
Van Dyk, Robert *health care center executive*

NEW MEXICO

Albuquerque
Clark, Teresa Watkins *psychotherapist, mental health counselor*
Dencoff, John Edgar *research technologist*

NEW YORK

Albany
Paravati, Michael Peter *medico-legal investigator, forensic sciences program administrator, correction specialist*

Bronx
Weiner, Richard Lenard *hospital administrator, educator, pediatrician*

Brooklyn
Astwood, William Peter *psychotherapist*
Bradford, Susan Kay *nutritionist*
Dobrin, Raymond Allen *psychometrician*
Peters, Mercedes *psychoanalyst*

Buffalo
Hare, Daphne Kean *medical association director, educator*

Clinton
McKee, Francis John *medical association executive, lawyer*

Cold Spring Harbor
Chmelev, Sandra D'Arcangelo *laboratory administrator*

East Hampton
Hellman, Harriet Louise *pediatric nurse practitioner*

Hampton Bays
Hoberman, Shirley E. *speech pathologist, audiologist*

Hicksville
Hallak, Joseph *optometrist*
Hart, Dean Evan *research optometrist*

Howard Beach
Istrico, Richard Arthur *physician*

Huntington
Grossfeld, Michael L. *hospital administrator*

Hurley
Soltanoff, Jack *nutritionist, chiropractor*

Livingston Manor
Zagoren, Joy Carroll *health facility director, researcher*

Malverne
Ryan, Suzanne Irene *nursing educator*

Massapequa
Margulies, Andrew Michael *chiropractor*

Nesconset
Feldman, Gary Marc *nutritionist, consultant*

New York
Billig, Robert Emmanuel *psychiatric social worker*
Clark, Lynne Wilson *speech and language pathology educator*
Goldberg, Harold Howard *psychologist, educator*
Hochberg, Irving *audiologist, educator*
Kirshenbaum, Richard Irving *public health physician*
Krasnow, Maurice *psychoanalyst, educator*
Manly, Carol Ann *speech pathologist*
Nauert, Roger Charles *health care executive*
Sharp, Victoria Lee *medical director*
Wiemer-Sumner, Anne-Marie *psychotherapist, educational administrator*
Witkin, Mildred Hope Fisher *psychotherapist, educator*

Norwich
Garzione, John Edward *physical therapist*

Ozone Park
Catalfo, Betty Marie *health service executive, nutritionist*

Pomona
Masters, Robert Edward Lee *neural re-education researcher, psychotherapist, human potential educator*

Rochester
Harris, Howard Alan *laboratory director*
Johnson, Jean Elaine *nursing educator*
Moore, Duncan Thomas *optics educator*
Snell, Karen Black *audiologist, educator*

Salt Point
Lackey, Mary Michele *physician assistant*

Staten Island
Sabido, Almeda Alice *mental health facility administrator*

Syracuse
Evans, Judy Anne *health center administrator*
Pelczer, István *spectroscopist*

Wantagh
Malden, Joan Williams *physical therapist*

Woodstock
Villchur, Edgar *audiology research scientist*

NORTH CAROLINA

Burlington
Mason, James Michael *biomedical laboratories executive*

Chapel Hill
Hackney, Anthony C. *nutrition and physiology educator, researcher*
Handler, Enid Irene *health care administrator, consultant*

Hickory
Kuehnert, Deborah Anne *medical center administrator*

Hope Mills
Baylor, John Patrick *nurse*

Raleigh
Ciraulo, Stephen Joseph *nurse, anesthetist*

Research Triangle Park
Olden, Kenneth *public health service administrator, researcher*

Winston Salem
Hutcherson, Karen Fulghum *nursing administrator*
Yeatts, Dorothy Elizabeth Freeman *nurse, retired county official, educator*

NORTH DAKOTA

Fargo
Hadley, Mary *nutritionist*

Grand Forks
Hunt, Janet Ross *research nutritionist*
Nielsen, Forrest Harold *research nutritionist*

OHIO

Cincinnati
Hensgen, Herbert Thomas *medical technologist*
Rubinstein, Jack Herbert *health center administrator, pediatrics educator*
Schubert, William Kuenneth *hospital medical center executive*

Columbus
Bai, Sungchul Charles *nutritionist*
Covault, Lloyd R., Jr. *hospital administrator, psychiatrist*
McNulty, Frank John *laboratory coordinator*
Richards, Ernest William *clinical nutritionist*

Dayton
Murphy, Martin Joseph, Jr. *cancer research center executive*
Nixon, Charles William *bioacoustician*

Elyria
Mahjoub, Elisabeth Mueller *health facility administrator*

Hamilton
Johnson, Pauline Benge *nurse, anesthetist*

Middletown
Easley, Michael Wayne *public health professional*

Oxford
Pfohl, Dawn Gertrude *laboratory executive*

Youngstown
Varma, Raj Narayan *nutrition educator, researcher*

OKLAHOMA

Bethany
Reinschmiedt, Anne Tierney *nurse, health care executive*

Norman
Dille, John Robert *physician*

Oklahoma City
Jones, Renee Kauerauf *health care administrator*
Thurman, William Gentry *medical research foundation executive, pediatric hematology and oncology physician, educator*

Tahlequah
Wickham, M(arvin) Gary *optometry educator*

Tulsa
Alexander, John Robert *hospital administrator, internist*
Quillin, Patrick *nutritionist, writer*

OREGON

Corvallis
Oldfield, James Edmund *nutrition educator*

Eugene
Watson, Mary Ellen *ophthalmic technologist*

Grants Pass
Selinger, Rosemary Celeste Lee *medical psychotherapist*

Medford
Linn, Carole Anne *dietitian*

Portland
Baker, Timothy Alan *healthcare administrator, educator, consultant*

Salem
Edge, James Edward *health care administrator*

PENNSYLVANIA

Bethlehem
Snyder, John Mendenhall *medical administrator, retired thoracic surgeon*

Blakeslee
Hayes, Alberta Phyllis Wildrick *retired health service executive*

Drexel Hill
Amoroso, Marie Dorothy *retired medical technologist*

Erie
Jones, William V(incent) *health center administrator*
Macosko, Paul John, II *psychotherapist*

Harrisburg
Gallaher, William Marshall *dental laboratory technician*

Landenberg
Aldrich, Nancy Armstrong *psychotherapist, clinical social worker*

New Castle
Flannery, Wilbur Eugene *health science association administrator, internist*

Philadelphia
Anyanwu, Chukwukre *alcohol and drug abuse facility administrator*
Doty, Richard Leroy *medical researcher*
Norwood, Carol Ruth *research laboratory administrator*
Sridhara, Channarayapatna Ramakrishna Setty *health facility administrator, consultant*

Pittsburgh
Abdelhak, Sherif Samy *health science executive*
Constantino-Bana, Rose Eva *nursing educator, researcher*
Omiros, George James *medical foundation executive*
Safar, Peter *emergency health care facility administrator, educator*

Reading
Bell, Frances Louise *medical technologist*

Slippery Rock
Fallon, L(ouis) Fleming, Jr. *public health educator, researcher*

State College
Moon, Marla Lynn *optometrist*

University Park
Achterberg, Cheryl Lynn *nutrition educator*

RHODE ISLAND

Providence
Petrucci, Jane Margaret *medical technologist, laboratory director*

SOUTH CAROLINA

Charleston
Stroud, Sally Dawley *nursing educator, researcher*

Clemson
Lee, Daniel Dixon, Jr. *nutritionist, educator*

Columbia
Cooper, William Allen, Jr. *audiologist*
Ramsey, Bonnie Jeanne *mental health facility administrator, psychiatrist*

Hartsville
Edson, Herbert Robbins *hospital executive*

Hilton Head Island
Stockard, Joe Lee *public health service officer, consultant*

Hopkins
Clarkson, Jocelyn Adrene *medical technologist*

Myrtle Beach
Madory, James Richard *hospital administrator, former air force officer*

TENNESSEE

Kingsport
Denton, David Lee *laboratory manager, chemical engineer*

Knoxville
Kidd, Janice Lee *nutritionist, consultant*

Maryville
Lucas, Melinda Ann *health facility director*

Memphis
Mendel, Maurice *audiologist, educator*
Ohman, Marianne *medical technologist*
Wolfson, Lawrence Aaron *hospital administrator*

Nashville
Cullen, Marion Permilla *nutritionist*

Oak Ridge
Snyder, Fred Leonard *health sciences administrator*

TEXAS

Austin
Nicholas, Nickie Lee *industrial hygienist*

Dallas
Barnett, Peter Ralph *health science facility administrator, dentist*
Hanratty, Carin Gale *pediatric nurse practitioner*
Kaufman, Tina Marie *physician recruiter, research consultant, presentation graphics consultant*
Metzler, Jerry Don *nursing administrator*
Roeser, Ross Joseph *audiologist, educator*
Withrow, Lucille Monnot *nursing home administrator*

El Paso
Mitchell, Paula Rae *nursing educator*
Seaman, Edwin Dwight *physician, laboratory director*

Galveston
Head, Elizabeth Spoor *retired medical technologist*
Terrebonne, Annie Marie *medical technologist, educator, clinical laboratory scientist*

Grand Prairie
Payne, Anthony Glen *clinical nutritionist, naturopathic physician*

Houston
Becker, Frederick Fenimore *cancer center administrator, pathologist*
Burns, Sally Ann *medical association administrator*
Frenger, Paul Fred *medical computer consultant, physician*
Kwok, Lance Stephen *optometry educator, vision researcher*
Lotze, Evie Daniel *psychodramatist*

Humble
Stevens, Elizabeth *psychotherapist*

Huntsville
Vick, Marie *retired health science educator*

Irving
Donnelly, Barbara Schettler *medical technologist*

Kilgore
Springer, Andrea Paulette Ryan *physical therapist, biology educator*

Mcallen
Farias, Fred, III *optometrist*

Nacogdoches
Brennan, Thomas George, Jr. *audiologist, speech-language pathologist*

Plano
Franklin, Thomas Doyal, Jr. *medical research administrator*

San Antonio
Barnes, Betty Rae *counselor*
Cook, Harold Rodney *military officer, medical facility administrator*

UTAH

Clearfield
Ashmead, Harve DeWayne *nutritionist, executive, educator*

Kaysville
Ashmead, Allez Morrill *speech-hearing-language pathologist, orofacial myologist, consultant*

Salt Lake City
Belliston, Edward Glen *medical facility administrator, consultant*

VIRGINIA

Annandale
Abdellah, Faye Glenn *retired public health service executive*

Falls Church
Adams, Nancy R. *nurse, military officer*
Fink, Charles Augustin *behavioral systems scientist*

Norfolk
Burtoft, John Nelson, Jr. *cardiovascular physician assistant*

Richmond
Freund, Emma Frances *medical technologist*
Hardage, Page Taylor *health care administrator*

Springfield
McMillan, Ronald Therow *optician*

WASHINGTON

Bellevue
Lipkin, Mary Castleman Davis (Mrs. Arthur Bennett Lipkin) *retired psychiatric social worker*

Port Orchard
Kammin, William Robert *environmental laboratory director*

Puyallup
Walize, Reuben Thompson, III *health research administrator*

Seattle
Day, Robert Winsor *research administrator*
Goodall, Frances Louise *nurse, production company assistant*

Stanwood
Lipscomb, David Milton *audiologist, consultant*

Tacoma
Ketchersid, Wayne Lester, Jr. *medical technologist*

WEST VIRGINIA

Rainelle
Scott, Pamela Moyers *physician assistant*

Ronceverte
Johnson, Bret Gossard *dietitian, consultant*

WISCONSIN

Hartford
Janzen, Norine Madelyn Quinlan *medical technologist*

Janesville
Morgan, Donna Jean *psychotherapist*

Madison
Berven, Norman Lee *counselor, psychologist, educator*

Menasha
Mahnke, Kurt Luther *psychotherapist, clergyman*

Menomonie
Seaborn, Carol Dean *nutrition researcher*

Milwaukee
Babcock, Janice Beatrice *health care coordinator*
Montgomery, Robert Renwick *medical association administrator, educator*
Puta, Diane Fay *medical staff services director*

Whitewater
Newman, Lisa Ann *speech pathologist, educator*

WYOMING

Cheyenne
Laycock, Anita Simon *psychotherapist*

TERRITORIES OF THE UNITED STATES

PUERTO RICO

Rio Piedras
Perez, Victor *medical technologist, laboratory director*

MILITARY ADDRESSES OF THE UNITED STATES

EUROPE

APO
Foster, Kirk Anthony *emergency medical service administrator, educator, consultant*

CANADA

ONTARIO

Kingston
McGeer, James Peter *research executive, consultant*

Toronto
Chasin, Marshall Lewis *audiologist, educator*
De Nil, Luc Frans *speech-language pathologist*
Ramakrishnan, Ramani *acoustician*

Windsor
Courtenay, Irene Doris *nursing consultant*

QUEBEC

Montreal
Bailar, John Christian, III *public health educator, physician, statistician*
Gallagher, Tanya Marie *speech pathologist, educator*

SASKATCHEWAN

Saskatoon
Patience, John Francis *nutritionist*

ARGENTINA

Buenos Aires
Nitka, Hermann Guillermo *hospital administrator*

AUSTRIA

Vienna
Leibetseder, Josef Leopold *nutritionist, educator*

BANGLADESH

Dhaka
Abeyesundere, Nihal Anton Aelian *health organization representative*

BOLIVIA

La Paz
Hartmann, Luis Felipe *health science association administrator, endocrinologist*

CHINA

Beijing
Zhang, Li-Xing *physician, medical facility executive*

Nanjing
Du, Gonghuan *acoustics educator*

Shanghai
Wang, Ji-Qing (Chi-Ching Wong) *acoustician, educator*

DENMARK

Lyngby
Bjørnø, Leif *industrial acoustics educator*

ENGLAND

Cambridge
Barlow, Horace Basil *physiologist*

GERMANY

Bochum
Blauert, Jens Peter *acoustician, educator*

Bonn
Schwerdtfeger, Walter Kurt *public health official, researcher*

Jülich
Feinendegen, Ludwig Emil *retired hospital and research institute director*

Marburg
Rienhoff, Otto *physician, medical informatics educator*

Wiesbaden
Sachse, Guenther *health facility administrator, medical educator*

INDONESIA

Bandung
Wijaya, Andi *clinical laboratory executive, clinical chemistry educator*

ISRAEL

Jerusalem
Berns, Donald Sheldon *research scientist*

ITALY

Novara
Cerofolini, Gianfranco *laboratory administrator*

Pianezza
Badetti, Rolando Emilio *health science facility administrator*

JAPAN

Gotsu
Hirayama, Chisato *healthcare facility administrator, physician, educator*

Kamakura
Yamamoto, Mikio *public health and human ecology researcher, educator*

Osaka
Nakagawa, Yuzo *laboratory administrator*

THE NETHERLANDS

Amsterdam
Dreschler, Wouter Albert *audiologist, researcher*

REPUBLIC OF KOREA

Daejeon
Son, Ki Sub *health facility administrator, surgeon*

SEYCHELLES

Victoria
Shamlaye, Conrad Francois *health facility administrator, epidemiologist*

SINGAPORE

Singapore
Tong, Yit Chow *research scientist*

SWEDEN

Gothenburg
Wikman, Georg Karl *institution administrator*

Östersund
Jonsson, Lars Olov *hospital administrator*

Stockholm
Lilliehöök, J(ohan) Björn O(lof) *health science association administrator*
Sundberg, Johan Emil Fredrik *communications educator*

SWITZERLAND

Lausanne
Guillemin, Michel Pierre *occupational hygienist*

Zurich
Dechmann, Manfred *psychotherapist*

TANZANIA

Dar es Salaam
Paalman, Maria Elisabeth Monica *public health executive*

ADDRESS UNPUBLISHED

Abbott, Regina A. *neurodiagnostic technologist, consultant, business owner*
Alzofon, Julia *laboratory administrator*
Ball, Edna Marion *cardiac rehabilitation nurse educator*
Black, Patricia Jean *medical technologist*
Bois, Pierre *former medical research organization executive*
Bolton, Julia Gooden *hospital administrator*
Broadwell, Milton Edward *nurse, anesthetist, educator*
Brown, Geraldine *nurse, freelance writer*
Brown, Linda Joan *psychotherapist, psychoanalyst*
Clements, Michael Craig *health services consulting executive, retired renal dialysis technician*
Cowan, Mary Elizabeth *medical technologist*
Ellis, Wayne Enoch *nurse, anesthetist, air force officer*
Franks, Paul Todd *laboratory manager*
Herrmann, Walter *retired laboratory administrator*
Howe, John Prentice, III *health science center executive, physician*
Jahn, Billie Jane *nursing educator, consultant*
Keller, George Henry *research administrator, consulting biochemist*
Mangels, Ann Reed *nutrition educator, researcher*
Merenbloom, Robert Barry *hospital and medical school administrator*
Moffatt, Hugh McCulloch, Jr. *hospital administrator, physical therapist*
Moliere, Jeffrey Michael *cardio-pulmonary administrator*
Mooneyhan, Esther Louise *nurse, educator*
Oakes, Ellen Ruth *psychotherapist, health institute administrator*
Pepper, Dorothy Mae *nurse*
Sauvage, Lester Rosaire *health facility administrator, cardiovascular surgeon*
Scott, Amy Annette Holloway *nursing educator*
Sieloff, Christina Lyne *nurse*
Sovde-Pennell, Barbara Ann *sonographer*
Spencer, William Stewart *radiologic technologist*
Thomson, Grace Marie *nurse, minister*
Tourtillott, Eleanor Alice *nurse, educational consultant*
Vogel, H. Victoria *psychotherapist, educator*
Wharton, Thomas William *health administration executive*
Wong, Sue Siu-Wan *health educator*

LAW

UNITED STATES

ARIZONA

Scottsdale
Barbee, Joe Ed *lawyer*

CALIFORNIA

Irvine
Gess, Albin Horst *lawyer*

La Jolla
Seidman, Stephanie Lenore *lawyer*

Los Gatos
Dumas, Jeffrey Mack *lawyer*

COLORADO

Boulder
Forest, Carl Anthony *lawyer*

Colorado Springs
Kubida, William Joseph *lawyer*

DISTRICT OF COLUMBIA

Washington
Downey, Richard Morgan *lawyer*
Gross, Thomas Paul *lawyer*
Starrs, James Edward *law and forensics educator, consultant*

GEORGIA

Atlanta
Chisholm, Tommy *lawyer, utility company executive*
Fleming, Julian Denver, Jr. *lawyer*

ILLINOIS

Chicago
Long, Kevin Jay *medicolegal consultant*

INDIANA

Indianapolis
Sowers, Edward Eugene *lawyer*

MASSACHUSETTS

Cambridge
Weiler, Paul Cronin *law educator*

MISSOURI

Independence
Cady, Elwyn Loomis, Jr. *medicolegal consultant, educator*

Saint Louis
Lucchesi, Lionel Louis *lawyer*

NEW HAMPSHIRE

Concord
Rines, Robert Harvey *lawyer, inventor, law center executive, educator*

NEW MEXICO

Questa
Lamb, Margaret Weldon *lawyer*

NEW YORK

Commack
Frankenberger, Glenn F(rances) *lawyer*

Great Neck
Wachsman, Harvey Frederick *lawyer, neurosurgeon*

New York
Berle, Peter Adolf Augustus *lawyer, association executive*
Hoffberg, Steven Mark *lawyer*
Huettner, Richard Alfred *lawyer*

Poughkeepsie
McEnroe, Caroline Ann *legal assistant*

OHIO

Dayton
Muir, Herman Stanley, III *lawyer*

PENNSYLVANIA

King Of Prussia
Bramson, Robert Sherman *lawyer*

TEXAS

Dallas
Vestal, Tommy Ray *lawyer*

VIRGINIA

Charlottesville
Wadlington, Walter James *legal educator*

WISCONSIN

Madison
Woods, Thomas Fabian *lawyer*

CANADA

QUEBEC

Montreal
Somerville, Margaret Anne Ganley *law educator*

AUSTRALIA

Sydney
Hinde, John Gordon *lawyer, solicitor*

JAPAN

Tokyo
Ibayashi, Tsuguio *legal educator*
Iida, Yukisato *lawyer*

ADDRESS UNPUBLISHED

Casella, Peter F(iore) *patent and licensing executive*
Colodny, Edwin Irving *lawyer, retired airline executive*
Kasper, Horst Manfred *lawyer*
King, Patricia Ann *law educator*

LIFE SCIENCES: AGRICULTURE

UNITED STATES

ALABAMA

Maxwell AFB
Frank, Paul Sardo, Jr. *forester, air force officer*

ARIZONA

Flagstaff
Avery, Charles Carrington *forestry educator, researcher*

Tucson
Fritts, Harold Clark *dendrochronology educator, researcher*
Kaltenbach, Carl Colin *agriculturist, educator*

ARKANSAS

Monticello
Cain, Michael Dean *research forester*

CALIFORNIA

Berkeley
Vaux, Henry James *forest economist, educator*

Fresno
Harvey, John Marshall *agricultural scientist*

Pacific Palisades
Jennings, Marcella Grady *rancher, investor*

Palo Alto
Sandmeier, Ruedi Beat *agricultural research executive*

Saint Helena
Amerine, Maynard Andrew *enologist, educator*

San Diego
Stowell, Larry Joseph *agricultural consultant*

DISTRICT OF COLUMBIA

Washington
Mowbray, Robert Norman *forest ecologist, government agricultural and natural resource development officer*

FLORIDA

Arcadia
Durkin, John Charles *agriculturist*

Gainesville
Seale, James Lawrence, Jr. *agricultural economics educator, international trade researcher*

Ona
Pate, Findlay Moye *agriculture educator, university center director*

Port Charlotte
Parvin, Philip E. *retired agricultural researcher and educator*

Winter Haven
Grierson, William *retired agriculture educator, consultant*

GEORGIA

Atlanta
Clerke, William Henry III *forester*
Killorin, Edward Wylly *lawyer, tree farmer*

Griffin
Arkin, Gerald Franklin *agricultural research administrator, educator*

Thomasville
Buckner, James Lee *forester, biologist*

HAWAII

Honolulu
Ching, Chauncey Tai Kin *agricultural economics educator*

IDAHO

Boise
Burton, Lawrence DeVere *agriculturist, educator*

ILLINOIS

Chicago
Mundlak, Yair *agriculture and economics educator*

De Kalb
Robison, Norman Glenn *tropical research director*

Naperville
Harms, David Jacob *agricultural consultant*

Peoria
Ehmke, Dale William *agriculturist*

IOWA

Ames
Freeman, Albert E. *agricultural science educator*

Parkersburg
Boukerrou, Lakhdar *agricultural researcher*

KANSAS

Garden City
Dick, Gary Lowell *agricultural consultant, educator*

Parsons
Lomas, Lyle Wayne *agricultural research administrator, educator*

KENTUCKY

Lexington
Reed, Michael Robert *agricultural economist*

MARYLAND

Monkton
Doepkens, Frederick Henry *agriscience educator*

MASSACHUSETTS

Cambridge
Goldberg, Ray Allan *agribusiness educator*

Shutesbury
Smulski, Stephen John *wood scientist, consultant*

MICHIGAN

Midland
Davidson, John Hunter *agriculturist*

MINNESOTA

Elgin
Meyer, Robert Verner *farmer*

Saint Paul
Cheng, H(wei) H(sien) *agriculture and environmental science educator*

MISSISSIPPI

Starkville
Friend, Alexander Lloyd *forester educator*

NEW MEXICO

Carlsbad
Houghton, Woods Edward *agricultural science educator*

OKLAHOMA

Ada
Thompson, Rahmona Ann *plant taxonomist*

OREGON

Oregon City
Wall, Brian Raymond *forest economist, consultant*

Portland
Hagenstein, William David *consulting forester*

PENNSYLVANIA

Langhorne
Venable, Robert Ellis *crop scientist*

Shiremanstown
Willis, Donald J. *agricultural research professional*

University Park
Manbeck, Harvey B. *agriculturist, educator*

RHODE ISLAND

North Scituate
Dupree, Thomas Andrew *forester, state official*

SOUTH CAROLINA

Mullins
Rogers, Colonel Hoyt *agricultural consultant*

TEXAS

Bryan
Van Arsdel, Eugene Parr *tree pathologist, consultant meteorologist*

College Station
Jack, Steven Bruce *forest ecologist, educator*

San Angelo
Menzies, Carl Stephen *agricultural research administrator, ruminant nutritionist*

Valley View
Wallace, Donald John, III *rancher, former pest control company executive*

VIRGINIA

Dillwyn
Moseley, John Marshall *nurseryman*

WASHINGTON

West Richland
Cole, Charles R. *hydrologist, researcher*

WEST VIRGINIA

Morgantown
Gottschalk, Kurt William *research forester*

CANADA

ALBERTA

Beaverlodge
McElgunn, James Douglas *agriculturist, researcher*

Lethbridge
Sonntag, Bernard H. *agrologist, research executive*

BRITISH COLUMBIA

Burnaby
Runka, Gary G. *agricultural company executive*

Summerland
Dueck, John *agricultural researcher, plant pathologist*

Victoria
Wilkinson, R. L. *agriculturalist*

MANITOBA

Winnipeg
Bushuk, Walter *agricultural studies educator*
Storgaard, Anna K. *agriculturalist*

ONTARIO

Ottawa
Lister, E. Edward *animal science consultant*

PRINCE EDWARD ISLAND

Charlottetown
Gupta, Umesh Chandra *agriculturist, soil scientist*

QUEBEC

Laval
Frisque, Gilles *forestry engineer*

Lennoxville
Deschenes, Jean-Marie *agriculturist, researcher*

Montreal
Fortin, Joseph André *forestry educator, researcher*

SASKATCHEWAN

Saskatoon
Braidek, John George *agriculturist*
Storey, Gary Garfield *agricultural studies educator*

COLOMBIA

Cali
Cock, James Heywood *agricultural scientist*

ENGLAND

London
Kassam, Amirali Hassanali *agricultural scientist*

Wallington
Green, Maurice Berkeley *agrochemical research consultant*

FRANCE

Le Cannet des Maures
Gudin, Serge *plant breeder*

HUNGARY

Budapest
Stefanovits, Pál *agriculturalist*

ITALY

Bari
Nicastro, Francesco Vito Mario *agricultural science educator*

JAPAN

Tokyo
Kitani, Osamu *agriculture educator*

PANAMA

Panama
Tarte, Rodrigo *agriculture educator, researcher, consultant*

THE PHILIPPINES

Malabon Manila
Pizarro, Antonio Crisostomo *agricultural educator, researcher*

LIFE SCIENCES: AGRONOMY

UNITED STATES

ALABAMA

Normal
Mays, David Arthur *agronomy educator, small business owner*

ARKANSAS

Fayetteville
Beyrouty, Craig A. *agronomist, educator*

CALIFORNIA

Davis
Miller, Milton David *agronomist, educator*

Fresno
Rolfs, Kirk Alan *agronomist*

DISTRICT OF COLUMBIA

Washington
Plucknett, Donald Lovelle *scientific advisor*

FLORIDA

Gainesville
Edwardson, John Richard *agronomist*

Tallahassee
Onokpise, Oghenekome Ukrakpo *agronomist, educator, forest geneticist, agroforester*

GEORGIA

Athens
Boerma, Henry Roger *agronomist, educator*
Sumner, Malcom Edward *agronomist, educator*

HAWAII

Aiea
Heinz, Don J. *agronomist*

IDAHO

Pocatello
Smith, John Julian *agronomist*

ILLINOIS

Danville
Craig, Hurshel Eugene *agronomist*

INDIANA

Hobart
Seeley, Mark *agronomist*

West Lafayette
Johannsen, Chris Jakob *agronomist, educator, administrator*

IOWA

Ames
Bremner, John McColl *agronomy and biochemistry educator*
Owen, Michael *agronomist, educator*
Tabatabai, M. Ali *agronomist*

Sioux City
Petersen, Perry Marvin *agronomist*

KANSAS

Manhattan
Posler, Gerry Lynn *agronomist, educator*

LOUISIANA

Baton Rouge
Martin, Freddie Anthony *agronomist, educator*

MARYLAND

Adelphi
Miller, Raymond Jarvis *agronomist, college dean, university official*

Hyattsville
Pittarelli, George William *agronomist, researcher*

Princess Anne
Joshi, Jagmohan *agronomist, consultant*

MICHIGAN

Saginaw
Faubel, Gerald Lee *agronomist, golf course superintendent*

MINNESOTA

Saint Cloud
Kirick, Daniel John *agronomist*

Saint Paul
Hicks, Dale R. *agronomist, educator*
Rasmusson, Donald C. *agronomist, educator*
Schmitt, Michael A. *agronomist, educator*
Wyse, Donald L. *agronomist, educator*

MISSISSIPPI

Poplarville
Edwards, Ned Carmack, Jr. *agronomist, university program director*

Stoneville
Rutger, J. Neil *agronomy research administrator*

MISSOURI

Columbia
Coe, Edward Harold, Jr. *agronomist, educator, geneticist*
Mitchell, Roger Lowry *agronomy educator*
Poehlmann, Carl John *agronomist, researcher*

NEBRASKA

Clay Center
Roeth, Frederick Warren *agronomy educator*

Lincoln
Francis, Charles Andrew *agronomy educator, consultant*

NEW YORK

Ithaca
Alexander, Martin *educator, researcher*

NORTH CAROLINA

Raleigh
Dunphy, Edward James *crop science extension specialist*

NORTH DAKOTA

Fargo
Cross, Harold Zane *agronomist, educator*

OHIO

Columbus
Velagaleti, Ranga Rao *agronomist, environmental scientist*

PENNSYLVANIA

Kutztown
Janke, Rhonda Rae *agronomist, educator*

University Park
Berg, Clyde Clarence *research agronomist*

RHODE ISLAND

Cranston
Mruk, Charles Karzimer *agronomist*

SOUTH CAROLINA

Florence
Kittrell, Benjamin Upchurch *agronomist*

TENNESSEE

Nashville
Westbrook, Fred Emerson *agronomist, educator*

TEXAS

Beaumont
Bollich, Charles N. *agronomist*

Lubbock
Rummel, Don *agronomist*

VIRGINIA

Blacksburg
Alley, Marcus M. *agronomy educator*

Fredericksburg
Moorman, William Jacob *agronomist, consultant*

Leesburg
Mokhtarzadeh, Ahmad Agha *agronomist, consultant*

WISCONSIN

Madison
Barnes, Robert F. *agronomist*
Briggs, Rodney Arthur *agronomist, consultant*
Carter, Paul R. *agronomist consultant*

TERRITORIES OF THE UNITED STATES

PUERTO RICO

Corozal
Rodriguez Garcia, Jose A. *agronomist, investigator*

VIRGIN ISLANDS

Kingshill
Crossman, Stafford Mac Arthur *agronomist, researcher*

CANADA

SASKATCHEWAN

Regina
Webster, Alexander James *agrologist*

Saskatoon
Harvey, Bryan Laurence *crop science educator*

MEXICO

Aguascalientes
Santana-Garcia, Mario A. *plant physiologist*

COLOMBIA

Cali
Voysest, Oswaldo *agronomist, researcher*

GERMANY

Stuttgart
Pollmer, Wolfgang Gerhard *agronomy educator*

THE PHILIPPINES

Diffun
Temanel, Billy Estoque *agronomy research director, educator, consultant*

LIFE SCIENCES: BIOLOGY

UNITED STATES

ALABAMA

Auburn
Gandhi, Shailesh Ramesh *biotechnologist*

Birmingham
Brown, Ronnie Jeffrey *biologist, researcher*

Dauphin Island
Plakas, Steven Michael *biological research scientist*

Montevallo
Braid, Malcolm Ross *biology educator*

Tuscaloosa
Darden, William Howard, Jr. *biology educator*

ARIZONA

Tucson
Burgess, Kathryn Hoy *biologist*
Winfree, Arthur Taylor *biologist, educator*

ARKANSAS

Fayetteville
Etges, William James *biologist*

Little Rock
McAllister, Russell Benton *biologist, researcher*

CALIFORNIA

Alta Loma
Myers, Edward E. *retired biology and anthropology educator*

Berkeley
Penry, Deborah L. *biologist, educator*
Pfeiffer, Juergen Wolfgang *biologist*
Wake, David Burton *biology educator, researcher*

Chico
Kistner, David Harold *biology educator*

Davis
Armstrong, Peter Brownell *biologist*
Gifford, Ernest Milton *biologist, educator*
Grey, Robert Dean *biology educator*
Stemler, Alan James *plant biology educator*

Fresno
Gray, Jennifer Emily *biology educator*

Hayward
Baalman, Robert Joseph *biology educator*

Irvine
Bennett, Albert Farrell *biology educator*
Bryant, Peter James *biologist, educator*
Carpenter, F. Lynn *biology educator*
Graves, Joseph Lewis, Jr. *evolutionary biologist, educator*

La Jolla
Dulbecco, Renato *biologist, educator*

La Verne
Good, Harvey Frederick *biologist, educator*

Los Angeles
Lunt, Owen Raynal *biologist, educator*
McClure, William Owen *biologist*
Mohr, John Luther *biologist, environmental consultant*
Slavkin, Harold C. *biologist*
Stewart, Brent Kevin *radiological science educator*

Menlo Park
Newcomb, Robert Whitney *biotechnologist, neuroscience researcher*

Moreno Valley
Hamill, Carol *biologist, writing instructor*

Oakland
Whitsel, Richard Harry *biologist*

Oceanside
Hofmann, Frieder Karl *biotechnologist, consultant*

Pacific Grove
Beidleman, Richard Gooch *biologist, educator*

Palo Alto
Briggs, Winslow Russell *plant biologist, educator*

Pasadena
Lewis, Edward B. *biology educator*
Owen, Ray David *biology educator*
Rothenberg, Ellen *biologist*
Sperry, Roger Wolcott *neurobiologist, educator*
Sternberg, Paul Warren *biologist, educator*

Richmond
Balakrishnan, Krishna (Balki Balakrishnan) *biotechnologist, corporate executive*

Sacramento
Meral, Gerald Harvey *biologist*
Udvardy, Miklos Dezso Ferenc *biology educator*

San Diego
Archibald, James David *biology educator, paleontologist*
Crick, Francis Harry Compton *biologist, educator*
Helinski, Donald Raymond *biologist, educator*
Weinrich, James Donald *psychobiologist, educator*

San Francisco
Brown, Walter Creighton *biologist*
Cape, Ronald Elliot *biotechnology company executive*
Randall, Janet Ann *biology educator, researcher*
Wolff, Sheldon *radiobiologist, educator*

San Ramon
Montgomery, Elizabeth Ann *clinical research consultant*

Santa Cruz
Talamantes, Frank J. *biology educator*

Santa Rosa
Sibley, Charles Gald *biologist, educator*

Shaver Lake
Warren, Barbara Kathleen (Sue Warren) *wildlife biologist*

South San Francisco
Pitcher, Wayne Harold, Jr. *biotechnology company executive*

Stanford
Contag, Pamela Reilly *biologist, researcher*
Ehrlich, Paul Ralph *biology educator*
Liu, Guosong *neurobiologist*
Scott, Matthew Peter *biology educator*
Yanofsky, Charles *biology educator*

Woodland
Phan, Chuong Van *biotechnologist*

COLORADO

Carbondale
Cowgill, Ursula Moser *biologist, educator, environmental consultant*

Colorado State University
Moore, Janice Kay *biology educator*
Whicker, Floyd Ward *biology educator, ecologist*

Denver
Ferrell, Rebecca V. *biology educator*

Fort Collins
Ward, James Vernon *biologist, educator*

Ward
Benedict, Audrey DeLella *biologist, educator*

Westminster
Dotson, Gerald Richard *biology educator*

CONNECTICUT

Hartford
Child, Frank Malcolm *biologist educator*

Monroe
Turko, Alexander Anthony *biology educator*

New Haven
Altman, Sidney *biology educator*
Buss, Leo William *biologist, educator*
Goldsmith, Mary Helen M. *biology educator*
Seilacher, Adolf *biologist, educator*
Snyder, Michael *biology educator*

Norwalk
Reiss, Betti *biological and medical researcher, medical writer*

Storrs Mansfield
Hiestand, Nancy Laura *biology researcher*

DELAWARE

Lewes
Curtis, Lawrence Andrew *biologist, educator*

DISTRICT OF COLUMBIA

Washington
Alleva, John James *research biologist*
De Fabo, Edward Charles *photobiology and photoimmunology, research scientist, educator*
Harshbarger, John Carl *pathobiologist*
Lovejoy, Thomas Eugene *tropical and conservation biologist, association executive*
Nabholz, Joseph Vincent *biologist, ecologist*
Rao, Venigalla Basaveswara *biology educator*
Sparrowe, Rollin D. *wildlife biologist*
Suarez Quian, Carlos Andrés *biology educator*
Williams, Luther Steward *biologist, federal agency administrator*

FLORIDA

Coral Gables
Wilson, David Louis *biologist*

Fort Pierce
Rice, Mary Esther *biologist*

Gainesville
Buchholz, Richard *ethologist*
Drury, Kenneth Clayton *biological scientist*
Williams, Norris Hagan, Jr. *biologist, educator, curator*

Homestead
Ramos-Ledon, Leandro Juan *biologist*

Miami
Chen, Chun-fan *biology educator*

Palm Beach
Hopper, Arthur Frederick *biological science educator*

Pensacola
Mohrherr, Carl Joseph *biologist*

Punta Gorda
Beever, James William, III *biologist*

Saint Petersburg
Eldridge, Peter John *fishery biologist*
Ferguson, John Carruthers *biologist*

Tallahassee
Friedmann, E(merich) Imre *biologist, educator*

GEORGIA

Atlanta
Lucchesi, John C. *biology educator*
Spurlin, Lisa Turner *biologist*

Byron
Nyczepir, Andrew Peter *nematologist*

Demorest
Wainberg, Robert Howard *biology educator*

Kennesaw
McCoy, R. Wesley *biology educator*

Macon
Volpe, Erminio Peter *biologist, educator*

Milledgeville
Jones, Harold Charles *retired biologist*

Statesboro
Vives, Stephen Paul *biology educator, fish biologist*

Waycross
Nienow, James Anthony *biologist, educator*

HAWAII

Honolulu
Abbott, Isabella Aiona *biology educator*
Smith, Albert Charles *biologist, educator*

IDAHO

Moscow
Marshall, John David *forest biologist*
Scott, J(ames) Michael *research biologist*

ILLINOIS

Argonne
Drucker, Harvey *biologist*

Carlinville
Zalisko, Edward John *biology educator*

Chicago
Bieler, Rudiger *biologist, curator*
Castignetti, Domenic *microbiologist, biology educator*
Clevidence, Derek Edward *biologist*
Dumbacher, John Philip *evolutionary biologist*
Stark, Benjamin Chapman *biology educator, molecular biologist*

Dixon
Youker, David Eugene *biology educator*

Downers Grove
Mayer, Alejandro Miguel *biologist, educator*

Evanston
Galbreath, Gary John *biology educator, researcher*

Galesburg
Brodl, Mark Raymond *biology educator*

Libertyville
Munson, Norma Frances *biologist, ecologist, nutritionist, educator*

Macomb
Sather, J. Henry *biologist*
Stidd, Benton Maurice *biologist, educator*

Normal
Anderson, Roger Clark *biology educator*

Palos Heights
Wolff, Robert John *biology educator*

Urbana
Ducoff, Howard S. *radiation biologist*
Huang, Zhi-Yong *honey bee biologist*

Wilmette
Veneziano, Philip Paul *biologist, educator*

INDIANA

Butler
Ford, Lee Ellen *scientist, educator, retired lawyer*

Chesterton
Wiemann, Marion Russell, Jr. *retired executive, biologist, microscopist*

Fort Wayne
Gillespie, Robert Bruce *biology educator*

Indianapolis
Belagaje, Rama M. *biotechnology scientist*
Schaible, Robert Hilton *biologist*

Muncie
Mertens, Thomas Robert *biology educator*

Notre Dame
Bender, Harvey A. *biology educator*
Fraser, Malcolm James, Jr. *biological sciences educator*

Upland
Whipple, Andrew Powell *biology educator*

Valparaiso
Shipley-Phillips, Jeanette Kay *aquatic biologist*

West Lafayette
Sherman, Louis Allen *biologist, researcher*
Wasserman, Gerald Steward *psychobiology educator*

IOWA

Ames
Isely, Duane *biology and botany educator*
Wesley, Irene Varelas *research microbiologist*

Cedar Falls
Wiens, Darrell John *biologist, educator*

Iowa City
Solursh, Michael *biology educator, researcher*

KANSAS

Emporia
Schrock, John Richard *biology educator*

Kansas City
Smith, Donald Dean *biologist, educator*

Lawrence
Armitage, Kenneth Barclay *biology educator, ecologist*
Bennett, Stephen Christopher *biology educator*
Hersh, Robert Tweed *biology educator*

Manhattan
Barkley, Theodore Mitchell *biology educator*
Bechtel, Donald Bruce *biologist, educator, research chemist*
Fitch, Gregory Kent *biologist*
Johnson, Terry Charles *biologist, researcher*
Stalheim-Smith, Ann *biology educator*

Wichita
Zimniski, Stephen Joseph *biologist, educator*

KENTUCKY

Frankfort
Grubbs, Jeffrey Thomas *environmental biologist*

Lexington
Rodriguez, Lorraine Ditzler *biologist, consultant*
Sih, Andrew *biologist, educator*

LOUISIANA

Baton Rouge
Hansel, William *biology educator*
Killebrew, Charles Joseph *biologist*
Remsen, James Vanderbeek, Jr. *biologist, museum curator*

Lafayette
Deaton, Lewis Edward *biology educator*

Mandeville
Pollock, Jack Paden *biology and dental educator, consultant, free-lance writer, retired army officer*

Monroe
Baum, Lawrence Stephen *biologist, educator*

MAINE

Bar Harbor
Zoidis, Ann Margaret *biologist, researcher*

South Windham
DuBourdieu, Daniel John *biotechnologist, researcher*

MARYLAND

Baltimore
Brock, Mary Anne *research biologist, consultant*
Brown, Donald David *biology educator*
Coleman, Richard Walter *biology educator*
Hartman, Philip Emil *biology educator*
Nathans, Daniel *biologist*
Pollard, Thomas Dean *biologist, educator*
Saffer, Linda Diane *biology researcher*

Bethesda
Dawid, Igor Bert *biologist*
Elson, Hannah Friedman *research biologist*
Miller, Alexandra Cecile *radiation biologist, researcher, educator*
Myers, Lawrence Stanley, Jr. *radiation biologist*
Reid, Janet Warner *biologist consultant*
Robison, Wilbur Gerald, Jr. *research biologist*
Woolley, George Walter *biologist, geneticist, educator*

Columbia
Barton, William Elliott *biologist, administrator*
Beach, Robert Mark *biologist*

Frederick
Boyd, V(irginia) Ann Lewis *biology educator*
Moore, Sharon Pauline *biologist*

Laurel
Ellis, David H. *biologist, research behaviorist*

MASSACHUSETTS

Beverly
Keating, Carole Joanna *biotechnologist*

Boston
Call, Katherine Mary *biologist*
Carradini, Lawrence *comparative biologist, science administrator*
Gilmore, Thomas David *biologist*
Hagar, William Gardner, III *photobiology educator*
Jacobson, Gary Ronald *biology educator, researcher*
Murray, Mary Katherine *reproductive biologist, educator*
Smith, Wendy Anne *biologist, educator*

Cambridge
Barry, Brenda Elizabeth *respiratory biologist*
Bogorad, Lawrence *biologist*
Cumings, Edwin Harlan *biology educator*
Erikson, Raymond Leo *biology educator*
Farber, Neal Mark *biotechnologist, molecular biologist*
Holmes, Donna Jean *biologist*
Hynes, Richard Olding *biology educator*
Lynch, Harry James *biologist*
Miao, Shili *plant biology educator, plant ecology researcher*
Pfister, Donald Henry *biology educator*
Pierce, Naomi Ellen *biology educator, researcher*
Robbins, Phillips Wesley *biology educator*
Tonegawa, Susumu *biology educator*
Walker, Graham Charles *biology educator*
Wilson, Edward Osborne *biologist, educator*

Charlestown
Chung-Welch, Nancy Yuen Ming *biologist*

Duxbury
Hillman, Robert Edward *biologist*

Lexington
Gibbs, Martin *biologist, educator*

North Dartmouth
Read, Dorothy Louise *biology educator*

North Grafton
Rowan, Andrew Nicholas *biologist, educator*

Salem
Young, Alan M. *biologist, educator*

Shrewsbury
Pederson, Thoru Judd *biologist, research institute director*

Waltham
Tobet, Stuart Allen *neurobiologist*

Waquoit
Saunders, John Warren, Jr. *biology educator, consultant*

Westborough
Cardoza, James Ernest *wildlife biologist*

Westfield
Taylor, James Kenneth *biology educator, science teaching consultant*

Woods Hole
Burris, John Edward *biologist*
Ebert, James David *research biologist, educator*

Worcester
Boss, Michael Alan *biologist*
Theroux, Steven Joseph *biologist, educator*

MICHIGAN

Ann Arbor
Anderson, William R. *biologist, educator, curator, director*
Breck, James Edward *fisheries research biologist*
Easter, Stephen Sherman, Jr. *biology educator*
Shappirio, David Gordon *biologist, educator*

Big Rapids
Murnik, Mary Rengo *biology educator*

Dearborn
Zimmerman, William James *biologist, educator*

Detroit
Kopp, Monica *biologist, educator*

East Lansing
Bowerman, William Wesley, IV *biologist, researcher*

Flint
Adams, Paul Allison *biologist, educator*

Hickory Corners
Fitzstephens, Donna Marie *biologist*

Marquette
Twohey, Michael Brian *fishery biologist*

Mount Pleasant
Novitski, Charles Edward *biology educator*

MINNESOTA

Biddeford
Carter, Herbert Jacque *biologist, educator*

Duluth
Lindquist, Edward Lee *biological scientist, ecologist*

Grand Rapids
Garshelis, David Lance *wildlife biologist*

Minneapolis
Cornelissen-Guillaume, Germaine Gabrielle *chronobiologist, physicist*
Rahman, Yueh-Erh *biologist*

Saint Paul
Jones, Charles Weldon *biologist, educator, researcher*
Mech, Lucyan David *research biologist, conservationist*
Newman, Raymond Melvin *biologist, educator*
Tate, Jeffrey L. *biology institute administrator*

MISSISSIPPI

Hattiesburg
Bieslot, Patricia Marie *biology educator, researcher*

University
Kushlan, James A. *biology educator*
Parsons, Glenn Ray *biology educator*

MISSOURI

Columbia
Carrel, James Elliott *biologist*
Fahim, Mostafa Safwat *reproductive biologist, consultant*
Yanders, Armon Frederick *biological sciences educator, research administrator*

Eureka
Coles, Richard W(arren) *biology educator, research administrator*

Jefferson City
Weithman, Allan Stephen *fisheries biologist*

Kansas City
De Blas, Angel Luis *biologist, educator*

Kirksville
Twining, Linda Carol *biologist, educator*

New Bloomfield
Quay, Wilbur Brooks *biologist*

Saint Louis
Baile, Clifton A. *biologist, researcher*
Bourne, Carol Elizabeth Mulligan *biology educator, phycologist*
Sanes, Joshua Richard *biologist, researcher*
Thach, Robert Edwards *biology educator*
Varner, Joseph Elmer *biology educator, researcher*

Springfield
Havel, John Edward *biology educator*

Warrensburg
Voorhees, Frank Ray *biology educator*

MONTANA

Hamilton
Garon, Claude Francis *laboratory administrator, researcher*

Missoula
Craighead, John Johnson *wildlife biologist*

NEBRASKA

Hastings
Wilhelm, Dallas Eugene, Jr. *biology educator*

Kearney
Tillotson, Dwight Keith *biologist*

Omaha
Schalles, John Frederick *biology educator*

NEVADA

Reno
Hoelzer, Guy Andrew *biologist*

NEW HAMPSHIRE

Manchester
Dokla, Carl Phillip John *psychobiologist, educator*

Plymouth
Chabot, Christopher Cleaves *biology educator*

NEW JERSEY

Cape May Court House
Wood, Albert E(lmer) *biology educator*

Jersey City
Giuliani, Eleanor Regina *biology educator*

Kenilworth
Pellerito-Bessette, Frances *research biologist*

Lawrenceville
Campbell, David Bruce *biology educator, researcher*

New Brunswick
Ehrenfeld, David William *biology educator, author*
Vrijenhoek, Robert Charles *biologist*

Newark
Fadem, Barbara H. *psychobiologist, psychiatry educator*
Rosenblatt, Jay Seth *psychobiologist, educator*

Parsippany
Labriola, Joseph Arthur *environmental biologist*

Piscataway
Dornburg, Ralph Christoph *biology educator*

Princeton
Grant, Peter Raymond *biologist, researcher, educator*
Steinberg, Malcolm Saul *biologist, educator*
Tilghman, Shirley Marie *biology educator*

Sicklerville
Waldow, Stephen Michael *radiation biologist, educational consultant*

Somerset
Rosenstraus, Maurice Jay *biologist*

Sparta
Mott, Peter Andrew *biologist*

Teaneck
Rhodes, Rondell Horace *biology educator*

West Orange
Wiedl, Sheila Colleen *biologist*

NEW MEXICO

Socorro
Kieft, Thomas Lamar *biology educator*

NEW YORK

Albany
Mascarenhas, Joseph Peter *biologist, educator, researcher, consultant*
Sanchez de la Peña, Salvador Alfonso *biomedical chronobiologist*
Schmidt, John Thomas *neurobiologist*
Stewart, Margaret McBride *biology educator, researcher*
Tedeschi, Henry *bioscience educator*

Amherst
Vaughan, John Thomas *biology educator*

Annandale
Ferguson, John Barclay *biology educator*

Binghamton
Buckley, John Leo *retired environmental biologist*

Briarcliff Manor
Frair, Wayne Franklin *biologist, educator*

Bronx
Rothstein, Howard *biology educator*
Wharton, Danny Carroll *zoo biologist*

Brooklyn
Raffanielo, Robert Donald *research scientist, educator*

Buffalo
Duax, William Leo *biological researcher*
White, Thomas David *biology educator*

Cooperstown
Harman, Willard Nelson *malacologist, educator*

Cortland
Klotz, Richard Lawrence *biology educator*

Flushing
Commoner, Barry *biologist, educator*
Roze, Uldis *biologist, author*

Glen Cove
Grant, Anthony Victor *biology educator*

Hamilton
Kessler, Dietrich *biology educator*

Ithaca
Eisner, Thomas *biologist, educator*
Wasserman, Robert Harold *biology educator*

Lake Placid
Sato, Gordon Hisashi *retired biologist, researcher*

New York
Greenfield, Bruce Paul *investment analyst, biology researcher*
Hirano, Arlene Akiko *neurobiologist, research scientist*
Hof, Patrick Raymond *neurobiologist*
Hommes, Frits Aukustinus *biology educator*
Kopelman, Arthur Harold *biology educator, population ecologist*
Pestano, Gary Anthony *biologist*
Pollack, Robert Elliot *biological sciences educator, writer, scientist*
Qureshi, Sajjad Aslam *biologist, researcher*
Young, Helen Jamieson *biologist, educator*
Zuzolo, Ralph Carmine *biologist, educator, researcher, consultant*

Oswego
Seago, James Lynn *biologist, educator*
Tulve, Nicolle Suzanne *researcher*

Pearl River
Barrett, James Edward *biology educator, research*

Plattsburgh
Graziadei, William Daniel, III *biology educator, researcher*

Saranac Lake
North, Robert John *biologist*

Schenectady
Boyer, John Frederick *biology educator*

Southampton
Shumway, Sandra Elisabeth *shellfish biologist*

Sparkill
Rosko, John James *biology educator*

Stony Brook
Williams, George Christopher *biologist, ecology and evolution educator*

Syracuse
Cooper, John Edwin *fisheries biologist*
Phillips, Arthur William, Jr. *biology educator*

Upton
Laster, Brenda Hope *radiobiologist*

Westbury
Stalzer, John Francis *environmental biologist*

Woodbury
Gibbons, Edward Francis *psychobiologist*

NORTH CAROLINA

Boone
Shull, Julian Kenneth, Jr. *biology educator*

Chapel Hill
Kier, William McKee *biologist, educator*

Charlotte
Cornell, James Fraser, Jr. *biologist, educator*

Durham
Casseday, John Herbert *neurobiologist*

Emerald Isle
Hardy, Sally Maria *retired biological sciences educator*

Greensboro
O'Hara, Robert James *evolutionary biologist*

Hickory
Seaman, William Daniel *biology educator, clergyman*

Raleigh
Monteiro-Riviere, Nancy Ann *biologist, educator*

Research Triangle Park
Dibner, Mark Douglas *research executive, industry analyst*
Graves, Joan Page *biologist*

Wilmington
Merritt, James Francis *biological sciences educator*

Winston Salem
Allen, Nina Strömgren *biology educator*
Esch, Gerald Wisler *biology educator*

NORTH DAKOTA

Minot
Morgan, Rose Marie *biology educator, researcher*

OHIO

Bowling Green
Smith, Stan Lee *biology educator*
Walker, Daniel Jay *biologist*

Cincinnati
Kordenbrock, Douglas William *biomedical electronics technician*
Lacy, Mark Edward *computational biologist, systems scientist*

Cleveland
Canterbury, Ronald A. *biologist*
Caplan, Arnold I. *biology educator*
Ernsberger, Paul Roos *research biologist, neuropharmacologist*
Jarroll, Edward Lee *biology educator*

Columbus
Davis, Keith Robert *plant biology educator*
Whitford, Philip Clason *biology educator*

Dayton
Tsonis, Panagiotis Antonios *biologist, researcher*

Kent
Cooke, G. Dennis *biological science educator*
Stevenson, J. Ross *biological sciences educator, researcher*

Springfield
Hobbs, Horton Holcombe, III *biology educator*

Wooster
Madden, Laurence Vincent *plant pathology educator*

OKLAHOMA

Edmond
Bass, Thomas David *biology educator*
Radke, William John *biology educator*

Norman
Hutchison, Victor Hobbs *biologist, educator*

Oklahoma City
Barber, Susan Carrol *biology educator*
Branch, John Curtis *biology educator, lawyer*

Weatherford
Grant, Peter Michael *biologist, educator*

OREGON

Charleston
Shapiro, Lynda P. *biology educator, director*

Corvallis
Muir, Patricia Susan *biology educator, researcher*

Eugene
Wessells, Norman Keith *biologist, educator, university administrator*

La Grande
Betts, Burr Joseph *biology educator*
Thomas, Jack Ward *wildlife biologist*

Portland
Bortner, James Bradley *wildlife biologist*
Newell, Nanette *biotechnologist*
Ruben, Laurens Norman *biology educator*

PENNSYLVANIA

Altoona
Gannon, Michael Robert *biology educator*

Edinboro
Snyder, Donald Benjamin *biology educator*

Indiana
Lord, Thomas Reeves *biology educator*
Purdy, David Lawrence *biotechnical company executive*

Johnstown
McNair, Dennis Michael *biology educator*

Philadelphia
Franz, Craig Joseph *biology educator*
Greene, Joyce Marie *biology educator*
Hand, Peter James *neurobiologist, educator*
Hung, Paul Porwen *biotechnologist, educator, consultant*
Kirsch, Ted Michael *pathobiologist*
McEachron, Donald Lynn *biology educator, researcher*
Sheffield, Joel Bensen *biology educator*
Suntharalingam, Nagalingam *radiation therapy educator*

Pittsburgh
Claycamp, Henry Gregg *radiobiologist educator*
Kreithen, Melvin Louis *biologist, educator*
Murphy, Robert Francis *biology educator*

Spring House
Andrade-Gordon, Patricia *biological scientist*

Swiftwater
Anthony, Damon Sherman *biologist, researcher*
Lee, Chung Keel *biologist*

University Park
Cosgrove, Daniel Joseph *biology educator*
Dunson, William Albert *biology educator*
Proctor, Robert Neel *biologist, historian, educator*

Wilkes Barre
Hayes, Wilbur Frank *biology educator*

RHODE ISLAND

Providence
Hitti, Youssef Samir *biologist*

SOUTH CAROLINA

Aiken
Hernandez-Martich, Jose David *population and conservation biologist, educator*

Charleston
Arrigo, Salvatore Joseph *biologist*

Columbia
Lovell, Charles Rickey *biologist, educator*

Gaffney
Jones, Nancy Gale *retired biology educator*

Greenville
Cureton, Claudette Hazel Chapman *biology educator*

Rock Hill
Mitchell, Paula Levin *biology educator, editor*

SOUTH DAKOTA

Mitchell
Tatina, Robert Edward *biology educator*

Spearfish
Cox, Thomas Patrick *biological psychology educator*

TENNESSEE

Johnson City
Johnson, Dan Myron *biology educator*
Robbins, Charles Michael *teratologist, research developmental biologist*

Knoxville
Carroll, Roger Clinton *medical biology educator*
Chen, James Pai-fun *biology educator, researcher*
Trigiano, Robert Nicholas *biotechnologist, educator*

Maryville
Sievert, Lynnette Carlson *biologist, educator*

Memphis
Huggins, James Anthony *biology educator*
Monroe, Daniel Milton, Jr. *biologist, chemist*

Nashville
Richmond, Ann White *cell and molecular biologist, educator*

Oak Ridge
Gude, William D. *retired biologist*
Roop, Robert Dickinson *biologist*

Sewanee
Palisano, John Raymond *biologist, educator*

TEXAS

Austin
Albin, Leslie Owens *biology educator*
Arroyo-Vazquez, Bryan *wildlife biologist*
Biesele, John Julius *biologist, educator*
Norton, Jerry Don *biology educator*
Park, Thomas Joseph *biology researcher, educator*

Beaumont
Bianchi, Thomas Stephen *biology and oceanography educator*

College Station
Hall, Timothy C. *biology educator, consultant*

Commerce
Betts, James Gordon *biology educator*

Dallas
Fischer Lindahl, Kirsten *biologist, educator*

Denton
Lancaster, Francine Elaine *neurobiologist*
Schwalm, Fritz Ekkehardt *biology educator*

El Paso
Harris, Arthur Horne *biology educator*
Johnson, Jerry Douglas *biology educator, researcher*

Galveston
McTigue, Teresa Ann *biologist, researcher, educator*

Houston
Shakes, Diane Carol *developmental biologist, educator*
Tomasovic, Stephen Peter *radiobiologist, educator*
Weinstock, George Matthew *biology educator, researcher*

Kingsville
Perez, John Carlos *biology educator*

Lackland AFB
Dixon, Patricia Sue *medical biotechnologist, researcher*

Lubbock
McGlone, John James *biologist*

Odessa
Kurtz, Edwin Bernard *biology educator, researcher*

San Marcos
Longley, Glenn *biology educator, research director*

Stafford
Sparks, William Sidney *biologist*

Victoria
Wauer, Roland Horst *biologist*

Waco
Pierce, Benjamin Allen *biologist, educator*

UTAH

Ogden
Earley, Charles Willard *biologist*

Salt Lake City
Musci, Teresa Stella *developmental biologist, researcher*

Vernal
Folks, F(rancis) Neil *biologist, researcher*

VERMONT

Burlington
Banschbach, Valerie Suzanne *biologist*
Henson, Earl Bennette *biologist*

Johnson
Genter, Robert Brian *biology educator*

Northfield
Barnard, William Howard, Jr. *biologist, educator*

VIRGINIA

Arlington
Adams, Donald Edward *biology researcher*
Stone, Jacqueline Marie *biotechnology patent examiner*

Ashland
Falls, Elsa Queen *biologist, educator*

Charlottesville
Murray, Joseph James, Jr. *biology educator, zoologist*

Fairfax
Soyfer, Valery Nikolayevich *biology educator*

Hampden Sydney
Shear, William Albert *biology educator*

Herndon
Foster, Linda Ann *biomaterials research scientist*

Lorton
Koschny, Theresa Mary *environmental biologist*

Lynchburg
Peters, Ralph Irwin, Jr. *biology educator, researcher*

Norfolk
Stokes, Thomas Lane, Jr. *biologist, consultant*

Radford
McGraw, Katherine Annette *fisheries biologist, environmental consultant*

Richmond
Reynolds, Thomas Robert *scientist, biotechnology company executive*

Virginia Beach
Barranco, Sam Christopher *biologist, researcher*

WASHINGTON

Bothell
Taylor, Dean Perron *biotechnologist, researcher*

Pullman
Witmer, Gary William *research biologist, educator*

Richland
Bair, William J. *radiation biologist*

Seattle
Foe, Elizabeth *biologist, educator*
Hood, Leroy Edward *biologist*
Plisetskaya, Erika Michael *biology and physiology educator*
Ratner, Buddy Dennis *bioengineer, educator*
Rubel, Edwin W *neurobiologist*

Tacoma
Runde, Douglas Edward *wildlife biologist*

WEST VIRGINIA

Huntington
Harrison, Marcia Ann *biology educator*

WISCONSIN

Appleton
De Stasio, Elizabeth Ann *biology educator*

Madison
Fowler, John Francis *radiobiologist*
Lanher, Bertrand Simon *biological spectroscopist*

Sheboygan
Marr, Kathleen Mary *biologist, educator*

WYOMING

Rock Springs
Mitchell, Sandra Louise *biology educator, researcher*

MILITARY ADDRESSES OF THE UNITED STATES

ATLANTIC

APO
Knowlton, Nancy *biologist*
Rubinoff, Ira *biologist, research administrator, conservationist*

CANADA

ALBERTA

Edmonton
Cossins, Edwin Albert *biology educator, academic administrator*

BRITISH COLUMBIA

Bamfield
Druehl, Louis Dix *biology educator*

Burnaby
Baille, David L. *biologist, educator*
Brandhorst, Bruce Peter *biology educator*
Roitberg, Bernard David *biology educator*

Nanaimo
Ricker, William Edwin *biologist*

Vancouver
Hochachka, Peter William *biology educator*

ONTARIO

North York
Vafopoulou, Xanthe *biologist*

Ottawa
Armstrong, David William *biotechnologist, microbiologist*

Waterloo
Downer, Roger George H. *biologist*

Windsor
Sale, Peter Francis *biology educator, marine ecologist*

QUEBEC

Montreal
Chang, Thomas Ming Swi *biotechnologist, medical scientist*
Kalff, Jacob *biology educator*
Leggett, William C. *biology educator, educational administrator*

MEXICO

Mexico City
Ceballos, Gerardo *biology educator, researcher*

ARGENTINA

Buenos Aires
Diaz, Alberto *biotechnologist*
Florin-Christensen, Jorge *biologist*

AUSTRALIA

Prospect
Mayo, Oliver *biology researcher*

Rockhampton
Warner, Lesley Rae *biology educator*

AUSTRIA

Baden
Lukas, Elsa Victoria *radiobiologist, radiobiochemist*

CZECH REPUBLIC

Ceske Budejovice
Sláma, Karel *biologist, zoologist*

Prague
Říman, Josef *biology educator*

DENMARK

Aalborg
Zimmermann, Wolfgang Karl *biologist, researcher*

Beder
Fjerdingstad, Ejnar Jules *retired biological scientist and educator*

ENGLAND

Cambridge
Brenner, Sydney *biologist, educator*

Cambridgeshire
Ford, Brian J. *research biologist, author, broadcaster*

Cranfield
Turner, Anthony Peter Francis *biotechnologist, educator*

London
Pecorino, Lauren Teresa *biologist*

North Yorkshire
Martin, Carl Nigel *biotechnologist*

Norwich
Hopwood, David Alan *biotechnologist, geneticist, educator*

FRANCE

Aulnay-sous-Bois
Shahin, Majdi Musa *biologist*

Brest
Floch, Herve Alexander *medical biologist*

Paris
Courtois, Yves *biologist*
Jacob, François *biologist*
LeGoffic, Francois *biotechnology educator*

Strasbourg
Aunis, Dominique *neurobiologist*

Toulouse
Fonta, Caroline *biologist*

Villejuif
Hatzfeld, Jacques Alexandre *biologist, researcher*

GERMANY

Borstel
Loppnow, Harald *biologist*

Bremen
Valentine-Thon, Elizabeth Anne *biologist*

HONG KONG

Kowloon
Yang, Mildred Sze-ming *biologist, educator*

INDIA

New Delhi
Adholeya, Alok *biotechnologist, researcher*

IRELAND

Maynooth
Whittaker, Peter Anthony *biology educator*

ISRAEL

Nes Ziona
Schmell, Eli David *biotechnologist*

ITALY

Camerino
Miyake, Akio *biologist, educator*

Rome
Levi-Montalcini, Rita *neurobiologist, researcher*

JAPAN

Fukuoka
Aizawa, Keio *biology educator*

Ishinomaki
Hiwatashi, Koichi *biologist, educator*

Kobe
Baba, Yoshinobu *biotechnologist*

Kyoto
Kobayashi, Naomasa *biology educator*

Mishima
Nakatsuji, Norio *biologist*

Nagoya
Ando, Shigeru *biology educator*

Nara
Shigesada, Nanako *mathematical biology educator, researcher*

Obihiro
Goto, Ken *chronobiologist, educator*

Okubo
Nishimura, Susumu *biologist*

Otsu
Ohara, Akito *biology educator*

Sagamihara
Okui, Kazumitsu *biology educator*

Toyama
Hamada, Jin *biologist*

Yamaguchi
Chiba, Yoshihiko *biology educator*

NORWAY

Tromsø
El-Gewely, M. Raafat *biology educator*

PAKISTAN

Islamabad
Afzal, Mohammad *biologist*

POLAND

Poznan
Golab, Wlodzimierz Andrzej *biologist, geographer, librarian*

REPUBLIC OF KOREA

Iri
Kil, Bong-Seop *biology educator*

Seoul
Kang, Bin Goo *biologist*

Taejeon
Bok, Song Hae *biotechnologist, researcher*

RUSSIA

Moscow
Kefeli, Valentin Ilich *biologist*

SCOTLAND

Glasgow
Vickerman, Keith *biologist*

SLOVENIA

Novo Mesto
Sokolić, Milenko *biotechnologist, researcher*

SPAIN

Madrid
Ayala, Juan Alfonso *molecular biology research, biochemistry educator*
Gil-Loyzaga, Pablo Enrique *neurobiologist, researcher, educator*
Velasco Negueruela, Arturo *biology educator*

SWEDEN

Linköping
Andersson, Tommy Evert *researcher*

Lund
Nilsson, Kurt Gösta Ingemar *biotechnology and enzyme technology scientist*

SWITZERLAND

Carouge Geneva
Giordan, Andre Jean Pierre Henri *biologist researcher*

Hinterkappelen
Keller, Laurent *biologist, researcher*

TAIWAN

Taipei
Pan, Tzu-Ming *biotechnologist*

ADDRESS UNPUBLISHED

Alberts, Allison Christine *biologist*
Block, Barbara Ann *biology educator*
Bullock, Theodore Holmes *biologist, educator*
Burdett, Barbra Elaine *biology educator*
Camper, Jeffrey Douglas *biology educator*
Chang, Charles Shing *biology educator, researcher*
Darlington, Julian Trueheart *biology educator retired*
Dunski, Jonathan Frank *biologist, educator*
Fredine, C(larence) Gordon *biologist, former government agency official*
Gilbert, Charles D. *neurobiologist*
Goffman, Thomas Edward *radiobiologist, researcher*
Ishida, Andrew *neurobiologist*
Kathman, R. Deedee *biologist, educator*
McQueney, Patricia Ann *biologist, researcher*
Melia, Angela Therese *biologist, pharmacokineticist*
Mudar, M(arian) J(ean) *biologist, environmental scientist*
Post, Laura Cynthia *biologist*
Saneto, Russell Patrick *neurobiologist*
Sayigh, Laela Suad *biologist*
Sjostrand, Fritiof Stig *biologist, educator*
Tompkins, Laurie *biologist, educator*

LIFE SCIENCES: BOTANY & HORTICULTURE

UNITED STATES

ALABAMA

Auburn
Lemke, Paul Arenz *botany educator*

Normal
Pacumbaba, R.P. *plant pathologist, educator*
Sabota, Catherine Marie *horticulturist, educator*

Tuscaloosa
Xu, ZhaoRan *botanist*

ARIZONA

Flagstaff
Phillips, Arthur Morton, III *botanist, consultant*

Tucson
Yocum, Harrison Gerald *horticulturist, botanist, educator, researcher*

ARKANSAS

Fayetteville
Templeton, George Earl, II *plant pathologist*

CALIFORNIA

Berkeley
Baker, Herbert George *botany educator*
Silva, Paul Claude *botanist*

Claremont
Elias, Thomas Sam *botanist, author*

Davis
Addicott, Fredrick Taylor *retired botany educator*
Hess, Charles Edward *environmental horticulture educator*
Kester, Dale Emmert *pomologist, educator*
Marois, Jim *plant pathologist, educator*
Rappaport, Lawrence *plant physiology and horticulture educator*
Stebbins, George Ledyard *research botanist, retired educator*
Uyemoto, Jerry Kazumitsu *plant pathologist, educator*
Van Bruggen, Ariena Hendrika Cornelia *plant pathologist*

Gilroy
Barham, Warren S. *horticulturist*

Los Angeles
Mathias, Mildred Esther *botany educator*

Moffett Field
Wignarajah, Kanapathipillai *plant physiologist, researcher, educator*

Riverside
Embleton, Tom William *horticultural science educator*
Nothnagel, Eugene Alfred *plant cell biology educator*
Van Gundy, Seymour Dean *nematologist, plant pathologist, educator*
Zentmyer, George Aubrey *plant pathology educator*

Sacramento
Little, Robert John, Jr. *botanist, environmental consultant*
Rosenberg, Dan Yale *retired plant pathologist*

San Francisco
Peirce, Pamela Kay *horticulturist, writer*

COLORADO

University Of Colorado
Ranker, Tom A. *botanist, educator*

DISTRICT OF COLUMBIA

Washington
Feuillet, Christian Patrice *botanist*
Krauss, Robert Wallfar *botanist, university dean*
Peterson, Paul Michael *agrostologist*
Shetler, Stanwyn Gerald *botanist, museum official*

FLORIDA

Bradenton
Waters, Will Estel *horticulturist, researcher, educator*

Chuluota
Hatton, Thurman Timbrook, Jr. *retired horticulturist, consultant*

Gainesville
Agrios, George Nicholas *botanist*
Cantliffe, Daniel James *horticulture educator*
Childers, Norman Franklin *horticulture educator*
Christie, Richard G. *plant pathologist*
Hiebert, Ernest *plant pathologist, educator*

Orlando
Miller, Harvey Alfred *botanist, educator*

Venice
Hardenburg, Robert Earle *horticulturist*

GEORGIA

Athens
Jones, Samuel B., Jr. *botany educator*

HAWAII

Honolulu
Kamemoto, Haruyuki *horticulture educator*

IDAHO

Moscow
Roberts, Lorin Watson *botanist, educator*

ILLINOIS

Carbondale
Bozzola, John Joseph *botany educator, researcher*
Mohlenbrock, Robert Herman, Jr. *botanist, educator*

Glencoe
Taylor, Roy Lewis *botanist*

Urbana
Shurtleff, Malcolm C. *plant pathologist, consultant, educator, extension specialist*
Splittstoesser, Walter Emil *plant physiologist*

INDIANA

West Lafayette
Erickson, Homer Theodore *horticulture educator*
Janick, Jules *horticultural scientist, educator*
Larkins, Brian Allen *botany educator*

KANSAS

Lawrence
Haufler, Christopher Hardin *botany educator*
Lane, Meredith Anne *botany educator, museum director*

Manhattan
Jardine, Douglas Joseph *plant pathologist, educator*

LOUISIANA

Baton Rouge
Chapman, Russell Leonard *botany educator*
Tucker, Shirley Lois Cotter *botany educator, researcher*

Ruston
Walker, Harrell Lynn *plant pathologist, botany educator, researcher*

MARYLAND

Baltimore
Habermann, Helen Margaret *plant physiologist, educator*

Beltsville
Endo, Burton Yoshiaki *research plant pathologist*

Fort Detrick
Schaad, Norman Werth *plant pathologist*

Rockville
Barnes, John Maurice *plant pathologist*

Silver Spring
Romberger, John Albert *scientist*

Towson
Windler, Donald Richard *botanist, educator*

MASSACHUSETTS

Cambridge
Bazzaz, Fakhri A. *plant biology educator, administrator*
Schultes, Richard Evans *ethnobotanist, museum executive, educator, conservationist*

Holbrook
Noyes, Walter Omar *tree surgeon*

MICHIGAN

Ann Arbor
Knox, Eric *botanist, educator*
Reznicek, Anton Albert *plant systematist*
Wagner, Warren Herbert, Jr. *educator, botanist*

East Lansing
Fulbright, Dennis Wayne *plant pathologist, educator*
Hull, Jerome, Jr. *horticultural extension specialist*
Ohlrogge, John B. *botany and plant pathology educator*
Szerszen, Jedrzej Bogumil (Andrew Szerszen) *plant pathologist, educator*

MINNESOTA

Saint Paul
Christensen, Clyde Martin *plant pathology educator*
Kommedahl, Thor *plant pathology educator*
Leonard, Kurt John *plant pathologist, university program director*

MISSISSIPPI

Stoneville
Ranney, Carleton David *plant pathology researcher, administrator*

MISSOURI

Columbia
Niblack, Terry L. *nematologist, plant pathology educator*

Saint Louis
Armbruster, Barbara Louise *botanist*
Crosby, Marshall Robert *botanist, educator*
Hemming, Bruce Clark *plant pathologist*

NEW MEXICO

Las Vegas
Shaw, Mary Elizabeth *plant pathologist*

Mesilla
Cryder, Cathy M. *plant geneticist*

NEW YORK

Cobleskill
Ingels, Jack Edward *horticulture educator*

Geneva
Seem, Robert Charles *plant pathologist*
Wilcox, Wayne F. *plant pathologist, educator, researcher*

Highland
Rosenberger, David A. *research scientist, cooperative extension specialist*

Ithaca
Bates, David Martin *botanist, educator*
Crepet, William Louis *botanist, educator*
Davies, Peter John *plant physiology educator, researcher*
Jagendorf, Andre Tridon *plant physiologist*
Kingsbury, John Merriam *botanist, educator*
Seeley, John George *horticulture educator*
Uhl, Charles Harrison *botanist, plant cytologist*

New Paltz
Huth, Paul Curtis *ecosystem scientist, botanist*

NORTH CAROLINA

Archdale
Riddick, Douglas Smith *horticultural industrialist, industrial designer*

Durham
Abdel-Rahman, Mohamed *plant pathologist, physiologist*
Culberson, William Louis *botany educator*
Wilbur, Robert Lynch *botanist, educator*

Elm City
Parker, Josephus Derward *corporation executive*

Raleigh
Benson, D(avid) Michael *plant pathologist*
Campbell, Charles Lee *plant pathologist*
Daub, Margaret E. *plant pathologist, educator*
De Hertogh, August Albert *horticulture educator, researcher*
Hardin, James W. *botanist, herbarium curator, educator*

OHIO

Kent
Cooperrider, Tom Smith *botanist*

Oxford
Eshbaugh, W(illiam) Hardy *botanist, educator*
Powell, Martha Jane *botany educator*

Wooster
Ferree, David Curtis *horticultural researcher*

OKLAHOMA

Norman
Ortiz-Leduc, William *plant biologist, educator, researcher*

OREGON

Corvallis
Chambers, Kenton Lee *botany educator*
Fuchigami, Leslie Hirao *horticulturist, researcher*
Johnston, Larea Dennis *taxonomist*
Liston, Aaron Irving *botanist*

PENNSYLVANIA

California
Newhouse, Joseph Robert *plant pathologist, mycologist*

Philadelphia
Patrick, Ruth (Mrs. Charles Hodge) *limnologist, diatom taxonomist, educator*

University Park
Leath, Kenneth Thomas *research plant pathologist, educator*
Pennypacker, Barbara White *plant pathologist*
Tammen, James F. *plant pathologist, educator*

RHODE ISLAND

Kingston
Goos, Roger Delmon *mycologist*

TENNESSEE

Knoxville
Swingle, Homer Dale *horticulturist, educator*

TEXAS

Austin
Brown, Richard Malcolm, Jr. *botany educator*
Simpson, Beryl B. *botany educator*
Starr, Richard Cawthon *botany educator*
Turner, Billie Lee *botanist, educator*

Harlingen
Ryall, A(lbert) Lloyd *horticulturist, refrigeration engineer*

Weslaco
Amador, Jose Manuel *plant pathologist, research center administrator*

UTAH

Logan
Anderson, Jay LaMar *horticulture educator, researcher, consultant*

VIRGINIA

Dahlgren
Westbrook, Susan Elizabeth *horticulturist*

Farmville
Breil, David Allen *botany educator*

Meadowview
Hebard, Frederick V. *plant pathologist*

Richmond
Hayden, W(alter) John *botanist, educator*

WASHINGTON

Prosser
Proebsting, Edward Louis, Jr. *research horticulturist*

Pullman
Gurusiddaiah, Sarangamat *plant scientist*

Seattle
Tukey, Harold Bradford, Jr. *horticulture educator*

WISCONSIN

Madison
Daie, Jaleh *science educator, researcher, administrator*
Evert, Ray Franklin *botany educator*
Hopen, Herbert John *horticulture educator*
Iltis, Hugh Hellmut *plant taxonomist and evolutionist, educator*
Skoog, Folke Karl *botany educator*

TERRITORIES OF THE UNITED STATES

PUERTO RICO

San Juan
Lugo, Ariel E. *botanist, federal agency administrator*

CANADA

BRITISH COLUMBIA

Kaleden
Swales, John E. (Ted) *retired horticulturalist*

Sidney
Lanterman, William Stanley, III *plant pathologist, researcher, administrator*

Summerland
Looney, Norman Earl *pomologist, plant physiologist*

NEWFOUNDLAND

Saint John's
Sheath, Robert Gordon *botanist*

ONTARIO

Guelph
Lougheed, Everett Charles *retired horticulture educator, researcher*

Mississauga
Errampalli, Deena *molecular plant pathologist*

Ottawa
Baum, Bernard Rene *biosystematist*

Waterloo
Thompson, John Eveleigh *horticulturist, educator*

AUSTRALIA

Adelaide
Wiskich, Joseph Tony *botany educator, researcher*

Bundoora
Woelkerling, William J. *botanist educator*

Hobart
Murfet, Ian Campbell *botany educator*

BRAZIL

Passo Fundo
Baier, Augusto Carlos *plant researcher*

CHINA

Beijing
Yingqian, Qian *botanist*

ENGLAND

Bristol
Round, Frank E. *botanist, educator*

GERMANY

Bonn
Weiling, Franz Joseph Bernard *retired botany and biometry educator*

Bremerhaven
Crawford, Richard M. *botanist*

Jülich
Stengel, Eberhard Friedrich Otto *botanist*

ISRAEL

Rehovot
Halevy, Abraham Hayim *horticulturist, plant physiologist*

ITALY

Florence
Bennici, Andrea *botany educator*

JAPAN

Miki-lyo
Fujime, Yukihiro *horticultural science educator, researcher*

PERU

Lima
French, Edward Ronald *plant pathologist*

THE PHILIPPINES

Iloilo
Penecilla, Gerard Ledesma *botany educator*

POLAND

Lublin
Karczmarz, Kazimierz *botany educator*

RUSSIA

Saint Petersburg
Takhtadzhyan, Armen Leonovich *botanist*

SWITZERLAND

Basel
Reichstein, Tadeus *botanist, scientist, educator*

ADDRESS UNPUBLISHED

Fajardo, Julius Escalante *plant pathologist*
Kays, Stanley J. *horticulturist, educator*
Mosjidis, Cecilia O'Hara *botanist, researcher*

LIFE SCIENCES: CELL BIOLOGY

UNITED STATES

ARIZONA

Tucson
Payne, Claire Margaret *molecular and cellular biologist*

CALIFORNIA

Corona Del Mar
Brokaw, Charles Jacob *educator, cellular biologist*

Irvine
Berns, Michael W. *cell biologist, educator*
Bhalla, Deepak Kumar *cell biologist, toxicologist, educator*

Los Angeles
Price, Zane Herbert *cell biologist, research microscopist*

South San Francisco
Woodruff, Teresa K. *cell biologist*

COLORADO

Denver
Pfenninger, Karl H. *cell biology and neuroscience educator*

CONNECTICUT

Meriden
Merwin, June Rae *research scientist, cell biologist*

DISTRICT OF COLUMBIA

Washington
Vidic, Branislav *cell biologist*

FLORIDA

Gainesville
Khan, Saeed Rehman *cell biologist, researcher*

Miami
Bunge, Richard Paul *cell biologist, educator*

Tallahassee
Taylor, J(ames) Herbert *cell biology educator*

ILLINOIS

Urbana
Horwitz, Alan Fredrick *cell and molecular biology educator*

LOUISIANA

Baton Rouge
Head, Jonathan Frederick *cell biologist*

Covington
Cowden, Ronald Reed *biomedical educator, cell biologist*

New Orleans
Huot, Rachel Irene *cell biologist*

MARYLAND

Baltimore
Murphy, Douglas Blakeney *cell biology educator*

MASSACHUSETTS

Cambridge
Ladino, Cynthia Anne *cell biologist*

Shrewsbury
Aghajanian, John Gregory *electron microscopist, cell biologist*

Woods Hole
Inoué, Shinya *microscopy and cell biology scientist, educator*

MICHIGAN

Ann Arbor
Long, Michael William *cell/molecular biologist, educator*
Thall, Aron David *cell biologist*

Rochester
Nag, Asish Chandra *cell biology educator*

MISSOURI

Saint Louis
Symington, Janey Studt *cell and molecular biologist*

NEBRASKA

Omaha
Newton, Sean Curry *cell biologist*

NEW JERSEY

Belleville
Czirbik, Rudolf Joseph *cell biologist*

NEW MEXICO

Albuquerque
Carter, Charleata A. *cancer researcher, developmental biologist, cell biologist, toxicologist*

NEW YORK

Bronx
Frenz, Dorothy Ann *cell and developmental biologist*

Buffalo
Tomasi, Thomas B. *cell biologist, administrator*

Calverton
Racaniello, Lori Kuck *cellular and molecular biologist*

New York
Griff, Irene Carol *cell biology researcher*
Kleiman, Norman Jay *molecular biologist, biochemist*
Kromidas, Lambros *cell biologist, physical scientist*
Rothman, James Edward *cell biologist, educator*
Shelanski, Michael L. *cell biologist, educator*

Riverdale
Friedman, Ronald Marvin *cellular biologist*

Syracuse
Slepecky, Norma B. *cell biologist*

NORTH CAROLINA

Chapel Hill
Douglas, Michael Gilbert *cell biologist, educator*

Durham
Massaro, Edward Joseph *cell biology, biochemistry research scientist, experimental pathology educator*

Winston Salem
Jerome, Walter Gray *cell biologist*

OHIO

Mayfield Heights
Fatemi, Seyyed Hossein *cell biologist, physician*

PENNSYLVANIA

Malvern
Nardone, Robert Carmen *cell biologist*

Pittsburgh
Taylor, D. Lansing *cell biology educator*
Wang, Allan Zuwu *cell biologist*

SOUTH CAROLINA

West Columbia
Salthouse, Thomas Newton *cell biologist, biomaterial researcher*

TEXAS

Houston
O'Malley, Bert William *cell biologist, educator, physician*

VIRGINIA

Charlottesville
Herr, John Christian *cell biologist, educator*

WISCONSIN

Madison
Schatten, Gerald Phillip *cell biologist, educator*

CANADA

ONTARIO

Toronto
Kerbel, Robert Stephen *cell biologist, cancer researcher*

ENGLAND

London
Gahan, Peter Brian *cell biologist, researcher*

GERMANY

Heidelberg
Sakmann, Bert *physician, cell physiologist*

ITALY

Torino
Comoglio, Paolo Maria *cell and molecular biologist, educator*

JAPAN

Tsukuba
Imamura, Toru *molecular cell biologist*

NEW ZEALAND

Upper Hutt
Atkinson, Paul Henry *cell biologist*

ADDRESS UNPUBLISHED

Bauman, Jan Georgius Josef *cell biologist, histochemist*
Chegini, Nasser *cell biology educator, endocrinologist*
Palade, George Emil *cell biologist, educator*

LIFE SCIENCES: ECOLOGY

UNITED STATES

ALABAMA

Auburn
Dobson, F. Stephen *ecologist*

ARIZONA

Tucson
Rosenzweig, Michael Leo *ecology educator*

ARKANSAS

State University
Bednarz, James C. *wildlife ecologist educator*

CALIFORNIA

Bishop
Groeneveld, David Paul *plant ecologist*

Claremont
Eriksen, Clyde Hedman *ecology educator*

Eureka
Roberts, Robert Chadwick *ecologist, environmental scientist, consultant*

Los Angeles
Secor, Stephen Molyneux *physiological ecologist*

Tiburon
Pratt, Jeremy *human ecologist, researcher*

COLORADO

Fort Collins
Coffin, Debra Peters *ecologist, researcher*
Crist, Thomas Owen *ecologist*
Williamson, Samuel Chris *research ecologist*

CONNECTICUT

Norfolk
Egler, Frank Edwin *ecologist, administrator*

DISTRICT OF COLUMBIA

Washington
Filbin, Gerald Joseph *ecologist*
Lizotte, Michael Peter *aquatic ecologist, researcher*
Wickland, Diane Elizbeth *ecologist*

FLORIDA

Gainesville
Ewel, Katherine Carter *ecologist, educator*

Miami
Reark, John Benson *consulting ecologist, landscape architect*

Naples
McCollom, Jean Margaret *ecologist*

GEORGIA

Athens
Grossman, Gary David *ecological educator*
Plummer, Gayther L(ynn) *climatologist, ecologist, researcher*
Pulliam, Howard Ronald *ecology educator*

Macon
Harrison, James Ostelle *ecologist*

ILLINOIS

Champaign
Schwartz, Mark William *ecologist*

Chicago
Wadden, Richard Albert *environmental science educator, consultant, researcher*
Wade, Michael John *ecology and evolution educator, researcher*

Dundee
Burger, George Vanderkarr *wildlife ecologist, researcher*

Lanark
Gray, Gary Gene *ecologist, educator*

Macomb
Anderson, Richard Vernon *ecology educator, researcher*

INDIANA

Richmond
Sabine, Neil B. *ecology educator*

IOWA

Ames
Rieger, Phillip Warren *aquatic ecology educator, researcher*

Grinnell
Campbell, David George *ecologist*

KANSAS

Hays
Coyne, Patrick Ivan *physiological ecologist*

Manhattan
Kaufman, Donald Wayne *research ecologist*

Salina
Piper, Jon Kingsbury *ecologist, researcher*

KENTUCKY

Louisville
Thorp, James Harrison, III *aquatic ecologist*

MAINE

Bucks Harbor
Bandurski, Bruce Lord *ecological and environmental scientist*

MARYLAND

College Park
Jacobs, Dan *biometrician, ecologist*

MASSACHUSETTS

Cambridge
Liu, Jianguo *ecologist*

Groton
Kirshen, Paul Howard *water resources consultant*

Lincoln
Dunwiddie, Peter William *plant ecologist*

Manomet
Castro, Gonzalo *ecologist*

Woods Hole
Caswell, Hal *mathematical ecologist*
Woodwell, George Masters *ecologist, educator, author, lecturer*

MICHIGAN

East Lansing
Kevern, Niles Russell *aquatic ecologist, educator*
Petrides, George Athan *ecologist, educator*

MINNESOTA

Duluth
Johnston, Carol Arlene *ecological researcher*

Minneapolis
Gudmundson, Barbara Rohrke *ecologist*

Saint Paul
Davis, Mark Avery *ecologist, educator*

MONTANA

Bozeman
Maxwell, Bruce Dale *plant ecologist, educator*

Miles City
Heitschmidt, Rodney Keith *rangeland ecologist*

NEBRASKA

North Platte
Northup, Brian Keith *ecologist*

NEVADA

Las Vegas
Walker, Lawrence Reddeford *ecologist, educator*

Reno
Gifford, Gerald Frederic *environmental program director*

NEW HAMPSHIRE

Hanover
Boyce, Richard Lee *forest ecologist, researcher*

NEW YORK

Stony Brook
Futuyma, Douglas Joel *ecology educator*
Morgan, Steven Gaines *marine ecology researcher, educator*

Syracuse
Burgess, Robert Lewis *educator, ecologist*

NORTH CAROLINA

Chapel Hill
Hairston, Nelson George *animal ecologist*
Kuenzler, Edward Julian *ecologist and environmental biologist*

Durham
Billings, William Dwight *ecology educator*
Bush, Mark Bennett *ecologist, educator*
DeBusk, George Henry, Jr. *paleoecologist*
Qualls, Robert Gerald *ecologist*

Raleigh
Zimmerman, John Wayne *wildlife ecologist*

Research Triangle Park
Garner, Jasper Henry Barkdoll *ecologist*

OHIO

Columbus
Culver, David Alan *aquatic ecology educator*

OREGON

Bend
Riegel, Gregg Mason *ecologist, researcher*

Corvallis
Dixon, Robert Keith *plant physiologist, researcher*
Ford, Mary Spencer (Jesse) *ecologist, writer*
Ingham, Elaine Ruth *ecology educator*
Neilson, Ronald Price *ecology educator*
Phillips, Donald Lundahl *research ecologist*

La Grande
Tiedemann, Arthur Ralph *ecologist, researcher*

PENNSYLVANIA

Berwyn
Salvatore, Scott Richard *ecologist*

Erie
Czarnecki, Gregory James *ecologist*

Philadelphia
Sanders, Robert Walter *ecologist, researcher*

Plymouth Meeting
Parker, Jon Irving *ecologist*

SOUTH CAROLINA

Aiken
Smith, Michael Howard *ecologist*

TENNESSEE

Knoxville
Clark, Joseph Daniel *ecologist*
Hammer, Donald Arthur *ecologist*

Oak Ridge
Auerbach, Stanley Irving *ecologist, environmental scientist, educator*
Reed, Robert Marshall *ecologist*

TEXAS

Austin
Pulich, Warren Mark, Jr. *plant ecologist, coastal biologist*

College Station
Archer, Steven Ronald *ecology educator*

Fort Worth
Parker, Robert Hallett *ecologist*

Lubbock
Jorgensen, Eric Edward *wildlife ecologist, researcher*
Mathews, Nancy Ellen *wildlife ecologist, educator*
Winter, Jimmy Dale *ecology educator*

South Padre Island
Judd, Frank Wayne *population ecologist, physiological ecologist*

UTAH

Salt Lake City
Adler, Frederick Russell *mathematical ecologist*

VIRGINIA

Alexandria
Briggs, Jeffrey Lawrence *ecologist*

Charlottesville
Shugart, Herman Henry *environmental sciences educator, researcher*

Fairfax
Dietz, Thomas Michael *human ecology educator and researcher*

Luray
Tessler, Steven *ecologist*

WASHINGTON

Pullman
Thompson, John N. *ecology, evolutionary biology educator, researcher*

Seattle
Karr, James Richard *ecologist, researcher, educator*
Stanback, Mark Thomas *behavioral ecologist, researcher*

WEST VIRGINIA

Morgantown
Christiansen, Tim Alan *ecologist*

WISCONSIN

Madison
Rasmussen, Dennis Robert *behavioral ecologist*

WYOMING

Moose
Craighead, Frank Cooper, Jr. *ecologist*

CANADA

BRITISH COLUMBIA

Vancouver
Healey, Michael Charles *fishery ecologist, educator*

Victoria
Loring, Thomas Joseph *forest ecologist*

NEW BRUNSWICK

Moncton
Hanson, John Mark *ecologist, researcher*

Saint John
Thomas, Martin Lewis H. *marine ecologist, educator*

MEXICO

La Paz
Ortega-Rubio, Alfredo *ecologist, researcher*

Pabellon-Arteaga
De Alba-Avila, Abraham *plant ecologist*

AUSTRALIA

Adelaide
Possingham, Hugh Philip *mathematical ecologist*

EGYPT

Giza
Kassas, Mohamed *desert ecologist, environmental consultant*

GERMANY

Wuppertal
Von Weizsäcker, Ernst Ulrich *environmental scientist*

ISRAEL

Ramat Gan
Steinberger, Yosef *ecologist, biologist*

JAPAN

Osaka
Watanabe, Toshiharu *ecologist, educator*

Tsukuba
Takahashi, Masayuki *aquatic ecologist*

PANAMA

Balboa
Smith, Alan Paul *plant ecologist and physiologist*

REPUBLIC OF KOREA

Cheongju
Kang, Sang Joon *ecologist, educator*

SAUDI ARABIA

Riyadh
Chaudhary, Shaukat Ali *plant taxonomist, ecologist*

SPAIN

Bellaterra
Terradas, Jaume *ecologist*

Burjasot
Miracle, Maria Rosa *ecology educator*

Madrid
Martin de Agar, Pilar Maria *ecology educator*

SWEDEN

Lund
Malmberg, Torsten *human ecologist*

Umeå
Nilsson, Dan Christer *plant ecology educator, researcher*

ADDRESS UNPUBLISHED

Hanson, Hugh *ecologist, educator*
Johnson, Robert Andrew *ecologist*
Johnson, Warren Eliot *wildlife ecologist*

LIFE SCIENCES: ENTOMOLOGY

UNITED STATES

ARIZONA

Maricopa
Ellsworth, Peter Campbell *entomologist*

Tucson
Hagedorn, Henry Howard *entomology educator*

CALIFORNIA

Alameda
Schoeler, George Bernard *medical entomologist*

Berkeley
Poinar, George Orlo, Jr. *insect pathologist and paleontologist, educator*

Concord
Ivy, Edward Everett *entomologist, consultant*

Corona
Gjerde, Andrea Jo *entomologist*

Davis
Flint, Mary Louise *entomologist*
Laidlaw, Harry Hyde, Jr. *entomology educator*

Modesto
Steffan, Wallace Allan *entomologist, educator*

Riverside
Miller, Thomas Albert *entomology educator*

Westwood
Brydon, Harold Wesley *entomologist, writer*

DISTRICT OF COLUMBIA

Washington
Anderson, Donald Morgan *entomologist*
Perich, Michael Joseph *medical entomologist consultant*

FLORIDA

Gainesville
Focks, Dana Alan *medical entomologist, epidemiologist*
Hoy, Marjorie Ann *entomology educator, researcher*
Mead, Frank Waldreth *taxonomic entomologist*

Panama City
Cilek, James Edwin *medical and veterinary entomologist*

Saint Petersburg
Kormilev, Nicholas Alexander *retired entomologist*

Vero Beach
Hribar, Lawrence Joseph *entomologist, researcher*

GEORGIA

Griffin
Braman, S. Kristine *entomologist*

Savannah
Throne, James Edward *entomologist*

Tifton
Rogers, Charlie Ellic *entomologist*

Valdosta
Mares, Joseph Thomas *entomologist*

HAWAII

Honolulu
Vargas, Roger Irvin *entomologist, ecologist*

ILLINOIS

Belleville
Steffen, Alan Leslie *entomologist*

Champaign
Helm, Charles George *entomologist, researcher*

Chicago
Baumgartner, Donald Lawrence *entomologist, educator*

Harvey
Liem, Khian Kioe *medical entomologist*

Highland Park
Dobkin, Irving Bern *entomologist, sculptor*

INDIANA

Notre Dame
Craig, George Brownlee, Jr. *entomologist*

West Lafayette
Edwards, Charles Richard *entomology and pest management educator*

KANSAS

Lawrence
Alexander, Byron Allen *insect systematist*
Michener, Charles Duncan *entomologist, biologist, educator*

Manhattan
Stiefel, Vernon Leo *entomology educator*

KENTUCKY

Lexington
Knapp, Frederick Whiton *entomologist, educator*
Liu, Yong-Biao *entomologist*

LOUISIANA

Baton Rouge
Christian, Frederick Ade *entomologist, physiologist, biology educator*

Independence
Ryerson, Sunny Ann *entomologist*

MAINE

Greenville
Sommerman, Kathryn Martha *retired entomologist*

MARYLAND

Aberdeen Proving Ground
Evans, Edward Spencer, Jr. *entomologist*

Baltimore
Trpis, Milan *entomologist, scientist, educator*

Beltsville
Klun, Jerome Anthony *entomologist, researcher*

MASSACHUSETTS

Amherst
Cardé, Ring Richard Tomlinson *entomologist, educator*

Worcester
Andis, Michael D. *urban entomologist*

MICHIGAN

Ann Arbor
Cantrall, Irving J(ames) *entomologist, educator*

East Lansing
Haack, Robert Allen *forest entomologist*

MINNESOTA

Park Rapids
Tonn, Robert James *entomologist*

MISSISSIPPI

Jackson
Goddard, Jerome *medical entomologist*

MISSOURI

Columbia
Backus, Elaine Athene *entomologist, educator*

MONTANA

Bozeman
Onsager, Jerome Andrew *research entomologist*

NEBRASKA

North Platte
Seymour, Ronald Clement *entomologist, researcher*

NORTH CAROLINA

Oxford
Jackson, D. Michael *research entomologist, educator*

Raleigh
Kuhr, Ronald John *entomology and toxicology educator*

Research Triangle Park
Richmond, James Arthur *entomologist*

NORTH DAKOTA

Minot
Royer, Ronald Alan *entomologist, educator*

OKLAHOMA

Lane
Edelson, Jonathan Victor *entomologist*

OREGON

Corvallis
Krantz, Gerald William *entomology educator*

La Grande
Mason, Richard Randolph *entomologist*

SOUTH CAROLINA

Clemson
Fescemyer, Howard William *entomology educator, insect physiology reseacher*

TEXAS

College Station
Summers, Max (Duanne) *entomologist, scientist, educator*

Dallas
Reinert, James A. *entomologist, educator*

Kerrville
Drummond, Roger Otto *livestock entomologist*
Kunz, Sidney *entomologist*

Temple
Palmer, William Alan *entomologist*

UTAH

Logan
Dorst, Howard Earl *entomologist*
Vandenberg, John Donald *entomologist*

VIRGINIA

Arlington
Knipling, Edward Fred *retired research entomologist, agricultural administrator*

Big Stone Gap
Ogbonnaya, Chuks Alfred *entomologist, agronomist, educator*

Blacksburg
Ross, Mary Harvey *entomology educator, researcher*
Salom, Scott Michael *forest entomologist*

Halifax
Martin, Wesley Davis, Jr. *entomologist, research biologist*

Winchester
Horsburgh, Robert Laurie *entomologist*

WASHINGTON

Walla Walla
Gary, Walter J(oseph) *entomologist, educator*

Wenatchee
Hoyt, Stanley Charles *retired research administrator, entomologist*

WISCONSIN

Madison
Burkholder, Wendell Eugene *entomologist*
Jeanne, Robert Lawrence *entomologist, educator, researcher*

MILITARY ADDRESSES OF THE UNITED STATES

ATLANTIC

APO
Powers, Nelson Roger *entomologist*

CANADA

BRITISH COLUMBIA

Vancouver
Isman, Murray *entomology educator*

ONTARIO

Toronto
Wiggins, Glenn B. *entomologist*

HONG KONG

Sha Tin NT
Cheung, Wilkin Wai-Kuen *entomologist, educator*

ISRAEL

Bet Dagan
Halperin, Joseph *entomologist*

KENYA

Mbita
Khan, Zeyaur Rahman *entomologist*

SWITZERLAND

Zurich
Weber, Donald Charles *entomologist*

TAIWAN

Taichung
Chi, Hsin *entomology and ecology educator*

LIFE SCIENCES: FOOD SCIENCE

UNITED STATES

ARKANSAS

Fayetteville
Morris, Justin Roy *food scientist, enologist, consultant*

COLORADO

Fort Collins
Maga, Joseph Andrew *food science educator*
Smith, Gary Chester *meat scientist, researcher*

DISTRICT OF COLUMBIA

Washington
Falci, Kenneth J. *food and nutrition scientist*
Roberts, Howard Richard *food scientist, association administrator*

HAWAII

Honolulu
Smith-Kayode, Timi *food technologist*

INDIANA

Fort Wayne
Szuhaj, Bernard Francis *food research director*

West Lafayette
Nelson, Philip Edwin *food scientist, educator*

IOWA

Ames
Johnson, Lawrence Alan *cereal technologist, educator, researcher, administrator*

MAINE

Orono
Bushway, Rodney John *food science educator*

MICHIGAN

Zeeland
Hollingsworth, Cornelia Ann *food scientist*

MINNESOTA

Saint Paul
Caldwell, Elwood Fleming *food science educator, researcher, editor*
Padmanabhan, Mahesh *food and chemical engineer, researcher*

MISSOURI

Saint Louis
Graham, Donald James *food technologist*

NEBRASKA

Lincoln
Taylor, Steve Lloyd *food scientist, educator, consultant*

NEW JERSEY

East Brunswick
Chang, Stephen S. *food scientist, educator, researcher, inventor*

Maywood
Kennedy, Joanne Patricia *technical service representative*

New Brunswick
Hayakawa, Kan-Ichi *food science educator*
Solberg, Myron *food scientist, educator*

Somerset
Oh, Nam Hwan *food scientist*

NEW YORK

Baldwin
Lister, Bruce Alcott *food scientist, consultant*

Geneva
Siebert, Karl Joseph *food science educator, consultant*

Ithaca
Ledford, Richard Allison *food science educator, food microbiologist*

OREGON

Corvallis
Farkas, Daniel Frederick *food science and technology educator*

Hillsboro
Stewart, Kenneth Ray *food scientist*

PENNSYLVANIA

Hershey
Stumpf, David Michael *food scientist*

University Park
Kroger, Manfred *food science educator*

WISCONSIN

Madison
Mozdziak, Paul Edward *growth biologist*
Olson, Norman Fredrick *food science educator*
Pariza, Michael Willard *research institute executive, microbiology and toxicology educator*

DENMARK

Frederiksberg
Skibsted, Leif Horsfelt *food chemistry researcher*

ENGLAND

Oxford
Barnes, Peter John *food scientist*

GERMANY

Germersheim
Pollmer, Jost Udo *food chemist*

IRAN

Tehran
Vahabzadeh, Farzaneh *food scientist, educator, researcher*

JAPAN

Uji
Yasumoto, Kyoden *food science educator*

NEW ZEALAND

Palmerston North
Keen, Alan Robert *flavor scientist*

SPAIN

Madrid
Saura-Calixto, Fulgencio *food chemist, educator*

LIFE SCIENCES: GENETICS

UNITED STATES

ARIZONA

Tucson
McCluskey, Kevin *fungal molecular geneticist*

ARKANSAS

Little Rock
Bannon, Gary Anthony *molecular geneticist*

CALIFORNIA

Berkeley
King, Mary-Claire *epidemiologist, educator, geneticist*

Davis
Allard, Robert Wayne *geneticist, educator*
Bernoco, Domenico *immunogeneticist, educator*
Chetelat, Roger Topping *geneticist*
Rick, Charles Madeira, Jr. *geneticist, educator*
Syvanen, Michael *geneticist, educator*

Livermore
Amemiya, Chris Tsuyoshi *geneticist, biomedical scientist*
Batzer, Mark Andrew *molecular geneticist*

Los Angeles
Wu, David Li-Ming *cytogeneticist*

Pasadena
Celniker, Susan Elizabeth *geneticist, molecular biologist*

San Diego
Ryder, Oliver Allison *geneticist, conservation biologist, educator*

Stanford
Botstein, David *geneticist, educator*

COLORADO

Boulder
De Fries, John Clarence *behavioral genetics educator, institute administrator*

Denver
Puck, Theodore Thomas *geneticist, biophysicist, educator*

Fort Collins
Black, William Cormack, IV *insect geneticist, statistician*
Hecker, Richard Jacob *research geneticist*

CONNECTICUT

Storrs
Daniels, Stephen Bushnell *geneticist*

DELAWARE

Dover
Ofosu, Mildred Dean *immunogeneticist*

Wilmington
Irr, Joseph David *geneticist*

DISTRICT OF COLUMBIA

Washington
Chan, Wai-Yee *molecular geneticist*
Harriman, Philip Darling *geneticist, science foundation executive*

FLORIDA

Gainesville
Jones, David Alwyn *geneticist, botany educator*
Wilcox, Charles Julian *geneticist, educator*

GEORGIA

Atlanta
La Farge, Timothy *plant geneticist*

ILLINOIS

Bloomington
Weber, David Frederick *genetics educator*

Chicago
Shapiro, James A. *bacterial geneticist, educator*

Naperville
Yarger, James Gregory *technology company regulatory officer*

Urbana
Korban, Schuyler Safi *plant geneticist*

INDIANA

Bloomington
Schwartz, Drew *geneticist, educator*

Indianapolis
Christian, Joe Clark *medical genetics researcher, educator*
Kwon, Byoung Se *geneticist, educator*

IOWA

Ames
Hallauer, Arnel Roy *geneticist*

Iowa City
Milkman, Roger Dawson *genetics educator, molecular evolution researcher*

LOUISIANA

New Orleans
Cook, Julia Lea *geneticist*

MAINE

Bar Harbor
Davisson, Muriel Trask *geneticist*
Fox, Richard Romaine *geneticist, consultant*
Snell, George Davis *geneticist*

Presque Isle
Reeves, Alvin Frederick, II *genetics educator*

MARYLAND

Bethesda
Brady, Roscoe Owen *neurogeneticist, educator*
Di Paolo, Joseph Amadeo *geneticist*
Potter, Michael *genetics researcher, medical researcher*
Robbins, Keith Cranston *research scientist*

MASSACHUSETTS

Boston
Ruvkun, Gary B. *molecular geneticist*
Sager, Ruth *geneticist*

Cambridge
Hartl, Daniel Lee *genetics educator*

Chestnut Hill
Hoffman, Charles Stuart *molecular geneticist*

Worcester
Volkert, Michael Rudolf *molecular geneticist*

MICHIGAN

Ann Arbor
Gelehrter, Thomas David *medical and genetics educator, physician*
Markel, Dorene Samuels *geneticist*

University Center
Pelzer, Charles Francis *molecular geneticist, biology educator, cancer researcher*

MISSISSIPPI

Hattiesburg
Santangelo, George Michael *molecular geneticist*
Yarbrough, Karen Marguerite *genetics educator, university official*

NEW JERSEY

Cinnaminson
Evans, David A. *plant geneticist*

Mountainside
Buccini, Frank John *molecular geneticist*

New Brunswick
Day, Peter Rodney *geneticist, educator*

Nutley
Abbondanzo, Susan Jane *research geneticist*

Princeton
Seizinger, Bernd Robert *molecular geneticist, physician, researcher*

NEW YORK

Bronx
Waelsch, Salome Gluecksohn *geneticist, educator*

Brooklyn
Lee, Brendan *geneticist*
Luke, Sunny *medical geneticist, researcher*
Verma, Ram Sagar *geneticist, educator, author, administrator*

Ithaca
Altman, David Wayne *geneticist*
Last, Robert Louis *plant geneticist*

New York
Hirschhorn, Rochelle *medical educator*
Jagiello, Georgiana M. *geneticist, educator*
Lederberg, Joshua *geneticist, educator*
Terwilliger, Joseph Douglas *statistical geneticist*
Young, Michael Warren *geneticist, educator*

Stony Brook
Carlson, Elof Axel *genetics educator*

Syosset
Hershey, Alfred Day *geneticist*

NORTH CAROLINA

Chapel Hill
Maroni, Gustavo Primo *geneticist, educator*

Durham
Gross, Samson Richard *geneticist, biochemist, educator*

Greenville
Wiley, John Edwin *cytogeneticist*

Hendersonvlle
Kehr, August Ernest *geneticist, researcher*

Raleigh
Scandalios, John George *geneticist, educator*
Sederoff, Ronald Ross *geneticist*
Stuber, Charles William *genetics educator, researcher*

Research Triangle Park
Drake, John Walter *geneticist*
Judd, Burke Haycock *geneticist*

Winston Salem
Flory, Walter S., Jr. *geneticist, botanist, educator*

OHIO

Cincinnati
Tricoli, James Vincent *cancer genetics educator*

Columbus
Johnson, Lee Frederick *molecular geneticist*
Morrow, Grant, III *genetecist*

OREGON

Independence
Bhat, Bal Krishen *geneticist, plant breeder*

PENNSYLVANIA

Philadelphia
Knudson, Alfred George, Jr. *medical geneticist*

University Park
Buss, Edward George *geneticist*

SOUTH CAROLINA

Columbia
Dawson, Wallace Douglas, Jr. *geneticist*

TENNESSEE

Knoxville
Gresshoff, Peter Michael *molecular geneticist, educator*

Nashville
Phillips, John A(tlas), III *geneticist, educator*

TEXAS

Austin
Sutton, Harry Eldon *geneticist, educator*

College Station
Kohel, Russell James *geneticist*
Riggs, Penny Kaye *cytogeneticist*

Dallas
Brown, Michael Stuart *geneticist*

Denton
Wagers, William Delbert, Jr. *theoretical geneticist*

Galveston
McCombs, Jerome Lester *clinical cytogenetics laboratory director*

Houston
Greenberg, Frank *clinical geneticist, educator, academic administrator*
Horton, William Arnold *medical geneticist, educator*
Nelson, David Loren *geneticist, educator*
Schull, William J. *genetecist, educator*

Lubbock
Jackson, Raymond Carl *cytogeneticist*

Weslaco
Collins, Anita Marguerite *research geneticist*

UTAH

Provo
McArthur, Eldon Durant *geneticist, researcher*

Salt Lake City
Capecchi, Mario Renato *genetecist, educator*
Coon, Hilary Huntington *psychiatric genetics educator, researcher*

VERMONT

Burlington
Albertini, Richard Joseph *molecular geneticist, educator*

VIRGINIA

Blacksburg
Ha, Sam Bong *molecular geneticist*

Charlottesville
Gottesman, Irving Isadore *psychiatric genetics educator, consultant*

Richmond
Neale, Michael Churton *behavior geneticist*

WASHINGTON

Seattle
Hartwell, Leland Harrison *geneticist, educator*
Schellenberg, Gerard David *geneticist, researcher*
Weintraub, Harold M. *geneticist*

WISCONSIN

DeForest
Miller, Paul Dean *breeding company executive, geneticist*

Madison
Dentine, Margaret Raab *animal geneticist, educator*

CANADA

MANITOBA

Winnipeg
Hamerton, John Laurence *geneticist, educator*

ONTARIO

Guelph
Burnside, Edward Blair *geneticist, educator, administrator*

QUEBEC

Montreal
Pinsky, Leonard *geneticist*

SASKATCHEWAN

Saskatoon
Oelck, Michael M. *plant geneticist, researcher*

DENMARK

Aarhus
Pedersen, Bent Carl Christian *cytogeneticist*

ENGLAND

London
Bodmer, Walter Fred *cancer research administrator*

FRANCE

Montpellier
Pasteur, Nicole *population geneticist*

GERMANY

Braunschweig
Collins, John *molecular genetics educator, researcher*

GREECE

Athens
Tsakas, Spyros Christos *genetics educator*

ITALY

Ancona
Milani-Comparetti, Marco Severo *geneticist, bioethicist*

JAPAN

Hiroshima
Ban, Sadayuki *radiation geneticist*

Tokyo
Ishiwa, Sadao Chigusa *geneticist, educator*

THE NETHERLANDS

Utrecht
Van Zutphen, Lambertus F.M. (Bert Van Zutphen) *geneticist, educator*

THE PHILIPPINES

Manila
Khush, Gurdev Singh *geneticist*

RUSSIA

Moscow
Shestakov, Sergey Vasiliyevich *geneticist, biotechnologist*

SCOTLAND

Carnoustie
Harris, William Joseph *genetics educator*

Edinburgh
Chandley, Ann Chester *research scientist, cytogeneticist*

SPAIN

Palma Mallo
Petitpierre, Eduard *genetics educator*

SWEDEN

Stockholm
Daneholt, Per Bertil Edvard *molecular geneticist*

ADDRESS UNPUBLISHED

Leder, Philip *geneticist, educator*

LIFE SCIENCES: MARINE BIOLOGY

UNITED STATES

ALASKA

Fairbanks
Jewett, Stephen Carl *marine biologist, researcher, consultant*

CALIFORNIA

Bodega Bay
Hand, Cadet Hammond, Jr. *marine biologist, educator*

Fullerton
Murray, Steven Nelsen *marine biologist, educator*

Westminster
Allen, Merrill James *marine biologist*

DELAWARE

Lewes
Gaffney, Patrick Michael *marine biologist, aquacultural geneticist*

DISTRICT OF COLUMBIA

Washington
Homziak, Jurij *environment, aquaculture and marine resources specialist, consultant*

FLORIDA

Gainesville
Schelske, Claire L. *limnologist*

Jensen Beach
Martin, Roy Erik *marine biologist, consultant*

MAINE

West Boothbay Harbor
Field, John Douglas *marine biologist*

MASSACHUSETTS

Cambridge
Fell, Barry (Howard Barraclough Fell) *marine biologist, educator*

Nahant
Lavalli, Kari Lee *marine biologist*

MISSISSIPPI

Biloxi
Deegen, Uwe Frederick *marine biologist*

NEW HAMPSHIRE

Hanover
Gilbert, John Jouett *aquatic ecologist, educator*

NEW JERSEY

New Brunswick
Psuty, Norbert Phillip *marine sciences educator*
Wainright, Sam Chapman *marine scientist, educator*

NEW YORK

Stony Brook
Schubel, Jerry Robert *marine science educator, scientist, university dean and official*

NORTH CAROLINA

Chapel Hill
Frankenberg, Dirk *marine scientist*

Wilmington
McFall, Gregory Brennon *marine biologist*

OREGON

Corvallis
Lubchenco, Jane *marine biologist, educator*

PENNSYLVANIA

University Park
Fisher, Charles Raymond, Jr. *marine biologist*

RHODE ISLAND

Wakefield
Tarzwell, Clarence Matthew *aquatic biologist*

SOUTH CAROLINA

Charleston
Burrell, Victor Gregory, Jr. *marine scientist*

Columbia
Coull, Bruce Charles *marine ecology educator*
Vernberg, Frank John *marine and biological sciences educator*

TEXAS

Beaumont
Roller, Richard Allen *marine invertebrate physiologist*

Galveston
Giam, Choo-Seng *marine science educator*
Santschi, Peter Hans *marine sciences educator*

VIRGINIA

Gloucester Point
Penhale, Polly Ann *marine biologist, educator*

WASHINGTON

Anacortes
Sulkin, Stephen David *marine biology educator*

Olympia
Jamison, David W. *marine scientist*

Seattle
Edmondson, W(allace) Thomas *limnologist, educator*
Gunderson, Donald Raymond *fisheries educator and researcher*

WISCONSIN

Bayfield
Gallinat, Michael Paul *fisheries biologist*

Onalaska
Dukerschein, Jeanne Therese *aquatic biologist, educator, researcher*

CANADA

NOVA SCOTIA

Halifax
Scaratt, David J. *marine biologist*

AUSTRALIA

Queensland
Lawn, Ian David *marine biologist*

Townsville
Lucas, John Stewart *marine biologist*

GREECE

Athens
Jeftic, Ljubomir Mile *marine scientist*

JAPAN

Okinawa
Higa, Tatsuo *marine science educator*

NORWAY

Trondheim
Sundnes, Gunnar *retired marine biology educator*

THE PHILIPPINES

Iloilo
Tuburan, Isidra Bombeo *aquaculturist*

LIFE SCIENCES: MICROBIOLOGY

UNITED STATES

ALABAMA

Birmingham
Bhugra, Bindu *microbiologist, researcher*
Briles, David E(lwood) *microbiology educator*

ARIZONA

Tucson
Gerba, Charles Peter *microbiologist, educator*
Jeter, Wayburn Stewart *retired microbiology educator, microbiologist*

ARKANSAS

Fayetteville
Bhunia, Arun Kumar *microbiologist, immunologist, researcher*

Jefferson
Heflich, Robert Henry *microbiologist*
Sutherland, John Bruce, IV *microbiologist*

CALIFORNIA

Loma Linda
Taylor, Barry Llewellyn *microbiologist, educator*

Long Beach
Fung, Henry Chong *microbiologist, educator, administrator*
Wayne, Lawrence Gershon *microbiologist, researcher*

Los Angeles
Chen, Irvin Shao Yu *microbiologist, educator*
Esfandiari-Fard, Omid David *microbiologist*
Finegold, Sydney Martin *microbiology and immunology educator*

Mountain View
Couloures, Kevin Gottlieb *process development researcher*
Lu, Wuan-Tsun *microbiologist, immunologist*

Santa Clara
Klein, Harold Paul *microbiologist*

South San Francisco
Masover, Gerald Kenneth *microbiologist*

Stanford
Davis, Mark M. *microbiologist, educator*
Matin, Abdul *microbiology educator, consultant*

COLORADO

Denver
Barz, Richard L. *microbiologist*
Berens, Randolph Lee *microbiologist, educator*

Fort Collins
Ogg, James Elvis *microbiologist, educator*

Longmont
Ulrich, John August *microbiology educator*

CONNECTICUT

Hamden
Dubois, Normand Rene *microbiologist, researcher*

Lebanon
Feldman, Kathleen Ann *microbiologist, researcher*

Milford
Robohm, Richard Arthur *microbiologist, researcher*

New Haven
Dingman, Douglas Wayne *microbiologist*

Storrs Mansfield
Hinckley, Lynn Schellig *microbiologist*

West Haven
Huguenel, Edward David *microbiologist*

DELAWARE

Newark
Campbell, Linzy Leon *microbiologist, educator*

Rockland
Myoda, Toshio Timothy *microbiologist, consultant, educator*

DISTRICT OF COLUMBIA

Washington
Andrews, Wallace Henry *microbiologist*
Bellanti, Joseph A. *microbiologist, educator*
Day, Agnes Adeline *microbiology educator*
Kopecko, Dennis Jon *microbiologist, researcher*
Meyers, Wayne Marvin *microbiologist*

FLORIDA

Orlando
Trytek, Linda Faye *microbiologist*

Surfside
Polley, Richard Donald *microbiologist, polymer chemist*

GEORGIA

Alpharetta
Willis, Sharon White *microbiologist*

Athens
Lewis, David Lamar *research microbiologist*

Atlanta
Cavallaro, Joseph John *microbiologist*
Dowdle, Walter Reid *microbiologist, medical center administrator*
Lin, Jung-Chung *microbiologist, researcher*
Talkington, Deborah Frances *microbiologist*

Fort Gordon
Craft, David Walton *clinical microbiologist, army officer*

Griffin
Doyle, Michael Patrick *food microbiologist, educator, researcher*

HAWAII

Honolulu
Alicata, Joseph Everett *microbiology researcher, parasitologist*
Fujioka, Roger Sadao *research microbiology educator*

Kahului
Hughes, Arleigh Bruce *microbiologist, educator*

ILLINOIS

Abbott Park
Allen, Steven Paul *microbiologist*
Shipkowitz, Nathan L. *microbiologist*

Chicago
Corbett, Jules John *microbiology educator*
Lombardo, Janice Ellen *microbiologist*
Mullins, Obera *microbiologist*
Yamashiroya, Herbert Mitsugi *microbiologist, educator*

Forest Park
Orland, Frank Jay *oral microbiologist, educator*

Hinsdale
Morello, Josephine A. *microbiology educator, pathology educator*

Naperville
Wiatr, Christopher L. *microbiologist*

North Chicago
Schwartz, Robert David *fermentation microbiologist, bioengineer*

Peoria
Herrmann, Judith Ann *microbiologist*
Kurtzman, Cletus Paul *microbiologist*

Urbana
Isaacson, Richard Evan *microbiologist*

INDIANA

Bloomington
Gest, Howard *microbiologist, educator*

Indianapolis
Daily, William Allen *retired microbiologist*

IOWA

Cedar Rapids
Dvorak, Clarence Allen *microbiologist*

KENTUCKY

Louisville
Atlas, Ronald M. *microbiologist educator, ecologist*

LOUISIANA

Slidell
Stiffey, Arthur Van Buren *microbiologist*

MARYLAND

Baltimore
Suskind, Sigmund Richard *microbiology educator*
Walch, Marianne *microbiologist*

Bethesda
Broder, Christopher Charles *microbiologist*
Halula, Madelon Clair *microbiologist*
Varmus, Harold Eliot *microbiologist, educator*
Zierdt, Charles Henry *microbiologist*

Cockeysville Hunt Valley
Evans, George Leonard *microbiologist*
Marsik, Frederic John *microbiologist*

College Park
Colwell, Rita Rossi *microbiologist, molecular biologist*

Fort Washington
Godette, Stanley Rickford *microbiologist*

Frederick
Henchal, Erik Alexander *microbiologist*
Knisely, Ralph Franklin *retired microbiologist*

Rockville
Gougé, Susan Cornelia Jones *microbiologist*
Hackett, Joseph Leo *microbiologist, clinical pathologist*

Silver Spring
Lynt, Richard King *microbiologist*

Towson
Wubah, Daniel Asua *microbiologist*

MASSACHUSETTS

Bedford
Roche, Kerry Lee *microbiologist*

Boston
Fields, Bernard Nathan *microbiologist, physician*
Mekalanos, John J. *microbiology educator*
Prabhudas, Mercy Ratnavathy *microbiologist, immunologist*

Cambridge
Baltimore, David *microbiologist, educator*
Demain, Arnold Lester *microbiologist, educator*
Magasanik, Boris *microbiology educator*

Randolph
Jelley, Scott Allen *microbiologist*

Worcester
Escott, Shoolah Hope *microbiologist*

MICHIGAN

East Lansing
Reddy, Chilecampalli Adinarayana *microbiology educator*
Tiedje, James Michael *microbiology educator, ecologist*

Kalamazoo
Dietz, Alma *microbiologist*
Robertson, John Harvey *microbiologist*

Midland
McDade, Joseph John *microbiologist*

Rochester
Walia, Satish Kumar *microbiologist, educator*

MINNESOTA

Minneapolis
Dworkin, Martin *microbiologist, educator*
Haase, Ashley Thomson *microbiology educator, scientist*
McAloon, Todd Richard *food microbiologist*
Serstock, Doris Shay *retired microbiologist, educator, civic worker*

Rochester
Whelen, Andrew Christian *microbiologist*

MISSISSIPPI

Ridgeland
Evans, Wayne Edward *environmental microbiologist, researcher*

MISSOURI

Columbia
Brown, Olen Ray *medical microbiology research educator*
Eisenstark, Abraham *research director, microbiologist*
Finkelstein, Richard Alan *microbiologist*

Saint Louis
Murray, Patrick Robert *microbiologist, educator*

NEBRASKA

Omaha
Ehrhardt, Anton F. *medical microbiology educator*

NEVADA

Las Vegas
Pridham, Thomas Grenville *research microbiologist*

NEW HAMPSHIRE

Hanover
Jacobs, Nicholas Joseph *microbiology educator*

NEW JERSEY

Metuchen
Stapley, Edward Olley *retired microbiologist, research administrator*

Montvale
Bowman, Patricia Imig *microbiologist*

New Brunswick
Montville, Thomas Joseph *food microbiologist, educator*

North Branch
Kravec, Cynthia Vallen *microbiologist*

Rahway
Monaghan, Richard Leo *microbiologist*

Westfield
Stoudt, Thomas Henry *research microbiologist*

NEW MEXICO

Santa Fe
Stevenson, Robert Edwin *microbiologist, culture collection executive*

NEW YORK

Brooklyn
Bae, Ben Hee Chan *microbiologist*
Sultzer, Barnet Martin *microbiology and immunology educator*

Elmhurst
Kekatos, Deppie-Tinny Z. *microbiologist, researcher, lab technologist*

Floral Park
Dalto, Michael *medical microbiologist*

Flushing
Schnall, Edith Lea (Mrs. Herbert Schnall) *microbiologist, educator*

Ithaca
Mortlock, Robert Paul *microbiologist, educator*
Steinkraus, Keith Hartley *microbiology educator*

New York
Caroline, Leona Ruth *retired microbiologist*
Mayo, Joan Bradley *microbiologist, epidemiologist*
Moses, Johnnie, Jr. *microbiologist*
Philipson, Lennart Carl *microbiologist, science administrator*
Poshni, Iqbal Ahmed *microbiologist*
Racaniello, Vincent Raimondi *microbiologist, medical educator*
Sonnabend, Joseph Adolph *microbiologist*
Stotzky, Guenther *microbiologist, educator*

Nyack
Lee, Lillian Vanessa *microbiologist*

Oriskany Falls
Michels, Richard Steven *microbiologist*

Rochester
Iglewski, Barbara Hotham *microbiologist, educator*
McCormack, Grace *retired microbiology educator*

Syracuse
Lane, Michael Joseph *microbiology educator*

Troy
Soracco, Reginald John *microbiological biochemist*

Vestal
Lazaroff, Norman *microbiologist, researcher*

NORTH CAROLINA

Chapel Hill
Hutchison, Clyde Allen, III *microbiology educator*

Durham
Reller, L. Barth *microbiologist, educator*

Kinston
Arcino, Manuel Dagan *microbiologist, consultant*

Research Triangle Park
Sykes, Richard Brook *microbiologist*

OHIO

Bowling Green
Hann, William Donald *microbologist, biology educator*

Cincinnati
Hurst, Christon James *microbiologist*
Leusch, Mark Steven *microbiologist*
Reasoner, Donald J. *microbiologist*

Columbus
Kapral, Frank Albert *medical microbiology and immunology educator*
Olsen, Richard George *microbiology educator*

PENNSYLVANIA

Philadelphia
Kaji, Akira *microbiology scientist, educator*
Koprowski, Hilary *microbiology educator, medical scientist*
Rosan, Burton *microbiology educator*
Shockman, Gerald David *microbiologist, educator*
Tudor, John Julian *microbiologist*

University Park
Brenchley, Jean Elnora *microbiologist, researcher*

RHODE ISLAND

Kingston
Laux, David Charles *microbiologist, educator*

Warwick
Dickinson, Katherine Diana *microbiologist*

TENNESSEE

Knoxville
White, David Cleaveland *microbial ecologist, environmental toxicologist*

Memphis
Chung, King-Thom *microbiologist, educator*
Farrington, Joseph Kirby *microbiologist*
Howe, Martha Morgan *microbiologist, educator*

Nashville
Pincus, Theodore *microbiologist, educator*

TEXAS

Dallas
McCracken, George H. *microbiologist*
Norgard, Michael Vincent *microbiology educator, researcher*
Vitetta, Ellen S. *microbiologist educator, immunologist*

Irving
Fukui, George Masaaki *microbiology consultant*

Lubbock
Hentges, David John *microbiology educator*

San Antonio
Burno, John Gordon, Jr. *microbiologist*
Tokoly, Mary Andree *microbiologist*
Winters, Wendell Delos *microbiology educator, researcher, consultant*

UTAH

Logan
Tariq, Athar Mohammad *soil microbiologist*

VERMONT

Morrisville
Lechevalier, Mary Pfeil *retired microbiologist, educator*

VIRGINIA

Alexandria
Rafey, Larry Dean *microbiologist*

Blacksburg
Wilkins, Tracy Dale *microbiologist, educator*

Chantilly
Srivastava, Kailash Chandra *microbiologist*

Fairfax
Ware, Lawrence Leslie, Jr. *microbiologist*

Richmond
Bradley, Sterling Gaylen *microbiology and pharmacology educator*

WASHINGTON

Everett
Deboo, Behram Savakshaw *microbiologist*

Richland
Wiley, William Rodney *microbiologist, administrator*
Xun, Luying *microbiologist, educator*

Seattle
Koff, Andrew *microbiologist, biomedical researcher*
Nester, Eugene William *microbiology educator, immunology educator*

Tacoma
Tison, David Lawrence *microbiologist*

WISCONSIN

Milwaukee
Remsen, Charles Cornell, III *microbiologist, research administrator, educator*

Oconomowoc
Luedke, Patricia Georgianne *microbiologist*

Oshkosh
Rouf, Mohammed Abdur *microbiology educator*

TERRITORIES OF THE UNITED STATES

PUERTO RICO

Rio Piedras
Toranzos, Gary Antonio *microbiology educator*

San Juan
Rodriguez-del Valle, Nuri *microbiology educator*

CANADA

QUEBEC

Laval
Kluepfel, Dieter *microbiologist*

SASKATCHEWAN

Saskatoon
Khachatourians, George Gharadaghi *microbiology educator*

ENGLAND

Brighton
Spratt, Brian Geoffrey *microbiologist, educator, researcher*

FRANCE

Lille
Santoro, Ferrucio Fontes *microbiologist*

Nouzilly
Plommet, Michel Georges *microbiologist, researcher*

Paris
Chany, Charles *microbiology educator*
Lwoff, André Michel *retired microbiologist, virologist*

Strasbourg
Thierry, Robert Charles *microbiologist*

GERMANY

Frankfurt
Entian, Karl-Dieter *microbiology educator*

Marburg
Mannheim, Walter *medical microbiologist*

HUNGARY

Budapest
Deak, Tibor *microbiologist*

ISRAEL

Yavne
Fish, Falk *microbiologist, immunologist, researcher, inventor*

JAPAN

Kanagawa
Hankins, Raleigh Walter *microbiologist*

Kobe
Homma, Morio *microbiology educator*

Kokubunji
Shimahara, Kenzo *applied microbiology educator*

Osaka
Kitano, Kazuaki *microbiologist, researcher*
Sakaguchi, Genji *food microbiologist, educator*

MONGOLIAN PEOPLE'S REPUBLIC

Ulan Bator
Batbayar, Bat-Erdeniin *microbiologist*

NORWAY

Kristiansand
Csángó, Péter András *microbiologist*

SPAIN

Seville
Ventosa, Antonio *microbiologist, educator*

Valencia
Sentandreu, Rafael *microbiologist*

SWEDEN

Göteborg
Malmcrona-Friberg, Karin Elisabet *microbiologist*

Stockholm
Neilson, Alasdair Hewitt *microbiologist*

Umeå
Sandström, Gunnar Emanuel *microbiologist*

SWITZERLAND

Basel
Arber, Werner *microbiologist*

TAIWAN

Taichung
Wu, Wen Chuan *microbial geneticist, educator*

THAILAND

Bangkok
Punnapayak, Hunsa *microbiologist, educator*

ADDRESS UNPUBLISHED

Champlin, William Glen *clinical microbiologist-immunologist*
Crain, Danny B. *microbiologist*
Dombrowski, Anne Wesseling *microbiologist, researcher*
Dugan, Patrick Raymond *microbiologist, university dean*
Ellner, Paul D. *clinical microbiologist*
Flores-Lopez, Auremir *microbiologist*
Folkens, Alan Theodore *microbiologist*
Holden, Eric George *microbiologist*
Lichstein, Herman Carlton *microbiology educator emeritus*
Witte, Owen Neil *microbiologist, molecular biologist, educator*

LIFE SCIENCES: MOLECULAR BIOLOGY

UNITED STATES

ARKANSAS

Little Rock
Hardin, James Webb *molecular biologist, medical center administrator*

CALIFORNIA

Berkeley
Cline, Thomas Warren *molecular biologist, educator*
Cozzarelli, Nicholas R. *molecular biologist, educator*
Rubin, Gerald Mayer *molecular biologist, biochemistry educator*

Duarte
Smith, Steven Sidney *molecular biologist*

Los Angeles
Beamer, Lesa Jean *molecular biologist*

Menlo Park
Johnston, Brian Howard *molecular biologist*

Northridge
Taylor, Kent Douglas *molecular biologist*

Pasadena
Varshavsky, Alexander Jacob *molecular biologist*

San Diego
Eckhart, Walter *molecular biologist, educator*
Thomas, Charles Allen, Jr. *molecular biologist, educator*

San Francisco
Blackburn, Elizabeth Helen *molecular biologist*
McKnight, Steven Lanier *molecular biologist*

South San Francisco
Levinson, Arthur David *molecular biologist*
Pennica, Diane *molecular biologist*

Stanford
Baker, Bruce S. *molecular biologist*
Long, Sharon Rugel *molecular biologist, plant biology educator*

COLORADO

Boulder
Wood, William Barry, III *biologist, educator*

CONNECTICUT

New Haven
Coca-Prados, Miguel *molecular biologist*
Li, Jianming *molecular and cellular biologist*
Sigler, Paul Benjamin *molecular biology educator, protein crystallographer*
Squinto, Stephen Paul *molecular biologist, biochemist*

West Haven
Debeyssey, Mark Sammer *molecular and cellular biologist*
Pickle, Willian Neel, II *molecular biologist, researcher*

DISTRICT OF COLUMBIA

Washington
Holmes, George Edward *molecular biologist, researcher, educator*
Notario, Vicente *molecular biology educator, researcher*

FLORIDA

Gainesville
Boehm, Eric Walter Albert *molecular mycology researcher*

HAWAII

Honolulu
Mandel, Morton *molecular biologist*

ILLINOIS

Lombard
Velardo, Joseph Thomas *molecular biology and endocrinology educator*

Maywood
Amero, Sally Ann *molecular biologist, researcher*

Naperville
Hauptmann, Randal Mark *molecular biologist*

INDIANA

Indianapolis
Lahiri, Debomoy Kumar *molecular neurobiologist, educator*

Notre Dame
McLinden, James Hugh *molecular biologist*

West Lafayette
Ferreira, Paulo Alexandre *molecular biologist*

KENTUCKY

Lexington
Rangnekar, Vivek Mangesh *molecular biologist, researcher*

Richmond
Calie, Patrick Joseph *molecular biologist*

LOUISIANA

New Orleans
Wright, Maureen Smith *molecular biologist, microbiology educator*

MARYLAND

Baltimore
Clark, Patricia *molecular biologist*
Smith, Hamilton Othanel *molecular biologist, educator*

Beltsville
Zarlenga, Dante Sam, Jr. *molecular biologist*

Bethesda
Keith, Jerry M. *molecular biologist*
McKinney, Cynthia Eileen *molecular biologist*
Noguchi, Constance Tom *molecular biologist, researcher*
Pastan, Ira Harry *biomedical science researcher*

College Park
Kung, Shain-dow *molecular biologist, educator*

Gaithersburg
McClelland, Alan *molecular biologist, laboratory administrator*
Rollence, Michele Lynette *molecular biologist*

MASSACHUSETTS

Boston
Deresiewicz, Robert Leslie *molecular biologist, physician, educator*
Isberg, Ralph *molecular biologist, educator*
Struhl, Kevin *molecular biologist, educator*

Cambridge
Fox, Maurice Sanford *molecular biologist, educator*
Gilbert, Walter *molecular biologist, educator*
King, Jonathan Alan *molecular biology educator*
Mulligan, Richard C. *molecular biology educator*
Schwedock, Julie *molecular biologist*
Stewart, Sue Ellen *molecular biologist*

Kingston
Slot, Larry Lee *molecular biologist*

Waltham
Huxley, Hugh Esmor *molecular biologist, educator*

Worcester
Temsamani, Jamal *molecular biologist, researcher*

MICHIGAN

Ann Arbor
Jove, Richard *molecular biologist*
Lowe, John Burton *molecular biology educator, pathologist*

Detroit
Krawetz, Stephen Andrew *molecular biology and genetics educator*

Kalamazoo
Klein, Ronald Don *molecular biologist*
Marotti, Keith Richard *molecular biologist, researcher*

Rochester
Chaudhry, G. Rasul *molecular biologist, educator*

MINNESOTA

Saint Paul
Emeagwali, Dale Brown *molecular biologist*

MISSOURI

Saint Louis
Fleming, Timothy Peter *molecular biologist*
Green, Maurice *molecular biologist, virologist, educator*

NEW JERSEY

Nutley
Curran, Thomas *molecular biologist, educator*

Paramus
Bard, Jonathan Adam *molecular biologist*

Piscataway
Denhardt, David Tilton *molecular and cell biology educator*
Messing, Joachim Wilhelm *molecular biology educator*

Princeton
Fernandes, Prabhavathi Bhat *molecular biologist*
Levine, Arnold Jay *molecular biology educator, researcher*
Molloy, Christopher John *molecular cell biologist, pharmacist*
Shenk, Thomas Eugene *molecular biology educator*

Rahway
Van der Ploeg, Leonardus Harke Theresia *molecular biologist*

NEW YORK

Albany
Wulff, Daniel Lewis *molecular biologist, educator*

Buffalo
Belgrader, Phillip *molecular biologist*

Cold Spring Harbor
Roberts, Richard John *molecular biologist*
Watson, James Dewey *molecular biologist, educator*
Wigler, Michael H. *molecular biologist*

New York
Darnell, James Edwin, Jr. *molecular biologist, educator*
Horvath, Diana Meredith *plant molecular biologist*
Marks, Andrew Robert *molecular biologist*

NORTH CAROLINA

Chapel Hill
Bartlett, Jeffrey Stanton *molecular biologist*

Research Triangle Park
Huber, Brian Edward *molecular biologist*

OHIO

Toledo
Duran, Emilio *molecular biologist*

OKLAHOMA

Stillwater
Pennington, Rodney Edward *molecular biologist*

PENNSYLVANIA

Hershey
Stanley, Bruce Alan *molecular biologist*

Philadelphia
Chu, Mon-Li Hsiung *molecular biologist*
Davis, Alan Robert *molecular biologist*
Skalka, Anna Marie *molecular biologist, virologist*

Pittsburgh
McKnight, Jennifer Lee Cowles *molecular biologist*

RHODE ISLAND

Providence
Avissar, Yael Julia *molecular biology educator*

TEXAS

Austin
Brown, Dennis Taylor *molecular biology educator*

Dallas
Smagula, Cynthia Scott *molecular biologist*

Prairie View
Braithwaite, Cleantis Esewanu *molecular biologist*

The Woodlands
Maxwell, Steve A. *molecular biologist, researcher*
Ojwang, Joshua Odoyo *molecular biologist*

VERMONT

Burlington
Francklyn, Christopher Steward *molecular biologist*
Held, Paul G. *molecular biologist*

VIRGINIA

Falls Church
McEwan, Robert Neal *molecular biologist*

Richmond
Buck, Gregory Allen *molecular biology educator*

WASHINGTON

Kirkland
Pearson, Mark Landell *molecular biologist*

WISCONSIN

Madison
Borisy, Gary G. *molecular biology educator*

Middleton
Haynes, Joel Robert *molecular biologist*

CANADA

QUEBEC

Montreal
Skup, Daniel *molecular biologist, educator, researcher*
Stanners, Clifford Paul *molecular biologist, cell biologist, biochemistry educator*

Saint Jean-sur-Richelieu
Côté, Jean-Charles *research molecular biologist*

AUSTRALIA

Parkville
Foote, Simon James *molecular biologist*

ENGLAND

Cambridge
Klug, Aaron *molecular biologist*
Milstein, César *molecular biologist*
Perutz, Max Ferdinand *molecular biologist*
Sanger, Frederick *retired molecular biologist*

Linton
Kendrew, John Cowdery *molecular biologist, former college president*

FRANCE

Nice
Auwerx, Johan Henri *molecular biologist, medical educator*

GERMANY

Heidelberg
Papavassiliou, Athanasios George *molecular biologist, researcher*

Munich
Pääbo, Svante *molecular biologist, biochemist*

ITALY

Milan
Martegani, Enzo *molecular biology educator*
Soria, Marco Raffaello *molecular and cellular biologist*

JAPAN

Osaka
Nakazato, Hiroshi *molecular biologist*

REPUBLIC OF KOREA

Seoul
Park, Sang-Chul *molecular biologist, educator*

Suwon
Kim, Byung-Dong *molecular biology educator*

RUSSIA

Moscow
Mirzabekov, Andrey Daryevich *molecular biologist*

SWITZERLAND

Geneva
Rochaix, Jean-David *molecular biologist educator*

Zurich
Wüthrich, Kurt *molecular biologist, biophysical chemist*

YUGOSLAVIA

Belgrade
Kanazir, Dušan *molecular biologist, biochemist, educator*

ADDRESS UNPUBLISHED

Bryant, Donald Ashley *molecular biologist*
Galas, David John *molecular biology educator, researcher*
Hatano, Sadashi *molecular biology educator*
Huvos, Piroska Eva *molecular biology researcher, chemistry educator*
Rosario, Myra Odette *molecular biologist, pharmacist, educator*
Simon, Melvin I. *molecular biologist, educator*

LIFE SCIENCES: OTHER SPECIALTIES

UNITED STATES

ALABAMA

Huntsville
Marx, Richard Brian *forensic scientist*

ARIZONA

Tucson
Cogut, Theodore Louis *environmental specialist, meteorologist*
Foster, Kennith Earl *life sciences educator*

ARKANSAS

Fayetteville
Revels, Mia Renea *science educator*

CALIFORNIA

Fullerton
Silverman, Paul Hyman *parasitologist, former university official*

Irvine
Chicz-DeMet, Aleksandra *science educator, consultant*

La Jolla
Somerville, Richard Chapin James *science educator*

Manteca
Rainey, Barbara Ann *sensory evaluation consultant*

Merced
Olsen, David Magnor *science educator*

San Diego
Caulder, Jerry Dale *weed scientist*

COLORADO

Fort Collins
Ishimaru, Carol Anne *plant pathologist*

DISTRICT OF COLUMBIA

Washington
Begle, Douglas Pierce *ichthyologist*
West, Robert MacLellan *science education consultant*
Woods, Walter Ralph *animal scientist, research administrator*

FLORIDA

Homestead
Byrd, Mary Laager *life science researcher*

Jacksonville
Tardona, Daniel Richard *marine naturalist, park ranger, educator*

Jay
Brecke, Barry John *weed scientist, researcher*

Tallahassee
Collins, Angelo *science educator*
Williams, Theodore P. *biophysicist, biology educator*

Tavernier
Grove, Jack Stein *naturalist, marine biologist*

GEORGIA

Athens
Compton, Mark Melville *poultry science educator*

Bogart
Butts, David Phillip *science educator*

Columbus
Riggsby, Ernest Duward *science educator*

ILLINOIS

Champaign
Page, Lawrence Merle *ichthyologist, educator*

Chicago
Baker, M(ervin) Duane *healthcare, biotechnology marketing and business consultant*
Jeyendran, Rajasingam Sivaperagasam *andrologist, researcher*
Kass, Leon Richard *life sciences educator*

Hines
Trimble, John Leonard *sensor psychophysicist, biomedical engineer*

Park Ridge
Tomaszkiewicz, Francis Xavier *imaging technology educator*

Urbana
Chang, Ruey-Jang *life science researcher*

INDIANA

Warsaw
Cupp, Jon Michael *environmental scientist*

West Lafayette
McLaughlin, Gerald Lee *parasitology educator*

IOWA

Ames
Hammond, Earl Gullette *food science educator*

Larchwood
Onet, Virginia *veterinary parasitologist, researcher, educator*

LOUISIANA

New Orleans
Bundy, Kirk Jon *biomaterials educator, researcher, consultant*

MAINE

Orono
O'Connor, Raymond Joseph *wildlife educator*

MARYLAND

Baltimore
Maumenee, Irene Hussels *opthalmology educator*
Roberts, Randolph Wilson *health science educator*
Swift, David Leslie *environmental health educator, consultant*

Beltsville
Hoberg, Eric Paul *parasitologist*

College Park
Kuenzel, Wayne John *avian physiologist, neuroscientist*

Olney
Baker, Carl Gwin *science educator*

Princess Anne
Adams, James Alfred *natural science educator*

MASSACHUSETTS

Amherst
DeVries, Geert Jan *science educator*

Cambridge
Mendelsohn, Everett Irwin *history of science educator*
Wiegand, Thomas Edward von *psychophysicist, consultant*

MICHIGAN

Ann Arbor
Fink, William Lee *ichthyologist, systematist*
Hays, Paul B. *science educator, researcher*

Highland Park
Crittenden, Mary Lynne *science educator*

Kalamazoo
Conder, George Anthony *parasitologist*

MINNESOTA

Minneapolis
Gorham, Eville *scientist, educator*

Saint Paul
Wendt, Hans Werner *life scientist, educator*
Xie, Weiping *mass spectromist*

MISSOURI

Columbia
Merilan, Charles Preston *dairy husbandry scientist*

Saint Louis
Jakschik, Barbara A. *science educator, researcher*

NEBRASKA

Lincoln
Arumuganathan, Kathiravepillai *plant flow cytometrist, cell and molecular biologist*

NEW HAMPSHIRE

Durham
Burdick, David Maaloe *marine ecological researcher*

Hanover
Bzik, David John *parasitologist, researcher*

NEW JERSEY

Summit
Zelenakas, Ken Walter *clinical scientist*

NEW MEXICO

Tijeras
Saiz, Bernadette Louise *morphology technician*

NEW YORK

Chestertown
Wormwood, Richard Naughton *naturalist, theoretical field geologist, field research primatologist*

East Patchogue
Metz, Donald Joseph *scientist*

Fredonia
Tomlinson, Bruce Lloyd *biology educator, researcher*

Ithaca
Ballantyne, Joseph Merrill *science educator, program administrator, researcher*
Bowman, Dwight Douglas *parasitologist*
Isaacson, Michael Saul *physics educator, researcher*
Jones, Edward David *plant pathologist*

New York
Jacobson, Willard James *science educator*
Martin, Lenore Marie *bioorganic researcher, educator*

Riverhead
Senesac, Andrew Frederick *weed scientist*

White Plains
Theisz, Erwin Jan *scientist*

NORTH CAROLINA

Raleigh
Petitte, James Nicholas *poultry science educator*

Washington
Thomson, Stuart McGuire, Jr. *science educator*

OHIO

Athens
Skinner, Ray, Jr. *earth science educator, consultant*

Columbus
Zartman, David Lester *dairy science educator, researcher*

Oxford
Haley-Oliphant, Ann Elizabeth *science educator*

OKLAHOMA

Bartlesville
Rutledge, Kathleen Pillsbury *sensory scientist, researcher*

OREGON

Portland
Grimsbo, Raymond Allen *forensic scientist*

PENNSYLVANIA

Harrisburg
Wei, I-Yuan *research and development manager*

King Of Prussia
Bilofsky, Howard Steven *scientist*

Lancaster
Schwartz, John Howard *poultry science and extension educator*

Pittsburgh
Glencer, Suzanne Thomson *science educator*

Rheems
Adams, James Lee *poultry scientist*

RHODE ISLAND

Providence
Knopf, Paul Mark *immunoparasitologist*

Warwick
Dubois, Janice Ann *primatologist, educator*

SOUTH CAROLINA

Charleston
Cheng, Thomas Clement *parasitologist, immunologist, educator, author*

SOUTH DAKOTA

Brookings
Morgan, Walter *retired poultry science educator*

TENNESSEE

Dresden
Betz, Norman L. *science educator, consultant*

Memphis
Heitmeyer, Mickey E. *conservationist*

TEXAS

Arlington
Denny, Thomas Albert *product development specialist, researcher*

Dallas
Al-Hashimi, Ibtisam *oral scientist, educator*
Harbaugh, Lois Jensen *secondary science educator*

Denton
Goggin, Noreen Louise *kinesiology educator*

Galveston
Caillouet, Charles Wax, Jr. *fisheries scientist*

San Antonio
Quinn, Mary Ellen *science educator*

Tyler
Dodson, Ronald Franklin *electron microscopist, administrator*

UTAH

Logan
Sidle, Roy C. *research hydrologist*

Salt Lake City
Straight, Richard Coleman *photobiologist*

VIRGINIA

Arlington
Ratchford, Joseph Thomas *technical policy educator, consultant*

Manakin Sabot
Brickey, James Allan *owner environmental testing laboratory, consultant*

WASHINGTON

Olympia
Steiger, Gretchen Helene *marine mammalogist, research biologist*

WISCONSIN

Madison
Vailas, Arthur C. *biomechanics educator*

Onalaska
Soballe, David Michael *limnologist*

TERRITORIES OF THE UNITED STATES

PUERTO RICO

San German
Quintero, Héctor Enrique *science educator*

CANADA

ONTARIO

Kingston
Smol, John Paul *limnologist, educator*

ARGENTINA

Buenos Aires
Wais de Badgen, Irene Rut *limnologist*

AUSTRIA

Mondsee
Dokulil, Marin *limnologist*

CAYMAN ISLANDS

Georgetown
Husemann, Anthony James *science educator*

ENGLAND

Brighton
Kroto, Harold Walter *science educator*

Welwyn
Eckers, Christine *mass spectrometry scientist*

FRANCE

Castillon-du-Gard
Jerne, Niels Kaj *scientist*

Strasbourg
Danzin, Charles Marie *enzymologist*

GERMANY

Heidelberg
Abel, Ulrich Rainer *biometrician, researcher, financial consultant*

Munich
Berg, Jan Mikael *science educator*

HONG KONG

Hong Kong
Yoo, Kwong Mow *science educator*

INDONESIA

Pangkalanbun
Galdikas, Birute *primatologist*

JAPAN

Iwate
Kawauchi, Hiroshi *hormone science educator*

Okayama
Hamada, Hiroki *science educator*

Tochigi
Ishizaki, Tatsushi *physician, parasitologist*

Tokonaka
Tashiro, Kohji *macromolecular scientist*

Tokyo
Ishii, Akira *medical parasitologist, malariologist, allergist*
Kitazawa, Koichi *materials science educator*

SWEDEN

Lund
Wadsö, B. Ingemar *science educator*

TURKEY

Ankara
Tekelioglu, Meral *physician, educator*

UGANDA

Kampala
Kaddu, John Baptist *parasitologist, consultant*

ADDRESS UNPUBLISHED

Foster, Barbara Melanie *microscopist, consultant*

LIFE SCIENCES: PHYSIOLOGY

UNITED STATES

ALABAMA

Birmingham
Benos, Dale John *physiology educator*
Francis, Kennon Thompson *physiologist*

ARIZONA

Phoenix
Garcia, Richard Louis *plant physiologist/ micrometeorologist, researcher*

Tucson
Manciet, Lorraine Hanna *physiologist, educator*

ARKANSAS

Little Rock
Liu, Shi Jesse *physiologist, researcher*

CALIFORNIA

La Jolla
Guillemin, Roger C. L. *physiologist*

Loma Linda
Longo, Lawrence Daniel *physiologist, gynecologist*

Los Angeles
Langer, Glenn Arthur *cellular physiologist, educator*

Moffett Field
Greenleaf, John Edward *research physiologist*

Pasadena
Doupnik, Craig Allen *physiologist*

San Francisco
Borson, Daniel Benjamin *physiology educator, inventor, researcher*

Santa Barbara
Lennox Buchthal, Margaret Agnes *neurophysiologist*

Woodland Hills
Fox, Stuart Ira *physiologist*

COLORADO

Fort Collins
Bigiani, Albertino Roberto *electrophysiologist, researcher*
Niswender, Gordon Dean *physiologist, educator*
Turzillo, Adele Marie *reproductive physiologist, researcher*

CONNECTICUT

Branford
Sinton, Christopher Michael *neurophysiologist*

New Haven
DuBois, Arthur Brooks *physiologist, educator*

DISTRICT OF COLUMBIA

Washington
Kramer, Jay Harlan *physiologist, biochemist, researcher, educator*
Lorber, Mortimer *physiology educator, researcher*
Rabson, Robert *plant physiologist, administrator*

FLORIDA

Gainesville
Hansen, Peter James *reproductive physiologist, researcher*
Jaeger, Marc Julius *physiology educator, researcher*

Homestead
Schaffer, Bruce Alan *plant physiologist, educator*

GEORGIA

Augusta
Nosek, Thomas Michael *physiologist, educator*

Statesboro
Parrish, John Wesley, Jr. *physiologist, biology educator*

HAWAII

Honolulu
Uyehara, Catherine Fay Takako (Yamauchi) *physiologist, educator, pharmacologist*

ILLINOIS

Chicago
Ernest, J. Terry *ocular physiologist, educator*
Marotta, Sabath Fred *physiology educator*

Normal
Preston, Robert Leslie *cell physiologist, educator*

Skokie
Lavenda, Nathan *physiology educator*

Urbana
Bahr, Janice Mary *reproductive physiologist*
Rebeiz, Constantin Anis *plant physiology educator*

INDIANA

Bloomington
Hammel, Harold Theodore *physiology and biophysics educator, researcher*

Muncie
Costill, David Lee *physiologist, educator*

IOWA

Iowa City
Shibata, Erwin Fumio *cardiovascular physiologist*

KANSAS

Pratt
Rice, Ramona Gail *physiologist, psychologist, educator, consultant*

KENTUCKY

Lexington
Diana, John Nicholas *physiologist*
Dutt, Ray Horn *reproduction physiologist, educator*

Louisville
Harris, Patrick Donald *physiology educator*

LOUISIANA

Baton Rouge
Pezeshki, Sadrodin Reza *plant ecophysiology educator*

New Orleans
Barbee, Robert Wayne *cardiovascular physiologist*
Cairo, Jimmy Michael *physiologist*
Kreisman, Norman Richard *physiologist*
Roheim, Paul Samuel *physiology educator*

MAINE

Orono
Clapham, William Montgomery *plant physiologist*

MARYLAND

Baltimore
Littlefield, John Walley *physiology educator, geneticist, cell biologist, pediatrician*

Bethesda
Bunger, Rolf *physiology educator*
Burg, Maurice Benjamin *renal physiologist, physician*
Robinson, David Mason *physiologist*
Sokoloff, Louis *physiologist, neurochemist*

Frederick
Wellner, Robert Brian *physiologist*

Laurel
Hoffman, David John *physiologist*

MASSACHUSETTS

Boston
Hubel, David Hunter *physiologist, educator*
Marcum, James Arthur *physiology educator*

MICHIGAN

Ann Arbor
Brock, Thomas Gregory *plant physiologist*
Kluger, Matthew Jay *physiologist, educator*
Ning, Xue-Han (Hsueh-Han Ning) *physiologist, researcher*

Detroit
Phillis, John Whitfield *physiologist, educator*
Woodbury, Dixon John *physiology educator, researcher*

East Lansing
Dilley, David Ross *plant physiologist, researcher*

Kalamazoo
Tsai, Ti-Dao *electrophysiologist*

MINNESOTA

Duluth
Haller, Edwin Wolfgang *physiologist, educator*

Minneapolis
Miller, Robert Francis *physiologist, educator*

Rochester
Lockhart, John Campbell *bioengineer, physiologist*

Saint Paul
Hunter, Alan Graham *reproductive physiologist, educator*

MISSISSIPPI

Stoneville
Duke, Stephen Oscar *physiologist, researcher*

MISSOURI

Saint Louis
Montague, Michael James *plant physiologist*
Partridge, Nicola Chennell *physiology educator*

NEBRASKA

Chadron
Hardy, Joyce Margaret Phillips *plant physiologist, educator*

Lincoln
Sohaili, Aspi Isfandiar *physiologist, educator*

Omaha
Badeer, Henry Sarkis *physiology educator*

NEW JERSEY

Newark
Beyer-Mears, Annette *physiologist*

Woodbridge
Fee, Geraldine Julia *psychophysiologist*

NEW YORK

Brooklyn
Altura, Bella T. *physiologist, educator*
Altura, Burton Myron *physiologist, educator*
Pagala, Murali Krishna *physiologist*

Ithaca
Halpern, Bruce Peter *physiologist, consultant*
Nathanielsz, Peter William *physiologist*

New York
Cunningham, Dorothy Jane *physiology educator*
Windhager, Erich Ernst *physiologist, educator*

Old Westbury
Andrews, Mark Anthony William *physiologist, educator*

Rochester
Begenisich, Ted Bert *physiology educator*
Zhang, Guo He *physiologist*

Stony Brook
Smaldone, Gerald Christopher *physiologist*

NORTH CAROLINA

Chapel Hill
Warren, Donald William *physiology educator, dentistry educator*

Durham
Diamond, Irving T. *physiology educator*
Johnson, Edward A. *physician, educator*
Schmidt-Nielson, Knut *physiologist, educator*

Greenville
Lust, Robert Maurice, Jr. *physiologist, educator, researcher*

Raleigh
Grossfeld, Robert Michael *physiologist, zoologist, educator, neurobiologist*
Olson, Neil Chester *physiologist, educator*

Winston Salem
Simonsen, John Charles *exercise physiologist*

NORTH DAKOTA

Grand Forks
Bolonchuk, William Walter *physical educator*
Long, William McMurray *physiology educator*

OHIO

Cincinnati
Behnke, Erica Jean *physiologist*
Sperelakis, Nicholas *physiology and biophysics educator, researcher*

Cleveland
Lakshmanan, Mark Chandrakant *physiologist, physician*

Columbus
Evans, Michael Leigh *physiologist*
Sherman, William Michael *physiology educator, researcher*

Rootstown
Maron, Michael Brent *physiologist*

Springfield
Ryu, Kyoo-Hai Lee *physiologist*

OKLAHOMA

Norman
Bemben, Michael George *exercise physiologist*

Tulsa
Johnson, Gerald, III *cardiovascular physiologist, researcher*

OREGON

Pendleton
Klepper, Elizabeth Lee *physiologist*

PENNSYLVANIA

Philadelphia
Altamirano, Anibal Alberto *physiologist*
Beauchamp, Gary Keith *physiologist*
Cox, Robert Harold *physiology educator*
Fisher, Aron Baer *physiology and medicine educator*
Goldman, Yale E. *physiologist, educator*
Lefer, Allan Mark *physiologist*
Niewiarowski, Stefan *physiology educator, biomedical research scientist*
Xin, Li *physiologist*

Pittsburgh
Redgate, Edward Stewart *physiologist, educator*

Slippery Rock
Hart, Robert Gerald *physiology educator*

University Park
Buskirk, Elsworth Robert *physiologist, educator*
Hagen, Daniel Russell *physiologist*

Wellsboro
Rottiers, Donald Victor *physiologist*

SOUTH CAROLINA

Charleston
Pharr, Pamela Northington *physiology educator, researcher*

TEXAS

College Station
Granger, Harris Joseph *physiologist, educator*
Willard, Scott Thomas *reproductive physiologist*

Dallas
Mitchell, Jere Holloway *physiologist, researcher, medical educator*
Wilson, Lee Britt *physiology educator*

Houston
Brown, Jack Harold Upton *physiology educator, university official, biomedical engineer*
Nichols, Buford Lee, Jr. *physiologist*
Schultz, Stanley George *physiologist, educator*

Temple
Malone, Stephen Robert *plant physiologist*

VERMONT

Burlington
Low, Robert B. *physiology educator*

VIRGINIA

Charlottesville
Murphy, Richard Alan *physiology educator*

Richmond
Walsh, Scott Wesley *reproductive physiologist, researcher*

WASHINGTON

Seattle
Calvin, William Howard *neurophysiologist*
Deyrup-Olsen, Ingrith Johnson *physiologist, educator*
Klepper, John Richard *physiologist, educator*
Schiffrin, Milton Julius *physiologist*

WEST VIRGINIA

Huntington
Green, Todd Lachlan *physiologist, educator*
Mallory, David Stanton *physiology educator*

Morgantown
Brown, Paul B. *physiologist educator*
Gladfelter, Wilbert Eugene *physiology educator*

WISCONSIN

La Crosse
Meinertz, Jeffery Robert *physiologist*

Madison
Graham, James Miller *physiology researcher*
Helgeson, John Paul *physiologist, researcher*
Kemnitz, Joseph William *physiologist, researcher*
Peterson, David Maurice *plant physiologist, researcher*
Sheffield, Lewis Glosson *physiologist*

Milwaukee
Hutz, Reinhold Josef *physiologist*

TERRITORIES OF THE UNITED STATES

PUERTO RICO

San Juan
Opava-Stitzer, Susan Catherine *physiologist, researcher*

CANADA

ALBERTA

Calgary
Jones, Geoffrey Melvill *physiology research educator*
Wallace, John Lawrence *immunophysiologist, educator*

BRITISH COLUMBIA

Agassiz
Molnar, Joseph Michael *plant physiologist, research director*

Vancouver
Owen, Bruce Douglas *animal physiologist*
Randall, David John *physiologist, zoologist, educator*

MANITOBA

Winnipeg
Anthonisen, Nicholas R. *respiratory physiologist*

NOVA SCOTIA

Halifax
O'Dor, Ron *physiologist, marine biologist*

ONTARIO

North York
Buick, Fred J.R. *physiologist, researcher*

Toronto
Harrison, Robert Victor *auditory physiologist*

AUSTRALIA

Queensland
Pickles, James Oliver *physiologist*

DENMARK

Roskilde
Engvild, Kjeld Christensen *plant physiologist*

EGYPT

Monsura
El Nahass, Mohammed Refat Ahmed *physiology educator, researcher*

ENGLAND

Cambridge
Huxley, Sir Andrew (Fielding) *physiologist, educator*

London
Katz, Sir Bernard *physiologist*

FRANCE

Strasbourg
Petrovic, Alexandre Gabriel *physician, physiology educator, medical research director*

GERMANY

Waldbrunn
Goedde, Josef *physiologist*

GREECE

Thessaloniki
Smokovitis, Athanassios A. *physiologist, educator*

ITALY

Messina
Nigro, Aldo *physiology and psychology educator*

Rome
Tao, Kar-Ling James *physiologist, researcher*

THE NETHERLANDS

Haren
Den Otter, Cornelis Johannes *animal physiologist, educator*

POLAND

Lodz
Guzek, Jan Wojciech *physiology educator*

Poznan
Knapowski, Jan Boleslaw *pathophysiology educator, physician, researcher*

Zabrze
Gwóźdź, Bolesław Michael *physiologist*

RUSSIA

Saint Petersburg
Svidersky, Vladimir Leonidovich *neurophysiologist*

SWEDEN

Göteborg
Wallin, B(engt) Gunnar *neurophysiology educator*

SWITZERLAND

Bern
Fleisch, Herbert André *pathophysiologist*

Contra
Eccles, Sir John Carew *physiologist*

TAIWAN

Kaohsiung
Hsu, Zuey-Shin *physiology educator*

Taipei
Ma, Chueng-Shyang (Robert Ma) *reproductive physiology educator, geneticist*

UKRAINE

Kiev
Kostyuk, Platon Grigorevich *physiologist*

VENEZUELA

Caripe
Pereira, Jose Francisco *plant physiologist*

ADDRESS UNPUBLISHED

Dionigi, Christopher Paul *plant physiologist*
Griffith, Steven Lee *research physiologist*
Rose, William Cudebec *electrophysiologist*
Skinner, James Stanford *physiologist, educator*
Tenney, Stephen Marsh *physiologist, educator*
Thews, Gerhard *physiology educator*

LIFE SCIENCES: RESEARCH

UNITED STATES

CALIFORNIA

Berkeley
Rock, Irvin *research scientist*

Davis
Horowitz, Isaac M. *control research consultant, writer*

Moffett Field
Strawa, Anthony Walter *researcher*

Pleasanton
Choy, Clement Kin-Man *research scientist*

San Diego
Edwards, Jack Elmer *personnel researcher*

Tustin
Sinnette, John Townsend, Jr. *research scientist, consultant*

COLORADO

Boulder
Meier, Mark F. *research scientist, glaciologist, educator*

Fort Collins
Thurman, Pamela Jumper *research scientist*

DELAWARE

Newark
DeCherney, George Stephen *research scientist, research facility administrator*

DISTRICT OF COLUMBIA

Washington
Vandiver, Pamela Bowren *research scientist*

FLORIDA

Gainesville
Mishra, Vishnu S. *research scientist*

GEORGIA

Sapelo Island
Alberts, James Joseph *scientist, researcher*

HAWAII

Kamuela
Stillings, Dennis Otto *research director*

IOWA

Iowa City
Tye-Murray, Nancy *research scientist*

KENTUCKY

Wickliffe
Ford, Victor Lavann *research scientist, forest biologist*

MARYLAND

Bethesda
Collins, Francis S. *medical research scientist*
Fraumeni, Joseph F., Jr. *scientific researcher, medical educator, physician, military officer*
Gallo, Robert Charles *research scientist*

MASSACHUSETTS

Boston
Aldoori, Walid Hamid *researcher*
Kunkel, Louis Martens *research scientist, educator*

MICHIGAN

Ann Arbor
Losada, Marcial Francisco *research scientist, psychologist*

Midland
Morgan, Roger John *research scientist*

MISSOURI

Columbia
Chang, Jian Cherng *research analyst*

NEBRASKA

Omaha
St. John, Margaret Kay *research coordinator*

NEW JERSEY

Elizabeth
Trawick, Lafayette James, Jr. *research scientist*

Kenilworth
DiGiacomo, Ruth Ann *research scientist*

New Brunswick
Sachs, Clifford Jay *research scientist*

Piscataway
Gaffar, Abdul *research scientist, administrator*

NEW MEXICO

Las Cruces
Kilmer, Neal Harold *physical scientist*

NEW YORK

Albany
Davies, Kelvin James Anthony *research scientist, medical educator, consultant*

East Setauket
Duff, Ronald G. *research scientist*

Hawthorne
Mitchell, Joan LaVerne *research scientist*

New York
Anderson, Samuel Wentworth *research scientist*

Rensselaer
LaBrie, Teresa Kathleen *research scientist*

Yorktown Heights
Doganata, Yurdaer Nezihi *research scientist*

OHIO

Cincinnati
Adams, Donald Scott *research scientist, pharmacist*

OKLAHOMA

Ponca City
Wann, Laymond Doyle *retired petroleum research scientist*

RHODE ISLAND

Newport
Koch, Robert Michael *research scientist, consultant*

SOUTH CAROLINA

Spartanburg
Fundenberg, Herman Hugh *research scientist*

TEXAS

Fort Worth
Szal, Grace Rowan *research scientist*

Houston
Patterson, Donald Eugene *research scientist*
Sun, Yanyi *research scientist*

VIRGINIA

Alexandria
Bui, James *defense industry researcher*

WASHINGTON

Seattle
Kantowitz, Barry Howard *ergonomist, researcher*

WEST VIRGINIA

South Charleston
Britton, Laurence George *research scientist*

CANADA

ONTARIO

Mississauga
Evans, Essi H. *research scientist*

JAPAN

Kyoto
Kunugi, Shigeru *scientist, researcher, educator*

NIGERIA

Kano
Otokpa, Augustine Emmanuel Ogaba, Jr. *research scientist, consultant*

ADDRESS UNPUBLISHED

LaMunyon, Craig Willis *biology researcher*
Oprandy, John Jay *research scientist, molecular biologist*

LIFE SCIENCES: SOIL SCIENCE

UNITED STATES

ALABAMA

Auburn
Gill, William Robert *soil scientist*

Normal
Coleman, Tommy Lee *soil science educator, researcher, laboratory director*

Tuskegee
Datiri, Benjamin Chumang *soil scientist*

CALIFORNIA

Berkeley
DePaolo, Donald James *earth science educator*
Wohletz, Leonard Ralph *soil scientist, consultant*

COLORADO

Akron
Halvorson, Ardell David *research leader, soil scientist*

FLORIDA

Fort Pierce
Calvert, David Victor *soil science educator*

Gainesville
Olila, Oscar Gesta *soil and water scientist*

IDAHO

Kimberly
Carter, David LaVere *soil scientist, researcher, consultant*

Moscow
McGeehan, Steven Lewis *soil scientist*

ILLINOIS

Carbondale
Bates, Sharon Ann *plant and soil scientist, educator*

Urbana
Jones, Robert Lewis *soil mineralogy and ecology educator*

INDIANA

Carbon
Robinson, Glenn Hugh *soil scientist*

Indianapolis
Thomas, Jerry Arthur *soil scientist*

West Lafayette
Cochrane, Thomas Thurston *tropical soil scientist, agronomist*

IOWA

Ames
Keeney, Dennis Raymond *soil science educator*

KANSAS

Manhattan
Ham, George Eldon *soil microbiologist, educator*

LOUISIANA

Baton Rouge
Willis, Guye Henry, Jr. *soil chemist*

Haughton
De Ment, James Alderson *soil scientist*

MARYLAND

Beltsville
Starr, James LeRoy *soil scientist*

Silver Spring
Lunin, Jesse *retired soil scientist*

MASSACHUSETTS

Amherst
Hillel, Daniel *soil physics and hydrology educator, researcher, consultant*

MICHIGAN

Farmington Hills
Dragun, James *soil chemist*

MISSOURI

Manchester
Purdy, Donald Gilbert, Jr. *soil scientist*

NEW HAMPSHIRE

Durham
Pilgrim, Sidney Alfred Leslie *soil science educator*

Hanover
Brar, Gurdarshan Singh *soil scientist, researcher*

NEW MEXICO

Las Cruces
McCaslin, Bobby D. *soil scientist, educator*

Peralta
Diebold, Charles Harbou *seed company executive*

NORTH CAROLINA

Raleigh
Davey, Charles Bingham *soil science educator*

NORTH DAKOTA

Mandan
Halvorson, Gary Alfred *soil scientist*
Reichman, George Albert *soil scientist, educator, consultant*

OREGON

Corvallis
Liegel, Leon Herman *soil scientist, forester*

Portland
Jarrell, Wesley Michael *soil and ecosystem science educator, researcher, consultant*

PENNSYLVANIA

Monroeville
Cremeens, David Lynn *soil scientist, consultant*

TENNESSEE

Cookeville
Boswell, Fred C. *retired soil science educator, researcher*

TEXAS

Dallas
Senkayi, Abu Lwanga *environmental soil scientist*

UTAH

Provo
Thorup, Richard Maxwell *soil scientist*

Vernal
Remington, Delwin Woolley *soil conservationist*

VIRGINIA

Blacksburg
De Datta, Surajit Kumar *soil scientist, agronomist, educator*

WISCONSIN

Madison
Whitmyer, Robert Wayne *soil scientist, consultant, researcher*

WYOMING

Jackson
Davis, Randy L. *soil scientist*

CANADA

ALBERTA

Edmonton
Kalra, Yash Pal *soil chemist*

BRITISH COLUMBIA

Vancouver
Lavkulich, Leslie Michael *soil science educator*

ONTARIO

Guelph
Miller, Murray Henry *soil science educator*

Ottawa
Halstead, Ronald Lawrence *soil scientist*

SASKATCHEWAN

Saskatoon
Beaton, James Duncan *soil scientist*
Huang, Pan Ming *soil science educator*

BRAZIL

Rio de Janeiro
Döbereiner, Johanna *soil biology scientist*

INDIA

Varanasi
Srivastava, Om Prakash *soil science and agricultural chemistry educator*

JAPAN

Sendai
Shoji, Sadao *soil scientist*

KENYA

Nairobi
Sanchez, Pedro Antonio, Jr. *soil scientist, administrator*

SCOTLAND

Aberdeen
Cresser, Malcolm Stewart *soil scientist, educator*

ADDRESS UNPUBLISHED

Frere, Maurice Herbert *soil scientist*

LIFE SCIENCES: TOXICOLOGY

UNITED STATES

ALABAMA

Birmingham
Prejean, J. David *toxicologist*

ARIZONA

Phoenix
Adler, Eugene Victor *forensic toxicologist, consultant*
Meinhart, Robert David *analytical toxicologist*

ARKANSAS

Little Rock
Hinson, Jack Allsbrook *research toxicologist, educator*

CALIFORNIA

Alameda
Paustenbach, Dennis James *environmental toxicologist*

Colton
Halstead, Bruce Walter *biotoxicologist*

Davis
Klasing, Susan Allen *environmental toxicologist, consultant*

Foster City
Baselt, Randall Clint *toxicologist*

Menlo Park
MacGregor, James Thomas *toxicologist*

Pasadena
Damji, Karim Sadrudin *environmental toxicologist, consultant*

Sacramento
Rubin, Andrew Lawrence *toxicologist, cell biologist*

San Jose
Adamovics, Andris *toxicologist, consultant*

Santa Barbara
Hatherill, John Robert *toxicologist, educator*

Westlake Village
Huestis, Marilyn Ann *toxicologist, clinical chemist*

CONNECTICUT

Farmington
Sunderman, F(rederick) William, Jr. *toxicologist, educator, pathologist*

Groton
Katz, Lori Susan *toxicologist*

DELAWARE

Newark
Reinhardt, Charles Francis *toxicologist*

DISTRICT OF COLUMBIA

Washington
Benz, Robert Daniel *toxicologist*
Maciorowski, Anthony Francis *ecological toxicologist*
Thomas, Richard Dean *toxicologist, pathologist*

FLORIDA

Mount Dora
Staats, Dee Ann *toxicologist*

Tampa
Richards, Ira Steven *toxicology educator*

West Palm Beach
Freudenthal, Ralph Ira *toxicology consultant*

GEORGIA

Atlanta
Berg, George G. *toxicologist*
Grissom, Raymond Earl, Jr. *toxicologist*
Mumtaz, Mohammad Moizuddin *toxicologist, researcher*

ILLINOIS

Springfield
Henebry, Michael Stevens *toxicologist*

INDIANA

Greenfield
Kuyatt, Brian Lee *toxicologist, scientific systems analyst*

IOWA

Ames
Stahr, Henry Michael *analytical toxicology*

KANSAS

Kansas City
Pierce, John Thomas *industrial hygienist, toxicologist*

KENTUCKY

Lexington
Robertson, Larry Wayne *toxicologist, educator*

LOUISIANA

Baton Rouge
Iyer, Poorni Ramchandran *toxicologist*
Parent, Richard Alfred *toxicologist, consultant*

Monroe
Pope, Carey Nat *neurotoxicologist*

MARYLAND

Aberdeen Proving Ground
Armstrong, Robert Don *toxicologist*

Baltimore
Caplan, Yale Howard *toxicologist, consultant*

Bethesda
Amouzadeh, Hamid R. *toxicologist*
Malone, Winfred Francis *toxicologist*

Cockeysville Hunt Valley
Gift, James J. *toxicologist*

Frederick
Creasia, Donald Anthony *toxicologist, researcher*
Waalkes, Michael Phillip *toxicologist*

Rockville
Jacobson-Kram, David *toxicologist*

Silver Spring
Baker, Scott Ralph *toxicologist*
Rodgers, Imogene Sevin *toxicologist*

MASSACHUSETTS

Boston
Hayes, A(ndrew) Wallace *toxicologist*

MICHIGAN

Ann Arbor
Monteith, David Keith Brisson *toxicologist, researcher*

East Lansing
Fischer, Lawrence Joseph *toxicologist, educator*

MINNESOTA

Minneapolis
Garry, Vincent Ferrer *environmental toxicology researcher, educator*

MISSISSIPPI

University
Benson, William Hazlehurst *environmental toxicologist*

MISSOURI

Kansas City
von Kehl, Inge *toxicologist-pharmacologist*

Saint Louis
Long, Christopher *toxicologist*

NEBRASKA

Omaha
Alsharif, Naser Zaki *toxicologist, educator, researcher*

NEW JERSEY

East Millstone
Smith, Jacqueline Hagan *toxicologist*

Morristown
Derelanko, Michael Joseph *toxicologist*

Princeton
Schreiner, Ceinwen Ann *mammalian and genetic toxicologist*
Weiner, Myra Lee *toxicologist*

Somerset
Singh, Krishna Deo *toxicologist*

Somerville
Paulson, John Daniel *toxicologist*

Summit
Amemiya, Kenjie *toxicologist*
Yau, Edward Tintai *toxicologist, pharmacologist*

NEW MEXICO

Albuquerque
Robb, Jeffery Michael *forensic toxicologist*

NEW YORK

Brooklyn
Tanacredi, John T(homas) *ecotoxicologist*

Buffalo
Kostyniak, Paul John *toxicology educator*

Ithaca
Thompson, Larry Joseph *veterinary toxicologist, consultant*

Jamaica
Ford, Sue Marie *toxicology educator, researcher*

White Plains
Maslansky, Carol Jeanne *toxicologist*

NORTH CAROLINA

Chapel Hill
McBay, Arthur John *toxicologist, consultant*

Durham
McClellan, Roger Orville *toxicologist*

Greensboro
Heck, Jonathan Daniel *toxicologist*

Research Triangle Park
Chadwick, Robert William *toxicologist*
Lange, Robert William *immunotoxicologist*
Mayes, Mark Edward *molecular toxicologist*

OHIO

Dayton
Byczkowski, Janusz Zbigniew *toxicologist*

OKLAHOMA

Oklahoma City
Rohrig, Timothy Patrick *toxicologist, educator*

OREGON

Portland
Spencer, Peter Simner *neurotoxicologist*

PENNSYLVANIA

Philadelphia
Johnson, Elmer Marshall *toxicologist, teratologist*

Spring House
Frederick, Clay Bruce *toxicologist, researcher*

Willow Grove
Spikes, John Jefferson, Sr. *forensic toxicologist, pharmacologist*

TENNESSEE

Memphis
Lyman, Beverly Ann *biochemical toxicologist*

Nashville
Casillas, Robert Patrick *research toxicologist*

TEXAS

Dallas
Kurt, Thomas Lee *medical toxicologist*

Galveston
Bhat, Hari Krishen *biochemical toxicologist*

Lackland AFB
Cody, John Thomas *forensic toxicologist, biological chemist researcher*

San Antonio
Doane, Thomas Roy *environmental toxicologist*

VIRGINIA

Fairfax
Peters, Esther Caroline *aquatic toxicologist, pathobiologist, consultant*

Mc Lean
Dedrick, Robert Lyle *toxicologist, biomedical engineer*

WASHINGTON

Bellingham
Landis, Wayne G. *environmental toxicologist*

Puyallup
Stark, John David *pesticide toxicologist*

Redmond
Kelman, Bruce Jerry *toxicologist, consultant*

Spokane
Delistraty, Damon Andrew *toxicologist*

TERRITORIES OF THE UNITED STATES

PUERTO RICO

San Juan
Jiménez, Braulio Dueño *toxicologist*

CANADA

QUEBEC

Montreal
Plaa, Gabriel Leon *toxicologist, educator*

GERMANY

Leipzig
Mueller, Rudhard Klaus *toxicologist*

JAPAN

Nagasaki
Ariyoshi, Toshihiko *toxicologist, educator*

Tsukuba
Yamamoto, Hiro-Aki *toxicology and pharmacology educator*

UNITED ARAB EMIRATES

Dubai
Beg, Mirza Umair *toxicologist*

ADDRESS UNPUBLISHED

Dalderup, Louise Maria *medical and chemical toxicologist, nutritionist*
de Wolff, Frederik Albert *toxicology educator*
McMartin, Kenneth Esler *toxicology educator*

LIFE SCIENCES: VETERINARY SCIENCE

UNITED STATES

ALABAMA

Auburn
Clark, Terrence Patrick *veterinarian, researcher*

Montgomery
Parmer, Dan Gerald *veterinarian*

CALIFORNIA

Riverside
Hinkle, Nancy C. *veterinary entomologist*

Salinas
Reichenbach, Thomas *veterinarian*

San Juan Capistrano
Burrows, Barbara Ann *veterinarian*

COLORADO

Fort Collins
Benjamin, Stephen Alfred *veterinary medicine educator, environmental pathologist, researcher*

DISTRICT OF COLUMBIA

Washington
Crawford, Lester Mills, Jr. *veterinarian*
Norcross, Marvin Augustus *veterinarian, government agency official*

FLORIDA

Gainesville
Burridge, Michael John *veterinarian, educator, academic administrator*

GEORGIA

Alpharetta
Rettig, Terry *veterinarian, wildlife consultant*

Athens
Bounous, Denise Ida *veterinary pathology educator*

Atlanta
McClure, Harold Monroe *veterinary pathologist*

ILLINOIS

Chicago
Surgi, Elizabeth Benson *veterinarian*

Savoy
Ridgway, Marcella Davies *veterinarian*

INDIANA

Pendleton
Leonard, Elizabeth Ann *veterinarian*

IOWA

Ames
Greve, John Henry *veterinary parasitologist, educator*
Moon, Harley William *veterinarian*
Seaton, Vaughn Allen *veterinary pathology educator*

KANSAS

Manhattan
Troyer, Deryl Lee *life sciences educator*
Vorhies, Mahlon Wesley *veterinary pathologist, educator*

LOUISIANA

Baton Rouge
Besch, Everett Dickman *veterinarian, university dean emeritus*

MARYLAND

Bethesda
Donahue, Robert Edward *veterinarian, researcher*
Vickers, James Hudson *veterinarian, research pathologist*

Frederick
Bolon, Brad Newland *veterinary pathologist*

Hyattsville
Kohn, Barbara Ann *veterinarian*

Laurel
Ramos, Angel Salvador *veterinarian*

Rockville
Guest, Gerald Bentley *veterinarian*
Teske, Richard Henry *veterinarian*
Whitney, Robert A., Jr. *veterinarian, government public health executive*

MASSACHUSETTS

North Grafton
Loew, Franklin Martin *veterinary medical and biological scientist, university dean*

MICHIGAN

Ann Arbor
Ringler, Daniel Howard *lab animal medicine educator*

East Lansing
Tasker, John Baker *veterinary medical educator, college dean*
Witter, Richard Lawrence *veterinarian, educator*

Plainwell
Kleckner, Marlin Dallas *veterinarian*

MINNESOTA

Minneapolis
Atluru, Durgaprasadarao *veterinarian, educator*

MISSOURI

Columbia
Beckwith, Catherine S. *veterinarian*
Morehouse, Lawrence Glen *veterinarian, educational administrator*
Wagner, Joseph Edward *veterinarian, educator*

Saint Louis
Frazier, Kimberlee Gonterman *veterinarian*

NEW YORK

Albany
Csiza, Charles Karoly *veterinarian, microbiologist*

Ithaca
Gilbert, Robert Owen *veterinary educator, researcher*
Lopez, Jorge Washington *veterinary virologist*

New York
Garvey, Michael Steven *veterinarian, educator*

Owego
Kemp, Eugene Thomas *veterinarian*

NORTH CAROLINA

Chapel Hill
Chang, Jerjang *veterinarian, educator*

OHIO

Columbus
Capen, Charles Chabert *veterinary pathology educator, researcher*

OKLAHOMA

Stillwater
Faulkner, Lloyd C. *veterinary medicine educator*
Fox, Joseph Carl *veterinary medicine educator, researcher, parasitologist*
Kocan, Katherine Mautz *veterinary educator, researcher*
Qualls, Charles Wayne, Jr. *veterinary pathology educator*

PENNSYLVANIA

Birdsboro
Moyer, David Lee *veterinarian*

Flourtown
Kissileff, Alfred *veterinarian, researcher*

Philadelphia
Morrison, Adrian Russell *veterinarian educator*

Villanova
Melby, Edward Carlos, Jr. *veterinarian*

TEXAS

College Station
Carter, Craig Nash *veterinary epidemiologist, educator, researcher, software developer*

Fort Worth
Jensen, Harlan Ellsworth *veterinarian, educator*

WASHINGTON

Pullman
Davis, William Charles *veterinary microbiology educator*

Seattle
Van Hoosier, Gerald Leonard *veterinary science educator*

WISCONSIN

Beldenville
Mullenax, Charles Howard *veterinarian, researcher*

Hudson
Fahning, Melvyn Luverne *veterinary educator*

Madison
Easterday, Bernard Carlyle *veterinary medicine educator*

TERRITORIES OF THE UNITED STATES

PUERTO RICO

Luquillo
Arnizaut de Mattos, Ana Beatriz *veterinarian*

ECUADOR

Quito
Bedoya, Michael Julian *veterinarian*

FINLAND

Hautjärvi
Alanko, Matti Lauri Juhani *animal reproduction educator*

FRANCE

Corseul
Guerin, Patrick Gerard *veterinary surgeon, consultant*

GERMANY

Hannover
Liebler, Elisabeth M. *veterinary pathologist*

JAPAN

Wako
Ogawa, Tomoya *chemist, veterinary medical science educator*

NEW ZEALAND

Palmerston North
Parton, Kathleen *veterinarian*

ADDRESS UNPUBLISHED

Fox, Michael Wilson *veterinarian, animal behaviorist*
Hunt, Ronald Duncan *veterinarian, educator, pathologist*
Markle, Douglas Frank *ichthyologist, educator*
Morishita, Teresa Yukiko *veterinarian, consultant, researcher*
Roseig, Esther Marian *veterinary researcher*
Segre, Diego *veterinary pathology educator, retired*
Simeón Negrín, Rosa Elena *veterinary educator*

LIFE SCIENCES: VIROLOGY

UNITED STATES

COLORADO

Fort Collins
Smith, Ralph Earl *virologist*

DISTRICT OF COLUMBIA

Washington
Colberg-Poley, Anamaris Martha *virologist*

FLORIDA

Gainesville
Purcifull, Dan Elwood *plant virologist, educator*

Saint Petersburg
Bradley, William Guy *molecular virologist*

Tampa
Dunigan, David Deeds *biochemist, virologist, educator*

GEORGIA

Atlanta
Tamin, Azaibi *molecular virologist, researcher*

MARYLAND

Bethesda
Gruber, Jack *medical virologist, biomedical research administrator*
Moss, Bernard *virologist, researcher*
Purcell, Robert Harry *virologist*

MASSACHUSETTS

Boston
Robertson, Erle Shervinton *virologist, molecular biologist*
Schaffer, Priscilla Ann *virologist*

NEW YORK

Albany
Stellrecht Burns, Kathleen Anne *virologist*

New York
Hanafusa, Hidesaburo *virologist*
Morse, Stephen Scott *virologist, immunologist*

Rochester
Blumberg, Benjamin Mautner *virologist*

Stony Brook
Steigbigel, Roy Theodore *infectious disease physician and scientist, educator*

Troy
Tartaglia, James Joseph *molecular virologist*

NORTH CAROLINA

Chapel Hill
Huang, Eng Shang *virology educator, biomedical engineer*

Durham
Lambert, Dennis Michael *virologist, researcher*

OHIO

Cincinnati
Schiff, Gilbert Martin *virologist, microbiologist, medical educator*

PENNSYLVANIA

Hershey
Rapp, Fred *virologist*

West Point
Hilleman, Maurice Ralph *virus research scientist*
Tomassini, Joanne Elizabeth *virologist, researcher*

TEXAS

Galveston
Fons, Michael Patrick *virologist*

WISCONSIN

Madison
Dasgupta, Ranjit Kumar *virologist*

CANADA

ALBERTA

Calgary
Yoon, Ji-Won *virology, immunology and diabetes educator, research administrator*

Edmonton
Hiruki, Chuji *plant virologist, science educator*

QUEBEC

Laval
Trudel, Michel *virologist*

Ville de Laval
Tijssen, Peter H. T. *molecular virology educator, researcher*

SASKATCHEWAN

Saskatoon
Babiuk, Lorne Alan *virologist, immunologist, research administrator*

AUSTRIA

Vienna
Moreno-Lopez, Jorge *virologist, educator*

COSTA RICA

Turrialba
Lastra, Jose Ramon *plant virologist*

CUBA

Havana
Kouri, Gustavo Pedro *virologist*

ENGLAND

Leeds
Hambling, Milton Herbert *medical virologist consultant, lecturer*

FRANCE

Tours
Coursaget, Pierre Louis *virologist*

GERMANY

Hamburg
Bruns, Michael Willi Erich *virologist, veterinarian*

ITALY

Naples
Tarro, Giulio *virologist*

JAPAN

Chiba
Nishimura, Chiaki *molecular virologist educator*

THE NETHERLANDS

Leiden
Spaan, Willy Josephus *molecular virologist*

SPAIN

Madrid
Najera, Rafael *virologist*

SWITZERLAND

Geneva
Wunderli, Werner Hans Karl *virologist, researcher*

LIFE SCIENCES: ZOOLOGY

UNITED STATES

ALASKA

Fairbanks
Kessel, Brina *educator, ornithologist*

CALIFORNIA

Berkeley
Johnson, Ned Keith *ornithologist, educator*
Lidicker, William Zander, Jr. *zoologist, educator*
Pitelka, Frank Alois *zoologist, educator*

Clovis
Ensminger, Marion Eugene *animal science educator, author*

Davis
Baldwin, Ransom Leland *animal science educator*
O'Donnell, Sean *zoologist*

Hollywood
Bratcher, Twila Langdon *conchologist, malacologist*

San Diego
Brusca, Richard C. *zoologist, museum curator, researcher, educator*

DISTRICT OF COLUMBIA

Washington
Beehler, Bruce McPherson *research zoologist, ornithologist*
de Queiroz, Kevin *zoologist*
Kleiman, Devra Gail *zoologist, zoological park administrator*

FLORIDA

Dania
Messing, Charles Garrett *zoologist*

Gainesville
Kiltie, Richard Alan *zoology educator*
Meredith, Julia Alice *nematologist, biologist, researcher*

Tallahassee
James, Francis Crews *zoology educator*

GEORGIA

Athens
Willis, Judith H. *zoology educator, researcher*

Statesboro
Lefcort, Hugh George *zoologist*

HAWAII

Honolulu
Kay, Elizabeth Alison *zoology educator*

IDAHO

Hayden Lake
Lehrer, William Peter, Jr. *animal scientist*

ILLINOIS

Brookfield
Rabb, George Bernard *zoologist*

Carbondale
Burr, Brooks Milo *zoology educator*

Chicago
Altmann, Jeanne *zoologist, educator*

Springfield
Munyer, Edward A. *zoologist, museum adminstrator*

Urbana
Gaskins, H. Rex *animal sciences educator*

IOWA

Ames
Anderson, Lloyd Lee *animal science educator*
Young, Jerry Wesley *animal nutrition educator*

MAINE

Orono
Tyler, Seth *zoology educator*

MARYLAND

College Park
Imberski, Richard Bernard *zoology educator*
Wilkinson, Gerald Stewart *zoology educator*

MASSACHUSETTS

Cambridge
Mayr, Ernst *emeritus zoology educator, author*

Northampton
Hayssen, Virginia *mammalogist, educator*

Vineyard Haven
Billingham, Rupert Everett *zoologist, educator*

MICHIGAN

Ann Arbor
Dawson, William Ryan *zoology educator*

East Lansing
Hackel, Emanuel *science educator*

MINNESOTA

Saint Paul
McKinnell, Robert Gilmore *zoology, genetics and cell biology educator*
White, Michael Ernest *animal scientist*

MISSISSIPPI

Lorman
Williams, Richard, Jr. *animal scientist*

Ocean Springs
Gunter, Gordon *zoologist*

MISSOURI

Kansas City
Hagsten, Ib *animal scientist, educator*

Warrensburg
Belshe, John Francis *zoology and ecology educator*

MONTANA

Missoula
Weisel, George Ferdinand *retired zoology educator*

NEW MEXICO

Albuquerque
Findley, James Smith *biology and zoology educator, museum director*

NEW YORK

Bronx
Schaller, George Beals *zoologist*

Chenango Bridge
Fisher, Dale Dunbar *animal scientist, dairy nutritionist*

Syracuse
Russell-Hunter, W(illiam) D(evigne) *zoology educator, research biologist, writer*

NORTH CAROLINA

Durham
Brandon, Robert Norton *zoology and philosophy educator*
Livingstone, Daniel Archibald *zoology educator*
Wainwright, Stephen A. *zoology educator, design consultant*
Wilbur, Karl Milton *zoologist, educator*

OHIO

Athens
Reed, Michael Alan *scientist*

Columbus
Peterle, Tony John *zoologist, educator*

Delaware
Burtt, Edward Howland, Jr. *ornithologist, natural history educator*

OKLAHOMA

Stillwater
Campbell, John Roy *animal scientist educator, academic administrator*

OREGON

Corvallis
Mason, Robert Thomas *zoologist*

TEXAS

Austin
Hubbs, Clark *zoologist, researcher*
Kalthoff, Klaus Otto *zoology educator*

Dallas
Shepard, Mark Louis *animal scientist*

Galveston
Budelmann, Bernd Ulrich *zoologist, educator*

UTAH

Provo
Shiozawa, Dennis Kenji *zoology educator*
Smith, Howard Duane *zoology educator*

WASHINGTON

Friday Harbor
Willows, Arthur Owen Dennis *neurobiologist, zoology educator*

WISCONSIN

Madison
Hailman, Jack Parker *zoology educator*

WYOMING

Laramie
Hinds, Frank Crossman *animal science educator*

CANADA

BRITISH COLUMBIA

Vancouver
Jones, David Robert *zoology educator*
Larkin, Peter Anthony *zoology educator, university dean and official*

ONTARIO

Guelph
Stevens, (Ernest) Donald *zoology educator*

Toronto
Masui, Yoshio *zoology educator*
Tobe, Stephen Solomon *zoology educator*

QUEBEC

Lac Beauport
Lane, Peter *ornithologist*

AUSTRALIA

Sydney
Anderson, Donald Thomas *zoologist, educator*

ENGLAND

Oxford
Hamilton, William Donald *zoologist, educator*

FINLAND

Helsinki
Pulliainen, Erkki Ossi Olavi *zoology educator, legislator*

FRANCE

Paris
Ferrando, Raymond *animal nutrition scientist, educator*

GERMANY

Bochum
Mergner, Hans Konrad *zoology educator*

Münster
Thurm, Ulrich *zoology educator*

JAPAN

Tokyo
Takahashi, Keiichi *zoology educator*

TAIWAN

Nankang
Lin, Fei-Jann *zoology educator, researcher*

ADDRESS UNPUBLISHED

Holldobler, Berthold Karl *zoologist, educator*
Inouye, David William *zoology educator*
MacDonald, Stewart Dixon *ornithologist, ecologist, biologist*
Southwick, Charles Henry *zoologist, educator*
Wright, Philip Lincoln *zoologist, educator*

MATHEMATICS & STATISTICS

UNITED STATES

ALABAMA

Huntsville
McAuley, Van Alfon *aerospace mathematician*
Pruitt, Alice Fay *mathematician, engineer*

Mobile
Taylor, Washington Theophilus *mathematics educator*

Pelham
Turner, Malcolm Elijah *biomathematician, educator*

ALASKA

Anchorage
Mann, Lester Perry *mathematics educator*

Tuntutuliak
Bond, Ward C. *mathematics and computer educator*

ARIZONA

Fort Huachuca
Clark, Brian Thomas *mathematical statistician, operations research analyst*

Phoenix
Smarandache, Florentin *mathematics researcher, writer*

Tempe
Ahmed, Kazem Uddin *technical staff member, consultant*
Downs, Floyd L. *mathematics educator*
Ihrig, Edwin Charles, Jr. *mathematics educator*
Johnson, Ross Jeffrey *statistician*

Tucson
Merilan, Jean Elizabeth *statistics educator*

ARKANSAS

Conway
Spatz, Kenneth Chris *statistics educator*

El Dorado
Dlabach, Gregory Wayne *mathematics educator*

CALIFORNIA

Bellflower
Richardson, Bryan Kevin *mathematics educator, naval aviator*

Berkeley
Bickel, Peter John *statistician, educator*
Casson, Andrew J. *mathematician educator*
Chern, Shiing-Shen *mathematics educator*
Marsden, Jerrold Eldon *mathematician, educator*
Ratner, Marina *mathemetician, educator, researcher*
Rice, John Andrew *statistician*
Vojta, Paul Alan *mathematics educator*

Carlsbad
Halberg, Charles John August, Jr. *mathematics educator*

Claremont
Chang, Y(uan)-F(eng) *mathematics and computer science educator*

Davis
Mulase, Motohico *mathematics educator*
Puckett, Elbridge Gerry *mathematician, educator*
Utts, Jessica Marie *statistics educator*

Harbor City
Bornino-Glusac, Anna Maria *mathematics educator*

Imperial
Angelo, Gayle-Jean *mathematics and physical sciences educator*

Irvine
Juberg, Richard Kent *mathematician, educator*

La Jolla
Freedman, Michael Hartley *mathematician, educator*
Reissner, Eric (Max Erich Reissner) *applied mechanics educator*
Rosenblatt, Murray *mathematics educator*

Livermore
Verry, William Robert *mathematics researcher*

Los Angeles
Redheffer, Raymond Moos *mathematician, educator*
Re Velle, Jack B(oyer) *consulting statistician*
Thomas, Duncan Campbell *biostatistics educator*
Young, Lai-Sang *mathematician educator*

Monterey
Faulkner, Frank David *mathematics educator*

Pasadena
Sweetser, Theodore Higgins *mathematician*

Rancho Palos Verdes
Morizumi, Shigenori James *applied mathematician*

Riverside
Ghosh, Subir *statistician*

San Francisco
Christensen, David William *mathematician, engineer*
Farrell, Edward Joseph *mathematics educator*

San Jose
Smith, Harry James *mathematician*

Saratoga
Mihnea, Tatiana *mathematics educator*

Stanford
Dantzig, George Bernard *applied mathematics educator*
Efron, Bradley *mathematics educator*
Karlin, Samuel *mathematics educator, researcher*
Keller, Joseph Bishop *mathematician, educator*
Lieberman, Gerald J. *statistics educator*

Thousand Oaks
Andrews, Angus Percy *mathematician, writer*

Torrance
Malhotra, Vijay Kumar *mathematics educator*

COLORADO

Denver
Mendez, Celestino Galo *mathematics educator*

Durango
Spencer, Donald Clayton *mathematician*

Fort Collins
Mielke, Paul William, Jr. *statistician*

Greeley
Searls, Donald Turner *statistician, educator*

Loveland
Rosander, Arlyn Custer *mathematical statistician, management consultant*

University Of Colorado
Beylkin, Gregory *mathematician*

CONNECTICUT

Mystic
Parberry, Edward Allen *mathematician, consultant*

New Haven
Mostow, George Daniel *mathematics educator*
Piatetski-Shapiro, Ilya *mathematics educator*
Singer, Burton Herbert *statistics educator*

Ridgefield
Wahl, Martha Stoessel *mathematics educator*

DELAWARE

Newark
Olagunju, David Olarewaju *mathematician, educator*
Weinacht, Richard Jay *mathematician*

DISTRICT OF COLUMBIA

Washington
Childs, Sadie L. *mathematician, chemist, patent agent*
Deming, W(illiam) Edwards *statistics educator, consultant*
Gray, Mary Wheat *statistician, lawyer*
Jernigan, Robert Wayne *statistics educator*
Miller, Allen Richard *mathematician*
Phillips, Gary W. *psychometrician*
Shen, Yuan-Yuan *mathematics educator*
Tortora, Robert D. *mathematician*

FLORIDA

Cassadaga
Boland, Lois Walker *retired mathematician and computer systems analyst*

Cocoa
Kendrick, Pamela Ann *mathematics educator*

Cocoa Beach
Strickland, Christopher Alan *statistician, systems analyst*

Fort Myers
Kastenbaum, Marvin Aaron *statistician*

Gainesville
Wilson, David Clifford *mathematician, educator*

Indialantic
Carroll, Charles Lemuel, Jr. *mathematician*

Jacksonville
Robinson, Christine Marie *mathematics educator*

Miami
Cherepanov, Genady Petrovich *mathematician, mechanical engineer*

Ocala
Johnson, Winston Conrad *mathematics educator*

Orlando
Debnath, Lokenath *mathematician, educator*

Saint Petersburg
Kazor, Walter Robert *statistical process control and quality assurance consultant*

Sarasota
Petrie, George Whitefield, III *retired mathematics educator*

Tallahassee
Gilmer, Robert *mathematics educator*
Jones, Gladys Hurt *retired mathematics educator*
Nichols, Eugene Douglas *mathematics educator*

Tampa
Saff, Edward Barry *mathematics educator*

West Palm Beach
Bower, Ruth Lawther *mathematics educator*
Still, Mary Jane (M. J. Still) *mathematics educator*

GEORGIA

Atlanta
Ehrlich, Margaret Isabella Gorley *systems engineer, mathematics educator, consultant*

Fort Valley
Lewis, Larry *mathematics educator*

Marietta
Fox, Marian Cavender *mathematics educator*

HAWAII

Honolulu
Duncan, John Wiley *mathematics and computer educator, retired air force officer*
Swanson, Richard William *statistician*
Wheeler, Carl *mathematics educator*
Zhou, Chiping *mathematician, educator*

ILLINOIS

Champaign
Portnoy, Stephen Lane *statistician*
Wasserman, Stanley *statistician, educator*

Chicago
Calderon, Alberto P. *mathematician, educator*
Chen, Gui-Qiang *mathematician, educator, researcher*
Hanson, Floyd Bliss *applied mathematician, computational scientist, mathematical biologist*
McCullagh, Peter *statistician, mathematician*
Meier, Paul *statistician, mathematics educator*
Pavelka, Elaine Blanche *mathematics educator*
Stigler, Stephen Mack *statistician, educator*

Downers Grove
LaRocca, Patricia Darlene McAleer *mathematics educator*

Evanston
Ionescu Tulcea, Cassius *research mathematician, educator*
Matalon, Moshe *applied mathematician, educator*
Matkowsky, Bernard Judah *applied mathematician, educator*
Olmstead, William Edward *mathematics educator*

Fox Lake
Diprizio, Rosario Peter *mathematics educator*

Godfrey
McDaniels, John Louis *mathematics educator*

Libertyville
Nichols, Thomas Robert *biostatistician, consultant*

Springfield
Stonecipher, Larry Dale *mathematics educator*

Urbana
Azzi, Daniel W. *mathematician*
Burkholder, Donald Lyman *mathematician, educator*
Fossum, Robert Merle *mathematician, educator*
Henson, C. Ward *mathematician, educator*

INDIANA

Evansville
Knott, John Robert *mathematics educator*

Fort Wayne
Beineke, Lowell Wayne *mathematics educator*

Indianapolis
Cliff, Johnnie Marie *mathematics and chemistry educator*
Watt, Jeffrey Xavier *mathematics sciences educator, researcher*

IOWA

Ames
Dahiya, Rajbir Singh *mathematics educator, researcher*

Center Junction
Antons, Pauline Marie *mathematics educator*

Iowa City
Johnson, Eugene Walter *mathematician*

KENTUCKY

Lexington
Man, Chi-Sing *mathematician, educator*

Louisville
Greaver, Joanne Hutchins *mathematics educator, author*

Morehead
Mann, James Darwin *mathematics educator*

Richmond
Howard, Aughtum Luciel Smith *retired mathematics educator*

LOUISIANA

Baton Rouge
Marx, Brian *statistician*

Ruston
Dorsett, Charles Irvin *mathematics educator*

MARYLAND

Aberdeen
Anderson, Alfred Oliver *mathematician, consultant*

Baltimore
Jensen, Soren Stistrup *mathematics educator*
Krushat, William Mark *mathematical statistician*
Wierman, John Charles *mathematician, educator*
Winstead, Carol Jackson *mathematics educator*

Bethesda
Aldroubi, Akram *mathematician, researcher*
Everstine, Gordon Carl *mathematician, educator*
Freedman, Laurence Stuart *statistician*
Lange, Nicholas Theodore *statistician*
Weiss, George Herbert *mathematician, consultant*

Bowie
Reichmann, Péter Iván *mathematician*

College Park
Cohen, Michael Lee *statistics educator*
Embody, Daniel Robert *biometrician*

Gaithersburg
Carasso, Alfred Sam *mathematician*
Rosenblatt, Joan Raup *mathematical statistician*
Witzgall, Christoph Johann *mathematician*

Hyattsville
Piccinino, Linda Jeanne *statistician*
Pickle, Linda Williams *biostatistician*

Laurel
Farrell, William James, Jr. *mathematician*

Potomac
Peters, Carol Beattie Taylor (Mrs. Frank Albert Peters) *mathematician*

Silver Spring
Dalton, Robert Edgar *mathematician, computer scientist*
Shelton, William Chastain *retired statistician*

Upper Marlboro
Lisle, Martha Oglesby *mathematics educator*

MASSACHUSETTS

Boston
D'Agostino, Ralph Benedict *mathematician, statistician, educator, consultant*
Gilmore, Maurice Eugene *mathematics educator*

Cambridge
Bott, Raoul *mathematician, educator*
Carrier, George Francis *applied mathematics educator*
Diaconis, Persi W. *mathematical statistician, educator*
Elkies, Noam D. *mathematics educator*
Fang, Yue *statistician*
Helgason, Sigurdur *mathematician, educator*
Mackey, George Whitelaw *educator, mathematician*
MacPherson, Robert Duncan *mathematician, educator*
Minsky, Marvin Lee *mathematician, educator*
Mosteller, Frederick *mathematical statistician, educator*
Mumford, David Bryant *mathematics educator*
Papert, Seymour Aubrey *mathematician, educator, writer*
Rabin, Michael O. *computer scientist, mathematician*
Segal, Irving Ezra *mathematics educator*
Singer, Isadore Manuel *mathematician, educator*
Storch, Joel Abraham *mathematician*
Taubes, Clifford H. *mathematician educator*
Toomre, Alar *mathematics educator, theoretical astronomer*
Yau, Shing-Tung *mathematics educator*

Fitchburg
Nomishan, Daniel Apesuur *science and mathematics educator*

Lexington
Guivens, Norman Roy, Jr. *mathematician, engineer*

Medway
Yonda, Alfred William *mathematician*

Whitman
Thompson, Andrew Ernest *mathematics educator*

MICHIGAN

Ann Arbor
Hess, Ida Irene *statistician*
Hill, Bruce Marvin *statistician, scientist, educator*
Hochster, Melvin *mathematician, educator*

East Lansing
Hoppensteadt, Frank Charles *mathematician, university dean*

Southfield
Arlinghaus, William Charles *mathematics educator*

Ypsilanti
Randolph, Linda Jane *mathematics educator*

MINNESOTA

Duluth
Gallian, Joseph Anthony *mathematics educator*

Minneapolis
Carlson, Richard Raymond *statistician, consultant*
Du, Ding-Zhu *mathematician, educator*
Friedman, Avner *mathematician, educator*
Fryd, David Steven *biostatistician, consultant*
Jackson, Robert Loring *science and mathematics educator, academic administrator*
Nitsche, Johannes Carl Christian *mathematics educator*

Rochester
Naessens, James Michael *biostatistician*

MISSISSIPPI

Cleveland
Strahan, Jimmie Rose *mathematics educator*

Mississippi State
Zhu, Jianping *mathematics educator*

MISSOURI

Cape Girardeau
Young, John Edward *mathematics educator, consultant, writer*

Columbia
Beem, John Kelly *mathematician, educator*

Mexico
Kessler, Donna Kay Ens *mathematics educator*

Rolla
Ingram, William Thomas, III *mathematics educator*

Saint Louis
Haimo, Deborah Tepper *mathematics educator*

NEW HAMPSHIRE

Hanover
Baumgartner, James Earl *mathematics educator*
Stukel, Therese Anne *biostatistician, educator*

NEW JERSEY

Jersey City
Mastro, Victor John *mathematician, educator*

Murray Hill
Graham, Ronald Lewis *mathematician*

New Brunswick
Daubechies, Ingrid *mathematics educator*
Kruskal, Martin David *mathematical physicist, educator*
Strawderman, William E. *statistics educator*
Treves, Jean-François *mathematician educator*

New Providence
Shepp, Lawrence Alan *mathematician, educator*

Newark
Garfinkle, Devra *mathematician, educator*
Porter, Michael Blair *applied mathematics educator*
Skurnick, Joan Hardy *biostatistician, educator*

Oradell
Tong, Mary Powderly *mathematician educator, retired*

Princeton
Aizenman, Michael *mathematics and physics educator*
Borel, Armand *mathematics educator*
Caffarelli, Luis Angel *mathematician, educator*
Christodoulou, Demetrios *mathematics educator*
Deligné, Pierre R. *mathematician*
Fefferman, Charles Louis *mathematics educator*
Griffiths, Phillip A. *mathematician, academic administrator*
Jagerman, David Lewis *mathematician*
Langlands, Robert Phelan *mathematician*
Majda, Andrew J. *mathematician, educator*
Orszag, Steven Alan *applied mathematician, educator*
Sinai, Yakov G. *theoretical mathematician, educator*
Spencer, Thomas C. *mathmatician*
Wiles, Andrew J. *mathematican, educator*

Vineland
Steward, Mollie Aileen *mathematics and computer science educator*

NEW MEXICO

Las Cruces
Harary, Frank *mathematician, computer science educator*

Los Alamos
Kellner, Richard George *mathematician, computer scientist*

Portales
Lyon, Betty Clayton *mathematics educator*

Socorro
Arterburn, David Roe *mathematics educator*

White Sands Missile Range
Norman, James Harold *mathematician, researcher*

NEW YORK

Bronx
Farley, Rosemary Carroll *mathematics and computer science educator*

Brooklyn
Bolonkin, Alexander Alexandrovich *mathematician*

Buffalo
Hauptman, Herbert Aaron *mathematician, educator, researcher*
Tsekanovskii, Eduard Ruvimovich *mathematician, educator*
Wiesenberg, Russel John *statistician*

Corning
Johnson, Janet LeAnn Moe *statistician, project manager*

East Aurora
Young, William Lewis *retired mathematics educator*

Flushing
Artzt, Alice Feldman *mathematics educator*

Hempstead
Hastings, Harold Morris *mathematics educator, researcher, author*

Ithaca
Billera, Louis J(oseph) *mathematics educator*
Bramble, James Henry *mathematician, educator*
Dynkin, Eugene B. *mathematics educator*
Guckenheimer, John *mathematician*
Nerode, Anil *mathematician, educator*

Middletown
Edelhertz, Helaine Wolfson *mathematics educator*

Morrisville
Rouse, Robert Moorefield *mathematician, educator*

New York
Buttke, Thomas Frederick *mathematics educator*
Eisele, Carolyn *mathematician*
Gelbart, Abe *mathematician, educator*
Godbold, James Homer, Jr. *biostatistician, educator*
Gomory, Ralph Edward *mathematician, manufacturing company executive, foundation executive*
Kao, Richard Juichang *biostatistician*
Kulkarni, Ravi Shripad *mathematics educator, researcher*
Lax, Peter David *mathematician*
Lucchesi, Arsete Joseph *mathematician, university dean*
McKean, Henry P. *mathematics institute administrator*
Morawetz, Cathleen Synge *mathematician*
Nirenberg, Louis *mathematician, educator*
Posamentier, Alfred Steven *mathematics educator, university administrator*
Tyrl, Paul *mathematics educator, researcher, consultant*

Plattsburgh
Helinger, Michael Green *mathematics educator*

Schenectady
Stone, William C. *mathematics educator*

Southampton
Melter, Robert Alan *mathematics educator, researcher*

Stony Brook
Glimm, James Gilbert *mathematician*
Michelsohn, Marie-Louise *mathematician, educator*
Tewarson, Reginald Prabhakar *mathematics educator, consultant*

Syracuse
Graver, Jack Edward *mathematics educator*
Waterman, Daniel *mathematician, educator*

Troy
Cole, Julian D. *mathematician, educator*

West Point
Arkin, Joseph *mathematician, lecturer*

Yorktown Heights
Hamaguchi, Satoshi *mathematician*
Hoffman, Alan Jerome *mathematician, educator*
Mandelbrot, Benoit B. *mathematician, scientist, educator*
Winograd, Shmuel *mathematician*

NORTH CAROLINA

Chapel Hill
Coulter, Elizabeth Jackson *biostatistician, educator*
Kallianpur, Gopinath *statistician*

Cullowhee
Willis, Ralph Houston *mathematics educator*

Durham
Bryant, Robert Leamon *mathematics educator*
Winkler, Robert Lewis *statistics educator, researcher, author, consultant*

Greensboro
Blanchet-Sadri, Francine *mathematician*
Kurepa, Alexandra *mathematician*

Murfreesboro
McLawhorn, Rebecca Lawrence *mathematics educator*

Raleigh
Pollock, Kenneth Hugh *statistics educator*

Research Triangle Park
Hajian, Gerald *biostatistician, engineer*
Kramer, David Alan *biomathematician*
Smith, Luther A. *statistician*

OHIO

Berea
Little, Richard Allen *mathematics and computer science educator*

Bowling Green
Newman, Elsie Louise *mathematics educator*

Cleveland
Coroneos, Rula Mavrakis *mathematician*
Ellis, Brenda Lee *mathematician, computer scientist, consultant, educator*
Goffman, William *mathematician, educator*
Woyczynski, Wojbor Andrzej *mathematician, educator*

Columbus
Baker, Gregory Richard *mathematician*
Rubin, Karl Cooper *mathematician*

Dayton
Khalimsky, Efim *mathematics and computer science educator*

Lima
Johnson, Patricia Lyn *mathematics educator*

Mansfield
Gregory, Thomas Bradford *mathematics educator*

Portsmouth
Hamilton, Virginia Mae *mathematics educator, consultant*

Westerville
Bilisoly, Roger Sessa *statistician*

OKLAHOMA

Norman
Lyberopoulos, Athanasios Nikolaos *mathematician, engineer*

Tulsa
Thompson, Carla Jo Horn *mathematics educator*

Weatherford
Woods, John Merle *mathematics educator, chairman*

OREGON

Corvallis
Parks, Harold Raymond *mathematician, educator*

Portland
Hall, Howard Pickering *mathematics educator*
Lambert, Richard William *mathematics educator*

PENNSYLVANIA

Abington
Ayoub, Ayoub Barsoum *mathematician, educator*

Allentown
Sheesley, John Henry *statistician*

Allison Park
Sartori, David Ezio *statistician, consultant*

Harrisburg
Shatto, Ellen Latherow *mathematics educator*

Lehman
Nouri-Moghadam, Mohamad Reza *mathematics and physics educator*

Meadville
Cable, Charles Allen *mathematician*

Philadelphia
Calabi, Eugenio *mathematician, educator*
Kadison, Richard Vincent *mathematician, educator*
Pickands, James, III *mathematical statistician, educator*
Scedrov, Andre *mathematics and computer science researcher, educator*

Pittsburgh
Burbea, Jacob N. *mathematics educator*
Frieze, Alan Michael *mathematician, educator*
Hall, Charles Allan *numerical analyst, educator*
Kannan, Ravi *mathematician educator*

Schwenksville
Frech, Bruce *mathematician*

Uniontown
Dimitric, Ivko Milan *mathematician, educator*

University Park
Baum, Paul Frank *mathematics educator*

West Chester
Skeath, Ann Regina *mathematics educator*

West Mifflin
Starmack, John Robert *mathematics educator*

Williamsport
Owens, Edwin Geynet *mathematics educator*

RHODE ISLAND

Kingston
Driver, Rodney David *mathematics educator, state legislator*

Providence
Dafermos, Constantine Michael *applied mathematics educator*
Freiberger, Walter Frederick *mathematics educator, actuarial science consultant, educator*
Gottschalk, Walter Helbig *mathematician, educator*
Tabenkin, Alexander Nathan *metrologist*

SOUTH CAROLINA

Charleston
Hoel, David Gerhard *state administrator, statistician, scientist*

Columbia
Dilworth, Stephen James *mathematics educator*
Pumariega, JoAnne Buttacavoli *mathematics educator*

SOUTH DAKOTA

Aberdeen
Markanda, Raj Kumar *mathematics educator*

TENNESSEE

Brownsville
Kalin, Robert *retired mathematics educator*

Chattanooga
Johnson, Joseph Erle *mathematician*
Powell, Patricia Ann *mathematics and business educator*

Kingsport
Lee, Kwan Rim *statistician*

Nashville
Sloan, Paula Rackoff *mathematics educator*

TEXAS

Austin
Bledsoe, Woodrow Wilson *mathematics and computer sciences educator*
Gillman, Leonard *mathematician, educator*
Uhlenbeck, Karen Keskulla *mathematician, educator*

Dallas
Browne, Richard Harold *statistician, consultant*
Tiernan, J(anice) Carter Matheney *mathematics specialist, consultant*

El Paso
Dalton, Oren Navarro *mathematician*

Fort Worth
Doran, Robert Stuart *mathematics educator*

Galveston
Freeman, Daniel Herbert, Jr. *biostatistician*

Houston
Barnes, Ronald Francis *mathematics educator*
Dennis, John Emory, Jr. *mathematics educator*
Goerss, James Malcolm *statistician, electrical engineer*
Golubitsky, Martin Aaron *mathematician, educator*
Hardt, Robert Miller *mathematics educator*
Kimmel, Marek *biomathematician, educator*
Mifflin, Richard Thomas *applied mathematician*
Wang, Chao-Cheng *mathematician, engineer*
Zimmerman, Stuart ODell *biomathematician, educator*

Kingsville
Morey, Philip Stockton, Jr. *mathematics educator*

Kingwood
Burghduff, John Brian *mathematics educator*

Lubbock
Amir-Moez, Ali Reza *mathematician, educator*

Richardson
Wiorkowski, John James *mathematics educator*

San Antonio
Blaylock, Neil Wingfield, Jr. *applied statistician, educator*
Trench, William Frederick *mathematics educator*

The Woodlands
Savir, Etan *mathematics educator*

Waco
Rolf, Howard Leroy *mathematician, educator*

UTAH

Provo
Lang, William Edward *mathematics educator*

VERMONT

Norwich
Snapper, Ernst *mathematics educator*

VIRGINIA

Arlington
Waring, John Alfred *retired research writer, lecturer, consultant*

Blacksburg
Hinkelmann, Klaus Heinrich *statistician, educator*

Charlottesville
Mansfield, Lois Edna *mathematics educator, researcher*
Parshall, Karen Virginia Hunger *mathematics and science historian*
Rosenblum, Marvin *mathematics educator*
Scott, Leonard Lewy, Jr. *mathematician, educator*
Taylor, Samuel James *mathematics educator*

Clifton
Hoffman, Karla Leigh *mathematician*

Hampton
Hunt, Patricia Jacqueline *mathematician, system manager, graphics programmer*
Voigt, Robert G. *numerical analyst*

Lexington
Tierney, Michael John *mathematics and computer science educator*

Mc Lean
Stendahl, Steven James *mathematician, system engineer*

Norfolk
Hou, Jiashi *mathematician, educator*

Richlands
Witten, Thomas Jefferson, Jr. *mathematics educator*

Sterling
Bergeman, George William *mathematics educator, software author*

Sweet Briar
Wassell, Stephen Robert *mathematics educator, researcher*

Wise
Low, Emmet Francis, Jr. *mathematics educator*

WASHINGTON

Redmond
Yagi, Fumio *mathematician, systems engineer*

Seattle
Criminale, William Oliver, Jr. *applied mathematics educator*
Klee, Victor La Rue *mathematician, educator*
Osborne, Mason Scott *mathematician, educator*

WISCONSIN

Green Bay
Davis, Gregory John *mathematician, educator*

Madison
Askey, Richard Allen *mathematician*
de Boor, Carl *mathematician*
Kleene, Stephen Cole *retired mathematician, educator*

Milwaukee
Simms, John Carson *logic, mathematics and computer science educator*

Muskego
Brown, Serena Marie *mathematics and home economics educator*

TERRITORIES OF THE UNITED STATES

PUERTO RICO

Mayaguez
Collins, Dennis Glenn *mathematics educator*

MILITARY ADDRESSES OF THE UNITED STATES

EUROPE

APO
Huffman, Kenneth Alan *operations researcher, mathematician*

CANADA

BRITISH COLUMBIA

Vancouver
Seymour, Brian Richard *mathematician*

NEW BRUNSWICK

Sackville
Dekster, Boris Veniamin *mathematician, educator*

ONTARIO

Chatham
Shakhmundes, Lev *mathematician*

Ottawa
Dlab, Vlastimil *mathematics educator, researcher*

Toronto
Lehma, Alfred Baker *mathematician, educator*

Waterloo
Kalbfleisch, John David *statistics educator, dean*

Windsor
Barron, Ronald Michael *applied mathematician, educator, researcher*

QUEBEC

Rimouski
Jean, Roger V. *mathematician, educator*

ARGENTINA

Bahía Blanca
Panzone, Rafael *mathematics educator*

AUSTRALIA

Ascot Vale
Bish, Robert Leonard *applied mathematician, metallurgist, researcher*

Canberra
Neumann, Bernhard Hermann *mathematician*

East Saint Kilda
Zehnwirth, Ben Zion *mathematician consultant*

Melbourne
Tran-Cong, Ton *applied mathematician, researcher*

AUSTRIA

Linz
Pilz, Günter Franz *mathematics educator*

Vienna
Karigl, Günther *mathematician*
Niederreiter, Harald Guenther *mathematician, researcher*

BANGLADESH

Chittagong
Islam, Jamal Nazrul *mathematics and physics educator, director*

BELGIUM

Antwerp
Kuyk, Willem *mathematics educator*

Namur
Cornelis, Eric Rene *mathematician*

BENIN

Cotonou
Ezin, Jean-Pierre Onvêhoun *mathematician*

BRAZIL

Curitiba
Berman, Marcelo Samuel *mathematics and physics educator, cosmology researcher*

São Paulo
da Costa, Newton Carneiro Affonso *mathematics educator*

CHINA

Beijing
Chen, Gong Ning *mathematics educator*

Chengdu
Xu, Daoyi *mathematician, educator*

Huhot
Chen, Jian Ning *mathematics educator*

Linfen
Hou, Jin Chuan *mathematics educator*

Nanjing
Ton, Dao-Rong *mathematics educator*
Yang, Run Sheng *mathematics educator*
Zou, Yun *mathematician*

Nanning
Guo, Xin Kang *mathematics educator*

Shanghai
Cao, Jia Ding *mathematics educator*
Cheng, Ansheng *mathematics educator, researcher*
Ding, Hai *mathematician*
Ge, Guang Ping *mathematics educator, statistician*
Wei, Musheng *mathematics educator*

Wuhu
Mo, Jiaqi *mathematics educator*

COLOMBIA

Armenia
Pareja-Heredia, Diego *mathematics educator, bookseller consultant*

COSTA RICA

San Jose
Gongora-Trejos, Enrique *mathematician, educator*

DENMARK

Copenhagen
Olesen, Mogens Norgaard *mathematics educator*

Gentofte
Larsen, Jesper Kampmann *applied mathematician, educator*

ENGLAND

Cathay Park
Temperley, H. N. V. *mathematician educator*

Leicester
Light, William Allan *mathematics educator*

Nottingham
Soldatos, Kostas P. *mathematician*

Oxford
Penrose, Roger *mathematics educator*

FINLAND

Helsinki
Illman, Sören Arnold *mathematician, educator*

Oulu
Arjas, Elja *statistician*

Tampere
Merikoski, Jorma Kaarlo *mathematics educator*

FRANCE

Amiens
Chacron, Joseph *mathematics educator*

Paris
Giraudet, Michele *mathematics educator, researcher*
Yuechiming, Roger Yue Yuen Shing *mathematics educator*

Saint-Cloud
Mallet, Michel Francois *numerical analyst*

GERMANY

Bochum
Ponosov, Arcady Vladimirovitch *mathematician*

Bonn
Korte, Bernhard Hermann *mathematician, educator*

Dresden
Reinschke, Kurt Johannes *mathematician, educator*

Halle
Benker, Hans Otto *mathematics educator*
Goldschmidt, Bernd *mathematics educator*

Hamburg
Scheurle, Jurgen Karl *mathematician, educator*
Wolfinger, Bernd Emil *mathematics and computer science educator*

Hennef
Wette, Eduard Wilhelm *mathematician*

Jena
Krätzel, Ekkehard *mathematics educator*

Kaiserslautern
Fan, Tian-You *applied mathematician, educator*

Kassel
Dosbach, Bruno *mathematics educator*

Mannheim
Nürnberger, Günther *mathematician*

Munich
Reviczky, Janos *mathematician, researcher*
Schlee, Walter *mathematician*
Stiegler, Karl Drago *mathematician*

Saarbrücken
König, Heinz Johannes Erdmann *mathematics educator*

Steinfurt
Niederdrenk, Klaus *mathematician, educational administrator, educator*

Witten
an der Heiden, Wulf-Uwe *mathematics educator*

Wuerzburg
Buntrock, Gerhard Friedrich Richard *mathematician*

GHANA

Kumasi
Owusu-Ansah, Twum *mathematics educator*

GREECE

Athens
Demetriou, Ioannes Constantine *mathematics educator, researcher*
Panaretos, John *mathematics and statistics educator*

Patras
Dassios, George Theodore *mathematician, educator, researcher*

HONG KONG

Hong Kong
Chan, Raymond Honfu *mathematics educator*

Sha Tin
Lu, Wudu *mathematics educator*

Wanchai
Chen, Concordia Chao *mathematician*

HUNGARY

Debrecen
Sztrik, János *mathematics educator, researcher*

INDIA

Bangalore
Chakrabartty, Sunil Kumar *mathematician*

Pune
Purohit, Sharad Chandra *mathematician*

ISRAEL

Haifa
Shpilrain, Vladimir Evald *mathematician, educator*

ITALY

Cagliari
Arca, Giuseppe *mathematician, educator*

Padua
Rosati, Mario *mathematician, educator*

Parma
Bertolini, Fernando *mathematics educator, Italian embassy cultural administrator*

Pisa
Cimatti, Giovanni Ermanno *rational mechanics educator*
De Giorgi, Ennio *mathematics educator*

Potenza
Korchmaros, Gabor Gabriele *mathematics educator*

Torino
Elia, Michele *mathematics educator*
Rossi, Guido A(ntonio) *applied mathematics educator, researcher*

Udine
Dikranjan, Dikran Nishan *mathematics educator*

JAPAN

Chiba
Yamada, Shinichi *mathematician, computer scientist, educator*

Gifu
Hatada, Kazuyuki *mathematician, educator*

Hachioji
Shimoji, Sadao *applied mathematics educator, engineer*

Hiroshima
Shohoji, Takao *statistician*

Ibaraki
Kishimoto, Kazuo *mathematical engineering educator*

Kyoto
Hayashi, Takao *mathematics educator, historian, Indologist*
Mori, Shigejumi *mathematician educator*
Murazawa, Tadashi *mathematics educator*

Matsue
Kamiya, Noriaki *research mathematician*

Nagoya
Hida, Takeyuki *mathematics researcher, educator*

Sapporo
Giga, Yoshikazu *mathematician*

Yamazaki
Shiraiwa, Kenichi *mathematician*

MACAU

Taipa
Li, Yiping (Y.P.) *applied mathematics educator*

THE NETHERLANDS

Amsterdam
Hazewinkel, Michiel *mathematician, educator*

Delft
Koekoek, Roelof *mathematics educator*

Utrecht
Inoué, Takao *logician, philosopher*

NORWAY

Oslo
Fenstad, Jens Erik *mathematics educator*

PAKISTAN

Lahore
Chawla, Lal Muhammad *mathematics educator*
Majeed, Abdul *mathematics educator*

POLAND

Lublin
Kubacki, Krzysztof Stefan *mathematics educator*

Warsaw
Semadeni, Zbigniew Wladyslaw *mathematician, educator*

PORTUGAL

Lisbon
Campos, Luís Manuel Braga da Costa *mathematics, physics, acoustics and aeronautics educator*

REPUBLIC OF KOREA

Chinju
Nam, Jung Wan *mathematics educator*

ROMANIA

Bucharest
Cristescu, Romulus *mathematician, educator, science administrator*

RUSSIA

Barnaul
Karakozov, Sergei Dmitrievich *mathematics educator*

Moscow
Fursikov, Andrei Vladimirovich *mathematics educator*
Hohlov, Yuri Eugenievich *mathematician*
Litvinchev, Igor Semionovich *mathematician, educator*
Maslov, Viktor Pavlovich *mathematician, educator*
Mokhov, Oleg Ivanovich *mathematician*
Razborov, Alexander A. *mathematician*
Vladimirov, Vasiliy Sergeyevich *mathematician*

Novosibirsk
Koshelev, Yuriy Grigoryevich *mathematics educator*

Saint Petersburg
Faddeev, Ludwig D. *theoretical mathematician*
Faddeyev, Ludvig Dmitriyevich *mathematician, educator*
Sokolsky, Andrej Georgiyevich *mathematician, academic administrator*

Voronezh
Kostin, Vladimir Alexeevich *mathematics educator*

SAUDI ARABIA

Dhahran
Warne, Ronson Joseph *mathematics educator*

Unizah
Konsowa, Mokhtar Hassan *mathematics educator*

SCOTLAND

Glasgow
Cohen, Stephen Douglas *mathematician, educator*

SPAIN

Burjasot
Garcia, Domingo *mathematics educator*

Madrid
Pedregal, Pablo *mathematician, educator*

Santiago
Nieto-Roig, Juan Jose *mathematician, educator, researcher*

Valencia
Bonet, Jose *mathematics educator, researcher*

Vitoria
Gracia, Juan Miguel *mathematics educator*

SRI LANKA

Nugegoda
Somadasa, Hettiwatte *mathematician, educator*

SWEDEN

Linköping
af Ekenstam, Adolf W. *mathematics educator*

Stockholm
Fuchs, Laszlo Jehoshua *mathematician*
Petermann, Gotz Eike *mathematics educator, researcher*

Uppsala
Taubé, Adam A.S. *biometrician, educator*

SWITZERLAND

Locarno
Moresi, Remo P. *mathematician*

Zurich
Baladi, Viviane *mathematician*
Groh, Gabor Gyula *mathematician, educator*
Kalman, Rudolf Emil *research mathematician, systems scientist*
Klatte, Diethard W. *mathematician*

TAIWAN

Taichung
Chen, Chih-Ying *mathematics and computer science educator, researcher*

UKRAINE

Kiev
Martynyuk, Anatoly Andreevich *mathematician*

UNITED ARAB EMIRATES

Al-Ain
Zahreddine, Ziad Nassib *mathematician, educator, researcher*

UZBEKISTAN

Toshkent
Salakhitdinov, Makhmud *mathematics educator*

WALES

Swansea
Weatherill, Nigel Peter *mathematician, researcher*

ADDRESS UNPUBLISHED

Browder, Felix Earl *mathematician, educator*
Clogg, Clifford Collier *statistician, educator*
Császár, Ákos *mathematician*
Ferguson, James Clarke *mathematician, algorithmist*
Harris, Shari Lea *mathematics educator*
Horton, Wilfred Henry *mathematics educator*
Johnson, Deborah Crosland Wright *mathematics educator*
Lasry, Jean-Michel *mathematics educator*
Lindquist, Anders Gunnar *applied mathematician, educator*
Main, Myrna Joan *mathematics educator*
Mehta Malani, Hina *biostatistician, educator*
Patterson, Patricia Lynn *applied mathematician, geophysicist, engineer*
Paul, Vera Maxine *mathematics educator*
Somes, Grant William *statistician, biomedical researcher*
Suppes, Patrick *statistics, education, philosophy and psychology educator*
Temam, Roger M. *mathematician*
Varvak, Mark *mathematician, researcher*
Wylie, Clarence Raymond, Jr. *mathematics educator*

MEDICAL EDUCATION & RESEARCH

UNITED STATES

ALABAMA

Birmingham
Balschi, James Alvin *medical educator*
Finley, Wayne House *medical educator*
Sanders, Michael Kevin *hypertension researcher*

Mobile
Parmley, Loren Francis, Jr. *medical educator*

ARIZONA

Tucson
Witten, Mark Lee *lung injury research scientist, educator*

ARKANSAS

Little Rock
Chang, Louis Wai-wah *medical educator, researcher*

CALIFORNIA

Berkeley
Fuhs, G(eorg) Wolfgang *medical research scientist*

Profet, Margie *biomedical researcher*
Suzuki, Yuichiro Justin *biomedical scientist*

Chula Vista
Allen, Henry Wesley *biomedical researcher*

Davis
Schenker, Marc Benet *medical educator*

Encinitas
Rummerfield, Philip Sheridan *medical physicist*

La Jolla
Campbell, Iain Leslie *biomedical scientist*

La Puente
Goldberg, David Bryan *biomedical researcher*

Livermore
Swiger, Roy Raymond *biomedical scientist*

Los Angeles
Horwitz, David A. *medicine and microbiology educator*
Tran, Johan-Chanh Minh *research scientist*
Walsh, John Harley *medical educator*
Watanabe, Richard Megumi *medical research assistant*

Palo Alto
Cooper, Allen David *research scientist, educator*
Holman, Halsted Reid *medical educator*

Redwood City
Sharma, Kuldeepak Bhardwaj *pharmaceutical scientist*

San Francisco
Bishop, John Michael *biomedical research scientist, educator*
Lee, Philip Randolph *medical educator*

San Ramon
Brennan, Sean Michael *cancer research scientist, educator*

Santa Barbara
Peterson, Charles Marquis *medical educator*

Stanford
Baylor, Denis Aristide *neurobiology educator*
Jardetzky, Oleg *medical educator, scientist*

COLORADO

Denver
Dunn, Andrea Lee *biomedical researcher*
Henson, Peter Mitchell *physician, immunology and respiratory medicine executive*

CONNECTICUT

Hartford
Cheng, Wing-Tai Savio *medical technology consultant*

New Haven
Lentz, Thomas Lawrence *biomedical educator, dean*

DISTRICT OF COLUMBIA

Washington
Cole, Dean Allen *biomedical researcher*
Hicks, Jocelyn Muriel *laboratory medicine specialist*
Kelly, Douglas Elliott *biomedical researcher, association administrator*

FLORIDA

Miami
Byrnes, John Joseph *medical educator*

Pensacola
Gudry, Frederick E., Jr. *aerospace medical researcher*

GEORGIA

Augusta
Karp, Warren Bill *medical and dental educator, researcher*

Clarkston
Saint-Come, Claude Marc *science educator, consultant*

Marietta
Houser, Vincent Paul *medical research director*

ILLINOIS

Chicago
Datta, Syamal Kumar *medical educator, researcher*
Harris, Jules Eli *medical educator, physician, clinical scientist, administrator*
Koch, Alisa Erika *biomedical researcher, rheumatologist*
Kornel, Ludwig *medical educator, physician, scientist*

Maywood
Kovacs, Elizabeth J. *medical educator*

Springfield
Amador, Armando Gerardo *medical educator*

INDIANA

Gary
Echtenkamp, Stephen Frederick *biomedical researcher*

Indianapolis
Broxmeyer, Hal Edward *medical educator*
Weinberger, Myron Hilmar *medical educator*

West Lafayette
Kessler, Wayne Vincent *health sciences educator, researcher, consultant*

IOWA

Iowa City
Fellows, Robert Ellis *medical educator, medical scientist*
Husted, Russell Forest *research scientist*

KANSAS

Wichita
Cho, Sechin *medical geneticist*
Davis, John Stewart *medical educator, endocrinology researcher*

KENTUCKY

Lexington
Kraman, Steve Seth *physician, educator*

Louisville
Blumenreich, Martin Sigvart *medical educator*
Syed, Ibrahim Bijli *medical educator, theologist*
Thind, Gurdarshan S. *medical educator*

LOUISIANA

Hammond
Hejtmancik, Milton Rudolph *medical educator*

New Orleans
Bertrand, William Ellis *public health educator, international health center administrator*
Gottlieb, A(braham) Arthur *medical educator, biotechnology corporate executive*
Granger, Wesley Miles *medical educator*
Weill, Hans *physician, educator*

MARYLAND

Baltimore
Kwiterovich, Peter Oscar, Jr. *medical science educator, researcher, physician*
Myslinski, Norbert Raymond *medical educator*
Schnaar, Ronald Lee *biomedical researcher, educator*
Sherwin, Roger William *medical educator*

Beltsville
Lincicome, David Richard *biomedical and animal scientist*

Bethesda
Gottesman, Michael Marc *biomedical researcher*
Kaufmann, Peter G. *research biomedical scientist, psychologist, educator*
Moss, Joel *medical researcher*
Tabor, Edward *physician, researcher*
Templeton, Nancy Valentine Smyth *research scientist*
Waldmann, Thomas Alexander *medical research scientist, physician*

Havre De Grace
Wiegand, Gordon William *flow cytometrist, electro-optic consultant*

Rockville
Banfield, William Gethin *physician*
Zmudzka, Barbara Zofia *researcher*

Silver Spring
Lippman, Muriel Marianne *biomedical scientist*

MASSACHUSETTS

Boston
Abou-Samra, Abdul Badi *biomedical researcher, physician*
Alexander-Bridges, Maria Carmalita *medical researcher*
Bieber, Frederick Robert *medical educator*
Rosenberg, Irwin Harold *physician, educator*
Selker, Harry Paul *medical educator*
Swartz, Morton Norman *medical educator*

Cambridge
Manoharan, Ramasamy *biomedical scientist*

Shrewsbury
Paredes, Eduardo *medical physicist*

Wellesley
Reif, Arnold E. *medical educator*

MICHIGAN

Ann Arbor
Goldstein, Irwin Joseph *medical research executive*
Goldstein, Steven Alan *medical educator, engineering educator*

Detroit
Mayes, Maureen Davidica *physician, educator*
Tamimi, Nasser Taher *educator, medical physicist*

East Lansing
Grant, Rhoda *biomedical researcher, educator, medical physiologist*

Grosse Pointe
Wayland, Marilyn Ticknor *medical researcher, evaluator, educator*

MISSISSIPPI

Jackson
Cai, Zhengwei *biomedical researcher*

MISSOURI

Saint Louis
Houston, Devin Burl *biomedical scientist, educator*
Kurtz, Michael E. *medical educator*
Santiago, Julio V. *medical educator, medical association administrator*
Schonfeld, Gustav *medical educator*

NEBRASKA

Omaha
Ertl, Ronald Frank *research coordinator*
Godfrey, Maurice *biomedical scientist*

NEW HAMPSHIRE

Lebanon
Spencer-Green, George Thomas *medical educator*

NEW JERSEY

Browns Mills
Gu, Jiang *biomedical scientist*

Newark
Mistry, Kishorkumar Purushottamdas *biomedical scientist, educator*
Scherzer, Norman Alan *medical educator*

Paterson
DeBari, Vincent Anthony *medical researcher, educator*

Piscataway
Keller, Mark *medical educator*

Princeton Junction
Lee, Lihsyng Stanford *medical researcher, biotechnical consultant*

NEW YORK

Bellmore
Harris, Ira Stephen *medical and health sciences educator*

Bronx
Burk, Robert David *physician, medical educator*
Herbert, Victor Daniel *medical educator*
Rogler, Charles Edward *medical educator*

Manhasset
Hashimoto, Shiori *biomedical scientist*
Powell, Saul Reuben *research scientist, surgical educator*

New York
Bartlett, Elsa Jaffe *neuropsychologist, educator*
Bekesi, Julis George *medical researcher*
Ding, Aihao *medical educator, researcher*
Dole, Vincent Paul *medical research executive, educator*
Field, Steven Philip *medical educator*
Grumet, Martin *biomedical researcher*
McCarty, Maclyn *medical scientist*
Roberts, 2ames Lewis *medical sciences educator*
Roman, Stanford Augustus, Jr. *medical educator, dean*
Rosenfield, Richard Ernest *emeritus medical educator*
Sacks, Henry S. *medical researcher, infectious disease physician*
Saha, Dhanonjoy Chandra *biomedical research scientist*
Sorota, Steve *biomedical researcher*
Sun, Tung-Tien *medical science educator*
Tamm, Igor *biomedical scientist, educator*
Vilcek, Jan Tomas *medical educator*

Rochester
Goldsmith, Lowell Alan *medical educator*

Stony Brook
Rohlf, F. James *biometrist, educator*

Syracuse
Bellanger, Barbara Doris Hoysak *biomedical research technologist*

Valhalla
Ehrenkrantz, David *medical researcher*

White Plains
Blass, John Paul *medical educator, physician*

NORTH CAROLINA

Chapel Hill
Peters, Robert William *speech and hearing sciences educator*
Suzuki, Kunihiko *biomedical educator, researcher*

Durham
Bast, Robert Clinton, Jr. *research scientist, medical educator*
Kurlander, Roger Jay *medical educator, researcher*
Snyderman, Ralph *medical educator, physician*
Williams, Redford Brown *medical educator*

Greensboro
Robinson, Edward Norwood, Jr. *physician, educator*

Greenville
Lieberman, Edward Marvin *biomedical scientist, educator*
Waugh, William Howard *biomedical educator*

Research Triangle Park
Barry, David Walter *infectious diseases physician, researcher*
Lee, Paul Huk-Kai *biomedical research scientist*

Winston Salem
Clarkson, Thomas Boston *comparative medicine educator*
Li, Linxi *biomedical scientist*
Toole, James Francis *medical educator*

OHIO

Cincinnati
Albert, Roy Ernest *environmental health educator, researcher*

Cleveland
Scarpa, Antonio *medicine educator, biomedical scientist*

Columbus
Tzagournis, Manuel *physician, educator, university dean and official*

Rootstown
Westerman, Philip William *biomedical researcher, medical educator*

Wooster
Geho, Walter Blair *biomedical research executive*

OREGON

Beaverton
Haluska, George Joseph *biomedical scientist*

PENNSYLVANIA

Bangor
Wolf, Stewart George, Jr. *physician, medical educator*

Collegeville
Farmar, Robert Melville *medical scientist, educator*

Harrisburg
Grandon, Raymond Charles *physician, educator*

Philadelphia
Buerk, Donald Gene *medical educator*
Madaio, Michael Peter *medical educator*
Yassin, Rihab R. *biomedical researcher, educator*
Zweidler, Alfred *medical scientist, educator*

Pittsburgh
Blackburn, Ruth Elizabeth *biomedical research scientist*
Stanko, Ronald Thomas *physician, medical educator, clinical nutritionist*

Plymouth Meeting
Nobel, Joel J. *physician*

West Point
Callahan Graham, Pia Laaster *medical researcher, virology researcher*

RHODE ISLAND

Providence
DePetrillo, Paolo Bartolomeo *medical educator*
Galletti, Pierre Marie *artificial organ scientist, medical science educator*

TENNESSEE

Nashville
Newman, John Hughes *medical educator*

TEXAS

Dallas
Bowcock, Anne Mary *medical educator*
Goldstein, Joseph Leonard *physician, medical educator, molecular genetics scientist*

Galveston
Anderson, Karl Elmo *educator*
Yallampalli, Chandrasekhar *medical educator, researcher*
Yannariello-Brown, Judith I. *biomedical researcher, educator*

Houston
Brody, Baruch Alter *medical educator, academic center administrator*
Vassilopoulou-Sellin, Rena *medical educator*

Lubbock
Doris, Peter A. *biomedical scientist*

UTAH

Salt Lake City
Schoeff, Larry *educator*

VERMONT

Burlington
Tracy, Russell Peter *biomedical researcher, educator*

VIRGINIA

Charlottesville
Barrett, Eugene J. *researcher, medical educator, physician*
Woode, Moses Kwamena *medical educator*

WASHINGTON

Bothell
Daigle, Ronald Elvin *medical imaging scientist, researcher*

Hansville
Strahilevitz, Meir *inventor, researcher, psychiatry educator*

Seattle
Klebanoff, Seymour Joseph *medical educator*
Young, Nancy Lee *medical educator*
Zhao, Lue Ping *biostatistician, genetic epidemiologist*

WISCONSIN

Madison
Muller, Daniel *biomedical scientist*

Milwaukee
Story, Michael Thomas *biomedical researcher, educator*

TERRITORIES OF THE UNITED STATES

PUERTO RICO

Ponce
Torres-Aybar, Francisco Gualberto *medical educator*

San Juan
Sahai, Hardeo *medical educator*

CANADA

BRITISH COLUMBIA

Vancouver
Clement, Douglas Bruce *medical educator*
Silver, Hulbert K.B. *physician, educator*

ONTARIO

London
Valberg, Leslie Stephen *medical educator, physician, researcher*

Toronto
Paul, Leendert Cornelis *medical educator*

QUEBEC

Montreal
Carriere, Serge *physiologist, physician, educator*
Roberge, Fernand Adrien *biomedical researcher*

ARGENTINA

Tucuman
Valentinuzzi, Max Eugene *bioengineering and physiology educator, researcher*

BELGIUM

Ghent
de Thibault de Boesinghe, Léopold Baron *physician*
Ringoir, Severin Maria Ghislenus *medical educator, physician*

COSTA RICA

San Jose
Bergold, Orm *medical educator*

FRANCE

Boulogne
Dubourg, Olivier Jean *cardiology educator, researcher*

Lyons
Meunier, Pierre Jean *medical educator*

Montpellier
Michel, Henri Marie *medical educator*

GERMANY

Düsseldorf
Schadewaldt, Hans *medical educator*

Freiburg
Häussinger, Dieter Lothar *medical educator*

Munich
Cleve, Hartwig Karl *medical educator*
Porkert, Manfred (Bruno) *medical sciences educator, author*

Tübingen
Nüsslein-Volhard, Christiane *medical researcher*

GREECE

Athens
Mantas, John *health informatics educator*

IRELAND

Galway
Lavelle, Seān Marius *clinical informatics educator*

ITALY

Milan
Pontiroli, Antonio Ettore *medical educator*

Udine
Sorrentino, Dario Rosario *medical educator*

JAPAN

Fukuoka
Omura, Tsuneo *medical educator*

Kobe
Yamabe, Shigeru *medical educator*

Kumamoto
Maeda, Hiroshi *medical educator*

Sapporo
Fukuda, Morimichi *medical educator*

NEW ZEALAND

Dunedin
Adams, Duncan Dartrey *medical researcher, physician*

SPAIN

Arrecife
Jerndal, Jens *holistic medicine educator, health center promoter*

SWEDEN

Malmö
Manthorpe, Rolf *medical educator*

TAIWAN

Taipei
Wang, Ling Danny *research scientist*

ADDRESS UNPUBLISHED

Asch, David Kent *medical educator*
Loftin, Karin Christiane *biomedical specialist, researcher*
Lolis, Elias *biomedical researcher*
Okuda, Kunio *emeritus medical educator*
Warfel, John Hiatt *medical educator, retired*

MEDICINE: ALLERGY & IMMUNOLOGY

UNITED STATES

ALABAMA

Birmingham
Koopman, William James *medical educator, internist, immunologist*

ARIZONA

Tucson
Enriquez, Francisco Javier *immunologist*
Marchalonis, John Jacob *immunologist, educator*

CALIFORNIA

Berkeley
Koshland, Marian Elliott *immunologist, educator*

Irvine
Gupta, Sudhir *immunologist, educator*

La Jolla
Fischer, Peter *research immunologist*
Ishizaka, Kimishige *immunologist, educator*
Morrow, William John Woodroofe *immunologist*

Los Angeles
Fahey, John Leslie *immunologist*

Palo Alto
Zlotnik, Albert *immunologist, researcher*

San Diego
Yang, Hsin-Ming *immunologist*

Stanford
McDevitt, Hugh O'Neill *immunology educator, physician*

COLORADO

Denver
Talmage, David Wilson *physician, microbiology and medical educator, former university administrator*

Englewood
Cowen, Donald Eugene *physician*

CONNECTICUT

New Haven
Smith, Brian Richard *hematologist, oncologist, immunologist*

FLORIDA

Melbourne
Minor, Mark William *allergist*

Miami
Beck, Morris *allergist*

Tampa
Swan, Charles Wesley *psychoneuroimmunologist*

GEORGIA

Atlanta
Ibegbu, Chris Chidozie *immunologist, virologist*
Kiser, Don Curtis *immunologist*

ILLINOIS

Chicago
Miller, Stephen Douglas *immunologist, educator*
Norton, Steven David *immunologist, researcher*

North Chicago
Kim, Yoon Berm *immunologist, educator*

INDIANA

Indianapolis
Faulk, Ward Page *immunologist*
Gregory, Richard Lee *immunologist*

IOWA

Ames
Stevens, Mark Gregory *immunologist*

KENTUCKY

Lexington
Kang, Bann C. *immunologist*

LOUISIANA

New Orleans
Bautista, Abraham Parana *immunologist*
Sizemore, Robert Carlen *immunologist*

Shreveport
Griffith, Robert Charles *allergist, educator*

MARYLAND

Baltimore
Bochner, Bruce Scott *immunologist, educator*
Rose, Noel Richard *immunologist, microbiologist, educator*

Bethesda
Abrams, Scott Irwin *immunologist*
Grant, Angus John *immunologist*
Kaliner, Michael Aron *immunologist, allergist, medical association administrator*
Kozak, Robert William *immunologist*
Kresina, Thomas Francis *immunologist, educator*
Malech, Harry Lewis *immunologist, researcher*
Paul, William Erwin *immunologist, researcher*
Yarchoan, Robert *clinical immunologist, researcher*

Frederick
Ochoa, Augusto Carlos *immunologist, researcher*

Rockville
Cannon, Grace Bert *immunologist*
Lightfoote, Marilyn Madry *molecular immunologist*

MASSACHUSETTS

Boston
Weber, Georg Franz *immunologist*

Cambridge
Vercelli, Donata *immunologist, educator*

Fall River
Weltman, Joel Kenneth *immunologist*

Springfield
Andrzejewski, Chester, Jr. *immunologist, research scientist*

Worcester
Stavnezer, Janet Marie *immunologist*

MICHIGAN

Detroit
Swanborg, Robert Harry *immunology educator*

MINNESOTA

Rochester
Gleich, Gerald Joseph *immunologist, medical scientist*

MISSOURI

Columbia
Misfeldt, Michael Lee *immunologist, educator*

Saint Louis
Takes, Peter Arthur *immunologist*

NEVADA

Las Vegas
Seggev, Joram Simon *allergist, clinical immumologist*

NEW JERSEY

Monmouth Junction
Molnar-Kimber, Katherine Lu *immunologist, molecular biologist*

Nutley
Gately, Maurice Kent *research immunologist*

Princeton
McKearn, Thomas Joseph *immunology and pathology educator, scientist*

Yardville
Grace, Michael Judd *immunologist*

NEW MEXICO

Rodeo
Scholes, Robert Thornton *allergist, research administrator*

NEW YORK

Bay Shore
Pinsker, Walter *allergist, immunologist*

Bronx
Scharff, Matthew Daniel *immunologist, cell biologist, educator*

Buffalo
Milgrom, Felix *immunologist, educator*
Nair, Madhavan Puthiya Veethil *immunologist, consultant*
Perl, Andras *immunologist, educator*
Shanahan, Thomas Cornelius *immunologist*
Zaleski, Marek Bohdan *immunologist*

Manhasset
Chirmule, Narendra Bhalchandra *immunologist, educator*
McCloskey, Thomas Warren *flow cytometrist, immunologist*

New York
Abdelnoor, Alexander Michael *immunologist, educator*
Chase, Merrill Wallace *immunologist, educator*
Epstein, Seth Paul *immunologist, researcher*
Giancotti, Francesca Romana *immunologist, cancer research scientist*
Hashim, George A. *immunologist, biomedical researcher, educator*
Kabat, Elvin Abraham *immunochemist, biochemist, educator*
Tilley, Shermaine Ann *molecular immunologist, educator*
Tsuji, Moriya *immunologist*
Vivera, Arsenio Bondoc *allergist*

Stony Brook
Volkman, David J. *immunology educator*

Tuxedo Park
Flescher, Eliezer *immunologist*

NORTH CAROLINA

Burlington
Golden, Carole Ann *immunologist, microbiologist*

Durham
Amos, Dennis B. *immunologist*
Buckley, Rebecca Hatcher *physician*
Dawson, Jeffrey Robert *immunology educator*

OHIO

Cincinnati
Gallagher, Joan Shodder *research immunologist*
Newman, Simon Louis *immunologist, educator*

Marietta
Tipton, Jon Paul *allergist*

PENNSYLVANIA

Bryn Mawr
Pettit, Horace *allergist, consultant*

Danville
Schuller, Diane Ethel *allergist, immunologist, educator*

Monroeville
Lin, Ming Shek *allergist, immunologist*

Philadelphia
Winegrad, Albert Irvin *immunologist, educator*
Zweiman, Burton *physician, scientist, educator*

Pittsburgh
Green, Mayer Albert *physician*
Whiteside, Theresa Listowski *immunologist, educator*
Woo, Jacky *immunologist, educator, researcher*

Sayre
Beezhold, Donald Harry *immunologist*

West Point
Donnelly, John James, III *immunologist, blood banker*

RHODE ISLAND

Providence
Biron, Christine Anne *medical science educator, researcher*

TENNESSEE

Memphis
Pabst, Michael John *immunologist, dental researcher*

Nashville
Hollemweguer, Enoc Juan *immunologist*

TEXAS

Dallas
Stone, Marvin Jules *immunologist, educator*

Galveston
Suzuki, Fujio *immunologist, educator, researcher*

Houston
Couch, Robert Barnard *physician, educator*
Lotzová, Eva *immunologist, researcher, educator*

Lackland AFB
Charlesworth, Ernest Neal *immunologist, educator*
Ward, William Wade *clinical immunologist*

San Antonio
Murthy, Krishna Kesava *infectious desease and immunology scientist*

UTAH

Salt Lake City
Daynes, Raymond Austin *immunology educator*

VIRGINIA

Chantilly
Lobel, Steven Alan *immunologist*

WISCONSIN

Madison
Hong, Richard *pediatric immunologist, educator*
Sedgwick, Julie Beth *immunologist*

Marshfield
Marx, James John *immunologist*

Racine
Kim, Zaezeung *allergist, immunologist, educator*

MILITARY ADDRESSES OF THE UNITED STATES

PACIFIC

APO
Heppner, Donald Gray, Jr. *immunology research physician*

ARGENTINA

Rosario
Rotolo, Vilma Stolfi *immunology researcher*

FINLAND

Turku
Toivanen, Paavo Uuras *immunologist, microbiologist, educator*

FRANCE

Paris
Dausset, Jean *immunologist*

Tours
Renoux, Gerard Eugene *immunologist*

GERMANY

Marburg
Gemsa, Diethard *immunologist*

Tübingen
Pawelec, Graham Peter *immunologist*

ITALY

Verona
Tridente, Giuseppe *immunoligist educator*

JAPAN

Osaka
Takino, Masuichi *physician*

Tokyo
Kitahara, Shizuo *allergist*
Yakura, Hidetaka *immunologist*

THE NETHERLANDS

Maastricht
De Baets, Marc Hubert *immunologist, internist*

NORWAY

Oslo
Fagerhol, Magne Kristoffer *immunologist*

SAUDI ARABIA

Riyadh
Weheba, Abdulsalam Mohamad *immunologist, allergist, consultant*

SWEDEN

Stockholm
Möller, Göran *immunology educator*

SWITZERLAND

Zurich
Zinkernagel, Rolf Martin *immunologist educator*

ADDRESS UNPUBLISHED

Chorpenning, Frank Winslow *immunology educator, researcher*
Critchfield, Jeffrey Moore *immunology researcher*
Ein, Daniel *allergist*
Garvey, Justine Spring *immunochemistry educator, biology educator*
Köhler, Georges J. F. *scientist, immunologist*
Welsh, Elizabeth Ann *immunologist, research scientist*

MEDICINE: ANATOMY

UNITED STATES

CALIFORNIA

Davis
Enders, Allen Coffin *anatomy educator*
Hendrickx, Andrew George *anatomy educator*

Los Angeles
Clemente, Carmine Domenic *anatomist, educator*

FLORIDA

Gainesville
Lawless, John Joseph *anatomy educator*

GEORGIA

Augusta
Colborn, Gene Louis *anatomy educator, researcher*

ILLINOIS

Maywood
McNulty, John Alexander *anatomy educator*

INDIANA

Terre Haute
Duong, Taihung *anatomist*

IOWA

Des Moines
Canby, Craig Allen *anatomy educator*
Meetz, Gerald David *anatomist*

KANSAS

Kansas City
Hung, Kuen-Shan *anatomy educator*

KENTUCKY

Lexington
Hauser, Kurt Francis *anatomy and neurobiology educator*

LOUISIANA

New Orleans
Low, Frank Norman *anatomist, educator*

MASSACHUSETTS

Bedford
Siwek, Donald Fancher *neuroanatomist*

Boston
Raviola, Elio *anatomist, neurobiologist*

Roxbury
Peters, Alan *anatomy educator*

MICHIGAN

Ann Arbor
Huelke, Donald Fred *anatomy and cell biology educator, research scientist*

Detroit
Schaffler, Mitchell Barry *research scientist, anatomist, educator*

East Lansing
Walker, Bruce Edward *anatomy educator*

MINNESOTA

Rochester
Carmichael, Stephen Webb *anatomist, educator*

MISSISSIPPI

Jackson
Sinning, Allan Ray *anatomy educator, researcher*

NEW YORK

Bronx
Scharrer, Berta Vogel *anatomy and neuroscience educator*

NORTH CAROLINA

Durham
Merchenthaler, Istvan Jozsef *anatomist*

OHIO

Athens
Palmer, Brent David *microanatomy educator, reproductive biologist*

Cleveland
Bloch, Edward Henry *scientist, retired anatomy educator*

Toledo
Pansky, Ben *anatomy educator, science researcher*

OKLAHOMA

Oklahoma City
Dones, Maria Margarita *anatomist, educator*

RHODE ISLAND

Providence
Erikson, George Emil (Erik Erikson) *anatomist, archivist, historian, educator, information specialist*

TEXAS

Fort Worth
Agarwal, Neeraj *anatomy and biology educator*

Houston
Gibson, Kathleen Rita *anatomy and anthropology educator*

WISCONSIN

Milwaukee
Carroll, Edward William *anatomist, educator*

AUSTRALIA

Nedlands
Oxnard, Charles Ernest *anatomist, anthropologist, human biologist, educator*

JAPAN

Chuo-ku
Yamadori, Takashi *anatomy educator*

Hiroshima
Yasuda, Mineo *anatomy educator*

SWITZERLAND

Bern
Burri, Peter Hermann *anatomy, histology and embryology educator*

MEDICINE: ANESTHESIOLOGY

UNITED STATES

ALABAMA

Montgomery
Greene, Ernest Rinaldo, Jr. *anesthesiologist, chemical engineer*

CALIFORNIA

Chico
Welter, Lee Orrin *anesthesiologist, biomedical engineer*

Los Angeles
Ashley, Sharon Anita *pediatric anesthesiologist*

Napa
Wycoff, Charles Coleman *retired anesthesiologist*

CONNECTICUT

Hartford
Kang, Juliana Haeng-Cha *anesthesiologist*

Trumbull
Vaidya, Kirit Rameshchandra *anesthesiologist, physician*

FLORIDA

Fort Lauderdale
Bas Csatary, Laszlo Kalman *anesthesiologist, cancer researcher*

Tampa
Varlotta, David *anesthesiologist*

GEORGIA

Atlanta
Waller, John Louis *anesthesiology educator*

Augusta
Merin, Robert Gillespie *anesthesiology educator*

ILLINOIS

Chicago
Feingold, Daniel Leon *physician, consultant*

IOWA

Iowa City
Dexter, Franklin *anesthesiologist*

KANSAS

Kansas City
Mathewson, Hugh Spalding *anesthesiologist, educator*

MARYLAND

Ruxton
Duer, Ellen Ann Dagon *anesthesiologist*

MASSACHUSETTS

Boston
Carr, Daniel Barry *anesthesiologist, endocrinologist, medical researcher*
Vandam, Leroy David *anesthesiologist, educator*

Canton
Yerby, Joel Talbert *anesthesiologist, pain management consultant*

Weymouth
Johnson, Mark David *anesthesiologist, educator*

MICHIGAN

East Lansing
Beckmeyer, Henry Ernest *anesthesiologist, medical educator*

MISSOURI

Saint Louis
Thomas, Lewis Jones, Jr. *anesthesiology educator, biomedical researcher*

NEW YORK

Bronx
Foldes, Francis Ferenc *anesthesiologist*

Brooklyn
Gotta, Alexander Walter *anesthesiologist, educator*

New York
Bendixen, Henrik Holt *physician, educator, dean*

Rochester
Crino, Marjanne Helen *anesthesiologist*

Sands Point
Lear, Erwin *anesthesiologist, educator*

Stony Brook
Poppers, Paul Jules *anesthesiologist, educator*

NORTH CAROLINA

Durham
Bennett, Peter Brian *researcher, anesthesiology educator*
Murray, William James *anesthesiology educator, clinical pharmacologist*

OHIO

Toledo
Barrett, Michael John *anesthesiologist*

PENNSYLVANIA

Hershey
Marshall, Wayne Keith *anesthesiology educator*

Lancaster
Falk, Robert Barclay, Jr. *anesthesiologist, educator*

Norristown
Tornetta, Frank Joseph *anesthesiologist, educator, consultant*

TENNESSEE

Jackson
Bearb, Michael Edwin *anesthesiologist*

TEXAS

Richmond
Hay, Richard Carman *anesthesiologist*

San Antonio
Sloan, Tod Burns *anesthesiologist, researcher*
Smith, Reginald Brian Furness *anesthesiologist, educator*

WASHINGTON

Seattle
Hornbein, Thomas Frederic *anesthesiologist*

TERRITORIES OF THE UNITED STATES

PUERTO RICO

San Juan
De Jesús, Nydia Rosa *physician, anesthesiologist*

DENMARK

Kalundborg
Bitsch-Larsen, Lars Kristian *anaesthesia and intensive care specialist*

GERMANY

Bielefeld
Lauven, Peter Michael *anesthesiologist*

JAPAN

Oita
Honda, Natsuo *anesthesiologist, educator*

PORTUGAL

Lisbon
Soares, Eusebio Lopes *anesthesiologist*

SWEDEN

Lund
Larsson, Anders Lars *anesthesiologist*

Uppsala
Wiklund, K. Lars C. *anesthesiologist, scientist, educator*

ADDRESS UNPUBLISHED

Cork, Randall Charles *anesthesiology educator*

MEDICINE: CARDIOLOGY

UNITED STATES

ALABAMA

Birmingham
Pohost, Gerald M. *cardiologist, medical educator*

Gadsden
Hanson, Ronald Windell *cardiologist*

ARIZONA

Peoria
McKee, Margaret Crile *pulmonary medicine and critical care physician*

CALIFORNIA

Claremont
Johnson, Jerome Linné *cardiologist*

La Mesa
Wohl, Armand Jeffrey *cardiologist*

Los Angeles
Bernstein, Sol *cardiologist, medical services administrator*
Jelliffe, Roger Woodham *cardiologist, clinical pharmacologist*
Kaplan, Samuel *pediatric cardiologist*
Winsor, Travis Walter *cardiologist, educator*

National City
Morgan, Jacob Richard *cardiologist*

Oakland
Killebrew, Ellen Jane (Mrs. Edward S. Graves) *cardiologist*

Rancho Mirage
Bolton, Merle Ray, Jr. *cardiologist*

Sacramento
Sharma, Arjun Dutta *cardiologist*

San Diego
Jamieson, Stuart William *cardiologist, educator*

San Francisco
Bernstein, Harold Seth *pediatric cardiologist, molecular geneticist*
Havel, Richard Joseph *physician, educator*
Mason, Dean Towle *cardiologist*

Sylmar
Levine, Paul Allan *cardiologist*

Torrance
French, William J. *cardiologist, educator*

COLORADO

Denver
Washington, Reginald Louis *pediatric cardiologist*

DISTRICT OF COLUMBIA

Washington
Shine, Kenneth I. *cardiologist, educator*

FLORIDA

Miami
Slonim, Ralph Joseph, Jr. *cardiologist*

Palm Coast
Van Dusen, James *cardiologist*

GEORGIA

Augusta
Pallas, Christopher William *cardiologist*

ILLINOIS

Chicago
Arnsdorf, Morton Frank *cardiologist, educator*
Stamler, Jeremiah *physician, educator*

Glen Ellyn
Agruss, Neil Stuart *cardiologist*

Hazel Crest
Prentice, Robert Craig *cardiologist*

Hines
Zvetina, James Raymond *pulmonary physician*

Rock Island
Forlini, Frank John, Jr. *cardiologist*

INDIANA

Indianapolis
Ross, Edward *cardiologist*

IOWA

Iowa City
Eckstein, John William *physician, educator*

KANSAS

Kansas City
Dunn, Marvin Irvin *physician*

LOUISIANA

New Orleans
Hyman, Albert Lewis *cardiologist*

MARYLAND

Baltimore
Baltazar, Romulo Flores *cardiologist*
Becker, Lewis Charles *cardiology educator*

Silver Spring
Grossberg, David Burton *cardiologist*

MASSACHUSETTS

Boston
Chobanian, Aram Van *physician*
Hilkert, Robert Joseph *cardiologist*
Joyce-Brady, Martin Francis *medical educator, physician, researcher*
Lown, Bernard *cardiologist, educator*
Naimi, Shapur *cardiologist, educator*

MICHIGAN

Ann Arbor
Pitt, Bertram *cardiologist, consultant*

Bloomfield Hills
Cohen, Alberto *cardiologist*

Lansing
Ip, John H. *cardiologist*

MINNESOTA

Minneapolis
Weir, Edward Kenneth *cardiologist*

MISSOURI

Ballwin
López-Candales, Angel *cardiologist, researcher*

NEBRASKA

Omaha
Sketch, Michael Hugh *cardiologist, educator*

NEW JERSEY

West Trenton
Roman, Cecelia Florence *cardiologist*

NEW YORK

Albany
Hoffmeister, Jana Marie *cardiologist*

Bronx
Visco, Ferdinand Joseph *cardiologist, educator*

Brooklyn
Friedman, Howard Samuel *cardiologist, educator*
Lichstein, Edgar *cardiologist*

Buffalo
Naughton, John Patrick *cardiologist, medical school administrator*

Far Rockaway
Neches, Richard Brooks *cardiologist, educator*

Hempstead
Laano, Archie Bienvenido Maaño *cardiologist*

Hewlett
Lowenstein, Alfred Samuel *cardiologist*

Jamaica
Dubroff, Jerome M. *cardiologist*

Manhasset
Nelson, Roy Leslie *cardiac surgeon, researcher, educator*

Monticello
Lauterstein, Joseph *cardiologist*

New York
Cannon, Paul Jude *physician, educator*
Cornell, James S. *critical care and pulmonary physician*
Graf, Jeffrey Howard *cardiologist*
Katz, Jose *cardiologist, theoretical physicist*
Keefe, Deborah Lynn *cardiologist, educator*
Macken, Daniel Loos *cardiologist, educator*
Rosendorff, Clive *cardiologist*
Stutman, Leonard Jay *research scientist, cardiologist*
Zucker, Howard Alan *pediatric cardiologist, intensivist, anesthesiologist*

Valhalla
McClung, John Arthur *cardiologist*

OHIO

Cleveland
Healy, Bernadine P. *physician, educator, federal agency administrator*

Marion
Lim, Shun Ping *cardiologist*

OKLAHOMA

Oklahoma City
Oehlert, William Herbert, Jr. *cardiologist, educator*

Tulsa
Kalbfleisch, John McDowell *cardiologist, educator*
Okada, Robert Dean *cardiologist*

PENNSYLVANIA

Hershey
Davis, Dwight *cardiologist, educator*

Philadelphia
Frankl, William Stewart *cardiologist, educator*
Iskandrian, Ami Simon (Edward) *cardiologist, educator*
Kresh, J. Yasha *cardiovascular researcher, educator*
Nimoityn, Philip *cardiologist*
Walinsky, Paul *cardiology educator*

Pittsburgh
Joyner, Claude Reuben, Jr. *physician, medical educator*
Reichek, Nathaniel *cardiologist*

Somerset
Nair, Velupillai Krishnan *cardiologist*

Thorndale
Hodess, Arthur Bart *cardiologist*

SOUTH CAROLINA

Columbia
Brooker, Jeff Zeigler *cardiologist*

TENNESSEE

Memphis
Kossmann, Charles Edward *cardiologist*
Sullivan, Jay Michael *medical educator*
Watson, Donald Charles *cardiothoracic surgeon, educator*

Nashville
Hondeghem, Luc M. *cardiovascular and pharmacology educator*
Ross, Joseph Comer *physician, educator, academic administrator*

TEXAS

College Station
Rohack, John James *cardiologist*

El Paso
Pearl, William Richard Emden *pediatric cardiologist*

Galveston
James, Thomas Naum *cardiologist, educator*

Houston
Karim, Amin H. *cardiologist*
Kuo, Peter Te *cardiologist*

San Antonio
Schnitzler, Robert Neil *cardiologist*

UTAH

Salt Lake City
Abildskov, J. A. *cardiologist, educator*
Anderson, Jeffrey Lance *cardiologist, educator*

WEST VIRGINIA

Morgantown
Einzig, Stanley *pediatric cardiologist, researcher*

CANADA

ONTARIO

Ottawa
Keon, Wilbert Joseph *cardiologist, surgeon, educator*

AUSTRALIA

Woden
Nikolic, George *cardiologist, consultant*

BELGIUM

Brussels
Pouleur, Hubert Gustave *cardiologist*

Jemeppe-Sur-Sambre
Carlier, Jean Joachim *cardiologist, educator, administrator*

ENGLAND

London
Sapsford, Ralph Neville *cardiothoracic surgeon*

FINLAND

Helsinki
Siltanen, Pentti Kustaa Pietari *cardiologist*

FRANCE

Nice
Nicolay, Jean Honoré *cardiologist*

Nord
Lablanche, Jean-Marc Andre *cardiologist, educator*

GERMANY

Berlin
Eichstädt, Hermann Werner *cardiology educator*

Hannover
Kallfelz, Hans Carlo *cardiologist, pediatrics and pediatric cardiology educator*

Heidelberg
Bode, Christoph Albert-Maria *cardiology educator, researcher*

GREECE

Athens
Gotzoyannis, Stavros Eleutherios *cardiologist*

ICELAND

Reykjavik
Thorgeirsson, Gudmundur *physician, cardiologist*

INDIA

Bombay
Shah, Natverlal Jagjivandas *cardiologist*

INDONESIA

Jakarta
Lembong, Johannes Tarcicius *cardiologist*

ITALY

Milan
Colombo, Antonio *cardiologist*
Pelosi, Giancarlo *cardiologist*

Pisa
Simonetti, Ignazio *cardiologist*

Rome
Aiello, Pietro *cardiologist*

JAPAN

Iyomishima
Honda, Toshio *cardiologist*

Komagane-shi
Endoh, Ryohei *cardiologist*

Suita
Shimazaki, Yasuhisa *cardiac surgeon*

Yokohama
Kaneko, Yoshihiro *cardiologist, researcher*

PANAMA

Panama
Rognoni, Paulina Amelia *cardiologist*

SINGAPORE

Singapore
Mak, Koon Hou *cardiologist, researcher*
Ng, Kheng Siang *cardiologist, educator, researcher*

SPAIN

Madrid
Coma-Canella, Isabel *cardiologist*

SWITZERLAND

Basel
Odavic, Ranko *physician, pharmaceutical company executive*

Geneva
Bloch, Antoine *cardiologist*
Perrenoud, Jean Jacques *cardiologist, educator*

Zurich
von Segesser, Ludwig Karl *cardiovascular surgeon*

VENEZUELA

Cumana
Velasquez Perez, Jose R. *cardiologist*

ADDRESS UNPUBLISHED

Alexander, Jonathan *cardiologist, consultant*
Knapp, William Bernard *cardiologist*
St. Cyr, John Albert, II *cardiovascular and thoracic surgeon*
Terris, Susan *physician, cardiologist*

MEDICINE: DENTISTRY

UNITED STATES

ALABAMA

Birmingham
Birkedal-Hansen, Henning *dentist, educator*

Mobile
Nettles, Joseph Lee *dentist*

CALIFORNIA

Fairfield
Edwards, Richard Charles *oral and maxillofacial surgeon*

Long Beach
Gehring, George Joseph, Jr. *dentist*

San Francisco
Khosla, Ved Mitter *oral and maxillofacial surgeon, educator*

Torrance
Leake, Donald Lewis *oral and maxillofacial surgeon, oboist*

Woodland Hills
Merin, Robert Lynn *periodontist*

CONNECTICUT

Avon
Weiss, Robert Michael *dentist*

Farmington
Coykendall, Alan Littlefield *dentist, educator*
Waknine, Samuel *dental materials scientist, researcher, educator*

New Canaan
Gottlieb, Arnold *dentist*

DISTRICT OF COLUMBIA

Washington
Clarkson, John J. *dentist, dental association administrator*

FLORIDA

Gainesville
Clark, William Burton, IV *dentist, educator*
Widmer, Charles Glenn *dentist, researcher*

Jupiter
Nessmith, H(erbert) Alva *dentist*

Miami
Iver, Robert Drew *dentist*

Naples
Rehak, James Richard *orthodontist*

Pensacola
Noland, Robert Edgar *dentist*

Tampa
Pasetti, Louis Oscar *dentist*
Perret, Gerard Anthony, Jr. *orthodontist*
Wiest, John Andrew *dentist*

GEORGIA

Atlanta
Freedman, Louis Martin *dentist*

ILLINOIS

Chicago
Driskell, Claude Evans *dentist*

Northbrook
Williams, David Allan *dentist, educator*

INDIANA

Carmel
Roche, James Richard *pediatric dentist, university dean*

Gary
Stephens, Paul Alfred *dentist*

Indianapolis
Marlin, Donnell Charles *dental educator*
Roberts, Wilbur Eugene *dental educator, research scientist*

New Albany
Baxter, Joseph Diedrich *dentist*

IOWA

Ankeny
Weigel, Ollie J. *dentist, mayor*

Clinton
Martin, Dennis Charles *dentist*

MARYLAND

Bethesda
Löe, Harald *dentist, educator, researcher*
Rosenberg, Jacob Joseph *orthodontist*

MASSACHUSETTS

Boston
Hein, John William *dentist, educator*
Mandell, Robert Lindsay *periodontist, researcher*
White, George Edward *pedodontist, educator*

Brewster
Gumpright, Herbert Lawrence, Jr. *dentist*

Brockton
Hodge-Spencer, Cheryl Ann *orthodontist*

Quincy
Shalit, Bernard Lawrence *dentist*

Rockport
Gavelis, Jonas Rimvydas *dentist, educator*

MICHIGAN

Ann Arbor
Ash, Major McKinley, Jr. *dentist, educator*
Brooks, Sharon Lynn *dentist, educator*
Craig, Robert George *dental science educator*

Bay City
Pearsall, Harry James *dentist*

MISSISSIPPI

Hattiesburg
Nicholson, James Allen *orthodontist, inventor*
West, Michael Howard *dentist*

NEW JERSEY

Clifton
Swystun-Rives, Bohdana Alexandra *dentist*

Hammonton
Stephanick, Carol Ann *dentist, consultant*

Leonia
Armstrong, Edward Bradford, Jr. *oral and maxillofacial surgeon, educator*

Morristown
Baron, Hazen Jay *dental scientist*

Mountainside
Ricciardi, Antonio *prosthodontist, educator*

Newark
Jandinski, John Joseph *dentist, immunologist*
Kantor, Mel Lewis *dental educator, researcher*

Westfield
Feret, Adam Edward, Jr. *dentist*

NEW YORK

Brooklyn
Schweikert, Edgar Oskar *dentist*

Buffalo
Gogan, Catherine Mary *dental educator*

Flushing
Furst, George *forensic dentist*
Weiss, George Arthur *orthodontist*

Great Neck
Elkowitz, Lloyd Kent *dental anesthesiologist, dentist, pharmacist*

Middletown
Anderman, Irving Ingersoll *dentist*

New York
Beube, Frank Edward *periodontist, educator*
Joskow, Renee W. *dentist, educator*
Scopp, Irwin Walter *periodontist, educator*
Sendax, Victor Irven *dentist, educator, dental implant researcher*

Rochester
Bowen, William Henry *dental researcher, dental educator*
McHugh, William Dennis *dental educator, researcher*

Rockville Centre
Epel, Lidia Marmurek *dentist*

Syracuse
Marshall, David *orthodontist*

Tuckahoe
Reyes-Guerra, Antonio *dental surgeon*

Valley Stream
Tartell, Robert Morris *dentist*

Wappingers Falls
Engelman, Melvin Alkon *retired dentist, business executive, scientist*

Woodbury
Kitzis, Gary David *periodontics educator, periodontist*

NORTH CAROLINA

Charlotte
Owen, Kenneth Dale *orthodontist*

Raleigh
Clifton, Marcella Dawn *dentist*

OHIO

Cleveland
Goodman, Donald Joseph *dentist*
Neuger, Sanford *orthodontics educator*

Columbus
Jolly, Daniel Ehs *dental educator*

Ross
Nelson, Dennis George Anthony *dental researcher, life scientist*

Westchester
Kosti, Carl Michael *dentist, researcher*

OKLAHOMA

Claremore
Kelly, Vincent Michael, Jr. *orthodontist*

Durant
Craige, Danny Dwaine *dentist*

OREGON

Milwaukie
Schafer, Walter Warren *dentist*

PENNSYLVANIA

Danville
Kleponis, Jerome Albert *dentist*

Philadelphia
Breitman, Joseph B. *prosthodontist, dental educator*

Pittsburgh
Sniderman, Marvin *dentist*
Suzuki, Jon Byron *periodontist, educator*

York
Jacobs, Donald Warren *dentist*

RHODE ISLAND

Providence
Mehlman, Edwin Stephen *endodontist*

SOUTH CAROLINA

Charleston
Johnson, Dewey E(dward) *dentist*

Greenville
Mitchell, William Avery, Jr. *orthodontist*

Lake City
TruLuck, James Paul, Jr. *dentist, vintner*

Winnsboro
Williams, Charles Oliver, Jr. *dentist*

TENNESSEE

Knoxville
McGuire, John Albert *dentist*

TEXAS

Dallas
McWhorter, Kathleen *orthodontist*

De Soto
Marsh, Herbert Rhea, Jr. *dentist*

El Paso
Torres, Israel *oral and maxillofacial surgeon*

Houston
Newman, Paul Wayne *dentist, cattle rancher*

Lubbock
Illner-Canizaro, Hana *physician, oral surgeon, researcher*

UTAH

Salt Lake City
Thompson, Elbert Orson *retired dentist, consultant*

VIRGINIA

Great Falls
Dillon, Kathleen Gereaux *dentist*

Pearisburg
Morse, F. D., Jr. *dentist*

Portsmouth
Cox, William Walter *dentist*

Virginia Beach
Lowe, Cameron Anderson *dentist, endodontist, educator*

WASHINGTON

Bellevue
Carlson, Curtis Eugene *orthodontis, periodontist*
Randish, Joan Marie *dentist*

Seattle
Page, Roy Christopher *periodontist, educator*

Spokane
Steadman, Robert Kempton *oral and maxillofacial surgeon*

WISCONSIN

Green Bay
Swetlik, William Philip *orthodontist*

AUSTRALIA

Sydney
Barnard, Peter Deane *dentist*

CHINA

Shanghai
Xue, Miao *dentistry educator, biomaterial scientist*

GERMANY

Stuttgart
Luckenbach, Alexander Heinrich *dentist*

GREECE

Athens
Rapidis, Alexander Demetrius *maxillofacial surgeon*

JAPAN

Chiyoda-ku
Satsumabayashi, Sadayoshi *dental educator*

THE PHILIPPINES

Manila
Lim, Joseph Dy *oral surgeon*

SOUTH AFRICA

Johannesburg
Preston, Charles Brian *orthodontist, school administrator*

SWEDEN

Gothenburg
Bona, Christian Maximilian *dentist, psychotherapist*

TAIWAN

Kaohsiung
Lin, Li-Min *dentistry educator*

ADDRESS UNPUBLISHED

Archibald, David William *virologist, dentist*
Brooke, Ralph Ian *dental educator, vice provost, university dean*
Ligotti, Eugene Ferdinand *retired dentist*
Makins, James Edward *retired dentist, dental educator, educational administrator*
McHugh, Earl Stephen *dentist*
Paris, David Andrew *dentist*
Park, Jon Keith *dentist, educator*
Slaughter, Freeman Cluff *retired dentist*

MEDICINE: DERMATOLOGY

UNITED STATES

ARKANSAS

North Little Rock
Biondo, Raymond Vitus *dermatologist*

CALIFORNIA

Beverly Hills
Klein, Arnold William *dermatologist*

Castro Valley
Charney, Philip *dermatologist*

Corte Madera
Epstein, William Louis *dermatologist, educator*

Los Angeles
Moy, Ronald Leonard *dermatologist, surgeon*

Palo Alto
Farber, Eugene Mark *psoriasis research institute administrator*

San Diego
Gigli, Irma *physician, educator, academic administrator*

San Francisco
Shumate, Charles Albert *retired dermatologist*

Stanford
Bauer, Eugene Andrew *dermatologist, educator*

CONNECTICUT

Stamford
Sheard, Charles, III *dermatologist*

FLORIDA

Miami
Mertz, Patricia Mann *dermatology educator*

Tampa
Shenefelt, Philip David *dermatologist*
Trunnell, Thomas Newton *dermatologist*

GEORGIA

Atlanta
Dobes, William Lamar, Jr. *dermatologist*
Willis, Isaac *dermatologist, educator*

INDIANA

Elkhart
Arlook, Theodore David *dermatologist*

Indianapolis
Norins, Arthur Leonard *physician, educator*

LOUISIANA

New Orleans
Millikan, Larry Edward *dermatologist*

MASSACHUSETTS

Boston
Fitzpatrick, Thomas Bernard *dermatologist, educator*
Parrish, John Albert *dermatologist, research administrator*

MICHIGAN

Detroit
Krull, Edward Alexander *dermatologist*

MINNESOTA

Rochester
Pittelkow, Mark Robert *dermatology educator, researcher*

MISSOURI

Kansas City
Sauer, Gordon Chenoweth *physician, educator*

NEW JERSEY

West Orange
Brodkin, Roger Harrison *dermatologist, educator*

NEW YORK

Brooklyn
Shalita, Alan Remi *dermatologist*

New York
Baer, Rudolf Lewis *dermatologist, educator*
Carter, David Martin *dermatologist*
Lipkin, George *dermatologist, researcher*
Sanchez, Miguel Ramon *dermatologist, educator*
Weinberg, Samuel *pediatric dermatologist*

White Plains
Glassman, George Morton *dermatologist*

NORTH CAROLINA

Winston Salem
Howell, Charles Maitland *dermatologist*

OHIO

Cincinnati
Kitzmiller, Karl William *dermatologist*

TEXAS

Dallas
Menter, M(artin) Alan *dermatologist*

Houston
Rudolph, Andrew Henry *dermatologist, educator*

Irving
Garcia, Raymond Lloyd *dermatologist*

Lubbock
Zemtsov, Alexander *dermatology and biochemistry educator*

VIRGINIA

Norfolk
Pariser, Robert Jay *dermatologist*

TERRITORIES OF THE UNITED STATES

PUERTO RICO

Hato Rey
Rosario-Guardiola, Reinaldo *dermatologist*

AUSTRIA

Vienna
Tappeiner, Gerhard *dermatologist, educator*

FINLAND

Helsinki
Mustakallio, Kimmo Kalervo *dermatologist*

GERMANY

Münster
Voss, Werner *dermatologist*

Tübingen
Schaumburg-Lever, Gundula Maria *dermatologist*

HONG KONG

Hong Kong
Lee, Edward King Pang *dermatologist, public health service officer*

JAPAN

Tokyo
Shimada, Shinji *dermatology educator, researcher*

MEDICINE: EMERGENCY MEDICINE

UNITED STATES

CALIFORNIA

Los Angeles
Caldwell, Allan Blair *emergency health services company executive*

Salinas
Stubblefield, James Irvin *physician, surgeon, army officer*

CONNECTICUT

Hartford
Margulies, Robert Allan *physician, educator, emergency medical service director*

ILLINOIS

Harvey
Heilicser, Bernard Jay *emergency physician*

Springfield
Albright, Deborah Elaine *emergency physician*

KENTUCKY

Louisville
Danzl, Daniel Frank *emergency physician*

MAINE

Skowhegan
Hornstein, Louis Sidney *retired emergency room physician*

MICHIGAN

Three Rivers
Johnson, William Herbert *emergency medicine physician, aerospace physician, retired air force officer*

MINNESOTA

Robbinsdale
Dannewitz, Stephen Richard *emergency physician, consultant, toxicologist*

NEW MEXICO

Placitas
Silk, Marshall Bruce *emergency physician*

NEW YORK

Rochester
Papadakos, Peter John *critical care physician*

TEXAS

Lufkin
Perry, Lewis Charles *emergency medicine physician, osteopath*

WISCONSIN

Marshfield
Stueland, Dean Theodore *emergency physician*

SPAIN

Manresa
Castella, Xavier *critical care physician, researcher*

ADDRESS UNPUBLISHED

Hendricks, Leonard D. *emergency medicine physician, consultant*
Schreiner, Christina Maria *emergency physician*
Woytowitz, Donald Vincent *emergency physician*

MEDICINE: ENDOCRINOLOGY

UNITED STATES

ALABAMA

Birmingham
Pittman, James Allen, Jr. *endocrinologist, dean, educator*

ARIZONA

Mesa
Boren, Kenneth Ray *endocrinologist*

CALIFORNIA

Irvine
Charles, M. Arthur *endocrinologist, educator*

Los Angeles
Solomon, David Harris *physician, educator*

San Francisco
Baxter, John Darling *physician, educator, health facility administrator*

Stanford
Piquette, Gary Norman *reproductive endocrinologist*

CONNECTICUT

Waterbury
Fica, Juan *endocrinologist*

FLORIDA

Miami
Mintz, Daniel Harvey *endocrinologist, educator, academic administrator*

GEORGIA

Atlanta
Beasley, Ernest William, Jr. *endocrinologist*

Augusta
Gambrell, Richard Donald, Jr. *endocrinologist, educator*
Mahesh, Virendra Bhushan *endocrinologist*
Plouffe, Leo, Jr. *endocrinologist*

ILLINOIS

Carbondale
Chandrashekar, Varadaraj *endocrinologist*

Chicago
Landsberg, Lewis *endocrinologist, medical researcher*
Rubenstein, Arthur Harold *physician, educator*

IOWA

Iowa City
Bar, Robert S. *endocrinologist*

LOUISIANA

New Orleans
Re, Richard N. *endocrinologist*
Riddick, Frank Adams, Jr. *physician, health care facility administrator*

MARYLAND

Baltimore
Wilber, John Franklin *endocrinologist, educator*

Bethesda
Nelson, Lawrence Merle *reproductive endocrinologist*
Rodbard, David *endocrinologist, biophysicist*

Silver Spring
Levy, William Joel *endocrinologist*
Sharon, Michael *endocrinologist*

MASSACHUSETTS

Cambridge
Wurtman, Richard Jay *physician, educator*

MICHIGAN

Ann Arbor
Midgley, A(lvin) Rees, Jr. *reproductive endocrinology educator, researcher*
Payne, Anita Hart *reproductive endocrinologist, researcher*

Detroit
Lupulescu, Aurel Peter *medical educator, researcher, physician*

MISSOURI

Saint Louis
Nowak, Felicia Veronika *endocrinologist, molecular biologist, educator*

NEW JERSEY

West Orange
Feigenbaum, Abraham Samuel *nutritional biochemist*

NEW YORK

Buffalo
Dandona, Paresh *endocrinologist*

Flushing
Lorber, Daniel Louis *endocrinologist, educator*

Huntington
Trager, Gary Alan *endocrinologist, diabetologist*

Jamaica
Bradlow, Herbert Leon *endocrinologist, educator*

New Rochelle
Rivlin, Richard Saul *physician, educator*

New York
Ferin, Michel Jacques *reproductive endocrinologist, educator*
Ginsberg-Fellner, Fredda *pediatric endocrinologist, researcher*
Werner, Andrew Joseph *physician, endocrinologist, musicologist*

Woodbury
Bleicher, Sheldon Joseph *endocrinologist, medical educator*

OHIO

Cincinnati
Chin, NeeOo Wong *reproductive endocrinologist*

Columbus
Zipf, William Byron *pediatric endocrinologist, educator*

OKLAHOMA

Oklahoma City
Muchmore, John Stephen *endocrinologist*

RHODE ISLAND

Providence
Ocrant, Ian *pediatric endocrinologist*

SOUTH CAROLINA

Columbia
Lin, Tu *endocrinologist, educator, researcher, academic administrator*

TEXAS

Dallas
Breslau, Neil Art *endocrinologist, researcher, educator*
Marks, James Frederic *pediatric endocrinologist, educator*
Prihoda, James Sheldon *endocrinologist*

UTAH

Salt Lake City
Odell, William Douglas *physician, scientist, educator*

VIRGINIA

Charlottesville
Hartman, Mark Leopold *internist, endocrinologist*

WASHINGTON

Seattle
Porte, Daniel, Jr. *physician, educator, health facility administrator*

WEST VIRGINIA

Huntington
Driscoll, Henry Keane *endocrinologist, researcher*

WISCONSIN

La Crosse
Silva, Paul Douglas *reproductive endocrinologist*

CANADA

ONTARIO

Ottawa
Friesen, Henry George *endocrinologist, educator*

Toronto
Volpé, Robert *endocrinologist*

FRANCE

Bicêtre
Baulieu, Etienne-Emile *endocrinologist*

Marseilles
Vague, Jean Marie *endocrinologist*

GERMANY

Munich
Weisweiler, Peter *physician, educator*

GREECE

Athens
Koutras, Demetrios A. *physician, endocrinology investigator, educator*
Singhellakis, Panagiotis Nicolaos *endocrinologist, educator*

ITALY

Florence
Brandi, Maria Luisa *endocrinologist, educator*

JAPAN

Suita
Amino, Nobuyuki *endocrinologist, educator*

SPAIN

Madrid
Leon-Sanz, Miguel *physician, endocrinologist, nutritionist*

SWEDEN

Stockholm
Lindqvist, Jens Harry *physician*

THAILAND

Bangkok
Himathongkam, Thep *endocrinologist*

ADDRESS UNPUBLISHED

Guéritée, Nicolas *endocrinologist*

MEDICINE: EPIDEMIOLOGY & PREVENTIVE MEDICINE

UNITED STATES

ALABAMA

Birmingham
Kahlon, Jasbir Brar *viral epidemiologist, researcher*

CALIFORNIA

Los Angeles
Detels, Roger *epidemiologist, physician, former university dean*
Henderson, Brian Edmond *physician, educator*
Moss, Charles Norman *physician*

Novato
Reed, Dwayne Milton *medical epidemiologist, educator*

Oakland
Friedman, Gary David *epidemiologist, health facility administrator*

San Diego
Lockwood Hourani, Laurel Lee *epidemiologist*
Reed, Sharon Lee *infectious disease consultant*

San Francisco
Schmidt, Robert Milton *physician, scientist, educator*

CONNECTICUT

Hartford
Dembek, Zygmunt Francis *epidemiologist*

New Haven
Ryder, Robert Winsor *medical epidemiologist*
Shope, Robert Ellis *epidemiology educator*

DISTRICT OF COLUMBIA

Washington
Carlo, George Louis *epidemiologist*
Laukaran, Virginia Hight *epidemiologist*
Schwartz, Joel *epidemiologist*

GEORGIA

Atlanta
Schultz, Linda Jane *epidemiologist*
Tyler, Carl Walter, Jr. *physician, health research administrator*

ILLINOIS

Chicago
Wolinsky, Steven Mark *infectious diseases physician, educator*

KENTUCKY

Louisville
Hutchison, George Barkley *epidemiologist, educator*

MAINE

Augusta
Gensheimer, Kathleen Friend *epidemiologist*

Union
Buchan, Ronald Forbes *preventive medicine physician*

MARYLAND

Baltimore
Silbergeld, Ellen Kovner *environmental epidemiologist and toxicologist*

Bethesda
Greenwald, Peter *physician, government medical research director*
Lipton, James Abbott *epidemiologist, researcher*
Miller, Barry Alan *epidemiologist, cancer researcher*

Rockville
Seltser, Raymond *epidemiologist, educator*

MASSACHUSETTS

Boston
Tucker, Katherine Louise *nutritional epidemiologist, educator*

Cambridge
Wilson, Mary Elizabeth *physician*

Framingham
Castelli, William *cardiovascular epidemiologist, educator*

MICHIGAN

Ann Arbor
Schottenfeld, David *epidemiologist, educator*

Detroit
Saravolatz, Louis Donald *epidemiologist, educator*

MINNESOTA

Minneapolis
Luepker, Russell Vincent *epidemiology educator*

MISSOURI

Saint Louis
Cottler, Linda Bauer *epidemiologist*
Rice, Treva Kay *genetic epidemiologist*

NEW JERSEY

Hackensack
Gross, Peter Alan *epidemiologist, researcher*

NEW YORK

Buffalo
Freudenheim, Jo L. *social and preventive medicine educator*
Trevisan, Maurizio *epidemiologist, researcher*

Cooperstown
Pearson, Thomas Arthur *epidemiologist, educator*

New York
Harlap, Susan *epidemiologist, educator*
Stellman, Steven Dale *epidemiologist*
Taylor, Patricia Elsie *epidemiologist*
Whelan, Elizabeth Ann Murphy *epidemiologist*

Stony Brook
Hyman, Leslie Gaye *epidemiologist*
Schoenfeld, Elinor Randi *epidemiologist*

NORTH CAROLINA

Cary
Amtoft-Nielsen, Joan Theresa *physician, educator, researcher*

Chapel Hill
Popkin, Carol Lederhaus *epidemiologist*

OHIO

Athens
Hedges, Richard H. *epidemiologist*

Cincinnati
Buncher, Charles Ralph *epidemiologist, educator*

Columbus
Fass, Robert J. *epidemiologist, academic administrator*

OKLAHOMA

Oklahoma City
Gavaler, Judith Ann Stohr Van Thiel *epidemiologist*

SOUTH CAROLINA

Charleston
Schuman, Stanley *epidemiologist, educator*

TEXAS

Houston
Page, Valda Denise *epidemologist, researcher, nutritionist*
Schachtel, Barbara Harriet Levin *epidemiologist, educator*

San Antonio
Patterson, Jan Evans *epidemiologist, educator*

VIRGINIA

Arlington
Staffa, Judy Anne *epidemiologist*

Fairfax
Payne, Fred J. *physician, educator*

WASHINGTON

Seattle
Kukull, Walter Anthony *epidemiologist, educator*

WISCONSIN

La Crosse
Lindesmith, Larry Alan *physician, administrator*

Milwaukee
Morris, Robert DuBois *epidemiologist*

CANADA

QUEBEC

Ville de Laval
Siemiatycki, Jack *epidemiologist, biostatistician, educator*

SWEDEN

Malmö
Cronberg, Stig *infectious diseases educator*

Nässjö
Schultz, Per-Olov *occupational health physician, consultant*

SWITZERLAND

Geneva
Piot, Peter *medical microbiologist, epidemiologist*

MEDICINE: FAMILY & GENERAL PRACTICE

UNITED STATES

ALABAMA

Birmingham
Allen, James Madison *family practice physician, consultant*
Bueschen, Anton Joslyn *physician, educator*
Strickler, Howard Martin *physician*

Tuscaloosa
Jones, Jerry Edward *family physician, educator*
Moody, Maxwell, Jr. *retired physician*

ALASKA

Juneau
Mala, Theodore Anthony *physician, state official*

ARIZONA

Paradise Valley
Butler, Byron Clinton *physician, cosmologist, gemologist, scientist*

Scottsdale
DeHaven, Kenneth Le Moyne *retired physician*

Sun City West
Calderwood, William Arthur *physician*

Tempe
Anand, Suresh Chandra *physician*

ARKANSAS

De Valls Bluff
Jones, Robert Eugene *physician*

CALIFORNIA

Berkeley
Kubler-Ross, Elisabeth *physician*
Sheen, Portia Yunn-ling *retired physician*

Carmichael
Wagner, Carruth John *physician*

Covina
Schneider, Calvin *physician*

Downey
Redeker, Allan Grant *physician, medical educator*

La Canada Flintridge
Byrne, George Melvin *physician*

Los Angeles
Rimoin, David Lawrence *physician, geneticist*
Shacks, Samuel James *physician*
Shlian, Deborah Matchar *physician*

Orange
Appelbaum, Bruce David *physician*

Palo Alto
Lane, William Kenneth *physician*
Zunich, Kathryn Margaret *physician*

San Diego
Moser, Kenneth Miles *physician*
Oliphant, Charles Romig *physician*

San Francisco
Gottfried, Eugene Leslie *physician, educator*

Santa Ana Heights
George, Kattunilathu Oommen *homoeopathic physician, educator*

Stanford
Small, Peter McMichael *physician, researcher*

Sunnyvale
Altamura, Michael Victor *physician*

Whittier
Prickett, David Clinton *physician*

COLORADO

Denver
Rumack, Barry H. *physician, toxicologist, pediatrician*

CONNECTICUT

Farmington
Rothfield, Naomi Fox *physician*

DISTRICT OF COLUMBIA

Washington
Cowdry, Rex William *physician, researcher*
Gay-Bryant, Claudine Moss *physician*
Peterson, William Frank *physician*
Short, Elizabeth M. *physician, educator, federal agency administrator*

FLORIDA

Boca Raton
Gagliardi, Raymond Alfred *physician*

Clearwater
Loewenstein, George Wolfgang *retired physician, UN consultant*

Jacksonville
Halley, James Alfred *physician*

Largo
Brown, Warren Joseph *physician*

Melbourne
Baney, Richard Neil *physician, internist*
Broussard, William Joseph *physician, cattleman, environmentalist*

Miami
Ryan, James Walter *physician, medical researcher*
Sichewski, Vernon Roger *physician*

Orlando
Touffaire, Pierre Julien *physician*

Saint Petersburg
Willey, Edward Norburn *physician*

Tampa
Calderon, Eduardo *general practice physician, researcher*

HAWAII

Honolulu
Chock, Clifford Yet-Chong *family practice physician*

Koloa
Donohugh, Donald Lee *physician*

Wailuku
Savona, Michael Richard *physician*

IDAHO

Priest River
Freibott, George August *physician, chemist, priest*

ILLINOIS

Chicago
Lopez, Carolyn Catherine *physician*
Patterson, Roy *physician, educator*

Lake Bluff
Kelly, Daniel John *physician*

Lena
Vickery, Eugene Livingstone *retired physician, writer*

McHenry
Sturm, Richard E. *occupational physician*

North Chicago
Sapienza, Anthony Rosario *physician, educator, dean ambulatory facilities*
Weil, Max Harry *physician, medical educator, medical scientist*

Wheaton
Bogdonoff, Maurice Lambert *physician*

INDIANA

Avilla
Sneary, Max Eugene *physician*

Indianapolis
Hathaway, David Roger *physician, medical educator*

Marion
Green, Robert Frederick *physician, photographer*

Rockville
Swaim, John Franklin *physician, health care executive*

IOWA

Iowa City
Abboud, Francois Mitry *physician, educator*
Bertolatus, John Andrew *physician, educator*

KANSAS

Kansas City
Arakawa, Kasumi *physician, educator*

Paola
Banks, Robert Earl *family practice physician*

Shawnee Mission
Dockhorn, Robert John *physician*

Wichita
Haynes, Deborah Gene *physician*

KENTUCKY

Elkton
Manthey, Frank Anthony *physician, director*

Louisville
Uhde, George Irvin *physician*

Paducah
Bassi, Joseph Arthur *physician*

LOUISIANA

Baton Rouge
McLaury, Ralph Leon *physician*

Eunice
Landreneau, Rodney Edmund, Jr. *physician*

New Orleans
Berenson, Gerald Sanders *physician*

Pineville
Swearingen, David Clarke *general practice physician*

MARYLAND

Baltimore
De Hoff, John Burling *physician, consultant*
Friedman, Marion *internist, family physician, medical administrator*
Wagner, Henry Nicholas, Jr. *physician*

Bethesda
Danforth, David Newton, Jr. *physician, scientist*
Kirschstein, Ruth Lillian *physician*
Lenfant, Claude Jean-Marie *physician*
Maher, John Francis *physician, educator*
Neva, Franklin Allen *physician, educator*
Plaut, Marshall *physician, medical administrator, educator*
Saville, Michael Wayne *physician, scientist*
Yanovski, Susan Zelitch *physician, eating disorders specialist*

Gaithersburg
Top, Franklin Henry, Jr. *physician, researcher*

Greenbelt
Chan, Clara Suet-Phang *physician*

Rockville
Nora, James Jackson *physician, author, educator*

MASSACHUSETTS

Amherst
Ralph, James R. *physician*

Boston
Adelstein, S(tanley) James *physician, educator*
Kaye, Kenneth Marc *physician, educator*
Leaf, Alexander *physician, educator*
Potts, John Thomas, Jr. *physician, educator*
Rice, Peter Alan *physician, scientist*
Schlossman, Stuart Franklin *physician, educator, researcher*
Warth, James Arthur *physician, researcher*

Cambridge
Brusch, John Lynch *physician*

Jamaica Plain
Arbeit, Robert David *physician*

Natick
Jones, Bruce Hovey *physician, researcher*

Needham
Weller, Thomas Huckle *physician, emeritus educator*

Norwood
Seder, Richard Henry *physician*

Oxford
Schur, Walter Robert *physician*

MICHIGAN

Adrian
Haddad, Inad *physician*

Ann Arbor
Julius, Stevo *physician, educator, physiologist*
Weg, John Gerard *physician*

Birmingham
Miley, Hugh Howard *retired physician*

Okemos
Gillespie, Gary Don *physician*

West Bloomfield
Sarwer-Foner, Gerald Jacob *physician, educator*

MINNESOTA

Minneapolis
Peterson, Douglas Arthur *physician*

Saint Paul
Heuer, Marvin Arthur *physician, science foundation executive*

MISSISSIPPI

Jackson
Kliesch, William Frank *physician*

Meridian
Mutziger, John Charles *physician*

MISSOURI

Columbia
Allen, William Cecil *physician, educator*
Barbero, Giulio John *physician, educator*
Stonnington, Henry Herbert *physician, medical executive, educator*

NEBRASKA

Elkhorn
Graves, Harris Breiner *physician, hospital administrator*

Omaha
Bouda, David William *insurance medical officer*

NEVADA

Elko
Moren, Leslie Arthur *physician*

Las Vegas
Hattem, Albert Worth *physician*

NEW JERSEY

Allenhurst
Calabro, Joseph John, III *physician*

New Brunswick
Zawadsky, Joseph Peter *physician*

Passaic
Baum, Howard Barry *physician*

Plainsboro
Royds, Robert Bruce *physician*

Princeton
Kane, Michael Joel *physician*

Rumson
Pflum, William John *physician*

South Plainfield
Friedman, Mark *physician, consultant*

NEW MEXICO

Santa Fe
Schwartz, George R. *physician*

NEW YORK

Albany
Robinson, David Ashley *family physician*

Amityville
Upadhyay, Yogendra Nath *physician, educator*

Bronx
Goldberg, Myron Allen *physician, psychiatrist*
Shafritz, David Andrew *physician, research scientist*
Stein, Ruth Elizabeth Klein *physician*

Brooklyn
Benes, Solomon *biomedical scientist, physician*
Namba, Tatsuji *physician, researcher*
Ravitz, Leonard J., Jr. *physician, scientist, consultant*
Viswanathan, Ramaswamy *physician, educator*

Castle Point
Mehta, Rakesh Kumar *physician, consultant*

Chestnut Ridge
Day, Stacey Biswas *physician, educator*

East Meadow
Kurian, Pius *physician*

Edmeston
Price, James Melford *physician*

Flushing
Weiss, Joseph *physician*

New Hyde Park
Prisco, Douglas Louis *physician*

New York
Asanuma, Hiroshi *physician, educator*
Berns, Kenneth Ira *physician*
Bertino, Joseph Rocco *physician, educator*
Friedman, Robert Jay *physician*
Gertler, Menard M. *physician, educator*
Gilbert, Richard Michael *physician, educator*
Gotschlich, Emil Claus *physician, educator*
Hirsch, Jules *physician, biochemistry educator*
Michnovicz, Jon Joseph *physician, research endocrinologist*
Morishima, Akira *physician, director, educator, consultant*
Overweg, Norbert Ido Albert *physician*
Spielberger, Lawrence *physician, educator*
Stoopler, Mark Benjamin *physician*
Zinberg, Stanley *physician, educator*

Poughkeepsie
Golden, Reynold Stephen *family practice physician, educator*

Stony Brook
Kaplan, Allen P. *physician, educator, academic administrator*

Upton
Hamilton, Leonard Derwent *physician, molecular biologist*

Valhalla
Levere, Richard David *physician, academic administrator, educator*

West Haverstraw
Cochran, George Van Brunt *physician, surgery educator, researcher*

White Plains
Samii, Abdol Hossein *physician, educator*

NORTH CAROLINA

Durham
Cohen, Harvey Jay *physician, educator*
Littman, Susan Joy *physician*
Miller, David Edmond *physician*

Greenville
Metzger, Walter James, Jr. *physician, educator*

Winston Salem
Bowman, Marjorie Ann *physician, academic administrator*

OHIO

Ada
Elliott, Robert Betzel *physician*

Canal Winchester
Burrier, Gail Warren *physician*

Cleveland
Robbins, Frederick Chapman *physician, medical school dean emeritus*

Columbus
Stephens, Sheryl Lynne *family practice physician*

Springfield
Orhon, Necdet Kadri *physician*

OKLAHOMA

Lawton
Webb, O(rville) Lynn *physician, pharmacologist, educator*

Muskogee
Kent, Bartis Milton *physician*

Oklahoma City
Watson, Steven Edward *family physician*

OREGON

Portland
Swank, Roy Laver *physician, educator, inventor*

PENNSYLVANIA

Berwyn
Dickson, Brian *physician, researcher, educator*

Coatesville
Gehring, David Austin *physician, adminstrator, cardiologist*

Erie
Clark, Gordon Hostetter, Jr. *physician*
Kalkhof, Thomas Corrigan *physician*

Gibsonia
Cauna, Nikolajs *physician, medical educator*

Philadelphia
Holmes, Edward W. *physician, educator*
Kissick, William Lee *physician, educator*
Marino, Paul Lawrence *physician, researcher*
Mastroianni, Luigi, Jr. *physician, educator*
Owens, Gary Mitchell *physician, educator*
Yun, Daniel Duwhan *physician, foundation administrator*

Reading
Lusch, Charles Jack *physician*

Wilkes Barre
Hernandez, Wilbert Eduardo *physician*

RHODE ISLAND

North Providence
Stankiewicz, Andrzej Jerzy *physician, biochemistry educator*

Providence
Lonks, John Richard *physician*

SOUTH CAROLINA

North Augusta
McRee, John Browning, Jr. *physician*

Spartanburg
Guthrie, John Robert *physician, health science facility administrator*

TENNESSEE

Memphis
Godsey, William Cole *physician*
Nienhuis, Arthur Wesley *physician, researcher*

Nashville
Quinn, Robert William *physician, educator*

TEXAS

Austin
Elequin, Cleto, Jr. *retired physician*

Camp Wood
Triplett, William Carryl *physician, researcher*

Dallas
Berbary, Maurice Shehadeh *physician, military officer, hospital administrator, educator*
Frenkel, Eugene Phillip *physician*
Wheeler, Clarence Joseph, Jr. *physician*

Fort Worth
Hines, Dwight Allen, II *family practice physician*

Houston
Aslam, Muhammed Javed *physician*
Davis, Barry Robert *biometry educator, physician*
Kerwin, Joseph Peter *physician, former astronaut*
Wonnacott, James Brian *physician*

Humble
Trowbridge, John Parks *physician, nutritional medicine specialist, joint treatment specialist*

Lubbock
Winter, Mark Lee *physician*

Temple
Brasher, George Walter *physician*

UTAH

Salt Lake City
Moser, Royce, Jr. *physician, medical educator*

VERMONT

Stowe
Eisenberg, Howard Edward *physician, psychotherapist, educator, consultant*

VIRGINIA

Mechanicsville
McCahill, Thomas Day *physician*

WASHINGTON

Bellevue
Hackett, Carol Ann Hedden *physician*

Olympia
Flemming, Stanley Lalit Kumar *family practice physician, state legislator*

Seattle
Butler, John Ben, III *physician, computer specialist*
Robertson, William Osborne *physician*
Thomas, E(dward) Donnall *physician, researcher*

WEST VIRGINIA

Man
Bofill, Rano Solidum *physician*

WISCONSIN

La Crosse
Smith, Martin Jay *physician, biomedical research scientist*

Milwaukee
Atlee, John Light, III *physician*
Effros, Richard Matthew *medical educator, physician*

Neenah
Zimmerman, Delano Elmer *physician*

WYOMING

Laramie
Cronkleton, Thomas Eugene *physician*

TERRITORIES OF THE UNITED STATES

PUERTO RICO

Santurce
Fernandez-Martinez, Jose *physician*

CANADA

BRITISH COLUMBIA

Vancouver
Chow, Anthony Wei-Chik *physician*

MANITOBA

Winnipeg
Angel, Aubie *physician, academic administrator*

ONTARIO

Hamilton
Bienenstock, John *physician, educator*

QUEBEC

Montreal
Chretien, Michel *physician, educator, administrator*
Milic-Emili, Joseph *physician, educator*

AUSTRALIA

Fremantle
Flacks, Louis Michael *consulting physician*

West Perth
Brine, John Alfred Seymour *physician, consultant*

Woden
Sinnett, Peter Frank *physician, geriatrics educator*

BELGIUM

Aalst
De Loof, Jef Emiel Elodie *general physician*

Brussels
Mosselmans, Jean-Marc *physician*

Ghent
Colardyn, Francis Achille *physician*

BRAZIL

Rio de Janeiro
Leite, Carlos Alberto *physician*

DENMARK

Charlottenlund
Langer, Jerk Wang *physician, medical journalist*

Copenhagen
Andersen, Leif Percival *physician*
Langer, Seppo Wang *physician, registrar, medical journalist*

Nakskov
Christensen, Ole *general practice physician*

DOMINICA

Portsmouth
Cooles, Philip Edward *physician*

ENGLAND

Blackpool
Avasthi, Ram Bandhu *physician, general practitioner*

London
Asfoury, Zakaria Mohammed *physician*
Rubens, Robert David *physician, educator*

Manchester
Wilkinson, Peter Maurice *physician consultant, pharmacologist*

Newcastle upon Tyne
Gibson, Gerald John *physician*

FRANCE

Angers
Monroche, André Victor Jacques *physician*

Eure et Loir
Adjamah, Kokouvi Michel *physician*

Limoges
Moreau, Hugues Andre *physician*

Marcq en Baroeul
Choain, Jean Georges *physician, acupuncturist*

Nanterre
Nguyen-Trong, Hoang *physician, consultant*

Nyons
Bottero, Philippe Bernard *general practitioner*

Paris
Lefebure, Alain Paul *family physician*

Rognac
Castel, Gérard Joseph *physician*

GERMANY

Berlin
Witzel, Lothar Gustav *physician, gastroenterologist*

Essen
Erbel, Raimund *physician, educator*

Fulda
Stegmann, Thomas Joseph *physician*

Hamburg
Comberg, Hans-Ulrich *physician*

Heppenheim
Singer, Peter *physician, researcher, consultant*

Mainz
Schmitt, Heinz-Josef *physician*

Rostock
Loehr, Johannes-Matthias *physician*

HONG KONG

Kowloon
Chow, Stephen Heung Wing *physician*

JAPAN

Fukuoka
Shirai, Takeshi *physician*

Kurashiki
Itami, Jinroh *physician*

Osaka
Horiuchi, Atsushi *physician, educator*
Umeki, Shigenobu *physician, researcher*

Sendai
Okuyama, Shinichi *physician*

Suita
Iwatani, Yoshinori *physician, educator, researcher*

NORWAY

Oslo
Waaler, Bjarne Arentz *physician, educator*

POLAND

Katowice
Kokot, Franciszek Józef *physician*

Poznan
Szulc, Roman Władysław *physician*

RWANDA

Kigali
Fox, Emile *physician*

SPAIN

Barcelona
Gatell, Jose Maria *physician*

SWEDEN

Lund
Abdulla, Mohamed *physician, educator*

Stockholm
Lundquist, Per Birger *physician, medical illustrator, artist, surgeon*

Uppsala
Öberg, Kjell Erik *physician, educator, researcher*

THAILAND

Bangkok
Petchclai, Bencha *physician, researcher, inventor*

WALES

Aberystwyth Dyfe
Roberts, William James Cynfab *physician*

ADDRESS UNPUBLISHED

Alter, Blanche Pearl *physician, educator*
Becker, Bruce Carl, II *physician, educator*
de Séguin des Hons, Luc Donald *physician, medical biologist*
Dryden, Richie Sloan *physician*
Glassman, Armand Barry *physician, pathologist, scientist, educator, administrator*
Griffitts, James John *physician*
Hauben, Manfred *physician*
Hochenegg, Leonhard *physician*
Kruus, Harri Kullervo *physician*
Logan, Bruce David *physician*
MacDuffee, Robert Colton *family physician, pathologist*
Materson, Richard Stephen *physician, educator*
Schneider, Eleonora Frey *physician*
Vernon, Sidney *physician*
Wilmore, Douglas Wayne *physician, surgeon*

MEDICINE: GASTROENTEROLOGY

UNITED STATES

ALABAMA

Huntsville
Tietke, Wilhelm *gastroenterologist*

CALIFORNIA

Burlingame
Bender, Michael David *gastroenterologist*

Martinez
Geokas, Michael C. *gastroenterologist*

COLORADO

Aurora
Lewey, Scot Michael *gastroenterologist*

CONNECTICUT

Waterbury
Garsten, Joel Jay *gastroenterologist*

ILLINOIS

Chicago
Jilhewar, Ashok *gastroenterologist*
Ostrow, Jay Donald *gastroenterology educator, researcher*

MARYLAND

Baltimore
Wilson, Donald Edward *physician, educator*

MASSACHUSETTS

Boston
Belkind-Gerson, Jaime *gastroenterologist, nutritionist, researcher*
Carr-Locke, David Leslie *gastroenterologist*
Cave, David Ralph *gastroenterologist, educator*
Friedman, Lawrence Samuel *gastroenterologist, educator*

Haverhill
Niccolini, Drew George *gastroenterologist*

Springfield
Farkas, Paul Stephen *gastroenterologist*

MINNESOTA

Rochester
LaRusso, Nicholas F. *gastroenterologist, educator*

NEW HAMPSHIRE

Lebanon
Kelley, Maurice Leslie, Jr. *gastroenterologist, educator*

NEW YORK

Bronx
Bloom, Alan Arthur *gastroenterologist, educator*

Brooklyn
El Kodsi, Baroukh *gastroenterologist, educator*
Guy, Matthew Joel *gastroenterologist*
Rabinowitz, Simon S. *physician, scientist, pediatric gastroenterologist*

Manhasset
Mullin, Gerard Emmanuel *physician, educator, researcher*

Mount Kisco
Gutstein, Sidney *gastroenterologist*

New York
Dieterich, Douglas Thomas *gastroenterologist, researcher*
Lieberman, Harvey Michael *hepatologist, gastroenterologist, educator*
Miskovitz, Paul Frederick *gastroenterologist*
Sachar, David Bernard *gastroenterologist, medical educator*

NORTH CAROLINA

Chapel Hill
Ulshen, Martin Howard *pediatric gastroenterologist, researcher*

OHIO

Cleveland
Holzbach, Raymond Thomas *gastroenterologist, author, educator*

PENNSYLVANIA

Bala Cynwyd
Katz, Julian *gastroenterologist, educator*

Carlisle
Lewis, Gregory Lee *gastroenterologist*

Philadelphia
Frucht, Harold *physician*

TEXAS

Dallas
Fordtran, John Satterfield *physician*

Houston
McKechnie, John Charles *gastroenterologist, educator*

UTAH

Salt Lake City
Bjorkman, David Jess *gastroenterologist, educator*

GERMANY

Leipzig
Mössner, Joachim *internist, gastroenterologist*

SWITZERLAND

Lausanne
Borel, Georges Antoine *gastroenterologist, consultant*

ADDRESS UNPUBLISHED

Carey, Martin Conrad *gastroenterologist, molecular biophysicist, educator*
Rudert, Cynthia Sue *gastroenterologist*

MEDICINE: HEMATOLOGY

UNITED STATES

ALABAMA

Birmingham
Barton, James Clyde, Jr. *hematologist, medical oncologist*

ARIZONA

Tucson
Katakkar, Suresh Balaji *hematologist, oncologist*

ARKANSAS

Little Rock
Hauser, Simon Petrus *hematologist*

CALIFORNIA

Duarte
Zaia, John Anthony *hematologist*

Los Angeles
Garratty, George *immunohematologist*
Johnson, Cage Saul *hematologist, educator*
Rosenblatt, Joseph David *hematologist, oncologist*

San Francisco
Kan, Yuet Wai *physician, investigator*

CONNECTICUT

Fairfield
Burd, Robert Meyer *hematologist, educator*

DISTRICT OF COLUMBIA

Washington
Reaman, Gregory Harold *pediatric hematologist, oncologist*

FLORIDA

Jacksonville
Colon-Otero, Gerardo *hematologist, oncologist*

GEORGIA

Atlanta
Lollar, John Sherman, III (Pete Lollar) *hematologist*

Decatur
Kann, Herbert Ellis, Jr. *hematologist, oncologist*

ILLINOIS

Hines
Kanofsky, Jeffrey Ronald *physician, educator*

LOUISIANA

New Orleans
Veith, Robert Woody *hematologist*

MARYLAND

Bethesda
Hoak, John Charles *physician, educator*

Rockville
Monaghan, W(illiam) Patrick *immunohematologist, retired naval officer, health educator, consultant*

MASSACHUSETTS

Boston
Bern, Murray Morris *hematologist, oncologist*
Grous, John Joseph *hematologist, oncologist, educator*
Wright, Daniel Godwin *academic physician*

Burlington
Rabinowitz, Arthur Philip *hematologist*

Watertown
Kim, Byung Kyu *hematologist, consultant*

Worcester
Michelson, Alan David *pediatric hematologist*

MICHIGAN

Royal Oak
Wilner, Freeman Marvin *hematologist, oncologist*

MINNESOTA

Minneapolis
Miller, Jeffrey Steven *hematologist, researcher*

MISSOURI

Saint Louis
Kornfeld, Stuart A. *hematology educator*

NEW JERSEY

Piscataway
Nilsson, Bo Ingvar *hematologist*

NEW YORK

Brooklyn
Feldman, Felix *pediatric hematologist, oncologist*

Liverpool
DiFino, Santo Michael *hematologist*

New York
Adamson, John William *hematologist*
Straus, David Jeremy *hematologist, educator*
Weksler, Babette Barbash *hematologist*

NORTH CAROLINA

Chapel Hill
Roberts, Harold Ross *medical educator, hematologist*

OKLAHOMA

Oklahoma City
Hampton, James Wilburn *hematologist, medical oncologist*

PENNSYLVANIA

Philadelphia
Beacham, Dorothy Ann *medical research scientist*
Shapiro, Sandor Solomon *hematologist*

TEXAS

San Antonio
Williams, Thomas Eugene *pediatric hematologist and oncologist*

VIRGINIA

Richmond
Goldman, Israel David *hematologist, oncologist*

WISCONSIN

Milwaukee
Libnoch, Joseph Anthony *physician, educator*

CANADA

BRITISH COLUMBIA

Vancouver
Eaves, Allen Charles Edward *hematologist, medical agency administrator*

FRANCE

Paris
Wautier, Jean Luc *hematologist*

Rennes
Genetet, Bernard *hematologist, immunologist, educator*

JAPAN

Tokyo
Ozawa, Keiya *hematologist, researcher*

THE NETHERLANDS

Groningen
van der Meer, Jan *hematologist*

SWEDEN

Danderyd
Hast, Robert *hematologist*

Stockholm
Reizenstein, Peter Georg *hematologist*

SWITZERLAND

Le Mont
Bachmann, Fedor Wolfgang *hematology educator, laboratory director*

THAILAND

Bangkok
Issaragrisil, Surapol *hematologist*

ADDRESS UNPUBLISHED

Balaban, Edward Paul *physician*

MEDICINE: INTERNAL MEDICINE

UNITED STATES

ALABAMA

Birmingham
Hirschowitz, Basil Isaac *physician*
Pittman, Constance Shen *physician, educator*

Mobile
Esham, Richard Henry *internal medicine and geriatrics educator*

ARIZONA

Sun City
Cannady, Edward Wyatt, Jr. *retired physician*

Tempe
Oscherwitz, Steven Lee *internist, infectious disease physician*

Tucson
Burrows, Benjamin *physician, educator*
Gall, Eric Papineau *physician educator*

CALIFORNIA

Barstow
Sutterby, Larry Quentin *internist*

Belvedere Tiburon
Hinshaw, Horton Corwin *physician*

Carmel
Doe, Richard Philip *physician, educator*

Fresno
Chandler, Bruce Frederick *internist*

La Jolla
Hostetler, Karl Yoder *internist, endocrinologist, educator*

Los Angeles
Maronde, Robert Francis *internist, clinical pharmacologist, educator*

Oakland
Reitz, Richard Elmer *physician*

Orange
Hodges, Robert Edgar *physician, educator*

Pacific Palisades
Claes, Daniel John *physician*

Piedmont
Collen, Morris Frank *physician*

San Diego
Ross, John, Jr. *physician, educator*
Salk, Jonas Edward *physician, scientist*

San Francisco
Peterlin, Boris Matija *physician*

Stanford
Melmon, Kenneth Lloyd *physician, biologist, pharmacologist, consultant*

COLORADO

Denver
Aikawa, Jerry Kazuo *physician, educator*
Sussman, Karl Edgar *physician*

CONNECTICUT

Bridgeport
Skowron, Tadeusz Adam *physician*

Farmington
Jones, Thomas Gordon *internist*

New Haven
Elias, Jack Angel *physician, educator*
Feinstein, Alvan Richard *physician*

Ridgefield
Sobol, Bruce J. *internist, educator*

Torrington
Artushenia, Marilyn Joanne *internist, educator*

DISTRICT OF COLUMBIA

Washington
Arling, Bryan Jeremy *internist*
Grossman, Nathan *physician*
Hodes, Richard Michael *internist, educator*
LaRosa, John Charles *internist, educator, researcher*
Mann, Oscar *physician, internist, educator*
Oler, Wesley Marion, III *physician, educator*
Simon, Gary Leonard *internist, educator*
Thonnard, Ernst *internist, researcher*

FLORIDA

Clearwater
Rinde, John Jacques *internist*

Fort Lauderdale
Morris, James Bruce *internist*
Swiller, Randolph Jacob *internist*

Gainesville
Anderson, Richard McLemore *internist*
Cluff, Leighton Eggertsen *physician*

Miami
Getz, Morton Ernest *internist, gastroenterologist*

Tampa
Behnke, Roy Herbert *physician, educator*
Cho, Jai Hang *internist, hematologist, educator*
Young, Lawrence Eugene *internist*

West Palm Beach
Roberts, Hyman Jacob *internist, researcher, author, publisher*

GEORGIA

Atlanta
Barnett, Crawford Fannin, Jr. *internist, educator, cardiologist*
Evans, Edwin Curtis *internist, educator, geriatrician*

ILLINOIS

Chicago
Cassel, Christine Karen *physician*
Diamond, Seymour *physician*
Knospe, William Herbert *medical educator*
Siegler, Mark *internist, educator*

INDIANA

Fort Wayne
Richardson, Joseph Hill *physician, educator*

Indianapolis
Chernish, Stanley Michael *physician*
Dere, Willard Honglen *internist, educator*
Rohn, Robert Jones *internist, educator*

Walton
Chu, Johnson Chin Sheng *physician*

IOWA

Iowa City
Weiner, George Jay *internist*

Sioux City
Spellman, George Geneser, Sr. *internist*

LOUISIANA

Baton Rouge
Osterberger, James Sheldon, Jr. *internist, educator*

New Orleans
Incaprera, Frank Philip *internist*
Salvaggio, John Edmond *physician, educator*

MARYLAND

Baltimore
Adjei, Alex Asiedu *internist, pharmacologist*
Friedrich, Christopher Andrew *internist, geneticist*
Horn, Janet *physician*
Noar, Mark David *internist, gastroenterologist, therapeutic endoscopist, consultant, inventor*
Rayson, Glendon Ennes *internist, preventive medicine specialist, writer*

Bethesda
Barber, Ann McDonald *internist*
Gibson, Sam Thompson *internist, educator*
Mojcik, Christopher Francis *internist*
Ognibene, Frederick Peter *internist*

MASSACHUSETTS

Bedford
Alarcon, Rogelio Alfonso *physician, researcher*

Boston
Black, Paul Henry *medical educator, researcher*
Cohen, Alan Seymour *internist*
Crowley, William Francis, Jr. *medical educator*
Isselbacher, Kurt Julius *physician, educator*
Kazemi, Homayoun *physician, medical educator*
Levinsky, Norman George *physician, educator*
Moellering, Robert Charles, Jr. *internist, educator*

Brighton
Stanton, Joseph Robert *physician*

Cambridge
Augerson, William Sinclair *internist*
Wacker, Warren Ernest Clyde *physician, educator*

Chestnut Hill
Stanbury, John Bruton *physician, educator*

Waban
Aisner, Mark *internist*

MICHIGAN

Ann Arbor
Greene, Douglas A. *internist, educator*
Wiggins, Roger C. *internist, educator*

Copemish
Wells, Herschel James *physician, former hospital administrator*

Detroit
Iverson, Robert Louis, Jr. *internist, physician, intensive care administrator, medical educator*

Saint Clair Shores
Petz, Thomas Joseph *internist*

MINNESOTA

Minneapolis
Stenwick, Michael William *internist, geriatric medicine consultant*
Weber, Lowell Wyckoff *internist*

Rochester
Kyle, Robert Arthur *medical educator, oncologist*
Mayberry, William Eugene *retired physician*

MISSISSIPPI

Jackson
Finley, Richard Wade *internist, educator*

NEBRASKA

Lincoln
Koszewski, Bohdan Julius *internist, medical educator*
Weaver, Arthur Lawrence *physician*

Omaha
Bierman, Philip Jay *physician, reseracher, educator*
Eilts, Susanne Elizabeth *physician*
Nair, Chandra Kunju Pillai *internist, educator*

NEW HAMPSHIRE

Lebanon
Sox, Harold Carleton, Jr. *physician, educator*

NEW JERSEY

Hazlet
Chudzik, Douglas Walter *internist*

Plainfield
Yood, Harold Stanley *internist*

Teaneck
Friedhoff, Lawrence Tim *internist, bio-medical researcher*

Tenafly
Cosgriff, Stuart Worcester *internist, consultant*

Trenton
Evers, Martin Louis *internist*

NEW MEXICO

Albuquerque
Abrums, John Denise *internist*

NEW YORK

Astoria
Atkinson, Holly Gail *physician, journalist, author, lecturer*

Bronx
Abbey, Leland Russell *internist*

Brooklyn
Nurhussein, Mohammed Alamin *internist, geriatrician, educator*

Buffalo
Maloney, Milford Charles *internal medicine educator*

Manhasset
Scherr, Lawrence *physician, educator*

New York
Brown, Arthur Edward *physician*
Fins, Joseph Jack *internist, medical ethicist*
Gershengorn, Marvin Carl *physician, educator*
Gitlow, Stanley Edward *internist, educator*
Jackson, Sherry Diane *internist*
Jacobs, Jonathan Lewis *physician*
Kappas, Attallah *physician, medical scientist*
Kreek, Mary Jeanne *physician*
Laragh, John Henry *physician, scientist, educator*
Lipkin, Martin *physician, scientist*
Manger, William Muir *internist*
Schwimmer, David *physician, educator*
Spingarn, Clifford Leroy *internist, educator*
Sternlieb, Cheryl Marcia *internist*
Tichenor, Wellington Shelton *physician*
Urso, Charles Joseph *physician*
Verdesca, Arthur Salvatore *internist, corporate medical director*
Weinsaft, Paul Phineas *retired physician, administrator*
Weinstein, I. Bernard *physician*

Rochester
Willey, James Campbell *internist, educator*
Williams, Thomas Franklin *physician, educator*

Rye
Reader, George Gordon *physician, educator*

Staten Island
Berger, Herbert *retired internist, educator*

Tarrytown
Salop, Arnold *internist*

NORTH CAROLINA

Chapel Hill
Pagano, Joseph Stephen *physician, researcher, educator*

Winston Salem
Kaufman, William *internist*

OHIO

Cincinnati
Rouan, Gregory W. *internal medicine physician, educator*

Cleveland
Mc Henry, Martin Christopher *physician*

Columbus
Balcerzak, Stanley Paul *physician, educator*

Dayton
Arn, Kenneth Dale *physician, city official*
Chang, Jae Chan *internist, educator*

OKLAHOMA

Tulsa
Lewis, Ceylon Smith, Jr. *physician*

OREGON

Portland
Stalnaker, John Hulbert *physician*

PENNSYLVANIA

Erie
Michaelides, Doros Nikita *internist, educator*

Malvern
Langrall, Harrison Morton, Jr. *internist*

Palmyra
Moyer, John Henry, III *physician, educator*

Philadelphia
Austrian, Robert *physician, educator*
Gozum, Marvin Enriquez *internist*
Kelley, William Nimmons *physician, educator*
Levy, Robert Isaac *physician, educator, research director*
London, William Thomas *internist*
Mayock, Robert Lee *internist*
Schumacher, H(arry) Ralph *internist, researcher, medical educator*

Pittsburgh
Owens, Gregory Randolph *physician, medical educator*
Pyeritz, Reed Edwin *internist*

RHODE ISLAND

Providence
Parks, Robert Emmett, Jr. *medical science educator*

SOUTH DAKOTA

Sioux Falls
Zawada, Edward Thaddeus, Jr. *physician, educator*

TENNESSEE

Nashville
Crofford, Oscar Bledsoe, Jr. *internist, medical educator*

TEXAS

Austin
Schless, James Murray *internist*

Dallas
Dees, Tom Moore *internist*

Galveston
Koeppe, Patsy Poduska *internist, educator*

Houston
Marcus, Donald Martin *internist, educator*
Rensimer, Edward R. *internist, educator*

San Antonio
Hausheer, Frederick Herman *internist, cancer researcher, pharmaceutical company officer*

Webster
Rappaport, Martin Paul *internist, nephrologist, educator*

VERMONT

White River Junction
Colice, Gene Leslie *physician*

VIRGINIA

Arlington
Nguyen-Dinh, Thanh *internist, geriatrician*

Charlottesville
Davis, John Staige, IV *physician*
Gwaltney, Jack Merrit, Jr. *physician, educator, scientist*
Scheld, William Michael *internist, educator*

Richmond
Owen, Duncan Shaw, Jr. *physician, medical educator*

Wytheville
McConnell, James Joseph *internist*

WEST VIRGINIA

Huntington
Mufson, Maurice Albert *physician, educator*

WISCONSIN

Fort McCoy
Truthan, Charles Edwin *physician*

Madison
Maki, Dennis G. *medical educator, researcher, clinician*

Milwaukee
Dorff, Gerald J. *physician*
Funahashi, Akira *physician, educator*

TERRITORIES OF THE UNITED STATES

PUERTO RICO

Gurabo
Curet-Ramos, José Antonio *internist*

San Juan
Ramírez-Ronda, Carlos Héctor *physician*

CANADA

ONTARIO

Toronto
Mc Culloch, Ernest Armstrong *physician, educator*

QUEBEC

Montreal
Gold, Phil *physician, educator*

AUSTRALIA

West Perth
Woods, Thomas Brian *physician*

BRAZIL

Belo Horizonte
Pena, Sergio Danilo Junho *physician*

DENMARK

Odense
Kemp, Ejvind *internist*

ENGLAND

London
Kakati, Dinesh Chandra *physician*

FRANCE

Beaujon
Pariente, René Guillaume *physician, educator*

Bondy
Beaugrand, Michel *physician, educator*

Chartres
Benoit, Jean-Pierre Robert *pneumologist, consultant*

Paris
Gontier, Jean Roger *internist, physiology educator, consultant*

GERMANY

Bad Nauheim
Hueting, Juergen *internist*

Cologne
Merten, Utz Peter *physician*

Gladbeck
Geisler, Linus Sebastian *physician, educator*

Hannover
Von Zur Mühlen, Alexander Meinhard *physician, internal medicine educator*

GREECE

Elefsing
El-Husban, Tayseer Khalaf *internist, consultant*

ICELAND

Reykjavik
Gudjonsson, Birgir *physician*

JAPAN

Niigata
Asakura, Hitoshi *internal medicine educator*

Osaka
Ueda, Einosuke *physician*

Tokyo
Masuda, Gohta *physician, educator*
Terao, Toshio *physician, educator*

Yamanashi-ken
Onaya, Toshimasa *internal medicine educator*

NORWAY

Hjelset
Hartmann-Johnsen, Olaf Johan *internist*

PAKISTAN

Sheikhupura
Ahmad, Khalil *medical practitioner*

THE PHILIPPINES

Batangas
Malabanan, Ernesto Herella *internist*

Binondo
Gan, Felisa So *physician*

Pangasinah
Posadas, Martin Posadas *physician, educator, businessman*

SPAIN

Galdácano
Collazos Gonzalez, Julio *internist, researcher*

Madrid
Dominguez Ortega, Luis *medical educator, health facility administrator*

SWEDEN

Södertälje
Kock, Lars Anders Wolfram *physician, educator*

SWITZERLAND

Nyon
Jung, André *internist*

TAIWAN

Taipei
Chiang, Cheng-Wen *internal medicine educator, physician*

ADDRESS UNPUBLISHED

Abela, George Samih *medical educator, internist, cardiologist*
Bynes, Frank Howard, Jr. *physician*
Daley, George Quentin *internist, biomedical research scientist*
Doane, Woolson Whitney *internist*
Engel, Joanne Netter *internist, educator*
Frank, Sanders Thalheimer *physician, educator*
Hessen, Margaret Trexler *internist, educator*
Jones, Walton Linton *internist, former government official*
Kornfeld, Peter *internist*
Lowy, Israel *internist, educator*

MEDICINE: NEPHROLOGY

UNITED STATES

ALABAMA

Birmingham
Warnock, David Gene *nephrologist, pharmacology educator*

ARKANSAS

Fort Smith
Coleman, Michael Dortch *nephrologist*

FLORIDA

Panama City
Walker, Richard, Jr. *nephrologist, internist*

GEORGIA

Atlanta
Lowry, Robin Pearce *nephrologist, educator*
Neylan, John Francis, III *nephrologist, educator*

ILLINOIS

Chicago
Batlle, Daniel Campi *nephrologist*

Evanston
Quintanilla, Antonio Paulet *physician, educator*

KANSAS

Kansas City
Grantham, Jared James *nephrologist, educator*

MASSACHUSETTS

Boston
Brenner, Barry Morton *physician*

MICHIGAN

Ann Arbor
Humes, H(arvey) David *nephrologist, educator*

MINNESOTA

Minneapolis
Keane, William Francis *nephrology educator, research foundation executive*
Michael, Alfred Frederick, Jr. *physician, medical educator*

MISSOURI

Saint Louis
Slatopolsky, Eduardo *nephrologist, educator*

NEW YORK

New York
Birbari, Adil Elias *physician, educator*

Orangeburg
Squires, Richard Felt *research scientist*

OHIO

Cleveland
Wish, Jay Barry *nephrologist, specialist*

OKLAHOMA

Oklahoma City
Bourdeau, James Edward *nephrologist, researcher*

PENNSYLVANIA

Philadelphia
Ortiz-Arduan, Alberto *nephrologist*

FRANCE

Toulouse
Conte, Jean Jacques *medical educator, nephrologist*

GERMANY

Wuerzburg
Schaefer, Roland Michael *nephrologist, consultant*

ITALY

Bari
Schena, Francesco Paolo *nephrology educator*

Milan
D'Amico, Giuseppe *nephrologist*

JAPAN

Iizuka City
Nishimura, Manabu *nephrologist*

REPUBLIC OF KOREA

Chonju
Kang, Sung Kyew *medical educator*

SPAIN

Lleida
Montoliu, Jesus *nephrologist, educator*

ADDRESS UNPUBLISHED

Duarte, Cristobal G. *nephrologist, educator*
Glassock, Richard James *nephrologist*
Malluche, Hartmut Horst *nephrologist, medical educator*

MEDICINE: NEUROLOGY

UNITED STATES

ARIZONA

Phoenix
Flitman, Stephen Samuel *neurologist*

CALIFORNIA

Capo Beach
Roemer, Edward Pier *neurologist*

Davis
Fowler, William Mayo, Jr. *rehabilitation medicine physician*

Los Angeles
Baumhefner, Robert Walter *neurologist*
Engel, Jerome, Jr. *neurologist, neuroscientist, educator*
Treiman, David Murray *neurology educator*
Van Der Meulen, Joseph Pierre *neurologist*

Martinez
Efron, Robert *neurology educator, research institute administrator*

Orange
Lott, Ira Totz *pediatric neurologist*

San Diego
Smith, Richard Alan *neurologist, medical association administrator*

San Francisco
Aird, Robert Burns *neurologist, educator*
Fishman, Robert Allen *educator, neurologist*

Santa Barbara
Fisher, Steven Kay *neurobiology eductor*

CONNECTICUT

Farmington
Donaldson, James Oswell, III *neurology educator*

New Haven
Ransom, Bruce Robert *neurologist, neurophysiologist, educator*
Seigel, Arthur Michael *neurologist, educator*
Sontheimer, Harald Wolfgang *scientist, cell biology researcher*

Norwalk
Twist-Rudolph, Donna Joy *neurophysiology and psychology researcher*

Stratford
Sena, Kanaga Nitchinga *neurologist*

West Haven
Scheyer, Richard David *neurologist, researcher*

DISTRICT OF COLUMBIA

Washington
Henkin, Robert Irwin *neurobiologist, internal medicine, nutrition and neurology educator, scientific products company executive*
Laureno, Robert *neurologist*

FLORIDA

Gainesville
Palovcik, Reinhard Anton *research neurophysiologist*

Miami
Ginsberg, Myron David *neurologist*
Mettinger, Karl Lennart *neurologist*
Sánchez-Ramos, Juan Ramon *neurologist, researcher*
Weiner, William Jerrold *neurologist, educator*

Saint Augustine
Gerling, Gerard Michael *neurologist*

Tampa
Olanow, C(harles) Warren *neurologist, educator*
Werner, Mark Henry *neurologist, researcher*

GEORGIA

Atlanta
Bakay, Roy Arpad Earle *neurosurgeon, educator*

Augusta
Loring, David William *neuropsychologist, researcher*
Zamrini, Edward Youssef *behavioral neurologist*

Marietta
Holtz, Noel *neurologist*

ILLINOIS

Chicago
Arnason, Barry Gilbert Wyatt *neurologist, educator*
Hughes, John Russell *physician, educator*

Evanston
Dau, Peter Caine *neurologist, immunologist, educator*

Waukegan
Weisz, Reuben R. *neurology educator*

INDIANA

Bloomington
Conneally, P. Michael *medical educator*

KENTUCKY

Lexington
Markesbery, William R. *neurology and pathology educator, physician*

Louisville
Olson, William Henry *neurology educator, administrator*

LOUISIANA

Baton Rouge
Romero, Jorge Antonio *neurologist, educator*

MARYLAND

Baltimore
Drachman, Daniel Bruce *neurologist*
Johnson, Kenneth Peter *neurologist, medical researcher*
Lesser, Ronald Peter *neurologist*
Moses, Hamilton, III *neurology educator, hospital executive*
Price, Thomas Ransone *neurologist, educator*
Schuster, Frank Feist *neurologist, social services administrator*
Zee, David Samuel *neurologist*

Bethesda
Hallett, Mark *physician, neurologist, health research institute administrator*
Leventhal, Carl M. *neurologist*
Reese, Thomas Sargent *neurobiology educator and researcher*

Poolesville
Newman, John Dennis *neuroethologist, biomedical researcher*

MASSACHUSETTS

Boston
Caplan, David Norman *neurology educator*
Nathanson, James A *neurologist*
Oas, John Gilbert *neurologist, researcher*

Concord
Palay, Sanford Louis *retired scientist, educator*

Medford
Kinsbourne, Marcel *neurologist, behavioral neuroscientist*

Roxbury
Berman, Marlene Oscar *neuropsychologist, educator*

West Roxbury
Charness, Michael Edward *neurologist, neuroscientist*

MICHIGAN

Detroit
Benjamins, Joyce Ann *neurology educator*

Escanaba
Cooper, Janelle Lunette *neurologist*

Lansing
Vincent, Frederick Michael *neurologist, electromyographer, educational administrator*

Mancelona
Whelan, Joseph L. *neurologist*

MINNESOTA

Minneapolis
Swaiman, Kenneth Fred *pediatric neurologist, educator*

Rochester
Engel, Andrew George *neurologist*

Saint Paul
Witek, John James *neurologist*

MISSISSIPPI

Jackson
Currier, Robert David *neurologist*
Snodgrass, Samuel Robert *neurologist*

MISSOURI

Saint Louis
Hsu, Chung Yi *neurologist*
Landau, William Milton *neurologist*
Pestronk, Alan *neurologist*
Thach, William Thomas, Jr. *neurobiology and neurology educator*

Springfield
Hackett, Earl Randolph *neurologist*

NEBRASKA

Omaha
Futrell, Nancy Nielson *neurologist*

NEW JERSEY

Elizabeth
Sananman, Michael Lawrence *neurologist*

Englewood
Rabin, Aaron *neurologist*

New Brunswick
Duvoisin, Roger Clair *physician, medical educator*

NEW YORK

Brooklyn
Gintautas, Jonas *physician, scientist, administrator*

Manhasset
Halperin, John Jacob *neurology educator, researcher*

New Hyde Park
Biddle, David *neurologist*

New York
Brown, Jason Walter *neurologist, educator, researcher*
Brust, John Calvin Morrison *neurology educator*
DeAngelis, Lisa Marie *neurologist, educator*
Foley, Kathleen M. *neurologist, educator, researcher*
Jonas, Saran *neurologist, educator*
Maiese, Kenneth *neurologist*
Pedley, Timothy Asbury, IV *neurologist, educator, researcher*
Posner, Jerome Beebe *neurologist, educator*
Reis, Donald Jeffery *neurologist, neurobiologist, educator*
Scheinberg, Labe Charles *physician, educator*

Rochester
Pettee, Daniel Starr *neurologist*

Syracuse
Haas, David Colton *neurologist, educator*

Tarrytown
Cedarbaum, Jesse Michael *neurologist*

NORTH CAROLINA

Fort Bragg
Moss, Kenneth Wayne *neurologist*

Wilmington
Gillen, Howard William *neurologist, medical historian*

Wilson
Kushner, Michael James *neurologist, consultant*

Winston Salem
Penry, James Kiffin *physician, neurology educator*

OHIO

Cleveland
Ruff, Robert Louis *neurologist, physiology researcher*

Columbus
Platika, Doros *neurologist*

OREGON

Portland
Lezak, Muriel Deutsch *psychology, neurology and psychiatry educator*
Zimmerman, Earl Abram *physician, scientist, educator, neuroendocrinology researcher*

PENNSYLVANIA

Greensburg
Catalano, Louis William, Jr. *neurologist*

Philadelphia
Asbury, Arthur Knight *neurologist, educator*
Black, Perry *neurological surgeon, educator*
Kollros, Peter Richard *child neurology educator, researcher*
Mancall, Elliott Lee *neurologist, educator*
Molino-Bonagura, Lory Jean *neurobiologist*
Selzer, Michael Edgar *neurologist*
Shipkin, Paul M. *neurologist*
Silberberg, Donald H. *neurologist*

Upland
Green, Lawrence *neurologist, educator*

RHODE ISLAND

Providence
Dowben, Robert Morris *physician, scientist*

TEXAS

Corpus Christi
Lim, Alexander Rufasta *neurologist*

Galveston
McKendall, Robert Roland *neurologist, virologist, educator*

Houston
Appel, Stanley Hersh *neurologist*
Jankovic, Joseph *neurologist, educator, scientist*
Meyer, John Stirling *neurologist, educator*
Patten, Bernard Michael *neurologist, educator*

Lubbock
Green, Joseph Barnet *neurologist*

VIRGINIA

Charlottesville
Phillips, Larry H., II *neurologist*

Mc Lean
Kratz, Ruediger *neurologist, researcher*

WEST VIRGINIA

Wheeling
Zyznewsky, Wladimir A. *neurologist*

WISCONSIN

Madison
Sufit, Robert Louis *neurologist, educator*

WYOMING

Casper
Cole, Malvin *neurologist, educator*

CANADA

ONTARIO

London
Barnett, Henry Joseph Macaulay *neurologist*

QUEBEC

Montreal
Gjedde, Albert Hellmut *neuroscientist, neurology educator*
Karpati, George *neurologist*

AUSTRALIA

Randwick
Lance, James Waldo *neurologist*

BRAZIL

São Paulo
Gabbai, Alberto Alain *neurology educator, researcher*

FINLAND

Helsinki
Partinen, Markku Mikael *neurologist*

GERMANY

Berlin
Mauritz, Karl Heinz *neurology educator*

Frankfurt
Duus, Peter *neurology educator*

Wuerzburg
Reiners, Karlheinz *neurologist, educator*

ITALY

Milan
Boeri, Renato Raimondo *neurologist*

Rome
Manfredi, Mario Erminio *neurologist educator*

JAPAN

Kanagawa
Takeoka, Tsuneyuki *neurology educator*

Tokyo
Fukuyama, Yukio *child neurologist, pediatrics educator*
Sakuta, Manabu *neurologist, educator*

THE NETHERLANDS

Amsterdam
Koetsier, Johan Carel *clinical neurology educator*

Utrecht
Gooskens, Robert Henricus Johannus *pediatric neurologist*

VENEZUELA

Caracas
Vanegas, Horacio *neurobiology educator, director*

ADDRESS UNPUBLISHED

Belman, Anita Leggold *pediatric neurologist*
Bercel, Nicholas Anthony *neurologist, neurophysiologist*
Hunter, Richard Grant, Jr. *neurologist, executive*
Kaufmann, Walter Erwin *neurologist, neuropathologist, educator*
Prusiner, Stanley Ben *neurology and biochemistry educator, researcher*

MEDICINE: NEUROSCIENCE

UNITED STATES

ALABAMA

Auburn
Mehta, Jagjivan Ram *research scientist*

Birmingham
Friedlander, Michael J. *neuroscientist, animal physiologist, medical educator*

ARKANSAS

Little Rock
Garcia-Rill, Edgar Enrique *neuroscientist*

CALIFORNIA

Alameda
Mandel, Ronald James *neuroscientist*

Berkeley
Goodman, Corey Scott *neurobiology educator, researcher*

Irvine
Wong, Patrick Tin-Choi *neuroscientist*

La Jolla
Alvarez, Pablo *neuroscientist*

Los Angeles
Arnold, Arthur Palmer *neurobiologist*
Cohen, Randy Wade *neuroscientist*

Grinnell, Alan Dale *neurobiologist, educator, researcher*
Woolf, Nancy Jean *neuroscientist, educator*

Martinez
Nielsen-Bohlman, Lynn Tracy *neuroscientist*

Pasadena
Niebur, Ernst Dietrich *computational neuroscientist*

San Diego
Lewis, Gregory Williams *neuroscientist*

San Marino
Benzer, Seymour *neurosciences educator*

Stanford
Honig, Lawrence Sterling *neuroscientist, neurologist*

CONNECTICUT

Tolland
Roberge, Lawrence Francis *neuroscientist, biotechnology consultant*

DELAWARE

Newark
Plata-Salamán, Carlos Ramon *neuroscientist*

DISTRICT OF COLUMBIA

Washington
Brown, James Harvey *neuroscientist, government research administrator*
Olsen, Kathie Lynn *neuroscientist, administrator*
Werbos, Paul John *neural research director*

FLORIDA

Gainesville
Semple-Rowland, Susan Lynn *neuroscientist*

GEORGIA

Atlanta
King, Frederick Alexander *neuroscientist, educator*
Plotsky, Paul Mitchell *neuroscientist, educator*
Siegler, Melody Victoria Stephanie *neuroscientist*

ILLINOIS

North Chicago
McCandless, David Wayne *neuroscientist, anatomy educator*

Skokie
Stittsworth, James Dale *neuroscientist*

Urbana
Anastasio, Thomas Joseph *neuroscientist, educator, researcher*

Western Springs
Swiatek, Kenneth Robert *neuroscientist*

INDIANA

West Lafayette
Hall, Stephen Grow *research neuroscientist*

IOWA

Iowa City
Lim, Ramon (Khe-Siong) *neuroscience educator*

KANSAS

Kansas City
Samson, Frederick Eugene, Jr. *neuroscientist, educator*
Smith, Peter Guy *neuroscience educator, researcher*

LOUISIANA

New Orleans
Prasad, Chandan *neuroscientist*

MAINE

Bar Harbor
Johnson, Eric Walter *neuroscientist*

MARYLAND

Baltimore
Jastreboff, Pawel Jerzy *neuroscientist, educator*
Koliatsos, Vassilis Eleftherios *neurobiologist*
Markowska, Alicja Lidia *neuroscientist, researcher*
Mouton, Peter Randolph *neuroscientist, biologist*

Bethesda
Gold, Philip William *neurobiologist*
King, Ron Glen *neuroscientist*
Sirén, Anna-Leena Kaarina *neuroscientist*
Webster, Henry deForest *experimental neuropathologist*

Chevy Chase
Cowan, William Maxwell *neurobiologist*

Rockville
Clark, William Anthony *neuropharmacologist*

MASSACHUSETTS

Boston
Bullock, Daniel Hugh *computational neuroscience educator, psychologist*
Foote, Warren Edgar *neuroscientist, psychologist, educator*
Leeman, Susan Epstein *neuroscientist*

Natick
Stillman, Michael James *neuroscientist*

Southborough
Madras, Bertha Kalifon *neuroscientist, consultant*

Williamstown
Williams, Heather *neuroethologist, educator*

MICHIGAN

East Lansing
Johnson, John Irwin, Jr. *neuroscientist*

MINNESOTA

Minneapolis
Galeazza, Marc Thomas *neuroscientist*
Santi, Peter Alan *neuroanatomist, educator*

MISSISSIPPI

Jackson
Hutchins, James Blair *neuroscientist*

NEW JERSEY

New Brunswick
Walsh, Thomas Joseph *neuroscientist, educator*

Nutley
Connor, John Arthur *neuroscientist*

Princeton
Pothos, Emmanuel *neuroscientist*

NEW YORK

Bronx
Purpura, Dominick P. *neuroscientist, university dean*
Sircar, Ratna *neurobiology educator, researcher*
Van De Water, Thomas Roger *neuroscientist, educator*

Hamilton
Tierney, Ann Jane *neuroscientist*

New York
Goodrich, James Tait *neuroscientist, pediatric neurosurgeon*
Green, Maurice Richard *neuropsychiatrist*
Greengard, Paul *neuroscientist*
Hegde, Ashok Narayan *neuroscientist*
Reeke, George Norman, Jr. *neuroscientist, crystallographer, educator*
Wiesel, Torsten Nils *neurobiologist, educator*

Purchase
Santucci, Anthony Charles *neuroscientist, educator*

Rochester
Harris, Eric William *neuroscientist, pharmaceutical executive*

Stony Brook
Strecker, Robert Edwin *neuroscientist, educator*
Yang, Chen-yu *neuroscientist*

Syracuse
Verrillo, Ronald Thomas *neuroscientist*

White Plains
Greenberg, Sharon Gail *neuroscientist, researcher*

NORTH CAROLINA

Davidson
Ramirez, Julio Jesus *neuroscientist*

OHIO

Cleveland
Herrup, Karl *neurobiologist*

Rootstown
Chopko, Bohdan Wolodymyr *neuroscientist*

OKLAHOMA

Oklahoma City
Thompson, Ann Marie *neuroscientist, researcher, educator*

Tulsa
Kemp, Sarah (Sally Leech) *neurodevelopment specialist*

OREGON

Portland
Barmack, Neal Herbert *neuroscientist*

PENNSYLVANIA

Hershey
Norgren, Ralph *neuroscientist*

Philadelphia
Barchi, Robert Lawrence *neuroscience educator, clinical neurologist, neuroscientist*
Heyman, Julius Scott *neuroscientist*
Johnson, Bonnie Jean *neuroscientist*
Rawson, Nancy Ellen *neurobiology researcher*
Wysocki, Charles Joseph *neuroscientist*

Pittsburgh
Palmer, Alan Michael *neuroscientist*

RHODE ISLAND

Providence
Stopa, Edward Gregory *neuropathologist*

Wakefield
Fair, Charles Maitland *neuroscientist, author*

TENNESSEE

Nashville
Charlton, Clivel George *neuroscientist, educator*

TEXAS

Dallas
Chapman, Sandra Bond *neurolinguist, researcher*
Sanchez, Dorothea Yialamas *neuroscientist*
Suppes, Trisha *neuroscientist*

Denton
Schafer, Rollie Randolph, Jr. *neuroscientist*

Galveston
Puzdrowski, Richard Leo *neuroscientist*
Willis, William Darrell, Jr. *neurophysiologist, educator*

Houston
Kellaway, Peter *neurophysiologist, researcher*

VIRGINIA

Charlottesville
Koshiya, Naihiro *neuroscientist*

Palmyra
Weiss-Wunder, Linda Teresa *neuroscience research consultant*

Radford
Hudspeth, William Jean *neuroscientist*

Richmond
Kinsley, Craig Howard *neuroscientist*

WASHINGTON

Pullman
Barnes, Charles D. *neuroscientist, educator*

WISCONSIN

Madison
Fettiplace, Robert *neurophysiologist*
Whitlon, Donna Sue *neuroscientist, researcher*

CANADA

ALBERTA

Calgary
Stell, William Kenyon *neuroscientist, educator*

BELGIUM

Brussels
Goffinet, Serge *neuropsychiatrist, researcher*

GERMANY

Bochum
Kaernbach, Christian *psychophysicist*

Bonn
Eckmiller, Rolf Eberhard *neuroscientist, educator*

Magdeburg
Sabel, Bernhard August Maria *research neuroscientist, psychologist*

ITALY

Milan
Silani, Vincenzo *neurology and neuroscience educator*

Parma
Rizzolatti, Giacomo *neuroscientist*

POLAND

Warsaw
Tarnecki, Remigiusz Leszek *neurophysiology educator, laboratory director*

SAUDI ARABIA

Riyadh
Mousa, Alyaa Mohammed Ali *neurobiologist, researcher*

ADDRESS UNPUBLISHED

Hopfield, Jessica F. *neuroscientist, researcher, administrator*
Livingston, Robert Burr *neuroscientist, educator*
Michaelis, Elias K. *neurochemist*

MEDICINE: NEUROSURGERY

UNITED STATES

ARIZONA

Tucson
Carter, L. Philip *neurosurgeon, consultant*

CALIFORNIA

Los Angeles
Couldwell, William Tupper *neurosurgeon*

San Francisco
Wilson, Charles B. *neurosurgeon, educator*

San Jose
Lippe, Philipp Maria *neurosurgeon, educator*

CONNECTICUT

Hartford
Roberts, Melville Parker, Jr. *neurosurgeon, educator*

Norwalk
Needham, Charles William *neurosurgeon*

DISTRICT OF COLUMBIA

Washington
Curfman, David Ralph *neurological surgeon, musician*

FLORIDA

Fort Lauderdale
Gomez, Jaime G. *neurosurgeon*

Jacksonville
Bremer, Alfonso M. *neurosurgeon, neuro-oncologist*

ILLINOIS

Hinsdale
Kazan, Robert Peter *neurosurgeon*

Moline
Milas, Robert Wayne *neurosurgeon*

Quincy
Del Castillo, Julio Cesar *neurosurgeon*

KENTUCKY

Lexington
Hodes, Jonathan Ezra *neurosurgeon, educator*

Louisville
Garretson, Henry David *neurosurgeon*

LOUISIANA

New Orleans
Kline, David Gellinger *neurosurgery educator*

MASSACHUSETTS

Boston
Zervas, Nicholas Themistocles *neurosurgeon*

MICHIGAN

Ann Arbor
Hoff, Julian Theodore *physician, educator*

Detroit
Diaz, Fernando Gustavo *neurosurgeon*

MISSOURI

Kansas City
Schoolman, Arnold *neurological surgeon*

NEW MEXICO

Albuquerque
Ottensmeyer, David Joseph *neurosurgeon, health care executive*

NEW YORK

Bronx
Michelsen, W(olfgang) Jost *neurosurgeon, educator*
Waltz, Joseph McKendree *neurosurgeon, educator*

Brooklyn
Lohmann, George Young, Jr. *neurosurgeon, hospital executive*
Milhorat, Thomas Herrick *neurosurgeon*

Freeport
Burstein, Stephen David *neurosurgeon*

Nanuet
Savitz, Martin Harold *neurosurgeon*

New Hyde Park
Epstein, Joseph Allen *neurosurgeon*

New York
Fodstad, Harald *neurosurgeon*
Mc Murtry, James Gilmer, III *neurosurgeon*
Sachdev, Ved Parkash *neurosurgeon*

Rochester
Brzustowicz, Richard John *neurosurgeon, educator*
Wiley, Jason LaRue, Jr. *neurosurgeon*

Stony Brook
Kuchner, Eugene Frederick *neurosurgeon, educator*

NORTH CAROLINA

Greenville
Lee, Kenneth Stuart *neurosurgeon*

High Point
Schwarz, Saul Samuel *neurosurgeon*

PENNSYLVANIA

Pittsburgh
Vidovich, Danko Victor *neurosurgeon, researcher*

Rockledge
Polakoff, Pedro Paul, II *neurosurgeon*

TENNESSEE

Knoxville
Kliefoth, A(rthur) Bernhard, III *neurosurgeon*

VIRGINIA

Charlottesville
Bleck, Thomas Pritchett *neurologist, neuroscientist, educator*

WASHINGTON

Seattle
Haglund, Michael Martin *neurosurgeon*

WISCONSIN

Marshfield
Kelman, Donald Brian *neurosurgeon*

GERMANY

Frankfurt am Main
Lorenz, Ruediger *neurosurgeon*

Greifswald
Gaab, Michael Robert *neurosurgery educator, consultant*

JAPAN

Ota-ku
Sano, Keiji *neurosurgeon*

REPUBLIC OF KOREA

Seoul
Chung, Hwan Yung *neurosurgeon*

SAUDI ARABIA

Riyadh
Moutaery, Khalaf Reden *neurosurgeon*

MEDICINE: NUCLEAR MEDICINE

UNITED STATES

CALIFORNIA

Los Angeles
Siegel, Michael Elliot *nuclear medicine physician, educator*

Rocklin
Vande Streek, Penny Robillard *nuclear medicine physician, researcher, educator*

Walnut Creek
Farr, Lee Edward *physician*

FLORIDA

Tampa
Muroff, Lawrence Ross *nuclear medicine physician*

HAWAII

Honolulu
Gilbert, Fred Ivan, Jr. *physician, researcher*

ILLINOIS

Chicago
Beck, Robert N. *nuclear medicine educator*

MICHIGAN

Grosse Pointe
Beierwaltes, William Henry *physician, educator*

MINNESOTA

Minneapolis
Boudreau, Robert James *nuclear medicine physician, researcher*

NEW JERSEY

Long Branch
Makhija, Mohan *nuclear medicine physician*

NEW YORK

Roslyn
Silverstein, Seth *physician*

OHIO

Cincinnati
Maxon, Harry Russell, III *nuclear medicine physician*

Kettering
Mantil, Joseph Chacko *nuclear medicine physician, researcher*

TEXAS

Dallas
Parkey, Robert Wayne *radiology and nuclear medicine educator, research radiologist*
Simon, Theodore Ronald *physician, medical educator*

Houston
Holmquest, Donald Lee *physician, astronaut, lawyer*

San Antonio
Horton, Granville Eugene *nuclear medicine physician, retired air force officer*
Phillips, William Thomas *nuclear medicine physician, educator*

GERMANY

Volklingen
Reinhardt, Kurt *retired radiological and nuclear physician*

SWITZERLAND

Lausanne
Delaloye, Bernard *nuclear medicine physician*

MEDICINE: OBSTETRICS & GYNECOLOGY

UNITED STATES

ALABAMA

Birmingham
Wideman, Gilder LeVaugh *obstetrician/gynecologist*

ARIZONA

Tucson
Ricke, P. Scott *obstetrician/gynecologist*

ARKANSAS

Texarkana
Harrison, James Wilburn *gynecologist*

CALIFORNIA

Davis
Overstreet, James Wilkins *obstetrics and gynecology educator, administrator*

Los Angeles
Hobel, Calvin John *obstetrician/gynecologist*

Mountain View
Warren, Richard Wayne *obstetrician, gynecologist*

Orange
DiSaia, Philip John *gynecologist, obstetrician, radiology educator*
Thompson, William Benbow, Jr. *obstetrician/gynecologist, educator*

San Bernardino
Russo, Alvin Leon *obstetrician/gynecologist*

San Francisco
Jaffe, Robert Benton *obstetrician-gynecologist, reproductive endocrinologist*

Santa Barbara
Mathews, Barbara Edith *gynecologist*

Stanford
Yang, Zeren *obstetrician/gynecologist, researcher*

Torrance
Chao, Conrad Russell *obstetrician*

COLORADO

Snowmass Village
Diamond, Edward *gynecologist, infertility specialist, clinician*

CONNECTICUT

Stamford
Goodhue, Peter Ames *obstetrician/gynecologist, educator*

FLORIDA

Clearwater
Fromhagen, Carl, Jr. *obstetrician/gynecologist*

Naples
Gahagan, Thomas Gail *obstetrician/gynecologist*

South Miami
Remmer, Harry Thomas, Jr. *obstetrician/gynecologist*

Tampa
Jacobson, Howard Newman *obstetrics/gynecology educator, researcher*
Shephard, Bruce Dennis *obstetrician, medical writer*
Spellacy, William Nelson *obstetrician-gynecologist, educator*

GEORGIA

Augusta
Devoe, Lawrence Daniel *obstetrician/gynecologist educator*

Lithia Springs
Hebel, Gail Suzette *obstetrician/gynecologist*

ILLINOIS

Chicago
Bachicha, Joseph Alfred *obstetrician/gynecologist, educator*
Cohen, Melvin R. *physician, educator*
Coopersmith, Bernard Ira *obstetrician/gynecologist, educator*
Moawad, Atef *obstetrician-gynecologist, educator*
Osiyoye, Adekunle *obstetrician/gynecologist, educator*
Wied, George Ludwig *physician*
Williams, Philip Copelain *obstetrician, gynecologist*

Joliet
Nazos, Demetri Eleftherios *obstetrician, gynecologist, medical facility executive*

INDIANA

Huntingburg
Rossmann, Charles Boris *obstetrician/gynecologist*

Indianapolis
Cleary, Robert Emmet *gynecologist, infertility specialist*

IOWA

Mount Pleasant
Sandy, Edward Allen *obstetrician/gynecologist*

KENTUCKY

Bowling Green
Bryson, Keith *obstetrician/gynecologist*

LOUISIANA

Slidell
Muller, Robert Joseph *gynecologist*

MARYLAND

Baltimore
Nagey, David Augustus *physician, researcher*

Bethesda
Haseltine, Florence Pat *research administrator, obstetrician, gynecologist*

MASSACHUSETTS

Boston
De Cherney, Alan Hersh *obstetrics and gynecology educator*
Ryan, Kenneth John *physician, educator*

Chestnut Hill
Kosasky, Harold Jack *gynecologist*

MICHIGAN

East Lansing
Sauer, Harold John *physician, educator*

MINNESOTA

Duluth
Sebastian, James Albert *obstetrician/gynecologist, educator*

MISSOURI

Columbia
Hess, Leonard Wayne *obstetrician/gynecologist, perinatologist*

NEBRASKA

Omaha
Casey, Murray Joseph *obstetrician/gynecologist, educator*
DeJonge, Christopher John *obstetrics/gynecology educator*

NEW JERSEY

Camden
Ances, I. G(eorge) *obstetrician/gynecologist, educator*

Florham Park
Nash, Lillian Dorothy *gynecologist, reproductive endocrinologist*

Millburn
Heistein, Robert Kenneth *obstetrician/gynecologist*

Newark
Iffy, Leslie *medical educator*

Roselle Park
Wilchins, Sidney A. *gynecologist*

NEW MEXICO

Albuquerque
Guardia, David King *obstetrics/gynecology educator*

Las Cruces
Reeves, Billy Dean *obstetrics/gynecology educator emeritus*

Santa Fe
Mendez, C. Beatriz *obstetrician/gynecologist*

NEW YORK

Albany
Posner, Norman Ames *medical educator*

Brooklyn
Schwarz, Richard Howard *obstetrician, gynecologist, educator*

Buffalo
Enhorning, Goran *obstetrician/gynecologist, educator*

Geneva
Dickson, James Edwin, II *obstetrician/gynecologist*

Manhasset
Gal, David *gynecologic oncologist, obstetrician, gynecologist*

New York
Dantuono, Louise Mildred *obstetrician/gynecologist*
Hoskins, William John *obstetrician/gynecologist, educator*
Jewelewicz, Raphael *obstetrician/gynecologist, educator*
Mukherjee, Trishit *reproductive infertility specialist*
Ordorica, Steven Anthony *obstetrician/gynecologist, educator*

Poughkeepsie
Milano, Charles Thomas *obstetrician, legal medicine consultant*

Rochester
Abramowicz, Jacques Sylvain *obstetrician, perinatologist*

Sayville
Pagano, Alphonse Frederick *obstetrician/gynecologist*

NORTH CAROLINA

Durham
Hammond, Charles Bessellieu *obstetrician-gynecologist, educator*

Kernersville
Farrer-Meschan, Rachel (Mrs. Isadore Meschan) *obstetrics/gynecology educator*

Research Triangle Park
King, Theodore M. *obstetrician, gynecologist, educator*

OHIO

Columbus
Zuspan, Frederick Paul *obstetrician/gynecologist, educator*

OKLAHOMA

Oklahoma City
Everett, Royice Bert *obstetrician/gynecologist*
Rossavik, Ivar Kristian *obstetrician/gynecologist*

PENNSYLVANIA

Philadelphia
Clifford, Maurice Cecil *physician, former college president, foundation executive*

Pittsburgh
Allen, Thomas E. *obstetrician/gynecologist*
Silverstein, Alan Jay *physician*

SOUTH CAROLINA

Florence
Hunter, Nancy Quintero *obstetrician/gynecologist*

TENNESSEE

Jonesborough
Weaver, Kenneth *gynecologist, researcher*

Memphis
Buster, John Edmond *gynecologist, medical researcher*

TEXAS

Arlington
Keller, Ben Robert, Jr. *gynecologist*

Fort Worth
Suba, Steven Antonio *obstetrician/gynecologist*

Galveston
Dawson, Earl Bliss *obstetrics and gynecology educator*

Houston
Dawood, Mohamed Yusoff *obstetrician/gynecologist*
Ross, Patti Jayne *obstetrics and gynecology educator*

Katy
Tal, Jacob *obstetrician/gynecologist*

San Angelo
Dunham, Gregory Mark *obstetrician/gynecologist*

San Antonio
Jensen, Andrew Oden *obstetrician/gynecologist*

VIRGINIA

Norfolk
Anderson, Freedolph Deryl *gynecologist*

WISCONSIN

Madison
Jackson, Carl Robert *obstetrician/gynecologist*

AUSTRALIA

Sydney
McBride, William Griffith *research gynecologist*

FINLAND

Hämeenlinna
Kivinen, Seppo Tapio *obstetrician/gynecologist, hospital administrator*

Helsinki
Sipinen, Seppo Antero *obstetrician/gynecologist*

FRANCE

Paris
Atlan, Paul *gynecologist*
Ben Amor, Ismäil *obstetrician/gynecologist*

GERMANY

Munich
Zander, Josef *gynecologist*

Tübingen
Hirnle, Peter *gynecologist, obstetrician, radiation oncologist, cancer researcher*

GREECE

Thessaloniki
Agorastos, Theodoros *obstetrics and gynecology educator*

IRELAND

Dublin
Bonnar, John *obstetrics/gynecology educator, consultant*

ITALY

Rome
Cosmi, Ermelando Vinicio *obstetrics/gynecology educator, consultant*

JAPAN

Ibaragi
Usuki, Satoshi *physician, educator*

Kishiwada
Tateyama, Ichiro *gynecologist*

THE PHILIPPINES

Manila
Benitez, Isidro Basa *obstetrician/gynecologist, oncologist*

POLAND

Kraków
Kowalczyk, Maciej Stanislaw *obstetrician/gynecologist*

THAILAND

Khon Kaen
Lumbiganon, Pisake *obstetrician/gynecologist*

ADDRESS UNPUBLISHED

Clemendor, Anthony Arnold *obstetrician, gynecologist, educator*
Gusdon, John Paul, Jr. *obstetrics/gynecology educator, physician*
Hakim-Elahi, Enayat *obstetrician/gynecologist, educator*
Kent, Howard Lees *obstetrician/gynecologist*
Nicholls, Richard Aurelius *obstetrician, gynecologist*

MEDICINE: ONCOLOGY

UNITED STATES

ALABAMA

Birmingham
Durant, John Ridgeway *physician*

CALIFORNIA

Berkeley
Castro, Joseph Ronald *physician, oncology researcher, educator*

Loma Linda
Slater, James Munro *radiation oncologist*

Los Angeles
Haskell, Charles Mortimer *medical oncologist, educator*

San Bernardino
Skoog, William Arthur *oncologist*

Stanford
Brown, J. Martin *oncologist, educator*
Levy, Ronald *medical educator, researcher*
Rosenberg, Saul Allen *oncologist, educator*

CONNECTICUT

New Haven
DeVita, Vincent Theodore, Jr. *oncologist*

DISTRICT OF COLUMBIA

Washington
Goldson, Alfred Lloyd *oncologist educator*
Perry, Seymour Monroe *physician*

FLORIDA

Tampa
del Regato, Juan Angel *radio-therapeutist and oncologist, educator*

GEORGIA

Atlanta
Davis, Lawrence William *radiation oncologist*

ILLINOIS

Arlington Heights
Shetty, Mulki Radhakrishna *oncologist, consultant*

Chicago
Rosen, Steven Terry *oncologist, hematologist*
Weichselbaum, Ralph R. *oncologist chairman*

Springfield
Fields, Joseph Newton, III *oncologist*

IOWA

West Des Moines
Brunk, Samuel Frederick *oncologist*

KANSAS

Mission
Van Veldhuizen, Peter Jay *oncologist, hematologist*

KENTUCKY

Louisville
La Rocca, Renato V. *oncologist, researcher*

MARYLAND

Baltimore
Aisner, Joseph *oncologist, physician*
Owens, Albert Henry, Jr. *oncologist, educator*
Vogelstein, Bert *oncology educator*

Bethesda
Okunieff, Paul *radiation oncologist, physician*
Sausville, Edward Anthony *medical oncologist*
Schlom, Jeffrey Bert *research scientist*
Sporn, Michael Benjamin *cancer etiologist*
Yuspa, Stuart H. *cancer etiologist*

MASSACHUSETTS

Boston
Steele, Glenn Daniel, Jr. *surgical oncologist*

Shrewsbury
Zamecnik, Paul Charles *oncologist, medical research scientist*

MICHIGAN

Ann Arbor
Wicha, Max S. *oncologist, educator*

Detroit
Baker, Laurence Howard *oncology educator*
Porter, Arthur T. *oncologist, educator*

Livonia
Gordon, Craig Jeffrey *oncologist*

MISSOURI

Columbia
Khojasteh, Ali *medical oncologist, hematologist*

NEBRASKA

Omaha
Korbitz, Bernard Carl *oncologist, hematologist, educator, consultant*

NEVADA

Reno
MacKintosh, Frederick Roy *oncologist*

NEW HAMPSHIRE

Manchester
Khazei, Amir Mohsen *surgeon, oncologist*

NEW JERSEY

Newark
Kazem, Ismail *radiation oncologist, educator, health science facility administrator*
Zirvi, Karimullah Abd *experimental oncologist, educator*

Woodbury
Stambaugh, John Edgar *oncologist, hematologist, pharmacologist, educator*

NEW YORK

Bronx
Dutcher, Janice Jean Phillips *medical oncologist*
Wiernik, Peter Harris *oncologist, educator*

Buffalo
Horoszewicz, Juliusz Stanislaw *oncologist, cancer researcher, laboratory administrator*
Mayhew, Eric George *cancer researcher, educator*
Piver, M. Steven *gynecologic oncologist*

Mount Kisco
Schneider, Robert Jay *oncologist*

New York
Biedler, June L. *oncologist*
Cammarata, Angelo *surgical oncologist*
Goldsmith, Michael Allen *medical oncologist, educator*
Kushner, Brian Harris *pediatric oncologist*
Marks, Paul Alan *oncologist, cell biologist*
Mendelsohn, John *oncologist, hematologist, educator*

Pomona
Jaffrey, Ira *oncologist, educator*

Syracuse
Gold, Joseph *medical researcher*
Poiesz, Bernard Joseph *oncologist*

Valhalla
Chung, Fung-Lung *cancer research scientist*

NORTH CAROLINA

Chapel Hill
Rosenman, Julian Gary *radiation oncologist*

PENNSYLVANIA

Philadelphia
Comis, Robert Leo *oncologist, educator*
Glick, John H. *oncologist, medical educator*
Rubin, Stephen Curtis *gynecologic oncologist, educator*

TEXAS

Houston
Dreizen, Samuel *oncologist*
Hong, Waun Ki *medical oncologist, clinical investigator*

Tyler
D'Andrea, Mark *radiation oncologist*

WISCONSIN

Madison
Carbone, Paul Peter *oncologist, educator, administrator*
Temin, Howard Martin *scientist, educator*

TERRITORIES OF THE UNITED STATES

PUERTO RICO

San Juan
Rodriguez Arroyo, Jesus *gynecologic oncologist*

CANADA

ONTARIO

Toronto
Ling, Victor *oncologist, educator*
Till, James Edgar *scientist*

MEXICO

Juarez
Torres Medina, Emilio *oncologist, consultant*

AUSTRIA

Vienna
Wrba, Heinrich *oncologist, research institute administrator*

BELGIUM

Antwerp
Uyttenbroeck, Frans Joseph *gynecologic oncologist*

CHINA

Tianjin
Zhang, Theodore Tian-ze *oncologist, health association administrator*

EGYPT

Maadi-Cairo
Amer, Magid Hashim *oncologist, physician*

ENGLAND

London
Retsas, Spyros *oncologist*

GERMANY

Baden
Musshoff, Karl Albert *radiation oncologist*

ITALY

Milan
Bonadonna, Gianni *oncologist*

Rome
Frati, Luigi *oncologist, pathologist*

THE NETHERLANDS

Tilburg
Maat, Benjamin *radiation oncologist*

SWEDEN

Stockholm
Mellstedt, HÅkan SÖren Thure *oncologist, medical facility administrator*

ADDRESS UNPUBLISHED

Chin, Hong Woo *oncologist, educator, researcher*
Markoe, Arnold Michael *radiation oncologist*
Thuning-Robinson, Claire *oncologist*

MEDICINE: OPHTHALMOLOGY

UNITED STATES

ALABAMA

Birmingham
Skalka, Harold Walter *ophthalmologist, educator*

ARIZONA

Phoenix
Lorenzen, Robert Frederick *ophthalmologist*

Tucson
Potts, Albert M. *ophthalmologist*

CALIFORNIA

Beverly Hills
Fein, William *ophthalmologist*

Brawley
Jaquith, George Oakes *ophthalmologist*

La Jolla
Goldbaum, Michael Henry *ophthalmologist*

Los Angeles
Ryan, Stephen Joseph, Jr. *ophthalmology educator, university dean*
Straatsma, Bradley Ralph *ophthalmologist, educator*

Salinas
Lee, Gilbert Brooks *retired ophthalmology engineer*

San Francisco
Dawson, Chandler R. *opthalmologist, educator*
Jampolsky, Arthur *ophthalmologist*

Visalia
Riegel, Byron William *ophthalmologist*

COLORADO

Denver
Lubeck, Marvin Jay *ophthalmologist*

CONNECTICUT

Manchester
Milewski, Stanislaw Antoni *ophthalmologist, educator*

New Haven
Silverstone, David Edward *ophthalmologist*

Stamford
Walsh, Thomas Joseph *neuro-ophthalmologist*

FLORIDA

Atlantis
Newmark, Emanuel *ophthalmologist*

Gainesville
Rubin, Melvin Lynne *ophthalmologist, educator*

New Port Richey
Hauber, Frederick August *ophthalmologist*

GEORGIA

Atlanta
Edelhauser, Henry F. *physiologist, ophthalmologist, medical educator*

HAWAII

Honolulu
Pang, Herbert George *ophthalmologist*
Sugiki, Shigemi *ophthalmologist, educator*

MARYLAND

Baltimore
Goldberg, Morton Falk *ophthalmologist, educator*
Knox, David LaLonde *ophthalmologist*

Bethesda
Datiles, Manuel Bernaldes, III *ophthalmologist*
Kupfer, Carl *ophthalmologist, science administrator*

MASSACHUSETTS

Boston
Berson, Eliot Lawrence *ophthalmologist, medical educator*
Puliafito, Carmen Anthony *ophthalmologist, laser researcher*

Brookline
Kraut, Joel Arthur *ophthalmologist*

Northampton
Mintzer, Paul *ophthalmologist, educator*

MICHIGAN

Ann Arbor
Lichter, Paul Richard *ophthalmology educator*

Bad Axe
Rosenfeld, Joel *ophthalmologist*

Dearborn
Coburn, Ronald Murray *ophthalmic surgeon, researcher*

Detroit
Jampel, Robert Steven *ophthalmologist, educator*

Rochester
Reddy, Venkat Narsimha *ophthalmalogist, researcher*

MINNESOTA

Minneapolis
Wirtschafter, Jonathan Dine *neuro-ophthalmology educator, scientist*

MISSISSIPPI

Jackson
Russell, Robert Pritchard *ophthalmologist*

Laurel
Lindstrom, Eric Everett *ophthalmologist*

NEVADA

Carson City
Fischer, Michael John *ophthalmologist, physician*

Las Vegas
Buzard, Kurt Andre *ophthalmologist*

NEW JERSEY

Newark
Materna, Thomas Walter *ophthalmologist*

NEW YORK

Bronx
Lubkin, Virginia Leila *ophthalmologist*

Fishkill
Brocks, Eric Randy *ophthalmologist, surgeon*

Katonah
Mooney, Robert Michael *ophthalmologist*

New York
Candia, Oscar A. *ophthalmologist, physiology educator*
Haddad, Heskel Marshall *ophthalmologist*
Hyman, Bruce Malcolm *ophthalmologist*
Kelman, Charles D. *ophthalmologist, educator*
Lipton, Lester *ophthalmologist, entrepreneur*
Muchnick, Richard Stuart *ophthalmologist, educator*

Olean
Catalano, Robert Anthony *ophthalmologist, physician, hospital administrator, writer*

Staten Island
Greenfield, Val Shea *ophthalmologist*

NORTH CAROLINA

Greensboro
Cotter, John Burley *ophthalmologist, corneal specialist*

Research Triangle Park
Friedland, Beth Rena *ophthalmologist*

OHIO

Columbus
Fryczkowski, Andrzej Witold *ophthalmologist, educator, business executive*

OKLAHOMA

Oklahoma City
Bradford, Reagan Howard, Jr. *ophthalmology educator*
Parke, David W., II *ophthalmologist, educator, healthcare executive*

OREGON

Portland
Prendergast, William John *ophthalmologist*

PENNSYLVANIA

Philadelphia
DellaVecchia, Michael Anthony *ophthalmologist*

Waynesboro
Cryer, Theodore Hudson *ophthalmologist, educator*

SOUTH CAROLINA

Aiken
Gleichauf, John George *ophthalmologist*

Charleston
Apple, David Joseph *ophthalmology educator*

Columbia
Schwarz, Ferdinand (Fred Schwarz) *ophthalmologist, ophthalmic plastic surgeon*

TEXAS

Fort Worth
Smith, Thomas Hunter *ophthalmologist, ophthalmic plastic and orbital surgeon*

Galveston
Gold, Daniel Howard *ophthalmologist, educator*

Houston
Daily, Louis *ophthalmologist*
Hollyfield, Joe G. *ophthalmology educator*

VIRGINIA

Arlington
Sheridan, Andrew James, III *ophthalmologist*

Falls Church
Evans, Peter Yoshio *ophthalmologist, educator*

WEST VIRGINIA

Bluefield
Blaydes, James Elliott *ophthalmologist*

Morgantown
Weinstein, George William *ophthalmology educator*

WISCONSIN

Milwaukee
Gonnering, Russell Stephen *ophthalmic plastic surgeon*

CANADA

BRITISH COLUMBIA

Vancouver
Drance, Stephen Michael *ophthalmologist, educator*

AUSTRALIA

Brisbane
English, Francis Peter *ophthalmologist, educator*

Melbourne
Taylor, Hugh Ringland *ophthalmologist, educator*

ENGLAND

Dewsbury
Mishra, Arun Kumar *ophthalmologist*

Herefordshire
Sandhu, Bachittar Singh *ophthalmologist*

London
Arnott, Eric John *ophthalmologist*
Choyce, David Peter *ophthalmologist*

FINLAND

Helsinki
Tervo, Timo Martti *ophthalmologist*

FRANCE

Creteil
Coscas, Gabriel Josue *ophthalmologist, educator*

Landivisiau
Floch-Baillet, Daniele Luce *ophthalmologist*

GERMANY

Sulzbach
Mester, Ulrich *ophthalmologist*

ISRAEL

Tel Hashomer
Belkin, Michael *ophthalmologist, educator, researcher*

JAPAN

Miyagi
Nakazawa, Mitsuru *ophthalmologist, educator*

Osaka
Kinoshita, Shigeru *ophthalmologist*

PAKISTAN

Larkana
Soomro, Akbar Haider *ophthalmologist, educator*

PORTUGAL

Coimbra
Cunha-Vaz, Jose Guilherme Fernandes *ophthalmologist*

ADDRESS UNPUBLISHED

Esterman, Benjamin *ophthalmologist*
Monninger, Robert Harold George *ophthalmologist, educator*
Woodhouse, Derrick Fergus *ophthalmologist*

MEDICINE: ORTHOPAEDICS

UNITED STATES

ARIZONA

Sun City
Sabanas-Wells, Alvina Olga *orthopedic surgeon*

ARKANSAS

Conway
McCarron, Robert Frederick, II *orthopedic surgeon*

CALIFORNIA

Los Angeles
Bao, Joseph Yue-Se *orthopaedist, microsurgeon, educator*
Urist, Marshall Raymond *orthopedic surgeon, researcher*

CONNECTICUT

New Haven
Friedlaender, Gary Elliott *orthopedist, educator*

Wolcott
Hillsman, Regina Onie *orthopedic surgeon*

DELAWARE

Milford
Quinn, Edward Francis, III *orthopedic surgeon*

FLORIDA

Gainesville
Strates, Basil Stavros *biomedical educator, researcher*

Key West
Benavides, Jaime Miguel *orthopedist*

Miami Beach
Lehrman, David *orthopedic surgeon*

Stuart
Jaller, Michael M. *retired orthopaedic surgeon*

West Palm Beach
Whitfield, Graham Frank *orthopedic surgeon*

ILLINOIS

Chicago
Meyer, Paul Reims, Jr. *orthopaedic surgeon*
Singh, Manmohan *orthopedic surgeon, educator*

Oak Park
Brackett, Edward Boone, III *orthopedic surgeon*

INDIANA

Indianapolis
Lindseth, Richard Emil *orthopaedic surgeon*

IOWA

Iowa City
Cooper, Reginald Rudyard *orthopaedic surgeon, educator*

MARYLAND

Easton
Thompson, Robert Campbell *orthopaedic surgeon*

Silver Spring
Schonholtz, George Jerome *orthopaedic surgeon*

MASSACHUSETTS

Boston
Glimcher, Melvin Jacob *orthopedic surgeon*
Lavine, Leroy Stanley *orthopedist, surgeon, consultant*
Leach, Robert Ellis *physician, educator*
Shields, Lawrence Thornton *orthopedic surgeon, educator*
Sledge, Clement Blount *orthopedic surgeon, educator*

Brighton
Cohen, Jonathan *orthopedic surgery educator, researcher*

Lunenburg
Rhodin, Anders G.J. *orthopaedic surgeon, chelonian researcher and herpetologist*

MICHIGAN

Detroit
Fitzgerald, Robert Hannon, Jr. *orthopedic surgeon*

MISSOURI

Columbia
Corcoran, Michael John *orthopedic surgeon*

Saint Louis
Kuhlman, Robert E. *orthopedic surgeon*

NEVADA

Las Vegas
Rask, Michael Raymond *orthopaedist*

NEW MEXICO

Albuquerque
Omer, George Elbert, Jr. *orthopaedic surgeon, educator*

NEW YORK

Albany
Dougherty, James *orthopedic surgeon, educator*

Bronx
Bassett, C(harles) Andrew L(oockerman) *orthopaedic surgeon, educator*

New York
Ergas, Enrique *orthopedic surgeon*
McClelland, Shearwood Junior *orthopaedic surgeon*
Rubin, Gustav *orthopedic surgeon, consultant, researcher*

Syracuse
Baker, Bruce Edward *orthopedic surgeon, consultant*

NORTH CAROLINA

Durham
Baker, Lenox Dial *orthopaedist, genealogist*

OKLAHOMA

Muskogee
Dandridge, William Shelton *orthopedic surgeon*

Tulsa
Nebergall, Robert William *orthopedic surgeon, educator*

Vinita
Neer, Charles Sumner, II *orthopaedic surgeon, educator*

PENNSYLVANIA

Pittsburgh
Herndon, James Henry *orthopedic surgeon, educator*

RHODE ISLAND

Providence
Merlino, Anthony Frank *orthopedic surgeon*

SOUTH CAROLINA

Clemson
DeVault, William Leonard *orthopedic surgeon*

SOUTH DAKOTA

Sioux Falls
Van Demark, Robert Eugene, Sr. *orthopedic surgeon*

TEXAS

San Antonio
Hall, Brad Bailey *orthopaedic surgeon, educator*

VERMONT

Burlington
Wilder, David Gould *orthopaedic biomechanics researcher*

VIRGINIA

Richmond
Elmore, Stanley McDowell *orthopaedic surgeon*

WASHINGTON

Moses Lake
Leadbetter, Mark Renton, Jr. *orthopaedic surgeon*

WISCONSIN

Madison
Nordby, Eugene Jorgen *orthopedic surgeon*

CANADA

ONTARIO

Sault Sainte Marie
Banerjee, Samarendranath *orthopaedic surgeon*

FRANCE

Marseilles
Poitout, Dominique Gilbert M. *orthopedic surgeon, educator*

GERMANY

Munich
Toft, Jürgen Herbert *orthopedic surgeon*

JAPAN

Toyama
Tsuji, Haruo *orthopaedics educator*

SWEDEN

Lund
Wingstrand, Hans Anders *orthopedic surgeon*

ADDRESS UNPUBLISHED

Popp, Dale D. *orthopedic surgeon*
Worrell, Richard Vernon *orthopedic surgeon, educator*

MEDICINE: OSTEOPATHY

UNITED STATES

COLORADO

Granby
McGrath, Richard William *osteopathic physician*

DISTRICT OF COLUMBIA

Washington
Petersdorf, Robert George *medical educator, association executive*

FLORIDA

Dade City
McBath, Donald Linus *osteopathic physician*

Fort Lauderdale
Price, Alexander *retired osteopathic physician*

Seminole
Schwartzberg, Roger Kerry *osteopath, internist*

ILLINOIS

Chicago
Ray, Richard Eugene *osteopath, psychiatrist*

MICHIGAN

Battle Creek
Waite, Lawrence Wesley *osteopathic physician*

MISSISSIPPI

Jackson
Forks, Thomas Paul *osteopath*

MISSOURI

Florissant
Schwarze, Robert Francis *osteopath, dermatologist*

NORTH CAROLINA

Fort Bragg
Sears, Catherine Marie *osteopath, radiologist*

OHIO

Cincinnati
Cole, Theodore John *osteopathic physician*

Columbus
Hilliard, Kirk Loveland, Jr. *osteopathic physician, educator*
Hom, Theresa Maria *osteopathic physician*

Greenfield
Jenkins, James William *osteopath*

OKLAHOMA

Ada
Van Burkleo, Bill Ben *osteopath, emergency physician*

Fort Sill
Evans, Paul *osteopath*

Jenks
Wootan, Gerald Don *osteopathic physician, educator*

Tulsa
McCullough, Robert Dale, II *osteopath*
Schmidt, Gregory Martin *osteopathic physician, independent oil producer*

PENNSYLVANIA

Philadelphia
Spector, Harvey M. *osteopathic physician*

TEXAS

Fort Worth
Brooks, Lloyd William, Jr. *osteopath, interventional cardiologist, educator*
Oakford, Lawrence Xavier *electron microscopist, laboratory administrator*

Granbury
Wilson, Robert Storey *osteopathic physician*

Raymondville
Montgomery-Davis, Joseph *osteopathic physician*

Rockwall
Sparks, Sherman Paul *osteopathic physician*

VIRGINIA

Virginia Beach
Kornylak, Harold John *osteopathic physician*

ADDRESS UNPUBLISHED

Linz, Anthony James *osteopathic physician, consultant, educator*
Maurer, Robert (Stanley) *osteopathic physician*

MEDICINE: OTHER SPECIALTIES

UNITED STATES

ALABAMA

Alexander City
Powers, Runas, Jr. *rheumatologist*

Birmingham
Stover, Samuel Landis *physiatrist*

ARIZONA

Scottsdale
Pomeroy, Kent Lytle *physical medicine and rehabilitation physician*

Tucson
Kischer, Clayton Ward *embryologist, educator*
Verdery, Roy Burton, III *gerontologist, consultant*

ARKANSAS

Little Rock
Maloney, Francis Patrick *physiatrist*

CALIFORNIA

Campbell
Wu, William Lung-Shen (You-Ming Wu) *aerospace medical engineering design specialist*

Long Beach
Tabrisky, Phyllis Page *physiatrist, educator*

San Francisco
Arieff, Allen Ives *physician*
Engleman, Ephraim Philip *physician*

COLORADO

Denver
Kassan, Stuart S. *rheumatologist*

CONNECTICUT

New Haven
Jacoby, Robert Ottinger *comparative medicine educator*

DISTRICT OF COLUMBIA

Washington
Haber, Paul Adrian Life *geriatrician*
Herman, Barbara Helen *pediatric psychiatrist, educator*
Holloway, Harry *aerospace medical doctor*

FLORIDA

Miami
Chang, Hsuan Hung *dosimetrist*

GEORGIA

Augusta
Oleskowicz, Jeanette *physician*

ILLINOIS

Chicago
Oshiyoye, Adekunle Emmanuel *physician, realtor*
Pope, Richard M. *rheumatologist*
Schwartzberg, Joanne Gilbert *physician*

INDIANA

Huntingburg
Von Taaffe-Rossmann, Cosima T. *physician, writer, inventor*

KANSAS

Manhattan
Oehme, Frederick Wolfgang *medical researcher and educator*

Parsons
Spradlin, Joseph E. *embryologist, psychologist, medical educator*

KENTUCKY

Louisville
Hasselbacher, Peter *rheumatologist, educator*

Whitesville
Robertson, Clifford Houston *physician*

MARYLAND

Bethesda
Krause, Richard Michael *medical scientist, government official, educator*
Lockshin, Michael Dan *rheumatologist*
Smith, Thomas Graves, Jr. *image processing scientist, neurophysiologist*
Wolffe, Alan Paul *molecular embryologist, molecular biologist*

Rockville
Harkonen, Wesley Scott *physician*

MASSACHUSETTS

Boston
Hay, Elizabeth Dexter *embryology researcher, educator*

Springfield
Kottamasu, Mohan Rao *physician*

Watertown
Moskowitz, Richard *physician*

MICHIGAN

Detroit
Roth, Thomas *psychiatry educator*

Southfield
Morales, Raul Hector *physician*

MINNESOTA

Minneapolis
Craig, James Lynn *physician, consumer products company executive*
Greaves, Ian Alexander *occupational physician*

MISSOURI

Saint Louis
Mattison, Richard *psychiatry educator*

NEBRASKA

Omaha
Klassen, Lynell W. *rheumatologist, transplant immunologist*

NEW YORK

Bath
Huang, Edwin I-Chuen *physician, environmental researcher*

Bronx
Horwitz, Susan Band *molecular pharmacologist*

Buffalo
Calkins, Evan *physician, educator*

New York
Davis, Kenneth Leon *psychiatrist, pharmacologist, medical educator*
Dayanim, Farangis *occupational and environmental medicine consultant*
Lee, Mathew Hung Mun *physiatrist*

Roosevelt Island
Montemayor, Jesus Samson *physician*

Syracuse
Hyla, James Franklin *rheumatologist, educator*

Valhalla
Williams, Gary Murray *medical researcher, pathology educator*

NORTH CAROLINA

Asheville
Von Viczay, Marika (Ilona) *naturopathic medical doctor, physician*

Chapel Hill
Williams, Mark Edward *geriatrician*
Winfield, John Buckner *rheumatologist, educator*

Monroe
Smith, Jeffrey Alan *occupational medicine physician, toxicologist*

OHIO

Columbus
Kaplan, Paul Elias *physiatrist, educator*

OKLAHOMA

McLoud
Whinery, Michael Albert *physician*

Oklahoma City
Parker, John R. *physician, radiologist*

OREGON

Portland
Rosenbaum, James Todd *rheumatologst, educator*

PENNSYLVANIA

Erie
Makarowski, William Stephen *rheumatologist*

Philadelphia
DeHoratius, Raphael Joseph *rheumatologist*
Ridenour, Marcella V. *motor development educator*

SOUTH CAROLINA

Charleston
LeRoy, Edward Carwile *rheumatologist*

TEXAS

Houston
Fisher, Anna Lee *physician, astronaut*

San Antonio
Yu, Byung Pal *gerontological educator*

VIRGINIA

Vienna
Austin, Frank Hutches, Jr. *aerospace physician, educator*

WASHINGTON

Seattle
Cardenas, Diana Delia *physician, educator*

WISCONSIN

Milwaukee
Gorelick, Jeffrey Bruce *physician, educator*

CANADA

ONTARIO

Hamilton
Basmajian, John Varoujan *medical scientist, educator, physician*

North York
Wyatt, Philip Richard *geneticist, physician, researcher*

DENMARK

Bronshoj
Skylv, Grethe Krogh *rheumatologist, anthropologist*

ENGLAND

Farnborough
Benson, Alan James *avaiation medical doctor*

FRANCE

Aix Les Bains
Tabau, Robert Louis *rheumatologist, researcher*

Périgueux
Delluc, Gilles *physician, researcher*

GERMANY

Munich
Paumgartner, Gustav *hepatologist, educator*

Paderborn
Frank, Helmar Gunter *educational cyberneticist*

INDIA

Madras
Chandra Sekharan, Pakkirisamy *forensic scientist*

ITALY

Rome
Rotondo, Gaetano Mario *aerospace medicine physician, retired military officer*

THE NETHERLANDS

Maastricht
Verheyen, Marcel Mathieu *homoeopathist, consultant*

POLAND

Lodz
Kmieć, Bogumil Leon *embryologist, educator, histologist*

SWEDEN

Lund
Sundler, Frank Eskil Georg *histology educator, biomedical scientist*

Strömstad
Vigmo, Josef *retired geriatrician*

ADDRESS UNPUBLISHED

Goodman, Alan Noel *physician*
Lacerna, Leocadio Valderrama *research physician*
Wong, Dennis Ka-Cheong *physician, physical therapist*

MEDICINE: OTOLARYNGOLOGY

UNITED STATES

CALIFORNIA

Covina
Takei, Toshihisa *otolaryngologist*

Los Angeles
House, John W. *otologist*

Newport Beach
Zalta, Edward *otorhinolaryngologist, utilization review physician*

CONNECTICUT

Stamford
Rosenberg, Charles Harvey *otorhinolaryngologist*

DISTRICT OF COLUMBIA

Washington
Feldman, Bruce Allen *otolaryngologist*

FLORIDA

Palm Beach Gardens
Jacobson, Alan Leonard *otolaryngologist*

GEORGIA

Atlanta
Turner, John Sidney, Jr. *otolaryngologist, educator*

Savannah
Zoller, Michael *otolaryngologist*

ILLINOIS

Elmhurst
Fornatto, Elio Joseph *otolaryngologist, educator*

INDIANA

Indianapolis
Miyamoto, Richard Takashi *otolaryngologist*

LOUISIANA

New Orleans
Berlin, Charles I. *otolaryngologist, educator*

MARYLAND

Baltimore
Johns, Michael Marieb Edward *otolaryngologist, university dean*

Bethesda
Snow, James Byron, Jr. *physician, research administrator*

MASSACHUSETTS

Boston
Kimura, Robert Shigetsugu *otologic researcher*

MICHIGAN

Ann Arbor
Arts, Henry Alexander *otolaryngologist*
Hawkins, Joseph Elmer, Jr. *acoustic physiologist*
Miller, Josef M. *otolaryngologist, educator*

MISSOURI

Saint Louis
Hirsch, Ira J. *otolaryngologist, educator*

NEW YORK

Bronx
Ruben, Robert Joel *physician, educator*

New York
Aviv, Jonathan Enoch *otolaryngologist, educator*
Blitzer, Andrew *otolaryngologist, educator*
Brookler, Kenneth Haskell *otolaryngologist, educator*
Cohen, Noel Lee *otolaryngologist, educator*
Coleman, Lester Laudy *otolaryngologist*
Conley, John Joseph *otolaryngologist*
Komisar, Arnold *otolaryngologist, educator*

NORTH CAROLINA

Chapel Hill
Henson, Anna Miriam *otolaryngology researcher, medical educator*

OHIO

Canton
Maioriello, Richard Patrick *otolaryngologist*

Chillicothe
Chen, Wen Fu *otolaryngologist*

Zanesville
Ray, John Walker *otolaryngologist, educator*

OKLAHOMA

Oklahoma City
Hough, Jack Van Doren *otologist*

OREGON

Portland
Vernon, Jack Allen *otolaryngology educator, laboratory administrator*

PENNSYLVANIA

Wynnewood
Harkins, Herbert Perrin *otolaryngologist, educator*

TENNESSEE

Memphis
Lazar, Rande Harris *otolaryngologist*

VIRGINIA

Chesapeake
Crane, Richard Turner *otolaryngologist*

TERRITORIES OF THE UNITED STATES

PUERTO RICO

Bayamon
Juarbe, Charles *otolaryngologist, neck surgeon*

GERMANY

Tübingen
Zenner, Hans Peter *otolaryngologist*

JAPAN

Okinawa
Noda, Yutaka *physician, otolaryngologist*

Ube
Sekitani, Toru *otolaryngologist, educator*

MEDICINE: PATHOLOGY

UNITED STATES

ALABAMA

Birmingham
Anderson, Peter Glennie *research pathologist, educator*
Kelly, David Reid *pathologist*
Mowry, Robert Wilbur *pathologist, educator*

Florence
Burford, Alexander Mitchell, Jr. *physician, pathologist*

CALIFORNIA

Glendale
Dent, Ernest DuBose, Jr. *pathologist*

La Jolla
Terry, Robert Davis *neuropathologist, educator*

Los Angeles
Lewin, Klaus J. *pathologist, educator*
Nathwani, Bharat Narottam *pathologist, consultant*

Reseda
Ross, Amy Ann *experimental pathologist*

San Francisco
Jeffrey, Robert Asahel, Jr. *pathologist*

Stanford
Cleary, Michael *pathologist, educator*

COLORADO

Denver
Krikos, George Alexander *pathologist, educator*

CONNECTICUT

Bridgeport
Bernstein, Larry Howard *clinical pathologist*

Hartford
Pastuszak, William Theodore *hematopathologist*

New Haven
Manuelidis, Laura *pathologist, neuropathologist, experimentalist*
Pober, Jordan S. *pathologist, educator*

DISTRICT OF COLUMBIA

Washington
Redman, Robert Shelton *pathologist, dentist*
Wagner, Glenn Norman *pathologist*

FLORIDA

Miami
Reik, Rita Ann Fitzpatrick *pathologist*

Sarasota
Klutzow, Friedrich Wilhelm *neuropathologist*

GEORGIA

Atlanta
Cooper, Gerald Rice *clinical pathologist*
Huber, Douglas Crawford *pathologist*
Schwartz, David Alan *infectious disease and placental pathologist*

Augusta
Krauss, Jonathan Seth *pathologist*

Griffin
Carter, Edward Fenton, III *pathologist, medical examiner*
Gillaspie, Athey Graves, Jr. *pathologist, researcher*

HAWAII

Honolulu
McCarthy, Laurence James *physician, pathologist*

ILLINOIS

Chicago
Boggs, Joseph Dodridge *pediatric pathologist, educator*
Hinojosa, Raul *physician, ear pathology researcher*
McLawhon, Ronald William *pathology educator, biochemist*
Ramsey, Glenn Eugene *pathologist, blood bank physician*
Swerdlow, Martin Abraham *physician, pathologist*

Hinsdale
Robertson, Abel L., Jr. *pathologist*

Olympia Fields
Kasimos, John Nicholas *pathologist*

INDIANA

Indianapolis
Allen, Stephen D(ean) *pathologist, microbiologist*

IOWA

Iowa City
Hammond, Harold Logan *pathology educator, oral pathologist*

KANSAS

Kansas City
Cuppage, Francis Edward *physician, educator*

LOUISIANA

New Orleans
Gerber, Michael Albert *pathologist, researcher*

MARYLAND

Baltimore
Crain, Barbara Jean *pathologist, educator*
Prendergast, Robert A. *pathologist educator*
Racusen, Lorraine Claire *pathologist, researcher*

Beltsville
Hackett, Kevin James *insect pathologist*

Bethesda
Cheever, Allen Williams *pathologist*
Elin, Ronald John *pathologist*
Friedman, Robert Morris *pathologist, molecular biologist*
Liotta, Lance A. *pathologist*
Rabson, Alan Saul *physician, educator*
Saffiotti, Umberto *pathologist*

MASSACHUSETTS

Boston
Benacerraf, Baruj *pathologist, educator*
Ehrmann, Robert Lincoln *pathologist*
Gottlieb, Leonard Solomon *pathology educator*
Hutchinson, Martha LuClare *pathologist*
Ingber, Donald Elliot *pathology and cell biology educator*
Karnovsky, Morris John *pathologist, biologist*
Schneeberger, Eveline Elsa *pathologist, cell biologist, educator*

Falmouth
Sato, Kazuyoshi *pathologist*

MICHIGAN

Allen Park
Kaldor, George *pathologist, educator*

Detroit
Riser, Bruce L. *research pathologist*

Flint
Himes, George Elliott *pathologist*

Royal Oak
Robbins, Thomas Owen *pathologist, educator*

Saint Clair Shores
Walker, Frank Banghart *pathologist*

MISSISSIPPI

Jackson
Dunsford, Harold Atkinson *pathologist, researcher*

MISSOURI

Kansas City
Welling, Larry Wayne *pathologist, educator, physiologist*

MONTANA

Billings
Mueller, Kenneth Howard *pathologist*

NEBRASKA

Omaha
Chan, Wing-Chung *pathologist, educator*

NEW JERSEY

Alpine
Sommers, Sheldon Charles *pathologist*

Livingston
Caballes, Romeo Lopez *pathologist, bone tumor researcher*

Newark
Goldenberg, David Milton *experimental pathologist, oncologist*

Teaneck
Churg, Jacob *pathologist*

NEW YORK

Buffalo
Brooks, John Samuel Joseph *pathologist, researcher*

Ithaca
Hajek, Ann Elizabeth *insect pathologist*
Quimby, Fred William *pathology educator, veterinarian*

New York
Baden, Michael M. *pathologist, educator*
Ellis, John Taylor *pathologist, educator*
Kaunitz, Hans *physician, pathologist*
Liu, Si-kwang *veterinary pathologist*
Upton, Arthur Canfield *retired experimental pathologist*
Yee, Herman Terence *pathologist*

Oneida
Muschenheim, Frederick *pathologist*

Port Jefferson
Hirschl, Simon *pathologist*

Williamsville
Hertzog, Robert William *pathologist, consultant, educator*

NORTH CAROLINA

Chapel Hill
Brinkhous, Kenneth Merle *pathologist, educator*
Grisham, Joe Wheeler *pathologist, educator*
Smithies, Oliver *pathologist, educator*

Durham
Adams, Dolph O. *pathologist, educator*
Jennings, Robert Burgess *experimental pathologist, medical educator*

Greensboro
Baird, Haynes Wallace *pathologist*

Greenville
Volkman, Alvin *pathologist, educator*

Hampstead
Solomon, Robert Douglas *pathology educator*

Oteen
Chapman, William Edward *pathologist*

OHIO

Cincinnati
Fenoglio-Preiser, Cecilia Mettler *pathologist, educator*
Fody, Edward Paul *pathologist*
Pavelic, Zlatko P. *physician, pathologist*

Girard
German, Norton Isaiah *pathologist, educator*

Strongsville
Opplt, Jan Jiri *pathologist, educator*

Toledo
Harris, James Herman *pathologist*
Stoner, Gary David *pathology educator*

PENNSYLVANIA

Chalfont
Mendlowski, Bronislaw *retired pathologist*

Philadelphia
Colman, Robert Wolf *physician, medical educator*
Conn, Rex Boland, Jr. *physician, educator*
De La Cadena, Raul Alvarez *physician, pathology and thrombosis educator*
Lewis, Paul Le Roy *pathology educator*
Ming, Si-Chun *pathologist, educator*
Taichman, Norton Stanley *pathology educator*
Young, Donald Stirling *clinical pathology educator*
Zimmerman, Michael Raymond *pathology educator, anthropologist*

Pittsburgh
Perper, Joshua Arte *forensic pathologist*

RHODE ISLAND

Providence
Martin, Horace Feleciano *pathologist, law educator*

SOUTH CAROLINA

Charleston
Sens, Mary Ann *pathology educator*

Greenville
Kilgore, Donald Gibson, Jr. *pathologist*

TENNESSEE

Jackson
Harwood, Thomas Riegel *pathologist*

Johnson City
Youngberg, George Anthony *pathology educator*

Memphis
Shanklin, Douglas Radford *physician*

Nashville
Pribor, Hugo Casimer *physician*

TEXAS

Dallas
McCracken, Alexander Walker *pathologist*
Montgomery, Philip O'Bryan, Jr. *pathologist*

Houston
Bruner, Janet M. *neuropathologist*
Mc Bride, Raymond Andrew *pathologist, physician, educator*
Milam, John Daniel *pathologist, educator*
Ordonez, Nelson Gonzalo *pathologist*
Popek, Edwina Jane *pathologist*
Sell, Stewart *pathologist, immunologist, educator*
Stimson, Paul Gary *pathologist*

San Antonio
Di Maio, Vincent Joseph Martin *forensic pathologist*
Jorgensen, James H. *pathologist, educator, microbiologist*
Townsend, Frank Marion *pathology educator*

UTAH

Salt Lake City
Matsen, John Martin *pathology educator, microbiologist*

VERMONT

Burlington
Morrow, Paul Lowell *forensic pathologist*

VIRGINIA

Lynchburg
Cresson, David Homer, Jr. *pathologist*

Reston
Jaffe, Russell Merritt *pathologist, research director*

Richmond
Hadfield, M. Gary *neuropathologist, educator*
Kornstein, Michael Jeffrey *pathologist*
Sirica, Alphonse Eugene *pathology educator*

WASHINGTON

Longview
Sandstrom, Robert Edward *physician, pathologist*

Seattle
Martin, George M. *pathologist, gerontologist*
Ross, Russell *pathologist, educator*
Todaro, George Joseph *pathologist*

WEST VIRGINIA

Huntington
Leppla, David Charles *pathology educator*
Simmons, Harry Dady *pathologist*

WISCONSIN

Madison
Laessig, Ronald Harold *pathology educator, state official*
Pitot, Henry Clement, III *physician, educator*

Sheboygan
Golubski, Joseph Frank *pathologist, physician*

CANADA

ONTARIO

Ottawa
de Bold, Adolfo J. *pathology and physiology educator, research scientist*

Toronto
Broder, Irvin *pathologist, educator*

AUSTRALIA

Rockhampton
Lynch, Thomas Brendan *pathologist*

DENMARK

Hellerup
Jacobsen, Grete Krag *pathologist*

FINLAND

Turku
Collan, Yrjö Urho *pathologist, medical educator, physician, toxicopathology consultant*

FRANCE

Garches
Durigon, Michel Louis *pathologist, forensic medicine educator*

GERMANY

Freiburg
Schaefer, Hans-Eckart *pathologist*

Mainz
Schirmacher, Peter *molecular pathologist, educator*

Munich
Weis, Serge *neuropathologist*

Wuppertal
Schubert, Guenther Erich *pathologist*

GREECE

Athens
Barbatis, Calypso *histopathologist*
Sarantopoulos, Theodore *physician, cardiologist, pathologist*

GRENADA

Saint George's
Brunson, Joel Garrett *pathologist, educator*

JAPAN

Chiba
Maruyama, Koshi *pathologist, educator*

Hiroshima
Tahara, Eiichi *pathologist, educator*

Mitaka
Fukuzumi, Naoyoshi (Hai-chin Chen) *pathology educator*

Okayama
Okada, Shigeru *pathology educator*

Saitama
Sugawara, Isamu *molecular pathology educator*

MALAYSIA

Kuala Lumpur
Rajadurai, Pathmanathan *pathologist, educator, consultant*

MARTINIQUE

Fort-de-France
Bucher, Bernard Jean-Marie *immunopathologist, researcher, consultant*

PAKISTAN

Faisalabad
Irfan, Muhammad *pathology educator*

THE PHILIPPINES

Quezon City
Javier, Aileen Riego *pathologist*

SOUTH AFRICA

Johannesburg
Mendelsohn, Dennis *chemical pathology educator, consultant*

SWEDEN

Stockholm
Collins, Vincent Peter *pathologist*

SWITZERLAND

Geneva
Rabinowicz, Théodore *neuropathology educator*

ADDRESS UNPUBLISHED

Bjornsson, Johannes *pathologist*
Kemnitz, Josef Blazek *pathologist, scientist*

MEDICINE: PEDIATRICS

UNITED STATES

ALABAMA

Birmingham
Cooper, Max Dale *physician, medical educator, researcher*
Palmisano, Paul Anthony *pediatrician, educator*

ARIZONA

Phoenix
Charlton, John Kipp *pediatrician*

ARKANSAS

Little Rock
Sotomora-von Ahn, Ricardo Federico *pediatrician, educator*

CALIFORNIA

Anaheim
Liem, Annie *pediatrician*

La Jolla
Nyhan, William Leo *pediatrician, educator*

Loma Linda
Mace, John Weldon *pediatrician*

Los Angeles
Sinatra, Frank Raymond *pediatric gastroenterologist*

Orange
Silverman, Benjamin K. *pediatrician, educator*

San Diego
Harwood, Ivan Richmond *pediatric pulmonologist*

San Francisco
Heyman, Melvin Bernard *pediatric gastroenterologist*

San Jose
Stein, Arthur Oscar *pediatrician*

San Juan Capistrano
Fisher, Delbert Arthur *physician, educator*

Wilton
Shapero, Harris Joel *pediatrician*

COLORADO

Denver
Nelson, Nancy Eleanor *pediatrician, educator*
Repine, John E. *pediatrician, educator*

CONNECTICUT

New Haven
Cohen, Donald Jay *pediatrics, psychiatry and psychology educator, administrator*
Dolan, Thomas F., Jr. *pediatrician, educator*
Genel, Myron *pediatrician, educator*
Solnit, Albert Jay *commissioner, physician, educator*

Wallingford
Dunkle, Lisa Marie *pediatrics educator*

DISTRICT OF COLUMBIA

Washington
Catoe, Bette Lorrina *physician, health educator*
MacDonald, Mhairi Graham *neonatologist*
Pollack, Murray Michael *physician*
Scott, Roland Boyd *pediatrician*
Sivasubramanian, Kolinjavadi Nagarajan *neonatologist, educator*
Todd, Richard Henry *retired physican, investor*

FLORIDA

Miami
Howell, Ralph Rodney *pediatrician, educator*

Saint Petersburg
Good, Robert Alan *physician, educator*

Tallahassee
Maguire, Charlotte Edwards *retired physician*

Tampa
Barness, Lewis Abraham *physician*

Vero Beach
Cooke, Robert Edmond *physician, educator, former college president*

HAWAII

Waipahu
Caldwell, Peter Derek *pediatrician, pediatric cardiologist*

ILLINOIS

Aurora
Ball, William James *pediatrician*

Chicago
Goodman, Harold *pediatrician, consultant*
Shulman, Stanford Taylor *pediatrics educator, infectious disease researcher*
Vasa, Rohitkumar Bhupatrai *pediatrician, neonatologist*

Elk Grove Village
DeAngelis, Catherine D. *pediatrics educator*

Oak Lawn
Rathi, Manohar *neonatologist, pediatrician*

INDIANA

Indianapolis
Brady, Mary Sue *pediatric dietitian, educator*
Eigen, Howard *pediatrician, educator*
Escobar, Luis Fernando *pediatrician, geneticist*
Merritt, Doris Honig *pediatrics educator*

KANSAS

Kansas City
Moore, Wayne V. *pediatrician, educator, endocrinologist*

KENTUCKY

Louisville
Marshall, Gary Scott *pediatrician, educator*

LOUISIANA

New Orleans
Beckerman, Robert Cy *pediatrician, educator*
Corrigan, James John, Jr. *pediatrician*

MAINE

Camden
Spock, Benjamin McLane *physician, educator*

MARYLAND

Baltimore
Bosma, James Frederick *pediatrician*

Bethesda
Alexander, Duane Frederick *pediatrician, research administrator*
Levine, Arthur Samuel *physician, scientist*
Sheridan, Philp Henry *pediatrician, neurologist*

MASSACHUSETTS

Boston
Avery, Mary Ellen *pediatrician, educator*
Crocker, Allen Carrol *pediatrician*
Nathan, David Gordon *physician, educator*
Schaller, Jane Green *pediatrician*

Peabody
Lipman, Richard Paul *pediatrician*

Sharon
Honikman, Larry Howard *pediatrician*

MICHIGAN

Ann Arbor
Kurnit, David Martin *pediatrician, educator*

Detroit
Hsu, Julie Man-ching *pediatric pulmonologist*
Kaplan, Joseph *pediatrician*

MINNESOTA

Minneapolis
Baisch, Steven Dale *pediatrician*
Warwick, Warren J. *pediatrics educator*

MISSOURI

Saint Louis
Tolan, Robert Warren *pediatric infectious disease specialist*

Saint Peters
Warren, Joan Leigh *pediatrician*

NEBRASKA

Grand Island
Bosley, Warren Guy *pediatrician*

Omaha
McIntire, Matilda Stewart *pediatrician, educator, retired*

NEVADA

Las Vegas
Kurlinski, John Parker *physician*

NEW JERSEY

East Orange
Brundage, Gertrude Barnes *pediatrician*

Elizabeth
Poch, Herbert Edward *pediatrician, educator*

Old Bridge
Brennan, George Gerard *pediatrician*

Phillipsburg
Kim, Ih Chin *pediatrician*

Westfield
Schrager, Gloria Ogur *pediatrician*

NEW YORK

Albany
Winnie, Glenna Barbara *pediatric pulmonologist*

Bronx
Lopez, Rafael *pediatrician*
Neuspiel, Daniel Robert *pediatrician, epidemiologist*
Rubinstein, Arye *pediatrician, microbiology and immunology educator*

Brooklyn
Greenberg, Bernard *pediatrician*
Kravath, Richard Elliot *pediatrician, educator*
Mendez, Hermann Armando *pediatrician, educator*

Buffalo
Azizkhan, Richard George *pediatric surgeon*

Great Neck
Ratner, Harold *pediatrician, educator*

Monroe
Werzberger, Alan *pediatrician*

New York
Breslow, Jan Leslie *scientist, educator, physician*
Dell, Ralph Bishop *pediatrician, researcher*
Gaerlan, Pureza Flor Monzon *pediatrician*
Gordon, Ronnie Roslyn *pediatrics educator, consultant*
Hajjar, Katherine Amberson *physician, pediatrician*
Hirschhorn, Kurt *pediatrics educator*

Rochester
Forbes, Gilbert Burnett *physician, educator*

NORTH CAROLINA

Durham
Falletta, John Matthew *pediatrician*

Gastonia
Prince, George Edward *pediatrician*

Salisbury
Kiser, Glenn Augustus *pediatrician*

OHIO

Cleveland
Berzins, Erna Marija *physician*
Davis, Pamela Bowes *pediatric pulmonologist*
Moore, John James Cunningham *neonatologist*

Youngstown
Bearer, Cynthia Frances *neonatologist*

OKLAHOMA

Tulsa
Smith, Vernon Soruix *neonatologist, pediatrician, educator*

OREGON

Portland
Pillers, De-Ann Margaret *neonatologist*
Rosenfeld, Ron Gershon *pediatrics educator*

PENNSYLVANIA

Huntingdon
Schock, William Wallace *pediatrician*

Philadelphia
Sato, Takami *pediatrician, medical oncologist*
Schidlow, Daniel *pediatrician, medical association administrator*
Schwartz, Elias *pediatrician*

SOUTH CAROLINA

Charleston
Favaro, Mary Kaye Asperheim (Mrs. Biagino Philip Favaro) *pediatrician*
Ohning, Bryan Lawrence *neonatologist, educator*

TENNESSEE

Memphis
Mauer, Alvin Marx *physician, medical educator*

Nashville
Russell, William Evans *pediatric endocrinologist*

TEXAS

Dallas
Mize, Charles Edward *academic pediatrician*

Galveston
Richardson, Carol Joan *pediatrician*

Houston
Heird, William Carroll *pediatrician, educator*
Shearer, William T. *pediatrician, educator*

VIRGINIA

Falls Church
Ho, Hien Van *pediatrician*

Norfolk
Oelberg, David George *neonatologist, biomedical researcher*

Richmond
Kendig, Edwin Lawrence, Jr. *physician, educator*
Roth, Karl Sebastian *pediatrician*

WASHINGTON

Seattle
Clarren, Sterling Keith *pediatrician*

WEST VIRGINIA

Charleston
Zangeneh, Fereydoun *pediatrics educator, pediatric endocrinologist*

WISCONSIN

La Crosse
Costakos, Dennis Theodore *neonatologist, researcher*

CANADA

BRITISH COLUMBIA

Vancouver
Schultz, Kirk R. *pediatric hematology-oncology educator*

MANITOBA

Winnipeg
Greenberg, Arnold H. *pediatrics educator, cell biologist*

ONTARIO

Toronto
Templeton, John Marks, Jr. *pediatric surgeon, financial service executive*

AUSTRALIA

Parkville
Shann, Frank Athol *paediatrician*

BELGIUM

Liège
Battisti, Oreste Guerino *pediatrician*

BRAZIL

Campinas
Brandalise, Silvia Regina *pediatrician*

ENGLAND

London
Ross, Euan Macdonald *pediatrician, educator*

FINLAND

Helsinki
Perheentupa, Jaakko Pentti *pediatrician*

Kuopio
Matilainen, Riitta Marja *pediatrician, pediatric neurologist*

GERMANY

Mainz
Schönberger, Winfried Josef *pediatrics educator*

HONG KONG

Hong Kong
Tam, Alfred Yat-Cheung *pediatrician, consultant*

Kowloon
Yang Ko, Lillian Yang *pediatrician*

ITALY

Milan
Boehm, Günther *pediatrician*

JAPAN

Kurume
Yamashita, Fumio *pediatrics educator*

Sendai
Niitu, Yasutaka *pediatrician, educator*

Shimotsuga
Ichimura, Tohju *pediatrician, educator*

SINGAPORE

Singapore
Tan, Kim Leong *pediatrician, neonatologist, medical educator*

SWEDEN

Uppsala
Gustavson, Karl-Henrik *physician*

ADDRESS UNPUBLISHED

Brown, Elizabeth Ruth *neonatologist*
Gajdusek, Daniel Carleton *pediatrician, research virologist*
Kowlessar, Muriel *retired pediatric educator*
Richmond, Julius Benjamin *retired physician, health policy educator emeritus*
Rosemberg, Eugenia *physician, scientist, educator, medical research administrator*
Taylor, Lesli Ann *pediatric surgery educator*
Tisdale, Patrick David *retired pediatrician*
Wessel, Morris Arthur *pediatrician*
Winter, Harland Steven *pediatric gastroenterologist*

MEDICINE: PHARMACY & PHARMACOLOGY

UNITED STATES

ALABAMA

Auburn
Parsons, Daniel Lankester *pharmaceutics educator*

Birmingham
Sekar, M. Chandra *pharmacology researcher, educator*

ALASKA

Fairbanks
Stragier, Cynthia Andreas *pharmacist*

ARIZONA

Phoenix
Drea, Edward Joseph *pharmacist*

Tucson
Regan, John Ward *molecular pharmacologist*
Slack, Marion Kimball *pharmacy educator*

ARKANSAS

Little Rock
Wessinger, William David *pharmacologist*

CALIFORNIA

Davis
Hollinger, Mannfred Alan *pharmacologist, educator, toxicologist*

La Jolla
Walker, Sydney, III *pharmacologist, psychiatric administrator*

Los Angeles
Jenden, Donald James *pharmacologist, educator*
Wincor, Michael Z. *psychopharmacology educator, clinician, researcher*

San Francisco
Benet, Leslie Zachary *pharmacokineticist*
Katzung, Bertram George *pharmacologist*
Kuck, Marie Elizabeth Bukovsky *retired pharmacist*

San Rafael
Danse, Ilene Homnick Raisfeld *physician, educator, toxicologist*

South San Francisco
Gonda, Igor *pharmaceutical scientist*

COLORADO

Denver
Tabakoff, Boris *pharmacologist educator*
Weiner, Norman *pharmacology educator*

CONNECTICUT

New Haven
Sartorelli, Alan Clayton *pharmacology educator*

Ridgefield
Letts, Lindsay Gordon *pharmacologist, educator*

Storrs Mansfield
Kelleher, William Joseph *pharmaceutical consultant*

Suffield
Chung, Douglas Chu *pharmacist, consultant*

DELAWARE

Newark
Uricheck-Holzapfel, Maryanne *pharmacist*

Wilmington
Howe, Burton Brower *pharmacologist*
Pan, Henry Yue-Ming *clinical pharmacologist*
Rudolph, Jeffrey Stewart *pharmacist, chemist*
Sands, Howard *pharmacologist, biochemist, research scientist*
Zinkand, William Collier *neuropharmacologist*

DISTRICT OF COLUMBIA

Washington
Carson, Regina Edwards *pharmacy administrator, educator*
Gregory, Robert Scott *pharmacist*
Lippman, Marc Estes *pharmacology educator*
Solimando, Dominic Anthony, Jr. *pharmacist*
Udeinya, Iroka Joseph *pharmacologist, researcher*

FLORIDA

Bay Pines
Laven, David Lawrence *nuclear and radiologic pharmacist, consultant*

Daytona Beach
Wehner, Henry Otto, III *pharmacist, consultant*

Dunnellon
Lapp, Roger James *consulting pharmacist*

Mango
Spencer, Francis Montgomery James *pharmacist*

Miami
Haynes, Duncan Harold *pharmacology educator*
Itzhak, Yossef *neuropharmacologist*
Marcus, Joy John *pharmacist, educator, consultant*
Monsalve, Martha Eugenia *pharmacist*
Ugwu, Martin Cornelius *pharmacist*

Pinellas Park
Tower, Alton G., Jr. *pharmacist*

GEORGIA

Athens
Boudinot, Frank Douglas *pharmaceutics educator*
Bowen, John Metcalf *pharmacologist, toxicologist, educator*

Augusta
Daniell, Laura Christine *pharmacology and toxicology educator*

Griffin
Trimmer, Brenda Kay *pharmacist*

La Grange
Toth, Danny Andrew *pharmacist*

Norcross
Buchanan, Diane Kay *pharmacist*

Valdosta
Lankford, Mary Angeline Gruver *pharmacist*

IDAHO

Boise
Olson, Richard Dean *researcher, pharmacology educator*

ILLINOIS

Chicago
Chiou, Win Loung *pharmacokinetics educator, director*
Giannopoulos, Joanne *pharmacist, consultant*
Monyak, Wendell Peter *pharmacist*
Narahashi, Toshio *pharmacology educator*

Deerfield
Schechter, Paul J. *pharmacologist*

Downers Grove
Currie, Bruce LaMonte *pharmaceutical sciences educator, medicinal chemistry researcher*

Lincolnshire
West, Dennis Paul *pharmacologist, pharmacist, educator*

Long Grove
Dajani, Esam Zapher *pharmacologist*

Maywood
Van De Kar, Louis David *pharmacologist, educator*

North Chicago
Bush, Eugene Nyle *pharmacologist, research scientist*
Nair, Velayudhan *pharmacologist, medical educator*

Round Lake
Yeung, Tin-Chuen *pharmacologist*

Springfield
Somani, Satu Motilal *pharmacologist, toxicologist, educator*

INDIANA

Elkhart
Byrd, William Garlen *clinical pharmacist, medical researcher*

Indianapolis
Ashmore, Robert Winston *computational pharmacologist*
Besch, Henry Roland, Jr. *pharmacologist, educator*
Bonate, Peter Lawrence *pharmacologist*
Gehlert, Donald Richard *pharmacologist*
Kauffman, Raymond Francis *biochemical pharmacologist*
Svoboda, Gordon Howard *pharmacognosist, consultant*
Weber, George *oncology and pharmacology researcher, educator*

IOWA

Iowa City
Berg, Mary Jaylene *pharmacy educator, researcher*
Wurster, Dale Eric *pharmacy educator*
Wurster, Dale Erwin *pharmacy educator, university dean emeritus*

Mason City
Hughes, Mark Lee *pharmacist*

KENTUCKY

Ashland
Kovar, Dan Rada *pharmacist*

Lexington
DeLuca, Patrick Phillip *pharmaceutical scientist, educator, administrator*

Louisville
Aronoff, George Rodger *medicine and pharmacology educator*
Bhattacherjee, Parimal *pharmacologist*
Hurst, Harrell Emerson *pharmacology and toxicology educator*

LOUISIANA

Harvey
Broussard, Malcolm Joseph *pharmacist, consultant*

MAINE

Bangor
Fiori, Michael J. *pharmacist*

MARYLAND

Baltimore
Funk Orsini, Paula Ann *pharmaceutical administration educator*
Khazan, Naim *pharmacology educator*
Lee, Carlton K. K. *clinical pharmacist, consultant, educator*

Bethesda
Axelrod, Julius *biochemist, pharmacologist*
DeFilippes, Mary Wolpert *pharmacologist*
Kandasamy, Sathasiva Balakrishna *pharmacologist*
Onufrock, Richard Shade *pharmacist, researcher*
Thorgeirsson, Snorri Sveinn *physician, pharmacologist*

Rockville
Lewis, Benjamin Pershing, Jr. *pharmacist, public health service officer*
Tabibi, S. Esmail *pharmaceutical researcher, educator*
Van Arsdel, William Campbell, III *pharmacologist*

Saint Michaels
Young, Donald Roy *pharmacist*

Salisbury
May, Everette Lee, Jr. *pharmacologist, educator*

MASSACHUSETTS

Boston
Lam, Bing Kit *pharmacologist*

Burlington
Anaebonam, Aloysius Onyeabo *pharmacist*

South Dennis
Svikla, Alius Julius *pharmacist*

MICHIGAN

Ann Arbor
La Du, Bert Nichols, Jr. *physician, educator*
Trujillo, Keith Arnold *psychopharmacologist*
Yang, Victor Chi-Min *pharmacy educator*

Detroit
Sloane, Bonnie Fiedorek *pharmacology and cancer biology educator, researcher*

East Lansing
Atchison, William David *pharmacology educator*

Kalamazoo
Stiver, James Frederick *pharmacist, health physicist, administrator, scientist*

MISSISSIPPI

Brandon
King, Kenneth Vernon, Jr. *pharmacist*

University
McChesney, James Dewey *pharmaceutical scientist*

MISSOURI

Independence
Sturges, Sidney James *pharmacist, educator, investment and development company executive*

Kansas City
Yourtee, David Merle *pharmaceutical science educator, molecular toxicologist*

Kirksville
Martin, John Richard *pharmacology educator, researcher*

Saint Louis
McIntosh, Helen Horton *research scientist*

NEW HAMPSHIRE

Hanover
Gosselin, Robert Edmond *pharmacologist, educator*

Weare
Dombrowski, Frank Paul, Jr. *pharmacist*

NEW JERSEY

Cranbury
Sofia, R. D. *pharmacologist*

Hillside
Patell, Mahesh *pharmacist, researcher*

Kenilworth
Colucci, Robert Dominick *cardiovascular pharmacologist*
Filippone Steinbrick, Gay *pharmacist, educator*

Kinnelon
Preston, Andrew Joseph *pharmacist, drug company executive*

Morris Plains
Fielding, Stuart *psychopharmacologist*

Nutley
Dennin, Robert Aloysius, Jr. *pharmaceutical research scientist*
Drews, Jürgen *pharmaceutical researcher*
Enthoven, Dirk *clinical pharmacologist, researcher*

Piscataway
Chien, Yie Wen *pharmaceutics educator*
Sinko, Patrick J. *pharmacist, educator*

Princeton Junction
Marchisotto, Robert *pharmacologist*

Rahway
Monek, Donna Marie *pharmacist*
Tobert, Jonathan Andrew *clinical pharmacologist*

Somerville
O'Neill, Patrick J. *pharmacologist*

South Plainfield
Borah, Kripanath *pharmacist*

South River
Geczik, Ronald Joseph *pharmaceutical researcher*

NEW MEXICO

Albuquerque
Davis, Robin L. *pharmacy educator*

Farmington
MacCallum, (Edythe) Lorene *pharmacist*

NEW YORK

Albany
DeNuzzo, Rinaldo Vincent *pharmacy educator*

Amherst
Halvorsen, Stanley Warren *neuropharmacologist, researcher*

Ardsley
Kochak, Gregory Michael *biophysical pharmacy/ pharmacokinetics researcher*

Bronx
Alexander, Christina Lillian *pharmacist*

Brooklyn
Baird, Rosemarie Annette *pharmacist*
Mesiha, Mounir Sobhy *industrial pharmacy educator, consultant*
Rumore, Martha Mary *pharmacist, educator*

Buffalo
Mihich, Enrico *medical researcher*

New Rochelle
Resnick, Henry Roy *pharmacist*

New York
Cranefield, Paul Frederic *pharmacology educator, physician, scientist*
Schlessinger, Joseph *pharmacology educator*
Torigian, Puzant Crossley *clinical research pharmacist, pharmaceutical company executive*
Watanabe, Kyoichi A(loysius) *chemist, researcher, pharmacology educator*

Norwich
Brooks, Robert Raymond *pharmacologist*

Orangeburg
Reilly, Margaret Anne *pharmacologist, educator*

Pearl River
Zisa, David Anthony *pharmaceutical researcher*

Rochester
Borch, Richard Frederic *pharmacology and chemistry educator*

Staten Island
Gokarn, Vijay Murlidhar *pharmacist, consultant*

Tarrytown
Davis-Bruno, Karen L. *pharmacologist, researcher*

Williamsville
Paladino, Joseph Anthony *clinical pharmacist*

NORTH CAROLINA

Cary
Chignell, Colin Francis *pharmacologist*

Chapel Hill
Diliberto, Pamela Allen *pharmacologist, researcher*

Durham
Fouts, James Ralph *pharmacologist, educator, clergyman*

Gorner
Jones, Stephen Yates *pharmacist*

Greenville
Dar, Mohammad Saeed *pharmacologist, educator*
Jain, Sunil *pharmaceutical scientist*

Morrisville
Voisin-Lestringant, Emmanuelle Marie *pharmacologist, consultant*

Raleigh
Berry, Joni Ingram *hospice pharmacist, educator*

Research Triangle Park
Elion, Gertrude Belle *research scientist, pharmacology educator*

NORTH DAKOTA

Grand Forks
Messiha, Fathy S *pharmacologist, toxicologist, educator*

OHIO

Cincinnati
Albrecht, Helmut Heinrich *medical director*

Columbus
Bianchine, Joseph Raymond *pharmacologist*
Feller, Dennis Rudolph *pharmacology educator*
Hayton, William Leroy *pharmacology educator*

OKLAHOMA

Healdton
Eck, Kenneth Frank *pharmacist*

Oklahoma City
Wallis, Robert Joe *pharmacist, retail executive*

Tulsa
Brenner, George Marvin *pharmacologist*

OREGON

Portland
North, Richard Alan *neuropharmacologist*
Sklovsky, Robert Joel *pharmacology educator*

PENNSYLVANIA

Hanover
Davis, Ruth C. *pharmacy educator*

King Of Prussia
Del Tito, Benjamin John, Jr. *pharmaceuticals researcher*

Philadelphia
Johnson, Mark Dee *pharmacologist, researcher*
Lehmann, John Charles *neuropharmacologist*
Malis, Bernard Jay *pharmacologist*
Price, Karen Overstreet *pharmacist, medical editor*

Pittsburgh
Collins, Charles Curtis *pharmacist, educator*
deGroat, William Chesney *pharmacology educator*

West Point
Gross, Dennis Michael *pharmacologist*
Johnson, Robert Gahagen, Jr. *medical researcher, educator*

SOUTH CAROLINA

Charleston
Margolius, Harry Stephen *pharmacologist, physician*

Cheraw
Williams, Morgan Lewis *pharmacist*

Columbia
Madden, Arthur Allen *nuclear pharmacist*

TENNESSEE

Johnson City
Ferslew, Kenneth Emil *pharmacology and toxicology educator*

Knoxville
Sherrill, Ronald Nolan *pharmacist, consultant*

Memphis
Patel, Tarun R. *pharmaceutical scientist*

Nashville
Brase, David Arthur *neuropharmacologist*
Byron, Joseph Winston *pharmacologist*
Robertson, David *clinical pharmacologist, physician, educator*
Wilson, Sheryl A. *pharmacist*

Ripley
Nunn, Jenny Wren *pharmacist*

TEXAS

Austin
Hurley, Laurence Harold *medicinal chemistry educator*

College Station
Chiou, George Chung-Yih *pharmacologist, educator*

Dallas
Gilman, Alfred Goodman *pharmacologist, educator*
Klein, Edward Robert *pharmacist*
Margolin, Solomon Begelfor *pharmacologist*

Fort Worth
Estes, Jacob Thomas, Jr. *pharmacist, consultant*

Galveston
Liehr, Joachim Georg *pharmacology educator, cancer researcher*

Houston
Haddox, Mari Kristine *biomedical scientist*
Housholder, Glenn Tholen *pharmacology educator*

Lubbock
Elder, Bessie Ruth *pharmacist*
Young, Teri Ann Butler *pharmacist*

San Antonio
Keeler, Jill Rolf *pharmacologist, army officer, consultant*

UTAH

Salt Lake City
Kim, Sung Wan *pharmacology educator*
Oakeson, Ralph Willard *pharmaceutical and materials scientist*

VIRGINIA

Alexandria
Huffman, D. C., Jr *pharmacology educator, health science association administrator*

Berryville
White, Eugene Vaden *pharmacist*

Richmond
Harris, Louis Selig *pharmacologist, researcher*
Robinson, Susan Estes *pharmacology educator*
Ward, John Wesley *pharmacologist*

WASHINGTON

Pullman
Klavano, Paul Arthur *veterinary pharmacologist, educator, anesthesiologist*

Redmond
Ransdell, Tod Elliot *pharmaceutical, in vitro diagnostics biotechnologist*

Seattle
Wang, Ji-Ping *pharmacist, researcher*

WEST VIRGINIA

Morgantown
Ponte, Charles Dennis *pharmacist, educator*

TERRITORIES OF THE UNITED STATES

PUERTO RICO

Caguas
Tulenko, Maria Josefina *pharmacist*

San Juan
Fernandez-Repollet, Emma D. *pharmacology educator*

CANADA

MANITOBA

Winnipeg
Steele, John Wiseman *pharmacy educator*

ONTARIO

Toronto
Seeman, Philip *pharmacology educator, neurochemistry researcher*

QUEBEC

Dorval
El-Duweini, Aadel Khalaf *clinical pharmacologist, information scientist*

Montreal
Cuello, Augusto Claudio Guillermo *medical research scientist, author*

AUSTRALIA

Melbourne
Story, David Frederick *pharmacologist, educator*

BELGIUM

Brussels
Godfraind, Theophile Joseph *pharmacologist educator*

Wilrijk
Van Ooteghem, Marc Michel Martin *pharmacology educator*

BRAZIL

São Paulo
Fernicola, Nilda Alicia Gallego Gándara de *pharmacist, biochemist*
Korolkovas, Andrejus *pharmaceutical chemistry educator*

DENMARK

Copenhagen
Christensen, Søren Brøgger *medicinal chemist*

ENGLAND

London
Vane, John Robert *pharmacologist*

FRANCE

Ballan Mire
Delbarre, Bernard *pharmacologist, consultant*

GERMANY

Frankfurt am Main
Vogel, Gerhard, Hans *pharmacologist, toxicologist*

Greifswald
Teuscher, Eberhard *pharmacist*

Hamburg
Maurer, Hans Hilarius *pharmacology educator, researcher*
Sprecher, Gustav Ewald *pharmacy educator*

ICELAND

Reykjavik
Thorarensen, Oddur C.S. *pharmacist*

ITALY

Brescia
Bosio, Angelo *pharmacologist, psychiatrist*

Siena
Ghiara, Paolo *immunopharmacologist*

Trieste
Nistri, Andrea *pharmacology educator*

JAPAN

Chiba
Fujii, Akira *pharmacology educator*

Hirakata
Nakanishi, Tsutomu *pharmaceutical science educator*
Yoshioka, Masanori *pharmaceutical sciences educator*

Isehara
Oka, Tetsuo *medical educator*

Kanagawa
Saitoh, Tamotsu *pharmacology educator*

Kobe
Tani, Shohei *pharmacy educator*

Nagoya
Shioiri, Takayuki *pharmaceutical science educator*

Noda
Kwon, Glenn S. *pharmaceutical scientist*

Okinawa
Sakanashi, Matao *pharmacology educator*

Sendai
Oikawa, Atsushi *pharmacology educator*

Tokushima
Nishimoto, Nobushige *pharmacognosy educator*

Tokyo
Nagai, Tsuneji *pharmaceutics educator*
Watanabe, Kouichi *pharmacologist, educator*

SCOTLAND

Dundee
Black, Sir James (Whyte) *pharmacologist*

SINGAPORE

Singapore
Adaikan, Ganesan Periannan *pharmacologist, medical scientist*

SWITZERLAND

Bern
Reichen, Jürg *pharmacology educator*

TAIWAN

Taipei
Chern, Ji-Wang *pharmacy educator*

ADDRESS UNPUBLISHED

De Salva, Salvatore Joseph *pharmacologist, toxicologist*
Gordon, Helmut Albert *biomedical researcher, pharmacology educator*
Gordon, Jacqueline Regina *pharmacist*
Hancock, John C. *pharmacologist*
Solomon, Julius Oscar Lee *pharmacist, hypnotherapist*

MEDICINE: PODIATRY

UNITED STATES

CALIFORNIA

Fresno
Hickey, Rosemary Becker *retired podiatrist, lecturer, writer*

San Bruno
Bradley, Charles William *podiatrist*

Sunnyvale
Saxena, Amol *podiatrist, consultant*

CONNECTICUT

Ansonia
Yale, Jeffrey Franklin *podiatrist*

Darien
Weiss, Robert Franklin *podiatrist*

West Hartford
Rimiller, Ronald Wayne *podiatrist*

DISTRICT OF COLUMBIA

Washington
Gottlieb, H. David *podiatrist*

FLORIDA

Hallandale
Haspel, Arthur Carl *podiatrist, surgeon*

Orlando
Rafetto, John *podiatrist*

ILLINOIS

Oak Lawn
Byrnes, Michael Francis *podiatrist*
Karr, Joseph Peter *podiatrist*

MARYLAND

Darnestown
Gottlieb, Julius Judah *podiatrist*

MASSACHUSETTS

Quincy
Bernardone, Jeffrey John *podiatrist*

MICHIGAN

Oak Park
Borovoy, Marc Allen *podiatrist*

NEW JERSEY

Belleville
Caputo, Wayne James *surgeon, podiatrist*

Brick
Kowalski, Lynn Mary *podiatrist*

Middletown
Anania, William Christian *podiatrist*

Union
Davison, Glenn Alan *podiatric surgeon*

NEW YORK

Binghamton
Hogan, Joseph Thomas *podiatrist*

Brooklyn
Saphire, Gary Steven *podiatrist*

Deer Park
Wernick, Justin *podiatrist, educator*

Miller Place
Gresser, Mark Geoffrey *podiatrist*

Mount Vernon
Brenner, Amy Rebecca *podiatrist*

New York
Kauth, Benjamin *podiatric consultant*
Leshnower, Alan Lee *podiatrist*

NORTH CAROLINA

Chapel Hill
Baerg, Richard Henry *podiatrist, educator, consultant*

OREGON

Portland
Mozena, John Daniel *podiatrist*

PENNSYLVANIA

Philadelphia
Freed, Edmond Lee *podiatrist*

CANADA

ONTARIO

Toronto
Nesbitt, Lloyd Ivan *podiatrist*

ADDRESS UNPUBLISHED

Commito, Richard William *podiatrist*
Kaczanowski, Carl Henry *podiatrist, educator*

MEDICINE: PSYCHIATRY

UNITED STATES

ALABAMA

Birmingham
Nuckols, Frank Joseph *psychiatrist*

ARIZONA

Tempe
Noce, Robert Henry *neuropsychiatrist, educator*

ARKANSAS

Calico Rock
Grasse, John M., Jr. *physician, missionary*

North Little Rock
Lawson, William Bradford *psychiatrist*

CALIFORNIA

Fairfield
Martin, Clyde Verne *psychiatrist*

Fresno
Leigh, Hoyle *psychiatrist, educator*

Los Angeles
Green, Richard *psychiatrist, lawyer, educator*
Liberman, Robert Paul *psychiatry educator, researcher, writer*
Scheibel, Arnold Bernard *psychiatrist, educator, researcher*

Mill Valley
Wallerstein, Robert Solomon *psychiatrist*

Pasadena
Barnard, William Marion *psychiatrist*
Shalack, Joan Helen *psychiatrist*

Sacramento
Zil, John Stephen *psychiatrist, physiologist*

San Francisco
Barondes, Samuel Herbert *psychiatrist, educator*
David, George *psychiatrist, economic theory lecturer*

Stockton
Renson, Jean Felix *psychiatry educator*

Walnut Creek
McKnight, Lenore Ravin *child psychiatrist*

West Hollywood
Wilson, Myron Robert, Jr. *former psychiatrist*

COLORADO

Denver
Cullum, Colin Munro *psychiatry and neurology educator*
Wamboldt, Marianne Zdeblick *psychiatrist*

CONNECTICUT

Bridgeport
Dolan, John Patrick *psychiatrist*

Farmington
Viner, Mark William *psychiatrist*

Greenwich
Foraste, Roland *psychiatrist*

Hamden
Leckman, James Frederick *psychiatry and pediatrics educator*

Hartford
Brauer, Rima Lois *psychiatrist*

New Haven
Bunney, Benjamin Stephenson *psychiatrist*
Comer, James Pierpont *psychiatrist*
Heninger, George R. *psychiatry educator, researcher*
Price, Lawrence H(oward) *psychiatrist, researcher, educator*
Pruett, Kyle Dean *psychiatrist, writer, educator*

Norwalk
Rose, Gilbert Jacob *psychiatrist, writer, psychoanalyst*

Westport
Lopker, Anita Mae *psychiatrist*

DELAWARE

Georgetown
Yossif, George *psychiatrist*

Wilmington
Kaye, Neil S. *psychiatry educator*

DISTRICT OF COLUMBIA

Washington
Greenberg, Richard Alan *psychiatrist, educator*
Natsukari, Naoki *psychiatrist, neurochemist*
Robinowitz, Carolyn Bauer *psychiatrist, educator*
Sabshin, Melvin *psychiatrist, educator, medical association administrator*
Wyatt, Richard Jed *psychiatrist, educator*

FLORIDA

Bay Pines
Keskiner, Ali *psychiatrist*

Dania
Apter, Nathaniel Stanley *psychiatrist, educator, retired, researcher*

Gainesville
Evans, Dwight Landis *psychiatrist, educator*

Inverness
Esquibel, Edward Valdez *psychiatrist, clinical medical program developer*

Miami
Goodnick, Paul Joel *psychiatrist*
Mandri, Daniel Francisco *psychiatrist*

Ocala
Corwin, William *psychiatrist*

Winter Park
Pollack, Robert William *psychiatrist*

GEORGIA

Atlanta
Clements, James David *psychiatry educator, physician*
Zumpe, Doris *psychiatry researcher, educator*

Augusta
Loomis, Earl Alfred, Jr. *psychiatrist*
Rausch, Jeffrey Lynn *psychiatrist, psychopharmacologist*

Decatur
Alderete, Joseph Frank *psychiatrist, medical service adminstrator*

HAWAII

Honolulu
Stevens, Stephen Edward *psychiatrist*

Kaneohe
Ahmed, Iqbal *psychiatrist, consultant*

ILLINOIS

Chicago
Barter, James T. *psychiatrist, educator*
Basch, Michael Franz *psychiatrist, psychoanalyst, educator*
Giovacchini, Peter Louis *psychoanalyst*
Goldberg, Arnold Irving *psychoanalyst, educator*
Visotsky, Harold Meryle *psychiatrist, educator*

Springfield
Feldman, Bruce Alan *psychiatrist*

INDIANA

Indianapolis
Campbell, Judith Lowe *child psychiatrist*
Nurnberger, John I., Jr. *psychiatrist, educator*
Sullivan, John Lawrence, III *psychiatrist*

IOWA

Iowa City
Andreasen, Nancy Coover *psychiatrist, educator*

KANSAS

Shawnee Mission
Sternberg, David Edward *psychiatrist*

Topeka
Menninger, William Walter *psychiatrist*

Wichita
Dyck, George *psychiatry educator*

KENTUCKY

Bowling Green
Marshall, Willis Henry *psychiatrist*

LOUISIANA

Houma
Conrad, Harold Theodore *psychiatrist*

Lake Charles
Yadalam, Kashinath Gangadhara *psychiatrist*

New Orleans
Svenson, Ernest Olander *psychiatrist, psychoanalyst*

MARYLAND

Baltimore
Fishman, Jacob Robert *psychiatrist, educator, corporate executive, investor*
Jani, Sushma Niranjan *child and adolescent psychiatrist*
Mc Hugh, Paul R. *psychiatrist, neurologist, educator*
Snyder, Solomon Halbert *psychiatrist, pharmacologist*

Bethesda
Cath, Stanley Howard *psychiatrist, psychoanalyst*
Johnson, Joyce Marie *psychiatrist, epidemiologist*
Rapoport, Judith *psychiatrist*

Rockville
Bridge, T(homas) Peter *psychiatrist, researcher*
Fenton, Wayne S. *psychiatrist*
Goodwin, Frederick King *psychiatrist*
Paul, Steven M. *psychiatrist*

Tantallon
Dickens, Doris Lee *psychiatrist*

MASSACHUSETTS

Belmont
Pope, Harrison Graham, Jr. *psychiatrist, educator*
Popper, Charles William *child and adolescent psychiatrist*

Boston
Burns, Padraic *physician, psychiatrist, psychoanalyst, educator*
Surman, Owen Stanley *psychiatrist*

Brookline
Jakab, Irene *psychiatrist*

Jamaica Plain
Pierce, Chester Middlebrook *psychiatrist, educator*

Stockbridge
Rothenberg, Albert *psychiatrist, educator*

Weston
Wells, Lionelle Dudley *psychiatrist*

Worcester
Och, Mohamad Rachid *psychiatrist, consultant*

MICHIGAN

Ann Arbor
Greden, John Francis *psychiatrist, educator*
Tandon, Rajiv *psychiatrist, educator*

Northville
Abbasi, Tariq Afzal *psychiatrist, educator*

Southfield
Rosenzweig, Norman *psychiatry educator*

MINNESOTA

Minneapolis
Clayton, Paula Jean *psychiatry educator*
Lentz, Richard David *psychiatrist*

MISSOURI

Saint Louis
Bird, Harrie Waldo, Jr. *psychiatrist, educator*
Robins, Eli *psychiatrist, biochemist, educator*

NEW HAMPSHIRE

Manchester
Emery, Paul Emile *psychiatrist*

NEW JERSEY

Englewood
Wuhl, Charles Michael *psychiatrist*

Flemington
Scasta, David Lynn *psychiatrist*

Maplewood
Shuttleworth, Anne Margaret *psychiatrist*

Paramus
Greenberg, William Michael *psychiatrist*

Piscataway
Sugerman, Abraham Arthur *psychiatrist*

Princeton
Mueller, Peter Sterling *psychiatrist, educator*

Springfield
Kim, Myunghee *psychiatrist, child psychiatrist, psychoanalyst*

Vineland
Clinton, Lawrence Paul *psychiatrist*

West Orange
Ghali, Anwar Youssef *psychiatrist, educator*

NEW YORK

Amityville
Liang, Vera Beh-Yuin Tsai *psychiatrist, educator*
Sodaro, Edward Richard *psychiatrist*

Bath
Sandt, John Joseph *psychiatrist*

Bronx
Davidson, Michael *psychiatrist, neuroscientist*
Fernandez-Pol, Blanca Dora *psychiatrist, researcher*
Kanofsky, Jacob Daniel *psychiatrist, educator*

Brooklyn
Begleiter, Henri *psychiatry educator*
Crum, Albert Byrd *psychiatrist, consultant*
Macroe-Wiegand, Viola Lucille (Countess Des Escherolles) *psychiatrist, psychoanalyst*
Norstrand, Iris Fletcher *psychiatrist, neurologist, educator*

Buffalo
Levy, Harold James *physician, psychiatrist*

Canaan
Bell, James Milton *psychiatrist*

East Meadow
Rachlin, Stephen Leonard *psychiatrist*

Elmhurst
Barron, Charles Thomas *psychiatrist*

Great Neck
Ratner, Lillian Gross *psychiatrist*

Manhasset
Spater-Zimmerman, Susan *psychiatrist, educator*

Mount Vernon
Zucker, Arnold Harris *psychiatrist*

New York
Boksay, Istvan Janos Endre *geriatric psychiatrist*
Bryt, Albert *psychiatrist*
Cancro, Robert *psychiatrist*
DeFlorio, Mary Lucy *physician, psychiatrist*
Druss, Richard George *psychiatrist, educator*
Freedman, Alfred Mordecai *pscyhiatrist, educator*
Friedhoff, Arnold J. *psychiatrist, medical scientist*
Hamburg, David A. *psychiatrist, foundation executive*
Hogan, Charles Carlton *psychiatrist*
Katz, Steven Edward *psychiatrist, state health official*
Marcus, Eric Robert *psychiatrist*
Millman, Robert Barnet *psychiatry and public health educator*
Newbold, Herbert Leon, Jr. *psychiatrist, writer*
Pardes, Herbert *psychiatrist, educator*
Reisner, Milton *psychiatrist*
Ristich, Miodrag *psychiatrist*
Schiavi, Raul Constante *psychiatrist, educator, researcher*
Schneck, Jerome M. *psychiatrist, medical historian, educator*
Sorrel, William Edwin *psychiatrist, educator, psychoanalyst*
Spiegel, Herbert *psychiatrist, educator*
Steinglass, Peter Joseph *psychiatrist, educator*
Wharton, Ralph Nathaniel *psychiatrist, educator*
Witenberg, Earl George *psychiatrist, educator*

Nyack
Esser, Aristide Henri *psychiatrist*

Rochester
Barton, Russell William *psychiatrist, author*

Schenectady
Pasamanick, Benjamin *psychiatrist, educator*

Stony Brook
Henn, Fritz Albert *psychiatrist*

Syracuse
Clausen, Jerry Lee *psychiatrist*
Szasz, Thomas Stephen *psychiatrist, educator, writer*

Valhalla
Brook, Judith Suzanne *psychiatry educator*
Levy, Norman B. *psychiatrist, educator*

White Plains
Pfeffer, Cynthia Roberta *psychiatrist, educator*
Seelye, Edward Eggleston *psychiatrist*

Woodmere
Erdberg, Mindel Ruth *psychiatrist*

NORTH CAROLINA

Chapel Hill
Hawkins, David Rollo, Sr. *psychiatrist, educator*
Peterson, Gary *child psychiatrist*

Durham
Krystal, Andrew Darrell *psychiatrist, biomedical engineer*

Research Triangle Park
DeVeaugh-Geiss, Joseph *psychiatrist*
Metz, Alan *psychiatrist*

OHIO

Boardman
Price, William Anthony *psychiatrist*

Cleveland
Denko, Joanne D. *psychiatrist, writer*
Resnick, Phillip Jacob *psychiatrist*

Columbus
Berntson, Gary Glen *psychiatry, psychology and pediatrics educator*
Goodman, Hubert Thorman *psychiatrist, consultant*

OKLAHOMA

Oklahoma City
Kimerer, Neil Banard, Sr. *psychiatrist, educator*

Tulsa
King, Joseph Willet *child psychiatrist*

PENNSYLVANIA

Coatesville
Nocks, James Jay *psychiatrist*

Haverford
Goppelt, John Walter *physician, psychiatrist*

Hershey
Tan, Tjiauw-Ling *psychiatrist, educator*

Philadelphia
Adom, Edwin Nii Amalai *psychiatrist*
Amsterdam, Jay D. *psychiatry educator, researcher*
Beck, Aaron Temkin *psychiatrist*
Comer, Nathan Lawrence *psychiatrist, educator*
Lief, Harold Isaiah *psychiatrist*
Stunkard, Albert James *physician, educator*

Pittsburgh
Rigatti, Brian Walter *psychiatric researcher*

University Park
Rolls, Barbara Jean *biobehavioral health educator, laboratory director*

RHODE ISLAND

Providence
Capone, Antonio *psychiatrist*

SOUTH CAROLINA

Charleston
Adinoff, Bryon Harlen *psychiatrist*
Zealberg, Joseph James *psychiatry and behavioral sciences educator*

Summerville
Orvin, George Henry *psychiatrist*

TEXAS

Bellaire
Pokorny, Alex Daniel *psychiatrist*

Galveston
Felthous, Alan Robert *psychiatrist*

Houston
Dilsaver, Steven Charles *psychiatry educator*
Karacan, Ismet *psychiatrist, educator*

San Antonio
Romero, Emilio Felipe *psychiatry educator, psychotherapist, hospital administrator*

VERMONT

Burlington
Bickel, Warren Kurt *psychiatry and psychology educator*

VIRGINIA

Charlottesville
Abse, David Wilfred *psychiatrist, psychologist*

Richmond
Pandurangi, Ananda Krishna *psychiatrist*

Woodstock
Vachher, Prehlad Singh *psychiatrist*

WASHINGTON

Seattle
Neppe, Vernon Michael *neuropsychiatrist, author, educator*

WISCONSIN

Madison
Jefferson, James Walter *psychiatry educator*
Roberts, Leigh Milton *psychiatrist*

Milwaukee
Lord, Guy Russell, Jr. *psychiatry educator*

CANADA

ALBERTA

Edmonton
Dewhurst, William George *physician, psychiatrist, educator, researcher*

BRITISH COLUMBIA

Vancouver
Roy, Chunilal *psychiatrist*

ONTARIO

Toronto
San, Nguyen Duy *psychiatrist*

FINLAND

Kuopio
Hakola, Hannu Panu Aukusti *psychiatry educator*

FRANCE

Fontainebleau
Saillon, Alfred *psychiatrist*

Nanterre
Berquez, Gérard Paul *psychiatrist, psychoanalyst*

Paris
Meyer, Jean-Pierre *psychiatrist*

Strasbourg
Meyer, Richard *psychiatrist*

GERMANY

Berlin
Priebe, Stefan *psychiatrist*

Haina
Müller-Isberner, Joachim Rüdiger *psychiatrist, educator*

HONG KONG

Hong Kong
Chung, Cho Man *psychiatrist*

ISRAEL

Jerusalem
Lipman, Daniel Gordon *neuropsychiatrist*

ITALY

Ancona
Marchesi, Gian Franco *psychiatry educator*

Rome
Ferracuti, Stefano Eugenio *forensic psychiatrist*

JAPAN

Chiba
Kitamura, Toshinori *psychiatrist*

Minamata-shi
Inoue, Takeshi *psychiatrist*

Sapporo
Okada, Fumihiko *psychiatrist*

Tokyo
Nagasawa, Yuko *aerospace psychiatrist*

THE NETHERLANDS

Amsterdam
Geerlings, Peter Johannes *psychiatrist, psychoanalyst*

NORWAY

Oslo
Retterstol, Nils *psychiatrist*

REPUBLIC OF KOREA

Taegu
Lee, Zuk-Nae *psychiatry educator, psychotherapist*

ADDRESS UNPUBLISHED

Borg, Stefan Lennart *psychiatrist, educator*
Dziewanowska, Zofia Elizabeth *neuropsychiatrist, researcher, physician*
Giannini, A. James *psychiatrist, educator, researcher*
Gordon, Richard Edwards *psychiatrist*
Marinas, Manuel Guillermo, Jr. *psychiatrist*
Werbach, Melvyn Roy *physician, writer*
Zukin, Stephen Randolph *psychiatrist*

MEDICINE: RADIOLOGY

UNITED STATES

ALABAMA

Mobile
Raider, Louis *physician, radiologist*

ARIZONA

Green Valley
Fischer, Harry William *radiologist, educator*

ARKANSAS

Fort Smith
Snider, James Rhodes *radiologist*

CALIFORNIA

Berkeley
Budinger, Thomas Francis *radiologist, educator*

Inglewood
Sukov, Richard Joel *radiologist*

Los Angeles
Steckel, Richard J. *radiologist, academic administrator*

Sacramento
Bogren, Hugo Gunnar *radiology educator*

Stanford
Glazer, Gary Mark *radiology educator*

FLORIDA

Fort Lauderdale
Lodwick, Gwilym Savage *radiologist, educator*

Jacksonville
Paryani, Shyam Bhojraj *radiologist*
Prempree, Thongbliew *oncology radiologist*

Miami
Ganz, William I. *radiology educator*

Saint Petersburg
Clarke, Kit Hansen *radiologist*

GEORGIA

Atlanta
Olkowski, Zbigniew Lech *physician, educator*

Stone Mountain
Rogers, James Virgil, Jr. *radiologist, educator*

ILLINOIS

Chicago
Hendrix, Ronald Wayne *physician, radiologist*

Normal
Yin, Raymond Wah *radiologist*

North Chicago
Hindo, Walid Afram *radiology educator, researcher*

INDIANA

Michigan City
Mothkur, Sridhar Rao *radiologist*

KANSAS

Kansas City
Lee, Kyo Rak *radiologist*

KENTUCKY

Lexington
Maruyama, Yosh *physician, educator*

LOUISIANA

Metairie
Harell, George S. *radiologist*

New Orleans
Ochsner, Seymour Fiske *radiologist, editor*

MASSACHUSETTS

Boston
LeMay, Marjorie Jeannette *neuroradiologist*
Little, John Bertram *physician, radiobiology educator, researcher*

Gardner
Wagenknecht, Walter Chappell *radiologist*

MICHIGAN

Ann Arbor
Kuhl, David Edmund *physician, radiology educator*

Detroit
McCarroll, Kathleen Ann *radiologist, educator*

Grand Rapids
Bartek, Gordon Luke *radiologist*

MISSOURI

Saint Louis
Fernandez-Pol, Jose Alberto *physician, radiology and nuclear medicine educator*
Reh, Thomas Edward *radiologist, educator*
Ter-Pogossian, Michel Mathew *radiation sciences educator*

NEBRASKA

Omaha
Dalrymple, Glenn Vogt *radiologist*
Harter, David John *radiation oncologist*

NEW YORK

New Hyde Park
Lee, Won Jay *radiologist*

New Rochelle
Parmer, Edgar Alan *radiologist, musician*

New York
Goldfarb, Richard Charles *radiologist*
Hirschy, James Conrad *radiologist*
Noz, Marilyn Eileen *radiology educator*

Rochester
Paterson, Eileen *radiation oncologist, educator*

Yonkers
Arnone, Mary Grace *radiology technologist*

NORTH CAROLINA

Durham
Tien, Robert Deryang *radiologist, educator*

Kernersville
Meschan, Isadore *radiologist, educator*

OHIO

Cleveland
Meaney, Thomas Francis *radiologist*

Columbus
Christoforidis, A. John *radiologist, educator*

Norwalk
Gutowicz, Matthew Francis, Jr. *radiologist*

OKLAHOMA

Ponca City
Oster, Pamela Ann *radiologic technologist*

PENNSYLVANIA

Allentown
Levin, Ken *radiologist*

Bryn Mawr
Friedman, Arnold Carl *diagnostic radiologist*

Norristown
Weiner, Harold M. *radiologist*

Philadelphia
Axel, Leon *radiologist, educator*
Kundel, Harold Louis *radiologist, educator*
Kurtz, Alfred Bernard *radiologist*
Pollack, Howard Martin *radiologist, teacher, author, researcher*

Pittsburgh
Kanal, Emanuel *radiologist*
Oh, Kook Sang *diagnostic radiologist, pediatric radiologist*

Scranton
Rhiew, Francis Changnam *radiologist*

TEXAS

Dallas
Dance, William Elijah *industrial neutron radiologist, researcher*

Houston
Libshitz, Herman I. *radiologist, educator*

VIRGINIA

Roanoke
Enright, Michael Joseph *radiologist*

Vienna
Koutrouvelis, Panos George *radiologist*

WASHINGTON

Seattle
Krohn, Kenneth Albert *radiology educator*
Nelson, James Alonzo *radiologist, educator*

WEST VIRGINIA

Lewisburg
Moore, Joan L. *radiology educator, physician*

CANADA

BRITISH COLUMBIA

Vancouver
Burhenne, Hans Joachim *physician, radiology educator*

AUSTRIA

Vienna
Mosser, Hans Matthias *radiologist*

BELGIUM

Antwerp
Metdepenninghen, Carlos Maurits W. *radiologist*

CROATIA

Split
Boschi, Srdjan *radiologist, nephrologist*

GERMANY

Berlin
Fobbe, Franz Caspar *radiologist*

Munich
Dühmke, Eckhart *radiation oncology educator*

Tübingen
Frommhold, Walter *radiologist*

THE NETHERLANDS

Maastricht
Janevski, Blagoja Kame *radiologist, educator*

ADDRESS UNPUBLISHED

Brent, Robert Leonard *physician, educator*
Calvert, William Preston *radiologist*
Steinberg, Fred Lyle *radiologist*
Weissmann, Heidi Seitelblum *radiologist, educator*

MEDICINE: SURGERY

UNITED STATES

ARIZONA

Cortaro
Lindsey, Douglas *trauma surgeon*

Mesa
Bunchman, Herbert Harry, II *plastic surgeon*

Scottsdale
Friedman, Shelly Arnold *cosmetic surgeon*

Sedona
D'Javid, Ismail Faridoon *surgeon*

ARKANSAS

Fort Smith
Still, Eugene Fontaine, II *plastic surgeon, educator*

CALIFORNIA

Agoura Hills
Bleiberg, Leon William *surgical podiatrist*

Irvine
Connolly, John Earle *surgeon, educator*

Loma Linda
Bailey, Leonard Lee *surgeon*

Los Angeles
Rosenthal, John Thomas *surgeon, transplantation surgeon*
Sprague, Norman Frederick, Jr. *surgeon, educator*

Orange
Furnas, David William *plastic surgeon*

Palo Alto
Shumway, Norman Edward *surgeon, educator*

Rolling Hills Estates
Bellis, Carroll Joseph *surgeon*

San Diego
Chambers, Henry George *surgeon*
Hansbrough, John Fenwick *surgery educator*
Lerner, Sheldon *plastic surgeon*

San Francisco
Cline, Carolyn Joan *plastic and reconstructive surgeon*
Debas, Haile T. *gastrointestinal surgeon, physiologist, educator*

San Jose
Shatney, Clayton Henry *surgeon*

Upland
Robinson, Hurley *surgeon*

COLORADO

Denver
Meldrum, Daniel Richard *general surgeon, physician*
Perreten, Frank Arnold *surgeon, ophthalmologist*

Golden
Tegtmeier, Ronald Eugene *physician, surgeon*

CONNECTICUT

New Haven
Baldwin, John Charles *surgeon, researcher*
Letsou, George Vasilios *cardiothoracic surgeon*
Ravikumar, Thanjavur Subramaniam *surgical oncologist*

Norwalk
Greenberg, Sheldon Burt *plastic and reconstructive surgeon*

Stamford
Cottle, Robert Duquemin *plastic surgeon, otolaryngologist*

Stratford
Beres, Milan *surgeon*

DELAWARE

Lewes
Saliba, Anis Khalil *surgeon*

DISTRICT OF COLUMBIA

Washington
Callender, Clive Orville *surgeon*
Feller, William Frank *surgery educator*

FLORIDA

Bradenton
Ambrusko, John Stephen *retired surgeon, county official*

Clearwater
Wheat, Myron William, Jr. *cardiothoracic surgeon*

Daytona Beach
Rubin, Mark Stephen *ophthalmic surgeon*

Gainesville
Morse, Martin A. *surgeon*

Miami
Freshwater, Michael Felix *hand surgeon*
Peck, Michael Dickens *burn surgeon*
Ricordi, Camillo *transplant surgeon, diabetes researcher, educator*

Orlando
Norris, Franklin Gray *thoracic and cardiovascular surgeon*

Ponte Vedra Beach
ReMine, William Hervey, Jr. *surgeon*

Vero Beach
Small, Wilfred Thomas *surgeon, educator*

GEORGIA

Atlanta
Foster, Roger Sherman, Jr. *surgeon, educator, health facility administrator*
Jones, Mark Mitchell *plastic surgeon*

Augusta
Allen, Marshall Bonner, Jr. *neurosurgeon, consultant*
Rubin, Joseph William *surgeon, educator*
Smith, Randolph Relihan *plastic surgeon*

La Grange
West, John Thomas *surgeon*

HAWAII

Hilo
Taniguchi, Tokuso *surgeon*

ILLINOIS

Aurora
Bleck, Phyllis Claire *surgeon, musician*

Berwyn
Misurec, Rudolf *physician, surgeon*

Chicago
Chmell, Samuel Jay *orthopedic surgeon*
Deorio, Anthony Joseph *surgeon*
Griffith, B(ezaleel) Herold *physician, educator, plastic surgeon*
Huggins, Charles Brenton *surgical educator*
Todd, James S. *surgeon, educator, medical association administrator*

Maywood
Gamelli, Richard L. *surgeon, educator*

Peoria
Watkins, George M. *surgeon, educator*

INDIANA

Indianapolis
Manders, Karl Lee *neurosurgeon*

Scottsburg
Kho, Eusebio *surgeon*

Zionsville
Heck, David Alan *orthopaedic surgery educator, mechanical engineering educator*

IOWA

Des Moines
Pandeya, Nirmalendu Kumar *plastic and flight surgeon, military officer*

Iowa City
Corson, John Duncan *vascular surgeon*

Mason City
Chanco, Amado Garcia *surgeon*

KENTUCKY

Louisville
Jacob, Robert Allen *surgeon*

LOUISIANA

New Orleans
Carter, Rebecca Davilene *surgical oncology educator*
Martin, Louis Frank *surgery and physiology educator*
Nichols, Ronald Lee *surgeon, educator*

Shreveport
Mancini, Mary Catherine *cardiothoracic surgeon, researcher*
Shelby, James Stanford *cardiovascular surgeon*

MAINE

Togus
Sensenig, David Martin *surgeon*

MARYLAND

Baltimore
Turney, Stephen Zachary *cardiothoracic surgeon, educator*

Bethesda
Rosenberg, Steven Aaron *surgeon, medical researcher*
Theuer, Charles Philip *surgeon*
Thom, Arleen Kaye *surgeon*

MASSACHUSETTS

Boston
Bougas, James Andrew *physician, surgeon*
Burke, John Francis *surgeon, educator, researcher*
Doody, Daniel Patrick *pediatric surgeon*
Egdahl, Richard Harrison *surgeon, medical educator, health science administrator*
Folkman, Moses Judah *surgeon*
Kim, Samuel Homer *pediatric surgeon*
Mannick, John Anthony *surgeon*
Moore, Francis Daniels, Jr. *surgeon*
Rodriguez, Agustin Antonio *surgeon*

Springfield
Frankel, Kenneth Mark *thoracic surgeon*
Friedmann, Paul *surgeon, educator*
Reed, William Piper, Jr. *surgeon, educator*

Wellesley
Murray, Joseph Edward *plastic surgeon*

MICHIGAN

Ann Arbor
Burke, Robert Harry *surgeon, educator*
Coran, Arnold Gerald *pediatric surgeon, educator*

Grand Blanc
Wasfie, Tarik Jawad *surgeon, educator*

Southfield
Zubroff, Leonard Saul *surgeon*

MINNESOTA

Duluth
Kubista, Theodore Paul *surgeon*

Minneapolis
Caldwell, Michael DeFoix *surgeon, educator*
Najarian, John Sarkis *surgeon, educator*
Shumway, Sara J. *cardiothoracic surgeon*

Rochester
Beahrs, Oliver Howard *surgeon*

MISSISSIPPI

Jackson
Das, Suman Kumar *plastic surgeon, researcher*
Didlake, Ralph Hunter, Jr. *surgeon*

MISSOURI

Saint Louis
McFadden, James Frederick, Jr. *surgeon*
Spray, Thomas L. *surgeon, educator*

MONTANA

Missoula
Newman, Jan Bristow *surgeon*

Wolf Point
Listerud, Mark Boyd *surgeon*

NEBRASKA

Lincoln
Hirai, Denitsu *surgeon*

NEW HAMPSHIRE

Hanover
Koop, Charles Everett *surgeon, government official*

Manchester
DesRochers, Gerard Camille *surgeon*

NEW JERSEY

Cherry Hill
Olearchyk, Andrew S. *cardiothoracic surgeon, educator*

Fair Lawn
Nizin, Joel Scott *surgeon*

Linwood
Godfrey, George Cheeseman, II *surgeon*

New Brunswick
Borah, Gregory Louis *plastic and reconstructive surgeon*
Greco, Ralph Steven *surgeon, researcher, medical educator*

Newark
Hochberg, Mark S. *cardiac surgeon*

Paramus
Sergi, Anthony Robert *physician, surgeon*

NEW MEXICO

Alamogordo
Stapp, John Paul *surgeon, former air force officer*

Albuquerque
Morris, Don Melvin *surgical oncologist*

NEW YORK

Binghamton
Bethje, Robert *general surgeon, retired*

Brooklyn
Edemeka, Udo Edemeka *surgeon*

Buffalo
Kurlan, Marvin Zeft *surgeon*
Shedd, Donald Pomroy *surgeon*

New Hyde Park
Wise, Leslie *surgeon, educator*

New York
Eaton, Richard Gillette *surgeon, educator*
Edmunds, Robert Thomas *retired surgeon*
Ego-Aguirre, Ernesto *surgeon*
Friedlander, Ralph *thoracic and vascular surgeon*
Gelernt, Irwin M. *surgeon, educator*
Golomb, Frederick Martin *surgeon, educator*
Lane, Joseph M. *orthopaedic surgeon, oncologist*
LaQuaglia, Michael Patrick *pediatric surgeon, neuroblastoma researcher*
Lewis, Jonathan Joseph *surgeon, molecular biologist*
Lewis, Stuart Weslie *surgeon*
McCabe, John Cordell *surgeon*
McCarthy, Joseph Gerald *plastic surgeon, educator*
Momtaheni, Mohsen *oral and maxillofacial surgeon, academician, clinician*
Osborne, Michael Piers *surgeon, health facility administrator*
Rothenberg, Robert Edward *physician, surgeon, author*
Schaefer, Steven David *head and neck surgeon, physiologist*
Tolete-Velcek, Francisca Agatep *pediatric surgeon, surgery educator*

Pomona
Glassman, Lawrence S. *plastic surgeon*

Rochester
Chan, Donald Pin-Kwan *orthopaedic surgeon, educator*
Lanzafame, Raymond Joseph *surgeon, researcher*

Saratoga Springs
Dorsey, James Baker *surgeon, lawyer*

NORTH CAROLINA

Chapel Hill
Bowman, Frederick Oscar, Jr. *cardiothoracic surgeon, retired*

Durham
Davis, James Evans *general and thoracic surgeon, parliamentarian, author*
Sabiston, David Coston, Jr. *educator, surgeon*

Greensboro
Truesdale, Gerald Lynn *plastic and reconstructive surgeon*

OHIO

Cincinnati
Heimlich, Henry Jay *physician, surgeon*
Meese, Ernest Harold *thoracic and cardiovascular surgeon*
Warden, Glenn Donald *burn surgeon*

Columbus
Furste, Wesley Leonard, II *surgeon, educator*

Toledo
Rubin, Allan Maier *physician, surgeon*

OKLAHOMA

Oklahoma City
Robison, Clarence, Jr. *surgeon*
Zuhdi, Nazih *surgeon*

OREGON

Portland
Jacob, Stanley Wallace *surgeon, educator*
Swan, Kenneth Carl *physician, surgeon*

PENNSYLVANIA

Allentown
Gaylor, Donald Hughes *surgeon, educator*
Kratzer, Guy Livingston *surgeon*

Bala Cynwyd
Kirschner, Ronald Allen *osteopathic plastic surgeon, otolaryngologist, educator*

Boothwyn
McLaughlin, Edward David *surgeon, educator*

Gaines
Beller, Martin Leonard *retired orthopaedic surgeon*

Greensburg
Austin, George Lynn *surgeon*

Hershey
Pierce, William Schuler *cardiac surgeon, educator*

Langhorne
Shah, Mubarik Ahmad *surgeon*

Media
Behbehanian, Mahin Fazeli *surgeon*

Monessen
Lementowski, Michal *surgeon*

Philadelphia
Barker, Clyde Frederick *surgeon, educator*
Brayman, Kenneth Lewis *surgeon*
Rhoads, Jonathan Evans *surgeon*
Schwartz, Gordon Francis *surgeon, educator*

Pittsburgh
Fisher, Bernard *surgeon, researcher, educator*
Marino, Ignazio Roberto *transplant surgeon, researcher*

Sayre
Thomas, John Melvin *surgeon*

Springfield
Mirman, Merrill Jay *physician, surgeon*

RHODE ISLAND

Providence
Amaral, Joseph Ferreira *surgeon*
Vezeridis, Michael Panagiotis *surgeon, researcher, educator*

TENNESSEE

Oak Ridge
Spray, Paul *surgeon*

TEXAS

Corpus Christi
Cox, William Andrew *cardiovascular thoracic surgeon*
Hammer, John Morgan *surgeon*

Dallas
Rice, Charles Lane *surgical educator*
Rohrich, Rodney James *plastic surgeon, educator*

Galveston
Desai, Manubhai Haribhai *surgeon*
Heggers, John Paul *surgery and microbiology educator, microbiologist, retired army officer*
Herndon, David N. *surgeon*

Houston
Barrett, Bernard Morris, Jr. *plastic and reconstructive surgeon*
Cooley, Denton Arthur *surgeon, educator*
DeBakey, Michael Ellis *cardiovascular surgeon, educator*
Dudrick, Stanley John *surgeon, educator*
Mountain, Clifton Fletcher *surgeon, educator*
NosE, Yukihiko *surgeon, educator*
Pollock, Raphael Etomar *surgeon, educator*
Rubio, Pedro A. *cardiovascular surgeon*
Stehlin, John Sebastian, Jr. *surgeon*

Lubbock
Shires, George Thomas *surgeon, physician, educator*

San Antonio
Smith, John Marvin, III *surgeon, educator*

UTAH

Salt Lake City
Bauer, A(ugust) Robert, Jr. *surgeon, educator*

VIRGINIA

Annandale
Simonian, Simon John *surgeon, scientist, educator*

Charlottesville
Nolan, Stanton Peelle *surgeon, educator*

Leesburg
McDow, Russell Edward, Jr. *surgeon*

Richmond
Christie, Laurence Glenn, Jr. *surgeon*
Neifeld, James Paul *surgical oncologist*

Virginia Beach
Oswaks, Roy Michael *surgeon, educator*

Woodbridge
Reha, William Christopher *urologic surgeon*

WASHINGTON

Seattle
Hutchinson, William Burke *surgeon, research center director*

Spokane
Mackay, Alexander Russell *surgeon*

WEST VIRGINIA

Huntington
Cocke, William Marvin, Jr. *plastic surgeon, educator*

Morgantown
Hill, Ronald Charles *surgeon, educator*

WISCONSIN

Green Bay
von Heimburg, Roger Lyle *surgeon*

Milwaukee
Frantzides, Constantine Themis *general surgeon*
Kloehn, Ralph Anthony *plastic surgeon*
Namdari, Bahram *surgeon*
Wagner, Marvin *general and vascular surgeon, educator*

WYOMING

Cheyenne
Flick, William Fredrick *surgeon*

CANADA

ALBERTA

Calgary
Kimberley, Barry Paull *ear surgeon*

QUEBEC

Montreal
Cruess, Richard Leigh *surgeon, university dean*

AUSTRALIA

Sydney
Ehrlich, Frederick *surgery consultant, orthopedist, rehabilitation specialist*

AUSTRIA

Vienna
Sammad, Mohamed Abdel *cardiovascular surgeon, consultant*

BELGIUM

Liège
Alexandre, Gilbert Fernand A.E. *surgeon*

BRAZIL

Salvador
Silva, Benedicto Alves de Castro *surgeon, educator*

ENGLAND

Bedford
Hadfield, James Irvine Havelock *surgeon*

Northwood
Smith, Paul John *plastic and reconstructive surgeon, consultant*

Preston
Gaze, Nigel Raymond *plastic surgeon*

FRANCE

Besançon
Pelissier, Edouard-Pierre *surgeon*

Laval
Sauvé, Georges *surgeon*

Longjumeau
Kapandji, Adalbert Ibrahim *orthopedic surgeon*

Maisons-Lafitte
Obadia, Andre Isaac *surgeon*

Metz
Tran-Viet, Tu *cardiovascular and thoracic surgeon*

Paris
Levy, Etienne Paul Louis *surgical department administrator*
Malherbe, Bernard *surgeon*
Mitz, Vladimir *plastic surgeon*
Rosenschein, Guy Raoul *pediatric and visceral surgeon, airline pilot*

GERMANY

Allgau
Lichtenheld, Frank Robert *physician, plastic and reconstructive surgeon, urologist*

Böblingen
Mühe, Erich *surgical educator*

GREECE

Athens
Antoniou, Panayotis A. *surgeon*

HONG KONG

Kowloon
Chin, James Kee-Hong *surgeon*

Sha Tin
King, Walter Wing-Keung *surgeon, head and neck surgery consultant*

ISLE OF MAN

Douglas
Lamming, Robert Love *retired surgeon*

JAPAN

Itogun
Sakurai, Takeo *surgery educator*

Sambu-Gun
Morishige, Fukumi *surgeon*

Tenri
Koizumi, Shunzo *surgeon*

Tokyo
Hatanaka, Hiroshi *neurosurgeon*
Shikata, Jun-ichi *surgery educator*

NORTHERN IRELAND

Dungannon
Peyton, James William Rodney *consultant surgeon*

NORWAY

Oslo
Ekeland, Arne Erling *surgeon, educator*

THE PHILIPPINES

Manila
Tangco, Ambrosio Flores *health administrator, surgeon, orthopedist*

POLAND

Wroclaw
Staniszewski, Andrzej Marek *surgeon, educator*

PORTUGAL

Lisbon
De Almeida, Antonio Castro Mendes *surgery educator*

REPUBLIC OF KOREA

Seoul
Baek, Se-Min *plastic surgeon*

SAUDI ARABIA

Riyadh
Isbister, William Hugh *surgeon*

SCOTLAND

Edinburgh
Boulter, Patrick Stewart *surgery educator*

SPAIN

Oviedo
Garcia-Moran, Manuel *surgeon*

SWEDEN

Linköping
Borch, Kurt Esben *surgeon*

Uppsala
Dubiel, Thomas Wieslaw *cardiothoracic surgeon, educator*

SWITZERLAND

Fribourg
Bydzovsky, Viktor *surgeon*

ADDRESS UNPUBLISHED

Amiel, David *orthopaedic surgery educator*
Chaikof, Elliot Lorne *vascular surgeon*
DePalma, Ralph George *surgeon, educator*
Fitchett, Vernon Harold *retired physician, surgeon, educator*
Glass, Thomas Graham, Jr. *retired general surgeon, educator*
Greene, Laurence Whitridge, Jr. *surgical educator*
Groves, Sheridon Hale *orthopedic surgeon*
Halliday, William Ross *retired physician, speleologist, writer*
Heckadon, Robert Gordon *plastic surgeon*
Hirose, Teruo Terry *surgeon*
Stone, James Robert *surgeon*
Toledo-Pereyra, Luis Horacio *transplant surgeon, researcher, educator*

MEDICINE: UROLOGY

UNITED STATES

CALIFORNIA

Gold River
Forbes, Kenneth Albert Faucher *urological surgeon*

La Jolla
Gittes, Ruben Foster *urological surgeon*

Los Angeles
Madlang, Rodolfo Mojica *urologic surgeon*
Martinez, Miguel Acevedo *urologist, consultant, lecturer*

San Diego
Schmidt, Joseph David *urologist*

CONNECTICUT

New Haven
Weiss, Robert M. *urologist, educator*

FLORIDA

Miami
Suarez, George Michael *urologist*

HAWAII

Honolulu
Edwards, John Wesley, Jr. *urologist*

ILLINOIS

Maywood
Albala, David Mois *urologist, educator*

IOWA

Sioux City
Vaught, Richard Loren *urologist*

KENTUCKY

Lexington
Gilliam, M(elvin) Randolph *urologist, educator*

MARYLAND

Baltimore
Berger, Bruce Warren *physician, urologist*

Grasonville
Prout, George Russell, Jr. *medical educator, urologist*

MASSACHUSETTS

Boston
Loughlin, Kevin Raymond *urologic surgeon*
Morgentaler, Abraham *urologist, researcher*

MICHIGAN

Detroit
Cerny, Joseph Charles *urologist, educator*

NEBRASKA

Omaha
Mardis, Hal Kennedy *urological surgeon, educator, researcher*

NEW HAMPSHIRE

Lebanon
Rous, Stephen Norman *urologist, educator, editor*

NEW YORK

New York
McGovern, John Hugh *urologist, educator*
Mininberg, David T. *pediatric urology surgeon, educator*

PENNSYLVANIA

Philadelphia
Seidmon, E. James *urologist*
Wein, Alan Jerome *urologist, educator, researcher*

SOUTH CAROLINA

Charleston
Bissada, Nabil Kaddis *urologist, educator, researcher, author*

UTAH

Salt Lake City
Middleton, Anthony Wayne, Jr. *urologist, educator*

VIRGINIA

Charlottesville
Jenkins, Alan Deloss *urologic surgeon, educator*

Danville
Hoffman, Allan Augustus *retired urologist, consultant*

WASHINGTON

Tacoma
Grenley, Philip *urologist*

WEST VIRGINIA

Morgantown
Lamm, Donald Lee *urologist*

BRAZIL

São Paulo
Sadi, Marcus Vinicius *urologist*

Uberlandia
Prado, Neilton Gonçalves *urologist*

ENGLAND

Pontefract
Robinson, Melvyn Roland *urologist*

JAPAN

Tochigi
Takasaki, Etsuji *urology educator*

Tokorozawa
Nakamura, Hiroshi *urology educator*

THE NETHERLANDS

Amsterdam
Newling, Donald William *urological surgeon*

SAUDI ARABIA

Dhahran
Milad, Moheb Fawzy *consultant, urologist*

TRINIDAD AND TOBAGO

San Fernando
Sawh, Lall Ramnath *urologist*

PHYSICAL SCIENCES: ASTRONOMY

UNITED STATES

ALABAMA

Huntsville
McCollough, Michael Leon *astronomer*

ARIZONA

Flagstaff
Millis, Robert Lowell *astronomer*

Tempe
Chaffee, Frederic H., Jr. *astronomer*

Tucson
Angel, James Roger Prior *astronomer*
Howard, Robert Franklin *observatory administrator, astronomer*
Jefferies, John Trevor *astronomer, astrophysicist, observatory administrator*
Leibacher, John William *astronomer*
Roemer, Elizabeth *astronomer, educator*
Strittmatter, Peter Albert *astronomer, educator*
Tifft, William Grant *astronomer*
Wolff, Sidney Carne *astronomer, observatory administrator*

ARKANSAS

Fayetteville
Lacy, Claud H. Sandberg *astronomer*

Russellville
Cook, Stephen Patterson *physical sciences and astronomy educator, researcher*

CALIFORNIA

Berkeley
Spinrad, Hyron *astronomer*
Welch, William John *astronomer, educator*

Big Bear City
Zirin, Harold *astronomer, educator*

Los Altos
Fraknoi, Andrew *astronomy educator, astronomical society executive*

Los Angeles
Krupp, Edwin Charles *astronomer*
McLean, Ian Small *astronomer, physics educator*

Moffett Field
O'Handley, Douglas Alexander *astronomer*
Seiff, Alvin *planetary scientist, atmosphere physics and aerodynamics consultant*

Pasadena
Babcock, Horace W. *astronomer*
Kulkarni, Shrinivas R. *astronomy educator*
Libbrecht, Kenneth *astronomy educator*
Sandage, Allan Rex *astronomer*
Sargent, Wallace Leslie William *astronomer, educator*
Schmidt, Maarten *astronomy educator*
Searle, Leonard *astronomer, researcher*
Weymann, Ray J. *astronomy educator*

Santa Barbara
Grunke, Andrew Frederick *astronomy educator, instrument designer*

Santa Cruz
Drake, Frank Donald *astronomy educator*
Faber, Sandra Moore *astronomer, educator*
Kraft, Robert Paul *astronomer, educator*
Osterbrock, Donald E(dward) *astronomy educator*

COLORADO

Boulder
Pryor, Wayne Robert *astronomer*

Denver
Peregrine, David Seymour *astronomer, consultant*

CONNECTICUT

New Haven
Hoffleit, Ellen Dorrit *astronomer*
Oemler, Augustus, Jr. *astronomy educator*

Rocky Hill
Griesé, John William, III *astronomer*

DISTRICT OF COLUMBIA

Washington
Boyce, Peter Bradford *astronomer, professional association executive*
Gergely, Tomas *astronomer*
Strand, Kaj Aage *astronomer*
Weiler, Kurt Walter *radio astronomer*

FLORIDA

Gainesville
Smith, Alexander Goudy *physics and astronomy educator*
Wood, Frank Bradshaw *retired astronomy educator*

Marco Island
Jacobson, Linda S(ue) *astronomy educator*

HAWAII

Honolulu
Chambers, Kenneth Carter *astronomer*
Hall, Donald Norman Blake *astronomer*
Meech, Karen Jean *astronomer*

Kamuela
Hamilton, John Carl *astronomer, telescope operator*

ILLINOIS

Peoria
Chamberlain, Joseph Miles *astronomer, educator*

Urbana
Lo, Kwok-Yung *astronomer*

MARYLAND

Baltimore
Baum, Stefi Alison *astronomer*
Hart, Helen Mavis *planetary astronomer*
Kowal, Charles Thomas *astronomer*
Westerhout, Gart *retired astronomer*

College Park
Kerr, Frank John *astronomer, educator*
Kissell, Kenneth Eugene *astronomer*

Greenbelt
Haxton, Donovan Merle, Jr. *astronomer*
Liu, Han-Shou *space scientist, researcher*
Maran, Stephen Paul *astronomer*
Palmer, David Michael Oliver *astronomer*

MASSACHUSETTS

Amherst
Strom, Stephen Eric *astronomer*

Cambridge
Dame, Thomas Michael *radio astronomer*
Lada, Elizabeth A. *astronomer*
Luu, Jane *astronomer*
Marsden, Brian Geoffrey *astronomer*
Narayan, Ramesh *astronomy educator*

MICHIGAN

Ann Arbor
Haddock, Fred T. *astronomer, educator*

MINNESOTA

Minneapolis
Kuhi, Leonard Vello *astronomer, university administrator*

NEW HAMPSHIRE

Hanover
Wegner, Gary Alan *astronomer*

NEW JERSEY

Holmdel
Wilson, Robert Woodrow *radio astronomer*

Princeton
Taylor, Joseph Hooton, Jr. *radio astronomer, physicist*

NEW MEXICO

Mesilla Park
Tombaugh, Clyde William *astronomer, educator*

NEW YORK

Ithaca
Sagan, Carl Edward *astronomer, educator, author*

Schenectady
Philip, A. G. Davis *astronomer, editor, educator*

Stony Brook
Lissauer, Jack Jonathan *astronomy educator*

PENNSYLVANIA

Philadelphia
Levitt, Israel Monroe *astronomer*

University Park
Usher, Peter Denis *astronomy educator*

RHODE ISLAND

Warwick
Jacoby, Margaret Mary *astronomer, educator*

TEXAS

Austin
Bash, Frank Ness *astronomer, educator*
Duncombe, Raynor Lockwood *astronomer*
Lambert, David L. *astronomy educator*
Vishniac, Ethan Tecumseh *astronomy educator*

Houston
Chamberlain, Joseph Wyan *astronomer, educator*
Dessler, Alexander Jack *space physics and astronomy educator, scientist*

VIRGINIA

Charlottesville
Chevalier, Roger Alan *astronomy educator, consultant*
Kellermann, Kenneth Irwin *astronomer*
Vanden Bout, Paul Adrian *astronomer, physicist, educator*

Dahlgren
Knowles, Stephen Howard *space scientist*

WASHINGTON

Seattle
Nelson, George Driver *astronomy and education educator, former astronaut*

WISCONSIN

Madison
Churchwell, Edward Bruce *astronomer, educator*

Williams Bay
Harper, Doyal Alexander, Jr. *astronomer, educator*
Hobbs, Lewis Mankin *astronomer*

CANADA

ALBERTA

Calgary
Kwok, Sun *astronomer*
Milone, Eugene Frank *astronomer, educator*

BRITISH COLUMBIA

Sidney
van den Bergh, Sidney *astronomer*

Victoria
Hesser, James Edward *astronomy researcher*

ONTARIO

Ottawa
MacLeod, John Munroe *radio astronomer, academic administrator*
Morton, Donald Charles *astronomer*

Toronto
Seaquist, Ernest Raymond *astronomy educator*

QUEBEC

Montreal
Carignan, Claude *astronomer, educator*

ARGENTINA

Mendoza
Branham, Richard Lacy, Jr. *astronomer*

AUSTRALIA

Saint Lucia
Page, Arthur Anthony *astronomer*

BRAZIL

São Paulo
Barbuy, Beatriz *astronomy educator*

ENGLAND

Cambridge
Lynden-Bell, Donald *astronomer*
Rees, Martin John *astronomy educator*

Macclesfield
Cohen, Raymond James *radio astronomy educator*

FRANCE

Strasbourg
Heck, André *astronomer*

GERMANY

Katlenburg-Lindau
Hagfors, Tor *national astronomy center director*

IRELAND

Wicklow
Wayman, Patrick Arthur *astronomer*

JAPAN

Ishikawa
Fukuda, Ichiro *astronomer, researcher*

Mitaka
Kaifu, Norio *astronomer*

Nagoya
Fukui, Yasuo *astronomer*

Setagaya-ku
Iijima, Shigetaka *astronomy educator*

Tokyo
Kozai, Yoshihide *astronomer*

Toyonaka
Miyamoto, Sigenori *astronomy educator, researcher*

THE NETHERLANDS

Leiden
Miley, George Kildare *astronomy educator*
Shane, William Whitney *astronomer*

NORWAY

Honefoss
Pettersen, Bjørn Ragnvald *astronomer, researcher*

REPUBLIC OF KOREA

Kyunggi-Do
Minn, Young Key *astronomer*

Seoul
Lee, Sang-Gak *astronomy educator*

RUSSIA

Moscow
Boyarchuk, Alexander *astronomer*

Saint Petersburg
Abalakin, Viktor Kuz'mich *astronomer*
Finkelshtein, Andrey Michailovich *astronomy educator*

SOUTH AFRICA

Cape Town
Ellis, George Francis Rayner *astronomy educator*

Randburg
Dean, John Francis *astronomer, researcher*

SWEDEN

Saltsjöbaden
Elvius, Aina Margareta *retired astronomer*

Stockholm
Jörsäter, Steven Bertil *astronomer, educator*

ADDRESS UNPUBLISHED

Livengood, Timothy Austin *astronomer*
Rubin, Vera Cooper *research astronomer*
Schwarzschild, Martin *astronomer, educator*

PHYSICAL SCIENCES: ASTROPHYSICS

UNITED STATES

ALABAMA

Huntsville
Derrickson, James Harrison *astrophysicist*
Rubin, Bradley Craig *astrophysicist*

ARIZONA

Amado
Weekes, Trevor C. *astrophysicist*

Tucson
De Young, David Spencer *astrophysicist*
Hunten, Donald Mount *planetary scientist, educator*
Krider, E. Philip *atmospheric scientist, educator*
Levy, Eugene Howard *planetary sciences educator, researcher*

CALIFORNIA

Berkeley
Sadoulet, Bernard *astrophysicist, educator*

Inglewood
Lewis, Roy Roosevelt *space physicist*

La Jolla
Burbidge, Geoffrey *astrophysicist, educator*

Palo Alto
Datlowe, Dayton Wood *space scientist, physicist*

Pasadena
Goldreich, Peter Martin *astrophysics and planetary physics educator*
Neugebauer, Gerry *astrophysicist, educator*
Wang, Joseph Jiong *astrophysicist*

Santa Ana
Gross, Herbert Gerald *space physicist*

Santa Cruz
Woosley, Stanford Earl *astrophysicist*

Stanford
Sturrock, Peter Andrew *space science and astrophysics educator*
Wagoner, Robert Vernon *astrophysicist, educator*

COLORADO

Boulder
Begelman, Mitchell C. *astrophysicist, educator*
Cuntz, Manfred Adolf *astrophysicist, researcher*
Gebbie, Katharine Blodgett *astrophysicist*
McCray, Richard Alan *astrophysicist, educator*
Norcross, David Warren *physicist, researcher*
Snow, Theodore Peck, Jr. *astrophysicist, author*

DELAWARE

Newark
Ness, Norman Frederick *astrophysicist, educator, administrator*

DISTRICT OF COLUMBIA

Washington
Harwit, Martin Otto *astrophysicist, educator, museum director*
Kinzer, Robert Lee *astrophysicist*
Williams, Earl George *acoustics scientist*

GEORGIA

Decatur
Sadun, Alberto Carlo *astrophysicist, physics educator*

ILLINOIS

Chicago
Chandrasekhar, Subrahmanyan *theoretical astrophysicist, educator*
Schramm, David Norman *astrophysicist, educator*

Urbana
Iben, Icko, Jr. *astrophysicist, educator*

MARYLAND

Baltimore
Zhang, Cheng-Yue *astrophysicist*

Bethesda
Silberberg, Rein *nuclear astrophysicist, researcher*

Greenbelt
Blackburn, James Kent *astrophysicist*
England, Martin Nicholas *astro/geophysicist*
Mather, John Cromwell *astrophysicist*
Ozernoy, Leonid Moissey *astrophysicist*
Whitlock, Laura Alice *research scientist*
Wood, H(oward) John, III *astrophysicist, astronomer*

MASSACHUSETTS

Boston
Brecher, Kenneth *astrophysicist*

Cambridge
Canizares, Claude Roger *astrophysicist, educator*
Covert, Eugene Edzards *aerophysics educator*
Field, George Brooks *theoretical astrophysicist*
Geller, Margaret Joan *astrophysicist, educator*
Gregory, Bruce Nicholas *astrophysicist, educator*
Grindlay, Jonathan Ellis *astrophysics educator*
Kirshner, Robert P. *astrophysicist, educator*
Layzer, David *astrophysicist, educator*
Press, William Henry *astrophysicist, computer scientist*

MINNESOTA

Minneapolis
Woodward, Paul Ralph *computational astrophysicist, applied mathematician, educator*

MISSOURI

Saint Louis
Sorrell, Wilfred Henry *astrophysics educator*

NEW JERSEY

Princeton
Bahcall, Neta Assaf *astrophysicist*
Gott, J. Richard, III *astrophysicist*
Hulse, Russell Alan *astrophysicist, plasma physicist*
Hut, Piet *astrophysics educator*
Ostriker, Jeremiah Paul *astrophysicist, educator*
Paczynski, Bohdan *astrophysicist, educator*

NEW YORK

Oneonta
Merilan, Michael Preston *astrophysicist, educator*

Schenectady
Mead, Kathryn Nadia *astrophysicist, educator*

PENNSYLVANIA

Philadelphia
Shen, Benjamin Shih-Ping *scientist, engineer, educator*

University Park
Córdova, France Anne-Dominic *astrophysics educator*
Skinner, Mark Andrew *astrophysicist*

SOUTH CAROLINA

Clemson
Clayton, Donald Delbert *astrophysicist, nuclear physicist, educator*

TEXAS

Austin
Wheeler, John Craig *astrophysicist, writer*

Houston
Black, David Charles *astrophysicist*

The Woodlands
Stanford, Michael Francis *scientist, engineer*

VIRGINIA

Alexandria
Shapiro, Maurice Mandel *astrophysicist*

Hampton
Schutte, Paul Cameron *research scientist*

WISCONSIN

Williams Bay
Kron, Richard G. *astrophysicist, educator*

CANADA

ALBERTA

Calgary
Venkatesan, Doraswamy *astrophysicist, physics educator*

ONTARIO

Toronto
Tremaine, Scott Duncan *astrophysicist*

QUEBEC

Montreal
Michaud, Georges Joseph *astrophysics educator*

MEXICO

Puebla
Chatterjee, Tapan Kumar *astrophysics researcher*

AUSTRALIA

Sydney
Cram, Lawrence Edward *astrophysicist*

BELGIUM

Liège
Perdang, Jean Marcel *astrophysicist*

CHILE

La Serena
Eggen, Olin Jeuck *astrophysicist, administrator*

ENGLAND

Cambridge
Hawking, Stephen W. *astrophysicist, mathematician*

FRANCE

Marseilles
Azzopardi, Marc Antoine *astrophysicist, scientist*

Meudon
Mamon, Gary Allan *astrophysicist*

GERMANY

Bonn
Henkel, Christian Johann *astrophysicist*

Heidelberg
Duschl, Wolfgang Josef *astrophysicist*

Munich
Giacconi, Riccardo *astrophysicist, educator*

INDIA

Bombay
Tarafdar, Shankar Prosad *astrophysicist, educator*

JAPAN

Toyonaka
Ikeuchi, Satoru *astrophysicist, educator*

RUSSIA

Moscow
Bochkarev, Nikolai Gennadievich *astrophysics researcher*

ADDRESS UNPUBLISHED

Fu, Albert Joseph *astrophysicist*
Penzias, Arno Allan *astrophysicist, research scientist, information systems specialist*
Smoot, George Fitzgerald, III *astrophysicist*

PHYSICAL SCIENCES: BIOCHEMISTRY

UNITED STATES

ALABAMA

Auburn
Daron, Harlow H. *biochemist*

Birmingham
Bugg, Charles Edward *biochemistry educator, scientist*
Elgavish, Ada *biochemist*
Elgavish, Gabriel Andreas *physical biochemistry educator*
Moore, William Gower Innes *biochemist*
Shealy, Y. Fulmer *biochemist*
Singh, Raj Kumar *biochemist, researcher*

Mobile
Campbell, Naomi Flowers *biochemist*

Montgomery
Tan, Boen Hie *biochemist*

Tuskegee
Adeyeye, Samuel Oyewole *nutritional biochemist*
Almazan, Aurea Malabag *biochemist*

ALASKA

Fairbanks
Duffy, Lawrence Kevin *biochemist, educator*

CALIFORNIA

Berkeley
Ames, Bruce N(athan) *biochemistry and molecular biology educator*
Koshland, Daniel Edward, Jr. *biochemist, educator*

Brea
Wheeler, Annemarie Ruth *biochemist*

Davis
Feeney, Robert Earl *research biochemist*
Stumpf, Paul Karl *former biochemistry educator*
Yang, Shang Fa *biochemistry educator, plant physiologist*

Emeryville
Masri, Merle Sid *biochemist, consultant*

Foster City
Iovannisci, David Mark *biochemist researcher*
Wiktorowicz, John Edward *research biochemist*
Zaidi, Iqbal Mehdi *biochemist, scientist*

Irwindale
Saless, Fathieh Molaparast *biochemist*

La Jolla
Anel, Alberto *biochemist*
Benson, Andrew Alm *biochemistry educator*
Edelman, Gerald Maurice *biochemist, educator*
Kitada, Shinichi *biochemist*
Mullis, Kary Banks *biochemist*

Loma Linda
Hill, Kelvin Arthur Willoughby *biochemistry educator*

Los Angeles
Allerton, Samuel Ellsworth *biochemist*
Boado, Ruben Jose *biochemist*

North Hollywood
Thomson, John Ansel Armstrong *biochemist*

Oakland
Jukes, Thomas Hughes *biological chemist, educator*

Orange
Varsanyi-Nagy, Maria *biochemist*

Orinda
Heftmann, Erich *biochemist*

Pasadena
Story, Randall Mark *biochemist*

Redwood City
Nacht, Sergio *biochemist*

Riverside
Mudd, John Brian *biochemist*

San Diego
Klump, Wolfgang Manfred *biochemist*
Mestril, Ruben *biochemist, researcher*

San Francisco
Betlach, Mary Carolyn *biochemist, molecular biology researcher*
Boisvert, William Andrew *nutritional biochemist, researcher*
Boyer, Herbert Wayne *biochemist*
Dill, Kenneth Austin *pharmaceutical chemistry educator*
Kirschner, Marc Wallace *biochemist, cell biologist*
Rutter, William J. *biochemist, educator*

Santa Cruz
Noller, Harry Francis, Jr. *biochemist, educator*

Santa Monica
Smith, Roberts Angus *biochemist, educator*

South San Francisco
Gaertner, Alfred Ludwig *biochemist*

Stanford
Berg, Paul *biochemist, educator*
Kaiser, Armin Dale *biochemist, educator*
Kornberg, Arthur *biochemist*
McConnell, Harden Marsden *biophysical chemistry researcher, chemistry educator*
Stryer, Lubert *biochemist, educator*

Tustin
Cruzen, Matt Earl *research biochemist*

COLORADO

Boulder
Cech, Thomas Robert *chemistry and biochemistry educator*

Fort Collins
Bamburg, James Robert *biochemistry educator*
Booth, Karla Ann Smith *biochemist*

CONNECTICUT

Bloomfield
Cohen, Patricia Ann *biochemist*

New Haven
Cometto-Muñiz, Jorge Enrique *biochemist*
Crothers, Donald Morris *biochemist, educator*
Steitz, Joan Argetsinger *biochemistry educator*

DELAWARE

Newark
Wetlaufer, Donald Burton *biochemist, educator*

Wilmington
Kinney, Anthony John *biochemist, researcher*

DISTRICT OF COLUMBIA

Washington
Casterline, James Larkin, Jr. *research biochemist*
Goldstein, Allan Leonard *biochemist, educator*
Heineken, Frederick G. *biochemical engineer*
Singer, Maxine Frank *biochemist*

FLORIDA

Gainesville
Frost, Susan Cooke *biochemistry educator*

Miami
Hawkins, Pamela Leigh Huffman *biochemist*
Sapan, Christine Vogel *protein biochemist, researcher*

North Miami Beach
Bernstein, Sheldon *biochemist*

GEORGIA

Athens
Darvill, Alan G. *biochemist, botanist, educator*
Eriksson, Karl-Erik Lennart *biochemist, educator*

Atlanta
Coluccio, Lynne M. *biochemistry educator*
Doetsch, Paul William *biochemist, educator*

Augusta
Huisman, Titus Hendrik Jan *biochemist, educator*

Duluth
Mowrey-McKee, Mary Flowers *biochemist*

Statesboro
Hurst, Michael Owen *biochemistry educator*

HAWAII

Waimanalo
Divakaran, Subramaniam *biochemist*

ILLINOIS

Abbott Park
Jeng, Tzyy-Wen *biochemist*
Peterson, Bryan Charles *biochemist*
Wideburg, Norman Earl *biochemist*

Chicago
Chambers, Donald Arthur *biochemistry educator*
Chandler, John W. *biochemistry educator, ophthalmology educator*
Iqbal, Zafar Mohd *cancer researcher, biochemist, pharmacologist, toxicologist, consultant*
Moskal, Joseph Russell *biochemist*
Papatheofanis, Frank John *biochemist*
Robbins, Kenneth Carl *biochemist*
Steiner, Donald Frederick *biochemist, physician, educator*
Tao, Mariano *biochemistry educator*
Williams-Ashman, Howard Guy *biochemistry educator*

Evanston
Goldberg, Erwin *biochemistry educator*

Jacksonville
Hainline, Adrian, Jr. *biochemist*

Maywood
Schultz, Richard Michael *biochemistry educator*

North Chicago
Thompson, Richard Edward *biochemist*
Walters, D. Eric *biochemistry educator*
Winkler, Martin Alan *biochemist*

Urbana
Briskin, Donald Phillip *biochemist*
Hager, Lowell Paul *biochemistry educator*
Ngai, Ka-Leung *biochemist, researcher*

INDIANA

Indianapolis
DeLong, Allyn Frank *biochemist*
Henry, Matthew James *biochemist, plant pathologist*
Hurley, Thomas Daniel *biochemist*
Long, Eric Charles *biochemist*
Stephens, Thomas Wesley *biochemist*
Wong, David T. *biochemist*

Lafayette
Chandrasekaran, Rengaswami *biochemist, educator*
Mertz, Edwin Theodore *biochemist, emeritus educator*

Notre Dame
Huber, Paul William *biochemistry educator, researcher*

West Lafayette
Baird, William McKenzie *chemical carcinogenesis researcher, biochemistry educator*
Schneegurt, Mark Allen *biochemist, researcher*

IOWA

Clive
Iruvanti, Pran Rao *endocrine biochemist, researcher*

Iowa City
Fulton, Alice Bordwell *biochemist, educator*
Routh, Joseph Isaac *biochemist*
Stegink, Lewis Dale *biochemist, educator*
Yorek, Mark Anthony *biochemist*

KANSAS

Kansas City
Andrews, Glen K. *biochemist*
Grisolia, Santiago *biochemistry educator*

Lawrence
Sanders, Robert B. *biochemistry educator*

Manhattan
Krishnamoorthi, Ramaswamy *biochemistry educator*
Wang, Xuemin *biochemistry educator*

KENTUCKY

Lexington
Spearman, Terence Neil *biochemist*

Louisville
Kotwal, Girish Jayant *biochemist, educator*
Wheeler, Thomas Jay *biochemist*

LOUISIANA

Metairie
Sibley, Deborah Ellen Thurston *immunochemist*

New Orleans
Dalferes, Edward Roosevelt, Jr. *biochemical researcher*
Schally, Andrew Victor *biochemist, researcher*

MAINE

Scarborough
Barrantes, Denny Manny *biochemist*

MARYLAND

Aberdeen Proving Ground
Porter, Dale Wayne *biochemist*
Stopa, Peter Joseph *biochemist, microbiologist*

Baltimore
Anfinsen, Christian Boehmer *biochemist*
Dong, Dennis Long-Yu *biochemist*
Grossman, Lawrence *biochemist, educator*
Harrington, William Fields *biochemist, educator*
McCarty, Richard Earl *biochemist, biochemistry educator*
Raben, Daniel Max *biochemist*
Reddi, A. Hari *orthopaedics and biological chemistry educator*
Steiner, Robert Frank *biochemist*
Weber, David Joseph *biochemist, educator*

Beltsville
Kearney, Philip Charles *biochemist*
Soper, Thomas Sherwood *biochemist*

Benedict
Sanders, James Grady *biogeochemist*

Berlin
Passwater, Richard Albert *biochemist, writer*

Bethesda
Adamson, Richard Henry *biochemist*
Bernstein, Lori Robin *biochemist, molecular biologist, educator*
Cantoni, Giulio Leonardo *biochemist, government official*
Contois, David Francis *biochemist*
Gelboin, Harry Victor *biochemistry educator, researcher*
Hook, Vivian Yuan-Wen Ho *biochemist, neuroscientist*
Kaufman, Seymour *biochemist*
Korn, Edward David *biochemist*
Lee, Fang-Jen Scott *biochemist, molecular biologist*
Murayama, Makio *biochemist*
Nash, Howard Allen *biochemist, researcher*
Nelson, Thomas John *research biochemist*
Nirenberg, Marshall Warren *biochemist*
Poston, John Michael *biochemist*
Stadtman, Earl Reece *biochemist*
Stadtman, Thressa Campbell *biochemist*
Vaughan, Martha *biochemist*
Wu, Roy Shih Shyong *biochemist, health scientist administrator*

Columbia
Walter, James Frederic *biochemical engineer*

Rockville
O'Rangers, John Joseph *biochemist*
Zoon, Kathryn Egloff *biochemist*

MASSACHUSETTS

Amherst
Goldsby, Richard Allen *biochemistry educator*

Andover
Tangarone, Bruce Steven *biochemist*

Belmont
Hauser, George *biochemist, educator*

Boston
Baleja, James Donald *biochemist, educator*
Blout, Elkan Rogers *biological chemistry educator, university dean*
Chishti, Athar H. *biochemist*
Gergely, John *biochemistry educator*
Goldin, Barry Ralph *biochemist, researcher, educator*
Kennedy, Eugene Patrick *biochemist, educator*
Nugent, Matthew Alfred *biochemist*
Pardee, Arthur Beck *biochemist, educator*
Rashkovetsky, Leonid *biochemist*
Rühlmann, Andreas Carl-Erich Conrad *biochemist*
Strominger, Jack Leonard *biochemist*
Thomas, Peter *biochemistry educator*
Walsh, Christopher Thomas *biochemist, department chairman*

Cambridge
Bloch, Konrad Emil *biochemist*
Kim, Peter Sung-bai *biochemistry educator*
Meselson, Matthew Stanley *biochemist, educator*
Schimmel, Paul Reinhard *biochemist, biophysicist, educator*
Venkatesh, Yeldur Padmanabha *biochemist, researcher*
Wald, George *biochemist, educator*
Wang, James Chuo *biochemistry and molecular biology educator*
Weinberg, Robert Allan *biochemist, educator*
Wiley, Don Craig *biochemistry and biophysics educator*

Haverhill
DeSchuytner, Edward Alphonse *biochemist, educator*

Lexington
Buchanan, John Machlin *biochemistry educator*

Medfield
Slovacek, Rudolf Edward *biochemist*

Natick
Akkara, Joseph Augustine *biochemist, educator*

Newton
Adams, Onie H. Powers (Onie H. Powers) *retired biochemist*

Waltham
Fasman, Gerald David *biochemistry educator*
Jencks, William Platt *biochemist, educator*
Perez-Ramirez, Bernardo *biochemist, researcher, educator*

Worcester
Shalhoub, Victoria Aman *molecular biochemist, researcher*

MICHIGAN

Ann Arbor
Agranoff, Bernard William *biochemist, educator*
Feng, Hsien Wen *biochemistry educator, researcher*
Wang, Kevin Ka-Wang *pharmaceutical biochemist*

Dearborn
Reeve, Lorraine Ellen *biochemist, researcher*

Detroit
Yamazaki, Akio *biochemistry educator*

East Lansing
Preiss, Jack *biochemistry educator*

Houghton
Podila, Gopi Krishna *biochemistry and molecular biology educator*

Kalamazoo
Baker, Carolyn Ann *research biochemist*
Greenfield, John Charles *bio-organic chemist*
Stapleton, Susan Rebecca *biochemistry educator*

Lake Linden
Campbell, Wilbur Harold *research plant biochemist, educator*

Midland
Leng, Marguerite Lambert *regulatory consultant, biochemist*

Saint Clair Shores
Rownd, Robert Harvey *biochemistry and molecular biology educator*

MINNESOTA

Austin
Schmid, Harald Heinrich Otto *biochemistry educator, academic director*

Minneapolis
Wood, Wellington Gibson, III *biochemistry educator*

Rochester
Bodine, Peter Van Nest *biochemist*
Moyer, Thomas Phillip *biochemist*

Saint Paul
Karl, Daniel William *biochemist, researcher, consultant*
Thenen, Shirley Warnock *nutritional biochemistry educator*

MISSOURI

Chesterfield
Franz, John E. *bio-organic chemist, researcher*

Kansas City
Dileepan, Kottarappat Narayanan *biochemist, researcher, educator*
Waterborg, Jakob Harm *biochemistry educator*

Kirksville
Rearick, James Isaac *biochemist, educator*

Saint Louis
Francis, Faith Ellen *biochemist*
Howard, Susan Carol Pearcy *biochemist*
Sly, William S. *biochemist, educator*

NEVADA

Las Vegas
Carper, Stephen William *biochemist, researcher*
Chiang, Tom Chuan-Hsien *biochemist*

NEW JERSEY

Allendale
Macaya, Roman Federico *biochemist*

Camden
Beck, David Paul *biochemist*
Foglesong, Paul David *microbiology educator*

Edgewater
Schiltz, John Raymond *biochemist, researcher*

Jersey City
Nakhla, Atif Mounir *biochemist*

Kenilworth
Pachter, Jonathan Alan *biochemist, researcher*

Newark
Fu, Shou-Cheng Joseph *biomedicine educator*
Ryzlak, Maria Teresa *biochemist, educator*

Nutley
Hall, Clifford Charles *pharmaceutical biochemist*
Pruess, David Louis *biochemist*

Piscataway
Shatkin, Aaron Jeffrey *biochemistry educator*
Thomas, Thresia K. *biomedical researcher, biochemistry educator*

Princeton
Liu, Edward Chang-Kai *biochemist*
Shiber, Mary Claire *biochemist*

Rahway
Shapiro, Bennett Michaels *biochemist, educator*

Raritan
Bottenus, Ralph Edward *biochemist*

Wayne
White, Doris Gnauck *science educator, biochemical and biophysics researcher*

West Milford
Parent, Edward George *biochemist*

NEW MEXICO

Las Cruces
Cho, Michael Yongkook *biochemical engineer, educator, consultant*
Kemp, John Daniel *biochemist, educator*

NEW YORK

Albany
Grasso, Patricia Gaetana *biochemist, educator*
Reichert, Leo Edmund, Jr. *biochemist, endocrinologist*

Beacon
Robison, Peter Donald *biochemist*

Bronx
Murthy, Vadiraja Venkatesa *biochemist, researcher, educator*
Philipp, Manfred Hans Wilhelm *biochemist, educator*

Brooklyn
Nandivada, Nagendra Nath *biochemist, researcher*

Buffalo
Derechin, Moises *biochemistry educator*

Cooperstown
Peters, Theodore, Jr. *research biochemist, consultant*

Geneva
Roelofs, Wendell Lee *biochemistry educator, consultant*

Glen Cove
Weitzmann, Carl Joseph *biochemist*

Grand Island
Epstein, David Aaron *biochemist*

Ithaca
Barker, Robert *biochemistry educator*
Bauman, Dale Elton *nutritional biochemistry educator*
Hardy, Ralph W. F. *biochemist, biotechnology executive*

Lake Placid
Fearn, Jeffrey Charles *biochemist*

Manhasset
Cerami, Anthony *biochemistry educator*

Melville
Pelle, Edward Gerard *biochemist*

New Hyde Park
Stein, Theodore Anthony *biochemist, educator*

New York
Chargaff, Erwin *biochemistry educator emeritus, writer*
Cheng, Chuen Yan *biochemist*
Cross, George Alan Martin *biochemistry educator, researcher*
Furneaux, Henry Morrice *biochemist, educator*
Hajjar, David P. *biochemist, educator*
Hendrickson, Wayne A(rthur) *biochemist, educator*
Markiewicz, Leszek *research biochemist*
Merrifield, Robert Bruce *biochemist, educator*
Roepe, Paul David *biophysical chemist*
Spielman, Andrew Ian *biochemist*
Zakim, David *biochemist*

Norwich
Sietsema, William Kendall *biochemist*

Oneonta
Helser, Terry Lee *biochemistry educator*

Orangeburg
Lajtha, Abel *biochemist*

Plattsburgh
Heintz, Roger Lewis *biochemist, educator, researcher*

Port Washington
Zahnd, Hugo *emeritus biochemistry educator*

Rochester
Dumont, Mark Eliot *biochemist, educator*
Ewing, James Francis *biochemist, researcher*

Staten Island
Chauhan, Ved Pal Singh *biochemist, researcher*

Stony Brook
Scarlata, Suzanne Frances *biophysical chemist*

Valhalla
Rosenfeld, Louis *biochemist, educator*

NORTH CAROLINA

Beaufort
Bonaventura, Joseph *biochemist, educator, research center director*

Buies Creek
Nemecz, George *biochemist*

Chapel Hill
Krasny, Harvey Charles *research biochemist, medical research scientist*
Topal, Michael David *biochemistry educator*
Wilson, John Eric *biochemist*

Durham
Hill, Robert Lee *biochemistry educator, administrator*
Straub, Karl David *biochemist researcher*

Greenville
Johnson, Ronald Sanders *physical biochemist*
Marks, Richard Henry Lee *biochemist, educator*

Hickory
Steelman, Sanford Lewis *research scientist, biochemist*

Huntersville
Acheson, Scott Allen *research biochemist*

Raleigh
Hassan, Hosni Moustafa *biochemistry, toxicology and microbiology educator, biologist*
Tove, Samuel B. *biochemistry educator*

Research Triangle Park
Kohn, Michael Charles *theoretical biochemistry professional*
Miller, Wayne Howard *biochemist*

Salemburg
Baugh, Charles Milton *biochemistry educator, college dean*

Winston Salem
Sorci-Thomas, Mary Gay *biomedical researcher, educator*

NORTH DAKOTA

Grand Forks
Nordlie, Robert Conrad *biochemistry educator*

OHIO

Cincinnati
Ball, William James, Jr. *biochemistry educator*
Brand, Larry Milton *biochemist*
Brankamp, Robert George *research biochemist*
Fanger, Bradford Otto *biochemist*
Nelson, Sandra Lynn *biochemist*
Smith, Philip Luther *research biochemist*

Cleveland
Banerjee, Amiya Kumar *biochemist*

Columbus
Behrman, Edward Joseph *biochemistry educator*
Kolattukudy, Pappachan Ettoop *biochemist, educator*
Slonim, Arnold Robert *biochemist, physiologist*

Toledo
Jankun, Jerzy Witold *biochemist*

Yellow Springs
Spokane, Robert Bruce *biophysical chemist*

OKLAHOMA

Edmond
Hanson-Painton, Olivia Lou *biochemist, educator*

Oklahoma City
Alaupovic, Petar *biochemist, educator*

Stillwater
Davis, Gordon Dale, II *biochemist*
Spivey, Howard Olin *biochemistry and physical chemistry educator*

OREGON

Corvallis
Arp, Daniel J. *biochemistry educator*
Reed, Donald James *biochemistry educator*

Portland
Peyton, David Harold *biochemistry educator*

PARAGUAY

Asunción
Vera Garcia, Rafael *food and nutrition biochemist*

POLAND

Lublin
Rogalski, Jerzy Marian *biochemist, educator*

PORTUGAL

Lisbon
Salvador, Armindo Jose Alves Silva *biochemist*

REPUBLIC OF KOREA

Buk-gu
Han, Oksoo *biochemistry educator*

Seoul
Paik, Young-Ki *biochemist, nuclear biologist*

RUSSIA

Moscow
Chernyak, Boris Victor *biochemist, researcher*
Poglazov, Boris Fedorovich *biochemist, researcher, administrator*

SOUTH AFRICA

Cape Town
Berman, Mervyn Clive *biochemist*

SWEDEN

Gothenburg
Hansson, Gunnar Claes *medical biochemist*

Malmö
Andersson, Ulf Göran Christer *biochemist*

Stockholm
Bergström, K. Sune D. *biochemist*

Uppsala
Laurent, Torvard Claude *biochemist, educator*

SWITZERLAND

Zurich
Hauser, Helmut Otmar *biochemistry educator*

TAIWAN

Hsinchu
Yang, C. C. *biochemistry educator, researcher*

Taipei
Wei, Yau-Huei *biomedical research scientist*

VENEZUELA

Caracas
Rangel-Aldao, Rafael *biochemist*

ADDRESS UNPUBLISHED

Azaryan, Anahit Vazgenovna *biochemist, researcher*
Deisenhofer, Johann *biochemistry educator, researcher*
Kotha, Subbaramaiah *biochemist*
Medzihradsky, Fedor *biochemist, educator*
Ochoa, Severo *biochemist*
Sawlivich, Wayne Bradstreet *biochemist*
Snyder, Melissa Rosemary *biochemist*

PHYSICAL SCIENCES: BIOPHYSICS

UNITED STATES

ALABAMA

Auburn
Vodyanoy, Vitaly Jacob *biophysicist, educator*

CALIFORNIA

Berkeley
Alpen, Edward Lewis *biophysicist, educator*

Loma Linda
Tosk, Jeffrey Morton *biophysicist, educator*

Long Beach
Smith, William Ray *biophysicist, engineer*

Los Angeles
Greenfield, Moses A. *medical physicist, educator*

Orange
Nguyen, Quan A. *medical physicist*

Pasadena
Hopfield, John Joseph *biophysicist, educator*

San Francisco
Düzgünes, Nejat A. *biophysicist*

COLORADO

Evergreen
Pullen, Margaret I. *genetic physicist*

Fort Collins
Elkind, Mortimer Murray *biophysicist, educator*

CONNECTICUT

Farmington
Loew, Leslie Max *biophysicist*

New Haven
Brünger, Axel Thomas *biophysicist, researcher, educator*
Konigsberg, William Henry *molecular biophysics and biochemistry educator, administrator*

DISTRICT OF COLUMBIA

Washington
Ledley, Robert Steven *biophysicist*

FLORIDA

Jacksonville
Givens, Stephen Bruce *medical physicist*

Miami
Hornicek, Francis John, Jr. *cell biophysicist, orthopedic surgeon*

Sarasota
Liu, Suyi *biophysicist*

Tallahassee
Mandelkern, Leo *biophysics and chemistry educator*

GEORGIA

Atlanta
Garcia, Ernest Victor *medical physicist*
Stafford, Patrick Morgan *biophysicist*

Thomasville
Haynes, Harold Eugene, Jr. *medical physicist*

IDAHO

Lewiston
Heidorn, Douglas Bruce *medical physicist*

ILLINOIS

Argonne
Holtzman, Richard Beves *health physicist, chemist*

Evanston
Offner, Franklin Faller *biophysics educator*
Saxena, Renu *medical physicist*

Urbana
Crofts, Antony Richard *biophysics educator*
Govindjee *biophysics and biology educator*
Jakobsson, Eric Gunnar *biophysicist, educator*
Xu, Dong *biophysicist, researcher*

INDIANA

Indianapolis
Chern, Jiun-Der *medical physicist*
Ehringer, William Dennis *membrane biophysicist*

Lafayette
Speir, Jeffrey Alan *biophysicist*

IOWA

Ames
Moyer, James Wallace *biophysicist*

Iowa City
Pennington, Edward Charles *medical physicist*

LOUISIANA

Lafayette
Kavanaugh, Howard Van Zant *medical physicist*

MARYLAND

Baltimore
Cone, Richard Allen *biophysics educator*
Mahesh, Mahadevappa Mysore *medical physicist, researcher*
Tewari, Kirti Prakash *biophysicist*

Bethesda
Ehrenstein, Gerald *biophysicist*
Leikin, Sergey L. *biophysicist, researcher*
Xu, Sengen *biophysicist*

MASSACHUSETTS

Boston
Webster, Edward William *medical physicist*

Burlington
Shaikh, Naimuddin *medical physicist*

Hyannis
Dubuque, Gregory Lee *medical physicist*

MICHIGAN

Ann Arbor
Oncley, John Lawrence *biophysics educator, consultant*

Detroit
Shih, Jing-Luen Allen *medical physicist*

Royal Oak
Langer, Steve Gerhardt *biomedical physicist, consultant*

MINNESOTA

Minneapolis
Geise, Richard Allen *medical physicist*

Rochester
Robb, Richard Arlin *biophysics educator, scientist*
Silva, Norberto DeJesus *biophysicist*

MONTANA

Bigfork
Thomas, Robert Glenn *biophysicist*

NEW JERSEY

Dover
Dombroski, Lee Anne Zarger *medical physicist*

Princeton
Novotny, Jiri *biophysicist*

Ridgefield
Goldman, Arnold Ira *biophysicist, statistical analyst*

Short Hills
Gerstein, Mark Bender *biophysicist*

NEW MEXICO

Albuquerque
Hylko, James Mark *health physicist*
Majumder, Sabir Ahmed *biophysicist*

NEW YORK

Bronx
Yalow, Rosalyn Sussman *medical physicist*

Buffalo
Anbar, Michael *biophysics educator*
Loomis, Ronald Earl *biophysicist*
Okhi, Shinpei *biophysicist*

Mount Morris
Sala, Martin Andrew *biophysicist*

New York
Callender, Robert Howard *biophysics educator, research scientist*
Knowles, Richard James Robert *medical physicist, educator, consultant*

Rochester
Goldstein, David Arthur *biophysicist, educator*
La Celle, Paul Louis *biophysics educator*

Stony Brook
Baldo, George Jesse *biophysicist, physiology educator*

Upton
Setlow, Richard Burton *biophysicist*
Studier, Frederick William *biophysicist*

NORTH CAROLINA

Chapel Hill
Lockett, Stephen John *medical biophysicist*
Xu, Le *biophysicist, physicist*

OHIO

Columbus
Schmalbrock, Petra *medical physicist, researcher, educator*

Dayton
Jones, Hobert W *health physicist*

OKLAHOMA

Oklahoma City
D'Souza, Maximian Felix *medical physicist*

PENNSYLVANIA

Hershey
King, Steven Harold *health physicist*

King Of Prussia
Dolce, Kathleen Ann *health physicist, inspector*

Philadelphia
Chance, Britton *biophysics and physical chemistry educator emeritus*
Liebman, Paul Arno *biophysicist, educator*
Pfeffer, Philip Elliot *biophysicist*
Roder, Heinrich *biophysicist, educator*

Radnor
Ohnishi, Stanley Tsuyoshi *biomedical director, biophysicist*

State College
Ginoza, William *retired biophysics educator*

Wilkes Barre
Shukla, Kapil P. *medical physicist*

RHODE ISLAND

Narragansett
Jacob, Ninni Sarah *health physicist*

Providence
North, David Lee *medical physicist*

TENNESSEE

Nashville
Reinisch, Lou *medical physics researcher, educator*
Stone, Michael Paul *biophysical chemist, researcher*

Oak Ridge
Reichle, David Edward *ecologist, biophysicist*

TEXAS

Barker
Hranitzky, E. Burnell *medical physicist*

Galveston
Prakash, Louise *biophysics educator*

Houston
O'Neill, Michael James *medical physicist*
Trkula, David *biophysics educator*

VIRGINIA

Fairfax
Morowitz, Harold Joseph *biophysicist, educator*

WASHINGTON

Richland
Fisher, Darrell Reed *medical physicist, researcher*

Seattle
Roubal, William Theodore *biophysicist, educator*

Woodinville
Kaufman, William Carl *biophysicist*

WISCONSIN

Milwaukee
Kusumi, Akihiro *scientist, educator*
Yin, Jun-jie *biophysicist*

CANADA

ALBERTA

Calgary
Nigg, Benno M. *biomechanics educator*

MANITOBA

Winnipeg
Smith, Ian Cormack Palmer *biophysicist*

ONTARIO

Ottawa
Carey, Paul Richard *biophysicist, scientific administrator*

Toronto
Siminovitch, Louis *biophysics educator, scientist*

CHINA

Guangzhou
Xie, Nan-Zhu *medical physics educator*

ENGLAND

Cambridge
Hodgkin, Sir Alan Lloyd *biophysicist*

London
Wallace, Bonnie Ann *biophysics educator, researcher*

FRANCE

Laxou
Anghileri, Leopoldo José *researcher*

GERMANY

Göttingen
Neher, Erwin *biophysicist*

Kleinseebach
Kalender, Willi Alfred *medical physicist*

HONG KONG

Kowloon
Chang, Donald Choy *biophysicist*

JAPAN

Aichi
Sekimura, Toshio *theoretical biophysics educator*

Nagaoka
Matsuno, Koichiro *biophysics educator*

Osaka
Aoki, Ichiro *theoretical biophysics educator, researcher*

Toyonaka
Kishimoto, Uichiro *biophysicist*

THE NETHERLANDS

Soesterberg
Smoorenburg, Guido Franciscus *biophysicist, educator*

SWEDEN

Stockholm
Fernandez-Moran, Humberto *biophysicist*

SWITZERLAND

Bern
Shiner, John Stewart *biophysicist educator*

ADDRESS UNPUBLISHED

Coohill, Thomas Patrick *biophysicist, photobiologist*
Hedge, Jeanne Colleen *health physicist*
Langmore, John Preston *biophysicist, educator*
Muranaka, Ken-ichiro *biophysicist*
Parsegian, V. Adrian *biophysicist*
Stern, Robin Lauri *medical physicist*
Wilkins, Maurice Hugh Frederick *biophysicist*

PHYSICAL SCIENCES: CHEMISTRY

UNITED STATES

ALABAMA

Auburn
Rostohar, Raymond *chemist, chromatographer*

Birmingham
Lammertsma, Koop *chemist, consultant*
Rose, Lucy McCombs *chemist*

McIntosh
Von Tersch, Frances Knight *analytical chemist*

Mobile
Beeman, Curt Pletcher *research chemist*
Lambert, James LeBeau *chemistry educator*
Madura, Jeffry David *chemistry educator*

Montgomery
Bezoari, Massimo Daniel *chemistry educator, writer*

Selma
Collins, Eugene Boyd *chemist, molecular pathologist, consultant*

Tuskegee
Koons, Lawrence Franklin *chemistry educator*

ALASKA

Fairbanks
Hoskins, L. Claron *chemistry educator*

Kenai
Morse, Aaron Holt *chemist, consultant*

ARIZONA

Chandler
Hatch, Jeffrey Scott *chemist*

Claypool
Halstead, John Irvin, II *analytical chemist*
Lowe, Douglas George *analytical chemist*

Flagstaff
Venedam, Richard Joseph *chemist, educator*

Page
Leus McFarlen, Patricia Cheryl *water chemist*

Phoenix
Allen, John Rybolt L. *chemist, biochemist*

Tempe
Blankenship, Robert Eugene *chemistry educator*
Pettit, George Robert *chemistry educator, cancer researcher*
Swink, Laurence Nim *chemist, consultant*
Vandenberg, Edwin James *chemist, educator*

Tucson
Chau, Lai-Kwan *research chemist*
Cheddar, Donville Glen *chemist, educator*
Feeney, Craig Michael *chemist*
Hatfield, (David) Brooke *analytical chemist, spectroscopist*
Lebl, Michal *peptide chemist*
Milberg, Morton Edwin *chemist*
Mollova, Nevena Nikolova *chemist*

ARKANSAS

Bauxite
Watson, David Raymond *chemist*

Little Rock
Compadre, Cesar Manuel *chemistry educator*
Darsey, Jerome Anthony *chemistry educator, researcher*

Magnolia
Willis, Rebecca Lynn *chemist, educator*

Marianna
Carroll, Stephen Douglas *chemist, research specialist*

Russellville
Albert, Milton John *retired chemist, microbiologist*

CALIFORNIA

Arcata
Russell, John Blair *chemistry educator*

Berkeley
Bergman, Robert George *chemist, educator*
Calvin, Melvin *chemist, educator*
Dauben, William Garfield *chemist, educator*
Diamond, Richard Martin *nuclear chemist*
Fisher, Robert Amos *physical chemist*
Head-Gordon, Teresa Lyn *chemist*
Heathcock, Clayton Howell *chemistry educator, researcher*
Hoffman, Darleane Christian *chemistry educator*
Huff, Welcome Rex Anthony *chemical researcher*
Johnston, Harold S(ledge) *chemistry educator*
Kim, Sung-Hou *chemistry educator, biophysical and biological chemist*
Lee, Yuan T(seh) *chemistry educator*
Lucas, Donald *physical chemist*
Miller, William Hughes *theoretical chemist, educator*
Pines, Alexander *chemistry educator, researcher*
Pitzer, Kenneth Sanborn *chemist, educator*
Quesada, Mark Alejandro *physical chemist*
Rosenblatt, Gerd Matthew *chemist*
Saykally, Richard James *chemistry educator*
Schultz, Peter G. *chemistry educator*
Seaborg, Glenn Theodore *chemistry educator*
Somorjai, Gabor Arpad *chemist, educator*
Streitwieser, Andrew, Jr. *chemistry educator*
Tinoco, Ignacio, Jr. *chemist, educator*

Brea
Shen, Gene Giin-Yuan *organic chemist*

Cardiff by the Sea
Elsner, Frederick Harvey *chemical scientist*

Claremont
Grieman, Fred Joseph *chemist, educator*
Long, Franklin Asbury *chemistry educator*
Musson, Donald George *chemist*

Commerce
Olbrantz, Don Lee *chemist*

Concord
Cole, Timothy David *chemist*

Costa Mesa
Lorance, Elmer Donald *organic chemistry educator*

Davis
Wooten, Frederick (Oliver) *applied science educator*

El Cerrito
Gwinn, William Dulaney *physical chemist, educator, consultant*

Encino
Hawthorne, Marion Frederick *chemistry educator*

Fallbrook
Tess, Roy William Henry *chemist*

Fremont
Berry, Michael James *chemist*

Fullerton
Solomon, Allen Louis *chemist*

Hercules
Zabin, Burton Allen *chemist*

Irvine
Bradic, Zdravko *chemist*
Rentzepis, Peter M. *chemistry educator*
Rowland, Frank Sherwood *chemistry educator*

Kensington
Connick, Robert Elwell *chemistry educator*

La Jolla
Boger, Dale L. *chemistry educator*
Kearns, David Richard *chemistry educator*
Lerner, Richard Alan *chemistry educator, scientist*
Nicolaou, Kyriacos Costa *chemistry educator*
Sharpless, K. Barry *chemist*
Tara, (Tara Singh) *research chemist*
Tsien, Roger Yonchien *chemist and cell biologist*
Wall, Frederick Theodore *chemistry educator*
Wong, Chi-Huey *chemistry educator*
Yakovlev, Vladislav Victorovich *research chemist, consultant*

La Mesa
Schelar, Virginia Mae *chemistry educator, consultant*

La Verne
Hwang, Cordelia Jong *chemist*

Livermore
Batzel, Roger Elwood *chemist*
Hall, Howard Lewis *nuclear chemist*
Olsen, Clifford Wayne *physical chemist*
Ornellas, Donald Louis *chemist researcher*

Los Altos Hills
van Tamelen, Eugene Earle *chemist, educator*

Los Angeles
Bayes, Kyle David *chemistry educator*
Benson, Sidney William *chemistry researcher*
Biles, John Alexander *pharmaceutical chemistry educator*
Chapman, Orville Lamar *chemist, educator*
Cram, Donald James *chemistry educator*
Edwards, Kenneth Neil *chemist, consultant*
El-Sayed, Mostafa Amr *chemistry educator*
Felker, Peter *chemistry educator*
Laaly, Heshmat Ollah *research chemist, roofing consultant, author*
Lee, David Woon *chemist, lawyer*
Levine, Raphael David *chemistry educator*
Miller, Wendell Smith *chemist, consultant*
Olah, George Andrew *chemist, educator*
Perlmutter, Milton Manuel *chemist, accountant, financial consultant, property manager, real estate appraiser, home building inspector*
Razouk, Rashad Elias *retired chemistry educator*
Reiss, Howard *chemistry educator*
Ropchan, Jim R. *research chemist, administrator*
Schep, Raymond Albert *chemist*
Vasquez, Rodolfo Anthony *protein chemist, researcher*
West, Charles David *chemistry educator*
Wittig, Curt *chemist, educator*

Malibu
Yamagishi, Frederick George *chemist*

Menlo Park
Hem, John David *research chemist*
Meyers, Paul Allan *chemist*

Monterey
Shull, Harrison *chemist, educator*

Orangevale
Gibson, Gordon Ronald *chemist*

Pacific Palisades
Csendes, Ernest *chemist, corporate and financial executive*

Palo Alto
Fried, John H. *chemist*
Henry, Kent Douglas *analytical chemist, hardware engineer*
Pauling, Linus Carl *chemistry educator*
Schreiber, Everett Charles, Jr. *chemist, educator*
Theeuwes, Felix *physical chemist*
Ullman, Edwin Fisher *research chemist*

Palos Verdes Estates
Joshi, Satish Devdas *organic chemist*

Pasadena
Anson, Fred Colvig *chemistry educator*
Baldeschwieler, John Dickson *chemist, educator*
Beauchamp, Jesse Lee (Jack Beauchamp) *chemistry educator*
Bercaw, John Edward *chemistry educator, consultant*
Dervan, Peter Brendan *chemistry educator*
Goddard, William Andrew, III *chemist, applied physicist, educator*
Gray, Harry Barkus *chemistry educator*
Grubbs, Robert H. *chemistry educator*
Leonard, Nelson Jordan *chemistry educator*
Marcus, Rudolph Arthur *chemist, educator*
Myers, Andrew Gordon *chemistry educator*
Roberts, John D. *chemist, educator*
Sinha, Mahadeva Prasad *chemist, consultant*
Zewail, Ahmed Hassan *chemistry and chemical engineering educator, editor, consultant*

Penn Valley
Klohs, Murle William *chemist, consultant*

Petaluma
Belmares, Hector *chemist*

Pleasanton
Llenado, Ramon *chemist, chemicals executive*

Sacramento
Nussenbaum, Siegfried Fred *chemistry educator*

San Diego
Ewig, Carl Stephen *chemist*
Goldstein, Mark Kingston Levin *high technology company executive, researcher*
Youkharibache, Philippe Bijin *computational chemist, researcher*

San Francisco
Bigler, William Norman *chemistry educator, biochemist*
Burlingame, Alma Lyman *chemist, educator*
Nguyen, Ann Cac Khue *pharmaceutical and medicinal chemist*

San Jose
Kasai, Paul Haruo *chemist*

San Luis Obispo
Race, Lisa Anne *environmental chemist*

Santa Barbara
Blaskó, Andrei *chemist, researcher*
Butler, Alison *chemist, educator*
Ford, Peter C. *chemistry educator*
Metiu, Horia Ion *chemistry educator*
Schofield, Keith *research chemist*
Wudl, Fred *chemistry educator, consultant*

Santa Clarita
Buck, Douglas Earl *chemist*

Santa Cruz
Bunnett, Joseph Frederick *chemist, educator*
Goldbeck, Robert Arthur, Jr. *physical chemist*

Santa Fe Springs
Hovanec, B. Michael *chemist, researcher*

South San Francisco
Nuwaysir, Lydia Marie *analytical chemist*
Winkler, Marjorie Everett *protein chemist, researcher*

Stanford
Andersen, Hans Christian *chemistry educator*
Brauman, John I. *chemist, educator*
Djerassi, Carl *chemist, educator, writer*
Johnson, William Summer *chemistry educator*
Mosher, Harry Stone *chemistry educator*
Ross, John *physical chemist, educator*
Taube, Henry *chemistry educator*
Trost, Barry Martin *chemist, educator*
Zare, Richard Neil *chemistry educator*

Stockton
Gross, Paul Hans *chemistry educator*

Trona
Laire, Howard George *chemist*

Woodland Hills
Sharma, Brahama Datta *chemistry educator*

COLORADO

Boulder
Birks, John William *chemistry educator*
Cantrell, Christopher Allen *atmospheric chemist*
Chung, Young Chu *chemist*
Cristol, Stanley Jerome *chemistry educator*
Lieberman, James Lance *chemistry educator*
Lineberger, William Carl *chemistry educator*
Usman, Nassim *research chemist and biochemist*
Walton, Harold Frederic *retired chemistry educator*

Colorado Springs
Vorndam, Paul Eric *analytical and organic chemist*

Denver
Eaton, Gareth Richard *chemistry educator, university dean*

Fort Collins
Bernstein, Elliot Roy *chemistry educator*
Fixman, Marshall *chemist, educator*

Golden
Cowley, Scott West *chemist, educator*
Edwards, Kenneth Ward *chemistry educator*
Sunderman, Duane Neuman *chemist, research institute executive*

Hayden
Dunn, Steven Allen *chemist*

CONNECTICUT

Avon
Goodson, Richard Carle, Jr. *chemist, hazardous waste management consultant*

Bloomfield
Oszurek, Paul John *industrial chemist*

Bridgeport
Reed, Charles Eli *retired chemist, chemical engineer*

Danbury
Joyce, William H. *chemist*
Kuryla, William Collier *chemist, consultant*
Pastor, Stephen Daniel *chemistry educator, researecher*

Dayville
Violette, Carol Ann *chemist, environmental compliance consultant*

East Hampton
Goncarovs, Gunti *radio chemist*

East Haven
Wasserman, Harry Hershal *chemistry educator*

Groton
Kelly, Sarah Elizabeth *chemist*
Vinick, Fredric James *chemist*

Hartford
Moyer, Ralph Owen, Jr. *chemist, educator*

Middlebury
Davis, Robert Glenn *research scientist*

Jancis, Elmar Harry *chemist*

Mystic
Chiang, Albert Chinfa *polymer chemist*

New Haven
Berson, Jerome Abraham *chemistry educator*
Wiberg, Kenneth Berle *chemist, educator*
Zoghbi, Sami Spiridon *radiochemist*

New Milford
Lee, Eldon Chen-Hsiung *chemist*

New Preston
Duffis, Allen Jacobus *polymer chemistry extrusion specialist*

Newtown
Cullen, Ernest André *chemist, researcher*

Norwalk
Workman, Jerome James, Jr. *chemist*

Quaker Hill
Conover, Lloyd Hillyard *retired pharmaceutical research scientist and executive*

Southington
Barry, Richard William *chemist, consultant*

Stamford
Panzer, Hans Peter *chemist*

Trumbull
Kontos, Emmanuel George *polymer chemist*

West Haven
Wheeler, George Lawrence *chemist, educator*

DELAWARE

Greenville
Schroeder, Herman Elbert *scientific consultant*

Newark
Evans, Dennis Hyde *chemist, educator*
Gorski, Robert Alexander *chemist, consultant*
Henry, Norman Whitfield, III *research chemist*
Olson, Carl Marcus *chemist retired*

Wilmington
Banville, Debra Lee *research chemist*
Dax, Scott Louis *chemist, researcher*
Hoegger, Erhard Fritz *chemist, consultant*
Kissa, Erik *chemist*
Krape, Philip Joseph *chemist*
Levitt, George *retired chemist*
Marcali, Jean Gregory *chemist*
Moore, Carl Gordon *chemist, educator*
O'Bryan, Saundra M. *clinical chemist, biosafety officer*
Parshall, George William *research chemist*
Simmons, Howard Ensign, Jr. *chemist, research administrator*
Watts, Malcolm L. *chemist*
Webster, Owen Wright *chemist*
Wingate, Phillip Jerome *chemist*
Wu, Dan Qing *research chemist*
Zinck, Barbara Bareis *chemist*

DISTRICT OF COLUMBIA

Washington
Burke, Jerry Alan *retired chemist*
Cassidy, James Edward *chemistry consultant*
Dressick, Walter J. *chemist*
Dusold, Laurence Richard *chemist, computer specialist*
Firestone, David *chemist*
Frank, Richard Stephen *chemist*
Harrison, Edward Thomas, Jr. *chemist*
Karle, Isabella *chemist*
Karle, Jean Marianne *chemist*
Klein, Philipp Hillel *physical chemistry consultant*
McGrath, Kenneth James *chemist*
Shamaiengar, Muthu *chemist*
Soderberg, David Lawrence *chemist*
Treichler, Ray *agricultural chemist*
Wakelyn, Phillip Jeffrey *chemist, consultant*
Weiss, Richard Gerald *chemist educator*
Weisz, Adrian *chemist*
Williams, Frederick Wallace *research chemist*

FLORIDA

Bartow
McFarlin, Richard Francis *industrial chemist, researcher*

Boca Raton
Vijayabhaskar, Rajagopal Coimbatore *chemist, educator*

Coral Gables
Kaifer, Angel Emilio *chemistry educator, researcher*

Crystal River
Skramstad, Phillip James *chemist*

Fort Myers
March, Jacqueline Front *chemist*

Fort Pierce
Killday, K. Brian *organic chemist*

Gainesville
Bertholf, Roger Lloyd *chemist, toxicologist*
Bodor, Nicholas Stephen *medicinal chemistry researcher, educator, consultant*
Colgate, Samuel Oran *chemistry educator*
Dewar, Michael James Steuart *chemistry educator*
Harrison, Willard W. *chemist, educator*
Micha, David Allan *chemistry and physics educator*
Ohrn, Nils Yngve *chemistry and physics educator*
Pop, Emil *research chemist*

Gonzalez
Plischke, Le Moyne Wilfred *research chemist*

Jacksonville
Goldberg, William K. *chemist*
Reynolds, Ellis W. *chemist*
Ulrich, Alfred Daniel, III *chemist*

Lake Worth
Kline, Gordon Mabey *chemist, editor*

Longboat Key
Brown, Henry *chemist*

Marco Island
Hyde, James Franklin *industrial chemical consultant*

Melbourne
Nelson, Gordon Leigh *chemist, educator*

Miami
Cooper, William James *chemist*
Dewanjee, Mrinal Kanti *radiopharmaceutical chemist*
Fernandez Stigliano, Ariel *chemistry educator*
Parra-Diaz, Dennisse *biophysical chemist*
Vought, Franklin Kipling *pharmaceuticals chemist, researcher*

New Port Richey
Eldred, Nelson Richards *chemist, consultant*

Pensacola
Davis, Leslie Shannon *research chemist, chemistry educator*
Jones, Walter Harrison *chemist*

Riviera Beach
Dominick, Paul Scott *chemist, researcher*

Saint Petersburg
Castle, Raymond Nielson *chemist, educator*

Sarasota
Myerson, Albert Leon *physical chemist*

Tallahassee
Choppin, Gregory Robert *chemistry educator*

Tampa
Miller, Ronald Lewis *research hydrologist, chemist*
Wise, Roger M. *chemist*

GEORGIA

Albany
McManus, James William *chemist, researcher*

Alpharetta
Hung, William Mo-Wei *chemist*

Athens
Allinger, Norman Louis *chemistry educator*
Annis, Patricia Anne *textile scientist, educator, researcher*

Atlanta
Anderson, Gloria Long *chemistry educator*
Ashby, Eugene Christopher *chemistry educator, consultant*
Clark, Benjamin Cates, Jr. *flavor chemist, organic research chemist*
Dennison, Daniel B. *chemist*
Hill, Craig Livingston *chemistry educator, consultant*
Iacobucci, Guillermo Arturo *chemist*
Kobelski, Robert John *analytical chemist*
Queen, Brian Charles *chemist*
Smith, David Carr *organic chemist*
Underwood, Arthur Louis, Jr. *chemistry educator, researcher*
Zalkow, Leon H. *organic chemistry educator*

Columbus
Moussa, Khalil Mahmoud *polymer photochemist, researcher*

Decatur
Mills, Terry, III *forensic chemist*

Gordon
Young, Raymond H(inchcliffe), Jr. *chemist*

Sandersville
Malla, Prakash Babu *research materials chemist*

Savannah
Brown, Thomas Edward *chemist*
Simonaitis, Richard Ambrose *chemist*

Valdosta
Hoff, Edwin Frank, Jr. *research chemist*

HAWAII

Camp Smith
Surface, Stephen Walter *water treatment chemist, environmental protection specialist*

Honolulu
Scheuer, Paul Josef *chemistry educator*
Seifert, Josef *chemist, educator*

IDAHO

Idaho Falls
Zaccardi, Larry Bryan *spectrochemistry scientist, microbiologist*

Lewiston
Bjerke, Robert Keith *chemist*

ILLINOIS

Abbott Park
Boyd, Steven Armen *medicinal chemist*
Swift, Kerry Michael *physical chemist*
Trivedi, Jay Sanjay *chemist*

Argo
Totten, Venita Laverne *chemist*

Argonne
Demirgian, Jack Charles *analytical chemist*
Kini, Aravinda Mattar *materials chemist*
Thorn, Robert Jerome *chemist*

Aurora
Ika, Prasad Venkata *chemist*

Bellwood
Gregory, Vance Peter, Jr. *chemist*

Buffalo Grove
Kalvin, Douglas Mark *research chemist*

Burr Ridge
Bathina, Harinath Babu *chemist, researcher*

Champaign
Wood, Susanne Griffiths *analytical environmental chemist*

Chicago
Freed, Karl Frederick *chemistry educator*
Fried, Josef *chemist, educator*
Gomer, Robert *chemistry educator*
Greaves, William Webster *chemist, patent analyst*
Halpern, Jack *chemist, educator*
Harwood, John Simon *chemist, university official, consultant*
Krawetz, Arthur Altshuler *chemist, science administrator*
Levy, Donald Harris *chemistry educator*
Miller, Michael Carl *chemist, researcher*
Murphy, Thomas Joseph *chemistry educator*
Norris, James Rufus, Jr. *chemist, educator, consultant*
Pluth, Joseph John *chemist, consultant*
Trenary, Michael *chemistry educator*
Triplett, Kelly B. *chemist*
Turkevich, Anthony Leonid *chemist, educator*

Decatur
Weatherbee, Carl *retired chemistry educator, genealogist*

Des Plaines
Arena, Blaise Joseph *research chemist*
Laban, Saad Lotfy *medicinal chemist*
Surgi, Marion Rene *chemist*

Downers Grove
Boese, Mark Alan *forensic scientist, chemist, educator*
Boese, Robert Alan *forensic chemist*

Evanston
Basolo, Fred *chemistry educator*
Colton, Frank Benjamin *retired chemist*
Ibers, James Arthur *chemist, educator*
Klotz, Irving Myron *chemist, educator*
Lambert, Joseph Buckley *chemistry educator*
Letsinger, Robert Lewis *chemistry educator*
Marks, Tobin Jay *chemistry educator*
Pople, John Anthony *chemistry educator*
Sachtler, Wolfgang Max Hugo *chemistry educator*

Forest Park
Johnson, Calvin Keith *research executive, chemist*

Freeport
Trickel, Neal Edward *chemist, chemical researcher*

Galesburg
Kooser, Robert Galen *chemical educator*

Glen Ellyn
Kelada, Nabih Philobbos *chemist, consultant*

Kankakee
Armstrong, Douglas *organic chemist, educator*

Lemont
Katz, Joseph Jacob *chemist, educator*
Melnikov, Paul *analytical chemist, instrumentation engineer*
Williams, Jack Marvin *chemist*

Maywood
Haschke, Paul Charles *analytical chemist*

Naperville
Carrera, Martin Enrique *research scientist*
Dieterle, Robert *chemist*
Fields, Ellis Kirby *research chemist*
Meyer, Delbert Henry *organic chemist, researcher*
Narutis, Vytas *chemist, researcher*
Shannon, James Edward *water chemist, consultant*
Sherren, Anne Terry *chemistry educator*
Weinstein, David Ira *industrial chemist*

Normal
Morse, Philip Dexter, II *chemist, educator*

North Chicago
Carney, Ronald Eugene *chemist*

Oak Forest
Lekberg, Robert David *chemist*

Orland Park
Germino, Felix Joseph *chemist, research-development company executive*

Peoria
Cunningham, Raymond Leo *research chemist*
King, Jerry Wayne *research chemist*

Urbana
Beak, Peter Andrew *chemistry educator*
Brown, Theodore Lawrence *chemistry educator*
Curtin, David Yarrow *chemist, educator*
Devadoss, Chelladurai *physical chemist*
Drickamer, Harry George *retired chemistry educator*
Gruebele, Martin *chemistry educator*
Gutowsky, H. S. *chemistry educator*
Jonas, Jiri *chemistry educator*
Katzenellenbogen, John Albert *chemistry educator*
Klemperer, Walter George *chemistry educator, researcher*
Lauterbur, Paul C(hristian) *chemistry educator*
Minear, Roger Allan *chemist, educator*
Stork, Wilmer Dean, II *physical chemist, researcher*
Wolynes, Peter Guy *chemistry researcher, educator*
Zimmerman, Steven Charles *chemistry educator*

Westchester
Hernandez, Medardo Concepcion *chemist*

Woodstock
McKittrick, Philip Thomas, Jr. *analytical chemist*

INDIANA

Bloomington
Chisholm, Malcolm Harold *chemistry educator*
Davidson, Ernest Roy *chemist, educator*
Hieftje, Gary Martin *analytical chemist, educator*
Magnus, Philip Douglas *chemistry educator*
Novotny, Milos V. *chemistry educator*
Parmenter, Charles Stedman *chemistry educator*

Columbus
Totten, Gary Allen *spectroscopist*

Elkhart
Free, Alfred Henry *clinical chemist, consultant*
Free, Helen M. *chemist, consultant*

Evansville
Moody, Brian Wayne *chemist*

Gary
Meyerson, Seymour *retired chemist*

Granger
Chmiel, Chester T. *adhesive chemist, consultant*

Indianapolis
Childers, Richard Herbert, Jr. *chemist*
Cooley, Rick Eugene *chemist*
Davis, Robert Drummond, Sr. *chemist, researcher*
Desai, Mukund Ramanlal *research and development chemist*
Fife, Wilmer Krafft *chemistry educator*
Hildebrand, William Clayton *chemist*
Marshall, Frederick Joseph *retired research chemist*
Scott, William Leonard *research chemist, educator*

Lafayette
Bowers, Conrad Paul *chemist*
Pyer, John Clayton *analytical chemist*
Weaver, Michael John *chemist, educator*

Muncie
Harris, Joseph McAllister *chemist*
Lang, Patricia Louise *chemistry educator, vibrational spectroscopist*

Notre Dame
Fehlner, Thomas Patrick *chemistry educator*
Hayes, Robert Green *chemical educator, researcher*
Schuler, Robert Hugo *chemist, educator*
Su, Yali *chemist*
Thomas, John Kerry *chemistry educator*
Trozzolo, Anthony Marion *chemistry educator*

Terre Haute
Guthrie, Frank Albert *chemistry educator*

Valparaiso
Cook, Addison Gilbert *chemistry educator*

West Lafayette
Brown, Herbert Charles *chemistry educator*
Gorenstein, David G. *chemistry educator*
Hanks, Alan R. *chemistry educator*
Johnston, Clifford Thomas *soil and environmental chemistry educator*
Krockover, Gerald Howard *science educator*
Margerum, Dale William *chemistry educator*
Morrison, Harry *chemistry educator, university dean*

IOWA

Ames
Bastiaans, Glenn John *analytical chemist, researcher*
Cink, James Henry *chemical safety consultant, educator*
DePristo, Andrew E. *chemist, educator*
DeYong, Gregory Donald *chemist*
Franzen, Hugo Friedrich *chemistry educator, researcher*
Houk, Robert Samuel *chemistry educator*
Ruedenberg, Klaus *theoretical chemist, educator*
Yeung, Edward Szeshing *chemist*

Fairfield
Zsigo, Jozsef Mihaly *chemist*

Iowa City
Goff, Harold Milton *chemistry educator*
Grassian, Vicki Helene *chemistry educator*
Lee, Shyan Jer *physical chemist*

KANSAS

Atchinson
Chinnaswamy, Rangan *cereal chemist*

Kansas City
Gray, Donald Lee *chemist*

Lawrence
Harmony, Marlin Dale *chemistry educator, researcher*
Stella, Valentino John *pharmaceutical chemistry educator*

Manhattan
Hoseney, Russell Carl *chemistry educator*

North Newton
Lehman, Thomas Alan *chemistry educator*

Stilwell
Fought, Lorianne *chemist*

KENTUCKY

Ashland
Alexander, William Robert *chemist*

Bowling Green
Hartman, David Robert *chemistry educator*

Carrollton
Tatera, James Frank *chemist, process analysis specialist*

Georgetown
Tieman, Robert Scott *chemist*

Lexington
Bachas, Leonidas Gregory *chemistry educator*
Butterfield, D. Allan *chemistry educator, university administrator*
Porter, William Hudson *clinical chemist*
Thio, Alan PooAn *chemist*

Louisville
Yff, David Robert *chemist*

Murray
McCreary, Terry Wade *chemist, educator*

Pikeville
Christensen, Ralph J. *chemistry educator*

LOUISIANA

Baton Rouge
Brown, Leo Dale *inorganic chemist*
Gooden, Robert *chemist*
Imhoff, Donald Wilbur *chemist*
West, Philip William *chemistry educator*

Kenner
deMonsabert, Winston Russel *chemist*

Metairie
Horkowitz, Sylvester Peter *chemist*

New Orleans
Andrews, Bethlehem Kottes *research chemist*
French, Alfred Dexter *chemist*
Harper, Robert John, Jr. *chemist, researcher*
Striegel, Andre Michael *chemist*
Sumrell, Gene *research chemist*
Wiewiorowski, Tadeusz Karol *research chemist, consultant*

Saint Gabriel
Bayer, Arthur Craig *organic chemist*

MAINE

Damariscotta
Emerson, William Stevenson *retired chemist, consultant, writer*

South Portland
Tewhey, John David *geochemist, environmental consultant*

MARYLAND

Aberdeen Proving Ground
Arroyo, Carmen Milagros *research chemist*
Mackay, Raymond Arthur *chemist*
Steger, Ralph James *chemist*

Annapolis
Siatkowski, Ronald E. *chemistry educator*

Baltimore
Berg, Jeremy M. *chemistry educator*
Fenselau, Catherine Clarke *chemistry educator*
Posner, Gary Herbert *chemist, educator*
White, Emil Henry *chemistry educator*

Beltsville
Howe, Juliette Coupain *chemist*
Schmidt, Walter Friedrich *research chemist*
Stern, Michael *chemist*
Velasco, James *chemist, consultant*

Bethesda
Backus, Robert Coburn *biophysical chemist*
Bartky, Ian Robertson *physical chemist*
Creveling, Cyrus Robbins *chemist, neuroscientist*
Fales, Henry Marshall *chemist*
Jacobson, Arthur Eli *research chemist*
Pastor, Richard Walter *research chemist*
Regdos, Shane Lawrence *nuclear chemist*
Witkop, Bernhard *chemist*

Cockeysville Hunt Valley
Cohen, Kenneth Allan *chemist*

College Park
Andrews, Kevin Paul *physical chemist*
Moore, John Hays *chemistry educator*
Ponnamperuma, Cyril Andrew *chemist*
Weeks, John David *chemistry and physical science educator*

Columbia
Khare, Mohan *chemist*
Raksis, Joseph W. *chemist*
Shiue, Gong-Huey *chemist*

Crofton
Watson, Robert Tanner *physical scientist*

Easton
Hurley, Richard Keith *chemist*

Frederick
Cragg, Gordon Mitchell *government chemist*
Kappe, David Syme *environmental chemist*

Gaithersburg
Bonnell, David William *research chemist*
Brinckman, Frederick Edward, Jr. *research chemist, consultant*
Epler, Katherine Susan *chemist*
Frahm, Veryl Harvey, Jr. *laboratory manager*
Hall, Dale Edward *electrochemist*
Hamer, Walter Jay *chemical consultant, science writer*

Highland
MacFadden, Kenneth Orville *chemist*

Laurel
Benson, Richard Charles *chemist*
Linevsky, Milton Joshua *physical chemist*
Sundaresan, P. Ramnathan *research chemist, consultant*

Rockville
Grady, Lee Timothy *pharmaceutical chemist*
Grist, Clarence Richard *chemist, precious metals investor*
Ji, Xinhua *physical chemist, educator*
Kindt, Thomas James *chemist*
Schmuff, Norman Robert *organic chemist*

Sandy Spring
Kanarowski, Stanley Martin *chemist, chemical engineer, government official*

Woodbine
Herrick, Elbert Charles *chemist, consultant*

MASSACHUSETTS

Amherst
MacKnight, William John *chemist, educator*
Maroney, Michael James *chemistry educator*
Voigtman, Edward George, Jr. *chemist educator*

Arlington
Leonardos, Gregory *chemist, odor consultant*

Attleboro
Griffin, Edwin H., Jr. (Hank Griffin) *chemist*

Bedford
Lopatin, George *research chemist*

Billerica
Conoby, Joseph Francis *chemist*

Boston
Browning, William Elgar, Jr. *chemistry consultant*
Bucher, Gail Phillips *cosmetic chemist*
Levy, Boris *chemistry educator*
Turner, Raymond Edward *chemistry educator, researcher*

Bridgewater
Daley, Henry Owen, Jr. *chemist, educator*

Cambridge
Alberty, Robert Arnold *chemistry educator*
Anderson, James Gilbert *chemistry educator*
Ashford, Nicholas Askounes *technology and policy educator*
Buchi, George Hermann *chemistry educator*
Chen, Peter *chemistry educator*
Ciappenelli, Donald John *chemist, academic administrator*
Corey, Elias James *chemistry educator*
Doering, William von Eggers *organic chemist, educator*
Evans, David A(lbert) *chemistry educator*
Field, Robert Warren *chemistry educator*
Friend, Cynthia M. *chemist, educator*
Gordon, Roy Gerald *chemistry educator*
Herschbach, Dudley Robert *chemistry educator*
Holm, Richard Hadley *chemist, educator*
Karplus, Martin *chemistry educator*
Khorana, Har Gobind *chemist, educator*
Klibanov, Alexander Maxim *chemistry educator*
Knowles, Jeremy Randall *chemist, educator*
Lippard, Stephen James *chemist, educator*
Lipscomb, William Nunn, Jr. *retired physical chemistry educator*
Martinez, Luis Enrique, Jr. *chemist researcher*
Molina, Mario Jose *physical chemist, educator*
Schreiber, Stuart L. *chemist, educator*
Steinfeld, Jeffrey Irwin *chemistry educator, consultant, author*
Stout, Philip John *chemist*
Stubbe, JoAnne *chemistry educator*
Virgil, Scott Christopher *chemistry educator*
Waugh, John Stewart *chemist, educator*
Westheimer, Frank Henry *chemist, educator*
Whitesides, George McClelland *chemistry educator*

Chestnut Hill
Bennett, Ovell Francis *chemistry educator, researcher, consultant*

East Falmouth
Dixon, Brian Gilbert *chemist*

Falmouth
Allen, Duff Shederic, Jr. *retired chemist*

Fitchburg
Wickstrom, Carl Webster *analytical chemist*

Hanscom AFB
Ahmadjian, Mark *chemist*
Paulson, John Frederick *research chemist*

Holden
Gordon, Steven B. *chemist*

Hopkinton
MacLean, Mary Elise *chemist*

Lexington
Bartlett, Paul Doughty *chemist, educator*
Cathou, Renata Egone *chemist, consultant*
Jensen, Mona Dickson *chemist researcher*
Samour, Carlos M. *chemist*
Schonhorn, Harold *chemist, researcher*

Lowell
Kyriacou, Demetrios *electro-organic chemist, researcher, consultant*
McCabe, Douglas Raymond *analytical chemist*

Malden
Rollins, Scott Franklin *chemist*

Marlborough
Szmanda, Charles Raymond *chemist*

Medford
Blanco, Frances Rebecca Briones *chemist, medical technologist*

Natick
Spada, Marianne Rina *medicinal chemist*

Newton
Heyn, Arno Harry Albert *retired chemistry educator*
Quinlan, Kenneth Paul *chemist*

Northampton
Sommer, Alfred Hermann *retired physical chemist*
White, Virginia *chemistry educator*

Pittsfield
Wheelock, Kenneth Steven *chemist*

Somerville
Leventis, Nicholas *chemist, consultant*

South Hadley
Harrison, Anna Jane *chemist, educator*
Williamson, Kenneth Lee *chemistry educator*

Springfield
Cohen, Saul Mark *chemist, retired*

Waltham
Kustin, Kenneth *chemist*
Petit, William *chemist*
Timasheff, Serge Nicholas *chemist, educator*

Wayland
Kant, Arthur *scientist emeritus, physical chemist*

MICHIGAN

Adrian
Scioly, Anthony Joseph *chemistry and physics educator*

Ann Arbor
DeWitt, Sheila Hobbs *research chemist*
Lee, Stephen *chemist, educator*
McBee, Gary L. *chemist*
Morris, Michael David *chemistry educator*

Battle Creek
Robie, Donna Jean *chemist*

Dearborn
Carter, Roscoe Owen, III *chemist*
Kelly, Carol Johnson *chemist*

Detroit
Belfield, Kevin Donald *chemistry educator*
Johnson, Carl Randolph *chemist, educator*
Romano, Louis James *chemist, educator*
Ryntz, Rose Ann *chemist*
Tewari, Kewal Krishan *chemist, researcher*

East Lansing
Dye, James Louis *chemistry educator*
Hoganson, Curtis Wendell *physical chemist*
Kamdem, Donatien Pascal *chemistry educator*

Farmington
Koshy, Vettithara Cherian *chemistry educator, technical director/formulator*
Theodore, Ares Nicholas *research scientist, educator, entrepreneur*

Holland
Benko, James John *chemist*
Witzel, William Martin *analytical chemist, educator, consultant*

Kalamazoo
Elrod, David Wayne *computational chemist, information scientist*
Finzel, Barry Craig *research scientist*
Mizsak, Stephen Andrew *chemist*

Midland
Collard, Randle Scott *chemist*
Dreyfuss, Patricia *chemist, researcher*
Rudolf, Philip Reinhold *chemist, crystallographer*
Snow, Steven Ashley *chemist*
Speier, John Leo, Jr. *chemist*
Stull, Daniel Richard *research thermochemist, educator, consultant*
Weyenberg, Donald Richard *chemist*

Okemos
Loconto, Paul Ralph *chemist, consultant*

Rochester
Eliezer, Isaac *chemistry educator*

Saginaw
Namboodiri, Krishnan *chemist*

Southfield
Ramesh, Swaminathan *chemist*

Three Rivers
Boyer, Nicodemus Elijah *organic-polymer chemist, consultant*

Troy
Nagata, Isao *chemist*
Purcell, Jerry *chemist*

Warren
Deak, Charles Karol *chemist*
Furey, Robert L. *research chemist*
Mance, Andrew Mark *chemist*

Ypsilanti
Kennelly, William James *chemist*

MINNESOTA

Lake Elmo
Vivona, Daniel Nicholas *chemist*

Marshall
Carberry, Edward Andrew *chemistry educator*

Minneapolis
Carlson, Marvin *analytical chemist*
Moscowitz, Albert Joseph *chemist, educator*
Smyrl, William H. *chemistry educator*
Truhlar, Donald Gene *chemist, educator*

Moorhead
Samaraweera, Upasiri *research chemist*

Northfield
Hardgrove, George Lind, Jr. *chemistry educator*

Plymouth
Ding, Ni *chemist*

Saint Paul
Alm, Roger Russell *chemist*
Erickson, Brice Carl *chemist*
Stebbings, William Lee *chemist, researcher*

Saint Peter
Thompson, H. Bradford *chemist, educator*

White Bear Lake
Holmen, Reynold Emanuel *chemist*

MISSISSIPPI

Jackson
Adams-Hilliard, Beverly Lynn *chemist*
Zubkowski, Jeffrey Dennis *chemistry educator*

University
Jung, Mankil *organic synthetic chemist, educator*
Williamson, John Steven *medicinal chemist*

MISSOURI

Cape Girardeau
Hathaway, Ruth Ann *chemist*

Chesterfield
Bockserman, Robert Julian *chemist*
Schuermann, Mark Harry *chemist, educator*

Columbia
Adams, John Ewart *chemistry educator*
Stalling, David Laurence *research chemist*
Wong, Tuck Chuen *chemist, educator*

Independence
Johnson, Cecil Kirk, Jr. *cement chemist*

Joplin
Whittle, Philip Rodger *chemistry educator, crime laboratory director*

Kansas City
Gier, Audra May Calhoon *environmental chemist*
Heath, Timothy Gordon *bioanalytical chemist*
Wickham-St. Germain, Margaret Edna *mass spectrometrist*

Maryland Heights
Chinn, Rex Arlyn *chemist*
Dahiya, Jai Bhagwan *chemist*

Saint Louis
Cooper, Stephen Randolph *chemistry educator and industrial research manager*
Hamper, Bruce Cameron *organic chemist, agricultural chemist*
Henis, Jay Myls Stuart *research scientist, research director*
Kaufmann, Gary Bryan *chemist*
Moran, Sharon Joyce *chemist*
Murray, Robert Wallace *chemistry educator*
Quicksall, Carl Owen *chemist, researcher*
Redmore, Derek *chemist*
Sharp, Dexter Brian *organic chemist, consultant*
Shelke, Kantha *cereal chemist*
Taylor, Morris Anthony *chemistry educator*
Worley, Jimmy Weldon *chemist*

Springfield
Criswell, Charles Harrison *analytical chemist, evironmental and forensic consultant and executive*

MONTANA

Butte
Murray, Joseph *chemistry educator*

Rollins
Zelezny, William Francis *retired physical chemist*

Troy
Sherman, Signe Lidfeldt *securities analyst, former research chemist*

NEBRASKA

Lincoln
Gross, Michael Lawrence *chemistry educator*
Yoder, Bruce Alan *chemist*

Omaha
Vasiliades, John *chemist*

Wayne
Johar, Jogindar Singh *chemistry educator*

NEVADA

Reno
Pierson, William R. *chemist*

NEW HAMPSHIRE

Durham
Seitz, William Rudolf *chemistry educator*

Glen
Zager, Ronald I. *chemist, consultant*

Hanover
Naumann, Robert Bruno Alexander *chemist, physicist, educator*
Stockmayer, Walter H(ugo) *chemistry educator*

Jaffrey
Walling, Cheves Thomson *chemistry educator*

Lyme
Kelemen, Denis George *physical chemist, consultant*

North Sutton
Springsteen, Arthur William *organic chemist*

NEW JERSEY

Annandale
Gorbaty, Martin Leo *chemist, researcher*
Sinfelt, John Henry *chemist*
Varadaraj, Ramesh *research chemist*

Basking Ridge
McCall, David W. *chemist, administrator, materials consultant*

Bedminster
Beiman, Elliott *research pharmaceutical chemist*
Bovey, Frank Alden *research chemist*
Collins, George Joseph *chemist*

Berkeley Heights
Zaret, Efrem Herbert *chemist*

Blairstown
Martin, James Walter *chemist, technology executive*

Bloomfield
Kwon, Joon Taek *chemistry researcher*

Bridgewater
Twardowski, Thomas Edward, Jr. *development chemist*

Budd Lake
Krause, Lois Ruth Breur *chemistry educator, engineer*

Camden
Simon, Frederick Edward *chemist*

Chatham
KixMiller, Richard Wood *chemist, corporate executive*

Cranbury
Wang, Chih Chun *chemist, scientific administrator*

Deepwater
Millican, David Wayne *analytical chemist*

Dover
Capellos, Chris Spiridon *chemist*

East Brunswick
Irgon, Joseph *physical chemist*
Steiger, Fred Harold *chemist*

East Hanover
Klemann, Lawrence Paul *chemical scientist*

Edgewater
Spaltro, Suree Methmanus *chemist, researcher*

Edison
Farazdel, Abbas *chemist*
Lin, Yi-Hua *environmental chemist*
Parsons, William Hugh *chemist, researcher*

Englewood
Grier, Nathaniel *chemist*

Fair Lawn
Humiec, Frank Stanley *chemist*
Mills, Lester Stephen *chemist*

Flemington
Mitry, Mark *chemist*

Franklin Lakes
Hetzel, Donald Stanford *chemist*

Glassboro
Schultz, Charles William *chemistry educator*

Hillside
Tencza, Thomas Michael *chemist*

Hoboken
Bose, Ajay Kumar *chemistry educator*

Kenilworth
Evans, Charlie Anderson *chemist*
Ganguly, Ashit Kumar *organic chemist*

Linden
Dietz, Thomas Gordon *chemist*
Pink, Harry Stuard *petroleum chemist*

Livingston
Peterson, Glenn Stephen *chemist*

Maplewood
Leal, Joseph Rogers *chemist*

Milltown
Sacharow, Stanley *chemist, consultant*

Monmouth Junction
Mylonakis, Stamatios Gregory *chemist*

Morris Plains
Montana, Anthony James *analytical chemist*
Moroni, Antonio *chemist, consultant, international coordinator*

Morristown
Reimschuessel, Herbert Kurt *chemist, consultant*
Wu, Tse Cheng *research chemist*

Mount Holly
Worne, Howard Edward *biotechnology company executive*

Mount Laurel
Cazes, Jack *chemist, marketing consultant, editor*

Murray Hill
Baker, William Oliver *research chemist, educator*

New Brunswick
Ho, Chi-Tang *food chemistry educator*
Lerner, Henry Hyam *chemist*
Martin, Frank Scott *chemist*
Pandey, Ramesh Chandra *chemist*
Rosen, Joseph David *chemist, educator*
Rosen, Robert Thomas *analytical and food chemist*
Strauss, Ulrich Paul *educator, chemist*
Winnett, A(sa) George *analytical research chemist, educator*

New Providence
Chandross, Edwin A. *chemist, polymer researcher*
Stillinger, Frank Henry *chemist, educator*

Newark
Ledeen, Robert Wagner *neurochemist, educator*
Suchow, Lawrence *chemistry educator, researcher, consultant*
Waelde, Lawrence Richard *chemist*

Nutley
Amornmarn, Lina *chemist*
Douvan-Kulesha, Irina *chemist*

Oakland
Pepe, Teri-Anne *development chemist*

Paramus
Klarreich, Susan Rae *chemistry educator*

Parsippany
Boulos, Atef Zekry *chemist*

Paulsboro
Domingue, Raymond Pierre *chemist, consultant, educator*
Garwood, William Everett *chemist researcher*

Pennsville
Ibrahim, Fayez Barsoum *chemist*
Ryan, Timothy William *analytical chemist*

Piscataway
Hara, Masanori *chemist*
Neishlos, Arye Leon *chemist*

Port Newark
McKenna, James Emmet *chemist*

Pottersville
Konecky, Milton Stuart *chemist*

Princeton
Bentz, Bryan Lloyd *chemist*
Green, Joseph *chemist*
Groves, John Taylor, III *chemist, educator*
Jones, Maitland, Jr. *chemistry educator*
Kauzmann, Walter Joseph *chemistry educator*
Kilian, Robert Joseph *chemist*
Kyin, Saw William *chemist, consultant*
Lawrence, Robert Michael *research chemist*
Liao, Hsiang Peng *chemist*
Little, Dorothy Marion Sheila *chemist*
Los, Marinus *agrochemical researcher*
Mc Clure, Donald Stuart *physical chemist, educator*
Ondetti, Miguel Angel *chemist, consultant*
Plummer, Ernest Lockhart *industrial research chemist*
Ramaprasad, Kackadasam Raghavachar *physical chemist*
Rebenfeld, Ludwig *chemist*
Scoles, Giacinto *chemistry educator*
Spiro, Thomas George *chemistry educator*
Strike, Donald Peter *pharmaceutical research director, research chemist*
Taylor, Edward Curtis *chemistry educator*
Weigmann, Hans-Dietrich H. *chemist*
Wong, Ching-Ping *chemist*
Zask, Arie *chemist, researcher*

Rahway
Patchett, Arthur Allan *medicinal chemist, pharmaceutical executive*

Randolph
Price, Elizabeth Cain *environmental chemist*

Raritan
Haller, William Paul *analytical chemist, robotics specialist*
Johnson, Dana Lee *biophysical chemist*
Stuting, Hans Helmuth *chemist, researcher, consultant*

Ridgefield Park
Eul, Wilfried Ludwig *chemist*

Ringoes
Overton, Santford Vance *applications chemist*

Rochelle Park
Long, Timothy Scott *chemist, consultant*

Skillman
Ma, Tony Yong *chemist, international business consultant*

Somerset
Avolio, John *chemist, inventor*
Heller, Donald Franklin *chemical and laser physicist*

Stewartsville
Eckert, John Andrew *chemist, technical consultant*

Summit
Charbonneau, Larry Francis *research chemist*
Zachary, Louis George *chemical company consultant*
Ziegler, John Benjamin *chemist, lepidopterist*

Trenton
Barclay, Robert, Jr. *chemist*
Parsa, Bahman *nuclear chemist*
Rau, Eric *chemist*

Warren
Ropp, Richard Claude *chemist*

West Caldwell
Nittoli, Thomas *chemist, consultant*

West Long Branch
Aguiar, Adam Martin *chemist, educator*

NEW MEXICO

Albuquerque
Cahill, Paul Augustine *chemistry researcher*
Dorko, Ernest Alexander *chemist, researcher*
Ortiz, Joseph Vincent *chemistry educator*
Santillanes, Simon Paul *analytical chemist, biotoxicologist*

Farmington
Norvelle, Norman Reese *environmental and industrial chemist*

Las Cruces
Naser, Najih A. *chemistry educator, researcher*

Los Alamos
Briesmeister, Richard Arthur *chemist*
Kubas, Gregory Joseph *research chemist*
Mullen, Ken Ian *chemist*
Onstott, Edward Irvin *research chemist*

Santa Fe
Cowan, George Arthur *chemist, bank executive, director*
Marcy, Willard *chemist, chemical engineer, retired*
Sheehan, William Francis *chemist educator*

Socorro
Gullapalli, Pratap *chemist, researcher, educator*

NEW YORK

Albany
Azam, Farooq *chemist, researcher*
Demerjian, Kenneth L. *atmospheric science educator, research center director*
Frisch, Harry Lloyd *chemist, educator*

Binghamton
Doetschman, David Charles *chemistry educator*
Eisch, John Joseph *chemist, educator*
Huie, Carmen Wah-Kit *chemistry educator*
Whittingham, M(ichael) Stanley *chemist*

Bronx
Gottfried, David Scott *chemist researcher*
Peterson, Eric Scott *physical chemist*

Brooklyn
Banks, Ephraim *chemistry educator, consultant*
Hardy, Major Preston, Jr. *analytical chemist*
Hirsch, Warren Mitchell *chemistry educator*
Ma, Tsu Sheng *chemist, educator, consultant*
Nolan, Robert Patrick *chemistry educator*
Pearce, Eli M. *chemistry educator, administrator*
Vogl, Otto *polymer science and engineering educator*
Yablonsky, Harvey Allen *chemistry educator*

Buffalo
Garvey, James Francis *physical chemist*
Ohrt, Jean Marie *chemist*
Patel, Suresh *chemist, researcher*
Sharma, Minoti *chemist, researcher*
Steward, A(lma) Ruth *chemistry educator, researcher*
Thompson, John James *chemist, researcher, consultant*
Williams, James Samuel, Jr. *chemist*

Carmel
Strojny, Norman *analytical chemist*

Chestnut Ridge
Huntoon, Robert Brian *chemist, food industry consultant*

Corning
Williams, Jimmie Lewis *research chemist*

Dobbs Ferry
Hoey, Michael Dennis *organic chemist*

Endicott
Creasy, William Russel *chemist, writer*

Farmingdale
Purandare, Yeshwant K. *chemistry educator, consultant*

Garden City
Kirsch, Robert *director analytical research and development*
Singer, Jeffrey Michael *organic analytical chemist*

Geneva
Gardner, Audrey V. *chemist, researcher*

Glens Falls
Elton, Richard Kenneth *polymer chemist*

Hastings On Hudson
Weil, Edward David *chemist*

Inwood
Fine, Sidney Gilbert *chemist*

Ithaca
Di Salvo, Francis Joseph *chemistry educator*
Fay, Robert Clinton *chemist, educator*
Harrison, Aidan Timothy *chemist*
McLafferty, Fred Warren *chemist, educator*
Meinwald, Jerrold *chemist, educator*
Morrison, George Harold *chemist, educator*
Scheraga, Harold Abraham *physical chemistry educator*
Van Campen, Darrell Robert *chemist*
Widom, Benjamin *chemistry educator*

Jamaica
Sun, Siao Fang *chemistry educator*
Testa, Anthony Carmine *chemistry educator*

Latham
Sciabica, Vincent Samuel *chemist, researcher*

New York
Ashen, Philip *chemist*
Breslow, Ronald Charles *chemist, educator*
de Duve, Christian René *chemist, biologist, educator*
Guo, Chu *chemistry educator*
Karouna, Kir George *chemist, consultant*
Keller, Stephen *chemist educator*
Nakanishi, Koji *chemistry educator, research institute administrator*
Parkin, Gerard Francis Ralph *chemistry educator, researcher*
Rosenfeld, Louis *biochemist*
Safier, Lenore Beryl *research chemist*
Sapse, Anne-Marie *chemistry educator*
Stork, Gilbert (Josse) *chemistry educator, investigator*
Turro, Nicholas John *chemistry educator*
Valentini, James Joseph *chemistry educator*
Wazneh, Leila Hussein *organic chemist*

Oakdale
Panzarella, John Edward *water quality chemist*

Painted Post
Hammond, George Simms *chemist*
Stookey, Stanley Donald *chemist*

Pearl River
Citardi, Mattio H. *chemist, researcher, system manager*
Kolor, Michael Garrett *research chemist*
Trust, Ronald Irving *organic chemist*

Pleasantville
Nabirahni, David M.A. *chemist, educator*

Pomona
Ciaccio, Leonard Louis *chemist researcher, science administrator*

Potsdam
Matijevic, Egon *chemistry educator, researcher, consultant*

Poughkeepsie
Rossi, Miriam *chemistry educator, researcher*

Rensselaer
Krasney, Ethel Levin *research chemist*

Rochester
Bambury, Ronald Edward *polymer chemist*
Cain, B(urton) Edward *chemistry educator*
Factor, Ronda Ellen *research chemist*
Gates, Marshall DeMotte, Jr. *chemistry educator*
Hailstone, Richard Kenneth *chemist, educator*
Huizenga, John Robert *nuclear chemist, educator*
Jansen, Kathryn Lynn *chemist*
Krogh-Jespersen, Mary-Beth *chemist, educator*
Luckey, George William *research chemist*

Martic, Peter Ante *research chemist*
Paz-Pujalt, Gustavo Roberto *physical chemist*
Rao, Joseph Michael *chemist*
Saunders, William Hundley, Jr. *chemist, educator*
Thomas, Telfer Lawson *chemist, researcher*
Whitten, David George *chemistry educator, researcher*

Rouses Point
Al-Hakim, Ali Hussein *chemist*

Saint James
Bigeleisen, Jacob *chemist, educator*

Saratoga Springs
Walter, Paul Hermann Lawrence *chemistry educator*

Scarsdale
Cox, Robert Hames *chemist, scientific consultant*

Schenectady
Billmeyer, Fred Wallace, Jr. *chemist, educator*

Spencerport
Astill, Bernard Douglas *environmental health and safety consultant*

Spring Valley
McCormick, Jerry Robert Daniel *chemistry consultant*

Syracuse
Birge, Robert Richards *chemistry educator*
LaLonde, Robert Thomas *chemistry educator*
Pearse, George Ancell, Jr. *chemistry educator, researcher*
Schenerman, Mark Allen *research chemist, educator*
Sleezer, Paul David *organic chemist*

Tarrytown
Belanger, Ronald Louis *chemist*
Margoshes, Marvin *chemist, consultant*

Troy
Krause, Sonja *chemistry educator, researcher*

Tuxedo Park
Hall, Frederick Keith *chemist*
Noel, Dale Leon *chemist*

Upton
Friedlander, Gerhart *nuclear chemist*
Sutin, Norman *chemistry educator, scientist*
Wolf, Alfred Peter *chemist, educator*

Utica
Pulliam, Curtis Richard *chemistry educator*

Webster
O'Neill, James Francis *chemist*

Wolcott
Wilt, John Robert *chemist*

Yonkers
Anderson, Bror Ernest *chemist*

Yorktown Heights
Buchwalter, Stephen L. *chemist, researcher*
Haller, Ivan *chemist*
Holtzberg, Frederic *chemist, solid state researcher*
Kanicki, Jerzy *chemist, researcher*

NORTH CAROLINA

Asheville
Squibb, Samuel Dexter *chemistry educator*
Stevens, John Gehret *chemistry educator*
Van Engelen, Debra Lynn *chemistry educator*

Chapel Hill
Driscol, Jeffrey William *chemist, researcher*
Eliel, Ernest Ludwig *chemist, educator*
Jorgenson, James Wallace *chromatographer, educator*
Murray, Royce Wilton *chemistry educator*
Parr, Robert Ghormley *chemistry educator*

Charlotte
Jones, Daniel Silas *chemistry educator*
Monroe, Frederick Leroy *chemist*

Davidson
Schuh, Merlyn Duane *chemist, educator*

Duck
Majewski, Theodore Eugene *chemist*

Durham
Coury, Louis Albert, Jr. *chemistry educator, researcher*
Fraser-Reid, Bertram Oliver *chemistry educator*
Hammes, Gordon G. *chemistry educator*
Pirrung, Michael Craig *chemistry educator, consultant*

Greenville
Blackmon, Margaret Lee *pharmaceutical chemist*

Linwood
Barnes, Melver Raymond *chemist*

Mount Holly
Roberts, Warren Hoyle, Jr. *chemist*

New Hill
Wilson, Robert Thaniel, Jr. *environmental chemist*

Raleigh
Owens, Tyler Benjamin *chemist*
Tyczkowska, Krystyna Liszewska *chemist*
Whitten, Jerry Lynn *chemistry educator*

Research Triangle Park
Chao, James Lee *chemist*
Huang, Jim Jay *chemist*

Lewin, Anita Hana *research chemist*
Profeta, Salvatore, Jr. *chemist*

Warrenton
Padgett, Bobby Lee, II *chemist*

Winston Salem
Dobbins, James Talmage, Jr. *analytical chemist, researcher*

NORTH DAKOTA

Fargo
Fatland, Charlotte Lee *chemist*
Urban, Marek Wojciech *chemist educator, consultant*

Grand Forks
Willson, Warrack G. *physical chemist*

OHIO

Akron
Cheng, Stephen Zheng Di *chemistry educator, polymeric material researcher*
Galiatsatos, Vassilios *chemist, educator*
Holtman, Mark Steven *chemist*
Huckstep, April Yvette *chemist*
Livigni, Russell A. *polymer chemist*
Scott, Mary Ellen Ann *chemist*
Uscheek, David Petrovich *chemist*

Alliance
Clark, Gregory Alton *research chemist*

Avon Lake
Farkas, Julius *chemist*
Martin, Christine Kaler *chemist*

Bowling Green
Midden, William Robert *chemist*

Canton
Arora, Sardari Lal *chemistry educator*

Cincinnati
Ashley, Kevin Edward *research chemist*
Burrows, Richard Steven *chemist*
Carr, Albert Anthony *organic chemist*
Heineman, William Richard *chemistry educator*
Hubbard, Arthur Thornton *chemistry educator, electro-surface chemist*
Hutchison, Robert B. *chemist*
Kawahara, Fred Katsumi *research chemist*
Kiser, Thelma Kay *analytical chemist*
Kupper, Philip Lloyd *chemist*
Lakes, Stephen Charles *research chemist, educator*
Liang, Nong *chemist, researcher*
McCarthy, James Ray *organic chemist*
Meal, Larie *chemistry educator, consultant*
Rickabaugh, Janet Fraley *environmental chemistry educator*
Swaine, Robert Leslie, Jr. *chemist*

Cleveland
Abraham, Tonson *chemist, researcher*
Berridge, Marc Sheldon *chemist, educator*
Cook, William R., Jr. *chemist*
Dunbar, Robert Copeland *chemist, educator*
Herrington, Daniel Robert *chemist*
Koenig, Jack L. *chemist, educator*
Krieger, Irvin Mitchell *chemistry educator, consultant*
Maximovich, Michael Joseph *chemist, consultant*
Myers, Ronald Eugene *research chemist*
Wolfe, Lowell Emerson *space scientist*

Columbus
Bernays, Peter Michael *retired chemical editor*
Crenshaw, Michael Douglas *chemist, researcher*
Golightly, Danold Wayne *chemist*
Greenlee, Kenneth William *chemical consultant*
Marshall, Alan George *chemistry and biochemistry educator*
Meites, Samuel *clinical chemist, educator*
Merritt, Joy Ellen *chemist, editor*
Miller, Terry Alan *chemistry educator*
Pfau, Richard Olin *forensic chemist, forensic science educator*
Relle, Ferenc Matyas *chemist*
Shupe, Lloyd Merle *chemist, consultant*
Singer, Sherwin Jeffrey *theoretical chemist, chemistry educator*
Weeks, Thomas J. *chemist*

Dayton
Emrick, Donald Day *chemist, consultant*
Loughran, Gerard Andrew *chemistry consultant, polymer scientist*
Mehta, Rajendra *chemist, researcher*
Standley, Paul Melvin *chemist*

Dublin
Blakley, Brent Alan *polymer chemist, computer programmer*
Mueller, Donald Scott *chemist*

Fairborn
Workman, John Mitchell *chemist*

Findlay
Zeyen, Richard Leo, III *analytical chemist, educator*

Granville
Carr, Thomas Michael *analytical chemist*

Hinckley
Sarbach, Donald Victor *retired chemist*

Ironton
Mitchell, Maurice McClellan, Jr. *chemist*

Loveland
Masters, Ron Anthony *research chemist*

Marietta
Putnam, Robert Ervin *chemistry educator*

Marion
Youll, Peter Jerome *environmental chemist*

Melmore
Cox, James Grady *chemist*

Miamisburg
Attalla, Albert *chemist*
Nease, Allan Bruce *research chemist*

Milford
Green, David Richard *development chemist*
Kinstle, James Francis *polymer scientist*

Newark
Swope, Robert J. *physical science laboratory administrator*
Tiburcio, Astrophel Castillo *polymer chemist*

Newbury
Kirman, Lyle Edward *chemist, engineer, consultant*

North Olmsted
Puhk, Heino *chemist, researcher*

Oxford
Danielson, Neil David *chemistry educator*

Painesville
Scozzie, James Anthony *chemist*

Parma
Laughlin, Ethelreda Ross *chemistry educator*

Piketon
Patton, Finis S., Jr. *nuclear chemist*

Piqua
Anderson, Christine Lee *analytical chemist*

Springboro
Dawson, Brian Robert *chemist, environmentalist*

Strongsville
Schroeder, Stanley Brian *chemist, coating application engineer*

Tipp City
Klimkowski, Robert John *photo reproduction process technical executive*

Toledo
Chrysochoos, John *chemistry educator*
Edwards, Jimmie Garvin *chemistry educator, consultant*
Guo, Hua *chemist, educator*

Wakeman
Arhar, Joseph Ronald *chemist*

Warren
Zimmerman, Doris Lucile *chemist*

Wickliffe
Dunn, Horton, Jr. *organic chemist*
Kornbrekke, Ralph Erik *colloid chemist*

Willoughby
Sieglaff, Charles Lewis *chemist*

OKLAHOMA

Bartlesville
Beever, William Herbert *chemist*
Byers, Jim Don *research chemist*
Guillory, Jack Paul *chemist, researcher*
Moczygemba, George Anthony *research chemist*

Fort Towson
Pike, Thomas Harrison *plant chemist*

Muskogee
Washington, Allen Reed *chemist*

Norman
Allen, Jonathan Dean *chemist, researcher*
Ciereszko, Leon Stanley *chemistry educator*

Nowata
Osborn, Ann George *retired chemist*

Stillwater
Holt, Elizabeth Manners *chemistry educator*
Whaley, Max Weldon *chemist*

Tulsa
Rotenberg, Don Harris *chemist*

OREGON

Aloha
Willis, Lawrence Jack *chemist*

Ashland
Grover, James Robb *retired chemist, editor*

Beaverton
Cole, Samuel Joseph *computational chemist*
Purvis, George Dewey, III *computational chemist*

Corvallis
Laver, Murray Lane *chemist, educator*
Shoemaker, Clara Brink *retired chemistry educator*

Eugene
Boekelheide, Virgil Carl *chemistry educator*
Noyes, Richard Macy *physical chemist, educator*
Schellman, John A. *chemistry educator*

Portland
Loehr, Thomas Michael *chemist, educator*

Selma
Roy, Harold Edward *research chemist*

Springfield
Detlefsen, William David, Jr. *chemist, administrator*

PENNSYLVANIA

Alcoa Center
Auses, John Paul (Jay) *technical specialist*
Bonewitz, Robert Allen *chemist, manufacturing executive*
Dobbs, Charles Luther *analytical chemist*

Allentown
Orphanides, Gus George *chemical company official*
Schweighardt, Frank Kenneth *chemist*

Allison Park
Xu, Zhifu *chemist, researcher*

Carlisle
Egolf, Kenneth Lee *chemistry educator*
Jones, Randall Marvin *chemist*

Doylestown
Weber, Charles Walton *chemistry educator*

East Stroudsburg
Bergo, Conrad Hunter *chemistry educator*

Fleetwood
Pangrazi, Ronald Joseph *chemist*

Gettysburg
Holland, Koren Alayne *chemistry educator*

Haverford
Lazar, Anna *chemist*

Hazleton
Smith, David *chemistry educator*

Holtwood
Liebman, Shirley Anne *analytical research scientist*

Huntingdon Valley
Godfrey, John Carl *medicinal chemist*

Johnstown
Kintner, Elisabeth Turner *chemistry educator*

King Of Prussia
Gitlitz, Melvin Hyman *chemist, researcher*
Peerce-Landers, Pamela Jane *chemical researcher*

Lancaster
Hess, Earl Hollinger *laboratory executive, chemist*

Lansdale
Schnable, George Luther *chemist*

Lehigh Valley
Golden, Timothy Christopher *chemist, researcher*

Lewisburg
Veening, Hans *chemistry educator*

Lock Haven
Sweeny, Charles David *chemist*

Malvern
Fisher, Sallie Ann *chemist*

McMurray
Mortimer, James Winslow *analytical chemist*

Media
Voltz, Sterling Ernest *physical chemist, researcher*

Millersville
Greco, Thomas G. *chemist, educator*

Monroeville
Fedak, Mitchel George *chemistry educator*
Parker, James Roger *chemist*

Moylan
Eberl, James Joseph *physical chemist, consultant*

New Kensington
Dando, Neal Richard *chemist*

Newtown
Long, Harry (On-Yuen Eng) *chemist, rubber science and technology consultant*
Marton, Joseph *paper chemistry consultant, educator*

Oakdale
Dean, Frank Warren, Jr. *chemist, pet food company executive*

Philadelphia
Dai, Hai-Lung *physical chemist, researcher*
Davis, Raymond, Jr. *chemist, researcher, educator*
Dymicky, Michael *retired chemist*
Hameka, Hendrik Frederik *chemist, educator*
Hopson, Kevin Mathew *chemist*
Kauffman, Joel Mervin *chemistry educator, researcher, consultant*
Otvos, Laszlo Istvan, Jr. *organic chemist*
Raftery, M. Daniel *chemistry researcher*
Smith, Amos Brittain, III *chemist, educator*

Pittsburgh
Hercules, David Michael *chemistry educator, consultant*
MacDonald, Hubert Clarence *analytical chemist*
Schonhardt, Carl Mario *analytical chemist*
Schultz, Hyman *analytical chemist*
Wipf, Peter *chemist*

Plumsteadville
Vaughan, Stephen Owens *project chemist*

Point Pleasant
Moss, Herbert Irwin *chemist*

Pottstown
Diana, Guy Dominic *chemist*
Weathington, Billy Christopher *analytical chemist*

Quarryville
Schuck, Terry Karl *chemist, environmental consultant*

Reading
Dulski, Thomas R. *chemist, writer*
Feeman, James Frederic *chemist, consultant*
Richart, Douglas Stephen *chemist*
Rowe, Jay E., Jr. *research and development director*

Spring Grove
Gleim, Jeffrey Eugene *research chemist*

Spring House
Caldwell, Gary Wayne *chemist, researcher*
De Jong, Gary Joel *chemist*
Emmons, William David *chemist*
Greer, Edward Cooper *chemist*
Klotz, Wendy Lynnett *analytical chemist*

State College
Baker, Dale Eugene *soil chemist*

Swarthmore
Hiltz, Arnold Aubrey *former chemist*
Pasternack, Robert Francis *chemistry educator*
Voet, Judith Greenwald *chemistry educator*

Tamaqua
Cusatis, John Anthony *chemist*

University Park
Jurs, Peter Christian *chemistry educator*
Winograd, Nicholas *chemist*

Valley Forge
Erb, Doretta Louise Barker *polymer applications scientist*

Villanova
Edwards, John Ralph *chemist, educator*

Waynesburg
Maguire, Mildred May *chemistry educator, magnetic resonance researcher*

West Mifflin
Smith, Stewart Edward *physical chemist*

West Point
Kalejta, Paul Edward *chemist*
Katrinak, Thomas Paul *analytical chemist*
Woolf, Eric Joel *analytical chemist*

Wheatland
Gruber, Jack Alan *chemist*

Williamsport
Leisey, April Louise Snyder *chemist*

RHODE ISLAND

Bristol
Von Riesen, Daniel D. *chemistry educator*

Cranston
Watt, Norman Ramsay *chemistry educator, researcher*

Fiskeville
Yang, Sen *chemist*

Kingston
Freeman, David Laurence *chemist, educator*

Narragansett
Arimoto, Richard *atmospheric chemist*

Providence
Rieger, Philip Henri *chemistry educator, researcher*

SOUTH CAROLINA

Aiken
Coleman, Jerry Todd *chemist*
Hyder, Monte Lee *chemist*

Charleston
Delli Colli, Humbert Thomas *chemist, product development specialist*

Columbia
Coleman, Robert Samuel *chemistry researcher, educator*

Eastover
Williams, Anthony M. *chemist*

SOUTH DAKOTA

Brookings
Hecht, Harry George *chemistry educator*

TENNESSEE

Chattanooga
Scrudder, Eugene Owen *chemist, environmental specialist*

Dyersburg
Bell, Helen Cherry *chemistry educator*

Kingsport
Embree, Norris Dean *chemist, consultant*
Gray, T(heodore) Flint, Jr. *chemist*
Kashdan, David Stuart *chemist*
Sharma, Mahendra Kumar *chemist*

Knoxville
Copper, Christine Leigh *chemist*
Dean, John Aurie *chemist, author, chemistry educator emeritus*
Kovac, Jeffrey Dean *chemistry educator*
Schweitzer, George Keene *chemistry educator*

Memphis
Fang, Chunchang *physical chemist, chemical engineer*
Kress, Albert Otto, Jr. *polymer chemist*
Lasslo, Andrew *medicinal chemist, educator*

Morristown
Culvern, Julian Brewer *chemist, educator, writer-naturalist*

Nashville
Martin, James Cullen *chemistry educator*
Tarbell, Dean Stanley *chemistry educator*

Oak Ridge
Barkley, Linda Kay *chemical analyst, spectroscopist*
Burtis, Carl A., Jr. *chemist*
Carlsmith, Roger Snedden *chemistry and energy conservation researcher*
Marshall, William Leitch *chemist*
Miller, John Cameron *research chemist*
Morrow, Roy Wayne *chemist*
Toth, Louis McKenna *chemist*

TEXAS

Austin
Bard, Allen Joseph *chemist, educator*
Folkers, Karl August *chemistry educator*
Fonken, Gerhard Joseph *chemistry educator, university administrator*
Hammond, Charles Earl *chemist, researcher*
Lewis, Richard Van *chemist*
Nguyen, Truc Chinh *analytical chemist*
Posey, Daniel Earl *analytical chemist*
Pradzynski, Andrzej Henryk *chemist*

Baytown
Lander, Deborah Rosemary *chemist*

Beaumont
Streeper, Robert William *environmental chemist*

College Station
Barton, Derek Harold Richard *chemist*
Cotton, Frank Albert *chemist, educator*
Goodman, David Wayne *research chemist*
Latimer, George Webster, Jr. *chemist*
Lin, Guang Hai *research scientist, consultant*
Natowitz, Joseph B. *chemistry educator, administrator*
Stipanovic, Robert Douglas *chemist, researcher*

Corpus Christi
Smith, Jeremy Owen *chemist*

Dallas
Frank, Steven Neil *chemist*
Hosmane, Narayan Sadashiv *chemistry educator*
Ravichandran, Kurumbail Gopalakrishnan *chemist*

Deer Park
Taggart, Austin Dale, II *chemist*

Denton
Braterman, Paul Sydney *chemistry educator*
Hurdis, Everett Cushing *chemistry educator*

Fort Worth
Bonakdar, Mojtaba *chemistry educator*
Gutsche, Carl David *chemistry educator*
Minter, David Edward *chemistry educator*

Freeport
Mercer, William Edward, II *chemical research technician*

Fulshear
Lurix, Paul Leslie, Jr. *chemist*

Horseshoe Bay
Ramey, James Melton *chemist*

Houston
Askew, William Earl *chemist, educator*
Bonchev, Danail Georgiev *chemist, educator*
Brooks, Philip Russell *chemistry educator, researcher*
Coleman, Samuel Ebow *chemist, engineer*
Curl, Robert Floyd, Jr. *chemistry educator*
Downs, Hartley H., III *chemist*
Fukuyama, Tohru *organic chemistry educator*
Havrilek, Christopher Moore *technical specialist*
Herz, Josef Edward *chemist*
Kinsey, James Lloyd *chemist, educator*
Kochi, Jay Kazuo *chemist, educator*
Margrave, John Lee *chemist, educator, university administrator*
McCown, Shaun Michael Patrick *chemist, consultant*
Smalley, Richard Errett *chemistry and physics educator, researcher*
Zlatkis, Albert *chemistry educator*

Irving
Holdar, Robert Martin *chemist*

La Porte
Young, George Hansen *chemist*

Longview
Robinson, Alfred G. *petroleum chemist*

Lubbock
Dasgupta, Purnendu Kumar *chemist, educator*

Marshall
Ford, Clyde Gilpin *chemistry educator*

Mcallen
Julian, Elmo Clayton *analytical chemist*

Mont Belvieu
Raczkowski, Cynthia Lea *chemist*

Pasadena
Griffin, John Joseph, Jr. *chemist, video producer*

Port Lavaca
Anfosso, Christian Lorenz *analytical chemist*

Richardson
Ramasamy, Ravichandran *chemistry educator*

Rockport
Jones, Lawrence Ryman *retired research scientist*

San Antonio
Bach, Stephan Bruno Heinrich *chemistry educator*
Budalur, Thyagarajan Subbanarayan *chemistry educator*
Fodor, George Emeric *chemist*
Lyle, Robert Edward *chemist*
Panda, Markandeswar *chemistry researcher*

Spring
Forester, David Roger *research scientist*
Kust, Roger Nayland *chemist*

Sugar Land
Mata, Zoila *chemist*

Temple
Gaa, Peter Charles *organic chemist, researcher*

Texas City
Fuchs, Owen George *chemist*

Tyler
Walsh, Kenneth Albert *chemist*

UTAH

Logan
Scouten, William Henry *chemistry educator, academic administrator*

Salt Lake City
Boyd, Richard Hays *chemistry educator*
Conceicao, Josie *chemistry researcher*
Eyring, Edward Marcus *chemical educator*
Giddings, J. Calvin *chemistry educator*
Grissom, Charles Buell *chemistry educator*
Miller, Joel Steven *solid state scientist*
Myers, Marcus Norville *research educator*
Parry, Robert Walter *chemistry educator*
Stang, Peter John *organic chemist*
Strickley, Robert Gordon *pharmaceutical chemist*

Sandy
Lambert, James Michael *chemist, researcher*

VERMONT

Burlington
Houston, John F. *chemist*
Sentell, Karen Belinda *chemist*

Essex Junction
Linde, Harold George *chemist*

VIRGINIA

Annandale
Berg, Lillian Douglas *chemistry educator*
Matuszko, Anthony Joseph *research chemist, administrator*

Arlington
Bikales, Norbert M. *chemist, science administrator*
Watt, William Stewart *physical chemist*
Wodarczyk, Francis John *chemist*

Blacksburg
Graybeal, Jack Daniel *chemist, educator*
Samaranayake, Gamini Saratchandra *chemist, researcher*

Charlottesville
Averill, Bruce Alan *chemistry educator*
Chapman, Martin Dudley *immunochemist*
MacDonald, Timothy Lee *chemistry educator*
Martin, Robert Bruce *chemistry educator*

Danville
Wright, Donald Lee *chemist*

Fairfax
Ney, Ronald Ellroy, Jr. *chemist*

Falls Church
Feldmann, Edward George *pharmaceutical chemist*
Spindel, William *chemistry educator, scientist, educational administrator*

Gainesville
Steger, Edward Herman *chemist*

Hampden Sydney
Porterfield, William Wendell *chemist, educator*

Herndon
Stirewalt, Edward Neale *chemist, scientific analyst*

Locust Grove
Stein, Richard Louis *chemist, educator*

Lynchburg
Morgan, Evan *chemist*

Mc Lean
Marinenko, George *chemist*

Norfolk
Brown, Kenneth Gerald *chemistry educator*
Overby, Veriti Page *chemist, environmental protection specialist*

Richmond
Chakravorty, Krishna Pada *chemist, spectroscopist*
Kinsley, Homan Benjamin, Jr. *chemist, chemical engineer*
Rutan, Sarah Cooper *chemistry educator*
Safo, Martin Kwasi *chemist*
Thomas, Charles Edwin *chemist, researcher*
Zimmermann, Michael Louis *chemist*

Salem
Fisher, Charles Harold *chemistry educator, researcher*

Springfield
Elbarbary, Ibrahim Abdel Tawab *chemist*

Surry
Johnson, Keith Edward *chemist*

Williamsburg
Muller, Julius Frederick *chemist, business administrator*
Starnes, William Herbert, Jr. *chemist, educator*

Winchester
Murtagh, John Edward *chemist, consultant*

WASHINGTON

Federal Way
Hansen, Michael Roy *chemist*

Kalama
Liang, Jason Chia *research chemist*

Manchester
Fearon, Lee Charles *chemist*

Richland
Rebagay, Teofila Velasco *chemist, chemical engineer*
Smith, Richard Dale *chemist, researcher*

Seattle
Brotherton, Vince Morgan *research chemist*
Brown, Craig William *chemist*
Gelb, Michael H. *chemistry educator*
Gruger, Edward Hart, Jr. *retired chemist*
Hol, Wim Gerardus Jozef *biophysical chemist*
Pocker, Yeshayau *chemistry, biochemistry educator*
Schomaker, Verner *chemist, educator*

Spokane
Benson, Allen B. *chemist, educator, consultant*

Tacoma
Harding, Karen Elaine *chemistry educator and department chair*

Vancouver
Dietze, Gerald Roger *chemist*

WEST VIRGINIA

Lewisburg
Cardis, Thomas Michael *chemist*

Morgantown
Das, Kamalendu *chemist*

Parkersburg
Lindstrom, Timothy Rhea *chemist*

Rocket Center
Hartman, Kenneth Owen *chemist*

WISCONSIN

Eau Claire
St. Louis, Robert Vincent *chemist, educator*

Madison
Dahl, Lawrence Frederick *chemistry educator, researcher*
Ediger, Mark D. *chemistry educator*
Ellis, Arthur Baron *chemist, educator*
Farrar, Thomas C. *chemist, educator*
Fennema, Owen Richard *food chemistry educator*
Ferry, John Douglass *chemist*
Johnson, Richard Warren *chemist*
Kuzmic, Petr *chemist*
Morton, Stephen Dana *chemist*
Rich, Daniel Hulbert *chemist*
Sih, Charles John *pharmaceutical chemistry educator*
Skinner, James Lauriston *chemist, educator*
Smith, Matthew Jay *chemist*
West, Robert Culbertson *chemistry educator*
Wright, John Curtis *chemist, educator*
Yu, Hyuk *chemist, educator*
Zimmerman, Howard Elliot *chemist, educator*

Milwaukee
Doumas, Basil Thomas *chemist, researcher, educator*
McKinney, Bryan Lee *chemist*
Petering, David Harold *chemistry educator*
Petersen, Ralph Allen *chemist*

Schofield
Adams, James William *retired chemist*

TERRITORIES OF THE UNITED STATES

PUERTO RICO

Mayaguez
Souto Bachiller, Fernando Alberto *chemistry educator*

CANADA

ALBERTA

Calgary
Codding, Penelope Wixson *chemistry educator*

Edmonton
Freeman, Gordon Russel *chemistry educator*
Khan, Abdul Quasim *chemistry researcher*
Lemieux, Raymond Urgel *chemistry educator*

BRITISH COLUMBIA

Burnaby
Einstein, Frederick William Boldt *chemistry educator*

Vancouver
Fryzuk, Michael Daniel *chemistry educator*

MANITOBA

Winnipeg
Secco, Anthony Silvio *chemistry educator*

NOVA SCOTIA

Wolfville
Ogilvie, Kelvin Kenneth *chemistry educator*

ONTARIO

Delhi
Court, William Arthur *chemist, researcher*

Gloucester
Mykytiuk, Alex P. *chemist*

Hamilton
Kenney-Wallace, Geraldine *chemistry and physics educator*
Thode, Henry George *chemistry educator*

Ottawa
Dubey, Ram Janam *inorganic and environmental chemist*
Ingold, Keith Usherwood *chemist, educator*
Ramsay, Donald Allan *physical chemist*
Siebrand, Willem *theoretical chemist, science editor*

Petawawa
Elchuk, Steve *chemist*

Toronto
Polanyi, John Charles *chemist, educator*
Riordan, John Richard *chemist*
Tang, You-Zhi *chemist, researcher*
Yates, Keith *chemistry educator*

Waterloo
Rudin, Alfred *chemistry educator emeritus*

QUEBEC

Laval
Bellini, Francesco *chemist*

Montreal
Chan, Tak Hang *chemist, educator*
Eisenberg, Adi *chemist*
Stern, Eric Petru *chemist*

Quebec
Engel, Charles Robert *chemist, educator*

Trois Rivières
Leblanc, Roger Maurice *chemistry educator*

Varennes
Vijh, Ashok Kumar *chemistry educator, researcher*

SASKATCHEWAN

Regina
Kybett, Brian David *chemist*

Saskatoon
Quail, John Wilson *chemist, educator*

MEXICO

Mexico City
Tamariz, Joaquin *chemist, educator*

ARGENTINA

La Plata
Vilche, Jorge Roberto *physical chemistry educator*

AUSTRALIA

Brisbane
Knight, Alan Edward Whitmarsh *physical chemistry educator*
Wentrup, Curt *chemist, educator*

Bundoora
James, Bruce David *chemistry educator*

Canberra
Bennett, Martin Arthur *chemist, educator*

Caulfield East
Hart, Barry Thomas *environmental chemist*

Clayton
Bray, Andrew Malcolm *chemist*
Evans, Richard Alexander *research chemist*
Meijs, Gordon Francis *chemist, research scientist*

Kensington
Bradley, Susan M. *chemistry educator*

Kwinana
Grocott, Stephen Charles *industrial research chemist*

Lane Cove
Jackson, Peter Edward *chemist*

Melbourne
Banwell, Martin Gerhardt *chemistry researcher*
Kemp, Bruce E. *protein chemist*

North Ryde
Wrigley, Colin Walter *cereal chemist*

Parkville
Lim, Kieran Fergus *chemistry educator*

Queensland
Connell, Desley William *chemist, educator, administrator*

Saint Albans
Orbell, John Donald *chemist, educator*

Sydney
Lay, Peter Andrew *chemistry educator*
Mackie, John Charles *chemistry educator*

Tamworth
Cook, John Bell *chemist*

AUSTRIA

Innsbruck
Kräutler, Bernhard *chemistry educator*

Lenzing
Gamerith, Gernot *chemist, researcher*

Leoben
Lederer, Klaus *macromolecular chemistry educator*

Leonding
Wechsberg, Manfred Ingo *chemical researcher*

Linz
Lell, Eberhard *retired inorganic chemist*

Vienna
Margarétha, Herbert Moriz Paul Maria *chemical consultant*

BAHRAIN

Isa Town
Ali Mohamed, Ahmed Yusuf *chemistry educator, researcher*

BARBADOS

Bridgetown
Headley, Oliver St. Clair *chemistry educator*

BELGIUM

Braine L'Alleud
Wulfert, Ernst Arne *research and development director, educator*

Brussels
Luxen, Andre Jules Marie *chemist*
Scheelen, André Joannes *chemical researcher*

Ghent
Goethals, Eric Jozef *chemistry educator*

Heverlee
L'abbe, Gerrit Karel *chemist*

Leuven
De Ranter, Camiel Joseph *chemistry educator*

Rixensart
Thrower, Keith James *chemist*

Turnhout
Caruso, Nancy Jean *chemist*

BRAZIL

Belo Horizonte
Piló-Veloso, Dorila *chemistry educator, researcher*

Itaguai
Braz-Filho, Raimundo *organic chemistry educator*

Rio de Janeiro
Barreiro, Eliezer Jesus *medicinal chemistry educator*
Costa Neto, Adelina *chemistry educator, consultant*
Felcman, Judith *chemistry educator*

São Paulo
Zinner, Léa Barbieri *inorganic chemistry educator, researcher*

Salvador
Andrade, Francisco Alvaro Conceicao *chemistry educator*

Sao Carlos
Motheo, Artur de Jesus *chemist, educator*

CHILE

Santiago
Arratia-Perez, Ramiro *chemistry educator*
Loeb, Barbara L. *chemistry educator*

CHINA

Beijing
Hua, Tong-Wen *chemistry educator, researcher*
Leng, Xin-Fu *chemist, educator*
Liu, Yuanfang *chemistry educator*
Wei, Gaoyuan *chemistry educator*
Zhang, Jun Yi *chemist, educator*

Changsha
Yu, Ru-Qin *chemistry educator*

Shanghai
Cheng, J. S. (Zhong-Zhi Zheng) *nuclear astrophysical chemist, educator*

Wuhan
Cha, Chuan-sin *chemistry educator*

Xiamen
Zhou, Shao-min *chemist, educator*

CROATIA

Zagreb
Ćosović, M. Božena *chemist, researcher*
Kojić-Prodić, Biserka *chemist*

CZECH REPUBLIC

Brno
Vesely, Karel *chemist, educator*

Olomouc
Šimánek, Vilím *chemist, educator*

Prague
Macek, Karel *analytical chemistry educator*
Procházka, Karel *chemistry educator*

DENMARK

Copenhagen
Wilkins, Cornelius Kendall *chemist, researcher*

Gentofte
Jaroszewski, Jerzy W. *chemist*

Odense
Oddershede, Jens Norgaard *chemistry educator*

ENGLAND

Bath
Buchanan, John Grant *chemistry educator*
Wilson, Derek Edward *chemist*

Bocking
Axon, Kenneth Stuart *chemist*

Brighton
Cornforth, Sir John Warcup *chemist*

Bristol
Albone, Eric Stephen *chemist, educator*

Cambridge
Martin, Archer John Porter *retired chemistry educator*
Rees, David Charles *chemist, researcher*
Todd, Alexander Robertus (Baron Todd of Trumpington) *chemistry educator*

Durham
Feast, William James *polymer chemist*
Parker, David *chemistry educator*
Richards, Randal William *chemist, educator*

Essex
Kneen, Geoffrey *chemist*

Huddersfield
Jewsbury, Roger Alan *chemistry educator*

Hull
Shorter, John *chemistry lecturer*

Liverpool
Bulman Page, Philip Charles *chemistry researcher*

London
Drake, Simon Robert *inorganic chemistry*
Eddo, James Ekundayo *industrial chemist, chemicals executive*
Kathirgamanathan, Poopathy *chemist*
Mason, Stephen Finney *chemistry educator*
Porter, Lord George *chemist, educator*
Reese, Colin Bernard *chemistry educator*
Thomas, John Meurig *chemist, scientist*
Wilkinson, Geoffrey *chemist, educator*

Loughborough
Arvanitoyannis, Ioannis *polymer chemistry researcher*

Manchester
Banks, Ronald Eric *chemistry educator*

Nottingham
Logan, Norman *chemistry researcher, educator*

Oxford
Atkins, Peter William *chemistry educator*
Hodgkin, Dorothy Crowfoot *chemist*

Reading
Almond, Matthew John *chemistry educator, researcher, writer*

Sheffield
Henshilwood, James Andrew *chemist, researcher*

Southhampton
Fleischmann, Martin *chemistry educator*

Sunbury on Thames
Packer, Kenneth John *spectroscopist*

Sutton
Wilman, Derry Edward Vincent *medicinal chemist*

York
Waddington, David James *chemistry educator*

FINLAND

Oulu
Koskinen, Ari Mauri Petri *organic chemistry educator*

FRANCE

Antony
Jolles, Georges Edgar René *scientist*
Mazzo, David Joseph *analytical chemist*

Bagneux
Jegham, Samir *chemist*

Meudon
Sculfort, Jean-Lou *chemistry and physics educator*

Mulhouse
Decker, Christian Lucien *research chemist*
Fouassier, Jean-Pierre *chemist, educator*
Riess, Henri Gerard *chemist, educator*

Orsay
Guillaumont, Robert *chemist, educator*
Seyden-Penne, Jacqueline *research chemist*

Paris
Aaron, Jean-Jacques *chemist, educator*
Delahay, Paul *chemistry educator*
Genet, Jean Pierre *chemistry educator, researcher*
Lehn, Jean-Marie Pierre *chemistry educator*
Pfau, Michel Alexandre *chemist, researcher, consultant*
Rips, Richard Maurice *chemist, pharmacologist*

Port-Marly
Dégeilh, Robert *retired chemist*

Reims
Mathlouthi, Mohamed *chemistry educator*

Saint Etienne
Soustelle, Michel Marcel Philippe *chemistry educator, researcher*

Strasbourg
Drain, Charles Michael *biophysical chemistry educator*

Toulouse
Lattes, Armand *educator, researcher*

Villeurbanne
Roche, Alain Andre *research chemist*

Villiers Le Bel
Verillon, Francis Charles *chemist*

Wissembourg
Grisar, Johann Martin *research chemist*

GERMANY

Bamberg
Krauss, Hans Ludwig *emeritus educator*

Berlin
Brätter, Peter *chemist, educator*
Henglein, Friedrich Arnim *chemistry educator*
Klapoetke, Thomas Matthias *chemist*
Möller, Detlev *atmospheric chemist*
Schulze, Matthias Michael *chemist*
Vorbrueggen, Helmut Ferdinand *organic chemist*

Bielefeld
Muller, Achim *chemistry educator*

Bochum
Wieghardt, Karl Ernst *chemistry educator*

Bonn
Giannis, Athanassios *chemist, physician*

Braunschweig
Boldt, Peter *chemistry educator, researcher*
Schomburg, Dietmar *chemist, researcher*

Bremen
Baykut, Mehmet Gökhan *chemist*

Cologne
Stuhl, Oskar Paul *organic chemist*

Darmstadt
Clausen, Thomas Hans Wilhelm *chemist*
Fetting, Fritz *chemistry educator*
Lichtenthaler, Frieder Wilhelm *chemist, educator*

Dessaŭ
Bach, Günther *organic chemist, researcher*

Dortmund
Neumann, Wilhelm Paul *chemistry educator*

Düsseldorf
Strehblow, Hans-Henning Steffen *chemistry educator*

Emden
Gombler, Willy Hans *chemistry educator*

Essen
Hoffmann, Günter Georg *chemist*

Frankfurt
Engel, Juergen Kurt *chemist, researcher*
Hilgenfeld, Rolf *chemist*
Ilten, David Frederick *chemist*
Klöpffer, Walter *chemist, educator*

Garching
Schlag, Edward William *chemistry educator*

Giessen
Hoppe, Rudolf Reinhold Otto *chemist, educator*

Göttingen
Gandhi, Suketu Ramesh *chemist*
Oellerich, Michael *chemistry educator, chemical pathologist*
Roesky, Herbert Walter *chemistry educator*
Sheldrick, George Michael *chemistry educator, crystallographer*
Tietze, Lutz Friedjan *chemist, educator*

Hagen
Struecker, Gerhard *analytical chemist*

Hamburg
Voss, Jürgen *chemistry educator*

Hannover
Habermehl, Gerhard Georg *chemist, educator*

Heidelberg
Oberdorfer, Franz *chemist*
Staab, Heinz A. *chemist*

Iserlohn
Paradies, Hasko Henrich *chemistry educator*

Jülich
Qaim, Syed Muhammad *nuclear chemist, researcher, educator*

Köln
Funken, Karl-Heinz *chemist*

Kruft
Lekim, Dac *chemist*

Leipzig
Zimmermann, (Arthur) Gerhard *chemist, educator, researcher*

Ludwigshafen
Hibst, Hartmut *scientist, chemistry educator*

Mainz
Gütlich, Philipp *chemistry educator*

Marburg
Brandt, Reinhard *chemist*
Patzelt, Paul *nuclear chemist*

Mûlhein
Wilke, Gunther *chemistry educator*

Münster
Jeitschko, Wolfgang Karl *chemistry educator*
Rüter, Ingo *research chemist, toxicology consultant*

Munich
Fischer, Ernst Otto *chemist, educator*
Wessjohann, Ludger Aloisius *chemistry educator, consultant*

Paderborn
Krohn, Karsten *chemistry educator*

Saarbrücken
Veith, Michael *chemist*

Steinfôrde
Kolditz, Lothar *chemistry educator*

Stuttgart
Kramer, Horst Emil Adolf *physical chemist*

Wiesbaden
Elben, Ulrich *chemist*

GREECE

Athens
Koupparis, Michael Andreas *analytical chemistry educator*
Paleos, Constantinos Marcos *chemist*
Screttas, Constantinos George *chemistry educator*

Ioannina
Albanis, Triadafillos Athanasios *chemistry educator, researcher*

HONG KONG

Clear Water Bay
Che, Chun-Tao *chemistry educator*

Kowloon
Tin, Kam Chung *industrial chemist, educator*

Tai Wai
Barbalas, Lorina Cheng *chemist*

HUNGARY

Budapest
Braun, Tibor *chemist*
Hajos, Zoltan George *chemist*
Kis-Tamás, Attila *chemist*
Markó, László *chemist*
Pungor, Erno *chemist, educator*

Szeged
Dékány, Imre Lajos *chemistry educator*

Tiszavasvari
Timar, Tibor *chemist*

ICELAND

Reykjavik
Kvaran, Agust *chemistry educator, research scientist*

INDIA

Bhavnagar
Natarajan, Paramasivam *chemistry educator*

Bombay
Holla, Kadambar Seetharam *chemist, educator*
Khopkar, Shripad Moreshwer *chemistry educator*
Krishnamurthy, Suresh Kumar *chemist, researcher*

Chandigarh
Chawla, Amrik Singh *chemist, educator*
Singh, Harkishan *chemist, educator*

Gorakhpur
Das, Ishwar *chemistry educator*

Madurai
Raju, Perumal Reddy *chemist*

Tiruchirapalli
Jeyaraman, Ramasubbu *chemist, educator*

IRAN

Babolsar
Mohanazadeh, Farajollah Bakhtiari *chemist, educator*

Tehran
Ahmadinejad, Behrouz *chemist, educator*
Sharifi, Iraj Alagha *organic chemistry educator*
Zohourian Mashmoul, Mohammad Jalal-od-din *chemistry educator*

IRELAND

Cork
Burke, Laurence Declan *chemistry educator*

ISRAEL

Haifa
Apeloig, Yitzhak *chemistry educator, researcher*

Jerusalem
Isaacs, Philip Klein *retired chemist*
Marcus, Yizhak *chemistry educator*
Metzger, Gershon *chemist, patent attorney*
Shaik, Sason Sabakh *chemistry educator*

Migdal
Selivansky, Dror *chemistry educator, resrarcher*

Ramallar
Laila, Abduhameed Abdelrahman *chemistry educator*

Ramat Gan
Hassner, Alfred *chemistry educator*

Rehovot
Cahen, David *materials chemist*

Technion City
Halevi, Emil Amitai *chemistry educator*

Tel Aviv
Jortner, Joshua *physical chemistry scientist, educator*
Kaldor, Uzi *chemistry educator*

ITALY

Bari
Pizzoli, Elsa Maria *chemistry educator*

Bologna
Susi, Enrichetta *chemist, researcher*

Brescia
Villa, Roberto Riccardo *chemist*

Catania
Montaudo, Giorgio *chemistry educator, researcher*

Ferrara
Manfredini, Stefano *medicinal chemistry educator*

Florence
Giolitti, Alessandro *chemist*

Frascati
Paparazzo, Ernesto *chemist*

Milan
Bellobono, Ignazio Renato *chemist, educator*
Benfenati, Emilio *chemist, researcher*
Chan, Ah Wing Edith *chemist*
Gavezzotti, Angelo *chemistry educator*
Resnati, Giuseppe Paolo *chemistry researcher, chemistry educator*
Trasatti, Sergio *chemistry educator*
Valcavi, Umberto *chemistry educator*

Novara
Pernicone, Nicola *catalyst consultant*

Padua
Mammi, Mario *chemist, educator*

Perugia
Laganà, Antonio *chemical kinetics educator*

Potenza
Battaglia, Franco *chemistry educator*

Povo-Trento
Pietra, Francesco *chemist*

Rome
Giomini, Marcello *chemistry educator*
Marini Bettolo, Giovanni Battista *chemistry educator, researcher*

San Donato
Bellussi, Giuseppe Carlo *chemical research manager*
Roggero, Arnaldo *polymer chemistry executive*

Torino
Gasco, Alberto *medicinal chemistry educator*

Vittorio Veneto
Albrizio, Francesco *chemical consultant, chemistry educator*

JAMAICA

Kingston
Szentpály, László Von *chemistry educator*

JAPAN

Aichi
Okada, Akane *chemist*
Tsuda, Takao *chemistry educator*

Amagasaki
Yamada, Shoichiro *chemistry educator*

Bunkyo
Kaneko, Masao *chemist*

Chiba
Kuroda, Rokuro *chemist, educator*

Daito
Sakai, Shogo *theoretical chemist*

Fukuoka
Hirokawa, Shoji *chemistry educator*

Hiroshima
Otsuka, Hideaki *chemistry educator*
Yamashita, Kazuo *chemistry educator*

Hokkaido
Suzuki, Akira *chemistry educator*

Ibaraki
Murao, Kenji *chemist*
Takahashi, Tsutomu *chemist*

Iizuka
Fujii, Masayuki *chemistry educator*

Ise
Hasegawa, Akinori *chemistry educator*

Kagawa
Kobayashi, Yoshinari *polymer chemist*

Kagoshima
Itahara, Toshio *chemistry educator*

Kamigori
Terabe, Shigeru *chemistry educator*

Kawagoe-Shi
Endo, Hajime *research chemist*

Kawasaki
Yoshizaki, Shiro *medicinal chemist*

Kobe
Takao, Hama *physiological chemistry educator*

Kochi
Hojo, Masashi *chemistry educator*
Kotsuki, Hiyoshizo *chemist, educator*

Koganei
Akiyama, Masayasu *chemistry educator*
Hasegawa, Tadashi *chemical educator*

Kumamoto
Kida, Sigeo *chemistry educator*

Kurashiki
Masamoto, Junzo *chemist, researcher*

Kyoto
Araki, Takeo *chemistry educator*
Einaga, Yoshiyuki *chemist*
Fukui, Kenichi *chemist*
Hanai, Toshihiko *chemist*
Kawabata, Nariyoshi *chemistry educator*
Kitagawa, Toshikazu *chemistry educator*
Saito, Isao *chemist*
Tachibana, Akitomo *chemistry educator*
Tachiwaki, Tokumatsu *chemistry educator*
Watanabe, Yoshihito *chemistry educator*
Yamakawa, Hiromi *polymer chemist, educator*
Yamana, Shukichi *chemistry educator*

Miyakonojo
Eto, Morifusa *chemistry educator*

Nagasaki
Hamada, Keinosuke *chemistry educator emeritus*

Nagoya
Aoki, Keizo *chemistry educator*
Esaki, Toshiyuki *pharmaceutical chemist*
Kato, Michinobu *chemistry educator*
Murakami, Edahiko *chemistry educator*
Okamoto, Yoshio *chemistry educator, researcher*
Shoji, Eguchi *organic chemistry educator*
Takagi, Shigeru *chemistry educator*
Tsuge, Shin *chemistry educator*

Niigata
Satsumabayashi, Koko *chemistry educator*

Niimi-shi
Tanabe, Yo *chemistry researcher*

Nishiku
Mataga, Noboru *scientist*

Nishinomiya
Takeda, Yasuhiro *chemistry educator*

Okayama
Torii, Sigeru *chemistry educator*

Okazaki
Mitsuke, Koichiro *chemistry educator*
Yoshihara, Keitaro *chemistry educator*

Okinawa
Tako, Masakuni *chemistry researcher and educator*

Osaka
Fueno, Takayuki *chemistry educator*
Kobayashi, Mitsue *chemistry educator*
Kuwata, Kazuhiro *chemist*
Murai, Shinji *chemistry educator*
Oka, Kunio *chemistry educator*
Sonogashira, Kenkichi *chemistry educator*
Suga, Hiroshi *chemistry educator*
Sugawara, Tamio *chemist*
Susumu, Kamata *organic and medicinal chemist*
Yonetani, Kaoru *chemist, consultant*
Yoneyama, Hiroshi *chemistry educator*
Yoshida, Kunihisa *chemistry educator, researcher*

Otsu
Ando, Takashi *chemistry educator*
Matsuura, Teruo *chemistry educator*
Takemoto, Kiichi *chemistry educator*

Sagamihara
Sakai, Kunikazu *chemistry researcher*

Sakai-Gun
Ise, Norio *chemistry educator*

Sapporo
Iwamoto, Masakazu *chemistry educator*
Mizutani, Junya *chemist, educator*

Sendai
Abiko, Takashi *peptide chemist*
Takahashi, Kazuko *organic chemistry researcher, educator*
Yasumoto, Takeshi *chemistry educator*

Shinagawa-ku
Kittaka, Atsushi *chemist*
Shiozaki, Masao *synthetic and organic chemist*

Shinjyuku-ku
Takeoka, Shinji *chemist*

Shiso-Gun
Domae, Takashi *cereal chemist*

Showa
Einaga, Hisahiko *chemistry educator*

Suita
Masuhara, Hiroshi *chemist, educator*

Takaishi
Yamamura, Kazuo *chemist*

Tokushima
Kaneshina, Shoji *biophysical chemistry educator*
Oshiro, Yasuo *medicinal chemist*
Tori, Motoo *chemist, educator*

Tokyo
Chang, Chia-Wun *chemist, researcher*
Chiba, Kiyoshi *chemist*
Crist, B. Vincent *chemist*
Harada, Yoshiya *chemistry educator*
Imai, Noriyoshi *chemist, researcher*
Iwakura, Yoshio *chemistry educator*
Kamiya, Yoshio *research chemist, educator*
Kurusu, Yasuhiko *chemistry educator*
Matsumoto, Kazuko *chemistry educator*
Muramatsu, Ichiro *chemistry educator*
Nakagaki, Masayuki *chemist*
Nakahara, Masayoshi *chemist*
Noguchi, Shun *chemistry educator*
Nozoe, Tetsuo *organic chemist, research consultant*
Okitani, Akihiro *food chemistry educator*
Sakurada, Yutaka *chemist*
Shima, Mikiko *polymer scientist*
Suami, Tetsuo *chemistry educator*
Tsuchida, Eishun *chemistry educator*

Toyama
Matsugo, Seiichi *chemistry educator, chemistry researcher*

Tsukuba
Araki, Hiroshi *chemist*

Urawa
Narasaki, Hisatake *analytical chemist*

Utsunomiya
Gunzo, Izawa *chemistry educator*

Yamaguchi
Suzuki, Nobutaka *chemistry educator*

Yamakita
Mita, Itura *polymer chemist*

Yokohama
Aika, Ken-ichi *chemistry educator*
Asawa, Tatsurō *chemistry researcher*
Hirota, Minoru *chemistry educator*
Ibata, Koichi *chemistry research manager*
Kunieda, Hironobu *physical chemistry educator*
Matsuzaki, Takao *chemist*
Momoki, Kozo *analytical chemist, educator*
Niki, Katsumi *chemist, educator*
Ogawa, Seiichiro *chemistry educator*
Oi, Ryu *chemist*

KENYA

Eldoret
Patel, Kiritkumar Natwerbhai *researcher, chemist*

NAURU

Nauru
Tun, Maung Myint Thein *analytical chemist*

THE NETHERLANDS

Amsterdam
Van Os, Nico Maria *research chemist*
Wiegman, Lenore Ho *chemist*

Arnhem
Sikkema, Doetze Jakob *chemist*

Delft
Moulijn, Jacob A. *chemical technology educator*

Eindhoven
De Bokx, Pieter Klaas *research chemist*
Van Santen, Rutger Anthony *catalysis educator*

Enschede
Seshan, Kulathu Iyer *chemistry educator, researcher*

Leiden
Bedeaux, Dick *chemist*
Lugtenburg, Johan *chemistry educator*
Maassen, Johannes Antonie *chemistry educator, consultant*

NEW ZEALAND

Auckland
Schwerdtfeger, Peter Adolf *research chemist*

NIGERIA

Ilorin
Adediran, Suara Adedeji *chemistry educator*

NORWAY

Bergen
Nerdal, Willy *research chemist*

Stavanger
Ruoff, Heinz Peter *chemistry educator*

OMAN

Muscat
Khair, Abul *chemistry educator*

PAKISTAN

Badamibagh
Murtaza, Ghulam *chemist, consultant*

Hyderabad
Ali, Syed Wajahat *physical chemist, researcher*

Islamabad
Qureshi, Iqbal Hussain *nuclear chemist*

PAPUA NEW GUINEA

Port Moresby
Koyyalamudi, Sundarrao *chemistry educator*

THE PHILIPPINES

Makati
Juliano, Petronilo Ochoa *chemist*

POLAND

Bydgoszcz
Paczkowski, Jerzy *chemistry educator*

Kraków
Jaskula, Marian Józef *chemist*

Lodz
Penczek, Stanislaw *chemistry educator*

Warsaw
Ziabicki, Andrzej Jozef *chemist*

Wroclaw
Wilgocki, Michal *electrochemist, chemistry educator*

PORTUGAL

Coimbra
Varandas, Antonio Joaquim De Campos *chemistry educator*

Lisbon
Pimenta, Gervásio Manuel *chemist, researcher*
Reis, João Carlos Ribeiro *chemistry educator*

Porto
Machado, Adelio Alcino Sampaio Castro *chemistry educator*

REPUBLIC OF KOREA

Andong
Park, Byung-Soo *chemistry educator, dean*

Chinju
Yoon, Yong-Jin *chemistry educator*

Daejeon
Kim, Wan Joo *medicinal chemist*

Daejon
Cho, Chang Gi *polymer scientist*
Kim, Heesook Park *chemist*

Inchon
Kim, Jong Soo *polymer scientist*

Jochiwon
Lee, Chi-Woo *chemistry educator*

Pohang
Chung, Sung-Kee *chemistry educator*

Pusan
Sung, Dae Dong *chemistry educator*

Seoul
Kim, Myung Soo *chemist, educator*
Lee, Young Moo *polymer chemist*

Taegu
Park, Yong-Tae *chemistry educator*

Taejon
Kim, Jae Nyoung *chemist*

ROMANIA

Bucharest
Spacu, Petru George *chemistry educator*

Cluj-Napoca
Farkas, György-Miklós *chemist, researcher*

RUSSIA

Moscow
Platé, Nicolai A. *polymer chemist*
Tartakovskiy, Vladimir Alexandrovich *chemist, researcher*
Zolotov, Yurii Alexander *chemist*

Saint Petersburg
Shevchenko, Sergey Markovich *organic chemist*

SAUDI ARABIA

Alkhobar
Balasubramanian, Krishna *chemistry educator, consultant*

Dhahran
Al-Afaleq, Eljazi *organic chemistry educator, administrator*
Ali, Mohammad Farhat *chemistry educator*
Fayad, Nabil Mohamed *chemist, researcher*
Harruff, Lewis Gregory *industrial chemist*
Raju, Krishnam *chemist*

Jubail
Gray, Kenneth Wayne *chemist, researcher*

Makkah
Muathen, Hussni Ahmad *chemistry educator*

SCOTLAND

Glasgow
Courtney, James McNiven *chemist*

SINGAPORE

Singapore
Chapman, Mark St. John *chemist*
Gan, Leong-Huat *chemist, educator*
Xu, Guo-Qin *chemist, educator*

SLOVAK REPUBLIC

Bratislava
Daněk, Vladimir *chemistry scientist*
Ertl, Peter *chemist, researcher*

SOUTH AFRICA

Bellville
Gihwala, Dherendra Isver *chemist*

Johannesburg
Staskun, Benjamin *retired chemistry educator*

Umtata
Rutherford, John Stewart *chemistry educator*

SPAIN

Alcala de Henares
Saiz, Enrique *chemist, educator*

Badajoz
Olivares Del Valle, Francisco Javier *physical chemistry educator*

Barcelona
Casañ-Pastor, Nieves *chemist, researcher*
Geijo, Fernando Antonio *chemist*
Subirana, Juan Antonio *polymer chemist, educator*

Girona
Carbo, Ramon *chemistry educator, researcher*

La Coruna
Nunez-Centella, Ramon Antonio *chemist, science educator, journalist*

Madrid
Garrido Perez, Mercedes *chemist*
González Ureña, Angel *chemistry researcher and educator*
Martinez Sanz, Antonio F. *chemist*
Rasines, Isidoro *chemist*

Santiago
Alonso-Fernandez, José Ramón *chemist*

Valencia
Corma, Avelino *chemist, researcher*

SRI LANKA

Colombo
Arambewela, Lakshmi Sriyani Rajapakse *chemist*

SWEDEN

Gothenburg
Nordén, Bengt Johan Fredrik *chemist*

Lund
Shinoda, Kōzō *chemistry educator, researcher*

Malmö
Wistrand, Lars Goran *chemist*

Stockholm
Arbin, Astrid Valborg *chemist*
Garegg, Per Johan *chemist, educator*
Magneli, Arne *chemist*
Olsson, Nils Urban *research manager*
Samuelsson, Bengt Ingemar *medical chemist*
Sjoberg, Berndt Olof Harald *chemist, research administrator*
Wranglén, Karl Gustaf (Gösta) *chemist*

Uppsala
Båckvall, Jan-Erling *chemist*

SWITZERLAND

Basel
Le-Van, Ngo *research chemist*
von Sprecher, Andreas *chemist*
Wolf, Romain Mathias *theoretical chemist*

Dubendorf
Hug, Rudolf Peter *chemist*

Geneva
Berclaz, Theodore M. *chemistry educator*
Kündig, Ernst Peter *chemist, researcher, educator*
Mueller, Paul *chemist, educator*

Neuchâtel
Neier, Reinhard Werner *chemistry educator*

Uetikon
Pfenninger, Armin *chemist, researcher*

Villigen
Reddy, Guvvala Nagabhushana *neurochemist*

Winterthur
Bührer, Heiner Georg *chemist, educator*

Zurich
Dunitz, Jack David *retired chemistry educator, researcher*
Ernst, Richard Robert *chemist, educator*
Geier, Gerhard *chemistry educator*
Huston, Rima *chemist*
Laube, Thomas *chemist, researcher*
Prelog, Vladimir *chemist*

TAIWAN

Chang-Hwa
Tai, Dar Fu *chemist*

Hsinchu
Chen, Chu-Chin *food chemist*
Hwu, Reuben Jih-Ru *chemistry educator*

Kaohsiung
Chen, Chen-Tung Arthur *chemistry and oceanography educator*

Kwei-San
Wu, Albert M. *glyco-immunochemistry educator*

Lungtan
Cheng, Sheng-San *chemistry research scientist, consultant*

Tainan
Low, Chow-Eng *chemistry educator*

Taipei
Lin, Ying-Chih *chemistry educator*
Su, Yuh-Long *chemistry educator, researcher*
Yang, Chin-Ping *chemist, engineering educator*

TURKEY

Maslak-Istanbul
Bekâroğlu, özer *chemist, educator*

UNITED ARAB EMIRATES

Al-Ain
Kiwan, Abdul Mageed Metwally *chemistry educator*

URUGUAY

Montevideo
Ventura, Oscar Nestor *chemistry educator, researcher*

VENEZUELA

Caracas
Nakano, Tatsuhiko *chemist, researcher, educator*

YUGOSLAVIA

Belgrade
Solaja, Bogdan Aleksandar *chemist, educator*

Novi Sad
Dokic, Petar *chemist, educator*

ADDRESS UNPUBLISHED

Armstrong, Andrew Thurman *chemist*
Arnett, Edward McCollin *chemistry educator, researcher*
Bailey, Arthur Emery *natural products chemist*
Batson, Susan Catherine *chemistry educator*
Berry, Richard Stephen *chemist*
Brown, Rhonda Rochelle *chemist, health facility administrator*
Butenandt, Adolf Friedrich Johann *physiological chemist*
Cairns, Theodore LeSueur *chemist*
Chen, Chaozong *chemistry educator*
Chou, David Chih-Kuang *analytical chemist*
Chu, Benjamin Thomas Peng-Nien *chemistry educator*
Claessens, Pierre *chemist, researcher*
Cox, James Carl, Jr. *chemist, researcher, consultant*
Crabtree, Robert Howard *chemistry educator, consultant*
DeAngelis, Thomas P. *chemist*
Denton, M. Bonner *research chemistry educator*
Fang, Joseph Pe Yong *chemistry educator*
Flor, Loy Lorenz *chemist, corrosion engineer, consultant*
Gass, Arthur Edward *chemist*
Gerber, Arthur Mitchell *chemist*
Hanton, Scott Douglas *physical chemist*
Hasha, Dennis Lloyd *research chemist, spectroscopist*
Hertz, Patricia Ritenour *analytical chemist*
Ho, John Wing-Shing *chemistry educator, researcher*
Hoffmann, Roald *chemist, educator*
Knowles, David Eugene *chemist*
Knowles, Richard Norris *chemist*
Kravitz, Rubin *chemist*
Kwan, Henry King-Hong *pharmaceutical chemist*

Lee, Sung Jai *medicinal chemist*
Lewis, Graham Thomas *analytical inorganic chemist*
Martin, Charles Raymond *chemist, educator*
Musmanni, Sergio *chemist, researcher*
Odink, Debra Alida *chemist, researcher*
Oppenheimer, Larry Eric *physical chemist*
Page, Philip Ronald *chemist*
Pearson, Ralph Gottfrid *chemistry educator*
Peterson, Dwight Malcolm *chemist*
Pinkert, Dorothy Minna *chemist*
Rakutis, Ruta *chemical economist*
Rendina, George *chemistry educator*
Rice, Stuart Alan *chemist, educator*
Roberts, Earl John *carbohydrate chemist*
Root, M. Belinda *chemist*
Rüetschi, Paul *electrochemist*
Rutstrom, Dante Joseph *chemist*
Solomon, Susan *chemist, scientist*
Spurr, Paul Raymond *organic chemist*
Starek, Rodger William *chemist*
Strier, Murray Paul *chemist, consultant*
Stuart, James Davies *analytical chemist, educator*
Tokue, Ikuo *chemist, researcher*
Vigler, Mildred Sceiford *retired chemist*
Watson, Stuart Lansing *chemist*
Weiss, Michael James *chemistry educator*
Whistler, Roy Lester *chemist, educator, industrialist*
Wickman, Herbert Hollis *physical chemist, condensed matter physicist*
Wong, Kwee Chang *chemist*
Zimm, Bruno Hasbrouck *physical chemistry educator*

PHYSICAL SCIENCES: EARTH & ENVIRONMENT

UNITED STATES

ARIZONA

Tucson
Brusseau, Mark Lewis *environmental educator, researcher*
Lane, Leonard J. *hydrologist*
Long, Austin *geosciences educator*

CALIFORNIA

Fontana
Poulsen, Dennis Robert *environmentalist*

Irvine
Phalan, Robert F. *environmental scientist*

Pacific Grove
Lindstrom, Kris Peter *environmental consultant*

Palo Alto
Warne, William Elmo *irrigationist*

Torrance
Krueger, Kurt Edward *environmental management company official*

COLORADO

Fort Collins
Connell, James Roger *atmospheric turbulence researcher, educator*

Golden
McNeill, William *environmental scientist*

Greeley
Dingeman, Thomas Edward *wastewater treatment plant administrator*

CONNECTICUT

Hartford
Casale, Joseph Wilbert *environmental organic chemist, researcher*

Waterford
Johnson, Gary William *environmental scientist, consultant*

DELAWARE

Newark
Hutton, David Glenn *environmental consultant, chemical engineer*

DISTRICT OF COLUMBIA

Washington
Kostka, Madonna Lou (Donna) *naturalist, environmental scientist, ecologist*
Krug, Edward Charles *environmental scientist*
Sexton, Ken *environmental health scientist*

FLORIDA

Coral Gables
Walsh-McGehee, Martha Bosse *conservationist*

Panama City
Leitheiser, James Victor *environmental specialist*

Punta Gorda
Beever, Lisa Britt-Dodd *environmental planner*

Tallahassee
Means, Donald Bruce *environmental educator, research ecologist*

Tampa
Kalmaz, Ekrem Errol *environmental scientist*

GEORGIA

Athens
Straw, William Russell *environmental scientist*

ILLINOIS

Argonne
Barrett, Gregory Lawrence *environmental scientist*
Bhatti, Neeloo *environmental scientist*

Breese
Anderson, Donald Thomas, Jr. *environmental consultant*

Champaign
Hodge, Winifred *environmental scientist, researcher*

Deerfield
Runkle, Robert Scott *environmental company executive*

Urbana
Aref, Hassan *fluid mechanics educator*

INDIANA

Schererville
Ontto, Donald Edward *environmental analytical chemist, wastewater treatment consultant*

IOWA

Iowa City
Folk, George Edgar, Jr. *environmental physiology educator*

LOUISIANA

Baton Rouge
Dartez, Charles Bennett *environmental consultant*
Lynam, Donald Ray *environmental health scientist*

MARYLAND

Aberdeen
Fought, Sheryl Kristine *environmental scientist, engineer*

Cockeysville Hunt Valley
Morhardt, Josef Emil, IV *environmentalist, engineering company executive*
Morhardt, Sia S. *environmental scientist*

Edgewood
Russell, William Alexander, Jr. *environmental scientist*

Gaithersburg
Sobers, David George *environmentalist*

Mount Savage
Warren, D. Elayne *environmental sanitarian*

MASSACHUSETTS

Amherst
Walker, Robert Wyman *environmental sciences educator*

Belmont
McHenry, Douglas Bruce *naturalist*

Boston
Scuderi, Louis Anthony *climatology educator*

Rockland
Doucette, Paul Stanislaus *environmental scientist*

MICHIGAN

Ann Arbor
DeYoung, Raymond *conservation behavior educator*

Howell
Dombkowski, Joseph John *water treatment specialist*

Lansing
Chazell, Russell Earl *environmental chemist*

Mount Pleasant
VanHouten, Jacob Wesley *environmental project manager, consultant*

Olivet
Seabrook, Barry Steven *environmental scientist, consultant*

MINNESOTA

Saint Paul
Swain, Edward Balcom *environmental research scientist*

MISSISSIPPI

Gulfport
Doudrick, Robert Lawrence *research plant pathologist*

MISSOURI

Kansas City
Kinsey, John Scott *environmental scientist*
Simmons, Robert Marvin *environmental scientist, consultant*

Neosho
Jefferson, Michael L *environmental educator*

Saint Louis
Churchill, Ralph John *environmental chemist*

NEVADA

Las Vegas
Barth, Delbert Sylvester *environmental studies educator*

NEW HAMPSHIRE

Durham
Bowden, William Breckenridge *natural resources educator*

NEW JERSEY

East Orange
Astor, Peter H. *environmental consultant, mathematician*

Hackensack
Skovronek, Herbert Samuel *environmental scientist*

Lyndhurst
Moese, Mark Douglas *environmental consultant*

Piscataway
Mehlman, Myron A. *environmental and occupational medicine educator, environmental toxicologist*

Princeton
Williams, Robert H. *environmental scientist*

Spring Lake
Burke, J(ohn) Michael *environmental company executive*

Trenton
Tucker, Robert Keith *environmental scientist, research administrator*

Wall
Montgomery, John Harold *environmentalist*

NEW MEXICO

Albuquerque
Mauzy, Michael Philip *environmental consultant, chemical engineer*

NEW YORK

Bronx
Hart, John Amasa *wildlife biologist, conservationist*

Ithaca
Evans, Gary William *human ecology educator*
Smith, Charles Robert *conservationist, naturalist, ornithologist, educator*

Millbrook
Cole, Jonathan Jay *aquatic scientist, researcher*

New York
Koestler, Robert John *conservation research scientist, biologist*
Rampino, Michael R. *earth scientist*

Ray Brook
Roy, Karen Mary *limnologist, state government regulator*

Syracuse
Elligott, Linda A. *environmental scientist*

Upton
Lewin, Keith Frederic *environmental science research associate*
Morris, Samuel Cary *environmental scientist, consultant, educator*

Valhalla
Gross, Stanislaw *environmental sciences educator, activist*

NORTH CAROLINA

Cary
Robertson, William Bell, Jr. *environmental scientist*

Raleigh
Rose, Thoma Hadley *environmental consultant*

Research Triangle Park
Hyatt, David Eric *environmental scientist*

OHIO

Cincinnati
Young, Daniel Lee *environmental scientist*

Columbus
Wali, Mohan Kishen *environmental science and natural resources educator*

Tiffin
Baker, David B. *environmental scientist*

OREGON

Portland
Raniere, Lawrence Charles *environmental scientist*

PENNSYLVANIA

East Stroudsburg
Rymon, Larry Maring *environmental scientist, educator*

Harrisburg
Willis, David Paul *environmental scientist*

Indiana
Cale, William Graham, Jr. *environmental sciences educator, university administrator, researcher*

Johnstown
Brezovec, Paul John *environmental specialist*

Philadelphia
Peck, Robert McCracken *naturalist, science historian, writer*

Pittsburgh
Leney, George Willard *environmental administrator, consulting geologist*

Sharon
Anttila, Samuel David *environmental scientist*

SOUTH CAROLINA

Aiken
Miller, Phillip Edward *environmental scientist*

Columbia
Shafer, John Milton *hydrologist, consultant, software developer*

Greenville
Holland, David Lee *environmental scientist, consultant*

TENNESSEE

Murfreesboro
Mitchell, Jerry Calvin *environmental company executive*

Oak Ridge
Cawley, Charles Nash *enviromental scientist*

TEXAS

Abilene
Pickens, Jimmy Burton *earth and life science educator, military officer*

Beaumont
Walker, John Michael *consulting environmental hydrogeologist*

College Station
Anderson, Duwayne Marlo *earth and polar scientist, university administrator*

Dallas
Urquhart, Sally Ann *environmental scientist, chemist*

Houston
Jackson, Douglas Webster *environmental scientist, consultant*

UTAH

Logan
Sigler, William Franklin *environmental consultant*

VIRGINIA

Arlington
Leibowitz, Alan Jay *environmental scientist*

Blacksburg
Cairns, John, Jr. *environmental science educator, researcher*

Mc Lean
Bizzigotti, George Ora *environmental chemist, consultant*

Norfolk
Corl, William Edward *environmental chemist*

Palmyra
Mulckhuyse, Jacob John *energy conservation and environmental consultant*

Reston
Miller, Lynne Marie *environmental company executive*

WASHINGTON

Richland
Konkel, R(ichard) Steven *environmental and social science consultant*
Link, Steven Otto *environmental scientist, statistician*
Onishi, Yasuo *environmental researcher*

WYOMING

Cheyenne
Beaven, Thornton Ray *physical scientist*

Laramie
Meyer, Edmond Gerald *energy and natural resources educator, resources scientist, entrepreneur, former chemistry educator*

CANADA

ALBERTA

Lethbridge
Johnson, Daniel Lloyd *biogeographer*

QUEBEC

Montreal
Conrad, Bruce R. *earth scientist*
Dubé, Ghyslain *earth scientist*
Fyffe, Les *earth scientist*
Healey, Chris M. *earth scientist*
Imorde, Henry K. *earth scientist*
Kerby, R. C. *earth scientist*
Leitch, Craig H. B. *earth scientist*
Savard, G. S. *earth scientist*
Swinden, H. Scott *earth scientist*

Otterburn Park
Roth, Annemarie *conservationalist*

MEXICO

Tehuacán
Romero, Miguel A. *animal nutrition director*

ARGENTINA

Salta
Barbarán, Francisco Ramón *educator, researcher*

AUSTRALIA

Milton
Brown, Trevor Ernest *environment risk consultant*

BELGIUM

Brussels
Bourdeau, Philippe *environmental scientist*

CZECH REPUBLIC

Ceske Budejovice
Dolejs, Petr *aquatic scientist, consultant*

ENGLAND

Burton-on-Trent
Page, Dennis *coal scientist, consultant*

FRANCE

Nanterre
Chambolle, Thierry Jean-Francois *environmental scientist*

ITALY

Trieste
Panza, Giuliano Francesco *seismologist, educator*

NAMIBIA

Kalkfeld
Oelofse, Jan Harm *game rancher, wildlife management consultant*

THE NETHERLANDS

Apeldoorn
Cramer, Jacqueline Marian *environmental scientist, researcher*

SWEDEN

Lund
Akselsson, Kjell Roland *environment technology educator, researcher*

Stockholm
Håfors, Aina Birgitta *wood conservation chemist*

ADDRESS UNPUBLISHED

Arisman, Ruth Kathleen *environmental manager*
Brown, Barbara S. *environmental scientist*
Davis, Kenneth Earl, Sr. *environmentalist*
Goldman, Charles Remington *environmental scientist, educator*
La Rivière, Jan Willem Maurits *environmental biology educator*
Nagys, Elizabeth Ann *environmental issues educator*
Sadusky, Maria Christine *environmental scientist*

PHYSICAL SCIENCES: GEOLOGY

UNITED STATES

ALABAMA

Auburn
Molz, Fred John, III *hydrologist, educator*

Tuscaloosa
LaMoreaux, Philip Elmer *geologist, hydrogeologist, consultant*
Mancini, Ernest Anthony *geologist, educator, researcher*

ARIZONA

Flagstaff
Shoemaker, Eugene Merle *geologist*

Sun City
Dapples, Edward Charles *geologist, educator*

Tempe
Mock, Peter Allen *hydrogeologist*
Péwé, Troy Lewis *geologist, educator*

Tucson
Kamilli, Robert Joseph *geologist*
Kiersch, George Alfred *geological consultant, educator emeritus*
More, Syver Wakeman *geologist*
Willis, Clifford Leon *geologist*

ARKANSAS

Russellville
Meeks, Lisa Kaye *hydrogeologist, researcher*

CALIFORNIA

Arcadia
Proctor, Richard J. *geologist, consultant*

Berkeley
Berry, William Benjamin Newell *geologist, educator, former museum administrator*
Carmichael, Ian Stuart Edward *geologist, educator*
Leopold, Luna Bergere *geology educator*

Big Bear Lake
Mac Iver, Douglas Yaney *geologist*

Canyon Lake
Schilling, Frederick Augustus, Jr. *geologist, consultant*

Davis
Moores, Eldridge Morton *geology educator*

Emeryville
McKereghan, Peter Fleming *hydrogeologist, consultant*

Livermore
Crow, Neil Byrne *geologist*

Los Angeles
Bottjer, David John *geological sciences educator*
Fischer, Alfred George *geology educator*

Menlo Park
Bukry, John David *geologist*
Carr, Michael Harold *geologist*
Dalrymple, Gary Brent *research geologist*
Wallace, Robert Earl *geologist*

Monrovia
Rudnyk, Marian E. *planetary photogeologist*

Pasadena
Albee, Arden Leroy *geologist, educator*
Sharp, Robert Phillip *geology educator, researcher*
Taylor, Hugh Pettingill, Jr. *geologist, educator*
Wasserburg, Gerald Joseph *geology and geophysics educator*

Sacramento
Maurath, Garry Caldwell *hydrogeologist*

San Juan Capistrano
Testa, Stephen Michael *geologist, consultant*

Santa Barbara
Kennett, James Peter *geology and zoology educator*

Santa Cruz
Lay, Thorne *geosciences educator*

Scotts Valley
Snyder, Charles Theodore *geologist*

Stanford
Coleman, Robert Griffin *geology educator*
Ernst, Wallace Gary *geology educator, dean*

Walnut Creek
Oakeshott, Gordon B(laisdell) *geologist*

COLORADO

Boulder
Kauffman, Erle Galen *geologist, paleontologist*

Colorado Springs
Henrickson, Eiler Leonard *geologist, educator*

Denver
Garske, Jay Toring *geologist, oil and minerals consultant*
Kellogg, Karl Stuart *geologist*
Landon, Susan Melinda *petroleum geologist*
Sherman, David Michael *geologist*
Todd, Donald Frederick *geologist*

Durango
Lauth, Robert Edward *geologist*
Osterhoudt, Walter Jabez *geophysical and geological exploration consultant*

Evergreen
Heyl, Allen Van, Jr. *geologist*
Phillips, Adran Abner (Abe Phillips) *geologist, oil and gas exploration consultant*

Fort Collins
Erslev, Eric Allan *geologist, educator*
Johnson, Robert Britten *geology educator*

Frisco
Power, Walter Robert *geologist*

Golden
Halstead, Philip Hubert *geologist, consultant*
Lachel, Dennis John *geologist, engineer, consultant*
Morrison, Roger Barron *geologist, executive*

CONNECTICUT

Middletown
Horne, Gregory Stuart *geologist, educator*

New Britain
Baskerville, Charles Alexander *geologist, educator*

Winchester
Carter, John Angus *geologist, geochemist, environmental engineer*

DELAWARE

Newark
Jordan, Robert R. *geologist, educator*

DISTRICT OF COLUMBIA

Washington
Eghbal, Morad *geologist, lawyer*
Felsher, Murray *geologist, publisher*
Fiske, Richard Sewell *geologist*
Morehouse, David Frank *geologist*
Price, Jonathan G. *geologist*
Wallace, Jane House *geologist*
Watters, Thomas Robert *geologist, museum administrator*

FLORIDA

Boca Raton
Finkl, Charles William, II *geologist, educator*

Live Oak
Exley, Sheck *geologist*

West Palm Beach
Witt, Gerhardt Meyer *hydrogeologist*

GEORGIA

Atlanta
Cramer, Howard Ross *geologist, environmental consultant*

HAWAII

Honolulu
Helsley, Charles Everett *geologist, geophysicist*
Keil, Klaus *geology educator, consultant*
Kwong, James Kin-Ping *geological engineer*
Wessel, Paul *geology and geophysics educator*

IDAHO

Moscow
Miller, Maynard Malcolm *geologist, educator, research foundation director, explorer, state legislator*

ILLINOIS

Champaign
Cartwright, Keros *hydrogeologist, researcher*

Chicago
Koster van Groos, August Ferdinand *geology educator*

Skokie
Nisperos, Arturo Galvez *engineering geologist, petrographer*

Urbana
White, W(illiam) Arthur *geologist*

INDIANA

Notre Dame
Gutschick, Raymond Charles *geology educator, researcher, micropaleontologist*

West Lafayette
Leap, Darrell Ivan *hydrogeologist*

IOWA

Iowa City
Koch, Donald LeRoy *geologist, state agency administrator*

North Liberty
Glenister, Brian Frederick *geologist, educator*

KANSAS

Lawrence
Enos, Paul *geologist, educator*
Gerhard, Lee Clarence *geologist, educator*
Merriam, Daniel F(rancis) *geologist*
Olea, Ricardo Antonio *geological researcher, engineer*

Pittsburg
Foresman, James Buckey *geologist, geochemist, industrial hygienist*

LOUISIANA

Alexandria
Rogers, James Edwin *geology and hydrology consultant*

Baton Rouge
Van Lopik, Jack Richard *geologist, educator*

Metairie
Hartman, James Austin *geologist*

Monroe
Glawe, Lloyd Neil *geology educator*

MAINE

Freeport
Hebson, Charles Stephan *hydrogeologist, civil engineer*

Winthrop
O'Brien, Ellen K. *hydrologist*

MARYLAND

Baltimore
Weaver, Kenneth Newcomer *geologist, state official*

Bethesda
Osterman, Lisa Ellen *geologist, researcher*

College Park
Zen, E-an *research geologist*

MASSACHUSETTS

Amherst
Bromery, Randolph Wilson *geologist, educator*

Cambridge
Burchfiel, Burrell Clark *geology educator*
Siever, Raymond *geology educator*

Sudbury
Blackey, Edwin Arthur, Jr. *geologist*

Woods Hole
Berggren, William Alfred *geologist, research micropaleontologist, educator*
Stone, Thomas Alan *geologist*
Uchupi, Elazar *geologist, researcher*

MICHIGAN

Ann Arbor
Gurnis, Michael *geological sciences educator*
Kelly, William Crowley *geological sciences educator*

Temperance
Piniewski, Robert James *geologist*

Warren
Abdul, Abdul Shaheed *hydrogeologist*

MINNESOTA

Minneapolis
Wright, Herbert E(dgar), Jr. *geologist*

MISSISSIPPI

Picayune
Lowrie, Allen *geologist, oceanographer*

MISSOURI

Cape Girardeau
Close, Edward Roy *hydrogeologist, environmental engineer, physicist*

Columbia
Shelton, Kevin L. *geology educator*

Kansas City
Hilpman, Paul Lorenz *geology educator*

Rolla
Hatheway, Allen Wayne *geological engineer, educator*

Springfield
Miller, James Frederick *geologist, educator*

MONTANA

Billings
Darrow, George F. *natural resources company owner, consultant*

Butte
Ruppel, Edward Thompson *geologist*

Florence
Campbell, Charles *geologist*

Missoula
Peterson, James Algert *geologist, educator*

NEVADA

Las Vegas
Harpster, Robert Eugene *engineering geologist*
Levich, Robert Alan *geologist*

Reno
Ritter, Dale F. *geologist, research association administrator*
Taranik, James Vladimir *geologist, educator*

NEW HAMPSHIRE

Berlin
Cabaup, Joseph John *geology educator*

NEW JERSEY

Fort Hancock
Klein, George D. *geologist, business executive*

Morristown
de Mauret, Kevin John *geologist*

NEW MEXICO

Albuquerque
Papike, James Joseph *geology educator, science institute director*
Rhodes, Doug *geologist*
Wengerd, Sherman Alexander *geologist, educator*

Socorro
Kottlowski, Frank Edward *geologist*

NEW YORK

Brooklyn
Friedman, Gerald Manfred *geologist, educator*

Canton
Romey, William Dowden *geologist, educator*

Nyack
Ryan, William B. F. *geologist*

Palisades
Eaton, Gordon Pryor *geologist, research director*

Sea Cliff
Rich, Charles Anthony *geologist*

Stony Brook
Schoonen, Martin Adrianus Arnoldus *geology educator*

Syracuse
Gass, Tyler Evan *hydrogeologist*

Troy
Miller, Donald Spencer *geologist, educator*

Watertown
Miller, David William *geologist*

NORTH CAROLINA

Chapel Hill
Miller, Daniel Newton, Jr. *geologist, consultant*
Rogers, John James William *geology educator*

OHIO

Athens
Nance, Richard Damian *geologist*

Cleveland
Elliott, W(illiam) Crawford *geology researcher*

Dayton
Gregor, Clunie Bryan *geology educator*

OKLAHOMA

Ardmore
Olsen, Thomas William *geologist*

Norman
Mankin, Charles John *geology educator*

Tulsa
Bennison, Allan Parnell *geological consultant*
Hall, George Joseph, Jr. *geologist, educator, consultant, geotechnical engineer*
Meyerhoff, Arthur Augustus *geologist, consultant*

OREGON

Ashland
Ferrero, Thomas Paul *engineering geologist*

Roseburg
Pendleton, Verne H., Jr. *geologist*

Seaside
See, Paul DeWitt *geology educator*

PENNSYLVANIA

Berwyn
Triegel, Elly Kirsten *geologist*

Bryn Mawr
Crawford, Maria Luisa Buse *geology educator*

Kutztown
Zei, Robert William *geology educator, consultant*

Mechanicsburg
Beard, Robert Douglas *geologist*

Philadelphia
Wagner, Mary Emma *geologist educator, researcher*

University Park
Biederman, Edwin Williams, Jr. *petroleum geologist*

RHODE ISLAND

Providence
Mustard, John Fraser *geologist, educator*

SOUTH CAROLINA

Columbia
Kanes, William Henry *geology educator, research center administrator*

SOUTH DAKOTA

Vermillion
Chadima, Sarah Anne *geologist*
Hammond, Richard Horace *geologist*

TENNESSEE

Oak Ridge
Grimes, James Gordon *geologist*
Rasor, Elizabeth Ann *hydrogeologist, environmental scientist*

TEXAS

Arlington
Damuth, John Erwin *marine geologist*

Austin
Baghai, Nina Lucille *geology researcher*
Boyer, Robert Ernst *geologist, educator*
Fisher, William Lawrence *geologist, educator*
Hilburn, John Charles *geologist, geophysicist*
McKenna, Thomas Edward *hydrogeologist*
Sharp, John Malcolm, Jr. *geology educator*
Young, Keith Preston *geologist*

Bowie
Reynolds, Don William *geologist*

College Station
Berg, Robert Raymond *geologist, educator*
Rezak, Richard *geology and oceanography educator*
Wiltschko, David Vilander *geology educator, tectophysics director*

Dallas
Brooks, James Elwood *geologist, educator*
Gibbs, James Alanson *geologist*
Marshall, John Harris, Jr. *geologist, oil company executive*
Oppenheim, Victor Eduard *consulting geologist*
Ries, Edward Richard *petroleum geologist, consultant*

Fort Worth
Caldwell, Billy Ray *geologist*

Houston
Bally, Albert W. *geology educator*
Davids, Robert Norman *petroleum exploration geologist*
Farley, Martin Birtell *geologist*
Golbraykh, Isaak German *geologist*
Phinney, William Charles *geologist*
Sullivan, Kathryn D. *geologist, astronaut*
Vail, Peter Robbins *geologist*

Lubbock
Laing, Malcolm Brian *geologist, consultant*

New Braunfels
Wilson, James Lee *retired geology educator, consultant*

Odessa
Reeves, Robert Grier LeFevre *geology educator, scientist*

San Angelo
Cline, David Christopher *geologist*

San Antonio
Jones, James Ogden *geologist, educator*

UTAH

Salt Lake City
Allison, Merle Lee *geologist*
Hunt, Charles Butler *geologist*

VIRGINIA

Arlington
Wayland, Russell Gibson, Jr. *geology consultant, retired government official*

Bluefield
Mullennex, Ronald Hale *geologist, consultant*

Herndon
Peck, Dallas Lynn *geologist*

Lexington
Spencer, Edgar Winston *geology educator*

Reston
Cohen, Philip *hydrogeologist*
Guptill, Stephen Charles *physical scientist*
Huebner, John Stephen *geologist*
Sato, Motoaki *geologist, researcher*
Wood, Warren Wilbur *hydrologist*

WASHINGTON

Bellingham
Ross, Charles Alexander *geologist*

Seattle
Creager, Joe Scott *geology and oceanography educator*
Owens, Edward Henry *geologist, consultant*

Vancouver
Iverson, Richard Matthew *earth scientist*

WISCONSIN

Madison
Clark, David Leigh *marine geologist, educator*
Dott, Robert Henry, Jr. *geologist, educator*
Maher, Louis James, Jr. *geologist, educator*
Pray, Lloyd Charles *geologist, educator*

Milwaukee
Cronin, Vincent Sean *geologist*

WYOMING

Cheyenne
Rust, Lynn Eugene *geologist*

CANADA

ALBERTA

Calgary
Nowlan, Godfrey S. *geologist*

BRITISH COLUMBIA

Sidney
Weichert, Dieter Horst *seismologist, researcher*

Vancouver
Chase, Richard Lionel St. Lucian *geology and oceanography educator*
Kieffer, Susan Werner *geology educator*
Mathews, William Henry *geologist, educator*

Victoria
Hoffman, Paul Felix *geologist, educator*

ONTARIO

Hamilton
Kingwood, Alfred E. *geologist, educator*
McNutt, R. H. *geologist, geochemist, educator*
Walker, Roger Geoffrey *geology educator, consultant*

Ottawa
Babcock, Elkanah Andrew *geologist*
McLaren, Digby Johns *geologist, educator*
Veizer, Ján *geology educator*

Saint Catharines
Jolly, Wayne Travis *geologist, educator*

Toronto
Scott, Steven Donald *geology educator, researcher*

Waterloo
Morgan, Alan Vivian *geologist, educator*

AUSTRIA

Vienna
Aulitzky, Herbert *retired erosion and avalanche control educator*

BELGIUM

Ghent
Walschot, Leopold Gustave *conservator*

BRAZIL

Rio Claro
Potter, Paul Edwin *geologist, educator, consultant*

CHINA

Beijing
Shu, Sun *geologist*

DENMARK

Copenhagen
Rose-Hansen, John *geologist*

ENGLAND

Durham
Robson, Geoffrey Robert *geologist, seismologist, consultant*

Oxford
Dewey, John F. *geologist, educator*
Reading, Harold G. *geology educator*

GERMANY

Braunschweig
Hollmann, Rudolf Werner *geologist, palaeontology researcher*

Jülich
Mann, Ulrich *petroleum geologist*

Neckargemuend
Kirchmayer-Hilprecht, Martin *geologist*

GREECE

Patras
Anagnostopoulos, Stavros Aristidou *earthquake engineer, educator*

ITALY

Macerata
Fruzzetti, Oreste Giorgio *geologist*

Parma
Cita, Maria *geology educator*

JAPAN

Funabashi
Katayama, Tetsuya *geologist, materials research petrographer*

Ishikawa
Konishi, Kenji *geology educator*

Kumamoto
Tsusue, Akio *geology educator*

Yokosuka
Sekioka, Mitsuru *geoscience educator*

NORWAY

Trondheim
Roaldset, Elen *geologist*

RUSSIA

Moscow
Bogdanov, Nikita Alexeevich *geology educator*
Knipper, Andrei Lvovich *geologist, administrator, researcher*

SWITZERLAND

Gais
Langenegger, Otto *hydrogeologist*

WALES

Clwyd
Nichol, Douglas *geologist*

ADDRESS UNPUBLISHED

Dickinson, William Richard *geologist, educator*
Dodson, James Noland *geologist*
Foster, Norman Holland *geologist*
Gault, Donald Eiker *planetary geologist*
Ingle, James Chesney, Jr. *geology educator*
Joseph, Lura Ellen *geologist*
Metsger, Robert William *geologist*
Middleton, Gerard Viner *geology educator*
Saines, Marvin *hydrogeologist*
Shariff, Asghar J. *geologist*
Sharp, William Wheeler *geologist*
West, Jack Henry *petroleum geologist*
Wilhelms, Don Edward *geologist*
Wilson, Leonard Richard *geologist, consultant*
Woods, Arnold Martin *geologist*

PHYSICAL SCIENCES: GEOPHYSICS

UNITED STATES

ARIZONA

Tucson
Fink, James Brewster *geophysicist, consultant*
Wallace, Terry Charles, Jr. *geophysicist, educator*

CALIFORNIA

Berkeley
Bolt, Bruce Alan *seismologist, educator*
Jeanloz, Raymond *geophysicist, educator*

Bonita
Wood, Fergus James *geophysicist, consultant*

La Jolla
Gilbert, James Freeman *geophysics educator*
Munk, Walter Heinrich *geophysics educator*
Sclater, John George *geophysics educator*

Menlo Park
Lindh, Allan Goddard *seismologist*

Pasadena
Anderson, Don Lynn *geophysicist, educator*
Smith, Edward John *geophysicist, physicist*

Ridgecrest
St. Amand, Pierre *geophysicist*

Santa Barbara
Atwater, Tanya Maria *marine geophysicist, educator*

Santa Cruz
Williams, Quentin Christopher *geophysicist*

Stanford
Thompson, George Albert *geophysics educator*

Sunnyvale
Breiner, Sheldon *geophysics educator, business executive*

COLORADO

Boulder
Kisslinger, Carl *geophysicist, educator*
Little, Charles Gordon *geophysicist*

Denver
Johnson, Walter Earl *geophysicist*

CONNECTICUT

New Haven
Clark, Sydney Procter *geophysics educator*

DISTRICT OF COLUMBIA

Washington
Greenewalt, David *geophysicist*
Mao, Ho-kwang *geophysicist, educator*
Press, Frank *geophysicist, educator*
Wetherill, George West *geophysicist, planetary scientist*
Whitcomb, James Hall *geophysicist, foundation administrator*
Yoder, Hatten Schuyler, Jr. *petrologist*

FLORIDA

Naples
Long, James Alvin *exploration geophysicist*

Tallahassee
Pfeffer, Richard Lawrence *geophysics educator*

HAWAII

Honolulu
Edwards, Margo H. *marine geophysicist, researcher*
Khan, Mohammad Asad *geophysicist, educator, former energy minister and senator of Pakistan*

MARYLAND

Baltimore
Salisbury, John William *research geophysicist, consultant*
Sibeck, David G. *geophysicist*

Gaithersburg
Kolstad, George Andrew *physicist, geoscientist*

MASSACHUSETTS

Hanscom AFB
Van Tassel, Roger Alan *infrared phenomenologist, geophysicist*

Worcester
Bell, Peter Mayo *geophysicist*

MICHIGAN

Ann Arbor
Pollack, Henry Nathan *geophysics educator*

MINNESOTA

Minneapolis
Karato, Shun-ichiro *geophysicist*

NEW JERSEY

Princeton
Navrotsky, Alexandra *geophysicist*

NEW MEXICO

Albuquerque
Cabaniss, Gerry Henderson *geophysicist, analyst*

NEW YORK

New York
Broecker, Wallace S. *geophysics educator*

Palisades
Kent, Dennis Vladimir *geophysicist, researcher*

Stony Brook
Weidner, Donald J. *geophysicist educator*

NORTH CAROLINA

Chapel Hill
Droessler, Earl G. *geophysicist educator*

Raleigh
Janowitz, Gerald Saul *geophysicist, educator*
Won, Ihn-Jae *geophysicist*

OHIO

Columbus
Jezek, Kenneth Charles *geophysicist, educator, researcher*

OKLAHOMA

Norman
Gal-Chen, Tzvi *geophysicist*

Tulsa
Robinson, Enders Anthony *geophysics educator, writer*

PENNSYLVANIA

University Park
Hardy, Henry Reginald, Jr. *geophysicist, educator*

SOUTH DAKOTA

Rapid City
Johnson, L. Ronald *geophysicist*
Smith, Paul Letton, Jr. *research scientist*

TENNESSEE

Memphis
Johnston, Archibald Currie *geophysics educator, research director*

TEXAS

Arlington
Ellwood, Brooks Beresford *geophysicist, educator*

Austin
Stoffa, Paul L. *geophysicist, educator*

Houston
Barlow, Nadine Gail *planetary geoscientist*
Cantwell, Thomas *geophysicist, electrical engineer*
Crook, Troy Norman *geophysicist, consultant*
Jordan, Neal Francis *geophysicist, researcher*
Mateker, Emil Joseph, Jr. *geophysicist*
Rosenthal, Alan Irwin *geophysicist*
Talwani, Manik *geophysicist, educator*

Plano
Broyles, Michael Lee *geophysics and physics educator*

Richardson
Ward-McLemore, Ethel *research geophysicist, mathematician*

UTAH

Provo
Benson, Alvin K. *geophysicist, consultant, educator*

VIRGINIA

Reston
Hanna, William Francis *geophysicist*

WASHINGTON

Lynnwood
Olsen, Kenneth Harold *geophysicist, astrophysicist*

Seattle
Merrill, Ronald Thomas *geophysicist, educator*

CANADA

BRITISH COLUMBIA

Sidney
Best, Melvyn Edward *geophysicist, researcher*

ONTARIO

Manotick
Hobson, George Donald *retired geophysicist*

AUSTRALIA

Hobart
Tilbrook, Bronte David *research scientist*

BARBADOS

Christchurch
Ramsahoye, Lyttleton Estil *geophysicist, consultant, educator*

ENGLAND

Crawley
Van Schuyver, Connie Jo *geophysicist*

FRANCE

Paris
Tarantola, Albert *geophysicist, educator*

GERMANY

Kiel
Meissner, Rudolf Otto *geophysicist, educator*

JAPAN

Hiratsuka
Asada, Toshi *seismologist, educator*

Sapporo
Maeno, Norikazu *geophysics educator*

Tokyo
Ishii, Yoshinori *geophysics educator*

PORTUGAL

Lisbon
De Aguiar, Ricardo Jorge Frutuoso *research geophysicist*

RUSSIA

Moscow
Ostrovsky, Alexey *geophysicist, researcher*
Strakhov, Vladimir Nikolayevich *geophysics educator*

SWITZERLAND

Zurich
Mueller, Stephan *geophysicist, educator*

ADDRESS UNPUBLISHED

Akasofu, Syun-Ichi *geophysicist*
Haskin, Larry Allen *earth and planetary scientist, educator*
Mitchell, Neil Charles *geophysicist*

PHYSICAL SCIENCES: MATERIALS

UNITED STATES

CALIFORNIA

Berkeley
Haller, Eugene Ernest *materials scientist, educator*
Ritchie, Robert Oliver *materials science educator*
Yao, Xiang Yu *materials scientist*

Cupertino
Schmid, Anthony Peter *materials scientist, engineer*

Livermore
Lindner, Duane Lee *materials science management professional*

Los Angeles
Dunn, Bruce Sidney *materials science educator*
Shapiro, Isadore *materials scientist, consultant*

Morgan Hill
Gehman, Bruce Lawrence *materials scientist*

Palo Alto
Stringer, John *materials scientist*

San Jose
Nimmagadda, Rao Rajagopala *materials scientist, researcher*
Parker, Michael Andrew *materials scientist*

Santa Clara
Flinn, Paul Anthony *materials scientist*

Stanford
Hagstrom, Stig Bernt *materials science and engineering educator*
Nix, William Dale *materials scientist, educator*

Valencia
Golijanin, Danilo M. *materials scientist, physics educator*

COLORADO

Crested Butte
Green, Walter Verney *materials scientist*

CONNECTICUT

Bethany
Bergen, Robert Ludlum, Jr. *materials scientist*

Windsor
Garde, Anand Madhav *materials scientist*

DELAWARE

Newark
Ravi, Vilupanur Alwar *materials scientist*
Schultz, Jerold Marvin *materials scientist, educator*

Wilmington
Nair, K. Manikantan *materials scientist*

DISTRICT OF COLUMBIA

Washington
Imam, M. Ashraf *materials scientist, educator*
Wang, Franklin Fu Yen *materials scientist, educator*

FLORIDA

Gainesville
Abbaschian, Reza *materials science and engineering educator*
Ernsberger, Fred Martin *former materials scientist*
Holloway, Paul Howard *materials science educator*

GEORGIA

Marietta
Gerhardt, Rosario Alejandrina *materials scientist*

ILLINOIS

Argonne
Alexander, Dale Edward *materials scientist*
Erdemir, Ali *materials scientist*

Evanston
Chang, R. P. H. *materials science educator*
Cohen, Jerome Bernard *materials science educator*
Meshii, Masahiro *materials science educator*
Teng, Mao-Hua *materials scientist, researcher*
Vaynman, Semyon *materials scientist*
Weertman, Johannes *materials science educator*

Lisle
Wouch, Gerald *materials scientist*

Morris
Mirabella, Francis Michael, Jr. *polymer scientist*

Urbana
Rockett, Angus Alexander *materials science educator*
Veeraraghavan, Dharmaraj Tharuvai *materials scientist, researcher*

Willowbrook
Rothman, Alan Bernard *consultant, materials and components technologist*

INDIANA

West Lafayette
Sato, Hiroshi *materials science educator*

MARYLAND

Baltimore
Djordjevic, Borislav Boro *materials scientist, researcher*
Kruger, Jerome *materials science educator, consultant*

College Park
Wang, Liqin *materials scientist*

Gaithersburg
Nguyen, Tinh *materials scientist*
Schwartz, Lyle H. *materials scientist, government official*

MASSACHUSETTS

Billerica
Bathey, Balakrishnan R. *materials scientist*

Cambridge
Wang, Jian-Sheng *materials scientist*

Lexington
Tustison, Randal Wayne *materials scientist*

Salem
Khattak, Chandra Prakash *materials scientist*

Worcester
Apelian, Diran *materials scientist, provost*

MICHIGAN

East Lansing
Case, Eldon Darrel *materials science educator*

Midland
Lipowitz, Jonathan *materials scientist*
Serrano, Myrna *materials scientist, chemical engineer*
Shastri, Ranganath Krishna *materials scientist*

Warren
Meng, Wen Jin *materials scientist*
Smith, John Robert *materials scientist*

MINNESOTA

Saint Paul
Rzepecki, Edward Louis *packaging management educator*

NEW JERSEY

Glassboro
Otooni, Monde A. *materials scientist, physicist*

Monmouth Junction
Drzewinski, Michael Anthony *materials scientist*

Murray Hill
Johnson, David W., Jr. *ceramic scientist, researcher*

New Providence
MacChesney, John Burnette *materials scientist, researcher*

Parsippany
Hasegawa, Ryusuke *materials scientist*

Piscataway
Kear, Bernard Henry *materials scientist*

Summit
Sawyer, Linda Claire *materials scientist*

NEW MEXICO

Los Alamos
Kung, Pang-Jen *materials scientist, electrical engineer*

NEW YORK

Alfred
Fukuda, Steven Ken *materials science educator*
Shelby, James Elbert *materials scientist, educator*
Snyder, Robert Lyman *ceramic scientist, educator*

Briarcliff Manor
Mehrotra, Vivek *materials scientist*

Buffalo
Shui, Xiaoping *materials scientist*

Corning
Whitney, William Percy, II *materials scientist*

Ithaca
Ast, Dieter Gerhard *materials science educator*

Middletown
Zheng, Maggie (Xiaoci) *materials scientist*

New York
Osbourn, Gordon Cecil *materials scientist*
Pye, Lenwood David *materials science educator, researcher, consultant*

Rochester
Li, James Chen Min *materials science educator*

Tarrytown
Flanigen, Edith Marie *materials scientist*

Troy
Fradkov, Valery Eugene *materials scientist*

Webster
Bluhm, Terry Lee *materials scientist, chemist*
Pan, David Han-Kuang *polymer scientist*

West Valley
Palmer, Ronald Alan *materials scientist, consultant*

Yorktown Heights
Colgan, Evan George *materials scientist*
Cuomo, Jerome John *materials scientist*

NORTH CAROLINA

Raleigh
Narayan, Jagdish *materials science educator*

OHIO

Cleveland
Blackwell, John *polymer scientist, educator*
DeGuire, Mark Robert *materials scientist, educator*

Columbus
Daehn, Glenn Steven *materials scientist*
Gupta, Prabhat Kumar *materials scientist*

Dayton
Chuck, Leon *materials scientist*

Toledo
Bagley, Brian G. *materials science educator, researcher*

PENNSYLVANIA

Bethlehem
Lyman, Charles Edson *materials scientist, educator*

Breinigsville
Reynolds, C(laude) Lewis, Jr. *materials scientist, researcher*

Pittsburgh
Cao, You Sheng *materials science and engineering researcher*
Laughlin, David Eugene *materials science educator, metallurgical consultant*
Matway, Roy Joseph *material scientist*

University Park
Amateau, Maurice Francis *materials scientist, educator*
German, Randall Michael *materials science educator, consultant*
Newnham, Robert Everest *materials scientist, department chairman*
Roy, Rustum *interdisciplinary materials researcher, educator*

SOUTH CAROLINA

Clemson
Nevitt, Michael Vogt *materials scientist, educator*

Greer
Roldan, Luis Gonzalez *materials scientist*

TENNESSEE

Memphis
Ramey, Harmon Hobson, Jr. *materials scientist*

Nashville
Grugel, Richard Nelson *materials scientist*

Tullahoma
Dahotre, Narendra Bapurao *materials scientist, researcher, educator*

TEXAS

El Paso
McClure, John Casper *materials scientist, educator*

San Antonio
Page, Richard Allen *materials scientist*

UTAH

Salt Lake City
Speck, Kenneth Richard *materials scientist*

VERMONT

Saint Johnsbury
Taylor, Leland Alan *chemical and materials scientist, researcher*

VIRGINIA

Arlington
Vasudevan, Asuri Krishnaswami *materials scientist*

Charlottesville
Johnson, Robert Alan *materials science educator*

Petersburg
Correale, Steven Thomas *materials scientist*

Richmond
Deevi, Seetharama C. *materials scientist*

WASHINGTON

Kirkland
Williford, John Frederic, Jr. *materials research and development scientist*

Richland
Ramesh, Kalahasti Subrahmanyam *materials scientist*

WISCONSIN

Milwaukee
Tekkanat, Bora *materials scientist*

CANADA

ONTARIO

Hamilton
Purdey, Gary Rush *materials science and engineering educator, dean*

AUSTRALIA

Lucas Heights
Nowotny, Janusz *materials scientist*

BRAZIL

Rio de Janeiro
de Biasi, Ronaldo Sergio *materials science educator*

CHINA

Beijing
Liu, Bai-Xin *materials scientist, educator*

ENGLAND

Cambridge
Cottrell, Alan *materials scientist*
Hall, Christopher *materials scientist, researcher*

London
McLean, Malcolm *materials scientist, educator, researcher*

Ormskirk
Machlachlan, Julia Bronwyn *materials scientist, researcher*

FINLAND

Turku
Suoninen, Eero Juhani *materials science educator*

FRANCE

Paris
Bathias, Claude *materials science educator, consultant*

GERMANY

Bochum
Kneller, Eckart Friedrich *materials scientist, electrical engineer, educator, researcher*

Hannover
Phan-Tan, Tai *scientist, researcher, educator*

Stuttgart
Konuma, Mitsuharu *materials scientist*

JAPAN

Hitachi
Hashimoto, Tsuneyuki *materials scientist*

Kyoto
Shibayama, Mitsuhiro *materials science educator*

Sendai
Oikawa, Hiroshi *materials science educator*

Toyohashi
Kobayashi, Toshiro *materials science educator*

Yokohama
Tsuruta, Yutaka *building materials researcher*

THE NETHERLANDS

Delft
Katgerman, Laurens *materials scientist*

SWEDEN

Fagersta
Vuorinen Ruppi, Sakari Antero *materials scientist*

TAIWAN

Hsinchu
Chen, Lih-Juann *materials science educator*
Lee, Sanboh *materials scientist*

ADDRESS UNPUBLISHED

Coffey, Kevin Robert *materials scientist, researcher*
Jia, Quanxi *electrical and material scientist, researcher*
Lin, Otto Chui Chau *materials scientist, educator*
Vanderwalker, Diane Mary *materials scientist*

PHYSICAL SCIENCES: METALLURGY

UNITED STATES

CALIFORNIA

Berkeley
Thomas, Gareth *metallurgy educator*

Carson
Gittleman, Morris *consultant, metallurgist*

Livermore
Price, Clifford Warren *metallurgist, researcher*

San Diego
Klimowicz, Thomas F. *metallurgist*

San Jose
Bryhan, Anthony James *metallurgist*

COLORADO

Golden
Yarar, Baki *metallurgical educator*

CONNECTICUT

Storrs
Devereux, Owen Francis *metallurgy educator*

DISTRICT OF COLUMBIA

Washington
Cooper, Khershed Pessie *metallurgist*

FLORIDA

De Land
Liberman, Michael *metallurgist, researcher*

INDIANA

Hammond
Vojcak, Edward Daniel *metallurgist*

IOWA

Ames
Han, Sang Hyun *metallurgist*

MARYLAND

Gaithersburg
Cahn, John Werner *metallurgist, educator*

Rockville
Murray, Peter *metallurgist, manufacturing company executive*

MASSACHUSETTS

Cambridge
Eagar, Thomas Waddy *metallurgist, educator*

NEW JERSEY

Bridgewater
Albrethsen, Adrian Edysel *metallurgist, consultant*

NEW MEXICO

Albuquerque
Romig, Alton Dale, Jr. *metallurgist, educator*
Vianco, Paul Thomas *metallurgist*

Los Alamos
Hecker, Siegfried Stephen *metallurgist*

NEW YORK

Brooklyn
Whang, Sung H. *metallurgical science educator*

Hopewell Junction
Marcotte, Vincent Charles *metallurgist*

PENNSYLVANIA

Bethlehem
Stephenson, Edward Thomas *consulting metallurgist*

Brackenridge
Houze, Gerald Lucian, Jr. *metallurgist*

Philadelphia
MacDiarmid, Alan Graham *metallurgist, educator*

Wheatland
Bolt, Michael Gerald *metallurgist*

TENNESSEE

Kingston
Manly, William Donald *metallurgist*

TEXAS

Conroe
Robbins, Jessie Earl *metallurgist*

Houston
Chaku, Pran Nath *metallurgist*

UTAH

Salt Lake City
Miller, Jan Dean *metallurgy educator*
Olson, Ferron Allred *metallurgist, educator*

VIRGINIA

Washington
Ayers, Jack Duane *metallurgist*

WISCONSIN

Madison
Hesse, Thurman Dale *welding metallurgy educator, consultant*

ENGLAND

Oxford
Hirsch, Peter Bernhard *metallurgist*

FINLAND

Espoo
Korhonen, Antti Samuli *metallurgist, educator*

FRANCE

Orsay
Reich, Robert Claude *metallurgist, physicist*

JAPAN

Hiroshima
Ebara, Ryuichiro *metallurgist, researcher*

Tondabayashi
Nozato, Ryoichi *metallurgy educator, researcher*

Tsukuba
Hirano, Ken-ichi *metallurgist, educator*

RUSSIA

Moscow
Liakishev, Nikolai Pavlovich *metallurgist, materials scientist*

SWEDEN

Nykoping
Grounes, Mikael *metallurgist*

ADDRESS UNPUBLISHED

Brown, David Edwin *retired metallurgist, consultant*
Budinski, Kenneth Gerard *metallurgist*

PHYSICAL SCIENCES: METEOROLOGY

UNITED STATES

ALASKA

Fairbanks
Fathauer, Theodore Frederick *meteorologist*

CALIFORNIA

Davis
Stewart, James Ian *agrometeorologist*

La Jolla
Barnett, Tim P. *meteorologist*

Livermore
Ellsaesser, Hugh Walter *retired atmospheric scientist*

Northridge
Court, Arnold *climatologist*

Palm Springs
Krick, Irving Parkhurst *meteorologist*

Pasadena
Chahine, Moustafa Toufic *atmospheric scientist*

Santa Maria
Geise, Harry Fremont *retired meteorologist*

Thousand Oaks
Wang, I-Tung *atmospheric scientist*

Vacaville
Coulson, Kinsell Leroy *meteorologist*

COLORADO

Boulder
Albritton, Daniel L. *aeronomist*
Kellogg, William Welch *meteorologist*
Meehl, Gerald Allen *research climatologist*
Morris, Alvin Lee *meteorologist, retired consulting corporation executive*
Rotunno, Richard *meteorologist*
Schneider, Stephen Henry *climatologist, researcher*

CONNECTICUT

New Haven
Saltzman, Barry *meteorologist, educator*

DISTRICT OF COLUMBIA

Washington
Hallgren, Richard Edwin *meteorologist*
Johnson, David Simonds *meteorologist*
Rodenhuis, David Roy *meteorologist, educator*

FLORIDA

Miami
Dorst, Neal Martin *meteorologist, computer programmer*
Sheets, Robert Chester *meteorologist*

Tallahassee
O'Brien, James Joseph *meteorology and oceanography educator*

Tampa
Johnson, Anthony O'Leary (Andy Johnson) *meteorologist, consultant*

ILLINOIS

Chicago
Fujita, Tetsuya Theodore *educator, meteorologist*

O'Fallon
Jenner, William Alexander *meteorologist, educator*

INDIANA

West Lafayette
McClelland, Thomas Melville *meteorologist, researcher*

MARYLAND

College Park
Rasmusson, Eugene Martin *meteorology researcher*

Greenbelt
Simpson, Joanne Malkus *meteorologist*

Laurel
Meyer, James Henry *meteorologist*

MASSACHUSETTS

Beverly
Harris, Miles Fitzgerald *meteorologist*

Billerica
Miller, Dawn Marie *meteorologist, product marketing specialist*

Boston
Spengler, Kenneth C. *meteorologist, professional society administrator*

Cambridge
Emanuel, Kerry Andrew *earth sciences educator*
Lindzen, Richard Siegmund *meteorologist, educator*
Lorenz, Edward Norton *meteorologist, educator*

Lexington
Halberstam, Isidore Meir *meteorologist*

MISSOURI

Kansas City
McNulty, Richard Paul *meteorologist*
Ostby, Frederick Paul, Jr. *meteorologist, government official*

NEBRASKA

Lincoln
Hubbard, Kenneth Gene *climatologist*

Omaha
Silberberg, Steven Richard *meteorology educator*

NEW JERSEY

Princeton
Mahlman, Jerry David *research meteorologist*
Manabe, Syukuro *climatologist*
Oort, Abraham Hans *meteorologist, researcher, educator*
Smagorinsky, Joseph *meteorologist*

Union
Zois, Constantine Nicholas Athanasios *meteorology educator*

NEW YORK

Albany
Bosart, Lance F. *meteorology educator*

Utica
Harney, Patrick Joseph Dwyer *meteorologist, consultant*

NORTH CAROLINA

Asheville
Haggard, William Henry *meteorologist*

Raleigh
Arya, Satya Pal Singh *meteorology educator*

OKLAHOMA

Norman
Eilts, Michael Dean *research meteorologist, manager*
Lamb, Peter James *meteorology educator, researcher, consultant*
Maddox, Robert Alan *atmospheric scientist*

Oklahoma City
England, Gary Alan *television meteorologist*

PENNSYLVANIA

Coraopolis
Manuel, Phillip Earnest *meteorologist*

Harrisburg
Zook, Merlin Wayne *meteorologist*

SOUTH CAROLINA

Columbia
Gandy, James Thomas *meteorologist*

TEXAS

Dallas
Blattner, Wolfram Georg Michael *meteorologist*

Houston
Ledley, Tamara Shapiro *earth system scientist, climatologist*

VERMONT

Pittsford
Betts, Alan Keith *atmospheric scientist*

VIRGINIA

Alexandria
Perry, John Stephen *meteorologist*

Hampton
Deepak, Adarsh *meteorologist, atmospheric optician*
Smith, Mary-Ann Hrivnak *meteorologist, geophysicist*

Mc Lean
McConathy, Donald Reed, Jr. *meteorologist, remote sensing program manager, systems engineer*

WASHINGTON

Seattle
Wallace, John Michael *meteorology educator*

WISCONSIN

Madison
Eloranta, Edwin Walter *meteorologist, researcher*
Smith, William Leo *meteorologist, researcher, educator*

CANADA

ONTARIO

North York
Godson, Warren Lehman *meteorologist*

QUEBEC

Sainte Anne de Bellevue
Davies, Roger *geoscience educator*

HUNGARY

Budapest
Mészáros, Ernö *meteorologist, researcher, science administrator*

ADDRESS UNPUBLISHED

Jeck, Richard Kahr *research meteorologist*
Thuillier, Richard Howard *meteorologist*

PHYSICAL SCIENCES: OCEANOGRAPHY

UNITED STATES

CALIFORNIA

Del Mar
Reid, Joseph Lee *physical oceanographer, educator*

La Jolla
Buckingham, Michael John *oceanography educator*
Cox, Charles Shipley *oceanography researcher, educator*
Huntley, Mark Edward *biological oceanographer*
Knox, Robert Arthur *oceanographer, academic director*

Menlo Park
Nichols, Frederic Hone *oceanographer*

Oakland
Mikalow, Alfred Alexander, II *deep sea diver, marine surveyor, marine diving consultant*

Santa Cruz
Griggs, Gary Bruce *earth sciences educator, oceanographer, geologist, consultant*

Sherman Oaks
Pothitt, Kathleen Marie *physical oceanography researcher*

COLORADO

Boulder
Butler, James Hall *oceanographer, atmospheric chemist*

CONNECTICUT

Bloomfield
Brooks, Douglas Lee *retired oceanographic and atmospheric policy analyst, environmental policy consultant*

Groton
Cooper, Richard Arthur *oceanographer*
Fitzgerald, William F. *chemical oceanographer, educator*

Milford
Koch, Evamaria Wysk *oceanographer, educator*

DISTRICT OF COLUMBIA

Washington
Pyle, Thomas Edward *oceanographer, academic director*
Spilhaus, Athelstan Frederick, Jr. *oceanographer, association executive*

FLORIDA

Coral Gables
Stewart, Harris Bates, Jr. *oceanographer*

Fort Pierce
Mooney, John Bradford, Jr. *oceanographer, engineer, consultant*

Miami
Mooers, Christopher Northrup Kennard *physical oceanographer, educator*

GEORGIA

Savannah
Menzel, David Washington *oceanographer*

HAWAII

Honolulu
Wyrtki, Klaus *oceanography educator*

MAINE

West Boothbay Harbor
Sieracki, Michael Edward *biological oceanographer*

MARYLAND

Silver Spring
Baer, Ledolph *oceanographer, meteorologist*
Briscoe, Melbourne G. *oceanographer, administrator*
Ostenso, Ned Allen *oceanographer, government official*
Tokar, John Michael *oceanographer, ocean engineer*
Wilson, William Stanley *oceanographer*

MASSACHUSETTS

Cambridge
McCarthy, James Joseph *oceanography educator*

Falmouth
Hollister, Charles Davis *oceanographer*

Woods Hole
Anderson, Donald Mark *biological oceanographer*
Ballard, Robert Duane *marine scientist*
Dorman, Craig Emery *oceanographer, academic administrator*
Milliman, John D. *oceanographer, geologist*
Steele, John Hyslop *marine scientist, oceanographic institute administrator*
Stewart, William Kenneth, Jr. *ocean engineer*

MISSISSIPPI

Bay Saint Louis
Skramstad, Robert Allen *oceanographer*

Stennis Space Center
Hurlburt, Harley Ernest *oceanographer*
Lewando, Alfred Gerard, Jr. *oceanographer*
Sprague, Vance Glover, Jr. *oceanography executive, naval reserve officer*

NEW YORK

Palisades
Cane, Mark Alan *oceanography and climate researcher*

Stony Brook
Pritchard, Donald William *oceanographer*
Scranton, Mary Isabelle *oceanographer*

NORTH CAROLINA

Chapel Hill
Neumann, Andrew Conrad *geological oceanography educator*

OREGON

Corvallis
Caldwell, Douglas Ray *oceanographer, educator*

RHODE ISLAND

Kingston
Sigurdsson, Haraldur *oceanography educator, researcher*

Narragansett
Leinen, Margaret Sandra *oceanographic researcher*
Pilson, Michael Edward Quinton *oceanography educator*

TEXAS

Austin
Maxwell, Arthur Eugene *oceanographer, marine geophysicist, educator*

College Station
Rowe, Gilbert Thomas *oceanography educator*

VIRGINIA

Gloucester Point
Schaffner, Linda Carol *biological oceanography educator*

Norfolk
Csanady, Gabriel Tibor *oceanographer, meteorologist, environmental engineer*

WASHINGTON

Seattle
Bernard, Eddie Nolan *oceanographer*
Kunze, Eric *physical oceanographer, educator*
Miles, Edward Lancelot *marine studies educator, consultant, director*
Quay, Paul Douglas *oceanography educator*

WISCONSIN

Milwaukee
Strickler, John Rudi *biological oceanographer*

CANADA

BRITISH COLUMBIA

Sidney
Wong, Chi-Shing *chemical oceanographer*
Xie, Yunbo *oceanographer*

QUEBEC

Rimouski
Walton, Alan *oceanographer*

Westmount
Dunbar, Maxwell John *oceanographer, educator*

FRANCE

Paris
Cousteau, Jacques-Yves *marine explorer, film producer, writer*

GREECE

Athens
Stergiou, Konstantinos *biological oceanographer, researcher*

JAPAN

Osaka
Sakamoto, Ichitaro *oceanologist, consultant*

TAIWAN

Keelung
Chien, Yew-Hu *aquaculture educator*

ADDRESS UNPUBLISHED

Fish, John Perry *oceanographic company executive, historian*
Gilmartin, Malvern *oceanographer*

PHYSICAL SCIENCES: OPTICS

UNITED STATES

ARIZONA

Tucson
Peyghambarian, Nasser *optical science educator*

CALIFORNIA

Concord
Edgerton, Robert Frank *optical scientist*

Pasadena
Terhune, Robert William *optics scientist*

San Jose
Jutamulia, Suganda *electro-optic scientist*
Rabolt, John Francis *optics scientist*

Santa Ana
Nelson, Richard David *electro-optics professional*

Santa Barbara
Meinel, Aden Baker *scientist*

Torrance
Shirk, Kevin William *fiber optic product manager*

FLORIDA

Gainesville
Hope, George Marion *vision scientist*

Orlando
Li-Kam-Wa, Patrick *research optics scientist, consultant*

ILLINOIS

Chicago
Levine, Michael William *vision researcher, educator*

LOUISIANA

New Orleans
Smolek, Michael Kevin *optics scientist*

MARYLAND

Beltsville
Norris, Karl Howard *optics scientist, agricultural engineer*

MASSACHUSETTS

Cambridge
Baker, James Gilbert *optics scientist*

Lexington
Hardy, John W. *optics scientist*

MICHIGAN

Ann Arbor
Islam, Mohammed N. *optics scientist*

NEW JERSEY

Edison
Pagdon, William Harry *optical systems technician*

Holmdel
Stolen, Rogers Hall *optics scientist*

Murray Hill
Gordon, James Power *optics scientist*

NEW MEXICO

Los Alamos
Donohoe, Robert James *spectroscopist*

NEW YORK

Great Neck
Wang, Jian-Ming *research optics scientist*

Irvington
Elbaum, Marek *electro-optical sciences executive, researcher*

Rochester
Burns, Peter David *imaging scientist*
Kingslake, Rudolf *retired optical designer*
Margevich, Douglas Edward *spectroscopist*

Wappingers Falls
McCamy, Calvin Samuel *optics scientist*

TEXAS

Dallas
Anderson, Douglas Warren *optics scientist*

WASHINGTON

Seattle
Ray, Sankar *opto-electroic device scientist*

AUSTRALIA

Mulgrave
Porter, Colin Andrew *optics scientist*

ENGLAND

Colchester
Loudon, Rodney *optics scientist, educator*

Highfield
Hanna, David *optics scientist, educator*

Southhampton
Payne, David N. *optics scientist*

FRANCE

Paris
Haroche, Serge *optics scientist*

HONG KONG

Sha Tin NT
Cheung, Kwok-wai *optical network researcher*

PHYSICAL SCIENCES: OTHER SPECIALTIES

UNITED STATES

ALABAMA

Birmingham
Pritchard, David Graham *research scientist, educator*

Huntsville
Chappell, Charles Richard *space scientist*
Lee, Thomas J. *aerospace scientist*

Redstone Arsenal
Miller, Walter Edward *physical scientist, researcher*

ALASKA

Fairbanks
Guthrie, Russell Dale *vertebrate paleontologist*

ARIZONA

Flagstaff
Colbert, Edwin Harris *paleontologist, museum curator*

Riviera
Jones, Vernon Quentin *surveyor*

Tempe
Moore, Carleton Bryant *geochemistry educator*
Skibitzke, Herbert Ernst, Jr. *hydrologist*

ARKANSAS

Rogers
Parris, Luther Allen *gemologist, goldsmith*

CALIFORNIA

El Segundo
Radys, Raymond George *laser scientist*

Gardena
Zhou, Simon Zhengzhuo *laser scientist*

Livermore
Fortner, Richard J. *physical scientist*
Orel, Ann Elizabeth *applied science educator, researcher*
Shotts, Wayne J. *nuclear scientist, federal agency administrator*
Van Devender, J. Pace *physical scientist*

Los Angeles
Loeblich, Helen Nina Tappan *paleontologist, educator*
Wu, Robert Chung Yung *space sciences educator*

Manhattan Beach
Pringle, Ronald Sandy Alexander *seismic inspector*

Moffett Field
Ellis, Stephen Roger *research scientist*

North Hollywood
McGee, Sam *laser scientist*

Pacific Palisades
Gregor, Eduard *laser physicist*

Pacifica
Hitz, C. Breck *optics and laser scientist*

Pasadena
Crisp, Joy Anne *research scientist*

San Carlos
Zanoni, Michael McNeal *forensic scientist*

San Francisco
Dawdy, David Russell *hydrologist*

Santa Rosa
de Wys, Egbert Christiaan *geochemist*

Stanford
Olson, Darin S. *research scientist*

Sunnyvale
Devgan, Onkar Dave N. *technologist, consultant*

Woodside
Ashley, Holt *aerospace scientist, educator*

COLORADO

Boulder
Miller, Harold William *nuclear geochemist*

Denver
Cobban, William Aubrey *paleontologist*

Lakewood
Hill, Walter Edward, Jr. *geochemist, extractive metallurgist*

Littleton
Ross, Reuben James, Jr. *paleontologist*

CONNECTICUT

New Haven
Skinner, Helen Catherine Wild *biomineralogist*

Norwalk
Hannah, Robert Wesley *infrared spectroscopist*

DISTRICT OF COLUMBIA

Washington
Finger, Larry Wayne *crystallographer, mineralogist*
Johnston, Gordon Innes *aerospace program manager*
Prewitt, Charles Thompson *geochemist*
Weitzman, Stanley Howard *ichthyologist*
Yochelson, Ellis L(eon) *paleontologist*

FLORIDA

Daytona Beach
Simatos, Nicholas Jerry *aerospace sciences educator, consultant*

Gainesville
Halperin, Joseph *nuclear chemist*

Tallahassee
Miller, John Richard *scientist*

GEORGIA

Alpharetta
Clay, Stanton Tower *technical consultant*

Atlanta
Bragdon, Clifford Richardson *city planner, educator*
Nash, William Wray, Jr. *retired city planning educator*

IDAHO

Pocatello
Dykes, Fred William *retired nuclear scientist*

ILLINOIS

Chicago
Sereno, Paul C. *paleontologist, educator*

Elgin
Alvarez, Guillermo A. *research scientist*

Springfield
Henry, Theodore Lynn *nuclear scientist, educator*

LOUISIANA

Baton Rouge
Lambremont, Edward Nelson, Jr. *nuclear science educator*

MARYLAND

Aberdeen Proving Ground
Stuempfle, Arthur Karl *physical science manager*

Bethesda
Blush, Steven Michael *nuclear scientist, safety consultant*
Hennet, Remy Jean-Claude *geochemist, environmental consultant*
Snitch, Thomas Harold *science educator, consultant*
Zwanzig, Robert Walter *chemist, physical science educator*

Gaithersburg
Masters, Larry William *physical science administrator*

Greenbelt
McGee, Thomas Joseph *atmospheric scientist*
Parkinson, Claire Lucille *climatologist*

Laurel
Babin, Steven Michael *atmospheric scientist, researcher*

Rockville
Beattie, Donald A. *energy scientist, consultant*
Rinkenberger, Richard Krug *physical scientist*

Silver Spring
Whitten, Charles Arthur *geodetic consultant*

Solomons
Baranowski, Paul Joseph *nuclear instrumentation technician*

MASSACHUSETTS

Boston
Dolnikowski, Gregory Gordon *scientist*

Cambridge
Burns, Roger George *mineralogist, educator*
Holmberg, Eva Birgitta *research scientist*
Howitt, Andrew Wilson *scientist*
Sanders, John Lyell, Jr. *educator, researcher*
Witt, August Ferdinand *aerospace scientist, educator*

Lexington
Kestigian, Michael *scientist*

North Billerica
Wong, David Chungyao *scientist*

Taunton
Shastry, Shambhu Kadhambiny *scientist, engineering executive*

Woods Hole
Hart, Stanley Robert *geochemist, educator*

MICHIGAN

Ann Arbor
Gingerich, Philip Derstine *paleontologist, evolutionary biologist, educator*

Dearborn
Morris, John Michael *energy technology educator*

East Lansing
Wilson, Ronald Wayne *natural science educator*

Holland
McConnell, Anthony *polymer scientist, consultant*

MINNESOTA

Minneapolis
Chornenky, Victor Ivan *laser scientist, researcher*

MISSISSIPPI

Mississippi State
Monts, David Lee *scientist, educator*

Vicksburg
Lashlee, Jon David *physical scientist*

MISSOURI

Kansas City
Bergman, Carla Elaine *hydrologist, consultant*

Rolla
Shrestha, Bijaya *nuclear scientist*

MONTANA

Bozeman
Horner, John Robert *paleontologist, researcher*

Columbia Falls
Spade, George Lawrence *scientist*

NEVADA

Las Vegas
Peck, Gaillard Ray, Jr. *aerospace business and healthcare consultant*

NEW HAMPSHIRE

Hanover
Oreskes, Naomi *earth sciences educator, historian*

NEW JERSEY

Bridgewater
Ramharack, Roopram *research scientist*

Murray Hill
Lemaire, Paul Joseph *scientist*

New Providence
Sharp, Louis *scientist*

Princeton
Lemelson, Jerome H. *inventor*
Silver, Lee Merrill *science educator*

Sea Girt
Moore, Penelope Ann *forensic scientist*

Summit
Burbank, Robinson Derry *crystallographer*

Union Beach
Schreck, Lisa Tanzman *sensory analyst*

NEW MEXICO

Albuquerque
Ewing, Rodney Charles *mineralogist, geology educator, materials scientist*

Los Alamos
Bish, David Lee *mineralogist*
Fisk, Zachary *physical scientist*
Hakkila, Eero Arnold *nuclear safeguards technology chemist*
Tait, Carleton Drew *geochemist*
Trocki, Linda Katherine *geoscientist, natural resource economist*

NEW YORK

Brooklyn
Langer, Arthur Mark *mineralogist*
Mendelson, Sol *physical science educator, consultant*
Teraoka, Iwao *polymer scientist*

Buffalo
Guo, Dongyao *crystallographer, researcher*

Corning
Gulati, Suresh Thakurdas *glass scientist, researcher*

Flushing
Finks, Robert Melvin *paleontologist, educator*

Ithaca
Salpeter, Edwin Ernest *physical sciences educator*

New York
Lynch, John Joseph *city and regional planner, consultant*
Mader, Bryn John *vertebrate paleontologist*
Mc Kenna, Malcolm Carnegie *vertebrate paleontologist, curator, educator*
Norell, Mark Allen *paleontology educator*
Worgul, Basil Vladimir *radiation scientist*

Rochester
Greener, Jehuda *polymer scientist*

Stony Brook
McLennan, Scott Mellin *geochemist, educator*

Troy
Tham, Fook Siong *crystallographer*

Upton
Pogorelsky, Igor Vladislav *laser physicist*

Webster
Drumheller, Jerry Paul *laser scientist, development engineer*

White Plains
Marshall, L(uther) Gerald *acoustic consultant*

OHIO

Bellefontaine
Reames, Spencer Eugene *science educator*

Dayton
Heneghan, Shawn Patrick *research scientist, educator*
von Ohain, Hans Joachim P. *aerospace scientist*

Kettering
Haas, Trice Walter *researcher*

Sylvania
Kurek, Dolores Bodnar *physical science and mathematics educator*

Toledo
Skrzypczak-Jankun, Ewa *crystallographer*

OREGON

Ashland
Goddard, Kenneth William *forensic scientist*

PENNSYLVANIA

Hershey
Demers, Laurence Maurice *science educator*

Mechanicsburg
Windsor, Ted Bacon *feed microscopy consultant*

Pittsburgh
Hillebrand, Julie Ann *biotechnology executive*
Sax, Martin *crystallographer*

University Park
Barnes, Hubert Lloyd *geochemistry educator*
Cuffey, Roger J. *paleontology educator*
Smith, Deane Kingsley, Jr. *mineralogy educator*
White, William Blaine *geochemist, educator*

TENNESSEE

Knoxville
Panjehpour, Masoud *research scientist, educator*

Oak Ridge
Mehrotra, Subhash Chandra *hydrologist, consultant*
Wesolowski, David Jude *geochemist*

Tullahoma
Wu, Jie-Zhi *fluid mechanist*

TEXAS

Austin
Langston, Wann, Jr. *paleontologist, educator, researcher emeritus*
Lundelius, Ernest Luther, Jr. *vertebrate paleontologist, educator*
Sprinkle, James Thomas *paleontologist, educator*

College Station
Frenz, Bertram Anton *crystallographer*

Dallas
Gerken, George Manz *neuroscientist, educator*

Houston
Clark, Carolyn Archer *technologist, scientist*
Gibson, Robert Lee *astronaut*
Soileau, Kerry Michael *aerospace technologist, researcher*

Univ Of Texas At Arlington
Wang, Shu-Shaw (Peter Wang) *educator*

VIRGINIA

Arlington
Roy, Gabriel Delvis *scientific researcher, research manager*

Herndon
Marshall, Larry Ronald *laser physicist*

Reston
Riehle, James Ronald *volcanologist*

Surry
Enroughty, Christopher James *nuclear chemistry technician*

WASHINGTON

Issaquah
Ford, Dennis Harcourt *laser systems consultant*

Richland
Austin, Inez J. *nuclear scientist*

WEST VIRGINIA

Masontown
Wilson, Dale William, Jr. *laser technician*

WISCONSIN

Madison
Radwin, Robert Gerry *science educator, researcher, consultant*

Racine
Neumiller, Phillip Joseph, III *research scientist*

WYOMING

Casper
Rasmussen, Niels Lee *geologist, environmental chemist*

CANADA

ONTARIO

Downsview
Tennyson, Roderick C. *aerospace scientist*

London
Fyfe, William Sefton *geochemist, educator*

Willowdale
Rhodes, Wayne Robert *ergonomist, consultant*

QUEBEC

Pointe Claire
Sherry, Cameron William *occupational hygienist, consultant*

AUSTRALIA

Canberra
Evans, Denis James *research scientist*

Melbourne
Allardice, David John *fuel technologist*

Ryde
Caffin, Roger Neil *research scientist, consultant*

Wembley
Graham, James *mineralogist*

BELGIUM

Brussels
Nicolis, Gregoire *science educator*

CHINA

Beijing
Chen, Naixing *thermal science educator, researcher*

DENMARK

Hoersholm
Jensen, Ole *energy researcher*

ENGLAND

Barnwood
Collier, John Gordon *nuclear scientist*

Hayes
Hounsfield, Godfrey Newbold *scientist*

London
Kolb, Noel Joseph *research scientist*

FRANCE

Gif sur Yvette
Jouzel, Jean *researcher*

Paris
Curien, Hubert *mineralogy educator*

Saclay
Teixier, Annie Mireille J. *research scientist*

GERMANY

Aachen
Nastase, Adriana *aerospace scientist, educator, researcher*

Berlin
Saenger, Wolfram Heinrich Edmund *crystallography educator*

Frankfurt am Main
Baur, Werner Heinz *mineralogist, educator*

Königswinter
Ternyik, Stephen *polytechnology researcher*

Ludwigshafen am Rhein
Laun, Hans Martin *rheology researcher*

Munich
Wess, Julius *nuclear scientist*

Stuttgart
Dittrich, Herbert *mineralogist, researcher*

Wiesbaden
Braun, Michael Walter *technology consultant*

INDIA

Bombay
Shikarkhane, Naren Shriram *laser scientist*

JAPAN

Ichihara
Niu, Keishiro *science educator*

Otsu
Wada, Eitaro *biogeochemist*

Tokyo
Kigoshi, Kunihiko *geochemistry educator*
Torii, Tetsuya *retired science educator*

THE NETHERLANDS

Delft
Peters, Charles Martin *research and development scientist, consultant*

NORWAY

Grimstad
Conway, John Thomas *computational fluid dynamacist, educator*

Oslo
Østrem, Gunnar Muldrup *glaciologist*

Trondheim
Saether, Ola Magne *geochemist*

REPUBLIC OF KOREA

Seoul
Yun-Choi, Hye Sook *natural products science educator*

ROMANIA

Bucharest
Mihaileanu, Andrei Calin *energy researcher*

RUSSIA

Moscow
Anodina, Tatyana Grigoryevna *aviation expert*
Feschenko, Alexander *nuclear scientist*
Kliouev, Vladimir Vladimirovitch *control systems scientist*

SAUDI ARABIA

Riyadh
Vora, Manhar Morarji *radiopharmaceutical scientist*

SWITZERLAND

Geneva
Parthé, Erwin *crystallographer, educator*

Ruschlikon Zurich
Bednorz, J. Georg *crystallographer*

Zurich
Gruen, Armin *photogrammetry educator*

ADDRESS UNPUBLISHED

Holub, Robert Frantisek *nuclear chemist, physicist*
Madueme, Godswill C. *nuclear scientist, international safeguards agency administrator*
Parkinson, William Quillian *paleontologist*
Peters, Randy Alan *scientist*
Ramachandran, Narayanan *aerospace scientist*
Tate, Manford Ben *guided missile scientist, investor*
Zhou, Ming De *aeronautical scientist, educator*

PHYSICAL SCIENCES: PHYSICS

UNITED STATES

ALABAMA

Birmingham
Sobol, Wlad Theodore *physicist*

Huntsville
Campbell, Jonathan Wesley *astrophysicist, aerospace engineer*
Carter, Regina Roberts *physicist*
Decher, Rudolf *physicist*
Eagles, David M. *physicist*
Hadaway, James Benjamin *physicist*
Jaenisch, Holger Marcel *physicist*
Montgomery, Willard Wayne *physicist*
Musielak, Zdzislaw Edward *physicist, educator*
Perkins, James Francis *physicist*
Reiss, Donald Andrew *physicist*
Roberts, Thomas George *retired physicist*
Stone, Richard John *physicist*
Witherow, William Kenneth *physicist*
Zwiener, James Milton *physicist*

Madison
Rosenberger, Franz Ernst *physics educator*

Mobile
Harpen, Michael Dennis *physicist, educator*

Normal
Caulfield, Henry John *physics educator*
Wang, Jai-Ching *physics educator*

Tuscaloosa
Clavelli, Louis John *physicist*

ARIZONA

Coolidge
Hiller, William Clark *physics educator, engineering educator, consultant*

Sedona
Stamm, Robert Franz *research physicist*

Tempe
Cowley, John Maxwell *physics educator*
Marcus, David Alan *physicist*
Nigam, Bishan Perkash *physics educator*
Tillery, Bill W. *physics educator*

Tucson
Burrows, Adam Seth *physicist*
Carruthers, Peter Ambler *physicist, educator*
Hill, Henry Allen *physicist, educator*
Jackson, Kenneth Arthur *physicist, researcher*
Kessler, John Otto *physicist, educator*
Kilkson, Rein *physics educator*
Lamb, Lowell David *physicist*
Lamb, Willis Eugene, Jr. *physicist, educator*

ARKANSAS

Little Rock
Lyublinskaya, Irina E. *physicist, researcher*

CALIFORNIA

Anaheim
Aronowitz, Frederick *physicist*

Berkeley
Alonso, Jose Ramon *physicist*
Attwood, David Thomas *physicist, educator*
Chamberlain, Owen *nuclear physicist*
Daftari, Inder Krishen *physicist, researcher*
Fowler, Thomas Kenneth *physicist*
Gaillard, Mary Katharine *physics educator*
Gardner, Wilford Robert *physicist, educator*
Glaser, Donald A(rthur) *physicist*
Goldhaber, Gerson *physicist, educator*
Hahn, Erwin Louis *physicist, educator*
Helmholz, August Carl *physicist, educator emeritus*
Leemans, Wim Pieter *physicist*
Mandelstam, Stanley *physicist*
McKinney, Wayne Richard *physicist*
Shen, Yuen-Ron *physics educator*
Siri, William E. *physicist*
Symons, Timothy James McNeil *physicist*
Townes, Charles Hard *physics educator*
Weber, Eicke Richard *physicist*
Westphal, Andrew Jonathan *physicist, researcher*
Zumino, Bruno *physics educator, researcher*

Burlingame
Hotz, Henry Palmer *physicist*

Canoga Park
Carman, Robert Lincoln, Jr. *physicist*
Holmes, Richard Brooks *mathematical physicist*

Chula Vista
Smith, Peggy O'Doniel *physicist, educator*

Claremont
Brown, Robert James Sidford *physicist*

Danville
Peterson, Jack Milton *retired physicist*

Davis
Cahill, Thomas Andrew *physicist, educator*
Knox, William Jordan *physicist*

Duarte
DuBridge, Lee Alvin *physicist*

El Dorado Hills
Todd, Terry Ray *physicist*

El Segundo
Liaw, Haw-Ming (Charles) *physicist*
Lomheim, Terrence Scott *physicist*

Fountain Valley
Turkel, Solomon Henry *physicist*

Fremont
Khan, Mahbub R. *physicist*

Glendale
Farmer, Crofton Bernard *atmospheric physicist*

Goleta
Jones, Colin Elliott *physicist, educator*

Hayward
Pearce-Percy, Henry Thomas *physicist*

Irvine
Juhasz, Tibor *physicist, researcher*
Moe, Michael K. *physicist*
Reines, Frederick *physicist, educator*

La Jolla
Abarbanel, Henry Don Isaac *physicist, academic director*
Anderson, Victor Charles *applied physics educator*
Braun, Hans-Benjamin *physicist*
Deane, Grant Biden *physicist*
Driscoll, Charles F. *research physicist*
Feher, George *physics and biophysics scientist, educator*
Goodkind, John Morton *physics educator*
Kerr, Donald MacLean, Jr. *physicist*
Morikis, Dimitrios *physicist*
O'Neil, Thomas Michael *physicist, educator*
Ride, Sally Kristen *physics educator, scientist, former astronaut*
Rotenberg, Manuel *physics educator*
Sham, Lu Jeu *physics educator*
Stern, Martin O(scar) *physicist, consultant*
York, Herbert Frank *physics educator*

Laguna Hills
Batdorf, Samuel B(urbridge) *physicist*

Livermore
Alder, Berni Julian *physicist*
Baldwin, David E. *physicist*
Chapline, George Frederick, Jr. *theoretical physicist*
Correll, Donald Lee, Jr. *physicist*
Drake, Richard Paul *physicist, educator*
Erskine, David John *physicist*
Hammer, James Henry *physicist*
Kidder, Ray Edward *physicist, consultant*
Kulander, Kenneth Charles *physicist*
McMillan, Charles Frederick *physicist*
Milanovich, Fred Paul *physicist*
Nuckolls, John Hopkins *physicist, researcher*
Picraux, Samuel Thomas *physics researcher*
Reitze, David Howard *physicist*
Struble, Gordon Lee *physicist, researcher*
Warshaw, Stephen Isaac *physicist*

Los Angeles
Braginsky, Stanislav Iosifovich *physicist, geophysicist, researcher*
Burns, Marshall *physicist*
Byers, Nina *physics educator*
Campos, Joaquin Paul, III *chemical physicist, regulatory affairs specialist*
Chen, Francis F. *physics and engineering educator*
Cline, David Bruce *physicist, educator*
Coleman, Paul Jerome, Jr. *physicist, educator*
Conn, Robert William *applied physics educator*
Cornwall, John Michael *physics educator, consultant, researcher*
Dawson, John Myrick *plasma physics educator*
Decyk, Viktor Konstantyn *research physicist, consultant*
Domaradzki, Julian Andrzej *physics educator*
Gruntman, Michael A. *physicist, researcher, educator*
Jiang, Hongwen *physicist*
Johnson, Charles Erik *physics researcher*
Kennel, Charles Frederick *physicist, educator*
Kivelson, Margaret Galland *physicist*
Knize, Randall James *physics educator*
Kunc, Joseph Anthony *physics and engineering educator, consultant*
Maki, Kazumi *physicist, educator*
Schwinger, Julian *physicist, educator*
Senitzky, Israel Ralph *physicist*
Walker, Raymond John *physicist*
Wittry, David Beryle *physicist, educator*

Los Angeles AFB
Moe, Osborne Kenneth *physicist*

Los Gatos
Burnett, James Ray *physicist, consultant*

Malibu
Chester, Arthur Noble *physicist*
Hasenberg, Thomas Charles *physicist*

Menlo Park
Darrah, James Gore *physicist, financial executive, real estate developer*
Tokheim, Robert Edward *physicist*

Milpitas
Lee, Kenneth *physicist*

Montecito
Wheelon, Albert Dewell *physicist*

Monterey
Atchley, Anthony Armstrong *physicist, educator*
Rockower, Edward B. *physicist, operations researcher, consultant*

Moorpark
Monteiro, Sergio Lara *physics educator*

Orange
Talbott, George Robert *physicist, mathematician, educator*

Palo Alto
Aberth, William Henry *physicist*
Bienenstock, Arthur Irwin *physicist, educator*
Cutler, Leonard Samuel *physicist*
Drell, Sidney David *physicist, educator*
Fisher, Thornton Roberts *physicist*
Hartman, Keith Walter *physicist*
Ingham, David R. *physicist*
MacDonald, Alexander Daniel *physics consultant*
Panofsky, Wolfgang Kurt Hermann *physicist, educator*
Richter, Burton *physicist, educator*
Schneider, Thomas R(ichard) *physicist*
Taimuty, Samuel Isaac *physicist*
Taylor, Richard Edward *physicist, educator*
Van de Walle, Chris Gilbert *physicist*
Varney, Robert Nathan *retired physicist, researcher*
Verdonk, Edward Dennis *physicist*

Pasadena
Cagin, Tahir *physicist*
Culick, Fred Ellsworth Clow *physics and engineering educator*
Fowler, William Alfred *retired physics educator*
Frautschi, Steven Clark *physicist, educator*
Gaskell, Robert Weyand *physicist*
Gell-Mann, Murray *theoretical physicist, educator*
Hitlin, David George *physicist, educator*
Jastrow, Robert *physicist*
Kanamori, Hiroo *physics and astronomy educator*
Kannan, Rangaramanujam *polymer physicist, chemical engineer*
Kirby, Shaun Keven *physicist*
Liepmann, Hans Wolfgang *physicist, educator*
Lo, Shui-yin *physicist*
Man, Kin Fung *physicist, researcher*
McGill, Thomas Conley *physics educator*
Schwarz, John Henry *theoretical physicist, educator*
Stone, Edward Carroll *physicist, educator*
Thompson, Robert James *physicist*
Thorne, Kip Stephen *physicist, educator*
Vining, Cronin Beals *physicist*
Vogt, Rochus Eugen *physicist, educator*

Pico Rivera
Ling, Rung Tai *physicist*

Pine Valley
Liddiard, Glen Edwin *physicist*

Pomona
Aurilia, Antonio *physicist*
Eagleton, Robert Don *physics educator*

Redondo Beach
Ball, William Paul *physicist, engineer*
Foster, John Stuart, Jr. *physicist, former defense industry executive*
Roth, Thomas J. *physicist*
Schwarzbek, Stephen Mark *physicist*

Redwood City
Murthy, Sudha Akula *physicist, consultant*

Riverside
Fung, Sun-Yiu Samuel *physics educator*
Karlow, Edwin Anthony *physicist, educator*
White, Robert Stephen *physics educator*

Sacramento
Dedrick, Kent Gentry *retired physicist, researcher*

San Diego
Ennis, Joel Brian *physicist*
Greene, John M. *physicist*
Luce, Timothy Charles *physicist*
Mooradian, Gregory Charles *physicist*
Overskei, David *physicist*
Paz, Nils *physicist*
Waltz, Ronald Edward *physicist*

San Jose
Eigler, Donald Mark *physicist*
Gruber, John Balsbaugh *physics educator, university administrator*
Gunter, William Dayle, Jr. *physicist*
Lakkaraju, Harinarayana Sarma *physics educator, consultant*
Mizrah, Len Leonid *physicist*
Rudge, William Edwin, IV *computational physicist*
Scifres, Donald Ray *physics and engineering administrator*

San Luis Obispo
Schumann, Thomas Gerald *physics educator*

San Mateo
Burke, Richard James *optical physicist, consultant*

Santa Barbara
Awschalom, David Daniel *physicist*
Dudziak, Walter Francis *physicist*
Gutsche, Steven Lyle *physicist*
Heeger, Alan Jay *physicist*
Kohn, Walter *educator, physicist*
Luyendyk, Bruce Peter *geophysicist, educator, institution administrator*
Morrison, Rollin John *physicist, educator*
Peale, Stanton Jerrold *physics educator*
Witherell, Michael S. *physics educator*

Santa Clara
Gozani, Tsahi *nuclear physicist*
Mand, Ranjit Singh *device physicist, educator*
Nowak, Romuald *physicist*

Santa Monica
Intriligator, Devrie Shapiro *physicist*
Shipbaugh, Calvin LeRoy *physicist*

Veneklasen, Paul Schuelke physicist, acoustics consultant

Sherman Oaks
Wróblewski, Ronald John physicist

Solana Beach
Agnew, Harold Melvin physicist

Stanford
Bai, Taeil Albert research physicist
Bashaw, Matthew Charles physicist
Beasley, Malcolm Roy physics educator
Chang-Hasnain, Constance Jui-Hua educator
Chu, Steven physics educator
Cutler, Cassius Chapin physicist, educator
Gliner, Erast Boris theoretical physicist
Harrison, Walter Ashley physicist, educator
Herrmannsfeldt, William Bernard physicist
Lathrop, Kaye Don nuclear scientist, educator
Sa, Luiz Augusto Discher physicist
Schawlow, Arthur Leonard physicist, educator
Teller, Edward physicist
Winick, Herman physicist

Sunland
Karney, James Lynn physicist, optical engineering consultant

Sunnyvale
Bullis, W(illiam) Murray physicist, consultant
Herman, Michael Harry physicist, researcher
Martinez-Galarce, Dennis Stanley physicist

Thousand Oaks
Newman, Paul Richard physicist

Torrance
Sun, Zongjian physicist, researcher

COLORADO

Boulder
Archambeau, Charles Bruce physics educator, geophysics research scientist
Benz, Samuel Paul physicist
Cornman, Larry Bruce physicist
Dunn, Gordon Harold physicist
Goldfarb, Ronald B. research physicist
Hall, John Lewis physicist, researcher
Hildner, Ernest Gotthold, III solar physicist, science administrator
Kamper, Robert Andrew physicist
Low, Boon Chye physicist
Phelps, Arthur Van Rensselaer physicist, consultant
Smythe, William Rodman physicist, educator
Sparks, Larry Leon physicist
Wieman, Carl E. physics educator

Colorado Springs
Burciaga, Juan Ramon physics educator
Lee, Kotik Kai physicist

Denver
Brown, Mark Steven medical physicist
Creech-Eakman, Michelle Jeanne physicist, educator
Drăgoi, Dănut physicist
Liu, Chaoqun staff scientist
Murcray, Frank James physicist

Golden
Bertness, Kristine Ann physicist

Grand Junction
Misra, Prasanta Kumar physics educator

Lakewood
Pitts, John Roland physicist

Louisville
Schemmel, Terence Dean physicist

University Of Colorado
Greene, Chris H. physicist, educator

CONNECTICUT

Bridgeport
Chih, Chung-Ying physicist, consultant
Tucci, James Vincent physicist

Danbury
Tittman, Jay physicist, consultant

Farmington
Olson, David P. physicist

Glastonbury
Grubin, Harold Lewis physicist, researcher

Hartford
Lindsay, Robert physicist, educator

Higganum
Marcus, Jules Alexander physicist, educator

New Haven
Adair, Robert Kemp physicist, educator
Bromley, David Allan physicist, educator
Chubukov, Andrey Vadim physicist
Iachello, Francesco physicist educator
Ibbott, Geoffrey Stephen physicist
Klein, Martin Jesse physicist, educator, historian of science
Nath, Ravinder physicist
Parker, Peter D.M. physicist, educator, researcher
Rose, Marian Henrietta physics educator
Zhong, Jianhui physicist

New London
Browning, David Gunter physicist
Moffett, Mark Beyer physicist

Old Lyme
Anderson, Theodore Robert physicist

Ridgefield
Beck, Vernon David physicist, consultant
Sen, Pabitra N. physicist, researcher
Zeroug, Smaine electrophysicist

Shelton
Adank, James P. physicist, administrator
Zeller, Claude physicist, researcher

West Hartford
Goldstein, Rubin consulting physicist

West Redding
Foster, Edward John engineering physicist

DELAWARE

Newark
Böer, Karl Wolfgang physicist, educator
Bragagnolo, Julio Alfredo physicist
Daniels, William Burton physicist, educator
Eyler, Edward Eugene physicist, educator
Gangopadhyay, Sunita Bhardwaj physicist, researcher
Morgan, John Davis physicist
Smith, Charles William physicist

Wilmington
Blanchet-Fincher, Graciela Beatriz physicist
Kobsa, Henry research physicist

DISTRICT OF COLUMBIA

Washington
Abelson, Philip Hauge physicist
Abraham, George research physicist, engineer
Beall, James Howard physicist, educator
Blanc, Theodore Von Sickle physicist
Britt, Harold C. physicist
Brown, Louis physicist, researcher
Carrico, John Paul physicist
Chow, Gan-Moog physicist
Chubb, Scott Robinson research physicist
Coffey, Timothy physicist
Crandall, David Hugh physicist
Crannell, Hall Leinster physics educator
Culbertson, James Clifford physicist
Cunningham, LeMoine Julius physicist
Dickman, Robert Laurence physicist, researcher
Edelstein, Alan Shane physicist
Erb, Karl Albert physicist, government official
Fainberg, Anthony physicist
Feldman, Charles physicist
Fisk, Lennard Ayres physicist, educator
Fogleman, Guy Carroll physicist, mathematician, educator
Friedman, Herbert physicist
Friedman, Moshe research physicist
Gibbons, John Howard (Jack) physicist, government official
Haskins, Caryl Parker scientist, author
Karle, Jerome research physicist
Kirchhoff, William Hayes chemical physicist
Kirwan, Gayle M. physics educator
Lambert, James Morrison physics educator
Lord, Norman William physicist, consultant
Lozansky, Edward Dmitry physicist, consultant
Marlay, Robert Charles physicist, engineer
McGrory, Joseph Bennett physicist
Nelson, David Brian physicist
Obenauer, John Charles physicist
Oertel, Goetz K. H. physicist, professional association administrator
Pecora, Louis Michael physicist
Qadri, Syed Burhanullah physicist
Quinn, Jarus William physicist, association executive
Raab, Harry Frederick, Jr. physicist
Raine, William Alexis physicist
Rife, Jack Clark physicist
Rodgers, James Earl physicist, educator
Romanowski, Thomas Andrew physics educator
Sorrows, Howard Earle executive, physicist
Tousey, Richard physicist
White, Robert Marshall physicist, government official, educator
Youtcheff, John Sheldon physicist

FLORIDA

Boca Raton
Ding, Mingzhou physicist
Qiu, Shen Li physicist
Schroeck, Franklin Emmett, Jr. mathematical physicist

Cape Canaveral
McCluney, (William) Ross physics researcher, technical consultant

Gainesville
Hanson, Harold Palmer physicist, government official, editor, academic administrator
Seiberling, Lucy Elizabeth physicist
Sullivan, Neil Samuel physicist, researcher, educator
Tanner, David Burnham physics educator
Tavano, Frank physicist, electrical engineer
Trickey, Samuel Baldwin physics educator, researcher

Miami
Ashkenazi, Josef physicist
Blanco, Luciano-Nilo physicist
Chang, Francis medical physicist
Tappert, Frederick Drach physicist

Orlando
Blue, Joseph Edward physicist
Breakfield, Paul Thomas, III physicist
Eligon, Ann Marie Paula physicist
Mangold, Vernon Lee physicist
Piquette, Jean Conrad physicist
Ting, Robert Yen-ying physicist

Saint Petersburg
Rester, Alfred Carl, Jr. physicist

Tallahassee
Berg, Bernd Albert physics educator
Herndon, Roy Clifford physicist
Kemper, Kirby Wayne physics educator
Owens, Joseph Francis, III physics educator
Schrieffer, John Robert physics educator, science administrator
Walton, Jeffrey Howard physicist

Tampa
Hickman, Hugh Vernon physics educator

Winter Haven
Mandal, Krishna Pada radiation physicist

GEORGIA

Atlanta
Barnett, Robert Neal physicist, researcher
Chou, Mie-Yin physicist
Crowe, Devon George physicist, engineering consultant
Eichholz, Geoffrey G(unther) physics educator
Erbil, Ahmet physics educator
Fox, Ronald Forrest physics educator
Gersch, Harold Arthur physics educator
Martin, Kevin Paul physicist
McMillan, Robert Walker physicist, consultant
Rashford, David R. physicist, educator
Shen, Kangkang research physicist

Roswell
Arnold, Jack Waldo physicist

Statesboro
Dean, Cleon Eugene physicist
Mobley, Cleon Marion, Jr. (Chip Mobley) physics educator, real estate executive

IDAHO

Idaho Falls
Long, John Kelley nuclear reactor physicist, consultant
Ramshaw, John David chemical physicist, mechanical engineer

Moscow
Stumpf, Bernhard Josef physicist

ILLINOIS

Argonne
Blair, Robert Eugene physicist, researcher
Coffey, Howard Thomas physicist
Dunford, Robert Walter physicist
Jorgensen, James Douglas research physicist
Lawson, Robert Davis theoretical nuclear physicist
Routbort, Jules Lazar physicist, editor
Toppel, Bert Jack reactor physicist

Batavia
Cooper, Peter Semler physicist
Rapidis, Petros A. research physicist
Tollestrup, Alvin Virgil physicist

Bloomington
Jaggi, Narendra K. physics educator, researcher

Cahokia
Redmount, Ian H. physicist

Carbondale
Ali, Naushad physicist
Chen, Tian-Jie physicist, educator
Tao, Rongjia physicist, educator

Cary
White, William research physicist

Champaign
Slichter, Charles Pence physicist, educator

Chicago
Chakravarthy, Sreenathan Ramanna physicist
Chang, Wei medical physicist
Chu, James Chien-Hua medical physicist, educator
Cronin, James Watson physicist, educator
Fritzsche, Hellmut physics educator
Gofron, Kazimierz Jan physicist
Kadanoff, Leo Philip physicist
Lederman, Leon Max physicist, educator
Mazenko, Gene Francis physics educator
Meyer, Peter physicist, educator
Muller, Dietrich Alfred Helmut physicist, educator
Parker, Eugene Newman physicist, educator
Schutt, Jeffry Allen physicist
Shawhan, Peter Sven physicist
Simpson, John Alexander physicist
Thompson, Robert W. theoretical physicist
Zhao, Meishan chemical physics educator, researcher

De Kalb
Rosenmann, Daniel physicist, educator
Rossing, Thomas D. physics educator

Elmhurst
Betinis, Emanuel James physics and mathematics educator

Evanston
Dutta, Pulak physicist, educator
Garg, Anupam K. physicist
Udler, Dmitry physicist, educator

Glenview
Savic, Stanley Dimitrius physicist

Hinsdale
Karplus, Henry Berthold physicist, research engineer

Normal
Young, Robert Donald physicist, educator

Oak Park
Vandervoort, Kurt George physicist

Urbana
Chiang, Tai-Chang physics educator
Choe, Won-Ho (Wayne) plasma physicist
Ginsberg, Donald Maurice educator, physicist
Goldbart, Paul Mark theoretical physicist, educator
Greene, Laura Helen physicist
Klein, Miles Vincent physics educator
Nayfeh, Munir Hasan physicist
Schweizer, Kenneth Steven physics educator

Wheaton
Sloan, Michael Lee physics and computer science educator, author

INDIANA

Bloomington
Pollock, Robert Elwood nuclear physicist
Szymanski, John James physicist, educator

Indianapolis
Jones, Katharine Jean research physicist
Liu, Pingyu physicist, educator

Merrillville
Chang, Kai Siung medical physicist

Terre Haute
Western, Arthur Boyd, Jr. physics educator

West Lafayette
Overhauser, Albert Warner physicist

Westville
Das, Purna Chandra physics and mathematics educator

IOWA

Ames
Cao, Zhijun physicist
Clem, John Richard physicist, educator
Hill, John Christian physics educator
Kelly, William Harold physicist, physics educator
Knox, Ralph David physicist
Luban, Marshall physicist
Ostenson, Jerome Edward physicist

Dubuque
Schaefer, Joseph Albert physics and engineering educator, consultant

Fairfield
Hagelin, John Samuel theoretical physicist

Iowa City
Van Allen, James Alfred physicist, educator

KANSAS

Lawrence
Kwak, Nowhan physics educator
Ralston, John Peter theoretical physicist, educator

KENTUCKY

Hopkinsville
Spencer, Lewis VanClief retired physicist

Lexington
Ng, Kwok-Wai physics educator
Standler, Ronald B. physicist

Louisville
Park, Chang Hwan physicist, educator

Paducah
Walden, Robert Thomas physicist, consultant

LOUISIANA

Baton Rouge
Callaway, Joseph physics educator
Craft, Benjamin Cole, III accelerator physicist
Fan, J.D. (Jiangdi) physics educator
Luckett, John Paul, Jr. medical radiation physicist
Rupnik, Kresimir physicist, researcher

New Orleans
McGuire, James Horton physics educator
Rosensteel, George T. physics educator, nuclear physicist
Tang, Jinke physicist

Ruston
McCall, Richard Powell physics educator

MAINE

Castone
Otto, Fred Bishop physics educator

Chebeague Island
Allen, Clayton Hamilton physicist, acoustician

Lewiston
Semon, Mark David physicist, educator

Orono
Tarr, Charles Edwin physicist, educator

MARYLAND

Aberdeen Proving Ground
Anderson, William Robert physicist

Annapolis
Dickey, Joseph Waldo *physicist*
Heiner, Lee Francis *physicist*

Baltimore
Green, Robert Edward, Jr. *physicist, educator*
Inglehart, Lorretta Jeannette *physicist*
Jensen, Arthur Seigfried *consulting engineering physicist*
Kaplan, Alexander Efimovich *physics educator, engineering educator*
Lee, Yung-Keun *physicist, educator*
Rao, Gopala Utukuru *radiological physicist, educator*
Zukas, Jonas Algimantas *physicist*

Bel Air
Cash, (Cynthia) LaVerne *physicist*

Berlin
Brodsky, Allen *radiological and health physicist, consultant*

Bethesda
Hagen, Stephen James *physicist*
McLellan, Katharine Esther *health physicist, consultant*
Rich, Harry Louis *physicist, marine engineering consultant*
Sinclair, Warren Keith *radiation biophysicist, organization executive, consultant*
Stone, Philip M. *physicist, nuclear engineer*
Zoltick, Brad J. *physicist, mathematics educator*

Clarksburg
Townsend, John William, Jr. *physicist, retired federal aerospace agency executive*

College Park
Busby, Kenneth Owen *applied physicist*
De Souza-Machado, Sergio Guilherme *physics research assistant*
Dragt, Alexander James *physicist*
Fetter, Steve *physicist*
Fisher, Michael Ellis *mathematical physicist, chemist*
Gomez, Romel Del Rosario *physicist*
Griem, Hans Rudolf *physicist, educator*
Lau, Yun-Tung *physicist, consultant*
Peaslee, David Chase *physics educator*
Redish, Edward Frederick *physicist, educator*
Sagdeev, Roald Zinnurovi *physicist educator*
Sengers, Jan Vincent *physicist*
Skuja, Andris *physics educator*
Wang, Jian Guang *engineering physicist*

Columbia
Clark, Billy Pat *physicist*

Fort Meade
Madden, Robert William *physicist*

Gaithersburg
Bruening, Robert John *physicist, researcher*
Caplin, Jerrold Leon *health physicist*
Casella, Russell Carl *physicist*
Evenson, Kenneth M. *physicist*
Hoffer, James Brian *physicist, consultant*
Lu, Kwang-Tzu *physicist*
Nakatani, Alan Isamu *physicist*
Reader, Joseph *physicist*
Seiler, David George *physicist*
Simons, David Stuart *physicist*
Teague, Edgar Clayton *physicist*
Vorburger, Theodore Vincent *physicist, metrologist*
Weber, Alfons *physicist*
Wineland, David J. *physicist*

Germantown
Charlton, Gordon Randolph *physicist*

Glenelg
Williams, Donald John *research physicist*

Greenbelt
Alexander, Joseph Kunkle, Jr. *physicist*
Comiso, Josefino Cacas *physical scientist*
Day, John H. *physicist*
Degnan, John James, III *physicist*
Mumma, Michael Jon *physicist*
Petuchowski, Samuel Judah *physicist*
Schatten, Kenneth Howard *physicist*

Kensington
Aronson, Casper Jacob *physicist*

Landover
Zehl, Otis George *optical physicist*

Lanham
Nithianandam, Jeyasingh *physicist*

Lanham Seabrook
Lesikar, James Daniel, II *physicist*

Laurel
Blum, Norman Allen *physicist*
Bostrom, Carl Otto *physicist, laboratory director*
Krimigis, Stamatios Mike *physicist, researcher, space science/engineering manager, consultant*
Maurer, Richard Hornsby *physicist*
Moorjani, Kishin *physicist, researcher*
Voss, Paul Joseph *physicist*

Monrovia
Atanasoff, John Vincent *physicist*

Pasadena
Young, Russell Dawson *physics consultant*

Rockville
Corley, Daniel Martin *physicist*

Silver Spring
Forbes, Jerry Wayne *research physicist*
Gaunaurd, Guillermo C. *physicist, engineer, researcher*
Mathur, Veerendra Kumar *physicist, researcher, project manager*
Restorff, James Brian *physicist*
Riel, Gordon Kienzie *research physicist*

Suitland
Guss, Paul Phillip *physicist*
Vojtech, Richard Joseph *nuclear physicist*

MASSACHUSETTS

Acton
Richter, Edwin William *physicist, editor, consultant*

Amherst
Romer, Robert Horton *physicist, educator*
Scott, David Knight *physicist, university administrator*
Sokolik, Igor *physicist*

Bedford
Eldering, Herman George *physicist*
Frederickson, Arthur Robb *physicist*
Greenwald, Anton Carl *physicist, researcher*

Belmont
Chiu, Tak-Ming *magnetic resonance physicist*

Boston
Cohen, Robert Sonné *physicist, philosopher, educator*
Deutsch, Thomas Frederick *physicist*
Hoop, Bernard *physicist, researcher, educator*
Lu, Hsiao-ming *physicist*
Malenka, Bertram Julian *physicist, educator*
Sage, James Timothy *physicist, educator*
Stachel, John Jay *physicist, educator*
Stanley, H(arry) Eugene *physicist, educator*
Von Goeler, Eberhard *physics educator*

Brookline
Strauss, Bruce Paul *engineering physicist, consultant*

Cambridge
Barger, James Edwin *physicist*
Benedek, George Bernard *physicist, educator*
Bloembergen, Nicolaas *physicist, educator*
Branscomb, Lewis McAdory *physicist*
Burke, Bernard Flood *physicist, educator*
Chin, Aland Kwang-Yu *physicist*
Coleman, Sidney Richard *physicist, educator*
Cook, Andrew Robert *experimental chemical physicist*
Dresselhaus, Mildred Spiewak *physics and engineering educator*
Evans, Robley Dunglison *physicist*
Feshbach, Herman *physicist, educator*
Flatté, Michael Edward *physicist, researcher*
French, Anthony Philip *physicist, educator*
Friedman, Jerome Isaac *physics educator, researcher*
Goldstone, Jeffrey *physicist*
Golovchenko, Jene Andrew *physics and applied physics educator*
Gordon, George Stanwood, Jr. *physicist*
Guth, Alan Harvey *physicist, educator*
Hwa, Terence Tai-Li *physicist*
Jacobsen, Edward Hastings *physicist*
Jaffe, Arthur Michael *physicist, mathematician, educator*
Jones, Richard Victor *physics and engineering educator*
Keck, James Collyer *physicist, educator*
Kendall, Henry Way *physicist*
Kerman, Arthur Kent *physicist, educator*
King, Ronold Wyeth Percival *physics educator*
Kleppner, Daniel *physicist, educator*
Lee, Patrick A. *physics educator*
Li, Yonghong *physics researcher*
Litster, James David *physics educator, dean*
Lyon, Richard Harold *educator, physicist*
Martin, Paul Cecil *physicist, educator*
McElroy, Michael *physicist, researcher*
Moniz, Ernest Jeffrey *physics educator*
Negele, John William *physics educator, consultant*
Oppenheim, Irwin *chemical physicist, educator*
Paul, William *physicist, educator*
Poggio, Tomaso Armando *physicist, educator, computer scientist, researcher*
Porkolab, Miklos *physics educator, researcher*
Postol, Theodore A. *physicist, educator*
Pritchard, David Edward *physics educator*
Purcell, Edward Mills *physics educator*
Rahman, Anwarur *physicist*
Ramsey, Norman F. *physicist, educator*
Raymond, John Charles *physicist*
Revol, Jean-Pierre Charles *physicist*
Rhodes, William George, III *physicist*
Runge, Erich Karl Rainer *physicist*
Shapiro, Irwin Ira *physicist, educator*
Smith, Peter Lloyd *physicist*
Snipes, Joseph Allan *research scientist, physicist*
Spaepen, Frans August *applied physics researcher, educator*
Ting, Samuel Chao Chung *physicist, educator*
Wilson, Richard *physicist, educator*

Charlestown
Lizak, Martin James *physicist*

Chelmsford
Sepucha, Robert C. *chemical physicist, optics scientist*

Chestnut Hill
Bakshi, Pradip M. *physicist*
Mohanty, Udayan *chemical physicist, theoretical chemist*

Danvers
Newell, Philip Bruce *physicist, lighting engineer*

Hanscom AFB
Kirkwood, Robert Keith *applied physicist*
Lai, Shu Tim *physicist*
Shepherd, Freeman Daniel *physicist*

Lexington
Blanchard, Robert Lorne *engineering physicist*
Brown, Elliott Rowe *physicist*
Callerame, Joseph *physicist*
Dionne, Gerald Francis *research physicist*
Schloemann, Ernst Fritz (Rudolf August) *physicist, engineer*

Lowell
Jourjine, Alexander N. *theoretical physicist*
Sebastian, Kunnat Joseph *physics educator*

Marlborough
Pittack, Uwe Jens *engineer, physicist*

Medford
Cormack, Allan MacLeod *physicist, educator*
Kafka, Tomas *physicist*

Natick
Wang, Chia Ping *physicist, educator*

Newton
Wadzinski, Henry Teofil *physicist*
Weisskopf, Victor Frederick *physicist*

North Dartmouth
Sauro, Joseph Pio *physics educator*

Northampton
Decowski, Piotr *physicist*

Quincy
Giberson, Karl Willard *physics and philosophy educator*

Rockport
Hull, Gordon Ferrie *physicist*

Salem
Brown, Walter Redvers John *physicist*
Kaiser, Kurt Boye *physicist*

Sturbridge
Feller, Winthrop Bruce *physicist*

Waltham
Caspar, Donald Louis Dvorak *physics and structural biology educator*
Deser, Stanley *educator, physicist*

Watertown
Lin, Alice Lee Lan *physicist, researcher, educator*
Oakes, Carlton Elsworth *physicist*

Westborough
Oliver, David Edwin *physicist*

Williamstown
Strait, Jefferson *physicist, educator*

Woods Hole
Stanton, Timothy Kevin *physicist*

Worcester
Reilly, Judith Gladding *physics educator*

MICHIGAN

Alma
Reed, Bruce Cameron *physics educator, astronomy researcher*

Ann Arbor
Bucksbaum, Philip Howard *physicist*
Chupp, Timothy E. *physicist, educator, nuclear scientist, academic administrator*
Clarke, Roy *physicist, educator*
Crane, Horace Richard *educator, physicist*
Donahue, Thomas Michael *physics educator*
Filisko, Frank Edward *physicist, educator*
Peterson, Lauren Michael *physicist, educator*
Rand, Stephen Colby *physicist*
Rauchmiller, Robert Frank, Jr. *physicist*
Roe, Byron Paul *physics educator*
Sudijono, John Leonard *physicist*
Veltman, Martinus J. *physics educator*

Dearborn
Ginder, John Matthew *physicist*

Detroit
Gupta, Suraj Narayan *physicist, educator*
Thomas, Robert Leighton *physicist, researcher*
Wierzbicki, Jacek Gabriel *physicist, researcher*

East Lansing
Antaya, Timothy Allen *physicist*
Austin, Sam M. *physics educator*
Blosser, Henry Gabriel *physicist*
McIntyre, John Philip, Jr. *physics educator*
Pollack, Gerald Leslie *physicist, educator*
Yussouff, Mohammed *physicist, educator*

Ferndale
Hyder, Ghulam Muhammad Ali *physicist*

Grand Rapids
Van Zytveld, John Bos *physicist, educator*

Hemlock
Wheelock, Scott A. *physicist*

Houghton
Jaszczak, John Anthony *physicist*

Kalamazoo
Askew, Thomas Rendall *physics educator, researcher, consultant*

Madison Heights
Chapman, Gilbert Bryant *physicist*

Metamora
Blass, Gerhard Alois *physics educator*

Midland
Bernius, Mark T. *research physicist*

Warren
Franetovic, Vjekoslav *physicist*
Heremans, Joseph Pierre *physicist*
Sell, Jeffrey Alan *physicist*
Vaz, Nuno Artur *physicist*

MINNESOTA

Minneapolis
Cline, James Michael *physicist*
Giese, Clayton Frederick *physics educator, researcher*
Goldman, Allen Marshall *physics educator*
Hamermesh, Morton *physicist, educator*
Heller, Kenneth Jeffrey *physicist*
Hobbie, Russell Klyver *physicist*
Johnson, Robert Glenn *physics educator*
Marshak, Marvin Lloyd *physicist, educator*
Nier, Alfred Otto Carl *physicist*

Saint Paul
Fisch, Richard S. *physicist, psychophysicist*
Lee, Charles C. *physicist*
Mitchell, William Cobbey *physicist*
Schumer, Douglas Brian *physicist*
Watson, James Edwin *physicist*

Saint Peter
Fuller, Richard Milton *physics educator*

MISSISSIPPI

Bay Saint Louis
Quinn, John Michael *physicist, geophysicist*
Zingarelli, Robert Alan *research physicist*

Oxford
Breazeale, Mack Alfred *physics educator*

Pascagoula
Corben, Herbert Charles *physicist, educator*

MISSOURI

Cape Girardeau
Dahiya, Jai Narain *physics educator, researcher*

Columbia
Boley, Mark S. *physicist, mathematician*
Popovici, Galina *physicist*

Gladstone
Moffitt, Christopher Edward *physicist*

Kansas City
Ching, Wai Yim *physics educator, researcher*
Grosskreutz, Joseph Charles *physicist, engineering researcher, educator*
Murphy, Richard David *physics educator*
Shi, Zheng *radiological physicist*
Wegst, Audrey V. *physicist, consulting firm executive*
Wieliczka, David Michael *physics educator, researcher*

Liberty
Philpot, John Lee *physics educator*

Rolla
Schulz, Michael *physicist*
Sparlin, Don Merle *physicist*

Saint Louis
Binns, Walter Robert *physics researcher*
Carlsson, Anders Einar *physicist*
Conradi, Mark Stephen *physicist, educator*
Cowsik, Ramanath *physics educator*
Huddleston, Philip Lee *physicist*
Kelton, Kenneth Franklin *physicist, educator*
Norberg, Richard Edwin *physicist, educator*
Visser, Matthew Joseph *physicist*
Walker, Robert Mowbray *physicist, educator*

NEBRASKA

Lincoln
Sellmyer, David Julian *physicist, educator*

Omaha
Zepf, Thomas Herman *physics educator, researcher*

NEVADA

Carson City
Wadman, William Wood, III *health physicist, consulting company executive*

NEW HAMPSHIRE

Durham
Distelbrink, Jan Hendrik *physicist, researcher*

Hanover
Kantrowitz, Arthur *physicist, educator*
Sturge, Michael Dudley *physicist*

Weare
Pierce, John Alvin *physics researcher*

NEW JERSEY

Annandale
Flannery, Brian Paul *physicist, educator*

Atlantic City
Vansuetendael, Nancy Jean *physicist*

Cherry Hill
Hayasi, Nisiki *physicist*

Fort Monmouth
Kronenberg, Stanley *research physicist*

Kennedy, Douglas Wayne *physicist*
Martin, Joel Jerome *physics educator*

Tulsa
Banik, Niranjan Chandra-Dutta *physicist, researcher*
Blais, Roger Nathaniel *physics educator*
Duncan, Lewis Mannan, III *physicist, education administrator*

Yukon
Huynh, Nam Hoang *physics educator*

OREGON

Beaverton
Allen, Paul C. *physicist*
Murdock, Bruce *physicist*

Corvallis
Fontana, Peter Robert *physics educator*

Mcminnville
Deer, James William *physicist*

Medford
Kunkle, Donald Edward *physicist*

Salem
Gillette, (Philip) Roger *physicist, systems engineer*

PENNSYLVANIA

Bethlehem
Licini, Jerome Carl *physicist, educator*

Conshohocken
Shanbaky, Ivna Oliveira *physicist*

Downingtown
Tarpley, William Beverly, Jr. *physics and chemistry consultant*

Jenkintown
Mifsud, Lewis *electrical engineer, fire origin investigator, physicist*

Johnstown
Salem, Kenneth George *theoretical physicist*

Lehman
Alston, Steven Gail *physicist, educator*

Lock Haven
Smith, Augustine Joseph *physicist, educator*

Merion Station
Amado, Ralph David *physics educator*

New Kensington
Oder, Robin Roy *physicist*

Philadelphia
Frankel, Sherman *physicist*
Joseph, Peter Maron *physics educator*
Tyagi, Som Dev *physics educator*
Weisz, Paul B(urg) *physicist, chemical engineer*

Pittsburgh
Anderson, Russell Karl, Jr. *physicist, horse breeder*
Cutkosky, Richard Edwin *physicist, educator*
Dutt, David Alan *physicist*
Friedberg, Simeon Adlow *physicist, educator*
Griffiths, Robert Budington *physics educator*
Huang, Mei Qing *physics educator, researcher*
Page, Lorne Albert *physicist, educator*
Sinharoy, Samar *physicist, researcher*
Sorensen, Raymond Andrew *physics educator*
Taylor, Lyle H. *physicist*
Wolfenstein, Lincoln *physicist, educator*
Worthington, Charles Roy *physics educator*

State College
Lannin, Jeffrey S. *physicist, educator*

University Park
Badzian, Andrzej Ryszard *physicist*
Gu, Claire Xiang-Guang *physicist*
Kurtz, Stewart Kendall *physics educator, researcher*
McKenna, Mark Joseph *physicist, educator, researcher*
Olivero, John Joseph, Jr. *physics educator*
Thomas King, McCubbin, Jr. *physicist*

West Chester
Dionne, Ovila Joseph *physicist*
Vinokur, Roman Yudkovich *physicist, engineer*

Williamsport
Fisher, David George *physics educator*

RHODE ISLAND

Kingston
McCorkle, Richard Anthony *physicist*

Newport
Mellberg, Leonard Evert *physicist*

Providence
Cooper, Leon N. *physicist, educator*
Elbaum, Charles *physicist, educator, researcher*
Tauc, Jan *physics educator*
Walecki, Wojciech Jan *physicist, engineer*

SOUTH CAROLINA

Charleston
Fenn, Jimmy O'Neil *physicist*

Clemson
McNulty, Peter J. *physics educator*

Columbia
Datta, Timir *physicist, solid state/materials consultant*
Edge, Ronald Dovaston *physics educator*
Preedom, Barry Mason *physicist, educator*
Schuette, Oswald Francis *physics educator*
Wilson, Scott Roland *physicist*

Conway
Skinner, Samuel Ballou, III *physics educator, researcher*

Greenville
Cronemeyer, Donald Charles *physicist*

Hilton Head Island
Hamlin, Scott Jeffrey *physicist*

Spartanburg
Vassy, David Leon, Jr. *radiological physicist*

Sumter
Bryant, Wendy Sims *medical physicist*

SOUTH DAKOTA

Brookings
Duffey, George Henry *physics educator*
O'Brien, John Joseph *physicist*

TENNESSEE

Cookeville
Kumar, Krishna *physics educator*

Knoxville
Borie, Bernard Simon, Jr. *physicist, educator*
Deeds, William Edward *physicist, educator*
Parks, James Edgar *physicist, consultant*

Memphis
Jahan, Muhammad Shah *physicist*

Nashville
Pan, Zhengda *physicist, researcher*

Oak Ridge
Beecher, Stephen Clinton *physicist*
Bicehouse, Henry James *health physicist*
Chen, Chung-Hsuan *research physicist*
Chen, Gwo-Liang *physicist, researcher*
Kerchner, Harold Richard *physicist, researcher*
Larson, Bennett Charles *solid state physicist, researcher*
Postma, Herman *physicist, consultant*
Sauers, Isidor *physicist, researcher*
Specht, Eliot David *physicist*
Trivelpiece, Alvin William *physicist, corporate executive*
Weinberg, Alvin Martin *physicist*
Whealton, John H. *physicist*
Wong, Cheuk-Yin *physicist*

Tullahoma
Li, Liqiang *physicist*

TEXAS

Austin
Bohm, Arno Rudolf *physicist*
Cobb, John Winston *plasma physicist*
Curran, Dian Beard *physicist, consultant*
Downer, Michael C. *physicist*
Drummond, William Eckel *physics educator*
Gentle, Kenneth William *physicist*
Hazeltine, Richard Deimel *physics educator, university institute director*
Herman, Robert *physics educator*
Newberger, Barry Stephen *physicist, research scientist*
Oakes, Melvin Ervin Louis *physics educator*
Prigogine, Vicomte Ilya *physics educator*
Ray, Robert Landon *nuclear physicist*

Beaumont
Pizzo, Joe *physics educator*

Brooks AFB
Donovan, Lawrence *physicist*

Carrollton
Wang, Peter Zhenming *physicist*

Cleveland
Dolney, Tabatha Ann *physics educator*

College Station
Jaric, Marko Vukobrat *physicist, educator, researcher*
Kirk, Wiley Price, Jr. *physics and electrical engineering educator*
Magnuson, Charles Emil *physicist*
McIntyre, Peter Mastin *physicist, educator*
Mouchaty, Georges *physicist*
Pandey, Raghvendra Kumar *physicist, educator*

Dallas
Chang, Cheng-Hui (Karen) *medical physicist*
Chen, Zhan *physicist*
Schwitters, Roy Frederick *physicist, educator*
Toohig, Timothy E. *physicist*
Trahern, Charles Garrett *physicist*
Womersley, William John *physicist*

Denton
Kim, Yong-Dal *physicist*

El Paso
Wang, Paul Weily *physics educator*

Fort Worth
Miller, Bruce Neil *physicist*

Houston
Allman, Mark C. *physicist*
Dickens, Thomas Allen *physicist*
Mehra, Jagdish *physicist*
Moss, Simon Charles *physics educator*
Nordlander, Jan Peter Arne *physicist, educator*
Thornton, Kathryn C. *physicist, astronaut*
Weinstein, Roy *physics educator, researcher*
Wilson, Thomas Leon *physicist*

Kingsville
Kruse, Olan E. *physics educator*
Suson, Daniel Jeffrey *physicist*

Richardson
Hanson, William Bert *physics educator, science administrator*

San Antonio
Edlund, Carl E. *physicist*
Sablik, Martin John *research physicist*

Sugar Land
Phares, Lindsey James *consultant, retired physicist and engineer*

Tyler
Mattern, James Michael *physicist*

Waco
Wilcox, Walter Mark *elementary particle physicist, educator*

UTAH

Salt Lake City
Meyer, Frank Henry *physicist*
Newport, Brian John *physicist*

VERMONT

Middlebury
Dunham, Jeffrey Solon *physicist, educator*

VIRGINIA

Alexandria
Baker, George Harold, III *physicist*
Robson, Anthony Emerson *plasma physicist*
Straus, Leon Stephan *physicist*

Annandale
Neubauer, Werner George *physicist*

Arlington
Keeler, Roger Norris *physicist*
Zirkind, Ralph *physicist*

Blacksburg
Collins, George Briggs *retired physicist*

Charlottesville
Baragiola, Raul Antonio *physicist*
Song, Xiaotong *physicist*

Crystal City
Lintz, Paul Rodgers *physicist, patent examiner*

Dahlgren
Holt, William Henry *physicist, researcher*

Falls Church
Rosen, Coleman William *radiological physicist*

Fort Belvoir
Heberlein, David Craig *physicist*

Greenwood
Findlay, John Wilson *retired physicist*

Hampden Sydney
Joyner, Weyland Thomas *physicist, educator, business consultant*

Hampton
Sun, Keun Jenn *physicist*
Wang, Liang-guo *research scientist*

Harrisonburg
Giovanetti, Kevin Louis *physicist, educator*

Kinsale
Gould, Gordon *physicist, retired optical communications executive*

Newport News
Hartline, Beverly Karplus *physicist, science educator*
Neil, George Randall *physicist*
Yan, Chen *electromagnetic physicist*

Norfolk
Vušković, Leposava *physicist, educator*

Petersburg
Chandler, Paul Anderson *physicist, researcher*

Richmond
Kalen, Joseph David *physicist, researcher*

Roanoke
Al-Zubaidi, Amer Aziz *physicist, educator*

Springfield
Reed, Charles Kenneth *physicist*

Sweet Briar
Hyman, Scott David *physicist, educator*

Vienna
Edwards, John William *physicist*

Virginia Beach
Yu, Brian Bangwei *research physicist, executive*

Williamsburg
Krakauer, Henry *physics educator*

Winchester
Ludwig, George Harry *physicist*

WASHINGTON

Bothell
Hadjicostis, Andreas Nicholas *physicist*

Bremerton
Thovson, Brett Lorin *physicist*

Ellensburg
Mitchell, Robert Curtis *physicist, educator*
Rosell, Sharon Lynn *physics and chemistry educator, researcher*
Yu, Roger Hong *physics educator*

Pullman
Banas, Emil Mike *physicist, educator*
Pandey, Lakshmi Narayan *physicist, researcher*

Redmond
Wang, Lin *physicist, computer science educator, computer software consultant*

Richland
Brenden, Byron Byrne *physicist, optical engineer*
Lemon, Douglas Karl *physicist*

Seattle
Cramer, John Gleason, Jr. *physics educator, experimental physicist*
Dehmelt, Hans Georg *experimental physicist*
Henley, Ernest Mark *physics educator, university dean emeritus*
Johnston, Allan Hugh *physicist*
Mills, David Michael *physicist*
Rimbey, Peter Raymond *physicist, mathematical engineer*
Thouless, David James *physicist, educator*
Wilets, Lawrence *physics educator*

Spokane
McGrew, Stephen Paul *physicist, entrepreneur*

WEST VIRGINIA

Green Bank
Parker, David Hiram *physicist, electrical engineer*

WISCONSIN

Appleton
Van den Akker, Johannes Archibald *physicist*

Kenosha
Greenebaum, Ben *physicist*

Madison
Anderson, Frederic Simon B. *physicist*
Knutson, Lynn Douglas *physics educator*
Lagally, Max Gunter *physics educator*
Lawler, James Edward *physics educator*
Richards, Hugh Taylor *physics educator*

Middleton
Adney, James Richard *physicist*
Herb, Raymond G. *physicist, manufacturing company executive*

Milwaukee
Greenler, Robert George *physics educator, researcher*
Karkheck, John Peter *physics educator, researcher*
Levy, Moises *physics educator*

Stoughton
Huber, David Lawrence *physicist, educator*

Waukesha
Otu, Joseph Obi *mathematical physicist*

Whitewater
Stekel, Frank Donald *physics educator*

WYOMING

Laramie
Grandy, Walter Thomas, Jr. *physicist*

TERRITORIES OF THE UNITED STATES

PUERTO RICO

Humacao
Esteban, Ernesto Pedro *physicist*
Zypman-Niechonski, Fredy Ruben *physicist*

CANADA

ALBERTA

Calgary
Sreenivasan, Sreenivasa Ranga *physicist, educator*

Edmonton
Israel, Werner *physics educator*
Kanasewich, Ernest Roman *physics educator*

Khanna, Faqir Chand *physics educator*
Kitching, Peter *physics educator*
Page, Don Nelson *theoretical gravitational physics educator*
Rostoker, Gordon *physicist, educator*

BRITISH COLUMBIA

Vancouver
Affleck, Ian Keith *physics educator*
Hardy, Walter Newbold *physics educator, researcher*
Kiefl, Robert Frances *physicist, educator*
Vogt, Erich Wolfgang *physicist, university administrator*
Young, Margaret Elisabeth *physicist*

Victoria
Zakarauskas, Pierre *physicist*

MANITOBA

Winnipeg
McKee, James Stanley Colton *physics educator*

NOVA SCOTIA

Halifax
Andrew, John Wallace *medical physicist*

ONTARIO

Chalk River
Dolling, Gerald *physicist, research executive*

Guelph
Karl, Gabriel *physics educator*

Hamilton
Collins, Malcolm Frank *physicist, educator*

Kingston
Stewart, Alec Thompson *physicist*

London
Bancroft, George Michael *chemical physicist, educator*
Battista, Jerry Joseph *medical physicist*

North York
Nicholls, Ralph William *physicist, educator*

Orleans
Vanier, Jacques *physicist*

Ottawa
Herzberg, Gerhard *physicist*
Lees, Ron Milne *physicist, educator*
Lockwood, David John *physicist, researcher*
Stinson, Michael Roy *physicist*
Wood, Gordon Harvey *physicist*

Scarborough
Teitsma, Albert *physicist*

Toronto
Brumer, Paul William *chemical physicist, educator*
Kaiser, Nicholas *physicist, educator*
Zhu, Yunping *clinical physicist*

QUEBEC

Montreal
Barrette, Jean *physicist, researcher*
Kalman, Calvin Shea *physicist*
Leroy, Claude *physics educator, researcher*
MacFarlane, David B. *physicist, educator*
Wesemael, François *physics educator*

Outremont
Levesque, Rene Jules Albert *former physicist*

Quebec
Lessard, Roger A. *physicist, educator*

Sainte-Foy
Beaulieu, Jacques Alexandre *physicist*

Sherbrooke
Lecomte, Roger *physicist*

Vaudreuil
Webb, Paul *physicist*

MEXICO

Leon
Aboites, Vicente *physicist*

Mexico City
Asomoza, Rene *physicist*
Figueroa, Juan Manuel *physicist*
González Flores, Agustín Eduardo *physicist*

Puebla
Zehe, Alfred Fritz Karl *physics educator*

ARGENTINA

Bariloche
Barbero, José Alfredo *physics researcher*
Wio, Horacio Sergio *physicist*

Buenos Aires
Gaggioli, Nestor Gustavo *physicist, researcher, educator*
Paneth, Thomas *retired physicist*

La Plata
Scalise, Osvaldo Hector *physics researcher*

AUSTRALIA

Lindfield
Chiang, Kin Seng *optical physicist, engineer*

Lucas Heights
Collins, Richard Edward *physicist*

Melbourne
Przelozny, Zbigniew *physicist*

Murdoch
Bauchspiess, Karl Rudolf *physicist*

Pyrmont
Lawrence, Martin William *physicist*

Salisbury
Hermann, John Arthur *physicist*

AUSTRIA

Kapfenberg
Mitter, Werner Sepp *physicist, researcher, educator*

BRAZIL

Campinas
Brito Cruz, Carlos Henrique *physicist, science administrator*
Gonçalves da Silva, Cylon Eudóxio Tricot *physics educator*

Petrópolis
Gomide, Fernando de Mello *physics educator, researcher*

Recife
Coelho, Hélio Teixeira *physics researcher, consultant*

São Paulo
Arruda-Neto, João Dias de Toledo *nuclear physicist*

BULGARIA

Sofia
Khristov, Khristo Yankov *physicist*

CHILE

Santiago
Cabrera, Alejandro Leopoldo *physics researcher and educator*

Temuco
Vogel, Eugenio Emilio *physics educator*

CHINA

Beijing
Guangzhao, Zhou *theoretical physicist*
Wang, Yibing *physicist*
Zhang, Hong Tu *physics educator*

Lanzhou
Gao, Yi-Tian *physicist, educator, astronomer*

Nanjing
He, Duo-Min *physics educator*
Lin, You Ju *physicist*

Zhengzhou
Zhang, Dinglin *physics educator*
Zhang, Tao *physics educator, director*

CUBA

Havana
Fernández Miranda, Jorge *physico-mathematician, researcher*

CZECH REPUBLIC

Prague
Brdička, Miroslav *retired physicist*
Kočka, Jan Vilém *physicist*

DENMARK

Copenhagen
Bohr, Aage Niels *physicist*
Mottelson, Ben R. *physicist*

Lyngby
Kozhevnikova, Irina N. *physicist*

EGYPT

Cairo
El-Sayed, Karimat Mahmoud *physics and crystallography educator*

ENGLAND

Abingdon
Denne-Hinnov, Gerd Boël *physicist*

Amersham
Finlan, Martin Francis *physicist, consultant*

Bristol
Wilson, John Anthony *physicist*

Cambridge
Josephson, Brian David *physicist*
Tabor, David *physics educator*

Chelmsford
Cross, Trevor Arthur *physicist*

Colchester
Allen, Leslie *physics educator*
Boyd, Thomas James Morrow *physicist*

Coventry
Rowlands, George *physics educator*

Leeds
Slechta, Jiri *theoretical physicist*

Liverpool
Twin, Peter John *physics educator*

London
Hall, Trevor James *physicist, educator, consultant*
Seaton, Michael John *physicist*
Stelle, Kellogg Sheffield *physicist*

Nottingham
Choi, Kwing-So *physics and engineering educator*

Oxford
Binney, James Jeffrey *physicist*
Carlow, John Sydney *research physicist, consultant*

Southampton
Markvart, Tomas *physicist*

FRANCE

Bagneux
Wang, Zhao Zhong *physics researcher*

Besançon
Boillat, Guy Maurice Georges *mathematical physicist*

Bures-sur-Yvette
Ruelle, David Pierre *physicist*

Crolles
Reader, Alec Harold *physicist, researcher*

Dijon
Chanussot, Guy *physics educator*

Duvy
Brevignon, Jean-Pierre *physicist*

Gif sur Yvette
Cotton, Jean-Pierre Aimé *physicist*
Daoud, Mohamed *physicist*
Fournet, Gerard Lucien *physics educator*
Pierre, Françoise *physics educator*
Radvanyi, Pierre Charles *physicist*

Le Mans
Castagnede, Bernard Roger *physicist, educator*

Le Vesinet Yvelines
Hillion, Pierre Théodore Marie *mathematical physicist*

Marseilles
Favre, Alexandre Jean *physics educator*

Meudon-Bellevue
Neel, Louis Eugene Felix *physicist*

Nice
Sornette, Didier Paul Charles Robert *physicist*

Orsay
Deutsch, Claude David *physicist, educator*

Paris
Cohen-Tannoudji, Claude Nessim *physics educator*
de Gennes, Pierre Gilles *physicist, educator*
Lions, Jacques Louis *physicist*
Sirat, Gabriel Yeshoua *physics educator*

Paris La Defense
Farge, Yves Marie *physicist*

Saclay
Soukiassian, Patrick Gilles *physics educator, physicist*

Strasbourg
Broll, Norbert *physicist, consultant*
Elkomoss, Sabry Gobran *physicist*

Toulouse
Portal, Jean-Claude *physicist*

Villeurbanne
Campigotto, Corrado Marco *physicist*

GERMANY

Aachen
Meixner, Josef *emeritus physics educator*

Bayreuth
Esquinazi, Pablo David *physicist*

Berlin
Becker, Uwe Eugen *physicist*
Gobrecht, Heinrich Friedrich *physicist, educator*
Selle, Burkhardt Herbert Richard *physicist*

Blumberg
Wesley, James Paul *theoretical physicist, lecturer, consultant*

Böblingen
Fischer, Bernhard Franz *physicist*

Bochum
Kunze, Hans-Joachim Dieter *physics educator*

Bonn
Paul, Wolfgang *physics educator*
Wehrberger, Klaus Herbert *physics educator, research manager*

Bremen
Theile, Burkhard *physicist*

Clausthal
Bauer, Ernst Georg *physicist, educator*

Darmstadt
Kozhuharov, Christophor *physicist*
Theobald, Jürgen Peter *physics educator*

Duisburg
Franke, Hilmar *physics educator*

Erlangen
Hartmann, Werner *physicist*

Frankfurt
Greiner, Walter Albin Erhard *physicist*
Lüthi, Bruno *physicist, educator*

Freiburg
Gao, Hong-Bo *physics educator, researcher*
Wintgen, Dieter *physicist educator*

Garching
Grieger, Günter *physicist*
Mössbauer, Rudolf Ludwig *physicist, educator*
Scott, Bruce Douglas *physicist*
Walther, Herbert *physicist, educator*

Göttingen
Eigen, Manfred *physicist*
Faubel, Manfred *physicist*
Wedemeyer, Erich Hans *physicist*

Hamburg
Trinks, Hauke Gerhard *physicist, researcher*
Tscheuschner, Ralf Dietrich *theoretical physicist*
Wiik, Björn H. *physicist researcher, director*

Hanau
Vasak, David Jiři Jan *physicist, consultant*

Jülich
Krasser, Hans Wolfgang *physicist*
Rogister, Andre Lambert *physicist*

Karlsruhe
Schulz, Paul *physicist*

Konstanz
Jäger-Waldau, Arnulf Albert *physicist*

Luebeck
Gober, Hans Joachim *physics educator*

Mainz
Meisel, Werner Paul Ernst *physicist*
Rabe, Jürgen P. *chemical physicist*

Munich
Binnig, Gerd Karl *physicist*
Bohn, Horst-Ulrich *physicist*
Grünewald, Michael *physics researcher, research program coordinator*
Pfleiderer, Hans Markus *physicist*
Stroke, George Wilhelm *physicist, educator*

Neubiberg
Triftshäuser, Werner *physics educator*

Regensburg
Steinborn, E(rnst) Otto H. *physicist, educator*

Reutlingen
Hartmann, Jürgen Heinrich *physicist*

Schwaebisch
Khan, Hamid Raza *physicist*

Stuttgart
Klitzing, Klaus von *institute administrator, physicist*
Ning, Cun-Zheng *physicist*
Pfisterer, Fritz *physicist*
Zwicknagl, Gertrud *physicist*

Sulzbach
Deutsch, Hans-Peter Walter *physicist*

Ulm
Presting, Hartmut *physicist*
Reincker, Peter *physics educator*

Waldhilsbach
Zeh, Heinz-Dieter *retired theoretical physics educator*

GREECE

Athens
Petropoulos, Labros S. *physicist, researcher*

Crete
Economou, Eleftherios Nickolas *physics educator, researcher*

Ioannina
Alexandropoulos, Nikolaos *physics educator*
Mikropoulos, Anastassios (Tassos Mikropoulos) *physicist, researcher*

HONG KONG

Hong Kong
Ge, Weikun *physicist, educator*

Kowloon
Chang, Leroy L. *physicist*

HUNGARY

Budapest
Deák, Peter *physicist, educator*
Szasz, Andras István *physicist, educator, researcher*

INDIA

New Delhi
Menon, Mambillikalathil Govind Kumar *physicist*

IRELAND

Carlow
Kavanagh, Yvonne Marie *physicist*

Dublin
Walton, Ernest Thomas Sinton *physicist*

ISRAEL

Yavne
Caner, Marc *physicist*

ITALY

Florence
Verga Sheggi, Annamaria *physicist*

Ispra
Blaesser, Gerd *theoretical physics researcher*
Rickerby, David George *physicist, materials scientist*

Milan
Lugiato, Luigi Alberto *physics educator*
Mandorini, Vittorio *research physicist*
Sindoni, Elio *physics educator*

Naples
Covello, Aldo *physics educator*

Palermo
San Biagio, Pier Luigi *physicist*

Rome
Frova, Andrea Fausto *physicist, author*
Lavenda, Bernard Howard *chemical physics educator, scientist*
Mignani, Roberto *physics educator, researcher*
Parisi, Giorgio *physicist, educator*
Salvini, Giorgio *physicist, educator*
Vulpetti, Giovanni *physicist*
Zagara, Maurizio *physicist*

Trieste
Salam, Abdus *physicist, educator*

JAPAN

Aichi
Toyoda, Tadashi *physicist*

Bunkyo
Sasaki, Taizo *physicist*

Chiba
Kondo, Jun *physicist*
Yagi, Takashi *physicist, researcher*

Chikusa-ku
Iizuka, Jugoro *physics educator, researcher*

Fuchu
Takaki, Ryuji *physics educator*

Funabashi
Sasaki, Wataru *physics educator*

Hyogo
Terasawa, Mititaka *physics educator*

Ise
Hayashi, Takemi *physics educator*

Kagoshima
Tanaka, Toshijiro *physicist*

Kanagawa
Wada, Akiyoshi *physicist*

Kumamoto
Mitarai, Osamu *physics educator*

Matsuyama
Yano, Tadashi *physics educator*

Mito
Misawa, Susumu *physicist, educator*

Neyagawa
Motoba, Toshio *physics educator*

Nishinomiya
Imamura, Tsutomu *physicist, educator*

Noda
Suzuki, Taira *physics educator*

Okayama
Yasue, Kunio *physicist, applied mathematician*

Osaka
Ikeda, Kazuyosi *physicist, poet*
Nishiyama, Toshiyuki *physics educator*

Tochigi
Iida, Shuichi *educator, physicist*

Tokyo
Fukai, Yuh *physics educator*
Hirota, Jitsuya *reactor physicist*
Iwashita, Takeki *physics educator, researcher*
Miyake, Akira *physics educator*
Musha, Toshimitsu *physicist, educator*

Toshima-Ku
Furuichi, Susumu *physics researcher*

Toyama
Hayashi, Mitsuhiko *physics educator*

Tsukuba
Esaki, Leo *physicist*
Hara, Yasuo *physics educator and researcher*
Iguchi, Ienari *physicist, educator*
Ishimaru, Hajime *physics and engineering educator*
Onuki, Hideo *physicist*
Yatsu, Kiyoshi *plasma physicist*

Uji
Hata, Koichi *heat transfer researcher*

Yoshii
Hayashi, Shizuo *physics educator*

LATVIA

Riga
Silins, Andrejs Roberts *physics educator*

MALAYSIA

Penang
Ibrahim, Kamarulazizi *physics educator, researcher*

THE NETHERLANDS

Amsterdam
Kuyper, Paul *physicist*
Lodder, Adrianus *physics educator*
Seal, Michael *physicist, industrial diamond consultant*

Den Helder
Van Der Meij, Govert Pieter *physics educator*

Eindhoven
Cowern, Nicholas Edward Benedict *physical scientist*
Kroesen, Gerrit Maria Wilhelmus *physicist, educator*

Ra
Van Dishoect, Edwine *physicist educator*

Utrecht
Muradin-Szweykowska, Maria *physicist*

NEW ZEALAND

Auckland
Wu, Cheng Yi *physicist*

Dunedin
Dodd, John Newton *retired physics educator*

NIGERIA

Ikoyi
Madakson, Peter Bitrus *physicist, engineer*

Ilorin
Singh, Sardul *physicist*

NORWAY

As
Thue-Hansen, Vidar *physics educator*

Bergen
Stamnes, Jakob Johan *physicist educator*

Oslo
Flottorp, Gordon *audiophysicist*
Guttormsen, Magne Sveen *nuclear physicist*

Stavanger
Papatzacos, Paul George *mathematical physicist, educator*

Trondheim
Chao, Koung-An *physics educator*

THE PHILIPPINES

Quezon City
Alarcon, Minella Clutario *physics educator, researcher*

POLAND

Kraków
Broda, Rafal Jan *physicist*

Warsaw
Wiśniewski, Roland *physics educator*

Zielona Gora
Gil, Janusz Andrzej *physics educator*

REPUBLIC OF KOREA

Daejon
Lee, Elhang Howard *physicist, researcher, educator*

Taejon
Lee, Choochon *physics educator, researcher*

ROMANIA

Bucharest
Milu, Constantin Gheorghe *physicist*

RUSSIA

Moscow
Basov, Nikolai Gennadievich *physicist*
Cherenkov, Pavel Alexeyevich *physicist*
Frank, Ilya Mikhailovich *physicist*
Galeyev, Albert Abubakirovich *physicist*
Goldanskii, Vitalii Iosifovich *chemist, physicist*
Gribov, Vladimir N. *physicist*
Itskevich, Efim Solomonovich *physicist*
Keldysh, Leonid Veniaminovich *physics educator*
Khalatnikov, Isaac Markovich *theoretical physicist, educator*
Lobashev, Vladimir Mikhailovich *physicist*
Osipyan, Yuri Andreyevich *physicist*
Ossipyan Yuriy, Andrew *physicist, metallurgist, educator*
Prokhorov, Aleksandr Mikhailovich *radiophysicist*

Nizhniy Novgorod
Gaponov-Grekhov, Andrey Viktorovich *physicist*

Novosibirsk
Aleksandrov, Leonid Naumovitsh *physicist, educator, researcher*

Saint Petersburg
Amusia, Miron Ya *physics educator*
Andreev, Vacheslav Mikchaylovitch *physicist*
Golant, Victor Evgen'evich *physicist, researcher, educator*

Sverdlovsk
Oshtrakh, Michael Iosifovich *physicist, biophysicist*

Troitzk
Letokhov, Vladilen Stepanovich *physicist, educator*

SCOTLAND

Edinburgh
Simone, James Nicholas *physicist*

Peebles
Hooper, John Edward *retired physicist, researcher*

Stirling
Kleinpoppen, Hans Johann Willi *physics educator, researcher*

SLOVENIA

Ljubljana
Ribaric, Marijan *physicist, mathematician*
Zavrtanik, Danilo *physicist, researcher*

Škofja Loka
Roblek, Branko *physics educator*

SPAIN

Barcelona
Wagensberg, Jorge *physicist*

Madrid
Algora del Valle, Carlos *physics educator, researcher*
Bonilla, Luis Lopez *physics and mathematics educator*
Cárabe López, Julio *physicist*
Carbo-Fite, Rafael *physicist, researcher*
Delgado-Barrio, Gerardo *physics educator, researcher*
Gallego-Juarez, Juan Antonio *ultrasonics research scientist*
Moya de Guerra, Elvira *physics educator*

Salamanca
Cervero, Jose Maria *physics educator*

San Sebastian
Echenique, Pedro Miguel *physicist, educator*

SWEDEN

Göteborg
Andersson, Stig Ingvar *physicist*

Linköping
Steinvall, Kurt Ove *physicist*
Willander, Lars Magnus *physics educator*

Lund
Grimmeiss, Hermann Georg *physics educator, researcher*

Nykoping
Kivikas, Tõivelemb *physicist, executive*

Stockholm
Alfvén, Hannes Olof Gosta *physicist*
Carlson, Per J. *physics educator, academic administrator*
Enflo, Anita Margarita *physicist, chemist*

Ulricehamn
Alfredsson, Mats Lennart *physics researcher, administrator*

Uppsala
Ekelöf, Tord Johan Carl *elementary particle physicist*

SWITZERLAND

Basel
Fattinger, Christof Peter *physicist*

Bern
Gabutti, Alberto *physicist*

Geneva
Charpak, Georges *physicist*
Harigel, Gert Günter *physicist*
Héritier, Charles André *physicist, computer systems consultant, educator*
Hofmann, Albert Josef *physicist*
Jowett, John Martin *physicist*
Rubbia, Carlo *physicist*
Steinberger, Jack *physicist, educator*
Telegdi, Valentine Louis *physicist*
Winter, Klaus H. *physicist educator*

La Rippe
Johnsen, Kjell *accelerator physicist, educator*

Lausanne
Howling, Alan Arthur *physicist, researcher*
Schneider, Wolf-Dieter *physicist, educator*

Porrentruy
Chevalier, Jean *physics educator*

Rueschlikon
Müller, Karl Alexander *physicist, researcher*
Rohrer, Heinrich *physicist*

Villigen
Rehwald, Walther R. *physicist, researcher*

Zurich
Fröhlich, Jürg Martin *physicist, educator*
Quack, Martin *physical chemistry educator*

TAIWAN

Chung-Li
Tseng, Tien-Jiunn *physics educator*

Hsinchu
Liu, Ti Lang *physics educator*
Lu, Tian-Huey *physics educator*

Taipei
Chen, Yang-Fang *physicist, educator*
Hwang, Woei-Yann Pauchy *physics educator*

UKRAINE

Kiev
Gamarnik, Moisey Yankelevich *solid state physicist*

VENEZUELA

Caracas
Sáez, Alberto M. *physics educator*

YUGOSLAVIA

Belgrade
Miljevic, Vujo I(lija) *physicist, researcher*

ADDRESS UNPUBLISHED

Abella, Isaac David *physicist, educator*
Baym, Gordon Alan *physicist, educator*
Berry, Chester Ridlon *physicist*
Binder, Kurt *physicist, educator*
Black, Kristine Mary *physicist*
Bova, Michael Anthony *physicist*
Boyd, Edward Lee *physicist, computer scientist*
Cahn, Robert Nathan *physicist*
Castracane, James *physicist*
Compton, W. Dale *physicist*
Critoph, Eugene *retired physicist, nuclear research company executive*
Dicke, Robert Henry *educator, physicist*
Dickson, Paul Wesley, Jr. *physicist*
Ewen, H.I. *physicist*
Fellner-Feldegg, Hugo Robert *scientific consultant*
Foster, Lynn Irma *physicist*
Glashow, Sheldon Lee *physicist, educator*

Hall, Grace Rosalie *physicist, educator, literary scholar*
He, Xing-Fei *physicist*
Herzfeld, Charles Maria *physicist*
Jacobs, Ralph Raymond *physicist*
Jeffries, Robert Alan *physicist*
Jones, George Richard *physicist*
Jones, Robert William *physicist*
Kane, John Vincent, Jr. *nuclear physicist, researcher*
Lande, Alexander *physicist, educator*
Laor, Herzel *physicist*
Lloyd, Joseph Wesley *physicist, researcher*
MacQueen, Robert Moffat *solar physicist*
Matossian, Jesse Nerses *physicist*
Mattoussi, Hedi Mohamed *physicist*
Mazarakis, Michael Gerassimos *physicist, researcher*
McEnnan, James Judd *physicist*
McKinley, John McKeen *retired physics educator*
Metlay, Michael Peter *nuclear physicist*
Miller, Herman Lunden *retired physicist*
Mil'shtein, Samson *semiconductor physicist*
Mott, Sir Nevill (Francis Mott) *physicist, educator, author*
Myers, Eric Arthur *physicist*
Nagel, Max Richard *retired, applied optics physicist*
Narayan, K(avassey) Sureswaran *physicist*
Nedoluha, Alfred Karl Franz *physicist*
Oldham, Timothy Richard *physicist*
Oster, Ludwig Friedrich *physicist*
Polkosnik, Walter *physicist*
Pound, Robert Vivian *physics educator*
Prasad, Satish C(handra) *physicist*
Presley, Alice Ruth Weiss *physicist, researcher*
Pritzker, Andreas Eugen Max *physicist, administrator, author*
Pullin, Jorge Alfredo *physics researcher*
Pytlinski, Jerzy Teodor *physicist, research administrator, educator*
Redlich, Robert Walter *physicist, electrical engineer, consultant*
Richardson, Jasper Edgar *nuclear physicist*
Robinson, Bruce Butler *physicist*
Rockwell, Benjamin Allen *physicist*
Rosenkilde, Carl Edward *physicist*
Roychoudhuri, Chandrasekhar *physicist*
Rugge, Hugo Robert *physicist*
Sanchez Muñoz, Carlos Eduardo *physicist*
Siegbahn, Kai Manne Börje *physicist, educator*
Steinert, Leon Albert *mathematical physicist*
Synek, M. *physics educator, researcher*
Tamor, Stephen *physicist*
Teal, Edwin Earl *engineering physicist, consultant*
Tirkel, Anatol Zygmunt *physicist*
Tribble, Alan Charles *physicist*
Vanderford, Frank Josire *physicist, computer scientist consultant*
van der Meer, Simon *accelerator physicist*
Vary, James Patrick *physics educator*
Victor, Andrew Crost *physicist, consultant, small business owner*
Villarrubia, John Steven *physicist*
Wald, Francine Joy Weintraub (Mrs. Bernard J. Wald) *physicist, academic administrator*
Wang, Ruqing *research physicist*
Watson, Robert Barden *physicist*
Weinberg, Steven *physics educator*
Wildt, Daniel Ray *physicist*
Wilson, Kenneth Geddes *physics research administrator, educator*

SOCIAL SCIENCES: ECONOMICS

UNITED STATES

ALABAMA

Birmingham
McCarl, Henry N. *economics and geology educator*

Huntsville
Kestle, Wendell Russell *cost and economic analyst, consultant*

ARIZONA

Sedona
Eggert, Robert John, Sr. *economist*

Tempe
Metcalf, Virgil Alonzo *economics educator*

ARKANSAS

Conway
Hamblin, Daniel Morgan *economist*

CALIFORNIA

Berkeley
Hall, Bronwyn Hughes *economics educator*
Teece, David John *economics and management educator*

Fresno
O'Brien, John Conway *economist, educator, writer*

Lake Arrowhead
Beckman, James Wallace Bim *economist, marketing executive*

Los Angeles
Wong, James Bok *economist, engineer, technologist*

Menlo Park
Thiers, Eugene Andres *economist*

Monterey
Boger, Dan Calvin *economics educator, statistical and economic consultant*

Placerville
Craib, Kenneth Bryden *resource development executive, physicist, economist*

San Francisco
Brown, H. William *urban economist, private banker*
Warner, Rollin Miles, Jr. *economics educator, financial planner, real estate broker*

Santa Clara
Field, Alexander James *economics educator, dean*

Stanford
Anderson, Martin Carl *economist*
Arrow, Kenneth Joseph *economist, educator*
Huntington, Hillard Griswold *economist*
Pearson, Scott Roberts *economics educator*
Ricardo-Campbell, Rita *economist, educator*

COLORADO

Golden
Packey, Daniel J. *economist, researcher*

CONNECTICUT

New Britain
Charkiewicz, Mitchell Michael, Jr. *economics and finance educator*

New Haven
Lonergan, Brian Joseph *economist, planner*

Storrs Mansfield
McEachern, William Archibald *economics educator*

West Hartford
Giannaros, Demetrios Spiros *economist, educator*

DISTRICT OF COLUMBIA

Washington
Bailey, Norman Alishan *economist*
Bohi, Douglas Ray *economist*
Dailey, Victoria Ann *economist, policy analyst*
Gardner, Bruce Lynn *agricultural economist*
Johnson, Omotunde Evan George *economist*
Maudlin, Robert V. *economics and government affairs consultant*
Moran, Ricardo Julio *economist*
Munasinghe, Mohan *development economist*
Randall, Robert L(ee) *industrial economist*
Reynolds, Robert Joel *economist, consultant*
Rosenberg, Joel Barry *government economist*
Silverman, Lester Paul *economist, energy industry consultant*
Solomon, Elinor Harris *economics educator*
Turner, John Andrew *economist*
Wallis, W(ilson) Allen *economist, educator, statistician*
Weaver, Carolyn Leslie *economist, public policy researcher*
Wright, Charles Leslie *economist*

FLORIDA

Boynton Beach
Mittel, John J. *economist, corporate executive*

Bradenton
Balsley, Howard Lloyd *economist*

Miami
Clarkson, Kenneth Wright *economics educator*
Salazar-Carrillo, Jorge *economics educator*
Wheeler, Donald Keith *community and economic development specialist*

Naples
Halvorson, William Arthur *economic research consultant*

Ocala
Killian, Ruth Selvey *home economist*

Tallahassee
Ashler, Philip Frederic *international trade and development advisor*

GEORGIA

Athens
Allsbrook, Ogden Olmstead, Jr. *economics educator*

Atlanta
Levy, Daniel *economics educator*

Clarkston
Grimes, Richard Allen *economics educator*

HAWAII

Honolulu
Laney, Leroy Olan *economist, banker*
Staff, Robert James, Jr. *international economist, consultant*

ILLINOIS

Chicago
Becker, Gary Stanley *economist, educator*

Evanston
Braeutigam, Ronald Ray *economics educator*
Reiter, Stanley *economist, educator*

Macomb
Rao, Vaman *economics educator*

Urbana
Dovring, Folke *land economics educator, consultant*
Schmidt, Stephen Christopher *agricultural economist, educator*

INDIANA

Franklin
Launey, George Volney, III *economics educator*

South Bend
Apostolides, Anthony Demetrios *economist, educator*

West Lafayette
Connor, John Murray *agricultural economics educator*

IOWA

Ames
Fox, Karl August *economist, eco-behavioral scientist*
Johnson, Stanley R. *economist, educator*

Iowa City
Forsythe, Robert Elliott *economics educator*

KANSAS

Manhattan
Babcock, Michael Ward *economics educator*

LOUISIANA

Metairie
Huber, John Henry, III *economic scientist, researcher*

MAINE

Orono
Boyle, Kevin John *economics educator, consultant*

MARYLAND

Bethesda
Gschwindt de Gyor, Peter George *economist*

Chevy Chase
Wallerstein, Leibert Benet *economist*

Takoma Park
Yates, Renee Harris *economist*

Wheaton
Ghosh, Arun Kumar *economics, social sciences and accounting educator*

MASSACHUSETTS

Boston
Michaud, Richard Omer *financial economist, researcher*
Newhouse, Joseph Paul *economics educator*
Siegel, Richard Allen *economist*
Sinai, Allen Leo *economist, educator*

Boxford
Laderoute, Charles David *engineer, economist, consultant*

Cambridge
Greenberg, Paul Ernest *economics consultant*
Hsiao, William C. *economist, actuary educator*
Joskow, Paul Lewis *economist, educator*
Kennedy, Stephen Dandridge *economist, researcher*

Ipswich
Jennings, Frederic Beach, Jr. *economist, consultant*

Northampton
Carfora, John Michael *economics and political science educator*

MICHIGAN

Berrien Springs
Stokes, Charles Junius *economist, educator*

MINNESOTA

Minneapolis
Chipman, John Somerset *economist, educator*
Hurwicz, Leonid *economist, educator*

Moorhead
Sun, Li-Teh *economics educator*

MISSISSIPPI

Jackson
Lee, Daniel Kuhn *economist*

MISSOURI

Columbia
Bevins, Robert Jackson *agricultural economics educator*

Saint Louis
Kagan, Sioma *economics educator*

NEW HAMPSHIRE

Center Sandwich
Shoup, Carl Sumner *retired economist*

NEW JERSEY

Caldwell
Kapusinski, Albert Thomas *economist, educator*

Plainsboro
Pinninti, Krishna Rao *economist*

Princeton
Reinhardt, Uwe Ernst *economist, educator*

Randolph
Lichtig, Leo Kenneth *health economist*

NEW YORK

Bronx
Peck, Fred Neil *economist, educator*
Sun, Emily M. *economics educator*

Brooklyn
Lyubavina, Olga Samuilovna *economist*

Hamilton
Haines, Michael Robert *economist, educator*

Ithaca
Jones, Barclay Gibbs *regional economics researcher*
Tomek, William Goodrich *agricultural economist*

New York
Cochrane, James Louis *economist*
Grindea, Daniel *international economist*
Heal, Geoffrey Martin *economics educator*
Hyman, Leonard Stephen *finanical executive, economist, author*
Ivanovitch, Michael Stevo *economist*
Kalamotousakis, George John *economist*
Marlin, John Tepper *economist, writer, consultant*
Melamid, Alexander *economics educator, consultant*
Mishkin, Frederic Stanley *economics educator*
Sylla, Richard Eugene *economics educator*

Old Westbury
Ozelli, Tunch *economics educator, consultant*

Rochester
Hopkins, Thomas Duvall *economics educator*
Plosser, Charles Irving *economics educator*
Vernarelli, Michael Joseph *economics educator, consultant*

Scarsdale
Cohen, Irwin *economist*

Staten Island
Meltzer, Yale Leon *economist, educator*

NORTH CAROLINA

Chapel Hill
Behrman, Jack Newton *economist*
Pfouts, Ralph William *economist, consultant*

Elizabeth City
Oriaku, Ebere Agwu *economics educator*

Research Triangle Park
Franklin, David Lee *economist*

OHIO

Athens
Gallaway, Lowell Eugene *economist, educator*

Berea
Miller, Dennis Dixon *economics educator*

Columbus
Parsons, Donald Oscar *economics educator*

Dayton
Premus, Robert *economics educator*

Dublin
Viezer, Timothy Wayne *economist*

Kent
McKee, David Lannen *economics educator*

OKLAHOMA

Stillwater
Poole, Richard William *economics educator*

OREGON

Mcminnville
Blodgett, Forrest Clinton *economics educator*

PENNSYLVANIA

Allentown
Bannon, George *economics educator, department chairman*

Ambler
Sorrentino, John Anthony *environmental economics educator, consultant*

Bala Cynwyd
Burtle, James Lindley *economist, educator*

Media
Ataiifar, Ali Akbar *economist, educator*

Philadelphia
Rohrlich, George Friedrich *social economist*

Pittsburgh
Kenkel, James Lawrence *economics educator*

TENNESSEE

Chattanooga
Clark, Jeff Ray *economist*

Nashville
Klein, Christopher Carnahan *economist*

Oak Ridge
Das, Sujit *policy analyst*

TEXAS

Dallas
Maasoumi, Esfandiar *economics educator*
Murphy, John Carter *economics educator*

Georgetown
Camp, Thomas Harley *economist*

VIRGINIA

Arlington
Bandopadhyaya, Amitava (Amit Bando) *economist, consultant, educator*

Blacksburg
Morgan, George Emir, III *financial economics educator*

Burke
Uwujaren, Gilbert Patrick *economist, consultant*

Charlottesville
Snavely, William Pennington *economics educator*

Falls Church
Shriner, Robert Dale *economist, management consultant*
Weiss, Armand Berl *economist, association management executive*

WASHINGTON

Richland
Roop, Joseph McLeod *economist*

Tacoma
Hamner, Homer Howell *economist, educator*

Tumwater
Miles, Donald Geoffrey *economist*

WEST VIRGINIA

Bethany
Cooey, William Randolph *economics educator*

WYOMING

Laramie
Forster, Bruce Alexander *economics educator*

TERRITORIES OF THE UNITED STATES

GUAM

Tamuning
Mayer, Peter Conrad *economics educator*

CANADA

BRITISH COLUMBIA

Burnaby
Copes, Parzival *economist, researcher*

Vancouver
Nemetz, Peter Newman *policy analysis educator, economics researcher*

ONTARIO

Ottawa
Dagum, Camilo *economist, educator*

QUEBEC

Montreal
Matziorinis, Kenneth N. *economist*

ANDORRA

Andorra la Vella
Mestre, S(olana) Daniel *economics consultant*

ARGENTINA

Buenos Aires
Buscaglia, Adolfo Edgardo *economist, educator*
Macon, Jorge *fiscal economist*
Piccione, Nicolas Antonio *economist, educator*

AUSTRALIA

Sydney
Okabe, Mitsuaki *economist*

AUSTRIA

Laxenburg
Nakicenovic, Nebojsa *economist, interdisciplinary researcher*

Leoben
Schmidt, Walter J. *exploration and mineral economist*

Vienna
Lim, Youngil *economist*
Pichler, J(ohann) Hanns *economics educator*

BANGLADESH

Dhaka
Ahmed, Abu *economics educator*

BELGIUM

Asse
Lorijn, Johannes Albertus *economist*

Brussels
Ledic, Michèle *economist*
Vissol, Thierry-Louis *senior economist, researcher*

Heverlee
Van Assche, Frans Jan Maurits *economics educator*

Louvain
Van Rompuy, Paul Frans *economics educator*

BOLIVIA

Sucre
Larrazábal Antezana, Erik *economics educator, organization executive*

BRAZIL

São Paulo
Hersztajn Moldau, Juan *economist*

CHILE

Santiago
Jara Diaz, Sergio R. *transport economics educator*

CHINA

Beijing
Fan, Jiaxiang *economics educator, researcher*
Lin, Justin Yifu *economist, educator*
Mao, Yu-shi *economist, engineer, educator*

Dalian
Wu, Zhenshan *economics educator*

Shanghai
Xie (Hsieh), Shu-Sen *economics and finance educator, consultant*

CYPRUS

Nicosia
Chacholiades, Miltiades *economics educator*

DENMARK

Copenhagen
Jensen, Bjarne Sloth *economist*

ECUADOR

Quito
Casals, Juan Federico *economist, consultant*

ENGLAND

Leicester
Riach, Peter Andrew *economist*

FRANCE

Besançon
Olsem, Jean-Pierre *economics educator*

Lille
Gillet, Roland *financial economist*

Paris
Bénassy, Jean-Pascal *economist, researcher, educator*
Chahid-Nourai, Behrouz J.P. *economist, corporate executive*
Nême, Jacques *economist*

GERMANY

Berlin
Albach, Horst *economist*
Klinkmuller, Erich *economist*

Bonn
Cloes, Roger Arthur Josef *economist, consultant*

Bremen
Hansohm, Dirk Christian *economics educator, editor*

Giessen
Sell, Friedrich Leopold *economics educator, researcher*

Karlsruhe
Hohmeyer, Olav Hans *economist, researcher*

Lüneburg
Linde, Robert Hermann *economics educator*

Marburg
Zimmermann, Horst Ernst Friedrich *economics educator*

Mittelberg
Nanz, Claus Ernest *economist, consultant*

Münster
Bonus, Holger *economics educator*

GREECE

Athens
Deliyannis, Constantine Christos *economist, mathematician, educator*

Kalamaki
Cacouris, Elias Michael *economist, consultant*

HONG KONG

Hong Kong
Kwok, Raymond Hung Fai *economics educator*

IRELAND

Dublin
Norton, Desmond Anthony *economics educator*

ITALY

Milan
Montesano, Aldo Maria *economics educator*

Rome
Lopez-Portillo, José Ramon *economist, international government representative*
Rossmiller, George Eddie *agricultural economist*

JAMAICA

Kingston
Persaud, Bishnodat *sustainable development educator*

JAPAN

Fukuoka
El-Agraa, Ali M. *economics educator*
Hattori, Akira *economics educator*
Takahashi, Iichiro *economics educator*

Kawasaki
Asajima, Shoichi *economics educator*
Terada, Yoshinaga *economist*

Kitakyushu
Kuryu, Masao *economics educator*

Kobe
Hirooka, Masaaki *economics educator*

Nagoya
Mori, Shigeya *economist, educator*

Osaka
Iyoda, Mitsuhiko *economics educator*
Matsumura, Fumitake *economics educator*
Takagi, Shinji *economist, educator*

Seto
Rin, Zengi *economic history educator*

Suita
Murata, Yasuo *economist, educator*

Takahashi
Matsui, Eiichi *economics and sociology educator*

Tokyo
Honda, Hiroshi *engineer, energy economist*
Kinoshita, Tomio *economics educator*
Maki, Atsushi *economics educator*
Odawara, Ken'ichi *economist, educator*
Ohta, Hiroshi *economics educator*
Onishi, Akira *economics educator*
Sakurai, Kiyoshi *economics educator*
Wakimura, Yoshitaro *economics educator*
Yamada, Ryoji *economist*

Urawa
Rhodes, James Richard *economics educator*

MACAU

Macau
Leong, Mang Su *economist*

MALAYSIA

Kuala Lumpur
Oguntoye, Ferdinand Abayomi *economist, statistician, computer consultant*

NEPAL

Kathmandu
Poudyal, Sri Ram *economics educator, consultant*

THE NETHERLANDS

The Hague
Irvin, George William *economics educator*

Zoetermeer
Ritzen, Jozef Maria Mathias *economist*

NEW ZEALAND

Dunedin
Bairam, Erkin Ibrahim *economics educator*

NORWAY

Hvasser
Reinert, Erik Steenfeldt *economist, researcher, administrator*

PAKISTAN

Islamabad
Awan, Ghulam Mustafa *economist, political scientist, educator*

Lahore
Azhar, Barkat Ali *economic adviser, researcher*

PARAGUAY

Asunción
Ferreira Falcon, Magno *economist*

POLAND

Olsztyn
Gazinski, Benon *agricultural economics educator*

Szczecin
Flejterski, Stanislaw *economist, educator, consultant*

Warsaw
Wilczynski, Ryszard Leslaw *economist, educator*
Wrebiak, Andrzej *economic and financial consultant*

PORTUGAL

Lisbon
Martins, Ana Paula *economics educator*

Porto
Guedes-Silva, António Alberto Matos *economist*

REPUBLIC OF KOREA

Seoul
Cha, Dong Se *economist, research institute administrator*
Kang, Shin Il *economist*
Lee, Young Ki *economist, researcher*

RUSSIA

Moscow
Saltykov, Boris Georgievich *economist*

SAUDI ARABIA

Riyadh
Deeik, Khalil George *economist, financing company executive*

SINGAPORE

Singapore
Ohta, Hideaki *economics and finance researcher*

SLOVENIA

Ljubljana
Sicherl, Pavle *economics educator, consultant*

SPAIN

Barcelona
Mira Galiana, Jaime Jose Juan *economist, consultant*

Madrid
Galindo, Miguel Angel *economics educator*

Santiago
Díez, José Alberto *business educator*

SWEDEN

Linköping
Grubbström, Karl Robert William *economist*

Stockholm
Glader, Mats Lennart *economics educator, researcher*
Palmer, Edward Emery *economics educator*
Söderström, Hans Tson *economist*

Sundsvall
Sarafoglou, Nikias *economist, educator*

SWITZERLAND

Geneva
Sethuraman, Salem Venkataraman *economist*
Stahel, Walter Rudolf *industrial analyst, consultant*

Zurich
von Schuller-Goetzburg, Viktorin Wolfgang *economist, consultant*

TAIWAN

Taipei
Chou, Tein-chen *economics educator*

TRINIDAD AND TOBAGO

Port of Spain
Williams, Allan Nathaniel *agricultural economist*

ADDRESS UNPUBLISHED

Del Castillo, Jaime *economic planning consultant, educator*
Duffy, Martin Edward *management consultant, economist*
Dwyer, Gerald Paul, Jr. *economics educator, consultant*
Fowler, Nancy Crowley *government economist*
Fromlet, K. Hubert *banking economist*
Grow, Robert Theodore *economist, association executive*
Harrison, Barry *economics educator, researcher, author*
Hirsch, Walter *economist, researcher*
Knottenbelt, Hans Jorgen *economist*
Krappinger, Herbert Ernst *economist*
Ludden, John Franklin *financial economist*
Markovich-Treece, Patricia *economist, art consultant*
Robinson, Rudyard Livingstone *economist, financial analyst*
Siddayao, Corazon Morales *economist, educator*
Studness, Charles Michael *economist*

SOCIAL SCIENCES: HUMANITIES & LIBRARIES

UNITED STATES

ARIZONA

Prescott
Moses, Elbert Raymond, Jr. *speech and dramatic arts educator*

CALIFORNIA

Los Angeles
Black, Craig Call *museum administrator*

Santa Ana
Labbe, Armand Joseph *museum curator, anthropologist*

COLORADO

Fort Collins
Rolston, Holmes, III *theologian, educator, philosopher*

Littleton
Brown, Elizabeth Eleanor *retired librarian*

DELAWARE

Newark
Valdata, Patricia *English language educator, aviation writer*

DISTRICT OF COLUMBIA

Washington
Level, Allison Vickers *science librarian*
Stine, Jeffrey K. *science historian, curator*
Talbot, Frank Hamilton *museum director, marine researcher*
Veatch, Robert Marlin *medical ethics researcher, philosophy educator*
Wilkinson, Ronald Sterne *science administrator and historian*

FLORIDA

Gainesville
Bennett, Thomas Peter *museum director, educator, biologist*
Primack, Alice Lefler *librarian*

Tampa
Schuh, Sandra Anderson *ethics educator*

GEORGIA

Austell
Prather, Brenda Joyce *librarian*

Marietta
Sloan, Mary Jean *media specialist*

HAWAII

Honolulu
Duckworth, Walter Donald *museum executive, entomologist*
Grace, George William *linguistics educator*

ILLINOIS

Chicago
Lammers, Thomas Gerard *museum curator*

Naperville
Tucker, Beverly Sowers *information specialist*

INDIANA

Bloomington
Johnson, Sidney Malcolm *foreign language educator*

Notre Dame
Manier, August Edward *philosophy of biology educator*

KANSAS

Manhattan
Shanteau, James *psychology educator, researcher*

University Of Kansas
Humphrey, Philip Strong *university museum director*

MAINE

Surry
Pickett, James McPherson *speech scientist*

MARYLAND

Bethesda
Didsbury, Howard Francis *futurist educator, lecturer, consultant*
Lindberg, Donald Allan Bror *library administrator, pathologist, educator*

MASSACHUSETTS

Boston
Naeser, Margaret Ann *linguist, medical researcher*
Sienkiewicz, Frank Frederick *observatory curator*
Washburn, H. Bradford, Jr. *museum administrator, cartographer, photographer*

Cambridge
Watson, Joyce Margaret *observatory librarian*

MICHIGAN

Ann Arbor
Dougherty, Richard Martin *library and information science educator*
Moore, Thomas Edwin *museum director, biology educator*

Bloomfield Hills
Jacobowitz, Ellen Sue *museum administrator*

East Lansing
Dewhurst, Charles Kurt *museum director, curator, folklorist, English educator*

Grand Rapids
Tomlinson, Gary Earl *museum curator*

MISSOURI

Kansas City
Gale, George Daniel, Jr. *philosophy of science educator, researcher*

NEBRASKA

Lincoln
Genoways, Hugh Howard *museum director*

NEW JERSEY

Kenilworth
Wagman, Gerald Howard *library administrator*

Newark
Blount, Alice McDaniel *museum curator*

Ridgewood
Hartmann, Gregor Louis *technical translator, writer*

NEW YORK

Albany
Levine, Louis David *museum director, archaeologist*

Brooklyn
Benton, Peter Montgomery *business development and information science consultant*

East Fishkill
Poschmann, Andrew William *information systems and management consultant*

New York
Brandt, Kathleen Weil-Garris *art history educator*
Kidd, Julie Johnson *museum director*
Langdon, George Dorland, Jr. *museum administrator*
Rothman, David J. *history and medical educator*

Poughkeepsie
Henry, Charles Jay *library director*

Purchase
Lucas, Billy Joe *philosophy educator*

Rochester
Hayes, Charles Franklin, III *museum research director*
Paradowski, Robert John *history of science educator*

NORTH CAROLINA

Chapel Hill
Stephens, Laurence David, Jr. *linguist, consultant*

OHIO

Columbus
Adams, David Parrish *historian, educator*
Jackson, Michel Tah-Tung *phonetician, linguist*

Oberlin
Ricker, Alison Scott *science librarian*

OKLAHOMA

Norman
Nicewander, Walter Alan *psychology educator*

PENNSYLVANIA

Lancaster
Dunlap, Lawrence Hallowell *museum curator*

Philadelphia
Andes, Charles Lovett *museum executive, technology association executive*
Berg, Ivar Elis, Jr. *social science educator*
Bolt, Eugene Albert, Jr. *historic natural science museum curator, historian*
Hoenigswald, Henry Max *linguist, educator*
Spamer, Earle Edward *museum executive*

Pittsburgh
Yourison, Karola Maria *information specialist, librarian*

TEXAS

Sanger
Gervasi, Anne *language professional, English language educator*

VIRGINIA

Arlington
Tilton, James Joseph *research librarian*

Riner
Foster, Joy Via *library media specialist*

Virginia Beach
Lichtenberg, Byron K. *futurist, manufacturing executive, space flight consultant*

Woodbridge
Campbell, Robert P. *information scientist*

WASHINGTON

Olympia
Snow, Blaine Arlie *language educator*

WISCONSIN

Madison
Westphal, Klaus Wilhelm *university museum director*

Milwaukee
Rosen, Barry Howard *museum director, history educator*
Young, Allen Marcus *museum curator of zoology, educator, naturalist, consultant, writer*

CANADA

ONTARIO

Ottawa
Emery, Alan Roy *museum director*

QUEBEC

Boucherville
Venne, Louise Marguerite *librarian*

Montreal
Shea, William Rene *historian, philosopher of science, educator*

Pointe Claire
De Brouwer, Nathalie *librarian*

AUSTRALIA

Melbourne
Rich, Thomas Hewitt *curator*

AUSTRIA

Vienna
Welzig, Werner *philologist*

CHINA

Beijing
Shao, Wenjie *librarian*

GERMANY

Heidelberg
Dihle, Albrecht Gottfried Ferdinand *professional society administrator, classics educator*

HUNGARY

Budapest
Kosáry, Domokos *historian*

ICELAND

Reykjavik
Pind, Jörgen Leonhard *computational linguist, psycholinguist*

INDIA

Rasayani
Modak, Chintamani Krishna *information and documentation officer*

THE PHILIPPINES

Quezon City
David, Lourdes Tenmatay *librarian*

TUNISIA

Mahrajane
Ghazali, Salem *linguist, educator*

ADDRESS UNPUBLISHED

English, Bruce Vaughan *museum director and executive, environmental consultant*
Lorenzino, Gerardo Augusto *linguist*

SOCIAL SCIENCES: OTHER SPECIALTIES

UNITED STATES

ARIZONA

Tempe
Brazel, Anthony James *geographer, climatologist*

Schneller, Eugene S. *sociology educator*

Tucson
Goodall, Jane *ethnologist*
Homburg, Jeffrey Allan *geoarchaeologist*
Netting, Robert M. *anthropology educator*
Soren, David *archaeology educator, administrator*

CALIFORNIA

Berkeley
Gurgin, Vonnie Ann *social scientist*
Johanson, Donald Carl *physical anthropologist*
Smelser, Neil Joseph *sociologist*

Davis
Skinner, G(eorge) William *anthropologist, educator*

La Jolla
Spiro, Melford Elliot *anthropology educator*

Sacramento
Gilmore, Allan Emory *forensic consultant*
Moore, David Sumner *forensic document examiner*

San Clemente
Ditty, Marilyn Louise *gerontologist, educator*

San Jose
McDowell, Jennifer *sociologist, composer, playwright, publisher*

Santa Barbara
Goodchild, Michael *geographer, educator*

Stanford
Greenberg, Joseph H. *anthropologist*

COLORADO

Aurora
Zimmerman, Jeannine *crime laboratory specialist, researcher*

Boulder
Brues, Alice Mossie *physical anthropologist, educator*

Denver
Mooney, Dennis John *forensic document examiner*

Estes Park
Moore, Omar Khayyam *experimental sociologist*

CONNECTICUT

Bridgeport
van der Kroef, Justus Maria *political science educator*

New Haven
Conklin, Harold Colyer *anthropologist, educator*
Marmor, Theodore Richard *political science and public management educator*
Pospisil, Leopold Jaroslav *anthropology educator*
Rouse, Irving *anthropologist, emeritus educator*

Southbury
Wescott, Roger Williams *anthropology educator*

DELAWARE

Newark
Bilinsky, Yaroslav *political scientist*

DISTRICT OF COLUMBIA

Washington
Epstein, Gerald Lewis *technology policy analyst*
Grimm, Curt David *anthropologist*
Leggon, Cheryl Bernadette *sociologist, staff officer*
Lewis, Gwendolyn L. *sociologist, policy analyst*
Snyder, Jed Cobb *foreign affairs specialist*
Stanford, Dennis Joe *archaeologist, museum curator*

FLORIDA

Gainesville
Maples, William Ross *anthropology educator, consultant*

Lecanto
McLean-Wainwright, Pamela Lynne *educational consultant, college educator, counselor, program developer*

Saint Petersburg
Serrie, Hendrick *anthropology and international business educator*

Tampa
Wienker, Curtis Wakefield *physical anthropologist, educator*

GEORGIA

Athens
Field, Dorothy *gerontologist, educator*

Lilburn
Neumann, Thomas William *archaeologist*

Norcross
Nelson, Larry Keith *questioned document examiner*

HAWAII

Honolulu
Fuchs, Roland John *geography educator, university administrator*

ILLINOIS

Carbondale
Gumerman, George John *archaeologist*

Chicago
Harris, Chauncy Dennison *geographer, educator*
Nicholas, Ralph Wallace *anthropologist, educator*

Evanston
Brown, James Allison *anthropology educator*

Schaumburg
Kennedy, Patrick Michael *fire analyst*

Urbana
Linowes, David Francis *political economist, educator*
Rich, Robert F. *political sciences educator, science administrator*

INDIANA

Bloomington
Caldwell, Lynton Keith *social scientist, educator*
Mead, Sean Michael *anthropological researcher, consultant*

IOWA

Iowa City
Rajagopal, Rangaswamy *geography and engineering educator*
Reckase, Mark Daniel *psychometrician*

KANSAS

Overland Park
Burger, Henry G. *anthropologist, vocabulary scientist, publisher*

KENTUCKY

Lexington
Schmitt, Frederick Adrian *gerontologist, neuropsychologist*

Prestonsburg
Mc Aninch, Robert Danford *philosophy and government affairs educator*

MAINE

Phillips
Appell, George Nathan *social anthropologist*

MARYLAND

Baltimore
Soeken, Karen Lynne *research methods educator, researcher*
Stacy, Pheriba *archaeologist*

Bethesda
Baldwin, Wendy Harmer *social demographer*
Lisack, John, Jr. *cartographer, executive*

Hyattsville
Guadagno, Mary Ann Noecker *social scientist, consultant*

Potomac
Mc Bryde, Felix Webster *geographer, ecologist, consultant*

Temple Hills
Day, Mary Jane Thomas *cartographer*

Woodstock
Ballweber, Hettie Lou *archaeologist*

MASSACHUSETTS

Amherst
Woodbury, Richard Benjamin *anthropologist, educator*

Andover
Mac Neish, Richard Stockton *archaeologist, educator*

Belmont
Lex, Barbara Wendy *medical anthropologist*
Washburn, Barbara *cartography researcher*

Cambridge
Lamberg-Karlovsky, Clifford Charles *anthropologist, archaeologist*
Moore, Sally Falk *anthropology educator*
Pilbeam, David Roger *paleoanthropology educator*
Turkle, Sherry *sociologist, psychologist, educator*

Waltham
McBrearty, Sally Ann *archaeologist*

MICHIGAN

Ann Arbor
Flannery, Kent V. *anthropologist, educator*
Joscelyn, Kent Buckley *criminologist, research scientist, lawyer*

Houghton
Reynolds, Terry Scott *social science educator*

Kalamazoo
Sundick, Robert Ira *anthropologist, educator*

MISSOURI

Saint Louis
Beck, Lois Grant *anthropologist, educator*

MONTANA

Missoula
Gritzner, Jeffrey Allman *geographer, educator*

NEBRASKA

Omaha
Thorson, James Alden *gerontologist, author, consultant*

NEW JERSEY

Paramus
Aronson, Miriam Klausner *gerontologist, consultant, researcher, educator*

Princeton
Von Hippel, Frank Niels *public and international affairs educator*

NEW MEXICO

Las Vegas
Riley, Carroll Lavern *anthropology educator*

NEW YORK

Albany
Falk, Dean *anthropology educator*

Binghamton
Greiner, Thomas Moseley *physical anthropologist, archaeologist, consultant*

Brooklyn
Kaplan, Mitchell Alan *sociologist, researcher*
Vroman, Georgine Marie *medical anthropologist*

Buffalo
Batty, J. Michael *geographer, educator*
Pollock, Donald Kerr *anthropologist*
Tedlock, Dennis *anthropology and literature educator*

Hempstead
Krauze, Tadeusz Karol *sociologist, educator*

New Paltz
Schnell, George Adam *geographer, educator*
Sperber, Irwin *sociologist*

New York
Heyerdahl, Thor *anthropologist, explorer, author*
James, Gary Douglas *biological anthropologist, educator, researcher*
Nelkin, Dorothy *sociology and science policy educator, researcher*
Pavis, Jesse Andrew *sociology educator*
Rothschild, Nan Askin *archaeologist*

Yonkers
Varma, Baidya Nath *sociologist, broadcaster*

NORTH CAROLINA

Chapel Hill
Finkler, Kaja *anthropologist, educator*

Durham
Simons, Elwyn LaVerne *physical anthropologist, primatologist, paleontologist, educator*

Gastonia
Kiser, Clyde Vernon *retired demographer*

OHIO

Bowling Green
Perry, Robert Lee *sociologist, ethnologist*

Cincinnati
Trimpe, Michael Anthony *forensic scientist*

Columbus
Camboni, Silvana Maria *environmental sociologist*
Namboodiri, Krishnan *sociology educator*

OKLAHOMA

Oklahoma City
Snow, Clyde Collins *anthropologist*

PENNSYLVANIA

Bethlehem
Frankel, Barbara Brown *cultural anthropologist*

Kutztown
Dougherty, Percy H. *geographer, educator*

Philadelphia
Cutler, Neal Evan *gerontologist, educator*
Elesh, David Bert *sociology educator*

Pittsburgh
Ohlsson, Stellan *scientist, researcher*
Sabloff, Jeremy Arac *archaeologist*

University Park
Stoneking, Mark Allen *anthropologist*
Taylor, Alan Henry *geography educator, ecological consultant*

RHODE ISLAND

Providence
Kates, Robert William *geographer, educator*

SOUTH CAROLINA

Columbia
Rathbun, Ted Allan *anthropologist, educator*

TENNESSEE

Nashville
Abernethy, Virginia Deane *population and environment educator*

Oak Ridge
Tonn, Bruce Edward *social scientist, researcher*

TEXAS

Austin
Butzer, Karl W. *archaeology and geography educator*
Lopreato, Joseph *sociology educator, author*

College Station
Bass, George Fletcher *archaeology educator*
Gaston, Jerry Collins *sociology educator*
Steffy, John Richard *nautical archaeologist, educator*

Dallas
Free, Mary Moore *anthropologist*

UTAH

Logan
Ford, Robert Elden *natural resources educator, geographer, consultant*

VIRGINIA

Alexandria
Corson, Walter Harris *sociologist*

Blacksburg
Bryant, Clifton Dow *sociologist, educator*

Charlottesville
Hymes, Dell Hathaway *anthropologist*

Fairfax
Adamson, Sandra Lee *anthropologist*
Williams, Thomas Rhys *educator, anthropologist*

Reston
Wood, John Thurston *cartographer, jazz musician*

WASHINGTON

Port Townsend
Speser, Philip Lester *social scientist, consultant*

Pullman
Dunlap, Riley E. *sociologist*

Seattle
Quimby, George Irving *anthropologist, former museum director*

WEST VIRGINIA

Fairmont
Fulda, Michael *political science educator, space policy researcher*

WISCONSIN

Madison
Freudenburg, William R. *sociology educator*
Haller, Archibald Orben *sociologist, educator*

Milwaukee
Silverberg, James Mark *anthropology educator, researcher*

WYOMING

Laramie
Chai, Winberg *political science educator, foundation chair*

CANADA

ALBERTA

Edmonton
Krotki, Karol Jozef *sociology educator, demographer*

NOVA SCOTIA

Halifax
Borgese, Elisabeth Mann *political science educator, author*

ONTARIO

Hamilton
Woo, Ming-Ko *geographer, educator*

Kingston
Meisel, John *political scientist*

Ottawa
Tomlinson, Roger W. *geographer*

Toronto
Kerr, Peter Donald *geography educator emeritus*
Simmons, James *geography educator*

Waterloo
Hewitt, Kenneth *geography educator*
Warner, Barry Gregory *geographer*

QUEBEC

Montreal
Rowland, Helen *geographer*

Sainte-Croix
Grenier, Fernand *geographer, consultant*

Sainte-Foy
Denis, Paul-Yves *geography educator*

Sherbrooke
Bonn, Ferdinand J. *geography educator, environmental scientist*

MEXICO

Mexico City
Arizpe, Lourdes *anthropologist, researcher*

AUSTRALIA

Adelaide
Twidale, C(harles) R(owland) *geomorphologist, educator*

AUSTRIA

Salzburg
Rasssem, Mohammed Hassan *sociology and cultural science educator*

GERMANY

Berlin
Poehlmann, Gerhard Manfred *cartography educator*

Bremen
Heinz, Walter Richard *sociology educator*

ITALY

Rome
Mercurio, Antonino Marco *anthropologist, artist*

JAPAN

Kobe
Funaba, Masatomi *political economy educator*

Kyoto
Seki, Hiroharu *political science educator, researcher*

Otsu
Matsushita, Keiichiro *sociology educator*

Tokyo
Ori, Kan *political science educator*

KENYA

Nairobi
Leakey, Mary Douglas *archaeologist, anthropologist*

POLAND

Kielce
Ripinsky-Naxon, Michael *archaeologist, art historian, ethnologist*

RUSSIA

Moscow
Kotlyakov, Vladimir Michailovich *geographer, glaciologist researcher*

SWEDEN

Uppsala
Vedung, Evert Oskar *political science educator*

ADDRESS UNPUBLISHED

Adelman, Richard Charles *gerontology educator, researcher*
Bateson, Mary Catherine *anthropology educator*
Grusky, David Bryan *sociology educator*
Leakey, Richard Erskine *paleoanthropologist, museum director*
Lonergan, Thomas Francis, III *criminal justice consultant*
McGervey, Teresa Ann *cartographer*
Willey, Gordon Randolph *retired anthropologist, archaeologist, educator*
Woodrow-Lafield, Karen Ann *demographer*

SOCIAL SCIENCES: PSYCHOLOGY

UNITED STATES

ALABAMA

Birmingham
Ramey, Craig T. *pschology educator*

Mobile
Vitulli, William Francis *psychology educator*

Tuscaloosa
Curtner, Mary Elizabeth *psychologist*
Williams, Maurice *clinical psychologist*

ALASKA

Anchorage
Condy, Sylvia Robbins *psychologist*
Crawford, E(dwin) Ben *psychologist*
Gier, Karan Hancock *counseling psychologist*
Guinn, Janet Martin *psychologist, consultant*

ARIZONA

Flagstaff
Irving, Douglas Dorset *behavioral scientist, consultant*
Windes, James Dudley *psychology educator*

Florence
Girtman, Gregory Iverson *psychologist*

Phoenix
Archer, Gregory Alan *clinical psychologist, writer*
Hancock, Thomas Emerson *educational psychologist, educator*

Prescott
Davidson-Moore, Kathy Louise *psychologist*
Longfellow, Layne *psychologist*

Sun City
Dale, Martha Ericson *clinical psychologist, educator*

Tucson
Glisky, Elizabeth Louise *psychology educator*

ARKANSAS

Little Rock
Dykman, Roscoe Arnold *psychologist, educator*

CALIFORNIA

Arcata
Hu, Senqi *psychologist, educator*

Berkeley
Block, Jack *psychology educator*
Canfield, Judy Ohlbaum *psychologist*
Dake, Karl Manning *research psychologist*
Gough, Harrison Gould *psychologist, educator*
Grube, Joel William *psychologist, researcher*
Rosenzweig, Mark Richard *psychology educator*

Beverly Hills
Alhanati, Shelley *clinical psychologist*
Evans, Louise *psychologist*
Fox, David Peter *psychologist, educator*
Lee, Nancy Francine *psychologist*

Claremont
Gable, Robert S. *psychology educator*
Johnson, James Lawrence *clinical psychologist, writer*
Leeb, Charles Samuel *clinical psychologist*
Wents, Doris Roberta *psychologist*

Covina
Jameson, Julianne *clinical psychologist*

Davis
Natsoulas, Thomas *psychology educator*
Waller, Niels Gordon *psychologist*

Escondido
Damsbo, Ann Marie *psychologist*

Fresno
Botwin, Michael David *psychologist, researcher*
Mallory, Mary Edith *psychology educator*
Swanson, Steven Clifford *clinical psychologist*
Templer, Donald Irvin *psychologist, educator*

Glendale
De Santis, James Joseph *clinical psychologist*
Jernazian, Levon Noubar *psychologist*

Hayward
Kirkland, Shari Lynn *clinical psychologist*
Whalen, Thomas Earl *psychology educator*

Hermosa Beach
Wickwire, Patricia Joanne Nellor *psychologist, educator*

Irvine
Mc Gaugh, James Lafayette *psychobiologist*
Rauscher, Frances Helen *psychologist*
Russell, Roger Wolcott *psychobiologist, educator, researcher*
Sperling, George *cognitive scientist, educator*

La Jolla
Polich, John Michael *experimental psychologist*

La Mesa
Butler, Paul Clyde *psychologist*

Laguna Hills
Daily, Augustus Dee, Jr. *psychologist, retired*

Loma Linda
Betancourt, Hector Mainhard *psychology scientist, educator*

Long Beach
Seymour, Janet Martha *psychologist*

Los Altos
Carr, Jacquelyn B. *psychologist, educator*

Los Angeles
Burns, Marcelline *psychologist, researcher*
Edwards, Ward Dennis *psychology and industrial engineering educator*
Gross, Sharon Ruth *psychology educator, researcher*
Morales, Cynthia Torres *clinical psychologist, consultant*
O'Neil, Harold Francis, Jr. *psychologist, educator*
Raven, Bertram H(erbert) *psychology educator*
Sasao, Toshiaki *psychologist, researcher*
Sears, David O'Keefe *psychology educator*
Uijtdehaage, Sebastian Hendricus J. *psychophysiology researcher*
Wilcox, Rano Roger *psychology educator*
Williams, Bruce Warren *psychologist*
Wittrock, Merlin Carl *educational psychologist*
Ziskin, Jay Hersell *psychologist*

Malibu
Forer, Bertram Robin *psychologist, researcher*

Martinez
Miller, Nicole Gabrielle *clinical psychologist*

Menlo Park
Alexander, Theron *psychologist, writer*

Moffett Field
Cohen, Malcolm Martin *psychologist, researcher*

Nevada City
Vaughan-Kroeker, Nadine *psychologist*

Newport Beach
Whittemore, Paul Baxter *psychologist*

Northridge
Fidell, Linda Selzer *psychology educator, consultant*

Novato
Sacks, Colin Hamilton *psychologist educator*

Oakland
Cumming, Janice Dorothy *clinical psychologist*
Ellison, Carol Rinkleib *psychologist, educator*
Morier, Dean Michael *psychology educator*
Nebelkopf, Ethan *psychologist*
Solomon, Daniel *psychologist*

Orange
Becker, Juliette *psychologist, marriage and family therapist*
Knoth, Russell Laine *psychologist, educator*

Oroville
Shelton, Joel Edward *clinical psychologist*

Pasadena
Bjorck, Jeffrey Paul *psychology educator, clinical psychologist*
Court, John Hugh *psychology educator*
Gorsuch, Richard Lee *psychologist, educator*
Henninger, Polly *neuropsychologist, researcher*

Piedmont
Elgin, Gita *psychologist*

Pomona
Freeland, John Chester, III *neuropsychologist*

Rancho Mirage
Deiter, Newton Elliott *clinical psychologist*

Riverside
Parke, Ross Duke *psychology educator*

Sacramento
Morrow, Joseph Eugene *psychology educator*
Pryor, Douglas Keith *clinical psychologist, consultant*

Salinas
Eifler, Carl Frederick *retired psychologist*
Finnberg, Elaine Agnes *psychologist, editor*

San Carlos
Hoffman, Paul Jerome *psychologist*

San Diego
Edwards, Darrel *psychologist*
Grossberg, John Morris *psychologist, educator*
Kent, Theodore Charles *psychologist*
Kobus, David Allan *research psychologist, consultant*
Langlais, Philip Joseph *psychologist, educator*
Linnville, Steven Emory *psychologist, navy officer*
Schorr, Martin Mark *psychologist, educator, writer*
Simpson, Henry Kertan, Jr. *psychologist, educator*
Trejo, Leonard Joseph *psychologist*

San Francisco
Calvin, Allen David *psychologist, educator*
Harary, Keith *psychologist*

San Jose
Pellegrini, Robert J. *psychology educator*

San Rafael
Hudy, John Joseph *psychologist*

Santa Ana
Daniel, Ramon, Jr. *psychologist, consultant, bilingual educator*
London, Ray William *clinical and forensic psychologist*

Santa Barbara
Beutler, Larry Edward *psychology educator*
Casey, Steven Michael *ergonomist, human factors engineer*
Kendler, Tracy Seedman *psychology educator*

Santa Cruz
Machotka, Pavel *psychology and art educator*

Santa Monica
Coleman, William Eliah *psychologist*

Sherman Oaks
Rosen, Alexander Carl *psychologist, consultant*

Stanford
Korner, Anneliese F. *psychology research scientist*
Levine, Seymour *psychology educator, researcher*
Pratto, Felicia *psychologist*
Steele, Claude Mason *psychology educator*

Torrance
Black, Suzanne Alexandra *clinical psychologist, clinical neuropsychologist, researcher*

Ventura
Naurath, David Allison *engineering psychologist, researcher*

West Los Angeles
Van Zak, David Bruce *psychologist*

Woodland Hills
Blanchard, William Henry *psychologist*

COLORADO

Aurora
Bilett, Jane Louise *clinical psychologist*

Boulder
Haenggi, Dieter Christoph *psychologist, researcher*

Colorado Springs
Ginnett, Robert Charles *organizational psychologist*

Denver
Handelsman, Mitchell M. *psychologist, educator*
McShane, Eugene Mac *psychologist*
Rubin, Betsy Claire *clinical psychologist, researcher*
Zimet, Carl Norman *psychologist, educator*

Durango
Jones, Janet Lee *psychology educator, cognitive scientist*

Gunnison
Drake, Roger Allan *psychology educator*

Lakewood
Civish, Gayle Ann *psychologist*

LaSalle
Stevenson, James Ralph *school psychologist, author*

CONNECTICUT

Fairfield
Goodrich, David Charles *management psychologist*

Farmington
Kegeles, S. Stephen *behavioral science educator*
Mulhearn, Cynthia Ann *industrial/organizational psychology researcher*

Glastonbury
Magnavita, Jeffrey Joseph *psychologist*

Greenwich
Gargiulo, Gerald John *psychoanalyst, writer*

Guilford
Chatt, Allen Barrett *psychologist, neuroscientist*

Middletown
Wilkes-Gibbs, Deanna Lynn *psychologist, educator*

New Britain
Cotten-Huston, Annie Laura *psychologist, educator*

New Haven
Miller, Joan G. *psychology educator*
Salovey, Peter *psychology educator*
Zigler, Edward Frank *educator, psychologist*

Putnam
Epstein, Sandra Gail *psychologist*

Storrs Mansfield
Brown, Scott Wilson *educational psychology educator*

Waterbury
Phillips, Walter Mills, III *psychologist, educator*

West Haven
Cooney, Ned Lyhne *psychologist*

DELAWARE

Dover
Porterfield, Craig Allen *psychologist, consultant*

Wilmington
Obrinski, Virginia Wallin *retired school psychologist*

DISTRICT OF COLUMBIA

Washington
Fischl, Myron Arthur *psychologist*
Huey, Beverly Messick *psychologist*
Lauber, John K. *research psychologist*
Newhouse, Quentin, Jr. *social psychologist, educator, researcher*
O'Connell, Daniel Craig *psychology educator*
Prothro, Edwin Terry *psychologist educator*
Ross, Bruce Mitchell *psychology educator*
Viscuso, Susan Rice *psychologist, researcher*
Wallace, Joan S. *psychologist*
Ward, Wanda Elaine *psychologist*
Wepman, Barry Jay *psychologist*

FLORIDA

Bal Harbour
Radford, Linda Robertson *psychologist*

Boca Raton
Richardson, Deborah Ruth *psychology educator*

Brandon
Mussenden, Gerald *psychologist*

Clearwater
Davidson, Joan Gather *psychologist*

Coral Gables
Jacobson, Leonard I. *psychologist, educator*

Fort Lauderdale
Laursen, Brett Paul *psychology educator*
McGreevy, Mary *retired psychology educator*

Gainesville
Green, David Marvin *psychology educator, researcher, consultant*
Iwata, Brian Anthony *psychologist*
Mussenden, Maria Elisabeth *psychologist, substance abuse counselor*
Sorkin, Robert Daniel *psychologist, educator*
White, Susie Mae *school psychologist*

Hialeah
Fisher, Barbara Turk *school psychologist*

Lakeland
Campbell, Doris Klein *retired psychology educator*

Maitland
Von Hilsheimer, George Edwin, III *neuropsychologist*

Melbourne
Harrell, Thomas Hicks *psychologist*

Miami
Gibby, Mabel Enid Kunce *psychologist*
Humphries, Joan Ropes *psychologist, educator*
Huysman, Arlene Weiss *psychologist, educator*
Russell, Elbert Winslow *neuropsychologist*
Thornton, Thomas Elton *psychologist, consultant*
Westman, Wesley Charles *psychologist*

Orlando
Deaton, John Earl *aerospace experimental psychologist*
Kennedy, Robert Samuel *experimental psychologist, consultant*

Saint Augustine
Koger, Mildred Emmelene Nichols *educational psychologist*

Saint Petersburg
Eastridge, Michael Dwayne *clinical psychologist*

Sarasota
Hendon, Marvin Keith *psychologist*

Tallahassee
Brigham, John Carl *psychology educator*
Figley, Charles Ray *psychology educator*
Glenn, Rogers *psychologist, student advisor, consultant*
Whitney, Glayde Dennis *psychologist, educator, geneticist*

Tampa
Clark, Michael Earl *psychologist*
Phares, Vicky *psychology educator*

GEORGIA

Americus
Barnes, David Benton *school psychologist*

Athens
Martin, Leonard Louis *social psychologist, educator*
Torrance, Ellis Paul *psychologist, educator*

Atlanta
Boothe, Ronald George *psychology and ophthalmology educator*
Davis, Elizabeth Emily Louise Thorpe *vision psychophysicist, psychologist and computer scientist*
Rumbaugh, Duane M. *psychology educator*
Seifert, Alvin Ronald *psychologist*
Snarey, John Robert *psychologist, researcher, educator*
Zirps, Fotena Anatolia *psychologist, researcher*

Augusta
Zachert, Virginia *psychologist, educator*

Carrollton
Gustin, Ann Winifred *psychologist*

Fort Gordon
Latiff-Bolet, Ligia *psychologist*

Kennesaw
Ziegler, Christine Bernadette *psychology educator, consultant*

Kingsland
Shockley, Carol Frances *psychologist, psychotherapist*

Lawrenceville
Carter, Dale William *psychologist*

Roswell
Jensen, Mogens Reimer *psychologist*

HAWAII

Aiea
Brassfield, Patricia Ann *psychologist*

Honolulu
Cattell, Heather Birkett *psychologist*
Flannelly, Kevin J. *psychologist, research analyst*
Ishikawa-Fullmer, Janet Satomi *psychologist, educator*
Kop, Tim M. *psychologist*

ILLINOIS

Arlington Heights
Monti, Laura Anne *psychology researcher, educator*

Carbondale
DiLalla, Lisabeth Anne *developmental psychology researcher, educator*

Champaign
Donchin, Emanuel *psychologist, educator*
Eriksen, Charles Walter *psychologist, educator*
Hirsch, Jerry *psychology and biology educator*
Komorita, Samuel Shozo *psychology educator*
Laughlin, Patrick Ray *psychologist*

Chicago
Bryant, Fred Boyd *psychology educator*
Carney, Jean Kathryn *psychologist*
Crawford, Isiaah *clinical psychologist*
Gruba-McCallister, Frank Peter *psychologist*
Osowiec, Darlene Ann *psychologist, consultant, educator*
Rizzo, Thomas Anthony *psychologist*
Rosenberg, Sheldon *psychologist, educator*
Tipp, Karen Lynn Wagner *school psychologist*
Trapani, Catherine *special education and educational psychology educator*
Wall, Jacqueline Remondet *industrial psychologist, rehabilitation counselor*
Young, Kathleen *clinical psychologist*

Danville
Elghammer, Richard William *psychologist*

De Kalb
Kerr, Sandra Lee *psychology educator*

Decatur
Lambirth, Thomas Thurman *clinical psychologist*

Edwardsville
Skitka, Linda Jean *psychology educator*

Evanston
Howard, Kenneth Irwin *psychology educator*
Revelle, William Roger *psychology educator*

Macomb
Harris, Karen L. *psychologist*

Oak Brook
Maides-Keane, Shirley Allen *psychologist*

Rosemont
Martin, Scott Lawrence *psychologist*

South Holland
Poprick, Mary Ann *psychologist*

Springfield
Aylward, Glen Philip *psychologist*

Urbana
Larson, Reed William *psychologist, educator*
Taylor, Henry L. *aerospace psychologist, educator*

Westmont
Forbes, Bo Crosby *clinical psychologist*
Mendelsohn, Avrum Joseph *psychologist*

INDIANA

Bloomington
Anderson, Brenda Jean *biological psychologist*
Hofstadter, Douglas Richard *cognitive, computer scientist, educator*
Townsend, James Tarlton *psychologist*

Hanover
Krantz, John Howell *psychology educator*

Indianapolis
Jones, James Lamar *psychologist*
Reilly, Jeanette P. *clinical psychologist*
Traicoff, Ellen Braden *psychologist*

Kokomo
Kramer, Geoffrey Philip *psychology educator*

Morgantown
Jones, Barbara Ewer *school psychologist*

North Vernon
Karkut, Richard Theodore *clinical psychologist*

Notre Dame
Buyer, Linda Susan *psychologist, educator*

West Lafayette
Davidson, Terry Lee *experimental psychology educator*
Jagacinski, Carolyn Mary *psychology educator*
Schweickert, Richard Justus *psychologist, educator*
Tiffany, Stephen Thomas *psychologist*

Westville
Spores, John Michael *psychologist*

IOWA

Ames
Hanisch, Kathy Ann *psychologist*
Johnson, Willie Roy *industrial psychology educator*

Ankeny
Irwin, Donald Berl *psychology educator*

Decorah
Dengler, Madison Luther *psychologist, educator*

Iowa City
Block, Robert I. *psychologist, researcher, educator*

Mason City
Rosenberg, Dale Norman *psychology educator*

Tabor
Reese, William Albert, III *psychologist*

KANSAS

Emporia
Boor, Myron Vernon *psychologist, educator*

Lawrence
Binns, William Arthur *clinical psychologist*
Brehm, Jack Williams *social psychologist, educator*
Crandall, Christian Stuart *social psychology educator*

Manhattan
Samelson, Franz *psychologist*

Saint John
Robinson, Alexander Jacob *clinical psychologist*

Topeka
Spohn, Herbert Emil *psychologist*

Wichita
Clark, Susan Matthews *psychologist*
Schalon, Charles Lawrence *psychologist*

KENTUCKY

Crestview Hills
Boehm, Lawrence Edward *psychologist, educator*

Lexington
Partington, John Edwin *retired psychologist*

Louisville
Bloch, James Phillips *clinical psychologist*
Colley, Thomas Elbert, Jr. *psychologist*
Petry, Heywood Megson *psychology educator*

Wilmore
Nonneman, Arthur J. *psychology educator*

LOUISIANA

Baton Rouge
Steiner, Dirk Douglas *psychology educator*
Waters, William Frederick *psychology educator*

Hammond
Holt-Ochsner, Liana Kay *psychology educator, researcher*

Lake Charles
Inman, James Carlton, Jr. *psychological counselor*

New Orleans
Harwood, Robin Louise *psychologist*

Ruston
Livingston, Mary M. *psychology educator*

Slaughter
Gremillion, Curtis Lionel, Jr. *psychologist, hospital administrator, musician*

MAINE

Orono
Farthing, G. William *psychology educator*

MARYLAND

Aberdeen Proving Ground
Monty, Richard Arthur *experimental psychologist*

Baltimore
Bylsma, Frederick Wilburn *neuropsychologist*
Chapanis, Alphonse *human factors engineer, ergonomist*
Feldstein, Stanley *psychologist*
Gaber, Robert *psychologist*
Giambra, Leonard Michael *psychologist*
Gottfredson, Gary Don *psychologist*
Holland, John Lewis *psychologist*
Hutton, Larrie Van *cognitive scientist, neural networker*
Kimmel, Melvin Joel *psychologist*
Schwandt, Leslie Marion *psychologist, researcher*

Bethesda
Duncan, Constance Catharine *psychologist*
Mishkin, Mortimer *neuropsychologist*
Papaioannou, Evangelia-Lilly *psychologist, researcher*

Chevy Chase
Raifman, Irving *clinical psychologist*

College Park
Teglasi, Hedwig *psychologist, educator*

Columbia
Roomsburg, Judy Dennis *industrial/organizational psychologist*

Ellicott City
Elgort, Andrew Charles *school psychologist*

Laurel
Hamill, Bruce W. *psychologist*

Rockville
Brumback, Gary Bruce *industrial and organizational psychologist*
Donahue, Mary Rosenberg *psychologist*
Lewis, Paul Martin *engineering psychologist*

Salisbury
Flory, Charles David *retired psychologist, consultant*

MASSACHUSETTS

Amherst
O'Connor, Kevin Neal *psychologist*

Bedford
Nickerson, Raymond Stephen *psychologist*

Boston
Cristiani, Vincent Anthony *counseling psychology educator*
Harrison, Robert Hunter *psychology educator*
Wayland, Sarah Catherine *cognitive psychologist*

Brockton
O'Farrell, Timothy James *psychologist*

Cambridge
Gardner, Howard Earl *psychologist, author*
Holzman, Philip Seidman *psychologist, educator*
Piattelli-Palmarini, Massimo *cognitive scientist*
Pinker, Steven A. *cognitive science educator*
Rosenthal, Robert *psychology educator*
Swets, John Arthur *psychologist, scientist*

Feeding Hills
Norris, Pamela *school psychologist*

Georgetown
O'Brien, John Steininger *clinical psychologist*

Haverhill
Woo, Buck Hong *neuropsychologist*

Northampton
Tallent, Norman *psychologist*

Salem
Vaughan, Margaret Evelyn *psychologist, consultant*

Sharon
Cahn, Glenn Evan *psychologist*

Shrewsbury
Sassen, Georgia *psychologist*

Waltham
Zebrowitz, Leslie Ann *psychology educator*

West Springfield
McKenzie, Rita Lynn *psychologist*

MICHIGAN

Ann Arbor
Eby, David W. *research psychologist*
Gruppen, Larry Dale *psychologist, educational researcher*
Morrel-Samuels, Palmer *experimental social psychologist*
Nisbett, Richard Eugene *psychology educator*
Stevenson, Harold William *psychology educator*
Woronoff, Israel *retired psychology educator*

Big Rapids
Weinlander, Max Martin *retired psychologist*

Detroit
Cantoni, Louis Joseph *psychologist, poet, sculptor*
Rickel, Annette Urso *psychology educator*
Shantz, Carolyn Uhlinger *psychology educator*

East Lansing
Ilgen, Daniel Richard *psychology educator*
Kerr, Norbert Lee *experimental social psychologist, educator*
Leigh, Linda Diane *psychologist, clinical neuropsychologist*
Lindell, Michael Keith *psychology educator*

Grand Haven
Parmelee, Walker Michael *psychologist*

Grand Rapids
MacDonald, David Richard *industrial psychologist*

Lansing
Kozlowski, Steve W.J. *organizational psychologist*

Mount Pleasant
Colarelli, Stephen Michael *psychology educator, organizational psychologist*
Dunbar, Gary Leo *psychology educator*

Novi
Lewes, Kenneth Allen *clinical psychologist*

Rochester
Braunstein, Daniel Norman *management psychologist, educator, consultant*

Sandusky
Keeler, Lynne Livingston Mills *psychologist, educator, consultant*

Ypsilanti
Bonem, Elliott Jeffrey *psychology educator*

MINNESOTA

Bloomington
Seashore, Stanley E(manuel) *social and organizational psychology researcher*

Duluth
Hoffman, Richard George *psychologist*

Eden Prairie
Penn, Sherry Eve *communication psychologist, educator*

Mankato
Sachau, Daniel Arthur *psychology educator*
Zeller, Michael James *psychologist, educator*

Minneapolis
Barden, Robert Christopher *psychologist, educator, lawyer*
Born, David Omar *psychologist, educator*
Eberly, Raina Elaine *psychologist, educator*
Engdahl, Brian Edward *psychologist*
Jesness, Bradley L. *psychology educator, testing and professional selection consultant*
Johnson, David Wolcott *psychologist, educator*
Will, Thomas Eric *psychologist*

Northfield
Huff, Charles William *psychologist, educator*

Worthington
Carlson, Rolf Stanley *psychologist*

MISSISSIPPI

Biloxi
Wasserman, Karen Boling *clinical psychologist, nursing consultant*

Brandon
Cooley, Sheila Leanne *psychologist, consultant*

Hattiesburg
Noblin, Charles Donald *clinical psychologist, educator*

Mississippi State
McMillen, David L. *psychology educator*

University
Sufka, Kenneth Joseph *psychology educator*

MISSOURI

Columbia
Altomari, Mark G. *clinical psychologist*
Cowan, Nelson *cognitive psychologist, researcher*

Kansas City
Fry-Wendt, Sherri Diane *psychologist*
Winitz, Harris *psychology educator*
Woods, Karen Marguerite *psychologist, human development, institute executive*

Nevada
Kuchta, Steven Jerry *psychologist*

Rolla
Haemmerlie, Frances Montgomery *psychology educator, consultant*

Saint Louis
Clark, Carl Arthur *retired psychology educator, researcher*
Du Bois, Philip Hunter *psychologist, educator*
Gfeller, Donna Kvinge *clinical psychologist*
Hile, Matthew George *psychologist, researcher*
Patterson, Miles Lawrence *psychology educator*
Richards, Diana Lyn *psychologist*
Wisdom, Guyrena Knight *psychologist, educator*

MONTANA

Billings
Bütz, Michael Ray *psychologist*
Gumper, Lindell Lewis *psychologist*

Bozeman
Block, Richard Atten *psychology educator*
Gray, Philip Howard *psychologist, educator*

Miles City
Birk-Updyke, Dawn Marie *psychologist*

Missoula
Strobel, David Allen *psychology educator*

NEBRASKA

Kearney
Miller, Richard Lee *psychology educator*

NEVADA

Ely
Alderman, Minnis Amelia *psychologist, educator, small business owner*

Las Vegas
McWhirter, Joan Brighton *psychologist*

Reno
Cummings, Nicholas Andrew *psychologist*
Smith, Aaron *research director, clinical psychologist*

NEW HAMPSHIRE

Hanover
Kleck, Robert Eldon *psychology educator*

Lebanon
Emery, Virginia Olga Beattie *psychologist, researcher*

Newport
Gibbs, Elizabeth Dorothea *developmental psychologist*

Portsmouth
Powers, Eva Agoston *clinical psychologist*

Windham
Hurst, Michael William *psychologist*

NEW JERSEY

Clark
Apelian, Virginia Matosian *psychologist, assertiveness training instructor, lecturer, consultant*

Cranford
Dashevsky, Sidney George *clinical and industrial psychologist*

Englewood Cliffs
Farrell, Patricia Ann *psychologist, educator*

Florham Park
Perham, Roy Gates, III *industrial psychologist*

Glassboro
Kaufman, Denise Norma *psychologist, addictions counselor, educator*

Holmdel
Francis, Eulalie Marie *psychologist*

Lyons
Youngelman, David Roy *psychologist*

Millburn
Nicol, Marjorie Carmichael *research psychologist*

Morristown
Tenopyr, Mary Louise Welsh (Mrs. Joseph Tenopyr) *psychologist*

New Brunswick
Contrada, Richard J(ude) *psychologist*
Glass, Arnold Lewis *psychology educator*
Hadani, Itzhak *experimental psychologist, human factors engineer*

Newark
Greene, Clifford *psychologist*
Tallal, Paula *psychologist*

Pequannock
MacMurren, Harold Henry, Jr. *psychologist, lawyer*

Perth Amboy
Hall, Pamela Elizabeth *psychologist*

Piscataway
Julesz, Bela *experimental psychologist, educator, electrical engineer*
Pandina, Robert John *neuropsychologist*
Spence, Donald Pond *psychologist, psychoanalyst*
Williams, James Richard *human factors engineering psychologist*

Princeton
Breland, Hunter Mansfield *psychologist*
Harnad, Stevan Robert *cognitive scientist*
Hoebel, Bartley Gore *psychology educator*
Miller, George Armitage *psychologist, educator*
Shafir, Eldar *psychology educator*
Willingham, Warren Willcox *psychologist, testing service executive*

Scotch Plains
Sweeney, Lucy Graham *psychologist*

Teaneck
Shein, Samuel T. *clinical psychologist*

Trenton
Singh, Allan *psychology educator*

Upper Montclair
Townsend, David John *psychologist*

Wayne
Kressel, Neil Jeffrey *psychologist*

Westwood
Fabrikant, Craig Steven *psychologist*

NEW MEXICO

Las Cruces
Gregory, W. Larry *social psychology educator*
Hunt, Darwin Paul *psychology educator*
Wilkinson, William Kunkel *psychologist, educator*

Los Alamos
Oakley, Marta Tlapova *psychologist*
Thompson, Lois Jean Heidke Ore *industrial psychologist*

Univ Of New Mexico
Velk, Robert James *psychologist*

NEW YORK

Albany
Apostle, Christos Nicholas *social psychologist*
Israel, Allen Charles *psychology educator*

Alfred
Lovelace, Eugene Arthur *psychology educator*

Amityville
Bogorad, Barbara Ellen *psychologist*

Beechhurst Flushing
Biegen, Elaine Ruth *psychologist*

Binghamton
Burright, Richard George *psychology educator*
Levis, Donald James *psychologist, educator*

Brewster
Vigdor, Martin George *psychologist*

Brockport
Wallnau, Larry Brownstein *psychologist educator*

Bronx
Charry, Jonathan M. *psychologist*
Perez, Luz Lillian *psychologist*

Brooklyn
Feldmar, Gabriel Gabor *psychologist, educator*
Levy, Sidney *psychologist*
Weinstein, Marie Pastore *psychologist*

Brookville
Heimer, Walter Irwin *psychologist, educator*

Buffalo
Masling, Joseph Melvin *psychology educator*
Moran, Joseph John *psychology educator*
Suess, James Francis *clinical psychologist*

City Island
Ward, Joan Gaye *psychologist*

College Point
Judge, Joseph B. *clinical psychologist*

Deer Park
Rosenblum, Judith Barbara *psychologist*

East Meadow
Albert, Gerald *clinical psychologist*

Floral Park
Schwartz, Teri J(ean) *clinical psychologist*

Geneseo
Olczak, Paul Vincent *psychologist*

Glen Oaks
Snyder, Peter Rubin *neuropsychologist*

Great Neck
Shaw, Martin Andrew *clinical and research psychologist*

Hamilton
Dovidio, John Francis *psychology educator*

Irvington
Rembar, James Carlson *psychologist*

Ithaca
Beins, Bernard Charles *psychology educator*
Darlington, Richard Benjamin *psychology educator*
Krumhansl, Carol Lynne *psychologist, educator*
Williams, David Vandergrift *organizational psychologist*

New York
Allison, David Bradley *psychologist*
Barron, Susan *clinical psychologist*
Castro-Blanco, David Raphael *clinical psychologist, researcher*
Daniele, Joan O'Donnell *clinical psychologist*
Freidenbergs, Ingrid *psychologist*
Fried, Robert *psychology educator*
Glanzer, Murray *psychology educator*
Greenleaf, Marcia Diane *psychologist, writer, educator*
Hamilton, Linda Helen *clinical psychologist*
Handel, Yitzchak S. *psychologist, educator*
Haywood, H(erbert) Carl(ton) *psychologist*
Hirsch, Joseph Allen *psychology and pharmacology educator*
Jonas, Ruth Haber *psychologist*
Kline, Milton Vance *psychologist, educator*
Kothera, Lynne M. *clinical neuropsychologist*
Miller, Cate *psychologist, educator*
Mischel, Harriet Nerlove *psychologist, educator*
Pappas, John *clinical psychologist*
Sciacca, Kathleen *psychologist*
Shrout, Patrick Elliot *psychometrician, educator*
Tallent, Marc Andrew *clinical psychologist*
Tester, Leonard Wayne *psychology educator*
Weiss, Samuel Abraham *psychologist, psychoanalyst*
Yarris, Steve *child psychologist, educator*

Niagara University
Osberg, Timothy M. *psychologist, educator, researcher, clinician*

North Tarrytown
Schippa, Joseph Thomas, Jr. *school psychologist, educational consultant, hypnotherapist*

Oneonta
Trotti, Lisa Onorato *psychology educator*

Oswego
Gooding, Charles Thomas *psychology educator, college dean*

Plattsburgh
Smith, Noel Wilson *psychology educator*

Pomona
Gordon, Edmund Wyatt *psychologist, educator*

Potsdam
Herman, William Elsworth *psychology educator*

Rochester
Johnston, Frank C. *psychologist*
Levy, Harold David *psycholinguist*

Rye Brook
Aquino, Joseph Mario *clinical psychologist*

Schenectady
DeBono, Kenneth George *psychology educator*

Setauket
Gelinas, Paul Joseph *psychologist, author*

Stony Brook
Franklin, Nancy Jo *psychology educator*
Levine, Marvin *psychologist, author*
Rachlin, Howard *psychologist, educator*

Stuyvesant
Lincoln, Janet Elizabeth *experimental psychologist, consultant*

Syracuse
Fiske, Sandra Rappaport *psychologist, educator*
Miron, Murray Samuel *psychologist, educator*
Sprafkin, Robert Peter *psychologist, educator*
Wadden, Thomas Antony *psychology educator*

Troy
Kalsher, Michael John *psychology educator*

Valley Stream
Sapadin, Linda Alice *psychologist, writer*

White Plains
Dalessio, Anthony Thomas *industrial and organization psychologist*
Parker, Robert Andrew *psychologist, computer consultant*

Whitesboro
Serling, Joel Martin *educational psychologist*

Whitestone
Juszczak, Nicholas Mauro *psychology educator*

Williston Park
Drago, Robert John *psychologist*

Yonkers
Lupiani, Donald Anthony *psychologist*

NORTH CAROLINA

Cary
Krasner, Paul R. *psychologist*

Chapel Hill
Powell, Judith Carol *clinical psychologist*

Charlotte
Duffy, Sally M. *psychologist*
Van Wallendael, Lori Robinson *psychology educator*

Davidson
Barton, Cole *psychologist, educator*

Durham
Bunch, Michael Brannen *psychologist, educator*
Keefe, Francis Joseph *psychology educator*
Lockhead, Gregory Roger *psychology educator*
Staddon, John Eric Rayner *psychology, zoology, neurobiology educator*

Greensboro
Lange, Garrett Warren *psychology educator*

Greenville
Webster, Raymond Earl *psychology educator, psychotherapist*

Morganton
Faunce, William Dale *clinical psychologist, researcher*

Oriental
Rowell, John Thomas *psychologist, consultant*

Raleigh
Zande, Michael Dominic *clinical psychologist, consultant*

NORTH DAKOTA

Grand Forks
Honts, Charles Robert *psychology educator*

OHIO

Akron
Franck, Ardath Amond *psychologist*

Bowling Green
Guion, Robert Morgan *psychologist, educator*

Cincinnati
Graen, George B. *psychologist, researcher*
Lyman, Howard B(urbeck) *psychologist*
Rosenthal, Susan Leslie *psychologist*

Cleveland
Bate, Brian R. *psychologist*
Ferguson, Sheila Alease *psychologist, consultant, researcher*
Tarnowski, Kenneth J. *psychologist*

Columbus
Carlton, Robert L. *clinical psychologist*
Davis, June Leah *psychologist*
Janus, Mark David *priest, psychologist, researcher, consultant*
Petty, Richard Edward *psychologist, educator, researcher*

Dayton
Christensen, Julien Martin *psychologist, educator*
Shaffer, Jill *clinical psychologist*

Galion
Reisner, Andrew Douglas *psychologist, chief clinical officer*

Hamilton
Fulero, Solomon M. *psychologist, educator, lawyer*

Maumee
Mohler, Terence John *psychologist*

Oxford
Harris, Yvette Renee *psychologist, educator*

Powell
Manchester, Carol Ann Freshwater *psychologist*

Shaker Heights
Blue, Reginald C. *psychologist, educator*

Toledo
McSweeney, Austin John *psychology educator, researcher*
Riseley, Martha Suzannah Heater (Mrs. Charles Riseley) *psychologist, educator*

Wright Patterson AFB
Moore, Thomas Joseph *research psychologist*

Zanesville
Wilcox, Roger Clark *psychologist, researcher*

OKLAHOMA

Guthrie
Bell, Thomas Eugene *psychologist, educational administrator*

Norman
Gronlund, Scott Douglas *psychology educator, researcher*
McClanahan, Walter Val *psychologist*
Stoltenberg, Cal Dale *psychology educator*

Oklahoma City
Hamm, Robert MacGowan *psychologist*
Sanders, Gilbert Otis *educational and research psychologist, addictions treatment therapist, consultant, educator*

Tulsa
Härtel, Charmine Emma Jean *industrial/organizational psychology educator, consultant*
Young, William Dean *psychologist*

Weatherford
Nail, Paul Reid *psychology educator*

OREGON

Beaverton
Hughes, Laurel Ellen *psychologist, educator, writer*

Corvallis
Bernieri, Frank John *social psychology educator*
Gillis, John Simon *psychologist, educator*

Eugene
Goldberg, Lewis Robert *psychology educator, researcher*
Reed, Diane Marie *psychologist*
Slovic, Stewart Paul *psychologist*

PENNSYLVANIA

Allentown
Graham, Kenneth Robert *psychologist, educator*

Altoona
Arbitell, Michelle Reneé *clinical psychologist*

Annville
Cullari, Salvatore Santino *clinical psychologist, educator*

Bloomsburg
Waggoner, John Edward *psychology educator*

Doylestown
Dimond, Roberta Ralston *psychology and sociology educator*

Fairless Hills
Rosella, John Daniel *psychologist*

Haverford
Blumberg, Herbert Haskell *psychology educator*

Horsham
Logue, John Joseph *psychologist*

Meadville
Wharton, William Polk *psychologist, consultant*

Media
Harnish, Richard John *psychologist*

Middletown
Foxx, Richard Michael *psychology educator*

Philadelphia
Gibson, William Charles *neuropsychologist*
Kleban, Morton Harold *psychologist*
Mancini, Nicholas Angelo *psychologist*
Nezu, Christine Maguth *clinical psychologist, educator*
Rescorla, Robert Arthur *psychology educator*
Rosenberg, Robert Allen *psychologist, educator, optometrist*

Pittsburgh
Donovan, John Edward *psychologist*
Fichman, Mark *industrial/organizational psychologist*
Fischhoff, Baruch *psychologist, educator*
Geiwitz, (Peter) James *psychologist, writer, researcher*
Johnson, Mark Henry *psychology educator*
Kotovsky, Kenneth *psychology educator*
Orbison, David Vaillant *clinical psychologist, consultant*
Perloff, Robert *psychologist, educator*
Reder, Lynne Marie *cognitive science educator*

Plymouth Meeting
Friedman, Philip Harvey *psychologist*

Polk
Hall, Richard Clayton *psychologist, consultant, researcher*

Shippensburg
Clark, Mary Diane *psychologist, educator*

University Park
Stern, Robert Morris *psychology educator and psychophysiology researcher*

Villanova
Bush, David Frederic *psychologist, educator*

West Chester University
Renner, Michael John *psychologist, biologist, educator*

Wilkes Barre
Brooks, Charles Irving *psychology educator*

Williamsport
Sutliff, Kimberly Ann *psychologist*

York
Casteel, Mark Allen *psychology educator*

RHODE ISLAND

Kingston
Rossi, Joseph Stephen *research psychologist*

Providence
O'Keeffe, Mary Kathleen *psychology educator*

Smithfield
Morahan-Martin, Janet May *psychologist, educator*

Warwick
Solomon, Richard Strean *psychologist, educator*

SOUTH CAROLINA

Abbeville
Cellura, Angele Raymond *psychologist*

Charleston
Beidel, Deborah Casamassa *clinical psychologist, researcher*

Clinton
Buggie, Stephen Edward *psychology educator*

Columbia
Shea, Mary Elizabeth Craig *psychologist, educator*

Greenville
Simrall, Dorothy Van Winkle *psychologist*

Greer
Hawkins, Janet Lynn *school psychologist*

Rock Hill
Prus, Joseph Stanley *psychology educator, consultant*

Spartanburg
Armstrong, Joanne Marie *clinical psychologist, family mediator*

SOUTH DAKOTA

Sioux Falls
Sadler, James Bertram *psychologist, clergyman*

TENNESSEE

Bristol
Dirlam, David Kirk *psychology educator*

Chattanooga
Warren, Amye Richelle *psychologist, educator*

Harrison
Fisher, Paul Douglas *psychologist, program director*

Johnson City
Isaac, Walter Lon *psychology educator*

Knoxville
Bell, Corinne Reed *psychologist*

Maryville
Crisp, Polly Lenore *psychologist*

Memphis
Johnson, Johnny *research psychologist, consultant*
Kreuz, Roger James *psychology educator*

Millington
Melcher, Jerry William Cooper *clinical psychologist, army officer*

Murfreesboro
Schmidt, Constance Rojko *psychology educator, researcher*

Nashville
Thompson, Travis *psychology educator, administrator, researcher*

TEXAS

Abilene
Hennig, Charles William *psychology educator*

Amarillo
Ayad, Joseph Magdy *psychologist*

Arlington
Cox, Verne Caperton *psychology educator*
Ickes, William *psychologist, educator*

Austin
Diehl, Randy Lee *psychology educator, researcher*
Drake, Stephen Douglas *clinical psychologist, health facility administrator*
Eldredge-Thompson, Linda Gaile *psychologist*
Holtzman, Wayne Harold *psychologist, educator*
Lobb, Michael Louis *psychologist*
Pickhardt, Carl Emile, III *psychologist*
Schade, Mark Lynn *psychologist*

Bastrop
O'Connell, Walter Edward *psychologist*

Bellaire
Mayo, Clyde Calvin *organizational psychologist*

Brooks AFB
Goettl, Barry Patrick *personnel research psychologist*
Spector, Jonathan Michael *research psychologist, cognitive scientist*
Stamper, David Andrew *psychologist*
Tirre, William Charles *research psychologist*

College Station
Murphy, Kathleen Jane *psychologist*

Corpus Christi
Cutlip, Randall Brower *retired psychologist, former college president*

Dallas
Dudley, George William *behavioral scientist, writer*
Karlson, Kevin Wade *trial, forensic, and clinical psychologist, consultant*
Kress, Gerard Clayton, Jr. *psychologist, educator*
Peek, Leon Ashley *psychologist*
Wolf, Michael Ellis *clinical psychologist*

Denton
Kennelly, Kevin Joseph *psychology educator*

El Paso
Cuevas, David *psychologist*
Malpass, Roy Southwell *psychology educator*

Fort Worth
Demaree, Robert Glenn *psychologist, educator*
Knight, Kevin Kyle *research psychologist*
Papini, Mauricio Roberto *psychologist*
Simpson, Dennis Dwayne *psychologist, educator*

Galveston
Miller, Todd Q. *social psychology educator*

Houston
Cadwalder, Hugh Maurice *psychology educator*
Dougan, Deborah Rae *neuropsychology professional*
Garnes, Delbert Franklin *clinical and consulting psychologist, educator*
Justice, (David) Blair *psychology educator, author*
Kershaw, Carol Jean *psychologist*
Mariotto, Marco Jerome *psychology educator, researcher*
Miller, Janel Howell *psychologist*
Novy, Diane Marie *psychologist*
Pomerantz, James Robert *psychology educator, academic admininstrator*
Prather, Rita Catherine *psychology educator*
Roney, Lynn Karol *psychologist*
Williams, M. Wright *psychologist, educator*

Killeen
Roberts, Mary Lou *school psychologist*

Nacogdoches
Clagett, Arthur F(rank), Jr. *psychologist, sociologist, qualitative research writer, retired sociology educator*
Ludorf, Mark Robert *cognitive psychologist, educator*
Speer, James Ramsey *developmental psychologist*

Odessa
McCullough, Gary William *psychology educator, researcher*

Pasadena
Meyer, Kathleen Anne *school psychologist*

San Antonio
Bloom, Wallace *psychologist*
Mylar, J(ames) Lewis *psychologist, consultant*
Ribble, Ronald George *psychologist, educator, writer*
Sherman, James Owen *psychologist*
Skelton, John Goss, Jr. *psychologist*

San Marcos
Fletcher, John Lynn *psychology educator*

Tyler
Loughmiller, Grover Campbell *psychologist, consultant*
Rudd, Leo Slaton *psychology educator, minister*

Waco
Hynan, Linda Susan *psychology educator*

UTAH

Logan
Van Dusen, Lani Marie *psychologist*

Midvale
Morris, Stephen Blaine *clinical psychologist*

Salt Lake City
McCusker, Charles Frederick *psychologist, consultant*

VERMONT

Brattleboro
Kotkov, Benjamin *clinical psychologist*

Middlebury
Gibson, Eleanor Jack (Mrs. James J. Gibson) *psychology educator*

VIRGINIA

Alexandria
Chatelier, Paul Richard *aviation psychologist, training company officer*

Arlington
Stollnitz, Fred *comparative psychologist, government program administrator*

Blacksburg
Ash, Philip *psychologist*

Charlottesville
McCarty, Richard Charles *psychology educator*
Owens, Justine Elizabeth *cognitive psychology educator*
Reppucci, Nicholas Dickon *psychologist, educator*
Scarr, Sandra Wood *psychology educator, researcher*
Wilson, Melvin Nathaniel *psychology educator*

Dayton
Couch, James Vance *psychology educator*

Lynchburg
Looney, Thomas Albert *psychologist, educator*

Manassas
Carvalho, Julie Ann *psychologist*

Awards Index Listings

Awards Index

The Awards Index is organized by general field of scientific or engineering endeavor and, within each field, organized by award-granting agency. Each listing is further arranged alphabetically by award, in descending chronology, with the award recipients listed alphabetically.

In order for a person to be listed below, he or she must have a biographical sketch in Marquis *Who's Who in Science and Engineering*.

AERONAUTICS & ASTRONAUTICS

American Institute of Aeronautics and Astronautics *Washington, DC*

Aerodynamic Decelerator Systems Award
1990 Maurice Paul Gionfriddo

Aerodynamics Award
1992 Joe F. Thompson
1991 Norman David Malmuth

Aerospace Communications Award
1992 Tadahiro Sekimoto
1990 John E. Keigler

Aerospace Contribution to Society Award
1993 James T. Rose
1990 Arthur Kantrowitz

Aerospace Software Engineering Award
1991 Christine Marlene Anderson

Air Breathing Propulsion Award
1991 F. Blake Wallace

Aircraft Design Award
1992 Richard Allen Hardy
D. Randall Kent, Jr.
Sherman N. Mullin
1991 Antonio L. Elias
1990 Alan Charlton Brown

Hugh L. Dryden Lectureship in Research
1992 Frederick Billig
1991 Vijaya V. Shankar
1990 Seymour Moses Bogdonoff

Durand Lectureship for Public Service
1992 Eugene Edzards Covert
1990 Konrad K. Dannenberg

Engineer of the Year Award
1992 Antonio L. Elias
1991 Frank C. Gillette, Jr.
1990 Clarence J. Wesselski

Fluid and Plasmadynamics Award
1992 Julian D. Cole
1991 Dennis M. Bushnell
1990 Hans Wolfgang Liepmann

Robert H. Goddard Astronautics Award
1992 James W. Plummer
1991 Eberhardt Rechtin
1990 Richard H. Truly

Daniel Guggenheim Medal Award
1992 Bernard L. Koff
1991 Hans J. P. von Ohain
1990 Joseph F. Sutter

Haley Space Flight Award
1993 Daniel Charles Brandenstein
1991 Bruce McCandless II
Kathryn D. Sullivan

International Cooperation Award
1992 Irving Statler
1991 Herbert Friedman
1990 Robert F. Freitag

William Littlewood Memorial Lecture
1991 Bernard L. Koff

Losey Atmospheric Sciences Award
1993 Moustafa T. Chahine
1992 Charles Gordon Little

George M. Low Space Transportation Award
1992 Alan M. Lovelace
1990 Edward C. Aldridge, Jr.

Plasmadynamics and Laser Award
1992 Abraham Hertzberg

Sylvanus A. Reed Aeronautics Award
1992 James Norton Krebs
1991 Richard H. Petersen

Space Processing Award
1992 August Ferdinand Witt

Space Science Award
1993 John Cromwell Mather
1992 Peter Andrew Sturrock
1991 Michael B. Duke

Space Systems Award
1992 John Anthony Jamieson
Edward E. Noneman

Structures, Structural Dynamics and Materials Award
1992 Elbert L. Rutan
1991 Hyman Norman Abramson
1990 Richard Hugo Gallagher

Thermophysics Award
1992 Leroy Stevenson Fletcher
1990 John Reid Howell

Von Karmen Lectureship in Astronautics
1993 James A. Abrahamson
1992 Hans Michael Mark
1991 John R. Casani

Wright Brothers Lectureship in Aeronautics
1992 Sherman N. Mullin
1991 Benjamin A. Cosgrove
Irving T. Waaland
1990 Edwin Irving Colodny

Wyld Propulsion Award
1992 Robert L. Sackheim
1991 Thomas Ferguson Davidson

American Society of Mechanical Engineers
New York, NY

Spirit of St. Louis Medal
1992 Holt Ashley
1990 Charles Feltz

Aviation Week & Space Technology *New York, NY*

Laurels Awards
Aeronautics/Propulsion
1993 Daniel M. Tellep

(Laurels Awards, continued)
Space/Missiles
1993 John Cromwell Mather
George Fitzgerald Smoot III

National Academy of Sciences *Washington, DC*

NAS Award in Aeronautical Engineering
1990 Robert Thomas Jones

Smithsonian Institution *Washington, DC*

National Air and Space Museum Trophy: Lifetime Achievement Award
1992 Francis M. Rogallo
1991 Arthur E. Raymond

AGRICULTURAL ENGINEERING

American Society of Agricultural Engineers
Saint Joseph, MI

ASAE Fellow
1991 Walter J. Ochs
John Deere Medal
1990 William R. Gill
Kishida International Award
1990 Gajendra Singh
Massey-Ferguson Medal
1991 Gerald W. Isaacs
Cyrus Hall McCormick–Jerome Increase Case Medal
1990 Gordon L. Nelson

AGRICULTURE

Agricultural Institute of Canada *Ottawa, ON, Canada*

AIC Fellowship Award
1992 John George Braidek
Keith Russell Collver
Richard James Huggard
Everett C. Lougheed
Robert Arthur Simmons
Anna K. Storgaard
John E. Swales
R. L. Wilkinson
1991 Umesh Chandra Gupta
Murray Henry Miller
Bruce Douglas Owen
Gary Garfield Storey
Alexander James Webster
1990 James Duncan Beaton
Charles E. Broadwell
Walter Bushuk
Ronald Lawrence Halstead
Bryan L. Harvey
Gary G. Runka

American Association of Cereal Chemists
Saint Paul, MN

Thomas Burr Osborne Medal
1992 Colin W. Wrigley
1991 Russell Carl Hoseney

American Society of Agronomy *Madison, WI*

Agronomic Achievement Award — Crops
1992 Henry Roger Boerma
1991 Charles N. Bollich
1990 Donald C. Rasmusson
Agronomic Research Award
1992 Arnel R. Hallauer
1990 Surajit Kumar Dedatta
CIBA-GEIGY Award in Agronomy
1992 Michael A. Schmitt
1991 Craig A. Beyrouty
1990 Paul R. Carter
Laurence V. Madden
Don Rummel

Soil Science Award
1992 M. Ali Tabatabai
1991 Malcom E. Sumner
1990 James M. Tiedje

Robert E. Wagner Award for Efficient Agriculture
1992 Charles A. Francis
1991 Dale R. Hicks
1990 Marcus M. Alley

United States Department of Agriculture
Washington, DC

Distinguished Service Award
1992 Ronald T. Noyes

Wolf Foundation *Bet Herzlia, Israel*

Wolf Foundation Prize (in agriculture)
1991 Shang-Fa-Yang

ARCHAEOLOGY

American Anthropological Association
Washington, DC

Alfred Vincent Kidder Award
1992 Kent Flannery

ASTRONOMY

American Astronomical Society *Washington, DC*

Dick Brouwer Award
1993 Alar Toomre
1992 Stanton J. Peale
1991 Martin Schwarzschild

Annie Jump Cannon Award in Astronomy
1993 Stefi Alison Baum
1992 Elizabeth A. Lada
1991 Jane Luu

George Ellery Hale Award
1992 Richard Tousey
1990 Horace W. Babcock

Dannie Heineman Prize for Astrophysics
1993 John C. Mather
1992 Bohdan Paczynski
1991 Wallace L. W. Sargent

Gerard P. Kuiper Prize
1992 Peter Goldreich

Newton Lacy Pierce Prize in Astronomy
1991 Kenneth Libbrecht

Henry Norris Russell Lectureship
1991 Donald Osterbrock

Beatrice M. Tinsley Prize
1992 Robert H. Dicke

Harold C. Urey Prize
1992 Jack J. Lissauer

Helen B. Warner Prize for Astronomy
1990 Ethan T. Vishniac

Astronomical Society of the Pacific *San Francisco, CA*

Catherine Wolfe Bruce Medal
1991 Donald Osterbrock

Meteoritical Society *Piscataway, NJ*

Leonard Medal
1991 Donald Delbert Clayton

National Academy of Sciences *Washington, DC*

Henryk Arctowski Medal
1993 John Alexander Simpson
1990 Peter A. Sturrock

Henry Draper Medal
1993 Ralph Asher Alpher
Robert Herman

J. Lawrence Smith Medal
1991 Robert Mowbray Walker

James Craig Watson Medal
1991 Maarten Schmidt

BIOLOGY

American Ornithologists' Union *Washington, DC*

William Brewster Memorial Award
1992 Ned Keith Johnson

Eliot Coues Award
1992 Francis Crews James

American Society for Biochemistry and Molecular Biology *Bethesda, MD*

AMGEN Award
1992 Richard C. Mulligan

ASBMB-Merck Award
1992 William Platt Jencks
William C. Rose Award in Biochemistry
1993 Irving Myron Klotz
1992 Eugene Patrick Kennedy
1991 Robert Lee Hill

American Society for Cell Biology *Bethesda, MD*

E. B. Wilson Award
1992 Shinya Inoue

Canadian Society of Zoologists *Ottawa, ON, Canada*

Fry Medal
1993 David John Randall
1992 David R. Jones
1991 Roger G. H. Downer

CIBA-GEIGY Corporation *Greensboro, NC*

CIBA-GEIGY/Entomological Society of America Award
1992 Frederick Whiton Knapp
1991 Sidney Kunz

Columbia University *New York, NY*

Louisa Gross Horowitz Prize
1992 Edward B. Lewis
Christiane Nüsslein-Volhard
1991 Richard Ernst
Kurt Wuthrich

Crustacean Society *San Antonio, TX*

Excellence in Research
1990 Donald T. Anderson

Entomological Society of Canada *Ottawa, ON, Canada*

Entomological Society of Canada Gold Medal Award
1992 Glenn B. Wiggins

(Entomological Society of Canada Gold Medal Award, continued)
1991 Roger G. H. Downer
1990 Stephen Tobe
C. Gordon Hewitt Award
1992 Daniel L. Johnson
1991 Murray Isman
1990 Bernard Roitberg

Federation of American Societies for Experimental Biology *Bethesda, MD*

Women's Excellence in Sciences Award
1993 Susan E. Leeman
1992 Bettie Sue Siler Masters
1991 Ellen S. Vitetta

Genetics Society of America *Bethesda, MD*

Thomas Hunt Morgan Medal
1993 Ray D. Owen
1992 Edward H. Coe, Jr.
1991 Armin Dale Kaiser

National Academy of Sciences *Washington, DC*

Richard Lounsbery Award
1993 Stanley Ben Prusiner
Bert Vogelstein
1991 Marc Wallace Kirschner
Harold M. Weintraub
NAS Award in Molecular Biology
1993 Peter Sung-Bai Kim
1992 Bruce S. Baker
Thomas Warren Cline
1991 Steven L. McKnight
Selman A. Waksman Award in Microbiology
1993 Boris Magasanik
1991 Melvin I. Simon

Royal Society of Canada *Ottawa, ON, Canada*

Flavelle Medal
1990 Peter William Hochachka

BOTANY

American Phytopathological Society *Saint Paul, MN*

Ruth Allen Award
1993 Richard G. Christie
John R. Edwardson
Ernest Hiebert
Dan E. Purcifull
1991 George E. Templeton
Award of Distinction
1993 James F. Tammen
CIBA-GEIGY Award
1993 Wayne F. Wilcox
1992 Margaret E. Daub
1991 Jim Marois
Extension Award
1991 Malcolm C. Shurtleff
1990 Jose M. Amador
Lee M. Hutchins Award
1993 Jerry Kazumitsu Uyemoto

American Society of Plant Physiologists *Rockville, MD*

Charles Reid Barnes Life Membership Award
1992 Paul Karl Stumpf
1990 Frederick T. Addicott
Adolph E. Gude, Jr. Award
1992 Martin Gibbs

Stephen Hales Prize
1992 Clarence A. Ryan
1990 Joseph E. Varner

Charles F. Kettering Award
1992 Antony Crofts

Charles Albert Shull Award
1991 Daniel Cosgrove

American Society of Plant Taxonomists *Albany, NY*

George R. Cooley Award
1992 Eric Knox

Asa Gray Award
1992 Albert Charles Smith

Botanical Society of America *Storrs, CT*

BSA Merit Awards
1992 William Hardy Eshbaugh
Raymond Carl Jackson
Beryl B. Simpson
1990 Kenton L. Chambers

Darbaker Prize
1991 John P. Smol

Henry Allan Gleason Award
1992 Warren H. Wagner, Jr.

Edgar T. Wherry Award
1992 Tom A. Ranker

CIBA-GEIGY Corporation *Greensboro, NC*

CIBA-GEIGY/Weed Science Society of America Award
1992 Michael Owen
1991 Donald L. Wyse
1990 Stephen O. Duke

National Academy of Sciences *Washington, DC*

Gilbert Morgan Smith Medal
1991 Jean-David Rochaix

Phycological Society of America *Williamsburg, VA*

Gerald W. Prescott Award
1991 Richard M. Crawford
Frank E. Round
1989 William J. Woelkerling

Luigi Provasoli Award
1990 David L. Baille
1988 Louis D. Druehl

CHEMICAL ENGINEERING

American Chemical Society *Washington, DC*

E. V. Murphree Award in Industrial and Engineering Chemistry
1993 James J. Carberry

American Institute of Chemical Engineers *New York, NY*

Founders Award
1991 Donald F. Othmer

Robert E. Wilson Award in Nuclear Chemical Engineering
1992 Finis S. Patton, Jr.

Electrochemical Society *Pennington, NJ*

David C. Grahame Award
1993 Arthur T. Hubbard

Society of the Plastics Industry *Washington, DC*

Clare E. Bacon Person of the Year Award
1993 Robert Dieterle
1992 Francis W. Como
1991 Jerry Purcell

CHEMISTRY

American Association for Clinical Chemistry *Washington, DC*

International Fellowship
1992 Carl A. Burtis, Jr.

Outstanding Contributions to Clinical Chemistry
1993 Jocelyn Muriel Hicks

American Association of Textile Chemists and Colorists *Research Triangle Park, NC*

Olney Medal
1992 Bethlehem Kottes Andrews
1991 Robert J. Harper, Jr.
1990 Hans-Dietrich H. Weigmann

American Chemical Society *Washington, DC*

ACS Award for Creative Invention
1993 Albert A. Carr
ACS Award for Creative Work in Fluorine Chemistry
1993 Ronald Eric Banks
ACS Award for Creative Work in Synthetic Organic Chemistry
1993 Kyriacos C. Nicolaou
ACS Award for Distinguished Service in the Advancement of Inorganic Chemistry
1993 Theodore L. Brown
ACS Award for Nuclear Chemistry
1993 Richard M. Diamond
ACS Award in Analytical Chemistry
1993 Jeanette G. Grasselli
1991 Royce W. Murray
ACS Award in Applied Polymer Science
1993 Owen Wright Webster
ACS Award in Chromatography
1993 James W. Jorgenson
ACS Award in Colloid or Surface Chemistry
1993 David Wayne Goodman
ACS Award in Industrial Chemistry
1993 Larry Flack Thompson
ACS Award in Inorganic Chemistry
1993 Gregory Joseph Kubas
ACS Award in Organometallic Chemistry
1993 Robert H. Crabtree
ACS Award in Petroleum Chemistry
1993 Bruce Clark Gates
ACS Award in Polymer Chemistry
1993 Takeo Saegusa
ACS Award in Pure Chemistry
1993 Jeremy M. Berg
ACS Award in Separation Science and Technology
1993 James Rutherford Fair, Jr.
ACS Award in Theoretical Chemistry
1993 Martin Karplus
Roger Adams Award in Organic Chemistry
1993 Elias James Corey
1991 Gilbert Josse Stork
Arthur W. Adamson Award for Distinguished Service in the Advancement of Surface Chemistry
1993 David M. Hercules
Alfred R. Bader Award in Bioinorganic or Bioorganic Chemistry
1993 William Wallace Cleland
Earle B. Barnes Award for Leadership in Chemical Research Management
1993 William David Emmons
Arthur C. Cope Scholar Awards
1993 Peter Beak
Peter Chen
Peter B. Dervan
Tohru Fukuyama
Alexander M. Klibanov
Robert L. Letsinger
Andrew G. Myers
Joanne Stubbe
Chi-Huey Wong
Fred Wudl
Peter Debye Award in Physical Chemistry
1993 Frank Sherwood Rowland
Wilford T. Doherty Award
1993 Paul A. Srere
Gustavus John Esselen Award
1993 James G. Anderson
1992 Bruce N. Ames
Francis P. Garvan–John M. Olin Medal
1993 Edith Marie Flanigen
1991 Darleane C. Hoffman
1990 Cynthia M. Friend
Willard Gibbs Medal
1993 Peter B. Dervan
1992 Harry B. Gray
1991 Gunther Wilke
Charles Goodyear Medal
1993 Leo Mandelkern
Ernest Guenther Award in the Chemistry of Natural Products
1993 Amos B. Smith III
Charles H. Herty Medal
1993 Leon H. Zalkow
Joel Henry Hildebrand Award in the Theoretical and Experimental Chemistry of Liquids
1993 Jerome K. Percus
Ralph F. Hirschmann Award in Peptide Chemistry
1993 Daniel H. Rich
Harrison Howe Award
1992 Richard J. Saykally
Claude S. Hudson Award in Carbohydrate Chemistry
1993 Irwin J. Goldstein
Ralph K. Iler Award in the Chemistry of Colloidal Materials
1993 Egon Matijević
Victor K. LaMer Award
1993 Robert D. Tilton
Irving Langmuir Prize in Chemical Physics
1993 James David Litster
1992 John Ross
1991 Richard Errett Smalley
Eli Lilly Award in Biological Chemistry
1993 Stuart L. Schreiber
Herman F. Mark Award
1992 Edwin J. Vandenberg
Harry and Carol Mosher Award
1992 Carl Randolph Johnson
William H. Nichols Medal
1993 Richard Errett Smalley
1992 Koji Nakanishi
1991 J. Calvin Giddings
James Flack Norris Award in Physical Organic Chemistry
1993 Keith Usherwood Ingold
Linus Pauling Award
1993 Richard N. Zare
1992 Kenneth Wiberg
1991 Rudolph Arthur Marcus
Perkin Medal
1993 Lubomyr T. Romankiw
1992 Edith Marie Flanigen
1991 Miguel A. Ondetti
Pfizer Award in Enzyme Chemistry
1993 Michael H. Gelb
Priestley Medal
1994 Howard Ensign Simmons, Jr.
1993 Robert W. Parry
1992 Carl Djerassi

Repligen Corporation Award in Chemistry of Biological Processes
1993 Jeremy R. Knowles

Kenneth A. Spencer Award
1992 Clarence A. Ryan, Jr.

Charles H. Stone Award
1992 Craig L. Hill

Henry H. Storch Award in Fuel Chemistry
1993 Martin Leo Gorbaty

Roy W. Tess Award in Coatings
1993 Larry Flack Thompson

Richard C. Tolman Medal
1993 Peter C. Ford

American Institute of Chemists *Bethesda, MD*

Chemical Pioneer Award
1993 Derek Harold Richard Barton
Robert Bruce Merrifield
George Andrew Olah
Jule Anthony Rabo

Gold Medal
1993 Fred Basolo
1992 Roy Whistler
1991 Bruce Ames

Canadian Biochemical Society *London, ON, Canada*

Boehringer-Mannheim Canada Prize
1993 Michel Chretien
1991 Jack Riordan

Chemical Institute of Canada *Ottawa, ON, Canada*

Norman and Marion Bright Memorial Award
1993 Ray A. Cullen
1992 Steve Elchuk
1991 Alex P. Mykytiuk

Chemical Institute of Canada Medal
1993 Paul Brumer
1992 Donald A. Ramsay
1991 Keith Yates

Montreal Medal
1993 Henry G. Thode
1992 Geraldine A. Kenny-Wallace
1991 Robert H. Marchessault

Dickinson College *Carlisle, PA*

Joseph Priestley Award
1992 Solomon H. Snyder

(Joseph Priestley Award, continued)
1991 Harry B. Gray
1990 Wallace S. Broecker

National Academy of Sciences *Washington, DC*

NAS Award for Chemistry in Service to Society
1993 Harold Sledge Johnston
1991 Vladimir Haensel

NAS Award in Chemical Sciences
1993 Richard H. Holm
1992 Donald James Cram
1991 Richard N. Zare

Nobel Foundation *Stockholm, Sweden*

Nobel Prize in Chemistry
1993 Kary B. Mullis
Michael Smith
1992 Rudolph Arthur Marcus
1991 Richard R. Ernst

Society of Chemical Industry, American Section *New York, NY*

Chemical Industry Medal of the American Section
1993 H. Eugene McBrayer

University of Wisconsin, Theoretical Chemistry Institute *Madison, WI*

Joseph O. Hirschfelder Prize
1993 Rudolph Arthur Marcus
1991 Benjamin Widom

Robert A. Welch Foundation *Houston, TX*

Welch Award in Chemistry
1993 Gilbert Josse Stork
1992 Richard Errett Smalley
1991 Edwin G. Krebs
Earl R. Stadtman

Wolf Foundation *Bet Herzlia, Israel*

Wolf Prize (in chemistry)
1993 Ahmed Hassan Zewail
1992 John Anthony Pople
1991 Richard R. Ernst
Alexander Pines

CIVIL & STRUCTURAL ENGINEERING

American Concrete Institute *Detroit, MI*

Henry C. Turner Medal
1993 Norman L. Scott
1992 Clifford L. Freyermuth
1991 Daniel P. Jenny

American Institute of Architects *Washington, DC*

Gold Medal
1993 Eamonn Kevin Roche
1992 Benjamin Thompson
1991 Charles W. Moore

American Society of Civil Engineers *New York, NY*

Alfred M. Freudenthal Medal
1992 Pol D. Spanos
1990 James T. P. Yao

James Laurie Prize
1990 James Alexander Caywood, III

Nathan M. Newmark Medal
1990 Lawrence Goodman

Canadian Society of Civil Engineering
Montreal, PQ, Canada

Albert E. Berry Medal
1992 Gerhard William Heinke
1991 Donald S. Mavinic
1990 Steve E. Hrudey

Les Prix Camille A. Dagenais Award
1992 Michael Isaacson
1991 Bommanna Gounder Krishnappan
1990 James J. Sharp

Sir Casimir Gzowski Medal
1992 Anita Brattland
D. J. Laurie Kennedy
1991 Denis Mitchell
Patrick Paultre
1990 André Filiatrault

T. C. Keefer Medal
1992 S. J. Stanley
1991 J. A. Kells
C. D. Smith
1990 Peter S. McCreath

Horst Leipholz Medal
1992 N. R. Thomson
1991 Lorne W. Gold
1990 Anthony P. S. Selvadurai

Le Prix P. L. Pratley Award
1992 Fred M. P. Bartlett
Peter G. Buckland
D. J. Laurie Kennedy
1991 Peter G. Buckland
Brian D. Morgenstern
1990 Andrew Scanlon

Le Prix A. B. Sanderson Award
1992 Amin Ghali
1991 J. Springfield
1990 M. Saeed Mirza

Le Prix E. Whitman Wright Award
1992 Mehmet Baltacioglu
1991 Siegfried F. Stiemer
1990 Aftab A. Mufti

Franklin Institute *Philadelphia, PA*

Frank P. Brown Medal
1992 John W. Fisher

National Academy of Sciences *Washington, DC*

Gibbs Brothers Medal
1993 Olin James Stephens, Jr.
1991 Bruce Garfield Collipp

Pritzker Architecture Prize *New York, NY*

Pritzker Architecture Prize
1993 Fumihiko Maki

EARTH SCIENCES

American Association of Petroleum Geologists
Tulsa, OK

Michel T. Halbouty Human Needs Award
1993 Gordon B. Oakeshott
1992 Donald Frederick Todd

Sidney Powers Memorial Medal Award
1993 Robert R. Berg
1992 Sherman A. Wengerd

American Crystallographic Association *Buffalo, NY*

Martin J. Buerger Award
1991 Jack D. Dunitz

Fankuchen Memorial Award in X-Ray Crystallography
1992 Donald Louis Dvorak Caspar

A. L. Patterson Award
1993 George M. Sheldrick

American Geographical Society *New York, NY*

Charles P. Daly Medal
1991 Robert P. Sharp

Samuel Finely Breese Morse Medal
1991 Alexander Melamid

American Geophysical Union *Washington, DC*

William Bowie Medal
1992 Alfred O. C. Nier
1990 Don L. Anderson
Eugene N. Parker

Maurice Ewing Medal
1993 Charles S. Cox

John Adam Fleming Award
1993 Stanislov Iosifovich Braginsky

Robert E. Horton Medal
1993 Luna B. Leopold
James B. Macelwane Medal
1993 Eric Kunze
Terry Charles Wallace, Jr.
1992 David G. Sibeck
Roger Revelle Medal
1993 Edward N. Lorenz
Waldo E. Smith Medal
1993 Earl G. Droessler

American Institute of Professional Geologists
Arvada, CO

Ben H. Parker Memorial Medal
1993 Daniel N. Miller, Jr.
1992 Kenneth N. Weaver
1991 Susan M. Landon

American Meteorological Society *Boston, MA*

Cleveland Abbe Award for Distinguished Service to Atmospheric Sciences by an Individual
1993 Kenneth C. Spengler
Award for Outstanding Contributions to the Advance of Applied Meteorology
1993 Richard H. Thuillier
Jule G. Charney Award
1993 Abraham H. Oort
1992 Lance F. Bosart
Banner I. Miller Award
1992 Kerry Andrew Emanuel
Richard Rotunno
Carl-Gustaf Rossby Research Medal
1993 John Michael Wallace
1992 Syukuro Manabe
Harald Ulrick Sverdrup Gold Medal
1993 Tim P. Barnett
1992 Mark A. Cane
1991 Klaus Wyrtki

Canadian Association of Geographers
Montreal, PQ, Canada

CAG Award for Scholarly Distinction in Geography
1992 James Simmons
Ming-Ko Woo
1991 Kenneth Hewitt
1990 Paul-Yves Denis
Michael Goodchild
CAG Award for Service to the Profession of Geography
1993 Fernand Grenier
1992 Peter Donald Kerr
1991 Helen Rowland
Roger Tomlinson

Canadian Institute of Mining and Metallurgy
Montreal, PQ, Canada

H. T. Airey Award
1992 Ghyslain Dubé
1991 C. A. Sutherland
1990 D. Robert Weir
Barlow Medal
1992 H. Scott Swinden
1991 Craig H. B. Leitch
Selwyn G. Blaylock Medal
1992 Come Carbonneau
1991 Henry K. Imorde
1990 Norman Bell Keevil
Julian Boldy Memorial Award
1992 Les Fyffe
1990 Chris M. Healey
1989 Arthur E. Soregaroli
Falconbridge Innovation Award
1992 Robert Gum Hong Lee
G. S. Savard
Inco Medal
1992 Robert E. Hallbauer
1991 Alfred Powis
1990 Patrick C. MacCulloch
Sherritt Hydrometallurgy Award
1992 Bruce R. Conrad
1991 Pierre Claessens
1990 R. C. Kerby

Canadian Society of Petroleum Geologists
Calgary, AB, Canada

R. J. W. Douglas Memorial Medal
1992 Philip Simony
1991 Paul Felix Hoffman
1990 Roger G. Walker

Columbia University *New York, NY*

G. Unger Vetlesen Prize
1993 Walter Heinrich Munk

Geochemical Society *Washington, DC*

F. W. Clarke Medal
1991 David Michael Sherman
V. M. Goldschmidt Award
1992 Stanley Robert Hart
1991 Alfred E. Kingwood

Geological Society of America *Boulder, CO*

Arthur L. Day Medal
1993 Alfred George Fischer
1992 Susan W. Kieffer
1991 Ian Carmichael
Penrose Medal
1993 Hugh P. Taylor, Jr.
1992 John F. Dewey
1991 William R. Dickinson

Mineralogical Society of America *Washington, DC*

Roebling Medal
1992 Hatten Schuyler Yoder, Jr.
1991 E-an Zen

National Academy of Sciences *Washington, DC*

Alexander Agassiz Medal
1992 Joseph Lee Reid
Arthur L. Day Prize and Lectureship
1993 Hiroo Kanamori
1990 Ho-kwang Mao
G. K. Warren Prize
1990 John Rybolt L. Allen

National Speleological Society *Huntsville, AL*

William J. Stephenson Award for Outstanding Service
1992 Sheck Exley
1991 Doug Rhodes

Royal Canadian Geographical Society
Ottawa, ON, Canada

Massey Medal
1992 Stewart Dixon MacDonald
1991 George Donald Hobson

Royal Society of Canada *Ottawa, ON, Canada*

Bancroft Award
1992 Godfrey S. Nowlan
1990 Steven D. Scott
Willet G. Miller Medal
1991 Ján Veizer
1989 William H. Mathews

Society for Mining, Metallurgy, and Exploration
Littleton, CO

Daniel C. Jackling Award
1992 Milton H. Ward
1991 Hilbert Douglas Dahl

Society for Sedimentary Geology *Tulsa, OK*

Raymond C. Moore Medal for Paleontology
1993 Reuben J. Ross, Jr.
1992 Raymond C. Gutschick
1991 Erle Kauffman
Francis J. Pettijohn Medal for Sedimentology
1994 Gerard V. Middleton
1993 Charles Campbell
1992 Paul Potter
Francis P. Shepard Medal
1994 Maria Cita
1993 William B. F. Ryan
1992 John D. Milliman
William H. Twenhofel Medal
1994 Harold G. Reading
1993 Robert H. Dott, Jr.
1992 Peter Vail

Society of Economic Geologists *Golden, CO*

Lindgren Award
1993 Naomi Oreskes
1992 Mark Barton
1991 Kevin L. Shelton

ELECTRICAL ENGINEERING

Franklin Institute *Philadelphia, PA*

Stuart Ballantine Medal
1992 Rolf Landauer

Institute of Electrical and Electronics Engineers
Washington, DC

Thomas A. Edison Medal
1991 John Lewis Moll
1990 Archie W. Straiton
Founders Medal
1990 Erich Bloch
Irwin Dorros
IEEE Medal of Honor
1991 Leo Esaki
David Sarnoff Award
1990 Leroy L. Chang

ENGINEERING, GENERAL

American Association of Engineering Societies
Washington, DC

Chairman's Award
1993 George E. Brown, Jr.
1992 David T. Kearns
1991 James D. Watkins
National Engineering Award
1993 Ruben Frederick Mettler
1992 Bonnie J. Dunbar
1991 Norman R. Augustine

Association of Engineering Geologists *Sudbury, MA*

Claire P. Holdredge Award
1991 Richard W. Galster
1990 Robert B. Johnson

Canadian Council of Professional Engineers
Ottawa, ON, Canada

Canadian Engineers' Gold Medal Award
1992 Douglas T. Wright
1991 Robert D. Neill
1990 Gerald J. Maier

Young Engineer Achievement Award
1992 Katherine Crewe
1991 Phil Mundy

Columbia Engineering School Alumni Association
New York, NY

Pupin Medal
1992 Norman F. Ramsey

Engineering Institute of Canada *Ottawa, ON, Canada*

Sir John Kennedy Medal
1991 Russell Stafford Allison

Julian C. Smith Medal
1992 Angus A. Bruneau
Roland Doré
1991 Victor Milligan
Hugh C. Rynard
1990 James R. McFarlane
Robert Stollery

John B. Stirling Medal
1992 William A. Devereaux
Peter M. Wright
1991 Alan B. Connelly
1990 Vijay Bharghava
Michael Bozozuk

National Academy of Engineering *Washington, DC*

Arthur M. Bueche Award
1993 Ralph Edward Gomory
1992 Ruben Frederick Mettler
1991 Norman R. Augustine

Distinguished Honoree
1993 James H. Doolittle

Charles Stark Draper Prize
1993 John Backus
1992 Frank Whittle
1991 Hans J. P. von Ohain

Founders Award
1993 William Redington Hewlett
1992 George Harry Heilmeier
1991 George W. Housner

National Society of Professional Engineers
Alexandria, VA

Federal Engineer of the Year Award
1993 Randy L. Haupt
1992 Jonathan P. Deason
1991 M. Y. Shahin

NSPE Award
1993 William R. Kimel
1992 John W. Reh
1991 James Johnson Duderstadt

Young Engineer of the Year Award
1993 William R. Toole
1992 Josephine L. Emerick
1991 Roddy Rogers

Optical Society of America *Washington, DC*

Engineering Excellence Award
1992 Bertrand H. Johnson
Louis A. Koszi
Anthony Phillips

Joseph Fraunhofer Award
1992 James C. Wyant
1991 James G. Baker

David Richardson Medal
1991 Gary Keith Starkweather

John Tyndall Award
1991 David N. Payne
1990 Thomas G. Giallorenzi

Royal Society of Canada *Ottawa, ON, Canada*

Thomas W. Eadie Medal
1991 Ernest A. McCulloch

Society for Mining, Metallurgy, and Exploration
Littleton, CO

Howard N. Evenson Award
1992 Howard L. Hartman
1991 John L. Schroder, Jr.

Antoine M. Gaudin Award
1992 Jan D. Miller
1991 Frank F. Aplan

Rock Mechanics Award
1992 J. F. T. Agapito
1991 Steven L. Crouch

ENVIRONMENT

American Association of Engineering Societies
Washington, DC

Joan Hodges Queneau Award
1992 Hsieh Wen Shen
1990 Joseph Tso-Ti Ling

American Association of Zoological Parks and Aquariums *Wheeling, WV*

Edward H. Bean Memorial Award
1992 Don Farst
Louis Garibaldi

American Chemical Society *Washington, DC*

ACS Award in Creative Advances in Environmental Science and Technology

1993 John H. Seinfeld

American Society for Microbiology *Washington, DC*

ASM Award in Applied and Environmental Microbiology

1993 David C. White
1992 James M. Tiedje
1991 Ronald M. Atlas

Ecological Society of America *Tempe, AZ*

Eminent Ecologist

1992 Frank A. Pitelka
1991 W. Dwight Billings
Nelson G. Hairston, Sr.
1990 William Ricker

Québec Zoological Gardens *Charlesbourg, PQ, Canada*

Snowy Owl Conservation Award

1992 Peter Lane
1991 Annemarie Roth

United States Department of Energy *Washington, DC*

Sadi Carnot Award in Energy Conservation

1991 Robert H. Williams

John Ericsson Award in Renewable Energy

1991 Melvin Calvin

University of Southern California *Los Angeles, CA*

Tyler Prize for Environmental Achievement

1992 Robert Marshall White
1991 Charles Everett Koop
1990 Thomas Eisner
Jerrold Meinwald

Wildlife Society *Bethesda, MD*

Aldo Leopold Memorial Award

1993 L. David Mech
1992 Glen C. Sanderson
1991 Jack Ward Thomas

INFORMATION SCIENCE

American Society for Information Science
Silver Spring, MD

ASIS Research Award

1991 Abraham Bookstein

Association for Computing Machinery *New York, NY*

A. M. Turing Award

1992 Butler Wright Lampson
1990 Fernando J. Corbato

Cambridge Center for Behavioral Studies
Cambridge, MA

Loebner Prize (Turing Test Award)

1992 Joseph Weintraub

Data Processing Management Association
Park Ridge, IL

Distinguished Information Sciences Award

1992 Audrey Kathleen Hennessey
1991 Robert P. Campbell
1990 Mitchell Kapor

MANUFACTURING ENGINEERING

American Society of Mechanical Engineers
New York, NY

William T. Ennor Manufacturing Technology Award

1992 Bei Tse Chao
Kenneth J. Trigger
1991 Kuo-King Wang

M. Eugene Merchant Manufacturing Medal of ASME/SME

1992 Günter Spur
1991 Edson I. Gaylord
1990 Stephen Dale Antolovich

American Institute of Industrial Engineers
Atlanta, GA

Baker Distinguished Researcher Award
1991 Donald B. Chaffin
Hugh Donald Ratliff
1990 John Austin White, Jr.

Gilbreth Industrial Engineering Award
1991 A. Alan B. Pritsker
1990 Joe H. Mize

Canadian Association of Physicists
Ottawa, ON, Canada

Outstanding Achievement in Industrial and Applied Physics
1993 Jacques J. A. Beaulieu
1991 Paul Webb

Industrial Research Institute *Washington, DC*

IRI Achievement Award
1992 Richard H. Frenkiel
1990 Leonard S. Cutler

IRI Medal
1992 John S. Mayo
1991 Mary L. Good
1990 Edward M. Scolnick

National Academy of Sciences *Washington, DC*

NAS Award for the Industrial Application of Science
1993 Nick Holonyak, Jr.
1990 Carl Djerassi

Society of Manufacturing Engineers *Dearborn, MI*

SME Gold Medal
1991 Keith E. McKee
1990 James L. Koontz

SME Albert M. Sargent Progress Award
1991 John A. Manoogian
1990 Inyong Ham

SME Frederick W. Taylor Research Medal
1991 Clayton Daniel Mote, Jr.
1990 Kazuaki Iwata

Eli Whitney Productivity Award
1991 Earl E. Walker
1990 Thage Berggren

MATERIALS SCIENCE

Acta Metallurgica *Bethesda, MD*

Acta Metallurgica Gold Medal
1993 William D. Nix
1992 Jerome B. Cohen
1991 Jun Kondo

Acta/Scripta Metallurgica Lecturer
1993 Masahiro Meshii
Alton D. Romig
1992 John Campbell
Malcolm McLean
1991 Robert A. Rapp
Brian Wilshire

J. Herbert Holloman Award
1993 Frederick Seitz
1991 Alan Cottrell

American Ceramic Society *Westerville, OH*

Albert Victor Bleininger Award
1993 Fred Martin Ernsberger
1992 Keith K. Kappmeyer
1991 Van Derck Frechette

Samuel Geijsbeek Award for Innovation in Ceramics
1993 George Halsey Beall

John Jeppson Medal
1993 George Halsey Beall
1992 John Burnette MacChesney
1991 Robert E. Newnham

American Chemical Society *Washington, DC*

ACS Award in the Chemistry of Materials
1993 Rustum Roy

Exxon Solid-State Chemistry Fellowship Award
1993 William S. Hammack

Frederic Stanley Kipping Award in Organosilicon Chemistry
1990 John L. Speier, Jr.

American Physical Society *New York, NY*

David Adler Lectureship Award in the Field of Materials Physics
1993 Simon C. Moss
1991 John R. Smith

International Prize for New Materials
1993 Gordon C. Osbourn
Richard Errett Smalley
1992 Robert F. Curl
Harold Kroto
1991 Francis J. DiSalvo, Jr.
Frederic Holtzberg

American Society of Mechanical Engineers
New York, NY

Arpard L. Nadai Award
1992 George J. Dvorak
1991 John W. Hutchinson

Franklin Institute *Philadelphia, PA*

Francis J. Clamer Medal
1993 Alan Graham MacDiarmid

Materials Research Society *Pittsburgh, PA*

Von Hippel Award
1992 Michael F. Ashby

Minerals, Metals, and Materials Society
Warrendale, PA

Robert Lansing Hardy Gold Medal Award
1993 Glenn S. Daehn
William Hume-Rothery Award
1993 John Werner Cahn
1992 James Charles Phillips
Champion H. Mathewson Gold Medal Award
1993 Hong Yong Sohn
1992 Diran Apelian

MATHEMATICS

American Mathematical Society *Providence, RI*

Frank Nelson Cole Prize in Algebra
1992 Shigejumi Mori
Frank Nelson Cole Prize in Number Theory
1992 Karl Rubin
Paul A. Vojta
Delbert Ray Fulkerson Fund
1992 Alan M. Frieze
Alfred B. Lehma
Ruth Lyttle Satter Prize in Mathematics
1993 Lai-Sang Young
Leroy P. Steele Prizes
1993 Peter D. Lax
1992 James Gilbert Glimm
Ravi Kannan
1991 Armand Borel
Eugenio Calabi
Jean-François Treves
Oswald Veblen Prize in Geometry
1991 Andrew J. Casson
Clifford H. Taubes
Norbert Wiener Prize in Applied Mathematics
1990 Michael Aizenman
Jerrold E. Marsden

American Physical Society *New York, NY*

Dannie N. Heineman Prize for Mathematical Physics
1992 Martin Charles Gutzwiller
Stanley Mandelstam
1991 Jürg Fröhlich
Thomas C. Spencer

International Mathematical Union
Helsinki, Finland

Rolf Nevanlinna Prize
1990 Alexander A. Razborov

National Academy of Sciences *Washington, DC*

NAS Award in Applied Mathematics and Numerical Analysis
1992 Andrew J. Majda
NAS Award in Mathematics
1992 Robert Duncan MacPherson

Wolf Foundation *Bet Herzlia, Israel*

Wolf Foundation Prize (in mathematics)
1990 Ennio de Giorgi
Ilya Piatetski-Shapiro

MECHANICAL ENGINEERING

American Society of Mechanical Engineers
New York, NY

ASME Medal
1992 Daniel C. Drucker
1990 Harley A. Wilhelm
Per Bruel Gold Medal for Noise Control and Acoustics
1992 Miguel C. Junger
1991 Alan Powell
Edwin F. Church Medal
1992 Stephen Juhasz
1991 Joseph A. Falcon
1990 James R. Welty
Codes and Standards Medal
1992 Walter R. Mikesell, Jr.
1991 Arthur R. Machell
1990 Robert J. Bosnak

Fluids Engineering Award
1992 Christopher E. Brennen
1991 Frank M. White
1990 Turgut Sarpkaya

Freeman Scholar Award
1992 William A. Sirignano
1990 Budugur Lakshminarayana

Gas Turbine Award
1992 James H. Leylek
David C. Wisler
1991 Thomas J. Hajek
Bruce V. Johnson
Joel H. Wagner
1990 Roger L. Davis
David E. Hobbs
Harris D. Weingold

Heat Transfer Memorial Award
1992 Vijay K. Dhir
Thomas F. Irvine, Jr.
Wataru Nakayama
1991 John Reid Howell
Suhas V. Patankar
1990 Michael M. Chen
Alfred Linden Crosbie

Mayo D. Hersey Award
1992 Maurice Godet
1991 Kenneth L. Johnson
1990 Herbert S. Cheng

Henry Hess Award
1992 Jeffrey S. Marshall
1990 Stephen E. Bechtel

Holley Medal
1991 James Robert Thompson, Jr.
1990 Roy J. Plunkett

Soichiro Honda Medal
1992 Hiroyuki Hiroyasu
1991 Hans C. List
1990 Charles M. Heinen

Honorary Members
1992 Richard Monson Christensen
Ferdinand Freudenstein
Richard J. Goldstein
Bernard L. Koff
1991 Simon Ostrach
Eric Reissner
George N. Sandor
1990 William E. Cooper
George Herrmann

Internal Combustion Engine Award
1992 John A. Kimberley
1991 Fred S. Schaub
1990 Daniel C. Garvey

James N. Landis Medal
1992 George V. McGowan
1991 William States Lee
1990 John E. Dolan

Bernard F. Langer Nuclear Codes and Standards Award
1992 Marcus N. Bressler
1991 T. Eugene Northup
1990 Edward L. Williamson

Gustus L. Larson Memorial Award
1992 George P. Peterson

(Gustus L. Larson Memorial Award, continued)
1991 Pol D. Spanos
1990 Dale E. Klein

H. R. Lissner Award
1992 John C. Chato
1991 Savio Lau-Yuen Woo
1990 Albert B. Schultz

Machine Design Award
1992 Edward J. Haug, Jr.
1991 Francis Rendel Erskine Crossley
1990 Charles R. Mischke

Melville Medal
1992 Bharat Bhushan
Arunava Majumdar
1991 Koichi Hata
Akira Sakurai
Masahiro Shiotsu
1990 Shu Chien
Cheng Dong
Geert Wilfried Schmid-Schoenbein
Richard Skalak
Kuo-Li Paul Sung

Burt L. Newkirk Award
1992 Thomas N. Farris
1991 Farshid Sadeghi
1990 Michael M. Khonsari

Rufus Oldenburger Medal
1992 Isaac M. Horowitz
1991 John G. Truxal
1990 Harold Chestnut

Pressure Vessel and Piping Award
1992 Donald S. Griffin
1990 Robert J. Cepluch

James Harry Potter Gold Medal
1992 David Japikse
1991 James B. Jones
1990 Adrian Bejan

R. Tom Sawyer Award
1992 William R. Hawthorne
1991 Gerhard Neumann
1990 Hans J. P. von Ohain

Timoshenko Medal
1992 Jan D. Achenbach
1991 Yuang-Cheng B. Fung
1990 Stephen H. Crandall

Worcester Reed Warner Medal
1992 J. Narasimh Reddy
1991 Bruno A. Boley
1990 John Tinsley Oden

George Westinghouse Gold Medal
1992 Daniel R. Wilkins
1991 Ralph J. Ortolano
1990 John J. Taylor

George Westinghouse Silver Medal
1991 John B. Kitto, Jr.
1990 Atambir Singh Rao

Franklin Institute *Philadelphia, PA*

Edward Longstreth Medal
1991 Slobodan Cuk
Robert David Middlebrook

MEDICINE & PHYSIOLOGY

Abbott Laboratories, Pharmaceutical Division
Abbott Park, IL

Abbott Distinguished Investigator Award
1993 Tohru Fukuyama

Aerospace Medical Association *Alexandria, VA*

Eric Liljencrantz Award
1993 Fred Buick
1992 Alan J. Benson
1991 Frederick E. Gudry, Jr.

American Academy of Allergy & Immunology
Milwaukee, WI

Burroughs Wellcome Developing Investigator Award: Immunopharmacology of Allergic Diseases
1992 Bruce S. Bochner
1991 Donata Vercelli

American Association for Cancer Research
Philadelphia, PA

B. F. Cain Memorial Award
1993 Victor Ling
1992 Susan Band Horwitz
1991 Michael B. Sporn

G. H. A. Clowes Memorial Award
1993 Stuart H. Yuspa
1992 June L. Biedler
1991 Michael H. Wigler

Richard and Hinda Rosenthal Foundation Award
1993 Waun Ki Hong
1992 Michael M. Gottesman
1991 Owen N. Witte

American Association of Anatomists
New Orleans, LA

Henry Gray Award
1990 Sanford Palay

American Association of Pathologists
Bethesda, MD

Warner-Lambert/Parke-Davis Award
1993 Michael Cleary
1991 David P. Hajjar

American Association of Physicists in Medicine
New York, NY

William D. Coolidge Award
1992 Nagalingam Suntharalingam

American Cancer Society *Atlanta, GA*

ACS Medal of Honor
1992 Joseph R. Bertino
Bert Vogelstein

(ACS Medal of Honor, continued)
1991 Gianni Bonadonna
Louis Wade Sullivan

Distinguished Service Award
1992 Kathleen M. Foley

American Chemical Society *Washington, DC*

Alfred Burger Award in Medicinal Chemistry
1992 Everette Lee May

E. B. Hershberg Award for Important Discoveries in Medicinally Active Substances
1993 Arthur A. Patchett

Marvin J. Johnson Award in Microbial and Biochemical Technology
1993 Thomas Kent Kirk

American College of Radiology *Reston, VA*

Gold Medal
1991 Thomas F. Meaney
Edward W. Webster

American Institute of Aeronautics and Astronautics *Washington, DC*

Jeffries Medical Research Award
1991 Frank Hutches Austin, Jr.

American Heart Association *Dallas, TX*

James B. Herrick Award
1990 John Ross, Jr.

American Medical Association *Chicago, IL*

AMA Distinguished Service Award
1992 John J. Conley
Seymour Ochsner

Scientific Achievement Award
1992 Juan Angel Del Regato
1991 Henry Nicholas Wagner, Jr.

American Pediatric Society *Elk Grove Village, IL*

John Howland Award
1993 Lewis A. Barness
1992 Gilbert B. Forbes
1991 Robert E. Cooke
Roland B. Scott

American Society for Microbiology
Washington, DC

Abbott Laboratories Award in Clinical and Diagnostic Immunology
1993 Noel R. Rose
1992 Ellen S. Vitetta

Chiron Corporation Biotechnology Research Award
1993 David A. Hopwood
1992 Kary B. Mullis
1991 Eugene W. Nester

Becton Dickinson and Company Award in Clinical Microbiology
1993 Patrick R. Murray
1992 James H. Jorgensen
1991 L. Barth Reller

Hoechst-Roussel Award
1993 Brian G. Spratt
1992 Sydney M. Finegold
1991 George H. McCracken

Eli Lilly and Company Research Award in Microbiology and Immunology
1993 Ralph Isberg
1992 Vincent Racaniello
1991 John J. Mekalanos

J. Roger Porter Award
1993 Alma Dietz
1992 Mary Lechevalier
1991 Lawrence G. Wayne

Sonnenwirth Memorial Award
1993 John M. Matsen
1992 Paul D. Ellner
1991 Josephine A. Morello

Bristol-Myers Squibb Foundation *New York, NY*

Bristol-Myers Squibb Award for Distinguished Achievement in Infectious Disease Research
1992 Bernard N. Fields

Bristol-Myers Squibb Company Unrestricted Cancer Research Grants Program
1993 Gianni Bonadonna
Bernard Fisher

Chemical Industry Institute of Toxicology
Research Triangle Park, NC

Founders' Award
1993 Arthur B. Pardee
Henry C. Pitot III
1992 Kenneth Bruce Bischoff
Robert Lyle Dedrick
1991 Frederick P. Guengerich

Endocrine Society *Bethesda, MD*

Fred Conrad Koch Award
1991 John T. Potts
1990 Donald F. Steiner

Gairdner Foundation *Willowdale, ON, Canada*

Gairdner Foundation International Awards
1993 Mario Renato Capecchi
Alvan R. Feinstein
Stanley Ben Prusiner
Oliver Smithies
Michel Mathew Ter-Pogossian
1992 Leland H. Hartwell
Yoshio Masui
Bert Vogelstein
Robert A. Weinberg
1991 Sydney Brenner
M. Judah Folkman
David MacLennan
Kary Mullis

Gairdner Foundation Wightman Award
1992 John R. Evans

Institute of Medicine *Washington, DC*

Gustav O. Lienhard Award
1992 Faye Glenn Abdellah
Charles Everett Koop
1990 Loretta C. Ford

Albert and Mary Lasker Foundation
New York, NY

Albert Lasker Basic Medical Research Award
1991 Edward B. Lewis
Christiane Nüsslein-Volhard

Albert Lasker Clinical Medical Research Award
1991 Yuet Wai Kan

National Academy of Sciences *Washington, DC*

Jessie Stevenson Kovalenko Medal
1991 Roscoe Owen Brady

NAS Award in the Neurosciences
1991 Paul Greengard

Nobel Foundation *Stockholm, Sweden*

Nobel Prize for Medicine or Physiology
1993 Richard John Roberts
Phillip Allen Sharp
1992 Edmond Henri Fischer
Edwin G. Krebs
1991 Erwin Neher
Bert Sakmann

Optical Society of America *Washington, DC*

Edward D. Tillyer Award
1992 Horace B. Barlow

Passano Foundation *Baltimore, MD*

Passano Foundation Laureates
1993 Jack L. Strominger
Don C. Wiley
1992 Thomas Curran
Charles Yanofsky
1991 Stuart Kornfeld
William S. Sly
Roger Y. Tsien

Passano Senior Award
1990 Alfred G. Gilman

Royal Society of Canada *Ottawa, ON, Canada*

McLaughlin Medal
1991 Geoffrey Melvill Jones

Swedish Academy of Pharmaceutical Sciences
Stockholm, Sweden

Scheele-priset
1992 Koji Nakanishi

Wolf Foundation *Bet Herzlia, Israel*

Wolf Foundation Prize (in medicine)
1991 Seymour Benzer
1990 Maclyn McCarty

NATURAL SCIENCES

Academy of Natural Sciences of Philadelphia
Philadelphia, PA

Richard Hopper Day Memorial Award
1991 Robert McCracken Peck

Gold Medal for Distinction in Natural History Art
1992 William T. Cooper

National Academy of Sciences *Washington, DC*

Daniel Giraud Elliot Medal
1992 George Christopher Williams

PETROLEUM ENGINEERING

Society of Petroleum Engineers *Richardson, TX*

John Franklin Carll Award
1990 Robert Charles Earlougher, Sr.

DeGolyer Distinguished Service Medal
1991 James Roy Jorden
1990 Frederick Heinz Poettmann

Anthony F. Lucas Gold Medal
1990 George Wheeler Govier

Uren Award
1990 Necmettin Mungan

PHYSICS

Academy of Sciences of the EUR *Moscow, Russia*

Faraday Cup
1992 Alexander Feschenko

Acoustical Society *Woodbury, NY*

Gold Medal
1992 Ira J. Hirsch

R. Bruce Lindsay Award
1992 Anthony Armstrong Atchley

Silver Medal
1993 Victor C. Anderson
Alan Powell
1992 George Croswell Maling, Jr.
Thomas D. Rossing

American Academy of Arts and Sciences
Cambridge, MA

Rumford Medal
1992 George Feher
Joseph J. Katz
James Rufus Norris, Jr.

American Chemical Society *Washington, DC*

Frank H. Field and Joe L. Franklin Award for Outstanding Achievement in Mass Spectrometry
1993 Seymour Meyerson

American Crystallographic Association
Buffalo, NY

Bertram Eugene Warren Diffraction Physics Award
1991 James D. Jorgensen

American Institute of Physics *New York, NY*

Karl Taylor Compton Medal
1992 Victor Weisskopf

Germont Award
1990 Jeremy Bernstein

John T. Tate International Award
1992 Roald Z. Sagdeev

American Nuclear Society *LaGrange Park, IL*

Eugene P. Wigner Award
1992 Julius Wess
Bruno Zumino

American Physical Society *New York, NY*

Will Allis Prize
1992 James E. Lawler
1990 Arthur V. Phelps

Award for Excellence in Plasma Physics Research
1992 Nathaniel J. Fisch
John M. Greene
1991 Charles F. Driscoll
Thomas M. O'Neil

Biological Physics Prize
1992 Hans Frauenfelder

Tom W. Bonner Prize in Nuclear Physics
1993 Akito Arima
Francesco Iachello
1992 Henry G. Blosser
Robert E. Pollock
1991 Peter J. Twin

Herbert Broida Prize in Atomic, Molecular, or Chemical Physics
1993 Curt Wittig
1991 David E. Pritchard

Oliver E. Buckley Condensed-Matter Physics Prize
1993 Frederick Duncan M. Haldane
1992 Richard A. Webb
1991 Patrick A. Lee

Davisson-Germer Prize in Atomic or Surface Physics
1992 Joseph DeMuth
Larry Spruch
1991 Neville V. Smith
1990 David J. Wineland

John H. Dillon Medal for Research in Polymer Physics
1992 Mark D. Ediger
1991 Kenneth S. Schweizer

Fluid Dynamics Prize
1992 William R. Sears
1991 Andreas Acrivos
1990 John L. Lumley

Forum Award for Promoting Public Understanding of the Relation of Physics to Society
1991 Victor Weisskopf
1990 Richard Wilson

Maria Goeppert-Mayer Award
1993 Edwine Van Dishoect

High-Polymer Physics Prize
1992 Benjamin Thomas Peng-Nien Chu
Philip A. Pincus
1991 Edwin L. Thomas

Frank Isakson Prize for Optical Effects in Solids
1992 Paul A. Fleury

Otto Laporte Award for Research in Fluid Dynamics
1991 Steven A. Orszag
1990 Tony Maxworthy

Julius Edgar Lilienfeld Prize
1993 David N. Schramm
1992 Claude N. Cohen-Tannoudji
Alan H. Guth
1991 Daniel Kleppner

James Clerk Maxwell Prize for Plasma Physics
1992 John M. Greene
1991 Hans R. Griem

George E. Pake Prize
1993 Roland W. Schmitt
1992 Alan G. Chynoweth
1991 Albert Narath

W. K. H. Panofsky Prize
1992 Raymond Davis, Jr.
Frederick Reines
1991 Gerson Goldhaber
François Pierre
1990 Michael S. Witherell

Earle K. Plyler Prize
1993 Ahmed H. Zewail
1991 Kenneth M. Evenson

I. I. Rabi Prize
1993 Timothy E. Chupp
1991 Chris H. Greene

Simon Ramo Prize
1992 Wim Pieter Leemans

J. J. Sakurai Prize for Theoretical Particle Physics
1993 Mary Katherine Gaillard
1992 Lincoln Wolfenstein
1991 Vladimir N. Gribov

Leo Szilard Award for Physics in the Public Interest
1991 John H. Gibbons
1990 Theodore A. Postol

Robert R. Wilson Prize for Achievement in the Physics of Particle Accelerators
1993 John P. Blewett
1990 Kjell Johnsen

American Vacuum Society *New York, NY*

Gaede-Langmuir Award
1992 Russell D. Young

John A. Thornton Memorial Award
1992 Thomas R. Anthony

Medard W. Welch Award in Vacuum Science
1992 Ernst Bauer

King Abdul Aziz Research Centre
Riyadh, Saudi Arabia

King Faisal International Prize in Science (physics)
1993 Steven Chu
Herbert Walther

Canadian Association of Physicists
Ottawa, ON, Canada

CAP Medal for Achievement in Physics
1993 Walter N. Hardy
1992 Alec T. Stewart
1991 Gabriel Karl

Gerhard Herzberg Medal
1993 Nicholas Kaiser
1992 Robert Kiefl
1991 David B. MacFarlane

Coblenz Society *Norwalk, CT*

Coblenz Award
1993 Peter Felker

Deutsche Physikalische Gesellschaft
Bad Honnef, Germany

Gustav Hertz Preis
1993 Dieter Wintgen

Max-Planck-Medaille
1993 Kurt Binder

Robert-Wichard-Pohl-Preis
1993 Bruno Lüthi

Walter-Schottky-Preis für Festkorperforschung
1993 Gertrud Zwicknagl

Stern-Gerlach-Preis für Physik
1993 Klaus H. Winter

Albert Einstein Peace Prize Foundation
Northbrook, IL

Einstein Peace Prize
1993 Hans A. Bethe

Franklin Institute *Philadelphia, PA*

Albert A. Michelson Medal
1993 Serge Haroche
Herbert Walther
1992 John W. Hardy

International Centre for Theoretical Physics
Trieste, Italy

Paul Adrian Maurice Dirac Medal
1992 Yakov G. Sinai
1991 Stanley Mandelstam
1990 Sidney R. Coleman
Ludwig D. Faddeev

International Union of Pure and Applied Physics
Göteborg, Sweden

Boltzmann Medal
1992 Joel L. Lebowitz
Giorgio Parisi
1990 Leo Philip Kadanoff

National Academy of Sciences *Washington, DC*

Comstock Prize
1993 Erwin Louis Hahn
Charles Pence Slichter

Nobel Foundation *Stockholm, Sweden*

Nobel Prize in Physics
1993 Russell Alan Hulse
Joseph Hooton Taylor, Jr.
1992 Georges Charpak
1991 Pierre-Gilles de Gennes

Optical Society of America *Washington, DC*

Max Born Award
1993 David Hanna
1992 Rodney Loudon
1991 James P. Gordon

Frederic Ives Medal
1992 Robert William Terhune
1991 John Lewis Hall
1990 Joseph Wilfred Goodman

Ellis R. Lippincott Award
1993 John F. Rabolt
1992 Richard J. Saykally
1991 Alfons Weber

Adolf Lomb Medal
1992 Mohammed N. Islam

William F. Meggars Award
1992 Joseph Reader
1991 Daniel Kleppner

Charles Hard Townes Award
1991 Elias Snitzer
1990 Herbert Walther

R. W. Wood Prize
1991 Thomas F. Deutsch
Daniel J. Ehrlich
Richard M. Osgood, Jr.
1990 Rogers H. Stolen

Palacký University *Olomouc, Czech Republic*

Memorial Gold Medal
1992 Malvin C. Teich

Royal Physical Society *Edinburgh, Scotland*

Massey Award
1992 Herbert Friedman

Royal Society *London, England*

Hughes Medal
1992 Michael John Seaton

Rumford Medal
1992 H. N. V. Temperley
1991 Ian Affleck

Royal Society of Canada *Ottawa, ON, Canada*

Rutherford Memorial Medals
1992 François Wesemael
1990 Michael Daniel Fryzuk

Society for Applied Spectroscopy *Frederick, MD*

Maurice F. Hasler Award
1993 Robert Samuel Houk
1991 Karl Howard Norris

Lester W. Strock Award
1993 Willard W. Harrison
1992 Gary M. Hieftje
1991 M. Bonner Denton

United States Department of Energy
Washington, DC

Enrico Fermi Award
1992 Harold Brown
John Foster, Jr.
Leon Lederman
1990 George A. Cowan
Robley Dunglison Evans
1988 Richard Setlow
Victor Weisskopf

Ernest Orlando Lawrence Memorial Award
1991 Zachary Fisk
Richard J. Fortner
Rulon Kesler Linford
Peter G. Schultz
Richard Errett Smalley
Pace Van Devender

(Ernest Orlando Lawrence Memorial Award, continued)

1990 John J. Dorning
James Rufus Norris, Jr.
Samuel Thomas Picraux
Wayne J. Shotts
Maurice Tigner
Floyd Ward Whicker

Wolf Foundation *Bet Herzlia, Israel*

Wolf Prize (in physics)

1993 Benoit B. Mandelbrot
1992 Joseph Hooton Taylor, Jr.
1991 Maurice Goldhaber
Valentine Louis Telegdi

PSYCHOLOGY & PSYCHIATRY

Aerospace Medical Association *Alexandria, VA*

Raymond F. Longacre Award

1993 Robert S. Kennedy
1992 Henry L. Taylor
1991 Yuko Nagasawa

American Psychiatric Association *Washington, DC*

Distinguished Service Award

1993 James P. Comer
Herbert Pardes
1992 Eli Robins
Albert J. Solnit
1991 David A. Hamburg
Carolyn B. Robinowitz

SCIENCE, GENERAL

Adhesion Society *Hampton, VA*

Adhesion Society Award for Excellence in Adhesion Science

1993 Louis Sharp

American Association for the Advancement of Science *Washington, DC*

AAAS Hilliard Roderick Prize in Science, Arms Control and International Security

1993 Sidney D. Drell
1991 Wolfgang K. H. Panofsky

AAAS Newcomb Cleveland Prize

1991-92 Paul Douglas Quay
Bronte David Tilbrook
Chi-Shing Wong
1990-91 Michael C. Pirrung
Lubert Stryer

AAAS Scientific Freedom and Responsibility Award

1993 Daniel L. Albritton
Robert Barden Watson
1992 Inez Austin
1991 Adrian Russell Morrison

Franklin Institute *Philadelphia, PA*

Bower Award and Prize for Achievement in Science

1991 Solomon H. Snyder
1990 Paul Christian Lauterbur

Delmar S. Fahrney Medal

1993 John Bruton Stanbury
1992 Harold Paul Furth
1991 Timothy Coffey

Franklin Medal

1992 Frederick Reines

John Price Wetherill Medal

1992 Gerald Edward Brown
1991 Peter J. Twin

Laser Institute of America *Orlando, FL*

Arthur L. Schawlow Award

1990 Herbert M. Dwight, Jr.

John D. and Katherine T. MacArthur Foundation *Chicago, IL*

MacArthur Fellows

1993 Demetrios Christodoulou
Maria Luisa Buse Crawford
Ingrid Daubechies
Elizabeth Foe
Stephen Lee
Jane Lubchenco
Margie Profet
Stephen H. Schneider
Ellen Kovner Silbergeld
Frank Von Hippel
Heather Williams
Robert H. Williams

National Academy of Sciences *Washington, DC*

John J. Carty Medal for the Advancement of Science

1991 Joseph Hooton Taylor, Jr.

NAS Award for Initiatives in Research

1992 Alice P. Gast
Sangtae Kim
1991 Noam D. Elkies

NAS Public Welfare Medal
1993 Jerome B. Wiesner
1992 Philip Hauge Abelson
Robertson Memorial Lecture of the NAS
1993 Frank Sherwood Rowland
Rosenblith Lectures in Science and Technology
1993 Jose Goldemberg

National Geographic Society *Washington, DC*

Centennial Awards
1988 Robert D. Ballard
George F. Bass
Jacques-Yves Cousteau
John J. Craighead
Frank C. Craighead, Jr.
John Herschel Glenn, Jr.
Jane Goodall
Edmund Hillary
Mary D. Leakey
Richard E. Leakey
Thayer Soule
Barbara Washburn
H. Bradford Washburn, Jr.

National Science Foundation *Washington, DC*

Vannevar Bush Award
1993 Norman Hackerman
1992 Jerome B. Wiesner
1991 James A. Van Allen
National Medal of Science
1993 Alfred Y. Cho
Donald J. Cram
Val L. Fitch
Norman Hackerman
Martin D. Kruskal
Daniel Nathans
Vera C. Rubin
Salome G. Waelsch
1992 Eleanor J. Gibson
Calvin Forrest Quate
Eugene Merle Shoemaker
Howard E. Simmons, Jr.
Maxine F. Singer
Howard M. Temin
John Roy Whinnery
1991 Mary Ellen Avery
Ronald Breslow
Alberto P. Calderon
Gertrude B. Elion
George H. Heilmeier
Dudley R. Herschbach
Elvin A. Kabat
Robert W. Kates
Luna B. Leopold
Paul A. Marks
George A. Miller
Arthur L. Schawlow
Glenn T. Seaborg
Folke K. Skoog
H. Guyford Stever

(National Medal of Science, continued)
Edward C. Stone
Steven Weinberg
Paul C. Zamecnik
Alan T. Waterman Award
1993 Deborah L. Penry
1992 Shrinivas R. Kulkarni
1991 Herbert Edelsbrunner

Royal Society *London, England*

Copley Medal
1992 George Porter
Royal Medal
1992 David Tabor

Royal Society of Canada *Ottawa, ON, Canada*

Henry Marshall Tory Medal
1991 Willem Siebrand
Alice Wilson Award
1992 Susan M. Bradley

Royal Swedish Academy of Sciences
Stockholm, Sweden

Crafoord Prize
1993 Seymour Benzer
William Donald Hamilton
1992 Adolf Seilacher
1990 Paul Ralph Ehrlich
Edward O. Wilson

Science and Technology Foundation of Japan
Tokyo, Japan

Japan Prize
1993 Kary Banks Mullis
Frank Press
1992 Knut Schmidt-Nielson

Smithsonian Institution *Washington, DC*

Joseph Henry Medal
1991 Joseph J. Bonsignore
1990 August Heckscher
Smithson Medal
1991 Julie Johnson Kidd

Technion–Israel Institute of Technology
Haifa, Israel

Harvey Prize (in technology)
1992 Amnon Yariv

United States Department of Commerce Technology Administration *Washington, DC*

National Medal of Technology
1993 Amos E. Joel, Jr.
William H. Joyce

(National Medal of Technology, continued)

George Kozmetsky
George Levitt
Hans W. Liepmann
Marinus Los
William D. Manly
Kenneth H. Olsen
Walter L. Robb
1992 William H. Gates III
Joseph M. Juran
Charles D. Kelman
Delbert H. Meyer
Paul Burg Weisz
N. Joseph Woodland
1991 Stephen D. Bechtel, Jr.
Chester Gordon Bell
Geoffrey Boothroyd
John Cocke

(National Medal of Technology, continued)

Peter Dewhurst
Carl Djerassi
James Johnson Duderstadt
Antonio L. Elias
Robert W. Galvin
David S. Hollingsworth
Francis Kenneth Iverson
Robert R. Lovell
Charles E. Reed
John Paul Stapp
David W. Thompson

Alexander von Humboldt-Stiftung *Bonn, Germany*

Max-Planck-Forschungs-Preis

1992 Tomaso Arnardo Poggio